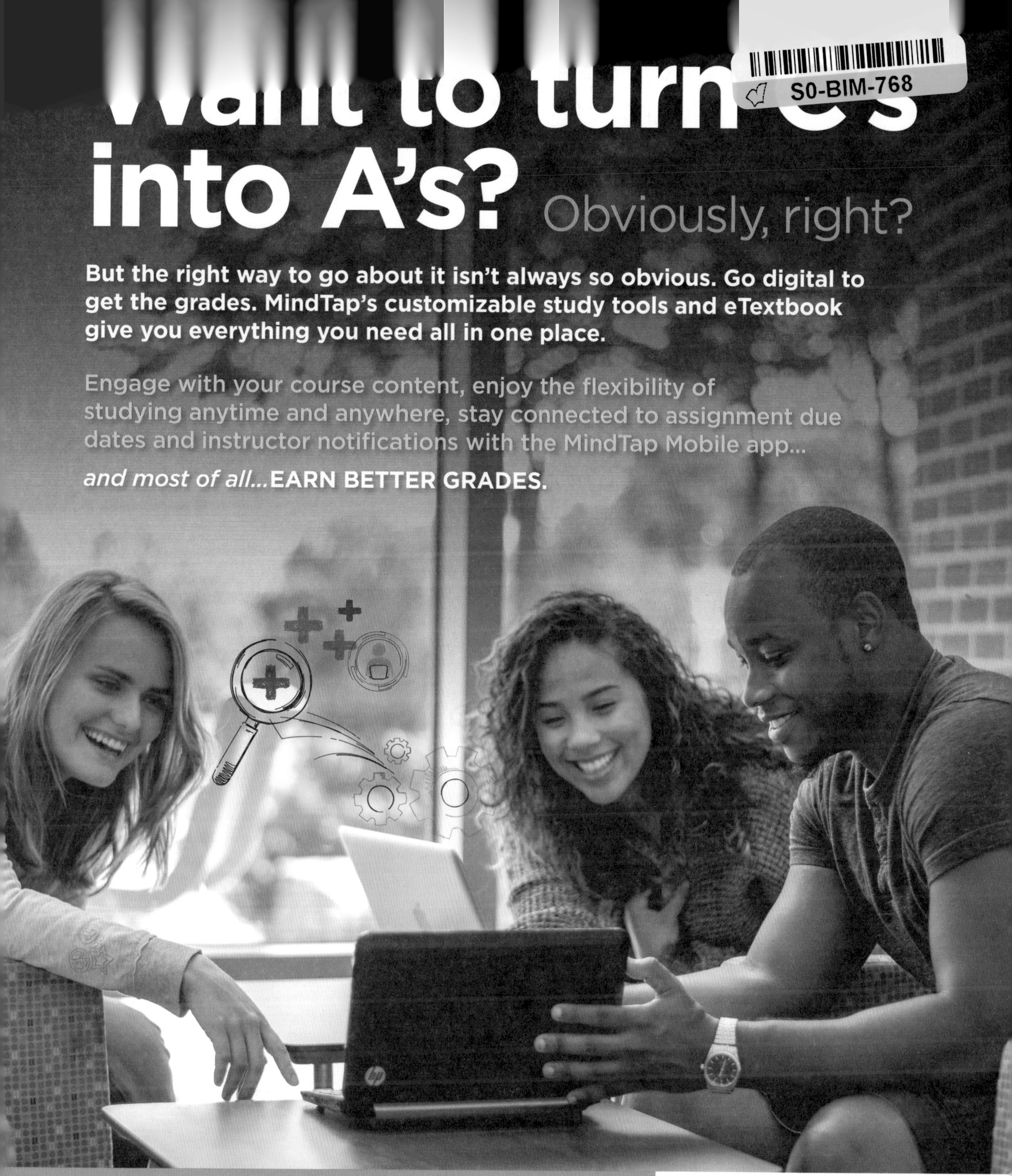

S0-BIM-768
Want to turn
into A's? Obviously, right?
But the right way to go about it isn't always so obvious. Go digital to get the grades. MindTap's customizable study tools and eTextbook give you everything you need all in one place.
Engage with your course content, enjoy the flexibility of studying anytime and anywhere, stay connected to assignment due dates and instructor notifications with the MindTap Mobile app...
and most of all...EARN BETTER GRADES.

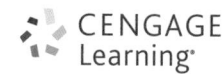

Biology
The Dynamic Science
Fourth Edition

Russell   Hertz   McMillan

Australia • Brazil • Mexico • Singapore • United Kingdom • United States

CENGAGE
Learning®

Biology: The Dynamic Science, **Fourth Edition**
Peter J. Russell, Paul E. Hertz,
Beverly McMillan

Product Director: Dawn Giovanniello

Product Manager: Kristen Sutton

Senior Content Developer: Jake Warde

Content Developers: Lauren Oliveira,
 Elesha Hyde, Susannah Alexander

Associate Content Developer: Kellie Petruzzelli

Product Assistant: Victor Luu

Executive Marketing Manager: Tom Ziolkowski

Product Development Specialist: Nicole Hurst

Content Project Manager: Hal Humphrey

Managing Art Director: Andrei N. Pasternak

Manufacturing Planner: Karen Hunt

Production Service: Megan Knight,
 Graphic World Inc.

Photo Researcher: Cheryl DuBois,
 Lumina Datamatics

Text Researcher: Manjula Subramanian,
 Lumina Datamatics

Copy Editor: Graphic World Inc.

Illustrators: Dragonfly Media Group,
 Steve McEntee, Graphic World Inc.

Text and Cover Designer: Jean Calabrese

Cover Image: ImageMore/Age Fotostock

Compositor: Graphic World Inc.

For product information and technology assistance, contact us at
Cengage Learning Customer & Sales Support, 1-800-354-9706.

For permission to use material from this text or product,
submit all requests online at **www.cengage.com/permissions.**
Further permissions questions can be e-mailed to
permissionrequest@cengage.com.

Library of Congress Control Number: 2015946817

Student Edition:

ISBN: 978-1-305-38989-2

Loose-leaf Edition:

ISBN: 978-1-305-65591-1

Cengage Learning
20 Channel Center Street
Boston, MA 02210
USA

Cengage Learning is a leading provider of customized learning solutions with employees residing in nearly 40 different countries and sales in more than 125 countries around the world. Find your local representative at **www.cengage.com.**

Cengage Learning products are represented in Canada by Nelson Education, Ltd.

To learn more about Cengage Learning Solutions, visit **www.cengage.com.**

Purchase any of our products at your local college store or at our preferred online store **www.cengagebrain.com.**

Printed in the United States of America
Print Number: 02   Print Year: 2017

Brief Contents

Peter J. Russell received a B.Sc. in Biology from the University of Sussex, England, in 1968 and a Ph.D. in Genetics from Cornell University in 1972. He has been a member of the Biology faculty of Reed College since 1972 and is currently a Professor of Biology, Emeritus. Peter taught a section of the introductory biology course, a genetics course, and a research literature course on molecular virology. In 1987 he received the Burlington Northern Faculty Achievement Award from Reed College in recognition of his excellence in teaching. Since 1986, he has been the author of a successful genetics textbook; current editions are *iGenetics: A Molecular Approach, iGenetics: A Mendelian Approach,* and *Essential iGenetics.* Peter's research was in the area of molecular genetics, with a specific interest in characterizing the role of host genes in the replication of the RNA genome of a pathogenic plant virus, and the expression of the genes of the virus; yeast was used as the model host. His research has been funded by agencies including the National Institutes of Health, the National Science Foundation, the American Cancer Society, the Department of Defense, the Medical Research Foundation of Oregon, and the Murdoch Foundation. He has published his research results in a variety of journals, including *Genetics, Journal of Bacteriology, Molecular and General Genetics, Nucleic Acids Research, Plasmid,* and *Molecular and Cellular Biology.* Peter has a long history of encouraging faculty research involving undergraduates, including cofounding the biology division of the Council on Undergraduate Research in 1985. He was Principal Investigator/Program Director of a National Science Foundation Award for the Integration of Research and Education (NSF–AIRE) to Reed College, 1998 to 2002.

Aaron Kinard

Paul E. Hertz was born and raised in New York City. He received a B.S. in Biology from Stanford University in 1972, an A.M. in Biology from Harvard University in 1973, and a Ph.D. in Biology from Harvard University in 1977. While completing field research for the doctorate, he served on the Biology faculty of the University of Puerto Rico at Rio Piedras. After spending two years as an Isaac Walton Killam Postdoctoral Fellow at Dalhousie University, Paul accepted a teaching position at Barnard College, where he has taught since 1979. He was named Ann Whitney Olin Professor of Biology in 2000, and he received The Barnard Award for Excellence in Teaching in 2007. In addition to serving on numerous college committees, Paul chaired Barnard's Biology Department for eight years and served as Acting Provost and Dean of the Faculty from 2011 to 2012. He is the founding Program Director of the Hughes Science Pipeline Project at Barnard, an undergraduate curriculum and research program that has been funded continuously by the Howard Hughes Medical Institute since 1992. The Pipeline Project includes the Intercollegiate Partnership, a program for local community college students that facilitates their transfer to four-year colleges and universities. He teaches one semester of the introductory sequence for Biology majors and pre-professional students, lecture and laboratory courses in vertebrate zoology and ecology, and a year-long seminar that introduces first-year students to scientific research. Paul is an animal physiological ecologist with a specific research interest in the thermal biology of lizards. He has conducted fieldwork in the West Indies since the mid-1970s, most recently focusing on the lizards of Cuba. His work has been funded by the NSF, and he has published his research in *The American Naturalist, Ecology, Nature, Oecologia,* and *Proceedings of the Royal Society.* In 2010, he and his colleagues at three other universities received funding from NSF for a project designed to detect the effects of global climate warming on the biology of *Anolis* lizards in Puerto Rico.

Courtesy of Beverly McMillan

Beverly McMillan has been a science writer for more than 25 years. She holds undergraduate and graduate degrees from the University of California, Berkeley, and is coauthor of a college text in human biology, now in its eleventh edition. She has also written or coauthored numerous trade books on scientific subjects and has worked extensively in educational and commercial publishing, including eight years in editorial management positions in the college divisions of Random House and McGraw-Hill.

Preface

Welcome to the fourth edition of *Biology: The Dynamic Science.* The book's title reflects the speed with which our knowledge of biology is growing. Although biologists have made enormous progress in solving the riddles posed by the living world, every discovery raises new questions and provides new opportunities for further research. As in previous editions, we have encapsulated the dynamic nature of biology in the fourth edition by explaining biological concepts—and the data from which they are derived—in the historical context of each discovery and by describing what we know now and what new discoveries will be likely to advance the field in the future.

Building on a strong foundation . . .

The first three editions of this book provided students with the tools they need to learn fundamental biological concepts and processes. More important, the previous editions encouraged students to *think like scientists* by applying the process of science. Our approach encourages students to think about biological questions and hypotheses through clear examples of hypothesis development, observational and experimental tests of hypotheses, and the conclusions that scientists draw from data. The many instructors and students who have used previous editions have generously provided valuable feedback that has allowed us to strengthen the elements that enhance student learning. We have also received comments from expert reviewers. As a result of these inputs, every chapter has been revised and updated with recent studies, including many based on genomic and proteomic analyses. In addition, the chapters in Unit One (Molecules and Cells) have been reorganized. This edition also includes new or modified illustrations and photos, as well as some new features.

The fourth edition of *Biology: The Dynamic Science* represents a fully integrated package of print and media that will appeal to today's students. Although the traditional format of the printed text can stand alone for both instructors and students, MindTap, the most engaging and easily personalized online solution in biology, enables instructors to deliver what they know is best for their students. MindTap offers an online version of the text, as well as before-class and in-class exercises, assignable and gradable homework exercises drawn from the book's content, and other resources and features that allow students to assess their learning as they progress through their study of biology.

Emphasizing the big picture . . .

In this textbook, we have applied our collective experience as teachers, researchers, and writers to create a readable and understandable foundation for students who choose to enroll in more advanced biology courses in the future. Where appropriate, we provide straightforward explanations of fundamental concepts from the evolutionary perspective that bind together all of the biological sciences. Recognizing that students in an introductory biology course face a potentially daunting quantity of ideas and information, we strive to provide an appropriate balance between factual and conceptual material, taking great care to provide clear explanations of how scientists draw conclusions from empirical data. Our approach helps students understand how we achieved our present knowledge. Having watched our students struggle to navigate the many arcane details of college-level introductory biology, we constantly remind ourselves and each other to "include fewer facts, provide better explanations, and maintain the narrative flow," thereby enabling students to see the big picture. Clarity of presentation, thoughtful organization, a logical and seamless flow of topics within chapters, and carefully designed illustrations are key to our approach. With this edition, full integration with MindTap engages students with appealing and useful exercises that encourage them to learn biology by thinking like scientists.

Focusing on research to help students engage the living world as scientists . . .

A primary goal of this book is to sharpen and sustain students' curiosity about biology, rather than dulling it with a mountain of disconnected facts. We can help students develop the mental habits of scientists and a fascination with the living world by conveying our passion for biological research. We want to excite students not only with *what biologists know* about the living world but also with *how they know it* and *what they still need to learn.* In doing so, we can encourage some students to accept the challenge and become biologists themselves, posing and answering important new questions through their own innovative research. For students who pursue other careers, we hope that they will leave their introductory—and perhaps only—biology course armed with intellectual skills that will enable them to evaluate future knowledge with a critical eye.

In this book, we introduce students to a biologist's "ways of learning." Research biologists constantly integrate new observations, hypotheses, questions, experiments, and insights with existing knowledge and ideas. To help students engage the world as biologists do, we must not simply introduce them to the current state of knowledge. We must also foster an appreciation of the historical context within which those ideas developed, and identify the future directions that biological research is likely to take.

To achieve these goals, our explanations are rooted in the research that established the basic facts and principles of biology. Thus, a substantial proportion of each chapter focuses on studies that define the state of biological knowledge today. When describing research, we first identify the hypothesis or question that inspired the work and then relate it to the broader topic under discussion. Our research-oriented theme teaches students, through example, how to ask scientific questions and pose hypotheses, two key elements of the scientific process.

Because advances in science occur against a background of research, we also give students a feeling for how biologists of the past formulated basic knowledge in the field. By fostering an appreciation of such discoveries, given the information and theories available to scientists in their own time, we can help students understand the successes and limitations of what we consider cutting edge today. This historical perspective also encourages students to view biology as a dynamic intellectual enterprise, not just a collection of facts and generalities to be memorized.

We have endeavored to make the science of biology come alive by describing how biologists formulate hypotheses and evaluate them using hard-won data; how data sometimes tell only part of a story; and how the results of studies often end up posing more questions than they answer. Although students might prefer simply to learn the "right" answer to a question, they must be encouraged to embrace "the unknown," those gaps in knowledge that create opportunities for further research. An appreciation of what biologists do *not* yet know will draw more students into the field. And by defining *why* scientists do not understand interesting phenomena, we encourage students to think critically about possible solutions and to follow paths dictated by their own curiosity. We hope that this approach will encourage students to make biology a part of their daily lives by having informal discussions and debates about new scientific discoveries.

Presenting the story line of the research process . . .

In preparing this book, we developed several special features, all of which are included in MindTap, to help students broaden their understanding of the material presented and of the research process itself. A Visual Tour of these features and more begins on page xiv.

- The chapter openers, titled *Why it matters . . .,* are engaging, short vignettes designed to capture students' imaginations and whet their appetites for the topic that the chapter addresses. In many cases, this feature tells the story of how a researcher or researchers arrived at a key insight or how biological research solved a major societal problem, explained a fundamental process, or elucidated a phenomenon. The *Why it matters . . .* feature also provides a brief summary of the contents of the chapter.

- To complement this historical or practical perspective, each chapter closes with a brief essay titled *Unanswered Questions,* prepared by an expert or experts in the field. These essays identify important unresolved issues relating to the chapter topic and describe cutting-edge research that will advance our knowledge in the future.

- Most chapters include a short, boxed essay titled *Molecular Insights* (formerly called *Insights from the Molecular Revolution*), which describes how molecular tools allow scientists to answer questions that they could not have posed even 30 years ago. Most *Molecular Insights* focus on a single study and include sufficient detail for their content to stand alone.

- Many chapters are further supplemented with one or more short, boxed essays called *Focus on Research.* Each essay focuses on one of three different aspects of research. *Focus on Research: Basic Research* essays describe how research has provided understanding of basic biological principles. *Focus on Research: Applied Research* essays describe research designed to solve practical problems in the world, such as those relating to health or the environment. *Focus on Research: Model Organisms* essays introduce model research organisms—such as *Escherichia coli, Drosophila, Arabidopsis, Caenorhabditis,* the mouse, and *Anolis*—and explain why they are used as subjects for in-depth analysis.

- Three types of specially designed *research figures* provide more detailed information about how biologists formulate specific hypotheses and test them by gathering and interpreting data. The research figures are listed on the endpapers at the back of the book. *Experimental Research* figures describe specific studies in which researchers used both experimental and control treatments—either in the laboratory or in the field—to test hypotheses or answer research questions by manipulating the system they studied. *Observational Research* figures describe specific studies in which biologists have tested hypotheses by comparing systems under varying natural circumstances. *Research Method* figures provide examples of important techniques, such as light and electron microscopy, the polymerase chain reaction, making a knockout mouse, DNA microarray analysis, plant cell culture, producing monoclonal antibodies, radiometric dating, and cladistic analysis. Each *Research Method* figure leads a student through the purpose of the technique and protocol and describes how scientists interpret the data it generates.

Integrating effective, high-quality visuals into the narrative . . .

Today's students are accustomed to receiving ideas and information visually, making the illustrations and photographs in a textbook and the fully integrated online resources critically important. From the first edition, our illustration program has provided an exceptionally clear supplement to the narrative in a style that is consistent throughout the book. Graphs and anatomical drawings are annotated with interpretive explanations that lead students, step by step, through the major points they convey.

Over subsequent editions, we have enhanced the illustration program, focusing on features that reviewers and users of the book identified as the most useful pedagogical tools. For this most recent edition, we focused explicitly on helping students to think like scientists. A revised Figure 1.14 illuminates the intellectual steps that collectively lead researchers to new scientific discoveries. These steps—observation, hypothesis, prediction, experiment, and interpretation—represent the fundamentals of our "think like a scientist" theme in this book.

In revising the text, we reevaluated each illustration and photograph and made appropriate changes to improve their utility as teaching tools. New illustrations for the fourth edition were created in the same style as existing ones. In addition, some illustrations of key biological processes were recast as *Closer Look* figures in which a Summary and a concluding *Think Like a Scientist* question enhance student learning.

Organizing chapters around important concepts...

As authors and college teachers, we understand how easily students can get lost within a chapter. When students request advice about how to read a chapter and learn the material in it, we usually suggest that, after reading each section, they pause and quiz themselves on the material they have just encountered. After completing all of the sections in a chapter, they should quiz themselves again, even more rigorously, on the individual sections and, most important, on how the concepts developed in the different sections fit together. Accordingly, we have adopted a structure for each chapter to help students review concepts as they learn them.

- The organization within chapters presents material in digestible sections, building on students' knowledge and understanding as they acquire it. Each major section covers one broad topic. Each subsection, titled with a declarative sentence that summarizes the main idea of its content, explores a narrower range of material.
- Whenever possible, we include the derivation of unfamiliar terms so that students will see connections between words that share etymological roots. Mastery of the technical language of biology will allow students to discuss ideas and processes precisely. At the same time, we have minimized the use of unnecessary jargon.
- *Study Break* questions follow every major section. These questions encourage students to pause at the end of a section and review what they have learned before going on to the next topic within the chapter. Short answers to these questions appear in an appendix.

Encouraging active learning, critical thinking, and self-assessment of learning outcomes...

In the third edition we introduced an active learning feature, *Think Like a Scientist,* which is designed to help students think analytically and critically about research presented in the chapter. *Think Like a Scientist* questions appear at the end of *Experimental Research* figures, *Observational Research* figures, *Closer Look* figures, *Molecular Insights* boxes, and *Unanswered Questions*. In this new edition, *Experimental Research* figures and *Observational Research* figures include a new icon that identifies the particular step in the process of science that the *Think Like a Scientist* question addresses: Observe, Hypothesize, Predict, Experiment, or Interpret. (Sample shown.)

This edition also continues the popular *Think Outside the Book* active learning feature. *Think Outside the Book* activities have been designed to encourage students to explore biology directly or through electronic resources, working either individually or collaboratively.

Supplementary materials at the end of each chapter—all of which are fully integrated into MindTap—help students review the material they have learned, assess their understanding, and think analytically as they apply the principles developed in the chapter to novel situations. Many end-of-chapter questions also serve as good starting points for class discussions or out-of-class assignments.

Review Key Concepts provides a summary of important ideas developed in the chapter, referencing specific figures and tables in the chapter. These *Reviews* are no substitute for reading the chapter, but students may use them as a valuable outline of the material, filling in the details on their own.

Test Your Knowledge includes four types of end-of-chapter questions and problems that focus on the chapter's factual content while encouraging students to apply what they have learned: (1) Multiple-choice questions (with answers in an appendix) focus on factual material; (2) *Discuss Concepts* questions involve open-ended issues that emphasize key ideas, the interpretation of data, and practical applications of the material; (3) *Design an Experiment* questions help students hone their critical thinking skills by asking them to test hypotheses that relate to the chapter's main topic; and (4) *Apply Evolutionary Thinking* questions ask students to answer a question in relation to the principles of evolutionary biology.

In this edition, *Test Your Knowledge* questions are organized according to Bloom's taxonomy into three sections: Remember/Understand, Apply/Analyze, and Evaluate/Create. This structure allows students to review the material in a sequence that moves from the basic knowledge of factual material to more challenging and sophisticated applications of that knowledge to novel situations.

Interpret the Data questions, highlighted in a distinctive format, help students develop analytical and quantitative skills by asking them to interpret graphical or tabular results of experimental or observational research studies for which the hypotheses and methods of analysis are presented.

Effectively introducing digital solutions into your classroom—online or in class—is now easier than ever . . .

The fourth edition of *Biology: The Dynamic Science* represents a fully integrated package of print and media, providing comprehensive learning tools and flexible delivery options. In preparing this edition we conducted extensive research to determine how instructors prefer to present online learning opportunities. The result of this research is a new MindTap course organized around the instructor's preferred workflow. Instructors can now select just the content they want to assign, chosen from a comprehensive set of learning materials provided with the course for each chapter. Many types of learning activities are assignable and offer students immediate feedback and automated instructor assessment.

Research also indicates that online content is most effective when it enhances conceptual understanding through the use of relevant applications. In this edition, we have developed three new assessable, online learning activities that align with important book features and provide students the opportunity to explore and practice biology the way scientists practice biology:

- The *Interpret the Data* feature at the end of every chapter is enhanced by an additional online exercise to further develop student quantitative analysis and mathematical reasoning skills.

altrendo images/Getty Images.

- Many *Experimental Research* and *Observational Research* figures now include an additional activity to help students engage with the scientific process.
- The end-of-chapter *Design an Experiment* feature is presented online as a guided learning activity that takes the student through the process of designing an experiment.

The *Instructor Resource Center* provides everything you need for your course in one place. This collection of lecture and class tools is available online for instructors only via www.cengage.com/login. There you can access and download PowerPoint presentations, images, the *Instructor's Manual*, videos, and more.

We hope you agree that we have developed a clear, fresh, and well-integrated introduction to biology as it is understood by researchers today. Just as important, we hope that our efforts will excite students about the research process and the biological discoveries it generates.

Peter J. Russell   Paul E. Hertz   Beverly McMillan

New to This Edition

The enhancements we have made in the fourth edition of *Biology: The Dynamic Science* reflect our commitment to provide a text that introduces students to new developments in biology while fostering active learning and critical thinking. The new icon for *Think Like a Scientist* questions not only helps readers identify those questions easily, but it also helps students reinforce their understanding of the process of science.

We have also made important changes in coverage to follow recent scientific advances. In the third edition, we added a new Chapter 19, Genomes and Proteomes, that discusses methods of genomics and proteomics along with examples of new discoveries and insights. To reflect the importance of these approaches, genomics and proteomics coverage has been integrated throughout the book. In addition, 35 *Molecular Insights* boxes have been revised to include recently published research, and many of them focus on genomics or proteomics. Beyond these changes, we have made numerous improvements to update and clarify scientific information and to engage students as interested readers and active learners, as well as responsive scientific thinkers. The following sections highlight some of the new content and organizational changes in this edition.

Chapter 1

In the introductory chapter, Section 1.1 (What is Life? Characteristics of Living Organisms) has been revised to provide a more detailed discussion of the function and transmission of genetic information in living systems; and the accompanying figure (Figure 1.4) provides a more accurate portrayal of nucleic acids and the flow of information from DNA to RNA to proteins. In addition, Section 1.4 (Biological Research) now has a more explicit discussion of the process of science, including revisions to Figure 1.14 (now a *Closer Look* figure), which introduces the new icon for *Think Like a Scientist* questions in *Experimental Research* figures and *Observational Research* figures. Finally, the introduction of genomics and proteomics has been refined, and systems biology is introduced (with a new Figure 1.17) as an inclusive approach for exploring complex biological phenomena.

Unit One: Molecules and Cells

Unit One has been reorganized. The former Chapter 4 (Energy, Enzymes, and Biological Reactions) is now Chapter 6, preceding Chapter 7 (Cellular Respiration: Harvesting Chemical Energy) and Chapter 8 (Photosynthesis) to provide a more logical conceptual flow of metabolism. The former Chapter 7 (Cell Communication) is now Chapter 9.

Chapter 2 (Life, Chemistry, and Water) has a new *Why it matters* . . . relating biology to underlying chemical reactions, a new subsection on molecule geometry and function in the cell, and a new *Unanswered Questions* essay on the effect of climate change on marine ecosystems.

Chapter 3 (Biological Molecules: The Carbon Compounds of Life) has molecular models added to Table 3.1 on functional groups and to Table 3.2 on major protein functions.

In Chapter 4 (Cells), Section 4.2 on Prokaryotic Cells has been reorganized into subsections on structure and organization, and evolutionary divergence of bacteria and archaea, and Section 4.3 on Eukaryotic Cells has added proteomics content in the endomembrane system discussion, updated discussion of the Golgi complex and vesicle traffic, and enhanced discussion of the movement of organelles and vesicles. A new *Molecular Insights* describes a genome-wide analysis study identifying human proteins that regulate secretion.

Chapter 5 (Membranes and Transport) adds the kiss-and-run model of exocytosis.

Chapter 7 (Cellular Respiration: Harvesting Chemical Energy) adds more examples of human disorders related to cellular respiration, discussion of evolution of mitochondria, and discussion of anaerobic respiration. A new Section 7.6 on Interrelationships of Catabolic and Anabolic Pathways combines material previously in earlier sections with new material to discuss how food substances feed into glycolysis, how biosynthetic pathways link to glycolysis and the citric acid cycle (with a new figure), and the regulation of cellular respiration (with a new figure). A new *Molecular Insights* describes experiments studying the mitochondrial proteome and its variation among organs and organisms.

In Chapter 8 (Photosynthesis), a new *Experimental Research* figure describes the Engelmann experiment. Discussion of the evolution of photosynthesis and cellular respiration has been enhanced. A new *Molecular Insights* describes proteomics experiments on the effect of water deficit on growth and photosynthesis in a C_3 plant.

Chapter 9 (Cell Communication) has expanded discussion of the different types of signaling along with an updated figure. Sections on receptor tyrosine kinases and G-protein–coupled receptors are now combined with new material on ligand-gated ion channels into a single section on signaling pathways triggered by surface receptors. Nitric oxide is added to the internal receptor section.

Chapter 10 (Cell Division and Mitosis) has enhanced discussion of chromosomes now including presentation of chromosome structure and levels of chromosome organization (including a figure) moved from Chapter 14 (DNA Structure and Replication). A new *Molecular Insights* describes proteomics experiments on the regulation of proteins in the cell cycle of humans.

Unit Two: Genetics

In Chapter 12 (Mendel, Genes, and Inheritance), the terms *gene marker, genetic marker,* and *DNA marker* are now introduced to relate to discussions later in the unit.

Chapter 13 (Genes, Chromosomes, and Human Genetics) adds discussions of pedigrees and pedigree analysis with four new figures, and of the usefulness of mitochondrial DNA analysis in genealogy and forensics. A new *Molecular Insights* describes experiments on the involvement of noncoding RNAs in X-chromosome inactivation.

Chapter 14 (DNA Structure and Replication) adds discussion of the loading and unloading of the sliding clamp in human DNA replication, a new figure and description of the replisome complex for DNA replication, discussion of the role of telomeres as chromosome caps, and a description of how newly replicated DNA is assembled into nucleosomes. Section 14.4 is retitled Repair of Errors in DNA to reflect the addition of new material on excision repair mechanisms, and a new figure on thymine dimers and their repair.

Chapter 15 (From DNA to Protein) now discusses Garrod's classic experiment in *Why it matters*... adds discussion of the coupling of transcription, pre-mRNA processing and export of mRNA from the nucleus, and adds new material on spontaneous and induced mutations and mutagens, transposable element content of genomes, and the types of changes in gene expression eukaryotic transposable elements can cause. A new *Molecular Insights* describes genomics/proteomics experiments on the effect of sleep–wake timing on gene expression.

In Chapter 16 (Regulation of Gene Expression), Section 16.2 on Regulation of Transcription in Eukaryotes is reorganized to present chromatin modification and methylation before the molecular details of transcription initiation at promoters. Discussion of DNA methylation and gene regulation is expanded, including analysis of DNA methylomes. New material is added on the interference of transcription by long noncoding RNAs (lncRNAs), on the cataloging of human promoters and enhancers by genome analysis, and on the role of alternative splicing in adjusting gene output to match physiological requirements. In Section 16.4, Genetic and Molecular Regulation of Development, the discussion of *Hox* genes is rewritten and expanded, and lncRNAs are added to the discussion of noncoding RNAs and their roles in development. Section 16.5, Genetics of Cancer, is changed to The Genetics and Genomics of Cancer to reflect added content. Discussion of genes involved in cancer is rewritten to include new knowledge from genomic approaches on the types of genes involved in cancer, and the cellular processes affected by mutations associated with cancer. Also added is discussion of cancer therapy and the prospect of personalized medicine, and how cancer genomic analysis identifies cancer subtypes. A new *Molecular Insights* describes experiments on the role of a key long noncoding RNA in cardiac development.

In Chapter 17 (Bacterial and Viral Genetics), a new *Molecular Insights* describes genomics/proteomics experiments that identified the largest known giant virus.

Chapter 18 is retitled DNA Technologies: Making and Using Genetically Altered Organisms, and Other Applications, to reflect a reorganization of the chapter around the theme of making and using genetically altered organisms for basic and applied research. The three main sections of the chapter are now Key DNA Technologies for Making Genetically Altered Organisms, Applications of Genetically Altered Organisms, and Other Applications of DNA Technologies (not involving genetically altered organisms). PCR is now emphasized more as a widely used technique, and Southern and northern blotting are no longer discussed. A new figure is added on making knockout mice by gene targeting, and a new *Focus on Research* box describes programmable RNA-guided genome editing based on CRISPR-Cas. The *Why it matters*... is new, discussing the relationship between historical genetics approaches to modern-day genomics approaches. A new *Molecular Insights* describes metabolomic experiments on the nutritional quality of genetically modified food.

In Chapter 19 (Genomes and Proteomes), lncRNAs are added to the discussions of important genome sequences, the ENCODE project is added to the discussion of the profile of the human genome, and tables are updated.

Unit Three: Evolutionary Biology

Chapter 20 (The Development of Evolutionary Thinking) has been revised to contextualize the study of evolution within the science process described in Chapter 1, and Figure 20.10 presenting Darwin's observations, hypotheses, and predictions has been restructured to reflect the process of science theme we have developed. In addition, the discussion of the fossil record has been updated to reflect new discoveries about the relationships of birds to non-avian dinosaurs, and Figure 20.13 now includes a phylogenetic tree to highlight when key adaptations arose. Chapter 20 also includes a new *Molecular Insights* on the genetics of dog domestication.

In Chapter 21 (Microevolution: Genetic Changes within Populations), the discussion of variation in DNA sequences now includes the importance of single-nucleotide polymorphisms in biological research. The *Focus on Research* box about the Hardy–Weinberg genetic equilibrium has been subdivided to make the steps in the analysis more explicit. The figure on the distribution of color and striping patterns in European garden snails (Figure 21.15) has been revised to emphasize the associations between snail phenotypes and habitat types. The chapter includes a new example of frequency-dependent selection, using flower color in a European orchid (Figure 21.16). In Table 21.2, we have added a column about the fitness consequences of each of the agents of evolution. Finally, Chapter 21 includes a new *Molecular Insights* about the genetic and physiological mechanisms that produce exaggerated horns in male rhinoceros beetles.

In Chapter 22 (Speciation), a new introduction to the discussion of reproductive isolation in Section 22.2 provides a broader context for the discussion of the mechanisms of reproductive isolation. The discussion of parapatric speciation was deleted from Section 22.3 on the geography of speciation in favor of a more extended discussion of the possible consequences of secondary contact after a period of allopatry. The example of a hybrid zone between oriole *(Icterus)* species has been enhanced with additional text and a revised figure (22.12). The discussion of allopolyploidy and speciation in wheat has been updated with new genomic studies and a greatly revised figure (22.16). Chapter 22 includes a revised figure in the *Focus on Research* box about speciation in Hawaiian *Drosophila,* as well as a new *Molecular Insights* about the genetics of ecological and behavioral isolation between two *Drosophila* species.

In Chapter 23 (Paleobiology and Macroevolution), the table (Table 23.1) outlining geological time and the history of life has been consolidated and now includes more references to major geological events. The chapter also includes a revised discussion of biogeographical realms with a revised figure (23.10) based on a study published in 2013. The discussions of both adaptive radiations and mass extinctions have also been updated along with a new figure (23.14) showing the pattern of mass extinctions more explicitly. The discussion of feathers in non-avian dinosaurs and birds has been updated with new information, and a new figure (23.24) illustrates age-related changes in dinosaur plumage.

In Chapter 24 (Systematics and Phylogenetics: Revealing the Tree of Life), the taxonomic hierarchy is outlined in a new table (Table 24.1) that cross-references information in Figure 1.10. The discussion about reading a phylogenetic tree has been consolidated, and the accompanying figures have been combined into a *Closer Look* figure (24.2). The discussion of homology and homoplasy has been clarified. Chapter 24 also includes a new *Molecular Insights* (with accompanying phylogenetic tree) about the evolution of electric organs in fishes.

Unit Four: Biodiversity

In Chapter 25 (The Origin of Life), Section 25.1 on the formation of molecules necessary for life was revised to add material about what is needed to understand the origin of life, about questions asked in the scientific study of the origin of life, and about the possible role of alkaline hydrothermal vents in the origin of life. Section 25.2 on the origin of cells was largely rewritten, now to discuss the evolution of molecular replicators, the evolution of cellular membranes, and the evolution of biological energy sources, as well as adding to the discussion of prokaryotic cells as the first living cells.

Chapter 26 (Prokaryotes: Bacteria and Archaea) adds a discussion of microbiomes.

Chapter 27 (Protists) adds a discussion of the apicoplast and its origin. A new *Molecular Insights* describes experiments on two different types of bacteria associated with a cellular slime mold.

Chapter 28 (Seedless Plants) includes examples of comparative genomics that shed light on the evolution of vascular tissue in land plants and of new fossil finds that expand our understanding of land plant evolution. The chapter's new *Molecular Insights, Comparative Genomics Probes Plant Evolution,* reinforces these ideas. The chapter now concludes with a new section that surveys the ecological, economic, and research importance of seedless plants. New figures accompany chapter discussions of early adaptation for water transport and the evolutionary shift from homospory to heterospory.

Chapter 29 (Seed Plants) begins with a new *Why it matters . . .* that introduces students to the core concept that the evolution of the seed, together with pollen and pollination, was crucial in the radiation of vascular plants into nearly every land environment. A new Section 29.1, The Rise of Seed Plants, examines key innovations in the evolution of the seed; new line art illustrates key stages hypothesized for one of those innovations, the evolution of the ovule. This new section provides the conceptual foundation for subsequent sections on gymnosperms and angiosperms, both of which have been reorganized and expanded. Section 29.3, on angiosperms, includes new subsections that underscore the adaptive roles of flowers, double fertilization, fruits and seeds. Expanded coverage in Section 29.4 (previously 29.3), Insights from Plant Genome Research, includes the hypothesized role of whole genome duplication in plant polyploidy and examples of how genome sequencing is advancing research on the biology and evolutionary relationships of various seed plant lineages. A new, concise Section 29.5, Seed Plants and People, reminds students of the major roles of seed plants in human affairs. A new *Molecular Insights* describes insights that are emerging from the sequencing of the loblolly pine genome.

Revisions to Chapter 30 (Fungi) include a discussion of the proposed new fungal phylum Cryptomycota and a new *Molecular Insights, Researching Relationships of "Hidden Fungi."* Much of the third edition's discussion of mycorrhizae now appears in Chapter 34 on plant nutrition. A new concluding section discusses the effects of fungi on ecosystems and human endeavors.

Changes to Chapter 31 (Animal Phylogeny, Acoelomates, and Protostomes) include an updated description of how the molecular phylogeny for animals was constructed. The chapter now also includes descriptions of a recently discovered predatory sponge, all-female species of bdelloid rotifers, and a revised analysis of annelid systematics to reflect recent research. The discussion of insects includes a new description of the evo–devo origin of insect wings. Although *Molecular Insights* addresses the same topic (relationships among arthropods) as in prior editions, it is now based on research published in 2010 and includes a revised phylogenetic tree for arthropods.

In Chapter 32 (Deuterostomes: Vertebrates and Their Closest Relatives), revisions include an expanded discussion of conodont elements and of placoderms as a paraphyletic group. The chapter also includes a greatly expanded discussion of the origin of limbs with a completely revised figure (32.21). The discussion of amniote relationships, including the phylogenetic tree (Figure 32.24), reflects new insights into the relationships of turtles to archosaurs and recognizes the newly defined group

Archelosauromorpha. The phylogenetic tree for primates (Figure 32.34) now includes characters that distinguish clades, and the discussion of human evolution now includes a description of the Denisovans and an enhanced hominin timeline (Figure 32.39).

Unit Five: Plant Structure and Function

This edition's chapters on plant anatomy and physiology also have been significantly revised, updated, and reorganized. In Chapter 33 (The Plant Body), a new *Why it matters . . .* introduces the main chapter topic using the historical domestication of grasses (rice, wheat, corn) to underscore the relevance of plant parts to human concerns. Sections on root and shoot systems are reordered, with root systems considered first (Section 33.3) followed by primary shoot systems (Section 33.4). A new *Molecular Insights* features current research on the complex genetic events governing the formation of secondary cell walls.

In Chapter 34 (Transport in Plants), a new *Why it matters . . .* uses the example of the centuries-old Angel Oak on Johns Island in South Carolina to introduce transport in plants. The discussion of water potential is revised for clarity, with new subsections providing a more straightforward presentation on solute and pressure potential, turgor pressure, and wilting. Section 34.2, formerly Transport in Roots, is now titled Roots: Moving Water and Minerals into the Plant. A new Figure (34.7) provides an enhanced visual to support the text discussion of the Casparian strip. Section 34.3, Transport of Water and Minerals in the Xylem, now includes a subsection on effects of humidity, temperature, and wind on transpiration. The functioning and regulation of stomata are now considered in a separate section (Section 34.4). A new *Molecular Insights, Going with the Phloem,* looks at recent experiments exploring the role of the transcription factors called NAC proteins in shaping the development of sieve-tube elements.

Enhancements to Chapter 35 (Plant Nutrition) include new photographs (Figure 35.2) to illustrate a range of nutrient deficiency symptoms in plants. A new subsection discusses the use of fertilizers to remedy nutrient deficiencies in soil. Section 35.2 on soil characteristics now includes the role of weathering in soil formation. New subsections clearly distinguish the effects of organic and inorganic components of soils and discuss the role of proper soil management (including no-till farming) in sustainable agriculture. Section 35.3, Root Adaptations for Obtaining and Absorbing Nutrients, now presents much of the discussion of mycorrhizae that previously was in Chapter 30 (Fungi). The subsection on nitrogen fixation now includes a concise description of denitrification.

In Chapter 36 (Reproduction and Development in Flowering Plants), Section 36.2 on the formation of flowers and gametes now includes a new subsection on evolutionary trends in flower structure with accompanying photographs that allow students to readily visualize and compare key differences in flower structure. Two other new subsections expand the chapter's coverage of fruit diversity and seed dispersal mechanisms. Section 36.4, Asexual Reproduction of Flowering Plants, provides expanded examples of vegetative reproduction. The discussion of tissue culture methods emphasizes somatic embryogenesis, including the creation of artificial seeds.

Chapter 37 (Plant Signals and Responses to the Environment) also features important revisions and additions. The introduction to plant hormones in Section 37.1 is reorganized to consider auxins, cytokinins, and gibberellins, in that order. It also includes strigolactones (SLs) as a major plant hormone family, with accompanying line art reinforcing the varied functions of SLs in enhancing nutrient access and optimizing the growth of shoot and root parts. The subsection on ethylene now includes the triple response in seedling growth. A brief comparison of constitutive and inducible defenses is added to Section 37.2, Plant Chemical Defenses. One new subsection, Defensive Chemicals Reflect the Coevolution of Plants, Pathogens, and Herbivores, discusses bioactive specialized compounds including phytoalexins, plant alkaloids, terpenes, and phenolics. Another new subsection on inducible responses to specific threats introduces pattern recognition receptors and their role in detecting pathogen-associated molecular patterns (PAMPs). A new *Molecular Insights* discusses the recent work of Chunyang Wang and colleagues in exploring the evolution of hormone signaling in plants.

Unit Six: Animal Structure and Function

Chapter 38 (Introduction to Animal Organization and Physiology) expands discussion of anchoring junctions, tight junctions, and gap junctions, and of the basics of epithelial structure and structure, including new discussion and a figure on pseudostratified columnar epithelium. A new Molecular Insights describes miRNA regulation of epithelial cell differentiation in the lung and its relationship to lung cancer. In the presentation of homeostasis, a new subsection discusses how set points can change due to biorhythms or altered environmental conditions.

Chapter 39 (Information Flow and the Neuron) adds information about the proteomic analysis of chemical synapses, and the discussion of equilibrium potential now includes presentation of the Nernst equation.

Chapter 40 (Nervous Systems) adds discussion of the evolutionary changes in brain regions, brain size, and relative size of brain regions in vertebrate brains, and expanded description of the knee-jerk reflex. A new *Molecular Insights* describes experiments on sex differences in the neural connections of the human brain.

Chapter 41 (Sensory Systems) adds a new subsection on the evolutionary history of olfactory receptor genes revealed by bioinformatics analysis of genomes. A new *Molecular Insights* describes experiments on taste neuron changes associated with the emergence of an adaptive behavior in cockroaches.

Chapter 42 (The Endocrine System) adds discussion of nongenomic action of some steroid hormones. A new *Molecular Insights* describes experiments on the fear-enhancing effects of some oxytocin receptors in mice.

Chapter 43 (Muscles, Bones, and Body Movements) expands the discussion of synovial joints. A new *Molecular Insights* describes aspects of the genetics of bone formation learned from analysis of the elephant shark genome.

Chapter 44 (The Circulatory System) adds a figure on the control of red blood cell production.

Chapter 45 (Defenses against Disease) adds a new flow-chart figure of antibody-mediated and cell-mediated immune responses, and adds discussion of microbiome composition in preventing pathogen attack. A new *Molecular Insights* describes genomics experiments revealing the unique immune system of the Atlantic cod.

In Chapter 46 (Gas Exchange: The Respiratory System), a new *Molecular Insights* describes genomics-based experiments on the evolution of altitude adaptation in Tibetans.

Chapter 47 (Animal Nutrition) adds discussion of gut microbiomes and their roles in digestion and nutrition. A new *Molecular Insights* describes experiments on the association of intestinal bacterial populations with obesity in humans.

Chapter 48 (Regulating the Internal Environment) expands discussion of temperature regulation in endotherms. A new *Molecular Insights* describes experiments on the involvement of miRNAs with the development of polycystic kidney disease.

In Chapter 49 (Animal Reproduction), discussion of hormonal regulation of male reproductive functions is enhanced and includes a new replacement figure.

In Chapter 50 (Animal Development), a new *Molecular Insights* describes experiments showing an essential role of protein O-mannosylation in embryonic development.

Unit Seven: Ecology and Behavior

In Chapter 51 (Ecology and the Biosphere), the chapter opener is now more focused on natural climate cycles and less on disastrous weather events. The chapter also includes expanded discussions of positive feedback loops between climate warming and melting of permafrost in tundra, as well as more detailed descriptions of deep-sea environments. A new *Molecular Insights* describes the genetic basis of adaptation to extreme cold in polar octopuses.

In Chapter 52 (Population Ecology), we have updated Figures 52.22 and 52.23 on human population growth. We have subdivided the *Focus on Research* box into sections to clarify the take-home message about the effects of predation on guppy life histories. The chapter also includes a new *Molecular Insights* on the construction of a life table for ant colonies.

Chapter 53 (Population Interactions and Community Ecology) includes new examples of Batesian mimicry (between birds and insects) and fundamental and realized niches (cane toads) with dramatic new figures. In addition, the figure about primary succession (53.28) has been converted into the *Closer Look* format.

Chapter 54 (Ecosystems and Global Change) has undergone substantial revision. The distinction between detrital and grazing food webs has been eliminated, and the discussion of biological magnification (previously a *Focus on Research* box) has been tightened and incorporated into the text. Ecological pyramids are now represented in a single figure (54.8) instead of three; and new figures on the effects of temperature and precipitation on primary productivity (54.4), seasonal changes in primary productivity (54.5), the greenhouse effect (54.15), rising carbon dioxide levels and global temperature (54.16), and anthropogenic nitrogen fixation (54.17) have been added. In addition, Section 54.4 (now titled Human Activities and Anthropogenic Global Change) includes expanded and updated discussions of disruptions to the carbon and nitrogen cycles and a completely new discussion of the impact of global change on ecosystems, including ocean acidification, declining primary productivity, and dead zones in shallow marine environments. The *Molecular Insights* box has been condensed.

Chapter 55 (Biodiversity and Conservation Biology) includes updated and expanded discussions of overfishing, invasive species (updated Figure 55.10 on hemlock woolly adelgids), dam removals, amphibian declines, and vulture mortality in South Asia. It also features a new example of habitat fragmentation with a new figure (55.6). Section 55.3 (now titled Ecosystem Services That Biodiversity Provides) includes discussions of provisioning, regulating, and support services. Section 55.4 (Which Species and Ecosystems Are Most Threatened by Human Activities?) now includes a discussion of the *IUCN Red List of Threatened Species* and an accompanying new figure (55.13). Section 55.5 includes a new discussion on the cost of preserving biodiversity. The *Molecular Insights* box on DNA barcoding has been updated and includes a new figure, and the *Focus on Research* box about population viability analysis has been tightened and condensed.

Chapter 56 (Animal Behavior) has added emphasis on cost/benefit analyses in the chapter opener. The discussion of honeybee communication has been updated to reflect new research on the waggle dance (with revisions to Figure 56.19). We have also substantially revised the discussion of altruism, inclusive fitness, and kin selection to include Hamilton's inequality as a means of predicting altruistic behavior; and the discussion of haplodiploidy and eusociality has been expanded to include alternative hypotheses for the evolution of this complex behavior.

Welcome to *Biology: The Dynamic Science* 4e

Russell/Hertz/McMillan, *Biology: The Dynamic Science* 4e and MindTap engage students so they learn not only *WHAT* scientists know, but *HOW* they know it, and what they still need to learn.

Customize your students' learning experience with unparalleled content options presented in easily edited folders. ▼

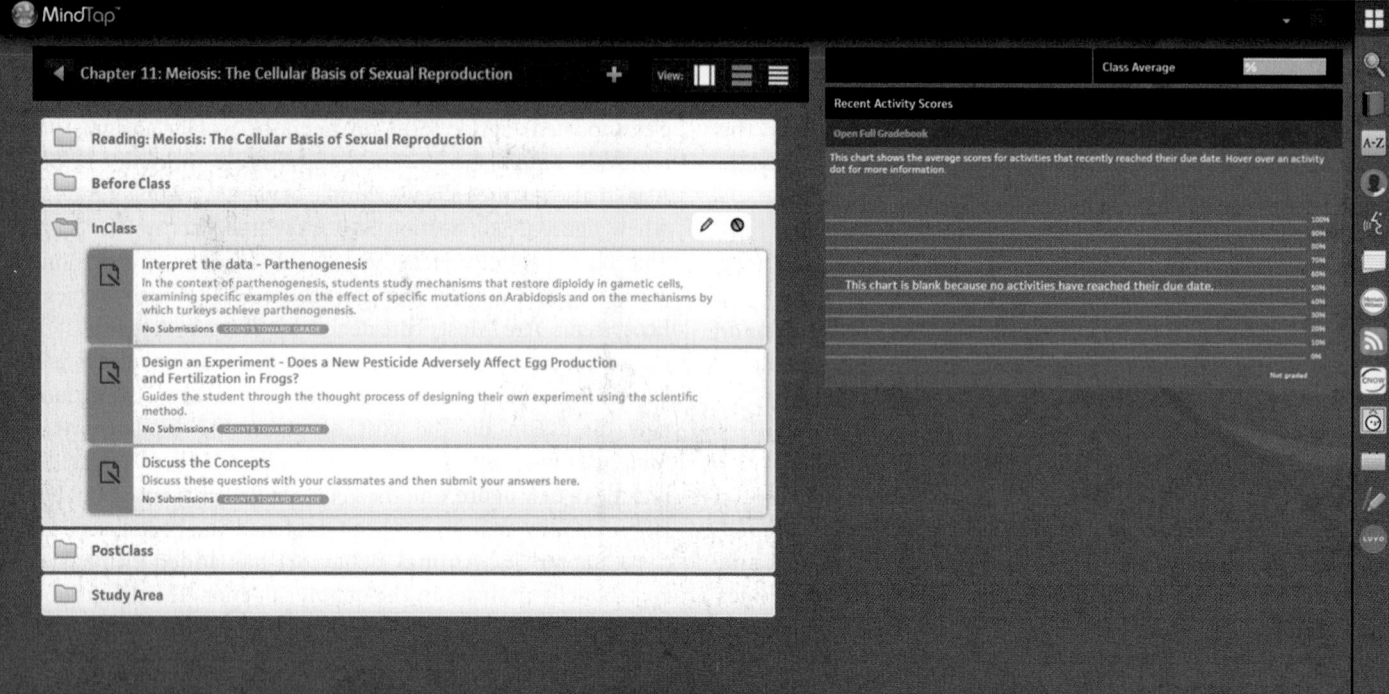

▲ Elevate learning and conceptual understanding with assignable and gradeable exercises that build on material presented in the text. Fully editable content delivery options are organized into categories to match course workflow. The InClass folder is expanded to show the pre-loaded content in this category.

Science as a Process

Immerse students in the process of doing biology, while building skills students need to succeed in more advanced courses.

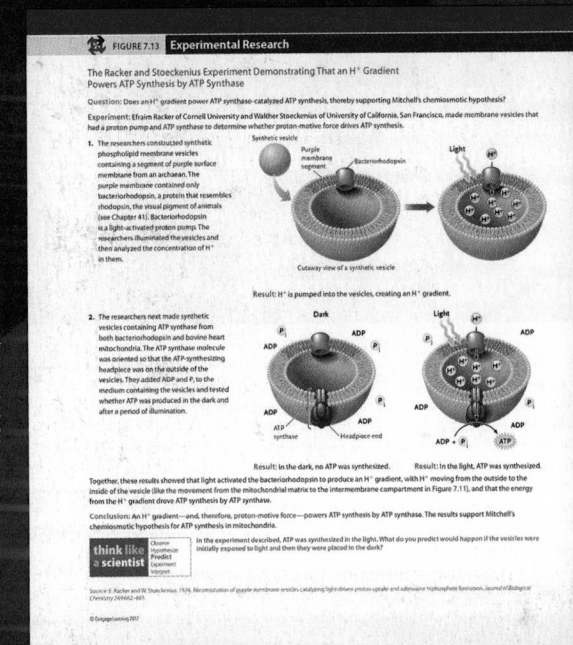

Science as a Process is emphasized throughout the ▶ text. *Research Figures* provide information about how biologists formulate and test specific hypotheses by gathering and interpreting data.

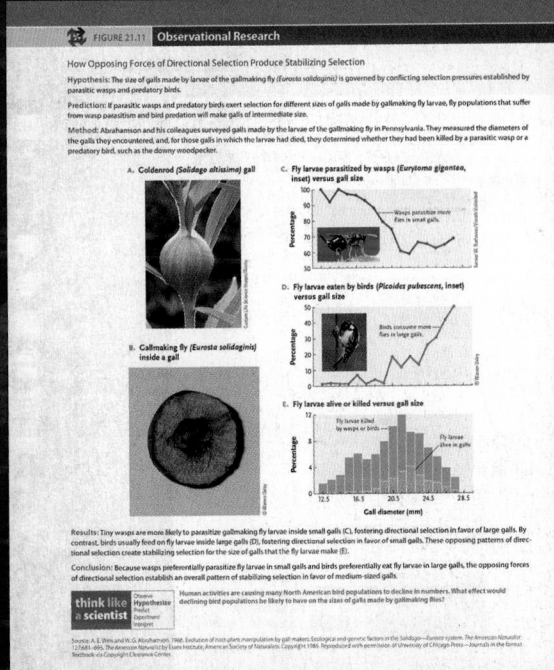

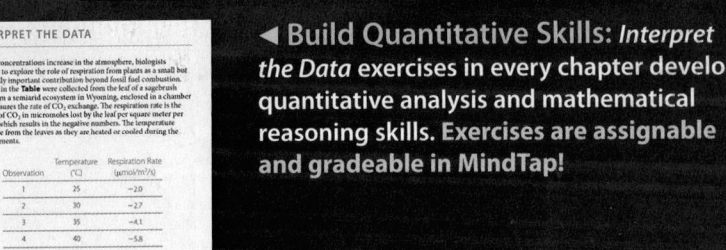

◀ **Build Quantitative Skills:** *Interpret the Data* exercises in every chapter develop quantitative analysis and mathematical reasoning skills. Exercises are assignable and gradeable in MindTap!

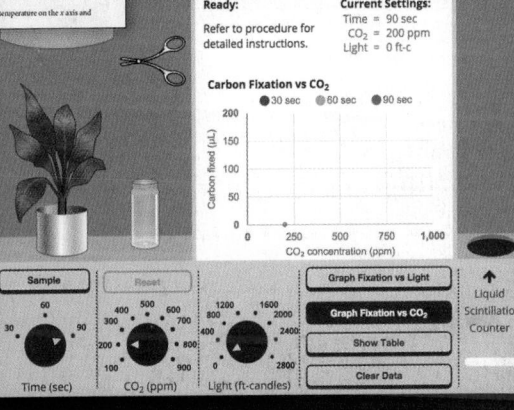

▲ **Apply the Process of Science:** *Think Like a Scientist* questions throughout the text ask students to apply what they have learned beyond the material presented in the book. New icon on selected figures relates the questions to steps in the scientific process. Selected *Think Like a Scientist* exercises are assignable and gradeable in MindTap!

▲ **Virtual Biology Laboratory (VBL)** enables students to "do" science by acquiring data, performing experiments, and using that data to explain biological concepts or phenomena.

Engagement Taken to the Next Level

Comprehensive content and flexible course delivery options...because you know what is best for your students.

Engage with your students before class. We provide a collection of learning exercises and activities including high quality videos for you to choose from.

Or insert your own content such as PowerPoints, videos, and animations! ▼

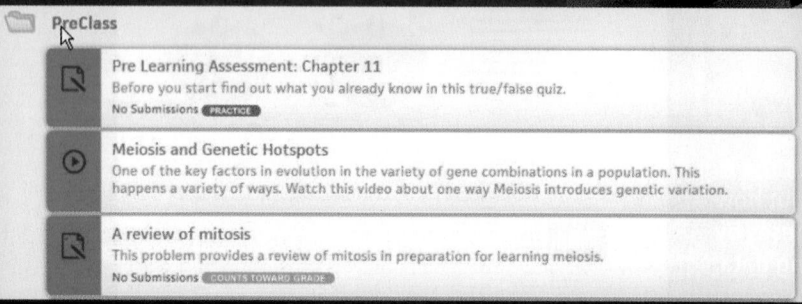

Straightforward explanations and carefully developed illustrations are followed by extensive opportunity

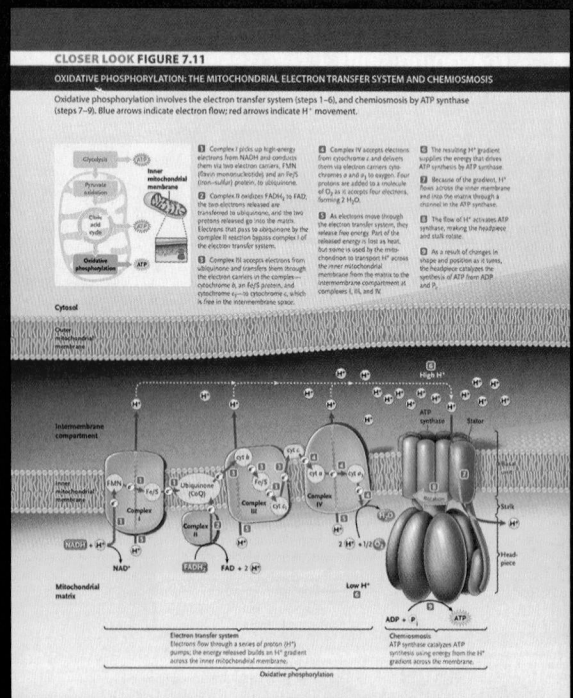

▲ **Spectacular illustrations** such as this *Closer Look* figure help students visualize complex processes. Numbered step-by-step explanations lead students through all the major concepts.

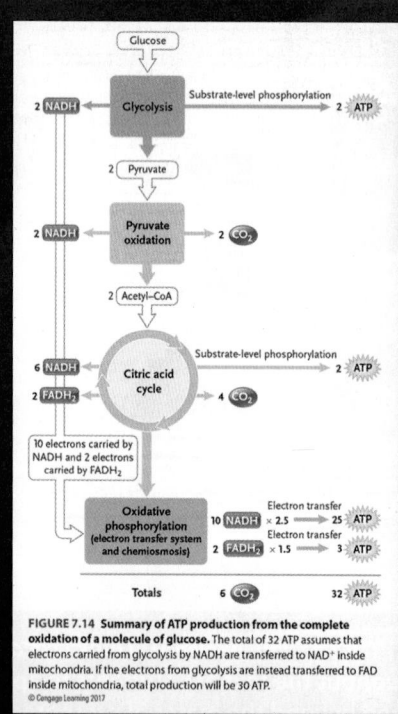

FIGURE 7.14 **Summary of ATP production from the complete oxidation of a molecule of glucose.** The total of 32 ATP assumes that electrons carried from glycolysis by NADH are transferred to NAD⁺ inside mitochondria. If the electrons from glycolysis are instead transferred to FAD inside mitochondria, total production will be 30 ATP.
© Cengage Learning 2017

▲ **Summary figures** help students see the big picture and understand important connections.

Full student engagement. On these two pages, take a tour of how you can design a dynamic learning path for students to help them learn important concepts. Our case example is from Chapter 7: Cellular Respiration: Harvesting Chemical Energy.

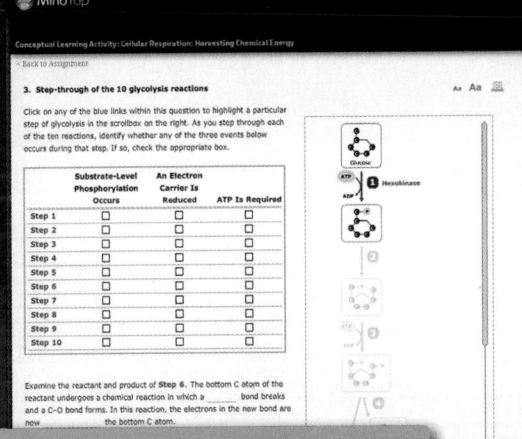

Engage with your students during class. We provide a collection of dynamic activities. Or (of course!) insert your own content. Assess student understanding on a topic during class with interactive figures and immediate feedback with explanations. ▶

"The explanations when you receive the wrong answer on a homework question were really good and helped me understand many concepts that I didn't originally understand."

—**Timothy, California State University of Northridge**

for reinforcement and practice. Sample content for learning about Cellular Respiration is shown below.

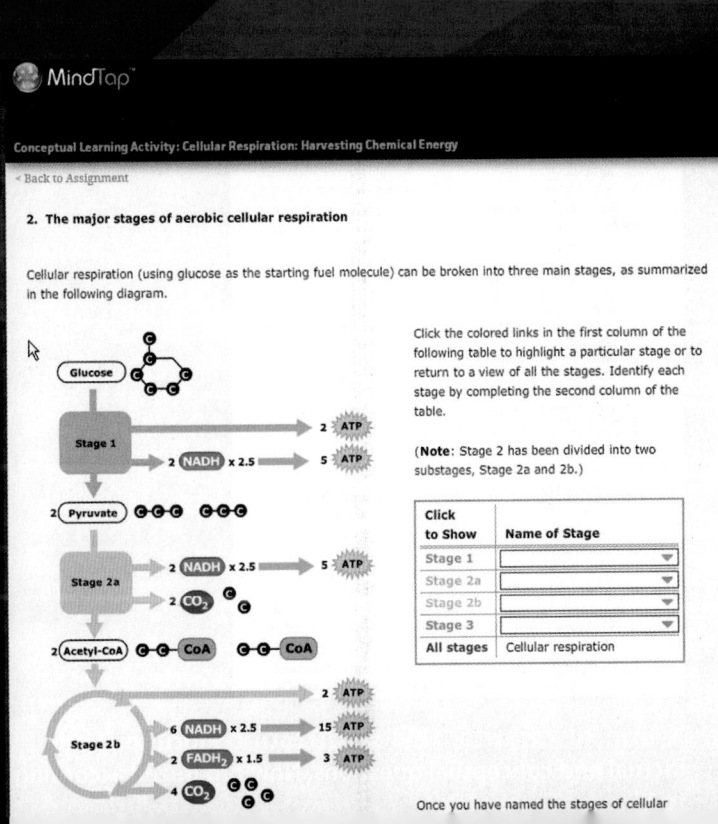

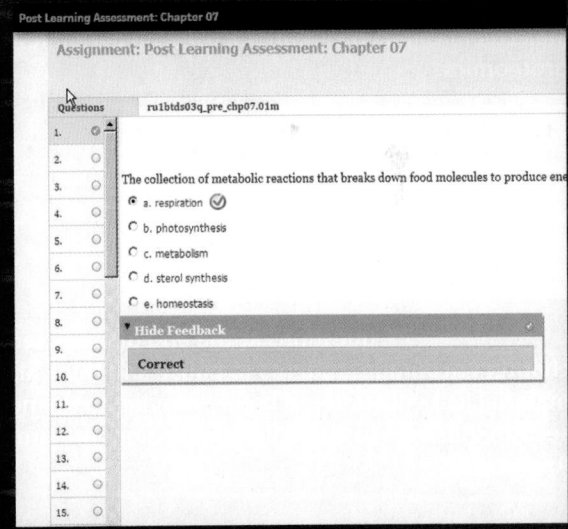

▲ **Post Learning Assessments** offer students opportunities for more practice before a full chapter test. **Practice Tests are assignable and gradeable in MindTap!**

◀ **Conceptual exercises** in MindTap, repeatable in alternate versions, help students learn the material. **Exercises are assignable and gradeable in MindTap!**

Active Learning

Features that engage your students in the process of learning because an engaged student is a successful student.

Molecular Insights

Genomics of Breast Cancer: How does binding of estrogen receptor β to the genome affect breast cancer progression?

The **Figure** shows an X-ray mammogram of a breast with cancer. Breast cancer is the leading cause of death among women in the United States and globally. About 50–70% of breast cancers are hormone-responsive, meaning that they need the female steroid hormone estrogen for the involved cells to grow and proliferate. Estrogens play a crucial role in regulating cell growth and differentiation in the mammary gland. In both normal and mammary gland cancer cells, estrogens act by binding to two specific steroid hormone receptors, estrogen receptors α and β (ERα and ERβ). Once activated by the binding of estrogen, these receptors are transcription factors, regulatory molecules that bind to the control regions of particular genes in the genome to alter those genes' expression, by turning them on or off, or increasing or decreasing their expression. (Expression of genes in this context means transcription, the copying of a gene's DNA sequence into a messenger RNA [mRNA] sequence; see Chapter 1 for an overview, and see Chapter 15 for detailed discussion.) Subsequent translation of the mRNA generates the protein encoded by the gene (see

Chapters 1 and 15). Both ERs have similar DNA-binding domains (see Figure 9.15), and both can bind to the conserved DNA control sequence, 5′-GGTCAnnnTGACC-3′, where "n" is any DNA nucleotide.

The two ERs may be present in the same cell. They have specific, often opposite functions, with ERα typically stimulating cell proliferation, and ERβ being inhibitory to cell proliferation. In breast cancer, ERβ often is lost, which removes its inhibitory effects, thereby allowing progression of the cancer via the effects of estrogen-bound ERα on gene expression. An important goal, therefore, is to understand where activated ERβ binds in the genome so as to determine the roles of this receptor in estrogen signaling and breast cancer.

Research Question

What are the genomic targets of activated ERβ in hormone-responsive breast cancer cells?

Experiments

Collaborating researchers in Italy, Germany, and the United States used genomics approaches to answer the question. They started with a human breast cancer cell line that is hormone-responsive and expresses ERα but not ERβ (ERα+ ERβ−). They genetically engineered those cells engineered to produce a cell line that expresses both ERα and ERβ (ERα+ ERβ+).

the ER–DNA complex precipitates it. After removing the proteins, the sequences of the DNA they were bound to the ER is determined using next-generation sequencing, a technique in which many short segments of DNA are sequenced in parallel. (Next-generation sequencing is discussed in Chapter 19.) The researchers also analyzed gene expression (transcription) on a genomics scale (discussed in Chapter 19) to determine the similarities and differences in the two cell lines.

Results

The effect of estrogen on growth of hormone-responsive ERα+ ERβ+ and ERα+ ERβ− cells was compared. The results showed that the expression of ERβ in the ERα+ ERβ+ cells brought about a large reduction in cell proliferation compared to ERα+ ERβ−, which were not expressing ERβ.

ChIP-Seq analysis of estrogen-stimulated cells identified 5,196 binding sites in the genome for ERβ, 1,516 binding sites for ERα, and 4,506 binding sites to which both ERβ and ERα bind. Summing, there are 9,702 genomic binding sites ERβ, and 6,024 genomic binding sites for ERα.

The researchers compared estrogen-stimulated expression from genes to which ERβ can bind in ERα+ ERβ+ vs. ERα+ ERβ− cells. They identified 921 genes whose expression was differentially regulated in the two cell types, meaning that their expression was

▲ **Molecular Insights** boxes in each chapter describe how molecular tools allow scientists to answer questions that they could not have posed even 30 years ago. Most *Molecular Insights* focus on a single study and include sufficient detail for its content to stand alone. 25 new *Molecular Insights* boxes are focused on genomics or proteomics.

STUDY BREAK 8.2

1. What is the difference in function between the chlorophyll *a* molecules in the antenna complexes and the chlorophyll *a* molecules in the reaction centers of the photosystems?
2. How is NADPH made in the linear electron flow pathway?
3. What is the difference between the linear electron flow pathway and the cyclic electron flow pathway?

▲ **Study Break** questions at the end of every section in a chapter encourage students to pause and think about the material just encountered before moving to the next section. Answers to Study Break questions are provided in Appendix A.

THINK OUTSIDE THE BOOK

A number of human genetic diseases result from mutations that affect mitochondrial function. Collaboratively or individually, find an example of such a disease and research how the genetic mutation disrupts mitochondrial function and leads to the disease symptoms.

▲ **Think Outside the Book activities** help students think analytically and critically as they explore the biological

Unanswered Questions explore important unresolved issues identified by experts in the field and describe cutting-edge research that will advance knowledge in the future. *Think Like a Scientist* questions encourage students to think critically about the research projects described. ▼

Unanswered Questions

What is the role of gene duplication in the evolution of plant diversity?

What is the genetic basis of the origin of new and complex structures, such as the flower, during the course of evolution? One important factor is gene duplication, thought to be one of the driving forces behind the increase in organismal complexity that we see with evolution. When a plant gene is duplicated, initially there are two identical copies with identical functions (called "redundancy"); often over time one of the copies either is eliminated or becomes nonfunctional, so that the original condition of one gene is restored. But sometimes, through the process of mutation and sequence divergence, the two copies take on different functions. They may divide the functions of the original single gene between them ("subfunctionalization"), or one of the copies may take on entirely new functions, leaving the other copy to perform the original function ("neofunctionalization"). It is this last possibility—the origin of new gene functions after duplication—that is thought to provide the raw material for the origin of new plant structures such as the flower.

Like all gymnosperms, the ancestors of angiosperms produced separate cones with male and female reproductive structures, whereas flowers produce both male and female reproductive organs surrounded by a novel structure unique to flowers, the sterile *perianth* (sepals and petals). What genetic changes occurred in proto-angiosperms that allowed for the development of the complex bisexual flower and the perianth? Of course we cannot look at the genomes of the extinct angiosperm ancestors. However, by comparing the genomes of extant gymnosperms and angiosperms we can identify key flower development genes that are found only in angiosperms. In *Arabidopsis thaliana*, *APETALA1 (AP1)* and *SEPALLATA (SEP)* genes are required for proper flower formation. If *AP1* genes are eliminated, the plants will not form flowers. If *SEP* genes are eliminated, the plants form structures similar to flowers but with all the floral organs replaced by tiny leaves. *AP1* and *SEP* genes are found only in flowering plants; combined with the observation that these genes are required to form flowers, this suggests that they may have played a role in the evolution of the flower.

AP1 and *SEP* genes appear to have arisen by way of two duplications from a third gene group *(AGL6)* that is found in both gymnosperms and angiosperms. Along with the B- and C-function genes of the ABC model, these genes belong to the MADS-box family of transcription factors, members of which play key roles throughout plant development. Repeated duplication during the course of plant evolution has led to an increase in the number of these regulatory genes from 1 in algae to over 100 in *Arabidopsis;* the proliferation of these key developmental regulators may be one of the driving forces behind the increase in complexity from algae to angiosperms, and is likely to have played a role in the origin of the flower. Thus, duplications in these genes led to the origin in gymnosperms of the B- and C-function genes, which in gymnosperms appear to play a role in specifying reproductive organ identity that is carried into the angiosperms. In contrast, later duplications, coinciding with the origin of the angiosperms, led to the establishment of the *AP1* and *SEP* gene lineages. These angiosperm-specific genes are required for flower formation, and may have been critical, in particular, in the origin of the flower-specific perianth.

think like a scientist If the A function of the ABC model is not found in any species outside the mustard family, explain how the "BCE" model can still account for the formation of four different types of floral organs.

Courtesy of Amy Litt

Amy Litt is on the faculty of the Department of Botany and Plant Sciences at the University of California, Riverside. Her main interests lie in the evolution of plant form and how changes in gene function during the course of plant evolution have produced novel plant forms and functions—particularly new flower and fruit morphologies.

Apply/Analyze

9. Which of the following statements is *false*? Imagine that you ingested three chocolate bars just before sitting down to study this chapter. Most likely:
   a. your brain cells are using ATP.
   b. there is no deficit of the initial substrate to begin glycolysis.
   c. the respiratory processes in your brain cells are moving atoms from glycolysis through the citric acid cycle to the electron transfer system.
   d. after a couple of hours, you change position and stretch to rest certain muscle cells, which removes lactate from these muscles.
   e. after 2 hours, your brain cells are oxygen-deficient.
10. In the 1950s, a diet pill that had the effect of "poisoning" ATP synthase was tried. The person taking it could not use glucose and "lost weight"—and ultimately his or her life. Today, we know that the immediate effect of poisoning ATP synthase is:
    a. ATP would not be made in the electron transfer system.
    b. H^+ movement across the inner mitochondrial membrane would increase.
    c. more than 32 ATP could be produced from a molecule of glucose.
    d. ADP would be united with phosphate more readily in the mitochondria.
    e. ATP would react with oxygen.
11. **Discuss Concepts** Why do you think nucleic acids are not oxidized extensively as a cellular energy source?

Evaluate/Create

12. **Discuss Concepts** A hospital patient was regularly found to be intoxicated. He denied that he was drinking alcoholic beverages. The doctors and nurses made a special point to eliminate the possibility that the patient or his friends were smuggling alcohol into his room, but he was still regularly intoxicated. Then, one of the doctors had an idea that turned out to be correct and cured the patient of his intoxication. The idea involved the patient's digestive system and one of the oxidative reactions covered in this chapter. What was the doctor's idea?
13. **Design an Experiment** There are several ways to measure cellular respiration experimentally. For example, CO_2 and O_2

▲ **End-of-Chapter review questions** focus on both factual and conceptual questions. Now organized according

MindTap Course Development: Simple & Powerful

Auto-graded learning activities are easily assignable by instructors and offer students immediate feedback and coaching. Request a demo today: www.cengage.com/learningsolutions.

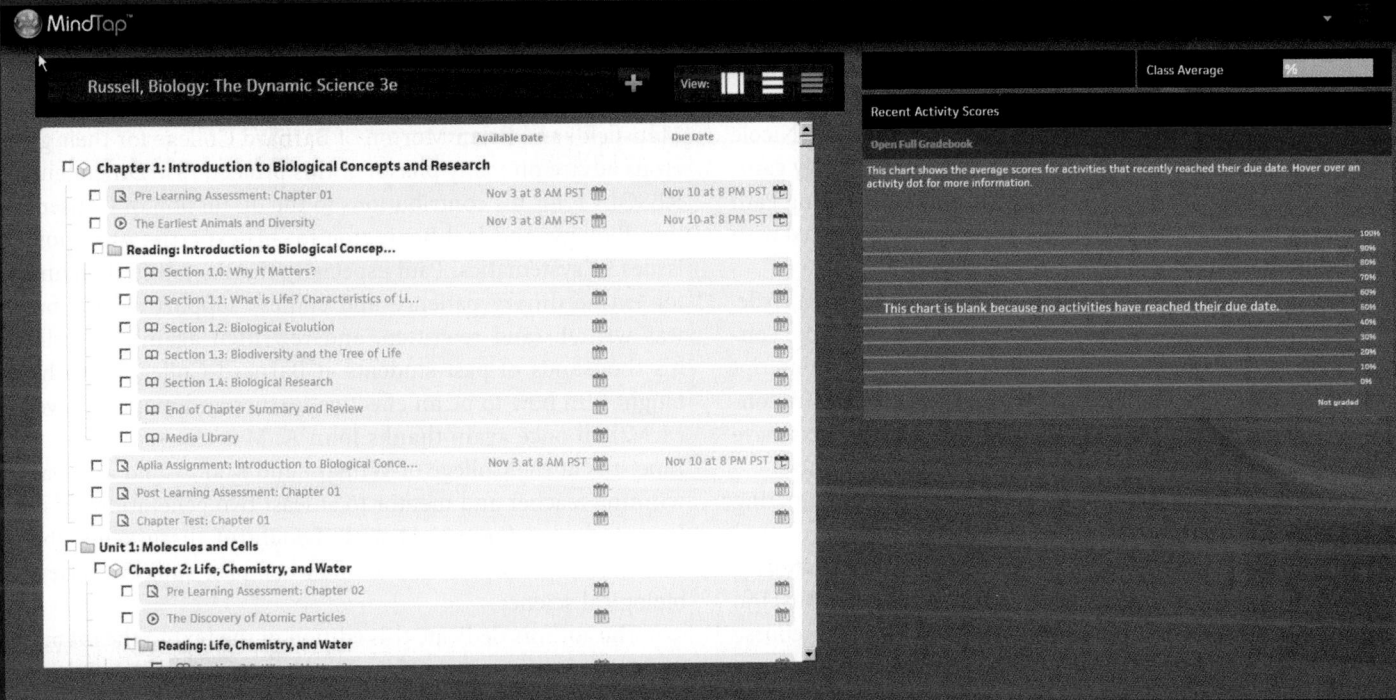

▲ Easily assign content and assignments with due dates.

▲ Comprehensive course analytics to assess student learning and understanding in real time.

Effectively introducing digital solutions into your classroom — online or on-ground — is now easier than ever. We're with you every step of the way.

◄ Our Digital Course Support Team
When you adopt from Cengage Learning, you have a dedicated team of Digital Course Support professionals, who will provide hands-on, start-to-finish support, making digital course delivery a success for you and your students.

"The technical help provided by Cengage staff in using MindTap was superb. The Digital Support Team was proactive in training and prompt in answering follow-up questions for their excellent online learning resources."

—K. Sata Sathasivan, The University of Texas at Austin

Acknowledgments

Revising a text—both the printed version and its wholly integrated digital applications—from one edition to the next is a complex, though exciting and rewarding, project. The helpful assistance of many people enabled us to accomplish the task in a timely manner.

Kristen Sutton, Life Sciences Product Manager, provided the essential vision, leadership, and support needed to bring our unified course solution to fruition. Kristen and Nicole Hurst, Product Development Specialist, worked tirelessly gaining instructor input to ensure that the MindTap Learning Path reflects the current state of biological knowledge conveyed with cutting edge and effective pedagogical approaches.

Our Content Developers, Suzannah Alexander, Elesha Hyde, and Jake Warde, served as pilots for the generation of core content. They provided very helpful guidance as the manuscript matured. They compiled, interpreted, and sometimes deconstructed reviewer comments; their analyses and insights have helped us tighten the narrative and maintain a steady course.

We are grateful to Kellie Petruzelli for coordinating the supplements and our Product Assistants Victor Luu and Kristina Cannon for managing all our reviewer information.

We offer many thanks to John Anderson, Lauren Oliveira, and Colleen O'Rourke, who spearheaded the MindTap course development. Their collective efforts allowed us to create a set of tools that support students in learning and instructors in teaching.

We appreciate the help of the production staff led by Content Project Manager Hal Humphrey and Art Director Andrei Pasternak at Cengage and Dan Fitzgerald and Megan Knight at Graphic World.

We also wish to acknowledge Tom Ziolkowski, our Marketing Manager, whose expertise ensured that all of you would know about this new book.

Our outstanding art program is the result of the collaborative talent, hard work, and dedication of a select group of people. The meticulous styling and planning of the program are credited to Steve McEntee and to Dragonfly Media Group (DMG), led by Mike Demaray. The DMG group created hundreds of complex, vibrant art pieces. Steve's role was crucial in overseeing the development and consistency of the art program; he was the original designer for the *Experimental Research, Observational Research,* and *Research Methods* figures.

We also want to extend our sincere gratitude to the following dedicated faculty advisers who provided expert guidance throughout the MindTap Learning Path development process:

Deborah Dardis, *Southeastern Louisiana University*
Mark Mort, *University of Kansas*
Marcia Shofner, *University of Maryland*
Justin Ronald St. Juliana, *Ivy Tech Community College*

Peter Russell thanks Joel Benington of St. Bonaventure University for his extensive reviewing and contributions to Chapter 19, Genomes and Proteomes, and Chapter 25, The Origin of Life; and Sharon Thoma of the University of Wisconsin–Madison for her expert input and advice during the revision of the Unit Six chapters on Animal Structure and Function. Paul Hertz thanks Hilary Callahan, John Glendinning, Jennifer Mansfield, and Brian Morton of Barnard College for their generous advice on many phases of this project; Eric Dinerstein of RESOLVE for his contributions to the discussion of Conservation Biology; and Joel Benington for his expertise on genomic issues in systematics. Paul especially thanks Jamie Rauchman for extraordinary patience and endless support as this book was written (and rewritten, and rewritten again), as well as his thousands of past students at Barnard College, who have taught him how to be an effective teacher and mentor. Beverly McMillan once again thanks John A. Musick, Acuff Professor Emeritus at the College of William and Mary—and an award-winning teacher and mentor to at least two generations of college students—for patient and thoughtful discussions about effective ways to present the often complex subject matter of biological science.

The authors dedicate this edition of the text to the memory of Mary Arbogast, our first Developmental Editor, without whose superhuman efforts the first edition would have never seen the light of day.

We would also like to thank our advisors and contributors:

Supplements Authors

David Asch, *Youngstown State University*
Carolyn Bunde, *Idaho State University*
Shelli Carter, *University of Alabama*
Karie Cherwin, *Chimborazo Publishing*
Albia Duggar, *Miami Dade College*
Frederick B. Essig, *University of South Florida*
Brent Ewers, *University of Wyoming*
Anne Galbraith, *University of Wisconsin–La Crosse*
Alan Hecht, *Hofstra University*
Kathleen Hecht, *Nassau Community College*
Qinzi Ji, *Instructional Curriculum Specialist*
Sarah Kohler, *Chimborazo Publishing*
William Kroll, *Loyola University Chicago–Lake Shore*
Shari Laprise, *Chimborazo Publishing*
Julianna Lemieux, *Mercy College*
Todd Osmundson, *University of California–Berkeley*
Mark Paternostro, *West Virginia University*
Debra Pires, *University of California–Los Angeles*
Elena Pravosudova, *University of Reno–Nevada*
Jeff Roth-Vinson, *Cottage Grove High School*

Mark Sheridan, *North Dakota State University*
Gary Shin, *California State University–Long Beach*
Michael Silva, *El Paso Community College*
Michelle Taliaferro, *Auburn University–Montgomery*
Jeffrey Taylor, *State University of New York–Canton*
Catherine Anne Ueckert, *Northern Arizona University*
Jyoti Wagle, *Houston Community College, Central College*
Alexander Wait, *Missouri State University*

MindTap and Media Reviewers

Thomas Abbott, *University of Connecticut*
David Asch, *Youngstown State University*
John Bell, *Brigham Young University*
Anne Bergey, *Truman State University*
Gerald Bergtrom, *University of Wisconsin–Milwaukee*
Scott Bowling, *Auburn University*
Joi Braxton-Sanders, *Northwest Vista College*
Carolyn Bunde, *Idaho State University*
Jung H. Choi, *Georgia Institute of Technology*
Tim W. Christensen, *East Carolina University*
Patricia J. S. Colberg, *University of Wyoming*
Robin Cooper, *University of Kentucky*
Karen Curto, *University of Pittsburgh*
Joe Demasi, *Massachusetts College of Pharmacy and Health Science*
Nicholas Downey, *University of Wisconsin–LaCrosse*
Albia Dugger, *Miami-Dade College*
Natalie Dussourd, *Illinois State University*
Lisa Elfring, *University of Arizona*
Bert Ely, *University of South Carolina*
Kathleen Engelmann, *University of Bridgeport*
Helene Engler, *Science Writer*
Monika Espinasa, *State University of New York at Ulster*
Michael Ferrari, *University of Missouri–Kansas City*
David Fitch, *New York University*
Paul Fitzgerald, *Northern Virginia Community College*
Steven Francoeur, *Eastern Michigan University*
Daria Hekmat-Scafe, *Stanford University*
Jutta Heller, *Loyola University Chicago–Lake Shore*
Ed Himelblau, *California Polytechnic State University–San Luis Obispo*
Justin Hoffman, *McNeese State University*
Kelly Howe, *University of New Mexico*
Carrie Hughes, *San Jacinto College (Central Campus)*
Ashok Jain, *Albany State University*

Susan Jorstad, *University of Arizona*
Judy Kaufman, *Monroe Community College*
David Kiewlich, *Research Biologist*
Christopher Kirkhoff, *McNeese State University*
Richard Knapp, *University of Houston*
William Kroll, *Loyola University Chicago–Lake Shore*
Nathan Lents, *John Jay College*
Janet Loxterman, *Idaho State University*
Susan McRae, *East Carolina University*
Brad Mehrtens, *University of Illinois at Urbana–Champaign*
Jennifer Metzler, *Ball State University*
Bruce Mobarry, *University of Idaho*
Jennifer Moon, *The University of Texas at Austin*
Robert Osuna, *State University of New York at Albany*
Matt Palmer, *Columbia University*
Roger Persell, *Hunter College*
Michael Reagan, *College of Saint Benedict and Saint John's University*
Ann Rushing, *Baylor University*
Jeanne Serb, *Iowa State University*
Leah Sheridan, *University of Northern Colorado*
Mark Sheridan, *North Dakota State University*
Nancy N. Shontz, *Grand Valley State University*
Michael Silva, *El Paso Community College*
Julia Snyder, *Syracuse University*
Linda Stabler, *University of Central Oklahoma*
Mark Staves, *Grand Valley State University*
Eric Strauss, *University of Wisconsin–LaCrosse*
Mark Sturtevant, *Oakland University*
Mark Sugalski, *Southern Polytechnic State University*
David Tam, *University of North Texas*
Salvatore Tavormina, *Austin Community College*
Rebecca Thomas, *Montgomery College*
David H. Townson, *University of New Hampshire*
David Vleck, *Iowa State University*
Neal Voelz, *St. Cloud State University*
Camille Wagner, *San Jacinto College (Central Campus)*
Miryam Wahrman, *William Paterson University*
Alexander Wait, *Missouri State University*
Suzanne Wakim, *Butte Community College*
Johanna Weiss, *Northern Virginia Community College*
Lisa Williams, *Northern Virginia Community College*
Marilyn Yoder, *University of Missouri–Kansas City*
Martin Zahn, *Thomas Nelson Community College*

Reviewers and Contributors

Thomas D. Abbott, *University of Connecticut*
Sheena Abernathy, *College of the Mainland*
Lori Adams, *University of Iowa*
Heather Addy, *The University of Calgary*
Adrienne Alaie-Petrillo, *Hunter College-CUNY*
Richard Allison, *Michigan State University*
Terry Allison, *The University of Texas–Pan American*

Phil Allman, *Gulf Coast University*
Tracey M. Anderson, *University of Minnesota Morris*
Deborah Anderson, *Saint Norbert College*
Robert C. Anderson, *Idaho State University*
Andrew Andres, *University of Nevada, Las Vegas*

Steven M. Aquilani, *Delaware County Community College*
Stephen Arch, *Reed College*
Jonathan W. Armbruster, *Auburn University*
Peter Armstrong, *University of California, Davis*
John N. Aronson, *The University of Arizona*
Joe Arruda, *Pittsburgh State University*

Karl Aufderheide, *Texas A&M University*

Charles Baer, *University of Florida*

Lee Baines, *University of Wisconsin, La Crosse*

Gary I. Baird, *Brigham Young University*

Aimee Bakken, *University of Washington*

Marica Bakovic, *University of Guelph*

Mitchell F. Balish, *Miami University*

Michael Baranski, *Catawba College*

W. Brad Barbazuk, *University of Florida*

Michael Barbour, *University of California, Davis*

Gerry Barclay, *Highline Community College*

Timothy J. Baroni, *State University of New York at Cortland*

Edward M. Barrows, *Georgetown University*

Anton Baudoin, *Virginia Polytechnic Institute and State University*

Penelope H. Bauer, *Colorado State University*

Erwin A. Bautista, *University of California, Davis*

Kevin Beach, *The University of Tampa*

Mike Beach, *Southern Polytechnic State University*

Ruth Beattie, *University of Kentucky*

Robert Beckmann, *North Carolina State University*

Jane Beiswenger, *University of Wyoming*

Asim Bej, *University of Alabama at Birmingham*

Michael C. Bell, *Richland College*

Andrew Bendall, *University of Guelph*

Joel H. Benington, *St. Bonaventure University*

Anne Bergey, *Truman State University*

William L. Bischoff, *The University of Toledo*

Catherine Black, *Idaho State University*

Andrew Blaustein, *Oregon State University*

Jeffrey Blaustein, *University of Massachusetts Amherst*

Anthony H. Bledsoe, *University of Pittsburgh*

Harriette Howard-Lee Block, *Prairie View A&M University*

Dennis Bogyo, *Valdosta State University*

David Bohr, *University of Michigan*

Emily Boone, *University of Richmond*

Hessel Bouma III, *Calvin College*

Nancy Boury, *Iowa State University*

Scott Bowling, *Auburn University*

Robert S. Boyd, *Auburn University*

Laurie Bradley, *Hudson Valley Community College*

William Bradshaw, *Brigham Young University*

J. D. Brammer, *North Dakota State University*

Heather Brasher, *College of the Mainland*

G. L. Brengelmann, *University of Washington*

Randy Brewton, *University of Tennessee-Knoxville*

Bob Brick, *Blinn College-Bryan*

Mirjana Brockett, *Georgia Institute of Technology*

William Bromer, *University of Saint Francis*

William Randy Brooks, *Florida Atlantic University-Boca Raton*

Mark Browning, *Purdue University*

Gary Brusca, *Humboldt State University*

Alan H. Brush, *University of Connecticut*

Arthur L. Buikema, Jr., *Virginia Polytechnic Institute and State University*

Carolyn Bunde, *Idaho State University*

E. Robert Burns, *University of Arkansas for Medical Sciences*

Ruth Buskirk, *The University of Texas at Austin*

David Byres, *Florida Community College at Jacksonville*

Christopher S. Campbell, *The University of Maine*

Angelo Capparella, *Illinois State University*

Marcella D. Carabelli, *Broward Community College-North*

Jeffrey Carmichael, *University of North Dakota*

Bruce Carroll, *North Harris Montgomery Community College*

Robert Carroll, *East Carolina University*

Patrick Carter, *Washington State University*

Christine Case, *Skyline College*

Domenic Castignetti, *Loyola University Chicago-Lake Shore*

Peter Chen, *College of DuPage*

Jung H. Choi, *Georgia Institute of Technology*

Kent Christensen, *University of Michigan Medical School*

James W. Clack, *Indiana University –Purdue University Indianapolis*

John Cogan, *Ohio State University*

Patricia J. S. Colberg, *University of Wyoming*

Linda T. Collins, *University of Tennessee-Chattanooga*

Lewis Coons, *University of Memphis*

Robin Cooper, *University of Kentucky*

Joe Cowles, *Virginia Polytechnic Institute and State University*

George W. Cox, *San Diego State University*

David Crews, *The University of Texas at Austin*

Paul V. Cupp, Jr., *Eastern Kentucky University*

Karen Curto, *University of Pittsburgh*

Anne M. Cusic, *The University of Alabama at Birmingham*

David Dalton, *Reed College*

Frank Damiani, *Monmouth University*

Melody Danley, *University of Kentucky*

Deborah Athas Dardis, *Southeastern Louisiana University*

Rebekka Darner, *University of Florida*

Shawn Dash, *University of Texas at El Paso*

Peter J. Davies, *Cornell University*

Jonathan Davis, *Florida State University*

Fred Delcomyn, *University of Illinois at Urbana-Champaign*

Jerome Dempsey, *University of Wisconsin-Madison*

Philias Denette, *Delgado Community College-City Park*

Nancy G. Dengler, *University of Toronto*

Jonathan J. Dennis, *University of Alberta*

Daniel DerVartanian, *University of Georgia*

Donald Deters, *Bowling Green State University*

Kathryn Dickson, *California State University, Fullerton*

Eric Dinerstein, *World Wildlife Fund*

Kevin Dixon, *University of Illinois at Urbana-Champaign*

Nick Downey, *University of Wisconsin-LaCrosse*

Stella Doyungan, *Texas A & M University, Corpus Christi*

Gordon Patrick Duffie, *Loyola University Chicago-Lake Shore*

Charles Duggins, *University of South Carolina*

Carolyn S. Dunn, *University of North Carolina-Wilmington*

Kathryn A. Durham, *Luzerne County Community College*

Roland R. Dute, *Auburn University*

Melinda Dwinell, *Medical College of Wisconsin*

Gerald Eck, *University of Washington*

Gordon Edlin, *University of Hawaii*

William Eickmeier, *Vanderbilt University*

Jamin Eisenbach, *Eastern Michigan University*

Ingeborg Eley, *Hudson Valley Community College*

Paul R. Elliott, *Florida State University*

John A. Endler, *University of Exeter*

Kathleen Engelmann, *University of Bridgeport*

Helene Engler, *Science Consultant and Lecturer*

Robert B. Erdman, *Florida Gulf Coast University*

Jose Luis Ergemy, *Northwest Vista College*

Joseph Esdin, *University of California, Los Angeles*

Frederick B. Essig, *University of South Florida*

Brent Ewers, *University of Wyoming*

Daniel J. Fairbanks, *Utah Valley University*

Piotr G. Fajer, *Florida State University*

Richard H. Falk, *University of California, Davis*

Ibrahim Farah, *Jackson State University*

Mark A. Farmer, *University of Georgia*

Jacqueline Fern, *Lane Community College*

Michael B. Ferrari, *University of Missouri-Kansas City*

Victor Fet, *Marshall University*

David H. A. Fitch, *New York University*

Daniel P. Fitzsimons, *University of Wisconsin-Madison*

Daniel Flisser, *Camden County College*

Paul Florence, *Kentucky Community and Technical College*

R. G. Foster, *University of Virginia*

Austin W. Francis Jr., *Armstrong Atlantic University*

Dan Friderici, *Michigan State University*

J. W. Froehlich, *The University of New Mexico*

Anne M. Galbraith, *University of Wisconsin-LaCrosse*

Paul Garcia, *Houston Community College-Southwest*

E. Eileen Gardner, *William Paterson University*

Umadevi Garimella, *University of Central Arkansas*

David W. Garton, *Georgia Institute of Technology*

John R. Geiser, *Western Michigan University*

Robert P. George, *University of Wyoming*

Stephen George, *Amherst College*

Tim Gerber, *University of Wisconsin-LaCrosse*

John Giannini, *St. Olaf College*

Joseph Glass, *Camden County College*

Florence Gleason, *University of Minnesota Twin Cities*

Scott Gleeson, *University of Kentucky*

John Glendinning, *Barnard College*

Elizabeth Godrick, *Boston University*

Andres Gomez, *American Museum of Natural History*

Judith Goodenough, *University of Massachusetts Amherst*

H. Maurice Goodman, *University of Massachusetts Medical School*

Bruce Grant, *College of William and Mary*

Becky Green-Marroquin, *Los Angeles Valley College*

Christopher Gregg, *Louisiana State University*

Katharine B. Gregg, *West Virginia Wesleyan College*

John Griffin, *College of William and Mary*

Erich Grotewold, *Ohio State University*

Samuel Hammer, *Boston University*

Aslam Hassan, *University of Illinois at Urbana-Champaign*

Albert Herrera, *University of Southern California*

Wilford M. Hess, *Brigham Young University*

Martinez J. Hewlett, *The University of Arizona*

R. James Hickey, *Miami University*

Christopher Higgins, *Tarleton State University*

Phyllis C. Hirsch, *East Los Angeles College*

Carl Hoagstrom, *Ohio Northern University*

Stanton F. Hoegerman, *College of William and Mary*

Kelly Hogan, *University of North Carolina*

Ronald W. Hoham, *Colgate University*

Jill A. Holliday, *University of Florida*

Margaret Hollyday, *Bryn Mawr College*

John E. Hoover, *Millersville University*

Howard Hosick, *Washington State University*

Carrie Hughes, *San Jacinto College*

William Irby, *Georgia Southern University*

John Ivy, *Texas A&M University*

Alice Jacklet, *University at Albany, State University of New York*

John D. Jackson, *North Hennepin Community College*

Jennifer Jeffery, *Wharton County Junior College*

Eric Jellen, *Brigham Young University*

Rick Jellen, *Brigham Young University*

John Jenkin, *Blinn College-Bryan*

Dianne Jennings, *Virginia Commonwealth University*

Leonard R. Johnson, *The University of Tennessee College of Medicine*

Walter Judd, *University of Florida*

Prem S. Kahlon, *Tennessee State University*

Thomas C. Kane, *University of Cincinnati*

Peter Kareiva, *University of Washington*

Gordon I. Kaye, *Albany Medical College*

Greg Keller, *Eastern New Mexico University-Roswell*

Stephen Kelso, *University of Illinois at Chicago*

Bryce Kendrick, *University of Waterloo*

Bretton Kent, *University of Maryland*

Jack L. Keyes, *Linfield College Portland Campus*

David Kiewlich, *Science Consultant and Research Biologist*

Scott L. Kight, *Montclair State University*

John Kimball, *Tufts University*

Hillar Klandorf, *West Virginia University*

Barrett Klein, *University of Wisconsin, Lacrosse*

Michael Klymkowsky, *University of Colorado at Boulder*

Loren Knapp, *University of South Carolina*

Richard Knapp, *University of Houston*

David Kooyman, *Brigham Young University*

Olga Ruiz Kopp, *Utah Valley State University*

Ana Koshy, *Houston Community College-Northwest*

Donna Koslowsky, *Michigan State University*

Kari Beth Krieger, *University of Wisconsin-Green Bay*

David T. Krohne, *Wabash College*

William Kroll, *Loyola University Chicago-Lake Shore*

Josepha Kurdziel, *University of Michigan*

Allen Kurta, *Eastern Michigan University*

Howard Kutchai, *University of Virginia*

Paul K. Lago, *The University of Mississippi*

John Lammert, *Gustavus Adolphus College*

William L'Amoreaux, *College of Staten Island-CUNY*

Brian Larkins, *The University of Arizona*

William E. Lassiter, *University of North Carolina-Chapel Hill*

Kary Latham, *Victory University*

Shannon Lee, *California State University, Northridge*

Lissa Leege, *Georgia Southern University*

Matthew Levy, *Case Western Reserve University*

Harvey Liftin, *Broward Community College-Central*

Hsin Lin, *University of Texas at El Paso*

Tom Lonergan, *University of New Orleans*

Lynn Mahaffy, *University of Delaware*

Charly Mallery, *University of Miami*

Alan Mann, *University of Pennsylvania*

Paul Manos, *Duke University*

Kathleen Marrs, *Indiana University-Purdue University Indianapolis*

Robert Martinez, *Quinnipiac University*

Patricia Matthews, *Grand Valley State University*

Joyce B. Maxwell, *California State University, Northridge*

Jeffrey D. May, *Marshall University*

Geri Mayer, *Florida Atlantic University*

Jerry W. McClure, *Miami University*

Andrew G. McCubbin, *Washington State University*

Mark McGinley, *Texas Tech University*

Jacqueline S. McLaughlin, *Penn State University-Lehigh Valley*

F. M. Anne McNabb, *Virginia Polytechnic Institute and State University*

Mitch McVeigh, *Tufts University*

Mark Meade, *Jacksonville State University*

Bradley Mehrtens, *University of Illinois at Urbana-Champaign*

Amee Mehta, *Seminole State University*

Michael Meighan, *University of California, Berkeley*

Catherine Merovich, *West Virginia University*

Richard Merritt, *Houston Community College*

Jennifer Metzler, *Ball State University*

Ralph Meyer, *University of Cincinnati*

Melissa Michael, *University of Illinois at Urbana-Champaign*

James E. "Jim" Mickle,
North Carolina State University

Hector C. Miranda, Jr.,
Texas Southern University

Jasleen Mishra,
Houston Community College–Southwest

Jeanne M. Mitchell, *Truman State University*

David Mohrman,
University of Minnesota Medical School Duluth

John M. Moore, *Taylor University*

Roderick M. Morgan,
Grand Valley State University

Mark Mort, *University of Kansas*

David Morton, *Frostburg State University*

Alexander Motten, *Duke University*

Alan Muchlinski,
California State University–Los Angeles

Michael Muller, *University of Illinois at Chicago*

Richard Murphy, *University of Virginia*

Darrel L. Murray,
University of Illinois at Chicago

Allan Nelson, *Tarleton State University*

David H. Nelson, *University of South Alabama*

Jacalyn Newman, *University of Pittsburgh*

David O. Norris, *The University of Colorado*

Bette Nybakken, *Hartnell College*

Victoria Ochoa, *El Paso Community
College–Rio Grande Campus*

Tom Oeltmann, *Vanderbilt University*

Bruce F. O'Hara,
University of Kentucky

Diana Oliveras,
The University of Colorado at Boulder

Alexander E. Olvido,
Virginia State University

Todd W. Osmundson,
University of California, Berkeley

Robert Osuna,
State University of New York, Albany

John Osterman, *University of Nebraska*

Karen Otto, *The University of Tampa*

William W. Parson,
University of Washington School of Medicine

James F. Payne, *The University of Memphis*

Craig Peebles, *University of Pittsburgh*

Joe Pelliccia, *Bates College*

Kathryn Perez,
University of Wisconsin–LaCrosse

Vinnie Peters, *Indiana University–Purdue
University Fort Wayne*

Susan Petro, *Ramapo College of New Jersey*

Debra Pires, *University of California, Los Angeles*

Jarmila Pittermann,
University of California, Santa Cruz

Thomas Pitzer, *Florida International University*

Roberta Pollock, *Occidental College*

Steve Vincent Pollock,
Louisiana State University

Elena Pravosudova, *University of Nevada, Reno*

Jerry Purcell, *San Antonio College*

Jason M. Rauceo,
John Jay College of Criminal Justice

Kim Raun, *Wharton County Junior College*

Michael Reagan, *College of
Saint Benedict and Saint John's University*

Tara Reed, *University of Wisconsin-Green Bay*

Melissa Murray Reedy,
University of Illinois at Urbana-Champaign

Sean Rice, *Texas Tech University*

Lynn Robbins, *Missouri State University*

Carolyn Roberson,
Roane State Community College

Laurel Roberts, *University of Pittsburgh*

George R. Robinson,
State University of New York, Albany

Kenneth Robinson, *Purdue University*

Frank A. Romano, *Jacksonville State University*

Michael R. Rose, *University of California, Irvine*

Michael S. Rosenzweig, *Virginia Polytechnic
Institute and State University*

Linda S. Ross, *Ohio University*

Ann Rushing, *Baylor University*

Scott D. Russell, *University of Oklahoma*

Christine Russin, *Northwestern University*

Linda Sabatino, *Suffolk Community College*

Tyson Sacco, *Cornell University*

Peter Sakaris,
Southern Polytechnic State University

Frank B. Salisbury, *Utah State University*

Mark F. Sanders, *University of California, Davis*

Stephen G. Saupe, *College of Saint Benedict
and Saint John's University*

Andrew Scala, *Dutchess Community College*

John Schiefelbein, *University of Michigan*

Deemah Schirf,
The University of Texas at San Antonio

Kathryn J. Schneider,
Hudson Valley Community College

Jurgen Schnermann,
University of Michigan Medical School

Thomas W. Schoener,
University of California, Davis

Brian Shea, *Northwestern University*

Mark Sheridan, *North Dakota State University*

Dennis Shevlin, *The College of New Jersey*

Rebecca F. Shipe,
University of California, Los Angeles

Nancy N. Shontz,
Grand Valley State University

Richard Showman, *University of South Carolina*

Jennifer L. Siemantel, *Cedar Valley College*

Michael Silva, *El Paso Community College*

Bill Simcik, *Lone Star College-Tomball*

Robert Simons,
University of California, Los Angeles

Roger Sloboda, *Dartmouth College*

Jerry W. Smith, *St. Petersburg College*

Val Smith, *University of Kansas*

Nancy Solomon, *Miami University*

Christine C. Spencer,
Georgia Institute of Technology

Bruce Stallsmith,
The University of Alabama in Huntsville

Richard Stalter,
College of St. Benedict and St. John's University

Sonja Stampfler, *Kellogg Community College*

Karl Sternberg, *Western New England College*

Pat Steubing, *University of Nevada, Las Vegas*

Karen Steudel,
University of Wisconsin–Madison

Tom Stidham, *Texas A&M University*

Richard D. Storey, *The Colorado College*

Tara Stoulig, *Southeastern Louisiana University*

Brian Stout, *Northwest Vista College*

Gregory W. Stunz,
Texas A&M University

Mark T. Sugalski,
Southern Polytechnic State University

Michael A. Sulzinski,
The University of Scranton

Marshall Sundberg, *Emporia State University*

David Tam, *University of North Texas*

David Tauck, *Santa Clara University*

Salvatore Tavormina,
Austin Community College

Jeffrey Taylor,
Slippery Rock University of Pennsylvania

Franklyn Te, *Miami Dade College*

Roger E. Thibault,
Bowling Green State University

Ken Thomas,
Northern Essex Community College

Megan Thomas,
University of Nevada, Las Vegas

Patrick Thorpe, *Grand Valley State University*

Ian Tizard, *Texas A&M University*

Terry M. Trier, *Grand Valley State University*

Robert Turner, *Western Oregon University*

Joe Vanable, *Purdue University*

William Velhagen, *New York University*

Linda H. Vick, *North Park University*

J. Robert Waaland, *University of Washington*

Alexander Wait, *Missouri State University*

Douglas Walker,
Wharton County Junior College

James Bruce Walsh, *The University of Arizona*

Fred Wasserman, *Boston University*

R. Douglas Watson, *The University of Alabama at Birmingham*

Chad M. Wayne, *University of Houston*

Cindy Wedig, *The University of Texas–Pan American*

Michael N. Weintraub, *The University of Toledo*

Edward Weiss, *Christopher Newport University*

Mark Weiss, *Wayne State University*

Adrian M. Wenner, *University of California, Santa Barbara*

Sue Simon Westendorf, *Ohio University*

Ward Wheeler, *American Museum of Natural History, Division of Invertebrate Zoology*

Adrienne Williams, *University of California, Irvine*

Elizabeth Willott, *The University of Arizona*

Mary Wise, *Northern Virginia Community College*

Robert Wise, *University of Wisconsin, Oshkosh*

Shawn Wright, *Central New Mexico Community College*

Charles R. Wyttenbach, *The University of Kansas*

Robert Yost, *Indiana University–Purdue University Indianapolis*

Xiaoning Zhang, *St. Bonaventure University*

Yunde Zhao, *University of California, San Diego*

Heping Zhou, *Seton Hall University*

Xinsheng Zhu, *University of Wisconsin–Madison*

Adrienne Zihlman, *University of California–Santa Cruz*

Unanswered Questions Contributors

Chapter 2
David I. Kline, *Scripps Institution of Oceanography*

Chapter 3
Michael S. Brown and Joseph L. Goldstein, *University of Texas Southwestern Medical Center*

Chapter 4
Ulrich Müller, *University of California–San Diego*

Chapter 5
Matthew Welch, *University of California–Berkeley*

Chapter 6
Peter Agre, *Johns Hopkins Malaria Research Institute*

Chapter 7
Jeffrey Blaustein, *University of Massachusetts Amherst*

Chapter 8
Gail A. Breen, *University of Texas at Dallas*

Chapter 9
David Kramer, *Washington State University*

Chapter 10
Raymond Deshaies, *California Institute of Technology*

Chapter 11
Monica Colaiácovo, *Harvard Medical School*

Chapter 12
Nicholas Katsanis, *Duke University*

Chapter 13
Michelle Le Beau and Angela Stoddart, *The University of Chicago*

Chapter 14
Janis Shampay, *Reed College*

Chapter 15
Harry Noller, *University of California–Santa Cruz*

Chapter 16
Mark A. Kay, *Stanford University School of Medicine*

Chapter 17
Gerald Baron, *Rocky Mountain Laboratories*

Chapter 18
John F. Engelhardt and Tom Lynch, *University of Iowa*

Chapter 19
Larisa H. Cavallari, *University of Illinois at Chicago College of Pharmacy*

Chapter 20
Douglas J. Futuyma, *Stony Brook University*

Chapter 21
Peter Grant and Rosemary Grant, *Princeton University*

Chapter 22
Jerry Coyne, *University of Chicago*

Chapter 23
Elena M. Kramer, *Harvard University*

Chapter 24
Richard Glor, *University of Kansas*

Chapter 25
Andrew Pohorille, *National Aeronautics and Space Administration (NASA)*

Chapter 26
Stephen D. Bell and Rachel Y. Samson, *Oxford University*

Chapter 27
Geoff McFadden, *University of Melbourne*

Contents

Unit Four Biodiversity

Unit Seven   Ecology and Behavior

Introduction to Biological Concepts and Research

1

Earth, a planet teeming with life, is seen here in a satellite photograph.

Why it matters . . . Life abounds in almost every nook and cranny on our planet Earth. A lion creeps across an African plain, ready to spring at a zebra. The leaves of a sunflower in Kansas turn slowly through the day, keeping their surfaces fully exposed to rays of sunlight. Fungi and bacteria in the soil of a Canadian forest obtain nutrients by decomposing dead organisms. A child plays in a park in Madrid, laughing happily as his dog chases a tennis ball. In one room of a nearby hospital, a mother hears the first cry of her newborn baby; in another room, an elderly man sighs away his last breath. All over the world, countless organisms are born, live, and die every moment of every day. How did life originate, how does it persist, and how is it changing? Biology, the science of life, provides scientific answers to these questions.

What *is* life? Offhandedly, you might say that although you cannot define it, you know it when you see it. The question has no simple answer, because life has been unfolding for billions of years, ever since nonliving materials assembled into the first organized, living cells. Clearly, any list of criteria for the living state only hints at the meaning of "life." Deeper scientific insight requires a wide-ranging examination of the characteristics of life, which is what this book is all about.

Over the next semester or two, you will encounter examples of how organisms are constructed, how they function, where they live, and what they do. The examples provide evidence in support of concepts that will greatly enhance your appreciation and understanding of the living world, including its fundamental unity and striking diversity. This chapter provides a brief overview of these basic concepts. It also describes some of the ways in which biologists conduct research, the process by which they observe nature, formulate explanations of their observations, and test their ideas.

FIGURE 1.1 Living organisms and inanimate objects. Living organisms, such as this Texas brown tarantula (*Aphonopelma hentzi*) have characteristics that are fundamentally different from those of inanimate objects, like the rock on which it is sitting.

1.1 What Is Life? Characteristics of Living Organisms

Picture a tarantula on a rock, waiting patiently for a food item to wander within its reach **(Figure 1.1).** You know that the tarantula is alive and that the rock is not. At the atomic and molecular levels, however, the differences between them blur. Tarantulas, rocks, and all other matter are composed of atoms and molecules, which behave according to the same physical laws. Nevertheless, living organisms share a set of characteristics that collectively set them apart from nonliving matter.

The differences between a tarantula and a rock depend not only on the kinds of atoms and molecules present, but also on their organization and their interactions. Individual organisms are at the middle of a hierarchy that ranges from the atoms and molecules within their bodies to the diverse assemblages of organisms that occupy Earth's environments. Within every individual, certain biological molecules contain instructions for building other molecules, which, in turn, are assembled into complex structures. Living organisms must gather energy and materials from their surroundings to build new biological molecules, grow in size, maintain and repair their parts, and produce offspring. They must also respond to environmental changes by altering their chemistry and activity in ways that allow them to survive. Finally, aspects of their structure and function may change from one generation to the next.

Life on Earth Exists at Several Levels of Organization, Each with Its Own Emergent Properties

The organization of life extends through several levels of a hierarchy **(Figure 1.2).** Complex biological molecules exist at the lowest level of organization, but by themselves, these molecules are not alive. The properties of life do not appear until they are arranged into cells. A **cell** is an organized chemical system that includes many specialized molecules surrounded by a membrane. A cell is the lowest level of biological organization that can survive and reproduce—as long as it has access to a usable energy source, the necessary raw materials, and appropriate environmental conditions. However, a cell is alive only as long as it is organized as a cell; if broken into its component parts, a cell is no longer alive even if the parts themselves are unchanged. Characteristics that depend on the level of organization of matter, but do not exist at lower levels of organization, are called **emergent properties.** Life is thus an emergent property of the organization of matter into cells.

Many single cells, such as bacteria and protozoans, exist as **unicellular organisms.** By contrast, plants and animals are **multicellular organisms.** Their cells live in tightly coordinated groups and are so interdependent that they cannot survive on their own. For example, human cells cannot live by themselves in nature because they must be bathed in body fluids and supported by the activities of other cells. Like individual cells, multicellular organisms have emergent properties that their individual components lack; for example, humans can learn biology.

The next, more inclusive level of organization is the **population,** a group of organisms of the same kind that live together in the same place. The humans who occupy the island of Tahiti and a group of sea urchins living together on the coast of Washington State are examples of populations. Like multicellular organisms, populations have emergent properties that do not exist at lower levels of organization. For example, a population has characteristics such as its birth or death rate—that is, the number of individual organisms who are born or die over a period of time—that do not exist for single cells or individual organisms.

Working our way up the biological hierarchy, all the populations of different organisms that live in the same place form a **community.** The algae, snails, sea urchins, and other organisms that live along the coast of Washington State, taken together, make up a community. The next higher level, the **ecosystem,** includes the community *and* the nonliving environmental factors with which it interacts. For example, a coastal ecosystem comprises a community of living organisms, as well as rocks, air, seawater, minerals, and sunlight. The highest level, the **biosphere,** encompasses all the ecosystems of Earth's waters, crust, and atmosphere. Communities, ecosystems, and the biosphere also have emergent properties. For example, communities can be described in terms of their *diversity*—the number and types of different populations they contain—and their *stability*—the degree to which the populations within the community remain the same through time.

Living Organisms Contain Genetic Information That Governs Their Structure and Function

The most fundamental and important molecule that distinguishes living organisms from nonliving matter is **deoxyribonucleic acid (DNA).** DNA is a large, double-stranded, helical

Biosphere
All regions of Earth's crust, waters, and atmosphere that sustain life

Ecosystem
Communities interacting with their shared physical environment

Community
Populations of all species that occupy the same area

Population
Group of individuals of the same species living in the same area

Multicellular organism
Individual consisting of interdependent cells

Cell
Smallest unit with the capacity to live and reproduce, independently or as part of a multicellular organism

FIGURE 1.2 The hierarchy of life. Each level in the hierarchy of life exhibits emergent properties that do not exist at lower levels. The middle four photos depict a rocky intertidal zone on the coast of Washington State.
© Cengage Learning 2017

molecule that contains instructions for assembling a living organism from simpler molecules **(Figure 1.3A).** The two strands in a molecule of DNA each consist of a chain of chemical building blocks called **nucleotides.** The four different nucleotides present in DNA are commonly identified by the first letters of their chemical names: A, T, G, and C. (The structures of nucleotides and how they are organized in DNA are discussed in Section 3.5.) Nucleotides in one DNA strand chemically bond with nucleotides in the other DNA strand to form the double-helical structure of the molecule **(Figure 1.3B).** All of the DNA in the cells of a living organism constitutes its **genome,** which contains the genetic information that makes each organism unique. Genetic information is encoded in the sequence of nucleotides in an organism's DNA, just as this book conveys information in sequences of letters that make up words and sentences.

Genes are particular regions of the genome where specific nucleotide sequences encode instructions that cells use to build **ribonucleic acid (RNA)** molecules and **proteins.** The molecules produced from these instructions fold into specific three-dimensional shapes, and the shape of each protein or RNA molecule determines how it will function within the cell.

The process by which information encoded in genes guides the production of RNA molecules and proteins is called **gene expression.** The expression of a protein-encoding gene involves two steps **(Figure 1.4).** First, information in the nucleotide sequence of one of the gene's DNA strands is copied into a

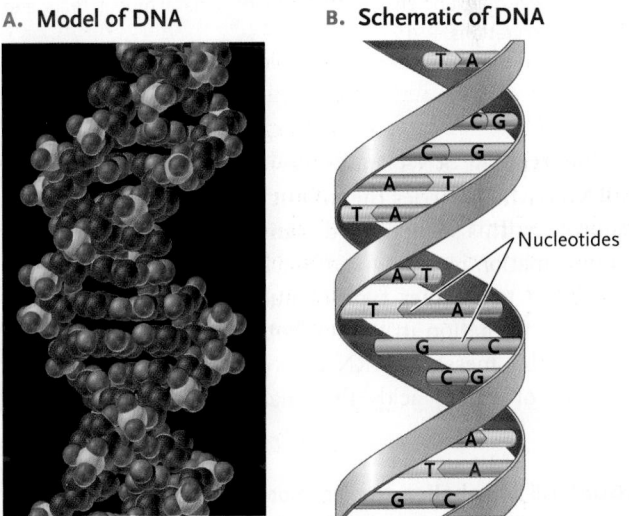

A. Model of DNA

B. Schematic of DNA

Nucleotides

FIGURE 1.3 Deoxyribonucleic acid (DNA). (A) A computer-generated model of DNA illustrates that it is made up of two strands twisted into a double helix. **(B)** A schematic diagram shows how nucleotides on the two strands bind to each other.
© Cengage Learning 2017

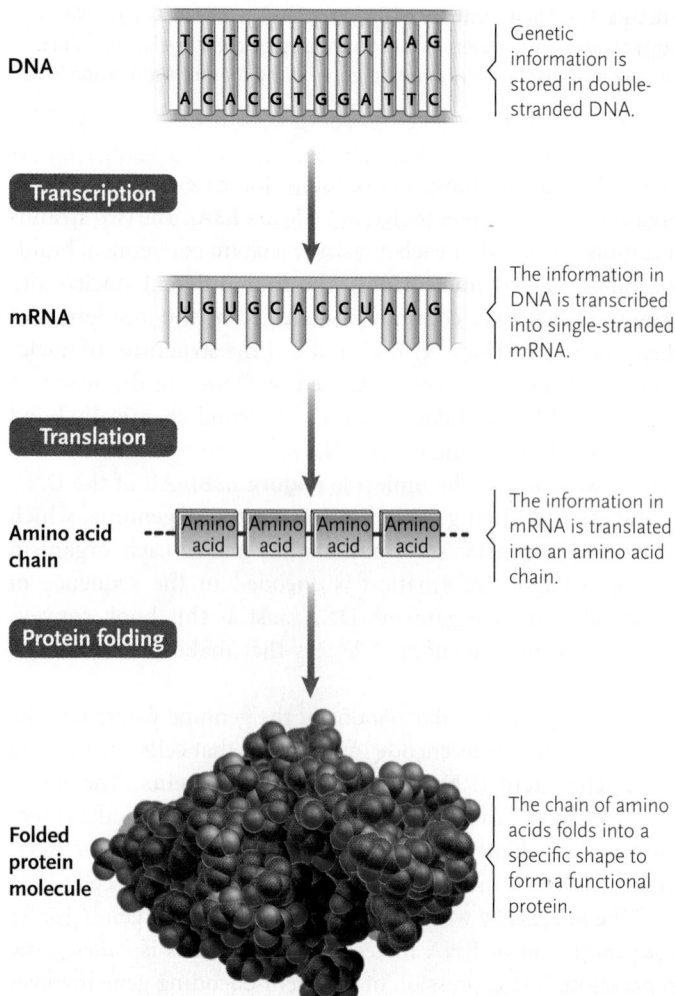

DNA T G T G C A C C T A A G / A C A C G T G G A T T C

Genetic information is stored in double-stranded DNA.

Transcription

mRNA U G U G C A C C U A A G

The information in DNA is transcribed into single-stranded mRNA.

Translation

Amino acid chain - - - Amino acid | Amino acid | Amino acid | Amino acid - - -

The information in mRNA is translated into an amino acid chain.

Protein folding

Folded protein molecule

The chain of amino acids folds into a specific shape to form a functional protein.

FIGURE 1.4 The pathway of information flow in living organisms. Information stored in DNA is transcribed to mRNA, which is then translated into a chain of amino acids. The amino acid chain folds to assume the functional structure of the protein. The protein shown here is lysozyme, a bacterial-wall digesting enzyme that is found in nasal mucus, tears, and other body secretions.
© Cengage Learning 2017

specific type of RNA molecule, **messenger RNA (mRNA),** which carries the instructions for building the protein. This step is called **transcription** because the information in one type of nucleic acid (DNA) is *transcribed* to another type of nucleic acid (RNA). Second, information in the nucleotide sequence carried by the messenger RNA is converted into a sequence of amino acids that makes up a protein

molecule. This step is called **translation** because the nucleotide sequence in the mRNA is *translated* into a sequence of amino acids, producing a protein, an entirely different type of molecule. Translation is carried out on **ribosomes,** roughly spherical particles that are abundant in the cytoplasm. Ribosomes act as molecular machines that catalyze the assembly of amino acid chains.

The amino acid sequence of a protein determines how it folds into its functional structure. Thus, the different nucleotide sequences of genes produce mRNA molecules with different nucleotide sequences, which, in turn, guide the production of proteins with different amino acid sequences. (Transcription and translation are discussed in detail in Chapter 15.) Genes that do not encode proteins are transcribed to produce RNA molecules, but rather than being translated to produce a protein, those RNA molecules themselves fold into shapes that enable them to perform functions like regulating the expression of other, protein-encoding genes.

We can think of the genes within a genome as blueprints for the life of the organism: each protein or RNA product of a gene is a molecular tool that the organism uses to stay alive, grow, and reproduce. All of the ways that frogs differ from oak trees or from human beings result from differences in the genetic instructions encoded in the unique DNA sequences of these very different organisms, which in turn cause their cells to produce different proteins and RNA molecules. The human genome contains about 20,500 genes that encode proteins, and at least 20,000 other genes that encode RNA molecules. Together, these genes encode all of the molecular tools that perform all of the functions of human cells—and they determine how those cells combine together to form larger, multicellular structures like the tissues and organs of the human body.

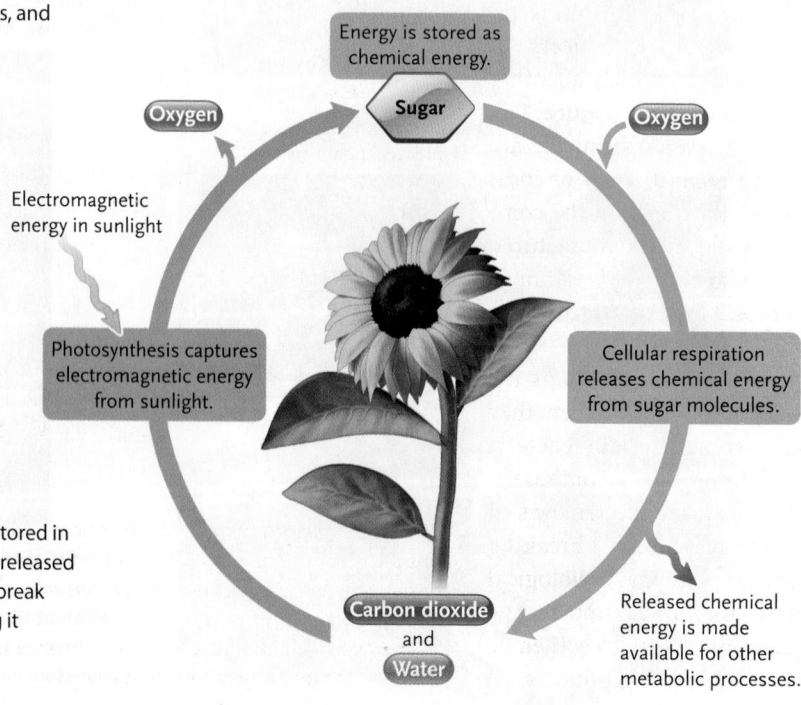

Energy is stored as chemical energy.

Oxygen

Sugar

Oxygen

Electromagnetic energy in sunlight

Photosynthesis captures electromagnetic energy from sunlight.

Cellular respiration releases chemical energy from sugar molecules.

Carbon dioxide and Water

Released chemical energy is made available for other metabolic processes.

FIGURE 1.5 Metabolic activities. Photosynthesis converts the electromagnetic energy in sunlight into chemical energy, which is stored in sugars and starches built from carbon dioxide and water; oxygen is released as a by-product of the reaction. Cellular respiration uses oxygen to break down sugar molecules, releasing their chemical energy and making it available for other metabolic processes.
© Cengage Learning 2017

Living Organisms Engage in Metabolic Activities

Some genes in all organisms code for molecules responsible for **metabolism** (described in Chapters 7 and 8), the ability of a cell or organism to extract energy from its surroundings and use that energy to maintain itself, grow, and reproduce. As a part of metabolism, cells carry out chemical reactions that assemble, alter, and disassemble molecules **(Figure 1.5).** For example, a growing sunflower plant carries out **photosynthesis,** in which the electromagnetic energy in sunlight is absorbed and converted into chemical energy. The cells of the plant store some chemical energy in sugar and starch molecules, and they use the rest to manufacture other biological molecules from simple raw materials obtained from the environment.

Sunflowers concentrate some of their energy reserves in seeds from which more sunflower plants may grow. The chemical energy stored in the seeds also supports other organisms, such as insects, birds, and humans, that eat them. Most organisms, including sunflower plants, tap stored chemical energy through another metabolic process, **cellular respiration.** In cellular respiration complex biological molecules are broken down with oxygen, releasing some of their energy content for cellular activities.

Energy Flows and Matter Cycles through Living Organisms

With few exceptions, energy from sunlight supports life on Earth. Plants and other photosynthetic organisms absorb energy from sunlight and convert it into chemical energy. They use this chemical energy to assemble complex molecules, such as sugars, from simple raw materials, such as water and carbon dioxide. As such, photosynthetic organisms are the **primary producers** of the food on which all other organisms rely **(Figure 1.6).** By contrast, animals are **consumers:** directly or indirectly, they feed on the complex molecules manufactured by plants. For example, zebras tap directly into the molecules of plants when they eat grass, and lions tap into it indirectly when they eat zebras. Certain bacteria and fungi are decomposers: they feed on the remains of dead organisms, breaking down complex biological molecules into simpler raw materials, which may then be recycled by the producers.

As you will see in Chapter 54, much of the energy that photosynthetic organisms trap from sunlight *flows* within and between populations, communities, and ecosystems. But because the transfer of energy from one organism to another is not 100% efficient, a portion of that energy is lost as heat. Although some animals can use this form of energy to maintain body temperature, it cannot sustain other life processes. By contrast, matter—nutrients such as carbon and nitrogen—*cycles* between living organisms and the nonliving components of the biosphere, to be used again and again (see Figure 1.6).

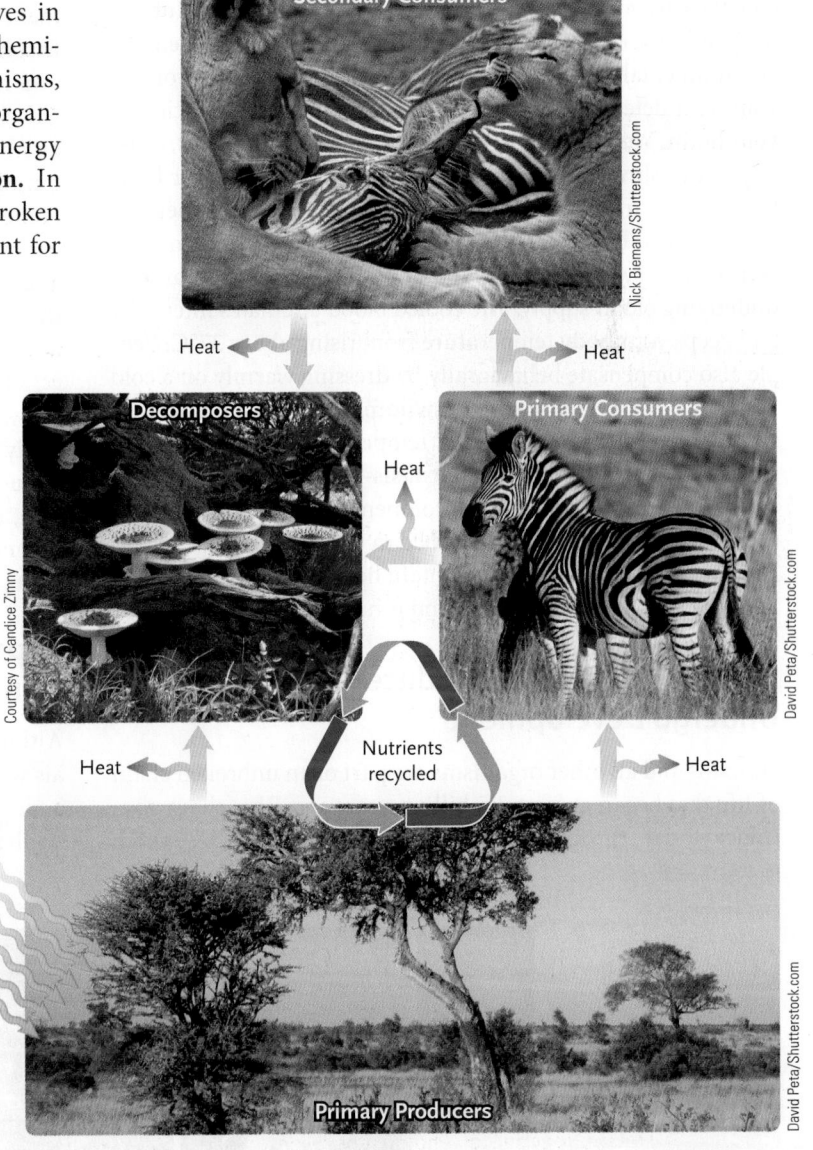

KEY

→ Energy transfer

⤳ Energy ultimately lost as heat

FIGURE 1.6 Energy flow and nutrient recycling. In most ecosystems, energy flows from the Sun to producers to consumers to decomposers. On the African savanna, the Sun provides energy to grasses (producers); zebras (primary consumers) then feed on the grasses before being eaten by lions (secondary consumers); fungi (decomposers) absorb nutrients and energy from the digestive wastes of animals and from the remains of dead animals and plants. All of the energy that enters an ecosystem is ultimately lost from the system as heat. Nutrients move through the same pathways, but they are conserved and recycled.

Living Organisms Compensate for Changes in the External Environment

All objects, whether living or nonliving, respond to changes in the environment; for example, a rock warms up on a sunny day and cools at night. But only living organisms have the capacity to detect environmental changes and *compensate* for them through controlled responses. Diverse and varied *receptors*—molecules or larger structures located on individual cells and body surfaces—can detect changes in external and internal conditions. When stimulated, the receptors trigger reactions that produce a compensating response.

For example, your internal body temperature remains reasonably constant, even though the environment in which you live is usually either cooler or warmer than you are. Your body compensates for these environmental variations and maintains its internal temperature at about 37° Celsius (C). When the environmental temperature drops significantly, receptors in your skin detect the change and transmit that information to your brain. Your brain may send a signal to your muscles, causing you to shiver, thereby releasing heat that keeps your body temperature from dropping below its optimal level. When the environmental temperature rises significantly, glands in your skin secrete sweat, which evaporates, cooling the skin and its underlying blood supply. The cooled blood circulates internally and keeps your body temperature from rising above 37°C. People also compensate behaviorally by dressing warmly on a cold winter day or jumping into a swimming pool in the heat of summer. Keeping your internal temperature within a narrow range is one example of **homeostasis**—a steady internal state maintained by responses that compensate for changes in the external environment. As described in Units 5 and 6, all organisms have mechanisms that maintain homeostasis in relation to temperature, blood chemistry, and other important factors.

Living Organisms Reproduce and Many Undergo Development

Humans and all other organisms are part of an unbroken chain of life that began at least 3.5 billion years ago. This chain continues today through **reproduction,** the process by which parents produce offspring. Offspring generally resemble their parents because the parents pass copies of their DNA—with all the accompanying instructions for virtually every life process—to their offspring. The transmission of DNA (that is, genetic information) from one generation to the next is called **inheritance.** For example, the eggs produced by storks hatch into little storks, not into pelicans, because they inherited stork DNA, which is different from pelican DNA.

Multicellular organisms also undergo a process of **development,** a series of programmed changes encoded in DNA, through which a fertilized egg divides into many cells that ultimately are transformed into an adult, which is itself capable of reproduction. As an example, consider the development of a moth **(Figure 1.7).** This insect begins its life as a tiny egg that contains all the instructions necessary for its development into an adult moth. Following these instructions, the egg first hatches into a caterpillar, a larval form adapted for feeding and rapid growth. The caterpillar increases in size until internal chemical signals indicate that it is time to spin a cocoon and become a pupa. Inside its cocoon, the pupa undergoes profound developmental changes that remodel its body completely. Some cells die; others multiply and become organized in different patterns. When these transformations are complete, the adult moth emerges from the cocoon. It is equipped with structures and behaviors, quite different from those of the caterpillar, that enable it to reproduce.

The sequential stages through which individuals develop, grow, maintain themselves, and reproduce are known collectively as the **life cycle** of an organism. The moth's life cycle includes egg, larva, pupa, and adult stages. Through reproduction, adult moths continue the cycle by producing the sperm and eggs that unite to form the fertilized egg, which starts the next generation.

Populations of Living Organisms Change from One Generation to the Next

Although offspring generally resemble their parents, individuals with unusual characteristics sometimes suddenly appear in a population. Moreover, the features that distinguish these oddballs are often inherited by their offspring. Our awareness

A. Egg   **B. Larva**   **C. Pupa**   **D. Adult**

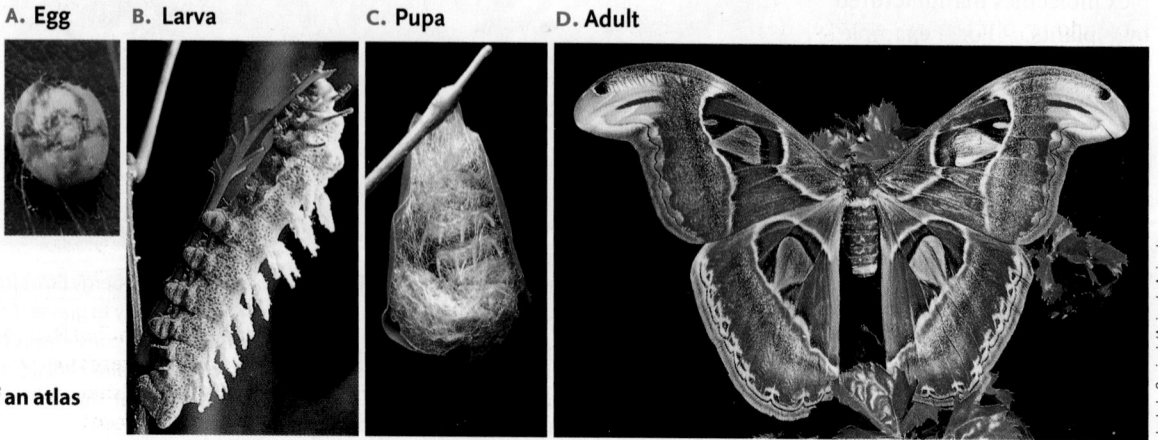

FIGURE 1.7 Life cycle of an atlas moth (Attacus atlas).

Jack de Coningh/Animals Animals

of the inheritance of unusual characteristics has had an enormous impact on human history because it has allowed plant and animal breeders to produce crops and domesticated animals with especially desirable characteristics.

Biologists have observed that similar changes also take place under natural conditions. In other words, populations of all organisms change from one generation to the next, because some individuals experience changes in their DNA and they pass those modified instructions along to their offspring. We introduce this fundamental process, **biological evolution,** in the next section. Although we explore biological evolution in great detail in Unit 3, every chapter in this book—indeed, every idea in biology—references our understanding that all biological systems are the products of evolutionary change.

STUDY BREAK 1.1

1. List the major levels in the hierarchy of life, and identify one emergent property of each level.
2. What do living organisms do with the energy they collect from the external environment?
3. What is a life cycle?

1.2 Biological Evolution

All research in biology—ranging from analyses of the precise structure of biological molecules to energy flow through the biosphere—is undertaken with the knowledge that biological evolution has shaped life on Earth. Our understanding of the evolutionary process reveals several truths about the living world: (1) all populations change through time, (2) all organisms are descended from a common ancestor that lived in the distant past, and (3) evolution has produced the spectacular diversity of life that we see around us. Evolution is the unifying theme that links all the subfields of the biological sciences, and it provides cohesion to our treatment of the many topics discussed in this book.

Darwin and Wallace Explained How Populations of Organisms Change through Time

How do evolutionary changes take place? One important mechanism was first explained in the mid-nineteenth century by two British naturalists, Charles Darwin and Alfred Russel Wallace. On a five-year voyage around the world, Darwin observed many "strange and wondrous" organisms. He also found fossils of species that are now extinct (that is, all members of the species are dead). The extinct forms often resembled living species in some traits but differed in others. Darwin originally believed in special creation—the idea that living organisms were placed on Earth in their present numbers and kinds and have not changed since their creation. But he became convinced that species do not remain constant with the passage of time: instead, they change from one form to another over

generations. Wallace came to the same conclusion through his observations of the great variety of plants and animals in the jungles of South America and Southeast Asia.

Darwin also studied the process of evolution through observations and experiments on domesticated animals. Pigeons were among his favorite experimental subjects. Domesticated pigeons exist in a variety of sizes, colors, and shapes, but all of them are descended from the wild rock dove **(Figure 1.8).** Darwin noted that pigeon breeders who wished to promote a certain characteristic, such as elaborately curled tail feathers, selected individuals with the most curl in their feathers as parents for the next generation. By permitting only these birds to mate, the breeders fostered the desired characteristic and gradually eliminated or reduced other traits. The same practice is still used today to increase the frequency of desirable traits in tomatoes, dogs, and other domesticated plants and animals. Darwin called this practice **artificial selection.** He termed the equivalent process that occurs in nature **natural selection.**

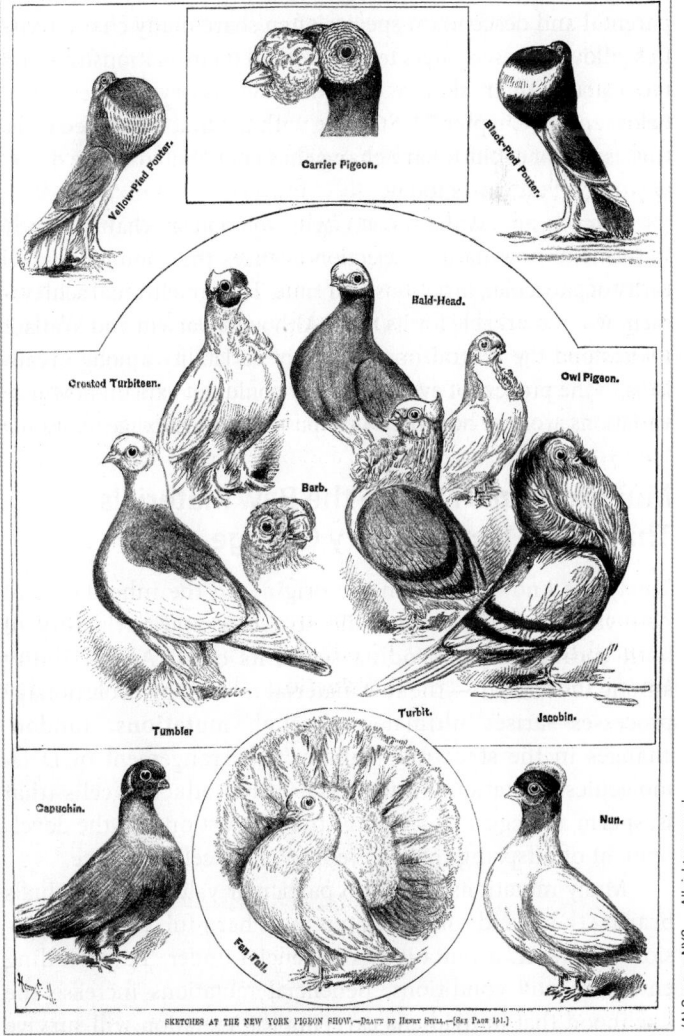

FIGURE 1.8 Artificial selection. This lithograph, published in an American newspaper, illustrates breeds that were exhibited at the New York Pigeon Show in 1879. Darwin studied the inheritance of strange new characteristics in similar pigeon breeds.

In 1858, Darwin and Wallace formally summarized their observations and conclusions explaining biological evolution:

1. Most organisms can produce numerous offspring, but environmental factors limit the number that actually survive and reproduce.
2. Heritable variations allow some individuals to compete more successfully for space, food, and mates.
3. These successful individuals somehow pass the favorable characteristics to their offspring.
4. As a result, the favorable traits become more common in the next generation, and less successful traits become less common.

This process of natural selection results in evolutionary change. Today, evolutionary biologists recognize that natural selection is just one of several potent evolutionary processes, as described in Chapter 21.

Over many generations, the evolutionary changes in a population may become extensive enough to produce a population of organisms that is distinct from its ancestors. Nevertheless, parental and descendant species often share many characteristics, allowing researchers to understand their relationships and reconstruct their shared evolutionary history, as described below and in Chapter 24. Starting with the first organized cells, this aspect of evolutionary change has contributed to the diversity of life that exists today.

Darwin and Wallace described evolutionary change largely in terms of how natural selection changes the commonness or rarity of particular variations over time. Their intellectual achievement was remarkable for its time. Although Darwin and Wallace understood the central importance of variability among organisms to the process of evolution, they could not explain how new variations arose or how they were passed to the next generation.

Mutations in DNA Are the Raw Materials That Allow Evolutionary Change

Today, we know that both the origin and the inheritance of new variations arise from the structure and variability of both coding and noncoding segments of DNA. Variability among individuals—the raw material molded by evolutionary processes—arises ultimately through **mutations,** random changes in the structure, number, or arrangement of DNA molecules. Mutations in the DNA of reproductive cells (that is, sperm and eggs) may change the instructions for the development of offspring that the reproductive cells produce.

Many mutations are of no particular value to individuals bearing them, and some turn out to be harmful. On rare occasions, however, a mutation is beneficial under the prevailing environmental conditions. Beneficial mutations increase the likelihood that individuals carrying the mutation will survive and reproduce. Thus, through the persistence and spread of beneficial mutations among individuals and their descendants, the genetic makeup of a population will change from one generation to the next.

Adaptations Enable Organisms to Survive and Reproduce in the Environments Where They Live

Favorable mutations may produce **adaptations,** characteristics that help an organism survive longer or reproduce more under a particular set of environmental conditions. To understand how organisms benefit from adaptations, consider an example from the recent literature on *cryptic coloration* (camouflage) in animals. Many animals have skin, scales, feathers, or fur that matches the color and appearance of the background in their environment, enabling them to blend into their surroundings. Camouflage makes it harder for predators to identify and then catch them—an obvious advantage to survival. Animals that are not camouflaged are often just sitting ducks.

The rock pocket mouse *(Chaetodipus intermedius),* which lives in the deserts of the southwestern United States, is mostly nocturnal (that is, active at night). At most desert localities, the rocks are pale brown, and rock pocket mice have sandy-colored fur on their backs. However, at several sites, the rocks—remnants of lava flows from now-extinct volcanoes—are black; here, the rock pocket mice have black fur on their backs. Thus, like the sandy-colored mice in other areas, they are camouflaged in their **habitats,** the types of areas in which they live **(Figure 1.9A).** Camouflage appears to be important to these mice because owls, which locate prey using their exceptionally keen eyesight, frequently eat nocturnal desert mice.

Examples of cryptic coloration are well documented in scientific literature, and biologists generally interpret them as adaptations that reduce the likelihood of being captured by a predator. Michael W. Nachman, Hopi E. Hoekstra, and their colleagues at the University of Arizona explored the genetic and evolutionary basis for the color difference between rock pocket mice that live on light and dark backgrounds. In an article published in 2003, they reported the results of an analysis of mice sampled at six sites in southern Arizona and New Mexico. In two regions (Pinacate, AZ, and Armendaris, NM), both light and dark rocks were present, allowing the researchers to compare mice that lived on differently colored backgrounds. Two other sites had only light rocks and sandy-colored mice.

Nachman and his colleagues found that nearly all of the mice they captured on dark rocks had dark fur and that nearly all of the mice they captured on light rocks had light fur **(Figure 1.9B).** The researchers then studied the structure of the melanocortin 1 receptor gene *(Mc1r),* which influences fur color in laboratory mice; random mutations in this gene can produce fur colors ranging from light to dark in any population of mice, regardless of the habitat it occupies. (Variations in this gene are also responsible for differences in hair and skin color in humans, and many are associated with an increased risk of developing skin cancer.) The 17 black mice from Pinacate all shared certain mutations in their *Mc1r* gene, which established four specific changes in the structure of the Mc1r protein. However, none of the 12 sandy-colored mice from

A. Camouflage in rock pocket mice *(Chaetodipus intermedius)*

Sandy-colored mice are well camouflaged on pale rocks, and black mice are well camouflaged on dark rocks (top); but mice with fur that does not match their backgrounds (bottom) are easy to see.

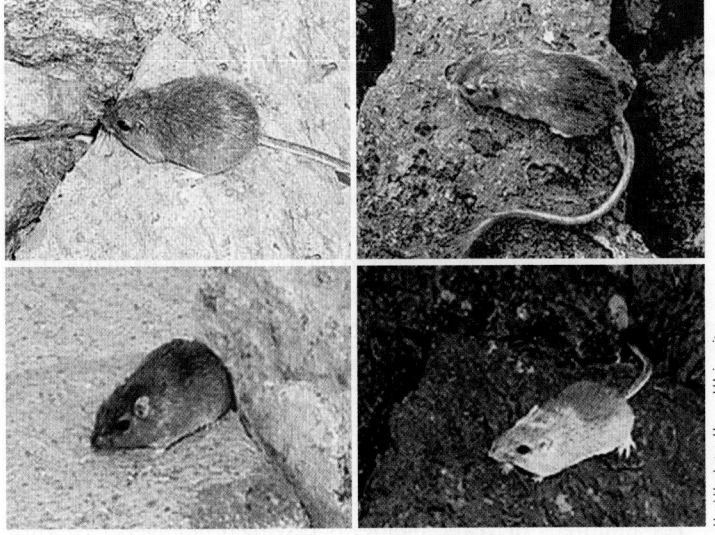

FIGURE 1.9 Adaptive coloration in rock pocket mice *(Chaetodipus intermedius).*

B. Distributions of rock pocket mice with light and dark fur

At sites in Arizona and New Mexico, mouse fur color closely matched the color of the rocks where they lived. The pie charts show the proportion of mice with sandy-colored or black fur, N = the number of mice sampled at each site. The bars beneath the pie charts indicate the rock color.

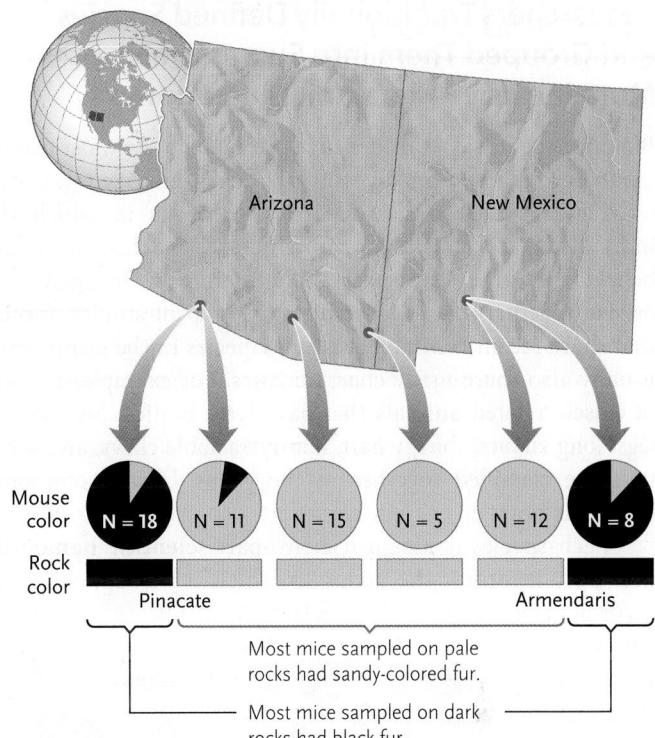

Pinacate carried these mutations. The exact match between the presence of the mutations and the color of the mice strongly suggests that these mutations in the *Mc1r* gene are responsible for the dark fur in the mice from Pinacate. These data on the distributions of light and dark mice, coupled with analyses of their DNA, suggest that the color difference is the product of specific mutations that were favored by natural selection. In other words, natural selection *conserved* random mutations that produced black fur in mice that live on black rocks.

Nachman's team then analyzed the *Mc1r* gene in the dark and light mice from Armendaris and in the light mice at two intermediate sites. Because the mice in these regions also closely matched the color of their environments, the researchers expected to find the *Mc1r* mutations in the dark mice but not in the light mice. However, none of the mice from Armendaris shared any of the mutations that apparently contribute to the dark color of mice from Pinacate. Thus, mutations in some other gene or genes, which the researchers have not yet identified, must be responsible for the camouflaging black coloration of mice that live on black rocks in Armendaris.

The example of an adaptation provided by the rock pocket mice illustrates the observation that genetic differences often develop between populations. Sometimes these differences become so great that the organisms develop different appearances and adopt different ways of life. If they become different enough, biologists may regard them as distinct types, as

described in Chapter 22. Over immense spans of time, evolutionary processes have produced many types of organisms, which constitute the diversity of life on Earth. In the next section, we survey this diversity and consider how it is studied.

STUDY BREAK 1.2

1. What is the difference between artificial selection and natural selection?
2. How do random changes in the structure of DNA affect the characteristics of organisms?
3. What is the usefulness of being camouflaged in natural environments?

1.3 Biodiversity and the Tree of Life

Millions of different kinds of organisms live on Earth today, and many millions more existed in the past and became extinct. This mind-boggling biodiversity, the product of evolution, represents the many ways in which the common elements of life have combined to survive and reproduce. To make sense of the past and present diversity of life on Earth, biologists analyze the evolutionary relationships of these organisms and use classification systems to keep track of them. As described in Chapter 24, the task is daunting, and there is no clear consensus on the numbers and kinds of divisions and categories to use. Moreover, our understanding of evolutionary relationships is constantly changing as researchers develop new analytical techniques and learn more about extinct and living organisms.

Researchers Traditionally Defined Species and Grouped Them into Successively More Inclusive Hierarchical Categories

Biologists generally consider the species to be the most fundamental grouping in the diversity of life. As described in Chapter 22, a **species** is a group of populations in which the individuals are so similar in structure, biochemistry, and behavior that they can successfully interbreed. Biologists recognize a **genus** (plural, *genera*) as a group of similar species that share recent common ancestry. Species in the same genus usually also share many characteristics. For example, a group of closely related animals that have large bodies, four stocky legs, long snouts, shaggy hair, non-retractable claws, and short tails are classified together in the genus *Ursus,* commonly known as bears.

Each species is assigned a two-part **scientific name:** the first part identifies the genus to which it belongs, and the second part designates a particular species within that genus. In the genus *Ursus,* for example, *Ursus americanus* is the scientific name of the American black bear; *Ursus maritimus,* the polar bear, and *Ursus arctos,* the brown bear, are two other species in the same genus. Scientific names are always written in italics, and only the first letter of the genus name is capitalized. After its first mention in a discussion, the genus name is frequently abbreviated to its first letter, as in *U. americanus.*

In a traditional classification, biologists first identified species and then grouped them into successively more inclusive categories **(Figure 1.10)**: related genera are placed in the same **family,** related families in the same **order,** and related orders in the same **class.** Related classes are grouped into a **phylum** (plural, *phyla*), and related phyla are assigned to a **kingdom.** In recent years, biologists have added the **domain** as the most inclusive group.

Today Biologists Identify the Trunks, Branches, and Twigs on the Tree of Life

For hundreds of years, biologists classified biodiversity within the hierarchical scheme described above, mostly using structural similarities and differences as clues to evolutionary relationships.

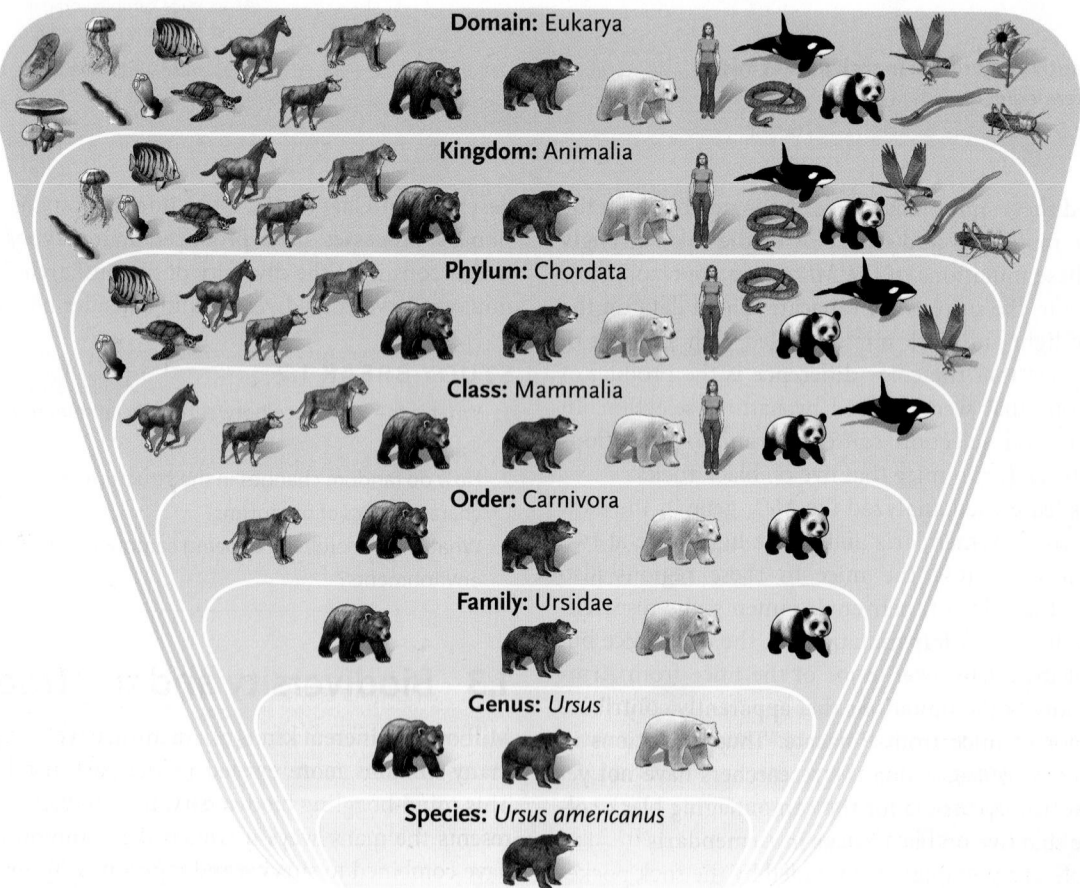

FIGURE 1.10 Traditional hierarchical classification. The classification of the American black bear *(Ursus americanus)* illustrates how each species fits into a nested hierarchy of ever more inclusive categories. The following sentence can help you remember the order of categories in a classification, from *Domain* to *Species:* Diligent Kindly Professors Cannot Often Fail Good Students.

© Cengage Learning 2017

An overview of the Tree of Life illustrates the relationships between the three domains. Branches and twigs are not included for Bacteria and Archaea. The branches of the Eukarya include three well-defined kingdoms (Plantae, Fungi, and Animalia), as well as five groups of organisms that were once collectively described as protists (marked with *); biologists have not yet clarified their evolutionary relationships.

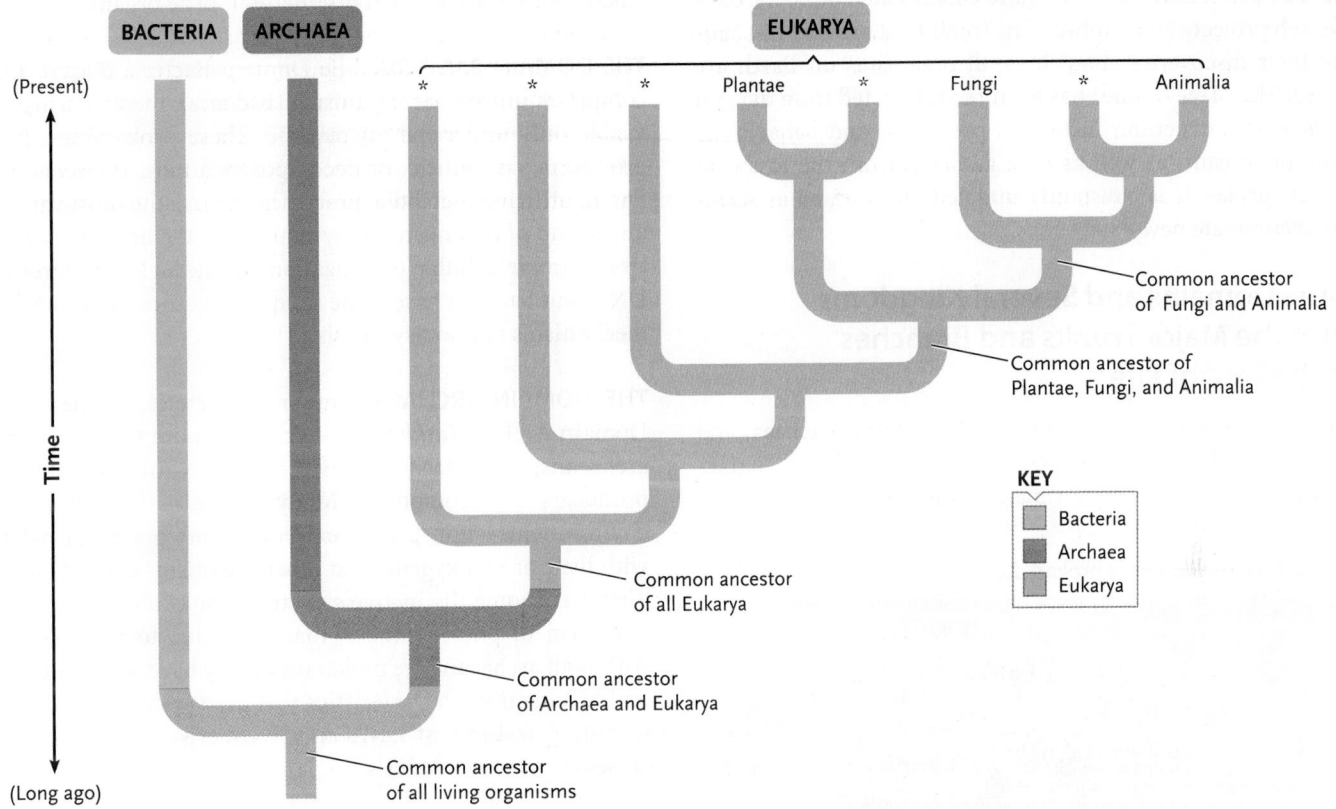

SUMMARY Phylogenetic trees contain more information than simple hierarchical classifications do because the trees illustrate which ancestors gave rise to which descendants, as well as when those evolutionary events occurred. Each fork between trunks, branches, and twigs on the phylogenetic tree represents an evolutionary event in which one ancestral species gave rise to two descendant species. Detailed phylogenetic trees illustrate how, over time, descendant species gave rise to their own descendants, producing the great diversity of life.

think like a scientist Given the structure of the Tree of Life represented in this figure, do you think that animals are more closely related to fungi or to plants?

© Cengage Learning 2017

With the development of new techniques late in the twentieth century, biologists began to use the precise sequence of nucleotides in DNA and other biological molecules to trace the evolutionary pathways through which biodiversity evolved. This approach allows the comparison of species as different as bacteria and humans because all living organisms share the same genetic code. It also provides so much data that biologists are now able to construct very detailed **phylogenetic trees** (*phylon* = race; *genetikos* = origin)—illustrations of the evolutionary pathways through which species and more inclusive groups appeared—for all organisms **(Figure 1.11)**. Phylogenetic trees, described further in Chapter 24, are like family genealogies

spanning the many millions of years that evolution has been occurring. In most of the phylogenetic trees you will encounter in Units 3 and 4 of this book, time is usually represented vertically, with forks closer to the base of the tree representing evolutionary events in the distant past and those near the top representing more recent evolutionary events.

In many ways, information in a phylogenetic tree parallels the traditional hierarchical classification, because organisms on the same branch share the common ancestor that is represented at the base of their branch. If the base of a branch that includes two species is near the bottom of the tree, biologists would judge the species to be only distant relatives because their ancestries separated very long ago. By contrast, if the base of the branch containing two species is close to the top of the tree, we would describe the species as being close relatives. Major branches on the tree are therefore roughly equivalent to kingdoms and phyla; progressively smaller branches represent classes, orders, families, and genera. The twigs represent species or the individual populations they comprise.

Since 1994, with substantial support from the National Science Foundation, biologists have collaborated on the Tree of Life web project (http://tolweb.org/tree/) to share and disseminate their discoveries about how all organisms on Earth are related. The "Tree of Life" has been reconstructed from data on the genetics, structure, metabolic processes, and behavior of living organisms, as well as data gathered from the fossils of extinct species. It is constantly updated and revised as scientists accumulate new data.

Three Domains and Several Kingdoms Form the Major Trunks and Branches on the Tree of Life

Biologists distinguish three domains—Bacteria, Archaea, and Eukarya—each of which is a group of organisms with characteristics that set it apart as a major trunk on the Tree of Life.

A. *Escherichia coli*, a prokaryote

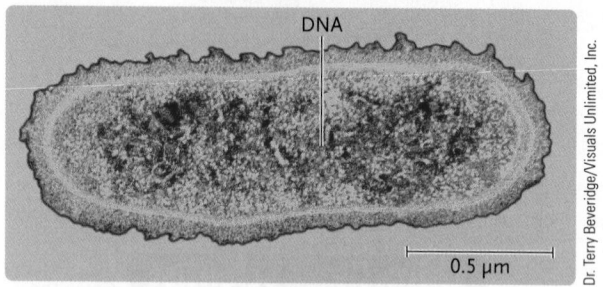

DNA

0.5 μm

Dr. Terry Beveridge/Visuals Unlimited, Inc.

B. *Paramecium aurelia*, a eukaryote

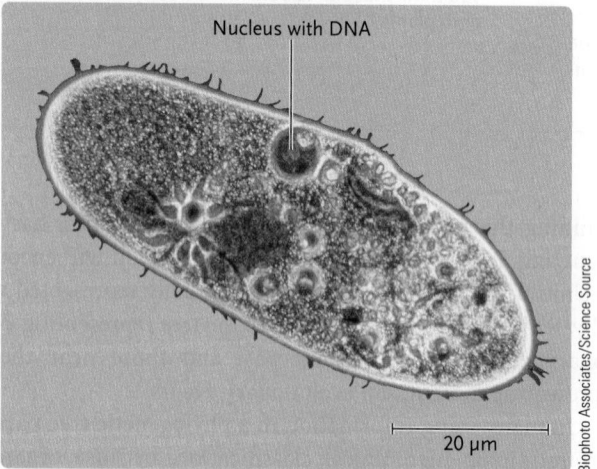

Nucleus with DNA

20 μm

Biophoto Associates/Science Source

FIGURE 1.12 Prokaryotic and eukaryotic cells. (A) *Escherichia coli*, a prokaryote, lacks the complex internal structures apparent in **(B)** *Paramecium aurelia*, a eukaryote. Color coding of the domains follows the key in Figure 1.11.

Species in two of the three domains, Bacteria and Archaea, are described as **prokaryotes** (*pro* = before; *karyon* = nucleus). Their DNA is suspended inside the cell without being separated from other cellular components **(Figure 1.12A)**. By contrast, the domain Eukarya comprises organisms that are described as **eukaryotes** (*eu* = true) because their DNA is enclosed in a nucleus, a separate structure within the cells **(Figure 1.12B)**. The nucleus and other specialized internal compartments of eukaryotic cells are called **organelles** ("little organs").

THE DOMAIN BACTERIA The Domain Bacteria **(Figure 1.13A)** comprises unicellular organisms (bacteria), most of which are visible only under the microscope. These prokaryotes live as producers, consumers, or decomposers almost everywhere on Earth, utilizing metabolic processes that are the most varied of any group of organisms. They share with the archaeans a relatively simple cellular organization of internal structures and DNA, but bacteria have some unique structural molecules and mechanisms of photosynthesis.

THE DOMAIN ARCHAEA Similar to bacteria, species in the Domain Archaea (*arkhaios* = ancient) **(Figure 1.13B),** known as archaeans, are unicellular, microscopic organisms that live as producers or decomposers. Many archaeans inhabit extreme environments—hot springs, extremely salty ponds, or habitats with little or no oxygen—that other organisms cannot tolerate. They have some distinctive structural molecules and a primitive form of photosynthesis that is unique to their domain. Although archaeans are prokaryotic, they have some molecular and biochemical characteristics that are typical of eukaryotes, including features of DNA and RNA organization and processes of protein synthesis.

THE DOMAIN EUKARYA All the remaining organisms on Earth, including the familiar plants and animals, are members of the Domain Eukarya **(Figure 1.13C)**. Organisms with eukaryotic cell structure are currently described as protists or classified as members of one of three well-defined kingdoms: Plantae, Fungi, and Animalia.

The Protists The term *protists* describes a diverse set of single-celled and multicellular eukaryotic species. They do not constitute a kingdom, because they do not share a unique common ancestry (see asterisks in Figure 1.11). The most familiar protists are protozoans, which are primarily unicellular, and algae, which range from single-celled, microscopic species to large, multicellular seaweeds. Protozoans are consumers and decomposers, but almost all algae are photosynthetic producers.

The Kingdom Plantae Members of the Kingdom Plantae are multicellular organisms that, with few exceptions, carry out photosynthesis; they, therefore, function as producers in ecosystems. Except for the reproductive cells (pollen) and seeds of some species, plants do not move from place to place. The

A. Domain Bacteria

1 μm

A. Barry Dowsett/Science Source

B. Domain Archaea

1 μm

Eye of Science/Science Source

C. Domain Eukarya

Protists

50 μm

Jubal Harshaw/Shutterstock.com

Kingdom Fungi

© Ann & Rob Simpson

Kingdom Plantae

Col Swain/Shutterstock.com

Kingdom Animalia

James Carmichael Jr./NHPA/Photoshot

FIGURE 1.13 Three domains of life. (A) This member of the Domain Bacteria *(Proteus mirabilis)* is commonly found in the digestive systems of humans. **(B)** These organisms from the Domain Archaea *(Pyrococcus furiosus)* live in hot ocean sediments near an active volcano. **(C)** The Domain Eukarya includes the protists and three kingdoms in this book. The protists are represented by a stalked predatory protozoan *(Vorticella* species) that is common in ponds. Coast redwoods *(Sequoia sempervirens)* are among the largest members of the Kingdom Plantae; the picture shows a young tree amidst a grove of mature redwoods. The Kingdom Fungi includes the big laughing mushroom *(Gymnopilus* species), which lives on the forest floor. Members of the Kingdom Animalia are consumers, as illustrated by the fishing spider *(Dolomedes* species), which is feasting on a minnow it has captured. Color coding of the domains follows the key in Figure 1.11. The photos of the redwood tree, the mushroom, and the spider are not at the same scale.

kingdom includes the familiar flowering plants, conifers, and mosses.

The Kingdom Fungi The Kingdom Fungi includes a highly varied group of unicellular and multicellular species, among them the yeasts and molds. Most fungi live as decomposers by breaking down and then absorbing biological molecules from dead organisms. Fungi do not carry out photosynthesis.

The Kingdom Animalia Members of the Kingdom Animalia are multicellular organisms that live as consumers by ingesting protists and organisms from all three domains. The ability to move actively from one place to another during some stage of their life cycles is a distinguishing feature of animals. The kingdom encompasses a great range of organisms, including groups as varied as sponges, worms, insects, fishes, amphibians, reptiles, birds, and mammals.

Biologists Often Use Model Organisms to Study Fundamental Biological Processes

Although biologists strive to learn as much as possible about the diversity of life, certain species or groups of organisms have become favorite research subjects because their characteristics make them relatively easy to study. In most cases, biologists began working with these **model organisms** because they have rapid development, short life cycles, and small adult size. Thus, researchers can rear and house large numbers of them in the laboratory. Their appeal as research subjects grows as fuller portraits of their biology emerge.

Many forms of life share similar molecules, structures, and processes; thus, research on these small and often simple organisms provides insights into biological processes that operate in larger and more complex ones. Moreover, by strategically adopting model organisms that represent different branches on the Tree of Life, biologists can extrapolate their findings broadly to the millions of species that they have not yet studied in detail. For example, research in the mid-twentieth century on gene expression in the bacterium *Escherichia coli* formed the intellectual and technical foundation that now allows scientists to make and clone (that is, produce multiples copies of) DNA molecules. Similarly, early analyses of inheritance in a fruit fly (*Drosophila melanogaster*) established our basic understanding of genetics in all eukaryotic organisms. Finally, research on a tiny mustard plant (*Arabidopsis thaliana*) is providing information about the genetic and molecular control of development in all plants, including important agricultural crops. Other model organisms facilitate research in ecology and evolution. You will read about eight of the organisms most frequently used in research in *Focus on Research: Model Research Organisms* boxes distributed throughout this book.

Now that we have introduced the characteristics of living organisms, basic concepts of evolution, and biological diversity, we turn our attention to the ways in which biologists examine the living world to make new discoveries and gain new insights about life on Earth.

STUDY BREAK 1.3

1. What is a major difference between prokaryotic and eukaryotic organisms?
2. In which domain and kingdom are humans classified?
3. Why do biologists often use model organisms in their research?

THINK OUTSIDE THE BOOK

Learn more about the Tree of Life web project by visiting this web site: http://tolweb.org/tree/.

1.4 Biological Research

The entire content of this book—every observation, experimental result, and generality—is the product of **biological research,** the collective effort of countless individuals who have worked to understand every aspect of the living world. This section describes how biologists working today pose questions and find answers to them. The narrative of this book and many of its special features (*Focus on Research; Molecular Insights; Closer Look* figures; *Experimental Research, Observational Research,* and *Research Methods* figures) highlight particularly elegant or insightful research methods and studies.

Biologists Confront the Unknown by Conducting Basic and Applied Research

As you read this book, you may at first be surprised to discover how many fundamental questions in biology have not yet been answered. How and where did life begin? How exactly do genes govern the growth and development of an organism? What triggers the signs of aging? Scientists embrace these "unknowns" as opportunities to apply creative thinking to important problems. To show you how exciting it can be to venture into unknown territory, most chapters close with an essay that addresses *Unanswered Questions* about the topic of the chapter. Although the concepts and facts that you will learn are profoundly interesting, you will discover that unanswered questions are even more exciting. In many cases, we do not even know *how* you and other scientists of your generation will answer these questions.

Research science is often broken down into two complementary activities—basic research and applied research—that constantly inform one another. Biologists who conduct **basic research** often seek explanations about natural phenomena to satisfy their own curiosity and to advance our collective knowledge. Sometimes, they may not have a specific practical goal in mind. For example, some biologists study how lizards control their body temperatures in different environments. At other times, basic research is inspired by specific practical concerns. For example, understanding how certain bacteria attack the cells of larger organisms might someday prove useful for the development of a new antibiotic (that is, a bacteria-killing agent).

Other scientists conduct **applied research,** with the goal of solving specific practical problems. For example, biomedical scientists conduct applied research to develop new drugs and to learn how illnesses spread from animals to humans or through human populations. Similarly, agricultural scientists try to develop varieties of important crop plants that are more productive and more pest-resistant than the varieties currently in use.

The Scientific Method Helps Researchers Crystallize and Test Their Ideas

People have been adding to our knowledge of biology ever since our distant ancestors first thought about gathering food or hunting game. However, beginning about 500 years ago in Europe, inquisitive people began to understand that direct observation is the most reliable and productive way to study natural phenomena. By the nineteenth century, researchers were using the **scientific method,** an investigative approach to acquiring knowledge in which scientists make observations about the natural world, develop working explanations about what they observe, and then test those explanations by collecting more information.

Grade school teachers often describe the scientific method as a stepwise, linear procedure for observing and explaining the world around us **(Figure 1.14).** But scientific research is both an intellectual and a technical process, and because scientists

The process of science allows researchers to crystallize their thoughts about a topic and devise a formal way to test their ideas by making observations and collecting measurable data. Although the scientific method is often described as a stepwise process, scientists think faster than they can act, and they may engage in some steps of the process simultaneously.

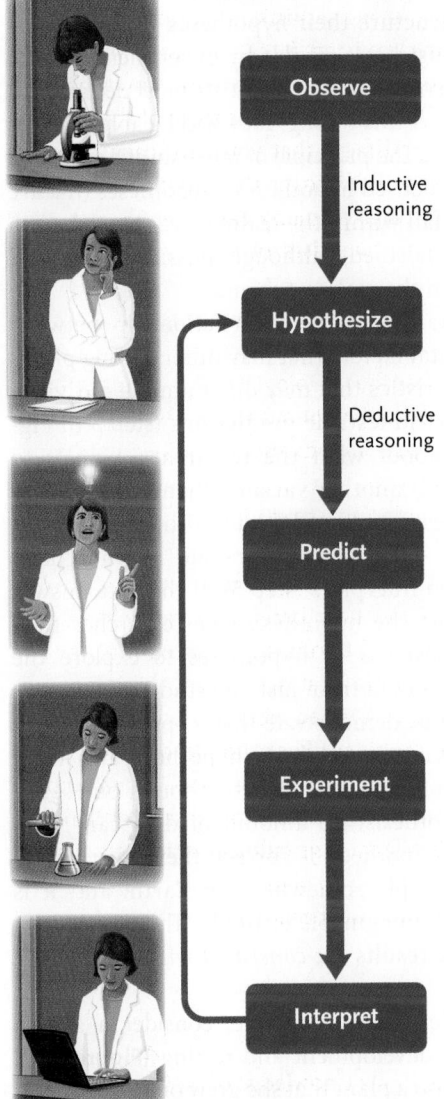

Observe

Inductive reasoning

Hypothesize

Deductive reasoning

Predict

Experiment

Interpret

1. Make detailed observations about a phenomenon of interest.

2. Devise a basic question for your research to answer. In some instances, you can use inductive reasoning to create a testable hypothesis that provides a working explanation of the observations you made. Hypotheses may be expressed in words or in mathematical equations. Many scientists also formulate alternative hypotheses (that is, alternative explanations) at the same time.

3. Use deductive reasoning to make predictions about what you would observe if the hypothesis were applied to a novel situation.

4. Design and conduct a controlled experiment (or new observational study) to answer the question or to test the predictions of the hypothesis. The experiment must be clearly defined so that it can be repeated in future studies. It must also lead to the collection of measurable data that other researchers can evaluate and reproduce if they choose to repeat the experiment themselves.

5. For question-based research, determine whether or not the data provide a clear answer to your question. For hypothesis-based research, compare the results of the experiment or new observations with those predicted by the hypothesis. Scientists often use formal statistical tests to determine whether the results match the predictions of the hypothesis.

SUMMARY After (1) making initial observations, (2) posing a question or defining a hypothesis, (3) making predictions, (4) answering the question or testing the predictions of the hypothesis with experimental or observational data, and (5) interpreting the data you collected, you may be able to answer your original question or evaluate the correctness of your hypothesis.

If the results do not answer the original question, you may try to answer it with a different approach. For hypothesis-based research, if the results do not match the predictions of the hypothesis, the hypothesis is refuted, and it must be rejected or revised.

For question-based research, if the data provide a clear answer to your original question, define a follow-up question that, when answered, provides deeper insight. For hypothesis-based research, if the statistical tests suggest that the prediction was correct, the hypothesis is confirmed—at least until new data refute it in the future.

Figures 1.15 and 1.16 provide examples of this process, using experimental and observational approaches, respectively.

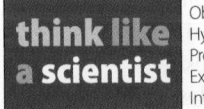

think like a scientist

Observe
Hypothesize
Predict
Experiment
Interpret

This Closer Look has introduced you to the basic steps in the process of science. Throughout this book you will encounter questions labeled "Think Like a Scientist," each of which is directed toward one of the steps in the scientific process. For the "Think Like a Scientist" questions in Experimental Research and Observational Research figures, the particular step each question addresses is identified. All of the "Think Like a Scientist" questions are good opportunities to engage in the science process.

© Cengage Learning 2017

usually think faster than they can work, they often undertake the different steps simultaneously. Application of the scientific method requires both curiosity and skepticism: successful scientists question the current state of our knowledge and challenge old concepts with new ideas and new observations. Scientists like to be shown *why* an idea is correct, rather than simply being told that it is: explanations of natural phenomena must be backed up by objective evidence rooted in observation and measurement.

Most important, scientists share their ideas and results through the publication of their work. Publications typically include careful descriptions of the methods employed and details of the results obtained so that other researchers can repeat and verify the findings at a later time.

Biologists Conduct Research by Collecting Observational and Experimental Data

Biologists generally use one of two complementary approaches—*descriptive science* and *experimental science*—or a combination of the two to advance our knowledge. In many cases, they collect **observational data,** basic information on biological structures or the details of biological processes. This approach, which is sometimes called *descriptive science,* provides information about systems that have not yet been well studied. For example, biologists are now collecting observational data about the precise nucleotide sequences of the genomes of different species of organisms for the purpose of understanding their evolutionary relationships.

In other cases, researchers collect **experimental data,** information that describes the result of a careful manipulation of the system under study. This approach, which is known as *experimental science,* often answers questions about why or how systems work as they do. For example, a biologist who wonders whether a particular snail species influences the distribution of algae on a rocky shoreline might remove the snail from some enclosed patches of shoreline and examine whether the distribution of algae changes as a result. Similarly, a geneticist who wants to understand the role of a particular gene in the functioning of an organism might make mutations in the gene and examine the consequences.

Research Often Begins with Basic Questions That Allow the Development of Hypotheses and Controlled Experiments

Research on a previously unexplored system usually starts with basic observations (step 1 in Figure 1.14). For example, a biologist may notice a novel structure in a cell or an intact organism, or she may observe many more species of trees living in a tropical forest than in a temperate forest. Once the facts have been carefully observed and described, scientists may pose a basic question (step 2), such as "What is the structure's function?" Direct observation or laboratory manipulations—such as the removal of the structure—may provide preliminary answers.

Scientists frequently develop a **hypothesis** to explain their observations (step 2). In a report from the National Academy of Sciences, a *hypothesis* is defined as "a tentative explanation for an observation, phenomenon, or scientific problem that can be tested by further investigation." And whenever scientists create a hypothesis, they simultaneously define—either explicitly or implicitly—a **null hypothesis,** a statement of what they would see if the hypothesis being tested is not correct.

Many scientists structure their hypotheses with one crucial requirement: it must be *falsifiable* by experimentation or further observation. In other words, scientists must describe an idea in such a way that, if it is wrong, they will be able to demonstrate that it is wrong. The principle of falsifiability helps scientists define testable, focused hypotheses. Hypotheses that are testable and falsifiable fall within the realm of science, whereas those that cannot be falsified—although possibly valid and true—do not fall within the realm of science.

Hypotheses generally explain the relationship between **variables,** environmental factors that may differ among places or organismal characteristics that may differ among individuals. Thus, hypotheses yield testable **predictions** (step 3 in Figure 1.14), statements about what the researcher expects to happen to one variable if another variable changes. Scientists then test their hypotheses and predictions with experimental or observational tests (step 4) that generate relevant data, which the scientist then interprets (step 5). If the data answer the question or support the hypothesis, the researcher then develops follow-up questions or hypotheses to explore the topic more deeply. But if data from just one study refute a scientific hypothesis (that is, demonstrate that its predictions are incorrect), the scientist must modify the hypothesis and test it again or abandon it altogether. Note that although one set of data can refute a hypothesis, no amount of data can *prove* beyond a doubt that a hypothesis is correct; there may always be a contradictory example somewhere on Earth, and it is impossible to test every imaginable example. That is why scientists say that positive results *are consistent with, support,* or *confirm* a hypothesis.

To make these ideas more concrete, consider a simple example of hypothesis development and testing **(Figure 1.15).** Say that a friend gives you a plant that she grew on her windowsill. She watered it and fertilized it regularly, and under her loving care, the plant always flowered. You place the plant on your windowsill and water it regularly, but the plant never blooms—your observation (step 1). You wonder if fertilizing the plant would make it flower; in other words, you create a hypothesis (step 2) with the specific prediction that this type of plant will flower if it receives fertilizer as well as water (step 3). This is a good scientific hypothesis because it is falsifiable. To test the hypothesis, you would simply give the plant fertilizer (step 4). If it blooms, you would interpret the data (step 5) and say that it confirms your hypothesis. If it does not bloom, the data suggest that you should reject or revise your hypothesis.

One problem with this experiment is that the hypothesis does not address other possible reasons that the plant did not

FIGURE 1.15 **Experimental Research**

Hypothetical Experiment Illustrating the Use of Control Treatment and Replicates

Question: Your friend fertilizes a plant that she grows on her windowsill, and it flowers. After she gives you the plant, you put it on your windowsill, but you do not give it any fertilizer and it does not flower. Will giving the plant fertilizer induce it to flower?

Friend added fertilizer · You did not add fertilizer

Experiment: Establish six replicates of an experimental treatment (identical plants grown with fertilizer) and six replicates of a control treatment (identical plants grown without fertilizer).

Experimental Treatment · **Control Treatment**

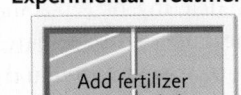 

Add fertilizer · No fertilizer

Possible Result 1: Neither experimental nor control plants flower.

Experimentals · **Controls**

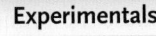  

Conclusion: Fertilizer alone does not cause the plants to flower. Consider alternative hypotheses and conduct additional experiments, each testing a different experimental treatment, such as the amount of water or sunlight the plant receives or the temperature to which it is exposed.

Possible Result 2: Plants in the experimental group flower, but plants in the control group do not.

Experimentals · **Controls**

Conclusion: The application of fertilizer induces flowering in this type of plant, confirming your original hypothesis. Pat yourself on the back and apply to graduate school in plant biology.

think like a scientist
Observe
Hypothesize
Predict
Experiment
Interpret

Suppose you were interested in studying the effects of both fertilizer and water on the flowering of your plants. How would you design an experiment that addressed the two experimental variables at the same time?

© Cengage Learning 2017

flower. Maybe it received too little water. Maybe it did not get enough sunlight. Maybe your windowsill was too cold. All of these explanations could be the basis of **alternative hypotheses,** which a conscientious scientist always considers when designing experiments. You could easily test any of these hypotheses—or some combination of them—by providing more water, more hours of sunlight, or warmer temperatures to the plant.

But even if you provide each of these necessities in turn, your efforts will not definitively confirm or refute your hypothesis unless you introduce a control treatment. The **control,** as it is often called, represents a null hypothesis; it tells us what we would see in the absence of the experimental manipulation. For example, your experiment would need to compare plants that received fertilizer (the experimental treatment) with plants grown without fertilizer (the control treatment). The presence or absence of fertilizer is the **experimental variable,** and in a controlled experiment, everything except the experimental variable—the flower pots, the soil, the amount of water, and exposure to sunlight—is exactly the same, or as close to exactly the same as possible. Thus, if your experiment is well controlled, any difference in flowering pattern observed between plants that receive the experimental treatment (fertilizer) and those that receive the control treatment (no fertilizer) can be attributed to the experimental variable. If the plants that receive fertilizer did not flower more than the control plants, you would reject your initial hypothesis. The elements of a typical experimental approach, as well as our hypothetical experiment, are summarized in Figure 1.15. Figures that present observational and experimental research using this basic format are provided throughout this book.

Notice that in the preceding discussion we discussed plants (plural) that received fertilizer and plants that did not. Nearly all experiments in biology include **replicates,** multiple subjects that receive either the same experimental treatment or the same control treatment. Scientists use replicates in experiments because individuals typically vary in genetic makeup, size, health, or other characteristics—and because accidents may disrupt a few replicates. By exposing multiple subjects to both treatments, we can use a statistical test to compare the average result of the experimental treatment with the average result of the control treatment, giving us more confidence in the overall findings. Thus, in the fertilizer experiment we described, we might expose six or more individual plants to each treatment and compare the results obtained for the experimental group with those obtained for the control group. We would also try to ensure that the individuals included in the experiment were as similar as possible. For example, we might specify that they all must be the same age or size.

When Controlled Experiments Are Unfeasible, Researchers Employ Null Hypotheses to Evaluate Observational Data

In some fields of biology, especially ecology and evolution, the systems under study may be too large or complex for experimental manipulation. In such cases, biologists can use a null hypothesis to evaluate observational data. For example, Paul E. Hertz of Barnard College studies temperature regulation in lizards. As in many other animals, a lizard's body temperature can vary substantially as environmental temperatures change. Research on many lizard species has demonstrated that they often compensate for fluctuations in environmental temperature—that is, maintain thermal homeostasis—by perching in the Sun to warm up or in the shade when they feel hot.

Hertz hypothesized that the crested anole, *Anolis cristatellus*, a lizard species in Puerto Rico, regulates its body temperature by perching in patches of Sun when environmental temperatures are low and seeking shade when environmental temperatures are high. To test this hypothesis, Hertz needed to determine what he would see if lizards were *not* trying to control their body temperatures. In other words, he needed to know the predictions of the null hypothesis: "Lizards do *not* regulate their body temperature, and they select perching sites at random with respect to factors that influence body temperature" **(Figure 1.16)**. Of course, it would be impossible to force a natural population of lizards to perch in places that define the null hypothesis. Instead, he and his students created a population of artificial lizards, copper models that served as lizard-sized, lizard-shaped, and lizard-colored thermometers. Each hollow copper model was equipped with a built-in temperature-sensing wire that can be plugged into an electronic thermometer. After constructing the copper models, Hertz and his students verified that the models reached the same internal temperatures as live lizards under various laboratory conditions. They then traveled to Puerto Rico and hung 60 models at randomly selected positions in each of the habitats where this lizard species lives.

How did the copper models allow Hertz and his students to interpret their data? Because the researchers placed these inanimate objects at random positions in the lizards' habitats, the percentages of models observed in Sun and in shade provided a measure of how sunny or shady a particular habitat was. In other words, the copper models established the null hypothesis about the percentage of lizards that would perch in sunlit spots just by chance. Similarly, the temperatures of the models provided a null hypothesis about what the temperatures of lizards would be if they perched at random in their habitats. Hertz and

his students gathered data on the use of sunny perching places and temperatures from both the copper models and live lizards. By comparing the behavior and temperatures of live lizards with the random "behavior" and random temperatures of the copper models, they demonstrated that *A. cristatellus* did, in fact, regulate its body temperature (see Figure 1.16).

Molecular Tools Allow Researchers to Explore Genes, Genomes, and Proteomes

Our knowledge of DNA and the molecules for which it codes has grown exponentially since its molecular structure was first described by James Watson and Francis Crick in 1953. Technical advances in the second half of the twentieth century enabled scientists to isolate individual genes, study them in detail, and manipulate them in the laboratory. Today, we can modify organisms by deleting or replacing some genes or by adding others. In the late 1970s, researchers began to identify the nucleotide sequences of individual genes. The sequencing process has become so much faster and cost-effective that biologists have found themselves inundated with data describing the entire genome sequences of thousands of organisms ranging from archaeans to important agricultural plants and from disease-causing microorganisms to humans.

Scientists mine this treasure trove of data using the techniques of **genomics,** the branch of biology that characterizes entire genomes. Research in genomics includes three main areas of study: (1) sequencing genomes and identifying likely protein coding and noncoding segments of DNA; (2) determining the specific functions of protein-coding genes and noncoding DNA sequences; and (3) comparing the genomes of different organisms to see how the genomes—and, by extension, the organisms themselves—have evolved. Indeed, genomics has become an important tool for elucidating the Tree of Life.

Genomics is closely associated with **proteomics,** the study of which protein molecules are produced by a cell; what functions those proteins perform; where they are localized in the cell; and to which other molecules they bind to form larger cellular structures. The complete inventory of proteins produced by each cell is called its **proteome.** For the most part, an individual's genome is the same in each of its cells and does not change through its life cycle. However, each cell's proteome is unique, and it constantly changes as a cell responds to the ever-varying environment in the individual's body. Different specialized cell types, like skin cells or muscle cells, are shaped and function differently because their proteomes contain different molecular tools. Thus, studying an organism's diverse proteomes presents a much greater challenge than sequencing and analyzing its genome. (Genomics and proteomics are discussed in Chapter 19.)

The study of genomics and proteomics has required the development of new computational and mathematical tools, collectively called **bioinformatics.** The huge volume of data on DNA or protein sequences is stored in databases accessible to the public. The tools of bioinformatics allow researchers to

FIGURE 1.16 **Observational Research**

A Field Study Using a Null Hypothesis

Anolis cristatellus

Alejandro Sanchez

Hypothesis: *Anolis cristatellus,* the crested anole, uses patches of Sun and shade to regulate its body temperature.

Null Hypothesis: If these lizards do not regulate their body temperature, individuals would select perching sites at random with respect to environmental factors that influence body temperature.

Method: The researchers created a set of hollow, copper lizard models, each equipped with a temperature-sensing wire. At study sites where the lizards live in Puerto Rico, the researchers hung 60 models at random positions in trees. They observed how often live lizards and the randomly positioned copper models were perched in patches of Sun or shade, and they measured the temperatures of live lizards and the copper models. Data from the randomly positioned copper models define the predictions of the null hypothesis.

Results: The researchers compared the frequency with which live lizards and the copper models perched in Sun or shade, as well as the temperatures of live lizards and the copper models. The data revealed that the behavior and temperatures of *A. cristatellus* were different from those of the randomly positioned models, therefore confirming the original hypothesis.

Copper *Anolis* model

Kevin de Queiroz, National Museum of Natural History, Smithsonian Institution

Percentage of models and lizards perched in Sun or shade

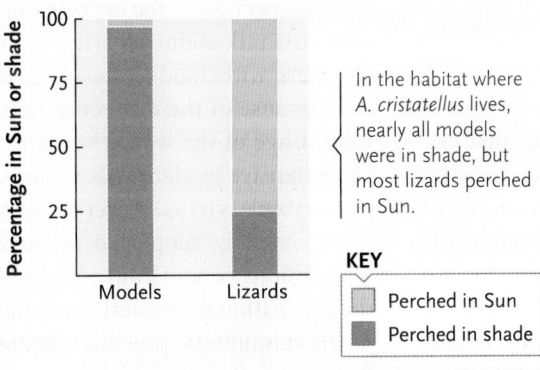

In the habitat where *A. cristatellus* lives, nearly all models were in shade, but most lizards perched in Sun.

KEY
- Perched in Sun
- Perched in shade

Temperatures of models and lizards

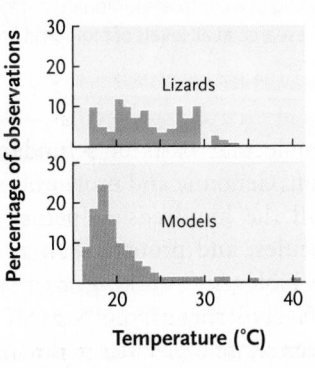

Body temperatures of *A. cristatellus* were significantly higher than those of the randomly placed models.

Conclusion: *A. cristatellus* uses patches of Sun and shade to regulate its body temperature.

think like a scientist

Observe
Hypothesize
Predict
Experiment
Interpret

Based on the data presented in the two graphs, how would you describe the variability in temperature among shaded sites in this environment?

Source: P. E. Hertz. 1992. Temperature regulation in Puerto Rican *Anolis* lizards: A field test using null hypotheses. *Ecology* 73:1405–1417.

compare newly acquired sequences to those documented in other organisms, to predict the structure and function of gene products, and to create hypotheses about how particular DNA sequences have evolved. To give you a sense of the exciting impact of molecular, genomic, and proteomic research on all areas of biology, nearly every chapter in this book includes an essay titled *Molecular Insights.*

Systems biology explains the emergent properties of biological systems by studying the interactions of their parts. Many scientists use a *reductionist* approach in their work: they focus on the small, individual parts of living system, such as identifying the DNA sequence in a gene or determining the temperature at which a digestive enzyme in a lizard's gut functions most efficiently. Other scientists, including systems biologists, adopt a *holistic* approach: they integrate discoveries from many fields to understand a complete biological entity, such as a living cell. Recent advances in genomics, proteomics, and bioinformatics now allow systems biologists to learn how molecules in cells interact with each other, helping us to identify the functions of proteins and other molecules *in relation to* the functions of the molecules with which they work. Understanding this dynamic network of interactions allows biologists to pose new questions and hypotheses.

Research in systems biology is necessarily a collaborative effort that begins with questions posed by a biologist (**Figure 1.17**). These questions are answered with data generated by genomics and proteomics, coupled with bioinformatic analyses. For example, biologists might ask a question about

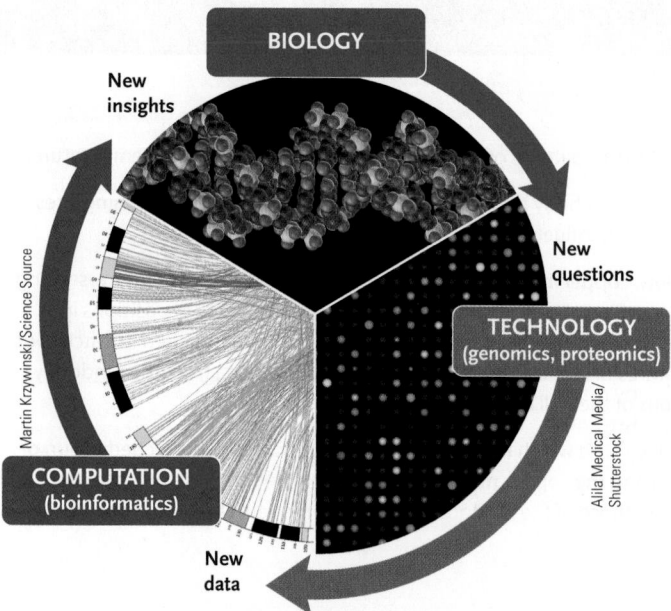

New insights

BIOLOGY

New questions

TECHNOLOGY (genomics, proteomics)

COMPUTATION (bioinformatics)

New data

Martin Krzywinski/Science Source

Ailia Medical Media/Shutterstock

FIGURE 1.17 Systems biology. Systems biology is a holistic approach to science in which researchers integrate information from many fields to develop an understanding of a complete biological entity. The systems approach is useful for research at all levels of biological organization, from the cell to the biosphere.

the genetic and molecular basis of a fundamental process, such as cell division. Genomic and proteomic studies identify the "parts list," all the hundreds or perhaps thousands of genes, RNA molecules, and proteins that participate in the process. Then the biologists, working with statisticians and computer scientists, construct a hypothesis or model of all the interactions between those molecules to describe the complex sequence of events that allows one cell to divide into two. (Cell division is described in detail in Chapter 10.)

The systems approach—that is, understanding the "whole" by studying the interactions of its "parts"—is useful to scientists working at every level of biological organization, and it contributes to many applications across the biological sciences. For example, genomic analyses of humans can identify their susceptibility to certain diseases, such as cancer or Alzheimer disease. And by modeling the body's complex and dynamic responses to both the disease and medications, a systems biologist can ask what effect the administration of a preventive or therapeutic drug might have on the affected organ and on other organs throughout the body. In another application, systems biology is helping neuroscientists understand the complex interactions among neurons, the cells that carry information in the brains of animals; these processes enable us to learn and remember new information, solve problems, and experience emotional responses. As a final example, systems biology is also helping ecologists understand how bacteria and other tiny organisms, many of which cannot yet be cultured in the laboratory, affect the health of ecosystems and how they might be useful in decontamination efforts that reduce the harm caused by pollutants that humans increasingly release into the environment.

Scientific Theories Are Grand Ideas That Have Withstood the Test of Time

When a hypothesis stands up to repeated experimental tests, it is gradually accepted as an accurate explanation of natural events. This acceptance may take many years, and it usually involves repeated experimental confirmations. When many different tests have consistently confirmed a hypothesis that addresses many broad questions, it may become regarded as a **scientific theory.** In the report from the National Academy of Sciences, a *scientific theory* is defined as "a plausible or scientifically acceptable, well-substantiated explanation of some aspect of the natural world. . . that applies in a variety of circumstances to explain a specific set of phenomena and predict the characteristics of as yet unobserved phenomena."

Most scientific theories are supported by exhaustive experimentation; thus, scientists usually regard them as established truths that are unlikely to be contradicted by future research. Note that this use of the word *theory* is quite different from its informal meaning in everyday life. In common usage, the word *theory* most often labels an idea as either speculative or downright suspect, as in the expression "It's only a theory." But when scientists talk about theories, they do so with respect for ideas that have withstood the test of many experiments.

Because of the difference between the scientific and common usage of the word *theory,* many people fail to appreciate the extensive evidence that supports most scientific theories. For example, virtually every scientist accepts the theory of evolution as fully supported scientific truth: all species change with time, new species are formed, and older species eventually die off. Although evolutionary biologists debate the details of how evolutionary processes bring about these changes, very few scientists doubt that the theory of evolution is essentially correct. Moreover, *no scientist who has tried to cast doubt on the theory of evolution has ever devised or conducted a study that disproves any part of it.* Unfortunately, the confusion between the scientific and common usage of the word *theory* has led, in part, to endless public debate about supposed faults and inadequacies in the theory of evolution.

Curiosity and the Joy of Discovery Motivate Scientific Research

What drives scientists in their quest for knowledge? The motivations of scientists are as complex as those driving people toward any goal. Intense curiosity about ourselves, our fellow creatures, and the chemical and physical objects of the world and their interactions is a basic ingredient of scientific research. The discovery of information that no one knew before is as exciting to a scientist as finding buried treasure. There is also an element of play in science, a joy in the manipulation of scientific ideas and apparatus, and the chase toward a scientific goal. Biological research also has practical motivations—for example, to cure disease or improve agricultural productivity. In all of this research, one strict requirement of science is honesty—without honesty in the gathering and reporting of results, the work of science is meaningless. Dishonesty is actually rare in science, not least because repetition of experiments by others soon exposes any funny business.

Whatever the level of investigation or the motivation, the work of every scientist adds to the fund of knowledge about us and our world. For better or worse, the scientific method—that inquiring and skeptical approach—has provided knowledge and technology that have revolutionized the world and improved the quality of human life immeasurably. This book presents the fruits of biologists' labors in the most important and fundamental areas of biological science—cell and molecular biology, genetics, evolution, systematics, physiology, developmental biology, ecology, and behavioral science.

STUDY BREAK 1.4

1. In your own words, explain the most important requirement of a scientific hypothesis.
2. What information did the copper lizard models provide in the study of temperature regulation described earlier?
3. How would you respond to a nonscientist who told you that Darwin's ideas about evolution were "just a theory"?

THINK OUTSIDE THE BOOK

Learn more about the scientific method and the richness of the scientific enterprise by visiting this web site, maintained by researchers at the University of California, Berkeley: http://undsci.berkeley.edu/.

REVIEW KEY CONCEPTS

For access to MindTap and additional study materials visit www.cengagebrain.com.

1.1 What Is Life? Characteristics of Living Organisms

- Living systems are organized in a hierarchy, each level having its own emergent properties (Figure 1.2): cells, the lowest level of organization that is alive, are organized into unicellular or multicellular organisms; populations are groups of organisms of the same kind that live together in the same area; an ecological community comprises all the populations living in an area, and ecosystems include communities that interact through their shared physical environment; the biosphere includes all of Earth's ecosystems.

- Living organisms have complex structures established by instructions coded in their DNA (Figure 1.3). The instructions for many characteristics are coded by segments of DNA called genes, which are passed from parents to offspring in reproduction. The information in genes is transcribed into RNA, some of which is then translated into protein molecules (Figure 1.4). Proteins carry out most of the activities of life. All of the DNA in the cells of an organism constitutes its genome.

- Living cells and organisms engage in metabolism, obtaining energy and using it to maintain themselves, grow, and reproduce. The two primary metabolic processes are photosynthesis and cellular respiration (Figure 1.5).

- Energy that flows through the hierarchy of life is eventually released as heat. By contrast, matter is recycled within the biosphere (Figure 1.6).

- Cells and organisms use receptors to detect environmental changes and trigger a compensating reaction that allows the organism to survive.

- Organisms reproduce, and their offspring develop into mature, reproductive adults (Figure 1.7).

- Populations of living organisms undergo evolutionary change as generations replace one another over time.

1.2 Biological Evolution

- The structure, function, and types of organisms in populations change with time. According to the theory of evolution by natural selection, certain characteristics allow some organisms to survive better and reproduce more than others in their population. If the instructions that produce those characteristics are coded in DNA, successful characteristics will become more common in later generations. As a result, the characteristics of the offspring generation will differ from those of the parent generation (Figure 1.8).

- Mutations—changes in the structure, number, or arrangement of DNA molecules—create variability among individuals. Variability is the raw material of natural selection and other processes that cause biological evolution.

- Over many generations, the accumulation of favorable characteristics may produce adaptations, which enable individuals to survive longer or reproduce more (Figure 1.9).

- Over long spans of time, the accumulation of different adaptations and other genetic differences between populations has produced the diversity of life on Earth.

1.3 Biodiversity and the Tree of Life

- Scientists classify organisms in a hierarchy of categories. The species is the most fundamental category, followed by genus, family, order, class, phylum, and kingdom as increasingly inclusive categories (Figure 1.10). Analyses of DNA structure now allow biologists to construct the Tree of Life, a model of the evolutionary relationships among all known living organisms (Figure 1.11).

- Most biologists recognize three domains—Bacteria, Archaea, and Eukarya—based on fundamental characteristics of cell structure and molecular analysis. The Bacteria and Archaea each include one kingdom; the Eukarya is divided into the protists and three kingdoms: Plantae, Fungi, and Animalia (Figures 1.12 and 1.13).

- Model organisms, many of which are easy to maintain in the laboratory, have been the subject of much research.

1.4 Biological Research

- Biologists conduct basic research to advance our knowledge of living organisms and applied research to solve practical problems.

- The scientific method allows researchers to crystallize and test their ideas. Scientists ask questions and develop hypotheses, which are working explanations about the relationships between variables. Scientific hypotheses must be falsifiable (Figure 1.14).

- Scientists may collect observational data, which describe particular organisms or the details of biological processes, or experimental data, which describe the results of an experimental manipulation.

- A well-designed experiment considers alternative hypotheses and includes control treatments and replicates (Figure 1.15). When experiments are unfeasible, biologists often use null hypotheses, explanations of what they would see if their hypothesis was wrong, to evaluate data (Figure 1.16).
- Genomics is the branch of biology that characterizes entire genomes, allowing researchers to study the protein coding and noncoding segments of DNA; predict the structure and function of DNA products; and develop and test hypotheses about the evolution of the Tree of Life. The science of proteomics is the study of all the proteins coded in a genome and the interactions of those molecules within cells. Systems biology is a holistic approach that integrates research from many fields to understand the functioning of entire biological entities (Figure 1.17).

- A scientific theory is a set of broadly applicable hypotheses that have been supported by repeated tests under many conditions and in many different situations. The theory of evolution by natural selection is of central importance to biology because it explains how life evolved through natural processes.

TEST YOUR KNOWLEDGE

Remember/Understand

1. What is the lowest level of biological organization that biologists consider to be alive?
   a. a protein
   b. DNA
   c. a cell
   d. a multicellular organism
   e. a population of organisms

2. Which category falls immediately below "class" in the systematic hierarchy?
   a. species
   b. order
   c. family
   d. genus
   e. phylum

3. Houseflies develop through a series of programmed stages from egg, to larva, to pupa, to flying adult. This series of stages is called:
   a. artificial selection.
   b. respiration.
   c. homeostasis.
   d. a life cycle.
   e. metabolism.

4. Which structure allows living organisms to detect changes in the environment?
   a. a protein
   b. a receptor
   c. a gene
   d. RNA
   e. a nucleus

5. Which of the following characteristics would *not* qualify an animal as a model organism?
   a. It has rapid development.
   b. It has a small adult size.
   c. It has a rapid life cycle.
   d. It has unique genes and unusual cells.
   e. It is easy to raise in the laboratory.

6. **Discuss Concepts** Explain why control treatments are a necessary component of well-designed experiments.

Apply/Analyze

7. Which of the following represents the application of the "scientific method"?
   a. comparing one experimental subject to one control subject
   b. believing an explanation that is too complex to be tested
   c. using controlled experiments to test falsifiable hypotheses
   d. developing one testable hypothesis to explain a natural phenomenon
   e. observing a once-in-a-lifetime event under natural conditions

8. Which of the following is *not* a component of Darwin's theory as he understood it?
   a. Some individuals in a population survive longer than others.
   b. Some individuals in a population reproduce more than others.
   c. Heritable variations allow some individuals to compete more successfully for resources.
   d. Mutations in genes produce new variations in a population.
   e. Some new variations are passed to the next generation.

9. Which of the following questions best exemplifies basic research?
   a. How did life begin?
   b. How does alcohol intake affect aging?
   c. How fast does H1N1 flu spread among humans?
   d. How can we reduce hereditary problems in purebred dogs?
   e. How does the consumption of soft drinks promote obesity?

10. When researchers say that a scientific hypothesis must be falsifiable, they mean that:
    a. the hypothesis must be proved correct before it is accepted as truth.
    b. the hypothesis has already withstood many experimental tests.
    c. they have an idea about what will happen to one variable if another variable changes.
    d. appropriate data can prove without question that the hypothesis is correct.
    e. if the hypothesis is wrong, scientists must be able to demonstrate that it is wrong.

11. **Discuss Concepts** Many viruses are infectious agents that contain DNA surrounded by a protein coat. They cannot reproduce on their own, but they can take over the cells of the organisms they infect and force those cells to produce more virus particles. Based on the characteristics of living organisms described in this chapter, should viruses be considered living organisms?

Evaluate/Create

12. What role did the copper lizard models play in the field study of temperature regulation?
    a. They attracted live lizards to the study site.
    b. They measured the temperatures of live lizards.
    c. They established null hypotheses about basking behavior and temperatures.
    d. They scared predators away from the study site.
    e. They allowed researchers to practice taking lizard temperatures.

13. **Discuss Concepts** While walking in the woods, you discover a large rock covered with a gelatinous, sticky substance. What tests could you perform to determine whether the substance is inanimate, alive, or the product of a living organism?

14. **Design an Experiment** Design an experiment to test the hypothesis that the color of farmed salmon is produced by pigments in their food.

15. **Apply Evolutionary Thinking** When a biologist first tested a new pesticide on a population of insects, she found that only 1% of the insects survived their exposure to the poison. She allowed the survivors to reproduce and discovered that 10% of the offspring survived exposure to the same concentration of pesticide. One generation later, 50% of the insects survived this experimental treatment. What is a likely explanation for the increasing survival rate of these insects over time?

For selected answers, see Appendix A.

INTERPRET THE DATA

While working in Puerto Rico, Paul E. Hertz and his students studied a second species of lizard, the yellow-chinned anole, *Anolis gundlachi,* using the procedures described in Figure 1.16. Their results for copper models and living lizards are presented in the **Figure.** Based on these data, do you think that *A. gundlachi* regulates its body temperature? Why or why not?

Percentage of models and lizards perched in Sun or shade

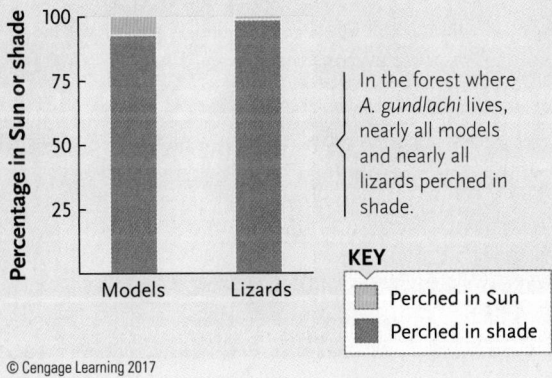

In the forest where *A. gundlachi* lives, nearly all models and nearly all lizards perched in shade.

KEY

Perched in Sun
Perched in shade

© Cengage Learning 2017

Temperatures of models and lizards

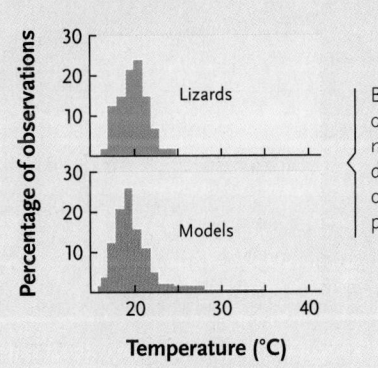

Body temperatures of *A. gundlachi* were not significantly different from those of the randomly placed models.

© Cengage Learning 2017

Anolis gundlachi

2 Life, Chemistry, and Water

Source: Kuttelvaserova Stuchelova/shutterstock.com

Life as we know it would be impossible without water, a small inorganic compound with unique properties.

Why it matters . . . Biology is the study of biological organisms and the viruses that infect them, and of the biological processes they exhibit. We show wonder at the wide range of those processes: A bacterium infects a human and causes an infection. A fungus grows on a downed tree and obtains nutrients from it. A spider spins a web and uses it to capture prey. An embryo grows within a mother's womb into a baby that is born into the world. To understand these processes completely, we need to understand the chemistry of organisms. That is, biological organisms are collections of atoms and molecules linked together by chemical bonds. Molecules are synthesized and broken down by chemical reactions. The pathways involved developed over evolutionary time, meaning that modern-day organisms utilize the same basic chemical reactions.

Chemical reactions take place mostly within cells, and cells make up tissues, organs, and the whole organism itself. In fact, many thousands of chemical reactions take place inside living organisms. Decades of research have taught us much about these reactions and have confirmed that the same laws of chemistry and physics govern both living and nonliving things. Therefore, we can apply with confidence information obtained from chemical experiments in the laboratory to the processes inside living organisms. An understanding of the relationship between the structure of chemical substances and their behavior is the first step in learning biology.

2.1 The Organization of Matter: Elements and Atoms

All **matter** of the universe—anything that occupies space and has mass—is composed of *elements* and combinations of elements. An **element** is a pure substance that cannot be broken down into simpler substances by ordinary chemical or physical techniques. Ninety-two different elements occur naturally on Earth, and more than fifteen artificial elements have been synthesized in the laboratory.

Living Organisms Are Composed of about 25 Key Elements

Four elements—carbon, hydrogen, oxygen, and nitrogen—make up more than 96% of the weight of living organisms. Seven other elements—calcium, phosphorus, potassium, sulfur, sodium, chlorine, and magnesium—contribute most of the remaining 4%. Several other elements occur in organisms in quantities so small (less than 0.01%) that they are known as **trace elements. Figure 2.1** compares the relative proportions of different elements in a human, a plant, Earth's crust, and seawater, and lists the most important trace elements in a human. The proportions of elements in living organisms, as represented by the human and the plant, differ markedly from those of Earth's crust and seawater; these differences reflect the highly ordered chemical structure of living organisms.

Trace elements are vital for normal biological functions. For example, iodine makes up only about 0.0004% of a human's weight. However, a lack of iodine in the human diet severely impairs the function of the thyroid gland, which produces hormones that regulate metabolism and growth (see Chapter 42). Symptoms of iodine deficiency include lethargy, apathy, and sensitivity to cold temperatures. Prolonged iodine deficiency causes a *goiter*, a condition in which the thyroid gland enlarges so much that the front of the neck swells significantly. Once a common condition, goiter has almost been eliminated by adding iodine to table salt.

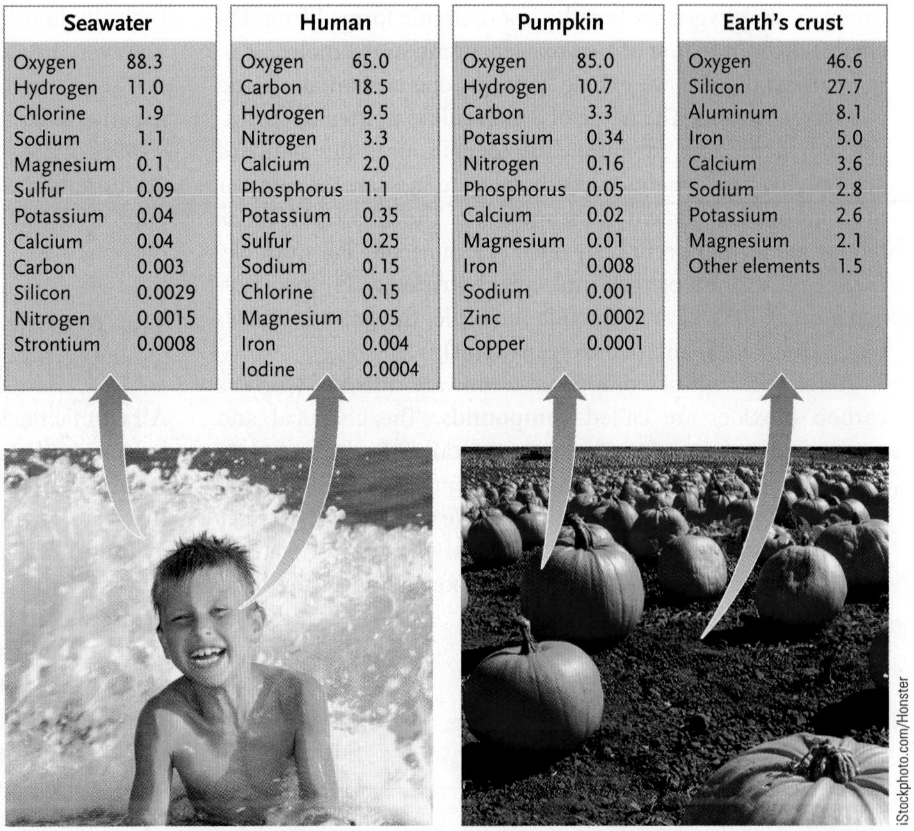

Elements Are Composed of Atoms, Which Combine to Form Molecules

Elements are composed of individual **atoms**—the smallest units that retain the chemical and physical properties of an element. Any given element has only one type of atom. Several million atoms arranged side by side would be needed to equal the width of the period at the end of this sentence.

Atoms are identified by a standard one- or two-letter symbol. For example, the element carbon is identified by the single letter C, which stands for both the carbon atom and the element, and iron is identified by the two-letter symbol Fe (*ferrum* = iron). **Table 2.1** lists the chemical symbols of these and other atoms common in living organisms.

Atoms, combined chemically in fixed numbers and ratios, form the **molecules** of living and nonliving matter. For

Seawater	
Oxygen	88.3
Hydrogen	11.0
Chlorine	1.9
Sodium	1.1
Magnesium	0.1
Sulfur	0.09
Potassium	0.04
Calcium	0.04
Carbon	0.003
Silicon	0.0029
Nitrogen	0.0015
Strontium	0.0008

Human	
Oxygen	65.0
Carbon	18.5
Hydrogen	9.5
Nitrogen	3.3
Calcium	2.0
Phosphorus	1.1
Potassium	0.35
Sulfur	0.25
Sodium	0.15
Chlorine	0.15
Magnesium	0.05
Iron	0.004
Iodine	0.0004

Pumpkin	
Oxygen	85.0
Hydrogen	10.7
Carbon	3.3
Potassium	0.34
Nitrogen	0.16
Phosphorus	0.05
Calcium	0.02
Magnesium	0.01
Iron	0.008
Sodium	0.001
Zinc	0.0002
Copper	0.0001

Earth's crust	
Oxygen	46.6
Silicon	27.7
Aluminum	8.1
Iron	5.0
Calcium	3.6
Sodium	2.8
Potassium	2.6
Magnesium	2.1
Other elements	1.5

FIGURE 2.1 The proportions by mass of different elements in seawater, the human body, a fruit, and Earth's crust. Trace elements in humans include boron, chromium, cobalt, copper, fluorine, iodine, iron, manganese, molybdenum, selenium, tin, vanadium, and zinc, as well as variable traces of other elements.
© Cengage Learning 2017

TABLE 2.1	Atomic Number and Mass Number of the Most Common Elements in Living Organisms		
Element	Symbol	Atomic Number	Mass Number of the Most Common Form
Hydrogen	H	1	1
Carbon	C	6	12
Nitrogen	N	7	14
Oxygen	O	8	16
Sodium	Na	11	23
Magnesium	Mg	12	24
Phosphorus	P	15	31
Sulfur	S	16	32
Chlorine	Cl	17	35
Potassium	K	19	39
Calcium	Ca	20	40
Iron	Fe	26	56
Iodine	I	53	127

example, the oxygen we breathe is a molecule formed from the chemical combination of two oxygen atoms, and a molecule of the carbon dioxide we exhale contains one carbon atom and two oxygen atoms. The name of a molecule is written in chemical shorthand as a **formula.** The formula uses the standard symbols for the elements and subscripts to indicate the number of atoms of each element in the molecule. The subscript is omitted for atoms that occur only once in a molecule. For example, the formula for an oxygen molecule is written as O_2 (two oxygen atoms); for a carbon dioxide molecule, the formula is CO_2 (one carbon atom and two oxygen atoms).

Molecules whose component atoms are different (such as carbon dioxide) are called **compounds.** The chemical and physical properties of compounds typically are distinct from those of their atoms or elements. For example, we all know that water is a liquid at room temperature. We also know that water does not burn. However, the individual elements of water—hydrogen and oxygen—are gases at room temperature, and both are highly reactive.

STUDY BREAK 2.1

Distinguish between an element and an atom, and between a molecule and a compound.

2.2 Atomic Structure

Each element consists of one type of atom and all atoms share the same basic structure **(Figure 2.2).** Located at the center of an atom is the **atomic nucleus,** which consists of subatomic particles called **protons** and **neutrons.** A proton carries one unit of positive electrical charge, whereas a neutron is uncharged. The atomic nucleus is surrounded by one or more smaller, fast-moving particles called **electrons.** An electron carries one unit of negative electrical charge. In figures, electrons are shown

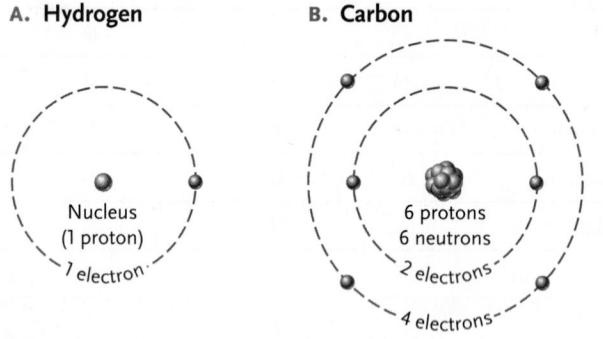

A. Hydrogen   **B. Carbon**

FIGURE 2.2 Atomic structure. The nucleus of an atom contains one or more positively charged protons and, except for the most common form of hydrogen, a similar number of uncharged neutrons. Fast-moving negatively charged electrons, in numbers equal to the protons, surround the nucleus. The most common form of hydrogen, the simplest atom, has a single proton in its nucleus and a single electron **(A).** Carbon, a more complex atom, has a nucleus surrounded by electrons at two levels **(B).** The electrons in the outer level follow more complex pathways than shown here.

© Cengage Learning 2017

typically in orbital paths around the atomic nucleus. In reality, however, the electrons move rapidly around the atomic nucleus and they can be in any location, essentially forming a cloud of negative charge. Therefore, the electrons occupy more than 99.99% of the space of an atom, even though the nucleus makes up more than 99.99% of its total mass. The attraction between opposite electrical charges keeps the electrons near the atomic nucleus.

The Atomic Nucleus Contains Protons and Neutrons

All atomic nuclei contain one or more protons. The number of protons in the nucleus of each kind of atom is referred to as the **atomic number** (see Table 2.1). This number does not vary and, therefore, identifies the atom specifically. The smallest atom, hydrogen, has a single proton in its nucleus, so its atomic number is 1. The heaviest naturally occurring atom, uranium, has 92 protons in its nucleus and therefore has an atomic number of 92. Carbon with six protons, nitrogen with seven protons, and oxygen with eight protons have atomic numbers of 6, 7, and 8, respectively.

With one exception, the nuclei of all atoms also contain neutrons. Neutrons occur in variable numbers approximately equal to the number of protons. The lone exception is the most common form of hydrogen, which has a single proton in its nucleus. There are two less common forms of hydrogen: one, deuterium, has one neutron and one proton in its nucleus; the other, tritium, has two neutrons and one proton.

Other atoms also have common and less common forms with different numbers of neutrons. For example, the most common form of the carbon atom has six protons and six neutrons in its nucleus, but about 1% of carbon atoms have six protons and seven neutrons in their nuclei and an even smaller percentage of carbon atoms have six protons and eight neutrons.

The distinct forms of the atoms of an element, all with the same number of protons but different numbers of neutrons, are called **isotopes (Figure 2.3).** The various isotopes of an atom differ in mass and other physical characteristics, but all have essentially the same chemical properties. Therefore, organisms can use any hydrogen or carbon isotope, for example, without a change in their chemical reactions.

A neutron and a proton have almost the same mass, about 1.66×10^{-24} grams (g). This mass is defined as a standard unit, the **dalton** (Da), named after John Dalton, a nineteenth-century English scientist who contributed to the development of atomic theory. Atoms are assigned a **mass number** based on the total number of protons and neutrons in the atomic nucleus (see Table 2.1). Electrons are ignored in determinations of atomic mass because the mass of an electron, at only 1/1,800th of the mass of a proton or neutron, does not contribute significantly to the mass of an atom. Thus, the mass number of the hydrogen isotope with one proton in its nucleus is 1, and its mass is 1 Da. The mass number of the hydrogen isotope

Applied Research: Using Radioisotopes in Medicine

Radioisotopes are widely used in medicine to diagnose and cure disease, to produce images of diseased body organs, and to trace the locations and routes followed by individual substances marked for identification by radioactivity. One example of their use is in the diagnosis of thyroid gland disease. The thyroid is the only structure in the body that absorbs iodine in quantity. The size and shape of the thyroid, which reflect its health, are measured by injecting a small amount of a radioactive iodine isotope into the patient's bloodstream. After the isotope is concentrated in the thyroid, the gland is then scanned by an apparatus that uses the radioactivity to produce an image of the gland on X-ray film (the energy of the radioisotope decay exposes the film). Examples of what the scans may show are presented in the **Figure.**

Another application uses the fact that radioactive thallium is not taken up by regions of the heart muscle with poor circulation to detect coronary artery disease. Other isotopes are used to detect bone injuries and defects, including injured, arthritic, or abnormally growing segments of bone.

Treatment of disease with radioisotopes takes advantage of the fact that radioactivity in large doses can kill cells (radiation generates highly reactive chemical groups that break and disrupt biological molecules). For instance, dangerously overactive thyroid glands are treated by giving patients a dose of radioactive iodine calculated to destroy just enough thyroid cells to reduce activity of the gland to normal levels. In radiation therapy, cancer cells are killed by bombarding them with radiation emitted by radium-226 or cobalt-60. As much as is possible, the radiation is focused on the tumor to avoid destroying nearby healthy tissues.

Normal   Enlarged   Cancerous

FIGURE Scans of human thyroid glands after iodine-123 (^{123}I) was injected into the bloodstream. The radioactive iodine becomes concentrated in the thyroid gland.

---

deuterium is 2, and the mass number of tritium is 3. These hydrogen mass numbers are written as ^{2}H and ^{3}H, respectively. The carbon isotope with six protons and six neutrons in its nucleus has a mass number of 12; the isotope with six protons and seven neutrons has a mass number of 13, and the isotope with six protons and eight neutrons has a mass number of 14 (see Figure 2.3). These carbon mass numbers are written as ^{12}C, ^{13}C, and ^{14}C, or carbon-12, carbon-13, and carbon-14, respectively. However, all carbon isotopes have the same atomic number of 6, because this number reflects only the number of protons in the nucleus.

What is the meaning of *mass* as compared to *weight*? **Mass** is the amount of matter in an object, whereas **weight** is a measure of the pull of gravity on an object. Mass is constant, but the weight of an object may vary because of differences in gravity. For example, a piece of lead that weighs 1 kilogram (kg) on Earth is weightless in an orbiting spacecraft, even though its mass is the same in both places. However, on Earth's surface, an object has the same mass and weight. Thus, the weight of an object in the laboratory accurately reflects its mass.

The Nuclei of Some Atoms Are Unstable and Tend to Break Down to Form Simpler Atoms

The nuclei of some isotopes are unstable and break down, or *decay,* giving off particles of matter and energy that can be detected as **radioactivity.** The decay transforms the unstable, radioactive isotope—called a **radioisotope**—into an atom of another element. The decay continues at a steady, clocklike rate, with a constant proportion of the radioisotope breaking down at any instant. The rate of decay is not affected by chemical reactions or environmental conditions such as temperature or pressure. For example, the carbon isotope ^{14}C is unstable and undergoes radioactive decay in which one of its neutrons splits into a proton and an electron. The electron is ejected from the nucleus, but the proton is retained, giving a new total of seven

Isotopes of hydrogen

^{1}H ^{2}H (deuterium) ^{3}H (tritium)
1 proton   1 proton   1 proton
                         1 neutron   2 neutrons
Atomic number = 1   Atomic number = 1   Atomic number = 1
Mass number = 1   Mass number = 2   Mass number = 3

Isotopes of carbon

^{12}C ^{13}C ^{14}C
6 protons   6 protons   6 protons
6 neutrons   7 neutrons   8 neutrons
Atomic number = 6   Atomic number = 6   Atomic number = 6
Mass number = 12   Mass number = 13   Mass number = 14

FIGURE 2.3 The atomic nuclei of hydrogen and carbon isotopes. Note that isotopes of an atom have the same atomic number but different mass numbers.

protons and seven neutrons, which is characteristic of the most common form of nitrogen. Thus, the decay transforms the carbon atom into an atom of nitrogen.

A technique called **radiometric dating** uses the clocklike decay of unstable isotopes to estimate the age of organic material, rocks, or fossils that contain them. With respect to evolution, radiometric dating is a particularly important technique for tracing evolutionary lineages through analysis of fossils (see Chapter 23). Isotopes are also used in biological research as **tracers** to label molecules so that they can be tracked as they pass through biochemical reactions. Radioactive isotopes of carbon (^{14}C), phosphorus (^{32}P), and sulfur (^{35}S) can be traced easily by their radioactivity. A number of stable, nonradioactive isotopes, such as ^{15}N (called heavy nitrogen), can be detected by their mass differences and have also proved valuable as tracers in biological experiments. *Focus on Research: Applied Research* describes some applications of radioisotopes in medicine.

The Electrons of an Atom Occupy Orbitals around the Nucleus

In an atom, the number of electrons surrounding the nucleus is equal to the number of protons in the nucleus. An electron carries a negative charge that is exactly equal and opposite to the positive charge of a proton. The equality of numbers of electrons and protons in an atom balances the positive and negative charges and makes the total structure of an atom electrically neutral.

As mentioned in the introduction to this section, an atom is often drawn with electrons following a pathway around the nucleus similar to planets orbiting a Sun. The reality is different. The speed of electrons in motion around the nucleus approaches the speed of light. At any instant, an electron may be in any location with respect to its nucleus, from the immediate vicinity of the nucleus to practically infinite space. An electron moves so fast that it almost occupies all the locations at the same time. However, the electron passes through some locations much more frequently than others. The locations where an electron occurs most frequently around the atomic nucleus define a path called an *orbital*. An **orbital** is the region of space where the electron "lives" most of the time. Although either one or two electrons may occupy an orbital, the most stable and balanced condition occurs when an orbital contains a pair of electrons.

Electrons are maintained in their orbitals by a combination of attraction to the positively charged nucleus and mutual repulsion because of their negative charge. The orbitals take different shapes depending on their distance from the nucleus and their degree of repulsion by electrons in other orbitals.

Under certain conditions, electrons may pass from one orbital to another within an atom, enter orbitals shared by two or more atoms, or pass completely from orbitals in one atom to orbitals in another. As discussed later in this chapter, the ability of electrons to move from one orbital to another underlies the chemical reactions that combine atoms into molecules.

Orbitals Occur in Discrete Layers around an Atomic Nucleus

Within an atom, electrons are found in regions of space called **energy levels,** or more simply, **shells.** Within each energy level, electrons are grouped into orbitals. The lowest energy level of an atom, the one nearest the nucleus, may be occupied by a maximum of two electrons in a single orbital **(Figure 2.4A).** This orbital, which has a spherical shape, is called the 1s orbital. (The "1" signifies that the orbital is in the energy level closest to the nucleus, and the "s" signifies the shape of the orbital, in this case, spherically symmetric around the nucleus.) Hydrogen has one electron in this orbital, and helium has two.

Atoms with atomic numbers between 3 (lithium) and 10 (neon) have two energy levels, with two electrons in the 1s orbital and one to eight electrons in orbitals at the next highest energy level. The electrons at this second energy level occupy one spherical orbital, called the 2s orbital **(Figure 2.4B),** and as

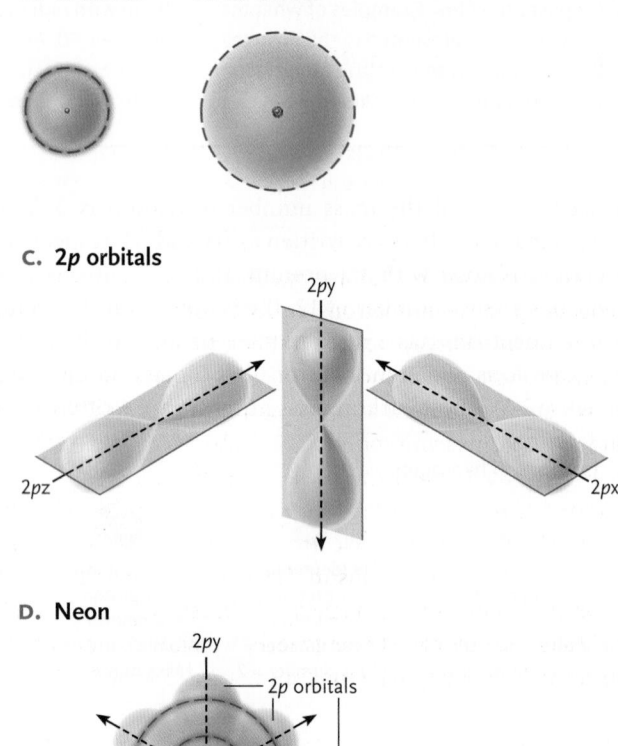

A. 1s orbital   **B. 2s orbital**

C. 2p orbitals

2py

2pz

2px

D. Neon

2py

2p orbitals

1s orbital
2s orbital

2pz   2px

FIGURE 2.4 Electron orbitals. (A) The single 1s orbital of hydrogen and helium approximates a sphere centered on the nucleus. **(B)** The 2s orbital. **(C)** The 2p orbitals lie in the three planes x, y, and z, each at right angles to the others. **(D)** In atoms with two energy levels, such as neon, the lowest energy level is occupied by a single 1s orbital as in hydrogen and helium. The second, higher energy level is occupied by a maximum of four orbitals—a spherical 2s orbital and three dumbbell-shaped 2p orbitals.
© Cengage Learning 2017

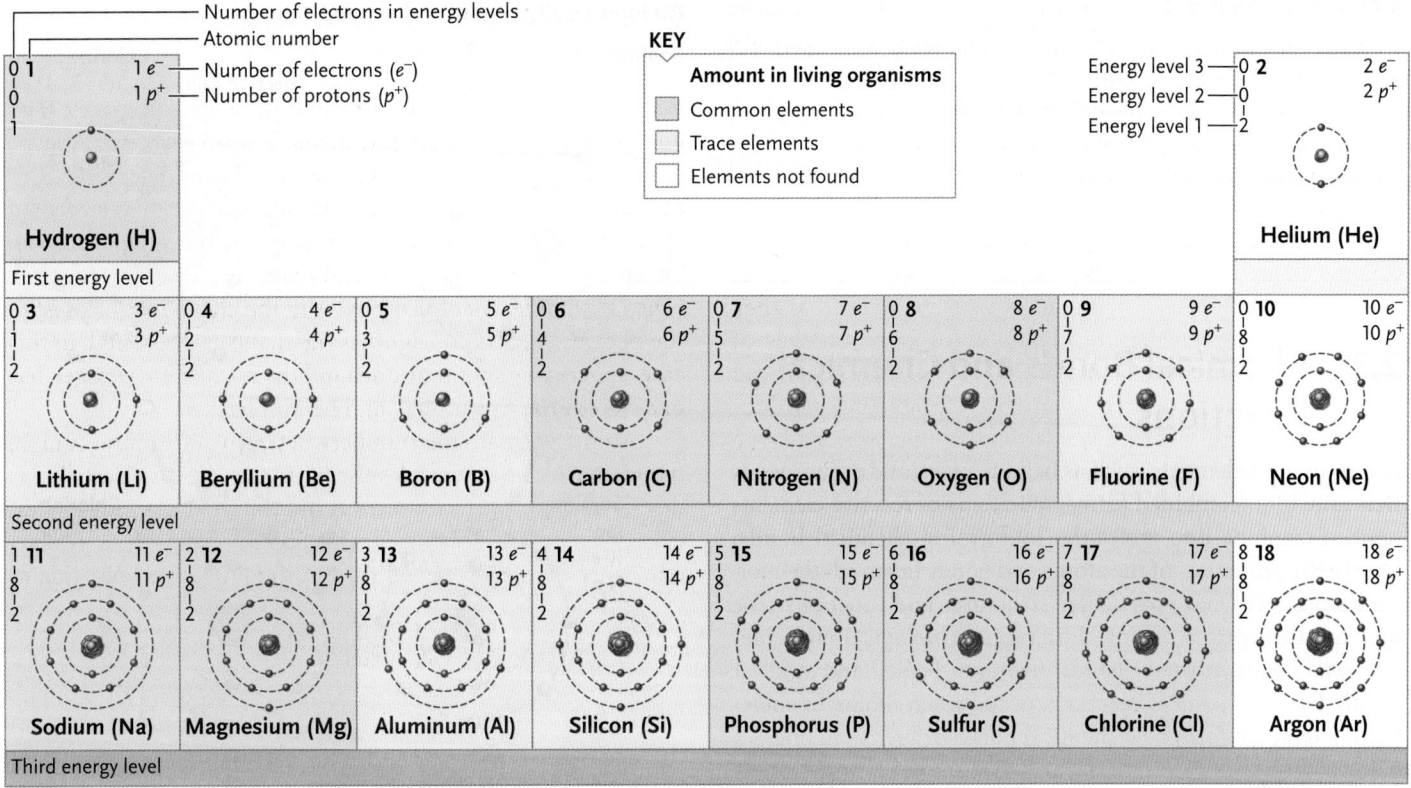

FIGURE 2.5 The atoms with electrons distributed in one, two, or three energy levels. The atomic number of each element (shown in boldface in each panel) is equivalent to the number of protons in its nucleus.

© Cengage Learning 2017

many as three orbitals that are pushed into a dumbbell shape by repulsions between electrons, called *2p* orbitals **(Figure 2.4C)**. The orbitals for neon are shown in **Figure 2.4D.**

Larger atoms have more energy levels. The third energy level, which may contain as many as 18 electrons in 9 orbitals, includes the atoms from sodium (11 electrons) to argon (18 electrons). (**Figure 2.5** shows the 18 elements that have electrons in the lowest three energy levels only.) The fourth energy level may contain as many as 32 electrons in 16 orbitals. In all cases, the total number of electrons in the orbitals is matched by the number of protons in the nucleus. However, whatever the size of an atom, the outermost energy level typically contains one to eight electrons occupying a maximum of four orbitals.

The Number of Electrons in the Outermost Energy Level of an Atom Determines Its Chemical Activity

The electrons in an atom's outermost energy level are known as **valence electrons** (*valentia* = power or capacity). Atoms in which the outermost energy level is not completely filled with electrons tend to be chemically reactive, whereas those with a completely filled outermost energy level are nonreactive, or inert. For example, hydrogen has a single, unpaired electron in its outermost and only energy level, and it is highly reactive; helium has two valence electrons filling its single orbital, and it is inert. For atoms with two or more energy levels, only those

with unfilled outer energy levels are reactive. Those with eight electrons completely filling the four orbitals of the outer energy level, such as neon and argon, are stable and chemically unreactive (see Figure 2.5).

Atoms with outer energy levels that contain electrons near the stable numbers tend to gain or lose electrons to reach the stable configuration. For example, sodium has two electrons in its first energy level, eight in the second, and one in the third and outermost level (see Figure 2.5). The outermost electron is readily lost to another atom, giving the sodium atom a stable second energy level (now the outermost level) with eight electrons. Chlorine, with seven electrons in its outermost energy level, tends to take up an electron from another atom to attain the stable number of eight electrons.

Atoms that differ from the stable configuration by more than one or two electrons tend to attain stability by *sharing* electrons in joint orbitals with other atoms rather than by gaining or losing electrons completely. Among the atoms that form biological molecules, electron sharing is most characteristic of carbon, which has four electrons in its outer energy level and thus falls at the midpoint between the tendency to gain or lose electrons. Oxygen, with six electrons in its outer level, and nitrogen, with five electrons in its outer level, also share electrons readily. Hydrogen may either share or lose its single electron. The relative tendency to gain, share, or lose valence electrons underlies the chemical bonds and forces that hold the atoms of molecules together.

STUDY BREAK 2.2

1. Where are protons, electrons, and neutrons found in an atom?
2. The isotopes carbon-11 and oxygen-15 do not occur in nature, but they can be made in the laboratory. Both are used in a medical imaging procedure called positron emission tomography. How many protons and neutrons are in carbon-11 and in oxygen-15?
3. What determines the chemical reactivity of an atom?

2.3 Chemical Bonds and Chemical Reactions

Atoms of inert elements, such as helium, neon, and argon, occur naturally in uncombined forms, but atoms of reactive elements tend to combine into molecules by forming **chemical bonds.** Due to the properties of the atoms and bonds involved, the molecules produced have particular structures and functions. The four most important chemical linkages in biological molecules are *ionic bonds, covalent bonds, hydrogen bonds,* and *van der Waals forces.* Chemical reactions occur when atoms or molecules interact to form new chemical bonds or break old ones.

Ionic Bonds Are Multidirectional and Vary in Strength

Ionic bonds result from electrical attractions between atoms that gain or lose valence electrons completely. A sodium atom (Na) readily loses a single electron to achieve a stable outer energy level, and chlorine (Cl) readily gains an electron:

$$Na\cdot \; + \; \cdot \ddot{\underset{..}{Cl}}: \; \rightarrow \; Na^+ \; :\ddot{\underset{..}{Cl}}:^-$$

(The dots in the preceding formula represent the electrons in the outermost energy level.) After the transfer, the sodium atom, now with 11 protons and 10 electrons, carries a single, unit positive charge. The chlorine atom, now with 17 protons and 18 electrons, carries a single, unit negative charge. In this charged condition, the sodium and chlorine atoms are called **ions** instead of atoms and are written as Na^+ and Cl^- **(Figure 2.6).** A positively charged ion such as Na^+ is called a **cation,** and a negatively charged ion such as Cl^- is called an **anion.** The difference in charge between cations and anions creates an attraction—the ionic bond—that holds the ions together in solid NaCl (sodium chloride).

Many other atoms that differ from stable outer energy levels by one electron, including hydrogen, can gain or lose electrons completely to form ions and ionic bonds. When a hydrogen atom loses its single electron to form a hydrogen ion (H^+), it consists of only a proton and is often simply called a proton to reflect this fact. A number of atoms with outer energy levels that differ from the stable number by two or three electrons, particularly metallic atoms such as calcium (Ca^{2+}), magnesium (Mg^{2+}), and iron (Fe^{2+} or Fe^{3+}), also lose their electrons readily to form cations and to join in ionic bonds with anions.

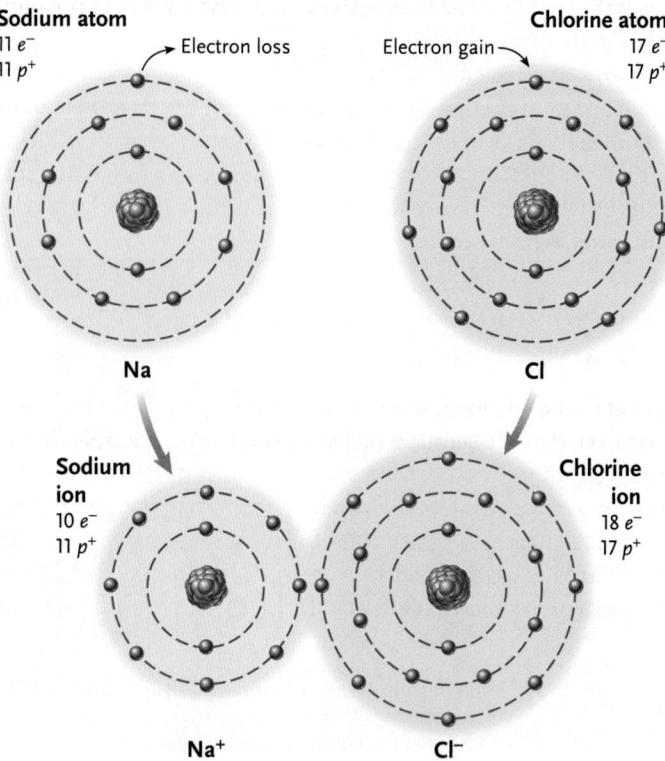

A. Ionic bond between sodium and chlorine

B. Combination of sodium and chlorine in sodium chloride

FIGURE 2.6 Formation of an ionic bond. Sodium, with one electron in its outermost energy level, readily loses that electron to attain a stable state in which its second energy level, with eight electrons, becomes the outer level. Chlorine, with seven electrons in its outer energy level, readily gains an electron to attain the stable number of eight. The transfer creates the ions Na^+ and Cl^-. The combination of Na^+ and Cl^- forms sodium chloride (NaCl), common table salt.
© Cengage Learning 2017

Ionic bonds are common among the forces that hold ions, atoms, and molecules together in living organisms because these bonds have three key features: (1) they exert an attractive force over greater distances than any other chemical bond; (2) their attractive force extends in all directions; and (3) they vary in strength depending on the presence of other charged substances. That is, in some systems, ionic bonds form in locations that exclude other charged substances, setting up strong and stable attractions that are not easily disturbed. For example, iron ions in the large biological molecule hemoglobin are stabilized by ionic bonds; these iron ions are key to the molecule's distinctive chemical properties. In other systems, particularly at molecular surfaces exposed to water molecules, ionic bonds are relatively weak, allowing ionic attractions to be established or broken quickly. For example, as part of their activity in speeding biological reactions, many enzymatic proteins bind and release molecules by forming and breaking relatively weak ionic bonds.

Covalent Bonds Are Formed by Electrons in Shared Orbitals

Covalent bonds form when atoms share a pair of valence electrons rather than gaining or losing them. For example, if two hydrogen atoms collide, the single electron of each atom may join in a new, combined two-electron orbital that surrounds both nuclei. The two electrons fill the orbital and, therefore, the hydrogen atoms tend to remain linked stably together in the form of molecular hydrogen, H_2. The linkage formed by the shared orbital is a covalent bond.

In molecular diagrams, a covalent bond is designated by a pair of dots or a single line that represents a pair of shared electrons. For example, in H_2, the covalent bond that holds the molecule together is represented as H:H or H—H.

Unlike ionic bonds, which extend their attractive force in all directions, the shared orbitals that form covalent bonds extend between atoms at discrete angles and directions, giving covalently bound molecules distinct, three-dimensional shapes. For biological molecules such as proteins, which are held together primarily by covalent bonds, the three-dimensional structure imparted by these bonds is critical to their functions.

An example of a molecule that contains covalent bonds is methane, CH_4, the main component of natural gas. Carbon, with four unpaired outer electrons, typically forms four covalent bonds to complete its outermost energy level, here with four atoms of hydrogen. **Figure 2.7A** shows a shared-orbital model of methane. Each of the four shared orbitals has a shape like a water droplet, with the narrow end of each at the atomic nucleus. The four shared orbitals—the four covalent bonds—are fixed at an angle of 109.5° from each other. If the centers of the larger ends of the shared orbitals are connected by lines, the shape formed is a tetrahedron. The tetrahedral shape of methane can also be seen in two other ways of showing molecules that you will encounter throughout this book, the ball-and-stick model **(Figure 2.7B)** and the space-filling model **(Figure 2.7C)**. The tetrahedral arrangement of the bonds allows carbon "building blocks" **(Figure 2.7D)** to link to each other in both branched and unbranched chains and rings **(Figure 2.7E)**. Such structures form the backbones of an almost unlimited variety of molecules. Carbon can also form double bonds, in which atoms share two pairs of electrons, and triple bonds, in which atoms share three pairs of electrons. The carbon compounds of life are the subjects of Chapter 3.

A. Shared-orbital model of methane (CH₄)

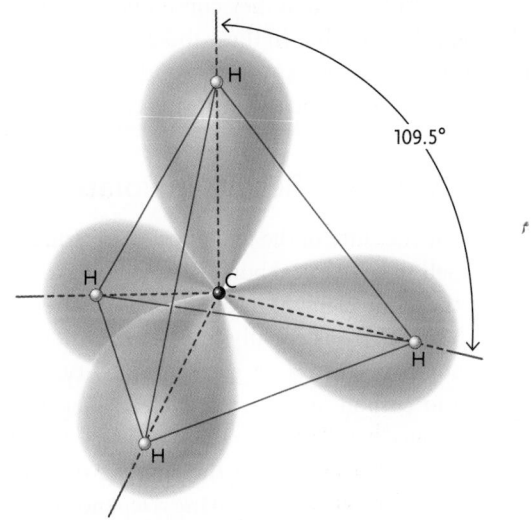

109.5°

B. Ball-and-stick model of methane

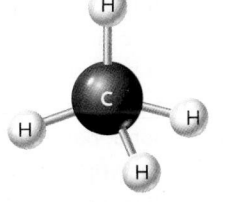

C. Space-filling model of methane

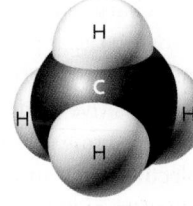

D. A carbon "building block" used to make molecular models

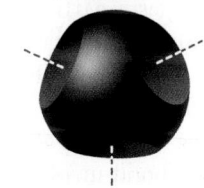

E. Cholesterol

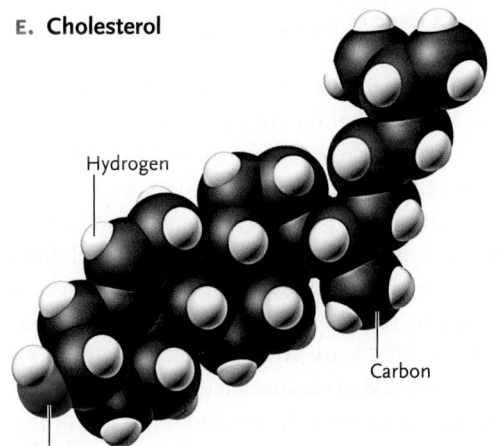

Hydrogen

Carbon

Oxygen

FIGURE 2.7 Covalent bonds shared by carbon. (A) Shared-orbital model of methane (CH_4) showing the four covalent bonds. The bonds extend outward from the carbon nucleus at angles of 109.5° from each other (dashed lines). The red lines connecting the hydrogen nuclei form a regular tetrahedron with four triangular faces. **(B)** Ball-and-stick model of methane. **(C)** In the space-filling model of methane, the diameter of the sphere representing an atom shows the approximate limit of its electron orbitals. **(D)** A tetrahedral carbon "building block." One of the four faces of the block is not visible. **(E)** Carbon atoms assembled into rings and chains forming a complex molecule.

© Cengage Learning 2017.

Oxygen, hydrogen, nitrogen, and sulfur also share electrons readily to form covalent linkages, and they commonly combine with carbon in biological molecules. In these linkages with carbon, oxygen typically forms two covalent bonds; hydrogen forms one; nitrogen forms three; and sulfur forms two.

Unequal Electron Sharing Results in Polarity

Electronegativity is the measure of the tendency of an atom to attract electrons to itself in a chemical bond (that is, to become negative). Atoms of the same element have equal electronegativity; that is, they share electrons equally in a bond. Different elements often do not have the same electronegativity. For example, hydrogen and carbon are much less electronegative than nitrogen and oxygen.

Although all covalent bonds involve the sharing of electrons, they differ widely in the degree of sharing. Depending on the difference in electronegativity between the bonded atoms, covalent bonds are classified as *nonpolar covalent bonds* or *polar covalent bonds*. In a **nonpolar covalent bond,** electrons are shared equally or nearly equally; the atoms involved have no charge. In a **polar covalent bond,** electrons are shared unequally; the atom that attracts the electrons more strongly carries a partial negative charge, δ^- ("delta minus"), and the atom deprived of electrons carries a partial positive charge, δ^+ ("delta plus"). The atoms carrying partial charges may give the whole molecule partially positive and negative ends making the molecule *polar,* hence the name given to the bond. Based on the bond involved, molecules vary in electron sharing with a continuum of electron sharing seen from equal sharing in nonpolar covalent bonds, to partial sharing in polar covalent bonds, to no sharing in ionic bonds.

Nonpolar covalent bonds are characteristic of molecules that contain atoms of one element, such as elemental hydrogen (H_2) and oxygen (O_2), although there are some exceptions. Polar covalent bonds are characteristic of molecules that contain atoms of different elements.

For example, in water—the primary biological example of a polar molecule—an oxygen atom forms polar covalent bonds with two hydrogen atoms. Because the oxygen nucleus with its eight protons attracts electrons much more strongly than the hydrogen nuclei do, the bonds are strongly polar **(Figure 2.8).** In addition, the water molecule is asymmetric, with the oxygen atom located on one side and the hydrogen atoms on the other. This arrangement gives the entire molecule an unequal charge distribution, with the hydrogen end partially positive and the oxygen end partially negative, and makes water molecules strongly polar.

Oxygen, nitrogen, and sulfur, which all share electrons unequally with hydrogen, are located asymmetrically in many biological molecules. Therefore, the presence of —OH, —NH, or —SH groups tends to make regions in biological molecules containing them polar.

Although carbon and hydrogen share electrons somewhat unequally, these atoms tend to be arranged symmetrically in

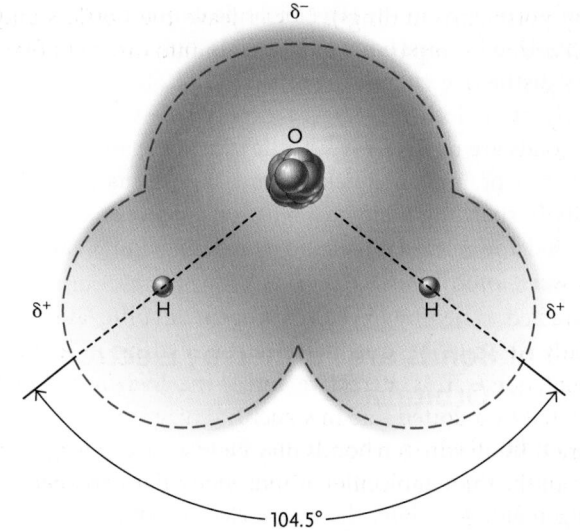

FIGURE 2.8 Polarity in the water molecule, created by unequal electron sharing between the two hydrogen atoms and the oxygen atom and the asymmetric shape of the molecule. The unequal electron sharing gives the hydrogen end of the molecule a partial positive charge, δ^+ ("delta plus"), and the oxygen end of the molecule a partial negative charge, δ^- ("delta minus"). Regions of deepest color indicate the most frequent locations of the shared electrons. The orbitals occupied by the electrons are more complex than the spherical forms shown here.
© Cengage Learning 2017

biological molecules. Thus, regions that contain only carbon–hydrogen chains are typically nonpolar. For example, the C—H bonds in methane are located symmetrically around the carbon atom (see Figure 2.7), so their partial charges cancel each other and the molecule as a whole is nonpolar.

Polar Molecules Tend to Associate with Each Other and Exclude Nonpolar Molecules

Polar molecules attract and align themselves with other polar molecules and with charged ions and molecules. These **polar associations** create environments that tend to exclude nonpolar molecules. When present in quantity, the excluded nonpolar molecules tend to clump together in arrangements called **nonpolar associations;** these nonpolar associations reduce the surface area exposed to the surrounding polar environment. Polar molecules that associate readily with water are identified as **hydrophilic** (*hydro* = water; *philic* = having an affinity or preference for). Nonpolar substances that are excluded by water and other polar molecules are identified as **hydrophobic** (*phobic* = having an aversion to or fear of).

Polar and nonpolar associations can be demonstrated with an apparatus as simple as a bottle containing water and vegetable oil. If the bottle has been standing for some time, the nonpolar oil and polar water form separate layers, with the oil on top. If you shake the bottle, the oil becomes suspended as spherical droplets in the water; the harder you shake, the smaller the oil droplets become (the spherical form of the oil droplets exposes the least surface area per unit volume to the watery polar surroundings). If you leave the bottle standing, the oil and water separate quickly again into distinct polar and nonpolar layers.

Hydrogen Bonds Also Involve Unequal Electron Sharing

When hydrogen atoms are made partially positive by sharing electrons unequally with oxygen, nitrogen, or sulfur, they may be attracted to nearby oxygen, nitrogen, or sulfur atoms made partially negative by unequal electron sharing in a different covalent bond. This attractive force, the **hydrogen bond,** is illustrated by a dotted line in structural diagrams of molecules **(Figure 2.9A).** Hydrogen bonds may be *intramolecular* (between atoms in the same molecule) or *intermolecular* (between atoms in different molecules).

A. Representation of hydrogen bond
A hydrogen bond (dotted line) between the hydrogen of an —OH group and a nearby nitrogen atom, which also shares electrons unequally with another hydrogen. Regions of deepest blue indicate the most likely locations of electrons.

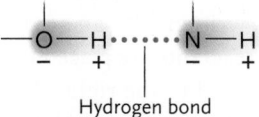

Hydrogen bond

B. Stabilizing effect of hydrogen bonds
Multiple hydrogen bonds stabilize the backbone chain of a protein molecule into a spiral called the alpha helix. The spheres labeled *R* represent chemical groups of different kinds.

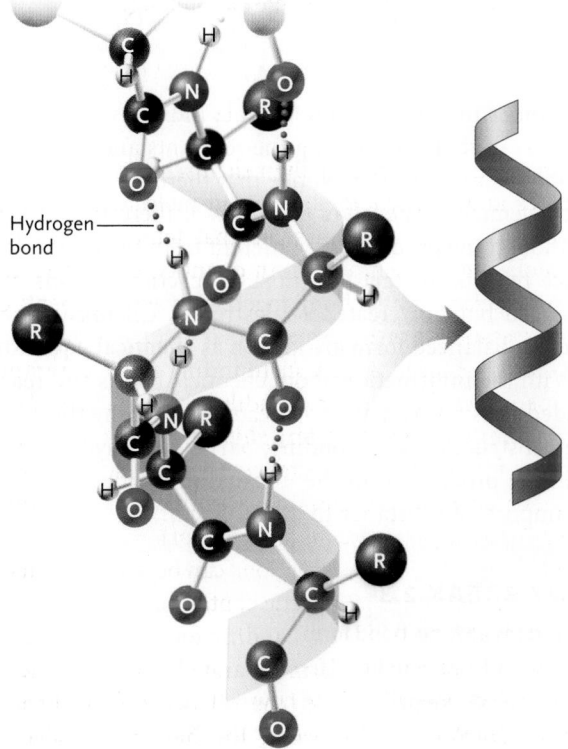

Hydrogen bond

FIGURE 2.9 Hydrogen bonds.
© Cengage Learning 2017

Individual hydrogen bonds are weak compared with ionic and covalent bonds. However, large biological molecules may offer many opportunities for hydrogen bonding, both within and between molecules. When numerous, hydrogen bonds are collectively strong and lend stability to the three-dimensional structure of molecules such as proteins **(Figure 2.9B).** Hydrogen bonds between water molecules are responsible for many of the properties that make water uniquely important to life (see Section 2.4 for a more detailed discussion).

The weak attractive force of hydrogen bonds makes them much easier to break than covalent and ionic bonds, particularly when elevated temperature increases the movements of molecules. Hydrogen bonds begin to break extensively as temperatures rise above 45°C and become practically nonexistent at 100°C. The disruption of hydrogen bonds by heat—for instance, the bonds in proteins—is one of the primary reasons most organisms cannot survive temperatures much greater than 45°C. Thermophilic (temperature-loving) organisms (see Chapter 26), which live at temperatures higher than 45°C, some at 120°C or more, have different molecules from those of organisms that live at lower temperatures. The proteins in thermophiles are stabilized at high temperatures by van der Waals forces and other noncovalent interactions.

Van der Waals Forces Are Weak Attractions over Very Short Distances

Van der Waals forces are even weaker than hydrogen bonds. These forces develop over very short distances between nonpolar molecules or regions of molecules when, through their constant motion, electrons accumulate by chance in one part of a molecule or another. This process leads to zones of positive and negative charge, making the molecule polar. If they are oriented in the right way, the polar parts of the molecules are attracted electrically to one another and cause the molecules to stick together briefly. Although an individual bond formed with van der Waals forces is weak and transient, the formation of many bonds of this type can stabilize the shape of a large molecule, such as a protein.

A striking example of the collective power of van der Waals forces concerns the ability of geckos, a group of tropical lizard species, to cling to and walk up vertical smooth surfaces **(Figure 2.10).** The toes of the gecko are covered with millions of hairs, called *setae* (singular, *seta*). At the tip of each seta are hundreds of thousands of pads, called *spatulae*. Each pad forms a weak interaction—using van der Waals forces—with molecules on the smooth surface. Magnified by the huge number of pads involved, the attractive forces are 1,000 times greater than necessary for the gecko to hang on a vertical wall or even from a ceiling. To climb a wall, the animal rolls the setae onto the surface and then peels them off like a piece of tape. Understanding the gecko's remarkable ability to climb has led to the development of Geckskin™, a prototype adhesive capable of supporting 318 kg (700 pounds) from glass with an index-card–sized piece.

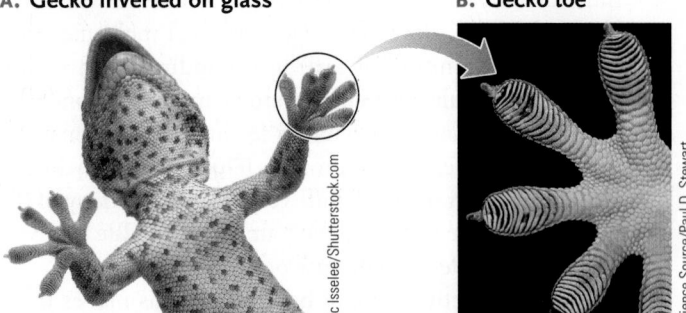

A. Gecko inverted on glass   **B. Gecko toe**

Eric Isselee/Shutterstock.com

Science Source/Paul D. Stewart

FIGURE 2.10 An example of van der Waals forces in biology.
(A) A Tokay gecko *(Gekko gecko)* climbing while inverted on a glass plate.
(B) Gecko toe.

Molecules Have Characteristic Geometries That Determine Their Functions in the Cell

Molecular geometry is the three-dimensional arrangement of the atoms in a molecule; more simply, we refer to the geometry of a molecule as its shape. Every molecule has a characteristic size and three-dimensional shape, properties that are determined by the atoms it contains and the positions of the orbitals of those atoms.

The simplest molecules, those with two atoms, are linear. Examples are hydrogen (H_2), oxygen (O_2), and nitrogen (N_2). Molecules with more than two atoms have more complex geometries. Methane, discussed earlier, is an example; it has a tetrahedral shape.

Molecular shape is crucial in biology because it determines the function of a molecule. To understand this more completely, consider the analogy of a lock and key. A lock fits into a key and can open what the lock protects. A cell's function—and, therefore, an organism's function—depends on interactions between molecules. For instance, hormones circulating in the blood bind specifically to receptors on the surfaces of or within cells to trigger a change in the cell's functions (see Chapters 9 and 42), and specific proteins bind to DNA sequences adjacent to genes to control their expression (see Chapters 15 and 16). Those interactions are lock-and-key interactions at a molecular level; one molecule interacts with another because their shapes facilitate it. The precise recognition that occurs between the molecules triggers the cellular event controlled by the interaction. Therefore, if the two molecules cannot interact, that cellular event cannot take place. Such a change in cellular function can occur, for example, when mutations in a gene affect the shape of its encoded protein (see Chapter 15).

A number of drugs have been developed that work by mimicking the shape of natural molecules so that they block or interfere with the interactions of, and therefore the functions of, those natural molecules. For example, acetylcholine ($C_7H_{16}NO_2$) is a molecule that, when it binds in a lock-and-key fashion to a specific receptor, brings about contraction of the circular pupillary sphincter muscle to contract the pupil in the eye. The drug atropine ($C_{17}H_{23}NO_3$) has a three-dimensional shape, part of which can bind to the same receptor to which acetylcholine binds, thereby blocking the ability for acetylcholine to bind. When atropine is administered in drops to the eye, the consequence is that acetylcholine cannot bring about the contraction of the pupillary sphincter muscle. As a result, the radial pupillary dilator muscle can contract and the pupil dilates, facilitating examination of the eye by an ophthalmologist.

Bonds Form and Break in Chemical Reactions

In **chemical reactions,** atoms or molecules interact to form new chemical bonds or break old ones. As a result of bond formation or breakage, atoms are added to or removed from molecules, or the linkages of atoms in molecules are rearranged. When any of these alterations occur, molecules change from one type to another, usually with different chemical and physical properties. In biological systems, chemical reactions are accelerated by molecules called *enzymes* (discussed in more detail in Chapter 6).

The atoms or molecules entering a chemical reaction are called the **reactants,** and those leaving a reaction are the **products.** A chemical reaction is written with an arrow showing the direction of the reaction; reactants are placed to the left of the arrow, and products are placed to the right. Both reactants and products are usually written in chemical shorthand as formulas.

For example, the overall reaction of photosynthesis, in which carbon dioxide and water are combined to produce sugars and oxygen (see Chapter 8), is written as follows:

$$6\,CO_2 + 6\,H_2O \rightarrow C_6H_{12}O_6 + 6\,O_2$$

carbon   water   a sugar   molecular
dioxide   oxygen

The number in front of each formula indicates the number of molecules of that type among the reactants and products (the number 1 is not written). Notice that there are as many atoms of each element to the left of the arrow as there are to the right, even though the products are different from the reactants. This balance reflects the fact that in such reactions, atoms may be rearranged but not created or destroyed. Chemical reactions written in balanced form are known as **chemical equations.**

With the information about chemical bonds and reactions provided thus far, you are now ready to examine the effects of chemical structure and bonding, particularly hydrogen bonding, in the production of the unusual properties of water, the most important substance to life on Earth.

STUDY BREAK 2.3
1. How does an ionic bond form?
2. How does a covalent bond form?
3. What is electronegativity, and how does it relate to nonpolar covalent bonds and polar covalent bonds?
4. What is a chemical reaction?

2.4 Hydrogen Bonds and the Properties of Water

All living organisms, whether they live in water or in dry environments, contain water. Indeed, water is the main component of most cells. The water inside organisms is crucial for life: it is required for many important biochemical reactions and plays major roles in maintaining the shape and organization of cells and tissues. Key properties of water molecules that make them so important to life include the following:

- Hydrogen bonds between water molecules produce a *water lattice*. (A lattice is a cross-linked structure.) The lattice arrangement makes water denser than ice and gives water several other properties that make it a highly suitable medium for the molecules and reactions of life. For example, *water absorbs or releases relatively large amounts of energy as heat without undergoing extreme changes in temperature.* This property stabilizes both living organisms and their environments. Also, the water lattice has an unusually high internal *cohesion* (resistance of water molecules to separate). This property plays an important role, for example, in water transport from the roots to the leaves of plants (see Chapter 34). Further, water molecules at surfaces facing air are even more resistant to separation, producing the force called *surface tension,* which, for example, allows droplets of water to form and small insects and arachnids to walk on water.
- The polarity of water molecules in the hydrogen-bond lattice contributes to the formation of distinct polar and nonpolar environments that are critical to the organization of cells.

- Water is a *solvent* for charged or polar molecules, meaning that it is a solution in which such molecules can dissolve. Water's solvent properties made life possible because it enabled the chemical reactions needed for life to evolve. The chemical reactions in all organisms take place in aqueous solutions.
- Water molecules separate into ions. Those ions are important for maintaining an environment within cells that is optimal for the chemical reactions that occur there.

A Lattice of Hydrogen Bonds Gives Water Several Unusual, Life-Sustaining Properties

Hydrogen bonds form readily between water molecules in both liquid water and ice. In liquid water, each water molecule establishes an average of 3.4 hydrogen bonds with its neighbors, forming an arrangement known as the **water lattice (Figure 2.11A)**. In liquid water, the hydrogen bonds that hold the lattice together constantly break and reform, allowing the water molecules to break loose from the lattice, slip past one another, and reform the lattice in new positions.

THE DIFFERING DENSITIES OF WATER AND ICE In ice, the water lattice is a rigid, crystalline structure in which each water molecule forms four hydrogen bonds with neighboring molecules **(Figure 2.11B)**. The rigid **ice lattice** spaces the water molecules farther apart than the water lattice. Because of this greater spacing, water has the unusual property of being about 10% less dense when solid than when liquid. (Almost all other substances are denser in solid form than in liquid form.) Hence, water filling a closed glass vessel will expand, breaking the

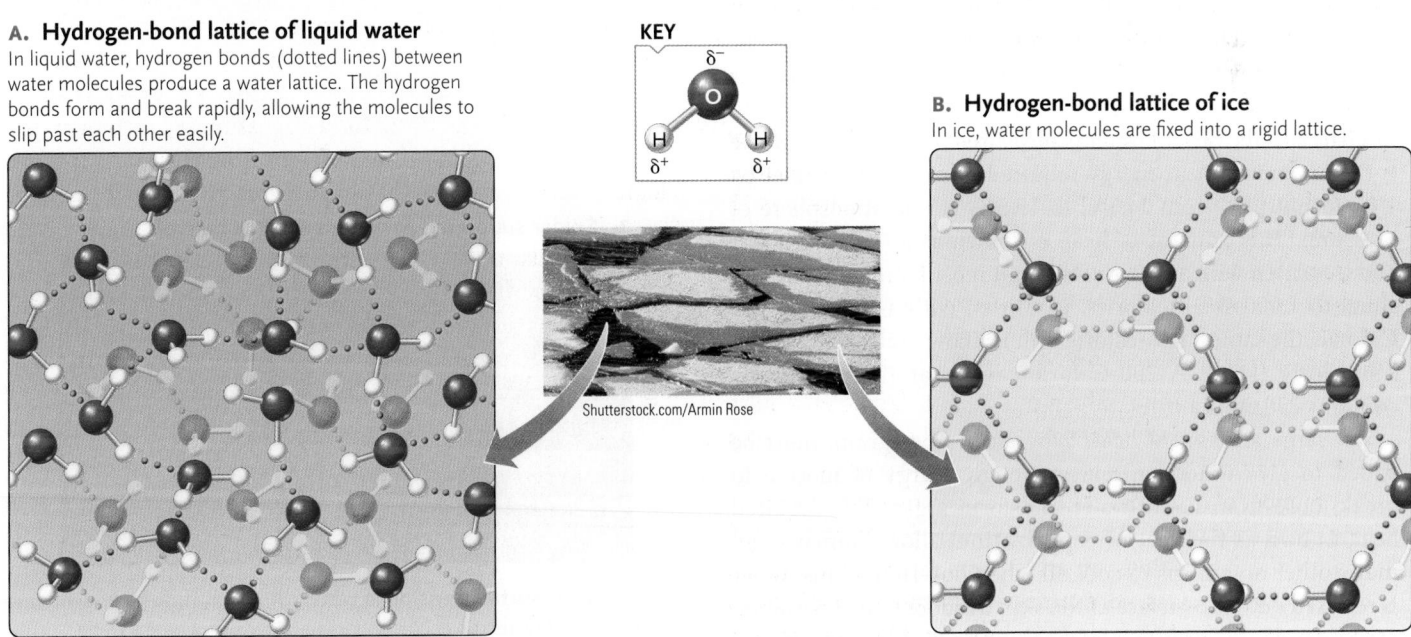

A. Hydrogen-bond lattice of liquid water
In liquid water, hydrogen bonds (dotted lines) between water molecules produce a water lattice. The hydrogen bonds form and break rapidly, allowing the molecules to slip past each other easily.

KEY

B. Hydrogen-bond lattice of ice
In ice, water molecules are fixed into a rigid lattice.

Shutterstock.com/Armin Rose

FIGURE 2.11 Hydrogen bonds and water.
© Cengage Learning 2017

vessel when the water freezes. At atmospheric pressure, water reaches its greatest density at a temperature of 4°C, while it is still a liquid.

Because it is less dense than liquid water, ice forms at the surface of a body of water and remains floating at the surface. The ice creates an insulating layer that helps keep the water below from freezing. If ice were denser than liquid water, a body of water would freeze from the bottom up, killing most aquatic plants and animals in it.

THE BOILING POINT AND TEMPERATURE-STABILIZING EFFECTS OF WATER The hydrogen-bond lattice of liquid water retards the escape of individual water molecules as the water is heated. As a result, relatively high temperatures and the addition of considerable heat are required to break enough hydrogen bonds to make water boil. Its high boiling point maintains water as a liquid over the wide temperature range of 0°C to 100°C (at sea level). Similar molecules that do not form an extended hydrogen-bond lattice, such as H_2S (hydrogen sulfide), have much lower boiling points and are gases rather than liquids at room temperature. Without its hydrogen-bond lattice, water would boil at −81°C. If this were the case, most of the water on Earth would be in gaseous form and life as described in this book would not have developed and evolved.

As a result of water's stabilizing hydrogen-bond lattice, it also has a relatively high **specific heat,** the amount of energy as heat required to increase the temperature of a given quantity of water. As heat energy flows into water, much of it is absorbed in the breakage of hydrogen bonds. As a result, the temperature of water increases relatively slowly as heat energy is added. For example, a given amount of heat energy increases the temperature of water by only half as much as that of an equal quantity of ethyl alcohol. High specific heat allows water to absorb or release relatively large quantities of heat energy without undergoing extreme changes in temperature; this gives it a moderating and stabilizing effect on both living organisms and their environments.

The specific heat of water is measured in *calories.* A **calorie** is the amount of heat energy required to raise 1 g of water by 1°C (technically, from 14.5°C to 15.5°C at one atmosphere of pressure). This amount of heat is known as a "small" calorie and is written with a small *c*. The unit most familiar to dieters, equal to 1,000 small calories, is written with a capital *C* as a **Calorie;** the same 1,000-calorie unit is known scientifically as a **kilocalorie (kcal).** A 250-Calorie candy bar therefore really contains 250,000 calories.

A large amount of heat, 586 calories per gram, must be added to give water molecules enough energy of motion to break loose from liquid water and form a gas. This required heat, known as the **heat of vaporization,** allows humans and many other organisms to cool off when hot. In humans, water is released onto the surface of the skin by more than 2.5 million sweat glands. The heat energy absorbed by the water in sweat as the sweat evaporates cools the skin and the underlying blood vessels. The heat loss helps keep body temperature from increasing when environmental temperatures are high. Plants use a similar cooling mechanism as water evaporates from their leaves (see Chapter 34).

COHESION AND SURFACE TENSION The high resistance of water molecules to separation, provided by the hydrogen-bond lattice, is known as internal **cohesion.** For example, in land plants, cohesion holds water molecules in unbroken columns in microscopic conducting tubes that extend from the roots to the highest leaves. As water evaporates from the leaves, water molecules in the columns, held together by cohesion, move upward through the tubes to replace the lost water. This movement raises water from roots to the tops of the tallest trees (see discussion in Chapter 34). Maintenance of the long columns of water in the tubes is aided by *adhesion,* in which molecules "stick" to the walls of the tubes by forming hydrogen bonds with charged and polar groups in molecules that form the walls of the tubes.

Water molecules at surfaces facing air can form hydrogen bonds with water molecules beside and below them but not on the sides that face the air. This unbalanced bonding produces a force that places the surface water molecules under tension, making them more resistant to separation than the underlying water molecules **(Figure 2.12A).** The force, called **surface tension,** is strong enough to allow small insects or arachnids

A. Creation of surface tension by unbalanced hydrogen bonding

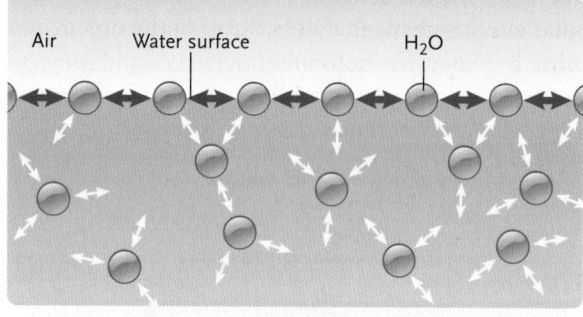

B. Spider supported by water's surface tension

FIGURE 2.12 Surface tension in water. (A) Unbalanced hydrogen bonding places water molecules under lateral tension where a water surface faces the air. **(B)** A raft spider (*Dolomedes fimbriatus*) is supported by the surface tension of water.

such as raft spiders to walk on water **(Figure 2.12B).** Surface tension also causes water to form water droplets (as seen in the chapter-opening photo); the surface tension pulls the water in around itself to produce the smallest possible area, which is a spherical bead or droplet.

The Polarity of Water Molecules in the Hydrogen-Bond Lattice Contributes to Polar and Nonpolar Environments in and around Cells

In liquid water, the hydrogen-bond lattice resists invasion by other molecules unless the invading molecule also contains polar or charged regions that can form competing attractions with water molecules. If present, the competing attractions open the water lattice, creating a cavity into which the polar or charged molecule can move. By contrast, nonpolar molecules are unable to disturb the water lattice. The lattice thus excludes nonpolar substances, forcing them to form the nonpolar associations that expose the least surface area to the surrounding water—such as the spherical droplets of oil that form when oil and water are mixed together and shaken.

The distinct polar and nonpolar environments created by water are critical to the organization of cells. For example, biological membranes, which form boundaries around and inside cells, consist of lipid molecules with dual polarity: one end of each molecule is polar, and the other end is nonpolar. (Lipids are described in more detail in Chapter 3.) The membranes are surrounded on both sides by strongly polar water molecules. Exclusion by the water molecules forces the lipid molecules to associate into a double layer, a **bilayer,** in which only the polar ends of the surface molecules are exposed to the water **(Figure 2.13).** The nonpolar ends of the molecules associate in the interior of the bilayer, where they are not exposed to the water. Exclusion of their nonpolar regions by water is all that holds membranes together.

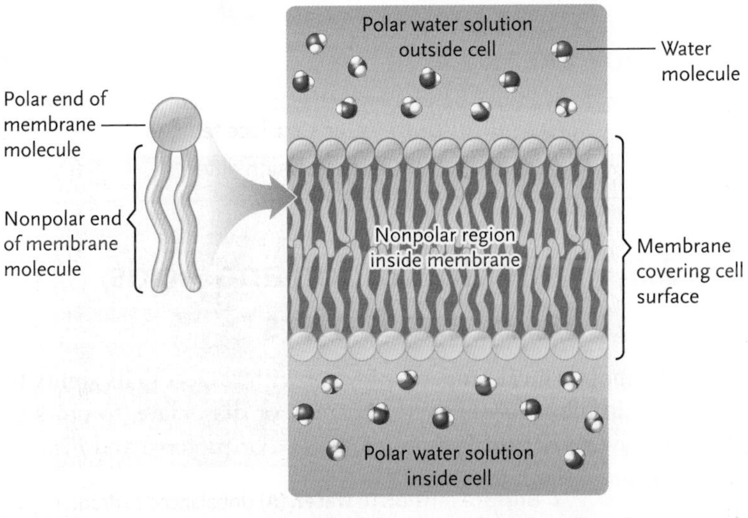

The membrane at the surface of cells prevents the watery solution inside the cell from mixing directly with the watery solution outside the cell. By doing so, the surface membrane, kept intact by nonpolar exclusion by water, maintains the internal environment and organization necessary for cellular life and its evolution.

The Small Size and Polarity of Its Molecules Makes Water a Good Solvent

Because water molecules are small and strongly polar, they can penetrate or coat the surfaces of other polar and charged molecules and ions. The surface coat, called a **hydration layer,** reduces the attraction between the molecules or ions and promotes their separation and entry into a **solution,** where they are suspended individually, surrounded by water molecules. Once in solution, the hydration layer prevents the polar molecules or ions from reassociating. In such a solution, water is called the **solvent,** and the molecules of a substance dissolved in water are called the **solute.**

For example, when a teaspoon of table salt is added to water, water molecules quickly form hydration layers around the Na^+ and Cl^- ions in the salt crystals, reducing the attraction between the ions so much that they separate from the crystal and enter the surrounding water lattice as individual ions **(Figure 2.14).** If the water evaporates, the hydration layer is eliminated, exposing the strong positive and negative charges of the ions. The opposite charges attract and reestablish the ionic bonds that hold the ions in salt crystals. As the last of the water evaporates, all of the Na^+ and Cl^- ions reassociate, reestablishing the solid, crystalline form.

In a Cell, Chemical Reactions Involve Solutes Dissolved in Aqueous Solutions

In a cell, chemical reactions depend on solutes dissolved in aqueous solutions. To understand these reactions, you need to know the number of atoms and molecules involved. **Concentration** is the number of molecules or ions of a substance in a unit volume of space, such as a milliliter (mL) or liter (L). The number of molecules or ions in a unit volume cannot be counted directly but can be calculated indirectly by using the mass number of atoms as the starting point. The same method is used to prepare a solution with a known number of molecules per unit volume.

The mass number of an atom is equivalent to the number of protons and neutrons in its nucleus. From the mass number, and the fact that neutrons and protons are approximately the same weight (that is, 1.66×10^{-24} g), you can calculate the weight of an atom of any substance. For an atom of the most common form of carbon, with 6 protons and 6 neutrons in its nucleus, the total weight is calculated as follows:

$$12 \times (1.66 \times 10^{-24} \text{ g}) = 1.992 \times 10^{-23} \text{ g}$$

FIGURE 2.13 Formation of the membrane covering the cell surface by lipid molecules. Exclusion by polar water molecules forces the nonpolar ends of lipid molecules to associate into the bilayer that forms the membrane.
© Cengage Learning 2017

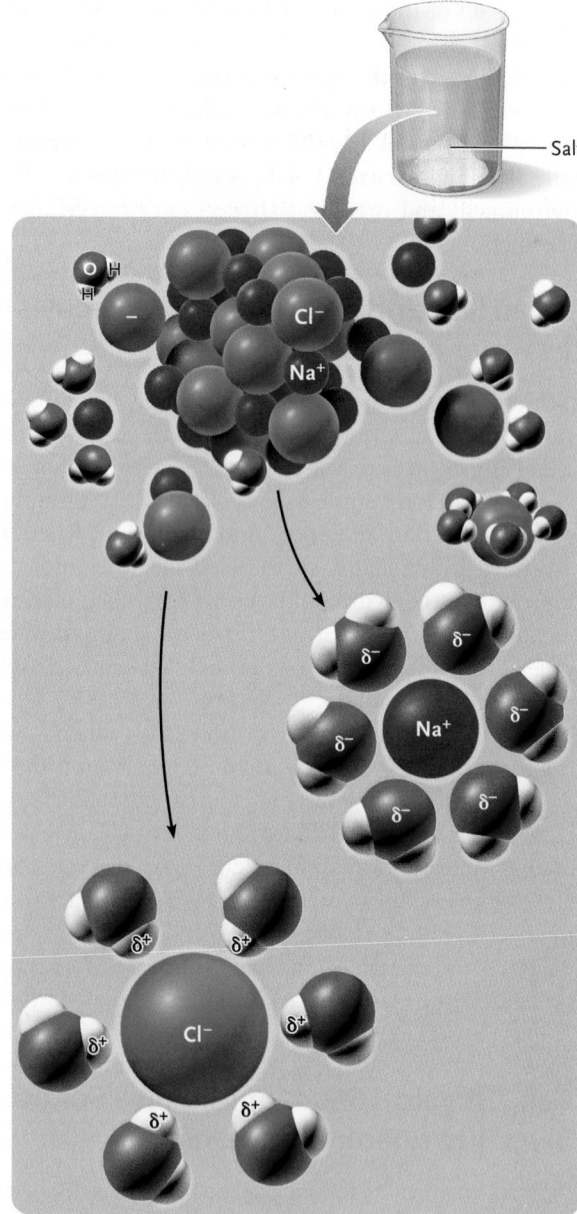

FIGURE 2.14 Water molecules forming a hydration layer around Na^+ and Cl^- ions, which promotes their separation and entry into solution.
© Cengage Learning 2017

For an oxygen atom, with 8 protons and 8 neutrons in its nucleus, the total weight is calculated as follows:

$$16 \times (1.66 \times 10^{-24}\,g) = 2.656 \times 10^{-23}\,g$$

Dividing the total weight of a sample of an element by the weight of a single atom gives the number of atoms in the sample. Suppose you have a carbon sample that weighs 12 g—a weight in grams equal to the atom's mass number. (A weight in grams equal to the mass number is known as an **atomic weight** of an element.) Dividing 12 g by the weight of one carbon atom gives the following result:

$$\frac{12}{(1.992 \times 10^{-23}\,g)} = 6.022 \times 10^{23}\,atoms$$

If you divide the atomic weight of oxygen (16 g) by the weight of one oxygen atom, you get the same result:

$$\frac{16}{(2.656 \times 10^{-23}\,g)} = 6.022 \times 10^{23}\,atoms$$

Dividing the atomic weight of any element by the weight of an atom of that element always produces the same number: 6.022×10^{23}. This number is called **Avogadro's number** after Amedeo Avogadro, the nineteenth-century Italian chemist who first discovered the relationship.

The same relationship holds for molecules. The **molecular weight** of any molecule is the sum of the atomic weights of all of the atoms in the molecule. For NaCl, the total mass number is $23 + 35 = 58$ (a sodium atom has 11 protons and 12 neutrons, and a chlorine atom has 17 protons and 18 neutrons). The weight of an NaCl molecule is therefore:

$$58 \times (1.66 \times 10^{-24}\,g) = 9.628 \times 10^{-23}\,g$$

Dividing a molecular weight of NaCl (58 g) by the weight of a single NaCl molecule gives:

$$\frac{58}{(9.628 \times 10^{-23}\,g)} = 6.022 \times 10^{23}\,molecules$$

When concentrations are described, the atomic weight of an element or the molecular weight of a compound—the amount that contains 6.022×10^{23} atoms or molecules—is known as a **mole** (abbreviated **mol**). More strictly, chemists define the mole as the amount of a substance that contains as many atoms or molecules as there are atoms in exactly 12 g of carbon-12. As we saw above, the number of atoms in 12 g of carbon-12 is 6.022×10^{23}. The number of moles of a substance dissolved in 1 L of solution is known as the **molarity** (abbreviated *M*) of the solution. This relationship is highly useful in chemistry and biology because we know that two solutions having the same volume and molarity but composed of different substances will contain the same number of molecules of the substances.

STUDY BREAK 2.4

1. How do hydrogen bonds between water molecules contribute to the properties of water?
2. How do a solute, a solvent, and a solution differ?

2.5 Water Ionization and Acids, Bases, and Buffers

The most critical property of water with respect to its ability to sustain life is its ability to separate, or **dissociate,** to produce positively charged *hydrogen ions* (H^+, or protons) and *hydroxide ions* (OH^-):

$$H_2O \rightleftharpoons H^+ + OH^-$$

(The double arrow means that the reaction is **reversible**—depending on conditions, it may go from left to right or from right to left.) The proportion of water molecules that dissociates to release protons and hydroxide ions is small. However, because of the dissociation, water always contains some H^+ and OH^- ions. These ions help maintain the environment inside living organisms that promotes the chemical reactions of life.

Substances Act as Acids or Bases by Altering the Concentrations of H^+ and OH^- Ions in Water

In pure water, the concentrations of H^+ and OH^- ions are equal. However, adding other substances may alter the relative concentrations of H^+ and OH^-, making them unequal. Some substances, called **acids,** are proton donors that release H^+ (and anions) when they are dissolved in water, effectively increasing the H^+ concentration. For example, hydrochloric acid (HCl) dissociates into H^+ and Cl^- when dissolved in water:

$$HCl \rightleftharpoons H^+ + Cl^-$$

Other substances, called **bases,** are proton acceptors that reduce the H^+ concentration of a solution. Most bases dissociate in water into a hydroxide ion (OH^-) and a cation. The hydroxide ion can act as a base by accepting a proton (H^+) to produce water. For example, sodium hydroxide (NaOH) separates into Na^+ and OH^- ions when dissolved in water:

$$NaOH \rightarrow Na^+ + OH^-$$

The excess OH^- combines with H^+ to produce water:

$$OH^- + H^+ \rightarrow H_2O$$

thereby reducing the H^+ concentration.

Other bases do not dissociate to produce hydroxide ions directly. For example, ammonia (NH_3), a poisonous gas, acts as a base when dissolved in water by accepting a proton from water to produce an ammonium ion and releasing a hydroxide ion:

$$NH_3 + H_2O \rightarrow NH_4 + OH^-$$

The concentration of H^+ ions in a water solution, as compared with the concentration of OH^- ions, determines the **acidity** of the solution. Scientists measure acidity using a numerical scale from 0 to 14, called the **pH scale.** Because the number of H^+ ions in solution increases exponentially as the acidity increases, the scale is based on logarithms of this number to make the values manageable:

$$pH = -\log_{10}[H^+]$$

In this formula, the brackets indicate concentration in moles per liter of the substance within them. The negative of the logarithm is used to give a positive number for the pH value. For example, a pure water solution is *neutral*—neither acidic nor basic. That is, in a neutral water solution, the concentration of *both* H^+ and OH^- ions is 1×10^{-7} M (0.0000001 M), with the product of the two ion concentrations being constant in an aqueous solution at 25°C and given by $[H^+][OH^-] = (1 \times 10^{-7}) \times (1 \times 10^{-7}) = 1 \times 10^{-14}$. The $\log_{10}$ of 1×10^{-7} is -7. The negative of the logarithm -7 is 7. Thus, a neutral

water solution with an H^+ concentration of 1×10^{-7} M has a pH of 7. *Acidic* solutions have pH values less than 7, with pH 0 being the value for the highly acidic 1 M hydrochloric acid (HCl). *Basic* solutions have pH values greater than 7, with pH 14 being the value for the highly basic 1 M sodium hydroxide (NaOH) (basic solutions are also called *alkaline* solutions). Each whole number on the pH scale represents a value 10 times greater or less than the next number. Thus, a solution with a pH of 4 is 10 times more acidic than one with a pH of 5, and a solution with a pH of 6 is 100 times more acidic than a solution with a pH of 8. (The pH of many familiar solutions is shown in **Figure 2.15.**)

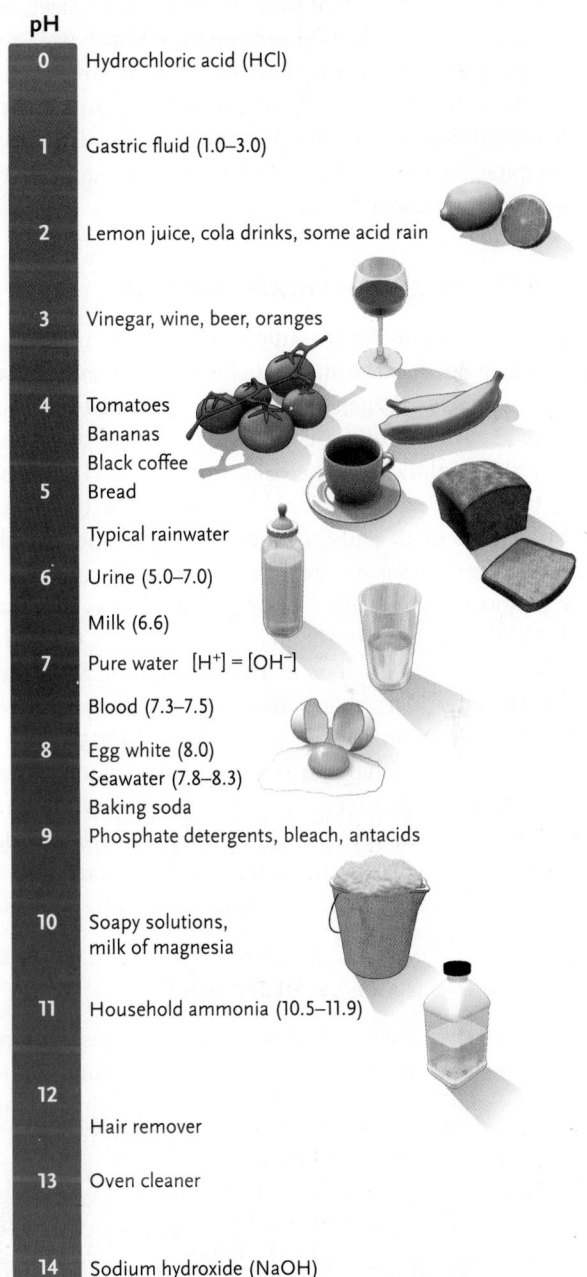

pH	
0	Hydrochloric acid (HCl)
1	Gastric fluid (1.0–3.0)
2	Lemon juice, cola drinks, some acid rain
3	Vinegar, wine, beer, oranges
4	Tomatoes
	Bananas
	Black coffee
5	Bread
	Typical rainwater
6	Urine (5.0–7.0)
	Milk (6.6)
7	Pure water $[H^+] = [OH^-]$
	Blood (7.3–7.5)
8	Egg white (8.0)
	Seawater (7.8–8.3)
	Baking soda
9	Phosphate detergents, bleach, antacids
10	Soapy solutions, milk of magnesia
11	Household ammonia (10.5–11.9)
12	
	Hair remover
13	Oven cleaner
14	Sodium hydroxide (NaOH)

FIGURE 2.15 The pH scale, showing the pH of substances commonly encountered in the environment.
© Cengage Learning 2017

Acidity is important to cells because even small changes, on the order of 0.1 or even 0.01 pH unit, can drastically affect biological reactions. In large part, this effect reflects changes in the structure of proteins that occur when the water solution surrounding the proteins has too few or too many hydrogen ions. Consequently, all living organisms have elaborate systems that control their internal acidity by regulating H^+ concentration near the neutral value of pH 7.

Acidity is also important to the environment in which we live. Where the air is unpolluted, rainwater is only slightly acidic. However, in regions where certain pollutants are released into the air in large quantities by industry and automobile exhaust, the polluting chemicals combine with atmospheric water to produce **acid precipitation**—rain, sleet, snow, or fog with a pH below 6. Acid precipitation may have a pH as low as 3, about the same pH as that of vinegar. It can sicken and kill wildlife such as fishes and birds, as well as plants and trees (see also discussion in Chapter 55). Humans are also affected; acid precipitation can contribute to human respiratory diseases such as bronchitis and asthma.

Buffers Help Keep pH under Control

Living organisms control the internal pH of their cells with **buffers,** substances that compensate for pH changes by absorbing or releasing H^+. When H^+ ions are released in excess by biological reactions, buffers combine with them and remove them from the solution. If the concentration of H^+ decreases greatly, buffers release additional H^+ to restore the balance. Most buffers are weak acids or bases, or combinations of these substances that dissociate reversibly in water solutions to release or absorb H^+ or OH^-. (Weak acids, such as acetic acid, or weak bases, such as ammonia, are substances that release relatively few H^+ or OH^- ions in a water solution. Strong acids or bases are substances that dissociate extensively in a water solution. HCl is a strong acid; NaOH is a strong base.)

The buffering mechanism that maintains blood pH near neutral values is a primary example. In humans and many other animals, blood pH is buffered by a *carbonic acid–bicarbonate buffer system*. In water solutions, carbonic acid (H_2CO_3), which is a weak acid, dissociates readily into bicarbonate ions (HCO_3^-) and H^+:

$$H_2CO_3 \rightleftharpoons HCO_3^- + H^+$$

The reaction is reversible. If H^+ is present in excess, the reaction is pushed to the left—the excess H^+ ions combine with bicarbonate ions to form H_2CO_3. If the H^+ concentration declines below normal levels, the reaction is pushed to the right—H_2CO_3 dissociates into HCO_3^- and H^+, restoring the H^+ concentration. **Figure 2.16** graphs the pH of a solution (such as blood) as its relative proportions of H_2CO_3 and HCO_3^- change. The colored zone of the graph indicates the range of greatest *buffering capacity,* that is, the range in which changes in the relative proportions of H_2CO_3 and HCO_3^- produce *little*

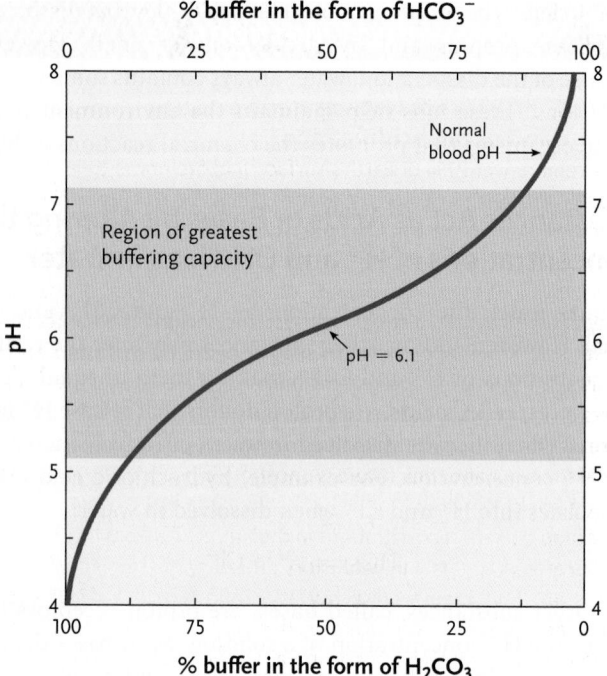

FIGURE 2.16 Properties of the carbonic acid–bicarbonate buffer system. The colored zone is the pH range of greatest buffering capacity for this buffer system.
© Cengage Learning 2017

change in the pH of the solution. Note that at pH values lower than 5.1 and higher than 7.1, the slope of the curve is much greater. Here, changes in the relative proportions of H_2CO_3 and HCO_3^- produce a *large* change in the pH of the solution. Interestingly, the normal pH of blood is 7.4, which, as Figure 2.16 shows, is outside the region of greatest buffering capacity for this buffer system. In the body, other mechanisms help keep the blood pH relatively constant. (More on blood pH appears in Section 46.4.)

All buffers have curves similar to that of Figure 2.16. Each buffer has a specific range of greatest buffering capacity.

This chapter examined the basic structure of atoms and molecules and discussed the unusual properties of water that make it ideal for supporting life. The next chapter looks more closely at the structure and properties of carbon and at the great multitude of molecules based on this element.

STUDY BREAK 2.5

1. Distinguish between acids and bases. What are their properties?
2. Why are buffers important for living organisms?

THINK OUTSIDE THE BOOK

The H_2CO_3–HCO_3^- buffering system is only one of the mechanisms by which the human body maintains blood pH at a relatively constant level. Collaboratively, or on your own, research what happens to the body when blood pH becomes abnormal.

How will marine ecosystems be affected by climate change and ocean acidification?

The oceans cover 71% of Earth's surface and are home to hundreds of thousands and likely millions of species of marine organisms. The majority of these species are microscopic, from bacteria and viruses to microscopic algae and animals (called plankton), but the oceans are also home to the largest animal on the planet—the blue whale. The oceans include many colorful and charismatic ecosystems, including coral reefs, kelp forests, and deep-sea ecosystems such as hydrothermal vents and seamounts. These marine ecosystems harbor some of the highest marine diversity and provide billions of dollars in food and resources. However, these marine ecosystems are increasingly threatened from both local and global stressors, and their future is uncertain in a world of higher carbon dioxide levels.

Carbon dioxide concentrations in the atmosphere are increasing at a rate faster than any time in geohistory due to rapidly growing anthropogenic inputs from industry, transportation, and deforestation. The oceans absorb approximately one third of this carbon dioxide, resulting in a change in the oceans' chemistry and a decline in pH in a process called *ocean acidification.* Ocean acidification and global warming affect even the most remote marine ecosystems, and, in combination with other local stressors such as overfishing, pollution, and development, many marine organisms and ecosystems will be imperiled in a high CO_2 future. Although ocean acidification has become a major research focus over the last decade, most of our current understanding is based on relatively short-term experiments with single species or strains in the laboratory or aquarium isolated from their natural ecosystems. Our understanding of climate change and ocean acidification effects at the ecosystem level is relatively limited, but critical for providing data for policy and management.

An array of experimental approaches is increasingly being used to study ocean acidification and climate change effects, including laboratory studies, aquarium studies, mesocosm studies (where large-volume tanks are used to create a mini-ecosystem), studies in environments with naturally low pH such as gradients, vents, and seeps, and *in situ* experimental approaches (where future levels of CO_2 are produced in the environment itself in a controlled manner). Each of these approaches has it strengths and limitations, and only through collaborative studies that involve a diversity of approaches can our understanding of ecosystem-level effects grow.

Coral reefs harbor the highest level of biodiversity in the oceans yet are one of the marine ecosystems most susceptible to ocean acidification. In my research I focus on controlled experimental studies that are as close to natural as possible to gain a deeper understanding of ocean acidification effects from the molecular to the ecosystem level. Coastal ecosystems such as coral reefs often have large daily, monthly, and seasonal changes in temperature and chemistry, yet few studies have incorporated this natural variability into their experimental treatments. At the Heron Island Research Station on the Great Barrier Reef with the Ove Hoegh-Guldberg and Sophie Dove laboratory at the University of Queensland, we developed an aquarium and mesocosm experimental system that carefully produces both CO_2 and temperature levels predicted for different future scenarios with the conditions produced as an offset from those measured in real time on the reef. These tightly controlled future conditions are maintained in 72 replicate aquaria and 12 mesocosms so that we can conduct experiments on both individual organisms and communities for months to years.

Another approach to studying ocean acidification is to actually move the experiments into the environment itself. Peter Brewer's group at the Monterey Bay Aquarium Research Institute (MBARI) first developed the Free Ocean Carbon Enrichment (FOCE) system for *in situ* ocean acidification experiments in the deep ocean. We collaborated with MBARI, the University of Queensland, Stanford, and others to develop a replicated, shallow-water FOCE system for coral reefs. The coral proto–FOCE (CP-FOCE) uses pH and flow sensors for feedback-controlled addition of low-pH seawater to produce future predicted levels of pH as an offset from levels measured in the environment. We have run experiments in the CP-FOCE for 8 months on the Great Barrier Reef. Our research suggests that in a high-CO_2 future, reef calcification will decline while reef dissolution will increase, leading to net dissolution of the reef.

We are currently working on developing multistressor FOCE systems that can test the effects of ocean acidification in combination with other stressors such as elevated temperature, pollution, or low oxygen. We are also planning to deploy FOCE systems along natural gradients, vents, and seeps. As new techniques are developed and applied at a range of spatial and temporal scales we hope to provide the data needed to better protect increasingly endangered marine ecosystems.

think like a scientist Describe both local and global strategies for protecting marine ecosystems from ocean acidification and climate change. What experiments and studies could be performed to track the success of these strategies and maximize their effectiveness?

Erik Jepsen, UC San Diego

David I. Kline is a project scientist at the Scripps Institution of Oceanography, University of California, San Diego, in the integrative oceanography division. His main research interests include coral reef ecology, automating the analysis of reef imagery using computer vision techniques, and determining climate change effects on coral reefs and other shallow coastal benthic ecosystems. To learn more about his research, go to http://scrippsscholars.ucsd.edu/dkline.

REVIEW KEY CONCEPTS

For access to MindTap and additional study materials visit www.cengagebrain.com.

2.1 The Organization of Matter: Elements and Atoms

- Matter is anything that occupies space and has mass. Matter is composed of elements, each consisting of atoms of the same kind.
- Atoms combine chemically in fixed numbers and ratios to form the molecules of living and nonliving matter. Compounds are molecules in which the component atoms are different.

2.2 Atomic Structure

- Atoms consist of an atomic nucleus that contains protons and neutrons surrounded by one or more electrons traveling in orbitals. Each orbital can hold a maximum of two electrons (Figure 2.2).
- All atoms of an element have the same number of protons, but the number of neutrons is variable. The number of protons in an atom is designated by its atomic number; the number of protons plus neutrons is designated by the mass number (Figure 2.3 and Table 2.1).
- Isotopes are atoms of an element with differing numbers of neutrons. The isotopes of an atom differ in physical but not chemical properties (Figure 2.3).
- Electrons surround an atomic nucleus in orbitals occupying energy levels that increase in discrete steps (Figures 2.4 and 2.5).
- The chemical activities of atoms are determined largely by the number of electrons in the outermost energy level. Atoms that have the outermost level filled with electrons are nonreactive, whereas atoms in which that level is not completely filled with electrons are reactive. Atoms tend to lose, gain, or share electrons to fill the outermost energy level.

2.3 Chemical Bonds and Chemical Reactions

- An ionic bond forms between atoms that gain or lose electrons in the outermost energy level completely, that is, between a positively charged cation and a negatively charged anion (Figure 2.6).
- A covalent bond is established by a pair of electrons shared between two atoms. If the electrons are shared equally, the covalent bond is nonpolar (Figure 2.7).
- If electrons are shared unequally in a covalent bond, the atoms carry partial positive and negative charges and the bond is polar (Figure 2.8).
- Polar molecules tend to associate with other polar molecules and to exclude nonpolar molecules. Polar molecules that associate readily with water are hydrophilic; nonpolar molecules excluded by water are hydrophobic.

- A hydrogen bond is a weak attraction between a hydrogen atom made partially positive by unequal electron sharing and another atom—usually oxygen, nitrogen, or sulfur—made partially negative by unequal electron sharing (Figure 2.9).
- Van der Waals forces, bonds even weaker than hydrogen bonds, can form when natural changes in the electron density of molecules produce regions of positive and negative charge, which cause the molecules to stick together briefly.
- The three-dimensional arrangement of the atoms in a molecule—its molecular geometry or shape—is characteristic of the molecule and determines the function of the molecule.
- Chemical reactions occur when molecules form or break chemical bonds. The atoms or molecules entering into a chemical reaction are the reactants, and those leaving a reaction are the products.

2.4 Hydrogen Bonds and the Properties of Water

- The hydrogen-bond lattice gives water unusual properties that are vital to living organisms, including high specific heat, boiling point, cohesion, and surface tension (Figures 2.11 and 2.12).
- The polarity of the water molecules in the hydrogen-bond lattice makes it difficult for nonpolar substances to penetrate the lattice. The distinct polar and nonpolar environments created by water are critical to the organization of cells (Figure 2.13).
- The polar properties of water allow it to form a hydration layer over the surfaces of polar and charged biological molecules, particularly proteins. Many chemical reactions depend on the special molecular conditions created by the hydration layer (Figure 2.14).
- The polarity of water allows ions and polar molecules to dissolve readily in water, making it a good solvent.

2.5 Water Ionization and Acids, Bases, and Buffers

- Acids are substances that increase the H^+ concentration by releasing additional H^+ as they dissolve in water; bases are substances that decrease the H^+ concentration by gathering H^+ or releasing OH^- as they dissolve.
- The relative concentrations of H^+ and OH^- in a water solution determine the acidity of the solution, which is expressed quantitatively as pH on a number scale ranging from 0 to 14. Neutral solutions, in which the concentrations of H^+ and OH^- are equal, have a pH of 7. Solutions with pH less than 7 have H^+ in excess and are acidic; solutions with pH greater than 7 have OH^- in excess and are basic or alkaline (Figure 2.15).
- The pH of living cells is regulated by buffers, which absorb or release H^+ to compensate for changes in H^+ concentration (Figure 2.16).

TEST YOUR KNOWLEDGE

Remember/Understand

1. Which of the following statements about the mass number of an atom is *incorrect*?
   a. It has a unit defined as a dalton.
   b. On Earth, it equals the atomic weight.
   c. Unlike the atomic weight of an atom, it does not change when gravitational forces change.
   d. It equals the number of electrons in an atom.
   e. It is the sum of the protons and neutrons in the atomic nucleus.

2. Oxygen (O) is a(n) _____; the oxygen we breathe (O_2) is a(n) _____; and the carbon dioxide we exhale is a(n) _____ .
   a. compound; molecule; element
   b. atom; compound; element
   c. element; atom; molecule
   d. atom; element; molecule
   e. element; molecule; compound

3. The chemical activity of an atom:
   a. depends on the electrons in the outermost energy level.
   b. is increased when the outermost energy level is filled with electrons.
   c. depends on its $1s$ but not its $2s$ or $2p$ orbitals.
   d. is increased when valence electrons completely fill the outer orbitals.
   e. of oxygen prevents it from sharing its electrons with other atoms.

4. When electrons are shared equally between atoms, they form:
   a. a polar covalent bond.
   b. a nonpolar covalent bond.
   c. an ionic bond.
   d. a hydrogen bond.
   e. a van der Waals force.

5. Which of the following is *not* a property of water?
   a. It has a low boiling point compared with other molecules.
   b. It has a high heat of vaporization.
   c. Its molecules resist separation, a property called cohesion.
   d. It has the property of adhesion, the ability to stick to charged and polar groups in molecules.
   e. It can form hydrogen bonds to molecules below but not above its surface.

6. The water lattice:
   a. is formed from hydrophobic bonds.
   b. causes ice to be denser than water.
   c. causes water to have a relatively low specific heat.
   d. excludes nonpolar substances.
   e. is held together by hydrogen bonds that are permanent; that is, they never break and reform.

7. A hydrogen bond is:
   a. a strong attraction between hydrogen and another atom.
   b. a bond between a hydrogen atom already covalently bound to one atom and made partially negative by unequal electron sharing with another atom.
   c. a bond between a hydrogen atom already covalently bound to one atom and made partially positive by unequal electron sharing with another atom.
   d. weaker than van der Waals forces.
   e. exemplified by the two hydrogens covalently bound to oxygen in the water molecule.

8. **Apply Evolutionary Thinking** What properties of water made the evolution of life possible?

Apply/Analyze

9. If pond water has a pH of 5, the hydroxide concentration would be:
   a. $10^{-5} M$.
   b. $10^{-10} M$.
   c. $10^5 M$.
   d. $10^9 M$.
   e. $10^{-9} M$.

10. Due to a sudden hormonal imbalance, a patient's blood was tested and shown to have a pH of 7.5. What does this pH value mean?
    a. This is more acidic than normal blood.
    b. It represents a weak alkaline fluid.
    c. This is caused by a release of large amounts of hydrogen ions into the system.
    d. The reaction $H_2CO_3 \rightleftharpoons HCO_3^- + H^+$ is pushed to the left.
    e. This is probably caused by excess CO_2 in the blood.

11. **Discuss Concepts** Detergents allow particles of oil to mix with water. From the information presented in this chapter, how do you think detergents work?

12. **Discuss Concepts** What would living conditions be like on Earth if ice were denser than liquid water?

Evaluate/Create

13. **Discuss Concepts** You place a metal pan full of water on the stove and turn on the heat. After a few minutes, the handle is too hot to touch but the water is only warm. How do you explain this observation?

14. **Discuss Concepts** You are studying a chemical reaction accelerated by an enzyme. H^+ forms during the reaction, but the enzyme's activity is lost at low pH. What could you include in the reaction mix to keep the enzyme's activity at high levels? Explain how your suggestion might solve the problem.

15. **Design an Experiment** You know that adding NaOH to HCl results in the formation of common table salt, NaCl. You have a $0.5\ M$ HCl solution. What weight of NaOH would you need to add to convert all of the HCl to NaCl? (Note: Chemical reactions have the potential to be dangerous. Please do not attempt to perform this reaction.)

For selected answers, see Appendix A.

INTERPRET THE DATA

The pH of human stomach acid ranges from 1.0 to 3.0, whereas a healthy esophagus has a pH of approximately 7.0. In gastroesophageal reflux disease (GERD), often called acid reflux, stomach acid flows backward from the stomach into the esophagus. Repeated episodes in which esophageal pH goes below 4.0, considered clinical acid reflux, can result in bleeding ulcers and damage to the esophageal lining. The data in the **Figure,** from a patient with GERD, show esophageal pH during a sleeping reflux event.

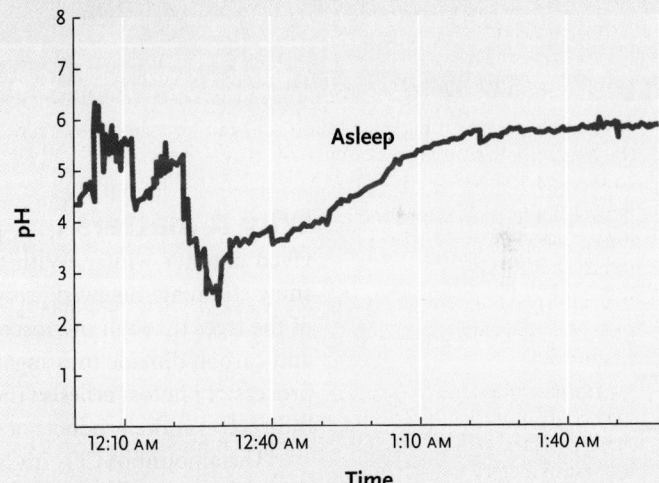

© Cengage Learning 2017

1. How many minutes does it take to go from the peak of the reflux event (when pH is most acidic) to when the reflux event is over?

2. What is the molar concentration of H^+ and OH^- ions: (a) during sleep after the reflux event; and (b) during the peak of the reflux event? Be sure to include the correct concentration units in your answer.

3. What is the change in concentration of H^+ and OH^- during the peak of the reflux attack compared to the clinical value of acid reflux? Be sure to include the correct concentration units in your answer.

Source: Based on T. Demeester et al. 1976. Patterns of gastroesophageal reflux in health and disease. *Annals of Surgery* 184:459–469.

3

Biological Molecules: The Carbon Compounds of Life

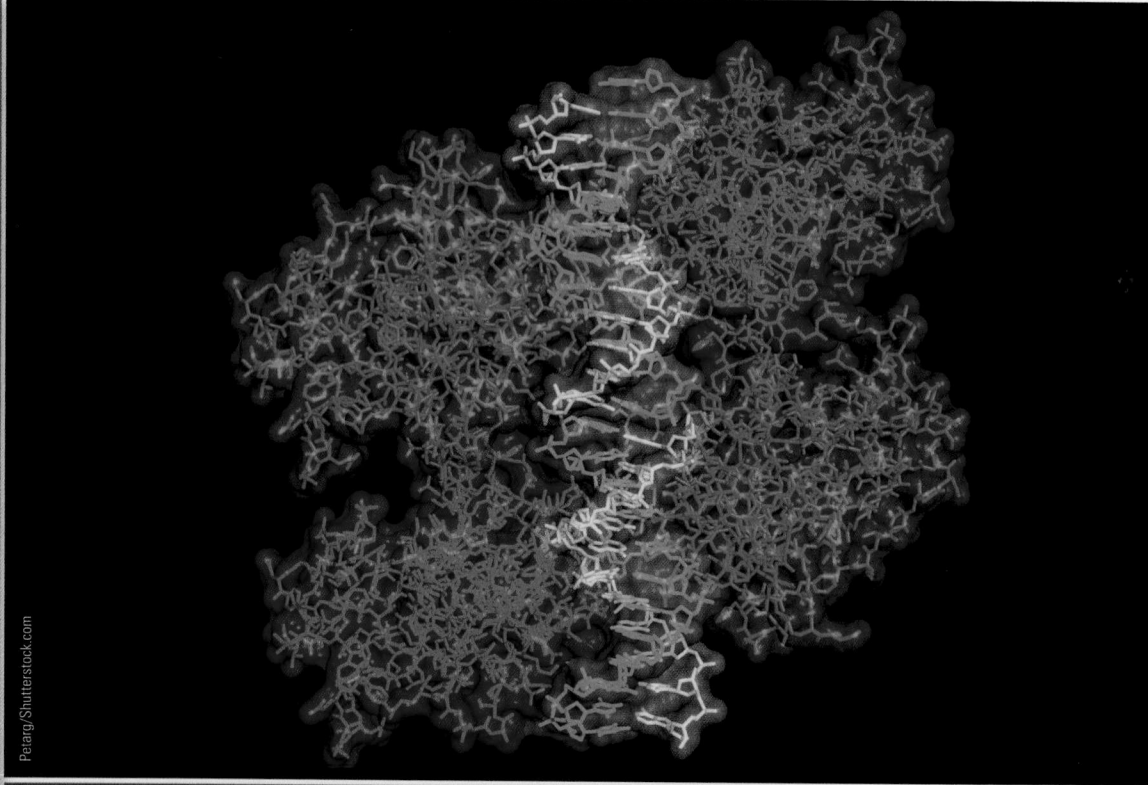

Petarg/Shutterstock.com

A regulatory protein (red and blue) bound to a DNA molecule (helix of yellow and green).

Why it matters . . . In the Pacific Northwest, vast forests of coniferous trees have survived another winter. With the arrival of spring, rising temperatures and water from melting snow stimulate renewed growth. Carbon dioxide (CO_2) from the air enters the needlelike leaves of the trees through microscopic pores. Using energy from sunlight, the trees combine the water and carbon dioxide into sugars such as glucose and other carbon-based compounds through the process of photosynthesis. The lives of plants, and almost all other organisms, depend directly or indirectly on the products of photosynthesis.

The amount of CO_2 in the atmosphere is critical to photosynthesis. Researchers have found that CO_2 concentration changes with the seasons. It declines during spring and summer, when plants and other photosynthetic organisms withdraw large amounts of the gas from the air and convert it into sugars and other complex carbon compounds. It increases during fall and winter, when photosynthesis decreases and decomposers that release the gas as a metabolic by-product increase. Great quantities of CO_2 are also added to the atmosphere by forest fires and by the burning of coal, oil, gasoline, and other fossil fuels in automobiles, aircraft, trains, power plants, and other industries.

The importance of atmospheric CO_2 to food production and world climate is just one example of how carbon and its compounds are fundamental to the living world. Carbon compounds form the structures of all living organisms on Earth and take part in all biological reactions. They also serve as sources of energy for living organisms and as an energy resource for much of the world's industry—for example, coal and oil are the fossil remains of long-dead organisms. This chapter outlines the structures and functions of biological carbon compounds.

3.1 Formation and Modification of Biological Molecules

The bonding properties of carbon (described and shown in Section 2.3 and Figure 2.7) enable it to form an astounding variety of chain and ring structures that are the backbones of all biological molecules. The wide variety of carbon-based molecules has been responsible for the wide diversity of organisms that have evolved.

Carbon Chains and Rings Form the Backbones of All Biological Molecules

Collectively, molecules based on carbon are known as **organic molecules.** All other substances—those without carbon atoms in their structures—are **inorganic molecules.** A few of the smallest carbon-containing molecules that occur in the environment as minerals or atmospheric gases, such as CO_2, are also considered inorganic molecules. Outside of water, the four major classes of organic molecules—*carbohydrates, lipids, proteins,* and *nucleic acids*—form almost the entire substance of living organisms. They are discussed in turn in subsequent sections in this chapter.

In organic molecules, carbon atoms bond covalently to each other and to other atoms (chiefly hydrogen, oxygen, nitrogen, and sulfur) in molecules that range in size from a few atoms to thousands, or even millions, of atoms. Molecules consisting of carbon linked only to hydrogen atoms are called **hydrocarbons** (*hydro-* refers to hydrogen, not to water).

As discussed in Section 2.3, carbon has four unpaired outer electrons that it readily shares to complete its outermost energy level, forming four covalent bonds. The simplest hydrocarbon, CH_4 (methane), consists of a single carbon atom bonded to four hydrogen atoms (see Figure 2.7A–C). More complex hydrocarbons vary in length, the bonds involved, branching, double-bond position, and whether or not one or more ring is present. **Figure 3.1A** illustrates two-carbon hydrocarbons with single, double, and triple bonds. **Figure 3.1B** shows longer hydrocarbons, one branched and one unbranched. **Figure 3.1C** shows a hydrocarbon ring structure with double bonds. The number of bonds between neighboring carbon atoms diversifies the structures. A triple bond can only occur in a two-carbon hydrocarbon (see Figure 3.1A), but single and double bonds are found in both linear and ring hydrocarbons (see Figures 3.1B–C). There is almost no limit to the number of different hydrocarbon structures that carbon and hydrogen can form. The structures of the hydrocarbon molecules are responsible for their functions.

Organic Molecules Formed by Chemical Evolution

Formation of the organic molecules that allowed the first forms of life on Earth to originate is termed **chemical evolution.** Chemical evolution occurred as a result of reactions involving

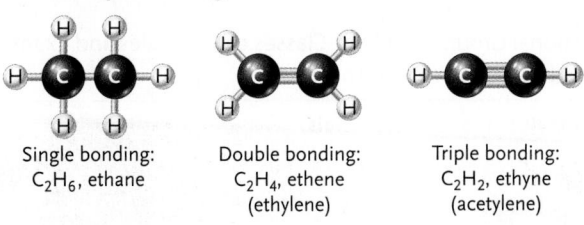

A. Two-carbon hydrocarbons with single, double, and triple bonding

Single bonding: C_2H_6, ethane

Double bonding: C_2H_4, ethene (ethylene)

Triple bonding: C_2H_2, ethyne (acetylene)

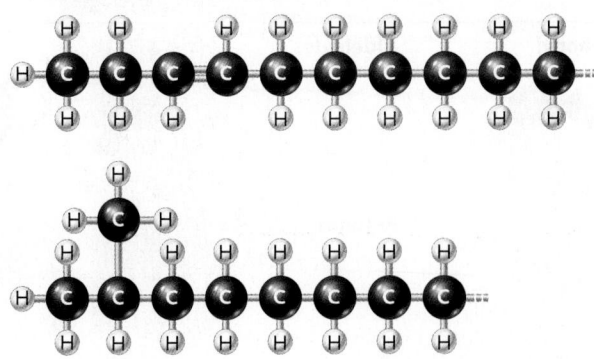

B. Linear and branched hydrocarbon chains

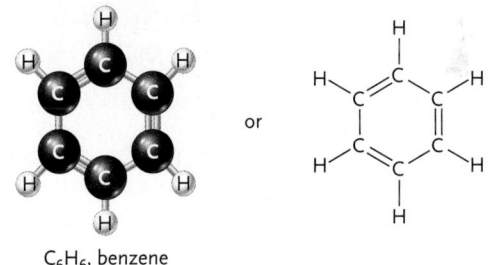

C. Hydrocarbon ring, in this case with double bonds

or

C_6H_6, benzene

FIGURE 3.1 Examples of hydrocarbon structures.
© Cengage Learning 2017

inorganic molecules present on primordial Earth, and the conditions of the Earth at the time. Those conditions were a lot different from present-day conditions, importantly involving an atmosphere that lacked oxygen and that contained hydrogen, methane, ammonia, and water. Energy for the chemical evolution reactions came from solar energy and other natural sources, such as lightning and volcanic activity. A classic laboratory simulation experiment of chemical evolution was performed in 1953 by Stanley Miller and Harold Urey. Using an apparatus they designed, they achieved the synthesis of complex organic molecules from methane, ammonia, water, and hydrogen. The Miller–Urey experiment is described in more detail in Chapter 25: The Origin of Life.

Functional Groups Confer Specific Functions to Biological Molecules

Carbohydrates, lipids, proteins, and nucleic acids contain particular small, reactive groups of atoms called **functional groups.** Each of those groups has specific chemical properties,

TABLE 3.1 | Common Functional Groups of Organic Molecules

Functional Group (boxed in blue)	Major Classes of Molecules and Examples of Them		Properties

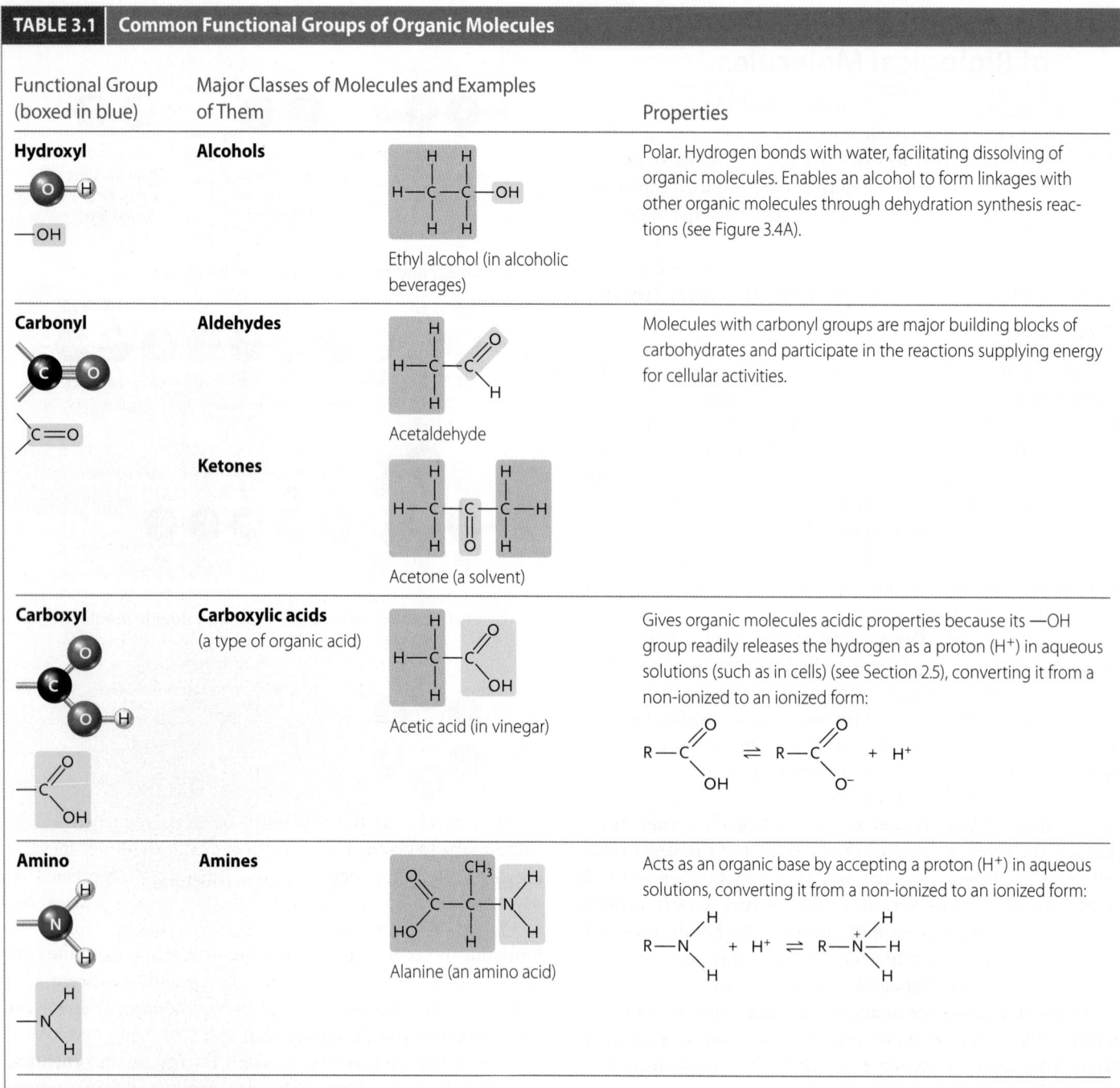

Hydroxyl

Alcohols

Ethyl alcohol (in alcoholic beverages)

Polar. Hydrogen bonds with water, facilitating dissolving of organic molecules. Enables an alcohol to form linkages with other organic molecules through dehydration synthesis reactions (see Figure 3.4A).

Carbonyl

Aldehydes

Acetaldehyde

Ketones

Acetone (a solvent)

Molecules with carbonyl groups are major building blocks of carbohydrates and participate in the reactions supplying energy for cellular activities.

Carboxyl

Carboxylic acids (a type of organic acid)

Acetic acid (in vinegar)

Gives organic molecules acidic properties because its —OH group readily releases the hydrogen as a proton (H^+) in aqueous solutions (such as in cells) (see Section 2.5), converting it from a non-ionized to an ionized form:

Amino

Amines

Alanine (an amino acid)

Acts as an organic base by accepting a proton (H^+) in aqueous solutions, converting it from a non-ionized to an ionized form:

(Continued)

which are then also found in the larger molecules containing them. Thus, the number and arrangement of functional groups in a larger molecule give that molecule its particular function.

Functional groups can participate in biological reactions. The functional groups that enter most frequently into biological reactions are the **hydroxyl group** (—OH), **carbonyl group** ($>C=O$), **carboxyl group** (—COOH), **amino group** (—NH_2), **phosphate group** (—OPO_3^{2-}), and **sulfhydryl group** (—SH) **(Table 3.1).** The functional groups (boxed in blue in the table) are linked by covalent bonds to other atoms in biological molecules, usually carbon atoms. Varying chemical groups (boxed in brown in the table) are attached to a functional group, and

more than one functional group can be present in a molecule. A chemical group attached to a functional group is often symbolized by *R* as a shorthand way with which to represent it. A double bond, such as between the C and O in a carbonyl group, indicates that two pairs of electrons are shared between the carbon and oxygen atoms.

Isomers Have the Same Chemical Formula but Different Molecular Structures

Isomers are two or more molecules with the same chemical formula but different molecular structures. Molecules that are

| TABLE 3.1 | Common Functional Groups of Organic Molecules *(Continued)* |

Functional Group (boxed in blue)	Major Classes of Molecules and Examples of Them	Properties
Phosphate	**Organic phosphates**	Molecules that contain phosphate groups react as weak acids because one or both —OH groups readily release their hydrogens as H^+ in aqueous solutions, converting them from a non-ionized to an ionized form:

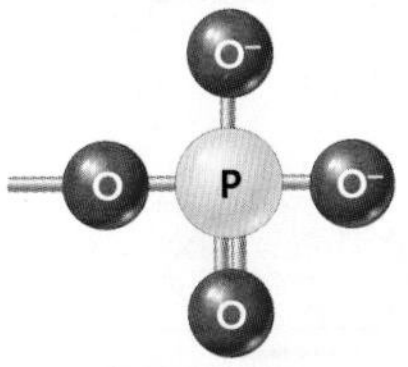

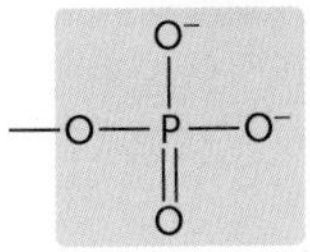

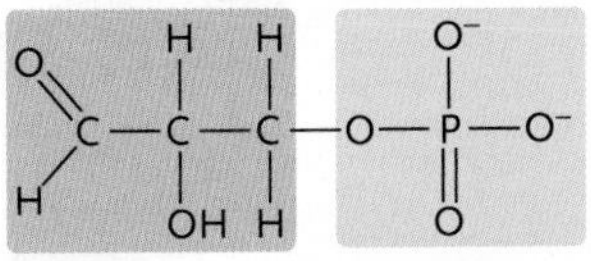

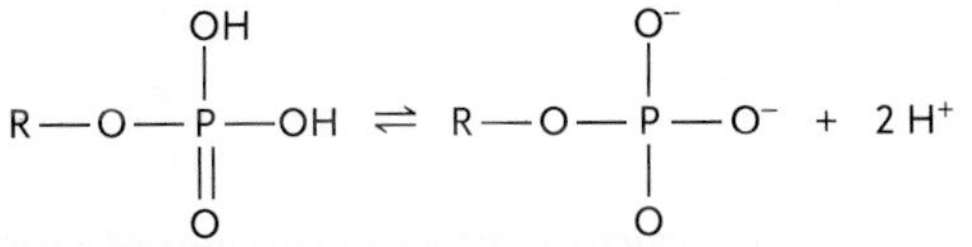

Glyceraldehyde-3-phosphate (product of photosynthesis). Nucleotides and nucleic acids are also examples.

$$R-O-\overset{\overset{OH}{|}}{\underset{\underset{O}{\|}}{P}}-OH \;\rightleftharpoons\; R-O-\overset{\overset{O^-}{|}}{\underset{\underset{O}{\|}}{P}}-O^- \;+\; 2\,H^+$$

A phosphate group can bridge two organic building blocks to form a larger structure, for example, DNA:

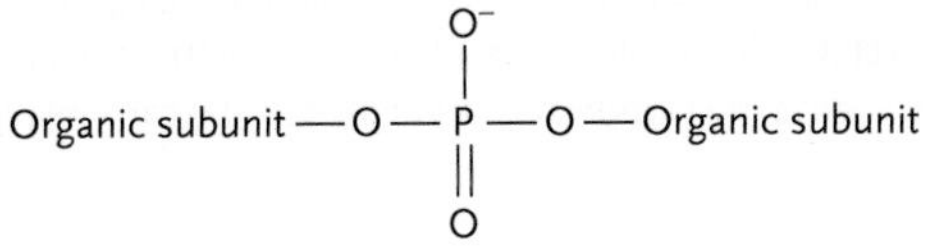

$$\text{Organic subunit}-O-\overset{\overset{O^-}{|}}{\underset{\underset{O}{\|}}{P}}-O-\text{Organic subunit}$$

Phosphate groups are added to or removed from biological molecules as part of reactions that conserve or release energy, or, for many proteins, to alter activity.

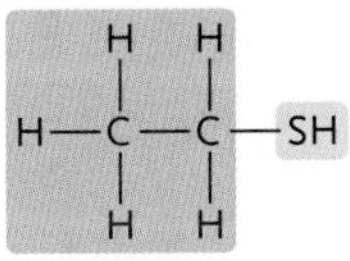

| **Sulfhydryl** | **Thiols** | Easily converted into a covalent linkage, in which it loses its hydrogen atom as it binds. In many linking reactions, two sulfhydryl groups form a **disulfide linkage** (—S—S—): |

Mercaptoethanol

$$R-SH + HS-R \rightarrow R\underset{\substack{\text{disulfide}\\\text{linkage}}}{-S-S-}R + 2\,H^+ + 2 \text{ electrons}$$

---

mirror images of one another are an example of **stereoisomers.** That is, often one or more of the carbon atoms in an organic molecule link to four different atoms or functional groups. A carbon linked in this way is called an *asymmetric carbon.* Asymmetric carbons have important effects on the molecule because they can take either of two fixed positions in space with respect to other carbons in a carbon chain.

Figure 3.2 shows the stereoisomers of an amino acid. The chemical formula of each, as well as the connections between atoms and groups, is the same. The difference between the two

forms is similar to the difference between your two hands. Although both hands have four fingers and a thumb, they are not identical; rather, they are mirror images of each other. That is, when you hold your right hand in front of a mirror, the reflection looks like your left hand and vice versa. One of the stereoisomers is designated the L isomer (L for *laevus* = left) (see left side of Figure 3.2 with the left hand). The other stereoisomer is called the D isomer (D for *dexter* = right) (see right side of Figure 3.2 with the right hand).

The difference between L and D stereoisomers is critical to biological function. Typically, one of the two forms enters much more readily into reactions within a cell; just as your left hand does not fit readily into a right-hand glove, enzymes (proteins that accelerate chemical reactions in living organisms) fit best to one of the two forms of a stereoisomer. For example, most of the enzymes that catalyze the biochemical reactions involving amino acids recognize the L stereoisomer. Many other kinds of biological molecules besides amino acids form stereoisomers. Most of the enzymes that catalyze the biochemical reactions involving sugars recognize the D stereoisomer, the more common form found among cellular carbohydrates.

Structural isomers are two molecules with the same chemical formula but atoms that are connected in different ways. The sugars glucose and fructose are examples of structural isomers **(Figure 3.3).**

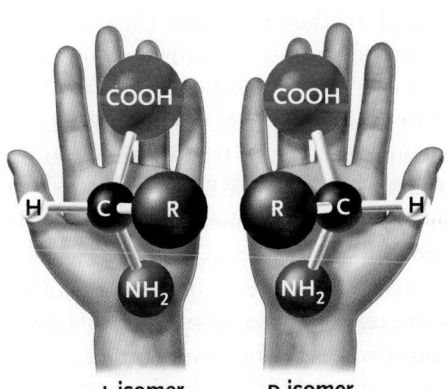

FIGURE 3.2 Stereoisomers of an amino acid.

Adapted from http://creationwiki.org/File:Chirality.jpg

A. Glucose
(an aldehyde)

B. Fructose
(a ketone)

FIGURE 3.3 Glucose and fructose, structural isomers of a six-carbon sugar with the chemical formula $C_6H_{12}O_6$. **(A)** In glucose, the aldehyde isomer, the carbonyl group (shaded region) is located at the end of the carbon chain. **(B)** In fructose, the ketone isomer, the carbonyl group is located inside the carbon chain. For convenience, the carbons of the sugars are numbered, with 1 being the carbon at the end nearest the carbonyl group.
© Cengage Learning 2017

A Water Molecule Is Added or Removed in Many Reactions Involving Functional Groups

In many of the reactions that involve functional groups, the components of a water molecule, —H and —OH, are removed from or added to the groups as they interact. In a **dehydration synthesis reaction** or **condensation reaction (Figure 3.4A),** the components of a water molecule are *removed* during a reaction (usually as part of the assembly of a larger molecule from smaller subunits). For example, this type of reaction occurs when individual sugar molecules combine to form a starch molecule. In **hydrolysis,** the reverse reaction, the components of a water molecule are *added* to functional groups as molecules are broken into smaller subunits **(Figure 3.4B).** For example, the breakdown of a protein molecule into individual amino acids occurs by hydrolysis in the digestive processes of animals.

Of the functional groups, hydroxyl groups readily enter dehydration synthesis reactions and are formed as part of hydrolysis reactions. Carboxyl groups readily enter into dehydration synthesis reactions, giving up hydroxyl groups as organic molecules combine into larger assemblies. Amino groups also readily enter dehydration synthesis reactions, releasing H^+ as it links subunits into larger molecules. For example, the joining of amino acids in the synthesis of proteins involves a dehydration reaction involving the carboxyl group of one amino acid and the amino group of another amino acid (see Figure 3.16).

Many Carbohydrates, Proteins, and Nucleic Acids Are Macromolecules

Many carbohydrates, proteins, and nucleic acids are large polymers (*poly* = many; *mer* = unit). A **polymer** is a molecule assembled from subunit molecules called *monomers* into a

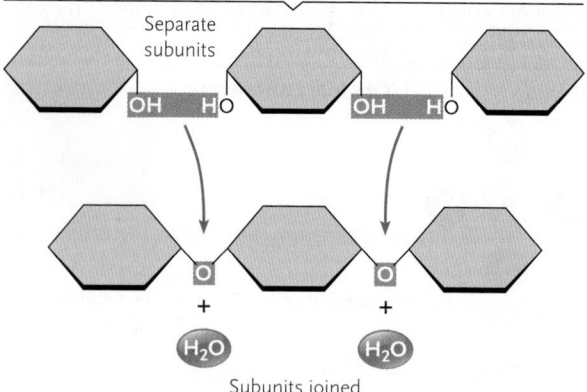

A. Dehydration synthesis reaction
Dehydration synthesis reactions remove the components of a water molecule as new covalent bonds join subunits into a larger molecule.

B. Hydrolysis reaction
Hydrolysis reactions add the components of a water molecule as covalent bonds are broken, splitting a molecule into smaller subunits.

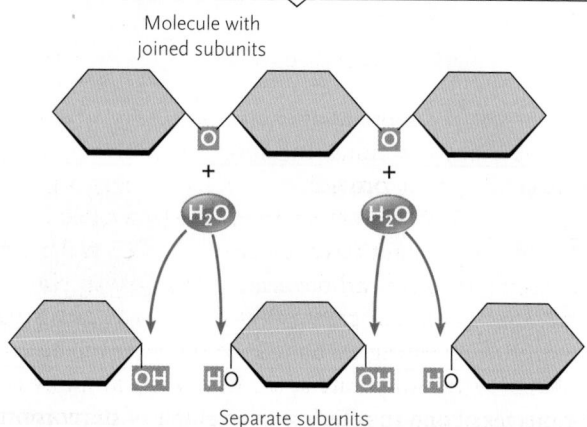

FIGURE 3.4 Dehydration synthesis and hydrolysis reactions.

chain by covalent bonds. The process of assembly of a polymer from monomers is called **polymerization.** The polymerization reactions are dehydration synthesis reactions. The breakdown of polymers into monomers—the opposite reactions—occur by hydrolysis.

Each type of polymeric biological molecule contains one type of monomer. Depending on the molecule, the monomers may be identical, or they may have chemical variations. The variations among monomer structures are responsible for the highly diverse and varied biological molecules found in living organisms. For instance, proteins are polymers consisting of amino acid monomers. There are 20 different amino acids, each with identical amino and carboxyl functional groups that enable them to undergo polymerization (see Figure 3.14), but each also with a different R group. The proteins assembled from these amino acids vary in number and organization in the polymer chains, resulting in a huge variety of proteins with different structures and, therefore, functions.

Somewhat arbitrarily, a single polymer molecule with a mass of 1,000 Da (daltons) or more is called a **macromolecule** (*macro* = large). By that criterion, many representatives of carbohydrates, proteins, and nucleic acids are macromolecules. In a number of instances, macromolecules interact to form even larger functional molecular structures in cells. For example, the ribosome, the cellular structure that plays the central role in the polymerization of amino acids into a protein chain, consists of several RNA macromolecules and many protein macromolecules.

STUDY BREAK 3.1

1. What is the difference between hydrocarbons and other organic molecules?
2. What is the maximum number of bonds that a carbon atom can form?
3. Do carboxyl groups, amino groups, and phosphate groups act as acids or bases?
4. What is the difference between a dehydration synthesis reaction (condensation reaction) and hydrolysis?

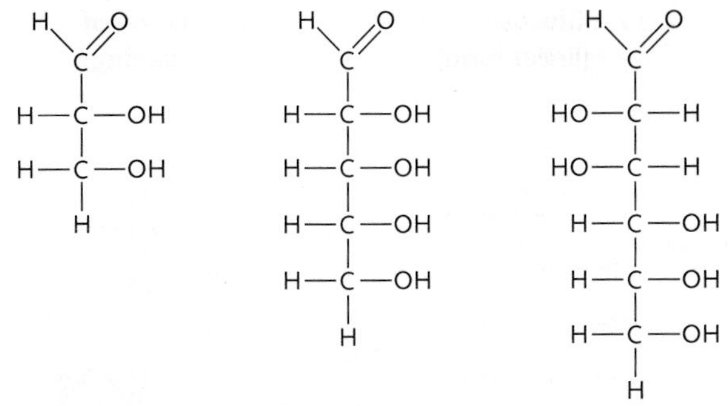

Glyceraldehyde
(3 carbons;
a triose)

Ribose
(5 carbons;
a pentose)

Mannose
(6 carbons;
a hexose)

FIGURE 3.5 Some representative monosaccharides. The triose, glyceraldehyde, takes part in energy-yielding reactions and photosynthesis. The pentose, ribose, is a component of RNA and of molecules that carry energy. The hexose, mannose, is a fuel substance and a component of glycolipids and glycoproteins.
© Cengage Learning 2017

3.2 Carbohydrates

Carbohydrates, the most abundant organic molecules in the world, serve many functions. Together with fats, they act as the major fuel substances providing chemical energy for cellular activities. The carbohydrate sucrose, common table sugar, is consumed in large quantities as an energy source in the human diet. Energy-providing carbohydrates are stored in plant cells as **starch** and in animal cells as **glycogen,** both consisting of long chains of repeating carbohydrate subunits linked end to end. Chains of carbohydrate subunits also form many structural molecules, such as **cellulose,** one of the primary constituents of plant cell walls.

Carbohydrates contain only carbon, hydrogen, and oxygen atoms, in an approximate ratio of 1 carbon:2 hydrogens:1 oxygen (CH_2O). The names of many carbohydrates end in -ose. The smallest carbohydrates, the **monosaccharides** (mono = one; saccharum = sugar), contain three to seven carbon atoms. For example, the monosaccharide glucose consists of a chain of six carbons and has the molecular formula $C_6H_{12}O_6$. Two monosaccharides polymerize to form a **disaccharide** such as sucrose. Carbohydrate polymers with more than 10 linked monosaccharide monomers are called **polysaccharides.** Starch, glycogen, and cellulose are common polysaccharides.

Monosaccharides Are the Structural Units of Carbohydrates

Carbohydrates occur either as monosaccharides or as polymers of monosaccharide units linked together. Monosaccharides are soluble in water, and most have a sweet taste. Of the monosaccharides, those that contain three carbons (trioses), five carbons (pentoses), and six carbons (hexoses) are most common in living organisms **(Figure 3.5).**

All monosaccharides can occur in the linear form shown in Figure 3.5. In this form, each carbon atom in the chain

except one has both an —H and an —OH group attached to it. The remaining carbon is part of a carbonyl group, which may be located at the end of the carbon chain in the aldehyde position, or inside the chain in the ketone position, resulting in structural isomers (see Figure 3.3).

Monosaccharides with four or more carbons can fold back on themselves to assume a ring form. Folding into a ring occurs through a reaction between two functional groups in the same monosaccharide, as occurs in glucose **(Figure 3.6).** The ring form of most five- and six-carbon sugars is much more common in cells than the linear form.

In the ring form of many five- or six-carbon monosaccharides, including glucose, the carbon at the 1 position of the ring is asymmetric because its four bonds link to different groups of atoms. This asymmetry allows monosaccharides such as glucose to exist as two different stereoisomers. The glucose stereoisomer with an —OH group pointing below the plane of the ring is known as *alpha-glucose,* or *α-glucose*; the stereoisomer with an —OH group pointing above the plane of the ring is known as *beta-glucose,* or *β-glucose* (see Figure 3.6B). Other five- and six-carbon monosaccharide rings have similar α- and β-configurations.

The α- and β-rings of monosaccharides can give the polysaccharides assembled from them vastly different chemical properties. For example, starches, which are assembled from α-glucose units, are biologically reactive polysaccharides easily digested by animals; cellulose, which is assembled from β-glucose units, is relatively unreactive and, for most animals, indigestible.

Two Monosaccharides Link to Form a Disaccharide

Disaccharides typically are assembled from two monosaccharides covalently joined by a dehydration synthesis reaction. For example, the disaccharide maltose is formed by the linkage of

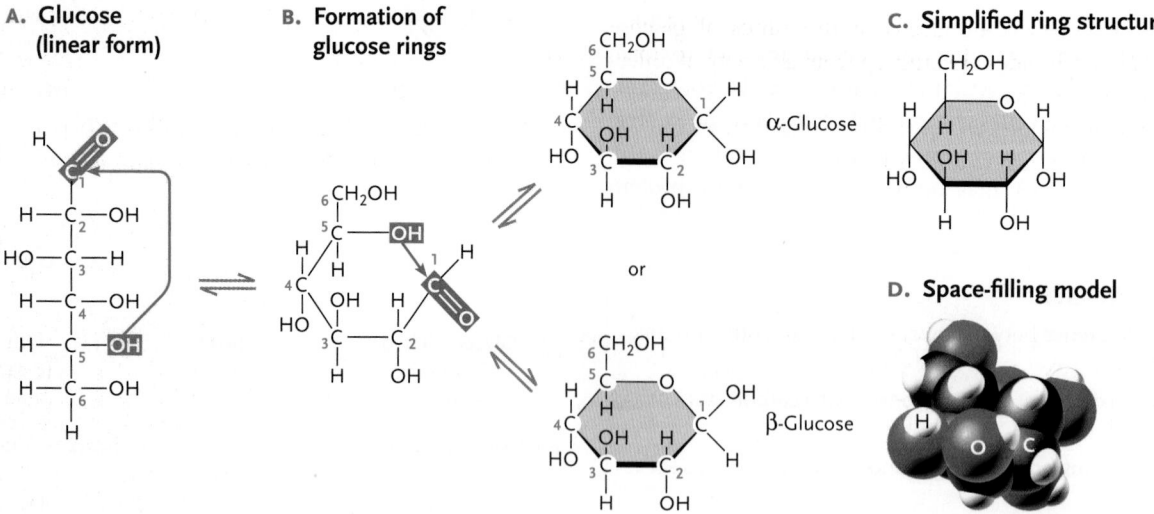

FIGURE 3.6 Ring formation by glucose. (A) Glucose in linear form. **(B)** The ring form of glucose is produced by a reaction between the aldehyde group at the 1 carbon and the hydroxyl group at the 5 carbon. The reaction produces two glucose stereoisomers, α- and β-glucose. If the ring is considered to lie in the plane of the page, the —OH group points below the page in α-glucose and upward from the page in β-glucose. In the ring form, the thicker lines along one side indicate that you are viewing the ring edge on. For simplicity, the group at the 6 carbon is shown as CH_2OH in this and later diagrams. **(C)** A commonly used, simplified representation of the glucose ring (in this case α-glucose), in which the Cs designating carbons of the ring are omitted. Other sugar rings similarly are drawn this way. **(D)** A space-filling model of glucose, showing the volumes occupied by the atoms. Carbon atoms are black, oxygen atoms are red, and hydrogen atoms are white.
© Cengage Learning 2017

two α-glucose molecules **(Figure 3.7A)** with oxygen as a bridge between the number 1 carbon of the first glucose unit and the 4 carbon of the second glucose unit. Bonds of this type, which commonly link monosaccharides into chains, are known as **glycosidic bonds.** A glycosidic bond between a 1 carbon and a 4 carbon is written in chemical shorthand as a 1→4 linkage; 1→2, 1→3, and 1→6 linkages are also common in carbohydrate chains. The linkages are designated as α or β depending on the orientation of the —OH group at the 1 carbon that forms the bond. In maltose, the —OH group is in the α position. Therefore, the link between the two glucose subunits of maltose is written as an α(1→4) linkage.

Maltose, sucrose, and lactose are common disaccharides. Maltose (see Figure 3.7A) is present in germinating seeds and is a major sugar used in the brewing industry. Sucrose, which contains a glucose and a fructose unit **(Figure 3.7B),** is transported to and from different parts of leafy plants. It is probably the most plentiful sugar in nature. Table sugar is made by extracting and crystallizing sucrose from plants, such as sugar cane and sugar beets. Lactose, assembled from a glucose and a galactose unit **(Figure 3.7C),** is the primary sugar of milk.

Monosaccharides Link in Longer Chains to Form Polysaccharides

Polysaccharides are the macromolecules formed by polymerization of monosaccharide monomers through dehydration synthesis reactions. The most common polysaccharides—the plant starches, glycogen, and

A. Formation of maltose
Maltose is assembled from two glucose molecules.

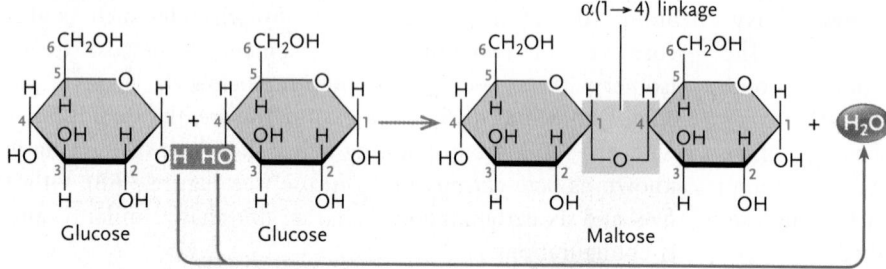

B. Sucrose
Sucrose is assembled from glucose and fructose.

C. Lactose
Lactose is assembled from galactose and glucose.

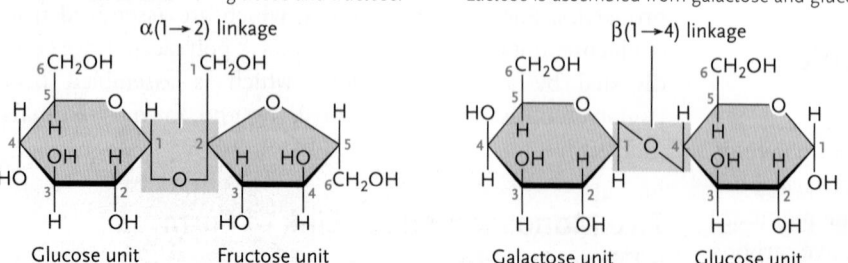

FIGURE 3.7 Disaccharides. A dehydration synthesis reaction involving two monosaccharides produces a disaccharide. The components of a water molecule (in blue) are removed from the monosaccharides as they join.
© Cengage Learning 2017

cellulose—are polymers of hundreds or thousands of glucose units. Other polysaccharides are built up from a variety of different sugar units. Polysaccharides may be linear, unbranched molecules, or they may contain one or more branches in which side chains of sugar units attach to a main chain.

Figure 3.8 shows four common polysaccharides. Plant starches include both linear, unbranched forms such as amylose **(Figure 3.8A)** and branched forms such as amylopectin. Glycogen **(Figure 3.8B),** a more highly branched polysaccharide than amylopectin, can be assembled or disassembled readily to

A. Amylose, a plant starch

Amylose, formed from α-glucose units joined end to end in α(1→4) linkages. The coiled structures are induced by the bond angles in the α-linkages.

α(1→4) linkage

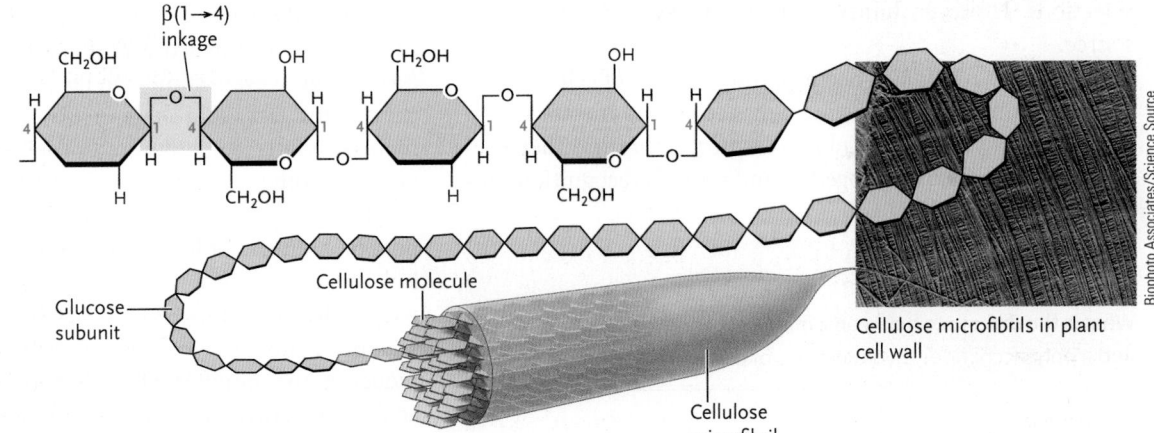

Amylose grains (purple) in plant root tissue

Ed Reschke/Getty Images

B. Glycogen, found in animal tissues

Glycogen, formed from glucose units joined in chains by α(1→4) linkages; side branches are linked to the chains by α(1→6) linkages (boxed in blue).

α(1→4) linkage

α(1→6) linkage

Glycogen particles (magenta) in liver cell

DeAgostini/Getty Images

C. Cellulose, the primary fiber in plant cell walls

Cellulose, formed from glucose units joined end to end by β(1→4) linkages. Hundreds to thousands of cellulose chains line up side by side, in an arrangement reinforced by hydrogen bonds between the chains, to form cellulose microfibrils in plant cells.

β(1→4) linkage

Glucose subunit

Cellulose molecule

Cellulose microfibril

Cellulose microfibrils in plant cell wall

Biophoto Associates/Science Source

D. Chitin, a reinforcing fiber in the external skeleton of arthropods and the cell walls of some fungi

Chitin, formed from β(1→4) linkages joining glucose units modified by the addition of nitrogen-containing groups. The external body armor of the tick is reinforced by chitin fibers.

β(1→4) linkage

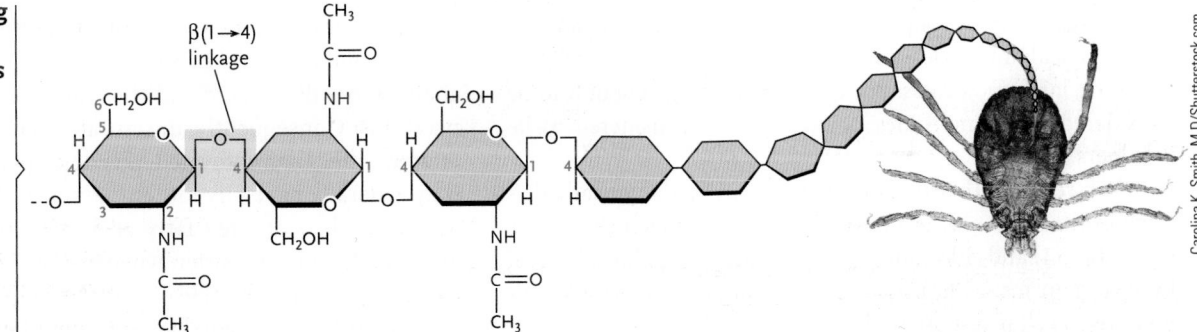

Carolina K. Smith, M.D/Shutterstock.com

FIGURE 3.8 Four common polysaccharides.

© Cengage Learning 2017

take up or release glucose; it is stored in large quantities in the liver and muscle tissues of many animals.

Cellulose **(Figure 3.8C)**, probably the most abundant carbohydrate on Earth, is an unbranched polysaccharide assembled from glucose monomers bound together by β-linkages. It is the primary structural fiber of plant cell walls; in this role, cellulose has been likened to the steel rods in reinforced concrete. Its tough fibers enable the cell walls of plants to withstand enormous weight and stress. Fabrics such as cotton and linen are made from cellulose fibers extracted from plant cell walls. Animals such as mollusks, crustaceans, and insects synthesize an enzyme that digests the cellulose they eat. In ruminant mammals, such as cows, microorganisms in the digestive tract break down cellulose. Cellulose passes unchanged through the human digestive tract as indigestible fibrous matter. Many nutritionists maintain that the bulk provided by cellulose fibers helps maintain healthy digestive function.

Chitin **(Figure 3.8D)**, another tough and resilient polysaccharide, is assembled from glucose units modified by the addition of nitrogen-containing groups. The modified glucose monomers of chitin are held together by β-linkages. Chitin is the main structural fiber in the external skeletons and other hard body parts of arthropods such as insects, crabs, and spiders. It is also a structural material in the cell walls of fungi such as mushrooms, bread molds, and yeasts. Unlike cellulose, chitin is digested by enzymes that are widespread among microorganisms, plants, and many animals. In plants and animals, including humans and other mammals, chitin-digesting enzymes occur primarily as part of defenses against fungal infections. However, humans cannot digest chitin as a food source.

Polysaccharides also occur on the surfaces of cells, particularly in animals. These surface polysaccharides are attached to both the protein and lipid molecules in membranes. They help hold the cells of animals together and serve as recognition sites between cells.

STUDY BREAK 3.2

What is the difference between a monosaccharide, a disaccharide, and a polysaccharide? Give examples of each.

3.3 Lipids

Lipids are a diverse group of water-insoluble, primarily nonpolar biological molecules composed mostly of hydrocarbons. Some are large molecules, but they are not large enough to be considered macromolecules. Lipids are not considered to be polymers.

As a result of their nonpolar character, lipids typically dissolve much more readily in nonpolar solvents, such as acetone (see Table 3.1) and chloroform, than in water, the polar solvent of living organisms. Their insolubility in water underlies their ability to form cell membranes, the thin molecular films that create boundaries between and within cells (see Chapters 4 and 5).

In addition to forming membranes, some lipids are stored and used in cells as an energy source. Other lipids serve as hormones that regulate cellular activities. Three types of lipid molecules—*neutral lipids, phospholipids,* and *steroids*—occur most commonly in living organisms.

Neutral Lipids Are Familiar as Fats and Oils

Neutral lipids, commonly found in cells as energy-storage molecules, are called "neutral" because at cellular pH they have no charged groups; they are, therefore, nonpolar. **Oils** and **fats** are the two types of neutral lipids. Oils are liquid at biological temperatures, and fats are semisolid. Generally, neutral lipids are insoluble in water.

Almost all neutral lipids are formed by dehydration synthesis reactions involving glycerol and three *fatty acids* **(Figure 3.9)**. The product is also called a **triglyceride.** Glycerol is a three-carbon alcohol with an —OH attached to each carbon. In its free state, glycerol is a polar, water-soluble, sweet-tasting alcohol. The glycerol forms the backbone of the triglyceride. A **fatty acid** contains a single hydrocarbon chain with a carboxyl group (—COOH) at one end. The carboxyl group gives the fatty acid its acidic properties. In the synthesis of a triglyceride, three dehydration synthesis reactions occur, each involving the carboxyl group of one fatty acid and one of the hydroxyl groups of glycerol. A covalent bond formed between a carboxyl group and a hydroxyl group, as here, is called an **ester linkage.** In the formation of the three ester linkages, the polar groups of glycerol are eliminated, resulting in the nonpolar triglyceride.

The fatty acids in living organisms contain four or more carbons in their hydrocarbon chain, with the most common forms having even-numbered chains of 14 to 22 carbons. Only the shortest fatty acid chains are water-soluble. As chain length increases, fatty acids become progressively less water-soluble and more oily.

If the hydrocarbon chain of a fatty acid binds the maximum possible number of hydrogen atoms, so that only single bonds link the carbon atoms, the fatty acid is said to be **saturated** with hydrogen atoms (as in stearic acid in **Figure 3.10A**). If one or more double bonds link the carbons (**Figure 3.10B,** arrow), reducing the number of hydrogen atoms bound, the fatty acid is **unsaturated.** Fatty acids with one double bond are **monounsaturated;** those with more than one double bond are **polyunsaturated.**

Unsaturated fatty acid chains tend to bend or "kink" at a double bond (see Figures 3.10B and 3.11C). The kink makes the chains more disordered and thus more fluid at biological temperatures. Consequently, unsaturated fatty acids melt at lower temperatures than saturated fatty acids of the same length, and they generally have oily rather than fatty characteristics.

In foods, saturated fatty acids are usually found in solid animal fats, such as butter, whereas unsaturated fatty acids are usually found in vegetable oils, such as liquid canola oil. Nonetheless, both solid animal fats and liquid vegetable oils contain some saturated and some unsaturated fatty acids.

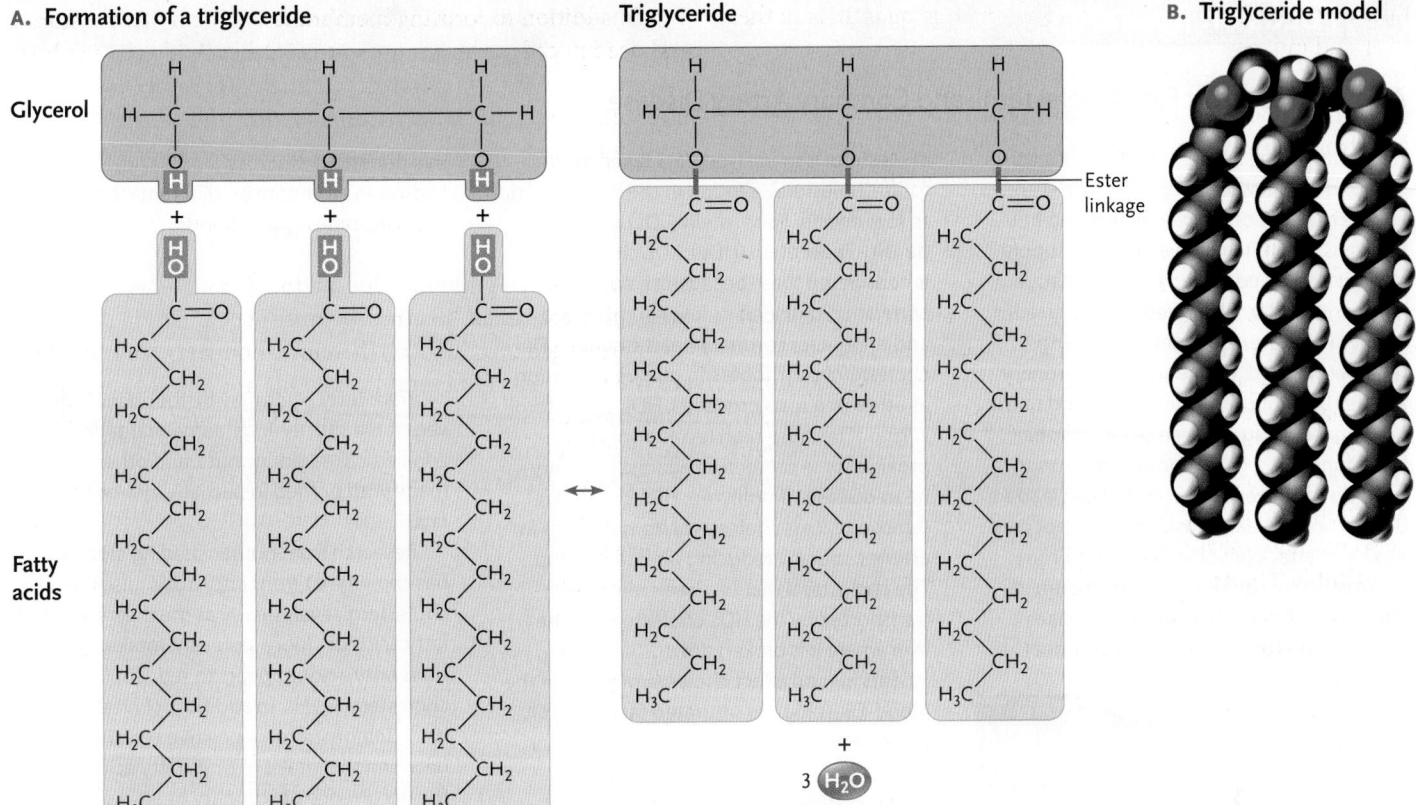

A. Formation of a triglyceride

Glycerol

Fatty acids

Triglyceride

Ester linkage

3 H_2O

B. Triglyceride model

FIGURE 3.9 Triglycerides. (A) Formation of a triglyceride by dehydration synthesis of glycerol with three fatty acids. The fatty acid shown is palmitic acid, and the triglyceride product is glyceryl palmitate. The components of a water molecule (in blue) are removed from the glycerol and fatty acids in each of the three bonds formed. **(B)** Space-filling model of the triglyceride, glyceryl palmitate.

© Cengage Learning 2017

The fatty acids linked to a glycerol may be different or the same. Different organisms usually have distinctive combinations of fatty acids in their triglycerides. As with individual fatty acids, triglycerides generally become less fluid as the length of their fatty acid chains increases; those with shorter chains remain liquid as oils at biological temperatures, and those with longer chains solidify as fats. The degree of saturation of the fatty acid chains also affects the fluidity of triglycerides—the more saturated, the less fluid the triglyceride. Plant oils are converted commercially to fats by *hydrogenation*—that is, adding hydrogen atoms to increase the degree of saturation, as in the conversion of vegetable oils to margarines and shortening.

Triglycerides are used widely as stored energy in animals. Gram for gram, they yield more than twice as much energy as carbohydrates do (see Chapter 47). Therefore, fats are an excellent source of energy in the diet. Storing the equivalent amount of energy as carbohydrates rather than fats would add more than 100 pounds to the weight of an average man or woman. A layer of fatty tissue just under the skin also serves as an insulating blanket in humans, other mammals, and birds. Triglycerides secreted from special glands in waterfowl and other birds help make their feathers water repellent.

Unsaturated fats are considered healthier than saturated fats in the human diet. Saturated fats have been implicated in

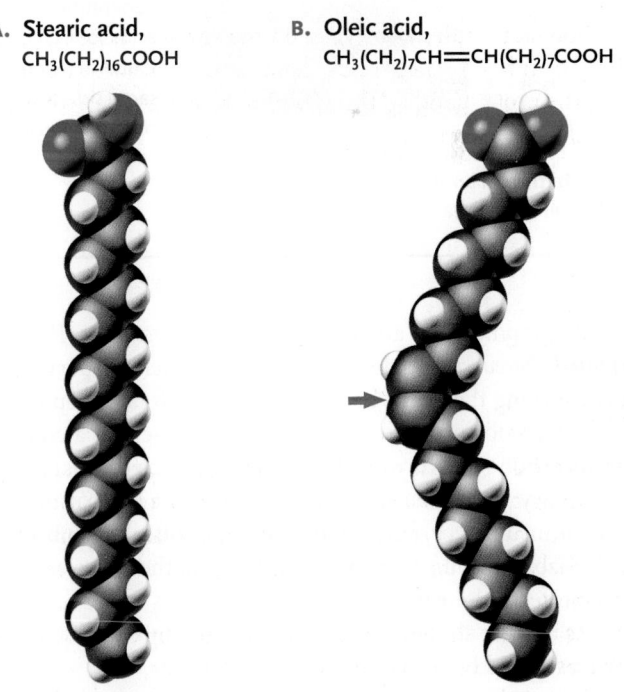

A. Stearic acid, $CH_3(CH_2)_{16}COOH$

B. Oleic acid, $CH_3(CH_2)_7CH=CH(CH_2)_7COOH$

FIGURE 3.10 Fatty acids. (A) Stearic acid, a saturated fatty acid. **(B)** Oleic acid, an unsaturated fatty acid. An arrow marks the "kink" introduced by the double bond.

© Cengage Learning 2017

Focus on Research

Applied Research: Fats, Cholesterol, and Coronary Artery Disease

Hardening of the arteries, or *atherosclerosis,* is a condition in which deposits of lipid and fibrous material called *plaque* build up in the walls of arteries, the vessels that supply oxygenated blood to body tissues. Plaque reduces the internal diameter of the arteries, restricting or even completely blocking the flow of blood. Blockage of the coronary arteries that supply oxygenated blood to the heart muscle **(Figure)** can severely impair heart function, a condition called *coronary heart disease.* In extreme cases, it can lead to destruction of heart muscle tissue, as occurs in a heart attack (myocardial infarction).

Your body requires a certain amount of cholesterol, but the liver normally makes enough to meet this demand. Additional

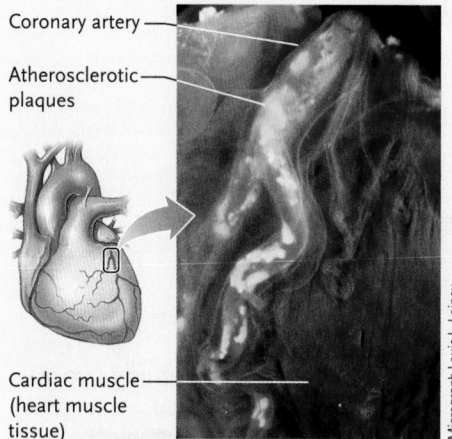

Coronary artery ——

Atherosclerotic —— plaques

Cardiac muscle —— (heart muscle tissue)

Micrograph Louis L. Lainey

FIGURE **Atherosclerotic plaques (bright areas) in the coronary arteries of a patient with heart disease.**

cholesterol is made from fats taken in as food. Cholesterol is found in the blood bound to low-density lipoprotein (LDL) and high-density lipoprotein (HDL). LDL cholesterol is considered "bad" because of a positive correlation between its level in the blood and the risk for coronary heart disease. LDL cholesterol contributes to plaque formation as atherosclerosis proceeds. HDL cholesterol is "good" because high levels appear to provide some protection against coronary heart disease. Simplifying, HDL cholesterol removes excess cholesterol from plaques in arteries, thereby reducing plaque buildup. The cholesterol that has been removed is transported by the HDL cholesterol to the liver where it is broken down.

Fats in food affect cholesterol levels in the blood. Diets high in saturated fats raise LDL cholesterol levels, but not HDL cholesterol levels. Foods of animal origin typically contain saturated fats, and foods of plant origin typically contain unsaturated fats.

In the food industry, unsaturated vegetable oils are often processed to solidify the fats. The process, partial hydrogenation, adds hydrogen atoms to unsaturated sites, eliminating many double bonds and generating substances known as *trans* fatty acids (or *trans* fats). Usually the hydrogen atoms at a double bond are positioned on the same side of the carbon chain, producing a *cis* (Latin, "on the same side") fatty acid:

$$-\overset{\overset{\displaystyle H}{|}}{C}=\overset{\overset{\displaystyle H}{|}}{C}-$$

In a *trans* (Latin, "across") fatty acid, the hydrogen atoms are on different sides of the chain at some double bonds:

$$-\overset{\overset{\displaystyle H}{|}}{C}=\overset{\underset{\displaystyle H}{|}}{C}-$$

Trans fatty acids may be found in many vegetable shortenings, some margarines, cookies, cakes, doughnuts, and other foods made with or fried in partially hydrogenated fats.

Research from human feeding studies has shown that *trans* fatty acids raise LDL cholesterol levels nearly as much as saturated fatty acids do. More seriously, intake of *trans* fatty acids appears to reduce HDL cholesterol levels. In addition, clinical studies have demonstrated a positive correlation between the intake of *trans* fatty acids and the occurrence of coronary heart disease. In the United States, the Food and Drug Administration (FDA) required trans fats to be listed on food labels as of January 2008. However, the regulations allow 0.5 g per serving to be listed as 0 g *trans* fats on the food label. Then, in November 2013, the FDA made a preliminary determination that *trans* fats are not *generally recognized as safe* (GRAS). Potentially this means that, in the future, *trans* fats will be reclassified as food additives. If that is done, *trans* fats would require specific authorization to be used in foods. This would likely almost eliminate *trans* fats from food in the United States.

© Cengage Learning 2017

the development of atherosclerosis (see *Focus on Research: Applied Research*), a disease in which arteries, particularly those serving the heart, become clogged with fatty deposits.

Fatty acids may also combine with long-chain alcohols or hydrocarbon structures to form **waxes,** which are harder and less greasy than fats. Insoluble in water, waxy coatings help keep skin, hair, or feathers of animals protected, lubricated, and pliable. In humans, earwax lubricates the outer ear canal and protects the eardrum. Honeybees use a wax secreted by glands in their abdomen to construct the comb in which larvae are raised and honey is stored.

Many plants secrete waxes that form a protective exterior layer, which greatly reduces water loss from the plants and resists invasion by infective agents such as bacteria and viruses.

This waxy covering gives cherries, apples, and many other fruits their shiny appearance.

Phospholipids Provide the Framework of Biological Membranes

Phosphate-containing lipids called **phospholipids** are the primary lipids of cell membranes. In the most common phospholipids, glycerol forms the backbone of the molecule as in triglycerides, but only two of its binding sites are linked to fatty acids **(Figure 3.11).** The third site is linked to a polar phosphate group, which binds to yet another polar unit. The end of the molecule with the fatty acids is nonpolar and hydrophobic, and the end with the phosphate group is polar and hydrophilic.

In polar environments, such as a water solution, phospholipids assume arrangements in which only their polar ends are exposed to the water; their nonpolar ends collect together in a region that excludes water. One of these arrangements, the *bilayer,* is the structural basis of membranes, the organizing boundaries of all living cells (see Figure 2.13 and Chapter 5). In a bilayer, formed by a film of phospholipids just two molecules thick, the phospholipid molecules are aligned so that the polar groups face the surrounding water molecules at the surfaces of the bilayer. The hydrocarbon chains of the phospholipids are packed together in the interior of the bilayer, where they form a nonpolar, hydrophobic region that excludes water. The bilayer remains stable because, if disturbed, the hydrophobic, nonpolar

hydrocarbon chains of the phospholipids become exposed to the surrounding watery solution, and the molecule returns to its normal bilayer arrangement.

A Boundary Lipid Membrane around Organic Molecules Was Key to the Formation of Cells and the Origin of Life

In the origin of cells billions of years ago, researchers hypothesize that organic molecules assembled into aggregates that became bounded by lipid membranes to form primitive protocells. Protocells are considered key to the origin of life because life depends on reactions occurring in a controlled and

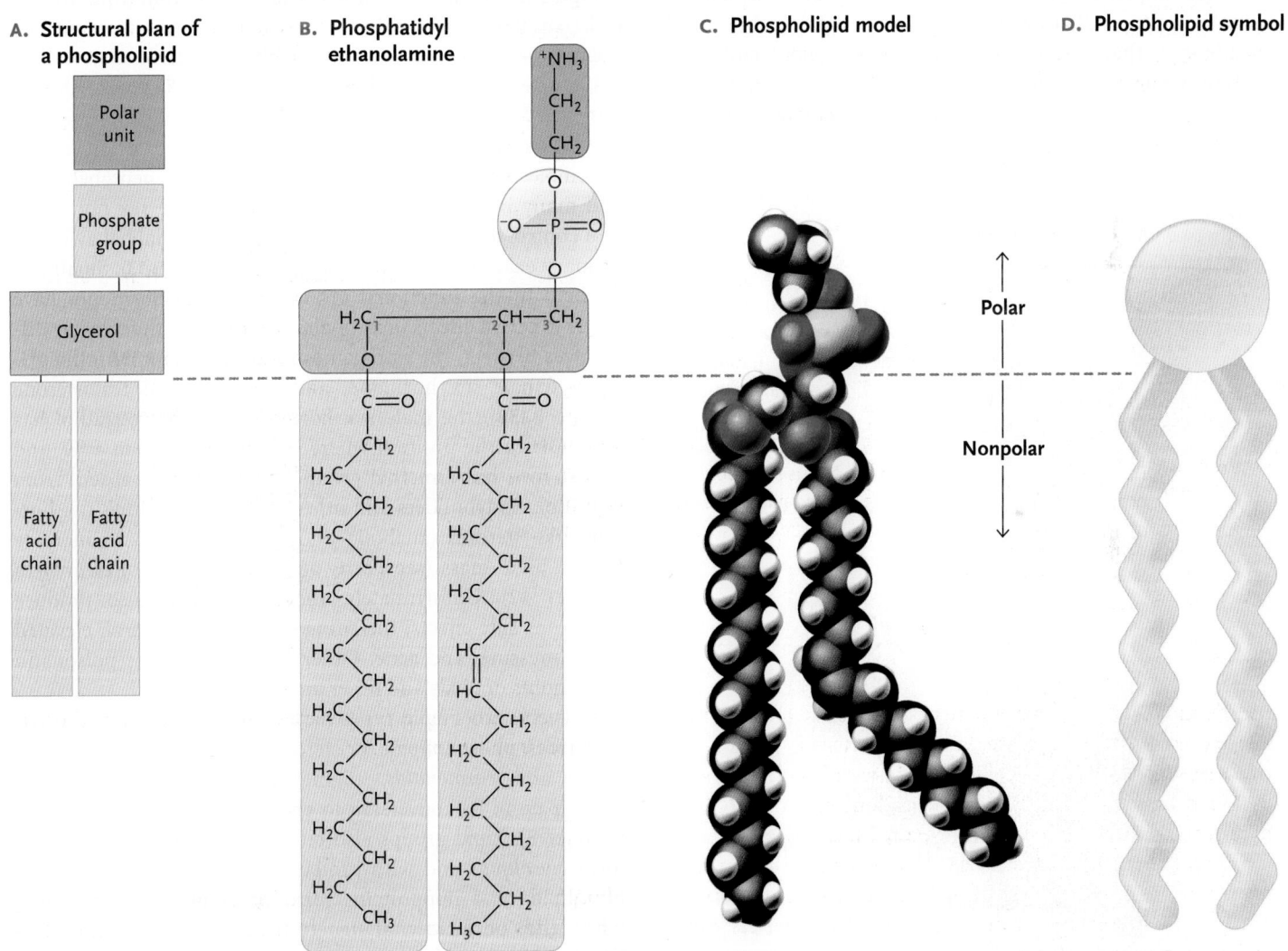

FIGURE 3.11 Phospholipid structure. (A) The arrangement of components in phospholipids. **(B)** Phosphatidyl ethanolamine, a common membrane phospholipid. **(C)** Space-filling model of phosphatidyl ethanolamine. The kink in the fatty acid chain on the right reflects a double bond at this position. **(D)** Diagram widely used to depict a phospholipid molecule in cell membrane diagrams. The sphere represents the polar end of the molecule, and the zigzag lines represent the nonpolar fatty acid chains. (The kink in the fatty acid chain is not shown.)

sequestered environment; namely, the cell. Protocells are the presumed precursors of cells. (The origin of cells is discussed in more detail in Chapter 25.)

Steroids Contribute to Membrane Structure and Work as Hormones

Steroids are a group of lipids with structures based on a framework of four carbon rings **(Figure 3.12A)**. Small differences in the side groups attached to the rings distinguish one steroid from another. The most abundant steroids, the **sterols,** have a single polar —OH group linked to one end of the ring framework and a complex, nonpolar hydrocarbon chain at the other end **(Figure 3.12B)**. Although sterols are almost completely hydrophobic, the single hydroxyl group gives one end of the molecule a slightly polar, hydrophilic character. As a result, sterols also have dual solubility properties and, like phospholipids, tend to assume positions that satisfy these properties. In biological membranes, they line up beside the phospholipid molecules with their polar —OH group facing the membrane surface and their nonpolar ends buried in the nonpolar membrane interior.

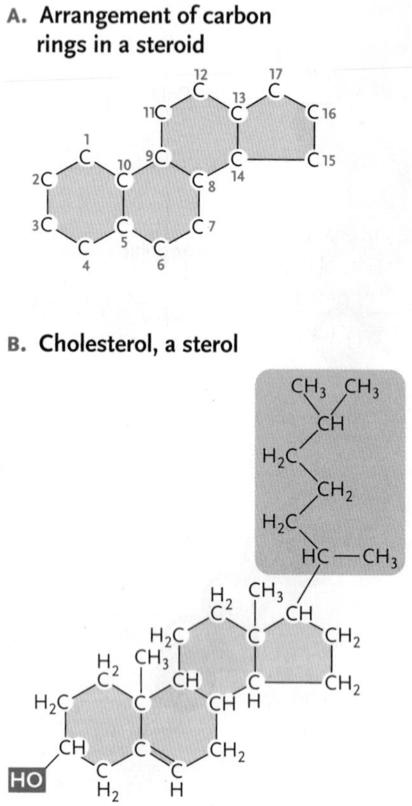

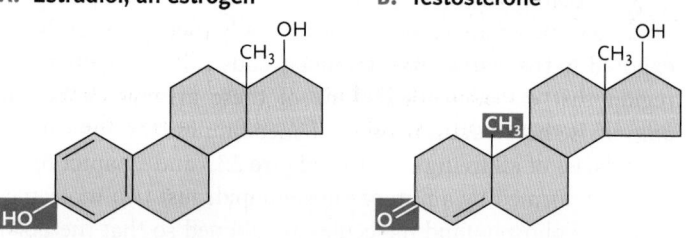

A. Estradiol, an estrogen

B. Testosterone

FIGURE 3.13 Steroid sex hormones and their effects. The female sex hormone, estradiol **(A)**, and the male sex hormone, testosterone **(B)**, differ only in substitution of an —OH group for an oxygen and the absence of one methyl group (—CH₃) in the estrogen. Although small, these differences greatly alter sexual structures and behavior in animals, such as humans.
© Cengage Learning 2017

A. Arrangement of carbon rings in a steroid

B. Cholesterol, a sterol

FIGURE 3.12 Steroids. (A) Typical arrangement of four carbon rings in a steroid molecule. **(B)** A sterol, cholesterol. Sterols have a hydrocarbon side chain linked to the ring structure at one end and a single —OH group at the other end (boxed in red). The —OH group makes its end of a sterol slightly polar. The rest of the molecule is nonpolar.
© Cengage Learning 2017

Cholesterol (see Figures 3.12B and 2.7E) is an important component of the boundary membrane surrounding animal cells; similar sterols, called phytosterols, occur in plant cell membranes. Deposits derived from cholesterol also collect inside arteries in atherosclerosis (see *Focus on Research: Applied Research*).

Other steroids, the *steroid hormones,* are important regulatory molecules in animals; they control development, behavior, and many internal biochemical processes. The sex hormones that control differentiation of the sexes and sexual behavior are primary examples of steroid hormones **(Figure 3.13)**. Small differences in the functional groups of steroid hormones have vastly different effects in animals. For instance, the two key differences between the estrogen estradiol **(Figure 3.13A)**, the primary female sex hormone, and the androgen testosterone **(Figure 3.13B)**, the male sex hormone, are that estradiol has an —OH in the position where testosterone has an =O, and testosterone has a methyl group (—CH₃) that is absent from estradiol. Bodybuilders and other athletes sometimes use hormonelike steroids (anabolic–androgenic steroids) to increase their muscle mass (see *Focus on Research: Basic Research* in Chapter 42). Unfortunately, these substances also produce numerous side effects, including elevated cholesterol, elevated blood pressure, and acne. Other steroids occur as poisons in the venoms of toads and other animals.

Several other lipid types have structures unrelated to triglycerides, phospholipids, or steroids. Among these are *chlorophylls* and *carotenoids,* pigments that absorb light and participate in its conversion to chemical energy in plants (see Chapter 8). Lipid groups also combine with carbohydrates to form *glycolipids* and with proteins to form *lipoproteins.* Both glycolipids and lipoproteins form parts of cell membranes, where they perform vital structural and functional roles (see Chapter 5).

STUDY BREAK 3.3

What are the three most common lipids in living organisms? How do their structures differ?

3.4 Proteins

Proteins are macromolecules that perform many vital functions in living organisms **(Table 3.2)**. Some provide structural support for cells; others, called **enzymes,** increase the rate of cellular reactions; still others impart movement to cells and cellular structures. Proteins also transport substances across biological membranes, serve as recognition and receptor molecules at cell surfaces, or regulate the activity of other proteins and DNA. Some proteins work as hormones or defend against

TABLE 3.2	Major Protein Functions		
Protein Type	Function	Examples	
Structural proteins	Support	Microtubule fibers and microfilament fibers (red in **Figure**) inside cells; collagen and other proteins of animal cells; cell wall proteins of plant cells.	Courtesy of Dr. Vincenzo Cirulli, Department of Medicine, University of Washington, Seattle, WA
Enzymatic proteins	Increase the rate of biological reactions	DNA polymerase in DNA replication; RuBP (ribulose 1,5-bisphosphate) carboxylase/oxygenase (rubisco) in photosynthesis **(Figure);** digestive enzymes.	
Membrane transport proteins	Speed up movement of substances across biological membranes	Ion transporters move ions such as Na^+, K^+, and Ca^{2+} across membranes **(Figure);** glucose transporters move glucose into cells; aquaporins allow water molecules to move across membranes.	
Motile proteins	Produce cellular movements	Myosin acts on microfilaments to produce muscle movements; kinesin **(Figure)** acts on microtubules involved in cell division and in movement of some materials within the cell.	
Regulatory proteins	Promote or inhibit the activity of other cellular molecules	Nuclear regulatory proteins (red and blue in **Figure** and chapter-opening photo) turn genes on or off to control the activity of DNA (yellow and green in Figure and chapter-opening photo); protein kinases add phosphate groups to other proteins to modify their activity.	Petarg/Shutterstock.com
Receptor proteins	Bind molecules at cell surface or within cell; some trigger internal cellular responses	Hormone receptors bind hormones at the cell surface or within cells and trigger cellular responses; cellular adhesion molecules help hold cells together by binding molecules on other cells **(Figure:** β-catenin).	Molekuul.be/Shutterstock.com
Hormonal proteins	Carry regulatory signals between cells	Insulin **(Figure)** regulates sugar levels in the bloodstream; growth hormone regulates cellular growth and division.	Lculig/Shutterstock.com
Defensive proteins	Defend against invading molecules and organisms	Antibodies **(Figure)** recognize, bind, and help eliminate essentially any protein of infecting bacteria and viruses, and many other types of molecules, both natural and artificial.	molekuul.be/Shutterstock.com
Storage proteins	Hold amino acids and other substances in stored form	Ovalbumin is a storage protein of eggs; apolipoproteins **(Figure)** hold cholesterol in stored form for transport through the bloodstream.	Molekuul.be/Shutterstock.com

foreign substances, such as infectious microorganisms. In addition, many toxins and venoms are proteins.

All of the protein molecules that carry out these and other functions are fundamentally similar in structure. All are polymers consisting of one or more unbranched chains of monomers called *amino acids*. An **amino acid** is a molecule that contains both an amino and a carboxyl group. Although the most common proteins contain 50 to 1,000 amino acids, some proteins found in nature have as few as 3 or as many as 50,000 amino acid units. Proteins range in shape from globular or spherical forms to elongated fibers, and they vary from soluble to completely insoluble in water solutions. The amino acid composition and sequence of a protein determines its structure and its function. Some proteins have single functions, whereas others have multiple functions.

Cells Assemble 20 Kinds of Amino Acids into Proteins by Forming Peptide Bonds

The cells of all organisms use 20 different amino acids as the building blocks of proteins **(Figure 3.14)**. Of these 20 amino acids, 19 have the same structural plan— a central carbon atom is attached to an amino group ($-NH_2$), a carboxyl group ($-COOH$), and a hydrogen atom:

$$H_2N-\overset{\overset{\displaystyle R}{|}}{\underset{\underset{\displaystyle H}{|}}{C}}-COOH$$

The remaining bond of the central carbon is linked to 1 of 19 different side groups represented by the *R* (see shaded regions in Figure 3.14); its usage for

FIGURE 3.14 The 20 amino acids used by cells to make proteins. The side group of each amino acid is boxed in light brown. The amino acids are shown in the ionic forms in which they are found at the pH within the cell. Three-letter and one-letter abbreviations commonly used for the amino acids appear below each diagram. All amino acids assembled into proteins are in the L-form, one of two possible stereoisomers.
© Cengage Learning 2017

A. Nonpolar amino acids

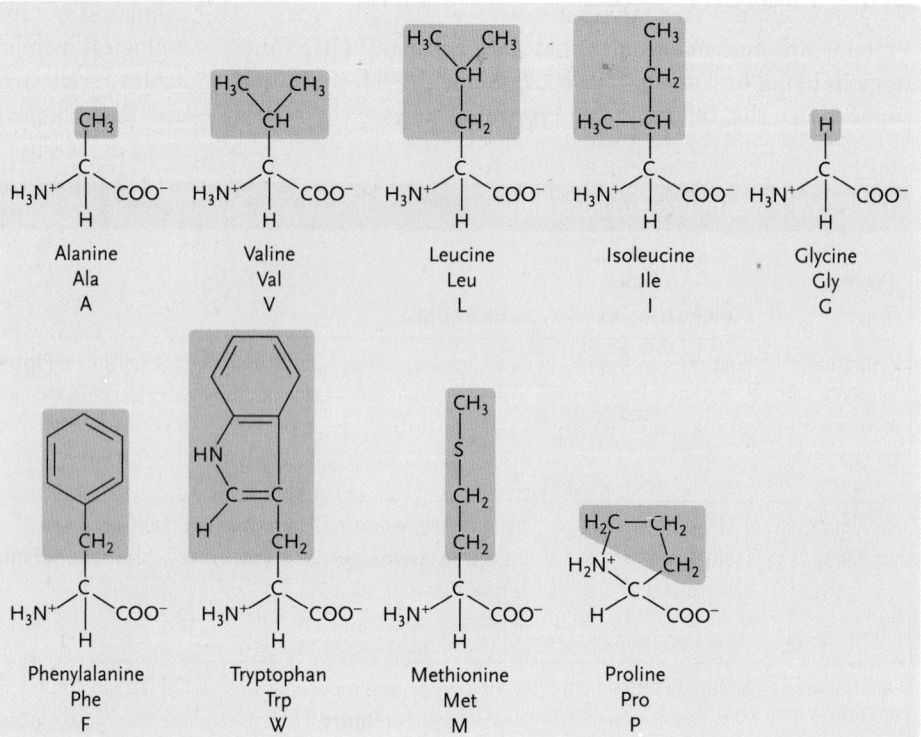

B. Uncharged polar amino acids

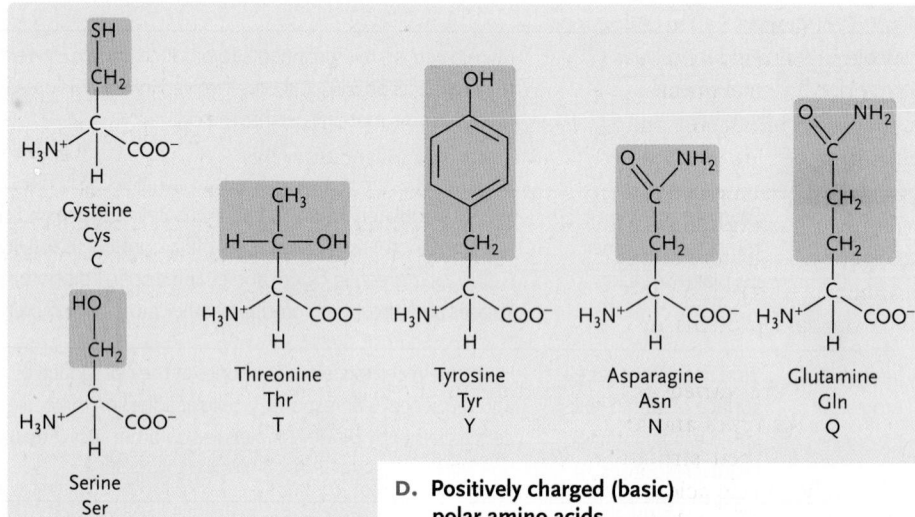

C. Negatively charged (acidic) polar amino acids

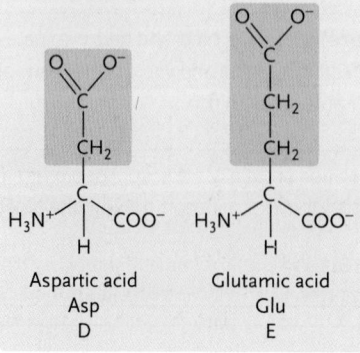

D. Positively charged (basic) polar amino acids

amino acids refers to a range from a single hydrogen atom to complex carbon-containing chains or rings. The twentieth amino acid, proline, has a ring structure that includes the central carbon atom; that carbon bonds to a —COOH group on one side and to an =NH (imino) group that forms part of the ring at the other side (see Figure 3.14). (Strictly speaking, proline is an imino acid.) At the pH of the cell, the amino acids assume ionic forms: the amino group becomes —NH$_3^+$ (—NH$_2^+$ in the case of proline), and the carboxyl group becomes —COO$^-$.

Differences in the side groups give the amino acids their individual properties. Some amino acids are nonpolar (see Figure 3.14A), and some are polar (see Figure 3.14B–D). Among the polar amino acids, some are uncharged (see Figure 3.14B), some are negatively charged (acidic) (see Figure 3.14C), and some are positively charged (basic) (see Figure 3.14D). Many of the side groups contain reactive functional groups, such as —NH$_2$, —OH, —COOH, or —SH, which may interact with atoms located elsewhere in the same protein or with molecules and ions outside the protein.

The sulfhydryl group (—SH) in the amino acid cysteine is particularly important in protein structure. The sulfhydryl groups in the side groups of two cysteines located in different regions of the same protein, or in different proteins, can react to produce disulfide linkages (—S—S—) (see Table 3.1). The linkages fasten different regions of the same amino acid chain together or different amino acid chains together **(Figure 3.15)** and help hold proteins in their three-dimensional functional shape.

Overall, the varied properties and functions of proteins depend on the types and locations of the different amino acid side groups in their structures. The variations in the number and types of amino acids mean that the total number of possible proteins is extremely large.

Covalent bonds link amino acids into the chains of subunits that make proteins. The link, a **peptide bond,** is formed by a dehydration synthesis reaction between the amino group of one amino acid and the carboxyl group of a second **(Figure 3.16)**. In the cell, an amino acid chain always has an —NH$_3^+$ group at one end, called the **N-terminal end,** and a —COO$^-$ group at the other end, called the **C-terminal end.** Ignoring the R groups of the amino acids, the repeating structure of an amino acid chain is referred to as the *backbone* of the chain.

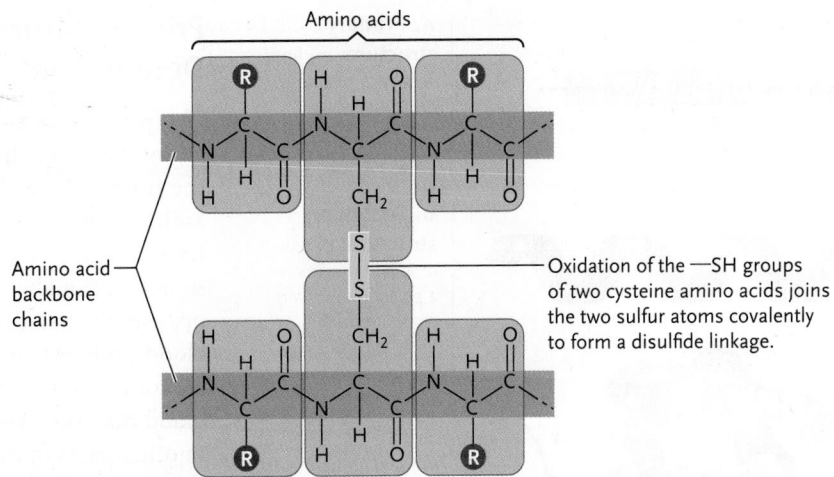

FIGURE 3.15 A disulfide linkage between two amino acid chains or two regions of the same chain. The linkage is formed by a reaction between the sulfhydryl groups (—SH) of cysteines. The circled Rs indicate the side groups of other amino acids in the chains. Figure 3.21 shows how disulfide linkages help maintain a protein's conformation.
© Cengage Learning 2017

The chain of amino acids formed by sequential peptide bonds is a **polypeptide,** and it is only part of the complex structure of proteins. That is, once assembled, an amino acid chain may fold in various patterns, and more than one chain may combine to form a finished protein, adding to the structural and functional variability of proteins.

Proteins Have as Many as Four Levels of Structure

Proteins have up to four levels of structure, with each level imparting different characteristics and degrees of structural complexity to the molecule **(Figure 3.17)**. **Primary structure** is the particular and unique linear sequence of amino acids linked to each other by peptide bonds to form a polypeptide. **Secondary structure** is produced by hydrogen bonding between different amino acids in a segment of amino acids within a polypeptide chain. The result is coiling or folding of the segment. Most proteins include coils and many also include folds. **Tertiary structure** is the folding of the complete amino

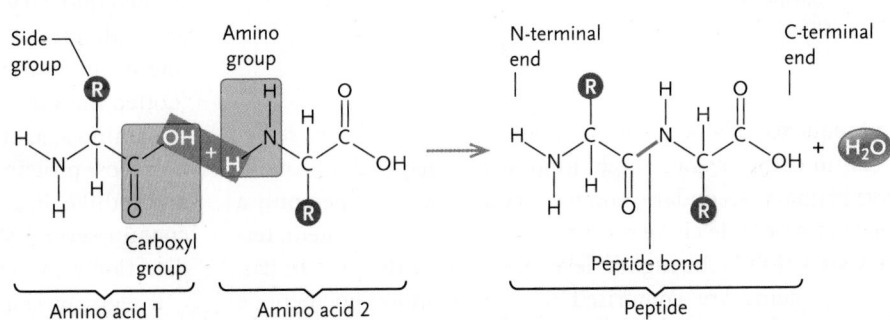

FIGURE 3.16 A peptide bond formed by reaction of the carboxyl group of one amino acid with the amino group of a second amino acid. The reaction is a typical dehydration synthesis reaction.
© Cengage Learning 2017

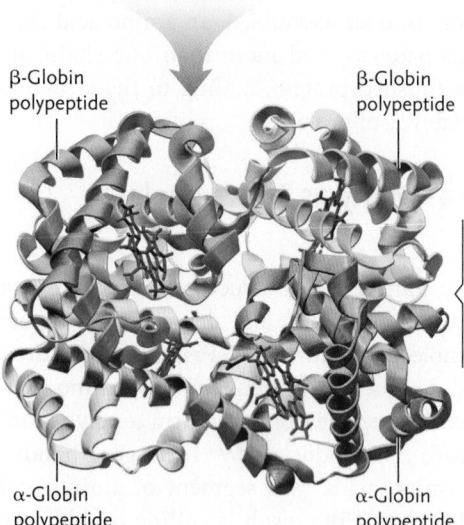

A. Primary structure: the linear sequence of amino acids joined by peptide bonds in a polypeptide chain

B. Secondary structure: coiled or folded regions in a polypeptide chain resulting from hydrogen bonding within particular sequences of amino acids

C. Tertiary structure: overall three-dimensional folding of a polypeptide chain

Heme group

β-Globin polypeptide

β-Globin polypeptide

D. Quaternary structure: the arrangement of polypeptide chains in a protein that contains more than one chain

α-Globin polypeptide

α-Globin polypeptide

FIGURE 3.17 The four levels of protein structure. The protein shown in **(C)** is one of the subunits of a hemoglobin molecule; the heme group (in red) is an iron-containing group that binds oxygen. **(D)** A complete hemoglobin molecule.

© Cengage Learning 2017

acid sequence of a polypeptide chain, with its secondary structures, into the overall three-dimensional shape. All proteins have primary, secondary, and tertiary structures. When only a single polypeptide chain comprises the functional protein, tertiary structure is the highest level of structure the protein has. Some proteins are comprised of two or more polypeptide chains bonded together. Each of those polypeptide chains has a tertiary structure. The combined arrangement of the bonded polypeptide chains in a protein formed from more than one chain is its **quaternary structure.**

Primary Structure Is the Fundamental Determinant of Protein Form and Function

The primary structure of a protein—the sequence in which amino acids are linked together by peptide bonds—underlies the other, higher levels of structure. Changing even a single amino acid of the primary structure alters the secondary, tertiary, and quaternary structures to at least some degree and, by so doing, can alter or even destroy the biological functions of a protein. For example, substitution of a single amino acid in the blood protein hemoglobin produces an altered form responsible for sickle-cell anemia (see Chapter 15); a number of other blood disorders are caused by single amino acid substitutions in other parts of the protein.

Because primary structure is so fundamentally important, many years of research have been devoted to determining the amino acid sequence of proteins. Initial success came in 1953, when the English biochemist Frederick Sanger of the University of Cambridge, UK, deduced the 51-amino acid sequence of the two-polypeptide chain protein, insulin (a protein hormone: see Table 3.2), using samples obtained from cows. Sanger was awarded a Nobel Prize in 1958 "for his work on the structure of proteins, especially that of insulin." Now, the amino acid sequences of thousands of proteins have been determined. Knowledge of the primary structure of proteins can allow their three-dimensional structure and functions to be predicted and reveal relationships among proteins.

Coils and Folds of the Amino Acid Chain Form the Secondary Structure of a Protein

The amino acid chain of a protein, rather than being stretched out in linear form, is folded into arrangements that form the protein's secondary structure. Two highly regular secondary structures, the *alpha helix* and the *beta strand*, are particularly stable and make an amino acid chain resistant to bending. Most proteins have segments of both secondary structures.

THE ALPHA HELIX In the **alpha (α) helix,** first identified by Linus Pauling and Robert Corey at the California Institute of Technology in 1951, the backbone of the amino acid chain is twisted into a regular, right-hand spiral **(Figure 3.18).** The amino acid side groups extend outward from the twisted backbone. The structure is stabilized by regularly spaced hydrogen bonds (dotted lines in Figure 3.18) between atoms in the backbone of the amino acid chain.

Most proteins contain segments of α helix, which are rigid and rodlike, in at least some regions. Globular proteins usually contain several short α-helical segments that run in different directions, connected by segments of *random coil*. A random coil segment is a sequence of amino acids that has neither an α-helical nor a β-strand structure. It provides flexible sites that allow α-helical or β strand segments to bend or fold back on themselves. Segments of random coil also commonly act as "hinges" that allow major parts of proteins to move with respect

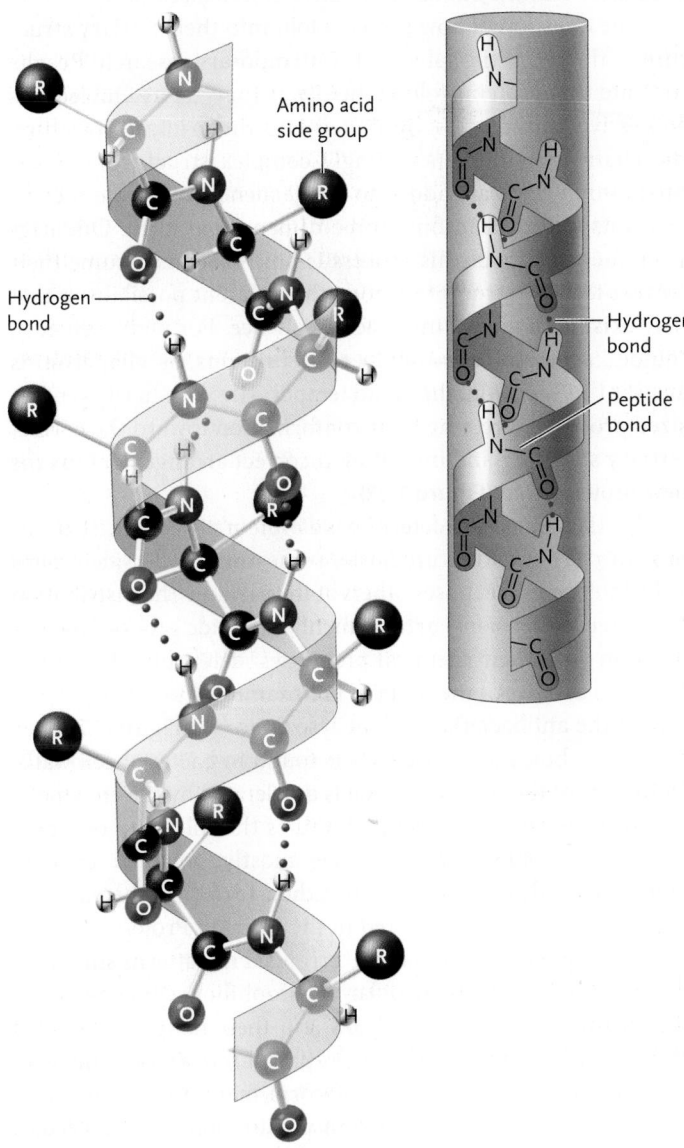

A. Ball-and-stick model of α helix

Amino acid side group

Hydrogen bond

B. Cylinder representation of α helix

Hydrogen bond

Peptide bond

FIGURE 3.18 The α helix, a type of secondary structure in proteins.
(A) Ball-and-stick model of the α helix. The backbone of the amino acid chain is held in a spiral by hydrogen bonds formed at regular intervals between backbone atoms. **(B)** The cylinder often is used to depict an α helix in protein diagrams; peptide and hydrogen bonds may also be shown.
© Cengage Learning 2017

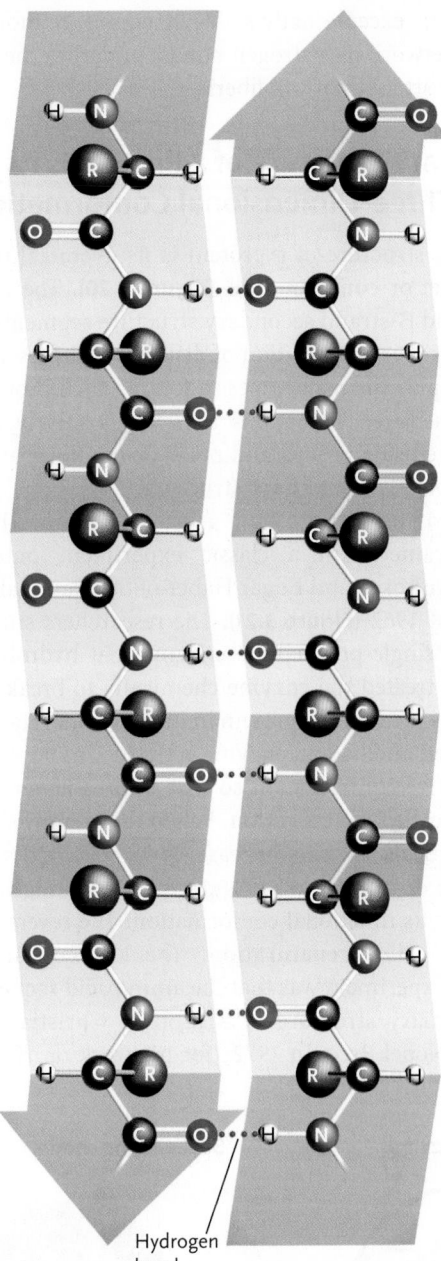

Hydrogen bond

FIGURE 3.19 A β sheet formed by side-by-side alignment of two β strands, a type of secondary structure in proteins. The β strands are held together stably by hydrogen bonds. In this sheet, the β strands run in opposite directions, as shown by the arrows, which point in the direction of the C-terminal end of each polypeptide chain. Strands may also run in the same direction in a β sheet. Arrows alone often are used to represent β strands in protein diagrams.
© Cengage Learning 2017

to one another. Fibrous proteins, such as the collagens, a major component of tendons, bone, and other extracellular structures in animals, typically contain one or more α-helical segments that run the length of the molecule, with few or no bendable regions of random coil.

THE BETA STRAND Pauling and Corey were also the first to identify the beta (β) strand as a major secondary protein structure. In a β strand, the amino acid chain is folded into zigzags in a flat plane rather than being twisted into a coil. In many proteins, β strands are aligned side by side in the same or

opposite directions to form a structure known as a **beta (β) sheet (Figure 3.19).** Hydrogen bonds between adjacent β strands stabilize the sheet, making it a highly rigid structure. Beta sheets may lie in a flat plane or may twist into propeller- or barrel-like structures.

Beta strands and sheets occur in many proteins, usually in combination with α-helical segments. One notable exception is in the silk protein secreted by silk worms, which contains only β sheets. This exceptionally stable structure, reinforced by an extensive network of hydrogen bonds, underlies the unusually high tensile strength of silk fibers.

The Tertiary Structure of a Protein Is Its Overall Three-Dimensional Conformation

The tertiary structure of a protein is its overall three-dimensional shape, or **conformation (Figure 3.20).** The contents of α-helical and β-strand secondary structure segments, together with the number and position of disulfide linkages and hydrogen bonds, play the major roles in folding each protein into its tertiary structure. Attractions between positively and negatively charged side groups and polar or nonpolar associations also contribute to the tertiary structure.

The first insight into how a protein assumes its tertiary structure came from a classic experiment published by Christian Anfinsen and Edgar Haber of the National Institutes of Health in 1962 **(Figure 3.21).** The researchers studied ribonuclease, a single-polypeptide enzyme that hydrolyzes RNA. When they treated the enzyme chemically to break the disulfide linkages holding the protein in its functional state, the protein unfolded and had no enzyme activity. Unfolding a protein from its active conformation so that it loses its structure and function is called **denaturation.** When they removed the denaturing chemicals, the ribonuclease slowly regained full activity because the disulfide linkages reformed, enabling the protein to reassume its functional conformation. The reversal of denaturation is called **renaturation.** The key conclusion from Anfinsen's experiment was that the amino acid sequence specifies the tertiary structure of a protein. Christian Anfinsen received a Nobel Prize in 1972 "for his work on ribonuclease,

especially concerning the connection between the amino acid sequences and the biologically active conformation."

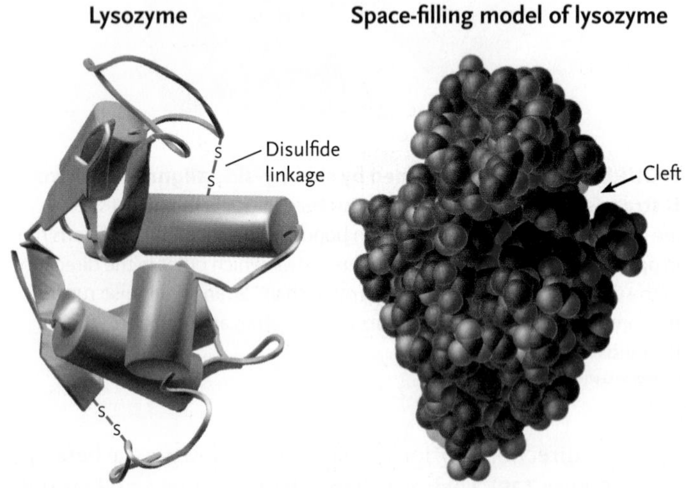

The question of how proteins fold into their tertiary structure in the cell is the subject of contemporary research. Results indicate that proteins fold gradually as they are assembled—as successive amino acids are linked into the primary structure, the chain folds into increasingly complex structures. As the final amino acids are added to the sequence, the protein completes its folding into final three-dimensional form. One nagging question about this process is how proteins assume their correct tertiary structure among the different possibilities that may exist for a given amino acid sequence. For many proteins, "guide" proteins called **chaperone proteins** or **chaperonins** answer this question; they bind temporarily with newly synthesized proteins, directing their conformation toward the correct tertiary structure and inhibiting incorrect arrangements as the new proteins fold **(Figure 3.22).**

Tertiary structure determines a protein's function. That is, a protein's tertiary structure buries some amino acid side groups in its interior and exposes others at the surface. The distribution and three-dimensional arrangement of the side groups, in combination with their chemical properties, determine the overall chemical activity of the protein. For example, the tertiary structure of the antibacterial enzyme lysozyme (see Figure 3.20) has a cleft that binds a polysaccharide found in bacterial cell walls; hydrolysis of the polysaccharide is accelerated by the enzyme.

Tertiary structure also determines the solubility of a protein. Water-soluble proteins have mostly polar or charged amino acid side groups exposed at their surfaces, whereas nonpolar side groups are clustered in the interior. Proteins embedded in nonpolar membranes are arranged in patterns similar to phospholipids, with their polar (hydrophilic) segments facing the surrounding watery solution and their nonpolar surfaces embedded in the nonpolar (hydrophobic) membrane interior. These dual-solubility proteins perform many important functions in membranes, such as transporting ions and molecules into and out of cells.

The tertiary structure of most proteins is flexible, allowing them to undergo limited alterations in three-dimensional shape known as **conformational changes.** These changes contribute to the function of many proteins, particularly those working as enzymes, in cellular movements, or in the transport of substances across cell membranes.

As you learned earlier, chemical treatment can denature a protein in the test tube (see Figure 3.21). Excessive heat can also break the hydrogen bonds holding a protein in its natural conformation, causing it to denature and lose its biological function. Denaturation is one of the major reasons few living organisms can tolerate temperatures greater than 45°C. Extreme changes in pH, which alter the charge of amino acid side groups and weaken or destroy ionic bonds, can also cause protein denaturation.

For some proteins, denaturation is permanent. A familiar example of a permanently denatured protein is a cooked egg white. In its natural form, the egg white protein albumin

Lysozyme **Space-filling model of lysozyme**

Disulfide linkage

Cleft

FIGURE 3.20 Tertiary structure of the protein lysozyme, with α helices shown as cylinders, β strands as arrows, and random coils as ropes. Lysozyme is an enzyme found in nasal mucus, tears, and other body secretions; it destroys the cell walls of bacteria by breaking down molecules in the wall. Disulfide bonds are shown in red. A space-filling model of lysozyme is shown for comparison.
© Cengage Learning 2017

FIGURE 3.21 | **Experimental Research**

Anfinsen's Experiment Demonstrating That the Amino Acid Sequence of a Protein Specifies Its Tertiary Structure

Question: What is the relationship between the amino acid sequence of a protein and its conformation?

Experiment: Anfinsen and Haber studied the 124–amino acid enzyme ribonuclease in the test tube. They knew that the native (functional) enzyme has four disulfide linkages between amino acids 26 and 84, 40 and 95, 58 and 110, and 65 and 72 (see figure). They treated the active enzyme with a mixture of urea and β-mercaptoethanol, which breaks disulfide linkages. Then they removed the two chemicals and left the enzyme solution in air.

Results: The chemical treatment broke the four disulfide linkages, which caused the protein to denature and lose its enzyme activity. After the chemicals were removed and the enzyme solution exposed to air, the researchers observed that the protein renatured, slowly regaining enzyme activity. Ultimately the solution showed 90% of the activity of the native enzyme.

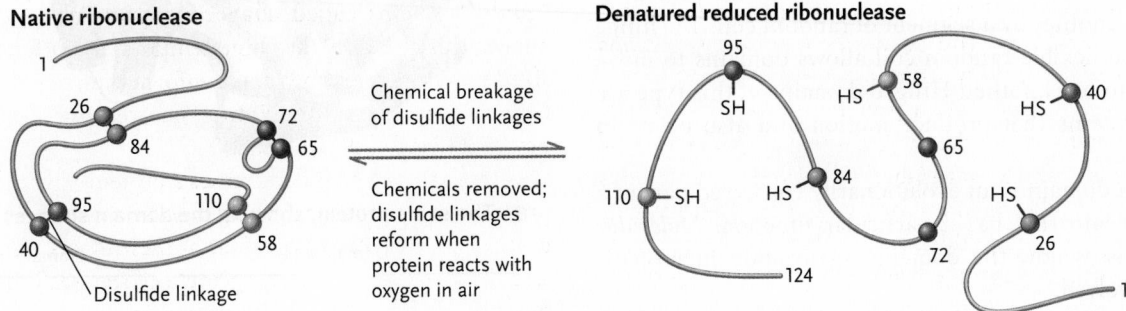

Anfinsen realized that oxygen from the air had reacted with the —SH groups of the denatured enzyme causing disulfide linkages to reform, and that the enzyme had spontaneously refolded into its native, active conformation. All physical and chemical properties of the refolded enzyme the researchers measured were the same as those of the native enzyme, confirming that the same disulfide bridges had formed as in the native enzyme.

Conclusion: Anfinsen concluded that the information for determining the three-dimensional shape of ribonuclease is in its amino acid sequence.

Observe
Hypothesize
Predict
Experiment
Interpret

If denatured ribonuclease renatures in the presence of a high concentration of urea, the renatured enzyme has physical and chemical properties similar to those of the native enzyme indicating that refolding had occurred, but enzyme activity is less than 1% of that of the native enzyme. Interpret this result.

Source: E. Haber and C. Anfinsen. 1962. Side-chain interactions governing the pairing of half-cystine residues in ribonuclease. *Journal of Biological Chemistry* 237:1839–1844.

dissolves in water to form a clear solution. The heat of cooking denatures it permanently into an insoluble, whitish mass. For other proteins, such as ribonuclease (mentioned previously), denaturation is reversible; the proteins can renature and return to their functional form if the temperature or pH returns to normal values.

Multiple Polypeptide Chains Form Quaternary Structure

Some complex proteins, such as hemoglobin and antibody molecules, have *quaternary structure*—that is, the presence and arrangement of two or more polypeptide chains (see Figure 3.17D). The same bonds and forces that fold single polypeptide chains into tertiary structures, including hydrogen bonds, polar and nonpolar attractions, and disulfide

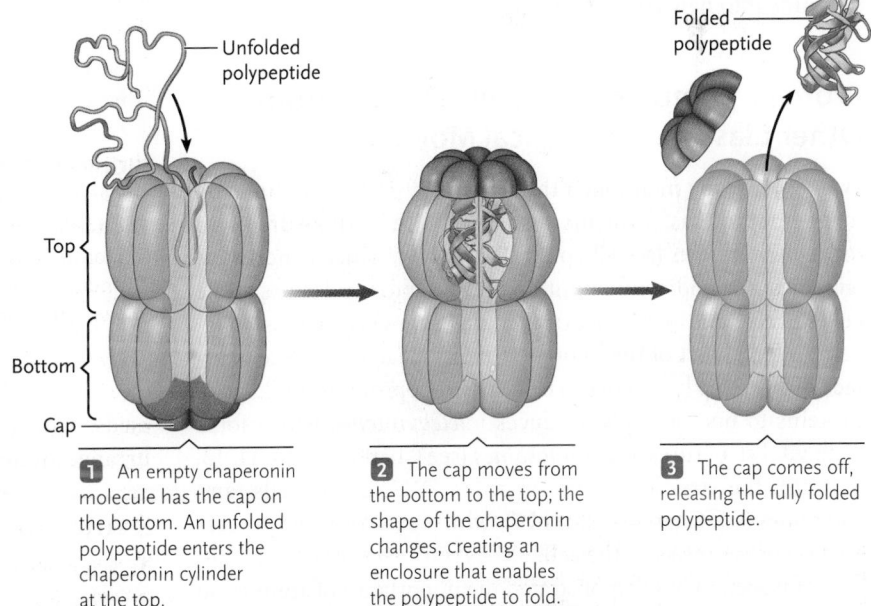

1 An empty chaperonin molecule has the cap on the bottom. An unfolded polypeptide enters the chaperonin cylinder at the top.

2 The cap moves from the bottom to the top; the shape of the chaperonin changes, creating an enclosure that enables the polypeptide to fold.

3 The cap comes off, releasing the fully folded polypeptide.

FIGURE 3.22 Role of a chaperonin in folding a polypeptide. The three parts of the chaperonin are the top and bottom, which form a cylinder, and the cap.

© Cengage Learning 2017

linkages, also hold the multiple polypeptide chains together. During the assembly of multichain proteins, chaperonins also promote correct association of the individual amino acid chains and inhibit incorrect formations.

Combinations of Secondary, Tertiary, and Quaternary Structure Form Functional Domains in Many Proteins

In many proteins, folding of the amino acid chain (or chains) produces distinct, large structural subdivisions called **domains (Figure 3.23A).** Often, one domain of a protein is connected to another by a segment of random coil. The hinge formed by the flexible random coil allows domains to move with respect to one another. Hinged domains of this type are typical of proteins that produce motion and also occur in many enzymes.

A protein domain is an evolutionarily conserved part of a polypeptide chain that has a particular function. *Molecular Insights* discusses how the domain organization in proteins might have evolved.

Many proteins have multiple functions. For instance, the sperm surface protein SPAM1 (sperm adhesion molecule 1) plays multiple roles in mammalian fertilization. In proteins with multiple functions, individual functions are often located in different domains **(Figure 3.23B),** meaning that domains are functional as well as structural subdivisions. Different proteins often share one or more domains with particular functions. For example, a type of domain that releases energy to power biological reactions appears in similar form in many enzymes and motile proteins. The appearance of similar domains in different proteins suggests that the proteins may have evolved through a mechanism that mixes existing domains into new combinations.

Proteins Combine with Units Derived from Other Classes of Biological Molecules

We have already mentioned the linkage of proteins to lipids to form lipoproteins. Proteins also link with carbohydrates to form *glycoproteins* (see Chapters 4, 5, and 9), which function as enzymes, antibodies, recognition and receptor molecules at the cell surface, and parts of extracellular supports such as collagen. In fact, most of the known proteins located at the cell surface or in the spaces between cells are glycoproteins. Linkage of proteins to nucleic acids produces *nucleoproteins,* which form such vital structures as *chromosomes* (see Chapters 4, 10, 11, 13, and 16), the structures that organize DNA inside cells, and *ribosomes* (see Chapters 4 and 15), which carry out the process of protein synthesis in the cell.

This section has demonstrated the importance of amino acid sequence to the structure and function of proteins and highlighted the great variability in proteins produced by differences in their amino acid sequence. The next section considers the nucleic

A. Two domains in an enzyme that assembles DNA molecules

Domain a   Domain b

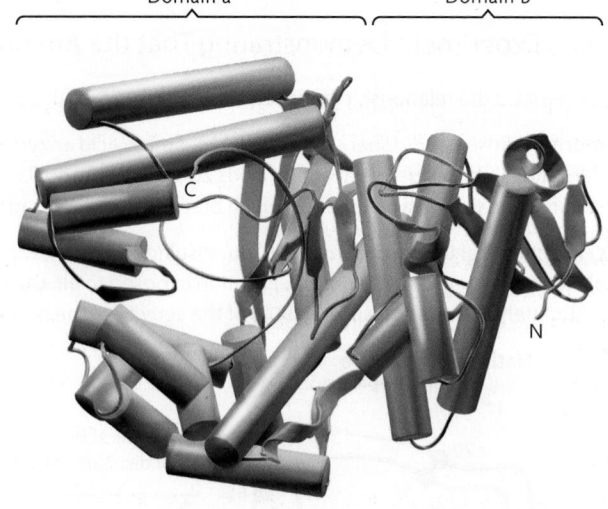

B. The same protein, showing the domain surfaces

Domain a   Domain b

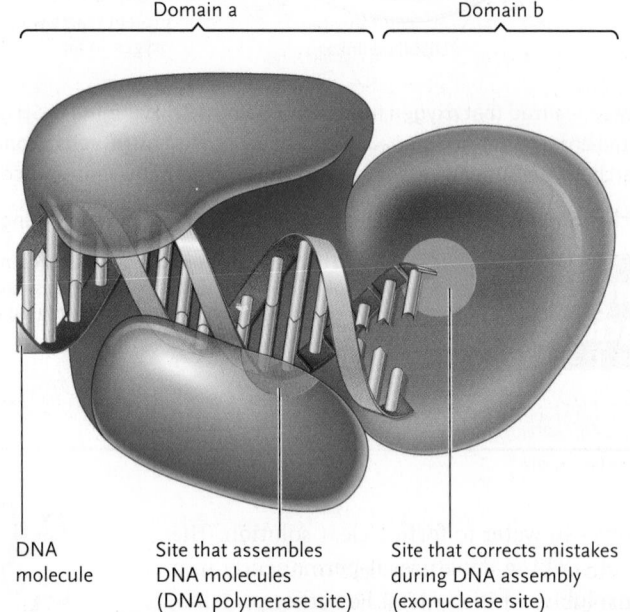

DNA molecule   Site that assembles DNA molecules (DNA polymerase site)   Site that corrects mistakes during DNA assembly (exonuclease site)

FIGURE 3.23 Domains in proteins. (A) Two domains in part of an enzyme that assembles DNA molecules in the bacterium *Escherichia coli*. The α helices are shown as cylinders, the β strands as arrows, and the random coils as ropes. "N" indicates the N-terminal end of the protein, and "C" indicates the C-terminal end. **(B)** The same view of the protein as in **(A)**, showing the domain surfaces and functional sites.
© Cengage Learning 2017

acids, which store and transmit the information required to arrange amino acids into particular sequences in proteins.

STUDY BREAK 3.4

1. What gives amino acids their individual properties?
2. What is a peptide bond, and what type of reaction forms it?
3. What are functional domains of proteins, and how are they formed?

A Big Bang in Protein Structure Evolution: How did the domain organization in proteins evolve?

Many proteins have distinct, large structural regions called *domains* (see Figure 3.23). In multifunctional proteins, different domains are responsible for the various functions. Further, proteins that have similar functions will have similar domains responsible for those functions. In other words, protein domains may be considered as units of structure and function that have combined to produce a wide variety of complex domain arrangements.

Research Question

How did the domain organization in proteins evolve?

Method of Analysis

Gustavo Caetano-Anollés and Minglei Wang at the University of Illinois at Urbana-Champaign used a bioinformatics approach to answer the question. Bioinformatics is a fusion of biology with mathematics and computer science used to manage and analyze biological data (see Chapter 19). Typically, bioinformatics involves the analysis of DNA sequences and/or protein sequences in comprehensive databases. The researchers analyzed domain structure and organization in proteins encoded in hundreds of fully sequenced genomes. From the data they reconstructed phylogenetic trees of protein structures. These phylogenetic trees laid out a rough timeline for, and details of, the evolution of domain organization in proteins.

Conclusion

The researchers discovered that, before the emergence of the three taxonomic domains—Bacteria, Archaea, and Eukarya (see Chapter 1)—most proteins contained only single domains that performed multiple tasks. With time, these protein domains began to combine with one another, becoming more specialized. After a long period of gradual evolution, there was an explosion ("big bang") of change in which protein domains combined with each other or split apart to produce a wide range of novel domains, each typically more specialized than its ancestors. The explosion in the evolution of domain organization is of particular interest because it coincided with the rapidly increasing diversity of the three taxonomic domains. Thereafter, the protein domains diverged markedly from each other as the diversification of life continued. Diversification of protein domains was most extensive in the Eukarya, allowing eukaryotic organisms to exhibit functions of their proteins not possible in other organisms.

think like a scientist

1. Give an example of a multiple-domain protein that is mentioned in the text. What are its domains?

2. What sort of bioinformatics evidence would indicate that a domain in a protein had diverged?

Source: M. Wang and G. Caetano-Anollés. 2009. The evolutionary mechanics of domain organization in proteomes and the rise of modularity in the protein world. *Structure* 17:66–78.

THINK OUTSIDE THE BOOK

Collaboratively or on your own, investigate whether chaperonins are the same in bacteria, archaeans, and eukaryotes.

3.5 Nucleotides and Nucleic Acids

Nucleic acids are another class of macromolecules, in this case, long polymers assembled from repeating monomers called *nucleotides.* The two types of nucleic acids are DNA and RNA. **DNA (deoxyribonucleic acid)** stores the hereditary information responsible for inherited traits in all eukaryotes and prokaryotes and in a large group of viruses. In terms of information flow, genes in the DNA are expressed to produce proteins and RNAs that, together, specify and control the functions of cells (introduced in Section 1.1 and Figure 1.4, and described in more detail in Chapters 15 and 16). **RNA (ribonucleic acid)** is the hereditary molecule of another large group of viruses. In all organisms, one major type of RNA carries the instructions for assembling proteins from DNA to the sites where the proteins are made inside cells (see Chapter 15). Another major type of RNA forms part of ribosomes, the structural units that assemble proteins, and a third major type of RNA brings amino acids to the ribosomes for their assembly into proteins (see Chapter 15). Some other types of RNA are involved in regulating gene expression (see Chapter 16).

Nucleotides Consist of a Nitrogenous Base, a Five-Carbon Sugar, and One or More Phosphate Groups

A **nucleotide,** the monomer of nucleic acids, consists of three parts linked together by covalent bonds: (1) a **nitrogenous base** (a nitrogen-containing molecule that accepts protons), formed from rings of carbon and nitrogen atoms; (2) a five-carbon, ring-shaped sugar; and (3) one to three phosphate groups **(Figure 3.24).** The two types of nitrogenous bases are **pyrimidines,** with one carbon–nitrogen ring, and **purines,** with two carbon–nitrogen rings **(Figure 3.25).** Three pyrimidines—uracil (U), thymine (T), and cytosine (C)—and two purines—adenine (A) and guanine (G)—form parts of nucleic acids in cells.

In nucleotides, the nitrogenous bases link covalently to **deoxyribose** in DNA and to **ribose** in RNA. Nucleotides containing deoxyribose are called **deoxyribonucleotides** and nucleotides containing ribose are called **ribonucleotides.** Deoxyribose and ribose are sugars with five carbons. The carbons are numbered with a prime symbol—1′, 2′, 3′, 4′, and 5′—to distinguish them from the unnumbered carbons and nitrogens in the nitrogenous bases (see Figure 3.24). The two sugars differ only in the chemical group bound to the 2′ carbon (boxed in red in Figure 3.24B): deoxyribose has an —H at this position and ribose has an —OH group. The prefix *deoxy-* in deoxyribose indicates that oxygen is absent at this position. In

A. Overall structural plan of a nucleotide

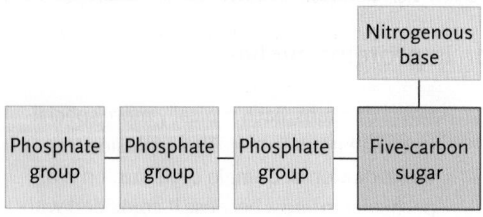

B. Chemical structures of nucleotides

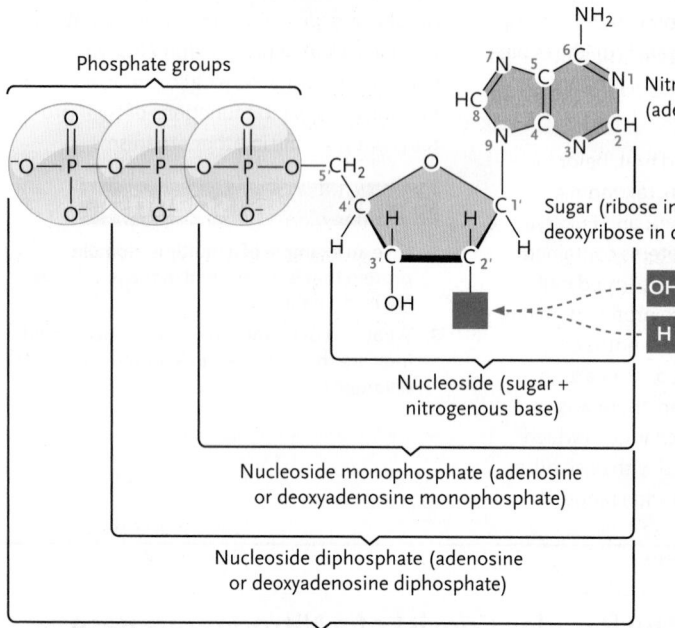

Other nucleotides:

Containing guanine: Guanosine or deoxyguanosine monophosphate, diphosphate, or triphosphate

Containing cytosine: Cytidine or deoxycytidine monophosphate, diphosphate, or triphosphate

Containing thymine: Thymidine monophosphate, diphosphate, or triphosphate

Containing uracil: Uridine monophosphate, diphosphate, or triphosphate

FIGURE 3.24 Nucleotide structure.
© Cengage Learning 2017

individual, unlinked nucleotides, a chain of one, two, or three phosphate groups is bonded covalently to the ribose or deoxyribose sugar typically at the 5′ carbon. Nucleotides are called monophosphates, diphosphates, or triphosphates according to the number of phosphates.

A structure containing only a nitrogenous base and a five-carbon sugar is called a **nucleoside** (see Figure 3.24B). Thus, nucleotides are *nucleoside phosphates*. For example, the nucleoside containing adenine and ribose is called *adenosine*. Adding one phosphate group to this structure produces *adenosine monophosphate (AMP)*, adding two phosphate groups produces

adenosine diphosphate (ADP), and adding three produces adenosine triphosphate (ATP). The corresponding adenine-deoxyribose complexes are named *deoxyadenosine monophosphate (dAMP), deoxyadenosine diphosphate (dADP)*, and *deoxyadenosine triphosphate (dATP)*. The lowercase *d* in the abbreviations indicates that the nucleoside contains the deoxyribose form of the sugar. Equivalent names and abbreviations are used for the other nucleotides (see Figure 3.24B). Whether a nucleotide is a monophosphate, diphosphate, or triphosphate has fundamentally important effects on its activities.

Nucleotides perform many functions in cells in addition to serving as the building blocks of nucleic acids. Two ribose-containing nucleotides in particular, adenosine triphosphate (ATP) and guanosine triphosphate (GTP), are the primary molecules that transport chemical energy from one reaction system to another. The same nucleotides regulate and adjust cellular activity. Molecules derived from nucleotides play important roles in biochemical reactions by delivering reactants or electrons from one system to another.

DNA and RNA Consist of Chains of Nucleotides and Are the Informational Molecules in Cells

DNA and RNA consist of *polynucleotide chains,* that is, chains of nucleotides with one nucleotide linked to the next by a bridging phosphate group between the 5′ carbon of one sugar and the 3′ carbon of the next sugar in line. This linkage is called

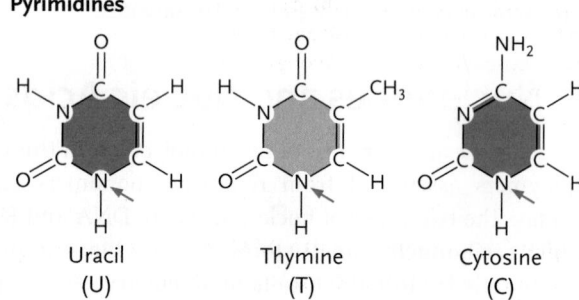

FIGURE 3.25 Pyrimidine and purine bases of nucleotides and nucleic acids. Red arrows indicate where the bases link to ribose or deoxyribose sugars in the formation of nucleotides.
© Cengage Learning 2017

A. DNA

B. RNA

FIGURE 3.26 caption area labels: Phosphate groups, Phosphodiester bond, Bases, Bases

replaced by a hydrogen (see Figure 3.25). The differences in sugar and pyrimidine bases between DNA and RNA account for important differences in the structure and functions of these nucleic acids inside cells.

DNA Molecules in Cells Consist of Two Nucleotide Chains Wound Together

In cells, DNA takes the form of a **double helix,** first discovered by James D. Watson and Francis H. C. Crick in 1953, in collaboration with Maurice Wilkins and Rosalind Franklin (see Chapter 14 for details of their discovery). In the double helix, two nucleotide chains are wrapped around each other in a spiral that resembles a twisted ladder **(Figure 3.27)**. The sides of the ladder are the sugar–phosphate backbones of the two chains, which twist around each other in a right-handed direction to form the double spiral. The rungs of the ladder are the nitrogenous bases, which extend inward from the sugars toward the center of the helix. Each rung consists of a pair of nitrogenous bases held in a flat plane roughly perpendicular to the long axis of the helix. The two nucleotide chains of a DNA double helix are held together primarily by hydrogen bonds between the base pairs. Slightly more than 10 base pairs are packed into

FIGURE 3.26 Linkage of nucleotides to form the nucleic acids DNA and RNA. P is a phosphate group (see Figure 3.27). **(A)** In DNA, the bases adenine (A), thymine (T), cytosine (C), or guanine (G) may be bound at the base positions. The lilac zones are the sugar (deoxyribose)–phosphate backbones of the two polynucleotide strands of the DNA molecule. Dotted lines designate hydrogen bonds. **(B)** In RNA, A, G, C, or uracil (U) may be bound at the base positions. The light green zone is the sugar (ribose)–phosphate backbone of the polynucleotide strand of the RNA molecule.

© Cengage Learning 2017

a **phosphodiester bond (Figure 3.26).** This arrangement of alternating sugar and phosphate groups forms the backbone of a nucleic acid chain, which is a constant chemical structure in the molecule. The nitrogenous bases of the nucleotides project from this backbone. Different DNA molecules have different sequences of bases, as do different RNA molecules. The particular sequences of bases represent the informational content of the molecules, as outlined in Section 1.1.

Each nucleotide of a DNA chain contains deoxyribose and one of the four bases A, T, G, or C. Each nucleotide of an RNA chain contains ribose and one of the four bases A, U, G, or C. Thymine and uracil differ only in a single functional group: in T a methyl (—CH_3) group is linked to the ring, but in U it is

A. DNA double helix, showing arrangement of sugars, phosphate groups, and bases

B. Space-filling model of DNA double helix

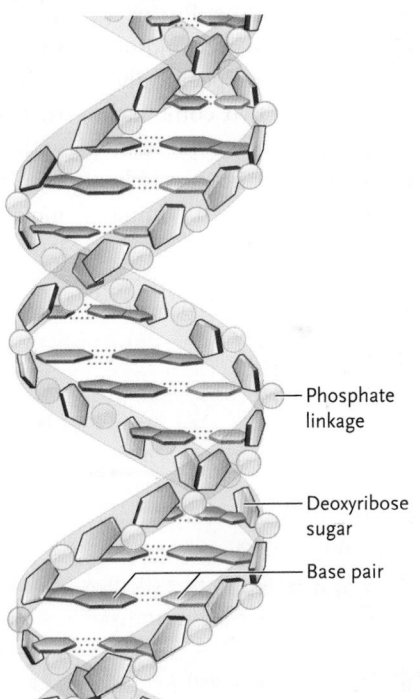
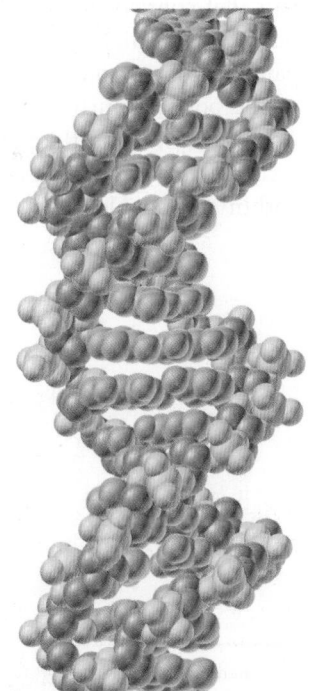

Phosphate linkage

Deoxyribose sugar

Base pair

FIGURE 3.27 The DNA double helix. (A) Arrangement of sugars, phosphate groups, and bases in the DNA double helix. The dotted lines between the bases designate hydrogen bonds. **(B)** Space-filling model of the DNA double helix. The paired bases, which lie in flat planes, are seen edge-on in this view.

© Cengage Learning 2017

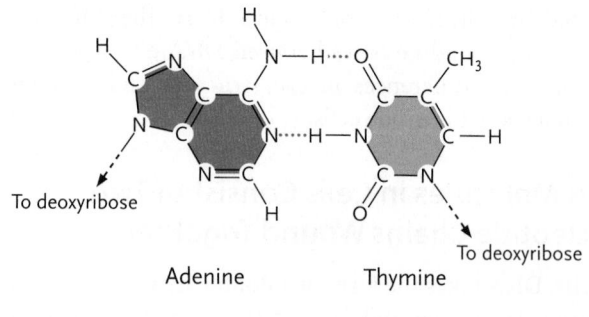

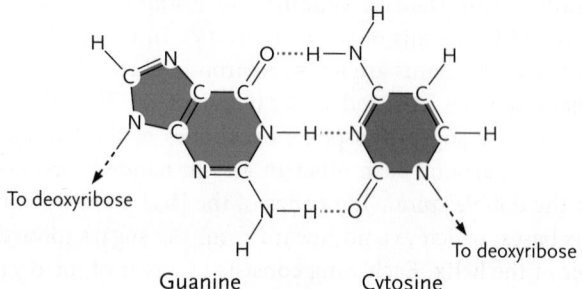

FIGURE 3.28 The DNA base pairs A–T (adenine–thymine) and G–C (guanine–cytosine), as seen from one end of a DNA molecule. Dotted lines between the bases designate hydrogen bonds.

© Cengage Learning 2017

each turn of the double helix. A DNA double-helix molecule is also referred to as double-stranded DNA.

The space separating the sugar–phosphate backbones of a DNA double helix is just wide enough to accommodate a base pair that consists of one purine and one pyrimidine. Purine–purine base pairs are too wide and pyrimidine–pyrimidine pairs are too narrow to fit this space exactly. More specifically, of the possible purine–pyrimidine pairs, only two combinations, adenine with thymine, and guanine with cytosine, can form stable hydrogen bonds so that the base pair fits precisely within the double helix **(Figure 3.28)**. An adenine–thymine

(A–T) pair forms two stabilizing hydrogen bonds; a guanine–cytosine (G–C) pair forms three.

As Watson and Crick pointed out in the initial report of their discovery, the formation of A–T and G–C pairs allows the sequence of one nucleotide chain to determine the sequence of its partner in the double helix. That is, wherever a T occurs on one chain of a DNA double helix, an A occurs opposite it on the other chain; wherever a C occurs on one chain, a G occurs on the other side (see Figure 3.26). Because of this, the nucleotide sequence of one chain is said to be *complementary* to the nucleotide sequence of the other chain. The complementary nature of the two chains underlies the processes when DNA molecules are copied—replicated—to pass hereditary information from parents to offspring (see Chapter 14), and when RNA copies are made of DNA molecules to transmit information within cells (see Chapter 15). In DNA replication, one nucleotide chain is used as a **template** for the assembly of a complementary chain according to the A–T and G–C base-pairing rules **(Figure 3.29)**.

The sequence of base pairs distinguishes one DNA molecule or a segment of a DNA molecule from another. Hence, as indicated previously, the informational content of DNA is located in its sequence of base pairs. A gene is a specific sequence of base pairs in a section of a long DNA chain. "Reading" that sequence at the molecular level to produce the product specified by the gene is what we mean by gene expression (see Section 1.1 and Figure 1.4).

RNA Molecules Are Usually Single Nucleotide Chains

In contrast to DNA, RNA molecules exist largely as single, rather than double, polynucleotide chains in living cells. That is, RNA is typically single-stranded. (An exception is found with some viruses which have double-stranded RNA genomes.) Similar to DNA, the sequence of bases distinguishes one RNA molecule from another. With the exception of viruses that have

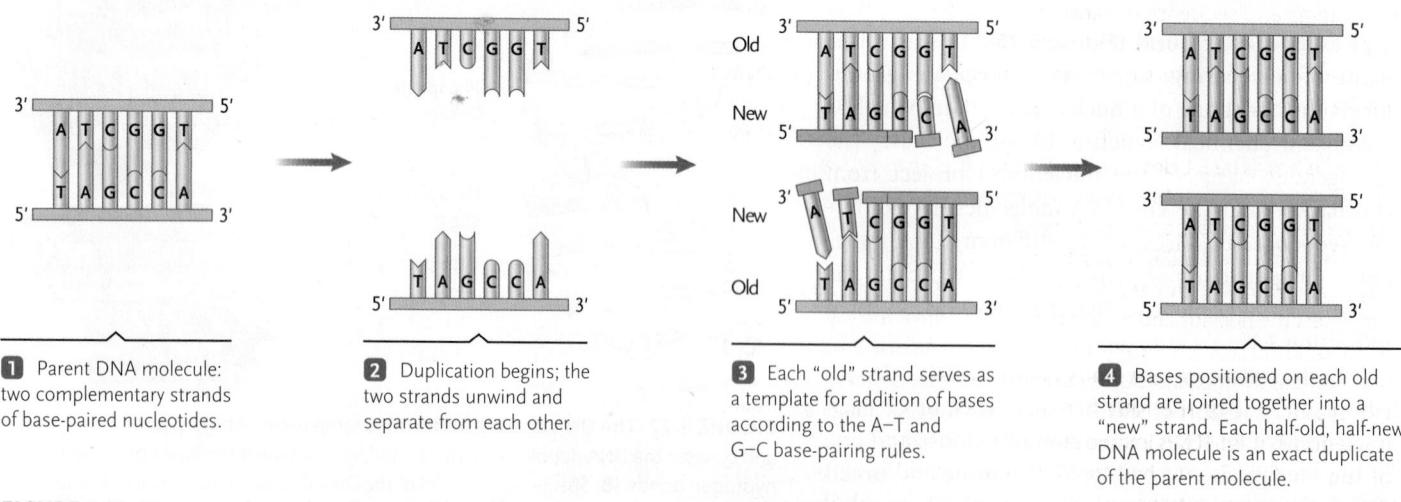

1 Parent DNA molecule: two complementary strands of base-paired nucleotides.

2 Duplication begins; the two strands unwind and separate from each other.

3 Each "old" strand serves as a template for addition of bases according to the A–T and G–C base-pairing rules.

4 Bases positioned on each old strand are joined together into a "new" strand. Each half-old, half-new DNA molecule is an exact duplicate of the parent molecule.

FIGURE 3.29 How complementary base pairing allows DNA molecules to be replicated precisely.

© Cengage Learning 2017

RNA as their genetic material, all RNAs are expressed from genes in DNA (see Figure 1.4 and Chapter 15). Once synthesized, the linear RNA molecules can fold and twist back on themselves to form double-helical regions. The patterns of these fold-back double helices are as vital to RNA function as the folding of amino acid chains is to protein function. "Hybrid" double helices, which consist of an RNA chain paired with a DNA chain, are formed temporarily when RNA copies are made of DNA chains (see Figure 1.4 and Chapter 15). In the RNA–RNA or hybrid RNA–DNA helices, U in RNA takes over the pairing functions of T, forming A–U rather than A–T base pairs.

The description of nucleic acid molecules in this section, with the discussions of carbohydrates, lipids, and proteins in earlier sections, completes our survey of the major classes of organic molecules found in living organisms. In the next chapter, we survey the structure of cells, the fundamental units into which biological molecules are organized and where molecules interact to produce the characteristics of life.

STUDY BREAK 3.5

1. What is the monomer of a nucleic acid macromolecule?
2. What are the chemical differences between DNA and RNA?

 Unanswered Questions

How are the enzymes that synthesize fatty acids and cholesterol regulated in the cells?

The cells of humans and animals contain thousands of different lipids that form the lipid bilayer membranes that bound cells and subcellular organelles. Several of the signaling molecules or hormones that regulate metabolic processes also derive from lipids. Despite their ubiquitous functions, we know very little about how the body controls the level of each lipid substance to ensure a concentration that is adequate for function but guards against overaccumulation. Overaccumulation can produce disorders ranging from brain degeneration to heart attacks.

One area of recent progress concerns the mechanism that controls the body's production of two of the most abundant lipids, cholesterol and fatty acids. Much of this work has been carried out in a laboratory that is led jointly by Joseph L. Goldstein and myself at the University of Texas Southwestern Medical Center in Dallas, Texas. Several years ago our group discovered sterol regulatory element binding proteins (SREBPs), which are transcription factors. (Transcription factors are regulatory proteins [see Table 3.2] that bind to specific sequences in DNA and activate transcription of nearby genes. Transcription is the first step in the process of gene expression, in which the DNA sequence of a gene is copied into an RNA molecule. This RNA sequence later directs assembly of the protein's amino acid sequence in translation. [See Section 1.1 and Chapters 15 and 16.]) The SREBPs selectively activate several dozen genes, including all of the ones necessary for the production of cholesterol or its uptake from plasma.

SREBPs are synthesized initially on membranes of the endoplasmic reticulum (ER). (The ER, described in Chapter 4, is an extensive interconnected network of membranes in eukaryotic cells that is responsible for the synthesis, transport, and modification of lipids and some proteins.) To reach the nucleus, the SREBPs must be transported from the ER to the cell's Golgi complex where they are cleaved by proteases to release a fragment that enters the nucleus and activates transcription. (The Golgi complex, described in Chapter 4, is the eukaryotic organelle responsible for the final modification, sorting, and distribution of proteins and lipids that arrive from the ER.) Transport of SREBPs from ER to Golgi requires the protein Scap, which binds SREBPs immediately after their synthesis.

Why are SREBPs attached to ER membranes, and why do cells go through this elaborate transport process to release the active fragments?

The reason is simple: it allows the synthesis of membrane cholesterol to be regulated by the concentration of cholesterol in the ER membrane. When cholesterol builds up in ER membranes, the cholesterol binds to Scap, changing its conformation and blocking the transport of SREBPs to the Golgi. The SREBPs are no longer processed by the Golgi proteases, and the amount of the fragment that goes to the nucleus decreases. Synthesis of cholesterol is diminished, and it remains low until the cholesterol content of ER membranes falls, upon which the SREBPs are again transported to the Golgi where they activate the genes required for cholesterol synthesis.

The elucidation of the SREBP processing pathway is a step forward, but many questions remain unanswered. Many of these center on the precise way in which Scap senses the cholesterol level in membranes. Also, we must determine the mechanisms that control the other constituents of cell membranes, primarily phospholipids. Only then will we understand how our cells manage to create membranes with the precise chemical and physical properties that allow them to perform their vital functions.

think like a scientist Defects in SREBP regulation contribute to common diseases, ranging from heart attacks to obesity and diabetes. How might better understanding of the SREBP regulatory mechanism contribute to treatment of diabetes?

Dr. Michael S. Brown and **Dr. Joseph L. Goldstein,** both of the University of Texas Southwestern Medical Center, discovered the low-density lipoprotein (LDL) receptor. They shared many awards for this work, including the Nobel Prize for Medicine or Physiology. Dr. Brown is Paul J. Thomas Professor of Molecular Genetics and Director of the Jonsson Center for Molecular Genetics; Dr. Goldstein is chairman of the Department of Molecular Genetics. To learn more about their research, go to http://www4.utsouthwestern.edu/.

REVIEW KEY CONCEPTS

For access to MindTap and additional study materials visit www.cengagebrain.com.

3.1 Formation and Modification of Biological Molecules

- Carbon atoms readily share electrons, allowing each carbon atom to form four covalent bonds with other carbon atoms or atoms of other elements. The resulting extensive chain and ring structures form the backbones of diverse organic compounds (Figure 3.1).

- The structure and behavior of organic molecules, as well as their linkage into larger units, depend on the chemical properties of functional groups (Table 3.1).

- Isomers have the same chemical formula but different molecular structures. Molecules that are mirror images of one another are an example of stereoisomers. Structural isomers are two molecules with atoms arranged in different ways (Figures 3.2 and 3.3).

- In a dehydration synthesis reaction, the components of a water molecule are removed as subunits assemble. In hydrolysis, the components of a water molecule are added as subunits are broken apart (Figure 3.4).

- Many carbohydrates, proteins, and nucleic acids are large polymers of monomer subunits. Polymerization reactions assemble monomers into polymers. Polymers larger than 1,000 daltons in mass are considered to be macromolecules.

3.2 Carbohydrates

- Carbohydrates are molecules in which carbon, hydrogen, and oxygen occur in the approximate ratio 1:2:1.

- Monosaccharides are carbohydrate subunits that contain three to seven carbons (Figures 3.5–3.6).

- Monosaccharides have D and L stereoisomers (Figure 3.6B). Typically, one of the two forms is used in cellular reactions because it has a molecular shape that can be recognized by the enzyme accelerating the reaction, whereas the other form does not.

- Two monosaccharides join to form a disaccharide; greater numbers form polysaccharides (Figures 3.7 and 3.8).

3.3 Lipids

- Lipids are hydrocarbon-based, water-insoluble, nonpolar molecules. Biological lipids include neutral lipids, phospholipids, and steroids.

- Neutral lipids, which are primarily energy-storing molecules, have a glycerol backbone and three fatty acid chains (Figures 3.9 and 3.10).

- Phospholipids are similar to neutral lipids except that a phosphate group and a polar organic unit substitute for one of the fatty acids (Figure 3.11). In polar environments (such as water), phospholipids orient with their polar end facing the water and their nonpolar ends clustered in a region that excludes water. This orientation underlies the formation of bilayers, the structural framework of biological membranes.

- Steroids, which consist of four carbon rings carrying primarily nonpolar groups, function chiefly as components of membranes and as hormones in animals (Figures 3.12 and 3.13).

- Lipids link with carbohydrates to form glycolipids and with proteins to form lipoproteins, both of which play important roles in cell membranes.

3.4 Proteins

- Proteins are assembled from 20 different amino acids. Amino acids have a central carbon to which is attached an amino group, a carboxyl group, a hydrogen atom, and a side group (R) that differs in each amino acid (Figure 3.14).

- Sulfhydryl groups join different regions of the same amino acid chain together or different amino acid chains together (Figures 3.15 and 3.21).

- Peptide bonds between the amino group of one amino acid and the carboxyl group of another amino acid link amino acids into chains (Figure 3.16).

- A protein may have four levels of structure. Its primary structure is the linear sequence of amino acids in a polypeptide chain; secondary structure is the arrangement of the amino acid chain into α helices or β strands and sheets brought about by hydrogen bonding; and tertiary structure is the protein's overall conformation. Quaternary structure is the number and arrangement of polypeptide chains in a protein (Figures 3.17–3.20).

- Chaperonins facilitate folding newly synthesized polypeptides into their correct tertiary structure (Figure 3.22).

- In many proteins, combinations of secondary, tertiary, and quaternary structure form functional domains (Figure 3.23).

- Proteins combine with lipids to produce lipoproteins, with carbohydrates to produce glycoproteins, and with nucleic acids to form nucleoproteins.

3.5 Nucleotides and Nucleic Acids

- A nucleotide consists of a nitrogenous base, a five-carbon sugar, and one to three phosphate groups (Figures 3.24 and 3.25).

- DNA and RNA consist of chains of nucleotides linked by covalent bonds between their sugar and phosphate groups to form polynucleotide chains. Alternating sugar and phosphate groups form the backbone of a nucleic acid chain (Figure 3.26).

- DNA contains nucleotides with the nitrogenous bases adenine (A), thymine (T), guanine (G), or cytosine (C) linked to the sugar deoxyribose. RNA contains nucleotides with the nitrogenous bases adenine, uracil (U), guanine, or cytosine linked to the sugar ribose (Figures 3.24–3.26).

- In a DNA double helix, two nucleotide chains wind around each other like a twisted ladder, with the sugar–phosphate backbones of the two chains forming the sides of the ladder and the nitrogenous bases forming the rungs of the ladder (Figure 3.27). The sequence of bases pairs in DNA represents the informational content of the molecule.

- A–T and G–C base pairs mean that the sequences of the two nucleotide chains of a DNA double helix are complements of each other. Complementary pairing underlies the processes that replicate DNA and copy RNA from DNA (Figures 3.28 and 3.29).

TEST YOUR KNOWLEDGE

Remember/Understand

1. Which functional group has a double bond and forms organic acids?
   a. carboxyl
   b. amino
   c. hydroxyl
   d. carbonyl
   e. sulfhydryl

2. Which of the following characteristics is *not* common to carbohydrates, lipids, and proteins?
   a. They are composed of a carbon backbone with functional groups attached.
   b. Monomers of these molecules undergo dehydration synthesis to form polymers.
   c. Their polymers are broken apart by hydrolysis.
   d. The backbones of the polymers are primarily polar molecules.
   e. The molecules are held together by covalent bonding.

3. Cellulose is to carbohydrate as:
   a. amino acid is to protein.
   b. lipid is to fat.
   c. collagen is to protein.
   d. nucleic acid is to DNA.
   e. nucleic acid is to RNA.

4. Maltose, sucrose, and lactose differ from one another:
   a. because not all contain glucose.
   b. because not all of them exist in ring form.
   c. in the number of carbons in the sugar.
   d. in the number of hexose monomers involved.
   e. by the linkage of the monomers.

5. Lipids that are liquid at room temperature:
   a. are fats.
   b. contain more hydrogen atoms than lipids that are solids at room temperature.
   c. if polyunsaturated, contain several double bonds in their fatty acid chains.
   d. lack glycerol.
   e. are not stored in cells as triglycerides.

6. Which of the following statements about steroids is *false*?
   a. They are classified as lipids because, like lipids, they are nonpolar.
   b. They can act as regulatory molecules in animals.
   c. They are composed of four carbon rings.
   d. They are highly soluble in water.
   e. Their most abundant form is as sterols.

7. The term *secondary structure* refers to a protein's:
   a. sequence of amino acids.
   b. structure that results from local interactions between different amino acids in the chain.
   c. interactions with a second protein chain.
   d. interaction with a chaperonin.
   e. interactions with carbohydrates.

8. The first and major effect in denaturation of proteins is that:
   a. peptide bonds break.
   b. α helices unwind.
   c. β sheet structures unfold.
   d. tertiary structure is changed.
   e. quaternary structures disassemble.

9. In living systems:
   a. proteins rarely combine with other macromolecules.
   b. enzymes are always proteins.
   c. proteins are composed of 24 amino acids.
   d. chaperonins inhibit protein movement.
   e. a protein domain refers to the place in the cell where proteins are synthesized and function.

10. RNA differs from DNA because:
    a. RNA may contain the pyrimidine uracil, and DNA does not.
    b. RNA is always single-stranded when functioning, and DNA is always double-stranded.
    c. the pentose sugar in RNA has one less O atom than the pentose sugar in DNA.
    d. RNA is more stable and is broken down by enzymes less easily than DNA.
    e. RNA is a much larger molecule than DNA.

Apply/Analyze

11. **Discuss Concepts** Identify the following structures as a carbohydrate, fatty acid, amino acid, or polypeptide:

    a. $H_3N^+ — \overset{\overset{R}{|}}{\underset{\underset{H}{|}}{C}} — COO^-$ (The R indicates an organic group.)

    b. $C_6H_{12}O_6$
    c. (glycine)$_{20}$
    d. $CH_3(CH_2)_{16}COOH$

12. **Discuss Concepts** Lipoproteins are relatively large, spherical clumps of protein and lipid molecules that circulate in the blood of mammals. They are like suitcases that move cholesterol, fatty acid remnants, triglycerides, and phospholipids from one place to another in the body. Given what you know about the insolubility of lipids in water, which of the three kinds of lipids would you predict to be on the outside of a lipoprotein clump, bathed in the fluid portion of blood?

13. **Discuss Concepts** Explain how polar and nonpolar groups are important in the structure and functions of lipids, proteins, and nucleic acids.

Evaluate/Create

14. **Design an Experiment** A clerk in a health food store tells you that natural vitamin C extracted from rose hips is better for you than synthetic vitamin C. Given your understanding of the structure of organic molecules, how would you respond? Design an experiment to test whether the rose hips and synthetic vitamin C preparations differ in their effects.

15. **Apply Evolutionary Thinking** How do you think the primary structure (amino acid) sequence of proteins could inform us about the evolutionary relationships of proteins?

For selected answers, see Appendix A.

INTERPRET THE DATA

Cholesterol does not dissolve in blood. It is carried through the bloodstream by lipoproteins, as described in *Focus on Research: Applied Research*. Low-density lipoprotein (LDL) carries cholesterol to body tissues, such as artery walls, where it can form health-endangering deposits. LDL is often called "bad" cholesterol. High-density lipoprotein (HDL) carries cholesterol away from tissues to the liver for disposal; it is often called "good" cholesterol.

Ronald Mensink and Martijn Katan tested the effects of different dietary fats on blood lipoprotein levels. They placed 59 men and women on a diet in which 10% of their daily energy intake consisted of *cis* fatty acids, *trans* fatty acids, or saturated fats. (All subjects were tested on each of the diets.) Blood LDL and HDL levels of the subjects were measured after three weeks on the diet. The **Table** displays the averaged results, shown in milligrams per deciliter (mg/dL) of blood.

Effect of Diet on Lipoprotein Levels (mg/dL)

	cis Fatty Acids	*trans* Fatty Acids	Saturated Fats	Optimal Level
LDL	103	117	121	<100
HDL	55	48	55	>40
LDL-to-HDL ratio	1.87	2.43	2.2	<2

© Cengage Learning 2017

1. In which group was the level of LDL ("bad" cholesterol) highest and in which group was the level of HDL ("good" cholesterol) lowest?

2. An elevated risk of heart disease has been correlated with increasing LDL-to-HDL ratios. Which group had the highest LDL-to-HDL ratio?

3. Rank the three diets from best to worst according to their potential effect on cardiovascular health.

Source: R. P. Mensink and M. B. Katan. 1990. Effect of dietary *trans* fatty acids on high-density and low-density lipoprotein cholesterol levels in healthy subjects. *New England Journal of Medicine* 323:439–445.

Cells

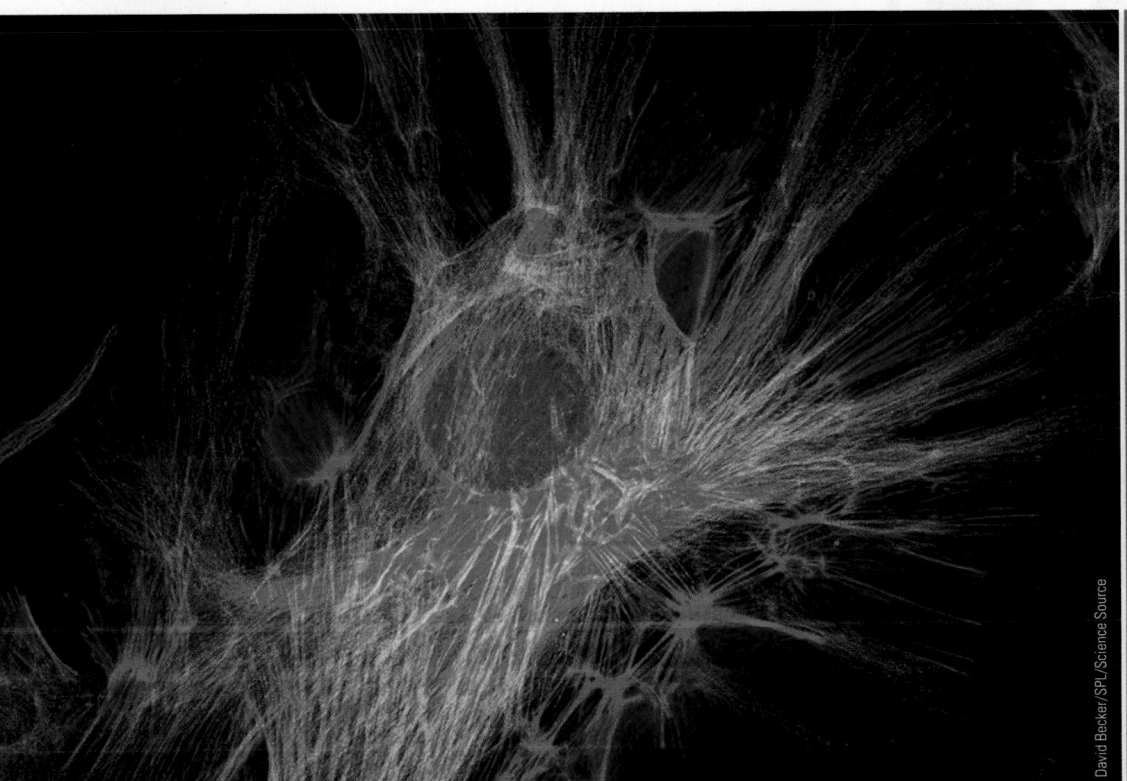

David Becker/SPL/Science Source

Cells labeled fluorescently to visualize their internal structure (confocal light micrograph). Cell nuclei are shown in blue and parts of the cytoskeleton in red and green.

Why it matters . . . In the mid-1600s, Robert Hooke, Curator of Instruments for the Royal Society of England, used the newly invented light microscopes to study biological materials. When Hooke looked at thinly sliced cork from a mature tree through a microscope, he observed tiny compartments **(Figure 4.1).** He gave them the Latin name *cellulae,* meaning "small rooms"—hence, the origin of the biological term *cell.* Cork consists of the walls of dead cells, which is what Hooke was observing.

Reports of cells also came from other sources. By the late 1600s, Anton van Leeuwenhoek, a Dutch shopkeeper, observed "many very little animalcules, very prettily a-moving," using a single-lens microscope of his own construction. Leeuwenhoek discovered and described diverse protists (see Chapter 27), sperm cells, and even bacteria, organisms so small that they would not be seen by others for another two centuries.

In the 1820s, improvements in microscopes brought cells into sharper focus. Robert Brown, an English botanist, noticed a discrete, spherical body inside some cells. He called the body a *nucleus.* In 1838, a German botanist, Matthias Schleiden, speculated that the nucleus had something to do with the development of a cell. The following year, zoologist Theodor Schwann of Germany expanded Schleiden's idea to propose that all animals and plants consist of cells that contain a nucleus. He also proposed that even when a cell forms part of a larger organism, it has an individual life of its own. However, an important question remained: Where do cells come from? A decade later, from his studies of cell growth and reproduction, the German physiologist Rudolf Virchow proposed that cells arise only from preexisting cells by a process of division.

FIGURE 4.1 The cork cells drawn by Robert Hooke and the compound microscope he used to examine them.

Thus, by the middle of the nineteenth century, microscopic observations had yielded three profound generalizations, which together constitute what is now known as the **cell theory:**

1. All organisms are composed of one or more cells.
2. The cell is the basic structural and functional unit of all living organisms.
3. Cells arise only from the division of preexisting cells.

These tenets were fundamental to the development of biological science.

This chapter describes the structure and functions of cells, emphasizing both the similarities among all cells and some of the most basic differences among cells of various organisms.

4.1 Basic Features of Cell Structure and Function

As the basic structural and functional units of all living organisms, cells carry out the essential processes of life. They contain highly organized systems of molecules, including the nucleic acids DNA and RNA, which carry hereditary information and direct the manufacture of cellular molecules. Cells use chemical molecules or light as energy sources for their activities. Cells also respond to changes in their external environment by altering their internal reactions. Further, cells duplicate and pass on their hereditary information as part of cellular reproduction. All these activities occur in cells that, in most cases, are invisible to the naked eye.

Some types of organisms are unicellular. Unicellular organisms include almost all bacteria and archaeans; some protists, such as amoebas; and some fungi, such as yeasts. Each of

these cells is a functionally independent organism capable of carrying out all activities necessary for its life. In more complex multicellular organisms, including plants and animals, the activities of life are divided among varying numbers of specialized cells. However, individual cells of multicellular organisms are potentially capable of surviving by themselves if placed in a chemical medium that can sustain them.

If cells are broken open, the property of life is lost: they are unable to grow, reproduce, or respond to outside stimuli in a coordinated, potentially independent fashion. This fact confirms the second tenet of the cell theory: life as we know it does not exist in units more simple than individual cells. *Viruses,* which consist only of a nucleic acid molecule surrounded by a protein coat, cannot carry out most of the activities of life. Their only capacity is to infect living cells and direct them to make more virus particles of the same kind. (Viruses are discussed in Chapter 17.)

Cells Are Visualized Using a Microscope

Cells assume a wide variety of forms in different prokaryotes and eukaryotes **(Figure 4.2).** Individual cells range in size from tiny bacteria to an egg yolk, a single cell that can be several centimeters in diameter. Yet, all cells are organized according to the same basic plan, and all have structures that perform similar activities.

Most cells are too small to be seen by the unaided eye: humans cannot see objects smaller than about 0.1 mm in diameter. The smallest bacteria have diameters of about 0.5 μm (a micrometer is 1,000 times smaller than a millimeter). Most animal and plant cells range from 10 to 100 μm in diameter. Your red blood cells are 7 to 8 μm across—a string of 2,500 of these cells is needed to span the width of your thumbnail. (**Figure 4.3** explains the units of measurement used in biology to study molecules and cells.)

To see cells and the structures within them biologists use **microscopy,** a technique for producing visible images of objects, biological or otherwise, that are too small to be seen by the human eye **(Figure 4.4).** The instrument of microscopy is the **microscope.** The two common types of microscopes are **light microscopes,** which use light to illuminate the specimen (the object being viewed), and **electron microscopes,** which use electrons to illuminate the specimen. Different types of microscopes give different *magnification* and *resolution* of the specimen. Just as for a camera or a pair of binoculars, **magnification** is the ratio of the object as viewed to its real size, usually given as something like 1,200✕. **Resolution** is the minimum distance two points in the specimen can be separated and still be seen as two points. Resolution depends primarily on the wavelength of light or electrons used to illuminate the specimen; the shorter the wavelength, the better the resolution. Hence, electron microscopes have higher resolution than light microscopes. Biologists choose the type of microscopy technique based on what they need to see in the specimen; selected examples are shown in Figure 4.4.

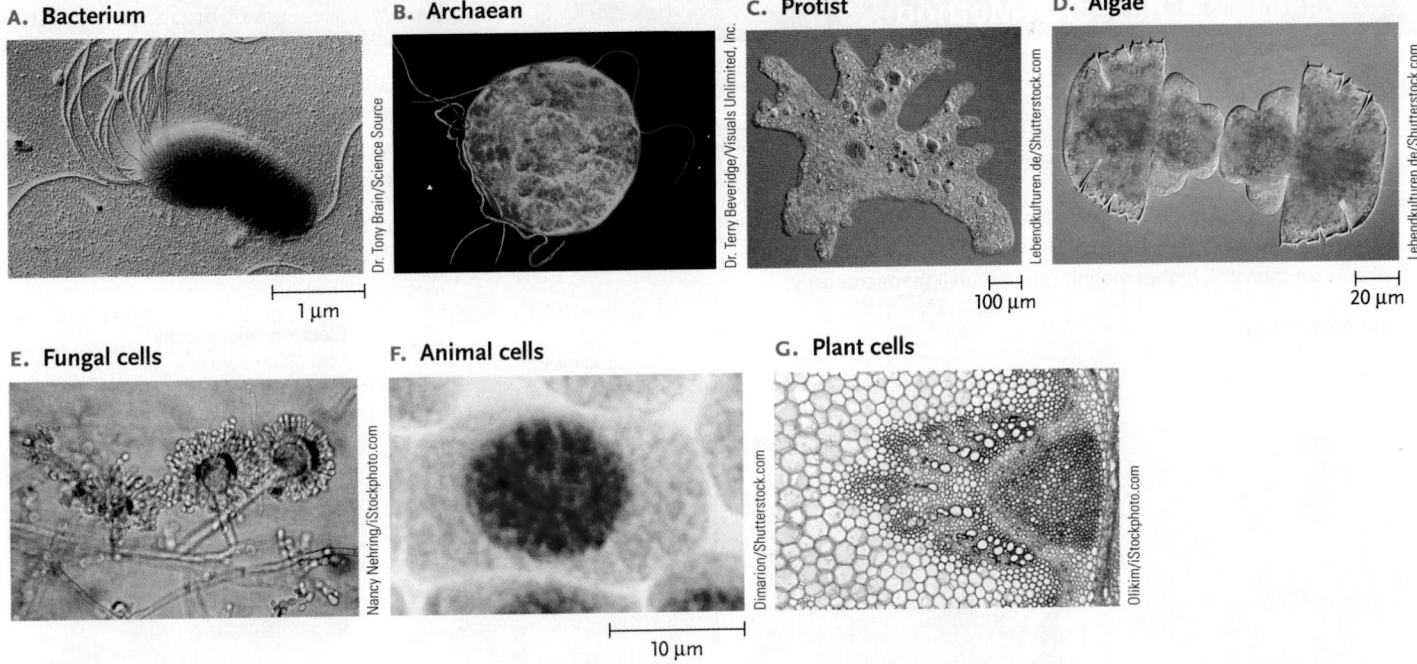

A. Bacterium
Dr. Tony Brain/Science Source

B. Archaean
Dr. Terry Beveridge/Visuals Unlimited, Inc.

C. Protist
Lebendkulturen.de/Shutterstock.com

D. Algae
Lebendkulturen.de/Shutterstock.com
20 µm

1 µm

100 µm

E. Fungal cells
Nancy Nehring/iStockphoto.com

F. Animal cells
Dimarion/Shutterstock.com

G. Plant cells
Olikim/iStockphoto.com

10 µm

FIGURE 4.2 Examples of the varied kinds of cells: (A) and (B) are prokaryotes, the others are eukaryotes. (A) A bacterial cell with flagella, *Pseudomonas fluoresceins* (color-enhanced). (B) An archaean, the extremophile *Sulfolobus acidocaldarius* (color-enhanced). (C) *Amoeba proteus*, a protist. (D) *Micrasterias*, an algal protist, in the process of dividing into two cells. (E) Fungal cells of the bread mold *Aspergillus*. (F) Animal cells. (G) Cells in the stem of a sunflower, *Helianthus annuus*.

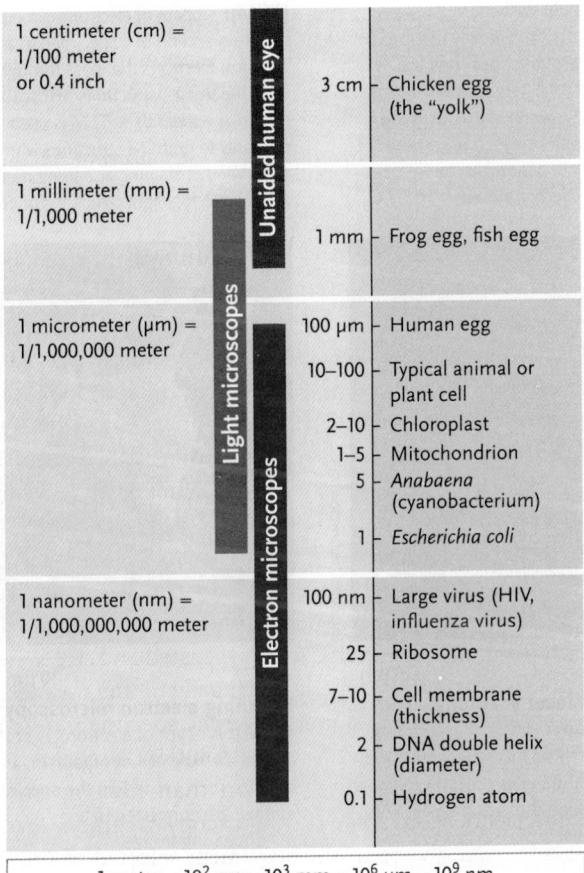

1 centimeter (cm) = 1/100 meter or 0.4 inch	**Unaided human eye**	3 cm – Chicken egg (the "yolk")
1 millimeter (mm) = 1/1,000 meter		1 mm – Frog egg, fish egg
1 micrometer (µm) = 1/1,000,000 meter	**Light microscopes**	100 µm – Human egg
		10–100 – Typical animal or plant cell
		2–10 – Chloroplast
		1–5 – Mitochondrion
		5 – *Anabaena* (cyanobacterium)
		1 – *Escherichia coli*
1 nanometer (nm) = 1/1,000,000,000 meter	**Electron microscopes**	100 nm – Large virus (HIV, influenza virus)
		25 – Ribosome
		7–10 – Cell membrane (thickness)
		2 – DNA double helix (diameter)
		0.1 – Hydrogen atom

1 meter = 10^2 cm = 10^3 mm = 10^6 µm = 10^9 nm

Evolutionary Adaptations Circumvent Cell Size Limits

There is an upper limit to cell size due to the change in the surface area-to-volume ratio of an object as its size increases **(Figure 4.5).** For example, doubling the diameter of a cell increases its volume by eight times but increases its surface area by only four times. The significance of this relationship is that the volume of a cell determines the amount of chemical activity that can take place within it, whereas the surface area determines the amount of substances that can be exchanged between the inside of the cell and the outside environment. Nutrients must enter cells constantly, and wastes must leave constantly. However, past a certain point, increasing the diameter of a cell gives a surface area that is insufficient to maintain an adequate nutrient–waste exchange for its entire volume. At that point cell growth must stop or the cell must divide to begin anew with a functional surface area-to-volume ratio.

Some cells have adaptations that allow them to circumvent the surface area limitation just described. For instance, eggs of some species, such as birds and frogs, are much larger than

FIGURE 4.3 Units of measure and the ranges in which they are used in the study of molecules and cells. The vertical scale in each box is logarithmic.
© Cengage Learning 2017

Light and Electron Microscopy

Purpose: In biology, microscopy is used to view organisms, cells, and structures within cells in their natural state or after being treated (stained) so that specific structures can be seen more clearly. All of the photographs of cells and cell structures in this book were made using microscopy.

Protocol: A light microscope uses a beam of light to illuminate the specimen and forms a magnified image of the specimen with glass lenses. An electron microscope uses a beam of electrons to illuminate the specimen and forms an image with magnetic fields. Electron microscopy provides higher resolution and higher magnification than light microscopy.

Light microscopy

Bright field, dark field, and phase-contrast micrographs are of a human epithelial cell; the Nomarski micrograph is of the protist *Paramecium;* and the fluorescent and confocal laser scanning micrographs are of a human fibroblast cell.

Electron microscopy

Micrographs are of a cell of the human immune system (top), and a human fibroblast cell (bottom).

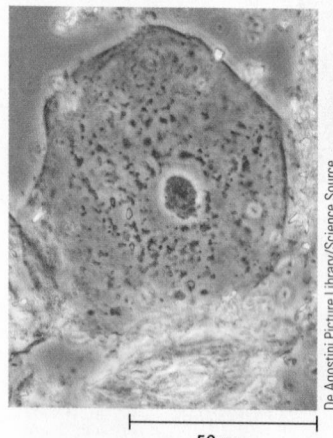

De Agostini Picture Library/Science Source

50 μm

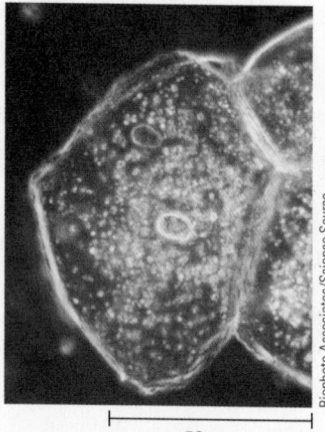

Biophoto Associates/Science Source

50 μm

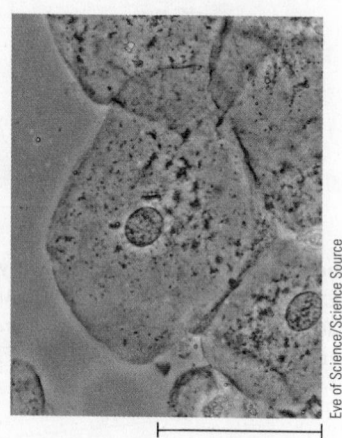

Eye of Science/Science Source

50 μm

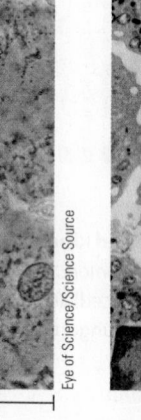

Dlumen/Shutterstock.com

2 μm

Bright field microscopy:
Light passes directly through the specimen. Many cell structures have insufficient contrast to be discerned. Staining with a dye is used to enhance contrast in a specimen, as shown here, but this treatment usually fixes and kills the cells.

Dark field microscopy:
Light illuminates the specimen at an angle, and only light scattered by the specimen reaches the viewing lens of the microscope. This gives a bright image of the cell against a black background.

Phase-contrast microscopy:
Differences in refraction (the way light is bent) caused by variations in the density of the specimen are visualized as differences in contrast. Otherwise invisible structures are revealed with this technique, and living cells in action can be photographed or filmed.

Transmission electron microscopy (TEM): A beam of electrons is focused on a thin section of a specimen in a vacuum. Electrons that pass through form the image; structures that scatter electrons appear dark. TEM is used primarily to examine structures within cells. Various staining and fixing methods are used to highlight structures of interest.

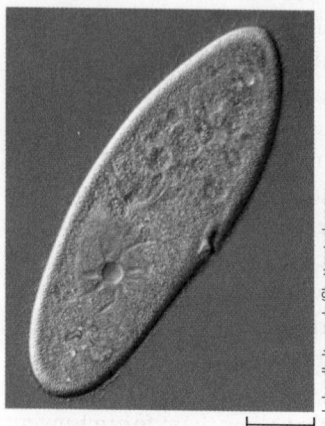

Lebendkulturen.de/Shutterstock.com

50 μm

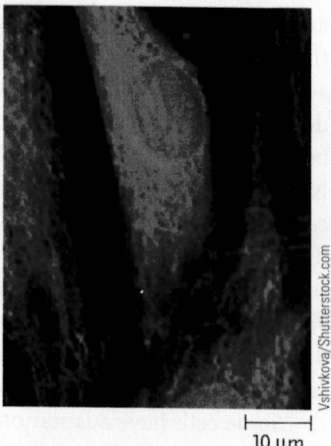

Vshivkova/Shutterstock.com

10 μm

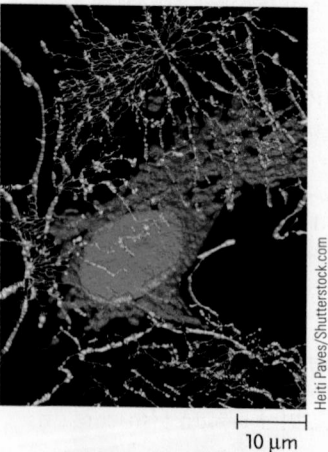

Heiti Paves/Shutterstock.com

10 μm

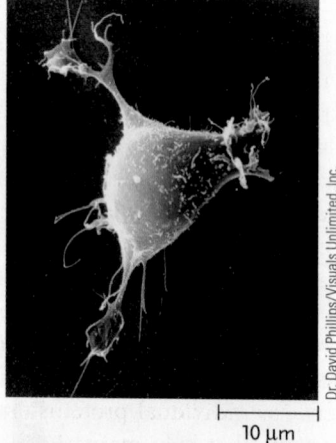

Dr. David Phillips/Visuals Unlimited, Inc.

10 μm

Nomarski (differential interference contrast): Similar to phase-contrast microscopy, special lenses enhance differences in density, giving a cell a 3D appearance.

Fluorescence microscopy:
Different structures or molecules in cells are stained with specific fluorescent dyes. The stained structures or molecules fluoresce when the microscope illuminates them with ultraviolet light, and their locations are seen by viewing the emitted visible light.

Confocal laser scanning microscopy: Lasers scan across a fluorescently stained specimen, and a computer focuses the light to show a single plane through the cell. This provides a sharper 3D image than other light microscopy techniques.

Scanning electron microscopy (SEM): A beam of electrons is scanned across a whole cell or organism, and the electrons excited on the specimen surface are converted to a 3D-appearing image.

Interpreting the Results: Different techniques of light and electron microscopy produce images that reveal different structures or functions of the specimen. A micrograph is a photograph of an image formed by a microscope.

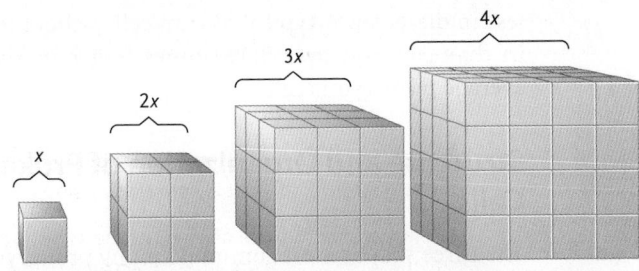

Total surface area	$6x^2$	$6(2x)^2 = 24x^2$	$6(3x)^2 = 54x^2$	$6(4x)^2 = 96x^2$
Total volume	x^3	$(2x)^3 = 8x^3$	$(3x)^3 = 27x^3$	$(4x)^3 = 64x^3$
Surface area/ volume ratio	6:1	3:1	2:1	1.5:1

FIGURE 4.5 Relationship between surface area and volume. The surface area of an object increases as a square of the linear dimension, whereas the volume increases as a cube of that dimension.
© Cengage Learning 2017

typical cells, meaning that they have a low surface area-to-volume ratio. In this case the eggs contain a large store of nutrients so no nutrients need to be brought into the cells. In addition, once fertilized, the eggs divide rapidly to produce a multicelled embryo with each cell of typical cell size (see Chapter 50). Another adaptation to circumvent the surface area limitation is for cells to be long and thin, or skinny and flat, both of which increase surface area. Examples are nerve cells (neurons; see Chapter 39) and muscle cells (see Chapter 43). Yet another adaptation is seen in human intestinal cells, which have closely packed, fingerlike extensions that increase their surface area (see Chapter 47).

Cells Have a DNA-Containing Central Region That Is Surrounded by Cytoplasm

All cells are bounded by the **plasma membrane,** a lipid bilayer with embedded protein molecules **(Figure 4.6).** (The chemical structure of the lipid bilayer was described in Section 3.3.) The plasma membrane is a selective barrier for the passage of water, ions, nutrients, and waste. Overall, it has the important role of maintaining the specialized internal ionic and molecular environments required for cellular life. (Membrane structure and functions are discussed further in Chapter 5.)

A central region of all cells contains DNA molecules, which store hereditary information. The hereditary information is organized in the form of *genes*—segments of DNA that code for individual proteins. The central region also contains proteins that help maintain the DNA structure and enzymes that duplicate DNA and copy its information into RNA. (*Enzymes* are molecules that speed up the rates of a reaction; they are discussed in detail in Chapter 6.)

All the parts of the cell between the plasma membrane and the central region comprise the **cytoplasm.** The cytoplasm

contains the *organelles,* the *cytosol,* and the *cytoskeleton.* The **organelles** ("little organs") are small, organized structures important for cell function. The **cytosol** is an aqueous (water) solution containing ions and various organic molecules. The **cytoskeleton** is a protein-based framework of filamentous structures that, among other things, helps maintain proper cell shape and plays key roles in cell division and chromosome segregation from cell generation to cell generation. The cytoskeleton was once thought to be specific to eukaryotes, but research has shown that all major eukaryotic cytoskeletal proteins have functional equivalents in prokaryotes.

Many of the cell's vital activities occur in the cytoplasm, including the synthesis and assembly of most of the molecules required for growth and reproduction (except those made in the central region) and the conversion of chemical and light energy into forms that can be used by cells. The cytoplasm also conducts stimulatory signals from the outside into the cell interior and carries out chemical reactions that respond to these signals.

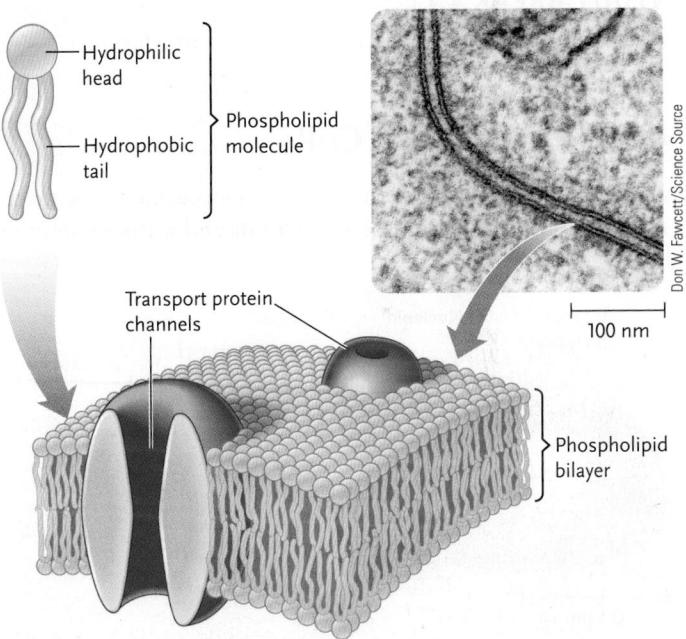

FIGURE 4.6 The plasma membrane, the outer boundary of a cell's cytoplasm. The plasma membrane consists of a lipid bilayer, an arrangement of phospholipids two molecules thick, which provides the framework of all biological membranes. The lipid bilayer is a hydrophobic barrier to the passage of water-soluble substances, but selected water-soluble substances can penetrate cell membranes through transport protein channels. Two proteins that transport substances across the membrane are shown. Other types of proteins are also associated with the plasma membrane. *(Inset)* Electron micrograph showing the plasma membranes of two adjacent animal cells.
© Cengage Learning 2017

Cells Occur in Prokaryotic and Eukaryotic Forms, Each with Distinctive Structures and Organization

Organisms fall into two fundamental groups, prokaryotes and eukaryotes, based on the organization of their cells. **Prokaryotes** (*pro* = before; *karyon* = nucleus) make up two domains of organisms, the Bacteria and the Archaea (discussed in Chapter 26). The DNA-containing central region of prokaryotic cells, the **nucleoid,** has no boundary membrane separating it from the cytoplasm. Many species of bacteria contain few if any internal membranes, whereas some bacterial species contain extensive internal membranes.

The **eukaryotes** (*eu* = true) make up the Domain Eukarya, which includes all the remaining organisms. The DNA-containing central region of eukaryotic cells, a true **nucleus,** is separated by membranes from the surrounding cytoplasm. The cytoplasm of eukaryotic cells typically contains extensive membrane systems that form organelles with their own distinct environments and specialized functions. As in prokaryotes, a plasma membrane surrounds eukaryotic cells as the outer limit of the cytoplasm.

The remainder of this chapter surveys the components of prokaryotic and eukaryotic cells in more detail.

STUDY BREAK 4.1

What is the plasma membrane, and what are its main functions?

4.2 Prokaryotic Cells

Most prokaryotic cells are relatively small, usually not much more than a few micrometers in length and a micrometer or less in diameter. A typical human cell is about ten times larger in diameter and over 8,000 times larger in volume than an average prokaryotic cell.

Structure and Organization of Prokaryotic Cells

The three shapes most common among prokaryotes are spherical, rodlike, and spiral. *Escherichia coli (E. coli),* a normal inhabitant of the mammalian intestine that has been studied extensively as a model organism in genetics, molecular biology, and genomics research, is rodlike in shape. **Figure 4.7** shows an electron micrograph (EM) and diagram of *E. coli* to illustrate the basic features of prokaryotic cell structure. Chapter 26 presents more detail about prokaryotic cell structure and function, as well as about the diversity of prokaryotic organisms.

The genetic material of prokaryotes is located in the nucleoid. In an EM, the nucleoid is seen to contain a highly folded mass of DNA (see Figure 4.7). For most species, the DNA is a single, circular molecule that unfolds when released from the cell. This DNA molecule is called the **prokaryotic chromosome.** (Chapter 17 discusses the genetics of prokaryotes.)

Individual genes in the DNA molecule encode the information required to make proteins. This information is copied into a type of RNA molecule called *messenger RNA (mRNA)* (see Section 1.1, Figure 1.4, and Chapter 15). Small, roughly spherical particles in the cytoplasm, the **ribosomes,** use the information in the mRNA to assemble amino acids into proteins (see Chapter 15). A prokaryotic ribosome consists of a large and a small subunit, each formed from a combination of *ribosomal RNA (rRNA)* and protein molecules. Each prokaryotic ribosome

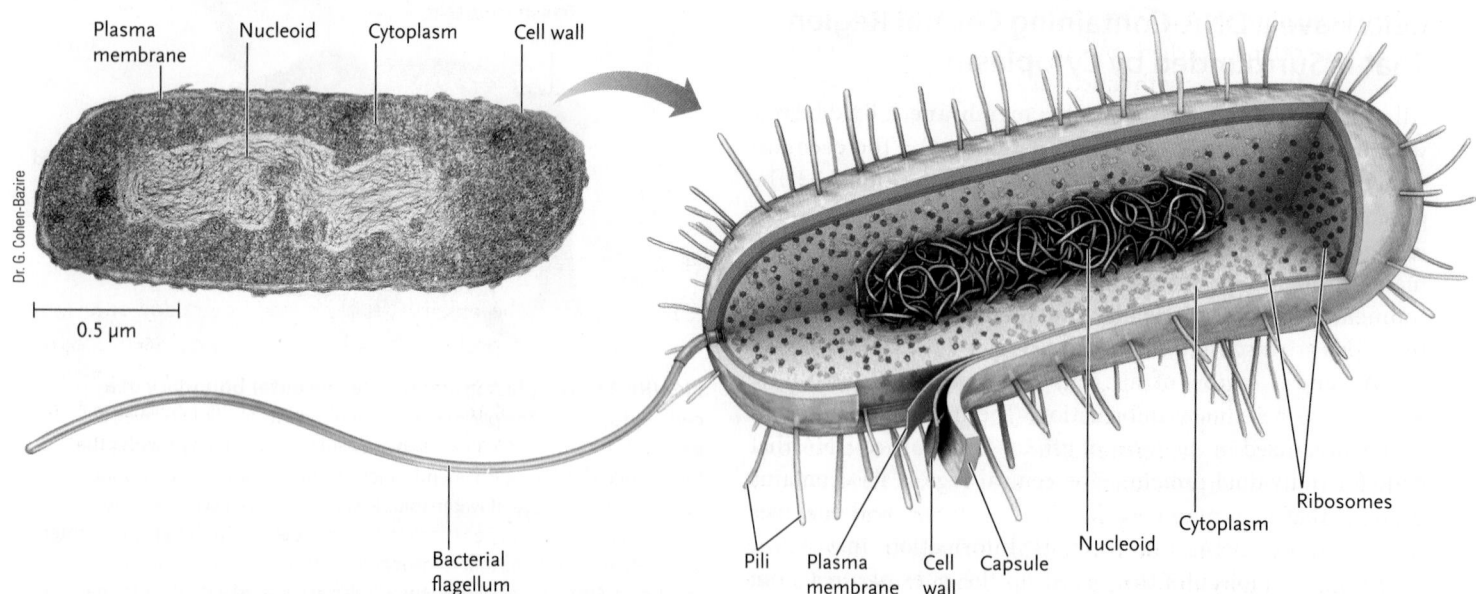

Dr. G. Cohen-Bazire

Plasma membrane • Nucleoid • Cytoplasm • Cell wall

0.5 μm

Bacterial flagellum

Pili • Plasma membrane • Cell wall • Capsule • Nucleoid • Cytoplasm • Ribosomes

FIGURE 4.7 Prokaryotic cell structure. An electron micrograph (left) and a diagram (right) of the bacterium *Escherichia coli.* The pili extending from the cell wall attach bacterial cells to other cells of the same species or to eukaryotic cells as a part of infection. A typical *E. coli* has four flagella.
© Cengage Learning 2017

contains three types of rRNA molecules, which are also copied from the DNA, and more than 50 proteins.

In almost all prokaryotes, the plasma membrane is surrounded by a rigid external layer of material, the **cell wall,** which ranges in thickness from 15 to 100 nm or more (a nanometer is one-billionth [10^{-9}] of a meter). The cell wall provides rigidity to prokaryotic cells and helps protect the cell from physical damage. In many prokaryotic cells, the wall is coated with an external layer of polysaccharides called the **glycocalyx** (a "sugar coating" from *glykys* = sweet; *calyx* = cup or vessel). When the glycocalyx is diffuse and loosely associated with the cells, it is a **slime layer;** when it is gelatinous and attached more firmly to cells, it is a **capsule.** The glycocalyx helps protect prokaryotic cells from physical damage and desiccation, and may enable a cell to attach to a surface, such as other prokaryotic cells (as in forming a colony), eukaryotic cells (as in *Streptococcus pneumoniae* attaching to lung cells), or nonliving substrates (such as rocks).

The plasma membrane itself performs several vital functions in prokaryotes. Besides transporting materials into and out of the cells, it contains most of the molecular systems that metabolize food molecules into the chemical energy of ATP. In photosynthetic prokaryotes, the molecules that absorb light energy and convert it to the chemical energy of ATP are also associated with the plasma membrane or with internal, saclike membranes derived from the plasma membrane.

Many prokaryotic species contain few if any internal membranes; in such cells, most cellular functions occur either on the plasma membrane or in the cytoplasm. But some prokaryotes have more extensive internal membrane structures. For example, photosynthetic bacteria and archaeans have complex layers of intracellular membranes formed by invaginations of the plasma membrane on which photosynthesis takes place. And members of the bacterial phylum Planctomycetes have complex internal membranes that form distinct compartments.

As mentioned earlier, prokaryotic cells have filamentous cytoskeletal structures with functions similar to those in eukaryotes. Prokaryotic cytoskeletons play important roles in creating and maintaining the proper shape of cells, in cell division, and, for certain prokaryotes, in determining polarity of the cells.

Many bacteria and archaeans can move through liquids and across wet surfaces. Most commonly they do so using long, threadlike protein fibers called **flagella** (singular, *flagellum,* meaning whip), which extend from the cell surface (see Figure 4.2A). The **bacterial flagellum,** which is helically shaped, rotates in a socket in the plasma membrane and cell wall to push the cell through a liquid medium (see Chapter 26). In *E. coli,* for instance, rotating bundles of flagella propel the bacterium. Archaeal flagella function similarly to bacterial flagella, but the two types differ significantly in their structures and mechanisms of action. Both types of prokaryotic flagella are also fundamentally different from the much larger and more complex flagella of eukaryotic cells (see Section 4.3).

Some bacteria and archaeans have hairlike shafts of protein called **pili** (singular, *pilus*) extending from their cell walls. There are two types of pili. *Common pili* (also known as

fimbriae), as their name suggests, are common among bacteria. They are relatively short, there are many per cell, and there are many subvarieties. Common pili are important for attaching a bacterial cell to surfaces or other cells, biofilm formation (biofilms are discussed in Chapter 26), cell motility, and the transport of protein and DNA across membranes. *Sex pili* are specialized structures on a bacterium that attach that cell to another bacterium lacking sex pili during conjugation. In conjugation, DNA is transferred in one direction from the cell with the sex pili to the cell without (see Chapter 17). Sex pili are long structures, there are only two or three of them per cell, and only particular bacterial cells can produce them.

Although prokaryotic cells appear relatively simple, their simplicity is deceptive. Most can use a variety of substances as energy and carbon sources, and they are able to synthesize almost all of their required organic molecules from simple inorganic raw materials. In many respects, prokaryotes are more versatile biochemically than eukaryotes. Their small size and metabolic versatility are reflected in their abundance; prokaryotes vastly outnumber all other types of organisms and live successfully in almost all regions of Earth's surface.

Evolutionary Divergence of Bacteria and Archaea

The two domains of prokaryotes, the Bacteria and the Archaea, share many biochemical and molecular features. However, the archaeans also share some features with eukaryotes and have other characteristics that are unique to their group. As shown in Figure 1.11, evolutionary divergence from the common ancestor of all living organisms resulted in the ancestral lineage of present-day members of the Domain Bacteria and the common ancestral lineage of present-day members of the Domain Archaea and members of the Domain Eukarya.

STUDY BREAK 4.2

Where in a prokaryotic cell is DNA found? How is that DNA organized?

4.3 Eukaryotic Cells

The domain of the eukaryotes, Eukarya, is divided into four major groups: the protists (see Chapter 27), fungi (see Chapter 30), animals (see Chapters 31 and 32), and plants (see Chapters 28 and 29). The rest of the chapter focuses on the cell components that are common to all or large groups of eukaryotic organisms.

Eukaryotic Cells Have a True Nucleus and Cytoplasmic Organelles Enclosed within a Plasma Membrane

The cells of all eukaryotes have a true nucleus enclosed by membranes. The cytoplasm surrounding the nucleus contains a system of membranous organelles, each specialized to carry

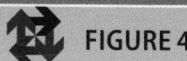 **FIGURE 4.8** **Research Method**

Cell Fractionation

Purpose: Cell fractionation partitions cells into fractions containing a single cell component, such as mitochondria or ribosomes. Once isolated, the cell component can be disassembled by the same general techniques to analyze its structure and function.

Protocol:

1. Break open intact cells by sonication (high-frequency sound waves), grinding in fine glass beads, or exposure to detergents that disrupt plasma membranes.

2. Use sequential centrifugations at increasing speeds to separate and purify cell structures. The spinning centrifuge drives cellular structures to the bottom of tube at a rate that depends on their shape and density. With each centrifugation, the largest and densest components are isolated and concentrated into a pellet; the remaining solution, the supernatant, is drawn off and can be centrifuged again at higher speed.

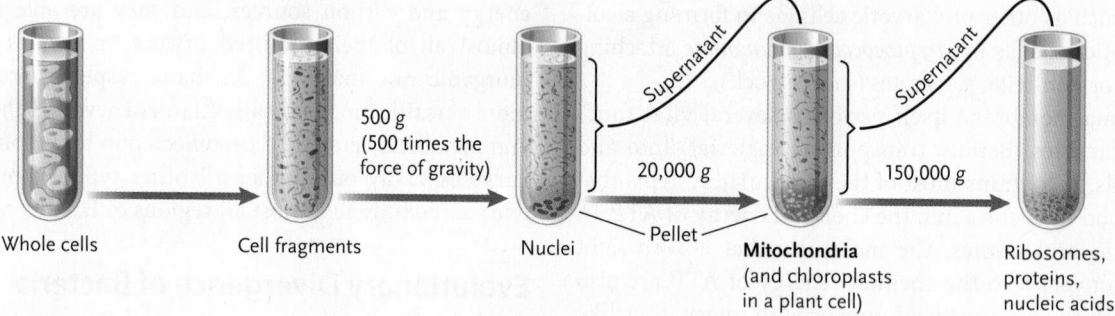

Whole cells Cell fragments 500 g (500 times the force of gravity) Supernatant Nuclei Pellet 20,000 g Supernatant **Mitochondria** (and chloroplasts in a plant cell) 150,000 g Ribosomes, proteins, nucleic acids

3. Resuspend the pellet containing the isolated cell components and subfractionate using the same general techniques to examine the components of organelles.

Interpreting the Results: Many of the cell or organelle subfractions generated by cell fractionation retain their biological activity, making them useful in studies of various cellular processes. For example, cell fractionation is used to determine the cellular location of a protein or biological reaction, such as whether it is free in the cytosol or associated with a membrane.

out one or more major functions of energy metabolism and molecular synthesis, storage, and transport. The cytosol, the cytoplasmic solution surrounding the organelles, participates in energy metabolism and molecular synthesis and performs specialized functions in support and motility. Researchers have discovered much about the structures and functions of various cellular organelles by using the research method of **cell fractionation** to isolate them and purify them **(Figure 4.8).**

The eukaryotic plasma membrane carries out various functions through several types of embedded proteins. Some of these proteins form channels through the plasma membrane that transport substances into and out of the cell (see Chapter 5). Other proteins in the plasma membrane act as receptors; they recognize and bind specific signal molecules in the cellular environment and trigger internal responses (see Chapter 9). In some eukaryotes, particularly animals, plasma membrane proteins recognize and adhere to molecules on the surfaces of other cells. Yet other plasma membrane proteins are important markers in the immune system, labeling cells as "self," meaning belonging to the organism. Therefore, the immune system can identify cells without those markers as being foreign, most likely *pathogens* (disease-causing organisms or viruses) (see Chapter 45).

A supportive cell wall surrounds the plasma membrane of fungal, plant, and many protist cells. (The cell wall makes it harder to break open the cells in those organisms than is the case with animal cells.) The cell wall is an *extracellular* structure (*extra* = outside) because it lies outside the plasma membrane. Although animal cells do not have cell walls, they also form extracellular material with supportive and other functions (see Section 4.5).

Figure 4.9A presents a diagram of a representative animal cell and **Figure 4.9B** presents a diagram of a representative plant cell to show where the nucleus, cytoplasmic organelles, and other structures are located. The following sections discuss the structure and function of eukaryotic cell parts in more detail, beginning with the nucleus.

The Eukaryotic Nucleus Is Much More Complex Than the Prokaryotic Nucleoid

The nucleus (see Figures 4.9A–B) is separated from the cytoplasm by the **nuclear envelope,** which consists of two lipid bilayer membranes, one just inside the other and separated by a narrow space **(Figure 4.10).** A network of protein filaments

A. Diagram of an animal cell, highlighting the major organelles and their primary locations (Not shown are flagellae and cilia, which are found in some cells.)

Present in animal cells but not plant cells
Centrosomes with centrioles
Lysosomes
Cilia

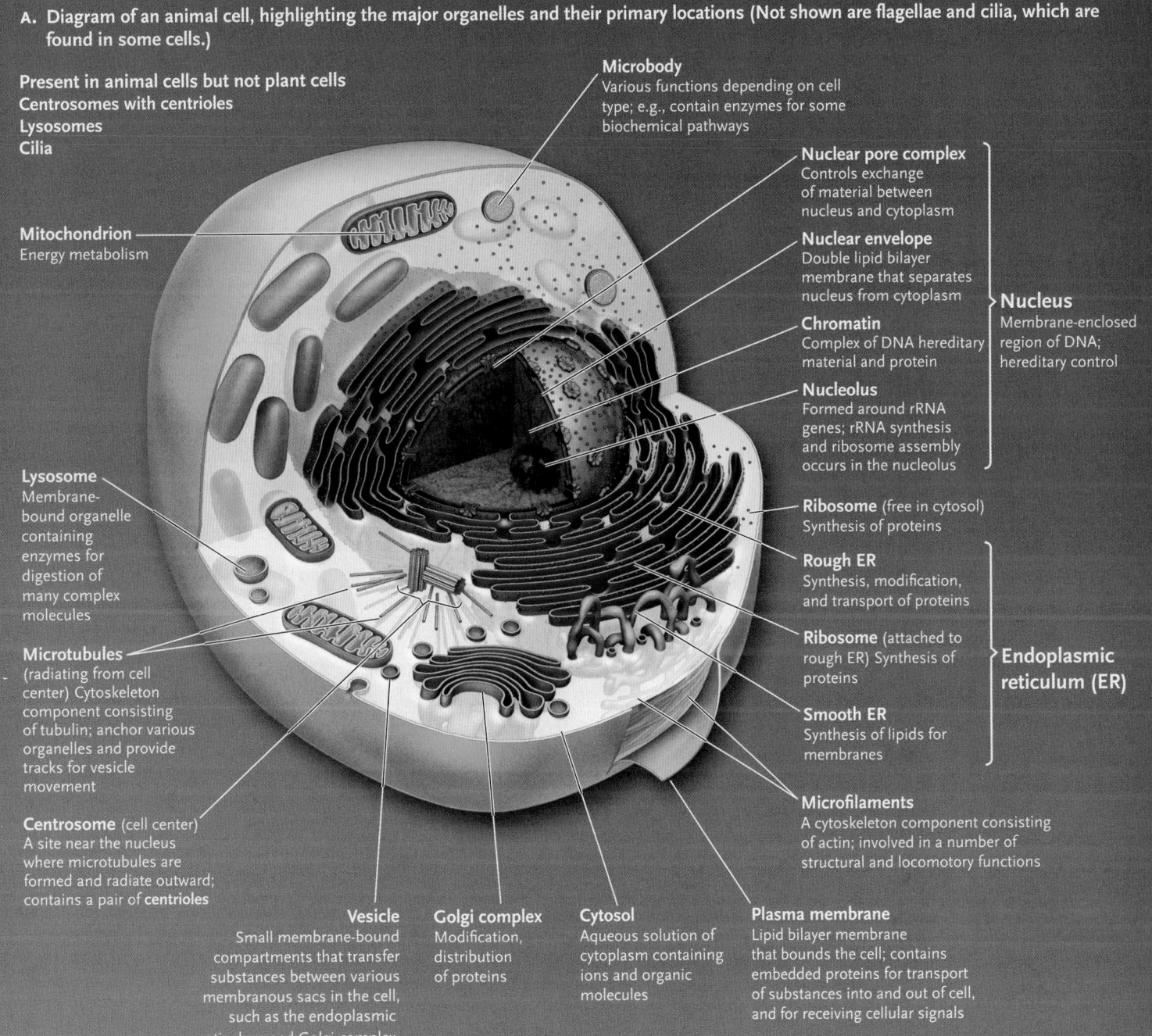

Microbody
Various functions depending on cell type; e.g., contain enzymes for some biochemical pathways

Mitochondrion
Energy metabolism

Nuclear pore complex
Controls exchange of material between nucleus and cytoplasm

Nuclear envelope
Double lipid bilayer membrane that separates nucleus from cytoplasm

Chromatin
Complex of DNA hereditary material and protein

Nucleus
Membrane-enclosed region of DNA; hereditary control

Nucleolus
Formed around rRNA genes; rRNA synthesis and ribosome assembly occurs in the nucleolus

Ribosome (free in cytosol)
Synthesis of proteins

Rough ER
Synthesis, modification, and transport of proteins

Ribosome (attached to rough ER) Synthesis of proteins

Smooth ER
Synthesis of lipids for membranes

Endoplasmic reticulum (ER)

Lysosome
Membrane-bound organelle containing enzymes for digestion of many complex molecules

Microtubules
(radiating from cell center) Cytoskeleton component consisting of tubulin; anchor various organelles and provide tracks for vesicle movement

Centrosome (cell center)
A site near the nucleus where microtubules are formed and radiate outward; contains a pair of **centrioles**

Microfilaments
A cytoskeleton component consisting of actin; involved in a number of structural and locomotory functions

Vesicle
Small membrane-bound compartments that transfer substances between various membranous sacs in the cell, such as the endoplasmic reticulum and Golgi complex

Golgi complex
Modification, distribution of proteins

Cytosol
Aqueous solution of cytoplasm containing ions and organic molecules

Plasma membrane
Lipid bilayer membrane that bounds the cell; contains embedded proteins for transport of substances into and out of cell, and for receiving cellular signals

FIGURE 4.9 Eukaryotic cells, highlighting the major organelles and their primary locations.

(Continued)

© Cengage Learning 2017

called *lamins* lines and reinforces the inner surface of the nuclear envelope in animal cells. Lamins are a type of intermediate filament (see later in this section). Evolutionarily unrelated proteins line the inner surface of the nuclear envelope in protists, fungi, and plants and carry out the same function.

Embedded in the nuclear envelope are many hundreds of nuclear pore complexes. A **nuclear pore complex** is a large, octagonally symmetric, cylindrical structure formed of many types of proteins, called **nucleoporins.** Probably the largest

protein complex in the cell, it exchanges components between the nucleus and cytoplasm and prevents the transport of material not meant to cross the nuclear membrane. A **nuclear pore**—a channel through the nuclear pore complex—is the path for the assisted exchange of large molecules such as proteins and RNA molecules with the cytoplasm, whereas small molecules simply pass through unassisted. A protein or RNA molecule (called the *cargo*) associates with a transport protein acting as a chaperone to shuttle the cargo through the pore.

B. Diagram of a plant cell, highlighting the major organelles and their primary locations

Present in plant cells but not animal cells
Cell wall (with plamodesmata)
Chloroplasts
Central vacuole

Cytosol

Mitochondrion

Golgi complex

Vesicle

Nuclear pore complex

Nuclear envelope

Chromatin

Nucleolus

Nucleus

Tonoplast
(central
vacuole
membrane)

Central vacuole
Cell growth, support,
and storage; contains
enzymes for digestion
of many complex
molecules

Plasmodesmata
Channels through
cell wall

Chloroplast
Photosynthesis;
some starch storage

Rough ER

Ribosome
(attached to
rough ER)

Endoplasmic reticulum

Microtubules

Smooth ER

Cell wall
Protection,
structural
support

Plasma membrane

Ribosome (free
in cytosol)

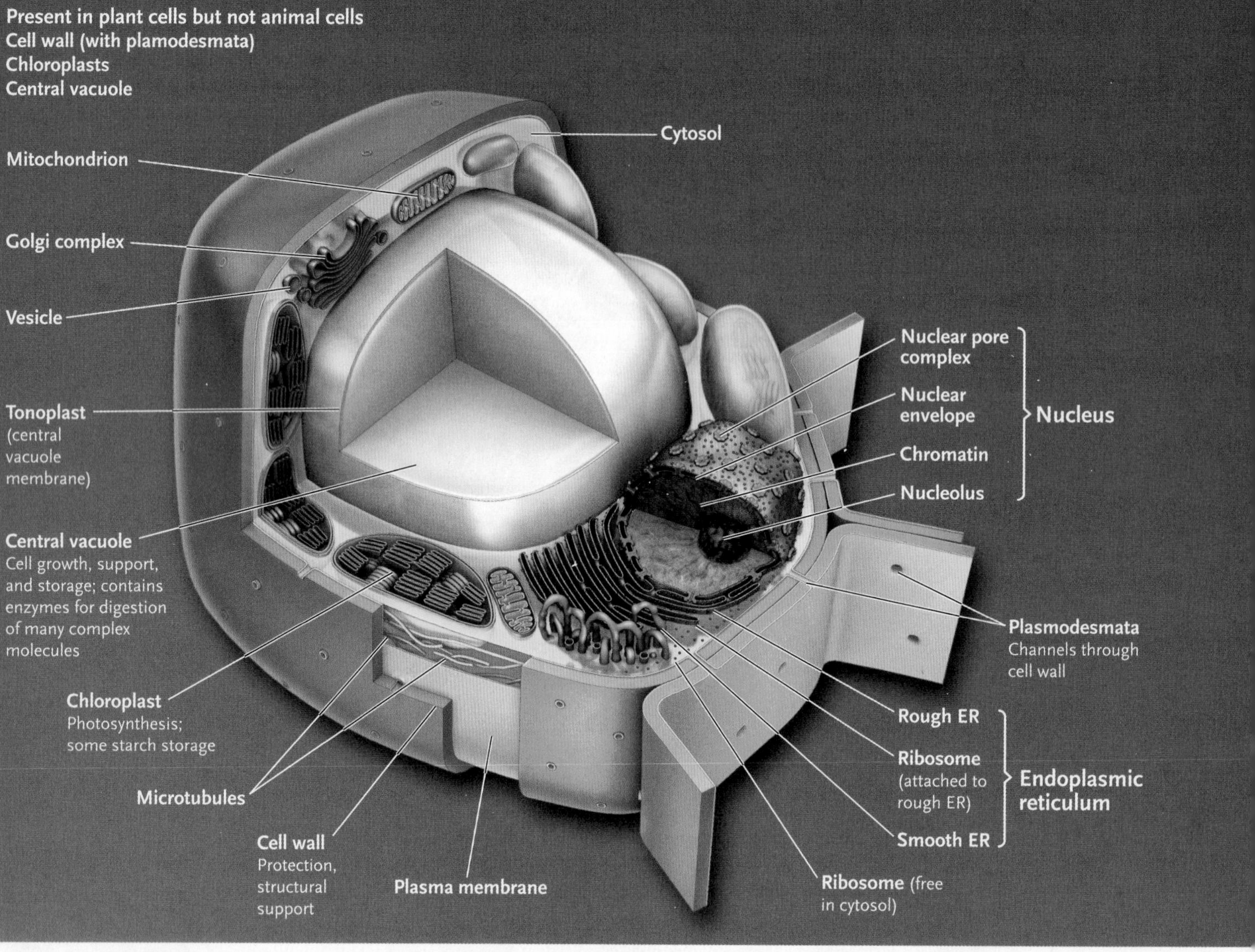

FIGURE 4.9 *(Continued)* **Eukaryotic cells, highlighting the major organelles and their primary locations.**
© Cengage Learning 2017

Some proteins—for instance, the enzymes for replicating and repairing DNA—must be imported into the nucleus to carry out their functions. Proteins to be imported into the nucleus are distinguished from those that function in the cytosol by the presence of a special, short amino acid sequence called a **nuclear localization signal.** A specific protein in the cytosol recognizes and binds to the signal and moves the protein containing it to the nuclear pore complex where it is then transported through the pore into the nucleus. **Figure 4.11** shows how researchers discovered the nuclear localization signal.

The liquid or semiliquid substance within the nucleus is the **nucleoplasm.** Most of the space inside the nucleus is filled with **chromatin,** a complex of DNA and proteins. By contrast with most prokaryotes, most of the nuclear DNA of a eukaryote is distributed among several to many linear molecules in the nucleus. Each individual DNA molecule with its associated

proteins is a **eukaryotic chromosome.** The terms *chromatin* and *chromosome* are similar but have distinct meanings. *Chromatin* refers to any collection of eukaryotic DNA molecules with their associated proteins. *Chromosome* refers to one complete DNA molecule with its associated proteins.

Eukaryotic nuclei contain much more DNA than do prokaryotic nucleoids. For example, the entire complement of 46 chromosomes in the nucleus of a human cell has a total DNA length of about 2 meters (m), compared with about 1,500 μm in prokaryotic cells with the most DNA. Some eukaryotic cells contain even more DNA; for example, a single frog nucleus, although of microscopic diameter, is packed with about 10 m of DNA.

A eukaryotic nucleus also contains one or more **nucleoli** (singular, *nucleolus*), which look like irregular masses of small fibers and granules (see Figure 4.9). These structures form

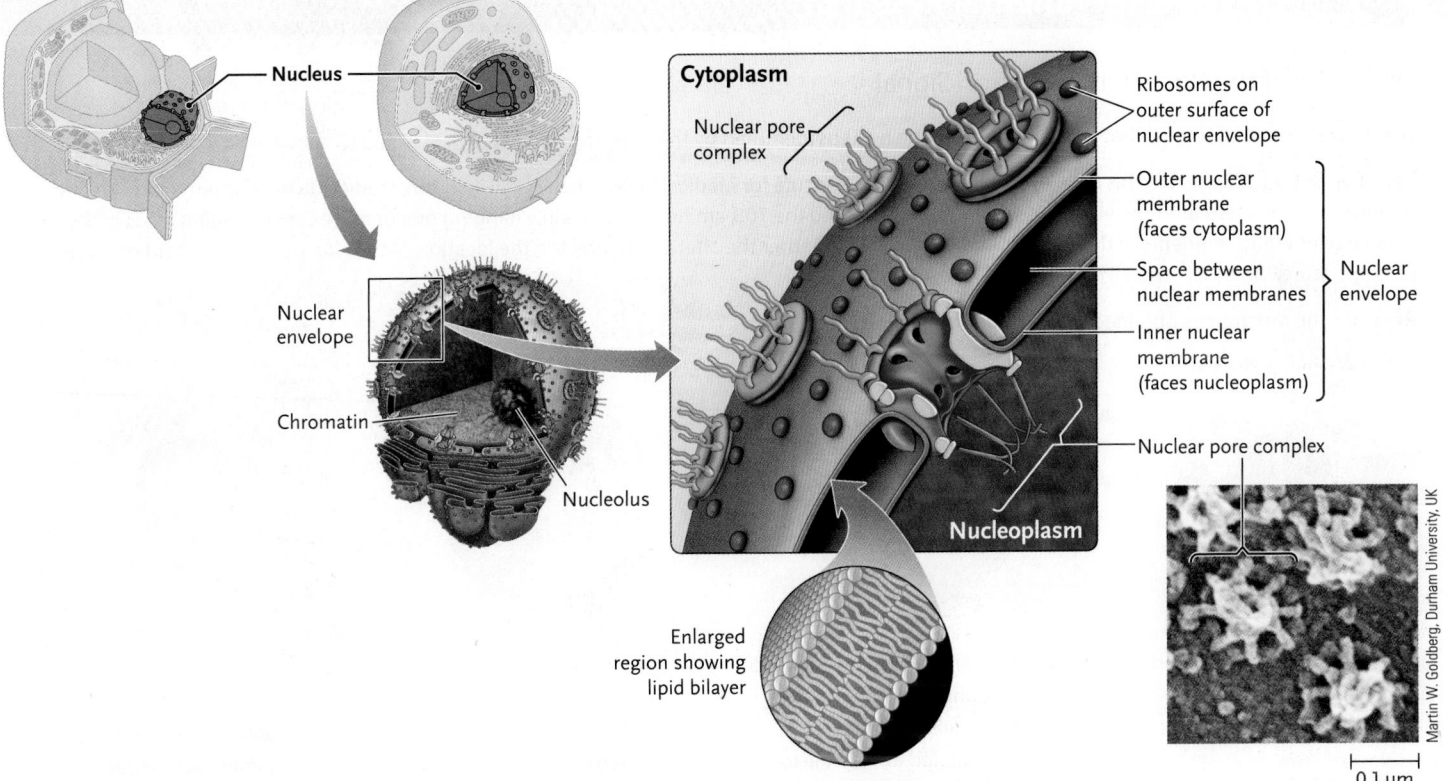

FIGURE 4.10 The nuclear envelope, which consists of a system of two concentric membranes with nuclear pore complexes embedded. Nuclear pore complexes are octagonally symmetric protein structures with a channel—the nuclear pore—through the center. They control the transport of molecules between the nucleus and cytoplasm.

© Cengage Learning 2017

around the genes coding for the rRNA molecules of ribosomes. Within the nucleolus, the information in rRNA genes is copied into rRNA molecules, which combine with proteins to form ribosomal subunits. The ribosomal subunits then leave the nucleoli and exit the nucleus through the nuclear pore complexes to enter the cytoplasm, where they join to form complete ribosomes on mRNAs (see Chapter 15).

The genes for most of the proteins that the organism can make are found within the chromatin, as are the genes for specialized RNA molecules such as rRNA molecules. Expression of these genes is carefully controlled as required for the function of each cell. The other proteins in the cell are specified by genes in the DNA of mitochondria and chloroplasts. Mitochondria and chloroplasts are discussed later in the chapter.

Eukaryotic Ribosomes Are Either Free in the Cytosol or Attached to Membranes

Like prokaryotic ribosomes, a eukaryotic ribosome consists of a large and a small subunit **(Figure 4.12)**. However, the structures of bacterial, archaeal, and eukaryotic ribosomes, although similar, are not identical. In general, eukaryotic ribosomes are larger than either bacterial or archaeal ribosomes; they contain four types of rRNA molecules and more than 80 proteins. Their

function is identical to that of prokaryotic ribosomes: They use the information in mRNA to assemble amino acids into proteins.

Some eukaryotic ribosomes are suspended freely in the cytosol; others are attached to membranes. Proteins made on free ribosomes in the cytosol may remain in the cytosol, pass through the nuclear pores into the nucleus, or become parts of mitochondria, chloroplasts, the cytoskeleton, or other cytoplasmic structures. Proteins that enter the nucleus become part of chromatin, line the nuclear envelope (the lamins), or remain in solution in the nucleoplasm.

Many ribosomes are attached to membranes. Some ribosomes are attached to the nuclear envelope, but most are attached to a network of membranes in the cytosol called the *endoplasmic reticulum* (ER) (described in more detail next). The proteins made on ribosomes attached to the ER follow a special path to other organelles within the cell (see Chapter 15).

An Endomembrane System Divides the Cytoplasm into Functional and Structural Compartments

Eukaryotic cells are characterized by an **endomembrane system** (*endo* = within), a collection of interrelated internal

FIGURE 4.11 | **Experimental Research**

Discovery of the Nuclear Localization Signal

Question: How are proteins that are imported into the nucleus identified by the import machinery?

Experiment: Alan Smith and his colleagues at the National Institute for Medical Research, Mill Hill, London, studied a viral protein that normally is found in the nucleus after the virus infects a cell. They mutated the 708-amino-acid protein, changing one or more specific amino acids in the protein or deleting segments of the protein, and determined whether the alterations affected the location of the viral protein in rodent or monkey cells in culture.

Results: The researchers obtained the following results:

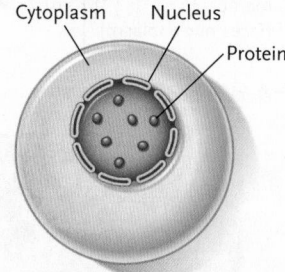 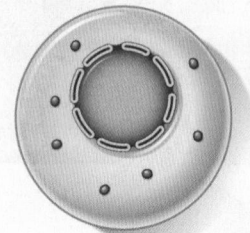 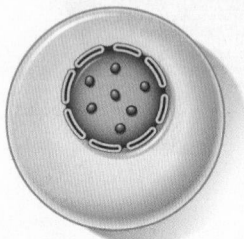 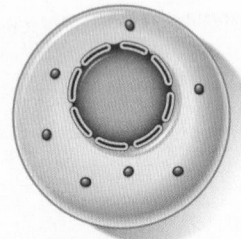

1. Normal protein: Localized to nucleus.

2. The amino acid at position 128 (thought to be important for the protein to bind to DNA) was mutated from lysine to threonine: Mutated protein localized in cytoplasm. The researchers interpreted this result to mean that the mutated amino acid was important for localizing the protein to the nucleus. Mutating other amino acids in the same region of the protein impaired import of the protein into the nucleus, but did not abolish it.

3. Deleting amino acids 1 to 126 (or any part of that region) or 136 to 708 (or any part of that region): Protein localized to nucleus, meaning that amino acids in those regions are not important for nuclear localization.

4. Deleting amino acids 127 to 133: Protein localized to cytoplasm, meaning that this amino acid sequence is necessary for nuclear localization of the viral protein. Other deletions involving parts of this region gave the same result.

Conclusion: By mutating the viral protein sequence, the researchers identified a seven-amino-acid segment of the protein, amino acids 127 to 133, that is necessary for localization of the protein to the nucleus. In follow-up experiments, they added this amino acid sequence to a cellular enzyme protein normally found only in the cytoplasm and showed that the modified protein localized to the nucleus. Therefore, the seven-amino-acid sequence is a nuclear localization signal. Continuing research has shown that similar sequences occur in other nuclear proteins. The identification of nuclear localization signals was a key step toward understanding the import of proteins into the nucleus.

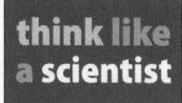 **think like a scientist** | Observe **Hypothesize** Predict Experiment Interpret | At the time of the discovery of the nuclear localization signal, the nature of the molecular system for the transport of nuclear proteins was not known. Imagine you are at that time and present a hypothesis for the molecular nature of the transport system.

Sources: D. Kalderon, W. D. Richardson, A. F. Markham, and A. E. Smith. 1984. Sequence requirements for nuclear location of simian virus 40 large-T antigen. *Nature* 31 1:33–38; D. Kalderon, B. L. Roberts, W. D. Richardson, and A. E. Smith. 1984. A short amino acid sequence able to specify nuclear location. *Cell* 39:499–509.

membranous sacs that divide the cell into functional and structural compartments. The membranes of the endomembrane system are lipid bilayers.

The endomembrane system has a number of functions, including the synthesis and modification of proteins and their transport into membranes and organelles or to the outside of the cell, the synthesis of lipids, and the detoxification of some toxins. The membranes of the system are connected either directly in the physical sense or indirectly by **vesicles,** which are small membrane-bound compartments that transfer substances between parts of the system (described later in Figure 4.17). Researchers refer to the substances as *cargo*.

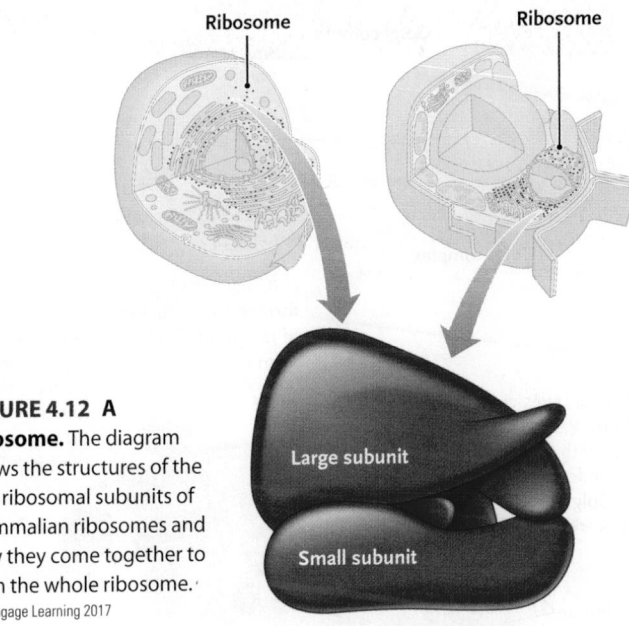

Ribosome   Ribosome

FIGURE 4.12 A ribosome. The diagram shows the structures of the two ribosomal subunits of mammalian ribosomes and how they come together to form the whole ribosome.
© Cengage Learning 2017

Large subunit

Small subunit

The components of the endomembrane system include the nuclear envelope, endoplasmic reticulum, Golgi complex, lysosomes, vesicles, and plasma membrane. The plasma membrane and the nuclear envelope are discussed earlier in this chapter. The functions of the other organelles are described in the following sections.

ENDOPLASMIC RETICULUM The **endoplasmic reticulum** (ER) is an extensive interconnected network (*reticulum* = little net) of membranous channels and vesicles called **cisternae** (singular, *cisterna*). Each cisterna is formed by a single membrane that surrounds an enclosed space called the **ER lumen (Figure 4.13).** The ER occurs in two forms:

1. The **rough ER** (see Figure 4.13A) gets its name from the ribosomes that stud its outer surface. The synthesis, processing, and trafficking (movement) of a large variety of proteins occurs on the rough ER. Those proteins represent about one third of the proteins encoded in the

A. Rough ER

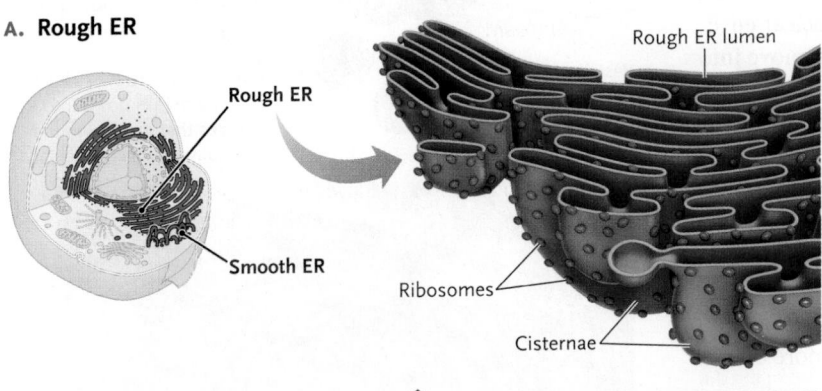

Rough ER

Smooth ER

Rough ER lumen

Ribosomes

Cisternae

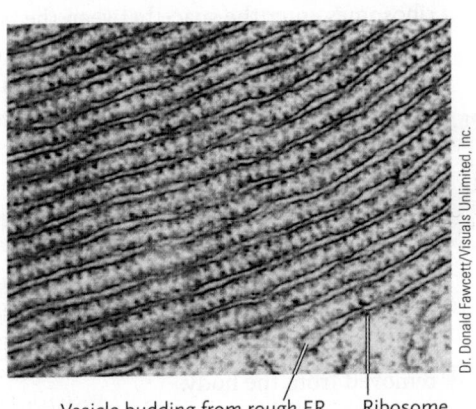

Vesicle budding from rough ER   Ribosome

Dr. Donald Fawcett/Visuals Unlimited, Inc.

Ribosomes stud the membrane surfaces facing the cytoplasm. Proteins synthesized on these ribosomes enter the lumen of the rough ER where they are modified chemically and then begin their path to their final destinations in the cell.

B. Smooth ER

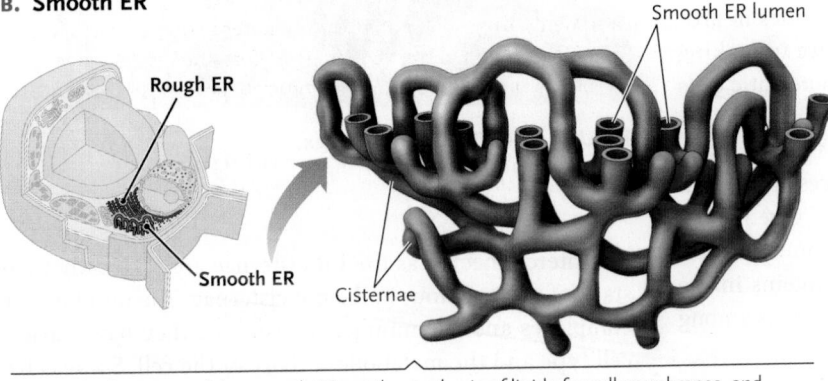

Rough ER

Smooth ER

Cisternae

Smooth ER lumen

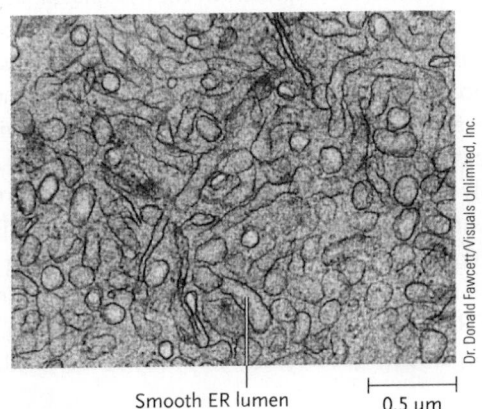

Smooth ER lumen   0.5 µm

Dr. Donald Fawcett/Visuals Unlimited, Inc.

Among the functions of the smooth ER are the synthesis of lipids for cell membranes, and enzymatic conversion of certain toxic molecules to safer molecules.

FIGURE 4.13 The endoplasmic reticulum. (A) Rough ER, showing the ribosomes that stud the membrane surfaces facing the cytoplasm. Proteins synthesized on these ribosomes enter the lumen of the rough ER where they are modified chemically and then begin their path to their final destinations in the cell. **(B)** Smooth ER membranes. Among their functions are the synthesis of lipids for cell membranes, and enzymatic conversion of certain toxic molecules to safer molecules.
© Cengage Learning 2017

human genome and include enzymes (see Chapter 6), receptors (see Chapter 9), ion channels and transporters (see Chapter 5), and hormones (see Chapter 42). The proteins are synthesized on the ribosomes attached to the ER and then enter the ER lumen. Processing then occurs in which the proteins are folded into their final form, and then chemically modified, such as by the addition of carbohydrate groups to produce glycoproteins. In trafficking, the proteins are then delivered to other regions of the cell within small vesicles that pinch off from the ER, travel through the cytosol, and join with the organelle that performs the next steps in their modification and distribution. For most of the proteins made on the rough ER, the next destination is the Golgi complex, which packages and sorts them for delivery to their final destinations.

The outer membrane of the nuclear envelope is closely related in structure and function to the rough ER, to which it is connected. This membrane is also a "rough" membrane, studded with ribosomes attached to the surface facing the cytoplasm. The proteins made on these ribosomes enter the space between the two nuclear envelope membranes. From there, the proteins can move into the ER and on to other cellular locations.

2. The **smooth ER** (see Figure 4.13B) is so called because its membranes have no ribosomes attached to their surfaces. The smooth ER has various functions in the cytoplasm, including synthesis of lipids that become part of cell membranes. In some cells, such as those of the liver, smooth ER membranes contain enzymes that convert drugs, poisons, and toxic by-products of cellular metabolism into substances that can be tolerated or more easily removed from the body.

The rough and smooth ER membranes are often connected, making the entire ER system a continuous network of interconnected channels in the cytoplasm. The relative proportions of rough and smooth ER reflect cellular activities in protein and lipid synthesis. Cells that are highly active in making proteins to be released outside the cell, such as pancreatic cells that make digestive enzymes, are packed with rough ER but have relatively little smooth ER. By contrast, cells that primarily synthesize lipids or break down toxic substances are packed with smooth ER but contain little rough ER. Current proteomics research is directed at characterizing the number, similarities, and differences in the populations of proteins in the ER of different cell types within an organism, as well as among organisms. (Proteomics was introduced in Chapter 1.)

GOLGI COMPLEX Camillo Golgi, a late-nineteenth-century Italian neuroscientist and Nobel laureate, discovered the **Golgi complex.** The Golgi complex consists of a stack of flattened, membranous sacs (without attached ribosomes) known as *cisternae* **(Figure 4.14).** In most cells, the complex looks like a stack of cupped pancakes, and like pancakes, they are separate sacs,

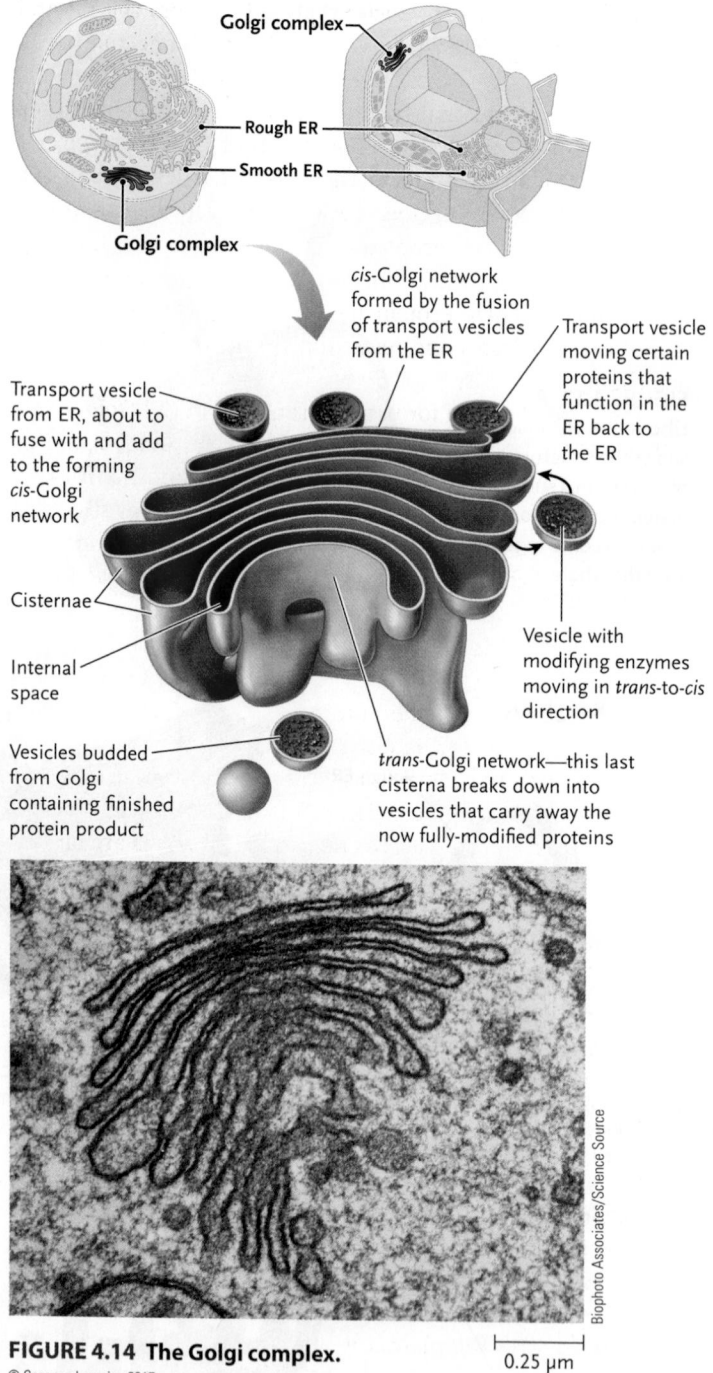

FIGURE 4.14 **The Golgi complex.**
© Cengage Learning 2017

0.25 μm

not interconnected as the ER cisternae are. Typically there is a stack of between four and eight cisternae. The number of Golgi complexes and the number of cisternae they have varies with cell type and the metabolic activity of the cell. Some cells have a single complex, whereas cells highly active in secreting proteins from the cell can have hundreds of complexes. Golgi complexes are usually located near concentrations of rough ER membranes, between the ER and the plasma membrane.

The Golgi complex receives proteins from the ER, then modifies and releases them. Newly synthesized proteins made in the ER that are destined to go to the Golgi complex are packaged into transport vesicles that bud off from the ER and move to the Golgi. Transport vesicles fuse with one another to form the *cis*-Golgi network, the cisterna that is closest to the ER. Certain proteins that function in the ER are moved to the ER in transport vesicles that bud off from the *cis*-Golgi network. The remaining proteins are transported across the stack of cisternae to the *trans*-Golgi network, the last cisterna, which faces the plasma membrane. During that transport, the proteins are modified to their final functional forms by enzymes within the cisternae. Protein modification includes the addition of carbohydrates and removing segments of the polypeptide chain.

The favored model for protein transport in the Golgi complex is cisternal maturation. In this model, the Golgi cisternae form anew from fusion of transport vesicles from the ER (see above), and then they mature progressively until they dissipate in the form of vesicles formed from the *trans*-Golgi network (see Figure 4.14). In other words, a new cisterna that forms gradually becomes the next cisterna in the stack as yet another cisterna is formed and the last *(trans)* cisterna breaks down into vesicles. The contents of the vesicles are kept separate from the cytosol by the vesicle membrane.

Vesicles also bud from cisternae that are internally located within the stack and move to and fuse with cisternae closer to the *cis* side. Their cargo consists of modifying enzymes of the Golgi complex. In this way, the modifying enzymes are moved constantly to their locations within the maturing Golgi complex where they are needed to modify the ER-derived proteins that are passing from the *cis* to the *trans* side.

The Golgi complex regulates the movement of several types of proteins. Some proteins are secreted from the cell, some become embedded in the plasma membrane as integral membrane proteins (see Chapter 5), some are transported back to the ER (see above), and some are placed in lysosomes (see below). The modifications of the proteins within the Golgi complex include adding "zip codes" to the proteins, which tags them for sorting to their final destinations. For instance, proteins secreted from the cell are transported to the plasma membrane in **secretory vesicles,** which release their contents to the exterior by **exocytosis (Figure 4.15A).** In this process, a secretory vesicle fuses with the plasma membrane and the vesicle contents are released to the outside. The contents of secretory vesicles vary, including signaling molecules such as peptide hormones (see Chapters 9 and 42) and neurotransmitters (see Chapters 9 and 39), waste products or toxic substances, and enzymes (such as from cells lining the intestine). The vesicle membrane may fuse with and become part of the plasma membrane. *Molecular Insights* describes a genome-wide research study that characterized the proteins involved in regulating the secretory pathway in human cells.

Vesicles in the cell also may form by the reverse process, called **endocytosis,** which brings molecules into the cell from the exterior **(Figure 4.15B).** In this process, the plasma membrane forms a pocket, which bulges inward and pinches off into the cytoplasm as an **endocytic vesicle.** Once in the

A. Exocytosis: A secretory vesicle fuses with the plasma membrane, releasing the vesicle contents to the cell exterior. The vesicle membrane may fuse with and become part of the plasma membrane.

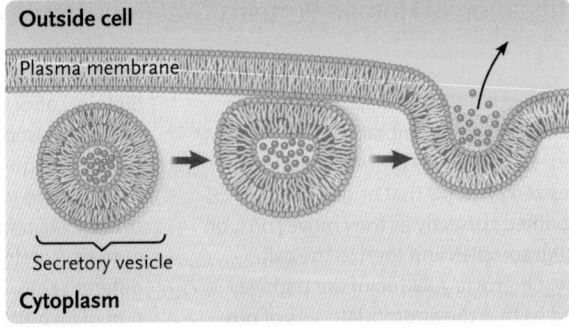

B. Endocytosis: Materials from the cell exterior are enclosed in a segment of the plasma membrane that pockets inward and pinches off as an endocytic vesicle.

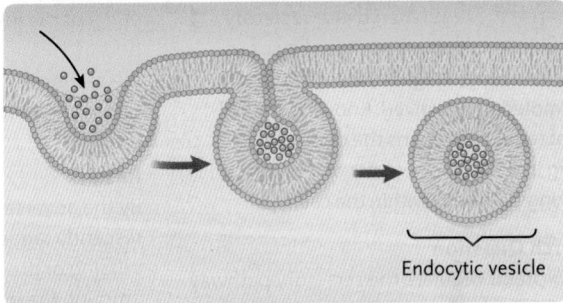

FIGURE 4.15 Exocytosis and endocytosis.
© Cengage Learning 2017

cytoplasm, endocytic vesicles, which contain segments of the plasma membrane as well as proteins and other molecules, are carried to the Golgi complex or to other destinations such as lysosomes in animal cells. The substances carried to the Golgi complex are sorted and placed into vesicles for routing to other locations, which may include lysosomes. Those routed to lysosomes are digested into molecular subunits that may be recycled as building blocks for the biological molecules of the cell. Exocytosis and endocytosis are discussed in more detail in Chapter 5.

LYSOSOMES Lysosomes (*lys* = breakdown; *some* = body) are small, membrane-bound vesicles that contain more than 30 enzymes for the hydrolysis (digestion) of many complex molecules, including proteins, lipids, nucleic acids, and polysaccharides **(Figure 4.16).** The cell recycles the subunits of these molecules. Lysosomes are found in animals and protists. A number of researchers support the existence of lysosomes in plants, although the issue is still a topic of debate. Those who deny the existence of lysosomes in plants believe that the functions of lysosomes in plants are carried out by the central vacuole (see Section 4.4). Yeasts have a lysosome-like vacuole that functions similarly to an animal cell lysosome. Depending on the contents they are digesting, lysosomes assume a variety of sizes and shapes instead of a uniform structure as is characteristic of other organelles. Most commonly, lysosomes are small

Identification of Human Proteins That Regulate Secretion Using a Genome-Wide Analysis

The secretory pathway has evolved to transfer proteins and lipids to internal membranes and to cell surface membranes. Functionally, the secretory pathway ensures that newly synthesized proteins that originate at the ER are modified correctly as they move through the Golgi complex and then to the cell surface. Careful regulation of the pathway is needed so that the appropriate sets of proteins and lipids are located in the various membrane-bound organelles. Without such tight regulation, the activities of the cell cannot proceed smoothly.

Because the regulation of the secretory pathway is so important, for many years researchers have sought to identify the regulatory molecules involved. Knowing the full array of regulators opens the door to investigating in detail the network of membrane trafficking pathways within the cell.

Research Question

What proteins regulate the secretory pathway in humans?

Experiment

Jeremy Simpson and colleagues at University College Dublin, Ireland, and Rainer Pepperkok and colleagues at the European Molecular Biology Laboratory (EMBL) in Heidelberg, Germany, used genome-wide analysis to answer the question. They used a molecular technique to interfere with the translation of the mRNAs transcribed from the ~19,900 human protein-coding genes individually in different cells. By interfering with the translation of an mRNA, its encoded protein could not be synthesized or could only be synthesized in much-lower-than-normal amounts. The researchers then used fluorescence microscopy (see Figure 4.4) to determine the effect of the loss of that protein on the transport of a fluorescently tagged protein through the secretory pathway from the ER to the plasma membrane.

Results

By the adverse effect on secretion of the fluorescently tagged protein when a protein was missing or present in a lower-than-normal amount, the investigators identified 554 proteins that influence secretion. Of those proteins, 143 either influence an early stage of the secretory pathway or morphology of the Golgi complex.

Conclusion

The research identified a large number of proteins involved in the secretory pathway of human cells. In addition, the study revealed some previously unknown links between the secretory network and some other cell systems, such as cytoskeleton organization. (The cytoskeleton is described later in the chapter.) The researchers conclude that the work "provides an important resource for an integrative understanding of global cellular organization and regulation of the secretory pathway in mammalian cells."

think like a scientist

Similar genome-wide analyses in the fruit fly, *Drosophila*, identified a number of genes that regulate secretion in that organism. Approximately 60% of those genes have sequences that are similar to the sequences of genes in the human genome, whereas the other approximately 40% have sequences that are not similar to the sequences of any human genes. Interpret this observation from an evolutionary standpoint.

Source: J. Simpson et al. 2012. Genome-wide RNAi screening identifies human proteins with a regulatory function in the early secretory pathway. *Nature Cell Biology* 14:764–774.

© Cengage Learning 2017

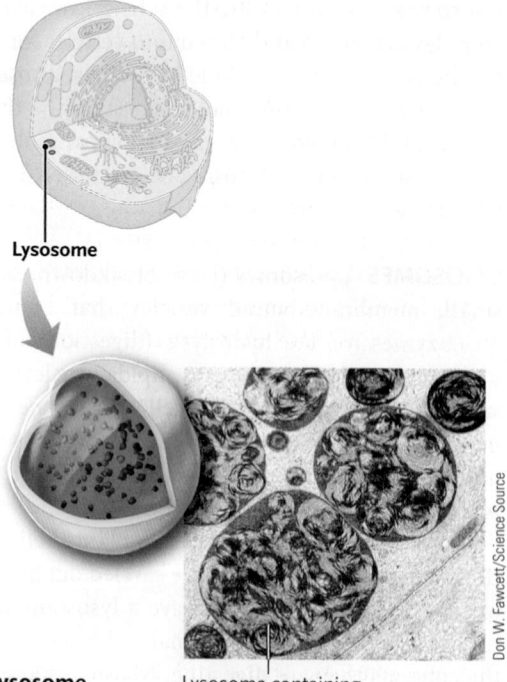

Lysosome

Lysosome containing ingested material

FIGURE 4.16 A lysosome.
© Cengage Learning 2017

Don W. Fawcett/Science Source

(0.1–0.5 μm in diameter) oval or spherical bodies. A human cell contains about 300 lysosomes.

Lysosomes are formed by budding from the Golgi complex. Their hydrolytic enzymes are synthesized in the rough ER, modified in the lumen of the ER to identify them as being bound for a lysosome, moved to the Golgi complex in a vesicle, and then packaged in the budding lysosome.

The pH within lysosomes is acidic (pH ~5) and is significantly lower than the pH of the cytosol (pH ~7.2). The hydrolytic enzymes in the lysosomes function optimally at the acidic pH within the organelle, but they do not function well at the pH of the cytosol; this difference reduces the risk to the viability of the cell should the enzymes be released from the vesicle.

Lysosomal enzymes can digest several types of materials. They digest food molecules entering the cell by endocytosis when an endocytic vesicle fuses with a lysosome. In a process called *autophagy*, they digest organelles that are not functioning correctly. A membrane surrounds the defective organelle, forming a large vesicle that fuses with one or more lysosomes; the organelle then is degraded by the hydrolytic enzymes. They also play a role in **phagocytosis**, a process in which some types of cells engulf bacteria or other cellular debris to break them down.

These cells include the white blood cells known as *phagocytes*, which play an important role in the immune system (see Chapter 45). Phagocytosis produces a large vesicle that contains the engulfed materials until lysosomes fuse with the vesicle and release the hydrolytic enzymes necessary for degrading them.

In certain human genetic diseases known as *lysosomal storage diseases*, one of the hydrolytic enzymes normally found in the lysosome is absent. As a result, the substrate of that enzyme accumulates in the lysosomes, and this accumulation eventually interferes with normal cellular activities. An example is Tay–Sachs disease, which is a fatal disease of the central nervous system caused by the failure to synthesize the enzyme needed for hydrolysis of fatty acid derivatives found in brain and nerve cells.

We have completed our survey of the components of the endomembrane system, including their interrelated structures and functions. **Figure 4.17** illustrates the vesicle traffic in the cytoplasm that involves the endomembrane system.

Mitochondria Are the Organelles in Which Some Reactions of Cellular Respiration Occur

MITOCHONDRIAL FUNCTION Mitochondria (singular, *mitochondrion*), located in the cells of all eukaryotic groups of organisms, are the membrane-bound organelles in which some of the reactions of cellular respiration occurs. *Cellular respiration* is the process by which energy-rich molecules such as sugars, fats, and other fuels are broken down to water and carbon dioxide by mitochondrial reactions, with the release of energy. Much of the energy released by the breakdown is captured in ATP. In fact, mitochondria generate most of the ATP of the cell. Mitochondria require oxygen for cellular respiration—when you breathe, you are taking in oxygen primarily for your mitochondrial reactions (see Chapter 7).

Chloroplasts are the sites of photosynthesis in plant cells and algae. Like mitochondria, chloroplasts convert energy, in this case solar energy into chemical energy. Here, too, energy is captured in ATP. The structure of chloroplasts is introduced in the next section, and photosynthesis (including its similarities to and differences from cellular respiration) is discussed in Chapter 8.

MITOCHONDRIAL STRUCTURE Mitochondria are enclosed by two lipid bilayer membranes (**Figure 4.18**). The **outer mitochondrial membrane** is smooth and covers the outside of the organelle. The surface area of the **inner mitochondrial membrane** is expanded by folds called **cristae** (singular, *crista*). Both membranes surround the innermost compartment of the mitochondrion, called the **mitochondrial matrix.** The ATP-generating reactions of mitochondria occur in the cristae and matrix.

The mitochondrial matrix also contains DNA, ribosomes, and other molecular machinery. Proteins encoded by the DNA are components of the enzyme machinery for the reactions of cellular respiration carried out by mitochondria. Other protein components of the enzyme machinery are encoded by nuclear genes and are imported into the organelle.

EVOLUTIONARY ORIGIN OF MITOCHONDRIA Mitochondrial DNA, ribosomes, and other molecular machinery resemble the equivalent structures in bacteria. These and other similarities led to the *endosymbiotic theory*, which proposes that mitochondria may have originated from a mutually advantageous relationship between an ingested prokaryotic cell and the prokaryotic cell that ingested it. The ingested prokaryotic cell evolved over time to become mitochondria. Chapter 25 discusses the endosymbiotic theory and the evolution of mitochondria (and chloroplasts).

Microbodies Carry Out Vital Reactions That Link Metabolic Pathways

Microbodies are small, relatively simple membrane-bound organelles found in various forms in essentially all eukaryotic cells. They consist of a single boundary lipid bilayer membrane that encloses a collection of enzymes and other proteins (**Figure 4.19**). Research has shown that the ER is involved in microbody production. Proteins and phospholipids are continuously imported into microbodies. The phospholipids are used for new membrane synthesis, leading to growth of the microbody. Division of a microbody then produces new microbodies.

Microbodies have various functions that are often specific to an organism or cell type. Commonly, they contain enzymes that conduct preparatory or intermediate reactions linking major biochemical pathways. For example, the series of reactions that allows cells to use fats as an energy source begins in microbodies and continues in mitochondria. Beginning or intermediate steps in the breakdown of some amino acids and alcohols also take place in microbodies, including about half of the ethyl alcohol that a human may consume. Many types of microbodies produce as a by-product the toxic substance hydrogen peroxide (H_2O_2), which is broken down into water and oxygen by the enzyme *catalase*. Microbodies with this reaction are often termed **peroxisomes.**

Microbodies in plants convert oils or fats to sugars that can be used directly for energy-releasing reactions in mitochondria or for reactions that require sugars as chemical building blocks. These microbody reactions are particularly important in plant embryos that develop from oily seeds, such as those of the peanut or soybean. Depending on the particular reaction pathways they carry out, plant microbodies are called *peroxisomes, glyoxysomes,* or *glycosomes.*

The Cytoskeleton Supports and Moves Cell Structures

The characteristic shape and internal organization of each type of cell is maintained in part by its cytoskeleton, the

The ER and Golgi complex are part of the endomembrane system, which releases proteins and other substances to the cell exterior and gathers materials from outside the cell.

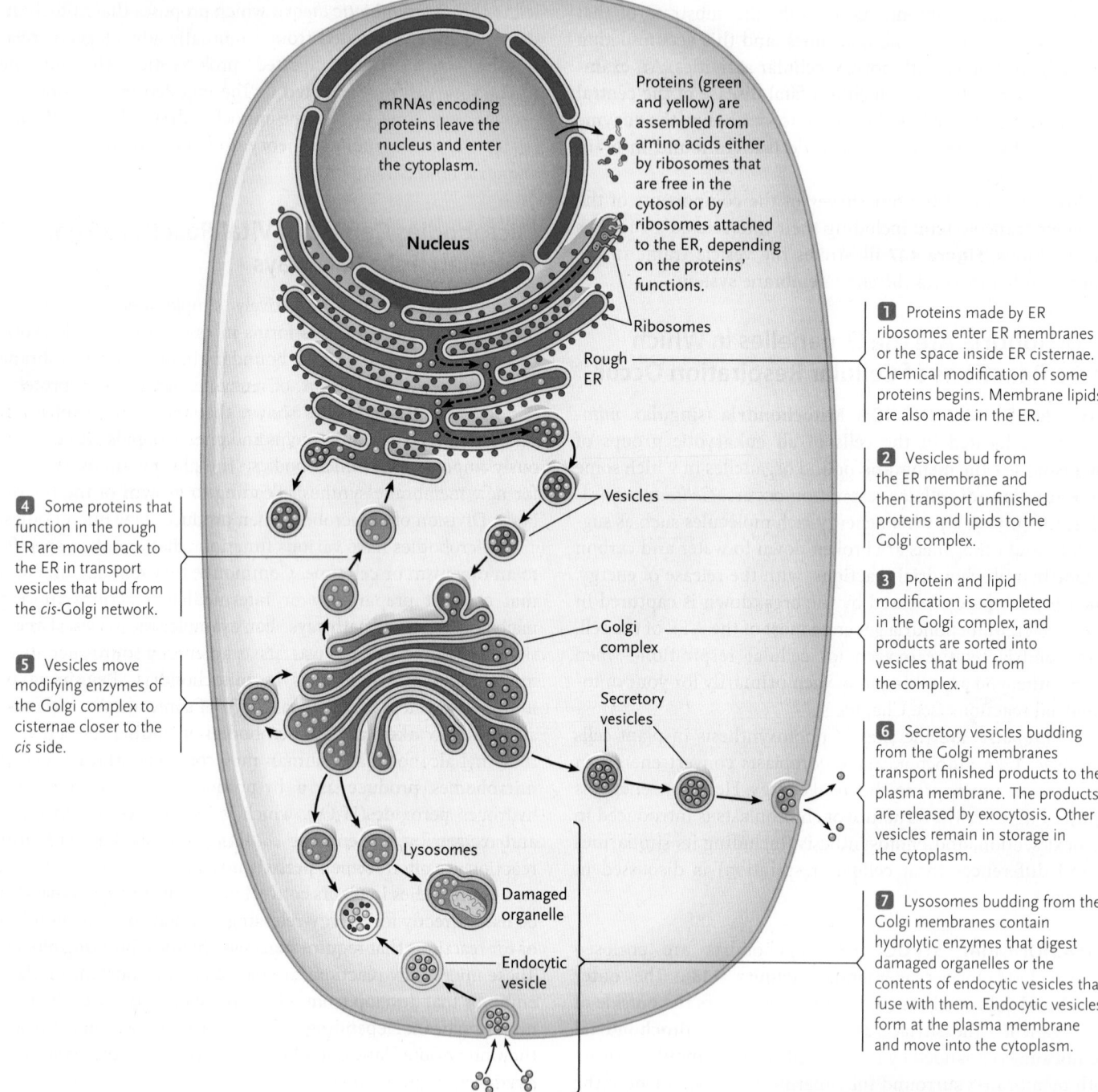

mRNAs encoding proteins leave the nucleus and enter the cytoplasm.

Proteins (green and yellow) are assembled from amino acids either by ribosomes that are free in the cytosol or by ribosomes attached to the ER, depending on the proteins' functions.

Nucleus

Ribosomes

Rough ER

Vesicles

Golgi complex

Secretory vesicles

Lysosomes

Damaged organelle

Endocytic vesicle

4 Some proteins that function in the rough ER are moved back to the ER in transport vesicles that bud from the *cis*-Golgi network.

5 Vesicles move modifying enzymes of the Golgi complex to cisternae closer to the *cis* side.

1 Proteins made by ER ribosomes enter ER membranes or the space inside ER cisternae. Chemical modification of some proteins begins. Membrane lipids are also made in the ER.

2 Vesicles bud from the ER membrane and then transport unfinished proteins and lipids to the Golgi complex.

3 Protein and lipid modification is completed in the Golgi complex, and products are sorted into vesicles that bud from the complex.

6 Secretory vesicles budding from the Golgi membranes transport finished products to the plasma membrane. The products are released by exocytosis. Other vesicles remain in storage in the cytoplasm.

7 Lysosomes budding from the Golgi membranes contain hydrolytic enzymes that digest damaged organelles or the contents of endocytic vesicles that fuse with them. Endocytic vesicles form at the plasma membrane and move into the cytoplasm.

SUMMARY The endomembrane system is a major traffic network for proteins and other substances within the cell. The Golgi complex in particular is a key distribution station for membranes and proteins. From the Golgi complex, lipids and proteins may move to storage or secretory vesicles, and from the secretory vesicles, they may move to the cell exterior by exocytosis. Membranes and proteins may also move between the nuclear envelope and the endomembrane system. Proteins and other materials that enter cells by endocytosis also enter the endomembrane system to travel to the Golgi complex for sorting and distribution to other locations. Lysosomes, originating as vesicles from the Golgi complex, digest damaged organelles or the contents of endocytic vesicles.

think like a scientist

Consider an animal cell that is secreting a particular protein. You treat cells with a chemical and find that the protein no longer is secreted. What possible steps could be affected by the chemical?

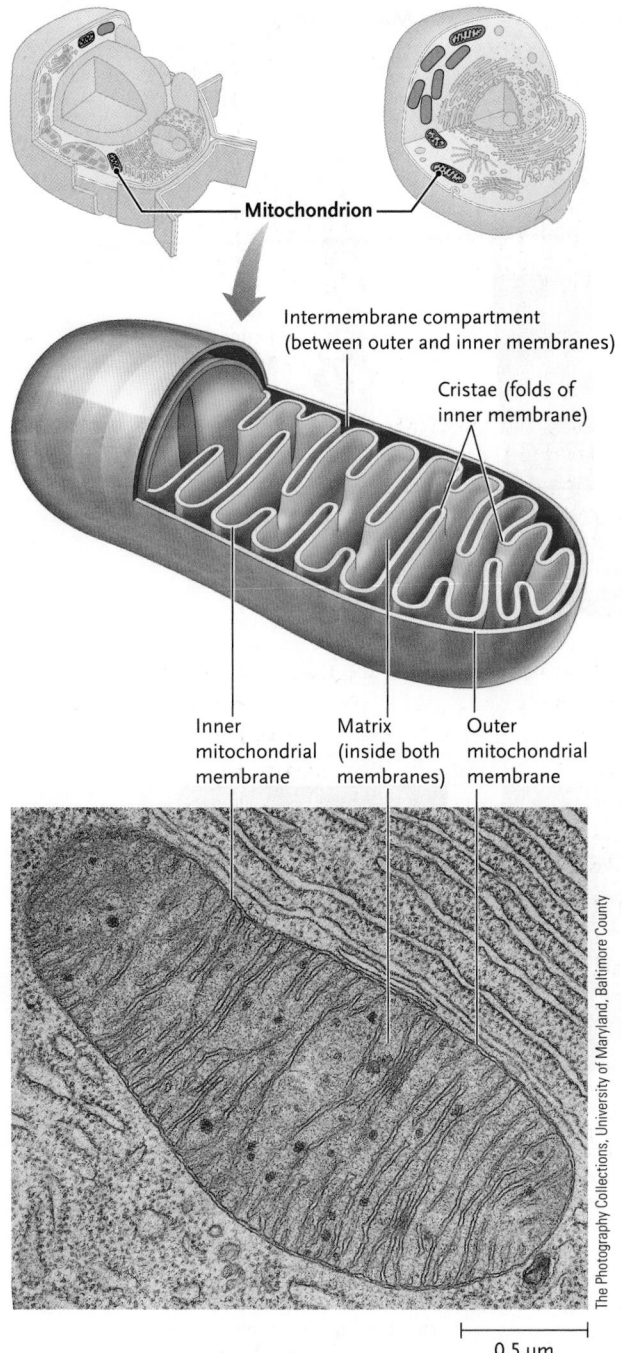

Mitochondrion

Intermembrane compartment
(between outer and inner membranes)

Cristae (folds of
inner membrane)

Inner
mitochondrial
membrane

Matrix
(inside both
membranes)

Outer
mitochondrial
membrane

0.5 μm

FIGURE 4.18 Mitochondrion. The EM shows a mitochondrion from bat pancreas, surrounded by cytoplasm containing rough ER. Cristae extend into the interior of the mitochondrion as folds from the inner mitochondrial membrane.

© Cengage Learning 2017

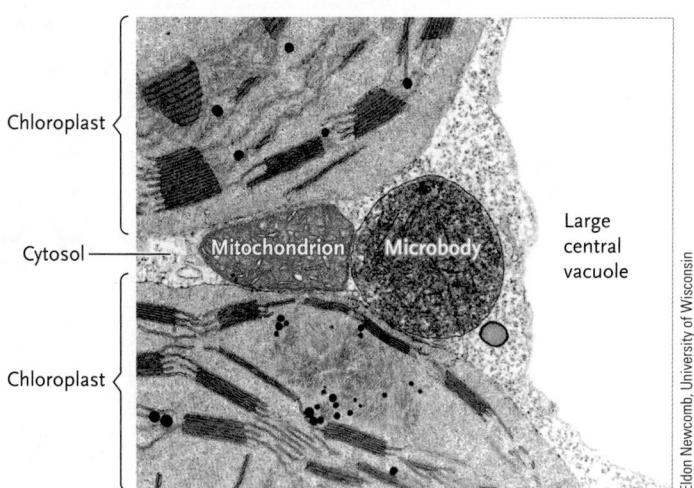

Chloroplast

Cytosol

Mitochondrion

Microbody

Large
central
vacuole

Chloroplast

Eldon Newcomb, University of Wisconsin

FIGURE 4.19 Microbody. EM of a microbody in the cytoplasm of a tobacco leaf. The EM has been color-enhanced to make the structures easier to identify.

cellular support in plants is provided by the cell wall and a large central vacuole (described in Section 4.4).

The cytoskeleton of animal and plant cells contains structural elements of three major types: *microtubules, intermediate filaments,* and *microfilaments* **(Figure 4.20).** Microtubules are the largest cytoskeletal elements, and microfilaments are the smallest. Each cytoskeletal element is assembled from proteins—microtubules from *tubulins,* intermediate filaments from a large and varied group of *intermediate filament proteins,* and microfilaments from *actins.*

Microtubules (see Figure 4.20A) are tubes with an outer diameter of about 25 nm and an inner diameter of about 15 nm. They vary widely in length from less than 200 nm to several micrometers. The wall of the microtubule consists of 13 protein filaments arranged side by side. Each filament is a linear polymer of tubulin dimers organized head-to-tail in each filament, each dimer consisting of one α-tubulin and one β-tubulin subunit. The end of a microtubule filament with α-tubulin subunits is the − (minus) end; the other end with the β-tubulin subunits is the + (plus) end, at the ends of the filaments. Microtubules change their lengths as required by their functions, such as when animal cells are changing shape. The length change occurs by the asymmetric addition or removal of tubulin dimers, with dimers adding or detaching more rapidly at the + end than at the − end. The lengths of microtubules are tightly regulated in the cell.

Many of the cytoskeletal microtubules in animal cells are formed and radiate outward from a site near the nucleus termed the **cell center** or **centrosome** (see Figure 4.9). At its midpoint are two short, barrel-shaped structures also formed from microtubules called the **centrioles** (see Figure 4.24). The + ends of microtubules are distal to the centrosome, and the − ends are proximal. Often, intermediate filaments also extend from the cell center, apparently held in the same radiating pattern by linkage to microtubules.

interconnected system of protein tubes and fibers that extends throughout the cytoplasm. The cytoskeleton also reinforces the plasma membrane and functions in movement, both of structures within the cell and of the cell as a whole. It is most highly developed in animal cells, in which it fills and supports the cytoplasm from the plasma membrane to the nuclear envelope. Although cytoskeletal structures are also present in plant cells, the fibers and tubes of the system are less prominent; much of

A. Microtubule

Structure

β-tubulin
α-tubulin

Tubulin dimers

+ end

− end

Thirteen filaments side by side in a microtubule

15 nm

25 nm

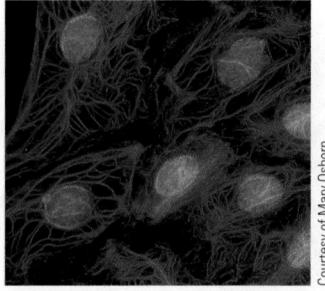

Microtubules (yellow)

Jennifer C. Waters/Science Source

- 25-nm diameter tubes formed from tubulin dimers arranged in 13 side-by-side filaments.

Function

- Anchor various membranous organelles (e.g., ER, Golgi)
- Provide tracks for vesicle movement
- Separating and moving chromosomes during cell division
- Maintaining animal cell shape
- Moving animal cells
- Determining orientation for growth of new cell wall during plant cell division

B. Intermediate filament

Structure

Each green line is an intermediate filament protein

8–12 nm

Courtesy of Mary Osborn

Intermediate filaments (red); nucleus is stained blue

- 8- to 12-nm diameter fibers formed from intermediate filament proteins with tissue-specific protein composition.

Function

- Structural support in many cells and tissues

C. Microfilament

Structure

+ end

Actin subunit

− end

5–7 nm

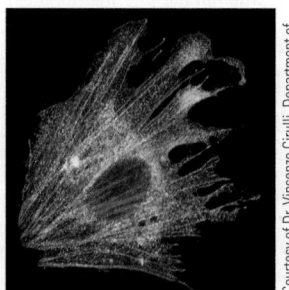

Microfilaments (red)

Courtesy of Dr. Vincenzo Cirulli, Department of Medicine, University of Washington, Seattle, WA

- 5- to 7-nm fibers consisting of two polymers of actin wound around each other in a helical spiral.

Function

- Involved in a number of structural and locomotor functions such as muscle contraction, cytoplasmic streaming, and dividing the cytoplasm during animal cell division

FIGURE 4.20 The major components of the cytoskeleton. The cells are visualized using fluorescence microscopy (see Figure 4.4). (Also see chapter-opening confocal light micrograph.)

Microtubules have various functions, as listed in Figure 4.20A. One of those functions is moving animal cells, which is particularly important in animal development (see Chapter 50). Animal cell movements are generated by "motor" proteins that push or pull against microtubules or microfilaments, much as our muscles produce body movements by acting on bones of the skeleton. One end of a motor protein is firmly fixed to a cell structure such as a vesicle or to a microtubule or microfilament. Using energy from ATP hydrolysis, the other end "walks" along another microtubule or microfilament by making an attachment, forcefully swiveling a short distance, and then releasing (**Figure 4.21**). The motor proteins that walk along microtubules are *dyneins* and *kinesins,* and the ones that walk along microfilaments are *myosins*. Some cell movements, such as the whipping motions of sperm tails, depend entirely on microtubules and their motor proteins (see later discussion of flagella and cilia).

Dyneins and kinesins transport organelles throughout an animal cell in a centrosome-dependent fashion. The orientation of microtubules is set by the centrosome, so transport away from or toward the centrosome depends on the motor protein. For example, the ER is transported through the cytoplasm by kinesin, which moves toward the + ends of microtubules, which are distal from the centrosome. As a result, the ER becomes distributed throughout the cytoplasm. Similarly, secretory vesicles budded off of the *trans* side of the Golgi complex and destined to fuse with the plasma membrane are moved along microtubules by kinesin. By contrast, dynein transports the Golgi complex toward the − ends of microtubules so that the organelle becomes located near the centrosome.

Intermediate filaments (see Figure 4.20B) are fibers with diameters of about 8 to 12 nm. ("Intermediate" signifies, in fact, that these filaments are intermediate in size between microtubules and microfilaments.) Found only in multicellular organisms, intermediate filaments occur singly, in parallel bundles, and in interlinked networks, either alone or in combination with microtubules, microfilaments, or both. Intermediate filaments consist of proteins that are tissue-specific in their composition. Despite the molecular diversity of intermediate filaments, however, they all play similar roles in the cell, providing structural support in many cells and tissues. For example, the nucleus in epithelial cells is held within the cell by a basketlike network of intermediate filaments made of keratins. Keratins are also found in animal hair, nails, and claws.

Microfilaments (see Figure 4.20C) are thin protein fibers 5 to 7 nm in diameter that consist of two polymers of actin subunits wound around each other in a long helical spiral. The actin subunits are asymmetric in shape, and they are all oriented in the same way in the polymer chains of a microfilament. Thus, as for microtubules, the two ends are designated + (plus) and − (minus). And as for microtubules, growth and disassembly occur more rapidly at the + end than at the − end.

Microfilaments occur in almost all eukaryotic cells and are involved in a number of structural and locomotor processes (see Figure 4.20C). They are best known as one of the two components of the contractile elements in muscle fibers of vertebrates (the roles of myosin and microfilaments in muscle contraction are discussed in Chapter 43). Microfilaments are involved in the actively flowing motion of cytoplasm called *cytoplasmic streaming,* which can transport nutrients, proteins, and organelles in both animal and plant cells, and which is responsible for amoeboid movement. When animal cells divide, microfilaments are responsible for dividing the cytoplasm (see Chapter 10).

FIGURE 4.21 The microtubule motor protein kinesin.
(A) Structure of the end of a kinesin molecule that "walks" along a microtubule, with α-helical segments shown as spirals and β strands as flat ribbons.
(B) How a kinesin molecule walks along the surface of a molecule by alternately attaching and releasing its "feet."
© Cengage Learning 2017

A. "Walking" end of a kinesin molecule

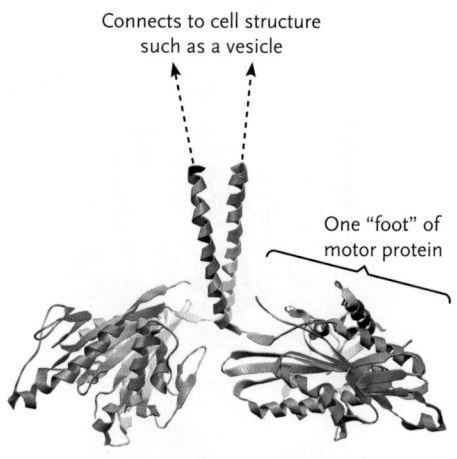

Connects to cell structure such as a vesicle

One "foot" of motor protein

B. How a kinesin molecule "walks"

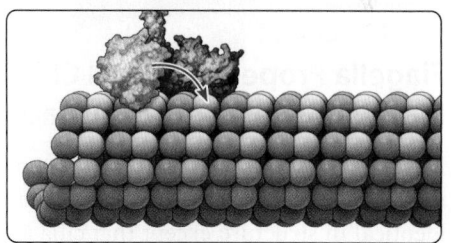

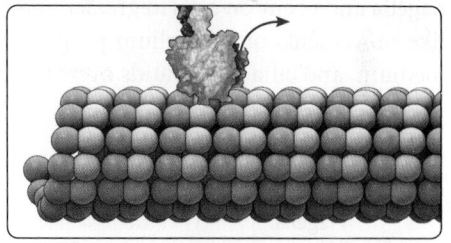

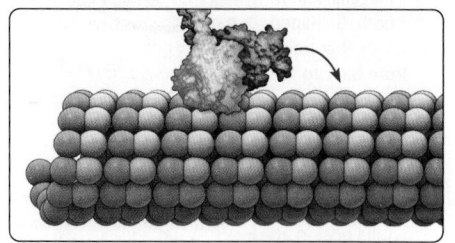

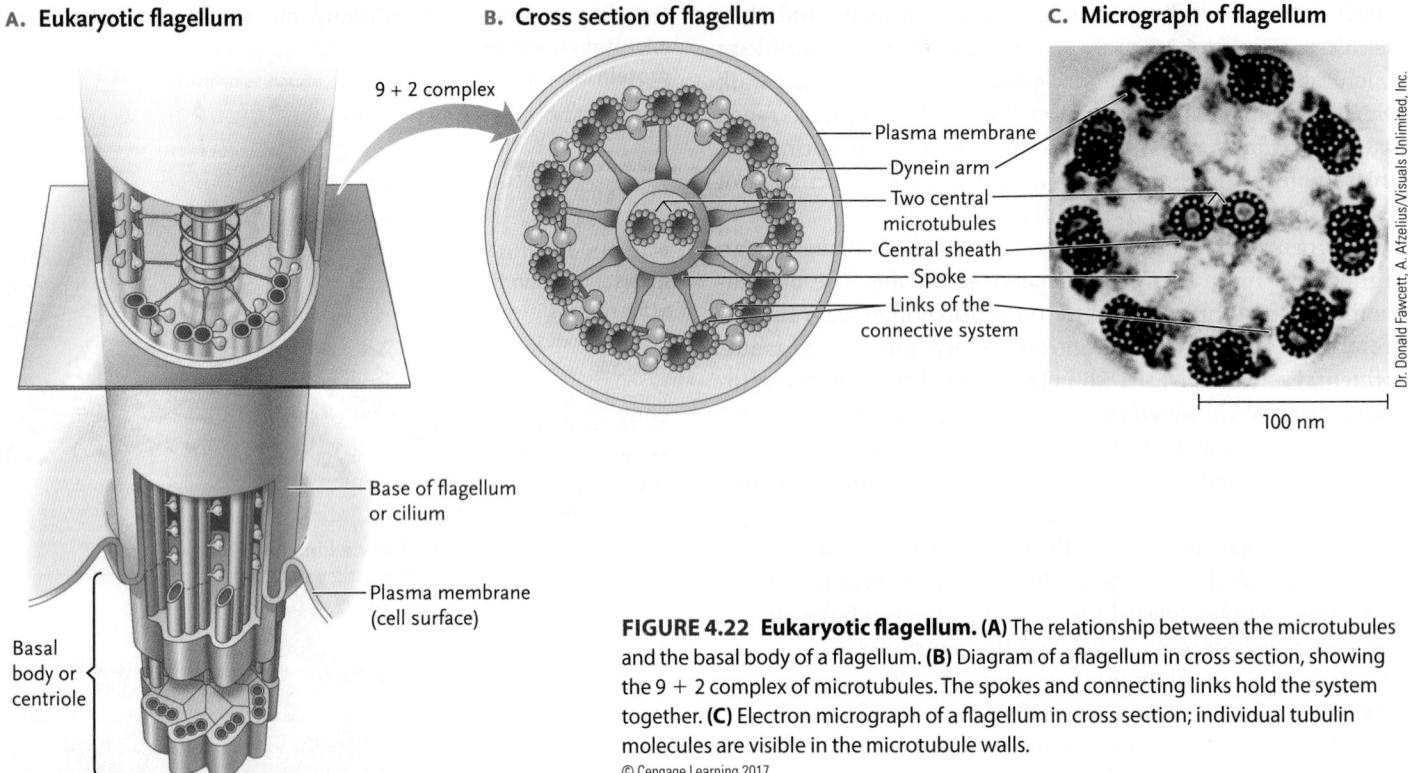

A. Eukaryotic flagellum

9 + 2 complex

Base of flagellum or cilium

Plasma membrane (cell surface)

Basal body or centriole

B. Cross section of flagellum

Plasma membrane
Dynein arm
Two central microtubules
Central sheath
Spoke
Links of the connective system

C. Micrograph of flagellum

100 nm

Dr. Donald Fawcett, A. Afzelius/Visuals Unlimited, Inc.

FIGURE 4.22 Eukaryotic flagellum. (A) The relationship between the microtubules and the basal body of a flagellum. **(B)** Diagram of a flagellum in cross section, showing the 9 + 2 complex of microtubules. The spokes and connecting links hold the system together. **(C)** Electron micrograph of a flagellum in cross section; individual tubulin molecules are visible in the microtubule walls.
© Cengage Learning 2017

Flagella Propel Cells, and Cilia Move Materials over the Cell Surface

Flagella and **cilia** (singular, *cilium*) are elongated, slender, motile structures that extend from the cell surface. They are identical in structure except that cilia are usually shorter than flagella and occur on cells in greater numbers. Whiplike or oarlike movements of a flagellum propel a cell through a watery medium, and cilia move fluids over the cell surface.

A bundle of microtubules extends from the base to the tip of a flagellum or cilium **(Figure 4.22)**. In the bundle, a circle of nine double microtubules surrounds a central pair of single microtubules, forming the 9 + 2 complex. Dynein motor proteins slide the microtubules of the 9 + 2 complex over each other to produce the movements of a flagellum or cilium **(Figure 4.23)**.

Flagella and cilia arise from the centrioles. The barrel-shaped centrioles contain a bundle of microtubules similar to

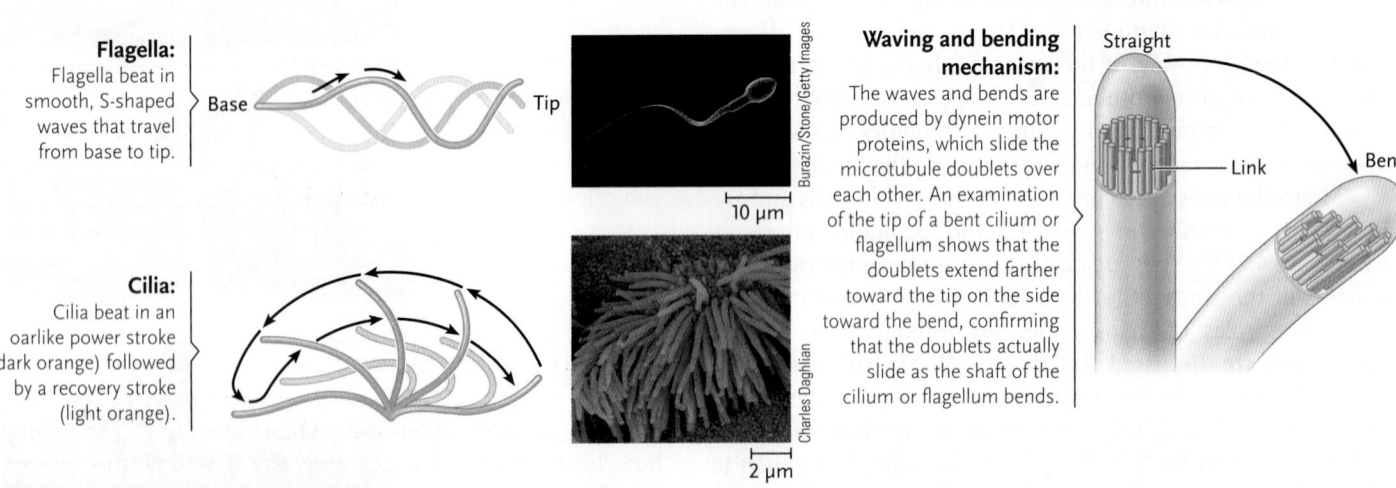

Flagella:
Flagella beat in smooth, S-shaped waves that travel from base to tip.

Base

Tip

Burazin/Stone/Getty Images

10 μm

Cilia:
Cilia beat in an oarlike power stroke (dark orange) followed by a recovery stroke (light orange).

Charles Daghlian

2 μm

Waving and bending mechanism:
The waves and bends are produced by dynein motor proteins, which slide the microtubule doublets over each other. An examination of the tip of a bent cilium or flagellum shows that the doublets extend farther toward the tip on the side toward the bend, confirming that the doublets actually slide as the shaft of the cilium or flagellum bends.

Straight

Link

Bent

FIGURE 4.23 Flagellar and ciliary beating patterns. The micrographs show a few human sperm, each with a flagellum (top), and cilia from the lining of an airway in the lungs (bottom).
© Cengage Learning 2017

the 9 + 2 complex, except that the central pair of microtubules is missing and the outer circle is formed from a ring of nine triple rather than double microtubules (compare Figure 4.23 and **Figure 4.24**). During the formation of a flagellum or cilium, a centriole moves to a position just under the plasma membrane. Then two of the three microtubules of each triplet grow outward from one end of the centriole to form the ring of nine double microtubules. The two central microtubules of the 9 + 2 complex also grow from the end of the centriole, but without direct connection to any centriole microtubules. The centriole remains at the innermost end of a flagellum or cilium when its development is complete as the **basal body** of the structure (see Figure 4.22).

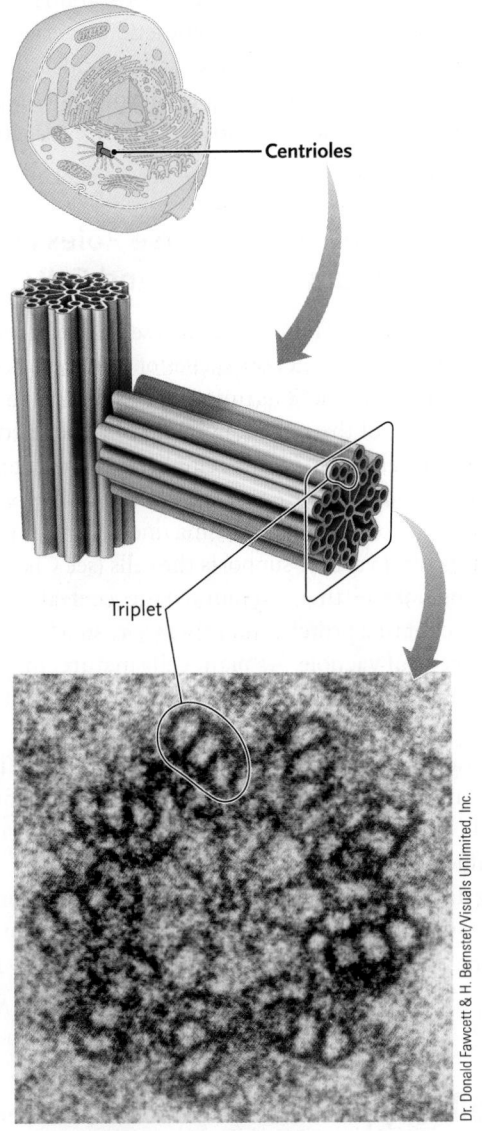

FIGURE 4.24 Centrioles. The two centrioles of the pair at the cell center usually lie at right angles to each other as shown. The electron micrograph shows a centriole from a mouse cell in cross section. A centriole gives rise to the 9 + 2 complex of a flagellum and persists as the basal body at the inner end of the flagellum.

© Cengage Learning 2017

Cilia and flagella are found in protozoa and algae, and many types of animal cells have flagella—the tail of a sperm cell is a flagellum—as do the reproductive cells of some plants. In humans, cilia cover the surfaces of cells lining cavities or tubes in some parts of the body. For example, cilia on cells lining the ventricles (cavities) of the brain circulate fluid through the brain, and cilia in the oviducts conduct eggs from the ovaries to the uterus. Cilia covering cells that line the air passages of the lungs sweep out mucus containing bacteria, dust particles, and other contaminants.

Although the purpose of the eukaryotic flagellum is the same as that of prokaryotic flagella, the genes that encode the components of the flagellar apparatus of cells of Domains Bacteria, Archaea, and Eukarya are different in each case. Thus, as mentioned earlier in the chapter, the three types of flagella are analogous, not homologous, structures, and they must have evolved independently.

STUDY BREAK 4.3

1. Where in a eukaryotic cell is DNA found? How is that DNA organized?
2. What is the nucleolus, and what is its function?
3. Explain the structure and function of the endomembrane system.
4. What is the structure and function of a mitochondrion?
5. What is the structure and function of the cytoskeleton?

THINK OUTSIDE THE BOOK

On your own or collaboratively, explore the Internet and the research literature to develop an outline of the molecular steps that a protein with a nuclear localization signal follows for nuclear import (that is, being transported through a nuclear pore complex).

4.4 Specialized Structures of Plant Cells

Chloroplasts, a large and highly specialized central vacuole, and cell walls are three structures that give plant cells their distinctive characteristics. Two of these structures also occur in some other eukaryotes—chloroplasts in algal protists (see Chapter 27) and cell walls in algal protists and fungi (see Chapter 30).

Chloroplasts Are Biochemical Factories Powered by Sunlight

Chloroplasts (*chloro* = yellow-green), the sites of photosynthesis in plant cells, are members of a family of plant organelles known collectively as **plastids.** Other members of the family include amyloplasts and chromoplasts. **Amyloplasts** (*amylo* = starch) are colorless plastids that store starch, a product of photosynthesis. They occur in great numbers in the roots or tubers of some plants, such as the potato. **Chromoplasts** (*chromo* =

color) contain red and yellow pigments and are responsible for the colors of ripening fruits or autumn leaves.

All plastids contain DNA genomes and molecular machinery for gene expression and the synthesis of proteins on ribosomes. For example, chloroplast-encoded proteins are components of the enzyme machinery for photosynthesis and for some of the proteins of chloroplast ribosomes. The other protein components of the enzyme machinery and of the ribosomes are encoded by nuclear genes and are imported into the organelle.

Chloroplasts, like mitochondria, are usually lens or disc shaped and are surrounded by a smooth **outer boundary membrane** and an **inner boundary membrane,** which lies just inside the outer membrane **(Figure 4.25).** These two boundary membranes completely enclose an inner compartment, the stroma. Within the stroma is a third membrane system that consists of flattened, closed sacs called **thylakoids.** In higher plants, the thylakoids are stacked, one on top of another, forming structures called **grana** (singular, *granum*).

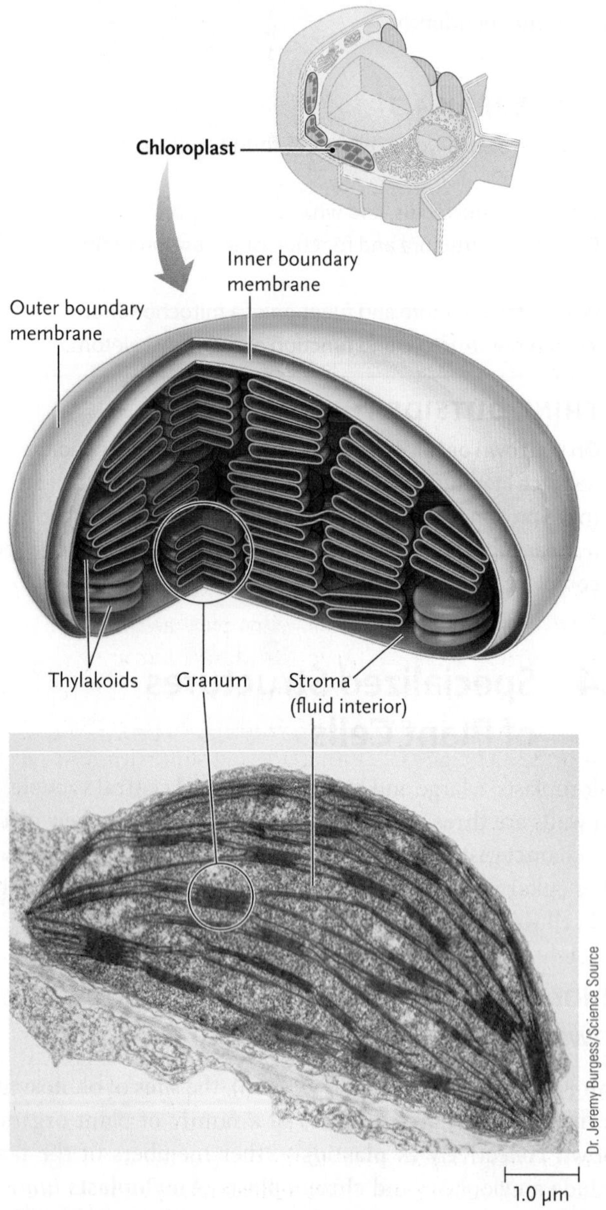

The thylakoid membranes contain molecules that absorb light energy and convert it to chemical energy in photosynthesis (see Chapter 8). The primary molecule absorbing light is *chlorophyll,* a green pigment that is present in all chloroplasts. The chemical energy is used by enzymes in the stroma to make carbohydrates and other complex organic molecules from water, carbon dioxide, and other simple inorganic precursors. The organic molecules produced in chloroplasts, or from biochemical building blocks made in chloroplasts, are the ultimate food source for most organisms.

The chloroplast stroma contains DNA and ribosomes that resemble those of certain photosynthetic bacteria. Because of these similarities, chloroplasts, like mitochondria, are believed to have originated from ancient prokaryotes that became permanent residents of the eukaryotic cells ancestral to the plant lineage (see Chapter 25).

Central Vacuoles Have Diverse Roles in Storage, Structural Support, and Cell Growth

Central vacuoles (see Figure 4.9B; also see Chapter 34) are large vesicles identified as distinct organelles of plant cells because they perform specialized functions unique to plants. In a mature plant cell, 90% or more of the cell's volume may be occupied by one or more large central vacuoles. The remainder of the cytoplasm and the nucleus of these cells are restricted to a narrow zone between the central vacuole and the plasma membrane. The pressure within the central vacuole supports the cells (see Chapter 5).

The membrane that surrounds the central vacuole, the **tonoplast,** contains proteins that transport substances into and out of the central vacuole. As plant cells mature, they grow primarily by increases in the pressure and volume of the central vacuole.

Central vacuoles have other vital functions. They store salts, organic acids, sugars, storage proteins, pigments, and, in some cells, waste products. Pigments concentrated in the vacuoles produce the colors of many flowers. Enzymes capable of breaking down biological molecules are present in some central vacuoles, supporting the view of some scientists that they may have some of the properties of animal lysosomes. Molecules that provide chemical defenses against pathogenic organisms also occur in the central vacuoles of some plants.

Cell Walls Support and Protect Plant Cells

The cell walls of plants are extracellular structures because they are located outside the plasma membrane **(Figure 4.26;** also see Chapter 33). Cell walls provide support to individual cells, contain the pressure produced in the central vacuole, and protect cells against invading bacteria and fungi.

FIGURE 4.25 Chloroplast structure. The EM shows a maize (corn) chloroplast.
© Cengage Learning 2017

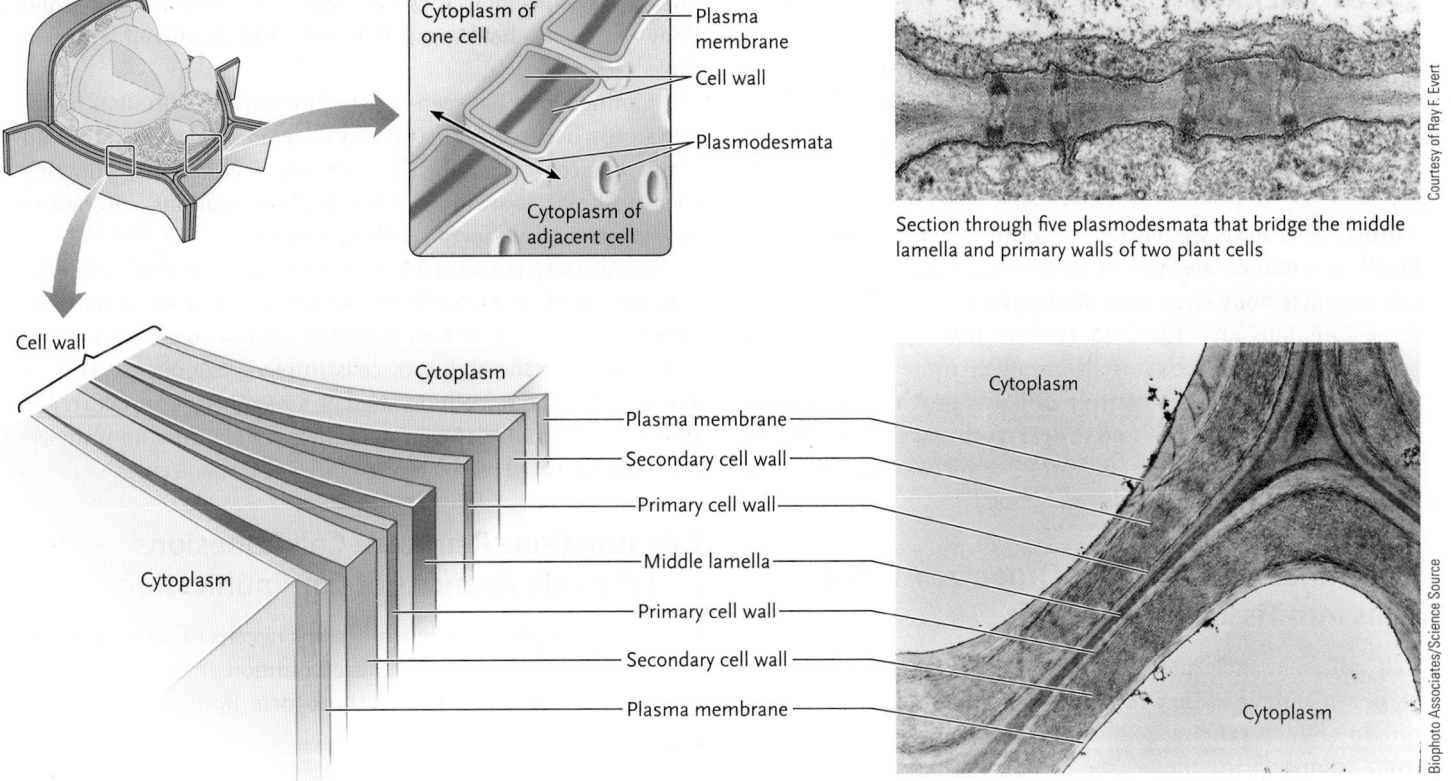

FIGURE 4.26 Cell wall structure in plants. The upper right diagram and EM show plasmodesmata, which form openings in the cell wall that directly connect the cytoplasm of adjacent cells. The lower diagram and EM show the successive layers in the cell wall between two plant cells that have laid down secondary wall material.

© Cengage Learning 2017

Cell walls consist of cellulose fibers (see Figure 3.8C), which give tensile strength to the walls, embedded in a network of highly branched carbohydrates. The initial cell wall laid down by a plant cell, the **primary cell wall,** is relatively soft and flexible. As the cell grows and matures, the primary wall expands and additional layers of cellulose fibers and branched carbohydrates are laid down between the primary wall and the plasma membrane. The added wall layer, which is more rigid and may become many times thicker than the primary wall, is the **secondary cell wall.** In woody plants and trees, secondary cell walls are reinforced by *lignin,* a hard, highly resistant substance assembled from complex alcohols, surrounding the cellulose fibers. Lignin-impregnated cell walls are actually stronger than reinforced concrete by weight; hence, trees can grow to substantial size, and the wood of trees is used extensively in human cultures to make many structures and objects, including houses, tables, and chairs. (Plant development is described in Chapter 36.)

The walls of adjacent cells are held together by a layer of gel-like polysaccharides called the **middle lamella,** which acts as an intercellular glue (see Figure 4.26). The polysaccharide material of the middle lamella, called *pectin,* is extracted from some plants and used to thicken jams and jellies.

Both primary and secondary cell walls are perforated by minute channels, the **plasmodesmata** (singular, *plasmodesma;*

see Figure 4.26 and Figure 4.9B). A typical plant cell has between 1,000 and 100,000 plasmodesmata connecting it to abutting cells. These cytosol-filled channels are lined by plasma membranes, so that connected cells essentially all have one continuous surface membrane. Most plasmodesmata also contain a narrow tubelike structure derived from the smooth endoplasmic reticulum of the connected cells. Plasmodesmata allow ions and small molecules to move directly from one cell to another through the connecting cytosol, without having to penetrate the plasma membranes or cell walls. Proteins and nucleic acids move through some plasmodesmata using energy-dependent processes.

Cell walls also surround the cells of fungi and algal protists. Carbohydrate molecules form the major framework of cell walls in most of these organisms, as they do in plants. In some, the wall fibers contain chitin (see Figure 3.8D) instead of cellulose. Details of cell wall structure in the algal protists and fungi, as well as in different subgroups of the plants, are presented in later chapters devoted to these organisms.

As noted earlier, animal cells do not form rigid, external, layered structures equivalent to the walls of plant cells. However, most animal cells secrete extracellular material and have other structures at the cell surface that play vital roles in the support and organization of animal body structures. The next section describes these and other surface structures of animal cells.

STUDY BREAK 4.4
1. What is the structure and function of a chloroplast?
2. What is the function of the central vacuole in plants?

4.5 The Animal Cell Surface

Animal cells have specialized structures that help hold cells together, produce avenues of communication between cells, and organize body structures. Molecular systems that perform these functions are organized at three levels: individual **cell adhesion molecules** bind cells together, more complex **cell junctions** seal the spaces between cells and provide direct communication between cells, and the **extracellular matrix (ECM)** supports and protects cells and provides mechanical linkages, such as those between muscles and bone.

Cell Adhesion Molecules Organize Animal Cells into Tissues and Organs

Cell adhesion molecules are glycoproteins embedded in the plasma membrane. They help maintain body form and structure in animals ranging from sponges to the most complex invertebrates and vertebrates. Cell adhesion molecules

bind to specific molecules on other cells. Most cells in solid body tissues are held together by many different cell adhesion molecules.

Connections between cells formed by cell adhesion molecules only become permanent as an embryo develops into an adult. Cancer cells typically lose these adhesions, allowing them to break loose from their original locations, migrate to new locations, and form additional tumors.

Some bacteria and viruses—such as the virus that causes the common cold—target cell adhesion molecules as attachment sites during infection. Cell adhesion molecules are also partly responsible for the ability of cells to recognize one another as being part of the same individual or foreign. For example, rejection of organ transplants in mammals results from an immune response triggered by the foreign cell surface molecules.

Cell Junctions Reinforce Cell Adhesions and Provide Avenues of Communication

Cell junctions reinforce cell adhesions in cells of adult animals. Three types of cell junctions are common in animal tissues **(Figure 4.27)**. **Anchoring junctions** form buttonlike spots, or belts, that run entirely around cells, "welding" adjacent cells together. For some anchoring junctions known as **desmosomes,** intermediate filaments anchor the junction in

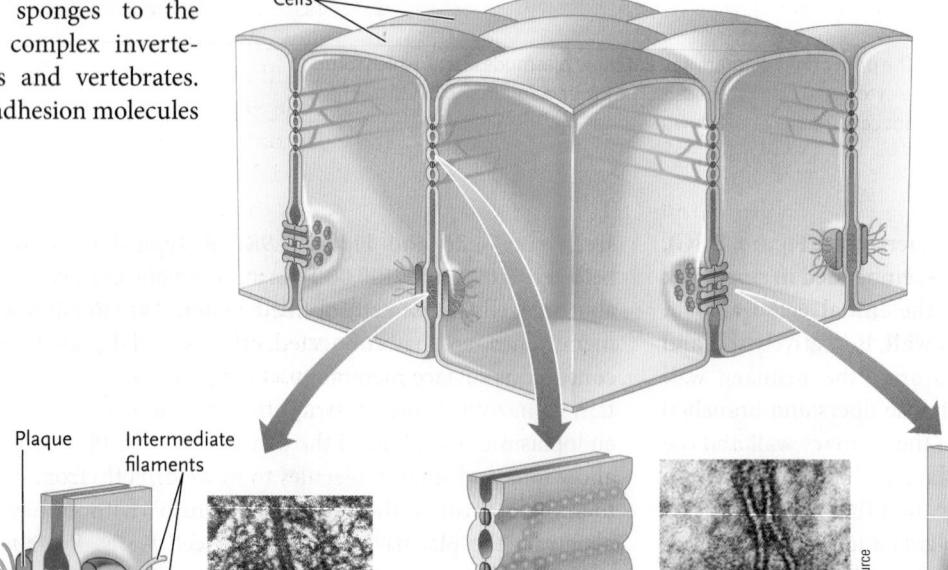

Cells

Anchoring junction:
Adjoining cells adhere at a mass of proteins (a plaque) anchored beneath their plasma membrane by many intermediate filaments (desmosome) or microfilaments (adherens junction) of the cytoskeleton.

Plaque   Intermediate filaments

SPL/Science Source

Tight junction:
Tight connections form between adjacent cells by fusion of plasma membrane proteins on their outer surfaces. A complex network of junction proteins makes a seal tight enough to prevent leaks of ions or molecules between cells.

Don W. Fawcett/Science Source

Gap junction:
Cylindrical arrays of proteins form direct channels that allow small molecules and ions to flow between the cytoplasm of adjacent cells.

Channel in a complex of proteins

Dr. Donald Fawcett/Visuals Unlimited, Inc.

FIGURE 4.27 Anchoring junctions, tight junctions, and gap junctions, which connect cells in animal tissues. Anchoring junctions reinforce the cell-to-cell connections made by cell adhesion molecules, tight junctions seal the spaces between cells, and gap junctions create direct channels of communication between animal cells.

© Cengage Learning 2017

the underlying cytoplasm; in other anchoring junctions known as **adherens junctions,** microfilaments are the anchoring cytoskeletal component. Anchoring junctions are most common in tissues that are subject to stretching, shear, or other mechanical forces—for example, heart muscle, skin, and the cell layers that cover organs or line body cavities and ducts.

Tight junctions, as the name indicates, are regions of tight connections between membranes of adjacent cells (see Figure 4.27). The connection is so tight that it can keep particles as small as ions from moving between the cells in the layers.

Tight junctions seal the spaces between cells in the cell layers that cover internal organs and the outer surface of the body, or the layers that line internal cavities and ducts. For example, tight junctions between cells that line the stomach, intestine, and bladder keep the contents of these body cavities from leaking into surrounding tissues.

A tight junction is formed by direct fusion of proteins on the outer surfaces of the two plasma membranes of adjacent cells. Strands of the tight junction proteins form a complex network that gives the appearance of stitch work holding the cells together. Within a tight junction, the plasma membrane is not joined continuously; instead, there are regions of intercellular space. Nonetheless, the network of junction proteins is sufficient to make the tight cell connections characteristic of these junctions.

Gap junctions open direct channels that allow ions and small molecules to pass directly from one cell to another (see Figure 4.27). Hollow protein cylinders embedded in the plasma membranes of adjacent cells line up and form a sort of pipeline that connects the cytoplasm of one cell with the cytoplasm of the next. The flow of ions and small molecules through the channels provides almost instantaneous communication between animal cells, similar to the communication that plasmodesmata provide between plant cells.

In vertebrates, gap junctions occur between cells within almost all body tissues, but not between cells of different tissues. These junctions are particularly important in heart muscle tissues and in the smooth muscle tissues that form the uterus, where their pathways of communication allow cells of the organ to operate as a coordinated unit. Although most nerve tissues do not have gap junctions, nerve cells in dental pulp are connected by gap junctions; they are responsible for the discomfort you feel if your teeth are disturbed or damaged, or when a dentist pokes a probe into a cavity.

The Extracellular Matrix Organizes the Cell Exterior

Many types of animal cells are embedded in an ECM that consists of proteins and polysaccharides secreted by the cells themselves **(Figure 4.28).** The primary function of the ECM is

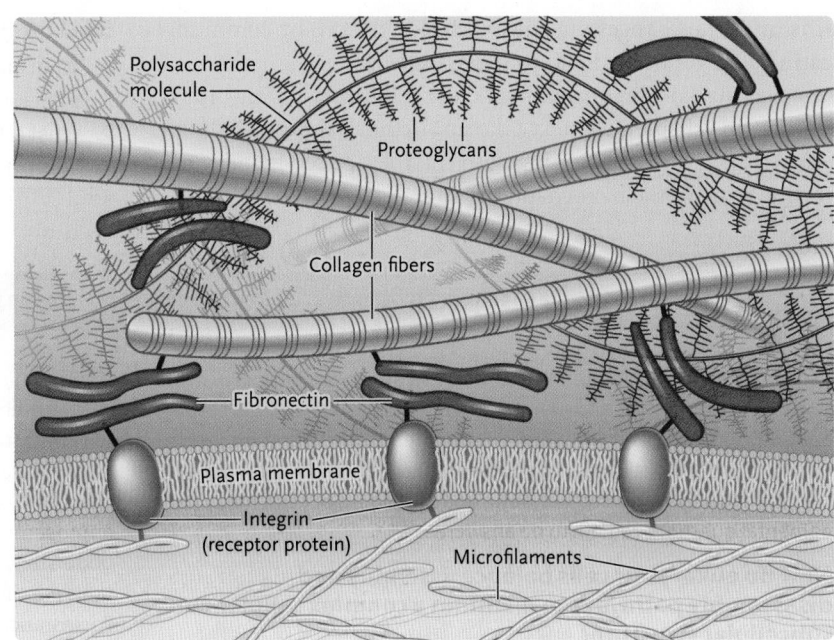

FIGURE 4.28 Components of the extracellular matrix in an animal cell.
© Cengage Learning 2017

protection and support. The ECM forms the mass of skin, bones, and tendons; it also forms many highly specialized extracellular structures such as the cornea of the eye and filtering networks in the kidney. The ECM also affects cell division, adhesion, motility, and embryonic development, and it takes part in reactions to wounds and disease.

Glycoproteins (see Chapter 3) are the main component of the ECM. In most animals, the most abundant ECM glycoprotein is **collagen,** which forms fibers with great tensile strength and elasticity. In vertebrates, the collagens of tendons, cartilage, and bone are the most abundant proteins of the body, making up about half of the total body protein by weight. (Collagens and their roles in body structures are described in further detail in Chapter 38.)

The consistency of the matrix, which may range from soft and jellylike to hard and elastic, depends on a network of proteoglycans that surrounds the collagen fibers. **Proteoglycans** are glycoproteins that consist of small proteins noncovalently attached to long polysaccharide molecules. Matrix consistency depends on the number of interlinks in this network, which determines how much water can be trapped in it. For example, cartilage, which contains a high proportion of interlinked glycoproteins, is relatively soft. Tendons, which are almost pure collagen, are tough and elastic. In bone, the glycoprotein network that surrounds collagen fibers is impregnated with mineral crystals, producing a dense and hard—but still elastic—structure that is about as strong as fiberglass or reinforced concrete.

Yet another class of glycoproteins is **fibronectins,** which aid in organizing the ECM and help cells attach to it. Fibronectins bind to receptor proteins called *integrins* that span the plasma membrane. On the cytoplasmic side of the plasma membrane, the integrins bind to microfilaments of the cytoskeleton. Integrins integrate changes outside and inside the cell by communicating changes in the ECM to the cytoskeleton.

Having laid the groundwork for cell structure and function in this chapter, we next take up further details of individual cell structures, beginning with the roles of cell membranes in transport in the next chapter.

STUDY BREAK 4.5

1. What are the structures and functions of anchoring junctions, tight junctions, and gap junctions?
2. What is the structure and function of the extracellular matrix?

 ## Unanswered Questions

Cell biologists seek to understand the properties and behaviors of cells, including their growth and division, shape and movement, subcellular organization and transport systems, and interactions and communication with each other and the environment. Although cell biology research has dramatically enhanced our basic understanding of cells, many fundamental questions remain to be answered.

How do eukaryotic cells balance assembly and disassembly pathways to maintain a complex cellular and subcellular structure?
Eukaryotic cells have a complex spatial organization, with vastly different cellular sizes and shapes and numerous subcellular organelles and structures. Maintenance of this complexity depends on a balance between the pathways that promote the formation and disassembly of each structure. How is this balance achieved? In my lab, we are working to answer this question in part by studying the interaction between bacterial and viral pathogens and their eukaryotic host cells. Many pathogens invade, move within, and spread between host cells by manipulating the assembly and disassembly of cytoskeletal polymers. Because pathogens exploit normal cellular pathways that regulate assembly and disassembly, they are very useful tools for identifying key molecular players and determining their mechanistic roles. For example, our studies of the intracellular actin-based movement of human bacterial pathogens in the genera *Listeria, Rickettsia, Burkholderia,* and *Mycobacterium* have helped define cellular molecules and pathways that control the assembly and organization of the actin cytoskeleton. Continued progress in addressing this question will come from studying pathogens and their host cells, and from reconstructing subcellular processes outside of the cell.

How do the complex properties of cells arise from the functions and interactions of their molecular components?
Over the past few decades, inspired by advances in molecular biology and genetics, most cell biologists have pursued a reductionist approach by determining the detailed role of individual genes and proteins in cell structure and behavior. This has been very successful, and together with advances in genome sequencing technology, it has enabled cell biologists to catalog many molecules that perform key roles in the cell. In the past few years, many cell biologists have embraced a more holistic approach called *systems biology* (see Chapter 1), which is focused on understanding how the complex properties of cells and subcellular systems arise from the properties and interactions of their individual parts. For instance, Jonathan Weismann's lab at the University of California, San Francisco, has studied the translation, localization, and abundance of thousands of proteins in the yeast *Saccharomyces cerevisiae,* revealing insights into global mechanisms that regulate protein abundance and function. Systems cell biologists often work with mathematicians and computer scientists to develop models of cellular processes that can be refined by experimental validation. Which approach, reductionist or holistic, represents the future of cell biology? Both do, as each relies on the other to generate insights and hypotheses that move the field forward.

think like a scientist There are several major approaches cell biologists use to investigate complex cell structures and processes. One is to perturb the activity of genes or proteins and assess the effect on cellular function. Another is to build or reconstitute the structure or process in a test tube. What are strengths and weaknesses of each approach? What types of questions can be answered using one or the other? Can you think of technical challenges that might arise for each one?

Matthew Welch is a professor of Molecular and Cell Biology at the University of California, Berkeley. His research interests include cytoskeleton dynamics and microbial pathogenesis. Learn more about his work at http://mcb.berkeley.edu/labs/welch.

Courtesy of Matthew Welch

REVIEW KEY CONCEPTS

For access to MindTap and additional study materials visit www.cengagebrain.com.

4.1 Basic Features of Cell Structure and Function

- According to the cell theory: (1) all living organisms are composed of cells; (2) cells are the structural and functional units of life; and (3) cells arise only from the division of preexisting cells.
- Cells of all kinds are divided internally into a central region containing the genetic material and the cytoplasm, which consists of the cytosol, the cytoskeleton, and organelles and is bounded by the plasma membrane.

- The plasma membrane is a lipid bilayer in which transport proteins are embedded (Figure 4.6).
- In the cytoplasm, proteins are made, most of the other molecules required for growth and reproduction are assembled, and energy absorbed from the surroundings is converted into energy usable by the cell.

4.2 Prokaryotic Cells

- Prokaryotic cells are surrounded by a plasma membrane and, in most groups, are enclosed by a cell wall. The genetic material, typically a single, circular DNA molecule, is located in the nucleoid. The cytoplasm contains masses of ribosomes (Figure 4.7).

4.3 Eukaryotic Cells

- Eukaryotic cells have a true nucleus, which is separated from the cytoplasm by the nuclear envelope perforated by nuclear pores. A plasma membrane forms the outer boundary of the cell. Other membrane systems enclose specialized compartments as organelles in the cytoplasm (Figure 4.9).

- The eukaryotic nucleus contains chromatin, a combination of DNA and proteins. A specialized segment of the chromatin forms the nucleolus, where ribosomal RNA molecules are made and combined with ribosomal proteins to make ribosomes. The nuclear envelope contains nuclear pore complexes with pores that allow passive or assisted transport of molecules between the nucleus and the cytoplasm. Proteins destined for the nucleus contain a short amino acid sequence called a nuclear localization signal (Figures 4.10 and 4.11).

- Eukaryotic cytoplasm contains ribosomes (Figure 4.12), an endomembrane system, mitochondria, microbodies, the cytoskeleton, and some organelles specific to certain organisms. The endomembrane system includes the nuclear envelope, ER, Golgi complex, lysosomes, vesicles, and plasma membrane.

- The endoplasmic reticulum (ER) occurs in two rough and smooth forms. The ribosome-studded rough ER makes proteins that become part of cell membranes or are released from the cell. Smooth ER synthesizes lipids and breaks down toxic substances (Figure 4.13).

- The Golgi complex chemically modifies proteins made in the rough ER and sorts finished proteins to be secreted from the cell, embedded in the plasma membrane, or included in lysosomes (Figures 4.14, 4.15, and 4.17).

- Lysosomes, specialized vesicles that contain hydrolytic enzymes, digest complex molecules such as food molecules that enter an animal cell by endocytosis, cellular organelles that are no longer functioning correctly, and engulfed bacteria and cell debris (Figure 4.16).

- Mitochondria carry out cellular respiration, the conversion of fuel molecules into the energy of ATP (Figure 4.18).

- Microbodies conduct the initial steps in fat breakdown and other reactions that link major biochemical pathways in the cytoplasm (Figure 4.19).

- The cytoskeleton is a supportive structure built from microtubules, intermediate filaments, and microfilaments. Motor proteins walking along microtubules and microfilaments produce most movements of animal cells, as well as the distribution of organelles in cells and the movements of vesicles (Figures 4.20 and 4.21).

- Motor protein-controlled sliding of microtubules generates the movements of flagella and cilia. Flagella and cilia arise from centrioles (Figures 4.22–4.24).

4.4 Specialized Structures of Plant Cells

- Plant cells contain all the eukaryotic structures found in animal cells. (Lysosomes are a matter of debate.) They also contain three structures not found in animal cells: chloroplasts, a central vacuole, and a cell wall (Figure 4.9B).

- Chloroplasts contain pigments and molecular systems that absorb light energy and convert it to chemical energy. The chemical energy is used inside the chloroplasts to assemble carbohydrates and other organic molecules from simple inorganic raw materials (Figure 4.25).

- The large central vacuole, which consists of a tonoplast enclosing an inner space, develops pressure that supports plant cells, accounts for much of cellular growth by enlarging as cells mature, and serves as a storage site for substances including waste materials (Figure 4.9B).

- A cellulose cell wall surrounds plant cells, providing support and protection. Plant cell walls are perforated by plasmodesmata, channels that provide direct pathways of communication between the cytoplasm of adjacent cells (Figure 4.26).

4.5 The Animal Cell Surface

- Animal cells have specialized surface molecules and structures that function in cell adhesion, communication, and support.

- Cell adhesion molecules bind to specific molecules on other cells. The adhesions organize and hold together cells of the same type in body tissues.

- Cell adhesions are reinforced by various junctions. Anchoring junctions hold cells together. Tight junctions seal together the plasma membranes of adjacent cells, preventing ions and molecules from moving between the cells. Gap junctions open direct channels between the cytoplasm of adjacent cells (Figure 4.27).

- The extracellular matrix, formed from collagen proteins embedded in a matrix of branched glycoproteins, functions primarily in cell and body protection and support but also affects cell division, motility, embryonic development, and wound healing (Figure 4.28).

TEST YOUR KNOWLEDGE

Remember/Understand

1. A prokaryote converts food energy into the chemical energy of ATP on/in its:
   a. chromosome.
   b. flagella.
   c. ribosomes.
   d. cell wall.
   e. plasma membrane.

2. Eukaryotic and prokaryotic ribosomes are similar in that:
   a. both contain a small subunit, but only eukaryotes contain a large subunit.
   b. both contain the same number of proteins.
   c. both use mRNA to assemble amino acids into proteins.
   d. both contain the same number of types of rRNA.
   e. both produce proteins that can pass through pores into the nucleus.

3. Which of the following structures does *not* require an immediate source of energy to function?
   a. central vacuoles
   b. cilia
   c. microtubules
   d. microfilaments
   e. microbodies

4. Which of the following structures is *not* used in eukaryotic protein manufacture and secretion?
   a. ribosome
   b. lysosome
   c. rough ER
   d. secretory vesicle
   e. Golgi complex

5. Which of the following are glycoproteins whose function is affected by the common cold virus?
   a. plasmodesmata
   b. desmosomes
   c. cell adhesion molecules
   d. flagella
   e. cilia

6. Which of the following contributes to the sealed lining of the digestive tract to keep food inside it?
   a. a central vacuole that stores proteins
   b. tight junctions formed by direct fusion of proteins
   c. gap junctions that communicate between cells of the stomach lining and its muscular wall
   d. desmosomes forming buttonlike spots or a belt to keep cells joined together
   e. plasmodesmata that help cells communicate their activities

7. Which of the following statements about proteins is correct?
   a. Proteins are transported to the rough ER for use within the cell.
   b. Lipids and carbohydrates are added to proteins by the Golgi complex.
   c. Proteins are transported directly into the cytosol for secretion from the cell.
   d. Proteins that are to be stored by the cell are moved to the rough ER.
   e. Proteins are synthesized in vesicles.

8. Which of the following is *not* a component of the cytoskeleton?
   a. microtubules
   b. actins
   c. microfilaments
   d. cilia
   e. cytokeratins

9. **Discuss Concepts** Explain why aliens invading Earth are not likely to be giant cells the size of humans.

Apply/Analyze

10. You are examining a cell from a crime scene using an electron microscope. It contains ribosomes, DNA, a plasma membrane, a cell wall, and mitochondria. What type of cell is it?
    a. lung cell
    b. plant cell
    c. prokaryotic cell
    d. cell from the surface of a human fingernail
    e. sperm cell

11. An electron micrograph shows that a cell has extensive amounts of rough ER throughout. One can deduce from this that the cell is:
    a. synthesizing and metabolizing carbohydrates.
    b. synthesizing and secreting proteins.
    c. synthesizing ATP.
    d. contracting.
    e. resting metabolically.

12. **Discuss Concepts** Many compound microscopes have a filter that eliminates all wavelengths except that of blue light, thereby allowing only blue light to pass through the microscope. Use the spectrum of visible light (see Figure 8.3) to explain why the filter improves the resolution of light microscopes.

13. **Discuss Concepts** An electron micrograph of a cell shows the cytoplasm packed with rough ER membranes, a Golgi complex, and mitochondria. What activities might this cell concentrate on? Why would large numbers of mitochondria be required for these activities?

14. **Apply Evolutionary Thinking** What aspects of cell structure suggest that prokaryotes and eukaryotes share a common ancestor in their evolutionary history?

Evaluate/Create

15. **Discuss Concepts** Assuming that mitochondria evolved from bacteria that entered cells by endocytosis, what are the likely origins of the outer and inner mitochondrial membranes?

16. **Discuss Concepts** Researchers have noticed that some men who were sterile because their sperm cells were unable to move also had chronic infections of the respiratory tract. What might be the connection between these two symptoms?

17. **Design an Experiment** The unicellular alga *Chlamydomonas reinhardtii* has two flagella assembled from tubulin proteins. If a researcher changes the pH from approximately neutral (their normal growing condition) to pH 4.5, *Chlamydomonas* cells spontaneously lose their flagella. After the cells are returned to neutral pH, they regrow the flagella—a process called reflagellation. Assuming that you have deflagellated *Chlamydomonas* cells, devise experiments to answer the following questions:

    1. Do new tubulin proteins need to be made for reflagellation to occur, or is there a reservoir of proteins in the cell?

    2. Is the production of new mRNA for the tubulin proteins necessary for reflagellation?

    3. What is the optimal pH for reflagellation?

For selected answers, see Appendix A.

INTERPRET THE DATA

Investigators studying protein changes during aging examined enzyme activity in cells extracted from the nematode worm *Caenorhabditis elegans*. The cell extracts were treated to conserve enzyme activity, although the investigators noted that some proteins were broken down by the extraction procedure. The extracts were centrifuged, and seven fractions were collected in sequence to isolate the location of activity by protease enzymes called cathepsins. Examine the activity profiles in the **Figure.** In which fraction and, hence, in which eukaryotic cellular structure are these enzymes most active?

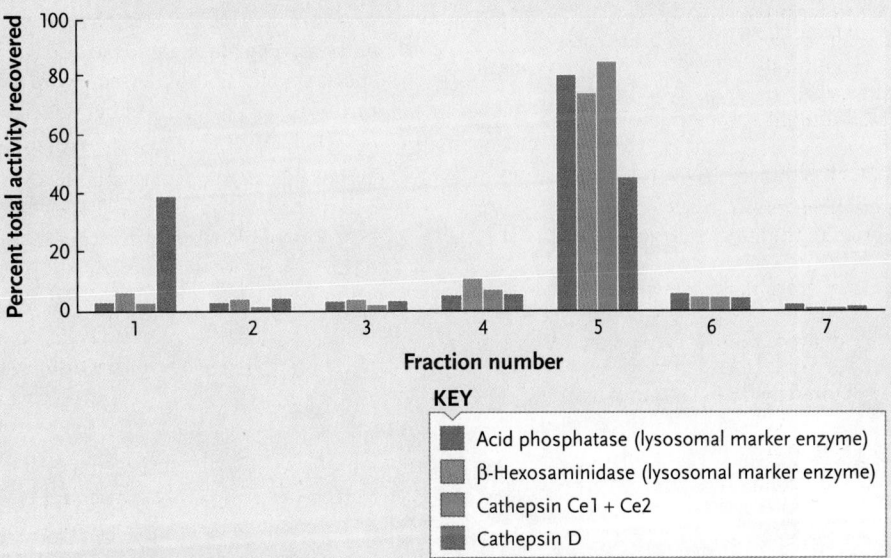

KEY

- ■ Acid phosphatase (lysosomal marker enzyme)
- ■ β-Hexosaminidase (lysosomal marker enzyme)
- ■ Cathepsin Ce1 + Ce2
- ■ Cathepsin D

FIGURE   Distribution of enzyme activity in fractions from centrifugation of an organelle pellet. The fractions are numbered 1 to 7 from the top to the bottom of the centrifuge tube. Fraction 1 contains cytosolic contents and is the supernatant, and fraction 7 contains cellular debris and membrane fragments.

Source: G. J. Sarkis et al. 1988. Decline in protease activities with age in the nematode *Caenorhabiditis elegans. Mechanisms of Ageing and Development* 45:191–201.

5

Membranes and Transport

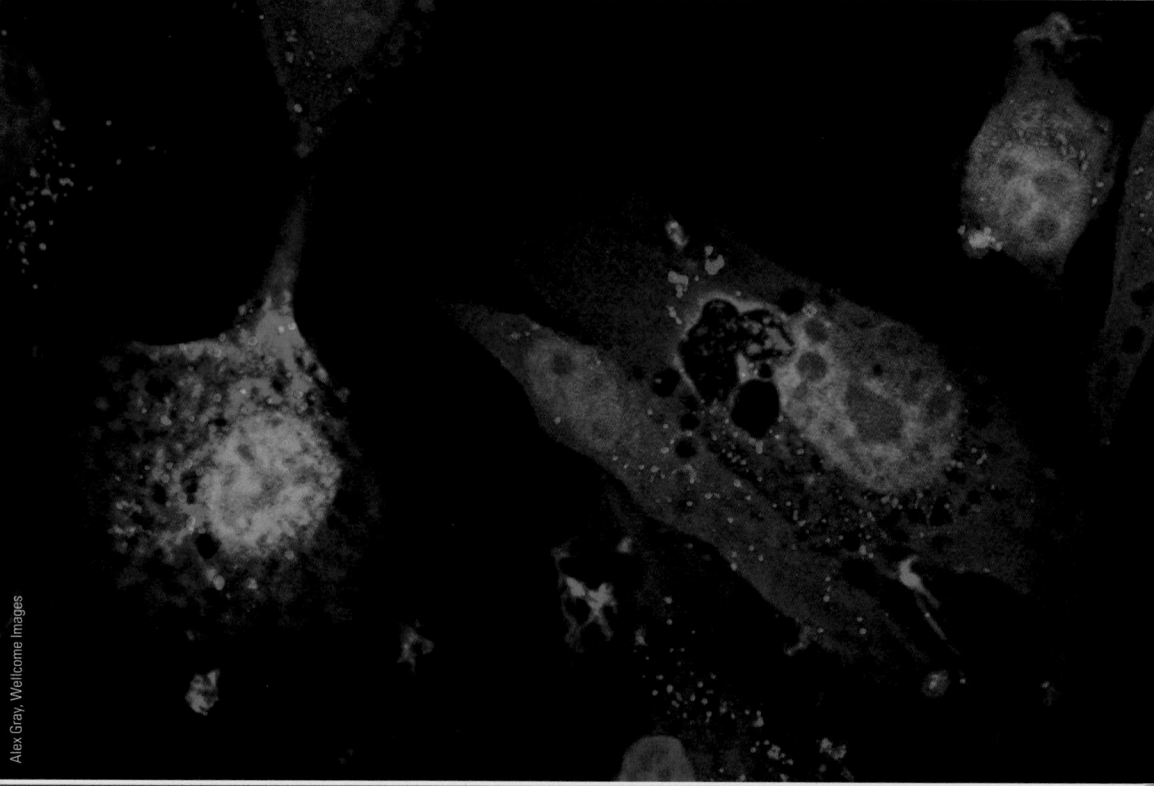

Endocytosis in cancer cells (confocal micrograph). The red spots are fluorescent spheres used to follow the process of endocytosis; some of the spheres have been taken up by cells.

Alex Gray, Wellcome Images

Why it matters . . . Cystic fibrosis (CF) is one of the most common genetic diseases. It affects approximately 1 in 4,000 children born in the United States. People with CF experience a progressive impairment of lung and gastrointestinal function. The average lifespan of CF patients is under 40 years. CF is caused by mutations in a single gene that codes for a protein so vital that a change in a particular amino acid has life-altering consequences. What is this protein, and what role does it play in the body?

To maintain their internal environment, the cells of all organisms must exchange molecules and ions constantly with the fluid environment that surrounds them. What makes this possible is the **plasma membrane,** the thin layer of lipids and proteins that separates a cell from its surroundings. Some of the proteins are transport proteins that move particular ions and molecules, including water, in a directed way across the membrane. Different transport proteins move different molecules or ions. One such transport protein is *cystic fibrosis transmembrane conductance regulator,* or CFTR, which is found in the plasma membrane of *epithelial cells* **(Figure 5.1).** Epithelial cells are organized into sheetlike layers that form coverings and linings, which are typically exposed to water, air, or fluids within the body (see Chapter 38). Epithelial cells line the passageways and ducts of the lungs, liver, pancreas, intestines, reproductive system, and skin. CFTR pumps chloride ions out of those cells, and water follows the ions, producing a thin, watery film over the epithelial tissue surface. Mucus can slide easily over the tissue because of the moist surface.

Particular mutations in the *CFTR* gene result in a CFTR molecule that transports chloride ions poorly or not at all. The most common mutation is a small deletion in the gene that removes one amino acid from the CFTR protein. If an individual has one normal and one mutant copy of the gene, there are enough working CFTR molecules for normal chloride transport. However, a CF

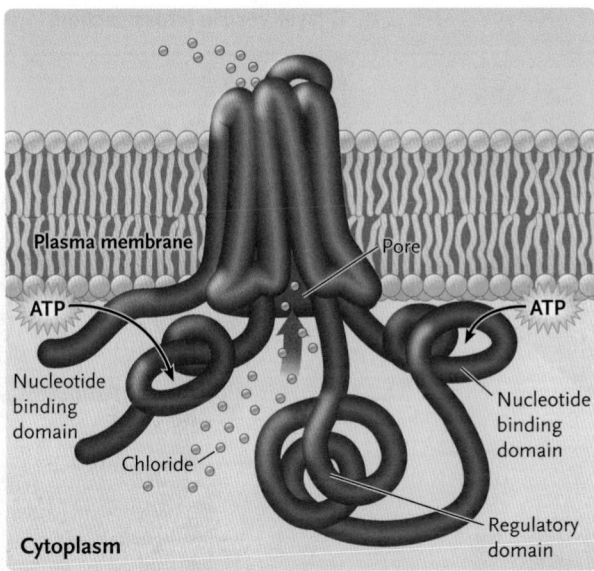

FIGURE 5.1 Molecular model of cystic fibrosis transmembrane conductance regulator (CFTR), a chloride ion transport protein embedded in the plasma membrane. The nucleotide binding domains bind ATP, which provides energy for ion transport by the protein.
© Cengage Learning 2017

patient has two mutant copies of the gene, one inherited from each parent, so all CFTR molecules are mutant, and chloride transport is defective. Not enough chloride ions are transported out of the cell, so not enough water leaves either. As a result, mucus sticks to the drier epithelial tissue, building up into a thick mass. In the respiratory tract, the abnormally thick mucus means much diminished protection from bacteria, which leads to infections. Chronic lung infections produce progressive damage to the respiratory tract. At some point, lung function becomes insufficient to support life and the CF patient dies.

Most treatments for CF center on managing the symptoms, and include pounding the back or chest to dislodge mucus and antibiotic or other pharmaceutical treatments to control infections. Research to develop more effective treatments includes gene therapy to repair the *CFTR* gene and developing pharmaceuticals that target the defective CFTR protein.

The plasma membrane, in which the CFTR transport protein is located, forms the outer boundary of every living cell and encloses the intracellular contents (see Chapter 4). The plasma membrane also regulates the passage of molecules and ions between the inside and outside of the cell, helping to determine the cell's composition. Primitive protocells, evolutionary precursors of cells, are hypothesized to have formed by the simultaneous evolution of a boundary membrane system (now a plasma membrane) and a genetic information system (discussed in Chapter 25). Within eukaryotic cells, membranes surrounding internal organelles play similar roles, creating environments that differ from the surrounding cytosol (see Chapter 4).

The structure and function of biological membranes are the focus of this chapter. We first consider the structure of membranes and then examine how membranes transport substances selectively in and out of cells and organelles. Other roles of membranes, including recognition of molecules on other cells, adherence to other cells or extracellular materials, and reception of molecular signals such as hormones, are the subjects of Chapters 9, 42, and 45.

5.1 Membrane Structure and Function

A watery fluid medium—or aqueous solution—bathes both surfaces of all biological membranes. The membranes are also fluid, but they are kept separate from their surroundings by the properties of the lipid and protein molecules from which they are formed.

Biological Membranes Contain Both Lipid and Protein Molecules

Biological membranes consist of lipids and proteins assembled into a thin film. The proportions of lipid and protein molecules in membranes vary, depending on the functions of the membranes in the cells.

MEMBRANE LIPIDS *Phospholipids* and *sterols* are the two major types of lipids in membranes (see Section 3.3). Phospholipids have a polar (electrically charged) end containing a phosphate group linked to one of several alcohols or amino acids, and a nonpolar (uncharged) end containing two nonpolar fatty-acid tails. **Figure 5.2A** shows an example of a phospholipid, phosphatidylcholine, in which the polar head contains choline. The polar end is hydrophilic, meaning that it "prefers" being in an aqueous environment, and the nonpolar end is hydrophobic, meaning that it "prefers" being in an environment from which water is excluded. In other words, phospholipids have dual solubility properties.

In an aqueous medium, phospholipid molecules satisfy their dual solubility properties by assembling into a **bilayer**—a layer two molecules thick **(Figure 5.2B)**. In a bilayer, the polar ends of the phospholipid molecules are located at the surfaces, where they face the aqueous media. The nonpolar fatty-acid chains arrange themselves end to end in the membrane interior, in a nonpolar region that excludes water. When a phospholipid bilayer sheet is shaken in water, it breaks and spontaneously forms small *vesicles* **(Figure 5.2C)**. Vesicles consist of a spherical shell of phospholipid bilayer enclosing a small droplet of water. At low temperatures, the phospholipid bilayer freezes into a semisolid, gel-like state **(Figure 5.2D)**.

Membrane sterols, which also have dual solubility properties, have nonpolar carbon rings with a nonpolar side chain at one end and a single polar group (an —OH group) at the other end (see Section 3.3). The sterols pack into membranes alongside the phospholipid hydrocarbon chains, with only the polar end extending into the polar membrane surface **(Figure 5.3)**. The predominant sterol of animal cell membranes is **cholesterol,** which is important for keeping the membranes fluid (see Figure 3.12).

A. Phospholipid molecule

Polar end (hydrophilic)

CH₃

H₃C—N⁺—CH₃

H₂C

CH₂ — Choline

O

⁻O—P=O — Phosphate group

O

H₂C—CH—CH₂ — Glycerol

C=O   C=O

Nonpolar end (hydrophobic)

H₂C   CH₂

CH₂   H₂C

H₂C   CH₂

CH₂   H₂C

H₂C   CH₂

CH₂   HC

H₂C   HC — Hydrophobic tail

CH₂   CH₂

H₂C   H₂C

CH₂   CH₂

H₂C   H₂C

CH₂   CH₂

H₃C   H₂C

CH₂

H₃C

B. Fluid bilayer

Aqueous solution

Aqueous solution

C. Bilayer vesicle (cross section)

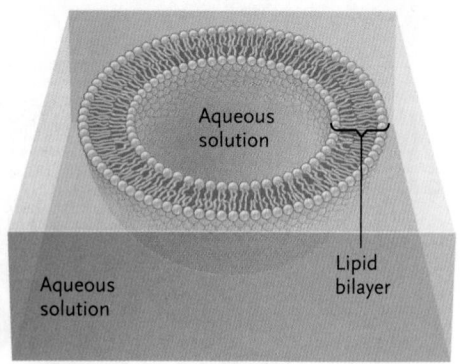

Aqueous solution

Aqueous solution

Lipid bilayer

D. "Frozen" bilayer

FIGURE 5.2 Phospholipid bilayers.
(A) Phosphatidylcholine, an example of a phospholipid molecule. Within the circle at the top representing the polar end of the molecule are choline (blue), the phosphate group (orange), and the glycerol unit (pink). **(B)** Phospholipid bilayer in the fluid state, in which individual molecules are free to flex, rotate, and exchange places. **(C)** Phospholipid bilayer forming a spherical vesicle. **(D)** A bilayer frozen in a semisolid, gel-like state; note the close alignment of the hydrophobic tails compared with part **(B)**.
© Cengage Learning 2017

A variety of sterols, called *phytosterols,* are found in plants. The high similarity of bilayer membranes in all prokaryotic and eukaryotic cells is evidence that the basic structure of membranes evolved during the earliest stages of life on Earth, and has been preserved ever since.

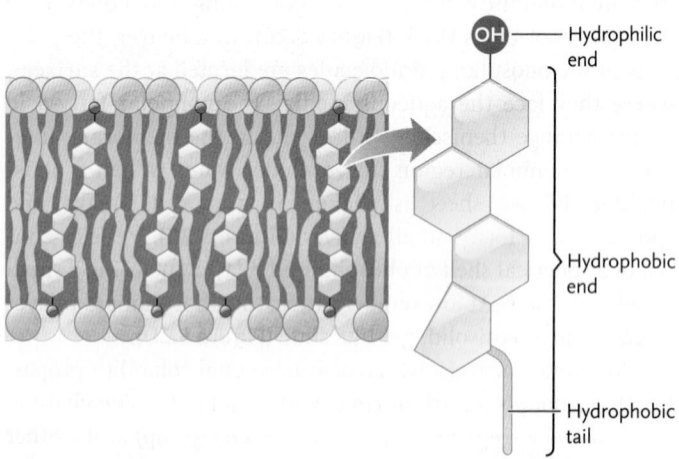

Cholesterol

OH — Hydrophilic end

Hydrophobic end

Hydrophobic tail

FIGURE 5.3 The position taken by cholesterol in bilayers. The hydrophilic —OH group at one end of the molecule extends into the polar regions of the bilayer; the ring structure extends into the nonpolar membrane interior.
© Cengage Learning 2017

MEMBRANE PROTEINS Membrane proteins also have hydrophilic and hydrophobic regions. The hydrophobic regions of membrane proteins are formed by segments of the amino acid chain with hydrophobic side groups. These segments are often wound into α helices, which span the membrane bilayer, and are connected by loops of hydrophilic amino acids that extend into the polar regions at the membrane surfaces **(Figure 5.4)**.

Each type of membrane has a characteristic group of proteins that is responsible for its specialized functions:

- **Transport proteins** form channels that allow selected polar molecules and ions to pass across a membrane.
- **Cell–cell recognition proteins** in the plasma membrane identify a cell as part of the same individual or as foreign, facilitate cell–cell linking, bind cells to the extracellular matrix (ECM), and link the ECM to the cytoskeleton.
- **Receptor proteins** recognize and bind molecules from other cells that act as chemical signals for altering cell activity, such as the peptide hormone insulin in animals.
- **Proteins in cell–cell junctions** bind cells tightly together, as exemplified in gap junctions and tight junctions (see Section 5.5).
- **Enzymatic proteins** confer specific properties on the membranes with them.

A. Typical membrane protein

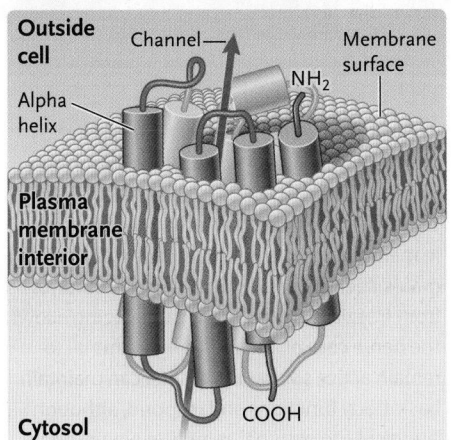

B. Hydrophilic and hydrophobic surfaces

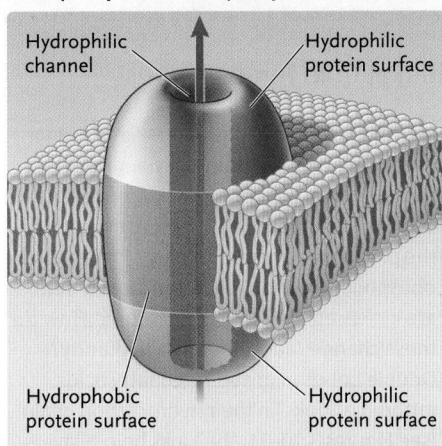

FIGURE 5.4 Structure of membrane proteins. (A) Typical membrane protein, bacteriorhodopsin, showing the membrane-spanning alpha-helical segments (blue cylinders), connected by flexible loops of the amino acid chain at the membrane surfaces. **(B)** The same protein as in **(A)** in a diagram that shows hydrophilic (blue) and hydrophobic (orange) surfaces and the membrane-spanning channel created by this protein. Bacteriorhodopsin absorbs light energy in plasma membranes of photosynthetic archaeans.
© Cengage Learning 2017

MEMBRANE GLYCOLIPIDS AND GLYCOPROTEINS Glyco-lipids—lipid molecules with carbohydrate groups attached—are another type of lipid found in membranes **(Figure 5.5)**. Glycolipids are found in the part of the membrane that faces the outside of the cell. The carbohydrate groups may form a linear or a branched chain. Carbohydrates are also attached to some of the proteins in the exterior-facing lipid layer, producing **glycoproteins** (see Figure 5.5). In many animal cells, such as intestinal epithelial cells, the carbohydrate groups of the cell surface glycolipids and glycoproteins form a surface coat called the **glycocalyx.** Part of the function of the glycocalyx is to protect cells against chemical and mechanical damage. (The prokaryotic glycocalyx was described in Section 4.2.)

The Fluid Mosaic Model Explains Membrane Structure

Membrane structure has been and continues to be a subject for research. The current view of membrane structure is based on the fluid mosaic model advanced by S. Jonathan Singer and Garth L. Nicolson at the University of California, San Diego, in 1972 (see Figure 5.5). The **fluid mosaic model** proposes that the membrane consists of a fluid phospholipid bilayer in which proteins are embedded and float freely. This

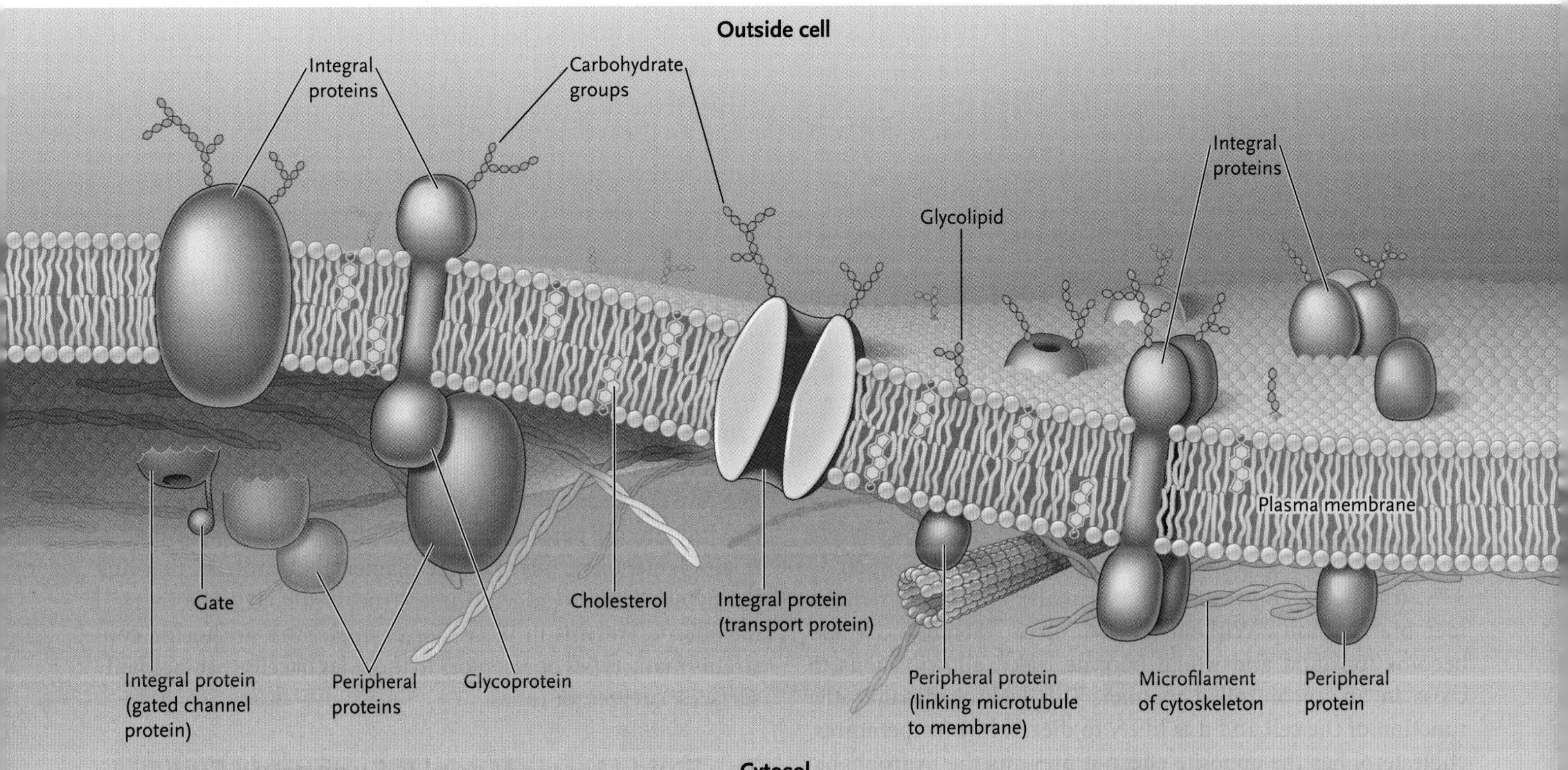

FIGURE 5.5 Membrane structure according to the fluid mosaic model, in which integral membrane proteins are suspended individually in a fluid bilayer. Peripheral proteins are attached to integral proteins or membrane lipids mostly on the cytoplasmic side of the membrane (shown only on the inner surface in the figure). In the plasma membrane, carbohydrate groups of membrane glycoproteins and glycolipids face the cell exterior.
© Cengage Learning 2017

Basic Research: Keeping Membranes Fluid at Cold Temperatures

Biological membranes of all organisms have been conserved throughout evolution. The fluid state of biological membranes is critical to membrane function and, therefore, vital to cellular life. When membranes freeze, the phospholipids form a semisolid gel in which they are unable to move (see Figure 5.2D), and proteins become locked in place. Freezing can kill cells by impeding vital membrane functions such as transport.

Many eukaryotic organisms, including higher plants, animals, and some protists, adapt to colder temperatures by changing membrane lipids. Experiments have shown that, in animals with body temperatures that fluctuate with environmental temperature, such as fish, amphibians, and reptiles, both the proportion of double bonds in membrane phospholipids and the cholesterol content are increased at lower temperatures. How do these changes affect membrane fluidity? Double bonds in unsaturated fatty acids introduce "kinks" in their hydrocarbon chain (see Figures 3.10–11; the kinks help bilayers stay fluid at lower temperatures by interfering with packing of the hydrocarbons.

Cholesterol depresses the freezing point by interfering with close packing of membrane phospholipids.

All of these membrane changes also occur in mammals that enter hibernation in cold climates, thereby preventing their membranes from freezing. The resistance to freezing allows the nerve cells of a hibernating mammal to remain active so that the animal can maintain basic body functions and respond, although sluggishly, to external stimuli. In active, nonhibernating mammals, membranes freeze into the gel state at about 15°C.

---

model revolutionized how scientists think about membrane structure and function.

The "fluid" part of the fluid mosaic model refers to the phospholipid molecules, which vibrate, flex back and forth, spin around their long axis, move sideways, and exchange places within the same bilayer half. Only rarely does a phospholipid flip-flop between the two layers. Phospholipids exchange places within a layer millions of times a second, making the phospholipid molecules in the membrane highly dynamic. Membrane fluidity is critical to the functions of membrane proteins and allows membranes to accommodate, for example, cell growth, motility, and surface stresses.

Lipid composition and temperature affect the fluidity of membranes. For example, shorter fatty acid chains or a greater proportion of unsaturated fatty acids reduces the ability of the hydrophobic tails to interact with each other. As a result, the membrane remains fluid at lower temperatures. This ability of membranes to remain fluid at lower temperatures is important for organisms whose temperature fluctuates with that of the environment. Such organisms adapt to the lower temperature by synthesizing fatty-acid tails that are shorter and with more double bonds.

Cholesterol also plays an important role in enabling membrane fluidity over a range of temperature. As you learned, cholesterol packs into the membrane alongside the phospholipid molecules (see Figure 5.3). At higher temperatures, such as 37°C (the body temperature of mammals), cholesterol reduces membrane fluidity. Without cholesterol, the membrane would become too fluid and would become leaky, allowing ions to cross in an uncontrolled manner. This leaking disrupts the function of the cell and it is likely to die. At low temperatures, cholesterol has the opposite effect: it prevents the hydrophobic tails from coming together and crystallizing, thereby maintaining membrane fluidity. (See *Focus on Research: Basic Research* for a description of other strategies that organisms use to keep their membranes from freezing at low temperatures.)

The "mosaic" part of the fluid mosaic model refers to the membrane proteins, most of which float individually in the fluid lipid bilayer, like icebergs in the sea. Membrane proteins are larger than membrane lipids, and those that move do so much more slowly than do lipids. A number of membrane proteins are attached to the cytoskeleton. These proteins either are immobile or move in a directed fashion, such as along cytoskeletal filaments.

Membrane proteins are oriented across the membrane so that particular functional groups and active sites face either the inside or the outside membrane surface. The inside and outside halves of the bilayer also contain different mixtures of phospholipids. These differences make biological membranes *asymmetric* and give their inside and outside surfaces different functions.

Proteins that are embedded in the phospholipid bilayer are termed **integral proteins** (see Figure 5.5). Essentially all transport, receptor, recognition, and cell adhesion proteins that give membranes their specific functions are integral membrane proteins.

Other proteins, called **peripheral proteins** (see Figure 5.5), are held to membrane surfaces by noncovalent bonds—hydrogen bonds and ionic bonds—formed with the polar parts of integral membrane proteins or membrane lipids. Most peripheral proteins are on the cytoplasmic side of the membrane. Some peripheral proteins are parts of the cytoskeleton, such as microtubules, microfilaments, or intermediate filaments, or proteins that link the cytoskeleton together. These structures hold some integral membrane proteins in place. For example, this anchoring constrains many types of receptors to the sides of cells that face body surfaces, cavities, or tubes.

The Fluid Mosaic Model Is Supported Fully by Experimental Evidence

The novel ideas of a fluid membrane and a flexible mosaic arrangement of proteins and lipids challenged an accepted

---

model in which a relatively rigid, stable membrane was coated on both sides with proteins arranged like jam on bread. The results of research experiments support every major hypothesis of the model completely: that membrane lipids are arranged in a bilayer; that the bilayer is fluid; that proteins are suspended individually in the bilayer; and that the arrangement of both membrane lipids and proteins is asymmetric.

Evidence that the membranes are fluid came from an experiment done by L. David Frye and Michael A. Edidin. They fused human cells and mouse cells and observed that the membrane-embedded proteins intermixed, demonstrating that the proteins were floating freely in a fluid membrane **(Figure 5.6)**. Based on the measured rates at which molecules mix in biological membranes, the membrane bilayer appears to be about as fluid as light machine oil, such as the lubricants you might use around the house to oil a door hinge or a bicycle.

Evidence that the membrane is a bilayer with proteins suspended in it individually and that the arrangement of membrane lipids and proteins is asymmetric came from an experiment using membranes prepared for electron microscopy by the **freeze–fracture technique (Figure 5.7)**. In the experiment, membrane bilayers of frozen cells were split into inner and outer halves, revealing the proteins embedded in the bilayers and their different patterns in the two halves (see Figure 5.7C).

Recent research has shown that lipids and proteins tend to form ordered arrangements in membranes, rather than freely mixed arrangements, and that their movements may be restricted. In fact, many cellular processes require a specific structural organization of membrane lipids and proteins. Cell communication, the topic of Chapter 9, is an example of such a process.

Membranes Have Several Functions

Membranes perform a diverse array of functions. As you will see, often the membrane protein defines the function of a membrane or membrane segment.

- Membranes define the boundaries of cells and, in eukaryotes, the boundaries of compartments (for example, the nucleus, mitochondria, and chloroplasts).
- Membranes are permeability barriers, permitting regulated control of the contents of cells compared with the extracellular environments.

 FIGURE 5.6 **Experimental Research**

The Frye–Edidin Experiment Demonstrating That the Phospholipid Bilayer Is Fluid

Question: Is the phospholipid bilayer fluid?

Experiment: Frye and Edidin grew human cells and mouse cells separately in tissue culture. Then they added antibodies that bound to either human or mouse membrane proteins. The anti-human antibodies were attached to dye molecules that fluoresce red under ultraviolet light, and the anti-mouse antibodies to molecules that fluoresce green. The researchers fused the two cells and followed the pattern of fluorescence under a microscope.

Results: After 40 minutes, the fluorescence pattern showed that the human and mouse membrane proteins had mixed completely.

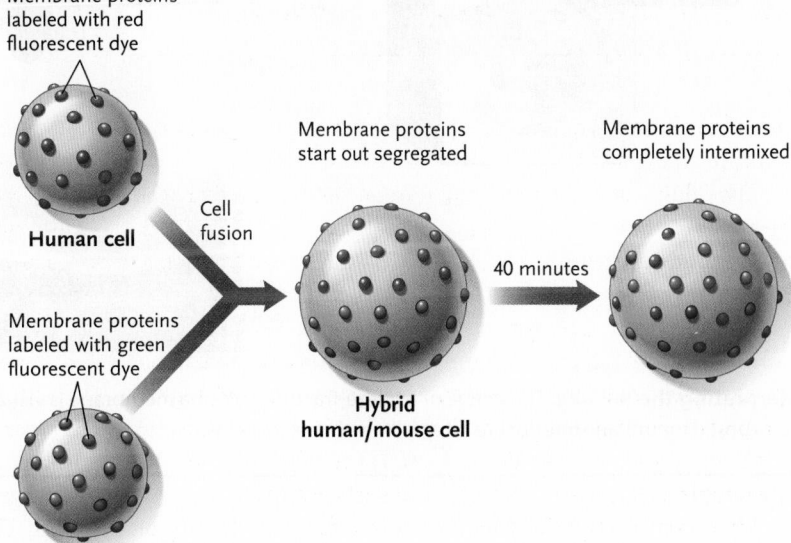

Membrane proteins labeled with red fluorescent dye

Human cell

Cell fusion

Membrane proteins labeled with green fluorescent dye

Mouse cell

Membrane proteins start out segregated

Hybrid human/mouse cell

40 minutes

Membrane proteins completely intermixed

Conclusion: The rapid mixing of membrane proteins in the fused human–mouse cells showed that membrane proteins move in the phospholipid bilayer, indicating that the membrane is fluid.

 **think like a scientist**
Observe
Hypothesize
Predict
Experiment
Interpret

An alternative explanation for the observed intermixing of membrane proteins is new synthesis of the proteins and the insertion of those proteins all over the plasma membrane to replace proteins that are broken down. Frye and Edidin did some experiments to determine whether the intermixing was due to the fluid nature of the phospholipid bilayer or to new protein synthesis. In one experiment, they incubated the hybrid cells with three different protein synthesis inhibitors. They observed that the membrane proteins were completely mixed after the incubation period. In another experiment, the researchers performed the experiment at a variety of temperatures below that of the initial experiment, which was performed at 37°C. They observed that the lower the temperature, the less intermixing of membrane proteins occurred. Essentially no intermixing occurred at 15°C and below. How do these results support the fluid phospholipid bilayer model?

Source: L. D. Frye and M. Edidin. 1970. The rapid intermixing of cell surface antigens after formation of mouse–human heterokaryons. *Journal of Cell Science* 7:319–335.

FIGURE 5.7 | Research Method

Freeze Fracture

Purpose: Quick-frozen cells are fractured to split apart lipid bilayers for analysis of the membrane interior.

Protocol:

1. The specimen—a block of cells—is frozen quickly in liquid nitrogen and then the block is fractured by a sharp blow from the sharp edge of a microscopic knife.

2. The fracture may travel over membrane surfaces as it passes through the specimen, or it may split membrane bilayers into inner and outer halves thereby exposing the interior, as shown here.

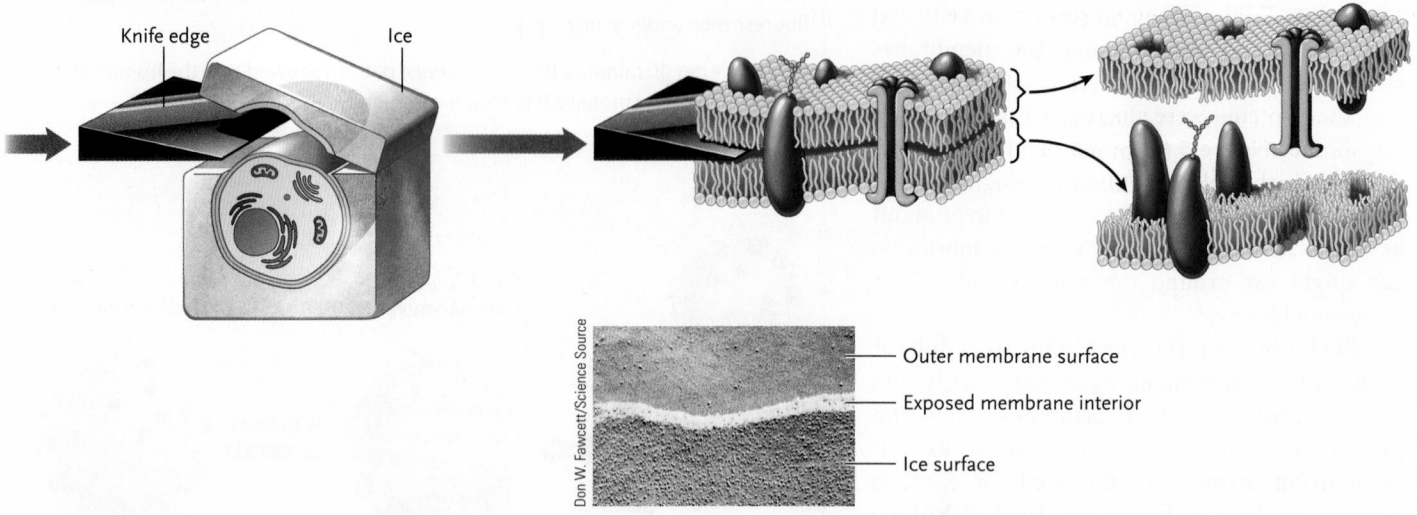

Knife edge   Ice

Outer membrane surface

Exposed membrane interior

Ice surface

Don W. Fawcett/Science Source

Interpreting the Results: The image of a freeze-fractured plasma membrane is visualized using the electron microscope. The particles visible in the exposed membrane interior are integral membrane proteins.

- Some membranes have enzymatic activities that are the properties of the membrane proteins. For instance, the particular enzymes found in or on their membranes define the specific properties of many eukaryotic organelles, such as the endoplasmic reticulum, Golgi complex, lysosome, mitochondrion, and chloroplast (see Sections 4.3 and 4.4). Enzymes on mitochondrial and chloroplast internal membranes, for example, play essential roles in converting the chemical energy in energy-rich nutrients to ATP in cellular respiration, and the conversion of light energy to the chemical energy of ATP in photosynthesis (see Sections 4.3 and 4.4 and Chapters 7 and 8).

- Membrane-spanning channel proteins form channels that transport specific ions or water selectively through the membrane. Proteins that form channels for ions may either allow the ions to pass freely, or may regulate their passage. For example, specific channel proteins control the passage of individual ions such as Na^+, K^+, Ca^{2+}, and Cl^- (see next section).

- Membrane-spanning carrier proteins bind to specific substances and transport them across the membrane. An example is the carrier protein for transporting glucose into cells (see next section).

- Some membrane proteins are receptors that recognize and bind specific molecules in the extracellular environment. Depending on the system, binding serves as the first step in bringing the substance into the cell, or it activates the receptor and triggers a series of molecular events within the cell that leads to a cellular response. For instance, some hormones bind to receptors in the plasma membrane and cause the cell to change its gene activity (discussed in Chapter 9).

- Membranes have electrical properties as a result of an uneven distribution of particular ions inside and outside of the cell. These electrical properties can serve as a mechanism of signal conduction when a cell receives an electrical, chemical, or mechanical stimulus. For instance, neurons (nerve cells) and muscle cells conduct electrical signals by using the electrical properties of membranes (discussed in Chapter 39).

- Some membrane proteins facilitate cell-to-cell communication. For example, in gap junctions particular plasma membrane proteins of adjacent cells line up and form pipelines between the two cells (see Section 4.5 and Figure 4.27).

- Surface membrane glycoproteins play an important role in cell–cell recognition and, hence, are vital molecules for the function of an organism. That is, cells recognize other cells by binding reactions involving molecules, often glycoproteins, that are on the extracellular surface of the plasma membrane. For instance, molecules known as *cell adhesion molecules (CAMs)* consist mostly of glyocoproteins. CAMs form various complexes and junctions to link cells to cells, cells to the extracellular matrix (ECM) (see Chapter 4), and the ECM to the cytoskeleton within the cell (see Chapter 4). In animal development, CAMs play an important role in organizing cells into tissues by their control of cell–cell adhesion and cell migration (see Chapter 50). As another example, the particular membrane glycoproteins on the surfaces of red blood cells determine human blood types (see Chapter 12). Matching blood types in transfusions is important because the immune system of an individual recognizes blood cells with different membrane glycoproteins than on its own blood cells as foreign, and initiates an immune response to remove the foreign cells from the circulation (see Chapter 45).

In the remainder of the chapter we focus on the functions of membranes in transporting substances into and out of the cell.

STUDY BREAK 5.1

1. Describe the fluid mosaic model for membrane structure.
2. Give two examples each of integral proteins and peripheral proteins.

5.2 Functions of Membranes in Transport: Passive Transport

Transport is the controlled movement of ions and molecules from one side of a membrane to the other. Membrane proteins are the molecules responsible for transport. Typically the movement is *directional* in that some ions and molecules move into cells, whereas others move out of cells. Transport is also *specific* in that only certain ions and molecules move directionally across membranes. Transport is critical to the ionic and molecular organization of cells and, with it, the maintenance of cellular life.

Transport occurs by two mechanisms:

1. **Passive transport** depends on concentration differences on the two sides of a membrane (concentration = number of molecules or ions per unit volume). A concentration difference is termed a **concentration gradient.** Passive transport moves ions and molecules across the membrane *with* the concentration gradient, meaning from the side with the higher concentration to the side with the lower concentration. The difference in concentration provides the energy for this form of transport.
2. **Active transport** moves ions or molecules *against* the concentration gradient, meaning from the side with the lower concentration to the side with the higher concentration. Active transport uses energy obtained directly or indirectly by breaking down ATP. **Table 5.1** compares the properties of passive and active transport.

TABLE 5.1	Characteristics of Transport Mechanisms		
	Passive Transport		
Characteristic	Simple Diffusion	Facilitated Diffusion	Active Transport
Membrane component responsible for transport	Lipids	Proteins	Proteins
Binding of transported substance	No	Yes	Yes
Energy source	Concentration gradients	Concentration gradients	ATP hydrolysis or concentration gradients
Direction of transport	With gradient of transported substance	With gradient of transported substance	Against gradient of transported substance
Specificity for molecules or molecular classes	Nonspecific	Specific	Specific
Saturation at high concentrations of transported molecules	No	Yes	Yes

In the figures depicting passive and active transport, a green arrow shows the concentration gradient with high concentration at the blunt end and low concentration at the arrow end, and a red arrow shows the direction of transport.

Passive Transport Is Based on Diffusion

Passive transport is a form of **diffusion,** the net movement of ions or molecules from a region of higher concentration to a region of lower concentration. Diffusion depends on the constant motion of ions or molecules at temperatures above absolute zero ($-273°C$). For instance, if you add a drop of food dye to a container of clear water, the dye molecules, and therefore the color, will spread or *diffuse* from their initial center of high concentration until they are distributed evenly. At this point, the water has an even color.

Diffusion involves a *net* movement of molecules or ions. Molecules and ions actually move in all directions at all times in a solution as a result of thermal (heat) energy. But when there is a concentration gradient, more molecules or ions move from the area of higher concentration to areas of lower concentration than in the opposite direction. Even after their concentration is the same in all regions, molecules or ions still move constantly from one space to another, but there is no net change in concentration on either side. This condition is an example of a *dynamic equilibrium* (*dynamic* with respect to the continuous movement, and *equilibrium* with respect to the exact balance between opposing forces).

Substances Move Passively through Membranes by Simple or Facilitated Diffusion

Hydrophobic (nonpolar) molecules are able to dissolve in the lipid bilayer of a membrane and move through it freely. By contrast, the hydrophobic core of the membrane blocks or slows the movement of hydrophilic molecules such as ions and polar molecules. Membranes that affect diffusion in this way are said to be **selectively permeable.**

TRANSPORT BY SIMPLE DIFFUSION A few small substances diffuse through the lipid part of a biological membrane. With one major exception—water—these substances are nonpolar inorganic gases such as O_2, N_2, and CO_2 and nonpolar organic molecules such as steroid hormones. This type of transport, which depends solely on molecular size and lipid solubility, is **simple diffusion** (see Table 5.1).

Water is a strongly polar molecule. Nevertheless, water molecules are small enough to slip through momentary spaces created between the hydrocarbon tails of phospholipid molecules as they flex and move in a fluid bilayer. This type of water movement across a membrane is relatively slow.

TRANSPORT BY FACILITATED DIFFUSION Many polar and charged molecules such as water, amino acids, sugars, and ions diffuse across membranes with the help of transport proteins, a mechanism termed **facilitated diffusion.** The transport proteins enable polar and charged molecules to avoid interaction with the hydrophobic lipid bilayer (see Table 5.1).

Facilitated diffusion is specific in that the membrane proteins involved transport particular polar and charged molecules, but not others. Facilitated diffusion is also dependent on concentration gradients. That is, proteins aid the transport of polar and charged molecules through membranes, but a favorable concentration gradient provides the energy for transport. Transport stops if the gradient falls to zero.

Two Groups of Transport Proteins Carry Out Facilitated Diffusion

The transport proteins that carry out facilitated diffusion are integral membrane proteins that extend entirely through the membrane. Two types are involved in facilitated diffusion:

1. **Channel proteins** form hydrophilic channels in the membrane through which water and ions can pass **(Figure 5.8A).** The channel "facilitates" the diffusion of molecules through the membrane by providing an avenue. For example, facilitated diffusion of water

through membranes occurs through specialized water channels called **aquaporins** (see Figure 5.8A[1]). A billion molecules of water per second can move through an aquaporin channel. How the molecules move is fascinating. Each water molecule is severed from its hydrogen-bonded neighbors as it is handed off to a succession of hydrogen-bonding sites on the aquaporin protein in the channel. Peter Agre at Johns Hopkins University in Baltimore, Maryland, received a Nobel Prize in 2003 for his discovery of aquaporins. (Aquaporins in plants are discussed in Chapter 34.)

Other channel proteins, **ion channels,** facilitate the transport of ions such as sodium (Na^+), potassium (K^+), calcium (Ca^{2+}), and chlorine (Cl^-). Ion channels occur in all eukaryotes. Most ion channels are **gated channels,** meaning that they switch between open, closed, or intermediate states. For instance, the gates may open or close in response to changes in voltage across the membrane, or by binding signal molecules. The opening or closing involves changes in the protein's three-dimensional shape. In animals, voltage-gated ion channels are used in nerve conduction and the control of muscle contraction (see Chapters 39 and 43); Figure 5.8A(2) shows a voltage-gated K^+ channel.

Gated ion channels perform functions that are vital to survival, as illustrated by the effects of hereditary defects in the channels. For example, as you learned in *Why it matters . . .*, the lethal genetic disease *cystic fibrosis* (CF) results from a fault in a Cl^- channel.

2. **Carrier proteins** also form passageways through the lipid bilayer **(Figure 5.8B).** Carrier proteins each bind a specific single solute, such as glucose or an amino acid, and transport it across the lipid bilayer. (Glucose is also transported by active transport, as described in the next section.) Because a single solute is transferred in this carrier-mediated fashion, the transfer is called *uniport transport.* In performing the transport step, the carrier protein undergoes conformational changes that progressively move the solute-binding site from one side of the membrane to the other, thereby transporting the solute. This property distinguishes carrier protein function from channel protein function.

Facilitated diffusion by carrier proteins can become *saturated* when there are too few transport proteins to handle all the solute molecules. For example, if glucose is added at higher and higher concentrations to the solution that surrounds an animal cell, the rate at which glucose passes through the membrane at first increases proportionately with the increase in concentration. However, at some point, as the glucose concentration is increased still further, the increase in the rate of transport slows. Eventually, further increases in concentration cause no additional rise in the rate of transport—the transport mechanism is saturated. By contrast, saturation does not occur for simple diffusion. Because the proteins that

A. Channel protein

Channel proteins form hydrophilic channels in the membrane through which water and ions can move.

1 Aquaporin

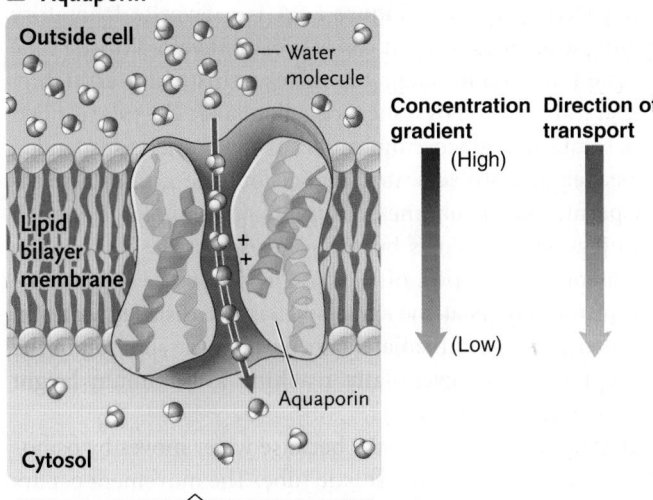

An aquaporin is a water channel. Water molecules move through the channel by being handed off to a succession of hydrogen-bonding sites on the channel in this protein.

2 K⁺ voltage-gated channel

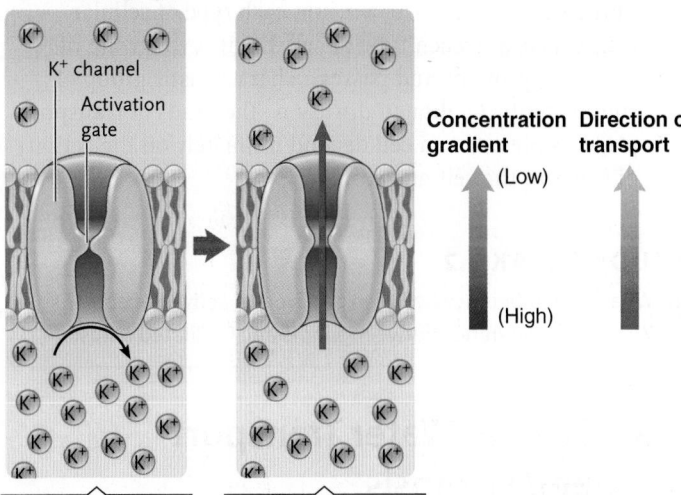

With normal voltage across the membrane, the activation gate of the K⁺ channel is closed and K⁺ cannot move across the membrane.

In response to a voltage change across the membrane, the activation gate of the K⁺ channel opens, and K⁺ moves with its concentration gradient from the cytoplasm to outside the cell.

B. Carrier protein

Carrier proteins each bind a single solute and transport it across the lipid bilayer. During the transport step, the carrier protein undergoes conformational changes that move the solute-binding site progressively from one side of the membrane to the other, thereby transporting the solute. Shown is the transport of glucose.

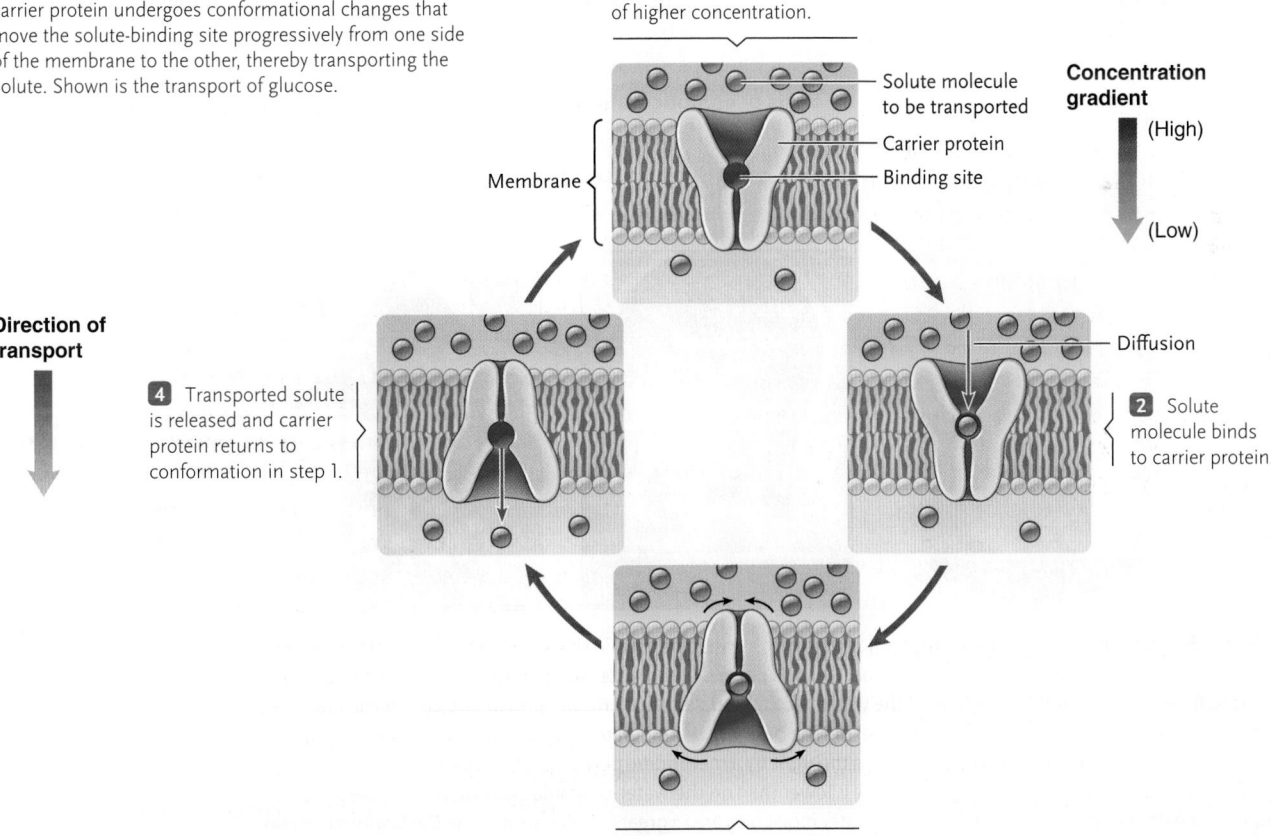

1 Carrier protein is in conformation so that binding site is exposed toward region of higher concentration.

2 Solute molecule binds to carrier protein.

3 In response to binding, carrier protein changes conformation so that binding site is exposed to region of lower concentration.

4 Transported solute is released and carrier protein returns to conformation in step 1.

FIGURE 5.8 Transport proteins for facilitated diffusion. (A) Channel protein: (1) aquaporin; (2) K⁺ voltage-gated channel. **(B)** Carrier protein.

© Cengage Learning 2017

perform facilitated diffusion are specific, cells can control the kinds of molecules and ions that pass through their membranes by regulating the types of transport proteins in their membranes. As a result, each type of cellular membrane, and each type of cell, has its own group of transport proteins and passes a characteristic group of substances by facilitated diffusion. The kinds of transport proteins present in a cell depend ultimately on the activity of genes in the cell nucleus.

STUDY BREAK 5.2

1. What is the difference between passive and active transport?
2. What is the difference between simple and facilitated diffusion?

5.3 Passive Water Transport and Osmosis

As discussed earlier, water can diffuse passively across membranes by following a concentration gradient. Water diffuses directly both through the membrane and aquaporins. The passive transport of water, called **osmosis,** occurs constantly in living cells. Inward or outward movement of water by osmosis develops forces that can cause cells to swell and burst or shrink and shrivel up. Much of the energy budget of many cell types, particularly in animals, is spent counteracting the inward or outward movement of water by osmosis.

Osmosis Can Be Demonstrated in a Purely Physical System

The apparatus shown in **Figure 5.9A** is a favorite laboratory demonstration of osmosis. It consists of an inverted thistle tube (so named because its shape resembles a thistle flower) tightly sealed at its lower end by a sheet of cellophane. The tube is filled with a solution of glucose molecules in water and is suspended in a beaker of distilled water. The cellophane acts as a selectively permeable membrane because its pores are large enough to admit water molecules but not glucose. At the start of the experiment, the position of the tube is set so the level of the liquid in the tube is at the same level as the distilled water in the beaker. Almost immediately, the level of the solution in the tube begins to rise, eventually reaching a maximum height above the liquid in the beaker.

The liquid rises in the tube because water moves by osmosis from the beaker into the thistle tube. The movement occurs passively, in response to a concentration gradient in which the water molecules are more concentrated in the beaker than inside the thistle tube. The basis for the gradient is shown in **Figure 5.9B.** The glucose molecules are more concentrated on one side of the selectively permeable membrane. On this side, association of water molecules with those solute molecules reduces the amount of water available to cross the membrane. Thus, although initially there is an equal apparent water concentration on each side of the membrane, there is a difference in the *free water* concentration—that is, the water available to

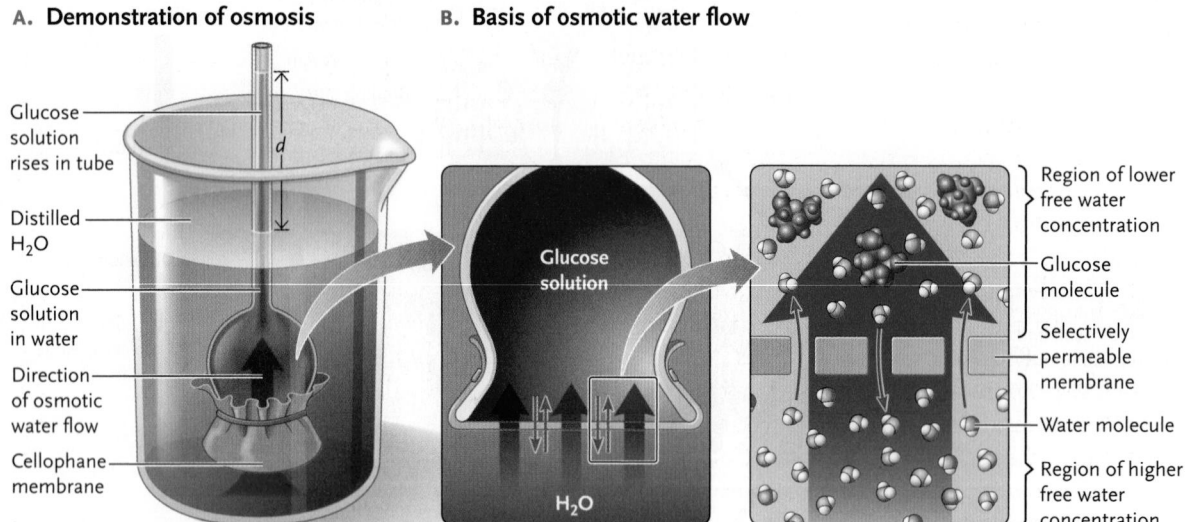

A. Demonstration of osmosis

Glucose solution rises in tube

Distilled H_2O

Glucose solution in water

Direction of osmotic water flow

Cellophane membrane

B. Basis of osmotic water flow

Glucose solution

H_2O

Region of lower free water concentration

Glucose molecule

Selectively permeable membrane

Water molecule

Region of higher free water concentration

FIGURE 5.9 Osmosis. (A) An apparatus demonstrating osmosis. The fluid in the tube rises because of the osmotic flow of water through the cellophane membrane, which is permeable to water but not to glucose molecules. Osmotic flow continues until the weight of the water in column *d* develops enough pressure to counterbalance the movement of water molecules into the tube. **(B)** The basis of osmotic water flow. The pure water solution on the left is separated from the glucose solution on the right by a membrane permeable to water but not to glucose. The free water concentration on the glucose side is lower than on the water-only side because water molecules are associated with the glucose molecules. That is, water molecules are in greater concentration on the bottom than on the top. Although water molecules move in both directions across the membrane (small red arrows), there is a net upward movement of water (blue arrows), with the water's concentration gradient.

move across the membrane. Specifically, the concentration of free water molecules is lower on the glucose side than on the pure water side. In response, more water molecules from the pure water side will hit the pores in the membrane than from the solute side, producing a net movement of water from the pure water side to the glucose solution side. Osmosis is the net diffusion of water molecules through a selectively permeable membrane in response to a gradient of this type.

The solution stops rising in the tube when the pressure created by the weight of the raised solution exactly balances the tendency of water molecules to move from the beaker into the tube in response to the concentration gradient. At this point, the system is in a state of dynamic equilibrium and no further net movement of water molecules occurs. **Osmotic pressure** is the term for the pressure that must be applied to a solution to prevent water movement across a membrane.

A formal definition for osmosis is *the net movement of water molecules across a selectively permeable membrane by passive diffusion, from a solution of lesser solute concentration to a solution of greater solute concentration* (the *solute* is the substance dissolved in water). For osmosis to occur, the selectively permeable membrane must allow water molecules, but not molecules of the solute, to pass. Pure water does not need to be on one side of the membrane; osmotic water movement also occurs if a solute is at different concentrations on the two sides.

Tonicity Is the Effect of the Concentration of Nonpenetrating Solutes in a Solution on Cell Volume

The solution surrounding a cell can affect cell volume—whether the cell remains the same size, shrinks, or swells—depending on the relative concentrations of nonpenetrating solutes surrounding the cell and within the cell. *Nonpenetrating solutes* are proteins and other molecules that cannot pass through a membrane that is impermeable to them but freely permeable to water. The effect a solution has on cell volume when the solution surrounds the cell is the **tonicity** of the solution (*tonos* = tension or tone). Tonicity is a property of a solution with respect to a particular membrane. That is, the same surrounding solution can have different effects on different cell types.

If the solution surrounding a cell contains nonpenetrating solutes at lower concentrations than in the cell, the solution is said to be **hypotonic** to the cell (*hypo* = under or below). When a cell is in a hypotonic solution, water enters by osmosis and the cell tends to swell **(Figure 5.10A)**. Animal cells (for instance, red blood cells) in a hypotonic solution may actually swell to the point of bursting. However, in most plant cells, strong walls prevent the cells from bursting in a hypotonic solution. In most land plants, the cells at the surfaces of roots are surrounded by almost pure water, which is hypotonic to the cells and tissues of the root. As a result, water flows from the surrounding soil into the root cells by osmosis. The osmotic pressure developed by the inward flow contributes part of the force required to raise water from the roots to the leaves of the plant. Osmosis

also drives water into cells of the stems and leaves of plants. The resulting osmotic pressure, called **turgor pressure,** pushes the cells tightly against their walls and supports the softer tissues against the force of gravity. (Turgor pressure in plants is described more in Chapter 34.)

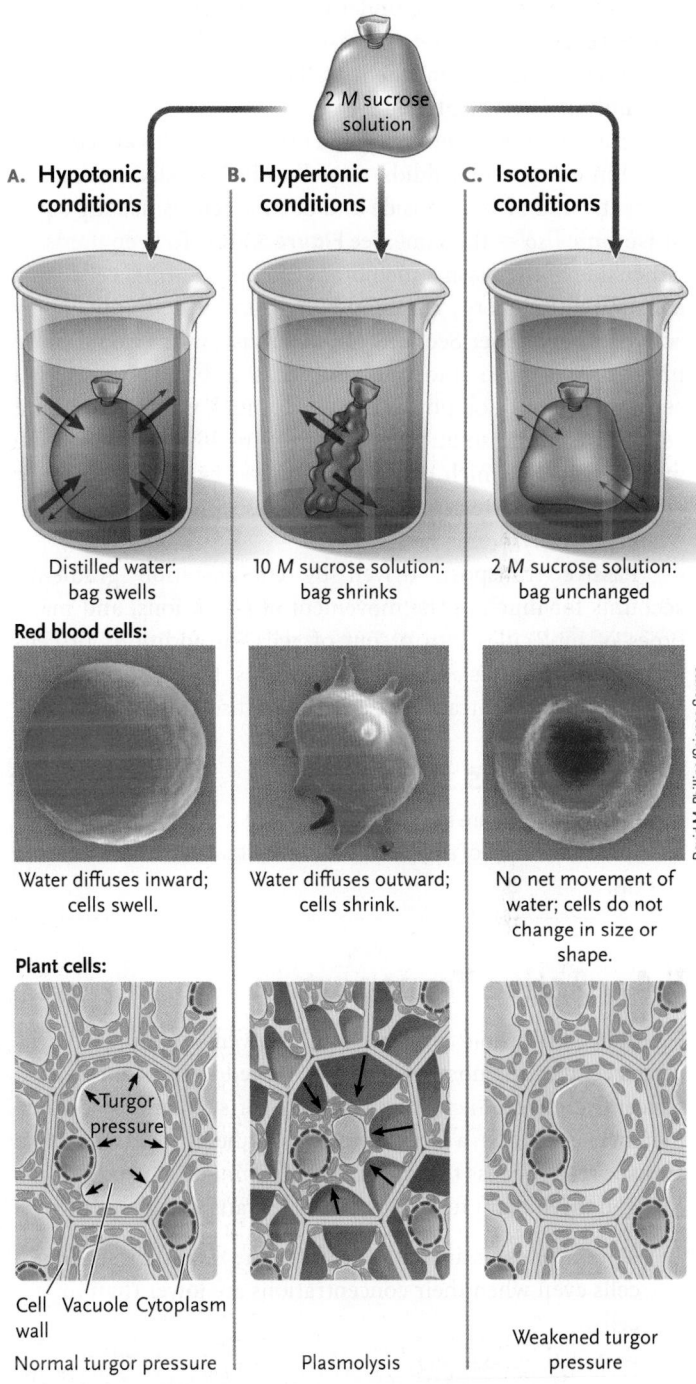

FIGURE 5.10 Tonicity and osmotic water movement. The diagrams show what happens when a cellophane bag filled with a 2 *M* sucrose solution is placed in a **(A)** hypotonic, **(B)** hypertonic, or **(C)** isotonic solution. They also show the corresponding effects of these three types of solutions on animal cells and plant cells. (See Chapter 34 and Figure 34.5 for more discussion of osmotic effects on plant cells.)

© Cengage Learning 2017

If the solution that surrounds a cell contains nonpenetrating solutes at higher concentrations than in the cell, the outside solution is said to be **hypertonic** to the cells (*hyper* = over or above). When a cell is in a hypertonic solution, water leaves by osmosis. If the outward osmotic movement exceeds the capacity of cells to replace the lost water, both animal and plant cells will shrink **(Figure 5.10B)**. In plants, the shrinkage and loss of internal osmotic pressure under these conditions causes stems and leaves to wilt. In extreme cases, plant cells shrink so much that they retract from their walls, a condition known as **plasmolysis** (see Chapter 34).

In animals, ions, proteins, and other molecules are concentrated in extracellular fluids, as well as inside cells, so that the concentration of water inside and outside cells is usually equal or **isotonic** (*iso* = the same; see **Figure 5.10C**). To keep fluids on either side of the plasma membrane isotonic, animal cells must constantly use energy to pump Na$^+$ from inside to outside by active transport (see Section 5.4); otherwise, water would move inward by osmosis and cause the cells to burst. For animal cells, an isotonic solution is usually optimal, whereas for plant cells, an isotonic solution results in some loss of turgor. The mechanisms by which plants and animals balance their water content by regulating osmosis are discussed in Chapters 34 and 48, respectively.

Passive transport, driven by concentration gradients, accounts for much of the movement of water, ions, and many types of molecules into or out of cells. In addition, all cells transport some ions and molecules against their concentration gradients by active transport (see the next section).

STUDY BREAK 5.3

1. What conditions are required for osmosis to occur?
2. Explain the effect of a hypertonic solution that surrounds animal cells.

5.4 Active Transport

Facilitated diffusion accelerates the movement of substances across cellular membranes, but it is limited to transport with a concentration gradient. The transport of substances across a membrane against a concentration gradient requires **active transport,** a process that requires energy input.

The three main functions of active transport are:

1. Uptake of essential nutrients from the fluid surrounding cells even when their concentrations are lower than in cells;
2. Removal of secretory or waste materials from cells or organelles even when the concentration of those materials is higher outside the cells or organelles; and
3. Maintenance of essentially constant intracellular concentrations of Na$^+$, K$^+$, H$^+$, and Ca^{2+}.

Because ions are charged molecules, active transport of ions may contribute to voltage—an electrical potential difference—across the plasma membrane, called a **membrane potential.** The unequal distribution of ions across the membrane created by passive transport also contributes to the voltage. Neurons and muscle cells use the membrane potential in a specialized way. That is, in response to electrical, chemical, mechanical, and certain other types of stimuli, their membrane potential changes rapidly and transiently. In nerve cells, for example, this type of transport is the basis for transmission of a nerve impulse (discussed in Chapter 39).

In short, active transport usually establishes differences in solute concentrations or of voltage across membranes that are important for cell or organelle function. By contrast, passive diffusion and facilitated diffusion act mostly to move substances across membranes in the direction toward equalizing their concentrations on each side.

Active Transport Requires a Direct or Indirect Input of Energy Derived from ATP Hydrolysis

There are two kinds of active transport:

1. In **primary active transport,** the same protein that transports a substance also hydrolyzes ATP to power the transport directly.
2. In **secondary active transport,** the transport is driven indirectly by ATP hydrolysis. That is, the transporter proteins do not break down ATP but use instead a favorable concentration gradient of ions, built up by primary active transport, as their energy source for active transport of a different ion or molecule.

Other features of active transport resemble facilitated diffusion (listed in Table 5.1). Both processes depend on membrane transport proteins, both are specific, and both can be saturated. In both, the transport proteins are *carrier proteins* that change their conformation as they function.

Primary Active Transport Moves Positively Charged Ions across Membranes

The primary active transport pumps all move positively charged ions—Na$^+$, K$^+$, H$^+$, and Ca^{2+}—across membranes. The gradients of positive ions established by primary active transport pumps underlie functions that are essential for cellular life. For example, the **Na$^+$/K$^+$ pump** (also known as the **sodium-potassium pump** or **Na$^+$/K$^+$-ATPase**), located in the plasma membrane of all animal cells, pushes three Na$^+$ out of the cell and two K$^+$ into the cell in the same pumping cycle **(Figure 5.11)**. As a result, positive charges accumulate in excess outside the membrane, and the inside of the cell becomes negatively charged with respect to the outside. This creates a membrane potential measuring from about −50 to −200 millivolts (mV; 1 millivolt = 1/1,000th of a volt), with the minus sign indicating that the charge inside the cell is negative versus the outside. That is, there is both a concentration difference (of the ions) and an electrical charge difference on the two sides of

ACTIVE TRANSPORT: THE Na$^+$/K$^+$ PUMP, AN ACTIVE TRANSPORT PROTEIN IN THE PLASMA MEMBRANE

Energy from the protein's hydrolysis of ATP directly transports Na$^+$ out of the cell and K$^+$ into the cell, each against its concentration gradient. The pump moves three Na$^+$ out and two K$^+$ in for each ATP molecule hydrolyzed.

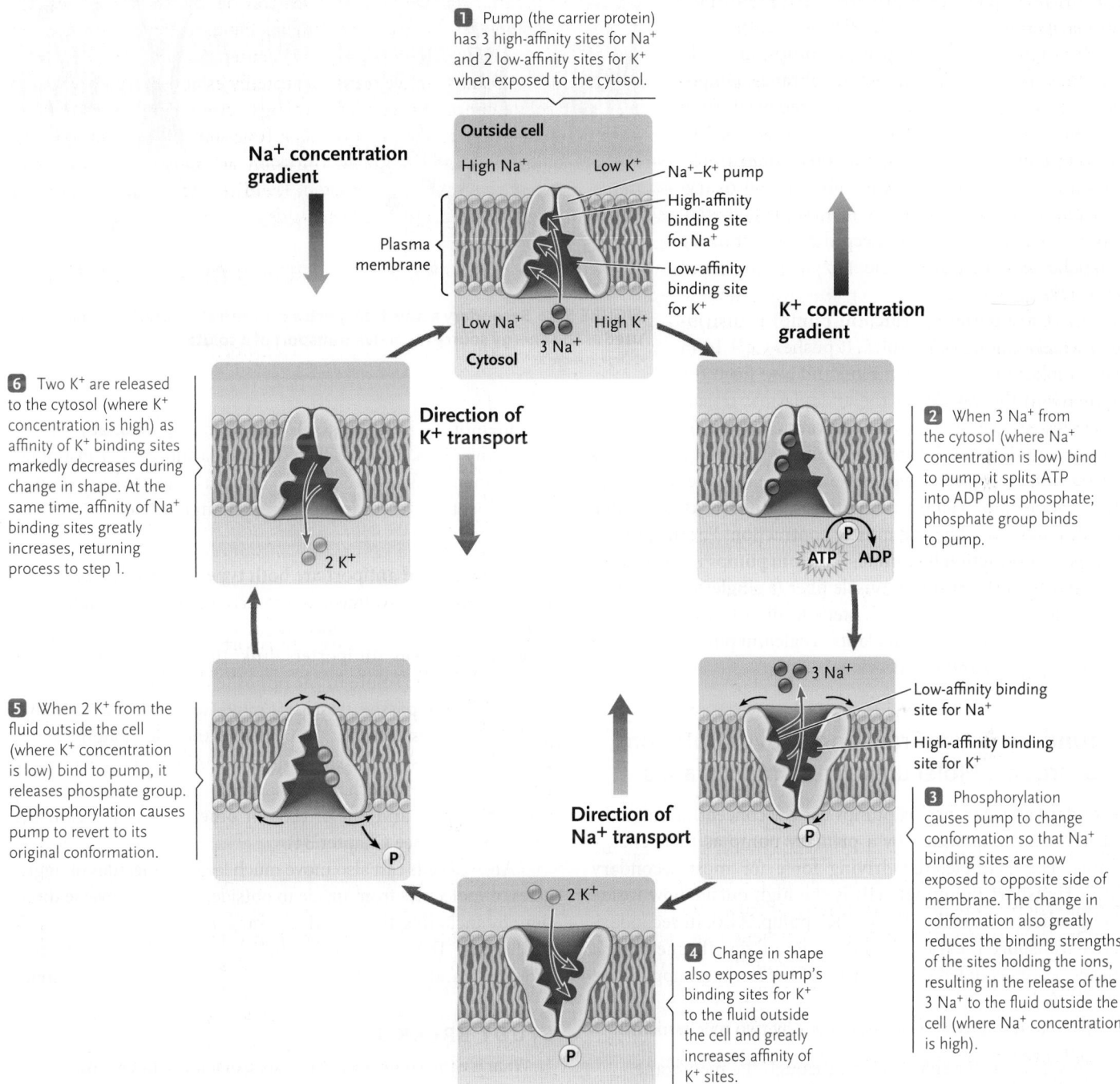

1 Pump (the carrier protein) has 3 high-affinity sites for Na$^+$ and 2 low-affinity sites for K$^+$ when exposed to the cytosol.

Na$^+$ concentration gradient

Outside cell

High Na$^+$ Low K$^+$

Na$^+$–K$^+$ pump

High-affinity binding site for Na$^+$

Plasma membrane

Low-affinity binding site for K$^+$

K$^+$ concentration gradient

Low Na$^+$ High K$^+$

3 Na$^+$

Cytosol

Direction of K$^+$ transport

6 Two K$^+$ are released to the cytosol (where K$^+$ concentration is high) as affinity of K$^+$ binding sites markedly decreases during change in shape. At the same time, affinity of Na$^+$ binding sites greatly increases, returning process to step 1.

2 K$^+$

2 When 3 Na$^+$ from the cytosol (where Na$^+$ concentration is low) bind to pump, it splits ATP into ADP plus phosphate; phosphate group binds to pump.

P

ATP   ADP

3 Na$^+$

Low-affinity binding site for Na$^+$

High-affinity binding site for K$^+$

5 When 2 K$^+$ from the fluid outside the cell (where K$^+$ concentration is low) bind to pump, it releases phosphate group. Dephosphorylation causes pump to revert to its original conformation.

P

3 Phosphorylation causes pump to change conformation so that Na$^+$ binding sites are now exposed to opposite side of membrane. The change in conformation also greatly reduces the binding strengths of the sites holding the ions, resulting in the release of the 3 Na$^+$ to the fluid outside the cell (where Na$^+$ concentration is high).

Direction of Na$^+$ transport

2 K$^+$

4 Change in shape also exposes pump's binding sites for K$^+$ to the fluid outside the cell and greatly increases affinity of K$^+$ sites.

P

SUMMARY In primary active transport, exemplified by the Na$^+$/K$^+$ pump, the carrier protein that transports a substance also hydrolyzes ATP to power the transport directly. Primary active transport pumps all move positively charged ions—Na$^+$, K$^+$, H$^+$, and Ca^{2+}—across membranes against their concentration gradients.

think like a scientist Suppose there was a mutation in the gene that encodes the channel protein, which reduces the ability of the protein to hydrolyze ATP drastically. What effect would that have on primary active transport?

the membrane, constituting what is called an **electrochemical gradient.** Electrochemical gradients store energy that is used for other transport mechanisms. For instance, an electrochemical gradient across a nerve cell membrane drives the movement of ions involved in nerve impulse transmission (see Chapter 39). Other primary active transport pumps operate similarly to the Na$^+$/K$^+$ pump.

H$^+$ pumps (also called **proton pumps**) move H$^+$ hydrogen ions (protons) across membranes, temporarily binding a phosphate group removed from ATP during the pumping cycle (see Figure 34.3A). Proton pumps have various functions. For example, proton pumps in the plasma membrane in prokaryotes, plants, and fungi generate membrane potential. Proton pumps in lysosomes keep the pH within the organelle low, activating the enzymes contained within them.

The **Ca^{2+} pump** (or **calcium pump**) is distributed widely among eukaryotes. It pushes Ca^{2+} from the cytoplasm to the cell exterior, and also from the cytosol into the vesicles of the endoplasmic reticulum (ER). As a result, Ca^{2+} concentration is typically high outside cells and inside ER vesicles and low in the cytosol. This Ca^{2+} gradient is used universally among eukaryotes as a regulatory control of cellular activities as diverse as secretion, microtubule assembly, and muscle contraction. For the process of muscle contraction in animals, calcium pumps release stored Ca^{2+} from vesicles inside a muscle fiber (a single cell in a muscle), which initiates a series of steps leading to contraction of the fiber (see Chapter 43). In plants, a calcium pump is involved in pollen growth and fertilization.

Secondary Active Transport Moves Both Ions and Organic Molecules across Membranes

Secondary active transport pumps use the concentration gradient of an ion established by a primary pump as their energy source. For example, the driving force for most secondary active transport in animal cells is the high outside/low inside Na$^+$ gradient created by the Na$^+$/K$^+$ pump. Also, in secondary active transport, the transfer of the solute across the membrane always occurs coupled with transfer of the ion that supplies the driving force.

Secondary active transport occurs by two mechanisms:

1. In **symport,** the solute moves through the membrane channel in the same direction as the driving ion **(Figure 5.12A).** Sugars, such as glucose, and amino acids are examples of molecules transported actively into cells by symporters.
2. In **antiport,** the driving ion moves through the membrane channel in one direction, providing the energy for the active transport of another molecule through the

A. Symport
The transported solute moves in the same direction as the gradient of the driving ion.

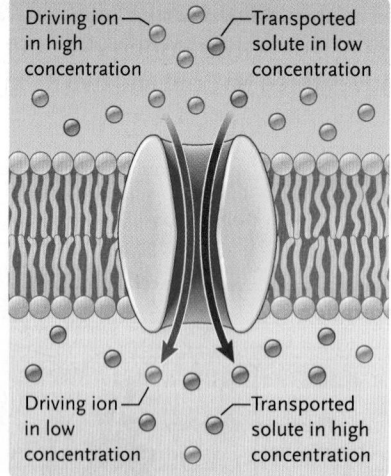

B. Antiport
The transported solute moves in the direction opposite from the gradient of the driving ion.

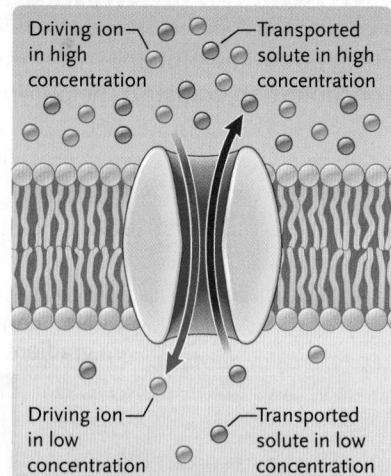

FIGURE 5.12 Secondary active transport: a concentration gradient of an ion is used as the energy source for active transport of a solute.
© Cengage Learning 2017

membrane in the opposite direction **(Figure 5.12B).** In many cases, ions are exchanged by antiport. For example, in many tissues, a Na$^+$/Ca^{2+} antiporter is used to remove cytosolic calcium from cells, exchanging one Ca^{2+} for three Na$^+$.

Symport and antiport are both types of **cotransport.** Symporters link the movement of the driving ion from high to low concentration with the movement of the solute in the same direction whereas antiporters link the ion movement with movement of the solute in the opposite direction.

Active and passive transport move ions and smaller hydrophilic molecules across cellular membranes. Although the discussion in this section has focused on animal cells, active and passive transport mechanisms are also important for moving substances into and out of cells of other types of organisms, such as plants (see Chapter 34).

Animal cells can also move much larger molecules or aggregates of molecules from inside to outside, or in the reverse direction, by including them in the inward or outward vesicle traffic of the cell. The mechanisms that carry out this movement—exocytosis and endocytosis—are discussed in the next section.

STUDY BREAK 5.4

1. What is active transport? What is the difference between primary and secondary active transport?
2. How is a membrane potential generated?

THINK OUTSIDE THE BOOK

Collaboratively or on your own, draw a diagram showing how ions and solutes exchange in a symport system and an antiport system in the intestinal tract.

5.5 Exocytosis and Endocytosis

The largest molecules transported through cellular membranes by passive and active transport are in the size range of amino acids or monosaccharides such as glucose. Eukaryotic cells import and export larger molecules by exocytosis and endocytosis (introduced in Section 4.3). The export of materials by exocytosis primarily carries secretory proteins and some waste materials from the cytoplasm to the cell exterior. Import by endocytosis may carry proteins, larger aggregates of molecules, or even whole cells from the outside into the cytoplasm. Exocytosis and endocytosis also contribute to the back-and-forth flow of membranes between the endomembrane system and the plasma membrane. Both exocytosis and endocytosis require energy; thus, both processes stop if the ability of a cell to make ATP is inhibited. The chapter-opening photo is a confocal micrograph of endocytosis in cancer cells.

Exocytosis Releases Molecules to the Outside of the Cell by Means of Secretory Vesicles

In exocytosis, secretory vesicles originated by budding from the Golgi complex move through the cytoplasm along microtubules of the cytoskeleton and contact the plasma membrane. In one major mechanism of exocytosis, the vesicle membrane fuses with the plasma membrane, releasing the contents of the vesicle to the cell exterior (**Figure 5.13A**). For some cells, such as particular nerve cells, a *kiss-and-run* mechanism occurs in which a vesicle fuses with the plasma membrane, the vesicle contents are released from the cell, and then the vesicle reforms and moves back into the cell. (The release of molecules from nerve cells and their role in conduction of signals in the nervous system is described in Chapter 39.)

All eukaryotic cells secrete materials to the outside through exocytosis. For example, in animals, glandular cells secrete peptide hormones or milk proteins, and cells that line the digestive tract secrete mucus and digestive enzymes. Plant cells, fungal cells, and bacterial cells use exocytosis to secrete proteins and other macromolecules associated with the cell wall, including enzymes, proteins, and carbohydrates. Fungi and bacteria also secrete enzymes by exocytosis to digest nutrients in their environments. Finally, all organisms use exocytosis to place integral membrane proteins in the plasma membrane.

Endocytosis Brings Materials into Cells in Endocytic Vesicles

In endocytosis, proteins and other substances are trapped in pitlike depressions that bulge inward from the plasma membrane. The depression then pinches off as an endocytic vesicle. Endocytosis occurs in most eukaryotic cells by one of two distinct but related pathways:

1. In **bulk endocytosis** (sometimes called **pinocytosis,** meaning "cell drinking") a drop of the aqueous fluid surrounding the cell—called the *extracellular fluid* (ECF)—is taken into the cell together with any molecules that happen to be in solution in the water (**Figure 5.13B,** top panel). The process is nonspecific in that it takes in any solutes present in the fluid because the membrane lacks surface receptors for specific molecules.

2. In **receptor-mediated endocytosis,** the target molecules to be taken in are bound to receptor proteins on the outer cell surface (**Figure 5.13B,** middle and bottom panels). The receptors, which are integral proteins of the plasma membrane, recognize and bind only certain molecules in the solution that surrounds the cell, which makes this type of endocytosis highly specific. After binding their target molecules, the receptors collect into a depression in the plasma membrane called a **coated pit** because a network of proteins (called **clathrin**) coat and reinforce the cytoplasmic side. With the target molecules attached, the pits deepen and pinch free of the plasma membrane to form endocytic vesicles. Once in the cytoplasm, an endocytic vesicle loses its clathrin coat rapidly and may fuse with a lysosome. The enzymes within the lysosome then digest the contents of the vesicle, breaking them down into smaller molecules useful to the cell. These molecular products—for example, amino acids and monosaccharides—enter the cytoplasm by crossing the vesicle membrane via transport proteins. The membrane proteins are recycled to the plasma membrane.

Mammalian cells take in many substances by receptor-mediated endocytosis, including peptide hormones such as insulin, growth factors, neurotransmitters, enzymes, antibodies, blood proteins, iron, and vitamin B_{12}. Some viruses exploit the receptor-mediated exocytosis mechanism to enter cells. For instance, HIV, the virus that causes AIDS, binds to membrane receptors that function normally to internalize a needed molecule. The receptors that bind these substances to the plasma membrane are present in thousands to hundreds of thousands of copies. For example, a mammalian cell plasma membrane has about 20,000 receptors for *low-density lipoprotein (LDL)*. LDL, a complex of lipids and proteins, is the way cholesterol moves through the bloodstream. When LDL binds to its receptor on the membrane, it is taken into the cell by receptor-mediated endocytosis. Then, by the steps described earlier, the LDL is broken down within the cell and cholesterol is released into the cytoplasm. *Molecular Insights* describes the discovery of receptor-mediated endocytosis.

Some specialized cells, such as certain white blood cells *(phagocytes)* in the bloodstream, or protists such as *Amoeba proteus,* can take in large aggregates of molecules, cell parts, or even whole cells by a process related to receptor-mediated endocytosis. The process, called **phagocytosis** (meaning "cell eating"), begins when surface receptors bind molecules on the substances to be taken in (**Figure 5.14**). Cytoplasmic lobes then extend, surround, and engulf the materials, forming a pit that

A. Exocytosis

Vesicle joins plasma membrane, releases contents.

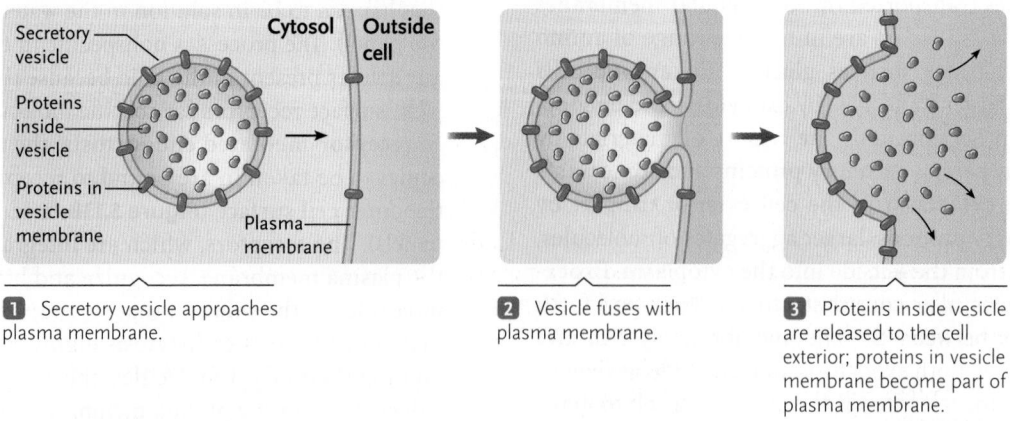

1 Secretory vesicle approaches plasma membrane.

2 Vesicle fuses with plasma membrane.

3 Proteins inside vesicle are released to the cell exterior; proteins in vesicle membrane become part of plasma membrane.

B. Endocytosis

Bulk endocytosis (pinocytosis): Vesicle imports water and other substances from outside cell.

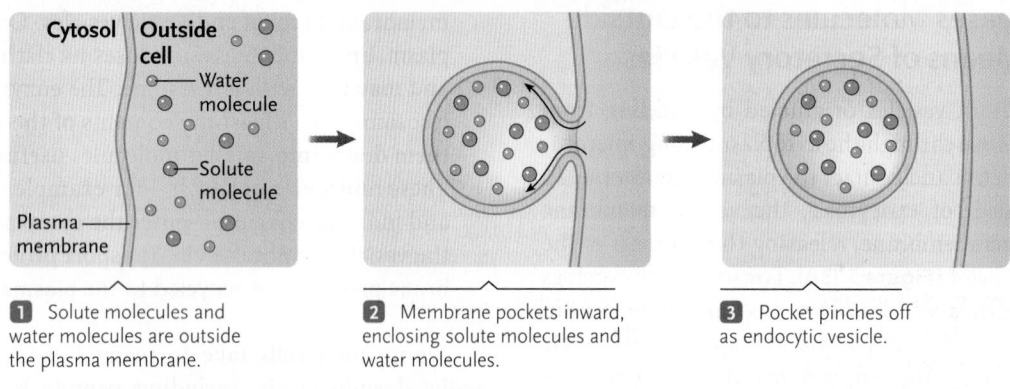

1 Solute molecules and water molecules are outside the plasma membrane.

2 Membrane pockets inward, enclosing solute molecules and water molecules.

3 Pocket pinches off as endocytic vesicle.

Receptor-mediated endocytosis: Vesicle imports specific molecules.

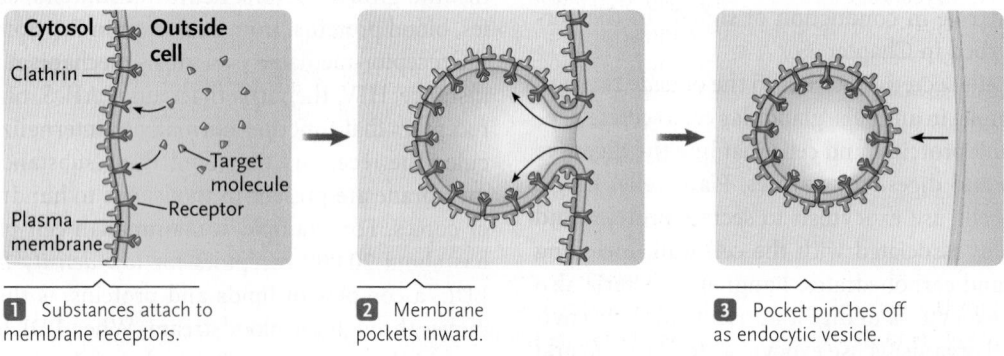

1 Substances attach to membrane receptors.

2 Membrane pockets inward.

3 Pocket pinches off as endocytic vesicle.

Micrographs of stages of receptor-mediated endocytosis shown above

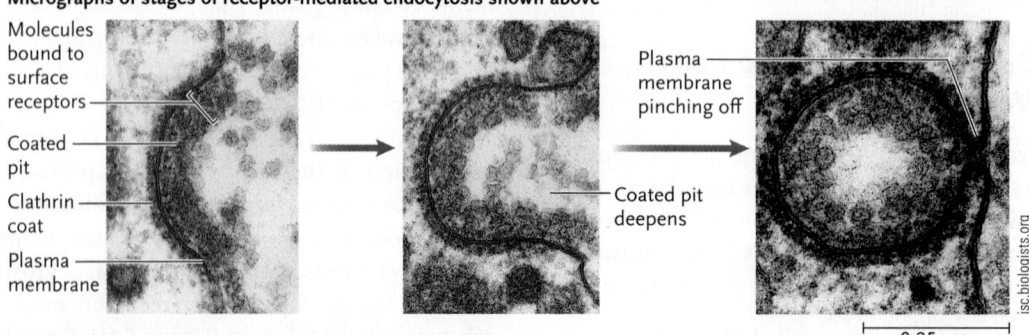

0.25 μm

jsc.biologists.org

FIGURE 5.13 Exocytosis (A) and endocytosis (B).

© Cengage Learning 2017

Research Serendipity: The discovery of receptor-mediated endocytosis

When they were still in training, two physicians, Michael Brown and Joseph Goldstein, treated two sisters, 6 and 8 years old. The children were dying of recurrent heart attacks brought on by extremely high cholesterol levels in their blood. Brown and Goldstein performed groundbreaking research to determine how such young children develop a condition that usually appears only in late middle or old age.

As you have learned, cholesterol is essential for keeping cell membranes fluid. The blood transports cholesterol to cells that need it. But blood cholesterol can cause *atherosclerosis* (thickening of the arteries as a result of the buildup of fatty materials such as cholesterol; see *Focus on Research: Applied Research* in Chapter 3), which can be lethal. To lessen this danger, the body packages cholesterol with proteins to form lipoproteins such as low-density lipoprotein (LDL).

Patients with familial hypercholesteremia (FH) have a higher than normal concentration of LDL in their blood and often experience atherosclerosis and heart attacks early in life. Individuals with one copy of the mutated gene responsible for the disease have about twice the normal level of LDL, and begin to have heart attacks at 30 to 40 years old. Individuals like the two sisters, with two copies of the mutated gene, have 6- to 10-fold higher than normal levels of LDL in their blood, and they often have heart attacks in childhood.

Research Question

What is the molecular basis of familial hypercholesteremia?

Experiments

Brown and Goldstein of the University of Texas Southwestern Medical Center, Dallas, Texas, carried out two key experiments:

1. They cultured human cells (skin fibroblasts) from normal individuals and found that radioactively labeled LDL bound strongly to the cells **(Figure).** They interpreted this result to mean that the cells had specific surface receptors for LDL. Once bound to the receptor, the LDL was taken into the cell with the cholesterol. In fibroblasts from patients with two mutant forms of the *FH* gene, very little radioactively labeled LDL bound to the cells **(Figure).** This experiment showed that FH occurs because patients either do not make the receptor or have defective receptor molecules. As a result, LDL accumulates in the blood because it cannot be taken into the cells. Further research has confirmed that a mutation in the gene for this receptor is responsible for FH.

2. To determine how the receptor–LDL complex enter cells, the two researchers, with Richard G. W. Anderson, used the same fibroblast system. In this case, they bound ferritin to the LDL. Ferritin is a protein that binds iron, which is electron dense, making it possible to detect the LDL visually as black dots under the electron microscope. When they incubated fibroblast cells from normal individuals with ferritin-labeled LDL, the investigators saw that the ferritin collected at short segments of the plasma membrane, which appeared to be indented and coated on both sides by "fuzzy material." These regions are now known to be clathrin-coated pits, and they are the cellular entry site for the receptor–LDL complex.

Conclusion

In a major step toward answering the question of what causes FH, Goldstein and Brown had discovered that LDL enters cells using a specific receptor on the cell surface, which is absent or reduced in FH patients. Serendipitously, they had also discovered the answer to broader question: How do cells take in specific molecules that cannot pass through the membrane? They had discovered receptor-mediated endocytosis (see Figures 5.13B, middle and bottom panels). The researchers received the Nobel Prize in 1985 "for their discoveries concerning the regulation of cholesterol metabolism."

think like a scientist

The researchers also carried out an important variation of their first key experiment. Specifically, they cultured fibroblast cells in the presence both of radioactively labeled LDL and of a 50-fold excess of unlabeled LDL. The result of this experiment was that only a limited amount of radioactivity became associated with the cells, far less than in the experiment described in the box. Interpret this result, and explain whether or not it supports their overall conclusions.

Sources: R. G. W. Anderson, J. L. Goldstein, and M. S. Brown. 1976. Localization of low density lipoprotein receptors on plasma membrane of normal human fibroblasts and their absence in cells from a familial hypercholesteremia homozygote. *Proceedings of the National Academy of Sciences USA* 73:2434–2438; R. G. W. Anderson, J. L. Goldstein, and M. S. Brown. 1977. A mutation that impairs the ability of lipoprotein receptors to localize in coated pits on the cell surface of human fibroblasts. *Nature* 270:695–699; M. S. Brown and J. L. Goldstein. 1974. Familial hypercholesteremia: Defective binding of lipoproteins to cultured fibroblasts associated with impaired regulation of 3-hydroxy-3-methylglutaryl coenzyme A reductase activity. *Proceedings of the National Academy of Sciences USA* 71:788–792.

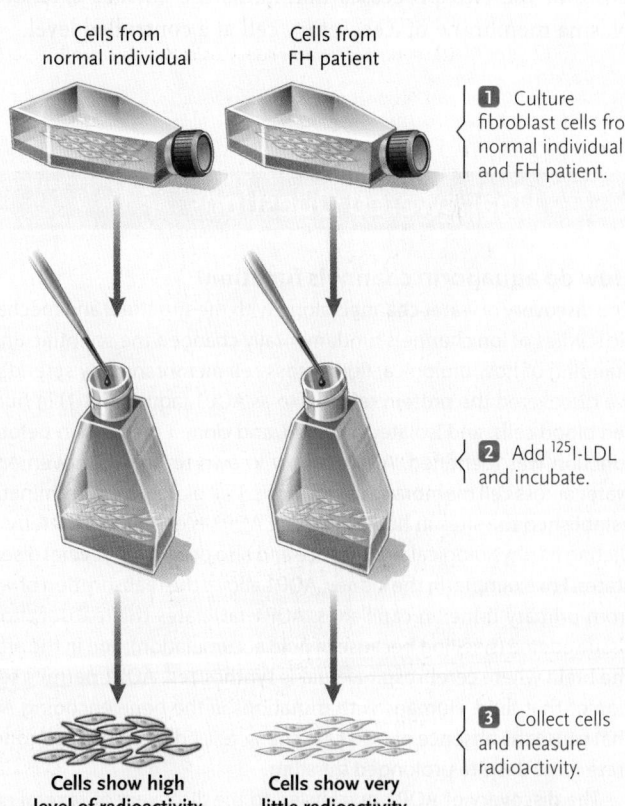

Cells from normal individual

Cells from FH patient

1 Culture fibroblast cells from normal individual and FH patient.

2 Add ^{125}I-LDL and incubate.

3 Collect cells and measure radioactivity.

Cells show high level of radioactivity

Cells show very little radioactivity

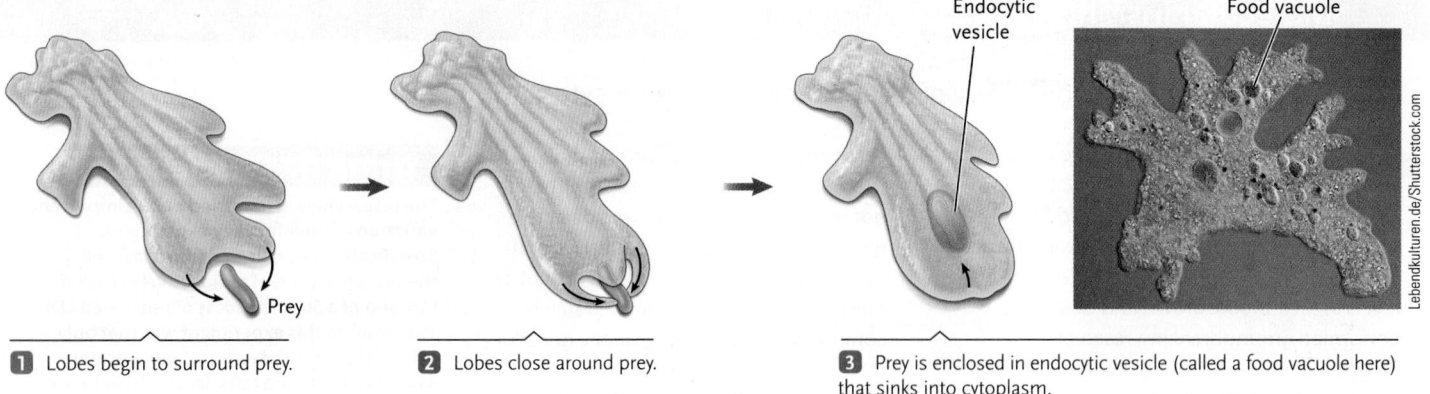

1 Lobes begin to surround prey.

2 Lobes close around prey.

3 Prey is enclosed in endocytic vesicle (called a food vacuole here) that sinks into cytoplasm.

Endocytic vesicle

Food vacuole

FIGURE 5.14 Phagocytosis, in which lobes of the cytoplasm extend outward and surround a cell targeted as prey. The differential interference contrast (Nomarski) micrograph shows the protist *Amoeba proteus* with a number of food vacuoles (corresponding to step 3 in the diagram) containing *Paramecium bursaria,* which has a symbiotic relationship with a green alga, and *Haemotococcus pluvialis,* a green alga; white blood cells called phagocytes carry out a similar process in mammals.
© Cengage Learning 2017

pinches off and sinks into the cytoplasm as a large endocytic vesicle. Enzymes then digest the materials as in receptor-mediated endocytosis, and the cell permanently sequesters any remaining residues into storage vesicles or expels them by exocytosis as wastes.

Exocytosis and endocytosis are coupled processes. Within seconds to minutes of exocytosis, endocytosis is initiated to retrieve vesicle membrane and its associated proteins. This coupling of the two processes maintains the surface area of the plasma membrane of a secretory cell at a controlled level.

In sum, through the combined mechanisms of passive transport, active transport, exocytosis, and endocytosis, cells maintain their internal concentrations of ions and molecules and exchange larger molecules such as proteins with their surroundings.

STUDY BREAK 5.5

1. What is the mechanism of exocytosis?
2. What is the difference between bulk endocytosis and receptor-mediated endocytosis?

 Unanswered Questions

How do aquaporin channels function?

The discovery of water channels along with the structural and mechanistic studies of ion channels fundamentally changed the scientific understanding of how biological fluids cross cell membranes. By serendipity, we discovered the protein referred to as AQP1 (aquaporin-1) in human red blood cells, and isolated, purified, and cloned the protein before its function was identified. AQP1 is now known to permit movement of water across cell membranes by osmosis. Cell biological determinations established the sites in humans where AQP1 is expressed, thereby predicting its physiological significance and also predicting several diseases states. For example, in the kidney, AQP1 allows the reabsorption of water from primary urine; in capillaries, AQP1 facilitates the reabsorption of tissue edema (swelling because of fluid accumulation); and in the area of the brain where cerebrospinal fluid is synthesized, AQP1 permits secretion of that fluid. Humans with mutations in the gene encoding AQP1 that cause the absence of AQP1 are characterized by inability to concentrate urine despite prolonged thirsting.

The discovery of AQP1 quickly led to the discovery of several other mammalian aquaporins. In humans, AQP2 resides in the final segment of the kidney where the neurohormone antidiuretic hormone (also known as vasopressin; see Chapter 48) regulates water reabsorption by controlling exocytosis and endocytosis of vesicles containing AQP2 channels. Genetic defects in AQP2 result in severe nephrogenic diabetes insipidus—a disease where children must drink gallons of fluid every day, because they make large volumes of dilute urine. AQP4 exists in multiple sites, including the perivascular membranes of astroglial cells in brain. This protein has been linked to epileptic seizures and brain edema after injury. Autoantibodies to AQP4 are diagnostic for neuromyelitis optic, an important disorder causing blindness and paralysis. AQP5 is present in secretory glands and is responsible for sweat, tears, and saliva. Other aquaporins are necessary for eye lens homeostasis; defects in AQP0 are linked to congenital cataracts.

The structures of the aquaporins are highly related. The narrowest span of the pore allows water to move rapidly in single file. All larger molecules, including hydrated ions, are blocked by their greater diameter. Fixed positively charged residues in the pore serve as barriers to movement of protons, thereby restricting passage of protons.

Thus, the mysterious process of transcellular water movement occurs through molecular water channels. Together these proteins form a "plumbing system" for cells. Research on the aquaporin family is ongoing to understand their structure and their known cellular and physiological functions in detail. Where defects in aquaporins are implicated in clinical disorders, that understanding will fuel new research into treatments or cures. Research is also being done to identify other cellular and physiological roles of individual aquaporins, as well as to search for new members of the family in humans and in other organisms.

Courtesy of Peter Agre

Dr. Peter Agre is director of the Johns Hopkins Malaria Research Institute. In 2003 he and Roderick McKinnon of Rockefeller University were awarded the Nobel Prize in Chemistry "for discoveries concerning channels in cell membranes." In 2009 Dr. Agre became president of the American Association for the Advancement of Science. To learn more about his research, visit http://www.jhsph.edu/faculty/directory/profile/4671/Agre/Peter.

think like a scientist Why do you think it is functional that the water channels in the cell membrane formed by aquaporins restrict the passage of protons?

REVIEW KEY CONCEPTS

For access to MindTap and additional study materials visit www.cengagebrain.com.

5.1 Membrane Structure and Function

- Both membrane phospholipids and membrane proteins have hydrophobic and hydrophilic regions, giving them dual solubility properties.

- Membranes are based on a fluid phospholipid bilayer, with the polar regions of the phospholipids at the surfaces of the bilayer and their nonpolar tails in the interior (Figures 5.2–5.5).

- Membrane proteins are suspended individually in the bilayer, with their hydrophilic regions at the membrane surfaces and their hydrophobic regions in the interior (Figures 5.4 and 5.5).

- The lipid bilayer forms the structural framework of membranes and is a barrier to the passage of most water-soluble molecules.

- Proteins embedded in the phospholipid bilayer perform most membrane functions, including transport of selected hydrophilic substances, recognition, signal reception, cell adhesion, and metabolism.

- Integral membrane proteins are embedded deeply in the bilayer, whereas peripheral membrane proteins associate with membrane surfaces (Figure 5.5).

- Membranes are asymmetric—different arrangements of membrane lipids and proteins occur in the two bilayer halves.

- Membranes have diverse functions, including defining the boundaries of cells and of internal compartments, acting as permeability barriers, and facilitating electric signal conduction. Membrane proteins also show diverse activities, acting as enzymes, channel proteins, carrier proteins, receptors, and cell adhesion molecules.

5.2 Functions of Membranes in Transport: Passive Transport

- Passive transport depends on diffusion, the net movement of molecules from a region of higher concentration to a region of lower concentration. It does not require cells to expend energy (Table 5.1).

- Simple diffusion is the passive transport of substances across the lipid portion of cellular membranes. It proceeds most rapidly for small molecules that are soluble in lipids (Table 5.1).

- Facilitated diffusion is the diffusion of polar and charged molecules through a membrane aided by transport proteins in the membrane. It follows concentration gradients, is specific for certain substances, and becomes saturated at high concentrations of the transported substance (Figure 5.8 and Table 5.1).

- Most proteins that carry out facilitated diffusion of ions are controlled by "gates" that open or close their transport channels (Figure 5.8).

5.3 Passive Water Transport and Osmosis

- Osmosis is the net diffusion of water molecules across a selectively permeable membrane in response to differences in the concentration of solute molecules (Figure 5.9). Water moves from hypotonic (lower solute concentrations) to hypertonic solutions (higher solute concentrations). When the solutions on each side are isotonic, net osmotic movement of water ceases (Figure 5.10).

5.4 Active Transport

- Active transport moves substances against their concentration gradients and requires cells to expend energy. It depends on membrane proteins, is specific for certain substances, and becomes saturated at high concentrations of the transported substance (Table 5.1).

- Active transport proteins are either primary transport pumps, which use ATP directly for energy, or secondary transport pumps, which use favorable concentration gradients of positively charged ions, created by primary transport pumps, as their energy source (Figures 5.11).

- Secondary active transport may occur by symport, in which the transported substance moves in the same direction as the concentration gradient that provides energy, or by antiport, in which the transported substance moves in the direction opposite to the concentration gradient that provides energy (Figure 5.12).

5.5 Exocytosis and Endocytosis

- Large molecules and particles are moved out of and into cells by exocytosis and endocytosis. The mechanisms allow substances to leave and enter cells without directly passing through the plasma membrane (Figure 5.13).

- In exocytosis, a vesicle carrying secreted materials contacts and fuses with the plasma membrane on its cytoplasmic side. The fused vesicle membrane releases the vesicle contents to the cell exterior (Figure 5.13A).

- In endocytosis, materials on the cell exterior are enclosed in a segment of the plasma membrane that pockets inward and pinches off on the cytoplasmic side as an endocytic vesicle. The two forms of endocytosis are bulk endocytosis (pinocytosis) and receptor-mediated endocytosis. Most of the materials that enter cells are digested into molecular subunits small enough to be transported across the vesicle membranes (Figures 5.13B).

- Exocytosis and endocytosis are coupled processes so as to control the plasma membrane surface area of a secretory cell.

- In phagocytosis, some protists and some specialized cells, take in aggregates of molecules, cell parts, or whole cells by a process related to receptor-mediated endocytosis (Figure 5.14).

TEST YOUR KNOWLEDGE

Remember/Understand

1. In the fluid mosaic model:
   a. plasma membrane proteins orient their hydrophilic sides toward the internal bilayer.
   b. phospholipids often flip-flop between the inner and outer layers.
   c. the mosaic refers to proteins attached to the underlying cytoskeleton.
   d. the fluid refers to the phospholipid bilayer.
   e. the mosaic refers to the symmetry of the internal membrane proteins and sterols.

2. Which of the following statements is false? Proteins in the plasma membrane can:
   a. transport ions.
   b. transport chloride ions when there are two mutant copies of the cystic fibrosis transmembrane conductance regulator gene.
   c. recognize self versus foreign molecules.
   d. allow adhesion between the same tissue cells or cells of different tissues.
   e. combine with lipids or sugars to form complex macromolecules.

3. The freeze–fracture technique demonstrated:
   a. that the plasma membrane is a bilayer with individual proteins suspended in it.
   b. that the plasma membrane is fluid.
   c. that the arrangement of membrane lipids and proteins is symmetric.
   d. that proteins are bound to the cytoplasmic side but not embedded in the lipid bilayer.
   e. the direction of movement of solutes through the membrane.

4. A characteristic of carrier molecules in a primary active transport pump is that:
   a. they cannot transport a substance and also hydrolyze ATP.
   b. they retain their same shape as they perform different roles.
   c. their primary role is to move negatively charged ions across membranes.
   d. they move Na^+ into a neural cell and K^+ out of the same cell.
   e. they act to establish an electrochemical gradient.

5. A driving ion moving through a membrane channel in one direction gives energy to actively transport another molecule in the opposite direction. What is this process called?
   a. facilitated diffusion
   b. antiport transport
   c. symport transport
   d. primary active transport pump
   e. cotransport

6. Phagocytosis illustrates which phenomenon?
   a. receptor-mediated endocytosis
   b. bulk endocytosis
   c. exocytosis
   d. pinocytosis
   e. cotransport

7. Place in order the following events of receptor-mediated endocytosis.
   (1) Clathrin coat disappears.
   (2) Receptors collect in a coated pit covered with clathrin on the cytoplasmic side.
   (3) Receptors recognize and bind specific molecules.
   (4) Endocytic vesicle may fuse with lysosome whereas receptors are recycled to the cell surface.
   (5) Pits deepen and pinch free of plasma membrane to form endocytic vesicles.
   a. 4, 1, 2, 5, 3
   b. 2, 1, 3, 5, 4
   c. 3, 2, 5, 1, 4
   d. 4, 1, 5, 2, 3
   e. 3, 1, 2, 4, 5

Apply/Analyze

8. In the following figure, assume that the setup was left unattended. Which of the following statements is correct?

Selectively permeable membrane

Inside a cell	Outside fluids
Solvent 95%	Solvent 98%
Solute 5%	Solute 2%

© Cengage Learning 2017

   a. The relation of the cell to its environment is isotonic.
   b. The cell is in a hypertonic environment.
   c. The net flow of solvent is into the cell.
   d. The cell will soon shrink.
   e. Diffusion can occur here but not osmosis.

9. Which of the following statements is true for the figure in question 8?
   a. The net movement of solutes is into the cell.
   b. There is no concentration gradient.
   c. There is a potential for plasmolysis.
   d. The solvent will move against its concentration gradient.
   e. If this were a plant cell, turgor pressure would be maintained.

10. Using the principle of diffusion, a dialysis machine removes waste solutes from a patient's blood. Imagine blood runs through a cylinder wherein diffusion can occur across an artificial selectively permeable membrane to a saline solution on the other side. Which of the following statements is correct?
    a. Solutes move from lower to higher concentration.
    b. The concentration gradient is lower in the patient's blood than in the saline solution wash.
    c. The solutes are transported through a symport in the blood cell membrane.
    d. The saline solution has a lower concentration gradient of solute than the blood.
    e. The waste solutes are actively transported from the blood.

11. **Discuss Concepts** The bacterium *Vibrio cholerae* causes cholera, a disease characterized by severe diarrhea that may cause infected people to lose up to 20 L of fluid in a day. The bacterium enters the body when someone drinks contaminated water. It adheres to the intestinal lining, where it causes cells of the lining to release sodium and chloride ions. Explain how this release is related to the massive fluid loss.

12. **Discuss Concepts** Irrigation is widely used in dryer areas of the United States to support agriculture. In those regions, the water evaporates and leaves behind deposits of salt. What problems might these salt deposits cause for plants?

Evaluate/Create

13. **Discuss Concepts** In hospitals, solutions of glucose with a concentration of 0.3 M can be introduced directly into the bloodstream of patients without tissue damage by osmotic water movement. The same is true of NaCl solutions, but these must be adjusted to 0.15 M to be introduced without damage. Explain why one solution is introduced at 0.3 M and the other at 0.15 M.

14. **Design an Experiment** Design an experiment to determine the concentration of NaCl (table salt) in water that is isotonic to potato cells. Use only the following materials: a knife, small cookie cutters, and a balance.

15. **Apply Evolutionary Thinking** What evidence would convince you that membranes and active transport mechanisms evolved from an ancestor common to both prokaryotes and eukaryotes?

For selected answers, see Appendix A.

INTERPRET THE DATA

Some cancer cells are insensitive to typical chemotherapy. Research into the mechanisms underlying this insensitivity uncovered an ability by these cells to "pump" the treatment drug out of the cell against its concentration gradient. Additional drugs have been developed that inhibit the pump, thus trapping the chemotherapeutic agent inside to promote cancer cell destruction. The **Figure** shows what happens when two types of cells are treated with a ³H-labeled anti-cancer drug, paclitaxel.

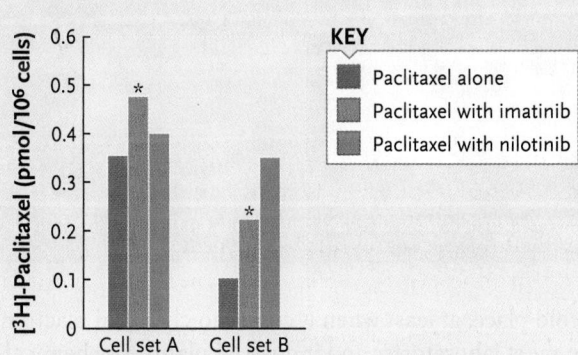

1. Which set of cells (A or B) would be described as resistant to the cancer treatment? Explain your answer. What type of transport are the resistant cells using?

2. Two additional drugs, imatinib and nilotinib, are evaluated for their ability to overcome the cancer cells' ability to "pump out" the chemotherapeutic agent. An asterisk (*) indicates a statistically significant difference from the cells receiving paclitaxel alone. Do the additional drugs seem to be effective in overcoming the pump? Which set of graphs (A or B) best supports your answer? Explain your answer.

Source: T. Shen et al. 2009. Imatinib and nilotinib reverse multidrug resistance in cancer cells by inhibiting the efflux activity of the MRP7 (ABCC10). *PLoS ONE* 4(10):e7520. doi:10.1371/journal.pone.0007520.

6 Energy, Enzymes, and Biological Reactions

Fly caught by a leaf of the English sundew *(Drosera anglica)*. Enzymes secreted by the hairs on the leaf digest trapped insects, providing nutrients to the plant.

Why it matters . . . Earth is a cold place, at least when it comes to chemical reactions. At the high temperatures routinely used in most laboratories and industrial plants for chemical synthesis, life cannot survive. Instead, life relies on substances called *catalysts*, which speed up the rates of reaction without the need for an increase in temperature. The acceleration of a reaction by a catalyst is called *catalysis*. Most of the catalysts in biological systems are proteins called *enzymes*.

How good are enzymes at increasing the rate of a reaction? Richard Wolfenden and his colleagues at the University of North Carolina experimentally measured the rates for a range of uncatalyzed and enzyme-catalyzed biological reactions. They found the greatest difference between the uncatalyzed rate and the enzyme-catalyzed rate for a reaction that removes a phosphate group from a molecule.

In the cell, a group of enzymes called *phosphatases* catalyze the removal of phosphate groups from a number of molecules, including proteins. The reversible addition and removal of a phosphate group from particular proteins is a central mechanism of intracellular communication in almost all cells (see Chapter 9). In a cell using a phosphatase enzyme, the phosphate removal reaction takes approximately 10 milliseconds (msec). Wolfenden's research group calculated that in an aqueous environment such as a cell, without an enzyme, the phosphate removal reaction would take over 1 trillion (10^{12}) years to occur. That is, the enzyme-catalyzed reaction is 21 orders of magnitude (10^{21}) faster.

For most reactions, the difference between the uncatalyzed rate and the enzyme-catalyzed rate is many millions of times. Because life requires temperatures that are relatively low (below 100°C), life as we know would not have evolved without enzymes to speed up the rates of reactions.

Enzymes are key players in **metabolism**—the biochemical modification and use of organic molecules and energy to support the activities of life. Metabolism, which occurs only in living organisms, comprises thousands of biochemical reactions that accomplish the special activities we associate with life, such as growth, reproduction, movement, and the ability to respond to stimuli. In other words, metabolism underlies essentially all life activities described in the chapters of this textbook.

Understanding metabolism means learning how biological reactions occur and how enzymes work. That requires knowledge of the basic laws of chemistry and physics. All reactions, whether they occur inside living organisms or in the outside, inanimate world, obey the same chemical and physical laws that operate everywhere in the universe. In this chapter, these fundamental laws are a starting point for an exploration of the nature of energy and how cells use it in metabolism.

6.1 Energy, Life, and the Laws of Thermodynamics

Life, like all chemical and physical activities, is an energy-driven process. Energy cannot be measured or weighed directly. Energy can be detected only through its effects on matter, including its ability to move objects against opposing forces, such as friction, gravity, or pressure, or to push chemical reactions toward completion. Therefore, energy is most conveniently defined as *the capacity to do work*. Even when you are asleep, cells of your muscles, brain, and other parts of your body are at work and using energy.

Energy Exists in Different Forms and States

Energy can exist in many different forms, including heat, chemical, electrical, mechanical, and radiant energy. Visible light, infrared and ultraviolet light, gamma rays, and X-rays are all types of radiant energy. Although the forms of energy are different, energy can be converted readily from one form to another. For example, chemical energy is transformed into electrical energy in a flashlight battery, and electrical energy is transformed into light and heat energy in the flashlight bulb. In green plants, the radiant energy of sunlight is transformed into chemical energy in the form of complex sugars and other organic molecules (see Chapter 8).

All forms of energy can exist in one of two states: kinetic and potential. **Kinetic energy** (*kinetikos* = putting in motion) is the energy of an object because it is in motion. Examples of everyday objects that possess kinetic energy are waves in the ocean, a hit baseball, and a falling rock. Examples from the natural world are electricity (which is a flow of electrons), photons of light, and heat. The movement present in kinetic energy is useful because it can perform work by making other objects move. **Potential energy** is stored energy, that is, the energy an object has because of its location or chemical structure. A boulder on the top of a hill has potential energy because of its

position in Earth's gravitational field. Chemical energy, nuclear energy, gravitational energy, and stored mechanical energy are forms of potential energy.

In biological systems, the potential energy that can be released in a chemical reaction is called **chemical energy.** Chemical energy is stored in the bonds between atoms.

Potential energy can be converted to kinetic energy and vice versa. Consider a roller coaster. When the train descends from its maximum height, its potential energy is converted into kinetic energy, and the coaster accelerates. On the next hill, kinetic energy is converted back to potential energy and the coaster slows. In the ups and downs of the ride, the coaster's potential energy and kinetic energy interchange continuously but, importantly, the sum of potential energy and kinetic energy remains constant. As a biological example, when you eat foodstuffs, enzymes in your digestive system help break down the molecules they contain. The digestion process converts the chemical energy to other forms of energy for the body's use, such as kinetic energy when you walk or run.

The Laws of Thermodynamics Describe the Energy Flow in Natural Systems

The study of energy and its transformations is called **thermodynamics.** Quantitative research by chemists and physicists in the nineteenth century regarding energy flow between systems and the surroundings led to the formulation of two fundamental laws of thermodynamics that apply equally to living cells and to stars and galaxies. These laws allow us to predict whether reactions of any kind, including biological reactions, can occur. That is, if particular groups of molecules are placed together, are they likely to react chemically and change into different groups of molecules? The laws also give us the information needed to trace energy flows in biological reactions: They allow us to estimate the amount of energy released or required as a reaction proceeds.

When discussing thermodynamics, scientists refer to a *system*, which is the object under study. A system is whatever we define it to be—a single molecule, a cell, a planet. Everything outside a system is its *surroundings*. The universe, in this context, is the total of the system and the surroundings. There are three types of systems:

1. An *isolated system* does not exchange matter or energy with its surroundings **(Figure 6.1A)**. A perfectly insulated Thermos flask is an example of an isolated system.
2. A *closed system* can exchange energy but not matter with its surroundings **(Figure 6.1B)**. Earth is a closed system. It takes in a great amount of energy from the Sun and releases heat, but essentially no matter is exchanged between Earth and the rest of the universe (barring the odd space probe).
3. An *open system* can exchange both energy and matter with its surroundings **(Figure 6.1C)**. All living organisms are open systems.

A. Isolated system: does not exchange matter or energy with its surroundings

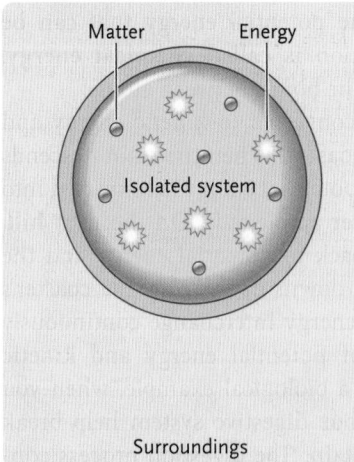

B. Closed system: exchanges energy with its surroundings

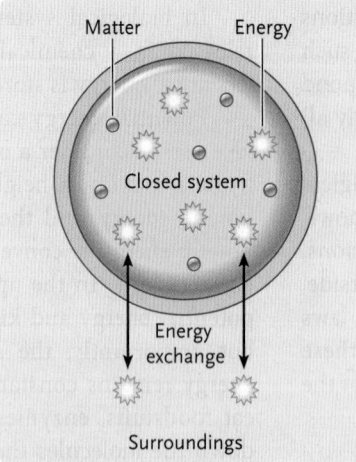

C. Open system: exchanges both energy and matter with its surroundings

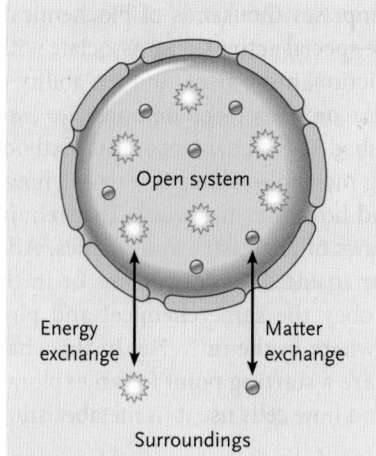

FIGURE 6.1 Isolated, closed, and open systems in thermodynamics.
© Cengage Learning 2017

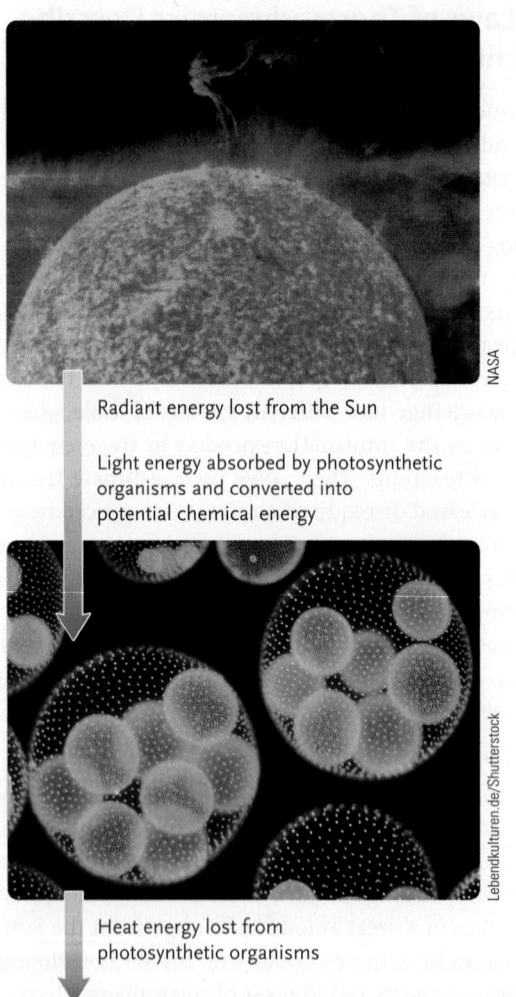

Radiant energy lost from the Sun

Light energy absorbed by photosynthetic organisms and converted into potential chemical energy

Heat energy lost from photosynthetic organisms

FIGURE 6.2 Energy flow from the Sun to photosynthetic organisms (here, colonies of the green alga *Volvox*), which capture the kinetic radiant energy of sunlight and convert it to potential chemical energy in the form of complex organic molecules.

The First Law of Thermodynamics Addresses the Energy Content of Systems and Their Surroundings

The **first law of thermodynamics** states that *energy can be transformed from one form to another or transferred from one place to another, but it cannot be created or destroyed.* That is, in any process that involves an energy change, the *total amount of energy in a system and its surroundings remains constant.* This law is also called the *principle of conservation of energy.*

If energy can be neither created nor destroyed, what is the ultimate source of the energy living organisms use? For almost all organisms, the ultimate source is the Sun **(Figure 6.2)**. Plants capture the kinetic energy of light radiating from the Sun by absorbing it and converting it to the chemical energy of complex organic molecules—primarily sugars, starches, and lipids (see Chapter 8). These substances are used as fuels by the plants themselves, by animals that feed on plants, and by organisms (such as fungi and bacteria) that break down the bodies of dead organisms. The chemical energy stored in sugars and other organic molecules is used for growth, reproduction, and other work of living organisms.

Eventually, most of the solar energy absorbed by green plants is converted into heat energy as the activities of life take place. Heat (a form of kinetic energy) is largely unusable by living organisms. As a result, most of the heat released by the reactions of living organisms radiates to their surroundings, and then from Earth into space.

How does the principle of conservation of energy apply to biochemical reactions? Molecules have both kinetic and potential energy. Kinetic energy for molecules above absolute zero (−273°C) is reflected in the constant motion of the molecules, whereas potential energy for molecules is chemical energy, which is the energy contained in the arrangement of atoms and chemical bonds. The energy content of reacting systems provides part of the information required to predict the likelihood and direction of chemical reactions. For chemical reactions in which a molecule is broken down to simpler molecules, the energy content of the reactants is larger than the energy content of the products. Thus, reactions usually progress to a state in which the products have *minimum energy content*. When this is the case, the difference in energy content between reactants and products in the reacting system is released to the surroundings.

The Second Law of Thermodynamics Considers Changes in the Degree of Order in Reacting Systems

The second law of thermodynamics explains why, as any energy change occurs, the objects (matter) involved in the change typically become more disordered. (Your room and the kitchen at home are probably the best examples of this phenomenon. You know from experience that it takes energy to straighten out—decrease—the disorder, such as when you clean up your room.)

The **second law of thermodynamics** states this tendency toward disorder formally: in any process in which a system changes from an initial to a final state, *the total disorder of a system and its surroundings always increases*. In thermodynamics, disorder is called **entropy.** If the system and its surroundings are defined as the entire universe, the second law means that as changes occur anywhere in the universe, the total disorder or entropy of the universe increases constantly. As the first law of thermodynamics asserts, however, the total energy in the universe does not change.

At first glance, living organisms—which are open systems—appear to violate the second law of thermodynamics. As a fertilized egg develops into an adult animal, it becomes more highly ordered (decreases its entropy) as it synthesizes organic molecules from less complex substances. However, the entropy of the whole system—the surroundings as well as the organism—must be considered as growth proceeds. For the fertilized egg—the initial state—its surroundings include all the carbohydrates, fats, and other complex organic molecules the animal will use as it develops into an adult. For the adult—the final state—the surroundings include the animal's waste products (water, carbon dioxide, and many relatively simple organic molecules), which are collectively much less complex than the organic molecules used as fuels. When the total reactants, including all the nutrients, and the total products, including all the waste materials, are tallied, the total change satisfies both laws of thermodynamics—the total energy content remains constant, and the entropy of the system and its surroundings increases.

STUDY BREAK 6.1

1. What are kinetic energy and potential energy?
2. In thermodynamics, what is meant by an isolated system, a closed system, and an open system?

6.2 Free Energy and Spontaneous Reactions

Applying the first and second laws of thermodynamics together allows us to predict whether any particular chemical or physical reaction will occur without an input of energy. Such reactions are called **spontaneous reactions** in thermodynamics. In this usage, the word *spontaneous* means only that a reaction will occur—it does not describe the rate of a reaction. Spontaneous reactions may proceed very slowly, such as the formation of rust on a nail, or very quickly, such as a match bursting into flame.

Energy Content and Entropy Contribute to Making a Reaction Spontaneous

Two factors related to the first and second laws of thermodynamics must be taken into account to determine whether a reaction is spontaneous: (1) the change in energy content of a system; and (2) its change in entropy.

1. *Reactions tend to be spontaneous if the products have less potential energy than the reactants.* The potential energy in a system is its **enthalpy** (*en* = to put into; *-thalpein* = to heat), symbolized by *H*. Reactions that release energy are **exothermic**—the products have less potential energy than the reactants. Reactions that absorb energy are **endothermic**—the products have more potential energy than the reactants.

   When natural gas burns, methane reacts spontaneously with oxygen producing carbon dioxide and water. This reaction is *exothermic,* producing a large quantity of heat, as the products have less potential energy than the reactants. In a system composed of a glass of ice cubes in water melting at room temperature, the system absorbs energy from its surroundings and the water has greater potential energy than the ice. The process of ice melting is *endothermic,* yet it is spontaneous. Clearly, some other factor besides potential energy is involved.

2. *Reactions tend to be spontaneous when the products are less ordered (more random) than the reactants.* Reactions tend to occur spontaneously if the entropy of the products is greater than the entropy of the reactants.

   Consider again the glass of ice in water. An increase in entropy makes the melting of the ice a spontaneous process at room temperature. Molecules of ice are far more ordered (possess lower entropy) than molecules of water moving around randomly (see Section 2.4).

We have now learned that a combination of factors, *energy content* and *entropy,* contributes to determining whether a chemical reaction proceeds spontaneously. These two concepts are used in the next section to explore quantitatively how to predict whether or not a reaction proceeds spontaneously.

The Change in Free Energy Indicates Whether a Reaction Is Spontaneous

From the second law of thermodynamics, energy transformations are not 100% efficient because some of the energy is lost as an increase in entropy. How much energy is available? The portion of a system's energy available to do work is called **free energy,** which is symbolized as G in recognition of the physicist Josiah Willard Gibbs, who developed the concept. In living organisms, free energy accomplishes the chemical and physical work involved in activities such as the synthesis of molecules, movement, and reproduction.

The change in free energy, ΔG ($\Delta G = G_{\text{final state}} - G_{\text{initial state}}$; Δ, pronounced delta, means change in), can be calculated for any chemical reaction from the following formula:

$$\Delta G = \Delta H - T\Delta S$$

where ΔH is the change in enthalpy, T is the absolute temperature in degrees Kelvin (K, where K = °C + 273.16), and ΔS is the change in entropy.

For a reaction to be spontaneous, ΔG must be negative. As the above formula tells us, both the entropy and the enthalpy of a reaction can influence the overall ΔG. In some processes, such as the combustion of methane, the large loss of potential energy, negative enthalpy (ΔH), dominates in making a reaction spontaneous. In other reactions, such as the melting of ice at room temperature, a decrease in order (ΔS increases) dominates. Once we know what the ΔG is for a reaction, we can determine if the reaction will proceed spontaneously.

Spontaneous Reactions Typically Reach an Equilibrium Point Rather Than Going to Completion

In many spontaneous biological reactions, the reactants may not convert completely to products even though the reactions have a negative ΔG. Instead, the reactions run in the direction of completion (toward reactants or toward products) until they reach the equilibrium point, a state of balance between the opposing factors pushing the reaction in either direction. At the equilibrium point, both reactants and products are present and the reactions typically are reversible.

Consider as an example the chemical reaction in which glucose-1-phosphate is converted into glucose-6-phosphate, starting with $0.02\ M$ glucose-1-phosphate as the reactant **(Figure 6.3)**. The reaction proceeds spontaneously until there is $0.019\ M$ glucose-6-phosphate (product) and $0.001\ M$ of glucose-1-phosphate (reactant) in the solution. In fact, regardless of the amounts of each you start with, the reaction reaches a point at which there is 95% glucose-6-phosphate and 5% glucose-1-phosphate. This is the point of *chemical equilibrium:* the reaction does not stop, but the rate of the forward reaction equals the rate of the reverse reaction. As a system moves toward equilibrium, its free energy becomes progressively lower and reaches its lowest point when the system achieves

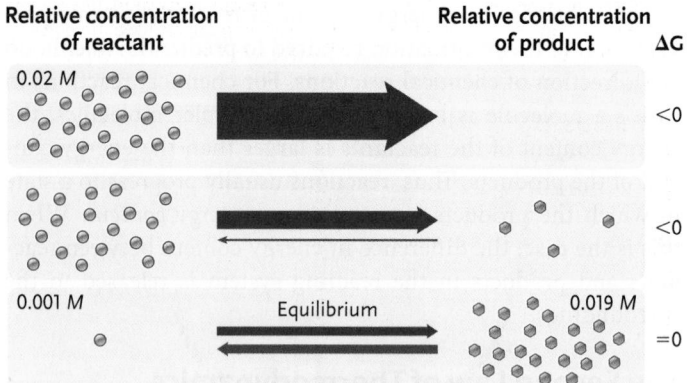

FIGURE 6.3 The equilibrium point of a reaction. No matter what quantities of glucose-1-phosphate and glucose-6-phosphate are dissolved in water, when equilibrium is reached, there is 95% glucose-6-phosphate (product) and 5% glucose-1-phosphate (reactant). At equilibrium, the number of reactant molecules being converted to product molecules equals the number of product molecules being converted back to reactant molecules. The reaction at the equilibrium point is reversible; it may be made to run to the right (forward) by adding more reactants, or to the left (reverse) by adding more products.

© Cengage Learning 2017

equilibrium ($\Delta G = 0$). You can think of a reaction as an energy valley with the equilibrium point being at the bottom. To move away from equilibrium requires free energy and thus will not be spontaneous.

The point of equilibrium of a reaction is related to its ΔG. The more negative the ΔG, the further toward completion the reaction will move before equilibrium is established. If the reaction shown in Figure 6.3 had a positive ΔG, the reaction would run in reverse toward glucose-1-phosphate. Many reactions have a ΔG that is near zero and, therefore, are readily **reversible** by adjusting the concentration of products and reactants slightly. Reversible reactions are written with a double arrow:

$$\underset{\text{reactants}}{A + B} \rightleftharpoons \underset{\text{products}}{C + D}$$

The reaction in Figure 6.3 is an isolated system and, over time, equilibrium is reached, with ΔG becoming zero. However, many individual reactions in living organisms never reach an equilibrium point because living systems are open. Therefore, the supply of reactants is constant and, as products are formed, they do not accumulate but become the reactants of another reaction. In fact, the ΔG of life is always negative as organisms constantly take in energy-rich molecules (or light, if they are photosynthetic) and use them to do work. Organisms reach equilibrium, with $\Delta G = 0$, only when they die.

Metabolic Pathways Consist of Exergonic and Endergonic Reactions

Based on the free energy of reactants and products, every reaction can be placed into one of two groups. An **exergonic reaction** (*ergon* = work) **(Figure 6.4A)** releases free energy, so

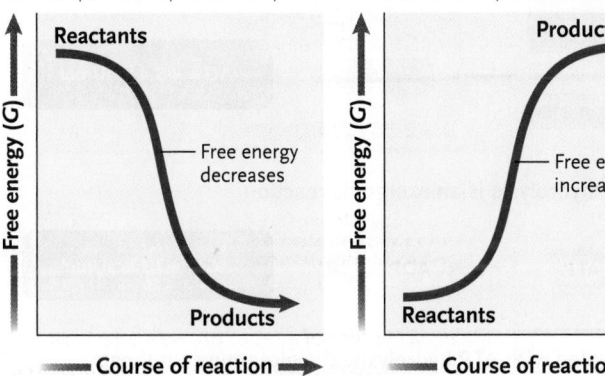

A. Exergonic reaction: free energy is released, products have less free energy than reactants, and the reaction proceeds spontaneously

B. Endergonic reaction: free energy is gained, products have more free energy than reactants, and the reaction is not spontaneous

FIGURE 6.4 Exergonic (A) and endergonic (B) reactions. An endergonic reaction proceeds only if energy is supplied by an exergonic reaction.

© Cengage Learning 2017

ΔG is negative because the products contain less free energy than the reactants. In an **endergonic reaction (Figure 6.4B),** the products contain more free energy than the reactants so ΔG is positive. The reactants involved in endergonic reactions need to gain free energy from the surroundings to form the products of the reaction.

In metabolism, individual reactions tend to be part of a *metabolic pathway,* a series of reactions in which the products of one reaction are used immediately as the reactants for the next reaction in the series. In one type of metabolic pathway, called a **catabolic pathway** (*cata* = downward, as in the sense of a rock releasing energy as it rolls down a hill), chemical energy is released by the breakdown of complex molecules to simpler compounds. (An individual reaction from which chemical energy is released is called a **catabolic reaction.**) An example of a catabolic pathway in metabolism is cellular respiration, the topic of Chapter 7, in which energy is extracted from the breakdown of food such as glucose. By contrast, in an **anabolic pathway** (*ana* = upward, as in the sense of using energy to push a rock up a hill), chemical energy is used to build complicated molecules from simpler ones. Anabolic pathways are often called *biosynthetic pathways.* (An individual reaction that requires chemical energy input is called an **anabolic reaction,** or a **biosynthetic reaction.**) Examples of anabolic pathways in metabolism include photosynthesis, the topic of Chapter 8, as well as the synthesis of macromolecules such as proteins and nucleic acids.

The overall ΔG of a catabolic pathway is negative, whereas the overall ΔG of an anabolic pathway is positive. Any one pathway consists of a number of individual reactions, each of which may have a positive or negative ΔG. However, when you sum the individual reaction ΔG values for a catabolic pathway, the overall free energy is negative, and for an anabolic reaction, the overall free energy is positive.

STUDY BREAK 6.2

1. What two factors must be taken into account to determine if a reaction will proceed spontaneously?
2. What is the relation between ΔG and the concentrations of reactants and products at the equilibrium point of a reaction?
3. Distinguish between exergonic and endergonic reactions, and between catabolic and anabolic reactions. How are the two categories of reactions related?

6.3 Adenosine Triphosphate (ATP): The Energy Currency of the Cell

Many reactions within cells involve the assembly of complex molecules from more simple components. As we learned in the previous section, these reactions have a positive ΔG and are called endergonic. They may be part of both catabolic and anabolic pathways. How the cell supplies chemical energy to drive these endergonic reactions is highly conserved among all forms of life and involves the nucleotide adenosine triphosphate (ATP).

ATP Hydrolysis Releases Free Energy

ATP is the best example of a molecule that contains large amounts of free energy because of its high-energy phosphate bonds. ATP consists of the five-carbon sugar ribose linked to the nitrogenous base adenine and a chain of three phosphate groups **(Figure 6.5A).** Much of the potential energy of ATP is associated with the three phosphate groups. Removal of one or two of the three phosphate groups is a spontaneous reaction that releases large amounts of free energy. (As you will learn in Chapter 15, ATP is also a building block in the synthesis of RNA.)

The breakdown of ATP is a hydrolysis reaction **(Figure 6.5B)** and results in the formation of adenosine diphosphate (ADP) and a molecule of inorganic phosphate (P_i).

$$ATP + H_2O \rightarrow ADP + P_i$$

$$\Delta G = -7.3 \text{ kcal/mol}$$

ADP can be hydrolyzed further to adenosine monophosphate (AMP), but this releases somewhat less free energy than the hydrolysis of ATP to ADP.

Phosphate Groups from ATP Hydrolysis Couple Reactions

The hydrolysis of ATP in water in a test tube is an exergonic reaction that releases free energy that simply warms up the water. How do living cells couple the hydrolysis of ATP to an endergonic reaction so that energy is not simply wasted as heat?

The answer is a process called **energy coupling,** the use of an exergonic reaction—ATP hydrolysis—to drive an endergonic reaction. In other words, the two reactions become coupled. In a **coupled reaction,** ATP is hydrolyzed and its terminal phosphate group is transferred to the reactant molecule of the

A. Chemical structures of AMP, ADP, and ATP

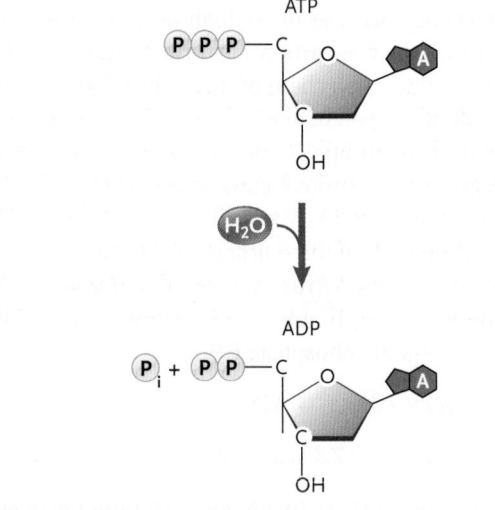

With one phosphate group, the molecule is known as AMP; with two phosphates, the molecule is called ADP. Each added phosphate packs additional potential chemical energy into the molecular structure.

B. Hydrolysis reaction removing a phosphate group from ATP

FIGURE 6.5 ATP, the primary molecule that couples energy-requiring reactions to energy-releasing reactions in living organisms. (P_i is the symbol used in this book for inorganic phosphate.)
© Cengage Learning 2017

endergonic reaction. The addition of a phosphate group to a molecule is called **phosphorylation,** and the modified molecule is said to have been *phosphorylated.* Phosphorylation makes a molecule less stable (more reactive) than when it is unphosphorylated. Energy coupling requires the action of an enzyme to bring the ATP and reactant molecule into close association. The enzyme has a specific site on it that binds both the ATP and the reactant molecule, allowing for transfer of the phosphate group.

A. Without ATP, reaction is not spontaneous because ΔG is positive

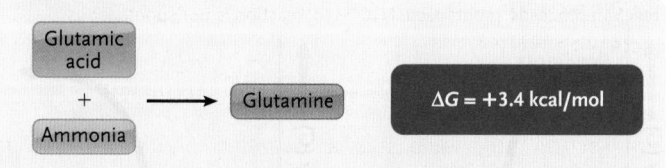

B. ATP hydrolysis is an exergonic reaction

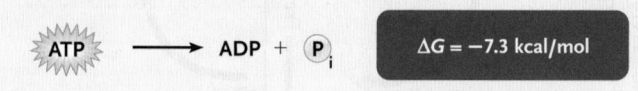

C. Coupled with ATP hydrolysis, the glutamine synthesis reaction is spontaneous because net ΔG is negative

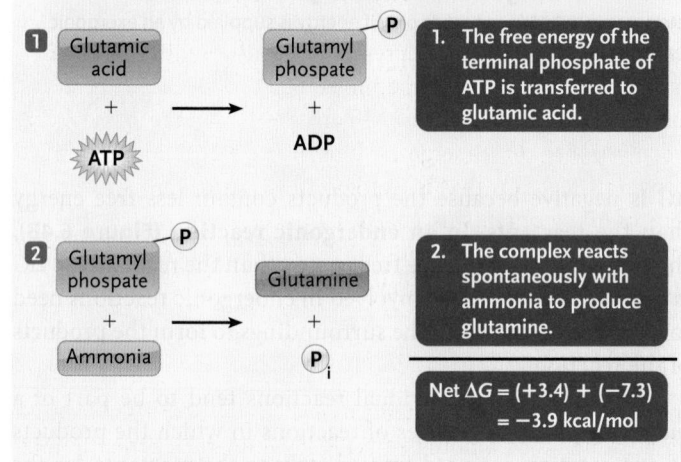

1. The free energy of the terminal phosphate of ATP is transferred to glutamic acid.

2. The complex reacts spontaneously with ammonia to produce glutamine.

Net ΔG = (+3.4) + (−7.3) = −3.9 kcal/mol

FIGURE 6.6 Energy coupling using ATP in the synthesis of glutamine from glutamic acid and ammonia.
© Cengage Learning 2017

Energy coupling is used, for example, in the reaction in which ammonia (NH_3) is added to glutamic acid, an amino acid with one amino group, to produce glutamine, an amino acid with two amino groups. The overall reaction is **(Figure 6.6A):**

$$\text{glutamic acid} + NH_3 \rightarrow \text{glutamine} + H_2O$$

$$\Delta G = +3.4 \text{ kcal/mol}$$

The positive value for ΔG indicates that the reaction cannot proceed spontaneously. Both glutamic acid and glutamine are used in the assembly of proteins. This reaction is common in most cells. The product, glutamine, is a donor of nitrogen for other reactions in the cell.

Cells carry out this endergonic reaction by using energy released from ATP hydrolysis **(Figure 6.6B).** The coupled reaction is shown in **Figure 6.6C.** As a first step, glutamic acid is phosphorylated using the phosphate group removed from ATP, forming glutamyl phosphate:

$$\text{glutamic acid} + ATP \rightarrow \text{glutamyl phosphate} + ADP$$

The ΔG for this step is negative, making the reaction spontaneous, but much less free energy is released than in the hydrolysis of ATP to ADP + P_i. In the second step, glutamyl phosphate reacts with NH_3:

$$\text{glutamyl phosphate} + NH_3 \rightarrow \text{glutamine} + P_i$$

This second step also has a negative ΔG value and is spontaneous.

Even though the reaction proceeds in two steps, it is usually written as one reaction, with a combined negative ΔG value:

$$\text{glutamic acid} + NH_3 + ATP \rightarrow \text{glutamine} + ADP + P_i$$

$$\Delta G = -3.9 \text{ kcal/mol}$$

Because ΔG is negative, the coupled reaction is spontaneous and releases energy. The difference between −3.9 kcal/mol and the −7.3 kcal/mol released by hydrolyzing ATP to ADP + P_i (that is, +3.4 kcal/mol) represents potential chemical energy transferred to the glutamine molecules produced by the reaction. That is, the coupling of an exergonic reaction—ATP hydrolysis—to an endergonic biosynthesis reaction produces an overall reaction that is exergonic. All the endergonic reactions of living organisms, including those of growth, reproduction, movement, and response to stimuli, are made possible by coupling reactions in this way.

Cells Also Couple Reactions to Regenerate ATP

Coupling reactions occur continuously in living cells, consuming a tremendous amount of ATP. How do cells generate that

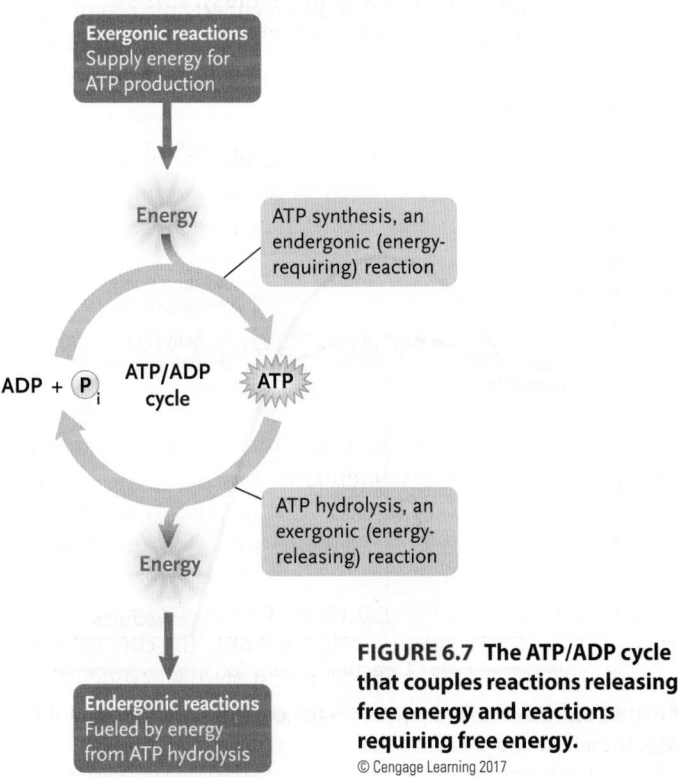

FIGURE 6.7 The ATP/ADP cycle that couples reactions releasing free energy and reactions requiring free energy.
© Cengage Learning 2017

ATP? ATP is a renewable resource that is synthesized by recombining ADP and P_i. Although ATP hydrolysis is an exergonic reaction, ATP synthesis from ADP and P_i is an energy-requiring, endergonic reaction. The energy for ATP synthesis comes from exergonic reactions that involve breakdown of complex molecules that contain an abundance of free energy: carbohydrates, proteins, and fats in food.

The continual hydrolysis and resynthesis of ATP is called the **ATP/ADP cycle (Figure 6.7).** Approximately 10 million ATP molecules are hydrolyzed and resynthesized each second in a typical cell, illustrating that this cycle operates at an astonishing rate. In fact, if ATP were not regenerated from ADP and P_i, the average human would use an estimated 75 kg of ATP per day.

6.4 Role of Enzymes in Biological Reactions

The laws of thermodynamics tell us if a process will occur spontaneously. However, the laws do not tell us anything about the rate of a reaction. For example, even though the breakdown of sucrose into glucose and fructose is a spontaneous reaction with a ΔG of −7 kcal/mol, a solution of sucrose will sit for years without any detectable glucose or fructose forming. That is, *spontaneous reactions do not necessarily proceed rapidly.* In this section, we discuss how the speed of a reaction can be altered by enzymes.

Activation Energy Represents a Kinetic Barrier for a Reaction

In our example above, what prevents sucrose from being converted rapidly into glucose and fructose? Chemical reactions require bonds to break and new bonds to form. For bonds to be broken, they must be strained or otherwise made less stable so that breakage can occur. To get reacting molecules into a less stable (more unstable) state requires a small input of energy. Thus, even though a reaction is spontaneous (negative ΔG), the reaction will not start unless a relatively small boost of energy is added **(Figure 6.8A).** This initial energy investment required to start a reaction is called the **activation energy,** symbolized E_a. Molecules that gain the necessary activation energy occupy what is called the *transition state,* where bonds are unstable and are ready to be broken.

A rock resting in a depression at the top of a hill provides a physical example of activation energy **(Figure 6.8B).** The rock

A. Activation energy barrier in the oxidation of glucose

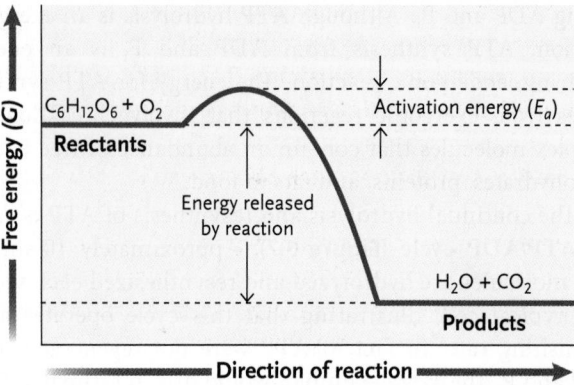

B. "Activation energy" barrier in the movement of a rock downhill

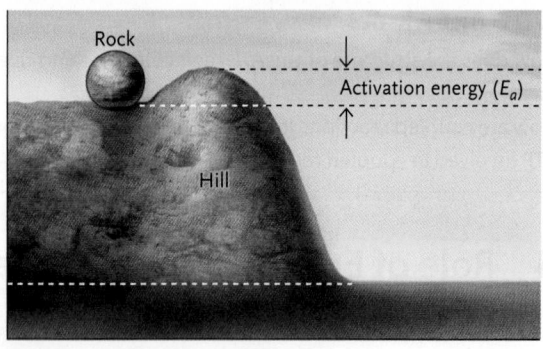

FIGURE 6.8 Activation energy (Eₐ).
© Cengage Learning 2017

will not roll downhill spontaneously, even though its position represents considerable potential energy and the total "reaction"—the downward movement of the rock—is spontaneous and releases free energy. In this example, the activation energy is the effort required to raise the rock over the rim of the depression and start its downhill roll.

What provides the activation energy in chemical reactions? The molecules taking part in chemical reactions are in constant motion at temperatures above absolute zero. Periodically, reacting molecules gain enough energy to reach the transition state. For a solution of sucrose, the number of molecules that reach the transition state at any one time is very small. However, if a significant number of reactant molecules reach the transition state, the free energy released may be enough to get the remaining reactants to the transition state. For example, in a chemistry laboratory, heat commonly provides the energy needed for reactant molecules to get to the transition state and, therefore, to speed up the rate of a reaction. In biology, however, using heat to speed up a reaction is problematic for two reasons: (1) high temperatures destroy the structural components of cells, particularly proteins; and (2) an increase in temperature would speed up all possible chemical reactions in a cell, not just the specific reactions that are part of metabolism.

Enzymes Accelerate Reactions by Reducing the Activation Energy

How can the rate of a reaction increase without a raise in the temperature? The answer is by the use of a **catalyst,** a chemical agent that accelerates the rate of a reaction without itself being changed by the reaction. The process of accelerating a reaction with a catalyst is called **catalysis,** and we say that the chemical agent responsible *catalyzes* the reaction. The most common biological catalysts are proteins called **enzymes** (*enzym* = in yeast).

The activation energy for a reaction represents a kinetic barrier that prevents spontaneous reactions from proceeding quickly. The greater the activation energy barrier, the slower the reaction will proceed. Enzymes increase the rate of reaction by lowering this barrier, that is, by lowering the activation energy of the reaction **(Figure 6.9).** Because the rate of a reaction is proportional to the number of reactant molecules that can acquire enough energy to get to the transition state, enzymes make it possible for a greater proportion of reactant molecules to attain the activation energy.

Although enzymes lower the activation energy of a reaction, as shown in Figure 6.9, they do not alter the change in free energy (ΔG) of the reaction. The free energy values of the reactants and products remain the same; the only difference is the path the reaction takes.

Let us be clear about what enzymes do and do not do with regard to biological reactions:

- By lowering the activation energy, enzymes DO speed up the rate of spontaneous (exergonic) reactions.
- Enzymes do NOT supply free energy to a reaction. Therefore, enzymes CANNOT make an endergonic reaction proceed spontaneously. ATP hydrolysis can be coupled to an endergonic reaction to make it proceed spontaneously but, alone, an enzyme cannot.
- Enzymes do NOT change the ΔG of a reaction.

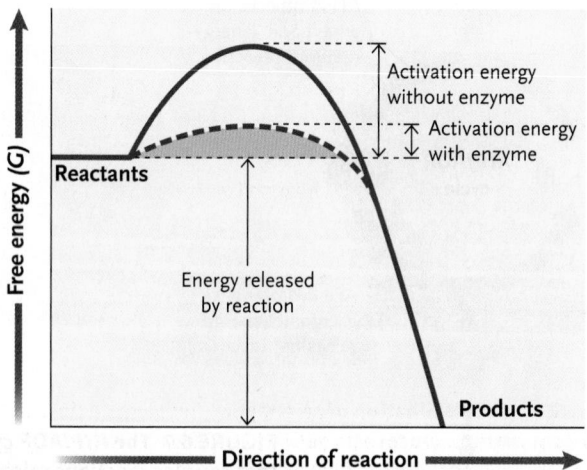

FIGURE 6.9 Effect of enzymes in reducing the activation energy (Eₐ). The reduction allows biological reactions to proceed rapidly at the relatively low temperatures that can be tolerated by living organisms.
© Cengage Learning 2017

Cells have thousands of different enzymes. They vary from relatively small molecules, with single polypeptide chains containing as few as 100 amino acids, to large complexes that include many polypeptide chains totaling thousands of amino acids. It is the three-dimensional structure of a protein—its *conformation*—that determines the protein's function. Conformation is determined by the number of polypeptides in the protein and the amino acid sequence of each polypeptide. (The three-dimensional structure of proteins is described in Section 3.4.) Each enzyme has a specific protein structure that has evolved to catalyze a specific reaction.

Different enzymes are found in all areas of the cell, from the aqueous cell solution to the cell membranes. Other enzymes are released to catalyze reactions outside the cell. For example, enzymes that catalyze reactions breaking down food molecules are released from cells into the digestive cavity in all animals.

The majority of enzymes have names ending in -*ase*. The rest of the name typically relates to the substrate of the enzyme or to the type of reaction with which the enzyme is associated. For example, enzymes that break down proteins are called *proteinases* or *proteases*.

Enzymes are not the only biological molecules capable of accelerating reaction rates. Some RNA molecules (see Section 6.6) also have this capacity. As we do in this book, most biologists reserve the term *enzyme* for protein molecules that can accelerate reaction rates and call the RNA molecules with this capacity *ribozymes*.

Enzymes Combine with Reactants and Are Released Unchanged

In enzymatic reactions, an enzyme combines briefly with reacting molecules and is released unchanged when the reaction is complete. For example, the enzyme in **Figure 6.10**, hexokinase, catalyzes the reaction:

$$\text{glucose} + \text{ATP} \xrightarrow{\text{hexokinase}} \text{glucose-6-phosphate} + \text{ADP}$$

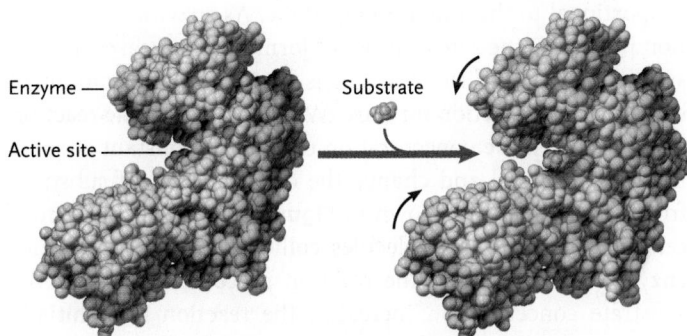

FIGURE 6.10 Combination of an enzyme, hexokinase (in blue), with its substrate, glucose (in yellow). Hexokinase catalyzes the phosphorylation of glucose to form glucose-6-phosphate. The phosphate group that is transferred to glucose is not shown. Note how the enzyme undergoes a conformational change (an induced fit), closing the active site more tightly as it binds the substrate.

© Cengage Learning 2017

Writing the enzyme name (here, hexokinase) above the reaction arrow indicates that it is required but not involved as a reactant or a product. The reactant on which an enzyme acts is called the **substrate,** or substrates if the enzyme binds two or more molecules. In this reaction, the substrates are glucose and ATP.

Each type of enzyme catalyzes the reaction of a single type of substrate molecule or a group of closely related molecules. This **enzyme specificity** explains why a typical cell needs about 4,000 different enzymes to function properly. Notice in Figure 6.10 that the enzyme is much larger than the substrate. Moreover, the substrate interacts with only a very small region of the enzyme called the **active site,** the place on the enzyme where catalysis occurs. The active site is usually a pocket or groove that is formed when the enzyme protein folds into its functional three-dimensional shape.

When the substrate binds initially at the active site of the enzyme, both the enzyme and substrate molecules are distorted, which stabilizes the substrate molecule in the transition state and makes its chemical bonds ready for reaction. Figure 6.10 illustrates the change in enzyme conformation on substrate binding, a phenomenon called *induced fit*.

Once an enzyme–substrate complex is formed, catalysis occurs, with the enzyme converting a single substrate molecule into one or more product molecules in a catabolic reaction, or one or more substrate molecules into one or more product molecules in an anabolic reaction. Because enzymes are released unchanged after a reaction, enzyme molecules cycle repeatedly through reactions, combining with reactants and releasing products **(Figure 6.11).** Depending on the enzyme, the rate at which reactants are bound and catalyzed and at which products are released varies from 100 times to 10 million times per second. These high rates of catalysis mean that a small number of enzyme molecules can catalyze large numbers of reactions.

Some enzymes consist of polypeptide chains only. However, many enzymes require a **cofactor,** a nonprotein group that binds precisely to the enzyme, for catalytic activity. The role of a cofactor in an enzyme's catalytic activity varies with the cofactor and the enzyme. Some cofactors are metallic ions, including iron, copper, magnesium, zinc, and manganese. Other cofactors are small organic molecules called **coenzymes,** which are often derived from vitamins. Some coenzymes bind loosely to enzymes; others—called *prosthetic groups*—bind tightly.

Enzymes Reduce the Activation Energy by Stabilizing the Transition State

How do enzymes reduce the activation energy of a reaction? Recall that substrate molecules need to be in the transition state for catalysis to occur. Enzymes stabilize the transition state, doing so through three major mechanisms:

1. Enzymes *bring the reacting molecules together*. Reacting molecules can assume the transition state only when they collide. Binding of reactant molecules to an enzyme's

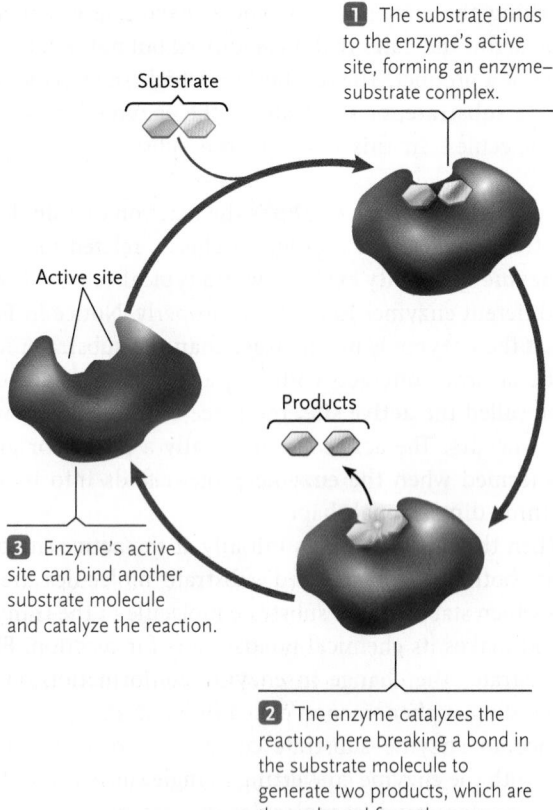

1 The substrate binds to the enzyme's active site, forming an enzyme–substrate complex.

Substrate

Active site

Products

3 Enzyme's active site can bind another substrate molecule and catalyze the reaction.

2 The enzyme catalyzes the reaction, here breaking a bond in the substrate molecule to generate two products, which are then released from the enzyme.

FIGURE 6.11 The catalytic cycle of enzymes.
© Cengage Learning 2017

active site brings them close together in the correct orientation for catalysis to take place.

2. Enzymes *expose the reactant molecules to altered charge environments that promote catalysis.* In some systems, the active site of the enzyme contains ionic groups whose positive or negative charges alter the substrate in a way that favors catalysis.

3. Enzymes *change the shape of the substrate molecules.* As mentioned previously, the active site can distort substrate molecules into a conformation that mimics the transition state.

Regardless of the mechanism, the binding of the substrate molecule(s) to the active site results in stabilization of the substrate in the transition state conformation. Substrate molecules do attain the transition state in the absence of an enzyme, but that is a rare event. The inclusion of an enzyme enables many more molecules to reach the transition state more rapidly. Fundamentally, this is why an enzyme speeds up the rate of a reaction.

STUDY BREAK 6.4

1. How do enzymes increase the rates of the reaction they catalyze?
2. Can enzymes alter the ΔG of a reaction?

6.5 Conditions and Factors That Affect Enzyme Activity

Several conditions can alter enzyme activity, including changes in the concentration of substrate and other molecules that can bind to enzymes. In addition, a number of control mechanisms modify enzyme activity, thereby adjusting reaction rates to meet a cell's requirements for chemical products. Changes in pH and temperature can also have a significant effect on enzyme activity.

Enzyme and Substrate Concentrations Influence the Rate of Catalysis

Biochemists use a wide range of approaches to study enzymes. These include molecular tools to study the structure and regulation of the gene encoding the enzyme and sophisticated computer algorithms to model the three-dimensional structure of the enzyme itself. The most fundamental and central approach has been to determine the rate of an enzyme-catalyzed reaction and how it changes in response to altering certain experimental parameters. Typically this requires isolating the enzyme from a cell, incubating it in an appropriate buffered solution, and supplying the reaction mixture with substrate. With these constituents, a researcher can then determine the rate of catalysis, usually by measuring the rate at which the product of the reaction is formed.

As shown in **Figure 6.12A,** in the presence of excess substrate (that is, at high concentrations), the rate of catalysis is proportional to the amount of enzyme. As enzyme concentration increases, the rate of product formation increases. In this system, the rate of the reaction is limited by the amount of enzyme in the reaction mixture. What happens to the reaction rate if we keep the concentration of enzyme constant at some intermediate level and change the concentration of substrate from low to high? As shown in **Figure 6.12B,** at very low concentrations, substrate molecules collide so infrequently with enzyme molecules that the reaction proceeds slowly. As the substrate concentration increases, the reaction rate initially increases as enzyme and substrate molecules collide more frequently. However, as the enzyme molecules approach the maximum rate at which they can combine with reactants and release products, increasing substrate concentration has a smaller and smaller effect, and the rate of reaction eventually levels off. When the enzymes are cycling as rapidly as possible, further increases in substrate concentration have no effect on

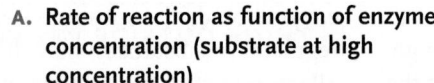

A. Rate of reaction as function of enzyme concentration (substrate at high concentration)

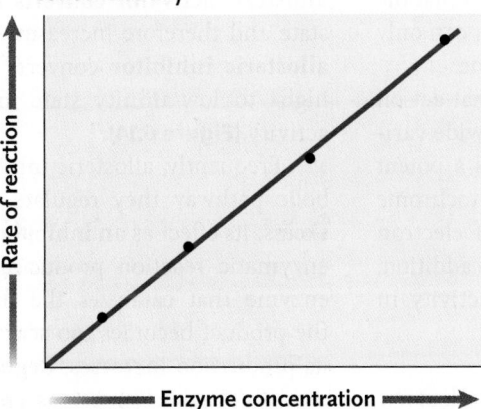

B. Rate of reaction as function of substrate concentration (enzyme amount constant)

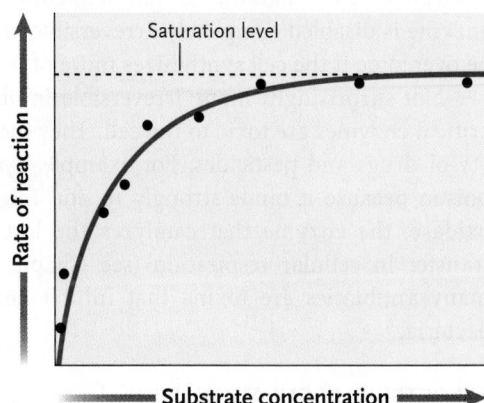

the reaction rate. At this point, the enzymes are said to be **saturated** with the substrate (the saturation level is shown by a horizontal dashed line in Figure 6.12B).

Cells Adjust Enzyme Activity to Meet Their Needs for Reaction Products

Cells adjust the activity of many enzymes upward or downward to meet their needs for reaction products. Several mechanisms are used in this regulation, including: (1) competitive and noncompetitive inhibition; (2) a form of noncompetitive control called *allosteric regulation;* and (3) covalent modification of enzyme structure by the addition or removal of chemical groups.

COMPETITIVE AND NONCOMPETITIVE INHIBITION Reversible inhibition of enzyme activity serves as an important mechanism of metabolism regulation. A typical cell contains thousands of enzymes, and for many enzymes that synthesize a specific molecule, usually another enzyme exists that catalyzes its breakdown. If both of these enzymes were active in the same cell compartment at the same time, the two metabolic pathways they catalyze would run simultaneously in opposite directions and have no overall effect other than using up energy. To prevent this, the cell is able to regulate enzyme activity in such a way that not all enzymes are active at the same time.

Many enzymes are regulated by natural inhibitors. These *enzyme inhibitors* are nonsubstrate molecules that bind to an enzyme and decrease its activity, thereby lowering the rate at which an enzyme can catalyze a reaction. Control by the inhibitors changes enzyme activity precisely to meet the needs of the cell for the products of the reaction catalyzed by the enzyme.

Regulation of enzyme activity by inhibitors may be competitive or noncompetitive depending on the inhibitor. In **competitive inhibition,** an inhibitor competes with the normal substrate for binding to an enzyme's active site **(Figure 6.13A).** That is, competitive inhibitors have shapes resembling the normal substrate closely enough to fit into and occupy the active

site, thereby blocking access for the normal substrate and slowing the reaction rate. If the concentration of the inhibitor is high enough, the reaction may stop completely. Competitive inhibitors are useful in enzyme research because their structure helps identify the region of a normal substrate that binds to an enzyme.

In **noncompetitive inhibition,** an inhibitor binds to an enzyme at a site other than its active site and changes the conformation of the enzyme so that the ability of the active site to bind substrate efficiently is reduced **(Figure 6.13B).**

Inhibitors (competitive or noncompetitive) differ with respect to how strongly they bind to enzymes. In *reversible inhibition,* the binding of an inhibitor to an enzyme is weak,

A. Competitive inhibition

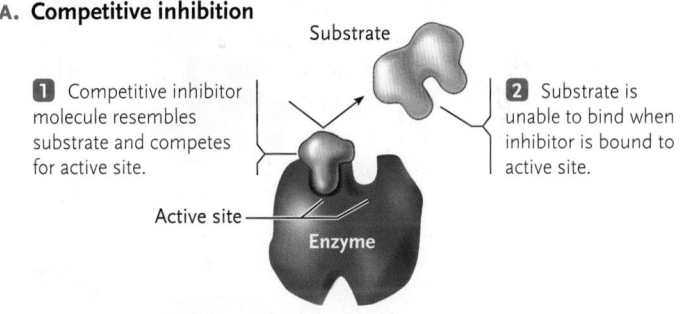

1 Competitive inhibitor molecule resembles substrate and competes for active site.

2 Substrate is unable to bind when inhibitor is bound to active site.

Substrate

Active site

Enzyme

B. Noncompetitive inhibition

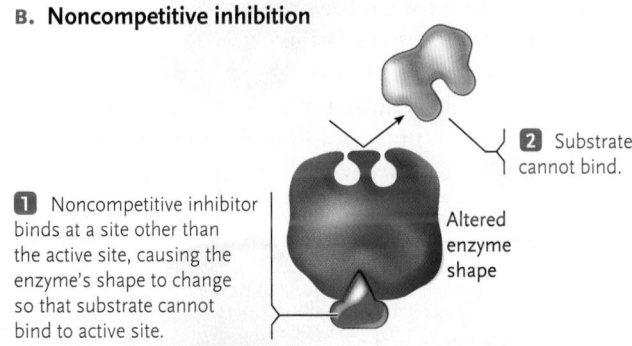

1 Noncompetitive inhibitor binds at a site other than the active site, causing the enzyme's shape to change so that substrate cannot bind to active site.

2 Substrate cannot bind.

Altered enzyme shape

FIGURE 6.13 How competitive (A) and noncompetitive (B) inhibitors reduce enzyme activity.
© Cengage Learning 2017

and when the inhibitor releases, enzyme activity returns to normal. In *irreversible inhibition,* an inhibitor binds so strongly to an enzyme through the formation of covalent bonds that the enzyme is disabled completely. Irreversible inhibition can only be overcome if the cell synthesizes more of the enzyme.

Not surprisingly, many irreversible inhibitors that act on critical enzymes are toxic to the cell. They include a wide variety of drugs and pesticides. For example, cyanide is a potent poison because it binds strongly to and inhibits cytochrome oxidase, the enzyme that catalyzes the last step of electron transfer in cellular respiration (see Chapter 7). In addition, many antibiotics are toxins that inhibit enzyme activity in bacteria.

ALLOSTERIC REGULATION In the mechanism of **allosteric regulation** (*allo* = different; *stereo* = shape), enzyme activity is controlled by the reversible binding of a regulatory molecule to the **allosteric site,** a location on the enzyme outside the active site. The mechanism may either increase or decrease enzyme activity. Because allosteric regulatory molecules work by binding to sites separate from the active site, their action is noncompetitive.

Enzymes controlled by allosteric regulation typically have two alternate conformations controlled from the allosteric site. In one conformation, called the *high-affinity state* (the active form), the enzyme binds strongly to its substrate; in the other conformation, *the low-affinity state* (the inactive form), the enzyme binds the substrate weakly or not at all. Binding with regulatory substances may induce either state: binding an **allosteric activator** converts it from the low- to high-affinity state and therefore increases enzyme activity, and binding an **allosteric inhibitor** converts an allosteric enzyme from the high- to low-affinity state and therefore decreases enzyme activity **(Figure 6.14)**.

Frequently, allosteric inhibitors are a product of the metabolic pathway they regulate. If the product accumulates in excess, its effect as an inhibitor automatically slows or stops the enzymatic reaction producing it, typically by inhibiting the enzyme that catalyzes the first reaction of the pathway. If the product becomes too scarce, the inhibition is reduced and its production increases. Regulation of this type, in which the product of a reaction acts as a regulator of the reaction, is termed **feedback inhibition** (also called **end-product inhibition**). Feedback inhibition prevents cellular resources from being wasted in the synthesis of molecules made at intermediate steps of the pathway.

The biochemical pathway that makes the amino acid isoleucine from threonine is an example of feedback inhibition. The pathway proceeds in five steps, each catalyzed by an enzyme **(Figure 6.15)**. The end product of the pathway, isoleucine, is an allosteric inhibitor of the first enzyme of the pathway, threonine deaminase. If the cell makes more isoleucine than it needs, isoleucine binds reversibly with threonine deaminase at the allosteric site, converting the enzyme to the

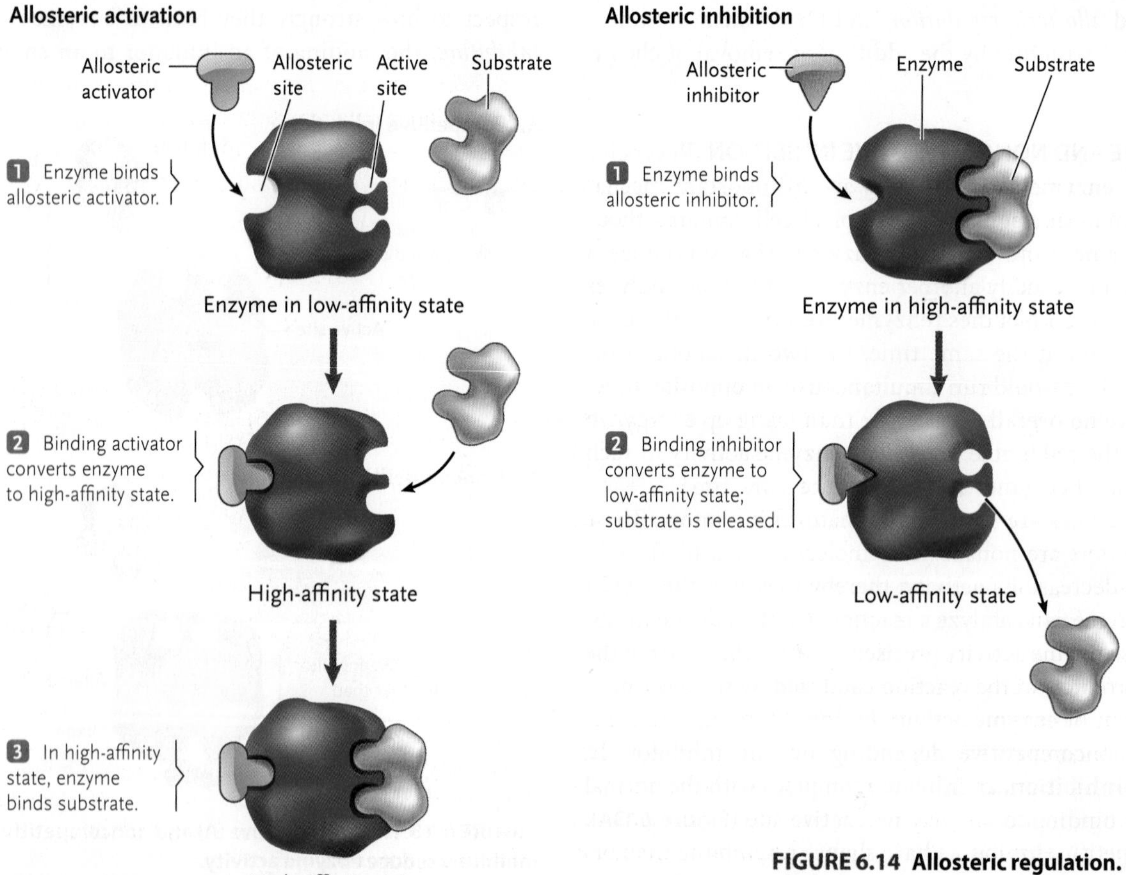

Allosteric activation

Allosteric activator — Allosteric site — Active site — Substrate

1 Enzyme binds allosteric activator.

Enzyme in low-affinity state

2 Binding activator converts enzyme to high-affinity state.

High-affinity state

3 In high-affinity state, enzyme binds substrate.

High-affinity state

Allosteric inhibition

Allosteric inhibitor — Enzyme — Substrate

1 Enzyme binds allosteric inhibitor.

Enzyme in high-affinity state

2 Binding inhibitor converts enzyme to low-affinity state; substrate is released.

Low-affinity state

FIGURE 6.14 Allosteric regulation.
© Cengage Learning 2017

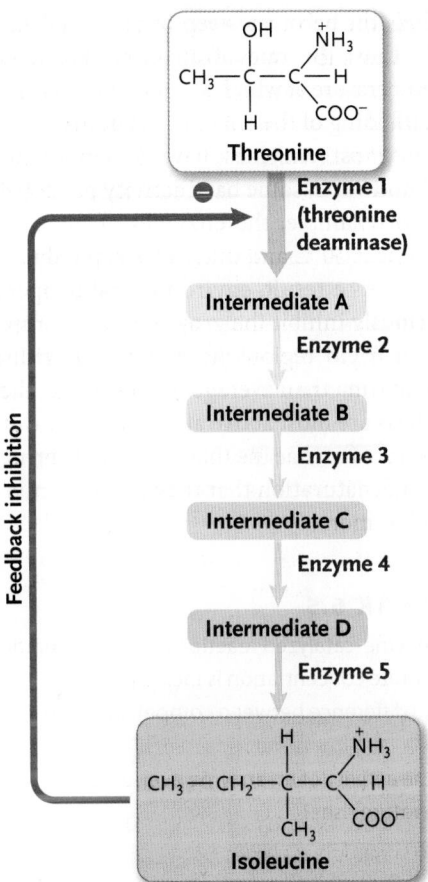

FIGURE 6.15 Feedback inhibition in the pathway that produces isoleucine from threonine. If the product of the pathway, isoleucine, accumulates in excess, it slows or stops the pathway by acting as an allosteric inhibitor of the enzyme that catalyzes the first step in the pathway.
© Cengage Learning 2017

low-affinity state and inhibiting its ability to combine with threonine, the substrate for the first reaction in the pathway. If isoleucine levels drop too low, the allosteric site of threonine deaminase is vacated, the enzyme is converted to the high-affinity state, and isoleucine production increases.

REGULATION BY CHEMICAL MODIFICATION Many key enzymes are regulated by chemical linkage to other substances, typically ions, functional groups such as phosphate or methyl groups, or units derived from nucleotides. The regulatory substances induce folding changes in the enzyme that increase or decrease its activity.

For example, chemical modification by the addition or removal of phosphate groups is a highly significant mechanism of cellular regulation that is used by all organisms from bacteria to humans. Typically, regulatory phosphate groups derived from ATP or other nucleotides are added to the regulated enzymes by other enzymes known as *protein kinases*. The addition of a phosphate group (phosphorylation, as you learned earlier) either increases or decreases enzyme activity or activates or deactivates the enzyme, depending on the particular enzyme and where the phosphate group is added to the enzyme.

Removal of phosphate groups—*dephosphorylation*—reverses the effects of phosphorylation. Dephosphorylation is carried out by a different group of enzymes called *protein phosphatases*. The balance between phosphorylation and dephosphorylation of the enzymes modified by the protein kinases and protein phosphatases closely regulates cellular activity, often as a part of the response to external signal molecules (see Chapter 9).

pH and Temperature Are Key Factors Affecting Enzyme Activity

The activity of most enzymes is altered strongly by changes in pH and temperature. Characteristically, enzymes reach maximal activity within a narrow range of pH or temperature; at levels outside this range, enzyme activity decreases. These effects produce a typically peaked curve when enzyme activity is plotted, with the peak where pH or temperature produces maximal activity.

EFFECTS OF pH CHANGES Typically, each enzyme has an optimal pH where it operates at peak efficiency in speeding the rate of its biochemical reaction **(Figure 6.16).** On either side of this pH optimum, the rate of the catalyzed reaction decreases. The effects become more extreme at pH values farther from the optimum, until the rate drops to zero. An enzyme's dependence on pH typically is caused by ionizable amino acids. A change in pH from the optimal value alters the charges of those amino acids, which modifies the conformation of the protein and eventually causes denaturation of the enzyme.

Most enzymes have an optimum of about pH 7, near the pH of the cellular contents. Enzymes that are secreted from cells may have pH optima farther from neutrality. For example, pepsin, a protein-digesting enzyme secreted into the stomach,

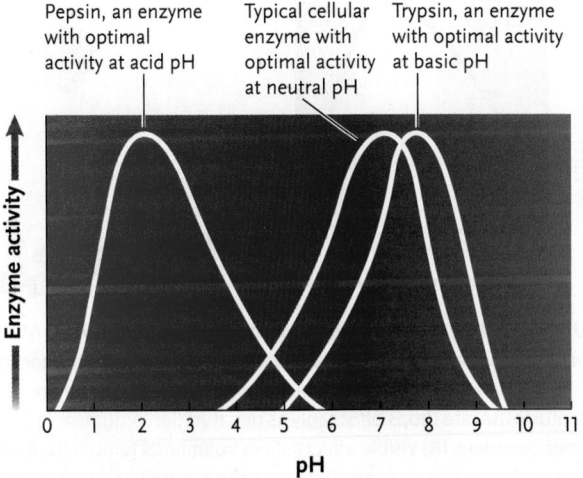

FIGURE 6.16 Effects of pH on enzyme activity. An enzyme typically has an optimal pH at which it is most active; at pH values above or below the optimum, the rate of enzyme activity drops off. At extreme pH values, the rate drops to zero.
© Cengage Learning 2017

has an optimum of pH 1.5, close to the acidity of stomach contents. Similarly, trypsin, also a protein-digesting enzyme, has an optimum of about pH 8, allowing it to function well in the somewhat alkaline contents of the intestine where it is secreted.

EFFECTS OF TEMPERATURE CHANGES The effects of temperature changes on enzyme activity reflect two distinct processes:

1. Temperature has a general effect on chemical reactions of all kinds. As the temperature rises, the rate of chemical reactions typically increases. This effect reflects increases in the kinetic motion of all molecules, with more frequent and stronger collisions as the temperature rises.
2. Temperature has an effect on all proteins, including enzymes. As the temperature rises, the kinetic motions of the amino acid chains of an enzyme increase, along with the strength and frequency of collisions between enzymes and surrounding molecules. At some point, these disturbances become strong enough to denature the enzyme: the hydrogen bonds and other forces that maintain its three-dimensional structure break, making the enzyme unfold and lose its function.

The two effects of temperature act in opposition to each other to produce characteristic changes in the rate of enzymatic catalysis **(Figure 6.17)**. In the range of 0° to about 40°C, the reaction rate doubles for every 10°C increase in temperature. Above 40°C, the increasing kinetic motion begins to unfold the enzyme, reducing the rate of increase in enzyme activity. At some point, as temperature continues to rise, the unfolding causes the reaction rate to level off at a peak. Further increases cause such extensive unfolding that the reaction rate decreases rapidly to zero. For most enzymes, the peak in activity lies between 40° and 50°C; the drop-off becomes steep at 55°C and falls to zero at about 60°C. Thus, the rate of an enzyme-catalyzed reaction peaks at a temperature at which kinetic motion is greatest but no significant unfolding of the enzyme has occurred.

Although most enzymes have a temperature optimum between 40° and 50°C, some have activity peaks below or above this range. For example, the enzymes of maize (corn) pollen function best near 30°C and undergo steep reductions in activity above 32°C. As a result, environmental temperatures above 32°C can seriously inhibit the growth of corn crops. Many animals living in frigid regions have enzymes with much lower temperature optima than average. For example, the enzymes of arctic snow fleas are most active at −10°C. At the other extreme are the enzymes of archaeans that live in hot springs, which are so resistant to denaturation that they remain active at temperatures of 85°C or more.

6.6 RNA-Based Biological Catalysts: Ribozymes

In 1981, biochemist Thomas R. Cech of the University of Colorado at Boulder discovered a group of RNA molecules that appeared to be capable of accelerating the rate of certain biological reactions without being changed by the reactions. This discovery was a great surprise to the scientific community. Further work demonstrated that these RNA-based catalysts, now called **ribozymes,** are part of the biochemical machinery of all cells. Cech and another scientist, Yale University biochemist Sidney Altman, received a Nobel Prize in 1989 "for their discovery of catalytic properties of RNA."

Most of the known ribozymes speed the cutting and splicing reactions that remove surplus segments from RNA molecules as part of their conversion into finished form. Some have other functions, however. For example, Harry F. Noller and his coworkers at the University of California, Santa Cruz, found that ribosomes, the cell structures that assemble amino acids into proteins, can still link amino acids together even if their proteins are removed. After the proteins are extracted, only RNA molecules are left in the ribosomes, indicating that a ribozyme catalyzes this central reaction of protein synthesis. (See Harry Noller's accounting of some of this work in the Unanswered Question of Chapter 15.) After Noller's discovery, Cech and his colleague, Biliang Zhang, confirmed that ribozymes can actually catalyze this reaction (see *Molecular Insights* for an outline of Cech and Zhang's experiment).

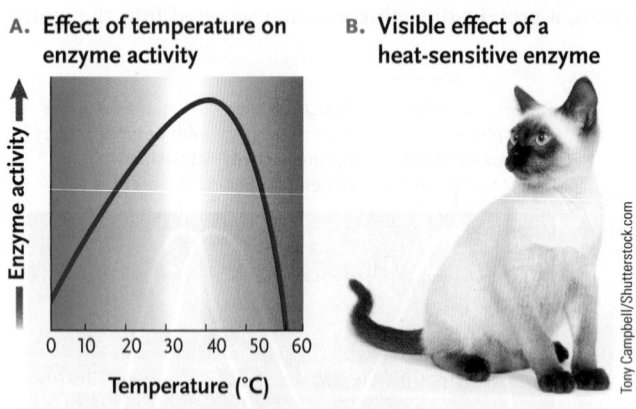

A. Effect of temperature on enzyme activity

Enzyme activity

Temperature (°C)
0 10 20 30 40 50 60

B. Visible effect of a heat-sensitive enzyme

Tony Campbell/Shutterstock.com

FIGURE 6.17 Effect of temperature on enzyme activity. (A) As the temperature rises, the rate of the catalyzed reaction increases proportionally until the temperature reaches the point at which the enzyme begins to denature. The rate drops off steeply as denaturation progresses and becomes complete. **(B)** Visible effects of environmental temperature on enzyme activity in Siamese cats. The fur on the extremities—ears, nose, paws, and tail—contains more dark brown pigment (melanin) than the rest of the body. A heat-sensitive enzyme controlling melanin production is denatured in warmer body regions, so dark pigment is not produced and fur color is lighter.
© Cengage Learning 2017

Ribozymes: Can RNA catalyze peptide bond formation in protein synthesis?

Harry Noller's experiment showed that if proteins were removed from ribosomes, the remaining ribosomal RNA (rRNA) molecules could still catalyze the central reaction of protein synthesis, linkage of amino acids into chains via peptide bonds. However, his work did not eliminate the possibility that undetectable small amounts of ribosomal proteins in the preparations might be catalyzing peptide bond formation.

Research Question

Can RNA catalyze peptide bond formation in protein synthesis?

Experiment

Research by Biliang Zhang and Thomas R. Cech of the University of Colorado Boulder showed that rRNA synthesized artificially and, therefore, never exposed to ribosomal proteins could catalyze the formation of peptide bonds between amino acids (see Figure 3.16).

1. The researchers first synthesized a large pool of RNA molecules in the test tube. Parts of the RNA sequence were the same in every molecule (blue in the **Figure**) and parts differed randomly from molecule to molecule (green in the **Figure**), but did not vary in length. The investigators linked the amino acid phenylalanine (Phe) to the 5' end of each RNA molecule by a disulfide (—S—S—) bond (**Figure,** left side).

2. To the pool of RNAs, they added the amino acid methionine (Met) linked to the nucleotide AMP (see Figure 3.24). (In the cell, single amino acids linked to AMP are used in the pathway that makes proteins.) The methionine–AMP was "tagged" by

linking the small molecule biotin to it (**Figure,** left side).

3. They allowed the molecules to react, hypothesizing that some of the RNA molecules would have the right sequence to act as a ribozyme and form a peptide bond between the methionine and the phenylalanine (**Figure,** right side).

To determine if peptide bonds had formed, Zhang and Cech poured the reaction mixture through a column packed with chemical beads that bind to biotin. Such binding trapped any RNA molecules that were able to catalyze the joining of the two amino acids, whereas unreactive RNA molecules flowed out the bottom of the column. The RNA molecules with the biotin tag were then washed from the column and separated from the linked amino acids by adding a reagent that breaks disulfide bonds.

4. The ribozyme RNA molecules washed from the column were analyzed and refined. Eventually, the researchers obtained ribozymes that catalyzed peptide bond formation at rates 100,000 times faster than the same reaction

occurring spontaneously without a catalyst.

Conclusion

Zhang and Cech's experiments confirmed a feature of ribozyme activity that is critical to the role proposed for these RNA-based catalysts in the primitive RNA world—their ability to catalyze formation of the fundamental linkage tying amino acids together in proteins. Thus, during the evolution of life, proteins could have been made first in quantity by RNA, with no requirement for either DNA or enzymatic proteins.

think like a scientist

Deletion of 20 nucleotides of the constant region from either the 5' or the 3' end of the active ribozymes identified resulted in a dramatic decrease in catalytic activity. Interpret this result.

Source: B. Zhang and T. R. Cech. 1997. Peptide bond formation by in vitro selected ribozymes. *Nature* 390:96–100.

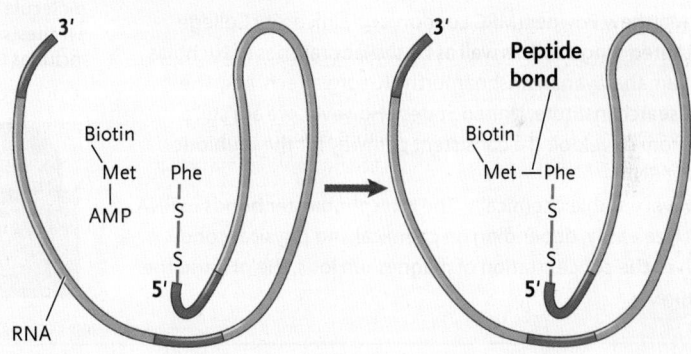

Ribozymes provide a possible solution to a long-standing "chicken-or-egg" paradox about the evolution of life: Did proteins or nucleic acids come first in evolution? It is difficult to understand how DNA could exist without the enzymatic proteins required for its duplication. At the same time, it is difficult to understand how enzymes could exist without nucleic acids, which contain the information required to make them. Ribozymes offer a way around this dilemma because they could have acted as *both* enzymes and informational molecules when cellular life first appeared. The earliest forms of life therefore might have inhabited an "RNA world" in which

neither DNA nor proteins played critical roles (see Chapter 25). If so, ribozymes—the most recently discovered biological catalysts—may have existed for the longest time.

In the next two chapters, we continue our discussion of metabolism with descriptions of cellular respiration (Chapter 7) and photosynthesis (Chapter 8).

STUDY BREAK 6.6

What is a ribozyme, and how does it fit the definition of an enzyme?

You learned in this chapter that RNA molecules can, like proteins, fold into complex three-dimensional structures, bind substrate molecules, and catalyze chemical reactions. You also learned the widely accepted *RNA world hypothesis,* which states that early forms of life had a stage in which RNA served both as genome and as the only genome-encoded catalyst. The RNA world hypothesis is based on several lines of evidence, one of which is that the concept of the RNA world solves the "chicken-or-egg" paradox described at the end of Section 6.6. Other pieces of evidence for the RNA world hypothesis are *molecular fossils*—molecules that exist in today's cells and whose existence is most easily explained by an earlier stage of life, in which RNA was the dominant type of molecule. For example, the ribosome, the machinery that synthesizes proteins by translation, is a catalytic RNA (ribozyme). A different confirmation of the RNA world hypothesis comes from *in vitro* selection experiments. For example, catalytic RNAs have been isolated from large pools of random RNA sequences. Such synthetic ribozymes are able to catalyze RNA polymerization, which means that it should be possible to form a self-replicating system of catalytic RNAs—an RNA world.

How did the RNA world originate?

There are at least three problems concerning how the first RNA world organism originated:

1. It is hard to explain how the components necessary for the synthesis of RNA could have been generated in sufficient purity and concentration in a prebiotic soup. However, our current understanding of chemistry is focused on pure systems; therefore, it is possible that we lack a good understanding of how reactions can proceed in mixtures with many different molecules. In recent years several advances were made to understand how nucleic acid building blocks could have been generated on early Earth, by the laboratories of John Sutherland and Matthew Powner (MRC London, and University College London, United Kingdom), as well as by the laboratories of Nicholas Hud and Ramanarayanan Krishnamurthy (Georgia Tech, and The Scripps Research Institute, United States). However, we are still a long way from developing a consistent pathway for the prebiotic synthesis of RNA.

2. RNA is not very stable chemically. The phosphodiester bonds in RNA can hydrolyze easily, depending on chemical and physical conditions such as the concentration of magnesium ions, the pH, and the temperature.

3. No catalytic RNA (ribozyme) has been made or found that could serve as the self-replicating core of an RNA world organism. The best human-made ribozymes that can catalyze RNA polymerization to facilitate self-replication have a size in the range of 200 nucleotides. However, there are 4^{200} possibilities to arrange 200 nucleotides in their sequence, which means that the probability that this ribozyme (or other, related sequences) could emerge from a prebiotic soup by chance is astronomically unlikely.

A possible solution to these three problems is that a different form of life existed before the RNA world. This pre-RNA world would not have used RNA but a related molecule with an easier prebiotic synthesis, a higher chemical stability, and a higher chance of finding catalytic molecules in a pool of random sequences. Possible candidates for such molecules are TNA (threose nucleic acid), GNA (glycerol nucleic acid), and PNA (peptide nucleic acid). These variants differ from RNA in their backbone, which is not composed of ribose and phosphate (as in the case of RNA) but of threose and phosphate (for TNA), glycerol and phosphate (for GNA), or amino acids (for PNA). Although none of these molecules has been found in cellular nuclei, they are called nucleic acids because they contain the same nucleobases as RNA. The molecular geometry of these molecules allows them to base pair with RNA, which would allow the transfer of genetic information from a pre-RNA world to the RNA world. One problem in this transition is that the genomic information of a pre-RNA world may not have been useful for an RNA world because the different chemistries lead to structural differences in three-dimensionally folded polymers. The next years of research will show whether these molecules have the potential to generate a pre-RNA world organism.

think like a scientist Could a single catalytic RNA molecule replicate itself? Keep in mind that it needs to template the synthesis of every nucleotide in its sequence by base-pairing, which includes the nucleotides of its catalytic center.

Ulrich Müller is Assistant Professor at the University of California, San Diego. He tries to generate an RNA world organism in the lab to find out how an RNA world could have looked like. To learn more about Dr. Müller's research go to http://www-chem.ucsd.edu/faculty.

REVIEW KEY CONCEPTS

For access to MindTap and additional study materials visit www.cengagebrain.com.

6.1 Energy, Life, and the Laws of Thermodynamics

- Energy is the capacity to do work. Kinetic energy is the energy of motion; potential energy is energy stored in an object because of its location or chemical structure. Energy may be readily converted between potential and kinetic states. Potential energy that can be released in a chemical reaction is chemical energy.

- Thermodynamics is the study of energy flow between a system and its surroundings during chemical and physical reactions. A system that does not exchange energy or matter with its surroundings is an isolated system. A system that exchanges energy but not matter with its surroundings is a closed system. A system that exchanges both energy and matter with its surroundings is an open system (Figure 6.1).

- The first law of thermodynamics states that the total amount of energy in a system and its surroundings remains constant. The second law states that in any process involving a spontaneous (possible) change from an initial to a final state, the total entropy (disorder) of the system and its surroundings always increases.

6.2 Free Energy and Spontaneous Reactions

- A spontaneous reaction is one that will occur without the input of energy from the surroundings. A spontaneous reaction releases free energy—energy that is available to do work.

- The free energy equation, $\Delta G = \Delta H - T\Delta S$, states that the free energy change, ΔG, is influenced by two factors: the change in enthalpy (potential energy in a system) and the change in entropy of the system as a reaction goes to completion.

- Factors that oppose the completion of spontaneous reactions, such as the relative concentrations of reactants and products, produce an equilibrium point at which reactants are converted to products and products are converted back to reactants, at equal rates (Figure 6.3).

- Organisms reach equilibrium ($\Delta G = 0$) only when they die.

- Reactions with a negative ΔG are spontaneous; they release free energy and are known as exergonic reactions. Reactions with a positive ΔG require free energy and are known as endergonic reactions (Figure 6.4).

- Metabolism is the biochemical modification and use of energy in the synthesis and breakdown of organic molecules. A catabolic reaction releases the potential energy of a molecule in breaking it down to a simpler molecule (ΔG is negative). An anabolic (biosynthetic) reaction uses energy to convert a simple molecule to a more complex molecule (ΔG is positive). Typically, individual reactions operate in metabolic pathways. Individual reactions in a particular pathway can be catabolic or anabolic; it is the sum of the reactions that makes the pathway catabolic or anabolic.

6.3 Adenosine Triphosphate (ATP): The Energy Currency of the Cell

- The hydrolysis of ATP releases free energy that can be used as a source of energy for the cell (Figure 6.5).

- Using enzymes, a cell can couple the exergonic reaction of ATP hydrolysis to make an otherwise endergonic (anabolic) reaction proceed spontaneously (Figure 6.6).

- The ATP used in coupling reactions is replenished by reactions that link ATP synthesis to catabolic reactions. ATP thus cycles between reactions that release free energy and reactions that require free energy (Figure 6.7).

6.4 Role of Enzymes in Biological Reactions

- What prevents many exergonic reactions from proceeding rapidly is that they need to overcome an energy barrier (the activation energy, E_a) to get to the transition state (Figure 6.8).

- Enzymes are catalysts that greatly speed the rate at which spontaneous reactions occur because they lower the activation energy (Figure 6.9).

- Enzymes usually are specific: they catalyze reactions of only a single type of molecule or a group of closely related molecules (Figure 6.10).

- Catalysis occurs at the active site, which is the site where the enzyme binds to the substrate (reactant molecule). After combining briefly with the substrate, the enzyme is released unchanged when the reaction is complete (Figure 6.11).

- Many enzymes require a cofactor, a nonprotein group that binds to the enzyme, for catalytic activity. Some cofactors are ions; others are small organic molecules called coenzymes. Some coenzymes bind loosely to enzymes whereas others, called prosthetic groups, bind tightly.

- Enzymes reduce the activation energy by inducing the transition state of the reaction, from which the reaction can move easily in the direction of either products or reactants.

- Three major mechanisms contribute to enzymatic catalysis by reducing the activation energy: (1) enzymes bring reacting molecules together; (2) enzymes expose reactant molecules to altered charge environments that promote catalysis; and (3) enzymes change the shape of a substrate molecule.

6.5 Conditions and Factors That Affect Enzyme Activity

- When substrate is abundant, the rate of a reaction is proportional to the amount of enzyme. At a fixed enzyme concentration, the rate of a reaction increases with substrate concentration until the enzyme becomes saturated with reactants. At that point, further increases in substrate concentration do not increase the rate of the reaction (Figure 6.12).

- Many cellular enzymes are regulated by nonsubstrate molecules called inhibitors. Competitive inhibitors interfere with reaction rates by combining with the active site of an enzyme; noncompetitive inhibitors combine with sites elsewhere on the enzyme (Figure 6.13).

- Allosteric regulation resembles noncompetitive inhibition except that regulatory molecules may either increase or decrease enzyme activity. Allosteric regulation often carries out feedback inhibition, in which a product of an enzyme-catalyzed pathway acts as an allosteric inhibitor of the first enzyme in the pathway (Figures 6.14 and 6.15).

- Many key enzymes are regulated by chemical modification, by substances such as ions and certain functional groups. The modifications change enzyme conformation resulting in increased or decreased activity.

- Typically, an enzyme has optimal activity at a certain pH and a certain temperature; at pH and temperature values above and below the optimum, the reaction rate falls off (Figures 6.16 and 6.17).

6.6 RNA-Based Biological Catalysts: Ribozymes

- RNA-based catalysts called ribozymes speed some types of biological reactions; these include cutting and splicing reactions in which surplus segments are removed from RNA molecules and linking reactions that combine amino acids into polypeptide chains.

TEST YOUR KNOWLEDGE

Remember/Understand

1. The capacity to do work best defines:
   a. a metabolic pathway.
   b. entropy.
   c. kinetic or potential energy.
   d. a chemical equilibrium.
   e. thermodynamics.

2. The assembly of proteins from amino acids is best described as:
   a. a conversion of kinetic energy to potential energy reaction.
   b. an entropy reaction.
   c. a catabolic reaction.
   d. an anabolic reaction.
   e. an energy-free reaction.

3. When two glucose molecules react to form maltose:
   a. the reaction represents a negative ΔG.
   b. free energy had to be available to allow the reaction to proceed.
   c. the reaction is exothermic.
   d. it supports the second law of thermodynamics, which states there is tendency of the universe toward disorder.
   e. the resulting product has less potential energy than the reactants.

4. When glucose reacts with ATP to form glucose-6-phosphate:
   a. the synthesis of glucose-6-phosphate is exergonic.
   b. ADP is at a higher energy level than ATP.
   c. glucose-6-phosphate is at a higher energy level than glucose.
   d. because ATP donates a phosphate to glucose, this is not a coupled reaction.
   e. the reaction is spontaneous.

5. In the following graph:

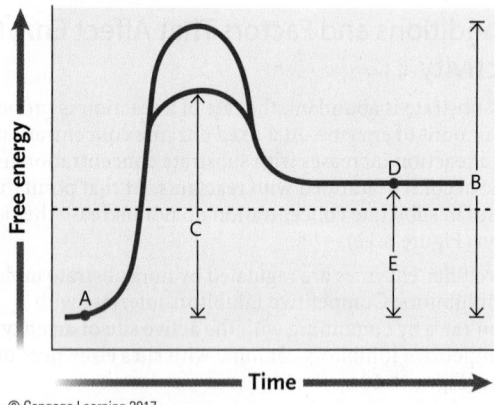

© Cengage Learning 2017

   a. A represents the product.
   b. B represents the energy of activation when enzymes are present.
   c. C is the free energy difference between A and D.
   d. C is the energy of activation without enzymes.
   e. E is the difference in free energy between the reactant and the products.

6. Which of the following methods is *not* used by enzymes to increase the rate of reactions?
   a. covalent bonding with the substrate at their active site
   b. bringing reacting molecules into close proximity
   c. orienting reactants into positions to favor transition states
   d. changing charges on reactants to hasten their reactivity
   e. increasing fit of enzyme and substrate that reduces the energy of activation

7. In an enzymatic reaction:
   a. the enzyme leaves the reaction chemically unchanged.
   b. if the enzyme molecules approach maximal rate, and the substrate is continually increased, the rate of the reaction does not reach saturation.
   c. in the stomach, enzymes would have an optimal activity at a neutral pH.
   d. increasing temperature above the optimal value slows the reaction rate.
   e. the least important level of organization for an enzyme is its tertiary structure.

8. Which of the following statements about the allosteric site is true?
   a. The allosteric site is a second active site on a substrate in a metabolic pathway.
   b. The allosteric site on an enzyme can allow the product of a metabolic pathway to inhibit that enzyme and stop the pathway.
   c. When the allosteric site of an enzyme is occupied, the reaction is irreversible and the enzyme cannot react again.
   d. An allosteric activator prevents binding at the active site.
   e. An enzyme that possesses allosteric sites does not possess an active site.

9. Which of the following statements about inhibition is true?
   a. Allosteric inhibitors and allosteric activators are competitive for a given enzyme.
   b. If an inhibitor binds the active site, it is considered noncompetitive.
   c. If an inhibitor binds to a site other than the active site, this is competitive inhibition.
   d. A noncompetitive inhibitor is believed to change the shape of the enzyme, making its active site inoperable.
   e. Competitive inhibition is usually not reversible.

10. Which of the following statements is *incorrect*?
    a. Ribozymes can link amino acids to form protein.
    b. Ribozymes can act as enzymes.
    c. Ribozymes can act as informational molecules.
    d. Ribozymes are suggested as the first molecules of life.
    e. Ribozymes are proteins.

Apply/Analyze

11. **Discuss Concepts** Trees become more complex as they develop spontaneously from seeds to adults. Does this process violate the second law of thermodynamics? Why or why not?

12. **Discuss Concepts** Trace the flow of energy through your body. What products increase the entropy of you and your surroundings?

13. **Discuss Concepts** You have found a molecular substance that accelerates the rate of a particular reaction. What kind of information would you need to demonstrate that this molecular substance is an enzyme?

14. **Discuss Concepts** The addition or removal of phosphate groups from ATP is a fully reversible reaction. In what way does this reversibility facilitate the use of ATP as a coupling agent for cellular reactions?

Evaluate/Create

15. **Discuss Concepts** Researchers once hypothesized that an enzyme and its substrate fit together like a lock and key but that the products do not fit the enzyme. Examine this idea with respect to reversible reactions.

16. **Design an Experiment** Succinate dehydrogenase is part of the cellular biochemical machinery for breaking down sugars, fatty acids, and amino acids into carbon dioxide and water, with the capture of their chemical energy as ATP. Suppose you are measuring the activity of this enzyme extracted from cells in test-tube reactions. You find that the rate of the reaction converting succinate to fumarate catalyzed by succinate dehydrogenase is inhibited by the addition of malonate to the reaction mixture. Design an experiment that will tell you whether malonate is acting as a competitive or a noncompetitive inhibitor.

17. **Apply Evolutionary Thinking** If RNA appeared first in evolution, establishing an RNA world, which do you think would evolve next: DNA or proteins? Why?

For selected answers, see Appendix A.

INTERPRET THE DATA

The postsynaptic density 95 (PSD-95) protein plays a key role in mammalian nervous system responses by concentrating and organizing receptors on the nerve cells. Future drugs that change these receptors could change learning and memory at the cellular level. A key part of the interaction between PSD-95 and the proposed drugs is the thermodynamics of binding between them.

The thermodynamics of binding between PSD-95 and two different small polypeptides (similar to those in the proposed drugs) are shown in the graphs. Graph A shows data for a parent polypeptide, and graph B shows the data for a mutated polypeptide (with some altered amino acids) of the same length. On the x axis is the ratio (in moles) between the amount of bound polypeptide and PSD-95. When the molar ratio is very near zero on the x axis in each graph, all of the PSD-95 molecules available are bound by the polypeptides. The molar ratio can be manipulated by injecting small amounts of polypeptide into the solution, and the change in energy content of the injectant (ΔH measured in kcal/mol) is measured with each accumulating injection.

1. If $T\Delta S$ for the parent polypeptide (graph A) is 4.0 kcal/mol and for the mutated polypeptide (graph B) is 3.5 kcal/mol, what is the ΔG for each polypeptide binding to the protein?

2. Using the ΔG values calculated in question 1, state whether each polypeptide binding is endergonic or exergonic, spontaneous or not spontaneous, and which would yield more free energy on binding to PSD-95.

3. If the mutated polypeptide represented a drug that needed to compete with the parent polypeptide in a nerve cell, would this mutated polypeptide be very effective?

Source: From D. Saro et al. 2007. A thermodynamic ligand binding study of the third PDZ domain (PDZ3) from the mammalian neuronal protein PSD-95. *Biochemistry* 46:6340–6352.

A. **Parent polypeptide**

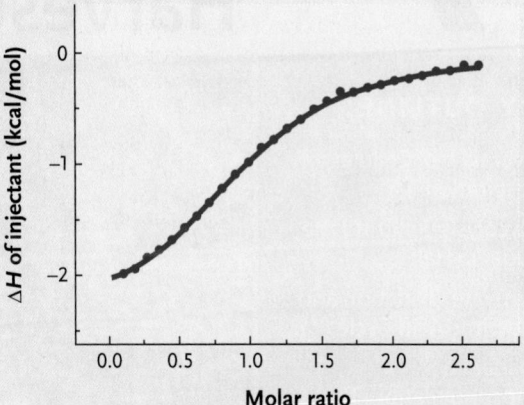

B. **Mutated polypeptide**

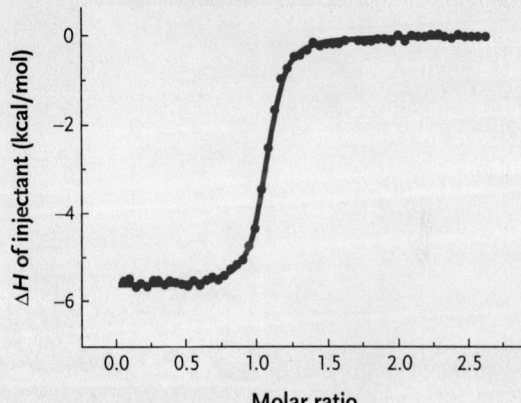

© Cengage Learning 2017

7 Cellular Respiration: Harvesting Chemical Energy

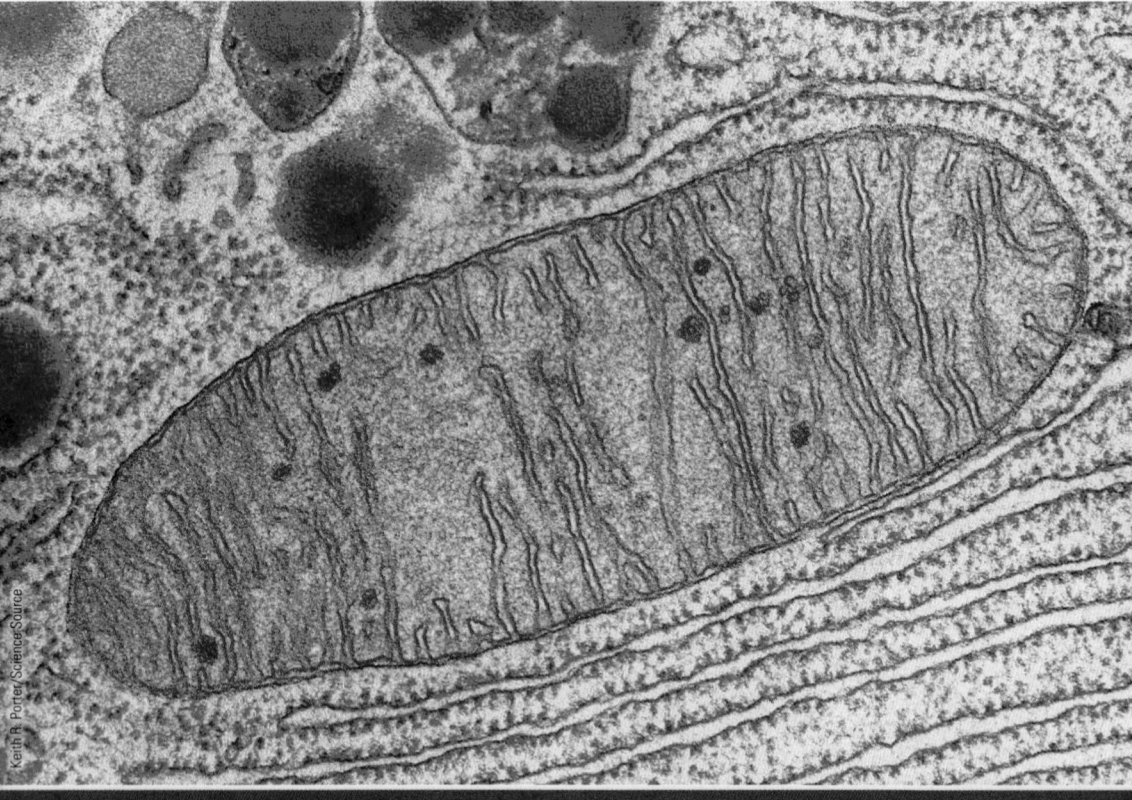

Keith R. Porter/Science Source

Mitochondrion (color-enhanced TEM). Mitochondria are the sites of cellular respiration.

Why it matters . . . In the early 1960s, Swedish physician Rolf Luft mulled over some odd symptoms of a patient. The young woman felt weak and too hot all the time (with a body temperature of up to 38.4°C). Even on the coldest winter days she never stopped perspiring, and her skin was always flushed. She was also underweight (40 kg), despite consuming about 3,500 calories per day.

Luft inferred that his patient's symptoms pointed to a metabolic disorder. Her cells were very active, but much of their activity was being dissipated as metabolic heat. Tests that measured her metabolic rate—the amount of energy her body was expending—showed the patient's oxygen consumption was the highest ever recorded, about twice the normal rate.

Luft also examined a tissue sample from the patient's skeletal muscles under a microscope. He found that her muscle cells contained many more mitochondria—the ATP-producing organelles of the cell—than are normally present in muscle cells. Other studies showed that the mitochondria were engaged in cellular respiration—their prime function—but little ATP was being generated.

The disorder, now called *Luft syndrome,* was the first disorder to be linked directly to a defective mitochondrion. By analogy, someone with this disorder functions like a city with half of its power plants shut down. Skeletal and heart muscles, the brain, and other hardworking body parts with high energy demands are hurt the most by the inability of mitochondria to provide enough energy for metabolic demands. To this day, the genetic defect responsible for Luft syndrome is not known.

We now know that mitochondrial dysfunction is involved in the development of a number of diseases, including cancer, diabetes, and heart failure, as well as in obesity and aging. Clearly, human health depends on mitochondria that are structurally sound and that function properly.

More broadly, every animal, plant, fungus, and most protists depend on mitochondria that are functioning correctly to grow and survive.

In mitochondria, ATP forms as part of the reactions of cellular respiration. **Cellular respiration** is the collection of metabolic reactions within cells that breaks down food molecules to produce energy in the form of ATP. ATP fuels nearly all the reactions that keep cells, and organisms, metabolically active. In eukaryotes and many prokaryotes, oxygen is a reactant in the ATP-producing process. This form of cellular respiration is **aerobic respiration** (*aero* = air, *bios* = life). In some prokaryotes, a molecule other than oxygen, such as sulfate or nitrate, is used in the ATP-producing process. This form of cellular respiration is **anaerobic respiration** (*an* = without). Commonly, cellular respiration is used synonymously with aerobic respiration, as we will do in this chapter.

Recall from our discussion of the laws of thermodynamics (see Section 6.1) that all living organisms are open systems, and that the ultimate source of energy they use is the Sun. That is, the primary source of the food molecules broken down in cellular respiration is *photosynthesis,* which is described in Chapter 8. Photosynthesis is the process in which light energy is captured and used to split water into hydrogen and oxygen. The hydrogen from the water is combined with carbon dioxide to synthesize carbohydrates and other organic molecules. The other product of the splitting of water is oxygen, a molecule needed for cellular respiration. Photosynthesis occurs in most plants, many protists, and some prokaryotes.

Together, cellular respiration and photosynthesis are the major biological steps of the carbon cycle, which is the global circulation of carbon atoms. The carbon cycle is described in Section 54.3. Atmospheric carbon in the carbon cycle is mostly in the form of CO_2, which is a product of cellular respiration.

7.1 Overview of Cellular Respiration

Electron-rich food molecules synthesized by photosynthetic organisms are used by the organisms themselves and by other organisms that ingest the photosynthetic organisms, or parts of them. The electrons are removed from fuel substances, such as sugars, and donated to other molecules, such as oxygen, that act as electron acceptors. In the process, some of the energy of the electrons is released and used to drive the synthesis of ATP. ATP provides energy for most of the energy-consuming activities in the cell. Thus, life and its systems are driven by a cycle of electron flow that is powered by light in photosynthesis and oxidation in cellular respiration (**Figure 7.1**).

Coupled Oxidation and Reduction Reactions Produce the Flow of Electrons for Energy Metabolism

The partial or full loss of electrons (e^-) from a substance is an **oxidation,** and the substance from which the electrons are lost—the *electron donor*—is said to be **oxidized.** The partial or full gain of electrons to a substance is a **reduction,** and the substance that gains the electrons—the *electron acceptor*—is said to be **reduced.** A simple mnemonic to remember the direction of electron transfer is OIL RIG: Oxidation Is Loss (of electrons), Reduction Is Gain (of electrons). The term *oxidation* was used originally to describe the reaction that occurs when fuel substances are burned in air, in which oxygen directly accepts electrons removed from the fuels. However, although the term *oxidation* suggests that oxygen is involved in electron loss, most cellular oxidations occur without the direct participation of oxygen. The term *reduction* refers to the decrease in positive electrical charge that occurs when electrons, which are

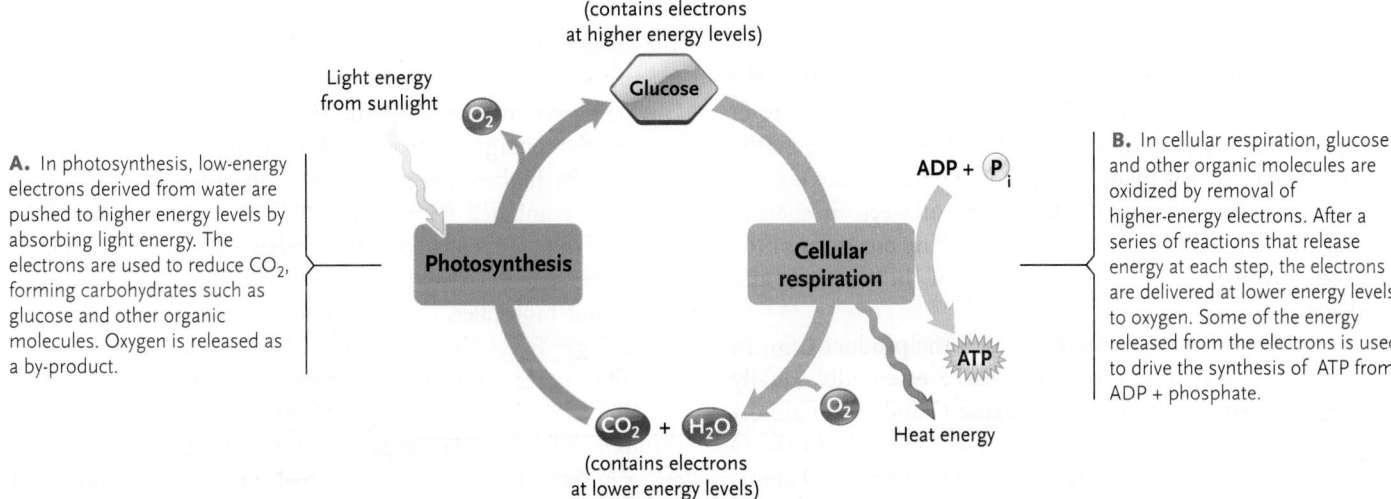

A. In photosynthesis, low-energy electrons derived from water are pushed to higher energy levels by absorbing light energy. The electrons are used to reduce CO_2, forming carbohydrates such as glucose and other organic molecules. Oxygen is released as a by-product.

B. In cellular respiration, glucose and other organic molecules are oxidized by removal of higher-energy electrons. After a series of reactions that release energy at each step, the electrons are delivered at lower energy levels to oxygen. Some of the energy released from the electrons is used to drive the synthesis of ATP from ADP + phosphate.

FIGURE 7.1 The flow of energy from sunlight to ATP. (A) Photosynthesis occurs in plants, many protists, and some prokaryotes; **(B)** cellular respiration (aerobic respiration) occurs in all eukaryotes, including plants, and in some prokaryotes.

© Cengage Learning 2017

negatively charged, are added to a substance. Although the term *reduction* suggests that the energy level of molecules is decreased when they accept electrons, molecules typically gain energy from added electrons.

Oxidation and reduction are coupled reactions that remove electrons from a donor molecule and simultaneously add them to an acceptor molecule. Such coupled oxidation–reduction reactions are also called **redox reactions;** a generalized redox reaction can be written as:

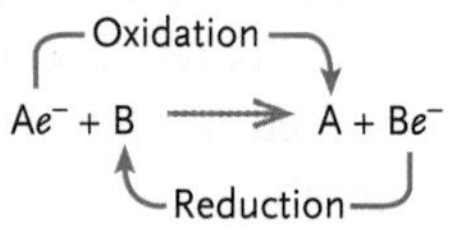

$$\overset{\overbrace{\qquad\text{Oxidation}\qquad}}{\underset{\underbrace{\qquad\text{Reduction}\qquad}}{Ae^- + B \longrightarrow A + Be^-}}$$

In a redox reaction, electrons release some of their energy as they pass from a donor molecule to an acceptor molecule. This free energy (see Section 6.2) is available for cellular work, such as ATP synthesis. ATP is the primary agent that couples exergonic and endergonic reactions in the cell to facilitate the synthesis of complex molecules (see Section 6.3). The harnessing of energy into a useful form such as ATP when electrons move from a higher-energy state to a lower-energy state is analogous to what happens when water moves through turbines at a dam and produces electricity.

As you have just learned, oxidation and reduction are defined with respect to the gain or loss of electrons. You will see in the reactions described later in the chapter that electron movement is associated with H atoms. Recall from Section 2.3 that a hydrogen atom, H, consists of a proton and an electron: $H = H^+ + e^-$. Therefore, the transfer of a hydrogen atom involves the transfer of an electron. As a result, when a molecule loses a hydrogen atom, it becomes oxidized.

The gain or loss of an electron in a redox reaction is not always complete. That is, in some redox reactions, electrons are transferred completely from one atom to another, whereas in others the degree to which electrons are shared between atoms changes. (Sharing of electrons in covalent bonds is described in Section 2.3.) The condition of electron sharing is said to involve a relative loss or gain of electrons. Most redox reactions in the electron transfer system discussed later in the chapter are of this type. The redox reaction between methane and oxygen (the burning of natural gas in air) that produces carbon dioxide and water illustrates a change in the degree of electron sharing **(Figure 7.2).** The dots in the figure indicate the positions of the electrons involved in the covalent bonds of the reactants and products.

Compare the reactant methane with the product CO_2. In methane, the covalent electrons are shared essentially equally between bonded C and H atoms because C and H are almost equally electronegative. In CO_2, electrons are closer to the O atoms than to the C atom in the C=O bonds because O atoms are highly electronegative. Overall, this means that the C atom has partially "lost" its shared electrons in the reaction— methane has been oxidized. Now compare the oxygen reactant with the product water. In the O_2 molecule, the two O atoms

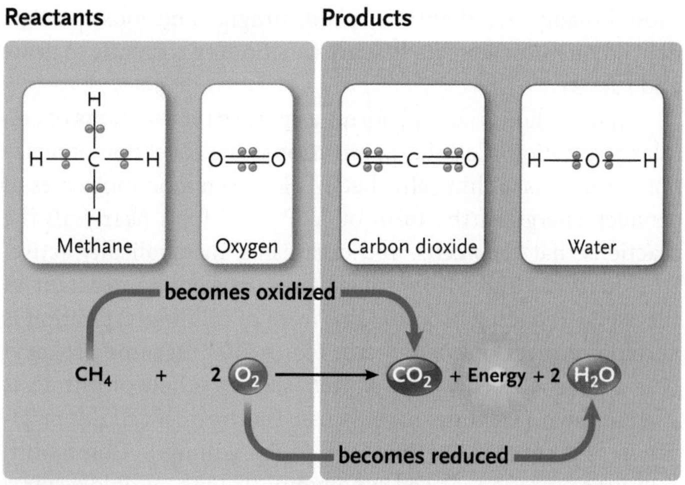

FIGURE 7.2 Relative loss and gain of electrons in a redox reaction, the burning of methane (natural gas) in oxygen. Compare the positions of the electrons in the covalent bonds of reactants and products. In this redox reaction, methane is oxidized because the carbon atom has partially lost its shared electrons, and oxygen is reduced because the oxygen atoms have partially gained electrons.
© Cengage Learning 2017

share their electrons equally. The oxygen reacts with the hydrogen from methane, producing water in which the electrons are closer to the O atom than to the H atoms. This means that each O atom has partially "gained" electrons—oxygen has been reduced. Because of this, the reaction between CH_4 and O_2 releases a lot of heat energy as the electrons in the C—H bonds of CH_4 move closer to the electronegative atoms that form CO_2. The more electronegative an atom is, the greater the force that holds the electrons to that atom and therefore the greater the energy required to lose an electron.

Electrons Flow from Fuel Substances to Final Electron Acceptors

The energy of the electrons removed during cellular oxidations originates in the reactions of photosynthesis (see Figure 7.1A). During photosynthesis, electrons derived from water are pushed to very high energy levels using energy from the absorption of light. These higher-energy electrons, together with H^+ from water, are combined with carbon dioxide to form sugar molecules and then are removed by the oxidative reactions that release energy for cellular activities (see Figure 7.1B). As electrons pass to acceptor molecules, they lose much of their energy. Some of this energy drives the synthesis of ATP from ADP and P_i (a phosphate group from an inorganic source) (see Section 6.3).

The total amount of energy obtained from electrons flowing through cellular oxidative pathways depends on the difference between their high energy level in fuel substances and the lower energy level in the molecule that acts as the *final acceptor* for electrons, that is, the last molecule reduced in cellular pathways. The lower the energy level in the final acceptor, the greater the yield of energy for cellular activities. Oxygen is the final acceptor

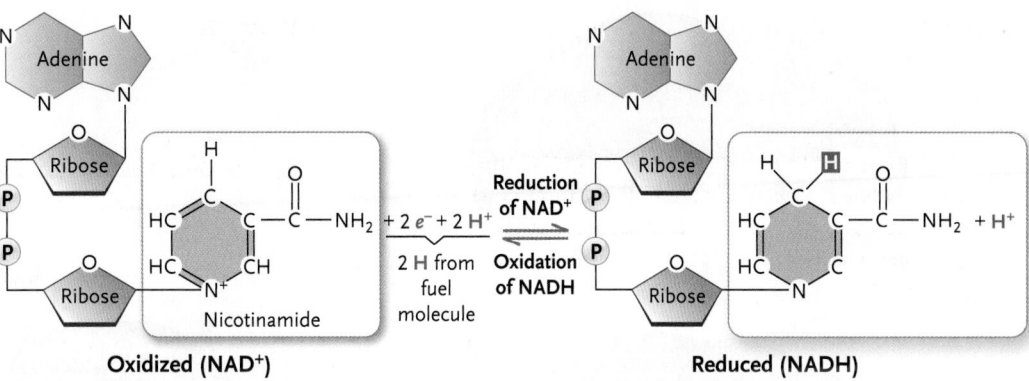

FIGURE 7.3 Electron carrier NAD⁺. When a fuel molecule is oxidized, releasing two hydrogen atoms, NAD⁺, the oxidized form of the carrier, accepts a proton (H^+) and two electrons and is transformed into NADH, the reduced form of the carrier. The nitrogenous base (blue) of NAD that adds and releases electrons and protons is nicotinamide, which is derived from the vitamin niacin (nicotinic acid).
© Cengage Learning 2017

Oxidized (NAD⁺)   **Reduced (NADH)**

in cellular respiration, the most efficient and highly developed form of cellular oxidation (see Figure 7.1B). The very low energy level of the electrons added to oxygen allows a maximum output of energy for ATP synthesis. As part of the final reduction, oxygen combines with protons and electrons to form water.

In Cellular Respiration, Cells Make ATP by Oxidative Phosphorylation

Cellular respiration includes both the reactions that transfer electrons from organic molecules (fuel molecules, commonly known as food) to oxygen and the reactions that make ATP. These reactions are often written in a summary form that uses glucose ($C_6H_{12}O_6$), the most common fuel molecule, as the initial reactant:

$$C_6H_{12}O_6 + 6\,O_2 + 32\,ADP + 32\,P_i \rightarrow 6\,H_2O + 6\,CO_2 + 32\,ATP$$

In this overall reaction, electrons and protons are transferred from glucose to oxygen, forming water, and the carbons left after this transfer are released as carbon dioxide. ATP synthesis by the addition of P_i to ADP is the key, and final, step of this reaction. *Phosphorylation* is the term for a reaction that adds a phosphate group to a substance such as ADP (see Section 6.3). How the 32 ATP molecules are derived is explained later in this chapter.

The oxidation of fuel molecules in cellular respiration is catalyzed by a number of *dehydrogenase* enzymes that transfer electrons from a fuel molecule to a molecule that acts as an *electron carrier*. The most common electron carrier is the coenzyme **nicotinamide adenine dinucleotide (NAD⁺) (Figure 7.3),** which is a type of nucleotide (see Section 3.5). (Coenzymes are discussed in Section 6.4.) In cellular respiration, dehydrogenases remove two H atoms from a substrate molecule and transfer the two electrons, but only one of the protons, to NAD⁺ (the oxidized form), resulting in its complete reduction to NADH (the reduced form). The other proton is released. Later you will learn that the potential energy carried in NADH is used in the synthesis of ATP.

The entire process of cellular respiration can be divided into three stages **(Figure 7.4):**

1. *Glycolysis:* In **glycolysis,** enzymes break a molecule of glucose (contains six carbon atoms) into two molecules

of pyruvate (an organic compound with a backbone of three carbon atoms). Some NADH is produced from NAD⁺, and some ATP is synthesized by **substrate-level phosphorylation,** an enzyme-catalyzed reaction that transfers a phosphate group from a substrate to ADP **(Figure 7.5).**

2. *Pyruvate oxidation and the citric acid cycle:* In **pyruvate oxidation,** enzymes convert the three-carbon pyruvate into a two-carbon acetyl group that enters the **citric acid**

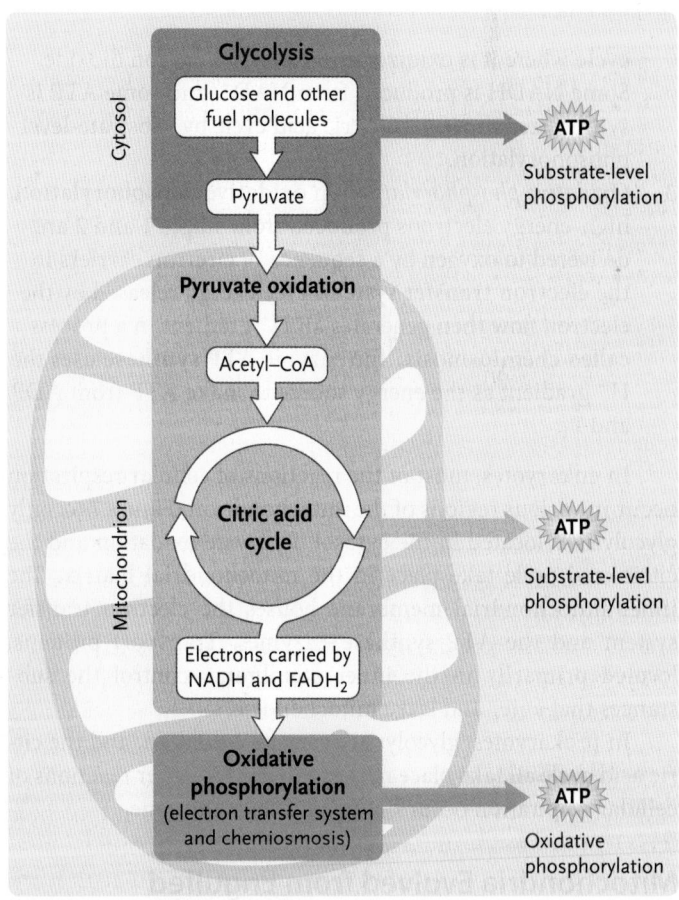

FIGURE 7.4 The three stages of cellular respiration: (1) glycolysis; (2) pyruvate oxidation and the citric acid cycle; and (3) oxidative phosphorylation, which includes the electron transfer system and chemiosmosis.
© Cengage Learning 2017

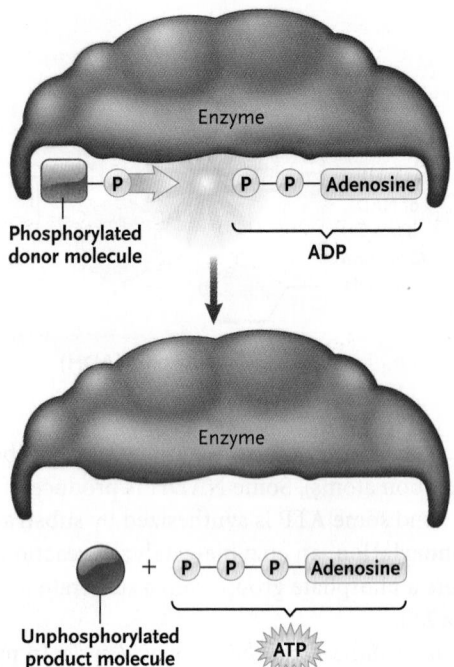

FIGURE 7.5 Substrate-level phosphorylation. A phosphate group is transferred from a high-energy donor directly to ADP, forming ATP.
© Cengage Learning 2017

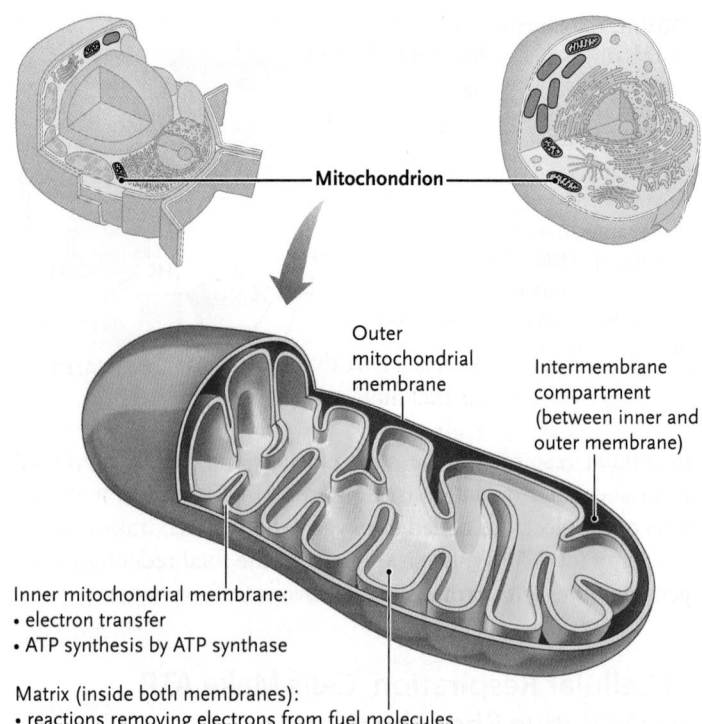

FIGURE 7.6 Membranes and compartments of a mitochondrion. Label lines that end in a dot indicate a compartment enclosed by the membranes.
© Cengage Learning 2017

cycle where it is oxidized completely to carbon dioxide. Some NADH is produced from NAD+, and some ATP is synthesized during the citric acid cycle by substrate-level phosphorylation.

3. *Oxidative phosphorylation:* In **oxidative phosphorylation,** high-energy electrons produced from stages 1 and 2 are delivered to oxygen by a sequence of electron carriers in the **electron transfer system.** Free energy released by the electron flow then generates an H+ gradient in a process called **chemiosmosis.** The enzyme **ATP synthase** uses the H+ gradient as the energy source to make ATP from ADP and P_i.

In eukaryotes, most of the reactions of cellular respiration occur in various regions of the mitochondrion **(Figure 7.6);** only glycolysis is located in the cytosol. Pyruvate oxidation and the citric acid cycle take place in the mitochondrial matrix. The inner mitochondrial membrane houses the electron transfer system and the ATP synthase enzymes. Transport proteins, located primarily in the inner membrane, control the substances that enter and leave mitochondria.

In prokaryotes, glycolysis, pyruvate oxidation, and the citric acid cycle all take place in the cytosol. The other reactions of cellular respiration occur on the plasma membrane.

Mitochondria Evolved from Engulfed Prokaryotic Cells

How did mitochondria evolve? These organelles began to develop when photosynthetic and nonphotosynthetic prokaryotes coexisted in an oxygen-rich atmosphere. An anaerobic

(nonphotosynthetic) prokaryote is proposed to have ingested an aerobic (photosynthetic) prokaryote. The aerobe became a symbiotic organism within the host cell. The cytoplasm of the host anaerobe, formerly limited to the use of organic molecules as final electron acceptors, was now home to an aerobe capable of carrying out the much more efficient transfer of electrons to oxygen. From an evolutionary perspective, cells that had engulfed prokaryotic cells capable of cellular respiration would have been at a considerable advantage compared to other cells, so natural selection would have favored their persistence in their ecosystem. They could ingest simple carbohydrates from their environment and metabolize them as a source of energy. As the two cells evolved together, the parasite adapted to doing one thing—powering the cell. As the parasite lost genes not needed for that function, it became simpler and more dependent on the host cell, and thus evolved into the mitochondrion. Researchers believe that the development of mitochondria may have occurred only once in the history of the Earth. (Chapter 25 discusses in more detail the evolution of mitochondria and chloroplasts from ingested prokaryotes.)

The next three sections examine the three stages of cellular respiration in turn.

STUDY BREAK 7.1

1. Distinguish between oxidation and reduction.
2. Distinguish between cellular respiration and oxidative phosphorylation.

7.2 Glycolysis: Splitting the Sugar in Half

Glycolysis, the first series of oxidative reactions that remove electrons from cellular fuel molecules, takes place in the cytosol of all organisms. In glycolysis (*glykys* = sweet, *lysis* = breakdown), glucose is oxidized partially and broken down into two smaller molecules, and a relatively small amount of ATP is produced. Glycolysis is also known as the Embden–Meyerhof pathway for Gustav Embden and Otto Meyerhof, two German physiological chemists who (separately) made the most important contributions to determining the sequence of reactions in the pathway. Meyerhof received a Nobel Prize in 1922 "for his discovery of the fixed relationship between the consumption of oxygen and the metabolism of lactic acid in the muscle."

Glycolysis starts with the six-carbon sugar glucose and produces two molecules of the three-carbon organic substance *pyruvate* or *pyruvic acid* in 10 sequential enzyme-catalyzed reactions. (The *-ate* suffix indicates the ionized form of an organic acid such as pyruvate, in which the carboxyl group —COOH dissociates to —COO⁻ + H⁺, which is usual under cellular conditions.) Pyruvate still contains many electrons that can be removed by oxidation, and it is the primary fuel substance for the second stage of cellular respiration.

The Reactions of Glycolysis Include Energy-Requiring and Energy-Releasing Steps

The initial steps of glycolysis are *energy-requiring reactions* (**Figure 7.7**, red arrow) in which 2 ATP are hydrolyzed. The steps convert a molecule of the six-carbon glucose into two molecules of the three-carbon phosphorylated compound, glyceraldehyde-3-phosphate (G3P). In the subsequent *energy-releasing reactions* of glycolysis (Figure 7.7, blue arrow), each G3P molecule is converted to a molecule of pyruvate. In these reactions, NAD⁺ is reduced to NADH, and 4 ATP molecules are produced by substrate-level phosphorylation. Energywise, glycolysis results in a net gain of energy per glucose molecule of 2 ATP and 2 NADH. The end product of the glycolysis pathway is two molecules of pyruvate for each input molecule of glucose. The two molecules of pyruvate contain all of the six carbon atoms of the glucose molecule.

The reactions of glycolysis are shown in detail in **Figure 7.8.** Two redox reactions occur in glycolysis, the first in reaction 6, and the second in reaction 9.

For each molecule of glucose that enters the pathway, the energy-requiring reactions 1 to 5 generate 2 molecules of G3P using 2 ATP, and the energy-releasing reactions 6 to 10 convert

the 2 molecules of G3P to 2 molecules of pyruvate, producing 4 ATP and 2 NADH. The net reactants and products of glycolysis in equation form are:

$$1 \text{ glucose} + 2 \text{ ADP} + 2 \text{ P}_i + 2 \text{ NAD}^+ + 4\,e^- + 4 \text{ H}^+ \rightarrow$$
$$2 \text{ pyruvate} + 2 \text{ ATP} + 2 \text{ NADH} + 2 \text{ H}^+ + 2 \text{ H}_2\text{O}$$

Each ATP molecule produced in the energy-releasing steps of glycolysis—steps 7 and 10 (see Figure 7.7)—results from

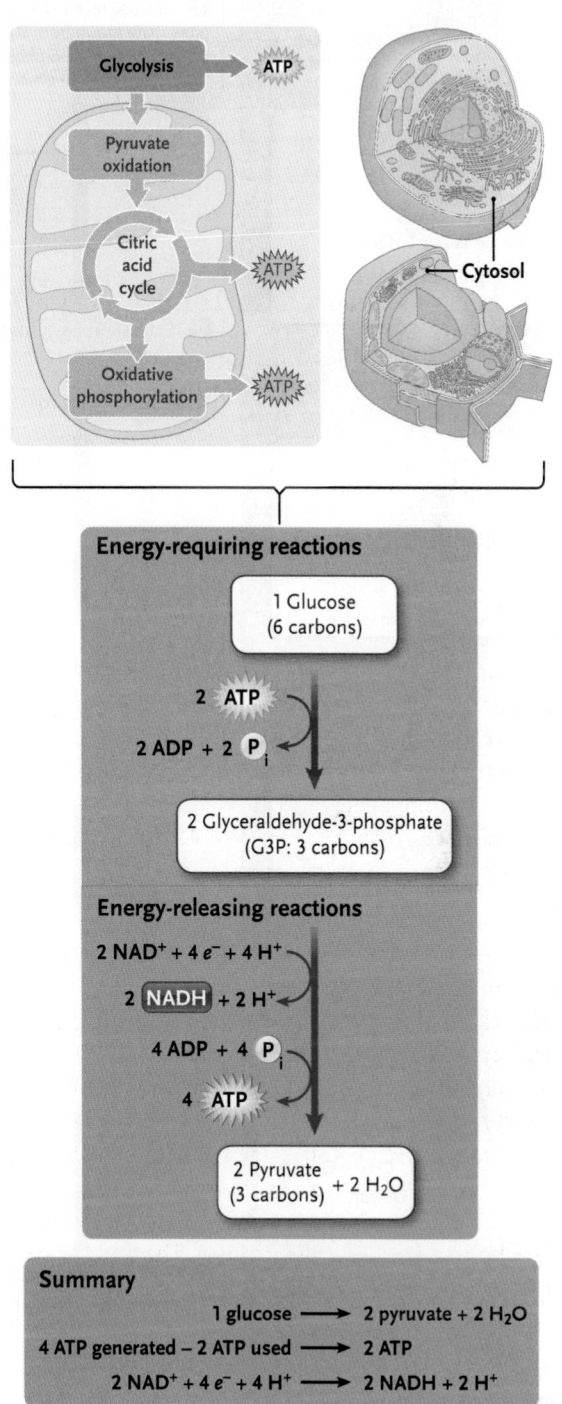

FIGURE 7.7 Overall reactions of glycolysis, which occur in the cytosol. Glycolysis splits glucose (six carbons) into pyruvate (three carbons) and yields ATP and NADH.
© Cengage Learning 2017

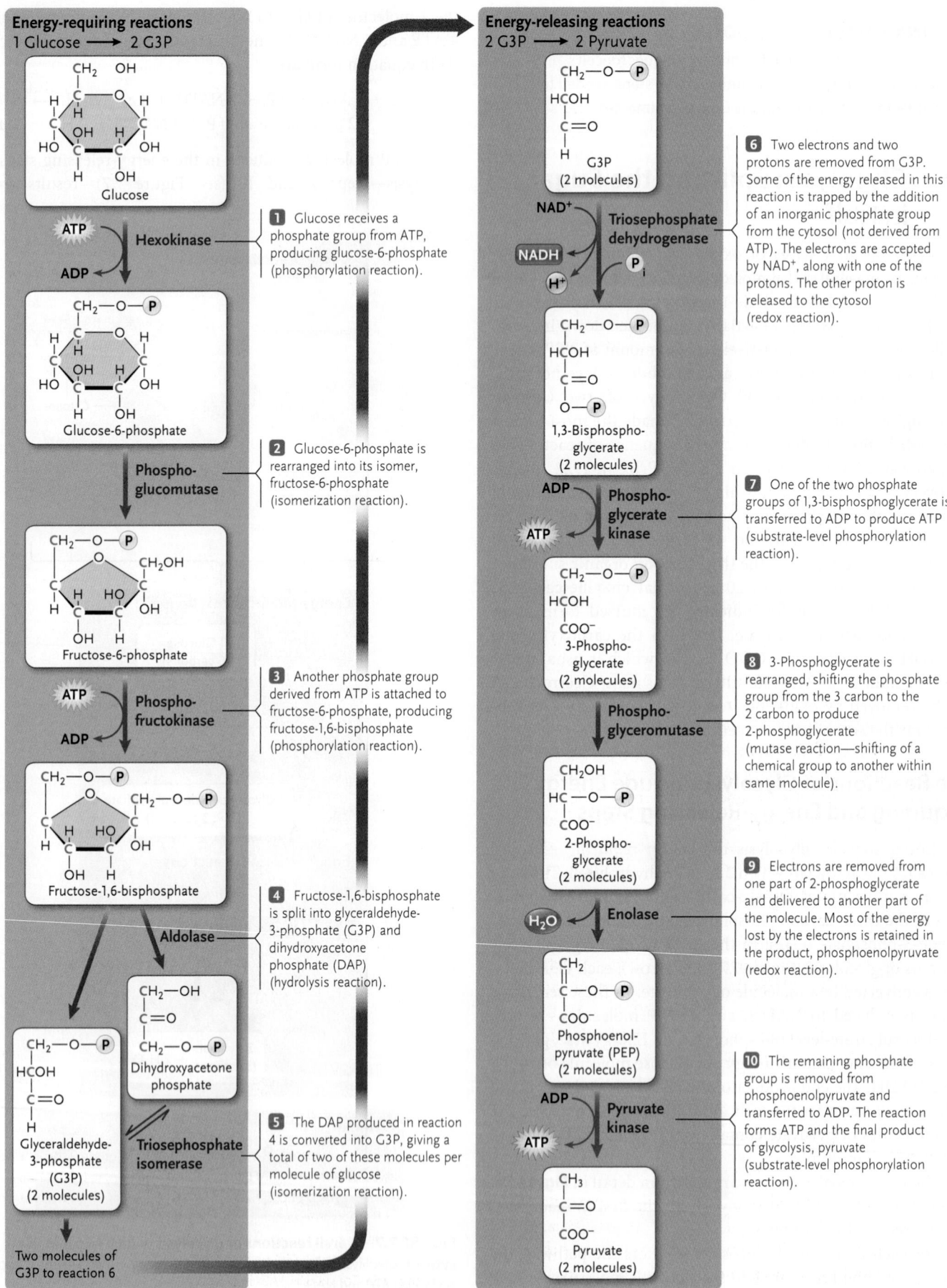

Energy-requiring reactions
1 Glucose ⟶ 2 G3P

Glucose

Hexokinase

1 Glucose receives a phosphate group from ATP, producing glucose-6-phosphate (phosphorylation reaction).

Glucose-6-phosphate

Phospho-glucomutase

2 Glucose-6-phosphate is rearranged into its isomer, fructose-6-phosphate (isomerization reaction).

Fructose-6-phosphate

Phospho-fructokinase

3 Another phosphate group derived from ATP is attached to fructose-6-phosphate, producing fructose-1,6-bisphosphate (phosphorylation reaction).

Fructose-1,6-bisphosphate

Aldolase

4 Fructose-1,6-bisphosphate is split into glyceraldehyde-3-phosphate (G3P) and dihydroxyacetone phosphate (DAP) (hydrolysis reaction).

Dihydroxyacetone phosphate

Glyceraldehyde-3-phosphate (G3P) (2 molecules)

Triosephosphate isomerase

5 The DAP produced in reaction 4 is converted into G3P, giving a total of two of these molecules per molecule of glucose (isomerization reaction).

Two molecules of G3P to reaction 6

Energy-releasing reactions
2 G3P ⟶ 2 Pyruvate

G3P (2 molecules)

NAD^+
Triosephosphate dehydrogenase
NADH
P_i
H^+

6 Two electrons and two protons are removed from G3P. Some of the energy released in this reaction is trapped by the addition of an inorganic phosphate group from the cytosol (not derived from ATP). The electrons are accepted by NAD^+, along with one of the protons. The other proton is released to the cytosol (redox reaction).

1,3-Bisphospho-glycerate (2 molecules)

ADP
Phospho-glycerate kinase
ATP

7 One of the two phosphate groups of 1,3-bisphosphoglycerate is transferred to ADP to produce ATP (substrate-level phosphorylation reaction).

3-Phospho-glycerate (2 molecules)

Phospho-glyceromutase

8 3-Phosphoglycerate is rearranged, shifting the phosphate group from the 3 carbon to the 2 carbon to produce 2-phosphoglycerate (mutase reaction—shifting of a chemical group to another within same molecule).

2-Phospho-glycerate (2 molecules)

H_2O
Enolase

9 Electrons are removed from one part of 2-phosphoglycerate and delivered to another part of the molecule. Most of the energy lost by the electrons is retained in the product, phosphoenolpyruvate (redox reaction).

Phosphoenol-pyruvate (PEP) (2 molecules)

ADP
Pyruvate kinase
ATP

10 The remaining phosphate group is removed from phosphoenolpyruvate and transferred to ADP. The reaction forms ATP and the final product of glycolysis, pyruvate (substrate-level phosphorylation reaction).

Pyruvate (2 molecules)

FIGURE 7.8 Reactions of glycolysis. Because two molecules of G3P are produced in reaction 5, all the reactions from 6 to 10 are doubled (not shown). The names of the enzymes that catalyze each reaction are in rust.

© Cengage Learning 2017

substrate-level phosphorylation (see Figure 7.5). In step 7 the molecule donating a phosphate group to ADP is 1,3-bisphosphoglycerate, and in step 10 the donor molecule is phosphoenolpyruvate (PEP).

Some human diseases are characterized by alterations in glycolysis. For example, genetic mutations that lead to a deficiency in pyruvate kinase, the last enzyme in the glycolytic pathway, cause hemolytic anemia, in which red blood cells are destroyed and removed from the bloodstream much earlier than normal.

Our discussion of the oxidative reactions that supply electrons now moves from the cytosol to mitochondria, the site of pyruvate oxidation and the citric acid cycle. These reactions complete the breakdown of fuel substances into carbon dioxide and provide most of the electrons that drive electron transfer and ATP synthesis.

STUDY BREAK 7.2

1. What are the energy-requiring and energy-releasing steps of glycolysis?
2. What is the redox reaction in glycolysis?
3. How is ATP synthesized in glycolysis?

7.3 Pyruvate Oxidation and the Citric Acid Cycle

Glycolysis produces pyruvate molecules in the cytosol, and an active transport mechanism moves them into the mitochondrial matrix where pyruvate oxidation and the citric acid cycle proceed. The overall reactions of these two processes are presented in **Figure 7.9.** *Pyruvate oxidation* generates CO_2, **acetyl–coenzyme A (acetyl–CoA),** and NADH in a redox reaction involving NAD^+. The acetyl group of acetyl–CoA enters the *citric acid cycle.* As the citric acid cycle turns, every available electron carried into the cycle from pyruvate oxidation is transferred to NAD^+ or to another nucleotide-based molecule, *flavin adenine dinucleotide* (FAD; the reduced form is $FADH_2$). With each turn of the cycle, substrate-level phosphorylation produces 1 ATP. The combined action of pyruvate oxidation and the citric acid cycle oxidizes the three-carbon products of glycolysis completely to carbon dioxide. The 3 NADH and 1 $FADH_2$ produced for each acetyl–CoA during this stage carry high-energy electrons to the electron transfer system in the mitochondrion.

Pyruvate Oxidation Produces the Two-Carbon Fuel of the Citric Acid Cycle

In **pyruvate oxidation** (also called **pyruvic acid oxidation**), a multienzyme complex called the pyruvate dehydrogenase complex removes the $—COO^-$ from pyruvate as CO_2 and then oxidizes the remaining two-carbon fragment of pyruvate to an acetyl group ($CH_3CO—$) (see Figure 7.9). Two electrons and

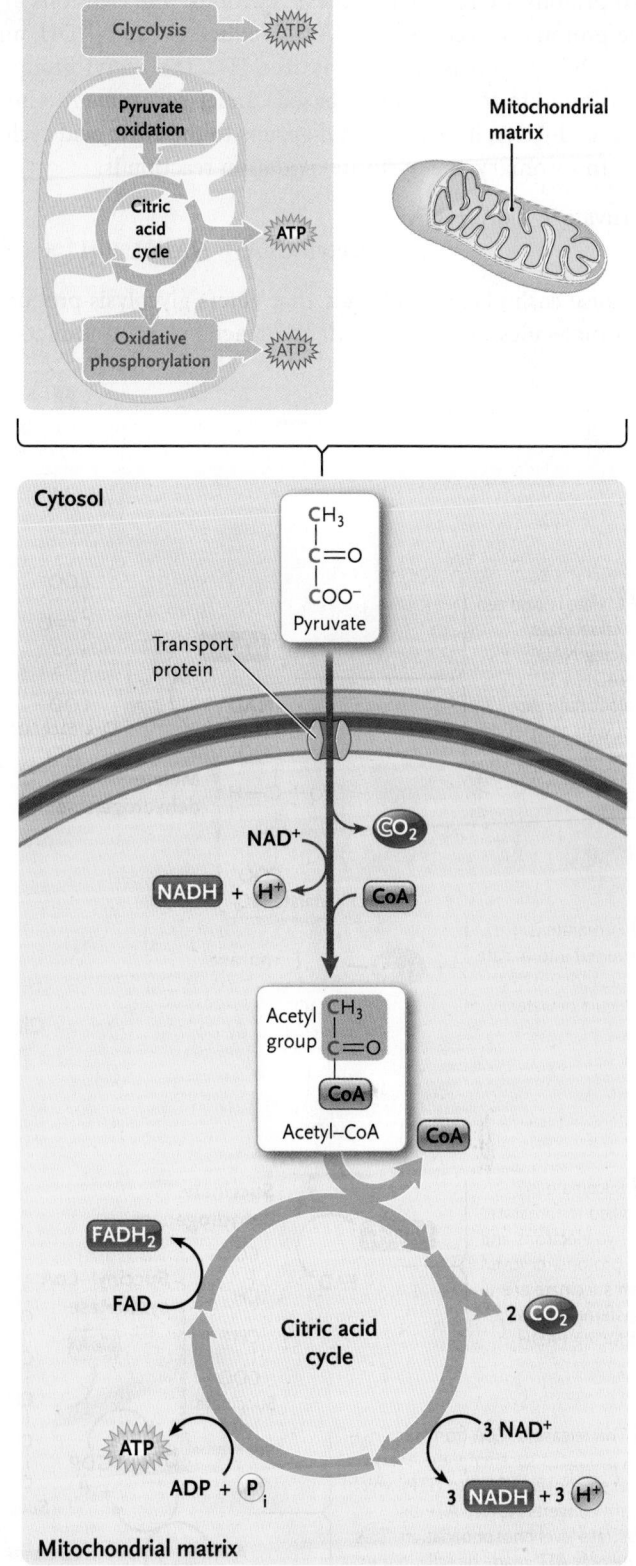

FIGURE 7.9 Overall reactions of pyruvate oxidation and the citric acid cycle, which occur in the mitochondrial matrix. Pyruvate (three carbons) is oxidized to an acetyl group (two carbons) and CO_2. NAD^+ accepts two electrons and one proton removed in the oxidation. The acetyl group, carried by CoA, is the fuel for the citric acid cycle. Each turn of the citric acid cycle oxidizes an acetyl group of acetyl–CoA to 2 CO_2. Acetyl–CoA, NAD^+, FAD, ADP, and P_i enter the cycle; CoA, NADH, $FADH_2$, ATP, and CO_2 are released as products.
© Cengage Learning 2017

two protons are released by these reactions. The electrons and one proton are accepted by NAD⁺, reducing it to NADH, and the other proton is released as free H⁺. The acetyl group is transferred to the nucleotide-based carrier *coenzyme A* (CoA). As acetyl–CoA, it carries acetyl groups to the citric acid cycle.

In summary, the pyruvate oxidation reaction is:

$$\text{pyruvate} + \text{CoA} + \text{NAD}^+ \rightarrow$$
$$\text{acetyl–CoA} + \text{NADH} + \text{H}^+ + \text{CO}_2$$

Because each glucose molecule that enters glycolysis produces two molecules of pyruvate, all the reactants and products in this equation are *doubled* when pyruvate oxidation is considered a continuation of glycolysis.

The Citric Acid Cycle Oxidizes Acetyl Groups Completely to CO₂

In the **citric acid cycle (Figure 7.10)**, enzymes oxidize acetyl groups completely to CO_2 and synthesize some ATP molecules by substrate-level phosphorylation (see Figure 7.5). The citric acid cycle gets its name from citrate, the product of the first reaction of the cycle. It is also called the **tricarboxylic acid**

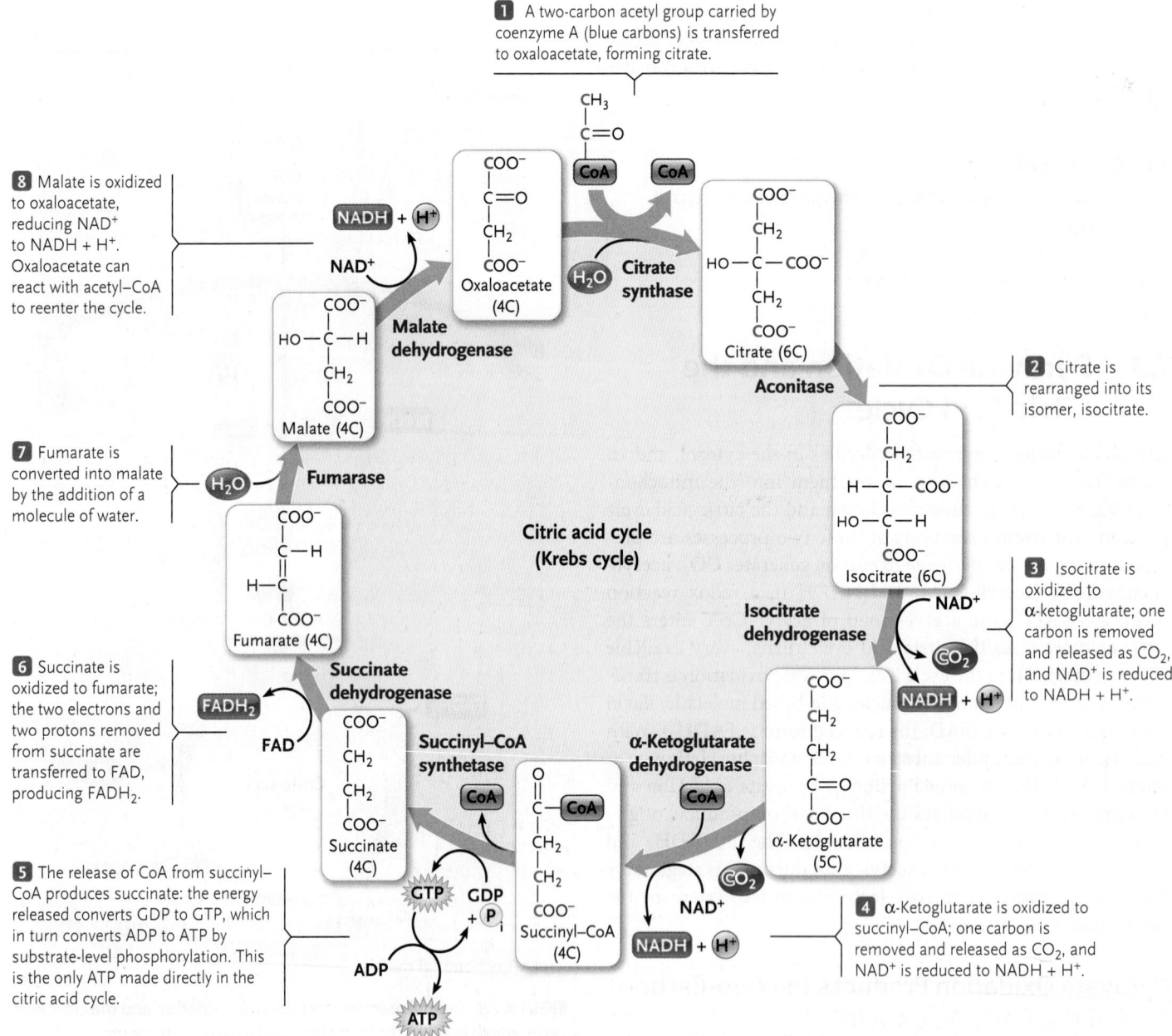

FIGURE 7.10 Reactions of the citric acid cycle. Acetyl–CoA, NAD⁺, FAD, ADP, and Pᵢ enter the cycle; CoA, NADH, FADH₂, ATP, and CO₂ are released as products. The CoA released in reaction 1 can cycle back for another turn of pyruvate oxidation. Enzyme names are in rust.

© Cengage Learning 2017

(TCA) **cycle** or the **Krebs cycle,** named after Hans Krebs, a German-born scientist who worked out the majority of the reactions in the cycle in research he conducted in England beginning in 1932. Krebs discovered the citric acid cycle in experiments using minced pigeon breast muscle (a tissue used in flight that has a very high rate of cellular respiration). He demonstrated that the four-carbon organic acids succinate, fumarate, malate, and oxaloacetate stimulated the consumption of oxygen by the tissue. He also showed that the oxidation of pyruvate by the muscle tissue is stimulated by the six-carbon organic acids citrate and isocitrate, and the five-carbon acid α-ketoglutarate. Although several other researchers pieced together segments of the reaction series, Krebs was the first to reason that the organic acids were linked into a cycle of reactions rather than a linear series. Krebs was awarded a Nobel Prize in 1953 "for his discovery of the citric acid cycle."

The citric acid cycle has eight reactions, each catalyzed by a specific enzyme. All of the enzymes are located in the mitochondrial matrix except the enzyme for reaction 6, succinate dehydrogenase, which is bound to the inner mitochondrial membrane on the matrix side. In a complete turn of the cycle, one two-carbon acetyl unit is consumed and two molecules of CO_2 are released (at reactions 3 and 4), thereby completing the conversion of all the C atoms originally in glucose to CO_2. The CoA molecule that carried the acetyl group to the cycle is released and participates again in pyruvate oxidation to pick up another acetyl group. Electron pairs are removed at each of four oxidations in the cycle (reactions 3, 4, 6, and 8). Three of the oxidations use NAD^+ as the electron acceptor, producing 3 NADH, and one uses FAD, producing 1 $FADH_2$. Substrate-level phosphorylation generates 1 ATP as part of reaction 5. Therefore, the net reactants and products of one turn of the citric acid cycle are:

$$1 \text{ acetyl–CoA} + 3\,NAD^+ + 1\,FAD + 1\,ADP + 1\,P_i + 2\,H_2O \rightarrow$$
$$2\,CO_2 + 3\,NADH + 1\,FADH_2 + 1\,ATP + 3\,H^+ + 1\,CoA$$

Because one molecule of glucose is converted to two molecules of pyruvate by glycolysis and each molecule of pyruvate is converted to one acetyl group, all the reactants and products in this equation are *doubled* when the citric acid cycle is considered a continuation of glycolysis and pyruvate oxidation.

Most of the energy released by the four oxidations of the cycle is associated with the high-energy electrons carried by the 3 NADH and 1 $FADH_2$. These high-energy electrons enter the electron transfer system, where their energy is used to make most of the ATP produced in cellular respiration.

Pyruvate oxidation is catalyzed by a cluster of three enzymes called the *pyruvate dehydrogenase complex* (PDC). The removal of carbon dioxide from pyruvate in pyruvate oxidation is catalyzed by the PDC. The rare genetic disorder, PDC deficiency, is the most common neurodegenerative disorder resulting from abnormal mitochondrial metabolism. PDC deficiency leads to malfunction of the citric acid cycle, leading to a major decrease in energy production by mitochondria. The age of onset and severity of the disease depend on how the mutation inherited affects the levels of the three different enzymes in the complex. Individuals with PDC deficiency symptoms beginning prenatally or early in infancy typically die in early childhood. Individuals who develop PDC deficiency later in childhood often have neurological problems and usually survive into adulthood.

STUDY BREAK 7.3

Summarize the fate of pyruvate molecules produced by glycolysis.

7.4 Oxidative Phosphorylation: The Electron Transfer System and Chemiosmosis

With regard to ATP synthesis, the most significant products of glycolysis, pyruvate oxidation, and the citric acid cycle are the many high-energy electrons removed from fuel molecules and picked up by the carrier molecules NAD^+ or FAD as a result of redox reactions. These electrons are released by the reduced form of these carriers into the electron transfer system of mitochondria (in eukaryotes).

The **mitochondrial electron transfer system** consists of a series of electron carriers that alternately pick up and release electrons, and ultimately transfer them to their final acceptor, oxygen. As the electrons flow through the system, they release free energy, which is used to build a gradient of H^+ across the inner mitochondrial membrane. The gradient goes from a high concentration of H^+ in the intermembrane compartment to a low concentration of H^+ in the matrix. The H^+ gradient supplies the energy that drives ATP synthesis by mitochondrial ATP synthase.

In the Electron Transfer System, Electrons Flow through Protein Complexes in the Inner Mitochondrial Membrane

The mitochondrial electron transfer system includes three major protein complexes, which serve as electron carriers **(Figure 7.11).** These protein complexes, numbered I, III, and IV, are integral membrane proteins located in the inner mitochondrial membrane. In addition, a smaller complex, complex II, is bound to the inner mitochondrial membrane on the matrix side. Associated with the system are two small, highly mobile electron carriers, *cytochrome c* and *ubiquinone* (also known as coenzyme Q, or CoQ), which shuttle electrons between the major complexes.

Cytochromes are proteins with a heme prosthetic group that contains an iron atom. (Heme is an iron-containing group that binds oxygen; see Figure 3.17. A prosthetic group is a

OXIDATIVE PHOSPHORYLATION: THE MITOCHONDRIAL ELECTRON TRANSFER SYSTEM AND CHEMIOSMOSIS

Oxidative phosphorylation involves the electron transfer system (steps 1–6), and chemiosmosis by ATP synthase (steps 7–9). Blue arrows indicate electron flow; red arrows indicate H^+ movement.

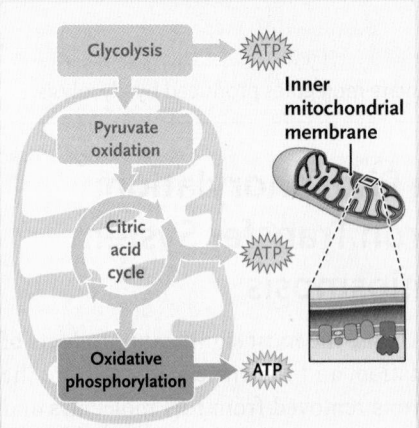

1 Complex I picks up high-energy electrons from NADH and conducts them via two electron carriers, FMN (flavin mononucleotide) and an Fe/S (iron–sulfur) protein, to ubiquinone.

2 Complex II oxidizes $FADH_2$ to FAD; the two electrons released are transferred to ubiquinone, and the two protons released go into the matrix. Electrons that pass to ubiquinone by the complex II reaction bypass complex I of the electron transfer system.

3 Complex III accepts electrons from ubiquinone and transfers them through the electron carriers in the complex— cytochrome b, an Fe/S protein, and cytochrome c_1—to cytochrome c, which is free in the intermembrane space.

4 Complex IV accepts electrons from cytochrome c and delivers them via electron carriers cytochromes a and a_3 to oxygen. Four protons are added to a molecule of O_2 as it accepts four electrons, forming 2 H_2O.

5 As electrons move through the electron transfer system, they release free energy. Part of the released energy is lost as heat, but some is used by the mitochondrion to transport H^+ across the inner mitochondrial membrane from the matrix to the intermembrane compartment at complexes I, III, and IV.

6 The resulting H^+ gradient supplies the energy that drives ATP synthesis by ATP synthase.

7 Because of the gradient, H^+ flows across the inner membrane and into the matrix through a channel in the ATP synthase.

8 The flow of H^+ activates ATP synthase, making the headpiece and stalk rotate.

9 As a result of changes in shape and position as it turns, the headpiece catalyzes the synthesis of ATP from ADP and P_i.

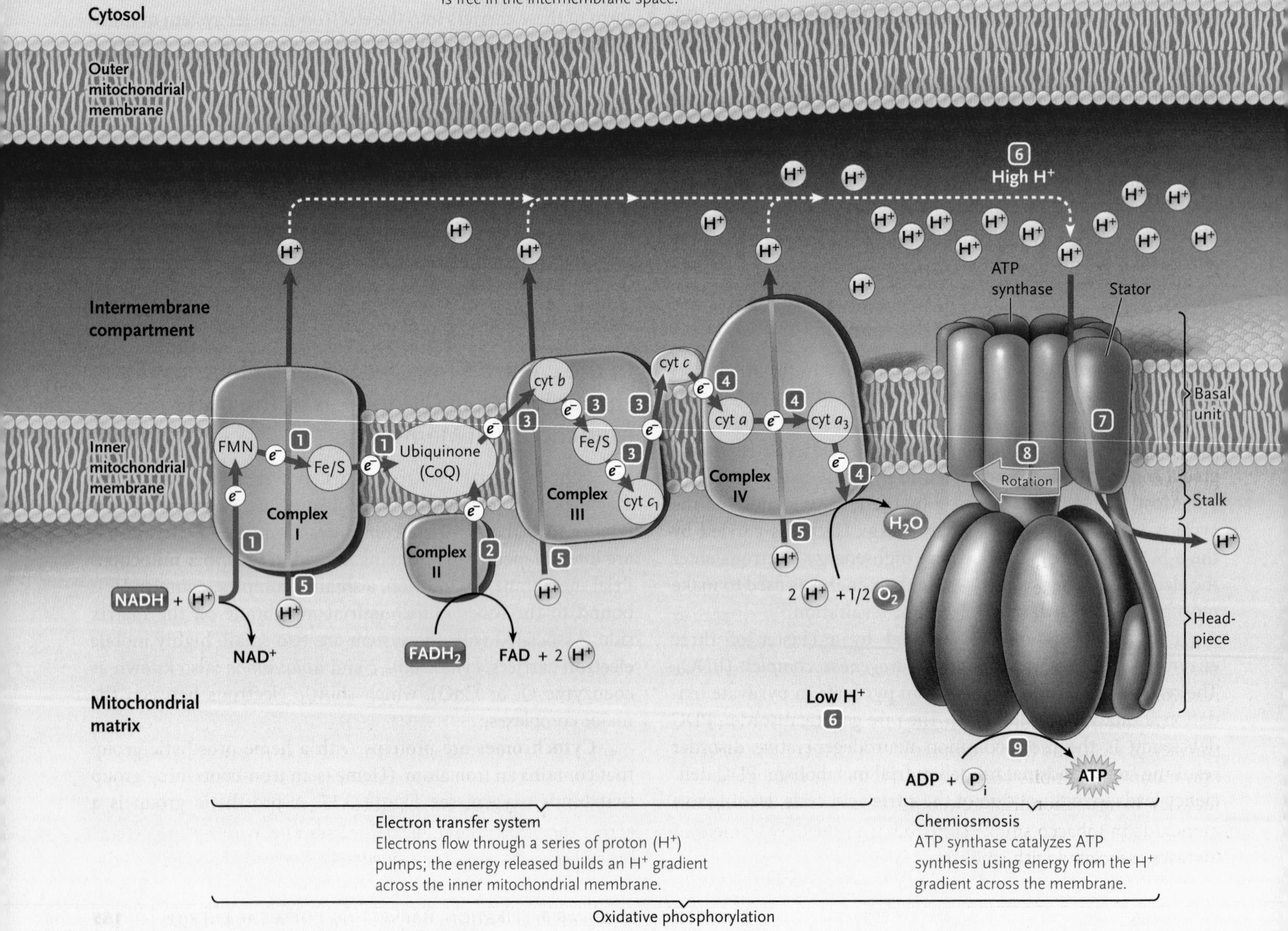

Cytosol

Outer mitochondrial membrane

Intermembrane compartment

Inner mitochondrial membrane

Mitochondrial matrix

ATP synthase **Stator**

High H^+

6

H^+

Basal unit

7

Rotation **8**

Stalk

H^+

Head-piece

cyt b **3** cyt c **4**

3 Fe/S **3**

FMN **1** Fe/S **1** Ubiquinone (CoQ) cyt a **4** cyt a_3

Complex I Complex III cyt c_1 Complex IV

Complex II **2**

NADH + H^+ **5** H^+

NAD$^+$ FADH$_2$ FAD + 2 H^+ H^+

5 H_2O

2 H^+ + 1/2 O_2

Low H^+ **6**

ADP + P_i **9** ATP

Electron transfer system
Electrons flow through a series of proton (H^+) pumps; the energy released builds an H^+ gradient across the inner mitochondrial membrane.

Chemiosmosis
ATP synthase catalyzes ATP synthesis using energy from the H^+ gradient across the membrane.

Oxidative phosphorylation

SUMMARY Three major protein complexes serve as electron carriers in the mitochondrial electron transfer system. Energy from electron flow through the complexes is used by the complexes to pump H$^+$ from the mitochondrial matrix into the intermembrane compartment. The resulting H$^+$ gradient supplies the energy that drives ATP synthesis by the membrane-embedded ATP synthase.

think like a scientist Using cell fractionation techniques (see Figure 4.8), you can isolate intact mitochondria from cells. Assume that the outer mitochondrial membrane, but not the inner mitochondrial membrane, is permeable to H$^+$. You place isolated mitochondria in a low-pH environment. The H$^+$ concentration will increase in which part of the mitochondrion? What effect will that increase have on ATP production?

© Cengage Learning 2017

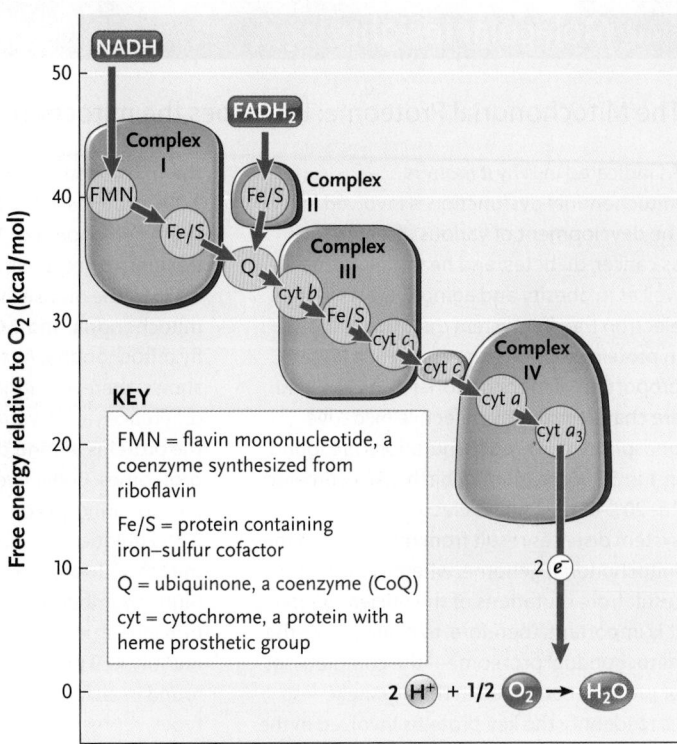

KEY

FMN = flavin mononucleotide, a coenzyme synthesized from riboflavin

Fe/S = protein containing iron–sulfur cofactor

Q = ubiquinone, a coenzyme (CoQ)

cyt = cytochrome, a protein with a heme prosthetic group

$2 \, H^+ + 1/2 \, O_2 \rightarrow H_2O$

Direction of electron flow →

FIGURE 7.12 Organization of the mitochondrial electron transfer system from high to low free energy. Electrons flow spontaneously from one molecule to the next in the series.
© Cengage Learning 2017

cofactor that binds tightly to a protein or enzyme; see Section 6.4.) The iron atom accepts and donates electrons. One cytochrome in particular, cytochrome *c*, was important historically in determining evolutionary relationships. Because the amino acid sequence of cytochrome *c* has been conserved by evolution, comparisons of cytochrome *c* amino acid sequences from a wide variety of eukaryotic species contributed to the construction of phylogenetic trees (see Figure 24.11). Present-day phylogeny studies typically use DNA or RNA sequences (see Chapter 24).

Thirteen proteins of the electron transfer system complexes are encoded by the 13 protein-coding genes of the mitochondrial genome. The remaining approximately 77 proteins of the electron transfer system are encoded by nuclear genes, synthesized in the cytosol, and imported into the mitochondria. Similarly, all of the proteins involved in the transcription, translation, and assembly of the 13 mitochondrial-encoded proteins, and proteins for pyruvate oxidation and the citric acid cycle, are nuclear-encoded. (*Molecular Insights* discusses the characteristics of the mitochondrial proteome—the complete set of proteins in the organelle—in different organs and organelles.)

Electrons flow through the major protein complexes as shown in Figure 7.11, steps 1 to 4. Note that complex II, a succinate dehydrogenase complex, catalyzes two reactions. One is reaction 6 of the citric acid cycle, the conversion of succinate to fumarate (see Figure 7.10). In that reaction, FAD accepts two protons and two electrons and is reduced to FADH$_2$. The other reaction is shown in Figure 7.11, step 2.

The poison cyanide does its deadly work by blocking the transfer of electrons from complex IV to oxygen. The gas carbon monoxide inhibits complex IV activity, leading to abnormalities in mitochondrial function. In this way, the carbon monoxide in tobacco smoke contributes to the development of diseases associated with smoking.

What is the driving force for electron flow through the protein complexes of the electron transfer system? **Figure 7.12** shows that the individual electron carriers of the system are organized specifically from high to low free energy. Any single component has a higher affinity for electrons than the preceding carrier in the series has. Overall, molecules such as NADH contain an abundance of free energy and can be oxidized readily, whereas O$_2$, the terminal electron acceptor of the series, can be reduced easily. As a consequence of this organization, electron movement through the system is spontaneous, releasing free energy.

Three Major Electron Transfer Complexes Pump H$^+$ across the Inner Mitochondrial Membrane

Using energy from electron flow, the proteins of complexes I, III, and IV pump (actively transport) H$^+$ (protons) from the matrix to the intermembrane compartment of the mitochondrion (see Figure 7.11, step 5). The result is an H$^+$ gradient with a high concentration of H$^+$ in the intermembrane compartment and a low concentration of H$^+$ in the matrix (step 6). Because protons carry a positive charge, the asymmetric distribution of protons generates an electrical and chemical gradient across the inner mitochondrial membrane, with the intermembrane compartment more positively charged than the matrix.

The Mitochondrial Proteome: How does the mitochondrial proteome vary among organs and organisms?

As indicated in *Why it matters . . .* , mitochondrial dysfunction is involved in the development of various diseases such as cancer, diabetes, and heart failure, as well as in obesity and aging. Defects in the electron transfer system due to mutations in protein components represent a large proportion of mitochondrial disorders, and are characterized by defective oxidative phosphorylation. Such mutations are found in 1 in 5,000 live human births. An estimated 15–20% of the known electron transfer system diseases result from mutations in the mitochondrial genome, whereas the rest result from mutations of the nuclear genome. It is important, therefore, to characterize the mitochondrial proteome—the complete set of proteins contained in the organelle—so as to identify the key proteins involved in the disorders. Prior to this study, the mammalian mitochondrial proteome was estimated to consist of 1,000 to 1,400 proteins.

Research Question

How does the mitochondrial proteome vary among organs and organisms?

Experiments

Collaborating researchers from the United States, the United Kingdom, and China isolated functional mitochondria from four model systems: human hearts, mouse hearts, mouse livers, and *Drosophila melanogaster* (fruit fly) and analyzed their proteomes qualitatively and quantitatively.

Results

The **Figure** shows the qualitative proteome data for the four mitochondrial types in Venn diagram form to show the number of common and unique proteins among

them. The researchers identified 1,398 proteins in the human heart mitochondria, 1,620 in the mouse heart mitochondria, 1,733 in the mouse liver mitochondria, and 1,015 in the fly mitochondria. As the Figure shows, there are significant differences and overlaps for the proteins among the four proteomes. Collectively, the data identify approximately 3,350 unique proteins in the mitochondrion, a significantly higher number than reported in previous studies. The data also identify 419 proteins that are found in all four mitochondrial types, representing a core conserved mitochondrial proteome (see the common area of overlap in the Venn diagram for the four mitochondrial types in the Figure).

Quantitative analysis showed that the core conserved proteins are highly abundant in the mitochondrial proteomes, representing 64% of proteins in human heart mitochondria, 64% in mouse heart mitochondria, 52% in mouse liver mitochondria, and 77% in fly mitochondria.

The researchers looked at the mitochondrial proteome core proteins with respect to diseases. They found that, as a group, about 54% of those proteins are involved in diseases, whereas about 82% of the conserved oxidative phosphorylation proteins among those proteins are involved in diseases. The authors conclude that "these data highlight the evolutionary prominence of these proteins."

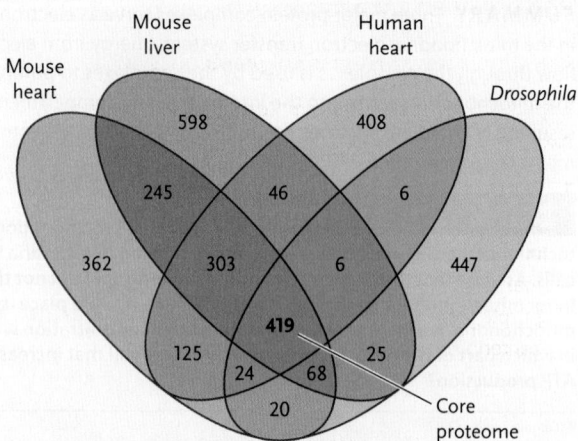

FIGURE Venn diagram showing the proteome compositions for the four mitochondrial model systems analyzed.

Discussion

The research advances our understanding of mitochondrial proteomes. It shows that there is a conserved set of proteins in the proteome, as well as large differences in proteome content between organs (heart and liver) and organisms (human, mouse, and fly). This means that the organization of a mitochondrial proteome in an organ or an organism is complex, and must be brought about in a highly regulated way.

think like a scientist

How might knowing the core proteome for mitochondria aid in detection and treatment of diseases associated with mitochondrial dysfunction?

Source: C. Lotz et al. 2014. Characterization, design, and function of the mitochondrial proteome: from organs to organisms. *Journal of Proteome Research* 13:433–446.

The combination of a proton gradient and voltage gradient across the membrane produces stored energy known as the **proton-motive force.** This force contributes energy for ATP synthesis, as well as for the cotransport of substances to and from mitochondria (see Section 5.4).

Chemiosmosis Powers ATP Synthesis by an H^+ Gradient

Within the mitochondrion, ATP is synthesized by ATP synthase, an enzyme embedded in the inner mitochondrial membrane. In 1961, British scientist Peter Mitchell of Glynn Research Laboratories proposed that mitochondrial electron transfer produces an H^+ gradient and that the gradient powers ATP synthesis by ATP synthase. He called this pioneering model the **chemiosmotic hypothesis;** the process is called *chemiosmosis* (see Figure 7.11). At the time, this hypothesis was a radical proposal because most researchers thought that the energy of electron transfer was stored as a high-energy chemical intermediate. No such intermediate was ever found, and eventually, Mitchell's hypothesis was supported by the results of many experiments, one of which is described in **Figure 7.13.**

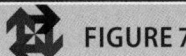 **FIGURE 7.13** | **Experimental Research**

The Racker and Stoeckenius Experiment Demonstrating That an H^+ Gradient Powers ATP Synthesis by ATP Synthase

Question: Does an H^+ gradient power ATP synthase-catalyzed ATP synthesis, thereby supporting Mitchell's chemiosmotic hypothesis?

Experiment: Efraim Racker of Cornell University and Walther Stoeckenius of University of California, San Francisco, made membrane vesicles that had a proton pump and ATP synthase to determine whether proton-motive force drives ATP synthesis.

1. The researchers constructed synthetic phospholipid membrane vesicles containing a segment of purple surface membrane from an archaean. The purple membrane contained only bacteriorhodopsin, a protein that resembles rhodopsin, the visual pigment of animals (see Chapter 41). Bacteriorhodopsin is a light-activated proton pump. The researchers illuminated the vesicles and then analyzed the concentration of H^+ in them.

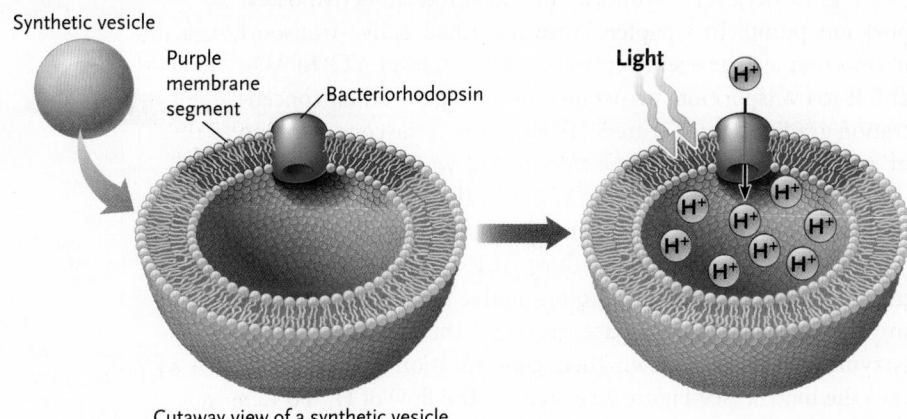

Cutaway view of a synthetic vesicle

Result: H^+ is pumped into the vesicles, creating an H^+ gradient.

2. The researchers next made synthetic vesicles containing ATP synthase from both bacteriorhodopsin and bovine heart mitochondria. The ATP synthase molecule was oriented so that the ATP-synthesizing headpiece was on the outside of the vesicles. They added ADP and P_i to the medium containing the vesicles and tested whether ATP was produced in the dark and after a period of illumination.

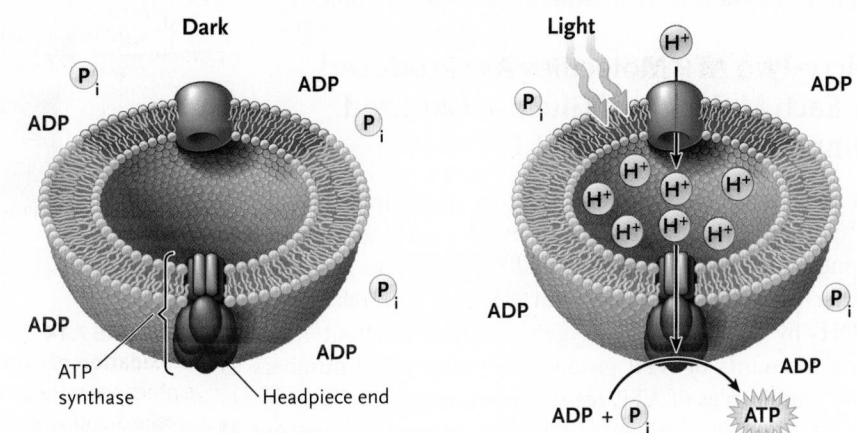

Result: In the dark, no ATP was synthesized.　　**Result:** In the light, ATP was synthesized.

Together, these results showed that light activated the bacteriorhodopsin to produce an H^+ gradient, with H^+ moving from the outside to the inside of the vesicle (like the movement from the mitochondrial matrix to the intermembrane compartment in Figure 7.11), and that the energy from the H^+ gradient drove ATP synthesis by ATP synthase.

Conclusion: An H^+ gradient—and, therefore, proton-motive force—powers ATP synthesis by ATP synthase. The results support Mitchell's chemiosmotic hypothesis for ATP synthesis in mitochondria.

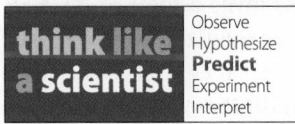

Observe
Hypothesize
Predict
Experiment
Interpret

In the experiment described, ATP was synthesized in the light. What do you predict would happen if the vesicles were initially exposed to light and then they were placed in the dark?

Source: E. Racker and W. Stoeckenius. 1974. Reconstitution of purple membrane vesicles catalyzing light-driven proton uptake and adenosine triphosphate formation. *Journal of Biological Chemistry* 249:662–663.

Mitchell received a Nobel Prize in 1978 "for his contribution to the understanding of biological energy transfer through the formulation of the chemiosmotic theory."

How does ATP synthase use the H^+ gradient to power ATP synthesis in chemiosmosis? ATP synthase consists of a *basal unit* embedded in the inner mitochondrial membrane connected by a *stalk* to a *headpiece* located in the matrix. A peripheral stalk called a *stator* bridges the basal unit and headpiece (see Figure 7.11). ATP synthase functions like an active transport ion pump. In Chapter 5, we described active transport pumps that use the energy created by hydrolysis of ATP to ADP and P_i to transport ions across membranes against their concentration gradients (see Figure 5.11). However, if the concentration of an ion is very high on the side toward which it is normally transported, the pump runs in reverse—the ion is transported backward through the pump, and the pump adds phosphate to ADP to generate ATP. That is how ATP synthase operates in mitochondrial membranes. Proton-motive force moves protons in the intermembrane space through the channel in the enzyme's basal unit, down their concentration gradient, and into the matrix (see Figure 7.11, step 7). The flow of H^+ powers ATP synthesis by the headpiece; this phosphorylation reaction is chemiosmosis (steps 8 and 9). ATP synthase occurs in similar form and works in the same way in mitochondria, chloroplasts, and prokaryotes capable of oxidative phosphorylation.

Thirty-Two ATP Molecules Are Produced for Each Molecule of Glucose Oxidized Completely to CO_2 and H_2O

The most recent research indicates that approximately 2.5 ATP are synthesized as a pair of electrons released by NADH travels through the entire electron transfer pathway to oxygen. The shorter pathway, followed by an electron pair released from $FADH_2$ by complex II to oxygen, synthesizes about 1.5 ATP. (Some accounts of ATP production round these numbers to 3 and 2 molecules of ATP, respectively.)

These numbers allow us to estimate the total amount of ATP that would be produced by the complete oxidation of glucose to CO_2 and H_2O if the entire H^+ gradient produced by electron transfer is used for ATP synthesis **(Figure 7.14)**. During glycolysis, substrate-level phosphorylation produces 2 ATP and 2 NADH. In pyruvate oxidation, 2 NADH are produced from the two molecules of pyruvate.

The subsequent citric acid cycle turns twice for each molecule of glucose that enters glycolysis, yielding a total of 2 ATP produced by substrate-level phosphorylation, as well as 6 NADH, 2 $FADH_2$, and 4 CO_2.

The combination of glycolysis, pyruvate oxidation, and the citric acid cycle has the following summary reaction:

glucose + 4 ADP + 4 P_i + 10 NAD^+ + 2 FAD →
$\qquad$ 4 ATP + 10 NADH + 10 H^+ + 2 $FADH_2$ + 6 CO_2

Therefore, high-energy electrons carried by 10 NADH molecules (2 from glycolysis, 2 from pyruvate oxidation, and 6

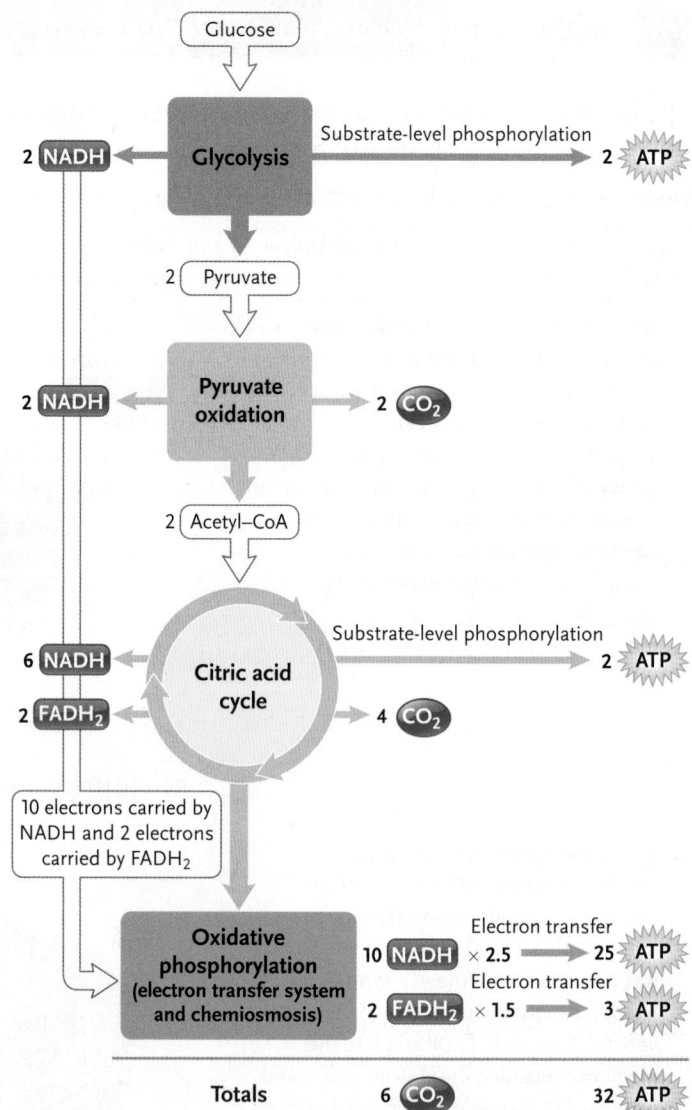

FIGURE 7.14 Summary of ATP production from the complete oxidation of a molecule of glucose. The total of 32 ATP assumes that electrons carried from glycolysis by NADH are transferred to NAD^+ inside mitochondria. If the electrons from glycolysis are instead transferred to FAD inside mitochondria, total production will be 30 ATP.

© Cengage Learning 2017

from the citric acid cycle) and 2 $FADH_2$ molecules (from the citric acid cycle) enter the electron transfer system of the mitochondrion (see Figure 7.14). Assuming 2.5 molecules of ATP generated per NADH molecule and 1.5 molecules of ATP generated per $FADH_2$ molecule, this gives a total of 32 ATP molecules for each molecule of glucose oxidized completely to CO_2 and H_2O.

The total of 32 ATP assumes that the two pairs of electrons carried by the 2 NADH reduced in glycolysis each drive the synthesis of 2.5 ATP when traversing the mitochondrial electron transfer system. However, because NADH cannot penetrate the mitochondrial membranes, its electrons are transferred into the mitochondrion by one of two shuttle systems. The more efficient shuttle mechanism transfers the electrons to

NAD$^+$ as the acceptor inside mitochondria. These electron pairs, when passed through the electron transfer system, result in the synthesis of 2.5 ATP each, producing the grand total of 32 ATP. The less efficient shuttle transfers the electrons to FAD as the acceptor inside mitochondria. These electron pairs, when passed through the electron transfer system, result in the synthesis of only 1.5 ATP each and produce a grand total of 30 ATP instead of 32.

Which shuttle system predominates depends on the particular species and the cell types involved. For example, heart, liver, and kidney cells in mammals use the more efficient shuttle system; skeletal muscle and brain cells use the less efficient shuttle system. Regardless, the numbers of ATP produced are idealized because mitochondria also use the H$^+$ gradient to drive cotransport. Any of the energy in the gradient used for this activity would reduce ATP production proportionately.

Defects in oxidative phosphorylation caused by mutations in mitochondrial or nuclear genes encoding proteins of the electron transfer system complexes reduce the amount of ATP produced in the cell. Human diseases with such defects are rare—about 1 in 10,000 live births—and symptoms often involve seriously impaired neurological conditions resulting from the consequences of ATP deficiency, such as loss of vision (Leber hereditary optic myophathy, described in Chapter 13) and kidney failure in childhood.

Cellular Respiration Conserves More Than 30% of the Chemical Energy of Glucose in ATP

The process of cellular respiration is not 100% efficient, in that it does not convert all the chemical energy of glucose to ATP. That is, under standard conditions, including neutral pH (pH = 7) and a temperature of 25°C, the hydrolysis of ATP to ADP yields about 7.0 kilocalories per mole (kcal/mol). Assuming that complete glucose oxidation produces 32 ATP, the total energy conserved in ATP production would be about 224 kcal/mol. By contrast, if glucose is burned in air, it releases 686 kcal/mol. On this basis, the efficiency of cellular glucose oxidation would be about 33% (224/686 × 100 = 32.7%). This value is considerably better than that of most devices designed by human engineers. For example, the engine of an automobile extracts only about 25% of the energy in the fuel it burns.

The chemical energy released by cellular oxidations that is not captured in ATP synthesis is released as heat. In mammals and birds, this source of heat maintains body temperature at a constant level. In certain mammalian tissues, including *brown fat* (see Chapter 48), the inner mitochondrial membranes contain *uncoupling proteins* (UCPs) that make the inner mitochondrial membrane "leaky" to H$^+$. As a result, electron transfer runs without building an H$^+$ gradient or synthesizing ATP and releases all the energy extracted from the electrons as heat. Brown fat with UCPs occurs in significant quantities in hibernating mammals and in very young mammals, including human infants.

STUDY BREAK 7.4

1. What distinguishes the four complexes of the mitochondrial electron transfer system?
2. Explain how the proton pumps of complexes I, III, and IV relate to ATP synthesis.

THINK OUTSIDE THE BOOK

A number of human genetic diseases result from mutations that affect mitochondrial function. Collaboratively or individually, find an example of such a disease and research how the genetic mutation disrupts mitochondrial function and leads to the disease symptoms.

7.5 Anaerobic Respiration and Fermentation

As you have learned, cellular respiration (aerobic respiration) is a process in which oxygen is the final electron acceptor in a collection of metabolic reactions within cells that catabolize food molecules to produce energy in the form of ATP. Two processes, *anaerobic respiration* and *fermentation*, have evolved that enable cells to produce ATP without using oxygen. Both of these processes are less efficient than aerobic respiration in producing ATP.

Anaerobic Respiration Uses a Molecule Other Than Oxygen as the Final Electron Acceptor of an Electron Transfer System

In **anaerobic respiration,** ATP is produced using an electron transfer system, but the final electron acceptor is an inorganic molecule other than oxygen. Anaerobic respiration is used by a number of types of prokaryotes living in oxygen-depleted environments. The electron transfer systems of these microorganisms are located in their plasma membranes. For instance, some bacteria and archaea use sulfate (SO_4^{2-}) as the final electron acceptor to form hydrogen sulfide (H_2S). Hydrogen sulfide has a bad-egg foul odor, thereby identifying sulfate-reducing microorganisms in anaerobic environments, such as groundwater and wells in some locations. Some other bacteria and archaea use nitrate (NO_3^-) as the final electron acceptor, reducing it to a nitrite ion (NO_2^-), nitrous oxide (N_2O), or nitrogen gas (N_2). Anaerobic respiration by microorganisms using sulfate and nitrate as final electron acceptors is essential for the sulfur and nitrogen cycles that occur in Earth's ecosystems (see Chapter 54). Other inorganic electron acceptors used for anaerobic respiration include CO_2, Fe^{3+}, and Mn^{4+}.

The amount of ATP produced by anaerobic respiration is limited by the fact that, under anaerobic conditions, only part of the citric acid cycle functions, and not all of the electron carriers of the electron transfer system are involved. The actual amount of ATP produced in anaerobic respiration depends on the organism and the biochemical pathway operating. It is always lower than that seen for aerobic respiration. Therefore, microorganisms using anaerobic respiration grow more slowly than those using aerobic respiration.

In Fermentation, Electrons Are Transferred to an Organic Molecule Rather Than to an Electron Transfer System

In anaerobic respiration, electrons are transferred to an electron transfer system. In **fermentation,** electrons carried by NADH are transferred to an organic acceptor molecule rather than to an electron transfer system. This transfer converts the NADH to NAD$^+$, which is required to accept electrons in reaction 6 of glycolysis (see Figure 7.8). As a result, glycolysis continues to supply ATP by substrate-level phosphorylation. The amount of ATP that is generated by fermentation is far less than is produced by cellular (aerobic) respiration.

Two types of fermentation reactions exist: *lactate fermentation* and *alcoholic fermentation* **(Figure 7.15).** **Lactate fermentation** converts pyruvate into lactate (Figure 7.15A). This reaction occurs in many bacteria, in some plant tissues, and in certain animal tissues such as skeletal muscle cells. When vigorous contraction of muscle cells calls for more oxygen than the circulation can supply, lactate fermentation takes place. For example, lactate accumulates in the leg muscles of a sprinter during a 100-meter race. The lactate temporarily stores electrons, and when the oxygen content of the muscle cells returns to normal levels, the reverse of the reaction in Figure 7.15A regenerates pyruvate and NADH. The pyruvate can then be used in the second stage of cellular respiration, and the NADH contributes its electron pair to the electron transfer system. Lactate is also the fermentation product of some bacteria. The sour taste of buttermilk, yogurt, and dill pickles is a sign of their activity.

Alcoholic fermentation (Figure 7.15B) occurs in some plant tissues, in certain invertebrates and protists, in certain bacteria, and in some single-celled fungi such as yeasts. In this reaction, pyruvate is converted into ethyl alcohol (which has two carbons) and carbon dioxide in a two-step series that also converts NADH into NAD$^+$. Alcoholic fermentation by yeasts has widespread commercial applications. Bakers use the yeast *Saccharomyces cerevisiae* to make bread dough rise. They mix the yeast with a small amount of sugar and blend the mixture into the dough, where oxygen levels are low. As the yeast cells convert the sugar into ethyl alcohol and carbon dioxide, the gaseous CO_2 expands and creates bubbles that cause the dough to rise. Oven heat evaporates the alcohol and causes further expansion of the bubbles, producing a light-textured product. Alcoholic fermentation is also the mainstay of beer and wine production. Fruits are a natural home to wild yeasts. The dusty appearance of the surface of grapes is yeast, for instance. In wine making, a mixture of wild and cultivated yeasts is used to produce wine. Alcoholic fermentation also occurs naturally in the environment. For example, overripe or rotting fruit frequently will start to ferment, and birds that eat the fruit may become too drunk to fly.

Fermentation is a lifestyle for some organisms. In bacteria and fungi that lack the electron transfer system to carry out oxidative phosphorylation, fermentation is the only source of ATP. These organisms are called **strict anaerobes.** In general, strict anaerobes *require* an oxygen-free environment because they cannot utilize oxygen as a final electron acceptor. Among these organisms are the bacteria that cause botulism, tetanus, and some other serious diseases. For example, the bacterium that causes botulism *(Clostridium botulinum)* thrives in the oxygen-free environment of improperly sterilized canned foods. That absence of oxygen in canned foods prevents the growth of most other microorganisms.

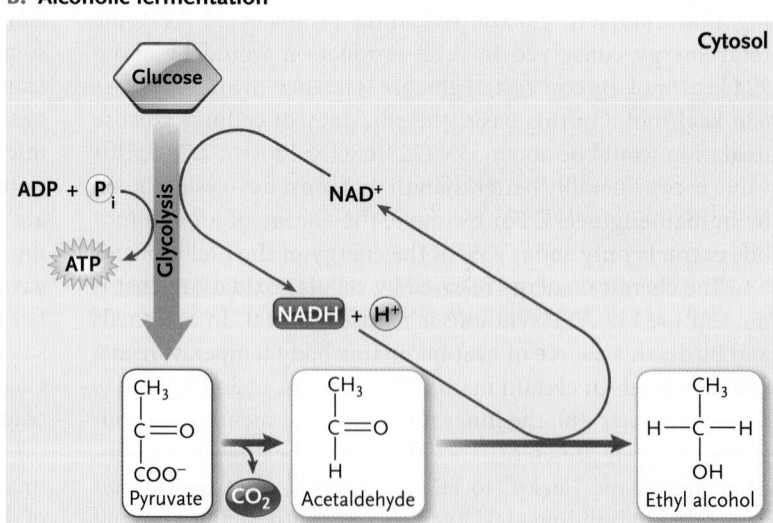

FIGURE 7.15 Fermentation reactions that produce (A) lactate and (B) ethyl alcohol. The fermentations, which occur in the cytosol, convert NADH to NAD$^+$, allowing the electron carrier to cycle back to glycolysis. This process keeps glycolysis running, with continued production of ATP.

© Cengage Learning 2017

Other organisms, called **facultative anaerobes,** can switch between fermentation and full oxidative pathways, depending on the oxygen supply. Facultative anaerobes include *Escherichia coli,* a bacterium that inhabits the digestive tract of humans; the *Lactobacillus* bacteria used to produce buttermilk and yogurt; and *S. cerevisiae,* the yeast used in brewing, wine making, and baking. Many cell types in higher organisms, including vertebrate muscle cells, also are facultative anaerobes. Some prokaryotic and eukaryotic cells are **strict aerobes,** meaning that they have an absolute requirement for oxygen to survive and are unable to live solely by fermentations. Vertebrate brain cells are key examples of strict aerobes.

Interestingly, energy metabolism is altered significantly in cancer cells. When oxygen is sufficient, normal cells use cellular respiration to produce ATP, and as you have just learned, only under anaerobic conditions do certain cells use anaerobic respiration to produce energy from glucose, thereby producing lactate in the process. However, most cancer cells depend on glycolysis for energy production. They have glycolysis rates much higher than those in normal cells, and use the substrate-level phosphorylation reactions of glycolysis to produce ATP regardless of the amount of oxygen present. Large amounts of lactate are generated because of the abnormal metabolism. This process is called **aerobic glycolysis,** and is also called the *Warburg effect* after Otto Warburg, who discovered the phenomenon in his studies of cancer cells. The Warburg effect is not a causative agent for cancer, but occurs as a result of the genetic mutations that convert normal cells to the cancerous state.

STUDY BREAK 7.5

1. What distinguishes the processes of anaerobic respiration and fermentation?
2. What are the two types of fermentation?

7.6 Interrelationships of Catabolic and Anabolic Pathways

Our discussion of cellular respiration used glucose as the starting point of the process. In this section, we discuss how we use other food substances as a source of energy, how the catabolic pathways of glycolysis and the citric acid cycle relate to biosynthetic (anabolic) pathways of the cell, and how the glycolysis and citric acid cycle stages of cellular respiration are regulated.

Carbohydrates, Fats, and Proteins Can Serve as Electron Sources for Oxidative Pathways

In addition to glucose (which is a relatively rare molecule in food) and other six-carbon sugars, reactions leading from glycolysis through pyruvate oxidation also oxidize a wide range of carbohydrates, lipids, and proteins. These molecules enter the

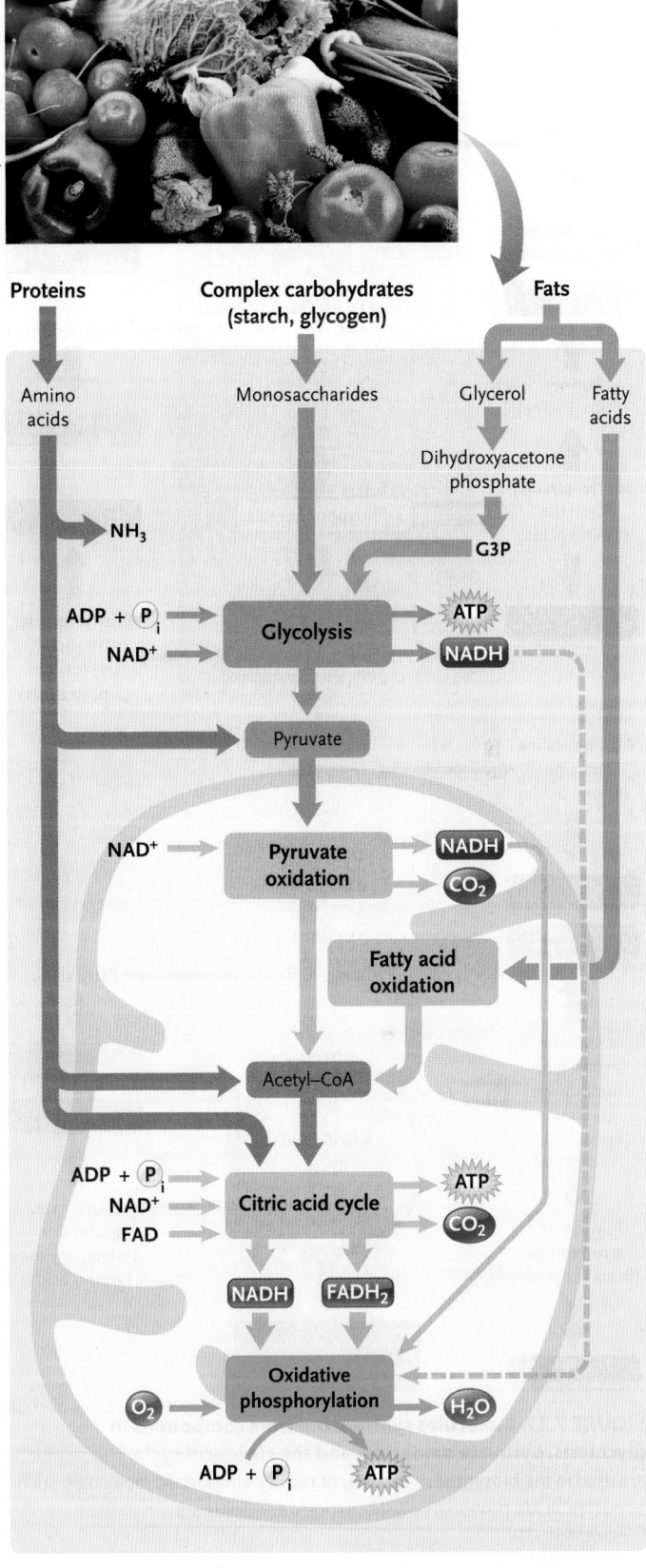

FIGURE 7.16 Major pathways that oxidize carbohydrates, fats, and proteins. Reactions that occur in the cytosol are shown against a tan background; reactions that occur in mitochondria are shown inside the organelle. CoA funnels the products of many oxidative pathways into the citric acid cycle.

© Cengage Learning 2017

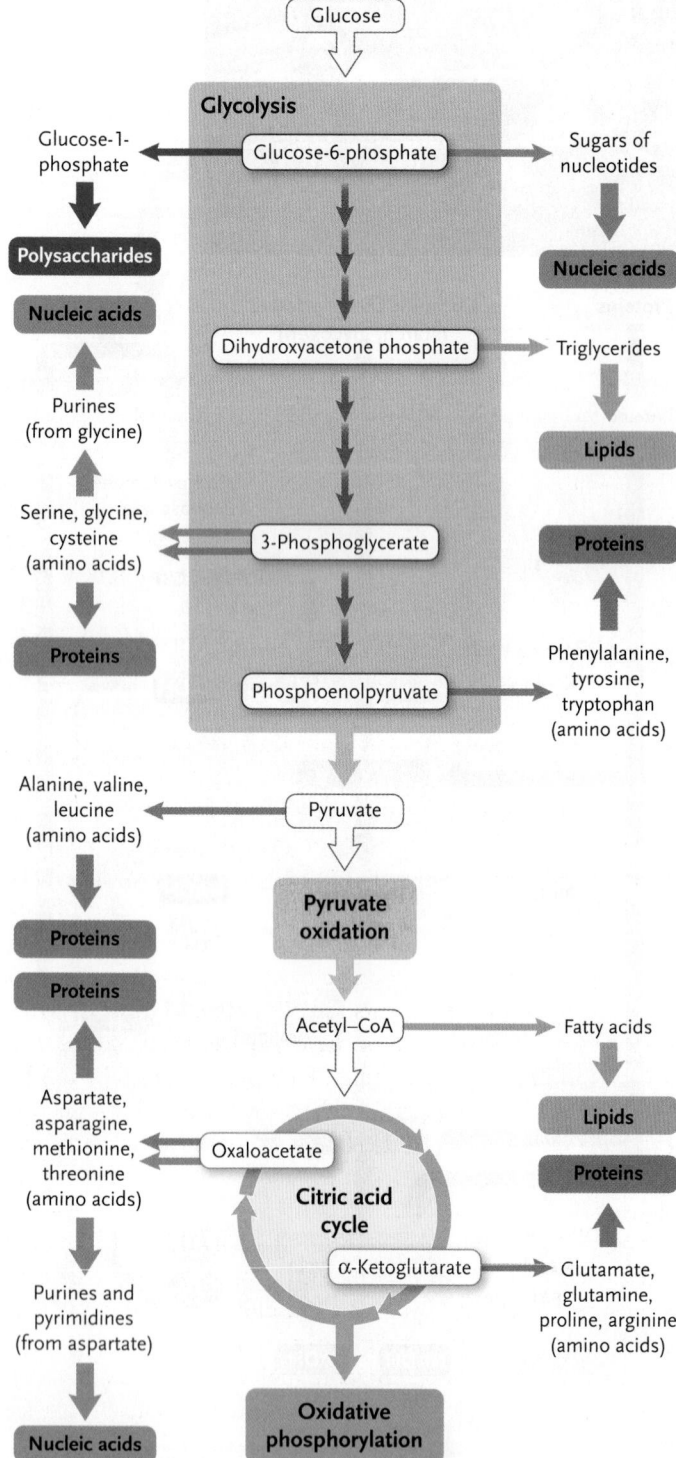

FIGURE 7.17 Molecules synthesized from compounds in glycolysis, pyruvate oxidation, and the citric acid cycle. The products are used in the biosynthesis of many of the cell's important molecules, such as nucleic acids, proteins, polysaccharides, and lipids.

© Cengage Learning 2017

reaction pathways at various points. **Figure 7.16** summarizes the cellular pathways involved. It shows the central role of CoA in funneling acetyl groups from different pathways into the citric acid cycle and of the mitochondrion as the site where most of these groups are oxidized.

Carbohydrates such as sucrose (table sugar) and other disaccharides are easily broken into monosaccharides such as glucose and fructose that enter glycolysis at early steps. Starch (see Figure 3.8A) is hydrolyzed by digestive enzymes into individual glucose molecules, which enter the first reaction of glycolysis. Glycogen, a more complex carbohydrate that consists of glucose subunits (see Figure 3.8B), is broken down and converted by enzymes into glucose-6-phosphate, which enters glycolysis at reaction 2 of Figure 7.8.

Among the fats, triglycerides (see Figure 3.9) are major sources of electrons for ATP synthesis. Before entering the oxidative reactions, the triglycerides are hydrolyzed into glycerol and individual fatty acids. The glycerol is converted to G3P and enters glycolysis at reaction 6 of Figure 7.8, in the ATP-producing portion of the pathway. The fatty acids, and many other types of lipids, are split into two-carbon fragments that enter the citric acid cycle as acetyl–CoA. The energy released by the oxidation of fats, by weight, is comparatively high—a little more than twice the energy of oxidation of carbohydrates or proteins. This fact explains why fats are an excellent source of energy in the diet.

Proteins are hydrolyzed to amino acids before oxidation. During oxidation, the amino group is removed, and the remainder of the molecule enters the pathway of carbohydrate oxidation as either pyruvate, acetyl units carried by CoA, or intermediates of the citric acid cycle. For example, the amino acid alanine is converted into pyruvate, leucine is converted into acetyl units, and phenylalanine is converted into fumarate. Fumarate enters the citric acid cycle at reaction 7 of Figure 7.9.

Many Biosynthetic Pathways Start from Glycolysis or the Citric Acid Cycle

The catabolic reactions of glycolysis and the citric acid cycle not only produce energy for the cell, but also supply molecules from which many other cellular molecules are synthesized. **Figure 7.17** shows some of the molecules that are synthesized from compounds in glycolysis, pyruvate oxidation, and the citric acid cycle. Those molecules, in turn, are used in the biosynthesis of nucleic acids, proteins, polysaccharides, and lipids.

In addition, when energy is not needed by the body, glucose can be synthesized from intermediates of the glycolysis and citric acid cycle pathways, as well as from molecules derived from those pathways. Examples are pyruvate, lactate, malate, oxaloacetate, and several amino acids. The glucose biosynthesis process is called **gluconeogenesis.** Its reactions are the reverse of those of glycolysis, involving many of the same enzymes. However, four enzymes are unique to gluconeogenesis, making the pathway distinct from that of glycolysis. Gluconeogenesis consumes ATP, rather than produces it. As you would expect, the regulation of glycolysis and gluconeogenesis is carefully controlled according to the energy needs of the body.

Glycolysis and Citric Acid Cycle Stages of Cellular Respiration Are Regulated by Feedback Mechanisms

Key enzymes of glycolysis and the citric acid cycle are regulated to match the cell's need for ATP **(Figure 7.18).** If ATP is

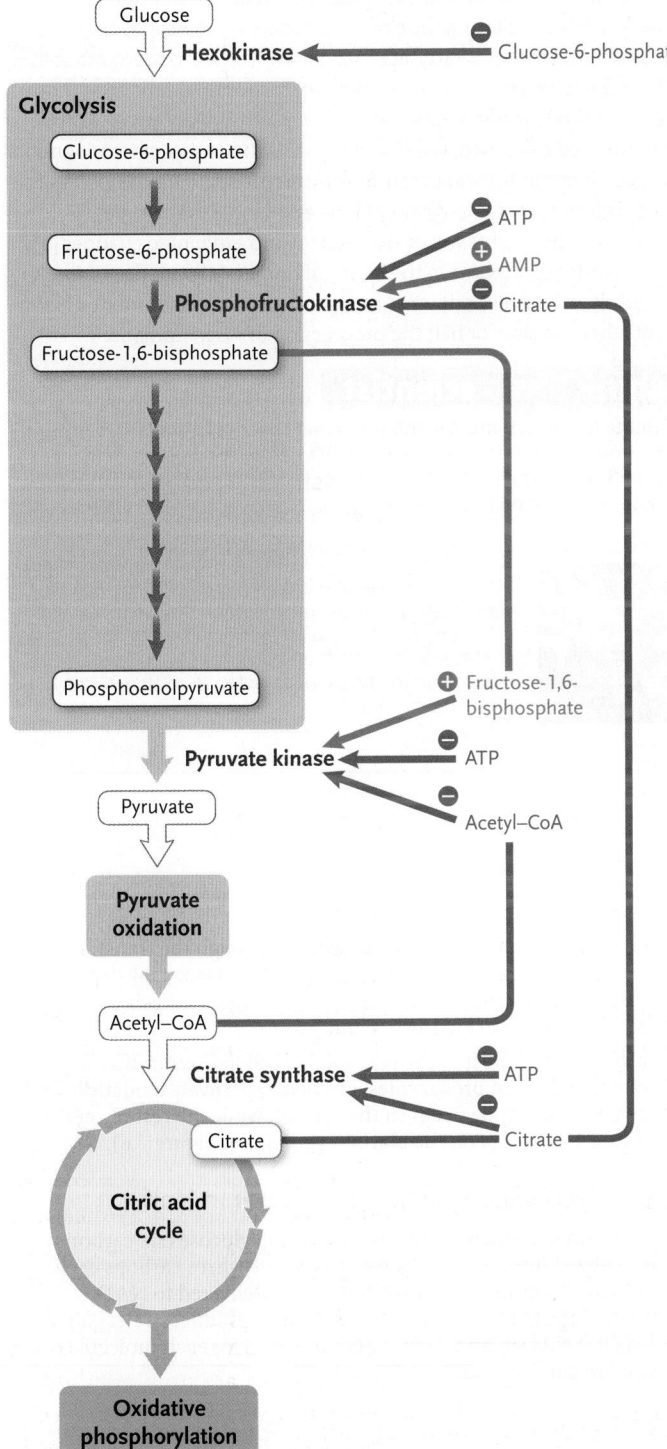

FIGURE 7.18 Feedback regulation of glycolysis and the citric acid cycle.
© Cengage Learning 2017

abundant, cellular respiration can slow down, conserving resources and allowing intermediates to be used in biosynthetic reactions. If ATP concentration is low, cellular respiration needs to increase to compensate. The regulatory mechanism involved is allosteric regulation by feedback inhibition. As you learned in Section 6.5, in allosteric regulation, enzyme activity is controlled by the reversible binding of a regulatory molecule to a site on the enzyme to either increase or decrease enzyme activity, and feedback inhibition is the process where a product of a reaction pathway acts as a regulator of a pathway (see Figure 6.15).

For glycolysis, the key enzyme regulated is phosphofructokinase; hexokinase and pyruvate kinase are also regulatory points (see Figure 7.18). Phosphofructokinase catalyzes reaction 3 of glycolysis (see Figure 7.8). If excess ATP is present in the cytosol, ATP binds to phosphofructokinase and inhibits it by feedback inhibition. Conversely, if the level of ATP is low, the level of AMP, a breakdown product of ATP, will be high. AMP stimulates phosphofructokinase, increasing the flow of intermediates through the pathways. Phosphofructokinase is also regulated by citrate. Citrate is the first product in the citric acid cycle. If citrate levels are high, some citrate diffuses out of the mitochondria and inhibits phosphofructokinase.

Hexokinase, which catalyzes the first reaction of glycolysis (see Figure 7.8), is inhibited by its product, glucose-6-phosphate. Pyruvate kinase, which catalyzes the tenth reaction of glycolysis (see Figure 7.8), is inhibited by ATP and acetyl–CoA, and stimulated by fructose-1,6-bisphosphate.

Like glycolysis, the citric acid cycle is regulated to match its rate to the cell's requirements for ATP. Here, the enzyme that catalyzes the first reaction of the citric acid cycle, *citrate synthase,* is inhibited by elevated ATP concentrations. The inhibitions automatically slow or stop the cycle when ATP production exceeds the demands of the cell and, by doing so, conserve cellular fuels. Citrate synthase is also inhibited by high concentrations of its product, citrate.

This chapter traced the flow of high-energy electrons from fuel molecules to ATP. As part of the process, the fuels are broken into molecules of carbon dioxide and water. The next chapter shows how photosynthetic organisms use these inorganic raw materials to produce organic molecules in a process that pushes the electrons back to high energy levels by absorbing the energy of sunlight.

STUDY BREAK 7.6

1. What molecule funnels the products of many oxidative pathways into the citric acid cycle?
2. What are the types of molecules that are the immediate products of biosynthetic pathways that start from glycolysis and the citric acid cycle?
3. Why is phosphofructokinase a target for inhibition by ATP?

Glycolysis and energy metabolism are crucial for the normal functioning of an animal. Research of many kinds is being conducted in this area, such as characterizing the molecular components of the reactions in detail and determining how they are regulated. The goal is to generate comprehensive models of cellular respiration and its regulation. Following is a specific example of ongoing research related to human disease caused by defects in cellular respiration.

How do mitochondrial proteins change in patients with Alzheimer disease?

Alzheimer disease (AD) is an age-dependent, irreversible, neurodegenerative disorder in humans. Symptoms include a progressive deterioration of cognitive functions and, in particular, a significant loss of memory. Neuropathologically, AD is characterized by the presence of extracellular amyloid plaques, intracellular neurofibrillary tangles, and synaptic and neuronal loss. Reduced brain metabolism occurs early in the onset of AD. One of the mechanisms for this physiological change appears to be damage to or reduction of key mitochondrial components, including enzymes of the citric acid cycle and the oxidative phosphorylation system. However, the complete scope of mitochondrial protein changes has not been established, nor have detailed comparisons been made of mitochondrial protein changes among AD patients.

To begin to address these questions, research is being carried out in my laboratory at The University of Texas at Dallas to analyze quantitatively the complete set of mitochondrial proteins in healthy versus AD brains. The results will show the changes that occur in the mitochondrial proteome between the two tissues. This approach is called *quantitative comparative proteomic profiling*.

We are using a transgenic mouse model of AD in the first stage of this research; that is, the mice have been genetically engineered with altered genes so that they develop AD. (The generation of transgenic organisms is described in Chapter 18.) The results of our experiments have demonstrated that the levels of many mitochondrial proteins are altered in the brains of transgenic AD mice. Interestingly, both down-regulated and up-regulated mitochondrial proteins were identified in AD brains. These dysregulated mitochondrial proteins participate in many different metabolic functions, including the citric acid cycle, oxidative phosphorylation, pyruvate metabolism, fatty acid oxidation, ketone body metabolism, metabolite transport, oxidative stress, mitochondrial protein synthesis, mitochondrial protein import, and cell growth and apoptosis (a type of programmed cell death; see Chapter 50). We have also determined that these changes in the mitochondrial proteome occur early in AD before the development of significant plaque and tangle pathologies.

Future experiments will be directed toward examining changes in the mitochondrial proteome in the brains of human AD patients. Ultimately, the results of our experiments may lead to the development of treatments that can slow or halt the progression of AD in humans.

think like a scientist What significance do you attribute to the statement in this essay that changes in levels of many mitochondrial proteins occurs early in AD, before the development of significant plaque and tangle pathologies? Develop a scientific hypothesis based on this statement.

Gail A. Breen is an associate professor in the Department of Biological Sciences at The University of Texas at Dallas. Her current research focuses on mitochondrial biogenesis and the role of mitochondria in neurodegenerative diseases, such as Alzheimer disease. To learn more about Dr. Breen's research, go to http://www.utdallas.edu/biology/faculty/research/breen.html.

REVIEW KEY CONCEPTS

For access to MindTap and additional study materials visit www.cengagebrain.com.

7.1 Overview of Cellular Respiration

- Plants and almost all other organisms obtain energy for cellular activities through cellular respiration, the process of transferring electrons from donor organic molecules to a final acceptor molecule such as oxygen; the energy that is released drives ATP synthesis (Figure 7.1).

- Oxidation–reduction reactions, called redox reactions, partially or completely transfer electrons from donor to acceptor atoms; the donor is oxidized as it releases electrons, and the acceptor is reduced (Figure 7.2).

- Cellular respiration occurs in three stages: (1) in glycolysis, glucose is converted to two molecules of pyruvate through a series of enzyme-catalyzed reactions. ATP is produced by substrate-level phosphorylation, an enzyme-catalyzed reaction that transfers a phosphate group from a substrate to ADP; (2) in pyruvate oxidation and the citric acid cycle, pyruvate is converted to an acetyl compound that is oxidized completely to carbon dioxide; and (3) in oxidative phosphorylation, which is comprised of the electron transfer system and chemiosmosis, high-energy electrons

produced from the first two stages pass through the transfer system, with much of their energy being used to establish an H^+ gradient across the membrane that drives the synthesis of ATP from ADP and P_i (Figures 7.4 and 7.5).

- In eukaryotes, most of the reactions of cellular respiration occur in mitochondria. In prokaryotes, glycolysis, pyruvate oxidation, and the citric acid cycle occur in the cytosol, while the rest of cellular respiration occurs on the plasma membrane (Figure 7.6).

7.2 Glycolysis: Splitting the Sugar in Half

- In glycolysis, which occurs in the cytosol, glucose (six carbons) is oxidized into two molecules of pyruvate (three carbons each). Electrons removed in the oxidations are delivered to NAD^+, producing NADH. The reaction sequence produces a net gain of 2 ATP, 2 NADH, and 2 pyruvate molecules for each molecule of glucose oxidized (Figures 7.7 and 7.8).

7.3 Pyruvate Oxidation and the Citric Acid Cycle

- In pyruvate oxidation, which occurs inside mitochondria, 1 pyruvate (three carbons) is oxidized to 1 acetyl group (two carbons) and 1 carbon dioxide (CO_2). Electrons removed in the oxidation are accepted by 1 NAD^+ to produce 1 NADH. The acetyl group is transferred to coenzyme A, which carries it to the citric acid cycle (Figure 7.9).

- In the citric acid cycle, acetyl groups are oxidized completely to CO_2. Electrons removed in the oxidations are accepted by NAD^+ or FAD, and substrate-level phosphorylation produces ATP. For each acetyl group oxidized by the cycle, 2 CO_2, 1 ATP, 3 NADH, and 1 $FADH_2$ are produced (Figure 7.10).

7.4 Oxidative Phosphorylation: The Electron Transfer System and Chemiosmosis

- Electrons are passed from NADH and $FADH_2$ to the electron transfer system, which consists of four protein complexes and two smaller shuttle carriers. As the electrons flow from one carrier to the next through the system, some of their energy is used by the complexes to pump protons across the inner mitochondrial membrane (Figures 7.11 and 7.12).
- The three major protein complexes (I, III, and IV) pump H^+ from the matrix to the intermembrane compartment, generating an H^+ gradient with a high concentration in the intermembrane compartment and a low concentration in the matrix (Figure 7.11).
- The H^+ gradient produced by the electron transfer system is used by ATP synthase as an energy source for synthesis of ATP from ADP and P_i. The ATP synthase is embedded in the inner mitochondrial membrane together with the electron transfer system (Figure 7.11).
- An estimated 2.5 ATP are synthesized as each electron pair travels from NADH to oxygen through the mitochondrial electron transfer system; about 1.5 ATP are synthesized as each electron pair travels through the system from $FADH_2$ to oxygen. Using these totals gives an efficiency of more than 30% for the utilization of energy released by glucose oxidation if the H^+ gradient is used only for ATP production (Figure 7.14).

7.5 Anaerobic Respiration and Fermentation

- Anaerobic respiration is used to produce ATP without using oxygen. Anaerobic respiration is similar to aerobic respiration in that an electron transfer system is used, but the final electron acceptor is an inorganic molecule other than oxygen, including sulfate, nitrate, CO_2, Fe^{3+}, and Mn^{4+}.
- Lactate fermentation and alcoholic fermentation are also processes used to produce ATP without using oxygen, but no electron transfer system is involved. These fermentation pathways deliver electrons carried from glycolysis by NADH to organic acceptor molecules, thereby converting NADH back to NAD^+. The NAD^+ is required to accept electrons generated by glycolysis, allowing glycolysis to supply ATP by substrate-level phosphorylation (Figure 7.15).

7.6 Interrelationships of Catabolic and Anabolic Pathways

- A wide range of carbohydrates, fats, and proteins can be used as electron sources, entering cellular respiration at various points in glycolysis through pyruvate oxidation (Figure 7.16).
- The catabolic reactions of glycolysis and the citric acid cycle also supply molecules that are used by biosynthetic pathways that stem from those stages of cellular respiration (Figure 7.17).
- Key enzymes of glycolysis and the citric acid cycle are regulated by feedback mechanisms to match the cell's need for ATP (Figure 7.18).

TEST YOUR KNOWLEDGE

Remember/Understand

1. What is the final acceptor for electrons in cellular respiration?
   a. oxygen
   b. ATP
   c. carbon dioxide
   d. hydrogen
   e. water

2. In glycolysis:
   a. free oxygen is required for the reactions to occur.
   b. ATP is used when glucose and fructose-6-phosphate are phosphorylated, and ATP is synthesized when 3-phosphoglycerate and pyruvate are formed.
   c. the enzymes that move phosphate groups on and off the molecules are uncoupling proteins.
   d. the product with the highest potential energy in the pathway is pyruvate.
   e. the end product of glycolysis moves to the electron transfer system.

3. Which of the following statements is *false*? In cellular respiration:
   a. one molecule of glucose can produce about 32 ATP.
   b. oxygen combines directly with glucose to form carbon dioxide.
   c. a series of energy-requiring reactions is coupled to a series of energy-releasing reactions.
   d. NADH and $FADH_2$ allow H^+ to be pumped across the inner mitochondrial membrane.
   e. the electron transfer system occurs in the inner mitochondrial membrane.

4. You are reading this text while breathing in oxygen and breathing out carbon dioxide. The carbon dioxide arises from:
   a. glucose in glycolysis.
   b. NAD^+ redox reactions in the mitochondrial matrix.
   c. NADH redox reactions on the inner mitochondrial membrane.
   d. $FADH_2$ in the electron transfer system.
   e. the oxidation of pyruvate, isocitrate, and α-ketoglutarate in the citric acid cycle.

5. In the citric acid cycle:
   a. NADH and H^+ are produced when α-ketoglutarate is both produced and metabolized.
   b. ATP is produced by oxidative phosphorylation.
   c. to progress from a four-carbon molecule to a six-carbon molecule, CO_2 enters the cycle.
   d. $FADH_2$ is formed when succinate is converted to oxaloacetate.
   e. the cycle "turns" once for each molecule of glucose metabolized.

6. For each NADH produced from the citric acid cycle, about how many ATP are formed?
   a. 38
   b. 36
   c. 32
   d. 2.5
   e. 2.0

7. Which of the following statements about phosphofructokinase is *false*?
   a. It is located and has its main activity in the inner mitochondrial membrane.
   b. It can be inhibited by NADH to slow glycolysis.
   c. It can be inactivated by ATP at an inhibitory site on its surface.
   d. It can be activated by ADP at an excitatory site on its surface.
   e. It can cause ADP to form.

8. If the level of ATP is low, and the level of AMP is high, the AMP will:
   a. bind glucose to turn off glycolysis.
   b. bind glucose-6-phosphate to turn off glycolysis.
   c. bind phosphofructokinase to turn on or keep glycolysis turned on.
   d. cause lactate to form.
   e. increase oxaloacetate binding to increase NAD^+ production.

Apply/Analyze

9. Which of the following statements is *false*? Imagine that you ingested three chocolate bars just before sitting down to study this chapter. Most likely:
   a. your brain cells are using ATP.
   b. there is no deficit of the initial substrate to begin glycolysis.
   c. the respiratory processes in your brain cells are moving atoms from glycolysis through the citric acid cycle to the electron transfer system.
   d. after a couple of hours, you change position and stretch to rest certain muscle cells, which removes lactate from these muscles.
   e. after 2 hours, your brain cells are oxygen-deficient.

10. In the 1950s, a diet pill that had the effect of "poisoning" ATP synthase was tried. The person taking it could not use glucose and "lost weight"—and ultimately his or her life. Today, we know that the immediate effect of poisoning ATP synthase is:
    a. ATP would not be made in the electron transfer system.
    b. H^+ movement across the inner mitochondrial membrane would increase.
    c. more than 32 ATP could be produced from a molecule of glucose.
    d. ADP would be united with phosphate more readily in the mitochondria.
    e. ATP would react with oxygen.

11. **Discuss Concepts** Why do you think nucleic acids are not oxidized extensively as a cellular energy source?

Evaluate/Create

12. **Discuss Concepts** A hospital patient was regularly found to be intoxicated. He denied that he was drinking alcoholic beverages. The doctors and nurses made a special point to eliminate the possibility that the patient or his friends were smuggling alcohol into his room, but he was still regularly intoxicated. Then, one of the doctors had an idea that turned out to be correct and cured the patient of his intoxication. The idea involved the patient's digestive system and one of the oxidative reactions covered in this chapter. What was the doctor's idea?

13. **Design an Experiment** There are several ways to measure cellular respiration experimentally. For example, CO_2 and O_2 gas sensors measure changes over time in the concentration of carbon dioxide or oxygen, respectively. Design two experiments to test the effects of changing two different variables or conditions (one per experiment) on the respiration of a research organism of your choice.

14. **Apply Evolutionary Thinking** Which of the two phosphorylation mechanisms, oxidative phosphorylation or substrate-level phosphorylation, is likely to have appeared first in evolution? Why?

For selected answers, see Appendix A.

INTERPRET THE DATA

As CO_2 concentrations increase in the atmosphere, biologists continue to explore the role of respiration from plants as a small but potentially important contribution beyond fossil fuel combustion. The data in the **Table** were collected from the leaf of a sagebrush plant from a semiarid ecosystem in Wyoming, enclosed in a chamber that measures the rate of CO_2 exchange. The respiration rate is the amount of CO_2 in micromoles lost by the leaf per square meter per second, which results in the negative numbers. The temperature values are from the leaves as they are heated or cooled during the measurements.

Observation	Temperature (°C)	Respiration Rate ($\mu mol/m^2/s$)
1	25	−2.0
2	30	−2.7
3	35	−4.1
4	40	−5.8
5	20	−1.3
6	15	−1.0
7	10	−0.7

© Cengage Learning 2017

1. Make a graph of the data, with temperature on the *x* axis and respiration rate on the *y* axis.

2. The Q10 value of respiration is the increase in respiration, expressed as the ratio of the higher rate to the lower rate, with a 10°C change in temperature. What is the approximate (whole number) Q10 of respiration for this sagebrush leaf?

3. What describes the relationship between temperature and respiration, a line or a curve? Does the Q10 that you calculated in 2 suggest a line or a curve?

4. How might the predicted increase in temperature due to elevated CO_2 concentrations affect respiration of sagebrush leaves? Do you think the data presented here are all that is needed to predict this impact?

Source: Data based on unpublished research by Brent Ewers, University of Wyoming.

Photosynthesis

Chloroplasts in the leaf of the pea plant *Pisum sativum* (color-enhanced TEM). The light-dependent reactions of photosynthesis take place within the thylakoids of the chloroplasts (thylakoid membranes are shown in yellow).

Dr. Kari Lounatmaa/Science Source

Why it matters . . . Plants, some protists (the algae), and some archaeans and bacteria, absorb the radiant energy of sunlight and convert it into chemical energy. These organisms use the chemical energy to convert simple inorganic raw materials—water, carbon dioxide (CO_2) from the air, and inorganic minerals from the soil—into organic molecules. The conversion of light energy to chemical energy in the form of sugar and other organic molecules is called **photosynthesis.** As part of their photosynthetic reactions, these organisms release oxygen.

Plants and other photosynthetic organisms are the most abundant and important *primary producers* of Earth. A **primary producer** is an organism that uses light energy or chemical energy to convert simple inorganic molecules into organic molecules. Plants and other photosynthetic organisms use some of the organic molecules they make as an energy source for their own activities, but they also serve—directly or indirectly—as a food source for *consumers,* the animals that live by eating plants or other animals. Eventually, the bodies of both primary producers and consumers provide chemical energy for bacteria, fungi, and other *decomposers.* In short, the entire food chain on the planet depends on photosynthetic organisms, and because photosynthesis releases oxygen as a product, animals rely on continuing photosynthetic activity to continue to breathe.

This chapter begins with an overview of the photosynthetic reactions. We then examine light and light absorption and the reactions that use absorbed energy to make organic molecules from inorganic substances. In this discussion, we focus on *oxygenic* (oxygen-generating) *photosynthesis* in plants and green algae; other eukaryotic photosynthesizers have individual variations on the process (see Chapter 27). Prokaryotic photosynthesis is described in Chapter 26.

8.1 Photosynthesis: An Overview

Photosynthesizers and other organisms that make all of their required organic molecules from CO_2 and other inorganic sources such as water are called **autotrophs** (*autos* = self; *trophos* = feeding). Autotrophs that use light as the energy source to make organic molecules by photosynthesis are called **photoautotrophs.** Consumers and decomposers, which need a source of organic molecules to survive, are called **heterotrophs** (*hetero* = different).

As the pathway of energy flows from the Sun through plants (primary producers) and animals to decomposers, the organic molecules made by photosynthesis are broken down into inorganic molecules again, and the chemical energy captured in photosynthesis is released as heat energy. Because the reactions capturing light energy are the first step in this pathway, photosynthesis is the vital link between the energy of sunlight and the vast majority of living organisms.

Electrons Play a Primary Role in Photosynthesis

Photosynthesis proceeds in two stages, each involving multiple reactions **(Figure 8.1).** In the first stage, the **light-dependent reactions,** the energy of sunlight is absorbed and converted into chemical energy in the form of ATP and NADPH. ATP is the main energy source for plant cells (as it is for all types of living cells), and **NADPH** (nicotinamide adenine dinucleotide phosphate) carries electrons that are pushed to high energy levels by absorbed light. In the second stage of photosynthesis, the **light-independent reactions** (also called the *Calvin cycle*), these electrons are used as a

source of energy to convert inorganic CO_2 to an organic form. The conversion process, called **CO_2 fixation,** is a reduction reaction, in which electrons are added to CO_2. As part of the reduction, protons are also added to CO_2 (reduction and oxidation are discussed in Section 7.1). With the added electrons and protons (H^+), CO_2 is converted to a carbohydrate that contains carbon, hydrogen, and oxygen atoms in the ratio 1 C:2 H:1 O.

$$CO_2 + H^+ + e^- \rightarrow (CH_2O)_n$$

Carbohydrate units are often symbolized as $(CH_2O)_n$, with the "*n*" indicating that different carbohydrates are formed from different multiples of the carbohydrate unit.

In plants, algae, and one group of photosynthetic bacteria (the cyanobacteria), the source of electrons and protons for CO_2 fixation is water (H_2O), the most abundant substance on Earth. Oxygen (O_2) generated from the splitting of the water molecule is released into the environment as a by-product of photosynthesis:

$$2\,H_2O \rightarrow 4\,H^+ + 4\,e^- + O_2$$

Thus, plants, algae, and cyanobacteria use three resources that are readily available—sunlight, water, and CO_2—to produce almost all the organic matter on Earth, and to supply the oxygen of our atmosphere.

In organisms that are able to split water, the two reactions shown above are combined and multiplied by 6 to produce a six-carbon carbohydrate such as glucose:

$$6\,CO_2 + 12\,H_2O \rightarrow C_6H_{12}O_6 + 6\,O_2 + 6\,H_2O$$

Note that water appears on both sides of the equation; it is both consumed as a reactant and generated as a product in photosynthesis.

Although glucose is the major product of photosynthesis, other monosaccharides, disaccharides, polysaccharides, lipids, and amino acids are also produced indirectly. In fact, all the organic molecules of plants are assembled as direct or indirect products of photosynthesis.

The relationships between the light-dependent and light-independent reactions are summarized in Figure 8.1. Notice that the ATP and NADPH produced by the light-dependent reactions, along with CO_2, are the reactants of the light-independent reactions. The ADP, inorganic phosphate (P_i), and **NADP$^+$** (the oxidized form of NADPH) produced by the light-independent reactions, along with H_2O, are the reactants for the light-dependent reactions. The light-dependent and light-independent reactions thus form a cycle in which the net inputs are H_2O and CO_2, and the net outputs are organic molecules and O_2.

Oxygen Released by Photosynthesis Derives from the Splitting of Water

Early investigators thought that the O_2 released by photosynthesis came from the CO_2 entering the process. The fact that it

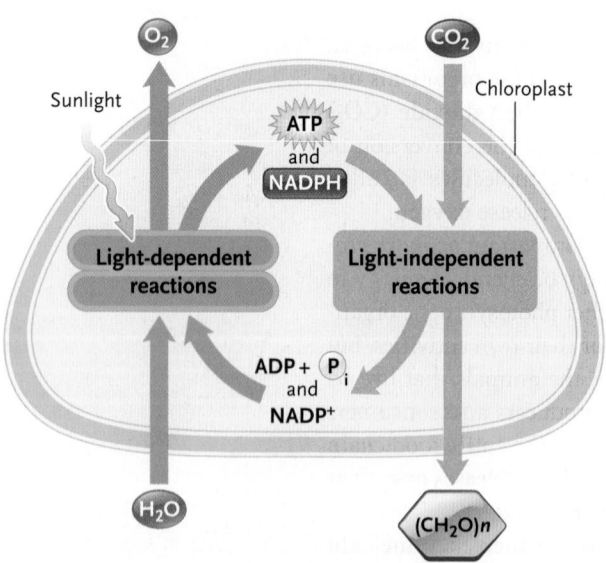

FIGURE 8.1 The light-dependent and light-independent reactions of photosynthesis, and their interlinking reactants and products. Both series of reactions occur in the chloroplasts of plants and algae.
© Cengage Learning 2017

comes from the splitting of water was demonstrated experimentally in 1941 when Samuel Ruben and Martin Kamen of the University of California, Berkeley, used a heavy isotope of oxygen, ^{18}O, to trace the pathways of the atoms through photosynthesis. A substance containing heavy ^{18}O can be distinguished readily from the same substance containing the normal isotope, ^{16}O. When a photosynthetic organism was supplied with water containing ^{18}O, the heavy isotope showed up in the O_2 given off in photosynthesis. However, if the organisms were supplied with carbon dioxide containing ^{18}O, the heavy isotope showed up in the carbohydrate and water molecules assembled during the reactions—but not in the oxygen gas. This experiment, and similar experiments using different isotopes, revealed where each atom of the reactants end up in products:

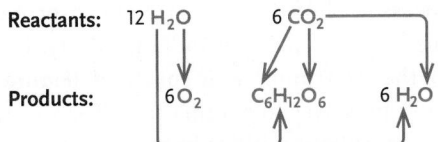

Reactants: $12\,H_2O$ $6\,CO_2$

Products: $6\,O_2$ $C_6H_{12}O_6$ $6\,H_2O$

The water-splitting reaction probably developed before oxygen-consuming organisms appeared, evolving first about 3 billion years ago in primitive photosynthetic bacteria that were ancestors to present-day cyanobacteria. The oxygen released by the reaction profoundly changed the atmosphere. It allowed for aerobic respiration in which oxygen serves as the final acceptor for electrons removed in cellular oxidations. The existence of all animals depends on the oxygen provided by the water-splitting reaction of photosynthesis.

In Eukaryotes, Photosynthesis Takes Place in Chloroplasts

In eukaryotes, the photosynthetic reactions take place in the chloroplasts of plants and algae. In cyanobacteria, the reactions are distributed between the plasma membrane and the cytosol.

Chloroplasts from individual algal and plant groups differ in structural details. The chloroplasts of plants and green algae are formed from three membranes that enclose three compartments inside the organelles **(Figure 8.2).** (Chloroplast structure is also described in Section 4.4.) An *outer membrane* covers the entire surface of the organelle. An *inner membrane* lies just inside the outer membrane. Between the outer and inner membranes is an *intermembrane compartment*. The fluid within the compartment formed by the inner membrane is the **stroma.** Within the stroma is the third membrane system, the *thylakoid membranes,* which form flattened, closed sacs called **thylakoids.** The space enclosed by a thylakoid is called the *thylakoid lumen.*

In green algae and higher plants, thylakoids are arranged into stacks called **grana** (singular, *granum;* shown in Figure 8.2). The grana are interconnected by flattened, tubular membranes called *stromal lamellae.* The stromal lamellae probably link the thylakoid lumens into a single continuous space within the stroma.

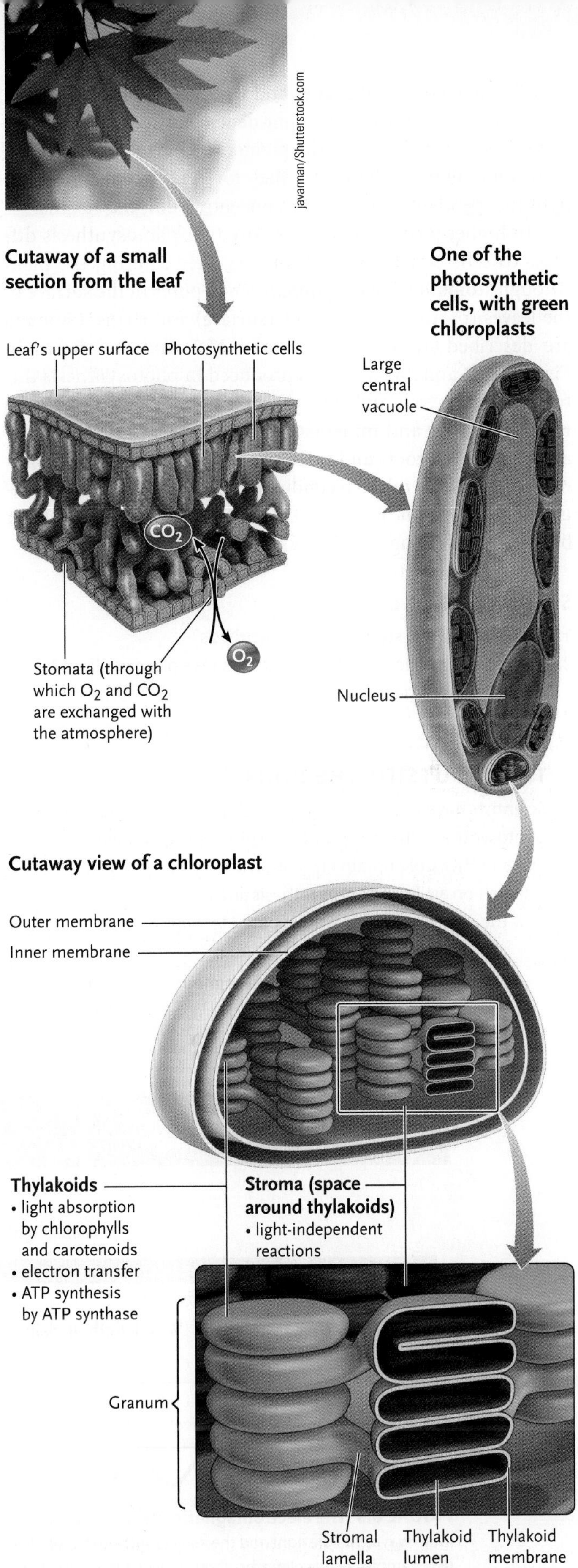

Cutaway of a small section from the leaf

Leaf's upper surface   Photosynthetic cells

CO_2

Stomata (through which O_2 and CO_2 are exchanged with the atmosphere)

O_2

One of the photosynthetic cells, with green chloroplasts

Large central vacuole

Nucleus

Cutaway view of a chloroplast

Outer membrane

Inner membrane

Thylakoids
- light absorption by chlorophylls and carotenoids
- electron transfer
- ATP synthesis by ATP synthase

Stroma (space around thylakoids)
- light-independent reactions

Granum

Stromal lamella   Thylakoid lumen   Thylakoid membrane

FIGURE 8.2 The membranes and compartments of chloroplasts.
© Cengage Learning 2017

The thylakoid membranes and stromal lamellae house the molecules that carry out the light-dependent reactions of photosynthesis, which include the pigments, electron transfer carriers, and ATP synthase enzymes for ATP production. The light-independent reactions are concentrated in the stroma.

In higher plants, the CO_2 required for photosynthesis diffuses to cells containing chloroplasts after entering the plant through **stomata** (singular, *stoma*), small pores in the surface of the leaves (particularly the undersurface) and stems. (Stomata are described in Sections 28.1 and 33.2, and are shown in Figures 8.13 and 28.3.) The O_2 produced in photosynthesis diffuses from the cells and exits through the stomata, as does the H_2O. The water and minerals required for photosynthesis are absorbed by the roots and transported to cells containing chloroplasts through tubular conducting cells. The organic products of photosynthesis are distributed to all parts of the plant by other conducting cells (see Chapter 34).

STUDY BREAK 8.1

1. What are the two stages of photosynthesis?
2. In which organelle does photosynthesis take place in plants? Where in that organelle are the two stages of photosynthesis carried out?

THINK OUTSIDE THE BOOK

Scientists have been working to develop an artificial version of photosynthesis that can be used to produce liquid fuels from CO_2 and H_2O. Collaboratively or individually, find an example of research on artificial photosynthesis and prepare an outline of how the system works or is anticipated to work.

8.2 The Light-Dependent Reactions of Photosynthesis

The light-dependent reactions (also referred to more simply as the light reactions), in which light energy is converted to chemical energy, involve two main processes: (1) light absorption; and (2) synthesis of NADPH and ATP. To keep the bigger picture in perspective as we discuss these processes, you may find it useful to refer periodically to the summary of photosynthesis shown in Figure 8.1.

Electrons in Pigment Molecules Absorb Light Energy in Photosynthesis

The first process in photosynthesis is light absorption. What is light? Visible light is a form of radiant energy. It makes up a small part of the **electromagnetic spectrum (Figure 8.3)**, which ranges from radio waves to gamma rays. The various forms of electromagnetic radiation differ in **wavelength**—the horizontal distance between the crests of successive waves. Radio waves have wavelengths in the range of 10 meters to hundreds of kilometers, and gamma rays have wavelengths in the range of one hundredth to one millionth of a nanometer. The average wavelength of radio waves for an FM radio station, for example, is 3 m.

The radiation humans detect as visible light has wavelengths between about 700 nm, seen as red light, and 400 nm, seen as blue light. We see the entire spectrum of wavelengths from 700 to 400 nm combined together as white light. The energy of light interacts with matter in elementary particles

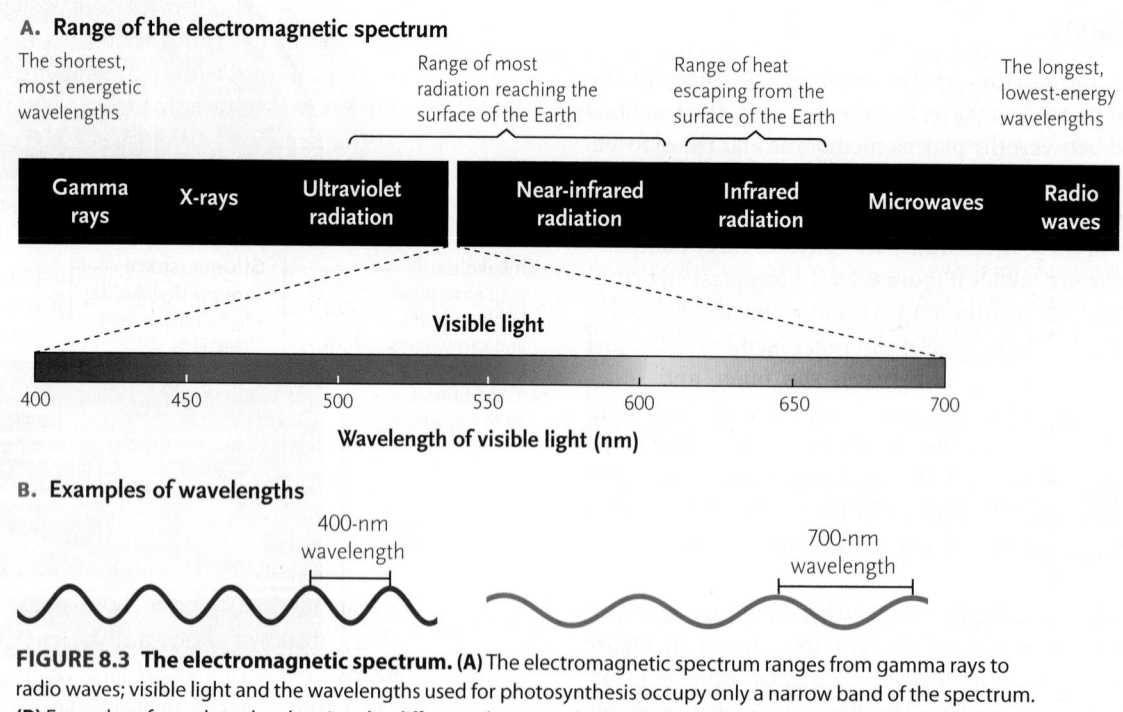

FIGURE 8.3 The electromagnetic spectrum. (A) The electromagnetic spectrum ranges from gamma rays to radio waves; visible light and the wavelengths used for photosynthesis occupy only a narrow band of the spectrum. **(B)** Examples of wavelengths, showing the difference between the longest and shortest wavelengths of visible light.
© Cengage Learning 2017

 **FIGURE 8.4** | **Experimental Research**

Engelmann's Experiment Showing the Action Spectrum of Light Used in Photosynthesis

Question: What wavelengths of light are used in photosynthesis?

Experiment: Engelmann used a light microscope and a glass prism to determine the most effective color of light for promoting photosynthesis. His experiment stands today as a classic one, both for the fundamental importance of his conclusion and for the simple but elegant methods he used to obtain it. Engelmann placed a strand of a filamentous green alga (see Chapter 27) on a glass microscope slide along with water containing motile bacteria that were strict aerobes (bacteria that require oxygen to survive: see Section 7.5). He adjusted the prism so that it split a beam of visible light into its separate colors, which spread like a rainbow across the strand. Under the microscope he observed where the bacteria moved.

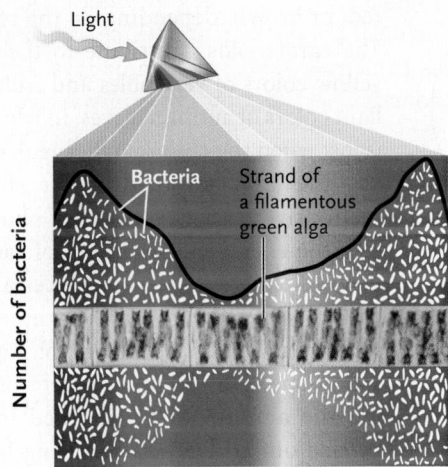

Results and Conclusions: After a short time, Engelmann observed that the bacteria had formed large clusters in the blue and violet light at one end of the strand, and in the red light at the other end. Very few bacteria were found in the green light. Evidently, violet, blue, and red light caused the most oxygen to be released, and Engelmann concluded that these colors of light—rather than green, the color of the algae—were used most effectively in photosynthesis.

Engelmann used the distribution of bacteria in the light to construct a curve called an *action spectrum* for the wavelengths of light that fell on the alga; it showed the relative effect of each color of light on photosynthesis (black curve in the figure). Engelmann's results were so accurate that an action spectrum obtained with modern equipment fits closely with his bacterial distribution. However, his results were controversial for some 60 years, until instruments that enabled direct measurements of the effects of specific wavelengths of light on photosynthesis became available.

 Observe
Hypothesize
Predict
Experiment
Interpret

What would the result be if the alga filament was illuminated with green light? With red light?

Source: T. W. Engelmann. 1882. Über Sauerstoffausscheidung von Pflanzenzellen im Mikrospectrum. *Botanische Zeitung* 40:419–426.

called *photons*. Each photon is a discrete unit that contains a fixed amount of energy. That energy is inversely proportional to its wavelength: the shorter the wavelength, the greater the energy of a photon.

What wavelengths of light are used in photosynthesis? This question was answered in an ingenious experiment performed in 1882 by a German scientist, Theodor Engelmann. **Figure 8.4** describes the experiment.

In photosynthesis, light is absorbed by molecules of green pigments called **chlorophylls** (*chloros* = yellow-green; *phyllon* = leaf) and yellow-orange pigments called **carotenoids** (*carota* = carrot). Pigment molecules such as chlorophyll appear colored to an observer because they absorb the energy of visible light at certain wavelengths and transmit or reflect other wavelengths. The color of a pigment is produced by the transmitted or reflected light. Plants look green because chlorophyll absorbs blue and red light most strongly, and transmits or reflects most of the wavelengths in between. We see the reflected light as green. This green light, as demonstrated by Engelmann's experiment, is the combination of wavelengths that are *not used* by the plants in photosynthesis.

Light is absorbed in a pigment molecule by excitable electrons occupying certain energy levels in the atoms of the pigments (see Section 2.2 for the discussion on energy levels). When not absorbing light, these electrons are at a relatively low energy level known as the *ground state*. If an electron in the pigment absorbs the energy of a photon, it jumps to a higher energy level that is farther from the atomic nucleus. This condition of the electron is called the *excited state*. The difference in energy level between the ground state and the excited state is equivalent to the energy of the photon of light that was absorbed.

One of three events then occurs, depending on both the pigment that is absorbing the light and other molecules in the vicinity of the pigment **(Figure 8.5).**

Chlorophylls Are the Main Light Receptor Pigments for Photosynthesis

Chlorophylls are the major photosynthetic pigments in plants, green algae, and cyanobacteria. They absorb photons and transfer excited electrons to nearby electron-accepting molecules, the **primary acceptor** molecules. In the transfer, the chlorophyll is oxidized because it loses an

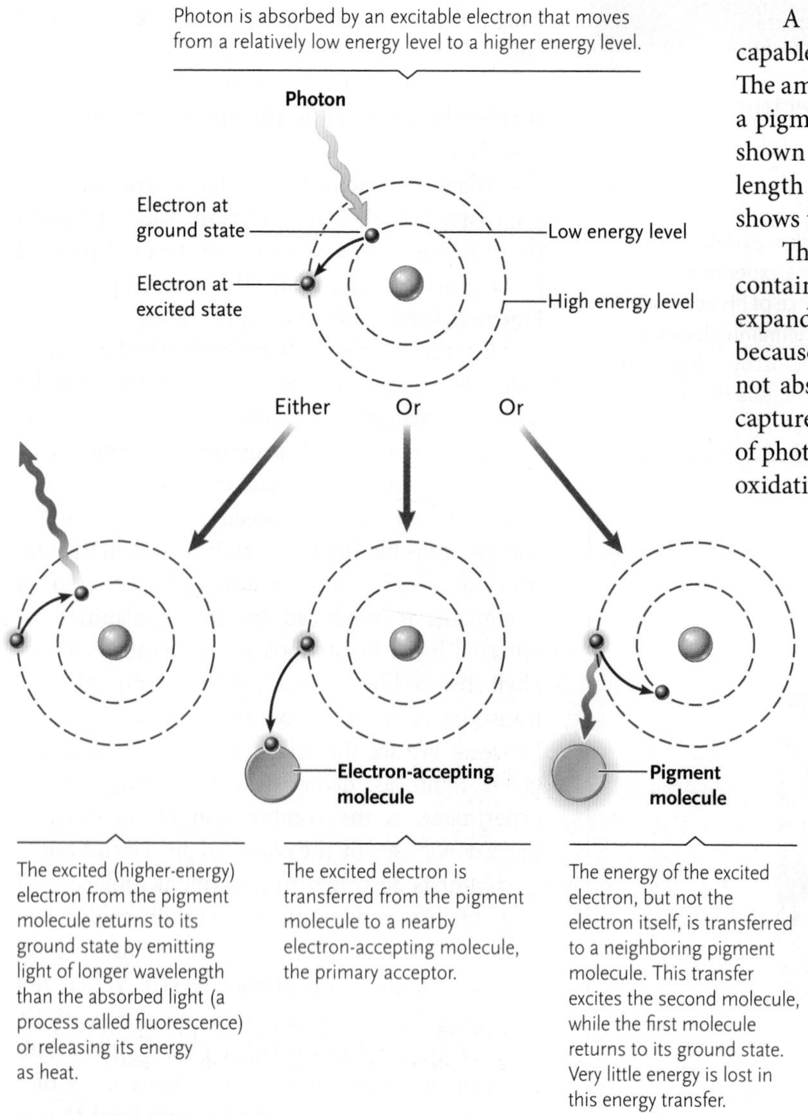

Photon is absorbed by an excitable electron that moves from a relatively low energy level to a higher energy level.

Photon

Electron at ground state ——— Low energy level

Electron at excited state

——— High energy level

Either   Or   Or

Electron-accepting molecule

Pigment molecule

The excited (higher-energy) electron from the pigment molecule returns to its ground state by emitting light of longer wavelength than the absorbed light (a process called fluorescence) or releasing its energy as heat.

The excited electron is transferred from the pigment molecule to a nearby electron-accepting molecule, the primary acceptor.

The energy of the excited electron, but not the electron itself, is transferred to a neighboring pigment molecule. This transfer excites the second molecule, while the first molecule returns to its ground state. Very little energy is lost in this energy transfer.

FIGURE 8.5 Alternative effects of light absorbed by a pigment molecule.
© Cengage Learning 2017

electron, and the primary acceptor is reduced because it gains an electron. Closely related molecules, the *bacteriochlorophylls,* carry out the same functions in other photosynthetic bacteria. Carotenoids are *accessory pigments* that absorb light energy at a different wavelength than those absorbed by chlorophylls.

Chlorophylls and carotenoids are bound to proteins that are embedded in photosynthetic membranes. In plants and green algae, they are located in the thylakoid membranes of chloroplasts. In photosynthetic bacteria, they are located in the plasma membrane.

Molecules of the chlorophyll family **(Figure 8.6A)** have a carbon ring structure with a magnesium atom bound at the center. The ring is attached to a long, hydrophobic side chain. The main types of chlorophyll are chlorophyll *a* and chlorophyll *b*—they differ only in one side group that is attached to a carbon of the ring structure (shown in Figure 8.6A).

A chlorophyll molecule contains a network of electrons capable of absorbing light (shaded in orange in Figure 8.6A). The amount of light of different wavelengths that is absorbed by a pigment is called an **absorption spectrum,** which is usually shown as a graph in which the height of the curve at any wavelength indicates the amount of light absorbed. **Figure 8.7A** shows the absorption spectra for chlorophylls *a* and *b*.

The carotenoids are built on a long backbone that typically contains 40 carbon atoms **(Figure 8.6B).** Carotenoids may expand the range of wavelengths used for photosynthesis because they absorb different wavelengths that chlorophyll does not absorb. A more important role for carotenoids than light capture for photosynthesis is in *photoprotection,* the protection of photosynthetic organisms against potentially harmful photo-oxidative processes. Carotenoids transmit or reflect other wavelengths in combinations that appear yellow, orange, red, or brown, depending on the type of carotenoid. The carotenoids contribute to the red, orange, and yellow colors of vegetables and fruits and to the brilliant colors of autumn leaves, in which the green color is lost when the chlorophylls break down.

The light absorbed by the chlorophylls is the main driver of the reactions of photosynthesis. Plotting the effectiveness of light of each wavelength in driving photosynthesis produces a graph called the **action spectrum** of photosynthesis. **Figure 8.7B** shows the action spectrum of higher plants. The action spectrum is usually determined by measuring the amount of O_2 released by photosynthesis carried out at different wavelengths of visible light, as Engelmann did indirectly in his experiment (compare Figures 8.4 and 8.7B).

In all eukaryotic photosynthesizers, a specialized chlorophyll *a* molecule passes excited electrons to the primary acceptor. Other chlorophyll molecules, along with carotenoids, act as accessory pigments that pass their energy to chlorophyll *a*. Light energy that is absorbed by the entire collection of chlorophyll and carotenoid molecules in chloroplasts is passed to the specialized chlorophyll *a* molecules that are directly involved in transforming light into chemical energy.

The Photosynthetic Pigments Are Organized into Photosystems in Chloroplasts

The light-absorbing pigments are organized with proteins and other molecules into large complexes called **photosystems (Figure 8.8),** which are embedded in thylakoid membranes and stromal lamellae. The photosystems are the sites at which light is absorbed and converted into chemical energy.

Plants, green algae, and cyanobacteria have two types of these complexes, called **photosystem II** and **photosystem I.** which carry out different parts of the light-dependent reactions. Photosystem II, in addition, is closely linked to a group of

A. Chlorophyll structure

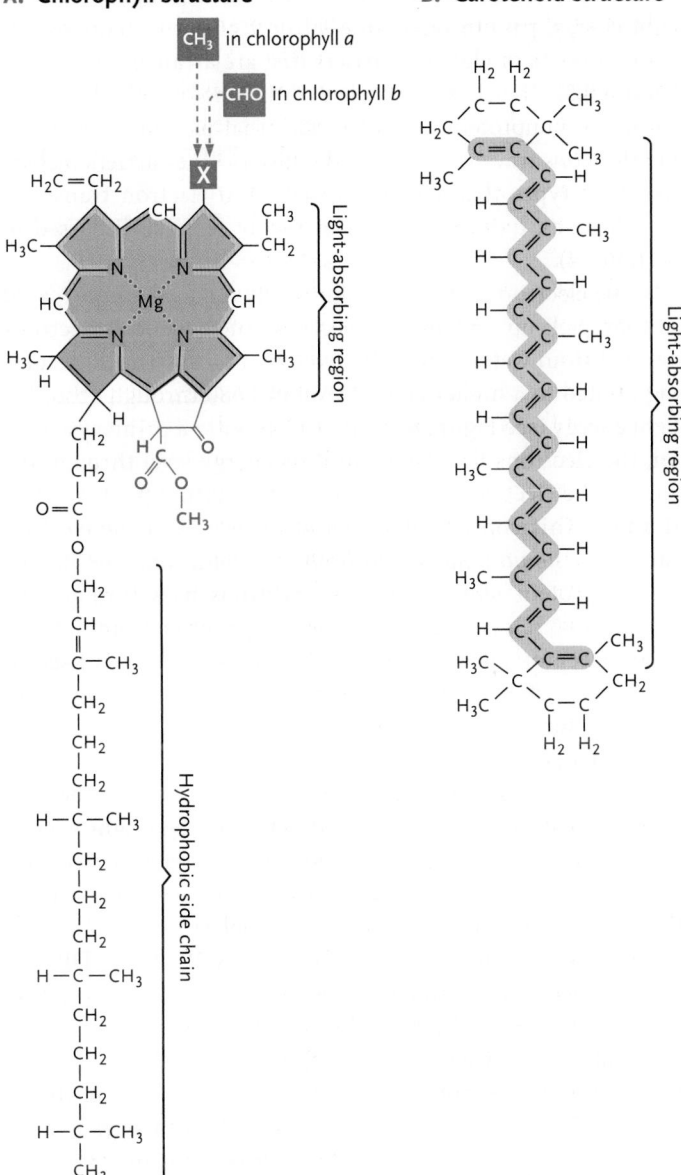

FIGURE 8.6 Pigment molecules used in photosynthesis.
(A) Chlorophylls *a* and *b*, which differ only in the side group attached at the X. Light-absorbing electrons are distributed among the bonds shaded in orange. The chlorophylls are similar in structure to the cytochromes, which occur in both the chloroplast and mitochondrial electron transfer systems. **(B)** In carotenoids, the light-absorbing electrons are distributed in a series of alternating double and single bonds in the backbone of these pigments.
© Cengage Learning 2017

B. Carotenoid structure

(image included in img_1 region: Light-absorbing region label)

A. The absorption spectra of chlorophylls *a* and *b*

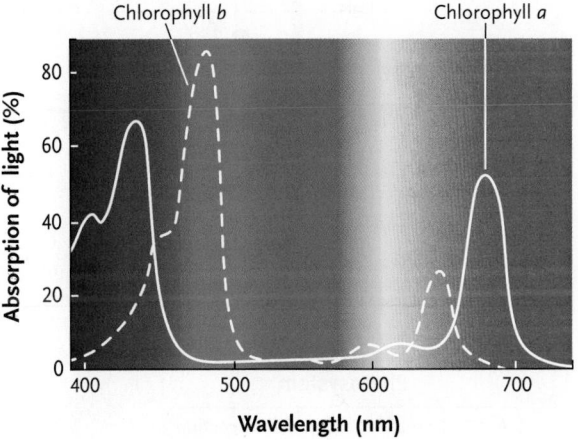

B. The action spectrum in higher plants

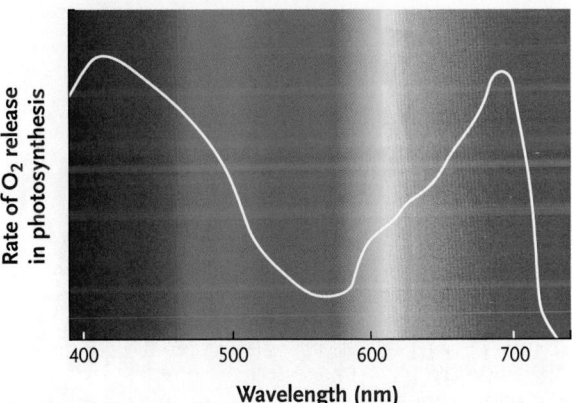

FIGURE 8.7 The absorption spectra of chlorophylls *a* and *b* (A) and the action spectrum of photosynthesis (B) in higher plants.
© Cengage Learning 2017

---

enzymes that carries out the initial reaction of splitting water into electrons, protons, and oxygen. (Note: As you will learn soon, photosystem II functions before photosystem I in the light-dependent reactions. However, photosystem I was named because it was discovered first and the systems were given their numbers before their order of use in the reactions was worked out.) Each photosystem consists of two closely associated components: an **antenna complex** (also called a *light-harvesting*

complex) and a **reaction center** (see Figure 8.8). The antenna complex contains an aggregate of some 300 chlorophyll pigments and 40 or so carotenoid pigments. The chlorophyll molecules are anchored in the complex by being bound to specific membrane proteins. In this form, they are arranged efficiently to optimize the capture of light energy.

The reaction center contains special subsets of chlorophyll *a* molecules complexed with proteins (two molecules are shown in Figure 8.8 for illustration). The chlorophyll *a* molecules in the reaction center of photosystem II are called *P680* (P = pigment) because they absorb light optimally at a wavelength of 680 nm. Those in the reaction center of photosystem I are called *P700* because they absorb light optimally at a wavelength of 700 nm.

Light energy in the form of photons is absorbed by the pigment molecules of the antenna complex. This absorbed light energy is conducted to the P680 or P700 molecules in the reaction center. The absorbed light is converted to chemical energy when an excited electron (*e*⁻ in the figure) from the special

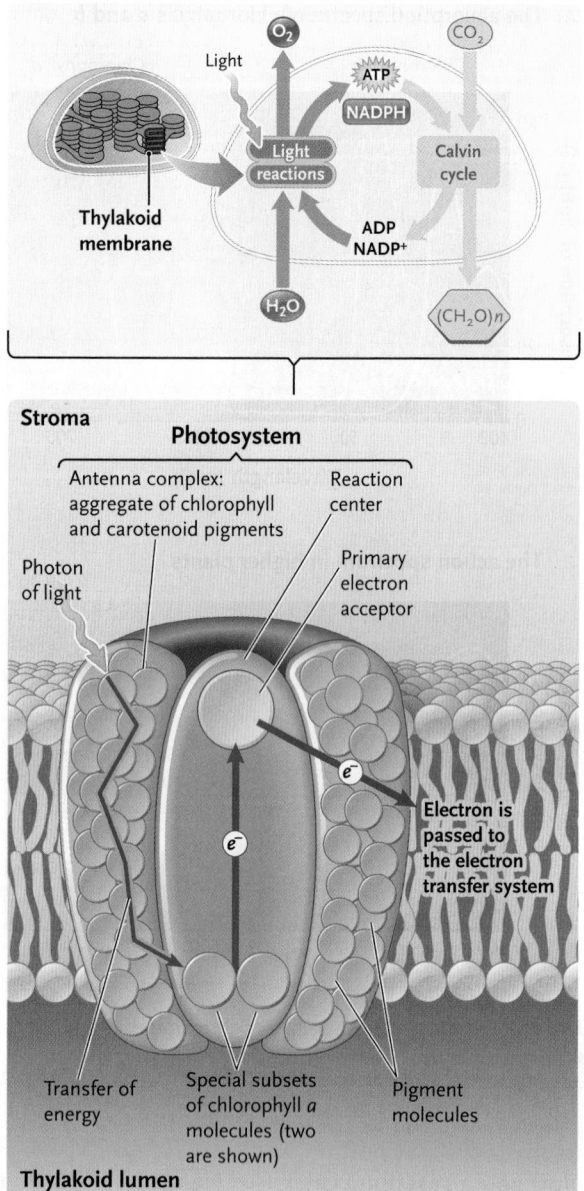

FIGURE 8.8 **Major components of a photosystem: a group of pigments forming an antenna complex (light-harvesting complex) and a reaction center.**

© Cengage Learning 2017

chlorophyll *a* molecule is transferred to a primary acceptor, which is also in the reaction center. That high-energy electron is passed out of the reaction center and out of the photosystem to the electron transfer system. Blue arrows in Figure 8.8 show the path of energy flow.

Electron Flow from Water to NADP+ Leads to the Synthesis of NADPH and ATP

In the second main process of the light-dependent reactions, the electrons obtained from the splitting of water (two electrons per molecule of water; see Section 8.1) are used for the synthesis of

NADPH. These electrons, which were pushed to higher levels by light energy, pass through an electron transfer system consisting of a series of electron carriers that are arranged in a chain **(Figure 8.9).** The electron carriers of the photosynthetic system consist of nonprotein organic groups that pick up and release the electrons traveling through the system. The carriers include the same types that act in mitochondrial electron transfer—cytochromes, quinones, and iron–sulfur centers (discussed in Section 7.4).

The electron carriers are alternately reduced and oxidized as they pick up and release electrons in sequence. Electrons derived from water first flow through photosystem II, becoming excited to a higher energy level in P680 through absorbed light energy (see Figure 8.9). After transfer to a primary acceptor, the electrons flow "downhill" in energy level through the part of the electron transfer system connecting photosystems II and I. This consists of a pool of molecules of the electron carrier *plastoquinone*, a *cytochrome complex*, and the mobile carrier protein *plastocyanin*. As electrons pass through the system they release free energy at each transfer from a donor to an acceptor molecule. Some of this free energy is used to create a gradient of H^+ across the membrane. The gradient provides the energy source for ATP synthesis, just as it does in mitochondria.

The electrons then pass to photosystem I, where they are excited a second time in P700 through absorbed light energy. After transfer to a primary acceptor, the high-energy electrons enter a short part of the electron transfer system that leads to the final electron acceptor of the chloroplast system, $NADP^+$. The enzyme $NADP^+$ reductase reduces $NADP^+$ to NADPH by using two electrons and two protons from the surrounding water solution, and by releasing one proton. This pathway is frequently called **linear electron flow** because electrons travel in a one-way direction from H_2O to $NADP^+$. It is sometimes called the *Z scheme* because of the zigzag-like changes in electron energy level (shown by the blue arrows in Figure 8.9).

Figure 8.10 shows how the electron transfer and ATP synthesis systems for the light-dependent reactions are organized in the thylakoid membrane and lays out the linear electron flow pathway for NADPH and ATP synthesis. The electron transfer system involves steps 1 to 9, and the chemiosmotic synthesis of ATP involves steps 10 to 12.

The flow of electrons through the electron transfer system leads to the generation of an H^+ gradient across the inner membrane (steps 1–8). That gradient is enhanced by the addition of two protons to the lumen for each water molecule split, and by the removal of one proton from the stroma for each NADPH molecule synthesized. Because protons carry a positive charge, an electrical gradient forms across the thylakoid membrane, with the lumen more positively charged than the stroma. The combination of a proton gradient and a voltage gradient across the membrane produces stored energy known as the *proton-motive force* (also discussed for cellular respiration in Section 7.4), which contributes energy for ATP synthesis by ATP synthase. Just as for the mitochondrial ATP synthase, the chloroplast ATP synthase is embedded in the same membranes as the electron transfer system. Protons flow through a membrane channel from the thylakoid lumen to the stroma along their concentration gradient (step 10). Free energy is released as H^+ moves through the channel; it powers synthesis of ATP from ADP and P_i by the ATP synthase (steps 11–12). This process of using an H^+ gradient to power ATP synthesis—*chemiosmosis*—is the same as that used for ATP synthesis in mitochondria (see Section 7.4).

The overall yield of the linear electron flow pathway is one molecule of NADPH and one molecule of ATP for each pair of electrons produced from the splitting of water. The synthesis of ATP coupled to the transfer of electrons energized by photons of light is called **photophosphorylation.** This process is analogous to oxidative phosphorylation in mitochondria (see Section 7.4), except that in chloroplasts light provides the energy for establishing the proton gradient.

Comparing the linear pathway with the mitochondrial electron transfer system (shown in Figure 7.11) reveals that the pathway from the plastoquinones through plastocyanin in chloroplasts is essentially the same as the pathway from the ubiquinones through cytochrome *c* in mitochondria. The similarities between the two pathways indicate that the electron transfer system is an ancient evolutionary development that became adapted to both photosynthesis and oxidative phosphorylation.

Electron Flow Can Also Drive ATP Synthesis by Flowing Cyclically around Photosystem I, but NADPH and O_2 Are Not Produced

Linear electron flow produces ATP and NADPH + H^+. In some cases, however, photosystem I works independently of photosystem II in a circular process called **cyclic electron flow (Figure 8.11).** In this process, electrons pass through the cytochrome complex and plastocyanin to the P700 chlorophyll *a* in the reaction center of photosystem I where they are excited by light energy. The electrons then flow from photosystem I to the mobile carrier ferredoxin, but rather than being used for $NADP^+$ reduction by $NADP^+$ reductase, they flow back to P700 in the cytochrome complex. The electrons again pass to plastocyanin and on to photosystem I where they receive another

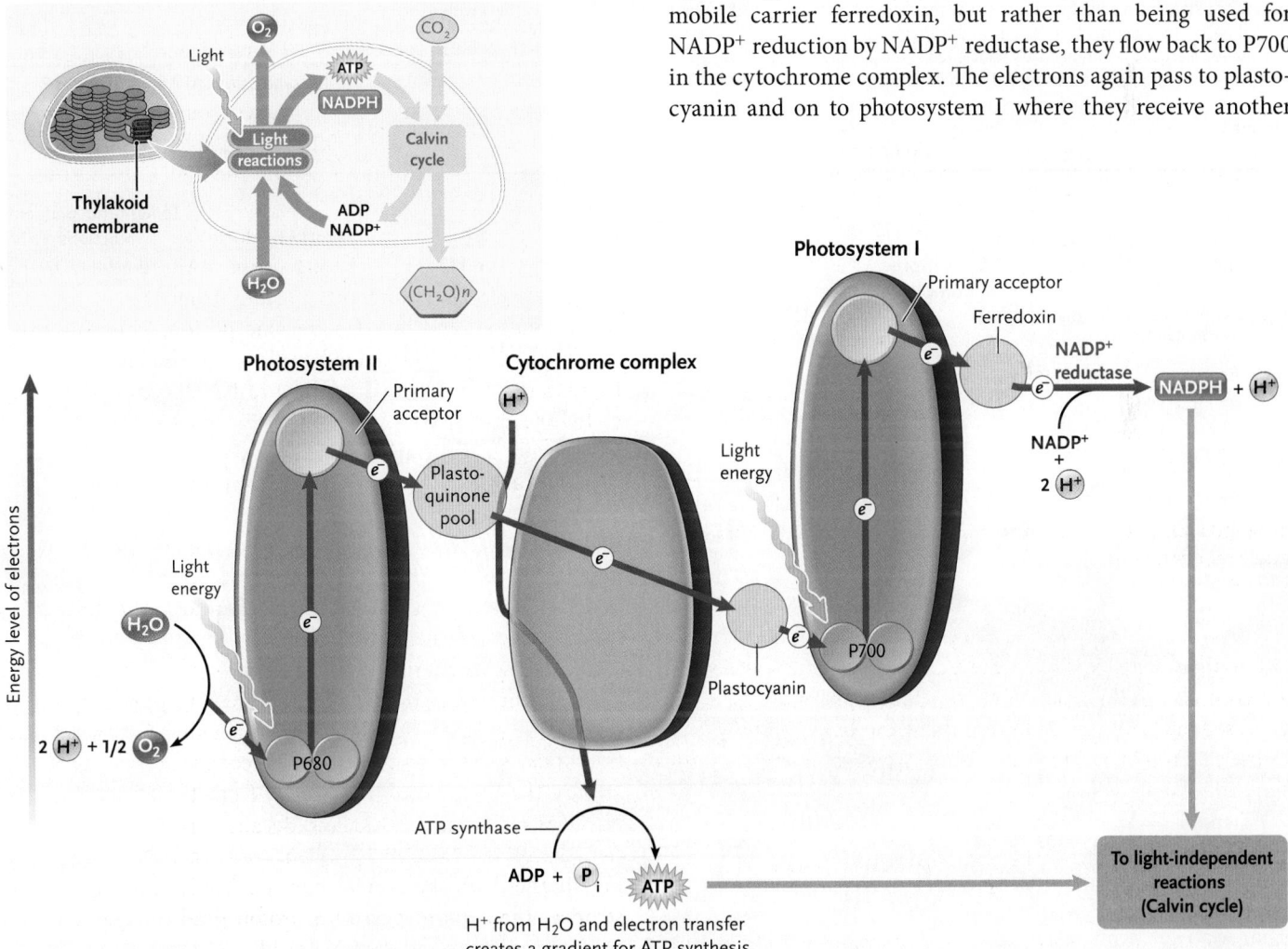

FIGURE 8.9 The pathway of the light-dependent reactions, linear electron flow.
© Cengage Learning 2017

THE CHLOROPLAST ELECTRON TRANSFER SYSTEM (STEPS 1–9) AND CHEMIOSMOSIS (STEPS 10–12), ILLUSTRATING THE SYNTHESIS OF NADPH AND ATP BY THE LINEAR (NONCYCLIC) ELECTRON FLOW PATHWAY

The components of the electron transfer system and chemiosmosis are located in the chloroplast thylakoid membrane.

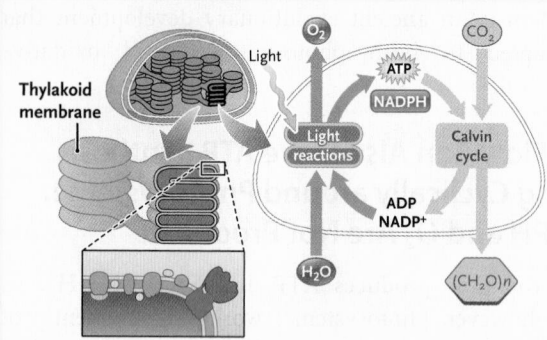

1 Electrons from the water-splitting reaction system are accepted one at a time by a P680 chlorophyll *a* in the reaction center of photosystem II. As P680 accepts the electrons, they are raised to the excited state, using energy passed to the reaction center from the light-absorbing pigment molecules in the antenna complex. The excited electrons are immediately transferred to the primary acceptor of photosystem II, a modified form of chlorophyll *a*.

2 The electrons flow through a short chain of carriers within photosystem II and then transfer to the mobile carrier plastoquinone. The plastoquinones form a "pool" of molecules within the thylakoid membranes.

3 Plastoquinones pass the electrons to the cytochrome complex. As it accepts and releases electrons, the cytochrome complex pumps H⁺ from the stroma into the thylakoid lumen. Those protons drive ATP synthesis (see step 8).

4 Electrons now pass to the mobile carrier *plastocyanin*, which shuttles electrons between the cytochrome complex and photosystem I.

5 Electrons flow to a P700 chlorophyll *a* in the reaction center of photosystem I, where they are excited to high energy levels again by absorbing more light energy. The excited electrons are transferred from P700 to the primary acceptor of photosystem I, a modified chlorophyll *a*.

6 After passage through carriers within photosystem I, the electrons are transferred to the iron–sulfur protein ferredoxin, which acts as a mobile carrier.

7 The ferredoxin transfers the electrons, still at very high energy levels, to NADP⁺, the final acceptor of the noncyclic pathway. NADP⁺ is reduced to NADPH by NADP⁺ reductase. Electron transfer is now complete.

8 Proton pumping by the plastoquinones and the cytochrome complex, as described in step 3, creates a concentration gradient of H⁺ with the high concentration within the thylakoid lumen and the low concentration in the stroma.

9 The H⁺ gradient supplies the energy that drives ATP synthesis by ATP synthase.

10 Due to the gradient, H⁺ ions flow across the inner membrane into the matrix through a channel in the ATP synthase.

11 The flow of H⁺ activates ATP synthase, making the headpiece and stalk rotate.

12 As a result of changes in shape and position as it turns, the headpiece catalyzes the synthesis of ATP from ADP and P$_i$.

Stroma

Electron transfer system | Chemiosmosis

Photosystem II | **Cytochrome complex** | **Photosystem I** | **9 Low H⁺**

To light-independent reactions (Calvin cycle)

Light energy — Antenna complex — Primary acceptor — Pigment molecules

Ferredoxin

2 H⁺ + NADP⁺

H⁺ + NADPH → ATP

2

1 P680

3

8

4 Plastocyanin

5 P700

NADP⁺ reductase

ADP + P$_i$

7

12

Plastoquinone

Water-splitting complex

H_2O 2 H⁺ + 1/2 O_2

Stator

10

Rotation

11

ATP synthase

High H⁺
9

Thylakoid lumen

Thylakoid membrane

SUMMARY Electrons derived from splitting water are used for the synthesis of NADPH and ATP. The electrons flow first through photosystem II, and then pass through part of the electron transfer system to photosystem I releasing energy that is used to create an H⁺ gradient across the membrane. ATP synthase uses the gradient to drive the synthesis of ATP.

think like a scientist

How many phospholipid bilayers separate the cytosol from the stroma?

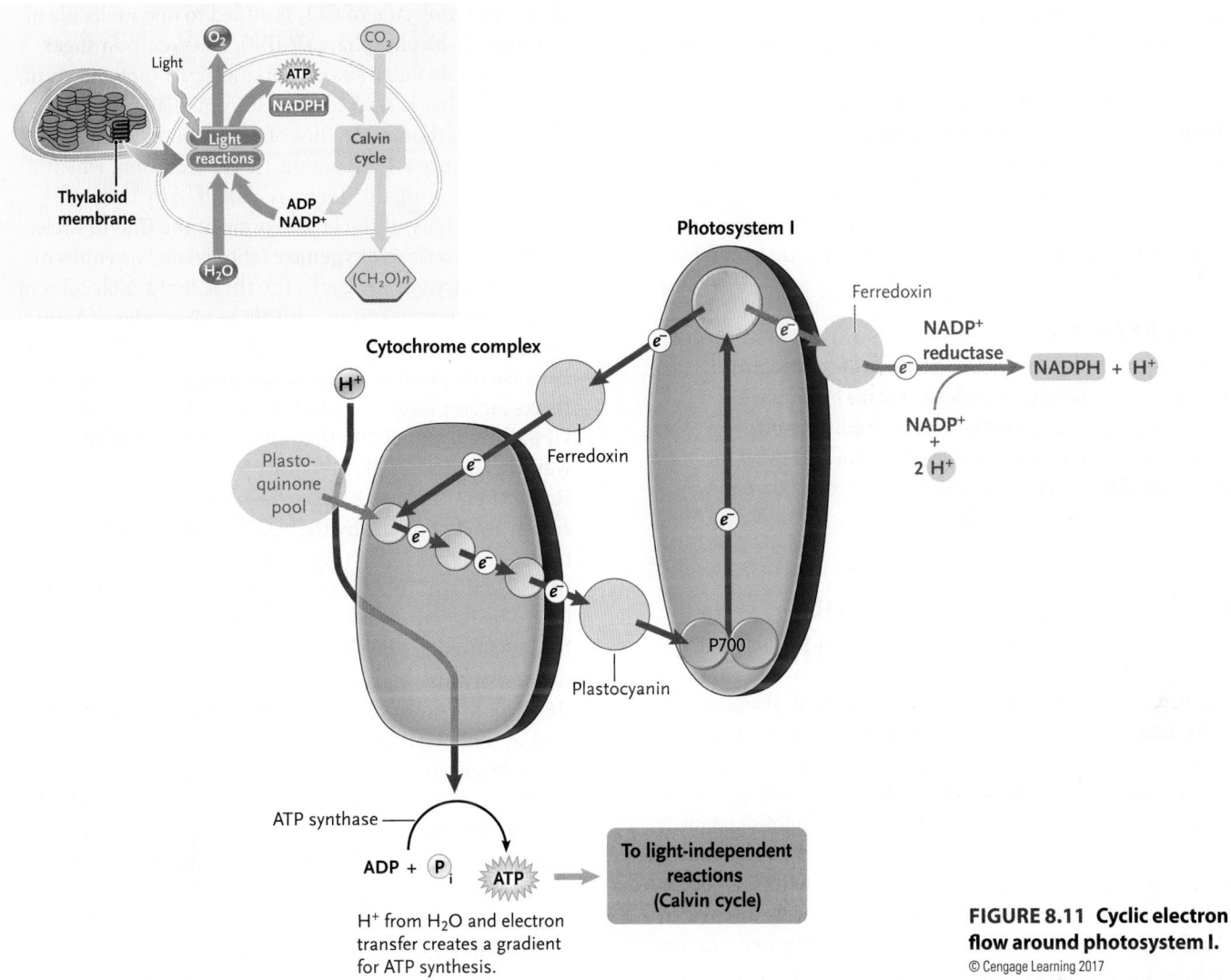

FIGURE 8.11 **Cyclic electron flow around photosystem I.**
© Cengage Learning 2017

Photosystem I

Ferredoxin

$NADP^+$ reductase

$NADPH + H^+$

$NADP^+ + 2 H^+$

Cytochrome complex

H^+

Plasto-quinone pool

Ferredoxin

P700

Plastocyanin

ATP synthase

$ADP + P_i$ → ATP

H^+ from H_2O and electron transfer creates a gradient for ATP synthesis.

To light-independent reactions (Calvin cycle)

Light

O_2 CO_2

ATP

NADPH

Light reactions

Calvin cycle

Thylakoid membrane

ADP $NADP^+$

H_2O $(CH_2O)n$

energy boost from light energy, and so the cycle continues. Each time electrons flow around the cycle, more H^+ is pumped across the thylakoid membranes, driving ATP synthesis in the way already described for the linear (noncyclic) electron flow pathway. The net result of cyclic electron flow is that light energy is converted into the chemical energy of ATP *without* the production of NADPH or O_2.

Cyclic electron flow was observed in higher plant chloroplasts over 50 years ago. However, researchers still debate whether it is a real physiological process. Recent experimental results, including studies of mutants that appear to lack the cyclic electron flow pathway, support the hypothesis that the pathway plays an important role in the responses of plants to stress. That is, because cyclic electron flow produces ATP but no NADPH, the plants avoid risky overproduction of reducing power when stressed. Excess reducing power can be damaging because of the potential for some of the extra electrons "leaking out" of the system to form reactive oxygen species such as superoxide. Reactive oxygen species are toxic and can lead to the oxidative destruction of cells.

Photosynthesis occurs also in some groups of bacteria. In those organisms the photosynthetic electron transfer system components are embedded in membranes, but there are no chloroplasts like those of plants. Cyanobacteria, the only prokaryotes that produce oxygen by photosynthesis, contain both photosystems II and I and normally carry out the light-dependent reactions using the linear electron flow pathway. They are also capable of using the cyclic electron flow pathway involving photosystem I. Photosynthesis is also carried out by some archaea.

Experiments with Chloroplasts Helped Confirm the Synthesis of ATP by Chemiosmosis

Our present understanding of the connection between electron transfer and ATP synthesis was first proposed for mitochondria in Mitchell's chemiosmotic hypothesis (discussed in Section 7.4). Several experiments have shown that the same mechanism operates in chloroplasts. In one experiment, André Jagendorf and Ernest Uribe isolated chloroplasts from cells, broke them open, and treated the broken chloroplasts to create an H^+

gradient across the thylakoid membrane. In the dark, ATP was made, indicating that the gradient, and not light-generated electron transfer, powers ATP synthesis.

Our description of photosynthesis to this point shows how the light-dependent reactions generate NADPH and ATP, which provide the reducing power and chemical energy required to produce organic molecules from CO_2. The next section follows NADPH and ATP through the light-independent reactions and shows how the organic molecules are produced.

8.3 The Light-Independent Reactions of Photosynthesis

The electrons carried from the light-dependent reactions by NADPH retain much of the energy absorbed from sunlight. These electrons provide the reducing power required to fix CO_2 into carbohydrates and other organic molecules in the light-independent reactions. The ATP generated in the light-dependent reactions supplies additional energy for the light-independent reactions. The reactions using NADPH and ATP to fix CO_2 occur in a circuit known as the **Calvin cycle,** named for its discoverer, Melvin Calvin. Calvin was awarded a Nobel Prize in 1961 "for his research on the carbon dioxide assimilation in plants." *Focus on Research: Basic Research* describes the experiments Calvin and his colleagues used to elucidate the light-independent reactions.

The Calvin Cycle Uses NADPH, ATP, and CO_2 to Generate Carbohydrates

The light-independent reactions of the Calvin cycle use CO_2, ATP, and NADPH as inputs. For three input molecules of CO_2, one of which is used in each of three turns of the cycle, the key product is one molecule of the three-carbon carbohydrate molecule glyceraldehyde-3-phosphate (G3P). The G3P is used in reactions to synthesize glucose and a number of other organic molecules. In plants, the Calvin cycle takes place entirely in the chloroplast stroma (see Figure 8.2). The Calvin cycle also occurs in most photosynthetic prokaryotes, where it takes place in the cytoplasm. We focus here on plants.

Figure 8.12 shows the Calvin cycle, summarizing the reactions of three turns of the cycle, starting with three molecules of CO_2 and resulting in the release of one molecule of the product, G3P. The Calvin cycle consists of three phases:

1. **Carbon fixation.** The first phase of the Calvin cycle, *carbon fixation,* involves the cycle's key reaction in which each input molecule of CO_2 is added to one molecule of ribulose 1,5-bisphosphate (RuBP), a five-carbon sugar, forming an unstable six-carbon molecule (not shown in the figure) that is cleaved almost immediately to produce two three-carbon molecules of 3-phosphoglycerate (3PGA) (Figure 8.12, reaction 1). This reaction, which fixes CO_2 into organic form, is catalyzed by the *carboxylase* activity of the key enzyme of the Calvin cycle, **RuBP carboxylase/oxygenase** (abbreviated as **rubisco**). For three turns of the cycle, the three input molecules of CO_2 (3 carbons) reacting with three molecules of RuBP (15 carbons) produce six molecules of 3PGA (18 carbons). Because the product of the carbon fixation reaction is a three-carbon molecule, the Calvin cycle is also called the C_3 **pathway,** and plants that initially fix carbon in this way are termed C_3 **plants.** Most plants are of this kind.

2. **Reduction.** In phase 2, *reduction,* reactions raise the energy level of 3PGA by the addition of a phosphate group transferred from ATP (Figure 8.12, reaction 2) and electrons from NADPH (Figure 8.12, reaction 3) to produce G3P, another three-carbon molecule. (The ATP and NADPH used are products of the light-dependent reactions.) For three turns of the cycle, six molecules of 3PGA (18 carbons) produce six molecules of G3P (18 carbons). One of these G3P molecule exits the cycle as a net product of the three turns and is used as the primary building block for reactions synthesizing the six-carbon glucose and many other organic molecules in chloroplasts. The other five molecules of G3P are used to regenerate RuBP in the next phase of the cycle.

3. **Regeneration.** In phase 3, *regeneration,* the G3P molecules generated by three turns of the cycle that do not exit the cycle are used to produce RuBP. First, G3P enters a complex series of reactions (Figure 8.12, reaction 4) that yields the five-carbon sugar ribulose 5-phosphate. Then, in the final reaction of the cycle (Figure 8.12, reaction 5), a phosphate group is transferred from ATP to regenerate the RuBP used in the first reaction. For three turns of the cycle, five molecules of G3P (15 carbons) produce three molecules of ribulose 5-phosphate (15 carbons) which then produce three molecules of RuBP (15 carbons).

In sum, for three turns of the Calvin cycle resulting in the net synthesis of one molecule of the product, G3P, 9 ATP and 6 NADPH are used. As mentioned earlier, those ATP and NADPH molecules derive from the light-dependent reactions of photosynthesis.

G3P Is the Starting Point for Synthesis of Many Other Organic Molecules

The net G3P formed by three turns of the Calvin cycle is the starting point for the production of a wide variety of organic molecules. More complex carbohydrates such as glucose and

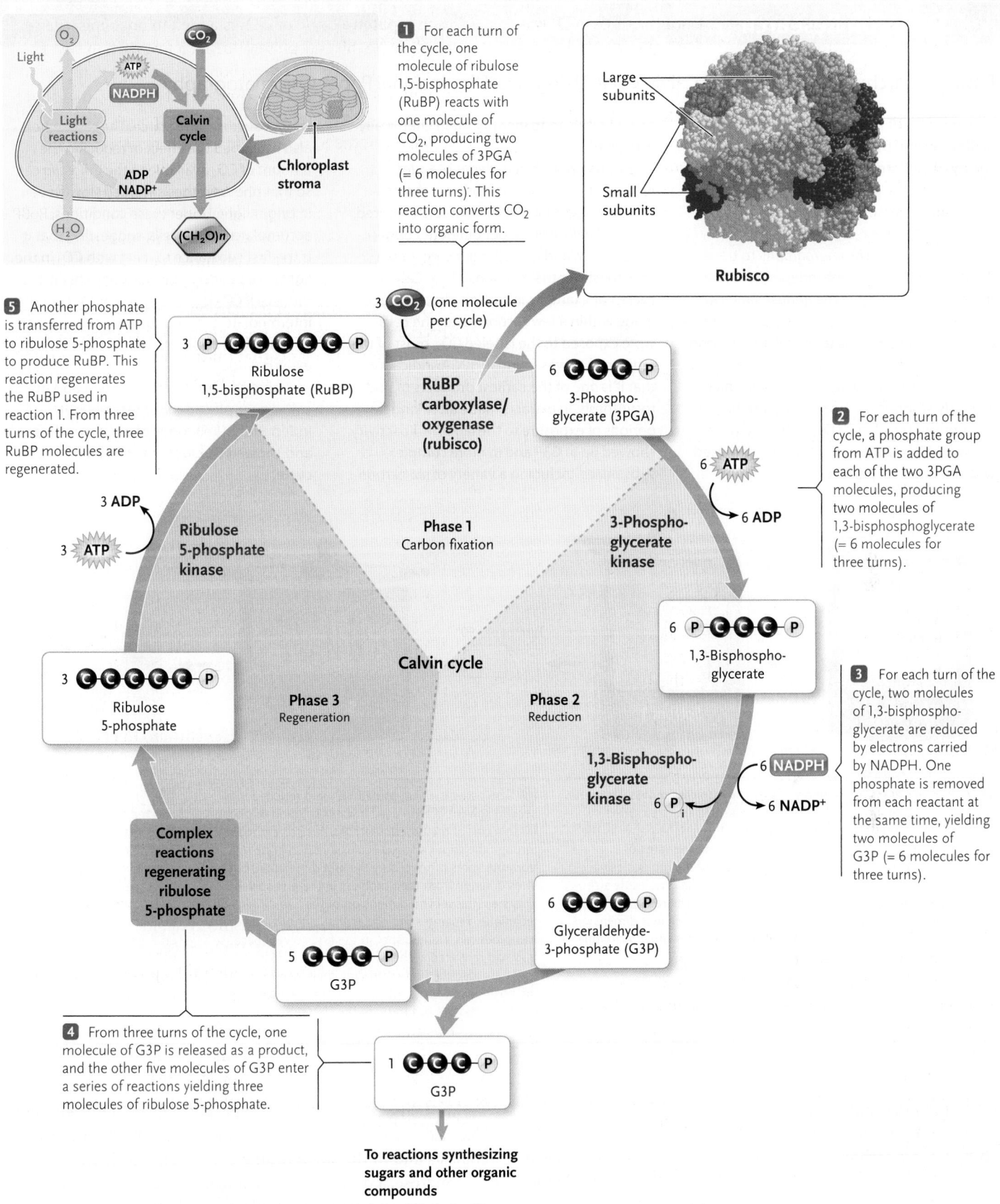

1 For each turn of the cycle, one molecule of ribulose 1,5-bisphosphate (RuBP) reacts with one molecule of CO_2, producing two molecules of 3PGA (= 6 molecules for three turns). This reaction converts CO_2 into organic form.

Large subunits

Small subunits

Rubisco

5 Another phosphate is transferred from ATP to ribulose 5-phosphate to produce RuBP. This reaction regenerates the RuBP used in reaction 1. From three turns of the cycle, three RuBP molecules are regenerated.

3 **P**–C–C–C–C–C–**P**
Ribulose
1,5-bisphosphate (RuBP)

3 **CO₂** (one molecule per cycle)

RuBP carboxylase/ oxygenase (rubisco)

6 C–C–C–**P**
3-Phospho-glycerate (3PGA)

2 For each turn of the cycle, a phosphate group from ATP is added to each of the two 3PGA molecules, producing two molecules of 1,3-bisphosphoglycerate (= 6 molecules for three turns).

6 **ATP**

6 **ADP**

3 **ADP**

Ribulose 5-phosphate kinase

Phase 1
Carbon fixation

3-Phospho-glycerate kinase

3 **ATP**

6 **P**–C–C–C–**P**
1,3-Bisphospho-glycerate

Calvin cycle

3 For each turn of the cycle, two molecules of 1,3-bisphospho-glycerate are reduced by electrons carried by NADPH. One phosphate is removed from each reactant at the same time, yielding two molecules of G3P (= 6 molecules for three turns).

3 C–C–C–C–C–**P**
Ribulose 5-phosphate

Phase 3
Regeneration

Phase 2
Reduction

1,3-Bisphospho-glycerate kinase

6 **P**ᵢ

6 **NADPH**

6 **NADP⁺**

Complex reactions regenerating ribulose 5-phosphate

6 C–C–C–**P**
Glyceraldehyde-3-phosphate (G3P)

5 C–C–C–**P**
G3P

4 From three turns of the cycle, one molecule of G3P is released as a product, and the other five molecules of G3P enter a series of reactions yielding three molecules of ribulose 5-phosphate.

1 C–C–C–**P**
G3P

To reactions synthesizing sugars and other organic compounds

FIGURE 8.12 The Calvin cycle. The figure tracks the carbon atoms in the molecules and summarizes the reactions of three turns of the cycle, starting with three molecules of CO_2. Enzymes are printed in red. Three turns of the cycle results in the synthesis of enough G3P molecules so that one can be released to be used in carbohydrate synthesis. The inset shows a model for the structure of rubisco in chloroplasts of higher plants. The large subunits are gray and white; the small subunits are blue and orange.

Basic Research: Elucidation of the Calvin Cycle Using Two-Dimensional Paper Chromatography

Beginning in 1945, Melvin Calvin, Andrew A. Benson, and their colleagues at the University of California, Berkeley, combined CO_2 labeled with the radioactive carbon isotope ^{14}C (radioactive isotopes are discussed in Section 2.2) with a technique called *two-dimensional paper chromatography* to trace the pathways of the light-independent reactions in a green alga of the genus *Chlorella*. The researchers exposed actively photosynthesizing *Chlorella* cells to the labeled carbon dioxide. Then, at various times, they removed cells and placed them in hot alcohol, which instantly stopped all the photosynthetic reactions of the algae. They extracted radioactive carbohydrates from the cells and used two-dimensional paper chromatography to separate and to identify them chemically **(Figure).**

By analyzing the labeled molecules revealed by the two-dimensional chromatography technique in the extracts prepared from *Chlorella* cells under different conditions, Calvin and his colleagues were able to reconstruct the reactions of the Calvin cycle. For example, in carbohydrate extracts made within a few seconds after the cells were exposed to the labeled CO_2, most of the radioactivity was found in 3PGA, indicating that it is one of the earliest products of photosynthesis. In extracts made after longer periods of exposure to the label, radioactivity showed up in G3P and in more complex substances, including a variety of six-carbon sugars, sucrose, and starch. The researchers also examined the effect of reducing the amount of CO_2 available to the *Chlorella* cells so that photosynthesis worked slowly even in bright light. Under these conditions, RuBP accumulated in the cells, suggesting that it is the first substance to react with CO_2 in the light-independent reactions and that it accumulates if CO_2 is in short supply. Most of the intermediate compounds between CO_2 and six-carbon sugars were identified in similar experiments.

Using this information, Calvin and his colleagues pieced together the light-independent reactions of photosynthesis and showed that they formed a continuous cycle.

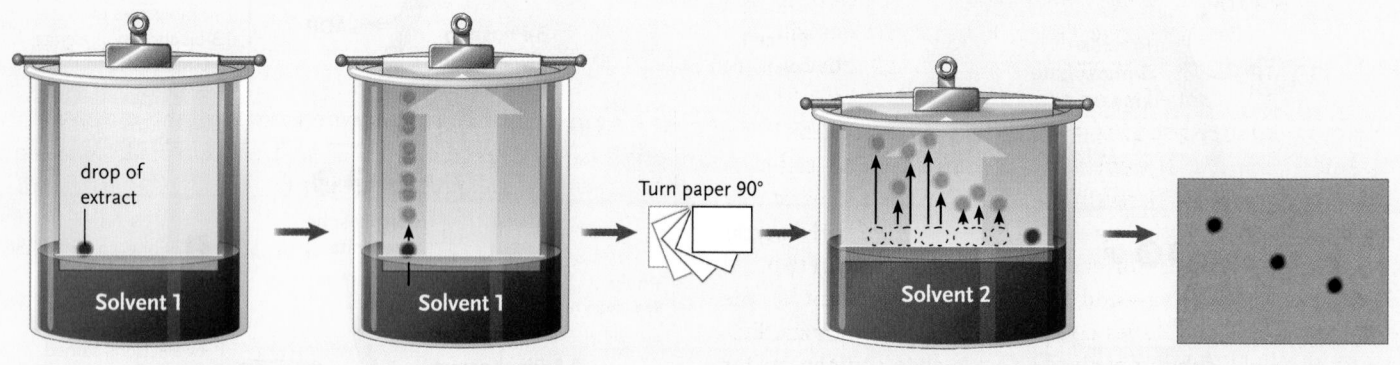

1 A drop of extract containing radioactive carbohydrates is placed at one corner of a piece of chromatography paper. The paper is placed in a jar with its edge touching a solvent. (Calvin used a water solution of butyl alcohol and propionic acid.)

2 The extracted molecules in the spot dissolve and are carried upward through the paper as the solvent rises. The rates of movement of the molecules vary according to their molecule size and solubility. The resulting vertical line of spots is the first dimension of the two-dimensional technique.

3 The paper is dried, turned 90°, and touched to a second solvent (Calvin used a water solution of phenol for this part of the experiment). As this solvent moves through the paper, the molecules again migrate upward from the spots produced by the first dimension, but at rates different from their mobility in the first solvent. This step, the second dimension of the two-dimensional technique, separates molecules that, although different, had produced a single spot in the first solvent because they had migrated at the same rate. The individual spots are identified by comparing their locations with the positions of spots made by known molecules when the "knowns" are run through the same procedure.

4 The paper is dried and covered with a sheet of photographic film. Radioactive molecules expose the film in spots over their locations in the paper. Developing the film reveals the locations of the radioactive spots. The spots on the film are compared with the spots on the paper to identify the molecules that were radioactive.

other monosaccharides are made from G3P by reactions that, in effect, reverse the first half of glycolysis (see Chapter 7). Glucose is a six-carbon sugar and G3P is a three-carbon sugar, which means that *six turns* of the Calvin cycle are needed for the synthesis of each molecule of glucose.

Once produced, the monosaccharides enter biochemical pathways that make disaccharides such as sucrose, polysaccharides such as starches and cellulose, and the other complex carbohydrates found in cell walls. Other pathways manufacture amino acids, fatty acids and lipids, proteins, and nucleic acids. The reactions forming these products occur both within chloroplasts and in the surrounding cytosol and nucleus.

Sucrose, a disaccharide of glucose linked to fructose, is the main form in which the products of photosynthesis circulate from cell to cell in vascular plants. Organic nutrients are stored in most vascular plants as sucrose or starch, or as a combination of the two in proportions that depend on the plant species. Sugarcane and sugar beets, which contain stored sucrose in high concentrations, are the main sources of the sucrose we use as table sugar.

Rubisco Is the Key Enzyme of the World's Food Economy

Rubisco, the enzyme that catalyzes the first reaction of the Calvin cycle, is unique to photosynthetic organisms. By catalyzing CO_2 fixation, it provides the source of organic molecules for most of the world's organisms—the enzyme converts about 100 billion tons of CO_2 into carbohydrates annually. There are so many rubisco molecules in chloroplasts that the enzyme may make up 50% or more of the total protein of plant leaves. As such, it is also the world's most abundant protein, estimated to total some 40 million tons worldwide, equivalent to about 6 kg for every human.

Rubisco has essentially the same overall structure in almost all photosynthetic organisms: eight copies each of a large and a small polypeptide, joined together in a 16-subunit structure (see inset in Figure 8.12). The large subunit contains all of the known active sites where substrates, including CO_2 and RuBP, can bind. (Active sites of enzymes are described in Section 6.4.) Although the small subunit has no active sites, it has a very significant effect on rubisco's rate of catalysis. That is, 99% of the catalytic activity of the enzyme is lost if the small subunit is removed. The effect of the small subunit is critically important when considered in the context of the comparatively slow reaction rate of the normal enzyme. The enzyme's multiple form—eight copies of each subunit, massed together, all doing the same thing—and the very large amount of the enzyme packed into leaves compensate for the slow rate. Evidently, the small subunit evolved as one way to compensate for the enzyme's slow action, by pushing the large subunit to do its job faster. Rubisco is also the key regulatory site of the Calvin cycle. The enzyme is stimulated by both NADPH and ATP; as long as these substances are available from the light-dependent reactions, the enzyme is active and the light-independent reactions proceed. During the daytime, when sunlight powers the light-dependent reactions, the abundant NADPH and ATP supplies keep the Calvin cycle running. In darkness, when NADPH and ATP become unavailable, the enzyme is inhibited and the Calvin cycle slows or stops. Similar controls based on the availability of ATP and NADPH also regulate the enzymes that catalyze other reactions of the Calvin cycle, including reactions 2 and 3 in Figure 8.12.

STUDY BREAK 8.3

1. What is the reaction that rubisco catalyzes? Why is rubisco the key enzyme for producing the world's food, and how is it the key regulatory site of the Calvin cycle?
2. How many molecules of carbon dioxide must enter the Calvin cycle for a plant to produce a sugar containing 12 carbon atoms? How many ATP and NADPH molecules would be required to make that molecule?

8.4 Photorespiration and Alternative Processes of Carbon Fixation

In this section we consider *photorespiration* in plants, a mechanism with some features similar to aerobic respiration that uses O_2 as the first step in a pathway to generate CO_2. We also describe two alternative processes of carbon fixation, the C_4 pathway and the CAM pathway.

Photorespiration Produces Carbon Dioxide That Is Used by the Calvin Cycle

Rubisco, RuBP carboxylase/oxygenase, is so named because it is both a carboxylase and an oxygenase. That is, the enzyme has an active site to which either CO_2 or O_2 can bind. When CO_2 binds, rubisco acts as a carboxylase fixing the carbon of CO_2 by combining it with RuBP to produce 3PGA in phase 1 of the Calvin cycle (see Figure 8.12):

$$RuBP + CO_2 \xrightarrow{\text{Rubisco}} 2 \text{ 3-phosphoglycerate (3PGA)}$$

Overall, the carboxylase reaction of rubisco results in *carbon gain*.

When O_2 binds, rubisco acts as an oxygenase converting RuBP to one molecule of 3PGA and one molecule of a two-carbon compound, phosphoglycolate:

$$RuBP + O_2 \xrightarrow{\text{Rubisco}} 1 \text{ 3PGA} + 1 \text{ Phosphoglycolate}$$

To reactions releasing CO_2

The 3PGA is used in the Calvin cycle within the chloroplast. The phosphoglycolate is hydrolyzed to its nonphosphorylated derivative, glycolate, which exits the chloroplasts and diffuses into peroxisomes (see Section 4.3). Glycolate is a toxic molecule, and it is broken down in the peroxisomes by reactions that release a molecule of CO_2, which can then be used for carbon fixation. In contrast to the carboxylase reaction, no carbon is fixed during the oxygenase reaction and there is no carbon gain. That is, in the oxygenase reaction, CO_2 is released, a *net loss of carbon*. But, in order to grow, all organisms must gain carbon.

The entire process from the oxygenase reaction of rubisco to the release of CO_2 is called **photorespiration.** The term was coined because the process occurs in the presence of light and, like aerobic respiration, it requires oxygen and produces carbon dioxide and water. However, unlike aerobic respiration, photorespiration does not generate ATP.

The extent to which photorespiration occurs in a plant depends on the relative concentrations of CO_2 and O_2 in the leaves. Gas exchange—CO_2 in and O_2 out—between the air and the cells within the leaf occurs through the stomata of the leaves and stem. If gas exchange occurs normally, the relative concentrations of CO_2 and O_2 within the leaf favor CO_2 fixation by the Calvin cycle. That is, even though the CO_2 concentration is lower than that of O_2, rubisco has a ten-times greater affinity for CO_2 than it does for O_2. However, at higher concentrations of O_2 within the leaf, oxygen acts as a competitive inhibitor of the enzyme, and this favors the reaction of RuBP with O_2 rather than with CO_2: photorespiration occurs.

Photorespiration activity occurs more as temperatures rise due to the mechanism used to limit water loss from leaves. That is, the surface of the leaf is covered by a waxy *cuticle* that prevents water loss (see Sections 28.1 and 33.2). The cuticle also prevents the rapid diffusion of gases, such as CO_2, into the leaf. Gas exchange occurs predominantly through the stomata (see earlier in this chapter and Sections 28.1 and 33.2). The plant regulates the size of stomata from fully closed to fully open so as to balance the demands for gas exchange with the need to minimize water loss. In particular, plants that are adapted to hot, dry climates are faced with a constant dilemma: their stomata must be open to let in CO_2 for the Calvin cycle, but their stomata should be closed to conserve water. A similar situation occurs for nonadapted plants when environmental conditions become hot and dry. With the stomata closed, photosynthesis rapidly consumes the CO_2 in the leaf and produces O_2, which accumulates in the chloroplasts and photorespiration occurs.

For a plant with high photorespiration rates at elevated temperatures, as much as 50% of the carbon fixed by the Calvin cycle may be lost because only one three-carbon 3PGA molecule is produced instead of two. Unfortunately, many economically important crop plants become seriously impaired by high photorespiration rates when temperatures rise. Among such plants are the C_3 plants rice, barley, wheat, soybeans, tomatoes, and potatoes.

Alternative Processes of Carbon Fixation Minimize Photorespiration

Many plant species that live in hot, dry environments have evolved alternative processes of carbon fixation that minimize photorespiration and, therefore, its negative effect on photosynthesis. C_4 plants use the C_4 *pathway* to fix CO_2 into a four-carbon molecule, *oxaloacetate,* while CAM plants use the *CAM pathway* at night to fix carbon in the form of oxaloacetate. These special pathways precede the C_3 pathway (Calvin cycle) rather than replacing it. (*Molecular Insights* describes experiments to determine the effects of water deficit on photosynthesis and metabolism in a C_3 plant.)

C_4 PLANTS AND THE C_4 PATHWAY At least 19 families of flowering plants are C_4 plants. They include crabgrass and several crops important agriculturally, including corn and sugarcane. C_4 plants have a characteristic leaf anatomy with photosynthetic mesophyll cells tightly associated with specialized, chloroplast-rich bundle sheath cells, which encircle the veins of the leaf **(Figure 8.13A)**. In a C_3 plant leaf, the bundle sheath

A. C_4 plant leaf cross section

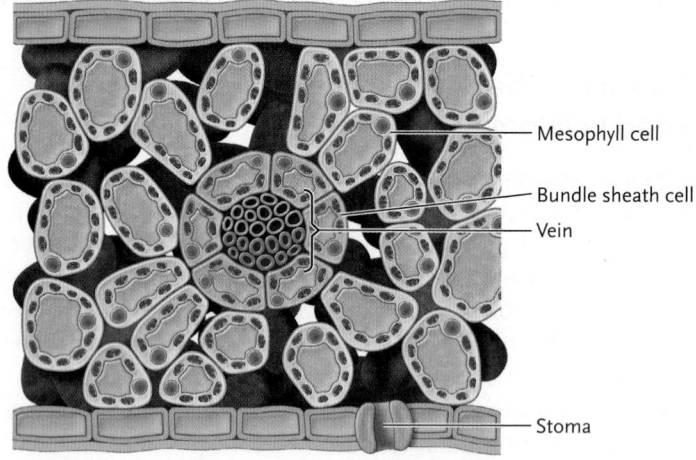

- Mesophyll cell
- Bundle sheath cell
- Vein
- Stoma

B. C_3 plant leaf cross section

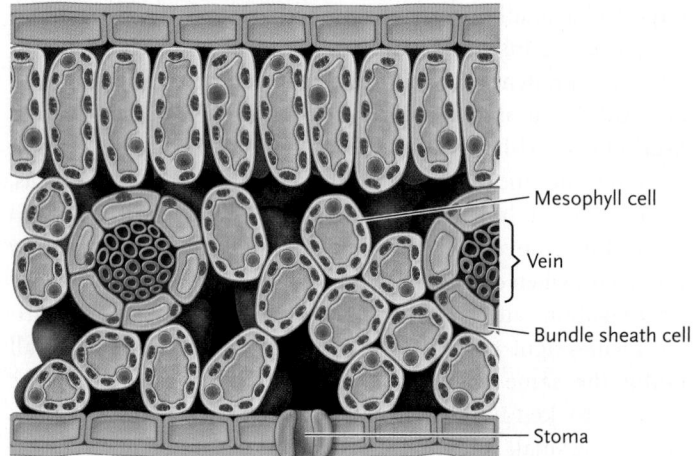

- Mesophyll cell
- Vein
- Bundle sheath cell
- Stoma

FIGURE 8.13 Leaf anatomy in (A) C_4 plants and (B) C_3 plants.
© Cengage Learning 2017

cells have few chloroplasts and low concentrations of rubisco **(Figure 8.13B)**. These cells generate ATP by cyclic electron flow (see Figure 8.11).

C_4 plants use the C_4 *pathway* for carbon fixation. In the **C_4 pathway,** CO_2 that has diffused into the leaf through the stomata initially is fixed in mesophyll cells by combining it with the three-carbon phosphoenolpyruvate (PEP) to produce the four-carbon oxaloacetate **(Figure 8.14A)**. The C_4 pathway gets its name because its first product is a four-carbon molecule rather than a three-carbon molecule, as in the C_3 pathway. The carbon fixation reaction producing oxaloacetate is catalyzed by **PEP carboxylase.** This enzyme has a much greater affinity for CO_2 than rubisco does and, unlike rubisco, it has no oxygenase activity.

As the C_4 pathway continues, the oxaloacetate is next reduced to the four-carbon *malate* by electrons transferred from NADPH. Malate diffuses into bundle sheath cells, the sites of the reactions of the Calvin cycle. (In C_3 plants, the

Growth and Photosynthesis in a C₃ Plant: What is the response to water deficit?

Drought is a significant issue for plants because the stress of water deficit can have deleterious effects on plant growth and physiological processes. With the tools now available to analyze an organism's proteome—the complete set of its proteins—experiments are possible to investigate how plant metabolism changes in response to the stress of water deficit.

Research Question

How does metabolism change in response to water deficit in grapevine, a C₃ plant?

Experiments

Grant Cramer and colleagues at the University of Nevada, Reno, used 2-year-old grapevines (*Vitis vinifera* L.) **(Figure)** grown in pots in a greenhouse for their experiments. Control (non-water-deficit) plants were watered daily with tap water, whereas experimental

yevgeniy11/Shutterstock.com

(water-deficit) plants were not watered. The researchers measured shoot elongation (as an indicator of growth) and photosynthesis over a 16-day period. During that time period, the scientists took shoot tip samples and extracted proteins for proteomic analysis to determine the effects of water deficit on protein abundance.

Results

After day 4 of water deficit, shoot elongation and photosynthesis decreased progressively. Among 2,777 proteins identified by proteomic analysis, 942 proteins were present in all samples. Approximately 50% of those proteins (472) were affected significantly by water deficit. Those 472 proteins fell into two groups with respect to time of response:

1. Early-responding proteins: 159 that increased significantly in abundance on day 4 and remained higher in abundance for the rest of the water-deficit treatment, and 52 proteins that decreased significantly in abundance on day 4 and remained lower in abundance.

2. Late-responding proteins: 58 proteins that increased progressively in abundance, especially on days 12 and 16, and 203 proteins that decreased progressively in abundance on those same sampling days.

Comparing the proteomic analysis results with the growth and photosynthesis results led the researchers to conclude that the experimental plants sensed the water deficit early and appeared to acclimate metabolically to the stress. That is, the abundance of many proteins (the early responders) changed (either increased or decreased) before

observed decreases in shoot elongation and photosynthesis. Among the early-responding proteins, the predominant ones in terms of function included those involved in photosynthesis, glycolysis, translation (the synthesis of proteins from mRNAs), antioxidant defense, and growth. Predominant among the late-responding proteins were those involved in photorespiration, amino acid metabolism, carbohydrate metabolism, and photorespiration.

Discussion

Large changes in the plant's proteome occurred in response to water deficit. The significant early changes in protein abundance took place before any detectable decreases in shoot elongation rate or photosynthesis, indicating an acclimation response to the water deficit. The authors state that the proteomics data indicate "massive and substantial changes in plant metabolism that appear to funnel carbon and energy into antioxidant defenses in the very early stages of plant response to water deficit before any significant injury."

think like a scientist

Excessive salts in the soil can cause inhibition of plant growth or plant death. How would you determine experimentally whether any of the same proteins that change in abundance in response to water deficit also change in abundance in response to stress from a high salt concentration?

Source: G. R. Cramer et al. 2013. Proteomic analysis indicates massive changes in metabolism prior to the inhibition of growth and photosynthesis of grapevine (*Vitis vinifera* L.) in response to water deficit. *BMC Plant Biology* 13:49.

© Cengage Learning 2017

Calvin cycle takes place in the mesophyll cells.) The malate is then oxidized to the three-carbon pyruvate, releasing CO_2, which is used in the rubisco-catalyzed first step of the Calvin cycle. The ability to use malate to donate CO_2 in this way, leads to a higher $CO_2{:}O_2$ ratio in the bundle sheath cells at the time when rubisco is functional in the Calvin cycle, namely in the daylight. The higher $CO_2{:}O_2$ ratio means that photorespiration is extremely low or negligible in C₄ plants. The pyruvate produced diffuses into mesophyll cells where it is converted back into PEP in a reaction that consumes ATP.

In sum, in C₄ plants the C₄ pathway takes place in the mesophyll cells, whereas the Calvin cycle occurs in the bundle sheath cells (see Figure 8.14A). That is, carbon fixation and the Calvin cycle are separated spatially in different cell types.

If the C₄ pathway is so effective at reducing photorespiration, why is the pathway not used by all plants? The answer is that the C₄ pathway has an additional energy requirement. For each turn of the C₄ pathway cycle, one ATP molecule is hydrolyzed to regenerate PEP from pyruvate. This adds an energy requirement of six ATP molecules for each G3P produced by

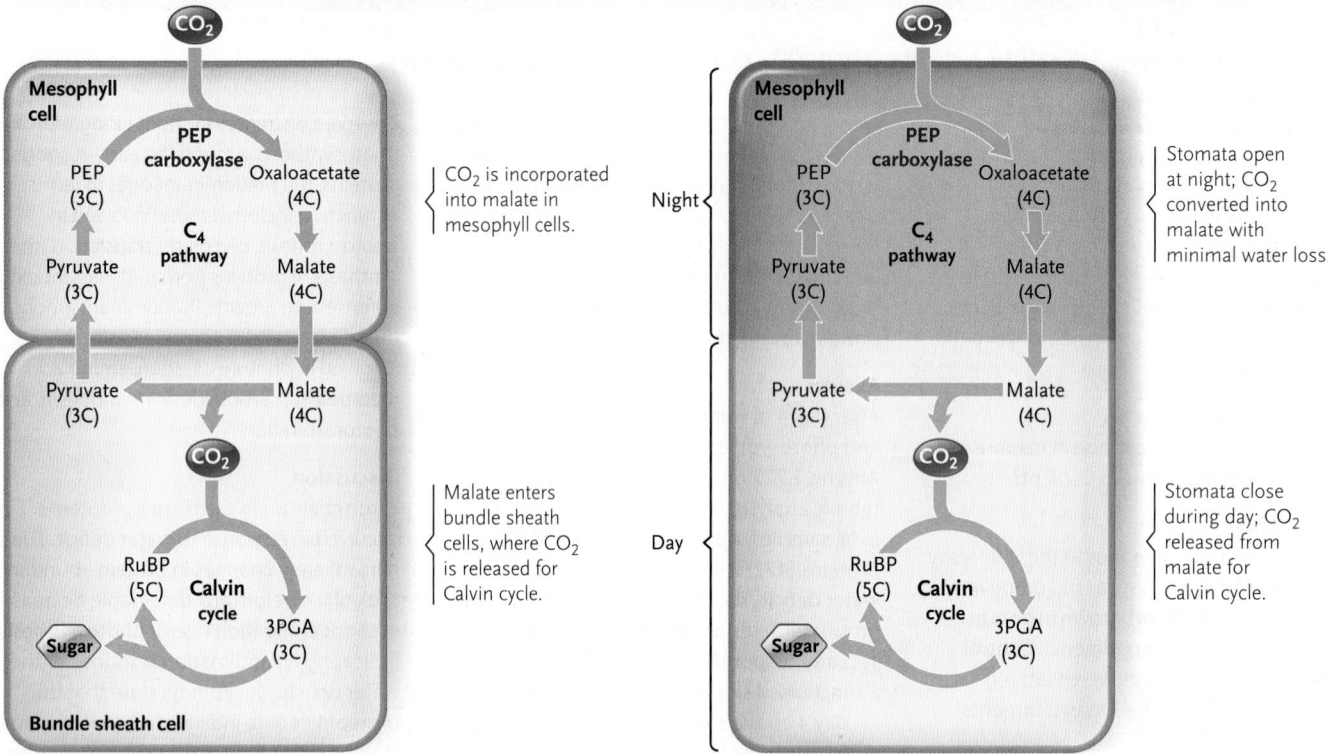

A. C_4 pathway in C_4 plants—spatial separation of steps

Mesophyll cell

PEP carboxylase

PEP (3C) → Oxaloacetate (4C)

C_4 pathway

Pyruvate (3C) ← Malate (4C)

CO_2 is incorporated into malate in mesophyll cells.

Pyruvate (3C) ← Malate (4C)

CO_2

RuBP (5C) Calvin cycle

Sugar → 3PGA (3C)

Malate enters bundle sheath cells, where CO_2 is released for Calvin cycle.

Bundle sheath cell

B. CAM pathway in CAM plants—temporal separation of steps

Mesophyll cell

PEP carboxylase

PEP (3C) → Oxaloacetate (4C)

C_4 pathway

Pyruvate (3C) ← Malate (4C)

Night

Stomata open at night; CO_2 converted into malate with minimal water loss.

Pyruvate (3C) ← Malate (4C)

CO_2

Day

RuBP (5C) Calvin cycle

Sugar → 3PGA (3C)

Stomata close during day; CO_2 released from malate for Calvin cycle.

FIGURE 8.14 Two alternative processes of carbon fixation to minimize photorespiration. In each case, carbon fixation produces the four-carbon oxaloacetate, which is processed to generate CO_2 that feeds into the Calvin (C_3) cycle. **(A)** In C_4 plants, carbon fixation and the Calvin cycle occur in different cell types: carbon fixation by the C_4 pathway takes place in mesophyll cells, while the Calvin cycle takes place in bundle sheath cells. **(B)** In CAM plants, carbon fixation and the Calvin cycle occur at different times in mesophyll cells: carbon fixation by the C_4 pathway takes place at night, while the Calvin cycle takes place during the day.

© Cengage Learning 2017

the Calvin cycle. However, as mentioned earlier, photorespiration potentially can decrease carbon fixation efficiency by as much as 50% in hot environments, so the additional ATP requirement is worthwhile. Moreover, hot environments typically receive a lot of sunshine. As a result, the additional ATP requirement can be met easily by increasing the output of the light-dependent reactions.

C_4 plants also perform better where it is dry. Because PEP carboxylase has a very high affinity only for CO_2, C_4 plants are more efficient at fixing CO_2 than are C_3 plants. Therefore, they do not have to keep their stomata open for as long as a C_3 plant does under the same conditions. Because this reduces water loss, C_4 plants are much better suited to arid conditions.

CAM PLANTS AND THE CAM PATHWAY The acronym CAM stands for **crassulacean acid metabolism.** The name comes from the Crassulaceae family (stonecrops), where the pathway was first described. CAM plants live in very dry environments and have several evolutionary adaptations that enable them to survive the arid conditions. Besides succulents (water-storing

plants) such as the stonecrops, CAM plants include some members of at least 25 plant families, including the cactus family, the lily family, and the orchid family.

Exactly as with the C_4 pathway in C_4 plants, in the **CAM pathway** of CAM plants CO_2 is initially fixed to oxaloacetate in a reaction catalyzed by PEP carboxylase **(Figure 8.14B).** The CO_2 produced by the oxidation of malate is used in the rubisco-catalyzed first step of the Calvin cycle. But, whereas carbon fixation and the Calvin cycle run in different cell types in C_4 plants, in CAM plants carbon fixation and the Calvin cycle *both* occur in mesophyll cells but are separated temporally. That is, they run at different times, initial carbon fixation at night and the Calvin cycle during the day.

CAM plants live in regions that are hot and dry during the day and cool at night. The stomata on their fleshy leaves or stems open only at night to minimize water loss. When they open, O_2 generated by photosynthesis is released, and CO_2 enters. The CO_2 is fixed into malate in the mesophyll cells by the CAM pathway. Malate accumulates throughout the night and is stored in large cell vacuoles. During the day, the stomata close to reduce water loss in the hot conditions, and this also cuts off the exchange of gases with the atmosphere. Malate now moves from the vacuole to the chloroplasts, where its oxidation to pyruvate generates CO_2 that is used by the Calvin cycle.

STUDY BREAK 8.4

1. When does photorespiration occur? What are the reactions of photorespiration, and what are the energetic consequences of the process?
2. What is the C_4 pathway, and how does it enable C_4 plants to circumvent photorespiration?
3. How are carbon fixation and the Calvin cycle different in C_4 plants and CAM plants?

8.5 Photosynthesis and Cellular Respiration Compared

This section addresses the similarities between photosynthesis and cellular respiration and the evolution of the two processes.

Photosynthesis and Cellular Respiration Are Similar Processes with Reactions That Essentially Are the Reverse of Each Other

A popular misconception is that photosynthesis occurs in plants, and cellular respiration occurs in animals only. In fact, both processes occur in plants, with photosynthesis confined to tissues containing chloroplasts and cellular respiration taking place in all cells. **Figure 8.15** presents side-by-side schematics of

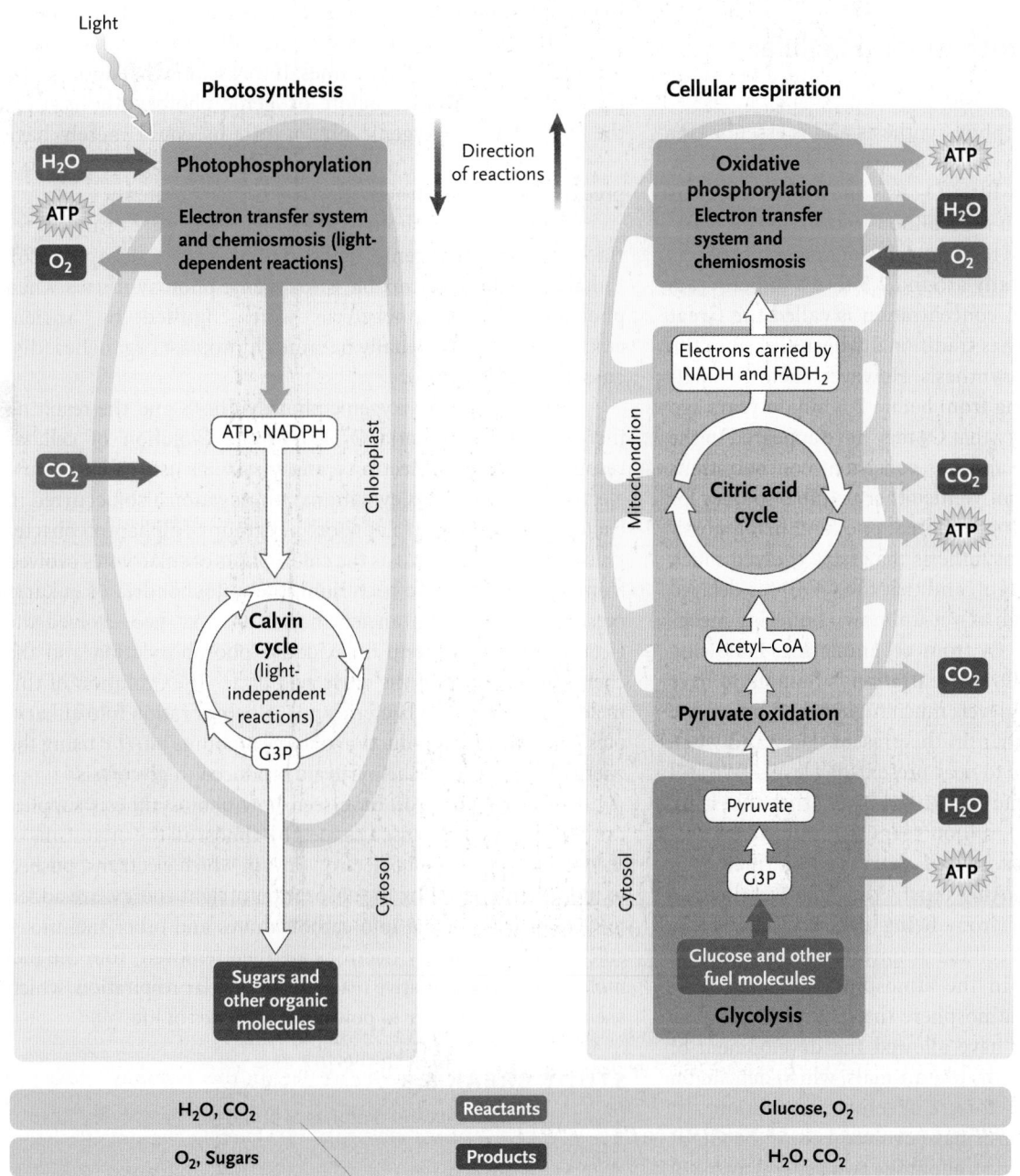

FIGURE 8.15 Schematic diagrams of the processes of photosynthesis (left) and cellular respiration (right). Cellular respiration is shown upside down with respect to the direction of reactions to help illustrate the similarities of the process with photosynthesis.

© Cengage Learning 2017

photosynthesis and cellular respiration to highlight their similarities and points of connection. Note that their overall reactions are essentially the reverse of each other. That is, the reactants of photosynthesis—CO_2 and H_2O—are the products of cellular respiration, and the reactants of cellular respiration—glucose and O_2—are the products of photosynthesis. Both processes have key phosphorylation reactions involving an electron transfer system—photophosphorylation in photosynthesis and oxidative phosphorylation in cellular respiration—followed by the chemiosmotic synthesis of ATP. Further, G3P is found in the pathways of both processes. In photosynthesis, it is a product of the Calvin cycle and is used for the synthesis of sugars and other organic fuel molecules. In cellular respiration, it is an intermediate generated in glycolysis in the conversion of glucose to pyruvate. Thus, G3P is used by anabolic pathways when it is generated by photosynthesis, and it is a product of a catabolic pathway in cellular respiration.

Evolution of Photosynthesis and Cellular Respiration

Oxygen (O_2) is a product of photosynthesis and is essential for cellular respiration. Oxygen makes up 21% of the present-day atmosphere. However, very little O_2 was present in approximately the first half of Earth's 4.5-billion-year existence. Various lines of evidence have led to the conclusion that atmospheric oxygen levels rose dramatically around 2.5 to 2.3 billion years ago. The increase in oxygen concentration is called the **Great Oxygen Event (GOE),** and has traditionally been linked to the evolution of oxygenic photosynthesis. However, chemical analyses of rocks and soils dating from before 2.5 billion years ago have suggested the possibility that O_2 may have appeared in the atmosphere in much lower and less consistent concentrations as early as 3 billion years ago. While other possible sources for this atmospheric O_2 are being explored, one tantalizing possibility is that oxygenic photosynthesis may have evolved much earlier than previously thought, and that the GOE was delayed by hundreds of years through as-yet-unknown buffering mechanisms that prevented free O_2 from accumulating. Following the GOE, the atmospheric O_2 concentration is thought to have remained relatively stable at levels much higher than before the GOE and yet much lower than in the present. How and when did atmospheric O_2 increase to near present-day levels? We still do not know for sure, but one hypothesis is that the most rapid increase happened 600–800 million years ago, most likely as a result of climate shifts that caused more organic matter to become trapped in the Earth's crust. The "fossilization" of organic matter prevents it from being decomposed by the O_2-consuming reactions of cellular respiration, which enables the unused O_2 to remain in the atmosphere. The dramatic accumulation of O_2 in the atmosphere that is thought to have occurred at that time may have allowed the development of complex multicellular life forms like animals, which depend on considerable quantities of O_2 for use in aerobic respiration.

How might photosynthesis have evolved before the GOE? To answer this, we need to think about the two general forms of

photosynthesis: *anoxygenic* (non-oxygen producing) *photosynthesis* and oxygenic photosynthesis. Like oxygenic photosynthesis, the most common forms of present-day anoxygenic photosynthesis use the excitation of electrons in pigment molecules to fuel electron transfer systems, but the electron donors are molecules other than H_2O, such as H_2S, and so molecular O_2 is not released. Electron transfer systems and ATP synthase enzymes driven by proton gradients are universal among living things, and are therefore thought to have provided energy for the earliest organisms. The evolution of the earliest form of photosynthesis may therefore have involved linking a photosynthetic reaction center (where the absorption of photons by a photosynthetic pigment excites electrons) to a preexisting electron transfer system.

Genomic and other molecular analysis of components of photosynthesis reactions in modern-day organisms suggest that all photosynthetic reaction centers have a single evolutionary origin, whereas antenna systems and carbon fixation pathways have evolved multiple times. If we assume that anoxygenic photosynthesis evolved before oxygenic photosynthesis, then the evolution of oxygenic photosynthesis could merely have involved addition of the oxygenic water-splitting complex to a preexisting photosynthetic electron transfer system.

Of present-day photosynthetic bacteria, only cyanobacteria carry out oxygenic photosynthesis (see Chapter 26). Eukaryotes became capable of oxygenic photosynthesis when photosynthetic cyanobacteria were engulfed by ancient eukaryotes, and eventually became chloroplast organelles (discussed in Chapter 25).

The evolution of oxygenic photosynthesis and the resulting increase in atmospheric O_2 led to the evolution of cellular respiration—that is, electron transfer systems using O_2 as a final electron acceptor. This evolutionary innovation also occurred in bacteria, most likely in a now-extinct group of alphaproteobacteria (see Chapter 26). Just as the chloroplasts of eukaryotes evolved from a photosynthetic bacterium, the mitochondria of eukaryotes evolved from an ancient bacterium that had evolved the electron transfer system of oxidative phosphorylation and the chemical reactions of the citric acid cycle. The evolution of this metabolic pathway in bacteria and its incorporation into eukaryotes permitted the production of large amounts of ATP using the energy released from the pyruvate produced in glycolysis.

In this chapter, you have seen how photosynthesis supplies the organic molecules used as fuels by almost all the organisms of the world. It is a story of electron flow in which electrons, pushed to high energy levels by the absorption of light energy, are added to CO_2, which is fixed into carbohydrates and other fuel molecules. The high-energy electrons are then removed from the fuel molecules by the oxidative reactions of cellular respiration, which use the released energy to power the activities of life.

STUDY BREAK 8.5

How are the reactants and products of photosynthesis and cellular respiration related?

Why is photosynthesis so inefficient, and what (if anything) can we do about it?

Photosynthesis is considered by many to be the most important biological process on Earth. Photosynthesis involves the highest energy processes of life; it is the process where (by far) most of the energy in our ecosystem is captured. All other biological processes are exergonic (they lose the energy captured by photosynthesis) and thus all other processes involve less energy than photosynthesis. It is also the process where, by far, the most energy in our ecosystem is *lost*. In fact, photosynthetic organisms dissipate (dump purposely) a large part of absorbed light energy to prevent the buildup of reactive oxygen species (intermediates of photosynthesis) that can damage the plant. At full sunlight, regulatory dissipation can involve more than 75% of absorbed light energy. Consequently, typical agricultural crops store only about 1% of their absorbed solar energy in the form of biomass. Interestingly, some plants and many green algae can store up to 10 times as much of this energy, at least under ideal conditions. Research is being done to understand why they are so efficient, and whether crop or biofuel plants can be engineered to produce more biomass.

How and why is the efficiency of photosynthesis regulated?

My laboratory at Michigan State University is interested in how plants set their photosynthetic "strategies" and thus how they control light capture and electron and proton transfer reactions to balance their competing needs for efficient photosynthesis while avoiding toxic side reactions. As you have learned, energy conversion by the chloroplast involves the capture of light energy and the channeling of that energy through an electron transfer system with the eventual synthesis of NADPH and ATP. At high concentrations, many of the intermediates produced in this energy conversion can interact with oxygen to produce reactive oxygen species (ROS) that can destroy the photosynthetic apparatus and the organism.

To prevent such damage, the efficiency of some of the photosystem components is decreased in a series of photoprotective mechanisms. Evidence from a range of studies indicates that the balance between protection against photoinhibition and photosynthetic efficiency is important in enabling plants to acclimate to environmental changes, but it can also limit their productivity.

In one area of our work, we found that the chloroplast ATP synthase acts as a major regulator of photosynthesis, sensing the metabolic status of the chloroplast, and in response, regulating the thylakoid proton-motive force by restricting proton flow out of the lumen. In turn, the proton-motive force controls the photoprotective dissipation of light energy and electron transfer. Among the major unanswered questions are the following: What controls the ATP synthase, what specifically does it sense, and how does this mechanism differ in high- and low-productivity plants? Can we adjust the regulation of the proton-motive force to improve the efficiency and robustness of photosynthesis? Answering these questions will illuminate how photosynthesis determines plant growth and survival. In addition, the technology developed as part of the research may lead to applications in plant breeding, particularly for improving energy storage by biofuels and food crops.

think like a scientist Find some specific examples of plants with higher- and lower-efficiency photosynthesis. Why do you think some plants have higher photosynthetic rates than others? What are some potential selective advantages of lower rates of photosynthesis?

David Kramer is the Hannah Distinguished Professor of Bioenergetics and Photosynthesis in the Biochemistry and Molecular Biology Department and the MSU–Department of Energy Plant Research Lab at Michigan State University. His research seeks to understand how plants convert light energy into forms usable for life, how these processes function at both molecular and physiological levels, how they are regulated and controlled, how they define the energy budget of plants and the ecosystem, and how they have adapted through evolution to support life in extreme environments. This work has led his research team to develop a series of novel spectroscopic tools for probing photosynthetic reactions *in vivo*. Some of these tools are available to support research projects initiated by scientists and citizens around the world, including the readers of this textbook. Learn more about his research at http://www.prl.msu.edu and http://photosynq.org.

REVIEW KEY CONCEPTS

For access to MindTap and additional study materials visit www.cengagebrain.com.

8.1 Photosynthesis: An Overview

- In photosynthesis, plants, algae, and photosynthetic prokaryotes use the energy of sunlight to drive synthesis of organic molecules from simple inorganic raw materials. The organic molecules are used by the photosynthesizers themselves as fuels; they also form the primary energy source for heterotrophs.

- The two overall stages of photosynthesis are the light-dependent and light-independent reactions. In eukaryotes, both stages take place inside chloroplasts (Figures 8.1 and 8.2).

- Photosynthesizers use light energy to push electrons to elevated energy levels. In eukaryotes and many prokaryotes, water is split as the source of the electrons for this process, and oxygen is released to the environment as a by-product.

- The high-energy electrons provide an indirect energy source for ATP synthesis and also for CO_2 fixation, in which CO_2 is fixed into organic substances by the addition of both electrons and protons.

8.2 The Light-Dependent Reactions of Photosynthesis

- In the light-dependent reactions of photosynthesis, light energy is converted to chemical energy when electrons, excited by absorption of light in a pigment molecule, are passed from the pigment to a stable orbital in a primary acceptor molecule (Figure 8.5).

- Chlorophylls are the major photosynthetic pigments in plants, green algae, and cyanobacteria. Carotenoids are accessory pigments that absorb light energy at a different wavelength than those absorbed by chlorophylls. Carotenoids may expand the range of wavelengths used for photosynthesis, but their more important role is in photoprotection (Figures 8.6 and 8.7).

- In organisms that split water as their electron source, the pigments are organized with proteins into two photosystems. Special subsets of chlorophyll *a* pass excited electrons to primary acceptor molecules in the photosystems (Figure 8.8).

- Electrons obtained from splitting water are used for the synthesis of NADPH and ATP. In the linear electron flow pathway, electrons first flow through photosystem II, becoming excited there to a higher energy level, and then pass through an electron transfer system to photosystem I releasing energy that is used to create an H^+ gradient across the membrane. The gradient is used by ATP synthase to drive synthesis of ATP. The net products of the light-dependent reactions are ATP, NADPH, and oxygen (Figures 8.9 and 8.10).

- Electrons can also flow cyclically around photosystem I and the electron transfer system, building the H^+ concentration and allowing extra ATP to be produced, but no NADPH (Figure 8.11).

8.3 The Light-Independent Reactions of Photosynthesis

- In the light-independent reactions of photosynthesis, CO_2 is reduced and converted into organic substances by the addition of electrons and hydrogen carried by the NADPH produced in the light-dependent reactions. ATP, also derived from the light-dependent reactions, provides additional energy. The key enzyme of the light-independent reactions is rubisco (RuBP carboxylase/oxygenase), which catalyzes the reaction that combines CO_2 into organic compounds (Figure 8.12).

- In the process, NADPH is oxidized to $NADP^+$, and ATP is hydrolyzed to ADP and phosphate. These products of the light-independent reactions cycle back as inputs to the light-dependent reactions.

- The Calvin cycle produces surplus molecules of G3P, which are the starting point for synthesis of glucose, sucrose, starches, and other organic molecules. The light-independent reactions take place in the chloroplast stroma in eukaryotes and in the cytoplasm of photosynthetic prokaryotes.

8.4 Photorespiration and Alternative Processes of Carbon Fixation

- When oxygen concentrations are high relative to CO_2 concentrations, rubisco acts as an oxygenase, catalyzing the combination of RuBP with O_2 rather than CO_2 and forming toxic products that cannot be used in photosynthesis. The toxic products are eliminated by reactions that release carbon as CO_2, greatly reducing the efficiency of photosynthesis. The entire process is called photorespiration because it uses oxygen and releases CO_2.

- Some plants use alternative processes of carbon fixation that minimize photorespiration. C_4 plants use the C_4 pathway to first fix CO_2 into the four-carbon oxaloacetate in mesophyll cells (the site of the Calvin cycle in C_3 plants), and then produce CO_2 for the Calvin cycle in bundle sheath cells. CAM plants use the CAM pathway also first to fix CO_2 into oxaloacetate and then to generate CO_2 for the Calvin cycle. Here both the carbon fixation and the Calvin cycle occur in mesophyll cells, but they are separated by time; initial carbon fixation occurs at night and the Calvin cycle occurs during the day (Figures 8.13 and 8.14).

8.5 Photosynthesis and Cellular Respiration Compared

- Photosynthesis occurs in the cells of plants that contain chloroplasts, whereas cellular respiration occurs in all cells. The overall reactions of the two processes are essentially the reverse of each other, with the reactants of one being the products of the other (Figure 8.15).

TEST YOUR KNOWLEDGE

Remember/Understand

1. An organism exists for long periods by using only CO_2 and H_2O. It could be classified as a(n):
   a. herbivore.
   b. carnivore.
   c. decomposer.
   d. autotroph.
   e. heterotroph.

2. During the light-dependent reactions:
   a. CO_2 is fixed.
   b. NADPH and ATP are synthesized using electrons derived from splitting water.
   c. glucose is synthesized.
   d. water is split and the electrons generated are used for glucose synthesis.
   e. photosystem I is unlinked from photosystem II.

3. Which of the following is a correct step in the light-dependent reactions of the Z system?
   a. Light is absorbed at P700, and electrons flow through a pathway to $NADP^+$, the final acceptor of the linear pathway.
   b. Electrons flow from photosystem II to water.
   c. $NADP^+$ is oxidized to NADPH as it accepts electrons.
   d. Water is degraded to activate P680.
   e. Electrons pass through a thylakoid membrane to create energy to pump H^+ through the cytochrome complex.

4. The light-dependent reactions of photosynthesis resemble aerobic respiration in that both:
   a. synthesize NADPH.
   b. synthesize NADH.
   c. require electron transfer systems to synthesize ATP.
   d. require oxygen as the final electron acceptor.
   e. have the same initial energy source.

5. The molecules that link the light-dependent and light-independent reactions are:
   a. ADP and H_2O.
   b. RuBP and CO_2.
   c. cytochromes and water.
   d. G3P and RuBP.
   e. ATP and NADPH.

6. Which of the following statements about the C_4 cycle is *incorrect*?
   a. CO_2 initially combines with PEP.
   b. PEP carboxylase catalyzes a reaction to produce oxaloacetate.
   c. Oxaloacetate transfers electrons from NADPH and is reduced to malate.
   d. Less ATP is used to run the C_4 cycle than the C_3 cycle.
   e. The cycle runs when O_2 concentration is high.

7. In one turn of the Calvin cycle, one molecule of CO_2 generates:
   a. 6 ATP.
   b. 6 NADH.
   c. 6 ATP and 6 NADPH.
   d. one (CH_2O) unit of carbohydrate.
   e. one molecule of glucose.

8. The oxygen released by photosynthesis comes from:
   a. CO_2.
   b. H_2O.
   c. light.
   d. NADPH.
   e. electrons.

Apply/Analyze

9. You bite into a spinach leaf. Which one of the following is true?
   a. You are getting 50% of the protein in the leaf in the form of ribulose 1,5-bisphosphate carboxylase.
   b. The major pigment you are ingesting is a carotenoid.
   c. The water in the leaf is a product of the light-independent reactions.
   d. Any energy from the leaf you can use directly is in the form of ATP.
   e. The spinach most likely was grown in an area with a low CO_2 concentration.

10. Animal metabolism and plant metabolism are related in that:
    a. plants carry out photosynthesis and animals carry out respiration.
    b. G3P is found in the metabolic pathways of both animals and plants.
    c. G3P is used by catabolic pathways when it is generated by photosynthesis, and it is a product of an anabolic pathway in cellular respiration.
    d. light drives electron excitation.
    e. the reactants of photosynthesis drive cellular respiration in animals.

11. **Discuss Concepts** Suppose a garden in your neighborhood is filled with red, white, and blue petunias. Explain the floral colors in terms of which wavelengths of light are absorbed and reflected by the petals.

12. **Discuss Concepts** About 200 years ago, Jan Baptista van Helmont tried to determine the source of raw materials for plant growth. To do so, he planted a young tree weighing 5 pounds in a barrel filled with 200 pounds of soil. He watered the tree regularly. After 5 years, he again weighed the tree and the soil. At that time the tree weighed 169 pounds, 3 ounces, and the soil weighed 199 pounds, 14 ounces. Because the tree's weight had increased so much, and the soil's weight had remained about the same, he concluded that the tree gained weight as a result of the water he had added to the barrel. Analyze his conclusion in terms of the information you have learned from this chapter.

13. **Discuss Concepts** Like other accessory pigments, the carotenoids extend the range of wavelengths absorbed in photosynthesis. They also protect plants from a potentially lethal process known as *photooxidation*. This process begins when excitation energy in chlorophylls drives the conversion of oxygen into free radicals, substances that can damage organic compounds and kill cells. When plants that cannot produce carotenoids are grown in light, they bleach white and die. Given this observation, what molecules in the plants are likely to be destroyed by photooxidation?

Evaluate/Create

14. **Discuss Concepts** What molecules would you have to provide a plant, theoretically speaking, for it to make glucose in the dark?

15. **Design an Experiment** Space travelers of the future land on a planet in a distant galaxy, where they find populations of a carbon-based life form. The beings on this planet are of a vibrantly purple color. The travelers suspect that the beings secure the energy necessary for survival by a process similar to photosynthesis on Earth. How might they go about testing this conclusion?

16. **Apply Evolutionary Thinking** If global warming raises the temperature of our climate significantly, will C_3 plants or C_4 plants be favored by natural selection? How will global warming change the geographical distributions of plants?

For selected answers, see Appendix A.

INTERPRET THE DATA

Photosynthesis directly opposes respiration in determining how plants influence atmospheric CO_2 concentrations. When a leaf is in the light, both photosynthesis and respiration are occurring simultaneously. The data in the **Table** were collected from the leaf of a sagebrush plant that was enclosed in a chamber that measures the rate of CO_2 exchange. The same leaf was used to collect the data in *Interpret the Data* in Chapter 7. Respiration is shown as a negative and photosynthesis as a positive rate of CO_2 exchange. The net photosynthesis rate is the amount of CO_2 (in micromoles per square meter per second) assimilated by the leaf while respiration is occurring; a positive value indicates more photosynthesis is occurring than respiration. The light exposed to the leaf is quantified as the number of photons in the 400 to 700 nm wavelength, the photosynthetic photon flux density (PPFD); 2,000 $\mu mol/m^2/s$ is equivalent to the amount of light occurring at midday in full Sun.

Observation	Photosynthetic Photon Flux Density (PPFD) ($\mu mol/m^2/s$)	Net Photosynthesis ($\mu mol/m^2/s$)
1	2,000	9.1
2	1,500	8.4
3	1,250	8.2
4	1,000	7.4
5	750	6.3
6	500	4.8
7	250	2.2
8	0	−2.0

1. Why is net photosynthesis negative when PPFD is zero? Looking at the respiration data from *Interpret the Data* in Chapter 7, at what temperature do you think these data were collected?

2. Make a graph of the data, with PPFD on the *x* axis and net photosynthesis rate on the *y* axis.

3. Is the relationship between PPFD and net photosynthesis best described as linear or curved? Why do you think the data have this type of relationship?

4. How might the predicted increase in temperature due to elevated CO_2 concentrations in the atmosphere affect net photosynthesis? Hint: Combine the data from this question with the data from the table in *Interpret the Data* in Chapter 7. Do you think the data presented in both of these questions are sufficient to explain this impact?

Source: Data based on unpublished research by Brent Ewers, University of Wyoming.

9

Cell Communication

Russell Kightley/Science Source

A B cell and a T cell communicating by direct contact in the human immune system (computer image). Cell communication coordinates the cellular defense against disease.

Why it matters . . . Hundreds of aircraft approach and leave airports traveling at various speeds, altitudes, and directions. These aircraft are kept separate and routed to and from their airports safely and efficiently by a highly organized system of controllers, signals, and receivers. As the aircraft arrive and depart, they follow directions issued by air traffic controllers. Although thousands of different messages are traveling through the airspace, each pilot has a radio receiver tuned to a frequency specific for only that aircraft. The flow of directing signals, followed individually by each aircraft in the vicinity, keeps the traffic unscrambled and moving safely.

An equivalent system of signals and tuned receivers evolved hundreds of millions of years ago, as one of the developments that made multicellular life possible. Within a multicellular organism, many of the activities of individual cells are directed by *signaling molecules* such as hormones that are released by certain controlling cells. Although the *controlling cells* release many signals, each receiving cell—the *target cell*—has *receptors* that are "tuned" to recognize only one or a few of the many signaling molecules that circulate in its vicinity. Other signaling molecules pass by without effect because the cell has no receptors for them.

When a target cell binds a signaling molecule, it modifies its internal activities in accordance with the signal, coordinating its functions with the activities of other cells of the organism. The responses of the target cell may include changes in gene activity, protein synthesis, transport of molecules across the plasma membrane, metabolic reactions, secretion, movement, division, or even "suicide"—that is, the programmed death of the receiving cell. As part of its response, a cell may itself become a controller by releasing signaling molecules that modify the activity of other cell types.

The effect of a signaling molecule on a target cell occurs by that molecule triggering a cellular pathway that results in the cell's response. The series of steps from signaling molecule to response is a **signaling pathway.** The total network of signaling pathways allows multicellular organisms to grow, develop, reproduce, and compensate for environmental changes in an internally coordinated fashion. Maintaining the internal environment within a narrow tolerable range is *homeostasis* (see Chapter 38). The system of communication between cells through signaling pathways is called **cell signaling.** Research in cell signaling is a highly important field of biology, motivated by the desire to understand the growth, development, and function of organisms.

This chapter describes the major pathways that form parts of the cell communication system based on both surface and internal receptors, including the links that tie the different response pathways into fully integrated networks.

9.1 Cell Communication: An Overview

Cell communication is essential to orchestrate the activities of cells in multicellular organisms, and also takes place among single-celled organisms. In this chapter we focus on the principles of cell communication in animals, and in later chapters you will see how similar principles apply to plants, fungi, and even bacteria and archaea.

Cell Communication in Animals May Involve Nearby or Distant Cells

Communication is critical for the function and survival of cells that compose a multicellular animal. For example, the ability of cells to communicate with one another in a regulated way is responsible for the controlled growth and development of an animal, as well as the integrated activities of its tissues and organs.

Cells communicate with one another in three general ways:

1. **By direct contact.** In communication by direct contact, adjacent cells have direct channels linking their cytoplasms. In this rapid means of communication, small molecules and ions exchange directly between the two cytoplasms without having to pass through the plasma membrane. In animal cells, the direct channels of communication are *gap junctions,* the specialized connections between the cytoplasms of adjacent cells (see Section 4.5 and Figure 4.27) **(Figure 9.1A[1]).** The main role of gap junctions is to synchronize metabolic activities or electrical signals between cells in a tissue. For example, gap junctions play a key role in spreading electrical signals from one cell to the next in cardiac (heart) muscle, and between certain neurons (nerve cells) in a nervous system. Regarding the latter, the communicating connection between neurons is called a **synapse (Figure 9.1A[2]),** and the communication across a synapse is called **synaptic signaling.** One type of synaptic signaling occurs across the **electrical synapse,** a synapse in which the plasma membranes of two connecting

neurons are in contact and communication across such a synapse occurs by the direct flow of the electrical signal. When an electrical signal in one neuron arrives at the synapse, gap junctions allow ions to flow directly between the two cells, leading to unbroken transmission of the electrical signal to the next neuron (described more in Chapter 39).

In plant cells, the direct channels of communication are plasmodesmata (see Section 4.4 and Chapter 34) **(Figure 9.1A[3]).** Small molecules moving between adjacent cells in plants include certain plant hormones that regulate growth (see Chapter 37). In this way, plant hormones are distributed to other cells.

Cells can also communicate directly through **cell–cell recognition.** In this process, animal cells with particular membrane-bound cell surface molecules dock with one another, initiating communication between the cells **(Figure 9.1A[4]).** For example, cell–cell recognition of this kind activates particular cells in a mammal's immune system in order to mount an immune response (see the chapter-opening figure and Figures 45.6 and 45.11).

2. **By local signaling.** In local signaling, a controlling (secreting) cell releases a signaling molecule that diffuses through the **extracellular fluid** (ECF: the aqueous fluid surrounding and between the cells; see Chapter 38) and causes a response in nearby target cells. Because the effect of cell signaling is local, the signaling molecule is called a *local regulator* and the process is called *paracrine regulation* **(Figure 9.1B[1],** and see Chapter 42).

A second type of *synaptic signaling* is a specialized type of paracrine regulation in which only one adjacent target cell rather than several is affected **(Figure 9.1B[2]).** Many neurons in animal nervous systems connect via **chemical synapses,** a narrow gap about 25 nm wide separating the plasma membranes of the two cells. An electrical signal arriving at the plasma membrane of one cell triggers the release of *neurotransmitters,* molecules that diffuse across the chemical synapse and bind to receptors in the plasma membrane of the connecting neuron, which then generates a new electrical signal. That electrical signal moves along the neuron to the next chemical synapse and the process repeats. Chemical synapses are also found between neurons and some other cells of the body, such as muscle cells. Chemical synapses and neurotransmitters are discussed in more detail in Chapter 39.

In some cases the local regulator acts on the same cell that produces it, a process called *autocrine regulation* **(Figure 9.1C)** (see Chapter 42). For example, many of the growth factors that regulate cell division are local regulators that act in both a paracrine and autocrine fashion.

3. **By long-distance signaling.** In this form of communication, a controlling cell secretes a long-distance signaling molecule called a **hormone** (*hormaein* = to excite), which produces a response in target cells that may be far

A. Cell communication by direct contact

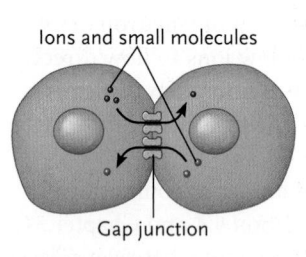

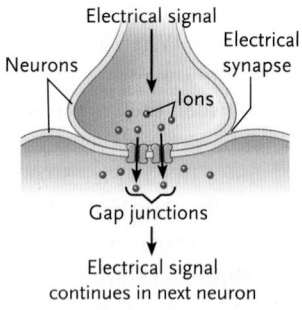

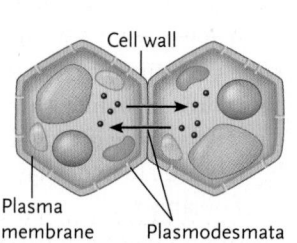

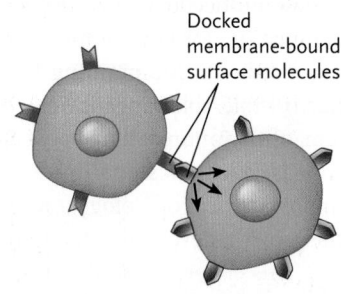

1. Communication between animal cells through gap junctions: Ions and small molecules move between adjacent cells through the gap junctions without going through the plasma membrane.

2. Communication through gap junctions at an electrical synapse: In this type of synaptic signaling, ion flow through the junctions allows transmission of an electrical signal through the synapse.

3. Communication between plant cells through plasmodesmata.

4. Communication through cell–cell recognition: Membrane-bound surface molecules dock, initiating communication between the cells.

B. Cell communication by local signaling: Paracrine regulation

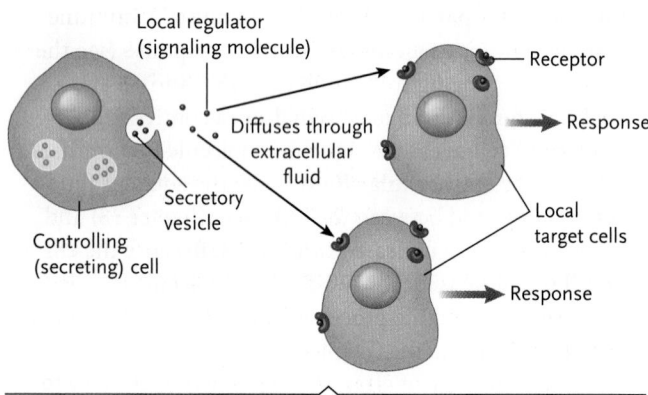

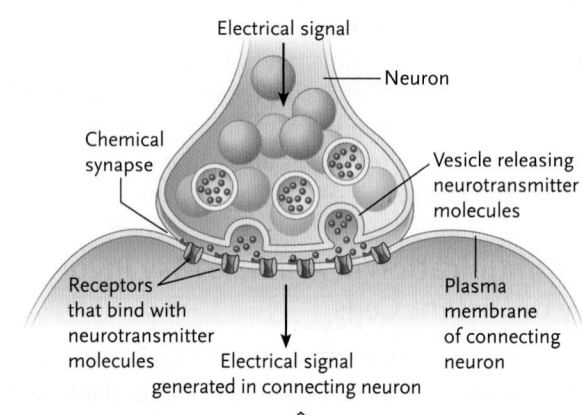

1. Signaling molecule released from a controlling (secreting) cell causes a response in nearby target cells.

2. Synaptic signaling at a chemical synapse: An electrical signal triggers the release of neurotransmitters, which diffuse across the synapse and bind to receptors on the adjacent neuron, triggering a new electrical signal in that cell.

C. Cell communication by local signaling: Autocrine regulation, in which a local regulator acts on the same cell that produces it

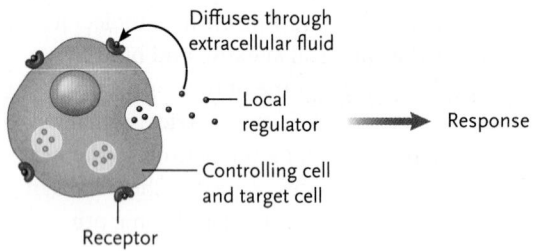

D. Cell communication by long-distance signaling: A secreted hormone triggers a response in target cells that are far away from the controlling cell

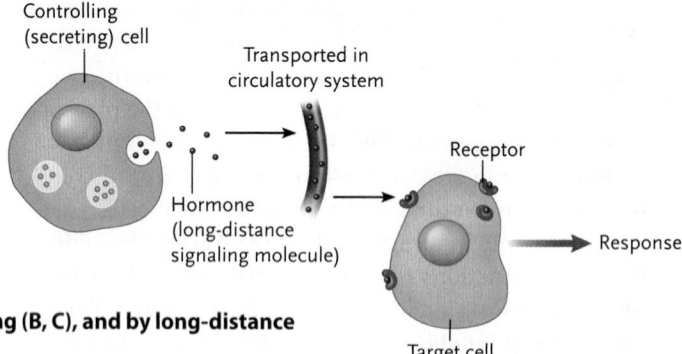

FIGURE 9.1 Cell communication by direct contact (A), by local signaling (B, C), and by long-distance signaling (D).

© Cengage Learning 2017

from the controlling cell **(Figure 9.1D).** This method is the most common means of cell communication. Hormones are found in both animals (see Chapter 42) and plants (see Chapter 37). In animals, hormones secreted by

controlling cells enter the circulatory system where they travel to target cells elsewhere in the body. For example, in response to stress, cells of a mammal's adrenal glands (located on top of the kidneys)—the controlling cells—secrete the hormone epinephrine (also known as *adrenaline*) into the bloodstream. Epinephrine acts on target cells to increase the amount of glucose in the blood. In plants, most hormones travel to target cells by moving through cells rather than by moving through vessels. Some plant hormones are gases that diffuse through the air to the target tissues.

Cell communication by long-distance signaling is the focus of this chapter, and we will use the epinephrine example to illustrate the principles involved. In the 1950s, Earl Sutherland and his research team at Case Western Reserve University, Cleveland, Ohio, wanted to understand how the hormone epinephrine activates the enzyme *glycogen phosphorylase*. In the liver, this enzyme catalyzes the breakdown of glycogen—a polymer of glucose molecules—into glucose molecules, which are then released into the bloodstream. The overall effect of this response to epinephrine secretion is to supply energy to the major muscles responsible for locomotion—the body is now ready for physical activity or to handle stress.

Sutherland's key experiments are shown in **Figure 9.2.** He demonstrated that enzyme activation did not involve epinephrine directly but required an unknown (at the time) cellular factor. Sutherland called the hormone the *first messenger* in the system and the unknown cellular factor the *second messenger*. He proposed the following chain of reactions: epinephrine (the first messenger) stimulates the membrane fraction of the cell to produce a second messenger molecule, which activates the glycogen phosphorylase for conversion of glycogen to glucose. Later in the chapter, we return to this topic and describe the natures and functions of second messenger molecules. Sutherland was awarded a Nobel Prize in 1971 "for his discoveries concerning the mechanisms of the action of hormones."

Sutherland's discovery was critical to understanding the mechanism of action of epinephrine and, in fact, of many other hormones. His work also illustrates how this type of long-distance cell signaling operates: a controlling cell releases a signaling molecule that causes a response (affects the function) in target cells. Target cells process the signal in three sequential steps **(Figure 9.3):**

1. **Reception.** A signaling molecule is an example of a **ligand** (Latin *ligandum* = binding), a molecule that binds to another molecule such as a protein. Reception is the binding of a signaling molecule with a specific receptor of target cells (Figure 9.3, step 1). Target cells have receptors that are specific for the signaling molecule, which distinguishes them from cells that do not respond to the signaling molecule. The signaling molecules themselves may be polar (charged, hydrophilic) molecules or nonpolar (hydrophobic) molecules, and their receptors have sites that are shaped three-dimensionally to recognize and bind them specifically. Receptors for polar signaling molecules are embedded in the plasma membrane with a binding site for the signaling molecule on the cell surface (as in Figure 9.3). Epinephrine, the first

messenger in Sutherland's research, is a polar hormone that is recognized by a surface receptor embedded in the plasma membrane of target cells. Receptors for nonpolar molecules are located within the cell (described later in the chapter; see Figure 9.15). In this case, the nonpolar signaling molecule passes freely through the plasma membrane and interacts with its receptor within the cell. Steroid hormones such as testosterone and estrogen are examples of nonpolar signaling molecules.

2. **Transduction.** Transduction is the process of changing the signal into the form necessary to cause the cellular response (Figure 9.3, step 2). As is characteristic of a ligand binding to a protein, binding of the initial signaling molecule to the receptor protein causes the protein to change shape. The change in shape leads to activation of the receptor protein, which then initiates transduction. Transduction typically involves a cascade of reactions that include several different molecules, referred to as a **signal transduction pathway.** For example, with respect to Sutherland's work, after epinephrine binds to its surface receptor, the signal is transmitted through the plasma membrane into the cell to another protein, which, in turn, causes the production of numerous small second messenger molecules. As we shall see later, both proteins and second messengers can be part of the signaling cascade that results in triggering a cellular response.

3. **Response.** Last, the transduced signal causes a specific cellular response (Figure 9.3, step 3). That response depends on the signal and the receptors of the target cell. In Sutherland's work, the response was the activation of the enzyme glycogen phosphorylase. The active enzyme catalyzes the conversion of stored glycogen to glucose, which is the response to the signal delivered by epinephrine.

The sequence of reception, transduction, and response is common to all the signaling pathways we will encounter in this chapter, although they vary greatly in detail. Importantly, all signaling pathways have "off switches" that serve to stop the cellular response when no such response is needed (see later in the chapter).

Cell Communication Is Evolutionarily Ancient

A number of cell communication properties are ancient, evolutionarily speaking. That is, mechanisms for one cell to signal to another cell and elicit a response most likely existed in unicellular organisms before the evolution of multicellularity. For instance, research with present-day bacteria has shown that a number of species alter their patterns of gene expression in response to changes in population density. In this process of *quorum sensing*, bacteria release signaling molecules in increasing concentration as cell density increases. The molecules are sensed by the cells in the population, and each cell then responds to adapt to the changing environment.

Sutherland's Experiments Discovering a Second Messenger Molecule

Question: How does epinephrine activate glycogen phosphorylase, the enzyme that breaks down glycogen into glucose in the liver?

Experiment: Sutherland had shown in one experiment that a homogenate (disrupted cells, consisting of cytoplasm, membranes, and other cell components) would activate glycogen phosphorylase if incubated with epinephrine, ATP, and magnesium ions. To learn more about the activation mechanism, in a second experiment, Sutherland prepared a liver cell homogenate and then centrifuged it, separating the cytoplasm from membranes and other cell debris.

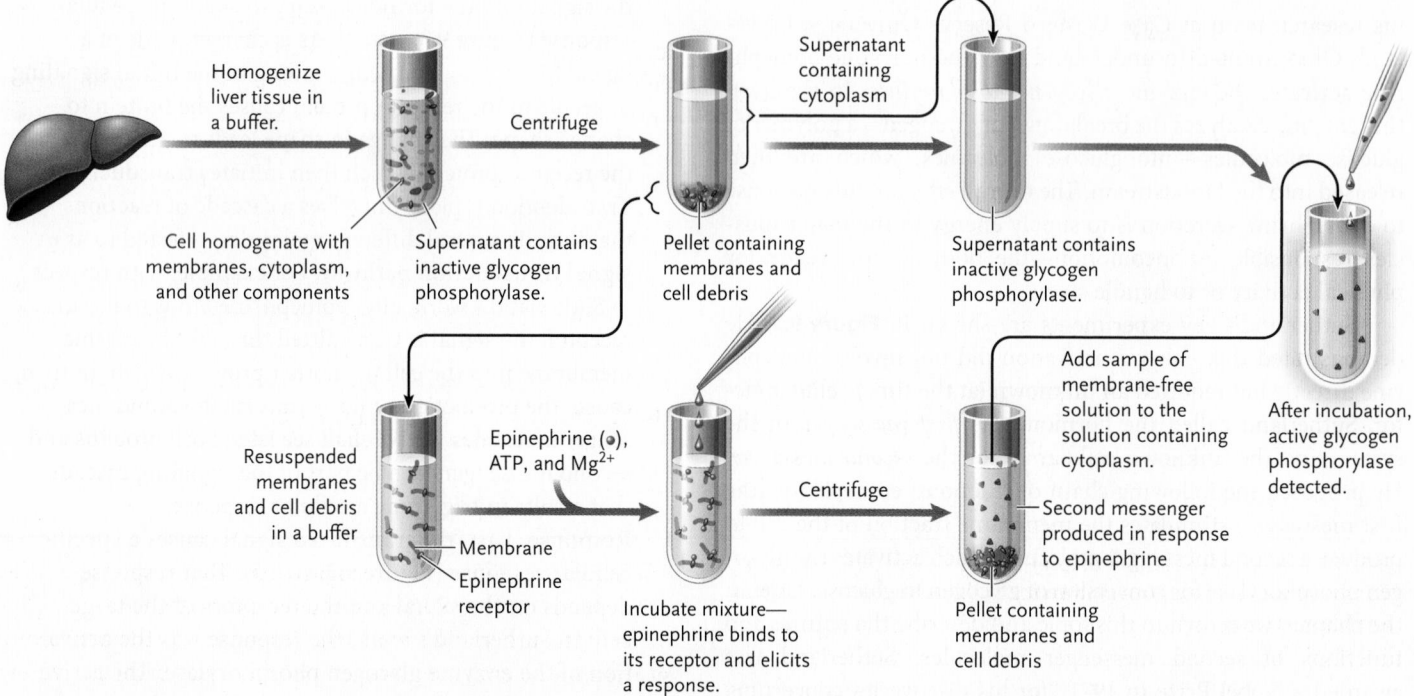

The cytoplasm was moved to a new tube, and the pellet with the membranes and cell debris was resuspended in a buffer. Neither the cytoplasm nor the membrane fractions had active glycogen phosphorylase. Next, he added epinephrine, ATP, and magnesium ions to the resuspended membranes and incubated the mixture. Centrifuging the mixture pelleted the membranes to the bottom of the tube. A sample of the supernatant (the membrane-free solution above the pellet) was added to the solution containing cytoplasm and the mixture was incubated.

Result: Active glycogen phosphorylase was detected in the mixture.

Conclusion: Sutherland had shown that the response to the hormone epinephrine—the activation of glycogen phosphorylase—does not involve epinephrine directly, but requires another cellular factor. He named the factor the *second messenger,* with the hormone itself being the *first messenger.*

Observe
Hypothesize
Predict
Experiment
Interpret

The researchers did some experiments to investigate the nature of the second messenger molecule. In one of their experiments, they showed that the second messenger remained active after heating in boiling water for 3 minutes. Interpret that result.

Source: T. W. Rail, E. W. Sutherland, and J. Berthet. 1957. The relationship of epinephrine and glucagon to liver phosphorylase. IV. Effect of epinephrine and glucagon on the reactivation of phosphorylase in liver homogenates. *Journal of Biological Chemistry* 224:463–475.

In the unicellular eukaryote, yeast, sexual mating begins when one cell secretes a hormone that is recognized by a cell of a different "sex," signaling that the two cells are compatible for mating. In multicellular eukaryotes, complex cell communication pathways coordinate the activities of multiple cell types. Some protein components of these pathways are found in both prokaryotes and eukaryotes, indicating that they are evolutionarily ancient. Other proteins in the pathways appeared only after eukaryotes evolved. For instance, protein kinases—enzymes that add phosphate groups to other proteins to control their activity—are one of the largest families of proteins in eukaryotes, yet they are absent in prokaryotes. Scientists believe

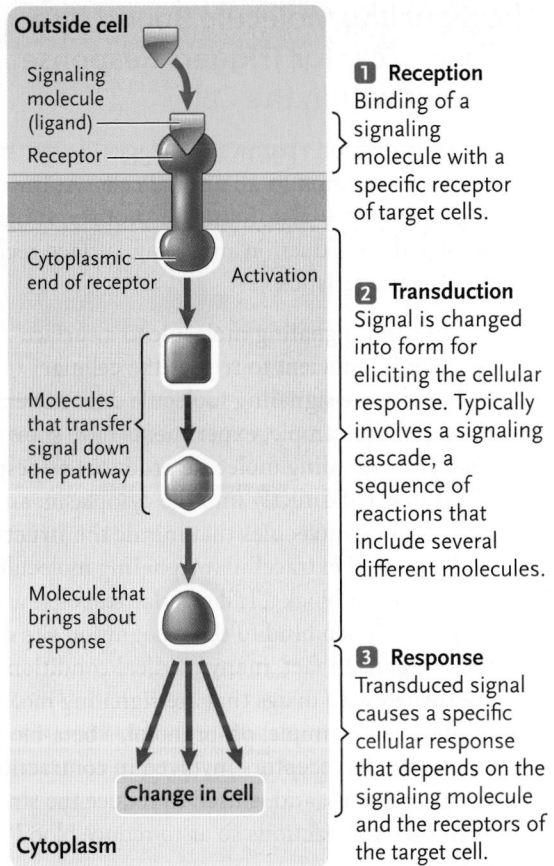

FIGURE 9.3 The three stages of long-distance signaling: reception, transduction, and response (shown for a signaling pathway using a surface receptor).
© Cengage Learning 2017

that the evolution of protein kinases was an important step in the development of multicellularity.

Beyond individual components, entire cell signaling pathways are conserved between organisms. For example, one pathway for cell growth control is conserved between *Drosophila* (the fruit fly) and humans, indicating that the pathway is at least 800 million years old. In short, the principles of cell communication are similar in unicellular and multicellular organisms, and some components are shared between them, but there is no single evolutionary root for the pathways involved. Many of the examples in this chapter focus on the systems working in animals, particularly in mammals, from which most of our knowledge of cell communication has been developed. (The plant communication and control systems are described in more detail in Chapter 37.) This discussion begins with a few fundamental principles that underlie the often complex networks of cell communication.

STUDY BREAK 9.1

What accounts for the specificity of a cellular response to a signal molecule?

9.2 Cell Communication Systems with Surface Receptors

Cell communication systems using surface receptors have three components: (1) the extracellular signaling molecules released by controlling cells; (2) the surface receptors on target cells that receive the signals; and (3) the internal response pathways triggered when receptors bind a signaling molecule. Surface receptors in mammals and other vertebrates recognize and bind polar, water-soluble signaling molecules. These molecules are released by controlling cells and enter the extracellular fluid, and then pass into the blood circulation (in animals with a circulatory system). Two major types of polar signaling molecules are polar hormones (hormones that are not steroids) and neurotransmitters. As you learned earlier, epinephrine is an example of a polar hormone, as are peptide hormones, which, as a group, affect all body systems (see Chapter 42). For example, insulin regulates sugar levels in blood. As you learned earlier, neurotransmitters are molecules released by neurons into chemical synapses; they trigger activity in other neurons or other cells in the body. Neurotransmitters include small peptides, individual amino acids or their derivatives, and other chemical substances (see Chapter 39).

Surface Receptors Are Integral Membrane Glycoproteins

The surface receptors that recognize and bind signaling molecules are all glycoproteins—proteins with attached carbohydrate chains (see Section 3.4). They are **transmembrane proteins**—integral membrane proteins that extend entirely through the plasma membrane **(Figure 9.4A)**. Typically the site of the receptor that binds a signaling molecule is the part of the protein that extends from the outer membrane surface, and which is folded in a way that closely fits that signaling molecule. The fit is specific, so a particular receptor binds only one type of signaling molecule or a group of signaling molecules that have closely related molecular structures.

A signaling molecule brings about specific changes in cells to which it binds. When a signaling molecule binds to a surface receptor, the receptor undergoes a conformation change (the molecular structure of the receptor changes) so that it transmits the signal through the plasma membrane, activating the cytoplasmic end of the receptor. The activated receptor then initiates the first step in a cascade of molecular events—the signaling cascade—that triggers the cellular response **(Figure 9.4B).**

Animal cells typically have hundreds to thousands of surface receptors that represent many receptor types. Receptors for a specific peptide hormone may number from 500 to as many as 100,000 or more per cell. Different cell types contain distinct combinations of receptors, allowing them to react individually to the polar signaling molecules in the extracellular fluid. The combination of surface receptors on particular cell types is not fixed but changes as cells develop. Changes also occur as normal cells are transformed into cancer cells.

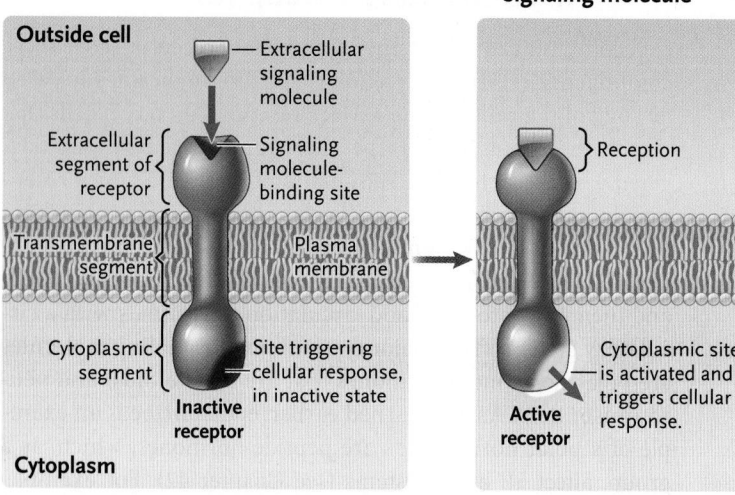

A. Surface receptor

Outside cell

Extracellular signaling molecule

Extracellular segment of receptor

Signaling molecule-binding site

Transmembrane segment

Plasma membrane

Cytoplasmic segment

Site triggering cellular response, in inactive state

Inactive receptor

Cytoplasm

A surface receptor has an extracellular segment with a site that recognizes and binds a particular signaling molecule.

B. Activation of receptor by binding of a specific signaling molecule

} Reception

Cytoplasmic site is activated and triggers cellular response.

Active receptor

When the signaling molecule is bound, a conformational change is transmitted through the transmembrane segment that activates a site on the cytoplasmic segment of receptor. The activation triggers a reaction pathway that results in the cellular response.

FIGURE 9.4 The mechanism by which a surface receptor responds when it binds a signal molecule.
© Cengage Learning 2017

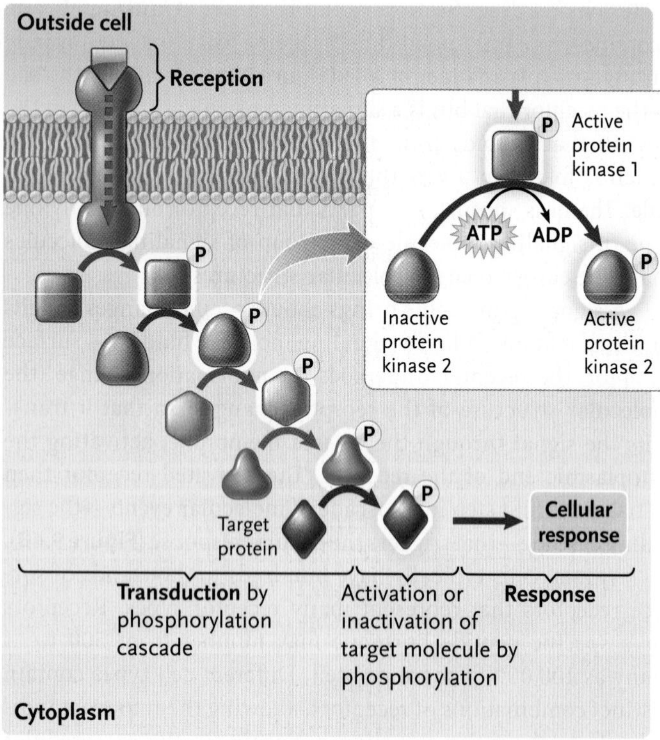

Outside cell

} Reception

P Active protein kinase 1

ATP ADP

Inactive protein kinase 2

Active protein kinase 2

Target protein

Cellular response

Transduction by phosphorylation cascade

Activation or inactivation of target molecule by phosphorylation

Response

Cytoplasm

FIGURE 9.5 Phosphorylation, a key reaction in many signaling pathways.
© Cengage Learning 2017

The Signaling Molecule Bound by a Surface Receptor Triggers Response Pathways within the Cell

Signal transduction pathways triggered by surface receptors are common to all animal cells. At least parts of the pathways are also found in protists, fungi, and plants. Signal transduction involving surface receptors has three characteristics:

1. Binding of a signaling molecule to a surface receptor is sufficient to trigger the cellular response—the signaling molecule does not enter the cell. For example, experiments have shown that: (1) a signaling molecule produces no response if it is injected directly into the cytoplasm; and (2) unrelated molecules that mimic the structure of the normal extracellular signaling molecule can trigger or block a full cellular response as long as they can bind to the recognition site of the receptor. In fact, many medical conditions are treated with drugs that are signaling molecule mimics. For example, propranolol, a beta-blocker drug, inhibits receptors involved in contractions of certain muscles, and is used to reduce the strength of cardiac contractions so as to reduce blood pressure.

2. The signal is relayed inside the cell by **protein kinases,** enzymes that transfer a phosphate group from ATP to one or more sites on particular proteins (see Section 6.5). The addition of a phosphate group to a molecule is called **phosphorylation** (see Chapter 6). As shown in **Figure 9.5,** protein kinases often act in a chain catalyzing a series of phosphorylation reactions called a **phosphorylation cascade,** to pass a signal along. The first kinase catalyzes phosphorylation of the second, which then becomes active and phosphorylates the third kinase, which then becomes active, and so on. The last protein in the cascade is the **target protein.** Phosphorylation of a target protein stimulates or inhibits its activity depending on the particular protein. This change in activity brings about the cellular response. For example, phosphorylating a target protein that regulates whether a set of genes are turned on or off could cause cells to start or stop producing the proteins the cell's genes encode. The change in this set of proteins then causes a response related to the functions of those proteins.

The effects of protein kinases in signal transduction pathways are balanced or reversed by another group of enzymes called **protein phosphatases,** which remove phosphate groups from proteins, in this case target proteins. Unlike the protein kinases, which are active only when a surface receptor binds a signal molecule, most of the protein phosphatases are active continuously in cells. By continually removing phosphate groups from

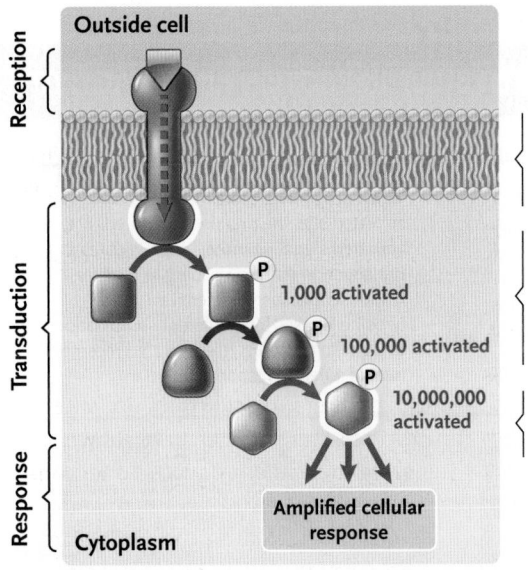

Outside cell

Reception

Transduction

Response

Cytoplasm

Signal enzyme activates 10 of 1st molecules in pathway.

1,000 activated

Each of these activates 100 of the 2nd enzyme in pathway, producing 1,000 activations.

100,000 activated

10,000,000 activated

Continued amplification of signal

Amplified cellular response

FIGURE 9.6 Amplification in signal transduction.
© Cengage Learning 2017

they are excreted by the kidneys. The removal process ensures that the signaling molecules are active only as long as controlling cells are secreting them.

As signal transduction runs its course, the receptors and their bound signaling molecules are removed from the target cell surface by endocytosis (see Section 5.5). Both the receptor and its bound signaling molecule may be degraded in lysosomes after entering the cell. Alternatively, the receptors may be separated from the signaling molecules and recycled to the cell surface, and the signaling molecules are degraded as above. Thus, surface receptors participate in an extremely lively cellular "conversation" with moment-to-moment shifts in the information.

target proteins, the protein phosphatases quickly shut off a signal transduction pathway if its signal molecule is no longer bound at the cell surface.

Two scientists, Edwin Krebs and Edmond Fischer at the University of Washington, Seattle, first discovered that protein kinases add phosphate groups to control the activities of key proteins in cells and obtained evidence showing that protein phosphatases reverse these phosphorylations. Krebs and Fischer, who began their experiments in the 1950s, received a Nobel Prize in 1992 "for their discoveries concerning reversible protein phosphorylation as a biological regulatory mechanism."

3. An increase in the magnitude of each step occurs as a signal transduction pathway proceeds, a phenomenon called **amplification (Figure 9.6).** Amplification occurs because many of the proteins that carry out individual steps in the pathways, including the protein kinases, are enzymes. Once activated, each enzyme in a signal transduction pathway can activate hundreds of proteins including other enzymes that enter the next step in the pathway. Generally, the more enzyme-catalyzed steps in a response pathway, the greater the amplification. As a result, just a few extracellular signal molecules binding to their receptors can produce a full internal response. For similar reasons, amplification also occurs for signal transduction pathways that involve internal receptors.

Signal Transduction Pathways Have "Off Switches"

Once signaling molecules are released into the body's circulation, they remain for only a certain time. Either they are broken down at a steady rate by enzymes in organs such as the liver, or

9.3 Signaling Pathways Triggered by Surface Receptors

In this section you will learn about three types of plasma membrane-embedded surface receptors: receptor tyrosine kinases, G-protein–coupled receptors, and ligand-gated ion channels.

Receptor Tyrosine Kinases Are Surface Receptors with Built-In Protein Kinase Activity

One major type of surface receptors, the **receptor tyrosine kinases (RTKs),** have their own protein kinase activity on the cytoplasmic end of the protein. Binding of a signaling molecule to this type of receptor turns on the receptor's built-in protein kinase, which leads to activation of the receptor **(Figure 9.7).** The activated receptor then initiates a signaling cascade, which results in a cellular response. Different RTKs can bind different combinations of signaling molecules, which explains how activated receptors can initiate different cellular responses.

RTKs are found in all multicellular animals, but not in plants or fungi. Fifty-eight genes in the human genome encode RTKs. In mammals, more generally, RTKs fall into about 20 different families, all related to one another in structure and amino acid sequence. Each family member is presumed to have diverged from a common ancestor, and research is being done to determine the mechanisms of divergence during their evolution.

The cellular responses triggered by RTKs are among the most important processes of animal cells. For example, the RTKs binding the peptide hormone *insulin,* a regulator of carbohydrate metabolism, trigger diverse cellular responses, including effects on glucose uptake, the rates of many metabolic reactions,

THE ACTION OF A RECEPTOR TYROSINE KINASE, A RECEPTOR TYPE WITH BUILT-IN PROTEIN KINASE ACTIVITY.

1 Signaling molecules bind and two receptor molecules assemble into a dimer

When no extracellular signaling molecules are bound, the two receptor monomers are separate and their cytoplasmic protein kinase sites are inactive. When each receptor monomer binds a signaling molecule, the monomers move together in the membrane and assemble into a dimer (a pair of monomers bonded together).

2 Activation of protein kinases and autophosphorylation of the receptor

Dimer formation triggers conformational changes in the receptor, which activate the protein kinases of each receptor monomer. The protein kinases then phosphorylate the partner monomer in the dimer, a process called autophosphorylation. The phosphorylation is of tyrosine amino acids. The multiple phosphorylations activate many different sites on the dimer.

3 Transduction and cellular responses

When a signaling protein binds to an activated site, the protein is activated. The activated signaling protein then initiates a transduction pathway leading to a cellular response. Because different receptor tyrosine kinases bind different combinations of signaling proteins, the activated receptors initiate different responses.

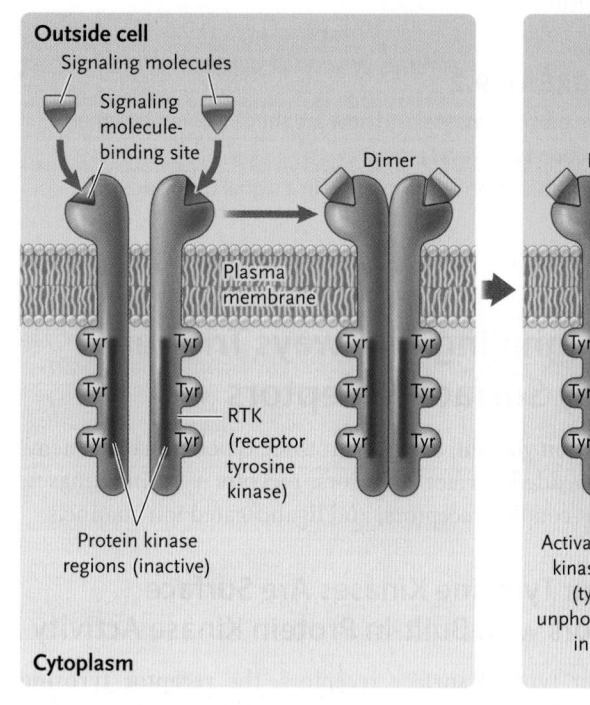

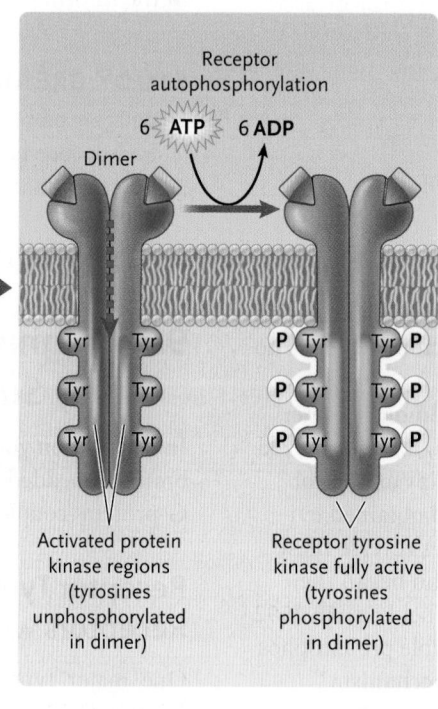

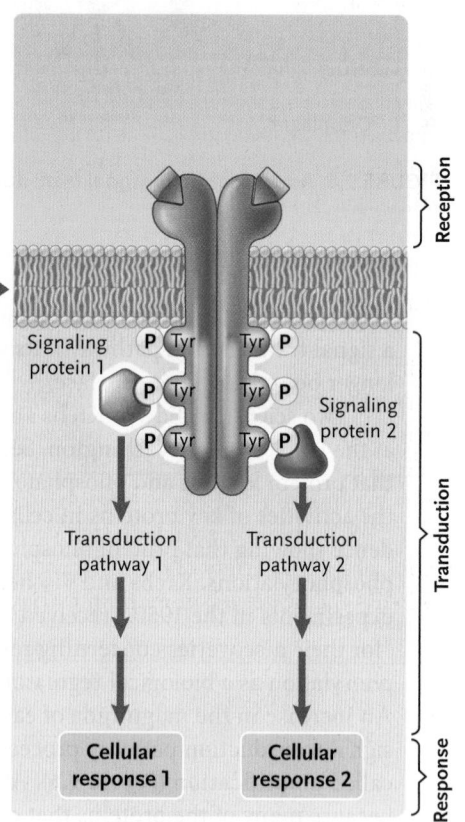

SUMMARY Binding of an extracellular signaling molecule causes receptor monomers to form a dimer that becomes active by autophosphorylation. Signaling proteins bind to the activated receptor and become activated, each initiating a transduction pathway that produces a cellular response. That is, binding of a signaling molecule to a receptor tyrosine kinase can initiate multiple signal transduction pathways and cause multiple cellular responses.

think like a scientist Suppose a mutation in a gene for a receptor tyrosine kinase receptor causes the receptor to form dimers and activate without having a signaling molecule bind. What would be the effect of that mutation?

and cell growth and division. (The insulin receptor is exceptional because it is permanently in a tetrameric [four-monomer] form.) Other RTKs bind growth factors (molecules that stimulate cell division of a target cell), including *epidermal growth factor, platelet-derived growth factor,* and *nerve growth factor,* which are important peptides.

Hereditary defects in the insulin receptor are responsible for some forms of *diabetes,* a disease in which glucose accumulates in the blood because it cannot be absorbed in sufficient quantity by body cells. The cells with faulty receptors do not respond to insulin's signal to add glucose receptors to take up glucose. (The role of insulin in glucose metabolism and diabetes is discussed further in Chapter 42.) Mutations in RTK genes are associated with a number of other inherited human diseases, including dwarfism and heritable cancer susceptibility. As just indicated, growth factors bind to RTKs. It is not surprising, therefore, that defective regulation of RTKs is found in a wide range of cancers where normal cell growth regulation is defective. Consequently, RTKs have become targets in research for therapeutic treatments of cancers.

G-Protein–Coupled Receptors Activate an Inner Membrane Protein and Lack Protein Kinase Activity

A second large family of surface receptors, known as the **G-protein–coupled receptors (GPCRs),** respond to a signal by activating an inner membrane protein called a G protein, which is closely associated with the cytoplasmic end of the receptor. Unlike receptor tyrosine kinases, GPCRs lack built-in protein kinase activity.

G proteins are so named because they bind the guanine nucleotides GDP (guanosine diphosphate) and GTP (guanosine triphosphate) (see Section 3.5). GPCRs are found in animals (both multicellular and unicellular forms), plants, fungi, and certain protists. Researchers have identified thousands of different GPCRs in mammals, including thousands involved in recognizing and binding odor molecules as part of the mammalian sense of smell, light-activated receptors in the eye, and many receptors for hormones and neurotransmitters. Humans have about 750 GPCRs, about 300 of which are not involved with sense of smell. Almost all of the receptors of this group are large transmembrane glycoproteins built up from a single polypeptide chain anchored in the plasma membrane by seven segments of the amino acid chain that zigzag back and forth across the membrane seven times **(Figure 9.8).**

GPCRs play key roles in cell communication events that control a wide variety of physiological functions. This has made GPCRs the target for more than 60% of all drugs prescribed at the present time. Mutations in GPCR genes in humans result in a number of inherited and acquired diseases such as hypothyroidism, hyperthyroidism, retinitis pigmentosa, some fertility disorders, short stature due to growth hormone deficiency, a form of diabetes, and some types of precocious puberty.

FUNCTION OF G PROTEINS IN SECOND-MESSENGER PATHWAYS Figure 9.9 illustrates a signal transduction pathway controlled by a G-protein–coupled receptor. In such a pathway, the extracellular signaling molecule is called the **first messenger** (recall Sutherland's experiment in Section 9.1). The binding of the first messenger to the receptor activates it. The activated receptor activates an intracellular G protein, which in turn activates a plasma membrane-associated **effector** molecule. The activated effector generates one or more internal, nonprotein signaling molecules called **second messengers,** which results in the activation of protein kinases. The activated protein kinases phosphorylate specific target proteins, thereby eliciting the cellular response.

The different protein kinases of these pathways all add phosphate groups to serine or threonine amino acids in their target proteins, which typically are enzymes that catalyze steps in metabolic pathways, ion channels in the plasma and other membranes, or regulatory proteins that control gene activity and cell division. The pathway from first messengers to target proteins is common to all GPCRs.

As long as a GPCR is bound to a first messenger, the receptor keeps the G protein active. The activated G protein, in turn, keeps the effector active in generating second messengers. If the first messenger is released from the receptor, or if the receptor is taken into the cell by endocytosis, GTP is hydrolyzed to GDP, which inactivates the G protein. As a result, the effector becomes inactive, turning "off" the response pathway.

Cells can make a variety of G proteins, with each type activating a different cellular response. Alfred G. Gilman at the University of Virginia, Charlottesville, and Martin Rodbell at the National Institutes of Health, Bethesda, Maryland, received a Nobel Prize in 1994 "for their discovery of G proteins and the role of these proteins in signal transduction in cells."

OPERATION OF TWO MAJOR G-PROTEIN–COUPLED RECEPTOR–RESPONSE PATHWAYS INVOLVING DIFFERENT SECOND MESSENGERS Activated G proteins bring about a cellular response through two major receptor–response pathways in which different effectors generate different second messengers. One pathway involves the second messenger **cyclic AMP (cAMP)**—cyclic 3′,5′-adenosine monophosphate, a relatively small, water-soluble molecule derived from ATP **(Figure 9.10).** The effector that produces cAMP is the enzyme *adenylyl cyclase,* which converts ATP to cAMP **(Figure 9.11).** cAMP diffuses through the cytoplasm and activates protein kinases that add phosphate groups to target proteins.

The other pathway involves two second messengers: **inositol triphosphate (IP₃)** and **diacylglycerol (DAG).** The effector of this pathway, an enzyme called *phospholipase C,*

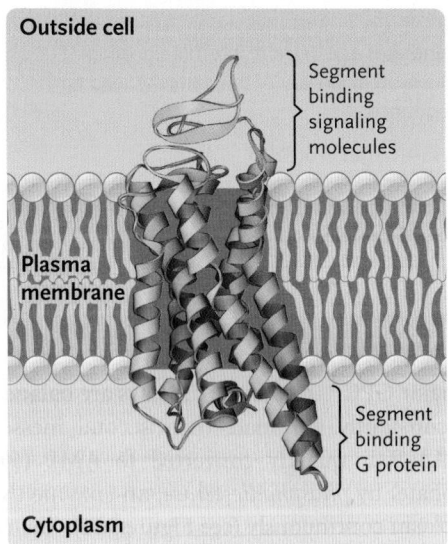

FIGURE 9.8 Structure of the G-protein–coupled receptors, which activate separate protein kinases. These receptors have seven transmembrane α-helical segments that zigzag across the plasma membrane. Binding of a signaling molecule at the cell surface, by inducing changes in the positions of some of the helices, activates the cytoplasmic end of the receptor.
© Cengage Learning 2017

Outside cell

Segment binding signaling molecules

Plasma membrane

Segment binding G protein

Cytoplasm

CLOSER LOOK FIGURE 9.9

RESPONSE PATHWAYS ACTIVATED BY G-PROTEIN–COUPLED RECEPTORS, IN WHICH PROTEIN KINASE ACTIVITY IS SEPARATE FROM THE RECEPTOR.

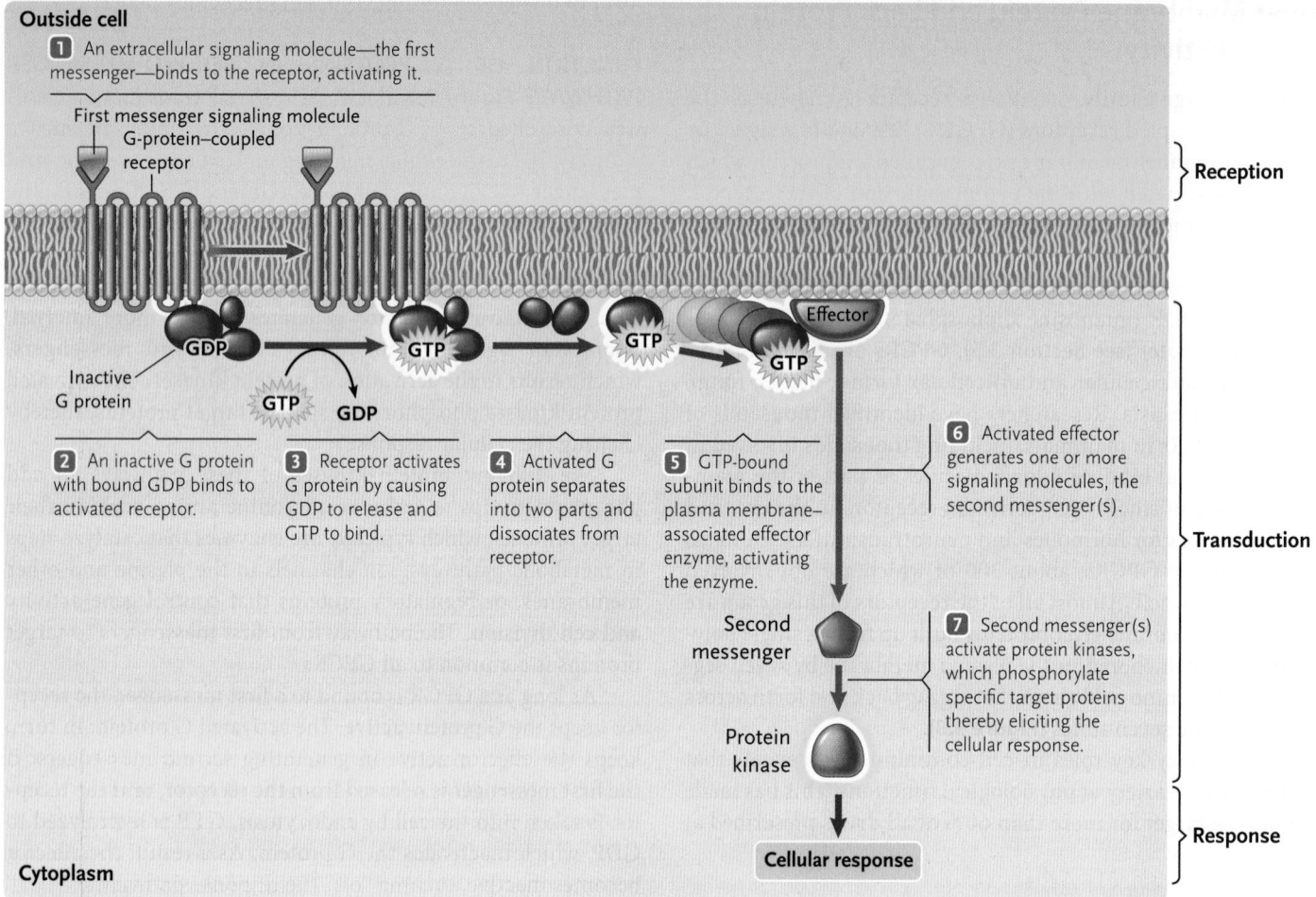

Outside cell

1 An extracellular signaling molecule—the first messenger—binds to the receptor, activating it.

First messenger signaling molecule
G-protein–coupled receptor

Reception

Inactive G protein

GDP · GTP · GTP · GTP · GDP

Effector

2 An inactive G protein with bound GDP binds to activated receptor.

3 Receptor activates G protein by causing GDP to release and GTP to bind.

4 Activated G protein separates into two parts and dissociates from receptor.

5 GTP-bound subunit binds to the plasma membrane–associated effector enzyme, activating the enzyme.

6 Activated effector generates one or more signaling molecules, the second messenger(s).

Transduction

Second messenger

7 Second messenger(s) activate protein kinases, which phosphorylate specific target proteins thereby eliciting the cellular response.

Protein kinase

Response

Cellular response

Cytoplasm

SUMMARY The signaling molecule is the first messenger. The effector is an enzyme that generates one or more internal signaling molecules called second messengers. The second messengers directly or indirectly activate the protein kinases of the pathway, leading to the cellular response. In sum, the entire control pathway operates through the following sequence:

first messenger → receptor → G proteins → effector → protein kinases → target proteins

think like a scientist What would be the possible effects of a malfunction of a G-protein–coupled receptor?

produces both of these second messengers by breaking down a membrane phospholipid called PIP_2 **(Figure 9.12)**. IP_3 is a small, water-soluble molecule that diffuses rapidly through the cytoplasm. DAG is hydrophobic; it remains and functions in the plasma membrane.

The primary effect of IP_3 in animal cells is to activate transport proteins in the endoplasmic reticulum (ER), which release Ca^{2+} stored in the ER into the cytoplasm. The released Ca^{2+}, either alone or in combination with DAG, activates a protein kinase cascade that brings about the cellular effect.

Both major GPCR–response pathways are balanced by reactions that constantly eliminate their second messengers. For example, cAMP is quickly converted to AMP (5'-adenosine monophosphate) by *phosphodiesterase,* an enzyme that is active in the cytoplasm continuously (see Figure 9.11). The rapid elimination of the second messengers provides another highly effective off switch for the pathways, ensuring that protein kinases are inactivated quickly if the receptor becomes inactive. Still another off switch is provided by protein phosphatases that remove the phosphate groups added to proteins by the protein kinases.

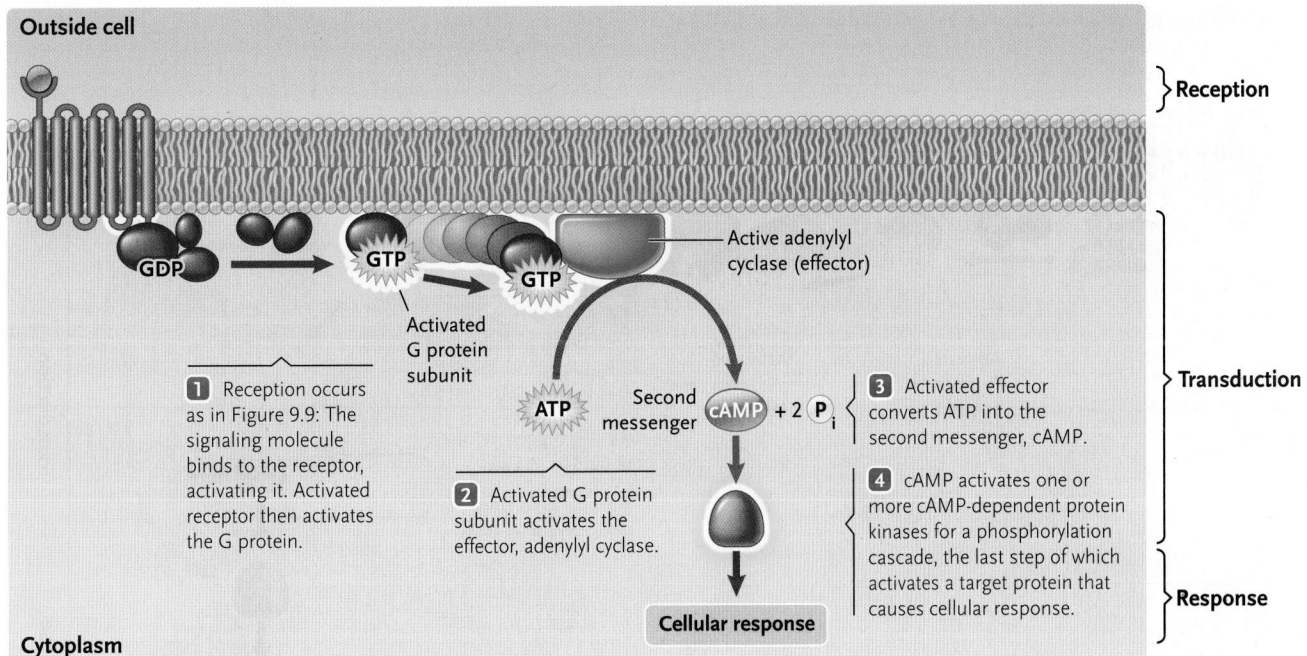

FIGURE 9.10 The operation of cAMP receptor–response pathways.
© Cengage Learning 2017

① Reception occurs as in Figure 9.9: The signaling molecule binds to the receptor, activating it. Activated receptor then activates the G protein.

② Activated G protein subunit activates the effector, adenylyl cyclase.

③ Activated effector converts ATP into the second messenger, cAMP.

④ cAMP activates one or more cAMP-dependent protein kinases for a phosphorylation cascade, the last step of which activates a target protein that causes cellular response.

As in the receptor tyrosine kinase pathways, the activities of the pathways controlled by cAMP and IP₃/DAG second messengers are also stopped by endocytosis of receptors and their bound extracellular signaling molecules. As with all cell signaling pathways, cells vary in their response to cAMP or IP₃/DAG pathways depending on the type of GPCRs on the cell surface and the kinds of protein kinases present in the cytoplasm.

The cAMP second-messenger pathway is found in animals and fungi. In plants, cAMP may be involved in germination and in some plant defensive responses (see Chapter 37), although the pathways are not well understood. The IP₃/DAG second-messenger pathway is universally distributed among eukaryotic organisms, including both vertebrate and invertebrate animals, fungi, and plants. In plants, IP₃ releases Ca^{2+} primarily from the large central vacuole rather than from the ER, and is involved in reactions that close plant stomata in response to a hormone signal (see Chapter 37).

Specific Examples of Cyclic AMP Pathways Many polar hormones act as first messengers for cAMP pathways in mammals and other vertebrates. The receptors that bind these hormones control such varied cellular responses as the uptake and oxidation of glucose, glycogen breakdown or synthesis, ion transport, the transport of amino acids into cells, and cell division.

For example, a cAMP pathway is involved in regulating the level of glucose, the fundamental fuel of cells. When the level of blood glucose falls too low in mammals, cells in the pancreas release the peptide hormone glucagon. Glucagon triggers a cAMP receptor–response pathway (see Figure 9.10) in liver cells, which stimulates them to break down glycogen into

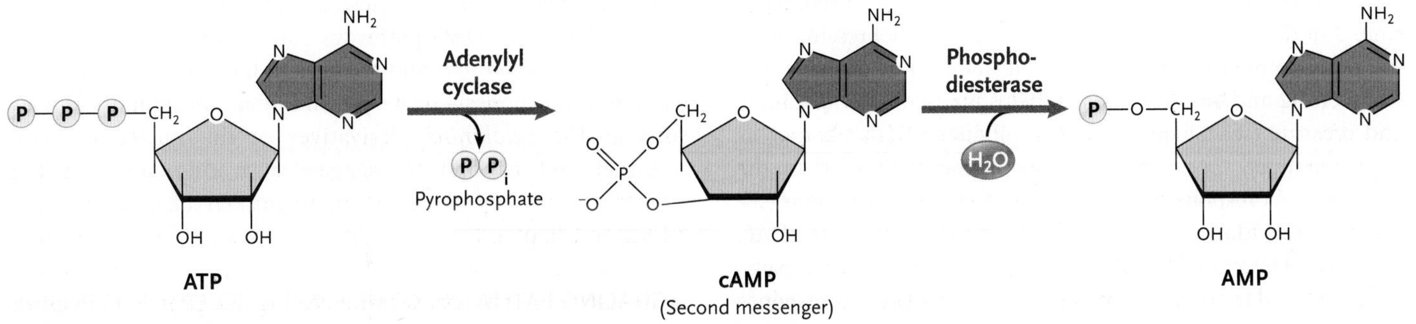

FIGURE 9.11 cAMP. The second messenger, cAMP, is made from ATP by adenylyl cyclase and is broken down to AMP by phosphodiesterase.
© Cengage Learning 2017

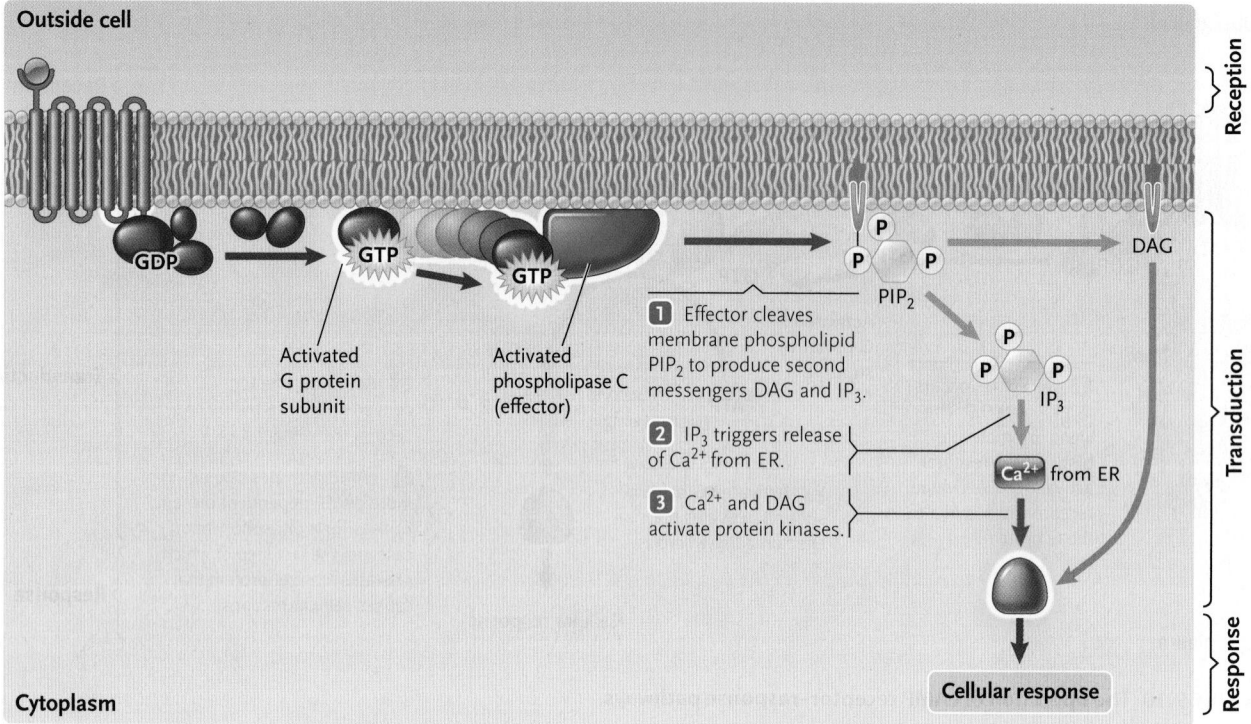

Outside cell

Reception

Transduction

Response

GDP

GTP

GTP

Activated
G protein
subunit

Activated
phospholipase C
(effector)

PIP₂

P P

P

P P

IP_3

1 Effector cleaves
membrane phospholipid
PIP_2 to produce second
messengers DAG and IP_3.

2 IP_3 triggers release
of Ca^{2+} from ER.

3 Ca^{2+} and DAG
activate protein kinases.

DAG

Ca^{2+} from ER

Cellular response

Cytoplasm

FIGURE 9.12 The operation of IP₃/DAG receptor–response pathways. Two second messengers, IP_3 and DAG, are produced by the pathway. IP_3 opens Ca^{2+} channels in ER membranes, releasing the ion into the cytoplasm. The Ca^{2+}, with DAG in some cases, directly or indirectly activates the protein kinases of the pathway, which add phosphate groups to target proteins to initiate the cellular response.
© Cengage Learning 2017

glucose units that pass from the liver cells into the bloodstream. When the level of blood glucose is excessive, the opposite occurs: the enzyme *glycogen synthase* joins glucose units to produce glycogen.

Specific Examples of IP₃/DAG Pathways The IP₃/DAG-response pathways are also activated by a large number of polar hormones (including growth factors) and neurotransmitters, leading to responses as varied as sugar and ion transport, glucose oxidation, cell growth and division, and movements such as smooth muscle contraction.

Among the mammalian hormones that activate the pathways are antidiuretic hormone, angiotensin, and norepinephrine. Antidiuretic hormone, also known as vasopressin, helps the body conserve water by reducing the output of urine (see Chapters 42 and 48). Angiotensin helps maintain blood volume and pressure (see Chapter 48). Norepinephrine (also known as noradrenaline), together with epinephrine, brings about the fight-or-flight response in threatening or stressful situations (see Chapter 42). Many growth factors operate through IP₃/DAG pathways. Defects in the receptors or other parts of the pathways that lead to higher-than-normal levels of DAG in response to growth factors are often associated with the progression of some forms of cancer. This is because DAG, in turn, causes an overactivity of the protein kinases responsible for stimulating

cell growth and division. Also, plant substances in a group called *phorbol esters* resemble DAG so closely that they can promote cancer in animals by activating the same protein kinases.

IP₃/DAG pathways have also been linked to mental disease, particularly *bipolar disorder* (previously called *manic depression*), in which patients experience periodic changes in mood. Lithium has been used for many years as a therapeutic agent for bipolar disorder. Research has shown that lithium reduces the activity of IP₃/DAG pathways that release neurotransmitters; among them are some neurotransmitters that are involved in brain function. Lithium also relieves cluster headaches and premenstrual tension, suggesting that IP₃/DAG pathways may be linked to these conditions as well.

In plants, IP₃/DAG pathways control responses to conditions such as water loss and changes in light intensity or salinity. Plant hormones—relatively small, nonprotein molecules such as the *cytokinins* (derivatives of the nucleotide base adenine)—act as first messengers activating some of the IP₃/DAG pathways of these organisms. Cytokinins are discussed in Chapter 37.

SIGNALING PATHWAYS COMBINING A RECEPTOR TYROSINE KINASE WITH THE G PROTEIN Ras Some pathways important in gene regulation link certain RTKs to a specific type of G protein called Ras **(Figure 9.13).** Activation of the RTK by a

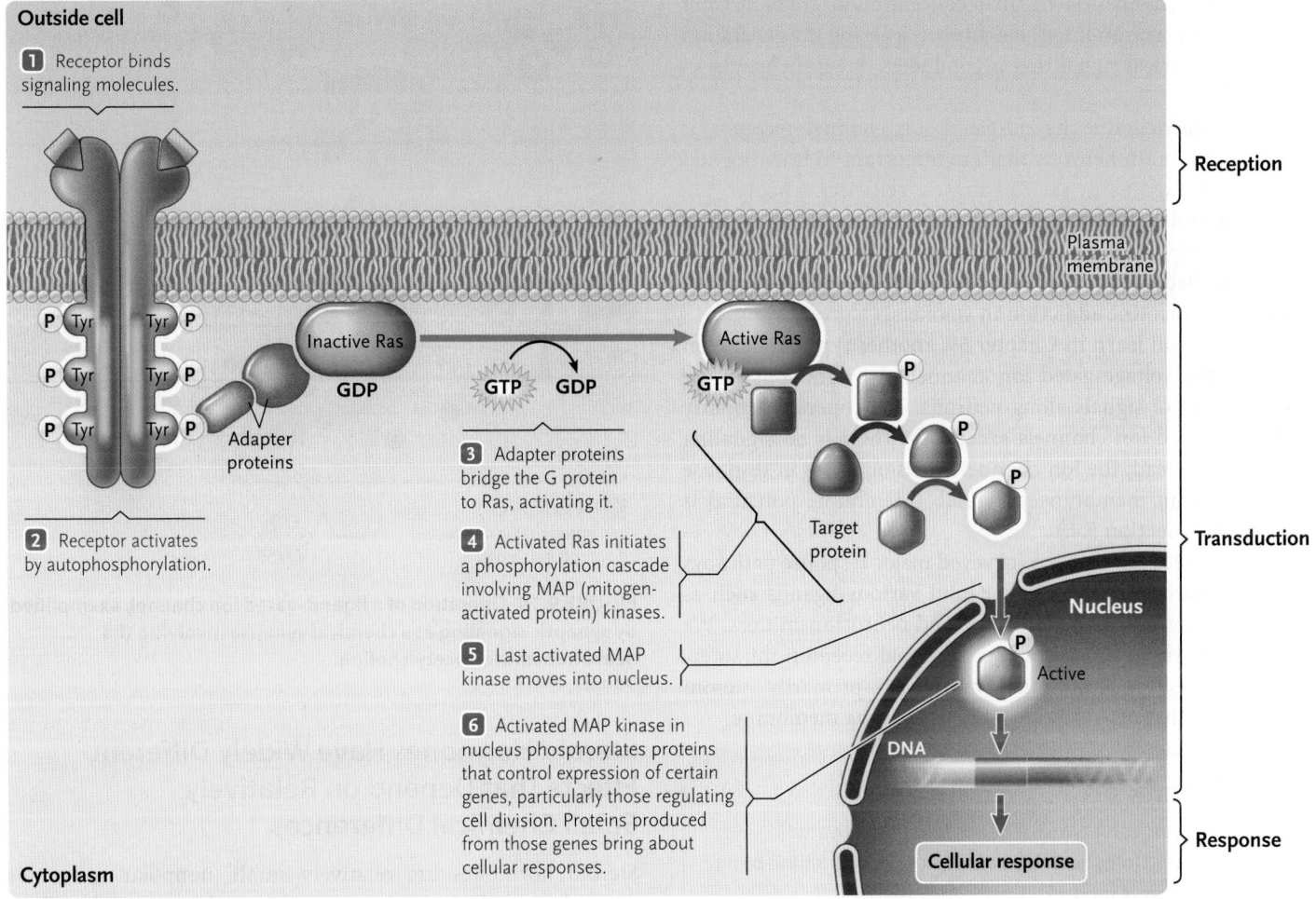

Outside cell

1 Receptor binds signaling molecules.

Reception

Plasma membrane

2 Receptor activates by autophosphorylation.

P Tyr   Tyr P
P Tyr   Tyr P
P Tyr   Tyr P

Adapter proteins

Inactive Ras
GDP

GTP → GDP

3 Adapter proteins bridge the G protein to Ras, activating it.

4 Activated Ras initiates a phosphorylation cascade involving MAP (mitogen-activated protein) kinases.

5 Last activated MAP kinase moves into nucleus.

6 Activated MAP kinase in nucleus phosphorylates proteins that control expression of certain genes, particularly those regulating cell division. Proteins produced from those genes bring about cellular responses.

Active Ras
GTP

Target protein

Transduction

Nucleus

P Active

DNA

Response

Cellular response

Cytoplasm

FIGURE 9.13 The pathway from receptor tyrosine kinases to gene regulation, including the G protein, Ras, and MAP kinase.
© Cengage Learning 2017

signaling molecule leads to activation of Ras. The activated Ras sets in motion a phosphorylation cascade that involves three *mitogen-activated protein kinase* (MAP kinase) enzymes. The last MAP kinase in the cascade, when activated, enters the nucleus and phosphorylates other proteins, which then change the expression of certain genes, particularly activating those involved in cell division. (A *mitogen* is a substance that controls cell division, hence the name of the kinases.)

Changes in gene expression can have far-reaching effects on the cell, such as determining whether a cell divides or how frequently it divides. The Ras proteins are of major interest to investigators because of their role in linking RTKs to gene regulation, as well as their major roles in the development of many types of cancer when their function is altered.

Both the Ras proteins and the MAP kinases are widely distributed among eukaryotes. Ras has been detected in eukaryotic organisms ranging from yeasts to humans and higher plants. Similarly, MAP kinases have been identified in eukaryotes as diverse as yeasts, roundworms, insects, humans, and plants.

Ligand-Gated Ion Channels Control Ion Flow in Response to Binding of a Ligand

Ligand-gated ion channels, another type of surface receptor, function in a simple way: In response to a ligand binding, the receptor changes conformation, opening or closing an ion channel, thereby allowing or blocking ion movement through the channel. These receptors are the ones involved in the type of neurotransmitter-based synaptic signaling that occurs at chemical synapses between a neuron and another neuron or another cell such as a muscle cell (see Section 9.1 and Figure 9.1B[2]). For the purposes of discussion, let us consider a chemical synapse between two neurons in the brain and the neurotransmitter, *acetylcholine* **(Figure 9.14)** (see Chapter 39). In response to an electrical signal arriving at the end of one neuron, acetylcholine is released from the cell. Acetylcholine (the ligand) diffuses across the chemical synapse and binds to a receptor—a ligand-gated ion channel—in the plasma membrane of the other neuron. Binding of acetylcholine to the receptor causes the ion channel part of the receptor to open. As a result, Na^+ ions move through the channel down their concentration gradient from outside of the neuron into the neuron, triggering the generation of a new electrical signal that is conducted down the length of the neuron.

The acetylcholine ligand-gated ion channel receptor is expressed in brain neurons that are important in learning and memory, as well as in skeletal muscle cells (see Chapter 43). Loss of acetylcholine receptors in the brain is seen in patients with epilepsy, Alzheimer disease, schizophrenia, and drug addiction. Because the receptor also binds nicotine, it might be involved in nicotine addiction in smokers.

As you will learn in Chapter 39, another type of gated ion channel, the voltage-gated ion channel, also functions in the transmission of signals along neurons. They operate similarly to ligand-gated ion channels except that there is no signaling molecule. Instead, the ion channel opens or closes in response to a change in membrane potential. (Membrane potential is explained in Section 5.4.)

In this section, we have surveyed major response pathways linked to surface receptors that bind various ligands such as peptide hormones, growth factors, and neurotransmitters. We now turn to the other major type of signal receptor: the internal receptors binding signal molecules—primarily steroid hormones—that penetrate through the plasma membrane.

STUDY BREAK 9.3

1. How does a receptor tyrosine kinase become activated?
2. Once activated fully, how does a receptor tyrosine kinase bring about a cellular response?
3. What is the role of the first messenger in a G-protein–coupled receptor-controlled pathway?
4. What is the role of the effector?
5. For a cAMP second-messenger pathway, how is the pathway turned off if no more signaling molecules are present in the extracellular fluids?
6. How does a ligand-gated ion channel function?

THINK OUTSIDE THE BOOK

Using a specific example, outline how an alteration in a receptor tyrosine kinase can contribute to the development of a cancer.

9.4 Signaling Pathways Triggered by Internal Receptors

Cells of many types have internal receptors that respond to signaling molecules arriving from the cell exterior. Unlike the signaling molecules that bind to surface receptors, these signaling molecules, primarily steroid hormones, penetrate through the plasma membrane to trigger response pathways inside the cells. In the case of steroid hormones, the internal receptors, called **steroid hormone receptors,** are typically control proteins that turn on (sometimes off) specific genes when they are activated by the binding of a signaling molecule.

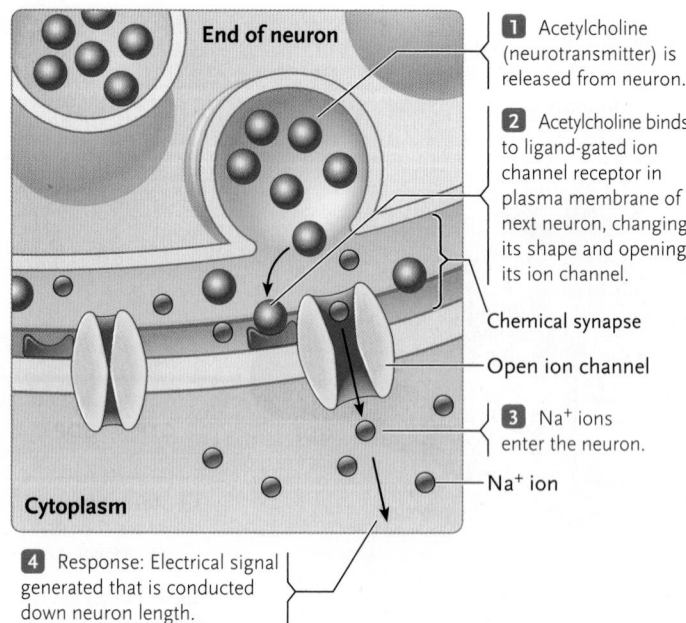

1. Acetylcholine (neurotransmitter) is released from neuron.
2. Acetylcholine binds to ligand-gated ion channel receptor in plasma membrane of next neuron, changing its shape and opening its ion channel.
3. Na⁺ ions enter the neuron.
4. Response: Electrical signal generated that is conducted down neuron length.

FIGURE 9.14 Operation of a ligand-gated ion channel, exemplified by synaptic signaling at a chemical synapse involving the neurotransmitter, acetylcholine.
© Cengage Learning 2017

Steroid Hormones Have Widely Different Effects That Depend on Relatively Small Chemical Differences

Steroid hormones are relatively small, nonpolar molecules derived from cholesterol, with a chemical structure based on four carbon rings (see Figure 3.12). Steroid hormones combine with hydrophilic carrier proteins that mask their hydrophobic groups and hold them in solution in the blood and extracellular fluids. When a steroid hormone–carrier protein complex contacts the surface of a cell, the hormone is released and penetrates directly through the plasma membrane. On the cytoplasmic side, the hormone binds to its internal receptor.

The various steroid hormones differ only in the side groups attached to their carbon rings. Although the differences are small, they are responsible for highly distinctive effects. For example, the male and female sex hormones of mammals, testosterone and estrogen, respectively, which are responsible for many of the structural and behavioral differences between male and female mammals, differ only in minor substitutions in side groups at two positions (see Figure 3.13). The differences cause the hormones to be recognized by different receptors, which activate specific group of genes leading to development of individuals as males or females.

The Response of a Cell to Steroid Hormones Depends on Its Internal Receptors and the Genes They Activate

Steroid hormone receptors are proteins with two major domains **(Figure 9.15).** The *hormone-binding domain* recognizes and

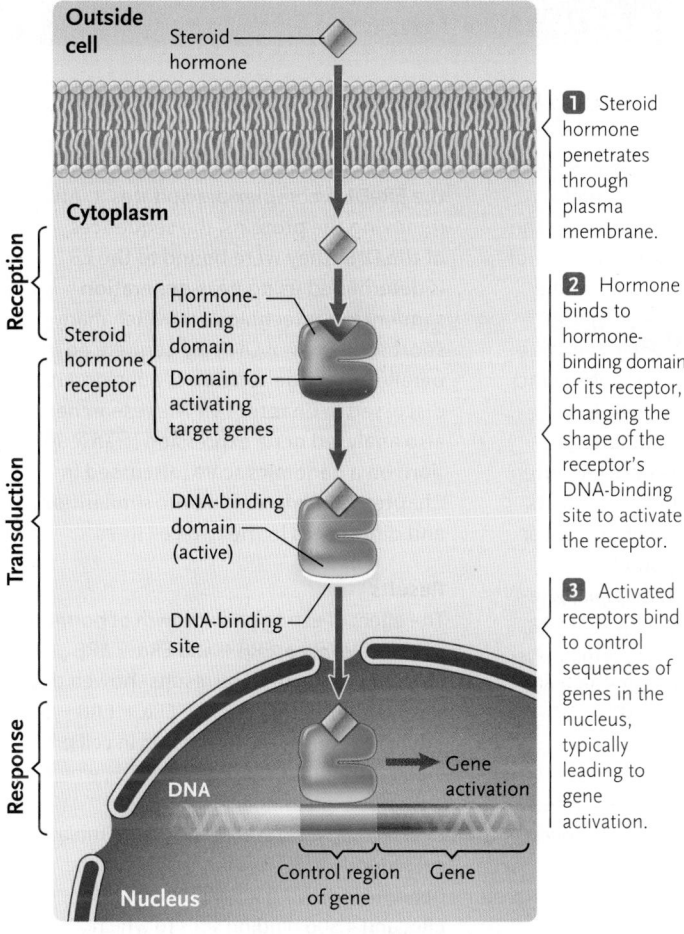

Outside cell — Steroid hormone

1 Steroid hormone penetrates through plasma membrane.

Cytoplasm

Reception

Steroid hormone receptor
- Hormone-binding domain
- Domain for activating target genes

2 Hormone binds to hormone-binding domain of its receptor, changing the shape of the receptor's DNA-binding site to activate the receptor.

Transduction

DNA-binding domain (active)

DNA-binding site

3 Activated receptors bind to control sequences of genes in the nucleus, typically leading to gene activation.

Response

DNA

Gene activation

Control region of gene — Gene

Nucleus

FIGURE 9.15 Pathway of gene activation by steroid hormone receptors.
© Cengage Learning 2017

binds a specific steroid hormone. The *DNA-binding domain* binds specifically with the regions of target genes that control the expression of those genes. When a steroid hormone enters a cell and combines with the hormone-binding domain of its receptor, the DNA-binding domain of that receptor changes shape. This shape change activates the receptor enabling the hormone–receptor complex to bind to the control regions of the target genes in the nucleus that the hormone affects. For most steroid hormone receptors, binding of the activated receptor to a gene control region activates that gene. *Molecular Insights* discusses the genomic targets of two activated estrogen receptors and their consequences to breast cancer.

Steroid hormones, like polar hormones, are released by cells in one part of an organism and are carried by the organism's circulation to other cells. Whether a cell responds to a steroid hormone depends on whether it has a receptor for the hormone within the cell. The type of response depends on the genes that are recognized and turned on (or off) by an activated receptor. Depending on the receptor type and the particular genes it recognizes, even the same steroid hormone can have highly varied effects on different cells. (The

effects of steroid hormones are described in more detail in Chapter 42.)

Nitric Oxide, a Gaseous Signaling Molecule, Functions via an Internal Receptor

In the body, the dissolved gas, nitric oxide (NO), can diffuse across the plasma membrane into a cell's cytosol and bind to an internal receptor to elicit a cellular response. NO is synthesized from the amino acid arginine (see Figure 3.14) and diffuses rapidly from its site of synthesis into nearby target cells. Thus, NO is a paracrine regulator. NO can only act locally because it is converted rapidly (in a few seconds) to nitrates and nitrites.

NO is synthesized in some neurons, in liver cells, in some cells of the immune cells, and in endothelial cells—flattened cells that line every blood vessel (see Chapter 44). For the purpose of discussion, let us consider the endothelial cells. NO is released from those cells in response to the neurotransmitter acetylcholine secreted from nearby neurons. Acting as a signaling molecule, NO causes the underlying smooth muscle cells in the adjacent blood vessel wall to relax (dilate). By this action, NO has an important role in controlling blood flow through tissues and maintaining blood pressure in arteries. Among other functions, NO helps regulate the contractions in the intestinal tract for moving digestive products forward, relaxes the smooth muscle cells in the lung airways to facilitate movement of air into and out of the lungs, and acts directly to control erection of the penis and the clitoris. In this latter function, for example, in response to NO signaling, the blood vessels in those reproductive organs dilate, allowing blood to flow into them to produce the erection.

Within a target cell, NO binds to and activates the enzyme guanylyl cyclase, which catalyzes the synthesis of cGMP (cyclic GMP) from GTP (see Section 3.5) in a reaction analogous to the synthesis of cAMP from ATP (see Figure 9.11). cGMP functions as a second messenger similar to cAMP, in this case triggering the cellular response specific for NO. cGMP is broken down to GMP by a specific phosphodiesterase. That enzyme is the target of the drug *sildenafil* (Viagra), which is prescribed for men with erectile dysfunction (ED). Sildenafil inhibits the phosphodiesterase, enabling cGMP to remain active longer so that blood vessel dilation in the penis continues and the erection can be sustained long enough to complete the sex act. Note that sildenafil does not have any effect on the release of NO and its subsequent activation of erection-producing cGMP; rather, it prolongs the triggered response.

NO is discussed further in Chapter 19 with respect to the evolution of a family of genes for the enzyme that synthesizes NO in cells, and in Chapter 44 with respect to its effect on blood vessel dilation.

Taken together, the various types of surface and internal receptors prime cells to respond to a stream of specific signals that fine-tune their function continuously. The next section shows how the various signal pathways are integrated within the cell into a coordinated response.

Genomics of Breast Cancer: How does binding of estrogen receptor β to the genome affect breast cancer progression?

The **Figure** shows an X-ray mammogram of a breast with cancer. Breast cancer is the leading cause of death among women in the United States and globally. About 50–70% of breast cancers are hormone-responsive, meaning that they need the female steroid hormone estrogen for the involved cells to grow and proliferate. Estrogens play a crucial role in regulating cell growth and differentiation in the mammary gland. In both normal and mammary gland cancer cells, estrogens act by binding to two specific steroid hormone receptors, estrogen receptors α and β (ERα and ERβ). Once activated by the binding of estrogen, these receptors are transcription factors, regulatory molecules that bind to the control regions of particular genes in the genome to alter those genes' expression, by turning them on or off, or increasing or decreasing their expression. (Expression of genes in this context means transcription, the copying of a gene's DNA sequence into a messenger RNA [mRNA] sequence; see Chapter 1 for an overview, and see Chapter 15 for detailed discussion.) Subsequent translation of the mRNA generates the protein encoded by the gene (see Chapters 1 and 15). Both ERs have similar DNA-binding domains (see Figure 9.15), and both can bind to the conserved DNA control sequence, 5'-GGTCAnnnTGACC-3', where "n" is any DNA nucleotide.

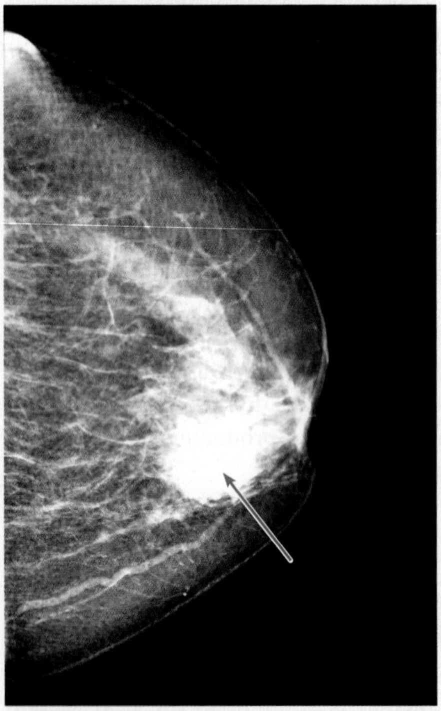

The two ERs may be present in the same cell. They have specific, often opposite functions, with ERα typically stimulating cell proliferation, and ERβ being inhibitory to cell proliferation. In breast cancer, ERβ often is lost, which removes its inhibitory effects, thereby allowing progression of the cancer via the effects of estrogen-bound ERα on gene expression. An important goal, therefore, is to understand where activated ERβ binds in the genome so as to determine the roles of this receptor in estrogen signaling and breast cancer.

Research Question

What are the genomic targets of activated ERβ in hormone-responsive breast cancer cells?

Experiments

Collaborating researchers in Italy, Germany, and the United States used genomics approaches to answer the question. They started with a human breast cancer cell line that is hormone-responsive and expresses ERα but not ERβ (ERα+ ERβ−). They genetically engineered those cells engineered to produce a cell line that expresses both ERα and ERβ (ERα+ ERβ+).

To identify DNA-binding sites for ERα and ERβ they used *ChIP-Seq (chromatin immunoprecipitation–DNA sequencing)*, a technique that analyzes specific interactions of proteins with DNA. In outline, cells were treated with estrogen, which activates the estrogen receptors and causes them to bind to their specific binding sites in DNA. The cells are treated to crosslink (bind strongly) proteins that are interacting with DNA in the genome. All proteins bound to DNA at the time of this step are now chemically bonded to the DNA. DNA is extracted from the cells and *sonicated* (bombarded with sound waves) to break the DNA into small segments, still with proteins attached. Next, DNA segments in the mix that have activated ER bound are collected by adding an antibody that is specific to the ERα or ERβ protein. Binding of the antibody to the ER–DNA complex precipitates it. After removing the proteins, the sequences of the DNA they were bound to the ER is determined using next-generation sequencing, a technique in which many short segments of DNA are sequenced in parallel. (Next-generation sequencing is discussed in Chapter 19.) The researchers also analyzed gene expression (transcription) on a genomics scale (discussed in Chapter 19) to determine the similarities and differences in the two cell lines.

Results

The effect of estrogen on growth of hormone-responsive ERα+ ERβ+ and ERα+ ERβ− cells was compared. The results showed that the expression of ERβ in the ERα+ ERβ+ cells brought about a large reduction in cell proliferation compared to ERα+ ERβ−, which were not expressing ERβ.

ChIP-Seq analysis of estrogen-stimulated cells identified 5,196 binding sites in the genome for ERβ, 1,516 binding sites for ERα, and 4,506 binding sites to which both ERβ and ERα bind. Summing, there are 9,702 genomic binding sites ERβ, and 6,024 genomic binding sites for ERα.

The researchers compared estrogen-stimulated expression from genes to which ERβ can bind in ERα+ ERβ+ vs. ERα+ ERβ− cells. They identified 921 genes whose expression was differentially regulated in the two cell types, meaning that their expression was increased or decreased. Of those 921 genes, 234 were expressed only in ERα+ ERβ− cells, 516 were expressed only in ERα+ ERβ+ cells, 154 had similar expression patterns in both cell types (up-regulated [more expression] or down-regulated [less expression] in all cases), and 17 which showed opposite responses in the two cell types (14 whose expression decreased in ERα+ ERβ− cells and increased in ERα+ ERβ+ cells, and 3 with the opposite pattern of increased in ERα+ ERβ− and decreased in ERα+ ERβ+). The genes in this set are known to control cell proliferation, cell differentiation, cell adhesion, cell motility, signal transduction, transcription, and cell death.

Conclusions

The results show that most of the genomic binding sites for ERβ can also bind ERα.

FIGURE X-ray mammogram of breast with cancer (arrow).

Tomas K/Shutterstock.com

The researchers concluded that the overall effect of ERβ on the genome of hormone-responsive breast cancer cells depends primarily on the relative concentrations of the two ERs in the cell. When both ERs are expressed in the same cell, ERα and ERβ compete for binding to many of the same DNA sequences in the genome. The researchers propose that "in hormone-responsive breast cancer, the final cellular response to estrogen is likely to depend upon the relative concentration of the two ERs in the cell, their activation status and DNA binding kinetics [rate of binding to DNA]."

think like a scientist

The investigators also found that ERβ binds to the mitochondrial genome. What might that mean?

Source: O. M. V. Grober et al. 2011. Global analysis of estrogen receptor beta binding to breast cancer cell genome reveals an extensive interplay with estrogen receptor alpha for target gene regulation. *BMC Genomics* 12:36.

© Cengage Learning 2017

STUDY BREAK 9.4

1. What distinguishes a steroid receptor from a receptor tyrosine kinase receptor or a G-protein–coupled receptor?
2. By what means does a specific steroid hormone result in a specific cellular response?

9.5 Integration of Cell Communication Pathways

Cells are under the continual influence of many simultaneous signal molecules. The cell signaling pathways may operate independently, or communicate with one another to integrate their responses to cellular signals coming from different controlling cells. The interpathway interaction is called **cross-talk;** a conceptual example that involves two second-messenger pathways is shown in **Figure 9.16.** For example, a protein kinase in one pathway might phosphorylate a site on a target protein in another signal transduction pathway, activating or inhibiting that protein, depending on the site of the phosphorylation. The cross-talk can be extensive, resulting in a complex network of interactions between cell communication pathways.

Cross-talk often leads to modifications of the cellular responses controlled by the pathways. Such modifications fine-tune the effects of combinations of signal molecules binding to the receptors of a cell. For example, cross-talk between second-messenger pathways is involved in particular types of olfactory (smell) signal transduction in rats and probably in many other animals, including humans. The two pathways involved are

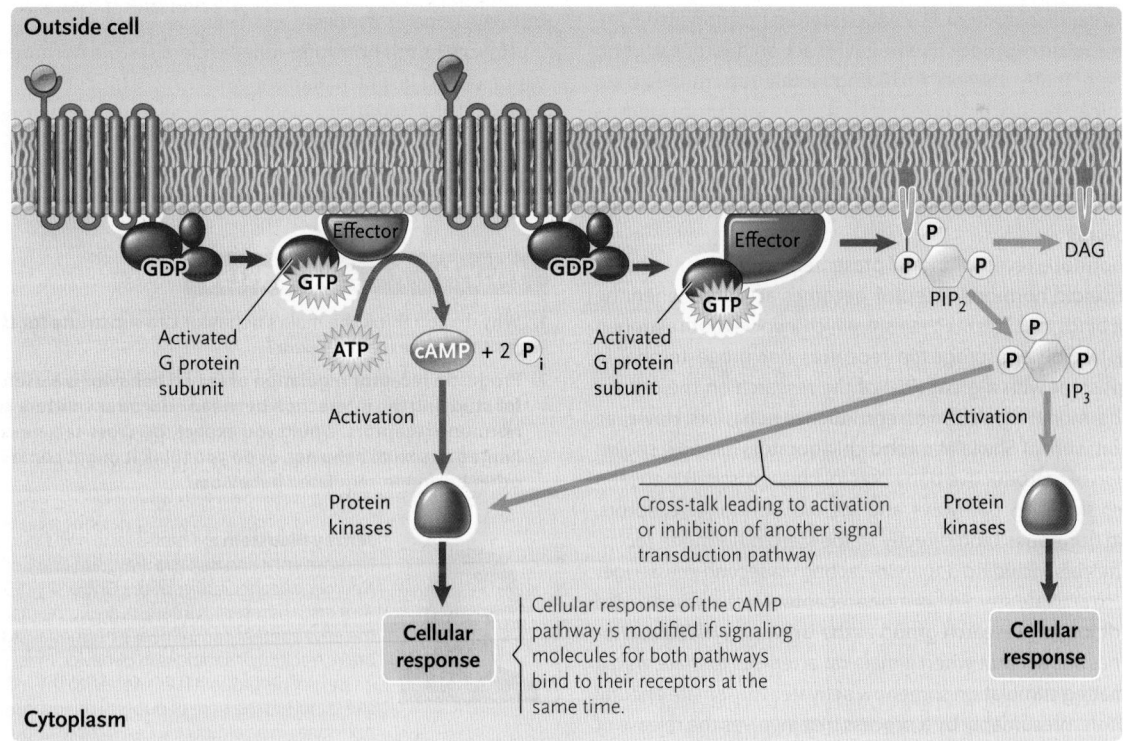

FIGURE 9.16 Cross-talk, the interaction between cell communication pathways to integrate the responses to signal molecules.
© Cengage Learning 2017

activated on stimulation with distinct odors. One pathway involves cAMP as the second messenger, and the other involves IP_3. However, the two olfactory second-messenger pathways do not work independently; rather, they operate in an antagonistic way. That is, experimentally blocking key enzymes of one signal transduction cascade inhibits that pathway, while simultaneously augmenting the activity of the other pathway. The cross-talk may be a way to refine an animal's olfactory sensory perception by helping discriminate different odor molecules more effectively.

Direct channels of communication may also be involved in a cross-talk network. For example, gap junctions between the cytoplasms of adjacent cells admit ions and small molecules, including the Ca^{2+}, cAMP, and IP_3 second messengers released by the receptor-response pathways. (Gap junctions are discussed in Section 4.5.) Thus, one cell that receives a signal through its surface receptors can transmit the signal to other cells in the same tissue via the connecting gap junctions, thereby coordinating the functions of those cells. For instance, cardiac muscle cells are connected by gap junctions, and the Ca^{2+} flow regulates coordinated muscle fiber contractions.

The entire system integrating cellular response mechanisms, tied together by many avenues of cross-talk between individual pathways, creates a sensitively balanced control mechanism that regulates and coordinates the activities of individual cells into the working unit of the organism.

Cross-talk is also discussed in the *Unanswered Questions* for this chapter.

STUDY BREAK 9.5

What cell communication pathways might be integrated in a cross-talk network?

Unanswered Questions

How does cross-talk between signaling pathways influence a behavior?

As you learned in this chapter, cell signaling by the sex hormones of animals is responsible for many of the structural and behavioral characteristics of males and females. Another type of cell signaling, called *neural signaling,* is also important in reproductive behavior. Currently, many laboratories are actively investigating possible cross-talk between these two major types of cell signaling. Specifically, researchers are investigating the cellular processes by which steroid hormones that are involved in mammalian reproductive behavior act on neurons. During the estrous cycle of many animals, including female rats, guinea pigs, hamsters, and mice, the ovarian hormones estradiol and progesterone regulate the expression of reproductive behaviors via cellular processes, including binding to steroid hormone receptors in neurons that are involved in the behaviors. But do neural signals also activate steroid hormone receptors?

The steroid hormone receptor model presented in this chapter is that an intracellular steroid hormone receptor becomes activated when the steroid hormone binds to it. This mechanism, which involves both estrogen receptors (for estradiol) and progestin receptors (for progesterone) in the brain, is consistent with a great deal of the research on the cellular mechanisms of hormonal regulation of reproductive behaviors. However, research based on work of Shaila Mani and collaborators has now shown that the regulation of reproductive behaviors involves cross-talk between neurotransmitter signaling pathways and steroid hormone receptors. Although steroid hormones bind directly to the steroid hormone receptors, neurotransmitters, including dopamine, acting via second-messenger pathways, can also activate steroid hormone receptors in the absence of a hormone. In addition, my research group at the University of Massachusetts, Amherst, has shown that when a male rat attempts to mate with a female rat, the mating stimulation somehow activates the female's neural progestin receptors, presumably by a process that involves the release of particular neurotransmitters onto neurons containing the receptors. In fact, although it had always been thought that progesterone is required to facilitate the full expression of sexual behaviors in female rats, stimulation by the male can substitute for progesterone. Mating stimulation induces reproductive behavior similar to that induced by the secretion of steroid hormones. That is, how a male behaves toward a female alters neurotransmitter release in her brain, presumably then activating steroid hormone receptors in some neurons. This activation results in neuronal changes, many of which are the same as those caused by the hormone secretions from the female's ovaries.

How does this hormone-independent steroid hormone receptor activation occur? In which neurons would you expect these events to occur, and what characteristics would you expect of the neurons (for example, inputs and outputs)? What might regulate the process? The results of experiments designed to answer these questions will give valuable insights into the mechanisms of steroid hormone action in the brain.

think like a scientist

1. Why do you think the male's behavior can substitute for the hormone in facilitating sexual behavior?

2. Progestin receptor regulation of sexual behavior is a useful model for studying the interaction between neurotransmitters and steroid hormone receptors. Would you expect this cross-talk mechanism to be limited to sexual behavior, or do you think it might come into play with other hormone-regulated behaviors?

Courtesy of Jeffrey Blaustein

Jeffrey Blaustein is a professor in and Director of the Neuroscience and Behavior Program and Center for Neuroendocrine Studies at the University of Massachusetts, Amherst. His research interests are in the many ways in which the environment can influence hormonal processes in the brain, resulting in changes in behavior. In recent years, the interest of his group has expanded to the influences of stress around the time of puberty on response to ovarian hormones in adulthood. To learn more about the work of his research group, go to http://www.umass.edu/cbd/people/jeffrey-d-blaustein.

REVIEW KEY CONCEPTS

For access to MindTap and additional study materials visit www.cengagebrain.com.

9.1 Cell Communication: An Overview

- Cells communicate with one another by direct contact, local signaling, and long-distance signaling (Figure 9.1).

- In long-distance signaling, a controlling cell releases a signaling molecule that causes a response by target cells. Target cells process the signal in three steps: reception, transduction, and response. This process is called signal transduction (Figure 9.3).

- Some cell communication properties are evolutionarily ancient. In some cases, entire cell signaling pathways are conserved between distantly related organisms.

9.2 Cell Communication Systems with Surface Receptors

- Cell communication systems based on surface receptors have three components: (1) extracellular signaling molecules; (2) surface receptors to which the signaling molecules bind; and (3) internal response pathways triggered when receptors bind a signaling molecule and are activated.

- The systems based on surface receptors respond to polar hormones and neurotransmitters. Polar hormones include peptide hormones and growth factors, which affect cell growth, division, and differentiation. Neurotransmitters include small peptides, individual amino acids or their derivatives, and other chemical substances.

- Surface receptors are integral membrane proteins that extend through the plasma membrane. Binding a signaling molecule induces a molecular change in the receptor that activates its cytoplasmic end (Figure 9.4).

- Many cellular response pathways operate by activating protein kinases, which add phosphate groups that stimulate or inhibit the activities of the target proteins, bringing about the cellular response (Figure 9.5). Protein phosphatases that remove phosphate groups from target proteins reverse the response. In addition, receptors are removed by endocytosis when signal transduction has run its course.

- Each step of a response pathway catalyzed by an enzyme is amplified, because each enzyme can activate many proteins that enter the next step in the pathway. Through amplification, a few signaling molecules can bring about a full cellular response (Figure 9.6).

9.3 Signaling Pathways Triggered by Surface Receptors

- When receptor tyrosine kinases (RTKs) bind an extracellular signaling molecule, the protein kinase site is activated and adds phosphate groups to tyrosines in the receptor itself activating those sites. When an intracellular signaling protein binds to an activated site, it initiates a transduction pathway leading to a cellular response. The binding of different combinations of signaling proteins to different tyrosine kinases produces different responses (Figure 9.7).

- In the pathways activated by G-protein–coupled receptors (GPCRs), binding of the extracellular signaling molecule (the first messenger) activates a site on the cytoplasmic end of the receptor (Figure 9.8). The activated receptor turns on a G protein. The active G protein switches on the effector, an enzyme that generates small internal signal molecules called second messengers. The second messengers activate the protein kinases of the pathway, which phosphorylate specific target proteins to elicit the cellular response (Figure 9.9).

- In one of the two major signaling pathways triggered by GPCRs, the effector, adenylyl cyclase, generates cAMP as second messenger. cAMP activates specific protein kinases (Figures 9.10 and 9.11).

- In the other major pathway, the activated effector, phospholipase C, generates two second messengers, IP_3 and DAG. IP_3 activates transport proteins in the ER, which release stored Ca^{2+}. The Ca^{2+} alone, or with DAG, activates specific protein kinases that phosphorylate their target proteins (Figure 9.12).

- Both the cAMP and IP_3/DAG pathways are balanced by reactions that constantly eliminate their second messengers. Both pathways are also stopped by protein phosphatases that remove phosphate groups from target proteins and by endocytosis of receptors and their bound signals.

- Some pathways important in gene regulation link certain RTKs to a specific G protein called Ras. When the receptor binds a signaling molecule, it phosphorylates itself, and adapter proteins then bind, bridging to Ras, activating it. Activated Ras turns on the MAP kinase cascade. The last MAP kinase in the cascade phosphorylates target proteins in the nucleus, which turn on specific genes (Figure 9.13). Many of those genes control cell division.

- When a signaling molecule (ligand) binds to a ligand-gated ion channel receptor, it triggers a change in the conformation of the receptor, opening or closing the receptor's ion channel. In this way, the signaling molecule controls the movement of ions into or out of the cell (Figure 9.14).

9.4 Signaling Pathways Triggered by Internal Receptors

- Steroid hormones penetrate through the plasma membrane to bind to receptors within the cell. Binding of the hormone to the hormone-binding domain of the receptor activates the receptor. The DNA-binding domain of the receptor then can interact with the control regions of target genes, typically activating them to produce the cellular response. A cell responds to a steroid hormone only if it has an internal receptor for the hormone, and the type of response depends on the genes that are turned on by an activated receptor (Figure 9.15).

- The gas, nitric oxide (NO), acts as a local (paracrine) regulator to cause smooth muscle cells to relax and to regulate nerve activity. To elicit these responses, NO diffuses across the plasma membrane into target cells and binds to an internal receptor.

9.5 Integration of Cell Communication Pathways

- In cross-talk, cell signaling pathways communicate with one another to integrate responses to cellular signals. Cross-talk may result in a complex network of interactions between cell communication pathways. Cross-talk often modifies the cellular responses controlled by the pathways, fine-tuning the effects of combinations of signal molecules binding to a cell (Figure 9.16).

- In animals, inputs from other cellular response systems, including cell adhesion molecules and molecules arriving through gap junctions, also can be involved in the cross-talk network.

TEST YOUR KNOWLEDGE

Remember/Understand

1. In signal transduction, which of the following is *not* a target protein?
   a. proteins that regulate gene activity
   b. hormones that activate the receptor
   c. enzymes of pathways
   d. transport proteins
   e. enzymes of cell reactions

2. Which of the following could *not* elicit a signal transduction response?
   a. a protein kinase
   b. a virus mimicking a normal signal molecule
   c. a peptide hormone
   d. a steroid hormone
   e. a neurotransmitter

3. A cell that responds to a signaling molecule is distinguished from a cell that does not respond by the fact that it has:
   a. a cell adhesion molecule.
   b. cAMP.
   c. a first messenger molecule.
   d. a receptor.
   e. a protein kinase.

4. The mechanism to activate an immune cell to make an antibody involves signal transduction using tyrosine kinases. Place in order the following series of steps to activate this function.
   (1) The activated receptor phosphorylates cytoplasmic proteins.
   (2) Conformational change occurs in the receptor tyrosine kinase.
   (3) Cytoplasmic protein crosses the nuclear membrane to activate genes.
   (4) An immune hormone signals the immune cell.
   (5) Activation of protein kinase site(s) adds phosphates to the receptor to activate it.
   a. 2, 1, 4, 3, 5
   b. 5, 3, 4, 2, 1
   c. 4, 1, 5, 2, 3
   d. 4, 2, 5, 1, 3
   e. 2, 5, 3, 4, 1

5. Which of the following describes the ability of enzymes, involving few surface receptors, to activate thousands of molecules in a stepwise pathway?
   a. autophosphorylation
   b. second-messenger enhancement
   c. amplification
   d. ion channel regulation
   e. G protein activation

6. Which of the following is *incorrect* about pathways activated by G-protein–coupled receptors?
   a. The extracellular signaling molecule is the first messenger.
   b. When activated, plasma membrane-bound G protein can switch on an effector.
   c. Second messengers enter the nucleus.
   d. ATP converts to cAMP to activate protein kinases.
   e. Protein kinases phosphorylate molecules to change cellular activity.

7. Which of the following would *not* inhibit signal transduction?
   a. Phosphate groups are removed from proteins.
   b. Endocytosis acts on receptors and their bound signals.
   c. Receptors and signaling molecules separate.
   d. Receptors and bound signaling molecules enter lysosomes.
   e. Autophosphorylation targets the cytoplasmic portion of the receptor.

8. An internal receptor binds both a signaling molecule and controlling region of a gene. What type of receptor is it?
   a. protein
   b. steroid
   c. IP_3/DAG
   d. receptor tyrosine kinase
   e. switch protein

9. Place in order the following steps for the normal activity of a Ras protein.
   (1) Ras turns on the MAP kinase cascade.
   (2) Adaptor proteins connect phosphorylated tyrosine on a receptor to Ras.
   (3) GTP activates Ras by binding to it, displacing GDP.
   (4) The last MAP kinase in the cascade phosphorylates proteins in the nucleus that activate genes.
   (5) Receptor tyrosine kinase binds a signaling molecule and is activated.
   a. 1, 2, 3, 4, 5
   b. 2, 3, 5, 1, 4
   c. 5, 2, 3, 1, 4
   d. 2, 3, 1, 5, 4
   e. 4, 1, 5, 3, 2

10. Which of the following does *not* exemplify cross-talk?
    a. a protein kinase in one pathway that phosphorylates a site on a target protein in another signal transduction pathway
    b. modifications of cellular responses controlled by pathways
    c. two second-messenger pathways interacting
    d. olfactory sensory perception
    e. signal transduction pathways controlled by G-protein–coupled receptors

Apply/Analyze

11. **Discuss Concepts** Describe the possible ways in which a G-protein–coupled receptor pathway could become defective and not trigger any cellular responses.

12. **Discuss Concepts** Is providing extra insulin an effective cure for an individual who has diabetes that is caused by a hereditary defect in the insulin receptor? Why or why not?

Evaluate/Create

13. **Discuss Concepts** There are molecules called GTP analogs that resemble GTP so closely that they can be bound by G proteins. However, they cannot be hydrolyzed by cellular GTPases. What differences in effect would you expect if you inject GTP or a nonhydrolyzable GTP analog into a liver cell that responds to glucagon?

14. **Discuss Concepts** Why do you suppose cells evolved internal response mechanisms using molecules that bind GTP instead of ATP?

15. **Design an Experiment** How would you set up an experiment to determine whether a hormone receptor is located on the cell surface or inside the cell?

16. **Apply Evolutionary Thinking** Based on their distributions among different groups of organisms, which signaling pathway is the oldest?

For selected answers, see Appendix A.

INTERPRET THE DATA

In most individuals with cystic fibrosis (CF), the 508th amino acid of the CFTR protein (a phenylalanine) is missing. A CFTR protein with this change is synthesized correctly, and it can transport ions correctly, but it never reaches the plasma membrane to do its job.

Sergei Bannykh and his coworkers developed a procedure to measure the relative amounts of the CFTR protein localized in different regions of the cell. They compared the pattern of CFTR distribution in normal cells with the pattern in CFTR-mutated cells. A summary of their results is shown in the **Figure.**

1. Which organelle contains the least amount of CFTR protein in normal cells? In CF cells? Which contains the most?

2. In which organelle is the amount of CFTR protein in CF cells closest to the amount in normal cells?

3. Where is the mutated CFTR protein getting held up?

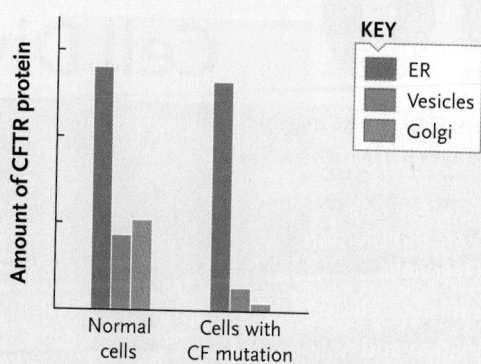

FIGURE **Comparison of the amounts of CFTR protein associated with endoplasmic reticulum (blue), vesicles traveling from ER to Golgi (green), and Golgi complexes (orange).** The patterns of CFTR distribution in normal cells, and the cells with the most common cystic fibrosis mutation, were compared.
© Cengage Learning 2017

10 Cell Division and Mitosis

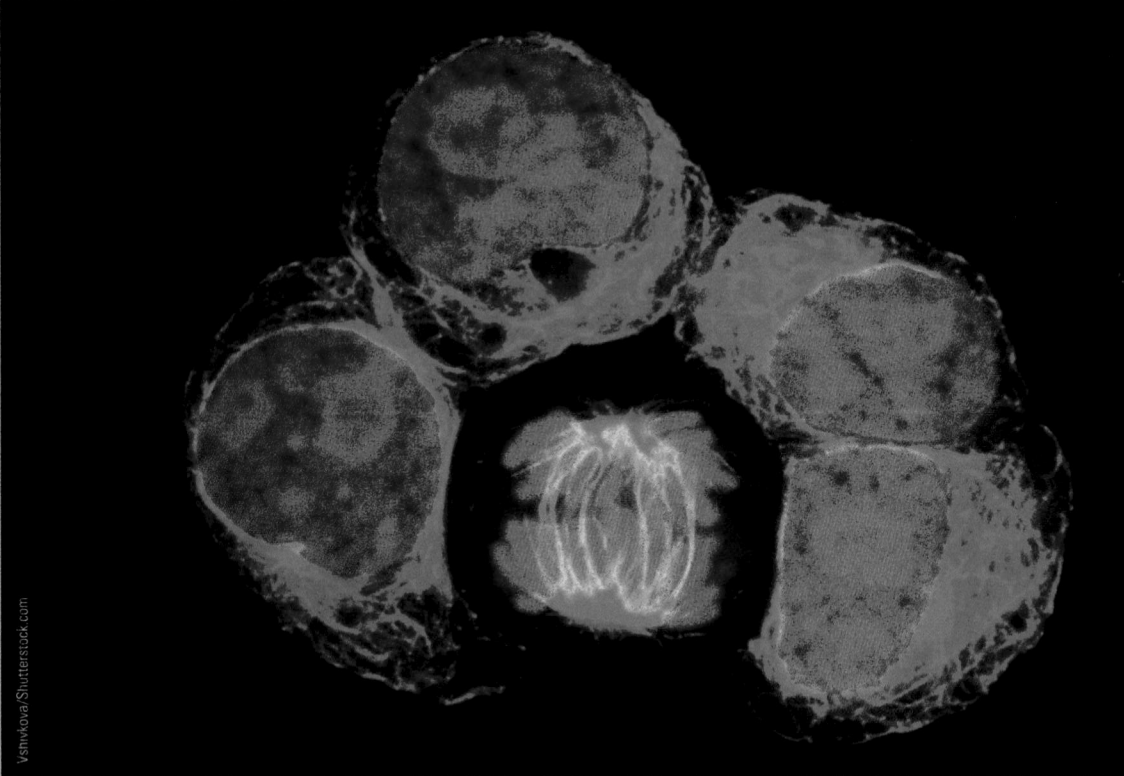

Cells labeled with fluorescent dyes (fluorescent micrograph). One of the cells is undergoing mitosis: the spindle in that cell (green) is separating copies of the cell's chromosomes (red) prior to cell division.

Why it matters . . . As the rainy season recedes in Northern India, rice paddies and other flooded areas begin to dry out. The resulting shallow, seasonal pools have provided an environment of slow-moving warm water for zebrafish *(Danio rerio)* to spawn **(Figure 10.1).** Over the past few months, many millions of cell divisions have changed the single-celled fertilized eggs into the complex multicellular tissues and organs of these small, boldly striped fish. Most cells in the adult fish have stopped dividing and are dedicated to particular functions.

Moving into the fast-running streams that feed the Ganges River, the young zebrafish often encounter larger predators such as knifefish *(Notopterus notopterus)*. Imagine that a knifefish attacks a zebrafish, tearing off part of a fin before its prey escapes. Within a week, the entire zebrafish fin will regenerate—skin, nerves, muscles, bones, and related tissues. The regeneration occurs because cells that were not growing and dividing are suddenly stimulated to grow and divide in a highly regulated way.

As a model organism for vertebrate development, the zebrafish is a popular tool for researchers to identify the stages of regeneration at the molecular level. In the first step of regeneration in injured zebrafish, existing skin cells migrate to close the wound and prevent bleeding. Then, cells just under the new skin transform into "regeneration cells" that form a temporary tissue called a *blastema*. The blastema cells begin to grow and divide. The large numbers of daughter cells produced are capable of maturing into new nerve, blood vessel, muscle, and bone cells in response to signal proteins that are produced by the skin. Once the regenerated fin has reached its normal size and shape, the new cells stop growing and dividing.

Since multicellular organisms are made almost entirely of cells and their products, understanding organismal development and structure at its most fundamental level by studying the regulation of cell division is necessary. Which conditions stimulate cells to divide? Which make

FIGURE 10.1 Zebrafish *(Danio rerio).*

214

them stop? This chapter focuses on how cell division occurs, and how that process is regulated.

10.1 The Cycle of Cell Growth and Division: An Overview

Cell division is the process whereby a preexisting cell divides to form two new cells. Cell division enables an organism to grow, reproduce, and repair damaged tissues and organs. For unicellular organisms such as the prokaryote *Escherichia coli* and the eukaryotic microorganism budding yeast *(Saccharomyces cerevisiae),* their life cycles consist of growth and cell division as long as environmental conditions allow. However, things are not as simple in multicellular eukaryotes in which cell division is under strict control to develop and maintain a mature body consisting of different subpopulations of cells. Most mature cells in multicellular organisms divide infrequently, if at all. However, the tissues of animals, plants, and other multicellular organisms also contain small populations of cells that are always actively dividing. The new cells are required, for instance, for growth (such as development of an animal or a plant), replacement of cells (such as blood cells and cells lining the intestine), and repair (such as for wounds, virus infections, and tissue damage [as in the regeneration of tissue in zebrafish discussed in *Why it matters . . .*]).

Before dividing, most cells enter a period of growth in which they synthesize proteins, lipids, and carbohydrates and, during one particular stage, replicate their nuclear DNA. After this growth period, the nuclei divide and, usually, **cytokinesis** *(cyto =* cell; *kinesis =* movement)—the division of the cytoplasm—follows, partitioning nuclei into daughter cells. Each daughter nucleus contains one copy of the replicated parent DNA. This sequence of events—the period of growth followed by nuclear division and cytokinesis—is known as the **cell cycle.** The nuclear division part of the cell cycle is **mitosis.**

The Products of Mitosis Are Genetic Duplicates of the Dividing Cell

Mitosis partitions the replicated DNA equally and precisely, generating daughter cells, which are genetic copies of the parent cell. This is accomplished by three interrelated systems:

1. A master program of molecular checks and balances that ensures an orderly and timely progression through the cell cycle
2. Within the overall regulation of the cell cycle, a process of DNA synthesis that replicates each DNA chromosome into two copies with almost perfect fidelity (see Chapter 14)
3. A structural and mechanical web of interwoven "cables" and "motors" of the mitotic cytoskeleton that separates the replicated DNA molecules precisely into the daughter cells

However, at a particular stage of the life cycle of sexually reproducing organisms, a cell division process called *meiosis* produces some cells that are genetically different from the parent cells. **Meiosis** produces daughter nuclei that are different in that they have one half the number of chromosomes the parental nucleus had. Also, the mechanisms involved in producing the daughter nuclei produce arrangements of genes on chromosomes that are different from those in the parent cell (see Chapter 11). The cells that are the products of meiosis may function as gametes in animals (fusing with other gametes to make a zygote) and as spores in plants and many fungi (dividing by mitosis).

This chapter concentrates on the mechanical and regulatory aspects of mitosis in eukaryotes (the next three sections) and cell division in prokaryotes (the last section). Meiosis and its role in eukaryotic sexual reproduction are addressed in Chapter 11.

Chromosomes Are the Genetic Units That Are Partitioned by Mitosis

In all eukaryotes, the hereditary information within the nucleus—its nuclear genome—is distributed among multiple, linear **chromosomes** *(chroma =* color; *soma =* body), each of which contains a single, linear, double-stranded DNA molecule. The number of chromosomes typically is characteristic of the species.

CHROMOSOME CONTENT OF CELLS Most eukaryotic cells have two copies of each type of chromosome in their nuclei, so their chromosome complement is said to be **diploid,** or $2n$. For example, human body cells have 23 pairs of different chromosomes, for a diploid number of 46 chromosomes ($2n = 46$). The two chromosomes of each pair in a diploid cell are called **homologous chromosomes,** meaning that they have the same genes, in the same order in the DNA of the chromosomes. One member of the pair derives from the maternal parent and the other from the paternal parent. Some eukaryotes, mostly fungi, have only one copy of each type of chromosome in their nuclei throughout much of their life cycle, so their chromosome complement is said to be **haploid,** or n. For example, cells of the orange bread mold *Neurospora crassa* are haploid ($n = 7$) throughout much of the organism's life cycle. Budding yeast *(S. cerevisiae),* a model organism that helped illuminate regulatory aspects of the cell cycle, is an example of an organism that can grow as haploid cells ($n = 16$) or as diploid cells ($2n = 32$). Still others, such as many plant species, have three, four, or even more complete sets of chromosomes in each cell. The number of chromosome sets is called the **ploidy** of a cell or species.

PACKING OF DNA IN CHROMOSOMES The length of DNA in a eukaryotic cell is far greater than the diameter of that cell's nucleus. For instance, the haploid genome of humans contains about 3.2 billion base pairs and, stretched out, the DNA in a diploid cell would be about 2 meters (2 m) long. A human cell is about 40–60 μm in diameter with a nucleus about 10 μm in diameter. DNA fits into a nucleus because it is packed into a shorter length by **histones,** a class of small, positively charged (basic) proteins that are complexed with DNA in eukaryotic chromosomes. The histones bind to DNA by an attraction between their positive charges and the negatively charged phosphate groups of the DNA. Five types of histones exist in most eukaryotic cells: H1, H2A, H2B, H3, and H4. The amino acid sequences of these proteins are highly similar among eukaryotes, indicating that they are evolutionarily conserved and suggesting that they perform the same functions in all eukaryotic organisms. Other proteins called **nonhistone proteins** are also associated with DNA; some of these proteins are also important for the structure of chromosomes, whereas others are involved in gene regulation. The complex of DNA and its associated proteins is termed **chromatin.**

Let us focus on the histones and how they pack DNA at several levels of chromatin structure. In the most fundamental structure, called a **nucleosome,** two molecules each of H2A, H2B, H3, and H4 combine to form a beadlike, eight-protein **nucleosome core particle** around which DNA winds for almost two turns **(Figure 10.2).** A short segment of DNA, the **linker,** extends between one nucleosome and the next. Under the electron microscope, this structure looks like beads on a string. The diameter of the beads (the nucleosomes) gives this structure its name—the **10-nm chromatin fiber** (see Figure 10.2). Each nucleosome and linker includes about 200 base pairs of DNA. Nucleosomes compact DNA by a factor of about 7; that is, a length of DNA becomes about 7 times shorter when it is wrapped into nucleosomes.

The fifth histone, H1, brings about the next level of chromatin packing. One H1 molecule binds both to the nucleosome (at the point where the DNA enters and leaves the core particle) and to the linker DNA. This binding causes the nucleosomes to package into a coiled structure 30 nm in diameter, called the **30-nm chromatin fiber.** One possible model for the 30-nm fiber is the **solenoid model,** with the nucleosomes spiraling helically with about six nucleosomes per turn (see Figure 10.2).

Yet higher levels of chromosome packing also occur. In cells that are not in mitosis (interphase cells—see next section), chromatin fibers are loosely packed in some regions and densely packed in others. The loosely packed regions are known as **euchromatin** (*eu* = true, regular, or typical), and the densely packed regions are called **heterochromatin** (*hetero* = different). Generally speaking, these chromosomes are not visible under a light microscope. During mitosis (and meiosis), chromatin fibers fold and pack further into thick, rodlike chromosomes visible under the light microscope. Experiments indicate that links formed between H1 histone molecules contribute to the packing of chromatin fibers, both into heterochromatin and into the chromosomes visible during nuclear division. However, the exact mechanism for the more complex folding and packing is not known.

STUDY BREAK 10.1

1. What are the three interrelated systems that contribute to the eukaryotic cell cycle?
2. What is the general composition of a eukaryotic chromosome?
3. What is the structure of the nucleosome?
4. What is the role of histone H1 in eukaryotic chromosome structure?

10.2 The Mitotic Cell Cycle

Growth and division of both diploid and haploid eukaryotic cells occurs in the mitotic cell cycle. Before a cell divides in mitosis, duplication of each chromosome produces two identical copies

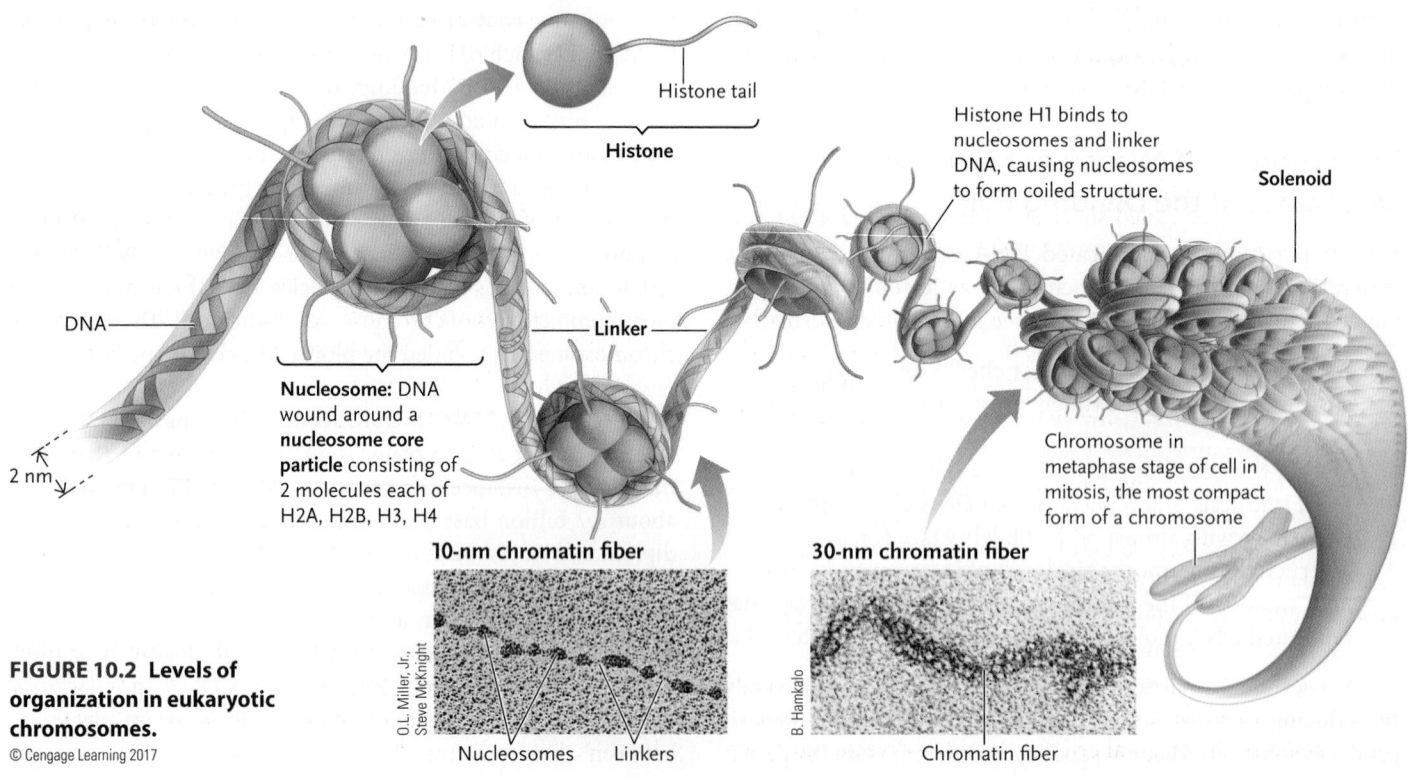

FIGURE 10.2 Levels of organization in eukaryotic chromosomes.
© Cengage Learning 2017

A. Cell cycle events

G_2 refers to the second gap in which there is no DNA synthesis. During G_2, the cell continues to synthesize RNAs and proteins, including those for mitosis, and it continues to grow. The end of G_2 marks the end of interphase; mitosis then begins.

If the cell is going to divide, DNA replication begins. During S phase, the cell duplicates each chromosome, including both the DNA and the chromosomal proteins, and it also continues synthesis of other cellular molecules.

G_1 phase is a period of growth before the DNA replicates. The cell makes various RNAs, proteins, and other types of cellular molecules but not DNA (the G in G_1 stands for *gap*, referring to the absence of DNA synthesis).

B. Chromosomes and DNA molecules at different stages of the cell cycle

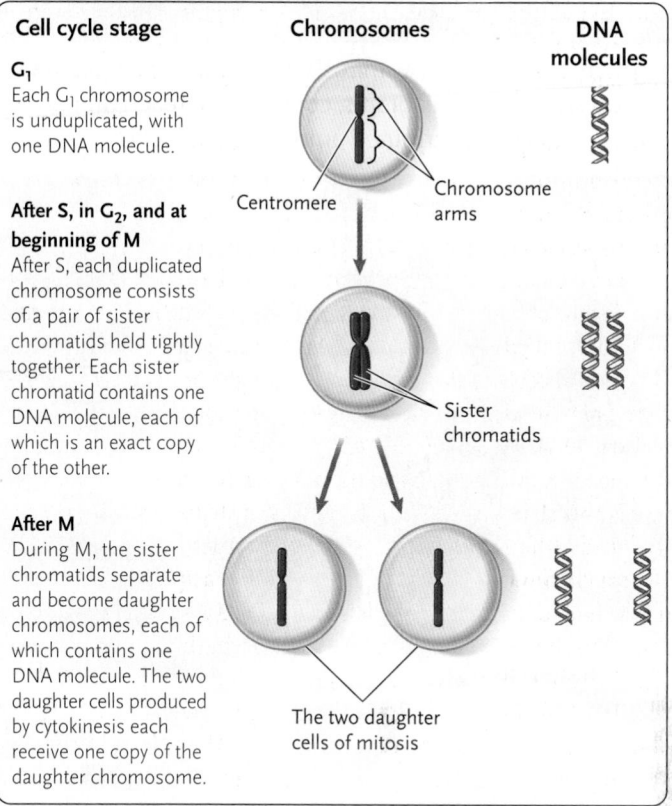

Cell cycle stage	Chromosomes	DNA molecules
G_1 Each G_1 chromosome is unduplicated, with one DNA molecule.		
After S, in G_2, and at beginning of M After S, each duplicated chromosome consists of a pair of sister chromatids held tightly together. Each sister chromatid contains one DNA molecule, each of which is an exact copy of the other.		
After M During M, the sister chromatids separate and become daughter chromosomes, each of which contains one DNA molecule. The two daughter cells produced by cytokinesis each receive one copy of the daughter chromosome.		

FIGURE 10.3 The cell cycle. (A) The length of G_1 varies, but for a given cell type, the timing of S, G_2, and mitosis is usually relatively uniform. Cytokinesis (segment at 2 o'clock in part A) usually begins while mitosis is in progress and reaches completion as mitosis ends. Cells in a state of division arrest enter a shunt from G_1 called G_0 (not shown). **(B)** Each chromosome has a centromere that is important for the segregation of chromosomes in mitosis. The position of the centromere defines the two arms of a chromosome. (Centromeres are discussed in more detail later in this section.)

© Cengage Learning 2017

of each chromosome called **sister chromatids.** Duplication of a chromosome involves replicating the DNA molecule it contains, plus doubling the proteins that are bound to the DNA. Newly formed sister chromatids are held together tightly by **sister chromatid cohesion,** in which proteins called **cohesins** encircle the sister chromatids along their length. During mitosis, the cohesins are removed and the sister chromatids are separated, with one of each pair going to each of the two daughter nuclei. *As a result of this precise division, each daughter nucleus receives exactly the same number and types of chromosomes, and contains the same genetic information, as the parent cell before its chromosomes were duplicated.* The equal distribution of daughter chromosomes into each of the two daughter cells that result from cell division is called **chromosome segregation.**

The accuracy of chromosome replication and segregation in the mitotic cell cycle creates a group of genetically identical cells—**clones** of the original cell. Since all the diverse cell types

of a complex multicellular organism arose by mitosis from a single zygote, they all contain the same genetic information. Forensic scientists rely on this feature of organisms when, for instance, they match the genetic profile of a small amount of tissue (for example, DNA from blood left at the scene of a crime) with a DNA sample from a suspect. In the laboratory, cells may be grown in **cell cultures,** living cells grown in laboratory vessels. Many types of prokaryotic and eukaryotic cells can be cultured in a growth medium optimized for the organism.

Interphase Extends from the End of One Mitosis to the Beginning of the Next Mitosis

If we set the formation of a new daughter cell as the beginning of the mitotic cell cycle, then the first and longest stage is **interphase (Figure 10.3A).** Three phases of the cell cycle comprise interphase:

THE STAGES OF MITOSIS

Light micrographs show mitosis in an animal cell. Diagrams show mitosis in an animal cell with two pairs of chromosomes. As you study these diagrams, consider that each diploid human cell, although only 40 to 60 μm in diameter, contains 2 meters of DNA distributed among 23 pairs of chromosomes.

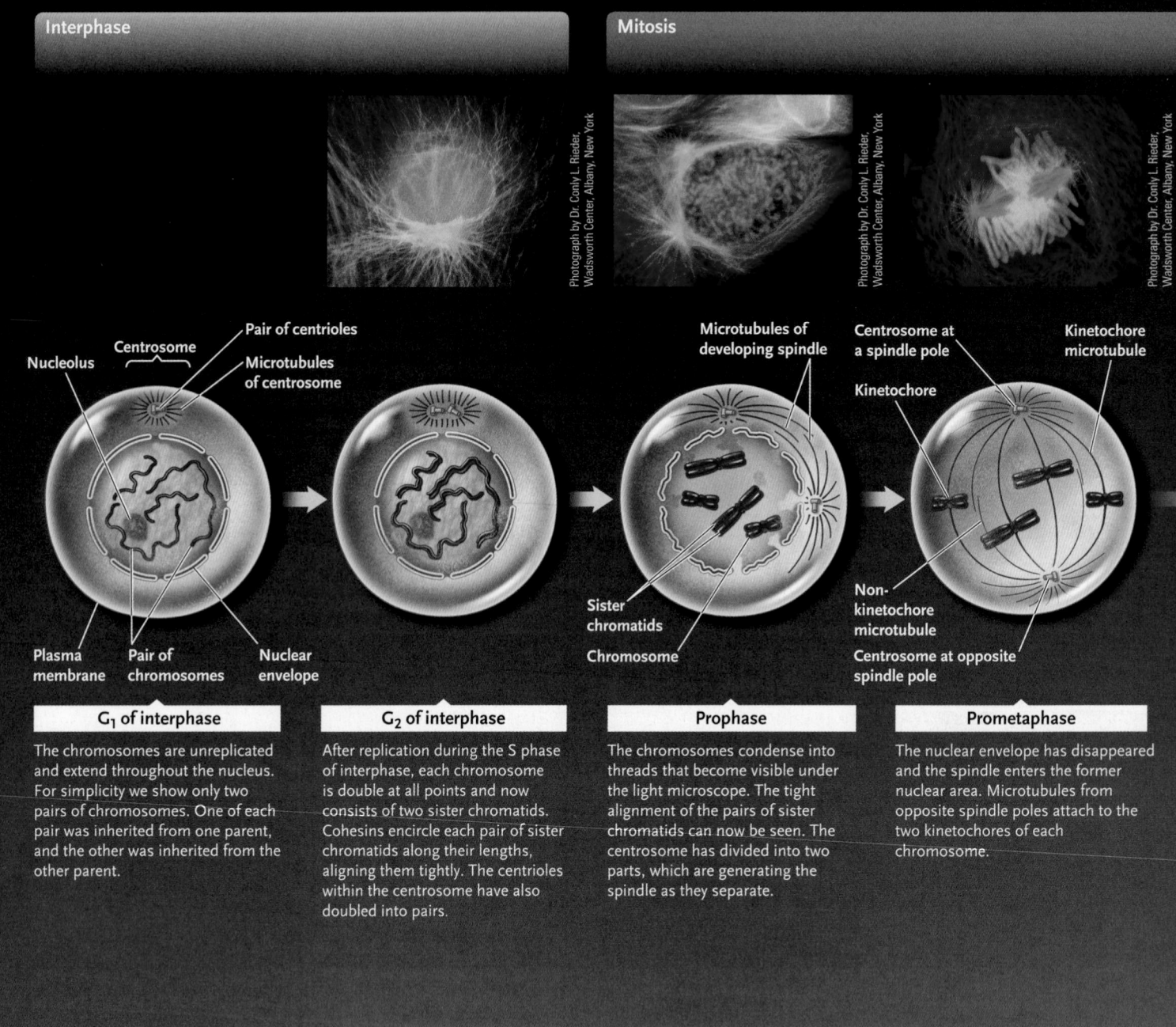

Interphase

Mitosis

Photograph by Dr. Conly L. Rieder, Wadsworth Center, Albany, New York

Nucleolus

Centrosome

Pair of centrioles

Microtubules of centrosome

Microtubules of developing spindle

Centrosome at a spindle pole

Kinetochore microtubule

Kinetochore

Plasma membrane

Pair of chromosomes

Nuclear envelope

Sister chromatids

Chromosome

Non-kinetochore microtubule

Centrosome at opposite spindle pole

G_1 of interphase	G_2 of interphase	Prophase	Prometaphase
The chromosomes are unreplicated and extend throughout the nucleus. For simplicity we show only two pairs of chromosomes. One of each pair was inherited from one parent, and the other was inherited from the other parent.	After replication during the S phase of interphase, each chromosome is double at all points and now consists of two sister chromatids. Cohesins encircle each pair of sister chromatids along their lengths, aligning them tightly. The centrioles within the centrosome have also doubled into pairs.	The chromosomes condense into threads that become visible under the light microscope. The tight alignment of the pairs of sister chromatids can now be seen. The centrosome has divided into two parts, which are generating the spindle as they separate.	The nuclear envelope has disappeared and the spindle enters the former nuclear area. Microtubules from opposite spindle poles attach to the two kinetochores of each chromosome.

© Cengage Learning 2017

1. **G_1 phase,** in which the cell carries out its function, and in some cases grows. In this phase, each chromosome is unduplicated with one DNA molecule **(Figure 10.3B).**
2. **S phase,** in which DNA replication and chromosome duplication occur. Duplication produces two sister chromatids

for each parental chromosome, each chromatid containing one DNA molecule (see Figure 10.3B). The sister chromatids remain together until separated during M phase.
3. **G_2 phase,** a brief gap in the cell cycle during which cell growth continues and the cell prepares for mitosis (the

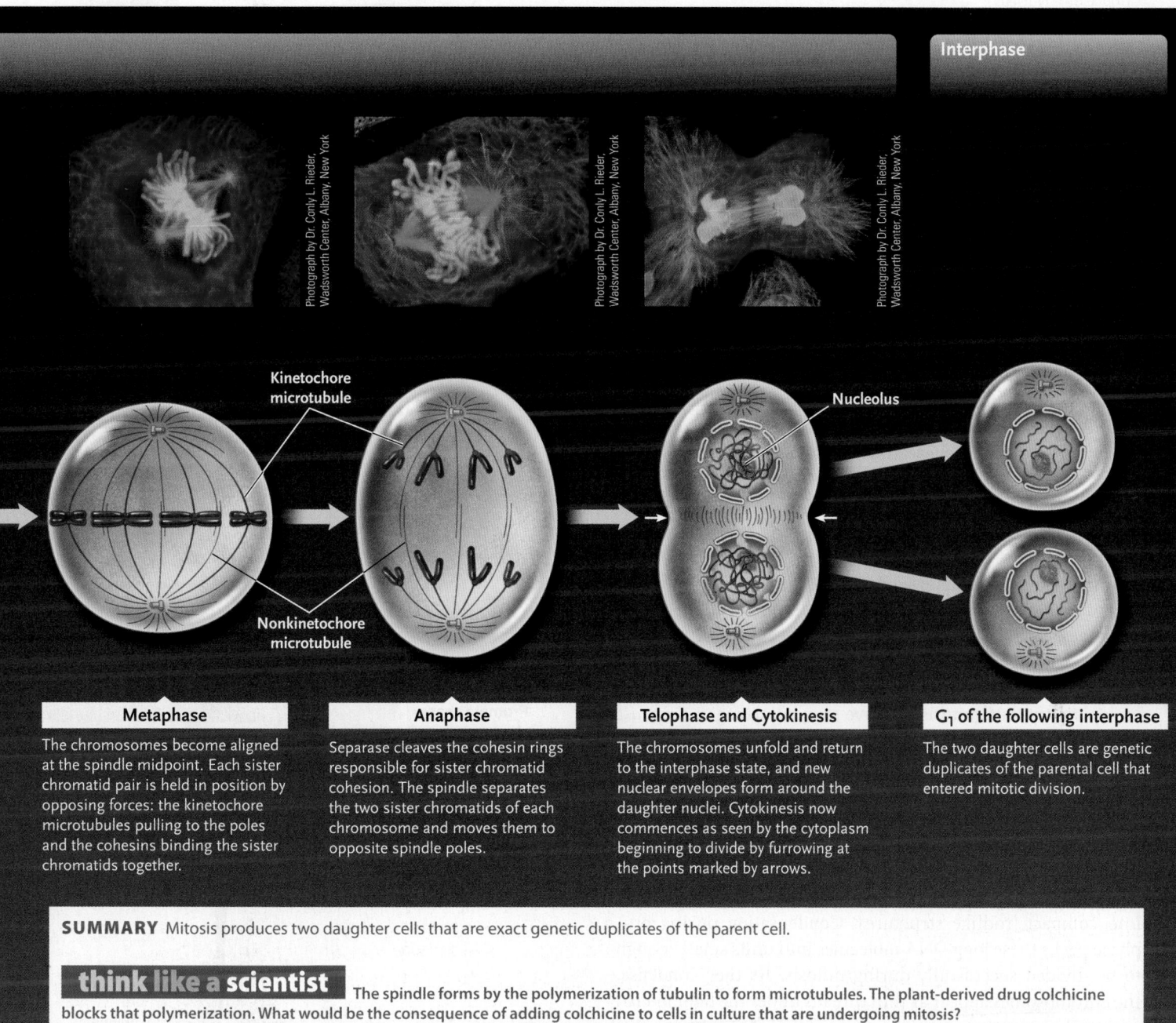

Photograph by Dr. Conly L. Rieder, Wadsworth Center, Albany, New York

Photograph by Dr. Conly L. Rieder, Wadsworth Center, Albany, New York

Photograph by Dr. Conly L. Rieder, Wadsworth Center, Albany, New York

Photograph by Dr. Conly L. Rieder, Wadsworth Center, Albany, New York

Kinetochore microtubule

Nonkinetochore microtubule

Nucleolus

Metaphase

The chromosomes become aligned at the spindle midpoint. Each sister chromatid pair is held in position by opposing forces: the kinetochore microtubules pulling to the poles and the cohesins binding the sister chromatids together.

Anaphase

Separase cleaves the cohesin rings responsible for sister chromatid cohesion. The spindle separates the two sister chromatids of each chromosome and moves them to opposite spindle poles.

Telophase and Cytokinesis

The chromosomes unfold and return to the interphase state, and new nuclear envelopes form around the daughter nuclei. Cytokinesis now commences as seen by the cytoplasm beginning to divide by furrowing at the points marked by arrows.

G_1 of the following interphase

The two daughter cells are genetic duplicates of the parental cell that entered mitotic division.

SUMMARY Mitosis produces two daughter cells that are exact genetic duplicates of the parent cell.

think like a scientist The spindle forms by the polymerization of tubulin to form microtubules. The plant-derived drug colchicine blocks that polymerization. What would be the consequence of adding colchicine to cells in culture that are undergoing mitosis?

fourth phase of the cell cycle; also called *M phase*) followed by cytokinesis.

Usually, G_1 is the only phase of the cell cycle that varies in length. The other phases are typically uniform in length within a species. Thus, whether cells divide rapidly or slowly depends primarily on the length of G_1. Once DNA replication begins, most mammalian cells take about 10 to 12 hours to proceed through the S phase, about 4 to 6 hours to go through G_2, and about 1 hour or less to complete mitosis.

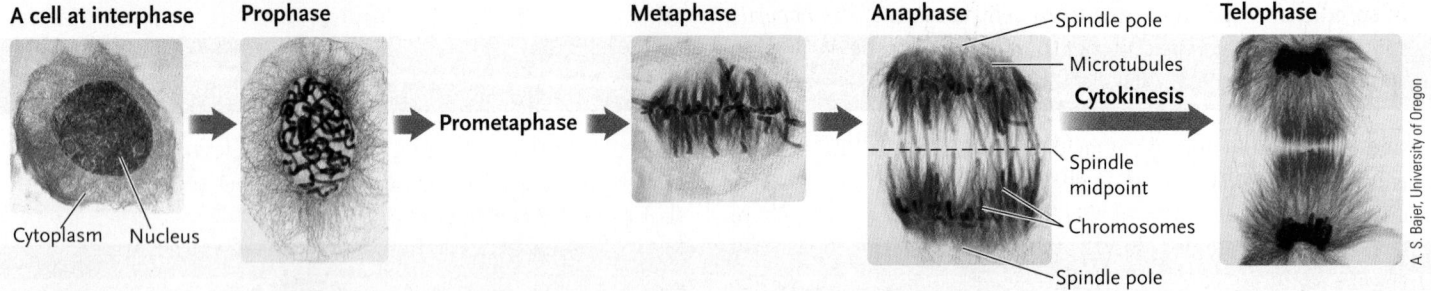

A cell at interphase | Prophase | Metaphase | Anaphase — Spindle pole, Microtubules, **Cytokinesis**, Spindle midpoint, Chromosomes, Spindle pole | Telophase

Prometaphase

Cytoplasm Nucleus

A. S. Bajer, University of Oregon

FIGURE 10.5 Mitosis in a plant cell. The chromosomes are stained blue and the spindle microtubules are stained red.

G_1 is also the stage in which many cell types stop dividing. Cells that are not destined to divide immediately enter a shunt from G_1 called the **G_0 phase.** In some cases, a cell in G_0 may start dividing again by reentering G_1. Some cells never resume the cell cycle; for example, most cells of the human nervous system stop dividing once they are fully mature.

Internal regulatory controls trigger each phase of the cell cycle, ensuring that the processes of one phase are completed successfully before the next phase can begin. Various internal mechanisms also regulate the overall number of cycles that a cell goes through. These internal controls may be subject to various external influences such as other cells or viruses, as well as signal molecules, including hormones, growth factors, and death signals.

After Interphase, Mitosis Proceeds in Five Stages

Mitosis follows interphase. During mitosis, the sister chromatids of each chromosome separate and become daughter chromosomes, each with one DNA molecule (see Figure 10.3B).

Once it begins, mitosis proceeds continuously, without significant pauses or breaks. However, for convenience in study, biologists separate mitosis into five sequential stages: *prophase* (*pro* = before), *prometaphase* (*meta* = between), *metaphase, anaphase* (*ana* = back), and *telophase* (*telo* = end). **Figure 10.4** and **Figure 10.5** show the process of mitosis in an animal cell and in a plant cell, respectively.

PROPHASE During **prophase,** the greatly extended chromosomes that were replicated during interphase begin to condense into compact, rodlike structures. Condensation during prophase packs these long DNA molecules into units small enough to be divided successfully during mitosis. As they condense, the chromosomes appear as thin threads under the light microscope. The word *mitosis* (*mitos* = thread) is derived from this threadlike appearance.

While condensation is in progress, the nucleolus becomes smaller and eventually disappears in most species. The disappearance reflects a substantial reduction in RNA synthesis, including the ribosomal RNA made in the nucleolus.

In the cytoplasm, the mitotic **spindle** (**Figure 10.6;** see also Figure 10.12 later in this chapter), the structure that will later separate the chromatids, begins to form between the two

centrosomes as they start migrating toward the opposite ends of the cell, where they will form the **spindle poles.** The spindle develops as two bundles of microtubules that radiate from the two spindle poles.

PROMETAPHASE At the end of prophase, the nuclear envelope breaks down, marking the beginning of **prometaphase.** Bundles

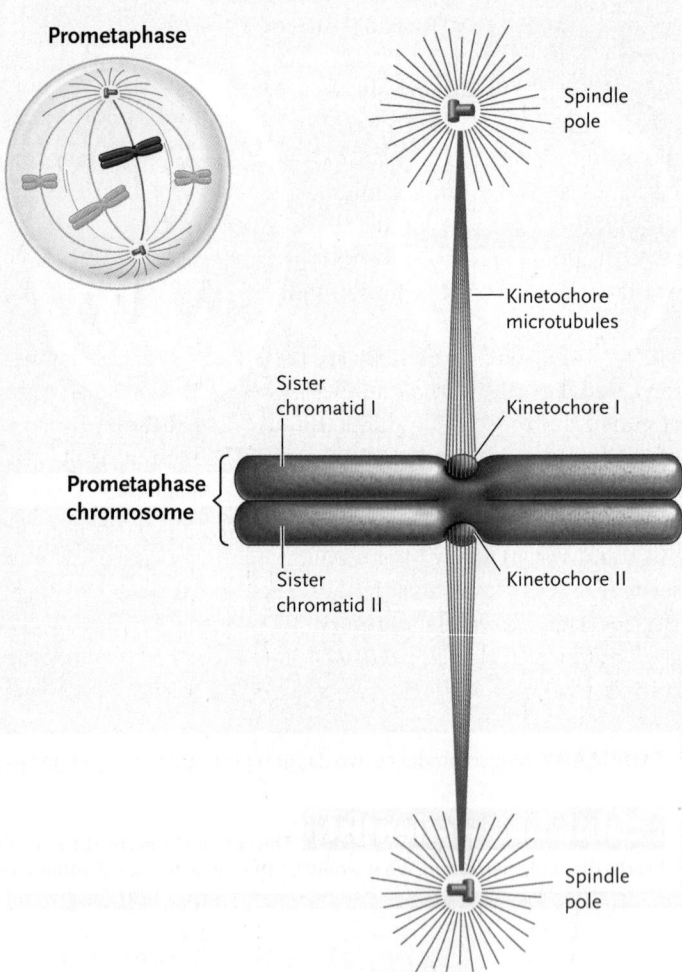

Prometaphase

Spindle pole

Kinetochore microtubules

Sister chromatid I

Kinetochore I

Prometaphase chromosome

Sister chromatid II

Kinetochore II

Spindle pole

FIGURE 10.6 Spindle connections made by chromosomes at mitotic prometaphase. The two kinetochores of the chromosome connect by kinetochore microtubules to opposite spindle poles, ensuring that the chromatids are separated and moved to opposite spindle poles during anaphase.

© Cengage Learning 2017

of spindle microtubules grow from centrosomes at the *opposite spindle poles* toward the center of the cell. Some of the developing spindle enters the former nuclear area.

Each chromosome is still in a duplicated state made of two identical sister chromatids held together throughout their entire lengths by sister chromatid cohesion. Each sister chromatid has a region of DNA called a **centromere** where the chromatid is more closely attached to its sister chromatid than elsewhere along the length of the chromosome. For condensed chromosomes, the paired centromeres of the sister chromatids are visible as a constriction along the length of the pair of sister chromatids. The position of the centromere is chromosome-specific: it may be central, towards one end or the other, or at or near a chromosome end. The position of a centromere gives the chromosome a particular morphology and biologists describe the segments of chromosomes to each side of the centromere as *chromosome arms* (see Figure 10.3B).

In prometaphase, a complex of several proteins, a **kinetochore,** forms on each chromatid at the centromere, a region located at a particular position in a given chromosome. The centromere is often narrower than the rest of the chromosome. **Kinetochore microtubules** originating from the spindle poles bind to the kinetochores. These connections determine the outcome of mitosis, because they attach the sister chromatids of each chromosome to microtubules that lead to the opposite spindle poles (see Figure 10.6). Microtubules that do not attach to kinetochores—the **nonkinetochore microtubules**— overlap those from the opposite spindle pole.

METAPHASE During **metaphase,** the spindle reaches its final form and the spindle microtubules move the chromosomes into alignment at the spindle midpoint, also called the *metaphase plate.* The chromosomes complete their condensation in this stage. The pattern of condensation gives each chromosome a characteristic shape, determined by the location of the centromere and the lengths of the chromatid arms (**Figure 10.7;** also see Figure 10.2). Remember that each chromosome at this stage still consists of two sister chromatids.

Only when the chromosomes are all assembled at the spindle midpoint, with the two sister chromatids of each one

attached to microtubules leading to opposite spindle poles, can metaphase give way to actual separation of chromatids in anaphase.

The complete set of metaphase chromosomes, arranged according to size and centromere position, forms the **karyotype** of a given species. In many cases, the karyotype is so distinctive that a species can be identified from this characteristic alone. **Figure 10.8** shows how human chromosomes are prepared for analysis as a karyotype.

ANAPHASE The proper alignment of chromosomes in metaphase triggers anaphase. An enzyme, *separase,* is activated and cleaves the cohesin rings around the pairs of sister chromatids. This cancels the force opposing the pull of sister chromatids to opposite poles. Thus, during **anaphase,** the spindle separates sister chromatids and pulls them to opposite spindle poles. The first signs of chromosome movement can be seen at the centromeres, where tension developed by the spindle pulls the kinetochores toward the poles. The movement continues until the separated chromatids, now called **daughter chromosomes,** have reached the two poles. At this point, chromosome segregation is complete.

TELOPHASE During **telophase,** the spindle disassembles and the chromosomes at each spindle pole decondense, returning to the extended state typical of interphase. As decondensation proceeds, the nucleolus reappears, RNA transcription resumes, and a new nuclear envelope forms around the chromosomes at each pole producing the two daughter nuclei. At this point, nuclear division is complete.

Mitosis produces two daughter nuclei, each with identical sets of chromosomes compared to the parental cell. Cytokinesis, the division of the cytoplasm, typically follows the nuclear division stage of mitosis, and produces two daughter cells each with one of the two daughter nuclei. Cytokinesis proceeds by different pathways in the various kingdoms of eukaryotic organisms. In animals, protists, and many fungi, a groove, the **furrow,** girdles the cell and gradually deepens until it cuts the cytoplasm into two parts (**Figure 10.9**). In plants, a new cell wall, called the **cell plate,** forms between the daughter nuclei and grows laterally until it divides the cytoplasm in two (**Figure 10.10**). In both cases, the plane of cytoplasmic division is determined by the layer of microtubules that persists at the former spindle midpoint.

The Mitotic Cell Cycle Is Significant for Both Development and Reproduction

The mitotic cycle of interphase, nuclear division, and cytokinesis accounts for the growth of multicellular eukaryotes from single initial cells, such as a fertilized egg, to fully developed adults. Mitosis also serves as a mechanism of organismal reproduction called **vegetative reproduction** or **asexual reproduction,** which occurs in many kinds of plants and protists and in some animals. In asexual reproduction, daughter

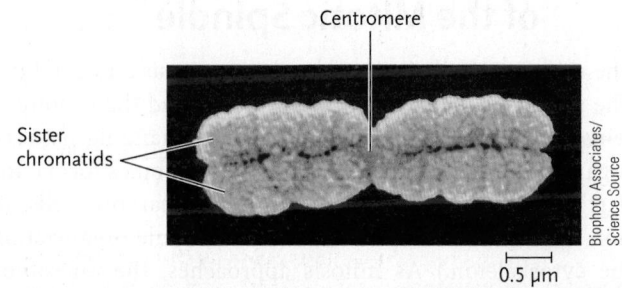

FIGURE 10.7 Metaphase chromosome, the most condensed form of a eukaryotic chromosome (scanning electron micrograph). At this stage of the cell cycle, the chromosome is duplicated; the shape results from the centromere position along the paired two sister chromatids.

FIGURE 10.8 | **Research Method**

Preparing a Human Karyotype

Purpose: A karyotype is a display of chromosomes of an organism arranged in pairs. A normal karyotype has a characteristic appearance for each species. Examination of the karyotype of the chromosomes from a particular individual indicates whether the individual has a normal set of chromosomes or whether there are abnormalities in number or appearance of individual chromosomes. A normal karyotype can be used to indicate the species.

Protocol:

1. Add sample to culture medium that has stimulator for growth and division of cells (white blood cells in the case of blood). Incubate at 37°C. Add colchicine, which blocks the formation of microtubules. As a result, the spindle does not form and this causes mitosis to arrest at metaphase.

2. Stain the cells so that the chromosomes are distinguished. Some stains produce chromosome-specific banding patterns, as shown in the photograph below.

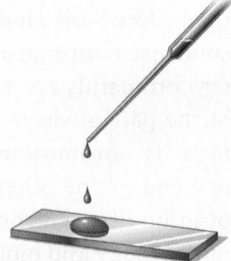

3. View the stained cells under a microscope equipped with a digital imaging system and take a digital photograph. A computer processes the photograph to arrange the chromosomes in pairs and numbers them according to size and shape, as shown in the photograph.

Pair of homologous chromosomes

Pair of sister chromatids closely aligned side-by-side by sister chromatid cohesion

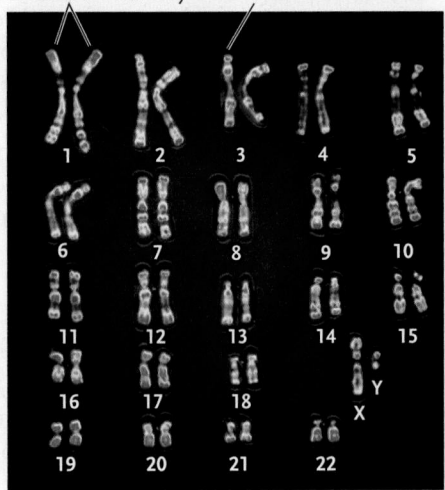

CNRI/Science Source

Interpreting the Results: The karyotype is evaluated with respect to the scientific question being asked. For example, it may identify a particular species, or it may indicate whether or not the chromosome set of a human (fetus, child, or adult) is normal or aberrant.

cells produced by mitotic cell division grow by further mitosis into complete individuals. For example, asexual reproduction occurs when a single-celled protist such as an amoeba divides by mitosis to produce two separate individuals, or when a stem cutting is used to propagate an entire new plant.

STUDY BREAK 10.2

1. Compare the chromosome content of daughter cells following mitosis with that of the parent cell before its chromosomes were duplicated.
2. In what order do the stages of mitosis occur?
3. What is the importance of centromeres to mitosis?

10.3 Formation and Action of the Mitotic Spindle

The mitotic spindle is central to both mitosis and cytokinesis. The spindle is made up of microtubules and their motor proteins, and its activities depend on their changing patterns of organization during the cell cycle. Microtubules form a major part of the interphase cytoskeleton of eukaryotic cells. (Section 4.3 outlines the patterns of microtubule organization in the cytoskeleton.) As mitosis approaches, the microtubules disassemble from their interphase arrangement and reorganize into the spindle, which grows until it fills almost the entire cell. This reorganization follows one of two pathways in

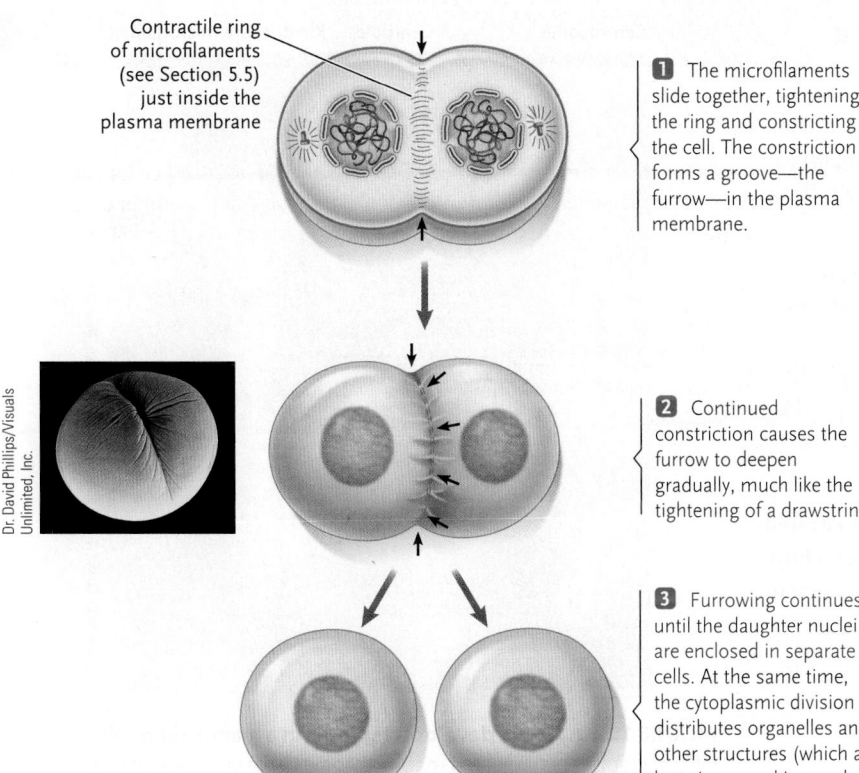

Contractile ring of microfilaments (see Section 5.5) just inside the plasma membrane

1 The microfilaments slide together, tightening the ring and constricting the cell. The constriction forms a groove—the furrow—in the plasma membrane.

2 Continued constriction causes the furrow to deepen gradually, much like the tightening of a drawstring.

3 Furrowing continues until the daughter nuclei are enclosed in separate cells. At the same time, the cytoplasmic division distributes organelles and other structures (which also have increased in number) approximately equally between the cells.

Dr. David Phillips/Visuals Unlimited, Inc.

FIGURE 10.9 Cytokinesis by furrowing. The micrograph shows a furrow developing in the first division of a fertilized egg cell.
© Cengage Learning 2017

Animals and Plants Form Spindles in Different Ways

Figure 10.11 shows the **centrosome,** a site near the nucleus from which microtubules radiate outward in all directions, and its role in spindle formation. The centrosome is the main **microtubule organizing center (MTOC)** of animal cells and many protists. The centrosome contains a pair of **centrioles,** arranged at right angles to each other.

When the nuclear envelope breaks down at the end of prophase, the spindle (see Figure 10.11, step 4) moves into the region formerly occupied by the nucleus and continues growing until it fills the cytoplasm. The microtubules that extend from the centrosomes also grow in length and extent, producing radiating arrays that appear starlike under the light microscope. Initially named by early microscopists, **asters** (*aster* = star) are the centrosomes at the spindle tips, which form the poles of the spindle. By separating the duplicated centrioles, the spindle ensures that, when the cytoplasm divides during cytokinesis, the daughter cells each receive a pair of centrioles. Angiosperms (flowering plants) and most gymnosperms, such as conifers, lack centrosomes and centrioles. In these organisms, the spindle forms from microtubules that assemble in all directions from multiple MTOCs surrounding the entire nucleus (see prophase in Figure 10.5). Then, when the nuclear envelope breaks down at the end of prophase, the spindle moves into the former nuclear region.

different organisms, depending on the presence or absence of a *centrosome* during interphase. However, once organized, the basic function of the spindle is the same, regardless of whether a centrosome is present or not.

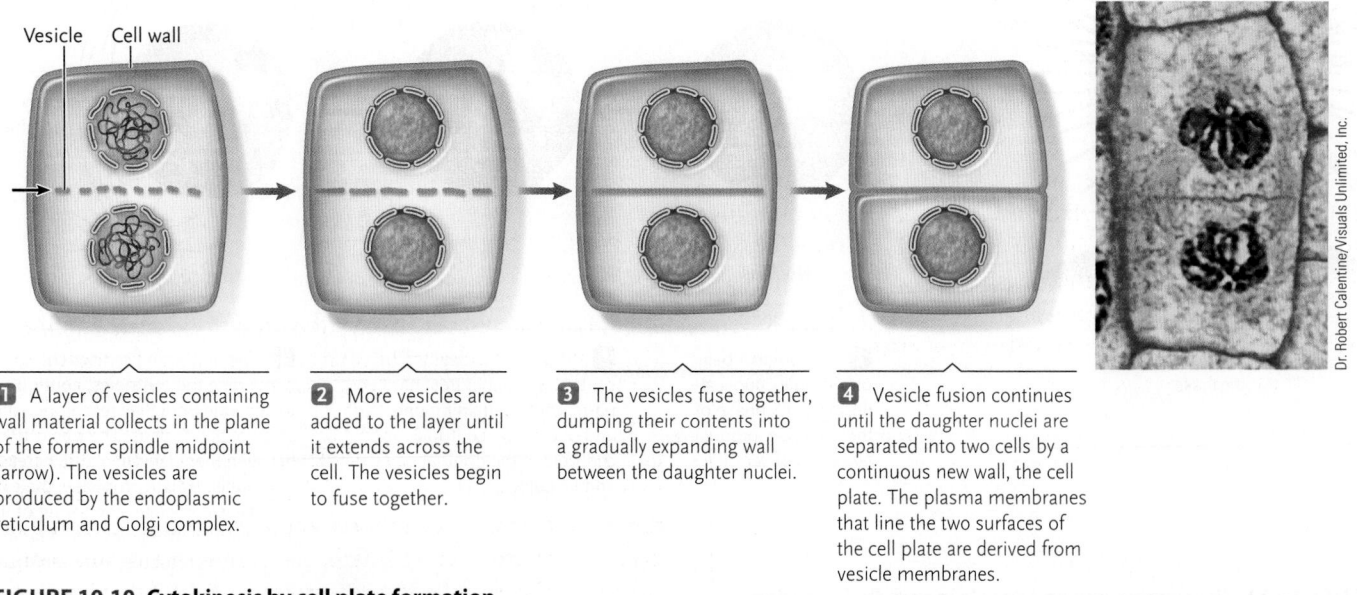

Vesicle   Cell wall

1 A layer of vesicles containing wall material collects in the plane of the former spindle midpoint (arrow). The vesicles are produced by the endoplasmic reticulum and Golgi complex.

2 More vesicles are added to the layer until it extends across the cell. The vesicles begin to fuse together.

3 The vesicles fuse together, dumping their contents into a gradually expanding wall between the daughter nuclei.

4 Vesicle fusion continues until the daughter nuclei are separated into two cells by a continuous new wall, the cell plate. The plasma membranes that line the two surfaces of the cell plate are derived from vesicle membranes.

Dr. Robert Calentine/Visuals Unlimited, Inc.

FIGURE 10.10 Cytokinesis by cell plate formation.
© Cengage Learning 2017

Mitotic Spindles Move Chromosomes by a Combination of Two Mechanisms

When fully formed at metaphase, the spindle may contain from hundreds to many thousands of microtubules, depending on the species **(Figure 10.12)**. As you learned earlier in the chapter, these microtubules are divided into two groups (see Figure 10.4):

1. *Kinetochore microtubules* connect the spindle poles to kinetochores, the complexes of proteins that form at centromeres during prometaphase.
2. *Nonkinetochore microtubules* extend between the spindle poles without connecting to chromosomes; at the spindle midpoint, the microtubules from one pole overlap with microtubules from the opposite pole.

Figure 10.13 describes an experiment that demonstrated that chromosomes can move by sliding along kinetochore microtubules toward the poles. Mechanistically this movement involves motor proteins in the kinetochores "walking" the

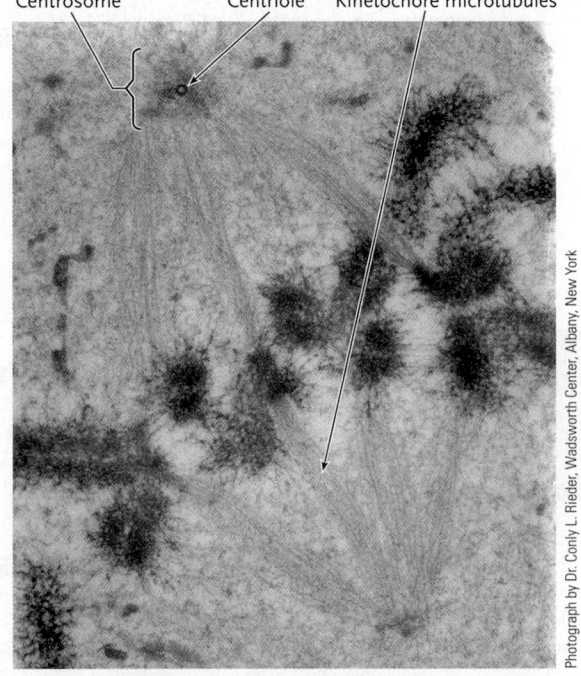

Centrosome   Centriole   Kinetochore microtubules

Photograph by Dr. Conly L. Rieder, Wadsworth Center, Albany, New York

FIGURE 10.12 A fully developed spindle in a mammalian cell. Only microtubules connected to chromosomes have been caught in the plane of this section. One of the centrioles is visible in cross section in the centrosome at the top of the micrograph.

Prophase

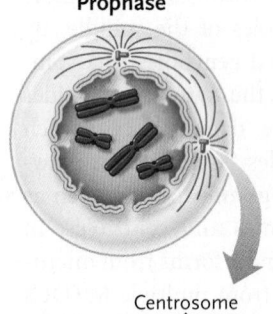

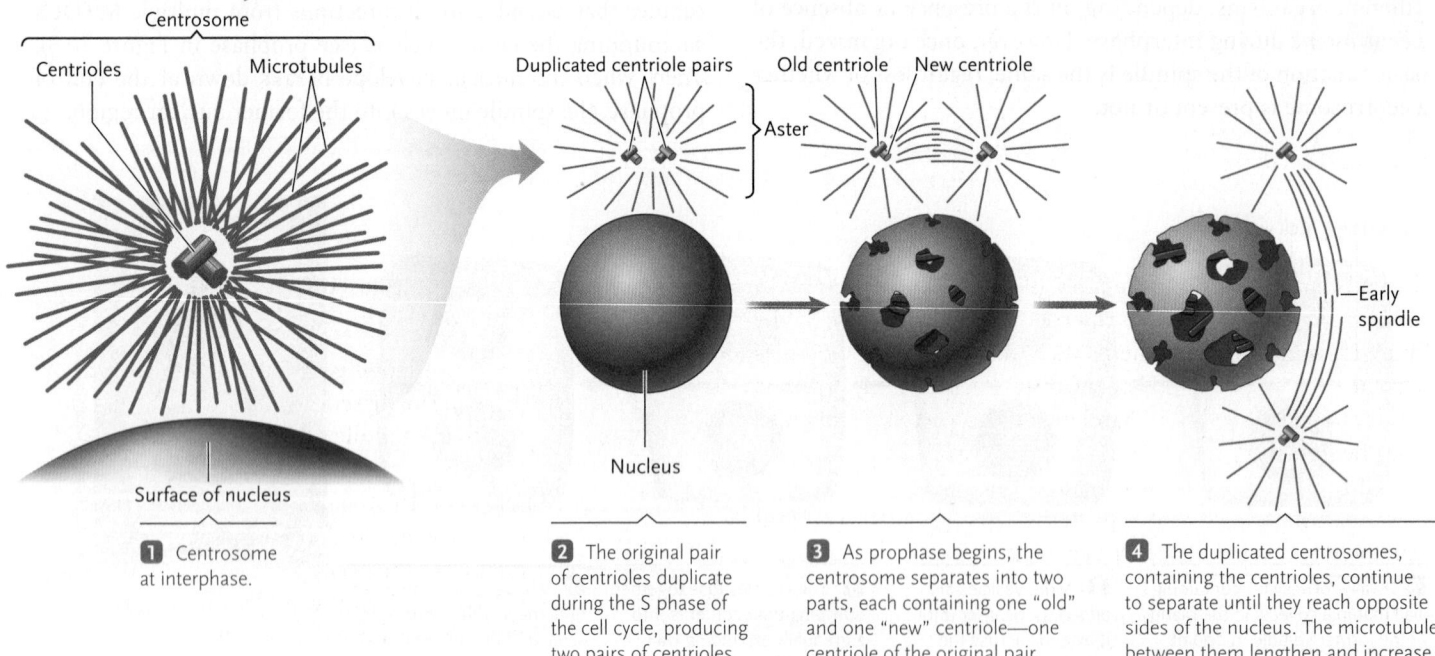

Centrosome

Centrioles   Microtubules

Surface of nucleus

Duplicated centriole pairs   Old centriole   New centriole

Aster

Nucleus

Early spindle

1 Centrosome at interphase.

2 The original pair of centrioles duplicate during the S phase of the cell cycle, producing two pairs of centrioles.

3 As prophase begins, the centrosome separates into two parts, each containing one "old" and one "new" centriole—one centriole of the original pair and its copy.

4 The duplicated centrosomes, containing the centrioles, continue to separate until they reach opposite sides of the nucleus. The microtubules between them lengthen and increase in number. By late prophase, the early spindle is complete, consisting of the separated centrosomes and a large mass of microtubules between them.

FIGURE 10.11 The centrosome and its role in spindle formation.
© Cengage Learning 2017

Movement of Chromosomes during Anaphase of Mitosis

Question: How do chromosomes move during anaphase of mitosis?

Experiment: One hypothesis for how chromosomes move during anaphase of mitosis was that the kinetochore microtubules moved, pulling chromosomes to the poles. An alternative hypothesis was that chromosomes move by sliding over or along kinetochore microtubules. To test the hypotheses, G. J. Gorbsky and his colleagues made regions of the kinetochore microtubules visibly distinct.

1. Kinetochore microtubules were combined with a dye molecule that bleaches when it is exposed to light.

2. The region of the spindle between the kinetochores and the poles was exposed to a microscopic beam of light that bleached a narrow stripe across the microtubules. The bleached region could be seen with a light microscope and analyzed as anaphase proceeded.

Results: The bleached region remained at the same distance from the pole as the chromosomes moved toward the pole.

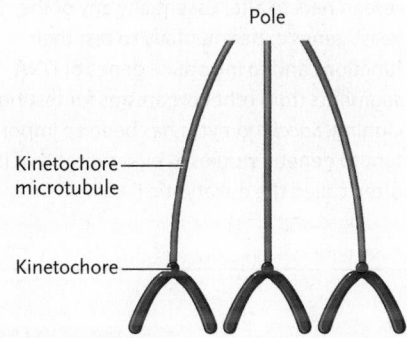

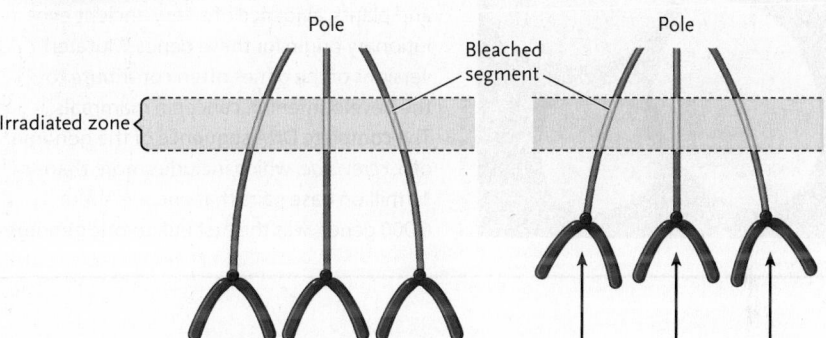

Conclusion: The results support the hypothesis that chromosomes move by sliding over or along kinetochore microtubules.

| Observe |
| Hypothesize |
| **Predict** |
| Experiment |
| Interpret |

What result would have supported the hypothesis that, during anaphase of mitosis, the kinetochore microtubules moved, pulling chromosomes to the poles?

Source: G. J. Gorbsky, P. J. Sammak, and G. G. Borisy. 1987. Chromosomes moved poleward in anaphase along stationary microtubules that coordinately disassemble from their kinetochore microtubules. *Journal of Cell Biology* 104:9–18.

chromosomes along microtubules **(Figure 10.14)**. The tubulin subunits of the kinetochore microtubules disassemble as the kinetochores pass along them. As a result, the microtubules become shorter as the movement progresses. The movement is similar to pulling yourself, hand over hand, up a rope as it falls apart behind you.

Chromosomes can also move toward the poles by a mechanism in which motor proteins at the spindle poles pull kinetochore microtubules polewards, disassembling those microtubules into tubulin subunits as that occurs. Both walking and pulling mechanisms are used in mitosis, though the relative contributions of the two mechanisms to chromosome movement varies among species and cell types. (The cell type used in the experiment of Figure 10.13 used walking predominantly.)

As a dividing cell goes through anaphase, it elongates (compare Metaphase and Anaphase in Figure 10.4). Nonkinetochore microtubules are responsible for this cell elongation in two ways:

1. Motor proteins on overlapping nonkinetochore microtubules walk in opposite directions, thereby reducing the extent of overlap at the spindle midpoint. This activity pushes the spindle poles further apart.

2. The nonkinetochore microtubules also push the poles apart by growing in length as they slide along. This activity also maintains the overlap of pairs of nonkinetochore microtubules at the spindle midpoint even as the motor proteins work to reduce the overlap.

STUDY BREAK 10.3

1. How does spindle formation differ in animals and plants?
2. How do mitotic spindles move chromosomes?

Model Organisms: The Yeast *Saccharomyces cerevisiae*

Saccharomyces cerevisiae, commonly known as budding yeast, baker's yeast, or brewer's yeast, was probably the first microorganism to have been grown and kept in cultures—a

SCIMAT/Science Source

beer-brewing vessel is basically an *S. cerevisiae* culture. The yeast has also been widely used in scientific research. Its microscopic size and relatively short generation time make it easy and inexpensive to culture in large numbers in the laboratory.

Genetic studies with *S. cerevisiae* led to the discovery of some of the genes that control the eukaryotic cell cycle. Many of these genes, after their first discovery in yeast cells, were found to have counterparts in animals and plants, evidence of a very ancient evolutionary origin for these genes. Mutated versions of the genes often contribute to the development of cancer in mammals. The complete DNA sequence of the genome of *S. cerevisiae,* which includes more than 12 million base pairs that encode about 6,000 genes, was the first eukaryotic genome

to be obtained. Analysis of the genome sequence revealed that yeast has many genes similar in sequence to counterpart genes in animals, including mammals, making this relatively simple microorganism an excellent subject for research that can be applied to the more complex species of interest, especially humans.

The genes of yeast can also be manipulated easily using genetic engineering techniques. This has made it possible for researchers to alter essentially any of the yeast genes experimentally to test their functions and to introduce genes or DNA segments from other organisms for testing or cloning. *Saccharomyces* has been so important to genetic studies in eukaryotes that it is often called the eukaryotic *E. coli.*

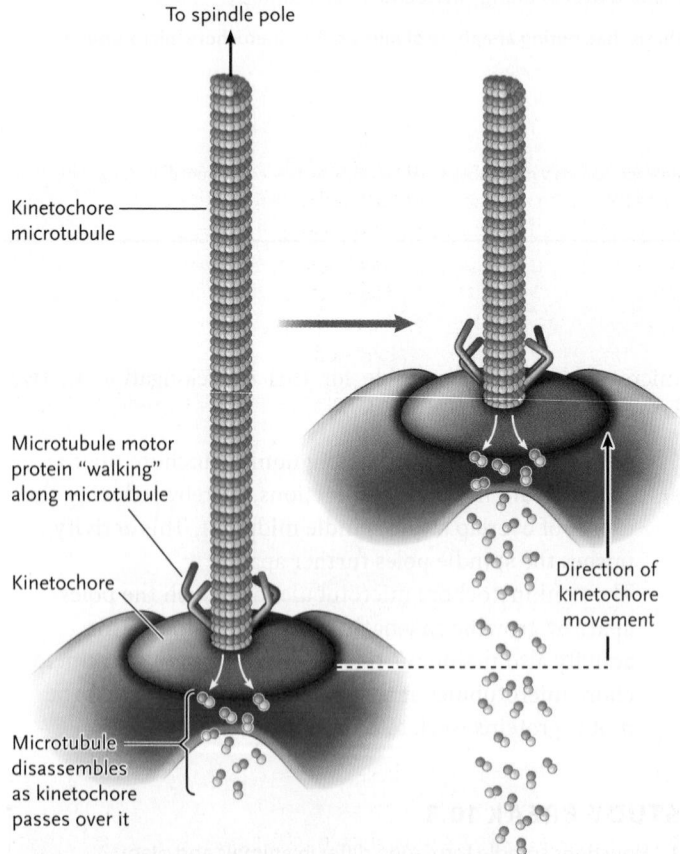

FIGURE 10.14 Microtubule motor proteins "walking" the kinetochore of a chromosome along a microtubule.
© Cengage Learning 2017

10.4 Cell Cycle Regulation

In this section we discuss experimental evidence for (and the operation of) regulatory mechanisms that control the mitotic cell cycle.

Cell Fusion Experiments and Studies of Yeast Mutants Identified Molecules That Control the Cell Cycle

The first insights into how the cell cycle is regulated came from experiments by Robert T. Johnson and Potu N. Rao at the University of Colorado Medical Center, Denver, published in 1970. They fused human HeLa cells (a type of cancer cell that can be grown in cell culture) that were in different stages of the cell cycle and determined whether one nucleus could influence the other **(Figure 10.15).** Their results suggested that specific molecules in the cytoplasm cause the progression of cells from G_1 to S, and from G_2 into M.

Some key research using budding yeast, *S. cerevisiae,* helped to identify these cell cycle control molecules and contributed to our general understanding of how the cell cycle is regulated. (*Focus on Research: Model Organisms* describes budding yeast and its role in research in more detail.) In particular, Leland Hartwell of the Fred Hutchinson Cancer Center, Seattle, investigated yeast mutants that become stuck at some point in the cell cycle, but only when they are cultured at a high temperature. By growing the mutant cells initially at the standard temperature, and then shifting the cells to the higher temperature, Hartwell

FIGURE 10.15 | **Experimental Research**

Demonstrating the Existence of Molecules Controlling the Cell Cycle by Cell Fusion

Question: Do molecules in the cytoplasm direct the progression through the cell cycle?

Experiment: Johnson and Rao fused human HeLa cells at different stages of the cell cycle. Cell fusion produces a single cell with two separate nuclei. The researchers allowed the fused cells to grow and determined whether one nucleus influenced the other in terms of progression through the cell cycle.

1. Fusion of cell in S phase with cell in G_1 phase.

2. Fusion of cell in M (mitosis) with cell in any other stage.

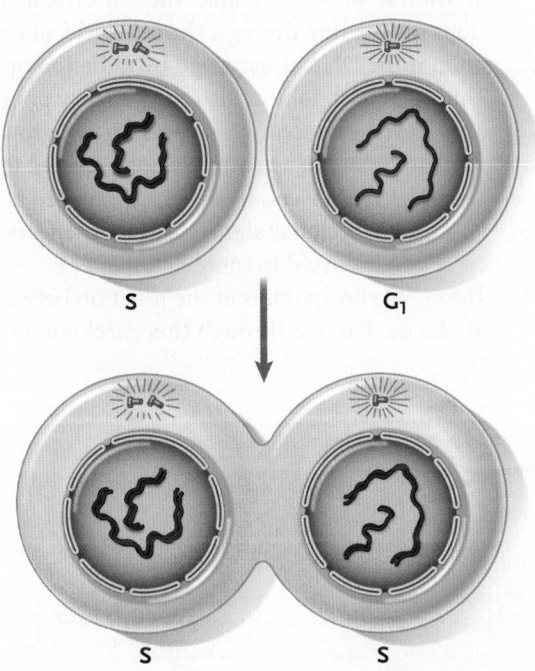

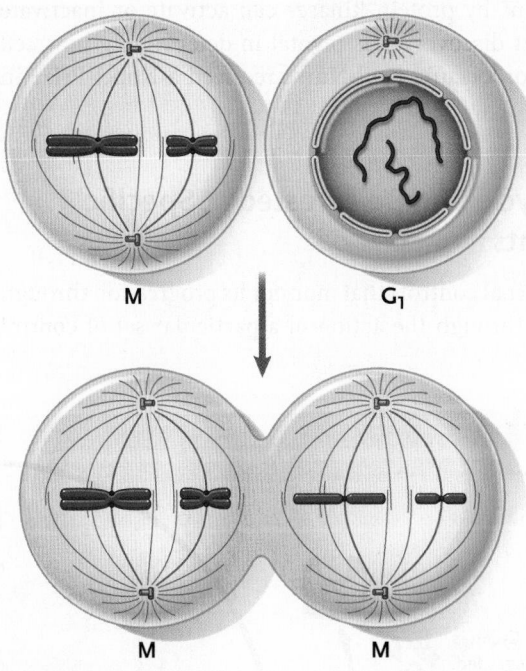

Result: DNA synthesis quickly began in the original G_1 nucleus. Normally, the G_1 nucleus would not have initiated DNA synthesis until it reached S phase itself, which could have been several hours later. The result suggested that one or more molecules that activate S phase are present in the cytoplasm of S phase cells.

Result: Regardless of the phase of the cell, the nucleus of the cell with which the M phase cell was fused immediately began the early stages of mitosis. This included condensation of the chromosomes, spindle formation, and breaking down of the nuclear envelope. For a cell in G_1 (shown in the diagram), the condensed chromosomes that appear have not replicated.

Conclusion: Taken together, the results showed that specific molecules in the cytoplasm direct the progression of cells from G_1 to S, and from G_2 to M in the cell cycle. Those molecules can move between the cytoplasm and the nucleus.

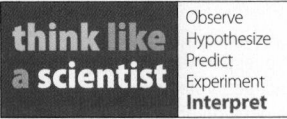
Observe
Hypothesize
Predict
Experiment
Interpret

Johnson and Rao also fused cells from other stages of the cell cycle. They observed that fused S and G_2 cells entered M phase earlier than did two fused S cells. What does this result indicate?

Source: R. T. Johnson and P. N. Rao. 1970. Mammalian cell fusion: induction of premature chromosome condensation in interphase nuclei. *Nature* 226:717–722.

used time-lapse photomicroscopy to see if and when growth and division were affected. In this way he isolated many *cell division cycle*, or *cdc* mutants. By examining the mutants, he could identify the stage in the cell cycle where each mutant type was blocked by noting whether nuclei had divided, chromosomes had condensed, the mitotic spindle had formed, cytokinesis had occurred, and so on. Using this approach, Hartwell identified many genes

that code for proteins involved in yeast's cell cycle and hypothesized where in the cycle these proteins operated. As might be expected, some of the proteins were involved in DNA replication, but a number of others were shown to function in cell cycle regulation. Hartwell received a Nobel Prize in 2001 for his discovery.

Paul Nurse of the Imperial Cancer Research Fund, London, carried out similar research with the fission yeast,

Schizosaccharomyces pombe, a species that divides by fission rather than budding. He identified a gene called *cdc2* that encodes a protein needed for the cell to progress from G$_2$ to M. Nurse also made the breakthrough discovery that all eukaryotic cells studied have counterparts of the yeast *cdc2* gene, implying that this gene originated early during eukaryotic evolution and has played an essential role in cell cycle regulation in all eukaryotes since that time. The protein product of *cdc2* is a protein kinase, an enzyme that catalyzes the phosphorylation of a target protein. (Recall from Section 9.2 that phosphorylation of proteins by protein kinases can activate or inactivate proteins.) That discovery was pivotal in determining how cell cycle regulation occurs. Paul Nurse received a Nobel Prize in 2001 for his discovery.

The Cell Cycle Can Be Arrested at Specific Checkpoints

A cell has internal controls that monitor its progression through the cell cycle through the action of a particular set of control proteins called *cyclins.* As part of the internal controls, the cell cycle has three key **checkpoints** to prevent critical phases from beginning until the previous phases are completed correctly **(Figure 10.16)**:

1. The *G$_1$/S checkpoint* is the main point in the cell cycle at which the mechanisms governing the cell cycle determine whether the cell will proceed through the rest of the cell cycle and divide. Once it passes this checkpoint, the cell is committed to continue the cell cycle through to cell division in M. For example, the cell cycle arrests (the cell stops proceeding through the cell cycle) at the G$_1$/S checkpoint if the DNA is damaged by radiation or chemicals. The G$_1$/S checkpoint is also the primary point at which cells "read" extracellular signals for cell growth and division. Therefore, if a growth factor required for stimulating cell growth is absent, the cells are arrested at this checkpoint. (Extracellular signals and their effects on the cell cycle are discussed in more detail later.)

2. The *G$_2$/M checkpoint* is at the junction between the G$_2$ and M phases. Passage through this checkpoint commits a

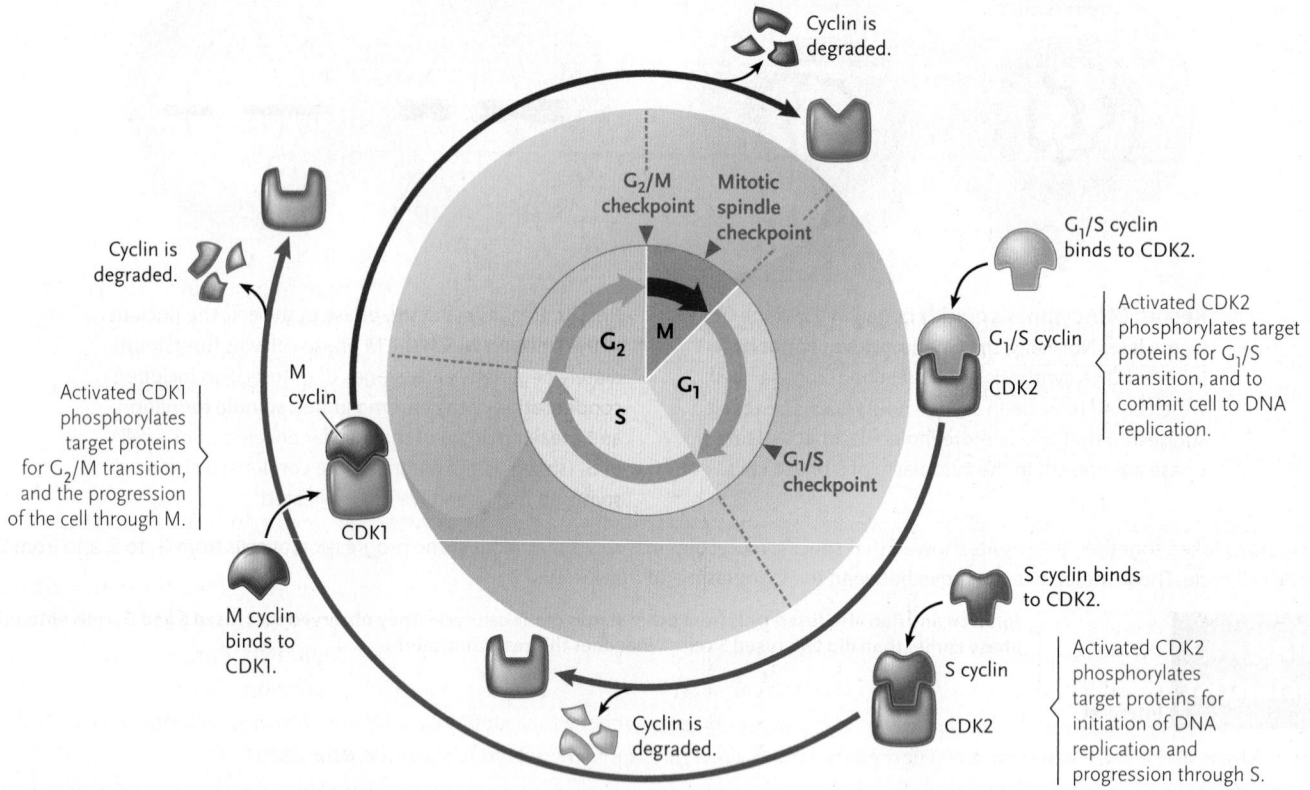

FIGURE 10.16 Regulation of the mitotic cell cycle by internal controls. Three key checkpoints for the G$_1$/S transition, for the G$_2$/M transition, and for the attachment of chromosomes to the mitotic spindle monitor cell cycle events to prevent crucial phases of the cell cycle from starting until previous phases are completed correctly. Complexes of cyclins and cyclin-dependent kinases (Cdks) regulate the progression of the cell through the cell cycle. The three cyclin–Cdks present in all eukaryotes are shown. The Cdks are present throughout the cell cycle, but they are active only when complexed with a cyclin (shown by the broad arrows in the figure). Each cyclin is synthesized and degraded in a regulated way so that it is present only for a particular phase of the cell cycle. During that phase, the Cdk to which it is bound phosphorylates and, thereby, regulates the activity of target proteins in the cell that are involved in initiating or regulating key events of the cell cycle.

© Cengage Learning 2017

cell to mitosis. Cells are arrested at the G_2/M checkpoint if DNA was not replicated fully in S, or if the DNA has been damaged by radiation or chemicals. Complete DNA replication is essential for producing genetically identical daughter cells, highlighting the importance of this checkpoint.

3. The *mitotic spindle checkpoint* is within the M phase before metaphase. This checkpoint assesses whether chromosomes are attached properly to the mitotic spindle so that they are aligned correctly at the metaphase plate. The checkpoint is essential for production of genetically identical daughter cells, which depends on separation of daughter chromosomes in anaphase. In turn, that separation depends on the correct alignment of the chromosomes on the spindle in metaphase. Once the cell begins anaphase, it is irreversibly committed to completing M, underlining the importance of the mitotic spindle checkpoint.

The control systems that operate at the checkpoints are signals to stop. Essentially, they are brakes. This becomes evident when a checkpoint is inactivated by mutation or chemical treatment. The consequence of inactivation of the checkpoints is that the cell cycle proceeds, even if DNA is damaged, DNA replication is incomplete, or the spindle did not assemble completely.

Cyclins and Cyclin-Dependent Kinases Are the Internal Controls That Directly Regulate Cell Division

The internal control system that acts at checkpoints causes cells to arrest in the cycle should proceeding through the cycle be harmful. The direct regulation of the cell cycle itself involves an internal control system consisting of proteins called **cyclins** and enzymes called **cyclin-dependent kinases (Cdks)** (see Figure 10.16). A Cdk is a *protein kinase,* which phosphorylates and thereby regulates the activity of target proteins. Cdk enzymes are "cyclin-dependent" because they are active *only* when bound to a cyclin molecule. Cyclins are named because their concentrations change at specific times in the cell cycle as they are synthesized and degraded. R. Timothy Hunt, Imperial Cancer Research Fund, London, received a Nobel Prize in 2001 for discovering cyclins. The basic control of the cell cycle by Cdks and cyclins is the same in all eukaryotes, but there are differences in the number and types of the molecules. We will focus on cell cycle regulation in vertebrates to explain how these proteins work.

The concentrations of the various Cdks remain constant throughout the cell cycle, whereas the concentrations of cyclins change as they are synthesized and degraded at specific stages of the cell cycle. Thus, a specific Cdk becomes active when the cell synthesizes the cyclin that binds to it and remains active until that cyclin is degraded. Each active Cdk phosphorylates particular target proteins, which play roles in initiating or regulating key events of the cell cycle. The phosphorylation regulates the activities of those proteins, and keeps the cycle operating in an orderly way. Those key events are DNA replication, mitosis, and cytokinesis. A succession of cyclin–Cdk complexes, each of which has specific regulatory effects, ensures that these stages follow in sequence somewhat like a clock passing through the sequence of hours. Regulation of the activity of cyclin–Cdk complexes is integrated with the regulatory events at the cell cycle checkpoints to ensure that daughter cells with damaged DNA or abnormal amounts of DNA are not generated.

Three classes of cyclins, each named for the stage of the cell cycle at which they bind and activate Cdks, operate in all eukaryotes (see Figure 10.16):

1. G_1/S cyclin binds to Cdk2 near the end of G_1 forming a complex required for the cell to make the transition from G_1 to S, and to commit the cell to DNA replication.
2. S cyclin binds to Cdk2 in the S phase forming a complex required for the initiation of DNA replication and the progression of the cell through S.
3. M cyclin binds to Cdk1 in G_2 forming a complex required for the transition from G_2 and M, and the progression of the cell through mitosis.

In most cells, a fourth class of cyclins, G_1 cyclin, binds to two additional Cdks, Cdk4 and Cdk6, before step 1 (the G_1/S transition) to form two cyclin–Cdk complexes. These complexes are needed to move the cell through the G_1 checkpoint stimulating it then to proceed from G_1 to S.

The M cyclin–Cdk1 complex is also called **M phase-promoting factor (MPF).** In addition to initiating mitosis, the M cyclin–Cdk1 complex (MPF) also orchestrates some of its key events. When all chromosomes are correctly attached to the mitotic spindle near the end of metaphase, the M cyclin–Cdk1 complex activates another enzyme complex, the **anaphase-promoting complex (APC).** Activated APC degrades an inhibitor of anaphase, and this leads to the separation of sister chromatids and the onset of daughter chromosome separation in anaphase. Later in anaphase, APC directs the degradation of the M cyclin, causing Cdk1 to lose its activity. The loss of Cdk1 activity then allows the separated chromosomes to become extended again, the nuclear envelope to reform around the two clusters of daughter chromosomes in telophase, and the cytoplasm then to divide in cytokinesis.

This discussion has focused on the main molecular players involved in cell cycle regulation. In reality, cell cycle regulation seems to be more complicated, as evidenced by research reports of large numbers of cell cycle-dependent proteins—proteins whose level or localization changes significantly through the cell cycle rather than remaining constant—in various organisms. *Molecular Insights* describes experiments that looked for cell cycle-dependent proteins in human cells.

The Cell Cycle: How do proteins vary in level and localization in the cell cycle?

Understanding the molecular details of the regulation of the cell cycle requires the identification and characterization of cell cycle-dependent proteins, the proteins whose function, level, or localization changes in the cell cycle. Proteins are produced by translation of mRNAs, which are generated by transcription of the DNA of protein-coding genes (see Chapter 15 for details of these processes). A number of experiments have studied mRNA levels through the cell cycle. Such experiments have shown that only 1–3% of human genes have cell cycle-dependent mRNA levels, meaning that those mRNA levels changed significantly through the cell cycle rather than staying constant. However, mRNA levels are only an indirect measure of protein levels in cells. It is not possible to determine the level of a protein in the cell from the measured level of the mRNA that is translated to produce that protein because each mRNA is translated many times and the production of proteins varies considerably from mRNAs of different genes. Therefore, a direct study of proteins is necessary to identify those proteins that are important for biological processes such as the cell cycle.

Research Question

What human proteins are cell cycle-dependent in level or localization?

Experiments

Researchers at the Weizmann Institute of Science in Rehovot, Israel, used a proteomics approach to answer the question. They studied 495 clones of a human cell line (cells that can be grown in culture). Each clone had a different protein-coding gene modified so that, when the gene was transcribed to produce mRNA and the mRNA was then translated to produce the protein, the protein had an extra segment of amino acids. That segment was a fluorescent tag. The tag allowed the specific protein (different in each clone) to be followed under the microscope by its fluorescence. To study proteins during the cell cycle, the researchers synchronized cells in cultures of each clone and quantified protein level and localization over a 24-hour growth period (about one cell cycle) by analyzing movies made using time-lapse fluorescent microscopy to follow the tagged proteins.

Results

Protein level: Eleven percent of the analyzed proteins (56/495) showed significant cell cycle-dependence in their levels. A number of the proteins in this set are known to be involved in cell cycle regulation and regulation of kinase activity. (A kinase is an enzyme that phosphorylates—adds a phosphate group—to a protein; see Chapter 6.) For example, the protein GMNN (Geminin) drops sharply in level at the beginning of the cell cycle (G_1), suggesting degradation of the protein immediately after mitosis **(Figure)**. Indeed, Geminin is known to be a substrate for the anaphase-promoting complex (APC), and is degraded by it after mitosis. Geminin shows a peak level at G_1/S.

Protein localization: By analyzing fluorescence in the nucleus compared with fluorescence in the entire cell, the researchers identified that 19% of proteins (96/495) showed cell cycle-dependent changes in nuclear localization. For example, PRC1 (*p*rotein *r*egulator of *c*ytokinesis) associates with the mitotic spindle during the M-to-G_1 transition when the spindle disappears and cytokinesis occurs, and then localizes to the reformed nucleus during S phase. Twenty-three percent of proteins (113/495) showed changes in localization other than moving to or from the nucleus.

Discussion

Considering the overlap in the groups of proteins identified, the researchers concluded that about 40% (199/495) of the proteins in the study showed cell cycle-dependent

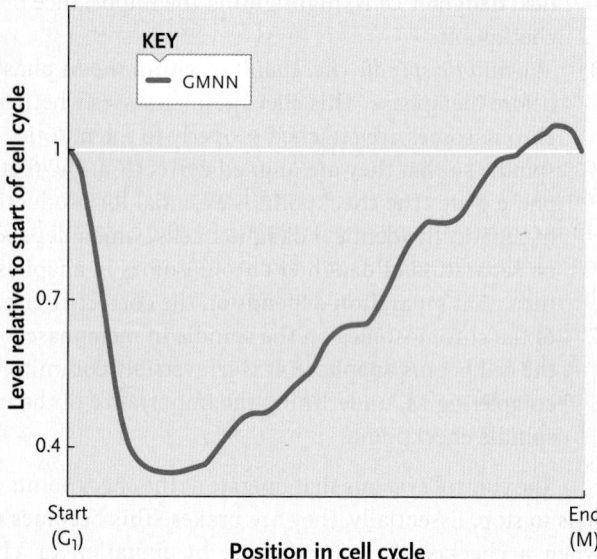

FIGURE Degradation and synthesis of protein GMNN through the cell cycle.

behavior in levels or localization. About 7% of the proteins (34/495) showed cell cycle-dependent behavior in both level and localization. Even considering the data for protein level alone, the 11% figure obtained is far higher than the few percent of mRNAs showing cell cycle-dependent behavior in earlier studies. This indicates that a much larger number of cell cycle-dependent proteins exists in human cells than was previously thought. Most of the cell cycle-dependent proteins identified in the study showed changes in cellular localization. In terms of regulating the cell cycle, changing a protein's location in the cell would be a more rapid and efficient way to regulate the cell cycle than a change in level, which would require synthesis or degradation activities.

think like a scientist

Would you expect the cell cycle-dependent proteins identified in the study to be conserved among eukaryotes or to be human-specific?

Source: S. Farkash-Amar et al. 2012. Dynamic proteomics of human protein level and localization across the cell cycle. *PLOS One* 7:e48722.

External Controls Coordinate the Mitotic Cell Cycle of Individual Cells with the Overall Activities of the Organism

The internal controls that regulate the cell cycle are modified by signaling molecules that originate from outside the dividing cells. In animals, these signaling molecules include the peptide hormones and similar proteins called *growth factors.*

The hormones and growth factors act on the cell by the reception–transduction–response pattern that applies to cell communication in general (see Chapter 9). The external factors bind to receptors at the cell surface, which respond by triggering reactions inside the cell. These reactions often include steps that add inhibiting or stimulating phosphate groups to the cyclin–Cdk complexes, particularly to the Cdks. The reactions triggered by the activated receptor may also directly affect the same proteins regulated by the cyclin–Cdk complexes. The overall effect is to speed, slow, or stop the progress of cell division, depending on the particular hormone or growth factor and the internal pathway that is stimulated.

Cell surface receptors in animals also recognize contact with other cells or with molecules of the extracellular matrix (see Section 4.5). The contact triggers internal reaction pathways that inhibit division by arresting the cell cycle, usually in the G_1 phase. The response, called **contact inhibition,** stabilizes cell growth in fully developed organs and tissues. As long as the cells of most tissues are in contact with one another or the extracellular matrix, they are shunted into the G_0 phase and prevented from dividing. If the contacts are broken, the freed cells often enter rounds of division.

Contact inhibition is easily observed in cultured mammalian cells grown on a glass or plastic surface. In such cultures, division proceeds until all the cells are in contact with their neighbors in a continuous, unbroken, single layer. At this point, division stops. If a researcher then scrapes some of the cells from the surface, cells at the edges of the "wound" are released from inhibition and divide until they form a continuous layer and all the cells are again in contact with their neighbors.

Cell Cycle Controls Are Lost in Cancer

Cancer occurs when cells lose the normal controls that determine when and how often they will divide. Cancer cells divide continuously and uncontrollably, producing a rapidly growing mass called a *tumor.* Cancer cells also typically lose their adhesions to other cells and often become actively mobile. As a result, in a process called *metastasis,* they break loose from an original tumor, spread throughout the body, and grow into new tumors in other body regions. Metastasis is promoted by changes that block contact inhibition and alter the cell surface molecules that link cells together or to the extracellular matrix.

Growing tumors damage surrounding normal tissues by compressing them and interfering with blood supply and nerve function. Tumors may also break through barriers such as the outer skin, internal cell layers, or the gut wall. The breakthroughs cause bleeding, open the body to infection by microorganisms, and destroy the separation of body compartments necessary for normal functioning. Both compression and breakthroughs can cause pain that, in advanced cases, may become extreme. As tumors increase in mass, the actively growing and dividing cancer cells may deprive normal cells of their required nutrients, leading to generally impaired body functions, muscular weakness, fatigue, and weight loss.

Cancer cells typically have a number of mutated genes of different types. The altered functions of those mutated genes in one way or another promote uncontrolled cell division or metastasis. In their normal (nonmutated) form, a number of these genes code for components of the cyclin/Cdk system that regulates cell division. Others encode proteins that regulate gene expression, form cell surface receptors, or make up elements of the systems controlled by the receptors. The mutated form of the genes, called **oncogenes** (*oncos* = bulk or mass), encodes altered versions of these products.

For example, a mutation in a gene that codes for a surface receptor might result in a protein that is constantly active, even without binding an extracellular signal molecule. As a result, the internal reaction pathways triggered by the receptor, which induce cell division, are continually stimulated. Another mutation, this time in a cyclin gene, could decrease the cyclin–Cdk binding that triggers DNA replication and the rest of the cell cycle. Increasingly, the genomes of cancer cells are being analyzed to identify the mutations they carry with the goal of understanding their molecular causes and then, hopefully, to indicate a direction of research for the development of treatments. The genetics of cancer is discussed further in Chapter 16.

The overview of the mitotic cell cycle and its regulation presented in this chapter only hints at the complexity of cell growth and division. The likelihood of any given cell dividing is determined by the interplay of various internal signals in the context of external cues from the environment. If a cell is destined to divide, then the problem of replicating and partitioning its DNA accurately requires a highly regulated, intricately interrelated series of mechanisms. It is a challenging operation even for male Australian Jack Jumper ants (*Myrmecia pilosula*), which have only two chromosomes ($2n = 2$); and think of the challenges faced by the fern species, *Ophioglossum pycnostichum*, which has 1,260 chromosomes in each cell.

STUDY BREAK 10.4

1. Why is a Cdk not active throughout the entire cell cycle?
2. How do cyclin–Cdk complexes typically trigger transitions in the cell cycle?
3. What is an oncogene? How might an oncogene affect the cell cycle?
4. What is metastasis?

10.5 Cell Division in Bacteria

Prokaryotes undergo a cycle of cytoplasmic growth, DNA replication, and cell division, producing two daughter cells from an original parent cell. The entire mechanism of prokaryotic cell division is called **binary fission**—that is, splitting or dividing into two parts.

DNA Is Organized More Simply in Prokaryotes Than in Eukaryotes

All prokaryotes use DNA as their genetic material. The vast majority of prokaryotes have a single, circular DNA molecule known as the **prokaryotic chromosome,** more specifically the **bacterial chromosome** for bacteria and the **archaeal chromosome** for archaeans. Here we focus on bacterial chromosomes; archaeal chromosomes are similar.

Several features of DNA organization in bacteria differ fundamentally from eukaryotic DNA organization. In contrast to the linear DNA in eukaryotes, the primary DNA molecule of most bacteria is circular, with only one copy per cell. The chromosome of the best-known bacterium, *E. coli,* includes about 1,460 μm of DNA, which is equivalent to 4.6 million base pairs. There are exceptions: some bacteria have two or more different chromosomes in the cell, and some bacterial chromosomes are linear.

Inside bacterial cells, the DNA circle is packed and folded into an irregularly shaped mass called the **nucleoid** (shown in Figure 4.7). The DNA of the nucleoid is suspended directly in the cytoplasm with no surrounding membrane.

Many bacterial cells also contain other DNA molecules, called **plasmids,** in addition to the main chromosome of the nucleoid. Most plasmids are circular, although some are linear. Plasmids have replication origins and are duplicated and distributed to daughter cells together with the bacterial chromosome during cell division.

Although bacterial DNA is not organized into nucleosomes, certain positively charged proteins do combine with bacterial DNA. Some of these proteins help organize the DNA into loops, thereby providing some compaction of the molecule. Bacterial DNA also combines with many types of genetic regulatory proteins that have functions similar to those of the nonhistone proteins of eukaryotes (see Chapter 16).

Replication Occupies Most of the Cell Cycle in Rapidly Dividing Prokaryotic Cells

When prokaryotic cells divide at the maximum rate, DNA replication occupies most of the period between cytoplasmic divisions. As soon as replication is complete, the cytoplasm divides to complete the cell cycle. For example, in *E. coli* cells, which are capable of dividing every 20 minutes, DNA replication occupies 19 minutes of the entire 20-minute division cycle.

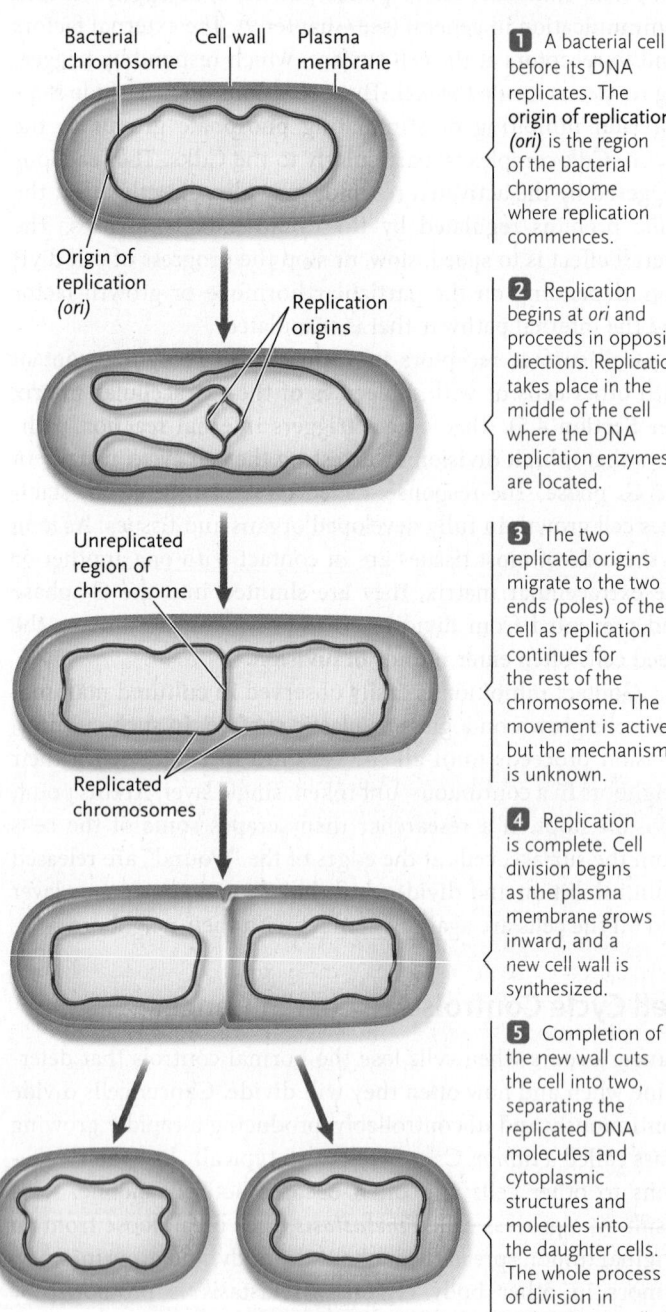

1 A bacterial cell before its DNA replicates. The **origin of replication** *(ori)* is the region of the bacterial chromosome where replication commences.

2 Replication begins at *ori* and proceeds in opposite directions. Replication takes place in the middle of the cell where the DNA replication enzymes are located.

3 The two replicated origins migrate to the two ends (poles) of the cell as replication continues for the rest of the chromosome. The movement is active, but the mechanism is unknown.

4 Replication is complete. Cell division begins as the plasma membrane grows inward, and a new cell wall is synthesized.

5 Completion of the new wall cuts the cell into two, separating the replicated DNA molecules and cytoplasmic structures and molecules into the daughter cells. The whole process of division in prokaryotes is called binary fission.

FIGURE 10.17 Model for the segregation of replicated bacterial chromosomes to daughter cells.
© Cengage Learning 2017

Replicated Chromosomes Are Distributed Actively to the Halves of the Bacterial Cell

In the 1960s, François Jacob of The Pasteur Institute, Paris, France, proposed a model for the segregation of bacterial chromosomes to the daughter cells in which the two chromosomes attach to the plasma membrane near the middle of the cell and separate as a new plasma membrane is added between the two sites during cell elongation. The essence of this model is that chromosome separation is passive. However, current research indicates that bacterial chromosomes rapidly separate in an active way that is linked to DNA replication events and that is independent of cell elongation. The new model is shown in **Figure 10.17.**

Mitosis Evolved from Binary Fission

The prokaryotic mechanism of replication works effectively because most prokaryotic cells have only a single chromosome. Thus, if a daughter cell receives at least one copy of the chromosome, its genetic information is complete. By contrast, in most cases the genetic information of eukaryotes is distributed among several chromosomes, with each chromosome containing a much greater length of DNA than a bacterial chromosome. If a daughter cell fails to receive a copy of even one chromosome, the effects may be lethal. The evolution of mitosis solved the mechanical problems associated with distributing long DNA molecules without breakage. Mitosis provided the level of precision required to ensure that each daughter cell receives a complete complement of the chromosomes the parent cell had.

Scientists believe that the ancestral division process was binary fission and that mitosis evolved from that process. Variations in the mitotic apparatus in modern-day organisms illuminate possible intermediates in this evolutionary pathway. For example, in many primitive eukaryotes, such as dinoflagellates (a type of single-celled protist [see Chapter 27]), the nuclear envelope remains intact during mitosis, and the chromosomes bind to the inner membrane of the nuclear membrane. When the nucleus divides, the chromosomes are segregated.

A more advanced form of the mitotic apparatus is seen in yeasts and diatoms (diatoms are another type of single-celled protist). In these organisms, the mitotic spindle forms and chromosomes segregate to daughter nuclei without the disassembly and reassembly of the nuclear envelope. Currently, scientists think that the types of mitosis seen in yeasts and diatoms, as well as in animals and higher plants, evolved separately from a common ancestral type.

Mitotic cell division, the subject of this chapter, produces two cells that have the same genetic information as the parental cell entering division. In the next chapter, you will learn about meiosis, a specialized form of cell division that produces gametes in animals and spores in plants and many fungi, which have half the number of chromosomes as that present in diploid cells.

STUDY BREAK 10.5

1. How do bacteria divide?
2. What processes involved in eukaryotic cell division are absent from bacterial cell division?

Unanswered Questions

Disrupted or defective control of cell growth and division can lead to diseases such as cancer. Complex, interacting molecular networks within the cell fine-tune the division of each cell in both unicellular and multicellular organisms. Identifying the genes and proteins involved in these networks is crucial both for a complete understanding of cell growth and division, and for developing models for diseases caused by cell cycle defects. Many researchers worldwide are working in this area of research.

How are transitions between phases of the cell cycle regulated?

Research in many labs has shown that transitions are important control points for progression through the cell cycle. If a cell in G_1 phase has damaged DNA, for instance, the cell pauses to repair the DNA before entering S phase, to ensure that any mutations are not passed on to progeny cells. My laboratory is studying how cells execute two transitions in the cell division program: the transition from G_1 phase to S phase and the transition from mitosis to G_1 phase. Most of our studies are focused on the budding yeast *Saccharomyces cerevisiae* because this organism is easy to manipulate and uses many of the same proteins as human cells do to carry out cell division.

The G_1-to-S and mitosis-to-G_1 transitions involve turnover—breakdown—of proteins that serve to maintain cells in the pretransition state. I focus here on the G_1-to-S transition. In the G_1 phase, the cyclin-dependent kinase (CDK) inhibitor Sic1 blocks the activity of the S phase cyclin–CDK complexes that promote DNA replication, and thereby delays the onset of S phase. Enzymes rapidly eliminate Sic1 at the end of the G_1 phase, thereby unmasking the activity of the S phase cyclin–CDK, which causes the rapid initiation of chromosome duplication. Although we now know much about how this transition works, a key unanswered question that is of interest to my laboratory is, How do the enzymes that eliminate Sic1 do their job? For example, we are trying to unravel the precise mechanism of action of the ubiquitin ligase enzyme that attaches ubiquitin to Sic1, which serves as a signal that activates Sic1 turnover. (Ubiquitin is a small protein added to proteins to designate those proteins for destruction; see Chapter 16. Ubiquitination of a particular protein is catalyzed by one of a family of ubiquitin ligases in the cell.) Another key question is how the timing of Sic1 turnover is controlled.

From our studies on both the G_1-to-S and mitosis-to-G_1 transitions, it is apparent that many of the key players have already been discovered. However, we still do not know how all of the proteins are organized and communicate with each other to bring about the transition in the cell division program at the right time, and we also do not know how the proteins operate as "nano-machines" to carry out a specific biological task.

Courtesy of Raymond Deshaies

Raymond Deshaies is professor of biology at CalTech and an investigator of Howard Hughes Medical Institute. The focus of his laboratory is investigation of the cellular machinery that mediates protein degradation by the ubiquitin–proteasome system, and how this machinery regulates cell division. Learn more about his work at www.deshaislab.com.

think like a scientist Why would researchers choose *Saccharomyces cerevisiae* as a model organism to address mitotic cell cycle defects over *Escherichia coli* or mammalian cells? How similar would you expect *S. cerevisiae* ubiquitin to be to human ubiquitin?

REVIEW KEY CONCEPTS

For access to MindTap and additional study materials visit www.cengagebrain.com.

10.1 The Cycle of Cell Growth and Division: An Overview

- In mitotic cell division, DNA replication is followed by the equal separation—that is, segregation—of the replicated DNA molecules and their delivery to daughter cells. The process ensures that the two cell products of a division end up with the same genetic information as the parent cell entering division.

- Mitosis is the basis for growth and maintenance of body mass in multicellular eukaryotes, and for the reproduction of many single-celled eukaryotes. Chromosomes are partitioned to daughter cells by mitosis.

- Eukaryotic chromosomes consist of DNA complexed with histone and nonhistone proteins.

- In eukaryotic chromosomes, DNA is wrapped around a core consisting of two molecules each of histones H2A, H2B, H3, and H4 to produce a nucleosome. Linker DNA connects adjacent nucleosomes. The chromosome structure in this form is the 10-nm chromatin fiber. The binding of histone H1 causes the nucleosomes to package into a coiled structure called the 30-nm chromatin fiber (Figure 10.2).

- Chromatin is distributed between euchromatin, a loosely packed region in which genes are active in RNA transcription, and heterochromatin, densely packed masses in which genes, if present, are inactive. Chromatin also folds and packs to form thick, rodlike chromosomes during nuclear division.

10.2 The Mitotic Cell Cycle

- The mitotic cell cycle consists of the G_1 phase, in which a cell may grow, then the S phase in which DNA replication and chromosome duplication occur to produce a pair of sister chromatids for each original chromosome, then G_2 in which cell growth continues and the cell prepares for mitosis, and finally mitosis (the M phase) in which chromosomes are partitioned to daughter cells (Figure 10.3). Together, G_1, S, and G_2 comprise interphase. Each daughter cell has the same set of chromosomes as the parental cell.

- Mitosis occurs in five stages. In prophase (stage 1), the chromosomes condense into short rods and the spindle forms in the cytoplasm (Figures 10.3 and 10.4).

- In prometaphase (stage 2), the nuclear envelope breaks down, the spindle enters the former nuclear area, and the sister chromatids of each chromosome make connections to opposite spindle poles. Each chromatid has a kinetochore that attaches to spindle microtubules (Figures 10.3, 10.4, and 10.6).

- In metaphase (stage 3), the spindle is fully formed and the chromosomes, moved by the spindle microtubules, become aligned at the metaphase plate (Figures 10.3, 10.4, and 10.7).

- In anaphase (stage 4), the spindle separates the sister chromatids and moves them to opposite spindle poles. At this point, chromosome segregation is complete (Figures 10.3 and 10.4).

- In telophase (stage 5), the chromosomes decondense and return to the extended state typical of interphase and a new nuclear envelope forms around the chromosomes (Figures 10.3 and 10.4).

- Cytokinesis, the division of the cytoplasm, completes cell division by producing two daughter cells, each containing a daughter nucleus produced by mitosis (Figures 10.3 and 10.4).

- Cytokinesis in animal cells proceeds by furrowing, in which a band of microfilaments just under the plasma membrane contracts, gradually separating the cytoplasm into two parts (Figure 10.9).

- In plant cytokinesis, cell wall material is deposited along the plane of the former spindle midpoint; the deposition continues until a continuous new wall, the cell plate, separates the daughter cells (Figure 10.10).

10.3 Formation and Action of the Mitotic Spindle

- In animal cells, the centrosome divides and the two parts move apart. As they do so, the microtubules of the spindle form between them. In plant cells, which do not have a centrosome, the spindle microtubules assemble around the nucleus (Figure 10.11).

- In the spindle, kinetochore microtubules run from the poles to the kinetochores of the chromosomes, and nonkinetochore microtubules run from the poles to a zone of overlap at the spindle midpoint without connecting to the chromosomes (Figure 10.4).

- During anaphase, chromosomes can move when motor proteins attached to kinetochores walk along the kinetochore microtubules, moving the chromosomes to the poles (Figures 10.13 and 10.14), or when motor proteins at the spindle poles pull kinetochore microtubules toward the poles. Both mechanisms are used, with their relative contributions being species-specific and cell type-specific.

- Also during anaphase, nonkinetochore microtubules reduce their overlap at the spindle midpoint and lengthen simultaneously, pushing the poles farther apart, thereby lengthening the cell.

10.4 Cell Cycle Regulation

- The internal controls that monitor progression through the cell cycle include checkpoints at key points to ensure that critical phases do not commence before previous phases are completed correctly (Figure 10.16).

- The internal control system that directly regulates cell division involves complexes of a cyclin and a cyclin-dependent protein kinase (Cdk). Cdk is activated when combined with a cyclin and then phosphorylates target proteins, regulating their activities. The altered target proteins then initiate or regulate key events of the cell cycle. Four classes of cyclins—G_1, G_1/S, S, and M—are distinguished by the stage of the cell cycle at which they activate Cdks (Figure 10.16).

- External controls are based primarily on surface receptors that recognize and bind signals such as peptide hormones and growth factors, surface groups on other cells, or molecules of the extracellular matrix. The binding triggers internal reactions that speed, slow, or stop cell division.

- In cancer, control of cell division is lost, and cells divide continuously and uncontrollably, forming a rapidly growing mass of cells that interferes with body functions. Cancer cells can also break loose from their original tumor (metastasize) to form additional tumors in other parts of the body.

10.5 Cell Division in Bacteria

- The bacterial chromosome is a closed, circular molecule of DNA; it is packed into the nucleoid region of the cell. Many bacteria also contain plasmids, which replicate independently of the host chromosome.

- Replication begins at the origin of replication of the bacterial chromosome in reactions catalyzed by enzymes located in the middle of the cell. Once the origin of replication is duplicated, the two origins migrate to the two ends of the cells. Division of the cytoplasm then occurs through a partition of cell wall material that grows inward until the cell is separated into two parts (Figure 10.17).

TEST YOUR KNOWLEDGE

Remember/Understand

1. During the cell cycle, the DNA mass of a cell:
   a. decreases during G_1.
   b. decreases during metaphase.
   c. increases during the S phase.
   d. increases during G_2.
   e. decreases during interphase.

2. A tumor suppressor protein, p21, inhibits Cdk1. The earliest effect of p21 on the cell cycle would be to stop the cell cycle at:
   a. early G_1.
   b. late G_1.
   c. the S phase.
   d. G_2.
   e. the mitotic prophase.

3. A major difference between hereditary information in eukaryotes and prokaryotes is:
   a. in prokaryotes, the hereditary information is distributed among individual, linear DNA molecules in the nucleus.
   b. in eukaryotes, the hereditary information is encoded in a single, circular DNA molecule.
   c. in prokaryotes, the hereditary information is usually distributed among multiple circular DNA molecules in the cytoplasm.
   d. in eukaryotes, the hereditary information is distributed among individual, linear DNA molecules in the cytoplasm.
   e. in eukaryotes, the hereditary information is distributed among individual, linear DNA molecules in the nucleus.

4. The major microtubule organizing center of the animal cell is:
   a. chromosomes, composed of chromatids.
   b. the centrosome, composed of centrioles.
   c. the chromatin, composed of chromatids.
   d. chromosomes, composed of centromere.
   e. centrioles, composed of centrosome.

5. The chromatids separate into chromosomes:
   a. during prophase.
   b. going from prophase to metaphase.
   c. going from anaphase to telophase.
   d. going from metaphase to anaphase.
   e. going from telophase to interphase.

6. Which of the following statements about mitosis is *incorrect*?
   a. Microtubules from the spindle poles attach to the kinetochores on the chromosomes.
   b. In anaphase, the spindle separates sister chromatids and pulls them apart.
   c. In metaphase, spindle microtubules align the chromosomes at the spindle midpoint.
   d. Cytokinesis describes the movement of chromosomes.
   e. Both the animal cell furrow and the plant cell plate form at their former spindle midpoint.

7. Mitomycin C is an anticancer drug that stops cell division by inserting itself into the strands of DNA and binding them together. This action is thought to have its major effect at:
   a. late G_1, early S phases.
   b. late G_2.
   c. prophase.
   d. metaphase.
   e. anaphase.

8. Which of the following statements about cell cycle regulators is *incorrect*?
   a. The concentrations of cyclins change throughout the cell cycle.
   b. Cyclins are present in all stages of the cell cycle except S.
   c. Cyclin–Cdk complexes phosphorylate target proteins.
   d. Cdks combine with cyclin to move the cycle into mitosis.
   e. During anaphase of mitosis, cyclin is degraded, allowing mitosis to end.

9. Which of the following is *not* a characteristic of cancer cells?
   a. less cytoplasmic volume than normal cells
   b. an absence of cyclin
   c. loss of adhesion to other cells
   d. loss of control of cell division
   e. loss of normal control of G_1/S phase transition

10. Which of the following does not accurately characterize bacterial DNA?
    a. The bacterial chromosome is a closed, circular molecule of DNA.
    b. Bacterial DNA is organized into nucleosomes.
    c. The primary DNA molecule of most bacteria is circular.
    d. DNA is organized more simply in bacteria than in eukaryotes.
    e. Positively charged proteins can combine with bacterial DNA.

11. In bacteria:
    a. several chromosomes undergo mitosis.
    b. binary fission produces four daughter cells.
    c. replication begins at the origin (*ori*), and the DNA strands separate.
    d. the plasma membrane plays an important role in separating the duplicated chromosomes into the two daughter cells.
    e. the daughter cells receive different genetic information from the parent cell.

Apply/Analyze

12. **Discuss Concepts** You have a means of measuring the amount of DNA in a single cell. You first measure the amount of DNA during G_1. At what points during the remainder of the cell cycle would you expect the amount of DNA per cell to change?

13. **Discuss Concepts** A cell has 38 chromosomes. After mitosis and cell division, one daughter cell has 39 chromosomes and the other has 37. What might have caused these abnormal chromosome numbers? What effects do you suppose this might have on cell function? Why?

Evaluate/Create

14. **Discuss Concepts** Paclitaxel (Taxol), a substance isolated from Pacific yew *(Taxus brevifolia)*, is effective in the treatment of breast and ovarian cancers. It works by stabilizing microtubules, thereby preventing them from disassembling. Why would this activity slow or stop the growth of cancer cells?

15. **Design an Experiment** Many chemicals in the food we eat have the potential to affect the growth of cancer cells. Chocolate, for example, contains a number of flavonoid compounds, which act as natural antioxidants. Design an experiment to determine whether any of the flavonoids in chocolate inhibit the cell cycle of breast cancer cells that are growing in culture.

16. **Apply Evolutionary Thinking** The genes and proteins involved in cell cycle regulation in prokaryotes and eukaryotes are very different. However, both types of organisms use similar molecular regulatory reactions to coordinate DNA synthesis with cell division. What does this observation mean from an evolutionary perspective?

For selected answers, see Appendix A.

INTERPRET THE DATA

Biologists have long been interested in the effects of radiation on cells. In one experiment, researchers examined the effect of radium on mitosis of chick embryo cells growing in culture. A population of experimental cells was examined under the microscope for the number of cells in telophase (as a measure of mitosis occurring) before, during, and after exposure to radium. The results are shown in the **Figure.**

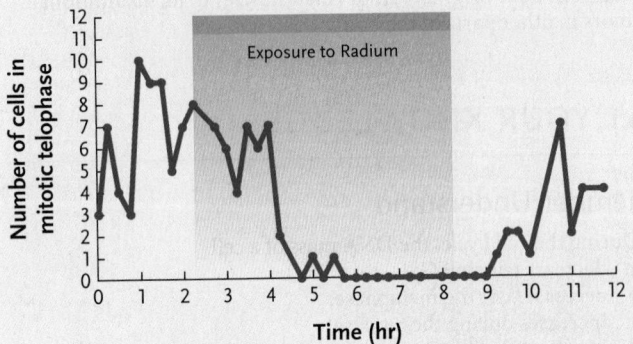

1. What is the effect of radium exposure on mitosis?

2. Was the effect of radium exposure permanent?

Source: R. G. Canti and M. Donaldson. 1926. The effect of radium on mitosis *in vitro. Proceedings of the Royal Society of London, Series B, Containing Papers of a Biological Character* 100:413–419.

Meiosis: The Cellular Basis of Sexual Reproduction

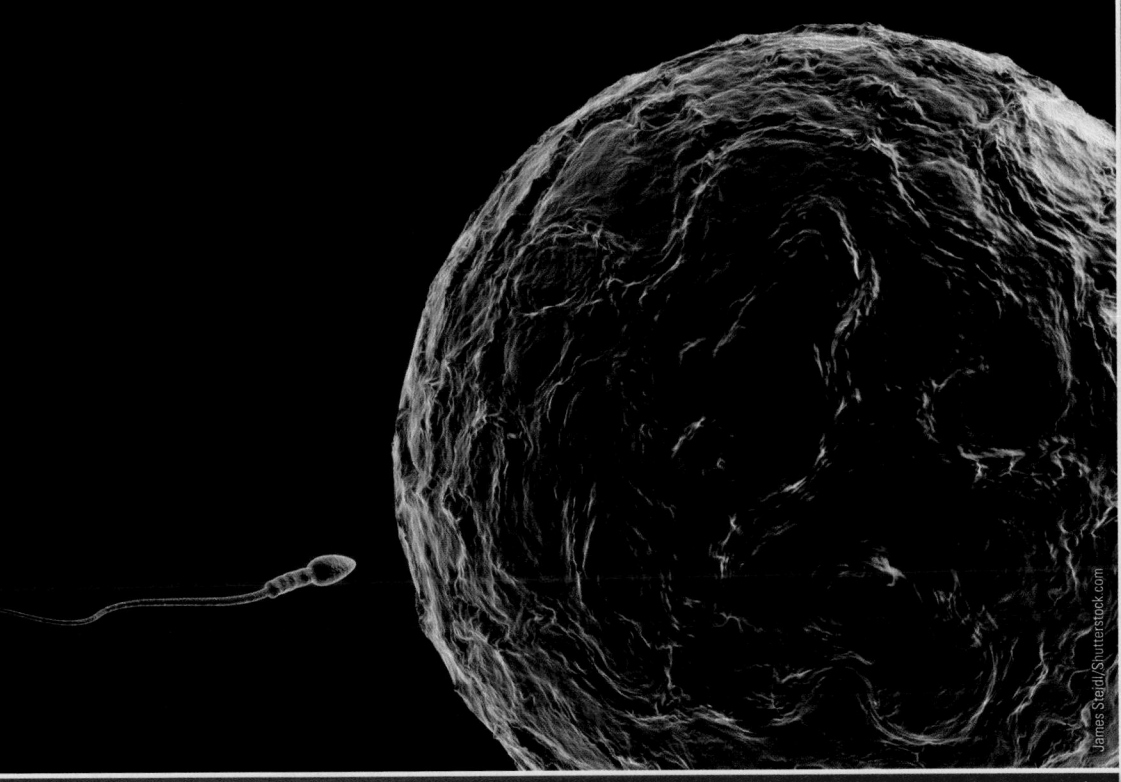

Sexual reproduction depends on meiosis, a specialized process of cell division that produces gametes such as eggs and sperm. Here a sperm is swimming toward an egg.

Why it matters . . . A couple clearly shows mutual interest. First, he caresses her with one arm, then another—then another, another, and another. She reciprocates. This interaction goes on for hours; a hug here, a squeeze there. At the climactic moment, the male reaches deftly under his mantle and removes a packet of sperm, which he inserts under the mantle of the female. For every one of his sperm that successfully performs its function, a fertilized egg can develop into a new octopus.

The octopuses are engaged in a form of **sexual reproduction,** the production of offspring through union of male and female **gametes**—for example, egg and sperm cells in animals. Sexual reproduction depends on **meiosis,** a specialized process of cell division that, in animals, produces gametes. Meiosis (*meioun* = to diminish) reduces the number of chromosomes, producing gametes with half the number of chromosomes present in the **somatic cells** (body cells) of a species. At **fertilization,** the nuclei of an egg and sperm cell fuse, producing a cell called the **zygote,** in which the chromosome number typical of the species is restored. Without the halving of chromosome number by the meiotic divisions, fertilization would double the number of chromosomes in each subsequent generation.

Both meiosis and fertilization also mix genetic information into new combinations allowing for genetic diversity in the offspring of a mating pair. By contrast, asexual reproduction generates genetically identical offspring because they are the products of mitotic divisions (asexual reproduction is discussed in Chapter 10). Sexual reproduction allows the genetic variability, which originally arose from mutations, to be shuffled and reshuffled throughout generations, resulting in a wide array of genetic combinations.

The halving of the chromosome number and shuffling of genetic information into new combinations—both by meiosis—and the restoration of the chromosome number by fertilization are the biological foundations of sexual reproduction. Intermingled tentacles in octopuses and the courting and complex mating rituals of humans have evolved as a means for promoting fertilization.

11.1 The Mechanisms of Meiosis

In humans, and other animals, meiosis takes place in the primary reproductive organs, the **gonads.** Meiosis in mature gonads of the male, the **testes,** produces **spermatozoa (sperm),** the gametes of the male. Meiosis in mature gonads of the female, the **ovaries,** produces **ova (eggs),** the gametes of the female. The cellular mechanisms of gamete formation—**gametogenesis**—are described in Chapter 49.

Meiosis Is Based on the Interactions and Distribution of Homologous Chromosome Pairs

To follow the steps of meiosis, you must understand the significance of the chromosome pairs in diploid organisms. As discussed in Section 10.1, the two representatives of each chromosome in a diploid cell constitute a *homologous pair*—they have the same genes, arranged in the same order in the DNA of the chromosomes. One chromosome of each homologous pair, the **paternal chromosome,** comes from the male parent of the organism, and the other chromosome, the **maternal chromosome,** comes from its female parent.

Although the genes of the two chromosomes of a homologous pair are arranged in the same order, the versions of each gene, called **alleles,** present in the members of the pair may be the same or different. For a gene that encodes a protein, the different alleles might encode different versions of the same protein, which have different structures, molecular properties, or both, or perhaps an allele may not encode a protein at all.

Recall from Section 10.1 that humans normally have 46 chromosomes in their diploid cells, which consist of 22 homologous pairs and a pair of sex chromosomes (see Figure 10.8). However, each individual (except for identical twins, identical triplets, and so forth) has a unique combination of the alleles in the two chromosomes of each homologous pair. The distinct set of alleles, arising from the mixing mechanisms of meiosis and fertilization, gives each individual his or her unique combination of inherited traits, including such attributes as height, hair and eye color, susceptibility to certain diseases, and even aspects of personality and intelligence.

Meiosis separates homologous pairs, thereby reducing the diploid or 2*n* number of chromosomes to the **haploid** or *n* number **(Figure 11.1).** Each gamete produced by meiosis receives only one member of each homologous pair. For example, a human egg produced in an ovary, or a sperm cell produced in a testis, contains 23 chromosomes, one of each pair.

When the egg and sperm combine in sexual reproduction to produce the *zygote*—the first cell of the new individual—the diploid number of 46 chromosomes (23 pairs) is regenerated. The processes of DNA replication and mitotic cell division ensure that this diploid number is maintained in the body cells as the zygote develops.

The Meiotic Cell Cycle Produces Four Genetically Different Daughter Cells with Half the Parental Number of Chromosomes

Meiosis follows a premeiotic interphase in which DNA replicates and the chromosomal proteins are duplicated. (This interphase passes through G_1, S, and G_2 stages as does a premitotic interphase: see Chapter 10 and Figure 10.3.) Meiosis consists of two cell divisions—**meiosis I** and **meiosis II**—in which duplicated chromosomes in the parental cell are distributed to four daughter cells,

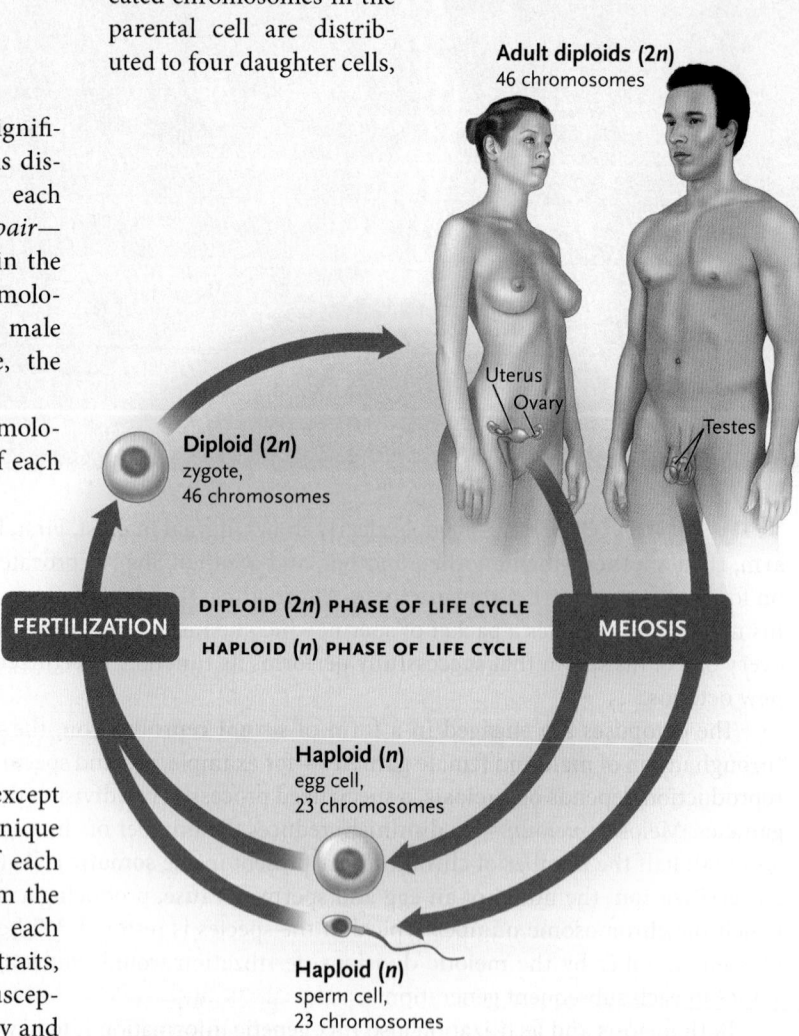

FIGURE 11.1 The cycle of meiosis and fertilization in animals, with humans as an example. Meiosis in animals produces gametes, spermatozoa (sperm) in the testes of the male, and ova (eggs) in the ovaries of the female. Meiosis reduces the chromosome number from the diploid level of two representatives of each chromosome to the haploid level of one representative of each chromosome. Fertilization restores the chromosome number to the diploid level.

© Cengage Learning 2017

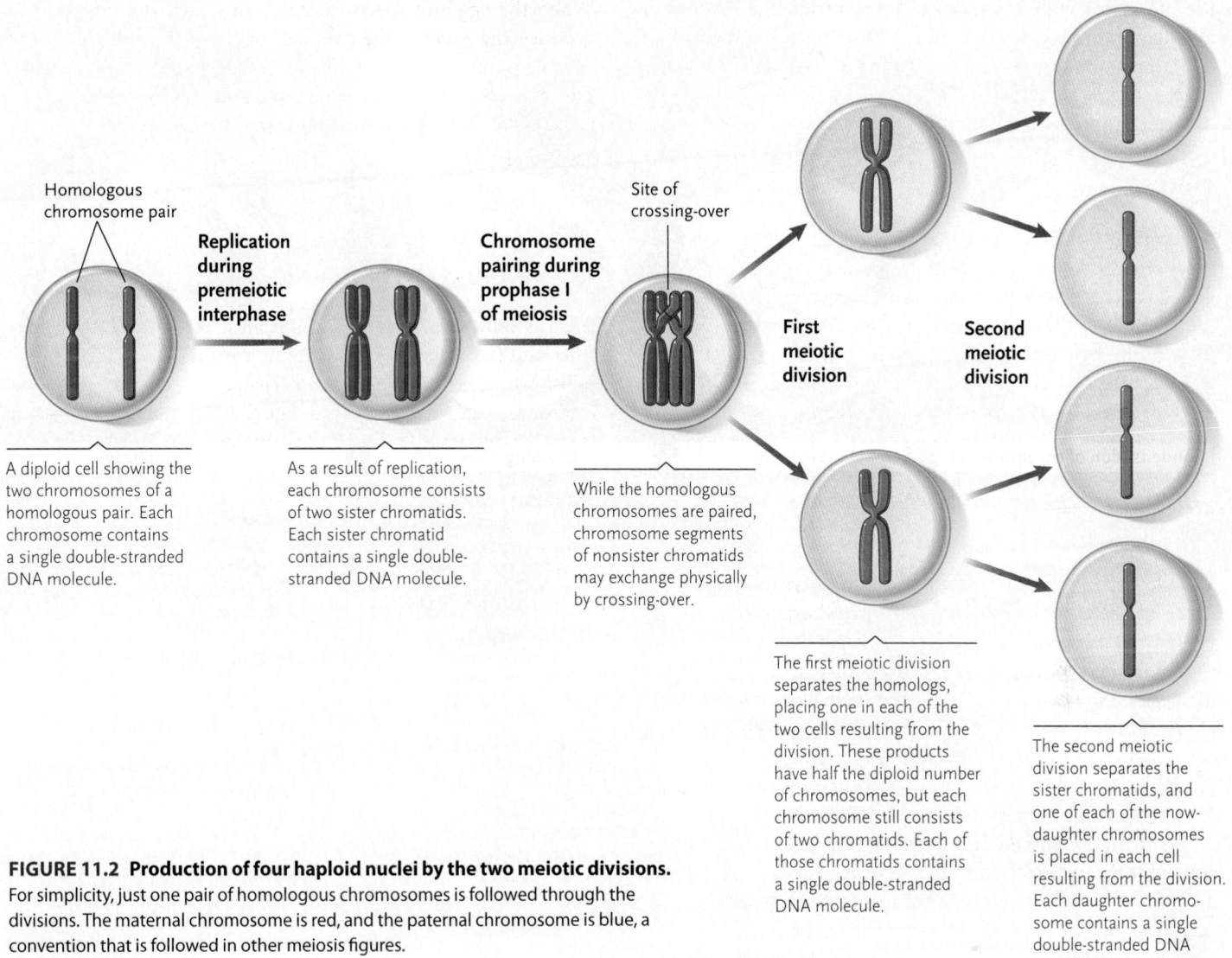

Homologous chromosome pair

Replication during premeiotic interphase

Chromosome pairing during prophase I of meiosis

Site of crossing-over

First meiotic division

Second meiotic division

A diploid cell showing the two chromosomes of a homologous pair. Each chromosome contains a single double-stranded DNA molecule.

As a result of replication, each chromosome consists of two sister chromatids. Each sister chromatid contains a single double-stranded DNA molecule.

While the homologous chromosomes are paired, chromosome segments of nonsister chromatids may exchange physically by crossing-over.

The first meiotic division separates the homologs, placing one in each of the two cells resulting from the division. These products have half the diploid number of chromosomes, but each chromosome still consists of two chromatids. Each of those chromatids contains a single double-stranded DNA molecule.

The second meiotic division separates the sister chromatids, and one of each of the now-daughter chromosomes is placed in each cell resulting from the division. Each daughter chromosome contains a single double-stranded DNA molecule.

FIGURE 11.2 Production of four haploid nuclei by the two meiotic divisions. For simplicity, just one pair of homologous chromosomes is followed through the divisions. The maternal chromosome is red, and the paternal chromosome is blue, a convention that is followed in other meiosis figures.
© Cengage Learning 2017

each of which, therefore, has half the number of chromosomes as does the parental cell (outlined in **Figure 11.2**). By contrast, in mitosis, each chromosome duplication is followed by a division. Consequently, the chromosome number remains constant from one cell generation to the next.

Figure 11.3 shows the steps of the two meiotic divisions in more detail. A brief interphase called **interkinesis** separates the two meiotic divisions, *but no DNA replication occurs during interkinesis*. **Figure 11.4** compares the two meiotic divisions with the single division of mitosis.

CHROMOSOME SEGREGATION FAILURE Rarely, chromosome segregation fails. That is, both chromosomes of a homologous pair may connect to a kinetochore microtubule from the same spindle pole in meiosis I. The result is **nondisjunction,** in which the spindle fails to separate the homologous chromosomes. As a result, one pole receives both chromosomes of the homologous pair, whereas the other pole has no copies of that chromosome. Nondisjunction can also occur in meiosis II. In this case, both chromatids of a sister-chromatid pair connect to a kinetochore

microtubule from the same spindle pole. Nondisjunction in this case produces a similar result: one pole receives both sister chromatids (as daughter chromosomes), whereas the other pole receives no copies of that chromosome. Zygotes that receive an extra chromosome because of nondisjunction have three copies of one chromosome instead of two. In humans, most zygotes of this kind do not result in live births. One exception is zygotes that have three copies of chromosome 21 instead of the normal two copies, which develop into individuals with Down syndrome. Chapter 13 discusses nondisjunction and Down syndrome in more detail.

SEX CHROMOSOMES IN MEIOSIS As you learned in Section 10.1, in many eukaryotes, including most animals, one or more pairs of chromosomes, called the **sex chromosomes,** are different in male and female individuals of the same species. For example, in humans, the cells of females contain a pair of sex chromosomes called the XX pair (the sex chromosomes are visible in Figure 10.8). Male humans contain a pair of sex chromosomes that consist of one X chromosome and a smaller

Prophase I

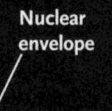

Plasma membrane · Duplicated centrioles · Nuclear envelope

Tetrad

Crossover

Condensation of chromosomes

1 • At the beginning of prophase I, the replicated chromosomes, each consisting of two identical sister chromatids, begin to condense into threadlike structures.

• As in mitosis, each pair of sister chromatids is held together tightly by sister chromatid cohesion, in which cohesin proteins encircle the sister chromatids along their length.

Synapsis

2 • The two chromosomes of each homologous pair come together and line up side-by-side in a zipperlike way in a process called **synapsis** or **pairing**. The tight association is facilitated by a protein framework called the **synaptonemal complex** (inset below).

• When fully paired, the homologs are called tetrads because each consists of four chromatids. No equivalent of chromosome pairing exists in mitosis.

Crossing-Over

3 • While they are paired, the chromatids of homologous chromosomes exchange segments by **crossing-over**. In crossing-over, enzymes break and rejoin DNA molecules of chromatids with great precision.

• The sites where crossing-over has occurred become visible under the light microscope when the chromosomes condense further as prophase I proceeds. The sites are called **crossovers** or **chiasmata** (singular, *chiasma* = crosspiece) (see enlarged circle).

• Once crossing-over is complete toward the end of prophase I, the synaptonemal complex disassembles and disappears.

Prometaphase I

4 • By the end of prophase I, a spindle has formed in the cytoplasm by the same mechanisms as described in Section 10.3. At the start of prometaphase I, the nuclear envelope breaks down, and the spindle moves into the former nuclear area.

• Kinetochore microtubules connect to the chromosomes—kinetochore microtubules from one pole attach to both sister kinetochores of one duplicated chromosome, and kinetochore microtubules from the other pole attach to both sister kinetochores of the other duplicated chromosome. That is, both sister chromatids of one homolog attach to microtubules leading to one spindle pole, whereas both sister chromatids of the other homolog attach to microtubules leading to the opposite pole.

• Nonkinetochore microtubules from the two poles overlap in the middle of the cell but do not attach to chromosomes.

Synaptonemal complex

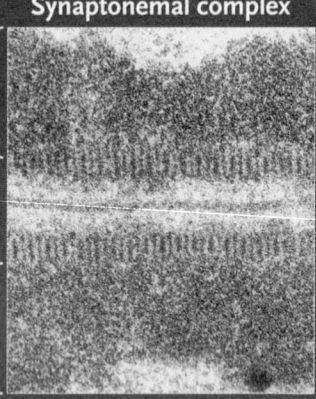

Sister chromatids of one of a homologous pair of chromosomes

Synaptonemal complex

Sister chromatids of the other of a homologous pair of chromosomes

INSET The synaptonemal complex as seen in a meiotic cell of the fungus *Neotiella*.
Courtesy of Diter Von Wettstein

Second meiotic division

Prophase II

8 • The chromosomes condense and a spindle forms.

Prometaphase II

9 • The nuclear envelope breaks down, and the spindle enters the former nuclear area.

• Kinetochore microtubules from the opposite spindle poles attach to the kinetochores of each chromosome.

FIGURE 11.3 **The meiotic divisions.** Two homologous pairs of chromosomes are shown. Maternal chromosomes are red; paternal chromosomes are blue.

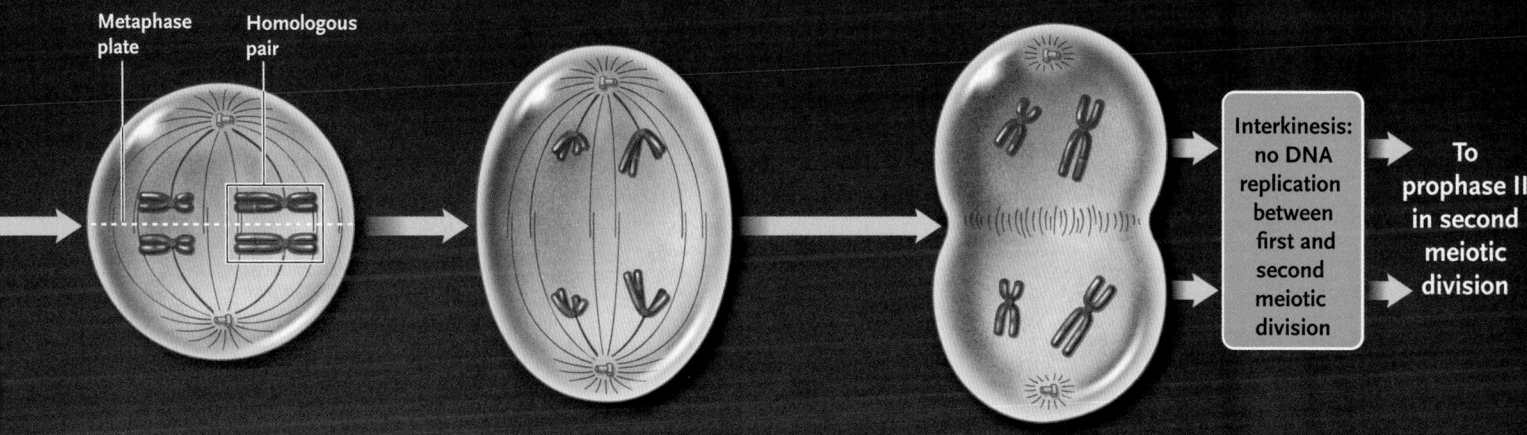

Metaphase I

5 • Movements of the kinetochore microtubules align the tetrads in the equatorial plane—meta-phase plate—between the two spindle poles.

Anaphase I

6 • Anaphase I is triggered when the enzyme separase (see Section 10.2) cleaves the cohesin rings just along the arms of the sister chromatids, leaving sister chromatid cohesion intact at the centromere region.
• The two chromosomes of each homologous pair segregate and move to opposite spindle poles. The movement delivers one-half the diploid number of chromosomes to each pole of the spindle. However, all the chromosomes at the poles are still double structures composed of two sister chromatids.

Telophase I

7 • Telophase I is a brief, transitory stage in which there is little or no change in the chromosomes except for limited decondensation or unfolding in some species.
• New nuclear envelopes form in some species but not in others.
• Telophase I is followed by an interkinesis in which the single spindle of the first meiotic division disassembles and the microtubules reassemble into two new spindles for the second division.

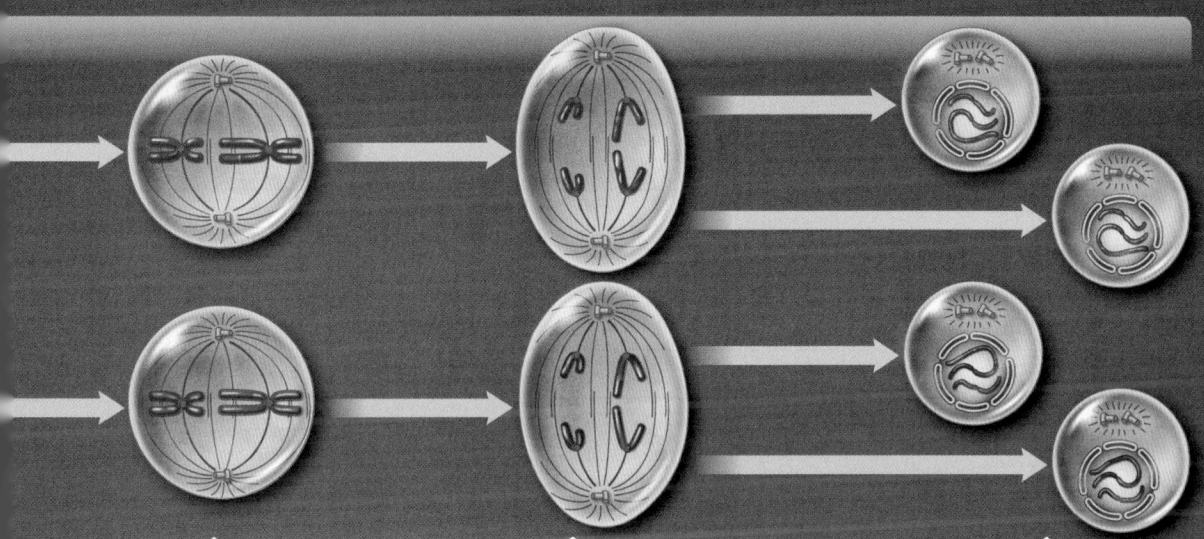

Metaphase II

10 • Movements of the spindle micro-tubules align the chromosomes on the metaphase plate.

Anaphase II

11 • Separase cleaves the remaining cohesin proteins that are holding the pairs of sister chromatids together in their centromere regions.
• Kinetochore microtubules separate the two chromatids of each chromosome and move them toward opposite spindle poles.
• At the completion of anaphase II, the chromatids—now called chromosomes—have been segregated to the two poles.

Telophase II

12 • The chromosomes begin decondensing, eventually reaching the extended interphase state.
• The spindles disassemble.
• New nuclear envelopes form around the masses of chromatin.
• Cytokinesis typically follows. The result is four haploid cells, each with a nucleus containing half the number of chromosomes present in a G_1 nucleus of the same species.

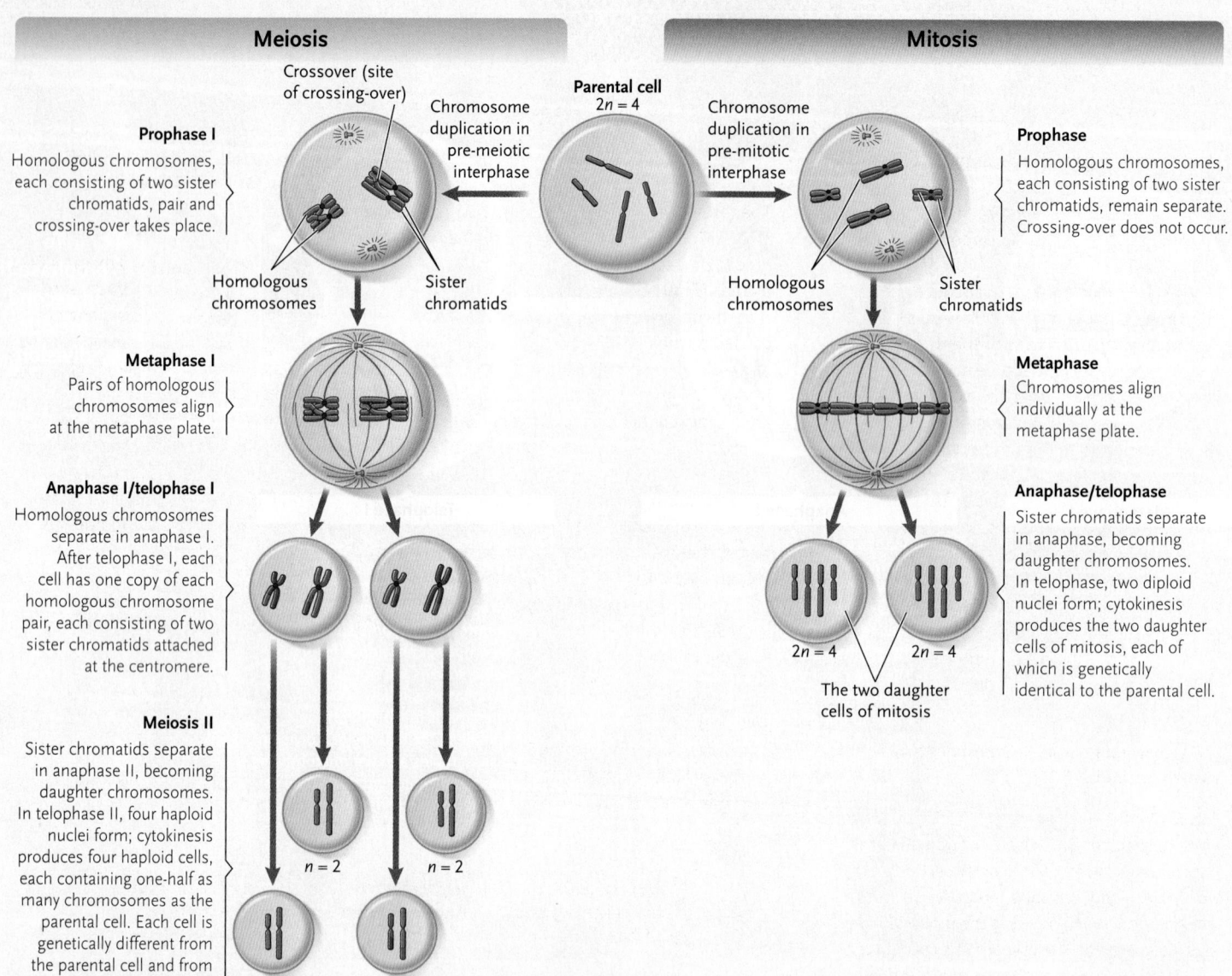

Meiosis

Prophase I
Homologous chromosomes, each consisting of two sister chromatids, pair and crossing-over takes place.

Crossover (site of crossing-over)

Chromosome duplication in pre-meiotic interphase

Homologous chromosomes

Sister chromatids

Metaphase I
Pairs of homologous chromosomes align at the metaphase plate.

Anaphase I/telophase I
Homologous chromosomes separate in anaphase I. After telophase I, each cell has one copy of each homologous chromosome pair, each consisting of two sister chromatids attached at the centromere.

Meiosis II
Sister chromatids separate in anaphase II, becoming daughter chromosomes. In telophase II, four haploid nuclei form; cytokinesis produces four haploid cells, each containing one-half as many chromosomes as the parental cell. Each cell is genetically different from the parental cell and from each other.

$n = 2$ $n = 2$

$n = 2$ $n = 2$

Parental cell
$2n = 4$

Mitosis

Chromosome duplication in pre-mitotic interphase

Prophase
Homologous chromosomes, each consisting of two sister chromatids, remain separate. Crossing-over does not occur.

Homologous chromosomes

Sister chromatids

Metaphase
Chromosomes align individually at the metaphase plate.

Anaphase/telophase
Sister chromatids separate in anaphase, becoming daughter chromosomes. In telophase, two diploid nuclei form; cytokinesis produces the two daughter cells of mitosis, each of which is genetically identical to the parental cell.

$2n = 4$ $2n = 4$

The two daughter cells of mitosis

Summary of Differences

Event or Feature	Meiosis	Mitosis
Cell divisions	Two; occurs only in diploid cells.	One; occurs in both haploid and diploid cells
Synapsis of homologous chromosomes	Yes	No
Crossing-over	Yes	No
Number of daughter cells	Four. Each cell has one-half the number of chromosomes vs the parental cell. The chromosomes are genetically different from the parent cell with various combinations of paternal and maternal chromosomes, and mixtures of paternal and maternal segments within chromosomes due to crossing-over.	Two; genetically identical to parental cell, with the same number of chromosomes.
Role in life cycle	Halving chromosome number in animal cells that produce gametes, in plant cells that produce spores, and in fungi and algae to produce spores. Generates genetic variability in gametes or spores.	Cell division for growth; asexual reproduction in some eukaryotes.

FIGURE 11.4 Meiosis and mitosis compared. Comparison of key steps in meiosis and mitosis. Both diagrams use an animal cell as an example. Maternal chromosomes are shown in red and paternal chromosomes are shown in blue.

Meiosis and Mammalian Gamete Formation: What determines whether an egg or a sperm will form?

The sex of a mammal is determined genetically. That is, the presence of a Y chromosome, and the specific male-determining gene it contains, directs the development of a male. In the absence of a Y chromosome, the default development of a female occurs. The gonads contain germ cells—cells that eventually turn into eggs or sperm. Interestingly, germ cells can become eggs or sperm, regardless of the genetic sex of the individual. Whether a germ cell develops into an egg or a sperm depends on the time at which meiosis begins. In females, meiosis commences in the fetus before birth and, as a result, eggs are produced in the ovaries. In males, meiosis begins after birth and sperm is produced in the testes. The working model was that germ cells in fetuses are programmed to enter meiosis and produce eggs unless prevented from doing so by a meiosis-inhibiting factor, in which case sperm is produced.

Research Question

What molecular signal determines germ cell fate in mammals?

Experiment

The research group of Peter Koopman at the Institute for Molecular Bioscience, University of Queensland, Brisbane, Australia, set out to answer the question using the mouse as the model mammal. The researchers used molecular tools to search for genes expressed in a sex-related fashion during gonad formation. They considered one gene, *Cyp26b1*, to be a strong candidate for the meiosis-inhibiting factor gene. *Cyp26b1* codes for an enzyme that breaks down retinoic acid, which is a natural derivative of vitamin A. Retinoic acid has many regulatory functions, one of which

is to cause germ cells in female embryos to begin meiosis.

The researchers obtained these results:

- The *Cyp26b1* gene is expressed in embryonic mouse testes but not in embryonic ovaries **(Figure)**. This result argued that the delay in meiosis in males was caused by the *Cyp26b1*-encoded enzyme degrading the retinoic acid.
- In *Cyp26b1*-knockout male mice, that is, mice in which the gene had been deleted, germ cells enter meiosis much earlier (see Figure).

The interpretation of the data is that, in the absence of the enzyme to degrade retinoic acid, meiosis was stimulated to occur at the same time as it does in the females. These two lines of evidence support the hypothesis that the product of the *Cyp26b1* gene is the meiosis-inhibiting factor in males.

Conclusion

The timing of the initiation of meiosis in mice (and probably in other mammals) is determined by the regulation of retinoic

acid presence by the *Cyp26b1*-encoded enzyme. In females, the gene is not active in embryonic ovaries, so the retinoic acid present in the fetus stimulates the initiation of meiosis before birth leading to the production of eggs. In males, the gene is active in embryonic testes, and the resulting enzyme degrades the retinoic acid. As a result, meiosis is delayed, initiating after birth, leading to the production of sperm.

Although potentially answering a very important biological question, the study does not provide all of the answers for how germ cell fate is regulated. Other questions remain, including how early meiosis favors egg formation over sperm formation.

think like a scientist

Sequence analysis has shown that the CYP26B1 protein is closely related in human, rat, mouse, and zebrafish. What do you think is the significance of this result with respect to germ cell development?

Source: J. Bowles et al. 2006. Retinoid signaling determines germ cell fate in mice. *Science* 312:596–600.

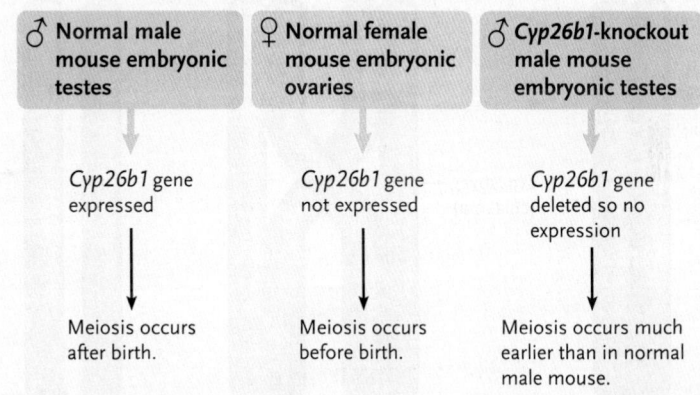

♂ Normal male mouse embryonic testes	♀ Normal female mouse embryonic ovaries	♂ *Cyp26b1*-knockout male mouse embryonic testes
Cyp26b1 gene expressed	*Cyp26b1* gene not expressed	*Cyp26b1* gene deleted so no expression
Meiosis occurs after birth.	Meiosis occurs before birth.	Meiosis occurs much earlier than in normal male mouse.

chromosome called the *Y chromosome*. Developing into a male, in fact, is directly determined by the presence of the Y chromosome because of a gene it contains (see Chapter 50). In the absence of a Y chromosome, as in an XX individual, a female is produced.

The two X chromosomes in females are fully homologous. In mammals, the X and Y chromosomes in males are homologous through a short region. This means that an X chromosome from the mother is able to pair up with either an X or a Y from

the father and follow the same pathways through the meiotic divisions as the other chromosome pairs.

As a result of meiosis, a gamete formed in a female (an egg) may receive either member of the X pair. A gamete formed in a male (a sperm) receives either an X or a Y chromosome.

The sequence of steps in the two meiotic divisions accomplishes the major outcomes of meiosis: the reduction of chromosome number and the generation of genetic variability. The latter is the subject of the next section.

1. How does the outcome of meiosis differ from that of mitosis?
2. What is recombination, and in what stage of meiosis does it occur?
3. Which of the two meiotic divisions is similar to a mitotic division?

11.2 Mechanisms That Generate Genetic Variability

The genetic variability due to the shuffling of alleles is a prime evolutionary advantage of sexual reproduction. The resulting variability increases the chance that at least some offspring will be successful in surviving and reproducing in changing environments.

The variability produced by sexual reproduction is apparent all around us, particularly in the human population. Except for identical twins (or identical triplets, identical quadruplets, and so forth), no two humans look exactly alike, act alike, or have identical biochemical and physiological characteristics, even if they are members of the same immediate family. Other species that reproduce sexually show the same type of variability arising from meiosis.

During meiosis and fertilization, genetic variability arises from three sources:

1. Crossing-over between paired homologous chromosomes during prophase I, which recombines alleles of genes on paired homologous chromosomes.
2. The independent assortment of chromosomes segregated to the poles during anaphase I, which recombines alleles of genes on nonhomologous chromosomes.
3. The particular sets of male and female gametes that unite in fertilization, which recombines the alleles of genes in the offspring of two parents.

Each of these sources of variability is discussed in further detail in the following sections. Genetic variation from an evolutionary perspective is discussed in Chapter 21.

Shuffling of Alleles on the Same Chromosome Depends on Chromosome Pairing and Crossing-Over Events between Homologous Chromosomes

If there are genetic differences between the homologs, crossing-over in prophase I can produce new allele combinations in a chromatid **(Figure 11.5).** Consider two genes on a homologous pair of chromosomes: the alleles for the two genes are *A* and *B* on the chromosome from one parent, and *a* and *b* on the chromosome

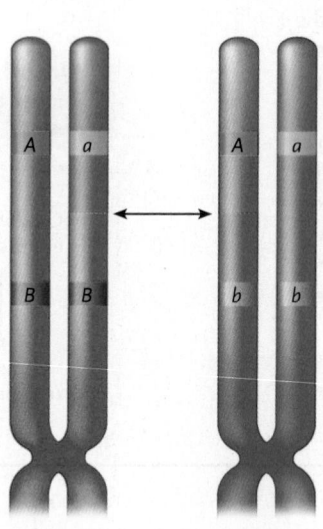
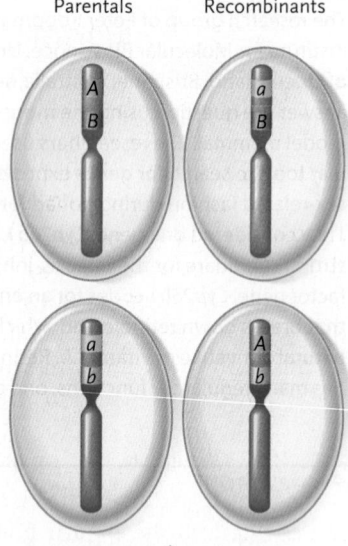

1 Homologous chromosomes pair. Each sister chromatid consists of a single double-stranded DNA molecule. The alleles for two genes on one chromosome (red) are *A* and *B*; the alleles for those genes on the other chromosome (blue) are *a* and *b*.

2 Crossing-over occurs between homologous chromatids, here between the two genes, resulting in the exchange of segments.

3 Homologous chromosomes separate at first meiotic division.

4 After meiosis II, two daughter chromosomes have parental combinations of alleles, *A–B* and *a–b*, and, as a result of crossing-over, the other two daughter chromosomes are recombinants for the alleles, *A–b* and *a–B*.

FIGURE 11.5 Recombination by crossing-over.
© Cengage Learning 2017

from the other parent. After chromosome duplication the homologous chromosomes, each consisting of two sister chromatids, pair in prophase I. Crossing-over exchanges segments of nonsister chromatids, producing new combinations of alleles. In our example, crossing-over occurred between the two genes, exchanging chromosome segments to produce two nonparental arrangements of alleles on two of the chromatids. Specifically, one sister chromatid of the left homologous chromosome now has an *a* allele and a *B* allele, and one sister chromatid of the right homologous chromosome now has an *A* allele and a *b* allele.

When meiosis is completed, there are four nuclei (see Figure 11.5). Two nuclei receive unchanged chromatids (now called chromosomes) with parental combinations of alleles, and two receive chromosomes that have new combinations of alleles resulting from crossing-over. Geneticists call the chromosomes with parental combinations of alleles **parental chromosomes,** and chromosomes with new, nonparental combinations of alleles **recombinant chromosomes.** Therefore, crossing-over is a mechanism for **genetic recombination**—it produces **genetic recombinants,** also called more simply **recombinants.**

Crossing-over can take place at almost any position along the chromosome arms, between any two of the four chromatids of a homologous pair. One or more additional crossing-over events may occur in the same chromosome pair and involve the same or different chromatids that exchanged segments in any single event. In most species, crossing-over occurs at two or three sites in each set of paired chromosomes.

Independent Assortment of Maternal and Paternal Chromosomes Is the Second Major Source of Genetic Variability in Meiosis

Recall that a diploid individual inherits one set of chromosomes from the mother and one from the father. The maternal and paternal members of a chromosome pair are homologous chromosomes. Before meiosis, each homologous chromosome duplicates to produce a pair of sister chromatids that remain connected at the centromere until they are separated in meiosis II and become daughter chromosomes.

Independent assortment of paired homologous chromosomes accounts for the second major mechanism of genetic recombination in meiosis. Recall that prometaphase I is the stage of meiosis in which the homologous pairs of chromosomes attach to the spindle poles. The maternal and paternal chromosomes of each homologous pair typically carry different alleles of many of the genes on that chromosome. For each homologous pair, one chromosome (at this stage consisting of a pair of sister chromatids) makes spindle connections leading to one pole and the other chromosome (also a pair of sister chromatids) connects to the opposite pole. In making these connections, the orientation of one chromosome pair—which member of a pair faces one pole and which faces the other—has no influence on the orientation of any other chromosome pair. How chromosomes orient themselves when aligning along the metaphase plate is entirely random. As a result, any combination of chromosomes of maternal and paternal origin may be segregated to the spindle poles **(Figure 11.6),** a phenomenon called **independent assortment** of chromosomes. The second meiotic division separates the chromatids containing these random combinations to gamete nuclei.

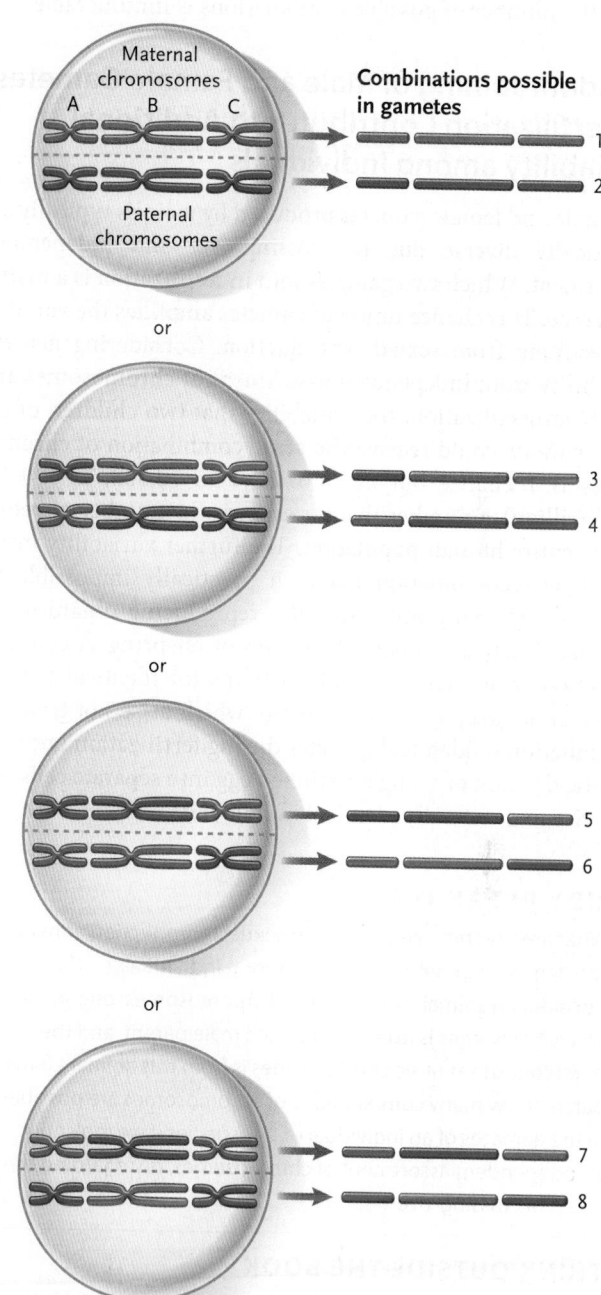

FIGURE 11.6 Possible outcomes of the random spindle connections of three pairs of chromosomes at metaphase I of meiosis. The three types of chromosomes are labeled A, B, and C, and for simplification, crossing-over is not considered. Maternal chromosomes are red; paternal chromosomes are blue. There are four possible patterns of connections, giving eight possible combinations of maternal and paternal chromosomes in gametes (labeled 1–8).
© Cengage Learning 2017

The number of combinations possible due to independent assortment depends on the number of chromosome pairs in a species. For example, the 23 chromosome pairs of humans allow 2^{23} different combinations of maternal and paternal chromosomes to be delivered to the poles, producing potentially 8,388,608 genetically different gametes from independent assortment alone. When independent assortment is combined with the rearrangement of alleles accomplished by crossing-over, the number of possible combinations is innumerable.

Random Joining of Male and Female Gametes in Fertilization Contributes to Additional Variability among Individuals

The male and female gametes produced by meiosis typically are genetically diverse due to crossing-over and independent assortment. Which two gametes join in fertilization is a matter of chance. This chance union of gametes amplifies the variability resulting from sexual reproduction. Considering just the variability from independent assortment of chromosomes and that from fertilization, the possibility that two children of the same parents could receive the same combination of chromosomes is 1 chance out of $(2^{23})^2$ or 1 in 70,368,744,000,000 (~70 trillion), a number that far exceeds the number of people in the entire human population. The further variability introduced by recombination makes it practically impossible for humans and many other sexually reproducing organisms to produce genetically identical gametes or offspring. A common exception in humans is identical twins (or identical triplets, identical quadruplets, and so forth), which arise not from the combination of identical gametes during fertilization but from mitotic division of a single fertilized egg into separate cells that give rise to genetically identical individuals.

STUDY BREAK 11.2

1. What are the three ways in which sexual reproduction enhances the degree of genetic variability among individuals?
2. Consider an animal with six pairs of chromosomes; one set of six chromosomes is from this animal's male parent, and the homologous set of six chromosomes is from this animal's female parent. How many combinations of chromosomes are possible in the gametes of an individual of this species if we look only at independent assortment of chromosomes, disregarding the effect of crossing-over?

THINK OUTSIDE THE BOOK

Mutations in several organisms have been identified that affect meiosis. Individually or collaboratively, use the Internet or research literature to find two examples of mutations that affect meiosis and outline: (1) how meiosis differs in the mutant cells of individuals compared with nonmutant cells of individuals; and (2) what we have learned about the molecular mechanisms of meiosis by studying these mutations.

11.3 The Time and Place of Meiosis in Organismal Life Cycles

The time and place at which meiosis occurs follow one of three major patterns in the life cycles of eukaryotes **(Figure 11.7)**. The differences reflect the portions of the life cycle spent in the haploid and diploid phases and whether mitotic divisions intervene between meiosis and the formation of gametes.

In Animals, the Diploid Phase Predominates, the Haploid Phase Is Reduced, and Meiosis Is Followed Directly by Gamete Formation

Animals follow the pattern **(Figure 11.7A)** in which the diploid phase predominates during the life cycle, the haploid phase is reduced, and meiosis is followed directly by gamete formation. In male animals, each of the four nuclei produced by meiosis is enclosed in a separate cell by cytoplasmic divisions, and each of the four cells differentiates into a functional sperm cell. In female animals, only one of the four nuclei becomes functional as an egg cell nucleus.

Fertilization restores the diploid phase of the life cycle. Thus, animals are haploids only as sperm or eggs, and no mitotic divisions occur during the haploid phase of the life cycle.

In Plants and Some Fungi, Generations Alternate between Haploid and Diploid Phases That Are Both Multicellular

Plants and some algae and fungi follow the life cycle pattern shown in **Figure 11.7B**. These organisms alternate between haploid and diploid generations in which, depending on the organism, either generation may dominate the life cycle, and mitotic divisions occur in both phases. In these organisms, fertilization produces the diploid generation, in which the individuals are called **sporophytes** (*spora* = seed; *phyta* = plant). After the sporophytes grow to maturity by mitotic divisions, some of their cells undergo meiosis, producing haploid, genetically different, reproductive cells called **spores.** The spores are not gametes; they germinate and grow directly by mitotic divisions into a generation of haploid individuals called **gametophytes** (*gameta* = gamete). At maturity, the nuclei of some cells in gametophytes develop into egg or sperm nuclei. All the egg or sperm nuclei produced by a particular gametophyte are genetically identical because they arise through mitosis; meiosis does not occur in gametophytes. Fusion of a haploid egg and sperm nucleus produces a diploid zygote nucleus that divides by mitosis to produce the diploid sporophyte generation again.

In many plants, including most bushes, shrubs, trees, and flowers, the diploid sporophyte generation is the most visible part of the plant. The gametophyte generation is reduced to a mostly microscopic stage that develops in the reproductive parts of the sporophytes—for flowering plants, in the structures of the flower. The female gametophyte remains in the flower; the male gametophyte is often released from flowers as microscopic pollen grains. When pollen contacts the stigma (sticky receptacle on the female portion of a flower) of the same species, it generates a pollen tube that penetrates the ovule and ultimately releases a haploid nucleus that fertilizes a haploid egg cell of a female gametophyte in the flower. The resulting cell reproduces by mitosis to form a sporophyte. (The reproduction of flowering plants is discussed in more detail in Chapter 36.)

In Some Fungi and Other Organisms, the Haploid Phase Is Dominant and the Diploid Phase Is Reduced to a Single Cell

The life cycle of some fungi and algae follows a third life cycle pattern (**Figure 11.7C**). In these organisms, the diploid phase is limited to a single cell, the zygote, produced by fertilization. Immediately after fertilization, the diploid zygote undergoes meiosis to produce the haploid phase. Mitotic divisions occur only in the haploid phase.

During fertilization, two haploid gametes fuse to form a diploid zygote nucleus. This nucleus immediately enters meiosis, producing four haploid cells. These cells develop directly or after one or more mitotic divisions into haploid spores. These spores germinate to produce haploid individuals, the gametophytes, which grow or increase in number by mitotic divisions. Eventually plus (+) and minus (−) mating types are formed in these individuals by differentiation of some of the cells produced by the mitotic divisions. Because the gametes ultimately are produced by mitosis from a single ancestral haploid cell, all the gametes of an individual are genetically identical.

In this chapter, we have seen that meiosis has three outcomes that are vital to sexual reproduction. Meiosis reduces the chromosomes to the haploid number so that the chromosome number does not double at fertilization. Through crossing-over and independent assortment of chromosomes, meiosis produces genetic variability; further variability is provided by the random combination of gametes in fertilization. The next chapter shows how the outcomes of meiosis and fertilization underlie the inheritance of traits in sexually reproducing organisms.

STUDY BREAK 11.3

How does the place of meiosis differ in the life cycles of animals and most plants?

FIGURE 11.7 Variations in the time and place of meiosis in eukaryotes. The diploid phase of the life cycles is shaded in blue; the haploid phase is shaded in yellow. n = haploid number of chromosomes; $2n$ = diploid number.
© Cengage Learning 2017

A. Animal life cycles
Diploid phase dominates the life cycle, the haploid phase is reduced, and meiosis is followed directly by gamete formation.

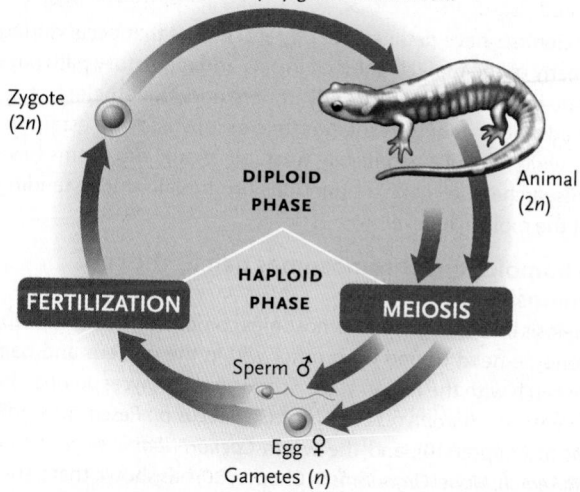

B. All plants and some fungi and algae (fern shown; relative length of the two phases varies widely in plants)
Generations alternate between haploid and diploid phases, each of which is multicellular.

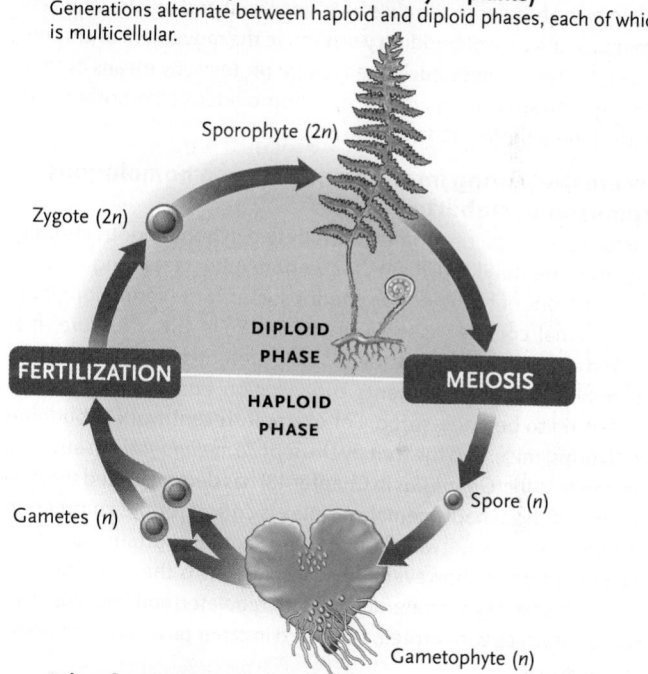

C. Other fungi and algae
Haploid phase is dominant and the diploid phase is reduced to a single cell.

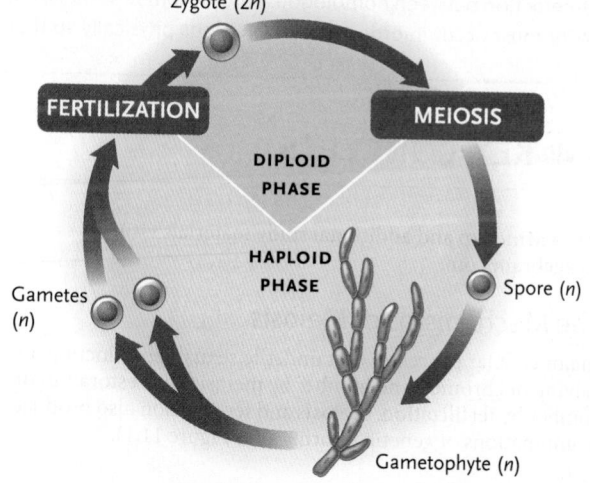

Whereas scientists have detailed the physical events that occur during meiosis, many of the molecular mechanisms and regulatory pathways that operate in meiosis are still poorly understood. Nonetheless, technological advances in high-resolution microscopy and the establishment of a wide range of genetically tractable model organisms have allowed researchers to make tremendous progress in understanding meiosis at the molecular level.

How do homologous chromosomes pair as the cell enters meiosis?

Early in meiosis I, homologous chromosomes comprising the newly replicated genome need to find each other within the nucleus and pair. Recent research with the fission yeast *Schizosaccharomyces pombe*, the budding yeast *Saccharomyces cerevisiae* (see *Focus on Research: Model Organisms* in Chapter 10), and the worm *Caenorhabditis elegans* (see *Focus on Research: Model Organisms* in Chapter 30) has shown that a specific family of nuclear envelope proteins connects cytoskeletal networks in the cytoplasm to chromosome ends in the nucleus. Different motors (dynein microtubules in the case of fission yeast and worms, and actin filaments in the case of budding yeast) drive the movement of chromosomes tethered to these nuclear envelope proteins. By means of these movements the matching sequences of homologous chromosomes are able to come together and become aligned.

How are the pairing interactions between homologous chromosomes stabilized?

Once formed, the pairing interactions between homologous chromosomes must be stabilized. A critical component for stabilizing the pairing interactions of homologous chromosomes is the formation of the synaptonemal complex (see Figure 11.5). While this structure, first observed over 50 years ago, is ubiquitously present during meiosis from yeast to humans, its components, organization, and function are only now starting to be understood. For example, recent work in budding yeast, worms, mice, and the fruit fly *Drosophila melanogaster* (see *Focus on Research: Model Organisms* in Chapter 13) has demonstrated that the formation of the synaptonemal complex is critical for the completion of crossover events between homologs. Some important questions remain unanswered, however. For example, how is the assembly and disassembly of the synaptonemal complex regulated and how does this structure interface with proteins involved in other processes occurring during meiosis?

How is the formation of crossovers ensured during meiosis?

Once the interaction between homologous chromosomes is stabilized, crossing-over must occur in order to link homologs physically so that

there will be sufficient mechanical tension between them as they align later at the metaphase I plate and are separated from each other by microtubules during the metaphase I to anaphase I transition. A highly conserved meiosis-specific protein known as Spo11 has been shown to be important in the formation of crossovers. That is, Spo11 causes the generation of physical breaks in the DNA along the paired homologous chromosomes. Repair of the breaks is initiated, which, at a subset of sites along the chromosomes, results in the formation of crossovers between nonsister chromatids. Here, an unanswered question is, How are both the frequency and distribution of crossing-over events regulated?

How do environmental chemicals affect meiosis?

Synthetic chemicals are everywhere in our environment, from those present in plastics to those in household cleaning products. One of the lines of research in our laboratory involves identifying which chemicals impair meiosis and understanding their mechanism of action. We are addressing this in the worm, *C. elegans*, which is amenable to various genetic, molecular, cytological, and biochemical approaches. As a proof-of-concept we demonstrated that key events in meiosis, namely the formation of the synaptonemal complex and the progression of meiotic recombination, were impaired following exposure to a commonly used plasticizer called bisphenol A (BPA). This result is of particular importance given that BPA has been associated with increased risk of miscarriages in humans. Moreover, we demonstrated that BPA exposure in worms, at doses that are physiologically relevant to humans, altered the germline expression of a subset of DNA break repair genes also present in humans. This simple biological system can therefore be extremely useful for the identification of the mechanism of action of these chemicals. We are currently designing various molecular approaches that will allow us to use the worm to study the effects on reproductive biology of hundreds of chemicals. The results from such experiments will provide key insights into the impact of environmental toxicants on human reproductive health.

think like a scientist If homologous chromosomes need to pair and synapse on entrance into meiosis, what happens when one of the homologs has undergone rearrangements resulting in large regions being either duplicated or inverted?

Courtesy of Monica Colaiácovo

Monica Colaiácovo is an Associate Professor in the Department of Genetics at Harvard Medical School. Her main research interests include understanding the mechanisms underlying germline maintenance and accurate chromosome inheritance during meiosis. To learn more about Dr. Colaiácovo's research go to http://genepath.med.harvard.edu/colaiacovo/index.php.

REVIEW KEY CONCEPTS

For access to MindTap and additional study materials visit www.cengagebrain.com.

11.1 The Mechanisms of Meiosis

- The major cellular processes that underlie sexual reproduction are the halving of chromosome number by meiosis and restoration of the number by fertilization. Meiosis and fertilization also produce new combinations of genetic information (Figure 11.1).

- Meiosis occurs in eukaryotes that reproduce sexually and typically in organisms that are at least diploid—that is, organisms that have at least two representatives of each chromosome.

- DNA replicates and the chromosomal proteins are duplicated during the premeiotic interphase, producing two copies, the sister chromatids, of each chromosome.

- During prophase I of the first meiotic division (meiosis I), the replicated chromosomes condense and come together and pair.

While they are paired, chromatids of homologous chromosomes exchange segments by crossing-over. While these events are in progress, the spindle forms in the cytoplasm. The crossing-over events become visible later as chiasmata (Figures 11.2–11.4).

- During prometaphase I, the nuclear envelope breaks down, the spindle enters the former nuclear area, and kinetochore microtubules leading to opposite spindle poles attach to one kinetochore of each set of sister chromatids of homologous chromosomes (Figure 11.3).
- At metaphase I, spindle microtubule movements have aligned the tetrads on the metaphase plate, the equatorial plane between the two spindle poles. The connections of kinetochore microtubules to opposite poles ensure that the homologous chromosomes of each pair segregate and move to opposite spindle poles during anaphase I, reducing the chromosome number to the haploid value. Each chromosome at the poles still contains two chromatids.
- Telophase I and interkinesis are brief and transitory stages; no DNA replication occurs during interkinesis. During these stages, the single spindle of the first meiotic division disassembles.
- During prophase II, the chromosomes condense and a spindle forms. During prometaphase II, the nuclear envelope breaks down, the spindle enters the former nuclear area, and spindle microtubules leading to opposite spindle poles attach to the two kinetochores of each chromosome. At metaphase II, the chromosomes align on the metaphase plate. The connections of kinetochore microtubules to opposite spindle poles ensure that during anaphase II, the chromatids of each chromosome are separated and migrate to those opposite spindle poles.
- During telophase II, the chromosomes decondense to their extended interphase state, the spindles disassemble, and new nuclear envelopes form. The result is four haploid cells, each containing half the number of chromosomes present in a G_1 nucleus of the same species.

11.2 Mechanisms That Generate Genetic Variability

- Crossing-over recombines genes on paired homologous chromosomes during meiosis (Figure 11.5). Chromatids acquire new combinations of alleles—recombinants—by physically exchanging segments in crossing-over. The exchange process involves precise breakage and exchange of DNA molecules through a complex mechanism. It is catalyzed by enzymes and occurs while the homologous chromosomes are held together tightly by the synaptonemal complex.
- The chiasmata visible between the chromosomes at late prophase I reflect the exchange of chromatid segments that occurred during the molecular steps of crossing-over.
- The independent assortment of homologous chromosomes is another mechanism through which meiosis produces genetic variability. The homologous pairs segregate at anaphase I of meiosis, the orientation and segregation of one pair having no influence on the orientation and segregation of any other pair (Figure 11.6).
- Random union of male and female gametes at fertilization is a third mechanism for enhancing genetic variability.

11.3 The Time and Place of Meiosis in Organismal Life Cycles

- The time and place of meiosis follow one of three major pathways in the life cycles of eukaryotes, which reflect the portions of the life cycle spent in the haploid and diploid phases and whether mitotic divisions intervene between meiosis and the formation of gametes (Figure 11.7).
- In animals, the diploid phase predominates during the life cycle; mitotic divisions occur only in this phase. Meiosis in the diploid phase gives rise to products that develop directly into egg and sperm cells without undergoing mitosis (Figure 11.7A).
- In all plants and some fungi, the life cycle alternates between haploid and diploid generations that both grow by mitotic divisions. Fertilization produces the diploid sporophyte generation; after growth by mitotic divisions, some cells of the sporophyte undergo meiosis and produce haploid spores. The spores germinate and grow by mitotic divisions into the gametophyte generation. After growth of the gametophyte, cells develop directly into egg or sperm nuclei, which fuse in fertilization to produce the diploid sporophyte generation again (Figure 11.7B).
- In some fungi and protists, meiosis occurs immediately after fertilization, producing a haploid phase, which predominates during the life cycle; mitosis occurs only in the haploid phase. At some point in the life cycle, haploid cells differentiate directly into gametes, which fuse to produce the brief diploid phase (Figure 11.7C).

TEST YOUR KNOWLEDGE

Remember/Understand

1. The chromosome constitution number of this individual is $2n = 6$.

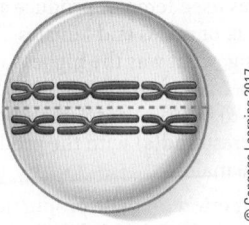

© Cengage Learning 2017

This drawing represents:
   a. mitotic metaphase.
   b. meiotic metaphase I.
   c. meiotic metaphase II.
   d. a gamete.
   e. six nonhomologous chromosomes.

2. Which of the following is *not* associated with sperm production?
   a. daughter cells identical to the parent cell
   b. variety in resulting cells
   c. chromosome number halved in resulting cells
   d. four daughter cells arising from one parent cell
   e. 23 chromosomes in the human sperm

3. Chiasmata:
   a. form during metaphase II of meiosis.
   b. occur between two nonhomologous chromosomes.
   c. represent chromosomes independently assorting.
   d. are sites of DNA exchange between homologous chromatids.
   e. ensure the resulting cells are identical to the parent cell.

4. If $2n$ is four, the number of possible combinations of chromosomes in the resulting gametes, excluding any crossing-over, is:
   a. 1.
   b. 2.
   c. 4.
   d. 8.
   e. 16.

5. In meiosis:
    a. homologous chromosomes pair at prophase II.
    b. chromosomes segregate from their homologous partners at anaphase I.
    c. the centromeres split at anaphase I.
    d. a female gamete has two X chromosomes.
    e. reduction of chromosome number occurs in meiosis II.

6. The DNA content in a diploid cell in G_2 is X. If that cell goes into meiosis at metaphase II, the DNA content will be:
    a. 0.1X.
    b. 0.5X.
    c. X.
    d. 2X.
    e. 4X.

7. Metaphase in mitosis is similar to what stage in meiosis?
    a. prophase I
    b. prophase II
    c. metaphase I
    d. metaphase II
    e. crossing-over

8. In the human sperm:
    a. there must be one chromosome of each type, except for the sex chromosomes, where either an X or a Y chromosome is present.
    b. a chromosome must be represented from each parent.
    c. there must be an unequal number of chromosomes from both parents.
    d. there must be representation of chromosomes from only one parent.
    e. there is the possibility of 2^{46} different combinations of maternal and paternal chromosomes.

9. In plants, the adult diploid individuals are called:
    a. spores.
    b. sporophytes.
    c. gametes.
    d. gametophytes.
    e. zygotes.

10. Which of the following sequences of events describes the general life cycle of an animal?
    a. $2n \rightarrow$ meiosis $\rightarrow 2n \rightarrow$ fertilization $\rightarrow 1n$
    b. $1n \rightarrow$ meiosis $\rightarrow 2n \rightarrow$ fertilization $\rightarrow 1n$
    c. $2n \rightarrow$ meiosis $\rightarrow 1n \rightarrow$ fertilization $\rightarrow 2n$
    d. $2n \rightarrow$ mitosis $\rightarrow 1n \rightarrow$ fertilization $\rightarrow 2n$
    e. $2n \rightarrow$ mitosis $\rightarrow 1n \rightarrow$ fertilization $\rightarrow 1n$

Apply/Analyze

11. **Discuss Concepts** You have a technique that allows you to measure the amount of DNA in a cell nucleus. You establish the amount of DNA in a sperm cell of an organism as your baseline. Which multiple of this amount would you expect to find in a nucleus of this organism at G_2 of premeiotic interphase? At telophase I of meiosis? During interkinesis? At telophase II of meiosis?

12. **Discuss Concepts** One of the human chromosome pairs carries a gene that influences eye color. In an individual human, one chromosome of this pair has an allele of this gene that contributes to the formation of blue eyes. The other chromosome of the pair has an allele that contributes to brown eye color (other genes also influence eye color in humans). After meiosis in the cells of this individual, what fraction of the nuclei will carry the allele that contributes to blue eyes? To brown eyes?

13. **Discuss Concepts** Mutations are changes in DNA sequence that can create new alleles. In which cells of an individual, somatic or meiotic cells, would mutations be of greatest significance to the survival of that individual? What about to the species to which the individual belongs?

Evaluate/Create

14. **Design an Experiment** Design experiments to determine whether a new pesticide on the market adversely affects egg production and fertilization in frogs.

15. **Apply Evolutionary Thinking** Explain aspects of the processes of mitosis and meiosis that would lead you to conclude that they are evolutionarily related processes. Do you think that mitosis evolved from meiosis, or did the opposite occur? Explain your conclusion.

For selected answers, see Appendix A.

INTERPRET THE DATA

A wingless female aphid can generate up to 100 almost-identical female offspring who themselves reproduce similarly, thereby creating a clone of offspring similar to the original female. In these cases, no eggs are laid, and the young are born as juveniles. However, if day length (photoperiod) and temperature change significantly (for example, in fall in temperate climates), the wingless female can switch and produce a combination of winged males, winged females that reproduce without laying eggs, or egg-laying females with wings. Depending on circumstances and aphid species, different proportions of each of these are produced.

The **Table** represents some data from an early study looking at the effect of temperature and day length (or night length) on the number of males produced by individual females of the species *Megoura viciae*.

Temperature	Photoperiod (hr light/24 hr)	Number of Mothers	Number of Males per Mother	
			Range	Mean
25°C	12	9	0–0	0
	16	9	0–0	0
20°C	12	27	5–21	13.7
	16	44	0–21	8.0
15°C	12	37	3–17	11.6
	16	25	1–20	10.4
11°C	12	19	0–5	2.5
	16	32	0–9	2.9

1. At the temperatures used females produce about 100 offspring each, typically a mix of males and females. Assuming this number of offspring, what was the highest and lowest percentage of male offspring per female parent at 20°C? What temperature range appears to be optimal for male production? Which temperature(s) correspond(s) with the production of no males? With less than 10% males?

2. The effect of photoperiod on male production was also studied. Various day lengths were simulated in the laboratory using artificial light. Citing examples from the data, what effect does photoperiod have on male production? How does it compare to the effect of temperature?

Source: A. D. Lees. 1959. The role of photoperiod and temperature in the determination of parthenogenetic and sexual forms in the aphid *Megoura viciae* Buckton—I: The influence of these factors on apterous virginoparae and their progeny. *Journal of Insect Physiology* 3:92–117.

Mendel, Genes, and Inheritance

Rabbits, showing genetic variation in coat color.

Biosphoto/Michel Gunther

Why it matters . . . On New Year's Eve, 1904, Ernest Irons, a medical intern, was examining a blood specimen from a new patient and was sketching what he saw through his microscope—peculiarly elongated red blood cells shaped like a sickle, a cutting tool with a crescent-shaped blade **(Figure 12.1).** He and his supervisor, James Herrick, had never seen anything like them.

The patient had complained of weakness, dizziness, shortness of breath, and pain. His father and two sisters had died from mysterious ailments that had damaged their lungs or kidneys. Did those deceased family members also have sickle-shaped red cells in their blood? Was there a connection between the abnormal cells and the ailments? How did the cells become sickled?

The medical problems that baffled Irons and Herrick killed their patient when he was only 32 years old. The patient's symptoms were characteristic of a genetic disorder now called *sickle-cell anemia*. This disease develops when a person has received two mutated copies of a gene (one from each parent) that codes for a polypeptide subunit of hemoglobin, the oxygen-transporting protein in red blood cells. The mutated gene codes for a slightly altered form of the hemoglobin subunit. When oxygen supplies are low, the altered hemoglobin forms long, fibrous, helical structures that distend red blood cells into the sickle shape. The mutant protein differs from the nonmutant protein by just a single amino acid.

A. A normal red blood cell

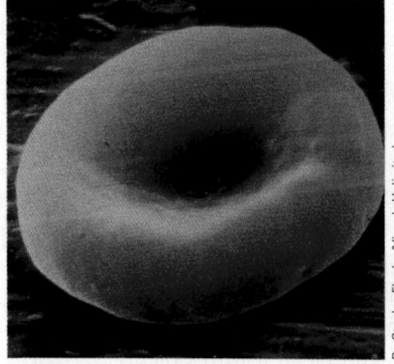

Dr. Stanley Flegler/Visuals Unlimited

B. A sickled red blood cell

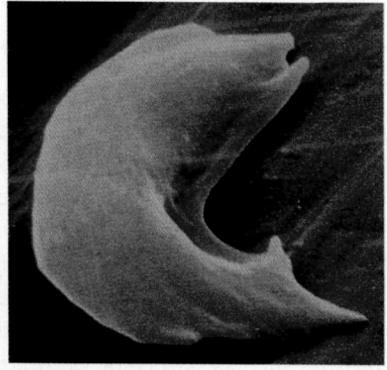

Dr. Stanley Flegler/Visuals Unlimited

FIGURE 12.1 Red blood cell shape in sickle-cell anemia.

FIGURE 12.2
Gregor Mendel (1822–1884), the founder of modern genetics.

The sickled red blood cells are too elongated and inflexible to pass freely through the capillaries, the smallest vessels in the circulatory system. As a result, the cells tend to increase in number and cluster at capillary openings, obstructing the flow of blood, when oxygen levels are low. Tissues become starved for oxygen and saturated with metabolic wastes, causing the symptoms experienced by Irons and Herrick's patient. The problem worsens as oxygen concentration falls in tissues and more red blood cells are pushed into the sickled form. (You will learn more about sickle-cell anemia later in this chapter and in Chapter 13.)

Researchers have studied sickle-cell anemia in great detail at both the molecular and the clinical levels. Interestingly, though, our understanding of sickle-cell anemia—and other inherited traits—actually began with studies of pea plants in a monastery garden.

Almost fifty years before Ernest Irons sketched sickled red blood cells, a scholarly monk named Gregor Mendel **(Figure 12.2)** used garden peas to study patterns of inheritance. He used the scientific method in his work (see Section 1.4 and Figure 1.14). To test his hypotheses about inheritance, Mendel bred generations of pea plants and carefully observed the patterns by which parents transmit traits to their offspring. Through his experiments and observations, Mendel discovered the fundamental principles that govern inheritance. His discoveries and conclusions founded the science of genetics.

12.1 The Beginnings of Genetics: Mendel's Garden Peas

Until about 1900, scientists and the general public believed in the **blending theory of inheritance,** which suggested that hereditary traits blend evenly in offspring through mixing of the parents' blood, much like the effect of mixing coffee and cream. Even today, many people assume that parental characteristics such as skin color, body size, and facial features blend evenly in their offspring, with the traits of the children

appearing about halfway between those of their parents. Yet if blending takes place, extremes, such as very tall and very short individuals, should disappear gradually over generations as repeated blending takes place, yet they do not. Also, why do children with blue eyes turn up among the offspring of brown-eyed parents?

Gregor Mendel's experiments with garden peas, performed in the 1860s, were not done to address directly the blending theory of inheritance, but their results provided answers to such questions and many more. Mendel was an Augustinian monk who lived in a monastery in Brno, now part of the Czech Republic. But he had an unusual education for a monk of his time. He had studied mathematics, chemistry, zoology, and botany at the University of Vienna. He had also been reared on a farm and was well aware of agricultural principles and their application. He kept abreast of breeding experiments published in scientific journals.

In his work with peas, Mendel studied a variety of *characters.* A **character** is a specific heritable attribute or property of an organism. It may be a visible attribute, or it may be one that can be detected only by biochemical, molecular, or physiological analysis. The characters Mendel studied included seed shape, seed color, and flower color. Mendel studied plants that had alternative forms of these characters, known as **character differences** or **traits.** For example, for the flower color character, the two traits studied—the alternative forms— were purple and white. In modern terms, we say that the alternative forms are two contrasting *phenotypes* (Greek *phaenin =* to appear or show; *typos =* mark). That is, a **phenotype** is a form taken by a character.

Mendel established that characters are passed to offspring in the form of discrete hereditary factors, which now are known as genes. Mendel observed that, rather than blending evenly, many parental traits appear unchanged in offspring, whereas others disappear in one generation to reappear unchanged in subsequent generations produced by particular breeding experiments. Although Mendel did not know it, the inheritance patterns he observed are the result of the segregation of chromosomes, on which the genes are located, to gametes in meiosis (see Chapter 11 and later in this chapter). Mendel's methods illustrate how rigorous scientific work is conducted: through observation, making hypotheses, and testing the hypotheses with experiments (see Section 1.4 and Figure 1.14). And, although others had studied inheritance patterns before him, Mendel's most important innovation was his quantitative approach to science, specifically his rigor and statistical analysis in an era when qualitative, descriptive science was the accepted practice.

Mendel Chose True-Breeding Garden Peas for His Experiments

Mendel used the garden pea *(Pisum sativum)* for his genetics experiments. **Figure 12.3** shows the structure of a pea flower. Normally, pea plants **self-fertilize** (also known as **self-pollinate,** or more simply, *self*): sperm nuclei in pollen produced by anthers

M. Hofer/National Library of Medicine

FIGURE 12.3 | Research Method

Making a Genetic Cross between Two Pea Plants

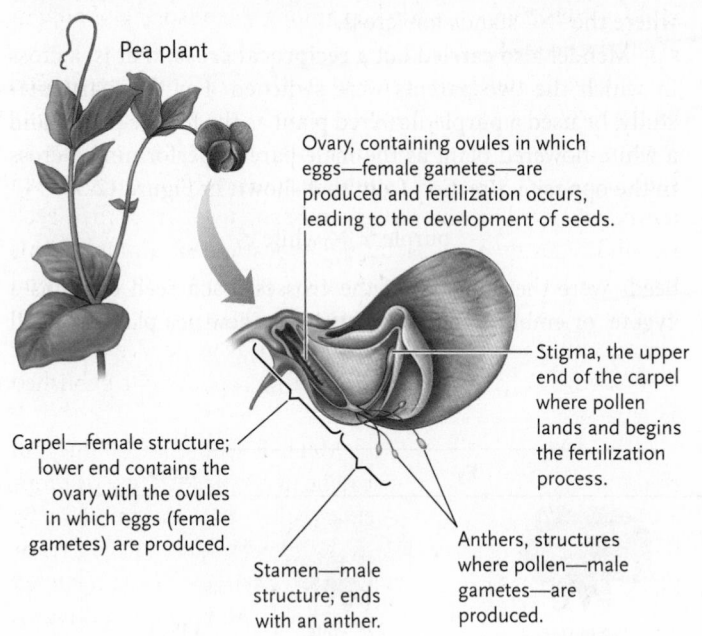

Pea plant

Ovary, containing ovules in which eggs—female gametes—are produced and fertilization occurs, leading to the development of seeds.

Stigma, the upper end of the carpel where pollen lands and begins the fertilization process.

Carpel—female structure; lower end contains the ovary with the ovules in which eggs (female gametes) are produced.

Stamen—male structure; ends with an anther.

Anthers, structures where pollen—male gametes—are produced.

Purpose: Mendel used the garden pea, *Pisum sativum,* for his genetic experiments. The goal of the experiments was to test various hypotheses about the patterns of inheritance by cross-breeding plants with easily observable characters, such as flower color and seed shape. He could then analyze whether the characters he observed and counted in the offspring supported the predictions made by a particular hypothesis.

In cross-breeding, the sperm and the egg must come from different plants. However, this type of flowering plant has both male and female structures within the same flower and is capable of self-fertilization, also called "selfing." The figure to the left shows a pea flower sectioned to show the location of the reproductive structures. (Details of plant fertilization are presented in Chapter 35).

The figure below shows how Mendel designed his experiments to prevent selfing and perform his crosses.

Protocol:

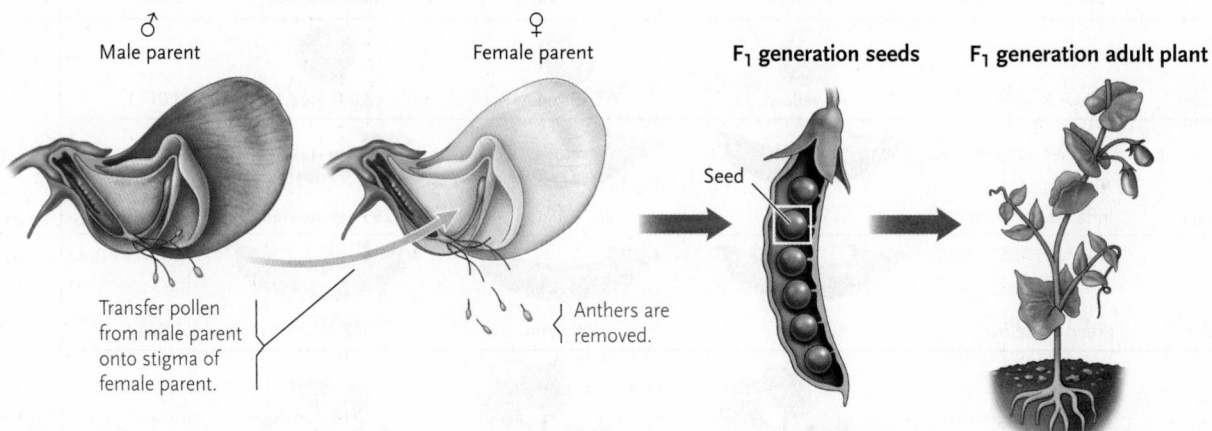

♂ Male parent

♀ Female parent

F₁ generation seeds

F₁ generation adult plant

Seed

Transfer pollen from male parent onto stigma of female parent.

Anthers are removed.

1. Remove the anthers from one of the parents (the white-flowered plant) to prevent self-fertilization. Transfer pollen from the male parent (the purple-flowered plant) onto the stigma of the white flower (the female parent). This results in cross-fertilization, the fertilization of one plant with pollen from another.

2. The cross-fertilized plant produces seeds. Seeds may be scored for seed traits, such as round vs. wrinkled shape. Seeds are grown into adult plants. Plants may be scored for adult traits, such as purple vs. white flower color.

fertilize egg cells housed in the ovule of the same flower. (Reproduction in flowering plants is described in Chapter 36.) However, for his experiments, Mendel prevented self-fertilization by removing the anthers (see Figure 12.3). Pollen to fertilize these flowers had to come from a different plant. This technique is called **cross-fertilization** or **cross-pollination,** or more simply,

a *cross.* Using this method, Mendel tested the effects of mating pea plants of different parental types.

To begin his experiments, Mendel chose pea plants that were known to be **true-breeding** (also called *pure-breeding*); that is, when self-fertilized—*selfed*—they passed traits without change from one generation to the next.

Mendel First Worked with Crosses of Plants Differing in One Character

Mendel selected seven characters for study **(Figure 12.4).** The alternative forms of each character could be readily seen visually. Although Mendel knew nothing about the molecular bases of character differences, keep in mind as you learn about his experiments that the visual differences result from underlying molecular differences.

Flower color was among the characters Mendel studied: one true-breeding variety of peas had purple flowers, and the other true-breeding variety had white flowers. Mendel fertilized a white flower with pollen from a purple flower as shown in Figure 12.3. In this cross, the white-flowered plant is the female parent and purple-flowered plant is the male parent. A simple way geneticists present crosses like this is:

$$\text{white} \; ♀ \times \text{purple} \; ♂$$

where the "×" stands for "cross."

Mendel also carried out a **reciprocal cross,** that is, a cross in which the two parents were switched. For this particular study, he used a purple-flowered plant as the female parent and a white-flowered plant as the male parent, performing a cross in the opposite direction from that shown in Figure 12.3:

$$\text{purple} \; ♀ \times \text{white} \; ♂$$

Seeds were the outcome of the crosses. Each seed contains a zygote, or embryo, that develops into a new pea plant. We call

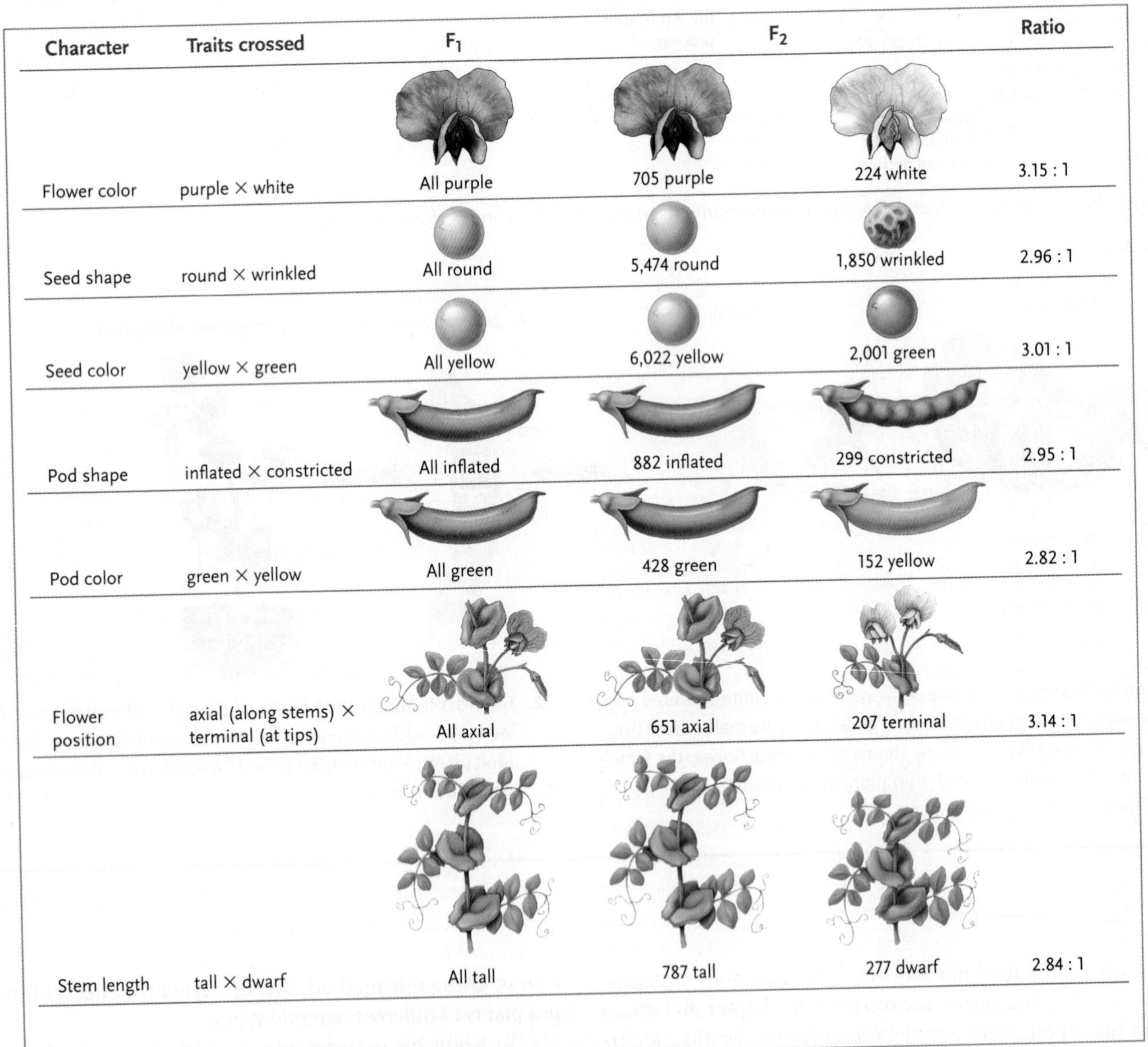

Character	Traits crossed	F₁	F₂		Ratio
Flower color	purple × white	All purple	705 purple	224 white	3.15 : 1
Seed shape	round × wrinkled	All round	5,474 round	1,850 wrinkled	2.96 : 1
Seed color	yellow × green	All yellow	6,022 yellow	2,001 green	3.01 : 1
Pod shape	inflated × constricted	All inflated	882 inflated	299 constricted	2.95 : 1
Pod color	green × yellow	All green	428 green	152 yellow	2.82 : 1
Flower position	axial (along stems) × terminal (at tips)	All axial	651 axial	207 terminal	3.14 : 1
Stem length	tall × dwarf	All tall	787 tall	277 dwarf	2.84 : 1

FIGURE 12.4 Mendel's crosses with seven characters in peas, including his results and the calculated ratios of offspring.

the plants used in an initial cross between two true-breeding parents the parental or **P generation.** We call the first generation of offspring from a cross between two true-breeding parents the **F₁ generation** (F stands for *filial; filius* = son).

From the white ♀ × purple ♂ cross, the plants that grew from the F₁ seeds all formed purple flowers, as if the trait for white flowers had disappeared. The flowers showed no evidence of blending. Exactly the same result was obtained for the purple ♀ × white ♂ reciprocal cross, showing that the sex of the parent did not affect the inheritance pattern of the traits.

Mendel then allowed purple-flowered F₁ plants to self, producing seeds that represented the **F₂ generation.** When he planted the F₂ seeds produced by this cross, plants with the white-flowered trait reappeared along with purple-flowered plants. Mendel counted 705 plants with purple flowers and 224 with white flowers, in a ratio that he noted was close to 3 purple : 1 white.

Mendel made similar crosses that involved six other characters with pairs of traits; for example, the character of seed color has the traits yellow and green (see Figure 12.4). In all cases, he observed a uniform F₁ generation, in which only one of the two traits was present. In the F₂ generation, the missing trait reappeared, and both traits were present among the offspring in a ratio that was also close to 3 : 1.

Single-Character Crosses Led Mendel to Propose the Principle of Segregation

Mendel made three important conclusions to explain the results of his crosses of plants differing in one character. The first two conclusions led to the third and most important conclusion, the *principle of segregation.*

Conclusion 1: *The adult plants carry a* pair *of factors that govern the inheritance of each trait.* Mendel correctly deduced that for each trait, an organism inherits one factor from each parent.

In modern terminology, Mendel's factors are *genes,* which are located on chromosomes. The different versions of a gene, producing different traits of a character, are **alleles** of the gene (see Section 11.1). Although Mendel did not use the terms *genes* and *alleles,* we use them in this chapter in our description of Mendel's work. Thus, there are two alleles of the gene that governs flower color in garden peas: one allele for purple flower color and another allele for white flower color. Organisms with two copies of each gene are known as *diploids* (see Section 11.1). The two alleles of a gene in a diploid individual may be identical or different. Alleles of genes are also known as **gene markers.** In turn, gene markers are one of two types of **genetic markers,** another name for a mutation or genetic variant that gives a distinguishable phenotype. The other type of genetic marker is a **DNA marker,** a site or region in the genome that is polymorphic (*poly* = many; *morphos* = form), meaning different sequences are found at the site or region in the population). DNA markers are detected by molecular analysis of DNA. (DNA markers are discussed more in later chapters.)

Conclusion 2: Mendel's second conclusion (in modern language) was: *If an individual's pair of genes consists of different alleles, one allele is dominant over the other, which is recessive.* Mendel had to explain why one of the traits, such as white flowers, disappears in the F₁ generation and then reappears in the F₂ generation. Mendel deduced that the trait that "disappeared" in the F₁ generation was actually present but was masked by the "stronger" allele. Mendel called the masking effect **dominance.** When a **dominant** allele of a gene is paired with a recessive allele of that gene, the dominant allele determines the trait that appears. By contrast, the trait determined by the **recessive** allele appears only when two copies of that allele are present. For example, the allele for purple flower color is dominant and the allele for white flower color is recessive.

Conclusion 3: Mendel's third conclusion was: *The pairs of alleles that control a character segregate (separate) as gametes are formed; half the gametes carry one allele, and the other half carry the other allele.* This conclusion is now known as Mendel's **Principle of Segregation.** During fertilization, fusion of the haploid maternal and paternal gametes produces a diploid nucleus called the *zygote nucleus.* The zygote nucleus receives one allele for the character from the male gamete and one allele for the same character from the female gamete, reuniting the pairs.

Figure 12.5 steps through the purple × white pea cross to illustrate the principle of segregation. Important genetic terms used in the figure are:

- **Homozygote** (*homo* = same): A true-breeding individual with both alleles of a gene the same. A homozygote produces only one type of gamete, which contains one copy of that allele.
- **Homozygous:** An individual that is a homozygote is said to be homozygous for the particular allele of the gene.
- **Heterozygote** (*hetero* = different): An individual with two different alleles of a gene. A heterozygote produces two types of gametes: one type has a copy of one allele, the other type has a copy of the other, different, allele.
- **Heterozygous:** An individual that is a heterozygote is said to be heterozygous for the pair of different alleles of a gene.
- **Monohybrid** (*mono* = one; *hybrid* = an offspring of parents with different traits): An F₁ heterozygote produced from a cross that involves a single character.
- **Monohybrid cross:** A cross between two individuals that are each heterozygous for the same pair of alleles.

Mendel's results explain how individuals may differ genetically but still look the same. The *PP* and *Pp* plants, although genetically different, both have purple flowers (see Figure 12.5). In modern terminology, **genotype** refers to the *genetic constitution of an organism in terms of genes and alleles,* and, as you learned earlier, *phenotype* refers to its *appearance.* In this case, the two different *genotypes PP* and *Pp* produce the same purple-flower *phenotype.*

How does the genotype relate to the phenotype for the flower-color character? Conceptually, at the molecular level,

FIGURE 12.5 **Experimental Research**

The Principle of Segregation: Inheritance of Flower Color in Garden Peas

Question: How is flower color in garden peas inherited?

Experiment: Mendel crossed a true-breeding purple-flowered plant with a true-breeding white-flowered plant and analyzed the progeny through the F_1 and F_2 generations. We explain this cross here in modern terms.

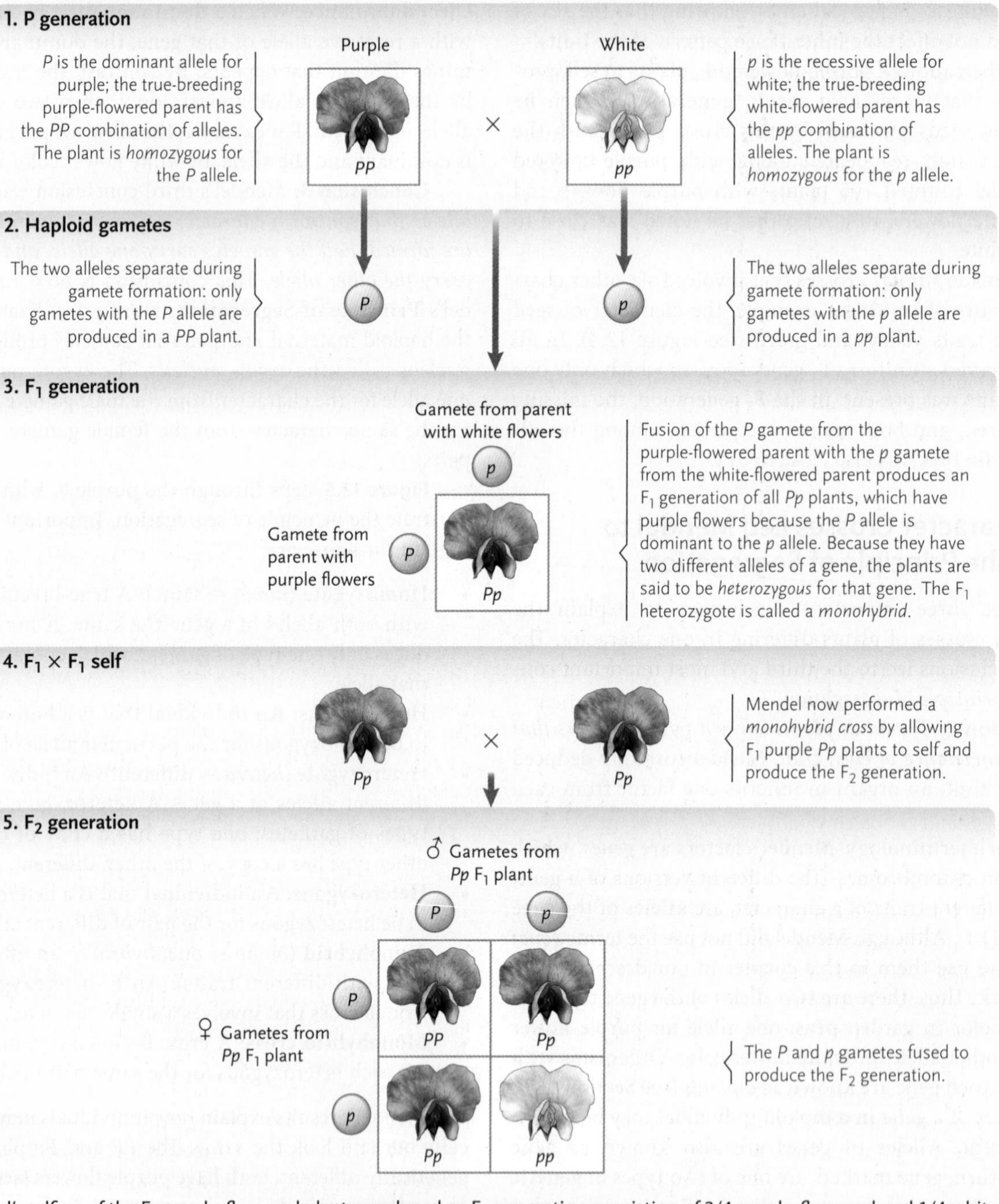

1. P generation

P is the dominant allele for purple; the true-breeding purple-flowered parent has the PP combination of alleles. The plant is *homozygous* for the P allele.

Purple

PP

×

White

pp

p is the recessive allele for white; the true-breeding white-flowered parent has the pp combination of alleles. The plant is *homozygous* for the p allele.

2. Haploid gametes

The two alleles separate during gamete formation: only gametes with the P allele are produced in a PP plant.

P

p

The two alleles separate during gamete formation: only gametes with the p allele are produced in a pp plant.

3. F_1 generation

Gamete from parent with white flowers

p

Gamete from parent with purple flowers

P

Pp

Fusion of the P gamete from the purple-flowered parent with the p gamete from the white-flowered parent produces an F_1 generation of all Pp plants, which have purple flowers because the P allele is dominant to the p allele. Because they have two different alleles of a gene, the plants are said to be *heterozygous* for that gene. The F_1 heterozygote is called a *monohybrid*.

4. $F_1 \times F_1$ self

Pp

×

Pp

Mendel now performed a *monohybrid cross* by allowing F_1 purple Pp plants to self and produce the F_2 generation.

5. F_2 generation

♂ Gametes from Pp F_1 plant

P p

♀ Gametes from Pp F_1 plant

P

p

PP Pp

Pp pp

The P and p gametes fused to produce the F_2 generation.

Results: Mendel's selfing of the F_1 purple-flowered plants produced an F_2 generation consisting of 3/4 purple-flowered and 1/4 white-flowered plants. White flowers were inherited as a recessive trait, disappearing in the F_1 and reappearing in the F_2.

Conclusion: The results supported Mendel's Principle of Segregation hypothesis that the pairs of alleles that control a character segregate as gametes are formed, with half of the gametes carrying one allele, and the other half carrying the other allele.

think like a scientist

Observe
Hypothesize
Predict
Experiment
Interpret

Suppose you pick at random one of the F_2 purple-flowered plants and allow it to self. What is the chance that the progeny will include both purple-flowered and white-flowered plants?

the *P* allele of the gene encodes a product that is needed for the synthesis of the purple pigment in the flower. Therefore, a *PP* plant has purple flowers. The *p* allele is a mutant form of the gene. The product of this allele is inactive or mostly inactive and therefore, in *pp* plants, purple pigment cannot be made. In the absence of purple pigment, the flower is white. The *Pp* heterozygote is purple, rather than intermediate between purple and white, because the amount and activity of the product of the *P* allele is sufficient to enable a normal amount of purple pigment to be synthesized. Similar molecular events underlie the genotype–phenotype relationships of the other six characters Mendel studied.

In the crosses Mendel analyzed, the phenotypic differences among individuals were caused directly by genetic differences. In populations of organisms in nature, phenotypic variation may be caused by genetic differences, by differences in environmental factors experienced by individuals, or by an interaction between genetics and the environment. That is, under some circumstances, organisms with different genotypes sometimes may have the same phenotype, or organisms with the same genotype sometimes may have the same phenotype. Importantly, only genetically based phenotypic variation is subject to evolutionary change. Phenotypic variation is discussed in more detail in Chapter 21 in the evolution unit.

Mendel Used Probability to Predict Both Classes and Proportions of Offspring from the Crosses He Made

Mendel could predict both which traits would appear in the offspring of a cross and their proportions. To understand this, let us review the mathematical rules that govern **probability**—that is, the possibility that an outcome will occur if it is a matter of chance, as in the random fertilization of an egg by a sperm cell that contains one allele or another.

In the mathematics of probability, we predict the likelihood of an outcome on a scale of 0 to 1. An outcome that is certain has a probability of 1, and an impossible outcome has a probability of 0. If two different outcomes are equally likely, as in getting heads or tails in tossing a coin, the probability of one of the outcomes is calculated by dividing that outcome by the total number of possible outcomes. The probability of obtaining a head in tossing a coin is 1 divided by 2, or 1/2. The probability of obtaining a tail is also 1 divided by 2, or 1/2. The probabilities of all the possible outcomes, when added together, must equal 1. Thus, a coin toss has only two possible outcomes, heads or tails, each with a probability of 1/2; the sum of these probabilities is: 1/2 + 1/2 = 1.

THE PRODUCT RULE IN PROBABILITY What is the chance of tossing two heads in succession? Because the outcome of one toss has no effect on the next one, the two successive tosses are *independent*. When two or more events are independent, the probability that they will occur in succession is calculated using the **product rule**—their individual probabilities are multiplied.

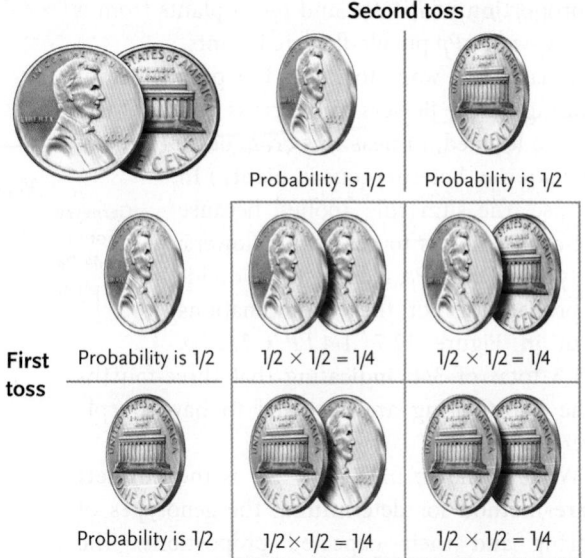

FIGURE 12.6 Rules of probability. For each coin toss, the probability of a head is 1/2; the probability of a tail is also 1/2. Because the outcome of the first toss is independent of the outcome of the second, the combined probabilities of the outcomes of successive tosses are calculated by multiplying their individual probabilities according to the product rule.
© Cengage Learning 2017

That is, the probability that events A and B *both* will occur equals the probability of event A *multiplied* by the probability of event B. For example, the probability of getting heads on the first toss is 1/2; the probability of heads on the second toss is also 1/2 **(Figure 12.6).** Because the events are independent, the probability of getting two heads in a row is 1/2 × 1/2 = 1/4. Applying the same principles, the probability of getting two tails is also 1/2 × 1/2 = 1/4 (see Figure 12.6). Similarly, because the sex of one child has no effect on the sex of the next child in a family, the probability of having four girls in a row is the product of their individual probabilities (very close to 1/2 for each birth): 1/2 × 1/2 × 1/2 × 1/2 = 1/16.

THE SUM RULE IN PROBABILITY We apply a different relationship, the **sum rule,** when there are two or more different ways of obtaining the same outcome. Returning to the coin toss example, we can determine the probability of getting a head and a tail in two tosses. We could toss the coin twice and get a head, then a tail. The probability that this will occur is 1/2 for the head × 1/2 for the tail = 1/4 (see Figure 12.6). However, we could toss the coin twice and get first a tail, then a head. The probability that this will occur is 1/2 for the tail × 1/2 for the head, which also = 1/4 (see Figure 12.6). Therefore, for the probability of tossing a head and a tail, we sum the individual probabilities to get the final probability: here, 1/4 + 1/4 = 1/2. (The other two probabilities for two coin tosses are 1/4 for two heads, and 1/4 for two tails.)

PROBABILITY IN MENDEL'S CROSSES The same rules of probability just discussed apply to Mendel's crosses. For example, **Figure 12.7** shows the rules of probability applied to calculating the proportion of *PP*, *Pp*, and *pp* F₂ plants from a cross of two F₁ *Pp* purple-flowered plants.

What if we want to know the probability of obtaining purple flowers in the cross *Pp* × *Pp*? (As you have learned, in peas this cross occurs by self-fertilization of heterozygous F₁ plants.) In this case, the sum rule applies, because there are two ways to get purple flowers: genotypes *PP* and *Pp*. Adding the individual probabilities of these combinations shown in Figure 12.7, 1/4 *PP* + 1/2 *Pp*, gives a total of 3/4, indicating that three-fourths of the F₂ offspring are expected to have purple flowers.

What is shown in Figure 12.7 is the **Punnett square** method for determining the genotypes of offspring and their expected proportions. The method was named for its originator, Reginald Punnett, a British geneticist who worked in the early part of the twentieth century. To use the Punnett square, write the probability of obtaining gametes with each type of allele from one parent at the top of the diagram and write the chance of obtaining each type of allele from the other parent on the left side. Then fill in the cells by combining the alleles from the top and from the left and multiply their individual probabilities.

Mendel Used a Testcross to Check the Validity of His Conclusions

Mendel realized that he could assess the validity of his conclusions by determining whether they could be used successfully to *predict* the outcome of a cross of a different type than he had tried so far. (Again, this is an example of Mendel using the scientific method.) Accordingly, he crossed an F₁ plant with purple flowers, assumed to have the heterozygous genotype *Pp*, with a true-breeding white-flowered plant, with the homozygous genotype *pp* (**Figure 12.8,** Experiment 1). There are two expected classes of offspring, *Pp* and *pp*, both with a probability of 1/2. Thus, the phenotypes of the offspring are expected to be 1 purple-flowered : 1 white-flowered.

Mendel's actual results closely approach that ratio. If you performed the same type of cross with all the other traits used in his study, including those traits affecting seed shape, seed color, and plant height, you would find the same 1 : 1 ratio.

A cross between an individual with the dominant phenotype and a homozygous recessive individual, such as the one just described, is called a **testcross**. Geneticists use a testcross to determine whether an individual with a dominant trait is a heterozygote or a homozygote, because their phenotypes are identical. If the offspring of the testcross are of two types, with half displaying the dominant trait and half the recessive trait,

Gametes from F₁ purple *Pp* plant

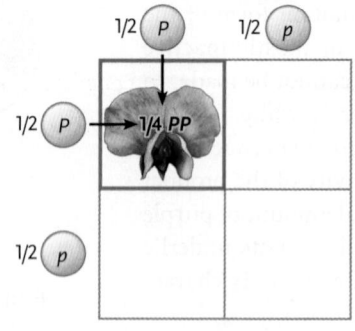

A. To produce an F₂ plant with the *PP* genotype, two *P* gametes must combine. The probability of selecting a *P* gamete from one F₁ parent is 1/2, and the probability of selecting a *P* gamete from the other F₁ parent is also 1/2. Using the product rule, the probability of producing a purple-flowered *PP* plant from a *Pp* × *Pp* cross is 1/2 × 1/2 = 1/4.

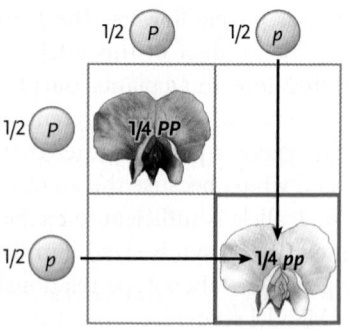

B. To produce an F₂ plant with the *pp* genotype, two *p* gametes must combine. The probability of selecting a *p* gamete from one F₁ parent is 1/2, and the probability of selecting a *p* gamete from the other F₁ parent is also 1/2. Using the product rule, the probability of producing a white-flowered *pp* plant from a *Pp* × *Pp* cross is 1/2 × 1/2 = 1/4.

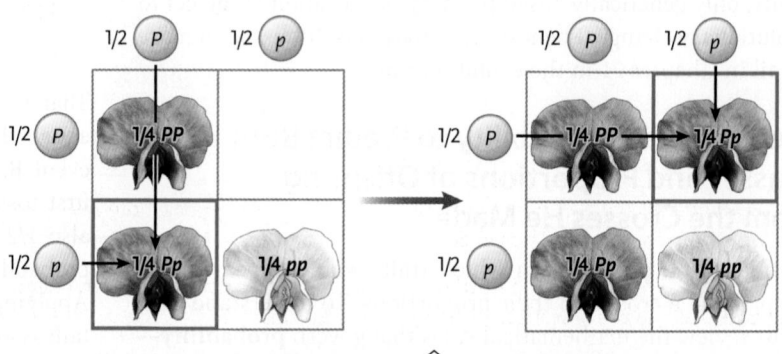

C. To produce an F₂ plant with the *Pp* genotype, a *P* gamete must combine with a *p* gamete. The cross *Pp* × *Pp* can produce *Pp* offspring in two different ways: (1) a *P* gamete from the first parent can combine with a *p* gamete from the second parent; or (2) a *p* gamete from the first parent can combine with a *P* gamete from the second parent. We apply the sum rule to obtain the combined probability: each of the ways to get *Pp* has an individual probability of 1/4, so the probability of *Pp*, purple-flowered offspring is 1/4 + 1/4 = 1/2.

FIGURE 12.7 Punnett square method for predicting offspring and their ratios in genetic crosses. The example is the F₁ × F₁ self of purple-flowered plants from Figure 12.5 to produce the F₂ generation. Each cell shows the genotype and proportion of one type of F₂ plant.
© Cengage Learning 2017

then the individual in question must be a heterozygote (see Figure 12.8, Experiment 1). If all the offspring display the dominant trait, the individual in question must be a homozygote. For example, the cross *PP* × *pp* gives all *Pp* progeny, which show the dominant purple flower phenotype (Figure 12.8, Experiment 2).

Obviously, the testcross method cannot be used for humans. However, it can be used in reverse, by noting the traits present in families over several generations and working backward to deduce whether a parent must have been a homozygote or a heterozygote (see also Chapter 13).

FIGURE 12.8 | Experimental Research

Testing the Predicted Outcomes of Genetic Crosses

Question: How can it be determined whether a plant with the dominant phenotype is a heterozygote or a homozygote?

Experiment 1: Mendel crossed an F_1 plant with purple flowers, predicted to have a Pp genotype, with a true-breeding white-flowered plant and analyzed the flower color phenotypes in the offspring.

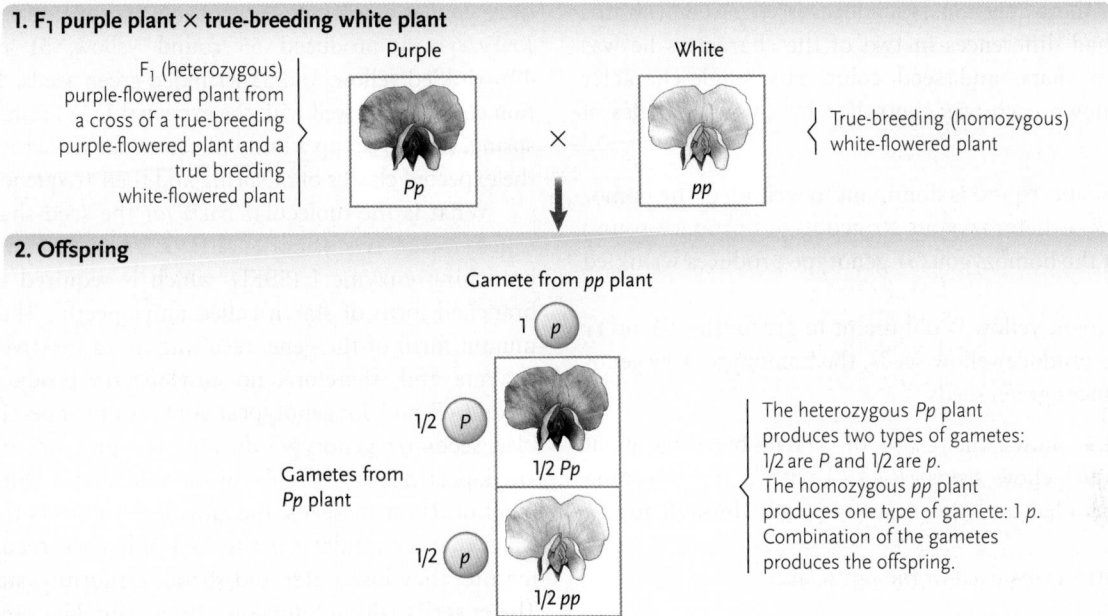

Results: Predicted progeny from a cross of a purple-flowered heterozygote with a true-breeding white-flowered plant is 1 Pp purple-flowered : 1 pp white-flowered. Mendel observed 85 purple-flowered and 81 white-flowered plants, close to the prediction.

Experiment 2: Mendel crossed a true-breeding plant with purple flowers, predicted to have a PP genotype, with a true-breeding white-flowered plant and analyzed the flower color phenotypes in the offspring.

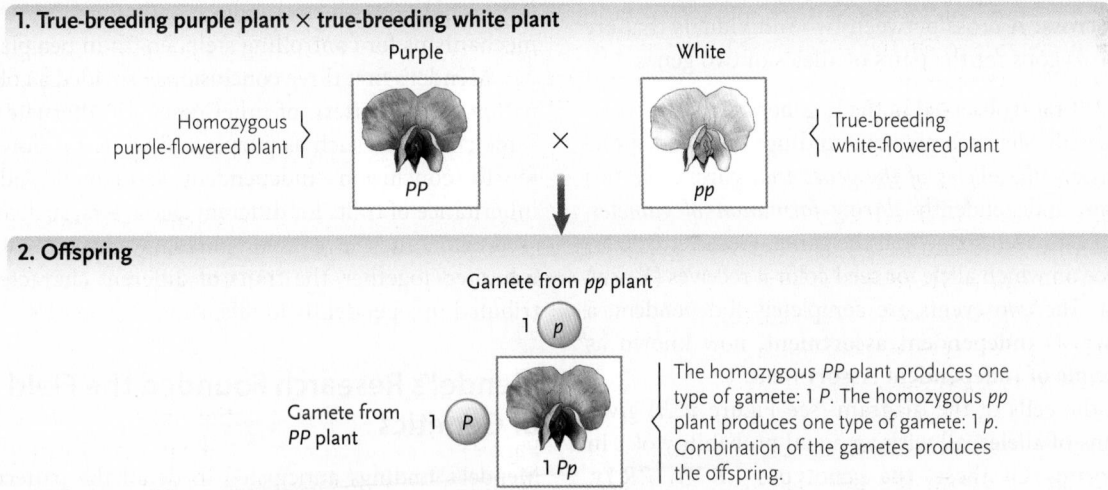

Results: The outcome of crossing a purple-flowered homozygote with a true-breeding white-flowered plant is all Pp plants, which have purple flowers.

Conclusion: The outcome of a cross between (here) a plant with a dominant phenotype and a plant with a recessive phenotype—a testcross—gives a different result depending on whether the plant with the dominant phenotype is a homozygote or a heterozygote. Therefore, a testcross is a useful way to determine the genotype for an individual with a dominant phenotype.

Observe
Hypothesize
Predict
Experiment
Interpret

In rabbits, white fat beneath the skin is dominant to yellow fat. A white-fat rabbit is bred to a yellow-fat rabbit for a season, resulting in 13 white-fat and 12 yellow-fat offspring. What are the genotypes of the parents and their offspring?

Crosses Involving Two Characters Led Mendel to Propose the Principle of Independent Assortment

Mendel next asked what happens in crosses when more than one character is involved. Would the alleles of the genes controlling the different characters be inherited independently, or would they be inherited together?

To answer these questions, Mendel crossed parental strains of peas that had differences in two of the characters he was studying: seed shape and seed color. His single-character crosses had shown each was controlled by a pair of alleles of one gene:

- For seed shape, round is dominant to wrinkled: the homozygous *RR* or heterozygous *Rr* genotypes produce round seeds and the homozygous *rr* genotype produces wrinkled seeds.
- For seed color, yellow is dominant to green: the *YY* or *Yy* genotypes produce yellow seeds; the homozygous *yy* genotype produces green seeds.

Figure 12.9 shows the cross of a true-breeding plant with round and yellow seeds *(RR YY)* with a true-breeding plant with wrinkled and green seeds *(rr yy)* through to the F_2 generation.

New genetic terms used in the figure are:

- **Dihybrid** (*di* = two): An F_1 that is produced from a cross that involves two characters and is heterozygous for each of the pairs of alleles of the two genes involved. An example would be *Aa Bb*, where genes *A* and *B* control different characters, and the uppercase and lowercase letters for the two genes are dominant and recessive alleles, respectively.
- **Dihybrid cross:** A cross between two individuals that are each heterozygous for the pairs of alleles of two genes.

This $9:3:3:1$ ratio observed in the F_2 generation of the cross was consistent with Mendel's previous findings if he added one further conclusion: *the alleles of the genes that govern the two characters assort* independently *during formation of gametes.* That is, the allele for seed shape that the gamete receives (*R* or *r*) has no influence on which allele for seed color it receives (*Y* or *y*) and vice versa. The two events are completely independent, a property known as **independent assortment,** now known as Mendel's **Principle of Independent Assortment.**

Filling in the cells of the diagram (see Figure 12.9) gives 16 combinations of alleles, all with an equal probability of 1 in every 16 offspring. Of these, the genotypes *RR YY, RR Yy, Rr YY,* and *Rr Yy* all have the same phenotype: round yellow seeds. These combinations occur in 9 of the 16 cells in the diagram, giving a total probability of 9/16. The genotypes *rr YY* and *rr Yy*, which produce the wrinkled yellow seeds, are found in three cells, giving a probability of 3/16 for this phenotype. Similarly, the genotypes *RR yy* and *Rr yy*, which yield round green seeds, occur in three cells, giving a probability of 3/16. Finally, the genotype *rr yy*, which produces wrinkled green

seeds, is found in only one cell and therefore has a probability of 1/16.

The actual results obtained by Mendel closely approximated the expected $9:3:3:1$ ratio of round yellow:round green:wrinkled yellow:wrinkled green seeds, as developed in Figure 12.9. Thus, Mendel's first three conclusions, with the added conclusion of independent assortment, explain the observed results of his dihybrid cross. Mendel's testcrosses confirmed his independent assortment conclusion. For example, the testcross *Rr Yy* $\times$ *rr yy* produced 55 round yellow, 51 round green, 49 wrinkled yellow, and 53 wrinkled green seeds. This distribution corresponds well with the expected $1:1:1:1$ ratio in the offspring. (Try to set up a Punnett square for this cross and predict the expected classes of offspring and their frequencies.)

What is the molecular basis for the seed-shape character differences? The normal *R* allele encodes an enzyme, starch-branching enzyme I (SBEI), which is required to produce a branched form of starch called amylopectin. The *r* allele is a mutant form of the gene, resulting in an inactive form of the enzyme and, therefore, no amylopectin production. Round seeds (*RR* and *Rr* genotypes) contain amylopectin, but wrinkled seeds (*rr* genotype) do not. The presence or absence of amylopectin is responsible for the seed shape. During development of *RR* or *Rr* seeds, the amylopectin limits the amount of water that accumulates in the seed. When the seeds dry as they mature, they lose water and shrink uniformly, staying round. The *rr* seeds without amylopectin accumulate excessive water. When they dry, they lose more water and the seeds collapse, giving them a wrinkled appearance.

As mentioned earlier, all of the phenotypic differences Mendel studied are the result of molecular differences resulting from normal and mutant alleles of the genes involved. *Molecular Insights* describes a molecular study that uncovered the mechanisms for controlling stem length in pea plants.

Mendel's first three conclusions provided a coherent explanation of the pattern of inheritance for alternate traits of the same character, such as purple and white for flower color. His fourth conclusion, independent assortment, addressed the inheritance of traits for different characters, such as seed shape, seed color, and flower color, and showed that, instead of being inherited together, the traits of different characters were distributed independently to offspring.

Mendel's Research Founded the Field of Genetics

Mendel's findings anticipated in detail the patterns by which genes and chromosomes determine inheritance. Yet, when Mendel first reported his findings, during the nineteenth century, the structure and function of chromosomes and the patterns by which they are separated and distributed to gametes were unknown; meiosis remained to be discovered; and, of course, nothing was known about the molecular nature of genes. In addition, his use of mathematical analysis was a new and radical departure from the usual biological techniques of his day.

FIGURE 12.9 | **Experimental Research**

The Principle of Independent Assortment

Question: Do alleles of genes for two different characters in garden peas assort independently in a cross?

Experiment: Mendel crossed a true-breeding plant with round and yellow seeds with a true-breeding plant with wrinkled and green seeds and analyzed the progeny through the F_1 and F_2 generations. We explain this cross here in modern terms.

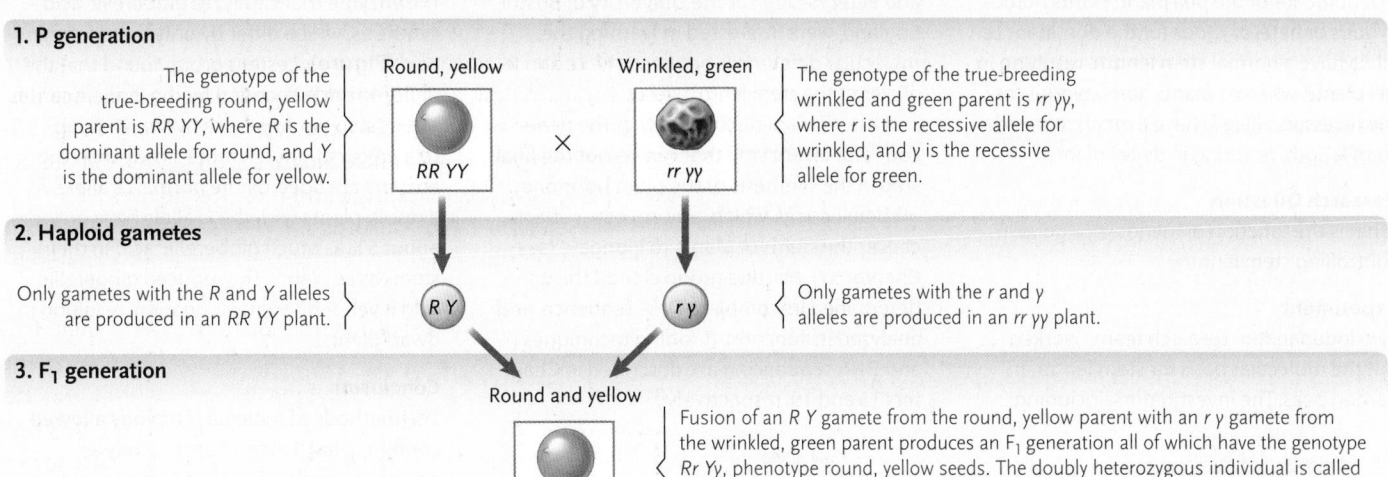

1. P generation

The genotype of the true-breeding round, yellow parent is *RR YY*, where *R* is the dominant allele for round, and *Y* is the dominant allele for yellow.

Round, yellow
RR YY

×

Wrinkled, green
rr yy

The genotype of the true-breeding wrinkled and green parent is *rr yy*, where *r* is the recessive allele for wrinkled, and *y* is the recessive allele for green.

2. Haploid gametes

Only gametes with the *R* and *Y* alleles are produced in an *RR YY* plant.

R Y

r y

Only gametes with the *r* and *y* alleles are produced in an *rr yy* plant.

3. F_1 generation

Round and yellow
Rr Yy

Fusion of an *R Y* gamete from the round, yellow parent with an *r y* gamete from the wrinkled, green parent produces an F_1 generation all of which have the genotype *Rr Yy*, phenotype round, yellow seeds. The doubly heterozygous individual is called a *dihybrid*. The seeds are round because the *R* allele is dominant to the *r* allele, and yellow because the *Y* allele is dominant to the *y* allele.

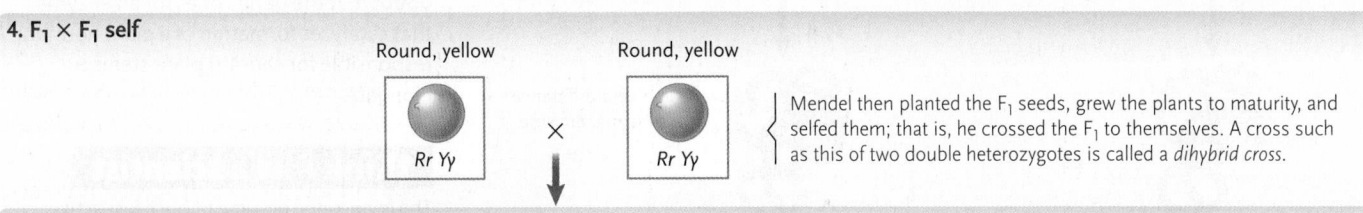

4. F_1 × F_1 self

Round, yellow
Rr Yy

×

Round, yellow
Rr Yy

Mendel then planted the F_1 seeds, grew the plants to maturity, and selfed them; that is, he crossed the F_1 to themselves. A cross such as this of two double heterozygotes is called a *dihybrid cross*.

5. F_2 generation

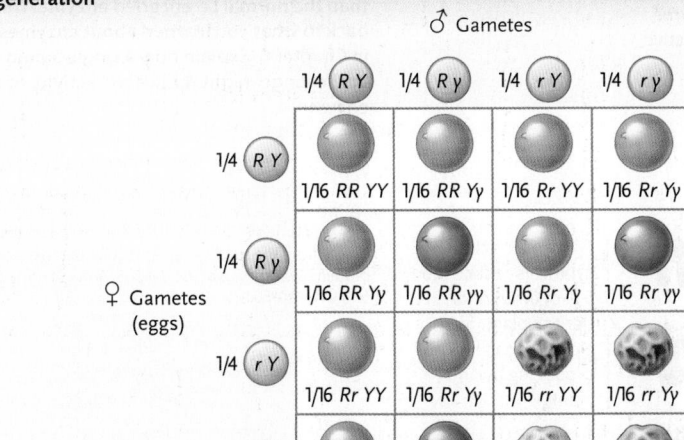

♂ Gametes

	1/4 *R Y*	1/4 *R y*	1/4 *r Y*	1/4 *r y*
1/4 *R Y*	1/16 *RR YY*	1/16 *RR Yy*	1/16 *Rr YY*	1/16 *Rr Yy*
1/4 *R y*	1/16 *RR Yy*	1/16 *RR yy*	1/16 *Rr Yy*	1/16 *Rr yy*
1/4 *r Y*	1/16 *Rr YY*	1/16 *Rr Yy*	1/16 *rr YY*	1/16 *rr Yy*
1/4 *r y*	1/16 *Rr Yy*	1/16 *Rr yy*	1/16 *rr Yy*	1/16 *rr yy*

♀ Gametes (eggs)

If the alleles that control seed shape and seed color assort independently, each F_1 plant grown from the seeds would produce four types of gametes: the *R* allele for seed shape would go to a gamete with either the *Y* or *y* allele for seed color, and similarly, the *r* allele would go to a gamete with either the *Y* or *y* allele. Thus, independent assortment of genes from the *Rr Yy* parents is expected to produce four types of gametes with equal probability: 1/4 *R Y*, 1/4 *R y*, 1/4 *r Y*, and 1/4 *r y*. Random fusion of the four different male gametes with the four different female gametes produces the F_2 generation.

Results: Filling in the cells of the Punnett square gives 16 combinations, each with an equal probability of 1 in every 16 offspring if the alleles of the two genes assort independently. The 16 combinations resulting from independent assortment give an expected F_2 phenotypic ratio of 9 round yellow : 3 round green : 3 wrinkled yellow : 1 wrinkled green. Mendel's selfing of the F_1 *Rr Yy* round yellow plants produced an F_2 generation with 315 round yellow : 108 round green : 101 wrinkled yellow : 32 wrinkled green, which is close to a 9 : 3 : 3 : 1 ratio (3 : 1 for round : wrinkled, and 3 : 1 for yellow : green).

Conclusion: The results indicate that the alleles of the genes for the two characters assort independently during the formation of gametes.

think like a scientist

Observe
Hypothesize
Predict
Experiment
Interpret

Suppose that, instead of the F_1 × F_1 cross (self-fertilization) shown in the figure, you cross an F_1 plant with a round, green plant of genotype *Rr yy*. What progeny would you get and in what ratio would they occur?

Mendel's Dwarf Pea Plants: How does a gene defect produce dwarfing?

One of the seven characters Mendel studied was stem length. The stem length gene *Le* controls the length of the stem between the leaf branches of the pea plant. Plants homozygous or heterozygous for the dominant *Le* allele have a normal stem length, resulting in tall plants, whereas plants homozygous for the recessive allele *le* have a much reduced stem length, resulting in dwarf plants.

Research Question

What is the function of Mendel's *Le* gene in controlling stem length?

Experiment

Two independent research teams worked out the molecular basis for stem length in garden peas. The investigators, including Diane Lester and her colleagues at the University of Tasmania in Australia, David Martin and his coworkers at Oregon State University, and Peter Hedden at the University of Bristol, England, were interested in learning the molecular differences between the *Le* and *le* alleles of the stem length gene.

Lester's team discovered that the gene codes for an enzyme that carries out the final step in the synthesis of the plant hormone gibberellic acid, which, among other effects, causes the stems of plants to elongate (see Chapter 37). Martin's group cloned the gene, determined its complete DNA sequence, and analyzed its function. (Cloning techniques and DNA sequencing are described in Chapters 18 and 19, respectively.)

Results

DNA sequencing showed that the *Le* and *le* alleles of the gene encode two versions of the enzyme that catalyzes gibberellic acid synthesis, which differ by only a single amino acid **(Figure)**. Lester's group found that the faulty enzyme encoded by the *le* allele carries out its step (addition of a hydroxyl group to a precursor) much more slowly than the enzyme encoded by the normal *Le* allele. As a result, plants with the *le* allele have only about 5% as much gibberellic acid in their stems as *Le* plants. The reduced gibberellic acid levels limit stem elongation, resulting in dwarf plants.

Conclusion

The methods of molecular biology allowed contemporary researchers to study a gene first studied genetically in the mid-nineteenth century. The findings leave little doubt that the gene codes for an enzyme that catalyzes formation of a plant hormone responsible for causing plant stems to elongate.

think like a scientist

The *le*-encoded enzyme is much less active than the normal *Le*-encoded enzyme. Think back to what you learned about enzymes in Chapter 6. Explain how a single amino acid change might reduce the activity of an enzyme.

Sources: D. R. Lester, J. J. Ross, P. J. Davies, and J. B. Reid. 1997. Mendel's stem length gene *(Le)* encodes a gibberellin 3β-hydroxylase. *Plant Cell* 9:1435–1443; D. N. Martin, W. M. Proebsting, and P. Hedden. 1997. Mendel's dwarfing gene: cDNAs from the *Le* alleles and function of the expressed proteins. *Proceedings of the National Academy of Sciences USA* 94:8907–8911.

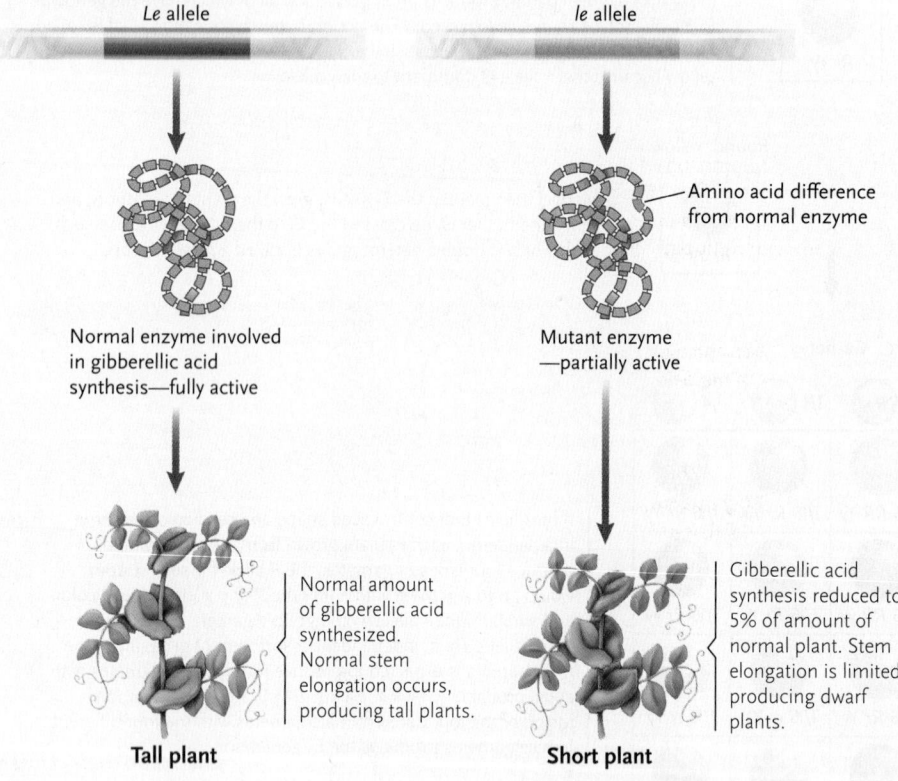

Le allele — Normal enzyme involved in gibberellic acid synthesis—fully active — Normal amount of gibberellic acid synthesized. Normal stem elongation occurs, producing tall plants. **Tall plant**

le allele — Amino acid difference from normal enzyme — Mutant enzyme —partially active — Gibberellic acid synthesis reduced to 5% of amount of normal plant. Stem elongation is limited, producing dwarf plants. **Short plant**

© Cengage Learning 2017

Mendel reported his results to a small group of fellow intellectuals in Brünn and presented his results in 1866 in a natural history journal published in the city. But Mendel's scientific conclusions were not immediately appreciated. His article received little notice outside of Brno, and those who read it did not appreciate the significance of his findings. His work was overlooked until the early 1900s, when three investigators—Hugo de Vries in Holland, Carl Correns in Germany, and Erich von Tschermak in Austria—independently performed a series of breeding experiments similar to Mendel's and reached the same conclusions. These investigators, in searching through previously published scientific articles, discovered to their surprise Mendel's article about his experiments conducted 34 years earlier. Each gave credit to Mendel's discoveries, and the quality and far-reaching implications of his work were at last realized. Mendel died in 1884, 16 years before the rediscovery of his experiments and conclusions, and thus he never received the recognition that he deserved during his lifetime.

Mendel's Results Are Explained by the Behavior of Chromosomes and Genes in Meiosis

By the time Mendel's results were rediscovered in the early 1900s, critical information from studies of meiosis was available. It was not long before Walter Sutton, a genetics graduate student at Columbia University in New York, recognized the similarities between the inheritance of the genes discovered by Mendel and the behavior of chromosomes in meiosis and fertilization (**Figure 12.10**).

In a historic article published in 1903, Sutton drew all the necessary parallels between genes and chromosomes:

- Chromosomes occur in pairs in sexually reproducing, diploid organisms, as do the alleles of each gene.
- The chromosomes of each pair are separated and delivered singly to gametes, as are the alleles of a gene.
- The separation of any pair of chromosomes in meiosis and gamete formation is independent of the separation of other

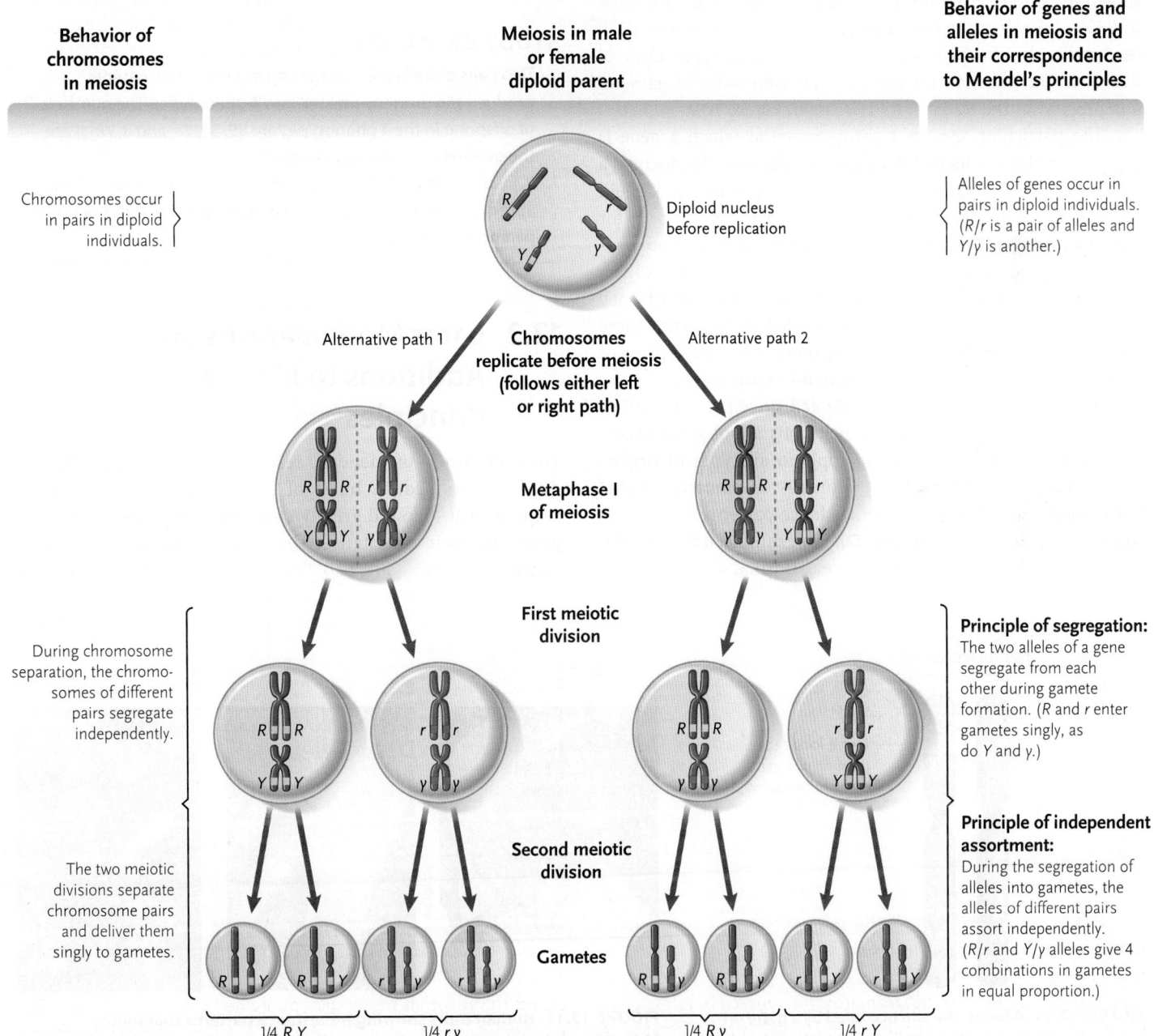

FIGURE 12.10 The parallels between the behavior of chromosomes and genes and alleles in meiosis. The gametes show four different combinations of alleles produced by independent segregation of chromosome pairs.

© Cengage Learning 2017

pairs (see Figure 12.10), as in the independent assortment of the alleles of different genes in Mendel's dihybrid crosses.

- One member of each chromosome pair is derived in fertilization from the male parent, and the other member is derived from the female parent, in an exact parallel with the two alleles of a gene.

From this total coincidence in behavior, Sutton correctly concluded that genes and their alleles are carried on the chromosomes, a conclusion known today as the **chromosome theory of inheritance.**

The exact parallel between the principles set forth by Mendel and the behavior of chromosomes and genes during meiosis is shown in Figure 12.10 for an *Rr Yy* diploid. For a dihybrid cross of *Rr Yy* × *Rr Yy*, when the gametes produced as in Figure 12.10 use randomly, the progeny will show a phenotypic ratio of 9 : 3 : 3 : 1. This mechanism explains the same ratio of gametes and progeny as the *Rr Yy* × *Ry Yy* cross in Figure 12.9.

The particular site on a chromosome at which a gene is located is called the **locus** (plural, *loci*) of the gene. The locus is a particular DNA sequence that encodes (typically) a protein responsible for the phenotype controlled by the gene. A locus for a gene with two alleles, *A* and *a*, on a homologous pair of chromosomes is shown in **Figure 12.11.** At the molecular level, different alleles of a gene have different DNA sequences, which may result in differences in the protein encoded by the gene. These differences are detected as distinct phenotypes in the offspring of a cross, such as was seen by Mendel in his experiments.

All the genetics research conducted since the early 1900s has confirmed Mendel's basic conclusions about inheritance. Moreover, Mendel's conclusions apply to all types of organisms, from yeast and fruit flies to humans. In humans, a number of easily seen traits show inheritance patterns that follow Mendelian principles **(Figure 12.12).** For example, *albinism,* the lack of normal skin color, is recessive to normal skin color, and

normally separated fingers are recessive to fingers with webs between them. Similarly, *achondroplasia,* the most frequent form of short-limb dwarfism, is a dominant trait that involves abnormal bone growth. Many human disorders that have no visibly obvious phenotypes also show simple inheritance patterns. For instance, *cystic fibrosis,* in which a defect in the membrane transport of chloride ions leads to pulmonary and digestive dysfunctions and eventually death, is a recessive trait.

Not all patterns of inheritance follow Mendel's principles discussed in this section. In the next section, we discuss patterns of inheritance that, in some circumstances, require modifications or additions to Mendel's principles.

STUDY BREAK 12.1

1. Two pairs of traits are segregating in a cross. Two parents produce 156 progeny that fall into 4 phenotypes. The numbers of offspring in the 4 phenotypes are 89, 31, 28, and 8. What are the genotypes of the two parents?

2. If, instead, the four phenotypes in question 1 occur in approximately equal numbers, what are the genotypes of the parents? What is this kind of cross called?

12.2 Later Modifications and Additions to Mendel's Principles

The rediscovery of Mendel's research in the early 1900s produced an immediate burst of interest in genetics, and the research that followed greatly expanded our understanding of genes and their inheritance. That research supported Mendel's conclusions fully, but also revealed many variations on the

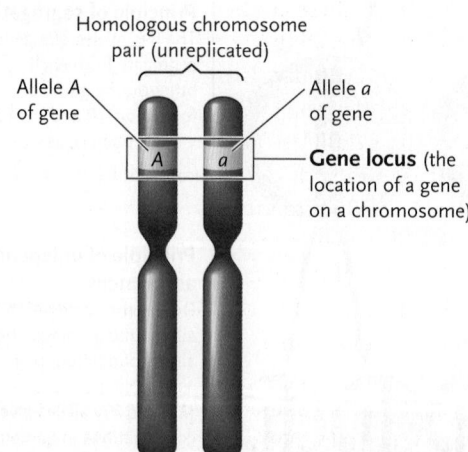

FIGURE 12.11 A locus, the site occupied by a gene on a pair of homologous chromosomes. Two alleles, *A* and *a*, of the gene are present at this locus in the homologous pair. These alleles have differences in the DNA sequence of the gene.
© Cengage Learning 2017

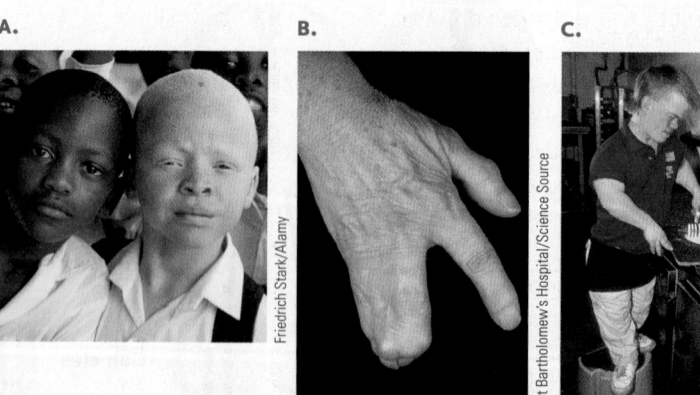

FIGURE 12.12 Human traits showing inheritance patterns that follow Mendelian principles. (A) Lack of normal skin color (albinism), a recessive trait. **(B)** Webbed fingers, a dominant trait. **(C)** Achondroplasia, or short-limbed dwarfism, a dominant trait.

In Incomplete Dominance, Dominant Alleles Do Not Mask Recessive Alleles Completely

When one allele of a gene is not completely dominant over another allele of the same gene, it is said to show **incomplete dominance.** With incomplete dominance, the phenotype of the heterozygote is somewhere between the phenotypes of individuals that are homozygous for either of the alleles. Flower color in snapdragons shows incomplete dominance. One gene controls the flower color character, with one allele for red and another allele for white. Because one allele is not completely dominant to the other in incomplete dominance, we use a different genetic symbolism: an italic letter is used that relates to the character, with superscripts for the different alleles. In this case, C signifies flower color and the superscript R is for red, and the superscript W is for white: C^R is the allele for red color and C^W is the allele for white color. We use these symbols in **Figure 12.13,** which follows a cross of a true-breeding red-flowered snapdragon with a true-breeding, white-flowered snapdragon through to the F_2 generation. The key differences from a cross in which complete dominance is the case are:

1. The F_1 phenotype is intermediate between the phenotypes of the two parents.
2. The phenotypes of F_2 individuals are directly determined by the different *genotypes* of those individuals, giving a $1:2:1$ ratio rather than a $3:1$ ratio as is characteristic for complete dominance.

In molecular terms, we can explain the flower colors as follows: the C^R allele encodes an enzyme that produces a red pigment, but two alleles ($C^R C^R$) are necessary to produce enough of the active form of the enzyme to produce fully red flowers. The enzyme is completely inactive in $C^W C^W$ plants, which produce colorless flowers that appear white because of the scattering of light by cell walls and other structures. With their single C^R allele, the $C^R C^W$ heterozygotes of the F_1 generation can produce only enough pigment to give the flowers a pink color. When pink $C^R C^W$ F_1 plants are crossed, the fully red and white colors reappear, together with the pink color, in exactly the same ratio as the ratio of genotypes produced from a cross of two heterozygotes in Mendel's experiments (for example, see Figure 12.7).

Some human disorders show incomplete dominance. For example, sickle-cell anemia (see *Why it matters . . .*) is characterized by an alteration in the hemoglobin molecule that changes the shape of red blood cells when oxygen levels are low. An individual with sickle-cell anemia is homozygous for a recessive allele that encodes a defective form of one of the polypeptides of the hemoglobin molecule. Individuals heterozygous for that recessive allele and the normal allele have a condition known as *sickle-cell trait,* which is a milder form of the disease

because the individuals produce some normal polypeptides from the normal allele as well as some abnormal ones. (Sickle-cell anemia and sickle-cell trait are two types of sickle-cell disease.)

In Codominance, the Effects of Different Alleles Are Equally Detectable in Heterozygotes

Codominance occurs when alleles have approximately equal effects in individuals, making the alleles equally detectable in heterozygotes. The inheritance of the human MN blood group presents an example of codominance. The L^M and L^N alleles of the MN blood group gene that control this character encode different forms of a glycoprotein molecule located on the surface of red blood cells. (Glycoproteins are discussed in Chapter 3.) If the genotype is $L^M L^M$, only the M form of the glycoprotein is present and the blood type is M; if it is $L^N L^N$, only the N form is present and the blood type is N. In heterozygotes with the $L^M L^N$ genotype, both glycoprotein types are present and can be detected, producing the blood type MN. Because each genotype has a different phenotype, the inheritance pattern for the MN blood group alleles is generally the same as for incompletely dominant alleles. The MN blood types do not affect blood transfusions and have relatively little medical importance.

In Multiple Alleles, More Than Two Alleles of a Gene Are Present in a Population

One of Mendel's major and most fundamental assumptions was that alleles (his factors) occur in pairs in individuals. In the pairs, the alleles may be the same or different. After the rediscovery of Mendel's principles, it soon became apparent that, although alleles do indeed occur in pairs in individuals, **multiple alleles** (more than two different alleles of a gene) may be present if all the individuals of a population are considered. For example, for a gene B, there could be the normal allele, B, and several alleles with alterations in this gene, for example, b_1, b_2, b_3, and so on. Some individuals in a population may have the B and b_1 alleles of a gene; others, the b_2 and b_3 alleles; still others, the b_3 and b_5 alleles; and so on, for all possible combinations. That is, although *any one individual can have only two alleles of the gene,* there are more than two alleles in the population as a whole. Genes may certainly occur in many more than the four alleles of the example. For instance, one of the genes that plays a part in the acceptance or rejection of organ transplants in humans has more than 200 different alleles.

The multiple alleles of a gene each contain differences at one or more points in their DNA sequences **(Figure 12.14)**, which cause detectable alterations in the structure and function of proteins encoded by the alleles. Multiple alleles present no real difficulty in genetic analysis because each diploid individual still has only two of the alleles, allowing gametes to be predicted and traced through crosses by the usual methods.

FIGURE 12.13 | **Experimental Research**

Experiment Showing Incomplete Dominance of a Trait

Question: How is flower color in snapdragons inherited?

Experiment: Cross a true-breeding red-flowered snapdragon with a true-breeding white-flowered snapdragon and analyze the progeny through the F_1 and F_2 generations.

1. P generation

The red-flowered snapdragon is homozygous for the C^R allele.

Homozygous red parent — Red $C^R C^R$

×

White $C^W C^W$ white parent — Homozygous

The white-flowered snapdragon is homozygous for the C^W allele.

2. F_1 generation

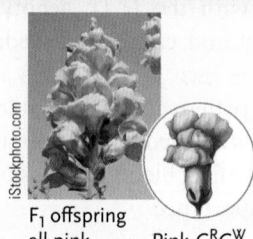

F_1 offspring all pink — Pink $C^R C^W$

Fusion of C^R gametes from the red-flowered plant and C^W gametes from the white-flowered plant produces $C^R C^W$ heterozygotes in the F_1. These plants have pink flowers, an intermediate phenotype between red and white. This phenotype is not that expected if one of the alleles shows complete dominance to the other allele. This phenotype is, however, consistent with incomplete dominance.

3. $F_1 \times F_1$ cross

 ×

Pink $C^R C^W$ Pink $C^R C^W$

F_1 pink-flowered plants are crossed to produce the F_2 generation.

4. F_2 generation

Gametes from one $C^R C^W$ F_1 pink-flowered plant

 C^R C^W

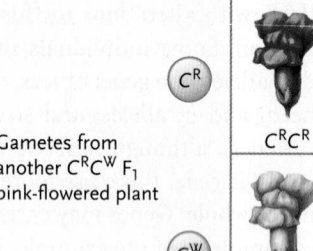

Gametes from another $C^R C^W$ F_1 pink-flowered plant

C^R

C^W

$C^R C^R$ $C^R C^W$

$C^R C^W$ $C^W C^W$

Each parent plant produces two types of gametes, C^R and C^W. Random fusion of the gametes from the two parents produces the F_2 generation.

Results: The F_2 phenotypic ratio is 1 red : 2 pink : 1 white. Each phenotype results from a distinct genotype, $C^R C^R$ for red flowers, $C^R C^W$ for pink flowers, and $C^W C^W$ for white flowers. The 1 : 2 : 1 phenotypic ratio is consistent with incomplete dominance.

Conclusion: In incomplete dominance, each genotype has a distinct phenotype. From a cross of two heterozygotes, the outcome is a phenotypic ratio of 1 : 2 : 1 rather than the 3 : 1 ratio characteristic of complete dominance.

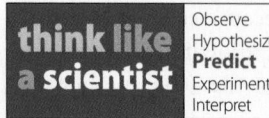

think like a scientist

Observe
Hypothesize
Predict
Experiment
Interpret

What progeny would you get if you testcrossed an F_1 pink plant to a red parent? To a white parent?

B allele	5'...ATGCAGATACCGATTACAGACCATAGG...3'	
	3'...TACGTCTATGGCTAATGTCTGGTATCC...5'	
b_1 allele	5'...ATGCAGAGACCGATTACAGACCATAGG...3'	
	3'...TACGTCTCTGGCTAATGTCTGGTATCC...5'	
b_2 allele	5'...ATGCAGATACCGACTACAGACCATAGG...3'	
	3'...TACGTCTATGGCTGATGTCTGGTATCC...5'	
b_3 allele	5'...ATGCAGATACCGATTACAGTCCATAGG...3'	
	3'...TACGTCTATGGCTAATGTCAGGTATCC...5'	

FIGURE 12.14 Multiple alleles. Multiple alleles consist of differences in the DNA sequence of a gene at one or more points, which result in detectable differences in the structure of the protein encoded by the gene. The differences shown here are single base-pair changes. The *B* allele is the normal allele, which encodes a protein with normal function. The three *b* alleles each have alterations of the normal protein-coding DNA sequence that may adversely affect the function of that protein.

© Cengage Learning 2017

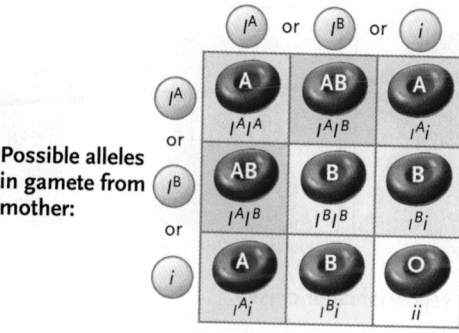

FIGURE 12.15 Inheritance of the blood types of the human ABO blood group.

© Cengage Learning 2017

HUMAN ABO BLOOD GROUP The human *ABO* blood group provides an example of multiple alleles, in a system that also exhibits both dominance and codominance. The ABO blood group was discovered in 1901 by Karl Landsteiner, an Austrian biochemist who was investigating the sometimes fatal outcome of attempts to transfer blood from one person to another. Landsteiner found that only certain combinations of four blood types, designated A, B, AB, and O, can be mixed safely in transfusions **(Table 12.1)**.

Landsteiner determined that, in the wrong combinations, red blood cells from one blood type are agglutinated (clumped) by an agent in the serum of another type (the serum is the fluid in which the blood cells are suspended). The clumping was later found to depend on the action of an antibody in the blood serum. (Antibodies, protein molecules that interact with specific substances called antigens, are discussed in Chapter 45.)

The antigens responsible for the blood types of the ABO blood group are the carbohydrate parts of glycoproteins located on the surfaces of red blood cells (unrelated to the glycoprotein carbohydrates responsible for the blood types of the MN blood group). For example, people with type A blood have *antigen A* on their red blood cells, and anti-B antibodies in their blood. If a person with type A blood receives a transfusion of type B blood, his or her anti-B antibodies will cause the blood to clump. Table 12.1 shows how the four blood types of the human ABO blood group determine compatibility in transfusions.

The four blood types—A, B, AB, and O—are produced by different combinations of multiple (three) alleles of a single gene *I* designated I^A, I^B, and *i* **(Figure 12.15)**. I^A and I^B are codominant alleles that are each dominant to the recessive *i* allele.

In Epistasis, Genes Interact, with the Activity of One Gene Influencing the Activity of Another Gene

The genetic characters discussed so far in this chapter, such as flower color, seed shape, and the blood types of the ABO group, all involve examples of genetic variation influenced by the alleles of single genes. This is not the case for every gene. In **epistasis** (*epi* = on or over; *stasis* = standing or stopping), two (or more) genes affect the same phenotype. The genes interact at the level of their products, with the phenotypic expression of one or more alleles of a gene at one locus inhibiting or masking the effects of the phenotypic expression of one or more alleles of a gene at a different locus. The result of epistasis is predictable ratios that can be explained by allele segregation patterns of the two (or more) genes involved.

Labrador retrievers (Labs) may have black, chocolate brown, or yellow fur **(Figure 12.16A–C)**. The different colors result from variations in the amount and distribution in hairs of a brownish black pigment called melanin. One gene, coding for an enzyme involved in melanin production, determines how much melanin is produced. The dominant *B* allele of this gene produces black fur color in *BB* or *Bb* Labs. Less pigment is produced in *bb* dogs, which are chocolate brown. Another gene at a different locus determines whether the black or chocolate color appears at all, by controlling the deposition of pigment in hairs. A dominant allele *E* of this second gene permits pigment deposition, so that the black color in *BB* or *Bb* individuals, or the chocolate color in *bb* individuals, actually appears in the fur. Pigment deposition is almost completely blocked in homozygous recessive *ee* individuals, so the fur lacks melanin and

TABLE 12.1	Blood Types of the Human ABO Blood Group		
Blood Type	Antigens	Antibodies	Blood Types Accepted in a Transfusion
A	A	Anti-B	A or O
B	B	Anti-A	B or O
AB	A and B	None	A, B, AB, or O
O	None	Anti-A, anti-B	O

© Cengage Learning 2017

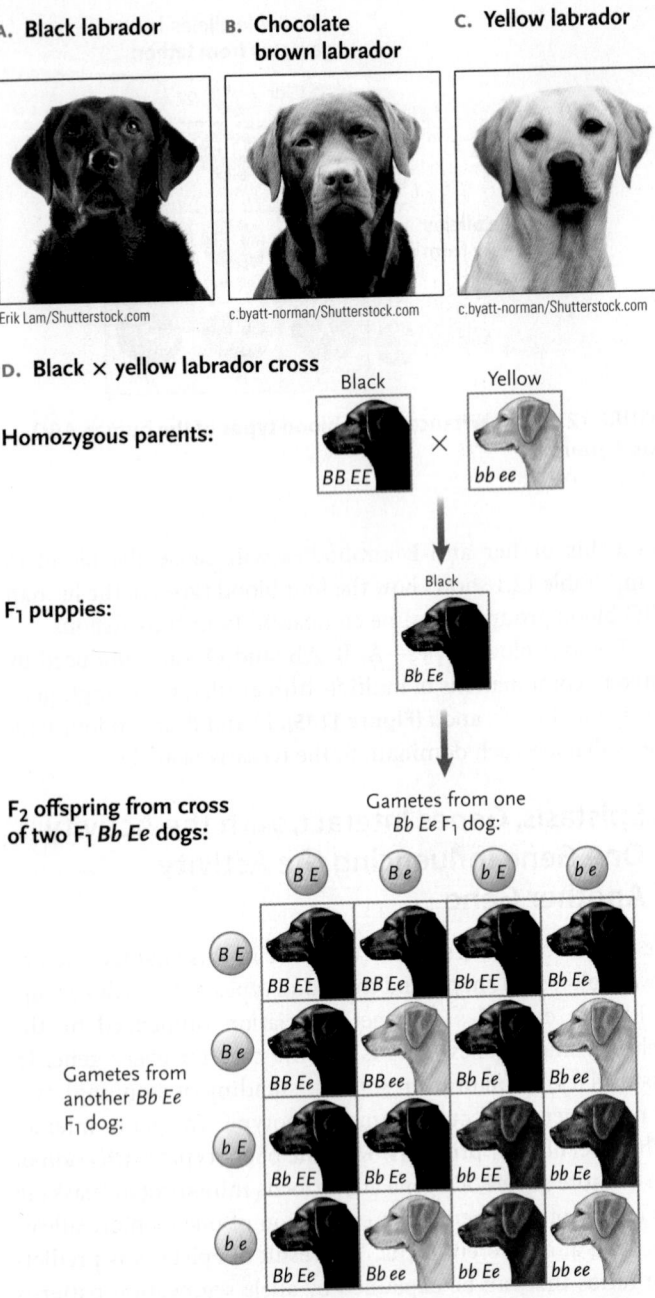

A. Black labrador

Erik Lam/Shutterstock.com

B. Chocolate brown labrador

c.byatt-norman/Shutterstock.com

C. Yellow labrador

c.byatt-norman/Shutterstock.com

D. Black × yellow labrador cross

Black   Yellow

Homozygous parents:

$BB\ EE$ × $bb\ ee$

F₁ puppies:

Black

$Bb\ Ee$

F₂ offspring from cross of two F₁ *Bb Ee* dogs:

Gametes from one $Bb\ Ee$ F₁ dog:

	B E	B e	b E	b e
B E	BB EE	BB Ee	Bb EE	Bb Ee
B e	BB Ee	BB ee	Bb Ee	Bb ee
b E	Bb EE	Bb Ee	bb EE	bb Ee
b e	Bb Ee	Bb ee	bb Ee	bb ee

Gametes from another *Bb Ee* F₁ dog:

F₂ phenotypic ratio is 9 black : 3 chocolate : 4 yellow

FIGURE 12.16 An example of epistasis: the inheritance of coat color in Labrador retrievers.

© Cengage Learning 2017

has a yellow color whether the genotype for the *B* gene is *BB*, *Bb*, or *bb*. Thus, the *E* gene is epistatic to the *B* gene.

Epistasis between the *B* and *E* genes combines two expected phenotypic classes into one in the progeny of crosses among Labs **(Figure 12.16D).** Rather than two separate classes, as would be expected from a dihybrid cross without epistasis, the *BB ee*, *Bb ee*, and *bb ee* genotypes produce a single yellow phenotype. Therefore, if we cross a true-breeding black Labrador with a

true-breeding yellow Labrador of genotype *bb ee*, the F₁ puppies are *Bb Ee* black heterozygotes (see Figure 12.16). F₂ progeny produced by crossing F₁ dogs have the distribution 9/16 black, 3/16 chocolate, and 4/16 yellow because of epistasis. That is, the ratio is 9:3:4 instead of the expected 9:3:3:1 ratio.

As you can see in this example, epistasis occurs when phenotypic expression of one or more alleles of a gene at one locus inhibits or masks the phenotypic expression of one or more alleles of a gene at a different locus. The result of epistasis is predictable ratios that can be explained by allele segregation patterns of the two (or more) genes involved. As in the Labrador coat color example, many other dihybrid crosses that involve epistatic interactions produce phenotypic distributions that differ from the expected 9:3:3:1 ratio.

In human biology, gene interactions and epistasis are common, and epistasis is an important factor in determining an individual's susceptibility to common human diseases. That is, different degrees of susceptibility are the result of different gene interactions in the individuals. A specific example is insulin resistance, a disorder in which muscle, fat, and liver cells do not use insulin correctly, with the result that glucose and insulin levels become high in the blood. This disorder is believed to be determined by several genes interacting with one another.

In Polygenic Inheritance, a Character Is Controlled by the Common Effects of Several Genes

Some characters follow a pattern of inheritance in which there is a more or less even gradation of types, forming a continuous distribution, rather than "on" or "off" (discontinuous) effects such as the production of purple or white flowers in pea plants. For example, in the human population, people range from short to tall, in a continuous distribution of gradations in height between limits of about 4 and 7 feet. Typically, a continuous distribution of this type is the result of **polygenic inheritance,** in which several to many different genes contribute to the *same* character. Other characters that exhibit a similar continuous distribution include skin color and body weight in humans, ear length in corn, seed color in wheat, and color spotting in mice. These characters are also known as **quantitative traits.** The individual genes that contribute to a quantitative trait are known as **quantitative trait loci** or QTLs. (Quantitative variation of characters are also discussed in Chapter 21.)

Polygenic inheritance can be detected by defining quantitative classes of a variation, such as human body height of 60 inches in one class, 61 inches in the next class, 62 inches in the next class, and so on **(Figure 12.17),** as opposed to qualitative classes such as tall and short. Variation for quantitative traits typically appears as continuous gradations among classes, often as a bell-shaped curve, with fewer individuals at the extremes and the greatest numbers clustered around the midpoint; this is a good indication that the trait is quantitative.

A. Students at Brigham Young University, arranged according to height

Dan Fairbanks

B. Actual distribution of individuals in the photo according to height

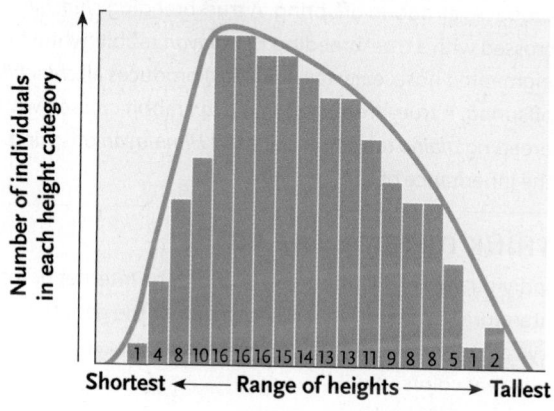

Number of individuals in each height category

1 4 8 10 16 16 16 15 14 13 13 11 9 8 8 5 1 2

Shortest ← Range of heights → Tallest

C. Idealized bell-shaped curve for a population that displays continuous variation in a trait

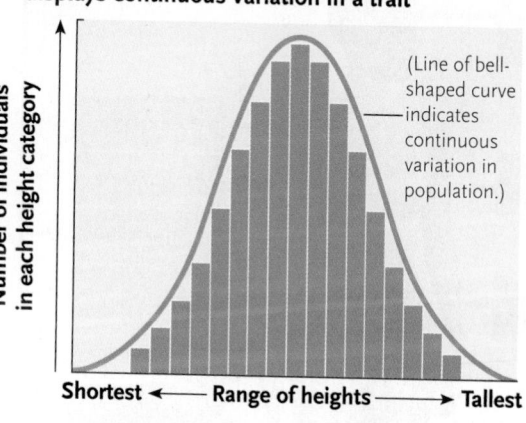

Number of individuals in each height category

(Line of bell-shaped curve indicates continuous variation in population.)

Shortest ← Range of heights → Tallest

If the sample in the photo included more individuals, the distribution would more closely approach this ideal.

FIGURE 12.17 Continuous variation in height due to polygenic inheritance.
© Cengage Learning 2017

Polygenic inheritance is often modified by the environment. For example, height in humans is not the result of genetics alone. Poor nutrition during infancy and childhood is one environmental factor that can limit growth and prevent individuals from reaching the height expected from genetic inheritance; good nutrition can have the opposite effect. Thus, the average young adult in Japan today is several inches taller than the average adult in the 1930s, when nutrition was poorer. Similarly, individuals who live in cloudy, northern or southern climates usually have lighter skin color than individuals with the same genotype who live in sunny climates.

At first glance, the effects of polygenic inheritance might appear to support the idea that characteristics of parents are blended in their offspring. Commonly, people believe that the children in a family with one tall and one short parent will be of intermediate height. Although the children of such parents are most likely to be of intermediate height, careful genetic analysis of the offspring in many such families shows that their offspring may range over a continuum from short to tall, forming a typical bell-shaped curve. Thus, genetic analysis does not support the idea of blending or even mixing of parental traits in polygenic characteristics such as body size or skin color; it is due to the combined effect of Mendelian inheritance for multiple genes influencing the same character.

In Pleiotropy, Two or More Characters Are Affected by a Single Gene

In **pleiotropy,** single genes affect more than one character of an organism. For example, sickle-cell anemia (see earlier discussion and *Why it matters . . .*) is caused by a recessive allele of a single gene that affects hemoglobin structure and function. However, the altered hemoglobin, the primary phenotypic change of the sickle-cell mutation, leads to blood vessel blockage, which can damage many tissues and organs in the body and affect many body functions, producing such wideranging symptoms as fatigue, abdominal pain, heart failure, paralysis, and pneumonia (**Figure 12.18**). Physicians recognize these wide-ranging pleiotropic effects as symptoms of sickle-cell anemia.

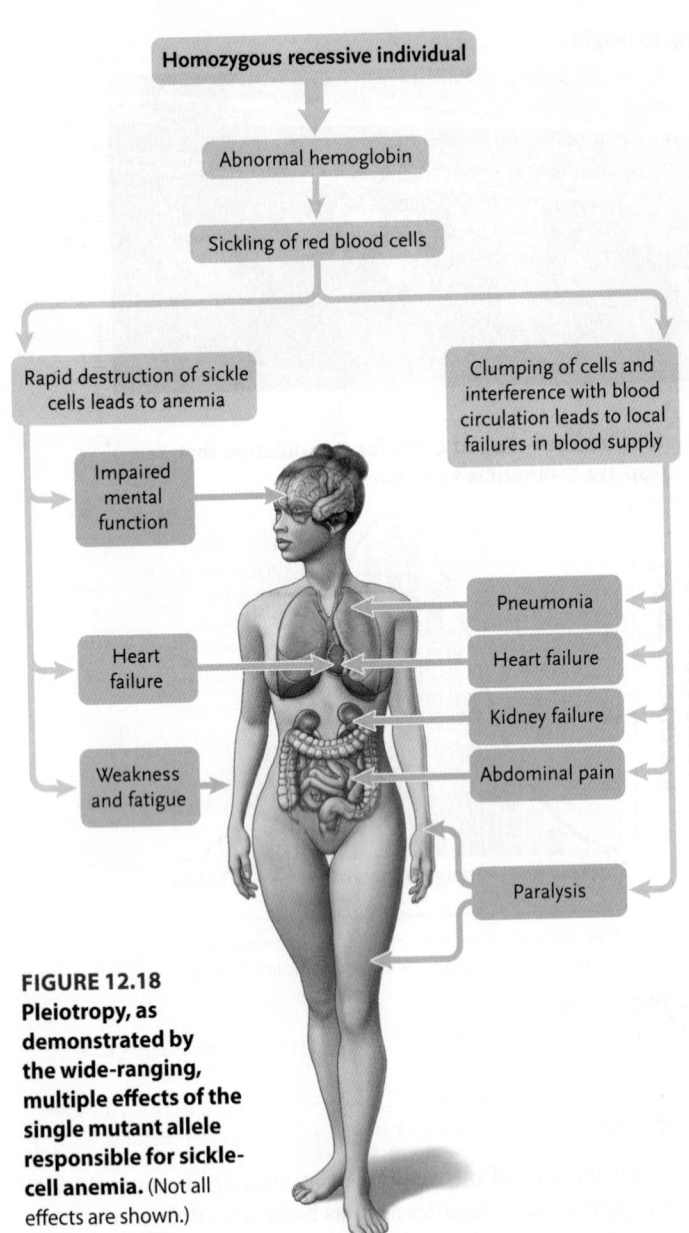

FIGURE 12.18
Pleiotropy, as demonstrated by the wide-ranging, multiple effects of the single mutant allele responsible for sickle-cell anemia. (Not all effects are shown.)
© Cengage Learning 2017

The next chapter describes additional patterns of inheritance that were not anticipated by Mendel, including the effects of recombination during meiosis. These additional patterns also extend, rather than contradict, Mendel's fundamental principles.

STUDY BREAK 12.2

1. Palomino horses have a golden coat color, with a white mane and tail. Palominos do not breed true. Instead, there is a 50% chance that a foal with two Palomino parents will be a Palomino. What is the explanation?

2. A true-breeding rabbit with *agouti* (mottled, grayish brown) fur crossed with a true-breeding rabbit with *chinchilla* (silver) fur produces all agouti offspring. A true-breeding *chinchilla* rabbit crossed with a true-breeding *Himalayan* rabbit (white fur with pigmented nose, ears, tail, and legs) produces all *chinchilla* offspring. A true-breeding *Himalayan* rabbit crossed with a true-breeding *albino* rabbit produces all *Himalayan* offspring. Explain the inheritance of the fur colors.

THINK OUTSIDE THE BOOK

Individually or collaboratively, explore the Internet or research literature to outline what is known about the genetics of human eye color, and about how certain individuals can have irises of different colors.

 ## Unanswered Questions

What is next for the genetics of human disease?

Technological and conceptual advances have propelled our understanding of the genetic basis of human disease to unprecedented levels of refinement. Comparing the *status quo* to that of only a decade ago, we now have detailed genetic and genomic information for humans, several primates, and a host of other organisms across all phyla, and our progress is set to accelerate. We have also witnessed the rapid identification of mutations that cause rare diseases, and we have recently discovered over 1,000 genomic segments that are associated with susceptibility to complex traits such as type I and type II diabetes, obesity, high cholesterol, and others.

Despite these advances, the most fundamental questions in genetics, raised a century ago, have yet to be addressed fully. Most fundamental of all is the question of how the genotype relates to the phenotype. Once

a genetic disorder has been diagnosed, we can often confirm the diagnosis by mutation analysis. However, our ability to use such analysis to predict whether or not a patient will develop a particular genetic disorder is still extremely limited. This is in part because we still evaluate the effects of genetic and genomic variations in isolation. We have not yet developed the tools to understand the effects of such variations in the context of an entire genome (as well as the environment). Examples from a wide variety of single-gene disorders have shown us that individuals can be genotypically affected but clinically normal. Likewise, the typical clinical experience is that patients with the same mutation can vary greatly in the presence and severity of particular symptoms.

One of the challenges ahead of us is to understand the effect of mutations in the context of an individual's total genomic variation; that is, the sum total of likely pathogenic mutations in the genome. Coupled to that goal, the availability of the total genome sequence poses both an opportunity and a problem. Given that each human has several thousand variants predicted to affect gene/protein function, we need to solve the problems of how to assign the effect of genetic variation for any particular phenotype and how to predict the frequency with which each allele results in disease symptoms.

An overlapping question is the century-old debate over the relative contribution of rare and common alleles to genetic disease (the so-called common-allele common disease versus rare-allele common disease argument). The premise that complex traits such as schizophrenia, hypertension, and diabetes must be caused by common mutations in the general population (otherwise they would not be so common) has also proven true of some disorders, such as age-related macular degeneration, the most frequent cause of blindness in the elderly. However, exhaustive analysis of large cohorts has also shown that common alleles can account for only a modest fraction of the genetic risk for any given complex trait and that most of that risk maps to regions of the genome that are not directly involved in coding protein sequences. For example, a common allele in a noncoding region of a gene called *Fused Toes (FTO)* that shows exceptionally significant association with obesity and diabetes accounts for only about 3 kg of increased weight. The questions ahead of us are therefore: (1) where are the genetic risk alleles that account for the majority of complex traits (the so-called genetic "dark matter"); (2) why is there a dearth of coding sequence changes in complex traits; and (3) how do common, probably mild, alleles (alleles that have small effects) interact with possibly rare alleles having strong effects to magnify the risk of a

complex trait? For example, work in our laboratory on Bardet–Biedl syndrome, a genetic disorder characterized by obesity and learning defects, and related disorders has shown that some alleles that essentially abolish protein function (and therefore have a strong effect) can interact with mutations in other genes that have a modest effect on protein function (mild alleles). The interaction between the two can significantly enhance the severity of the clinical phenotype, but until we are able to model the effect of these mutations, genetic analysis alone is insufficient to detect such disorders. It is likely that a combination of extensive sequencing, dense genotyping, and a new generation of computational and biological tools will be required to address these challenges.

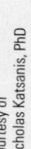 

think like a scientist Given significant resources, what would you choose to do as a means of understanding variation in the human genome and its involvement in disease: sequence 1,000 humans or sequence 1,000 other species? (*Hint:* One would give you access to a lot of human variation, the other would help you with evolutionary arguments.)

Nicholas Katsanis is professor of Cell Biology and director of the Center for Human Disease Modeling at Duke University. His research interests focus on the genetic basis of Bardet–Biedl syndrome, where his laboratory is engaged in the identification of causative genes. The Katsanis lab also pursues questions centered on the signaling roles of vertebrate cilia, the translation of signaling pathway defects on the causality of ciliary disorders, and the dissection of second-site modification phenomena as a consequence of genetic load in a functional system. To learn more, go to http://www.cellbio.duke.edu/nicholas-katsanis/.

REVIEW KEY CONCEPTS

For access to MindTap and additional study materials visit www.cengagebrain.com.

12.1 The Beginnings of Genetics: Mendel's Garden Peas

- Mendel was successful in his research because of his good choice of experimental organism, which had clearly defined characters, such as flower color or seed shape, and because he analyzed his results quantitatively (Figures 12.3 and 12.4).

- Mendel showed that traits are passed from parents to offspring as hereditary factors (now called genes and alleles) in predictable ratios and combinations, disproving the notion of blended inheritance (Figure 12.5).

- From the results of crosses that involve single characters (monohybrid crosses), Mendel made three conclusions: (1) the genes that govern genetic characters occur in pairs in individuals; (2) if different alleles of a gene are present in the pair of an individual, one allele is dominant over the other; and (3) the two alleles of a gene segregate and enter gametes singly (Figures 12.5 and 12.7).

- Mendel confirmed his conclusions by a testcross between an F_1 heterozygote with a homozygous recessive parent. This type of testcross is still used to determine whether an individual is homozygous or heterozygous for a dominant allele (Figure 12.8).

- To explain the results of his crosses with individuals showing differences in two characters—dihybrid crosses—Mendel made a fourth conclusion: the alleles of the genes that govern the two

characters segregate independently during formation of gametes (Figure 12.9).

- Walter Sutton was the first person to note the similarities between the inheritance of genes and the behavior of chromosomes in meiosis and fertilization. These parallels made it obvious that genes and alleles are carried on the chromosomes. Sutton's parallels are called the chromosome theory of inheritance (Figure 12.10).

- A locus is the site occupied by a gene on a chromosome (Figure 12.11).

12.2 Later Modifications and Additions to Mendel's Principles

- In incomplete dominance, some or all alleles of a gene are neither completely dominant nor recessive. In such cases, the phenotype of heterozygotes with different alleles of the gene can be distinguished from that of either homozygote (Figure 12.13).

- In codominance, different alleles of a gene have approximately equal effects in heterozygotes, also allowing heterozygotes to be distinguished from either homozygote.

- Many genes may have multiple alleles if all the individuals in a population are taken into account. However, any diploid individual in a population has only two alleles of these genes, which are inherited and passed on according to Mendel's principles (Figures 12.14 and 12.15).

- In epistasis, two (or more) genes affect the same phenotype; the genes interact at the level of their products, with the phenotypic expression of one or more alleles of one locus inhibiting or masking the phenotypic expression of one or more alleles at a different locus. The result is that some phenotypic classes among progeny may be combined into one (Figure 12.16).

- In polygenic inheritance, genes at several to many different loci control the same character, producing a more or less continuous variation in the character from one extreme to another. Plotting the distribution of such characters among individuals often produces a bell-shaped curve (Figure 12.17).

- In pleiotropy, one gene affects more than one character of an organism (Figure 12.18).

TEST YOUR KNOWLEDGE

Remember/Understand

1. The dominant *C* allele of a gene that controls color in corn produces kernels with color; plants homozygous for a recessive *c* allele of this gene have colorless or white kernels. What kinds of gametes, and in what proportions, would be produced by the plants in the following crosses? What seed color, and in what proportions, would be expected in the offspring of the crosses?
   a. *CC × Cc*
   b. *Cc × cc*
   c. *Cc × Cc*

2. In peas, the allele *Le* produces tall plants and the allele *le* produces dwarf plants. The *Le* allele is dominant to *le*. If a tall plant is crossed with a dwarf, the offspring are distributed about equally between tall and dwarf plants. What are the genotypes of the parents?

3. One gene has the alleles *A* and *a*; another gene has the alleles *B* and *b*. For each of the following genotypes, what types of gametes will be produced, and in what proportions, if the two gene pairs assort independently?
   a. *AA BB*
   b. *Aa BB*
   c. *Aa bb*
   d. *Aa Bb*

4. What genotypes, and in what frequencies, will be present in the offspring from the following matings?
   a. *AA BB × aa BB*
   b. *Aa Bb × aa bb*
   c. *Aa BB × AA Bb*
   d. *Aa Bb × Aa Bb*

5. In addition to the two genes in problem 4, assume you now study a third independently assorting gene that has the alleles *C* and *c*. For each of the following genotypes, indicate what types of gametes will be produced:
   a. *AA BB CC*
   b. *Aa BB Cc*
   c. *Aa BB cc*
   d. *Aa Bb Cc*

6. In garden peas, the genotypes *GG* or *Gg* produce green pods and *gg* produces yellow pods; *LeLe* or *Lele* plants are tall and *lele* plants are dwarfed; *RR* or *Rr* produce round seeds and *rr* produces wrinkled seeds. If a plant of a true-breeding, tall variety with green pods and round seeds is crossed with a plant of a true-breeding, dwarf variety with yellow pods and wrinkled seeds, what phenotypes are expected, and in what ratios, in the F$_1$ generation? What phenotypes, and in what ratios, are expected if F$_1$ individuals are crossed (allowed to self-fertilize)?

7. In cats, the genotype *AA* produces tabby fur color; *Aa* is also a tabby, and *aa* is black. Another gene at a different locus is epistatic to the gene for fur color. When present in its dominant *W* form (*WW* or *Ww*), this gene blocks the formation of fur color and all the offspring are white; *ww* individuals develop normal fur color. What fur colors, and in what proportions, would you expect from the cross *Aa Ww × Aa Ww*?

8. **Discuss Concepts** Explain how individuals of an organism that are phenotypically alike can produce different ratios of progeny phenotypes.

Apply/Analyze

9. The ability of humans to taste the bitter chemical phenylthiocarbamide (PTC) is a genetic trait. People with at least one copy of the normal, dominant allele of the *PTC* gene can taste PTC; those who are homozygous for a mutant, recessive allele cannot taste it. Could two parents able to taste PTC have a nontaster child? Could nontaster parents have a child able to taste PTC? A pair of taster parents, both of whom had one parent able to taste PTC and one nontaster parent, are expecting their first child. What is the chance that the child will be able to taste PTC? Unable to taste PTC? Suppose the first child is a nontaster. What is the chance that their second child will also be unable to taste PTC?

10. A man is homozygous dominant for alleles at 10 different genes that assort independently. How many genotypically different types of sperm cells can he produce? A woman is homozygous recessive for the alleles of 8 of these 10 genes, but she is heterozygous for the other 2 genes. How many genotypically different types of eggs can she produce? What hypothesis can you suggest to describe the relationship between the number of different possible gametes and the number of heterozygous and homozygous genes that are present?

11. In guinea pigs, an allele for rough fur *(R)* is dominant over an allele for smooth fur *(r)*; an allele for black coat *(B)* is dominant over that for white *(b)*. You have an animal with rough, black fur. What cross would you use to determine whether the animal is homozygous for these traits? What phenotype would you expect in the offspring if the animal is homozygous?

12. You cross a lima bean plant from a variety that breeds true for green pods with another lima bean from a variety that breeds true for yellow pods. You note that all the F$_1$ plants have green pods. These green-pod F$_1$ plants, when crossed, yield 675 plants with green pods and 217 with yellow pods. How many genes probably control pod color in this experiment? Give the alleles letter designations. Which is dominant?

13. Some recessive alleles have such a detrimental effect that they are lethal when present in both chromosomes of a pair. Homozygous recessives cannot survive and die at some point during embryonic development. Suppose that the allele *r* is lethal in the homozygous *rr* condition. What genotypic ratios would you expect among the living offspring of the following crosses?
   a. *RR × Rr*
   b. *Rr × Rr*

14. In chickens, feathered legs are produced by a dominant allele *F*. Another allele *f* of the same gene produces featherless legs. The dominant allele *P* of a gene at a different locus produces pea combs; a recessive allele *p* of this gene causes single combs. A breeder makes the following crosses with birds 1, 2, 3, and 4; all parents have both feathered legs and pea combs:

Cross	Offspring
1 × 2	All feathered, pea comb
1 × 3	3/4 feathered; 1/4 featherless, all pea comb
1 × 4	9/16 feathered, pea comb; 3/16 featherless, pea comb 3/16 feathered, single comb; 1/16 featherless, single comb

What are the genotypes of the four birds?

15. A mix-up in a hospital ward caused a mother with O and MN blood types to think that a baby given to her really belonged to someone else. Tests in the hospital showed that the doubting mother was able to taste PTC (see problem 9). The baby given to her had O and MN blood types and had no reaction when the bitter PTC chemical was placed on its tongue. The mother had four other children with the following blood types and tasting abilities for PTC:
    1. type A and MN blood, taster
    2. type B and N blood, nontaster
    3. type A and M blood, taster
    4. type A and N blood, taster

    Without knowing the father's blood types and tasting ability, can you determine whether the child is really hers? (Assume that all her children have the same father.)

16. Having malformed hands with shortened fingers is a dominant trait controlled by a single gene; people who are homozygous for the recessive allele have normal hands and fingers. Having woolly hair is a dominant trait controlled by a different gene; homozygous recessive individuals have normal, nonwoolly hair. Suppose a woman with normal hands and nonwoolly hair marries a man who has malformed hands and woolly hair. Their first child has normal hands and nonwoolly hair. What are the genotypes of the mother, the father, and the child? If this couple has a second child, what is the probability that it will have normal hands and woolly hair?

Evaluate/Create

17. **Discuss Concepts** ABO blood type tests can be used to exclude paternity. Suppose a defendant who is the alleged father of a child takes a blood type test and the results do not exclude him as the father. Do the results indicate that he is the father? What arguments could a lawyer make based on the test results to exclude the defendant from being the father? (Assume the tests were performed correctly.)

18. **Design an Experiment** Imagine that you are a breeder of Labrador retriever dogs. Labs can be black, chocolate brown, or yellow. Suppose that a yellow Lab is donated to you and you need to know its genotype. You have a range of dogs with known genotypes. What cross would you make to determine the genotype of the donated dog? Explain how the resulting puppies show you the Lab's genotype.

19. **Apply Evolutionary Thinking** How could an epistatic interaction shelter a harmful allele from the action of natural selection?

For selected answers, see Appendix A.

INTERPRET THE DATA

Half of the world's population eats rice at least twice a day. Much of this rice is grown in flooded conditions, and different strains of rice are tolerant (survive) or intolerant (die) under these conditions. Rice breeders used genetic crosses to test whether tolerance to flooding is a dominant trait. Researchers used three true-breeding flood-tolerant strains, FR143, BKNFR, and Kurk, and two true-breeding flood-intolerant strains, IR42 and NB, in the crosses. Results were obtained from three sets of crosses and are reported in the **Table** below:

1. F_2 results: Intolerant and tolerant strains were crossed and the resulting F_1 were interbred to produce the F_2.
2. Results of cross of F_1 to intolerant parent: F_1 plants were crossed with the intolerant parent of the cross.
3. Results of cross of F_1 to tolerant parent: F_1 plants were crossed with the tolerant parent of the cross.

Progeny Analyzed from Intolerant × Tolerant Cross	Number of Plants		
	Alive	Dead	Total
1. F_2 results of cross:			
IR42 × FR13A	187	77	264
IR42 × BKNFR	192	73	265
NB × Kurk	142	52	195
2. Results of cross of F_1 to intolerant parent:			
(F_1 of IR42 × FR13A) × IR42	14	17	31
(F_1 of IR42 × BKNFR) × IR42	15	10	25
(F_1 of NB × Kurk) × NB	21	35	56
3. Results of cross of F_1 to tolerant parent:			
(F_1 of IR42 × FR13A) × FR13A	31	0	31
(F_1 of IR42 × BKNFR) × BKNFR	28	0	28
(F_1 of NB × Kurk) × Kurk	40	0	40

Do the data support the hypothesis that the tolerance trait is dominant? Justify your conclusion by explaining the results from each of the three sets of crosses in terms of genotypes and phenotypic ratios.

Source: T. Setter et al. 1997. Physiology and genetics of submergence tolerance in rice. *Annals of Botany* 79:67–77.

13 Genes, Chromosomes, and Human Genetics

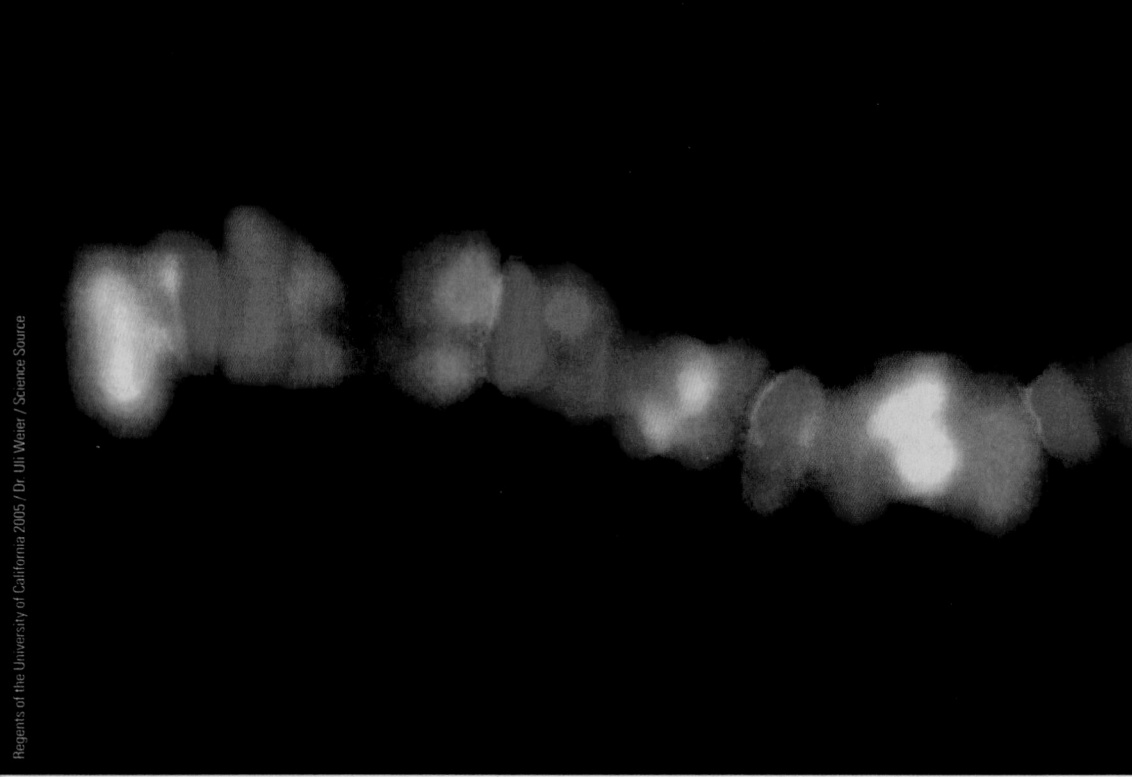

Regents of the University of California 2005 / Dr. Uli Weier / Science Source

Fluorescent probes bound to specific sequences along human chromosome 10 (light micrograph). New ways of mapping chromosome structure yield insights into the inheritance of normal and abnormal traits.

Why it matters . . . Imagine being 10 years old and trapped in a body that each day becomes more shriveled, frail, and old. You weigh less than 35 pounds, already you are bald, and you probably have only a few more years to live. But Mickey Hayes and Fransie Geringer **(Figure 13.1)** still have their courage and their childlike curiosity about life.

Progeria, the premature aging that afflicts Mickey and Fransie, is caused by a genetic mutation that is present in only 1 of every 4 to 8 million human births. The mutation is in the gene for lamin A, one of the lamin proteins that reinforces the inner surface of the nuclear envelope in animal cells (see Chapter 4). In some way not yet understood, the defective lamin A makes the nucleus unstable, leading to the premature aging and reduced life expectancy characteristic of progeria.

Progeria affects both sexes equally, and all races and ethnic groups. Usually, symptoms begin to appear between 18 and 24 months of age. The rate of body growth declines to abnormally low levels. Skin becomes thinner, muscles become flaccid, and limb bones start to degenerate. Children with progeria die from a stroke or heart attack brought on by hardening of the arteries, a condition typical of advanced age. Death occurs at an average age of 13, with a range of about 8 to 21 years.

The plight of Mickey and Fransie illustrates the dramatic effects that gene defects can have on living organisms. We are the products of our genes, and the characteristics of each individual, from humans to pine trees to protozoa, depend on the combination of genes, alleles, and chromosomes inherited from its parents, as well as on environmental effects. This chapter discusses extensions to Mendel's principles, particularly the effects of genetic recombination during meiosis, and develops further the relationship between genes and chromosomes in inheritance that was introduced in the discussion of the chromosome theory of inheritance in Section 12.1.

Model Organisms: The Marvelous Fruit Fly, *Drosophila melanogaster*

Hermann Eisenbeiss/
Science Source

The fruit fly was first described in 1830 by C. F. Fallén, who named it *Drosophila,* meaning "dew lover." The species identifier became *melanogaster,* which means "black belly."

The great geneticist Thomas Hunt Morgan began to culture *D. melanogaster* in 1909 in the famous "Fly Room" at Columbia University. Many important discoveries in genetics were made in the Fly Room, including sex-linked genes and sex linkage, and the first chromosome map. The subsequent development of methods to induce mutations in *Drosophila* led, through studies of the mutants produced, to many other discoveries that collectively established or confirmed essentially all the major principles and conclusions of eukaryotic genetics.

Among the many reasons for the success of *D. melanogaster* as a subject for genetics research are the ease of culturing these flies, their rapid cycle of growth and reproduction, easy identification of males and females, and

the many types of mutations that cause morphological differences, such as purple eyes or vestigial wings, which can be seen with the unaided eye or under a low-power binocular microscope.

The availability of a wide range of mutants, comprehensive linkage maps of each of its chromosomes, and the ability to manipulate genes readily by molecular techniques made the fruit fly one of the model organisms for genome sequencing in the Human Genome Project (see Chapter 19). The sequence of *Drosophila's* genome was published in 2000; there are approximately 13,500 protein-coding genes in its 144 million base-pair genome. Importantly, many fruit fly and human genes are similar, to the point that many human disease genes have counterparts in the fruit fly genome. This similarity is a consequence of the distant but still detectable evolutionary relationship between insects and vertebrates. It enables the fly genes to be studied as models of human disease genes in efforts to understand better the functions of those genes and how alterations in them lead to disease.

Drosophila has also become established as an excellent experimental model for neurobiology studies (the investigation of the structure and function of the nervous system). Scientists are investigating, for example, neural development, and analyzing behavior. Again, findings are helping us understand the human nervous system also.

The analysis of fruit fly embryonic development has also contributed significantly to the understanding of development in humans. For example, experiments on mutants that affect fly development have provided insights into the genetic basis of many human birth defects. Recognition of the importance of research with *Drosophila* developmental genetics to our understanding of development in general came in the form of the award of the Nobel Prize in 1995 to three scientists who pioneered the fruit fly work: Edward Lewis of the California Institute of Technology, Christiane Nusslein-Volhard of the Max Planck Institute for Developmental Biology in Tubingen, Germany, and Eric Wieschaus of Princeton University. Their work is discussed in Chapter 16.

13.1 Genetic Linkage and Recombination

In his historic experiments, Gregor Mendel found that each of the seven genes he studied assorted independently of the others in the formation of gametes (see Chapter 12). If

Mendel had extended his study to numerous characters, he would have found exceptions to this principle. This should not be surprising, because each chromosome of an organism contains many genes, with each gene at a particular locus. Genes located on different chromosomes assort independently in gamete formation because the two chromosomes behave independently of one another during meiosis (see Chapter 11). Genes located on the same chromosome may be inherited together in genetic crosses—that is, *not* assort independently—because the chromosome is inherited as a single physical entity in meiosis. Genes near each other on the same chromosome are known as **linked genes,** and the phenomenon is called **linkage.**

Linked Genes Show a Different Pattern of Inheritance Than Independently Assorting Genes

In the first part of the twentieth century, Thomas Hunt Morgan and his coworkers at Columbia University used the fruit fly, *Drosophila melanogaster,* as a model organism to investigate Mendel's principles in animals (see *Focus on Research: Model Organisms*).

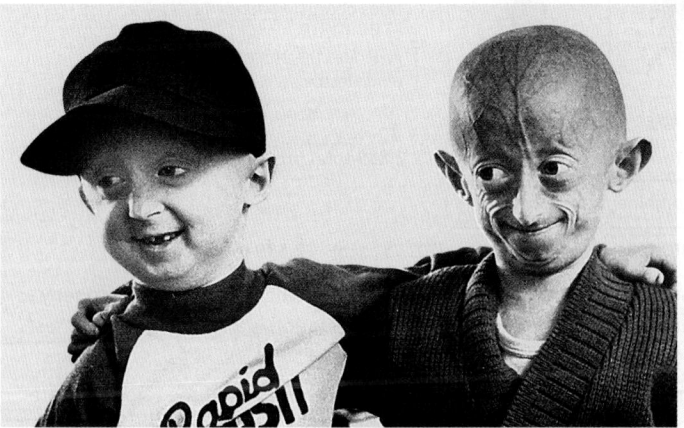

AP Images/Eddie Adams

FIGURE 13.1 Two boys, both younger than 10, who have progeria, a genetic disorder characterized by accelerated aging and extremely reduced life expectancy.

FIGURE 13.2 **Experimental Research**

Evidence for Gene Linkage

Question: Do the purple-eye and vestigial-wing genes of *Drosophila* assort independently?

Experiment: Morgan crossed true-breeding wild-type flies with red eyes and normal wings with true-breeding purple-eyed, vestigial-winged flies. He then testcrossed the F₁ flies, which were wild type in phenotype, and analyzed the distribution of phenotypes in the progeny.

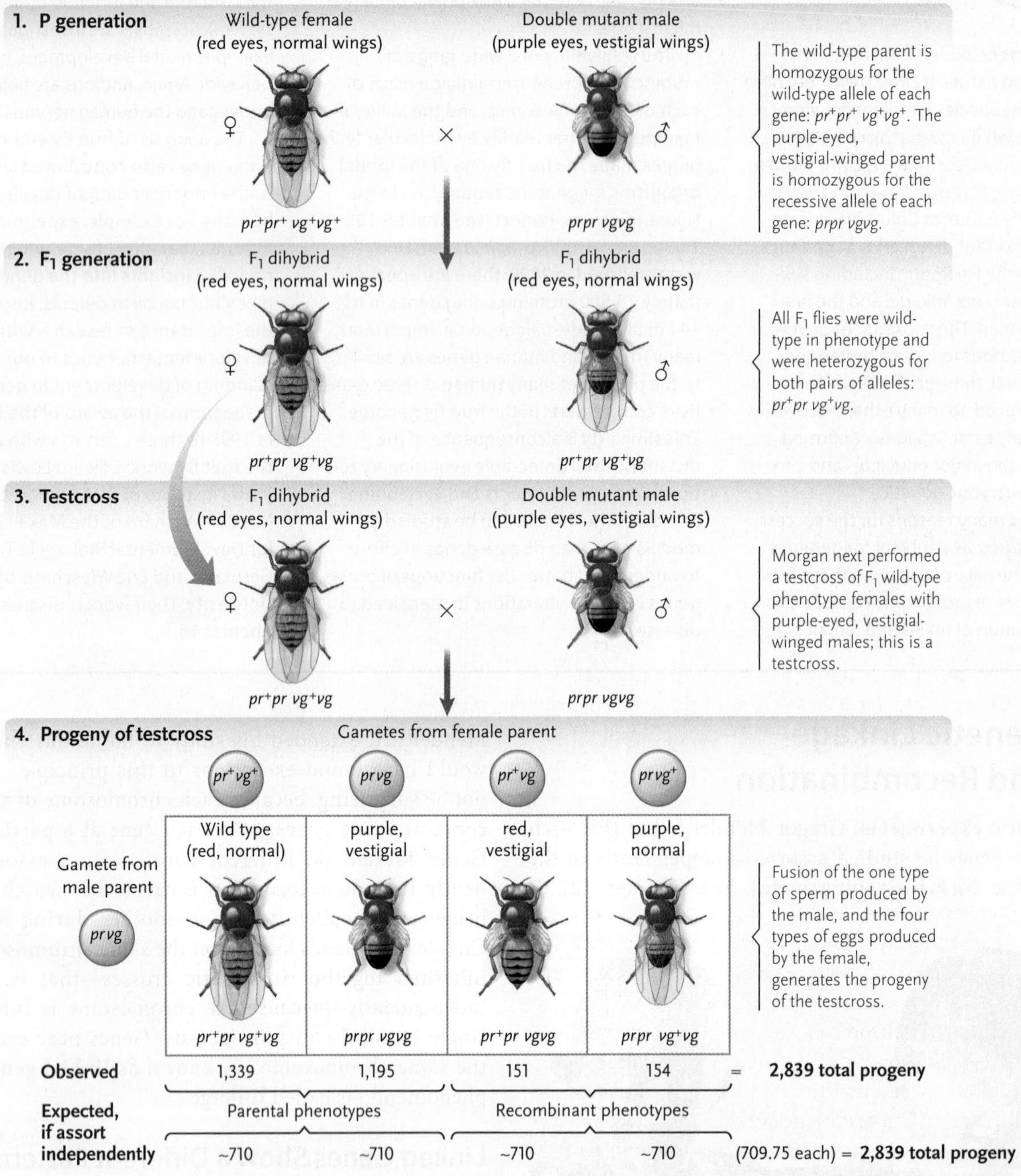

1. P generation

Wild-type female (red eyes, normal wings)

Double mutant male (purple eyes, vestigial wings)

$pr^+pr^+\ vg^+vg^+$ × $prpr\ vgvg$

The wild-type parent is homozygous for the wild-type allele of each gene: $pr^+pr^+\ vg^+vg^+$. The purple-eyed, vestigial-winged parent is homozygous for the recessive allele of each gene: $prpr\ vgvg$.

2. F₁ generation

F₁ dihybrid (red eyes, normal wings)

F₁ dihybrid (red eyes, normal wings)

$pr^+pr\ vg^+vg$ $pr^+pr\ vg^+vg$

All F₁ flies were wild-type in phenotype and were heterozygous for both pairs of alleles: $pr^+pr\ vg^+vg$.

3. Testcross

F₁ dihybrid (red eyes, normal wings)

Double mutant male (purple eyes, vestigial wings)

$pr^+pr\ vg^+vg$ × $prpr\ vgvg$

Morgan next performed a testcross of F₁ wild-type phenotype females with purple-eyed, vestigial-winged males; this is a testcross.

4. Progeny of testcross

Gametes from female parent

pr^+vg^+ $prvg$ pr^+vg $prvg^+$

	Wild type (red, normal)	purple, vestigial	red, vestigial	purple, normal
Gamete from male parent $prvg$	$pr^+pr\ vg^+vg$	$prpr\ vgvg$	$pr^+pr\ vgvg$	$prpr\ vg^+vg$

Fusion of the one type of sperm produced by the male, and the four types of eggs produced by the female, generates the progeny of the testcross.

Observed	1,339	1,195	151	154	= **2,839 total progeny**

Parental phenotypes   Recombinant phenotypes

Expected, if assort independently	~710	~710	~710	~710	(709.75 each) = **2,839 total progeny**

Results: 2,534 of the testcross progeny flies had parental phenotypes, wild-type (red, normal) and purple, vestigial, whereas 305 of the progeny had recombinant phenotypes of red, vestigial and purple, normal. If the genes assorted independently, the expectation is a 1 : 1 : 1 : 1 ratio for testcross progeny: approximately 1,420 of both parental and recombinant progeny.

Conclusion: The purple-eye and vestigial-wing genes do not assort independently. The simplest alternative hypothesis is that the two genes are linked on the same chromosome. The small number of recombinant phenotypes is explained by crossing-over.

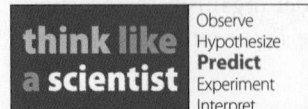
think like a scientist
Observe
Hypothesize
Predict
Experiment
Interpret

Suppose that instead of the cross shown, the P generation cross was between a true-breeding red-eyed, vestigial-winged female and a true-breeding purple-eyed, normal-winged male. What would be the two largest classes in the progeny of the testcross? What would be the two smallest classes?

Source: C. B. Bridges. 1919. The genetics of purple eye color in *Drosophila*. *The Journal of Experimental Zoology* 28:265–305.

Morgan crossed a true-breeding fruit fly with normal red eyes and normal wing length, genotype $pr^+pr^+ vg^+vg^+$, with a true-breeding fly with the recessive traits of purple eyes and vestigial (short and crumpled) wings, genotype $prpr\ vgvg$, to analyze the segregation of the two traits.

This gene symbolism is new to us. In this system, the superscript plus (+) symbol associated with a letter or letters indicates a wild-type—normal—allele of a gene. Typically, but not always, a *wild-type* allele is the most common allele found in a population. In most instances, the wild-type allele is dominant to mutant alleles, but there are exceptions. The letters for the gene are based on the phenotype caused by the *mutant* allele, for example, *pr* for *purple* eyes. Thus, we refer to the gene as the *purple* or *pr* gene; the dominant wild-type allele of the gene, pr^+, gives the wild-type red eye color. Using the terminology introduced in Section 12.1, the *pr* allele and the *vg* allele are gene markers, here being used to determine the genetic linkage relationship of the two genes.

Figure 13.2 steps through Morgan's cross of the two parents and his testcross of F_1 flies. Based on Mendel's principle of independent assortment (see Section 12.1), there should be four classes of phenotypes in the testcross offspring, in a $1:1:1:1$ ratio of red eyes, normal wings : purple, vestigial : red, vestigial : purple, normal. But, instead, Morgan analyzed 2,839 progeny flies and among them saw 1,339 red, normal and 1,195 purple, vestigial (step 4). These phenotypes are identical to the two original P generation flies and are called **parental** phenotypes. The remaining progeny flies consisted of 151 red, vestigial and 154 purple, normal. These phenotypes have different combinations of traits from those of the P generation flies and are called **recombinant** phenotypes. If the genes had shown independent assortment, there would have been 25% of each of the four classes, or 50% parental and 50% recombinant phenotypes. In numbers, there would have been 710 (approximately) of each of the 4 classes.

To explain the low frequency of recombinant phenotypes, Morgan hypothesized that *pr* and *vg* are linked genes—they are physically close to each other on the same chromosome. As you saw for this example, linked genes tend to be inherited together rather than independently. Morgan further hypothesized that the behavior of these linked genes in the testcross is explained by what he called *chromosome recombination,* a process in which two homologous chromosomes exchange segments with each other by crossing-over during meiosis (see Figure 11.5). Furthermore, he proposed that the frequency of this recombination is a function of the distance between linked genes. That is, the closer two genes are, the greater proportion of parental phenotypes compared with recombinant phenotypes will be seen in the progeny of a testcross.

Figure 13.3 shows how Morgan's thinking applies to the purple–vestigial cross. Cartoons of the chromosomes themselves allow us to follow pictorially the consequences of crossing-over during meiosis in the production of gametes, and then the fusion of parental and recombinant gametes in the female with the gamete from the male testcross parent.

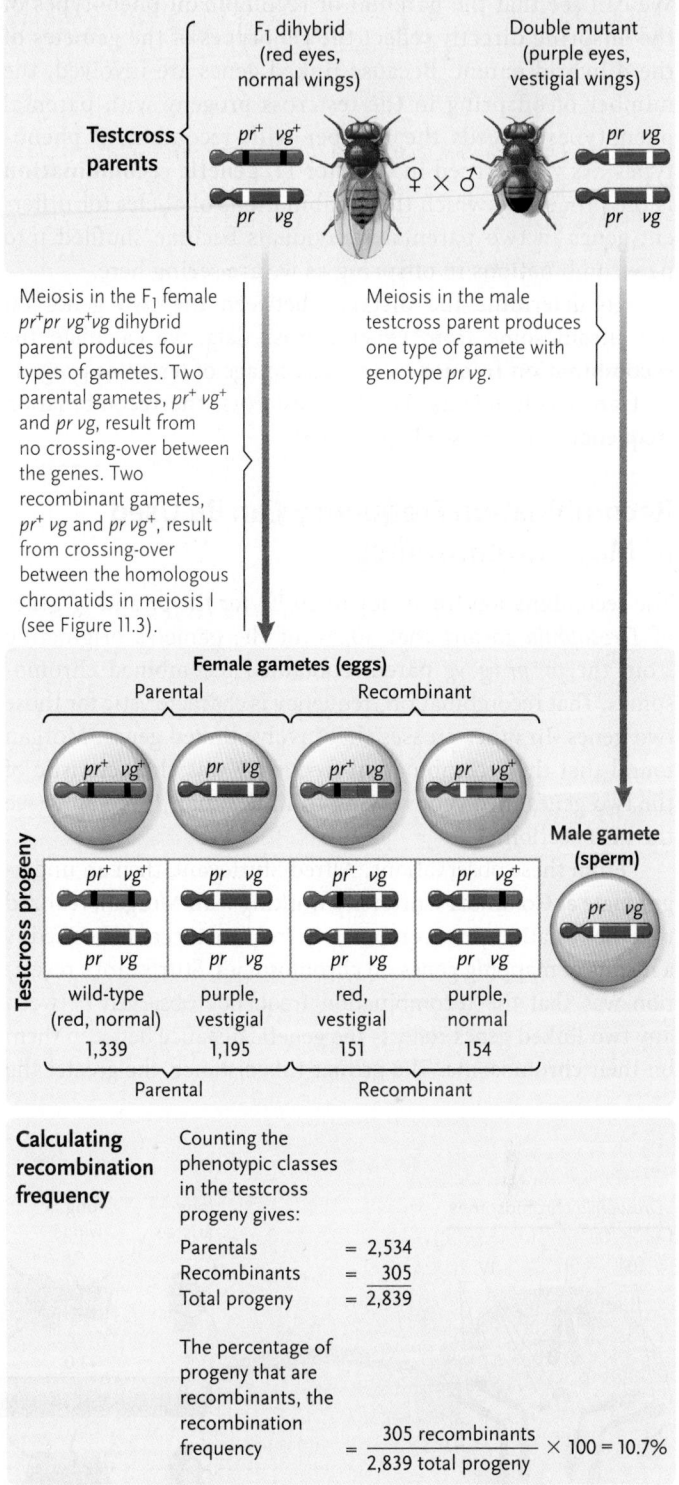

FIGURE 13.3 Recombination between the purple-eye gene and the vestigial-wing gene, resulting from crossing-over between homologous chromosomes. The testcross of Figure 13.2 is redrawn here showing the two linked genes on chromosomes. Chromosomes or chromosome segments with wild-type alleles are red, whereas chromosomes or segments with mutant alleles are blue. The parental phenotypes in the testcross progeny are generated by segregation of the parental chromosomes, whereas the recombinant phenotypes are generated by crossing-over between the two linked genes.

© Cengage Learning 2017

We can see that the parental or recombinant phenotypes of the offspring directly reflect the genotypes of the gametes of the dihybrid parent. Because linked genes are involved, the number of offspring in the testcross progeny with parental phenotypes exceeds the number with recombinant phenotypes. As you learned in Chapter 11, **genetic recombination** is the process by which the combinations of alleles for different genes in two parental individuals become shuffled into new combinations in offspring as we are seeing here.

To determine the distance between the two genes on the chromosome from genetic cross data, we calculate the **recombination frequency,** the percentage of testcross progeny that are recombinants. For this testcross, the recombination frequency is 10.7% (see Figure 13.3).

Recombination Frequency Can Be Used to Map Chromosomes

The recombination frequency of 10.7% for the *pr* and *vg* genes of *Drosophila* means that 10.7% of the gametes originating from the $pr^+pr\,vg^+vg$ parent contained recombined chromosomes. That recombination frequency is characteristic for those two genes. In other crosses that involve linked genes, Morgan found that the recombination frequency was characteristic of the two genes involved, varying from less than 1% to 50% (see the next section).

From these observations, Alfred Sturtevant, then an undergraduate at Columbia University working with Morgan, realized that the variations in recombination frequencies could be used as a means of mapping genes on chromosomes. Sturtevant's revelation was that the recombination frequency observed between any two linked genes reflects the genetic distance between them on their chromosome. The greater this distance, the greater the

chance that a crossover can form between the genes and the greater the recombination frequency.

Therefore, recombination frequencies can be used to make a **linkage map** of a chromosome showing the relative locations of genes. For example, assume that the three genes *a*, *b*, and *c* are carried together on the same chromosome. Crosses reveal a 9.6% recombination frequency for *a* and *b*, an 8% recombination frequency for *a* and *c*, and a 2% recombination frequency for *b* and *c*. These frequencies allow the genes to be arranged in only one sequence on the chromosomes as follows:

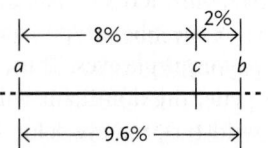

Note that the *a–b* recombination frequency does not exactly equal the sum of the *a–c* and *c–b* recombination frequencies. This is because genes farther apart on a chromosome are more likely to have more than one crossover occur between them. Whereas a single crossover between two genes gives recombinants, a double crossover (two single crossovers occurring in the same meiosis) between two genes gives parentals. You can see this simply by drawing single and double crossovers between two genes on a piece of paper. In our example, double crossovers that occur between *a* and *b* have slightly decreased the recombination frequency between these two genes.

Using this method, Sturtevant created the first linkage map showing the arrangement of six genes on the *Drosophila* X chromosome. (A partial linkage map of a *Drosophila* chromosome is shown in **Figure 13.4.**)

After the time of Morgan, many *Drosophila* genes and those of other eukaryotic organisms widely used for genetic

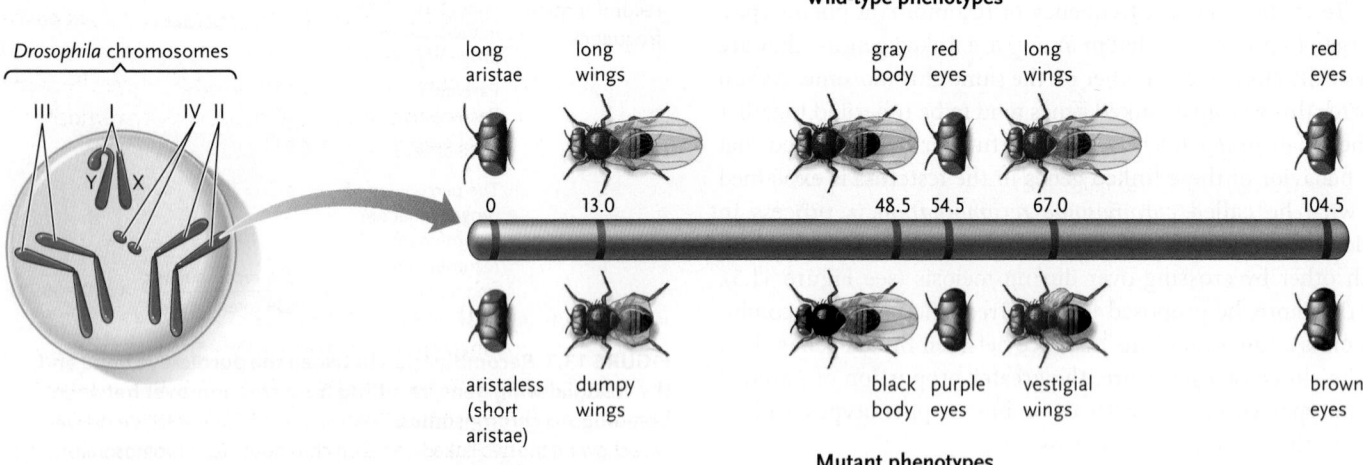

FIGURE 13.4 Relative map locations of several genes on chromosome 2 of *Drosophila,* as determined by recombination frequencies. For each gene, the diagram shows the normal or wild-type phenotype on the top and the mutant phenotype on the bottom. Mutant alleles at two different locations alter wing structure, one producing the dumpy wing and the other the vestigial wing phenotypes; the normal allele at these locations results in normal long-wing structure. Mutant alleles at two different locations also alter eye color.

research, including *Neurospora* (a fungus), yeast, maize (corn), and the mouse, were mapped using the same approach. Recombination frequencies, together with the results of other techniques, have also been used to create linkage maps of the locations of genes in the DNA of prokaryotes such as the human intestinal bacterium *Escherichia coli* (see Chapter 17).

The unit of a linkage map, called a **map unit** (abbreviated mu), is equivalent to a recombination frequency of 1%. The map unit is also called the **centimorgan** (cM) in honor of Morgan's discoveries of linkage and recombination. Map units are not absolute physical distances. Rather, they are *relative* distances due to the fact that the frequency of crossing-over varies to some extent from one position to another on chromosomes.

In recent years, the linkage maps of a number of species have been supplemented by DNA sequencing of whole genomes, which shows the precise physical locations of genes in the chromosomes. Moreover, nowadays it is not common to map genes in genomes using genetic crosses. Instead, typically such a map is created by first obtaining the sequence of a genome and then using computer algorithms to locate likely genes and predict their functions. Experiments are then done to characterize genes of potential interest to an investigator. An advantage of this approach is that it can be used with organisms that have no history of Mendelian genetic analysis in the laboratory.

Widely Separated Linked Genes Assort Independently

Genes can be so widely separated on a chromosome that recombination is likely to occur at some point between them in at least half of the cells undergoing meiosis. When this is the case, no linkage is detected and the genes assort independently. In other words, even though the alleles of the genes are carried on

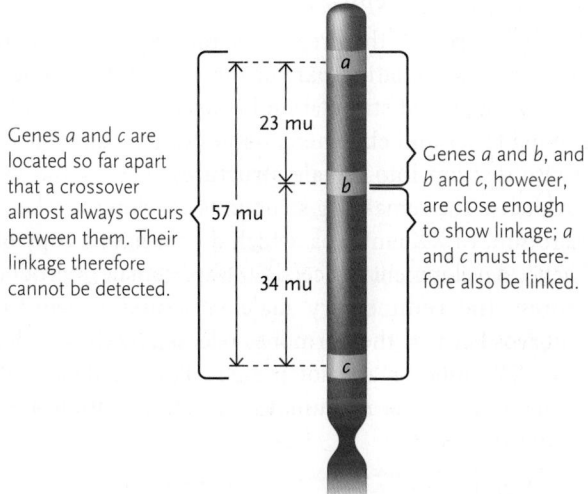

Genes *a* and *c* are located so far apart that a crossover almost always occurs between them. Their linkage therefore cannot be detected.

57 mu

23 mu
34 mu

Genes *a* and *b*, and *b* and *c*, however, are close enough to show linkage; *a* and *c* must therefore also be linked.

FIGURE 13.5 Genes far apart on the same chromosome. Genes *a* and *c* are far apart and will not show linkage, suggesting they are on different chromosomes. However, linkage between such genes can be established by noting their linkage to another gene or genes located between them—in this case, gene *b*.
© Cengage Learning 2017

the same chromosome, the 1:1:1:1 ratio of phenotypes is seen in the offspring of a dihybrid × double mutant testcross. That is, 50% of the progeny are parentals and 50% are recombinants. Linkage between such widely separated genes can still be detected, however, by testing their linkage to one or more genes that lie between them. For example, the genes *a* and *c* in **Figure 13.5** are located so far apart that they assort independently and show no linkage. However, crosses show that *a* and *b* are 23 map units apart (recombination frequency of 23%), and crosses that show *b* and *c* are 34 map units apart. Therefore, *a* and *c* must also be linked and carried on the same chromosome at 23 + 34 = 57 map units apart. We could not see a recombination frequency of 57% in testcross progeny because the maximum frequency of recombinants is 50%, which equals independent assortment.

We now know that some of the genes Mendel studied assort independently even though they are on the same chromosome, for example, the genes for flower color and seed color.

STUDY BREAK 13.1
You want to determine whether genes *a* and *b* are linked. What cross would you use and why? How would this cross tell you if they are linked?

13.2 Sex-Linked Genes

In many organisms, one or more pairs of chromosomes are different in males and females (see Section 11.1). Genes located on these chromosomes, the *sex chromosomes*, are called **sex-linked genes;** they are inherited differently in males and females. (Note that the word *linked* in *sex-linked gene* means that the gene is on a sex chromosome, whereas the use of the term *linked* when considering two or more genes means that the genes are on the same chromosome, not necessarily a sex chromosome.) Chromosomes other than the sex chromosomes are called **autosomes;** genes on these chromosomes have the same patterns of inheritance in both sexes. In humans, chromosomes 1 to 22 are the autosomes.

Females Are XX and Males Are XY in Both Humans and Fruit Flies

In most species with sex chromosomes, females have two copies of a chromosome known as the **X chromosome,** forming a homologous XX pair, whereas males have only one X chromosome. Another chromosome, the **Y chromosome,** occurs in males but not in females, giving males an XY combination. The human chromosome complement, including the sex chromosomes, is shown in the form of a karyotype (see Figure 10.8).

Because an XX female produces only one type of gamete with respect to the sex chromosomes, in this case ones with an X chromosome, she is called the **homogametic sex.** Because the XY male produces two types of gametes with respect to the sex chromosomes, in this case one with an X and one with a Y, the male is called the **heterogametic sex.** When a sperm cell carrying an X chromosome fertilizes an X-bearing egg cell, the new individual develops into an XX female; when a sperm cell carrying a Y chromosome fertilizes an X-bearing egg cell, the combination produces an XY male **(Figure 13.6).** Fertilization is expected to produce females and males with an equal frequency of 1/2. This expectation is closely matched in human and *Drosophila* populations.

Other sex chromosome arrangements occur in nature, as in some insects with XX females and XO males (the O means there is no Y chromosome). In some birds, butterflies, and some reptiles, the situation is reversed. Males are the homogametic sex with a homologous pair of sex chromosomes termed ZZ, and females are the heterogametic sex with ZW sex chromosomes. Researchers have compared the genes on sex chromosomes and have determined that the X and Y chromosomes of mammals are quite different from the Z and W chromosomes of birds. That is, mammalian X and Y chromosome genes typically are on bird autosomes, whereas bird Z and W chromosome genes are on mammalian autosomes. The

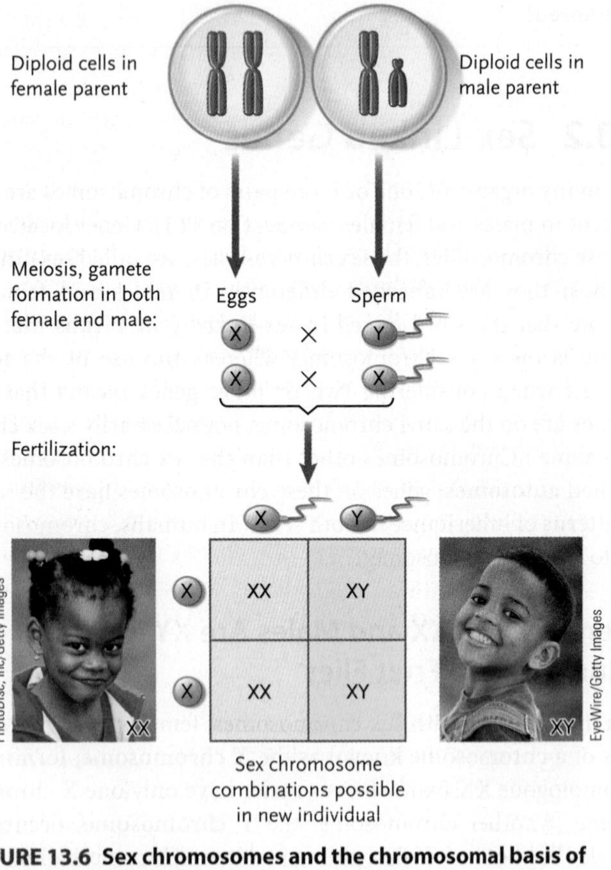

interpretation is that mammalian and bird sex chromosomes have evolved from different autosomal pairs.

In bees and wasps, and certain other arthropods, sex is determined not by sex chromosomes but whether the individual is haploid or diploid. Essentially this means that sex depends on the number of sets of chromosomes. In this system, an individual produced by fusion of an egg and a sperm is diploid and develops into a female, whereas an unfertilized egg, which is haploid, develops into a male.

A number of eukaryotic microorganisms do not have sex chromosomes but have a genic "sex" system specified by simple alleles of a gene. For example, budding yeast, *Saccharomyces cerevisiae* (see *Focus on Research: Model Organisms* in Chapter 10), is a haploid eukaryote with two *mating types* or sexes, designated **a** and α. The mating types are identical in appearance, but matings will only occur between two individuals of opposite type.

Human Sex Determination Depends on the Y Chromosome

The human X chromosome carries about 1,800 protein-coding and noncoding RNA genes. Although some of these genes are associated with sexual traits, such as differing distributions of body fat in males and females, most are concerned with nonsexual traits such as the ability to perceive color, metabolize certain sugars, or form blood clots when tissues are injured. Human sex determination depends on the Y chromosome, which contains the *SRY* gene (for *sex-determining region* of the *Y*) that switches development toward maleness at an early point in embryonic development.

For the first month or so of embryonic development in humans, the rudimentary structures that give rise to reproductive organs and tissues are the same in XX or XY embryos. After 6 to 8 weeks, the *SRY* gene becomes active in XY embryos, producing a protein that regulates the expression of other genes, thereby stimulating part of these structures to develop as testes. As a part of stimulation by hormones secreted in the developing testes and elsewhere, tissues degenerate that would otherwise develop into female structures such as the vagina and oviducts. The remaining structures develop into the penis and scrotum. In XX embryos, which do not have a copy of the *SRY* gene, development proceeds toward female reproductive structures. The rudimentary male structures degenerate in XX embryos because the hormones released by the developing testes in XY embryos are not present. Further details of the *SRY* gene and its role in human sex determination are presented in Chapter 50.

FIGURE 13.6 Sex chromosomes and the chromosomal basis of sex determination in humans. Females have two X chromosomes and produce gametes (eggs), all of which have the X sex chromosome. Males have one X and one Y chromosome and produce gametes, half with an X and half with a Y chromosome. Males transmit their Y chromosome to their sons, but not to their daughters. Males receive their X chromosome only from their mother.
© Cengage Learning 2017

Sex-Linked Genes Show a Different Pattern of Inheritance from Autosomal Genes

The different sets of sex chromosomes in males and females affect the inheritance of the alleles on these chromosomes in a distinct pattern known as *sex linkage.* Two features of the

A. Normal, red wild-type eye color

B. Mutant white eye color caused by recessive allele of a sex-linked gene on the X chromosome

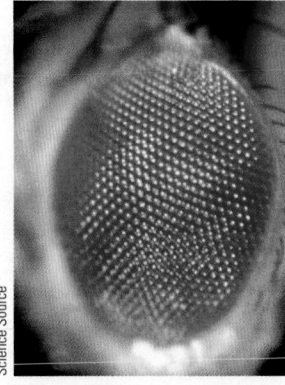
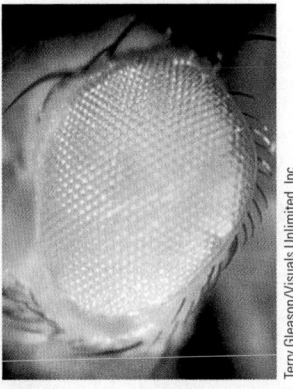

Science Source

Terry Gleason/Visuals Unlimited, Inc.

FIGURE 13.7 Eye color phenotypes in *Drosophila*. (A) Normal, red wild-type eye color. **(B)** Mutant white eye color caused by a recessive allele of a sex-linked gene carried on the X chromosome.

XX–XY arrangement cause sex linkage. One is that alleles carried on the X chromosome occur in two copies in females but in only one copy in males. The second feature is that alleles carried on the Y chromosome are present in males but not females. Those alleles do not correspond to alleles on the X chromosome.

Morgan discovered sex-linked genes and their pattern of sex linkage in 1910. It started when he found a male fly in his stocks with white eyes instead of the normal red eyes **(Figure 13.7)**. To determine how the white-eye allele is inherited, Morgan first crossed a true-breeding female with red eyes with the white-eyed male, and then interbred the F_1 generation to produce the F_2 generation. The F_1 flies of both sexes had red eyes, indicating that the white-eye trait is recessive. The F_2 flies showed a phenotypic ratio of 3 red-eyed : 1 white-eyed. But Morgan observed that, unexpectedly, the phenotypic ratio was not the same in males and females: all F_2 female flies had red eyes, whereas 1/2 the F_2 males had red eyes and the other 1/2 had white eyes.

Morgan hypothesized that the alleles segregating in the cross were of a gene located on the X chromosome and that the unexpected distribution of phenotypes in males and females in the F_2 generation could be accounted for by the inheritance pattern of X and Y chromosomes. A gene on a sex chromosome is a *sex-linked gene*, as introduced earlier. A gene on the X chromosome more precisely is called an **X-linked gene.** The pattern of inheritance of an X-linked gene is called **X-linked inheritance.** To designate X-linked genes and their alleles, we use a symbolism using the upper-case letter X for the X chromosome and superscript letter(s) for the gene. In this case, the white mutant allele is X^w, a white-eyed female is X^wX^w, and a white-eyed male is X^wY. The wild-type allele of the white gene is X^{w^+}; X^w is recessive to this allele.

Figure 13.8A follows the alleles in Morgan's cross of a true-breeding red-eyed female ($X^{w^+}X^{w^+}$) with the white-eyed male (X^wY) through to the F_2 generation. The transmission of the white-eye allele shown in this cross—from a male parent to a female offspring ("child") to a male "grandchild" is called **crisscross inheritance.**

Morgan also performed a **reciprocal cross,** meaning that he switched the phenotypes of the parents **(Figure 13.8B).** The reciprocal cross here was a true-breeding white-eyed female (X^wX^w) with a red-eyed male ($X^{w^+}Y$).

The results of the reciprocal crosses differed markedly in both the F_1 and F_2 generations. Morgan's experiments had shown a distinctive pattern in the phenotypic ratios for reciprocal crosses in which the gene involved is on the X chromosome. A key indicator of X-linked inheritance of a recessive trait is when all male offspring of a cross between a true-breeding mutant female and a wild-type male have the mutant phenotype. As we have seen, this occurs because a male receives his X chromosome from his female parent.

X-Linked Gene Inheritance in Humans Is Studied by Pedigree Analysis

The study of human genetics is complicated because controlled genetic matings of humans are not possible for ethical reasons. Instead, the inheritance patterns of human genetic traits are identified usually by examining the way a trait of interest occurs in family trees of individuals who exhibit the trait. Typically, the trees include several generations of phenotypes, and the associated genotypes must then be determined by interpretation. The family tree is called a **pedigree,** and the study of the pedigree is called **pedigree analysis.** Pedigree analysis has its own set of symbols: females are designated by a circle and males by a square; a solid circle or square indicates the presence of the trait; a horizontal line between a female and a male indicates a marriage or pairing; and a line down from that line leads to their offspring. For ease of referring to people in a pedigree, generations are numbered with Roman numerals, and individuals are numbered with Arabic numerals. You will see one pedigree in this section, and others will be discussed later in the chapter. Because the number of offspring in a generation in a pedigree is relatively small, typical Mendelian ratios that you might predict are not seen exactly, although they may be close. However, our confidence in determining a particular mode of inheritance for a human genetic trait from pedigree analysis comes from examining large numbers of pedigrees in which the same trait is found.

In humans, as in fruit flies, X-linked recessive traits appear more frequently among males than females because males need to receive only one copy of the allele on the X chromosome inherited from their mothers to develop the trait. Females must receive two copies of the recessive allele, one from each parent, to develop the trait. Two examples of human X-linked traits are red–green color blindness, a recessive trait in which the affected individual is unable to distinguish between the colors red and

 **FIGURE 13.8** | **Experimental Research**

Evidence for Sex-Linked Genes

Question: How is the white-eye gene of *Drosophila* inherited?

Experiment: Morgan crossed a white-eyed male *Drosophila* with a true-breeding female with red eyes and then interbred the F_1 flies to produce the F_2 generation. He also performed the reciprocal cross in which the phenotypes were switched in the parental flies—true-breeding white-eyed female $\times$ red-eyed male.

A. True-breeding red-eyed female $\times$ white-eyed male

P generation

Red eyes (wild type)   White eyes

F₁ generation

Red eyes   Red eyes

All F_1 flies have red eyes, indicating that the white-eye trait is recessive. The F_1 females inherit one X from each parent; their genotype is $X^{w^+}X^w$, and their phenotype is red eyes because the X^{w^+} allele is dominant. The F_1 males inherit their X chromosome from their mothers; their genotype is $X^{w^+}Y$, and their phenotype is red eyes.

F₂ generation

Sperm

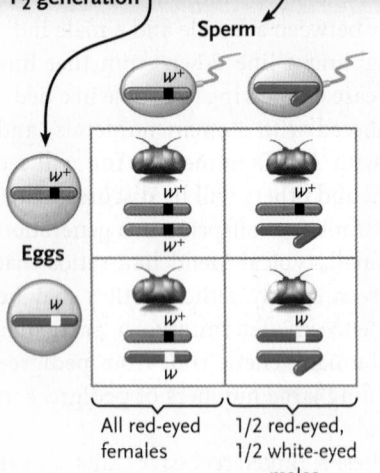

Eggs

All red-eyed females   1/2 red-eyed, 1/2 white-eyed males

3/4 red eyes : 1/4 white eyes

The F_2 females receive an X^{w^+} allele from the F_1 father and either an X^{w^+} or X^w allele from the F_1 mother; both these genotypes result in red eyes. The F_2 males inherit their one X chromosome from the F_1 mother whose genotype is $X^{w^+}X^w$. Therefore, F_2 males are half $X^{w^+}Y$ (red eyes) and half X^wY (white eyes). Females and males together show a phenotypic ratio of 3 red : 1 white.

B. White-eyed female $\times$ red-eyed male

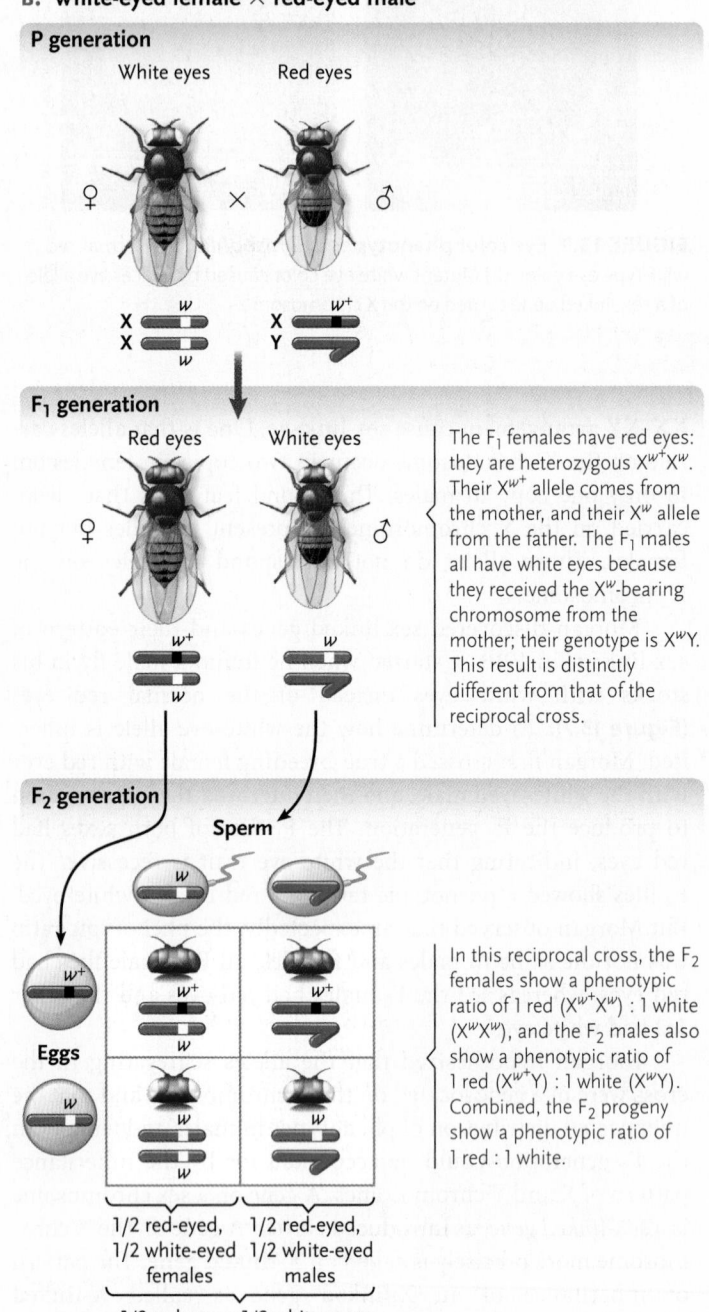

P generation

White eyes   Red eyes

F₁ generation

Red eyes   White eyes

The F_1 females have red eyes: they are heterozygous $X^{w^+}X^w$. Their X^{w^+} allele comes from the mother, and their X^w allele from the father. The F_1 males all have white eyes because they received the X^w-bearing chromosome from the mother; their genotype is X^wY. This result is distinctly different from that of the reciprocal cross.

F₂ generation

Sperm

Eggs

1/2 red-eyed, 1/2 white-eyed females   1/2 red-eyed, 1/2 white-eyed males

1/2 red eyes : 1/2 white eyes

In this reciprocal cross, the F_2 females show a phenotypic ratio of 1 red ($X^{w^+}X^w$) : 1 white (X^wX^w), and the F_2 males also show a phenotypic ratio of 1 red ($X^{w^+}Y$) : 1 white (X^wY). Combined, the F_2 progeny show a phenotypic ratio of 1 red : 1 white.

Results: Differences were seen in both the F_1 and F_2 generations for the red ♀ $\times$ white ♂ and white ♀ $\times$ red ♂ reciprocal crosses.

Conclusion: The segregation pattern for the white-eye trait showed that the white-eye gene is a sex-linked gene located on the X chromosome.

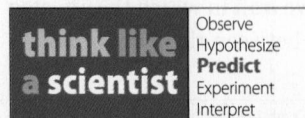

think like a scientist

Observe
Hypothesize
Predict
Experiment
Interpret

In humans, red-green color blindness *(c)* is recessive and X linked. The normal vision allele is c^+. A woman with normal vision whose father was color-blind marries a man with normal vision whose father was also color-blind. What possible children could they have with respect to color blindness, and in what proportion would they occur?

green because of a defect in light-sensing cells in the retina, and hemophilia, a recessive trait in which affected individuals have a defect in blood clotting.

Hemophiliacs—people with hemophilia—are "bleeders"; that is, if they are injured, they bleed much more than usual and are susceptible to severe bruising because a protein required for forming blood clots is not produced in functional form. Males are bleeders if they receive an X chromosome that carries the recessive allele, X^h. The disease also develops in females with the recessive allele on both of their X chromosomes, genotype X^hX^h—a rare combination. Although affected persons, with luck and good care, can reach maturity, they must be careful to avoid serious injury. The disease, which affects about 1 in 7,000 males, can be treated by injection of the required clotting molecules.

Hemophilia has had effects reaching far beyond individuals who inherit the disease. The most famous cases occurred in the royal families of Europe descended from Queen Victoria of England, the pedigree of whom is shown in **Figure 13.9.** Hemophilia was not recorded in Queen Victoria's ancestors, so the recessive allele for the trait probably appeared as a spontaneous mutation in the queen or one of her parents. Queen Victoria was heterozygous for the recessive hemophilia allele ($X^{h+}X^h$). That is, she was a **carrier,** meaning that she carried the mutant allele and could pass it on to her offspring but she did not have symptoms of the disease. In a pedigree, a carrier, where identi-

fied, may be indicated by a male or female symbol with a central dot.

Note in Queen Victoria's pedigree in Figure 13.9 that the trait alternates from generation to generation in males because a father does not pass his X chromosome to his sons. The X chromosome received by a male always comes from his mother.

At one time, 18 of Queen Victoria's 69 descendants were affected males or female carriers. Because so many sons of European royalty were affected, the trait influenced the course of history. In Russia, Crown Prince Alexis (highlighted in Figure 13.9) was one of Victoria's hemophiliac descendants. The hypnotic monk Rasputin manipulated the Czar Nicholas II and Czarina Alexandra to his advantage by convincing them that only he could control their son's bleeding. The situation helped trigger the Russian Revolution of 1917, which ended the Russian monarchy and led to the establishment of a Communist government in the former Soviet Union, a significant event in twentieth century history.

Hemophilia affected only sons in the royal lines but could have affected daughters if a hemophiliac son had married a carrier female. Because the disease is rare in the human population as a whole, the chance of such a mating is so low that only a few unions of this type have been recorded.

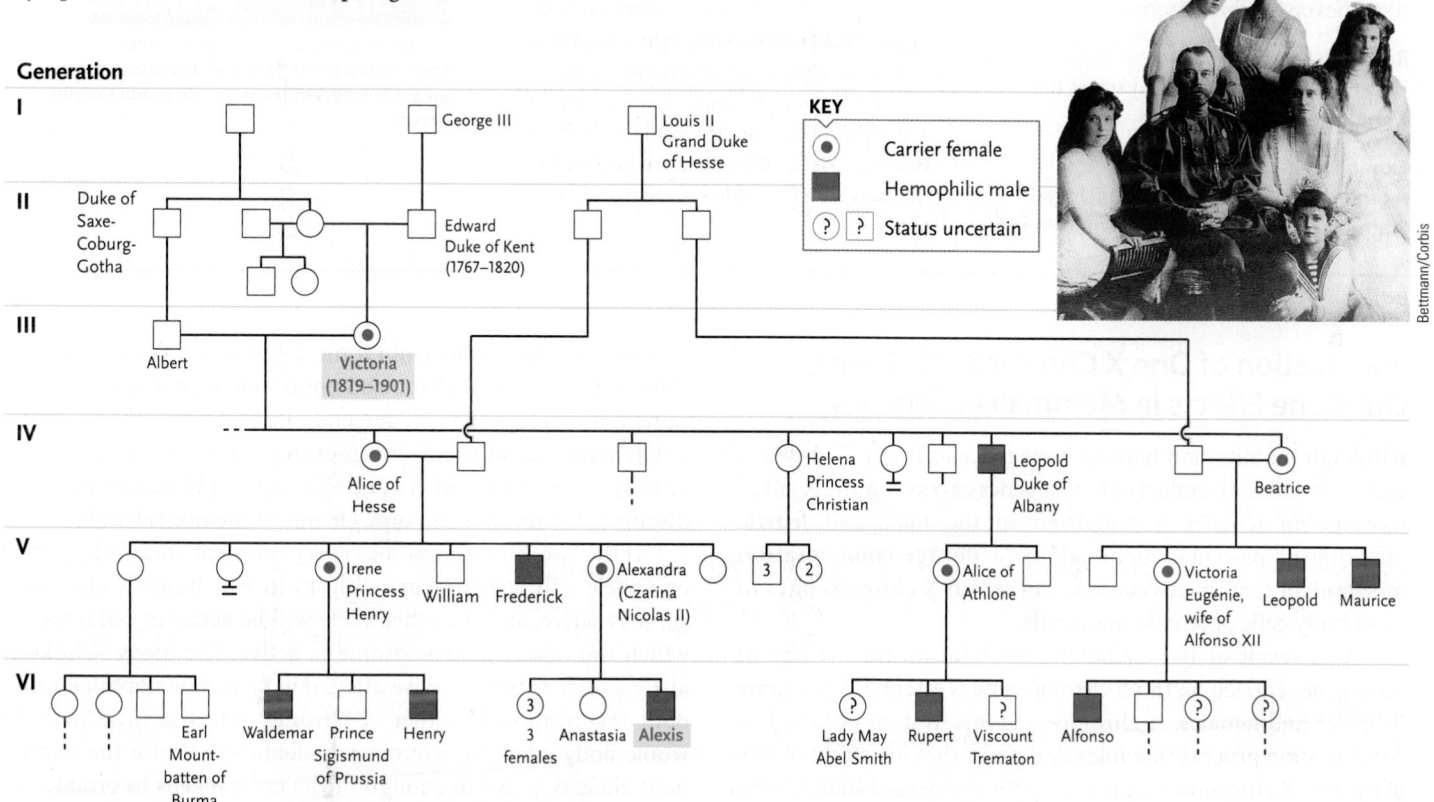

FIGURE 13.9 Inheritance of hemophilia in descendants of Queen Victoria of England. The photograph shows the Russian royal family in which the son, Crown Prince Alexis, had hemophilia. His mother was a carrier of the mutated gene. As is customary practice in many pedigrees, symbols for marriage partners who did not have the trait are omitted from the pedigree drawing.

© Cengage Learning 2017

X-Chromosome Inactivation in Mammals: What is the process of inactivation?

The X chromosome has a region called the X inactivation center (XIC). Two (or more) XICs must be present in a cell for X inactivation to occur and this is the case, of course, in females but not in males. Within XIC is a gene called *Xist*, for *X inactivation-specific transcripts*. *Xist* is expressed from the X chromosome that will be inactivated (Xi) but not from the X chromosome that will remain active. *Xist* is transcribed to produce an RNA that is not translated to produce a protein. As such, *Xist* is an example of a **noncoding RNA gene,** meaning that it is a gene that is expressed but does not encode a protein. (Some examples of noncoding RNA genes are discussed later in the book; see, for example, Chapter 16.) The *Xist*-encoded RNA is 17,000 bases (17 kb, where kb = 1 kilobase = 1,000 bases) long. Due to its length, it is considered to be a **long noncoding RNA,** or lncRNA for short. The Xist lncRNA spreads over the Xi—the chromosome from which it is transcribed—coating it and then recruiting protein complexes that silence the genes on the chromosome.

Research Question

How does Xist lncRNA spread across the X chromosome?

Experiments

A group of researchers at various United States universities and institutes developed a biochemical method that enabled them to map Xist lncRNA locations on the X chromosome with high resolution. This made it possible to study how Xist lncRNA localizes to the X chromosome initially and how it ends up coating the entire chromosome.

Results

First, the researchers examined Xist lncRNA localization in differentiated mouse cells in culture in which X chromosome inactivation was being maintained from cell generation to cell generation. Their results showed that Xist lncRNA is localized broadly across the entire X chromosome, appearing like a cloud over the chromosome.

Next, the researchers examined Xist lncRNA localization during the initiation of X-chromosome inactivation in mouse cells in culture. At 1 hour after initiation, Xist lncRNA appears as a strong focal point at the *Xist* locus, and grows to the characteristic cloud appearance as time progresses.

The scientists then demonstrated that Xist lncRNA localizes early in initiation to a number of sites distant from the *Xist* locus and spaced over the entire X chromosome. Xist lncRNA then spreads from those sites. However, Xist lncRNA does not localize to the early sites by binding to specific DNA sequences.

Next, the researchers investigated how Xist lncRNA localizes to the distant early sites during the initiation of X-chromosome inactivation by comparing available data on chromosome conformation with their Xist lncRNA results. Their results support a model in which the localization of Xist lncRNA to early sites occurs because those sites are physically close to the Xist lncRNA in the three-dimensional conformation of the chromosome. Xist lncRNA then spreads from those sites. Silencing of genes on the X chromosome then occurs through recruitment of proteins that lead to compaction of the chromosome.

Conclusion

The researchers concluded that their "findings suggest a model in which Xist coats the X chromosome by searching in three dimensions, modifying chromosome structure, and spreading to newly accessible locations."

think like a scientist

Suppose, using genetic engineering techniques, a *Xist* gene was expressed from an autosome. What would you predict might occur?

Source: J. M. Engreitz et al. 2013. The Xist lncRNA exploits three-dimensional genome architecture to spread across the X chromosome. *Science* 341:1237973.

Inactivation of One X Chromosome Evens Out Gene Effects in Mammalian Females

Although mammalian females have twice as many X chromosomes as males, the effects of most genes carried on the X chromosome in females is equalized in the male and female offspring of placental mammals by a **dosage compensation mechanism** that inactivates one of the two X chromosomes in most body cells of female mammals.

As a result of the equalizing mechanism, the activity of most genes carried on the X chromosome is essentially the same in males and females. X-chromosome inactivation occurs by a condensation process that folds and packs the chromatin of one of the two X chromosomes into a tightly condensed state. Under the light microscope, the inactive, condensed X chromosome can be seen at one side of the nucleus in cells of females as a mass of heterochromatin (see Section 10.1) called the **Barr body.**

X-chromosome inactivation occurs during early embryonic development. Which of the two X chromosomes becomes inactive in a particular embryonic cell line is a random event. But once one of the X chromosomes is inactivated in a cell, that same X is inactivated in all descendants of the cell. Thus, within one female, one of the X chromosomes is active in particular cells and inactive in others and vice versa. (*Molecular Insights* discusses the mechanism of X-chromosome inactivation.)

If the two X chromosomes carry different alleles of a gene, one allele will be active in cell lines in which one X chromosome is active, and the other allele will be active in cell lines in which the other X chromosome is active. For many X-linked alleles, such as the recessive allele that causes hemophilia, random inactivation of either X chromosome has little overall whole-body effect in heterozygous females because the dominant allele is active in enough of the critical cells to produce a normal phenotype. However, for some genes, the inactivation of either X chromosome in heterozygotes produces recognizably different effects in distinct regions of the body.

For example, the orange and black patches of fur in calico cats result from inactivation of one of the two X chromosomes in regions of the skin of heterozygous females **(Figure 13.10)**. Males, which get only one of the two alleles, normally have either black or orange fur.

An X-linked trait in humans has a similar, but less visible, phenotype. Called *anhidrotic ectodermal dysplasia*, the trait is characterized by the absence of sweat glands. Females heterozygous for the mutation that causes the trait may have a patchy distribution of skin areas with and without the glands.

The discovery of genetic linkage, recombination, and sex-linked genes led to the elaboration and expansion of Mendel's principles of inheritance. Next, we examine what happens when patterns of inheritance are modified by changes in the chromosomes.

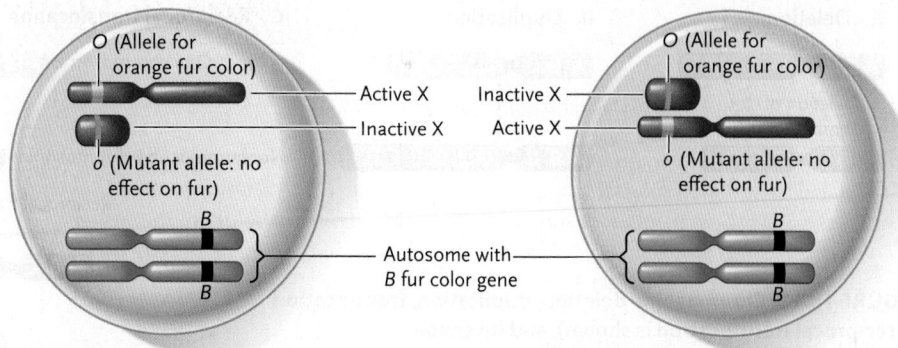

Orange fur: *O* allele is active, masking phenotypic expression of the *B* gene (an example of epistasis; see Section 12.2).

Black fur: *O* allele is inactive because the X chromosome it is on is inactivated; the mutant *o* allele on the active X chromosome does not mask the phenotypic expression of the *B* gene.

White patches result from interactions with a different, autosomal gene that blocks pigment deposition in the fur completely.

STUDY BREAK 13.2

You have a true-breeding strain of miniature-winged fruit flies, where this wing trait is recessive to the normal long wings. How would you show whether the miniature wing trait is sex-linked or autosomal?

THINK OUTSIDE THE BOOK

While *Drosophila* has X and Y chromosomes with XX flies being female and XY males being Y, this species does not have an *SRY* gene. Individually or collaboratively, explore the Internet or research papers to determine how sex determination occurs in *Drosophila*.

FIGURE 13.10 A female cat with the calico color pattern in which patches of orange and black fur are produced by random inactivation of one of the two X chromosomes. Two genes control the black and orange colors: the *O* allele on the X chromosome is for orange fur color, and the mutated *o* allele has no effect on color. The *B* gene on an autosome is for black fur color. A calico cat has the genotype *Oo BB* or *Oo Bb*; the former genotype is illustrated in the figure. In tortoiseshell cats, the same orange-black patching occurs as in calico cats but the gene for the white patching is not active.

© Cengage Learning 2017

13.3 Chromosomal Mutations That Affect Inheritance

Chromosomal mutations are variations from the normal condition in chromosome structure or chromosome number. Changes in chromosome structure occur when the DNA breaks, which can be generated by agents such as radiation or certain chemicals or by enzymes encoded in some infecting viruses. The broken chromosome fragments may be lost, or they may reattach to the same or different chromosomes. Chromosomal structure changes may have genetic consequences if alleles are eliminated, mixed in new combinations, duplicated, or placed in new locations by the alterations in cell lines that lead to the formation of gametes.

Genetic changes may also occur through changes in chromosome number, including addition or loss of one or more chromosomes or even entire sets of chromosomes. Both forms of chromosomal mutation, changes in chromosome structure and changes in chromosome number, can be a source of disease and disability.

Deletions, Duplications, Translocations, and Inversions Are the Most Common Chromosomal Mutations Affecting Chromosome Structure

Chromosomal mutations after breakages occur in four major forms **(Figure 13.11)**:

- A **deletion** occurs if a broken segment is lost from a chromosome.
- A **duplication** occurs if a segment is transferred from one chromosome and inserted into its homolog. In the receiving homolog, the alleles in the inserted fragment are added to the ones already there.
- A **translocation** occurs if a broken segment is attached to a different, nonhomologous chromosome.
- An **inversion** occurs if a broken segment reattaches to the same chromosome from which it was lost, but in reversed orientation, so that the order of genes is reversed.

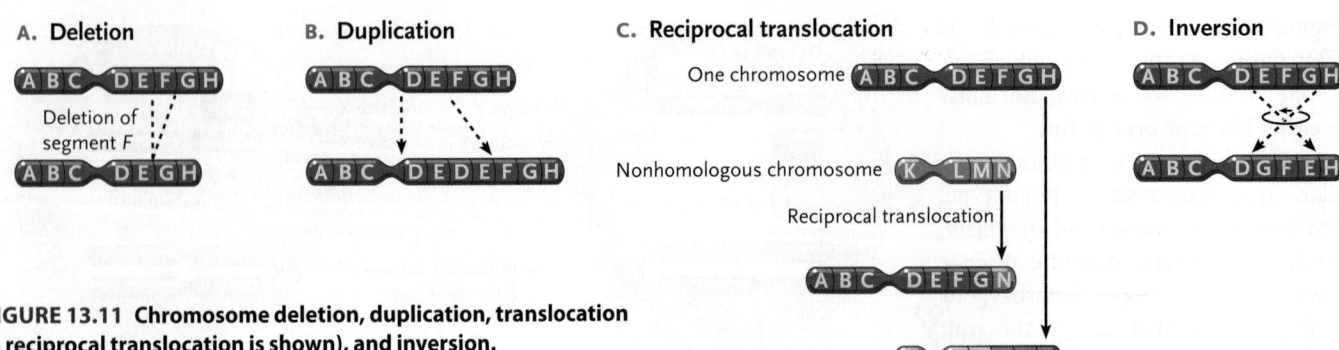

A. Deletion

A B C D E F G H

Deletion of segment F

A B C D E G H

B. Duplication

A B C D E F G H

A B C D E D E F G H

C. Reciprocal translocation

One chromosome A B C D E F G H

Nonhomologous chromosome K L M N

Reciprocal translocation

A B C D E F G N

K L M F G H

D. Inversion

A B C D E F G H

A B C D G F E H

FIGURE 13.11 Chromosome deletion, duplication, translocation (a reciprocal translocation is shown), and inversion.
© Cengage Learning 2017

To be inherited, chromosomal alterations must occur or be included in cells of the germ line leading to development of eggs or sperm.

DELETIONS AND DUPLICATIONS A deletion **(Figure 13.11A)** may cause severe problems if the missing segment contains genes that are essential for normal development or cellular functions. For example, an individual heterozygous for a deletion of part of human chromosome 5 typically has severe mental retardation, a variety of physical abnormalities, and a malformed larynx. The cries of an affected infant sound more like a meow than a human cry, hence the name of the disorder, *cri-du-chat* (meaning "cry of the cat").

A duplication **(Figure 13.11B)** may have effects that vary from harmful to beneficial, depending on the genes and alleles contained in the duplicated region. Although most duplications are likely to be detrimental, some have been important sources of evolutionary change. That is, because there are duplicate genes, one copy can mutate into new forms without seriously affecting the basic functions of the organism. For example, mammals have genes that encode several types of hemoglobin that are not present in vertebrates, such as sharks, which evolved earlier. The additional hemoglobin genes of mammals arose through evolutionary time through duplications, followed by mutations in the duplicates that created new and beneficial forms of hemoglobin as further evolution took place. Duplications sometimes arise if crossing-over occurs unequally during meiosis, so that a segment is deleted from one chromosome of a homologous pair and inserted in the other (see Section 19.4).

TRANSLOCATIONS AND INVERSIONS In a translocation, a segment breaks from one chromosome and attaches to another, nonhomologous chromosome. In many cases, a translocation is reciprocal, meaning that two nonhomologous chromosomes exchange segments **(Figure 13.11C)**. Reciprocal translocations resemble genetic recombination, except that the two chromosomes involved in the exchange do not contain the same genes.

In an inversion, a chromosome segment breaks and then reattaches to the same chromosome, but in reverse order **(Figure 13.11D)**. Inversions have essentially the same effects as

translocations—genes may be broken internally by the inversion, with loss of function, or they may be transferred intact to a new location within the same chromosome, producing effects that range from beneficial to harmful.

Many cancers have chromosomal mutations, and the most common type of chromosomal mutation involved is a translocation. For example, 90% of patients with chronic myelogenous leukemia (CML) have a chromosomal mutation called the *Philadelphia chromosome*. CML is a type of cancer of the blood involving the uncontrolled division of stem cells for white blood cells. The Philadelphia chromosome arises from a reciprocal translocation event involving chromosomes 9 and 22 **(Figure 13.12)**. The movement of the segment of chromosome 9 to chromosome 22 fuses together on the resulting Philadelphia chromosome, the *ABL* gene from 9 with the *BCR* gene on 22. The *ABL* gene is one of many genes that control cell growth and division. (Cell division control is described in Chapter 10.) Its product is a tyrosine kinase, an enzyme that adds phosphate to tyrosine amino acids in target proteins (see Section 9.3). In its new location on the Philadelphia chromosome, the *ABL* gene becomes much more active because of its fusion with the *BCR* gene and much more than normal of its tyrosine kinase product is made. As a result of its overactivity, normal cell cycle control breaks down and the cells are stimulated to growth and divide uncontrollably, becoming cancer cells. The drug Gleevec is used to treat CML patients. It works by inhibiting the tyrosine kinase enzyme so that the body stops, or at least reduces, the production of too many white blood cells.

CHROMOSOME STRUCTURE MUTATIONS AND GENOME EVOLUTION Chromosome structure mutations have played a role in the evolution of genomes. For instance, many new genes and gene families evolved as a result of duplication. Also, inversions and translocations have been important factors in the evolution of the genomes of plants and some animals, including insects and primates. As an example, nine of the chromosome pairs of humans show evidence of translocations and inversions that differ between humans and chimpanzees, and therefore must have occurred after the ancestral lineages leading to chimpanzees and humans split. (Genome evolution is discussed further in Chapter 19.)

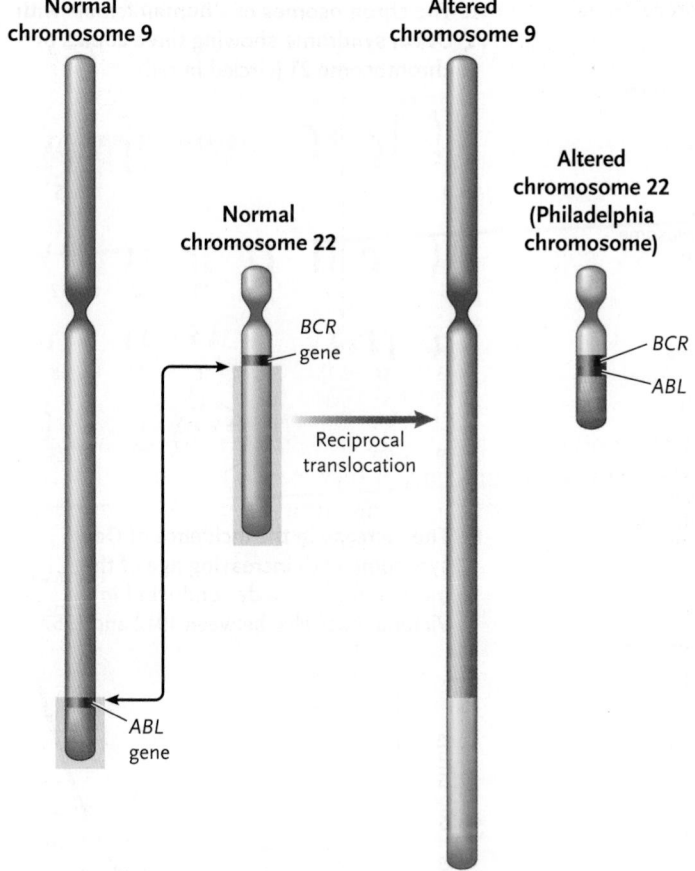

Normal chromosome 9

Normal chromosome 22

BCR gene

Reciprocal translocation

ABL gene

Altered chromosome 9

Altered chromosome 22 (Philadelphia chromosome)

BCR

ABL

FIGURE 13.12 Translocation found in many patients with a form of blood cancer called *chronic myelogenous leukemia* (CML). A reciprocal translocation involving chromosomes 9 and 22 produces a short chromosome named the Philadelphia. On this chromosome, the chromosome 9 *ABL* gene has become fused to the chromosome 22 *BCR*. The resulting overactivity of the *ABL* gene, which normally helps control cell division, causes the cell to convert to a cancer cell.

© Cengage Learning 2017

Some Chromosomal Mutations Involve Changes in the Number of Entire Chromosomes

At times, whole, single chromosomes are lost or gained from cells entering or undergoing meiosis, resulting in a change of chromosome number. Most often, these changes occur through **nondisjunction**—the failure of homologous pairs to separate during the first meiotic division or of chromatids to separate during the second meiotic division **(Figure 13.13)**. As a result, gametes are produced that lack one or more chromosomes or contain extra copies of one or more chromosomes. Fertilization by these gametes produces an individual with extra or missing chromosomes. Such individuals are called **aneuploids**, whereas individuals with a normal set of chromosomes are called **euploids.**

Changes in chromosome number can also occur through duplication or loss of entire sets, meaning individuals may receive fewer or more than the normal number of the entire haploid complement of chromosomes. Individuals with one set of chromosomes instead of the normal two are **monoploids.** Individuals with more than the two sets of chromosomes are called **polyploids.** *Triploids* have three copies of each chromosome instead of two; *tetraploids* have four copies of each chromosome; *hexaploids* have six copies of each chromosome. Multiples higher than hexaploids also occur.

ANEUPLOIDS The effects of addition or loss of whole chromosomes vary depending on the chromosome and the species. In animals, aneuploidy of autosomes usually produces debilitating or lethal developmental abnormalities. In humans, addition or loss of an autosomal chromosome causes embryos to develop so abnormally that generally they are aborted naturally. For reasons that are not understood, aneuploidy is as much as 10 times more frequent in humans than in other mammals. Of human fetuses that have been miscarried and examined, about 70% are aneuploids.

In some cases, autosomal aneuploids survive, for example humans who receive an extra copy of chromosome 21—the smallest human chromosome **(Figure 13.14A)**. Many of these individuals survive until young adulthood. The condition produced by the extra chromosome, called *Down syndrome* or *trisomy 21* (for "three chromosome 21s"), is characterized by short stature and moderate to severe mental retardation. About 40% of individuals with Down syndrome have heart defects, and skeletal development is slower than normal. Most do not mature sexually and remain infertile. However, with attentive care and special training, individuals with Down syndrome can participate with reasonable success in many activities.

Down syndrome arises from nondisjunction of chromosome 21 during the meiotic divisions, primarily in women (about 5% of nondisjunctions that lead to Down syndrome occur in men). The nondisjunction occurs more frequently as women age, increasing the chance that a child may be born with the syndrome **(Figure 13.14B)**. In the United States, 1 in every 1,000 children is born with Down syndrome, making it one of the most common serious human genetic disorders.

Aneuploidy of sex chromosomes can also arise by nondisjunction during meiosis **(Figure 13.15** and **Table 13.1)**. Unlike autosomal aneuploidy, which usually has drastic effects on survival, altered numbers of X and Y chromosomes are often tolerated in humans, producing individuals who progress through embryonic development and grow to adulthood. This is because, in the case of multiple X chromosomes, the X-chromosome inactivation mechanism converts all but one of the X chromosomes to a Barr body, so the dosage of active X-chromosome genes is the same as in normal XX females and XY males. Thus, XO (Turner syndrome) females have no Barr bodies, XXY (Klinefelter syndrome) males have one Barr body, and XXX (triple-X syndrome) females have two Barr bodies (see Table 13.1). However, X chromosomes are not inactivated until about 15 to 16 days after fertilization. Expression of the extra X chromosome genes

A. Nondisjunction during first meiotic division

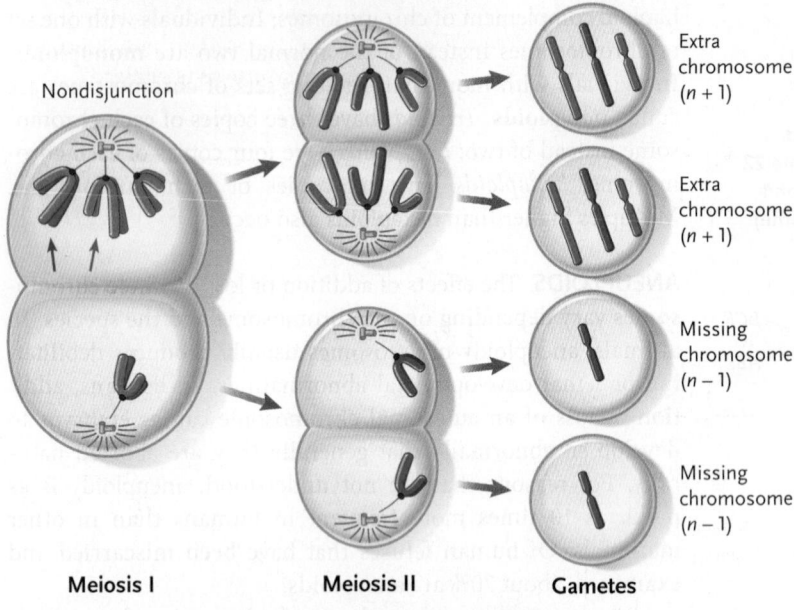

Nondisjunction during the first meiotic division causes both chromosomes of one pair to be delivered to the same pole of the spindle. The nondisjunction produces two gametes with an extra chromosome and two with a missing chromosome.

B. Nondisjunction during second meiotic division

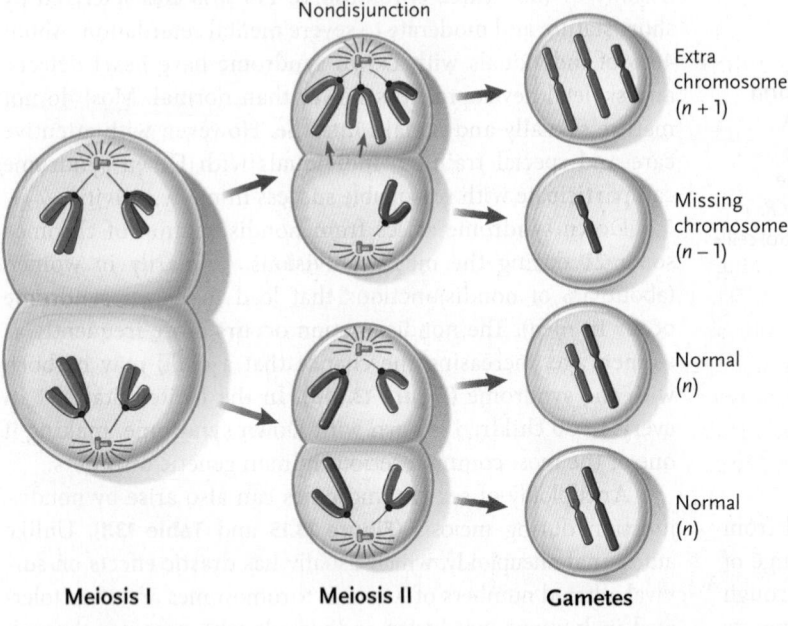

Nondisjunction during the second meiotic division produces two normal gametes, one gamete with an extra chromosome and one gamete with a missing chromosome.

FIGURE 13.13 Nondisjunction during (A) the first meiotic division and (B) the second meiotic division.
© Cengage Learning 2017

A. The chromosomes of a human female with Down syndrome showing three copies of chromosome 21 (circled in red)

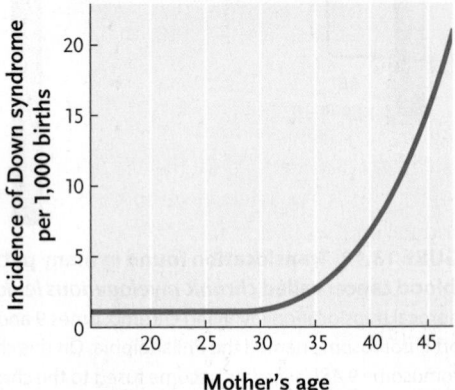

B. The increase in the incidence of Down syndrome with increasing age of the mother, from a study conducted in Victoria, Australia, between 1942 and 1957

C. Person with Down syndrome

FIGURE 13.14 Down syndrome.
© Cengage Learning 2017

early in development results in any deleterious effects associated with a particular sex chromosome aneuploidy.

Because sexual development in humans is pushed toward male or female reproductive organs primarily by the presence or absence of the *SRY* gene on the Y chromosome, people with a Y chromosome are externally malelike, no matter how many X chromosomes are present. If no Y chromosome is present, X chromosomes in various numbers give rise to femalelike individuals. (Table 13.1 lists the effects of some alterations in sex chromosome number.) Similar abnormal combinations of

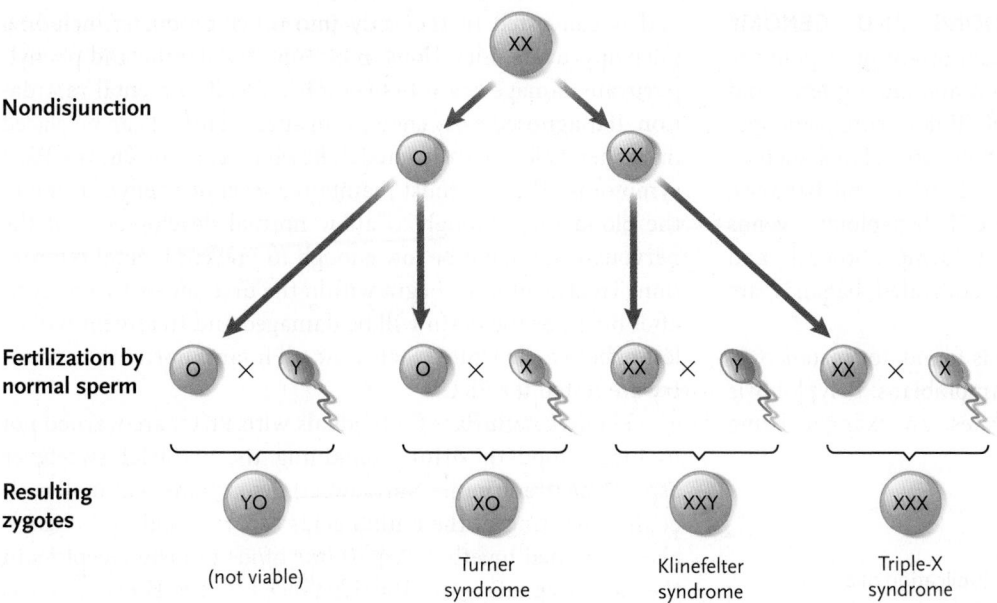

Nondisjunction

Fertilization by normal sperm

Resulting zygotes

YO (not viable)

XO Turner syndrome

XXY Klinefelter syndrome

XXX Triple-X syndrome

sex chromosomes also occur in other animals with varying effects on viability.

MONOPLOIDS A monoploid has only one set of chromosomes instead of the normal two. Monoploidy is lethal in most animal species, but is tolerated more in plants. Certain animal species produce monoploid organisms as a normal part of their life cycle. For instance, as mentioned earlier, some male wasps, ants, and bees are monoploid individuals that have developed from unfertilized eggs.

POLYPLOIDS Polyploidy often originates from failure of the spindle to function normally during mitosis in cell lines leading to germ-line cells. In these divisions, the spindle fails to separate the duplicated chromosomes, which are incorporated into a single nucleus with twice the usual number of chromosomes. Eventually, meiosis takes place and produces gametes with two copies of each chromosome instead of one. Fusion of one such gamete with a normal haploid gamete produces a triploid, and

fusion of two such gametes produces a tetraploid. Polyploidy may also occur when a single egg is fertilized by more than one sperm. For example, fertilization by two sperm produces a triploid with three sets of chromosomes. Such an event is rare in animals because of mechanisms that operate during fertilization of an egg with a sperm to prevent the subsequent fusion of other sperm with that egg (discussed more in Chapter 49).

The effects of polyploidy vary widely between plants and animals. In plants, polyploids are often hardier and more successful in growth and reproduction than the diploid plants from which they were derived. As a result, polyploidy is common in plants.

By contrast, among animals, polyploidy is relatively uncommon because it usually has lethal effects during embryonic development. For example, in humans, all but about 1% of polyploids die before birth, and the few who are born usually die within a month. The lethality is probably caused by disturbance of animal developmental pathways, which are typically much more complex than those of plants.

TABLE 13.1	**Effects of Unusual Combinations of Sex Chromosomes in Humans**		
Combination of Sex Chromosomes	Approximate Frequency	Barr Bodies	Effects
XO	1 in 5,000 births	0	Turner syndrome: females with underdeveloped ovaries; sterile; intelligence and external genitalia are normal; typically, individuals are short in stature with underdeveloped breasts
XXY	1 in 2,000 births	1	Klinefelter syndrome: male external genitalia with very small and underdeveloped testes; sterile; intelligence usually normal; sparse body hair and some development of the breasts; similar characteristics in XXXY and XXXXY individuals
XYY	1 in 1,000 births	0	XYY syndrome: apparently normal males but often taller than average
XXX	1 in 1,000 births	2	Triple-X syndrome: apparently normal female with normal or slightly retarded mental function

CHROMOSOME NUMBER MUTATIONS AND GENOME EVOLUTION Polyploidy has been an important process in the evolution of plant genomes and is common among ferns and flowering plants. That is, about half of all flowering plant species are polyploids, including important crop plants such as wheat and other cereals, cotton, strawberries, and bananas. Commercial bread wheat, for instance, is hexaploid; genus *Brassica* vegetables such as cabbage, cauliflower, turnips, broccoli, and Brussels sprouts are tetraploid; and cultivated bananas are triploid.

In vertebrate animals, polyploidy is found, for instance, in some fish species, and is common in amphibians. Polyploidy is relatively common among invertebrates, an example being flatworms.

STUDY BREAK 13.3

What mechanisms are responsible for: (a) duplication of a chromosome segment; (b) generation of a Down syndrome individual; (c) a chromosome translocation; and (d) polyploidy?

THINK OUTSIDE THE BOOK

Diagram the various ways a normal XX female and normal XY male could produce an XXY zygote.

13.4 Human Genetic Traits, Pedigree Analysis, and Genetic Counseling

We now turn to a description of the effects of altered alleles on human health and development. You have learned already about a number of human genetic traits and conditions caused by mutant alleles or chromosomal alterations. All these traits are of interest as examples of patterns of inheritance that amplify and extend Mendel's basic principles. Those with harmful effects are also important because of their impact on human life and society.

In Autosomal Recessive Inheritance, Heterozygotes Are Carriers and Homozygous Recessives Are Affected by the Trait

Phenylketonuria, sickle-cell anemia, and cystic fibrosis are examples of human disorders caused by recessive alleles on autosomes. Many other human genetic traits follow a similar pattern of inheritance. These traits are passed on according to the pattern known as **autosomal recessive inheritance,** in which individuals who are homozygous for the dominant allele are free of symptoms and are not carriers. Heterozygotes are usually symptom-free but are carriers. People who are homozygous for the recessive allele show the trait.

Phenylketonuria (PKU) appears in about 1 of every 15,000 births. Affected individuals cannot produce an enzyme that converts the amino acid phenylalanine to another amino acid, tyrosine. As a result, phenylalanine builds up in the blood and is converted in the body into other products, including phenylpyruvate. Elevations in both phenylalanine and phenylpyruvate damage brain tissue and can lead to mental retardation. If diagnosed early enough, an affected infant can be placed on a phenylalanine-restricted diet, which can prevent the PKU symptoms. The diet must maintain a level of phenylalanine in the blood high enough to allow normal development of the nervous system, but be low enough to prevent mental retardation. Treatment must begin within the first one or two months after birth, or the brain will be damaged and treatment will be less effective. By United States law, all infants born in the country are tested for PKU.

Phenylketonuriacs (individuals with PKU) are warned not to ingest food or drink containing the artificial sweetener aspartame (trade name NutraSweet). Aspartame is a small molecule consisting of the amino acids aspartic acid and phenylalanine joined together. Aspartame binds to taste receptors in the mouth signaling that the substance is sweet. However, it has essentially no calories. Once ingested, aspartame is broken down to its component amino acids. Phenylalanine released in this way could be harmful to an individual with PKU.

Figure 13.16A shows part of a pedigree for PKU. Analysis of the pedigree illustrates the general characteristics of autosomal recessive inheritance for a rare human genetic trait such as this:

- Most affected individuals have two normal parents, both of whom are heterozygotes. For example, III.3 and III.8 both have PKU, so their parents II.1 and II.2, and II.3 and II.4, respectively, all must have been heterozygotes.
- From a marriage of two heterozygotes, 1/4 of the offspring are expected to have the trait. From one pedigree, often there are not large enough numbers of individuals to see that ratio. However, the ratio would likely be seen when a large number of pedigrees are looked at together.

In sickle-cell anemia, the amino acid change in hemoglobin causes red blood cells to assume a sickle shape (see Figure 12.1, the Chapter 12 *Why it matters . . .* , and Section 12.2). The problems the sickled red blood cells have in passing through capillaries cause the serious symptoms of sickle-cell anemia. Between 10% and 15% of African Americans in the United States are carriers for this disorder—they have sickle-cell trait (see Section 12.2). Although carriers make enough normal hemoglobin through the activity of the dominant allele to be essentially unaffected, the mutant, sickle-cell form of the hemoglobin molecule is also present in their red blood cells. Carriers can be identified by a molecular test for the mutant hemoglobin. In countries where malaria is common, including several countries in Africa, carriers are less susceptible to malaria, which helps explain the increased proportions of the mutant allele in people whose ancestors originated in areas of the world where the malarial parasite is common.

Cystic fibrosis (CF), one of the most common genetic disorders among persons of Northern European descent, is another autosomal recessive trait (see Chapter 5 *Why it matters . . .*). About 1 in every 25 people from this line of descent is an unaffected carrier with one copy of the recessive allele. Approximately 1 per 4,000 children born in the United States has CF. Affected individuals have a mutated form of the transport protein called *cystic fibrosis transmembrane conductance regulator* (CFTR; see Figure 5.1). CFTR is embedded in the plasma membrane of epithelial cells, such as those of the passageways and ducts of the lungs, pancreas, and digestive tract. The mutant CFTR is deficient in the transport of Cl⁻ (chloride ions) out of the cells into the extracellular fluids. This alteration in chloride transport causes thick, sticky mucus to collect in airways of the lungs, in the ducts of glands such as the pancreas, and in the digestive tract. The accumulated mucus impairs body functions and, in the lungs, promotes pneumonia and other infections. With current management procedures, including daily chest thumps, back thumps, and repositioning to dislodge the thick mucus that collects in the lung airways, the life expectancy for a person with CF is about 40 years.

In Autosomal Dominant Inheritance, Only Homozygous Recessives Are Unaffected

Some human traits follow a pattern of **autosomal dominant inheritance.** In this case, the mutant allele that causes the trait is dominant, and people who are either homozygous or heterozygous for the dominant allele are affected. Individuals homozygous for the recessive nonmutant allele are unaffected.

Achondroplasia (see Figure 12.12C), a type of dwarfing that occurs in about 1 in 25,000 births worldwide, is caused by an autosomal dominant allele of a gene on chromosome 4. Of individuals with the dominant allele, only heterozygotes survive embryonic development; homozygous dominants are usually stillborn. When limb bones develop in heterozygous children, cartilage formation is defective, leading to disproportionately short arms and legs. They also have a relatively large head, but the trunk and torso are of normal size. Affected adults are usually not much more than 4 feet tall. Achondroplastic dwarfs are of normal intelligence, are fertile, and can have children.

The mutation responsible for achondroplasia maps to the *FGFR3* (fibroblast growth factor receptor 3) gene, which encodes a receptor tyrosine kinase (see Chapter 9). The normal role of FGFR3 is to repress the maturation of chondrocytes in the formation of bone in response to the signaling molecule that binds to the receptor. (Chondrocytes are cells that form cartilage and bone.) This negative effect on bone formation is regulated by opposing positive effects on bone growth by other pathways so that, in individuals homozygous for the normal *FGFR3* gene, bones of normal length are produced. In heterozygotes with the dominant mutation in *FGFR3*, the receptor is overly active all the time, resulting in a long-term negative effect on bone growth and, hence, the dwarfing symptoms of achondroplasia.

Figure 13.16B shows part of a pedigree for achondroplasia. Analysis of the pedigree illustrates the general characteristics of autosomal dominant inheritance for a rare trait such as this:

A. Phenylketonuria (PKU)
Generation

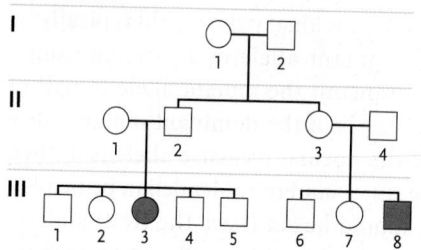

B. Achondroplasia
Generation

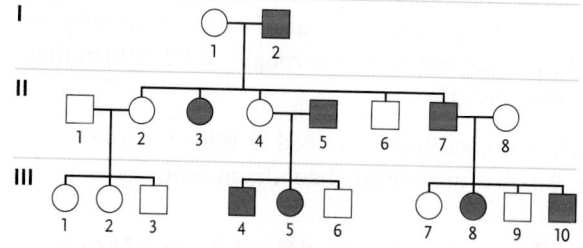

C. Duchenne muscular dystrophy
Generation

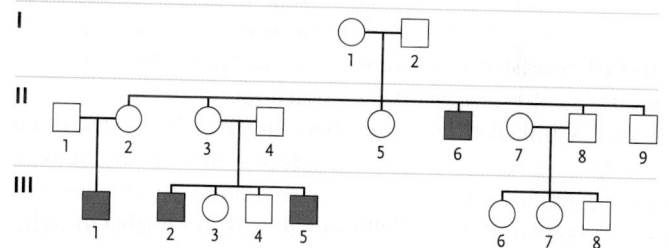

D. Hereditary enamel hypoplasia
Generation

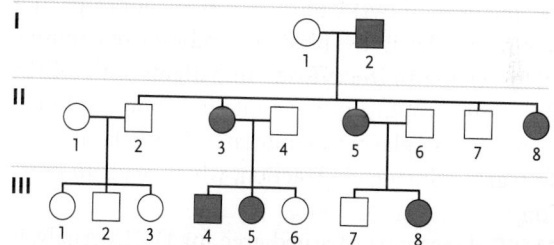

FIGURE 13.16 Pedigrees of human genetics traits showing different modes of inheritance. (A) Part of a pedigree for phenylketonuria (PKU), an autosomal recessive trait. **(B)** Part of a pedigree for achondroplasia, an autosomal dominant trait. **(C)** Part of a pedigree for Duchenne muscular dystrophy (DMD), an X-linked recessive trait. **(D)** Part of a pedigree for hereditary enamel hypoplasia, an X-linked dominant trait.
© Cengage Learning 2017

- Every affected person in a pedigree has at least one affected parent. That is readily seen in the pedigree. For rare traits, usually only one parent has the trait.
- The trait does not skip generations. Here, you can see that affected individuals are present in each generation of the pedigree. In contrast, for the autosomal recessive pedigree

shown in Figure 13.16A, affected individuals only appeared in generation III.

- On average, an affected individual (who would typically be heterozygous for the mutant allele of a rare autosomal dominant trait) will transmit the mutant allele to half of his or her offspring. That is, if the dominant mutant allele is designated *A*, and the normal recessive allele is *a*, then most pairings will be *Aa* × *aa*. From Mendelian principles, 1/2 of the offspring should be *Aa* (have the trait) and 1/2 should be *aa* (normal). Again, with small pedigrees there may be insufficient numbers of individuals to see such ratios. In this particular partial pedigree, you can see that three out of five generation II offspring have the trait, and, for the II.7–II.8 pairing, two out of four of their offspring have the trait, so overall we see roughly a 1/2 ratio in those parts of the pedigree.

Progeria, the condition described in *Why it matters . . .* in this chapter, is also an autosomal dominant trait.

X-Linked Recessive Traits Affect Males More Than Females

Red-green color blindness and hemophilia have already been presented as examples of human traits that demonstrate **X-linked recessive inheritance,** that is, traits resulting from inheritance of recessive alleles carried on the X chromosome. Another X-linked recessive human disease trait is Duchenne muscular dystrophy (DMD). In affected individuals, muscle tissue begins to degenerate late in childhood. By the onset of puberty, most individuals with this disease are unable to walk. Muscular weakness progresses, with later involvement of the heart muscle. The average life expectancy for individuals with DMD is 25 years. The nonmutant form of the gene that causes DMD encodes the protein dystrophin, which anchors a particular glycoprotein complex in the plasma membrane of a muscle fiber to the cytoskeleton in the cytosol. In patients with DMD, a mutation in the dystrophin gene results in a nonfunctional protein. As a result, the plasma membrane of muscle fibers is susceptible to tearing during contraction, which leads to muscle destruction.

Figure 13.16C shows part of a pedigree for DMD. Analysis of the pedigree illustrates some general characteristics of X-linked recessive inheritance:

- Many more males than females should exhibit the trait due to the different number of X chromosomes in the two sexes. We can see that property here in this pedigree, with only males showing the trait. This is also the case for Queen Victoria's pedigree showing hemophilia in Figure 13.9.
- All sons of an affected (homozygous mutant) mother should show the trait because males receive their only X chromosome from their mothers. There is no case of that in this pedigree or in Queen Victoria's.
- The sons of heterozygous (carrier) mothers should show an approximately 1:1 ratio of normal individuals to individuals

expressing the trait; that is, $X^{a+}X^a \times X^{a+}Y$ gives $1/2 \ X^{a+}Y$ and $1/2 \ X^aY$ sons. The male offspring of II.3 and II.4 are close to that ratio.

- From a pairing of a carrier female with a normal male, all daughters will be normal phenotypically, but 1/2 will be carriers; that is, $X^{a+}X^a \times X^{a+}Y$ gives $1/2 \ X^{a+}X^{a+}$ and $1/2 \ X^{a+}X^a$ females. In turn, 1/2 the sons of these carrier females will exhibit the trait. In the pedigree shown, the I.1 female must be a carrier because offspring II.6 has the trait. I.2 is a normal male because he does not show the trait. We see three female offspring, two of whom must be carriers because they produce some male offspring with the trait.
- A male expressing the trait, when paired with a homozygous normal female, will produce all normal children. But all the female progeny will be carriers; that is, $X^{a+}X^{a+} \times X^aY$ gives $X^{a+}X^a$ females and $X^{a+}Y$ (normal) males. This is seen in a limited way in Queen Victoria's pedigree of Figure 13.9 where hemophiliac Leopold in generation IV produces carrier offspring Alice in generation V who then produces hemophiliac Rupert in generation VI.

X-Linked Dominant Traits Are Rare

Only a few X-linked dominant traits have been identified in humans. One example is hereditary faulty enamel and dental discoloration, technical name *hereditary enamel hypoplasia* **(Figure 13.17)**. Another is a severe bleeding anomaly called *constitutional thrombopathy*.

Figure 13.16D shows part of a pedigree for hereditary enamel hypoplasia. Analysis of the pedigree illustrates some characteristics of X-linked dominant inheritance. Generally, X-linked dominant traits follow the same sort of inheritance rules as do X-linked recessives, except that heterozygous females express the trait.

- Because females have twice the number of X chromosomes as males, X-linked dominant traits are more frequent in females than in males, and we see that in the pedigree.
- Males with an X-linked dominant trait pass on the trait to all of their daughters and none of their sons. We see that in the pedigree where male I.2 with the trait passes on his X chromosome to each of his female offspring II.2, II.5,

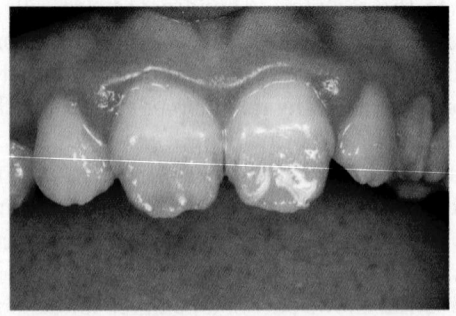

FIGURE 13.17 Teeth showing hereditary faulty enamel and dental discoloration, an X-linked dominant trait.

and II.8, and so they show the trait. However, the two male offspring, II.2 and II.4, received their X chromosomes from their mother who was normal; hence they do not show the trait.

Those females theoretically could have passed their X chromosome with the dominant mutant allele to male or female offspring who then would show the trait. In fact, we see that if the trait is rare, as is the case here, females with the trait are likely to be heterozygous. These females pass on the trait to 1/2 of their male progeny and 1/2 of their female progeny. We see this roughly in the pedigree where female II.3 produced both a male (III.4) and a female (III.5) child with the trait, plus a normal child (III.6).

Human Genetic Disorders Can Be Predicted, and Many Can Be Treated

Of all newborns, between 1% and 3% are homozygous for mutant alleles that encode defective forms of proteins required for normal functions. Possibly 1% have pronounced difficulties resulting from a chromosomal rearrangement or other aberration. Of all patients in children's hospitals, 10–25% are treated for problems arising from inherited disorders. Several approaches, which include genetic counseling, prenatal diagnosis, and genetic screening, can reduce the number of children born with genetic diseases.

Genetic counseling allows prospective parents to assess the possibility that they might have an affected child. For example, parents may seek counseling if they, a close relative, or one of their existing children has a genetic disorder. Genetic counseling begins with identification of parental genotypes through family pedigrees and often direct testing for an altered protein or DNA sequence. With this information in hand, counselors can often predict with a high degree of accuracy the chances of having a child with the trait in question. Couples can then make an informed decision about whether to have a child.

Genetic counseling is often combined with techniques of **prenatal diagnosis,** in which cells derived from a developing embryo or its surrounding tissues or fluids are tested for the presence of mutant alleles (by DNA testing or biochemical analysis) or for chromosomal mutations. In **amniocentesis,** cells are obtained from the amniotic fluid—the watery fluid surrounding the embryo or fetus in the mother's uterus **(Figure 13.18).** In **chorionic villus sampling,** cells are obtained from portions of the placenta that develop from tissues of the embryo. More than 100 genetic disorders can now be detected by these tests. If prenatal diagnosis detects a serious genetic disorder, the prospective parents can reach an informed decision about whether to continue the pregnancy, including religious and moral considerations, as well as genetic and medical advice.

Once a child is born, a number of inherited disorders can be identified by **genetic screening,** in which biochemical or

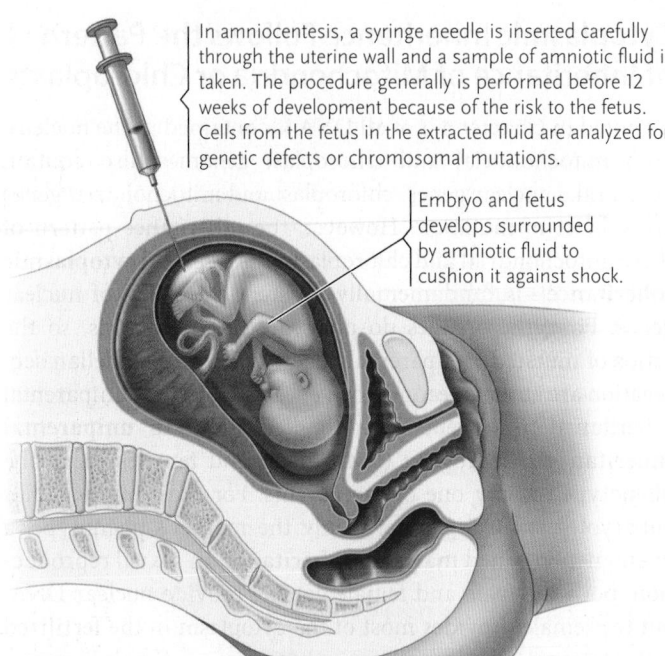

In amniocentesis, a syringe needle is inserted carefully through the uterine wall and a sample of amniotic fluid is taken. The procedure generally is performed before 12 weeks of development because of the risk to the fetus. Cells from the fetus in the extracted fluid are analyzed for genetic defects or chromosomal mutations.

Embryo and fetus develops surrounded by amniotic fluid to cushion it against shock.

FIGURE 13.18 Amniocentesis, a procedure used for prenatal diagnosis of genetic defects. The procedure is complicated and costly and, therefore, it is used primarily in high-risk cases.
© Cengage Learning 2017

molecular tests for disorders are routinely applied to children and adults or to newborn infants in hospitals. The tests can detect inherited disorders early enough to start any available preventive measures before symptoms develop (recall the earlier discussion of PKU, for instance).

In addition to the characters and traits described so far in this chapter, some patterns of inheritance depend on genes located not in the cell nucleus, but in mitochondria or chloroplasts in the cytoplasm, as discussed in the following section.

STUDY BREAK 13.4

1. A man has Simpson syndrome, an addiction to a certain television show. His wife does not have this syndrome. This couple has four children, two boys and two girls. One of the boys and one of the girls has this syndrome; the other children are normal. Can Simpson syndrome be an autosomal recessive trait? A sex-linked recessive trait?
2. In another family, a female child has wiggly ears, whereas her brother does not. Both parents are normal. Can the wiggly ear trait be an autosomal recessive trait? A sex-linked recessive trait?

13.5 Non-Mendelian Patterns of Inheritance

We consider two examples of patterns of inheritance in this section that do not follow the principles of Mendelian inheritance.

Cytoplasmic Inheritance Follows the Pattern of Inheritance of Mitochondria or Chloroplasts

As noted in Chapter 4, not all DNA is contained in the nucleus; both mitochondrial and chloroplast genomes also contain DNA. Like nuclear genes, chloroplast and mitochondrial genes are subject to mutation. However, the inheritance pattern of these mitochondrial and chloroplast genes—called **cytoplasmic inheritance**—is fundamentally different from that of nuclear genes. First, these genes do not segregate by meiosis, so the ratios of mutated and parental genes typical of Mendelian segregation are absent. Second, the genes usually show uniparental inheritance from generation to generation. In **uniparental inheritance,** all progeny (both males and females) have the phenotype of only one of the parents. For most multicellular eukaryotes, offspring inherit only the mother's phenotype, a phenomenon called **maternal inheritance.** In sexual reproduction, both the male and female gamete provide nuclear DNA, but the female provides most of the cytoplasm in the fertilized cell. Maternal inheritance occurs because, in animals, a zygote receives most of its cytoplasm, including mitochondria and (in plants) chloroplasts, from the female parent and little from the male parent. In plants, cytoplasmic inheritance varies depending on the species and the organelle (mitochondrion or chloroplast).

The first example of cytoplasmic inheritance of a mutant trait was found in 1909 by the German scientist Carl Correns. Correns studied variegated four-o'clock plants, *Mirabilis jalapa.* Variegation means that there are areas of pale green or white in otherwise normal green leaves. In some cases, whole branches may have green leaves, whereas others may have entirely pale green or white leaves. Correns made crosses of all combinations between these different types of branches on different plants. His results showed that the progeny always resembled the maternal parent (the parent producing the seed from which the offspring grew) and not the plant providing the pollen (the male parent). Correns was observing maternal inheritance. The explanation here is that the leaf colors are due to the plastids they contain. For example, green leaves contain normal chloroplasts, and white leaves contain chloroplasts that carry a mutation for the production of the green pigment chlorophyll (see Chapter 8). The maternal parent contributes essentially all plastids to the progeny, hence the progeny had the chloroplast phenotype of that parent and not the male parent.

Maternal inheritance of mutant traits involving the mitochondria have also been characterized in many eukaryotic species, including animals, plants, protists, and fungi. Similar to the mutant traits of chloroplasts, each mutant trait results from an alteration of a gene in the mitochondrial genome.

In humans, several inherited diseases have been traced to mutations in mitochondrial genes. Recall that the mitochondrion plays a critical role in synthesizing ATP, the energy source for many cellular reactions. Several maternally inherited diseases in humans involve mutations in mitochondrial genes that encode components of the ATP-generating system of the organelle. The resulting mitochondrial defects are especially destructive to the organ systems most dependent on mitochondrial reactions for energy: the central nervous system, skeletal and cardiac muscle, the liver, and the kidneys.

For example, *Leber's hereditary optic neuropathy* (LHON) is a maternally inherited human disease that affects midlife adults and is characterized by complete or partial blindness caused by optic nerve degeneration. Mutations in any one of the mitochondrial genes that encode eight electron transfer system proteins (see Chapter 7) all can lead to LHON. The electron transfer system is responsible for ATP synthesis by oxidative phosphorylation (see Chapter 7). Death of the optic nerve is a common result of defects in oxidative phosphorylation, which, in LHON, are caused by the inhibition of the electron transfer system.

The maternal inheritance of mitochondria in humans has some useful experimental applications, including genealogical studies to trace a maternal lineage back in time, and in forensics. For example, mitochondrial DNA testing has proved to be a valuable tool in missing persons investigations and in identifying remains in mass disasters.

In Genomic Imprinting, the Allele Inherited from One of the Parents Is Expressed Whereas the Other Allele Is Silent

Throughout our discussions of Mendelian inheritance, we have assumed that a particular allele has the same effect in an individual whether it was inherited from the mother or father. For the vast majority of genes, the assumption is correct. However, researchers have identified a number of genes whose effects do, in fact, depend on whether an allele is inherited from the mother or the father. For some of these genes, only the paternal, sperm-derived, allele is expressed. For others, only the maternal, egg-derived, allele is expressed. The phenomenon in which the expression of an allele of a gene depends on the parent that contributed it is called **genomic imprinting.** The gene involved is called an *imprinted gene.* The silent allele—the inherited allele that is not expressed—is called the *imprinted allele.* In humans and mice, at least 80 imprinted genes have been identified. This is an example of an *epigenetic phenomenon* where **epigenetics** is defined as the study of alterations of gene expression that occur without changes in DNA sequence (see also Chapter 16). Epigenetic changes such as this are reversible.

The first imprinted gene identified was *Igf2* in the mouse. *Igf2* encodes insulin-like growth factor 2, a protein that stimulates cell growth and division. The growth factor is needed for early embryos to develop normally. Researchers studying mice heterozygous for a deletion of the entire *Igf2* gene from the genome observed that if mice inherited the mutated chromosome from the father they were small, but if they had inherited the mutated chromosome from the mother they were normal size **(Figure 13.19A)**. It appeared that only the paternally inherited gene had an effect on size. The maternally inherited gene,

A. Phenotypes of mice heterozygous for a deletion of gene *Igf2*

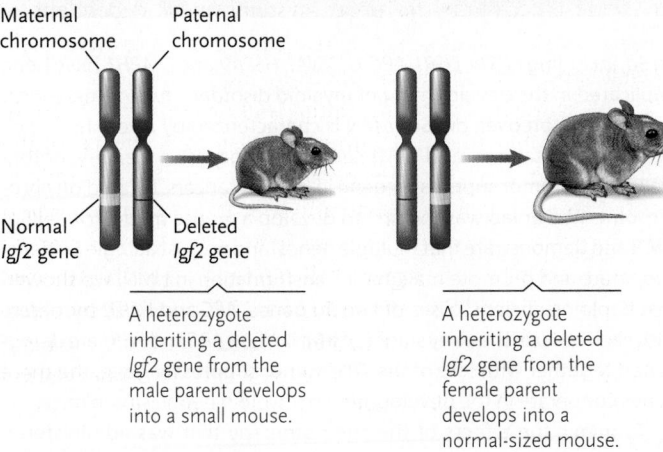

A heterozygote inheriting a deleted *Igf2* gene from the male parent develops into a small mouse.

A heterozygote inheriting a deleted *Igf2* gene from the female parent develops into a normal-sized mouse.

B. Phenotype of mice homozygous for the normal allele of *Igf2*

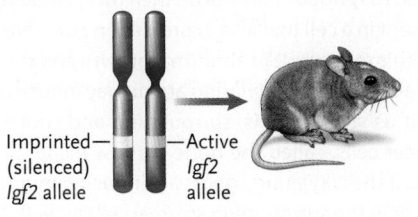

In a mouse heterozygous for the normal allele of *Igf2*, the paternal allele is active, and the maternal allele is imprinted (silenced). As long as a normal allele is inherited from the male parent, the mouse develops into a normal-sized adult.

FIGURE 13.19 Imprinting of the mouse *Igf2* (insulin-like growth factor 2) gene.
© Cengage Learning 2017

whether normal or not, had no effect. The researchers reasoned that, in normal mice homozygous for *Igf2*, the active form of the gene is the copy on the paternal chromosome, whereas the maternal copy of the gene is imprinted (silent) **(Figure 13.19B)**. So, if a heterozygote inherits the deletion from the father, that copy of the gene is inactive because of the deletion. Even if the maternal copy is nonmutant, it is not expressed because it is silenced by imprinting. As a result, normal development does not occur and the adult mouse produced is small.

The mechanism of genomic imprinting involves the modification of the DNA in the region that controls the expression of an allele by the addition of methyl ($-CH_3$) groups to cytosine (C) nucleotides. The methylation of the control region of a gene prevents it from being expressed. (You will learn more about the regulation of gene expression by methylation of DNA in Chapter 16.) Genomic imprinting occurs in the germ cells that develop into gametes. In those germ cells, the allele destined to be inactive in the new embryo after fertilization is methylated. That is, in the production of sperm, alleles for paternally imprinted genes are methylated, and in the production of eggs, alleles for maternally imprinted genes are methylated. That methylated (silenced) state of a gene is passed on cell generation to cell generation as the cells grow and divide to produce the somatic (body) cells of the organism.

Inherited imprints must first be erased in the germ cells before new imprinting occurs in the production of gametes. Consider a gene that is maternally imprinted, for example. In the adult, the maternal chromosome has an imprinted allele of that gene whereas the paternal chromosome has an active allele. When that adult produces gametes, it needs to imprint all alleles in a way appropriate to its sex—for example, if it is male, it must erase the imprint from the maternally inherited gene so that sperm the male produces has a nonimprinted copy of the gene. In the diploid cells that go through meiosis to produce the gametes, all imprints are first erased providing a clean slate for individuals to imprint alleles appropriately to their gender.

We must be clear that, although we are talking about the expression of alleles inherited from one or the other parent, genomic imprinting is a completely different phenomenon than sex linkage. For sex-linked genes, alleles inherited from the mother or father are both expressed. The phenotypic ratios are different than those for autosomal genes, but that is because of the difference in sex chromosome composition in males and females. And, in fact, most known imprinted genes are autosomal genes.

In this chapter, you have learned about genes and the role of chromosomes in inheritance. In the next chapter, you will learn about the molecular structure and function of the genetic material and about the molecular mechanism by which DNA is replicated.

STUDY BREAK 13.5

What key feature or features would suggest to you that a mutant trait shows cytoplasmic inheritance?

How do deletions of chromosome 5 cause myeloid neoplasms?

In this chapter, you learned that deletions are a type of chromosomal mutation that occurs if a broken segment is lost from a chromosome. Deletions may cause serious problems if the missing segment contains genes that are critical for normal cellular functions. Patients who have received chemotherapy and/or radiation for a primary cancer may subsequently develop acute myeloid leukemia (AML) or myelodysplastic syndrome (MDS), which are classified as therapy-related myeloid neoplasms (t-MNs). (A neoplasm is an abnormal growth of tissue.) A t-MN following chemotherapy with an alkylating agent typically develops after 3 to 7 years, and has characteristic loss or deletion of chromosomes 5 and/or 7. *Cytogeneticists* (geneticists who specialize in the study of chromosomes) refer to these patients as del(5q) and −7/del(7q), since either the entire chromosome is lost (−) or portions of the long or "q" arms of the chromosomes are deleted (del). The short or "petit" (p) arm of these chromosomes is typically intact in these patients.

Cytogenetic analysis of many cases of t-MN revealed that complex karyotypes are associated with abnormalities of chromosome 5, rather than 7. A karyotype is the number and appearance of chromosomes in the nucleus (see Figure 10.8). A complex karyotype means that three or more chromosomal aberrations, including deletions, duplications, translocations, and/or inversions, are detected. Recurring chromosomal abnormalities observed at a high frequency in patients with del(5q) included trisomy 8 and loss of portions of chromosomes: 13q, 16q, 17p, 18, and 20q. Our research has focused on the consequences of chromosome 5 deletions.

We and other groups of investigators mapped the deletions on chromosome 5 in t-MN patients with a del(5q) and defined a commonly deleted segment (CDS) that was predicted to contain one or more myeloid tumor suppressor genes. (Tumor suppressor genes are part of an elaborate regulatory system that controls cell division. As their name implies, they normally help suppress the growth of tumors or cancer. When they are not expressed normally, this can contribute to the development of cancer. They are discussed in Section 16.5.) The CDS contains 19 genes, which are deleted from one copy of chromosome 5 in every patient diagnosed with a del(5q). Additional genes on 5q, adjacent to the CDS, are also deleted frequently (up to 95% of cases), and are also likely to be involved in disease pathogenesis. The functions of these genes are diverse and include the regulation of mitosis, transcriptional control, and translational regulation.

Many genes require two copies to maintain normal expression levels and normal levels of protein. If one copy is lacking due to a chromosomal deletion and only one allele remains, the gene may be expressed at half the normal dose, and profound effects on cell growth and an abnormal phenotype may result. This phenomenon of a lowered gene dosage is referred to as "haploinsufficiency." In t-MN, a number of genes located on 5q, including *RPS14, EGR1, APC, CTNNA1, HSPA9,* and *DIAPH1,* have been implicated in the development of myeloid disorders due to a gene dosage effect. Moreover, del(5q) t-MN is characterized by a very high incidence of mutations and/or deletions of the *TP53* gene—another well-known tumor suppressor gene in human cancers, located on chromosome 17. Our lab was the first to develop a mouse model for del(5q) t-MN and demonstrate that multiple genes, and not just a single 5q gene, cooperate and promote malignant transformation in t-MN. We showed that haploinsufficient losses of two 5q genes, *APC* and *EGR1,* cooperate with each other in the early stages of t-MN. When *APC* and *EGR1* are deregulated together with loss of the *TP53* tumor suppressor gene, the three genes cooperate in the development of myeloid neoplasms in mice.

To mimic the effects of the chemotherapy that was administered before the development of t-MN, we are currently treating our mouse models with an alkylating agent. Alkylating agents may induce mutations in the DNA that may not be harmful on their own. However, if these mutations are present in a cell that also expresses one or more 5q genes at a lowered dose, this may result in abnormal growth and the development of cancer. In addition, the alkylating agent may induce mutations in the cells, known as stromal cells, surrounding and supporting the growth of the cancer cells, called the bone marrow niche. Our studies will help determine if the alkylating agents are inducing mutations only in the cancer cells, or in the surrounding stromal cells as well. Moreover, we can identify which genes are being mutated in mice and decipher if similar mutations occur in t-MN patients. Currently, it is very difficult to treat t-MN patients with a del(5q), and the patients' prognosis is poor. Development of mouse models of t-MN will help decipher the pathways that are destabilized in t-MN patients, and they will be of utmost importance in the development of more effective therapies.

think like a scientist In t-MN patients with a del(5q), are the genes in the deleted 5q segment completely inactive?

Michelle M. Le Beau and **Angela Stoddart** are researchers in the Section of Hematology/Oncology at the University of Chicago. Dr. Le Beau has had a long-standing interest in correlating specific chromosomal abnormalities with clinical features of the neoplastic disease. Dr. Stoddart joined Dr. Le Beau's group in 2004, and together they are characterizing the genes involved in therapy-related myeloid neoplasms with abnormalities of chromosome 5. Learn more about Drs. Le Beau and Stoddart's research at http://t-aml.uchicago.edu/scientific-leadership/michelle-le-beau/.

Courtesy of Michelle Le Beau

Courtesy of Angela Stoddart

REVIEW KEY CONCEPTS

For access to MindTap and additional study materials visit www.cengagebrain.com.

13.1 Genetic Linkage and Recombination

- Genes, consisting of sequences of nucleotides in DNA, are arranged linearly in chromosomes.

- Genes near each other on the same chromosome are linked together in their transmission from parent to offspring. Linked genes are inherited in patterns similar to those of single genes, except for changes in the linkage caused by recombination (Figures 13.2 and 13.3).

- In genetic recombination, alleles linked on the same chromosome are mixed into new combinations by exchange of segments between the chromosomes of a homologous pair. The exchanges occur by crossing-over while homologous chromosomes are paired during prophase I of meiosis.

- The recombination frequency between any two genes located on the same chromosome pair reflects the distance between them on the chromosome.

- The relationship between separation and recombination frequencies is used to produce chromosome maps in which genes are assigned relative locations with respect to each other (Figure 13.4).

13.2 Sex-Linked Genes

- Sex linkage is a pattern of inheritance produced by genes carried on sex chromosomes.

- Because males have only one X chromosome, they need to receive only one copy of a recessive allele from their mothers to develop the trait. Females must receive two copies of the recessive allele, one from each parent, to develop the trait (Figures 13.6–13.8).

- In placental mammals, inactivation of one of the two X chromosomes in cells of the female makes the dosage of X-linked genes the same in males and females (Figure 13.10).

13.3 Chromosomal Mutations That Affect Inheritance

- Inheritance is influenced by processes that delete, duplicate, or invert segments within chromosomes, or translocate segments between chromosomes (Figures 13.11 and 13.12).

- Chromosomes can change in number by addition or removal of individual chromosomes or entire sets. Changes in single chromosomes usually occur through nondisjunction, in which homologous pairs fail to separate during meiosis I, or sister chromatids fail to separate during meiosis II. As a result, one set of gametes receives an extra copy of a chromosome and the other set is deprived of the chromosome (Figures 13.13–13.15).

- Monoploids have only one set of chromosomes; monoploidy is lethal in most animal species. Polyploids have three or more copies of the entire chromosome set. Polyploids usually arise when the spindle fails to function during mitosis in cell lines leading to gamete formation, producing gametes that contain double the number of chromosomes typical for the species.

13.4 Human Genetic Traits, Pedigree Analysis, and Genetic Counseling

- Three modes of inheritance are most significant in human heredity: autosomal recessive, autosomal dominant, and X-linked recessive inheritance. X-linked dominant traits are rare. Pedigree analysis is used to identify the mode of inheritance for human genetic traits.

- In autosomal recessive inheritance, males or females carry a recessive allele on an autosome. Heterozygotes are carriers that are usually unaffected, but homozygous individuals show symptoms of the trait (Figure 13.16A).

- In autosomal dominant inheritance, a dominant gene is carried on an autosome. Individuals that are homozygous or heterozygous for the trait show symptoms of the trait; homozygous recessives are normal (Figure 13.16B).

- In X-linked recessive inheritance, a recessive allele for the trait is carried on the X chromosome. Male individuals with the recessive allele on their X chromosome or female individuals with the recessive allele on both X chromosomes show symptoms of the trait. Heterozygous females are carriers but usually show no symptoms of the trait (Figure 13.16C).

- In X-linked dominant inheritance, a dominant allele for the trait is carried on the X chromosome. The trait is expressed in males and females who receive such an X chromosome (Figures 13.16D and 13.17).

- Genetic counseling, based on identification of parental genotypes by constructing family pedigrees and prenatal diagnosis, allow prospective parents to reach an informed decision about whether to have a child or continue a pregnancy (Figure 13.18).

13.5 Non-Mendelian Patterns of Inheritance

- Cytoplasmic inheritance depends on genes carried on DNA in mitochondria or chloroplasts. Cytoplasmic inheritance follows the maternal line: it parallels the inheritance of the cytoplasm in fertilization, in which most or all of the cytoplasm of the zygote originates from the egg cell.

- Genomic imprinting is a phenomenon in which the expression of an allele of a gene is determined by the parent that contributed it. In some cases, the allele inherited from the father is expressed; in others, the allele from the mother is expressed. Commonly, the silencing of the other allele is the result of methylation of the region adjacent to the gene that is responsible for controlling the expression of that gene (Figure 13.19).

Remember/Understand

1. In humans, red–green color blindness is an X-linked recessive trait. If a man with normal vision and a color-blind woman have a son, what is the chance that the son will be color-blind? What is the chance that a daughter will be color-blind?

2. The following pedigree shows the pattern of inheritance of red–green color blindness in a family. Females are shown as circles and males as squares; the squares or circles of individuals affected by the trait are filled in red.

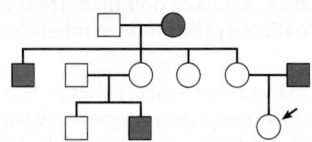

   What is the chance that a son of the third-generation female indicated by the arrow will be color-blind if the father is a normal man? If the father is color-blind?

3. Individuals affected by a condition known as polydactyly have extra fingers or toes. The following pedigree shows the pattern of inheritance of this trait in one family:

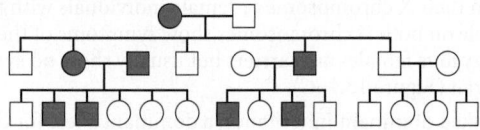

   From the pedigree, can you tell if polydactyly comes from a dominant or recessive allele? Is the trait sex-linked? As far as you can determine, what is the genotype of each person in the pedigree with respect to the trait?

4. A number of genes carried on the same chromosome are tested and show the following crossover frequencies. What is their sequence in the map of the chromosome?

Genes	Crossover Frequencies between Them
C and A	7%
B and D	3%
B and A	4%
C and D	6%
C and B	3%

5. In *Drosophila*, white eyes is an X-linked recessive trait. A white-eyed female is crossed with a male with normal red eyes. An F_1 female from this cross is mated with her father, and an F_1 male is mated with his mother. What will be the phenotypic ratios for eye color in the two sexes of the offspring of the last two crosses?

6. **Discuss Concepts** Can a linkage map be made for a haploid organism that reproduces sexually?

Apply/Analyze

7. In *Drosophila*, two genes, one for body color and one for eye color, are carried on the same chromosome. The wild-type gray body color is dominant to black body color, and wild-type red eyes are dominant to purple eyes. You make a cross between a fly with gray body and red eyes and a fly with black body and purple eyes. Among the offspring, about half have gray bodies and red eyes and half have black bodies and purple eyes. A small percentage have: (a) black bodies and red eyes; or (b) gray bodies and purple eyes. What alleles are carried together on the chromosomes in each of the flies used in the cross? What alleles are carried together on the chromosomes of the F_1 flies with black bodies and red eyes, and those with gray bodies and purple eyes?

8. Another gene in *Drosophila* determines wing length. The dominant wild-type allele of this gene produces long wings; a recessive allele produces vestigial (short) wings. A female that is true-breeding for red eyes and long wings is mated with a male that has purple eyes and vestigial wings. F_1 females are then crossed with purple-eyed, vestigial-winged males. From this second cross, a total of 600 offspring are obtained with the following combinations of traits:

   252 with red eyes and long wings

   276 with purple eyes and vestigial wings

   42 with red eyes and vestigial wings

   30 with purple eyes and long wings

   Are the genes linked, unlinked, or sex-linked? If they are linked, how many map units separate them on the chromosome?

9. **Discuss Concepts** Even though X inactivation occurs in XXY (Klinefelter syndrome) humans, they do not have the same phenotype as normal XY males. Similarly, even though X inactivation occurs in XX individuals, they do not have the same phenotype as XO (Turner syndrome) humans. Why might this be the case?

Evaluate/Create

10. You conduct a cross in *Drosophila* that produces only half as many male as female offspring. What might you suspect as a cause?

11. **Discuss Concepts** Crossing-over does not occur between any pair of homologous chromosomes during meiosis in male *Drosophila*. From what you have learned about meiosis and crossing-over, propose one hypothesis for why this might be the case.

12. **Discuss Concepts** All mammals have evolved from a common ancestor. However, the chromosome number varies among mammals. By what mechanism might this have occurred?

13. **Design an Experiment** Assume that genes *a*, *b*, *c*, *d*, *e*, and *f* are linked. Explain how you would construct a linkage map that shows the order of these six genes and the map units between them.

14. **Apply Evolutionary Thinking** How would the effects of natural selection differ on alleles that cause diseases fatal in childhood (such as progeria) and those that cause diseases that shorten life expectancy to 40 or 50 years (such as cystic fibrosis)?

For selected answers, see Appendix A.

INTERPRET THE DATA

Exposure to tobacco—whether by chewing it; by smoking cigars, cigarettes, or pipes; or by passive exposure to smoke—has been linked to cancer of the mouth and throat. Chemical compounds released during tobacco use form covalent bonds with the DNA in the oral cavity cells to generate structures called "adducts." These oral cavity cells replicate frequently, and the presence of DNA adducts is suspected to contribute to mutations, chromosomal alterations, and hence cellular defects passed onto offspring cells. The graph in the next column shows the incidence (attomol = 10^{-18} mol) of adduct formation in samples of *healthy* tissue from nonsmokers, smokers of various frequencies, and ex-smokers, all of whom required surgery to remove oral cancerous growths.

1. On the *y* axis, why is it necessary to express the data in terms of the quantity of adduct per μg of DNA, rather than simply noting the total quantity of adduct?

2. What do these data suggest about the relation between smoking and the formation of adducts?

3. What impact does smoking cessation have on the frequency of adduct formation?

4. Which data point(s) could suggest a relation between the potential to develop oral cancer and passive exposure to tobacco smoke?

Source: N. J. Jones, A. D. McGregor, and R. Waters. 1993. Detection of DNA adducts in human oral tissue: Correlation of adduct levels with tobacco smoking and differential enhancement of adducts using the butanol extraction and nuclease P1 versions of ^{32}P postlabeling. *Cancer Research* 53:1522–1528. Reprinted by permission of the American Association for Cancer Research.

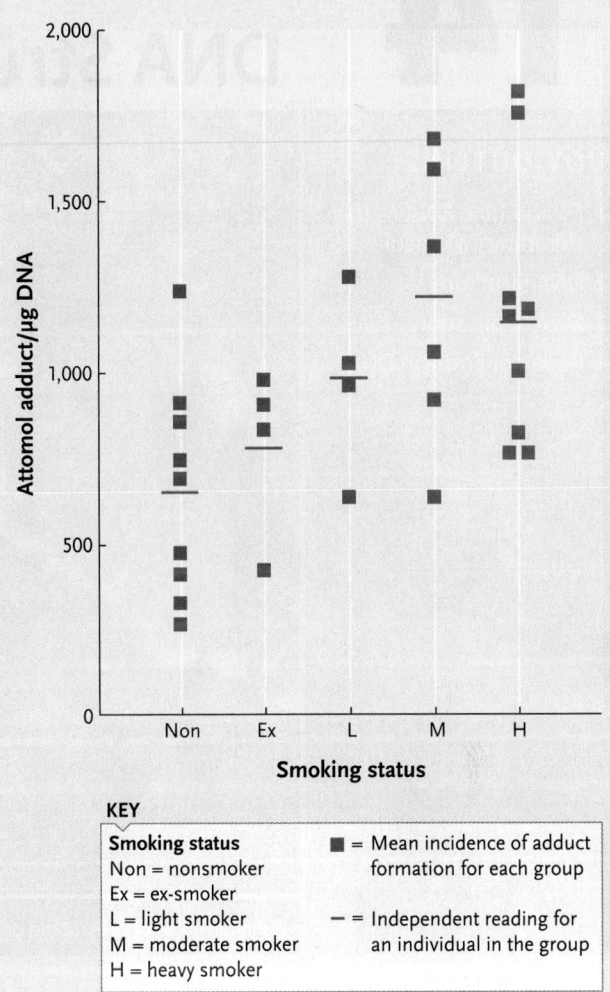

KEY

Smoking status	
Non = nonsmoker	■ = Mean incidence of adduct formation for each group
Ex = ex-smoker	
L = light smoker	— = Independent reading for an individual in the group
M = moderate smoker	
H = heavy smoker	

14

DNA Structure and Replication

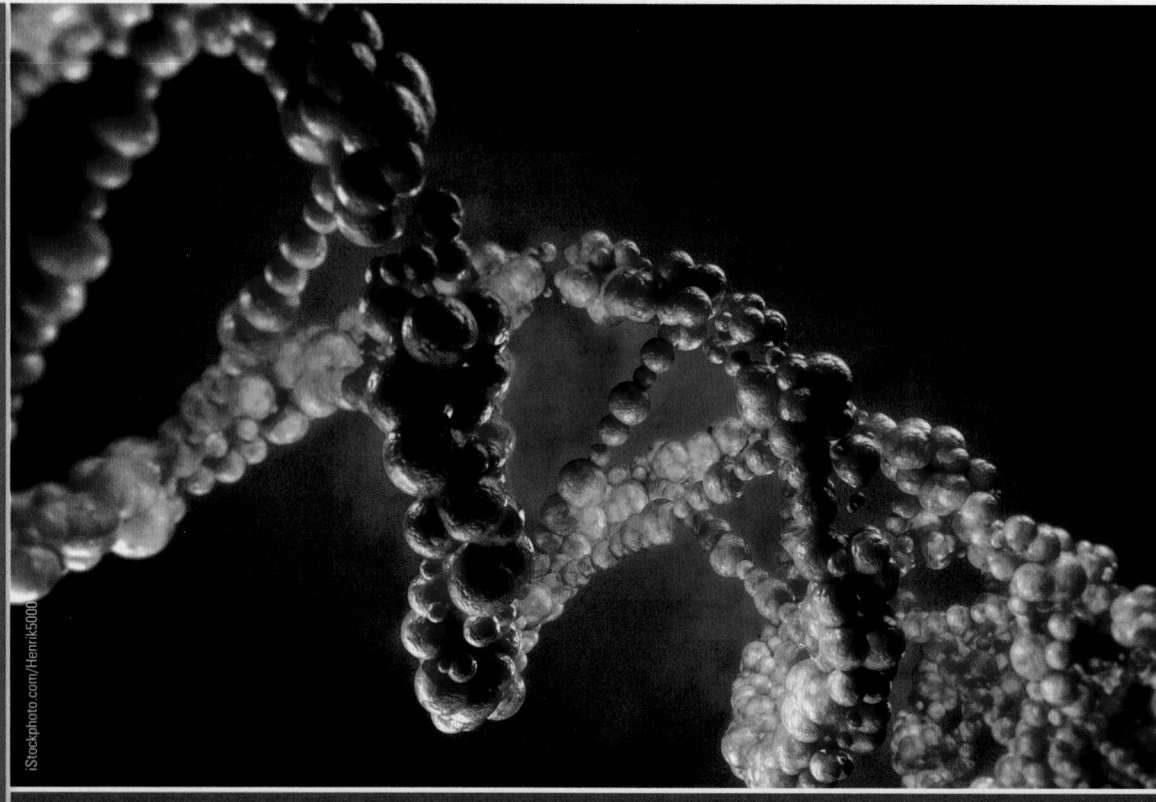

A digital model of DNA (based on data generated by X-ray crystallography).

Why it matters . . . In the spring of 1868, Johann Friedrich Miescher, a Swiss physician and physiological chemist, collected pus from discarded bandages. From the nuclei of the white blood cells found in the pus, Miescher extracted large quantities of an acidic substance with a high phosphorus content. He called the unusual substance "nuclein." His discovery is at the root of the development of our molecular understanding of life: nuclein is now known by its modern name, **deoxyribonucleic acid,** or **DNA,** the molecule that is the genetic material of all living organisms.

At the time of Miescher's discovery, scientists knew nothing about the molecular basis of heredity and very little about genetics. It was not known which chemical substance in cells actually carries the instructions for reproducing parental traits in offspring. Not until 1952 did scientists fully recognize that the hereditary molecule was DNA.

After DNA was established as the hereditary molecule, the focus of research changed to the three-dimensional structure of DNA. Among the scientists striving to work out the structure were James D. Watson, a young American postdoctoral student at Cambridge University in England, and the British scientist Francis H. C. Crick, then a graduate student at Cambridge University. Using chemical and physical information about DNA, the two investigators assembled molecular models from pieces of cardboard and bits of wire. Eventually they constructed a model for DNA that fit all the known data **(Figure 14.1).** Their discovery was of momentous importance in biology. The model immediately made it apparent how genetic information is stored and how it could be replicated faithfully. The discovery launched a molecular revolution within biology, making it possible for the first time to relate the genetic traits of living organisms to a universal molecular code present in the DNA of every cell. In addition, Watson and Crick's discovery

FIGURE 14.1 James D. Watson and Francis H. C. Crick in 1953, demonstrating their model for DNA structure, which revolutionized the biological sciences.

Barrington Brown/Science Source

opened the way for numerous advances in fields such as medicine, forensics, pharmacology, and agriculture, and eventually gave rise to the current rapid growth of the biotechnology industry.

14.1 Establishing DNA as the Hereditary Molecule

In the first half of the twentieth century, many scientists believed that proteins were the most likely candidates for the hereditary molecules because they appeared to offer greater opportunities for information coding than did nucleic acids. That is, proteins contain 20 different amino acids, whereas nucleic acids have only four different nitrogenous bases available for coding. In this section, we describe the experiments showing that DNA, and not protein, is the genetic material.

Griffith Found a Substance That Could Transform Pneumonia Bacteria Genetically

In 1928, Frederick Griffith, a British medical officer, experimented with the bacterium *Streptococcus pneumoniae* (also called pneumococcus), the species that can cause a severe form of pneumonia in mammals **(Figure 14.2).** He used two strains of the bacterium in his experiments. The smooth *(S)* strain is virulent (highly infective, or pathogenic), causing pneumonia and killing the mice in a day or two. The rough *(R)* strain is non-virulent (noninfective, or nonpathogenic), meaning that it does not affect mice.

The critical result of Griffith's experiment was that the mice died if he injected them with a mixture of living *R* bacteria and heat-killed *S* bacteria (Figure 14.2, experiment 4). He was able to isolate living *S* bacteria with polysaccharide capsules from those mice. In some way, living *R* bacteria had acquired the ability to make the polysaccharide capsule from the dead *S* bacteria, and they had changed—transformed—into virulent *S* cells. The smooth, virulence trait was stably inherited by the descendants of the transformed bacteria. Griffith called the conversion of *R* bacteria to *S* bacteria *transformation* and called the agent responsible the *transforming principle*.

Avery and His Coworkers Identified DNA as the Molecule That Transforms Nonvirulent *Streptococcus* Bacteria to the Virulent Form

In the 1940s, Oswald Avery at the Rockefeller Institute for Medical Research and his coworkers Colin MacLeod and Maclyn McCarty performed an experiment designed to identify the chemical nature of Griffith's transforming principle. Rather than working with mice, they reproduced the transformation using bacteria growing in culture tubes. They used heat to kill virulent *S* bacteria and then treated the macromolecules extracted from the *S* bacterial cells in turn with enzymes that break down each of the three main candidate molecules for the hereditary material—protein; DNA; and the other nucleic acid, RNA. When they destroyed proteins or RNA, the researchers saw no effect. The extract of *S* bacteria still transformed nonvirulent *R* bacteria into virulent *S* bacteria—the cells had polysaccharide capsules and produced smooth colonies on culture plates. When they destroyed DNA, however, no transformation occurred and no smooth colonies were seen on culture plates.

In 1944, Avery and his colleagues published their discovery that the transforming principle was DNA. Although their findings were revolutionary, Avery and his colleagues presented their conclusions in the paper cautiously. Some biologists accepted their results almost immediately. However, those who believed that the genetic material was protein argued that it was possible that not all protein was destroyed by the enzyme treatments and, as contaminants in their DNA transformation reaction, those remaining proteins were responsible for the transformation.

Hershey and Chase Obtained the Final Experimental Evidence Establishing DNA as the Hereditary Molecule

Experiments conducted in 1952 by bacteriologist Alfred D. Hershey and his laboratory assistant Martha Chase at the Cold Spring Harbor Laboratory removed any remaining doubts as to whether DNA or protein is the hereditary molecule. Hershey and Chase studied the infection of the bacterium *Escherichia coli* by bacteriophage T2. *E. coli* is a

FIGURE 14.2 | **Experimental Research**

Griffith's Experiment with Virulent and Nonvirulent Strains of *Streptococcus pneumoniae*

Question: What is the nature of the genetic material?

Experiment: Frederick Griffith studied the conversion of a nonvirulent (noninfective) *R* form of the bacterium *Streptococcus pneumoniae* to a virulent (infective) *S* form. The *S* form has a capsule surrounding the cell, giving colonies of it on a laboratory dish a smooth, shiny appearance. The *R* form has no capsule, so the colonies have a rough, nonshiny appearance. Griffith injected the bacteria into mice and determined how the mice were infected.

1. Mice injected with live *S* cells (control to show effect of *S* cells)

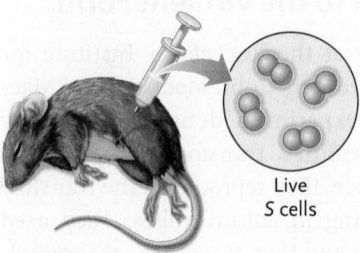

Live
S cells

Result: Mice die. Live *S* cells in their blood; shows that *S* cells are virulent.

2. Mice injected with live *R* cells (control to show effect of *R* cells)

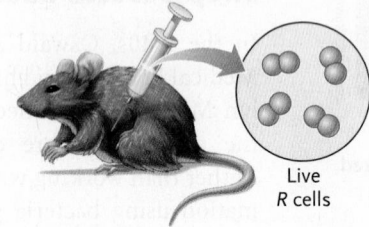

Live
R cells

Result: Mice live. No live *R* cells in their blood; shows that *R* cells are nonvirulent. Evidently the capsule is responsible for virulence of the *S* strain.

3. Mice injected with heat-killed *S* cells (control to show effect of dead *S* cells)

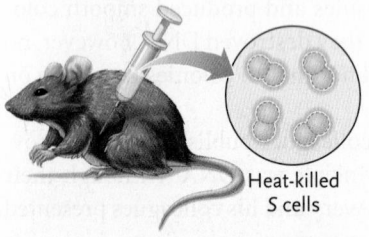

Heat-killed
S cells

Result: Mice live. No live *S* cells in their blood; shows that live *S* cells are necessary to be virulent to mice.

4. Mice injected with heat-killed *S* cells plus live *R* cells

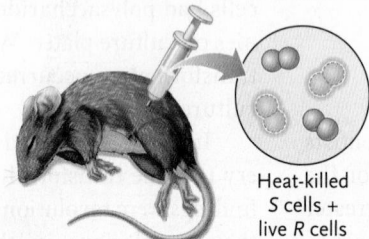

Heat-killed
S cells +
live *R* cells

Result: Mice die. Live *S* cells in their blood; shows that living *R* cells can be converted to virulent *S* cells with some factor from dead *S* cells.

Conclusion: Griffith concluded that some molecules released when *S* cells were killed could change living nonvirulent *R* cells genetically to the virulent *S* form. He called the molecule the *transforming principle* and the process of genetic change *transformation.*

think like a scientist

Observe
Hypothesize
Predict
Experiment
Interpret

Theoretically, could the existence of the transforming principle have been shown with the opposite setup, that is, by injecting mice with heat-killed *R* cells plus live *S* cells?

Source: F. Griffith. 1928. The significance of pneumococcal *types*. *Journal of Hygiene (London)* 27(1):13–159.

bacterium normally found in the intestines of mammals. **Bacteriophages** (or simply **phages;** see Chapter 17) are viruses that infect bacteria. A **virus** is an infectious agent that is made of either DNA or RNA surrounded by a protein coat. Viruses can reproduce only in a host cell. When a virus infects a cell, it can use the cell's resources to produce more virus particles.

The phage life cycle begins when a phage attaches to the surface of a bacterium and infects it. Phages such as T2 quickly stop the infected cell from producing its own molecules and instead use the cell's resources for making progeny phages. After about 100 to 200 phages are assembled inside the bacterial cell, a viral enzyme breaks down the cell wall, killing the cell and releasing the new phages. The whole life cycle takes approximately 30 minutes.

The T2 phage that Hershey and Chase studied consists of *only* a core of DNA surrounded by proteins. Therefore, either DNA or protein must be the genetic material that enters the bacterial cell and directs the infective cycle within. But which one? Hershey and Chase's definitive experiment to answer that question is presented in **Figure 14.3.** Their experimental approach was to label just the DNA or just the protein radioactively and then to use the label as a tag to follow the molecule through the phage life cycle. Because DNA contains phosphorus but not sulfur, they used radioactive phosphorus (^{32}P) to label only DNA, and because protein contains sulfur but not phosphorus, they used radioactive sulfur (^{35}S) to label only protein. Their results showed that most of the labeled DNA, but little of the labeled protein, entered the cell and appeared in progeny phages. Their results affirmed what most scientists at the time now believed—the genetic material is DNA, not protein.

When considered together, the experiments of Griffith, Avery and his coworkers, and Hershey and Chase established that DNA, not proteins, carries genetic information. The research also established the term *transformation,* which is still used in molecular biology. **Transformation** is the alteration of a cell's hereditary type by the uptake of DNA released by the breakdown of another cell, as in the Griffith and Avery experiments.

STUDY BREAK 14.1

Imagine that ^{35}S labeled *both* protein and DNA, whereas ^{32}P labeled only DNA. How would Hershey and Chase's results have been different?

FIGURE 14.3 **Experimental Research**

The Hershey and Chase Experiment Demonstrating That DNA Is the Hereditary Molecule

Question: Is DNA or protein the genetic material?

Experiment: Hershey and Chase performed a definitive experiment to show whether DNA or protein is the genetic material. They used phage T2 for their experiment; it consists only of DNA and protein. Because DNA contains phosphorus and not sulfur, they could label DNA selectively with radioactive ^{32}P. And, because protein contains sulfur and not phosphorus, they could label protein selectively with radioactive ^{35}S.

1. They infected *E. coli* growing in the presence of radioactive ^{32}P or ^{35}S with phage T2. The progeny phages were either labeled in their protein with ^{35}S (top), or in their DNA with ^{32}P (bottom).

2. Separate cultures of *E. coli* were infected with the radioactively labeled phages.

3. After a short period of time to allow the genetic material to enter the bacterial cell, the bacteria were mixed in a blender. The blending sheared from the cell surface the phage coats that did not enter the bacteria. The components were analyzed for radioactivity.

4. Progeny phages analyzed for radioactivity.

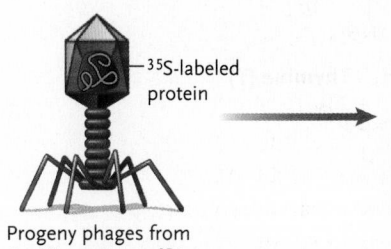

^{35}S-labeled protein

Progeny phages from *E. coli* growing in ^{35}S

E. coli

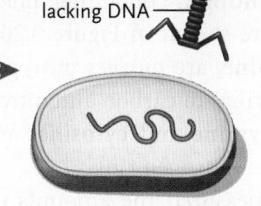

Phage coat lacking DNA

Result: Small amount of radioactivity within cell; most ^{35}S in phage coat

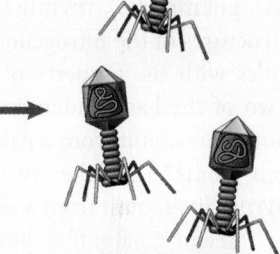

Result: Less than 1% of radioactivity in progeny phages

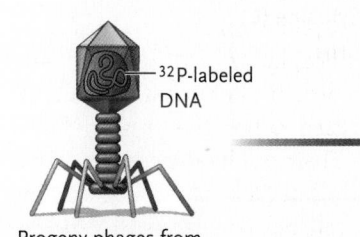

^{32}P-labeled DNA

Progeny phages from *E. coli* growing in ^{32}P

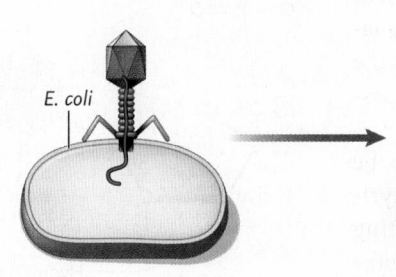

E. coli

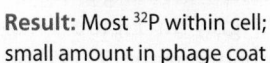

Phage coat lacking DNA

Result: Most ^{32}P within cell; small amount in phage coat

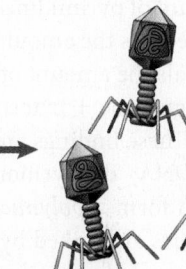

Result: Greater than 30% of ^{32}P in progeny phages

Conclusion: A significant amount of ^{32}P, the isotope used to label DNA, was found within phage-infected cells and in progeny phages, indicating that DNA is the genetic material. A significant amount of ^{35}S, the radioisotope used to label proteins, was found in phage coats after infection, but little was found in the infected cell or in progeny phages, showing that protein is not the genetic material.

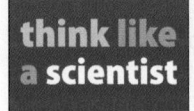

think like a scientist

Observe
Hypothesize
Predict
Experiment
Interpret

What isotope distribution would you expect to see if the phages used to infect *E. coli* were labeled with ^{14}C, the radioactive isotope of carbon?

Source: A. D. Hershey and M. Chase. 1952. Independent functions of viral protein and nucleic acid in growth of bacteriophage. *Journal of General Physiology* 36:39–56.

14.2 DNA Structure

The structure of DNA was elucidated by Watson and Crick in 1953.

Watson and Crick Brought Together Information from Several Sources to Develop a Model for DNA Structure

Before Watson and Crick began their research, other investigators had established that DNA contains four different nucleotides. Nucleotides were first described in Section 3.5 and are shown in detail in Figure 3.24 . Recall that the nucleotide in DNA is a deoxyribonucleotide because the sugar it contains is deoxyribose. As shown in **Figure 14.4,** each deoxyribonucleotide consists of the five-carbon sugar *deoxyribose* (carbon atoms on deoxyribose are numbered with primes from 1′ to 5′, a phosphate group, and one of the four *nitrogenous bases*—adenine (A), guanine (G), thymine (T), or cytosine (C). (The chemical structures of the nitrogenous bases—nitrogen-containing molecules with the property of a base—are shown in Figure 3.26.) Two of the bases, **adenine** and **guanine,** are *purines,* nitrogenous bases built from a pair of fused rings of carbon and nitrogen atoms. The other two bases, **thymine** and **cytosine,** are *pyrimidines,* built from a single carbon–nitrogen ring.

Erwin Chargaff, a biochemist, measured the amounts of nitrogenous bases in DNA and discovered that they occur in definite ratios. He observed that the amount of purines equals the amount of pyrimidines, but more specifically, the amount of adenine equals the amount of thymine, and the amount of guanine equals the amount of cytosine. He also discovered that the ratio of guanine + cytosine: adenine + thymine was species-specific. These findings are known as **Chargaff's rules.**

In DNA, deoxyribonucleotides had been shown to be joined to form a *polynucleotide chain* in which the deoxyribose sugars are linked by phosphate groups in an alternating sugar–phosphate–sugar–phosphate pattern, forming a **sugar–phosphate backbone** (highlighted in gray in Figure 14.4). Each phosphate group is a "bridge" between the 3′ carbon of one sugar and the 5′ carbon of the next sugar. The entire linkage, including the bridging phosphate group, is called a *phosphodiester bond,* also shown in Figure 14.4.

The polynucleotide chain of DNA has polarity—directionality. That is, the two ends of the chain are not the same. At one end, a phosphate group is bound to the 5′ carbon of a deoxyribose sugar, whereas at the other end, a hydroxyl group is bonded to the 3′ carbon of a deoxyribose sugar (see Figure 14.4). Consequently, the two ends are called the **5′ end** and **3′ end,** respectively.

Those were the known facts when Watson and Crick began their collaboration in the early 1950s. However, the number of polynucleotide chains in a DNA molecule and the manner in which they fold or twist in DNA were unknown. Watson and Crick themselves did not conduct experiments to study the structure of DNA. Instead, they used the research data of others

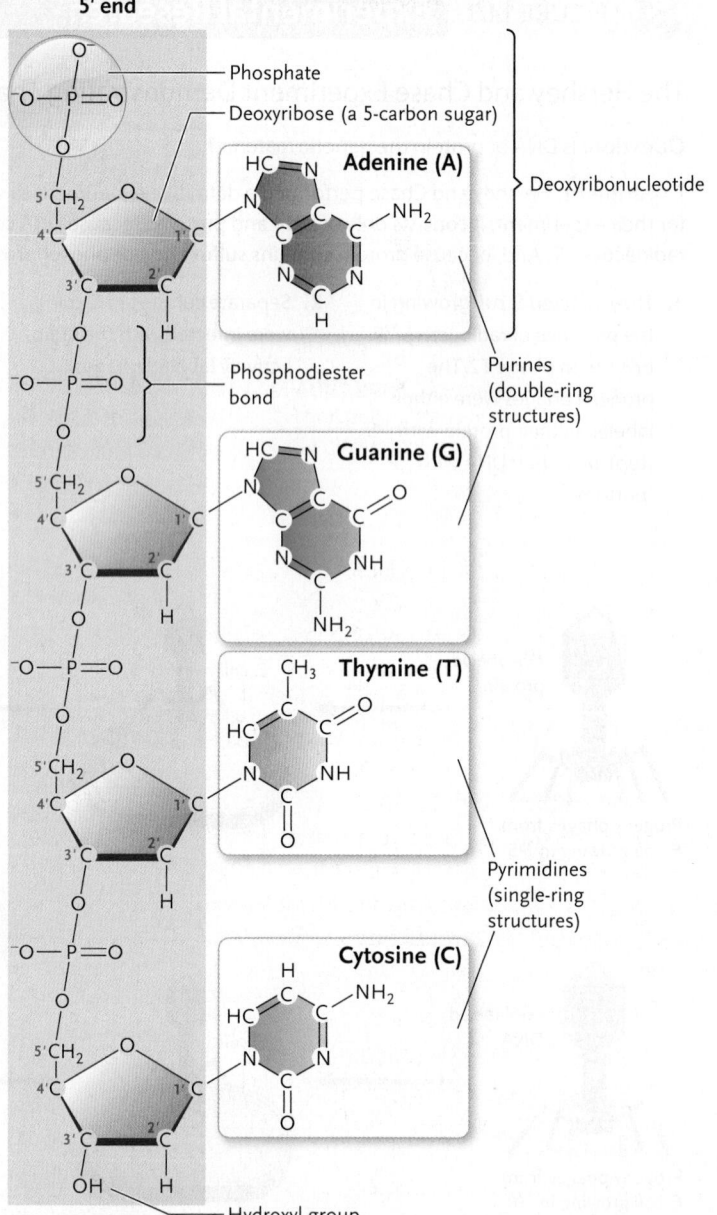

FIGURE 14.4 The four deoxyribonucleotide subunits of DNA, linked into a polynucleotide chain. The sugar–phosphate backbone of the chain is highlighted in gray. The connection between adjacent deoxyribose sugars is a phosphodiester bond. The polynucleotide chain has polarity; at one end, the 5′ end, a phosphate group is bound to the 5′ carbon of a deoxyribose sugar, whereas at the other end, the 3′ end, a hydroxyl group is bound to the 3′ carbon of a deoxyribose sugar.
© Cengage Learning 2017

for their analysis, relying heavily on data gathered by physicist Maurice H. F. Wilkins and his research colleague Rosalind Franklin at King's College, London. These researchers were using X-ray diffraction to study the structure of DNA **(Figure 14.5A).** In **X-ray diffraction,** an X-ray beam is directed at a molecule in the form of a regular solid, ideally in the form of a crystal. Within the crystal, regularly arranged rows and banks of atoms bend and reflect the X-rays into smaller beams that exit the crystal at definite angles determined by the arrangement of atoms in the

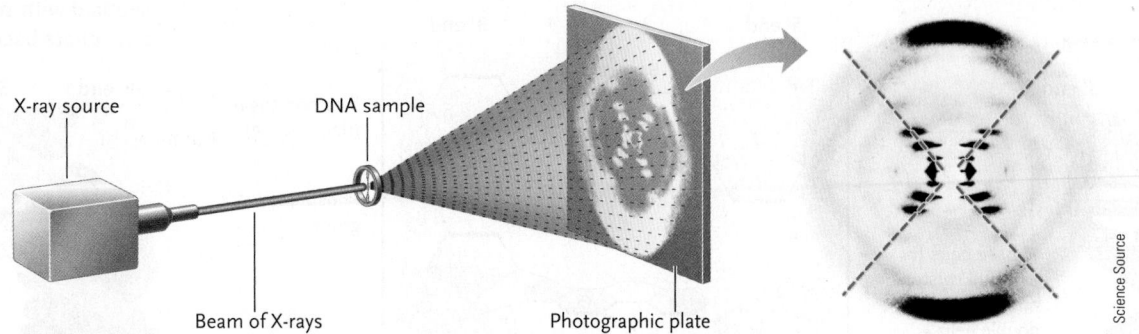

A. X-ray diffraction analysis of DNA

X-ray source

DNA sample

Beam of X-rays

Photographic plate

B. Franklin's DNA diffraction pattern

Science Source

FIGURE 14.5 X-ray diffraction analysis of DNA. (A) The X-ray diffraction method to study DNA. **(B)** The diffraction pattern Rosalind Franklin obtained. The X-shaped pattern of spots (dashed lines) was correctly interpreted by Franklin to indicate that DNA has a helical structure similar to a spiral staircase.
© Cengage Learning 2017

crystal. If an X-ray film is placed behind the crystal, the exiting beams produce a pattern of exposed spots. From that pattern, researchers can deduce the positions of the atoms in the crystal.

Wilkins and Franklin did not have DNA crystals, but they were able to obtain X-ray diffraction patterns from DNA molecules that had been pulled out into a fiber **(Figure 14.5B).** The patterns indicated that the DNA molecules within the fiber were cylindrical and about 2 nm in diameter. Separations between the spots showed that major patterns of atoms repeat at intervals of 0.34 and 3.4 nm within the DNA. Franklin interpreted an X-shaped distribution of spots in the diffraction pattern (see dashed lines in Figure 14.5B) to mean that DNA has a helical structure.

The New Model Proposed That Two Polynucleotide Chains Wind into a DNA Double Helix

Watson and Crick constructed scale models of the four DNA nucleotides and fitted them together in different ways until they arrived at an arrangement that satisfied both Wilkins' and Franklin's X-ray data and Chargaff's chemical analysis. Watson and Crick's trials led them to the **double-helix model** for DNA **(Figure 14.6).** The main features of the model are:

- *DNA is double stranded, consisting of two polynucleotide chains wound around each other in a right-handed double helix,* like a double spiral ladder (Figure 14.6A). The diameter of the double helix is 2 nm.
- *The sugar–phosphate backbones are on the outsides of the double helix, with the base pairs on the inside* (Figure 14.6A, B).
- *The two strands of a double helix are **antiparallel** (have opposite polarity),* meaning that they run in opposite directions (Figure 14.6B, arrows). In other words, the 3′ end of one strand is opposite the 5′ end of the other strand. This antiparallel arrangement is highly significant for the process of replication, which is discussed in the next section.

- *Connecting the two sugar–phosphate backbones are base pairs,* the rungs of the spiral ladder if you will. A purine and a pyrimidine, if paired together, are exactly wide enough to fill the space between the backbone chains in the double helix. However, two purines are too wide to fit the space exactly, and two pyrimidines are too narrow. The purine–pyrimidine base pairs in Watson and Crick's model are A–T and G–C pairs. Wherever an A occurs in one strand, a T must be opposite it in the other strand. Wherever a G occurs in one strand, a C must be opposite it. This feature of DNA is called **complementary base pairing,** and one strand is said to be *complementary* to the other. Complementary base pairing involving A–T and G–C base pairs fits Chargaff's rules. The base pairs, which fit together like pieces of a jigsaw puzzle, are stabilized by hydrogen bonds—two between A and T and three between G and C. (See Figures 14.6B and 3.28; hydrogen bonds are discussed in Section 2.3. A does not pair with C, and G does not pair with T because of the hydrogen bonding requirements.) The hydrogen bonds between the paired bases, repeated along the double helix, hold the two strands together in the helix.
- *The base pairs lie in flat planes almost perpendicular to the long axis of the DNA molecule.* In this state, each base pair occupies a length of 0.34 nm along the long axis of the double helix (Figure 14.6A). This spacing accounts for the repeating 0.34 nm pattern noted in the X-ray diffraction patterns. The larger 3.4 nm repeating pattern was interpreted to mean that each full turn of the double helix takes up 3.4 nm along the length of the molecule and therefore *10 base pairs are stacked in a full turn of DNA* (Figure 14.6A). Van der Waals forces (see Section 2.3) between stacked bases help stabilize the DNA double helix.
- *DNA has major and minor grooves* (Figure 14.6C). Within the grooves, the base pairs are accessible. Some proteins function by "reading" the base sequence in a major or minor groove. Examples are certain regulatory proteins that control gene transcription (see Chapter 16), and restriction enzymes that cut DNA at specific locations (see Chapter 18).

A. DNA double-helical structure

B. Chemical structure

C. Space-filling model overlaid with sugar–phosphate backbones

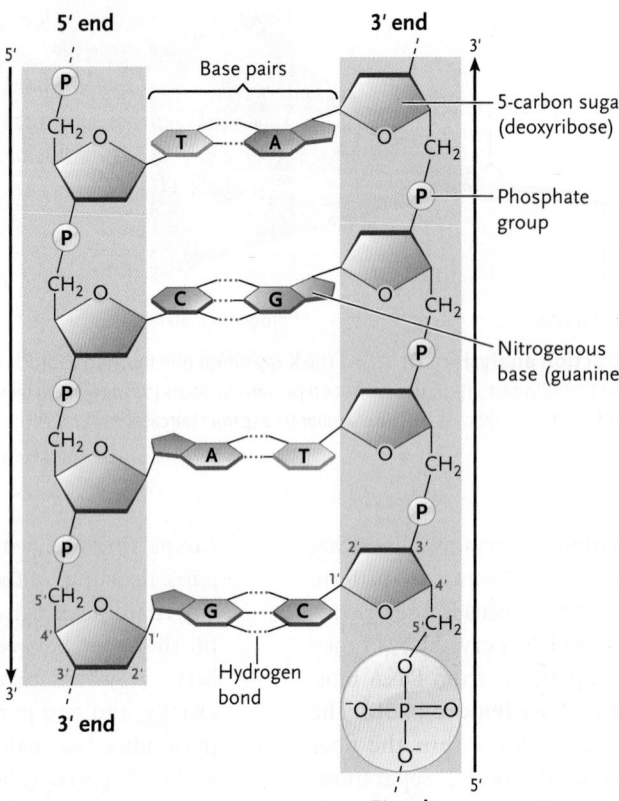
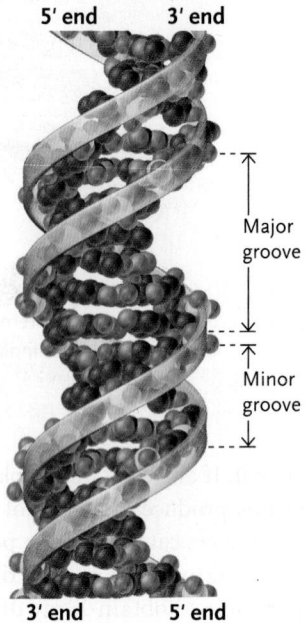

FIGURE 14.6 DNA double helix. (A) Space-filling model. **(B)** Schematic drawing. **(C)** Chemical structure drawing. Arrows and labeling of the ends show that the two polynucleotide chains of the double helix are antiparallel—that is, they have opposite polarity in that they run in opposite directions. In the space-filling model at the top, the spaces occupied by atoms are indicated by spheres. There are 10 base pairs per turn of the helix; only 8 base pairs are visible because the other 2 are obscured where the backbones pass over each other.

As hereditary material, DNA must faithfully store and transmit genetic information for the entire life cycle of an organism. Watson and Crick recognized that this information is coded into the DNA by the particular sequence of the four nucleotides. Although only four different kinds of nucleotides exist, combining them in groups allows an essentially infinite number of different sequences to be "written," just as the 26 letters of the alphabet can be combined in groups to create a virtually unlimited number of words. Chapter 15 shows how the four nucleotides form sequences of three nucleotides that form enough "words" to "spell out" the structure of any conceivable protein.

Watson and Crick announced their model for DNA structure in a brief but monumental paper published in the journal *Nature* in 1953. Watson and Crick shared a Nobel Prize with Wilkins in 1962 "for their discoveries concerning the molecular structure of nucleic acids and its significance for information transfer in living material." Rosalind Franklin might have been a candidate for a Nobel Prize had she not died of cancer at age 38 in 1958. (The Nobel Prize is given only to living investigators.) Watson and Crick's discovery of the structure of DNA opened the way to molecular studies of genetics and heredity, leading to our modern understanding of gene structure and action at the molecular level.

STUDY BREAK 14.2

1. Which bases in DNA are purines? Which are pyrimidines?
2. What bonds form between complementary base pairs? Between a base and the deoxyribose sugar?
3. Which features of the DNA molecule did Watson and Crick describe?
4. The percentage of A in a double-stranded DNA molecule is 20. What is the percentage of C in that DNA molecule?

14.3 DNA Replication

Once they had discovered the structure of DNA, Watson and Crick realized immediately that complementary base pairing between the two strands could explain how DNA replicates **(Figure 14.7)**. They imagined that, for replication, the hydrogen

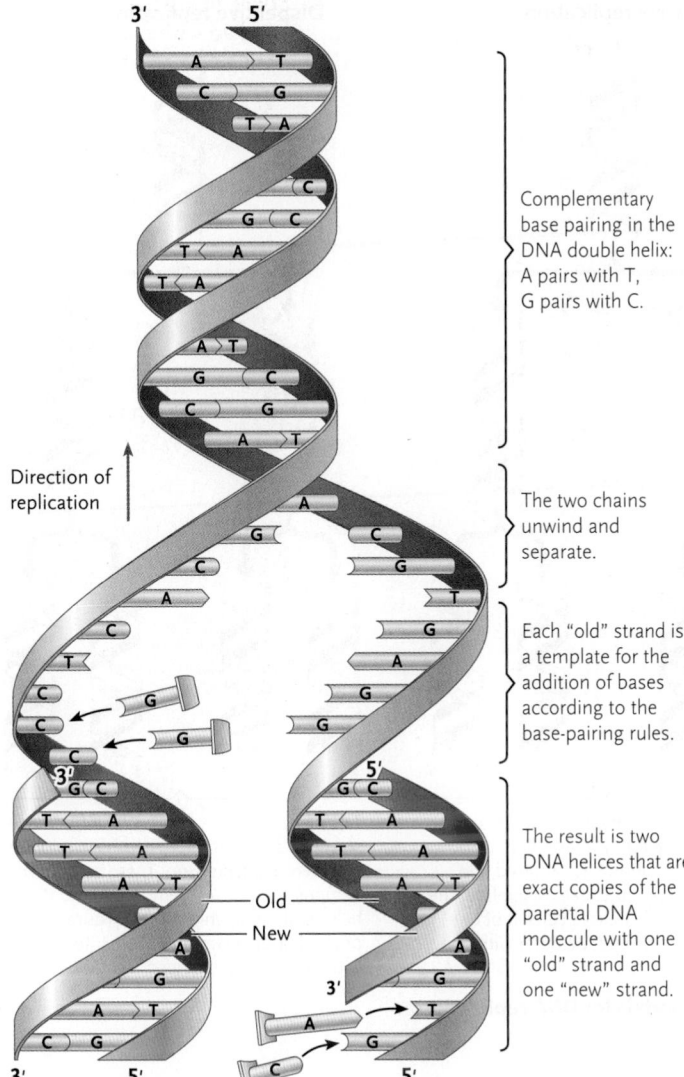

Direction of replication

Complementary base pairing in the DNA double helix: A pairs with T, G pairs with C.

The two chains unwind and separate.

Each "old" strand is a template for the addition of bases according to the base-pairing rules.

Old
New

The result is two DNA helices that are exact copies of the parental DNA molecule with one "old" strand and one "new" strand.

FIGURE 14.7 Watson and Crick's model for DNA replication. The original DNA molecule is shown in gray. A new polynucleotide chain (red) is assembled on each original chain as the two chains unwind. The template and complementary copy chains remain wound together when replication is complete, producing molecules that are half old and half new. The model is known as the semiconservative model for DNA replication.

© Cengage Learning 2017

bonds between the two strands break, and the two strands unwind and separate. Each strand then acts as a template for the synthesis of its partner strand. When replication is complete, there are two double helices, each of which has one strand derived from the parental DNA molecule base paired with a newly synthesized strand. Most important, each of the two new double helices has a base-pair sequence identical to that of the parental DNA molecule.

The model of replication Watson and Crick proposed is termed **semiconservative replication (Figure 14.8A).** Other scientists proposed two other models for replication. In the *conservative replication model,* the two strands of the original molecule serve as templates for the two strands of a new DNA

molecule, then rewind into an all "old" molecule **(Figure 14.8B).** After the two complementary copies separate from their templates, they wind together into an all "new" molecule. In the *dispersive replication model,* a variant of the conservative model, neither parental strand is conserved and both chains of each replicated molecule contain old and new segments **(Figure 14.8C).**

The Meselson and Stahl Experiment Showed That DNA Replication Is Semiconservative

A definitive experiment published in 1958 by Matthew Meselson and Franklin Stahl of the California Institute of Technology demonstrated that DNA replication is semiconservative **(Figure 14.9).** In their experiment, Meselson and Stahl distinguished parental DNA strands from newly synthesized DNA using a nonradioactive "heavy" nitrogen isotope to tag the parental DNA strands. The heavy isotope, ^{15}N, has one more neutron in its nucleus than the normal ^{14}N isotope. Molecules containing ^{15}N are measurably heavier (denser) than molecules of the same type containing ^{14}N. DNA molecules with different densities were distinguished by a special type of centrifugation.

DNA Polymerases Are the Primary Enzymes of DNA Replication

During replication, complementary polynucleotide chains are assembled from individual deoxyribonucleotides by enzymes known as DNA polymerases. More than one kind of DNA polymerase is required for DNA replication in bacteria, archaea, and eukaryotes. *Deoxyribonucleoside triphosphates* are the substrates for the polymerization reaction catalyzed by DNA polymerases **(Figure 14.10).** A nucleoside triphosphate is a nitrogenous base linked to a sugar, which is linked, in turn, to a chain of three phosphate groups (see Figure 3.24). You have encountered a nucleoside triphosphate before, namely ATP as the energy currency of the cell (see Section 6.3) and the ATP produced in cellular respiration (see Chapter 7). In that case, the sugar is ribose, making ATP a *ribonucleoside triphosphate.* The deoxyribonucleoside triphosphates used in DNA replication have the sugar *deoxyribose* rather than the sugar *ribose.* Because four different bases are found in DNA—adenine (A), guanine (G), cytosine (C), and thymine (T)—four different deoxyribonucleoside triphosphates are used for DNA replication. In keeping with the ATP naming convention, the deoxyribonucleoside triphosphates for DNA replication are given the short names dATP, dGTP, dCTP, and dTTP, where the "d" stands for "deoxyribose." The abbreviation dNTP will be used for *d*eoxyribo*n*ucleoside *t*ri-*p*hosphate in some of our discussions, where the "N" refers to a purine or pyrimidine base without specifying which one.

Figure 14.10 presents a section of a DNA polynucleotide chain being replicated, and shows how DNA polymerase catalyzes the assembly of a new DNA strand that is complementary

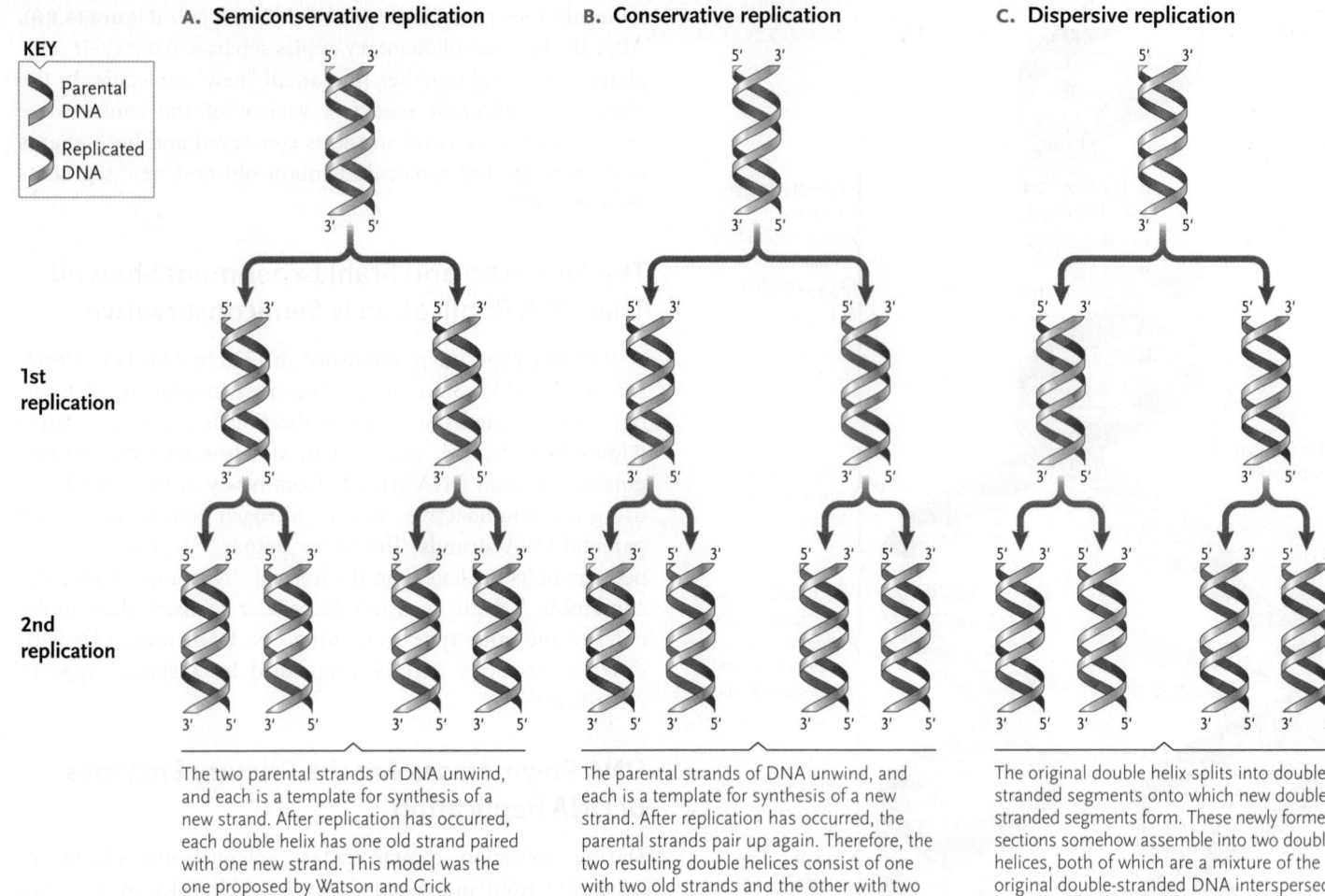

A. Semiconservative replication   **B. Conservative replication**   **C. Dispersive replication**

KEY

Parental DNA

Replicated DNA

1st replication

2nd replication

The two parental strands of DNA unwind, and each is a template for synthesis of a new strand. After replication has occurred, each double helix has one old strand paired with one new strand. This model was the one proposed by Watson and Crick themselves.

The parental strands of DNA unwind, and each is a template for synthesis of a new strand. After replication has occurred, the parental strands pair up again. Therefore, the two resulting double helices consist of one with two old strands and the other with two new strands.

The original double helix splits into double-stranded segments onto which new double-stranded segments form. These newly formed sections somehow assemble into two double helices, both of which are a mixture of the original double-stranded DNA interspersed with new double-stranded DNA.

FIGURE 14.8 Semiconservative (A), conservative (B), and dispersive (C) models for DNA replication.

© Cengage Learning 2017

to the template strand. To understand Figure 14.10, remember that the carbons in the deoxyriboses of nucleotides are numbered with primes. Each DNA strand has two distinct ends: the 5′ end has an exposed phosphate group attached to the 5′ carbon of the sugar, and the 3′ end has an exposed hydroxyl group attached to the 3′ carbon of the sugar. As you learned earlier, because of the antiparallel nature of the DNA double helix, the 5′ end of one strand is opposite the 3′ end of the other.

DNA polymerase can add a nucleotide *only to the* 3′ *end of* an existing nucleotide chain. As a new DNA strand is assembled, a 3′−OH group is always exposed at its "newest" end. The "oldest" end of the new chain has an exposed 5′ triphosphate. DNA polymerases are therefore said to assemble nucleotide chains in the 5′→3′ direction. Because of the antiparallel nature of DNA, DNA polymerase "reads" the template strand in the 3′→5′ direction for this new synthesis.

DNA polymerases of bacteria, archaea, and eukaryotes all consist of several polypeptide subunits arranged to form different domains (see Figure 3.23 and Section 3.4). The polymerases share a shape that is said to resemble a partially-closed human right hand in which the template DNA lies over the "palm" in a

groove formed by the "fingers" and "thumb" **(Figure 14.11A).** The palm domain is evolutionarily related among the polymerases of bacteria, archaea, and eukaryotes, while the finger and thumb domains are different sequences in each of those three types of organisms. The template strand does not pass through the tunnel formed by the thumb and finger domains, however. Instead, the template strand and the 3′−OH of the new strand meet at the active site for the polymerization reaction of DNA synthesis, located in the palm domain. A nucleotide is added to the new strand when an incoming dNTP enters the active site carrying a base complementary to the template strand base positioned in the active site. By moving along the template strand, one nucleotide at a time, DNA polymerase extends the new DNA strand as we saw in Figure 14.10.

Figure 14.11B shows the representation of DNA polymerase used in the following DNA replication figures, and it also shows a *sliding clamp*. The **sliding clamp** is a protein that encircles the DNA and binds to the rear of the DNA polymerase in terms of the enzyme's forward movement during replication. The sliding clamp anchors the DNA polymerase to the template strand. Without the sliding clamp, DNA polymerase detaches from the

FIGURE 14.9 | **Experimental Research**

The Meselson and Stahl Experiment Demonstrating the Semiconservative Model for DNA Replication to Be Correct

Question: Does DNA replicate semiconservatively?

Experiment: Matthew Meselson and Franklin Stahl proved that the semiconservative model of DNA replication is correct and that the conservative and dispersive models are incorrect.

1. Bacteria grown in ^{15}N (heavy) medium. The heavy isotope is incorporated into the bases of DNA, resulting in all the DNA being heavy, that is, labeled with ^{15}N.

2. Bacteria transferred to ^{14}N (light) medium and allowed to grow and divide for several generations. All new DNA is light.

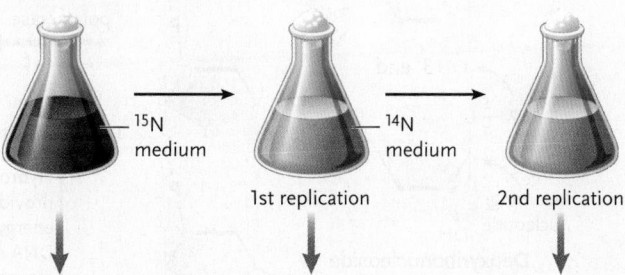

^{15}N medium

^{14}N medium
1st replication

2nd replication

3. DNA extracted from bacteria cultured in ^{15}N medium and after each generation in ^{14}N medium. Extracted DNA was centrifuged in a special solution to separate DNA of different densities.

Results: Meselson and Stahl obtained the following results:

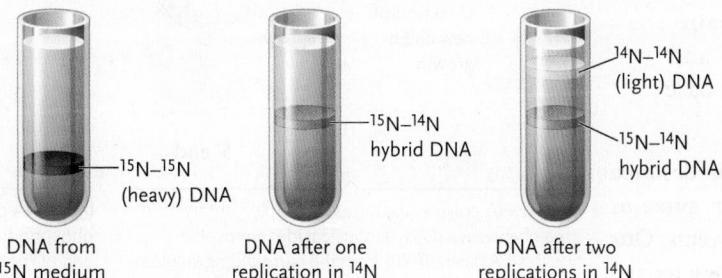

^{15}N–^{15}N (heavy) DNA

^{15}N–^{14}N hybrid DNA

^{14}N–^{14}N (light) DNA

^{15}N–^{14}N hybrid DNA

DNA from ^{15}N medium

DNA after one replication in ^{14}N

DNA after two replications in ^{14}N

Conclusion: The predicted DNA banding patterns for the three DNA replication models shown in Figure 14.8 were:

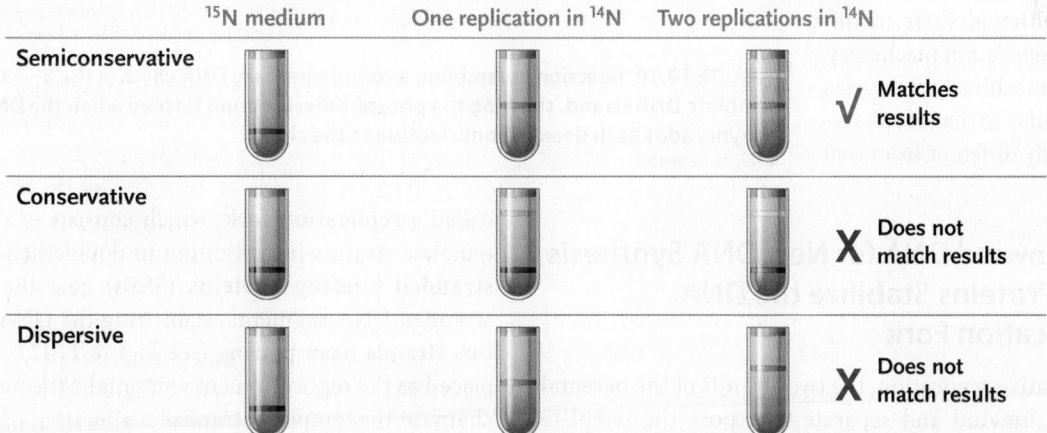

	^{15}N medium	One replication in ^{14}N	Two replications in ^{14}N	
Semiconservative				✓ Matches results
Conservative				✗ Does not match results
Dispersive				✗ Does not match results

The results support the semiconservative model.

think like a scientist

Observe
Hypothesize
Predict
Experiment
Interpret

For the semiconservative replication model, what proportion of ^{15}N–^{15}N, ^{15}N–^{14}N, and ^{14}N–^{14}N molecules would you expect after four and five replications in ^{14}N?

Source: M. Meselson and F. W. Stahl. 1958. The replication of DNA in *Escherichia coli. Proceedings of the National Academy of Sciences USA* 44:671–682.

template after only a few polymerizations. With the clamp, many thousands of polymerizations occur before the enzyme detaches. Overall, the rate of DNA synthesis is much faster because of the sliding clamp. *Molecular Insights* discusses how the sliding clamp is loaded onto replicating DNA in humans.

In sum, the key molecular events of DNA replication are as follows:

1. The two strands of the DNA molecule unwind for replication to occur.

2. DNA polymerase adds nucleotides to an existing chain using the parental template strand to determine which nucleotide to use.

3. The overall direction of new synthesis is in the 5′→3′ direction, which is a direction antiparallel to that of the template strand.

4. Nucleotides enter into a newly synthesized chain according to the A–T and G–C complementary base-pairing rules.

The following sections describe how enzymes and other proteins conduct these molecular events. Our focus is on the well-characterized replication system of *E. coli*. Replication in archaea and eukaryotes is highly similar, although there are differences in the replication machinery. The replication machinery of archaea is strikingly similar to that of eukaryotes and is clearly different from that of bacteria.

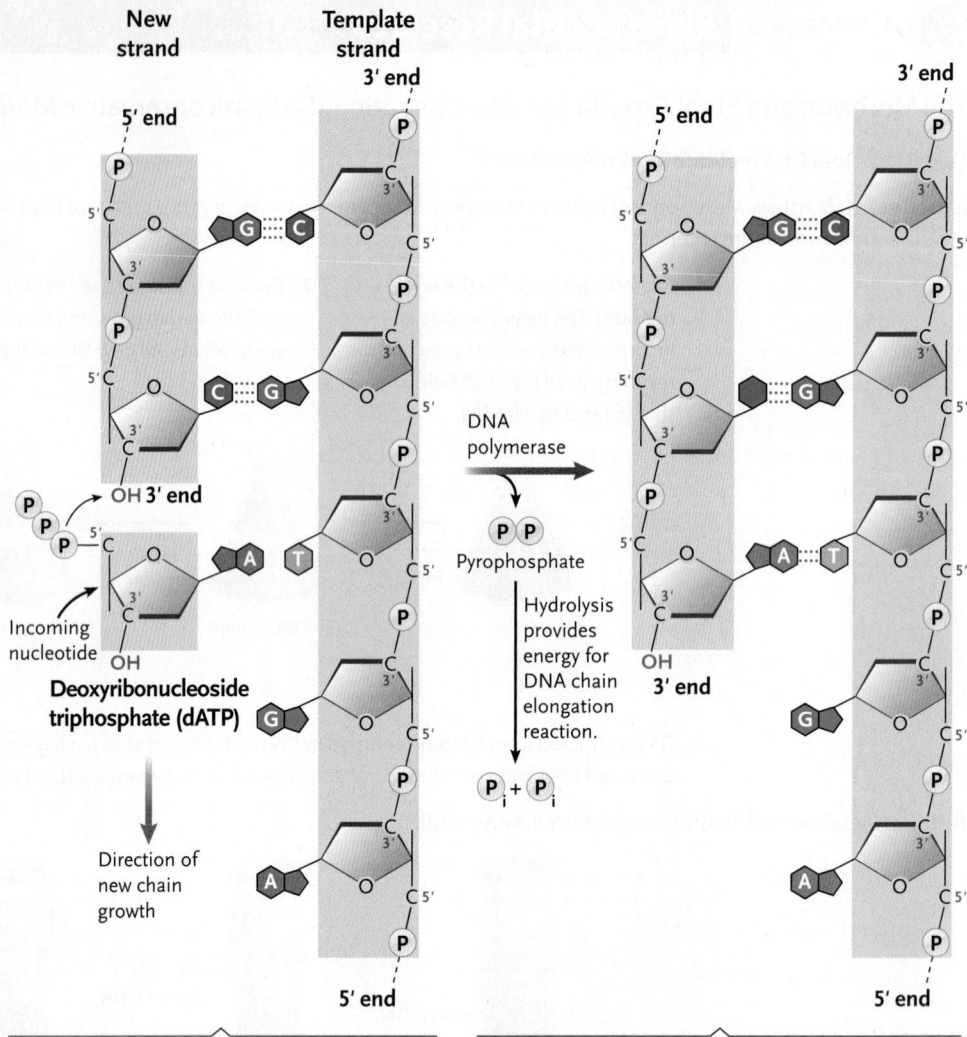

1 DNA polymerase forms a complementary base pair between a deoxyribonucleoside triphosphate with an A base (dATP) from the surrounding solution with the next, T, nucleotide of the template strand.

2 DNA polymerase catalyzes the formation of a phosphodiester bond involving the 3′–OH group at the end of the new chain and the innermost of the three phosphate groups of the dATP. The other two phosphates are released as a pyrophosphate molecule. The new chain has been lengthened by one nucleotide. The process continues, with DNA polymerase adding complementary nucleotides one by one to the growing DNA chain.

FIGURE 14.10 Reaction assembling a complementary DNA chain in the 5′→3′ direction on a template DNA strand, showing the phosphodiester bond formed when the DNA polymerase enzyme adds each deoxyribonucleotide to the chain.
© Cengage Learning 2017

Helicases Unwind DNA for New DNA Synthesis and Other Proteins Stabilize the DNA at the Replication Fork

In semiconservative replication, the two strands of the parental DNA molecule unwind and separate to expose the template strands for new DNA synthesis **(Figure 14.12)**. Unwinding of the DNA for replication occurs at a small, specific sequence in the bacterial chromosome known as an **origin of replication (ori)**. Specific proteins recognize an *ori* and recruit **DNA helicase,** which, using energy of ATP hydrolysis, unwinds the DNA strands. The unwinding produces a Y-shaped structure called a **replication fork,** which consists of the two unwound template strands transitioning to double-helical DNA. **Single-stranded binding proteins (SSBs)** coat the exposed single-stranded DNA segments, stabilizing the DNA and keeping the two strands from pairing (see Figure 14.12). The SSBs are displaced as the replication enzymes make the new polynucleotide chain on the template strands.

For circular chromosomes, such as the genomes of most bacteria, unwinding the DNA will eventually cause the still-wound DNA ahead of the unwinding to become highly twisted. You can visualize this phenomenon with some string. Take two equal lengths of string and twist them around each other. Now tie the two ends of each string together. You have created a model of a circular DNA double helix. Pick anywhere in the circle and pull apart the two pieces of string. The more you pull, the more the region where the two strings are still together becomes highly twisted. In the cell, the twisting of DNA during replication is prevented by **topoisomerase,** which cuts the DNA ahead of the replication fork, turns the DNA on one side of the break in the opposite direction of the twisting force, and rejoins the two strands again (see Figure 14.12).

RNA Primers Provide the Starting Point for DNA Polymerase to Begin Synthesizing a New DNA Chain

DNA polymerases can add nucleotides only to the 3′ end of an existing strand. How can a new strand begin when there is no existing strand in place? The answer lies in a short chain a few nucleotides long called a **primer** that is made of RNA instead of DNA **(Figure 14.13).** The primer is synthesized by the enzyme

A. Bacterial DNA polymerase

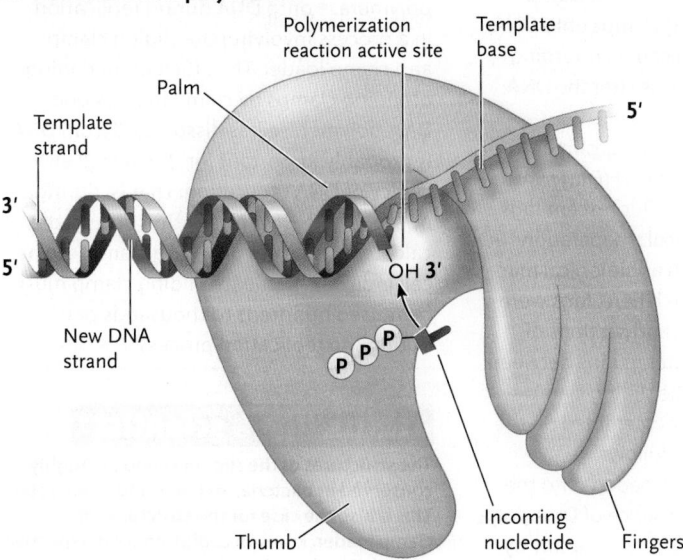

B. How a DNA polymerase and sliding clamp is shown in the book

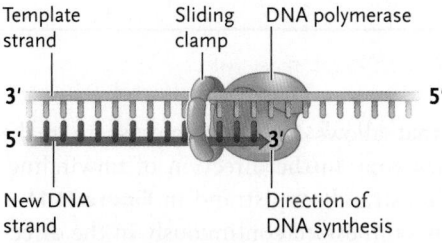

FIGURE 14.11 DNA polymerase structure. (A) Stylized drawing of a bacterial DNA polymerase. The enzyme viewed from the side resembles a human right hand. The polymerization reaction site lies on the palm. When the incoming nucleotide is added, the thumb and fingers close over the site to facilitate the reaction. **(B)** How DNA polymerase is shown in subsequent figures of DNA replication. The figure also shows a sliding clamp anchor the DNA polymerase to the template strand.
© Cengage Learning 2017

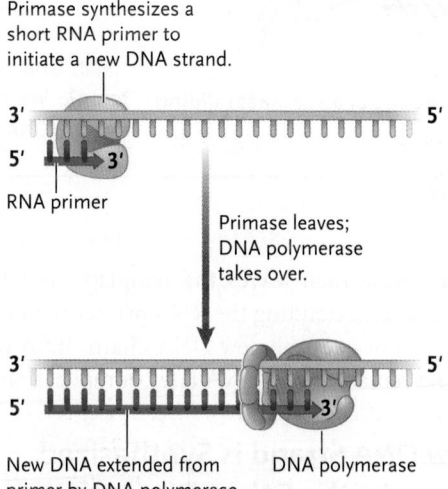

Single-stranded binding proteins (SSBs) coat and stabilize single-stranded DNA, preventing the two strands from reforming double-stranded DNA.

FIGURE 14.12 The roles of DNA helicase, single-stranded binding proteins (SSBs), and topoisomerase in DNA replication.
© Cengage Learning 2017

FIGURE 14.13 Initiation of a new DNA strand by synthesis of a short RNA primer by primase, and the extension of the primer as DNA by DNA polymerase.
© Cengage Learning 2017

DNA Replication in Humans: Loading and unloading the sliding clamp

For efficient DNA replication, DNA polymerases become anchored to ring-shaped sliding clamps (**Figure,** and see Figure 14.11B). A sliding clamp is added to the DNA in an ATP-dependent, enzyme-catalyzed reaction in which a protein called a *clamp loader* opens the clamp, loads it onto the DNA at the junction between the newly synthesized RNA primer and the DNA template, and closes the clamp around the DNA like a watchband. The sliding clamp can then anchor an incoming DNA polymerase to the 3′ end of the primer strand. The sliding clamp–DNA polymerase clamp complex then moves along the template as the enzyme synthesizes the new DNA strand. When DNA synthesis by DNA polymerase finishes, the sliding clamp is removed from the DNA by the clamp loader. Studies of the loading of the sliding clamp and anchoring of DNA polymerase in *E. coli*, T4 bacteriophage, and yeast have shown many similarities in the process, but some differences. Therefore, it was

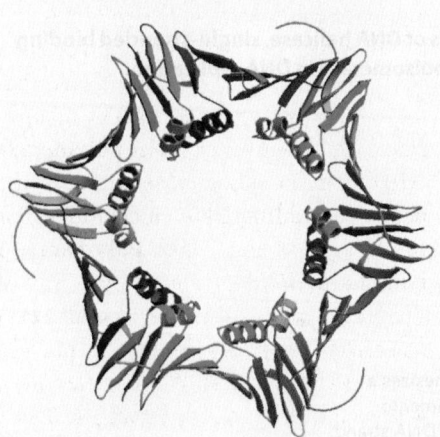

FIGURE Structure of an archaeal sliding clamp.

not possible to infer how the process occurs in humans, where there have been more limited studies.

Research Question
How is the sliding clamp loaded and unloaded onto replicating DNA in humans?

Experiments
Stephen Benkovic's research group at Pennsylvania State University set out to investigate how the sliding clamp and the clamp loader interact, and the timing of the latching and unlatching of the clamp on the DNA in humans. At the outset it was known that a clamp loader and a DNA polymerase cannot bind to a sliding clamp at the same time. The researchers hypothesized that clamp loaders latched the sliding clamps onto DNA, then left during DNA replication, returning to unlatch the sliding clamps after the DNA polymerases left the DNA.

To test their hypothesis, the group used *Förster resonance energy transfer (FRET)*. In outline, the technique uses fluorescent tags to examine the intermolecular separation between substances, such as biological macromolecules. In the research here, tags were added to human proteins and sections of DNA so as to follow the interactions between them, in particular how the sliding clamp–DNA polymerase complex assembled on the DNA. Their results were as follows:

1. When a sliding clamp is loaded onto the DNA template in the absence of DNA polymerase, the clamp is removed rapidly by the clamp loader. This action prevents the buildup of free sliding clamps on the DNA.
2. When a sliding clamp is loaded onto the DNA template in the presence of DNA

polymerase, the polymerase rapidly associates with the clamp and the clamp loader then dissociates from the DNA.
3. When the clamp loader places the closed sliding clamp on the DNA, the loader and clamp are not tightly associated. This allows the polymerase to associate with the clamp to complete the assembly of the complex that will synthesize new DNA. Once the clamp–polymerase complex forms, the clamp loader disengages from the DNA and DNA synthesis begins.

Conclusions
The researchers have advanced our understanding of the assembly of DNA polymerase onto DNA during replication in a process involving the sliding clamp and clamp loader. The efficient unloading of sliding clamps by clamp loaders once DNA polymerase has dissociated from DNA is probably important for the overall efficiency of DNA replication. That is, during S phase, where DNA replication occurs, the calculated number of sliding clamps in the cell indicates that each sliding clamp must be reused hundreds to thousands of times during the replication process.

think like a scientist
The structures of the sliding clamp are highly conserved in bacteria, archaea, and eukaryotes. This is also the case for the structures of the clamp loader. From an evolutionary perspective, what does this indicate?

Source: M. Hedglin et al. 2013. Stepwise assembly of the human replicative polymerase holoenzyme. *eLife* 2:e00278.

---

primase. Primase then leaves the template, and DNA polymerase takes over, extending the RNA primer with DNA nucleotides as it synthesizes the new DNA chain. RNA primers are removed and replaced with DNA later in replication.

One New DNA Strand Is Synthesized Continuously; the Other, Discontinuously

DNA polymerases synthesize a new DNA strand on a template strand in the 5′→3′ direction. Because the two strands of a DNA molecule are antiparallel, only one of the template strands

runs in a direction that allows DNA polymerase to make a 5′→3′ complementary copy in the direction of unwinding. That is, on this template strand—top strand in **Figure 14.14**—the new DNA strand is synthesized continuously in the direction of unwinding of the double helix. However, the other template strand—bottom strand in Figure 14.14—runs in the opposite direction. This means DNA polymerase has to copy it in the direction opposite to the unwinding direction.

How is the new DNA strand made in the opposite direction to the unwinding? The polymerases make this strand in short lengths that are synthesized in the direction opposite to

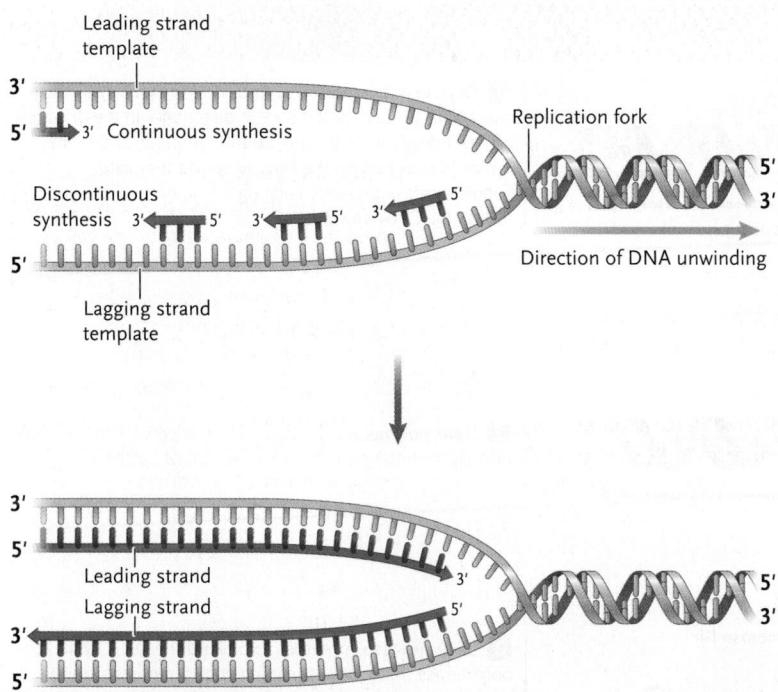

FIGURE 14.14 Replication of antiparallel template strands at a replication fork. Synthesis of the new DNA strand on the top template strand is continuous. Synthesis on the new DNA strand on the bottom template strand is discontinuous—short lengths of DNA are made, which are then joined into a continuous chain. The overall effect is synthesis of both strands in the direction of replication fork movement.

that of DNA unwinding (see Figure 14.14). The short lengths produced by this **discontinuous replication** are then covalently linked into a continuous polynucleotide chain. The short lengths are called **Okazaki fragments,** after Reiji Okazaki, the scientist who first detected them. The new DNA strand synthesized in the direction of DNA unwinding is called the **leading strand** of DNA replication. The template strand for that strand is the **leading strand template.** The strand synthesized discontinuously in the opposite direction is called the **lagging strand.** The template strand for that strand is the **lagging strand template.**

Multiple Enzymes Coordinate Their Activities in DNA Replication

Figure 14.15 shows how the enzymes and proteins we have introduced act in a coordinated way to replicate DNA. Primase initiates all new strands by synthesizing an RNA primer. **DNA polymerase III,** the main polymerase, extends the primer by adding DNA nucleotides. For the lagging strand, **DNA polymerase I** removes the RNA primer at the 5′ end of the previous newly synthesized Okazaki fragment, replacing the RNA nucleotides one by one with DNA nucleotides. RNA nucleotide removal uses the 5′→3′ exonuclease activity of the enzyme. (An exonuclease removes nucleotides from the end of a molecule.) DNA synthesis uses the 5′→3′ polymerization activity. **DNA ligase** (*ligare* = to tie) seals the nick left between the two fragments. The replication process continues in the same way until the entire DNA molecule is copied. **Table 14.1** summarizes the activities of the major enzymes and proteins of DNA replication.

Replication at a fork advances at a rate of about 1,000 nucleotides per second in *E. coli* and other bacteria, and at a rate of about 50 to 100 per second in eukaryotes. The entire process is so rapid that the RNA primers and nicks left by

TABLE 14.1	Major Enzymes and Proteins of DNA Replication	
Enzyme	**Symbol**	**Function**
Helicase		Unwinds DNA helix
Single-stranded binding proteins		Stabilize single-stranded DNA and prevent the two strands at the replication fork from reforming double-stranded DNA
Topoisomerase		Avoids twisting of the DNA ahead of replication fork (in circular DNA) by cutting the DNA, turning the DNA on one side of the break in the direction opposite to that of the twisting force, and rejoining the two strands again
Primase		Synthesizes RNA primer in the 5′→3′ direction to initiate a new DNA strand
DNA polymerase III		Main replication enzyme in *E. coli*. Extends the RNA primer by adding DNA nucleotides to it.
DNA polymerase I		*E. coli* enzyme that uses its 5′→3′ exonuclease activity to remove the RNA of the previously synthesized Okazaki fragment, and uses its 5′→3′ polymerization activity to replace the RNA nucleotides with DNA nucleotides.
Sliding clamp		Tethers DNA polymerase III to the DNA template, making replication more efficient.
DNA ligase		Seals nick left between adjacent fragments after RNA primers replaced with DNA

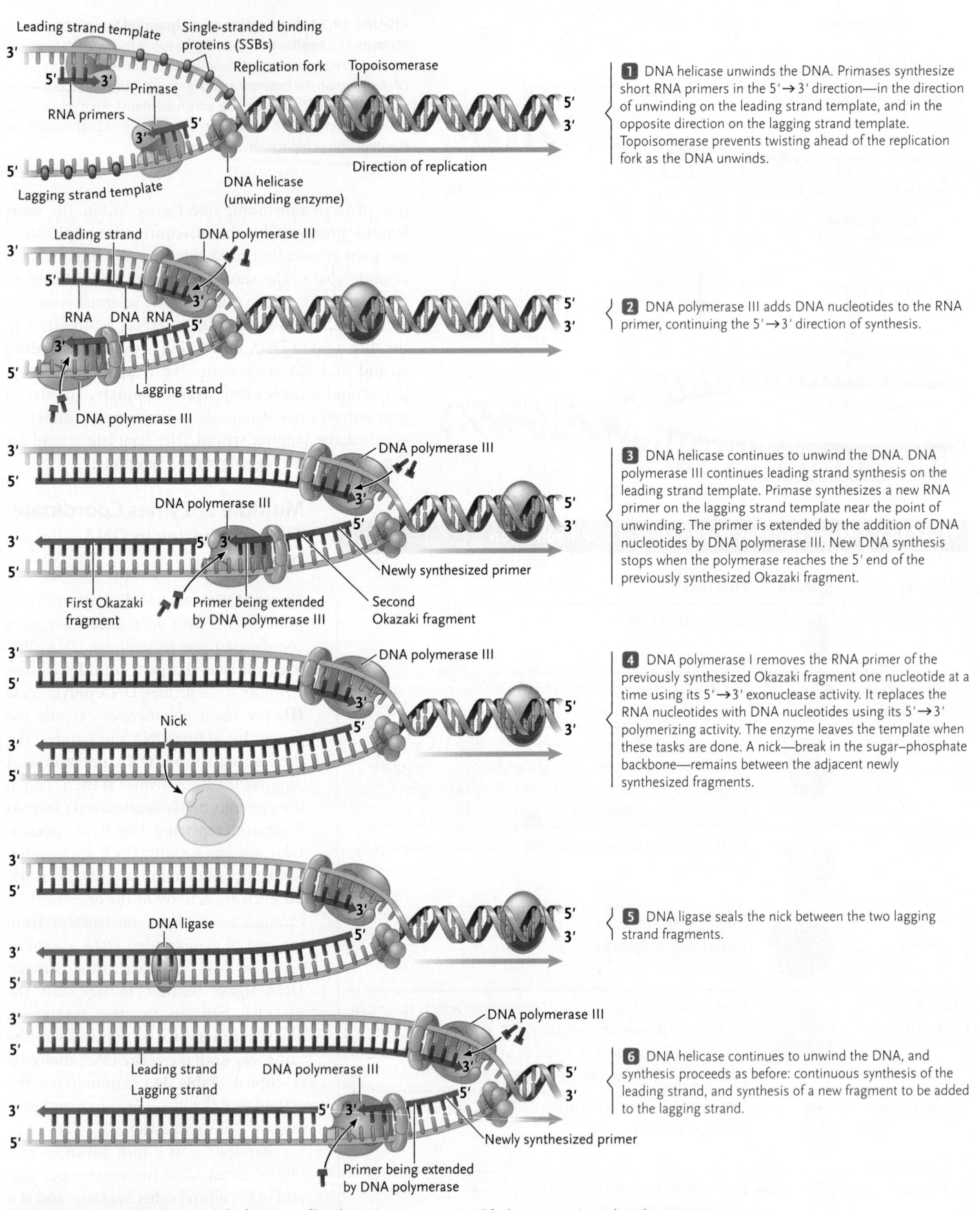

1 DNA helicase unwinds the DNA. Primases synthesize short RNA primers in the 5'→3' direction—in the direction of unwinding on the leading strand template, and in the opposite direction on the lagging strand template. Topoisomerase prevents twisting ahead of the replication fork as the DNA unwinds.

2 DNA polymerase III adds DNA nucleotides to the RNA primer, continuing the 5'→3' direction of synthesis.

3 DNA helicase continues to unwind the DNA. DNA polymerase III continues leading strand synthesis on the leading strand template. Primase synthesizes a new RNA primer on the lagging strand template near the point of unwinding. The primer is extended by the addition of DNA nucleotides by DNA polymerase III. New DNA synthesis stops when the polymerase reaches the 5' end of the previously synthesized Okazaki fragment.

4 DNA polymerase I removes the RNA primer of the previously synthesized Okazaki fragment one nucleotide at a time using its 5'→3' exonuclease activity. It replaces the RNA nucleotides with DNA nucleotides using its 5'→3' polymerizing activity. The enzyme leaves the template when these tasks are done. A nick—break in the sugar–phosphate backbone—remains between the adjacent newly synthesized fragments.

5 DNA ligase seals the nick between the two lagging strand fragments.

6 DNA helicase continues to unwind the DNA, and synthesis proceeds as before: continuous synthesis of the leading strand, and synthesis of a new fragment to be added to the lagging strand.

FIGURE 14.15 Molecular model of DNA replication. The drawings simplify the process. In reality, the enzymes assemble at the fork, replicating both strands from that position as the template strands fold and pass through the assembly.

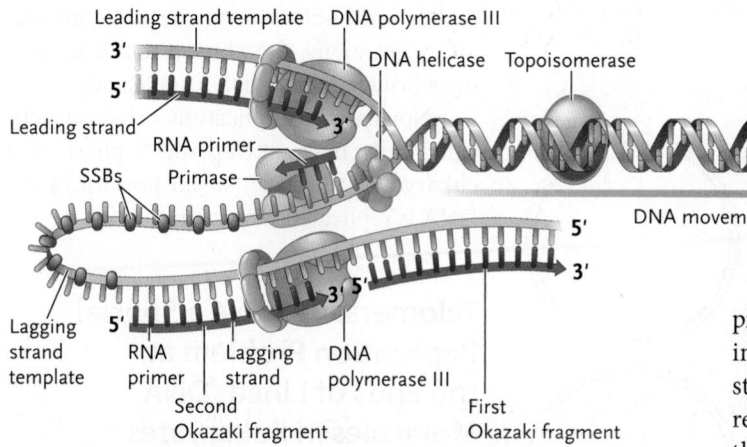

FIGURE 14.16 Simplified depiction of the *E. coli* replisome.
© Cengage Learning 2017

discontinuous synthesis persist for only seconds or fractions of a second. A short distance behind the fork, the new DNA chains are fully continuous and wound into complete DNA double helices. Each helix consists of one "old" and one "new" polynucleotide chain.

Researchers identified the enzymes that replicate DNA through experiments with a variety of bacteria and eukaryotes and with viruses that infect both types of cells. Experiments with the bacterium *E. coli* have provided the most complete information about DNA replication, particularly in the laboratory of Arthur Kornberg at Stanford University. Kornberg received a Nobel Prize in 1959 for his "discovery of the mechanisms in the biological synthesis of deoxyribonucleic acid."

Key Proteins and Enzymes for Replication Are Assembled into a Complex Called a Replisome

As just indicated, replication occurs at a rapid rate. At the same time, replication is an accurate process. Speed and accuracy are achieved by assembling the key proteins and enzymes for replication into a DNA replication complex called a *replication machine* or a **replisome.** That is, whereas we have presented drawings of the molecular steps of DNA replication with

somewhat scattered enzymes and proteins moving along tracks of DNA, in actuality, a replisome sits stationary at the fork and DNA moves through the machine as replication proceeds. In the process, the activities of the proteins and enzymes are integrated tightly. **Figure 14.16** shows a simplified view of the *E. coli* replisome. It contains one copy each of the helicase and the primase, and two copies of DNA polymerase III, one synthesizing the leading strand and the other synthesizing the lagging strand. The looping of the lagging strand template out from the replisome positions the 3′ end of that single-stranded DNA so that primase can synthesize the primer for the next Okazaki fragment. As replication proceeds, the loop becomes smaller and, after the Okazaki fragment is completed, a larger loop again forms for the next Okazaki fragment to be started.

Replisomes also carry out DNA replication in eukaryotic organisms. In this case, the replisomes are attached to specific locations on the nuclear matrix, a network of protein fibers within the nucleus. Multiple replisomes are thought to occur in assemblies known as *replication factories.* Each replisome remains stationary and, as in the *E. coli* replisome, the DNA to be replicated moves through it, entering as a double-stranded molecule and leaving as two identical double-stranded DNA molecules.

Bacterial Chromosomes Have a Single Replication Origin; Eukaryotic Chromosomes Have Multiple Replication Origins

Unwinding at an *ori* within a DNA molecule actually produces two replication forks: two Ys joined together at their tops to form a **replication bubble.** Typically, each of the replication forks moves away from the *ori* as DNA replication proceeds with the events at each fork mirroring those in the other **(Figure 14.17).**

For small circular genomes, such as those found in *E. coli,* and in many bacteria and archaea, there is a single *ori.* DNA replication begins from a single origin in the DNA circle,

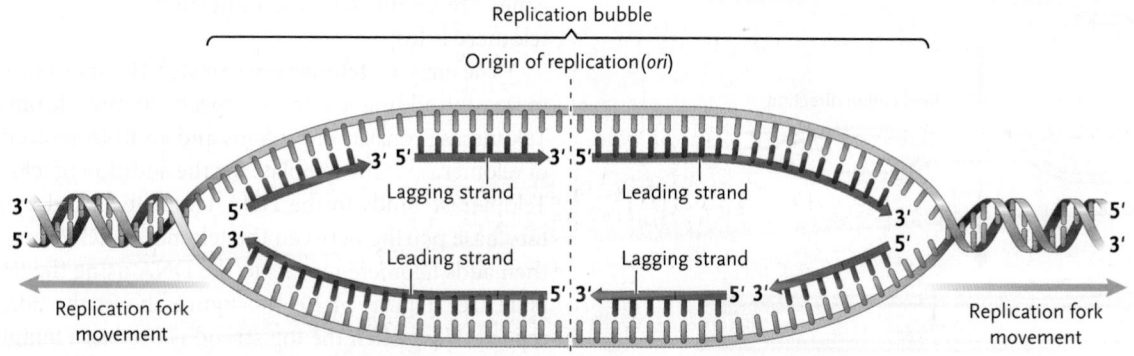

FIGURE 14.17 Synthesis of leading and lagging strands in the two replication forks of a replication bubble formed at an origin of replication.
© Cengage Learning 2017

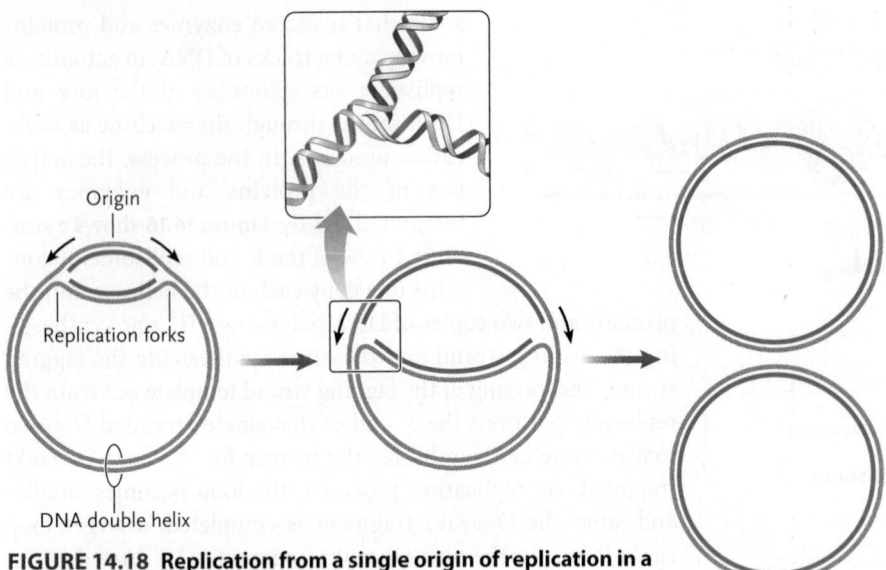

FIGURE 14.18 Replication from a single origin of replication in a circular bacterial chromosome.

© Cengage Learning 2017

forming two forks that travel around the circle in opposite directions. Eventually, the forks meet at the opposite side from the origin to complete replication **(Figure 14.18)**. The replicated chromosomes are distributed actively to the two halves of the bacterial cell. Subsequent binary fission of the cell produces two daughter bacterial cells, each with a copy of the chromosome (discussed in Section 10.5).

Eukaryotic genomes, by contrast, are distributed among several linear chromosomes, each of which can be very long. The average human chromosome, for instance, is about 25 times longer than the *E. coli* chromosome. Nonetheless, replication of long, eukaryotic chromosomes is relatively rapid—sometimes faster than the *E. coli* chromosome—because there are many, sometimes hundreds of origins of replications along eukaryotic chromosomes. Replication initiates at each origin, forming a replication bubble at each **(Figure 14.19)**. Movement of the two forks in opposite directions from each origin extends the

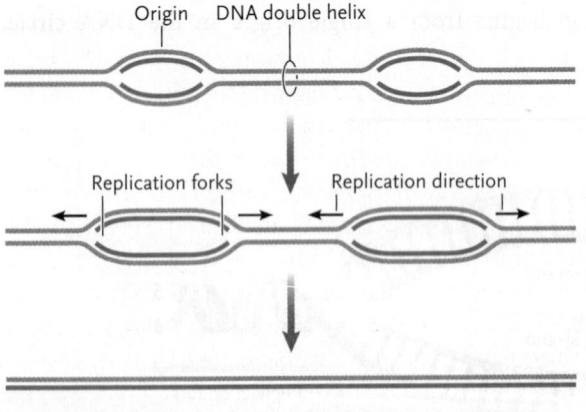

FIGURE 14.19 Replication from multiple origins in the linear chromosomes of eukaryotes.

© Cengage Learning 2017

replication bubbles until the forks eventually meet along the chromosomes to produce fully replicated DNA molecules.

Normally, a replication origin is activated only once during the S phase of a eukaryotic cell cycle, so no portion of the DNA is replicated more than once.

Telomerases Solve a Special Replication Problem at the Ends of Linear DNA Molecules in Eukaryotes

The RNA primer synthesized in DNA replication (see Figures 14.13 and 14.15) produces a problem for replicating the linear chromosomes of eukaryotes. Think about the end of a linear DNA molecule like a eukaryotic chromosome. New DNA synthesis on the 3′→5′ template strand starts with an RNA primer. That primer is removed subsequently, leaving a single-stranded region at the 5′ end of the new DNA strand **(Figure 14.20).** But, because there is no existing nucleotide chain that can be used, DNA polymerase cannot fill in that region with DNA nucleotides. In a similar way, a single-stranded region is produced at the 5′ end of the new strand made starting at the other end of the chromosome. When these new, now shortened DNA strands are used as a template for the next round of DNA replication, the new chromosome will be shorter. Indeed, when most somatic cells go through the cell cycle, the chromosomes shorten with each division. Deletion of genes by such shortening can eventually have lethal consequences for the cell.

In most eukaryotic chromosomes the genes near the ends of chromosomes are protected by a buffer of noncoding DNA. The region of noncoding DNA is called the **telomere** (*telo* = end, *mere* = segment). A telomere consists of short sequences repeated hundreds to thousands of times. In humans, the repeated sequence, the *telomere repeat*, is 5′-TTAGGG-3′ on the template strand (the top strand in Figure 14.20). With each replication, a fraction of the telomere repeats is lost but the genes are unaffected. The buffering fails only when the entire telomere is lost.

The enzyme **telomerase** can stop the shortening of the telomeres by adding telomere repeats to the chromosome ends. Telomerase consists of proteins and an RNA molecule. The RNA of telomerase is the template for the addition of telomere repeats. Telomerase binds to the DNA template strand by complementary base pairing between the telomerase RNA and the DNA. It then adds telomere repeats to the DNA using the RNA as a template (see Figure 14.20; the figure shows the addition of one repeat). Now when the top strand is used as a template for replication by primase and DNA polymerase and the RNA primer is removed, there will be a single-stranded region at the end of the chromosome as before. However, the chromosome has not shortened because of the extra telomere repeats added by the telomerase.

In humans, telomerase is expressed only in the early stages of embryo formation, in the male germ line, during development and differentiation of lymphocytes (white blood cells), and in stem cells. Telomerase is not expressed in somatic cells, meaning telomeres shorten when such cells divide. As a result, somatic cells are capable of only a certain number of mitotic divisions before they stop dividing and die. However, for many cancers, telomerase has become reactivated, preserving chromosome length during the rapid divisions characteristic of cancer. This explains how cancer cells can divide indefinitely and not be limited to a certain number of divisions as a result of telomere shortening.

Apart from the involvement of telomere length regulation in cancers, a lot of other evidence has been collected showing that telomeres are important factors in the regulation of cellular aging. For example, telomeres have also been causally linked to a number of human diseases, including ataxia-telangiectasia (A-T), Bloom syndrome, Fanconi anemia, and Werner syndrome. A common feature of these diseases is that patients have very short telomeres compared with healthy individuals of the same sex and age.

More fundamentally, telomeres also act as caps on the ends of chromosomes. As such, they prevent the staggered ends from being recognized by the cellular machinery that detects broken DNA in need of repair. Erroneously attempting to repair normal chromosome ends could lead to programmed cell death, that is, the purposeful destruction of the cell to avoid propagation of cells with DNA damage.

Elizabeth Blackburn, Carol Greider, and Jack Szostak were awarded a Nobel Prize in 2009 "for the discovery of how chromosomes are protected by telomeres and the enzyme telomerase."

In Eukaryotes, Newly Replicated DNA Is Assembled into Nucleosomes

Recall from Section 10.5 that bacterial chromosomes consist of DNA associated with few proteins. Also recall from Section 10.1 that, by contrast, eukaryotic chromosomes consist of DNA complexed with histones in a way that serves to compact the

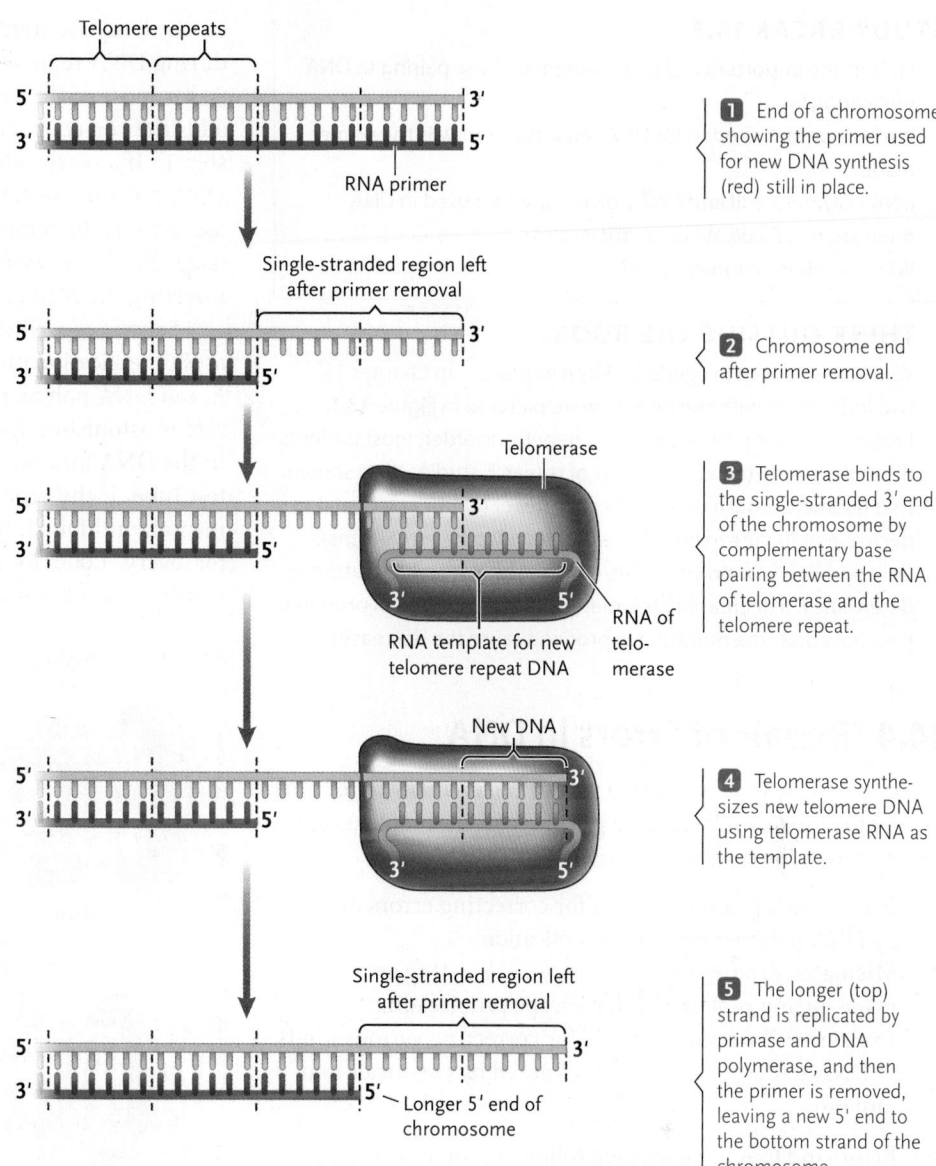

FIGURE 14.20 Addition of a telomere repeat to the 3′ end of a eukaryotic linear chromosome by telomerase.
© Cengage Learning 2017

1 End of a chromosome showing the primer used for new DNA synthesis (red) still in place.

2 Chromosome end after primer removal.

3 Telomerase binds to the single-stranded 3′ end of the chromosome by complementary base pairing between the RNA of telomerase and the telomere repeat.

4 Telomerase synthesizes new telomere DNA using telomerase RNA as the template.

5 The longer (top) strand is replicated by primase and DNA polymerase, and then the primer is removed, leaving a new 5′ end to the bottom strand of the chromosome.

chromatin in the nucleus (see Figure 10.2). The most fundamental structure of the eukaryotic chromosome is the nucleosome in which DNA wraps around an eight-protein histone complex called the nucleosome core particle. The nucleosome organization of eukaryotic chromosomes means that, in addition to DNA replication during the S phase of the cell cycle (see Section 10.2), the nucleosomes also must be duplicated so that, in the end, the entire chromosome structure is duplicated, ready to be partitioned to daughter cells in mitosis or meiosis. Conceptually, chromosome duplication in eukaryotes is straightforward. For replication to take place on a segment of DNA, first, nucleosomes disassemble from the chromosome as the replication fork passes. Then, almost immediately, the newly replicated DNA assembles into nucleosomes. Those nucleosomes are mixtures of parental histones—histones that were in nucleosomes before replication—and newly synthesized histones.

STUDY BREAK 14.3

1. What is the importance of complementary base pairing to DNA replication?
2. Why is a primer needed for DNA replication? How is the primer made?
3. DNA polymerase III and DNA polymerase I are used in DNA replication in *E. coli*. What are their roles?
4. Why are telomeres important?

THINK OUTSIDE THE BOOK

You learned about progeria in *Why it matters . . .* in Chapter 13; two individuals with the disease were pictured in Figure 13.1. Progeria is a rare premature aging genetic disorder; most patients die in their early teens. A number of research studies have shown that telomere length (that is, the number of telomere repeats) decreases with age in humans (and in many other organisms). Collaboratively, or on your own, explore the research literature to determine if a decrease in telomere length is involved in progeria. If so, how does the mutation in progeria cause the decrease?

14.4 Repair of Errors in DNA

Errors are made in DNA during replication or at any other time during the life of a cell. Three types of repair mechanisms operate to correct the errors:

1. **Proofreading,** a mechanism for correcting errors made by DNA polymerase during replication
2. **Mismatch repair,** a mechanism for correcting errors made during replication that escape proofreading
3. **Excision repair,** mechanisms for correcting various kinds of DNA damage, such as those caused by chemicals and radiation

Errors in DNA are corrected following three basic steps:

1. Recognition of the DNA error and its removal
2. Replacing the removed DNA by new DNA synthesis using a repair DNA polymerase
3. Sealing the new DNA to the old DNA using DNA ligase

Step 1 varies with the different error correction mechanisms, whereas steps 2 and 3 are essentially the same.

Base-pair mismatches are corrected, either by a proofreading mechanism carried out during replication by the DNA polymerases themselves or by a DNA repair mechanism that corrects mismatched base pairs after replication is complete.

Proofreading Depends on the Ability of DNA Polymerases to Reverse and Remove Mismatched Bases

DNA polymerases make very few errors as they assemble new nucleotide chains. Most of the mistakes made are **base-pair mismatches,** in which the new inserted nucleotide has a base that is the incorrect one to pair with the base of the nucleotide

on the template strand. A **proofreading** mechanism functions during DNA replication to correct base-pair mismatches. That is, for most of the polymerization reactions, DNA polymerase adds the correct nucleotide to the growing chain (**Figure 14.21,** step 1). If a newly added nucleotide is mismatched (step 2), the DNA polymerase can reverse, using a built-in 3′→5′ exonuclease activity to remove the newly added incorrect nucleotide (step 3). The enzyme then resumes forward synthesis, now inserting the correct nucleotide (step 4).

Several experiments showed that the major DNA polymerases of replication proofread their work. For example, when the *E. coli* DNA polymerase III is fully functional, its overall error rate is astonishingly low, with only about 1 mispair surviving in the DNA for every 1 million nucleotides polymerized in the test tube. If the proofreading activity of the enzyme is experimentally inhibited, the error rate increases to about 1 mistake for every 1,000 to 10,000 nucleotides polymerized. Experiments with eukaryotes have yielded similar results.

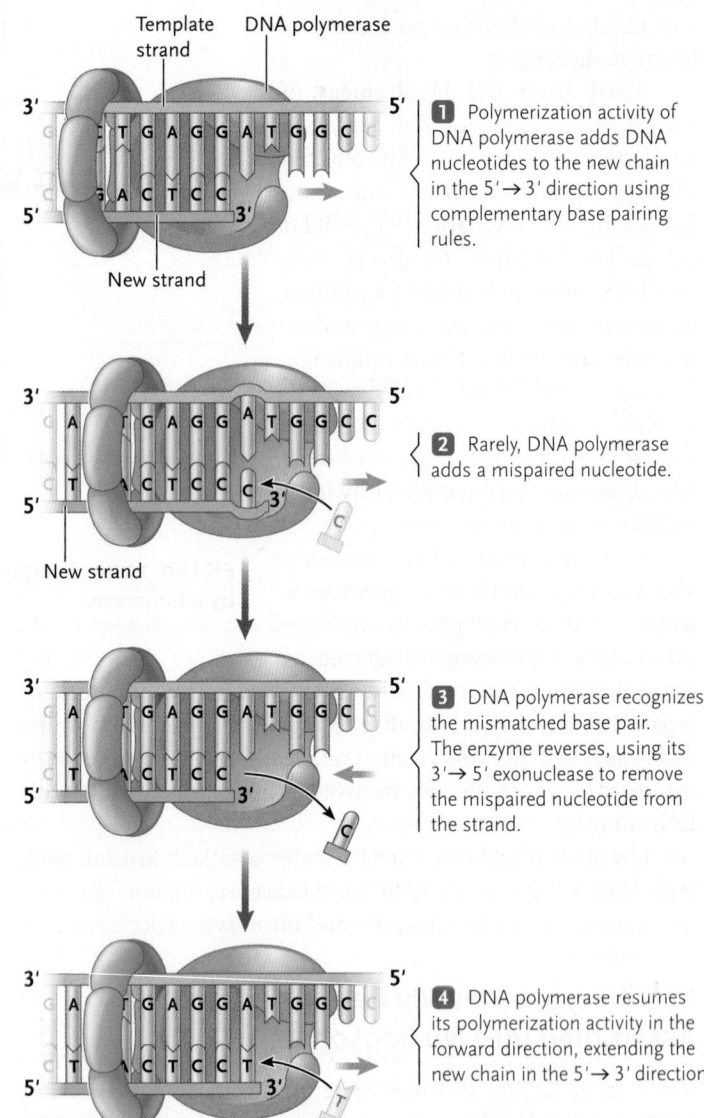

FIGURE 14.21 Proofreading by a DNA polymerase.
© Cengage Learning 2017

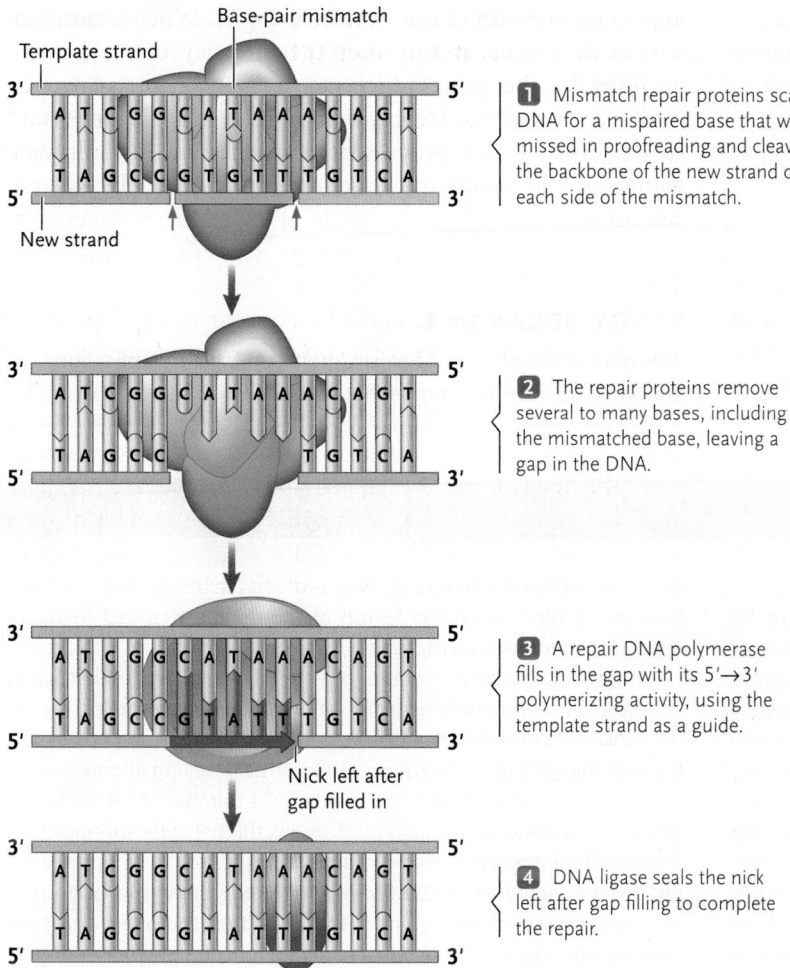

① Mismatch repair proteins scan DNA for a mispaired base that was missed in proofreading and cleave the backbone of the new strand on each side of the mismatch.

② The repair proteins remove several to many bases, including the mismatched base, leaving a gap in the DNA.

③ A repair DNA polymerase fills in the gap with its $5' \rightarrow 3'$ polymerizing activity, using the template strand as a guide.

④ DNA ligase seals the nick left after gap filling to complete the repair.

FIGURE 14.22 Repair of mismatched bases in replicated DNA.
© Cengage Learning 2017

A Mismatch Repair Mechanism Corrects Replication Errors That Escape Proofreading

Proofreading leaves about one error in 10^7 nucleotides copied. A **mismatch repair** mechanism corrects about 99% of those errors, resulting in an extremely high replication accuracy with only about one error in 10^9 nucleotides copied. Mismatch repair mechanisms operate similarly in all organisms and involve the removal of a segment of the DNA chain and its replacement with a newly synthesized segment complementary to the template strand. Evolutionary conservation of many of the mismatch repair proteins from bacteria to yeast to humans indicates that this mechanism is both ancient and vital to all living organisms.

To correct a postreplication error, mismatch repair proteins detect the mispaired base, cut the new DNA strand on each side of the mismatch, and remove a portion of the chain **(Figure 14.22)**. A repair DNA polymerase (DNA polymerase I in *E. coli*) fills in the gap with new DNA. The repair is completed by DNA ligase, which seals the nucleotide chain into a continuous DNA molecule.

Mismatch repair is important in the prevention of cancer. One type of colon cancer, for instance, is caused in part by a mutation in a genes that encodes a mismatch repair protein.

Excision Repair Mechanisms Correct Various Kinds of DNA Damage

DNA damage occurs all the time in cells as a result of chemical events that result in nonbulky damage to bases, such as the chemical modification of bases, or the loss of purine bases from the DNA. "Nonbulky" here means that there is no bulging of the DNA.

Repair of nonbulky damage to bases occurs by a **base-excision repair** mechanism that operates to remove the erroneous base and replace it with the correct one complementary to the base on the other DNA strand. Conceptually, this repair mechanism takes place similarly to mismatch repair (see Figure 14.22), with different proteins involved in the recognition step. After proofreading, base-excision repair is the most important mechanism to fix incorrect or damaged bases.

Some DNA damage is more extensive, causing bulky distortions in the DNA. For example, ultraviolet (UV) light causes thymine dimers to form in DNA in which adjacent thymine bases on one strand of the DNA form chemical bonds with each other **(Figure 14.23)**. Bulky DNA distortions such as thymine dimers, if unrepaired, have serious consequences because DNA polymerase cannot continue DNA synthesis past the distortion. As a result, replication stops, and a blocked replication fork can cause cell death. **Nucleotide-excision repair** is a mechanism that can repair DNA damage such as thymine dimers. This mechanism is similar to the others we have discussed (such as the mechanism shown in Figure 14.22), differing primarily in the recognition step. That is, proteins specific for nucleotide-excision repair recognize the bulky distortion in the DNA and remove a segment of the DNA strand containing the thymine dimer. Repair DNA polymerase and DNA ligase then replace the removed DNA with new DNA and seal it to the rest of the DNA strand.

The autosomal recessive disease in humans, *xeroderma pigmentosum* (XP), is caused by a genetic defect in nucleotide-excision repair. XP patients are extremely sensitive to UV light,

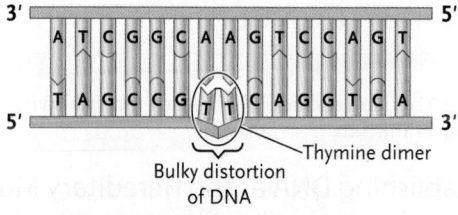

FIGURE 14.23 A thymine dimer in DNA, caused by UV light irradiation.
© Cengage Learning 2017

and typically show early development of cancers, particularly skin cancers. Skin in XP individuals usually shows intense pigmentation, freckling, and warty growths that can lead to malignancies.

Replication Errors and DNA Damage That Remain Unrepaired Are Mutations

Very few replication errors remain in DNA after proofreading and DNA repair. Many other kinds of DNA damage also are repaired routinely by the scavenging DNA repair systems. The errors that persist, although extremely rare, are a primary source of **mutations,** differences in DNA sequence that appear and remain in the replicated copies. When a mutation occurs in a gene, it can alter the property of the protein encoded by the gene, which, in turn, may alter how the organism functions. Hence, mutations are highly important to the evolutionary process because they are the ultimate source of the variability in offspring acted on by natural selection.

STUDY BREAK 14.4

Why is a proofreading mechanism important for DNA replication, and what are the mechanisms that correct errors?

Unanswered Questions

Does size matter?
In this chapter, you learned that the addition of DNA onto telomeres by the enzyme telomerase can counteract the shortening of chromosomes that is predicted by the end replication problem. In humans, telomerase is active in cells destined to become sperm and egg, and in stem cells (cells capable of cell division to replenish tissues such as bone marrow) and other highly proliferative cells. It is also active in most cancers and immortalized cell lines (cells that can grow and divide indefinitely in a culture dish). The presence of telomerase does not result in ever-growing telomeres, however. Cancer cells maintain a specific average telomere length, which can be shorter than the average telomere length for normal somatic cells. Furthermore, some species express high amounts of telomerase activity in all tissues, yet maintain species-specific average telomere lengths.

How do cells measure the length of the telomeric DNA tract?
Several proteins that assemble at the telomere have been identified, including three that bind specifically to the telomeric repeat DNA sequence, and are thought to regulate telomere length. Exactly how this is accomplished is an active area of research. The function of these proteins can be tested by manipulating the protein in question with contemporary genetic approaches and observing effects on the length of the telomeric DNA. For example, if a protein's role is to prevent telomerase access to the telomere, preventing the protein from being synthesized should result in telomere elongation; if it functions to recruit telomerase, telomeres should shorten. Proteins that bind to the double-stranded telomeric DNA sequences can somehow affect access of telomerase to the single-stranded 3′ end. Biochemical experiments and studies of the three-dimensional structure of these proteins are being used to dissect how information about the length of the double-stranded telomere region is communicated to the very terminus where telomerase acts.

Shortened telomeres, whether due to cell division in the absence of telomerase or experimental manipulations of telomere proteins, resemble broken chromosomes and can prevent the cell from progressing through the cell cycle. When normal human cells lacking telomerase are cultured in a dish, they undergo only a finite number of cell divisions. What kind of evidence would rigorously link this behavior specifically to shortened telomeres? A more complicated question is whether short telomeres are involved in organismal aging. Model organisms lacking the telomerase gene are an important tool to address this question, as are humans who have rare defects in telomerase function. Such cases display physiological signs of premature aging.

think like a scientist Contradictory results have been reported regarding the effect of exercise on telomerase and/or telomere length in renewable tissues like blood. In some instances, telomerase levels have increased; in others, no change in telomerase is detectable, but telomere length increases. How might telomere length change if telomerase levels remain modest?

Janis Shampay is a professor of biology at Reed College. Her research interests include the regulation of telomere metabolism and conservation of telomere protein function in nonmammalian model systems. Learn more about her work at http://academic.reed.edu/biology/professors/jshampay/index.html.

Courtesy of Janis Shampay

REVIEW KEY CONCEPTS

For access to MindTap and additional study materials visit www.cengagebrain.com.

14.1 Establishing DNA as the Hereditary Molecule
- Griffith found that a substance derived from killed virulent *Streptococcus pneumoniae* bacteria could transform nonvirulent living *S. pneumoniae* bacteria to the virulent type (Figure 14.2).

- Avery and his coworkers showed that DNA, and not protein or RNA, was the molecule responsible for transforming *S. pneumoniae* bacteria into the virulent form.
- Hershey and Chase showed that the DNA of a phage, not the protein, enters bacterial cells to direct the life cycle of the virus. Taken together, the experiments of Griffith, Avery and his coworkers, and Hershey and Chase established that DNA is the hereditary molecule (Figure 14.3).

14.2 DNA Structure

- Watson and Crick discovered that a DNA molecule consists of two polynucleotide chains twisted around each other into a right-handed double helix. Each nucleotide of the chains consists of deoxyribose, a phosphate group, and either adenine, thymine, guanine, or cytosine. The deoxyribose sugars are linked by phosphate groups to form an alternating sugar–phosphate backbone. The two strands are held together by adenine–thymine (A–T) and guanine–cytosine (G–C) base pairs. Each full turn of the double helix involves 10 base pairs (Figures 14.4 and 14.6).

- The two strands of the DNA double helix are antiparallel.

14.3 DNA Replication

- DNA is duplicated by semiconservative replication, in which the two strands of a parental DNA molecule unwind and each serves as a template for the synthesis of a complementary copy (Figures 14.7–14.9).

- DNA replication is catalyzed by several enzymes. Helicase unwinds the DNA; primase synthesizes an RNA primer used as a starting point for nucleotide assembly by DNA polymerases. DNA polymerases assemble nucleotides into a chain one at a time, in a sequence complementary to the sequence of bases in the template strand. After a DNA polymerase removes the primers and fills in the resulting gaps, DNA ligase closes the remaining single-strand nicks (Figures 14.10–14.13 and 14.15).

- As the DNA helix unwinds, only one template strand runs in a direction allowing the new DNA strand to be made continuously in the direction of unwinding. The other template strand is copied in short lengths that run in the direction opposite to unwinding. The short lengths produced by this discontinuous replication are then linked into a continuous strand (Figures 14.14 and 14.15).

- In both prokaryotes and eukaryotes, the key proteins and enzymes of replication are organized into a complex called a replisome in which the functions of those molecules are integrated tightly (Figure 14.16).

- DNA synthesis begins at sites that act as replication origins and proceeds from the origins as two replication forks moving in opposite directions (Figures 14.17–14.19).

- The ends of eukaryotic chromosomes consist of telomeres, short sequences repeated hundreds to thousands of times. These repeats provide a buffer against chromosome shortening during replication. Although most somatic cells show this chromosome shortening, some cell types do not because they have a telomerase enzyme that adds telomere repeats to the chromosome ends (Figure 14.20).

- The nucleosome organization of eukaryotic chromosomes is duplicated as replication forks move. Nucleosomes are disassembled to allow the replication fork to pass, and then new nucleosomes are assembled soon after the fork has passed.

14.4 Repair of Errors in DNA

- In proofreading, the DNA polymerase reverses and removes the most recently added base if it is mispaired as a result of a replication error. The enzyme then resumes DNA synthesis in the forward direction (Figure 14.21).

- A mismatch repair mechanism corrects most of the errors that escape proofreading. Mismatch repair involves removing a segment of a DNA chain and replacing it with new DNA (Figure 14.22).

- Excision repair mechanisms correct various nonbulky and bulky damage to DNA that occurs as a result of chemical events, such as UV-induced thymine dimers (Figure 14.23).

TEST YOUR KNOWLEDGE

Remember/Understand

1. Working on the Amazon River, a biologist isolated DNA from two unknown organisms, P and Q. He discovered that the adenine content of P was 15% and the cytosine content of Q was 42%. This means that:
   a. the amount of guanine in P is 15%.
   b. the amount of guanine and cytosine combined in P is 70%.
   c. the amount of adenine in Q is 42%.
   d. the amount of thymine in Q is 21%.
   e. it takes more energy to unwind the DNA of P than the DNA of Q.

2. The Hershey and Chase experiment showed that phage:
   a. ^{35}S entered bacterial cells.
   b. ^{32}P remained outside of bacterial cells.
   c. protein entered bacterial cells.
   d. DNA entered bacterial cells.
   e. DNA mutated in bacterial cells.

3. Pyrimidines built from a single carbon ring are:
   a. cytosine and thymine.
   b. adenine, cytosine, and guanine.
   c. adenine and thymine.
   d. cytosine and guanine.
   e. adenine and guanine.

4. Which of the following statements about DNA replication is *false*?
   a. Synthesis of the new DNA strand is from 3′ to 5′.
   b. Synthesis of the new DNA strand is from 5′ to 3′.
   c. DNA unwinds, primase adds RNA primer, and DNA polymerases synthesize the new strand and remove the RNA primer.
   d. Many initiation points exist in each eukaryotic chromosome.
   e. Okazaki fragments are synthesized in the opposite direction from the direction in which the replication fork moves.

5. Which of the following statements about DNA is *false*?
   a. Phosphate is linked to the 5′ and 3′ carbons of adjacent deoxyribose molecules.
   b. DNA is bidirectional in its synthesis.
   c. Each side of the helix is antiparallel to the other.
   d. The binding of adenine to thymine is through three hydrogen bonds.
   e. Avery identified DNA as the transforming factor in crosses between smooth and rough bacteria.

6. In the Meselson and Stahl experiment, the DNA in the parental generation was all ^{15}N^{15}N, and after one round of replication, the DNA was all ^{15}N^{14}N. What DNAs were seen after three rounds of replication, and in what ratio were they found?
   a. one ^{15}N^{14}N; one ^{14}N^{14}N
   b. one ^{15}N^{14}N; two ^{14}N^{14}N
   c. one ^{15}N^{14}N; three ^{14}N^{14}N
   d. one ^{15}N^{14}N; four ^{14}N^{14}N
   e. one ^{15}N^{14}N; seven ^{14}N^{14}N

7. During replication, DNA is synthesized in a 5'→3' direction. This implies that:
   a. the template is read in a 5'→3' direction.
   b. successive nucleotides are added to the 3'–OH end of the newly forming chain.
   c. because both strands are replicated nearly simultaneously, replication must be continuous on both.
   d. ligase unwinds DNA in a 5'→3' direction.
   e. primase acts on the 3' end of the replicating strand.

8. Telomerase:
   a. is active in many cancer cells.
   b. is more active in adult than embryonic cells.
   c. complexes with the ribosome to form telomeres.
   d. acts on unique genes called telomeres.
   e. shortens the ends of chromosomes.

9. Mismatch repair is the ability:
   a. to seal Okazaki fragments with ligase into a continual DNA strand.
   b. of primase to remove the RNA primer and replace it with the correct DNA.
   c. of some enzymes to sense the insertion of an incorrect nucleotide, remove it, and use a DNA polymerase to insert the correct one.
   d. to correct mispaired chromosomes in prophase I of meiosis.
   e. to remove worn-out DNA by telomerase and replace it with newly synthesized nucleotides.

10. **Discuss Concepts** Chargaff's data suggested that adenine pairs with thymine and guanine pairs with cytosine. What other data available to Watson and Crick suggested that adenine–guanine and cytosine–thymine pairs normally do not form?

Apply/Analyze

11. **Discuss Concepts** Eukaryotic chromosomes can be labeled by exposing cells to radioactive thymidine during the S phase of interphase. If cells are exposed to radioactive thymidine during the S phase, would you expect both or only one of the sister chromatids of a duplicated chromosome to be labeled at metaphase of the following mitosis (see Section 10.2)?

12. **Discuss Concepts** If the cells in question 11 finish division and then enter another round of DNA replication in a medium that has been washed free of radioactive label, would you expect both or only one of the sister chromatids of a duplicated chromosome to be labeled at metaphase of the following mitosis?

13. **Discuss Concepts** Strains of bacteria that are resistant to an antibiotic sometimes appear spontaneously among other bacteria of the same type that are killed by the antibiotic. In view of the information in this chapter about DNA replication, what might account for the appearance of this resistance?

Evaluate/Create

14. **Discuss Concepts** During replication, an error uncorrected by proofreading or mismatch repair produces a DNA molecule with a base mismatch at the indicated position:

```
AATTCCGACTCCTATGG
TTAAGGTTGAGGATACC
        ↑
```

The mismatch results in a mutation. This DNA molecule is received by one of the two daughter cells produced by mitosis. In the next round of replication and division, the mutation appears in only one of the two daughter cells. Develop a hypothesis to explain this observation.

15. **Design an Experiment** Design an experiment using radioactive isotopes to show that the process of bacterial transformation involves DNA and not protein.

16. **Apply Evolutionary Thinking** The amino acid sequences of the DNA polymerases found in bacteria show little similarity to those of the DNA polymerases found in eukaryotes and in archaea. By contrast, the amino acid sequences of the DNA polymerases of eukaryotes and archaea show a high degree of similarity. Interpret these observations from an evolutionary point of view.

For selected answers, see Appendix A.

INTERPRET THE DATA

Some cancer treatments target rapidly dividing cells while leaving nonproliferating cells undisturbed. The chemicals 5-fluorouracil (5-FU) and cisplatin (CDDP) are drugs that work in this way. 5-FU inhibits DNA replication, while CDDP binds to DNA causing changes that cannot be corrected by DNA repair enzymes so that programmed cell death (see Chapter 50) is triggered. Researchers suspected that these drugs might be useful in treating human gastric cancer and tested their effectiveness in gastric cancer cells growing in culture. They added 5-FU alone, or 5-FU and CDDP in various timed combinations (schedules) to cultured gastric cancer cells and measured the inhibitory effects of the drugs on cell proliferation compared with untreated cells. The results in the **Figure** show for each schedule the % cell proliferation, meaning the proliferation of treated cells /proliferation of control, untreated cells × 100%.

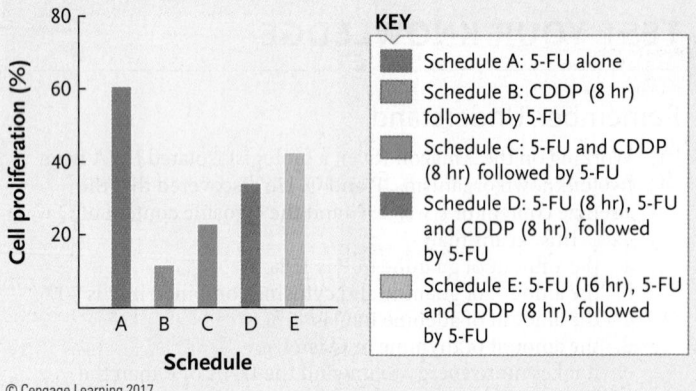

KEY
Schedule A: 5-FU alone
Schedule B: CDDP (8 hr) followed by 5-FU
Schedule C: 5-FU and CDDP (8 hr) followed by 5-FU
Schedule D: 5-FU (8 hr), 5-FU and CDDP (8 hr), followed by 5-FU
Schedule E: 5-FU (16 hr), 5-FU and CDDP (8 hr), followed by 5-FU

© Cengage Learning 2017

1. Which drug schedule was the most effective?
2. How did the drug schedule in the most effective treatment differ from the schedules in all the other treatments?

Source: H. Cho et al. 2002. In-vitro effect of a combination of 5-fluorouracil (5-FU) and cisplatin (CDDP) on human gastric cancer cell lines: Timing of cisplatin treatment. *Gastric Cancer* 5:43–46.

From DNA to Protein

LAGUNA DESIGN/Getty Images

Two transcription factors bound to DNA. In eukaryotes, transcription factors bound to the promoter of a gene recruit RNA polymerase to transcribe the gene.

Why it matters . . . At the end of the 19th century and the beginning of the 20th century, Sir Archibald Garrod, an English physician–biochemist, studied the human metabolic disease *alkaptonuria*. Alkaptonuria is characterized by a hardening and blackening of the cartilage and by urine that turns black when exposed to air. The disease is an autosomal recessive trait. The blackness is the result of the accumulation in the urine of homogentisic acid, a chemical derived from the amino acids phenylalanine and tyrosine. Garrod believed that people with alkaptonuria were unable to break down homogentisic acid as normal individuals can. He reasoned that this was due to the absence or inactivity of the enzyme that normally catalyzes that breakdown reaction and that, in turn, this resulted from the absence of the normal form of a specific gene. Garrod realized that the metabolic pathway for the formation of homogentisic acid could be studied experimentally. Thus, he fed people with alkaptonuria increasing quantities of the normal precursors of homogentisic acid, that is, the amino acids phenylalanine and tyrosine, and observed a corresponding increase in the amount of homogentisic acid excreted in their urine. These results led him to conclude in 1908 that alkaptonuria is an *inborn error of metabolism*. Garrod's work was the first evidence of a specific relationship between genes, enzymes, and metabolism.

The metabolic pathway involving homogentisic acid is now well understood. Homogentisic acid is an intermediate in the breakdown of phenylalanine and tyrosine. People with alkaptonuria have both copies of a gene mutated so that the enzyme required to break down homogentisic acid is not produced or is only produced at low levels. As a result, homogentisic acid accumulates causing the blackening of cartilage and the blackening of urine in air. Phenylketonuria (PKU), a genetic disease you learned about in Chapter 13, also results from a genetic defect that blocks or inhibits the same pathway, in this case an earlier step converting phenylalanine to tyrosine.

As you learned in Chapter 6, enzymes are proteins. One of life's universal truths is that *every protein is assembled on ribosomes according to instructions that are copied from DNA.* In this chapter we trace the reactions by which proteins are made, beginning with the instructions encoded in DNA and leading through RNA to the sequence of amino acids in a protein. Many enzymes and other proteins are players as well as products in this story, as are several kinds of RNA and the cell's protein-making molecular machines, the ribosomes. The same basic steps produce the proteins of all organisms. Our discussion begins with an overview of the entire process, starting with DNA and ending with a finished protein.

15.1 The Connection between DNA, RNA, and Protein

Genes that code for proteins are *protein-coding genes.* In this section you will learn how the DNA sequences of such genes *encode*—specify the amino acid sequences of—proteins. This section also presents an overview of the molecular steps from gene to protein: transcription and translation.

Proteins Are Specified by Genes

Garrod's research described in *Why it matters . . .* had little impact on geneticists at the time. The key research relating genes to enzymes and metabolism was done in the 1940s at Stanford University. Colleagues George Beadle and Edward Tatum used a genetic approach to study metabolism in the orange bread mold *Neurospora crassa,* a haploid fungus with simple nutritional needs. Wild-type *Neurospora*—the form of the mold found in nature—grows readily on a **minimal medium** (MM) consisting of a number of inorganic salts, a carbon source such as sucrose, and the vitamin biotin. Based on the work of Garrod and others, Beadle and Tatum assumed that *Neurospora* uses the simple chemicals in MM to synthesize all of the more complex molecules needed for growth and reproduction, including amino acids for proteins, and precursors for DNA and RNA, as well as most of the vitamins. They reasoned that mutations in genes involved in the biosynthesis of those molecules would result in new requirements for chemical nutrients in the medium in order for the mutant strains to grow.

Mutant strains that require a nutrient supplement in the MM to grow are called **auxotrophs** (*auxo* = increased; *troph* = eater), *auxotrophic mutants* or *nutritional mutants.* Beadle and Tatum realized that auxotrophs should not be able to grow in MM so they had to devise a method to find such mutants. First, they exposed spores of wild-type *Neurospora* to X-rays. An X-ray is a type of *mutagen,* an agent that causes mutations. Organisms treated experimentally with a mutagen are said to be *mutagenized.* Then they plated the mutagenized spores on a complete medium (CM), which is MM supplemented with the compounds that *Neurospora* normally synthesizes from MM components, including amino acids and vitamins. Next they

carried out a **genetic screen,** a technique to search through the mutagenized population of organisms to find individuals with the mutant phenotypes of interest, in this case auxotrophic mutants. For their genetic screen, they tested the colonies that grew on CM to see if they could grow on MM. In their initial experiment reported in 1941, approximately 2,000 colonies were tested. Of them, three were auxotrophs. Beadle and Tatum then tested each auxotroph for growth on MM with a variety of single nutrients added to it. In this way, they discovered that the three mutant strains required vitamin B_6 (pyridoxine), vitamin B_1 (thiamine), and para-aminobenzoic acid, respectively. Subsequent experiments led to the isolation of further auxotrophic mutants requiring a single nutrient in order to grow, including other vitamins, a variety of amino acids, and other compounds such as adenine. The researchers hypothesized that the nutrients needed by the mutant strains were synthesized by wild-type *Neurospora* but that synthesis was blocked in the mutant strains as a result of the genetic mutations.

In 1944, Adrian Srb and Norman Horowitz (a graduate student and postdoctoral fellow, respectively, in Beadle's laboratory), extended Beadle and Tatum's approach to study the steps in a biosynthesis pathway. The theory was that the synthesis of a compound from constituents of MM in many instances would involve a number of steps. If each step was under the control of a different gene, then mutations in any one of several different genes could result in the same nutrient requirement. The hypothesis was that those mutants could be distinguished by what intermediates in the biosynthesis pathway could satisfy the growth defect on MM. Srb and Horowitz tested this hypothesis using *arg* (arginine) auxotrophs isolated using the Beadle and Tatum method. These mutants require the amino acid arginine in order to grow. Their results showed that each of the three *arg* genes encoded an enzyme that controlled a different step in the arginine biosynthesis pathway **(Figure 15.1).** Based on this research, and research by others involving a number of different experimental organisms, Horowitz put forward the **one gene–one enzyme hypothesis:** "A large class of genes exists in which each gene controls the synthesis of, or the activity of, but a single enzyme." The central role of genes in controlling molecular reactions in cells had been demonstrated convincingly. Beadle and Tatum were awarded the Nobel Prize in 1958 "for their discovery that genes act by regulating definite chemical events."

As already mentioned, most enzymes are just one form of proteins, the amino acid-containing macromolecules that carry out many vital functions in living organisms. A functional protein consists of one or more subunits, called *polypeptides.* The protein hemoglobin, for instance, is made up of four polypeptides, two each of an α subunit and a β subunit. The ability of hemoglobin to transport oxygen is a functional property belonging only to the complete protein, and not to any of the polypeptides individually. A different gene encodes each distinct polypeptide, meaning that two different genes are needed to specify the hemoglobin protein: one for the α polypeptide

FIGURE 15.1   Experimental Research

The Gene–Enzyme Relationship

Question: What is the relationship between genes and enzymes?

Experiment: Adrian Srb and Norman Horowitz studied three *arg* auxotrophic mutants of *Neurospora crassa*. The mutants had been isolated because they did not grow on MM (minimal medium) but they did grow on MM + arginine. The researchers determined how each mutant grew in MM (minimal medium) with and without ornithine and citrulline intermediates involved in the arginine biosynthesis pathway in other organisms.

Results:

		Growth on MM +			
Strain		Nothing	Ornithine	Citrulline	Arginine
Wild type (control)	Grows on MM, and on all other supplemented media.	Growth			
***arg*-4 mutant**	Does not grow on MM; grows on all other supplemented media.	No growth			
***arg*-2 mutant**	Does not grow on MM; grows if citrulline or arginine is in the medium, but not if only ornithine is in the medium.				
***arg*-1 mutant**	Does not grow on MM; grows if arginine is in the medium, but not if only ornithine or citrulline is in the medium.				

Conclusion: Each of the three *arg* mutants showed a different pattern of growth on the supplemented MM. Srb and Horowitz concluded that the biosynthesis of arginine occurs in a series of steps, with each step controlled by a gene that encodes the enzyme for the step.

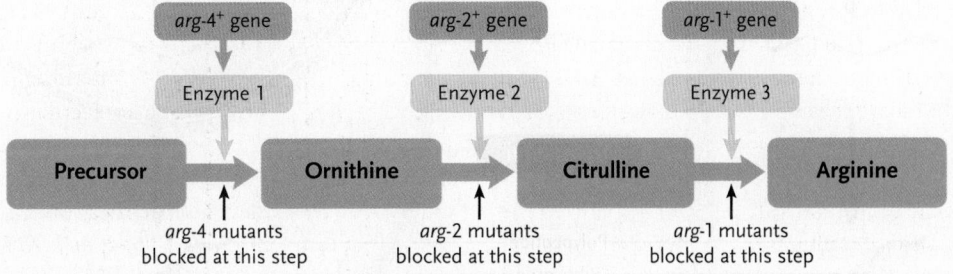

The logic is as follows, working from the end of the pathway back to its beginning:

- The *arg*-1 mutant grows on MM + arginine, but not on MM + citrulline or ornithine; this means that the mutant is blocked at the last step in the pathway that produces arginine.
- The *arg*-2 mutant grows on MM + arginine or citrulline, but not on MM + ornithine; this means that *arg*-2 is blocked in the pathway between ornithine and citrulline.
- The *arg*-4 mutant grows on MM + arginine, citrulline, or ornithine; this means that *arg*-1 is blocked in the pathway before ornithine is made.

Observe
Hypothesize
Predict
Experiment
Interpret

Consider two couples in which each partner has the recessive trait of albinism (lack of normal skin color; see Section 12.1). Each couple has four children. The children of the first couple all have albinism, whereas the children of the second couple all have normal skin color. How can you explain these results?

© Cengage Learning 2017

Source: A. M. Srb and N. H. Horowitz, 1944. The ornithine cycle in Neurospora and its genetic control. *Journal of Biological Chemistry* 154:129–139.

and one for the β polypeptide. Because some proteins consist of more than one polypeptide, and not all proteins are enzymes, the one gene–one enzyme hypothesis was modified later to the **one gene–one polypeptide hypothesis.** It is important to keep in mind the distinction between a protein, the functional molecule, and a polypeptide, the molecule specified by a gene, as we discuss transcription and translation in the rest of this chapter.

Beadle and Tatum's research approach exemplifies the classical genetics approach for identifying a gene or genes responsible for a phenotype of interest. That is, geneticists generate random mutations in an organism, and then they search for mutants with an altered phenotype of interest using a genetic screen. The mutants obtained are then analyzed in more detail by genetic, biochemical, and molecular methods.

The Pathway from Gene to Polypeptide Involves Transcription and Translation

The pathway from gene to polypeptide has two major steps, *transcription* and *translation*. **Transcription** is the mechanism by which the information encoded in DNA is made into a complementary RNA copy. It is called transcription because the information in one nucleic acid type is transferred to another nucleic acid type. **Translation** is the use of the information encoded in an RNA molecule to assemble amino acids into a polypeptide. It is called translation because the information in a nucleic acid, in the form of nucleotides, is converted into a different kind of molecule—amino acids. In 1956, Francis Crick gave the name **central dogma** to the information flow:

That flow of information more appropriately is:

DNA $\longrightarrow$ RNA $\longrightarrow$ Polypeptide

In transcription, the enzyme RNA polymerase copies the DNA sequence of a gene into an RNA sequence. The process is similar to DNA replication, except that only one of the two DNA strands—the **template strand**—is copied into an RNA strand, and only part of the DNA sequence of the genome is copied in any cell at any given time. A gene encoding a polypeptide is a **protein-coding gene,** and the RNA transcribed from it is called **messenger RNA (mRNA).**

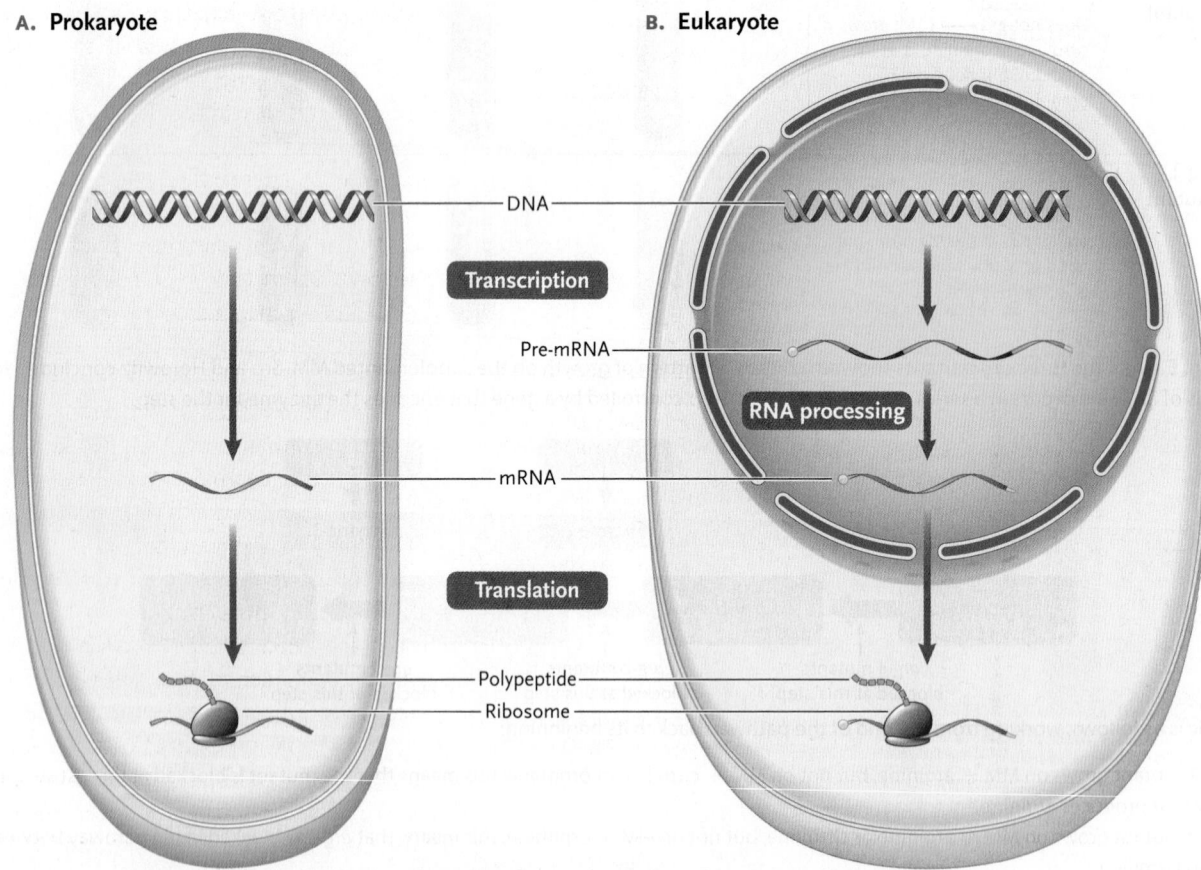

FIGURE 15.2 Transcription and translation in: (A) prokaryotes; and **(B)** eukaryotes. In prokaryotes, RNA polymerase synthesizes an mRNA molecule that is ready for translation on ribosomes. In eukaryotes, RNA polymerase synthesizes a precursor-mRNA (pre-mRNA molecule) that is processed to produce a translatable mRNA. That mRNA exits the nucleus through a nuclear pore and is translated on ribosomes in the cytoplasm.

© Cengage Learning 2017

In translation, an mRNA associates with a *ribosome*, a particle on which amino acids are linked into polypeptide chains. As the ribosome moves along the mRNA, the amino acids specified by the mRNA are joined one by one to form the polypeptide encoded by the gene.

Transcription and translation occur in all organisms. Both processes are similar, but not identical, in prokaryotes and eukaryotes **(Figure 15.2).** One key difference is that in eukaryotes, transcription in the nucleus produces a precursor-mRNA (pre-mRNA) that must be altered to generate the functional mRNA. Each end of the pre-mRNA is modified, and then extra segments within its sequence are removed by *RNA processing*. The result is the functional mRNA that exits the nucleus and is translated in the cytoplasm. In prokaryotes, transcription in the cytoplasm produces a functional mRNA directly, with no modifications.

Not all genes encode polypeptides; some encode **noncoding RNAs,** various RNA molecules that are not translated to produce polypeptides. They play roles in transcription and translation, and in some other processes in the cell such as the regulation of gene expression. You will learn about some of these important RNA molecules later in the chapter and in some other chapters (also see *Molecular Insights* in Chapter 13).

The Genetic Code Is Written in Three-Letter Words Using a Four-Letter Alphabet

Conceptually, the transcription of DNA into RNA is straightforward. The DNA "alphabet" consists of the four letters A, T, G, and C, representing the four bases of DNA nucleotides: adenine, thymine, guanine, and cytosine. The RNA "alphabet" consists of the four letters A, U, G, and C, representing the four RNA bases: adenine, uracil, guanine, and cytosine. In other words, the nucleic acids share three of the four bases but differ in the other one; T in DNA is equivalent to U in RNA. Translation of mRNA to form a polypeptide is more complex because, although there are four RNA bases, there are 20 amino acids. How is nucleotide information in an mRNA translated into the amino acid sequence of a polypeptide?

BREAKING THE GENETIC CODE The nucleotide information that specifies the amino acid sequence of a polypeptide is called the **genetic code.** Scientists realized that the four bases in an mRNA (A, U, G, and C) would have to be used in combinations of at least three to provide the capacity to code for 20 different amino acids. One- and two-letter words were eliminated because if the code used one-letter words, only four different amino acids could be specified (that is, 4^1); if two-letter words were used, only 16 different amino acids could be specified (that is, 4^2). But if the code used three-letter words, 64 different amino acids could be specified (that is, 4^3), more than enough to specify 20 amino acids. Experimental research showed that the genetic code is a three-letter code. Each three-letter word (triplet) of the code is called a **codon. Figure 15.3** illustrates the relationship between codons in a gene, codons in an mRNA, and the amino acid sequence of a polypeptide. The three-letter codons in DNA are first transcribed into complementary three-letter RNA codons. The process is similar to DNA replication except that in mRNA, the complement to adenine (A) in the template strand is uracil (U) instead of thymine (T) as in DNA replication.

How do the RNA codons correspond to the amino acids? The identity of most of the codons was established in 1964 by Marshall Nirenberg and Philip Leder of the National Institutes of Health (NIH). These researchers found that short, artificial mRNAs of codon length—three nucleotides—could bind to ribosomes in a test tube and cause a single transfer RNA (tRNA), with its linked amino acid, to bind to the ribosome. (As you will learn in Section 15.4, tRNAs are a special class of RNA molecules that bring amino acids to the ribosome for assembly into the polypeptide chain.) Nirenberg and Leder then made 64 of the short mRNAs, each

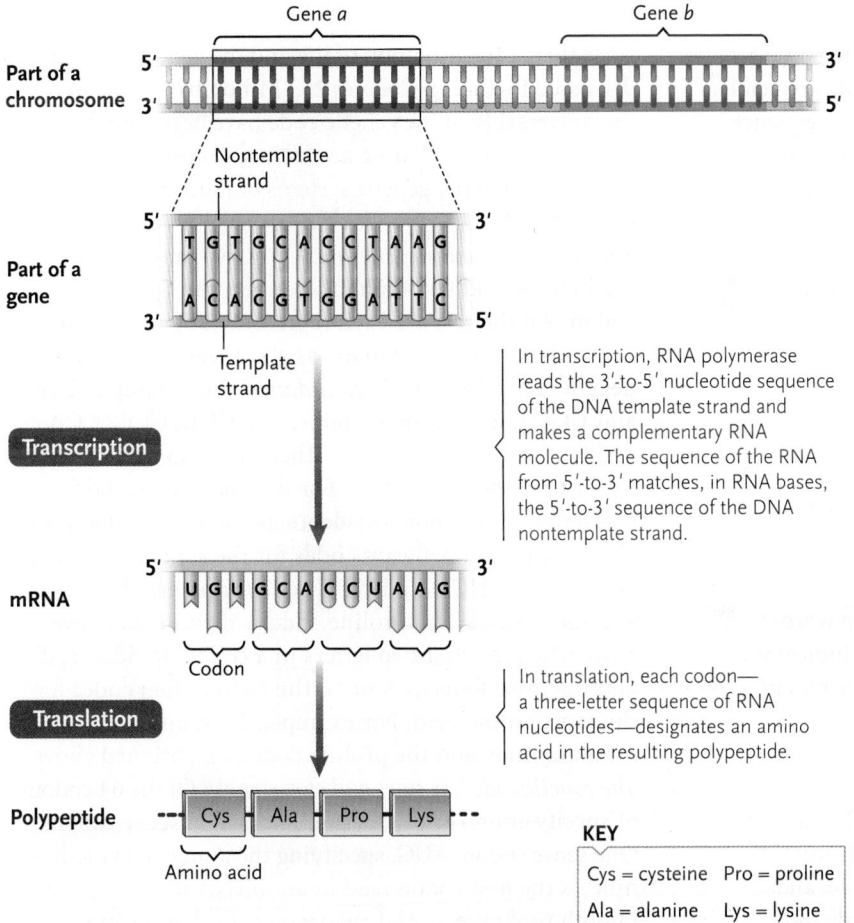

In transcription, RNA polymerase reads the 3'-to-5' nucleotide sequence of the DNA template strand and makes a complementary RNA molecule. The sequence of the RNA from 5'-to-3' matches, in RNA bases, the 5'-to-3' sequence of the DNA nontemplate strand.

In translation, each codon—a three-letter sequence of RNA nucleotides—designates an amino acid in the resulting polypeptide.

KEY

Cys = cysteine   Pro = proline
Ala = alanine   Lys = lysine

FIGURE 15.3 Relationship between a gene, codons in an mRNA, and the amino acid sequence of a polypeptide.

© Cengage Learning 2017

consisting of a different, single codon. They added the mRNAs, one at a time, to a test tube containing ribosomes and all the different tRNAs, each linked to its own amino acid. The idea was that each single-codon mRNA would link to the tRNA in the mixture that carried the amino acid corresponding to the codon. The experiment worked for 50 of the 64 codons, allowing those codons to be assigned to amino acids definitively.

Another approach, carried out in 1966 by H. Gobind Khorana and his coworkers at Massachusetts Institute of Technology, used long, artificial mRNA molecules containing only one nucleotide repeated continuously, or different nucleotides in repeating patterns. The researchers added each artificial mRNA to ribosomes in a test tube, and analyzed the sequence of amino acids in the polypeptide chain made by the ribosomes. For example, an artificial mRNA containing only uracil nucleotides in the sequence UUUUUU… resulted in a polypeptide containing only the amino acid phenylalanine: UUU must be a codon for phenylalanine. Khorana's approach, combined with the results of Nirenberg and Leder's experiments, identified the coding assignments of all the codons. Nirenberg and Khorana received a Nobel Prize in 1968 "for their interpretation of the genetic code and its function in protein synthesis."

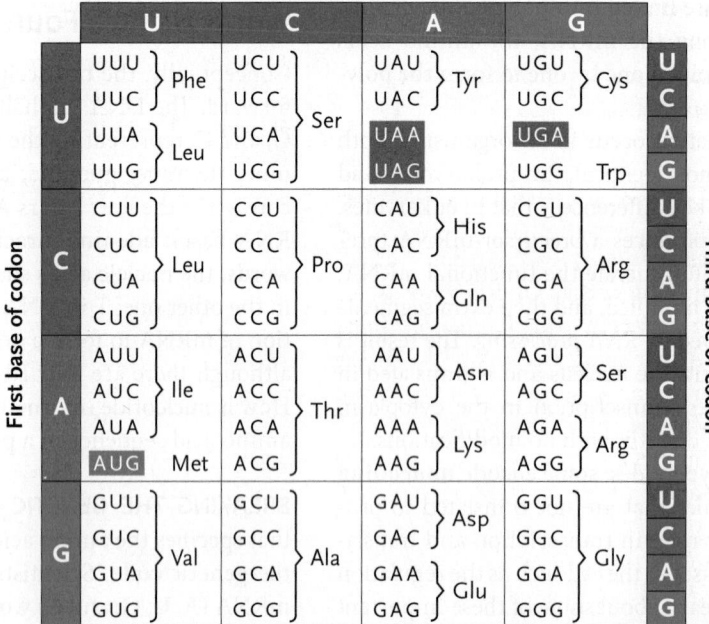

FIGURE 15.4 The genetic code, written in the form in which the codons appear in mRNA. The AUG initiator codon, which codes for methionine, is shown in green; the three terminator codons are boxed in red. The Key shows both the three-letter and one-letter abbreviations for each amino acid.
© Cengage Learning 2017

FEATURES OF THE GENETIC CODE Figure 15.4 shows the genetic code of the 64 possible codons. By convention, scientists write the codons in the 5′→3′ direction, as they appear in mRNAs. The features of the genetic code are:

1. *The genetic code is a three-letter code.*
2. *The genetic code is* **commaless,** meaning that the words of the nucleic acid code are sequential, with no indicators such as molecular commas or spaces to mark the end of one codon and the beginning of the next.
3. *The genetic code is* **universal.** With a few exceptions, the same codons specify the same amino acids in all living organisms, and also in viruses. In other words, the eukaryotic translation machinery can read a prokaryotic mRNA to make the same polypeptide as in the prokaryote, and vice versa. The universality of the nucleic acid code indicates that it was established in its present form very early in the evolution of life and has remained virtually unchanged

since then. (The evolution of life and the genetic code are discussed further in Chapter 25.) Minor exceptions to the universality of the genetic code have been found in a few organisms including a yeast, some protists, a prokaryote, and in the genetic systems of mitochondria and chloroplasts.

4. *The genetic code is degenerate.* Only two amino acids, methionine and tryptophan, are specified by a single codon. All the rest are each represented by more than one codon, some by as many as six, a feature known as **degeneracy** (also called *redundancy*). For example, UGU and UGC both specify cysteine, and CCU, CCC, CCA, and CCG all specify proline. There are also particular patterns in the degeneracy. For instance, when the first two letters in a codon are identical and the third letter is U or C, the codon always codes for the same amino acid. For example, UUU and UUC code for phenylalanine, and the cysteine and proline codons mentioned above. Also, when the first two letters in a codon are identical and the third letter is A or G, the codon often codes for the same amino acid. For example, CAA and CAG code for glutamine, and the proline codons mentioned above.

5. *The genetic code has start and stop signals.* Of the 64 codons, 61 specify amino acids. These are known as **sense codons.** One sense codon, AUG, specifying the amino acid methionine, is the first codon read in an mRNA in translation in both prokaryotes and eukaryotes. In that position, AUG is called a **start codon** or **initiator codon.** The three codons that do not specify amino acids—UAA, UAG, and

UGA—are **stop codons** (also called **nonsense codons** and **termination codons**) that act as "periods" indicating the end of a polypeptide-encoding "sentence." When a ribosome reaches one of the stop codons, polypeptide synthesis stops and the new polypeptide chain is released from the ribosome.

The genetic code can be read correctly only by starting at the right place—at the first base of the first three-letter codon at the beginning of a coded message—and reading three nucleotides at a time from this beginning codon. In other words, for each mRNA, there is only one correct **reading frame,** the linear sequence of codons in mRNA that specify amino acids during translation beginning at a particular start codon. By analogy, if you read the message SADMOMHASMOPCUTOFFBOYTOT three letters at a time, starting with the first letter of the first "codon," you would find that a mother reluctantly had her small child's hair cut. However, if you start incorrectly at the second letter of the first codon, you read the gibberish message ADM OMH ASM OPC UTO FFB OYT OT.

STUDY BREAK 15.1

1. On the basis of the work with auxotrophic mutants of the fungus *Neurospora crassa,* the one gene–one enzyme hypothesis was proposed. Why is it now known as the one gene–one polypeptide hypothesis?
2. If the codon were five bases long, how many different codons would exist in the genetic code?

THINK OUTSIDE THE BOOK

The section you have just read states that the genetic code is not completely universal. Use the Internet or research literature to determine what variants of the genetic code exist and in which organisms they occur.

15.2 Transcription: DNA-Directed RNA Synthesis

An organism's genome contains a large number of genes. For example, the human genome sequence has about 19,900 protein-coding genes. Transcription is the process of transferring the information coded in the DNA sequences of particular genes to complementary RNA copies. As you have learned, some of those genes are protein-coding genes that encode mRNAs that are translated; others are noncoding RNA genes that encode RNAs that are not translated, such as ribosomal RNAs (rRNAs) and transfer RNAs (tRNAs). Much of our initial understanding of transcription came from experimental research done with the model organism, *Escherichia coli.* We now know that transcription is generally similar, but not identical, in prokaryotes and eukaryotes. Throughout this section, our focus is eukaryotic transcription, and we will point out the important differences between eukaryotic and bacterial processes.

Transcription Proceeds in Three Stages

Figure 15.5 illustrates the general organization of a eukaryotic protein-coding gene and shows how it is transcribed. The gene consists of two main parts: a **promoter,** which is a control sequence for transcription, and a **transcription unit,** the section of the gene that is copied into an RNA molecule.

Transcription takes place in three stages:

1. In **initiation,** the molecular machinery that carries out transcription assembles at the promoter and begins synthesizing an RNA copy of the gene (Figure 15.5, steps 1 and 2). The molecular machinery includes **transcription factors (TFs),** proteins that bind to the promoter in the area of a special *initiator sequence,* and an **RNA polymerase,** an enzyme that catalyzes the assembly of RNA nucleotides into an RNA strand. In eukaryotes, RNA polymerase II is the enzyme that transcribes protein-coding genes. Unlike DNA polymerases, RNA polymerases do not require a primer to start a chain.

   The first initiator sequence identified, the **TATA box,** is shown in Figure 15.5. The TATA box has a 6-base pair *consensus sequence* (shown in the figure), where a **consensus sequence** is the series of nucleotides found most frequently at the particular sites in the different sequences which occur in nature. From analysis of eukaryotic genome sequences, it has now been shown to occur in only a minority of protein-coding gene promoters. In TATA-less promoters, other initiation sequences are present and function in the same way.

   In the initiation process, the DNA is unwound in the front of the RNA polymerase to expose the template strand. RNA polymerase II then begins RNA synthesis at the transcription start point (Figure 15.5, stage 3); the TFs are then released. As shown in Figure 15.3, only one of the two DNA strands is copied into an mRNA strand during transcription. The RNA strand is made in the $5' \rightarrow 3'$ direction using the $3' \rightarrow 5'$ DNA strand as template. Therefore, we refer to the beginning of the RNA strand as the *5' end,* and the other end as the *3' end.* You will recall that synthesis of new DNA strands also occurs in the $5' \rightarrow 3'$ direction.

   The sequence of the mRNA strand is determined by the DNA template strand and proceeds in a manner similar to DNA replication with new RNA nucleotides being added according to complementary base pairing rules. The one exception is that, when A (adenine) appears in the DNA template strand, a U (uracil) is paired with it in the RNA transcript (see Figure 15.4). Uracil has base-pairing properties similar to thymine.

2. In **elongation,** RNA polymerase II moves along the gene extending the RNA chain, with the DNA continuing to unwind ahead of the enzyme (Figure 15.5, step 4).

TRANSCRIPTION OF A EUKARYOTIC PROTEIN-CODING GENE

Transcription has three stages: initiation, elongation, and termination. RNA polymerase moves along the gene, separating the two DNA strands to allow RNA synthesis in the 5'→3' direction using the 3'→5' DNA strand as template.

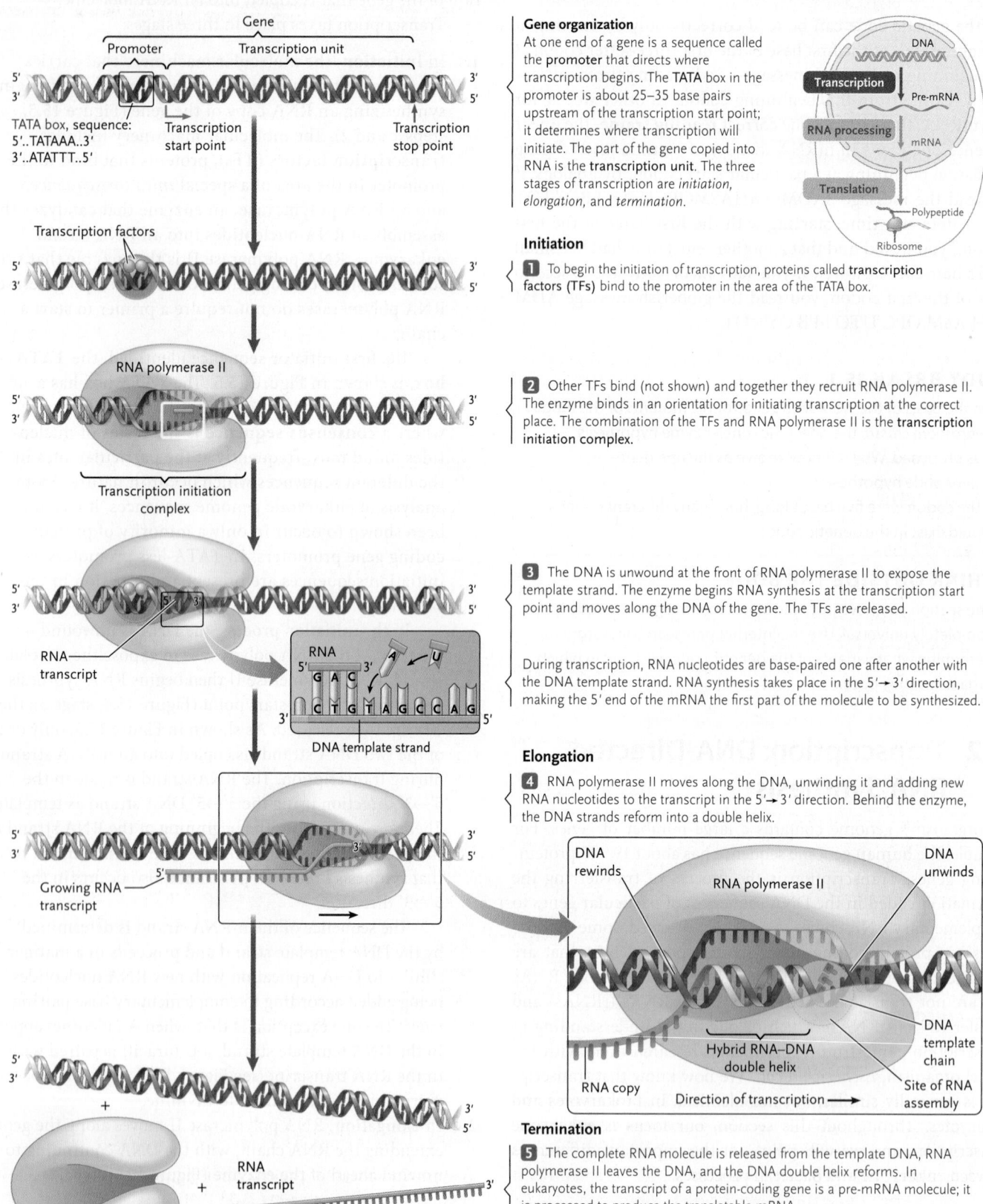

Gene organization

At one end of a gene is a sequence called the **promoter** that directs where transcription begins. The **TATA** box in the promoter is about 25–35 base pairs upstream of the transcription start point; it determines where transcription will initiate. The part of the gene copied into RNA is the **transcription unit**. The three stages of transcription are *initiation*, *elongation*, and *termination*.

Initiation

1 To begin the initiation of transcription, proteins called **transcription factors (TFs)** bind to the promoter in the area of the TATA box.

2 Other TFs bind (not shown) and together they recruit RNA polymerase II. The enzyme binds in an orientation for initiating transcription at the correct place. The combination of the TFs and RNA polymerase II is the **transcription initiation complex.**

3 The DNA is unwound at the front of RNA polymerase II to expose the template strand. The enzyme begins RNA synthesis at the transcription start point and moves along the DNA of the gene. The TFs are released.

During transcription, RNA nucleotides are base-paired one after another with the DNA template strand. RNA synthesis takes place in the 5'→3' direction, making the 5' end of the mRNA the first part of the molecule to be synthesized.

Elongation

4 RNA polymerase II moves along the DNA, unwinding it and adding new RNA nucleotides to the transcript in the 5'→3' direction. Behind the enzyme, the DNA strands reform into a double helix.

Termination

5 The complete RNA molecule is released from the template DNA, RNA polymerase II leaves the DNA, and the DNA double helix reforms. In eukaryotes, the transcript of a protein-coding gene is a pre-mRNA molecule; it is processed to produce the translatable mRNA.

think like a scientist Would you expect *E. coli* to be able to transcribe the gene shown in the figure? Explain your answer.

© Cengage Learning 2017

3. In **termination,** transcription ends and the RNA molecule—the transcript—and the RNA polymerase II are released from the DNA template (Figure 15.5, step 5).

Roger Kornberg of Stanford University received a Nobel Prize in 2006 "for his studies of the molecular basis of eukaryotic transcription."

Similarities and differences in transcription of eukaryotic and bacterial protein-coding genes are as follows:

- Gene organization is the same, although the specific sequences in the promoter where the transcription apparatus assembles differ.
- In eukaryotes, RNA polymerase II, the enzyme that transcribes protein-coding genes, cannot bind directly to DNA. It is recruited to the promoter once proteins called **transcription factors** have first bound. In bacteria, RNA polymerase binds directly to DNA at specific promoter sequences; the enzyme is directed to the promoter by a protein factor that is then released once transcription begins.
- Elongation is essentially identical in the two types of organisms.
- In bacteria, specific DNA sequences called **terminators** signal the end of transcription of the gene. For one type of terminator, a specific protein binds to the terminator, triggering the termination of transcription and the release of the RNA and RNA polymerase from the template. For another type of terminator, a protein factor is not involved in the termination mechanism. Instead, the terminator sequence is transcribed and it folds up into a shape that terminates transcription. Eukaryotic DNA has no equivalent sequences. Instead, the 3′ end of the mRNA is specified by a very different process, which is discussed in the next section.

Once an RNA polymerase molecule has started transcription and progressed past the beginning of a gene, another molecule of RNA polymerase may start transcribing the same gene. In most genes, this process continues until there are many RNA polymerase molecules spaced closely along a gene, each

making an RNA transcript. In this way, a large number of RNA transcripts can be produced from one gene.

Overall, transcription is similar to DNA replication. The main differences are that: (1) only one of the two DNA strands acts as a template for synthesis of the RNA transcript, instead of both for replication; and (2) only a relatively small part of a DNA molecule—the RNA-coding sequence of a gene—serves as a template, rather than all of both strands as in DNA replication.

Molecular Insights discusses how altering the sleep–wake cycle in humans affects the circadian rhythm of transcription on a genome-wide scale.

Transcription of Noncoding RNA Genes Occurs in a Similar Way

Noncoding RNA genes include those for tRNAs and rRNAs (ribosomal RNAs, the RNA components of ribosomes). In eukaryotes, whereas RNA polymerase II transcribes protein-coding genes, RNA polymerase III transcribes tRNA genes and the gene for one of the four rRNAs, and RNA polymerase I transcribes the genes for the three other rRNAs. The promoters for noncoding RNA genes differ from those of protein-coding genes, being specialized for the assembly of the transcription machinery that involves the correct RNA polymerase type. In bacteria, the promoters for all genes are essentially the same, so a single RNA polymerase transcribes them all.

STUDY BREAK 15.2

1. For the DNA template below, what would be the sequence of an RNA transcribed from it?

        3′-CAAATTGGCTTATTACCGGATG-5′

2. What is the role of the promoter in transcription?

15.3 Production of mRNAs in Eukaryotes

Both prokaryotic and eukaryotic mRNAs contain regions that code for proteins, along with noncoding regions that play key roles in the process of protein synthesis. In prokaryotic mRNAs, the coding region is flanked by untranslated ends, the **5′ untranslated region (5′ UTR)** and the **3′ untranslated region (3′ UTR)**. The same elements are present in eukaryotic mRNAs along with additional noncoding elements. The production of mRNA in eukaryotes is the focus of this section.

Eukaryotic Protein-Coding Genes Are Transcribed into Precursor-mRNAs That Are Modified in the Nucleus

A typical eukaryotic protein-coding gene is transcribed into a **precursor-mRNA (pre-mRNA)** that must be processed in the

Gene Expression and Sleep–Wake Timing: How does a mistimed sleep cycle affect transcription?

Circadian rhythms in a number of biological processes occur in many organisms. A circadian rhythm is characterized by repetitive oscillations that are very regular and cycle once every 24 hours. Rhythms are set by endogenous (built-in) oscillators. In mammals, circadian oscillators are present in the brain and in most peripheral tissues. A major influencing factor on rhythms in physiology and behavior is the sleep–wake cycle. If the sleep–wake cycle is moved in time (desynchronized) from the central circadian clock, those rhythms become significantly disrupted. Such desynchronization occurs, for instance, during jet lag and shift work.

Research Question

How does disruption of the human sleep–wake cycle affect genes that show a circadian rhythm of expression?

Experiments

Derk-Jan Dijk's research team at the University of Guildford, Surrey, United Kingdom, subjected 22 volunteers to a desynchronization treatment, in which the sleep–wake cycle (as well as feeding and light cycles) was shifted to 28 hours from the normal

Shutterstock.com/Joana Lopes

24 hours for three days. The researchers took blood samples at time intervals and analyzed the amount of melatonin and the transcriptome. Melatonin normally shows a circadian rhythm in blood, controlled by the central circadian pacemaker, which is located in the hypothalamus of the brain. A *transcriptome* is the complete set of transcripts in a cell, tissue, or organism; the study of a transcriptome qualitatively and quantitatively is *transcriptomics* (see Chapter 19). Transcriptomics analysis is a branch of genomics, the study of genes and their functions. The availability of an organism's genome sequence makes it possible to analyze the organism's transcriptome.

Results

Blood melatonin continued to show a normal 24-hour rhythm during sleep–wake cycle desynchronization. This result indicated that the central circadian pacemaker continued to keep time correctly.

Transcriptome analysis showed that, on a genome-wide scale, 6.4% of transcripts (1,502) exhibited circadian expression patterns when sleep was in phase with the melatonin rhythm. Approximately equal numbers showed peak expression levels at night and day.

Desynchronizing the sleep–wake cycle resulted in a significant change in the blood transcriptome. On a genome-wide scale, only 1.0% of transcripts (237) showed circadian rhythmicity when sleep was out of phase with the melatonin rhythm. This represents a greater than six-fold reduction compared with the sleeping in phase results. In comparison, in a previous study by the researchers, only a modest decrease in transcripts that

showed circadian rhythmicity was seen for volunteers subjected to sleep restriction (5.7 vs. 8.5 hours over 1 week; 22% of transcripts became arrhythmic).

Both day- and night-peaking transcripts showed disruption of circadian rhythmicity during desynchronization of the sleep–wake cycle. Genes that lost their rhythmic expression included many that regulate gene expression, genes involved in transcription such as RNA polymerase II, and genes involved in translation and targeting of proteins to the ER. Some important *clock genes*—genes that encode clock proteins, which are involved in setting rhythms—also showed disrupted circadian rhythms.

Conclusion

The study showed that humans whose sleep–wake cycle was altered to 28 hours for three days expressed far fewer genes with circadian rhythmicity. In other words, an altered sleep–wake cycle has significant effects at the molecular level. Genes affected included those involved in gene expression, as well as clock genes involved in circadian rhythm generation. The molecular mechanism(s) for the change in expression remain to be determined. The results may contribute to understanding the negative health effects of, for instance, jet lag and shift work.

think like a scientist

Sleep deprivation (sleep restriction) is a chronic issue for many college and university students. Propose a hypothesis based on the discussion in this *Molecular Insights* that could be tested using transcriptome analysis. Outline how you would test the hypothesis.

nucleus to produce the translatable mRNA (**Figure 15.6;** and see Figures 15.2 and 15.5, step 5). The mRNA exits the nucleus and is translated in the cytoplasm.

MODIFICATIONS OF PRE-mRNA AND mRNA ENDS At the 5′ end of the pre-mRNA is the **5′ cap,** consisting of a guanine-containing nucleotide that is reversed so that its 3′-OH group faces the beginning rather than the end of the molecule. A *capping enzyme* adds the 5′ cap to the pre-mRNA soon after RNA polymerase II begins transcription. The cap, which is connected to the rest of the chain by three phosphate groups,

remains when pre-mRNA is processed to mRNA. The cap is the site where ribosomes attach to mRNAs at the start of translation.

A eukaryotic protein-coding gene has no terminator sequence that signals RNA polymerase to stop transcription. Instead, near the 3′ end of the gene is a sequence called a **polyadenylation signal** that is transcribed into the pre-mRNA. Proteins bind to this *sequence,* and cleave the pre-mRNA just downstream of that point. Then, the enzyme **poly(A) polymerase** adds a chain of 50 to 250 adenine nucleotides, one nucleotide at a time, to that 3′ end of the pre-mRNA. This

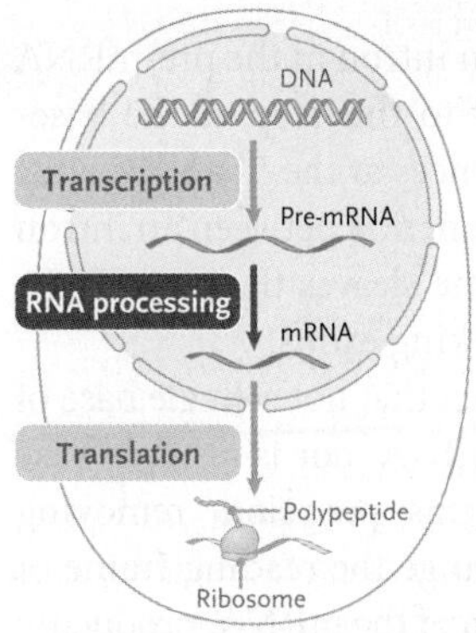

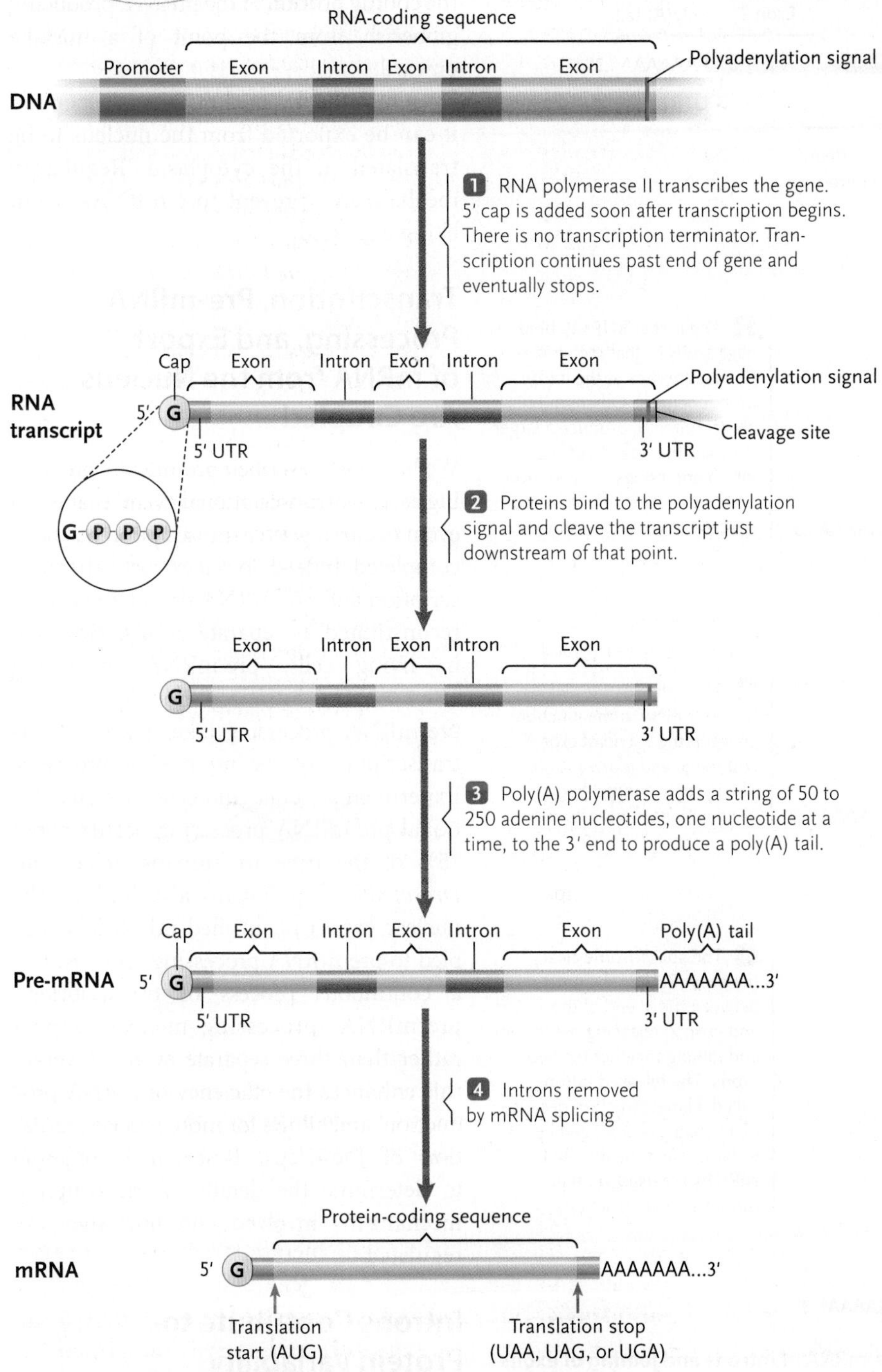

FIGURE 15.6 Relationship between a eukaryotic protein-coding gene, the pre-mRNA transcribed from it, and the mRNA processed from the pre-mRNA.

© Cengage Learning 2017

string of A nucleotides, called the **poly(A) tail,** enables the mRNA produced from the pre-mRNA to be translated efficiently, and protects it from attack by RNA-digesting enzymes in the cytoplasm. If the poly(A) tail of an mRNA is removed experimentally, the mRNA is degraded quickly inside cells.

SEQUENCES INTERRUPTING THE RNA-CODING SEQUENCE

The transcription unit of a typical eukaryotic protein-coding gene—the RNA-coding sequence—also contains one or more non-protein-coding sequences called **introns** that interrupt the protein-coding sequence (shown in Figure 15.6). The segments of the RNA-coding sequence interrupted by the introns are called **exons.** The introns are transcribed into pre-mRNA, but removed from pre-mRNA during processing in the nucleus, so that the amino acid-coding sequence in the finished mRNA is read continuously. As Figure 15.6 shows, the exons in finished mRNAs contain the protein-coding sequence of the gene. The exons at the two ends of the mRNA also contain the 5′ UTR and 3′ UTR sequences.

Introns were discovered by several methods, including direct comparisons between the nucleotide sequences of mature mRNAs and either pre-mRNAs or the genes encoding them. The majority of known eukaryotic genes contain at least one intron; some contain more than 60. The original discoverers of introns, Richard Roberts of New England Biolabs and Phillip Sharp of Massachusetts Institute of Technology, received a Nobel Prize in 1993 "for their discoveries of split genes."

Introns Are Removed During Pre-mRNA Processing to Produce the Translatable mRNA

A process called **mRNA splicing,** which occurs in the nucleus, removes introns from pre-mRNAs and joins exons together. As an illustration of mRNA splicing, **Figure 15.7** shows the processing of a pre-mRNA with a single intron to produce a mature mRNA. mRNA splicing takes place in a **spliceosome,** a complex formed between the pre-mRNA and a handful of **small nuclear ribonucleoprotein particles (snRNPs,** pronounced "snurps"). A *ribonucleoprotein particle* is a complex of RNA and proteins. Each of the snRNPs involved in mRNA splicing consists of a short *small nuclear RNA* (snRNA) bound to a number of proteins. The

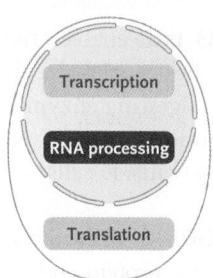

snRNPs bind in a particular order to an intron in the pre-mRNA and form the active spliceosome. Key to this step, is the base-pairing of snRNAs with specific sequences at the 5′ and 3′ ends of introns; those sequences define the junction between an intron and the adjacent exons. The spliceosome cleaves the pre-mRNA to release the intron, and joins the flanking exons.

The cutting and splicing are so exact that not a single base of an intron is retained in the finished mRNA, nor is a single base removed from the exons. Without this precision, removing introns would change the reading frame of the coding portion of the mRNA, producing gibberish from the point of a mistake onward.

Once the finished mRNA is generated it can be exported from the nucleus to be translated in the cytoplasm. Regulatory mechanisms prevent pre-mRNAs from being exported.

Transcription, Pre-mRNA Processing, and Export of mRNA from the Nucleus Are Coupled

We have just described pre-mRNA processing as a posttranscriptional event; that is, an event occurring after transcription has been completed. Indeed, in cell extracts, the transcription and pre-mRNA processing can be reconstituted as separate events. However, in living cells, pre-mRNA processing often occurs cotranscriptionally; that is, pre-mRNA processing takes place even as transcription of the pre-mRNA proceeds. Experimental data indicate cotranscriptional pre-mRNA processing occurs about 75% of the time in humans, mice, and *Drosophila*. Experiments also indicate the nuclear export of finished mRNAs is coupled to pre-mRNA processing. The result is a continuous process of transcription–pre-mRNA processing–nuclear export rather than three separate events. Overall, this enhances the efficiency of mRNA production, and allows for more precise regulation of the steps. Research is ongoing to determine the details of the coupling mechanisms involved and how they are regulated.

Introns Contribute to Protein Variability

Why are introns present in mRNA-encoding genes? Among a number of

Pre-mRNA

1 Pre-mRNA with an intron

2 The first snRNPs to bind have snRNAs that recognize RNA sequences at the intron–exon junctions. Other snRNPs are recruited to produce a larger complex that loops out the intron and brings the two exon ends close together. The active spliceosome has now formed.

Active spliceosome

Several snRNPs

Cut 5′ end of intron bonded to intron site near 3′ end

Cut 3′ end of exon 1

3 The spliceosome cleaves the pre-mRNA at the junction between the 3′ end of exon 1 and the 5′ end of the intron. The intron is looped back to bond with itself near its 3′ end.

Degraded

Released intron in lariat structure

Reused

4 The spliceosome cleaves the pre-mRNA at the junction between the 3′ end of the intron and exon 2, releasing the intron and joining together the two exons. The released intron, called a lariat structure because of its shape, is degraded by enzymes, and the released snRNPs are used in other mRNA splicing reactions.

Released snRNPs

FIGURE 15.7 mRNA splicing—the removal from pre-mRNA of introns and joining of exons in the spliceosome.
© Cengage Learning 2017

possibilities, introns may provide advantages by increasing the coding capacity of existing genes through a process called *alternative splicing* and by generating new proteins through a process called *exon shuffling*.

ALTERNATIVE SPLICING Many pre-mRNAs are processed by reactions that join exons in different combinations to produce different mRNAs from a single gene. The mechanism, called **alternative splicing,** greatly increases the number and variety of proteins encoded in the cell nucleus without increasing the size of the genome. The frequency of alternative splicing varies considerably with the species. For example, geneticists estimate that pre-mRNAs of more than 95% of human genes with multiple exons undergo alternative splicing, whereas only 25% of such genes in the worm, *C. elegans,* undergo alternative splicing. In each case, the different mRNAs produced from the "parent" pre-mRNA are translated to produce a family of related proteins with various combinations of amino acid sequences derived from the exons. Each protein in the family varies to a degree in its function. Alternative splicing helps us understand why humans have only about 20,500 genes. As a result of the alternative splicing process, the number of proteins produced far exceeds the number of genes, and it is proteins that mostly direct an organism's functions.

As an example, the pre-mRNA transcript of the mammalian α-tropomyosin gene, which encodes a protein component of muscle fibers, is alternatively spliced in smooth muscle (for example, muscles of the intestine and bladder), skeletal muscle (for example, biceps, glutes), fibroblast (connective tissue cell that makes collagen), liver, and brain. Alternative splicing results in different forms of the α-tropomyosin protein. **Figure 15.8** shows the alternative splicing of the α-tropomyosin pre-mRNA to the mRNAs found in smooth muscle and striated muscle. Exons 2 and 12 are exclusive to the smooth muscle mRNA, whereas exons 3, 10, and 11 are exclusive to the striated muscle mRNA.

The polypeptides made from the two mRNAs have some identical stretches of amino acids, along with others that differ. As you learned in Section 3.4, the primary structure of a protein—its amino acid structure—directs the folding of the chain into its three-dimensional shape. Therefore, the two forms of α-tropomyosin have related, but different shapes and, as a result, they participate in different types of muscle action. Smooth muscles perform squeezing actions in blood vessels and internal organs, whereas skeletal muscles pull on the bones of the skeleton to move body parts.

Alternative splicing forces us to reconsider the one gene–one polypeptide hypothesis introduced earlier in the chapter. We must now accept the fact that, for some genes at least, one gene may specify a number of polypeptides each of which has a related function.

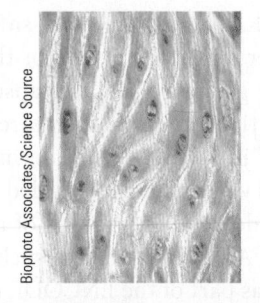

Smooth muscle
Found in walls of tubes and cavities of the body, including blood vessels, the stomach and intestine, the bladder, and the uterus. Contraction of smooth muscles typically produces a squeezing motion.

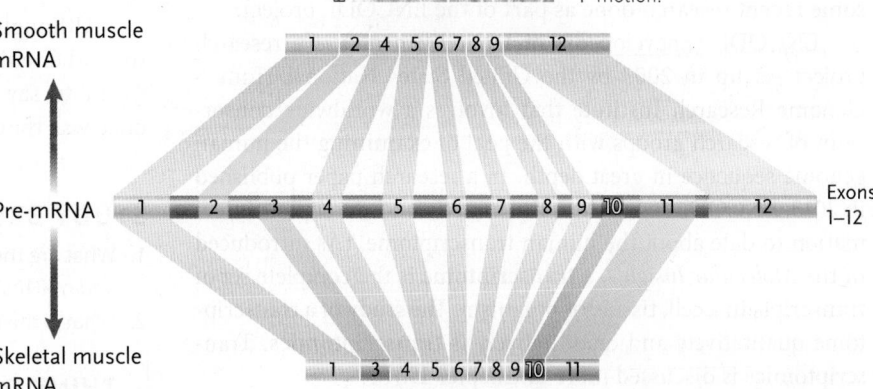

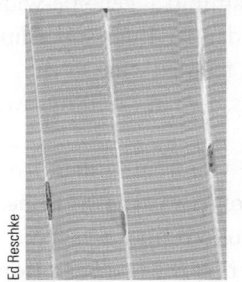

Skeletal muscle
Most muscles of this type are attached by tendons to the skeleton. Their function is locomotion and movement of body parts. The human body has more than 600 skeletal muscles, ranging in size from the small muscles that move the eyeballs, to the large muscles that move the legs.

FIGURE 15.8 Alternative splicing of the α-tropomyosin pre-mRNA to distinct mRNA forms found in smooth muscle and skeletal muscle. All of the introns are removed in both mRNA splicing pathways, but exons 3, 10, and 11 are also removed to produce the smooth muscle mRNA, and exons 2 and 12 are also removed to produce the skeletal muscle mRNA.
© Cengage Learning 2017

EXON SHUFFLING Intron–exon junctions often fall at points dividing major functional regions in encoded proteins, as they do in the genes for antibody proteins, hemoglobin blood proteins, and the peptide hormone insulin. The functional divisions may have allowed new proteins to evolve by **exon shuffling,** a process by which existing protein regions or domains, already selected for their functions by the evolutionary process, are mixed into novel combinations to create new proteins. Evolution of new proteins by this mechanism would produce changes much more quickly and efficiently than by alterations in individual amino acids at random points. (Exon shuffling as a mechanism for the evolution of new proteins is discussed in Section 19.4.)

Genome-Wide Analysis Has Revealed Unexpected Details of the Human Transcriptome

Only about 1.2% of the human genome encodes proteins (see Chapter 19). In the traditional view of the genome, noncoding genes such as those for tRNAs, rRNAs, and snRNAs were assumed to involve another small percentage of the genome, leaving the majority of the genome likely untranscribed and, therefore, of unknown function. Some old literature referred to the untranscribed DNA as "junk" DNA, considering it to have no function. The traditional view is being discarded as genome-wide technologies have made detailed analysis of the genome and its transcripts possible. An example of one such analysis is some recent research done as part of the ENCODE project.

ENCODE (Encyclopedia of DNA Elements) is a research project set up in 2003 by the United States National Human Genome Research Institute that involves a worldwide consortium of research groups with the goal of examining the human genome sequence in great depth. In a research paper published in 2013, ENCODE researchers reported the most complete information to date about the human transcriptome. (As introduced in the *Molecular Insights,* a **transcriptome** is the complete set of transcripts in a cell, tissue, or organism. The study of a transcriptome qualitatively and quantitatively is **transcriptomics.** Transcriptomics is discussed more in Chapter 19.)

With the goal of obtaining a genome-wide catalog of human transcripts and to identify their subcellular locations, the ENCODE researchers studied the RNAs expressed in 15 different human cell lines. Two key findings were:

- *Pervasive transcription:* Combining data from the 15 cell lines, 74.0% of the genome was shown to be transcribed. The average for the set of cell lines was 39%. These results show that there is pervasive transcription across the genome in humans, and likely in all eukaryotes. Most of the transcripts are noncoding RNAs. The noncoding RNAs are grouped into small noncoding RNAs and long noncoding RNAs (lncRNAs; greater than 200 nucleotides). Interestingly, the number of lncRNA genes likely equals or exceeds the number of protein-coding genes in the human genome. While the functions of most of the noncoding RNAs remain to be determined, it is clear that at least some small and long noncoding RNAs play a role in regulating gene expression (see Chapter 16).
- *Variable expression of isoforms of protein-coding genes:* Many protein-coding genes are transcribed to produce different forms of their mRNAs, called **gene isoforms.** You have already learned that alternative splicing is a mechanism that produces different mRNAs for a given gene. Those mRNAs are examples of gene isoforms. Other mechanisms that produce different mRNAs from a gene include the use of different transcription start sites and the generation of different 3′ ends. The ENCODE researchers studied gene isoforms within and between cell lines, discovering that:
  1. Genes typically express many isoforms simultaneously, with a plateau of about 10–12 isoforms per cell per gene per cell line.
  2. Isoform levels of a particular gene vary, with one usually predominating.

3. About three-quarters of protein-coding genes express different predominant isoforms depending on the cell line and condition.
4. Between cell lines, differences in gene expression levels of isoforms are more variable than differences in splicing. All in all, this adds up to a large amount of potential variation in gene expression for protein-coding genes.

Although this was a study of the human genome, it is likely that other eukaryotes have similar transcriptome properties. It is fair to say that genes are significantly more complex than once was thought.

STUDY BREAK 15.3

1. What are the similarities and differences between pre-mRNAs and mRNAs?
2. What is the role of snRNPs in mRNA splicing?

THINK OUTSIDE THE BOOK

Explore the Internet or research literature to find a molecular model that explains how an exon and its flanking introns are removed by alternative splicing.

15.4 Translation: mRNA-Directed Polypeptide Synthesis

Translation is the reading of an mRNA to assemble amino acids into a polypeptide **(Figure 15.9)**. In prokaryotes, the mRNA produced by transcription is immediately available for translation, which takes place throughout the cell. In eukaryotes, the mRNA produced by splicing of the pre-mRNA first exits the nucleus, and then is translated in the cytoplasm. (As you will see, a few specialized genes are transcribed and translated in mitochondria and chloroplasts.)

In translation, the mRNA associates with a ribosome, and tRNAs, another type of RNA, bring amino acids to the complex to be joined one by one into the polypeptide chain. The sequence of amino acids in the polypeptide chain is determined by the sequence of codons in the mRNA, whereas the ribosome facilitates the translation process. The mRNA is read from the 5′ end to the 3′ end, and the polypeptide is assembled from the N-terminal end to the C-terminal end.

In this section, we start by discussing the key players in the process, the tRNAs and ribosomes, and then walk through the translation process from a start codon to a stop codon.

tRNAs Are Small RNAs That Bring Amino Acids to the Ribosome

A **transfer RNA (tRNA)** brings an amino acid to the ribosome for addition to the polypeptide chain.

tRNA STRUCTURE tRNAs are small RNAs, 75–90 nucleotides long (mRNAs are typically hundreds of nucleotides long), with a highly distinctive structure that accomplishes their role in translation **(Figure 15.10).** All tRNAs can wind into four double-helical segments, forming in two dimensions a *cloverleaf* shape **(Figure 15.10A).** At the tip of one of the double-helical segments is the **anticodon,** the three-nucleotide segment that base pairs with a codon in mRNAs. Opposite the anticodon, at the other end of the cloverleaf, is a double-helical segment and the 3′ end of the molecule to which the amino acid corresponding to the anticodon binds covalently. For example, a tRNA that base pairs with the codon 5′-AGU-3′ has serine (Ser) linked to it (see Figure 15.9). The anticodon of the tRNA that pairs with this codon is 3′-UCA-5′. (The anticodon and codon pair in an antiparallel manner, as do the strands in DNA. We will write anticodons in the 3′→5′ direction to make it easy to see how they pair with codons.)

The tRNA cloverleaf folds in three dimensions into a structure resembling an upside-down L **(Figure 15.10B).** The anticodon and the 3′ end binding the amino acid are located at the opposite tips of the structure.

We learned earlier that 61 of the 64 codons of the genetic code specify an amino acid. But 61 different tRNAs are not needed to read the sense codons. Francis Crick's **wobble hypothesis** states that the 61 sense codons can be read by fewer than 61 distinct tRNAs because of particular pairing properties of the bases in the anticodons. That is, the pairing of the anticodon with the first two nucleotides of the codon is always precise, but the anticodon has more flexibility in pairing with the third nucleotide of the codon. Recall from the genetic code discussion that when the first two letters in a codon are identical and the third letter is U or C, the codon always codes for the same amino acid. In many cases, the same tRNA anticodon can read codons that have either U or C in the third position; for example, a tRNA carrying phenylalanine can read both codons 5′-UUU-3′ and 5′-UUC-3′. Similarly the same tRNA anticodon can read two codons with the same first two letters and A or G in the third position. For example, one tRNA carrying glutamine can read both codons 5′-CAA-3′ and 5′-CAG-3′. The special inosine purine in the alanine tRNA shown in Figure 15.10A allows even more

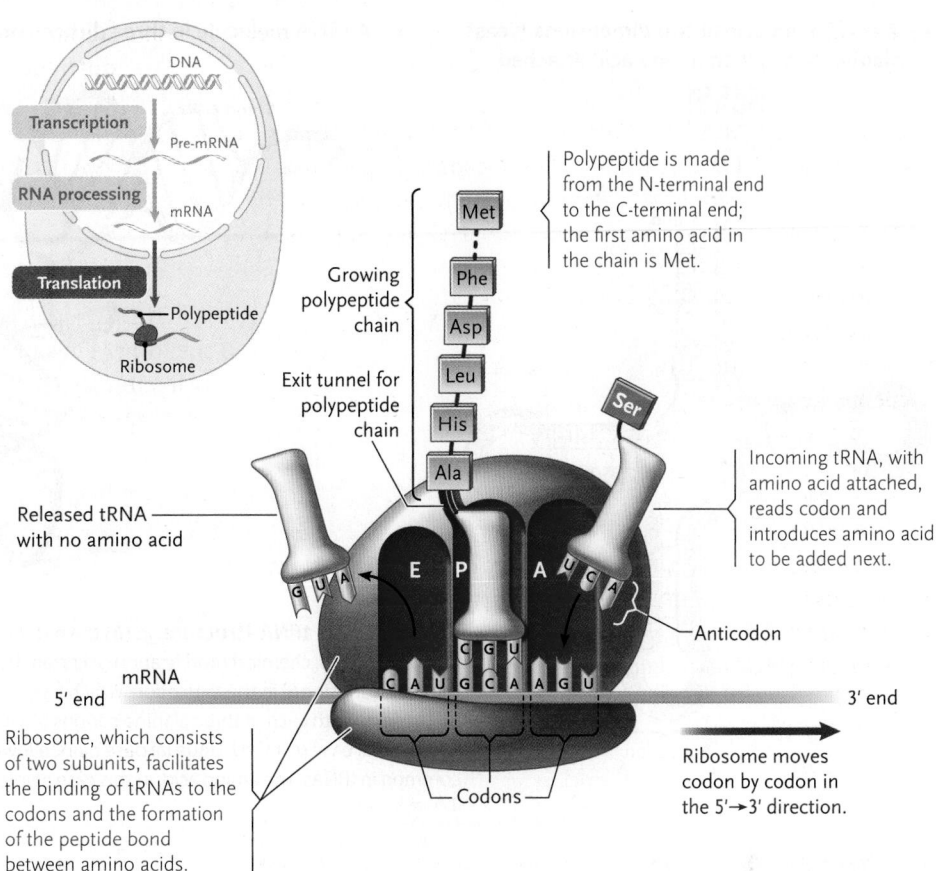

FIGURE 15.9 An overview of translation, in which ribosomes assemble amino acids into a polypeptide chain. The figure shows a ribosome in the process of translation. E, P, and A are tRNA binding sites. A tRNA molecule with an amino acid bound to it is entering the A site of the ribosome on the right. The anticodon on the tRNA will pair with the codon in the mRNA. Its amino acid will then be added to the growing polypeptide that is currently attached to the tRNA bound to the P site in the middle of the ribosome. As it assembles a polypeptide chain, the ribosome moves from one codon to the next along the mRNA in the 5′→3′ direction, and the tRNA with no amino acid is released from the E site.
© Cengage Learning 2017

extensive wobble by allowing the tRNA to pair with codons that have either U, C, or A in the third position. This apparent "looseness" in base pairing occurs because the base at the 5′ end of the anticodon can move in its chemical space—wobble—allowing it to form hydrogen bonds with different bases at the 3′ ends of codons.

ADDITION OF AMINO ACIDS TO THEIR CORRESPONDING tRNAs The correct amino acid must be present on a tRNA if translation is to be accurate. The process of adding an amino acid to a tRNA is called **aminoacylation** (literally, the addition of an amino acid) or **charging** (because the process adds free energy as the amino acid–tRNA combinations are formed). The finished product of charging, a tRNA linked to its "correct" amino acid, is called an **aminoacyl–tRNA. Figure 15.10C** shows how an aminoacyl–tRNA is shown in this book. Twenty different enzymes called **aminoacyl–tRNA synthetases**—one synthetase for each of the 20 amino acids—catalyze aminoacylation in the four steps shown in **Figure 15.11.**

A. A tRNA molecule in two dimensions (yeast alanine tRNA) with amino acid attached

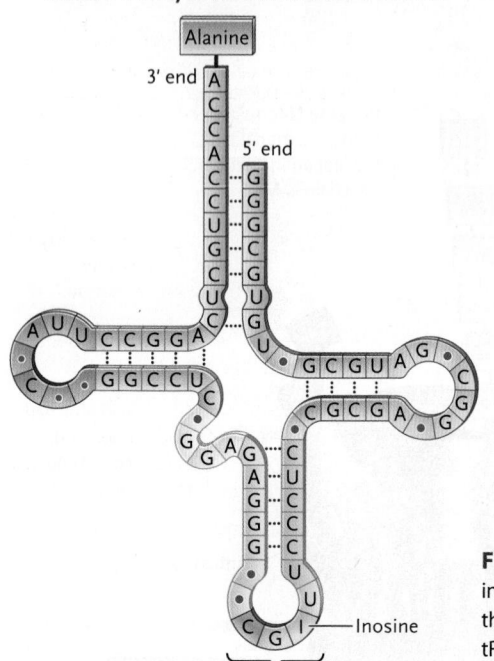

B. A tRNA molecule in three dimensions

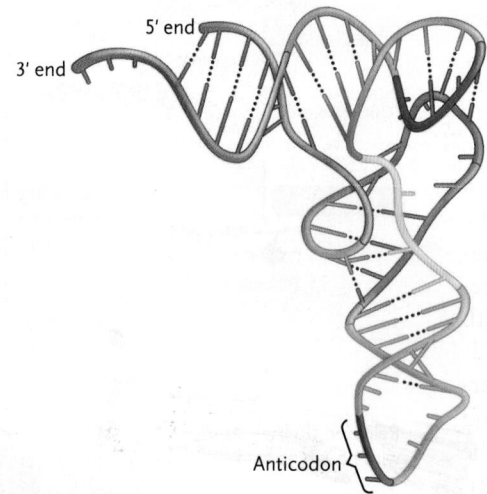

C. How a tRNA is shown in this book

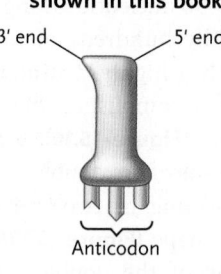

FIGURE 15.10 tRNA structure. In **(A)** the red dots show sites where bases are chemically modified into other forms; chemical modification of certain bases is typical of tRNAs. This yeast alanine tRNA has the purine inosine (I) in the anticodon, which has relatively loose base-pairing ability, allowing this single tRNA to pair with each of three alanine codons 5′-GCU-3′, 5′-GCC-3′, and 5′-GCA-3′. This tRNA also has the unusual base pair G–U. Unusual base pairs, allowed by the greater flexibility of short RNA chains, are common in tRNAs. The amino acid, in this case alanine, binds to the 3′ end of a tRNA molecule.
© Cengage Learning 2017

Ribosomes Are rRNA–Protein Complexes That Assemble Proteins

Ribosomes are ribonucleoprotein particles that carry out protein synthesis by translating mRNA into chains of amino acids. A ribosome reads the codons on an mRNA and joins the appropriate amino acids to make a polypeptide chain.

In prokaryotes, ribosomes carry out their assembly functions throughout the cell. In eukaryotes, ribosomes function in the cytoplasm, either suspended freely in the cytoplasmic solution, or attached to the membranes of the endoplasmic reticulum (ER), the system of tubular or flattened sacs in the cytoplasm (discussed in Section 4.3).

A ribosome is made up of two parts of dissimilar size, called the *large* and *small ribosomal subunits* **(Figure 15.12).** Each subunit is a combination of **ribosomal RNA (rRNA)** and ribosomal proteins.

Prokaryotic and eukaryotic ribosomes are similar in function, and quite similar in structure. However, certain differences in their molecular structure, particularly in the ribosomal proteins, give them distinguishable properties. For example, the antibiotics streptomycin and erythromycin are effective antibacterial agents because they inhibit the function of the bacterial ribosome, but not the eukaryotic ribosome.

In translation, the mRNA follows a bent path through a groove in the ribosome. The ribosome also has binding sites where tRNAs interact with the mRNA (see Figure 15.12 and refer also to Figure 15.9). The **A site** (aminoacyl site) is where the incoming aminoacyl–tRNA carrying the next amino acid

to be added to the polypeptide chain binds to the RNA. The **P site** (peptidyl site) is where the tRNA carrying the growing polypeptide chain is bound. The **E site** (exit site) is where a tRNA, now without an attached amino acid, binds before exiting the ribosome. These functional sites are discussed more later.

Translation Proceeds in Three Stages

Translation is similar in eukaryotes and bacteria. We will present translation from a eukaryotic perspective and indicate how it differs in bacteria.

The three major stages of translation are:

1. **Initiation,** in which the translation components assemble on the start codon of the mRNA.
2. **Elongation,** in which the assembled complex reads the string of codons in the mRNA one at a time while joining the specified amino acids into the polypeptide.
3. **Termination,** in which the complex disassembles after the last amino acid of the polypeptide specified by the mRNA has been added to the polypeptide.

The energy of GTP hydrolysis to GDP + P$_i$ fuels each of the three stages.

TRANSLATION INITIATION Figure 15.13 shows translation initiation in eukaryotes. Each initiation step is aided by proteins called **initiation factors (IFs).** The IFs are released when initiation is complete in step 3.

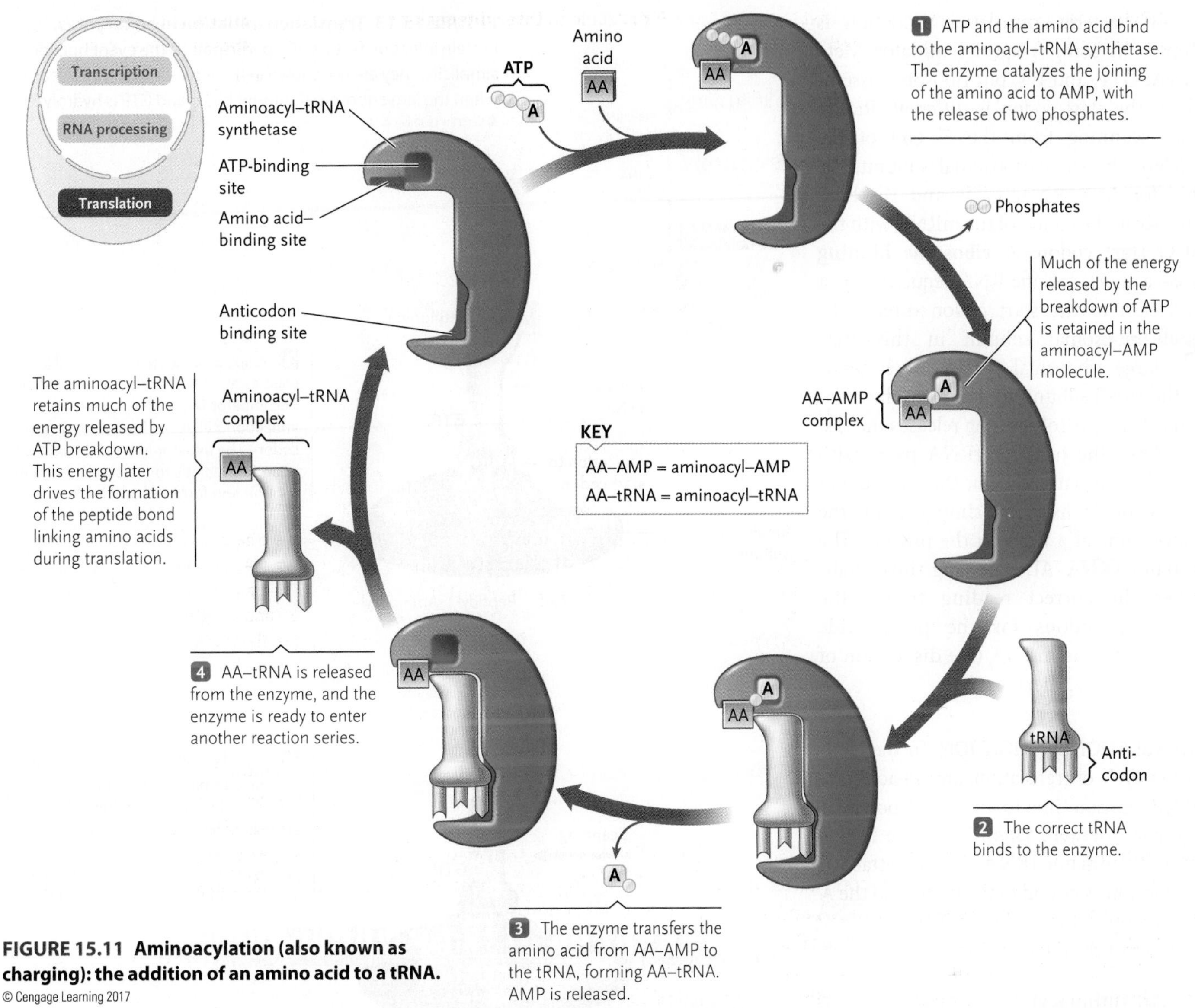

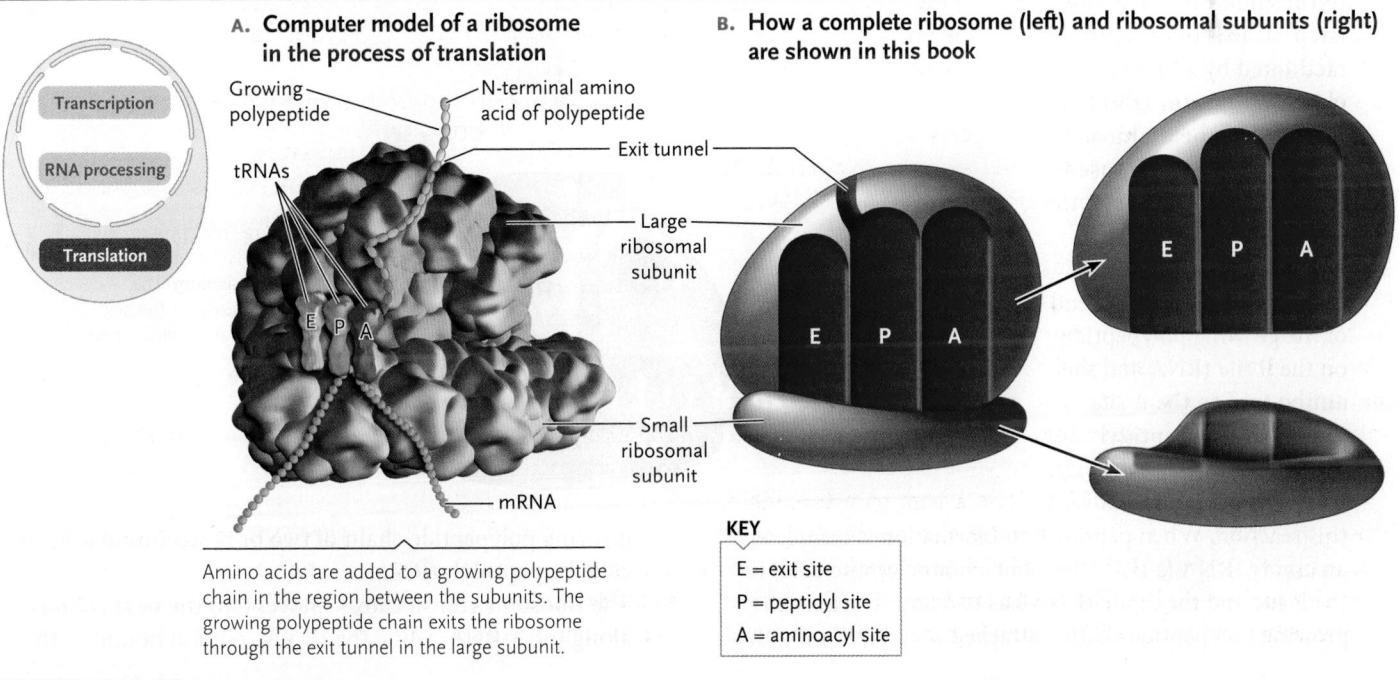

Transcription

RNA processing

Translation

Aminoacyl–tRNA
synthetase

ATP-binding
site

Amino acid–
binding site

Anticodon
binding site

ATP

Amino
acid

AA

AA

A

1 ATP and the amino acid bind
to the aminoacyl–tRNA synthetase.
The enzyme catalyzes the joining
of the amino acid to AMP, with
the release of two phosphates.

Phosphates

Much of the energy
released by the
breakdown of ATP
is retained in the
aminoacyl–AMP
molecule.

AA–AMP
complex

AA

A

The aminoacyl–tRNA
retains much of the
energy released by
ATP breakdown.
This energy later
drives the formation
of the peptide bond
linking amino acids
during translation.

Aminoacyl–tRNA
complex

AA

KEY

AA–AMP = aminoacyl–AMP

AA–tRNA = aminoacyl–tRNA

tRNA

Anti-
codon

4 AA–tRNA is released
from the enzyme, and the
enzyme is ready to enter
another reaction series.

AA

AA

A

2 The correct tRNA
binds to the enzyme.

A

3 The enzyme transfers the
amino acid from AA–AMP to
the tRNA, forming AA–tRNA.
AMP is released.

**FIGURE 15.11 Aminoacylation (also known as
charging): the addition of an amino acid to a tRNA.**

© Cengage Learning 2017

**A. Computer model of a ribosome
in the process of translation**

Transcription

RNA processing

Translation

Growing
polypeptide

N-terminal amino
acid of polypeptide

tRNAs

E P A

mRNA

Amino acids are added to a growing polypeptide
chain in the region between the subunits. The
growing polypeptide chain exits the ribosome
through the exit tunnel in the large subunit.

**B. How a complete ribosome (left) and ribosomal subunits (right)
are shown in this book**

Exit tunnel

Large
ribosomal
subunit

Small
ribosomal
subunit

E P A

E P A

KEY

E = exit site

P = peptidyl site

A = aminoacyl site

FIGURE 15.12 Ribosome structure.

© Cengage Learning 2017

In bacteria, translation initiation is similar in using a special initiator Met–tRNA, GTP, and IFs, but ribosome assembly at the start codon is different. Rather than scanning from the 5′ end of the mRNA, the small ribosomal subunit, the initiator Met–tRNA, GTP, and IFs bind directly to the region of the mRNA with the AUG start codon. A **ribosome binding site**—a short, specific RNA sequence—just upstream of the start codon directs the small ribosomal subunit in this step. The large ribosomal subunit then binds to the small subunit to complete the ribosome. GTP hydrolysis then releases the IFs.

After the initiator tRNA pairs with the AUG initiator codon, the subsequent stages of translation simply read the codons one at a time on the mRNA. The initiator tRNA–AUG pairing thus establishes the correct reading frame—the series of codons for the polypeptide encoded by the mRNA (see discussion of genetic code earlier).

TRANSLATION ELONGATION In the elongation cycle of translation, amino acids are added one at a time to a growing polypeptide chain (**Figure 15.14**). The cycle begins when an initiator tRNA with its attached methionine is bound to the P site, and the A site is empty (top of figure). Four steps occur during the cycle:

1. An aminoacyl–tRNA binds to the codon in the A site (step 1). This binding is facilitated by a protein **elongation factor (EF)** that is bound to the aminoacyl–tRNA and that is released once the tRNA binds to the codon.

2. A peptide bond is formed between the C-terminal end of the growing polypeptide on the P site tRNA and the amino acid on the A site tRNA (step 2). **Peptidyl transferase,** part of the large ribosomal subunit, catalyzes this reaction. When peptide bond formation is complete, an empty tRNA (a tRNA without a bound amino acid) in the P site and the A site tRNA has two amino acids—the growing polypeptide chain—attached to it. A tRNA with a

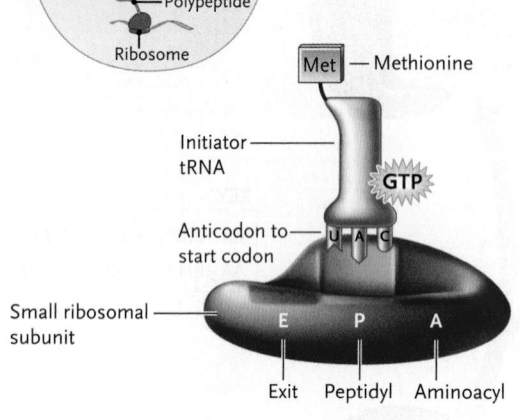

FIGURE 15.13 Translation initiation in eukaryotes.
Protein initiation factors (IFs) participate in the event but, for simplicity, they are not shown in the figure. The IFs are released when the large ribosomal subunit binds and GTP is hydrolyzed.
© Cengage Learning 2017

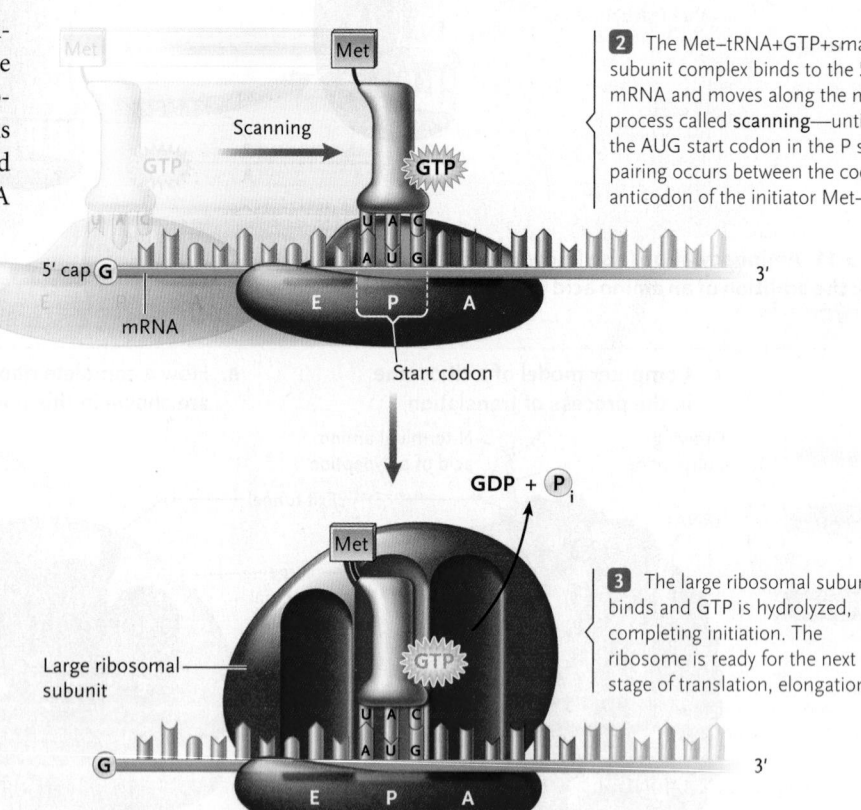

1 A specialized methionine–tRNA is used as an initiator tRNA in translation. The initiator Met–tRNA has an anticodon 3′-UAC-5′ for the AUG start codon. The initiator Met–tRNA with GTP bound to it binds to the small ribosomal subunit and forms a complex.

2 The Met–tRNA+GTP+small ribosomal subunit complex binds to the 5′ cap of the mRNA and moves along the mRNA—a process called **scanning**—until it reaches the AUG start codon in the P site. Base pairing occurs between the codon and the anticodon of the initiator Met–tRNA.

3 The large ribosomal subunit binds and GTP is hydrolyzed, completing initiation. The ribosome is ready for the next stage of translation, elongation.

growing polypeptide chain of two or more amino acids is called a **peptidyl–tRNA.**

3. The ribosome translocates—moves—to the next codon along the mRNA, while the tRNAs remain bound to the

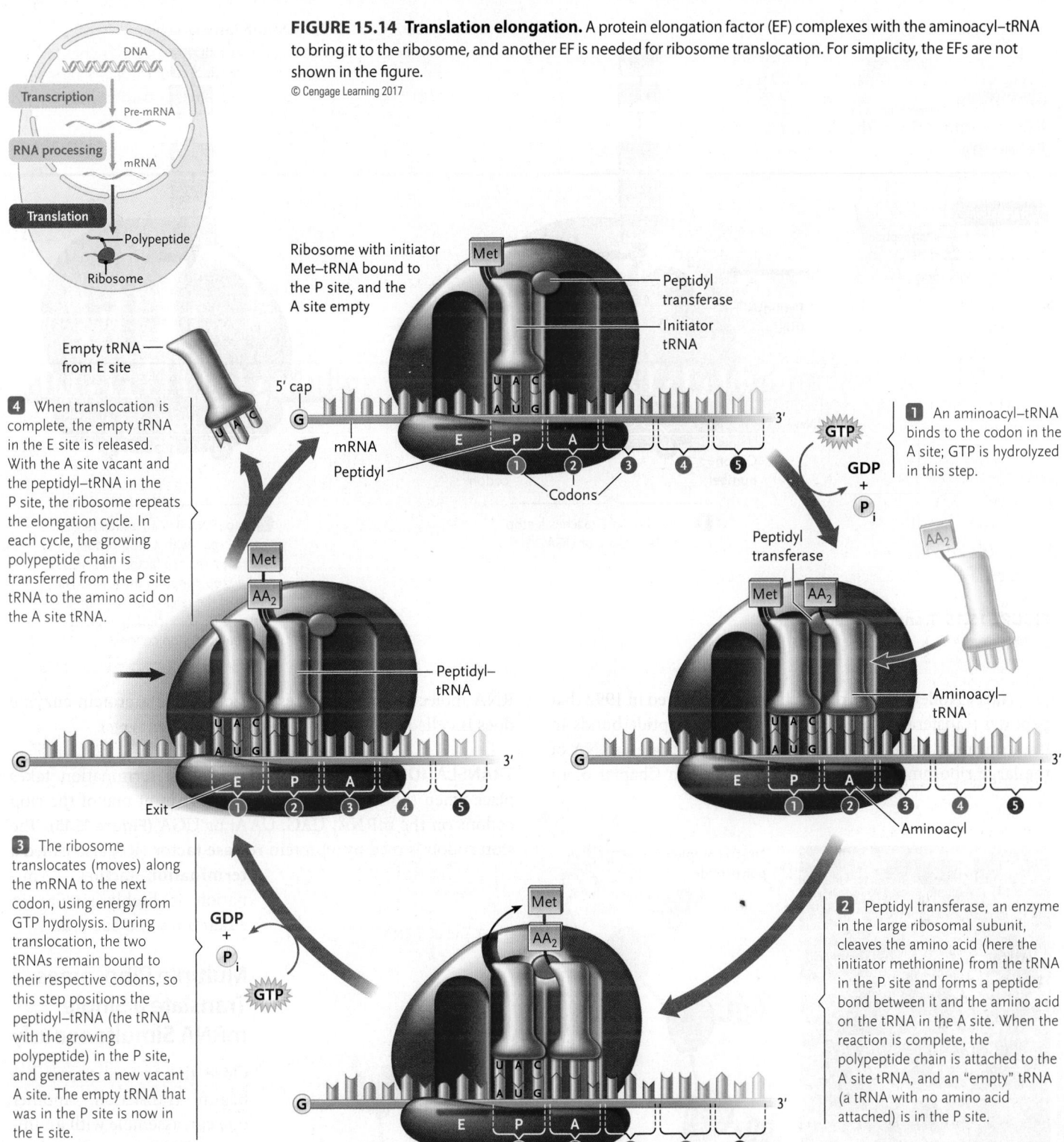

FIGURE 15.14 Translation elongation. A protein elongation factor (EF) complexes with the aminoacyl–tRNA to bring it to the ribosome, and another EF is needed for ribosome translocation. For simplicity, the EFs are not shown in the figure.
© Cengage Learning 2017

DNA
Transcription
Pre-mRNA
RNA processing
mRNA
Translation
Polypeptide
Ribosome

Ribosome with initiator Met–tRNA bound to the P site, and the A site empty

Met

Peptidyl transferase

Initiator tRNA

Empty tRNA from E site

4 When translocation is complete, the empty tRNA in the E site is released. With the A site vacant and the peptidyl–tRNA in the P site, the ribosome repeats the elongation cycle. In each cycle, the growing polypeptide chain is transferred from the P site tRNA to the amino acid on the A site tRNA.

5' cap
G
mRNA
Peptidyl
E P A
1 2 3 4 5
Codons
3'

GTP
GDP + P_i

1 An aminoacyl–tRNA binds to the codon in the A site; GTP is hydrolyzed in this step.

Peptidyl transferase

AA_2

Met
AA_2
Peptidyl–tRNA
G
Exit
E P A
1 2 3 4 5
3'

Met AA_2
Aminoacyl–tRNA
G
E P A
1 2 3 4 5
Aminoacyl
3'

2 Peptidyl transferase, an enzyme in the large ribosomal subunit, cleaves the amino acid (here the initiator methionine) from the tRNA in the P site and forms a peptide bond between it and the amino acid on the tRNA in the A site. When the reaction is complete, the polypeptide chain is attached to the A site tRNA, and an "empty" tRNA (a tRNA with no amino acid attached) is in the P site.

3 The ribosome translocates (moves) along the mRNA to the next codon, using energy from GTP hydrolysis. During translocation, the two tRNAs remain bound to their respective codons, so this step positions the peptidyl–tRNA (the tRNA with the growing polypeptide) in the P site, and generates a new vacant A site. The empty tRNA that was in the P site is now in the E site.

GDP + P_i
GTP

Met
AA_2
G
E P A
1 2 3 4 5
3'

mRNA in their same positions (step 3). An EF is used for this step and then is released. After translocation, the tRNA that was in the P site is now in the E site, the peptidyl–tRNA that was in the A site is now in the P site, and the A site is now empty.

4. The empty tRNA is released from the E site and the ribosome is ready to begin the next round of the elongation cycle.

Elongation is highly similar in eukaryotes and bacteria, with no significant conceptual differences. The elongation cycle turns at the rate of about one to three times per second in eukaryotes and 15 to 20 times per second in bacteria. Once it is long enough, the growing polypeptide chain extends from the ribosome through the exit tunnel (see Figure 15.12) as elongation continues.

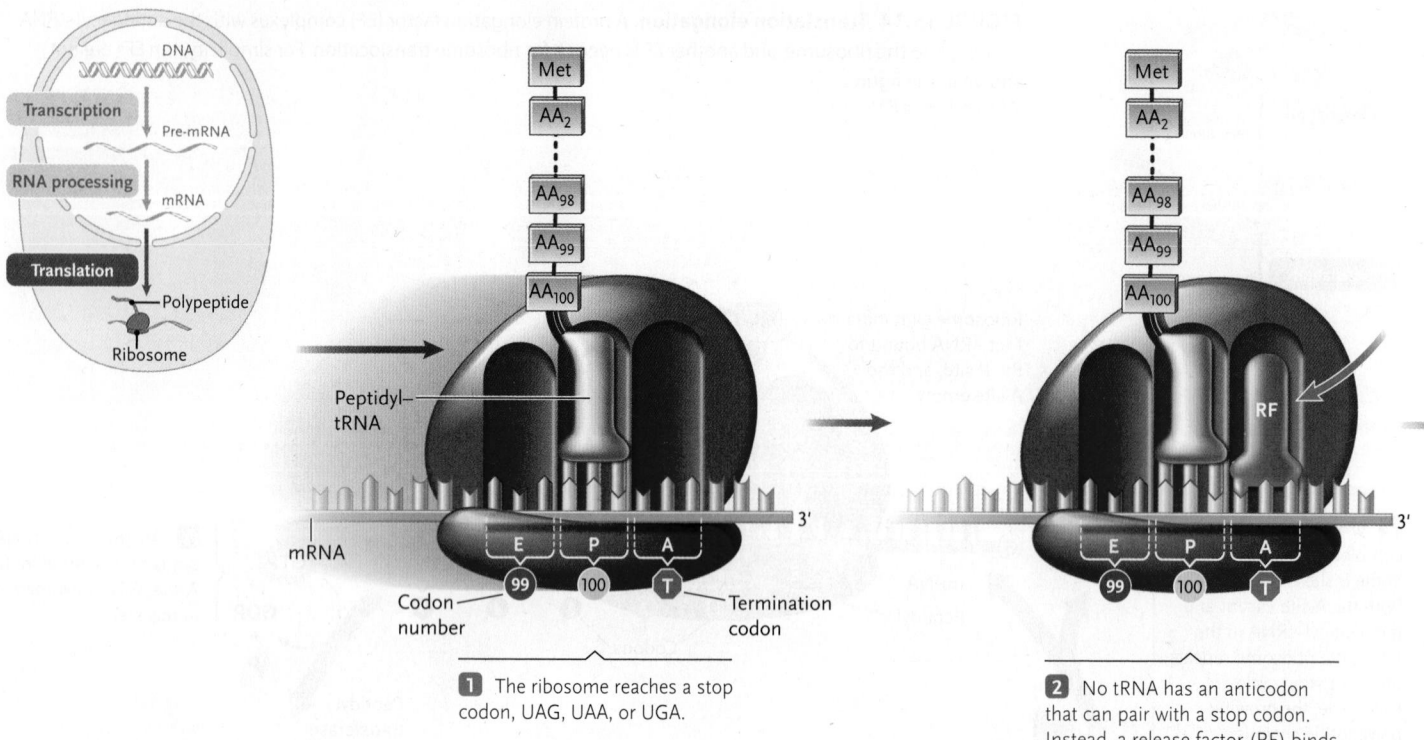

1 The ribosome reaches a stop codon, UAG, UAA, or UGA.

2 No tRNA has an anticodon that can pair with a stop codon. Instead, a release factor (RF) binds to the stop codon in the A site. The shape of the release factor mimics that of a tRNA, including regions that read the stop codons.

FIGURE 15.15 Translation termination.
© Cengage Learning 2017

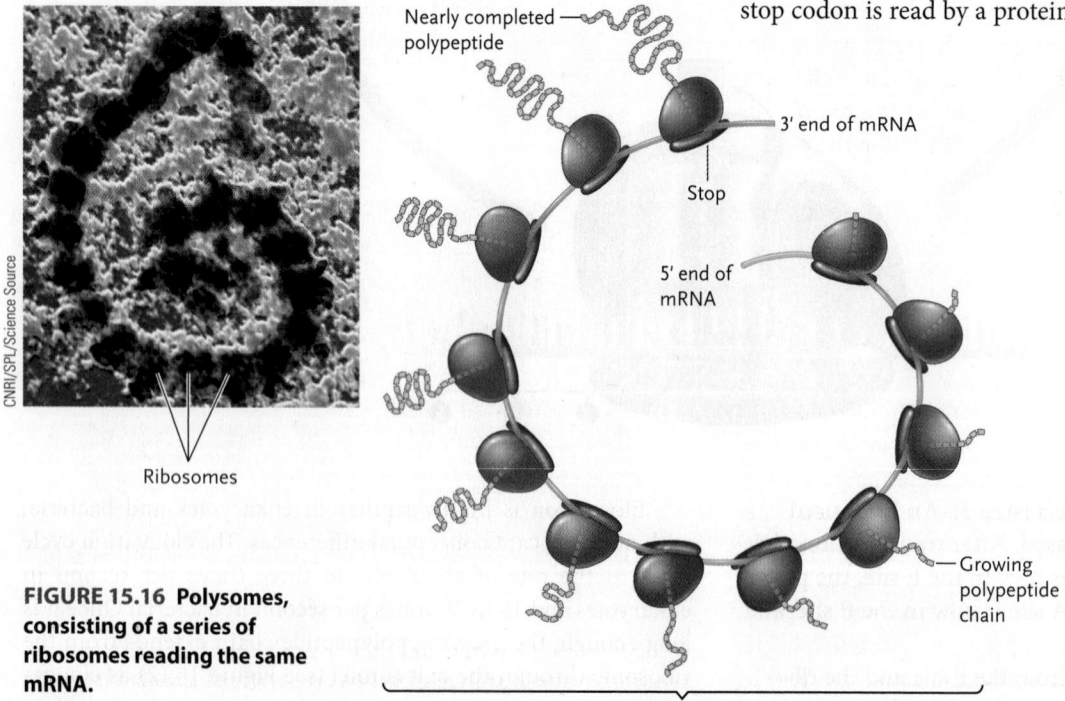

FIGURE 15.16 Polysomes, consisting of a series of ribosomes reading the same mRNA.
© Cengage Learning 2017

Harry Noller and his research team discovered in 1992 that peptidyl transferase, the enzyme that forms peptide bonds in the elongation cycle, is not a protein but a part of an rRNA of the large ribosomal subunit. As you learned in Chapter 6, an RNA molecule that catalyzes a reaction like a protein enzyme does is called a **ribozyme** (*ribo*nucleic acid en*zyme*).

TRANSLATION TERMINATION Translation termination takes place when the A site of a ribosome arrives at one of the stop codons on the mRNA, UAG, UAA, or UGA (**Figure 15.15**). The stop codon is read by a protein **release factor** (**RF;** also called a **termination factor**). Termination is highly similar in eukaryotes and bacteria.

Multiple Ribosomes Translate a Single mRNA Simultaneously

Once the first ribosome has begun translation, another one can assemble with an initiator tRNA as soon as there is room on the mRNA. Ribosomes continue to attach as translation continues and become spaced along the mRNA like beads on a string. The entire structure of an mRNA molecule and the multiple ribosomes attached to it is known as a **polysome**

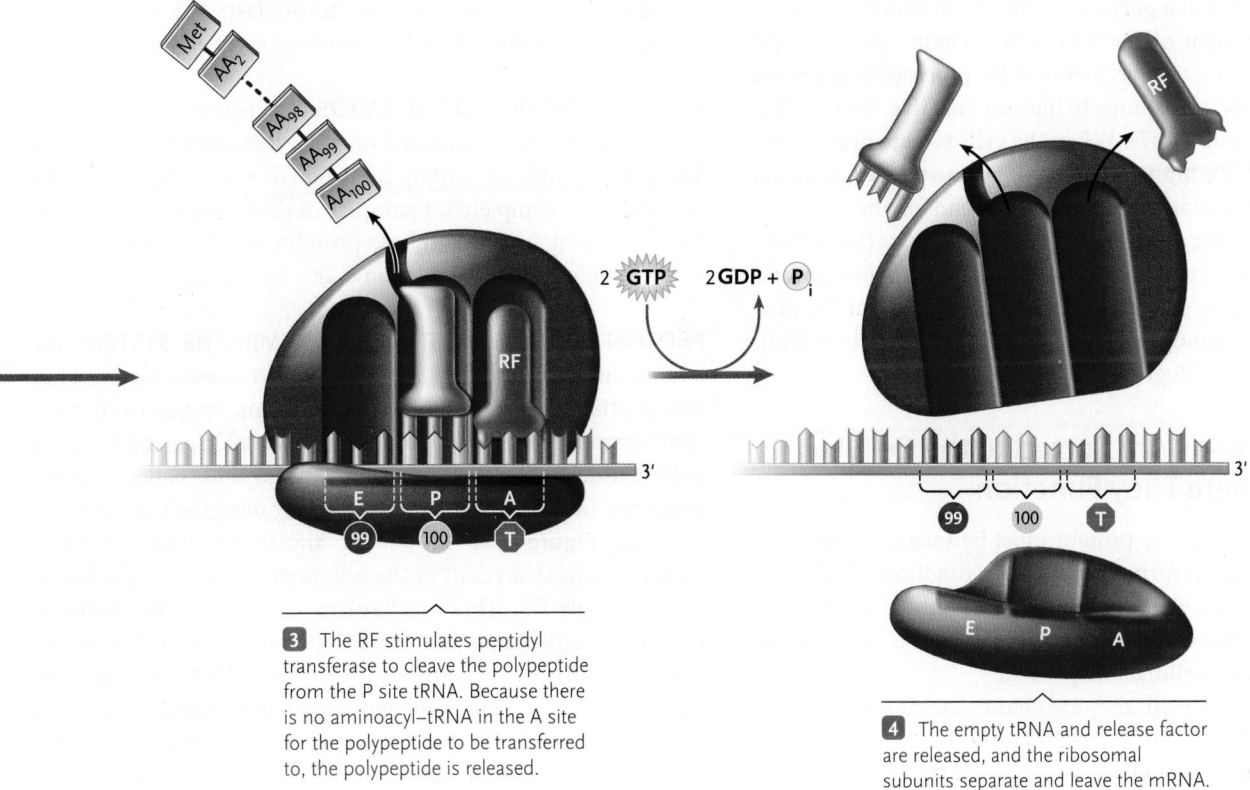

3 The RF stimulates peptidyl transferase to cleave the polypeptide from the P site tRNA. Because there is no aminoacyl–tRNA in the A site for the polypeptide to be transferred to, the polypeptide is released.

$2\ \textbf{GTP}$ $2\ \textbf{GDP} + P_i$

4 The empty tRNA and release factor are released, and the ribosomal subunits separate and leave the mRNA.

(a contraction of *polyribosome;* **Figure 15.16**). The multiple ribosomes greatly increase the overall rate of polypeptide synthesis from a single mRNA. The total number of ribosomes in a polysome depends on the length of the coding region of its mRNA molecule, ranging from a minimum of one or two ribosomes on the smallest mRNAs to as many as 100 on the longest mRNAs.

In prokaryotes, because of the absence of a nuclear envelope, transcription and translation typically are coupled. That is, as soon as the 5′ end of a new mRNA emerges from the RNA polymerase, ribosomal subunits attach and initiate translation **(Figure 15.17).** In essence, the polysome forms while the mRNA is still being made. By the time the mRNA is completely transcribed, it is covered with ribosomes from end to end, each assembling a copy of the encoded polypeptide.

Newly Synthesized Polypeptides Are Processed and Folded into Finished Form

Most eukaryotic proteins are in an inactive, unfinished form when ribosomes release them. Processing reactions that convert the new proteins into finished form include the removal of amino acids from the ends or interior of the polypeptide chain and the addition of larger organic groups, including carbohydrate or lipid structures.

Proteins fold into their final three-dimensional shapes as the processing reactions take place. For many proteins, helper proteins called **chaperones** or **chaperonins** assist the folding process by combining with the folding protein, promoting

"correct" three-dimensional structures, and inhibiting incorrect ones (see Section 3.4 and Figure 3.22).

In some cases, the same initial polypeptide may be processed by alternative pathways that produce different mature polypeptides, usually by removing different, long stretches of amino acids from the interior of the polypeptide chain. Alternative processing of a pre-mRNA is another mechanism that increases the number of polypeptides encoded by a single gene.

mRNAs with attached ribosomes

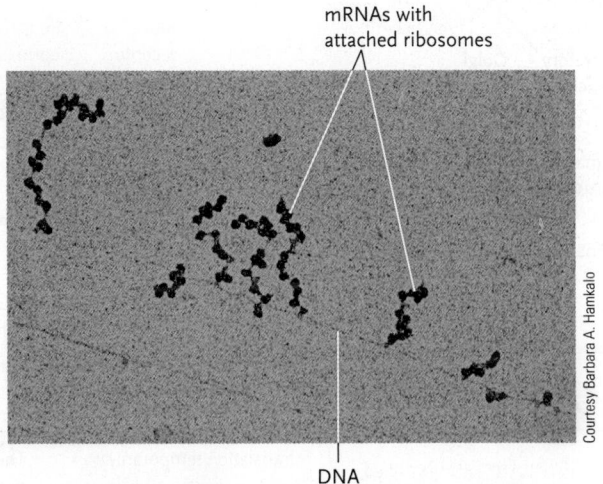

DNA

FIGURE 15.17 Simultaneous transcription and translation in progress in an electron microscope preparation extracted from *E. coli*. ×57,000.

Other proteins are processed into an initial, inactive form that is later activated at a particular time or location by removal of a covering segment of the amino acid chain. The digestive enzyme pepsin, for example, is made by processing reactions within cells lining the stomach into an inactive form called *pepsinogen* (see Chapter 47). When the cells secrete pepsinogen into the stomach, the high acidity of that organ triggers removal of a segment of amino acids from one end of the protein's amino acid chain. The amino acid removal converts the enzyme into the active form in which it rapidly breaks proteins in food particles into shorter pieces. The initial production of the protein as inactive pepsinogen protects the cells that make it from having their proteins degraded by the enzyme.

Finished Proteins Are Sorted to the Cellular Locations Where They Function

In a eukaryotic cell, every protein must be sorted to the compartment where it performs a necessary function. Without a sorting system, cells would wind up as a jumble of proteins floating about in the cytoplasm, with none of the spatial organization that makes cellular life possible.

As proteins are sorted, they are channeled into one of three compartments in the cell: (1) the cytosol; (2) the endomembrane system, which includes the endoplasmic reticulum (ER), Golgi complex, lysosomes, secretory vesicles, the nuclear envelope, and the plasma membrane (see Section 4.3); and (3) other

membrane-bound organelles distinct from the endomembrane system, including mitochondria, chloroplasts, microbodies (for example, peroxisomes), and the nucleus (see Section 4.3).

PROTEIN SORTING TO THE CYTOSOL Proteins that function in the cytosol are synthesized on *free ribosomes* in the cytosol. The polypeptides are simply released from the ribosomes once translation is completed. Examples of proteins that function in the cytoplasm are microtubule proteins and the enzymes that carry out glycolysis (see Section 7.2).

PROTEIN SORTING TO THE ENDOMEMBRANE SYSTEM The endomembrane system is a major traffic network for proteins. Polypeptides that sort to the endomembrane system begin their synthesis on free ribosomes in the cytosol. Specific to these polypeptides is a short segment of amino acids called a **signal sequence** (also called a **signal peptide**) near their N-terminal ends. As **Figure 15.18** shows, the signal sequence initiates a series of steps that result in the polypeptide entering the lumen of the rough ER. This mechanism is called **cotranslational import** because import of the polypeptide into the ER occurs simultaneously with translation of the mRNA encoding the polypeptide. The signal sequence was discovered in 1975 by Günter Blobel, B. Dobberstein, and colleagues at Rockefeller University, when they observed that proteins sorted through the endomembrane system initially contain extra amino acids at their N-terminal ends. Blobel received a Nobel Prize in 1999 "for the discovery that proteins have intrinsic signals that govern their transport and localization in the cell."

Once in the lumen of the rough ER, the proteins fold into their final form. They also have or attain a tag—a "zip code" if

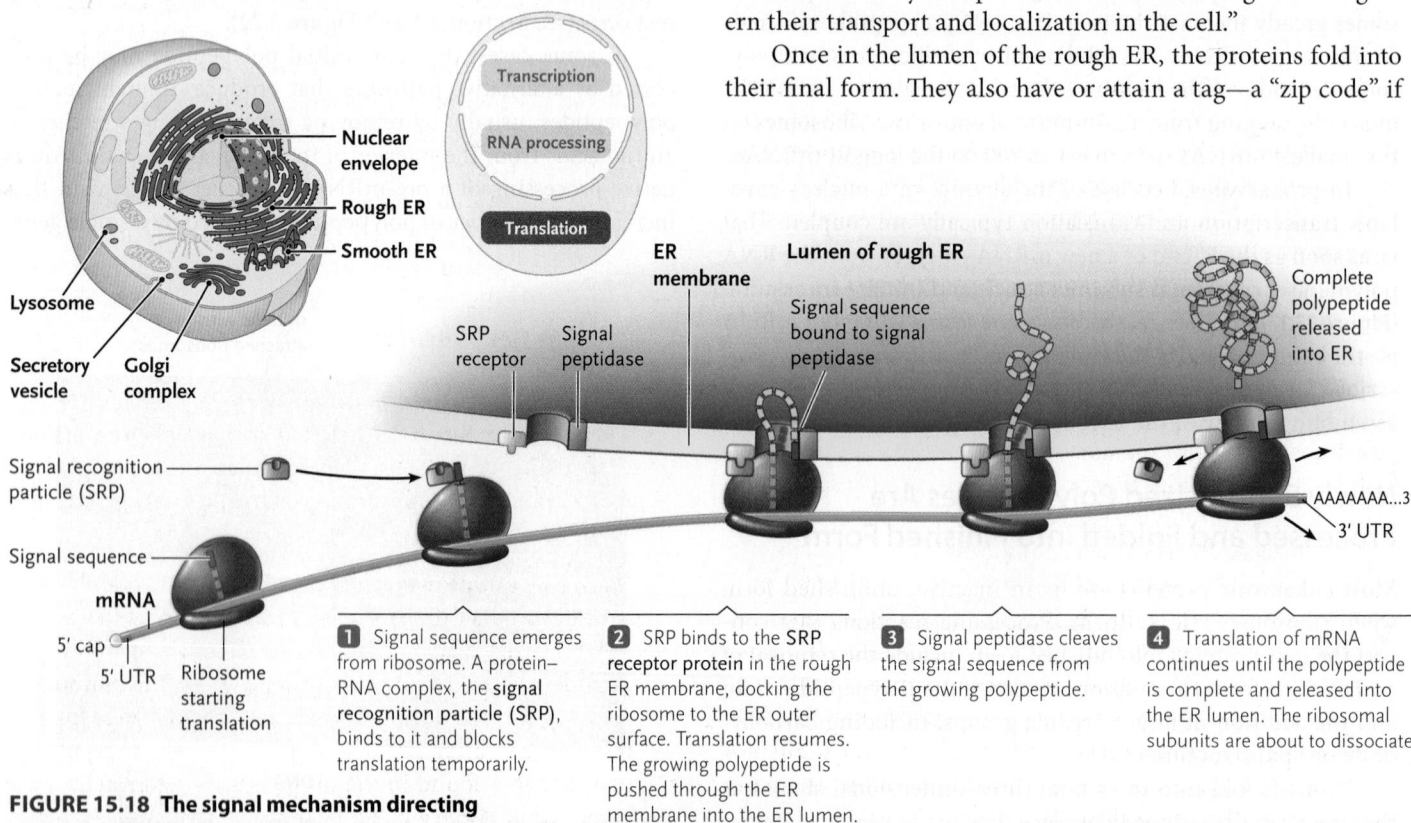

FIGURE 15.18 The signal mechanism directing proteins to the ER. The figure shows several ribosomes at different stages of translation of the mRNA.

© Cengage Learning 2017

1. Signal sequence emerges from ribosome. A protein–RNA complex, the **signal recognition particle (SRP)**, binds to it and blocks translation temporarily.

2. SRP binds to the **SRP receptor protein** in the rough ER membrane, docking the ribosome to the ER outer surface. Translation resumes. The growing polypeptide is pushed through the ER membrane into the ER lumen. The signal sequence binds to signal peptidase.

3. Signal peptidase cleaves the signal sequence from the growing polypeptide.

4. Translation of mRNA continues until the polypeptide is complete and released into the ER lumen. The ribosomal subunits are about to dissociate.

you will—that targets each protein for sorting to its final destination. Depending on the protein and its destination, the tag may be an amino acid sequence already in the protein, or a functional group or short sugar chain added in the lumen. Some proteins remain in the ER, whereas others are transported to the Golgi complex where they may be modified further. From the Golgi complex, proteins are packaged into vesicles, which may deliver them to lysosomes, secrete them from the cell (digestive enzymes, for example), or deposit them in the plasma membrane (cell surface receptors, for instance). Vesicle traffic in the cytoplasm involving the rough ER and Golgi complex is illustrated in Figure 4.14.

PROTEIN SORTING TO MITOCHONDRIA, CHLOROPLASTS, MICROBODIES, AND THE NUCLEUS Proteins are sorted to mitochondria, chloroplasts, microbodies, and the nucleus after they have been made on free ribosomes in the cytosol. This mechanism of sorting is called **posttranslational import.** Proteins destined for the mitochondria, chloroplasts, and microbodies have **targeting signals.** In some cases the targeting signal is an extension to an end of the protein; it is removed enzymatically once the protein enters the organelle. In other cases, the targeting signal is within the protein and is not removed once in the organelle. The targeting signals are recognized by specific receptors on the organelle surface triggering the import of the protein into the organelle by specific pathways.

A large number of proteins are imported by mitochondria and chloroplasts. For example, as you learned in the *Molecular Insights* of Chapter 7, a proteomics study identified approximately 1,400 proteins in human heart mitochondria. Yet, only 13 proteins are encoded by the human mitochondrial genome. The rest are encoded by nuclear genes and are imported into the organelle. Recent research has identified five main classes of imported proteins, each following a different import route into the mitochondrion.

Proteins sorted to the nucleus, such as the enzymes for DNA replication and RNA transcription, have short amino acid sequences called **nuclear localization signals.** A cytosolic protein binds to the signal and moves it to the nuclear pore complex (see Section 4.3 and Figures 4.10 and 4.11), where it is then transported into the nucleus through the pore. For these proteins, the nuclear localization signal remains because, if the cell divides, the proteins need to reenter the nucleus after the nuclear envelope breaks down and reforms.

The same basic system of sorting signals distributes proteins in prokaryotic cells, indicating that this mechanism probably evolved with the first cells. In bacteria, signals similar to the ER-directing signals of eukaryotes direct newly synthesized bacterial proteins to the plasma membrane (bacteria do not have ER membranes); further information built into the proteins keeps them in the plasma membrane or allows them to enter the cell wall or to be secreted outside the cell. Proteins without sorting signals remain in the cytoplasmic solution.

STUDY BREAK 15.4

1. How does translation initiation occur in eukaryotes versus prokaryotes?
2. Distinguish between the P, A, and E sites of the ribosome.
3. How are proteins directed to different compartments of a eukaryotic cell?

15.5 Genetic Changes That Affect Protein Structure and Function

We learned in Chapters 12 and 13 that a mutant allele of a gene can alter the phenotype controlled by the gene. In this section, we discuss two types of genetic change that can alter protein structure and function and, therefore, produce an altered phenotype. One type is mutation of one base pair to another in the DNA. The other is when genetic elements known as transposable elements (TEs) move from one location to another in the genome.

Base-Pair Mutations Can Alter Protein Structure and Function

Mutations, in general, are changes to the genetic material. Recall from Chapter 1 that variability among individuals—the raw material molded by evolutionary processes—arises ultimately through mutations. One type of mutation is the **base-pair mutation,** which involves a change of a single base pair in the genetic material. A base-pair mutation is also called a *point mutation,* although strictly speaking, a point mutation can refer to a change of one to a few adjacent base pairs. (Mutations involving more than one base pair, and chromosomal mutations [see Chapter 13] are not considered in this discussion.) If a base-pair mutation occurs in the protein-coding portion of a gene, it can change a base in a codon in the mRNA and thereby affect the structure and function of the encoded protein. More generally, mutations affecting the functions of genes are known as *gene mutations.*

TYPES OF BASE-PAIR MUTATIONS IN A PROTEIN-CODING GENE Consider a theoretical stretch of normal (unmutated) DNA encoding a string of amino acids in a polypeptide **(Figure 15.19A).** Four types of base-pair mutations affecting a protein-coding gene are:

1. **Missense mutation (Figure 15.19B):** A sense codon is changed to a different sense codon that specifies a different amino acid. Whether the function of a polypeptide is altered significantly depends on the amino acid change that occurs. Individuals homozygous for a missense mutation in the gene for one of the two polypeptide types found in the oxygen-carrying protein hemoglobin **(Figure 15.20)** have the genetic disease sickle-cell anemia, described in Chapter 12 (*Why it matters . . .* and Section 12.2). Many other human

A. Normal

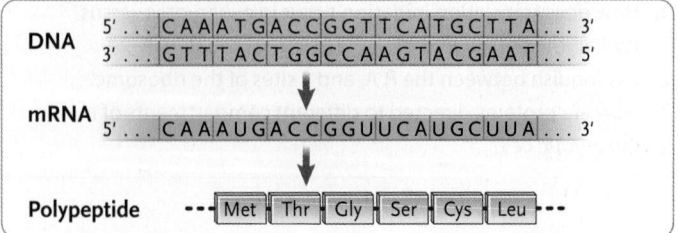

B. Missense mutation. Changes sense codon to a sense codon for a different amino acid.

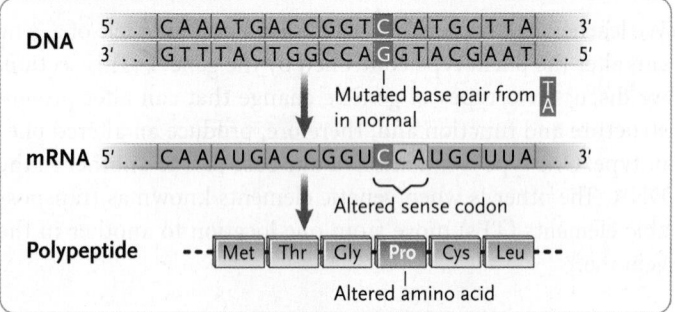

C. Nonsense mutation. Changes sense codon to a stop codon.

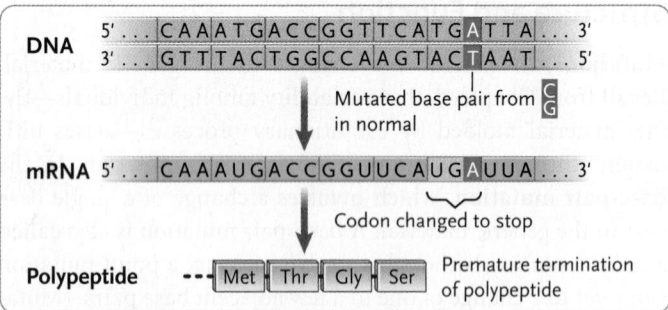

D. Silent mutation. Changes sense codon to another codon for the same amino acid.

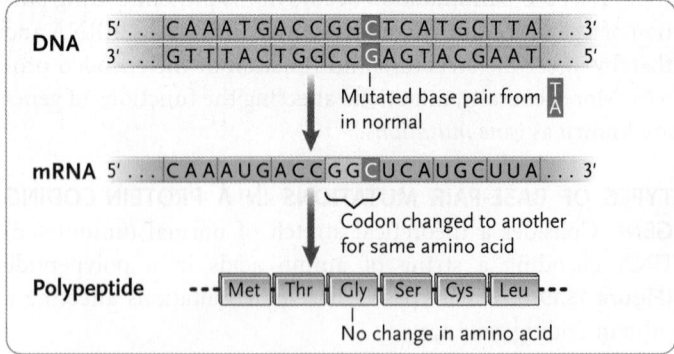

E. Frameshift mutation. Changes reading frame after mutation.

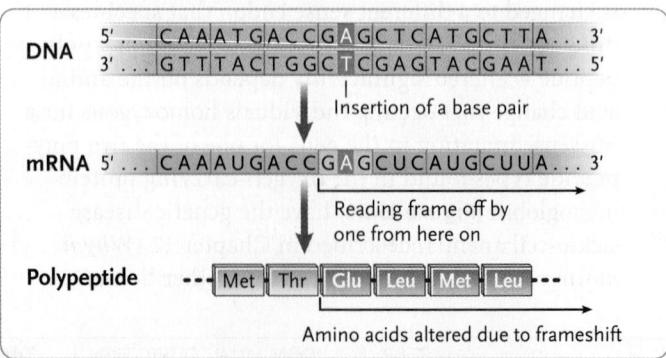

FIGURE 15.19 Effects of base-pair mutations in protein-coding genes on the amino acid sequence of the encoded polypeptide.
© Cengage Learning 2017

genetic diseases are caused by missense mutations, including albinism, hemophilia, and achondroplasia (see Section 13.4).

2. **Nonsense mutation (Figure 15.19C):** A sense codon is changed to a nonsense (stop) codon. Translation of an mRNA containing a nonsense mutation results in a shorter-than-normal polypeptide that, in many cases, will be only partially functional at best.

3. **Silent mutation (Figure 15.19D):** A sense codon is changed to a different sense codon, but that codon specifies the same amino acid as in the normal polypeptide, so the function of the polypeptide is unchanged.

4. **Frameshift mutation (Figure 15.19E):** A single base-pair deletion or insertion in the coding region of a gene alters the reading frame of the resulting mRNA (the figure shows an insertion). After the point of the mutation, the ribosome reads codons that are not the same as for the normal mRNA, producing a different amino acid sequence in the polypeptide from then on. The resulting polypeptide typically is nonfunctional because of the significantly altered amino acid sequence.

SPONTANEOUS MUTATIONS AND INDUCED MUTATIONS

Remember that mutations are the result of unrepaired damage to DNA. The generation of mutations can occur spontaneously or can be induced.

Spontaneous mutations are mutations that occur naturally within the cell. All types of base-pair mutations occur spontaneously and they can occur during DNA replication, as well as through cellular chemical activity during other stages of cell growth and division.

Induced mutations occur when an organism is exposed either deliberately or accidentally to a physical or chemical agent, called a **mutagen,** that interacts with DNA to cause a mutation. Production of mutations in a laboratory by exposure

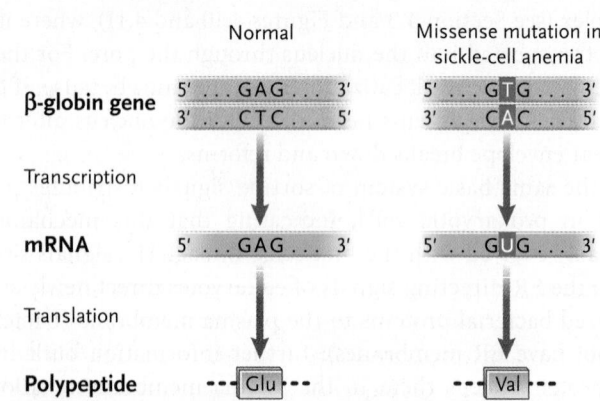

FIGURE 15.20 Missense mutation in a gene for one of the two polypeptides of hemoglobin that is the cause of sickle-cell anemia.
© Cengage Learning 2017

to a mutagen is called **mutagenesis,** and the treated organism is said to be *mutagenized.*

Induced mutations occur at a much higher frequency than do spontaneous mutations and, hence, have been useful in classical genetic studies of model organisms such as *E. coli,* yeast, *Neurospora* (see Section 15.1), and *Drosophila,* for identifying genes of interest. Mutagenesis in this way produces random mutations across the genome. (The classical methods are still used although, increasingly, genomics approaches are used to make specific mutations at targeted genomic locations.) Typically, experimental organisms are exposed to a mutagen and the mutagenized population and geneticists screen or select for a change in a phenotype of interest. Comparing mutants with normal individuals enables a researcher to identify the genes involved. Studying the structures and functions of those mutant genes in comparison with the corresponding normal genes provides information about the normal functions of the genes involved. You saw this approach in operation in the discussion of Beadle and Tatum's experiment in Section 15.1, where they used a mutagen to make auxotrophic mutants of *Neurospora* and study the genes involved.

Mutagens induce mutations by three main mechanisms: replacing a base in the DNA; altering a base so that it now mispairs specifically with a different base; or damaging a base so that it cannot pair with another base. The two types of mutagens are radiation and chemical mutagens. Radiation occurs in nonionizing and ionizing forms. Ionization occurs when energy is sufficient to knock an electron out of an atomic shell and hence break covalent bonds. The only nonionizing radiation that is mutagenic is UV light (see Figure 14.23). All forms of ionizing radiation, such as X-rays, cosmic rays, radon, and uranium, can induce mutations, which is why limiting exposure to those forms of radiation is important. Chemical mutagens include both naturally occurring chemicals (such as those ingested with food) and synthetic substances. Mutagenesis experiments use synthetic chemicals with known, specific effects on DNA. An example is base analogs, which can switch between two chemical forms. They incorporate into DNA in one form, with one type of pairing, say A–X, where X is the mutagen. If they switch to the other form at the next round of replication, they will now specify a G in the new strand to produce a G–X base pair. In the next round of replication, the G will specify a C on the new strand, resulting in a G–C base pair. The overall result is that the mutagen has caused an A–T base pair to mutate to a G–C base pair.

Outside of the laboratory, mutagenesis is of concern to all organisms. We, in the human population, are exposed every day to a wide variety of chemicals in our environment. The chemicals may be natural ones, such as those synthesized by the plants and animals we eat as food, or ones produced by humans, such as drugs, cosmetics, food additives, pesticides, and industrial compounds. Our exposure to chemicals occurs primarily through eating food, absorption through the skin, and inhalation. Some of these chemicals are mutagens. For a mutagenic chemical to cause DNA changes, it must enter cells and penetrate to the nucleus, which many chemicals cannot do.

Some chemicals are converted from nonmutagenic to mutagenic by our metabolism. That is, when these chemicals are directly tested for mutagenic activity on, say, a bacterial species, no mutations result. But after they are processed in the body, they become mutagens. For example, benzpyrene, a polycyclic aromatic hydrocarbon found in cigarette smoke, coal tar, automobile exhaust fumes, and charbroiled food, is nonmutagenic. But its metabolite, benzpyrene diol epoxide, which is both a mutagen and a *carcinogen* (an agent that increases the frequency with which cells become cancerous), can induce cancer. Because cancers are genetic diseases, requiring the accumulation of a number of genetic mutations to trigger the uncontrolled growth characteristic of the cancer (see Chapter 16), clearly it is important to limit our exposure to potential sources of environmental mutagens.

Transposable Elements Move from One Location to Another in the Genome and May Affect Gene Function

All organisms contain particular segments of DNA that can move from one place to another within a cell's genome. The movable sequences are called *transposable genetic elements,* or more simply, **transposable elements (TEs).**

The movement of TEs, called **transposition,** involves a type of genetic recombination process. However, the location in the DNA to which the TE moves—the **target site**—is not homologous with the TE. In this respect, transposition differs from genetic recombination in meiosis in eukaryotes, and in the processes that produce recombinants in bacteria (see Chapter 17), which involve crossing-over between homologous DNA molecules.

Transposition of a TE occurs at a low frequency. Depending on the TE, transposition occurs in one of two general ways: (1) a nonreplicative, cut-and-paste process, in which the TE leaves its original location and transposes to a new location **(Figure 15.21A);** or (2) a replicative, copy-and-paste process, in which a copy of a TE transposes to a new location, leaving the original TE behind **(Figure 15.21B).** For most TEs, transposition starts with contact between the TE and the target site. This also means that TEs do not exist free of the DNA. In both cases, the target site is duplicated as a result of the integration of the TE, with the copies flanking the TE.

TRANSPOSABLE ELEMENTS IN EUKARYOTES TEs were first discovered in maize (corn), in the 1940s by Barbara McClintock, a geneticist working at the Cold Spring Harbor Laboratory in New York. McClintock noted that some mutations affecting kernel and leaf color appeared and disappeared rapidly under certain conditions. Mapping the alleles by linkage studies produced a surprising result—the map positions changed frequently, indicating that the alleles could move from place to place in the corn chromosomes. Some of the movements were so frequent that changes in their effects could be noticed at different times in a single developing kernel **(Figure 15.22).**

A. Non-replicative, cut-and-paste transposition. The transposable element (TE) leaves one location in the DNA and moves to a new location.

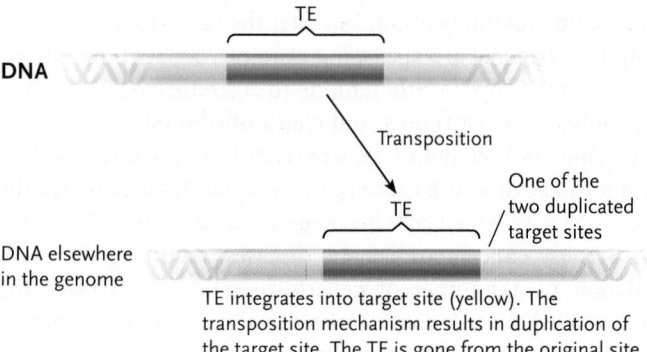

TE integrates into target site (yellow). The transposition mechanism results in duplication of the target site. The TE is gone from the original site.

B. Replicative, copy-and-paste transposition. A copy of the TE moves to a new location, leaving the original TE behind.

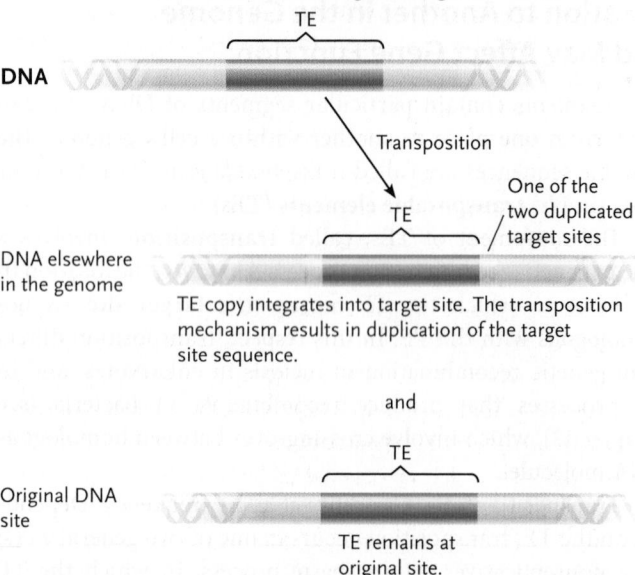

TE copy integrates into target site. The transposition mechanism results in duplication of the target site sequence.

and

TE remains at original site.

FIGURE 15.21 Two transposition mechanisms for transposable elements.
© Cengage Learning 2017

FIGURE 15.22 Barbara McClintock and corn kernels showing different color patterns because of the movement of transposable elements. As TEs move into or out of genes controlling pigment production in developing kernels, the ability of cells and their descendants to produce the dark pigment is destroyed or restored. The result is random patterns of pigmented and colorless (yellow) segments in individual kernels.

When McClintock first reported her results, her findings were regarded as an isolated curiosity, possibly applying only to corn. This was because the then-prevailing opinion among geneticists was that genes are fixed in the chromosomes and do not move to other locations. Her conclusions were widely accepted only after TEs were detected and characterized in bacteria in the 1960s. We now know that TEs are probably universally distributed among both prokaryotes and eukaryotes. McClintock was awarded a Nobel Prize in 1983 "for her discovery of mobile genetic elements."

Eukaryotic TEs fall into two major classes, *DNA transposons* and *retrotransposons,* distinguished by the way the TE sequence moves in the genome:

1. **DNA transposons** are TEs that transpose using a DNA intermediate. In most cases, movement of a DNA transposon is by a nonreplicative, cut-and-paste mechanism. A gene for **transposase,** an enzyme that catalyzes the reactions inserting or removing the TE from DNA, is in the central region of the transposon. Most DNA transposons have *inverted repeat sequences*—the same DNA sequence oriented in opposite directions—at their ends that enable the transposase to recognize the boundaries of the TE for transposition.

2. **Retrotransposons** are TEs that transpose using a replicative, copy-and-paste mechanism but, unlike DNA transposons, their transposition occurs via an intermediate RNA copy of the TE **(Figure 15.23)**. Some retrotransposons are bounded by sequences that are directly repeated rather than in inverted form; others have no repeated sequences at their ends.

Genomic analysis has shown that eukaryotic genomes vary considerably in the TEs they contain. For example, some research studies have revealed the extraordinary fact that about 42% of the human genome consists of retrotransposons, some of which are still active in transposition today. DNA transposons constitute about 3% of the human genome but, as is the case in other primates, none of them is capable of transposition. Transposition-competent DNA transposons are present in other eukaryotes, however. Some other research suggests that the proportion of the human genome derived from TEs may be closer to two-thirds. In maize (*Zea mays*, commonly referred to as corn), almost 85% of the genome consists of transposon sequences, the majority of which are retrotransposons. Interestingly, maize genes are clustered in the genome in small groups separated by long, uninterrupted segments of DNA made up of retrotransposons. In simpler eukaryotes, the percentage of transposons comprising the genome is much lower, for example about 5% in chicken, about 26% in zebrafish, about 10% in *Drosophila,* and about 9% in the worm, *C. elegans.* Interestingly, there is an approximate positive correlation between genome size and the percentage of the genome consisting of transposable element sequences. The inescapable conclusion is that the presence and activity of transposable elements in genomes have played important roles in generating the structures of genomes during genome evolution.

Active TEs can affect gene expression in various ways and, over evolutionary time, may have had a significant influence on generating transcript diversity. For example, insertion of a TE into a protein-coding gene could disrupt the gene and, therefore, prevent the synthesis of a protein. The insertion event, which occurs at random sites, is called **insertional mutagenesis.** A number of cases of human diseases have been shown to be caused by insertional mutagenesis like this involving retrotransposons, for example, cystic fibrosis, several types of cancer including breast cancer, hemophilia, and Duchenne muscular dystrophy.

TEs can also affect gene expression in other ways, such as providing an alternative promoter that may act in a tissue-specific manner; affecting the alternative splicing machinery; and providing an alternative polyadenylation site thereby affecting the 3′ end of a transcript. Overall, TE activity can have diverse effects on an organism's transcriptome and, therefore, contribute to genetic variability.

BACTERIAL TRANSPOSABLE ELEMENTS Bacterial TEs were discovered in the 1960s. They move from site to site within the bacterial chromosome, between the bacterial chromosome and plasmids, and between plasmids.

The two major types of bacterial TEs are *insertion sequences (ISs)* and *transposons* **(Figure 15.24). Insertion sequences** are the simpler of the two. They are relatively small and contain only the transposase gene for inserting or removing the TE from the DNA (see Figure 15.24A). At each of the two ends of an IS is a short inverted repeat sequence. As for eukaryotic TEs, the inverted repeat sequences enable the transposase enzyme to identify the ends of the TE when it catalyzes transposition.

The second type of bacterial TE, called a **transposon,** has an inverted repeat sequence at each end enclosing a central region with one or more genes (Figure 15.24B). In a number of bacterial transposons, the inverted repeat sequences are insertion sequences, one of which provides the transposase for movement of the element. Bacterial transposons without IS ends have short inverted repeat end sequences, and a transposase gene is within the central region. Additional genes in the central region of both types of transposons typically are for antibiotic resistance that originated from the main bacterial DNA circle or from plasmids. The additional genes are carried along as the TEs move from place to place within and between species.

Many antibiotics, such as penicillin, erythromycin, tetracycline, ampicillin, and streptomycin, that were once successful in curing bacterial infections have lost much of their effectiveness because of resistance genes carried in transposons. Movements of the transposons, particularly to plasmids that have been transferred between bacteria within the same species and between different species, greatly increase the spread of genes providing antibiotic resistance. Resistance genes have made many bacterial diseases difficult or impossible

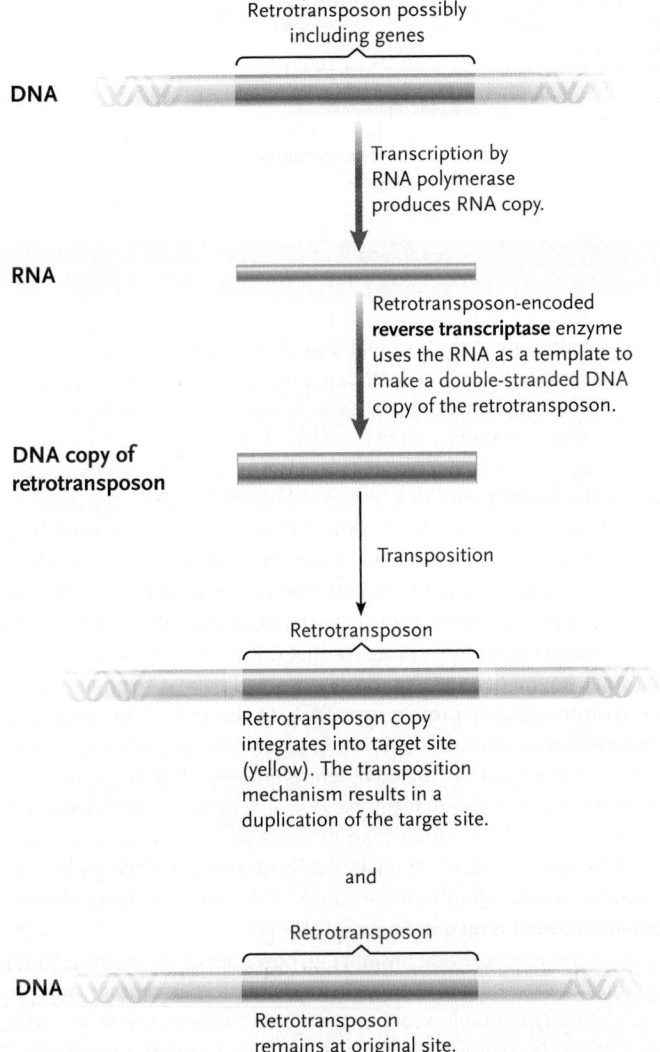

FIGURE 15.23 Transposition of a eukaryotic retrotransposon to a new location by means of an intermediate RNA copy.

© Cengage Learning 2017

to treat with standard antibiotics. (Chapter 26 discusses bacterial antibiotic resistance further.)

In this chapter you have learned how gene expression occurs by the processes of transcription and translation, and how gene expression may be changed by mutation or by the actions of transposable elements. In the next chapter, you will see how organisms and cells exert control over how their genes are expressed.

STUDY BREAK 15.5

1. How does a missense mutation differ from a silent mutation?
2. How do genetic recombination and TE transposition differ?
3. How do eukaryotic DNA transposons and retrotransposons differ?
4. In what ways are bacterial IS elements and transposons alike?

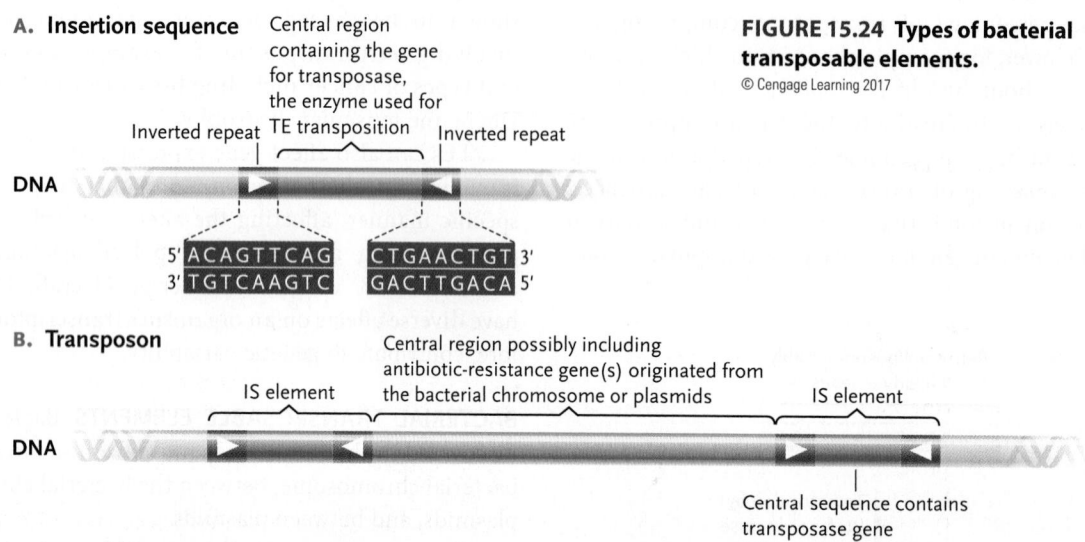

FIGURE 15.24 Types of bacterial transposable elements.
© Cengage Learning 2017

Unanswered Questions

How does the ribosome work?

When they were first discovered, ribosomes were thought to be passive "workbenches" on which proteins were synthesized, presumably by enzymes. But in the late 1960s, it was found that the ribosome itself catalyzes formation of peptide bonds. Binding and lining up the mRNA and the correct tRNAs were other things that appeared to be jobs carried out by the ribosome. But the ribosome is a huge macromolecular complex, containing more than 50 ribosomal proteins and large ribosomal RNAs containing many thousands of nucleotides.

In my laboratory, we wanted to identify the parts of the ribosome that carried out these important functions, as a step toward understanding *how* they manage to do it. Following the common wisdom that it is the proteins that carry out the biological functions of the cell, we assumed that ribosomal proteins were responsible for ribosome functions, but which ones? We decided to use *chemical modification* to approach this question.

Our plan was to inactivate the functions of the ribosome by attacking the ribosome with chemical reagents that were known to modify the reactive groups in proteins; we would then figure out which proteins were inactivated by rescuing the inactivated proteins with individual active ones, using *in vitro* reconstitution of a functional ribosome as the measure for a successful result. To our great surprise, it was very difficult to inactivate the ribosome using reagents that react with proteins. This disappointed me, because I was eager to use my skills as a protein chemist to understand how the ribosome works. As a sort of control experiment, we tried using a reagent that attacks RNA, and for the first time we observed rapid inactivation. Furthermore, ribosome activity could be rescued by reconstituting with active RNA, proving that inactivation of the ribosome was caused by modification of the ribosomal RNA. In 1972 we published

a paper proposing that ribosomal RNA plays a functional role in protein synthesis, and followed it with evidence from many other kinds of studies in the following years, but no one was ready to believe such a "crackpot idea" (as one prominent scientist was heard to describe our findings).

Twenty years after our original proposal, I was frustrated that people were still resistant to the idea, although less so because of the discovery of ribozymes (catalytic RNAs) in the 1980s. So in 1992 we carried out another set of experiments in which we subjected ribosomes to relatively brutal extraction procedures that are used to remove proteins from complexes with RNA and DNA. We washed the ribosomes with a strong detergent, treated them with a protease that is known to chew up proteins into small fragments, and vigorously extracted them with phenol, which is known to separate protein from RNA. At the end of this ordeal, the ribosomes were completely active in catalyzing formation of peptide bonds. Although we had not completely removed all of the protein from the ribosome, the idea that the biological activity of ribosomes might be based on its RNA, rather than its proteins, now became accepted widely by researchers. It was finally demonstrated convincingly by high-resolution crystallography of the ribosome that the site of peptide bond formation contains no protein, only RNA.

This experience provides insight into how science works; the scientific community is resistant to changes in fundamental paradigms ("Only proteins can carry out biological functions"), and acceptance of new ideas often depends on proposing them at "the right time." Also typical of science, the new insight leaves many questions still unanswered. Even after 50 years of research by hundreds of investigators, we are still trying to figure out the ribosome's fundamental mechanisms of action. And an even more daunting question is, How did the ribosome evolve?

Courtesy of Harry Noller

Harry Noller is Robert L. Sinsheimer Professor of Molecular Biology at the University of California, Santa Cruz, where he studies the structure and function of the ribosome, using such diverse approaches as biochemistry, molecular genetics, fluorescence spectroscopy, and X-ray crystallography. To learn more about his research, go to http://rna.ucsc.edu/rnacenter/noller_lab.html

REVIEW KEY CONCEPTS

For access to MindTap and additional study materials visit www.cengagebrain.com.

15.1 The Connection between DNA, RNA, and Protein

- In genetic experiments with *Neurospora crassa*, a direct correspondence was found between gene mutations and alterations of enzymes, which led to the proposal of the one gene–one enzyme hypothesis, later updated as the one gene–one polypeptide hypothesis (Figure 15.1).

- The pathway from genes to proteins involves transcription then translation. In transcription, a sequence of nucleotides in DNA is copied into a complementary sequence in an RNA molecule. In translation, the sequence of nucleotides in an mRNA molecule specifies an amino acid sequence in a polypeptide (Figure 15.2).

- The genetic code is a triplet code. AUG at the beginning of a coded message establishes a reading frame for reading the codons three nucleotides at a time until a stop codon is reached. The code is degenerate: most of the amino acids are specified by more than one codon (Figures 15.3 and 15.4).

- The genetic code is essentially universal.

15.2 Transcription: DNA-Directed RNA Synthesis

- Transcription begins when an RNA polymerase binds to a promoter sequence in the DNA and starts synthesizing an RNA molecule. The enzyme then adds RNA nucleotides in sequence according to the DNA template. At the end of the transcribed sequence, the enzyme and the completed RNA transcript release from the DNA template.

- Transcription occurs in three stages: initiation, elongation, and termination. Each stage is similar in eukaryotes and bacteria. For transcription of eukaryotic protein-coding genes, transcription factors bind to the promoter and recruit RNA polymerase II, which then begins transcription of the mRNA. Elongation of the mRNA occurs as the polymerase reads the coding sequence of the gene. The 3′ end of the mRNA is specified differently in prokaryotes and eukaryotes; termination of transcription in bacteria is determined by a terminator sequence in the gene (Figure 15.5).

- Transcription of noncoding RNA genes occurs in a similar way to transcription of protein-coding genes. In eukaryotes, special RNA polymerases are used to transcribe tRNA and rRNA genes, whereas in bacteria the same RNA polymerase that transcribes protein-coding genes transcribes those genes.

15.3 Production of mRNAs in Eukaryotes

- A gene encoding an mRNA molecule includes the promoter, which is recognized by the regulatory proteins and transcription factors that promote DNA unwinding and the initiation of transcription by an RNA polymerase. Transcription in eukaryotes produces a pre-mRNA molecule that consists of a 5′ cap, the 5′ untranslated region, interspersed exons (amino acid-coding segments) and introns, the 3′ untranslated region, and the 3′ poly(A) tail. The 5′ cap and 3′ poly(A) tail are not encoded in the DNA. The 5′ cap is added by a capping enzyme, and the 3′ poly(A) tail is added by poly(A) polymerase once a 3′ end of the pre-mRNA is generated by cleavage (Figure 15.6).

- Introns in pre-mRNAs are removed to produce functional mRNAs by splicing. snRNPs bind to the introns, loop them out of the pre-mRNA, clip the intron at each exon boundary, and join the adjacent exons together (Figure 15.7).

- In living cells, transcription, pre-mRNA processing, and nuclear export of finished mRNAs are often coupled resulting in a continuous process rather than separate events.

- Many pre-mRNAs are subjected to alternative splicing, a process that joins exons in different combinations to produce different mRNAs encoded by the same gene. Translation of each mRNA produced in this way generates a protein with different function (Figure 15.8).

- Analysis of the human transcriptome has shown that there is pervasive transcription of the genome, and that there is a significant amount of potential variation in gene expression for protein-coding genes because of gene isoforms, the different forms of mRNAs that can be transcribed from a gene.

15.4 Translation: mRNA-Directed Polypeptide Synthesis

- Translation is the assembly of amino acids into polypeptides. Translation occurs on ribosomes. The P, A, and E sites of the ribosome are used for the stepwise addition of amino acids to the polypeptide as directed by the mRNA (Figures 15.9 and 15.12).

- Amino acids are brought to the ribosome attached to specific tRNAs. Amino acids are added to their corresponding tRNAs by aminoacyl–tRNA synthetases (Figures 15.10 and 15.11).

- Translation proceeds through the stages of initiation, elongation, and termination. In initiation, a ribosome assembles with an mRNA molecule and an initiator methionine–tRNA. In elongation, amino acids linked to tRNAs add one at a time to the growing polypeptide chain. In termination, the new polypeptide is released from the ribosome, and the ribosomal subunits separate from the mRNA (Figures 15.13–15.15).

- Multiple ribosomes translate the same mRNA simultaneously, forming a polysome (Figures 15.16 and 15.17).

- After they are synthesized on ribosomes, many polypeptides in eukaryotes are converted into finished form by processing reactions, which include removal of one or more amino acids from the protein chains, addition of organic groups, and folding guided by chaperones.

- In eukaryotes, finished proteins are sorted to the cellular locations where they function. Proteins that function in the cytosol are synthesized on free ribosomes. Proteins are sorted to the endomembrane system by cotranslational import and proteins are sorted to the mitochondria, chloroplast, microbodies, and nucleus by posttranslational import. In each of these two cases, specific amino acid sequences in the polypeptides direct them to their destinations (Figure 15.18).

15.5 Genetic Changes That Affect Protein Structure and Function

- Base-pair mutations alter the mRNA and can lead to changes in the amino acid sequence of the encoded polypeptide. A missense mutation changes one sense codon to one that specifies a different amino acid, a nonsense mutation changes a sense codon to a stop codon, and a silent mutation changes one sense codon to another sense codon that specifies the same amino acid. A base-pair insertion or deletion is a frameshift mutation that alters the reading frame beyond the point of the mutation, leading to a different amino acid sequence from then on in the polypeptide (Figures 15.19 and 15.20).
- Mutations can occur spontaneously or they can be induced. Spontaneous mutations occur naturally within the cell, whereas induced mutations occur when an organism is exposed to a mutagen, a physical or chemical agent that interacts with DNA to cause a mutation. Induced mutations have been valuable tools in experimental organisms for determining gene function. The two types of mutagens are radiation and chemicals.
- Both eukaryotes and prokaryotes contain TEs (transposable elements)—DNA sequences that can move from place to place in the DNA. Transposition of a TE may be nonreplicative, in which the TE moves from one location in the DNA to another, or replicative, in which a duplicated copy of a TE inserts in a new location while leaving the "parent" copy in its original location (Figure 15.21).
- Eukaryotic TEs occur as DNA transposons, which typically transpose by a nonreplicative mechanism, and retrotransposons, which transpose replicatively by making an RNA copy which is then replicated by reverse transcriptase into a DNA copy that is inserted at a new location (Figures 15.22 and 15.23).
- Eukaryotic genomes vary considerably in the proportion made up of TE sequences. The presence and activity of TEs in genomes have played important roles in genome evolution.
- Bacterial TEs occur as insertion sequences and transposons. Both contain a gene for the transposase enzyme needed for transposition. The transposon may also contain genes, such as for antibiotic resistance, which originate in host DNA (Figure 15.24).

TEST YOUR KNOWLEDGE

Remember/Understand

1. Eukaryotic mRNA:
   a. uses snRNPs to cut out introns and seal together translatable exons.
   b. uses a spliceosome mechanism made of DNA to recognize consensus sequences to cut and splice.
   c. has a guanine cap on its 3′ end and a poly(A) tail on its 5′ end.
   d. is composed of adenine, thymine, guanine, and cytosine.
   e. codes the guanine cap and poly(A) tail from the DNA template.

2. A segment of a strand of DNA has a base sequence of 5′-GCATTAGAC-3′. What would be the sequence of an RNA molecule complementary to that sequence?
   a. 5′-GUCTAATGC-3′
   b. 5′-GCAUUAGAC-3′
   c. 5′-CGTAATCTG-3′
   d. 5′-GUCUAAUGC-3′
   e. 5′-CGUAAUCUG-3′

3. Which of the following statements about the initiation phase of translation is false?
   a. An initiation factor allows 5′ mRNA to attach to the small ribosomal subunit.
   b. Initiation factors complex with GTP to help Met–tRNA and AUG pair.
   c. mRNA attaches first to the small ribosomal subunit.
   d. GTP is synthesized.
   e. 3′-UAC-5′ on the tRNA binds 5′-AUG-3′ on mRNA.

4. Which of the following statements about aminoacylation is false?
   a. It precedes translation.
   b. It occurs in the ribosome.
   c. It requires ATP to bind an aminoacyl–tRNA synthetase.
   d. It joins the correct amino acid to a specific tRNA based on the tRNA's anticodon.
   e. It uses three binding sites on aminoacyl–tRNA synthetase.

5. Which of the following statements is false?
   a. GTP is an energy source during various stages of translation.
   b. In the ribosome, peptidyl transferase catalyzes peptide bond formation between amino acids.
   c. When the mRNA code UAA reaches the ribosome, there is no tRNA to bind to it.
   d. A long polypeptide is cut off the tRNA in the A site so its Met amino acid links to the amino acid in the P site.
   e. Forty-two amino acids of a protein are encoded by 126 nucleotides of the mRNA.

6. Which item binds to SRP receptor and to the signal sequence to guide a newly synthesized protein to be secreted to its proper "channel"?
   a. ribosome
   b. signal recognition particle
   c. endoplasmic reticulum
   d. signal peptidase
   e. receptor protein

7. A missense mutation cannot be:
   a. the code for the sickle-cell gene.
   b. caused by a frameshift.
   c. the deletion of a base in a coding sequence.
   d. the addition of two bases in a coding sequence.
   e. the same as a silent mutation.

8. Which of the following is *not* correct about transposable elements?
   a. They can be recognized by their ends of inverted transposable elements.
   b. They have an internal portion that can be transcribed.
   c. They encode a transposase enzyme.
   d. They have no harmful effects on cell function.
   e. They move by a cut-and-paste or copy-and-paste mechanism.

Apply/Analyze

9. Which statement about the following pathway is false?

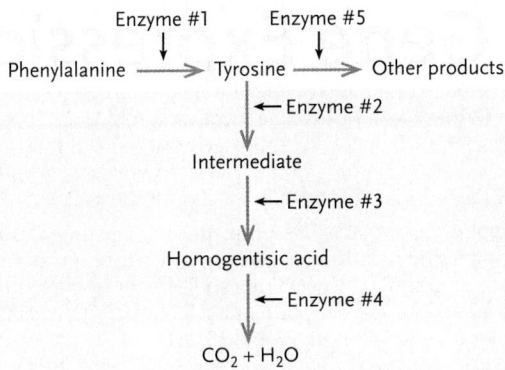

a. A mutation for enzyme #1 causes phenylalanine to build up.
b. A mutation for enzyme #2 prevents tyrosine from being synthesized.
c. A mutation at enzyme #3 prevents homogentisic acid from being synthesized.
d. A mutation for enzyme #2 could hide a mutation in enzyme #4.
e. Each step in a pathway such as this is catalyzed by an enzyme, which is coded by a gene.

10. A part of an mRNA molecule with the sequence 5′-UGC GCA-3′ is being translated by a ribosome. The following activated tRNA molecules are available.

tRNA Anticodon	Amino Acid
3′-GGC-5′	Proline
3′-CGU-5′	Alanine
3′-UGC-5′	Treonine
3′-CCG-5′	Glycine
3′-ACG-5′	Cysteine
3′-CGG-5′	Alanine

Which two of them can bind correctly to the mRNA so that a dipeptide can form?
a. cysteine–alanine
b. proline–cysteine
c. glycine–cysteine
d. alanine–alanine
e. threonine–glycine

11. **Discuss Concepts** A mutation occurs that alters an anticodon in a tRNA from 3′-AAU-5′ to 3′-AUU-5′. What effect will this mutation have on protein synthesis?

12. **Discuss Concepts** The normal form of a gene contains the nucleotide sequence:

5′-ATGCCCGCCTTTGCTACTTGGTAG-3′
3′-TACGGGCGGAAACGATGAACCATC-5′

When this gene is transcribed, the result is the following mRNA molecule:

5′-AUGCCCGCCUUUGCUACUUGGUAG-3′

In a mutated form of the gene, two extra base pairs (underlined) are inserted:

5′-ATGCCCGCCTAATTGCTACTTGGTAG-3′
3′-TACGGGCGGATTAACGATGAACCATC-5′

What effect will this particular mutation have on the structure of the protein encoded in the gene?

Evaluate/Create

13. **Discuss Concepts** Which do you think are more important to the accuracy by which amino acids are linked into proteins: nucleic acids or enzymatic proteins? Why?

14. **Discuss Concepts** A geneticist is attempting to isolate mutations in the genes for four enzymes acting in a metabolic pathway in the bacterium *Escherichia coli*. The end product *E* of the pathway is absolutely essential for life:

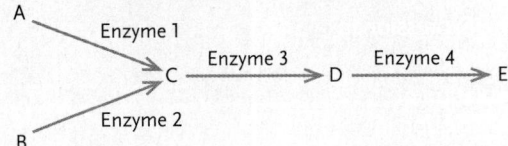

The geneticist has been able to isolate mutations in the genes for enzymes 1 and 2, but not for enzymes 3 and 4. Develop a hypothesis to explain why.

15. **Discuss Concepts** Experimental systems have been developed in which transposable elements can be induced to move under the control of a researcher. Following the induced transposition of a yeast TE element, two mutants were identified with altered activities of enzyme X. One of the mutants lacked enzyme activity completely, whereas the other had five times as much enzyme activity as normal cells did. Both mutants were found to have the TE inserted into the gene for enzyme X. Propose hypotheses for how the two different mutant phenotypes were produced.

16. **Design an Experiment** How could you show experimentally that the genetic code is universal; namely, that it is the same in bacteria as it is in eukaryotes such as fungi, plants, and animals?

17. **Apply Evolutionary Thinking** How might the process of alternative splicing and exon shuffling affect the rate at which new proteins evolve?

For selected answers, see Appendix A.

INTERPRET THE DATA

The **Figure** below shows amino acid changes that occurred at a particular position in a polypeptide as a result of mutations in the gene encoding the polypeptide. The amino acids connected by a line are specified by codons that differ in a single base. Using the genetic code shown in Figure 15.4, deduce the codons that specify the amino acids.

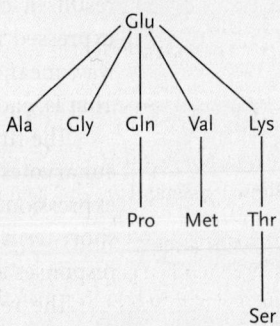

16

Regulation of Gene Expression

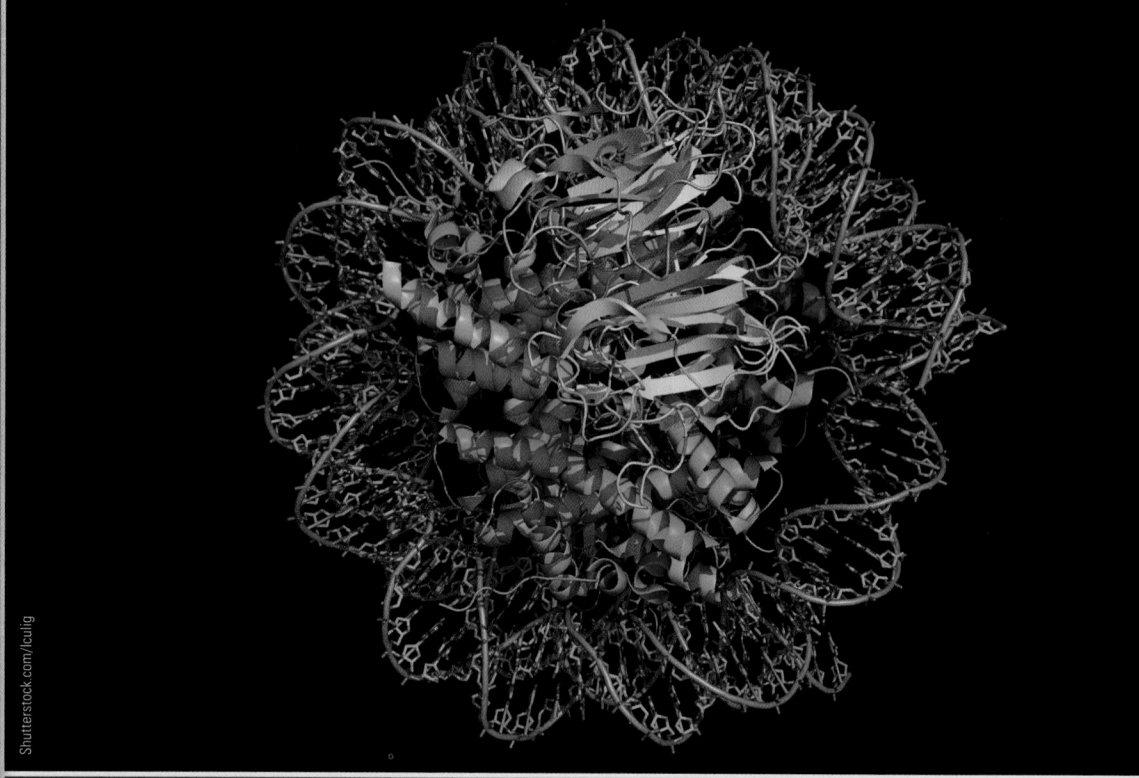

Shutterstock.com/Iculig

DNA wrapped around a core of eight histone proteins to form a nucleosome. A nucleosome is the basic structural unit of eukaryotic chromosomes.

Why it matters . . . A human egg cell is almost completely metabolically inactive when it is released from the ovary. In the oviduct leading from the ovary to the uterus, egg and sperm cells meet and embryonic development begins. Within seconds after the cells unite, the fertilized egg breaks its quiescent state and begins a series of divisions that continues as it moves to the uterus. Subsequent divisions produce specialized cells that *differentiate* into the distinct types tailored for specific functions in the body, such as muscle cells and cells of the nervous system.

All the nucleated cells in both the developing embryo and the adult retain the same set of genes. The structural and functional differences in the cell types are determined not by the presence or absence of certain genes but rather through differences in patterns of *gene expression* that result in cell type-specific sets of proteins. Some genes, known as **housekeeping genes,** are expressed in almost all cell types. **Regulated genes,** by contrast, are expressed in a controlled way, meaning that they may or may not be expressed at any given time or in a given cell type. That is, each differentiated cell is characterized by genes that are active in only that cell type.

The fundamental mechanisms that control gene expression are common to all multicellular eukaryotes. Even single-celled eukaryotes and prokaryotes have systems that regulate gene expression. With few exceptions, however, prokaryotic systems are limited almost exclusively to short-term responses to environmental changes; eukaryotic cells exhibit both short-term responses and long-term differentiation.

The regulation of gene expression occurs at several levels. **Transcriptional regulation,** the fundamental level of control, determines which genes are transcribed into mRNA. Additional controls fine-tune regulation by affecting the processing of mRNA (*posttranscriptional* regulation), its translation into proteins (*translational regulation*), and the lifespan and activity of the proteins themselves (*posttranslational regulation*).

These levels of regulation ultimately affect more than proteins, because among the proteins are enzymes that determine the types and kinds of all other molecules made in the developing cell. So, effectively, these regulatory mechanisms tailor the production of all cellular molecules. The entire spectrum of controls constitutes an exquisitely sensitive mechanism regulating when, where, and what kinds and numbers of cellular molecules are produced.

In this chapter we examine the regulation of gene expression first in prokaryotes and then in eukaryotes. Then we discuss the mechanisms by which genes regulate embryonic development, and finally how cancers develop when regulatory controls are lost.

16.1 Regulation of Gene Expression in Prokaryotes

Prokaryotes are relatively simple, single-celled organisms with generations that take a matter of minutes. Rather than the complex patterns of long-term cell differentiation and development typical of multicellular eukaryotes, prokaryotic cells typically undergo rapid and reversible alterations in biochemical pathways that allow them to adapt quickly to changes in their environment. These changes are the outcomes of regulatory events that control gene expression.

The human intestinal bacterium *Escherichia coli,* for example, can catabolize a number of sugars and other molecules to provide carbon and energy for the cell. One of those sugars is lactose (milk sugar). Lactose is not essential for *E. coli* growth but, when lactose is present, the regulated genes encoding three proteins for catabolizing the sugar are expressed. In the absence of lactose, those regulated genes are not expressed and the three proteins are not made.

More generally, when the environment in which a bacterium lives changes, some metabolic processes are stopped and others are started. The most rapid control occurs at the level of enzyme activity. For example, consider a biosynthetic pathway. If the end product of that pathway no longer needs to be synthesized because now it is available from the environment, *feedback inhibition* can block the pathway from operating. As described in Section 6.5, feedback inhibition operates typically when the end product of a biosynthetic pathway accumulates in the cell and inhibits the first enzyme in that pathway, blocking further synthesis. Another level of control involves turning off the genes for the metabolic processes not needed, and turning on the genes for the new metabolic processes. Each metabolic process involves a few to many genes, and the regulation of those genes is coordinated. The regulation of genes in this way conserves energy for the bacterium because gene products are made only when they are needed. In the following subsections

we learn about some examples of coordinated regulation of gene expression in prokaryotes.

Some Regulated Genes Occur in Clusters Called Operons

Some regulated genes in the prokaryotic genome occur singly, meaning that the gene is transcribed by RNA polymerase to produce an mRNA molecule that is translated to produce a single polypeptide. A useful term to introduce here is the **transcription unit,** which means the segment of DNA from the initiation point of transcription to the termination point of transcription (see Chapter 15). The transcription unit for a protein-coding gene that occurs singly in a genome corresponds to the mRNA-coding sequence of that gene.

Other regulated genes occur in clusters. Each cluster constitutes one transcription unit, meaning that the set of genes in the cluster is transcribed into a *single* mRNA molecule. Translation of the mRNA produces polypeptides corresponding to each of the genes in the transcription unit. The organization of genes in a cluster provides a means for efficient coordinate regulation of those genes, called the *operon model* for the control of gene expression. The operon model was proposed in 1961 by François Jacob and Jacques Monod of the Pasteur Institute in Paris to explain the control of the expression of genes for lactose catabolism in *E. coli* (described in the next two subsections). Subsequently, research has shown that the operon model is applicable to the regulation of expression of many genes in prokaryotes and their viruses. Jacob and Monod received the Nobel Prize in 1965 "for their discoveries concerning the genetic control of enzyme and virus synthesis."

An **operon** is a cluster of prokaryotic genes organized into a single transcription unit and its associated *regulatory sequences.* **Regulatory sequences** are DNA sequences to which specific proteins bind to control the transcription of the gene or genes. A protein that binds to a specific DNA sequence is a *DNA-binding protein.* A particular DNA-binding protein binds to a specific DNA sequence by interactions between particular amino acid regions in the three-dimensional shape of the protein and specific base pairs of the DNA accessed in the major or minor groove.

One regulatory DNA sequence is the **promoter,** which is the site to which RNA polymerase binds to begin transcription. The other regulatory DNA sequence in an operon is the **operator,** a short segment to which a *regulatory protein* binds to affect the expression of the operon. A **regulatory protein** is a DNA-binding protein that binds to a regulatory sequence and affects the expression of an associated gene or genes. For an operon, the gene for its regulatory protein is separate from the operon the protein controls. Two types of regulatory control systems are found:

1. **Negative control:** In negative control, genes are expressed unless they are switched *off* by a regulatory protein termed a **repressor.** In the absence of the repressor, or if the repressor is inactivated, the genes are *on.*

2. **Positive control:** In positive control, genes are switched *on* (expressed) only when an active regulatory protein called an **activator** is present. In the absence of the activator, or if the activator is inactivated, the genes are *off*.

In the rest of this section, we discuss two *E. coli* operons as examples of regulation of gene expression in bacteria. In the discussion, you will see examples of negative and positive control.

The *lac* Operon Is a Negatively Controlled Operon That Is Transcribed When an Inducer Inactivates a Repressor

Catabolism of the sugar lactose provides energy for the cell. Jacob and Monod used genetic and biochemical approaches to study the genetic control of lactose metabolism in *E. coli*. Their genetic studies showed that the protein products of three genes, *lacZ*, *lacY*, and *lacA*, are involved in lactose catabolism **(Figure 16.1).** The three genes constitute a single transcription unit with the order *Z–Y–A*. The promoter is upstream of *lacZ*, and the genes are transcribed into a single mRNA starting with the *lacZ* gene. The *lacZ* gene encodes the enzyme β-galactosidase, which has two catalytic activities. One activity is the hydrolysis of the disaccharide sugar, lactose, into the monosaccharide sugars, glucose and galactose. These sugars are then catabolized by other enzymes, producing energy for the cell. The other activity is converting lactose to *allolactose,* an isomer of lactose. (Isomers are discussed in Section 3.1.) As we will see, allolactose plays an important role in regulating the expression of the *lac* operon genes. The *lacY* gene encodes a permease (not an enzyme, despite its name), a protein that transports lactose actively into the cell. The *lacA* gene encodes a transacetylase enzyme, the function of which is unclear.

Jacob and Monod called the cluster of genes and adjacent sequences that control their expression the **lac operon** (see Figure 16.1). They coined the name *operon* from the *operator,* a key DNA sequence they discovered between the promoter and the *lacZ* gene that, through binding a regulatory protein, controls transcription of the operon.

The *lac* operon is a negatively regulated system controlled by a regulatory protein termed the **Lac repressor.** The Lac repressor is encoded by the *regulatory gene lacI,* which is nearby but separate from the *lac* operon, and is synthesized in *active* form (see Figure 16.1). In general, the term **structural gene** is used for a gene that encodes a protein that has a function other than gene regulation, such as the three *lac* operon genes. The term **regulatory gene** is used for a gene that encodes a protein that regulates the expression of structural genes, such as *lacI*.

Figure 16.2A shows how the Lac repressor inhibits transcription when lactose is *absent* from the medium. Here, the three-dimensional shape of the active Lac repressor positions amino acid regions that can recognize and bind specifically to base pairs in the operator. When lactose is added to the medium, expression of the *lac* operon increases about 100-fold **(Figure 16.2B).** The molecular switch for this change is set when some lactose molecules that enter the cell are converted by β-galactosidase in the cell to **allolactose.** Allolactose is the **inducer** for the *lac* operon, so called because it causes—induces—the transcription of the operon's structural genes. Allolactose works by inactivating the Lac repressor (see Figure 16.2B). It does so by binding to a specific site on the repressor protein, causing it to undergo an allosteric shift (change in shape; see Section 6.5). In this altered protein, the DNA-binding amino acid regions are no longer positioned appropriately to recognize the DNA sequence of the operator and, hence, the repressor cannot recognize and bind to the operator. Because an inducer molecule increases its expression, the *lac* operon is called an **inducible operon.**

When the lactose is used up from the medium, no allolactose inducer molecules are produced to inactivate the repressor. The now-active repressor binds to the operator, blocking

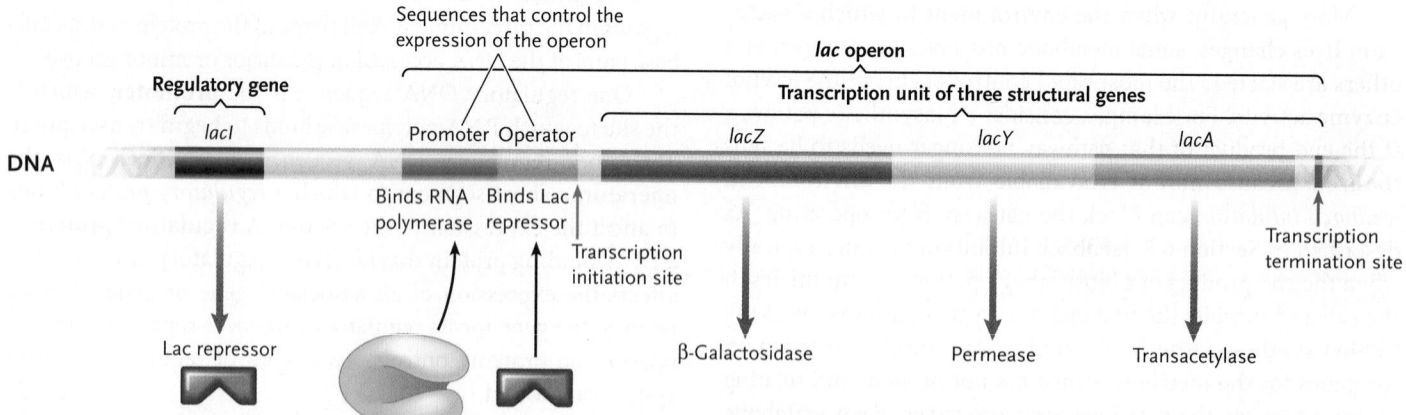

FIGURE 16.1 The *E. coli lac* operon. The *lacZ, lacY,* and *lacA* genes encode the proteins taking part in lactose metabolism. The separate regulatory gene, *lacI,* encodes the Lac repressor, which plays a pivotal role in the control of the operon. The promoter binds RNA polymerase, and the operator binds activated Lac repressor. The transcription unit, which extends from the transcription initiation site to the transcription termination site, contains the structural genes.

REGULATION OF THE INDUCIBLE *lac* OPERON BY THE LAC REPRESSOR IN THE ABSENCE (A) AND PRESENCE (B) OF LACTOSE

A. Lactose absent from medium: structural genes not transcribed

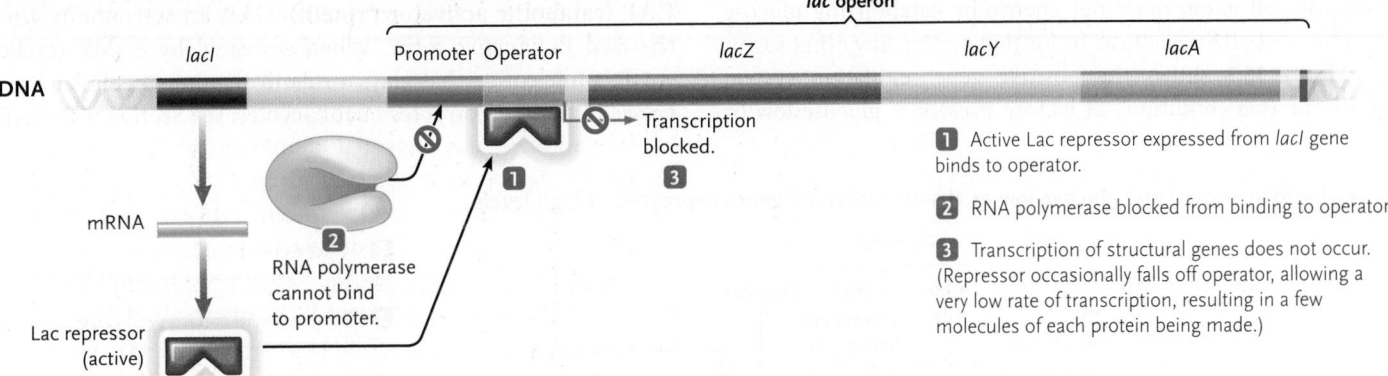

1 Active Lac repressor expressed from *lacI* gene binds to operator.

2 RNA polymerase blocked from binding to operator.

3 Transcription of structural genes does not occur. (Repressor occasionally falls off operator, allowing a very low rate of transcription, resulting in a few molecules of each protein being made.)

B. Lactose present in medium: structural genes transcribed

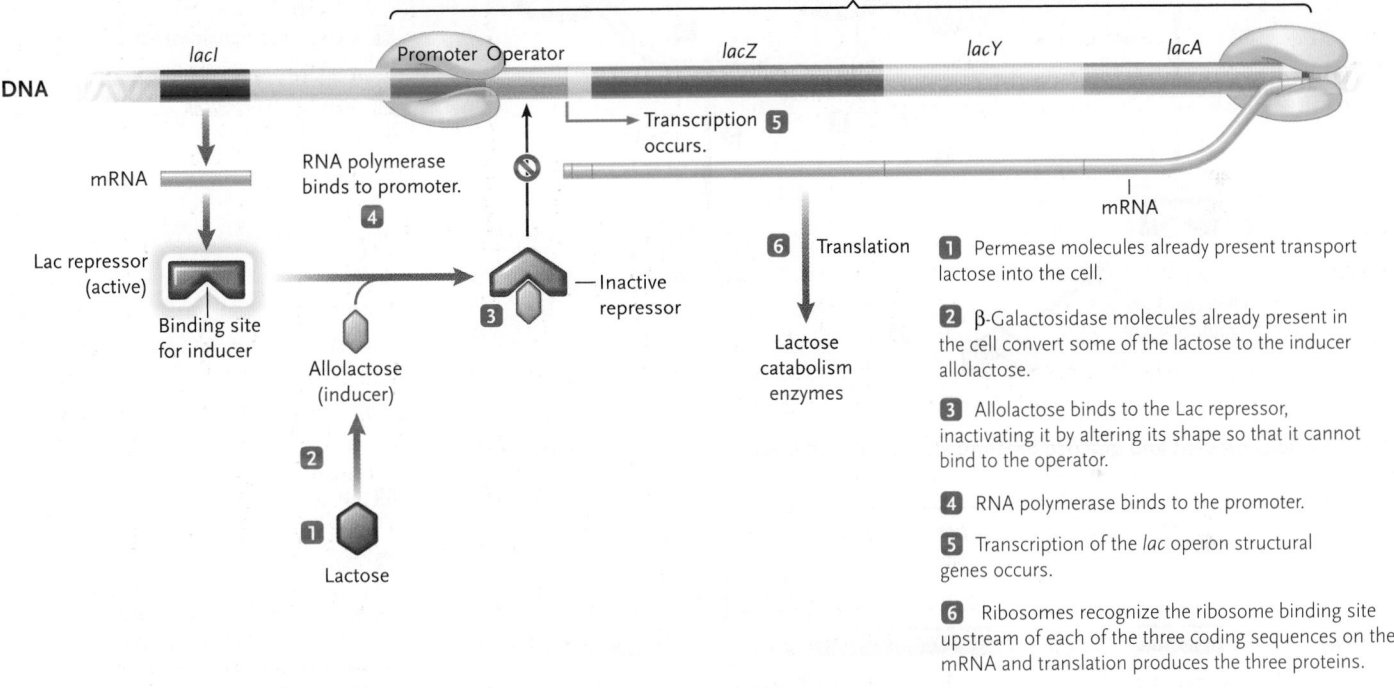

1 Permease molecules already present transport lactose into the cell.

2 β-Galactosidase molecules already present in the cell convert some of the lactose to the inducer allolactose.

3 Allolactose binds to the Lac repressor, inactivating it by altering its shape so that it cannot bind to the operator.

4 RNA polymerase binds to the promoter.

5 Transcription of the *lac* operon structural genes occurs.

6 Ribosomes recognize the ribosome binding site upstream of each of the three coding sequences on the mRNA and translation produces the three proteins.

SUMMARY In the absence of lactose, the *lac* operon is inactive: the active Lac repressor binds to the operator and blocks RNA polymerase from binding at the promoter so transcription of the structural genes does not occur. In the presence of lactose, the *lac* operon genes are expressed: the inducer allolactose binds to the Lac repressor inactivating it so that it does not bind to the operator, which allows RNA polymerase to bind to the promoter and transcribe the structural genes.

think like a scientist If there was a mutation in the *lacI* gene that results in a Lac repressor that could not bind to the operator, what effect would that have on the regulation of the *lac* operon?

transcription of the structural genes. This switching off of the operon is aided by the fact that bacterial mRNAs are very short-lived, about 3 minutes on the average. This quick turnover permits the cytoplasm to be cleared quickly of the mRNAs transcribed from an operon. The encoded proteins also have short lifetimes and are degraded quickly.

Transcription of the *lac* Operon Is Also Controlled by a Positive Regulatory System

A *positive gene regulation* system ensures that the *lac* operon is transcribed efficiently if lactose is provided as a sole energy source, but not if glucose, a more efficient energy source, is present in addition to lactose. Glucose can be used directly in the glycolysis pathway to produce energy for the cell (see Chapter 7). Lactose, on the other hand, first must be converted into glucose and galactose, and the galactose then converted into glucose. These conversions require energy from the cell. Thus the cell gains more net energy by catabolizing glucose than by catabolizing lactose, or for that matter, any other sugar.

Figure 16.3 shows the positive gene regulation system under the two conditions of lactose present + glucose low or absent (efficient transcription of *lac* operon genes), and lactose present + glucose present (very low level of transcription of *lac* operon genes). In essence, we are adding to the model shown in Figure 16.2B. The key regulatory molecule involved in positive gene regulation of the *lac* operon is the DNA-binding protein, **CAP (catabolite activator protein)**. CAP, an activator, is synthesized in *inactive* form. When activated by cAMP (cyclic AMP, a nucleotide that plays a role in regulating cellular processes in both prokaryotes in eukaryotes; see Section 9.4), CAP

A. Lactose present and glucose low or absent: structural genes expressed at high levels

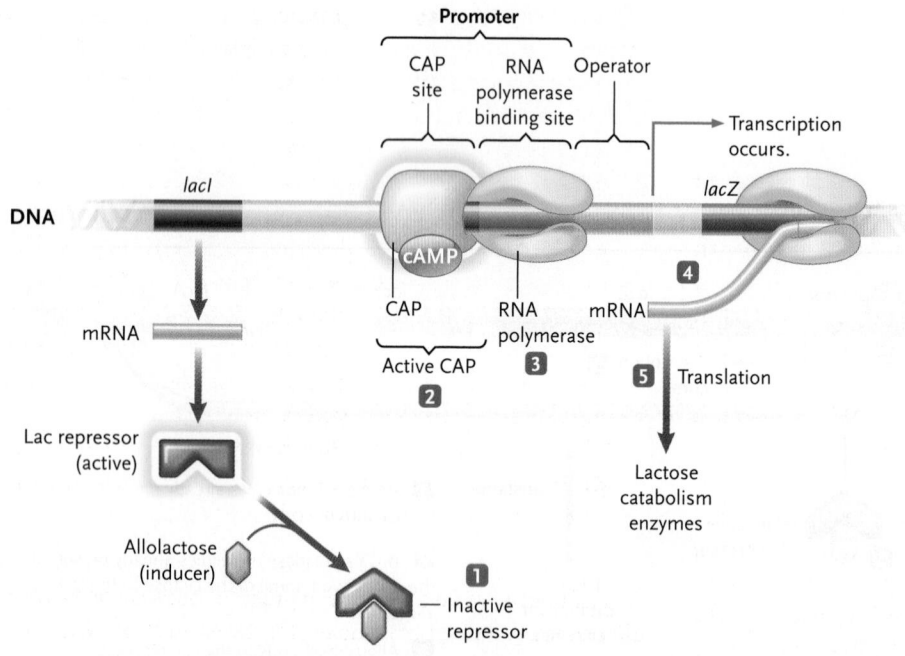

1 Lactose converted to the inducer, allolactose, which inactivates Lac repressor.

2 Active adenylyl cyclase synthesizes cAMP to high levels. cAMP binds to activator CAP, activating it. Activated CAP binds to CAP site in the promoter.

3 RNA polymerase binds efficiently to the promoter.

4 Genes of operon transcribed to high levels.

5 Translation produces high amounts of proteins.

B. Lactose present and glucose present: structural genes expressed at very low levels

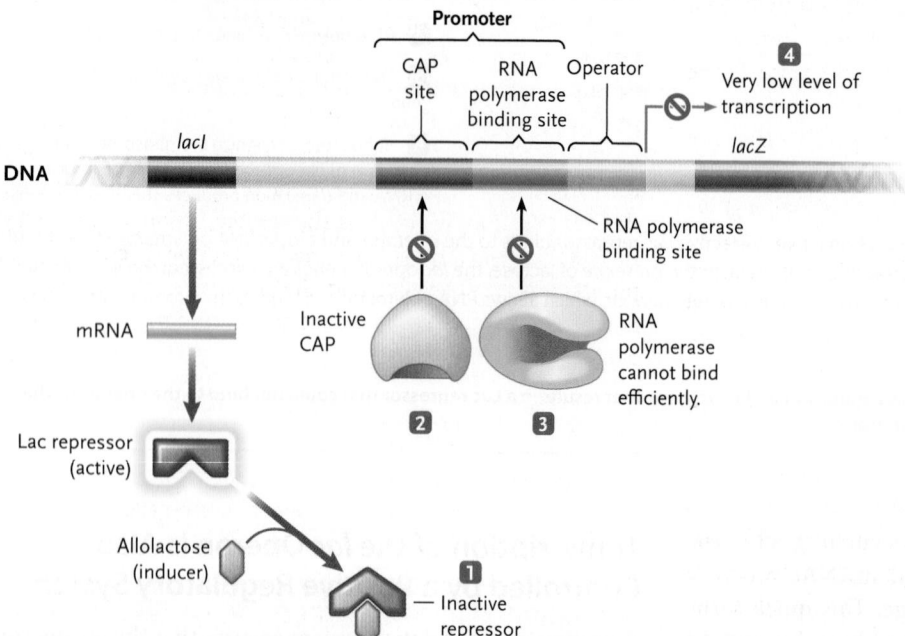

1 Lactose converted to the inducer, allolactose, which inactivates Lac repressor.

2 Catabolism of incoming glucose leads to inactivation of adenylyl cyclase, which causes the amount of cAMP in the cell to drop to a level too low to activate CAP. Inactive CAP cannot bind to the CAP site.

3 RNA polymerase is unable to bind to the promoter efficiently.

4 Transcription occurs at a very low level: Because the Lac repressor is not present to block RNA polymerase from binding to the promoter, the level of transcription is higher than when lactose is absent, but far lower than when lactose is present and glucose is absent.

FIGURE 16.3 Positive regulation of the *lac* operon by the CAP activator.
© Cengage Learning 2017

binds to the **CAP site** in the promoter and enables RNA polymerase to bind efficiently and transcribe the operon's genes. If glucose is absent, cAMP levels are high, resulting in active CAP. If glucose is present, cAMP levels are low, resulting in inactive CAP, which cannot bind to the CAP site.

The same positive gene regulation system using CAP and cAMP regulates a large number of other operons that control the catabolism of many sugars. In each case, the system functions so that glucose, if it is present in the growth medium, is catabolized first.

The *trp* Operon Gene Is a Negatively Controlled Operon, the Transcription of Which Is Blocked When Tryptophan Activates a Repressor

Tryptophan is an amino acid that is used in the synthesis of proteins, which are critical components for cell growth and survival. Therefore, if tryptophan is absent from the growth medium, *E. coli* must synthesize tryptophan so that it can make its proteins. If tryptophan is present in the medium, the cell will use that source of the amino acid rather than synthesizing its own.

Tryptophan biosynthesis involves the *trp* operon **(Figure 16.4)**. The five structural genes in this operon, *trpE–trpA*, encode the enzymes for the steps in the tryptophan biosynthesis pathway. Upstream of the *trpE* gene are the operon's promoter and operator sequences. Expression of the *trp* operon is controlled by the Trp repressor, a regulatory protein encoded by the regulatory gene, *trpR*, which is located elsewhere in the genome. In contrast to the Lac repressor, the Trp repressor is synthesized in an *inactive* form that cannot bind to the operator. That is, an inactive Trp repressor has a three-dimensional shape that *cannot* recognize and bind to the DNA sequence of the operator for which it is specific. The inactive state of the repressor leads to the default state of the operon, the expression of the *trp* operon structural genes (see Figure 16.4A). The repressor is activated when tryptophan is present in the medium (see Figure 16.4B). Activation occurs when tryptophan binds to a specific site on the Trp repressor and causes the DNA-binding protein to undergo an allosteric shift (change in shape) to a three-dimensional shape with amino acid regions that now can recognize and bind to the DNA sequence of the operator. Because the presence of tryptophan represses the expression of the tryptophan biosynthesis genes, this operon is an example of a **repressible operon**. Here, tryptophan acts as a **corepressor,** a regulatory molecule that activates the repressor to turn off expression of the operon. In addition, at the pathway level, tryptophan biosynthesis is regulated by feedback inhibition in which tryptophan, when it is abundant in the cell such as when it is present in the medium, inhibits the first enzyme of the pathway to block further synthesis.

Comparison of the *lac* and *trp* Operons

To compare and contrast the *lac* and *trp* operons:

- *lac* **operon:** the repressor is synthesized in an active form. The inducer allolactose inactivates the repressor. The structural genes are then transcribed. Specifically, when lactose is absent from the medium, the active repressor binds to the operator to block transcription, and when lactose is present in the medium, the allolactose made from lactose inactivates the repressor, thereby relieving the block on transcription. A positive control system also is involved. When lactose is present and glucose is low or absent, an activator binds to the promoter to facilitate RNA polymerase binding and transcription of the operon. When lactose is present and glucose is present, the activator is inactivated, RNA polymerase cannot bind to the promoter, and transcription of the operon is prevented.

- *trp* **operon:** the repressor is synthesized in an inactive form so the structural genes are transcribed. The corepressor (tryptophan from the growth medium) activates the repressor. Active repressor blocks transcription of the operon. Specifically, when tryptophan is absent from the medium, tryptophan must be made by the cell. In this case, the repressor is inactive and transcription of the operon occurs. When tryptophan is present in the medium, the cell uses that tryptophan rather than synthesizing it. In this case, the tryptophan activates the repressor, which then binds to the operator to block transcription.

These inducible and repressible operons illustrate two types of negative control gene expression because both are regulated by a repressor that turns off gene expression when it is in active form. Genes are expressed only when the repressor is in inactive form.

In sum, regulation of gene expression in prokaryotes occurs primarily at the transcription level. There are also some examples of regulation at the translation level. For example, some proteins can bind to the mRNAs that produce them and modulate their translation. This serves as a feedback mechanism to fine-tune the amounts of the proteins in the cell. In the next section, we discuss the regulation of gene expression in eukaryotes. You will see that regulation occurs at several points between the gene and the protein, and that regulatory mechanisms of eukaryotes are more complex than those in prokaryotes.

STUDY BREAK 16.1

1. Suppose the *lacI* gene is mutated so that the Lac repressor is not made. How does this mutation affect the regulation of the *lac* operon?
2. Answer the equivalent question for the *trp* operon: how would a mutation that prevents the Trp repressor from being made affect the regulation of the *trp* operon?

A. Tryptophan absent from medium: tryptophan must be made by the cell—structural genes transcribed

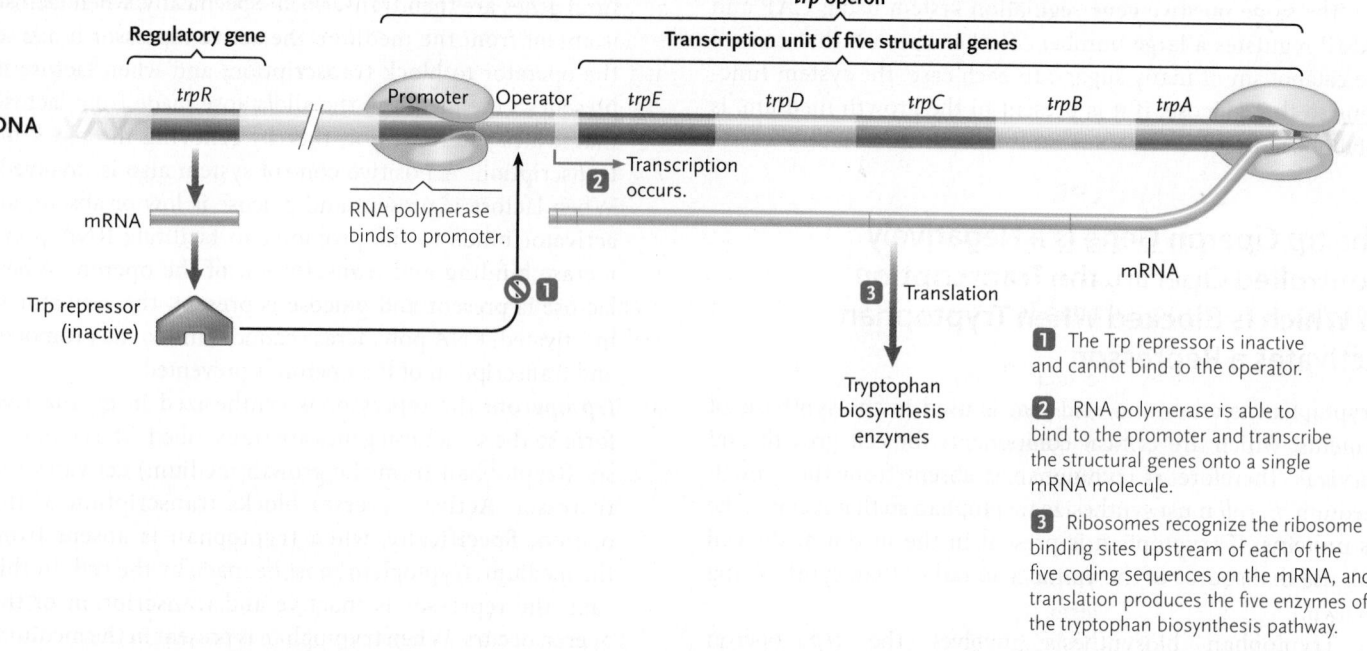

1. The Trp repressor is inactive and cannot bind to the operator.

2. RNA polymerase is able to bind to the promoter and transcribe the structural genes onto a single mRNA molecule.

3. Ribosomes recognize the ribosome binding sites upstream of each of the five coding sequences on the mRNA, and translation produces the five enzymes of the tryptophan biosynthesis pathway.

B. Tryptophan present in medium: cell uses tryptophan in medium rather than synthesizing it—structural genes not transcribed

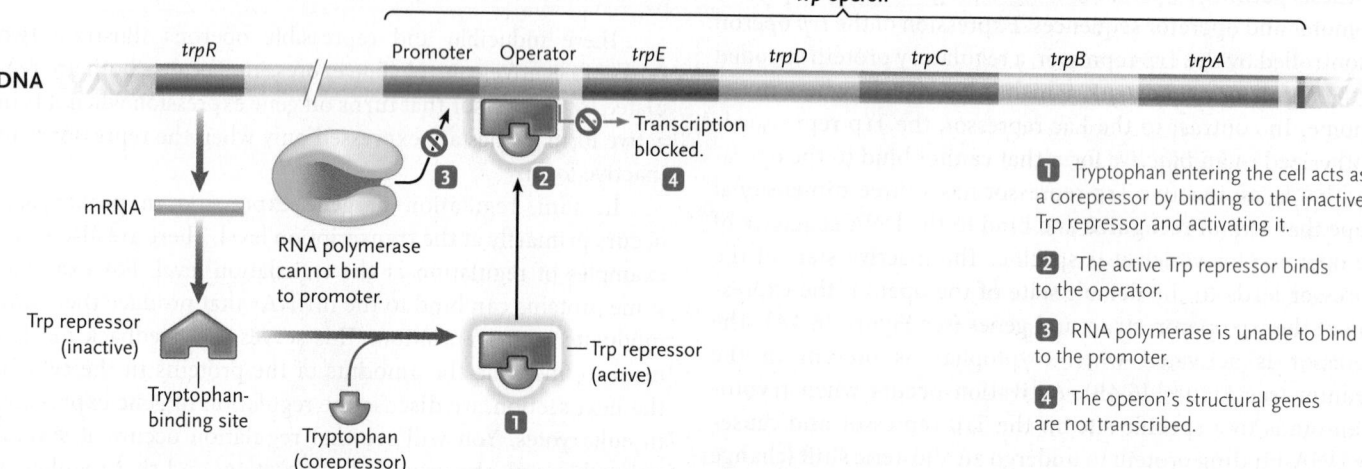

1. Tryptophan entering the cell acts as a corepressor by binding to the inactive Trp repressor and activating it.

2. The active Trp repressor binds to the operator.

3. RNA polymerase is unable to bind to the promoter.

4. The operon's structural genes are not transcribed.

SUMMARY In the absence of tryptophan in the medium, tryptophan must be made. In this scenario, the Trp repressor is inactive, which enables RNA polymerase to bind to the promoter and transcribe the structural genes for the tryptophan biosynthesis enzymes. In the presence of tryptophan in the medium, that tryptophan can be used for protein synthesis, so there is no need for the cell to make tryptophan. Tryptophan binds to the Trp repressor, activating it. The active repressor binds to the operator, thereby blocking RNA polymerase from binding to the promoter. Transcription of the structural genes does not occur.

think like a scientist If there was a mutation in the *trpR* gene that results in a Trp repressor that cannot bind tryptophan, what effect would that have on the regulation of the *trp* operon?

THINK OUTSIDE THE BOOK

In addition to the mechanism described, transcription of the *trp* operon is also regulated by another regulatory mechanism known as *attenuation*. Individually or collaboratively, use the Internet or the research literature to develop an outline of regulation of the *trp* operon by attenuation.

16.2 Regulation of Transcription in Eukaryotes

The molecular mechanisms of operon regulation in prokaryotes provide a simple means of coordinating synthesis of proteins that have related functions. In eukaryotes, the coordinated synthesis of proteins that have related functions also occurs, but the genes involved are usually scattered around the genomes and not organized into operons. Nonetheless, like operons, individual eukaryotic genes also consist of protein-coding sequences and their regulatory sequences.

There are two general types of eukaryotic gene regulation. Short-term regulation involves regulatory events in which gene sets are turned on or off quickly in response to changes in environmental or physiological conditions in the cell's or organism's environment. This type of regulation is most similar to prokaryotic gene regulation. Long-term gene regulation involves regulatory events required for an organism to develop and differentiate. Long-term gene regulation occurs in multicellular eukaryotes and not in simpler, unicellular eukaryotes. The mechanisms we discuss in this and the next section are applicable to both short-term and long-term regulation. The specific molecules and genes involved in short-term and long-term regulation are different and, of course, so is the outcome to the cell or organism.

In Eukaryotes, Regulation of Gene Expression Occurs at Several Levels

The regulation of gene expression is more complicated in eukaryotes than in prokaryotes because eukaryotic cells are more complex, because nuclear DNA is organized with histones into chromatin, and because multicellular eukaryotes produce large numbers and types of cells. Further, the eukaryotic nuclear envelope separates the processes of transcription and translation, whereas in prokaryotes translation can start on an mRNA that is still being synthesized. Consequently, regulation of gene expression in eukaryotes is regulated at the transcriptional, posttranscriptional, translational, and posttranslational levels **(Figure 16.5)**. The best characterized of these is transcriptional regulation. For most genes the initiation of transcription is a major control point and involves changes in the structure of the chromatin at the promoter, and regulatory events at a gene's promoter and regulatory sequences. We discuss these events in turn, focusing on protein-coding genes.

Chromatin Structure Plays an Important Role in Whether a Gene Is Active or Inactive

Eukaryotic DNA is organized into chromatin by combination with histone proteins (discussed in Section 10.1). Recall that DNA is wrapped around a core of two molecules each of

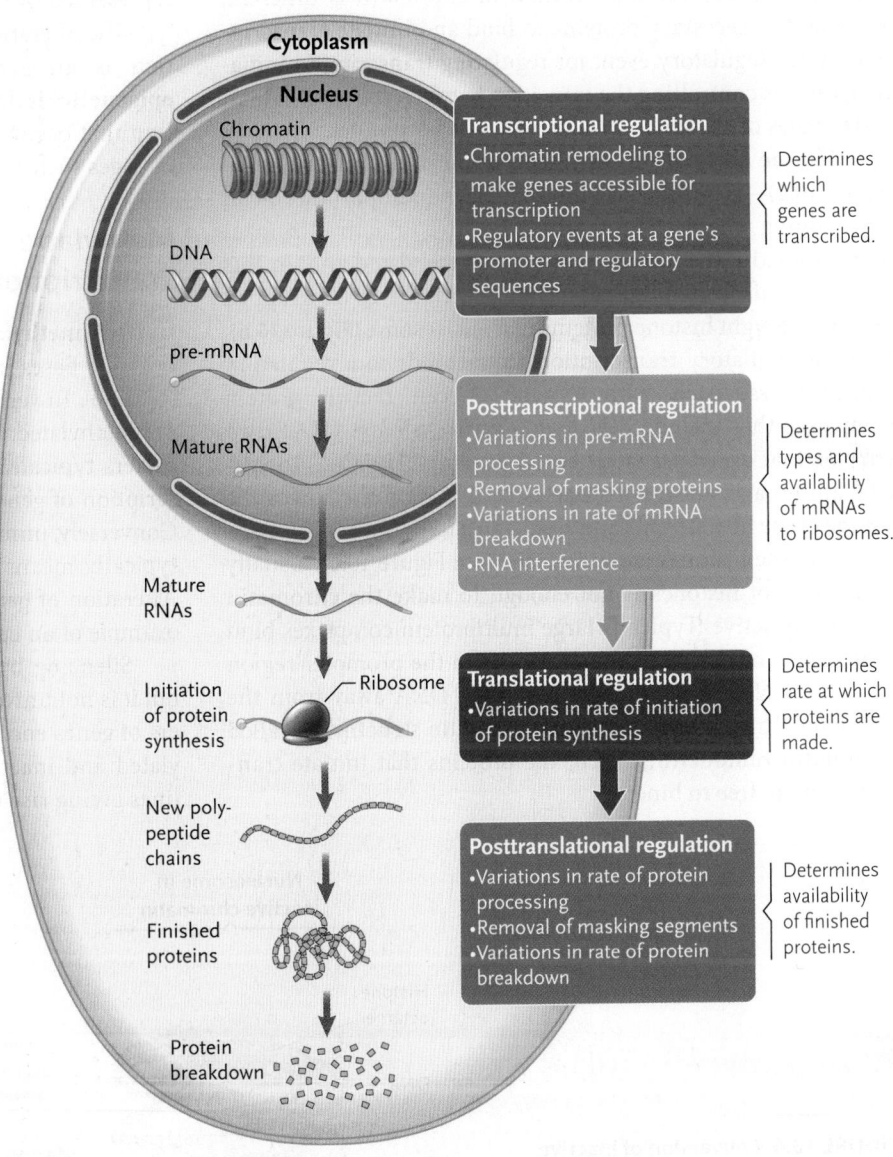

FIGURE 16.5 Steps in transcriptional, posttranscriptional, translational, and posttranslational regulation of gene expression in eukaryotes.

© Cengage Learning 2017

histones H2A, H2B, H3, and H4, forming the nucleosome (see Figure 10.2 and this chapter's chapter-opening figure). Higher levels of chromatin organization occur when histone H1 links adjacent nucleosomes.

Promoters in both bacteria and eukaryotes can exist in two states: active and inactive. For most bacterial promoters, the normal state of the associated gene(s) in the absence of regulatory proteins is "on." That is, in this situation, RNA polymerase usually can bind and initiate transcription. Conversely, initiation of transcription in bacteria is prevented or reduced if an active repressor regulatory protein binds to the DNA to prevent RNA polymerase from binding to the promoter. By contrast, the normal state of eukaryotic promoters in the absence of regulatory proteins is "off." In this inactive state, the nucleosomes in normal chromatin prevent the proteins that initiate transcription from binding so transcription does not occur. In the active state, the nucleosome organization of chromatin is different, allowing the necessary proteins to bind and initiate transcription. A key regulatory event for regulating transcription initiation, then, is controlling the transition between the inactive and active states of chromatin in the region of a promoter.

Histone acetylation—the acetylation of histone tails (see Figure 10.2)—is one mechanism that plays an important role in determining whether chromatin is inactive or active. In inactive chromatin, the histone tails are not acetylated and, in this form, the tails form a tight association with the DNA wrapped around the eight histone proteins of a nucleosome (Figure 16.6). When a regulatory transcription factor binds to a regulatory sequence associated with a gene, it can recruit protein complexes that include a *histone acetyltransferase* (HAT), an enzyme that *acetylates* (adds acetyl groups; CH₃CO—) to specific amino acids of the histone tails. Acetylation changes the charge of the histone tails and results in a loosening of the association of the histones with the DNA (see Figure 16.6). Usually acetylation of histones is not enough to make the chromatin completely active. Typically large multiprotein complexes bind to displace the acetylated nucleosomes in the promoter region from the DNA, or move them along the DNA away from the promoter. This type of change in chromatin structure is called **chromatin remodeling.** Then, the proteins that initiate transcription are free to bind.

Inactivation of an active gene involves essentially the opposite of this process. With respect to the histones, a *histone deacetylase* (HDAC) enzyme catalyzes the removal of acetyl groups from the histone tails restoring the inactive state of the chromatin in that region (see Figure 16.6).

The tails of histones can also be modified at specific positions by the enzyme-catalyzed covalent addition of methyl groups or phosphate groups. These chemical modifications also may affect chromatin structure and gene expression. Histone methylation, for instance, is associated with gene inactivation. Like acetylation, methylation and phosphorylation of histone tails is reversible. Overall, the patterns of modification of histone tails are important in determining chromatin structure and gene activity. This has led to the concept of the **histone code,** a regulatory mechanism for altering chromatin structure and, therefore, gene activity, based on signals in histone tails represented by chemical modification patterns.

The alteration of gene expression by chromatin modification is an example of an *epigenetic phenomenon* where **epigenetics** is defined as the study of alterations of gene expression that occur without changes in DNA sequence. Epigenetic changes such as this are reversible and may be heritable.

Methylation of DNA Can Control Gene Transcription

In **DNA methylation,** enzymes add a methyl group (—CH₃) to cytosine bases in the DNA that primarily are in the sequence 5′-CG-3′. In vertebrates, for example, 70–80% of CG sequences are methylated. Methylated cytosines in CG sequences of promoters typically are associated with **silencing,** in which transcription of genes controlled by those promoters is turned off. Conversely, unmethylated cytosines in promoter CG sequences typically means that the promoter is transcriptionally active. Alteration of gene expression by DNA methylation is another example of an epigenetic phenomenon.

Silencing by methylation is common among vertebrates, but it is not universal among eukaryotes. For example, promoters of genes encoding the blood protein hemoglobin are methylated and inactive in most vertebrate body cells. In the cell lines giving rise to red blood cells, enzymes remove the methyl

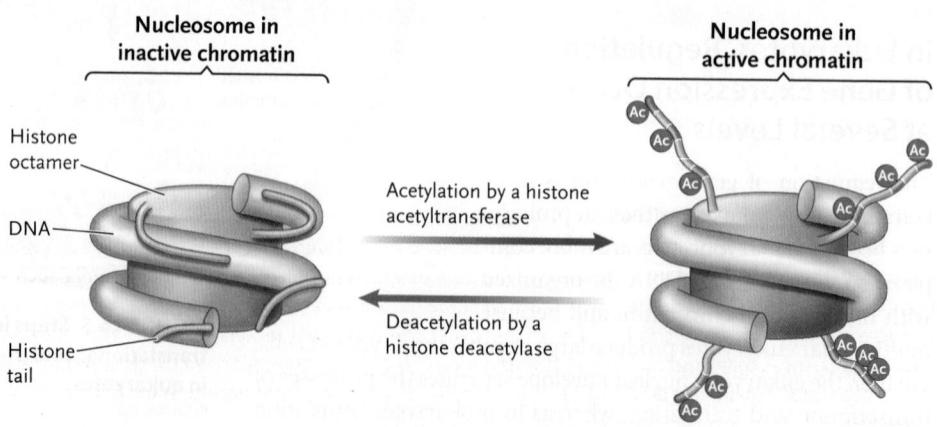

FIGURE 16.6 Conversion of inactive chromatin to active chromatin by acetylation of histone tails, and the reverse by deacetylation of histone tails.

© Cengage Learning 2017

groups from the promoters, and the hemoglobin genes are then transcribed.

DNA methylation is an important epigenetic phenomenon in human cancers. For example, cancer cells typically show a genome-wide and gene-specific loss of DNA methylation, indicating the significance of DNA methylation for gene expression in normal cells and for the development of particular diseases.

DNA methylation in some cases silences large blocks of genes, or even chromosomes. Recall from Section 13.2 that a dosage compensation mechanism inactivates one of the two X chromosomes in most body cells of female placental mammals, including humans. In X-chromosome inactivation, one of the two X chromosomes packs tightly into a heterochromatin mass known as a Barr body, in which most genes of the X chromosome are turned off. The inactivation occurs during embryonic development, and which X chromosome is inactivated in a particular embryonic cell line is a random event. Descendants of each cell with an inactivated X chromosome retain the setting for which parental X chromosome is inactivated, providing a heritable example of an epigenetic phenomenon. As part of X-chromosome inactivation, cytosines in CG sequences of promoters of genes that are silenced become methylated.

DNA methylation underlies **genomic imprinting,** an epigenetic phenomenon in which the expression of an allele is determined by the parent that contributed it (see Section 13.5). In genomic imprinting, methylation permanently silences transcription of either the inherited maternal or paternal allele of a particular gene. The methylation occurs during gametogenesis in a parent. An inherited methylated allele, the *imprinted allele,* is not expressed—it is silenced. The expression of the gene involved therefore depends on expression of the nonimprinted allele inherited from the other parent. The methylation of the parental allele is maintained as the DNA is replicated, so that the silenced allele remains inactive in progeny cells. Figure 13.19 shows an example of genomic imprinting.

The traditional view of DNA methylation and its effect on gene expression is as just described: methylation of CG sequences in promoters silences transcription. Genome analysis using high-throughput technologies has now uncovered new information about DNA methylation in the genome and its effect on gene expression. One exciting new discovery is that DNA methylation is present in the bodies (protein-coding regions) of active genes. In most cases examined, gene body DNA methylation is positively correlated with elevated expression of the genes. How gene body DNA methylation affects gene expression is not yet known. Interestingly, genome-wide methylation profiles (called *DNA methylomes*) for 20 eukaryotic genomes have shown that gene body methylation is ancient, evolutionarily speaking, in that it predates the divergence of plants and animals that occurred approximately 1.6 billion years ago. That is, gene body methylation appears to be highly conserved among diverse organisms.

Regulation of Transcription Initiation Involves the Effects of Transcription Factors Binding to a Gene's Promoter and Regulatory Sites

Transcription initiation is an important point at which regulation of gene expression of a eukaryotic protein-coding gene takes place. Just as in bacteria, the regulation of transcription initiation in eukaryotes involves regulatory proteins that bind to specific sequences in the DNA. Bacteria produce several hundred regulatory proteins, each with a specific DNA sequence partner with which to bind and, thereby, regulate a specific set of genes. By contrast, eukaryotes produce many more regulatory proteins, reflecting the greater complexity of gene regulation processes in their development and function.

Figure 16.7 shows a eukaryotic protein-coding gene, emphasizing the regulatory sequences involved in its expression. Eukaryotic protein-coding genes consist of single transcription units. Immediately upstream of the transcription unit is the promoter. The promoter in the figure contains an

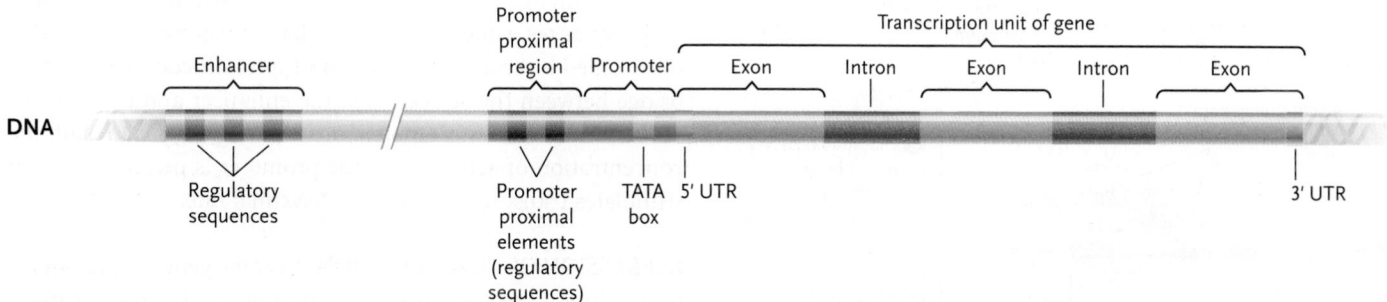

FIGURE 16.7 Organization of a eukaryotic gene. The transcription unit is the segment that is transcribed into the pre-mRNA; it contains exons and introns, the 5′ UTR (untranslated region) within the first exon, and the 3′ UTR within the last exon. Immediately upstream of the transcription unit is the promoter, which contains an initiator sequence such as the TATA box shown. Adjacent to the promoter and further upstream of the transcription unit is the promoter proximal region, which contains regulatory sequences called promoter proximal elements. More distant from the gene in some cases is the enhancer, which contains regulatory sequences that control the rate of transcription of the gene. Transcription of the gene produces a pre-mRNA molecule with a 5′ cap and 3′ poly(A) tail; processing of the pre-mRNA to remove introns generates the functional mRNA (see Chapter 15).
© Cengage Learning 2017

initiator sequence called a TATA box (see Chapter 15) that plays an important role in transcription initiation. The TATA box, historically the first initiator sequence identified, is found in some promoters and is located about 25–35 bp upstream of the start point for transcription. The TATA box has the 6-bp consensus sequence $\begin{smallmatrix}5'\text{-TATAAA-}3'\\3'\text{-ATATTT-}5'\end{smallmatrix}$. Promoters without TATA boxes have other initiator sequence elements that play a similar role. The following discussions involve a TATA box-containing promoter to parallel our earlier discussion of transcription in Chapter 15.

RNA polymerase II itself cannot recognize the promoter sequence. Instead, particular **transcription factors**—proteins required for RNA polymerase to initiate transcription or that regulate that process—recognize and bind to the TATA box and then recruit the polymerase. Once the RNA polymerase II–transcription factor complex forms, the polymerase unwinds the DNA and transcription begins. Adjacent to the promoter, further upstream, is the **promoter proximal region,** which contains regulatory sequences called **promoter proximal elements.** Regulatory proteins (types of transcription factors) that bind to promoter proximal elements may stimulate or inhibit the rate of transcription initiation. More distant from the beginning of some protein-coding genes is the *enhancer.* Other regulatory proteins (again, types of transcription factors) binding to regulatory sequences within an enhancer stimulate or inhibit the rate of transcription initiation, fine-tuning the regulation achieved at the promoter proximal elements.

ACTIVATION OF TRANSCRIPTION To initiate transcription, proteins called **general transcription factors** (also called *basal transcription factors*) bind to the promoter in the area of the TATA box **(Figure 16.8).** These factors recruit the enzyme RNA

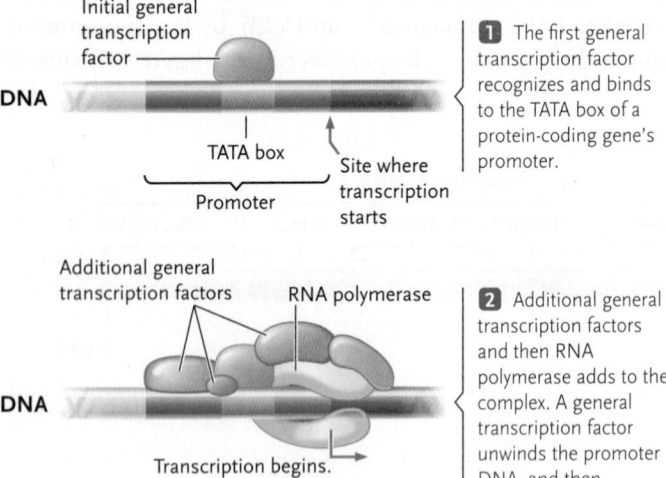

1 The first general transcription factor recognizes and binds to the TATA box of a protein-coding gene's promoter.

2 Additional general transcription factors and then RNA polymerase adds to the complex. A general transcription factor unwinds the promoter DNA, and then transcription begins.

FIGURE 16.8 Formation of the transcription complex on the promoter of a protein-coding gene by the combination of general transcription factors with RNA polymerase. The general transcription factors are needed for RNA polymerase to bind and initiate transcription at the correct place.

© Cengage Learning 2017

polymerase II to form the **transcription initiation complex,** which orients the enzyme to start transcription at the correct place. On its own, this complex brings about only a low rate of transcription initiation, which leads to just a few mRNA transcripts.

Activators, a type of regulatory protein, are transcription factors that stimulate transcription initiation. Activators that bind to the promoter proximal elements interact directly with the general transcription factors at the promoter to stimulate transcription initiation, so many more transcripts are synthesized in a given time. Housekeeping genes—genes that are expressed in all cell types for basic cellular functions such as glucose metabolism—have promoter proximal elements that are recognized by activators present in all cell types. By contrast, genes expressed only in particular cell types or at particular times have promoter proximal elements that are recognized by activators found only in those cell types, or at those times when transcription of these genes needs to be activated. To turn this around, the particular set of activators present within a cell at a given time is responsible for determining which genes in that cell are expressed to a significant level.

Many types of activators are found in eukaryotic cells. The DNA binding and activation functions of activators are properties of two distinct domains in the proteins. (Protein domains were introduced in Section 3.4.) The amino acid sequence of the DNA binding domain determines the DNA sequence to which it can bind. The chapter-opening figure for Chapter 15 shows transcription factors bound to DNA.

Some genes have enhancers associated with them that bind activators **(Figure 16.9).** By definition, an **enhancer** is a sequence that increases transcription of a gene independently of its position, orientation, and distance from a promoter. Orientation means that the regulatory effects of an enhancer are not dependent on the orientation of its sequence as compared with the orientation of the gene's coding sequence. Enhancers vary considerably in their location relative to the genes they control. They may be quite close, or as far away as hundreds of kilobase pairs. Some enhancers are located within introns.

The enhancers of different genes have specific sets of regulatory sequences, which bind particular activators. A **coactivator** (also called a *mediator*), a large multiprotein complex, forms a bridge between the activators at the enhancer and the proteins at the promoter and promoter proximal region. As a result, the concentration of activators at the promoter is increased, which stimulates transcription up to its maximal rate.

REPRESSION OF TRANSCRIPTION In some genes, repressors—transcription factors that are inhibitory to transcription initiation—oppose the effect of activators, reducing or blocking the rate of transcription. The final rate of transcription then depends on the "battle" between the activation signal and the repression signal.

Different kinds of repressors in eukaryotes work in different ways. Some repressors bind to the same regulatory sequence to which activators bind (often in the enhancer), thereby preventing

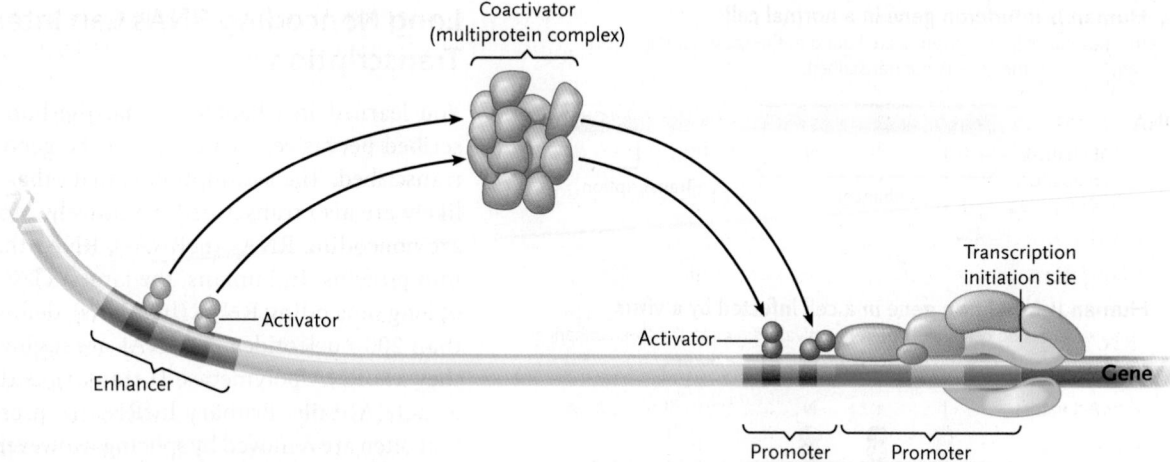

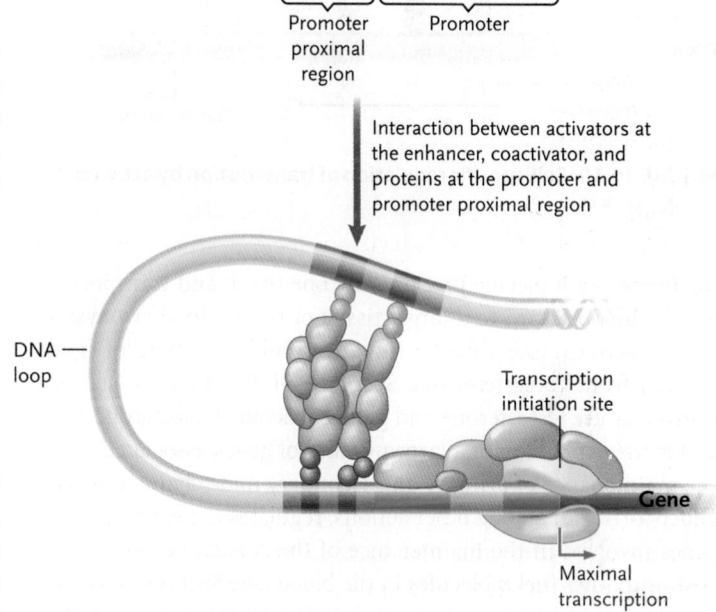

FIGURE 16.9 Interactions between activators at the enhancer, a coactivator, and general transcription factors at the promoter lead to maximal transcription of the gene. Interaction between activators at the enhancer, coactivator, and proteins at the promoter and promoter proximal region.
© Cengage Learning 2017

activators from binding to that site. Other repressors bind to their own specific site in the DNA near where the activator binds and interact with the activator so that it cannot interact with the coactivator. Yet other repressors bind to specific sites in the DNA and recruit **corepressors,** multiprotein complexes analogous to coactivators but which are negative regulators, inhibiting transcription initiation.

CONTROL OF TRANSCRIPTION BY COMBINATIONS OF TRANSCRIPTION FACTORS How is the binding of specific activators to regulatory sequences in promoter proximal elements and enhancers coordinated in regulating gene expression? Some genes may have one to a few regulatory sequences, but genes under complex regulatory control have many regulatory sequences. Each regulatory sequence binds a specific transcription factor. By groups of transcription factors functioning together, the transcription of a wide array of genes can be controlled. The process is called **combinatorial gene regulation.** Consider a theoretical example of two genes, each with four regulatory sequences in their enhancers. Gene *A* requires activators 2, 5, 7, and 8 to bind for transcription activation, whereas gene *B* requires activators 1, 5, 8, and 11 to bind for transcription activation. Looked at another way, both genes require activators 5 and 8 combined with other different activators for full activation.

The involvement of multiple transcription factors in controlling many eukaryotic genes helps us explain cell-specific gene expression. Because multiple activators are required to activate the transcription of a particular gene, the gene will be transcribed only in cells that contain those activators. **Figure 16.10** illustrates this for the human β-interferon gene, the product of which functions as part of the immune system to combat virus infections (see Chapter 45). In a normal (uninfected) cell, no activators are present to activate the gene so transcription does not occur **(Figure 16.10A).** In a virus-infected cell, activators J, I, and N are produced (the letters are shortened forms of their actual names) and bind to their specific regulatory sequences in the gene's enhancer, activating transcription of the gene **(Figure 16.10B).** Cell type-specific gene expression can be achieved in a similar way. That is, transcription of a gene will be activated only if the appropriate array of

activators is present for that gene in a cell. We can also use this example to illustrate combinatorial gene regulation. The N activator, for example, regulates many other genes including polypeptides that form part of immunoglobulin proteins (see Chapter 45).

COORDINATED REGULATION OF TRANSCRIPTION OF GENES WITH RELATED FUNCTIONS There are no operons in eukaryotes, yet the transcription of genes with related function is coordinately controlled. How is this accomplished? The answer is that all genes that are coordinately regulated have the same regulatory sequences associated with them. The set of genes can be switched on or off by having a single, key transcription factor bind to its regulatory sequence to complete the group of transcription factors controlling gene expression. Consider the control of gene expression by steroid hormones in mammals. A

A. Human β-interferon gene in a normal cell

In a normal cell, no activators are bound to the enhancer regulatory sequences so the gene is not transcribed.

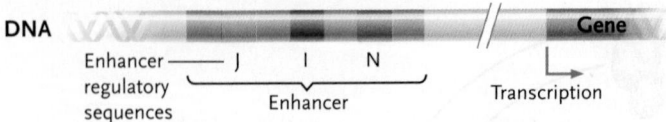

B. Human β-interferon gene in a cell infected by a virus

In a virus-infected cell, three different activators bind to the three enhancer regulatory sequences and transcription of the gene occurs.

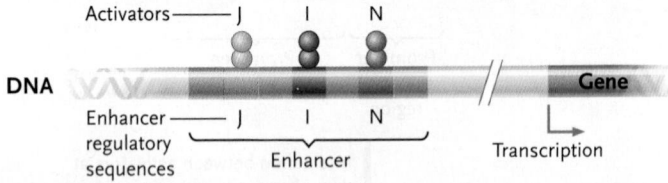

FIGURE 16.10 Cell-specific regulation of transcription by activators.
© Cengage Learning 2017

hormone is a molecule produced by one tissue and transported via the bloodstream to a target tissue or tissues to alter physiological activity (see Chapter 42). A **steroid** is a type of lipid derived from cholesterol (see Section 3.3). Examples of steroid hormones are testosterone and glucocorticoid. Testosterone regulates the expression of a large number of genes associated with the maintenance of primary and secondary male characteristics. Glucocorticoid, among other actions, regulates the expression of genes involved in the maintenance of the concentration of glucose and other fuel molecules in the blood (see Section 42.4).

A steroid hormone acts on specific target tissues in the body because only cells in those tissues have **steroid hormone receptors** in their cytoplasm that recognize and bind the hormone (see Section 9.4). The steroid hormone moves through the plasma membrane into the cytoplasm and binds to its specific receptor, activating it **(Figure 16.11)**. The hormone–receptor complex then enters the nucleus and, functioning as a transcription factor, binds to a specific regulatory sequence that is adjacent to the genes whose expression is controlled by the hormone. This binding completes the group of transcription factors needed for transcription activation, expression of the genes commences, and the proteins encoded by the genes are synthesized rapidly. If the cell no longer is exposed to the steroid hormone, the receptor (transcription factor) is inactivated and the genes it controls now switch off.

A single steroid hormone can regulate many different genes because all of the genes have an identical DNA sequence— a **steroid hormone response element**—to which the hormone–receptor complex binds. For example, all genes controlled by glucocorticoid have a glucocorticoid response element associated with them. Therefore, the release of glucocorticoid into the bloodstream coordinately activates the transcription of genes with that response element.

Long Noncoding RNAs Can Interfere with Transcription

You learned in Chapter 15 that the human genome is transcribed pervasively, with 74% of the genome capable of being transcribed. The assumption is that other eukaryotic genomes likely are also transcribed pervasively. Most of the transcripts are noncoding RNAs (ncRNAs), RNAs that are not translated into proteins. In humans, the largest class of ncRNAs consists of long noncoding RNAs (lncRNAs), defined as ncRNAs longer than 200 nucleotides. lncRNAs are highly similar to mRNAs: they are RNA polymerase II transcripts that have 5′ caps and 3′ poly(A) tails. Primary lncRNA transcripts contain introns that often are removed by splicing. However, like other ncRNAs, and unlike mRNAs, they do not encode proteins.

More than 10,000 lncRNA genes have been identified to date. Scientists estimate that the total number may equal or exceed the number of protein-coding genes. Many lncRNAs may be nonfunctional products of accidental transcription, but accumulating evidence indicates that particular lncRNAs may negatively or positively regulate the expression of protein-coding genes at the transcriptional or posttranscriptional level. For example, you learned in Chapter 13 of X-chromosome inactivation brought about by the lncRNA Xist. Xist lncRNA brings about silencing transcription of genes on the inactive X chromosome by recruiting epigenetic modifiers; that is, enzymes that methylate the promoters of those genes as described earlier.

Xist lncRNA interferes with the transcription of many genes. Particular lncRNAs also have been shown to regulate the expression of specific genes, including many known genomically imprinted genes. Diverse mechanisms of gene expression regulation have been seen, including guiding chromatin remodeling complexes to specific genomic loci so they can exert their effects, and binding to transcription factors to prevent them from binding to their target DNA sequence.

Research into lncRNA functions is at a relatively early stage. lncRNAs are thought to have a wide range of functions in cellular and developmental processes and future research will be focused on characterizing those functions and the mechanisms used by the lncRNAs involved.

Genome-Wide Analysis Has Cataloged Human Promoters and Enhancers

Researchers from the Functional Annotation of the Mammalian Genome (FANTOM) project and Japan's RIKEN Institute reported in 2014 on the results of a genome-wide analysis of promoters and enhancers in the human and mouse genomes. For promoters, the scientists searched for the beginnings of RNA transcripts and mapped them to the human genome sequence, then extrapolating the presence of promoters upstream of those transcription start sites (TSSs). To capture as many instances of gene expression as possible, the researchers analyzed TSSs in 975 human and 399 mouse samples, representing a wide range of cell types, tissues, and cancer cell lines. The result was the identification of an astonishing number of promoters: 184,476 in humans and 116,277 in the mouse. These numbers far exceed our present knowledge of genes in the two mammalian genomes, both protein-coding genes and noncoding genes. Many genes were shown to have multiple TSSs, with transcription beginning at different locations in different cell and tissue types, indicating the cell- and tissue-specific nature of promoters. Interestingly, transcription initiation as shown by the TSS locations is not highly evolutionarily conserved. In fact, 43% of the human TSSs could not be aligned, sequence-wise, to the mouse genome, and 39% of the mouse TSSs could not be aligned to the human genome. In order of TSS conservation, the highest conservation was shown for housekeeping TSSs, followed by protein-coding TSSs, then noncoding TSSs.

For the enhancer study, the investigators cataloged enhancers and enhancer activity in 808 different cell, tissue, and cancer cell line samples. They identified 43,011 likely enhancers, a data resource that will inform studies to link enhancers to the genes they control, and to investigate cell-type specific versus ubiquitous enhancers and their functions.

Overall, the data show that there are many more promoters and enhancers than the presently known genes. Some of the promoters can be accounted for by the presence of multiple promoters for particular genes. The dataset will be a valuable asset for experimental explorations of the complexities of gene expression regulation in humans.

The next section takes up the regulatory mechanisms operating at each of the steps from mRNA to proteins.

STUDY BREAK 16.2

1. What is the role of histones in gene expression? How does acetylation of the histones affect gene expression?
2. What are the roles of general transcription factors, activators, and coactivators in transcription of a protein-coding gene?

16.3 Posttranscriptional, Translational, and Posttranslational Regulation

Transcriptional regulation determines which genes are copied into mRNAs. Once mRNAs are transcribed from active genes, further regulation occurs at each of the major steps in the pathway from genes to proteins: during pre-mRNA processing and

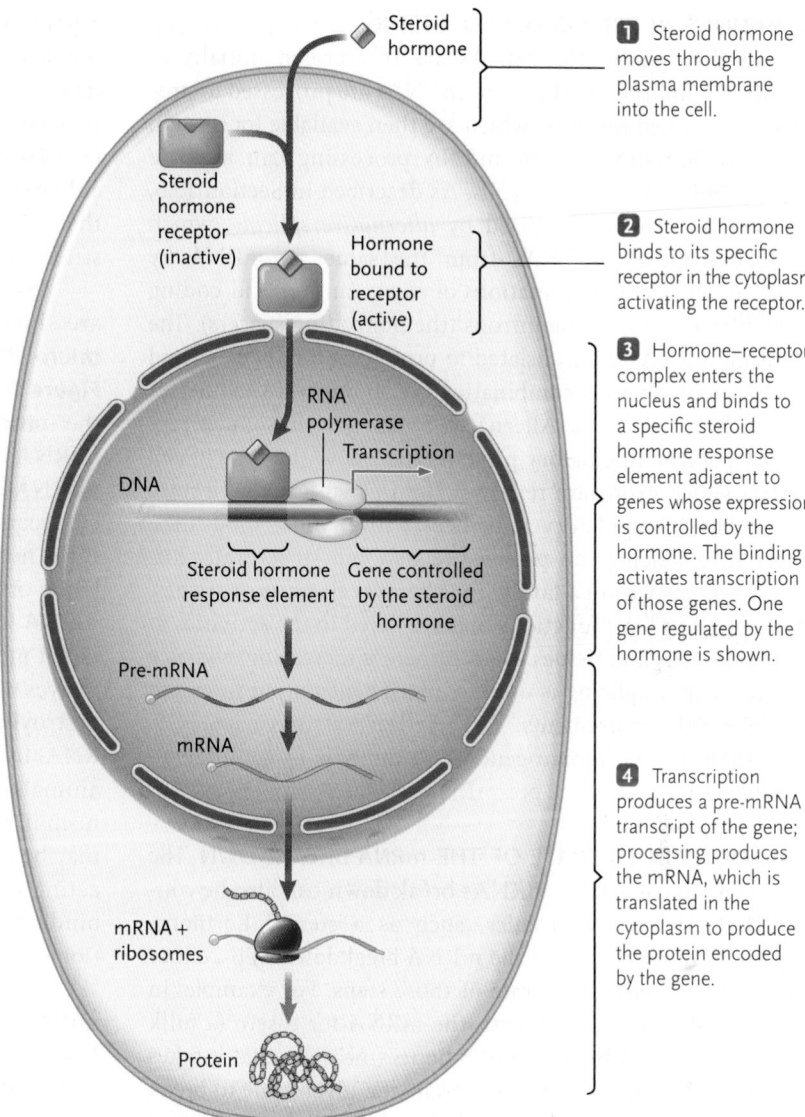

FIGURE 16.11 Steroid hormone regulation of gene expression. A steroid hormone enters the cell and forms a complex in the cytoplasm with a steroid hormone receptor that is specific to the hormone. Steroid hormone–receptor complexes migrate to the nucleus, bind to the steroid hormone response element next to each gene they control (one such gene is shown in the figure), and affect transcription of those genes.
© Cengage Learning 2017

the movement of finished mRNAs to the cytoplasm (posttranscriptional regulation), during protein synthesis (translational regulation), and after translation is complete (posttranslational regulation) (refer again to Figure 16.5).

Posttranscriptional Regulation Controls mRNA Availability

Posttranscriptional regulation involves several mechanisms, including changes in pre-mRNA processing and the rate at which mRNAs are degraded.

VARIATIONS IN PRE-mRNA PROCESSING In Chapter 15 you learned that eukaryotic mRNAs are transcribed initially as pre-mRNA molecules. These pre-mRNAs are processed to produce the finished mRNAs, which are then available for protein synthesis. Variations in pre-mRNA processing can regulate *which* proteins are made in cells. As described in Section 15.3, pre-mRNAs can be processed by *alternative splicing,* which produces different mRNAs from the same pre-mRNA by removing different combinations of exons (amino acid-coding segments) along with the introns (the noncoding spacers). The resulting mRNAs are translated to produce a family of related proteins with various combinations of amino acid sequences derived from the exons. Alternative splicing itself is under regulatory control. Regulatory proteins specific to the type of cell control which exons are removed from pre-mRNA molecules by binding to regulatory sequences within those molecules. Alternative splicing can produce related, but structurally different proteins within a family in different cell types or tissues. Those proteins have functions that are keyed to the activities of those cell types and tissues. In fact, there is increasing evidence that alternative splicing is used to adjust gene output to match physiological requirements. As you learned in Chapter 15, more than 95% of human genes with multiple exons undergo alternative splicing at the pre-mRNA level.

VARIATIONS IN THE RATE OF THE mRNA BREAKDOWN The rate at which eukaryotic mRNAs break down can also be controlled. Regulatory molecules, such as a steroid hormone, directly or indirectly affect the mRNA breakdown steps, either slowing or increasing the rate of those steps. For example, in the mammary gland of the rat, the mRNA for casein (a milk protein) has a half-life of about 5 hours (meaning that it takes 5 hours for half of the mRNA present at a given time to break down). The half-life of casein mRNA changes to about 92 hours if the peptide hormone prolactin is present. Prolactin (see Chapter 42) is synthesized in the brain and in other tissues, including the breast. Prolactin stimulates the mammary glands to produce milk. During milk production, a large amount of casein must be synthesized, and this is accomplished in part by decreasing radically the rate of breakdown of the casein mRNA.

Modulation of the stability of an mRNA typically involves a specific sequence or sequences in its 3′ UTR (untranslated region at the 3′ end of an mRNA following the stop codon; see Section 15.3). When specific proteins or regulatory RNAs (see next subsection) recognize and bind to the bases of the sequence, a degradation pathway is triggered for breaking down that RNA.

REGULATION OF GENE EXPRESSION IN EUKARYOTES BY SMALL NONCODING RNAs In 1998, Andrew Fire of Stanford University School of Medicine and Craig Mello of University of Massachusetts Medical School showed that RNA silenced the expression of a particular gene in the nematode worm, *Caenorhabditis elegans.* They called the phenomenon **RNA interference (RNAi).** Their discovery revolutionized the way

scientists thought about and studied gene regulation in eukaryotes because it showed that posttranscriptional regulation may involve not only regulatory proteins, but also single-stranded noncoding RNAs, which can bind to mRNAs and affect their translation. We now know that RNAi is widespread among eukaryotes. Fire and Mello received a Nobel Prize in 2006 "for their discovery of RNA interference - gene silencing by double-stranded RNA."

Two major groups of small noncoding regulatory RNAs are involved in RNAi at the posttranscriptional level, **microRNAs (miRNAs)** and **small interfering RNAs (siRNAs)**. **Figure 16.12** shows the transcription of an miRNA gene and the processing of its transcript to produce a functional miRNA molecule. The miRNA, in a protein complex called the **miRNA-induced silencing complex (miRISC),** binds to specific sequences in the 3′ UTRs of target mRNAs by base pairing. If the miRNA and target mRNA pair imperfectly, either translation of the mRNA is repressed (shown in Figure 16.12) or the mRNA is made less stable (shorter half-life). If the miRNA and target mRNA pair perfectly, an enzyme in the protein complex cleaves the target mRNA where the miRNA is bound to it, destroying the mRNA, which, of course, silences its expression. RNAi by imperfect pairing is the most common mechanism in animals. RNAi by perfect pairing is the most common mechanism in plants. Interestingly, research has shown that lncRNAs may be involved in regulating the action of miRNAs. By interacting directly with miRNAs, the lncRNAs block them from binding to their target mRNAs, thereby affecting their regulation of gene expression at the translation level.

Analysis of genome sequences has identified over 17,000 miRNAs in 142 species, including more than 1,900 in humans. Interestingly, many miRNAs have been implicated in common human diseases, including cancer (see later in the chapter). No longer can we consider proteins as the only regulators of gene expression. It is clear that miRNAs are key regulators of gene expression. Sequence comparisons show that miRNAs are conserved across species, and that they are expressed in different cell types. It is estimated that more than half of all mRNAs are targets of miRNAs, and that each miRNA may regulate hundreds of target mRNAs. Unquestionably miRNAs are involved in regulating many biological processes, including cell proliferation and development. We will discuss some examples in the next two sections.

The other major type of small noncoding regulatory RNAs that operates at the posttranscriptional level is the *small interfering RNA (siRNA).* Whereas miRNA is produced from RNA that is encoded in the cell's genome, siRNA is produced from double-stranded RNA that is *not* encoded by nuclear genes. For example, the life cycle and replication of many viruses with RNA genomes involves a double-stranded RNA stage. Cells attacked by such a virus can defend themselves using siRNA, which they produce from the virus' own RNA. The viral double-stranded RNA enters the cell's RNAi process in a way very similar to that described for miRNAs. Double-stranded RNA is cut into short double-stranded RNA molecules by an enzyme (the same Dicer enzyme shown in Figure 16.13), and a single-stranded siRNA is generated that is complexed with proteins to produce an **siRNA-induced silencing complex (siRISC)**. In the RNAi process, the siRNA in the siRISC acts similarly to miRNA in the miRISC. Single-stranded RNAs complementary to the siRNA are targeted and the target RNA is cleaved and the pieces are then degraded. In our viral example, the targeted RNAs would be viral mRNAs for proteins the virus uses to replicate itself, or a single-stranded RNA that is the viral genome itself, or that is produced from the viral genome during replication.

The expression of any gene can be knocked down to low levels experimentally using RNAi. Researchers are using this approach, for example, to study genes that have been detected by sequencing complete genomes, but whose functions are completely unknown (see Section 19.3). The expression of a gene of interest is knocked down and researchers look for a change in phenotype, such as properties relating to growth or metabolism. If such a change is seen, the researchers now have some insight into the gene's function.

Translational Regulation Controls the Rate of Protein Synthesis

Translational regulation controls the rate at which mRNAs are used in protein synthesis. It occurs in essentially all cell types and species. For example, translational regulation is involved in cell cycle control in all eukaryotes and in many processes during development in multicellular eukaryotes, such as red blood cell differentiation in animals. Many viruses exploit translational regulation to control their infection of cells and to shut off the host cell's own genes.

Consider the general role of translational regulation in animal development. During early development of most animals, little transcription occurs. The changes in protein synthesis patterns seen in developing cell types and tissues instead are the result of the activation, repression, or degradation of maternal mRNAs, the mRNAs that were in the mother's egg before fertilization. One important mechanism for translational regulation involves adjusting the length of the 3′ poly(A) tail of the mRNA (see Section 15.3). That is, enzymes in the cytoplasm can shorten or lengthen the poly(A) tail on an mRNA. Increases in poly(A) tail length result in increased translation. Decreases in length result in decreased translation. For example, during embryogenesis (the formation of the embryo) of the fruit fly, *Drosophila*, key proteins are synthesized when the poly(A) tails on the mRNAs for those proteins are lengthened in a regulated way. Evidence for this came from experiments in which poly(A) tail lengthening was blocked, resulting in inhibition of embryogenesis. The mechanism for controlling poly(A) tail length is not completely understood.

FIGURE 16.12 RNA interference—regulation of gene expression by microRNAs (miRNAs).

© Cengage Learning 2017

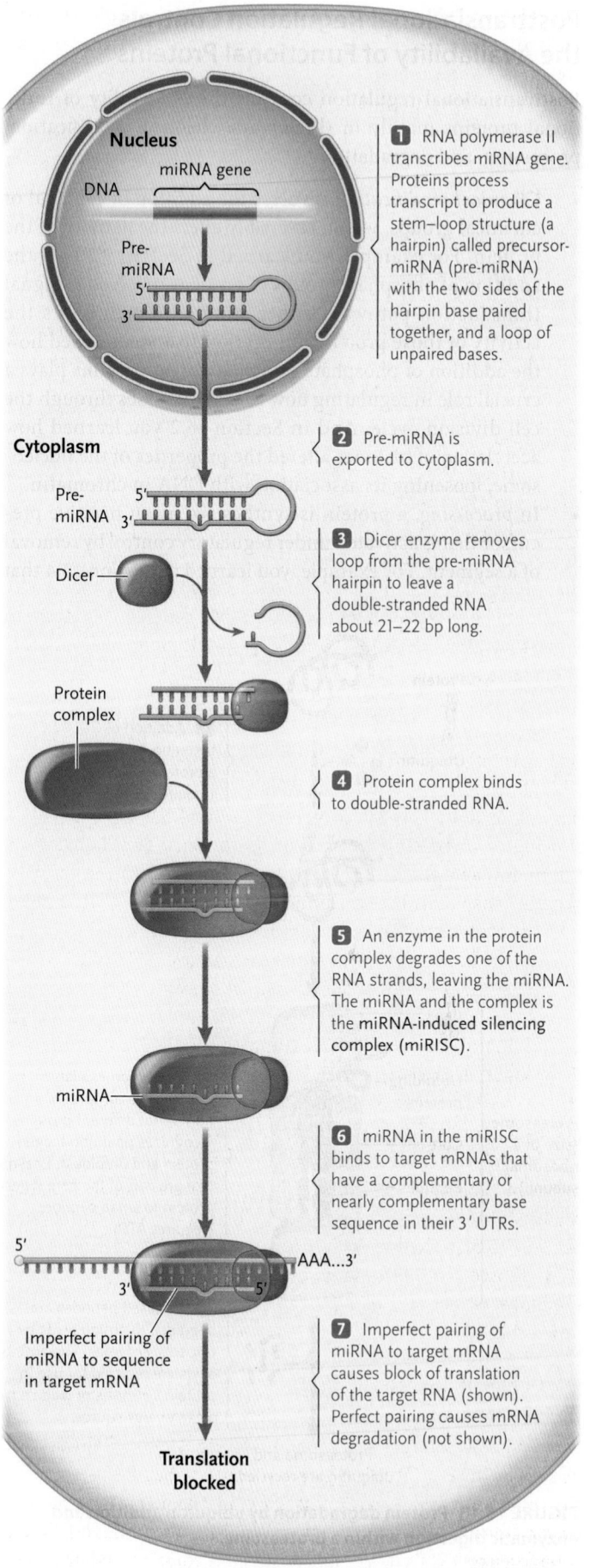

1. RNA polymerase II transcribes miRNA gene. Proteins process transcript to produce a stem–loop structure (a hairpin) called precursor-miRNA (pre-miRNA) with the two sides of the hairpin base paired together, and a loop of unpaired bases.

2. Pre-miRNA is exported to cytoplasm.

3. Dicer enzyme removes loop from the pre-miRNA hairpin to leave a double-stranded RNA about 21–22 bp long.

4. Protein complex binds to double-stranded RNA.

5. An enzyme in the protein complex degrades one of the RNA strands, leaving the miRNA. The miRNA and the complex is the **miRNA-induced silencing complex (miRISC)**.

6. miRNA in the miRISC binds to target mRNAs that have a complementary or nearly complementary base sequence in their 3′ UTRs.

7. Imperfect pairing of miRNA to target mRNA causes block of translation of the target RNA (shown). Perfect pairing causes mRNA degradation (not shown).

Nucleus

miRNA gene

DNA

Pre-miRNA

5′
3′

Cytoplasm

Pre-miRNA

5′
3′

Dicer

Protein complex

miRNA

5′

AAA...3′

3′

5′

Imperfect pairing of miRNA to sequence in target mRNA

Translation blocked

Posttranslational Regulation Controls the Availability of Functional Proteins

Posttranslational regulation controls the availability of functional proteins mainly in three ways: chemical modification, processing, and degradation.

- *Chemical modification* involves the addition or removal of chemical groups, which reversibly alters the activity of the protein. For example, you learned in Section 9.2 how the addition of phosphate groups to proteins involved in signal transduction pathways either stimulates or inhibits the activity of those proteins. In Section 10.4 you learned how the addition of phosphate groups to target proteins plays a crucial role in regulating how a cell progresses through the cell division cycle. And in Section 16.2 you learned how acetylation of histones altered the properties of the nucleosome, loosening its association with DNA in chromatin.
- In *processing*, a protein is synthesized as an inactive precursor that is activated under regulatory control by removal of a segment. For example, you learned in Section 15.4 that the digestive enzyme pepsin is synthesized as pepsinogen, an inactive precursor that activates by removal of a segment of amino acids (also see Figure 47.10). Similarly, the glucose-regulating hormone insulin (see Chapter 42) is synthesized as a precursor called *proinsulin;* processing of the precursor removes a central segment but leaves the insulin molecule, which consists of two polypeptide chains linked by disulfide bridges.

- The rate of *degradation* of proteins is also regulated. Some proteins in eukaryotic cells last for the lifetime of the individual, whereas others persist only for minutes. Proteins with relatively short cellular lives include many of the proteins regulating transcription. Typically, these short-lived proteins are marked for breakdown by enzymes that attach a "doom tag" consisting of a small protein called *ubiquitin.* The protein is given this name because it is indeed ubiquitous—present in almost the same form in essentially all eukaryotes. A ubiquinated protein is recognized by the **proteasome,** a large cytoplasmic complex of several different proteins, where degradation of the protein occurs **(Figure 16.13).** Aaron Ciechanover and Avram Hershko, both of the Israel Institute of Technology, Haifa, Israel, and Irwin Rose of the University of California, Irvine, received a Nobel Prize in 2004 "for the discovery of ubiquitin-mediated protein degradation."

STUDY BREAK 16.3

1. How does a microRNA silence gene expression?
2. If the poly(A) tail on an mRNA was removed, what would likely be the effect on the translation of that mRNA?

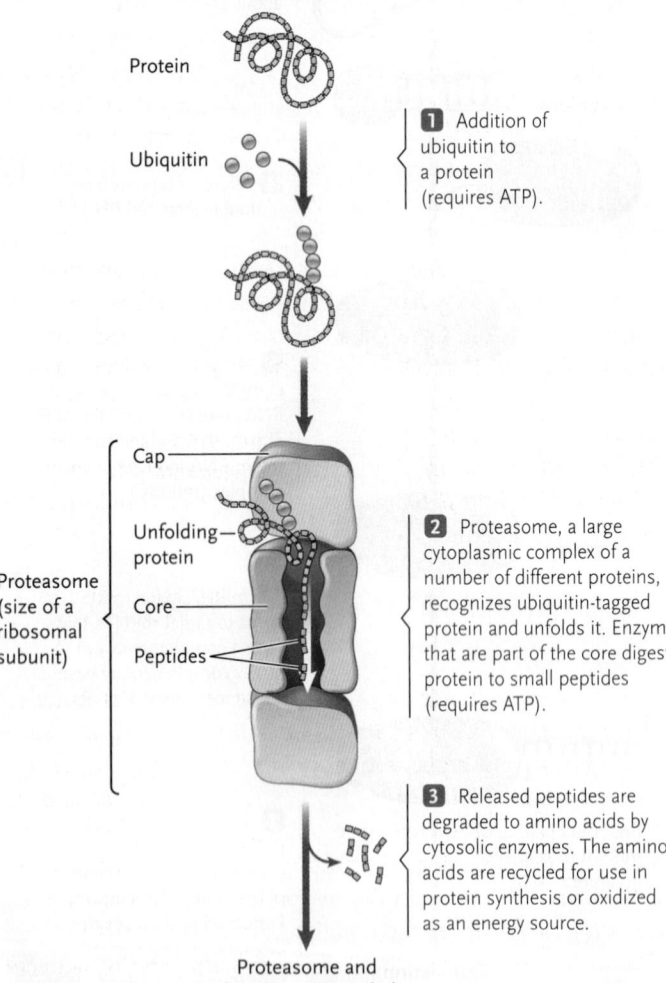

FIGURE 16.13 Protein degradation by ubiquitin addition and enzymatic digestion within a proteasome.

© Cengage Learning 2017

16.4 Genetic and Molecular Regulation of Development

In the development of multicellular eukaryotes, a fertilized egg divides into many cells that are ultimately transformed into an adult, which is itself capable of reproduction. The process is orchestrated by a series of programmed changes encoded in the genome, although the development of a multicellular organism is also influenced to some extent by the environment. The programmed changes do not alter the genome itself, which remains identical in virtually all of the cells of the developing organism. (Thus, the processes involved in development are all examples of epigenetic phenomena.) Instead, gene regulatory events direct sequential developments that are appropriate both in time—certain stages must unfold before others—and in place—new structures must arise in the correct location. An understanding of how genes are regulated therefore aids researchers in their genetic analysis of development. Earlier in this chapter you learned that regulatory proteins (transcription factors) and miRNAs are important players in the regulation of gene expression. So, too, are they important in development. For instance, many of the molecular mechanisms in development (specifically, determination and differentiation; see later) depend on **master regulatory genes,** genes encoding transcription factors that control genes for developmental events. The genes that are controlled may themselves be regulatory genes encoding transcription factors that, in turn, regulate the expression of other sets of genes. You also learned earlier in the chapter about combinatorial gene regulation, the involvement of multiple transcription factors in regulating gene expression. Not only is it used in regulating cell function on a continuing basis, combinatorial gene regulation is involved also in the generation of different cell types.

Key to understanding the genetic control of development is to identify and characterize the genes involved in development, and to understand how the products of the genes regulate and bring about the elaborate events that occur. One productive research approach has been to isolate mutants that affect developmental processes. Developmental geneticists can then identify the genes involved, and analyze them in detail to build models for the molecular functions of the gene products in development. A number of model organisms are used for these studies because of the relative ease with which mutants can be made and studied, and the ease of performing molecular analyses. These organisms include the fruit fly *(Drosophila melanogaster)* (see *Focus on Research: Model Organisms* in Chapter 13) and *Caenorhabditis elegans* (a nematode worm) (see *Focus on Research: Model Organisms* in Chapter 31) among invertebrates; the zebrafish *(Danio rerio)* (see Figure 10.1) and the house mouse *(Mus musculus)* (see *Focus on Research: Model Organisms* in Chapter 45) among vertebrates; and thale cress *(Arabidopsis thaliana)* among plants (see *Focus on Research: Model Organisms* in Chapter 36).

In this section we discuss some of the genetic and molecular mechanisms that regulate development in animals to illustrate the principles involved. In Chapter 50 you will learn about the cellular and morphological events that occur as an adult animal develops from a fertilized egg. The gene regulation of development in plants is described in Chapter 36.

Development in Animals Is Accomplished by Several Genetically Regulated Mechanisms

Fertilization of an egg by a sperm cell produces a zygote. Development begins at this point. Development in all animals is accomplished by a number of mechanisms that are under genetic control but are also influenced to some extent by the environment. The mechanisms include mitotic cell divisions, movements of cells, *induction, determination,* and *differentiation.*

Mitotic divisions produce the cells that are the subjects of the gene regulatory processes that create the various cell types in the adults. Cell movements during development are part of the program to create specific tissues and organs.

Induction is the process in which one group of cells (the *inducer cells*) causes or influences another nearby group of cells (the *responder cells*) to follow a particular developmental pathway.

Determination is the process by which the developmental fate of a cell is set. Before determination, a cell is **totipotent** (has the potential to become any cell type of the adult) but after determination, the cell commits to becoming a particular cell type. Induction is the major process responsible for determination.

Differentiation follows determination, and involves the establishment of a cell-specific developmental program in cells. Differentiation results in cell types with clearly defined structures and functions. Those features derive from specific patterns of gene expression in cells.

Developmental Information Is Located in Both the Nucleus and Cytoplasm of the Fertilized Egg

Part of the information that directs the initiation of development is stored in the zygote nucleus, in the DNA derived from the egg and sperm nuclei. This information directs development as individual genes are activated or turned off in a highly ordered manner. The rest of the information is stored in the egg cytoplasm, in the form of messenger RNA (mRNA) and protein molecules.

Because the fertilizing sperm contributes essentially no cytoplasm, nearly all the cytoplasmic information of the fertilized egg (zygote) is maternal in origin. Key mRNA and proteins stored in the egg cytoplasm that direct the early stages of development are called **cytoplasmic determinants.** They are distributed asymmetrically in the cell, rather than being evenly distributed. Therefore, when the zygote divides, the cytoplasmic determinants are distributed asymmetrically to the daughter cells, reflecting their distribution in the zygote. As a result, the two daughter cells resulting from division differ in the signals they have, and this leads to different patterns of gene expression. Cytoplasmic determinants direct the first stages of animal development before genes of the zygote become active. Depending on the animal group, the control of early development by cytoplasmic determinants may be limited to the first few divisions of the zygote, as in mammals, or it may last until the actual tissues of the embryo are formed, as in most invertebrates.

Induction Is the Major Process Responsible for Determination

Induction is the major process responsible for determination, which is the process by which the developmental fate of a cell is set. Induction is a highly selective process because only certain responder cells can respond to the signal from the inducer cells. Many experiments have shown that induction occurs through the interaction of signaling molecules of inducer cells with surface receptors on the responder cells. In some cases, a signaling molecule released by the inducer cell binds to a specific surface receptor of a responder cell. The binding activates the receptor, triggering a signaling pathway within the cell, which produces a developmental change, often involving a change in gene activity (the cellular response to the signal). (Surface receptors and their associated signaling pathways are discussed in Section 9.3.) In other cases, induction occurs by direct cell-to-cell contact involving interaction between a membrane-embedded protein on the inducer cell and a membrane-embedded receptor protein on the surface of the responder cell. The receptor is activated by the interaction, and in the same way as for the diffusible signal molecule example, the cell undergoes a developmental change.

Differentiation Produces Specialized Cells without Loss of Genes

Differentiation is the process by which cells that have been committed to a particular developmental fate by the determination process now develop into specialized cell types with distinct structures and functions, such as skin cells and nerve cells. As part of differentiation, cells produce molecules characteristic of the specific types. For example, in the lens cells of the eye, 80–90% of the total protein synthesized is crystallin, the protein responsible for the transparency of the lens.

Research into differentiation confirmed that, as cells specialize, the DNA in the genome remains constant, matching that of the original zygote. In other words, differentiation generally occurs as a result of differential gene activity, and not by a process in which DNA is lost, so that each type of differentiated cell retains only those genes required for that cell type. (There are just a few examples in particular organisms where gene loss occurs during differentiation.)

The most compelling evidence showing that the DNA in the genome remains constant during differentiation has come from experiments in which animals and plants have been cloned. Animal clones have been made by taking an egg produced by one animal, and then replacing its nucleus with a nucleus taken from a somatic cell of a different adult animal, showing that the adult nucleus is still totipotent. Different species of frogs and many types of mammals have been cloned in this way. The first mammal cloned was Dolly, a sheep (the cloning experiment is described in Chapter 18, and shown in Figure 18.10). Similarly, plants can be cloned from single cells isolated from a mature plant (see Chapter 36).

At the molecular level, combinatorial gene regulation (see Section 16.2) is important for generating the different cell types during development.

Genes Control Determination and Differentiation

Both determination and differentiation involve specific, regulated changes in gene expression. One well-studied example of the genetic control of determination and differentiation is the production of skeletal muscle cells from somites in mammals **(Figure 16.14).** Somites give rise to the vertebral column, the ribs, the repeating sets of skeletal muscles associated with the ribs and vertebral column, and the skeletal muscles of the limbs. Under the control of the master regulatory gene, *myoD*, particular cells of a somite differentiate into skeletal muscle

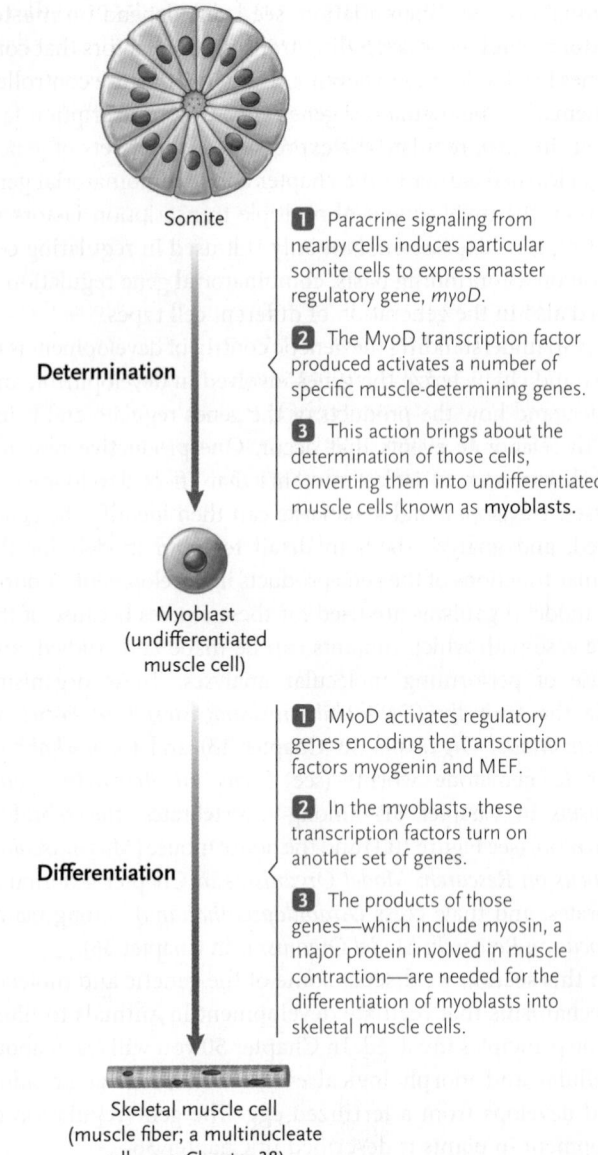

1 Paracrine signaling from nearby cells induces particular somite cells to express master regulatory gene, *myoD*.

2 The MyoD transcription factor produced activates a number of specific muscle-determining genes.

3 This action brings about the determination of those cells, converting them into undifferentiated muscle cells known as **myoblasts.**

1 MyoD activates regulatory genes encoding the transcription factors myogenin and MEF.

2 In the myoblasts, these transcription factors turn on another set of genes.

3 The products of those genes—which include myosin, a major protein involved in muscle contraction—are needed for the differentiation of myoblasts into skeletal muscle cells.

FIGURE 16.14 The genetic control of determination and differentiation involved in mammalian skeletal muscle cell formation.
© Cengage Learning 2017

cells. The product of *myoD* is the transcription factor MyoD, which acts to turn on muscle-specific genes coordinately.

Genes Regulate Pattern Formation during Development

As a part of the signals guiding differentiation, cells receive positional information that tells them where they are in the embryo. The positional information is vital to **pattern formation:** the arrangement of organs and body structures in their proper three-dimensional relationships. Positional information is laid down primarily in the form of concentration gradients of regulatory molecules produced under genetic control. In most cases, gradients of several different regulatory molecules interact to tell a cell, or a cell nucleus, where it is in the embryo. Below, we describe in brief the results of studies of the genetic control of pattern formation during the development of the fruit fly, *Drosophila melanogaster*. The developmental principles discovered from these studies apply to many other animal species, including humans.

THE LIFE CYCLE OF *DROSOPHILA* The production of an adult fruit fly from a fertilized egg occurs in a sequence of genetically controlled development events. The *Drosophila* life cycle is shown in **Figure 16.15.** As is typical of most insects, the life cycle proceeds from fertilized egg to embryo within the egg. The embryo then hatches from the egg and the life cycle proceeds through three larval stages to a pupal stage and then to the adult stage. The stages of development from fertilized egg to hatching collectively are called **embryogenesis.** As illustrated by the color usage in Figure 16.15, the segments of the embryo can be mapped to the segments of the adult fly. The development of vertebrates, including mammals, occurs quite differently, as you will learn in Chapter 50.

GENETIC ANALYSIS OF *DROSOPHILA* DEVELOPMENT The study of developmental mutants by a large number of researchers has given us important information about *Drosophila* development. Three researchers performed key, pioneering research with developmental mutants: Edward B. Lewis of the California Institute of Technology, Christiane Nüsslein-Volhard of the Max Planck Institute for Developmental Biology in Tübingen, Germany, and Eric Wieschaus of Princeton University. The three shared a Nobel Prize in 1995 "for their discoveries concerning the genetic control of early embryonic development."

Nüsslein-Volhard and Wieschaus studied early embryogenesis. They searched for *every* gene required for early pattern formation in the embryo. They did this by looking for recessive *embryonic lethal* mutations. These mutations, when homozygous, result in the death of the embryo during development. By examining the stage of development at which an embryo died, and how development was disrupted, they gained insights into the role of the particular genes in embryogenesis.

Lewis studied mutants that changed the fates of cells in particular regions in the embryo, producing structures in the adult that normally were produced by other regions. His work was the foundation of research identifying master regulatory genes that control the development of body regions in a wide range of organisms.

MATERNAL-EFFECT GENES AND SEGMENTATION GENES FOR ESTABLISHING THE BODY PLAN IN THE EMBRYO Two classes of genes—*maternal-effect genes* and *segmentation genes*—work sequentially to control the establishment of the embryo's body plan—how the organism is laid out in its anterior-to-posterior,

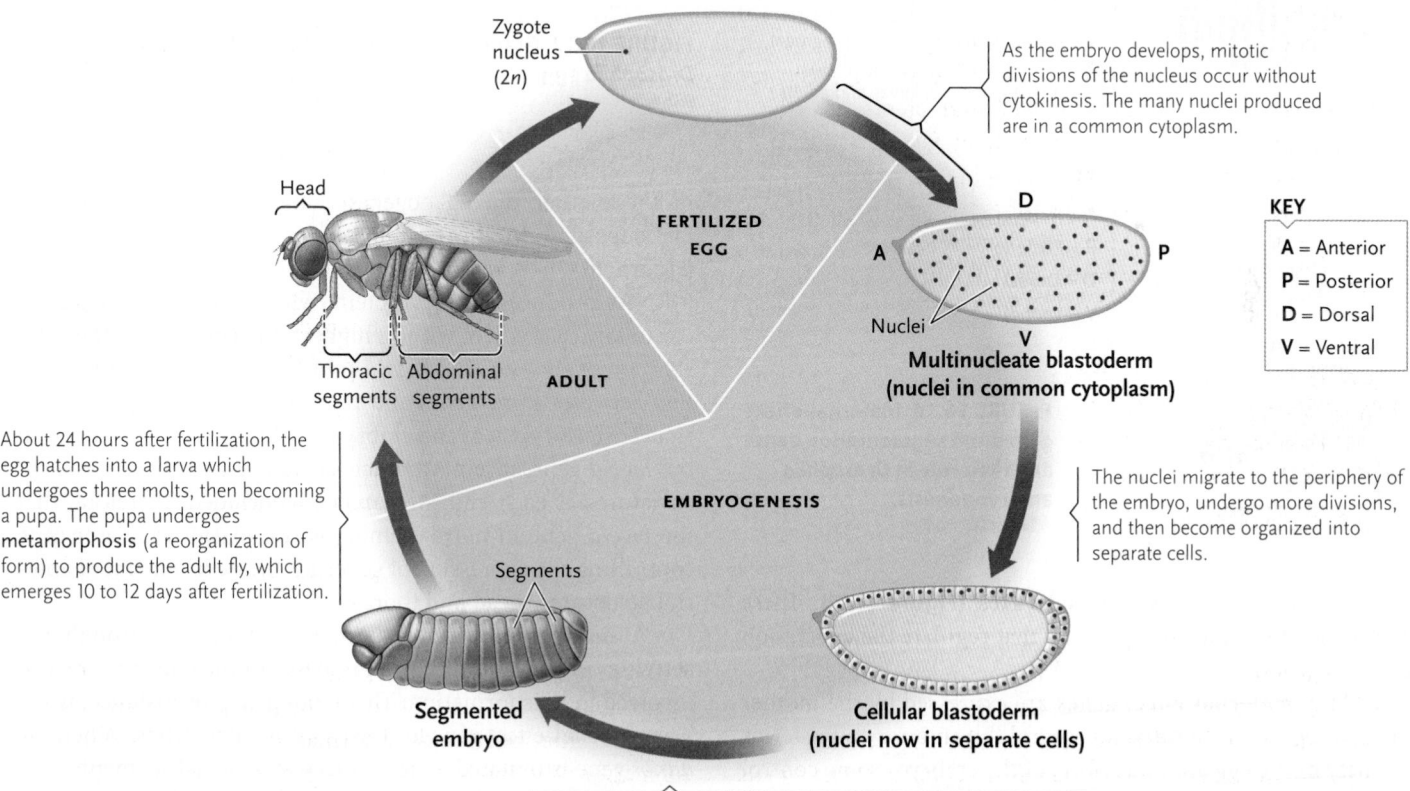

FIGURE 16.15 The life cycle of *Drosophila* and the relationship between segments of the embryo and segments of the adult.
© Cengage Learning 2017

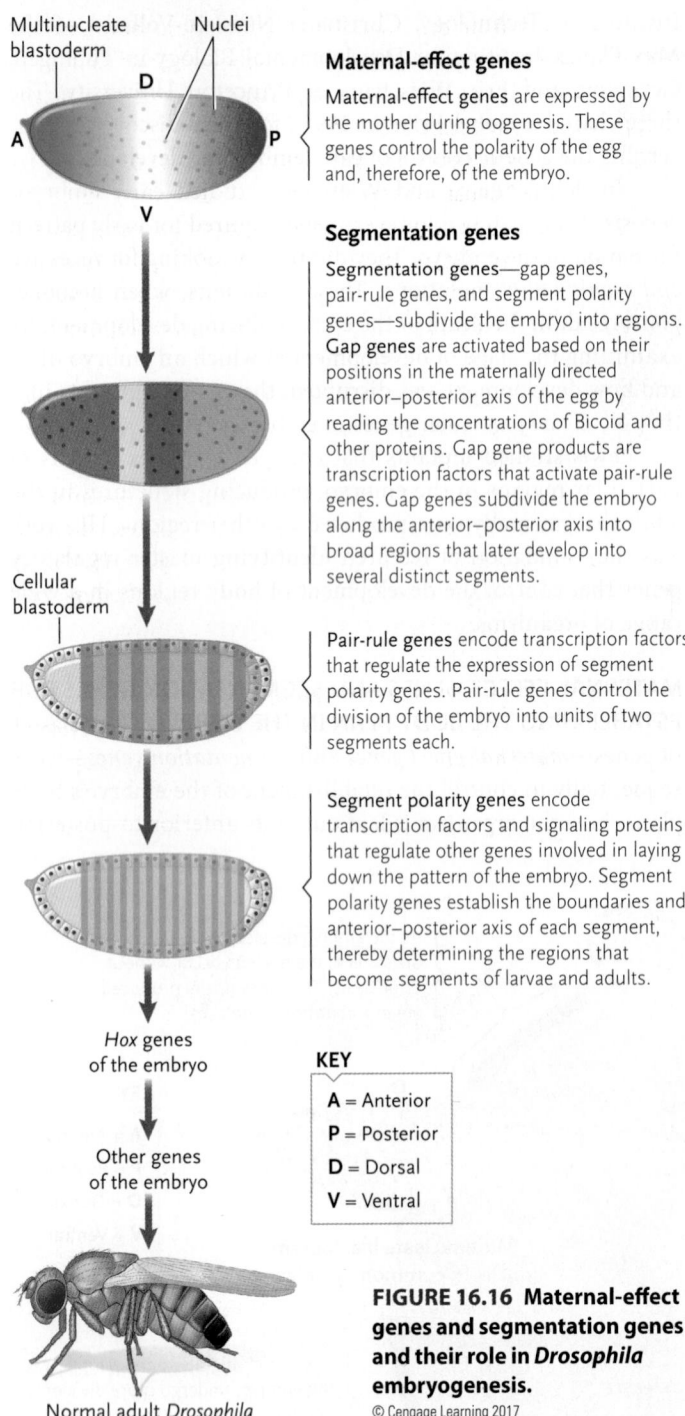

Maternal-effect genes

Maternal-effect genes are expressed by the mother during oogenesis. These genes control the polarity of the egg and, therefore, of the embryo.

Segmentation genes

Segmentation genes—gap genes, pair-rule genes, and segment polarity genes—subdivide the embryo into regions. **Gap genes** are activated based on their positions in the maternally directed anterior–posterior axis of the egg by reading the concentrations of Bicoid and other proteins. Gap gene products are transcription factors that activate pair-rule genes. Gap genes subdivide the embryo along the anterior–posterior axis into broad regions that later develop into several distinct segments.

Pair-rule genes encode transcription factors that regulate the expression of segment polarity genes. Pair-rule genes control the division of the embryo into units of two segments each.

Segment polarity genes encode transcription factors and signaling proteins that regulate other genes involved in laying down the pattern of the embryo. Segment polarity genes establish the boundaries and anterior–posterior axis of each segment, thereby determining the regions that become segments of larvae and adults.

KEY

A = Anterior
P = Posterior
D = Dorsal
V = Ventral

FIGURE 16.16 Maternal-effect genes and segmentation genes and their role in *Drosophila* embryogenesis.
© Cengage Learning 2017

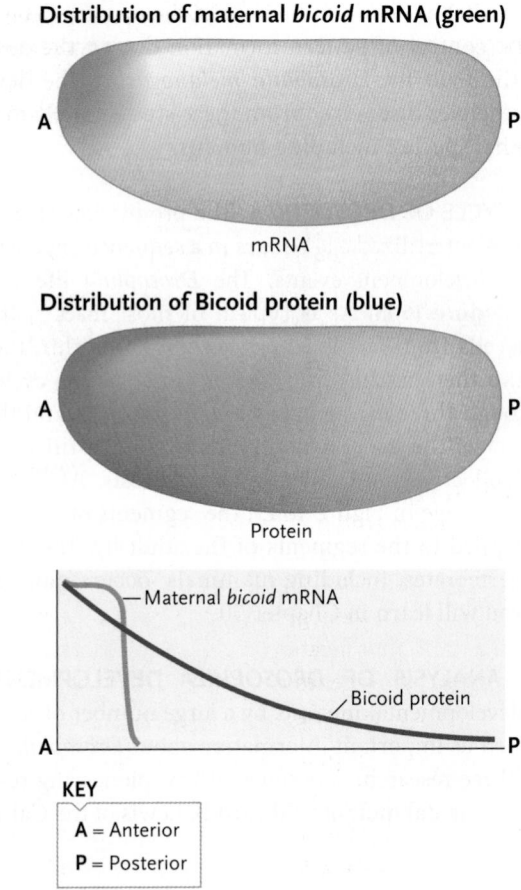

FIGURE 16.17 Gradients of *bicoid* mRNA and Bicoid protein in the *Drosophila* egg.
© Cengage Learning 2017

ventral-to-dorsal, and side-to-side axes **(Figure 16.16)**. These genes code for regulatory proteins that regulate the expression of other genes.

Many **maternal-effect genes** are expressed by the mother during oogenesis. These genes control the anterior-to-posterior polarity of the egg and, therefore, of the embryo. Some control the formation of the anterior structures of the embryo, others control the formation of the posterior structures, and yet others control the formation of the terminal end.

The *bicoid* gene is the key maternal-effect gene responsible for head and thorax development. The *bicoid* gene is transcribed in the mother during oogenesis, and the resulting mRNAs are deposited in the egg, localizing near the anterior end **(Figure 16.17)**. After the egg is fertilized, translation of the mRNAs produces Bicoid protein, which diffuses through the egg to form a gradient with its highest concentration at the anterior end of the egg. The Bicoid protein is a transcription factor that activates some genes and represses others along the anterior–posterior axis of the embryo. Embryos with mutations in the *bicoid* gene have no thoracic structures, but have posterior structures at each end. Researchers concluded, therefore, that the *bicoid* gene in normal embryos is a master regulator gene controlling the expression of genes for the development of anterior structures (head and thorax).

A number of other maternal-effect genes, through the activities of their products in gradients in the embryo, are also involved in axis formation. The *nanos* gene, for instance, is the key maternal-effect gene for the posterior structures. When the *nanos* gene is mutated, embryos lack abdominal segments.

Once the anterior–posterior axis of the embryo is set, the expression of at least 24 **segmentation genes** progressively subdivides the embryo into regions, determining the segments of the embryo and the adult (see Figure 16.16). Gradients of Bicoid and other proteins encoded by maternal-effect genes regulate

Normal   *Antennapedia* mutant

FIGURE 16.18 ***Antennapedia*, a *Hox* gene mutant of *Drosophila*, in which legs develop in place of antennae.**
© Cengage Learning 2017

expression of the embryo's segmentation genes differentially. That is, each segmentation gene is expressed at a particular time and in a particular location during embryogenesis. Three classes of segmentation genes act sequentially; their activities are regulated in a cascade of gene activations:

- **Gap genes,** through their activation of the next genes in the regulatory cascade, control the subdivision of the embryo along the anterior–posterior axis into several broad regions. Mutations in gap genes result in the loss of one or more body segments in the embryo.
- **Pair-rule genes,** through their activation of the next genes in the regulatory cascade, control the division of the embryo into units of two segments each. Mutations in pair-rule genes delete every other segment of the embryo.
- **Segment polarity genes** set the boundaries and anterior–posterior axis of each segment in the embryo. Mutations in segment polarity genes produce segments in which one part is missing and the other part is duplicated as a mirror image.

Hox GENES FOR SPECIFYING THE DEVELOPMENTAL FATE OF EACH SEGMENT

Once the segmentation pattern has been set, *Hox* genes of the embryo specify what each segment will become after metamorphosis. In normal flies, *Hox* genes are master regulatory genes that control the development of structures such as eyes, antennae, legs, and wings on particular segments (see Figure 16.15). Researchers discovered the role of *Hox* genes from the study of mutations in these genes. Such mutations alter the developmental fate of a segment in the embryo in a major way. For example, in flies with a mutation in the *Antennapedia* gene, legs develop in place of antennae **(Figure 16.18).** Such developmental fate changes are called **homeotic transformations.**

Hox genes are present in all major animal phyla. The genes encode a family of transcription factors that regulate developmental events along the anterior–posterior axis of the animals. Each *Hox* gene has a highly conserved 180-bp region called the **homeobox** which corresponds to a 60-amino-acid

section of the encoded transcription factor called the **homeodomain.** The homeodomain of each protein binds to regulatory sequences of the target genes whose transcription it regulates.

Hox genes are one class of **homeobox genes,** so called because they all contain the highly conserved homeobox sequence. Collectively, the several classes of homeobox genes regulate the formation of many body structures during early embryonic development. Humans, for example, have over 230 homeobox genes.

The genomic organization of *Hox* genes is varied and complex among animals. In *Drosophila*, there is a single cluster of eight *Hox* genes and, interestingly, they are organized along a chromosome in the same order as they are expressed along the anterior–posterior body axis **(Figure 16.19).** In vertebrates typically there are several *Hox* gene clusters; the mouse *Hox* gene organization is shown in the figure.

Hox genes are silent during the initial stages of development and then, as development proceeds, the genes become activated in a carefully controlled way. Specifically, individual *Hox* genes bring about their specific effects on the development

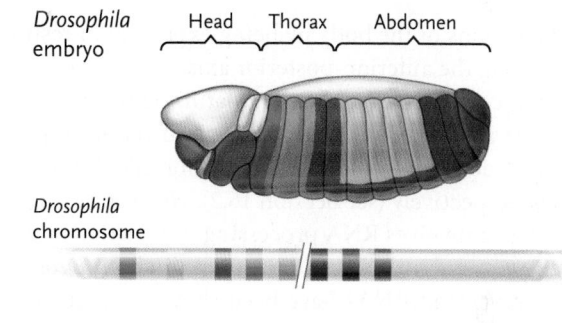

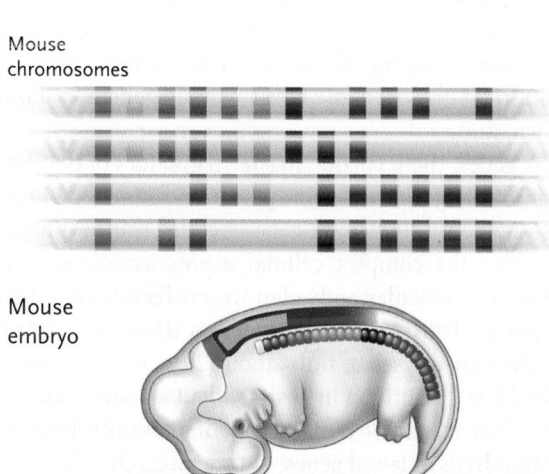

FIGURE 16.19 The *Hox* genes of the fruit fly and the corresponding regions of the embryo they affect. The mouse has four sets of *Hox* genes on four different chromosomes. Their relationship to the fruit fly genes is shown by the colors.
© Cengage Learning 2017

Have a Heart: Critical role for a long noncoding RNA in cardiac development

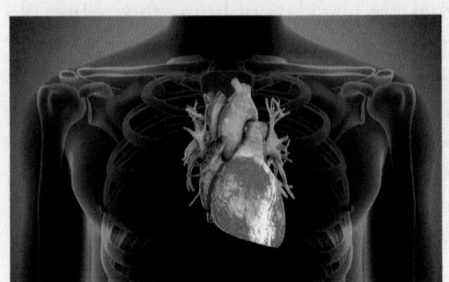

FIGURE Human heart anatomy.

In mammals, the development of the heart **(Figure)** involves the differentiation of embryonic stem cells (ESCs) in the embryo into cardiac cell types. (Stem cells are defined in Chapter 18.) The development of the cardiovascular system, including the heart, is known to involve multiple steps controlled by a network of transcription factors. Recent research has shown that long noncoding RNAs (lncRNAs) also may be involved in the regulation of processes that commit cells to particular developmental fates.

Research Question
Are lncRNAs involved in cardiac development in mammals?

Experiments
Researchers at MIT answered the question using mouse cells. Using genomics technologies, they analyzed lncRNA expression in ESCs and in differentiated tissues. One lncRNA they called *Braveheart (Bvht)* was expressed to a high level in ESCs and in differentiating heart cells. Experimentally they depleted *Bvht* levels in ESCs and observed that the cells did not differentiate into any of the three major types of heart cells of the cardiovascular system. The scientists then demonstrated that *Bvht* controls *MesP1*, a gene known to be a master regulator of cardiac cell differentiation in vertebrates. In normal cells, the *MesP1*-encoded protein initiates a cascade of hundreds of genes required for heart development. Without *Bvht*, heart development is blocked because *MesP1* is not expressed.

Conclusion
The lncRNA *Braveheart* plays a critical role in cardiac development in the mouse.

think like a scientist
What experimental questions might you ask next?

Source: C. A. Klattenhoff et al. 2013. *Braveheart,* a long noncoding RNA required for cardiovascular lineage commitment. *Cell* 152:570–583.

---

of specific regions of the body by being expressed in restricted locations along the anterior–posterior axis.

The molecular mechanisms regulating *Hox* gene expression are diverse and complex. Histone modifications typical of inactive and active chromatin are seen for silent and active *Hox* genes, respectively (see Section 16.2). *Hox* gene expression regulation also involves RNA processing, translational regulation, lncRNAs, and miRNAs. For instance, in both *Drosophila* and mice, several miRNAs have been shown to regulate *Hox* gene expression. As an example, inhibition of expression of the miRNA, miR-196, in chicks results in increased expression of *Hox*-encoded proteins (because there is less inhibition of translation of the *Hox* mRNAs), which generates homeotic transformations.

Hox genes, then, encode master regulators of developmental events along the anterior–posterior axis of animals. To understand how a single *Hox* gene-encoded transcription factor can regulate the complex cellular events associated with the generation of particular body plan structures and functions, it is important to identify the target genes for the various *Hox* genes. Before the genomics era, only about 20 candidate *Hox* target genes had been identified in *Drosophila*. Genome-wide analysis has now shown that *Hox*-encoded proteins control the expression of hundreds of target genes. Future research will be directed at how that regulation occurs and how it is coordinated.

Homeobox genes are also found in fungi and plants. For example, many mutations in homeobox genes that cause homeotic transformations in flower development have been identified and analyzed in *Arabidopsis* (see Section 36.5). Homeobox genes are also discussed in the Chapter 23 *Molecular Insights,* and in Sections 23.6 and 32.3.

Noncoding RNAs Play Important Roles in Development

Earlier in the chapter you learned how one type of noncoding RNA, the microRNAs (miRNAs), can regulate gene expression at the translational level. Studies of miRNA gene mutants, and of organisms with defective miRNA synthesis, have shown that miRNAs have critical roles in development, including embryogenesis, the formation of organs, and the development of the germline (the lineage of cells from which gametes are produced). For instance, in *Drosophila,* miRNAs are required for development of both somatic tissues and the germline, and in the maintenance of stem cells in the germline. (Stem cells are cells capable of differentiating into almost any adult cell type.) In zebrafish, miRNAs are essential for development. For example, a knockout of Dicer results in a developmental arrest at 7 to 10 days after fertilization. miRNAs are also involved in brain formation, somitogenesis (generation of somites), and heart development. In mice (and, by extrapolation, other mammals), miRNAs are essential for development. For example, a mouse with a knockout of Dicer dies at 7.5 days of gestation. Dicer is also required for embryonic stem cell differentiation *in vitro* and, therefore, probably *in vivo* also.

MicroRNAs are also important for regulating plant development. An example involves the regulation of plant growth and development by auxins, a class of plant hormones exemplified by indoleacetic acid (IAA) (see Section 37.1). Auxin affects development by modulating the expression of a number of genes that control cell division and cell elongation in specific parts of the plant and at specific stages during a plant's life cycle. The effects of auxin on plant development are controlled by several families of transcription factors. Among other results, researchers have shown that genetically engineered plants that express an altered mRNA for one of those transcription factors that cannot be cleaved as usual by an miRNA have significant growth defects. In another example, *Arabidopsis thaliana* (thale cress) mutants lacking a key, evolutionarily conserved, protein of the miRNA-induced silencing complex (miRISC; see Figure 16.13) exhibit severe developmental defects and are sterile.

You also learned earlier in this chapter about lncRNAs. Studies of lncRNAs are in their infancy compared with those of miRNAs. Nonetheless, the picture that is emerging is that, at least in animals, various lncRNAs are expressed in tissue-specific and developmental stage-specific patterns. It is thought that the lncRNAs may function along with other regulators to control gene expression during development. More research is needed before a clear understanding of their roles in development is in hand. (*Molecular Insights* describes the critical role of a lncRNA in mouse cardiac development.)

In short, regulatory proteins play critical roles in development and differentiation; however, they are but one player, and not the only player. Researchers must now discover how noncoding RNAs such as miRNAs and lncRNAs regulate the expression of protein-coding mRNAs to develop a more complete understanding of the regulatory circuits underlying development, as well as other biological processes.

STUDY BREAK 16.4

1. What are determination and differentiation and, in general, how are they controlled?
2. How do the segmentation genes and *Hox* genes of *Drosophila* differ in function?

16.5 The Genetics and Genomics of Cancer

You learned earlier in the book that the cell division cycle in all eukaryotes is carefully regulated by genes (see Section 10.4 and Figure 10.16). Complex signaling systems involving extracellular and cellular molecules are used in cell division regulation. The extracellular signaling molecules include polypeptide hormones and steroid hormones made in one tissue that influence the growth and division of cells in other tissues. The specific effects of the signaling molecules depend on the presence of receptors for those molecules on or in the cells they target. When a signaling molecule binds to its receptor, the receptor is activated, which triggers a signal transduction pathway in the target cell that brings about a cellular response. (Signaling and signal transduction pathways are discussed in Chapter 9.)

Figure 16.20 presents simplified presentations of the effects of two types of cell division-related signaling molecules on normal cells. Figure 16.20A shows a **growth factor**—a molecule that stimulates cell division of a target cell—binding to a surface receptor and triggering a signal transduction pathway. In the nucleus, the pathway results in a change in gene expression via its effect on transcription factors and a protein (or proteins) that stimulates cell division is produced. Figure 16.20B shows a **growth-inhibiting factor**—a molecule that inhibits cell division of a target cell by binding to a surface receptor and stimulating transcription of a gene for a protein that inhibits cell division. In this case the cellular response is the production of a protein (or proteins) that are inhibitory to cell division.

Cell division control depends on many stimulatory and inhibitory factors. For normal cells, the relationship between gene products that stimulate cell division and gene products that inhibit cell division governs whether the cell remains in a nondividing state or whether it grows and divides. Only when the balance shifts toward stimulatory signals does the cell grow and divide.

Occasionally, differentiated cells of complex multicellular organisms deviate from their normal genetic program and begin to grow and divide, giving rise to tissue masses called **tumors.** Those cells have lost their normal regulatory controls and have reverted partially or completely to an embryonic developmental state, in a process called **dedifferentiation.** If the altered cells stay together in a single mass, the tumor is *benign.* **Benign tumors** are not invasive, meaning that they cannot spread to other tissues. Such tumors typically are not life threatening, and their surgical removal generally results in a complete cure.

If the cells of a tumor invade and disrupt surrounding tissues, the tumor is a **malignant tumor** and is called a **cancer.** Sometimes, cells from malignant tumors break off and move through the blood system or lymphatic system, forming new tumors at other locations in the body. The spreading of a malignant tumor is called **metastasis** (meaning "change of state"). Malignant tumors can result in debilitation and death in various ways, including damage to critical organs, metabolic problems, hemorrhage, and secondary malignancies. In some cases, malignant tumors can be eliminated from the body by surgery or be destroyed by chemicals (*chemotherapy*) or radiation.

Cancers Are Genetic Diseases

Experimental evidence of various kinds shows that cancers are genetic diseases. That evidence includes:

1. Particular cancers can have a high incidence in some human families. Cancers that run in families are known as **familial (hereditary) cancers.** Cancers that are not inherited are known as **sporadic (nonhereditary) cancers.** Familial cancers are less frequent than sporadic cancers.
2. Descendants of cancer cells are all cancer cells. It is the cloned descendants of a cancer cell that form a tumor.

A. Stimulation of cell division by a growth factor

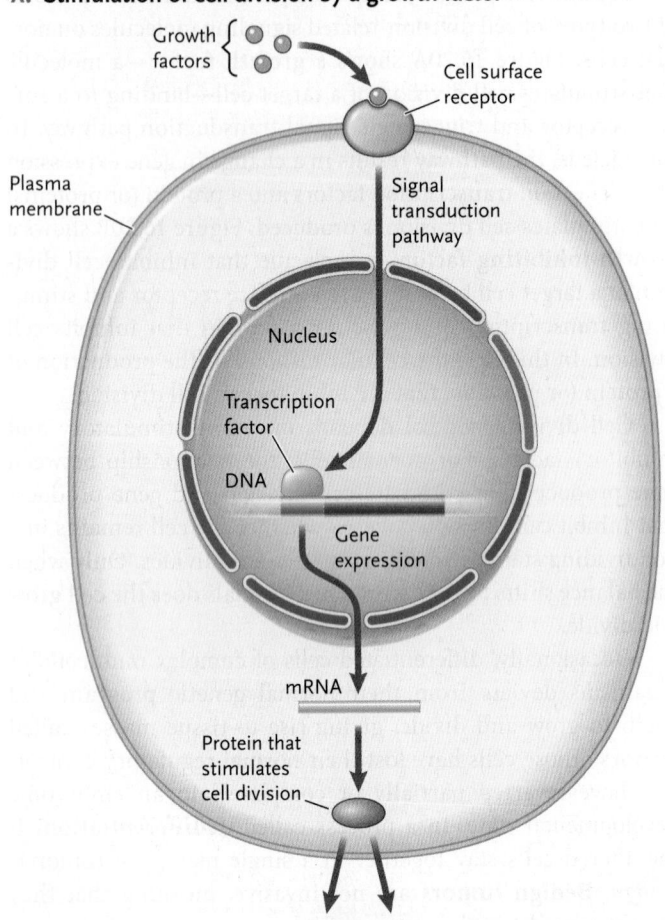

B. Inhibition of cell division by a growth-inhibiting factor

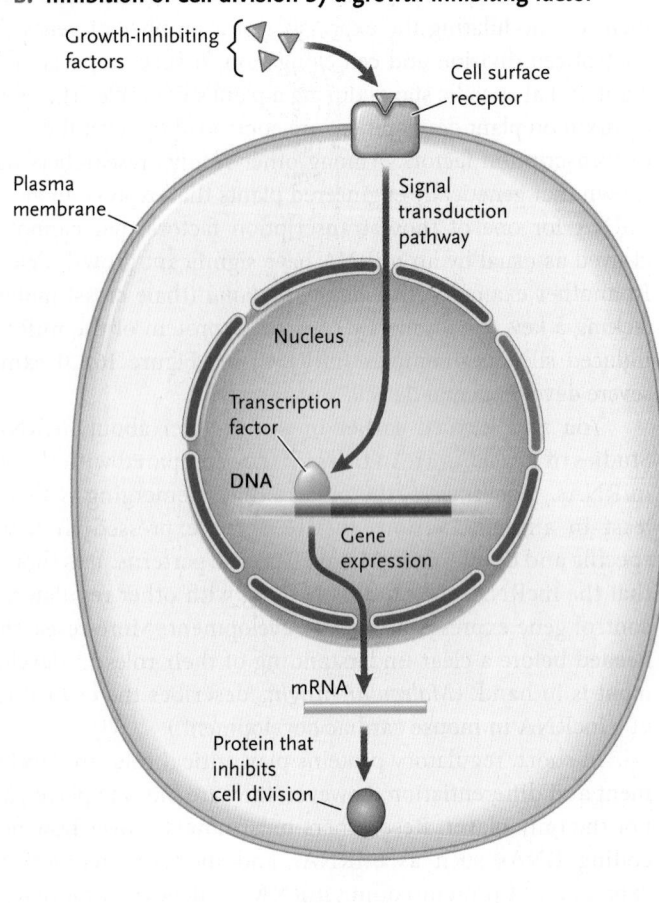

FIGURE 16.20 Stimulation of cell division by a growth factor (A) and inhibition of cell division by a growth-inhibiting factor (B) in a normal cell via signal transduction pathways.
© Cengage Learning 2017

3. The incidence of cancers increases on exposure to mutagens, agents that cause mutations in DNA. Particular chemicals and certain kinds of radiation are effective mutagens.

4. Some viruses can induce cancer. They do so by the expression of viral genes in the host, which disrupts normal cell cycle control.

All the characteristics of cancer cells that have been mentioned—dedifferentiation, uncontrolled division, and metastasis—reflect changes in gene activity.

Genetic and Genomic Approaches Have Identified Many Genes That Cause Cancer

Over several decades researchers have studied cancer cells to identify genes that cause the disease. Originally, conventional molecular genetics techniques were used, but in the past decade genome sequencing has been used for a more comprehensive analysis of genetic changes in cancer cells. The latter is part of **cancer genomics,** the study of the cancer genome to identify

genetic changes that cause a cell to become a cancer, and to distinguish one type or subtype of cancer from another. The rapid advances in cancer genomics are informing research to develop treatments for cancer patients.

DRIVER GENES, ONCOGENES, AND TUMOR SUPPRESSOR GENES A normal cell becomes a tumor cell when a particular mutation provides a selective growth advantage over surrounding cells. That growth advantage typically is very small, but enough for the original mutated cell to grow and divide and produce a clone—the benign tumor. Benign tumors can then progress to being malignant by the accumulation over time of a series of mutations that increase the selective growth advantage of the cells. Even though, like the original mutation, each subsequent mutation provides only a small selective growth advantage, collectively and over many years, the results can be a tumor mass containing billions of cells.

A mutation that confers a selective growth advantage to the cell in which it occurs is called a **driver mutation**—it "drives" tumor formation. A gene that contains a driver mutation or that is expressed abnormally so as to confer a selective growth advantage is called a **driver gene.** In a paper published in 2014, 168 driver genes were identified in a study in which the genomes of thousands of human tumors were sequenced. Several of the genes were already known from nongenomics experiments. A typical tumor contains between two and eight mutations in driver genes, although some cancer types can have many more driver gene mutations. A typical tumor also contains many other mutations, often more than 100, in other genes that do not have any effect on cancer progression. These mutations are called **passenger mutations.** Most tumors also have dozens of translocations (movements of chromosome segments to other locations in the genome; see Section 13.3), most of which are also passengers rather than drivers of tumor formation.

The identified driver genes fall into two major groups based on the role of the normal genes in cell growth and division. Approximately 40% of the genes are *oncogenes,* and the others are *tumor suppressor genes.* An **oncogene** (*onkos* = bulk or mass; cancer formation is also called *oncogenesis*) is a gene that, when *activated* by a mutation or when altered to increase its expression, has a selective growth advantage. Alterations to increase expression of the gene include amplification of the number of copies of the gene, a mutation in the gene's promoter or other regulatory sequence that increases transcription rate, or a translocation event that moves the gene so that it is now controlled by a stronger promoter or enhancer (see Figure 13.12).

An unmutated version of an oncogene in a normal cell is called a **proto-oncogene.** The products of proto-oncogenes stimulate growth and cell division; examples are growth factors and receptors on target cells that are activated by growth factors (see Figure 16.20A). From a genetics perspective, oncogenes are dominant because the mutant form of the gene determines the phenotype; the proto-oncogene allele of the oncogene is recessive.

A **tumor suppressor gene** is a gene that, when *inactivated* by mutation, results in a selective growth advantage of the cell. The normal alleles of tumor suppressor genes encode, for example, growth-inhibiting factors—proteins that inhibit cell growth and division (see Figure 16.20B). Tumor suppressor gene mutations are recessive; that is, both alleles of a tumor suppressor gene must be inactivated for inhibitory activity of the gene's product to be lost in cancer cells. **Figure 16.21** illustrates inactivation of the tumor suppressor gene *BRCA1 (breast cancer 1)* in sporadic and familial forms of breast cancer. Inactivating both alleles of *BRCA1* is not by itself sufficient for the development of breast cancer, but is one of the gene changes typically involved. Since sporadic breast cancer requires the mutational inactivation of two normal alleles of *BRCA1,* this form of the disease typically occurs later in life than the familial form. For familial breast cancer and other familial cancers, the term *predisposition* for the cancer is used. This term relates to the inactivation mechanism just described. That is, individuals are predisposed to develop a particular cancer if they inherit one mutant allele of an associated tumor suppressor disease because then a mutation inactivating the other allele is all that is needed to lose the growth inhibitory properties of the tumor suppressor gene's product.

A. Sporadic breast cancer
Two independent mutations of the *BRCA1* tumor suppressor.

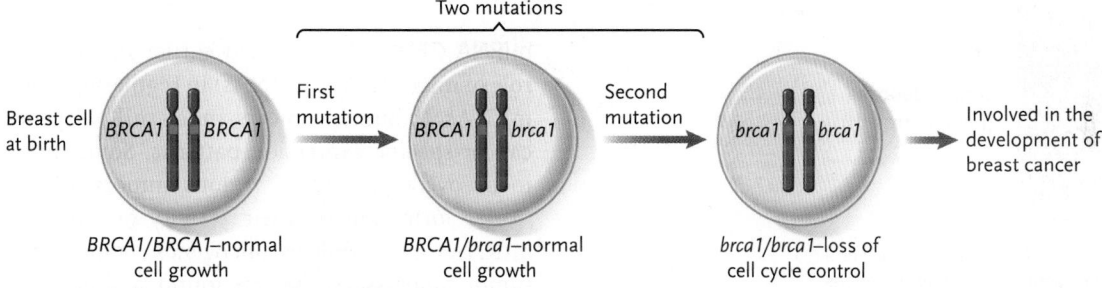

B. Familial breast cancer
An individual has a predisposition for breast cancer because of inheriting one mutated *brca1* allele; mutation of the other normal *BRCA1* allele then occurs.

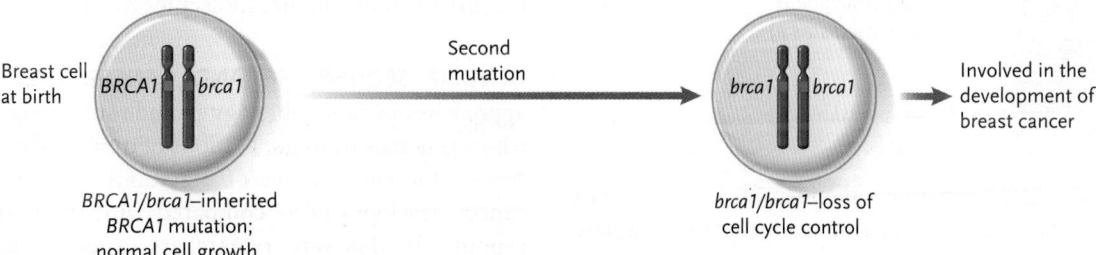

FIGURE 16.21 Mutational inactivation of tumor suppressor gene alleles in sporadic (A) and familial (B) cancers as exemplified by the *BRCA1* gene associated with breast cancer.

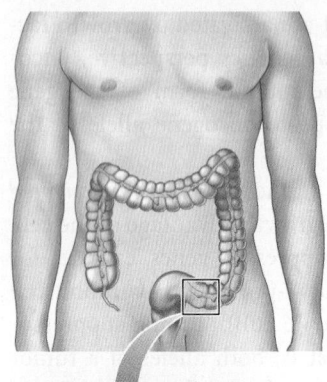

FIGURE 16.22 A multistep model for the development of a type of colorectal cancer.
© Cengage Learning 2017

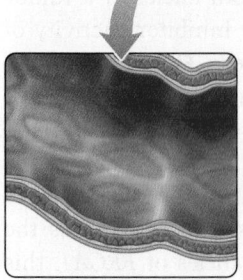

Normal colon cells

↓ Loss of the *APC* tumor suppressor gene activity, and other DNA changes

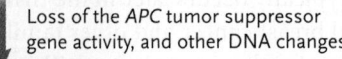

Small adenoma (benign growth)

↓ *ras* oncogene activation; loss of *DCC* tumor suppressor gene

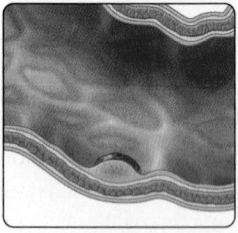

Large adenoma (benign growth)

↓ Loss of *TP53* tumor suppressor gene activity and other mutations

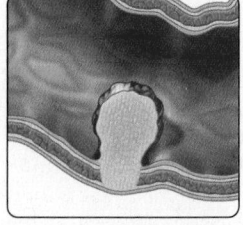

Carcinoma (malignant tumor with metastasis)

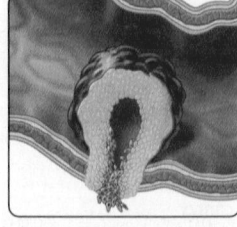

In terms of specific functions, the identified driver genes fall into one or more of twelve signaling pathways that regulate three core cellular processes:

- *Cell survival:* Cell survival genes include genes that promote cell division such as growth factors, the genes for

which in normal cells are proto-oncogenes. They also include genes that inhibit the normal death of damaged cells by apoptosis (programmed cell death; see Chapters 45 and 50); mutations of these genes allow damaged—mutated—cells to continue dividing.

- *Cell fate:* Cell fate genes determine how a cell will differentiate. Typically, differentiated cells do not grow and divide. Mutations in particular driver genes can perturb differentiation and allow continued cell division.

- *Genome maintenance:* Genome maintenance genes maintain the integrity of the genome. They include genes for the components of the DNA repair systems (see Section 14.4). An example is the tumor suppressor gene, *TP53,* so called because its encoded protein, p53, has a molecular weight of 53,000 daltons. The p53 protein is a transcription factor that turns on the expression of cell division inhibiting proteins. While p53 activity is important as part of the checks and balances involved in cell division of normal cells, it is also important if the cell has sustained genomic DNA damage. In normal cells, p53, along with other tumor suppressor gene-encoded proteins, arrests the cell cycle to give the cell time to repair the damage, or, if the damage cannot be repaired, to trigger the cell to undergo apoptosis. However, if both copies of the *TP53* gene are mutated so that the p53 protein is not produced or is produced in an inactive form, this may allow a damaged cell to continue through the cell cycle, passing its mutations on to progeny cells. The importance of p53 to cell division control is shown by the fact that inactive *TP53* genes are found in at least 50% of all cancers.

Gaining a better understanding of the twelve pathways is currently a major goal of cancer research.

miRNA GENES You learned earlier in this chapter about the role of microRNAs (miRNAs) in regulating expression of target mRNAs. In human cancers, many miRNA genes show altered, cancer-specific expression patterns. Some miRNAs are found to be overexpressed in various tumors, directly stimulating tumor formation by their activity on target mRNAs. These miRNAs are acting as oncogenes. Some other miRNAs are tumor suppressors; they are found at abnormally low levels or are absent in tumors. Rather than having defects in individual miRNAs, many tumors have deficiencies in many miRNAs, caused by the cell having diminished ability to process miRNA precursors to the mature molecules (see Figure 16.13).

GENOME SEQUENCING AND CANCER THERAPY Genomic approaches in cancer biology include surveying systematically whole genomes from populations of tumors of one type so as to catalog the entire range of mutations that have occurred in cancer development as compared with the host's unmutated genome. In this way, researchers expect to identify as-yet-undiscovered driver gene mutations involved in particular cancers. As scientists learn more about the functions of those genes and how mutations in them contribute to tumor progression, the paths to developing targeted therapies for particular cancers may become clearer. More broadly, research on the genes associated with cancer identified through genomics and other approaches will provide insights into the molecular genetics mechanisms of carcinogenesis and metastasis in general, and for specific cancer types. These research efforts should have future implications for cancer treatment.

Cancer Develops Gradually by Multiple Steps

As you learned earlier, most tumors contain several mutations in driver genes. That is, cancer rarely develops by alteration of, say, a single proto-oncogene to an oncogene, or inactivation of the two alleles of a single tumor-suppressor gene. Rather, in almost all cancers, successive alterations in several to many genes gradually accumulate over time to transform normal cells to cancer cells. This gradual mechanism is called the *multistep progression of cancer*. **Figure 16.22** shows one example of the steps that can occur, in this case for a form of colorectal cancer.

In the next chapter, you will learn about the molecular genetics of bacteria and their phages, and about DNA sequences in prokaryotic and eukaryotic genomes that have the ability to move to different chromosomal locations.

STUDY BREAK 16.5

1. What is a driver mutation?
2. What is the normal function of a tumor-suppressor gene? How do mutations in tumor-suppressor genes contribute to the onset of cancer?
3. What is the normal function of a proto-oncogene? How can mutations in proto-oncogenes contribute to the onset of cancer?
4. How can changes in expression of miRNA genes contribute to the onset of cancer?

THINK OUTSIDE THE BOOK

Individually or collaboratively, use the Internet and/or research papers to find two examples of how an miRNA potentially may be an effective therapeutic molecule to treat cancer. For the examples, outline the natures of the cancers and the state of the research on the particular miRNAs under investigation.

Unanswered Questions

Can RNA interference silence disease?

RNA interference (RNAi) is the process in which a single-stranded RNA such as small interfering RNA (siRNA) of about 21 to 22 nucleotides in length associated with a protein complex (RISC) inhibits gene expression by combining with an mRNA containing a totally or partially complementary sequence. The siRNA inhibits the mRNA by site-specific cleavage, enhanced degradation, or inducing a block to its translation. RNAi can be induced experimentally in cells by delivering either a synthetic double-stranded siRNA, or a DNA transcriptional template, which is transcribed to produce short-hairpin RNAs that are then converted into the mature siRNAs. Each of these approaches has advantages and disadvantages. Physicians and scientists are working together to develop RNAi for many clinical applications. Let us consider some examples of RNAi therapies that are in the works.

What are examples of diseases that are amenable to RNAi-based therapies?

Anti-viral treatment. Viruses are essentially exogenous genes that enter an organism and through the infection process can cause disease. Can RNAi treat viral infections? Our laboratory has had a long-standing interest in using RNAi to inhibit virus genes in order to short-circuit hepatitis virus infection. After we were able to establish that RNAi worked in whole mammals, we were successful in knocking down both hepatitis virus C gene sequences and hepatitis virus B replication in mice. Currently, scientists are using RNAi approaches to inactivate all types of viruses, including HIV, by directly attacking the viral genes and/or key cell proteins (for example, viral receptor) required for viral infection and spread. A number of different treatments for viral infections are in various phases of testing in animals and humans.

A rare (orphan) disease leads the way. The most advanced clinical trial is for treating a very rare disease. Amyloidosis caused by a mutation in the *transythyretin* gene results when the mutant protein made in the liver and secreted into the blood accumulates in many organ systems, disrupting their function. This can cause many clinical problems, especially when the nervous system is affected. RNAi delivery to the liver in humans has resulted in a 90% reduction in the protein. What is still needed with additional clinical trial data is the demonstration that the reduction in protein with continued administration of the drug will provide clinically meaningful benefits to patients who suffer from this disease, for which there are no other options.

Other disease targets. What still lags behind is very efficient and clinically relevant RNAi delivery into other organs. Cancer is an important target because it is now well established that various genes that are misregulated in cancer when reduced with RNAi will kill the tumor cells in the laboratory and animals. Early clinical studies are underway. Diseases that affect the central nervous system such as Huntington disease are another active pursuit of scientists and physicians. In this case, it may be possible to perform a one-time delivery of a gene therapy vector that makes the RNAi precursor to knock down the expression of the mutant Huntington protein that causes horrific neurological consequences in patients.

How long will it take before there are approved RNAi drugs?

It generally takes 10 to 15 years from the time a molecule is considered as a possible drug to its approval for use. With RNAi we are in the middle of that range, and it is possible that the first approval will occur in the next several years. What is clear is that over the last decade many improvements to the technology have resulted in an increase in the efficacy and decrease in toxicity responses, at least for some applications. These improvements come after human responses from small-scale clinical trials, as well as new information based on very basic molecular mechanistic studies on how RNAi functions in cells are collected. This exemplifies the need for both basic and applied research. Interestingly, the basic research by scientists has resulted in all sorts of new classes of small and very large noncoding RNAs. In fact, scientists have found that over 90% of the human genome is transcribed into RNA. This field is sometimes referred to as the dark matter of the genome. What are the functions of these RNAs? The biological importance of these noncoding RNAs is an intense area of fundamental research. While discovery is ongoing, the next intriguing question is whether these RNAs can be harnessed or manipulated in a manner similar to RNAi for therapeutic benefit.

Courtesy of Mark A. Kay, M.D., PhD

think like a scientist Researchers proceed cautiously with trials of RNAi disease therapies in animals before jumping into human trials. If you were working in Dr. Kay's laboratory, what would be your strategy to test for dangerous side effects of the drugs?

Mark A. Kay is the Dennis Farrey Family Professor in the Departments of Pediatrics and Genetics at the Stanford University School of Medicine. One major focus of his work is the role of small RNAs in mammalian gene regulation and their manipulation to produce new therapeutics. To learn more about his research, go to http://kaylab.stanford.edu.

REVIEW KEY CONCEPTS

For access to MindTap and additional study materials visit www.cengagebrain.com.

16.1 Regulation of Gene Expression in Prokaryotes

- Transcriptional control in prokaryotes involves short-term changes that turn specific genes on or off in response to changes in environmental conditions. The changes in gene activity are controlled by regulatory proteins that recognize operators of operons (Figure 16.1).

- Regulatory proteins may be repressors, which slow the rate of transcription of operons, or activators, which increase the rate of transcription.

- Some repressors are made in an active form, in which they bind to the operator of an operon and inhibit its transcription. Combination with an inducer blocks the activity of the repressor and allows the operon to be transcribed (Figure 16.2).

- Other repressors are made in an inactive form, in which they are unable to inhibit transcription of an operon unless they combine with a corepressor (Figure 16.3).

- Activators typically are made in inactive form, in which they cannot bind to their binding site next to an operon. Combining with another molecule, often a nucleotide, converts the activator into the form in which it binds with its binding site and recruits RNA polymerase, thereby stimulating transcription of the operon (Figure 16.4).

16.2 Regulation of Transcription in Eukaryotes

- Operons are not found in eukaryotes. Instead, genes that encode proteins with related functions typically are scattered through the genome, while being regulated in a coordinated manner.

- Two general types of gene regulation occur in eukaryotes. Short-term regulation involves relatively rapid changes in gene expression in response to changes in environmental or physiological conditions. Long-term regulation involves changes in gene expression that are associated with the development and differentiation of an organism.

- Gene expression in eukaryotes is regulated at the transcriptional level (where most regulation occurs) and at posttranscriptional, translational, and posttranslational levels (Figure 16.5).

- Transcriptionally active genes have a looser chromatin structure than transcriptionally inactive genes. The change in chromatin structure that accompanies the activation of transcription of a gene involves specific histone modifications, as well as chromatin remodeling, particularly in the region of a gene's

promoter (Figure 16.6). Chromatin remodeling is an epigenetic phenomenon.

- Methylation of cytosines in CG sequences of a promoter—an epigenetic phenomenon—is associated with silencing of transcription of the gene. Active genes have unmethylated promoter CG sequences. Methylation in the body of a gene is associated with elevated expression of the gene.

- A eukaryotic protein-coding gene contains a single transcription unit. Regulatory sequences are located upstream of the promoter and, for some genes, at a more distant site (Figure 16.7).

- Regulatory events at a gene's promoter and regulatory sequences control transcription initiation and involve the binding of an array of proteins. At the promoter, general transcription factors bind and recruit RNA polymerase II, giving a very low level of transcription. Activator proteins bind to promoter proximal elements and increase the rate of transcription. Other activators bind to the enhancer if one is present and, through interaction with a coactivator, which also binds to the proteins at the promoter, greatly stimulate the rate of transcription (Figures 16.8 and 16.9).

- The overall control of transcription of a gene depends on the particular regulatory transcription factors that bind to the regulatory sequences of genes. The regulatory transcription factors are cell-type specific and may be activators or repressors. This gene regulation is achieved by a relatively low number of regulatory proteins, acting in various combinations (Figure 16.10).

- The coordinate expression of genes with related functions is achieved by each of the related genes having the same regulatory sequences associated with them.

- Long noncoding RNAs (lncRNAs) can interfere with transcription in various ways, including interacting directly with RNA polymerase II, and bringing about methylation of a promoter, silencing the gene.

- The human genome contains far more promoters and enhancers than the presently known genes.

16.3 Posttranscriptional, Translational, and Posttranslational Regulation

- Posttranscriptional, translational, and posttranslational controls operate primarily to regulate the quantities of proteins synthesized in cells (Figure 16.5).

- Posttranscriptional controls regulate pre-mRNA processing, mRNA availability for translation, and the rate at which mRNAs are degraded. In alternative splicing, different mRNAs are derived from the same pre-mRNA. In another process, small single-stranded RNAs complexed with proteins bind to mRNAs that have complementary sequences, and either the mRNA is cleaved or translation is blocked (Figure 16.12).

- Translational regulation controls the rate at which mRNAs are used by ribosomes in protein synthesis.

- Posttranslational controls regulate the availability of functional proteins. Mechanisms of regulation include the alteration of protein activity by chemical modification, protein activation by processing of inactive precursors, and affecting the rate of degradation of a protein (Figure 16.13).

16.4 Genetic and Molecular Regulation of Development

- Development proceeds as a result of cell division, cell movements, induction, determination, and differentiation.

- Developmental information is stored in both the nucleus and cytoplasm of the fertilized egg. The mRNA and protein molecules that direct the first stages of development are the cytoplasmic determinants.

- In determination, the developmental fate of a cell is set. The major process responsible for determination is induction, which results from the effects of signaling molecules of inducing cells on responding cells.

- In differentiation, cells change from embryonic form to specialized types with distinct structures and functions. Differentiation produces specialized cells without loss of genes from the genome.

- Determination and differentiation both involve regulated changes in gene expression (Figure 16.14).

- Pattern formation derives from the positions of cells in the embryo. Typically, positional information is detected by the cells in the form of concentration gradients of regulatory molecules encoded by genes (Figures 16.15–16.17).

- *Hox* genes—a class of homeobox genes—are evolutionary conserved master regulatory genes that, during embryonic development, control the development of structures of the body along the anterior–posterior axis (Figures 16.18 and 16.19).

- Both miRNAs and lncRNAs play important roles in development.

16.5 The Genetics and Genomics of Cancer

- Complex signaling systems are involved in cell division regulation. Growth factors bind to cellular receptors of target cells and result in the production of proteins that stimulate cell division. Growth-inhibiting factors bind to cellular receptors of target cells and result in the production of proteins that inhibit cell division (Figure 16.20)

- In cancer, cells partially or completely dedifferentiate, divide rapidly and uncontrollably, and may break loose to form additional tumors in other parts of the body.

- Cancers are genetic diseases. Cancers that run in families are familial (hereditary) cancers; cancers that are not inherited are sporadic (nonhereditary) cancers.

- Many genes cause cancer. A normal cell becomes a tumor cell when a driver mutation provides a selective growth advantage to the cell. A gene with a driver mutation is a driver gene. Two major groups of driver genes are oncogenes (genes that, when activated by a mutation or when its expression is increased, result in a selective growth advantage for the cell), and tumor suppressor genes (genes that, when inactivated by mutation, result in a selective advantage for the cell) (Figure 16.21).

- The known driver genes fall into one or more of twelve signaling pathway that regulate three core cellular processes: cell survival, cell fate, and genome maintenance.

- In human cancers, many miRNAs exhibit altered, cancer-specific expression patterns. Some miRNAs act as oncogenes; other act as tumor suppressor genes.

- Most cancers develop by multistep progression involving the successive alteration of several to many genes (Figure 16.22).

TEST YOUR KNOWLEDGE

Remember/Understand

1. The control of the delivery of finished mRNAs to the cytoplasm is an example of:
   a. translational regulation.
   b. posttranslational regulation.
   c. transcriptional regulation.
   d. posttranscriptional regulation.
   e. deoxyribonucleic regulation.

2. For the *E. coli lac* operon, when lactose is present:
   a. and glucose is absent, cAMP binds and activates catabolic activator protein (CAP).
   b. and glucose is absent, the level of cAMP decreases.
   c. activated CAP binds the repressor protein to remove it from the operator gene.
   d. the cell prefers lactose over glucose.
   e. RNA polymerase cannot bind to the promoter.

3. For the *trp* operon:
   a. tryptophan is an inducer.
   b. when tryptophan binds to the Trp repressor, transcription is blocked.
   c. Trp repressor is synthesized in an active form.
   d. low levels of tryptophan bind to the *trp* operator and block transcription of the tryptophan biosynthesis genes.
   e. high levels of tryptophan activate RNA polymerase and induce transcription.

4. Transcriptional regulation is important because it:
   a. is the final and most important step in gene regulation.
   b. determines the availability of finished proteins.
   c. determines which genes are expressed.
   d. determines the rate at which proteins are made.
   e. removes masking proteins that block initiation of transcription.

5. Activation of a eukaryotic gene typically involves:
   a. release of transcription factors from the gene's promoter.
   b. acetylation of histone tails of nucleosomes in the region of the promoter.
   c. genomic imprinting.
   d. DNA methylation in the region of the promoter.
   e. binding of miRNAs to the DNA sequence of the promoter.

6. Which statement about activation of transcription is *not* correct?
   a. A transcription factor binds to the promoter in the area of the TATA box.
   b. A coactivator forms a bridge between the promoter and the gene to be transcribed.
   c. Transcription factors bind the promoter and RNA polymerase.
   d. Activators bind to the enhancer region on DNA.
   e. RNA is transcribed downstream from the promoter region.

7. Which of the following statements does *not* describe miRNA?
   a. miRNA is encoded by non-protein-coding genes.
   b. miRNA has a precursor that is folded and then cut by a Dicer enzyme.
   c. miRNA is an example of a molecule that induces RNA interference or gene silencing.
   d. miRNA is synthesized *in vitro* but probably not *in vivo*.
   e. miRNA has a similar function to that of small interfering RNAs.

8. In mammals, the nose is located at the anterior end of the embryo, and the heart at the center of the embryo. These positions are the result of activation of:
   a. *Hox* genes that are arranged along a number of chromosomes in the same order as they are expressed along the anterior–posterior body axis.
   b. maternal-effect genes, after somites differentiate into muscle.
   c. *Hox* genes that are scattered randomly among different chromosomes.
   d. a transcription factor called the homeobox.
   e. a homeodomain that binds ribosomes.

9. Which of the following is not a characteristic of cancer cells?
   a. proto-oncogenes altered to become oncogenes
   b. the mutation of a suppressor gene that results in normal genes becoming inactive
   c. the mutation of the *TP53* gene
   d. multistep progression
   e. amplification of growth factors and growth factor receptors

Apply/Analyze

10. Normal ears in a certain mammal are perky; mutants have droopy ears. In males of these mammals, the gene encoding perky ears is transcribed only from the female parent. This is because the gene from the male parent is silenced by methylation. If the maternal gene is mutated:
    a. male offspring have droopy ears.
    b. offspring have perky ears.
    c. male offspring have one droopy ear and one perky ear.
    d. the genetic mechanism is called alternative splicing.
    e. this is an example of posttranscriptional regulation.

11. **Discuss Concepts** In a mutant strain of *E. coli,* the CAP protein is unable to combine with its target region of the *lac* operon. How would you expect the mutation to affect transcription when cells of this strain are subjected to the following conditions?

    Lactose and glucose are both available.
    Lactose is available but glucose is not.
    Both lactose and glucose are unavailable.

12. **Apply Evolutionary Thinking** Fruit flies homozygous for a mutation in the tumor suppressor gene *HIPPO* develop tumors in every organ. Expression of the human gene *MST2* in flies homozygous for *HIPPO* show greatly reduced or no tumors. What does this result suggest about the evolution of tumor suppressor genes in animals?

Evaluate/Create

13. **Discuss Concepts** Duchenne muscular dystrophy, an inherited genetic disorder, affects boys almost exclusively. Early in childhood, muscle tissue begins to break down in affected individuals, who typically die in their teens or early twenties as a result of respiratory failure. Muscle samples from women who carry the mutation reveal some regions of degenerating muscle tissue adjacent to other regions that are normal. Develop a hypothesis explaining these observations.

14. **Discuss Concepts** Eukaryotic transcription is generally controlled by the binding of regulatory proteins to DNA sequences rather than by the modification of RNA polymerases. Develop a hypothesis explaining why this is so.

15. **Design an Experiment** Design an experiment using rats as the model organism to test the hypothesis that human chorionic gonadotropin (hCG), a hormone produced during pregnancy, leads to a significant protection against breast cancer.

For selected answers, see Appendix A.

INTERPRET THE DATA

Investigating a correlation between specific cancer-causing mutations and risk of mortality in humans is challenging, in part because each cancer patient is given the best treatment available at the time. There are no "untreated control" cancer patients, and the idea of what treatments are optimal changes quickly as new drugs become available and new discoveries are made.

The **Table** shows the results of a study in which 442 women who had been diagnosed with breast cancer were checked for mutations of the *BRCA1* tumor suppressor gene and of a second tumor suppressor gene involved with breast cancer, *BRCA2*. The treatments and progress of the women were followed over several years. All of the women in the study had at least two affected close relatives, so their risk of developing breast cancer due to an inherited factor was estimated to be greater than that of the general population.

BRCA Mutations in Women Diagnosed with Breast Cancer[1]

	BRCA1 Mutations	*BRCA2* Mutations	No *BRCA* Mutations	Total
Total number of patients	89	35	318.0	442.0
Average age at diagnosis	43.9	46.2	50.4	
Preventive mastectomy[2]	6	3	14	23
Preventive oophorectomy[2]	38	7	22	67
Number of deaths	16	1	21	38
Percent died	18.0	2.8	6.9	8.6

[1]Results from a 2007 study investigating *BRCA* mutations in women diagnosed with breast cancer. All women in the study had a family history of breast cancer.
[2]Some of the women underwent preventive mastectomy (removal of the noncancerous breast) during their course of treatment. Others had preventive oophorectomy (surgical removal of the ovaries).
© Cengage Learning 2017

1. According to this study, what is the woman's risk of dying of cancer if two of her close relatives have breast cancer?
2. What is her risk of dying of cancer if she carries a mutated *BRCA1* gene?
3. Is a *BRCA1* or *BRCA2* mutation more dangerous in breast cancer cases?
4. What other data would you have to see in order to make a conclusion about the effectiveness of preventive surgeries?

Bacterial and Viral Genetics

17

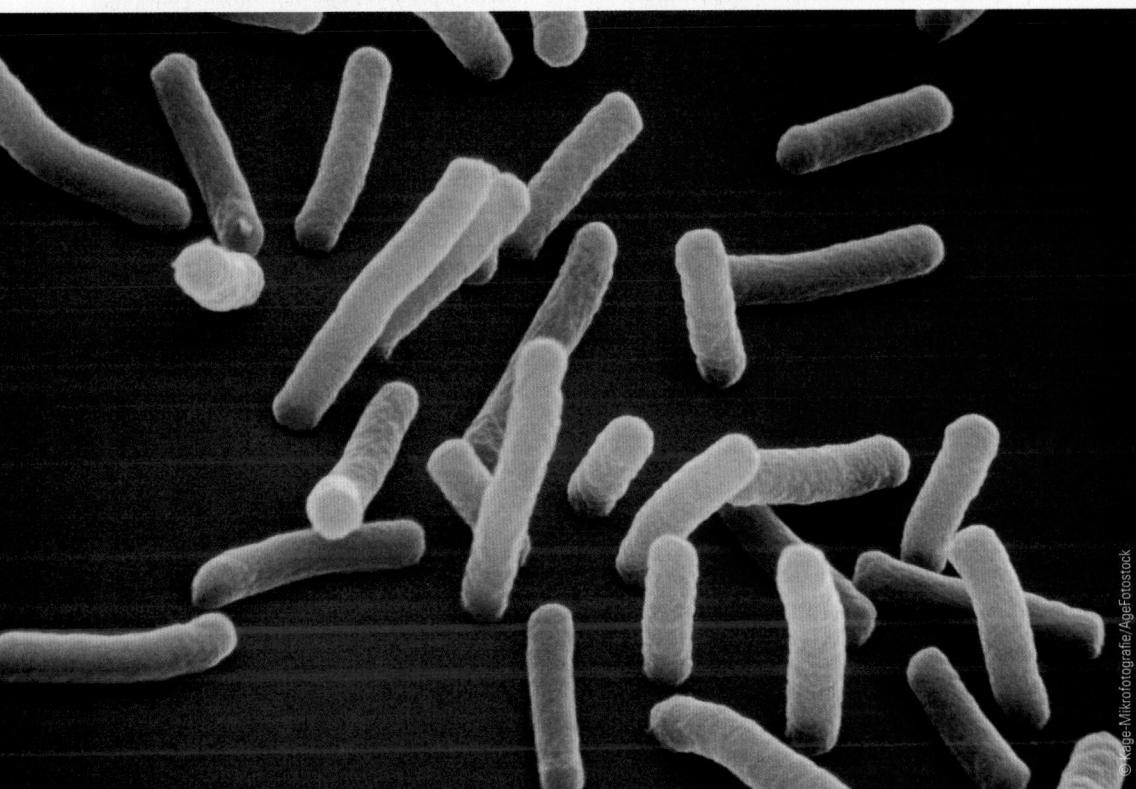

Escherichia coli, a model research organism for several types of biological studies, including bacterial genetics (colorized scanning electron micrograph).

Why it matters . . . In 1885, a German pediatrician, Theodor Escherich, identified a bacterium that caused severe diarrhea in infants. He named it *Bacterium coli*. Researchers were surprised to discover that *B. coli* is also present in healthy infants, and is a normal inhabitant of the human intestine. It was also discovered that only certain strains of *B. coli* cause human diseases.

As an organism that is readily available and easy to grow, this intestinal bacterium has been of central interest to scientists since its first discovery. Renamed *Escherichia coli* in honor of Escherich, the bacterium is an important model organism for scientific investigation. It can be grown quickly in huge numbers in nutrient solutions that are simple to prepare. And it can be infected with a group of viruses called **bacteriophages** (**phages** for short) that have been as valuable to scientists as *E. coli* has because phages can be grown by the billions in cultures of the host bacterium (see Section 14.1). (*Focus on Research: Model Organisms* tells more about *E. coli*'s use in the laboratory.)

Researchers naturally were curious about the genetics of bacteria and their viruses. Because they had constant properties from generation to generation, they must have hereditary systems. Geneticists discovered that *E. coli* can undergo genetic recombination and used the bacterium to analyze genetic crosses and their outcomes much more quickly than they could with eukaryotes. Geneticists also discovered that genetic recombination can occur between phage genomes in an infected bacterium. A number of experiments helped scientists to work out the structure, activity, and recombination of genes at the molecular level. For instance, recall from Section 14.1 that DNA was proved to be the genetic material in Hershey and Chase's experiment involving a bacteriophage infecting *E. coli*.

Focus on Research

Model Organisms: *Escherichia coli*

Escherichia coli is a cylindrically shaped bacterium (see chapter-opening photo). We probably know more about *E. coli* than any other organism. For example, the complete DNA sequence of the genome of a standard laboratory strain of *E. coli* was determined in 1997, identifying approximately 4,100 protein-coding genes (see Chapter 19). However, the functions of about one-third of these genes are still unidentified.

E. coli got its start in laboratory research because of the ease with which it can be grown in culture. *E. coli* cells divide about every 20 minutes under optimal conditions, producing a clone of 1 billion cells in a matter of hours, in only 10 mL of culture medium. The same amount of medium can accommodate as many as 10 billion cells before the growth rate of *E. coli* begins to slow. *E. coli* can be grown with minimal equipment, requiring little more than culture vessels in an incubator that is held at 37°C.

Early on, however, a major advantage of *E. coli* for research was the ability to use it for genetics experiments. The discovery that *E. coli* could conjugate, with DNA passing from one bacterium to the other, made it possible for geneticists to carry out genetic crosses with the bacterium, producing genetic recombinants that could map the positions of genes on the chromosome. The map showed that genes with related functions are clustered together, a fact that had significant implications for the regulation of expression of those genes. For example, François Jacob and Jacques Monod's work with the genes for lactose metabolism led to the pioneering operon model (described in Section 16.1). In their work, they used conjugation to map the genes and generated partial diploids by genetic means to help understand the details of the regulation of transcription of those genes.

The development of *E. coli* as a model organism for studying gene organization and the regulation of gene expression led to the field of molecular genetics. The study of naturally occurring plasmids in *E. coli* and of enzymes that cut DNA at specific sequences eventually resulted in techniques for combining DNA from different sources, such as inserting a gene from an organism into a plasmid. Today *E. coli* is used for amplifying (cloning) plasmids that contain inserted genes or other sequences.

In essence, the biotechnology industry has its foundation in molecular genetics studies of *E. coli*, and large-scale *E. coli* cultures are widely used as "factories" for production of desired proteins. For example, the human insulin hormone, required for treatment of certain forms of diabetes, is produced by *E. coli* factories. (Chapter 18 explains more about cloning and other types of DNA manipulation.)

Laboratory strains of *E. coli* are harmless to humans. Similarly, the natural *E. coli* cells in the colon of humans and other mammals are usually harmless. There are pathogenic strains of *E. coli*, though. Sometimes they make the news when humans who eat food that contains a pathogenic strain develop disease symptoms, notably colitis (inflammation of the colon) and bloody diarrhea. The genomes of several pathogenic *E. coli* strains have been sequenced. Each has more genes than either the laboratory strain or the strain that normally lives in the human colon. The extra genes of the pathogenic strains include the genes that make the bacteria pathogenic.

Bacteria are important to us because many diseases of humans and other organisms are caused by pathogenic strains of bacteria. Viruses are important to us for similar reasons. As a consequence, bacteria and viruses have been and are subjects of research to characterize the genetics and molecular biology of their *pathogenesis* (their mechanism to cause disease). This chapter outlines the basic findings of molecular genetics in bacteria and their viruses. It also describes more broadly the structure and genetic and molecular properties of viruses, viroids, and prions. We begin our discussion with genetic recombination mechanisms in bacteria.

17.1 Gene Transfer and Genetic Recombination in Bacteria

In the first half of the twentieth century, foundational genetic experiments with eukaryotes revealed the processes of genetic recombination during sexual reproduction, which led to the construction of genetic maps of chromosomes for a number of organisms (see Chapters 12 and 13). Bacteria became the subject of genetics research in the middle of the twentieth century. A key early question was whether genetic recombination can

occur in bacteria even though these organisms do not undergo meiosis. For particular bacteria, the answer to the question was yes—genes can be transferred from one bacterium to another by several different mechanisms, and the newly introduced DNA can recombine with DNA already present. Such genetic recombination performs the same function as it does in eukaryotes: it generates genetic variability through the exchange of alleles between homologous regions of DNA molecules from two different individuals. As you learned in Chapter 1, genetic variability is the raw material of natural selection and other processes that cause biological evolution.

E. coli is one of the bacteria in which genetic recombination has been studied. By the 1940s, geneticists knew that *E. coli* and many other bacteria could be grown in a minimal medium containing water, an organic carbon source such as glucose, and a selection of inorganic salts—including one that provides nitrogen—such as ammonium chloride. The growth medium can be in liquid form or in the form of a gel that is made by adding agar to the liquid medium. (Agar is a polysaccharide material, indigestible by most bacteria, that is extracted from red algae.)

Because it is not practical to study a single bacterium for most experiments, researchers soon developed techniques for

FIGURE 17.1 Experimental Research

Genetic Recombination in Bacteria

Question: Does genetic recombination occur in bacteria?

Experiment: To answer the question, Lederberg and Tatum used two mutant strains of *E. coli*: Mutant strain 1's genotype was *bio⁻ met⁻ leu⁺ thr⁺ thi⁺*, where the "+" means a normal allele and the "−" means a mutant allele. This strain required biotin and methionine to grow. Mutant strain 2's genotype was *bio⁺ met⁺ leu⁻ thr⁻ thi⁻*; it required leucine, threonine, and thiamine to grow.

Lederberg and Tatum plated about 100 million cells of a mixture of the two mutant strains on minimal medium, which lacked any of the nutrients the strains needed for growth. As controls, they also plated large numbers of the two mutant strains individually on minimal medium.

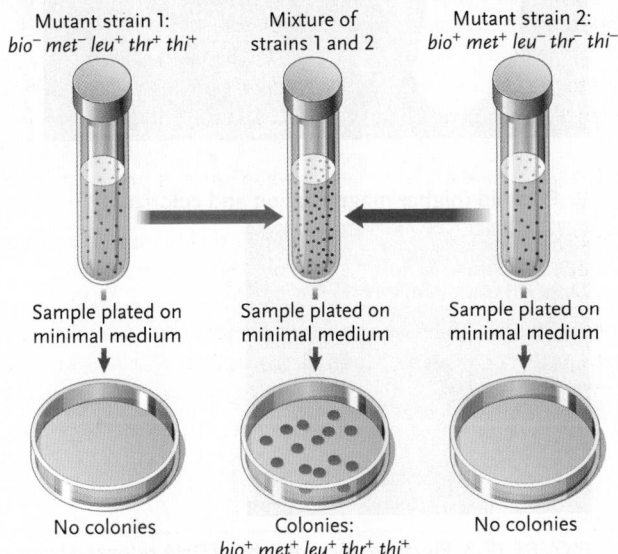

Results: No colonies grew on the control plates, meaning that the mutant alleles in the strains had not mutated back to normal alleles. However, for the mixture of mutant strain 1 and mutant strain 2, several hundred colonies grew on the minimal medium.

Conclusion: To grow on minimal medium, the bacteria must have been able to make biotin, methionine, leucine, threonine, and thiamine, meaning that they had the genotype: *bio⁺ met⁺ leu⁺ thr⁺ thi⁺*. Lederberg and Tatum concluded that the colonies on the plate must have resulted from genetic recombination between mutant strains 1 and 2.

Observe
Hypothesize
Predict
Experiment
Interpret

Suppose mutant strain 1 had the genotype *met⁻ leu⁺* and mutant strain 2 had the genotype *met⁺ leu⁻*. Would these have been better strains to use in the experiment? Explain why or why not.

Source: J. Lederberg and E. Tatum. 1946. Gene recombination in *Escherichia coli. Nature* 158:558.

© Cengage Learning 2017

starting bacterial cultures from a single cell, generating cultures with a large number of genetically identical cells. Cultures of this type are called **clones.** To start bacterial clones, a scientist spreads a drop of a diluted bacterial culture over a sterile agar gel in a culture dish. Each individual cell divides many times to produce a separate colony that is a clone of the

initial cell. Cells can be removed from a clone and introduced into liquid cultures or spread on agar and grown in essentially any quantity.

Genetic Recombination Occurs in *E. coli*

In 1946, Joshua Lederberg and Edward L. Tatum of Yale University performed an experiment to determine if genetic recombination occurs in bacteria **(Figure 17.1).** For this experiment, they used *auxotrophs* of *E. coli*, mutant strains that could not grow on minimal medium (see Section 15.1). When they mixed together cells of two mutant strains, each of which had different requirements for growth on minimal medium, they observed that some cells were able to grow on minimal medium. They interpreted the result to mean that *genetic recombination had occurred between the two strains.* Lederberg received a Nobel Prize in 1958 "for his discoveries concerning genetic recombination and the organization of the genetic material of bacteria."

Bacterial Conjugation Brings DNA of Two Cells Together, Allowing Genetic Recombination to Occur

How does genetic recombination occur in *E. coli*? Recombination in eukaryotes occurs in diploid cells undergoing meiosis by an exchange of segments between the chromatids of homologous chromosome pairs (see Section 13.1). Bacteria typically have a single, circular chromosome— they are haploid organisms. A different genetic recombination mechanism occurs in bacteria, as you will learn in this section.

CONJUGATION AND THE F FACTOR Rather than fusing together to produce the prokaryotic equivalent of a diploid zygote, bacterial cells *conjugate:* they contact each other, initially becoming connected by a long tubular structure on the cell surface called a **sex pilus (Figure 17.2A),** and then forming a cytoplasmic bridge that connects two cells **(Figure 17.2B).** During **conjugation,** DNA of one cell, the *donor* (the bristly cell in Figure 17.2A), moves through the cytoplasmic bridge into the other cell, the *recipient.*

The ability to conjugate depends on the presence within a donor cell of a plasmid called the *F factor* (F = fertility). Plasmids are small circles of DNA that occur in bacteria in addition to the main circular chromosomal DNA molecule **(Figure 17.3).** Plasmids contain several to many genes and a replication origin that permits them to replicate and be passed on during

A. Initial attachment of two cells by sex pilus

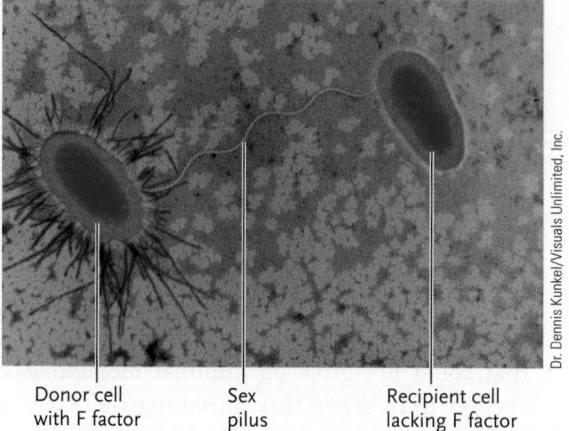

Donor cell with F factor | Sex pilus | Recipient cell lacking F factor

Dr. Dennis Kunkel/Visuals Unlimited, Inc.

B. Cytoplasmic bridge (arrow) formed between the cells through which DNA moves from one cell to the other

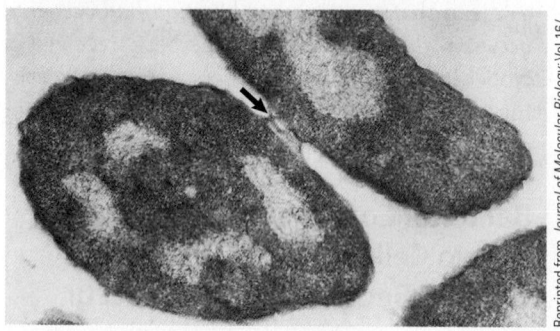

Reprinted from *Journal of Molecular Biology*; Vol 16/Issue 2; Julian D. Gross & Lucien G. Caro; DNA transfer in bacterial conjugation; Page 269; 1966; with permission from Elsevier.

FIGURE 17.2 Conjugating *E. coli* cells.

A. DNA released from bacterial cell showing plasmids (arrows) near the bacterial chromosome

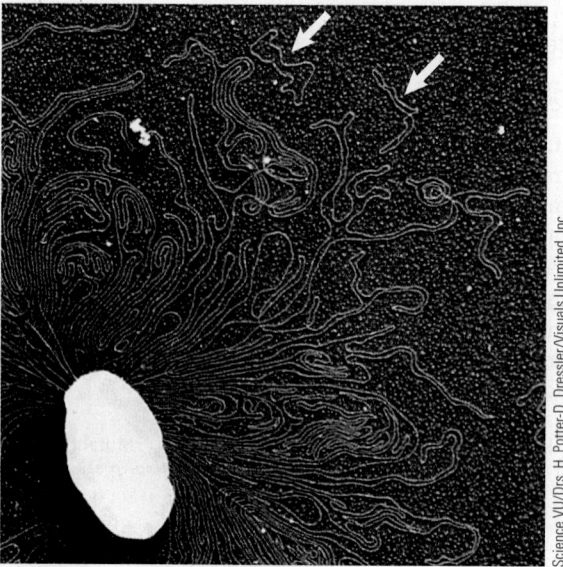

Science VU/Drs. H. Potter-D. Dressler/Visuals Unlimited, Inc.

B. Plasmid (higher magnification and colorized)

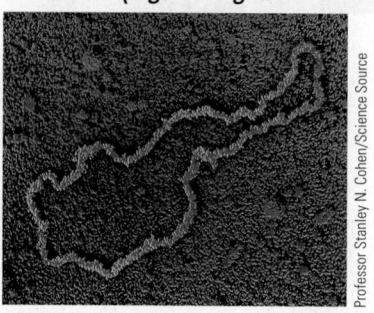

Professor Stanley N. Cohen/Science Source

FIGURE 17.3 Electron micrographs of DNA released from a bacterial cell.

bacterial division. Donor cells in conjugation are called **F⁺ cells** because they contain the F factor. They are able to conjugate (mate) with recipient cells but not with other donor cells. Recipient cells are called **F⁻ cells** because they lack the F factor and are unable to initiate conjugation.

The F factor contains genes that encode proteins of the sex pilus, also called here the **F pilus** (plural, *pili*). The sex pilus allows an F⁺ donor cell to attach to an F⁻ recipient (see Figure 17.2A). Once attached, the donor and recipient cells form a cytoplasmic bridge and conjugate **(Figure 17.4A)**. The outcome of F⁺ × F⁻ conjugation is that the F⁻ cell receives a copy of the F factor and, therefore, becomes an F⁺ cell. No chromosomal DNA is transferred between cells in F⁺ × F⁻ conjugation, however, so *no genetic recombination of bacterial genes occurs*.

Hfr CELLS AND GENETIC RECOMBINATION Genetic recombination of bacterial genes can occur if, in some F⁺ cells, the F factor integrates into the bacterial chromosome, producing a donor that can conjugate with and transfer genes on the bacterial chromosome to a recipient **(Figure 17.4B)**. These special donor cells are **Hfr cells** (Hfr = high-frequency recombination). The outcome of Hfr × F⁻ conjugation is that genetic recombination

of bacterial genes can occur because, as the F factor replicates and part of it enters the recipient cell, attached to it is a segment of replicated donor chromosomal DNA. Therefore, if donor and recipient have different alleles of genes, the recipient becomes a **partial diploid** for the donor DNA segment that has come through the conjugation bridge. Donor and recipient DNA can then pair and genetic recombination can occur. The recipient remains an F⁻ in this case, because the conjugating cells typically break apart long before the second part of the F factor is transferred to the recipient (it is the last DNA segment transferred).

In the example shown in Figure 17.4B, a b^+ recombinant is generated. In other pairs in the mating population, the a^+ gene could recombine with the homologous recipient gene, or both a^+ and b^+ genes could recombine. The genetic recombinants observed in Lederberg and Tatum's experiment (see Figure 17.1) were produced in this same general way, although those researchers did not know of the mechanism at the time.

Recombinants produced during conjugation can be detected only if the alleles of the genes in the DNA transferred from the donor differ from those in the recipient's chromosome. Following

TRANSFER OF GENETIC MATERIAL DURING CONJUGATION BETWEEN *E. coli* CELLS

A. Transfer of the F factor during conjugation between F⁺ and F⁻ cells

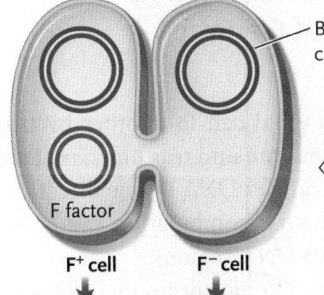

Bacterial chromosome

1 An F⁺ cell conjugates with an F⁻ cell. First, the two cells become attached by the sex pilus of the F⁺ cell. Then, the cells form a cytoplasmic bridge and conjugate.

F factor

F⁺ cell   F⁻ cell

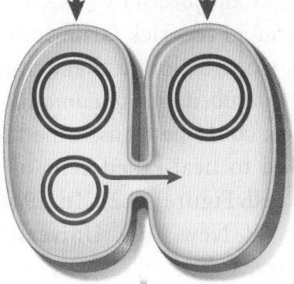

2 In a special type of DNA replication, one strand of the F factor breaks at a specific point and begins to move through the cytoplasmic bridge from the F⁺ (donor) to the F⁻ (recipient) cell as the F factor replicates.

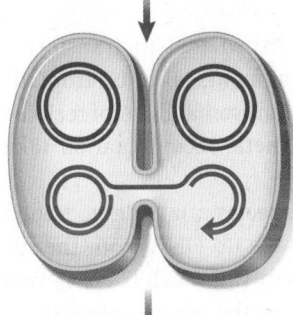

3 DNA replication of the F factor continues in the donor cell, and a complementary strand to the strand entering the recipient cell is synthesized.

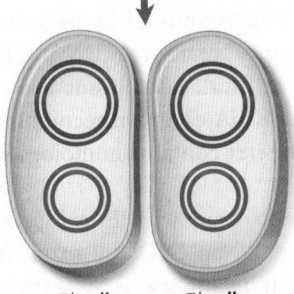

4 When transfer of the F factor is completed, replication has produced a copy of the F factor in both the donor and recipient cells; the recipient has become an F⁺. No chromosomal DNA is transferred in this mating.

F⁺ cell   F⁺ cell

SUMMARY F⁺ *E. coli* bacteria have an F factor free in the cytoplasm that is responsible for conjugation. An F⁺ donor cell can conjugate with an F⁻ recipient cell and transfer a copy of the F factor to the recipient, converting it to an F⁺ cell. No genetic recombination between donor and recipient genes occurs **(A).** Hfr bacteria have the F factor integrated into the chromosome. When an Hfr conjugates with an F⁻ cell, the transfer of DNA from donor to recipient includes a copy of some of the donor chromosomal genes, enabling genetic recombination to occur between donor and recipient genes in the recipient **(B).**

think like a scientist Why does transfer of chromosomal genes not occur from recipient to donor?

© Cengage Learning 2017

B. Transfer of bacterial genes and production of recombinants during conjugation between Hfr and F⁻ cells

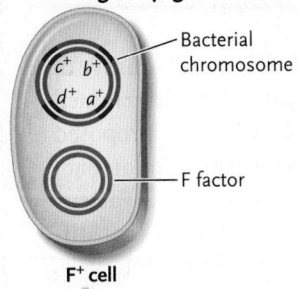

Bacterial chromosome

F factor

F⁺ cell

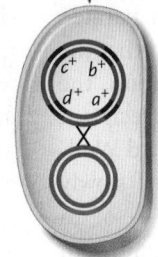

1 F factor integrates into the *E. coli* chromosome in a single crossover event producing an Hfr cell.

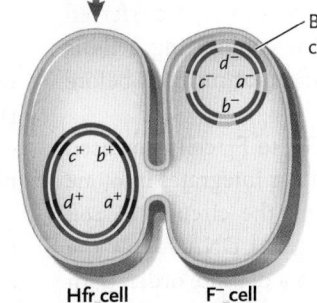

Bacterial chromosome

2 An Hfr cell and an F⁻ cell conjugate. These two cells differ in alleles: the Hfr is $a^+ b^+ c^+ d^+$, and the F⁻ cell is $a^- b^- c^- d^-$.

Hfr cell   F⁻ cell

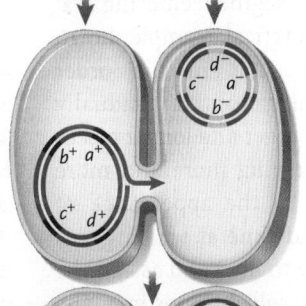

3 As with the F⁺ × F⁻ conjugation, one strand of the F factor breaks at a specific point and begins to move from the Hfr (donor) to the F⁻ (recipient) cell as replication takes place. The breakpoint is in the middle of the integrated F factor.

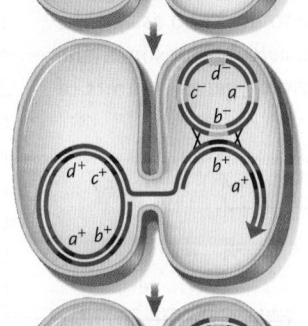

4 In the F⁻ cell the entering single-stranded F factor segment and the attached chromosomal DNA are replicated by synthesis of the complementary DNA strand. The recipient becomes a *partial diploid* for the donor DNA, here $a^+ b^+/a^- b^-$. The recipient's DNA and the homologous DNA segment from the donor can pair and recombine. Genetic recombination occurs by a double crossover event (see eukaryotic genetic recombination in Section 13.1).

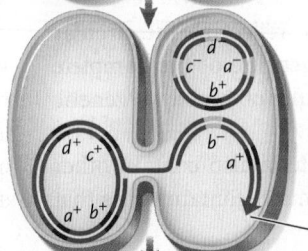

5 Here, a b^+ recombinant is produced by the double crossover of step 4. When the conjugating pair breaks apart, the linear piece of donor DNA is degraded and all descendants of the recipient will be b^+. The recipient remains F⁻ because not all the F factor has been transferred.

Hfr chromosome (part of F factor, followed by bacterial genes)

Conjugation bridge breaks, resulting in two cells. F⁻ is a b^+ recombinant.

recombination, the bacterial DNA replicates and the cell divides normally, producing a cell line with the new gene combination. Any remnants of the DNA fragment that originally entered the cell are degraded as division proceeds and do not contribute further to genetic recombination or cell heredity.

MAPPING GENES BY CONJUGATION The mechanism of genetic recombination in *E. coli* by conjugation was determined in 1957 by two scientists, François Jacob (who also proposed the operon model for the regulation of gene expression in bacteria; see Section 16.1) and Elie L. Wollman, at the Pasteur Institute in Paris. They conjugated Hfr and F⁻ cells that differed in a number of alleles and then, at regular time intervals, they removed some of the cells and agitated them in a blender to break apart attached cells. They then cultured the separated cells and analyzed them for recombinants. They found that the longer they allowed cells to conjugate before separation, the greater the number of donor genes that entered the recipient and produced recombinants. From this result, Jacob and Wollman concluded that, during conjugation, the Hfr cell slowly injects a copy of its DNA into the F⁻ cell. Full transfer of an entire DNA molecule to an F⁻ cell takes 90 to 100 minutes. In nature, however, the entire DNA molecule is rarely transferred because the cytoplasmic bridge between conjugating cells is fragile and easily broken by random molecular motions before transfer is complete.

The pattern of gene transfer from Hfr to F⁻ cells was used to map the *E. coli* chromosome. The F factor integrates into one of a few possible fixed positions around the circular *E. coli* DNA. As a result, the genes of the bacterial DNA follow the F factor segment into the recipient cell in a definite order, with the gene immediately behind the F factor segment entering first and the next genes following. In the theoretical example shown in Figure 17.4B, donor genes will enter in the order $a^+-b^+-c^+-d^+$. By breaking off conjugation at gradually increasing times, investigators allowed longer and longer pieces of DNA to enter the recipient cell, carrying more and more genes from the donor cell (detected by the appearance of recombinants). By noting the order and time at which genes were transferred, investigators were able to map and assign the relative positions of most genes in the *E. coli* chromosome. The resulting genetic map has distances between genes in units of minutes.

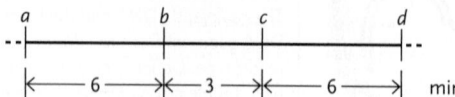

The *E. coli* genetic map is circular, reflecting the circular form of the *E. coli* chromosome. Sequencing of the complete *E. coli* genome has confirmed the results obtained by genetic mapping.

In addition to the F plasmid, bacteria also contain other types of plasmids. **R plasmids,** for example, contain genes that

provide resistance to unfavorable conditions, such as exposure to antibiotics. The competitive advantage provided by the genes in some plasmids may account for the wide distribution of plasmids of all kinds in prokaryotic cells.

In Transformation, DNA Taken Up by a Bacterium Is the Source of Genetic Recombination

DNA can transfer from one bacterial cell to another by two additional mechanisms, *transformation* and *transduction*. Like conjugation, these mechanisms transfer DNA in one direction, and create partial diploids in which recombination can occur between alleles in the homologous DNA regions.

In **transformation,** bacteria take up pieces of DNA that are released as other cells disintegrate. Frederick Griffith discovered this phenomenon in 1928, when he found that a non-virulent form of the bacterium *Streptococcus pneumoniae*, unable to cause pneumonia in mice, could be transformed to the virulent form if it was exposed to heat-killed cells of a virulent strain (see Section 14.1 and Figure 14.2). In 1944, Oswald Avery and his colleagues at New York University showed that the transforming substance most likely was DNA (see Section 14.1).

Subsequently, geneticists established that in transformation experiments the linear DNA fragments taken up from disrupted cells recombine with the chromosomal DNA of recipient cells, in much the same way as genetic recombination takes place in conjugation.

Approximately 1% of bacterial species can undergo *natural transformation*, in which bacteria take up DNA from the surrounding medium by natural mechanisms. Such bacteria typically have a DNA-binding protein on the outer surface of the cell wall. When DNA from the cell's surroundings binds to the protein, a deoxyribonuclease enzyme breaks the DNA into short pieces that pass through the cell wall and plasma membrane into the cytosol. The entering DNA can then recombine with the recipient cell's chromosome if it contains homologous regions.

Natural transformation does not occur with *E. coli* cells. However, they can be induced to take up DNA by *artificial transformation*. One transformation technique is to expose *E. coli* cells to calcium ions and the DNA of interest, incubate them on ice, and then give them a quick heat shock. This treatment alters the plasma membrane so that DNA can penetrate and enter. Another technique for artificial transformation, called *electroporation*, exposes cells briefly to rapid pulses of an electrical current. The electrical shock alters the plasma membrane so that DNA can enter. This method works well with many bacterial species, and also with many types of eukaryotic cells.

Artificial transformation is often used to insert plasmids containing DNA sequences of interest into *E. coli* cells as a part of DNA cloning techniques. After the cells are transformed, clones of the cells are grown in large numbers to

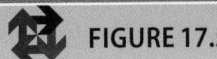 **FIGURE 17.5** **Research Method**

Replica Plating

Purpose: Replica plating is used to identify different strains of bacteria with respect to their growth requirements in a heterogeneous mixture of strains.

Protocol:

1. Press sterile velveteen gently onto the master plate of solid growth medium with bacterial colonies on it. Some of each colony transfers to the velveteen in the same pattern as the colonies on the plate. In the example, a mixture of colonies of normal and auxotrophic strains is on a plate of complete medium.

2. Press the velveteen gently onto a sterile replica plate to transfer some of each strain. In the example, the replica plate contains minimal medium. Incubate to allow colonies to grow, and compare the pattern of colonies on the replica plate with that on the master plate.

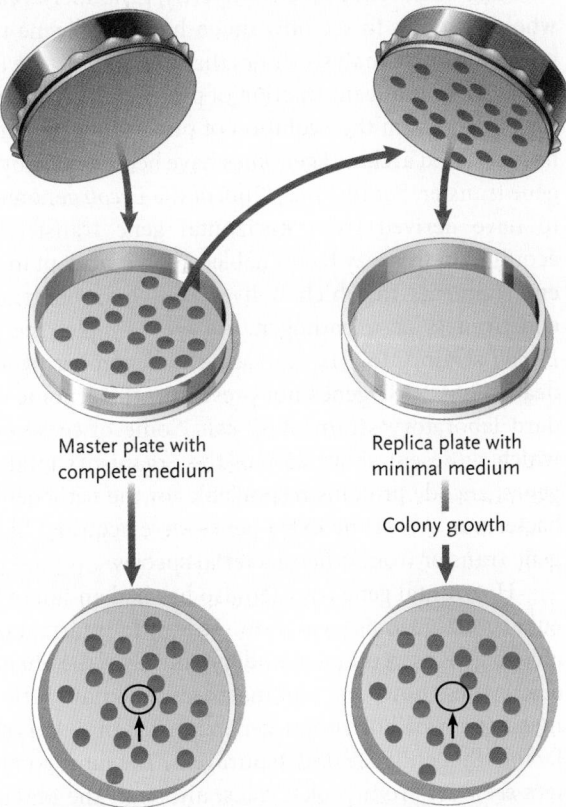

Master plate with complete medium

Replica plate with minimal medium

Colony growth

Interpreting the Results: A colony present on the master plate but not on the replica plate indicates that the strain requires some substance missing from the minimal medium in order to grow. In other words, the strain is an auxotroph. In actual experiments, the compositions of the master plate and replica plate media are chosen to be appropriate for the goals of the experiment.

Source: J. Lederberg and E. M. Lederberg. 1952. Replica plating and indirect selection of bacterial mutants. *Journal of Bacteriology* 63:399–406.

increase the quantity of the inserted DNA to the amounts necessary for sequencing or genetic engineering. (DNA cloning and genetic engineering are discussed further in Chapter 18.)

In Transduction, DNA Introduced into a Bacterium by Phage Infection Is the Source of Genetic Recombination

In **transduction,** DNA is transferred to recipient bacterial cells by an infecting phage (see Section 14.1). When new phages assemble in an infected bacterial cell, sometimes they incorporate fragments of the host cell DNA instead of viral DNA. After the phages are released from the host cell, they may attach to another cell and inject—*transduce*—the bacterial DNA into that cell. The introduction of this DNA, as in conjugation and transformation, makes the recipient cell a partial diploid and creates the potential for recombination to take place. And, because the phage carrying the donor bacterium's DNA does not have a phage genome, it cannot kill the recipient cell, allowing recombinant colonies to form. The DNA-containing capacity of a phage is limited, so the amount of donor DNA that can be transferred to a recipient by this mechanism is far less than is the case for conjugation. Joshua Lederberg and his graduate student, Norton Zinder, then at the University of Wisconsin–Madison, discovered transduction in 1952 in experiments with the bacterium *Salmonella typhimurium* and phage P22.

Replica Plating Allows Genetic Recombinants to Be Identified and Counted

Joshua Lederberg and Esther Lederberg developed a technique called **replica plating** for identifying and counting genetic recombinants in conjugation, transformation, and transduction experiments **(Figure 17.5).** In replica plating, researchers grow bacterial colonies on a master plate with a **complete medium,** that is, a medium containing a full complement of nutrients, including amino acids and other chemicals that normal strains can make for themselves. This master plate is pressed gently onto sterile velveteen, which transfers some of each colony to the velveteen in the same pattern as the colonies on the plate. The velveteen is then pressed onto new plates containing a minimal growth medium—the replica plates—thereby transferring some cells from each original colony to those plates. The replica plates are incubated to allow new colonies to grow.

Figure 17.5 shows the identification of auxotrophic mutants of *E. coli* by replica plating. Normal cells will grow in a minimal medium, but auxotrophic mutants will not, because they are unable to make one or more of the missing substances. Thus, the investigator takes for further study the colonies from the master plate that correspond to missing colonies on the minimal-medium plates.

In an actual experiment, the compositions of the media are appropriate for the goals of the experiment. For example, to identify a *met*⁺ recombinant in a conjugation experiment, the starting plate contains methionine and the colonies are replica plated to a plate lacking methionine. The *met*⁺ recombinants grow on the plate lacking methionine, whereas *met*⁻ parentals do not.

Horizontal Gene Transfer Is an Important Process for Genome Evolution

Our examples of gene inheritance patterns in earlier chapters all involved the movement of genetic material from generation to generation, whether that is from cell to cell or from parent to offspring. This is movement of genetic material by descent and, at its root, it is a feature of both asexual and sexual reproduction. By contrast, **horizontal gene transfer** is the movement of genetic material between organisms other than by descent. Horizontal gene transfer can occur between different species, introducing genes from one species into another. It is most common in prokaryotes, but there is also evidence that eukaryotes, including the ancestors of humans, obtained genes from viruses by horizontal transfer.

Conjugation, transformation, and transduction are three major mechanisms by which horizontal gene transfer occurs in bacterial species. Conjugation and transduction occur in archaea, and there is some evidence for natural transformation.

In conjugation, genetic material is transferred horizontally from donor to recipient under the control of a plasmid that facilitates cell–cell contact, and by the formation of a bridge through which DNA is transferred. Relatively long segments of DNA can be transferred in this process. The particular donor segment of DNA transferred depends on where the plasmid responsible for conjugation is integrated into the donor bacterium's chromosome. Our examples involved conjugation between donor and recipient cells of the same species, but the process can occur between bacteria that are not closely related, and it has also been shown to occur between bacteria and archaea.

In transformation, a DNA molecule in the environment surrounding a bacterial cell is taken up by that cell. In natural populations of bacteria, the bacteria taking up DNA typically are those capable of natural transformation. Any piece of DNA can be taken up in transformation, but usually only relatively short DNA molecules are involved.

In transduction, a phage transfers bacterial DNA from a donor to a recipient cell. The amount of DNA that can be transferred in this way is relatively small because it is limited by the capacity of the phage head in which the DNA must be packed. Horizontal gene transfer by transduction generally is limited to closely related bacteria because the phage carrying the donor

DNA has to bind to specific receptors on a recipient bacterium in order to inject the genetic material. Unless the bacteria are closely related, the necessary receptor will be absent.

In our examples of conjugation, transformation, and transduction in the same species, genetic recombination between donor and recipient DNA occurs because the sequences match, allowing recombination to occur. When different species are involved, the transferred DNA may have little sequence similarity to the recipient's DNA. However, if some stretches of sequences match sufficiently, recombination can lead to the integration of a segment of the transferred donor DNA. In this way, a recipient species can acquire new genes. If those genes confer some type of selective advantage to the recipient, the altered genome will be retained and potentially the old genome will be lost in the population.

Now that the genomes of many bacterial and archaeal species have been sequenced completely, genomicists can compare whole genomes to see how much horizontal gene transfer has occurred. Such analysis shows that horizontal gene transfer has involved a significant fraction of prokaryotic genes, making it a major process in the evolution of prokaryotic genomes. In fact, bacterial and archaeal genomes have been shaped by horizontal gene transfer. For instance, 20% of the *E. coli* genome is thought to have derived from horizontal gene transfer. The genes acquired in this way have enabled *E. coli* to adapt to the various environments in which it lives, and have contributed to its effectiveness as a pathogen. As an example, the genome of *E. coli* strain O157:H7, the cause of a number of food-related deaths, has 1,387 genes not present in the genome of the standard laboratory strain of *E. coli*. Some of these extra genes, which represent about 25% of the organism's total number of genes, encode proteins responsible for the pathogenicity of the bacterium. All of the extra genes were acquired by horizontal gene transfer from other bacterial species.

Horizontal gene transfer also has had an important role in eukaryotic genome evolution. Too little information is available to give us a complete understanding of the phenomenon in eukaryotes, however. For instance, we know little about the mechanism of horizontal gene transfer in eukaryotes, or the factors that enhance or discourage it. The most common transfers seen are from prokaryotes, although the amount of prokaryotic DNA found in eukaryotes varies widely. Most of these transfers are to the nuclear genomes, with only a few to organelle genomes. Eukaryote–eukaryote transfers of nuclear genes are also known, as are eukaryote–prokaryote transfers. Both of these types of transfer are rarer than prokaryote–prokaryote transfers, with eukaryote–prokaryote transfers being extremely rare.

STUDY BREAK 17.1

1. What are the properties of F⁺, F⁻, and Hfr cells of *E. coli*?
2. How is horizontal gene transfer in bacteria distinctive from gene segregation in sexually reproducing organisms?

A 30,000-Year-Old Pathogenic Giant Virus

Mimivirus was the first giant virus discovered. It consists of a roughly icosahedral capsid of diameter 0.4 μm surrounded by a fiber layer, and it is large enough to be seen with a light microscope. Mimivirus infects the amoeba *Acanthamoeba* (see Chapter 27). It has a genome size of 1.2 Mb (1 Mb = 1 million base pairs), which encodes an estimated 979 proteins. Recently, even larger giant viruses were discovered, the Pandoraviruses. These viruses also infect *Acanthamoeba;* they are roughly ovoid, about 1.0–1.2 μm long, with genomes up to 2.8 Mb and encoding

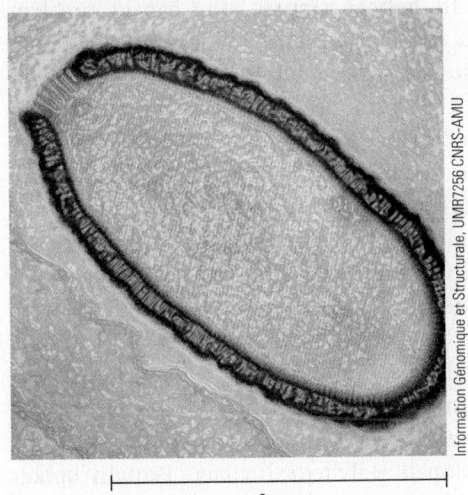

1 μm

FIGURE Pithovirus.

Information Génomique et Structurale, UMR7256 CNRS-AMU

up to 2,500 proteins. The proteins show little resemblance to those of Mimivirus.

Research Question

Do other types of giant viruses exist in nature?

Experiments

A group of researchers in France led by Jean-Michel Claverie and Chantal Abergel used a culture of *Acanthamoeba* as bait to search for giant viruses. They inoculated the culture with a sample of aseptically collected Siberian permafrost taken from a layer estimated to be more than 30,000 years old. Giant viruses initially were seen by light microscopy as ovoid particles within *Acanthamoeba* cells. Electron microscopic analysis showed the viruses to be similar morphologically to Pandoraviruses but slightly larger, approximately 1.5 μm long and 0.5 μm in diameter **(Figure)**. At one end is an apex structure like a protruding cork. The researchers named the giant virus Pithovirus.

The scientists followed a complete replication cycle of Pithovirus in *Acanthamoeba* starting from purified viral particles. The cycle starts with phagocytosis (see Section 5.5) of viral particles. The particles then lose the apical cork, enabling the viral contents to be released to begin the lytic cycle. Replication of the virus takes place in the cytoplasm,

and 10–20 hours after infection, hundreds of progeny viral particles are released upon host cell lysis.

The Pithovirus genome is 610,033 bp and is estimated to encode 467 proteins. The genome size and protein-coding content are surprisingly small given the size of the virus (compare with Pandoraviruses described earlier). Proteomic analysis of the Pithovirus particle showed the presence of 159 different gene products, two-thirds of which have unknown functions. The Pithovirus virus particle proteins are entirely different from those of Pandoraviruses.

Conclusion

As of early 2014, Pithovirus is the largest giant virus known. The genomic and proteomic data indicate that Pithovirus is distinct from Mimivirus and Pandoraviruses, indicating that it may be a prototype for a new virus family. Pithovirus is the most ancient eukaryote-infecting DNA virus revived to date.

think like a scientist

Why might there be a concern about thawing of ancient permafrost layers?

Source: M. Legendre et al. 2014. Thirty-thousand-year-old distant relative of giant icosahedral DNA viruses with a pandoravirus morphology. *Proceedings of the National Academy of Sciences USA.* 111:4274–4279.

© Cengage Learning 2017

17.2 Viruses and Viral Genetics

This section discusses the structures and infectious properties of viruses. Viruses infect all living organisms, the phages discussed in the previous section that infect bacteria being just one type of virus. Viruses infect cells and carry out their life cycles within those cells, releasing progeny viruses that infect other cells. Researchers study the virus–host cell interaction at the cellular and molecular levels so as to gain an understanding of *viral pathogenesis,* that is, the way viruses cause diseases. As you probably know, a number of serious and even lethal human diseases are caused by viruses. Examples are human immunodeficiency virus (HIV), influenza virus, and Ebola virus. Other viruses, by contrast, cause little or no symptoms of infection. Whereas all viruses are similar in being parasites of host cells they infect, viruses vary in their structure, the way they enter host cells, the way they replicate and express their genomes, the

way they assemble progeny viruses, and the way they evade, or attempt to evade, host defense systems.

Viruses in the Free Form Consist of a Nucleic Acid Core Surrounded by a Protein Coat

A **virus** (Latin *virus* = poison) is a biological entity that can infect the cells of a living organism. Viral infections usually have detrimental effects on their hosts. The study of viruses is called *virology,* and researchers studying viruses are known as *virologists.*

Viruses are the most abundant biological entities on Earth, and they are found everywhere. For instance, in soil and in the oceans, viruses outnumber cells by a factor of 10 to 100. And the genetic diversity of viruses, in terms of number of distinct genes, exceeds the genetic diversity of cellular organisms by a substantial amount.

All viruses consist of a DNA or RNA genome surrounded by a protein coat—a layer of proteins—called the **capsid.** The complete viral particle is also called a **virion.** Most, but not all, viruses are significantly smaller than bacteria. Among the few exceptions are viruses that are approximately 1.5 μm long and 0.5 μm in diameter. The smallest bacteria are 200 to 300 nm long. (*Molecular Insights* describes the discovery of the largest known giant virus in greater than 30,000-year-old permafrost.)

Are viruses living or nonliving? Many scientists consider viruses to be on the border of living and nonliving. That is, although they share some properties with living things, viruses are missing several important characteristics: (1) they cannot reproduce independently, but only within a host cell; (2) they are not made up of cells—they are merely pieces of DNA or RNA surrounded by a protein capsid and in mammals an additional membranous envelope; and (3) they do not grow, develop, or generate metabolic energy. Viruses can adapt very readily over time since they routinely mutate the glycoproteins in their capsids or their envelopes, which provide the lock-and-key fit for entry into their host cells. In summary, viruses cannot be defined as being alive at the cellular level. They do, however, share enough characteristics with organisms that they cannot be defined as nonliving, either.

Viral Structure Is the Minimum Necessary to Transmit Nucleic Acid Molecules from One Host Cell to Another

The genetic material of all organisms is double-stranded DNA, but the nucleic acid genome of a virus may be double-stranded DNA (dsDNA), singled-stranded DNA (ssDNA), double-stranded RNA (dsRNA), or single-stranded RNA (ssRNA), depending on the viral type. The genomes of some viruses, herpesvirus for example, consist of one dsDNA molecule. The genomes of other viruses are distributed among two or more nucleic acid molecules. For example, the influenza virus genome is distributed among eight (influenza A and B strains) or seven (influenza C strains) segments of ssRNA.

The simplest viruses contain only a few genes, whereas those of the most complex viruses contain a hundred or more. All viruses have genes encoding the proteins of their capsid. For some viruses, the capsid is assembled from protein molecules of a single type. More complex viruses have capsids assembled from several different proteins, including the recognition proteins that bind to cells of the specific host(s) they infect. All viruses also have a gene or genes encoding the protein(s) that make them pathogenic to their hosts. The particles of some viruses also contain the DNA or RNA polymerase enzymes required for viral genome replication and/or an enzyme that attacks cell walls or membranes.

Most viruses take one of two basic structural forms, helical or polyhedral. In **helical viruses,** the capsid proteins assemble in a rodlike spiral around the genome **(Figure 17.6A).** A number of viruses that infect plant cells are helical. In **polyhedral viruses,** the capsid proteins form triangular units that form an icosahedral structure **(Figure 17.6B).** The polyhedral viruses include forms that infect animals, plants, and bacteria. In some polyhedral viruses, protein spikes that provide host cell recognition extend from the corners where the facets fit together. Some viruses, the **enveloped viruses,** are covered by a surface membrane derived from the plasma membrane of their host cells; both enveloped helical and enveloped polyhedral viruses are known **(Figure 17.6C).** For example, the **human immunodeficiency virus (HIV),** the virus that causes acquired immunodeficiency syndrome (AIDS), is an enveloped polyhedral virus. Protein spikes extend through the membrane, giving the particle its recognition and adhesion functions.

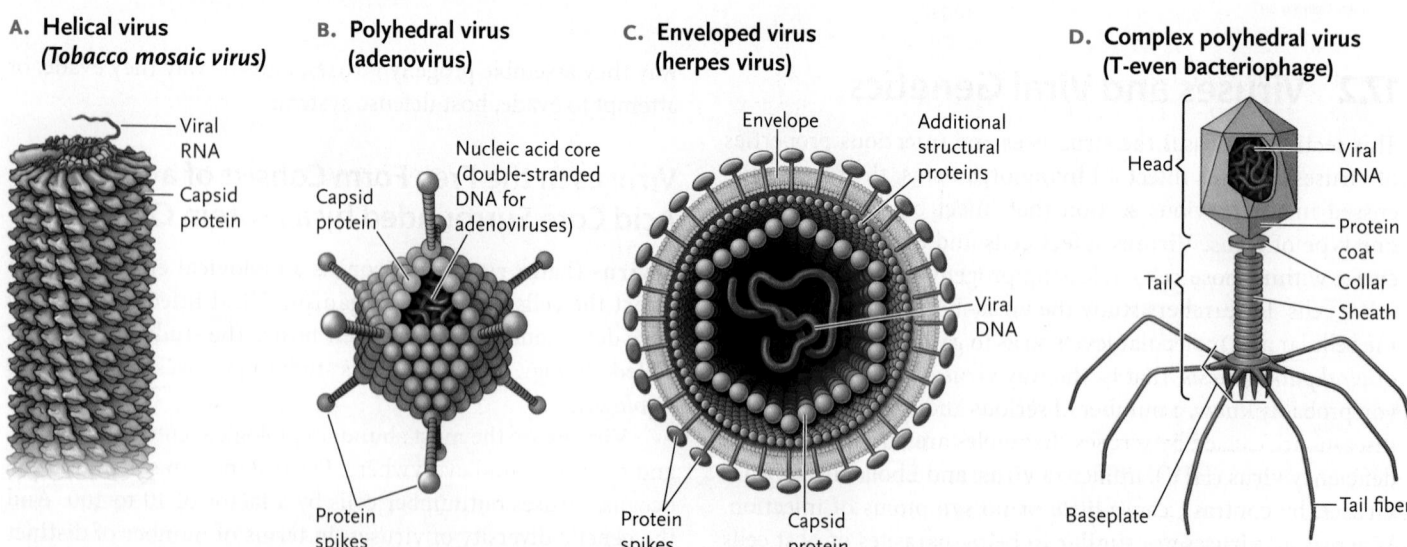

FIGURE 17.6 Basic viral structures. The tobacco mosaic virus in **(A)** assembles from more than 2,000 identical protein subunits. Protein spikes contain recognition proteins that allow the viral particle to bind to the surface of a host cell.
© Cengage Learning 2017

A number of bacteriophages with DNA genomes, such as T2 (see Section 14.1), have a **tail** attached at one side of a polyhedral head, forming what is known as a **complex virus** (**Figure 17.6D**). The genome is packed into the head and the tail is made up of proteins forming a collar, sheath, baseplate, and tail fibers. The tail has recognition proteins at its tip and, once attached to a host cell, functions as a sort of syringe that injects the DNA genome into the cell.

Viruses of Different Kinds Infect All Living Cells

Viruses are classified by the International Committee on Taxonomy of Viruses into orders, families, genera, and species using several criteria, including size and structure, type and number of nucleic acid molecules, method of replication of the nucleic acid molecules inside host cells, host range, and infective cycle. More than 4,000 species of viruses have been classified into more than 80 families according to these criteria.

One or more kinds of viruses probably infect all living organisms. Usually a virus infects only a single species or a few closely related species. A virus may even infect only one organ system, or a single tissue or cell type in its host. However, some viruses are able to infect unrelated species, either naturally or after mutating. For example, some humans have contracted bird flu from being infected with the natural avian flu virus as a result of contact with virus-infected birds. Of the viral families, 21 include viruses that cause human diseases. Viruses also cause diseases of wild and domestic animals. Plant viruses cause annual losses of millions of tons of crops, especially cereals, potatoes, sugar beets, and sugarcane. **Table 17.1** lists some virus families that infect animals, and **Table 17.2** lists some virus families that infect plants. DNA viruses replicate using a DNA-to-DNA mechanism; dsRNA and non-retroviral ssRNA viruses replicate using an RNA-to-RNA mechanism; and retroviruses, which have ssRNA genomes, replicate via a DNA intermediate (see later in the chapter).

The effects of viruses on the organisms they infect range from undetectable, to merely bothersome, to seriously debilitating or lethal diseases. For instance, some viral infections of humans, such as those causing cold sores, chickenpox, and the common cold, are usually little more than a nuisance to healthy adults. Others, including AIDS, influenza, encephalitis,

yellow fever, Ebola, and smallpox, are and/or have been among the most severe and deadly human diseases.

Although most viruses have detrimental effects, some may be considered beneficial. One of the primary reasons why

TABLE 17.1	Examples of Animal Viruses		
Virus Family	Envelope	Diseases and Infections in Humans and Other Animals	
Double-stranded DNA (dsDNA) Viruses			
Adenoviridae	No	Respiratory infections, tumors	
Baculoviridae	Yes	Gastrointestinal infections in arthropods	
Hepadnaviridae	Yes	Hepatitis B	
Herpesviridae	Yes		
Herpes simplex I		Oral herpes, cold sores	
Herpes simplex II		Genital herpes	
Varicella-zoster		Chickenpox, shingles	
Papillomaviridae	No	Benign and malignant warts	
Poxviridae	Yes	Smallpox, cowpox	
Single-stranded DNA (ssDNA) Viruses			
Parvoviridae	No	Fifth disease (humans), canine parvovirus disease (dogs)	
Double-stranded RNA (dsRNA) Viruses			
Reoviridae	No	Upper respiratory infections, enteritis, diarrhea	
Single-stranded RNA (ssRNA) Viruses			
Filoviridae	No	Ebola hemorrhagic fever, Marburg hemorrhagic fever	
Flaviviridae	Yes	Yellow fever, dengue, hepatitis C, West Nile fever	
Orthomyxoviridae	Yes	Influenza	
Paramyxoviridae	Yes	Measles, mumps, pneumonia	
Picornaviridae	No		
Enteroviruses		Polio, hemorrhagic eye disease, gastroenteritis	
Rhinoviruses		Common cold	
Hepatitis A virus		Hepatitis A	
Aphthovirus		Foot-and-mouth disease (livestock)	
Rhabdoviridae	Yes	Rabies, other animal diseases	
Retroviruses (ssRNA genome that replicates via a DNA intermediate)			
Retroviridae	Yes		
FeLV		Feline leukemia	
HTLV I, II		T-cell leukemia	
HIV		AIDS	

TABLE 17.2	Examples of Plant Viruses
Virus Family	Examples of Viruses (the name indicates a plant infected and its symptoms)
Double-stranded DNA (dsDNA) Viruses	
Caulimoviridae	Cauliflower mosaic virus, Blueberry red ringspot virus
Single-stranded DNA (ssDNA) Viruses	
Geminiviridae	Maize streak virus, Beet curly top virus
Double-stranded RNA (dsRNA) Viruses	
Reoviridae	Maize rough dwarf virus, Rice ragged stunt virus
Single-stranded RNA (ssRNA) Viruses	
Bromoviridae	Brome mosaic virus, Broad bean mottle virus
Luteoviridae	Barley yellow dwarf virus, Soybean dwarf virus
Rhabdoviridae	Broccoli necrotic yellow virus, Potato yellow dwarf virus
Tobamoviruses	Tobacco mosaic virus, Pepper mold mottle virus
Tombusviridae	Tomato bushy stunt virus, Carnation Italian ringspot virus

bacteria do not completely overrun the planet is that they are destroyed in incredibly huge numbers by bacteriophages. Viruses also provide a natural means to control some insect pests.

Viruses Infect Bacterial, Animal, and Plant Cells by Similar Pathways

Free viruses move by random molecular motions until they contact the surface of a host cell. For infection to occur, the virus must recognize the cell as its host, and then enter the cell. That is, viruses can only infect cells for which they are infectious, meaning that they have evolved the specialized properties to enter and reproduce within those cells. Inside the cell, typically the viral genes are expressed, leading to replication of the viral genome and assembly of progeny viruses. The viruses are then released from the host cell, a process that often ruptures the host cell, killing it.

INFECTION OF BACTERIAL CELLS Bacteriophages differ in how they infect and kill hosts cells. **Virulent bacteriophages** kill their host cells during each cycle of infection, whereas **temperate bacteriophages** may enter an inactive phase in which the host cell replicates and passes on the bacteriophage DNA for generations before the phage becomes active and kills the host. Bacteriophages of both types that infect E. coli are widely used in genetic research, and we will use examples of each type to illustrate their life cycles.

Virulent Bacteriophages Among the virulent bacteriophages infecting E. coli, the **T-even bacteriophages** T2, T4, and T6 have been most valuable in genetic studies (see Figure 17.6D). The coat of these phages is divided into a *head* and a *tail*. Packed into the head is a single linear molecule of double-stranded DNA. The tail, assembled from several different proteins, has recognition proteins at its tip that can bind to the surface of the host cell.

When a virulent DNA bacteriophage such as phage T2 infects E. coli, it enters the **lytic cycle (Figure 17.7),** in which the host cell is killed in each cycle of infection. The phage attaches to a host bacterial cell and injects its DNA genome into the cell. Within the cell, expression of phage genes directs the phage life cycle, leading to the production of progeny phages, which are released from the cell when a phage-encoded enzyme is synthesized that breaks open—*lyses*—the cell. Those phages can now infect other bacteria.

For some virulent phages (although not T-even phages), fragments of the host DNA may be included in the heads as the viral particles assemble, providing the basis for transduction of bacterial genes during the next cycle of infection. Because genes are randomly incorporated from essentially any DNA fragments, gene transfer by this mechanism is termed **generalized transduction.**

Temperate Bacteriophages Temperate bacteriophages alternate between a lytic cycle and a **lysogenic cycle,** in which the viral DNA inserts into the host cell DNA and production of new viral particles is delayed. During the lysogenic cycle, the integrated viral DNA, known as the **prophage,** remains partially or completely inactive, but is replicated and passed on with the host DNA to all descendants of the infected cell. In response to certain environmental signals, the prophage loops out of the chromosome and the lytic cycle of the phage proceeds.

A much-researched temperate phage that infects E. coli is lambda (λ). Phage λ infects E. coli in much the same way T-even phages do **(Figure 17.8).** The phage injects its linear double-stranded DNA chromosome into the bacterium. (Figure 17.8, step 1). Phage gene products then control whether the phage follows a lytic cycle (Figure 17.8, steps 7–9), or follows a lysogenic cycle (Figure 17.8, steps 3–5).

In the transition from the lysogenic cycle to the lytic cycle (Figure 17.8, step 6), excision of the λ chromosome from the E. coli DNA sometimes is not precise, resulting in the inclusion of one or more host cell genes. These genes are replicated with the viral DNA and packed into the coats, and may be carried to a new host cell in the next cycle of infection. Because of the mechanism involved, only genes that are adjacent to the integration site(s) of a temperate phage can be excised with the viral DNA, included in phage particles during the lytic stage, and undergo transduction. Accordingly, this mechanism of gene transfer is termed **specialized transduction.**

INFECTION OF ANIMAL CELLS In contrast to bacterial cells and plant cells, animal cells lack cell walls. As a result, viruses

FIGURE 17.7 The infective cycle of a T-even bacteriophage, an example of a virulent phage.
© Cengage Learning 2017

infecting animal cells follow a similar pattern of infection as for bacterial and plant cells, but a major difference is that both the viral capsid and the DNA or RNA genome enter a host cell.

As for bacteriophages, infection of animal cells requires interaction between specific proteins on the surface of the viral particle and specific cell surface proteins—receptors—which are located in the plasma membrane. The specificity of this interaction is responsible for the specific host range of each animal virus. Importantly, for the infection process, animal viruses exploit cell surface proteins that have specific cellular functions. For example, poliovirus attaches to a receptor for intracellular adhesion, and influenza viruses attach to the acetylcholine receptor, which is used in the transmission of neuronal impulses.

Viruses without an envelope, such as adenovirus (DNA genome), poliovirus (RNA genome), and Ebola virus (RNA genome), bind by their recognition proteins to the plasma membrane and are then taken into the host cell by receptor-mediated endocytosis (see Section 5.5). For some enveloped viruses, such as herpesviruses and poxviruses (DNA genome) and HIV (RNA genome), the viral capsid and its contained genome enter the host cell by fusion of their envelope with the host cell plasma membrane. In this case, the envelope does not enter the cell. For other enveloped viruses, such as orthomyxoviruses (RNA genome; for example, influenza virus) and rhabdoviruses (RNA genome), the entire virus, including the envelope, enters the cell by endocytosis and the viral capsid with its contained genome then is released into the cytoplasm.

Once inside the host cell, the viral capsid is disassembled to release the genome. The genome then directs the synthesis of additional viral particles by hijacking host cell machinery essentially as bacterial viruses do. Newly completed viruses that do not acquire an envelope are released by rupture of the cell's plasma membrane, which lyses and kills the cell. Most enveloped viruses receive their envelope as they pass through the plasma membrane, usually without breaking the membrane **(Figure 17.9).** This pattern of viral release typically does not injure the host cell.

Some animal viruses enter a **latent state** in which the virus remains in the cell in inactive form. The viral nucleic acid is present in the cytoplasm or nuclear DNA, but no complete viral particles or viral release can be detected. (The latent state is similar to the lysogenic cycle that is part of the life cycle of some bacteriophages.) At some point, the latent state may end as the viral DNA is replicated in quantity, capsid proteins are made, and completed viral particles are released from the cell. The herpesviruses that cause oral and genital ulcers in humans remain in a latent state in the cytoplasm of some body cells for the life of the individual. At

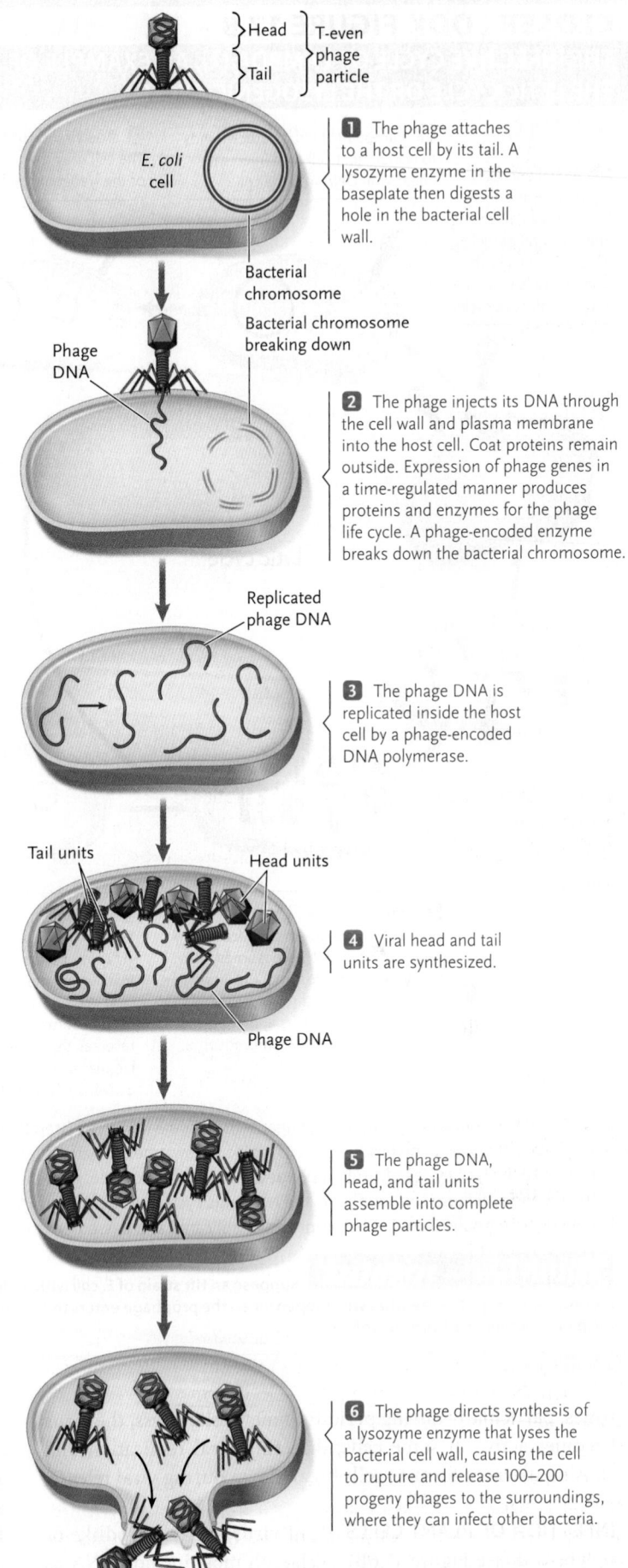

Head
Tail
} T-even phage particle

1 The phage attaches to a host cell by its tail. A lysozyme enzyme in the baseplate then digests a hole in the bacterial cell wall.

E. coli cell

Bacterial chromosome

Phage DNA

Bacterial chromosome breaking down

2 The phage injects its DNA through the cell wall and plasma membrane into the host cell. Coat proteins remain outside. Expression of phage genes in a time-regulated manner produces proteins and enzymes for the phage life cycle. A phage-encoded enzyme breaks down the bacterial chromosome.

Replicated phage DNA

3 The phage DNA is replicated inside the host cell by a phage-encoded DNA polymerase.

Tail units

Head units

4 Viral head and tail units are synthesized.

Phage DNA

5 The phage DNA, head, and tail units assemble into complete phage particles.

6 The phage directs synthesis of a lysozyme enzyme that lyses the bacterial cell wall, causing the cell to rupture and release 100–200 progeny phages to the surroundings, where they can infect other bacteria.

THE INFECTIVE CYCLE OF LAMBDA (λ), AN EXAMPLE OF A TEMPERATE PHAGE, WHICH CAN GO THROUGH THE LYTIC CYCLE OR THE LYSOGENIC CYCLE

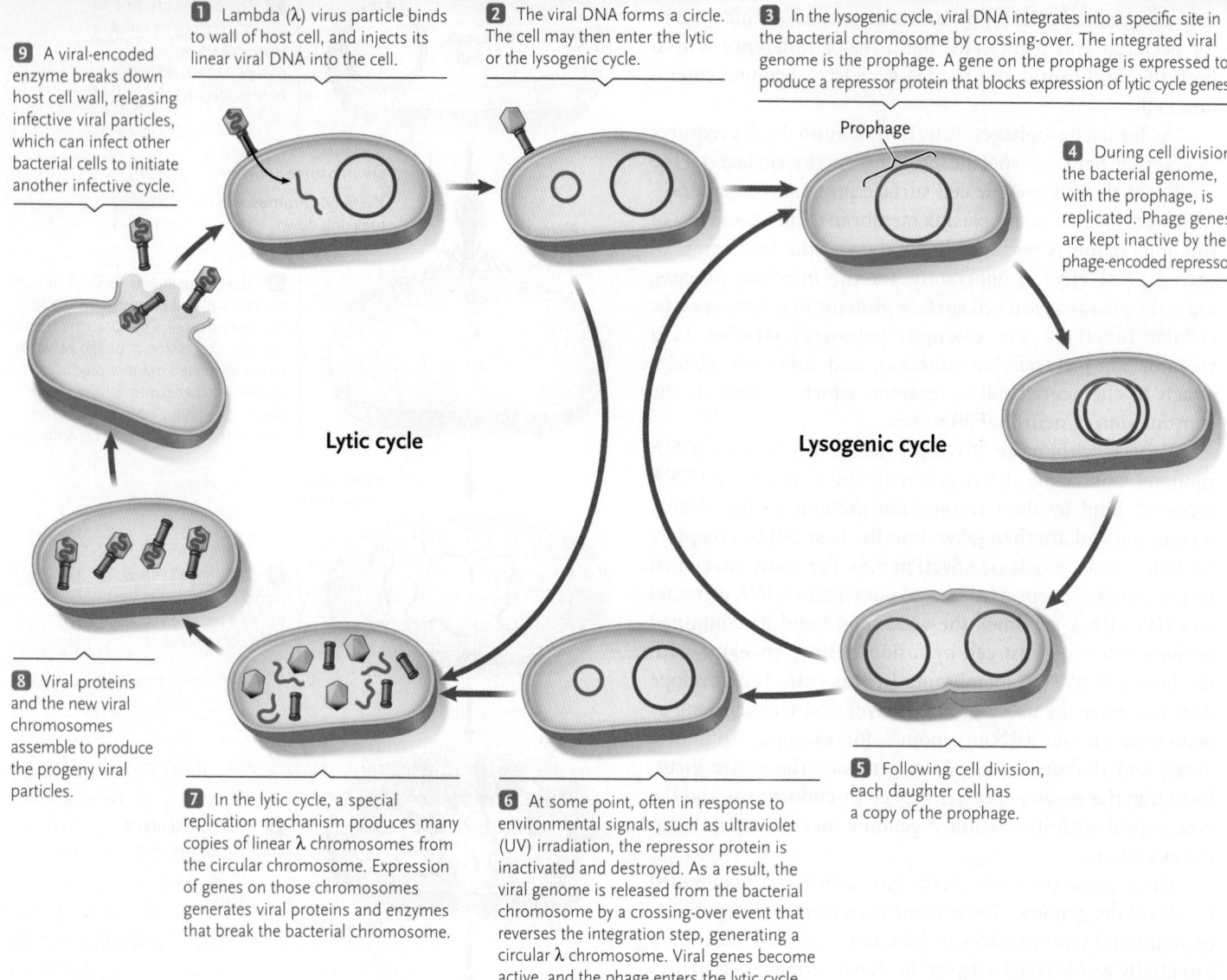

9 A viral-encoded enzyme breaks down host cell wall, releasing infective viral particles, which can infect other bacterial cells to initiate another infective cycle.

1 Lambda (λ) virus particle binds to wall of host cell, and injects its linear viral DNA into the cell.

2 The viral DNA forms a circle. The cell may then enter the lytic or the lysogenic cycle.

3 In the lysogenic cycle, viral DNA integrates into a specific site in the bacterial chromosome by crossing-over. The integrated viral genome is the prophage. A gene on the prophage is expressed to produce a repressor protein that blocks expression of lytic cycle genes.

Prophage

4 During cell division, the bacterial genome, with the prophage, is replicated. Phage genes are kept inactive by the phage-encoded repressor.

Lytic cycle

Lysogenic cycle

8 Viral proteins and the new viral chromosomes assemble to produce the progeny viral particles.

7 In the lytic cycle, a special replication mechanism produces many copies of linear λ chromosomes from the circular chromosome. Expression of genes on those chromosomes generates viral proteins and enzymes that break the bacterial chromosome.

6 At some point, often in response to environmental signals, such as ultraviolet (UV) irradiation, the repressor protein is inactivated and destroyed. As a result, the viral genome is released from the bacterial chromosome by a crossing-over event that reverses the integration step, generating a circular λ chromosome. Viral genes become active, and the phage enters the lytic cycle.

5 Following cell division, each daughter cell has a copy of the prophage.

SUMMARY Lambda phage injects its linear chromosome into a cell (step 1). Inside the cell, the chromosome forms a circle (step 2), and then follows one of two paths. Viral gene products act as molecular switches to determine which path is followed at the time of infection. One path is the lytic cycle, which is like the lytic cycles of virulent phages. During the lytic cycle, progeny phages are produced, which are then released from the cell (steps 7–9). In the second and more common path, the lysogenic cycle, the circular λ chromosome integrates into the host cell's chromosome by crossing-over, becoming a prophage (step 3). The prophage is replicated and passed on in division along with the host cell chromosome (steps 4–5).

think like a scientist Suppose an Hfr strain of *E. coli* with an integrated λ prophage conjugates with an F⁻ recipient that does not have a λ prophage. Hypothesize what will happen when the prophage enters the recipient during the conjugation process. Would your answer be different if the recipient contained a λ prophage?

times, particularly during periods of metabolic stress, the virus becomes active in some cells, directing viral replication and causing ulcers to form as cells break down during viral release.

INFECTION OF PLANT CELLS Plant viruses may be rodlike or polyhedral (see Figure 17.6B). Although most include RNA as

their nucleic acid, some contain DNA. None of the known plant viruses has an envelope. Plant viruses enter cells through mechanical injuries to leaves and stems or through transmission from plant to plant by biting and feeding insects such as leaf hoppers and aphids, by nematode worms, and by pollen during fertilization. Plant viruses can also be transmitted from

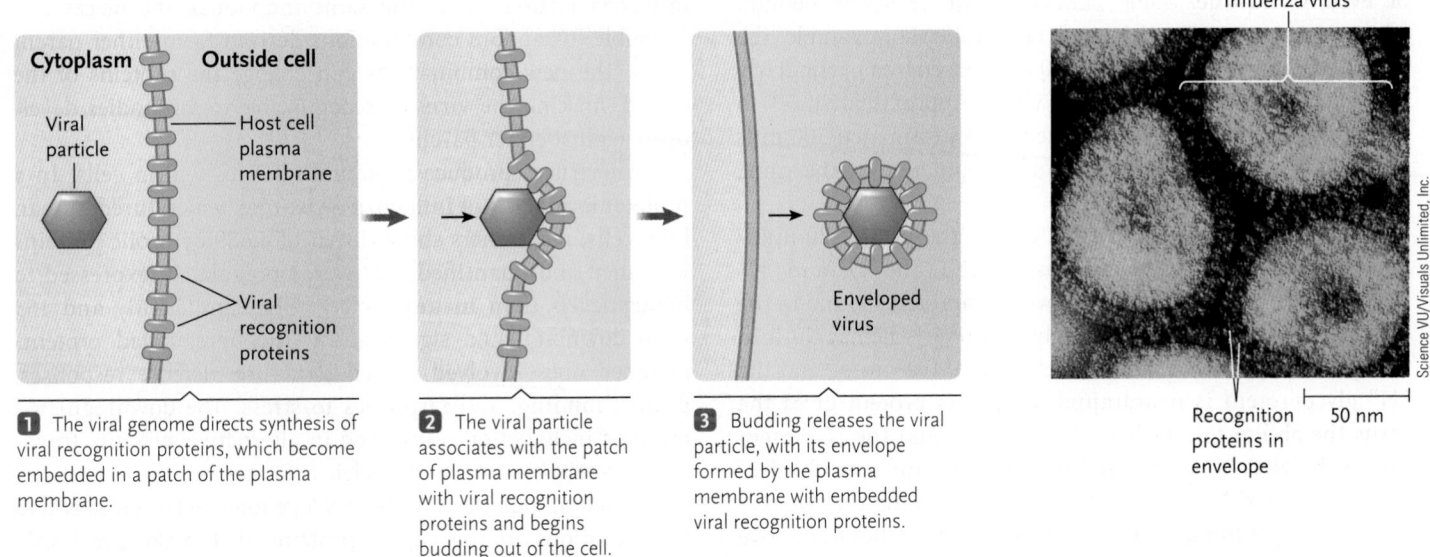

Cytoplasm   Outside cell

Viral particle

Host cell plasma membrane

Viral recognition proteins

1 The viral genome directs synthesis of viral recognition proteins, which become embedded in a patch of the plasma membrane.

2 The viral particle associates with the patch of plasma membrane with viral recognition proteins and begins budding out of the cell.

3 Budding releases the viral particle, with its envelope formed by the plasma membrane with embedded viral recognition proteins.

Enveloped virus

Influenza virus

Recognition proteins in envelope

50 nm

Science VU/Visuals Unlimited, Inc.

FIGURE 17.9 How enveloped viruses acquire their envelope. The micrograph shows the influenza virus with its envelope. Note the recognition proteins studding the envelope.

© Cengage Learning 2017

generation to generation in seeds. Once inside a cell, plant viruses replicate in the same patterns as animal viruses. Within plants, viral particles pass from infected to healthy cells through plasmodesmata, the openings in cell walls that directly connect plant cells (see Figure 4.26), and through the vascular system.

Plant viruses are generally named and classified by the type of plant they infect and their most visible effects. *Tomato bushy stunt virus,* for example, causes dwarfing and overgrowth of leaves and stems of tomato plants, and *Tobacco mosaic virus* causes a mosaic-like pattern of spots on leaves of tobacco plants. Most species of crop plants can be infected by at least one destructive virus.

Viral Infections Are Typically Difficult to Treat

The vast majority of animal viral infections are asymptomatic. However, there are many types of pathogenic viruses, and they employ a variety of ways to cause disease. In some instances, release of progeny viruses from the cell leads to massive cell death, destroying vital tissues such as nervous tissue or white or red blood cells, or causing lesions such as ulcers in skin and mucous membranes. Some other viruses release cellular molecules when infected cells break down, which can induce fever and inflammation. Yet other viruses alter gene function when they insert into the host cell DNA, leading to cancer and other abnormalities.

Viral infections are unaffected by the antibiotics and other treatments used for bacterial infections. As a result, many viral infections are allowed to run their course, with treatment limited to relieving the symptoms while the immune defenses of the patient attack the virus. Some viruses, however, cause serious and sometimes deadly symptoms on infection and,

consequently, researchers have spent considerable effort to develop antiviral drugs to treat them. Many of these drugs target a stage of the viral life cycle. For example, amantadine inhibits entry of hepatitis B and hepatitis C virus into cells, acyclovir (an analog of nucleosides; *analog* means it is chemically similar) inhibits replication of the genomes of herpesviruses, and zanamivir inhibits release of influenza virus particles from cells.

Let us consider influenza viruses in more detail. Influenza viruses (see Figure 17.9) are ssRNA viruses that infect a number of vertebrates, including birds (avian flu), pigs (swine flu), and humans and other mammals. In humans, influenza viruses are transmitted from person to person primarily by aerosols and droplets, and enter the host through the respiratory tract. The virus enters a host cell when the viral protein spike binds to a cell surface receptor and induces endocytosis. Release of progeny viruses lyses and kills the cell, which triggers inflammatory responses. The presence of viruses in the body also stimulates the immune system to make antibodies targeted at the virus. (Antibodies are highly specific protein molecules produced by the immune system that recognize and bind to foreign proteins originating from a pathogen; see Chapter 45.)

Virologists classify influenza viruses into three types, influenzavirus A, influenzavirus B, and influenzavirus C. Strains of each type can infect humans; the A and B types are highly contagious. Yearly epidemic outbreaks of flu are routine, even in developed countries, resulting in many deaths. However, each virus strain that causes an epidemic has a particular virulence (degree of pathogenicity) because of differences in the molecular properties of the viruses, so the death rate varies with each epidemic. Moreover, influenza viruses have the potential to cause *pandemics,* meaning the spread of a new strain through many human populations in a country, region, or even worldwide. Some pandemics are relatively benign, whereas others result in high mortality rates. For example, the so-called Spanish Flu of 1918–1920 was caused by a particularly virulent strain of the influenza virus. Approximately 50 to 100 million people died worldwide after an estimated 500 million people, one-third of the human population at the time, were infected.

The influenza virus type that caused the 1918–1920 influenza pandemic was influenzavirus A/H1N1. The "H" and "N" numbers refer to the subtypes of two glycoproteins found in the envelope of the virus. The "H" glycoprotein is hemagglutinin (the protein spike that binds to cell surface receptors), and the "N" glycoprotein is neuraminidase. The H protein gives the virus the ability to attach to the host cell, and the N protein allows the virus to be released from the cell and spread infection. The H and N proteins are *antigens,* meaning that we recognize them as foreign and then generate antibodies to remove the virus from the system (see Chapter 45). The reason that we continue to get influenza, even though we have produced antibodies against an infecting influenza virus, is that the virus mutates and infectious viruses are produced with new combinations of H and N proteins. Infection by a mutated virus means, therefore, that we see new antigens for which we now have to make new antibodies. In the meantime, the virus attacks cells and the symptoms of influenza develop.

The virus type of the 1918 influenza virus was determined experimentally in 2005, when researchers led by Jeffrey Taubenberger at the U.S. Armed Forces Institute of Pathology reconstructed the genome of the virus and produced infectious, pathogenic viruses in the laboratory. The team worked mainly with tissue from a 1918 flu victim found in permafrost in Alaska. Using modern DNA technology (see Chapter 18), they pieced together the sequences of the virus's 11 genes and characterized their protein products. They also transformed clones of the genes into animal cells and were able to produce complete viruses. The reconstructed 1918 viruses were about 50 times more virulent than most modern-day human influenza viruses; they killed a higher percentage of mice and killed them much more quickly, for instance. (All of these experiments were done with appropriate approval and under highly controlled experimental conditions.) By studying the 1918 virus genome and its pathogenicity, the researchers are learning how highly virulent viruses can be produced. So far, they have learned that the 1918 virus had mutations in polymerase genes for replicating the viral RNA genome in host cells, likely making this strain capable of replicating more efficiently. An influenza pandemic starting in 2009 involving a new strain of influenzavirus A/H1N1 was about 100 times less deadly than the 1918 flu virus (an estimated 579,000 people died worldwide). The molecular properties responsible for their differences in virulence are not known.

The influenza type A and B viruses have many unusual features that tend to keep them a step ahead of efforts to counteract their infections. One is the genome of the virus, which consists of eight segments of RNA. When two different influenza viruses infect the same individual, the pieces can assemble in random combinations derived from either parent virus. The new combinations can change the proteins of the capsid, making the virus unrecognizable to antibodies developed against either parent virus.

All viruses produce profound changes within cells. In a proteomic analysis of influenza virus-infected cultured human lung cells, researchers showed that of 4,689 cytosolic proteins identified and quantified, 127 were upregulated (expressed to higher levels than in uninfected cells) significantly and 153 were downregulated significantly. The upregulated proteins included ones involved in cell structure, defense responses, protein binding, and responses to stress. The downregulated proteins included ones involved in alternative splicing, transport, protein binding, and nucleic acid metabolism.

Random mutations in the RNA genome of the virus add to the variations in the capsid proteins that make previously formed antibodies ineffective, and also can alter the virulence of the virus.

Retroviruses Are Viruses That Replicate Their RNA Genomes via a DNA Intermediate

Most viruses with RNA genomes replicate those genomes directly to produce progeny RNA genomes. However, a **retrovirus** is an enveloped virus with an RNA genome that replicates via a DNA intermediate. In that respect, retroviruses resemble transposable elements known as retrotransposons (see Section 15.5). When a retrovirus infects a host cell, a *reverse transcriptase* enzyme carried in the viral particle (and encoded by the viral genome) is released and copies the single-stranded RNA genome into a double-stranded DNA copy. The viral DNA is then inserted into the host DNA, where it is replicated and passed to progeny cells during cell division. The inserted viral DNA is known as a **provirus (Figure 17.10).**

PROPERTIES OF RETROVIRUSES Retroviruses are found in a wide range of organisms, with most so far identified in vertebrates. You, as well as most other humans and mammals, probably contain from 1 to as many as 100 or more retroviruses in your genome as proviruses. Many of these retroviruses never produce viral particles. However, they may sometimes cause genetic disturbances of various kinds, including alterations of gene activity or DNA rearrangements such as deletions and translocations, some of which may be harmful to the host. HIV, the causal agent of acquired immunodeficiency syndrome (AIDS), does produce viral particles. (More discussion of AIDS is in Chapter 45.)

Some retroviruses, such as *avian sarcoma virus,* have been linked to cancer (in this case in chickens). Many of the cancer-causing retroviruses have picked up a host gene that triggers the entry of cells into uncontrolled DNA replication and cell division. When included in a retrovirus, the host gene comes under the influence of the highly active retroviral promoter, which makes the gene continually active and leads to the

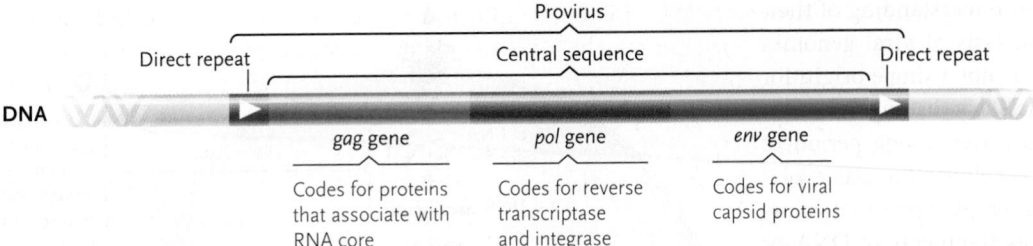

FIGURE 17.10 A mammalian retrovirus in the provirus form in which it is inserted into chromosomal DNA. The direct repeats at either end contain sequences capable of acting as enhancer, promoter, and termination signals for transcription. The central sequence contains genes coding for proteins, concentrated in the *gag, pol,* and *env* regions. The provirus of HIV, the virus that causes AIDS, takes this form.

© Cengage Learning 2017

uncontrolled cell division which is characteristic of cancer. In other words, the host gene has become an oncogene, a gene that promotes the development of cancer by stimulating cell division (see Section 16.5). Usually, the host gene replaces one or more retrovirus genes, making the virus unable to produce viral particles.

Retroviruses also may activate genes related to cell division by moving them to the vicinity of an active host cell promoter or enhancer, or by delivering an enhancer or active promoter to the vicinity of a host cell gene. In either case, the result may be uncontrolled cell division. Retroviruses that have been modified to make them harmless are used in genetic engineering to introduce genes into mammalian and other animal cells.

THE PROPERTIES AND LIFE CYCLE OF HIV HIV infects particular cells of the immune system. At the time of initial infection with HIV, many people suffer a mild fever and other symptoms that may be mistaken for the flu or the common cold. The symptoms disappear as antibodies against the viral proteins appear in the body, and the number of viral particles in the bloodstream drops. However, the genome of the virus is still present as a provirus, and the virus steadily spreads to infect other immune system cells. An infected person may remain apparently healthy for years, yet can transmit the virus to others. Both the transmitter and recipient of the virus may be unaware that the disease is present, making it difficult to control the spread of HIV infections. Ultimately, many important cells of the immune system are destroyed, wiping out the body's immune response and making the HIV-infected person susceptible to other infections. Steady debilitation may then occur, sometimes resulting in death.

Like all retroviruses, HIV is an enveloped virus with two copies of a single-stranded RNA genome contained within a capsid **(Figure 17.11)**. When HIV first infects a cell, a *gp120* glycoprotein of the capsid attaches the virus to a particular type of receptor in the cell's plasma membrane. Then, another viral protein triggers fusion of the viral envelope with the host cell's plasma membrane, releasing the virus into the cell; the life cycle then commences **(Figure 17.12)**. A key event of the life cycle is that a double-stranded DNA copy of the RNA genome made by the viral reverse transcriptase is integrated into the

host chromosome by the viral *integrase* to produce the provirus, which, like an integrated prophage, is dormant. When the dormancy of the provirus is broken (for example, when the host cell in which it is located is stimulated to grow and divide), the genes of the retrovirus are expressed leading to the production of infective HIV particles. Those viral particles are released from the host cell by budding. The viral particles may infect more body cells, or another person. Transmission to another person occurs when an infected person's body fluids, especially blood or semen, enter the blood or tissue fluids of another person's body.

Viruses May Have Originated and Evolved from Fragments of Cellular DNA or RNA

All viruses are *obligate intracellular parasites*, which means that they rely on host cell machinery for their life cycles. Viruses are known to infect all types of cellular organisms, which suggests there could be commonality of virus sequences and functions

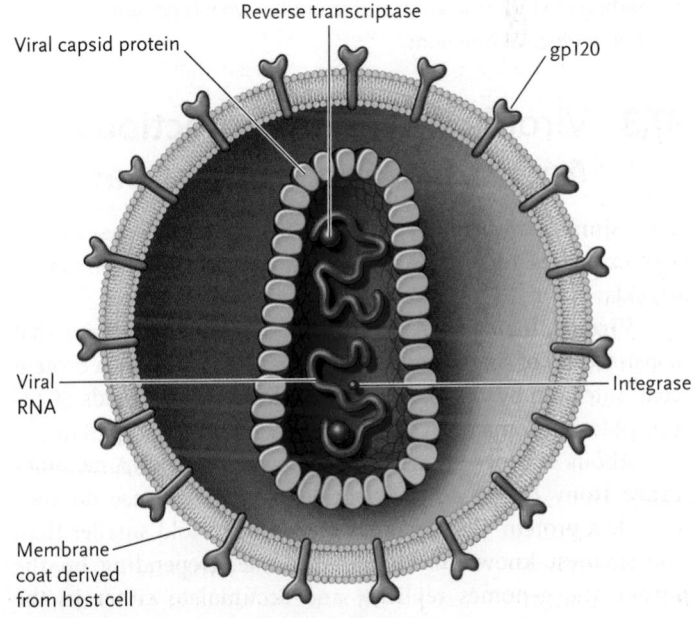

FIGURE 17.11 Structure of HIV, a retrovirus.

© Cengage Learning 2017

that could lead to an understanding of their origin. Sequence analysis of viral genomes indicates that there is not a single origin for viruses. Instead, it is likely that viruses originated multiple times over a long period of evolutionary time as cellular life was evolving. Researchers generally agree that many viruses originated as fragments of DNA or RNA, and then became independent entities when, in some way, the fragments became surrounded by a protective layer of protein that allowed them to escape from their parent cells. Over evolutionary time, the DNA or RNA segments have combined and mutated extensively, producing the present-day virus genomes. The genome changes have been so extensive that it is not possible to determine definitively whether or not some viruses diverged anciently from common ancestry or arose independently. Certainly it is clear that there were multiple and varied ancient origins for viruses, although those origins may not be understood entirely.

STUDY BREAK 17.2

1. What is the difference between a virulent phage and a temperate phage?
2. How does viral infection of an animal cell and a plant cell differ?
3. How are retroviruses distinctive among RNA viruses with respect to replication of their genome?

THINK OUTSIDE THE BOOK

Use the Internet or research literature to determine which gene(s) of the influenza virus genome is (are) responsible for the pathogenicity of the virus, and to outline what is known about their molecular functions.

17.3 Viroids and Prions, Infectious Agents Lacking Protein Coats

Even simpler and smaller infective agents than viruses exist. Two examples are *viroids* and *prions*. Both of these agents lack capsids.

Viroids, first discovered in 1971, are plant pathogens that consist solely of single-stranded, circular RNA without a protein coat. Infection by viroids can rapidly destroy entire fields of citrus, potatoes, tomatoes, coconut palms, and other crop plants.

About 33 types of viroids are known. Their RNA genomes range from 246 to 467 nucleotides and in no case do they encode a protein. The genomes are about 10-fold smaller than the smallest known viral RNA genome. Depending on the viroid, the genomes replicate and accumulate either in the nucleus or in the chloroplast of the host cell. Replication is catalyzed by host nuclear or chloroplast RNA polymerases.

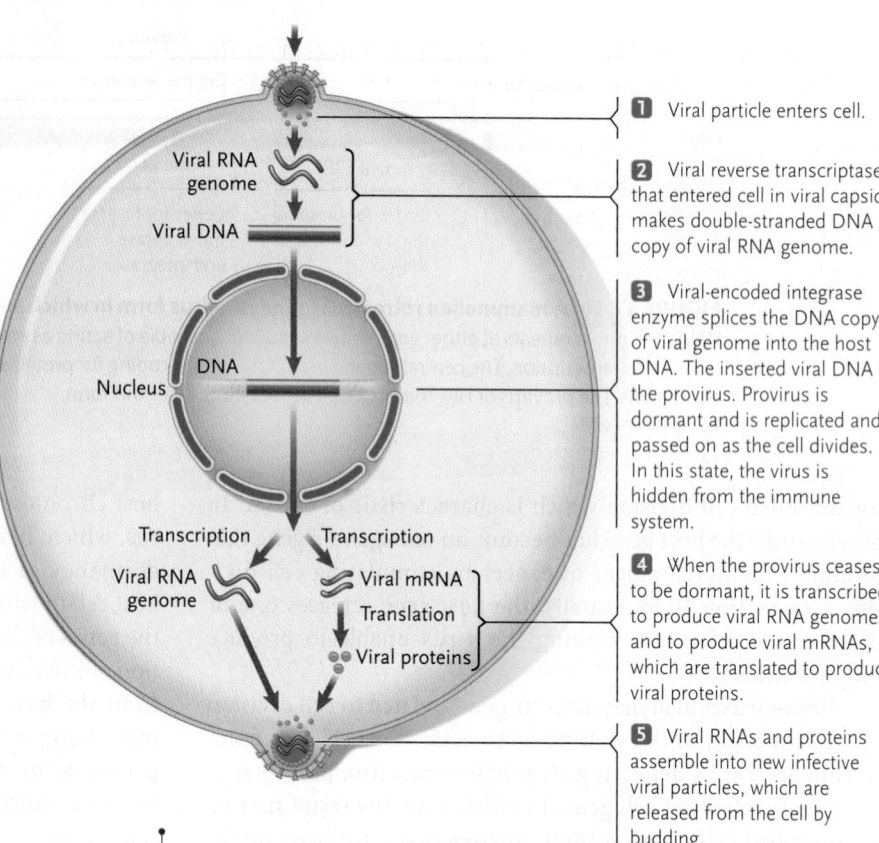

1 Viral particle enters cell.

2 Viral reverse transcriptase that entered cell in viral capsid makes double-stranded DNA copy of viral RNA genome.

3 Viral-encoded integrase enzyme splices the DNA copy of viral genome into the host DNA. The inserted viral DNA is the provirus. Provirus is dormant and is replicated and passed on as the cell divides. In this state, the virus is hidden from the immune system.

4 When the provirus ceases to be dormant, it is transcribed to produce viral RNA genomes and to produce viral mRNAs, which are translated to produce viral proteins.

5 Viral RNAs and proteins assemble into new infective viral particles, which are released from the cell by budding.

FIGURE 17.12 The steps in HIV infection of a host cell.
© Cengage Learning 2017

The tiny RNA genomes of viroids contain sufficient information to infect host plants and to induce the host to replicate the genomes. As a consequence of the replication, the viroids cause specific diseases, although the manner in which they do so is not well understood. In fact, researchers believe that there is more than one mechanism. In one known pathway, viroid RNA activates a protein kinase (an enzyme that adds phosphate groups to proteins; see Sections 6.5 and 9.2) in plants. This process leads to a reduction in protein synthesis and protein activity, and disease symptoms result.

Prions, named in 1982 by Stanley Prusiner of the University of California, San Francisco, for *proteinaceous infection,* are the only known infectious agents that do not include a nucleic acid molecule. Prions have been identified as the causal agents of certain diseases that degenerate the nervous system in mammals. One of these diseases is *scrapie,* a brain disease that causes sheep to rub against fences, rocks, or trees until they scrape off most of their wool. Another prion-based disease is *bovine spongiform encephalopathy (BSE),* also called *mad cow disease.* The disease produces spongy holes and deposits of proteinaceous material in brain tissue **(Figure 17.13).** In 1996, 150,000 cattle in Great Britain died from an outbreak of BSE, which was traced to cattle feed containing ground-up tissues of sheep that had died of scrapie. Humans are subject to a fatal prion infection called *Creutzfeldt–Jakob disease (CJD).* The symptoms of CJD include rapid mental deterioration, loss of vision and speech, and paralysis. Autopsies show spongy holes and deposits in brain tissue similar to those of cattle with BSE. Classic CJD occurs as a result of the spontaneous transformation of normal proteins into prion proteins. Fewer than 300 cases a year occur in the United States. Variant CJD is a form of the disease caused by eating meat or meat products containing nervous system tissue from cattle with BSE. Another prion-based disease of humans, *kuru*, is originally thought to have spread in a tribe in New Guinea when relatives ritualistically ate a deceased individual, including the brain, as a way (in their view) of returning that person's life force to the tribe.

STUDY BREAK 17.3

What distinguishes viroids and prions from viruses?

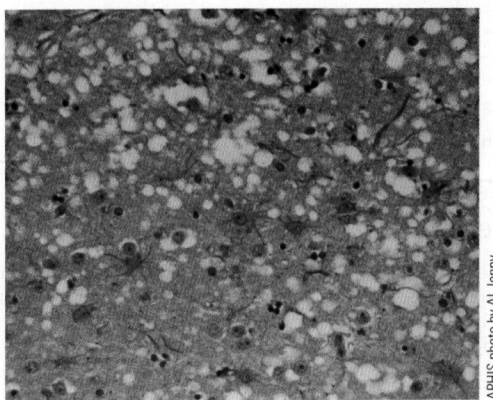

FIGURE 17.13 Bovine spongiform encephalopathy (BSE). The light-colored patches in this section from a brain damaged by BSE are areas where tissue has been destroyed.

Unanswered Questions

How do prions spread?

The brain-wasting diseases caused by prions are poorly understood despite intensive research, and the mechanisms by which prions spread between animals are under investigation. In scrapie, a prion disease of sheep, and chronic wasting disease (CWD), which affects deer and elk, infected animals shed prions into the environment. Remarkably, infectivity can persist in the soil for years. Under natural circumstances, prions enter the host at a peripheral site (for example, oral), invade the central nervous system (CNS), and spread within the CNS. These events require that prions infect new cells and spread between distant locations in the body. Prion propagation in the CNS follows the same circuits that neurons use to communicate with each other.

To understand mechanisms of prion infection, scientists have investigated how prions move through the CNS. Research led by our group at the Rocky Mountain Laboratories (RML), and Marco Prado at the University of Minas Gerais, followed prion aggregates as they invaded cultured mouse neurons. Using prion preparations labeled with a red fluorescent dye, we tracked the fate of the prion particles in living cells using fluorescence microscopy. Interestingly, the prion aggregates moved through neuronal projections to points of contact with other cells. In rare cases, there seemed to be transfer between cells. We proposed that newly described structures, called tunneling nanotubes (TNTs), mediate the intercellular transfer of prion aggregates. TNTs are delicate intercellular projections that can transport proteins between cells. Using methodologies developed by our team, another group later reported that prion aggregates can be transported between cells through TNTs.

Infection requires that prion aggregates interact with the normal prion protein in host cells and induce its conversion to the disease-associated form. We were unable to observe this process because the normal prion protein lacked a fluorescent tag. To permit simultaneous visualization of the normal and disease-associated aggregates of prion proteins, we developed a technique to label the normal prion protein in live cells with a green fluorescent dye. Importantly, we found that the green normal prion protein could convert to the aggregated prion disease-associated form. This technique, called IDEAL-labeling, provides the tools for visualizing the entire prion infection process in living cells by tracking green and red fluorescent prion proteins simultaneously. The advances in understanding how prions move through the nervous system were heralded as a significant step toward developing therapies to stop the spread of brain-wasting diseases by blocking pathways of prion replication and intercellular spreading.

Some important outstanding questions in the field include the following:

- What are the most relevant mechanisms of prion spreading in animals? Do they involve TNTs or perhaps prion-containing vesicles released from infected cells?
- How are prions shed into the environment?
- How does prion propagation cause the death of neurons? Does this involve corruption of the function of the normal prion protein?

think like a scientist One direction of research on prions is to determine if conversion to the infectious form causes corruption of the function of the normal prion protein. What are some approaches to understand further this process?

Gerald Baron is an investigator and head of the TSE/Prion cell biology section at Rocky Mountain Laboratories in Montana. To learn more about Dr. Baron's research, go to http://www3.niaid .nih.gov/labs/aboutlabs/lpvd/TSEPrionCellBiologySection /Pages/baron.aspx.

This research was supported in part by the Intramural Research Program of the NIH, NIAID.

REVIEW KEY CONCEPTS

For access to MindTap and additional study materials visit www.cengagebrain.com.

17.1 Gene Transfer and Genetic Recombination in Bacteria

- Recombination occurs in both bacteria and eukaryotes by exchange of segments between homologous DNA molecules. In bacteria, the DNA of the bacterial chromosome may recombine with DNA brought into the cell from outside.

- Three primary mechanisms bring DNA into bacterial cells from the outside: conjugation, transformation, and transduction.

- In conjugation, two bacterial cells form a cytoplasmic bridge, and part or all of the DNA of one cell moves into the other through the bridge. The donated DNA can then recombine with homologous sequences of the recipient cell's DNA (Figures 17.1, 17.2, and 17.4).

- *E. coli* bacteria that are able to act as DNA donors in conjugation have an F plasmid, making them F$^+$; recipients have no F plasmid and are F$^-$. In Hfr strains of *E. coli*, the F plasmid is integrated into the main chromosome. As a result, genes of the main chromosome can transfer into F$^-$ cells along with a portion of the F plasmid DNA. Genes on the *E. coli* chromosome are mapped by noting the order and timing in which they are transferred from Hfr to F$^-$ cells during conjugation (Figure 17.4).

- In transformation, intact cells take up pieces of DNA released from cells that have disintegrated. The entering DNA fragments can recombine with the recipient cell's chromosomal DNA.

- In transduction, bacterial DNA is transferred from one cell to another by an infecting phage.

- Conjugation, transformation, and transduction do not move genes from generation to generation (that is, by descent) but move genes between organisms by horizontal gene transfer.

17.2 Viruses and Viral Genetics

- A virus is a biological particle that can infect the cells of a living organism. A free viral particle consists of a nucleic acid core, either double-stranded or single-stranded DNA or double-stranded or single-stranded RNA, surrounded by a protein coat (the capsid). Some animal viruses also have a surrounding envelope derived from the host cell's plasma membrane (Figure 17.6). Many viruses have recognition proteins on the virus surface that enable the virus to attach to host cells.

- One or more kinds of viruses likely infect all living organisms. Viruses show specificity with respect to the host(s) they infect. Typically, a particular virus infects only a single species or a few closely related species.

- The cycle of viral infection begins when the nucleic acid molecule of a virus is introduced into a host cell. The virus then directs the cellular machinery to assemble viral capsid proteins with the new viral genomes into new viral particles.

- Virulent phages kill a host bacterial cell during a lytic cycle by releasing an enzyme that ruptures the plasma membrane and cell wall and releases the new viral particles (Figure 17.7).

- Temperate phages do not always kill their host bacterial cell. They may enter the lytic cycle, in which the phage DNA becomes active, exits the host DNA, and begins replication, or a lysogenic cycle. In the lysogenic cycle, the phage's DNA is integrated into the host cell's DNA to form a prophage, which may remain for many generations. At some point, the phage may enter the lytic cycle and begin replication. After production of viral capsids, the DNA is assembled into new progeny phages, which are released as the cell ruptures (Figure 17.8).

- During a cycle of viral infection with particular phages, one or more fragments of host cell DNA may be incorporated into phage particles. As an infected cell breaks down, it releases the viral particles containing host cell DNA. These phages, which form the basis of bacterial transduction, may infect a second cell and introduce the bacterial DNA segment into the new host, where it may recombine with the host DNA.

- The infection of animal cells by viruses depends on interaction between specific proteins on the viral particle surface and specific cell surface proteins. Viruses without envelopes enter the cell by receptor-mediated endocytosis. Enveloped viruses infect cells by endocytosis, or by release of the genome-containing capsid into the cell following fusion of the envelope with the plasma membrane.

- Progeny of nonenveloped animal viruses typically are released from a cell by rupture of the cell's plasma membrane, which lyses and kills the cell. Most enveloped viruses bud from the cell in a process that surrounds the protein capsid–nucleic acid genome with an envelope derived from the cell's plasma membrane and containing embedded viral encoded recognition proteins (Figure 17.9).

- Viruses are unaffected by antibiotics and most other treatment methods making infections caused by them difficult to treat.

- Retroviruses are viruses with single-stranded RNA genomes that replicate via a DNA intermediate that integrates into a nuclear chromosome to produce a provirus (Figures 17.10–17.12).

17.3 Viroids and Prions, Infectious Agents Lacking Protein Coats

- Viroids, which infect crop plants, consist only of a very small, single-stranded RNA molecule.

- Prions, which cause brain diseases in some animals, are infectious proteins with no associated nucleic acid. Prions are misfolded versions of normal cellular proteins that can induce other normal proteins to misfold (Figure 17.13).

TEST YOUR KNOWLEDGE

Remember/Understand

1. When studying the differences in the genes of bacteria, researchers:
   a. do not grow bacteria on a minimal medium as the medium lacks needed nutrients.
   b. use a bacterial clone, which is a group of cells from different bacteria of varying genetic makeup.
   c. use bacteria diploid for their full genome because they can grow on minimal medium.
   d. can study only one genetic trait in a single recombinant event.
   e. can measure the passage of genes between cells during conjugation, transduction, and transformation.

2. In conjugation, when a bacterial F factor is transferred:
   a. the donor cell becomes F^-.
   b. the recipient cell becomes F^+.
   c. the recipient cell becomes F^-.
   d. the donor cell turns into a recipient cell.
   e. chromosomal DNA is transferred in the mating.

3. Which of the following is *not* correct for bacterial conjugation?
   a. Both Hfr and F^+ bacteria have the ability to code for a sex pilus.
   b. After an F^- cell has conjugated with an F^+, its plasmid holds the F^+ factor.
   c. The recipient cell is Hfr following conjugation.
   d. In an Hfr $\times$ F^- conjugation, DNA of the main chromosome moves to a recipient cell.
   e. Genes on the F factor encode proteins of the sex pilus.

4. Which of the following is *not* correct for bacterial transformation?
   a. Artificial transformation is used in cloning procedures.
   b. Avery was able to transform live noninfective bacteria with DNA from dead infective bacteria.
   c. The cell wall and plasma membrane must be penetrated for transformation to proceed.
   d. A virus is required for the process.
   e. Electroporation is a form of artificial transformation used to introduce DNA into cells.

5. Transduction:
   a. may allow recombination of newly introduced DNA with host cell DNA.
   b. is the movement of DNA from one bacterial cell to another by means of a plasmid.
   c. can cause the DNA of the donor to change but not the DNA of the recipient.
   d. is the movement of viral DNA, but not bacterial DNA, into a recipient bacterium.
   e. requires a physical contact between two bacteria.

6. Viruses:
   a. have a protein core.
   b. have a nucleic acid coat.
   c. that infect and kill bacteria during a cell cycle are called virulent bacteriophages.
   d. were probably the first forms of life on Earth.
   e. if they are temperate bacteriophages, kill host cells immediately.

7. When a virus enters the lysogenic stage:
   a. the viral DNA is replicated outside the host cell.
   b. it enters the host cell and kills it immediately.
   c. it enters the host cell, picks up host DNA, and leaves the cell unharmed.
   d. it sits on the host cell plasma membrane with which it covers itself and then leaves the cell.
   e. the viral DNA integrates into the host genome.

8. An infectious material is isolated from a nerve cell. It contains protein with amino acid sequences identical to the host protein but no nucleic acids. It belongs to the group:
   a. prions.
   b. Archaea.
   c. toxin producers.
   d. viroids.
   e. spore formers.

9. Which is *not* correct about retroviruses?
   a. They are RNA viruses.
   b. They are believed to be the source of retrotransposons.
   c. They encode an enzyme for their insertion into host cell DNA.
   d. They encode single-stranded viral DNA from viral RNA.
   e. They encode a reverse transcriptase enzyme for RNA to DNA synthesis.

Apply/Analyze

10. **Discuss Concepts** If recombination occurred between two bacterial genomes as shown in the figure, the result would be:

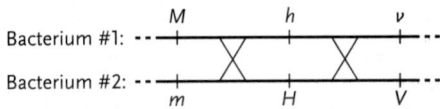

   a. *MHv* and *mhV*.
   b. *MHV* and *mhv*.
   c. *Mhv* and *mHv*.
   d. *MHV* and *mhV*.
   e. *mhv* and *MhV*.

11. **Discuss Concepts** You set up an experiment like the one carried out by Lederberg and Tatum, mixing millions of *E. coli* of two strains with the following genetic constitutions:

$$\text{Strain 1:} \quad bio^-\ met^-\ thr^+\ leu^+$$

$$\text{Strain 2:} \quad bio^+\ met^+\ thr^-\ leu^-$$

   Among the bacteria obtained after mixing, you find some cells that do not require threonine, leucine, or biotin to grow, but still need methionine. How might you explain this result?

12. **Discuss Concepts** As a control for their experiments with bacterial recombination, Lederberg and Tatum placed cells of either "parental" strain 1 or 2 on the surface of a minimal medium. If you set up this control and a few scattered colonies showed up, what might you propose as an explanation? How could you test your explanation?

Evaluate/Create

13. **Discuss Concepts** What rules would you suggest to prevent the spread of mad cow disease (BSE)?

14. **Design an Experiment** You have a culture of Hfr *E. coli* cells that cannot make biotin for themselves. To this culture you add some wild-type *E. coli* cells that have been heat killed, and then subject the culture to electroporation. After the addition, you find some cells that can grow on minimal medium. How could you establish whether the wild-type *bio*$^+$ allele was inserted in a plasmid or the chromosomal DNA of the Hfr cells?

15. **Apply Evolutionary Thinking** Are viruses evolutionarily derived from complex organisms that can reproduce themselves, or are they remnants of precellular "life"? Argue your case.

For selected answers, see Appendix A.

INTERPRET THE DATA

As indicated in the chapter, the F factor integrates into one of several positions around the circular *E. coli* chromosome, producing in each case a different Hfr strain. The various Hfr strains can be used to map the locations of genes around the bacterial chromosome. The **Table** presents the results of conjugation experiments with four different Hfr strains. For each Hfr strain, the order of transfer of the bacterial genes to the recipient is given. From the data, construct a map to show the arrangement of the genes on the circular *E. coli* chromosome.

Hfr strain	First	Order of Gene Transfer				Last
1	e	i	f	u	j	b
2	w	k	c	r	b	j
3	r	c	k	w	h	p
4	w	h	p	y	e	i

DNA Technologies: Making and Using Genetically Altered Organisms, and Other Applications

18

Scientist in a molecular biology laboratory carrying out an experiment using DNA manipulation techniques.

Shutterstock.com/anyaivanova

Why it matters . . . The activities of genes, and the influences of the environment on those activities, govern the structures and functions of organisms. The era of modern genetic analysis began with Gregor Mendel and his experiments in the nineteenth century to determine how particular flower traits are inherited (see Chapter 12). Since that time many researchers have focused their efforts on determining the structure and functions of genes of their particular interest, among them genes for biochemical pathways, genes for cell structure and function, genes for development, and genes for physiological processes in animals and plants. Many of these experiments were carried out using model organisms with which genetic and molecular studies are easy to do. The scope of the experiments, and hence of the questions that can be answered or the hypotheses that can be tested, has always been limited by the technologies available at the time. For instance, genetics studies in the first half of the twentieth century could not involve direct analysis of genes at the molecular level because techniques were not available to isolate and sequence DNA molecules. Nonetheless, geneticists could make mutants with agents that were known to make specific alterations in DNA and then study the effects of those mutations on the phenotype controlled by the gene. In that way, functions of many genes were elucidated. Then in the later part of the twentieth century the development of new techniques made it possible to study the molecular properties of genes directly. The new techniques included a method to isolate and clone (make many copies of) an individual gene in a bacterium, a method to amplify DNA sequences in the test tube, and a method to sequence DNA in the test tube. Together, the new techniques sparked a molecular revolution; they made it possible to ask questions about genes that were now much more molecular in nature. What is the sequence of the gene? What changes in the gene sequence are responsible for altering the phenotype in known mutants? How is the gene regulated at the DNA level? In addition, techniques were developed for manipulating the isolated genes and inserting them back into the same or a different organism.

Collectively, techniques to isolate, purify, analyze, and manipulate DNA sequences are known as **DNA technologies.** Not only are DNA technologies used in basic research, they are also used in applied research. The use of DNA technologies to alter genes for practical purposes is called **genetic engineering.** Genetic engineering is used in the field of applied biology known as **biotechnology,** which is any technique applied to biological systems or living organisms to make or modify products or processes for a specific purpose. Thus, biotechnology includes manipulations that do not involve DNA technologies, such as manipulation of the yeasts that brew beer and the manipulation of the bacteria that make yogurt using standard genetic techniques.

Although much research has been done with model research organisms, an obvious and important interest has always been understanding the structure and function of human genes, particularly those involved with genetic diseases. For ethical reasons, humans cannot be manipulated in the same way as can model research organisms. But the new techniques just described made it possible to clone individual human genes and study them, something that was not possible before.

As a result of continual advances in DNA sequencing technology, as well as development of sophisticated computational software, scientists realized that it was going to be possible to sequence and analyze entire genomes rather than just individual genes. Being able to sequence the complete genome of an organism enables a researcher to explore all aspects of that organism's life, including how many genes it has, how those genes interact, how those genes are regulated, how the genes control development, and what the genome contains other than genes. Importantly, with today's high-throughput DNA sequencing techniques, the genome of any organism can be sequenced rapidly once DNA is extracted from it. And that means that any gene within such an organism can be analyzed specifically, or all or many genes of the organism can be studied simultaneously.

Already in this book you have read about some examples of the outcomes of studies of genome sequences. Two key terms relate to studies of genomes: *genomics* and *bioinformatics.* **Genomics** is the characterization of whole genomes, including their structures (sequences), functions, and evolution. **Bioinformatics** is the application of mathematics and computer science to extract information from biological data, including those related to genome structure and function.

In this chapter, you will learn about the key DNA technologies that are used to create genetically altered organisms for basic and applied research, particular applications of genetically altered organisms, and the use of DNA technologies in biology that do not involve genetically altered organisms. Then, in the next chapter, you will learn about genomics and some of the information that has come from genomic analysis.

18.1 Key DNA Technologies for Making Genetically Altered Organisms

A **genetically altered organism** is an organism that has its genome altered to change a genetic traits or traits. For centuries, horticulturists and animal breeders cultivated genetically altered organisms by taking advantage of naturally occurring mutations, introducing mutations by chemical or radiation treatment, or by creating hybrids of existing organisms. DNA technologies have opened up a new and exciting era in biotechnology because now we can target very specific changes in a genome or introduce specific genes to make new genetically altered organisms tailored for particular purposes. Genetically altered organisms made using DNA technologies may be for either basic or applied research. In *basic research,* for example, a gene from a scientist's research organism may be introduced into *E. coli* for the purpose of **gene cloning,** that is, producing many identical copies of a gene in a host cell. (Recall from Chapter 17 that a *clone* is a line of genetically identical cells or individuals derived from a single ancestor.) We often refer to the gene as a *gene of interest,* defined as a gene that a researcher wants to study in detail to determine its structure and function, including how its expression is regulated, and the nature of the gene's product, or to manipulate it for experimental purposes, such as to help dissect the gene's function. In *applied research,* the interest in cloned genes is not in the structure and function of a gene or sequence; typically those are understood, at least to a significant degree, at the beginning of research projects. Rather, cloned genes or cloned DNA sequences are used, for instance, for medical, agricultural, or commercial applications. Examples include:

- Gene therapy to correct or treat genetic diseases. Diagnosis of genetic diseases, such as sickle-cell anemia.
- Production of pharmaceuticals, such as Humulin (human insulin) to treat diabetes, and tissue plasminogen activator to break down blood clots.
- Generation of genetically modified animals and plants, including animals that synthesize pharmaceuticals, and plants that are nutritionally enriched, insect-resistant, or herbicide-resistant.
- Modification of bacteria to use in cleanup of oil spills or toxic waste.

A term for a genetically altered organism whose genome has been engineered to introduce or change a genetically controlled trait is **genetically modified organism (GMO).** The last three bullet points involve examples of GMOs.

The same methods are used to clone important DNA sequences that are not genes. With our focus on creating genetically altered organisms, we will concentrate on cloning genes, particularly protein-coding genes.

An Overview of Gene Cloning

Most of the genetic alterations researchers do to create genetically altered organisms consist either of modifying a gene of the organism in a particular way to change its function, deleting a gene of the organism, or adding a new gene or regulatory sequence from another organism. In all of these cases, the genotype of the organism is changed as is its phenotype. Generally speaking, any of these alterations requires cloning a gene, which involves using methods for cutting DNA at particular points (using nucleases), joining two pieces of DNA together (using DNA ligase), and making many copies of the DNA (using DNA polymerases). Having many copies of the DNA facilitates whatever genetic manipulations researchers wish to make. You learned already about DNA polymerases and DNA ligase with respect to their role in DNA replication (see Section 14.3). In the remainder of this section we discuss in more detail these gene cloning tools.

An overview of one common method for cloning a gene of interest from a genome is shown in **Figure 18.1.** The method uses bacteria (commonly, *Escherichia coli*) and plasmids, small circular DNA molecules that replicate separately from the bacterial chromosome (see Section 17.1). The researcher isolates a gene of interest from the genome (the method is described later) and inserts it into a plasmid producing a *recombinant DNA molecule.* **Recombinant DNA** is DNA fragments from two or more different sources that have been joined together to form a single molecule, in this case a recombinant plasmid. The recombinant plasmid is introduced into a bacterium by artificial transformation (see Section 17.1), producing a genetically altered bacterium which then is cultured. As the bacterium grows and divides, the recombinant plasmid replicates. Through the replication of the plasmid, cloning of the gene inserted into the plasmid occurs.

Bacterial Enzymes Called Restriction Endonucleases Are the Basis of Gene Cloning

The key to gene cloning is the joining of two DNA molecules from different sources such as the gene of interest and a bacterial plasmid (see Figure 18.1). (Genomic DNA fragments that do not include genes are cloned similarly.) Bacterial enzymes called **restriction endonucleases** (also called **restriction enzymes**) can be used to generate DNA fragments that can be joined to produce recombinant DNA molecules. Restriction enzymes recognize *restriction sites,* specific DNA sequences that are typically 4 to 8 base pairs (bp) long; most restriction enzymes

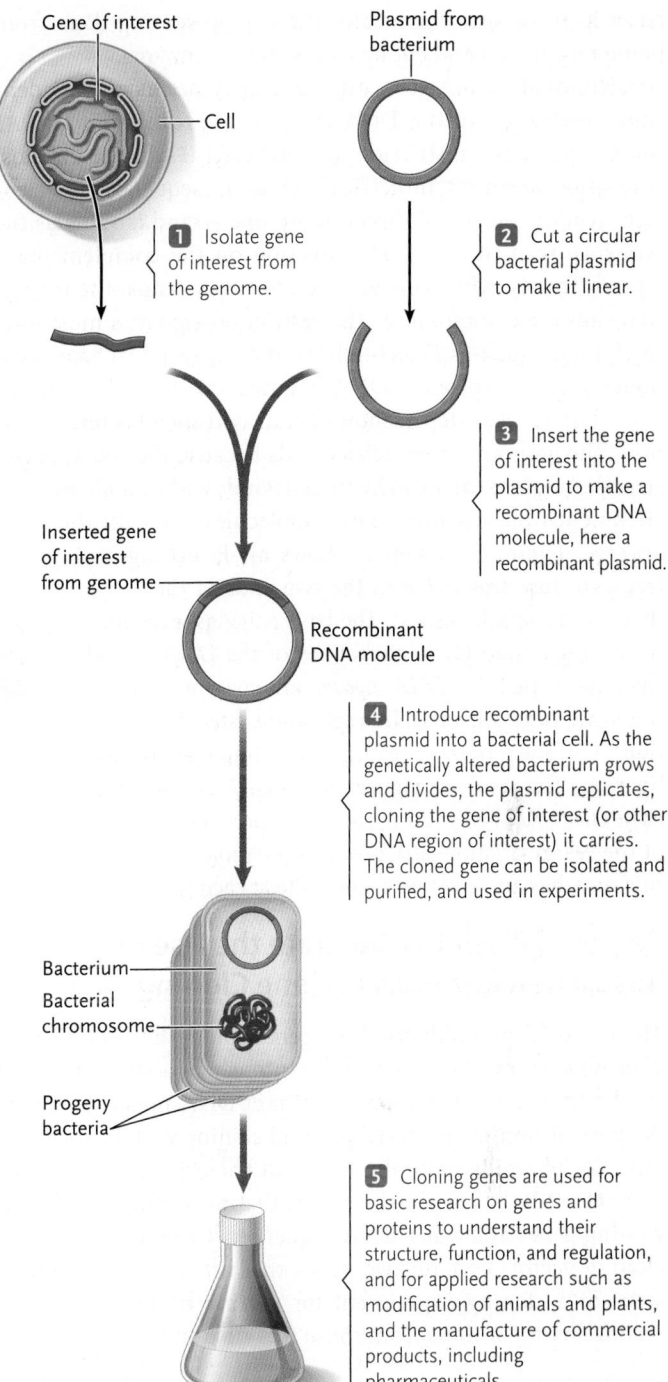

1 Isolate gene of interest from the genome.

2 Cut a circular bacterial plasmid to make it linear.

3 Insert the gene of interest into the plasmid to make a recombinant DNA molecule, here a recombinant plasmid.

4 Introduce recombinant plasmid into a bacterial cell. As the genetically altered bacterium grows and divides, the plasmid replicates, cloning the gene of interest (or other DNA region of interest) it carries. The cloned gene can be isolated and purified, and used in experiments.

5 Cloning genes are used for basic research on genes and proteins to understand their structure, function, and regulation, and for applied research such as modification of animals and plants, and the manufacture of commercial products, including pharmaceuticals.

FIGURE 18.1 Overview of cloning a gene in a bacterial plasmid.
© Cengage Learning 2017

cut the DNA at specific locations within those sites. The DNA fragments produced by a restriction enzyme are known as **restriction fragments.** By choosing a particular restriction enzyme or enzymes, DNA can be cut at specific places.

The "restriction" in the name of the enzymes refers to their normal role inside bacteria, in which the enzymes defend against viral attack by breaking down (restricting) the DNA molecules of infecting viruses. The bacterium protects the restriction sites in its own DNA from cutting by modifying bases in those sites chemically, thereby preventing them from being recognized and cut by its restriction enzyme.

Hundreds of different restriction enzymes have been identified, each one cutting DNA at a specific restriction site. As illustrated by the restriction site of *Eco*RI **(Figure 18.2)**, most restriction sites are symmetrical in that the sequence of nucleotides read in the 5′→3′ direction on one strand is same as the sequence read in the 5′→3′ direction on the complementary strand (that is, when considering the two strands, the restriction sites are *palindromes*). The restriction enzymes most used in cloning—such as *Eco*RI—cleave the sugar–phosphate backbones of DNA to produce DNA fragments with single-stranded ends. Figure 18.2, step 1, shows *Eco*RI cutting a bacterial plasmid. The ends are called **sticky ends** because the short single-stranded regions can form hydrogen bonds with complementary sticky ends on any other DNA molecules cut with the same enzyme. Figure 18.2, step 2, shows an *Eco*RI-digested fragment inserting between the two ends of the *Eco*RI-digested bacterial plasmid. The base pairings leave nicks in the sugar–phosphate backbones of the DNA strands that are sealed by *DNA ligase,* an enzyme that has the same function in DNA replication (step 3; see Section 14.3). The process of sealing with DNA ligase is called **ligation.** The result in Figure 18.2, step 3, is a recombinant plasmid. The ligation process restores the sequences of the two restriction sites producing a molecule that, in this case, has an *Eco*RI restriction site at each junction.

Bacterial Plasmids Illustrate the Use of Restriction Enzymes in Gene Cloning

The bacterial plasmids used for gene cloning are examples of **cloning vectors**—DNA molecules into which a DNA fragment can be inserted to form a recombinant DNA molecule for the purpose of cloning. Bacterial plasmid cloning vectors are derivatives of plasmids naturally found in bacteria, and have been engineered to have special features that make them useful for cloning genes and other DNA sequences. Commonly, plasmid cloning vectors contain two genes that are useful in the final steps of a cloning experiment for sorting bacteria that have recombinant plasmids from those that do not:

1. An antibiotic resistance gene. When a plasmid carrying an antibiotic resistance gene is introduced into an *E. coli* cell and the gene is expressed, the bacterium becomes resistant to the antibiotic. This allows researchers to select for bacteria that contain the plasmid. Hence, antibiotic resistance genes are examples of *selectable markers.* A common antibiotic resistance gene in bacterial plasmids is the *amp*^R gene, which encodes an enzyme (β-lactamase) that breaks down the antibiotic ampicillin.

2. The *lacZ*^+ gene encodes β-galactosidase (recall the *lac* operon from Section 16.1), which hydrolyzes the sugar lactose, as well as a number of synthetic substrates. A cluster of restriction sites is located within the *lacZ*^+ gene, but does not alter the gene's function. Each of

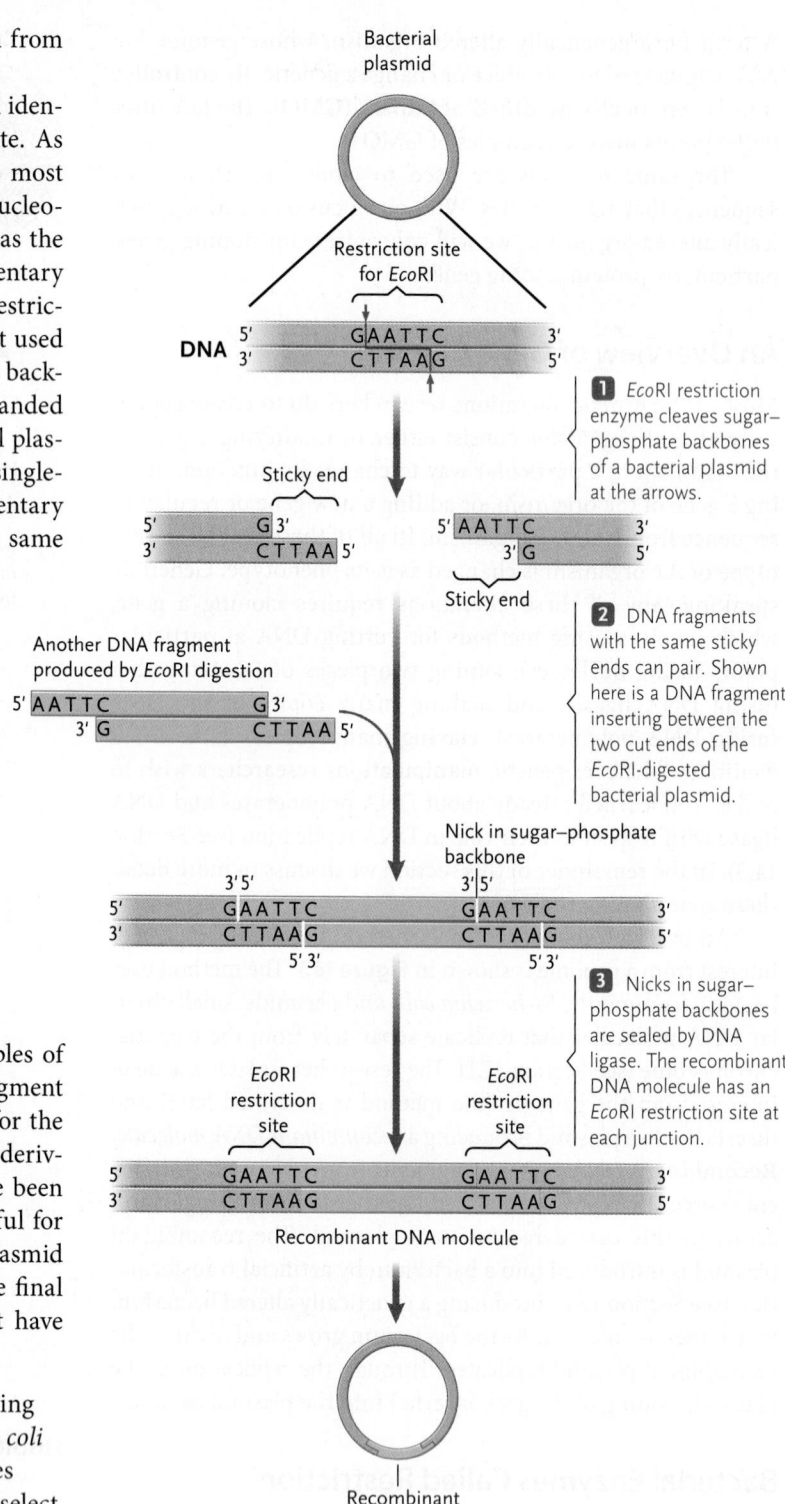

FIGURE 18.2 The restriction site for the restriction enzyme *Eco*RI, and the generation of a recombinant plasmid by complementary base pairing of a bacterial plasmid and a DNA fragment, each of which was digested with the *Eco*RI.

© Cengage Learning 2017

the restriction sites within the cluster is unique to the plasmid. For a given cloning experiment, one or two of these restriction sites is chosen. We will see a bit later how the *lacZ*^+ gene is useful in cloning.

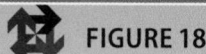 FIGURE 18.3   **Research Method**

Identifying a Recombinant Plasmid Containing a Gene of Interest

Purpose: To identify a recombinant plasmid containing a gene of interest from a ligation reaction mixture containing a bacterial cloning vector and a DNA fragment containing the gene of interest, each digested with the same restriction enzyme.

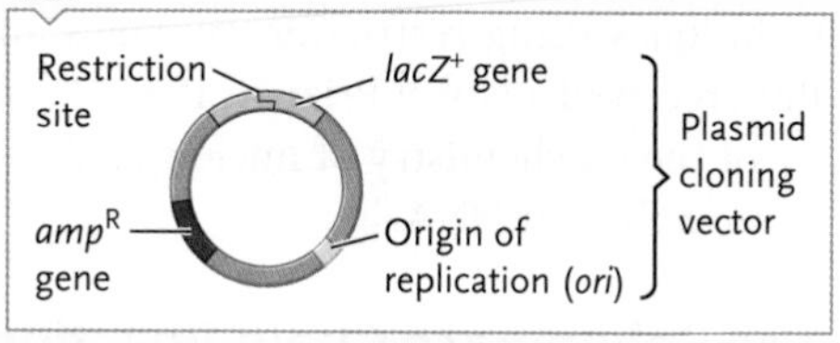

KEY

Restriction site · *lacZ⁺* gene · *amp^R* gene · Origin of replication (*ori*) · Plasmid cloning vector

Protocol:

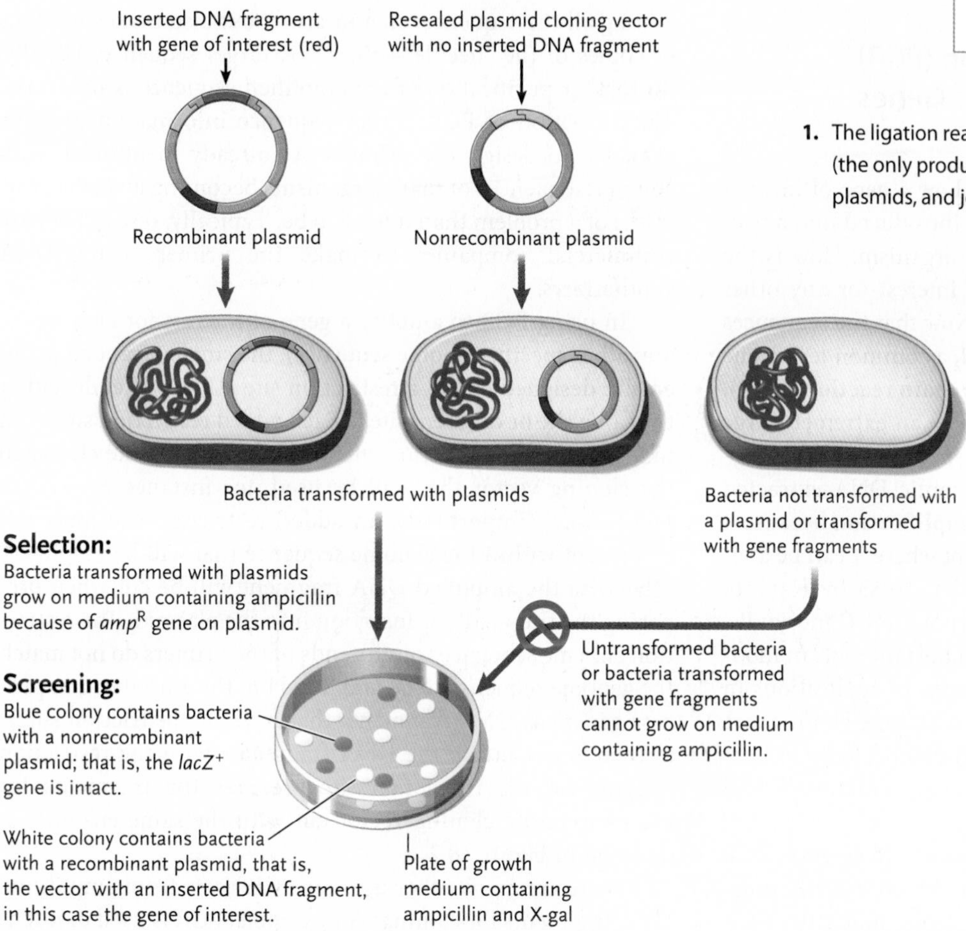

Inserted DNA fragment with gene of interest (red)

Resealed plasmid cloning vector with no inserted DNA fragment

Recombinant plasmid

Nonrecombinant plasmid

1. The ligation reaction produces recombinant plasmids (the only products with the gene of interest), nonrecombinant plasmids, and joined pieces of genomic DNA (not shown).

Bacteria transformed with plasmids

Bacteria not transformed with a plasmid or transformed with gene fragments

2. Transform ampicillin-sensitive, *lacZ⁻* E. coli (which cannot make β-galactosidase) with a sample of the ligation reaction. In this step, some bacteria will take up DNA whereas others will not.

Selection:
Bacteria transformed with plasmids grow on medium containing ampicillin because of *amp^R* gene on plasmid.

Screening:
Blue colony contains bacteria with a nonrecombinant plasmid; that is, the *lacZ⁺* gene is intact.

White colony contains bacteria with a recombinant plasmid, that is, the vector with an inserted DNA fragment, in this case the gene of interest.

Untransformed bacteria or bacteria transformed with gene fragments cannot grow on medium containing ampicillin.

Plate of growth medium containing ampicillin and X-gal

3. Spread the bacterial cells on a plate of growth medium containing ampicillin and X-gal, and incubate the plate until colonies appear.

Interpreting the Results: All of the colonies on the plate contain plasmids because the bacteria that form the colonies are resistant to the ampicillin present in the growth medium. Blue-white screening distinguishes bacterial colonies with nonrecombinant plasmids from those with recombinant plasmids. Bacteria in blue colonies contain nonrecombinant plasmids. These plasmids have intact *lacZ⁺* genes and produce β-galactosidase, which changes X-gal to a blue product. Bacteria in white colonies contain recombinant plasmids. Each recombinant plasmid has a DNA fragment (in this example, the gene of interest) inserted into the *lacZ⁺* gene, so β-galactosidase cannot be produced. As a result, bacteria with recombinant plasmids cannot convert X-gal to the blue product and the colonies are white. Culturing a white colony produces large quantities of the recombinant plasmid that can be isolated and purified for analysis and/or manipulation of the gene.

© Cengage Learning 2017

The plasmid also has an *origin of replication (ori)*. An origin of replication is the DNA region of the plasmid where DNA replication begins (see Section 14.3). The plasmid replicates independently of the bacterial chromosome so that many copies of the plasmid are produced in a cell.

Figure 18.1 outlined how a recombinant plasmid is made that includes a gene of interest. To construct such a plasmid in the laboratory, a researcher assembles a reaction mixture together in a tube containing molecules of the bacterial plasmid cloning vector and DNA fragments containing the gene of interest each digested with the same restriction enzyme, and DNA ligase. This produces a mixture of recombinant plasmids (plasmids with the gene inserted into the cloning vector), nonrecombinant plasmids (resealed cloning vectors with no gene), and joined-together copies of the DNA fragment containing the gene with no cloning vector involved. **Figure 18.3** shows how bacteria containing the desired recombinant plasmid with the gene of interest (in other words, the genetically altered bacteria that contain the cloned gene) are identified.

In 1973 three researchers, Paul Berg, Stanley N. Cohen, and Herbert Boyer, pioneered the development of DNA cloning techniques using restriction enzymes and bacterial plasmids. Berg received a Nobel Prize in 1980 "for his fundamental studies of the biochemistry of nucleic acids, with particular regard to recombinant DNA."

The Polymerase Chain Reaction (PCR) Amplifies DNA *in Vitro,* Such as Genes for Cloning

You have just learned the principles of how a gene of interest can be cloned using a bacterial plasmid introduced into a bacterium to produce a genetically altered organism. How is the DNA fragment containing the gene of interest (or any other DNA fragment) generated for cloning? Now that the sequences of many genomes have been determined, a common technique to use for this purpose is the **polymerase chain reaction (PCR).** PCR is an *in vitro* technique that produces an extremely large number of copies of a specific DNA sequence—such as a gene— from a DNA mixture, for example, genomic DNA extracted from a cell. The process is called **DNA amplification** because it increases the amount of DNA to the point where it can be analyzed or manipulated easily. Developed in 1983 by Kary B. Mullis and Fred Faloona at Cetus Corporation (Emeryville, CA), PCR has become one of the most important tools in modern molecular biology, having a wide range of applications in all areas of biology, some of which, such as gene cloning, are described in this chapter. Mullis received a Nobel Prize in 1993 "for his invention of the polymerase chain reaction (PCR) method."

Figure 18.4 shows how PCR is performed. In essence, PCR is a special case of DNA replication in which a DNA polymerase replicates only a portion of a DNA molecule. PCR takes advantage of a characteristic common to all DNA polymerases in living cells, namely, that these enzymes add nucleotides only to the end of an existing chain called the *primer* (see Section 14.3). For DNA amplification by PCR, a pair of primers must be base-paired to the template DNA. Each primer defines a point at which DNA replication can begin and, together, they define the segment of DNA to be amplified, called the *target sequence.* By cycling 20 to 30 times through a series of steps, PCR amplifies the target sequence, producing millions of copies—an amount that is sufficient for analysis and/or manipulation. Each of the steps is carried out at a specific temperature, with the changes in temperature controlled by the *thermocycler*—the machine in which the PCR is performed. Because a high temperature is used to denature double-stranded DNA to single strands (see Figure 18.4, step 1), a DNA polymerase isolated from most cells cannot be used because it would also be inactivated by the temperature. Instead, special heat-stable DNA polymerases are used that

have been isolated from microorganisms that live in high-temperature environments such as thermal pools or near deep-sea vents. An example is Taq polymerase, an enzyme from the bacterium *Thermus aquaticus* that was originally isolated from a hot spring in Yellowstone National Park.

The specificity of PCR lies in the two primers used to bracket the target sequence. The cycles of PCR replicate only this sequence from a mixture of essentially any DNA molecules. Thus PCR not only finds the "needle in the haystack" among all the sequences in a mixture, but also makes millions of copies of the "needle"—the DNA target sequence. Usually, no further purification of the amplified sequence is necessary. The limitation of PCR is that sequence information must be available to design the primers. As already mentioned, with genome sequences of many organisms becoming available, this is less of a problem than it used to be. Typically, researchers use commercial companies to make the primers using DNA synthesizers.

In using PCR to amplify a gene of interest for cloning (or any other specific genome sequence), the single-stranded primers are designed to add a restriction site at their 5' ends (either the same site or different sites). Each added restriction site must match a unique restriction site in the restriction site cluster in the cloning vector that will be used, for instance *Eco*RI (see Figure 18.2). Importantly, an added restriction site must not be present within the genome sequence that will be amplified; otherwise the amplified DNA fragment will be cut when preparing the fragment for insertion into the plasmid. The restriction enzyme sequences on the ends of the primers do not match the genome sequence to be amplified but the amplification process generates DNA fragments that have the restriction site(s) at each end. Cutting the DNA fragments with the restriction enzyme for the restriction site prepares the fragments for inserting into a cloning vector cut with the same enzyme, as outlined in Figure 18.2.

There are many other uses for PCR apart from amplifying DNA fragments for cloning. For example, PCR is used in forensics to produce enough DNA for analysis from the root of a single human hair, or from a small amount of blood, semen, or saliva, such as the traces left at the scene of a crime. It is also used to extract and amplify DNA sequences from skeletal remains; ancient sources such as mammoths, Neanderthals, and Egyptian mummies; and, in rare cases, from amber-entombed fossils, fossil bones, and fossil plant remains. Some specific examples of the uses of PCR are discussed later in the chapter.

A successful outcome of PCR is shown by analyzing a sample of the amplified DNA using **agarose gel electrophoresis** to see if the copies are the same length as the known length of the target DNA sequence **(Figure 18.5).** Gel electrophoresis is a technique by which DNA, RNA, or protein molecules are separated in a gel that is subjected to an electric field. The type of gel and the conditions used vary with the experiment, but in each case the gel functions as a molecular sieve to separate the macromolecules based on their size, electrical charge, or other

 FIGURE 18.4 | **Research Method**

The Polymerase Chain Reaction (PCR)

Purpose: To amplify—produce large numbers of copies of—a target DNA sequence (for example, a gene of interest) in the test tube without cloning.

Protocol: A polymerase chain reaction mixture has four key elements: (1) the DNA with the target sequence to be amplified, such as a gene of interest; (2) a pair of DNA primers, one complementary to one end of the target sequence and the other complementary to the other end of the target sequence; (3) the four nucleoside triphosphate precursors for DNA synthesis (dATP, dTTP, dGTP, and dCTP); and (4) DNA polymerase. Since PCR uses high temperatures that would break down normal DNA polymerases, a heat-stable DNA polymerase is used.

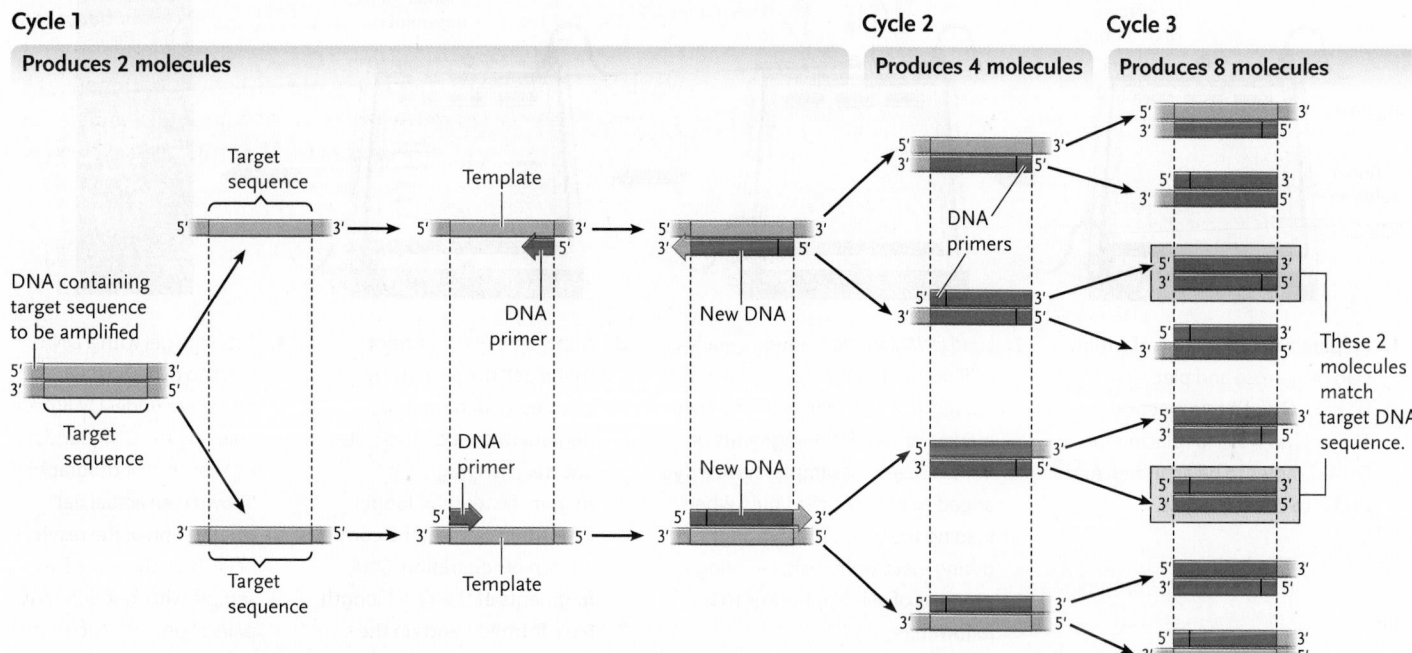

Cycle 1	**Cycle 2**	**Cycle 3**
Produces 2 molecules	Produces 4 molecules	Produces 8 molecules

1. **Denaturation:** heat DNA containing target sequence to 95°C to denature it to single strands.

2. **Annealing:** cool the mixture to 55 to 65°C (depending on the primers) to allow the two primers to anneal their complementary sequences at the two ends of the target sequence.

3. **Extension:** heat to 72°C, the optimal temperature for DNA polymerase to extend the primers, using the four nucleoside triphosphate precursors to make complementary copies of the two template strands. This completes cycle 1 of PCR; the end result is two molecules.

4. Repeat the same steps of denaturation, annealing of primers, and extension in cycle 2, producing a total of four molecules.

5. Repeat the same steps in cycle 3, producing a total of eight molecules. Two of the eight match the exact length of the target DNA sequence (highlighted in yellow).

Interpreting the Results: After three cycles, PCR produces a pair of molecules matching the target sequence. Subsequent cycles amplify these molecules to the point where they outnumber all other molecules in the reaction by many orders of magnitude.

properties. For separating large DNA molecules, such as those typically produced by PCR or by cutting with a restriction enzyme, a gel made of agarose (a natural molecule isolated from seaweed) is used because of its large pore size.

For PCR experiments, the size of the amplified DNA is determined by comparing the position of the DNA band with the positions of bands of a DNA marker ladder, that is, DNA fragments of known size that are separated on the gel at the same time. PCR is successful if the sample fragment size matches the predicted size for the target DNA fragment. In some cases, such as DNA from ancient sources, a size prediction may not be possible; here, agarose gel electrophoresis analysis simply indicates whether there was DNA in the sample that could be amplified. Further proof that the desired DNA fragment has been amplified correctly by PCR can come from restriction enzyme digestion of the DNA fragments to determine if the fragments are cut into restriction fragments of predicted size, or by sequencing the DNA fragment.

FIGURE 18.5 · Research Method

Separation of DNA Fragments by Agarose Gel Electrophoresis

Purpose: Gel electrophoresis separates DNA molecules, RNA molecules, or proteins according to their sizes, electrical charges, or other properties through a gel in an electric field. Different gel types and conditions are used for different molecules and types of applications. A common gel for separating large DNA fragments is made of agarose.

Protocol:

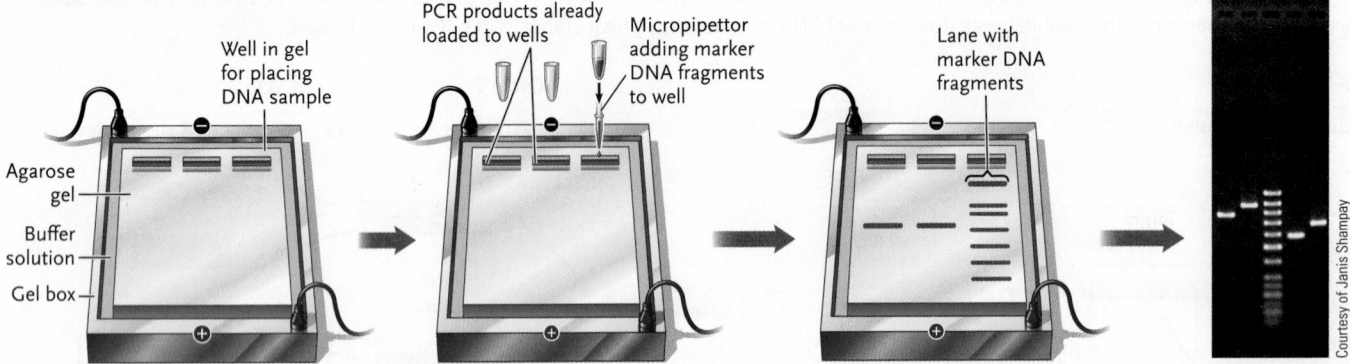

Courtesy of Janis Shampay

1. Prepare a gel consisting of a thin slab of agarose and place it in a gel box in between two electrodes. The gel has wells for placing the DNA samples to be analyzed. Add buffer to cover the gels.

2. Load DNA sample solutions, such as PCR products, into wells of the gel, alongside a well loaded with a DNA marker ladder (DNA fragments of known sizes). All samples have a dye added to help see the liquid when loading the wells. The dye migrates during electrophoresis, enabling the progress of electrophoresis to be followed.)

3. Apply an electric current to the gel; the negatively charged DNA fragments migrate to the positive pole. Shorter DNA fragments migrate faster than longer DNA fragments. At the completion of separation, DNA fragments of the same length have formed bands in the gel. At this point, the bands are invisible.

4. Stain the gel with a DNA-binding dye. The dye fluoresces under UV light, enabling the DNA bands to be seen and photographed. Shown is an actual gel photograph of the results of PCRs on the same DNA sample with four different pairs of primers, each with a different predicted size for the PCR products.

Interpreting the Results: Agarose gel electrophoresis separates DNA fragments according to their length. The lengths of the DNA fragments being analyzed are determined by measuring their migration distances and comparing those distances to a calibration curve of the migration distances of the bands of the DNA marker ladder, which have known lengths. For PCR, agarose gel electrophoresis shows whether DNA of the correct length was amplified. For restriction enzyme digests, this technique shows whether fragments are produced as expected.

Review of Some of the Materials, Concepts, and Techniques Introduced in This Section

More than any other chapter in the book, this chapter discusses a lot of research methods and contains a lot of new terms and techniques to learn. Here is a collection of a number of these terms and techniques and what they are or do:

- *Gene cloning.* A method for producing many copies of a gene by creating a genetically altered organism such as a bacterium in which the gene is replicated.

- *Recombinant DNA.* DNA fragments from two or more sources that have joined together.

- *Restriction enzyme (restriction endonuclease).* An enzyme that recognizes a specific DNA sequence and cuts the DNA within that sequence. Fragments produced by cutting DNA with a restriction enzyme are *restriction fragments*.

- *Ligation.* The process of joining two or more DNA fragments together to make one DNA molecule.

- *DNA ligase.* The enzyme that seals together DNA fragments generated by restriction enzyme digestion to produce a recombinant DNA molecule.

- *Cloning vector.* DNA molecules into which a DNA fragment can be inserted to form a recombinant DNA molecule that can be replicated in a host organism for the purpose of cloning the DNA fragment.

- *Polymerase chain reaction (PCR).* A DNA replication–based technique for amplifying DNA sequences, including genes, without cloning.

- *Agarose gel electrophoresis.* A technique in which an electric field passing through an agarose gel is used to separate DNA or RNA molecules on the basis of size.

STUDY BREAK 18.1

1. What features do restriction enzymes have in common? How do they differ?
2. Plasmid cloning vectors are one type of cloning vector that can be used with *E. coli* as a host organism. What features of a plasmid cloning vector make it useful for constructing and cloning recombinant DNA molecules?
3. What information and materials are needed to amplify a region of DNA using PCR?

18.2 Applications of Genetically Altered Organisms

The development of the ability to clone genes revolutionized biology. As discussed earlier, making genetically altered bacteria to clone genes has made it possible it isolate and purify genes to analyze them in detail, and to make specialized genetically altered organisms to use for the manufacture of commercial products, or for agricultural or medical applications. This section discusses selected examples of these applications.

Genetic Engineering Uses DNA Technologies to Alter the Genes of a Cell or Organism

Genetic engineering uses DNA technologies to alter genes of a cell or organism so as to change the genotype and, hence, the phenotype. Organisms that have been engineered to alter their genes are called **transgenic,** meaning that they have been modified to contain genetic information—the **transgene**—from an external source. Transgenic organism is synonymous with a genetically modified organism (GMO). The goals of genetic engineering include using prokaryotes, fungi, animals, and plants as factories for the production of proteins needed in medicine and scientific research; correcting hereditary disorders; and improving animals and crop plants of agricultural importance. In many of these areas, genetic engineering has already been highly successful.

The following sections discuss some applications of the genetic engineering of bacteria, animals, and plants.

GENETIC ENGINEERING OF BACTERIA TO PRODUCE PROTEINS
Transgenic bacteria have been made, for example, to synthesize proteins for medical applications, break down toxic wastes such as oil spills, produce industrial chemicals such as alcohols, and process minerals. *E. coli* is the organism of choice for many of these applications of DNA technologies.

Figure 18.6 shows how *E. coli* can be engineered to express a protein from a gene that comes from another organism. The protein-coding sequence of a gene is inserted into an **expression**

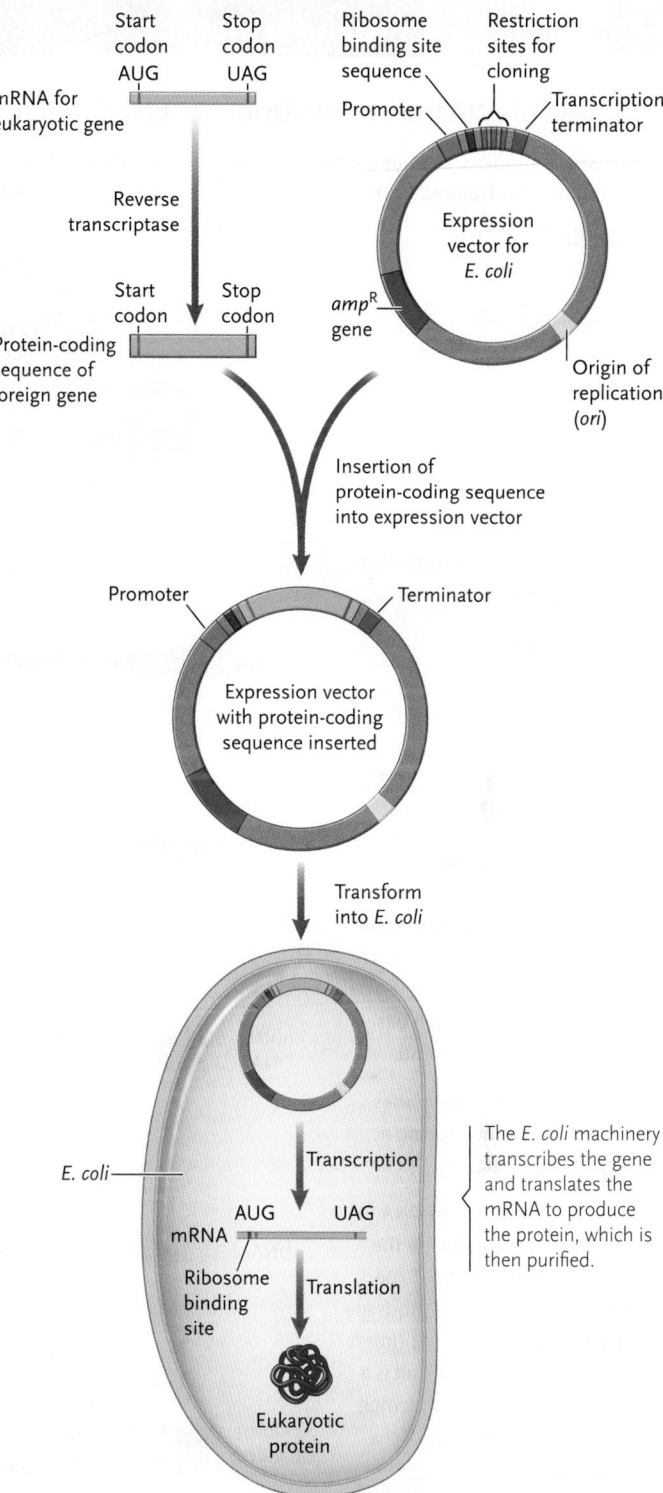

FIGURE 18.6 Using an expression vector to synthesize a foreign protein in *E. coli*. The *E. coli* machinery transcribes the coding sequence for the foreign protein and translates the mRNA to produce the protein, which is then purified.
© Cengage Learning 2017

vector, a cloning vector that has extra features, namely the regulatory sequences that allow transcription and translation of a gene. For a bacterial expression vector, this means having a promoter and a transcription terminator that are recognized by

FIGURE 18.7 | **Research Method**

Synthesis of DNA from mRNA Using Reverse Transcriptase

Purpose: To produce double-stranded, complementary DNA (cDNA) copies of mRNA molecules isolated from cells.

Protocol:

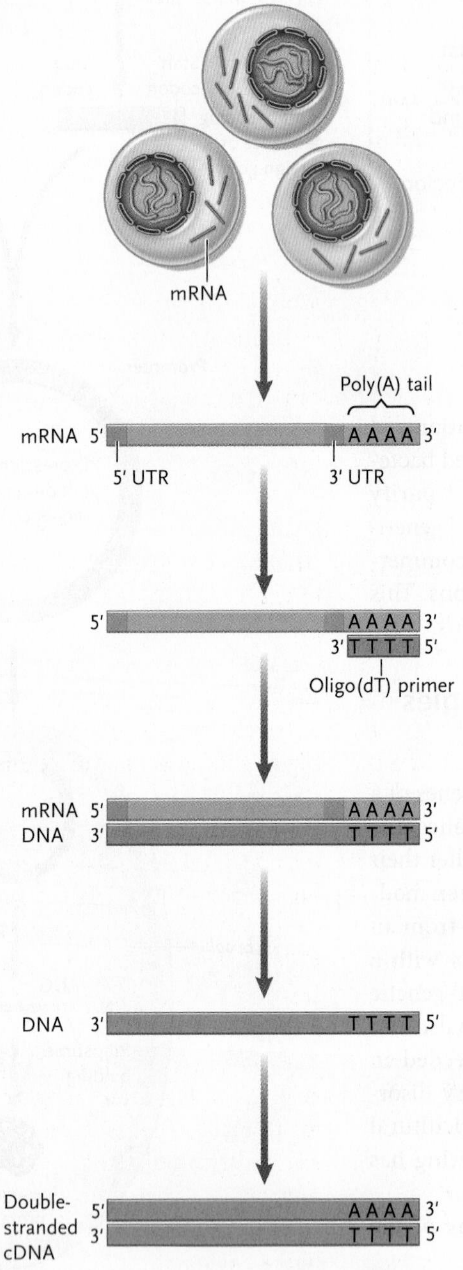

1. Isolate mRNAs from cells. One mRNA is shown.

2. Add primer of a short sequence of T DNA nucleotides (oligo(dT)). Primer base-pairs to poly(A) tail of mRNA.

3. Reverse transcriptase uses DNA precursors to synthesize a DNA copy of the mRNA in the 5′-to-3′ direction. The result is a hybrid nucleic acid molecule consisting of the mRNA base paired with a DNA strand.

4. An RNase enzyme degrades the mRNA strand, leaving a single strand of DNA.

5. DNA polymerase uses DNA precursors to synthesize the second strand of DNA. Experimentally different methods are available for the use of primers in this reaction. The result is a double-stranded complementary DNA (cDNA) copy of the starting mRNA.

Outcome: The outcome is a population of double-stranded, cDNA molecules that have base-pair sequences corresponding to the base sequences of the mRNA molecules isolated from the cell.

© Cengage Learning 2017

cloning so that the inserted gene is placed correctly for transcription and translation when the recombinant plasmid is transformed into *E. coli.*

Bacterial protein-coding genes and most archaeal protein-coding genes lack introns, so generating the protein-coding sequences of those genes for expression in a bacterial expression vector is straightforward. That is, PCR is used with primers that flank the protein-coding sequence and that have a restriction site added at their 5′ ends (see earlier discussion). However, most eukaryotic protein-coding genes have introns so a different strategy must be used. That strategy starts with mRNAs isolated from cells. Recall that eukaryotic mRNAs are produced by processing of pre-mRNAs during which introns are removed, and a 5′ cap and 3′ poly(A) tail are added (see Section 15.3). As shown in **Figure 18.7**, a single-stranded mRNA can be converted to a double-stranded DNA copy by **reverse transcription** followed by DNA synthesis. First the enzyme *reverse transcriptase* (isolated from retroviruses; see Section 17.2) is used to make a single-stranded DNA that is complementary to the mRNA. The primer for this reaction is a short sequence of T DNA nucleotides called a *oligo(dT) primer* ("d" = deoxy; "oligo" = few). Then the mRNA strand is degraded with an enzyme, and DNA polymerase is used to make a second DNA strand that is complementary to the first. The result is a double-stranded **complementary DNA (cDNA)** that, for the top strand in the figure, has the same sequence in DNA nucleotides as the RNA nucleotides in the mRNA.

Producing a gene-specific cDNA can be done by using **reverse transcriptase–PCR (RT–PCR)**, a method that couples reverse transcription with PCR. From genome sequence data, the DNA sequence equivalents of the ends of a particular gene's encoded mRNA can be predicted and primers can be made based on those sequences. As before, restriction sites are added to the two primers at their 5′ ends. The right-hand primer, which would correspond to the right end of the 3′ UTR in Figure 18.4, is used in the reverse transcriptase reaction with extracted RNA from the cell to produce a single-stranded cDNA copy of the specific mRNA of interest. In the PCR cycles that follow, the left-hand primer, which would correspond to the left end of the 5′ UTR in Figure 18.4, functions with

the *E. coli* transcriptional machinery, and having the ribosome binding site needed for the bacterial ribosome to recognize the start codon of the transgene (see Section 15.4). The regulatory sequences flank a cluster of restriction sites that are used for the right-hand primer to amplify the cDNA to produce many double-stranded copies. Restriction digestion of the copies prepares the DNA for insertion into the expression vector cut with the same enzyme.

Expression vectors are available for a number of organisms, both prokaryotes and eukaryotes. They vary in the regulatory sequences they contain and the selectable marker they carry so that the host organism transformed with the vector carrying a gene of interest can be detected and the host can express that gene.

For example, genetic engineering has been used to produce *E. coli* bacteria that make the human hormone insulin; the commercial product is called Humulin. Insulin is required by persons with some forms of diabetes. Humulin is a perfect copy of the human insulin hormone. Many other proteins, including human growth hormone to treat human growth disorders, tissue plasminogen activator to dissolve blood clots that cause heart attacks, and a vaccine against foot-and-mouth disease of cattle and other cloven-hoofed animals (a highly contagious and sometimes fatal viral disease), have been developed for commercial production in bacteria using similar methods.

GENETIC ENGINEERING OF ANIMALS Many animals, including fruit flies, fish, mice, pigs, sheep, goats, and cows, have been altered successfully by genetic engineering. There are many purposes for these alterations, including basic research, generating animal models of human diseases, correcting genetic disorders in humans and other mammals, and producing pharmaceutically important proteins. Some examples are discussed in this section.

Making a Knockout Mouse by Gene Targeting Gene **targeting** is the knocking out, replacement, or addition of a gene in a genome. Gene targeting methods have been developed for a number of model animals (as well as yeast and some plants). The specifics of the methods depend on the organism. Let us consider a method to knockout a gene in a mouse to create a **knockout mouse.** This method uses *stem cells,* so first we must learn about these important cells.

Stem cells are undifferentiated cells that proliferate indefinitely. When a stem cell divides, each progeny cell has a choice of remaining a stem cell, or changing into a *precursor cell* that follows a pathway to a fully differentiated state. The many specialized cell types of the body, such as skin cells and cells of the intestines, derive from stem cells in those cell types' tissues. Stem cells are found in both adults and embryos. In an adult mammal, tissues are made up of cells that have differentiated to be specific functional types and that are incapable of cell division. Tissues are renewed at a rate that is tissue specific; for example, red blood cells are replaced about every 120 days, and intestinal epithelial cells are replaced every 3–6 days. Small numbers of stem cells in the tissues divide to produce the precursor cells that differentiate into the cells needed for tissue renewal. Control mechanisms involving extracellular signaling molecules that regulate specific intracellular pathways (see Chapter 9) ensure that new cells are produced in the right places and in the right numbers. Stem cells not only are important for the normal replacement of cells in adult tissues, they are also used to repair damaged tissues.

Stem cells in adult tissues—*adult stem cells*—that are used for normal replacement or tissue cells or repairing damaged tissues are specialized for the tissues in which they are located. Hence, skin cell stem cells are responsible only for replacing the various cell types of the skin, and bone marrow stem cells are responsible only for replacing the various cell types of the blood. Stem cells like this are said to be *multipotent,* meaning that they have a restricted ability to produce cell types, specifically just those cell types in the tissues in which they are located. Another type of stem cell in mammals is the **embryonic stem cell (ES cell),** which is a *pluripotent* cell, meaning that it is capable of differentiating into many of the different cell types of the body, but not all of them. ES cells are found in a mass of cells inside an early-stage embryo (the blastocyst; see Figure 50.11). ES cells can be isolated from early mouse embryos and, under appropriate conditions, they can be proliferated using cell culture methods and still remain pluripotent. That is, if some of these cells are injected back into an early embryo, they can give rise to many of the tissues and cell types of the body, including germline cells. The cells can also be induced in culture to differentiate into different cell types.

Figure 18.8 shows a method using ES cells to make a knockout mouse. To make the method efficient, researchers use strains of mice that differ in coat color, and a selectable marker in the DNA construct used to replace the normal gene in the genome. Mice in which both alleles of the normal gene are knocked out are confirmed by DNA testing. In 2007, Mario Capecchi, Sir Martin Evans, and Oliver Smithies were awarded a Nobel Prize "for their discoveries of principles for introducing specific gene modifications in mice by the use of embryonic stem cells."

Because humans share many genes with yeast, knockout mice have proved to be a useful system for studying and modeling human diseases, including several different kinds of cancer, heart disease, arthritis, diabetes, Parkinson disease, and obesity. Knockout mice are also useful for testing pharmaceutical treatments and other therapies that might have efficacy in humans.

Making mice in which genes are replaced with mutated versions rather than being knocked out is also a valuable tool for research. That is, a gene can be isolated and cloned from a mouse, manipulated in the laboratory to introduce a specific mutation or mutations, and then used to replace the normal copies of the gene in the mouse. A mouse model for cystic fibrosis (CF) was made in this way. CF is caused by particular mutations in the gene for the cystic fibrosis transmembrane conductance regulator (CFTR) transport protein (see *Why it matters . . .* in Chapter 5). Researchers cloned the mouse homolog of the human CFTR gene, mutated it in the laboratory to introduce the mutations known to cause CF, and replaced the normal mouse CFTR genes with the mutated form. The mouse CF model is being used to study the disease in detail and for developing effective treatments.

Gene targeting methods for gene knockout and gene replacement are used in a variety of model organisms to create

FIGURE 18.8 | Research Method

Making a Knockout Mouse

Purpose: Make a transgenic mouse in which a specific gene has been knocked out (deleted) so that the function of that gene is lost.

Protocol:

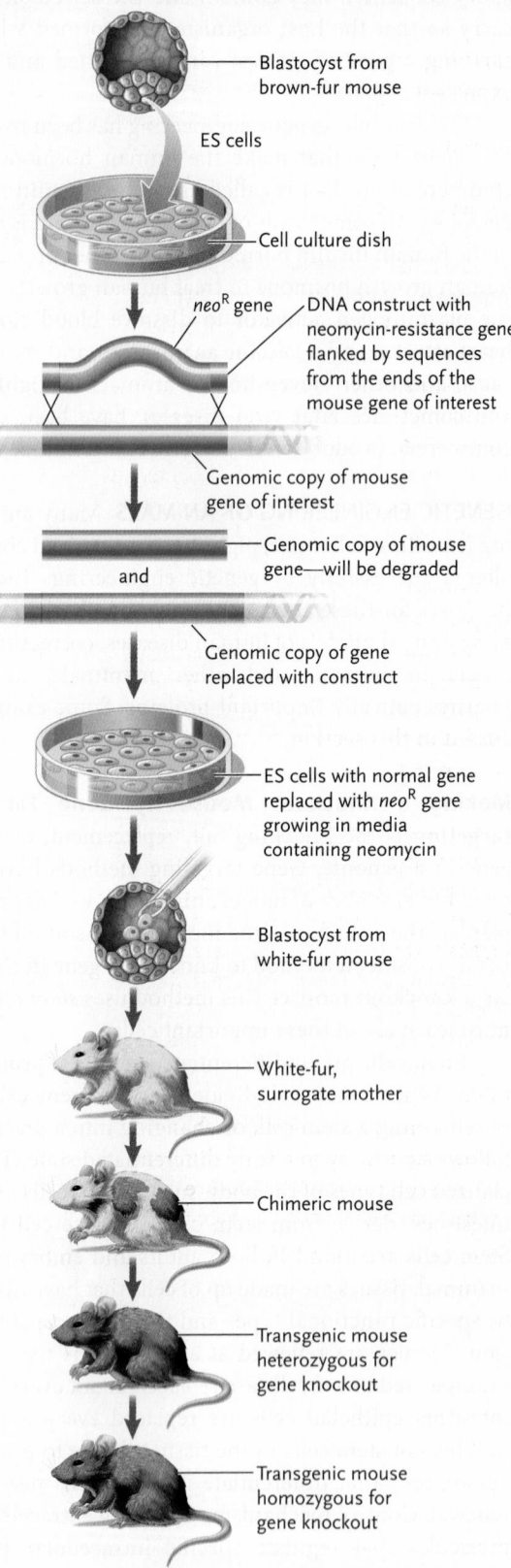

1. Extract ES cells from blastocyst of an agouti (mottled, grayish brown fur color) mouse and grow them in cell culture.

2a. Transform the ES cells with a DNA construct consisting of a selectable marker, here the neomycin-resistance (neo^R) gene (red), flanked by end sequences of the mouse gene of interest. The construct migrates to the nucleus and the mouse gene sequences align with the homologous sequences of a genomic copy of the mouse gene, allowing crossing-over to occur ...

2b. ... The outcome of crossing-over is the replacement of the genomic copy of the gene with the neo^R-containing DNA sequence. The chromosomal allele of the mouse gene is now nonfunctional.

3. Select for the ES cells in which a normal gene has been replaced by the neo^R-containing DNA sequence by growing cells in media containing neomycin.

4. Inject genetically engineered ES cells into blastocysts from white mice.

5. Implant the blastocysts into white, surrogate (foster) mother.

6. Some progeny mice will be white; others will be chimeric mice with patches of agouti fur. (Agouti fur is genetically dominant to white.) Chimeric mice have many cells derived from the original white mouse blastocyst, but some cells derived from the agouti mouse-derived genetically engineered ES cells that were introduced into the blastocyst in step 4.

7. Mate chimeric mice with white mice. If the chimeric mouse parent as gonads derived from the genetically engineered stem cells, all offspring will have agouti fur. The agouti offspring are heterozygous for the gene knockout.

8. Interbreed the heterozygous knockout mice. Perform DNA testing on agouti progeny to identify those homozygous for the gene knockout.

Labels in figure:
- Blastocyst from brown-fur mouse
- ES cells
- Cell culture dish
- neo^R gene
- DNA construct with neomycin-resistance gene flanked by sequences from the ends of the mouse gene of interest
- Genomic copy of mouse gene of interest
- Genomic copy of mouse gene—will be degraded
- and
- Genomic copy of gene replaced with construct
- ES cells with normal gene replaced with neo^R gene growing in media containing neomycin
- Blastocyst from white-fur mouse
- White-fur, surrogate mother
- Chimeric mouse
- Transgenic mouse heterozygous for gene knockout
- Transgenic mouse homozygous for gene knockout

Outcome: The result is a mouse in which both chromosomal alleles of a specific gene of interest have been knocked out. The effects of the loss of function of that gene can then be studied.

Basic and Applied Research: Programmable RNA-Guided Genome Editing System

Approximately 70% of sequenced bacterial genomes and 90% of sequenced archaeal genomes have CRISPR (pronounced "crisper") loci and *cas* genes that together encode an immune system against foreign bacteriophages and plasmids. In brief, the system works as follows: In the genomes are one or more **CRISPR (Clusters of Regularly Interspersed Short Palindromic Repeats)** loci **(Figure A)**. Each CRISPR locus consists of repeated sequences about 40 bp long that have palindromic regions, and that are interspersed with unique sequences of about the same length. The unique sequences have been "captured" from previous encounters with bacteriophages and plasmids and represent a signature of those invading DNA elements. Nearby each CRISPR locus are *cas* genes which encode DNA endonucleases, enzymes that cut DNA within the length of a strand (see Figure A). Transcription of the region containing the *cas* genes and the CRISPR locus generates a Cas protein and an RNA called *crisprRNA (crRNA)* that binds to the Cas protein, respectively. When a foreign DNA enters the cell that contains a DNA sequence represented by the RNA in crRNA, the crRNA/Cas complex binds to the DNA by DNA–RNA base pairing. The DNA endonuclease of the Cas protein is activated and cleaves both strands of the foreign DNA at the same position creating a double-strand break, thereby inactivating the invader. The crRNA/Cas complex in some ways is similar to the miRISC molecule used in RNA interference (see Section 16.3).

In this natural immune system, the key to inactivating foreign DNA is expressing an RNA molecule containing a sequence that can base pair with that DNA. Collaborators Jennifer Doudna (University of California, Berkeley) and Emmanuelle Charpentier (Helmholtz Centre for Infection Research, Germany) hypothesized that the natural system could be adapted to cutting within any DNA sequence by using custom-designed crRNA molecules. Support for their hypothesis came from in vitro experiments in which purified Cas protein and custom-designed crRNA molecules—called *single guide RNAs (sgRNAs)*—cut plasmid DNA to produce fragment sizes predicted by the crRNA sequences. The natural CRISPR–Cas system had been modified to be a **programmable RNA-guided genome editing system (Figure B).** This system is different from cutting DNA with a restriction enzyme because a restriction enzyme cuts DNA only at its recognition sequence. By contrast, the CRISPR–Cas system can cut any sequence of DNA simply by customizing the crRNA sequence.

CRISPR–Cas technology has brought about a revolution in genome engineering. It has been embraced rapidly by research groups around the world for both basic and applied research. In cell- or organism-based experiments, the Cas protein and the sgRNA are expressed from plasmids to initiate the biological effects of the programmed genome editing. Through various engineered modifications of the system, a wide range of programmed genome editing is possible, including gene knockouts, gene insertion, gene replacement, and altering specific sequences. Making gene knockouts with CRISPR–Cas is simpler and

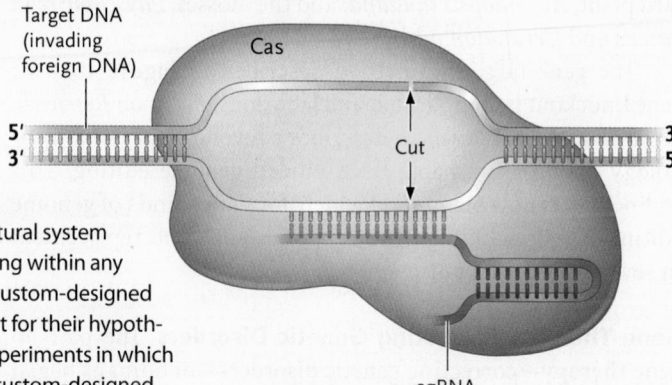

FIGURE B Model for programmable RNA-guided genome editing by Cas and sgRNA. The sgRNA/Cas complex binds to target DNA by complementary base pairing. Cas then cuts both strands of the target DNA.
© Cengage Learning 2017

more time-efficient than the traditional method shown in Figure 18.8. An example of altering specific sequences is introducing specific double-strand breaks into human cell lines that mimic double-strand breaks identified in specific cancers so as to model the cancers for research. Another example is correcting genetic mutations responsible for inherited disorders in cultured cells. CRISPR–Cas technology is being used in basic biological research in cell lines and model organisms such as fruit flies, nematodes, zebrafish, mice, frogs, monkeys, including genome modifications of their germ lines. CRISPR–Cas technology is also being used in biotechnology research, including manipulation of crop plants such as rice, wheat, and tobacco, and commercially useful fungi such as *Kluyveromyces*. In the future we can expect even more significant advances to be made using CRISPR–Cas technology, including benefitting research in human gene therapy.

In conclusion, the exciting CRISPR–Cas technology for genome editing came out of basic research with prokaryotes. The identification of unusual loci led to the discovery of an immune system in prokaryotes that responds to foreign DNA that had been encountered before. Once the molecular mechanisms of that process were understood, the process was adapted and developed for applications in genetics and molecular biology that is leading to revolutionary advances.

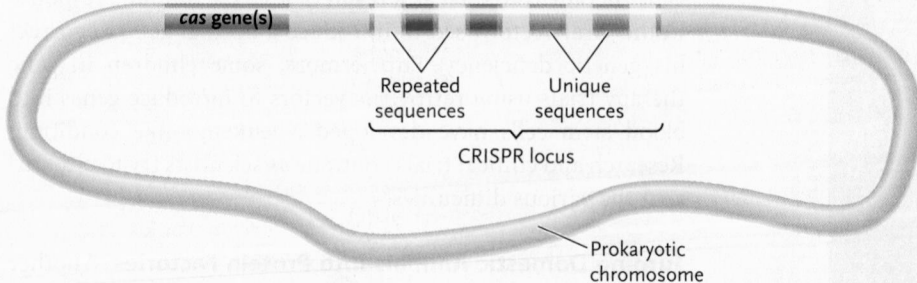

FIGURE A Organization of *cas* gene(s) and a CRISPR locus in a prokaryotic chromosome. The repeated sequences in the CRISPR locus are colored yellow and the unique sequences each have a different color.
© Cengage Learning 2017

specific transgenic organisms for biological or disease-related research. Those organisms include the yeast, *Saccharomyces cerevisiae;* the nematode, *Caenorhabditis elegans;* the fruit fly, *Drosophila melanogaster;* the rat, *Rattus norvegicus;* the mustard plant, *Arabidopsis thaliana;* and the mosses, *Physcomitrella patens* and *Ceratodon purpureus.*

The gene targeting method described in Figure 18.8 for gene knockout is complicated and laborious. *Focus on Research: Basic and Applied Research* describes a revolutionary new technology for programmable RNA-guided genome editing. This technology is now being used widely for many kinds of genome editing tasks, including gene knockout and gene replacement in several model organisms.

Gene Therapy: Correcting Genetic Disorders The path to **gene therapy**—correcting genetic disorders—in humans began with experiments using mice. In 1982, Richard Palmiter at the University of Washington, Ralph Brinster of the University of Pennsylvania, and their colleagues injected a growth hormone gene from rats into fertilized mouse eggs and implanted the eggs into a surrogate mother. Some normal-sized mouse pups were produced that grew faster than normal and became about twice the size of their normal litter mates: they were *giant mice* **(Figure 18.9).**

Palmiter and Brinster next attempted to cure a genetic disorder by gene therapy. In this experiment, they constructed transgenic mice by gene targeting, in this case adding a normal copy of the rat growth hormone gene to mutant mice known as little, genotype *lit/lit.* This mouse mutant is about one-half normal size due to decreased levels of growth hormone gene expression and of growth hormone itself. The transgenic mice grew faster than *lit/lit* and reached a size about three times that of mutant mice. In fact, because regulation of expression of the rat growth hormone in the transgenic mice is not under

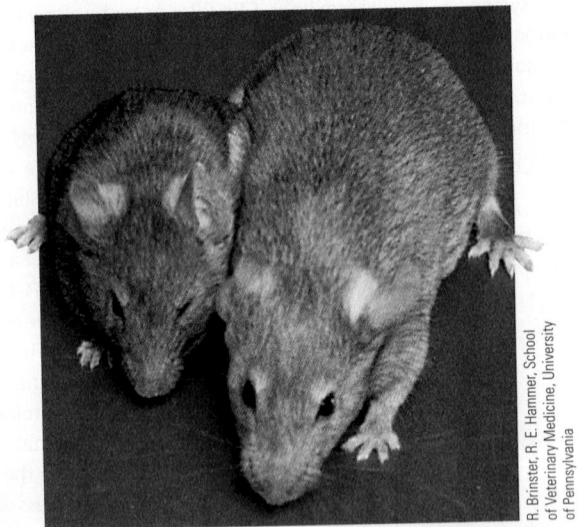

FIGURE 18.9 A genetically engineered giant mouse (right) produced by the introduction of a rat growth hormone gene into the animal. A mouse of normal size is on the left.

R. Brinster, R. E. Hammer, School of Veterinary Medicine, University of Pennsylvania

normal control, the transgenic mice grow to be slightly larger than even normal mice. Overall the results showed that the genetic defect in those mice had been corrected, at least partially.

This sort of experiment, in which a gene is introduced into germline cells of an animal to correct a genetic disorder, is **germline gene therapy.** For ethical reasons, germline gene therapy is not permitted with humans. Instead, humans are treated with **somatic gene therapy,** in which genes are introduced into somatic cells (as described in the previous section). In this method, somatic cells are removed from the body, cultured, and then transformed with an expression vector containing the transgene. The modified cells are then reintroduced into the body where the transgene functions. Because germ cells and their products are not involved, the transgene remains in the engineered individual and is not passed to offspring.

The first successful use of somatic gene therapy with a human subject who had a genetic disorder was carried out in the 1990s by W. French Anderson and his colleagues at the National Institutes of Health (NIH). The subject was a young girl with *adenosine deaminase deficiency (ADA).* Without the adenosine deaminase enzyme, white blood cells cannot mature (see Chapter 45). In the absence of normally functioning white blood cells, the body's immune response is so deficient that most children with ADA die of infections before reaching puberty. The researchers were successful in introducing a functional ADA gene into mature white blood cells isolated from the patient. Those cells were reintroduced into the girl, and expression of the ADA gene provided a temporary cure for her ADA deficiency. The cure was not permanent because mature white blood cells, produced by differentiation of stem cells in the bone marrow, are nondividing cells with a finite lifetime. Therefore, the somatic gene therapy procedure has to be repeated every few months. The subject of this example still receives periodic gene therapy to maintain the necessary levels of the ADA enzyme in her blood. In addition, she receives direct doses of the normal enzyme.

Despite enormous efforts, human somatic gene therapy has not been the panacea people expected. Relatively little progress has been made since the first gene therapy clinical trial for ADA deficiency, and, in fact, there have been major setbacks. In 1999, for example, a teenage patient in a somatic gene therapy trial died as a result of a severe immune response to the viral vector used to introduce a normal gene to correct his genetic deficiency. Furthermore, some children in gene therapy trials using retrovirus vectors to introduce genes into blood stem cells have developed a leukemia-like condition. Research and clinical trials continue as scientists try to circumvent the various difficulties.

Turning Domestic Animals into Protein Factories Another successful application of genetic engineering turns animals into pharmaceutical factories for the production of proteins required to treat human diseases or other medical conditions. Most of these *pharming* projects, as they are called, engineer

FIGURE 18.10 | **Experimental Research**

The First Cloning of a Mammal

Question: Does the nucleus of an adult mammal contain all the genetic information to specify a new organism? In other words, can mammals be cloned starting with adult cells?

Experiment: Ian Wilmut, Keith Campbell, and their colleagues fused a mammary gland cell from an adult sheep with an unfertilized egg cell from which the nucleus had been removed and tested whether that fused cell could produce a lamb.

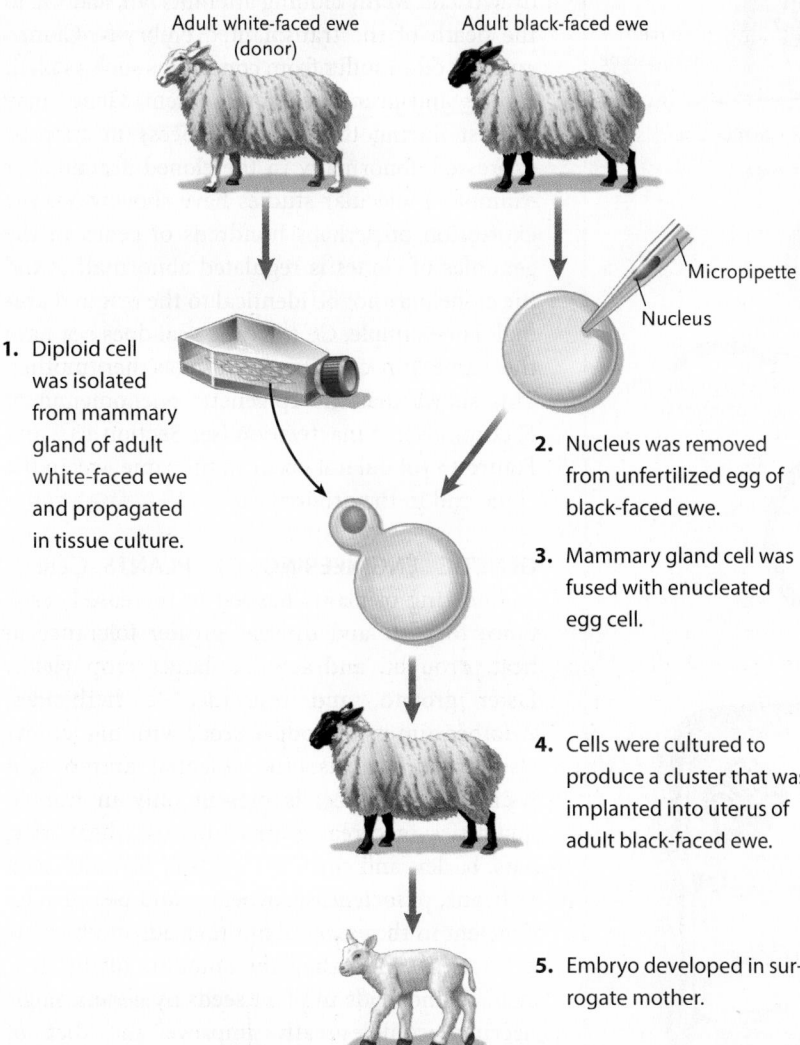

Adult white-faced ewe (donor)

Adult black-faced ewe

Micropipette

Nucleus

1. Diploid cell was isolated from mammary gland of adult white-faced ewe and propagated in tissue culture.

2. Nucleus was removed from unfertilized egg of black-faced ewe.

3. Mammary gland cell was fused with enucleated egg cell.

4. Cells were cultured to produce a cluster that was implanted into uterus of adult black-faced ewe.

5. Embryo developed in surrogate mother.

Result: Dolly was born and grew normally. She was white-faced—a clone of the donor ewe. DNA fingerprinting using STR loci showed her DNA matched that of the donor ewe and neither the ewe who donated the egg, nor the ewe who was the surrogate mother.

Conclusion: An adult nucleus of a mammal contains all the genetic material necessary to direct the development of a normal new organism, a clone of the original. Dolly was the first cloned mammal. The success rate for Wilmut and Campbell's experiment was very low—Dolly represented less than 0.4% of the fused cells they made—but its significance was huge.

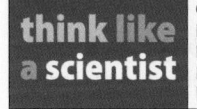
think like a scientist

Observe
Hypothesize
Predict
Experiment
Interpret

Why was it important that the donor ewe and surrogate mother be sheep of different kinds?

Source: I. Wilmut et al. 1997. Viable offspring derived from fetal and adult mammalian cells. *Nature* 385:810–813.

the animals to produce the desired proteins in milk, making the isolation and purification of the proteins easy, as well as harmless to the animals. ("Pharming" is a word made up from "pharmaceuticals" and "farming.")

One of the first successful applications of this approach was carried out with transgenic sheep that produce a protein required for normal blood clotting in humans. The protein, called a *clotting factor,* is deficient in persons with one form of hemophilia. (The genetics of hemophilia is described in Section 13.2.) These people require frequent injections of the factor to avoid bleeding to death from even minor injuries. Using DNA cloning techniques, researchers joined the gene encoding the normal form of the clotting factor to the promoter sequences of the β-lactoglobulin gene, which encodes a protein secreted in milk, and introduced it into fertilized eggs. Those cells were implanted into a surrogate mother, and the transgenic sheep born were allowed to mature. The β-lactoglobulin promoter controlling the clotting factor gene became activated in mammary gland cells of females. The clotting factor produced was secreted into the milk from which it could be collected and purified.

Other similar projects are being done to produce particular proteins in transgenic mammals. These include a protein to treat cystic fibrosis, collagen to correct scars and wrinkles, human milk proteins to be added to infant formulas, and normal hemoglobin for use as an additive to blood transfusions.

Producing Animal Clones Making transgenic mammals is expensive and inefficient. And, because only one copy of the transgene typically becomes incorporated into the treated cell, not all progeny of a transgenic animal inherit that gene. Scientists reasoned that an alternative to breeding a valuable transgenic mammal to produce progeny with the transgene would be to clone the mammal. Each clone would be identical to the original, including the expression of the transgene. That this is possible was shown when two Scottish scientists, Ian Wilmut of the Roslin Institute and Keith Campbell of PPL Therapeutics, Roslin, Scotland, announced in 1997 that they were successful in cloning a sheep using a single somatic cell taken from an adult sheep **(Figure 18.10)**—Dolly, the first cloned mammal.

After the successful cloning experiment that produced Dolly, many other mammals have been cloned, including mice, goats, pigs, monkeys,

FIGURE 18.11 **Research Method**

Using the Ti Plasmid of *Agrobacterium tumefaciens* to Produce Transgenic Plants

Purpose: To make transgenic plants. This technique is one way to introduce a transgene into a plant for genetic engineering purposes.

Protocol:

1. Isolate the Ti plasmid from *Agrobacterium tumefaciens*. The plasmid contains a segment called T DNA (T = transforming), which induces tumors in plants.

2. Digest the Ti plasmid with a restriction enzyme that cuts within the T DNA. Mix with a gene of interest on a DNA fragment that was produced by digesting with the same enzyme. Use DNA ligase to join the two DNA molecules together to produce a recombinant plasmid.

3. Transform the recombinant Ti plasmid into a disarmed *A. tumefaciens* that cannot induce tumors, and use the transformed bacterium to infect cells in plant fragments in a test tube. In infected cells, the T DNA with the inserted gene of interest excises from the Ti plasmid and integrates into the plant cell genome.

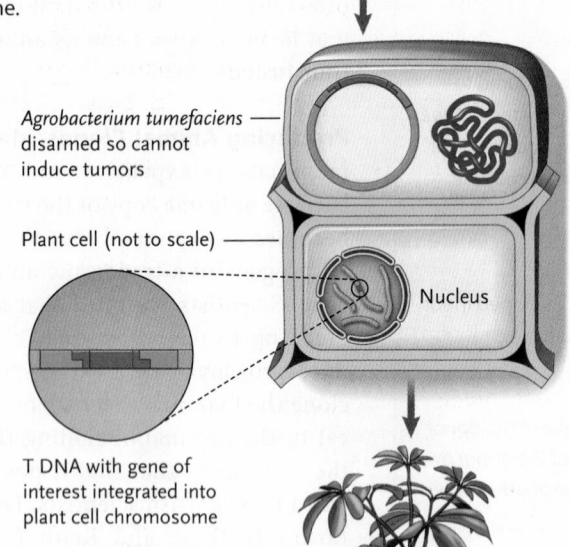

4. Culture the transgenic plant fragments to regenerate whole plants.

Outcome: The plant has been genetically engineered to contain a new gene. The transgenic plant will express a new trait based on that gene, perhaps resistance to an herbicide or production of an insect toxin, according to the goal of the experiment.

© Cengage Learning 2017

rabbits, dogs, a male calf appropriately named Gene, and a domestic calico cat called *CC* (for *Copy Cat*).

Cloning farm animals has been so successful that several commercial enterprises now provide cloned copies of champion animals. One example is a clone of an American Holstein cow, Zita, who was the U.S. national champion milk producer for many years.

The cloning of domestic animals has its drawbacks. Many cloning attempts fail, leading to the death of the transplanted embryos. Cloned animals often suffer from conditions such as birth defects and poor lung development. Genes may be lost during the cloning process or may be expressed abnormally in the cloned animal. For example, molecular studies have shown that the expression of perhaps hundreds of genes in the genomes of clones is regulated abnormally. And the clone may not be identical to the original animal. For example, *CC* the calico cat does not have the same fur coloring pattern as her mother. This shows that the epigenetic phenomenon of X-chromosome inactivation (see Section 13.2 and Figure 13.10) did not occur in the same way in the clone and in the original cat.

GENETIC ENGINEERING OF PLANTS Genetic engineering of plants has led to increased resistance to pests and disease; greater tolerance to heat, drought, and salinity; larger crop yields; faster growth; and resistance to herbicides. Another aim is to produce seeds with higher levels of amino acids. The essential amino acid lysine, for example, is present only in limited quantities in cereal grains such as wheat, rice, oats, barley, and corn; the seeds of legumes such as beans, peas, lentils, soybeans, and peanuts are deficient in the essential amino acids methionine or cysteine. Increasing the amounts of the deficient amino acids in plant seeds by genetic engineering would greatly improve the diet of domestic animals and human populations that rely on seeds as a primary food source.

Other applications for plant genetic engineering include plant pharming to produce pharmaceutical products. Plants are ideal for this purpose, because they are primary producers at the bottom rung of the food chain and can be grown in huge numbers with maximum conservation of the Sun's energy that is captured in photosynthesis.

Some plants, such as *Arabidopsis thaliana* (thale cress), tobacco, potato, cabbage, and carrot, have special advantages for genetic engineering because individual cells can be removed from an adult, altered by gene targeting to introduce a desired gene, and then grown in cultures into a multicellular mass of cloned cells called a *callus.* Subsequently, roots, stems, and leaves develop in the callus, forming a young plant that then can be cultivated by standard methods. Each cell in the plant contains the introduced gene. The gametes produced by the transgenic plants can then be used in crosses to produce offspring, some of which will have the transgene, as in the similar animal experiments.

Methods Used to Insert Genes into Plants Genes are inserted into plant cells by several techniques. A commonly used method takes advantage of properties of a bacterium, *Agrobacterium tumefaciens,* which causes crown gall disease. Crown gall disease is characterized by bulbous, irregular growths—tumors, essentially— that develop at wound sites on the trunks and limbs of deciduous trees. *A. tumefaciens* contains a large, circular plasmid called the **Ti (tumor-inducing) plasmid (Figure 18.11).** The interaction between the bacterium and a plant cell it infects stimulates the excision of a segment of the Ti plasmid called *T DNA* (for transforming DNA), which then integrates into the plant cell's genome. Genes on the T DNA are then expressed; the products stimulate the transformed cell to grow and divide, producing a tumor. The tumors provide essential nutrients for the bacterium. The Ti plasmid is used as a vector for making transgenic plants, in much the same way as cloning vectors introduce genes into bacteria.

Successful Plant Genetic Engineering Projects The most widespread application of genetic engineering of plants involves the production of transgenic crops. Thousands of such crops have been developed and field tested, and many have been approved for commercial use. If you examine the processed, plant-based foods sold at a national supermarket chain, you will find that at least two-thirds contain transgenic plants.

In many cases, plants are modified to make them resistant to insect pests, viruses, or herbicides. Crops that have been modified for insect resistance include corn, rice, tobacco, and potatoes. The most common approach to making plants resistant to insects is to introduce the gene from the bacterium *Bacillus thuringiensis* that encodes the *Bt* toxin, a natural pesticide. This toxin has been used in powder form to kill insects in agriculture for many years, and now transgenic plants making their own *Bt* toxin are resistant to specific groups of insects that feed on them. Millions of acres of crop plants planted in the United States, amounting to about 70% of the nation's agricultural acreage, are now *Bt*-engineered varieties.

Virus infections cause enormous crop losses worldwide. Virus-resistant transgenic crops would be highly valuable in agriculture. There is some promise in this area. By some unknown process, transgenic plants expressing certain viral proteins become resistant to infections by whole viruses that contain those same proteins. Engineered varieties of papaya, squash, and potatoes have been engineered to resist particular viral pathogens. Bacteria also cause significant crop losses and

genetic engineering approaches also are being used to combat those losses. For example, genetically modified oranges are in field trials to test resistance to citrus greening disease, which is caused by *Candidatus* bacteria introduced by insects.

Several crops have also been engineered to become resistant to herbicides. For example, *glyphosate* (commonly known by its brand name, Roundup) is a potent herbicide that is widely used in weed control. The herbicide works by inhibiting a particular enzyme in the chloroplast. Unfortunately, it also kills crops. But transgenic crops have been made in which a bacterial form of the chloroplast enzyme was added to the plants. The bacteria-derived enzyme is not affected by Roundup, and farmers who use these herbicide-resistant crops can spray fields of crops to kill weeds without killing the crops. Now most of the corn, soybean, and cotton plants grown in the United States and many other countries are glyphosate-resistant ("Roundup-ready") varieties that were produced by genetic engineering. Various crops are now commercially available that are resistant to the herbicide glufosinate, and tobacco plants have been modified to be resistant to the herbicide bromoxynil.

Crop plants are also being engineered to alter their nutritional qualities. For example, a strain of rice plants has been produced with seeds rich in β-carotene, a precursor of vitamin A **(Figure 18.12).** The new rice, which is given a yellow or golden color by its carotene content, may provide improved nutrition for the billions of people that depend on rice as a diet staple. In particular, the rice may help improve the nutrition of children younger than five years of age in southeast Asia, 70% of whom suffer from impaired vision because of vitamin A deficiency. Other examples of plants genetically modified to alter their nutritional qualities include a South African corn variety engineered to be enriched in vitamin C, folate, and β-carotene, and false flax (*Camelina sativa*) engineered to produce high levels of oils that are similar to fish oils. Neither plant is currently in production commercially. *Molecular Insights* describes an experiment which demonstrated that the metabolic fingerprint of genetically modified tomato fruits is not significantly different from nontransgenic fruits.

Plant pharming is also an active area in both university research labs and biotechnology companies. As with animal pharming, plant pharming involves the engineering of transgenic plants to produce medically valuable products. Products under development include vaccines for various bacterial and viral diseases, protease inhibitors to treat or prevent virus

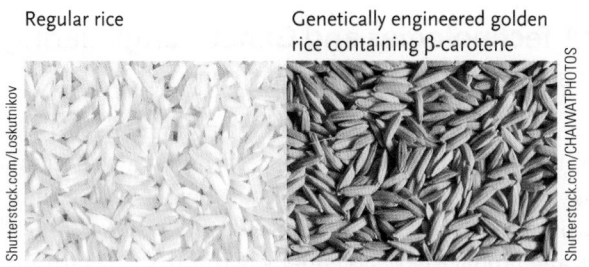

Regular rice   Genetically engineered golden rice containing β-carotene

FIGURE 18.12 Rice genetically engineered to contain β-carotene.

Nutritional Quality of Genetically Modified Food: Comparison of metabolomes in genetically modified and unmodified tomato fruits

Among the concerns people have is whether genetically modified foods are less tasty or nutritious than their unmodified versions. Nutrition and taste of food depend on the number and quantities of metabolites in the food. Metabolites are the intermediates and products of metabolism. The set of metabolites in a cell or organism is the metabolome, and the study of the metabolome is called *metabolomics*. Some people argue that the genetic manipulation of the food organism can cause unexpected changes in food quality and composition. Therefore, one could conclude that the genetic modification of the food is not detrimental to food quality if the metabolome is not significantly different from that of unmodified food.

Research Question

Are the metabolomes of genetically modified foods significantly different from their unmodified (nontransgenic) versions?

Experiment

Owen Hoekenga and his research team at Cornell University extracted approximately 1,000 metabolites of the metabolome

of normal (nontransgenic) tomato fruits **(Figure)** and of tomato fruits genetically engineered to delay fruit ripening. The engineered fruits remain fresher longer than normal fruits. The metabolites were analyzed using liquid chromatography–mass spectrometry to provide metabolic fingerprints of the different fruit types. The researchers used sophisticated statistical approaches to compare the large data sets for the various fruit samples to analyze for significant differences. The results showed that, with one exception, there were no significant differences between the genetically engineered fruits and a wide assortment of nontransgenic fruits that encompassed modern and heirloom varieties. The scientists did observe that the genetically engineered fruit had a metabolic fingerprint distinct from its nontransgenic parent, but that fingerprint fell within the range of fingerprints of the nontransgenic varieties. The exception was the metabolites associated with fruit ripening, and significant changes were expected there because that process was the target of the genetic modification.

FIGURE Different varieties of nontransgenic tomatoes.

Conclusion

Genetically modifying tomatoes to delay fruit ripening does not significantly affect the metabolic fingerprint of the tomato fruit other than for the fruit-ripening metabolites.

think like a scientist

Propose another research question that you could answer using metabolic fingerprinting.

Source: M. V. DiLeo et al. 2014. An assessment of the relative influences of genetic background, functional diversity at major regulatory genes, and transgenic constructs on the tomato fruit metabolome. *The Plant Genome* 7:1–16.

infections, collagen to treat scars and wrinkles, and the protein aprotinin, which is used to reduce bleeding and clotting during heart surgery. For example, a drug for treating Gaucher's disease is the first U.S. Food and Drug Administration (FDA)-approved plant-produced pharmaceutical. In the laboratory, bananas have been engineered to produce human vaccines for some infectious diseases, for instance, hepatitis B.

In contrast to modifying animals by genetic engineering techniques, genetically altered plants have been developed widely and appear to have become mainstays of agriculture. But, as the next section discusses, both animal and plant genetic engineering have not proceeded without concerns.

DNA Technologies and Genetic Engineering Are a Subject of Public Concern

When recombinant DNA technology was developed in the early 1970s, researchers were concerned that a bacterium carrying a recombinant DNA molecule might escape into the environment, transfer the recombinant molecule to other bacteria, and produce new, potentially harmful, strains. To address

this, and other concerns, the U.S. scientists who developed the technology drew up comprehensive safety guidelines for recombinant DNA research in the United States. That included growing the bacteria in laboratories that followed appropriate biosafety protocols, and using bacterial strains mutated so that they could not survive outside the growth media used in the laboratory. Since that time, countless thousands of experiments involving recombinant DNA molecules have shown that recombinant DNA manipulations can be done safely. Over time, therefore, the recombinant DNA guidelines have become more relaxed. Nonetheless, stringent regulations still exist for certain areas of recombinant DNA research that pose significant risk, such as cloning genes from highly pathogenic bacteria or viruses, or gene therapy experiments.

Guidelines for genetic engineering also extend to research in several areas that have been the subject of public concern and debate. While the public is concerned little about genetically engineered microorganisms, for example those cleaning up oil spills and hazardous chemicals, it is concerned about possible problems with genetically modified organisms (GMOs) used as food. The majority of GMOs are crop plants. Issues of concern include the safety of GMO-containing food, and the possible adverse effects of the GMOs on the environment, such as by interbreeding with natural species or by harming beneficial insect species. One example of the latter concern was whether *Bt*-expressing corn would have adverse effects on monarch butterflies that fed on the pollen of the plants. The most recent of a series of independent studies investigating this possibility has indicated that the risk to the butterflies from *Bt* toxin is extremely low.

More broadly, different countries have reacted to GMOs in different ways. In the United States, transgenic crops are planted widely. Before commercialization, such GMOs are evaluated for potential risk by appropriate government regulatory agencies, including the NIH, FDA, Department of Agriculture, and Environmental Protection Agency (EPA). Usually, the opposition to GMOs has come from particular activist and consumer groups.

Political opposition to GMOs has been greater in Europe, dampening the use of transgenic crop plants in the fields and GMOs in food. In 1999, the European Union (EU) imposed a 6-year moratorium on all GMOs. Now the EU has strict regulations for GMOs that involve extensive testing, labeling, the ability to trace them, and the monitoring of agricultural products derived from them.

Globally, an international agreement, the **Cartagena Protocol on Biosafety,** "promotes biosafety by establishing practical rules and procedures for the safe transfer [between countries], handling and use of GMOs." Separate procedures have been set up for GMOs that are to be introduced into the environment and those that are to be used as food or feed or for processing. As of March 2014, 166 countries and the European Union had ratified the Protocol; the United States was not one of them.

In sum, the use of DNA technologies in biotechnology has the potential for tremendous benefits to humankind. Such experimentation is not without risk, and so for each experiment, researchers must assess that risk and make a judgment about whether to proceed and, if so, how to do so safely. Furthermore, agreed-upon guidelines and protocols should ensure a level of biosafety that is acceptable to researchers, consumers, politicians, and governments.

STUDY BREAK 18.2

1. What is a transgenic organism?
2. What is the difference between using germline cells and somatic cells for gene therapy?

THINK OUTSIDE THE BOOK

Individually or collaboratively, investigate how it can be determined experimentally that processed foods that contain plant material contain transgenic plants. List the steps you would take to investigate whether a can of corn from a supermarket contains genetically modified corn.

18.3 Other Applications of DNA Technologies

In this section we discuss selected applications of DNA technologies that do not involve making or using genetically altered organisms.

DNA Technologies Are Used in Molecular Testing for Many Human Genetic Diseases

Many human genetic diseases are caused by defects in enzymes or other proteins that result from DNA mutations. Once scientists have identified the specific mutations responsible for human genetic diseases, they can often use DNA technologies to develop molecular tests for those diseases. One example is sickle-cell anemia (see *Why it matters . . .* in Chapter 12 and Sections 12.2 and 13.4). People with sickle-cell anemia are homozygous for a single base-pair mutation (see Section 15.5) that affects hemoglobin, the oxygen-carrying molecule of the blood. Hemoglobin consists of two copies each of the α-globin and β-globin polypeptides. The mutation, which is in the β-globin gene, changes one codon to another, resulting in a different amino acid at one place in the polypeptide. As a consequence of the different properties of the different amino acid, the function of hemoglobin is impaired significantly in individuals homozygous for the mutation (who have sickle-cell anemia), and mildly impaired in individuals heterozygous for the mutation (who have sickle-cell trait).

Three restriction sites for *Mst*II are associated with the normal β-globin gene, two within the coding sequence of the gene and one upstream of the gene. The sickle-cell mutation eliminates the middle site of the three **(Figure 18.13).** The human genome sequence is known, so the DNA sequence of the region containing the gene is known. Therefore, the region can be amplified by PCR **(Figure 18.14).** The resulting DNA fragments then can be cut with *Mst*II and the sizes of the fragment or fragments determined by agarose gel electrophoresis.

Cutting the amplified DNA fragment containing the β-globin gene with *Mst*II produces two DNA fragments from

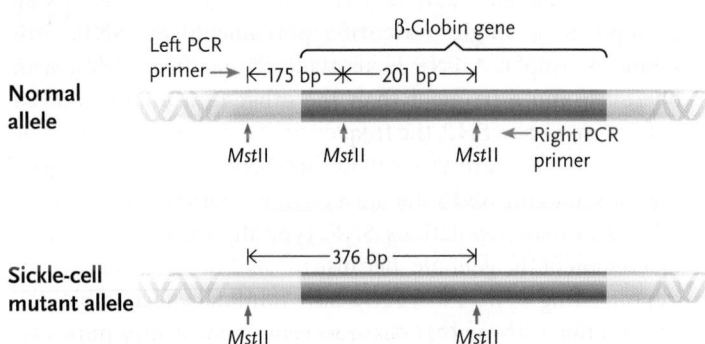

FIGURE 18.13 Restriction site differences between the normal and sickle-cell mutant alleles of the β-globin gene.
© Cengage Learning 2017

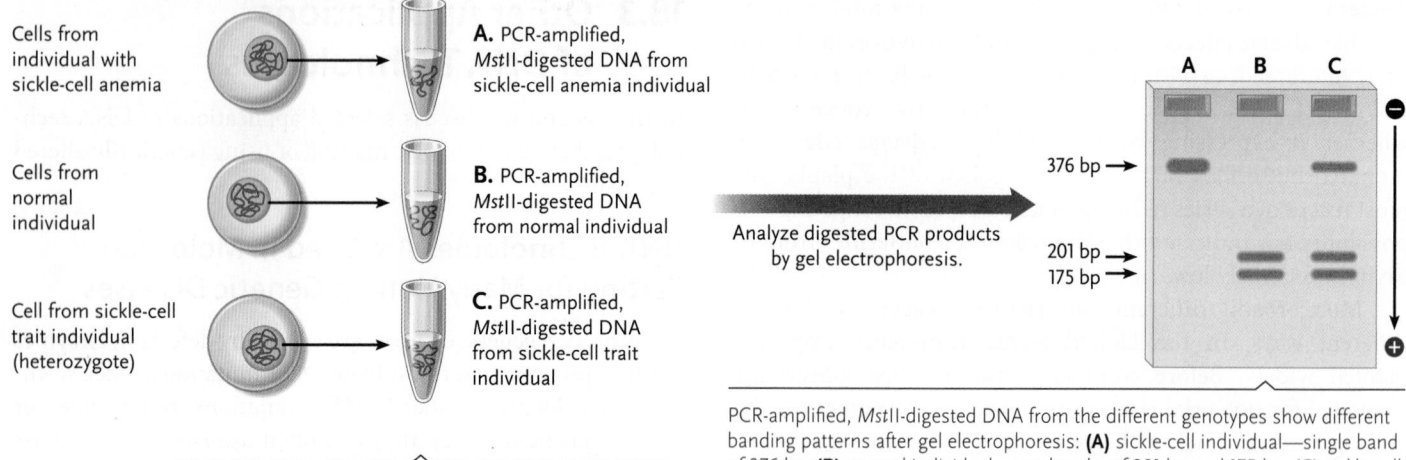

Cells from individual with sickle-cell anemia

A. PCR-amplified, *Mst*II-digested DNA from sickle-cell anemia individual

Cells from normal individual

B. PCR-amplified, *Mst*II-digested DNA from normal individual

Cell from sickle-cell trait individual (heterozygote)

C. PCR-amplified, *Mst*II-digested DNA from sickle-cell trait individual

Analyze digested PCR products by gel electrophoresis.

376 bp →
201 bp →
175 bp →

Isolate genomic DNA, amplify the β-globin gene region by PCR, and digest the resulting DNA fragments with *Mst*II.

PCR-amplified, *Mst*II-digested DNA from the different genotypes show different banding patterns after gel electrophoresis: **(A)** sickle-cell individual—single band of 376 bp; **(B)** normal individual—two bands, of 201 bp and 175 bp; **(C)** sickle-cell trait heterozygote—three bands, of 376 bp (from the sickle-cell mutant allele), and 201 bp and 175 bp (both from the normal allele).

FIGURE 18.14 Distinguishing normal and sickle-cell mutant alleles by analysis of *Mst*II restriction fragments of the PCR-amplified β-globin gene.
© Cengage Learning 2017

the normal gene and one fragment from the mutated gene. Restriction enzyme-generated DNA fragments of different lengths from the same region of the genome, such as in this example, are known as **restriction fragment length polymorphisms (RFLPs,** pronounced "riff-lips"). As illustrated in Figure 18.14 for this sickle-cell anemia example, analysis of RFLPs by agarose gel electrophoresis indicates the allele or alleles present in each tested individual.

PCR DNA amplification of a gene or region of a gene followed by digestion with a diagnostic restriction enzyme can be used only to test for human genetic diseases caused by a mutation (or mutations) that affects a restriction site. Many genetic diseases are caused by mutations involving a base pair change that does not affect a restriction site, or that affect more than one base pair, for example, deletions or insertions. Increasingly, due to the development of relatively inexpensive high-throughput DNA sequencing methods, DNA sequencing is being used as the technique of choice for screening for many genetic disease mutations.

The single base-pair mutation in the β-globin gene is an example of a **single-nucleotide polymorphism (SNP,** pronounced "snip"). A SNP locus typically has two alleles, with one allele more frequent than the other in a population. By definition, to be a SNP, the frequency of the rarer allele must be at least 1%. There are about 10 million SNPs in the human populations, making SNPs the most common form of genetic variation in human populations. SNPs typically are used as markers of a region of the genome. Because most of the genome consists of noncoding DNA, most SNPs have minimal effect on biological functions. Those that occur in genes may or may not cause a phenotypic change; the sickle-cell SNP is an example of one that does cause a phenotypic change. Further, most SNPs do not alter restriction sites; the sickle-cell SNP is an example of one that does.

We need to be clear about a mutation versus a SNP. As you have learned, both involve a single base-pair change. The frequency of the rare or mutant allele is the key to the distinction. Geneticists writing about their research typically use the term SNP for relatively common single base-pair changes (greater than 1% frequency in a population), a condition the sickle-cell anemia mutation meets. Geneticists typically use the term mutation for rare genetic base-pair changes (less than 1% frequency in a population). An example is cystic fibrosis (see *Why it matters . . .* in Chapter 5, and Section 13.4), which is a rare genetic disease.

DNA Fingerprinting Is Used to Identify Human Individuals, as well as Individuals of Other Species

Just as each human has a unique set of fingerprints, each also has unique combinations and variations of DNA sequences (with the exception of identical twins) known as *DNA fingerprints.* **DNA fingerprinting** (also called **DNA profiling**) is a technique used to distinguish between individuals of the same species using DNA samples. Invented by Sir Alec Jeffreys in 1984, DNA fingerprinting has become a mainstream technique for distinguishing human individuals, notably in forensics and paternity testing. The technique is applicable to all kinds of organisms. We focus on humans in the following discussion.

DNA FINGERPRINTING PRINCIPLES In DNA fingerprinting, molecular techniques, typically PCR, are used to analyze DNA variations at various loci in the genome. In the United States, the Federal Bureau of Investigation (FBI) has identified 13 core loci in noncoding regions of the genome as standards for PCR analysis. Each locus is an example of a **short tandem repeat**

A. Alleles at an STR locus

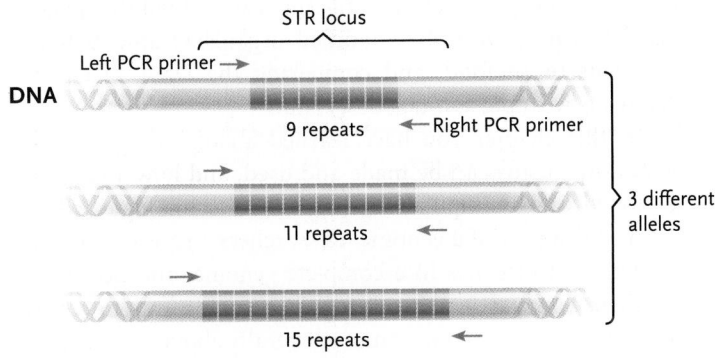

B. DNA fingerprint analysis of the STR locus by PCR

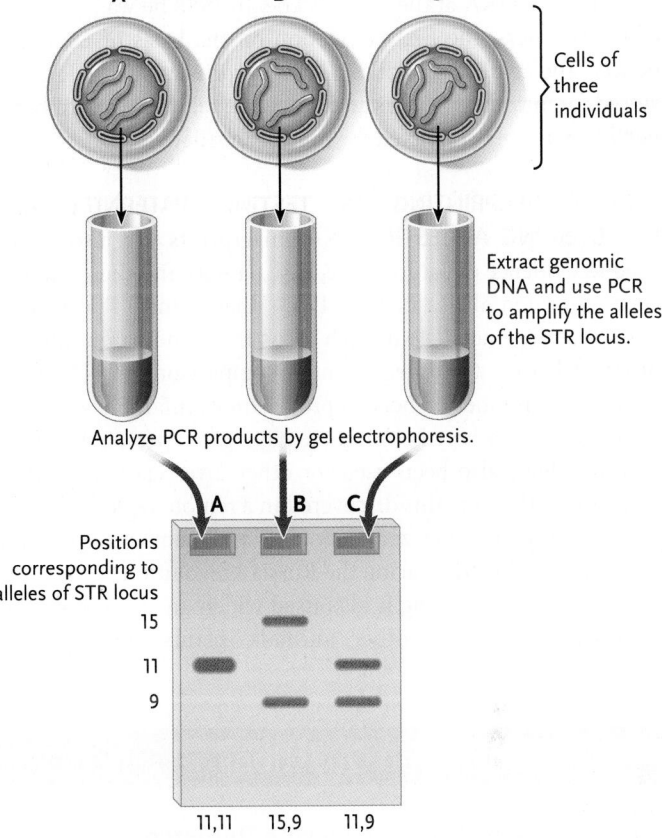

FIGURE 18.15 Using PCR to obtain a DNA fingerprint for an STR locus. (A) Three alleles of the STR locus with 9, 11, and 15 copies of the tandemly repeated sequence. The arrows indicate where left and right PCR primers bind to amplify the STR locus. **(B)** DNA fingerprint analysis of the STR locus by PCR. The number of bands on the gel and the sizes of the DNA in the bands show the STR alleles that were amplified. One band indicates that the individual was homozygous for an STR allele with a particular number of repeats, while two bands indicates the individual is heterozygous for two STR alleles with different numbers of repeats. Here, individual A is homozygous for an 11-repeat allele (designated 11,11), B is heterozygous for a 15-repeat allele and a 9-repeat allele (15,9), and C is heterozygous for the 11-repeat allele and the 9-repeat allele (11,9).

© Cengage Learning 2017

(STR) sequence (also called a **microsatellite**), meaning that it has a short 2–6 bp sequence of DNA repeated in series. Each locus has a different repeated sequence, and the number of repeats varies among individuals in a population. For example, one STR locus has the sequence 5′-AGAT-3′ repeated between 8 and 20 times. As a further source of variation, a given individual is either homozygous or heterozygous for an STR allele; perhaps you are homozygous for an 11-repeat allele, or heterozygous for a 9-repeat allele and a 15-repeat allele. Likely your DNA fingerprint for this locus is different from most of the others in your class. Because each individual has an essentially unique combination of alleles (identical twins are the exception), analysis of multiple STR loci can discriminate between DNA of different individuals and, therefore, potentially identify individuals on the basis of DNA analysis. **Figure 18.15** illustrates how PCR is used to obtain a DNA fingerprint for a theoretical STR locus with three alleles of 9, 11, and 15 tandem repeats.

DNA FINGERPRINTING IN FORENSICS DNA fingerprints are used routinely to identify criminals or eliminate innocent persons as suspects in legal proceedings. For example, a DNA fingerprint prepared from a hair found at the scene of a crime, or from a semen sample, might be compared with the DNA fingerprint of a suspect to link the suspect with the crime. Or a DNA fingerprint of blood found on a suspect's clothing or possessions might be compared with the DNA fingerprint of a

victim. Typically, the evidence is presented in terms of the probability that the particular DNA sample could have come from a random individual. Hence the media report probability values, such as one in several million, or in several billion, that a person other than the accused could have left his or her DNA at the crime scene.

Although courts initially met with legal challenges to the admissibility of DNA fingerprints, experience has shown that they are highly dependable as a line of evidence if DNA samples are collected and prepared with care, and if a sufficient number of loci which have variable numbers of DNA repeat sequences are examined. There is always concern, though, about the possibility of contamination of the sample with DNA from another source during the path from the crime scene to the forensic laboratory for analysis. Moreover, in some cases criminals themselves have planted fake DNA samples at crime scenes to confuse the investigation.

DNA fingerprinting has been used extensively to convict or exonerate a suspect. For example, in a case in England, the DNA fingerprints of more than 5,000 men were made using blood and saliva samples during an investigation of the rape and murder of two teenage girls in 1983 and 1986. The results led to the release of a man wrongly imprisoned for the crimes—the first person exonerated by DNA fingerprinting evidence. But no DNA fingerprints matched the DNA at the crime scene. Then a woman overheard a colleague at work saying that he had given his samples in the name of his friend, Colin Pitchfork. Pitchfork was arrested, and his DNA fingerprint subsequently was shown to match the DNA at the crime scene. In 1988 he was convicted of the murders, the first conviction on the basis of DNA evidence. The application of DNA fingerprinting techniques to stored forensic samples has also led to the release of a number of persons wrongly convicted for rape or murder.

DNA FINGERPRINTING IN TESTING PATERNITY AND ESTABLISHING ANCESTRY DNA fingerprints are also widely used as evidence of paternity because parents and their children share common alleles in their DNA fingerprints. That is, each child receives one allele of each locus from one parent and the other allele from the other parent. A comparison of DNA fingerprints for a number of loci can prove almost infallibly whether a child has been fathered or mothered by a given person. DNA fingerprints have also been used for other investigations, such as confirming that remains discovered in a remote region of Russia were actually those of Czar Nicholas II and members of his family, murdered in 1918 during the Russian revolution.

DNA fingerprinting is also used widely in studies of other organisms, including other animals, plants, and bacteria.

Examples include testing for pathogenic *E. coli* in food sources such as hamburger meat, investigating cases of wildlife poaching, detecting genetically modified organisms among living organisms or in food, and comparing the DNA of ancient organisms with present-day descendants.

In this chapter you have learned about how genetically altered organisms can be made and used, and how a gene can be cloned and manipulated using DNA technologies. But a gene is just a part of a genome. Researchers also want to know about the set of genes in a complete genome, and how genes and their gene products work together in networks to control life. They also want to know more generally about the organization of the genome with respect to both genes and nongene sequences. Genomes and proteomes (the complete sets of proteins expressed by a genomes) are the subjects of the next chapter.

STUDY BREAK 18.3

1. What is a restriction fragment length polymorphism (RFLP)?
2. What is a single-nucleotide polymorphism?
3. What are the principles of DNA fingerprinting?

 Unanswered Questions

What are scientists aiming for with "targeted gene therapy"?

The molecular basis of inherited disease resides in your genes. Thus, it is not surprising that gene therapy has evoked such excitement in the field of molecular medicine. *Gene therapy* is defined as the use of nucleic acid sequences (DNA or RNA) to treat inherited or acquired disease. Gene therapy for monogenic diseases (diseases caused by a single defective gene) has begun to demonstrate great promise. For example, gene therapy for a form of congenital blindness (called Leber congenital amaurosis) has restored vision to blind children. This shot in the arm for the field has renewed enthusiasm for gene therapy of other monogenic diseases such as cystic fibrosis (CF)—a devastating disease caused by a single gene defect in a chloride channel called the cystic fibrosis transmembrane conductance regulator (CFTR) that leads to chronic bacterial infections in the lung (see Chapter 5 *Why it matters* . . . and Section 13.4). How can scientists approach gene therapy of CF? What cells in the lung must scientists target to treat this disease? Can gene therapy cure CF? These and other questions about gene therapy for CF are common to a myriad of other genetic diseases, and patients' lives are depending on rapid answers. Our work at the University of Iowa focuses on obtaining answers to these questions needed to treat CF using gene therapy.

What are the cellular targets for gene therapy?

The cellular targets for gene therapy of any genetic disease are informed by a basic understanding of how the defective gene product (or protein) produces disease at the cellular and organ level. Using CF as an example, we know the defective gene encodes a chloride channel (CFTR; see Figure 5.1) that is expressed in a variety of cell types in the lung and airways. However, researchers still do not know which of these cell types must be targeted by gene therapy to treat the disease effectively. Defective innate immunity in CF may involve cells lining the surface of the

airways and/or mucous-producing cells underneath the airways. Do both of these cellular compartments need to be targeted by gene therapy to reverse CF lung disease? The answer to this question resides in research that bridges gene therapy and the mechanisms of disease progression.

Tissues within the body are made from a complex assortment of unique cell types. These cells perform unique differentiated functions and organize into distinct groups of specialized cell types that act together to perform unique functions in a given organ. Typically, each group has unique microenvironments that house stem cells (called *stem cell niches*), which replace differentiated cell types within the trophic unit following normal cell turnover or injury. Often, these stem cells lie dormant until a signal from the microenvironment tells them to divide. In the lung, four unique stem cell niches have been identified, each being potentially important targets for gene therapy of CF and other lung diseases. To achieve lasting disease correction, scientists must target the adult stem cells that replace terminally differentiated cell types throughout the patient's lifetime. Therapeutic genes that incorporate within the genome of tissue-specific stem cells will potentially persist indefinitely.

How do scientists deliver genes to treat disease?

Conceptually, gene therapy consists of two strategies: (1) *gene addition*, which incorporates an additional gene into target cells; and (2) *gene correction*, which corrects the gene mutation that exists in the target cells. Although more challenging, the second approach is considered the "Holy Grail" of gene therapy because it corrects the cause of disease—the DNA mutation itself. Viral gene delivery is generally the most efficient means of delivering genetic payloads to cells and thus is the approach used in most current clinical gene therapy trials. In this method, scientists remove most (or all) viral genes from the virus and replace them with the therapeutic gene of interest. Although viruses are generally efficient at gene addition, they can cause the host to mount an immune response. Thus, important areas of research are directed at generating recombinant viruses that can escape the immune response and/or improve efficiency of gene delivery through an increased understanding of viral infection. Certain viruses also have the capability of correcting a gene mutation through homologous recombination, whereas others insert randomly into the genome; both mechanisms allow therapeutic DNA to persist if stem cells are targeted.

think like a scientist Future gene therapy research must bridge the fields of stem cell biology, disease pathophysiology, virology, immunology, and homologous recombination. Tissue-specific stem cells are important targets of gene therapy, but they are often found in protective microenvironments with unique biological properties. Why do you think biological systems have evolved to sequester tissue-specific stem cells into constrained locations?

John F. Engelhardt is the Director of the Center for Gene Therapy, Professor and Head of the Department of Anatomy and Cell Biology at the University of Iowa Roy J. and Lucille A. Carver College of Medicine. Research in the Engelhardt laboratory focuses on the molecular basis of inherited and environmentally induced diseases, and on the development of gene therapies for these disorders.

Tom Lynch is a graduate student at the University of Iowa in Dr. Engelhardt's Laboratory. His graduate research focuses on the identification and isolation of tissue-specific stem cells in the airway epithelium. His research is particularly focused on signals that control stem cell dormancy and proliferation in the airway. To learn more about their research, go to: http://www.medicine.uiowa.edu/elab/.

REVIEW KEY CONCEPTS

For access to MindTap and additional study materials visit www.cengagebrain.com.

18.1 Key DNA Technologies for Making Genetically Altered Organisms

- Genetically altered organisms are made for both basic or applied research. Many key technologies are involved in making such organisms.

- Producing multiple copies of genes by cloning is a common first step for studying the structure and function of genes, or for manipulating genes. Cloning involves cutting genomic DNA and a cloning vector with the same restriction enzyme, joining the fragments to produce recombinant plasmids, and introducing those plasmids into a living cell such as a bacterium, where replication of the plasmid takes place (Figures 18.1–18.3).

- PCR amplifies a specific target sequence in DNA, such as a gene, which is defined by a pair of primers. PCR is an in vitro method that increases DNA quantities by successive cycles of denaturing the template DNA, annealing the primers, and extending the primers in a DNA synthesis reaction catalyzed by DNA polymerase. With each cycle of PCR, the amount of DNA doubles (Figure 18.4). PCR products are analyzed by agarose gel electrophoresis (Figure 18.5). PCR is a common method used to amplify copies of genes for use in gene cloning.

18.2 Applications of Genetically Altered Organisms

- Making genetically altered bacteria made it possible to isolate and purify genes for detailed analysis and for making specialized genetically altered organisms for basic and applied research applications.

- Genetic engineering uses DNA technologies to alter the genetic makeup of humans, other animals, plants, and microorganisms such as bacteria and yeast. Genetic engineering primarily aims to correct hereditary defects, improve domestic animals and crop plants, and provide proteins for medicine, research, and other applications (Figures 18.6–18.12).

- Genetic engineering has enormous potential for research and applications in medicine, agriculture, and industry. Potential risks include unintended damage to living organisms or to the environment.

18.3 Other Applications of DNA Technologies

- Recombinant DNA and PCR techniques are used in DNA molecular testing for human genetic disease mutations. One approach exploits restriction site differences between normal and mutant alleles of a gene that create restriction fragment length polymorphisms (RFLPs), which are detectable by PCR followed by restriction digestion and agarose gel electrophoresis (Figures 18.13 and 18.14).

- Human DNA fingerprints are produced from a number of loci in the genome characterized by short, tandemly repeated sequences that vary in number in all individuals (except identical twins). To produce a DNA fingerprint, PCR is used to amplify the region of genomic DNA for each locus, and the lengths of PCR products indicate the alleles an individual has for the repeated sequences at each locus. DNA fingerprints are widely used to establish paternity, ancestry, or criminal guilt (Figure 18.15).

Remember/Understand

1. The point at which a restriction enzyme cuts DNA is determined by:
   a. the sequence of nucleotides.
   b. the length of the DNA molecule.
   c. whether it is closer to the 5′ end or 3′ end of the DNA molecule.
   d. the number of copies of the DNA molecule in a bacterial cell.
   e. the location of a start codon in a gene.

2. Restriction endonucleases, ligases, plasmids, *E. coli,* electrophoretic gels, and a bacterial gene resistant to an antibiotic are all required for:
   a. dideoxyribonucleotide analysis.
   b. PCR.
   c. DNA cloning.
   d. DNA fingerprinting.
   e. DNA sequencing.

3. Why are antibiotic resistance markers such as amp^R important components of bacterial plasmid cloning vectors?
   a. The plasmid must have resistance to accept DNA inserts.
   b. They allow the detection of plasmids that contain an inserted DNA fragment.
   c. They ensure the presence of the *ori* site.
   d. They ensure that the plasmid can be cut by a restriction enzyme.
   e. They allow identification of bacteria that have taken up a plasmid.

4. After a polymerase chain reaction (PCR), agarose gel electrophoresis is often used to:
   a. amplify the DNA.
   b. convert cDNA into genomic DNA.
   c. convert cDNA into messenger RNA.
   d. verify that the desired DNA sequence has been amplified.
   e. synthesize primer DNA molecules.

5. A cDNA and a cloned fragment of genomic DNA share sequences from a mouse gene. What differences do you expect to see between the cDNA and genomic DNA sequences?
   a. None; they should be identical.
   b. The genomic DNA might have an intron or introns.
   c. The genomic DNA might have promoter sequences.
   d. The genomic DNA might have a poly(A) tail.
   e. Both b and c are correct.

6. Which of the following is needed both in using bacteria to produce proteins and in genetic engineering of human cells?
   a. DNA fingerprinting based on microsatellite sequences
   b. insertion of a transgene into an expression vector
   c. restriction fragment length polymorphism (RFLP)
   d. the Ti plasmid of *Agrobacterium*
   e. antibiotic resistance

7. Which of the following is *not* true of somatic cell gene therapy?
   a. White blood cells can be used.
   b. Somatic cells are cultured, and the desired DNA is introduced into them.
   c. Cells with the introduced DNA are returned to the body.
   d. The technique is still very experimental.
   e. The inserted genes are passed to the offspring.

8. Dolly, a sheep, was an example of reproductive (germline) cloning. Required to perform this process was:
   a. implantation of uterine cells from one strain into the mammary gland of another.
   b. fusion of the mammary cell from one strain with an enucleated egg of another strain.
   c. fusion of an egg from one strain with the egg of a different strain.
   d. fusion of an embryonic diploid cell with an adult haploid cell.
   e. fusion of two nucleated mammary cells from two different strains.

9. Restriction fragment length polymorphisms (RFLPs):
   a. are produced by reaction with restriction endonucleases and are detected by PCR and agarose gel electrophoresis.
   b. are of the same length for mutant and normal β-globin alleles.
   c. determine the sequence of bases in a DNA fragment.
   d. have in their middle short fragments of DNA that are palindromic.
   e. are used as vectors.

10. DNA fingerprinting, which is often used in forensics, paternity testing, and for establishing ancestry:
    a. compares coding regions of two or more people.
    b. measures different lengths of DNA from many repeating noncoding regions.
    c. requires the largest DNA lengths to run the greatest distance on a gel.
    d. requires amplification after the gels are run.
    e. can easily differentiate DNA between identical twins.

Apply/Analyze

11. **Discuss Concepts** What should juries know to be able to interpret DNA evidence? Why might juries sometimes ignore DNA evidence?

12. **Discuss Concepts** A forensic scientist obtained a small DNA sample from a crime scene. In order to examine the sample, he increased its quantity by cycling the sample through the polymerase chain reaction. He estimated that there were 50,000 copies of the DNA in his original sample. Derive a simple formula and calculate the number of copies he will have after 15 cycles of the PCR.

Evaluate/Create

13. **Design an Experiment** Suppose a biotechnology company has developed a GMO, a transgenic plant that expresses *Bt* toxin. The company sells its seeds to a farmer under the condition that the farmer may plant the seed, but not collect seed from the plants that grow and use it to produce crops in the subsequent season. The seeds are expensive, and the farmer buys seeds from the company only once. How could the company show experimentally that the farmer has violated the agreement and is using seeds collected from the first crop to grow the next crop?

14. **Apply Evolutionary Thinking** In PCR, researchers use a heat-stable form of DNA polymerase from microorganisms that are able to grow in extremely high temperatures. Given what you learned in Chapter 3 about protein folding, and in Chapter 6 about the effects of temperature on enzymes, would you predict that the amino acids of heat-stable DNA polymerase enzymes would have evolved so they can form stronger chemical attractions with each other, or weaker chemical attractions? Explain your answer.

For selected answers, see Appendix A.

INTERPRET THE DATA

You learned in the chapter that an STR locus is a locus where alleles differ in the number of copies of a short, tandemly repeated DNA sequence. PCR is used to determine the number of alleles present, as shown by the size of the DNA fragment amplified. In the **Figure** below are the results of PCR analysis for STR alleles at a locus where the repeat unit length is 9 bp, and alleles are known that have 5 to 11 copies of the repeat. Given the STR alleles present in the adults, state whether each of the four juveniles could or could not be an offspring of those two adults. Explain your answers.

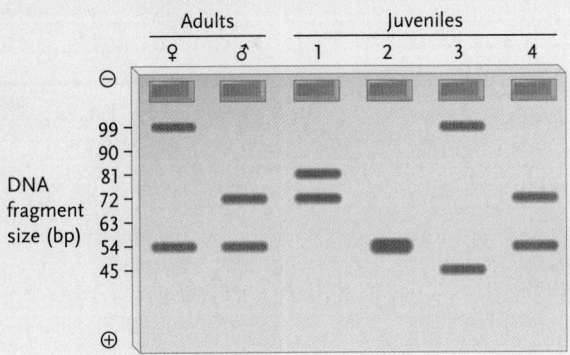

© Cengage Learning 2017

19

Genomes and Proteomes

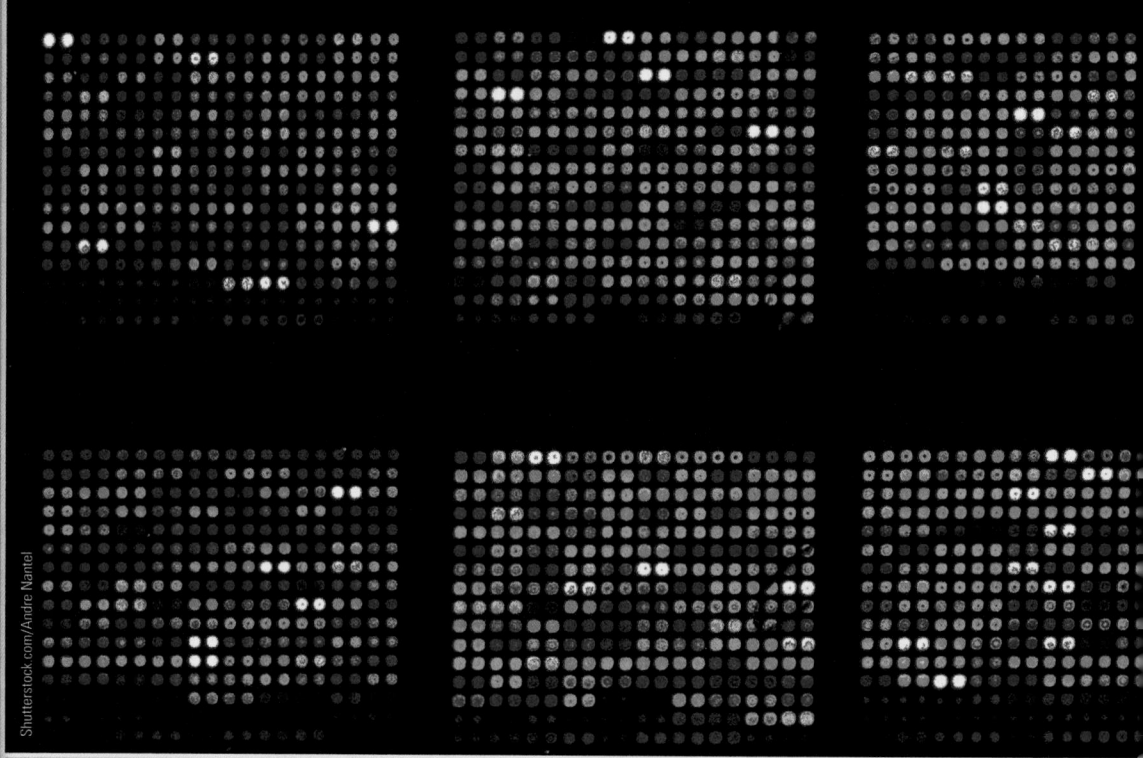

Shutterstock.com/Andre Nantel

Results of DNA microarray analysis. DNA microarrays can be used, at a genomic level, to study which protein-coding genes are being expressed and the relative levels of expression of those genes.

Why it matters . . . The 1000 Genomes Project (www.1000genomes.org), initiated in 2008, is an international collaborative research effort to find and catalog genetic variation in human genomes. As its name suggests, the Project intends to sequence the genomes of over 1,000 unidentified individuals from around the world. The sequencing data will be valuable in highly focused medical research, as well as in basic research. One interesting research study done as part of the Project analyzed the mutations each of us receives from our parents. As you learned in Chapters 12 and 13, mutations in the germline cells (sperm and egg) of a parent are inherited. Those mutations fall into two groups: mutations inherited by the parent, and new mutations that occur in the germline. In the study, researchers compared the genomes of two parent–offspring trios. They identified 49 and 35 mutations in the two offspring that must have originated as new mutations in their parents' germlines. Let that sink in: we have numerous mutations in our genomes that our parents do not have. Most interestingly, in one family, 92% of the new mutations were from the paternal germline, whereas in the other family 64% of the new mutations were from the maternal germline. These results mean that there is considerable variation in mutation rates within and between families. Although this is a relatively limited study (as it only can be at the moment, given the limited availability of complete genomes of families to study), nonetheless the results are provocative.

The 1000 Genomes Project research exemplifies the field of **genomics,** the characterization of whole genomes, including their structures (sequences), functions, and evolution. In this chapter, you will learn about genomics and some of the information that has come from genome analysis. You will also learn about **proteomics,** the study of the **proteome,** which is the complete set of proteins that can be produced by a genome. Proteomics involves characterizing the structures and functions of all expressed proteins of an organism, and the interactions among proteins in the cell.

The chapter also discusses how comparisons of genome sequences have contributed to our understanding of genome evolution.

19.1 Genomics: An Overview

As defined in *Why it matters* . . . genomics is the characterization of whole genomes, including their structures (sequences), functions, and evolution. Having the complete sequence of a genome makes it possible to study the complete set of genes in an organism or a virus, as well as other important sequences of their genomes. Understanding the structures and functions of all the genes in an organism or virus will give us a complete blueprint of the information for the development, structure, and function of that organism's life or for that virus's infection cycle. That is an exciting prospect, but we are not there yet.

Before we go further, let us be sure we are clear about genes. As you learned in Chapter 15, some genes encode proteins. Protein-coding genes are the main focus of our discussions. Other so-called noncoding RNA genes are transcribed to produce RNAs that function without being translated into proteins. They include genes for tRNAs, rRNAs, snRNAs, miRNAs, and long noncoding RNAs (lncRNAs).

Fundamental to genomics is determining the DNA sequences of organismal genomes and the DNA or RNA sequences of viral genomes. We focus here on the DNA genomes of organisms, though most of the techniques described in this chapter have also been applied to the study of viral genomes. Modern DNA sequencing techniques are advances on methods used to analyze the sequences of cloned DNA sequences and DNA sequences amplified by PCR (see Section 18.1). Having the complete sequence of a genome enables researchers to study the organization of genes in the genome as a whole, and to determine how genes function together in networks to control life.

Of natural interest to us is the human genome. The complete sequencing of the approximately 3 billion base-pair human genome—the Human Genome Project (HGP)—began in 1990. The task was completed in 2003 by an international consortium of researchers and by a private company, Celera Genomics (headed by J. Craig Venter). As part of the official HGP, the genomes of several important model organisms commonly used in genetic studies were sequenced for purposes of comparison: *E. coli* (representing prokaryotes); the yeast *Saccharomyces cerevisiae* (representing single-celled eukaryotes); *Drosophila melanogaster* and *Caenorhabditis elegans* (the fruit fly and a nematode worm, respectively, representing multicellular invertebrate animals); and *Mus musculus* (the mouse, representing nonhuman mammals). The sequences of the genomes of many organisms and viruses not part of the official Human Genome Project, including plants, have since been completed or are in progress.

A vast amount of DNA sequence data has been generated by genome sequencing projects. For those sequences to be useful, they need to be available centrally for access by all researchers. DNA sequences from genome sequencing projects are deposited into databases that are publicly available via the Internet. For example, GenBank is an "annotated collection of all publicly available DNA sequences" at the National Institutes of Health (NIH). The Internet link is http://www.ncbi.nlm.nih.gov/genbank/. Currently there are over 150 billion bases in the DNA sequence records at GenBank. Computational tools at NCBI GenBank enable researchers, and others, such as students like yourself, to perform various analyses with the sequence data.

Many other genomics databases are accessible using the Internet, with sequence data organized in different ways (perform an Internet search for "DNA sequence database"). For example, there are organism-specific sequence databases, as well as databases that include summaries of particular genomics studies. The databases are available for individual researchers to use and also for collaborative efforts involving researchers all over the world. One of the main benefits of collaborative research of this kind is that researchers with different specialties can tackle a particular question or questions at a genomic or multigenomic level.

Genomics consists of three main areas of study:

1. *Genome sequence determination and annotation,* which means obtaining the sequences of complete genomes and analyzing them to locate putative protein-coding and noncoding RNA genes and other functionally important sequences in the genome.

2. *Determining the functions of genes* (a somewhat outdated term for this is *functional genomics*), which means using genome sequence data as a basis to study and understand the functions of genes and other parts of the genome. With respect to genes, this includes developing an understanding of how their expression is regulated. For protein-coding genes it also includes determining what proteins they encode, and how those proteins function in the organism's metabolic processes.

3. *Studying how genomes have evolved,* which means comparing genome sequence data to develop an understanding of how genes, particularly protein-coding genes, originated and genes and genomes changed over evolutionary time. Studies of genome sequences or large parts of genome sequences for a number of organisms represent an area of genomics known as **comparative genomics.**

Advances in each of these areas of study are accelerating as techniques are developed and improved for automating experimental procedures and more sophisticated computer algorithms for data analysis are generated. Methods that facilitate the handling of many samples simultaneously, whether those samples are DNA molecules for sequencing or genes for analysis, are called **high-throughput techniques.** The next three sections of the chapter discuss each of the three areas of genomics in turn.

STUDY BREAK 19.1

What additional biological questions can be answered if one has the complete sequence of an organism's genome as compared with the sequences of individual genes?

FIGURE 19.1 **Research Method**

Whole-Genome Shotgun Sequencing

Purpose: Obtain the complete sequence of the genome of an organism.

Protocol:

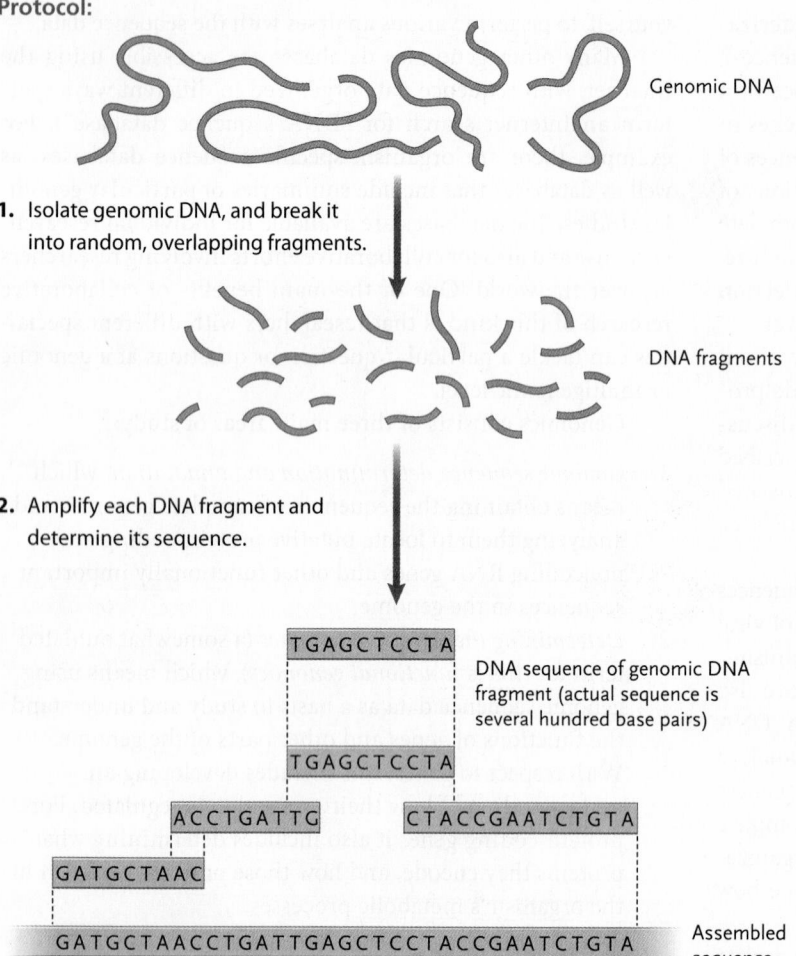

Genomic DNA

1. Isolate genomic DNA, and break it into random, overlapping fragments.

DNA fragments

2. Amplify each DNA fragment and determine its sequence.

TGAGCTCCTA

DNA sequence of genomic DNA fragment (actual sequence is several hundred base pairs)

TGAGCTCCTA

ACCTGATTG   CTACCGAATCTGTA

GATGCTAAC

GATGCTAACCTGATTGAGCTCCTACCGAATCTGTA   Assembled sequence

3. Enter the DNA sequences of the fragments into a computer, and use the computer to assemble overlapping sequences into the continuous sequence of each chromosome of the organism. This technique is analogous to taking 10 copies of a book that has been torn randomly into smaller sets of a few pages each and, by matching overlapping pages of the leaflets, assembling a complete copy of the book with the pages in the correct order.

Interpreting the Results: The method generates the complete sequence of the genome of an organism.

© Cengage Learning 2017

19.2 Genome Sequence Determination and Annotation

Genome sequence determination and annotation means obtaining the sequence of bases in a genome using DNA sequencing techniques, and then analyzing the sequence data using computer-based approaches to identify genes and other sequences of interest, which include gene regulatory sequences,

origins of replication, repetitive sequences, and transposable elements.

Genome Analysis Begins with DNA Sequencing

DNA sequencing was developed in the late 1970s by Allan M. Maxam, a graduate student, and his mentor, Walter Gilbert of Harvard University. A few years later, Frederick Sanger of Cambridge University designed a method that became the one commonly used in research. Gilbert and Sanger were awarded a Nobel Prize in 1980 "for their contributions concerning the determination of base sequences in nucleic acids." DNA sequencing technology has evolved since its development, and particularly rapidly in the past few years.

WHOLE-GENOME SHOTGUN SEQUENCING Before we discuss methods of DNA sequencing, let us consider the strategy generally used to determine the sequence of a genome, **whole-genome shotgun sequencing (Figure 19.1).** In this method, genomic DNA is isolated and purified, and that DNA is broken into thousands to millions of random, overlapping fragments. Each fragment is amplified to produce many copies, and then the sequence of the fragment is determined. The entire genome sequence is then assembled using computer algorithms that search for the sequence overlaps between fragments and stitch together the sequence reads to produce longer contiguous sequences.

DNA SEQUENCING METHODS All DNA sequencing methods have in common the following steps: (1) DNA purification; (2) DNA fragmentation; (3) amplification of fragments; (4) sequencing each fragment; and (5) assembly of fragment sequences into genome sequences. The methods differ in how the amplification is done, the lengths of the fragments, how many fragments are sequenced simultaneously, and how the sequencing reactions themselves are done.

For decades the method devised by Frederick Sanger was by far the most common DNA sequencing technique used. The Sanger method is a DNA synthesis-based method for DNA sequencing. It is based on the properties of nucleotides known as *dideoxyribonucleotides,* which have a —H on the 3′ carbon of the deoxyribose sugar instead of the —OH found in normal deoxyribonucleotides; therefore, the method, explained in **Figure 19.2,** is also called *dideoxy sequencing.*

In recent years the dideoxy sequencing method has been replaced largely, but not completely, by faster, cheaper, and

FIGURE 19.2 **Research Method**

Dideoxy (Sanger) Method for DNA Sequencing

Purpose: Obtain the sequence of a piece of DNA, such as in gene sequencing or genome sequencing. The method is shown here with a typical automated sequencing system.

Protocol:

1. A dideoxy sequencing reaction contains: (1) the fragment of DNA to be sequenced (denatured to single strands); (2) a DNA primer that will bind to the 3′ end of the sequence to be determined; (3) a mixture of the four deoxyribonucleotide precursors for DNA synthesis; (4) a mixture of the four dideoxyribonucleotide (dd) precursors, at about 1/100th of the concentration of the deoxyribonucleotides, each labeled with a different fluorescent molecule; and (5) DNA polymerase to catalyze the DNA synthesis reaction.

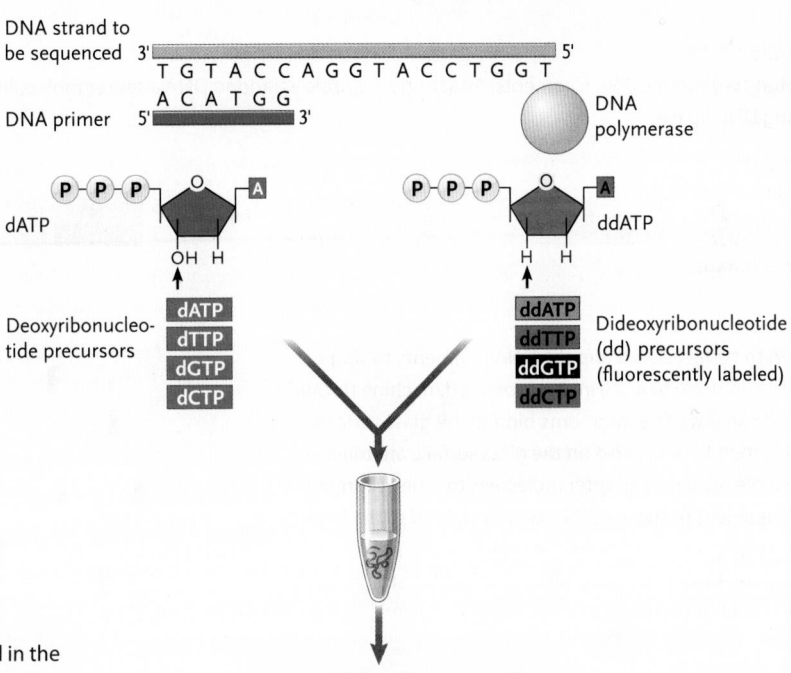

2. DNA polymerase synthesizes the new DNA strand in the 5′→3′ direction starting at the 3′ end of the primer. New synthesis continues until a dideoxyribonucleotide is incorporated randomly into the DNA. The dideoxyribonucleotide acts as a *terminator* for DNA synthesis because it has no 3′-OH group for the addition of the next base (see Section 14.3). For a large population of template DNA strands, the dideoxy sequencing reaction produces a series of new strands, with lengths from one on up. At the 3′ end of each new strand is the fluorescently labeled dideoxyribonucleotide that terminated the synthesis.

3. Separate the labeled strands by electrophoresis using a polyacrylamide gel prepared in a capillary tube. The principle of separation is the same as for agarose gel electrophoresis (see Figure 18.5), but this gel can discriminate between DNA strands that differ in length by one nucleotide. As the bands of DNA fragments move near the bottom of the tube, a laser beam shining through the gel excites the fluorescent labels on each DNA fragment. The fluorescence is registered by a detector, with the wavelength of the fluorescence indicating whether ddA, ddT, ddG or ddC is at the end of the fragment in each case.

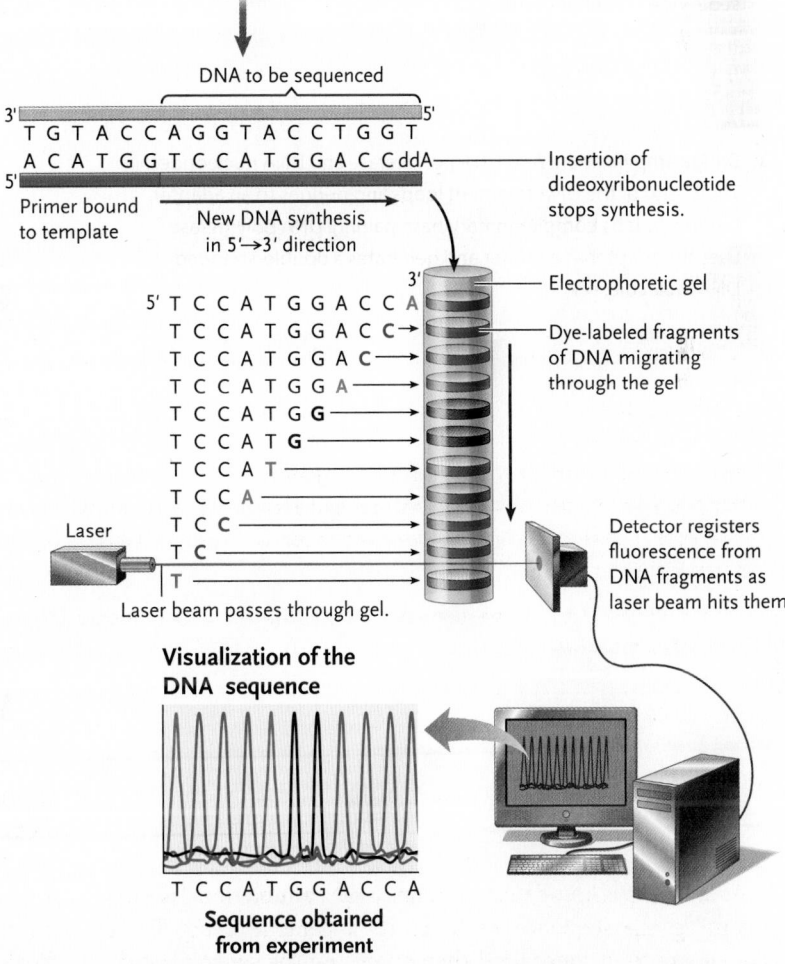

Visualization of the DNA sequence

T C C A T G G A C C A
Sequence obtained from experiment

Interpreting the Results: The data from the laser system are sent to a computer that interprets which of the four possible fluorescent labels is at the end of each DNA strand. The results show, on a computer screen or in printouts, colors for the labels as the DNA bands passed the detector. The sequence of the newly synthesized DNA, which is complementary to the template strand, is read from left (5′) to right (3′). (The sequence shown here begins after the primer.)

 FIGURE 19.3 | **Research Method**

Illumina/Solexa Method for DNA Sequencing

Purpose: Automated, massively parallel sequencing of up to a billion DNA fragments.

Protocol:

1. Add adapters to genomic DNA fragments: Attach short double-stranded DNA adapter molecules to each end of 100 to 300-bp genomic DNA fragments using DNA ligase.

2. Attach DNA to surface: Denature the DNA fragments to single strands and add them to a cell in an automated machine through which liquid can flow. The fragments bind to the glass surface of the cell at their 5′ ends. Also on the glass surface are trillions (10^{12}s) of single-stranded adapter molecules that have complementary sequences to those of the adapters added to the DNA in step 1; the adapters are also bound at their 5′ ends. The massive number of glass-bound adapters allows many DNA strands to be sequenced simultaneously.

3. Bridge amplification: Add DNA polymerase and DNA nucleotides. The 3′ end of the DNA fragment loops and bridges to an adapter, binding to it by complementary base pairing. DNA polymerase uses the adapter as a primer and generates a double-stranded DNA molecule.

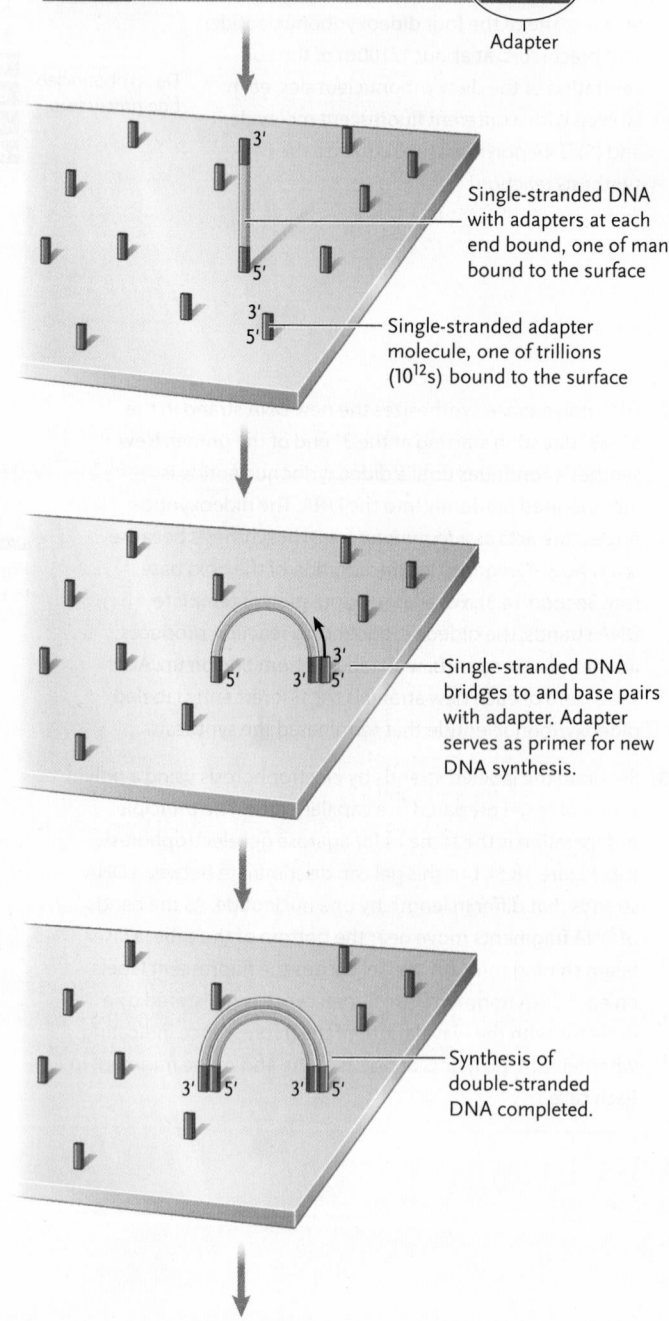

© Cengage Learning 2017

4. Denature the double-stranded DNA molecule. The result is two single-stranded DNA molecules, each bound to the surface.

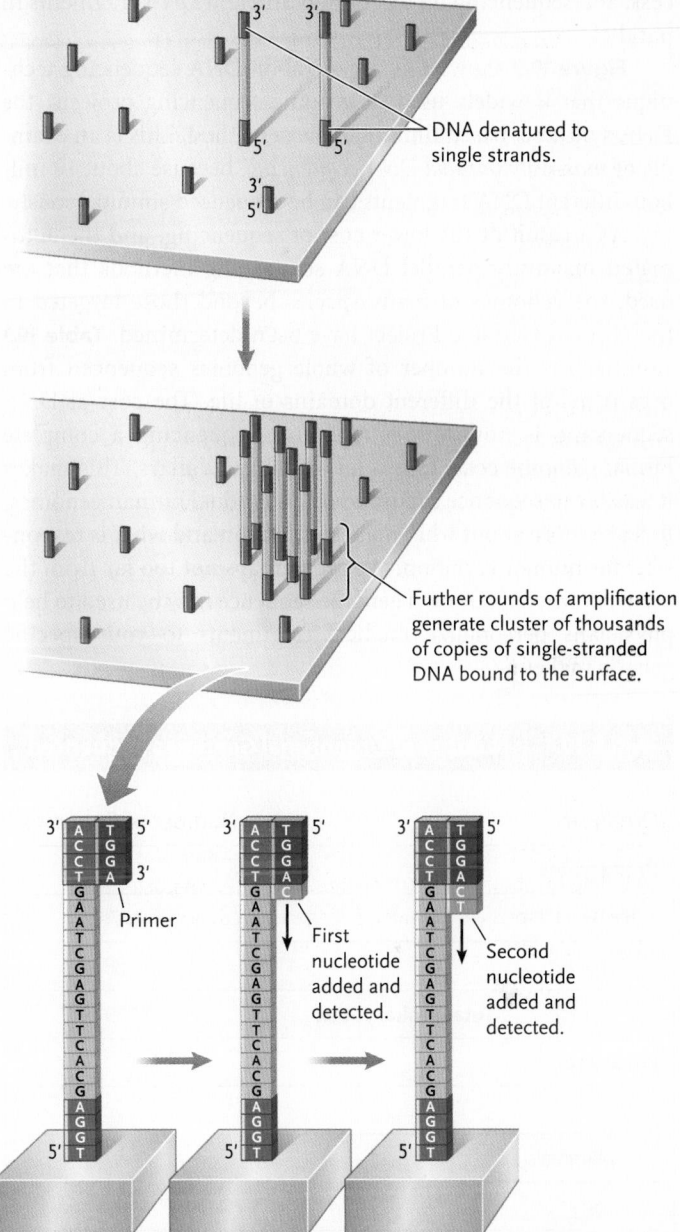

DNA denatured to single strands.

5. Carry out other rounds of amplification: The result is thousands of copies of each of the DNA fragments that was bound initially to one of the glass-bound adapters, clustered around the place on the cell where that DNA bound. Millions of different clusters can be produced in the sequencing cell simultaneously.

Further rounds of amplification generate cluster of thousands of copies of single-stranded DNA bound to the surface.

6. The DNA fragments are now ready for synthesis-based DNA sequencing. One fragment is shown for the sequencing steps.

7. DNA sequencing: Add DNA primers to the cell. A primer anneals to each DNA strand (it is the complementary strand to the adapter sequence at its end) and DNA synthesis is done in a cyclic manner, one nucleotide at a time, using four different fluorescently labeled DNA nucleotide precursors. Each time a labeled nucleotide is added, synthesis stops and the machine uses laser technology to measure the fluorescence, so as to identify the base added. That base is the same for all strands in a cluster. By repeated cycles of addition of a nucleotide and laser detection, the sequence of the strand is obtained. Up to about 100 bases of each fragment can be sequenced.

Primer

First nucleotide added and detected.

Second nucleotide added and detected.

Interpreting the Results: The DNA sequence obtained is complementary to the initial single-stranded DNA strand that bound to the surface. The DNA sequence data from all of the clusters of DNA fragments are analyzed by computer to determine overlaps between fragments and, by the principles described in Figure 19.2, the complete sequence of a genome is assembled.

more automated techniques. In general, these newer high-throughput techniques have decreased sequencing costs by reducing the preparatory steps, automating more of the process, and sequencing up to a billion different DNA fragments in parallel.

Figure 19.3 shows a next-generation DNA sequencing technique that is widely used in genome sequencing projects, the DNA synthesis-based Illumina/Solexa method. This is an example of *massively parallel DNA sequencing,* because about 10 million different DNA fragments can be sequenced simultaneously.

As a result of the lower cost of sequencing, and the automated massively parallel DNA sequencing methods that are used, the genomes of many species beyond those targeted in the Human Genome Project have been determined. **Table 19.1** summarizes the number of whole genomes sequenced from organisms of the different domains of life. The cost of DNA sequencing is now low enough that sequencing a complete human genome costs only a few thousand dollars. This makes it feasible to sequence thousands of individual human genomes, to learn more about what makes us human and what is responsible for human variation. We are perhaps not too far from the scenario where your own genome sequence may be used to help physicians personalize medical treatments to your specific genetic makeup.

TABLE 19.1	Number of Organismal Genomes Sequenced[a]
Organism	Number of Genomes
Prokaryotes	
Bacteria	4,471
Archaea	288
Total prokaryotes:	**4,759**
Eukaryotes	
Animals	
Mammals	97
Birds	60
Fishes	45
Insects	112
Other animals	101
Total animals:	**415**
Plants	139
Fungi	416
Protists	276
Total eukaryotes:	**1,126**
Total all organisms:	**5,885**

[a]As of October 2014

Genome Sequences Are Annotated to Identify Genes and Other Sequences of Importance

A raw genome sequence is simply a string of A, T, G, and C letters; it tells us practically nothing about the organism from which it derives, other than the total length of its genome. Therefore, once the complete sequence of a genome has been determined, the next step is *annotation,* the identification of functionally important features in the genome. These include:

- Protein-coding genes.
- Noncoding RNA genes. As mentioned earlier, "noncoding" means that the RNA transcript of the gene is not translated. Rather, the transcript is the functional product of the gene. Noncoding genes include genes for tRNAs, rRNAs, and snRNAs (see Chapter 15), and genes for miRNAs and lncRNAs (see Chapter 16).
- Regulatory sequences associated with genes (see Chapter 16).
- Origins of replication (see Chapter 14).
- Transposable elements and sequences related to them (see Chapter 15).
- Pseudogenes. A **pseudogene** is very similar to a functional gene at the DNA sequence level, but one or more inactivating mutations have changed the gene so that it can no longer produce a functional gene product. Most pseudogenes are derived from protein-coding genes and they are recognized by their sequence similarities to functional genes.
- Short repetitive sequences. These are sequences that are repeated a few too many times in the genome. The short tandem repeat (STR) sequences discussed in Section 18.3 are examples of short repetitive sequences.

Annotation is performed by researchers in the field of **bioinformatics,** which is the application of mathematics and computer science to extract information from biological data, including those related to genome structure and function. More broadly, the term *bioinformatics* may be used to include the experimental work that generates such data. For example, bioinformatics scientists predict the structure and function of gene products and postulate evolutionary relationships of sequences.

Some examples of genome annotation are illustrated in the following subsections. Protein-coding genes are the focus because they are of particular interest in genome analysis.

IDENTIFYING OPEN READING FRAMES BY COMPUTER SEARCH OF GENOME SEQUENCES Proteins are specified in mRNA molecules by a series of codons starting with the initiation codon AUG and ending with one of the three termination codons, UAG, UAA, or UGA. The span of codons from start to stop codon is called an **open reading frame (ORF),** and ORFs that are longer than 100 codons almost always indicate the presence of a protein-coding gene. Computer algorithms are used to identify possible protein-coding genes in a genome sequence by searching for ORFs. In a DNA sequence, this means searching

FIGURE 19.4 The six reading frames of double-stranded DNA. In this particular sequence, one of them is an open reading frame (ORF).
© Cengage Learning 2017

for ATG, separated from a stop codon (TAG, TAA, or TGA) by a multiple of three nucleotides. The search is complicated because which of the two DNA strands is the template strand for transcription is gene-specific. Theoretically, then, each DNA sequence has six reading frames for the three-letter genetic code, three on one strand and three on the other strand, and an ORF can be in any one of those frames. This is illustrated in **Figure 19.4** for a theoretical 30-nucleotide segment of DNA. Note that each single-stranded DNA sequence generated by DNA sequencing can be used to infer the sequence of the complementary DNA strand. If an ORF is present in a particular DNA sequence, it will be in one of those frames.

We can start looking for an ORF going in the 5′-to-3′ direction in the top strand of Figure 19.4, starting at the leftmost A nucleotide. In this case, reading in groups of three nucleotides will lead you to the TAG at the right end, which is the stop codon. We have found an ORF in the sequence, and it is coded by the entire length of top strand. However, if instead we start looking for an open reading frame in the top strand starting at either the second nucleotide (the T) or the third nucleotide (the G), we do not find a start codon so we have not found an ORF. For the bottom strand, none of the three frames has a start codon. Computer algorithms can search easily for ORFs in all six reading frames of a DNA sequence.

Searching for protein-coding ORFs is straightforward in prokaryotic genomes because few genes have introns. Eukaryotic protein-coding genes typically have introns and therefore more sophisticated algorithms must be used to identify such genes. For example, the algorithms may search for particular characteristics of protein-coding genes, such as junctions between exons and introns, sequences that are characteristic of eukaryotic promoters, and overrepresentation of certain three-base codons relative to others.

Computer identification typically is the first step in identifying protein-coding genes. However, other evidence is needed before biologists are confident that the sequence they have annotated is a functioning protein-coding gene. Two approaches to obtain such evidence are described in the next two subsections.

IDENTIFYING PROTEIN-CODING GENES BY SEQUENCE SIMILARITY SEARCHES One way of testing whether candidate protein-coding genes found by searching for ORFs are functioning protein-coding genes is by comparing their sequences with known, identified and verified genes in databases. This is a *sequence similarity search.* Such searches can be done using an Internet browser to access the computer programs and the databases. For example, to use the BLAST (Basic Local Alignment Search Tool) program at the National Center for Biotechnology Information (http://blast.ncbi.nlm.nih.gov), a researcher pastes the putative ORF DNA sequence, or the amino acid sequence of the protein it would encode, into a browser window and sets the program to begin searching. The BLAST program searches the databases of known sequences and returns the best matches, if any. The matches are listed in order, from the closest match to the least likely match. Finding a known gene's sequence that matches the putative ORF sequence closely would be good evidence that the ORF is in fact a protein-coding gene and that it encodes a protein functionally related to that of the matching gene sequence. The principle here is that genes of living organisms tend to be similar to each other because they have evolved from ancestral genes in ancestral organisms. Genes that have highly conserved sequences because they have evolved from a gene in a common ancestor are called **homologous genes.** For example, if a gene in the mouse has been characterized experimentally and its sequence is known, and that gene is evolutionarily conserved in mammals, there should be a match for that gene in the human genome sequence.

With sequence similarity searches, ORFs can be sorted into ones with known and unknown functions. For the latter, experiments are required to show whether they are real protein-coding genes and, if so, what their functions are. For example, analysis of the human genome sequence initially identified more than a thousand putative protein-coding genes with no sequence similarities to known genes. Most of these function-unknown genes have now been shown to be pseudogenes. Such uncertainty makes it difficult to determine the exact number of protein-coding genes in a genome just from its sequence.

IDENTIFYING PROTEIN-CODING GENES FROM SEQUENCES OF GENE TRANSCRIPTS The gold standard for identifying protein-coding genes in a genome sequence is the demonstration that the sequences are transcribed in cells to make mRNAs. The method involved is to sequence cDNAs that have been made from RNA transcripts. Recall from Section 18.2 that cDNAs are made starting with mRNA molecules isolated from a cell; that is, the cell's transcriptome. If the mRNA molecules are isolated under different conditions and from different cell types in a multicellular organism, they will represent the activity of many of the organism's protein-coding genes. However, protein-coding genes that are rarely transcribed or that produce very few mRNA molecules are likely to be missed by this approach.

The isolated single-stranded mRNA molecules are converted to double-stranded cDNA molecules using reverse transcriptase (see Figure 18.7) and then their sequences are obtained. The main way now used to collect the sequences of the cDNAs (and, hence, of the gene transcripts from which they derived) is by using **RNA-seq** ("RNA-seek"—whole-transcriptome sequencing). This technique uses high-throughput sequencing of cDNAs to identify and quantify RNA transcripts in a sample. About 30 to 400 nucleotides of each cDNA are sequenced using high-throughput techniques. The results are aligned with the genome sequence of the organism under study using computer algorithms to map the location of each sequence and, therefore, the protein-coding gene from which the original transcript was derived. A single RNA-seq study can identify over 100,000 sequence reads, each one of which indicates the presence of a specific mRNA in the cells being studied.

Sequencing cDNAs derived from gene transcripts is also useful for cataloging transcripts in humans and other eukaryotes to identify which genes are alternatively spliced. Recall from Section 16.3 that alternative splicing of pre-mRNA transcripts of protein-coding genes produces different mRNAs by using different splice sites that may remove different combinations of exons, or modify their lengths. The resulting mRNAs are translated to produce a family of related proteins having various combinations of amino acid sequences derived from the remaining exons.

Genome Landscapes Vary Markedly in Size, Gene Number, and Gene Density

With many genomes now sequenced, researchers can compare them to learn about genome sizes, the number of protein-coding genes, and the density of those genes (how widely spaced they are). A vast amount of new information is available about genome landscapes. Here we will present some broad brush strokes and then provide a more detailed description of the *E. coli* genome and the human genome, as examples of prokaryote and eukaryote genomes, respectively.

GENOME SIZES OF VIRUSES, BACTERIA, ARCHAEA, AND EUKARYA Figure 19.5 shows the ranges of genome sizes for viruses, bacteria, archaea, and different groups of eukaryotes; and **Table 19.2** gives examples of genome sizes and the number of protein-coding genes for some bacteria, archaea, and

eukaryotes. We can arrive at some general conclusions about the data. Members of both the Domain Bacteria and Domain Archaea have genomes that vary widely in size. In addition, their genes are densely packed in their genomes, with little noncoding space between them. Thus, the larger genomes of organisms in these two domains tend to reflect increased gene number. For example, in the Domain Bacteria, *Mycoplasma*

TABLE 19.2	Genome Sizes and Estimated Number of Protein-Coding Genes for Selected Members of Domains Bacteria, Archaea, and Eukarya		
Domain and Organism		Genome Size (Mb)	Protein-Coding Genes
Bacteria			
Mycoplasma genitalium		0.58	475
Escherichia coli		4.6	4,140
Archaea			
Thermoplasma acidophilum		1.56	1,484
Methanosarcina acetivorans		5.75	4,540
Eukarya			
Protists			
Tetrahymena thermophila (a ciliated protist)		103	24,725
Fungi			
Saccharomyces cerevisiae (budding yeast)		12.2	5,907
Neurospora crassa (orange bread mold)		39.2	9,822
Plants			
Arabidopsis thaliana (thale cress)		120	31,876
Oryza sativa (rice) (Japonica variety)		391	35,394
Invertebrates			
Caenorhabditis elegans (a nematode worm)		100	20,361
Drosophila melanogaster (fruit fly)		144	13,494
Vertebrates			
Takifugu rubripes (pufferfish)		391	19,448
Mus musculus (mouse)		2,799	19,952
Homo sapiens (human)		3,209	19,881

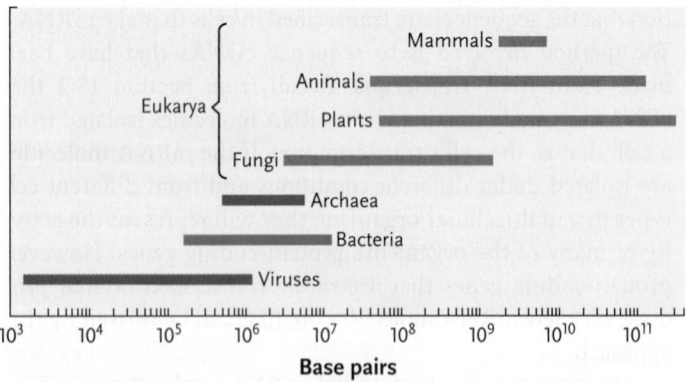

FIGURE 19.5 Ranges of genome sizes for viruses, bacteria, archaea, and eukaryotes. Note that the Domain Archaea has not been studied as extensively as the Domain Bacteria, so there may be representatives with substantially larger or smaller genomes than the range given.

© Cengage Learning 2017

genitalium has a genome size of 0.58 Mb with 475 protein-coding genes (and 43 noncoding RNA genes), and *E. coli* has a genome size of 4.6 Mb with 4,140 protein-coding genes (and 176 noncoding RNA genes). In the Domain Archaea, *Thermoplasma acidophilum* has a genome size of 1.56 Mb with 1,484 protein-coding genes, and *Methanosarcina thermophila* has a genome size of 5.75 Mb with 4,540 protein-coding genes. More broadly, bacteria and archaea have a fairly consistent density of about 900 genes for every Mb of DNA.

Members of the Domain Eukarya vary markedly in form and complexity, and their genomes also show great differences in size. For example, budding yeast, *Saccharomyces cerevisiae*, has a 12.2-Mb genome that is about 0.4% the size of the 3,200-Mb human genome, yet humans, with an identified 19,881 protein-coding genes, have only a little more than three times the number found in yeast, which has 5,907 protein-coding genes.

There are no clear rules relating organism complexity and genome size. For instance, the fruit fly, *Drosophila melanogaster*, and the locust, *Schistocerca gregaria*, have similar physiological complexity, but the genome of the locust is 9,300 Mb, which is 65 times the size of the 144-Mb fruit fly genome (and almost

three times the size of the human genome). Even within a genus there is not necessarily a consistency. For example, there is a 50-fold variation in the genome size of *Allium* species, which contains onions, leeks, shallots, and garlic. Even among vertebrates, there is great variation in genome size. The pufferfish, *Takifugu rubripes*, for example, has a 391-Mb genome, whereas the genomes of the mouse and humans are about seven times larger. Clearly the genes are spaced more closely in the pufferfish genome than they are in either the mouse or human genomes. But the human genome is by no means the largest among eukaryotes; the genomes of some amphibians and some ferns are about 200 times larger. In general, though, all genes are packed less densely in eukaryotes than they are in prokaryotes, with protein-coding gene density ranging from 500 per Mb (yeast) to fewer than 10 per Mb (humans and other mammals). However, there is no uniformity in the packing, as the pufferfish–mammal comparison shows.

It is important to think critically about the data presented for gene numbers in a genome. As you have learned, the determination of the protein-coding gene number involves both computer and experimental analysis. The outcomes of those analyses are therefore estimates of the number of genes present. Only when an entire genome has been studied experimentally to characterize every gene it contains can we be certain of an organism's exact gene number.

PROFILE OF THE *E. coli* GENOME *E. coli* is one of the most intensively studied model organisms, and the genome of laboratory strain K12 is one of the best annotated. In many ways, *E. coli* has a typical bacterial genome, with the vast majority of its genes on a single circular chromosome with one origin of replication **(Figure 19.6A)**, and the remainder of its genes on one or more plasmids, each of which is much smaller than the circular chromosome. With about 4.6 Mb and about 4,140 protein-coding genes, the *E. coli* K12 genome is in the middle range, sizewise, of bacterial genomes (see Figure 19.5). The noncoding genes are those for rRNAs and tRNAs. There are a small number of transposable elements and repetitive sequences.

Figure 19.6B shows a close up of a 10-kb segment of the *E. coli* genome containing a number of protein-coding genes to illustrate the following characteristics:

- The genes are close together, with little space in between. Promoters for the genes are located immediately upstream of each transcription unit (not shown in the figure).
- Some of the genes are transcribed in the left-to-right direction (using the bottom strand as the template), whereas the others are transcribed in the right-to-left direction (using the top strand as the template). (The two template strands are

A. Map of the circular *E. coli* K12 genome showing the genes transcribed clockwise (blue) and the genes transcribed counterclockwise (orange) and the location of the origin of replication

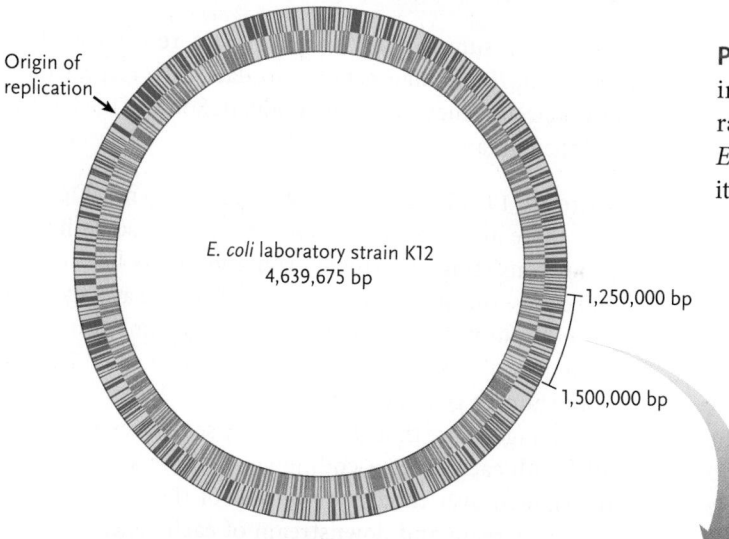

B. Detail of a 10-kb region of the *E. coli* K12 genome, from about 3:30 on the genome "clock"

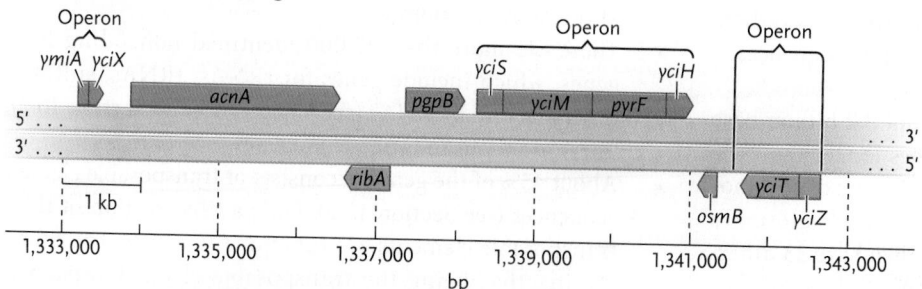

FIGURE 19.6 The genome of *E. coli*, laboratory strain K12.
© Cengage Learning 2017

TABLE 19.3 | Comparison of the *E. coli* K12 and Human Genomes

Property	*E. coli* K12 genome	*H. sapiens* genome
Chromosomes	1 circular (plus plasmids)	23 linear (pairs in diploid cells)[a]
Nucleotides	4.64 Mb	3,209 Mb
Protein-coding genes	4,140	19,881[b]
Noncoding RNA genes	178	25,411
% coding DNA	88%	1.2%
Protein-coding genes per Mb	892	6.2
Introns per average gene	0	8
Average polypeptide size	330 amino acids	430 amino acids

[a]There are 24 different human chromosomes: 22 autosomes, and the X and Y chromosomes. Each individual has 23 pairs of chromosomes.
[b]Through alternative splicing, 13,526 of the 19,881 protein-coding genes (68%) produce more that one different polypeptide. As a result, the 19,881 protein-coding genes produce a total of 54,420 full-length protein-coding transcripts, and 24,957 partial-length protein-coding transcripts.

transcribed in different directions because the two DNA molecules in a double helix are antiparallel; see Chapter 14.)

- Some genes are single transcription units, whereas others are organized into operons (see Chapter 16). In the genome as a whole, about one-half of protein-coding genes are organized into operons.
- The genes vary in length, reflecting the lengths of their encoded proteins.

Table 19.3 summarizes some of what has been learned about the *E. coli* K12 genome to date with respect to its physical aspects, genes, and gene products.

Other bacterial genomes may be larger or smaller than the *E. coli* K12 genome, but their genome landscapes are similar to that of *E. coli* in several ways. For example, typically there is one origin of replication, 85–92% of the DNA codes for proteins, there is a mixture of operons and single-gene transcription units, some genes are transcribed using one DNA strand as the template whereas others are transcribed using the other strand, and there are relatively few transposable elements or repetitive sequences.

PROFILE OF THE HUMAN GENOME At about 3.2 billion base pairs, the human genome is about 700 times longer than the *E. coli* genome. Each human individual has 23 pairs of chromosomes. Men have 24 different chromosomes, the 22 autosomes and the X and Y chromosomes, whereas women have 23 different chromosomes, the 22 autosomes and the X chromosome. **Figure 19.7A** displays the complete set of human chromosomes,

diagramming the banding patterns that help researchers identify regions of chromosomes (also see Figure 10.8). **Figure 19.7B** shows chromosome 6 in more detail and then a close-up of a 100-kb segment of the long arm of that chromosome to show the protein-coding genes it contains. Compare this figure with Figure 19.6B, which shows a 10-kb segment of the *E. coli* chromosome and note the following:

- Genes are relatively far apart, with a large amount of space in between. That is, even though the human genome segment shown is ten times longer than the *E. coli* segment shown in Figure 19.6.B, it contains only two protein-coding genes. Each of these genes consists of transcription units that are far longer than the genes in *E. coli*, largely because they consist of about 95% introns and 5% exons. The right-hand gene in Figure 19.7B illustrates at the DNA level the alternative splicing variants for that gene; note the different exons for the three gene drawings. (Alternative splicing is described in Section 16.3.)
- As in the genomes of other organisms, some genes are transcribed in the left-to-right direction and others in the right-to-left direction. For the particular segment shown in Figure 19.7B, both genes are transcribed in the right-to-left direction.
- All the genes are single transcription units. Eukaryotic genes are rarely organized in operons.

Table 19.3 summarizes some of what researchers have learned about the human genome to date with respect to its physical aspects, genes, and gene products. Some of the key features of the human genome are:

- There are 19,881 identified protein-coding genes. On average, there are 9 to 10 exons per gene, with some human genes consisting of a single exon and others having over 100 exons. Introns make up about 95% of the average transcription unit. Since about 2% of the genome consists of protein-coding sequences, introns represent about 20–25% of the human genome.
- On average, more than 20 regulatory sequences are associated with each protein-coding gene. Those sequences are distributed over thousands to tens of thousands of base pairs upstream and downstream of each gene, as well as being scattered within introns. The regulatory sequences are widely scattered within regions that encompass 15–25% of the human genome.
- There are more than 25,000 identified noncoding RNA genes, which include genes for rRNAs, tRNAs, snRNAs, miRNAs, and lncRNAs (see Chapter 15 and Section 16.3).
- There are thousands of pseudogenes.
- About 45% of the genome consists of transposable element sequences (see Section 15.5). Only a tiny fraction of those transposable elements are functionally active. The others are inactive, being the transposable element version of pseudogenes.

- The genome contains a variety of short, repeated sequences, including those at the centromeres and telomeres, as well as others scattered throughout the rest of the chromosome.
- Replication begins at hundreds to thousands of origins of replication per chromosome. However, there are no consistent, sequence-specific replication origins as is found in bacteria and yeast.

Up to this point, we have focused on the features of the enormous portion of the human genome encoded within the linear chromosomes of the cell's nucleus. It is also important to remember that in a eukaryote, each mitochondrion also contains a circular mitochondrial genome, or mtDNA. The human mtDNA is 16.6 kb, much smaller than even a prokaryotic genome. Its 37 genes (13 of them are protein-coding) perform essential functions related to cellular respiration. In addition, photosynthetic eukaryotes, including plants and algae, have a separate circular genome up to several hundred thousand bases in each of their chloroplasts: the cpDNA. Not surprisingly, this genome contains genes involved in photosynthesis.

Other mammalian genomes are very much like the human genome. But for other eukaryotes generally, particular features can vary considerably. For example, eukaryote genomes range

from having a very low percentage of protein-coding DNA as in mammals to almost as high a percentage as is seen for prokaryotes.

A more highly detailed annotation of the human genome is being generated as a result of research efforts of the ENCODE (Encyclopedia of DNA Elements) project (http://www.encodeproject.org), which was introduced in Section 15.3. The goals of the project include annotating the human genome in detail, including identifying all the genes (both protein-coding and noncoding), identifying all splicing variants of genes, and identifying all gene-regulatory sequences. This research is being done with over 100 different cell types, some of which are cancer cell types and others of which are normal cell types. The research data collected so far has helped refine details of genes, including finding new exons; revising transcription initiation and termination sites, and identifying alternative start and stop sites; and cataloging a wider range of splice variants. (The last two areas were discussed in Section 15.3.) The research have also identified chromatin modifications and other features of promoter and enhancer sites, thereby giving scientists further insights into how human genes are regulated. Using RNA-seq, research is underway with each of the chosen cell types to identify which genes are transcribed at higher or lower levels so as to give insights into gene expression patterns not only in cancerous vs normal cells, but in cells of various differentiated types.

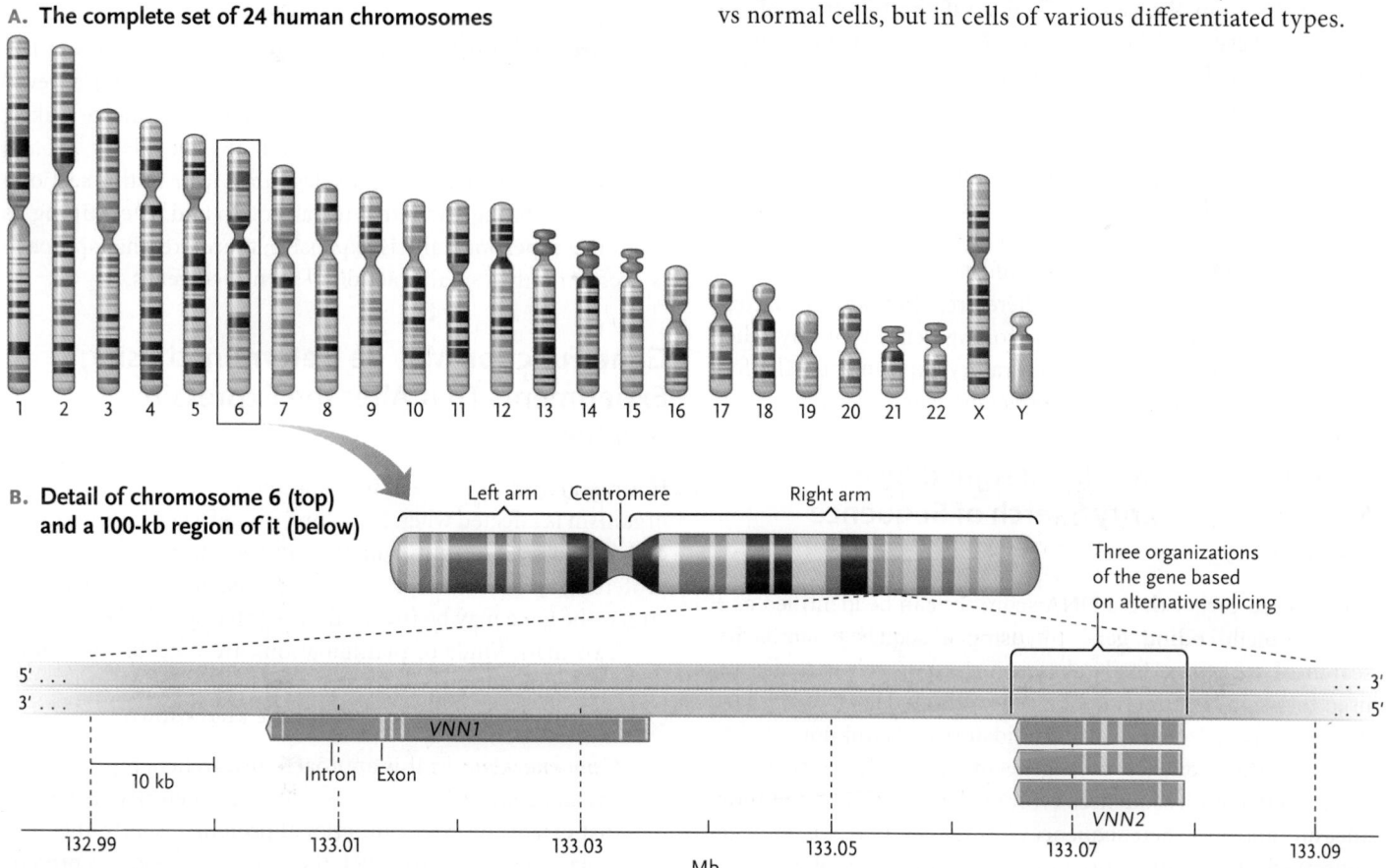

A. The complete set of 24 human chromosomes

1 2 3 4 5 6 7 8 9 10 11 12 13 14 15 16 17 18 19 20 21 22 X Y

B. Detail of chromosome 6 (top) and a 100-kb region of it (below)

Left arm   Centromere   Right arm

Three organizations of the gene based on alternative splicing

VNN1

Intron Exon

10 kb

VNN2

132.99   133.01   133.03   133.05   133.07   133.09

Mb

FIGURE 19.7 The human genome.
© Cengage Learning 2017

19.3 Determining the Functions of the Genes in a Genome

Once a genome is annotated, the next step is to use the genome sequence data to understand the functions of genes and other parts of the genome. "Gene function" is considered broadly here to include regulation of gene expression, the products genes encode, and the role of those products in the function of the organism. For protein-coding genes, the gene products are proteins. We study proteins to understand their structure and function, and to discover how they interact with other proteins and other nonprotein molecules in the cell and how those complexes are important functionally for the organism.

We also need to determine what genes there are for noncoding RNAs in the genome and what the functions of the RNAs are. Several of the noncoding RNA genes can be assigned functions based on evolutionary conservation principles; that is, by looking for sequence similarity with known gene sequences. An example is the rRNA genes. However, identifying and determining the functions of miRNA and lncRNA genes is more challenging in part because of their diversity of sequences. In this section, we focus on protein-coding genes, again because of the importance of their products in controlling the functions of cells and, therefore, of organisms. Determining the functions of protein-coding genes typically relies on computer analysis and on laboratory experiments. The following presents examples of those approaches.

Gene Function May Be Assigned by a Sequence Similarity Search of Sequence Databases

You learned earlier that a DNA sequence can be identified as a likely protein-coding gene by using a sequence similarity search of sequences in databases. This approach can also be used to assign the function of a gene. A high degree of similarity between the sequence of a candidate gene of unknown function and the sequence of a gene of known function likely indicates that both sequences evolved from a gene in a common ancestor and that their sequences in the present day have been conserved significantly because they code for proteins that have similar functions. As explained earlier, genes with highly conserved sequences as a result of divergence from a common ancestral gene are homologous genes.

Using sequence similarity searches to determine if a candidate gene and a known gene are homologous is by far the most common method for assigning the functions of genes. Experimental investigation of the functions of genes is considerably more expensive and time-consuming than DNA sequence comparisons; therefore, it is not feasible to repeat experiments in every species whose genome is sequenced. As a result, experimental data are available only for a small fraction of organisms. And because the functions of homologous protein-coding genes are so well conserved during the evolution of organisms, information about the function of a gene in one well-studied species very often applies to the homologous genes in another.

In some cases, the outcome of a sequence similarity search will indicate that the entire candidate gene's sequence is homologous to a known gene's sequence. In other cases, only part of the candidate gene sequence may match closely a sequence in a known gene. Typically this result indicates that the candidate gene encodes a protein with a domain that is related evolutionarily to a domain-encoding region of the known gene. (Protein domains are discussed in Section 3.4 and in the Genome Evolution section later in this chapter.)

Gene Function May Be Assigned Using Evidence from Protein Structure

Since the structure of a protein determines its function and the structures and functions of many proteins are known, similarity between the structure of a protein determined for a newly identified gene and a characterized protein will indicate a likely function of the protein product of that gene. (How protein structure may be determined is described later in this section.) However, isolating and purifying a protein and determining its structure experimentally is expensive; therefore, that approach is used for only a small fraction of identified genes.

Gene Function May Be Determined Using Experiments That Alter the Expression of a Gene

If a researcher can determine how the phenotype of a cell or organism is affected when the expression of a gene is turned off, or reduced significantly, functional properties of the encoded protein may be inferred. In a simple example, if cells grow larger, the gene may be involved in regulating cell size.

Two main kinds of manipulations are used to turn off or reduce significantly the expression of a gene in genome-scale experiments—gene knockout and gene knockdown:

1. *Gene knockout.* In this approach, researchers replace a normal gene on its chromosome with a defective gene that cannot express a functional protein. Usually, the replacement lacks the ORF that encodes the gene's protein product. In effect, this is a deletion mutation that has been engineered genetically. A deletion mutation is a *null mutation* because there is zero expression of the gene's

protein product. For a haploid organism, there is only one copy of each gene to knock out, whereas in diploid organisms both copies of each gene must be knocked out. On a genomic scale, experimental manipulations can be done to knockout each gene systematically one by one. The phenotypic consequences of zero expression of each gene can then be ascertained. Major projects have been done, or are being done, to knock out systematically the function of each gene in the genomes of several organisms, including yeast, the fruit fly, the nematode worm, and the mouse (the generation of knockout mice starting with embryonic stem cells was described in Section 18.2 and Figure 18.8). See *Focus on Research: Basic and Applied Research* in Chapter 18 for discussion of programmable RNA-guided genome editing based on the CRISPR–Cas system, a newer approach for making gene knockouts and other genome changes.

2. *Gene knockdown.* Knocking down a gene's expression can be done using RNA interference (RNAi). As discussed in Section 16.3, RNAi reduces the expression of a gene at the translation level. In RNAi, a small regulatory RNA (like a natural miRNA) is transcribed from an expression plasmid introduced into the cell. The sequence of that regulatory RNA forms complementary base pairs with the mRNA of a gene of interest. The base pairing triggers the RNAi molecular mechanisms (see Figure 16.13), which knocks down the expression of the gene by causing degradation of that gene's mRNA or by blocking its translation. For example, RNAi has been used to knock down gene expression of each of the approximately 20,000 genes of the nematode worm one by one. The advantage of RNAi in comparison to gene knockouts is that the decrease in function of a gene can be temporary. Gene knockdown can also be done at the transcriptional level using a newly developed technique called *CRISPRi* that is derived from the CRISPR–Cas system described in Chapter 18's *Focus on Research: Basic and Applied Research.*

Characterizing genes by studying the effects on phenotype of knockouts or knockdowns can be very expensive and time-consuming. In a genome-wide study, thousands of knockout or knockdown strains have to be engineered genetically, and then each one has to be screened for a battery of possible phenotypic changes. The most ambitious studies of this kind have examined hundreds of phenotypes for each gene, which is only a fraction of the phenotypes that could be characterized. To make this approach really productive in the future will require further development of high-throughput methods that automate the measurement of phenotypic changes.

The study of gene function by identifying changes in phenotypes is called **phenomics.** Knowing the phenotypic effects of knocking out or knocking down a gene usually only gives general information about the function of a gene. But that information is useful for follow-up experiments on the specific protein involved.

Transcriptomics Determines at the Genome Level When and Where Genes Are Transcribed

Some genes are transcribed in all cell types, whereas others are transcribed only when and where they are needed (see Chapter 16). Determining when and where genes are transcribed can shed light on their function. For instance, a researcher might be interested in determining at a genomic scale the gene expression patterns in different cell types, at different stages of embryonic development, at different points of the cell division cycle, or in response to mutation or changes in the environment. A medical example, for instance, would be identifying gene expression differences between normal cells and cells that have become cancerous. The experimental analysis itself may be qualitative—analyzing whether or not genes are expressed—or quantitative—analyzing how the level of expression of genes varies.

The complete set of transcripts in a cell, tissue, or organism is called a **transcriptome,** and the study of a transcriptome is called **transcriptomics.** Transcriptomics includes cataloging transcripts, and quantifying the changes in expression levels of each transcript during development, in different cell types, under different physiological conditions, and with other variations.

Analysis of transcriptomes is done using high-throughput hybridization or, increasingly, by sequence-based approaches. A hybridization-based approach uses **DNA microarrays,** also called **DNA chips.** The surface of a DNA microarray is divided into a microscopic grid of about 60,000 spaces. On each space of the grid, a computerized system deposits a microscopic spot containing about 10,000,000 copies of a DNA probe that is about 20 nucleotides long.

Studies of gene activity using DNA microarrays involve comparing gene expression under a defined experimental condition with expression under a reference (control) condition. As a theoretical example, **Figure 19.8** shows how a DNA microarray can be used to compare gene expression patterns in normal cells and cancer cells in humans. mRNAs are isolated from each cell type and cDNAs are made from them, incorporating different fluorescent labels: green for one cDNA, red for the other. The two cDNAs are mixed and added to the DNA microarray, where they hybridize with whichever spots on the microarray contain complementary DNA probes. A laser excites the fluorescent labels and the resulting green and red fluorescence is detected and quantified, enabling a researcher to see which genes are expressed in the cells. This technique is semiquantitative because it is also able to quantify differences in gene expression between the two cell types approximately (see *Interpreting the Results* in Figure 19.8).

In cancer research, for example, microarray analysis has been used to diagnose a particular cancer type, thereby potentially informing physicians about appropriate treatments. That is, many cancers can be identified based on **signature genes,** which are key genes whose expression changes in a way that correlates with a normal cell becoming a cancer cell. *Diffuse large B-cell lymphoma* (DLBCL) is the most common type of

FIGURE 19.8 **Research Method**

DNA Microarray Analysis of Gene Expression Levels

Purpose: DNA microarrays can be used in various experiments, including comparing the levels of gene expression in two different tissues, as illustrated here. The power of the technique is that the entire set of genes in a genome can be analyzed simultaneously.

Protocol:

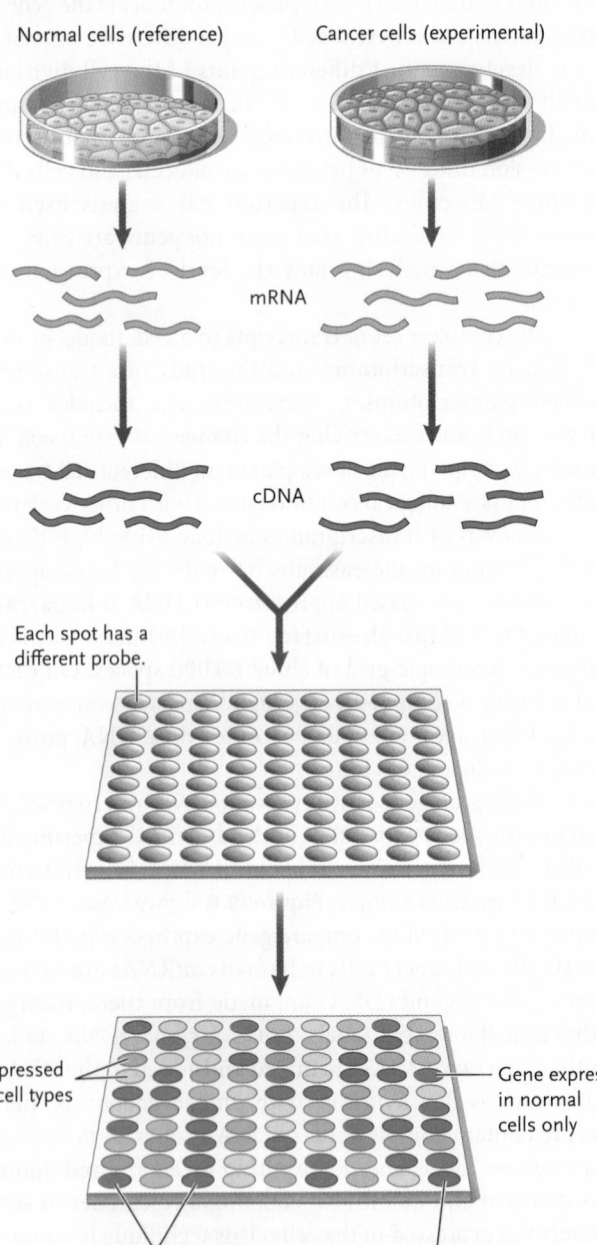

Normal cells (reference)   Cancer cells (experimental)

mRNA

cDNA

Each spot has a
different probe.

Gene expressed
in both cell types

Colored spots are
where labeled cDNAs
have hybridized.

Gene expressed
in normal
cells only

Gene expressed in
cancer cells only

1. Isolate mRNAs from a control cell type (here, normal human cells) and an experimental cell type (here, human cancer cells).

2. Prepare cDNA libraries from each mRNA sample. For the normal cell (control) library use nucleotides with a green fluorescent label, and for the cancer cell (experimental) library use nucleotides with a red fluorescent label.

3. Denature the cDNAs to single strands, mix them, and pump them across the surface of a DNA microarray containing a set of single-stranded probes representing every protein-coding gene in the human genome. The probes are spotted on the surface, with each spot containing a probe for a different gene. Allow the labeled cDNAs to hybridize with the gene probes on the surface of the chip, and then wash excess cDNAs off.

4. Locate and quantify the fluorescence of the labels on the hybridized cDNAs with a laser detection system.

Actual DNA microarray result

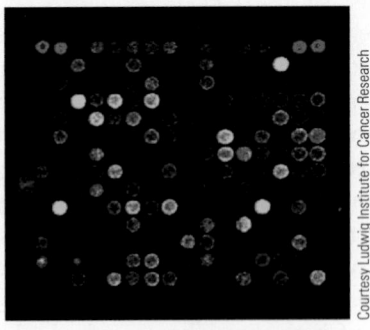

Courtesy Ludwig Institute for Cancer Research

Interpreting the Results: The colored spots on the microarray indicate where the labeled cDNAs have bound to the gene probes attached to the chip and, therefore, which genes were active in normal and/or cancer cells. Moreover, we can quantify the gene expression in the two cell types by the color detected. A purely green spot indicates the gene was active in the normal cell, but not in the cancer cell. A purely red spot indicates the gene was active in the cancer cell, but not in the normal cell. A yellow spot indicates the gene was equally active in the two cell types, and other colors tell us the relative levels of gene expression in the two cell types. For this particular experiment, we would be able to see how many genes have altered expression in the cancer cells, and exactly how their expression was changed.

non-Hodgkin's lymphoma, a type of cancer of the lymphatic system. Some patients respond to chemotherapy treatment, resulting in prolonged survival, whereas the remainder do not respond to such treatment and show no greater survival than untreated patients. DNA microarray-based transcriptome analysis of DLBCLs in a group of patients showed that the gene expression profile in chemotherapy-responsive tumors is different from that in chemotherapy-nonresponsive tumors. This result indicates that there are two subtypes of DLBCL. Identifying by DNA microarray analysis which type of DLBCL a patient has can, therefore, inform physicians as to the most effective treatment regimen. If the DLBCL is a responder, then chemotherapy is the best approach. If the DLBCL is a nonresponder based on the gene expression profile, then more aggressive treatments can be done.

DNA microarray is only one method that can be used to identify cancer subtypes. As you learned in Section 16.5, genome sequencing is now being used to identify subtypes of various cancers.

Other examples of DNA microarray analysis include screening individuals for mutations associated with genetic diseases (such as breast cancer) and studying changing gene expression profiles during *Drosophila* development, when a gingivitis-causing bacterium grows under different conditions (see *Molecular Insights* in Chapter 26), during flower development in plants (see *Focus on Research: Basic Research* in Chapter 37), and when sea urchin larvae are living under different pH conditions (see *Molecular Insights* in Chapter 54).

As you learned earlier in the section, RNA-seq can also be used to identify and quantify RNA transcripts in a sample. In fact, RNA-seq rapidly is becoming the replacement for DNA microarrays in transcriptomics, because of its decreasing cost and its greater precision for quantifying transcripts. RNA-seq is particularly useful in this regard in studies of non–model organisms for which microarrays would be very expensive to construct.

Proteomics Is the Characterization of All Expressed Proteins

Genome research also includes analysis of the proteins encoded by a genome, for proteins are largely responsible for cell function and, therefore, for most of the functions of an organism. The term *proteome* has been coined to refer to the complete set of proteins that can be expressed by an organism's genome. A *cellular proteome* is a subset of those proteins—the collection of proteins found in a particular cell type under a particular set of environmental conditions.

The study of the proteome is the field of *proteomics*. The number of possible proteins encoded by the genome is larger than the number of protein-coding genes in the genome, at least in eukaryotes. In eukaryotes, alternative splicing of gene transcripts and variation in protein processing means that expression of a gene may yield more than one protein product. The number of different proteins an organism can produce typically far exceeds the number of protein-coding genes.

Proteomics has three major goals: (1) to determine the structures and functions of all proteins; (2) to determine the location of each protein within or outside the cell; and (3) to identify physical interactions among proteins.

DETERMINING PROTEIN STRUCTURE Protein structure may be determined as follows:

Clone the coding sequence of the gene into an expression vector (see Section 18.2)
↓
Transform the cloned gene into a host to express the protein
↓
Purify the protein
↓
Determine the structure of the protein using X-ray crystallography or nuclear magnetic resonance (NMR)

Protein structure may also be predicted nonexperimentally using computer algorithms based on the known chemistry of amino acids and how they interact.

DETERMINING THE LOCATIONS OF PROTEINS IN CELLS The location of a protein in a cell is important because it is key to its function. The cellular location of a protein can be studied by tagging the protein in some way and then visualizing the location of the tag microscopically. Different tags are used for visualization using light microscopy or electron microscopy.

IDENTIFYING INTERACTIONS AMONG PROTEINS Many proteins function by interacting with other proteins. In some cases, proteins (actually polypeptides) interact to form the quaternary structure—and therefore the functional form—of a protein (see Section 3.4). Many multi-polypeptide proteins exist, and you have encountered several in this book, for example: (1) hemoglobin is a four-polypeptide protein consisting of two α-globin polypeptides and two (β-globin polypeptides and four associated heme groups (see Figure 3.17); (2) RuBP carboxylase/oxygenase (rubisco), the first enzyme of the light-independent reactions of photosynthesis (see Section 8.3), consists of eight copies of a large polypeptide and eight copies of a small polypeptide; and (3) the Lac repressor protein that controls the expression of the *lac* operon in *E. coli* (see Section 16.1) consists of four copies of the same polypeptide.

In other interactions among proteins, the interaction is not permanent, but instead serves to affect the function of one or other of the partners in the interaction. For example, in Chapter 6 you were introduced to protein kinases—enzymes that transfer a phosphate group from ATP to one or more sites on particular target proteins as part of a signal transduction pathway. The phosphorylation of the target proteins occurs as a result of the interaction between the enzymatic protein and each target protein. Once the target protein is phosphorylated,

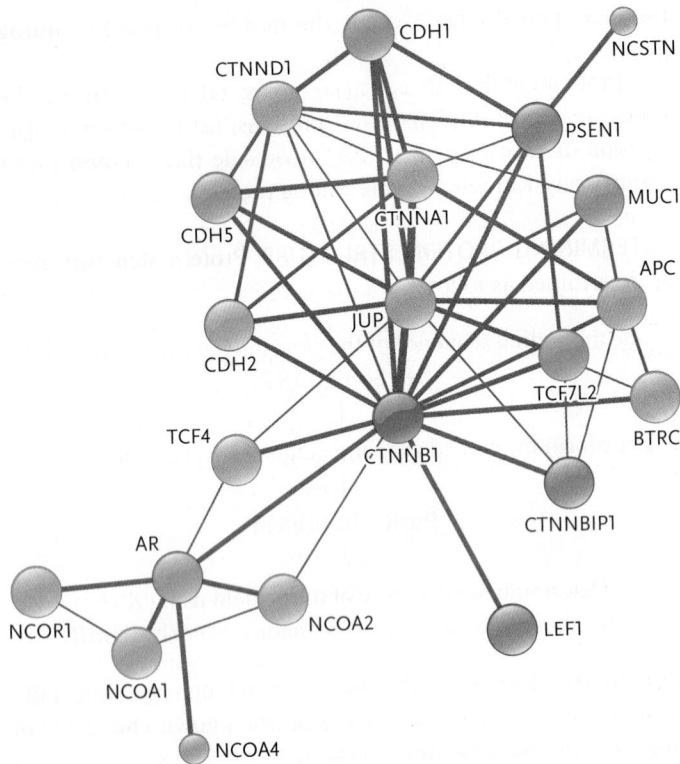

FIGURE 19.9 The protein-interaction network for human β-catenin (CTNNB1). Thicker lines show stronger associations between proteins.
© Cengage Learning 2017

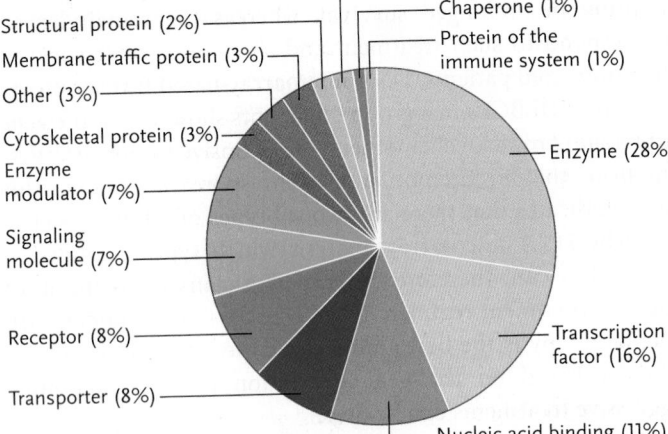

A. Protein classes

Transfer/carrier protein (2%)
Structural protein (2%)
Membrane traffic protein (3%)
Other (3%)
Cytoskeletal protein (3%)
Enzyme modulator (7%)
Signaling molecule (7%)
Receptor (8%)
Transporter (8%)
Chaperone (1%)
Protein of the immune system (1%)
Enzyme (28%)
Transcription factor (16%)
Nucleic acid binding (11%)

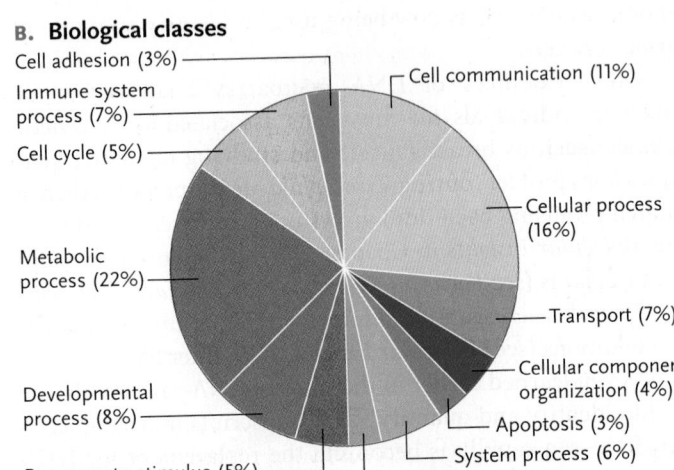

B. Biological classes

Cell adhesion (3%)
Immune system process (7%)
Cell cycle (5%)
Metabolic process (22%)
Developmental process (8%)
Response to stimulus (5%)
Cell communication (11%)
Cellular process (16%)
Transport (7%)
Cellular component organization (4%)
Apoptosis (3%)
System process (6%)
Reproduction (3%)

FIGURE 19.10 Functions of human protein-coding genes organized with respect to protein classes (A) and biological processes (B).
© Cengage Learning 2017

the two proteins no longer interact. Understanding the interactions among proteins is important, then, to help us understand how proteins work individually and together to determine the phenotype of a cell.

Thousands of interactions have been identified experimentally for a variety of organisms. The interaction data are assembled to produce protein-interaction networks, the analysis of which is informing us about the details and complexities of the functions of proteins in cells. **Figure 19.9** shows part of a protein-interaction network centered on the human protein, β-catenin (cadherin-associated protein; the central CTNNB1 sphere in the figure). β-Catenin is involved in the formation of adherens junctions (a type of cell junction; see Section 4.5) in epithelial cells where it links α-catenin (CTNNA1 in the figure) with E-cadherin (CDH1 in the figure; cadherins are discussed in Chapter 50). It also plays a key role in a signaling pathway that is important, for example, in regulating how cell fate is decided during development. Interactions with, for example, APC, LEF1, TCF4, and TCF7L2 occur as part of that signaling pathway's operation.

Just as for genomic DNA sequences, information about various properties of each protein are placed in databases, for example Entrez (http://www.ncbi.nlm.nih.gov) and UniProt (http://www.uniprot.org), to create a dossier of that protein that is available to researchers worldwide.

CHARACTERIZING PROTEIN FUNCTION Through various approaches, researchers learn about the functions of proteins. **Figure 19.10A** shows an example of what we have learned about the functions of human protein-coding genes with respect to protein classes, and **Figure 19.10B** shows the functions of human protein-coding genes with respect to the biological processes involving those proteins.

STUDY BREAK 19.3

1. What are the ways by which the function of a gene identified in a genome sequence may be assigned?
2. How would you determine how a steroid hormone affects gene expression in human tissue culture cells?
3. What is the proteome, and what are the major goals of proteomics?

19.4 Genome Evolution

DNA genomes with protein-coding genes are thought to have evolved over 3.5 billion years ago, by the time of the earliest fossil microorganisms that have been discovered (see Chapter 25). Those early cells probably had at most a few hundred protein-coding genes. New genes evolved as life evolved and became more complex, so that most present-day organisms have thousands or tens of thousands of protein-coding genes. In this section you will learn how genes and genomes have evolved, and how genome sequences inform us about the evolutionary history of life.

Comparative Genomics Reveals the Evolutionary History of Genes and Genomes

Understanding how genes evolved and how genomes evolved are major goals of the field of *comparative genomics*. Because the genes in present-day genomes evolved from ancestral genes that were in the genomes of organisms living millions to billions of years ago, we can trace the evolutionary history of genes by comparing the genomes of different groups of present-day organisms. From such comparisons, we can estimate when new genes first appeared in ancient organisms, describe how they changed over time, and gain insights into what molecular processes cause new genes to evolve in the first place.

Comparative genomics has shown that some genes are found in the genomes of almost all present-day organisms. Examples are genes involved in core biological processes such as protein synthesis, including genes for many of the proteins that make up part of the structure of a ribosome, and most of the aminoacyl–tRNA synthetase enzymes that attach amino acids to transfer RNA molecules. The proteins coded for by these genes not only perform the same function in every organism, but they are also related evolutionarily—they are homologous. This strongly suggests that the single-celled common ancestor of all living organisms had those genes in its genome, and that those genes have been passed down through the generations for billions of years. When genes in two or more different organisms are closely related to one another evolutionarily, and perform the same function in every organism, they are referred to as **orthologs.** For example, orthologs of the enzyme

that attaches valine amino acids to tRNAs are found in organisms from every kingdom of living things. Collectively, all of the orthologs of this gene in all species make up an **orthologous group** of genes. All members of an orthologous group have evolved from a single ancestral gene that was present in the common ancestor of all organisms whose genomes contain one or more genes in this orthologous group.

As an example, most genes do not appear in the genomes of all organisms but have a more restricted distribution. In the human genome, some genes are members of orthologous groups containing genes found only in other primates and not in any other mammals. This implies that the ancestral gene of these orthologs evolved first in a primate. Other human genes are members of larger orthologous groups representing a broader range of species: all vertebrates, for example, or all animals.

Analyzing the evolutionary history of genes provides valuable information about how life evolved on the molecular level. For example, by comparing the functions of almost 4,000 evolutionarily related groups of genes in 100 genomes of bacteria, archaea, and eukaryotes, researchers have identified a period about 3 billion years ago when many new genes evolved. By analyzing the functions of these new genes, the researchers concluded that many of the new genes evolved as adaptations to changes in the amount of oxygen in Earth's atmosphere, following the development of the oxygen-producing photosynthetic reactions (see Chapter 8).

Comparative genomics has also been applied to understanding human evolution. (Human evolution is discussed in Section 32.12.) The human genome was the first mammalian genome sequenced, and researchers now have the sequences of hundreds of human genomes to study and compare to discover what makes us human and what is responsible for human variation. We also have almost 100 other mammalian genomes to compare with the human genome. These include genomes of primate species that are closely related to humans, such as the common chimpanzee and the mountain gorilla, as well as less closely related mammals, such as the cow and the duck-billed platypus. Comparing the human genome with the genomes of other primates reveals which features are common to all of these primates and which are unique to humans. The human and chimpanzee genomes are strikingly similar, with 96% DNA sequence identity across the entire genome. The annotation of the chimpanzee genome is not yet complete, but it is likely that these two species share virtually all of their genes, so the genomic changes that occurred in human evolution probably involved only subtle mutations in the protein-coding sequences of genes, and mutations to regulatory sequences that determine how and when each gene is expressed (see Chapter 16). By contrast, comparisons of primate genomes with those of other mammals have identified new genes that evolved only in primates. Further studies of the functions of these genes may shed light on how primates evolved the characteristics that distinguish them from other mammals (see Chapter 32). And, interestingly, the primate-specific genes in the human genome contain the highest fraction of disease-related genes—19.4%—of any group of genes.

DNA microarrays (see Section 19.3) and RNA-seq (see Section 19.2) have been used to compare which genes are transcribed in which parts of the brain in humans, chimpanzees, and rhesus monkeys. Certain groups of genes are expressed in the brain only during embryonic and early postnatal development. In both chimpanzees and rhesus monkeys, many genes involved in the formation of new synapses (communicating junctions between neurons: see Chapter 39) are transcribed only in the first year after birth, whereas in humans expression of these genes continues up to age five. These differences are most prominent in the prefrontal cortex, which is an area of the brain involved in complex decision making (see Chapter 40). These comparative findings provide clues to how our species evolved enhanced learning abilities.

Comparative genomics also provides information about how the arrangement of genes on chromosomes has evolved. In Section 13.3 you learned about chromosomal mutations that occur when part of a chromosome is translocated to another chromosome, or inverted in place. Nondisjunction in meiosis can also cause entire chromosomes to be duplicated (see Figure 13.13). Such chromosomal mutations are uncommon, and usually they are harmful. But when a nonharmful chromosomal mutation occurs and spreads to all members of a species, the order of genes on the chromosomes of that species may then be different from the order in closely related species. Comparing the genomes of a range of related species reveals that, over the course of hundreds of millions of years of biological evolution, pieces of chromosomes have changed places repeatedly by translocation and inversion, rearranging the genes on chromosomes like shuffling a deck of cards (see Figure 22.17). Even so, the chromosomal arrangement of some genes is preserved after all this reshuffling, even in distantly related organisms. For example, comparisons of the human genome with the genomes of distantly related animals such as insects and sea anemones reveal blocks of homologous genes that are on the same chromosome and arranged in the same order in all of these species. This means that the order of these genes on the chromosome has been preserved from the time that all of these species evolved from a common ancestor, even though that common ancestor lived over 500 million years ago.

New Genes Evolve by Duplication and Exon Shuffling

The evolution of a new gene is a rare event—much less common than a mutation in an existing gene. But, nonetheless, over millions of years of biological evolution many new genes have been produced. For example, a comparison of genome sequences among four species in the *Drosophila* genus of fruit flies identified over 200 genes that had evolved in just the past 13 million years (a comparatively short time in evolutionary history).

Throughout the history of life on Earth, the evolution of new biological functions has almost always involved the evolution of new genes. For example, photosynthesis became possible only with the evolution of genes coding for proteins that could harness the energy in photons to synthesize ATP and electron carrier molecules. And comparative analysis of mammalian genomes has revealed genes involved in milk production and other biological functions found only in mammals. Evolutionary biologists have described a number of molecular mechanisms to explain where these new genes come from. The most common molecular mechanism to explain the origin of new genes is *gene duplication,* which produces *multigene families* after a series of duplications of genes that all derive from the same ancestral gene. New *types* of genes are produced by a process called *exon shuffling,* which combines parts of two or more genes.

GENE DUPLICATION Gene duplication is any process that produces two identical copies of a gene in an organism's genome. Genes can be duplicated by *unequal crossing-over* of homologous chromosomes during meiosis, or as a result of the replication of transposable elements (see Section 15.5).

Unequal crossing-over is the rare phenomenon in meiosis in which, instead of crossing-over occurring at the exact same point on each homolog of a homologous pair of chromosomes **(Figure 19.11A)**, crossing-over occurs at different points **(Figure 19.11B)**. The result of unequal crossing-over is that one of the recombinant chromosomes is missing one or more genes, whereas the other has duplicate copies of those genes. Unequal crossing-over produces **tandem duplication** of genes with the duplicate copies clustered together in the same region of the same chromosome.

Gene duplication may occur when a transposable element copies itself and splices the DNA copies elsewhere in a genome. (Transposable element movement is discussed in Section 15.5.) Rarely, transposable elements copy adjacent DNA in addition to their own, producing duplicate copies of any genes in that DNA. This produces **dispersed duplication** of genes, meaning that the copies of the gene are found in different places in the genome—often on two different chromosomes. Similarly, retrotransposons cause dispersed gene duplication when the enzymes produced by those retrotransposons make DNA copies of the mRNA molecules transcribed from other genes, and insert those DNA copies into a chromosome. Duplicate genes produced by this mechanism can be recognized because they lack the introns that had been removed in producing the mRNAs.

At first, the duplicate copies of a gene have the same protein-coding sequences and encode identical proteins. The two genes are functionally redundant, meaning that one could be eliminated from the genome with no loss of biological functionality. Often, one of the redundant copies is mutated into a pseudogene, or lost by deletion. But if both genes remain functional, they will evolve slowly in different ways, as different

A. Normal crossing-over

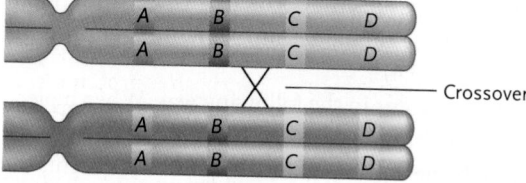

Crossover

Crossing-over occurs between homologous chromatids during prophase I of meiosis (see Figure 11.5). Normally crossing-over occurs at the exact same point on each homolog and results in recombinant chromosomes after meiosis that have the same number of genes in each homolog.

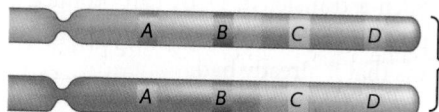

Recombinant chromosomes
(parental chromosomes
not shown)

B. Unequal crossing-over

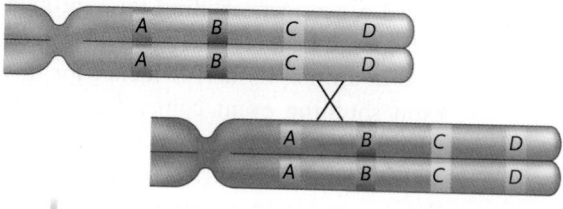

Unequal crossing-over results in recombinant chromosomes after meiosis that have a different number of genes. One (top) has duplicate genes, here B and C, whereas the other (bottom) has lost genes, here B and C.

FIGURE 19.11 Duplication of genes by unequal crossing-over.
© Cengage Learning 2017

mutations occur in each gene. Over many generations, this evolutionary process can produce two homologous genes with similar but distinct functions.

For example, nitric oxide synthase enzymes catalyze a reaction that produces nitric oxide (NO), a molecule that cells use to communicate with each other (see Section 44.5 for an example). In the human genome, there are three genes for different nitric oxide synthase enzymes. One gene is expressed only in neurons, another only in the endothelial cells lining blood vessels (see Chapter 44), and the third in the liver and in a type of white blood cell called a *macrophage* (see Chapter 45). Homologs of all three of these genes are found in other mammalian genomes, as well as in the genomes of birds and reptiles, so the gene duplications by which these genes evolved must have happened over 200 million years ago. Evolution of tissue-specific expression patterns for each gene must have involved mutations to regulatory sequences that control transcription. The evolution of these genes also involved mutations to the

protein-coding sequences, causing the proteins to have subtly different structures and functions. For example, the neuronal and endothelial nitric oxide synthase enzymes are regulated by Ca^{2+} ions, whereas the third enzyme is not. The three genes are found on three different chromosomes in the human genome, which suggests that they evolved by dispersed duplication.

PRODUCTION OF MULTIGENE FAMILIES Gene duplication is often followed by further duplication of one or both of the original duplicates. When this happens repeatedly over millions of years, a family of homologous genes called a **multigene family** evolves. The nitric oxide synthase enzymes described above are a small gene family, with three members in the human genome. Other multigene families contain tens to hundreds of genes. The members of a multigene family all evolve from one ancestral gene and therefore have similar DNA sequences and produce proteins with similar structures and functions. But because different mutations occur in each member of a multigene family, the genes gradually evolve subtly different characteristics.

Let us consider the *OPT/YSL* multigene family of the plant, *Arabidopsis thaliana,* as an example. The proteins coded for by these genes are oligopeptide transporters that shuttle short peptide molecules across cell membranes. This multigene family is found in other plant genomes as well as in the genomes of fungi and other eukaryotes. By comparing DNA sequences of *OPT/YSL* genes in the genomes of plants and other organisms, researchers have concluded that, hundreds of millions of years ago, and before the evolution of the plant kingdom, a gene duplication produced the ancestral *OPT* gene and the ancestral *YSL* gene. Mutations in each gene caused them to encode proteins specialized for transporting different oligopeptides. A series of more recent gene duplications occurred since the evolution of plants, but well before the evolution of *A. thaliana.* Each duplicate gene in this family accumulated different mutations, producing the functionally diverse set of *OPT* genes and *YSL* genes now found in *A. thaliana* and other plants.

Figure 19.12A illustrates the family relationships among the *OPT/YSL* genes using a phylogenetic tree, much as a family tree illustrates relationships in a human family (see Chapter 24 for more information on how phylogenetic trees are constructed). **Figure 19.12B** shows the distribution of the *OPT* genes on *A. thaliana* chromosomes and outlines the possible evolutionary history of some of those genes.

The oligopeptide transporter gene family in *A. thaliana* is larger than the nitric oxide synthase gene family in the human genome, but other multigene families are even larger. For example, some families of transcription factor proteins and membrane-bound receptor proteins include hundreds of genes. Some of the larger multigene families have members in all kingdoms of eukaryotes, or even in all three domains of living organisms. Such families each evolved from an ancestral gene that first appeared billions of years ago.

EXON SHUFFLING The new genes produced by gene duplication evolve distinct functions, but they retain the same general function as other members of the multigene family into which they have been "born," so to speak. Our examples have illustrated that. By contrast, **exon shuffling**—the duplication and rearrangement of exons—is a molecular evolutionary process that combines exons of two or more existing genes, to produce a gene that encodes a protein with an unprecedented function.

Remember from Section 15.3 that many protein-coding genes in eukaryotes contain introns, sequences that do not encode amino acids. The introns are present in the pre-mRNA transcripts of such genes but, by RNA processing, the introns are removed while the exons—the sequences encoding amino acids in the pre-mRNAs—are spliced together to make the mature mRNAs. In many genes, the junctions between exons fall at points within the protein-coding sequence between major functional regions in the protein. These functional regions correspond to the domains into which many proteins are divided (see Section 3.4).

Exon shuffling can occur in the following way. When a piece of DNA is cut out of a chromosome and reinserted elsewhere in the genome (through the activity of a transposable element, for example), the ends of the piece of DNA that moves may occur within the introns of a gene, causing one or more whole exons to be inserted somewhere else in a chromosome. If those exons are inserted into an intron in another gene, the amino acid sequence encoded by those exons may be added to the amino acid sequence of the encoded protein. Such a transfer of DNA can produce a new gene, coding for a protein that has one or more domains added to the other domains that it already had.

An exon shuffling event occurred very early in the evolution of animals (at least 700 million years ago) that produced a new gene coding for a protein that plays a key role in signaling between cells in animal tissues. Evidence for this exon shuffling event comes from comparing the genome sequence of the choanoflagellate *Monosiga brevicollis* with the sequences of a number of animal genomes, including *Homo sapiens*. Choanoflagellates (see Sections 27.2 and 31.1) are single-celled or colonial protists that are thought to be related evolutionarily to animals. The evolution of multicellularity in the first animals is thought to have involved molecular mechanisms that enabled choanoflagellate-like cells to attach to and communicate with one another, so they could then specialize in performing different functions.

Figure 19.13 shows the exon shuffling event. It involves the Notch family of proteins, which are multidomain, membrane-spanning proteins. The human Notch1 protein, encoded by the *NOTCH1* gene, contains a transmembrane (TM) region, 36 copies of an EGF domain, three copies of an NL domain, and six copies of an ankyrin domain. The TM region anchors the Notch1 protein in the plasma membrane. The three groups of

A. Family tree showing evolutionary relationships among *OPT* and *YSL* genes

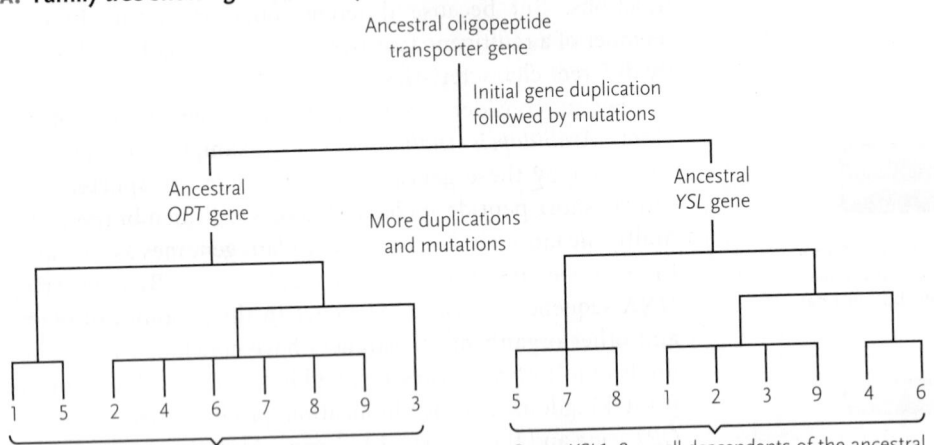

Genes *OPT1–OPT9* are all descendants of the ancestral *OPT* gene. An early duplication within this gene subfamily produced two genes, one an ancestor of *OPT1* and *OPT5* and the other an ancestor of the other seven *OPT* genes. A subsequent gene duplication in the latter group produced the ancestor of *OPT2, 4,* and *6–9*, and the ancestor of *OPT3*.

Genes *YSL1–9* are all descendants of the ancestral *YSL* gene. An early duplication within this gene subfamily produced two genes, one an ancestor of *YSL5, 7,* and *8* and the other an ancestor of the other six *YSL* genes. A subsequent gene duplication in the latter group produced the ancestor of *YSL1–3* and *9*, and the ancestor of *YSL4* and *6*.

B. Distribution of *OPT* genes on chromosomes in the *Arabidopsis thaliana* genome

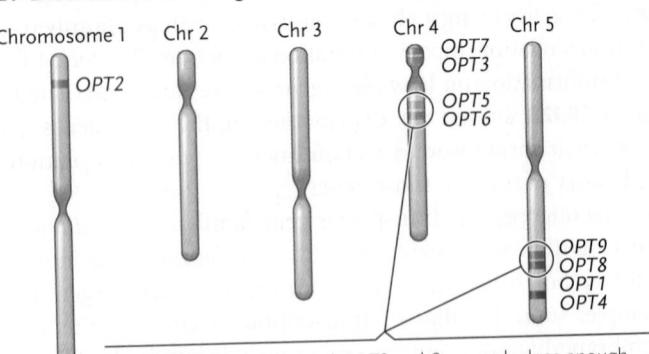

Gene pairs *OPT5* and *6* and *OPT8* and *9* are each close enough that they could have evolved by a recent tandem duplication. But only *OPT8* and *9* are near relatives in the *OPT* gene family tree (part **A** of figure). Therefore, we can hypothesize that *OPT8* and *9* likely resulted from a fairly recent tandem duplication, whereas *OPT6* is a dispersed duplicate of another *OPT* gene.

FIGURE 19.12 Evolution of the plant *OPT/YSL* multigene family. (A) Family tree showing evolutionary relationships among *OPT* and *YSL* genes. **(B)** Distribution of *OPT* genes on chromosomes in the *Arabidopsis thaliana* genome.
© Cengage Learning 2017

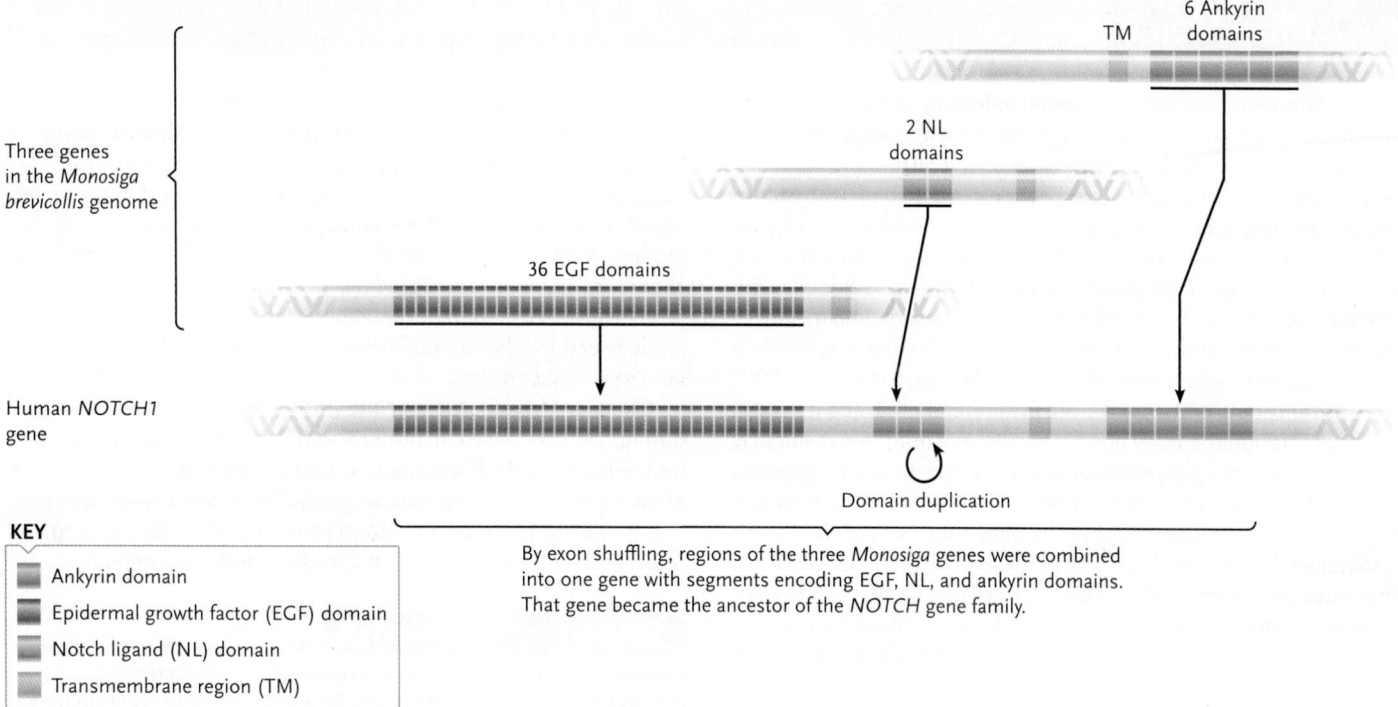

FIGURE 19.13 Evolution of Notch domains in animals by exon shuffling. At the top are three genes in the *Monosiga brevicollis* genome that encode three transmembrane (TM) region-containing proteins, one with epithelial growth factor (EGF) domains, a second with Notch ligand (NL) domains, and a third with ankyrin domains. At the bottom is the human *NOTCH1* gene, which encodes a protein with EFG, NL, and ankyrin domains.

© Cengage Learning 2017

domains are key to the protein's function. The part of the protein that is outside the cell membrane includes the EGF and NL domains, which enables a Notch protein in one cell to bind to other proteins in adjacent cells in a tissue. The ankyrin domains within the cell enable it to attach to the microfilaments that make up part of the cytoskeleton (see Section 4.3).

Figure 19.13 also shows three different genes in the *M. brevicollis* genome, one with a sequence encoding EGF domains, a second with a sequence encoding NL domains, and a third with a sequence encoding ankyrin domains. However, this organism lacks a gene homologous to the gene for the Notch protein. Researchers have hypothesized that, through exon shuffling early in animal evolution, the sequences coding for the EGF, NL, and ankyrin domains in the three genes were combined in one gene, producing the ancestor of the *NOTCH* gene family. At some point, a duplication of the sequence coding for one NL domain occurred, since the *M. brevicollis* gene has sequences coding for only two copies of that domain while animal genes for Notch proteins have sequences coding for three copies.

The genes produced by gene duplication typically encode functional proteins because they are duplicates of an existing functional gene. Most instances of exon shuffling theoretically should produce nonfunctional proteins, because an existing functional protein is interrupted by one or more domains from an unrelated protein. But in a small number of cases, proteins produced by exon shuffling combine the functions of two or more proteins in a new and useful way, such as the assembly of the gene for the Notch signaling protein (see Figure 19.13). Interestingly, exon shuffling is thought to account for perhaps one-third of newly evolved genes. The genes that are produced by exon shuffling have more novel functions than those produced by gene duplication. In cases such as the Notch protein, they become the ancestors of multigene families that evolve through subsequent gene duplications.

The domain structures of proteins in humans and other organisms provide evidence of how common exon shuffling has been in the evolution of proteins. The most widely used domains are found in thousands of different proteins, in dozens of different combinations with other domains.

STUDY BREAK 19.4

1. What molecular mechanisms cause tandem duplication of genes and dispersed duplication of genes?
2. Why do new genes produced by exon shuffling have more novel functions than new genes produced by gene duplication?

It is well known that individuals respond differently to medications. For example, a blood pressure–lowering medication may effectively lower blood pressure in one patient, have no effect on blood pressure in another patient, and cause intolerable adverse effects in a third patient. Factors such as age, weight, or liver and kidney function may influence response to medication. However, it remains difficult to predict how effective or safe a drug will be for a particular patient based on these factors alone. In clinical medicine, a trial-and-error approach is often used for drug therapy—several drugs or drug doses may be tried before finding the drug and dose that produce optimal response with acceptable tolerability.

Genes encoding proteins involved in the metabolism and transport of a drug through the body and proteins at the drug target site may greatly influence drug response. Genes involved in disease progression or phenotype may also influence how well a drug works. How can molecular approaches be used to personalize drug therapy? The answer to that question may well come from the field of *pharmacogenomics*. **Pharmacogenomics** is the study of how genes affect individual responses to drug therapy. The goal is to choose the most appropriate drug, drug dose, and treatment duration for a particular patient based on genetic information.

How can pharmacogenomics be used to optimize therapy for blood clots?

Prescribed to prevent the formation of a thrombus (clot) in the circulatory system, the drug warfarin ranks among the top 50 most commonly prescribed drugs in the United States. Even with newer agents available, warfarin remains the most commonly prescribed oral anticoagulant. Because it inhibits thrombus formation, warfarin also increases the chance of bleeding. The risk for bleeding increases when the warfarin dose is too high, and the risk for thrombosis increases when the dose is

too low. The dose of warfarin needed to prevent thrombosis without significantly increasing the chance for bleeding is highly variable among patients, with doses ranging by as much as 20-fold. Age, body size, other diseases the person may have, and medications the person may be taking all influence the dose of warfarin a person needs. However, considering age, body size, and clinical factors alone is often not enough to choose the right warfarin dose for a given patient. Genes encoding for proteins involved in warfarin metabolism (pharmacokinetics) and mechanism of action (pharmacodynamics) are now known to influence warfarin dose requirements.

Clinical trials have examined the effect of genotype-guided warfarin dosing, taking into account both genotype and clinical factors, but have had variable results. These trials focused on surrogate endpoints, such as time within the therapeutic anticoagulation range. Additional trials that may provide important information on the risk for bleeding and thrombosis with genotype-guided warfarin dosing are ongoing.

think like a scientist What are the ethical implications of pharmacogenomics if the genes in question are associated with disease risk as well as drug response? What are the ethical implications of pharmacogenomics if a patient possesses a genotype predictive of poor drug response and there are no alternative treatments available?

Courtesy of Larisa H. Cavallari

Larisa H. Cavallari is an Associate Professor and Director of the Center for Pharmacogenomics, College of Pharmacy at the University of Florida. She studies genetic contributions to cardiovascular drug response. To learn more about Dr. Cavallari's laboratory and research, go to http://pharmacy.ufl.edu/faculty/larisa-h-cavallari/.

REVIEW KEY CONCEPTS

For access to MindTap and additional study materials visit www.cengagebrain.com.

19.1 Genomics: An Overview

- Genomics is the characterization of whole genomes, including their structures (sequences), functions, and evolution.

- Genome sequence data are in databases that may be accessed by researchers worldwide.

- Genomics consists of three main areas of study: genome sequence determination and annotation, the determination of complete genome sequences and identification of putative genes and other important sequences; functional genomics, the study of the functions of genes and other parts of the genome; and comparative genomics, the comparison of entire genomes of parts of them to understand evolutionary relationships and basic biological similarities and differences among species.

19.2 Genome Sequence Determination and Annotation

- The whole-genome shotgun method of sequencing a genome involves breaking up the entire genome into random, overlapping fragments, cloning each fragment, determining the sequence of

the fragment in each clone, and using computer algorithms to assemble overlapping sequences into the sequence of the complete genome (Figure 19.1).

- DNA sequencing methods involve DNA purification; DNA fragmentation; amplification of fragments; sequencing each fragment; and assembly of fragment sequences into longer sequences, such as those of a genome (Figures 19.2 and 19.3).

- Once the complete sequence of a genome has been determined, it is annotated to identify key sequences, including protein-coding genes, noncoding RNA genes, regulatory sequences associated with genes, origins of replication, transposable elements, pseudogenes, and short repetitive sequences. Annotation of a genome is the task of researchers in bioinformatics.

- Identifying protein-coding genes in a genome sequence can be done by using a computer search for open reading frames, by searching databases for sequence similarity to genes of known function, and by studying gene transcripts (the gold standard).

- Complete genome sequences have been obtained for many viruses, a large number of prokaryotes, and many eukaryotes, including the human. For organismal genomes, those of bacteria and archaea are generally smaller than those of eukaryotes and their genes are densely packed in their genomes with little non-coding space in between them. Prokaryotic genes are organized either into single transcription units or into operons. Genomes of eukaryotes vary greatly in size, but there is no correlation between genome size and type of organism. Gene density also varies, but in general it is significantly less than is seen for prokaryotic genes. Eukaryotic genes are organized into single transcription units (Figures 19.5–19.7; Tables 19.2 and 19.3).

19.3 Determining the Functions of the Genes in a Genome

- The function of a protein-coding gene may be assigned by a sequence similarity search of sequence databases, by using evidence from protein structure, or by using knockout or knockdown experiments that alter the expression of a gene.

- Transcriptomics is the study at the genome level of when and where genes are transcribed. One experimental method for studying the transcription of all or many of the genes in a genome simultaneously is the DNA microarray (DNA chip); this technique can generate qualitative information about gene transcription, such as the similarities and differences in gene expression in two cell types or in two developmental stages, as well as quantitative information about the relative levels of gene transcription (Figure 19.8).

- Proteomics is the characterization of the complete set of proteins in an organism or in a particular cell type. Protein numbers, protein structures, protein functions, protein locations, and protein interactions are all topics of proteomics (Figures 19.9 and 19.10).

19.4 Genome Evolution

- Comparative genomics traces the evolution of genomes by analyzing similarities and differences in DNA sequences in the genomes of present-day organisms.

- Comparative analysis reveals how homologous protein-coding genes have evolved in groups of organisms, how regulation of gene expression has evolved through mutations in the regulatory sequences of genes, and how chromosome structure has evolved as parts of one chromosome have broken off and been attached to other chromosomes.

- New genes evolve by tandem duplication when chromosomes cross over unequally in meiosis, producing a chromosome containing two copies of the DNA coding for one or more genes (Figure 19.11). New genes evolve by dispersed duplication when transposable elements copy DNA coding for one or more genes, and insert it at another location in the genome.

- Duplicate copies of genes evolve distinct functions as different mutations occur in the two copies. Mutations in protein-coding sequences produce proteins with slightly different structures, while mutations in regulatory sequences cause the genes to be expressed in different cell types and in response to different stimuli.

- Repeated cycles of duplication followed by mutation produce multigene families, which are collections of homologous genes that code for similar but functionally distinct proteins. The largest multigene families comprise hundreds of different genes (Figure 19.12).

- Exon shuffling produces functionally novel proteins by combining parts of two or more different genes. When exons that code for one or more domains of a protein are copied from one gene and inserted into the protein-coding sequence of another gene, those domains are added to the structure of the protein coded for by that gene. Adding new domains to a protein gives the protein new molecular functionalities (Figure 19.13).

TEST YOUR KNOWLEDGE

Remember/Understand

1. Why is the Solexa/Illumina DNA sequencing method faster and less expensive than the Sanger method?
   a. It sequences longer fragments of DNA.
   b. It sequences more DNA fragments at the same time.
   c. It does not require the use of fluorescent markers.
   d. It does not require amplification of DNA fragments before sequencing.
   e. It does not require the use of computer algorithms to find places where sequence fragments overlap.

2. How do pseudogenes differ from genes?
   a. They are not transcribed.
   b. They contain longer open reading frames (ORFs).
   c. They do not have introns.
   d. They use a different genetic code.
   e. Their protein-coding sequence contains more than one start codon.

3. What is the main reason that searching for open reading frames (ORFs) is more useful for annotating prokaryote protein-coding genes than it is for annotating eukaryote protein-coding genes?
   a. Eukaryote protein-coding genes contain introns.
   b. The density of protein-coding genes is much higher in eukaryote genomes.
   c. In most prokaryotes, all of the protein-coding genes are located on a single circular chromosome.
   d. Prokaryotes use a different genetic code than eukaryotes.
   e. Prokaryotic protein-coding genes are much longer than eukaryotic protein-coding genes.

4. Which of the following is true about genome size?
   a. Bacteria have genomes that vary widely in size.
   b. The human genome is the largest among eukaryotes.
   c. Organisms with large genomes are always more complex than organisms with small genomes.
   d. As genome size increases in a lineage, the number of genes also always increases.
   e. The smallest known cellular genome is found in a species of Archaea.

5. Which of the following statements about the E. coli genome is false?
   a. Most of the genes are located on one circular chromosome.
   b. It has a much higher gene density than the human genome.
   c. It contains fewer genes than the human genome.
   d. All of the genes are transcribed from the same template strand of the DNA double helix.
   e. About half of the genes in the E. coli genome are grouped with other genes in operons.

6. Which of the following does not characterize the human genome?
   a. Introns occupy 20–25% of the genome.
   b. The protein-coding sequences occupy about 2% of the genome.
   c. About 45% of the genome consists of transposable element sequences.
   d. The genome sequence is comprised of approximately 30 million base pairs.
   e. Human cells have about 19,900 different protein-coding genes.

7. About 95% of the average human transcription unit consists of:
   a. short repeat sequences.
   b. protein-coding sequences.
   c. regulatory sequences.
   d. introns.
   e. origins of replication.

8. When the DNA sequences of two protein-coding genes are similar, but only for part of the protein-coding sequence, it suggests that:
   a. the two proteins have one or more domains in common.
   b. the two proteins were produced by duplication of an ancestral gene.
   c. the two proteins perform the same function.
   d. one of the two genes is actually a pseudogene.
   e. both genes are pseudogenes.

9. The proteins coded for by genes in a multigene family begin to evolve distinct functions when:
   a. gene duplication occurs.
   b. exon shuffling occurs.
   c. the genes are expressed by transcription and translation.
   d. different mutations occur in each protein-coding sequence.
   e. the two proteins evolve so they have the same three-dimensional structure.

Apply/Analyze

10. When two protein-coding genes have very similar nucleotide sequences and are located right next to each other on a chromosome, we can hypothesize that:
    a. one of them is a duplicate of the other, copied by a retrotransposon.
    b. they are nonhomologous.
    c. one of them is a pseudogene.
    d. they were produced by unequal crossing-over.
    e. they are transcribed in the same cell types.

11. **Discuss Concepts** Why are high-throughput techniques used so much in genomics research? Give examples from Chapter 19 of different uses of high-throughput techniques.

12. **Discuss Concepts** Why does the Sanger DNA sequencing method work best when the concentration of dideoxyribonucleotides is much less than the concentration of deoxyribonucleotides? If you wanted to adjust the reaction mixture to produce a greater number of very long complementary sequence fragments, how would you change the relative concentration of dideoxyribonucleotides, and why?

13. **Discuss Concepts** Which of the methods for annotating protein-coding genes would you expect to do the best job of distinguishing functioning genes from pseudogenes, and why?

Evaluate/Create

14. **Discuss Concepts** The genome of the yeast *Saccharomyces cerevisiae* is only about 0.4% the size of the human genome, yet it contains about 30% as many genes as are in the human genome. Given that, which of the features of the human genome would you expect to find many fewer of in the yeast genome?

15. **Discuss Concepts** How does sequencing the genomes of a greater number of animal species help in annotating and determining the functions of human protein-coding genes?

16. **Design an Experiment** You are studying the molecular mechanisms of sex determination in fruit flies (*Drosophila melanogaster*). How can you determine which genes are expressed at higher levels in male *Drosophila* embryos and which genes are expressed at higher levels in female *Drosophila* embryos? Once you have identified sex-specific genes, what further experiments could you do to test what effects they have on the development of *Drosophila* sex organs?

17. **Apply Evolutionary Thinking** You are studying a multigene family in the mouse genome, and would like to determine which genes evolved as a result of unequal crossing-over of chromosomes and which may have been copied through the action of transposable elements. What information would you use to help answer that question?

For selected answers, see Appendix A.

INTERPRET THE DATA

Below is a sequence of 540 bases from a genome. What information would you use to find the beginnings and ends of open reading frames? How many open reading frames can you find in this sequence? Which open reading frame is likely to represent a protein-coding sequence, and why? Which are probably not functioning protein-coding sequences, and why? Note: for simplicity's sake, analyze only this one strand of the DNA double helix, reading from left to right, so you will only be analyzing three of the six reading frames shown in Figure 19.4.

```
5'- AGTTTTATTTAAAAGAGTAGATTAAGAAAAGTAGTATTAGAATTTTATTGATTT
     ATGCAATTAGAGTACCTCAATCTTATTTCTCAAGCTAAAGTTATTGCAGAAAAA
     CAATTTAAAGCTAACCCTTTTTCTTTTGAAACAATTAGAAAGAAGTAGTTAAA
     CATTTCAAGATTTCAAAACAAGATGAACCAAGCTTAATTGGTCGTTTTTATCAA
     GATTTTCTTGAGGATCCTAACTTTGTCTATTAGGTGATAGAAAAGAAAACTT
     CGTGATTTTAGGAAGTTTGATAAATGGAACAAGATATCACAATCTATATTTGTT
     ACAAGGAGATTTTTGAAGAAGGTTATGAAGATCTTTCCAATAAAAAAGTAGAA
     CCTGAGGAAGGAGTTGGTGATTTCATTATGGGAAATGACGGTGCTGACACTGAA
     ACTGGCAGTGAAATAGTACAAGGTTTAATTAATAATTCATTCAGTGAGGAAAAT
     CAATAGTAGATACGCTTGTTAACTTTAAATTGACGCTTCAAAAAGCAAAGCTAG - 3'
```

Development of Evolutionary Thinking

20

Olaf Rahardt/The Image Works

An illustration of H.M.S. *Beagle,* the ship that carried Charles Darwin on his round-the-world journey of discovery.

Why it matters . . . On June 18, 1858, Charles Darwin received the shock of his life. Alfred Russel Wallace, a young naturalist working in the Asian tropics, solicited Darwin's opinion of a manuscript he had written about how species change through time. Darwin quickly realized that Wallace had independently described a mechanism for biological evolution that was nearly identical to the one he had been studying for more than 20 years, but had not yet described in print.

Like researchers today, scientists in the nineteenth century had to publish their work quickly to establish the "priority" on which their scientific reputations were made. Darwin's friend and colleague, the geologist Charles Lyell, had encouraged him to publish a preliminary essay on evolution two years before Wallace's letter arrived. But Darwin procrastinated, and because Wallace was the first to prepare his work for publication, Darwin feared that history would credit the younger man with these new ideas. Despite his anxiety, Darwin forwarded Wallace's manuscript to Lyell, who passed it along to the botanist Joseph Hooker. Lyell and Hooker engineered a solution that gave credit to both men **(Figure 20.1).** On July 1, 1858, papers by Darwin and Wallace were presented to the Linnaean Society of London, a prestigious scientific organization.

Darwin worked feverishly after this harrowing experience, and his now-famous book, *On the Origin of Species by Means of Natural Selection,* was published on November 24,

Charles Darwin

Historic England/Bridgeman Images

Alfred Russel Wallace

Historic England/Bridgeman Images

FIGURE 20.1 Pioneers of evolutionary theory. Charles Darwin (1809–1882) and Alfred Russel Wallace (1823–1913) independently discovered the mechanism of natural selection.

1859. The first printing of 1,250 copies sold out in one day. Today, we honor Darwin for developing the seminal idea about how biological evolution occurs and for accumulating and documenting vast quantities of evidence over decades of study.

In *The Origin*, Darwin proposed that natural mechanisms produce and transform the diversity of life on Earth. His concept of evolution still forms the unifying intellectual paradigm within which all biological research is undertaken. Even when researchers do not address explicitly evolutionary questions, their observations, theories, hypotheses, and experiments are formulated with the implicit knowledge that all forms of life are related and have evolved from ancestral forms.

Biological evolution occurs in populations when specific *processes* cause the genomes of organisms to differ from those of their ancestors. These genetic changes, and the phenotypic modifications they cause, are the *products* of evolution. By studying the products of evolution, biologists strive to understand the processes that cause evolutionary change.

The theory of evolution is so widely accepted that most people cannot think about the biological world in any other way. But the biological changes implied by Darwin's ideas and by modern evolutionary theory had not been included in earlier European worldviews.

20.1 Recognition of Evolutionary Change

The historical development of evolutionary theory is a fascinating tale of scientists struggling to reconcile evidence of change with a prevailing philosophy that change was impossible in a perfectly created universe.

Europeans Integrated Ideas from Ancient Greek Philosophy into Christian Doctrine

The Greek philosopher Aristotle (384–322 B.C.) was a keen observer of nature; he is generally considered to have been the first student of **natural history,** the branch of biology that examines the form and variety of organisms in their natural environments. Aristotle believed that both inanimate objects and living species had fixed characteristics. Careful study of their differences and similarities enabled him to create a ladderlike classification of nature, from the simplest to the most complex forms: minerals ranked below plants, plants below animals, animals below humans, and humans below the gods of the spiritual realm.

By the fourteenth century, Europeans had merged Aristotle's classification system with the biblical account of creation: each organism had been specially created by God; species could never change or become extinct; and new species could never arise. Biological research became dominated by **natural theology,** which sought to name and catalog all of God's creation. Careful study of each species would identify its position

and purpose in the *Scala Naturae,* or Great Chain of Being, as Aristotle's ladder of life was called. In the eighteenth century, the Swedish botanist Carolus Linnaeus (1707–1778), who developed the science of **taxonomy,** the branch of biology that classifies organisms (see Chapter 24), undertook this important work *ad majorem Dei gloriam* ("for the greater glory of God").

Scholars also used a literal interpretation of scripture to date the time of creation precisely. By tabulating the human generations described in the Bible, they determined that the creation had occurred around 4000 B.C., making Earth a bit less than 6,000 years old—hardly old enough for much change to have taken place.

Scientists Slowly Became Aware of Change in the Natural World

Modern science came of age in the fifteenth through eighteenth centuries. The English philosopher and statesman Sir Francis Bacon (1561–1626) established the importance of observation, experimentation, and inductive reasoning. Other scientists, notably Nicolaus Copernicus (1473–1543), Galileo Galilei (1564–1642), René Descartes (1596–1650), and Sir Isaac Newton (1643–1727), proposed mechanistic theories to explain physical events. In addition, three new disciplines—biogeography, comparative morphology, and geology—raised serious questions about natural theology's view of the history of life.

QUESTIONS ABOUT BIOGEOGRAPHY As long as naturalists encountered organisms only from Europe and surrounding lands, the task of understanding the *Scala Naturae* was manageable. But global explorations in the fifteenth through seventeenth centuries provided naturalists with thousands of unknown plants and animals from Asia, Africa, the Pacific Islands, and the Americas. Although some were similar to European species, others were new and very strange.

Studies of the world distribution of plants and animals, now called **biogeography,** raised puzzling questions. Was there no limit to the number of species created by God? Where did all these species fit in the *Scala Naturae*? If they had all been created in the Garden of Eden, why did some species have limited geographical distributions, whereas others were widespread? And why were some species found in Africa or Asia different from those found in Europe, whereas other species from far-flung locations had a similar appearance **(Figure 20.2)?**

QUESTIONS ABOUT COMPARATIVE MORPHOLOGY When biologists began to compare the **morphology** (anatomical structure) of various organisms, they discovered interesting similarities and differences. For example, the front legs of pigs, the flippers of dolphins, and the wings of bats differ markedly in size, shape, and function **(Figure 20.3).** But these appendages have similar locations in the animals' bodies; all are constructed of bones, muscles, and skin; and all develop similarly

African ostrich
(Struthio camelus)

South American rhea
(Rhea americana)

Australian emu
(Dromaius novaehollandiae)

FIGURE 20.2 Large, flightless birds. Three large bird species with greatly reduced wings occupy similar habitats in geographically separated regions.

in the animals' embryos. If these limbs were specially created for different means of locomotion, naturalists wondered, why didn't the Creator use entirely different materials and structures for walking, swimming, and flying?

Natural theologians countered this argument by stating that body plans were perfect, and there was no need to invent a new plan for every species. But a French scientist, George-Louis Leclerc (1707–1788), le Comte (Count) de Buffon, was still puzzled by the existence of body parts with no apparent function. For example, he noted that the feet of pigs and some other mammals have two toes that never touch the ground (pig digits 2 and 5 in Figure 20.3). If each species was anatomically perfect for its particular way of life, Buffon asked, why did useless structures exist?

Buffon proposed that some animals must have *changed* since their creation; he suggested that **vestigial structures,** the

useless parts we observe today, must have functioned in ancestral organisms. Buffon offered no explanation of how functional structures became vestigial, but he clearly recognized that some species were "conceived by Nature and produced by Time."

QUESTIONS ABOUT FOSSILS By the mid-eighteenth century, geologists were mapping the stratification, or horizontal layering, of sedimentary rocks beneath the soil surface (see Figure 23.2). Different layers held different kinds of **fossils** (*fossilis* = dug up). Relatively small and simple fossils appeared in the deepest layers. Fossils in the layers above them were more complex. Those in the uppermost layers often resembled living organisms. Moreover, fossils found in any particular layer were often similar, even if they were collected from geographically distant sites. What were these fossils, and why did they vary more from one layer of rock to another than from one geographical region to another?

Some scientists suggested that fossils were the remains of extinct organisms—but natural theology did not allow extinction. Thomas Jefferson, the third president of the United States and an amateur fossil hunter, thought that fossils were the remains of species that were now extremely rare; he believed that nature could not have "permitted any one race of her animals to become extinct" or "formed any link in her great works so weak [as] to be broken." He even asked Lewis and Clark to keep an eye out for giant ground sloths, now known to be extinct, during their exploration of the Pacific Northwest.

Georges Cuvier (1769–1832), a French zoologist and a founder of comparative morphology, as well as **paleobiology** (the study of ancient organisms), realized that the layers of fossils represented organisms that had lived at successive times in the past. He suggested that the abrupt changes between geological strata marked dramatic shifts in ancient environments. Cuvier and his followers developed the theory of **catastrophism,** reasoning that each layer of fossils represented the remains of organisms that had died in a local catastrophe, such as a flood. Somewhat different species then recolonized the area, and when another catastrophe struck, they formed a different set of fossils in the next, higher layer of rock.

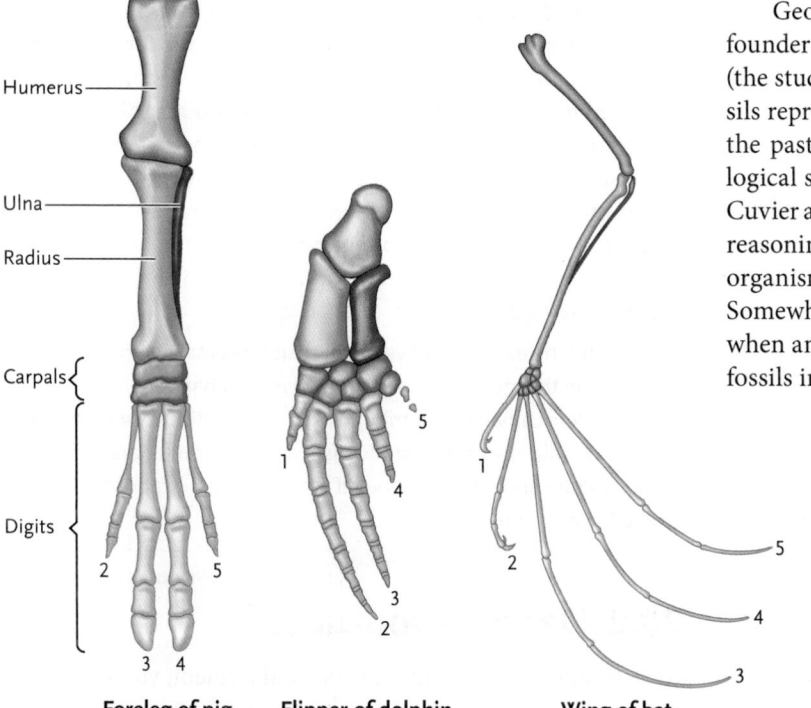

Humerus

Ulna

Radius

Carpals

Digits

Foreleg of pig   **Flipper of dolphin**   **Wing of bat**

FIGURE 20.3 Mammalian forelimbs and locomotion. Pigs use their legs to walk or run, dolphins use their flippers to swim, and bats use their wings to fly. Homologous (equivalent) bones are pictured in the same color, and digits (fingers) are numbered; pigs have lost the first digit over evolutionary time. (Limbs are not drawn to scale.)

© Cengage Learning 2017

Lamarck Developed an Early Theory of Biological Evolution

A contemporary of Cuvier and a student of Buffon, Jean Baptiste de Lamarck (1744–1829) proposed the first comprehensive theory of biological evolution that was based on specific mechanisms. He proposed that a metaphysical "perfecting principle" caused organisms to become better suited to their environments. Simple organisms evolved into more complex ones, moving up the ladder of life, and microscopic organisms were replaced at the bottom of the ladder by spontaneous generation.

Lamarck theorized that two mechanisms fostered evolutionary change. According to his *principle of use and disuse,* body parts grow in proportion to how much they are used, as anyone who "pumps iron" well knows. Conversely, structures that are not often used get weaker and shrink, as do the muscles of an arm immobilized in a cast. According to his second principle, the *inheritance of acquired characteristics,* changes that an animal acquires during its lifetime are inherited by its offspring. Thus, Lamarck argued that long-legged wading birds, such as herons **(Figure 20.4),** are descended from short-legged ancestors that stretched their legs to stay dry while feeding in shallow water. Their offspring inherited slightly longer legs, and after many generations, their legs became extremely long.

Today, we know that Lamarck's proposed mechanisms do not cause evolutionary change. Although muscles do grow larger through continued use, most structures do not respond in the way Lamarck predicted. Moreover, structural changes acquired during an organism's lifetime are not inherited by the next generation. Even in his own day, Lamarck's ideas were not widely accepted.

Despite the shortcomings of his theory, Lamarck made four tremendously important contributions to the development of an evolutionary worldview. First, he proposed that all species change through time. Second, he recognized that new characteristics are passed from one generation to the next. Third, he suggested that organisms change in response to their environments. And fourth, he hypothesized the existence of specific mechanisms that fostered evolutionary change. All four of these ideas became cornerstones of Darwin's evolutionary theory, although he proposed a different mechanism, natural selection, as the cause of evolutionary change. Perhaps Lamarck's most important contribution was that he fostered discussion. By the mid-nineteenth century, most educated Europeans were talking about evolutionary change, whether they believed in it or not.

Geologists Recognized That Earth Had Changed over Time

In 1795, the Scottish geologist James Hutton (1726–1797) argued that slow and continuous physical processes, *acting over long periods of time,* produced Earth's major geological features; for example, the movement of water in a river slowly erodes the land and deposits sediments near the mouth of the river. Given enough time, erosion creates deep canyons, and sedimentation creates thick topsoil on flood plains. Hutton's **gradualism,** the view that Earth changed *slowly* over its history, contrasted sharply with Cuvier's catastrophism.

The English geologist Charles Lyell (1797–1875) championed and extended Hutton's ideas in an influential series of books, *Principles of Geology.* Lyell argued that the geological processes that sculpted Earth's surface over long periods of time—such as volcanic eruptions, earthquakes, erosion, and the formation and movement of glaciers—are exactly the same as the processes observed today. This concept, called **uniformitarianism,** undermined any remaining notions of an unchanging Earth. Also, because geological processes proceed very slowly, it must have taken millions of years, not just a few thousand, to mold the landscape into its current configuration.

McPhoto/Blickwinkel/AGE Fotostock

FIGURE 20.4   A great blue heron (*Ardea herodias*). Like many other wading birds, herons have long, stiltlike legs. Lamarck hypothesized that as wading birds stretched their legs to keep their bodies dry while feeding, successive generations of their offspring would have progressively longer legs.

STUDY BREAK 20.1

1. Why did the existence of vestigial structures make Buffon question the idea that living systems never changed?
2. What were Lamarck's contributions to an evolutionary worldview?
3. How do the concepts of gradualism and uniformitarianism in geology undermine the belief that Earth is only about 6,000 years old?

20.2   Darwin's Journeys

In 1831, in the midst of this intellectual ferment, young Charles Darwin wondered what to do with his life. Raised in a wealthy English household, he had always collected shells and studied the habits of insects and birds; he preferred hunting and fishing to classical studies. Despite his lackluster performance as a student, Darwin was expected to continue the family tradition of practicing medicine. But after two years, he abandoned his medical studies at the University of Edinburgh. Instead, he followed his interest in natural history despite the objections of his father, who reputedly told him, "You care for nothing but shooting, dogs, and rat-catching and you will be a disgrace to yourself and all of your family."

At the suggestion of his father, Darwin studied for a career as a clergyman, earning a degree from Cambridge University. There, he found a mentor in the Reverend John Henslow, a leading botanist, who arranged for Darwin to travel as the captain's dining companion aboard H.M.S. *Beagle* (see chapter-opening illustration), a naval surveying ship. Darwin embarked on a sea voyage and an intellectual journey that altered the foundations of modern thought.

Darwin Traveled the World on the Voyage of the *Beagle*

The *Beagle* sailed southwest to map the coastline of South America and then circumnavigated the globe **(Figure 20.5)**. When the ship's naturalist quit his post mid-journey, Darwin replaced him in an unofficial capacity. For nearly five years Darwin toured the world, and because he suffered from seasickness, he seized every chance to go ashore. He collected plants and animals in Brazilian rainforests and fossils in Patagonia. He hiked the grasslands of the Pampas and climbed the Andes in Chile. Armed with Henslow's parting gift, the first volume of Lyell's *Principles of Geology,* Darwin was primed to apply gradualism and uniformitarianism to the living world.

WHAT DARWIN SAW When he began his travels, Darwin had no clue that biological evolution had produced the mind-boggling variety of species that he would encounter. Three broad sets of observations later helped him unravel the mystery of evolutionary change.

First, while exploring along the coast of Argentina, Darwin discovered fossils that often resembled organisms inhabiting the same region today. For example, despite an enormous size difference, living armadillos and fossilized glyptodonts had similar body armor, but they were unlike any other species known to science **(Figure 20.6)**. If both species had been created at the same time and both were found in South America, why didn't glyptodonts live alongside armadillos today? Darwin later wondered whether armadillos might be living descendants of the now-extinct glyptodonts.

Second, Darwin observed that the animals he encountered in different South American habitats clearly resembled each other, but they differed from species that occupied similar habitats in Europe. For example, he noted that nutria *(Myocastor coypus),* a semiaquatic rodent native to South America, bore a

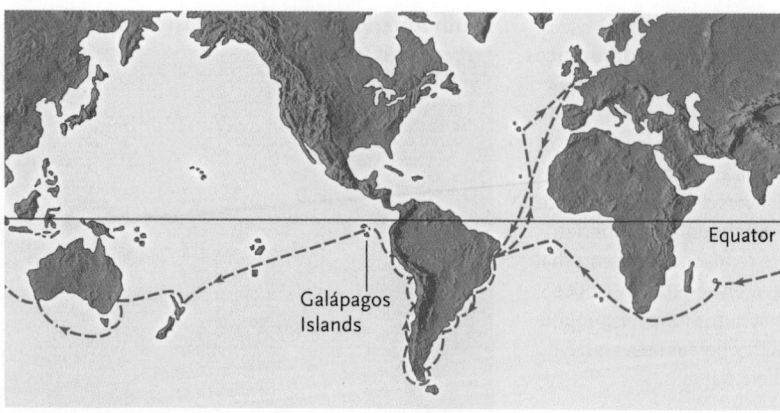

FIGURE 20.5 Darwin's voyage. H.M.S. *Beagle* circumnavigated the globe between 1831 and 1836.
© Cengage Learning 2017

closer resemblance to other rodents from the mountains or grasslands of that continent than it did to the beaver *(Castor fiber),* a semiaquatic rodent native to Europe **(Figure 20.7)**. Why did animals from markedly different South American environments resemble each other so closely? Why did animals that lived in similar environments on separate continents look so different? Darwin later understood that animals in South America resembled each other because they had inherited their similarities from a common ancestor.

Third, Darwin observed fascinating patterns in the distributions of species on the Galápagos Islands **(Figure 20.8)**. There he found strange and wonderful creatures, including giant tortoises and lizards that dove into the sea to feed on algae.

Charles R. Knight painting (negative CK21T), Field Museum of Natural History, Chicago

Steve Bower/Shutterstock.com

FIGURE 20.6 Ancestors and descendants. Darwin hypothesized that even though an extinct glyptodont (top) probably weighed 300 to 400 times as much as a living nine-banded armadillo *(Dasypus novemcinctus),* their obvious resemblance suggested that they are related.

FIGURE 20.7
Morphological differences in species from different continents. Darwin noted that the South American nutria and the European beavers differ in appearance, even though both species are semiaquatic rodents that feed on vegetation. Notice that nutria have long, round tails, whereas beavers have short, flat tails.

South American nutria (Myocastor coypus)

European beaver (Castor fiber)

Darwin quickly noted that the animals on different islands varied slightly in form. Indeed, experienced sailors could easily identify a tortoise's island of origin by the shape of its shell. Moreover, many species resembled those on the distant South American mainland. Why did so many different organisms occupy one small island cluster, and why did these species resemble others from the nearest continent? Darwin later hypothesized that the plants and animals of the Galápagos Islands were descended from South American ancestors, and that each species had changed after being isolated on a particular island.

DARWIN'S REFLECTIONS AFTER HIS VOYAGE The *Beagle* returned to England in 1836, and Darwin began his first notebook on the "transmutation of species" the following year. He realized that changes in species over time—*descent with modification*—provided the only plausible explanation for his observations.

A diverse group of finches from the Galápagos Islands **(Figure 20.9)** provided the single greatest spark for Darwin's work. He had noticed great variability in the shapes of their bills, but having incorrectly assumed that birds on different islands belonged to the same species, he had not recorded the

A. The Galápagos

B. Galápagos tortoise (*Geochelone elephantopus*)

C. Marine iguana (*Amblyrhynchus cristatus*)

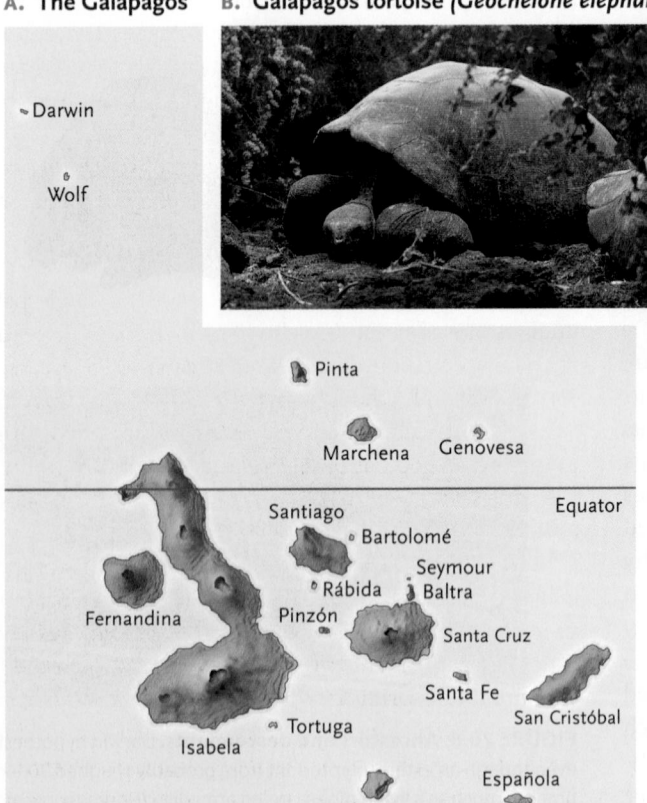

FIGURE 20.8 The Galápagos Islands and some of their unusual animal inhabitants. (A) Volcanic eruptions created the Galápagos archipelago (located 1,000 km west of Ecuador) between 3 and 5 million years ago. **(B)** The islands were named for the giant tortoises found there (in Spanish, *galápa* means tortoise); this tortoise is native to Isla Santa Cruz. **(C)** Marine iguanas dive into the Pacific Ocean to feed on algae.

© Cengage Learning 2017

A. Warbler finch
(Certhidea olivacea)

B. Common cactus finch
(Geospiza scandens)

C. Large ground finch
(Geospiza magnirostris)

D. Woodpecker finch
(Camarhynchus pallidus)

FIGURE 20.9 Bill shape and food habits. The 13 finch species that inhabit the Galápagos Islands are descended from a common ancestor, a seed-eating ground finch that migrated to the islands from South America. **(A)** The warbler finch uses its slender bill to probe for insects in vegetation. **(B)** The common cactus finch has a medium-sized bill suitable for eating cactus flowers and fruit. **(C)** The large ground finch uses its thick, strong bill to crush cactus seeds. **(D)** The woodpecker finch uses its bill to hammer at bark and to hold cactus spines, with which it probes for wood-boring insects, such as termites.

island where he had captured each specimen. Luckily, the *Beagle's* captain, Robert Fitzroy, had more thoroughly documented his own collection, allowing Darwin to study the relationships and geographical distributions of a dozen species. As Darwin reviewed the data, he began to focus on two aspects of a general problem. Why were the finches on a particular island slightly different from those on nearby islands, and how did all these different species arise?

Darwin Used Common Knowledge and Several Hypotheses to Develop His Theory

With a substantial inheritance, and burdened by chronic illness, Darwin led a reclusive life as he embarked on an intellectual journey every bit as exciting as his voyage on the *Beagle* (see *Focus on Research: Basic Research*). His lifetime goal was to accumulate evidence of evolutionary change and identify the mechanism that caused it.

SELECTIVE BREEDING AND HEREDITY Having grown up in the country, Darwin was well aware that "like begets like"; that is, offspring frequently resemble their parents. Plant and animal breeders had applied this basic truth of inheritance for thousands of years. By selectively breeding individuals with desired characteristics, they enhanced those traits in future generations.

Farmers use selective breeding to improve domesticated plants and animals. If one cow produces more milk than any other, the farmer selectively breeds her (rather than others), hoping that her offspring will also be good milk producers. Although the mechanism of heredity was not yet understood, this principle had been applied countless times to produce bigger beets, plumper pigs, and prize-winning pigeons (see

Figure 1.8). Darwin was well aware of this process, which he called **artificial selection,** but he puzzled over how it could operate in nature.

THE STRUGGLE FOR EXISTENCE Darwin had a revelation about how selective breeding could occur naturally when he read the famous publication by Thomas Malthus, *Essay on the Principles of Population.* Malthus, an English clergyman and economist, was worried about the fate of the nation's poor. England's population was growing much faster than its agricultural capacity, and with individuals competing for limited food resources, some would inevitably starve.

Darwin applied Malthus' argument to organisms in nature. Species typically produce many more offspring than are needed to replace the parent generation, yet the world is not overrun with sunflowers, tortoises, or bears. Darwin even calculated that, if its reproduction went unchecked, a single pair of elephants, the slowest breeding animal known, would leave roughly 19 million descendants after only 750 years. Happily for us (and all other species that might get underfoot), the world is not so crowded with elephants. Instead, some members of every population survive and reproduce, whereas others die without reproducing.

DARWIN'S HYPOTHESES AND PREDICTION Darwin's discovery of a mechanism for evolutionary change required him to predict the existence of a process that no one had envisioned, much less documented **(Figure 20.10)**. First, Darwin observed that not all offspring in a population survive to maturity. This observation led him to hypothesize that the individuals in a population compete for limited resources. Second, Darwin observed that individuals of every species vary in size, form, color, behavior, and other characteristics; moreover, he noted

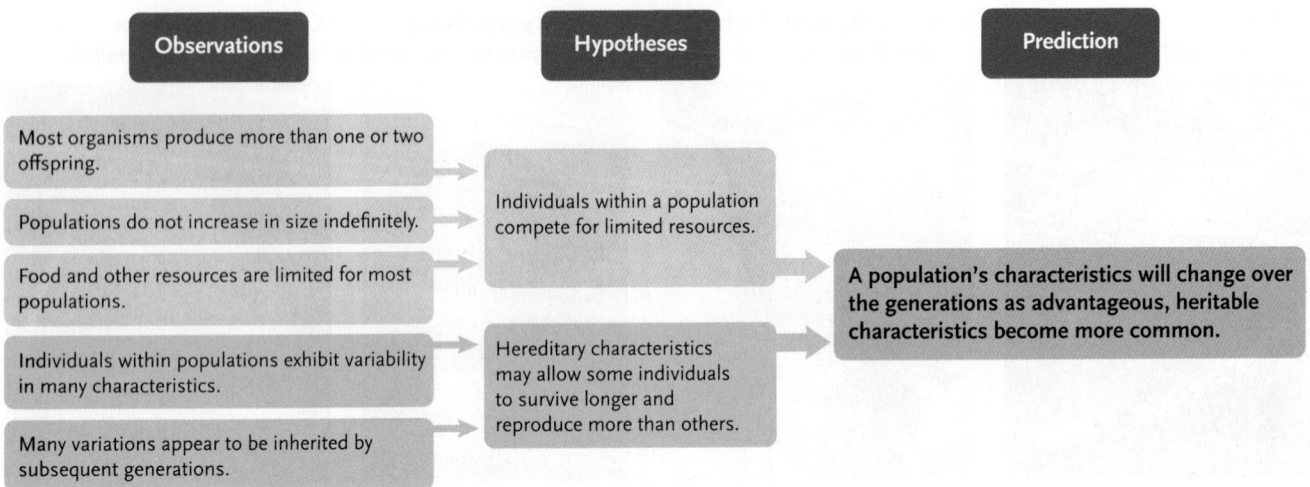

| Observations | Hypotheses | Prediction |

Most organisms produce more than one or two offspring.

Populations do not increase in size indefinitely.

Food and other resources are limited for most populations.

Individuals within populations exhibit variability in many characteristics.

Many variations appear to be inherited by subsequent generations.

Individuals within a population compete for limited resources.

Hereditary characteristics may allow some individuals to survive longer and reproduce more than others.

A population's characteristics will change over the generations as advantageous, heritable characteristics become more common.

FIGURE 20.10 Darwin's observations led him to develop two hypotheses that led to his prediction that natural selection caused evolutionary changes in species from one generation to the next.

© Cengage Learning 2017

that many of these variations are hereditary. He therefore hypothesized that variations in hereditary traits enabled some individuals to survive and reproduce more than others. Taken together, Darwin's hypotheses predicted that organisms with advantageous traits would leave many offspring, whereas those that lacked such traits would die leaving few, if any, descendants. Thus, advantageous hereditary traits would become more common in the next generation. If the next generation experienced the same process of selection, the traits would be even more common in the third generation. Because this process is analogous to artificial selection, Darwin called it **natural selection.**

As an evolutionary mechanism, natural selection favors **adaptive traits,** genetically based characteristics that make organisms more likely to survive and reproduce. And by favoring individuals that are well adapted to the environments in which they live, natural selection causes species to change through time. As shown in Figure 20.9, each species of Galápagos finch has a distinctive bill. Variations in bill size and shape make some birds better adapted for crushing seeds and others for capturing insects. Imagine an island where large seeds were the only food available; individuals with a stout bill would be more likely to survive and reproduce than would birds with slender bills. These favored individuals would pass the genes that produce stout bills to their descendants, and after many generations, their bills might resemble those of *Geospiza magnirostris* (see Figure 20.9C). Natural selection also changes nonmorphological characteristics of populations; for example, insect populations that are exposed to insecticides develop resistance to these toxic chemicals over time (see Figure 20.12).

Darwin realized that natural selection could also account for striking differences between populations and, given enough time, for the production of new species. For example, suppose that small insects were the only food available to finches on a different island. Birds with long, thin bills might be favored by natural selection, and the population of finches might eventually possess a bill shaped like that of *Certhidea olivacea* (see Figure 20.9A). If we apply parallel reasoning to the many characteristics that affect survival and reproduction, natural selection would cause the populations to become more different over time, a process called **evolutionary divergence.**

Darwin's Theory of Descent with Modification Revolutionized the Way We Think about the Living World

It would be hard to overestimate the impact of Darwin's theory on Western thought. In *The Origin,* Darwin proposed a logical mechanism for evolutionary change and provided enough supporting evidence to convince the educated public.

Darwin argued that all the organisms that have ever lived arose through **descent with modification,** the evolutionary alteration and diversification of ancestral species. In one of his notebooks, dated 1837, he sketched this pattern of descent as a tree **(Figure 20.11).** The trunk at the bottom (marked with a circled 1 in the figure) represents the ancestor of all organisms. Branching points above it identify the evolutionary divergence of ancestors into their descendants. Each limb represents a body plan suitable for a particular way of life, smaller branches more narrowly defined groups of organisms, and the twigs at the tips of the branches living species. Biologists still use this analogy today when analyzing the Tree of Life (see Figure 1.11 and Chapter 24).

Darwin proposed natural selection as the mechanism that drives evolutionary change. In fact, most of *The Origin* was an explanation of how natural selection acted on the variability within groups of organisms, preserving advantageous traits and eliminating disadvantageous ones.

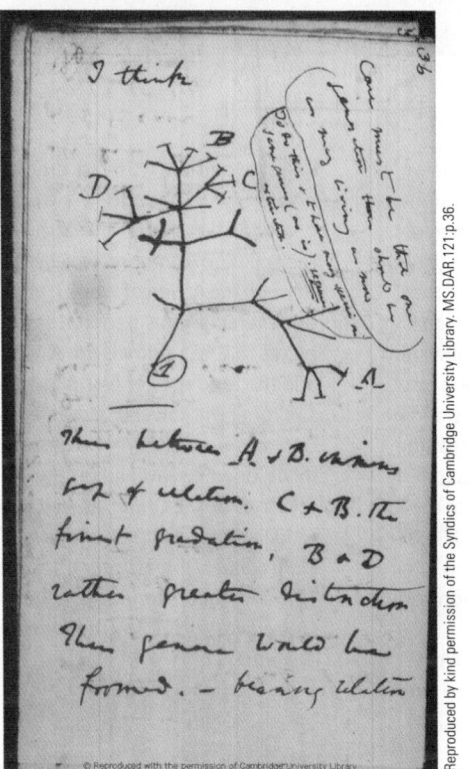

FIGURE 20.11 The Tree of Life. Darwin first envisioned the history of life as a tree more than 20 years before he published *On the Origin of Species*. In this sketch from his 1837 notebook on "the transmutation of species," living species are labeled A, B, C, and D. Forks in the tree represent the common ancestors of the branches that emerge from them. Unlabeled branches identify evolutionary lineages that became extinct.

Four characteristics distinguish Darwin's theory from earlier explanations of biological diversity and adaptive traits:

1. Darwin provided purely physical, rather than spiritual, explanations for the origins of biological diversity.
2. Darwin recognized that evolutionary change occurs in groups of organisms, rather than in individuals: some members of a group survive and reproduce more successfully than others.
3. Darwin described evolution as a multistage process: variations arise within groups, natural selection eliminates unsuccessful variations, and the next generation inherits successful variations.
4. Like Lamarck, Darwin understood that evolution occurs because some organisms function better than others *in a particular environment.*

What is most amazing about Darwin's intellectual achievement is that he knew nothing about genetics (see Chapter 12). Thus, he had no clear idea of how variation arose or how it was passed from one generation to the next.

Evolution was a popular topic in Victorian England, and Darwin's theory was both praised and ridiculed. Although he had not speculated about the evolution of humans in *The Origin,* many readers were quick to extrapolate Darwin's ideas to our own species. Needless to say, certain influential Victorians were not amused by the suggestion that humans and apes share a common ancestry.

Nevertheless, Darwin's painstaking logic and careful documentation convinced most readers that evolution really does take place. Thomas Huxley, so staunch an advocate that he was known as "Darwin's bulldog," summed up the reaction of many when he quipped that the theory was so obvious, once articulated, that he was surprised he had not thought of it himself. Darwin's vision of common ancestry quickly became the intellectual framework for nearly all biological research. Many nineteenth-century readers, however, did not readily accept the mechanism of natural selection. The major stumbling block was that Darwin had not provided any plausible theory of heredity.

STUDY BREAK 20.2

1. What observations that Darwin made on his round-the-world voyage influenced his later thoughts about evolution?
2. How did Darwin's understanding of artificial selection enable him to envision the process of natural selection?
3. What were the four great intellectual triumphs of Darwin's theory?

THINK OUTSIDE THE BOOK

Find a copy of *On the Origin of Species by Means of Natural Selection* in a library or bookstore or online, and read a chapter or two. How did Darwin's careful observation of organisms allow him to draw conclusions? Do you think that straightforward observation still has a role in twenty-first–century science?

20.3 Evolutionary Biology since Darwin

Although Gregor Mendel published his work on genetics in 1866, it was not well known in England until 1900. At that time, scientists perceived a fundamental conflict between Darwin's and Mendel's theories. One problem was that Darwin had used complex characteristics, such as the structure of bird bills, to illustrate how natural selection worked. We now know that several genes often control such traits (see Section 12.2). By contrast, Mendel had studied simpler characteristics, such as the height of pea plants. A single gene often controls simple traits, which is one reason Mendel could interpret his experimental results so clearly. Biologists had a hard time applying Mendel's straightforward experimental results to Darwin's complex examples.

A second problem arose because Darwin believed that biological evolution occurred gradually over many generations. However, early twentieth-century geneticists, focusing on simple traits such as those Mendel had studied, sometimes observed very rapid and dramatic changes in certain characteristics.

Basic Research: Charles Darwin's Life as a Scientist

Darwin's observations during the voyage of H.M.S. *Beagle* convinced him that species change through time, and that natural processes produced Earth's biodiversity. He spent the rest of his life gathering data to test his ideas and unravel the workings of natural selection.

Shortly after the *Beagle* returned to England in 1836, Darwin started to publish the geological and biological research that he had undertaken during the voyage. This task took him 10 years to complete—twice as long as the journey itself. The results of these efforts were numerous articles and several books, including the now famous *Journal of the Voyage of the* Beagle, published in 1839.

After preparing a sketch of his ideas about evolution in 1844, Darwin was puzzled by one species of barnacle, a small marine invertebrate, which he had collected in Chile. Over the next eight years he studied more than 10,000 barnacle specimens and revised their entire classification. Although colleagues viewed this study as a strange diversion from his work on evolution, the project honed Darwin's observational skills and provided a test case in which he could test his hypotheses about descent with modification to a large and diverse group of organisms. He published four volumes about barnacles in 1854.

While studying barnacles, Darwin continued to think about "the species question." He kept notebooks about variation in plants and animals, focusing on variation that was amplified by selective breeding. He was a tireless collector of facts, which he sought from every possible source. He badgered dog breeders, horse farmers, and horticulturists with long lists of questions about their work. His enthusiasm was infectious, and workers throughout the world supplied him with data and specimens. Darwin was also an eager and skilled experimentalist, and he took up pigeon breeding, marveling at the huge variety of morphological traits that he and other breeders could produce. In the late 1850s, a communication from Alfred Russel Wallace (see *Why it matters . . .*) forced him to complete *The Origin*, which revolutionized the study of biology.

Even after *The Origin* was published, Darwin continued to gather facts and write about evolution, working almost to the day he died in 1882 at age 74. He published a detailed analysis of how earthworms improve the soil *(The Formation of Vegetable Mould through the Action of Worms)* and wrote books on several botanical topics, among them plants that eat animals *(Insectivorous Plants)*, and the tendency of plants to grow toward sunlight *(The Power of Movement in Plants)*. Darwin's work always had an evolutionary focus, however, and he produced several revisions of *The Origin*, as well as books on artificial selection *(Variation of Animals and Plants under Domestication)*, human ancestry *(The Descent of Man)*, and animal behavior *(The Expression of the Emotions in Men and Animals)*.

FIGURE Darwin's study. Darwin undertook most of his life's work in this room at Down House. He hesitated to discard old papers and specimens, believing that he would find a use for them as soon as they were carried away in the trash.

A widely accepted theory, *mutationism,* suggested that evolution occurred in spurts, induced by the chance appearance of "hopeful monsters," rather than by gradual change.

The Modern Synthesis Created a Unified Theory of Evolution

In the early twentieth century, Thomas Hunt Morgan of Columbia University discerned that genes are carried on chromosomes. His experiments, described in Chapter 13, enabled geneticists and mathematicians to forge a critical link between Darwin's and Mendel's ideas. The new discipline, **population genetics,** recognized the importance of genetic variation as the raw material of evolution. Population geneticists constructed mathematical models, which applied equally well to simple and complex traits, to predict how natural selection and other processes influence a population's genetics.

In the 1930s and 1940s, a unified theory of evolution, the **modern synthesis,** integrated data from biogeography, comparative morphology, comparative embryology, paleontology, and taxonomy within an evolutionary framework. The authors of the modern synthesis focused on evolutionary change within populations, and although they considered natural selection the primary mechanism of evolution, they acknowledged the importance of other processes. Proponents of the modern synthesis also embraced Darwin's idea of gradualism and de-emphasized the significance of mutations that changed traits suddenly and dramatically.

The modern synthesis also tried to link the two levels of evolutionary change that Darwin had identified: microevolution and macroevolution. **Microevolution** describes the small-scale genetic changes that populations undergo, often in response to shifting environmental circumstances; a small evolutionary shift in the size of the bill of a finch species is an example of

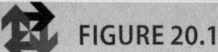

FIGURE 20.12 Experimental Research

How Exposure to Insecticide Fosters the Evolution of Insecticide Resistance

Question: Does exposure to insecticide foster the evolution of insecticide resistance in insect populations?

Experiment: Researchers studied samples of wild mosquitoes *(Anopheles culicifacies)* captured at a small village in India, where public health officials frequently sprayed the insecticide dichloro-diphenyl-trichloroethane (DDT) to control these pests. For each test, the researchers exposed samples of mosquitoes to a 4% concentration of DDT for 1 hour and then measured the percentage that died during the next 24 hours. Tests were repeated 12 months and 16 months after the first experiment.

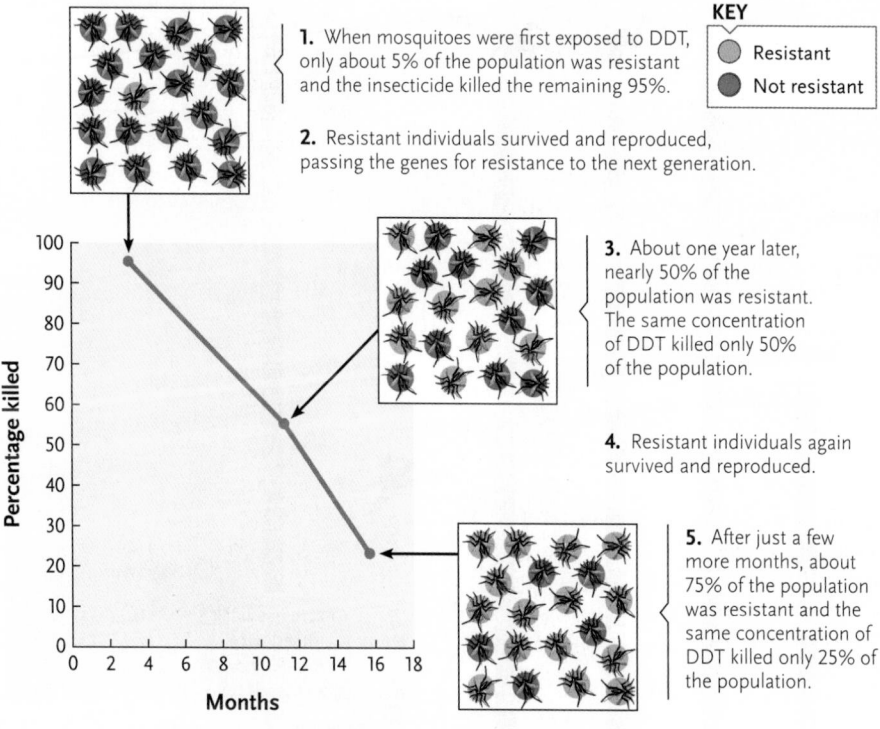

KEY
- ⬤ Resistant
- ⬤ Not resistant

1. When mosquitoes were first exposed to DDT, only about 5% of the population was resistant and the insecticide killed the remaining 95%.

2. Resistant individuals survived and reproduced, passing the genes for resistance to the next generation.

3. About one year later, nearly 50% of the population was resistant. The same concentration of DDT killed only 50% of the population.

4. Resistant individuals again survived and reproduced.

5. After just a few more months, about 75% of the population was resistant and the same concentration of DDT killed only 25% of the population.

Results: Over the course of the experiment, smaller and smaller percentages of the mosquitoes died after their exposure to the test concentration of the insecticide.

Conclusion: When public health officials used DDT indiscriminately, they created an environment in which natural selection favored DDT-resistant individuals. Exposure to DDT therefore fostered the evolution of an adaptive resistance to DDT in the mosquito population.

think like a scientist
Observe
Hypothesize
Predict
Experiment
Interpret

What effect would the continued use of DDT be likely to have on the mosquitoes' susceptibility to that chemical?

Source: A. M. Shalaby. 1968. Susceptibility studies on *Anopheles culicifacies* with DDT and Dieldrin in Gujarat state, India. *Journal of Economic Entomology* 61:533–541.

© Cengage Learning 2017

microevolution (see Chapter 21). **Macroevolution** describes larger-scale evolutionary changes observed in species and more inclusive groups. According to the modern synthesis, macroevolution results from the gradual accumulation of microevolutionary changes; researchers have recently begun to unravel the genetic mechanisms that establish a relationship between these two levels of evolutionary change (see Chapter 23).

Research in Many Fields Has Provided Consistent Evidence of Evolutionary Change

Since the emergence of the modern synthesis, scientists have assembled a huge and compelling body of evidence from many biological disciplines indicating that organisms living today are the products of descent with modification.

ADAPTATION BY NATURAL SELECTION Biologists interpret the products of natural selection as evolutionary adaptations. For example, the wings of birds, which have been modified by evolutionary processes over millions of years, have an obvious function that helps these animals survive and reproduce. In Units 5 and 6 of this book, you will encounter many examples of adaptive structures in plants and animals that have been modeled by natural selection. Sometimes, however, natural selection operates on a short time scale, as illustrated by the development of pesticide resistance in insects. When we first use a new pesticide, a low concentration often kills a large percentage of the pests. However, just by chance, a few insects may have genetic characteristics that confer resistance to the poison. These individuals survive and produce offspring, many of which inherit the resistance. As a result, a given concentration of the poison kills a smaller percentage of insects in the next generation, and over time, the entire population may become highly resistant **(Figure 20.12).**

THE FOSSIL RECORD Because evolution results from the modification of existing species, Darwin's theory proposes that all species that have ever lived are genetically related. The fossil record documents such continuity in morphological characteristics, providing clear evidence of ongoing change in **biological lineages,** evolutionary sequences of ancestral organisms and their descendants (see Chapter 23). For example, the evolution of modern birds can be traced from a nonflying dinosaur ancestor through fossils such as *Archaeopteryx lithographica* **(Figure 20.13).**

A. *Archaeopteryx* fossil

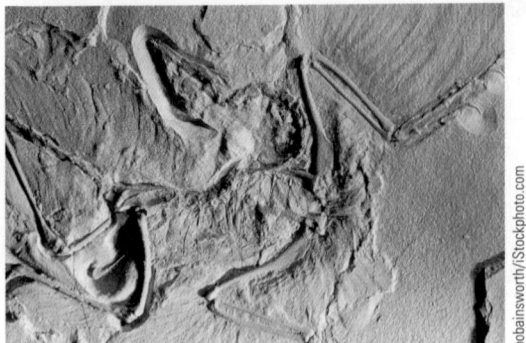

bobainsworth/iStockphoto.com

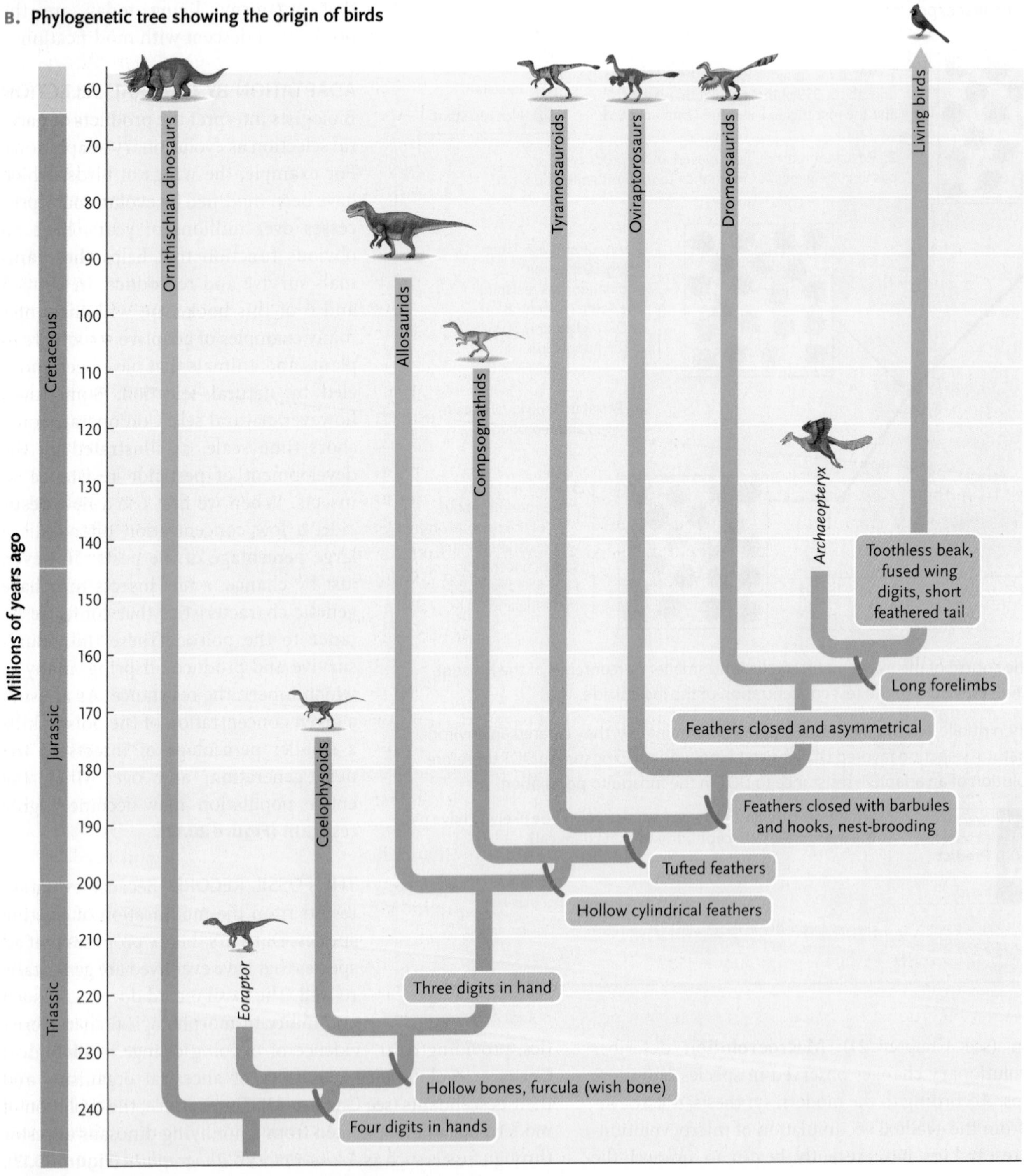

FIGURE 20.13 Bird ancestry. (A) One of the few known fossils of *Archaeopteryx lithographica,* from limestone deposits more than 140 million years old. **(B)** The evolutionary tree that includes modern birds illustrates their close relationship to small carnivorous dinosaurs. *Archaeopteryx* shared some characteristics with the Dromeosaurids (teeth, clawed fingers, and a long bony tail) and other traits with modern birds (long forelimbs).

© Cengage Learning 2017

Adaptation of phylogenic tree from http://evolution.berkeley.edu/evolibrary/article/evograms_06; credit on web page says: "Bird phylogeny adapted from/branched downy feather and symmetrical contour feather from The Tangled Bank.

B. Phylogenetic tree showing the origin of birds

Labels on the phylogenetic tree:

- Ornithischian dinosaurs
- Allosaurids
- Compsognathids
- Tyrannosauroids
- Oviraptorosaurs
- Dromeosaurids
- Living birds
- Coelophysoids
- *Eoraptor*
- *Archaeopteryx*

Trait labels:

- Toothless beak, fused wing digits, short feathered tail
- Long forelimbs
- Feathers closed and asymmetrical
- Feathers closed with barbules and hooks, nest-brooding
- Tufted feathers
- Hollow cylindrical feathers
- Three digits in hand
- Hollow bones, furcula (wish bone)
- Four digits in hands

Vertical axis: Millions of years ago (60–240)

Geologic periods: Cretaceous, Jurassic, Triassic

This species, first discovered only two years after *The Origin* was published, resembled both nonflying dinosaurs and modern birds. Like the small carnivorous Dromaeosaurids, *Archaeopteryx* walked on its hind legs and had teeth, long clawed fingers on its forelimbs, and a long, bony tail. Like modern birds, it had enlarged forelimbs with flight feathers. Recently discovered fossils reveal that many dinosaurian ancestors of birds had feathers (discussed further in Chapter 23).

HISTORICAL BIOGEOGRAPHY Analyses of **historical biogeography,** the study of the geographical distributions of plants and animals in relation to their evolutionary history, are generally consistent with Darwin's theory of evolution. Species on oceanic islands often closely resemble species on the nearest mainland, suggesting that the island and mainland species share a common ancestry. Moreover, species on a continental landmass are clearly related to one another and are often distinct from those on other continents. For example, monkeys in South America have long, prehensile tails and broad noses, traits that they inherited from a shared South American ancestor. By contrast, monkeys in Africa and Asia evolved from a different common ancestor in the Old World, and their shorter tails and narrower noses distinguish them from their American cousins.

COMPARATIVE MORPHOLOGY Other evidence of evolution comes from **comparative morphology,** analyses of the structure of living and extinct organisms. Such analyses are based on the comparison of **homologous traits,** characteristics that are similar in two species because they inherited the genetic basis of the trait from their common ancestor. For example, the forelimbs of all four-legged vertebrates are homologous because they evolved from a common ancestor with a forelimb composed of the same component parts (see Figure 20.3, which shows homologous bones in the same color). Even though the shapes of the bones are different in pigs, dolphins, and bats, similarities in the three limbs are apparent. The differences in structural details arose over evolutionary time, allowing pigs to walk, dolphins to swim, and bats to fly. The arms of humans and the wings of birds are constructed of comparable elements, suggesting that they, too, share a common ancestor with the three species that are illustrated.

Molecular Techniques Extend the Achievements of the Modern Synthesis

Molecular techniques provide biologists with powerful tools for exploring all aspects of life—and evolutionary biology is no exception. In this section, we describe two examples that illustrate how molecular techniques have furthered our understanding of evolution since the modern synthesis was articulated.

HOW SNAKES LOST THEIR LEGS From Darwin's time until the mid-twentieth century, biologists tried to discern the evolutionary history of animals by comparing their embryos and patterns of development **(Figure 20.14).** The early embryos of related species are often strikingly similar, but morphological differences appear as the embryos grow and develop their adult forms (see Chapter 50). For example, the early embryos of most four-legged vertebrates (such as lizards, mammals, and birds) develop "limb buds" from which the legs or wings typical of each species later grow. Forelimbs and their supporting structures grow at the base of the neck, just in front of the ribcage, and hindlimbs grow right behind the ribcage (see Figure 20.14A). Similarities in the limb buds and the positions of the limbs in these animals provide evidence of their descent from a shared ancestor. Differences in their adult structures are caused by additional genetic instructions that have evolved over time.

However, most snakes—which are very closely related to lizards—show no traces of limbs or necks; their ribcages are positioned right behind their heads (see Figure 20.14B). The fossil record shows us that snakes evolved from four-legged ancestors in stages: early snakes had small hindlimbs, and the most recently evolved snakes have no limbs at all. Only the most ancient living snakes, pythons and boas, have any traces of limbs—vestigial hindlimbs, which appear as a pair of tiny clawlike structures near the base of the tail (see Figure 20.14C). Observational studies of their embryos reveal that most living snakes never develop limb buds; by contrast, pythons and boas develop hind limb buds, which grow only slightly as the animal develops.

How did pythons, and presumably all other snakes, lose their necks and forelimbs? Although many genes control the differences in the adult forelimbs among species (see Figure 20.3), research reported in the 1990s revealed that two regulatory genes, *Hoxc6* and *Hoxc8*, determine whether forelimbs or ribs grow at a particular site along an animal's backbone. The *Hox* genes either activate or suppress other genes that direct the development of these structures. Forelimbs—but not ribs— grow just in front of the tissues where only *Hoxc6* is expressed. By contrast, ribs—but not forelimbs—grow where both *Hoxc6* and *Hoxc8* are expressed. In chickens and other vertebrates with four limbs, only *Hoxc6* is expressed just in front of the ribcage, and forelimbs grow nearby (see Figure 20.14D).

In 1999, Martin J. Cohn of the University of Reading and Cheryll Tickle of University College London reported that in pythons (primitive snakes), both *Hoxc6* and *Hoxc8* are expressed all along the backbone, beginning at the base of the skull; as a result, a python's ribcage develops right behind its head, and no limb buds or forelimbs develop (see Figure 20.14E). Thus, snakes have no forelimbs or necks because a mutation causes the expression of *Hoxc8* to extend into a more forward region of the animal's body. All descendants of the ancestor with that original mutation now lack necks and forelimbs. Cohn and Tickle's research also suggests that the second stage in snake evolution, the reduction or complete absence of hindlimbs, is caused by other genetic variations that appeared some time after the altered expression pattern of *Hoxc8*. Thus, molecular research has identified the genetic changes that caused snakes to lose their forelimbs and necks before losing their hindlimbs.

GENETICS OF LIMB LOSS IN SNAKES

A. Most lizards, like this monitor lizard (*Varanus* species), have four limbs attached to their backbones.

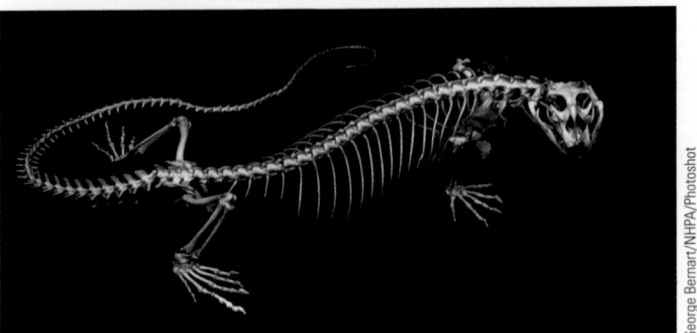

George Bernart/NHPA/Photoshot

B. Most snakes, like this grass snake (*Natrix* species), lack limbs altogether.

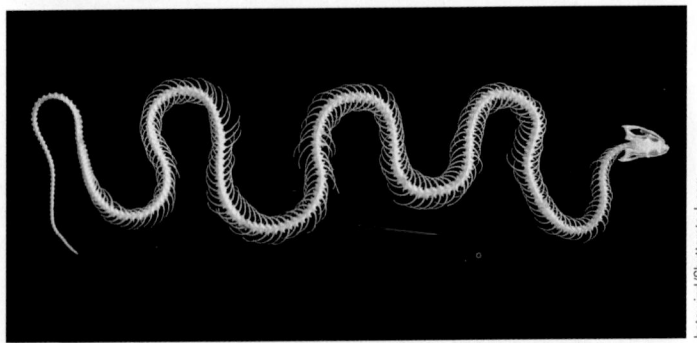

hotwind/Shutterstock.com

C. Some primitive snakes, like this ball python *(Python regius)*, have vestigial hindlimbs, visible as a pair of clawlike mating spurs near the base of the tail.

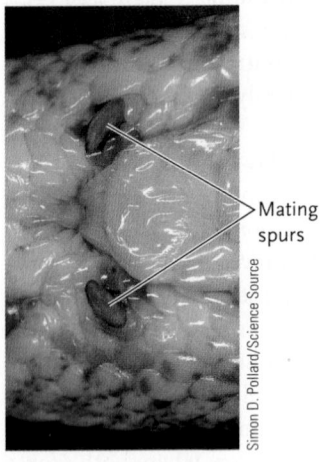

Mating spurs

Simon D. Pollard/Science Source

D. Molecular analyses reveal that the expression of *Hoxc6*, but not *Hoxc8*, at the base of the neck in a chick embryo (see arrow) causes limb buds and forelimbs to develop nearby.

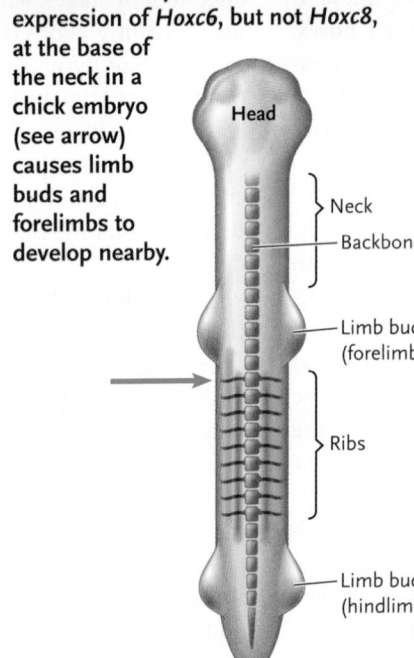

Head

Neck

Backbone

Limb bud (forelimb)

Ribs

Limb bud (hindlimb)

E. In pythons, both *Hoxc6* and *Hoxc8* are expressed alongside the backbone, beginning just behind the head (see arrow). The expression of these two genes appears to suppress the development of limb buds and forelimbs and promote the development of ribs at that location.

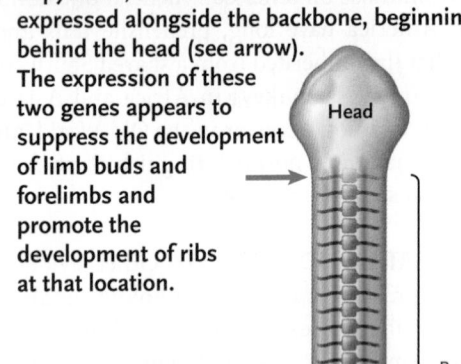

Head

Backbone

Ribs

Limb bud (hindlimb)

KEY

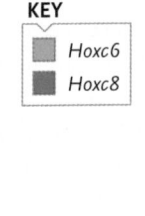

Hoxc6

Hoxc8

SUMMARY New molecular techniques from evolutionary developmental biology, coupled with traditional embryological observations, solved a long-standing evolutionary mystery. Evolutionary changes in the spatial pattern of gene expression caused the loss of forelimbs in snakes.

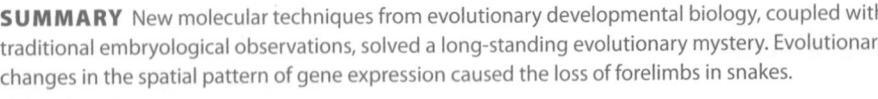 **think like a scientist** This research identifies the genetic mechanism underlying the loss of forelimbs in snakes, but why might natural selection have favored this anatomical change?

© Cengage Learning 2017

THE WOOLLY MAMMOTH'S CLOSEST LIVING RELATIVE Evolutionary biologists now routinely compare the genetic sequences of different species to determine their evolutionary relationships (see Chapter 24). The process is straightforward as long as researchers can collect small tissue samples from which they extract DNA. For many years, paleobiologists could not use these techniques to determine the evolutionary relationships between extinct organisms and those living today because they could not obtain the

Our Best Friends: Where on Earth and when were dogs domesticated?

Recent technical innovations now allow biologists to extract DNA from fossils of ancient organisms, providing answers to long-puzzling questions. Based on morphological evidence, paleontologists knew that domesticated dogs *(Canis familiaris)* are descended from grey wolves *(Canis lupus)*. Both species, along with foxes, coyotes, and jackals, are described as "canids" (that is, members of the taxonomic family Canidae). Indeed, dogs and wolves are so closely related that some biologists consider them to be of the same species, *C. lupus*. But deciphering the history of dog domestication on morphological grounds is complicated by the fact that early dogs and their wolf relatives frequently interbred. Wolf populations were abundant in most parts of the northern hemisphere until they were suppressed by human hunting, and dogs could have been domesticated anywhere within the wolf's geographic range. Although the oldest fossils of doglike animals are found in Europe, previous genetic research on the DNA of modern dogs and modern wolves suggested

that dogs were domesticated either in the Middle East or in East Asia. Critics suggested that comparing the DNA of modern dogs and modern wolves could not identify where dogs were first domesticated because of the subsequent interbreeding of dogs and wolves and the diversification of dog breeds by artificial selection. Thus, researchers compared DNA sequences of modern dogs to those of ancient wolves.

Research Question

To which populations of ancient wolves are ancient dogs and modern dogs most closely related and when did domestication take place?

Experiment

Olaf Thalmann, now at the University of Turku in Finland, Robert Wayne of the University of California at Los Angeles, and 29 collaborators at many institutions around the world used molecular techniques—specifically PCR amplification, cloning, and sequencing of

mitochondrial DNA (see Chapters 18 and 19)—to answer the question.

1. They extracted DNA from 18 fossilized canid samples (eight of them doglike and 10 of them wolflike) from sites in Alaska, Europe, Russia, South America, and the United States. The fossils ranged in age from 1,000 to 36,000 years.
2. The researchers generated complete and partial mitochondrial genomes for these animals after performing DNA capture and processing the samples with high-throughput sequencing. After filtering the data and assembling the sequences recovered, they used bioinformatic analyses to exclude sequences from modern DNA contamination.
3. They compared the DNA sequences extracted from the 18 fossils to equivalent sequences in 49 modern wolves, 77 modern dogs of varying breeds, and four coyotes.
4. Using several statistical approaches (see Chapter 24), they assembled a phylogenetic tree for wolves and dogs **(Figure).**

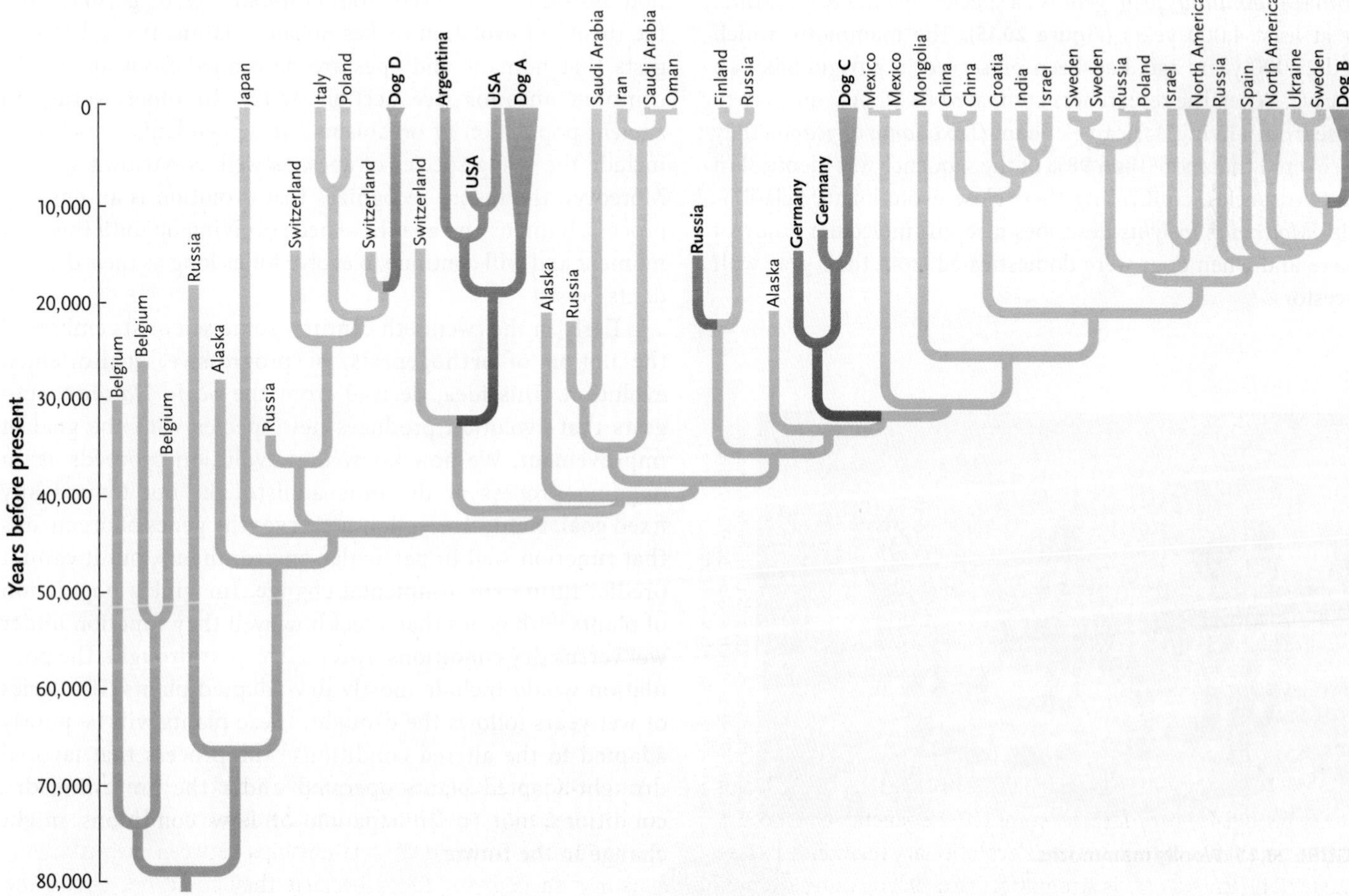

Evolutionary branches leading to dogs are illustrated in blue and those leading to wolves in green. Branches leading to modern canids extend to the top of the tree. Branches leading to ancient canids end at the time that their fossils formed, as indicated on the timeline on the vertical axis. The countries of origin of the ancient samples are listed at the ends of the branches.

Conclusions

The analysis suggests that living dogs fall within four evolutionary lineages (labeled Dog A, Dog B, Dog C, and Dog D in the Figure). Moreover, the closest ancient wolflike relatives of most modern dogs lived in Europe, and the domestication process began between 18,800 and 32,100 years ago.

The conclusions of this study are controversial, in part because the researchers were unable to analyze samples from ancient canids in the Middle East or East Asia. Previous research by other workers had suggested that dog domestication occurred after humans had developed agricultural practices 10,000 to 12,000 years ago, much later than the present results indicate. In response to this challenge, Thalman, Wayne, and their colleagues are trying to collect usable nuclear DNA sequence data from ancient dog and wolf fossils from a broader range of locations, but given how much dogs and wolves interbred during their shared history, the controversy may continue for many years.

Source: O. Thalmann et al. 2013. Complete mitochondrial genomes of ancient canids suggest a European origin of domestic dogs. *Science* 342:871–874.

necessary samples from fossilized material. Sequencing the DNA of extinct organisms was the "holy grail" of paleobiology.

Recent advances in molecular techniques now allow researchers to sequence DNA preserved in some fossils of extinct organisms. In a study published in 2006, Hendrick Poinar of McMaster University and colleagues at other institutions sequenced 13 million base pairs of mitochondrial and nuclear DNA extracted from the jawbone of a woolly mammoth (*Mammuthus primigenius),* a species that has been extinct for at least 4,000 years **(Figure 20.15)**. The mammoth, which died 27,000 years ago, had been preserved in a frigid Siberian ice cave. When the researchers compared its DNA sequences to those from a living African elephant (*Loxodonta africana),* they discovered that more than 98% of the sequence was identical in the two species, confirming their close evolutionary relationship. *Molecular Insights* describes a recent molecular study of where and when dogs were domesticated from their gray wolf ancestors.

Reproduced from Sedwick C (2008) What Killed the Woolly Mammoth? *PLoS Biol* 6(4): e99. doi:10.1371/journal.pbio.0060099 © 2008 Public Library of Science, illustration by Mauricio Anton

FIGURE 20.15 Woolly mammoths.

Evolution Is the Core Theory of Modern Biology

The theory of evolution has always been a contentious subject because it challenges deeply held traditional views of how living organisms originated. Many of Darwin's contemporaries were dismayed by the suggestion that all organisms share a common ancestry. Some people even misinterpreted this assertion as "humans evolved from chimpanzees or gorillas." But the theory of evolution makes no such claims. Instead, it suggests that humans and apes are descended from an apelike common ancestor (see Section 32.12). In other words, an ancient population of organisms left descendants, which now include the living species of apes, as well as our own species. Moreover, the theory recognizes that evolution is an ongoing process: humans and apes have been evolving up until this very moment and will continue to evolve for as long as their descendants persist.

Early in the twentieth century, some scientists embraced the notion of **orthogenesis,** or progressive, goal-oriented evolution. This idea, derived from the *Scala Naturae*, suggests that evolution produces new species with the goal of improvement. We now know that evolution proceeds as an ongoing process of dynamic adjustment, not toward any fixed goal. Natural selection preserves the genes of organisms that function well in particular environments, but it cannot predict future environmental change. Imagine a population of plants with genes that affect how well they function under wet versus dry conditions. After a five-year drought, the population would include mostly dry-adapted plants. If a series of wet years follows the drought, these plants will be poorly adapted to the altered conditions. The process that favored drought-adapted plants operated under the prevailing dry conditions, not in anticipation of how conditions might change in the future.

This species, first discovered only two years after *The Origin* was published, resembled both nonflying dinosaurs and modern birds. Like the small carnivorous Dromaeosaurids, *Archaeopteryx* walked on its hind legs and had teeth, long clawed fingers on its forelimbs, and a long, bony tail. Like modern birds, it had enlarged forelimbs with flight feathers. Recently discovered fossils reveal that many dinosaurian ancestors of birds had feathers (discussed further in Chapter 23).

HISTORICAL BIOGEOGRAPHY Analyses of **historical biogeography,** the study of the geographical distributions of plants and animals in relation to their evolutionary history, are generally consistent with Darwin's theory of evolution. Species on oceanic islands often closely resemble species on the nearest mainland, suggesting that the island and mainland species share a common ancestry. Moreover, species on a continental landmass are clearly related to one another and are often distinct from those on other continents. For example, monkeys in South America have long, prehensile tails and broad noses, traits that they inherited from a shared South American ancestor. By contrast, monkeys in Africa and Asia evolved from a different common ancestor in the Old World, and their shorter tails and narrower noses distinguish them from their American cousins.

COMPARATIVE MORPHOLOGY Other evidence of evolution comes from **comparative morphology,** analyses of the structure of living and extinct organisms. Such analyses are based on the comparison of **homologous traits,** characteristics that are similar in two species because they inherited the genetic basis of the trait from their common ancestor. For example, the forelimbs of all four-legged vertebrates are homologous because they evolved from a common ancestor with a forelimb composed of the same component parts (see Figure 20.3, which shows homologous bones in the same color). Even though the shapes of the bones are different in pigs, dolphins, and bats, similarities in the three limbs are apparent. The differences in structural details arose over evolutionary time, allowing pigs to walk, dolphins to swim, and bats to fly. The arms of humans and the wings of birds are constructed of comparable elements, suggesting that they, too, share a common ancestor with the three species that are illustrated.

Molecular Techniques Extend the Achievements of the Modern Synthesis

Molecular techniques provide biologists with powerful tools for exploring all aspects of life—and evolutionary biology is no exception. In this section, we describe two examples that illustrate how molecular techniques have furthered our understanding of evolution since the modern synthesis was articulated.

HOW SNAKES LOST THEIR LEGS From Darwin's time until the mid-twentieth century, biologists tried to discern the evolutionary history of animals by comparing their embryos and patterns of development **(Figure 20.14).** The early embryos of related species are often strikingly similar, but morphological differences appear as the embryos grow and develop their adult forms (see Chapter 50). For example, the early embryos of most four-legged vertebrates (such as lizards, mammals, and birds) develop "limb buds" from which the legs or wings typical of each species later grow. Forelimbs and their supporting structures grow at the base of the neck, just in front of the ribcage, and hindlimbs grow right behind the ribcage (see Figure 20.14A). Similarities in the limb buds and the positions of the limbs in these animals provide evidence of their descent from a shared ancestor. Differences in their adult structures are caused by additional genetic instructions that have evolved over time.

However, most snakes—which are very closely related to lizards—show no traces of limbs or necks; their ribcages are positioned right behind their heads (see Figure 20.14B). The fossil record shows us that snakes evolved from four-legged ancestors in stages: early snakes had small hindlimbs, and the most recently evolved snakes have no limbs at all. Only the most ancient living snakes, pythons and boas, have any traces of limbs—vestigial hindlimbs, which appear as a pair of tiny clawlike structures near the base of the tail (see Figure 20.14C). Observational studies of their embryos reveal that most living snakes never develop limb buds; by contrast, pythons and boas develop hind limb buds, which grow only slightly as the animal develops.

How did pythons, and presumably all other snakes, lose their necks and forelimbs? Although many genes control the differences in the adult forelimbs among species (see Figure 20.3), research reported in the 1990s revealed that two regulatory genes, *Hoxc6* and *Hoxc8,* determine whether forelimbs or ribs grow at a particular site along an animal's backbone. The *Hox* genes either activate or suppress other genes that direct the development of these structures. Forelimbs—but not ribs—grow just in front of the tissues where only *Hoxc6* is expressed. By contrast, ribs—but not forelimbs—grow where both *Hoxc6* and *Hoxc8* are expressed. In chickens and other vertebrates with four limbs, only *Hoxc6* is expressed just in front of the ribcage, and forelimbs grow nearby (see Figure 20.14D).

In 1999, Martin J. Cohn of the University of Reading and Cheryll Tickle of University College London reported that in pythons (primitive snakes), both *Hoxc6* and *Hoxc8* are expressed all along the backbone, beginning at the base of the skull; as a result, a python's ribcage develops right behind its head, and no limb buds or forelimbs develop (see Figure 20.14E). Thus, snakes have no forelimbs or necks because a mutation causes the expression of *Hoxc8* to extend into a more forward region of the animal's body. All descendants of the ancestor with that original mutation now lack necks and forelimbs. Cohn and Tickle's research also suggests that the second stage in snake evolution, the reduction or complete absence of hindlimbs, is caused by other genetic variations that appeared some time after the altered expression pattern of *Hoxc8*. Thus, molecular research has identified the genetic changes that caused snakes to lose their forelimbs and necks before losing their hindlimbs.

GENETICS OF LIMB LOSS IN SNAKES

A. Most lizards, like this monitor lizard (*Varanus* species), have four limbs attached to their backbones.

B. Most snakes, like this grass snake (*Natrix* species), lack limbs altogether.

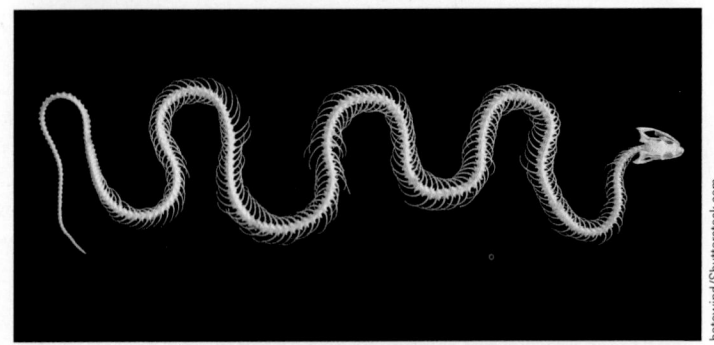

George Bernart/NHPA/Photoshot

hotowind/Shutterstock.com

C. Some primitive snakes, like this ball python *(Python regius)*, have vestigial hindlimbs, visible as a pair of clawlike mating spurs near the base of the tail.

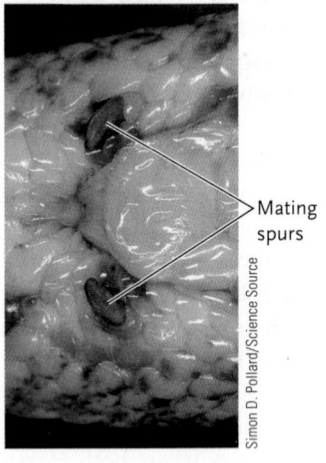

Mating spurs

Simon D. Pollard/Science Source

D. Molecular analyses reveal that the expression of *Hoxc6*, but not *Hoxc8*, at the base of the neck in a chick embryo (see arrow) causes limb buds and forelimbs to develop nearby.

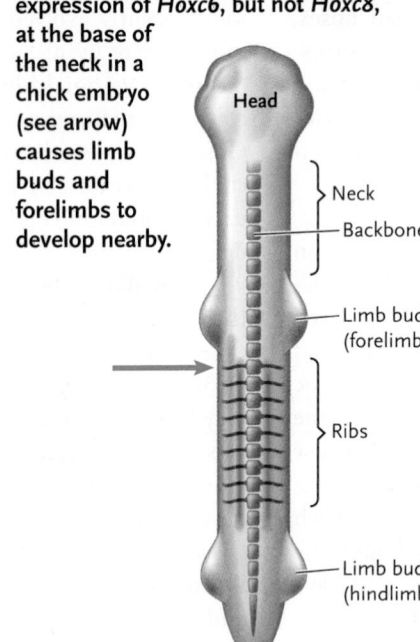

Head
Neck
Backbone
Limb bud (forelimb)
Ribs
Limb bud (hindlimb)

E. In pythons, both *Hoxc6* and *Hoxc8* are expressed alongside the backbone, beginning just behind the head (see arrow). The expression of these two genes appears to suppress the development of limb buds and forelimbs and promote the development of ribs at that location.

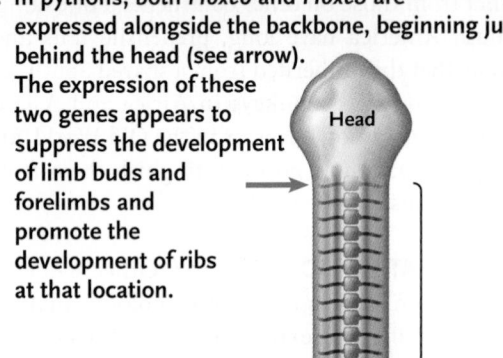
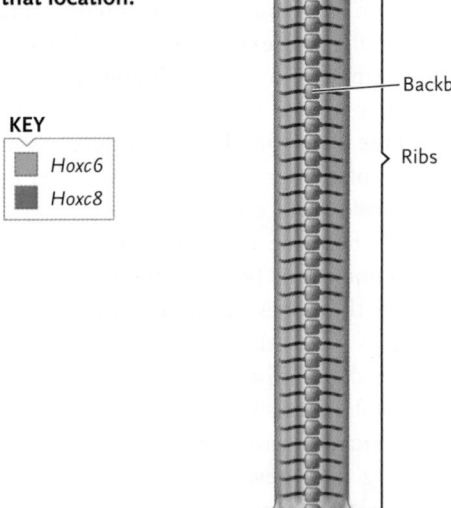

Head
Backbone
Ribs
Limb bud (hindlimb)

KEY

	Hoxc6
	Hoxc8

SUMMARY New molecular techniques from evolutionary developmental biology, coupled with traditional embryological observations, solved a long-standing evolutionary mystery. Evolutionary changes in the spatial pattern of gene expression caused the loss of forelimbs in snakes.

think like a scientist This research identifies the genetic mechanism underlying the loss of forelimbs in snakes, but why might natural selection have favored this anatomical change?

© Cengage Learning 2017

THE WOOLLY MAMMOTH'S CLOSEST LIVING RELATIVE Evolutionary biologists now routinely compare the genetic sequences of different species to determine their evolutionary relationships (see Chapter 24). The process is straightforward as long as researchers can collect small tissue samples from which they extract DNA. For many years, paleobiologists could not use these techniques to determine the evolutionary relationships between extinct organisms and those living today because they could not obtain the

Evolution is the core theory of modern biology because its explanatory power touches on every aspect of the living world. The application of molecular techniques to the study of evolutionary biology has greatly enhanced our knowledge. In the remaining chapters of this unit, you will discover how contemporary evolutionary theory explains changes at every level of biological organization, from adaptive modifications within populations (see Chapter 21), to the evolution of new species (see Chapter 22), major patterns in the history of life (see Chapter 23), and unraveling the evolutionary history of all organisms on Earth (see Chapter 24). Moreover, we will revisit many concepts from evolutionary biology in the discussions of biodiversity, the structure and function of plants and animals, and ecology and behavior in subsequent units.

STUDY BREAK 20.3

1. What two problems slowed the acceptance of Darwin's theory among scientists?
2. What is the difference between microevolution and macroevolution?
3. What types of data provide evidence that evolution has adapted organisms to their environments and promoted the diversification of species?

THINK OUTSIDE THE BOOK

Find examples from popular publications or advertisements for consumer products that misrepresent the theory of biological evolution. Explain how the theory is misrepresented.

Unanswered Questions

What determines whether a species adapts to a changing environment or becomes extinct?

Natural selection has produced marvelous adaptations in every species on Earth, and we know that evolutionary adaptation to certain environmental changes has allowed many species to persist. But we also know that more than 99% of the species that have ever lived became extinct, evidently because they failed to adapt to changes in climate, natural competitors or enemies, or other environmental factors. But what kinds of genetic variation are required for adaptation, and what kinds of characteristics must evolve to allow survival? This is a critical question today, because human activities are changing environments so rapidly and drastically that many species face the threat of extinction. Can aquatic species adapt to various kinds of water pollution? Can animals and plants that lived in prairies adapt to different habitats now that most prairies have been destroyed?

Is adaptation by natural selection responsible for most of the genetic differences between species?

New genetic variations sometimes become more common within populations or species because the proteins for which they code are advantageous and preserved by natural selection. But biologists who study molecular evolution have discovered that a large part of the genome in most organisms (about 98% of the human genome, for example) does not code for proteins and therefore appears to have no function. If this observation is generally correct, why do the noncoding parts of genomes exist? Are evolutionary changes in noncoding regions and the differences in noncoding sequences among species adaptive? For example, only about 1% of the DNA base pairs differ between human and chimpanzee genomes—but this amounts to about 34 million base-pair differences altogether, at least 60,000 of which alter the amino acid sequences of proteins. How can we determine which of these differences are adaptive and which differences underlie the unique characteristics of humans?

How do pathways of embryonic development evolve?

The characteristics of adult organisms are the product of developmental events, starting with the fertilized egg, that include growth in size, changes in the shape of various body parts, and the differentiation of cell types. These processes are largely controlled by genes, with input from the environment. Although biologists are beginning to learn how the genetic foundations of developmental processes evolve, many questions remain. For example, how do genetic changes induce differences in the branching patterns of antlers among species of deer, or differences in the length of the tails of monkeys and apes (including humans), or differences in the number and size of scales among species of lizards? We know that the proteins forming the lens of the eye are actually enzymes that play different roles in other cells, and that they have been "recruited" to form the lens, but what mechanisms induce them to assume this new role? And why do different enzymes form the lenses in eyes of birds and mammals? Evolutionary developmental biology, which is discussed in Chapter 23, is one of the most active, exciting fields in biology at this time.

think like a scientist As our production of carbon dioxide increases Earth's average temperature faster than ever before, will Arctic species be able to adapt to changes in climate?

Douglas J. Futuyma is Distinguished Professor of evolution and ecology at Stony Brook University. His research interests focus on speciation and the evolution of ecological interactions among species, and in particular on insect-plant interactions. Learn more about Dr. Futuyma's work at http://life.bio.sunysb.edu/ee/faculty.html.

REVIEW KEY CONCEPTS

For access to MindTap and additional study materials visit www.cengagebrain.com.

20.1 Recognition of Evolutionary Change

- Ancient Greek philosophers classified the natural world, ranking inanimate objects and living organisms from simple to complex.

- Natural theologians believed that all species were specially created and perfectly adapted; that existing species do not change or become extinct; and that new species do not arise. Studies in biogeography, comparative morphology, and paleontology led scientists to question whether species change through time (Figures 20.2 and 20.3).

- Lamarck proposed that species evolved into more complex forms that functioned better in their environments. He hypothesized that structures in an organism changed when they were used, and that those changes were inherited by the organism's offspring (Figure 20.4). Experiments have refuted Lamarck's proposed mechanisms.

- Two geologists, Hutton and Lyell, recognized that major features on Earth were created by the long-term action of the very slow geological processes that scientists observe today. Their insights suggested that Earth was much older than natural theologians had supposed.

20.2 Darwin's Journeys

- Darwin's observations during his voyage on the *Beagle* provided much of the data and inspiration for the development of his theory of evolution (Figures 20.5–20.8).

- Darwin based the theory of evolution by means of natural selection on two hypotheses and one prediction: (1) individuals within a population compete for limited resources, (2) hereditary characteristics allow some individuals to survive longer and reproduce more than others, and (3) a population's characteristics change over time as advantageous heritable characteristics become more common (Figure 20.10).

- Darwin also proposed that the accumulation of differences fostered by natural selection could cause populations to diverge over time. Such evolutionary divergence can lead to the production of new species, which can, in turn, give rise to new evolutionary lineages (Figures 20.9 and 20.11).

20.3 Evolutionary Biology since Darwin

- Scientists working in population genetics developed theories of evolutionary change by integrating Darwin's ideas with Mendel's research on genetics.

- In the 1930s and 1940s, the modern synthesis provided a unified view of evolution that drew on studies from many biological disciplines. It emphasized evolution within populations, the central role of variation in the evolutionary process, and the gradualism of evolutionary change.

- Studies of adaptation, the fossil record, historical biogeography, and comparative morphology provide compelling evidence of evolutionary change (Figures 20.12 and 20.13).

- Molecular techniques have extended the achievements of the modern synthesis, allowing precise analysis of the genetic basis of evolutionary change (Figure 20.14) and the genetic relatedness of living and extinct organisms (Figure 20.15).

TEST YOUR KNOWLEDGE

Remember/Understand

1. The "father" of taxonomy is:
   a. Charles Darwin.
   b. Charles Lyell.
   c. Alfred Wallace.
   d. Carolus Linnaeus.
   e. Jean Baptiste de Lamarck.

2. The wings of birds, the legs of pigs, and the flippers of dolphins provide examples of:
   a. vestigial structures.
   b. homologous structures.
   c. acquired characteristics.
   d. artificial selection.
   e. uniformitarianism.

3. Which of the following ideas proposed by Lamarck was *not* included in Darwin's theory?
   a. Organisms change in response to their environments.
   b. Changes that an organism acquires during its lifetime are passed to its offspring.
   c. All species change with time.
   d. New variations may be passed from one generation to the next.
   e. Specific mechanisms cause evolutionary change.

4. The belief that evolution is progressive or goal-oriented is called:
   a. gradualism.
   b. uniformitarianism.
   c. taxonomy.
   d. orthogenesis.
   e. the modern synthesis.

Apply/Analyze

5. Which of the following statements about evolutionary studies is *not* true?
   a. Biologists study the products of evolution to understand the processes causing it.
   b. Biologists design molecular experiments to examine evolutionary processes operating over short time periods.
   c. Biologists study the inheritance of characteristics that a parent acquired during its lifetime.
   d. Biologists study variation in homologous structures among related organisms.
   e. Biologists examine why a huge variety of species may inhabit a small island cluster.

6. Which of the following ideas is *not* included in Darwin's theory?
   a. All organisms that have ever existed arose through evolutionary modifications of ancestral species.
   b. The great variety of species alive today resulted from the diversification of ancestral species.
   c. Natural selection drives some evolutionary change.
   d. Natural selection preserves advantageous traits.
   e. Natural selection eliminates adaptive traits.

7. Which of the following could be an example of microevolution?
   a. a slight change in a bird population's color due to a small genetic change in the population
   b. large differences between fossils found near the ground surface and those found in deep rock layers
   c. the sudden disappearance of an entire genus
   d. the direct evolutionary link between living primates and humans
   e. a flood that drowns all members of a population

8. Medical advances now allow many people who suffer from genetic diseases to survive and reproduce. These advances:
   a. refute Darwin's theory.
   b. support Lamarck's theory.
   c. disprove descent with modification.
   d. reduce the effects of natural selection.
   e. eliminate adaptive traits

9. **Discuss Concepts** Explain why the characteristics we see in living organisms adapt them to the environments in which their ancestors lived rather than to the environments in which they live today.

Evaluate/Create

10. Which of the following statements is *not* compatible with Darwin's theory?
    a. All organisms have arisen by descent with modification.
    b. Evolution has altered and diversified ancestral species.
    c. Evolution occurs in individuals rather than in groups.
    d. Natural selection eliminates unsuccessful variations.
    e. Evolution occurs because some individuals function better than others in a particular environment.

11. Which of the following does *not* contribute to the study of evolution?
    a. population genetics
    b. inheritance of acquired characteristics
    c. the fossil record
    d. DNA sequencing
    e. comparative morphology

12. **Discuss Concepts** Imagine a population of mice that includes both brown and black individuals. They live in a habitat with brown soil, where predatory hawks can see black mice more easily than they can see brown ones. Design a study that would allow you to determine whether the brown mice are better adapted to this environment than the black mice.

13. **Design an Experiment** Design an experiment to test Lamarck's hypothesis that characteristics acquired during an organism's lifetime are inherited by their offspring. (You may wish to review the components of a well-designed experiment in Chapter 1 before formulating your answer.) Can you think of examples of acquired characteristics that are *not* inherited by offspring?

14. **Apply Evolutionary Thinking** Identify three discoveries or inventions that have changed how humans are affected by natural selection. Describe in detail how each discovery influences survival or reproduction in our species.

For selected answers, see Appendix A.

INTERPRET THE DATA

For centuries, animal breeders have used artificial selection to increase the speed of racehorses. The results of such efforts are illustrated in a graph that identifies how quickly winning horses ran the 1¼-mile track at the Kentucky Derby, a famous race held annually at Churchill Downs, Kentucky. (The time it takes a horse to run a track of fixed length is inversely proportional to the horse's speed.) Based on the data in the graph, has artificial selection increased the speed of horses consistently between 1900 and 2000? Do you think that artificial selection can increase the speed of racehorses in the future?

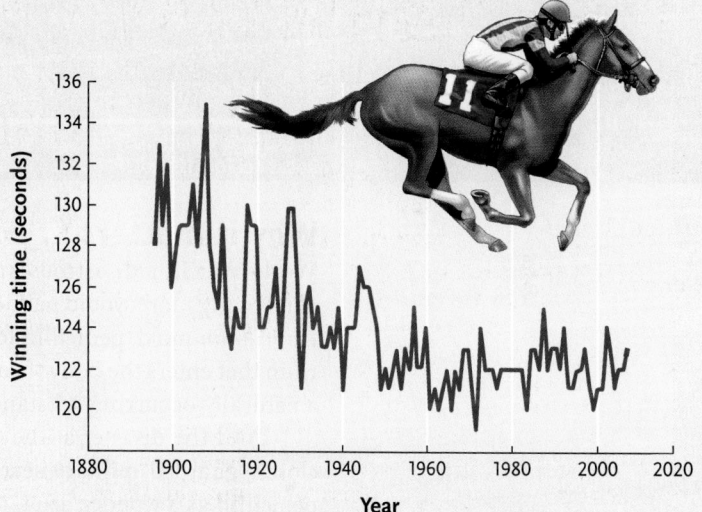

21 Microevolution: Genetic Changes within Populations

Phenotypic variation. The frog *Dendrobates pumilio* exhibits dramatic color variation in populations that inhabit the Bocas del Toro Islands, Panama.

Why it matters . . . On November 28, 1942, at the height of American involvement in World War II, a disastrous fire killed more than 400 people in Boston's Cocoanut Grove nightclub. Many more would have died later but for a new experimental drug, penicillin. A product of *Penicillium* mold, penicillin fought the usually fatal infections of *Staphylococcus aureus*, a bacterium that enters the body through damaged skin. Penicillin was the first antibiotic drug based on a naturally occurring substance that kills bacteria.

Until the disaster at the Cocoanut Grove, the production and use of penicillin had been a closely guarded military secret. But after its public debut, the pharmaceutical industry hailed penicillin as a wonder drug, promoting its use for the treatment of the many diseases caused by infectious microorganisms. Penicillin became widely available as an over-the-counter remedy, and Americans dosed themselves with it, hoping to cure all sorts of ills **(Figure 21.1)**. But in 1945, Sir Alexander Fleming, the scientist who discovered penicillin, predicted that some bacteria could survive low doses, and that the offspring of those germs would be more resistant to its effects. In 1946—just four years after penicillin's use in Boston—14% of the *Staphylococcus* strains isolated from patients in a London hospital were resistant. By 1950, more than half of the strains were resistant.

Scientists have discovered numerous antibiotics since the 1940s, and many strains of bacteria have developed resistance to these drugs. In fact, according to the Centers for Disease Control and Prevention, between 30,000 and 40,000 Americans die each year from infections caused by antibiotic-resistant bacteria.

How do bacteria become resistant to antibiotics? The genomes of bacteria—like those of all other organisms—vary among individuals, and some bacteria have genetic traits that allow them to withstand attack by antibiotics. When we administer antibiotics to an infected patient, we

FIGURE 21.1 **Selling penicillin.** This ad, from a 1944 issue of *Life* magazine, credits penicillin with saving the lives of wounded soldiers.

In this chapter, we first examine the extensive variation that exists within natural populations. We then take a detailed look at the most important processes that alter genetic variation within populations, causing microevolutionary change. Finally, we consider how microevolution can fine-tune the functioning of populations within their environments.

21.1 Variation in Natural Populations

In some species, individuals vary dramatically in appearance, but in most species, the members of a population look pretty much alike **(Figure 21.2).** However, even those that look alike, such as the *Cerion* snails in Figure 21.2B, are not identical. With a scale and ruler, you could detect differences in their weight as well as in the length and diameter of their shells. With suitable techniques, you could also document variations in their individual biochemistry, physiology, internal anatomy, and behavior. All of these are examples of **phenotypic variation,** differences in appearance or function that are passed from generation to generation.

Evolutionary Biologists Describe and Quantify Phenotypic Variation

Darwin's theory recognized the importance of heritable phenotypic variation, and today, microevolutionary studies often begin by assessing phenotypic variation within populations. Most characters exhibit **quantitative variation:** individuals differ in small, incremental ways. If you weighed everyone in your biology class, for example, you would see that weight varies almost continuously from your lightest to your heaviest classmate. Humans also exhibit quantitative variation in the length of their toes, the number of hairs on their heads, and their height, as discussed in Section 12.2.

We usually display data on quantitative variation in a bar graph or, if the sample is large enough, as a curve **(Figure 21.3).** The width of the curve is proportional to the variability—the amount of variation—among individuals, and the *mean* describes the average value of the character. As you will see shortly, natural selection often changes the mean value of a character or its variability within populations.

create an environment favoring bacteria that are even slightly resistant to the drug. The surviving bacteria reproduce, and resistant microorganisms—along with the genes that confer antibiotic resistance—become more common in later generations. In other words, bacterial strains adapt to the presence of antibiotics in their environment through the evolutionary process of selection. Our use of antibiotics is comparable to artificial selection by plant and animal breeders (see Chapter 20), but when we use antibiotics, we inadvertently select for the success of the organisms we are trying to eradicate.

The evolution of antibiotic resistance in bacteria is an example of **microevolution,** which is a heritable change in the genetics of a population. A **population** of organisms includes all the individuals of a single species that live together in the same place and time. Today, when scientists study microevolution, they analyze variations—the differences between individuals—in natural populations and determine how and why these variations are inherited. Darwin recognized the importance of heritable variation within populations; he also realized that natural selection can change the pattern of variation in a population from one generation to the next. Scientists have since learned that microevolutionary change results from several processes, not just natural selection, and that sometimes these processes counteract each other.

A. **European garden snails** *(Cepaea nemoralis)*

B. **Bahaman land snails** *(Cerion christophei)*

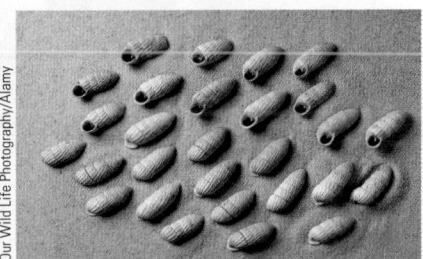

FIGURE 21.2 **Phenotypic variation. (A)** Shells of European garden snails from a population in Scotland vary considerably in appearance. **(B)** By contrast, shells of land snails from a population in the Bahamas look very similar.

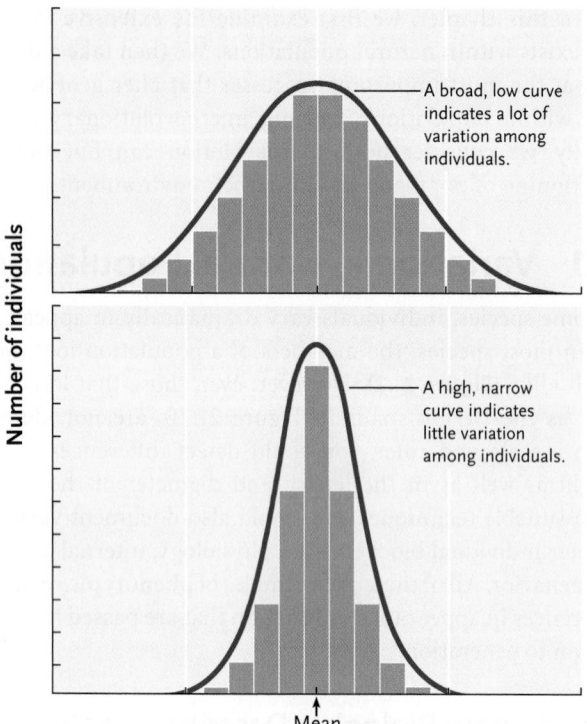

A broad, low curve indicates a lot of variation among individuals.

A high, narrow curve indicates little variation among individuals.

Mean

Measurement or value of trait

FIGURE 21.3 Quantitative variation. Many traits vary continuously among members of a population, and a bar graph of the data often approximates a bell-shaped curve. The mean defines the average value of the trait in the population, and the width of the curve is proportional to the variability among individuals.

© Cengage Learning 2017

Other characters, like those Mendel studied (see Section 12.1), exhibit **qualitative variation:** they exist in two or more discrete states, and intermediate forms are often absent. Snow geese, for example, have *either* blue *or* white feathers **(Figure 21.4).** The existence of discrete variants of a character is called a **polymorphism** (*poly* = many; *morphos* = form); we describe such traits as *polymorphic.* The *Cepaea nemoralis* snail shells in Figure 21.2A are polymorphic in background

FIGURE 21.4 Qualitative variation. Individual snow geese *(Chen caerulescens)* are either blue or white. Although both colors are present in many populations, geese tend to associate with others of the same color.

color, number of stripes, and color of stripes. Biochemical polymorphisms, like the human A, B, AB, and O blood types (described in Section 12.2), are also common.

We describe phenotypic polymorphisms quantitatively by calculating the percentage or *frequency* of each trait. For example, if you counted 123 blue snow geese and 369 white ones in a population of 492 geese, the frequency of the blue phenotype would be 123/492 or 0.25, and the frequency of the white phenotype would be 369/492 or 0.75.

Phenotypic Variation Can Have Genetic and Environmental Causes

Phenotypic variation within populations may be caused by genetic differences between individuals, by differences in the environmental factors that individuals experience, or by an interaction between genetics and the environment. As a result, genetic and phenotypic variations may not be perfectly correlated. Under some circumstances, organisms with different genotypes exhibit the same phenotype. For example, the black coloration of some rock pocket mice from Arizona is caused by certain mutations in the *Mc1r* gene (see Section 1.2); but black rock pocket mice from New Mexico do not share those mutations—that is, they have different genotypes—even though they exhibit the same phenotype. Conversely, organisms with the same genotype sometimes exhibit different phenotypes. For example, the acidity of soil influences flower color in some plants **(Figure 21.5).**

Knowing whether phenotypic variation is caused by genetic differences, environmental factors, or an interaction of the two is important because *only genetically based variation is subject to evolutionary change.* Moreover, knowing the causes of phenotypic variation has important practical applications. Suppose, for example, that one field of wheat produced more grain than another. If a difference in the availability of nutrients or water caused the difference in yield, a farmer might

FIGURE 21.5 Environmental effects on phenotype. Soil acidity affects the expression of the gene controlling flower color in the common garden plant *Hydrangea macrophylla.* When grown in acidic soil, it produces deep blue flowers. In neutral or alkaline soil, its flowers are bright pink.

FIGURE 21.6 Experimental Research

Using Artificial Selection to Demonstrate That Activity Level in Mice Has a Genetic Basis

Question: Do observed differences in activity level among house mice have a genetic basis?

Experiment: Swallow, Carter, and Garland knew that a phenotypic character responds to artificial selection only if it has a genetic, rather than an environmental, basis. In an experiment with house mice *(Mus domesticus),* they selected for the phenotypic character of increased wheel-running activity. In four experimental lines, they bred those mice that ran the most. Four other lines, in which breeders were selected at random with respect to activity level, served as controls.

Results: After 10 generations of artificial selection, mice in the experimental lines ran longer distances and ran faster than mice in the control lines. Thus, artificial selection on wheel-running activity in house mice increased **(A)** the distance that mice run per day and **(B)** their average speed. The data illustrate responses of females in four experimental lines and four control lines. Males showed similar responses.

A. Distance run

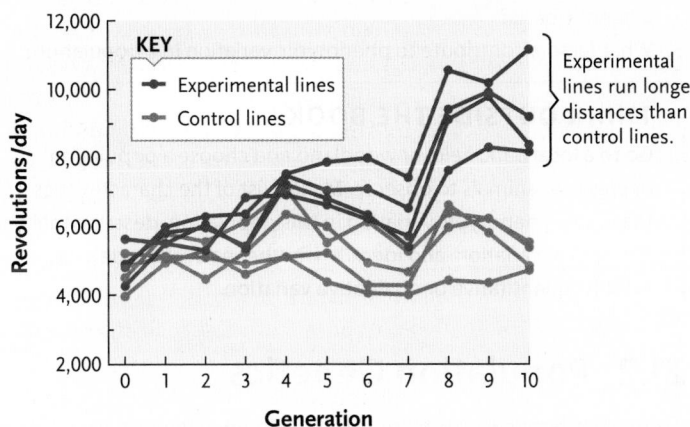

B. Average speed

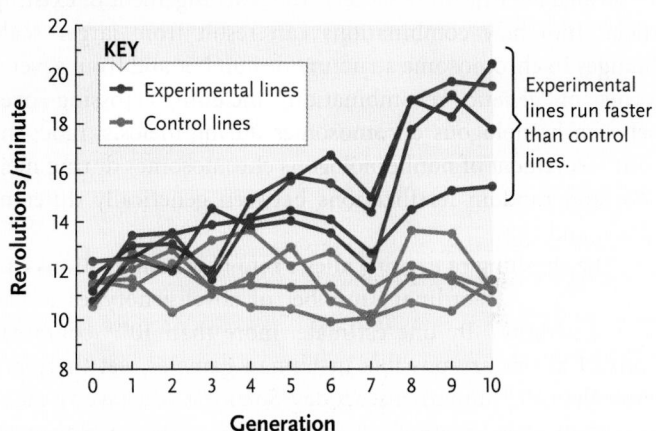

Conclusion: Because the two measures of activity level responded to artificial selection, researchers concluded that variation in this behavioral character has a genetic basis.

Observe
Hypothesize
Predict
Experiment
Interpret

Do the data presented in the graphs suggest that continued artificial selection would further increase the distance that the experimental lines run or their average speed?

Source: Adapted from Springer Science + Business Media: J. G. Swallow, P. A. Carter, and T. Garland, Jr. 1998. Artificial selection for increased wheel-running behavior in house mice. *Behavior Genetics* 28:227–237.

choose to fertilize or irrigate the less productive field. But if the difference in productivity resulted from genetic differences between plants in the two fields, a farmer might plant only the more productive genotype. Because environmental factors can influence the expression of genes, an organism's phenotype is frequently the product of an interaction between its genotype and its environment. In our hypothetical example, the farmer may maximize yield by fertilizing and irrigating the more productive genotype of wheat.

How can we determine whether phenotypic variation is caused by environmental factors or by genetic differences? We can test for an environmental cause experimentally by changing one environmental variable and measuring the effects on genetically similar subjects. You can try this yourself by growing some cuttings from an ivy plant in shade and other cuttings from the same plant in full sunlight. Although they all have the

same genotype, the cuttings grown in sunlight will produce smaller leaves and shorter stems.

Breeding experiments can demonstrate the genetic basis of phenotypic variation. For example, Mendel inferred the genetic basis of qualitative traits, such as flower color in peas, by crossing plants with different phenotypes. Although simple crosses will not reveal the genetic basis of variations in quantitative traits, these characteristics will respond to artificial selection if the variation has some genetic basis. For example, researchers observed that individual house mice *(Mus domesticus)* differ in activity levels, as measured by how much they use an exercise wheel and how fast they run. John G. Swallow and his colleagues at the University of Wisconsin, Madison, used artificial selection to produce lines of mice that exhibit increased wheel-running behavior, demonstrating that the observed differences in these two aspects of activity level have a genetic basis **(Figure 21.6).**

Breeding experiments are not always practical, however, particularly for organisms with long generation times. Ethical concerns also render these techniques unthinkable for humans. Instead, researchers sometimes study the inheritance of particular traits by analyzing genealogical pedigrees, as discussed in Section 13.4, but this approach often provides poor results for analyses of complex traits.

Several Processes Generate Genetic Variation

Genetic variation, the raw material molded by microevolutionary processes, has two potential sources: the production of new alleles and the rearrangement of existing alleles. Most new alleles probably arise from small-scale mutations in DNA (described later in this chapter). The rearrangement of existing alleles into new combinations can result from larger scale changes in chromosome structure or number and from several forms of genetic recombination, including crossing over between homologous chromosomes during meiosis, independent assortment of nonhomologous chromosomes during meiosis, and random fertilizations between genetically different sperm and eggs.

The shuffling of *existing* alleles into new combinations can produce an extraordinary number of novel genotypes in the next generation. By one estimate, more than 10^{600} combinations of alleles are possible in human gametes, yet there are fewer than 10^{10} humans alive today. So unless you have an identical twin, it is extremely unlikely that another person with your genotype has ever lived or ever will.

Populations Often Contain Substantial Genetic Variation

How much genetic variation actually exists within populations? In the 1960s, evolutionary biologists began to use gel electrophoresis (see Figure 18.7) to identify biochemical polymorphisms in organisms on many branches of the Tree of Life. This technique separates two or more forms of a given protein if they differ significantly in shape, mass, or net electrical charge. Thus, the identification of a protein polymorphism allowed researchers to infer genetic variation at the gene coding for that protein. Even though gel electrophoresis reveals only a portion of the existing genetic variation, it showed that nearly half the gene loci in many plant and animal populations were polymorphic, which was much more genetic variation in natural populations than anyone had anticipated.

DNA sequencing technologies now allow scientists to survey genetic variation directly and, as a result, researchers have accumulated an astounding knowledge of the sequences of genomes and the genes they contain. Research has shown that both coding and non-coding regions of DNA harbor extensive genetic variation. Indeed, *most* protein coding genes are polymorphic in their DNA sequences, having two or more alleles (variations in DNA sequences). This genetic variation may appear within heterozygous individuals as well as between

individuals in a single population, populations of the same species, and populations of related species. The study of DNA polymorphisms in humans, including single-nucleotide polymorphisms (SNPs, also called "snips") and short tandem repeat (STR) variants, is especially useful for understanding the genetic origins of certain diseases and in forensic applications (as described in Section 18.3).

21.2 Population Genetics

To predict how certain factors may influence genetic variation, population geneticists first describe the genetic structure of a population. They then create hypotheses, formalized in mathematical models, to describe how evolutionary processes may change the genetic structure under specified conditions. Finally, researchers test the predictions of these models to evaluate the ideas about evolution that are embodied within them.

All Populations Have a Genetic Structure

Populations are made up of individuals, each with its own genotype. In diploid organisms, which have pairs of homologous chromosomes, an individual's genotype includes two copies of every gene. The sum of all gene copies at all gene loci in all individuals is called the population's **gene pool.**

To describe the structure of a gene pool, scientists first identify the genotypes in a representative sample and calculate **genotype frequencies,** the percentages of individuals possessing each genotype. Knowing that each diploid organism has two copies of each gene (either two copies of the same allele or two different alleles), a scientist can then calculate **allele frequencies,** the relative abundances of the different alleles. For a gene with two alleles, scientists use the symbol p to identify the frequency of one allele, and q the frequency of the other.

The calculation of genotype and allele frequencies for the two alleles at the gene locus governing flower color in snapdragons (genus *Antirrhinum*) is straightforward **(Table 21.1).** This locus is easy to study because it exhibits incomplete

TABLE 21.1	Calculation of Genotype Frequencies and Allele Frequencies for the Snapdragon Flower Color Locus				
Because each diploid individual has two alleles at each gene locus, the entire sample of 1,000 individuals has a total of 2,000 alleles at the C locus.					
Flower Color Phenotype	Genotype	Number of Individuals	Genotype Frequency[1]	Total Number of C^R Alleles[2]	Total Number of C^W Alleles[2]
Red	C^RC^R	450	450/1,000 = 0.45	2 × 450 = 900	0 × 450 = 0
Pink	C^RC^W	500	500/1,000 = 0.50	1 × 500 = 500	1 × 500 = 500
White	C^WC^W	50	50/1,000 = 0.05	0 × 50 = 0	2 × 50 = 100
	Total	1,000	0.45 + 0.50 + 0.05 = 1.0	1,400	600

Calculate allele frequencies using the total of 1,400 + 600 = 2,000 alleles in the sample:

$$p = \text{frequency of } C^R \text{ allele} = 1,400/2,000 = 0.7$$
$$q = \text{frequency of } C^W \text{ allele} = 600/2,000 = 0.3$$
$$p + q = 0.7 + 0.3 = 1.0$$

[1]Genotype frequency = the number of individuals possessing a particular genotype divided by the total number of individuals in the sample.
[2]Total number of C^R or C^W alleles = the number of C^R or C^W alleles present in one individual with a particular genotype multiplied by the number of individuals with that genotype.

dominance (see Section 12.2). Individuals that are homozygous for the C^R allele (C^RC^R) have red flowers; those homozygous for the C^W allele (C^WC^W) have white flowers; and heterozygotes (C^RC^W) have pink flowers. Genotype frequencies represent how the C^R and C^W alleles are distributed among individuals. In this example, examination of the plants reveals that 45% of individuals have the C^RC^R genotype, 50% have the heterozygous C^RC^W genotype, and the remaining 5% have the C^WC^W genotype. Allele frequencies represent the commonness or rarity of each allele in the gene pool. As calculated in the table, 70% of the alleles in the population are C^R and 30% are C^W. Remember that for a gene locus with two alleles, there are three genotype frequencies, but only two allele frequencies (p and q). The sum of the three genotype frequencies must equal 1; so must the sum of the two allele frequencies.

The Hardy–Weinberg Principle Is a Null Model That Defines How Evolution Does Not Occur

When designing experiments, scientists often use control treatments to evaluate the effect of a particular factor: the control tells us what we would see if the experimental treatment had no effect. As you may recall from the hypothetical example presented in Chapter 1 (see Figure 1.15), to determine whether fertilizer has an effect on plant growth, you must compare the growth of fertilized plants (the experimental treatment) to the growth of plants that received no fertilizer (the control treatment). However, in studies that use observational rather than experimental data, there is often no suitable control. In such cases, investigators develop conceptual models, called **null models,** which predict what they would see if a particular factor had no effect. Null models serve as theoretical reference points against which observations can be evaluated.

Early in the twentieth century, geneticists were puzzled by the persistence of recessive traits because they assumed that natural selection replaced recessive or rare alleles with dominant or common ones. An English mathematician, G. H. Hardy, and a German physician, Wilhelm Weinberg, tackled this problem independently in 1908. Their analysis, now known as the **Hardy–Weinberg principle,** specifies the conditions under which a population of diploid organisms achieves **genetic equilibrium,** the point at which neither allele frequencies nor genotype frequencies change in succeeding generations. Their work also showed that dominant alleles need not replace recessive ones, and that the shuffling of genes in sexual reproduction does not in itself cause the gene pool to change.

The Hardy–Weinberg principle describes how genotype frequencies are established in sexually reproducing organisms. According to this mathematical model, genetic equilibrium is possible only if *all* of the following conditions are met:

1. No mutations are occurring.
2. The population is closed to migration from other populations.
3. The population is infinitely large.
4. All genotypes in the population survive and reproduce equally well.
5. Individuals in the population mate randomly with respect to genotypes.

If the conditions of the model are met, the allele frequencies of the population will never change, and the genotype frequencies will stop changing after one generation. In short, under these restrictive conditions, microevolution will *not* occur (see *Focus on Research: Basic Research* on pp. 482–483). The Hardy–Weinberg principle is thus a null model that serves as a reference point for evaluating the circumstances under which evolution *may* occur.

Basic Research: Using the Hardy–Weinberg Principle

To see how the Hardy–Weinberg principle can be applied, we will analyze the snapdragon flower color locus, using the hypothetical population of 1,000 plants described in Table 21.1. This locus includes two alleles—C^R (with its frequency designated as p) and C^W (with its frequency designated as q)—and three genotypes—homozygous $C^R C^R$, heterozygous $C^R C^W$, and homozygous $C^W C^W$. Table 21.1 lists the number of plants with each genotype: 450 have red flowers ($C^R C^R$), 500 have pink flowers ($C^R C^W$), and 50 have white flowers ($C^W C^W$). It also shows the calculation of both the genotype frequencies ($C^R C^R = 0.45$, $C^R C^W = 0.50$, and $C^W C^W = 0.05$) and the allele frequencies ($p = 0.7$ and $q = 0.3$) for the population.

1. Allele frequencies in parents and gametes

Let's assume for simplicity that each individual produces only two gametes and that both gametes contribute to the production of offspring. This assumption is unrealistic, of course, but it meets the Hardy–Weinberg requirement that all individuals in the population contribute equally to the next generation. In each parent, the two alleles segregate and end up in different gametes:

450 $C^R C^R$ individuals produce	$\rightarrow$	900 C^R gametes		
500 $C^R C^W$ individuals produce	$\rightarrow$	500 C^R gametes	+	500 C^W gametes
50 $C^W C^W$ individuals produce	$\rightarrow$	100 C^W gametes		

You can readily see that 1,400 of the 2,000 total gametes carry the C^R allele and 600 carry the C^W allele. The frequency of C^R gametes is 1,400/2,000 or 0.7, which is equal to p; the

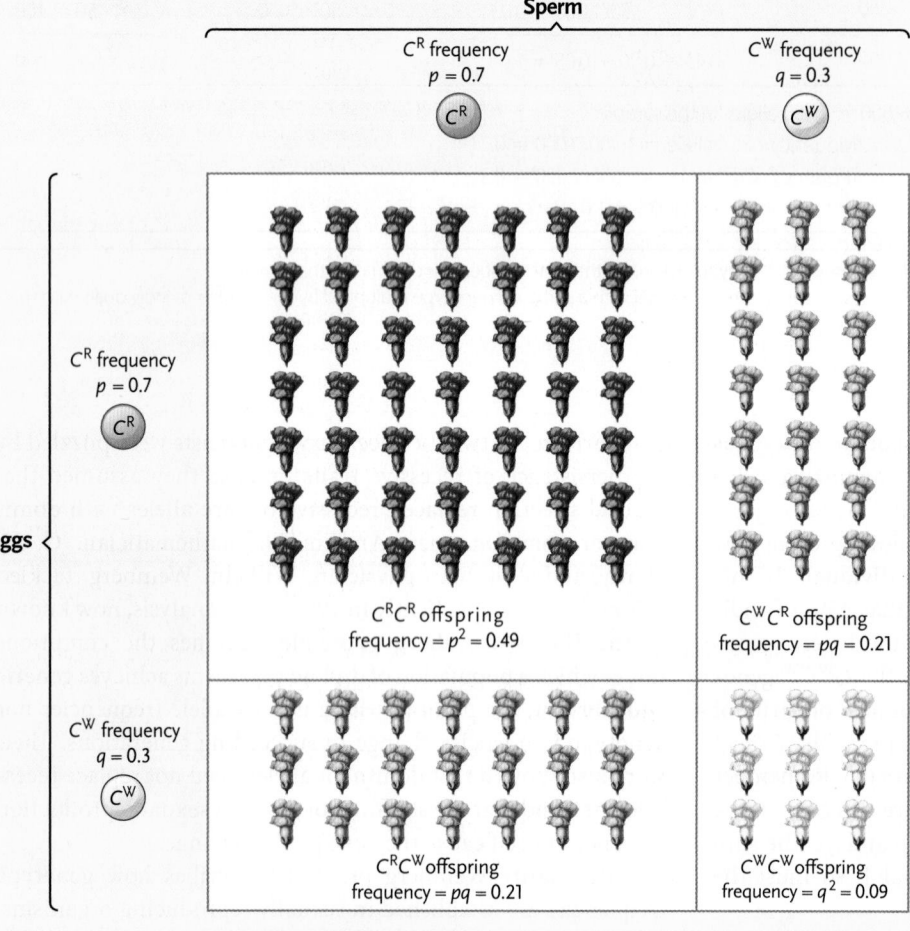

Sperm

C^R frequency $p = 0.7$ C^W frequency $q = 0.3$

Eggs

C^R frequency $p = 0.7$

C^W frequency $q = 0.3$

$C^R C^R$ offspring frequency $= p^2 = 0.49$

$C^W C^R$ offspring frequency $= pq = 0.21$

$C^R C^W$ offspring frequency $= pq = 0.21$

$C^W C^W$ offspring frequency $= q^2 = 0.09$

© Cengage Learning 2017

If a population's genotype frequencies do not match the predictions of this model or if its allele frequencies change over time, microevolution may be occurring. Determining which of the model's conditions are *not* met is a first step in understanding how and why the gene pool is changing.

STUDY BREAK 21.2

1. What is the difference between the genotype frequencies and the allele frequencies in a population?
2. Why is the Hardy–Weinberg principle considered a null model of evolution?
3. If the conditions of the Hardy–Weinberg principle are met, when will genotype frequencies stop changing?

21.3 The Agents of Microevolution

A population's allele frequencies will change over time if one or more conditions of the Hardy–Weinberg model are violated. In this section, we describe the processes that foster microevolutionary change—mutation, gene flow, genetic drift, natural selection, and nonrandom mating—which are summarized in **Table 21.2**.

Mutations Create New Genetic Variations

A **mutation** is a spontaneous and heritable change in DNA. Mutations are rare events: roughly one gamete in 100,000 to one in 1 million will include a new mutation at a particular gene locus. New mutations are so infrequent, in fact, that they

frequency of C^W gametes is 600/2,000 or 0.3, which is equal to q. Thus, the allele frequencies in the gametes are exactly the same as the allele frequencies in the parent generation—it could not be otherwise because each gamete carries one allele at each locus.

2. Genotype frequencies in offspring

Now assume that these gametes, both sperm and eggs, encounter each other at random. In other words, individuals reproduce without regard to the genotype of a potential mate. We can visualize the process of random mating in the mating table on the left.

We can also describe the consequences of random mating—$(p + q)$ sperm fertilizing $(p + q)$ eggs—with an equation that predicts the genotype frequencies in the offspring generation:

$$(p + q) \times (p + q) = p^2 + 2pq + q^2$$

If the population is at genetic equilibrium for this locus, p^2 is the predicted frequency of the $C^R C^R$ genotype; $2pq$, the predicted frequency of the $C^R C^W$ genotype; and q^2, the predicted frequency of the $C^W C^W$ genotype. Using the gamete frequencies determined above, we can calculate the predicted genotype frequencies in the next generation:

frequency of $C^R C^R$ = p^2 = (0.7×0.7) = 0.49
frequency of $C^R C^W$ =

$2pq = 2(0.7 \times 0.3) = 0.42$
frequency of $C^W C^W = q^2 = (0.3 \times 0.3) = 0.09$

Notice that the predicted genotype frequencies in the offspring generation have changed from the genotype frequencies in the parent generation: the frequency of heterozygous individuals has decreased, and the frequencies of both types of homozygous individuals have increased. This result occurred because the starting population was *not in equilibrium* at this gene locus. In other words, the distribution of parent genotypes did not conform to the predicted $p^2 + 2pq + q^2$ distribution.

The 2,000 gametes in our hypothetical population produced 1,000 offspring. Using the genotype frequencies we just calculated, we can predict how many offspring will carry each genotype:

490 red ($C^R C^R$)

420 pink ($C^R C^W$)

90 white ($C^W C^W$)

In a real study, we would examine the offspring to see how well their numbers match these predictions.

3. Allele frequencies in offspring

What about the allele frequencies in the offspring? The Hardy–Weinberg principle predicts that they did not change. Let's calculate them and see. Using the method shown in Table 21.1 and the prime symbol (') to indicate offspring allele frequencies:

$p' = ([2 \times 490] + 420)/2,000 =$

$1,400/2,000 = 0.7$
$q' = ([2 \times 90] + 420)/2,000 = 600/2,000 = 0.3$

You can see from this calculation that the allele frequencies did not change from one generation to the next, even though the alleles were rearranged to produce different proportions of the three genotypes.

4. Genetic equilibrium in future generations

The population is now at genetic equilibrium for the flower color locus; neither the genotype frequencies nor the allele frequencies will change in succeeding generations as long as the population meets the conditions specified in the Hardy–Weinberg model. To verify this, you can calculate the allele frequencies of the gametes for this offspring generation and predict the genotype frequencies and allele frequencies for a third generation. You could continue calculating until you ran out of either paper or patience, but these frequencies will not change.

Researchers use calculations like these to determine whether an actual population is near its predicted genetic equilibrium for one or more gene loci. When they discover that a population is not at equilibrium, they infer that microevolution is occurring and can investigate the factors that might be responsible.

exert little or no immediate effect on allele frequencies in most populations. But over evolutionary time scales, their numbers are significant—mutations have been accumulating in biological lineages for billions of years. And because it creates entirely new genetic variations, *mutation is a major source of heritable variation.*

For most animals, only mutations in the germ line (the cell lineage that produces gametes) are heritable; mutations in other cell lineages have no direct effect on the next generation. In plants, however, mutations may occur in meristem cells, which eventually produce flowers as well as nonreproductive structures (see Chapter 33); in such cases, a mutation may be passed to the next generation and ultimately influence the gene pool.

Deleterious mutations alter an individual's structure, function, or behavior in harmful ways. In mammals, for example, a

protein called collagen (see Chapter 38) is an essential component of most extracellular structures. Several simple mutations in humans cause forms of Ehlers–Danlos syndrome, a disruption of collagen synthesis that may result in loose skin, weak joints, or sudden death from the rupture of major blood vessels, the colon, or the uterus.

Lethal mutations can cause great harm to organisms carrying them. If a lethal allele is dominant, both homozygous and heterozygous carriers will, by definition, die from its effects; if recessive, it kills only homozygous recessive individuals. Any lethal mutation that causes an individual to die before it reproduces is eliminated from the population.

Neutral mutations are neither harmful nor helpful. Because of the degeneracy of the genetic code, several codons with different nucleotides in the third position may specify the same

TABLE 21.2	Agents of Microevolutionary Change			
Agent	Definition	Effect on Genetic Variation	Effect on Average Fitness
Mutation	Heritable change in DNA	Introduces new genetic variation into population; does not change allele frequencies quickly	Unpredictable effect on fitness; most mutations in protein-coding genes lower fitness
Gene flow	Change in allele frequencies as individuals join a population and reproduce	May introduce genetic variation from another population	Unpredictable effect on fitness; may introduce beneficial or harmful alleles
Genetic drift	Random changes in allele frequencies caused by chance events	Reduces genetic variation, especially in small populations; can eliminate rare alleles	Unpredictable effect on fitness; often harmful because of lost genetic diversity
Natural selection	Differential survivorship or reproduction of individuals with different genotypes	One allele can replace another or allelic variation can be preserved	Positive effect on fitness through evolution of adaptations
Nonrandom mating	Choice of mates based on their phenotypes and genotypes	Does not directly affect allele frequencies, but usually prevents genetic equilibrium	May have negative effect on fitness through the expression of recessive phenotypes

amino acid in the construction of a polypeptide chain (see Section 15.1). As a result, some DNA sequence changes—especially those in the third nucleotide of the codon—do not alter the amino acid sequence of the protein under construction. Not surprisingly, mutations at the third position appear to persist longer in populations than those at the first two positions. In other instances, mutations that change the amino acid sequence in a protein or even an organism's phenotype may have no influence on its survival and reproduction. A neutral mutation might even prove to be beneficial later if the environment changes.

Sometimes a change in DNA produces an *advantageous mutation,* which confers some benefit on an individual that carries it. However slight the advantage, natural selection may preserve the new allele and even increase its frequency over time. Once the mutation has been passed to a new generation, other agents of microevolution determine its long-term fate.

Gene Flow Introduces Novel Genetic Variants into Populations

Organisms or their gametes (for example, pollen) sometimes move from one population to another. If the immigrants reproduce, they may introduce novel alleles into a population, shifting its allele and genotype frequencies away from the values predicted by the Hardy–Weinberg model. This phenomenon, called **gene flow,** violates the Hardy–Weinberg requirement that populations must be closed to migration.

Gene flow is common in some animal species. For example, young male baboons typically move from one local population to another after experiencing aggressive behavior by older males. And many marine invertebrates disperse long distances as larvae carried by ocean currents.

Dispersal agents, such as pollen-carrying wind or seed-carrying animals, are responsible for gene flow in most plant populations. For example, blue jays foster gene flow among populations of oaks by carrying acorns from nut-bearing trees to their winter caches, which may be as much as a mile away **(Figure 21.7).** Transported acorns that go uneaten may germinate and contribute to the gene pool of a neighboring oak population.

The movement of individuals from one population to another is not sufficient to foster gene flow between two populations. The immigrants must also reproduce in the population they join, thereby contributing to its gene pool. In the San Francisco Bay area, for example, Bay checkerspot butterflies

FIGURE 21.7 Gene flow. Blue jays *(Cyanocitta cristata)* serve as agents of gene flow for oaks (genus *Quercus*) when the birds carry acorns from one oak population to another. An uneaten acorn may germinate and contribute to the gene pool of the population into which it was carried.

(*Euphydryas editha bayensis*) rarely move from one population to another because they are poor fliers (see Figure 52.19). But when adult females do change populations, it is often late in the breeding season, and their offspring have virtually no chance of finding enough food to mature. Thus, many immigrant females do not foster gene flow because they do not contribute to the gene pool of the population they join.

The evolutionary importance of gene flow depends on the degree of genetic differentiation between populations and the rate of gene flow between them. If two gene pools are very different, a little gene flow may increase genetic variability within the population that receives immigrants, and it will make the two populations more similar. But if populations are already genetically similar, even lots of gene flow will have little effect.

Genetic Drift Reduces Genetic Variability within Populations

Chance events sometimes cause the allele frequencies in a population to change unpredictably. This phenomenon, known as **genetic drift,** has especially dramatic effects on small populations, which clearly violate the Hardy–Weinberg assumption of infinitely large population size.

A simple analogy clarifies why genetic drift is more pronounced in small populations than in large ones. When individuals reproduce, male and female gametes often pair up randomly, as though the allele in any particular sperm or ovum was determined by a coin toss. Imagine that "heads" specifies the *R* allele and that "tails" specifies the *r* allele. If the two alleles are equally common (that is, their frequencies, *p* and *q*, are both equal to 0.5), heads should be as likely an outcome as tails. But if you toss the coin 20 or 30 times to simulate random mating in a small population, you won't often see a 50:50 ratio of heads and tails. Sometimes heads will predominate and sometimes tails will—just by chance. Tossing the coin 500 times to simulate random mating in a somewhat larger population is more likely to produce a 50:50 ratio of heads and tails. And if you tossed the coin 5,000 times, you would get even closer to a 50:50 ratio.

Chance deviations from expected results—which cause genetic drift—occur whenever organisms engage in sexual reproduction, simply because their population sizes are not infinitely large. But genetic drift is particularly common in small populations because only a few individuals contribute to the gene pool and because any given allele may be present in very few individuals.

Genetic drift generally leads to reduced genetic variability through the loss of rare alleles; it therefore causes genotype and allele frequencies to differ from those predicted by the Hardy–Weinberg model. Two general circumstances, population bottlenecks and founder effects, often foster genetic drift.

POPULATION BOTTLENECKS On occasion, a stressful factor such as disease, starvation, or drought kills a large proportion of the individuals in a population, producing a population bottleneck, a dramatic reduction in population size. This cause of genetic drift greatly reduces genetic variation even if the population numbers later rebound (**Figure 21.8**).

In the late nineteenth century, for example, hunters nearly wiped out northern elephant seals (*Mirounga angustirostris*) along the Pacific coast of North America. Since the 1880s, when the species received protected status, the population has increased to more than 30,000, all descended from a group of about 20 survivors. Today the population exhibits no variation in 24 proteins studied by gel electrophoresis. This low level of genetic variation, which is unique among seal species, is consistent with the hypothesis that genetic drift eliminated many alleles when the population experienced the bottleneck.

FOUNDER EFFECT When a few individuals colonize a distant locality and start a new population, they carry only a small sample of the parent population's genetic variation. By chance, some alleles may be totally missing from the new population, whereas other alleles that were rare "back home" might occur at relatively high frequencies. This change in the gene pool is called the **founder effect.**

The human medical literature provides some of the best-documented examples of the founder effect. The Old Order Amish, an essentially closed religious community in Lancaster County, Pennsylvania, have an exceptionally high incidence of Ellis–van Creveld syndrome, a genetic disorder caused by a recessive allele. In the homozygous state, the allele produces dwarfism, shortened limbs, and polydactyly (extra fingers). Genetic analysis suggests that, although this syndrome affects

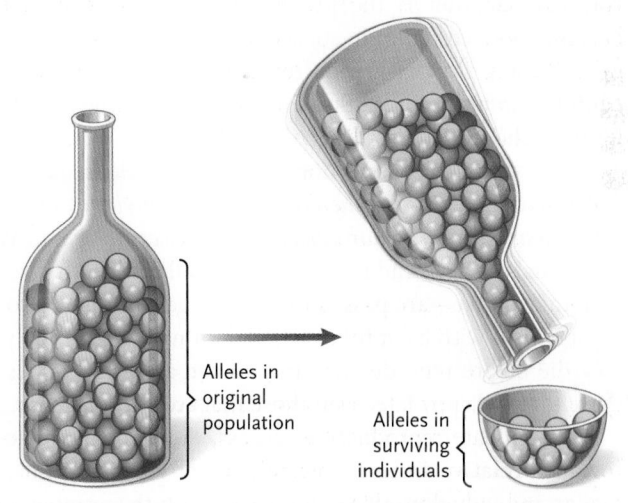

Alleles in original population

Alleles in surviving individuals

The gene pool of the original population, represented by a bottle filled with colored marbles, includes a locus with three alleles. Two of the alleles, represented by blue and green marbles, occur at high frequency; the third allele, represented by red marbles, occurs at low frequency.

A drastic reduction in population size is described as a population bottleneck. The process is analogous to shaking only a few of the marbles—the survivors—through the neck of the bottle. As a consequence of chance events associated with population bottlenecks, surviving individuals may not have the same allele frequencies as the original population. Rare alleles are inevitably lost.

FIGURE 21.8 Population bottlenecks, genetic drift, and the loss of genetic variability.
© Cengage Learning 2017

less than 1% of the Amish in Lancaster County, as many as 13% may be heterozygous carriers of the allele. All of the individuals exhibiting the syndrome are descended from one couple who helped found the community in the mid-1700s.

CONSERVATION IMPLICATIONS Genetic drift has important implications for conservation biology. By definition, endangered species experience severe population bottlenecks, which result in the loss of genetic variability. Moreover, the small number of individuals available for captive breeding programs may not fully represent a species' genetic diversity. Without such variation, no matter how large a population may become in the future, it will be less resistant to diseases or less able to cope with environmental change.

For example, scientists believe that an environmental catastrophe produced a bottleneck in the African cheetah (*Acinonyx jubatus*) population 10,000 years ago. Cheetahs today are remarkably uniform in genetic makeup. Their populations are highly susceptible to diseases; males also have a high proportion of sperm cell abnormalities and a reduced reproductive capacity. Thus, limited genetic variation, as well as small numbers, threatens populations of endangered species.

Natural Selection Shapes Genetic Variability by Favoring Some Traits over Others

The Hardy–Weinberg model requires all genotypes in a population to survive and reproduce equally well. But as you know from reading Section 20.2, heritable traits enable some individuals to survive better and reproduce more than others. **Natural selection** is the process by which successful traits become more common in subsequent generations. Thus, natural selection violates a requirement of the Hardy–Weinberg equilibrium and causes allele and genotype frequencies to differ from those predicted by the model.

Although natural selection can change allele frequencies in a population, *it is the phenotype of an individual organism, rather than any particular allele, that is successful or not.* When individuals survive and reproduce, their alleles—both favorable and unfavorable—are passed to the next generation. Of course, an organism with harmful or lethal dominant alleles will probably die before reproducing, and all the alleles it carries will share that unhappy fate, even those that are advantageous.

To evaluate reproductive success, evolutionary biologists consider **relative fitness,** the number of surviving offspring that an individual produces compared with the number left by others in the population. Thus, a particular allele will increase in frequency in the next generation if individuals carrying that allele leave *more* offspring than individuals carrying other alleles. A difference in the *relative* success of individuals is the essence of natural selection.

Natural selection tests fitness differences at nearly every stage of an organism's life cycle. One plant may be fitter than others in the population because its seeds survive colder conditions, because the arrangement of its leaves captures sunlight more efficiently, or because its flowers are more attractive to pollinators.

However, natural selection exerts little or no effect on traits that appear during an individual's postreproductive life. For example, Huntington disease, a dominant-allele disorder that first strikes humans after the age of 40, is not subject to strong selection. Carriers of the disease-causing allele can reproduce before the onset of the condition, passing it to the next generation.

Biologists measure the effects of natural selection on phenotypic variation by recording changes in the mean and variability of characters over time (see Figure 21.3). Three modes of natural selection have been identified: directional selection, stabilizing selection, and disruptive selection **(Figure 21.9).**

DIRECTIONAL SELECTION Traits undergo **directional selection** when individuals near one end of the phenotypic spectrum have the highest relative fitness. Directional selection shifts a trait away from the existing mean and toward the favored extreme (see Figure 21.9A). After selection, the trait's mean value is higher or lower than before, and variability in the trait may be reduced.

Directional selection is extremely common. For example, predatory fish promote directional selection for larger body size in guppies when they selectively feed on the smallest individuals in a guppy population (see *Focus on Research: Basic Research* in Chapter 52). And most cases of artificial selection, including the experiment on the activity levels of house mice described above, are directional, aimed at increasing or decreasing specific phenotypic traits. Humans routinely use directional selection to produce domestic animals and crops with desired characteristics, such as the small size of Chihuahuas and the intense "bite" of chili peppers.

STABILIZING SELECTION Traits undergo **stabilizing selection** when individuals expressing intermediate phenotypes have the highest relative fitness (see Figure 21.9B). By eliminating phenotypic extremes, stabilizing selection reduces genetic and phenotypic variation and increases the frequency of intermediate phenotypes. Stabilizing selection is probably the most common mode of natural selection, affecting many familiar traits. For example, very small and very large human newborns are less likely to survive than those born at an intermediate weight **(Figure 21.10).**

Warren G. Abrahamson and Arthur E. Weis of Bucknell University have shown that opposing forces of directional selection can sometimes produce an overall pattern of stabilizing selection **(Figure 21.11).**

The gallmaking fly (*Eurosta solidaginis*) is a small insect that feeds on the tall goldenrod plant (*Solidago altissima*). When a fly larva hatches from its egg, it bores into a goldenrod stem, and the plant responds by producing a spherical growth deformity called a gall. The larva feeds on plant tissues inside the gall. Galls vary dramatically in size; genetic experiments indicate that gall size is a heritable trait of the fly, although plant genotype also has an effect.

Fly larvae inside galls are subjected to two opposing patterns of directional selection. On one hand, a tiny wasp (*Eurytoma gigantea*) parasitizes gallmaking flies by laying eggs in fly

A hypothetical example using tail length of birds as the quantitative trait subject to selection.

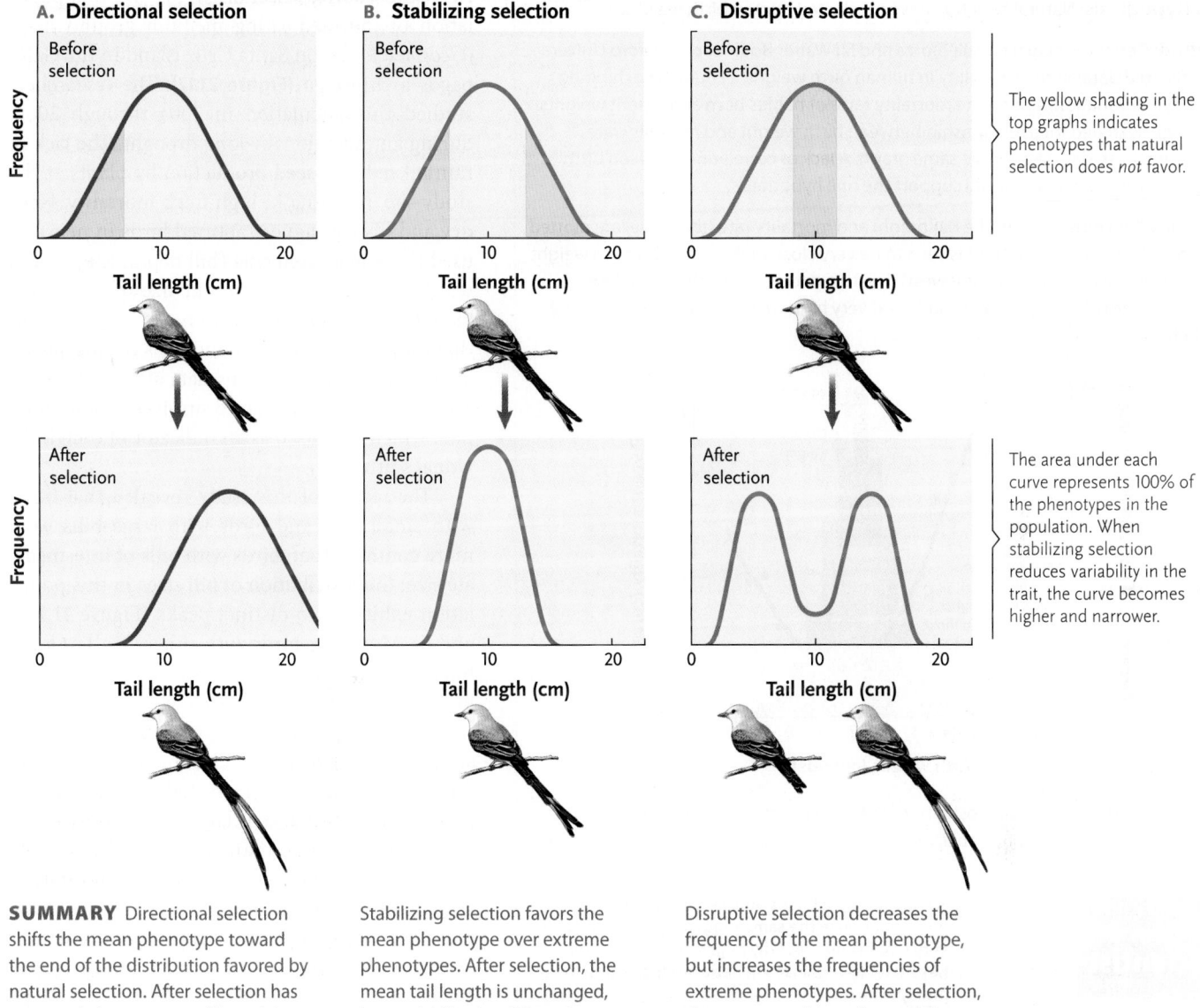

A. Directional selection

Before selection

Frequency / Tail length (cm)

B. Stabilizing selection

Before selection

Frequency / Tail length (cm)

C. Disruptive selection

Before selection

Frequency / Tail length (cm)

The yellow shading in the top graphs indicates phenotypes that natural selection does *not* favor.

After selection

Frequency / Tail length (cm)

After selection

Frequency / Tail length (cm)

After selection

Frequency / Tail length (cm)

The area under each curve represents 100% of the phenotypes in the population. When stabilizing selection reduces variability in the trait, the curve becomes higher and narrower.

SUMMARY Directional selection shifts the mean phenotype toward the end of the distribution favored by natural selection. After selection has operated, the average tail length in the population has increased.

Stabilizing selection favors the mean phenotype over extreme phenotypes. After selection, the mean tail length is unchanged, but variability in tail length among individuals in the population has decreased.

Disruptive selection decreases the frequency of the mean phenotype, but increases the frequencies of extreme phenotypes. After selection, the mean may be unchanged, but the variability among individuals has increased.

think like a scientist When people used artificial selection to produce the long body and short legs of a dachshund, which mode of selection did they use?

© Cengage Learning 2017

larvae inside their galls. After hatching, the young wasps feed on the fly larvae, killing them in the process. However, adult wasps are so small that they cannot easily penetrate the thick walls of a large gall; they generally lay eggs in fly larvae occupying small galls. Thus, wasps establish directional selection favoring flies that produce large galls, which are less likely to be parasitized. On the other hand, several bird species open galls to feed on mature fly larvae; these predators preferentially open large galls, fostering directional selection in favor of small galls.

In about one-third of the populations surveyed in central Pennsylvania, wasps and birds attacked galls with equal frequency, and flies producing galls of intermediate size had the

FIGURE 21.10 **Observational Research**

Do Humans Experience Stabilizing Selection?

Hypothesis: Human birth weight has been adjusted by natural selection.

Null Hypothesis: Natural selection has not affected human birth weight.

Method: Geneticists Luigi Cavalli-Sforza and Sir Walter Bodmer of Stanford University collected data on the variability in human birth weight, a character exhibiting quantitative variation, and on the mortality rates of babies born at different weights. They then searched for a relationship between birth weight and mortality rate by plotting both data sets on the same graph. A lack of correlation between birth weight and mortality rate would support the null hypothesis.

Results: When birth weight (the bar graph) and mortality rate (the curve) are plotted together, the mean birth weight is seen to be very close to the optimum birth weight (the weight at which mortality is lowest). The two data sets also show that few babies are born at the very low weights and very high weights associated with high mortality.

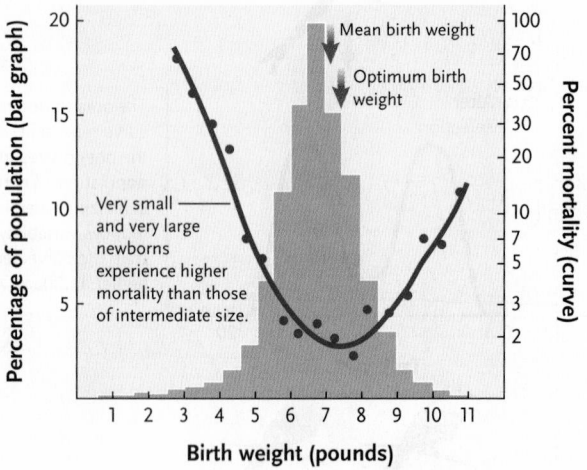

Conclusion: The shapes and positions of the birth weight bar graph and the mortality rate curve suggest that stabilizing selection has adjusted human birth weight to an average of 7 to 8 pounds.

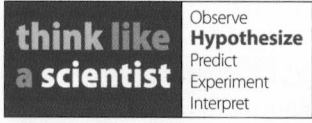

Observe	Having multiple births (that is, carrying three or
Hypothesize	more children in the same pregnancy) is historically
Predict	very rare in humans. Why might natural selection
Experiment	have selected against a genetic predisposition for
Interpret	multiple births?

Source: Based on L. L. Cavalli-Sforza and W. F. Bodmer. 1971. *The Genetics of Human Populations.* Freeman, San Francisco.

© Cengage Learning 2017

highest survival rate. The smallest and largest galls—as well as the genetic predisposition to make very small or very large galls—were eliminated from the population.

DISRUPTIVE SELECTION Traits undergo **disruptive selection** when extreme phenotypes have higher relative fitness than intermediate phenotypes (see Figure 21.9C). Thus, alleles producing extreme phenotypes become more common, promoting polymorphism. Under natural conditions, disruptive selection

is much less common than directional selection and stabilizing selection.

Andrew P. Hendry of McGill University and colleagues at the University of Massachusetts, Amherst and Harvard University analyzed a likely case of disruptive selection on bill size in a population of the seed-eating medium ground finch (*Geospiza fortis*) on Santa Cruz island in the Galápagos archipelago **(Figure 21.12).** The researchers studied this population in 2004 through 2006, during an exceptionally long drought. The lack of rainfall reduced seed production by plants at the study site, resulting in high finch mortality. Hendry and his colleagues captured birds in nets and used three measurements (bill depth, length, and width) to characterize the bill size of each individual. They also put rings on the birds' legs so that they could identify individuals during subsequent censuses. The census data allowed them to estimate how long each bird survived on the study plot, which they used as an indicator of each individual's fitness.

The results of this study revealed that birds with small bills and birds with large bills were more common than birds with bills of intermediate size: the distribution of bill sizes in this population exhibits two distinct peaks (Figure 21.12A and B). Moreover, birds with either small or large bills survived longer on the study plot, and thus had higher fitness, than birds with intermediate sized bills (Figure 21.12C). These results strongly suggest that disruptive selection is responsible for the polymorphism in bill size. Previous research had shown that large-billed individuals have a stronger bite than small-billed individuals, allowing them to feed on larger and harder seeds; thus, when the drought reduced food supplies, the two groups of birds probably relied on different resources.

Sexual Selection Often Exaggerates Showy Structures in Males

Darwin hypothesized that a special process, which he called **sexual selection,** has fostered the evolution of showy structures—such as brightly colored feathers, long tails, or impressive antlers—as well as elaborate courtship behavior in the males of many animal species. Sexual selection encompasses two related processes. As the result of *intersexual selection* (that is, selection based on the interactions between males and females), males produce these otherwise useless structures simply because females find them irresistibly attractive. Under *intrasexual selection* (that is, selection based on the interactions between members of the same sex), males

FIGURE 21.11   **Observational Research**

How Opposing Forces of Directional Selection Produce Stabilizing Selection

Hypothesis: The size of galls made by larvae of the gallmaking fly *(Eurosta solidaginis)* is governed by conflicting selection pressures established by parasitic wasps and predatory birds.

Prediction: If parasitic wasps and predatory birds exert selection for different sizes of galls made by gallmaking fly larvae, fly populations that suffer from wasp parasitism and bird predation will make galls of intermediate size.

Method: Abrahamson and his colleagues surveyed galls made by the larvae of the gallmaking fly in Pennsylvania. They measured the diameters of the galls they encountered, and, for those galls in which the larvae had died, they determined whether they had been killed by a parasitic wasp or a predatory bird, such as the downy woodpecker.

A. Goldenrod *(Solidago altissima)* gall

B. Gallmaking fly *(Eurosta solidaginis)* inside a gall

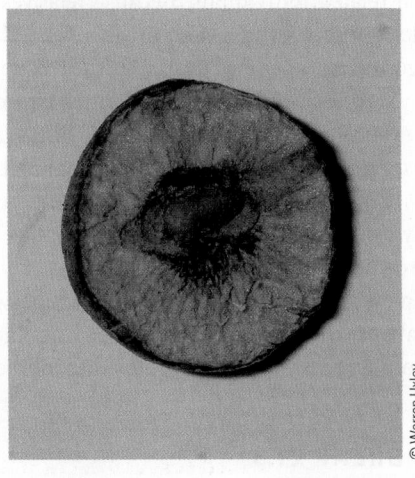

C. Fly larvae parasitized by wasps *(Eurytoma gigantea,* inset) versus gall size

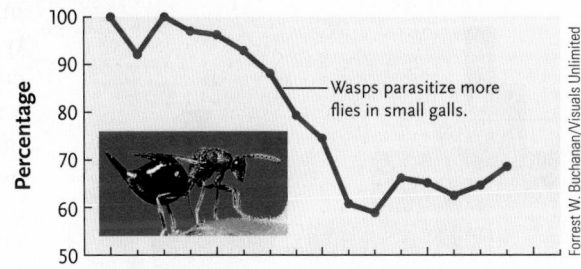

Wasps parasitize more flies in small galls.

D. Fly larvae eaten by birds *(Picoides pubescens,* inset) versus gall size

Birds consume more flies in large galls.

E. Fly larvae alive or killed versus gall size

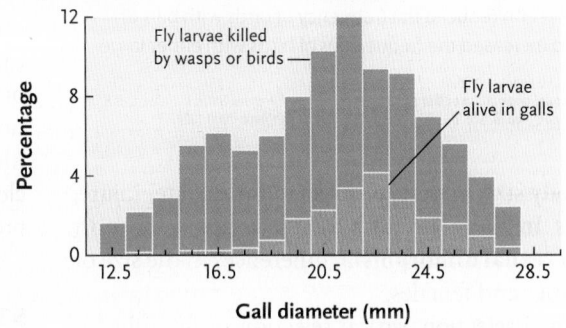

Fly larvae killed by wasps or birds

Fly larvae alive in galls

Gall diameter (mm)

Results: Tiny wasps are more likely to parasitize gallmaking fly larvae inside small galls (C), fostering directional selection in favor of large galls. By contrast, birds usually feed on fly larvae inside large galls (D), fostering directional selection in favor of small galls. These opposing patterns of directional selection create stabilizing selection for the size of galls that the fly larvae make (E).

Conclusion: Because wasps preferentially parasitize fly larvae in small galls and birds preferentially eat fly larvae in large galls, the opposing forces of directional selection establish an overall pattern of stabilizing selection in favor of medium-sized galls.

 Observe **Hypothesize** Predict Experiment Interpret

Human activities are causing many North American bird populations to decline in numbers. What effect would declining bird populations be likely to have on the sizes of galls made by gallmaking flies?

Source: A. E. Weis and W. G. Abrahamson. 1986. Evolution of host-plant manipulation by gall makers: Ecological and genetic factors in the *Solidago—Eurosta* system. *The American Naturalist* 127:681–695. *The American Naturalist* by Essex Institute; American Society of Naturalists. Copyright 1986. Reproduced with permission of University of Chicago Press—Journals in the format Textbook via Copyright Clearance Center.

**A. Small-billed and large-billed medium ground finches
(Geospiza fortis) from Santa Cruz island**

Dr. Andrew Hendry

B. Distribution of bill sizes for birds marked in 2004

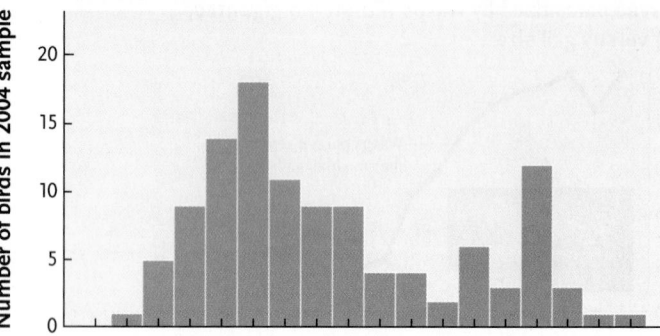

C. Fitness (survival on the study plot, 2004–2006)

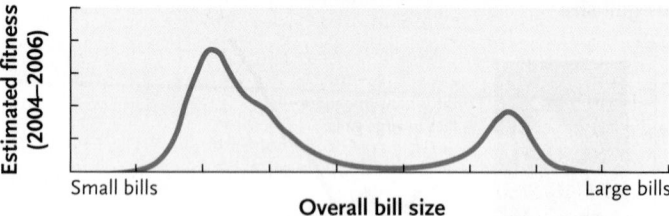

FIGURE 21.12 Disruptive selection. (A) Medium ground finches *(Geospiza fortis)* on Santa Cruz island have bills of varying sizes. In this bill size polymorphism, birds with small bills or large bills are more common **(B)** and have higher fitness **(C)** than birds with bills of intermediate size. Thus, natural selection reduced the frequency of birds with bills of intermediate size and increased the frequencies of birds with either large bills or small bills.

Disruptive selection in a bimodal population of Darwin's finches. Andrew P. Hendry, Sarah K. Huber, Luis F. De León, Anthony Herrel, Jeffrey Podos. Proc. R. Soc. B 2009 276 753–759; DOI:10.1098/rspb.2008.1321.

use their large body size, antlers, or tusks to intimidate, injure, or kill rival males. In many species, sexual selection is the most probable cause of **sexual dimorphism,** differences in the size or appearance of males and females.

Like directional selection, sexual selection pushes phenotypes toward one extreme. But the products of sexual selection are sometimes bizarre—such as the ridiculously long tail feathers of male African widowbirds. How could evolutionary processes favor such elaborate structures, which are costly to produce? Research by Malte Andersson of the University of Gothenburg, Sweden, suggests that the males' long tail feathers are a product of intersexual selection because females are more strongly attracted to males with long tails than to males with short tails, while tail length had no effect on a male's ability to compete with other males for space in the habitat **(Figure 21.13).**

Molecular Insights describes a study of the genetic and molecular mechanisms that foster the development of exaggerated horns in male rhinoceros beetles. Behavioral aspects of sexual selection are described further in Chapter 56.

Nonrandom Mating Can Influence Genotype Frequencies

The Hardy–Weinberg model requires individuals to select mates randomly with respect to their genotypes. This requirement is, in fact, often met; humans, for example, generally marry one another in total ignorance of their genotypes for digestive enzymes or blood types.

Nevertheless, many organisms mate nonrandomly, selecting a mate with a particular phenotype and underlying genotype. Snow geese, for example, usually select mates of their own color, and a tall woman is more likely to marry a tall man than a short man. If no one phenotype is preferred by all potential mates, nonrandom mating does not change allele frequencies by establishing selection for one phenotype over another. But because individuals with similar genetically based phenotypes mate with each other, the next generation will contain fewer heterozygous offspring—and more homozygous offspring—than the Hardy–Weinberg model predicts.

Inbreeding is a special form of nonrandom mating in which genetically related individuals mate with each other. Self-fertilization in plants (see Chapter 36) and a few animals (see Chapter 49) is an extreme example of inbreeding because offspring are produced from the gametes of a single parent. However, other organisms that live in small, relatively closed populations often mate with related individuals. Because relatives often carry the same alleles, inbreeding generally increases the frequency of homozygous genotypes and decreases the frequency of heterozygotes. Thus, recessive phenotypes are often expressed.

For example, the high incidence of Ellis–van Creveld syndrome among the Old Order Amish population, mentioned earlier, is caused by inbreeding. Although the founder effect originally established the disease-causing allele in this population, inbreeding increases the likelihood that it will be expressed. Most human societies discourage matings between genetically close relatives, thereby reducing inbreeding and the inevitable production of recessive homozygotes that inbreeding causes.

STUDY BREAK 21.3

1. Which agents of microevolution tend to increase genetic variation within populations, and which ones tend to decrease it?
2. Which mode of natural selection increases the representation of the average phenotype in a population?
3. In what way is sexual selection like directional selection?

THINK OUTSIDE THE BOOK

Search the Internet for examples of diseases or unusual phenotypes caused by inbreeding in specific human populations. Have human geneticists discovered the molecular basis of these conditions?

My, What a Big Horn You Have: Does the size of a male's ornaments provide a true indicator of his success and the relative quality of his genes?

In many animal species males have greatly enlarged structures—horns, antlers, or showy feathers—that they use as a signal to females, advertising their good health and vigor. In general, healthier males have disproportionately larger ornaments than unhealthy males, and females often mate with the male who displays the most exaggerated expression of the trait.

Research Question

Are the enlarged ornaments of males true indicators of their good health and vigor? In other words, does the choice of a mate based on the size of his ornament improve a female's chances of choosing a successful male among those available to father her offspring?

Experiment

Douglas J. Emlen at the University of Montana, working with colleagues at his institution and at Michigan State University and Washington State University, studied the growth of the horn on the Japanese rhinoceros beetle (*Trypoxylus dichotomus*) by manipulating the molecular pathway through which cells of a developing beetle respond to insulin and insulin-like growth factors (IGFs; see Chapter 42). Molecular signals in the insulin/IGF pathway stimulate tissue growth in many animal species by mediating the uptake of nutrients, and healthy, well-fed individuals have higher concentrations in their body fluids than do individuals that suffer from parasites, infections, or malnutrition.

Larger males of this beetle species have proportionately larger wings and genitalia than small males, but horn size is extremely variable among individuals. Some males have just a small bump on their heads, whereas others have elaborate, forked structures that are two-thirds the length of the animal's body (**Figure,** part A). The researchers wondered if this natural variation in horn size reflected differences in the health and overall condition of males and if the insulin/IGF pathway mediated variations in horn growth during the beetle's development.

To test whether rhinoceros beetle horns were more sensitive than wings or genitalia to insulin/IGF signaling during their development, they used RNA interference (see Section 16.3) to knock down (that is, disrupt transcription of) the insulin receptor gene. By disrupting the production of insulin receptors, the researchers effectively blocked the insulin/IGR pathway. They injected a double-stranded RNA copy of a 398-base-pair fragment of the beetle's insulin receptor gene into beetle larvae as they were about to begin metamorphosis. Because the beetles were already at their adult size, the knockdown of insulin receptors would affect only those structures that grow during metamorphosis (horns, wings, and genitalia). The researchers hypothesized that the exaggerated horn size and the variability in horn size in this species reflects its hypersensitivity to the insulin/IGF pathway.

Results

The administration of double-stranded RNA resulted in a substantial reduction in the synthesis of insulin receptor in developing wings, genitalia, and horns. After the beetles completed metamorphosis into adults, the researchers measured these structures and graphed their sizes against body size in both untreated controls and experimental animals. The knockdown of the synthesis of the insulin receptors had no effect on the size of genitalia (**Figure,** part B) and only a small effect on the size of wings (**Figure,** part C). But the treatment had a dramatic effect on horn size (**Figure,** part D): experimental animals had horns that were 16% smaller than those of controls.

Conclusion

The results suggest that the horns of rhinoceros beetles are highly sensitive to concentrations of insulin/IGF growth regulators, which mediate the uptake of nutrients by growing cells. In the wild, any male beetle with a large horn must have been in a healthy, well-fed state when his horn developed. Thus, the size of the horn is an honest indicator of its ability to harvest resources and resist disease and parasites. If those characteristics have a genetic basis, his offspring are likely to inherit them.

think like a scientist

On what developmental mechanism did sexual selection apparently act in the evolution of exaggerated horn size in male rhinoceros beetles?

Source: D. J. Emlen et al. 2012. A Mechanism of Extreme Growth and Reliable Signaling in Sexually Selected Ornaments and Weapons. *Science* 337:860–864.

A. Male rhinoceros beetle (*Trypoxylus dichotomus*)

© NH/Shutterstock.com

B. Genitalia

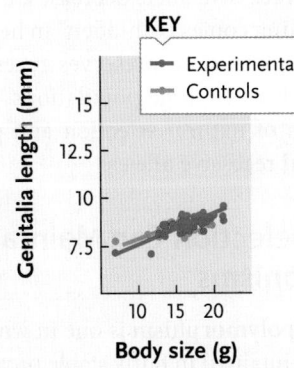

KEY
- Experimentals
- Controls

Genitalia length (mm) vs Body size (g)

C. Wings

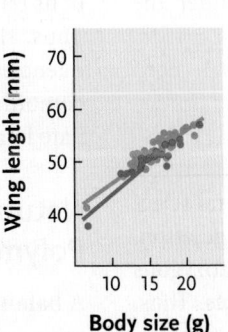

Wing length (mm) vs Body size (g)

D. Horns

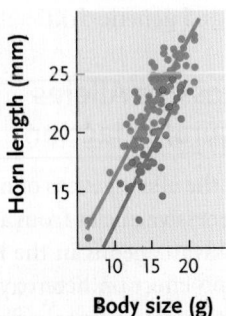

Horn length (mm) vs Body size (g)

E. Effect of treatment on horn size

Control Experimental

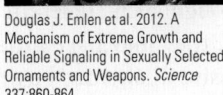

Douglas J. Emlen et al. 2012. A Mechanism of Extreme Growth and Reliable Signaling in Sexually Selected Ornaments and Weapons. *Science* 337:860-864.

© Cengage Learning 2017

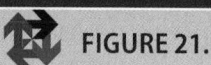 **FIGURE 21.13** | **Experimental Research**

Sexual Selection in Action

Question: Is the long tail of the male long-tailed widowbird *(Euplectes progne)* the product of intrasexual selection, intersexual selection, or both?

Experiment: Andersson counted the number of females that associated with individual male widowbirds in the grasslands of Kenya. He then shortened the tails of some males by cutting the feathers, lengthened the tails of others by gluing feather extensions to their tails, and left a third group essentially unaltered as a control. One month later, he again counted the number of females associating with each male and compared the results from the three groups.

Results: Males with experimentally lengthened tails attracted more than twice as many mates as males in the control group, and males with experimentally shortened tails attracted fewer. Andersson observed no differences in the ability of altered males and control group males to maintain their display areas.

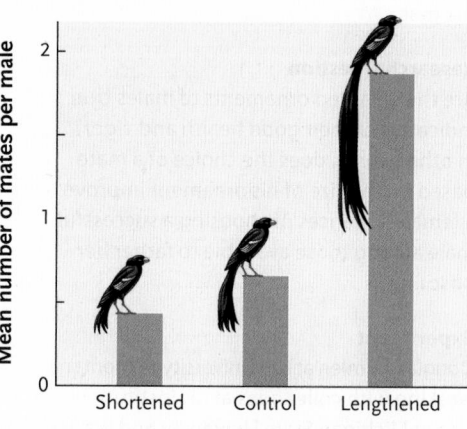

Conclusion: Female widowbirds clearly prefer males with experimentally lengthened tails to those with normal tails or experimentally shortened tails. Tail length had no obvious effect on the interactions between males. Thus, the long tail of male widowbirds is the product of intersexual selection.

 | Observe / Hypothesize / Predict / Experiment / **Interpret** | **Which experimental treatment had the greater effect on the number of females with which male widowbirds mated?**

Source: M. Andersson. 1982. Female choice selects for extreme tail length in a widowbird. *Nature* 299:818–820.

21.4 Maintaining Genetic and Phenotypic Variation

Evolutionary biologists continue to discover extraordinary amounts of genetic and phenotypic variation in most natural populations. How can so much variation persist under the action of stabilizing selection and genetic drift?

Diploidy Can Hide Recessive Alleles from the Action of Natural Selection

The diploid condition reduces the effectiveness of natural selection in eliminating harmful recessive alleles from a population. Although such alleles are disadvantageous in the homozygous state, they may have little or no effect on heterozygotes. Thus, recessive alleles can be protected from natural selection by the phenotypic expression of the dominant allele.

In most cases, the masking of recessive alleles in heterozygotes makes it almost impossible to eliminate them completely through selective breeding. Experimentally, we can prevent homozygous recessive organisms from mating. But, as the frequency of a recessive allele decreases, an increasing proportion of its remaining copies is "hidden" in heterozygotes **(Table 21.3).** Thus, the diploid state preserves recessive alleles at low frequencies, at least in large populations. In small populations, a combination of natural selection and genetic drift can eliminate harmful recessive alleles.

Natural Selection Can Maintain Balanced Polymorphisms

A **balanced polymorphism** is one in which two or more phenotypes are maintained in fairly stable proportions over many generations. Natural selection preserves balanced polymorphisms when heterozygotes have higher relative fitness, when different alleles are favored in different environments, and when the rarity of a phenotype provides a selective advantage.

HETEROZYGOTE ADVANTAGE A balanced polymorphism can be maintained by **heterozygote advantage,** when heterozygotes have higher relative fitness than either homozygote. As Darwin first discovered in his experiments on corn, the offspring of crosses between two homozygous strains of the same species often exhibit a robustness described as "hybrid vigor." Apparently, being heterozygous at many gene loci provides some advantage, perhaps by allowing organisms to respond effectively to environmental variation.

The best-documented example of heterozygote advantage with reference to a specific gene locus is the maintenance of the *HbS* (sickle) allele, which codes for a defective form of hemoglobin in humans. As you learned in Chapter 12, hemoglobin is an oxygen-transporting molecule in red blood cells. The hemoglobin produced by the *HbS* allele differs from normal hemoglobin (coded by the *HbA* allele) by just one amino acid (see Figure 15.20). In *HbS/HbS* homozygotes, the faulty hemoglobin forms long, fibrous chains under low oxygen conditions, causing red blood cells to assume a sickle shape (as shown in Figure 12.1). Homozygous *HbS/HbS* individuals often die of sickle-cell anemia before reproducing, yet in tropical and subtropical Africa, *HbS/HbA* heterozygotes make up nearly 25% of many populations.

Why is the harmful allele maintained at such high frequency in some populations? It turns out that sickle-cell anemia is common in regions where malarial parasites infect red blood cells in humans **(Figure 21.14)**. When heterozygous *HbA/HbS* individuals contract malaria, their infected red blood cells assume the same sickle shape as those of homozygous *HbS/HbS* individuals. The sickled cells lose potassium, killing the parasites, which limits their spread within the infected individual. Heterozygous individuals often survive malaria because the parasites do not multiply quickly inside them, their immune systems can effectively fight the infection, and they

A. Distribution of *HbS* allele

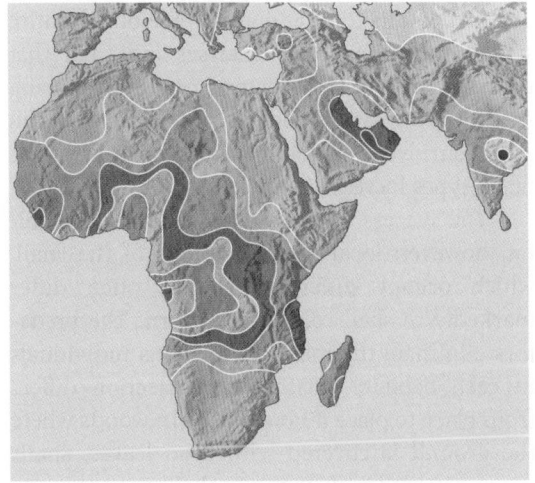

B. Distribution of malarial parasite

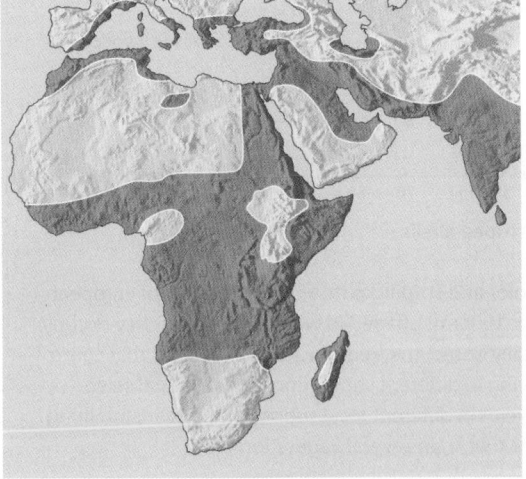

KEY

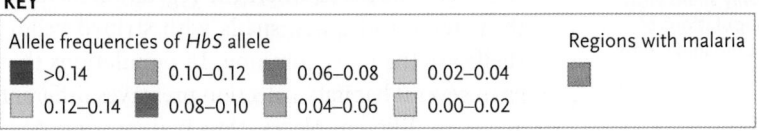

Allele frequencies of *HbS* allele

■ >0.14	▨ 0.10–0.12
▨ 0.12–0.14	■ 0.08–0.10
■ 0.06–0.08	▨ 0.02–0.04
▨ 0.04–0.06	▨ 0.00–0.02

Regions with malaria ▨

FIGURE 21.14 Heterozygote advantage. The distribution of the *HbS* allele **(A)**, which causes sickle-cell anemia in homozygotes, roughly matches the distribution of the malarial parasite *Plasmodium falciparum* **(B)** in southern Europe, Africa, the Middle East, and India. Gene flow among human populations has carried the *HbS* allele to some malaria-free regions.
© Cengage Learning 2017

FIGURE 21.15 | **Observational Research**

Habitat Variation in Color and Striping Patterns of European Garden Snails

Hypothesis: Birds and other visual predators are more likely to find snails that are not well camouflaged in their habitats than snails that blend in with their habitats.

Prediction: Genetically based variations in the shell color and striping patterns of the European garden snail *(Cepaea nemoralis)* will differ substantially between local subpopulations of snails that live in different types of vegetation.

Method: British researchers A. J. Cain and P. M. Shepard surveyed the distribution of color and striping patterns of snails in many local subpopulations. They plotted the data on a graph showing the percentage of snails with yellow shells versus the percentage of snails with striped shells, noting the vegetation type where each local population lived.

Results: The shell color and striping patterns of snails living in a particular vegetation type tend to be clustered on the graph, reflecting phenotypic differences that enable the snails to be camouflaged in different habitats. Thus, the alleles that control these characters vary from one local subpopulation to another.

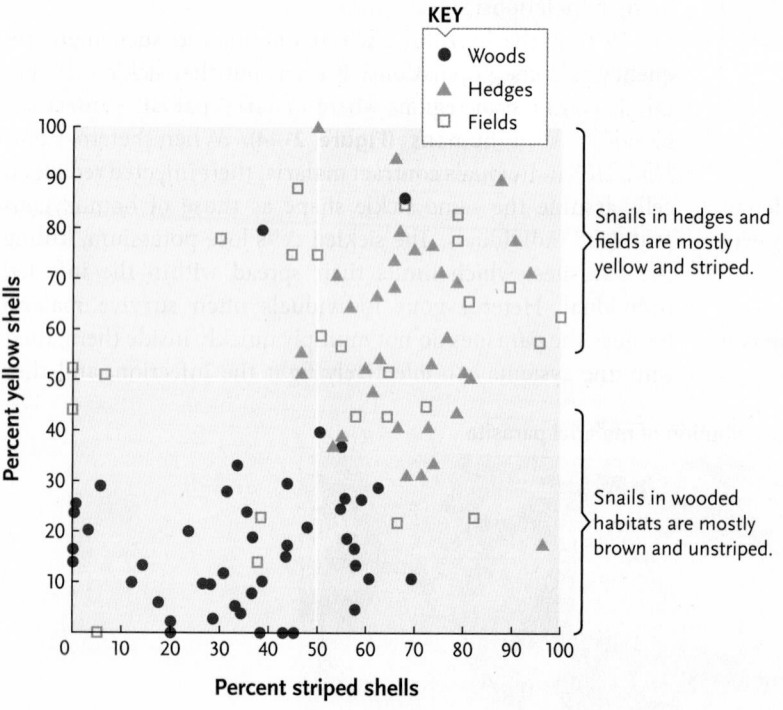

KEY
- ● Woods
- ▲ Hedges
- □ Fields

Snails in hedges and fields are mostly yellow and striped.

Snails in wooded habitats are mostly brown and unstriped.

Percent yellow shells (y-axis)

Percent striped shells (x-axis)

Conclusion: Variations in the color and striping patterns on the shells of European garden snails allow most snails to be camouflaged in whatever habitat they occupy. Because these traits are genetically based, the frequencies of the alleles that control them also differ among snails living in different vegetation types. Natural selection therefore favors different alleles in different local subpopulations, maintaining genetic variability in populations that span several vegetation types.

think like a scientist

Observe
Hypothesize
Predict
Experiment
Interpret

Do the data in the graph suggest that populations of snails in wooded habitats include *only* snails with brown, unstriped shells and that populations of snails in hedges and fields include *only* snails with yellow striped shells?

Source: A. J. Cain and P. M. Shepard. 1954. Natural selection in *Cepaea*. *Genetics* 39:89–116.

© Cengage Learning 2017

---

retain a large population of uninfected red blood cells. Homozygous *HbA/HbA* individuals are also subject to malarial infection, but because their infected cells do not sickle, the parasites multiply rapidly, causing a severe infection with a high mortality rate.

Therefore, *HbA/HbS* heterozygotes have greater resistance to malaria and are more likely to survive severe malarial infections in areas where the parasite is prevalent. Natural selection preserves the *HbS* allele in these populations because heterozygotes in malaria-prone areas have higher relative fitness than homozygotes for the normal *HbA* allele.

SELECTION IN VARYING ENVIRONMENTS

Genetic variability can also be maintained within a population when different alleles are favored in different places or at different times. For example, the shells of European garden snails range in color from nearly white to pink, yellow, or brown, and may be patterned by one to five stripes of varying color (see Figure 21.2A). This polymorphism, which is relatively stable through time, is controlled by several gene loci. The variability in color and striping pattern can be partially explained by selection for camouflage in different habitats.

Predation by song thrushes (*Turdus ericetorum*) is a major agent of selection on the color and pattern of these snails in England. When a thrush finds a snail, it smacks it against a rock to break the shell. The bird eats the snail, but leaves the shell near its "anvil." Researchers used the broken shells near an anvil to compare the phenotypes of captured snails to a random sample of the entire snail population. Their analyses indicated that thrushes are visual predators, usually capturing snails that are easy to find. Thus, well-camouflaged snails survive, and the alleles that specify their phenotypes increase in frequency.

The success of camouflage varies with habitat, however; local subpopulations of the snail, which occupy different habitats, often differ markedly in shell color and pattern. The predators eliminate the most conspicuous individuals in each habitat; thus, natural selection differs from place to place (**Figure 21.15**). In woods where the ground is covered with dead leaves, snails with unstriped pink or brown shells predominate. In hedges and fields, where the vegetation includes thin stems and grass, snails with striped yellow shells are the most common. In populations that span several habitats, selection preserves different alleles in different places, thus maintaining variability in the population as a whole.

FREQUENCY-DEPENDENT SELECTION Sometimes, genetic variability is maintained in a population simply because rare phenotypes—whatever they happen to be—have higher relative fitness than more common phenotypes. The rare phenotype will increase in frequency until it becomes so common that it loses its advantage. Such phenomena are examples of **frequency-dependent selection** because the selective advantage enjoyed by a particular phenotype depends on its frequency in the population.

In 2001, Luc D. B. Godard, Marc Macnair, and Ann Smithson of the University of Exter, United Kingdom, demonstrated that a flower color polymorphism in the European Elderflower orchid (*Dactylorhiza sambucina*) is maintained by the pattern of pollinator visitations to flowers of different colors. Populations of this orchid include both yellow-flowered and purple-flowered plants, with yellow-flowered plants slightly outnumbering purple-flowered plants. Elderflower orchids are pollinated by bumblebees (*Bombus lapidarius* and *B. terrestris*), which seek a nectar or pollen reward from the flowers they visit. (Plant-pollinator interactions are described in Section 29.3). But Elderflower orchids provide neither type of reward to pollinators; thus, neither color offers an advantage to the bees over the other. After visiting a yellow flower without gaining a reward, a bee will next visit a purple flower. After receiving no reward at a purple flower, it returns to a yellow flower, and then it alternates between the two colors until it eventually abandons both.

Yellow-flowered plants and purple-flowered plants produce equal numbers of flowers. Because the pollinators alternate their visits between yellow flowers and purple flowers, Godard and his colleagues hypothesized that a higher percentage of flowers with the less common color are visited by bees, simply because there are fewer of them available. The researchers conducted an experiment in which they manipulated the frequencies of plants with yellow and purple flowers and then measured how often bees removed pollen sacs (which include male gametes) and deposited them on the female reproductive parts of other plants; they also measured how many fruits each plant produced. Their results indicate that the reproductive success of plants with flowers of either color is higher when that color is rare and lower when it is common **(Figure 21.16)**. Thus, when yellow flowers are rare, yellow flowers are pollinated more frequently (as a percentage of the number available) and produce more fruits than purple flowers. When yellow-flowered plants eventually outnumber purple-flowered plants, the advantage of being less common shifts to purple flowers, and bees visit purple flowers more frequently (as a percentage of the number available) than yellow flowers. The activity of the pollinators thus provides an example of a polymorphism maintained by frequency-dependent selection.

Some Genetic Variations May Be Selectively Neutral

Many biologists believe that some genetic variations are neither preserved nor eliminated by natural selection. According to the **neutral variation hypothesis,** some of the genetic variation at loci coding for enzymes and other soluble proteins is **selectively neutral.** Even if various alleles code for slightly different amino acid sequences in proteins, the different forms of the proteins may function equally well. In those cases, natural selection would not favor some alleles over others.

Biologists who support the neutral variation hypothesis do not question the role of natural selection in producing complex anatomical structures or useful biochemical traits. They also recognize that selection reduces the frequency of harmful alleles. But they argue that we should not simply assume that every genetic variant that persists in a population has been preserved by natural selection. In practice, it is often very difficult to test the neutral variation hypothesis because the fitness effects of different alleles are often subtle and can vary with small changes in the environment.

The neutral variation hypothesis helps to explain why we see different levels of genetic variation in different populations. It proposes that genetic variation is directly proportional to a population's size and the length of time over which variations have accumulated. Small populations experience fewer mutations than large populations simply because they include fewer replicating genomes. Small populations also lose rare alleles more readily through genetic drift. Thus, small populations should exhibit less genetic variation than large ones, and a population, like the northern elephant seals, that has experienced a recent population bottleneck should exhibit an exceptionally low level of genetic variation. These predictions of the neutral variation hypothesis are generally supported by empirical data.

STUDY BREAK 21.4

1. How does the diploid condition protect harmful recessive alleles from natural selection?
2. What is a balanced polymorphism?
3. Why is the allele that causes sickle-cell anemia very rare in human populations that are native to northern Europe?

21.5 Adaptation and Evolutionary Constraints

Although natural selection preserves alleles that confer high relative fitness on the individuals that carry them, researchers are cautious about interpreting the benefits that particular traits may provide.

Scientists Construct Hypotheses about the Evolution of Adaptive Traits

An **adaptive trait** is any product of natural selection that increases the relative fitness of an organism in its environment. **Adaptation** is the accumulation of adaptive traits over time, and most chapters of this book describe examples that range across all levels of biological organization, from the

FIGURE 21.16 | Experimental Research

Demonstration of Frequency-Dependent Selection

Question: How does the frequency of a flower color influence the likelihood that it will be visited by pollinators?

Experiment: An Elderflower orchid *(Dactylorhiza sambucina)* produces either yellow flowers or purple flowers. Godard, Macnair, and Smithson established 20 experimental populations in which the frequency of yellow-flowered and purple-flowered plants varied from 0.1 to 0.9, and they allowed natural pollinators to visit the plants. The reproductive success of male plants was measured as the percentage of pollen sacs removed by pollinators, and the reproductive success of females was measured as the number of fruits produced per plant.

Result: The data are expressed as relative fitness of yellow-flowered plants compared with purple-flowered plants (the ratio of pollen sacs removed from yellow flowers versus purple flowers; the ratio of fruits produced by yellow flowers versus purple flowers). The horizontal line indicates equal reproductive success of yellow- and purple-flowered plants. The results indicate that both male and female reproductive success declined steadily as a flower color became more common.

A. Elderflower orchids *(Dactylorhiza sambucina)*

Paul Harcourt Davies/Nature Picture Library

B. Pollen sacs removed

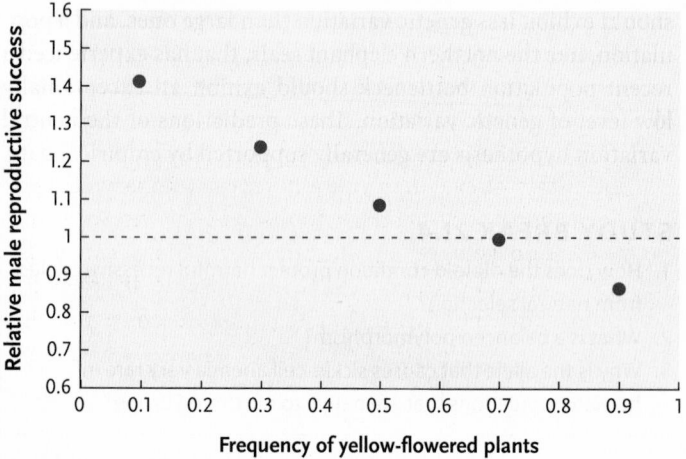

Frequency of yellow-flowered plants

C. Fruits produced

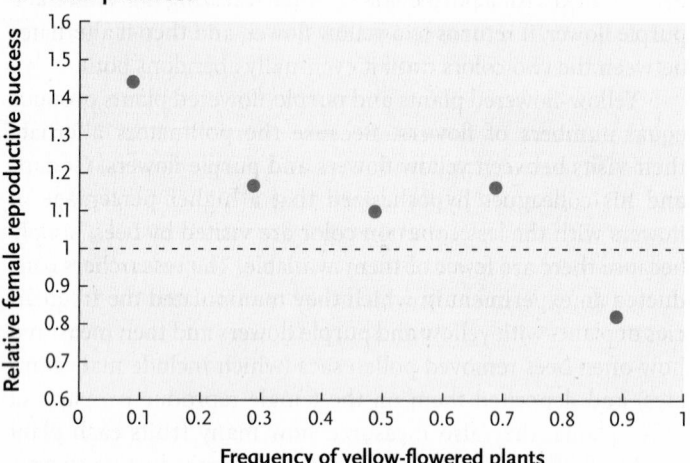

Frequency of yellow-flowered plants

Conclusion: The balanced polymorphism in flower color of the Elderflower orchid is maintained by frequency-dependent selection because pollinators visit flowers of the less common color more frequently.

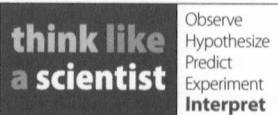

| Observe |
| Hypothesize |
| Predict |
| Experiment |
| **Interpret** |

At approximately what frequencies do yellow-flowered and purple-flowered plants have equal reproductive success?

Source: L. D. B. Gigord, M. R. Macnair, and A. Smithson. 2001. Negative frequency-dependent selection maintains a dramatic flower color polymorphism in the rewardless orchid *Dactylorhiza sambucina* (L.) Soò. *Proc Nat Acad Sci* 98:6253–6255.

molecular to the ecological. For example, the change in the oxygen-binding capacity of hemoglobin in response to carbon dioxide concentration, the water-retaining structures and special photosynthetic pathways of desert plants, and the warning coloration of poisonous animals can all be interpreted as adaptive traits.

In fact, we can concoct an adaptive explanation for almost any characteristic we observe in nature. But such explanations are just fanciful stories unless they are framed as testable hypotheses about the relative fitness of different phenotypes and genotypes. Unfortunately, evolutionary biologists cannot always conduct straightforward experiments because they sometimes study traits that do not vary much within a population or species. In such cases, they may compare variations of a trait in closely related species living in different environments. For example, one can test how the traits of desert plants are adaptive by comparing them to traits in related species from moister habitats.

When biologists try to unravel how and why a particular characteristic evolved, they must also remember that a trait they observe today may have had a different function in the past. For example, the structure of the shoulder joint in birds allows them to move their wings first upward and backward and then downward and forward during flapping flight. But analyses of the fossil record reveal that this adaptation, which is essential for flight, did not originate in birds: some predatory nonflying dinosaurs, including the ancestors of birds, had a similarly constructed shoulder joint. Researchers hypothesize that these fast-running predators may have struck at prey with a flapping motion similar to that used by modern birds. Thus, the structure of the shoulder may have evolved first as an adaptation for capturing prey, and only later proved useful for flapping flight. This hypothesis—however plausible it may be—cannot be tested by direct experimentation because the nonflying ancestors of birds have been extinct for millions of years. Instead, evolutionary biologists must use anatomical studies of birds and their ancestors, as well as theoretical models about the mechanics of movement, to challenge and refine the hypothesis.

Finally, although evolution has produced all the characteristics of organisms, not all are necessarily adaptive. Some traits may be the products of chance events and genetic drift. Others are produced by alleles that were selected for unrelated reasons. And still other characteristics result from the action of basic physical laws. For example, the seeds of many plants fall to the ground when they mature, reflecting the inevitable effect of gravity.

Several Factors Constrain Adaptive Evolution

When we analyze the structure and functions of an organism, we often marvel at how well adapted it is to its environment and mode of life. However, the adaptive traits of most organisms are compromises produced by competing selection pressures. Sea turtles, for example, must lay their eggs on beaches because their embryos cannot acquire oxygen under water. Although flippers allow females to crawl to nesting sites on beaches, they are not ideally suited for terrestrial locomotion. Their structure reflects their primary function, swimming.

Moreover, no organism can be perfectly adapted to its environment because environments change over time. When selection occurs in a population, it preserves alleles that are successful under the prevailing environmental conditions. Thus, each generation is adapted to the environmental conditions under which its parents lived. If the environment changes from one generation to the next, adaptation will always lag behind.

Another constraint on the evolution of adaptive traits is historical. Natural selection is not an engineer that designs new organisms from scratch. Instead, it acts on new mutations and existing genetic variation. Because new mutations are fairly rare, natural selection works primarily with alleles that have been present for many generations. Thus, adaptive changes in the morphology of an organism are often based on small modifications of existing structures. The bipedal (two-footed) posture of humans, for example, evolved from the quadrupedal (four-footed) posture of our ancestors. Natural selection did not produce an entirely new skeletal design to accompany this radical behavioral shift. Instead, existing characteristics of the spinal column and the musculature of the legs and back were modified, albeit imperfectly, for an upright stance.

The agents of evolution cause microevolutionary changes in the gene pools of populations. In the next chapter, we examine how microevolution in different populations can cause their gene pools to diverge. The extent of genetic divergence is sometimes sufficient to cause the populations to evolve into different species.

STUDY BREAK 21.5

1. How can a biologist test whether a trait is adaptive?
2. Why are most organisms adapted to the environments in which their parents lived?

Unanswered Questions

Can gene exchange between species facilitate the formation of a new lineage?

In this chapter you learned how gene flow between populations of a single species can increase the genetic variability of one or both populations. Let's now consider the implications of gene flow between two closely related species.

An exchange of genes between closely related species was once thought to be restricted to plants, but is now known to occur in all taxa from bacteria, to corals, arthropods, fish, birds, and mammals. Moreover, biologists now consider it to be one possible route to the formation of

a new lineage. Hybridization is always rare, because, as you will learn in Chapter 22, it requires the breakdown of a barrier preventing interbreeding between species. But when it does occur, it can lead to gene flow between species if a hybrid backcrosses (that is, mates with an individual from one of the parental species).

On the Galápagos island of Daphne Major there are three species of Darwin's ground finches: the medium ground finch (*Geospiza fortis*), the cactus finch (*Geospiza scandens*), and the large ground finch (*Geospiza magnirostris*). They differ in song, body size, and beak size and shape; all other traits, such as plumage, are similar. Only males sing, although young birds of both sexes learn their father's song in association with their parent's appearance during a short sensitive period early in life (described further in Chapter 56). As adults, females choose their mates according to these learned characteristics, a process named *sexual imprinting* by the famous behavioral biologist Konrad Lorenz. The consequence of this mate choice is a premating barrier to interbreeding. Because it is based on learning, the barrier may be incomplete, especially if a young bird learns the song of another species. This can happen, for example, when a chick hatches from an egg that remains in the nest after it has been taken over by another species: the leftover chick learns its foster father's song, not its biological father's song.

There are three possible consequences of hybridization. First, hybrids may die, due to either genetic incompatibility or inappropriate environmental conditions. Second, two species may fuse into one highly variable population (as described in Chapter 22). Third, natural selection can act on this genetic variation, and in a new environment, set the population on a new trajectory leading to a new lineage. We have seen all three outcomes in our forty-year study of Darwin's finches on Daphne Major.

First, *G. fortis* × *G. scandens* hybrids died during the first years of the study, because their intermediate beak size and shape left them unable to crack the large hard seeds that predominated in the food supply at that time.

Second, heavy rains in 1983 produced an abundance of small seeds, and these allowed hybrids to survive. Subsequent backcrossing resulted in a trickle of genes moving between the two species and a gradual genetic and phenotypic convergence over the next 30 years.

Third, the outcome that leads to a new lineage began when a bird immigrated to the island. An analysis of his DNA taken from a small drop of blood revealed he was a *G. fortis* × *G. scandens* backcross to *G. fortis* from the neighboring island of Santa Cruz. Weighing a hefty 28 grams, he was considerably larger than the 18 gram *G. fortis* on Daphne. He was homozygous for a rare allele (a useful genetic marker), and he sang a unique song. We banded him 5110. He eventually paired with a *G. fortis/ G. scandens/G. fortis* backcross female born on Daphne, and later with two *G. fortis* females. Subsequent offspring all bred among themselves. Between 2003 and 2005, a severe drought killed close to 90% of all finches on the island, including all birds from this lineage—except for two, a brother and sister. Both were large like 5110, both carried two copies of the rare allele as a result of previous inbreeding, and the male sang 5110's song. After the drought, the brother and sister bred with each other and produced 26 offspring, 17 of which subsequently bred.

One daughter bred with the father, one son with his mother, and the remainder with each other. This inbreeding continued for three more generations, currently six generations removed from their ancestor 5110.

Are these birds behaving as a separate species? All are large, all carry two copies of the rare allele, and all males sing 5110's unusual song. They hold contiguous territories, which they defend against each other (see Chapter 56). Their territories overlap those of the other three species, *G. fortis*, *G. scandens*, and *G. magnirostris*, who ignore them and who are ignored by them. Thus, in all respects they are behaving as a separate species. Will this lineage survive? We don't know. It could die out through the accumulation of harmful alleles through inbreeding or it could disappear through extensive interbreeding with another species. Although they are unusually flexible in their feeding behavior, encompassing the diet of all other finches, their supply of food could change.

Whether it survives or not, this new lineage gives insights into how a new species can arise and either persist or become extinct. As in all Darwin's finches, its premating barrier to interbreeding arose behaviorally through learning their parents' appearance and song early in life. These finches illustrate how genetic variation is generated through hybridization and backcrossing, and future gene flow could plausibly fuel a change in another direction in a different environment.

think like a scientist Try finding examples of two species (include plants) that look alike. Ask what keeps them from interbreeding? Could the barrier between them ever leak? What might be the causes and consequences of gene flow between them? What could be the evolutionary forces leading to a new lineage? How would you test your various hypotheses?

Denise Applewhite/Princeton University

Peter and Rosemary Grant have been studying Darwin's finches on the Galápagos islands since 1973. Their fieldwork is designed to understand the causes of an adaptive radiation. It combines analyses of archipelago-wide patterns of evolution with detailed investigations of population-level processes on two islands, Genovesa and Daphne. Their work is a blend of ecology, behavior, and genetics. Their most recent books are *How and Why Species Multiply* (2008) and *40 Years of Evolution* (2014), both published by Princeton University Press. They are both in the Department of Ecology and Evolution at Princeton University.

REVIEW KEY CONCEPTS

For access to MindTap and additional study materials visit www.cengagebrain.com.

21.1 Variation in Natural Populations

- Phenotypic traits exhibit either quantitative or qualitative variation within populations (Figures 21.2 and 21.3).

- Genetic variation, environmental factors, or an interaction between the two cause phenotypic variation within populations (Figures 21.4 and 21.5). Only genetically based phenotypic variation is heritable and subject to evolutionary change.

- Genetic variation arises within populations largely through mutation and genetic recombination. Artificial selection experiments (Figure 21.6) and analyses of protein and DNA

sequences reveal that most populations include significant genetic variation in both coding and noncoding areas of DNA.

21.2 Population Genetics

- All the gene copies in a population comprise its gene pool, which can be described in terms of allele frequencies and genotype frequencies (Table 21.1).

- The Hardy–Weinberg principle of genetic equilibrium is a null model that describes the conditions under which microevolution, a change in allele frequencies through time, will not take place. Microevolution occurs in populations when the restrictive requirements of the model are not met.

21.3 The Agents of Microevolution

- Several processes cause microevolution in populations (Table 21.2). Mutation introduces completely new genetic variation. Gene flow carries novel genetic variation into a population through the arrival and reproduction of immigrants (Figure 21.7). Genetic drift causes random changes in allele frequencies, especially in small populations (Figure 21.8). Natural selection occurs when the genotypes of some individuals enable them to survive and reproduce more than others.

- Natural selection alters phenotypic variation in three ways (Figure 21.9). Directional selection increases or decreases the mean value of a trait, shifting it toward a phenotypic extreme. Stabilizing selection increases the frequency of the mean phenotype and reduces variability in the trait (Figures 21.10 and 21.11). Disruptive selection increases the frequencies of extreme phenotypes and decreases the frequency of intermediate phenotypes (Figure 21.12).

- Sexual selection promotes the evolution of exaggerated structures and behaviors (Figure 21.13).

- Although nonrandom mating does not change allele frequencies, it can produce more homozygotes and fewer heterozygotes than the Hardy–Weinberg model predicts.

21.4 Maintaining Genetic and Phenotypic Variation

- Diploidy can maintain genetic variation in a population if recessive alleles are not expressed in heterozygotes and are thus hidden from natural selection (Table 21.3).

- Polymorphisms are maintained in populations when heterozygotes have higher relative fitness than both homozygotes (Figure 21.14), when natural selection occurs in variable environments (Figure 21.15), or when the relative fitness of a phenotype varies with its frequency in the population (Figure 21.16).

- The neutral variation hypothesis proposes that many genetic variations are selectively neutral, conferring neither advantages nor disadvantages on the individuals that carry them. It explains why large populations and those that have not experienced a recent population bottleneck exhibit the highest levels of genetic variation.

21.5 Adaptation and Evolutionary Constraints

- Adaptive traits increase the relative fitness of individuals carrying them. Adaptive explanations of traits must be framed as testable hypotheses.

- Natural selection cannot result in perfectly adapted organisms because most adaptive traits represent compromises among conflicting needs, because most environments change constantly, and because natural selection can affect only existing genetic variation.

TEST YOUR KNOWLEDGE

Remember/Understand

1. The reason spontaneous mutations do not have an immediate effect on allele frequencies in a large population is that:
   a. mutations are random events, and mutations may be either beneficial or harmful.
   b. mutations usually occur in males and have little effect on eggs.
   c. many mutations exert their effects after an organism has stopped reproducing.
   d. mutations are so rare that mutated alleles are greatly outnumbered by nonmutated alleles.
   e. most mutations do not change the amino acid sequence of a protein.

2. The phenomenon in which chance events cause unpredictable changes in allele frequencies is called:
   a. gene flow.
   b. genetic drift.
   c. inbreeding.
   d. balanced polymorphism.
   e. stabilizing selection.

3. Which of the following phenomena explains why the allele for sickle-cell hemoglobin is common in some tropical and subtropical areas where the malaria parasite is prevalent?
   a. genetic drift
   b. heterozygote advantage
   c. sexual dimorphism
   d. neutral selection
   e. stabilizing selection

4. The neutral variation hypothesis proposes that:
   a. complex structures in most organisms have not been fostered by natural selection.
   b. most mutations have a strongly harmful effect.
   c. some mutations are not affected by natural selection.
   d. natural selection cannot counteract the action of gene flow.
   e. large populations are subject to stronger natural selection than small populations.

5. Phenotypic characteristics that increase the fitness of individuals are called:
   a. mutations.
   b. founder effects.
   c. heterozygote advantages.
   d. adaptive traits.
   e. polymorphisms.

Apply/Analyze

6. Which of the following represents an example of qualitative phenotypic variation?
   a. the lengths of people's toes
   b. the body sizes of pigeons
   c. human ABO blood types
   d. the birth weights of humans
   e. the number of leaves on oak trees

7. A population of mice is at Hardy–Weinberg equilibrium at a gene locus that controls fur color. The locus has two alleles, M and m. A genetic analysis of one population reveals that 60% of its gametes carry the M allele. What percentage of mice contains both the M and m alleles?
   a. 60%
   b. 48%
   c. 40%
   d. 36%
   e. 16%

8. If the genotype frequencies in a population are 0.60 *AA*, 0.20 *Aa*, and 0.20 *aa*, and if the requirements of the Hardy–Weinberg principle apply, the genotype frequencies in the offspring generation will be:
   a. 0.60 *AA*, 0.20 *Aa*, 0.20 *aa*.
   b. 0.36 *AA*, 0.60 *Aa*, 0.04 *aa*.
   c. 0.49 *AA*, 0.42 *Aa*, 0.09 *aa*.
   d. 0.70 *AA*, 0.00 *Aa*, 0.30 *aa*.
   e. 0.64 *AA*, 0.32 *Aa*, 0.04 *aa*.

9. An Eastern European immigrant carrying the allele for Tay-Sachs disease settled in a small village on the St. Lawrence River. Many generations later, the frequency of the allele in that village is statistically higher than it is in the immigrant's homeland. The high frequency of the allele in the village probably provides an example of:
   a. natural selection.
   b. the concept of relative fitness.
   c. the Hardy–Weinberg genetic equilibrium.
   d. phenotypic variation.
   e. the founder effect.

10. If a storm kills many small sparrows in a population, but only a few medium-sized and large ones, which type of selection is probably operating?
    a. directional selection
    b. stabilizing selection
    c. disruptive selection
    d. intersexual selection
    e. intrasexual selection

11. **Discuss Concepts** Most large commercial farms routinely administer antibiotics to farm animals to prevent the rapid spread of diseases through a flock or herd. Explain why you think that this practice is either wise or unwise.

12. **Discuss Concepts** Many human diseases are caused by recessive alleles that are not expressed in heterozygotes. Some people think that eugenics—the selective breeding of humans to eliminate undesirable genetic traits—provides a way for us to rid our populations of such harmful alleles. Explain why eugenics cannot eliminate such genetic traits from human populations.

13. **Discuss Concepts** In what ways are the effects of sexual selection, disruptive selection, and nonrandom mating different? How are they similar?

14. **Apply Evolutionary Thinking** Captive breeding programs for endangered species often have access to a limited supply of animals for a breeding stock. As a result, their offspring are at risk of being highly inbred. Why and how might zoological gardens and conservation organizations avoid or minimize inbreeding?

Evaluate/Create

15. **Discuss Concepts** Using two types of beans to represent two alleles at the same gene locus, design an exercise to illustrate how population size affects genetic drift.

16. **Design an Experiment** Design an experiment to test the hypothesis that the differences in size among adult guppies are determined by the amount of food they eat rather than by genetic factors.

For selected answers, see Appendix A.

INTERPRET THE DATA

Peter and Rosemary Grant of Princeton University have studied the ecology and evolution of finches on the Galápagos Islands since the early 1970s. They have shown that finches with large bills (as measured by bill depth; see **Figure**) can eat both small seeds and large seeds, but finches with small bills can only eat small seeds. In 1977, a severe drought on the island of Daphne Major reduced seed production by plants. After the birds consumed whatever small seeds they found, only large seeds were still available. The resulting food shortage killed a majority of the medium ground finches *(Geospiza fortis)* on Daphne Major; their population plummeted from 751 in 1976 to just 90 in 1978. The Grants' research also documented a change in the distributions of bill depths in the birds from 1976 to 1978, as illustrated in the graphs to the right. In light of what you now know about the relationship between bill size and food size for these birds, interpret the change illustrated in the graph. What type of natural selection does this example illustrate?

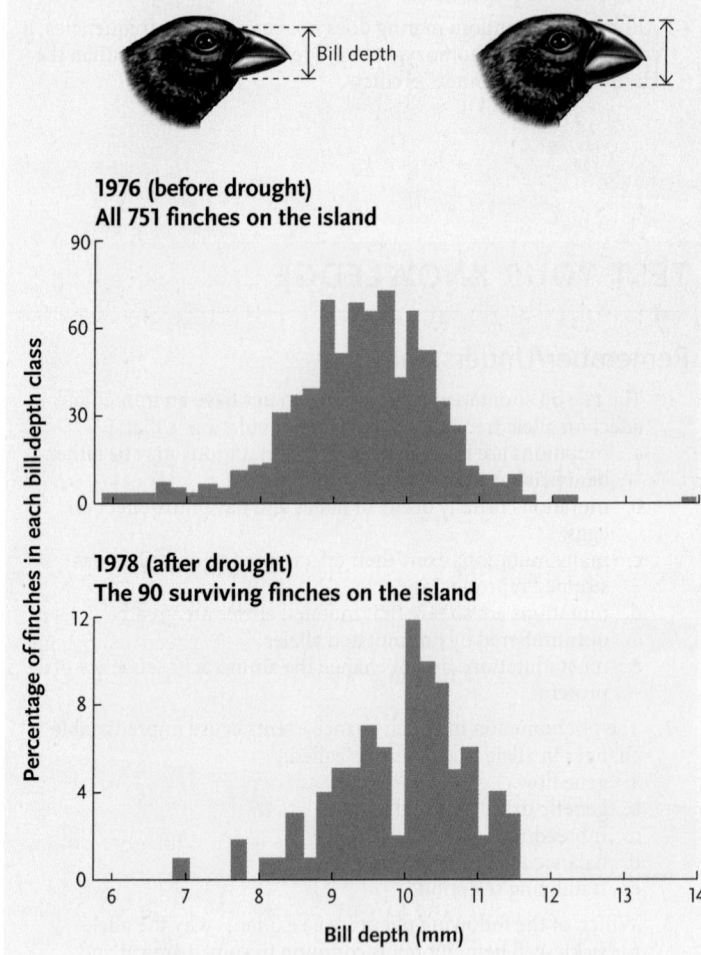

Source: P. R. Grant. 1986. *Ecology and Evolution of Darwin's Finches.* Princeton University Press.

Speciation

Two closely related bird species, purple martins *(Progne subis)* and tree swallows *(Tachycineta bicolor)* perching together on a branch in Crane Creek, Ohio.

Why it matters . . . In 1927, nearly 100 years after Darwin boarded the *Beagle,* a young German naturalist named Ernst Mayr embarked on his own journey, to the highlands of New Guinea. He was searching for rare "birds of paradise" **(Figure 22.1).** These birds were known in Europe only through their ornate and colorful feathers, which were used to decorate ladies' hats. On his trek through the remote Arfak Mountains, Mayr identified 137 bird species (including many birds of paradise) based on differences in their size, plumage, color, and other external characteristics.

To Mayr's surprise, the native Papuans—who were untrained in the ways of Western science, but who hunted these birds for food and feathers—had their own names for 136 of the 137 species he had identified. The close match between the two lists confirmed Mayr's belief that the *species* is a fundamental level of organization in nature. Each species has a unique combination of genes underlying its distinctive appearance and habits. Thus, people who observe them closely—whether indigenous hunters or Western scientists—can often distinguish one species from another.

Mayr also discovered some remarkable patterns in the geographical distributions of the bird species in New Guinea. For example, each mountain range he explored was home to some species that lived nowhere else. Closely related species often lived on different mountaintops, separated by deep valleys of unsuitable habitat. In 1942, Mayr published the book *Systematics and the Origin of Species,* in which he described the role of geography in the evolution of new

FIGURE 22.1 Birds of paradise. A male Count Raggi's bird of paradise *(Paradisaea raggiana)* tries to attract the attention of a female (not pictured) with his showy plumage and conspicuous display. There are 43 known bird of paradise species, 35 of them found only on the island of New Guinea.

species; the book quickly became a cornerstone of the modern synthesis (which was outlined in Section 20.3).

What mechanisms produce distinct species? As you discovered in Chapter 21, microevolutionary processes alter the pattern and extent of genetic and phenotypic variation within populations. When these processes differ between populations, the populations will diverge genetically, and they may eventually become so different that we recognize them as distinct species. Although Darwin's famous book was titled *On the Origin of Species*, he did not dwell on the question of *how* new species arise. But the concept of **speciation**—the process of species formation—was implicit in his insight that similar species often share inherited characteristics and a common ancestry. Darwin also recognized that "descent with modification" had generated the amazing diversity of organisms on Earth.

Today evolutionary biologists view speciation as a *process*, a series of events that occur through time. However, using classical, genomic, and bioinformatic approaches, they study the *products* of speciation, species that are alive today. Because they can rarely witness the process of speciation from start to finish, scientists make inferences about it by studying organisms in various stages of species formation. In this chapter, we consider four major topics: how biologists define and recognize species; how species maintain their genetic identity; how the geographical distributions of organisms influence speciation; and how different genetic mechanisms produce new species.

22.1 What Is a Species?

Like the hunters of the Arfak Mountains, most of us recognize the different species that we encounter every day. We can distinguish a cat from a dog and sunflowers from roses. The concept of species is based on our perception that Earth's biological diversity is packaged in discrete, recognizable units, and not as a continuum of forms grading into one another. As biologists have learned more about evolutionary processes—and the dazzling biodiversity those processes have produced—they have developed a variety of complementary species concepts.

The Morphological Species Concept Is a Practical Way to Identify Species

Biologists often describe new species on the basis of visible anatomical characteristics, a process that dates back to Linnaeus' classification of organisms in the eighteenth century (described in Chapter 24). This approach is based on the **morphological species concept,** the idea that all individuals of a species share measurable traits that distinguish them from individuals of other species.

The morphological species concept has many practical applications. For example, paleobiologists use morphological criteria to identify the species of fossilized organisms (see Chapter 23). And because we can observe the external traits of organisms in nature, field guides to plants and animals list

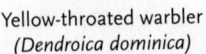

Yellow-throated warbler
(*Dendroica dominica*)

Myrtle warbler
(*Dendroica coronata*)

FIGURE 22.2 Diagnostic characters. Yellow-throated warblers and myrtle warblers can be distinguished by the color of feathers on the throat and rump.

© Cengage Learning 2017

diagnostic (that is, distinguishing) physical characters that allow us to recognize them **(Figure 22.2).**

Nevertheless, relying exclusively on morphology to identify species can present problems. Consider the variation in the shells of the European garden snail (*Cepaea nemoralis*; shown earlier, in Figure 21.2). How could anyone imagine that so variable a collection of shells represents just one species of snail? Conversely, morphology does not help us distinguish some closely related species that are nearly identical in appearance. Finally, morphological species definitions tell us little about the evolutionary processes that produce new species.

The Biological Species Concept Is Based on Reproductive Isolation

The **biological species concept** emphasizes the dynamic nature of species. Ernst Mayr defined *biological species* as "groups of . . . interbreeding natural populations that are reproductively isolated from [do not produce fertile offspring with] other such groups." The concept is based on reproductive criteria and is easy to apply, at least in principle: if the members of two populations interbreed and produce fertile offspring *under natural conditions*, they belong to the same species; their fertile offspring will, in turn, produce the next generation of that species. If two populations do not interbreed in nature, or fail to produce fertile offspring when they do, they belong to different species.

The biological species concept defines species in terms of population genetics and evolutionary theory. The first half of Mayr's definition notes the genetic *cohesiveness* of species: populations of the same species experience gene flow, which mixes their genetic material. Thus, we can think of a species as one large gene pool, which may be subdivided into local populations.

The second part of the biological species concept emphasizes the genetic *distinctness* of each species. Because populations of different species are reproductively isolated, they cannot exchange genetic information. In fact, the process of speciation is frequently defined as the evolution of reproductive isolation between populations.

The biological species concept also explains why individuals of a species generally look alike: members of the same gene pool share genetic traits that determine their appearance. Individuals of different species generally do not resemble one another as closely because they share fewer genetic characteristics. In practice, biologists often still use similarities or differences in morphological traits as convenient markers of genetic similarity or reproductive isolation.

However, the biological species concept does not apply to the many forms of life that reproduce asexually, including most bacteria; some protists, fungi, and plants; and a few animals. In these species, individuals do not breed, so it is pointless to ask whether members of different populations do. Similarly, we cannot use the biological species concept to study extinct organisms, because we have little or no data on their specific reproductive habits. These species must all be defined using morphological or biochemical criteria. Yet, despite its limitations, the biological species concept currently provides the best evolutionary definition of a sexually reproducing species.

The Phylogenetic Species Concept Focuses on Evolutionary History

Recognizing the limitations of the biological species concept, biologists have developed dozens of other ways to define a species. A widely accepted alternative is the **phylogenetic species concept.** Using both morphological and genetic sequence data, scientists first reconstruct the evolutionary tree for the organisms of interest. They then define a phylogenetic species as a cluster of populations—the tiniest twigs on this part of the Tree of Life—that emerge from the same small branch. Thus, a phylogenetic species comprises populations that share a recent evolutionary history. We will consider this approach for understanding the evolutionary relationships of organisms in Chapter 24.

One advantage of the phylogenetic species concept is that biologists can apply it to any group of organisms, including species that have long been extinct, as well as living organisms that reproduce asexually. Proponents of this approach also argue that the morphological and genetic distinctions between organisms on different branches of the Tree of Life reflect the absence of gene flow between them—one of the key requirements of the biological species definition. Nevertheless, because detailed evolutionary histories have been described for relatively few groups of organisms, biologists are not yet able to apply the phylogenetic species concept to all forms of life. As noted in *Unanswered Questions* for Chapter 24, continued research on the details of evolutionary relationships will increase this concept's usefulness in the future.

Many Species Exhibit Substantial Geographical Variation

Just as individuals within populations exhibit genotypic and phenotypic variation (see Section 21.1, Figures 21.2 and 21.15), populations within species

also differ both genetically and phenotypically. Neighboring populations often have shared characteristics because they live in similar environments, exchange individuals, and experience comparable patterns of natural selection. Widely separated populations, by contrast, may live under different conditions and experience different patterns of selection; because gene flow is less likely to occur between distant populations, their gene pools and phenotypes are often somewhat different.

When geographically separated populations of a species exhibit dramatic, easily recognized phenotypic variation, biologists may identify them as different **subspecies (Figure 22.3),** which are local variants of a species. Individuals from different subspecies usually interbreed where their geographical distributions meet, and their offspring often exhibit intermediate phenotypes. Biologists sometimes use the word "race" as shorthand for the term *subspecies.*

Various patterns of geographical variation—as well as analyses of how the variation may relate to climatic or habitat variation—have provided great insight into the speciation process. Two of the best-studied patterns are *ring species* and *clinal variation.*

RING SPECIES Some plant and animal species have a ring-shaped geographical distribution that surrounds uninhabitable

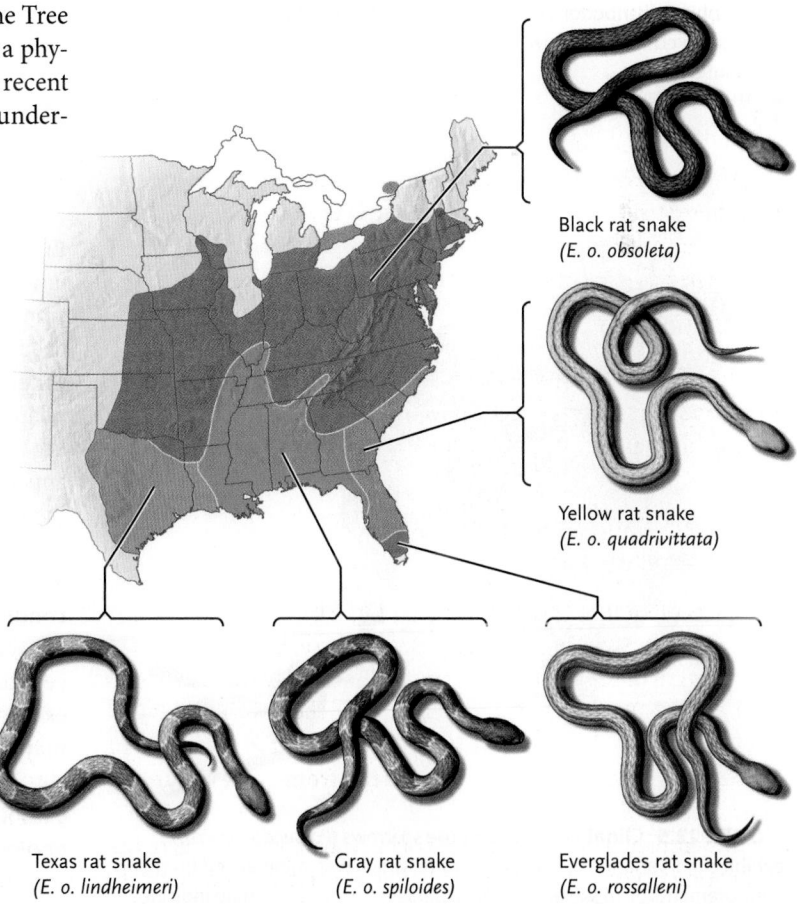

Black rat snake
(*E. o. obsoleta*)

Yellow rat snake
(*E. o. quadrivittata*)

Texas rat snake
(*E. o. lindheimeri*)

Gray rat snake
(*E. o. spiloides*)

Everglades rat snake
(*E. o. rossalleni*)

FIGURE 22.3 Subspecies. Five subspecies of rat snake *(Elaphe obsoleta)* in eastern North America differ in color and in the presence or absence of stripes or blotches.
© Cengage Learning 2017

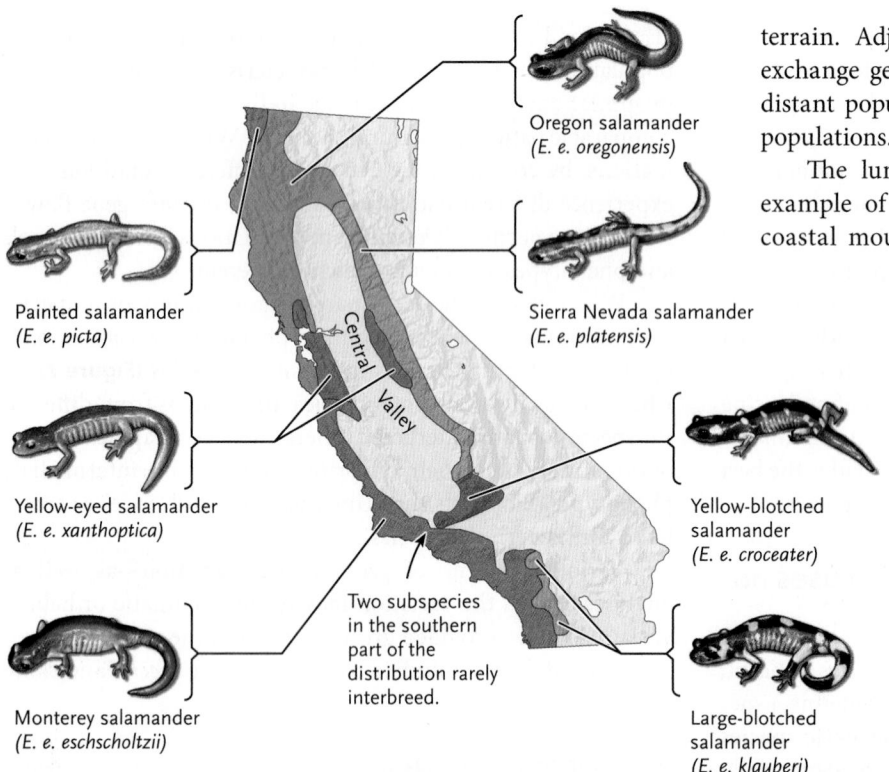

Oregon salamander
(E. e. oregonensis)

Painted salamander
(E. e. picta)

Sierra Nevada salamander
(E. e. platensis)

Central Valley

Yellow-eyed salamander
(E. e. xanthoptica)

Yellow-blotched
salamander
(E. e. croceater)

Two subspecies
in the southern
part of the
distribution rarely
interbreed.

Monterey salamander
(E. e. eschscholtzii)

Large-blotched
salamander
(E. e. klauberi)

FIGURE 22.4 Ring species. Six of the seven subspecies of the salamander *Ensatina eschscholtzii* are distributed in a ring around California's Central Valley. Subspecies often interbreed where their geographical distributions overlap. However, the two subspecies that nearly close the ring in the south (marked with an arrow), the Monterey salamander and the yellow-blotched salamander, rarely interbreed.
© Cengage Learning 2017

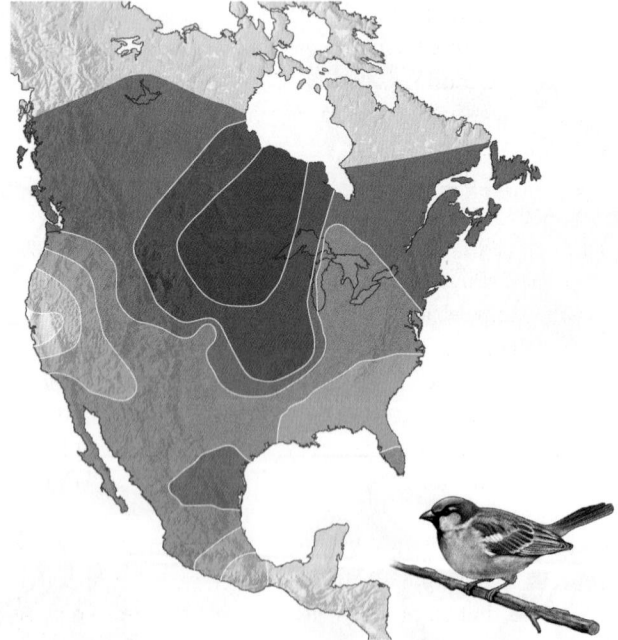

FIGURE 22.5 Clinal variation. House sparrows (*Passer domesticus*) exhibit clinal variation in overall body size, which was summarized from measurements of 16 skeletal features. Darker shading in the map indicates larger size.
© Cengage Learning 2017

terrain. Adjacent populations of these **ring species** can exchange genetic material directly, but gene flow between distant populations occurs only through the intermediary populations.

The lungless salamander *Ensatina eschscholtzii,* an example of a ring species, is widely distributed in the coastal mountains and the Sierra Nevada of California, but it cannot survive in the hot, dry Central Valley **(Figure 22.4).** Seven subspecies differ in biochemical traits, color, size, and ecology. Individuals from adjacent subspecies often interbreed where their geographical distributions overlap, and intermediate phenotypes are fairly common. But at the southern end of the Central Valley, adjacent subspecies rarely interbreed. Apparently, they have differentiated to such an extent that they can no longer exchange genetic material directly.

Are the southernmost populations of this salamander subspecies or different species? A biologist who saw *only* the southern populations, which coexist without interbreeding, might define them as separate species. However, they still have the potential to exchange genetic material through the intervening populations that form the ring. Hence, biologists recognize these populations as belonging to the same species. Most likely, the southern subspecies are in an intermediate stage of species formation.

CLINAL VARIATION When a species is distributed over a large, environmentally diverse area, some traits may exhibit a **cline,** a smooth pattern of variation across a geographical gradient. For example, many birds and mammals in the northern hemisphere show clinal variation in body size **(Figure 22.5)** and the relative length of their appendages. In general, populations living in colder environments have larger bodies and shorter appendages, a pattern that is usually interpreted as a mechanism to conserve body heat (see Chapter 48).

Clinal variation usually results from gene flow between adjacent populations that are each adapting to slightly different conditions. However, if populations at opposite ends of a cline are separated by great distances, they may exchange very little genetic material through reproduction. Thus, when a cline extends over a large geographical gradient, distant populations may be genetically and morphologically distinct.

Despite the geographical variation that many species exhibit, even closely related species are genetically and morphologically different from each other. In the next section, we consider the mechanisms that maintain the genetic distinctness of closely related species by preventing their gene pools from mixing.

THINK OUTSIDE THE BOOK

Search the biological literature and the Internet for a definition and application of the "ecological species concept." How does the ecological species concept differ from the three species concepts described above? Under what circumstances or for what purpose would the ecological species concept be useful?

22.2 Maintaining Reproductive Isolation

The biological species concept uses the criterion of reproductive isolation to define species of sexually reproducing organisms: most individuals produce offspring by mating with another individual of their own species. Indeed, a variety of biological characteristics, collectively described as **reproductive isolating mechanisms,** prevent individuals of different species from mating and producing successful progeny. Thus, by reducing the chance of interspecific (between-species) mating—and the production of **hybrid** offspring (that is, offspring with parents of different species)—these isolating mechanisms prevent the gene pools of distinct species from mixing.

Reproductive isolating mechanisms operate at different times during the reproductive process, and biologists categorize as occurring either before or after an egg is fertilized (summarized in **Table 22.1**). **Prezygotic isolating mechanisms** exert their effects before fertilization and the production of a zygote (a fertilized egg). **Postzygotic isolating mechanisms** operate after fertilization and zygote formation. These isolating mechanisms are not mutually exclusive, and two or more of them may affect the outcome of a between-species interaction.

Prezygotic Isolating Mechanisms Prevent the Production of Hybrid Offspring

Biologists have identified five mechanisms that can prevent the production of hybrid offspring. Four of these mechanisms limit the frequency of interspecific matings, whereas one blocks interspecific fertilizations. These five prezygotic mechanisms are *ecological, temporal, behavioral, mechanical,* and *gametic isolation.*

Species living in the same geographical region may experience **ecological isolation** if they live in different habitats. For example, lions and tigers were both common in India until the mid-nineteenth century, when hunters virtually exterminated the Asian lions. However, because lions live in open grasslands and tigers in dense forests, the two species did not encounter one another and did not interbreed. Lion-tiger hybrids are sometimes born in captivity, but they do not occur under natural conditions.

TABLE 22.1	Reproductive Isolating Mechanisms	
Timing Relative to Fertilization	Mechanism	Mode of Action
Prezygotic ("premating") mechanisms	Ecological isolation	Species live in different habitats
	Temporal isolation	Species breed at different times
	Behavioral isolation	Species cannot communicate
	Mechanical isolation	Species cannot physically mate
	Gametic isolation	Species have nonmatching receptors on gametes
Postzygotic ("postmating") mechanisms	Hybrid inviability	Hybrid offspring do not complete development
	Hybrid sterility	Hybrid offspring cannot produce gametes
	Hybrid breakdown	Hybrid offspring have reduced survival or fertility

Species living in the same habitat can experience **temporal isolation** if they mate at different times of day or different times of year. For example, the fruit flies *Drosophila persimilis* and *Drosophila pseudoobscura* overlap extensively in their geographical distributions, but they do not interbreed, in part because *D. persimilis* mates in the morning and *D. pseudoobscura* in the afternoon. Similarly, two species of pine in California are reproductively isolated where their geographical distributions overlap: even though both rely on the wind to carry male gametes (pollen grains) to female gametes (ova) in other cones, *Pinus radiata* releases pollen in February and *Pinus muricata* releases pollen in April.

Many animals rely on specific signals, which may differ dramatically between species, to identify the species of a potential mate. **Behavioral isolation** results when the signals used by one species are not recognized by another. For example, female birds rely on the song, color, and displays of males to identify members of their own species. Similarly, female fireflies identify males by their flashing patterns (**Figure 22.6**). These behaviors (collectively called *courtship displays*) are often so complicated that signals sent by one species are like a foreign language that another species simply does not understand.

Mate choice by females and sexual selection (discussed in Section 21.3) generally drive the evolution of mate recognition signals. Females often spend substantial energy in reproduction, and choosing an appropriate mate—that is, a male of her

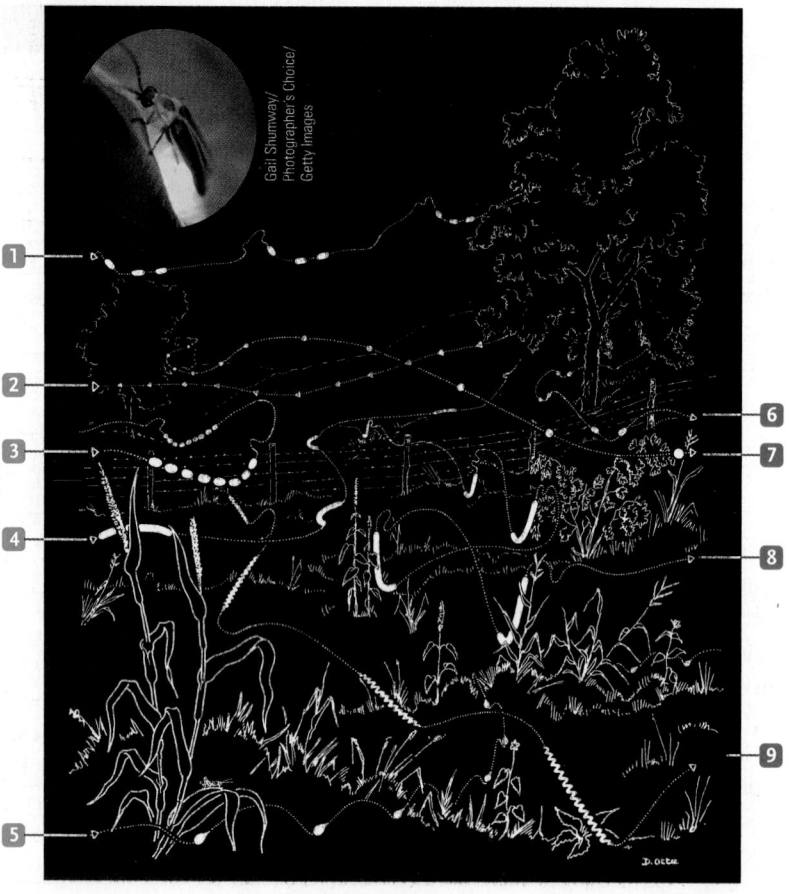

KEY

1 *P. consimilis*	**4** *P. collustrans*	**7** *P. ignitus*
2 *P. brimleyi*	**5** *P. marginellus*	**8** *P. pyralis*
3 *P. carolinus*	**6** *P. consanguineus*	**9** *P. granulatus*

FIGURE 22.6 Behavioral reproductive isolation. Male fireflies use bioluminescent signals to attract potential mates. The different flight paths and flashing patterns of males in nine North American *Photinus* species are represented here. Females respond only to the display given by males of their own species. The inset photo shows *P. pyralis*.

Illustration courtesy of James E. LLoyd. Miscellaneous Publications of the Museum of Zoology of The Univ. of Michigan, 130:1-100, 1966.

Purple monkey-flower
(Mimulus lewisii)

Scarlet monkey-flower
(Mimulus cardinalis)

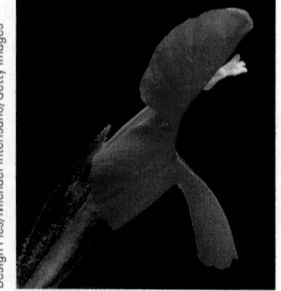

FIGURE 22.7 Mechanical reproductive isolation. Because of differences in floral structure, two species of monkey-flower attract different animal pollinators. *Mimulus lewisii* attracts bumblebees, and *Mimulus cardinalis* attracts hummingbirds.

own species—is critically important for the production of successful young. By contrast, a female that mates with a male from a different species is unlikely to leave any surviving offspring at all. Over time, the number of males with recognizable traits, as well as the number of females able to recognize the traits, increases in the population.

Differences in the structure of reproductive organs or other body parts—**mechanical isolation**—may prevent individuals of different species from interbreeding. In particular, many plants have anatomical features that allow only certain pollinators, usually particular bird or insect species, to collect and distribute pollen. For example, the flowers and nectar of two native California plants, the purple monkey-flower *(Mimulus lewisii)* and the scarlet monkey-flower *(Mimulus cardinalis)*, attract different animal pollinators **(Figure 22.7)**. *M. lewisii* is pollinated by bumblebees. It has shallow purple flowers with broad petals that provide a landing platform for the bees. Bright yellow streaks on the petals serve as "nectar guides," directing bumblebees to the short nectar tube and reproductive parts, which are located among the petals. Bees enter the flowers to drink their concentrated nectar, and they pick up and deliver pollen as their legs and bodies brush against the reproductive parts of the flowers. *M. cardinalis*, by contrast, is pollinated by hummingbirds. It has long red flowers with no yellow streaks, and the reproductive parts extend above the petals. The red color attracts hummingbirds but lies outside the color range detected by bumblebees. The nectar of *M. cardinalis* is more dilute than that of *M. lewisii* but is produced in much greater quantity, making it easier for hummingbirds to ingest. When a hummingbird visits *M. cardinalis* flowers, it pushes its long bill down the nectar tube, and its forehead touches the reproductive parts, picking up and delivering pollen. Recent research has demonstrated that where the two monkey-flower species grow side-by-side, animal pollinators restrict their visits to either one species or the other 98% of the time, providing nearly complete reproductive isolation.

Even when individuals of different species mate, **gametic isolation**, an incompatibility between the sperm of one species and the eggs of another, may prevent fertilization. Many marine invertebrates release gametes into the environment for external fertilization. The sperm and eggs of each species recognize one another's complementary surface proteins (see Chapter 49), but the surface proteins on the gametes of different species do not match. In animals with internal fertilization, sperm of one species may not survive and function within the reproductive tract of another. Interspecific matings between some *Drosophila* species, for example, induce a reaction in the female's reproductive tract that blocks "foreign" sperm from reaching eggs. Parallel physiological incompatibilities between a pollen tube and a stigma prevent interspecific fertilization in some plants.

Postzygotic Isolating Mechanisms Reduce the Success of Hybrid Individuals

If prezygotic isolating mechanisms between two closely related species are incomplete or ineffective, sperm from one species sometimes fertilizes an egg of the other species. In such cases the two species will be reproductively isolated if their offspring, interspecific hybrids, have lower fitness than the offspring of intraspecific (within species) matings. Three postzygotic isolating mechanisms—*hybrid inviability, hybrid sterility,* and *hybrid breakdown*—can reduce the fitness of hybrid individuals.

Many genes govern the complex processes that transform a zygote into a mature organism. Hybrid individuals have two sets of developmental instructions, one from each parent species, which may not interact properly for the successful completion of embryonic development. As a result, hybrid organisms frequently die as embryos or at an early age, a phenomenon called **hybrid inviability.** For example, domestic sheep and goats can mate and fertilize one another's ova, but the hybrid embryos always die before coming to term, presumably because the developmental programs of the two parent species are incompatible.

Although some hybrids between closely related species develop into healthy and vigorous adults, they may not produce functional gametes. This **hybrid sterility** often results when the parent species differ in the number or structure of their chromosomes, which cannot pair properly during meiosis. Such hybrids have zero fitness because they leave no descendants. The most familiar example is a mule, the product of mating between a female horse ($2n = 64$) and a male donkey ($2n = 62$). Zebroids, the offspring of matings between horses and zebras, are also usually sterile hybrids **(Figure 22.8).**

Some first-generation hybrids (F_1; see Section 12.1) are healthy and fully fertile. They can breed with other hybrids and with both parental species. However, the second generation (F_2), produced by matings between F_1 hybrids, or between F_1

hybrids and either parental species, may exhibit reduced survival or fertility, a phenomenon known as **hybrid breakdown.** For example, experimental crosses between *Drosophila* species may produce functional hybrids, but the offspring of hybrids experience high rates of chromosomal abnormalities and harmful types of genetic recombination. Thus, reproductive isolation is maintained between the species because there is little long-term mixing of their gene pools.

STUDY BREAK 22.2

1. What is the difference between prezygotic and postzygotic isolating mechanisms?
2. When a male duck of one species performed a courtship display to a female of another species, she interpreted his behavior as aggressive rather than amorous. What type of reproductive isolating mechanism does this scenario illustrate?

22.3 The Geography of Speciation

As Ernst Mayr recognized, geography has a huge impact on whether gene pools have the opportunity to mix. Biologists define two modes of speciation, based on the geographical relationship of populations as they become reproductively isolated: *allopatric speciation* (*allo* = different; *patria* = homeland) and *sympatric speciation* (*sym* = together).

Allopatric Speciation Occurs between Geographically Separated Populations

Allopatric speciation may take place when a physical barrier subdivides a large population or when a small population becomes separated from a species' main geographical distribution. Allopatric speciation occurs in two stages. First, two populations become *geographically* separated, preventing gene flow between them. Then, as the populations experience distinct mutations as well as different patterns of natural selection and genetic drift, they may accumulate genetic differences that isolate them *reproductively.* Allopatric speciation is probably the most common mode of speciation in large animals.

Geographical separation sometimes occurs when a barrier divides a large population into two or more units **(Figure 22.9).** For example, hurricanes may create new channels that divide low coastal islands and the populations inhabiting them. Uplifting mountains or landmasses as well as rivers or advancing glaciers can also produce barriers that subdivide populations. The uplift of the Isthmus of Panama, caused by movements of Earth's crust about 5 million years ago, separated a once-continuous shallow sea into the eastern tropical Pacific Ocean and the western tropical Atlantic Ocean. Populations of marine organisms were subdivided by this event. In the tropical Atlantic Ocean, populations experienced patterns of mutation, natural selection, and genetic drift that were

FIGURE 22.8 Interspecific hybrids. Horses and zebroids (hybrid offspring of horses and zebras) run in a mixed herd. Zebroids are usually sterile.

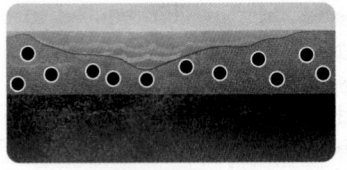

1 At first, a population is distributed over a large geographical area.

2 A geographical change separates the original population, creating a barrier to gene flow.

3 In the absence of gene flow, the separated populations evolve independently and diverge into different species.

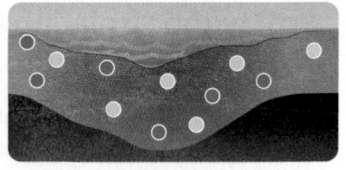

4 When another geographical change allows individuals of the two species to come into secondary contact, they do not interbreed.

FIGURE 22.9 The model of allopatric speciation and secondary contact.
© Cengage Learning 2017

different from those experienced by populations in the tropical Pacific Ocean. As a result, the populations diverged genetically, and pairs of closely related species now live on either side of this divide **(Figure 22.10).**

In other cases, small populations may become isolated at the edge of a species' geographical distribution. Such peripheral populations often differ genetically from the central population because they are adapted to somewhat different environments. Once a small population is isolated, founder effects and small population size may promote genetic drift (see Section 21.3), and natural selection may favor the evolution of distinctive traits. If the isolated population experiences limited gene flow from the parent population, these agents of evolution will foster genetic differentiation between them. In time, the accumulated genetic differences may lead to reproductive isolation.

Populations established by colonization of oceanic islands represent extreme examples of this phenomenon. The founder effect makes the populations genetically distinct. And on oceanic archipelagos, such as the Galápagos and Hawaiian islands, individuals from one island may colonize nearby islands, founding populations that differentiate into distinct species. Each island may experience multiple invasions, and the process may be repeated many times within the archipelago, leading to the evolution of a **species cluster,** a group of closely related species recently descended from a common ancestor **(Figure 22.11).** Sometimes a species cluster can evolve relatively quickly; for example, the nearly 800 species of fruit flies now living on the

Hawaiian Islands (described in *Focus on Research: Basic Research*) evolved in less than 5 million years, an average of just over 6,000 years per species.

Secondary Contact Provides a Test of Whether Allopatric Speciation Has Occurred

Allopatric populations may reestablish contact when a geographical barrier is eliminated or breached (see Figure 22.9, step 4). Such **secondary contact** (that is, contact after a period of geographic isolation) provides a test of whether or not the genes in the populations have diverged enough to make them reproductively isolated. If their gene pools did not differentiate much during geographical separation, the populations will interbreed and merge into one, a phenomenon described as **species fusion.** But if, during their separation, the populations accumulated enough genetic differences to be reproductively isolated on secondary contact, they will be separate species. (The ecological consequences of secondary contact are described in Chapter 53.)

During the early stages of secondary contact, prezygotic reproductive isolation may be weak or incomplete. Some members of each population may mate with individuals from the other, producing viable, fertile offspring in areas called **hybrid zones.** Although some hybrid zones may persist for hundreds or thousands of years, they are generally narrow, and ecological or geographical factors maintain the separation of the gene pools for the majority of individuals in both species.

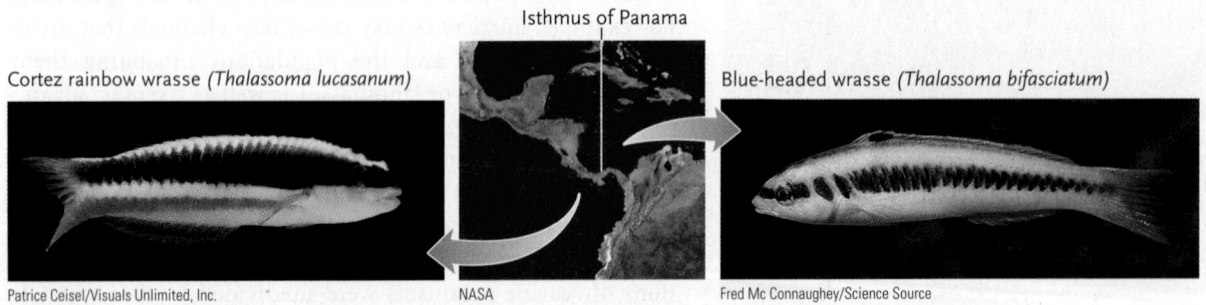

Cortez rainbow wrasse *(Thalassoma lucasanum)*

Isthmus of Panama

Blue-headed wrasse *(Thalassoma bifasciatum)*

Patrice Ceisel/Visuals Unlimited, Inc.　　　NASA　　　Fred Mc Connaughey/Science Source

FIGURE 22.10 Geographical separation. The uplift of the Isthmus of Panama divided an ancestral wrasse population. The Cortez rainbow wrasse now occupies the eastern Pacific Ocean, and the blue-headed wrasse now occupies the western Atlantic Ocean.

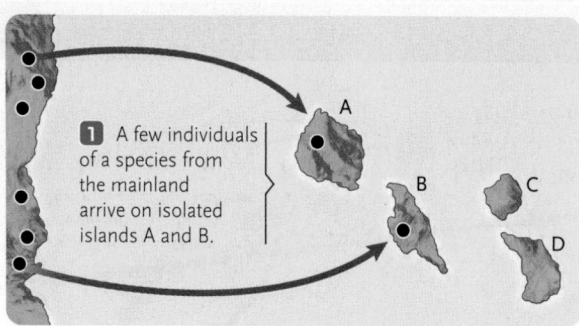

1 A few individuals of a species from the mainland arrive on isolated islands A and B.

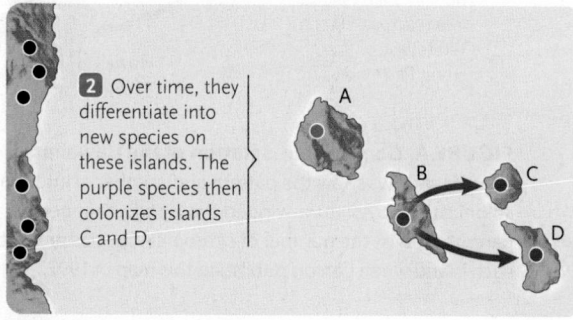

2 Over time, they differentiate into new species on these islands. The purple species then colonizes islands C and D.

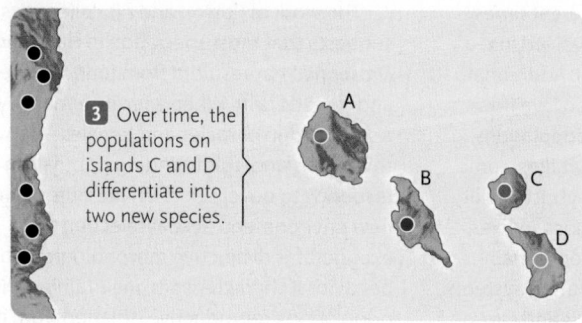

3 Over time, the populations on islands C and D differentiate into two new species.

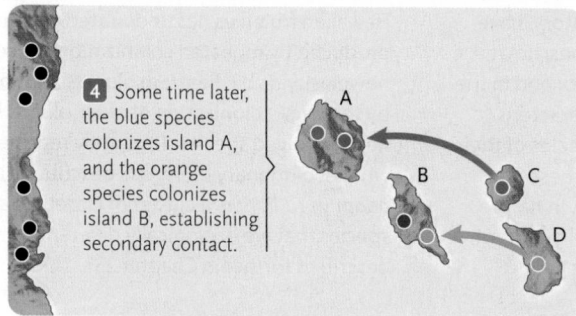

4 Some time later, the blue species colonizes island A, and the orange species colonizes island B, establishing secondary contact.

FIGURE 22.11 Hypothetical evolution of a species cluster on an archipelago. Letters identify four islands in an archipelago, and colored dots represent different species. The ancestor of all the species is represented by black dots on the mainland. At the end of the process, islands A and B are each occupied by two species, and islands C and D are each occupied by one species, all of which evolved on the islands.
© Cengage Learning 2017

For example, the breeding ranges of Bullock's oriole (*Icterus bullocki*) and Baltimore oriole (*Icterus galbula*) overlap in the Midwest of North America **(Figure 22.12)**. In 2011, Matthew D. Carling, Lindsay G. Serene, and Irby J. Lovette of Cornell University published a genetic analysis of orioles where the

geographical ranges overlap. Their research confirmed that the hybrid zone is narrow (only 325 km wide) and that it encompasses an area where two distinctive environments mix. Bullock's orioles live in the hotter and drier habitats to the west of the hybrid zone; Baltimore orioles occupy the cooler and moister habitats to the east. Researchers hypothesize that although hybrid individuals may survive and reproduce in the mixed habitat where the species' ranges overlap, hybrids may not be well adapted either to the dry habitats to the west or to the moist habitats to the east. Thus, if natural selection eliminates hybrids outside the narrow hybrid zone, the two species will remain reproductively isolated.

Postgygotic isolating mechanisms often cause hybrids to have lower fitness than either parent species. Under these circumstances, natural selection will favor individuals that choose mates of their own species, thus promoting the evolution of prezygotic isolating mechanisms. This phenomenon—the evolution of prezygotic barriers to reproduction after postzygotic barriers already exist—is called **reinforcement,** because the prezygotic mechanisms *reinforce* those postzygotic barriers. Studies of several *Drosophila* species suggest that reinforcement enhanced reproductive isolation that had begun to develop while the populations were geographically separated.

Sympatric Speciation Occurs within One Continuously Distributed Population

In **sympatric speciation,** reproductive isolation evolves between distinct subgroups that arise within one population. Models of sympatric speciation do not require that the populations be either geographically or environmentally separated as their gene pools diverge. We examine below general models of sympatric speciation in animals and plants; the genetic basis of sympatric speciation is considered in the next section.

Insects that feed on just one or two plant species are among the animals most likely to evolve by sympatric speciation. These insects generally carry out most important life cycle activities on or near their "host" plants. Adults mate on the host plant; females lay their eggs on it; and larvae feed on the host plant's tissues, eventually developing into adults, which initiate another round of the life cycle. Host plant choice is genetically determined in many insect species. In others, individuals associate with the host plant species they ate as larvae.

Theoretically, a genetic mutation could suddenly change some insects' choice of host plant. Mutant individuals would shift their life cycle activities to the new host, and then interact primarily with others preferring the same new host, an example of ecological isolation. These individuals would collectively form a separate subpopulation, called a **host race.** Reproductive isolation could evolve between different host races if the individuals of each host race are more likely to mate with members of their own host race than with members of another. Some biologists criticize this model, however, because it assumes that the genes controlling two traits, the insects' host

Basic Research: Explosive Speciation in Hawaiian Fruit Flies

The islands of the Hawaiian archipelago have been geographically isolated throughout their history, lying at least 3,200 km (approximately 2,000 miles) from the nearest continents or other islands **(Figure A).** Built by undersea volcanic eruptions over millions of years, they emerged from northwest to southeast: Kauai is at least 5 million years old, and Hawaii, the "Big Island," is less than 1 million years old. Individual islands differ in maximum elevation and include diverse habitats, from sparse, dry vegetation to lush, wet forests.

Resident species must have arrived from distant mainland localities or evolved on the islands from colonizing ancestors. The islands' isolation, different ages, and geographical and ecological complexity allowed repeated interisland colonizations followed by allopatric speciation events. Thus, it is not surprising that species clusters have evolved in several groups of organisms (including flowering plants, insects, and birds).

Nearly 800 species of Hawaiian fruit flies, most of which live on only one island, have been discovered. Biologists used many characters to identify the different species, including external and internal anatomy, cell structure, chromosome structure, ecology, and mating behavior. Their data suggest that the vast majority of native Hawaiian species arose from a single ancestral species that colonized the archipelago long ago, probably from eastern Asia. After repeated speciation events, the fruit flies of the Hawaiian Islands represent more than 25% of all known fruit fly species.

Hampton Carson, of the University of Hawaii, spearheaded studies on the evolutionary relationships of Hawaiian fruit flies. He and his colleagues gathered data on hundreds of fly species—a daunting task. Most species of fruit flies are sexually dimorphic. The females of different species may be similar in appearance, but the males of even closely related species differ in virtually every aspect of their external anatomy: body size, head shape, and the structure of their eyes, antennae, mouth-parts, bristles, legs, and wings. Their mating behavior and choice of mating sites also vary dramatically.

Nevertheless, closely related species on different islands occupy comparable habitats and associate with related plant species. Carson suggested that speciation in these flies resulted from the evolution of different genetically determined *mating systems*, the behaviors and morphological characteristics that males display when seeking a mate. The mating systems serve as prezygotic isolating mechanisms.

The 100 or more species of "picture-wing" *Drosophila,* relatively large flies with patterns on their wings, illustrate the evolution of a species cluster. Carson and his colleagues used similarities and differences in the banding patterns on the flies' giant salivary chromosomes (described in Figure 22.17) to trace the evolutionary origin of species on the younger islands by identifying their closest relatives on the older islands. Their analysis of 26 species on Hawaii, the youngest island, suggested that flies from the older islands colonized Hawaii at least 19 different times, and each founder population evolved into a new species there (see Figure A). Additional species apparently evolved when lava flows on Hawaii subdivided existing populations.

Among the picture-wing fruit flies, some interspecies matings result in hybrid sterility or hybrid breakdown. But for most species, prezygotic reproductive isolation is maintained by differences in their mating systems. For example, *Drosophila silvestris* and *Drosophila heteroneura,* which produce healthy and fertile hybrids in the laboratory, have similar geographical distributions; however, differences in courtship behavior and in the shape of the males' heads, a characteristic that females use to recognize males of their own species **(Figure B),** keep these two species reproductively isolated. In nature, they hybridize only in one small geographical area.

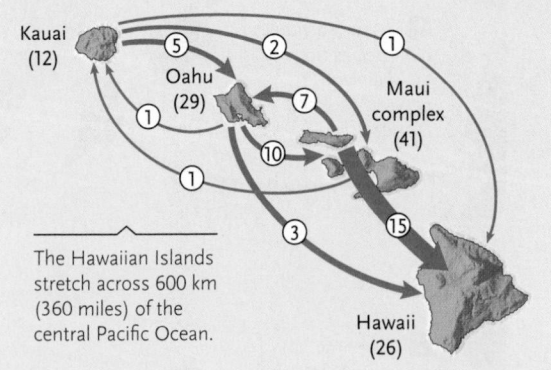

FIGURE A Geographic isolation of the Hawaiian Islands. Arrows show the pattern and number of interisland colonizations by picture-winged *Drosophila*. Numbers in parentheses list the number of *Drosophila* species present on each island when Carson published this map in 1992.
© Cengage Learning 2017

The work of Carson and his colleagues suggests that most speciation in Hawaiian *Drosophila* has resulted from founder effects and genetic drift. When a fertile female—or a small group of males and females—moves to a new island, this founding population responds to novel selection pressures in its new environment. Sexual selection then exaggerates distinctive morphological and behavioral characteristics, maintaining the population's reproductive isolation from its new neighbors. The tremendous variety of Hawaiian fruit flies has undoubtedly been produced by repeated colonizations of newer islands by flies from older islands and by the back-colonization of older islands by newly evolved species. Thus, they represent what evolutionary biologists describe as an *adaptive radiation,* a cluster of closely related species that are ecologically different (as described further in Chapter 23).

Drosophila heteroneura

Drosophila silvestris

FIGURE B Two *Drosophila* species in which the males' head shapes differ.

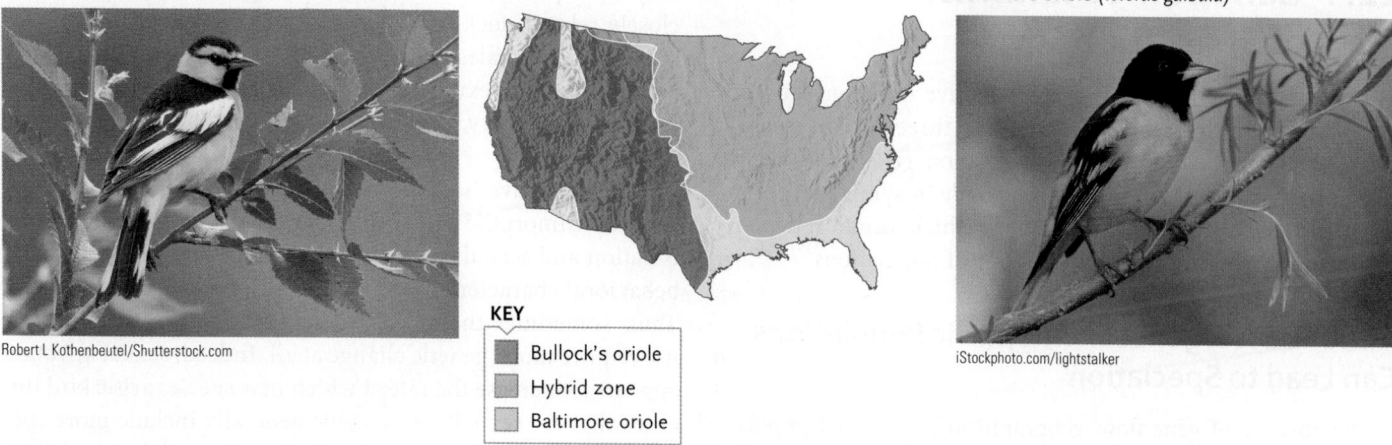

Bullock's oriole (*Icterus bullockii*)

Robert L Kothenbeutel/Shutterstock.com

Baltimore oriole (*Icterus galbula*)

iStockphoto.com/lightstalker

KEY
- Bullock's oriole
- Hybrid zone
- Baltimore oriole

FIGURE 22.12 Hybrid zones. Males of the Bullock's oriole and Baltimore oriole differ in color and courtship song, but in mixed habitats where their geographical ranges overlap, the two species produce hybrid offspring.
© Cengage Learning 2017

plant choice and their mating preferences, change simultaneously. Moreover, host plant choice is controlled by multiple gene loci in some insect species, and it is clearly influenced by prior experience in others.

The apple maggot (*Rhagoletis pomonella*) is the most thoroughly studied example of possible sympatric speciation in animals **(Figure 22.13).** This fly's natural host plant in eastern North America is the hawthorn (*Crataegus* species), but at least two host races have appeared in the past 150 years. The larvae of a new host race were first discovered feeding on apples in New York State in the 1860s. In the 1960s, a cherry-feeding host race appeared in Wisconsin.

Genetic analyses have shown that variations at just a few gene loci underlie differences in the feeding preferences of *Rhagoletis* host races. Other genetic differences cause the host races to develop at different rates, and adults of the three races mate during different summer months. Nevertheless, individuals show no particular preference for mates of their own host race, at least under simplified laboratory conditions. Thus, although behavioral isolation has not developed between races, ecological and temporal isolation may separate adults in nature. Researchers are still not certain that the different host races are reproductively isolated under natural conditions.

In 2010, Andrew P. Michel of the University of Notre Dame and colleagues elsewhere in the United States and Germany published a genomic analysis of the apple- and hawthorn-feeding races of *Rhagoletis*. Their results suggest that over the past 150 years, the two races have diverged at many loci in their genomes—not just at the loci that influence food choice and developmental rate—and that the divergence has largely been driven by disruptive selection, a diversifying form of natural selection (described in Section 21.3).

Sympatric speciation often occurs in plants through a genetic phenomenon, **polyploidy,** in which an individual has one or more *extra* copies of the entire haploid complement of chromosomes (see Section 13.3). Polyploidy can lead to speciation because these large-scale genetic changes may prevent polyploid individuals from breeding with individuals of the parent species. Nearly half of all flowering plant species are polyploid, including many important crops and ornamental species. The genetic mechanisms that produce polyploid individuals in plant populations are well understood; we describe them in the next section as part of a broader discussion of the genetics of speciation.

STUDY BREAK 22.3

1. What are the two stages required for allopatric speciation?
2. Why might insects from different host races be unlikely to mate with each other?

FIGURE 22.13 Sympatric speciation in animals. Male and female apple maggots (*Rhagoletis pomonella*) court on a hawthorn leaf. The female will later lay her eggs on the fruit, and the offspring will feed, mate, and lay their eggs on hawthorns as well.

Dr. Jim Smith, Michigan State University

THINK OUTSIDE THE BOOK

Figure 22.12 presents an example of a hybrid zone between two closely related bird species in North America. Search the biological literature and the Internet for three other examples of hybrid zones in animals. How large are the areas of hybridization? What prezygotic isolating mechanisms prevent the species from interbreeding outside the hybrid zones? Is the frequency of hybrid matings increasing or decreasing through time?

22.4 Genetic Mechanisms of Speciation

What genetic changes lead to reproductive isolation between populations? In this section we examine three genetic mechanisms that can lead to reproductive isolation: *genetic divergence* between allopatric populations, *polyploidy* in sympatric populations, and *chromosome alterations*, which occur independently of the geographical distributions of populations.

Genetic Divergence in Allopatric Populations Can Lead to Speciation

In the absence of gene flow, geographically separated populations inevitably accumulate genetic differences through the action of mutation, genetic drift, and natural selection.

How much genetic divergence is necessary for speciation to occur? To understand the genetic basis of speciation in closely related species, researchers first identify the specific causes of reproductive isolation. They then use standard techniques of genetic analysis along with new molecular, genomic, and bioinformatic approaches to analyze the genetic mechanisms that establish reproductive isolation. In cases of postzygotic reproductive isolation, mutations in just a few gene loci can establish reproductive isolation. For example, if two common aquarium fishes, swordtails (*Xiphophorus helleri*) and platys (*Xiphophorus maculatus*), mate, two genes induce the development of lethal tumors in their hybrid offspring. When hybrid sterility is the primary cause of reproductive isolation between *Drosophila* species, at least 5 gene loci are responsible. About 55 gene loci contribute to postzygotic reproductive isolation between the European fire-bellied toad (*Bombina bombina*) and the yellow-bellied toad (*Bombina variegata*).

In cases of prezygotic reproductive isolation, some mechanisms have a surprisingly simple genetic basis. For example, a single mutation reverses the direction of coiling in the shells of some snails (*Bradybaena* species): some individuals coil in a clockwise spiral and others coil in a counterclockwise spiral. Snails with shells that coil in opposite directions cannot approach each other closely enough to mate, making reproduction between them mechanically impossible. *Molecular Insights* provides another example of how a genetic change in one biochemical pathway contributes to reproductive isolation between two closely related fruit flies.

Many traits that now function as prezygotic isolating mechanisms may originally have evolved in response to sexual selection (described in Section 21.3). In sexually dimorphic species, this evolutionary process exaggerates showy structures and courtship behaviors in males, the traits that females use to identify appropriate mates. When two populations encounter one another on secondary contact, these traits

may also prevent interspecific mating. For example, many closely related duck species exhibit dramatic variation in the appearance of males, but not females **(Figure 22.14)**, an almost certain sign of sexual selection. Yet these species hybridize readily in captivity, producing offspring that are both viable and fertile.

Reproductive isolation and speciation in ducks and other sexually dimorphic birds probably results from geographical isolation and sexual selection on just a few morphological and behavioral characteristics that influence their mating behavior. Thus, sometimes the evolution of reproductive isolation may not require much genetic change at all. Indeed, sexual selection appears to increase the rate at which new species arise: bird lineages that are sexually dimorphic generally include more species than do related lineages in which males and females have a similar appearance.

Polyploidy Is a Common Mechanism of Sympatric Speciation in Plants

Polyploidy is common among plants, and it may be an important factor in the evolution of some fish, amphibian, and reptile species. Polyploid individuals can arise from chromosome duplications within a single species (autopolyploidy) or through hybridization of different species (allopolyploidy).

AUTOPOLYPLOIDY In **autopolyploidy (Figure 22.15A),** a diploid ($2n$) individual may produce, for example, tetraploid ($4n$) offspring, each of which has four complete chromosome sets. Auto-polyploidy often results when gametes, through an error in either mitosis or meiosis, spontaneously receive the same number of chromosomes as a somatic cell. Such gametes are called **unreduced gametes** because their chromosome number has not been reduced compared with that of somatic cells.

Diploid pollen can fertilize the diploid ovules of a self-fertilizing individual, or it may fertilize diploid ovules on another plant with unreduced gametes. The resulting tetraploid offspring can reproduce either by self-pollination or by breeding with other tetraploid individuals. However, a tetraploid plant cannot produce fertile offspring by hybridizing with its diploid parents. The fusion of a diploid gamete with a normal haploid gamete produces a triploid ($3n$) offspring, which is

Mallard ducks *(Anas platyrhynchos)*

Pintail ducks *(Anas acuta)*

FIGURE 22.14 Sexual selection and prezygotic isolation. In closely related species, such as mallard and pintail ducks, males have much more distinctive coloration than females, a sure sign of sexual selection at work.

A Dual-Function Gene: Ecological and behavioral isolation in two *Drosophila* species

The fruit fly species *Drosophila serrata* and *Drosophila birchii*, although closely related, occupy different environments along the eastern coast of Australia. *D. serrata*, which is resistant to dehydration, lives in a wide range of habitats; but *D. birchii*, which is sensitive to dehydration, lives only in moist rainforests. Research has shown that a specific class of molecules in fruit flies' exoskeletons—methyl-branched cuticular hydrocarbons (mbCHCs)—provide protection from evaporative water loss. Not surprisingly, *D. serrata* produces relatively large amounts of mbCHCs, whereas *D. birchii* produces much less. Moreover, mbCHCs also function as chemical signals that contribute to the choice of mates in *D. serrata*.

Research Question

Is the production of abundant mbCHC by *D. serrata* at least partially responsible for its ecological and behavioral isolation from *D. birchii*, a species that produces very little mbCHC?

Experiment

Henry Chung and colleagues at the University of Wisconsin–Madison and the Howard Hughes Medical Institute and a colleague at the University of California, Riverside, identified the gene—*CG3524*, part of a fatty acid synthase (FAS) pathway—responsible for the difference in the production of mbCHCs in the two species. Next, they used RNA interference (described in Section 16.3) to knock down the expression of *CG3524* in the structures that produce mbCHCs in both male and female *D. serrata*.

After producing the transgenic flies, the researchers compared the survival times of transgenic and wild-type *D. serrata* in a dehydrating environment. They also applied synthetic mbCHC to transgenic *D. serrata* and to wild-type *D. birchii* to determine whether the application of these molecules increased their resistance to dehydration.

The researchers then conducted a series of mate choice experiments in *D. serrata* to determine whether mbCHC level had an effect on their attractiveness to flies of the opposite sex. Males and females were each presented with a choice of either a wild-type or transgenic mate. They also tested whether applying mbCHC to *D. birchii*, which produces

very little of the compound, would induce *D. serrata* to mate with individuals of this closely related species.

Finally, the researchers explored the *D. birchii* genome to determine whether this species has the gene that codes for the production of mbCHCs. After discovering that the gene is present, they also conducted experiments to see if the gene is expressed in the structures that produce mbCHC.

Results

The results of the experiment on dehydration resistance demonstrate that mbCHCs contribute to dehydration resistance in these flies. Under dehydrating conditions, wild-type *D. serrata* survived at least three times longer than transgenic flies with reduced mbCHC levels **(Figure A)**. Moreover, transgenic *D. serrata* that received treatment with synthetic mbCHC survived longer than untreated individuals. Finally, wild-type *D. birchii* treated

KEY

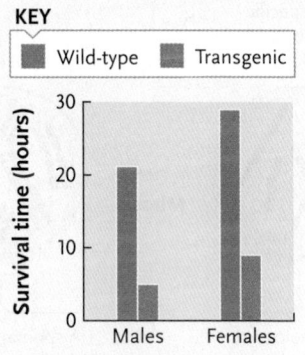

FIGURE A **Dehydration resistance of *D. serrata*.**

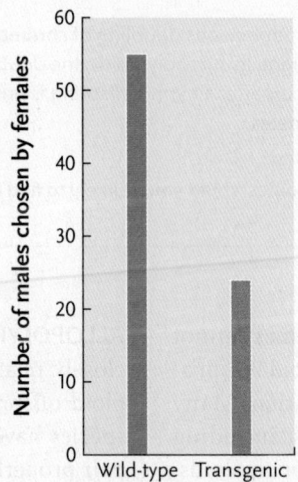

FIGURE B **Mate choice by female *D. serrata*.**

with synthetic mbCHCs survived longer than untreated flies; the latter result suggests that the low desiccation resistance of *D. birchii* results, at least in part, from their low levels of mbCHCs.

In the mate choice experiments, *D. serrata* males showed no preference for females based on the females' level of mbCHC. However, female *D. serrata* preferred males with normal mbCHC levels over transgenic males with reduced mbCHC levels **(Figure B)**.

The application of mbCHC to male and female *D. birchii* did not make them attractive to female and male *D. serrata*, respectively: no interspecific matings occurred. Although mbCHC production makes male *D. serrata* more attractive to female *D. serrata*, the compound alone is apparently not sufficient to override other prezygotic isolating mechanisms that prevent individuals of *D. serrata* from trying to mate with individuals of *D. birchii*.

The genetic and gene expression studies indicate that, although *D. birchii* still retains the gene coding for mbCHCs, the gene is not expressed in the structures that produce the molecule in adult flies.

Conclusion

The *CG3524* gene, which codes for the production of mbCHC, apparently serves a dual role in the reproductive isolation of *D. serrata* and *D. birchii*. The dehydration resistance conferred by mbCHC contributes to the ecological isolation of the two species, and the molecule also serves as a component of mate choice in *D. serrata*. The authors hypothesize that populations of a common ancestor of these two species diverged ecologically, with *D. birchii* becoming restricted to its moist rainforest habitat. Because resistance to dehydration is not critical for rainforest species, an ancestral *D. birchii* population lost the capacity to produce mbCHCs. As a result, flies of this species became less attractive as mates to *D. serrata*.

think like a scientist

How might natural selection have favored the function of mbCHC molecules as a sex attractant to female *D. serrata*?

Source: H. Chung et al. 2014. A single gene affects both ecological divergence and mate choice in *Drosophila*. *Science* 343:1148–1151.

A. Speciation by autopolyploidy in plants

A spontaneous doubling of chromosomes during meiosis produces diploid gametes. If the plant fertilizes itself, a tetraploid zygote will be produced.

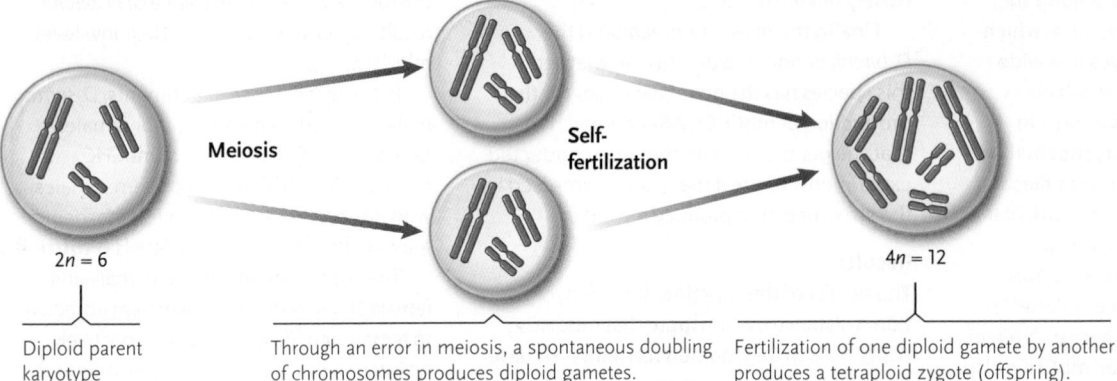

$2n = 6$ Meiosis   Self-fertilization $4n = 12$

Diploid parent karyotype	Through an error in meiosis, a spontaneous doubling of chromosomes produces diploid gametes.	Fertilization of one diploid gamete by another produces a tetraploid zygote (offspring).

B. Speciation by hybridization and allopolyploidy in plants

A hybrid mating between two species followed by a doubling of chromosomes during mitosis in gametes of the hybrid can instantly create sets of homologous chromosomes. Self-fertilization can then generate polyploid individuals that are reproductively isolated from both parent species.

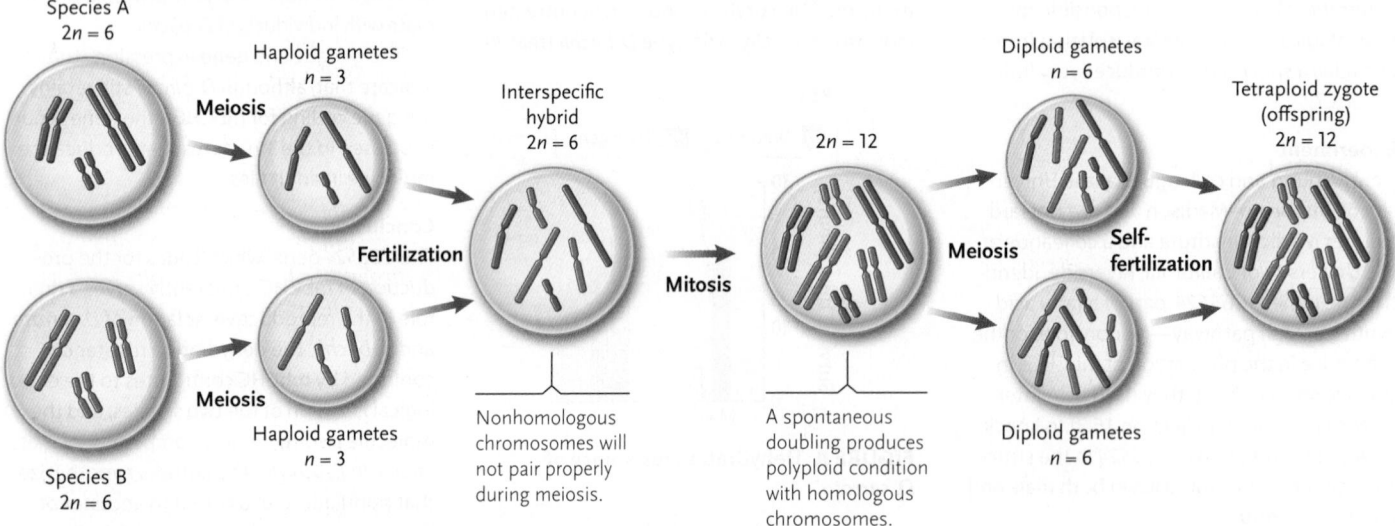

Species A
$2n = 6$

Meiosis → Haploid gametes $n = 3$

Fertilization → Interspecific hybrid $2n = 6$

Mitosis → $2n = 12$

Meiosis → Diploid gametes $n = 6$

Self-fertilization → Tetraploid zygote (offspring) $2n = 12$

Meiosis → Haploid gametes $n = 3$

Species B
$2n = 6$

Diploid gametes $n = 6$

Nonhomologous chromosomes will not pair properly during meiosis.

A spontaneous doubling produces polyploid condition with homologous chromosomes.

SUMMARY In both autopolyploidy and allopolyploidy, a spontaneous doubling of chromosome number produces gametes with twice the original number of chromosomes, but the timing of doubling is different. In autopolyploidy, the doubling occurs during a meiotic cell division that produces $2n$ gametes in the parent. In allopolyploidy, the doubling occurs after a hybrid offspring is produced, when some of its cells are undergoing mitosis; meiosis in the polyploid hybrid then produces polyploid gametes.

think like a scientist How many chromosomes would you be likely to find in the allopolyploid hybrid offspring of parents that had $2n = 8$ and $2n = 12$, respectively?

usually sterile because its odd number of chromosomes cannot segregate properly during meiosis. Thus, the tetraploid is reproductively isolated from the original diploid population. Many species of grasses, shrubs, and ornamental plants, including violets, chrysanthemums, and nasturtiums, are autopolyploids, having anywhere from 4 to 20 complete chromosome sets.

ALLOPOLYPLOIDY In **allopolyploidy** (**Figure 22.15B**), two closely related species hybridize and subsequently form polyploid offspring. Hybrid offspring are sterile if the two parent species have diverged enough that their chromosomes do not pair properly during meiosis. However, if the hybrid's chromosome number is doubled, the chromosome complement of the gametes is also doubled, producing homologous chromosomes that *can* pair during meiosis. The hybrid can then produce polyploid gametes and, through self-fertilization or fertilization with other doubled hybrids, establish a population of a new polyploid species. Compared with speciation by genetic divergence, speciation by allopolyploidy can be extremely rapid, causing a new species to arise in one generation without geographical isolation.

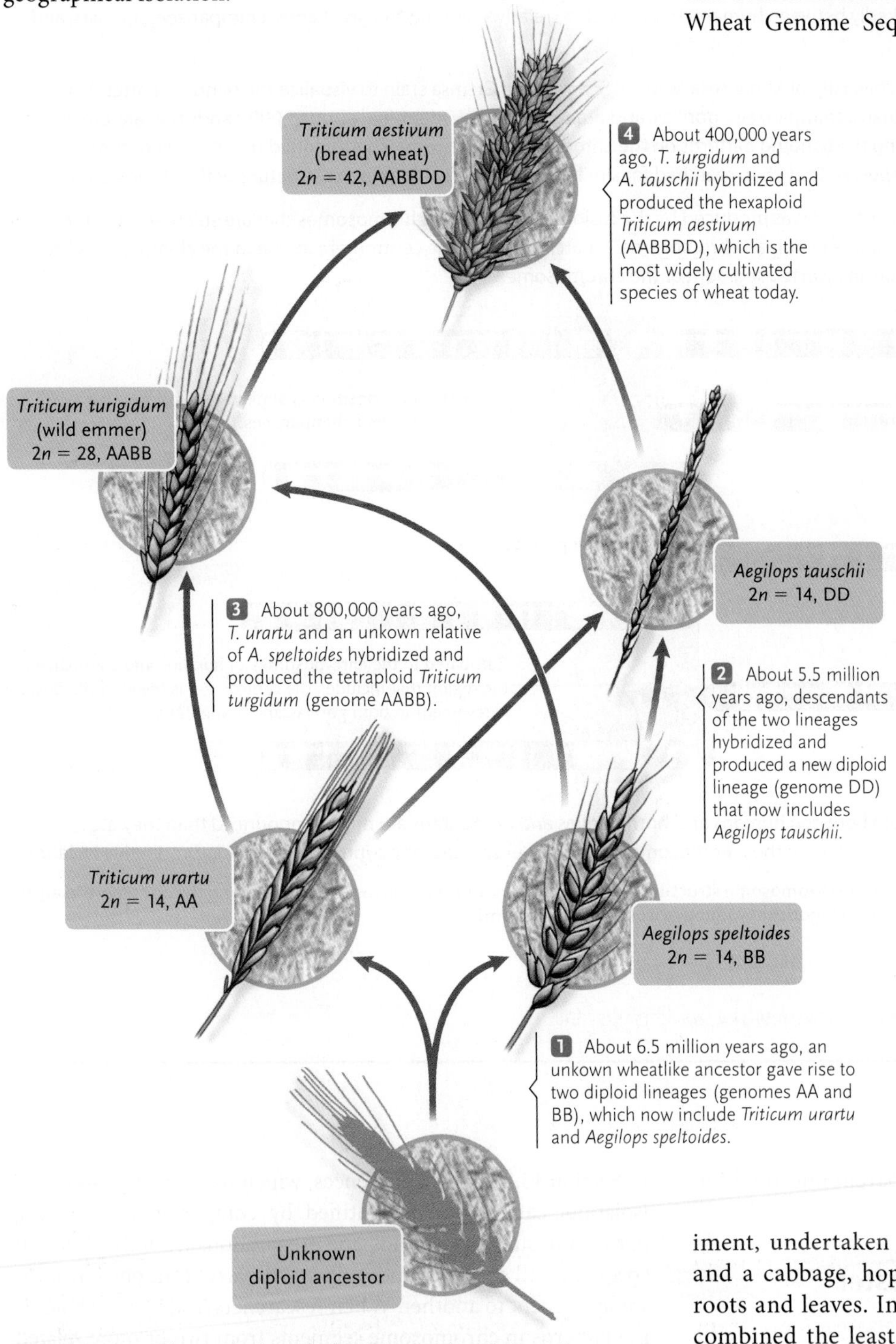

Even when sterile, polyploids are often robust, growing larger than either parent species. For that reason, both auto-polyploids and allopolyploids—including plantains (cooking bananas), coffee, cotton, potatoes, sugarcane, and tobacco—have been important to agriculture. For example, bread wheat *(Triticum aestivum),* a staple food for at least 30% of the world-wide human population, arose through a series of hybridization events. Recent research by members of the International Wheat Genome Sequencing Consortium, published in the journal *Science* in July 2014, has begun to reveal details of its genetics and ancestry **(Figure 22.16).** The bread wheat genome includes three diploid sub-genomes (identified as AA, BB, and DD) that originated in different ancestors. About 6.5 million years ago, divergence from a wheatlike ancestor produced lineages with subgenomes AA and BB. One million years later, hybridization between descendants of those lineages produced the lineage with subgenome DD. All three subgenomes are diploid, with two sets of seven chromosomes ($2n = 14$). Then, about 800,000 years ago, a hybridization between two species in lineage AA (*T. monococcum* and *T. urartu*) with one species from lineage BB (a close relative of *Aegilops speltoides*) produced a tetraploid ($2n = 28$, AABB) wheat, emmer *(Triticum turgidum),* which is still cultivated in the Middle East. About 400,000 years ago, *T. turgidum* hybridized with a diploid species in lineage DD *(A. tauschii),* producing the hexaploid ($2n = 42$, AABBDD) bread wheat *(T. aestivum)* that is widely grown today. Each of the three ancestors contributed two sets of seven chromosomes to bread wheat, making it a hexaploid with a total of 42 chromosomes.

Plant breeders often try to increase the probability of forming an allopolyploid by using chemicals that foster nondisjunction of chromosomes during mitosis. In the first such experiment, undertaken in the 1920s, scientists crossed a radish and a cabbage, hoping to develop a plant with both edible roots and leaves. Instead, the new species, *Raphanobrassica,* combined the least desirable characteristics of each parent, growing a cabbagelike root and radishlike leaves. Recent experiments have been more successful. For example, plant scientists have produced an allopolyploid grain, triticale,

FIGURE 22.16 The evolution of wheat. Researchers believe the evolution of common bread wheat *(Triticum aestivum)* resulted from a series of hybridizations, some of which produced allopolyploid species.
© Cengage Learning 2017

Triticum aestivum
(bread wheat)
$2n = 42$, AABBDD

4 About 400,000 years ago, *T. turgidum* and *A. tauschii* hybridized and produced the hexaploid *Triticum aestivum* (AABBDD), which is the most widely cultivated species of wheat today.

Triticum turigidum
(wild emmer)
$2n = 28$, AABB

Aegilops tauschii
$2n = 14$, DD

3 About 800,000 years ago, *T. urartu* and an unkown relative of *A. speltoides* hybridized and produced the tetraploid *Triticum turgidum* (genome AABB).

2 About 5.5 million years ago, descendants of the two lineages hybridized and produced a new diploid lineage (genome DD) that now includes *Aegilops tauschii.*

Triticum urartu
$2n = 14$, AA

Aegilops speltoides
$2n = 14$, BB

1 About 6.5 million years ago, an unkown wheatlike ancestor gave rise to two diploid lineages (genomes AA and BB), which now include *Triticum urartu* and *Aegilops speltoides.*

Unknown
diploid ancestor

FIGURE 22.17 | **Observational Research**

Chromosomal Similarities and Differences among Humans and the Great Apes

Question: Does chromosome structure differ between humans and their closest relatives among the apes?

Hypothesis: Large-scale chromosome rearrangements contributed to the development of reproductive isolation between species within the evolutionary lineage that includes humans and apes.

Prediction: Chromosome structure differs markedly between humans and their close relatives among the great apes: chimpanzees, gorillas, and orangutans.

Method: Jorge J. Yunis and Om Prakash of the University of Minnesota Medical School used Giemsa stain to visualize the banding patterns on metaphase chromosome preparations from humans, chimpanzees, gorillas, and orangutans. They identified about 1,000 bands that are present in humans and in the three ape species. By matching the banding patterns on the chromosomes, the researchers verified that they were comparing the same segments of the genomes in the four species. They then searched for similarities and differences in the structure of the chromosomes.

Results: Analysis of human chromosome 2 reveals that it was produced by the fusion of two smaller chromosomes that are still present in the other three species. Although the position of the centromere in human chromosome 2 matches that of the centromere in one of the chimpanzee chromosomes, in gorillas and orangutans it falls within an inverted segment of the chromosome.

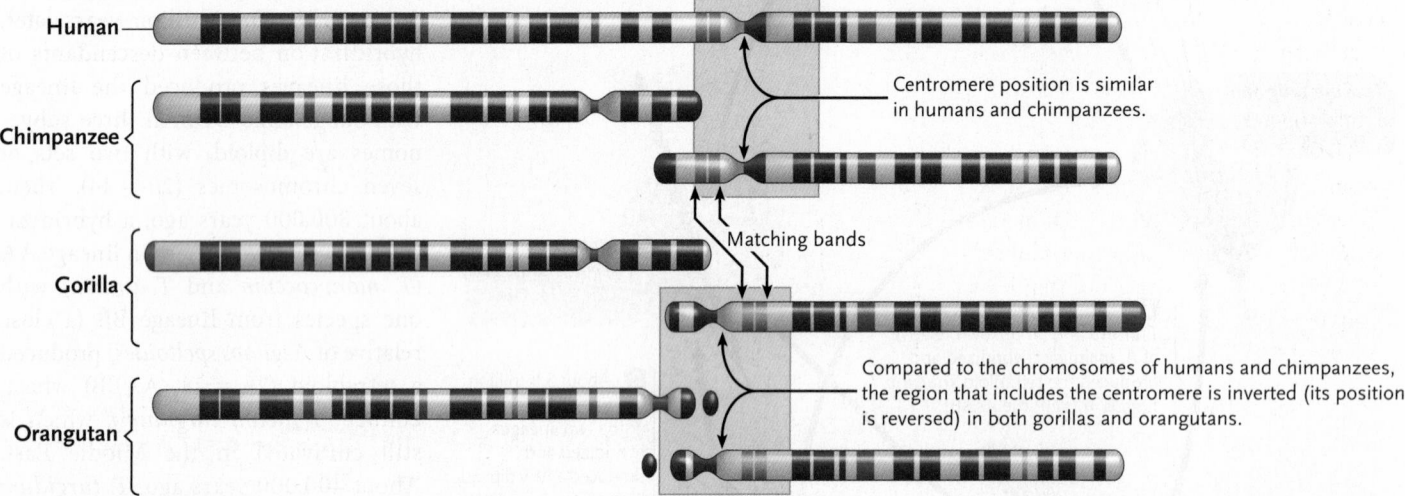

Centromere position is similar in humans and chimpanzees.

Matching bands

Compared to the chromosomes of humans and chimpanzees, the region that includes the centromere is inverted (its position is reversed) in both gorillas and orangutans.

Conclusion: Differences in chromosome structure between humans and both gorillas and orangutans are more pronounced than they are between humans and chimpanzees. Structural differences in the chromosomes of these four species may contribute to their reproductive isolation.

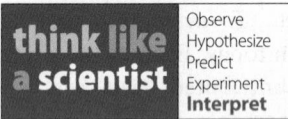
think like a scientist

Observe
Hypothesize
Predict
Experiment
Interpret

Are the differences in chromosome structure between humans and their closest relatives among the apes more likely to be prezygotic or postzygotic reproductive isolating mechanism?

Source: J. J. Yunis and O. Prakash. 1982. The origin of man: A chromosomal pictorial legacy. *Science* 215:1525–1530.

that has the disease resistance of its rye parent and the high productivity of its wheat parent.

Chromosome Alterations Can Foster Speciation

Other changes in chromosome structure or number may also foster speciation. Closely related species often have a substantial number of chromosome differences between them, including inversions, translocations, deletions, and duplications (described in Section 13.3). These differences, which may foster postzygotic isolation, can often be identified by comparing the *banding patterns* in stained chromosome preparations from the different species. In all species, banding patterns vary from one chromosome segment to another. When researchers find identical banding patterns in chromosome segments from two or more related species, they know that they are examining comparable portions of the species' genomes. Thus, the banding patterns allow scientists to identify specific chromosome segments and compare their positions in the chromosomes of different species.

The banding patterns of humans and their closest relatives among the apes—chimpanzees, gorillas, and orangutans—reveal that whole sections of chromosomes have been rearranged over evolutionary time (Figure 22.17). For example, humans have a diploid complement of 46 chromosomes, whereas chimpanzees, gorillas, and orangutans have 48 chromosomes. The difference can be traced to the fusion (that is, the joining together) of two ancestral chromosomes into chromosome 2 of humans; the ancestral chromosomes are separate in the other three species.

Moreover, banding patterns suggest that the position of the centromere in human chromosome 2 closely matches that of a centromere in one of the chimpanzee chromosomes, reflecting their close evolutionary relationship. But this centromere falls within an inverted region of the chromosome in gorillas and orangutans, reflecting their evolutionary divergence from chimpanzees and humans. (Recall from Section 13.3 that an inverted chromosome segment has a reversed orientation, so the order of genes on it is reversed relative to the order in a segment that is not inverted.) Nevertheless, humans and chimps differ from each other in centromeric inversions in six other chromosomes.

How might such chromosome rearrangements promote speciation? In a paper published in 2003, Arcadi Navarro of the Universitat Pompeu Fabra in Spain and Nick H. Barton of the University of Edinburgh in Scotland compared the rates of evolution in protein-coding genes that lie within rearranged chromosome segments of humans and chimpanzees to those in genes outside the rearranged segments. They discovered that proteins evolved more than twice as quickly in the rearranged chromosome segments. Navarro and Barton reasoned that because chromosome rearrangements inhibit chromosome pairing and recombination during meiosis, new genetic variations favored by natural selection would be conserved within the rearranged segments. These variations accumulate over time, contributing to genetic divergence between populations with the rearrangement and those without it. Thus, chromosome rearrangements can be a trigger for speciation: once a chromosome rearrangement becomes established within a population, it will diverge more rapidly from populations lacking the rearrangement. The genetic divergence eventually causes reproductive isolation.

Because speciation results from microevolutionary processes that divide one gene pool into two, it is sometimes the first step in the evolution of new biological lineages. In the next chapter we consider the effects of speciation over vast spans of time as we examine paleobiology and patterns of macroevolution. In Unit 4, we further explore the evolution of each of the major lineages in the Tree of Life.

STUDY BREAK 22.4

1. How can natural selection promote reproductive isolation in allopatric populations?
2. How does polyploidy promote speciation in plants?

Unanswered Questions

Do asexual organisms form species?

As you learned in this chapter, the biological species concept applies only to sexually reproducing organisms because only those organisms can evolve barriers to gene flow (asexual organisms reproduce more or less clonally). Nevertheless, research is starting to show that organisms whose reproduction is almost entirely asexual, such as bacteria, seem to form distinct and discrete clusters in nature. (These clusters could be considered "species.") That is, bacteria and other asexual forms may be as distinct as the species of birds described by Ernst Mayr in New Guinea. Workers are now studying the many species of bacteria in nature (only a small number of which have been discovered) to see if they indeed fall into distinct groups. If they do, scientists will need a special theory, independent of reproductive isolation, to explain this distinctness. Scientists are now working on theories of whether the existence of discrete ecological niches in nature might explain the possible discreteness of asexual "species."

How often does speciation occur allopatrically versus sympatrically?

Scientists do not know how often speciation occurs between populations that are completely isolated geographically (allopatric speciation), compared to how often it occurs in populations that exchange genes (sympatric speciation). The relative frequency of these modes of speciation in nature is an active area of research. The ongoing work includes studies on small, isolated islands: if an invading species divides into two or more species in this situation, it probably did so sympatrically, since geographical isolation of populations in small islands is unlikely. In addition, biologists are reconstructing the evolutionary history of speciation using molecular tools and correlating this history with the species' geographical distributions. If this research were to show, for example, that the most closely related pairs of species always had geographically isolated distributions, it would imply that speciation was usually allopatric. These lines of research should eventually answer the controversial question of the relative frequency of various forms of speciation.

What are the genetic changes underlying speciation?

Biologists know a great deal about the types of reproductive isolation that prevent gene flow between species, but almost nothing about their genetic underpinnings. Which genes control the difference between flower shape in monkey-flower species? Which genes lead to inviability and sterility of *Drosophila* hybrids? Which genes cause species of ducks to preferentially mate with members of their own species over members of other species? Do the genetic changes that lead to reproductive isolation tend to occur repeatedly at the same gene loci in a group of organisms, or at different gene loci? Do the changes occur mostly in protein-coding regions of genes, or in the noncoding regions that control the production of proteins? Were the changes produced by natural selection or by genetic drift? Biologists are now isolating "speciation genes" and sequencing their DNA. With only a handful of such genes known, and all of these causing hybrid sterility or inviability, there will undoubtedly be a lot to learn about the genetics of speciation in the next decade.

Jerry Coyne conducts research on speciation and teaches at the University of Chicago. To learn more about his research, go to http://pondside.uchicago.edu/ecol-evol/faculty/coyne.html.

think like a scientist Professor Coyne suggested that biologists will need to devise a special theory—or an additional type of species definition—to explain the distinctness of clusters of asexual organisms that are observed in nature. Do you think that this new explanation would be more similar to the morphological species concept or the phylogenetic species concept as described in this chapter?

REVIEW KEY CONCEPTS

For access to MindTap and additional study materials visit www.cengagebrain.com.

22.1 What Is a Species?

- In practice, most biologists describe, identify, and recognize species on the basis of morphological characteristics that serve as indicators of their genetic similarity to or divergence from other species (Figures 22.1 and 22.2).

- The biological species concept defines species as groups of interbreeding populations that are reproductively isolated from populations of other species in nature. A biological species thus represents a gene pool within which genetic material is potentially shared among populations. The biological species concept cannot be applied to organisms that reproduce only asexually, to those that are extinct, or to geographically separated populations.

- The phylogenetic species concept defines a species as a group of populations with a recently shared evolutionary history.

- Most species exhibit geographical variation of phenotypic and genetic traits. When marked geographical variation in phenotypes is discontinuous, biologists sometimes name subspecies (Figure 22.3). In ring species, populations are distributed in a ring around unsuitable habitat (Figure 22.4). Many species exhibit clinal variation of characteristics, which change smoothly over a geographical gradient (Figure 22.5).

22.2 Maintaining Reproductive Isolation

- Reproductive isolating mechanisms are characteristics that prevent two species from interbreeding (Table 22.1).

- Prezygotic isolating mechanisms either prevent individuals of different species from mating or prevent fertilization between their gametes. Prezygotic isolation occurs because species live in different habitats, breed at different times, use different courtship behavior (Figure 22.6), or differ anatomically (Figure 22.7). Prezygotic isolation can also result from genetic and physiological incompatibilities between male and female gametes.

- Postzygotic isolating mechanisms reduce the fitness of interspecific hybrids through hybrid inviability, hybrid sterility (Figure 22.8), or hybrid breakdown.

22.3 The Geography of Speciation

- The model of allopatric speciation proposes that speciation results from divergent evolution in geographically separated populations (Figures 22.9–22.11). If allopatric populations accumulate enough genetic differences, they will be reproductively isolated upon secondary contact. Nevertheless, some species hybridize over narrow areas of secondary contact (Figure 22.12).

- A model of sympatric speciation in insects suggests that reproductive isolation may evolve between host races that rarely contact one another under natural conditions (Figure 22.13). Sympatric speciation commonly occurs in flowering plants by allopolyploidy.

22.4 Genetic Mechanisms of Speciation

- Allopatric populations inevitably accumulate genetic differences, some of which may contribute to their reproductive isolation. Reproductive isolating mechanisms evolve as by-products of genetic changes that occur during divergence. Prezygotic isolating mechanisms may evolve in populations experiencing secondary contact (Figure 22.14).

- We cannot yet generalize about how many gene loci participate in the process of speciation, but at least several gene loci are usually involved.

- Speciation by polyploidy in flowering plants involves the duplication of an entire chromosome complement through nondisjunction of chromosomes during meiosis or mitosis. Polyploids can arise among the offspring of a single species (autopolyploidy; Figure 22.15A) or, more commonly, after hybridization between closely related species (allopolyploidy; Figures 22.15B and 22.16).

- Chromosome alterations can promote speciation by fostering the genetic divergence of, and reproductive isolation between, populations with different numbers of chromosomes or different chromosome structure (Figure 22.17).

Remember/Understand

1. The biological species concept defines species on the basis of:
   a. reproductive characteristics.
   b. biochemical characteristics.
   c. morphological characteristics.
   d. behavioral characteristics.
   e. all of the above.

2. Biologists can apply the biological species concept *only* to species that:
   a. reproduce asexually.
   b. lived in the past.
   c. are allopatric to each other.
   d. hybridize in captivity.
   e. reproduce sexually.

3. A characteristic that exhibits smooth changes in populations distributed along a geographical gradient is called a:
   a. ring species.
   b. hybrid.
   c. cline.
   d. hybrid breakdown.
   e. subspecies.

4. Prezygotic isolating mechanisms:
   a. reduce the fitness of hybrid offspring.
   b. generally prevent individuals of different species from producing zygotes.
   c. are found only in animals.
   d. are found only in plants.
   e. are observed only in organisms that reproduce asexually.

5. In the model of allopatric speciation, the geographical separation of two populations:
   a. is sufficient for speciation to occur.
   b. occurs only after speciation is complete.
   c. allows gene flow between them.
   d. reduces the relative fitness of hybrid offspring.
   e. inhibits gene flow between them.

6. Which of the following genetic characteristics is shared by humans and chimpanzees?
   a. They have the same number of chromosomes.
   b. The position of the centromere on human chromosome 2 matches the position of a centromere on a chimpanzee chromosome.
   c. A fusion of ancestral chromosomes formed chromosome 2.
   d. Centromeres on all of their chromosomes fall within inverted chromosome segments.
   e. all of the above

Apply/Analyze

7. If two species of holly (genus *Ilex*) flower during different months, their gene pools may be kept separate by:
   a. mechanical isolation.
   b. ecological isolation.
   c. gametic isolation.
   d. temporal isolation.
   e. behavioral isolation.

8. Geographically overlapping populations that produce hybrid offspring with low relative fitness may be undergoing:
   a. clinal isolation.
   b. sympatric speciation.
   c. allopatric speciation.
   d. reinforcement.
   e. geographical isolation.

9. An animal breeder, attempting to cross a llama with an alpaca for finer wool, found that the hybrid offspring rarely lived more than a few weeks. This outcome probably resulted from:
   a. genetic drift.
   b. prezygotic reproductive isolation.
   c. postzygotic reproductive isolation.
   d. sympatric speciation.
   e. polyploidy.

10. **Discuss Concepts** Human populations often differ dramatically in external morphological characteristics. On what basis are all human populations classified as a single species?

11. **Apply Evolutionary Thinking** How do human activities (such as destruction of natural habitats, diversion of rivers, and the construction of buildings) influence the chances that new species of plants and animals will evolve in the future? Frame your answer in terms of the geographical and genetic factors that foster speciation.

Evaluate/Create

12. Which of the following could be an example of allopolyploidy?
    a. One parent has 8 chromosomes, the other has 10, and their offspring have 36.
    b. Gametes and somatic cells have the same number of chromosomes.
    c. Chromosome number increases by one in a gamete and in the offspring it produces.
    d. Chromosome number decreases by one in a gamete and in the offspring it produces.
    e. Chromosome number in the offspring is exactly half of what it is in the parents.

13. **Discuss Concepts** All domestic dogs are classified as members of the species *Canis familiaris*. But it is hard to imagine how a tiny Chihuahua could breed with a gigantic Great Dane. Do you think that artificial selection for different breeds of dogs will eventually create different dog species?

14. **Discuss Concepts** If intermediate populations in a ring species go extinct, eliminating the possibility of gene flow between populations at the two ends of the ring, would you now identify those remaining populations as full species? Explain your answer.

15. **Design an Experiment** Design an experiment to test whether populations of birds on different islands belong to the same species.

For selected answers, see Appendix A.

INTERPRET THE DATA

David Hillis of Baylor University noted that three closely related species of leopard frog (genus *Rana*) exhibit substantial—but not complete—postzygotic reproductive isolation when crossed in the laboratory. Field surveys of numerous populations in Texas and surrounding states revealed that populations of the three species breed at various times during the year. Data on the breeding schedule of both allopatric and sympatric populations of these species are presented in the **Figure** below. Interpret the data in the figure and explain how they may demonstrate that these frogs experience prezygotic reproductive isolation in nature. What type of prezygotic reproductive isolation do the data suggest?

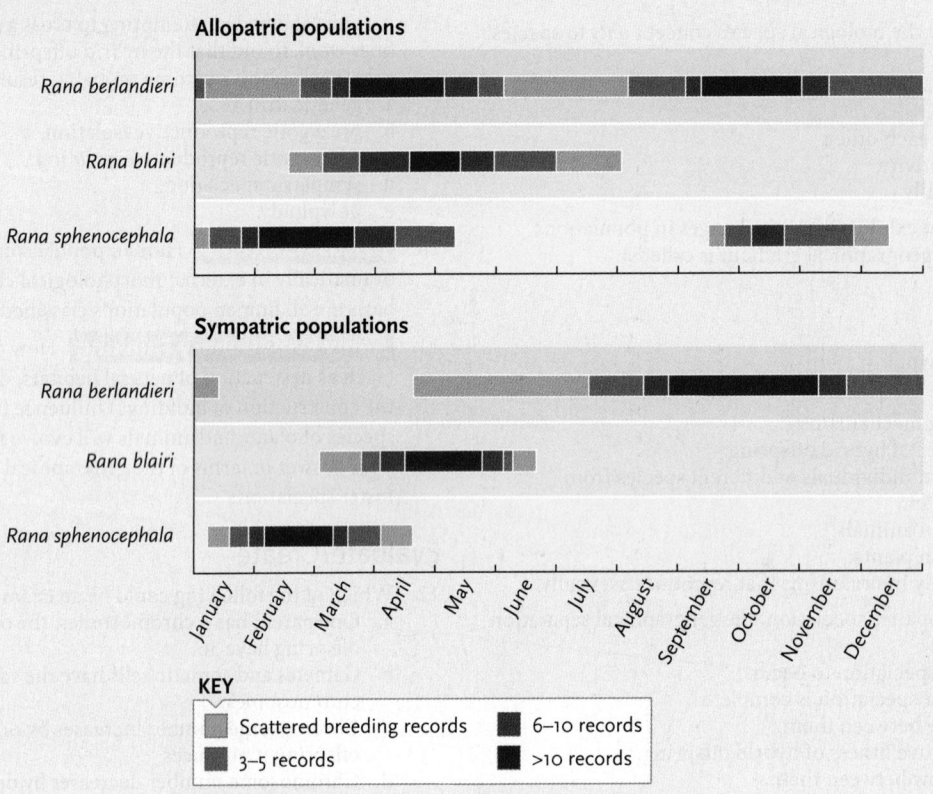

Source: D. M. Hillis. 1981. Premating isolating mechanisms among three species of the *Rana pipiens* complex in Texas and southern Oklahoma. *Copeia* 1981:312–319.

Paleobiology and Macroevolution

van Vdovin/AGE fotostock

Fossil of a trilobite *(Asaphus kowalewskii)* that lived 490 million years ago.

Why it matters . . . In a report on the fossil reptiles of Britain, published in April 1842, Sir Richard Owen described fragmentary fossils of three species that another researcher had unearthed in 1818. Recognizing that these gigantic creatures were distinct from lizards and snakes, Owen assigned them to a new group, which he named Dinosauria (*deinos* = terrible; *sauros* = lizards). In 1851, he brought the dinosaurs to the attention of a broader public in a display of the "inhabitants of the ancient world" at the Crystal Palace exhibition in London.

Owen's published accounts and Crystal Palace exhibition models led to a dinosaur craze that still grips scientists and the public today. Dinosaur discoveries continued apace, especially in North America, where many of the most complete fossils are found. In 1858, a nearly intact skeleton of the duck-billed dinosaur *Hadrosaurus foulkii* was uncovered in New Jersey. Analyses of its leg bones showed clearly that it was bipedal (that is, it stood on its hind legs rather than on all fours), a shocking hypothesis at the time. Ten years later, the specimen was exhibited at the Philadelphia Academy of Science, where it is still on display. The *Hadrosaurus* exhibit tripled attendance at the Philadelphia Academy, setting off a race to uncover other complete dinosaur skeletons. In the late 1870s, abundant fossils turned up in Colorado, Wyoming, and Alberta; and scientists and entrepreneurs scoured the region for fossils that would bring them fame and fortune. Their quest was not unlike the California gold rush of the mid-nineteenth century, with rivals reputedly hijacking one another's fossils and sometimes coming to blows over their ownership.

As natural history museums exhibited these discoveries, major questions plagued *preparators* (professionals who clean and reassemble fossils) as they mounted the specimens. What did dinosaurs really look like? How did these animals stand when they were alive? How did they hold their heads and tails?

FIGURE 23.1 Dinosaurs in living color. By examining the fine structure of pigment-containing organelles inside fossilized feathers, scientists were able to reconstruct the colors of *Anchiornis huxleyi,* a dinosaur that lived about 150 million years ago.
© Cengage Learning 2017

The continued study of dinosaur biology has answered many of these questions. In the 1960s, biologists hypothesized that dinosaurs were very active creatures, like mammals and birds are today. According to this hypothesis, many dinosaurs held their heads up and their tails off the ground; others adopted a bipedal stance. As our knowledge of dinosaurs increased, natural history museums repositioned their specimens and models into more active poses. But one question seemed impossible to answer. Did dinosaurs blend in with their surroundings, or were they brightly colored?

As you will discover later in this chapter, many dinosaurs were feathered, just as birds are today. In 2010, two groups of researchers published analyses of the colors of fossilized feathers from several species of small dinosaurs. Using scanning electron microscopes, the researchers compared the fine structure of *melanosomes* (pigment-bearing organelles) in the fossilized feathers to those of living birds, allowing them to reconstruct the feather colors of the dinosaurs. One analysis suggests that *Anchiornis huxleyi,* a dinosaur that lived about 150 million years ago in what is now northern China, had gray

or black feathers over most of its body, black and white feathers on its limbs, and a crown of red feathers on its head **(Figure 23.1).** In another study, published in 2012, researchers discovered that a feathered dinosaur (genus *Microraptor;* see Figure 23.23) was covered with iridescent black feathers, like those seen on male starlings *(Sternus vulgaris)* today. These animals must surely have used their colorful feathers in social displays.

This description of a truly remarkable discovery opens our chapter on paleobiology, the study of ancient organisms, and macroevolution, the large-scale changes in morphology and diversity observed over life's 3.8-billion-year history. Macroevolution has occurred over so vast a span of time and space that the evidence for it is fundamentally different from that for microevolution and speciation. In this chapter we consider what paleobiology and evolutionary developmental biology tell us about macroevolutionary patterns.

23.1 The Fossil Record

Paleobiologists discover, describe, and name new fossil species and analyze the morphology and ecology of extinct organisms. Because fossils provide physical evidence of life in the past, they are a primary source of data about the evolutionary history of many organisms.

Fossils Form When Organisms Are Buried by Sediments or Preserved in Oxygen-Poor Environments

Most fossils form in sedimentary rocks. Rain and runoff constantly erode the land, carrying fine particles of rock and soil downstream to a swamp, a lake, or the sea. Particles settle to the bottom as sediments, forming successive layers, called *strata,* over millions of years **(Figure 23.2).** The weight of newer sediments compresses the older layers beneath them into a

A. Sedimentation

Highest strata contain the most recent fossils.

Lowest strata contain the oldest fossils.

B. Geological strata in the Painted Desert, Arizona

Nick Greaves/Alamy

FIGURE 23.2 Sedimentation and geological strata. (A) Sedimentation deposits successive layers at the bottom of a lake or sea. **(B)** Over millions of years, the upper layers compress those below them into rock. When the rocks are later exposed by uplifting or erosion, the different layers are evident as geological strata.
© Cengage Learning 2017

solid matrix: sand into sandstone and silt or mud into shale. Fossils form within the layers when the remains of organisms are buried in the accumulating sediments. Because sedimentation superimposes new layers over old ones, the lowest strata in a sedimentary rock formation are usually the oldest and the highest layers are the newest.

The process of fossilization is a race against time because the soft remains of organisms are quickly consumed by scavengers or decomposed by microorganisms. Thus, fossils usually preserve the details of hard structures, such as the bones, teeth, and shells of animals and the wood, leaves, and pollen of plants. During fossilization, dissolved minerals replace some parts molecule by molecule, leaving a fossil made of stone **(Figure 23.3A)**; other fossils form as molds, casts, or impressions in material that is later transformed into solid rock **(Figure 23.3B).**

In some environments, the near absence of oxygen prevents decomposition, and even soft-bodied organisms are preserved. Some plants, insects, and tiny lizards and frogs are embedded in amber, the fossilized resin of coniferous trees **(Figure 23.3C)**. Other organisms are preserved in glacial ice, deeply frozen soil, coal, tar pits, or the highly acidic water of peat bogs **(Figure 23.3D)**. Sometimes organisms are so well preserved that researchers can examine their internal anatomy, cell structure, food in their digestive tracts, and even their DNA sequences.

The Fossil Record Provides an Incomplete Portrait of Life in the Past

The 300,000 described fossil species represent less than 1% of all the species that have ever lived. Several factors make the fossil record incomplete. First, soft-bodied organisms do not fossilize as readily as species with hard body parts. Moreover, we are unlikely to find fossilized remains of species that were rare or locally distributed. Finally, fossils rarely form in habitats where sediments do not accumulate, such as mountain forests. The most common fossils are those of hard-bodied, widespread, and abundant organisms that lived in swamps or shallow seas, where sedimentation is ongoing.

Although most fossils are composed of stone, they do not last forever. Many are deformed by pressure from overlying rocks or destroyed by geological disturbances such as volcanic eruptions and earthquakes. Once they are exposed on Earth's surface, where scientists are most likely to find them, rain and wind cause them to erode. Because the effects of these destructive processes are additive, old fossils are much less common than those formed more recently.

Scientists Assign Relative and Absolute Dates to Geological Strata and the Fossils They Contain

The sediments found in any one place form recognizable strata (layers) that differ in color, mineral composition, particle size, and thickness (see Figure 23.2B). If they have not been disturbed, the strata are arranged in the order in which they formed, with the youngest layers on top. However, strata are sometimes uplifted, warped, or even inverted by geological processes.

Geologists of the early nineteenth century deduced that the fossils discovered in a particular sedimentary stratum, no

A. Petrified wood (Araucariaceae)

B. *Sphenopteris*

C. Mosquitoes in amber

D. Mammoth *(Mammonteus)* in permafrost

FIGURE 23.3 Fossils. (A) Petrified wood in Arizona formed when minerals replaced the wood of dead trees, molecule by molecule. These forests lived during the late Triassic period, about 225 million years ago. **(B)** The remains of a fern *(Sphenopteris)* from the Carboniferous period, 300 million years ago, were preserved in coal. **(C)** These 10-million-year-old mosquitoes were trapped in the oozing resin of a coniferous tree and are now encased in amber. **(D)** A frozen baby mammoth that lived about 40,000 years ago was discovered embedded in Siberian permafrost in 1977.

matter where on Earth it is found, represent organisms that lived and died at roughly the same time in the past. Because each stratum formed at a specific time, the sequence of fossils in the lowest (oldest) to the highest (newest) strata reveals their *relative ages*. Geologists originally used the sequence of strata and their distinctive fossil assemblages to establish the geological time scale **(Table 23.1).**

Although the geological time scale provides a relative dating system for sedimentary strata, it does not tell us how old the rocks and fossils actually are. But many rocks contain unstable radioisotopes, which, from the moment they form, begin to break down into other, more stable elements. The breakdown proceeds at a steady rate that is unaffected by chemical reactions or environmental conditions such as temperature or pressure. Using a technique called **radiometric dating,** scientists can estimate the age of a rock by noting how much of an unstable "parent" isotope has decayed to another form. By measuring the relative amounts of the parent radioisotope and its breakdown products and comparing this ratio with the isotope's **half-life**—the time it takes for half of a given amount of radioisotope to decay—researchers can estimate the *absolute age* of the rock **(Figure 23.4).** Table 23.1 presents these age estimates along with the major geological and evolutionary events of each period.

Radiometric dating works best with volcanic rocks, which form when lava cools and solidifies. But most fossils are found in sedimentary rocks. To date a sedimentary fossil, scientists determine the age of volcanic rocks from the same stratum. Using this method, investigators have linked fossils to deposits that are hundreds of millions of years old.

Fossils that still contain organic matter, such as the remains of bones or wood, can be dated directly by measuring their content of the radioactive carbon isotope ^{14}C, which decays to ^{14}N. Living organisms absorb traces of ^{14}C and large quantities of ^{12}C, a stable carbon isotope, from the environment and incorporate them into biological molecules. As long as an organism is alive, its ^{14}C content remains constant because the ratio of ^{14}C to ^{12}C in its tissues is at equilibrium with the ratio in the environment. But as soon as the organism dies, ^{14}C begins its steady radioactive decay. Scientists use the ratio of ^{14}C to ^{12}C in a fossil to determine its age, as explained in Figure 23.4.

To develop a feeling for geological time, as revealed by radiometric data and other techniques, imagine the 4.5-billion-year history of Earth scaled onto a calendar year; each day represents a little over 12 million years. If the planet was formed on January 1, animal life originated in mid-November; terrestrial dinosaurs lived between December 14 and December 26; and the genus that includes modern humans appeared during the last 4 hours of December 31.

Fossils Provide Abundant Information about Life in the Past

Although imperfect, the fossil record provides our only direct information about life in the past. Fossilized skeletons, shells,

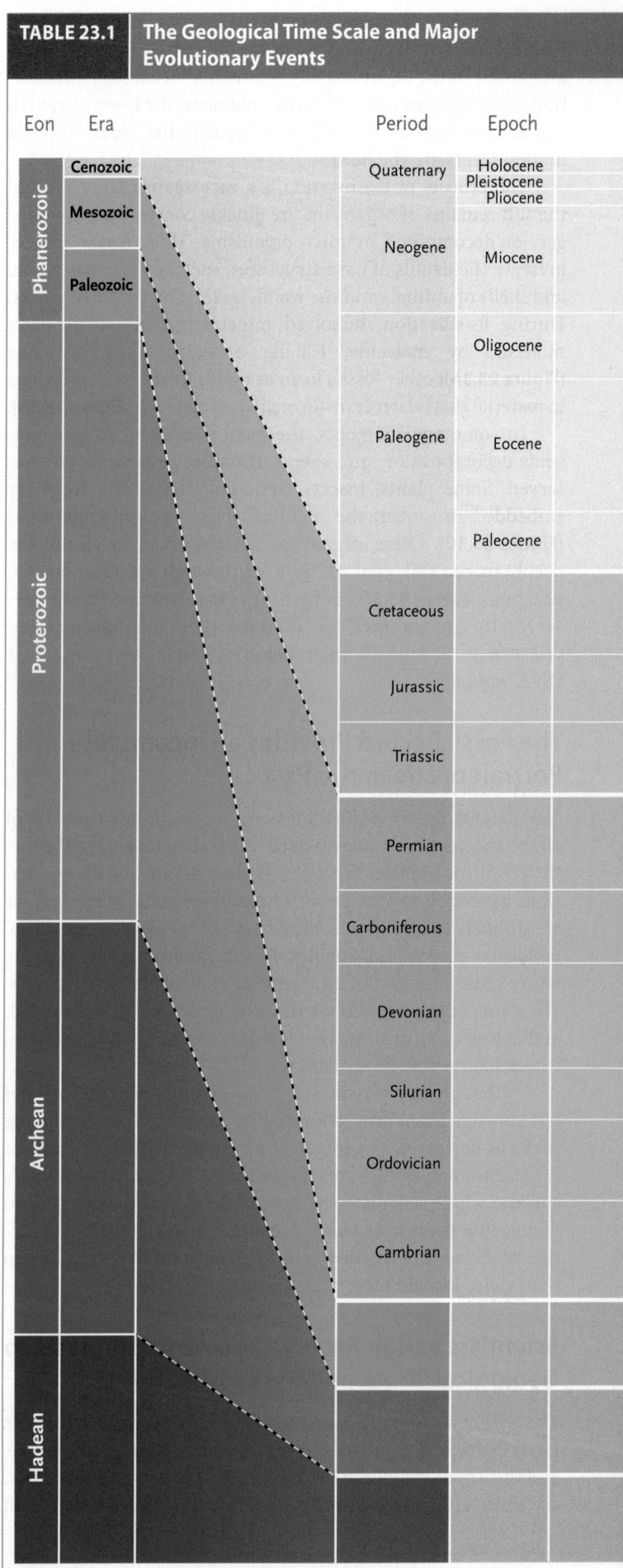

TABLE 23.1 The Geological Time Scale and Major Evolutionary Events

Eon	Era	Period	Epoch
Phanerozoic	Cenozoic	Quaternary	Holocene
			Pleistocene
			Pliocene
	Mesozoic	Neogene	Miocene
	Paleozoic		
		Paleogene	Oligocene
			Eocene
			Paleocene
Proterozoic		Cretaceous	
		Jurassic	
		Triassic	
		Permian	
		Carboniferous	
		Devonian	
Archean		Silurian	
		Ordovician	
		Cambrian	
Hadean			

Millions of Years Ago	Major Evolutionary Events	Mass Extinctions

0.01
2.6 — Origin of modern humans; major glaciations
5.3 — Origin of bipedal human ancestors

Angiosperms and mammals further diversify and dominate terrestrial habitats

23.0

Primates diversify; origin of apes

33.9

Angiosperms and insects diversify; modern orders of mammals differentiate

55.8

Grasslands and deciduous woodlands spread; modern birds, mammals, snakes, pollinating insects diversify; continents approach current positions

65.5 ┈┈┈┈┈┈┈┈┈┈┈┈┈┈┈┈┈┈┈┈┈┈┈┈┈┈┈┈┈┈┈┈┈► RIP **Cretaceous**

First angiosperms; insects, marine invertebrates, fishes, dinosaurs diversify; asteroid impact causes mass extinction at end of period, eliminating most dinosaurs and many other groups

145.5

Gymnosperms abundant in terrestrial habitats; modern fishes diversify; dinosaurs diversify and dominate terrestrial habitats; frogs, salamanders, lizards, birds, and placental mammals appear; continents continue to separate

201.6 ┈┈┈┈┈┈┈┈┈┈┈┈┈┈┈┈┈┈┈┈┈┈┈┈┈┈┈┈┈┈┈┈┈► RIP **Triassic**

Predatory fishes and reptiles dominate oceans; gymnosperms dominate terrestrial habitats; diversification of dinosaurs; early mammals; Pangaea starts to break up; mass extinction at end of period

251.0 ┈┈┈┈┈┈┈┈┈┈┈┈┈┈┈┈┈┈┈┈┈┈┈┈┈┈┈┈┈┈┈┈┈► RIP **Permian**

Insects and amniotes abundant and diverse in swamp forests; some amniotes colonize oceans; fishes colonize freshwater habitats; continents coalesce into Pangaea, causing glaciation and decline in sea level; huge volcanic eruptions cause mass extinction at end of period, eliminating 85% of species worldwide

299.0

Vascular plants form large swamp forests; first flying insects; amphibians diversify; first amniotes appear

359.0 ┈┈┈┈┈┈┈┈┈┈┈┈┈┈┈┈┈┈┈┈┈┈┈┈┈┈┈┈┈┈┈┈┈► RIP **Devonian**

Terrestrial vascular plants diversify; fungi, invertebrates, tetrapod vertebrates colonize land; first insects and seed plants; major glaciation at end of period; mass extinction, mostly of marine life

416.0

Jawless fishes diversify; first jawed fishes, first terrestrial arthropods and vascular plants

444.0 ┈┈┈┈┈┈┈┈┈┈┈┈┈┈┈┈┈┈┈┈┈┈┈┈┈┈┈┈┈┈┈┈┈► RIP **Ordovician**

Major radiations of marine invertebrates and jawless fishes; first terrestrial plants, fungi, and animals; major glaciation at end of period causes mass extinction of marine life

488.0 ┈┈┈┈┈┈┈┈┈┈┈┈┈┈┈┈┈┈┈┈┈┈┈┈┈┈┈┈┈┈┈┈┈► RIP **Cambrian**

Appearance of modern animal phyla, including vertebrates (Cambrian explosion); simple marine communities; mass extinctions eliminate many groups at end of period

542.0

High concentration of oxygen in atmosphere; origin of eukaryotic cells; evolution and diversification of protists, fungi, soft-bodied animals

2,500

Origin of life; evolution of prokaryotes, including anaerobic and photosynthetic bacteria; oxygen starts to accumulate in atmosphere; origin of aerobic respiration

3,850

Formation of Earth, including crust, atmosphere, and oceans

4,600

FIGURE 23.4 | **Research Method**

Radiometric Dating

Purpose: Radiometric dating allows researchers to estimate the absolute age of a rock sample or fossil.

Protocol:

1. Knowing the approximate age of a rock or fossil, select a radioisotope that has an appropriate half-life. Because different radioisotopes have half-lives ranging from seconds to billions of years, it is usually possible to choose one that brackets the estimated age of the sample under study. For example, if you think that your fossil is more than 10 million years old, you might use uranium-235. The half-life of ^{235}U, which decays into the lead isotope ^{207}Pb, is about 700 million years. Or if you think that your fossil is less than 70,000 years old, you might select carbon-14. The half-life of ^{14}C, which decays into the nitrogen isotope ^{14}N, is 5,730 years.

Radioisotopes Commonly Used in Radiometric Dating			
Radioisotope (Unstable)	More Stable Breakdown Product	Half-Life (Years)	Useful Range (Years)
Samarium-147 →	Neodymium-143	106 billion	>100 million
Rubidium-87 →	Strontium-87	48 billion	>10 million
Thorium-232 →	Lead-208	14 billion	>10 million
Uranium-238 →	Lead-206	4.5 billion	>10 million
Uranium-235 →	Lead-207	700 million	>10 million
Potassium-40 →	Argon-40	1.25 billion	>100,000
Carbon-14 →	Nitrogen-14	5,730	<70,000

2. Prepare a sample of the material and measure the quantities of the parent radioisotope and its more stable breakdown product.

Interpreting the Results: Compare the relative quantities of the parent radioisotope and its breakdown product (or some other stable isotope) to determine what percentage of the original parent radioisotope remains in the sample. Then use a graph of radioactive decay for that isotope to determine how many half-lives have passed since the sample formed.

Theory of radiometric dating

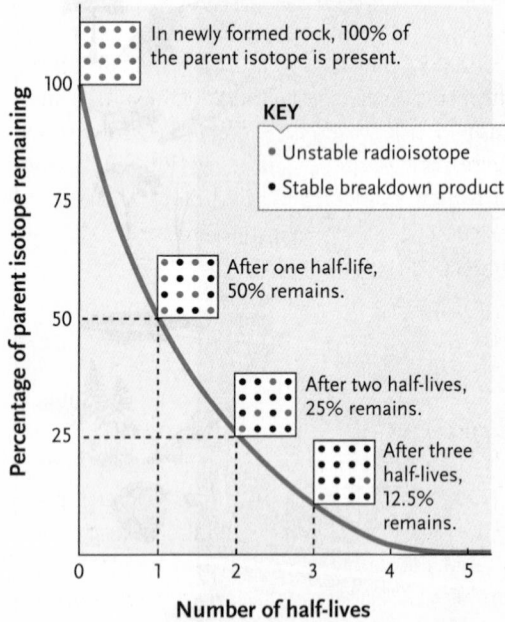

In newly formed rock, 100% of the parent isotope is present.

KEY
- Unstable radioisotope
- Stable breakdown product

After one half-life, 50% remains.

After two half-lives, 25% remains.

After three half-lives, 12.5% remains.

Percentage of parent isotope remaining

Number of half-lives

Knowing the number of half-lives that have passed allows you to estimate the age of the sample.

Radiometric dating of a fossilized mollusk

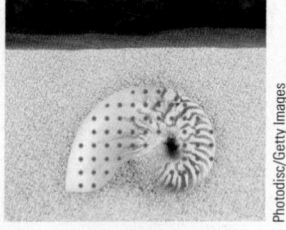

A living mollusk absorbed trace amounts of ^{14}C, a rare radioisotope of carbon, and large amounts of ^{12}C, which is the more stable and common isotope of carbon.

When the mollusk died, it was buried in silt and fossilized. From the moment of its death, the ratio of ^{14}C to ^{12}C began to decline through radioactive decay. Because the half-life of ^{14}C is 5,730 years, half of the original ^{14}C was eliminated from the fossil in 5,730 years and half of what remained was eliminated in another 5,730 years.

After the fossil was discovered, a scientist determined that its ^{14}C to ^{12}C ratio was one-eighth (12.5%) of the ^{14}C to ^{12}C ratio in living organisms. Thus, radioactive decay had proceeded for three half-lives—about 17,000 years—since the mollusk's death.

stems, leaves, and flowers tell us about the size and appearance of ancient animals and plants. The fossil record also allows scientists to see how structures were modified as they became adapted for specialized uses (see Figure 21.3). Moreover, fossils chronicle the proliferation and extinction of evolutionary lineages and provide data on their past geographical distributions.

Recent applications of new techniques allow paleobiologists to make truly remarkable discoveries about extinct organisms.

For example, researchers recently solved a long-standing puzzle about lambeosaurines, a group of huge duck-billed dinosaurs that lived in the swampy habitats of western North America in the late Cretaceous period. The heads of lambeosaurines sported bony crests with extensive nasal passages. Paleobiologists had proposed that the crests served as weapons in male combat, adornments that attracted mates, snorkels that facilitated breathing underwater, radiators that cooled the dinosaurs' bodies, structures that enhanced their sense of smell, or resonating chambers that produced honking vocalizations. Based on anatomical analyses, researchers accepted the vocalization hypothesis as the most probable: computerized acoustic models predicted that air flowing through the nasal passages would have produced low frequency sounds (30–375 hertz).

But could lambeosaurines hear sounds in that frequency range? In 2009, David Evans of the Royal Ontario Museum in Toronto, Canada, and colleagues in Ohio used computed tomography and 3-D visualization software to scan and reconstruct the interior anatomy of the skulls and brains of several lambeosaurine species **(Figure 23.5)**. Their findings indicate that the inner ears of lambeosaurines were attuned to hear low-frequency sounds that matched those predicted by the earlier research. The authors concluded that the elaborate nasal passages in lambeosaurine crests, as well as the structure of their inner ears, facilitated vocal communication, perhaps between parents and their offspring.

Fossils also provide indirect data about behavior, physiology, and ecology. For example, the fossilized footprints of some

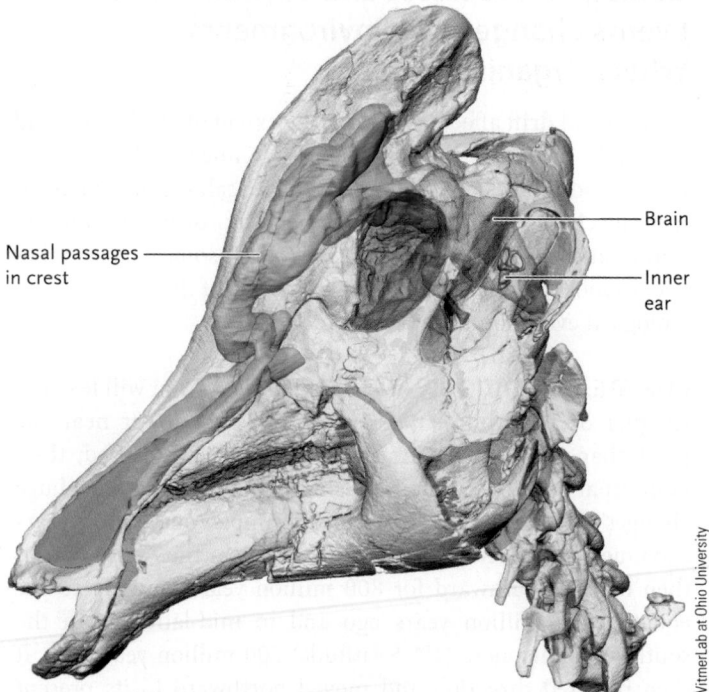

Nasal passages in crest — Brain

Inner ear

WitmerLab at Ohio University

FIGURE 23.5 Honking dinosaurs. Analysis of the sinuses and braincase of lambeosaurines (*Corythosaurus* species) revealed that the nasal passages in their crests served as resonating chambers for the production of low-frequency sounds that their inner ears could detect.

dinosaurs suggest that adults surrounded their young when the group moved, possibly to protect them from predators. Complex scrolls of bone in the nasal passages of early mammals suggest that they had a well-developed sense of smell, and fossilized teeth and dung provide data about the diets of extinct animals. As a final example, the study of fossilized pollen allows paleobiologists to reconstruct the vegetation and climate of ancient sites. The changing arrays of fossils that document biological evolution partly reflect large-scale shifts in Earth's physical environments, a topic that we explore in the next section.

STUDY BREAK 23.1

1. What biological materials are the most likely to fossilize?
2. Why does the fossil record provide an incomplete portrait of life in the past?
3. What sorts of information can paleobiologists discern from the fossil record?

23.2 Earth History

Organisms have interacted constantly with their environments over geological time. These interactions have triggered fundamental changes in Earth's physical environment, as well as in the organisms that occupy them. Perhaps the most striking effect of living systems on the physical environment was the development of an oxidizing atmosphere, discussed extensively in Chapter 25. Photosynthesis by prokaryotic organisms increased the concentration of oxygen in the atmosphere, facilitating the evolution of eukaryotic cells and later allowing animals to invade terrestrial habitats. In this section we focus on long-term shifts in geography and climate—as well as some brief but catastrophic events—that have significantly changed the environments where organisms live.

Continental Drift Has Altered the Configuration of Landmasses and Oceans

Major geological and climatic shifts occur because the planet's crust is constantly in motion. According to the theory of **plate tectonics,** Earth's crust is broken into irregularly shaped plates of rock—roughly 40 km thick—that float on its semisolid mantle **(Figure 23.6)**. Currents in the mantle cause the plates—and the continents embedded in them—to move, a phenomenon called **continental drift.**

Continent-sized landmasses have coalesced and broken apart several times in Earth's history. About 250 million years ago, they merged into a single supercontinent called Pangaea; continental drift later separated Pangaea into a northern continent, Laurasia, and a southern continent, Gondwana. Laurasia and Gondwana subsequently broke into the continents we know today **(Figure 23.7)**.

The breakup and separation of the continents had a huge impact on the evolution of terrestrial and aquatic organisms.

A. Earth's crustal plates

B. Model of plate tectonics

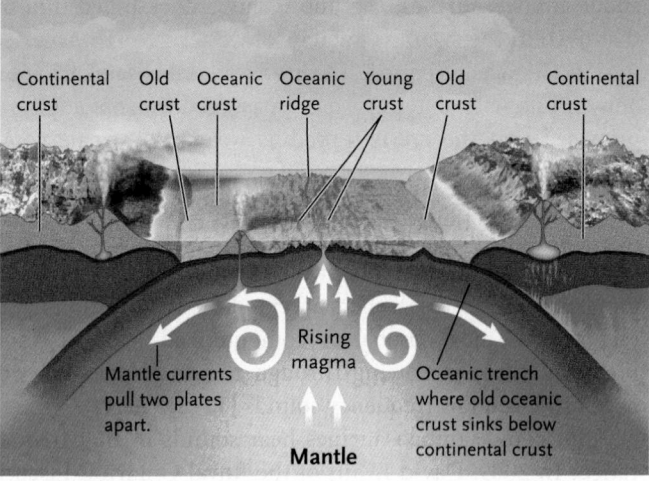

KEY

⎯⎯⎯ Oceanic ridge ‥‥‥‥ Oceanic trench

FIGURE 23.6 Plate tectonics. (A) Earth's crust is broken into large, rigid plates. New crust is added at oceanic ridges, and old crust is recycled into the mantle at oceanic trenches. **(B)** Oceanic ridges form where pressure in the mantle forces magma (molten rock) through fissures in the seafloor. Mantle currents pull the plates apart on either side of the ridge, forcing the seafloor to move laterally away from the ridge. This phenomenon, seafloor spreading, is widening the Atlantic Ocean about 3 cm per year. Oceanic trenches form where plates collide. The heavier oceanic crust sinks below the lighter continental crust, and it is recycled into the mantle, a process called *subduction*. The highest mountain ranges (including the Rockies, Himalayas, Alps, and Andes) formed where subduction uplifted continental crust. Earthquakes and volcanoes are also common near trenches.

© Cengage Learning 2017

For example, the widespread distribution of many terrestrial groups on Pangaea was a defining characteristic of the Paleozoic era. But continental drift later established large-scale patterns of geographical isolation. Each continent became a separate arena in which organisms evolved independently of those on other landmasses. Thus, the organisms that live on southern continents, the remnants of Gondwana, are very different from those that live on northern continents, which are the remnants of Laurasia. For example, the large flightless birds known as *ratites* (pictured in Figure 20.2) occur only in Africa, Australia, South America, and on nearby islands; no comparable or closely related species live in Eurasia or North America.

Moreover, the smaller landmasses produced by plate tectonics are surrounded by more extensive coastlines and shallow marine habitats than were the larger continents. Coastal waters have always harbored tremendous biodiversity, and as the continents became increasingly separated, the populations living in these shallow seas became geographically isolated from each other. As the continents assumed their present positions at different latitudes, locally distributed organisms diversified and differentiated from those in other regions. Today, shallow marine habitats in the temperate zone are often occupied by giant kelp beds, whereas those in the tropics harbor coral reefs (described further in Chapter 51).

Geological Processes and Unpredictable Events Changed the Environments Where Organisms Lived

Continental drift affected climate, the extent of glaciations, and sea levels, changing the physical environments where organisms live on local, regional, and global scales. Most environmental changes occurred incrementally over millions of years—and millions of generations—but some catastrophes had a sudden impact. The combined effects of these changes on biological evolution have been profound.

CLIMATE, GLACIATIONS, AND SEA LEVEL As you will learn in Chapter 51, environmental temperatures are lower near the poles than near the equator. As the continents drifted, their latitudinal positions shifted, causing what must have been huge changes in local temperatures. For example, one billion years ago Queensland, Australia, was situated near the North Pole; it then drifted southward for 800 million years, moving to the equator 440 million years ago and to mid-latitudes in the southern hemisphere (40° S latitude) 200 million years ago. It then reversed direction and moved northward to its present position near the equator (12° S).

On a regional level, the sizes of landmasses also have an effect on their climates. Coastal regions generally experience smaller daily and seasonal fluctuations in temperature than

FIGURE 23.7 Continental drift. Earth's many landmasses coalesced during the Permian period, forming the supercontinent Pangaea. About 160 million years ago, Pangaea began to separate into a southern continent, Gondwana, and a northern continent, Laurasia. Then Gondwana began to break apart. The landmasses that would form Africa and India pulled away first, opening the South Atlantic and Indian Oceans. Australia separated from Antarctica about 55 million years ago and slowly drifted northward. South America separated from Antarctica shortly thereafter. Laurasia remained nearly intact until 43 million years ago when North America and Greenland both separated from Europe and Asia. Movement of the continents also changed the shapes and the sizes of the oceans.

© Cengage Learning 2017

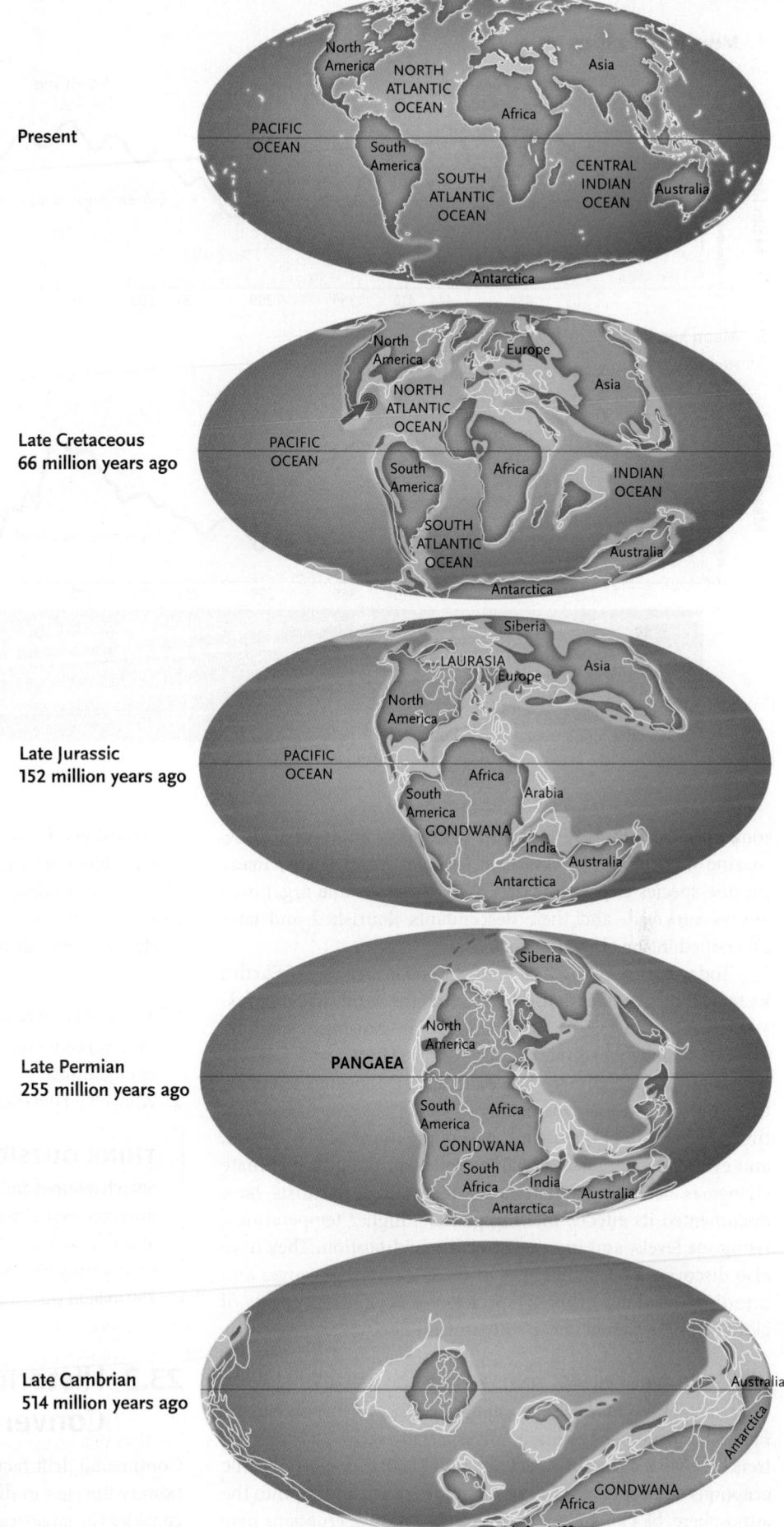

interior regions because their proximity to the sea moderates their climate. Whenever landmasses were joined into large continents, vast expanses of the interior landscape must have experienced frigid winters and hot summers. As the landmasses broke up into smaller continents, a larger fraction of the land was close to the sea and its moderating influence.

On a global scale, the overall climate has shifted from warm and wet to cool and dry several times in Earth's history **(Figure 23.8)**. The changing positions of continents altered the flow of ocean currents, which, along with small changes in Earth's orbit around the Sun, contributed to these climatic shifts. At times when Earth's climate was cooler than it is today, the polar ice caps grew larger. And because much of Earth's water was incorporated into these enormous glaciers, rainfall was greatly reduced and sea level fell dramatically. Between the cold spells, Earth's climate was warmer than it is today. Under those conditions, glaciers retreated, rainfall was luxuriant, temperature differences between equatorial and polar regions were reduced, and sea levels rose.

Changing climate patterns and the accompanying rise or fall of sea level influenced the evolution of living systems. Organisms adapted to the altered conditions, shifted their geographical ranges into suitable environments, or became extinct. Declining sea levels

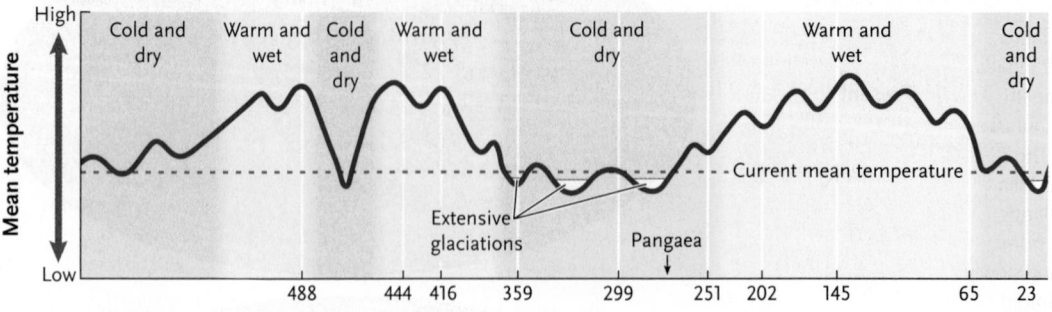

A. Mean temperature

High
Mean temperature
Cold and dry | Warm and wet | Cold and dry | Warm and wet | Cold and dry | Warm and wet | Cold and dry
Low

Extensive glaciations
Pangaea
Current mean temperature

488   444 416   359   299   251   202   145   65   23

B. Mean sea level

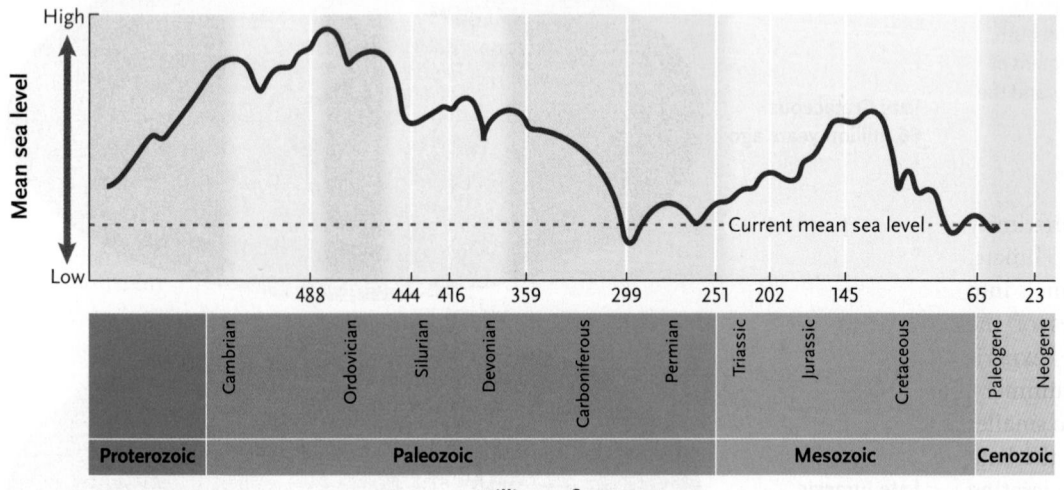

High
Mean sea level
Low

Current mean sea level

488   444 416   359   299   251   202   145   65   23

Cambrian | Ordovician | Silurian | Devonian | Carboniferous | Permian | Triassic | Jurassic | Cretaceous | Paleogene | Neogene

Proterozoic | Paleozoic | Mesozoic | Cenozoic

Millions of years ago

FIGURE 23.8 Variations in climate and sea level.
(A) Earth's climate has shifted from cold and dry to warm and wet multiple times in Earth's history. Particularly cold periods were accompanied by extensive glaciation. **(B)** As a result of climatic shifts and continental drift, sea level has often risen and fallen. Large drops in sea level eliminated many near-shore, shallow marine environments.
© Cengage Learning 2017

took an especially severe toll on organisms living in shallow marine environments. As their habitats disappeared, many marine species ceased to exist. Nevertheless, some organisms always survived, and their descendants flourished and later diversified in the altered environments.

Today we are living in one of the cooler periods of Earth's history. But human activities, including the combustion of fossil fuels and industrial-scale agriculture, are causing an extraordinarily rapid warming of the atmosphere and oceans. Although Earth's climate has experienced repeated temperature shifts, these changes occurred slowly relative to the lifetimes of individual organisms, and many populations, species, and evolutionary lineages adapted to them. But global climate change is occurring much faster today, and scientists have documented its effects: melting glaciers, higher temperatures, rising sea levels, and increasing ocean acidification. They have also discovered recent changes in the geographical ranges and breeding schedules of many species. We explore the effects of global climate change more fully in Unit Seven.

VOLCANIC ERUPTIONS AND ASTEROID IMPACTS Sudden, unpredictable events have also changed physical environments on Earth. For example, volcanoes are especially active near oceanic trenches where old crust is recycled into Earth's mantle. Volcanic eruptions often spew enormous quantities of ash and gas into the atmosphere, blocking incoming sunlight. Massive eruptions may

cause Earth's surface temperature to decrease several degrees for as long as a year. Other types of tectonic activity release molten rock on Earth's surface, sometimes with the opposite effect on climate, as you will discover in Section 23.4.

Asteroid strikes have also had devastating effects on living systems. In addition to obliterating the area around the site of impact, the collisions threw massive amounts of material into the atmosphere, blocking sunlight and potentially altering the planet's climate. As we describe in Section 23.4, these cataclysmic events have sometimes caused many forms of life to disappear over relatively short periods of geological time.

STUDY BREAK 23.2

1. How did continental drift affect the geographical distributions of organisms?
2. What effect did glaciations have on sea level?

THINK OUTSIDE THE BOOK

Search Internet and library resources for information about how continental drift and other major geological phenomena affected the area where you live. At what latitude was your home or university 65 million years ago, 100 million years ago, and 250 million years ago? Was it ever submerged below a shallow sea?

23.3 Historical Biogeography and Convergent Biotas

Continental drift facilitated the diversification of distinct evolutionary lineages in different regions on Earth. In this section we consider the large-scale geographical distributions of organisms.

A. Nothofagus

Geoff Bryant/Science Source

FIGURE 23.9 Distribution of southern beech trees (Nothofagus). The genus Nothofagus (A) exhibits a disjunct distribution (B); it includes species living in South America and in Australasia. (C) A reconstruction of the positions of Australia and South America in the late Cretaceous reveals that they were parts of a nearly continuous landmass at that time.

B. Current distribution of Nothofagus species

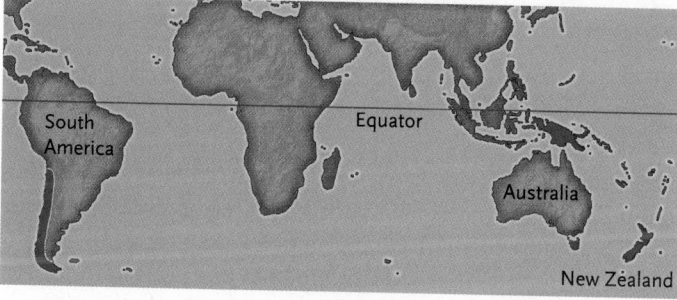

C. Position of southern continents in the late Cretaceous

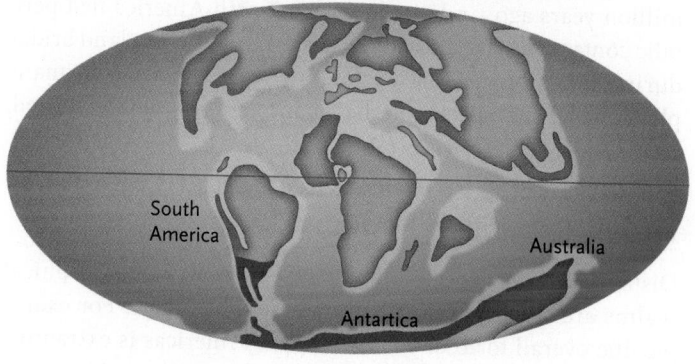

Historical Biogeography Explains the Broad Geographical Distributions of Organisms

More than a century after Darwin published his observations, the theory of plate tectonics refocused attention on biogeography. Historical biogeographers try to explain how organisms acquired their geographical distributions over evolutionary time.

CONTINUOUS AND DISJUNCT DISTRIBUTIONS Many species have a **continuous distribution:** they live in suitable habitats throughout a geographical area. For example, herring gulls (*Larus argentatus*) live along the coastlines of all northern continents. Continuous distributions, especially in mobile organisms such as birds, usually require no special historical explanation.

Other groups exhibit **disjunct distributions,** in which closely related species live in widely separated locations. Two

phenomena—dispersal and vicariance—can create disjunct distributions. **Dispersal** is the movement of organisms away from their place of origin; it can produce a disjunct distribution if a new population becomes established on the far side of a geographical barrier. **Vicariance** is the fragmentation of a once-continuous geographical distribution by external factors.

How can researchers determine whether a disjunct distribution is the product of dispersal or vicariance? Analysis of the fossil record for a group of organisms, interpreted in the context of both their evolutionary relationships and continental drift, can provide an answer. For example, southern beech trees (genus *Nothofagus*) are currently found in Australasia and South America **(Figure 23.9).** (Australasia is the region that includes Australia, New Guinea, New Zealand, and other islands of the South Pacific.) For many years, the genus' distribution only on southern continents and islands has been the classic example of vicariance: the group presumably originated on Gondwana, and the supercontinent's subsequent fragmentation apparently isolated ancestral *Nothofagus* populations on different continents and islands where their descendants live today. The fossil record generally supports that interpretation: researchers have identified fossilized *Nothofagus* pollen, dating from 55 to 34 million years ago, in locations that still formed a nearly contiguous landmass at that time (see Figure 23.9C).

However, Lyn G. Cook and Michael D. Crisp of the Australian National University recently undertook a molecular analysis of the relationships and geographical distributions of living *Nothofagus* species. Their research reveals that the genus had a much more complicated biogeographic history. It confirms that the distributions of some *Nothofagus* species in Australia and South America are the products of vicariance: their common ancestor originated on Gondwana *before* the two continents separated from each other, and their descendants now live on both landmasses. However, the analysis also suggests that the common ancestor of certain closely related *Nothofagus* species in New Zealand and Australia originated more than 30 million years *after* New Zealand had separated from Australia. Thus, vicariance cannot explain their current distribution. Instead, seeds of these trees must have dispersed across the open ocean from one place to the other. Therefore, both vicariance and dispersal have played important roles in establishing the present geographical distribution of *Nothofagus*.

BIOGEOGRAPHICAL REALMS For species that were widespread in the Mesozoic era, Pangaea's breakup was a powerful vicariant experience. The subsequent geographical isolation of continents fostered the evolution of distinctive regional **biotas.** (The word *biota* refers to all of the organisms living in a region; *fauna* to all of the animals; and *flora* to all of the plants.) In 1876, Alfred Russel Wallace, co-discoverer of the theory of evolution by natural selection, used differences in regional biotas to define six **biogeographical realms.** In 2013, Ben G. Holt of the University of Copenhagen, Denmark, and colleagues from multiple institutions in Europe and the United States published an updated analysis of these global patterns. Using data on the

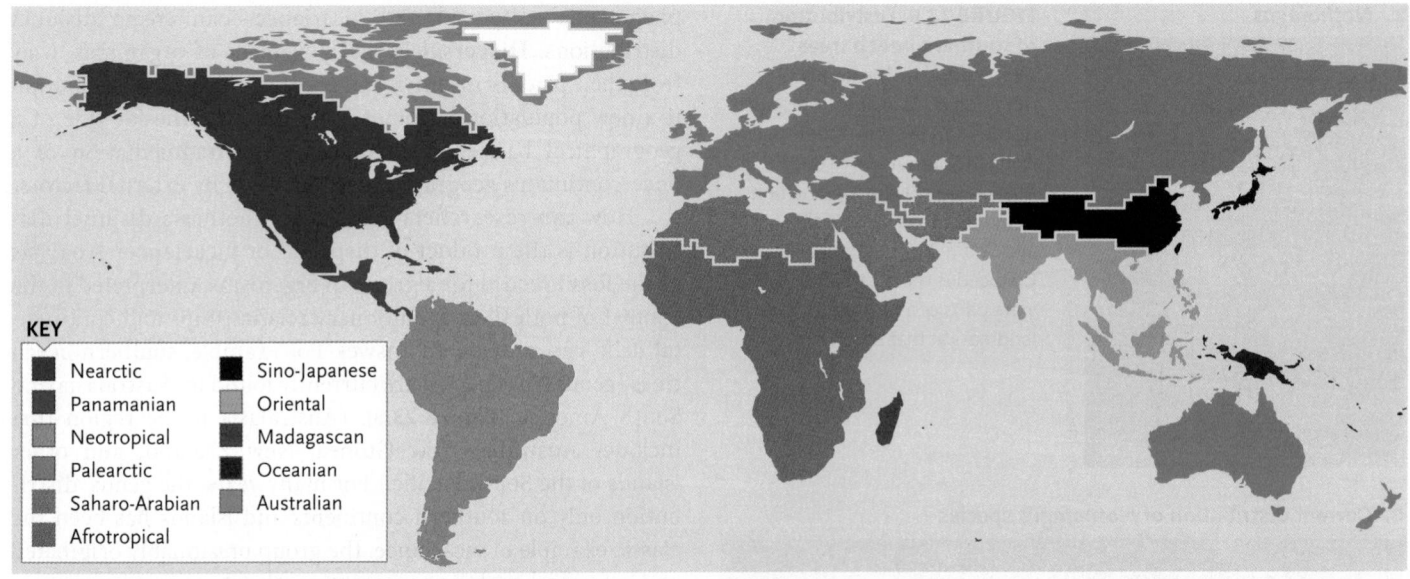

FIGURE 23.10 Zoogeographical realms. Each realm contains a distinctive fauna.
© Cengage Learning 2017

KEY

■	Nearctic	■	Sino-Japanese
■	Panamanian	■	Oriental
■	Neotropical	■	Madagascan
■	Palearctic	■	Oceanian
■	Saharo-Arabian	■	Australian
■	Afrotropical		

evolutionary histories of terrestrial vertebrates as well as their geographic distributions, the researchers identified 11 major **zoogeographical realms,** contiguous regions that are occupied by animals with a shared evolutionary history **(Figure 23.10).**

The Australian and Neotropical realms, which have been geographically isolated since the Mesozoic, contain many **endemic species** (those that occur nowhere else on Earth). The Australian realm, in particular, has had no complete land connection to any other continent for at least 55 million years. As a result, Australia's mammalian fauna is unique, made up almost entirely of endemic marsupials (mammals that give birth after a short gestation period and then carry their young in a pouch). Although Eocene fossils of placental mammals (mammals that do not rear their young in a pouch) have been found in Australia, all of them became extinct. The only native placental mammals that live there now are bats and rodents, which

apparently dispersed to Australia from either Asia or New Guinea after the marsupial fauna had already diversified there.

The biotas of the Nearctic and Palearctic realms are, by contrast, fairly similar. North America and Eurasia were frequently connected by land bridges: eastern North America was attached to Western Europe until the breakup of Laurasia 43 million years ago, and northwestern North America had periodic contact with northeastern Asia over the Bering land bridge during much of the past 60 million years. As a result, many plant and animal species on these continents are closely related.

Evolution Has Produced Convergent Biotas in Widely Separated Regions

Distantly related species living in different biogeographical realms are sometimes very similar in appearance. For example, the overall form of cactuses in the Americas is extraordinarily similar to that of spurges in Africa **(Figure 23.11).** But these lineages arose independently long after those continents had separated; thus, cactuses and spurges did not inherit their similarities from a shared ancestor. Their overall resemblance is the product of **convergent evolution,** the evolution of similar adaptations in distantly related organisms that occupy similar environments.

Convergent evolution has also fostered morphological similarities in distantly related animals that feed on similar foods and occupy similar habitats in widely separated geographic ranges. Sometimes, large portions of entire faunas develop convergent morphologies. For example, the marsupial mammals of Australia and the placental mammals of North America—groups that arose after the breakup of Pangaea—include many pairs of morphologically convergent species **(Figure 23.12).** Although the species in each pair differ in most of their anatomical details, their overall appearance and structure are

A. Cactus (*Echinocereus* species)

Edward S. Ross

FIGURE 23.11 Convergent evolution in plants. (A) North American cactuses (family Cactaceae) are strikingly similar to **(B)** African spurges (family Euphorbiaceae). Convergent evolution adapted both groups to desert environments with thick, water-storing stems, spiny structures that discourage animals from feeding on them, CAM photosynthesis (see Section 8.4), and stomata that open only at night.

B. Spurge (*Euphorbia* species)

Edward S. Ross

A. Mammal evolution and continental drift

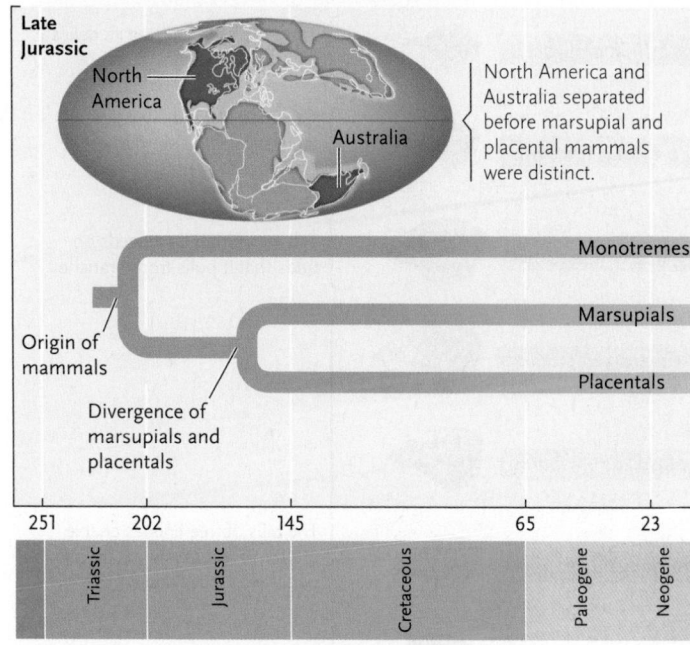

B. Convergence of placental and marsupial mammals

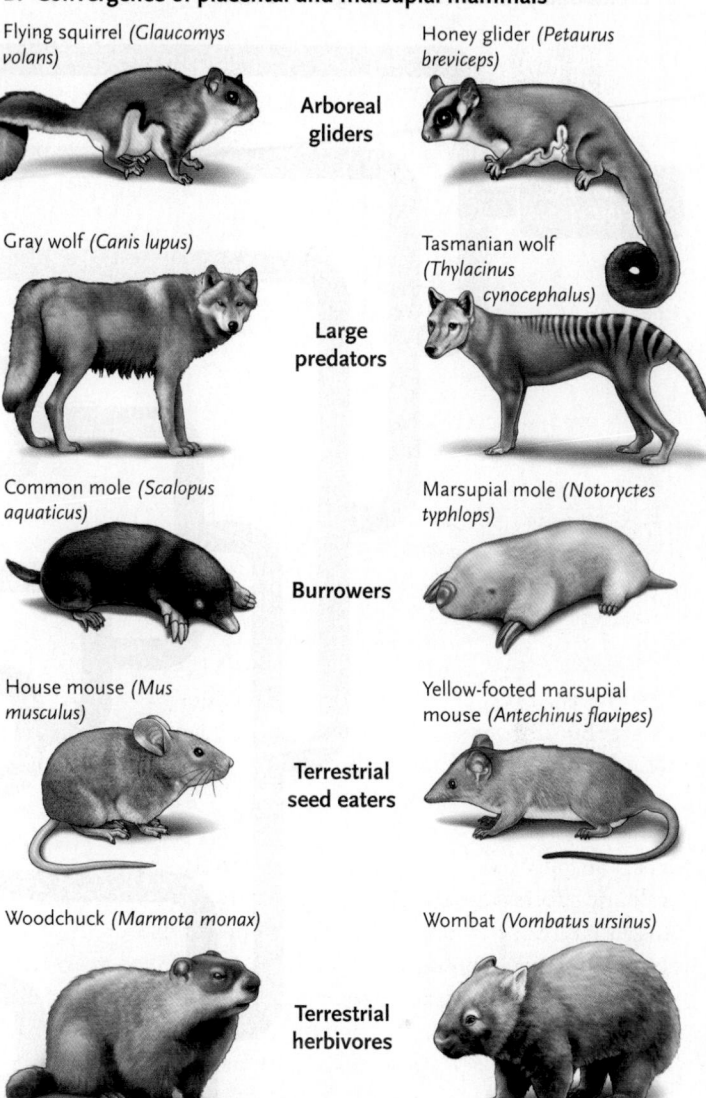

FIGURE 23.12 Convergent evolution in mammalian faunas.
(A) Two distinctive lineages of mammals, the placentals and the marsupials, diverged after Pangaea had already split into northern and southern continents. **(B)** Thus, biologists conclude that convergent evolution, rather than shared ancestry, produced similar body forms in North American placental mammals (on left) and Australian marsupial mammals (on right). The members of each pair differ in their anatomical details, but their overall forms are similar even though they are only distantly related. Biologists interpret these similarities as convergent adaptations in animals that fulfill similar ecological roles on the two continents.
© Cengage Learning 2017

remarkably similar. Biologists interpret these convergent morphologies as evolutionary responses to similar patterns of natural selection in the habitats where these animals live.

STUDY BREAK 23.3

1. Which type of geographical distribution requires no special explanation?
2. Why do distantly related species that live in different biogeographical realms sometimes resemble each other?

23.4 The History of Biodiversity

As organisms were isolated on different landmasses or in the seas that surrounded them, the number of species living on Earth—its overall **biodiversity**—changed over time as the result of adaptive radiations and extinctions.

Adaptive Radiations Are Clusters of Related Species with Diverse Ecological Adaptations

In some lineages, rapid speciation produces a cluster of closely related species that occupy different habitats or consume different foods; we describe such a lineage as an **adaptive radiation.** The Hawaiian fruit flies described in *Focus on Research: Basic Research* in Chapter 22 and the *Anolis* lizards described in *Focus on Research: Model Organisms* in Chapter 32 are examples of adaptive radiations in animals. Comparable adaptive radiations have also been identified in plants.

Adaptive radiation often occurs after an ancestral species moves to a new location and uses available resources in a unique way. In the case of the Galápagos finches, which inspired Darwin's theory of evolution by means of natural selection, an ancestral seed-eating finch arrived from the South American mainland between two and three million years ago. After becoming established on one island, individuals from

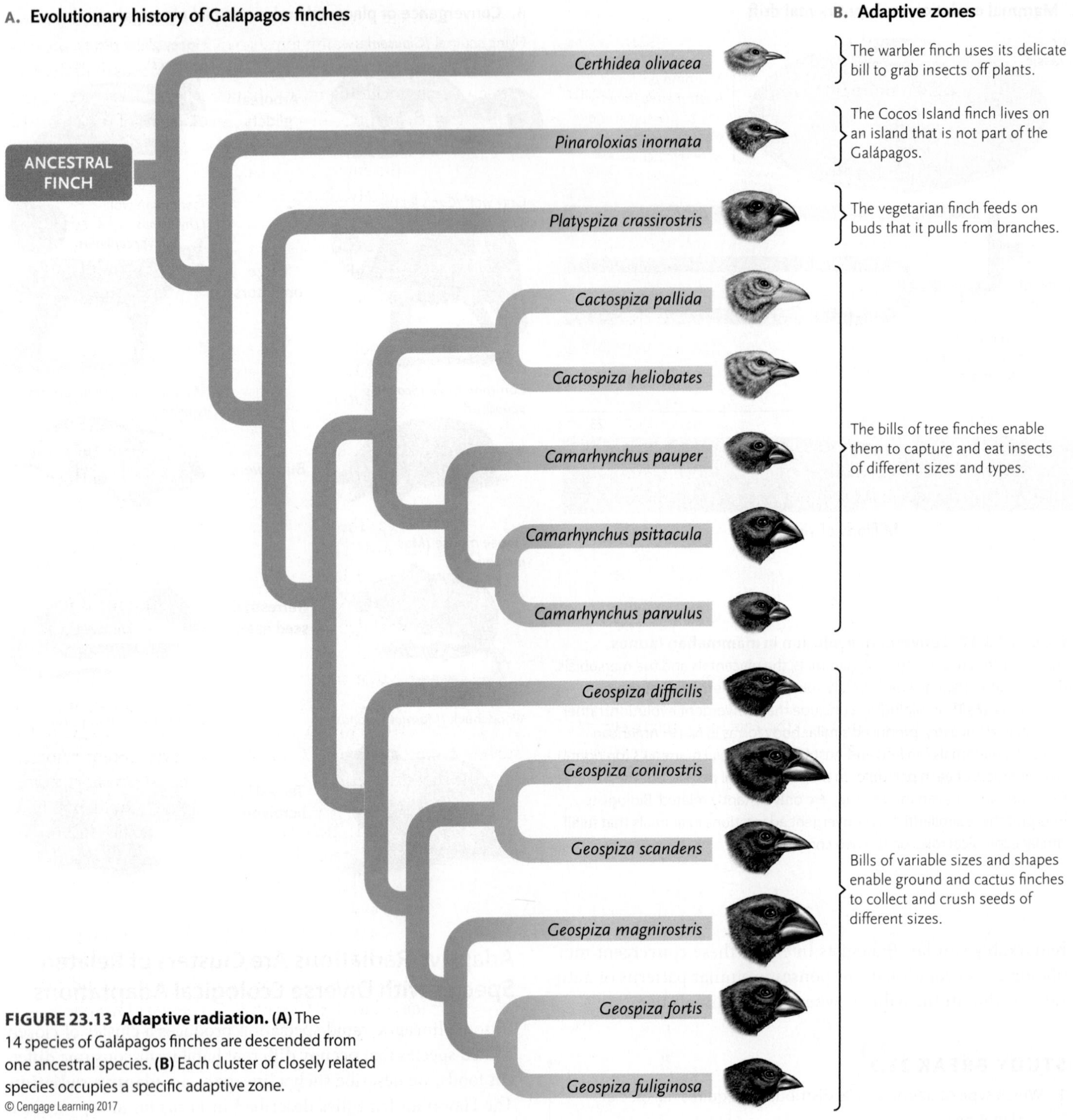

A. Evolutionary history of Galápagos finches

ANCESTRAL FINCH

Certhidea olivacea

Pinaroloxias inornata

Platyspiza crassirostris

Cactospiza pallida

Cactospiza heliobates

Camarhynchus pauper

Camarhynchus psittacula

Camarhynchus parvulus

Geospiza difficilis

Geospiza conirostris

Geospiza scandens

Geospiza magnirostris

Geospiza fortis

Geospiza fuliginosa

B. Adaptive zones

The warbler finch uses its delicate bill to grab insects off plants.

The Cocos Island finch lives on an island that is not part of the Galápagos.

The vegetarian finch feeds on buds that it pulls from branches.

The bills of tree finches enable them to capture and eat insects of different sizes and types.

Bills of variable sizes and shapes enable ground and cactus finches to collect and crush seeds of different sizes.

FIGURE 23.13 Adaptive radiation. (A) The 14 species of Galápagos finches are descended from one ancestral species. **(B)** Each cluster of closely related species occupies a specific adaptive zone.
© Cengage Learning 2017

this population colonized other islands in the archipelago (see Figures 20.9 and 22.11). Environments on the different islands vary from arid to moist and from flat to mountainous, and the allopatric finch populations diverged in their genetics, morphology, and food habits. Today, 14 descendant species occupy the Galápagos and neighboring Cocos Island, living in a range of habitats and feeding on a wide variety of foods **(Figure 23.13).** Different clusters of related species each occupy a somewhat different **adaptive zone,** a term that describes a general way of life. An organism's move into a new adaptive zone is sometimes

triggered by the chance evolution of a key structural or behavioral innovation that allows it to use the environment in a unique way. For example, the dehydration-resistant eggs of early reptiles enabled them to complete their life cycle on land, opening terrestrial habitats to them. Similarly, the evolution of flowers that attract insect pollinators was a key innovation in the history of flowering plants.

An adaptive radiation may also be triggered after the demise of a successful group. Mammals, for example, were relatively inconspicuous during their first 150 million years on Earth, presumably because dinosaurs dominated terrestrial habitats. But after dinosaurs declined in the late Mesozoic era, mammals underwent an explosive adaptive radiation. Today they are the dominant vertebrates in many terrestrial habitats.

Extinctions Have Been Common in the History of Life

Biodiversity has not always increased through the history of life. The additions to biodiversity caused by adaptive radiations and other speciation events are counteracted by **extinction,** the death of the last individual in a species or the last species in a lineage. Paleobiologists recognize two distinct patterns of extinction in the fossil record: background extinction and mass extinction.

Species and lineages have been going extinct since life first appeared. We should expect species to disappear at some low rate, the **background extinction rate;** as environments change, poorly adapted organisms will not survive and reproduce. In all likelihood, more than 99.9% of the species that have ever lived are now extinct. David Raup of the University of Chicago has suggested that, on average, as many as 10% of species go extinct every million years and that more than 50% go extinct every 100 million years. Thus, the history of life has been characterized by an ongoing turnover of species.

On at least six occasions, however, extinction rates rose well above the background rate. During these **mass extinctions,** large numbers of species and lineages died out over relatively short periods of geological time **(Figure 23.14).** The Permian extinction was the most severe: more than 85% of the species alive at that time—including all trilobites, many insects and amphibians, and the trees of the coal swamp forests—disappeared forever. During the last mass extinction, at the end of the Cretaceous, half the species on Earth, including most dinosaurs, became extinct. A seventh mass extinction, potentially the largest of all, may be occurring now as a result of global climate change and human degradation of the environment (see Unit Seven).

Scientists largely agree about the causes of three of the six historical mass extinctions. For example, the Ordovician extinction occurred after Gondwana moved toward the South Pole, triggering a glaciation that cooled the world's climate and lowered sea levels, eliminating many shallow marine environments.

Scientists hypothesize that the massive Permian extinction was triggered by a series of ongoing volcanic eruptions that released a flood of molten lava hundreds—or even thousands—of meters thick. Once solidified, the lava formed the "Siberian traps," a rock formation that covers 1.6 million km² of northeastern Asia. Several potential effects of this massive eruption may have sealed the fate of many marine and terrestrial species. The volcanic activity released enough sulfur dioxide to decrease the pH of rain to 2 (the equivalent of lemon juice), which would have devastated vegetation and the animals that fed on it. Volcanism may also have ignited massive underground coal deposits, and the resulting fires would have released toxic metal-containing ash into the atmosphere. Scientists estimate that the lava flood also released so much carbon dioxide, a greenhouse gas (see Section 54.4), into the atmosphere that it would have increased the average global surface temperature by 8°C to 10°C; however, recent research suggests that this period of global warming began shortly after the mass extinction—and, thus, could not have caused it. Regardless of which specific "kill mechanisms" caused the demise of so many organisms, 100 million years elapsed before global diversity returned to preextinction levels.

Most researchers agree that an asteroid impact initiated the mass extinction at the end of the Cretaceous period. The resulting dust cloud would have blocked the sunlight necessary for photosynthesis, setting up a chain reaction of extinctions that began with microscopic marine organisms. Geological evidence supports this hypothesis. Rocks dating to the end of the Cretaceous period (66 million years ago) contain a highly concentrated layer of iridium, a metal that is rare on Earth but common in asteroids. The impact from an iridium-laden asteroid only 10 km in diameter could have caused an explosion equivalent to that of 100 million megatons of TNT, scattering iridium dust around the world. Geologists have identified the Chicxulub crater, 180 km in diameter, on the edge of Mexico's Yucatán peninsula (see the red arrow in Figure 23.7) as the likely site of the impact.

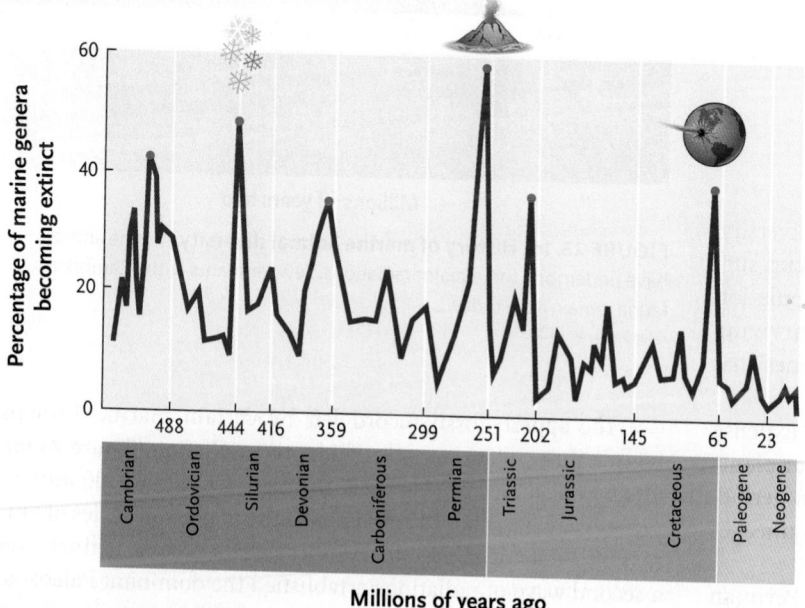

FIGURE 23.14 Mass extinctions. Extinction rates for marine animal genera spiked at least six times during the history of life. Genera are groups of species descended from a common ancestor.

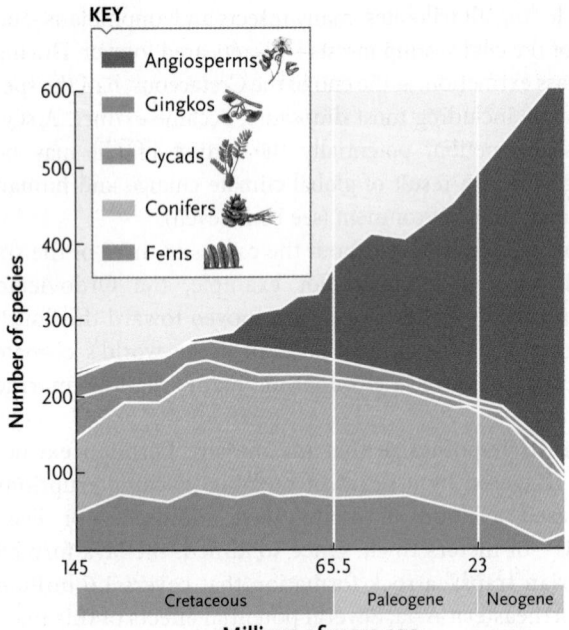

KEY
- Angiosperms
- Gingkos
- Cycads
- Conifers
- Ferns

FIGURE 23.15 History of vascular plant diversity. The diversity of angiosperms increased during the Cretaceous period as the diversity of other groups declined. Changes in the diversity of each group through time are indicated by changes in the height of the group's colored band in the graph.

© Cengage Learning 2017

The causes of the other three historical mass extinctions are less well understood. A series of large-scale extinctions marked the end of the Cambrian period; scientists hypothesize that a massive volcanic eruption, the release of sulfur-containing compounds into the oceans, and a rising sea level may have been responsible. The extinctions at the end of the Devonian and Triassic periods may have been initiated by a decrease in both atmospheric and oceanic temperatures.

Biodiversity Has Increased Repeatedly over Evolutionary History

Although mass extinctions temporarily reduce biodiversity, they also create evolutionary opportunities. Some species survive because they have highly adaptive traits, large population sizes, or widespread distributions. And some of the surviving species undergo adaptive radiation, filling adaptive zones that mass extinctions made available.

Sometimes, the success of one lineage comes at the expense of another. Although the diversity of terrestrial vascular plants has increased almost continuously since the Devonian period, this trend includes booms and busts in several lineages **(Figure 23.15)**.

Ferns and conifers recovered rapidly after the Permian extinction, maintaining their diversity into the early Cenozoic era. However, angiosperms, which arose and began to diversify in the early Cretaceous period, may have hastened the decline of these groups by replacing them in many environments.

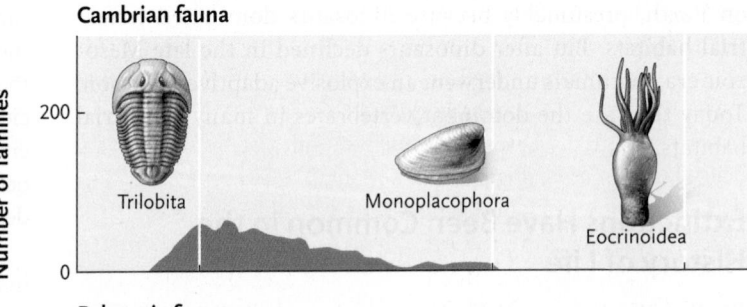

Cambrian fauna

Trilobita   Monoplacophora   Eocrinoidea

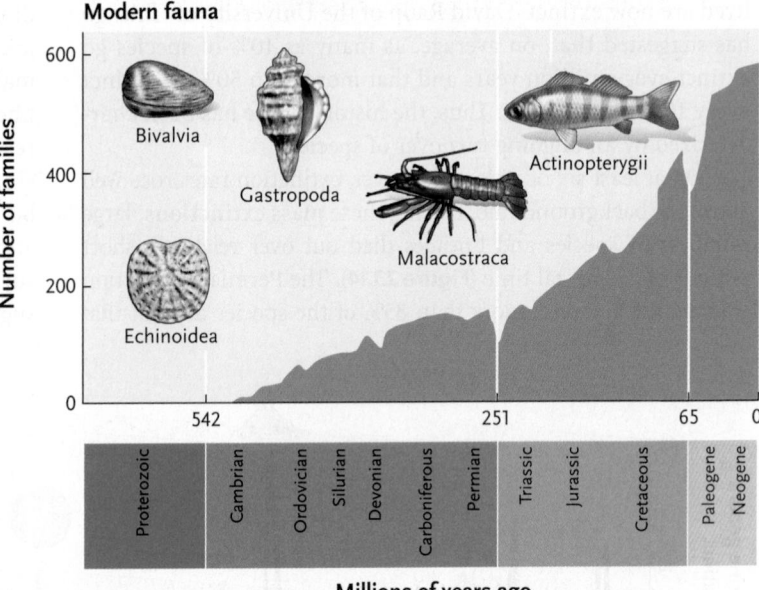

Paleozoic fauna

Articulata   Cephalopoda   Stelleroida

Modern fauna

Bivalvia   Gastropoda   Actinopterygii   Malacostraca   Echinoidea

Millions of years ago

Proterozoic | Cambrian | Ordovician | Silurian | Devonian | Carboniferous | Permian | Triassic | Jurassic | Cretaceous | Paleogene | Neogene

FIGURE 23.16 History of marine animal diversity. Marine animals have undergone three major radiations. Few remnants of the Cambrian fauna remain alive today.

© Cengage Learning 2017

The superb fossil record left by certain marine animals reveals three major periods of adaptive radiation **(Figure 23.16)**. The first occurred during the Cambrian, more than 500 million years ago, when many animal phyla, the major categories of animal life, first appeared. Most of these phyla became extinct, and a second wave of radiations established the dominant Paleozoic fauna during the Ordovician period. A third evolutionary fauna diversified in the Triassic period, right after the great Permian extinction; it produced the immediate ancestors of modern marine animals. The diversity of marine animals has increased consistently since the early Triassic, in large measure because of continental drift. As continents and shallow seas became increasingly isolated, regional biotas diversified independently of one another, increasing worldwide biodiversity.

Historical increases in biodiversity can also be attributed to the evolution of ecological interactions. For example, the number of plant species found *within* fossil assemblages has increased over time, suggesting the evolution of mechanisms that allow more species to coexist. In addition, insects continued to diversify dramatically in the Cretaceous period, possibly because the angiosperms created a new adaptive zone for them. New insect species then provided a novel set of pollinators that may have stimulated the radiation of angiosperms. Such long-term evolutionary interactions between ecologically intertwined lineages have played an important role in structuring ecological communities, which are described more fully in Chapter 53.

STUDY BREAK 23.4

1. What factors might allow a population of organisms to occupy a new adaptive zone?
2. What events apparently triggered the mass extinction at the end of the Permian period?
3. When did the first major adaptive radiation of animals occur?

THINK OUTSIDE THE BOOK

You have learned how the geography of the Galápagos and the Hawaiian Islands fostered adaptive radiations in finches and fruit flies. Search Internet or library sources for examples of adaptive radiations in other types of organisms on these archipelagos. Also look for information about the adaptive radiation of mammals on continental landmasses. Do adaptive radiations on archipelagos and on continental landmasses differ in substantive ways?

23.5 Interpreting Evolutionary Lineages

As newly discovered fossils demand the reinterpretation of old hypotheses, biologists constantly refine their ideas about the history of life. In this section we describe how our interpretation of evolutionary lineages is constantly changing.

Modern Horses Are Living Representatives of a Once-Diverse Lineage

The earliest known ancestors of modern horses were first identified by Othniel C. Marsh of Yale University just a year after Darwin published *On the Origin of Species*. These early horses, *Hyracotherium*, stood 25 to 50 cm high and weighed no more than 20 kg. Their toes (four on the front feet and three on the hind) were each capped with a tiny hoof, but the animals walked on soft pads as dogs do today. Their faces were short, their teeth were small, and they browsed on soft leaves in woodland habitats.

In 1879, Marsh published his analysis of 55 million years of horse family history. He described the evolution of this group of mammals as a sequence of stages from the tiny *Hyracotherium* through intermediates represented by *Mesohippus, Merychippus,* and *Pliohippus* to the modern *Equus* **(Figure 23.17A).** (Each of these names refers to a genus, a group of closely related species.) Marsh inferred a pattern of descent characterized by gradual, directional evolution in several skeletal features. Changes in the legs and feet allowed horses to run more quickly, and changes in the face and teeth accompanied a switch in diet from soft leaves to tough grasses.

The fossil record for horses is superb, and we now have fossils of more than 100 extinct species from five continents. These data reveal a macroevolutionary history very different from Marsh's interpretation. *Hyracotherium* was not gradually transformed into *Equus* along a linear track. Instead, the evolutionary tree for horses was highly branched **(Figure 23.17B)** and *Hyracotherium*'s descendants differed in size, number of toes, tooth structure, and other traits. In fact, in a paper published in 2012, Ross Secord of the University of Nebraska and colleagues at other institutions demonstrated that the size of the earliest horses changed rapidly in response to changing environmental temperatures during the Paleocene and Eocene epochs: horses became smaller at times of climate warming and larger during periods of climate cooling (see Figure 22.5 on the effects of environmental temperature on body size in birds and mammals). Although many branches of this lineage lived in the Miocene and Pliocene epochs, all but one are now extinct. The species of the genus *Equus* living today (horses, donkeys, and zebras) are the surviving tips of that one branch.

When we study extinct organisms, we tend to focus on traits that characterize modern species. Marsh, for example, assumed that the differences between *Hyracotherium* and *Equus* were typical of the changes that characterized the group's evolutionary history. But not all fossil horses were larger **(Figure 23.17C),** had fewer toes, or were better adapted to feed on grass than their ancestors.

The Tempo of Morphological Change Varies among Lineages

Analyses of the fossil record indicate that the tempo, or timing, of morphological change varies among lineages. The discovery of transitional fossils provides evidence in support of the **phyletic gradualism hypothesis** (*phylon* = race), which proposes that most morphological change occurs gradually over long periods of time. For example, a study of Ordovician trilobites revealed that the number of "ribs" in their tail region changed continuously over 3 million years. The change was so gradual that a sample from any given stratum was almost always intermediate between samples from the strata just above and below it. The changes in rib number probably evolved without the evolution of new, reproductively isolated species **(Figure 23.18).**

A. Marsh's reconstruction of horse evolution

Marsh's description of horse evolution focused on increased body size, reduced number of toes, and larger, grinding teeth at the back of the mouth.

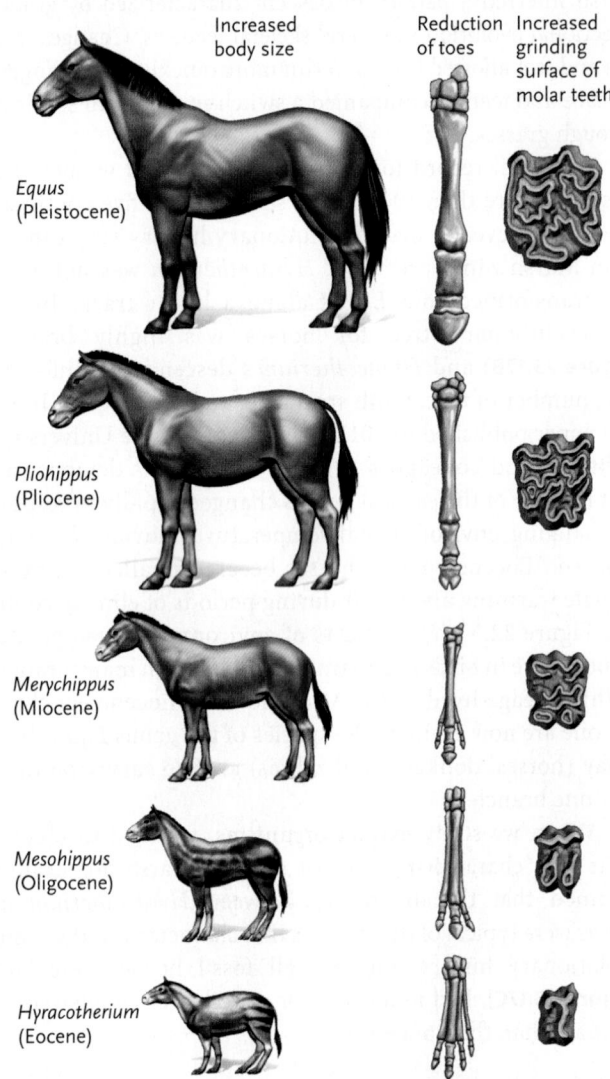

B. Modern reconstruction of horse evolution

Recent studies revealed that the horse family includes numerous evolutionary branches with variable morphology. The horses in Marsh's analysis are highlighted in green.

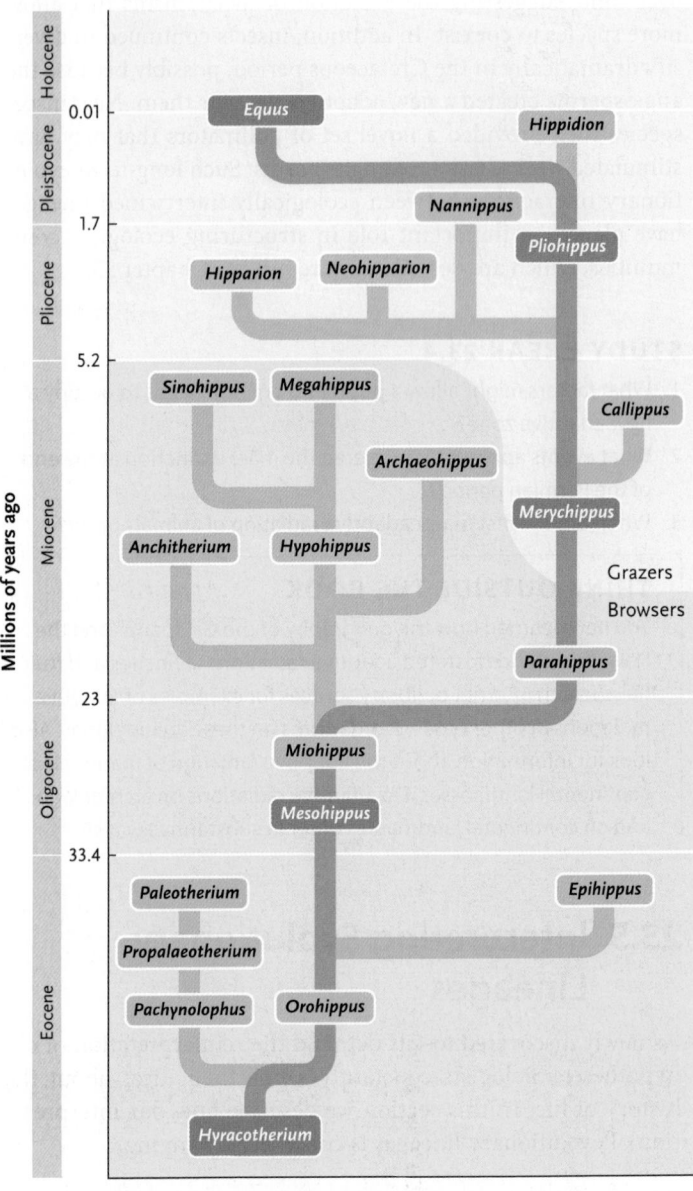

C. Changes in body size of horse species over time

Some branches of the lineage remained as small as the first horses.

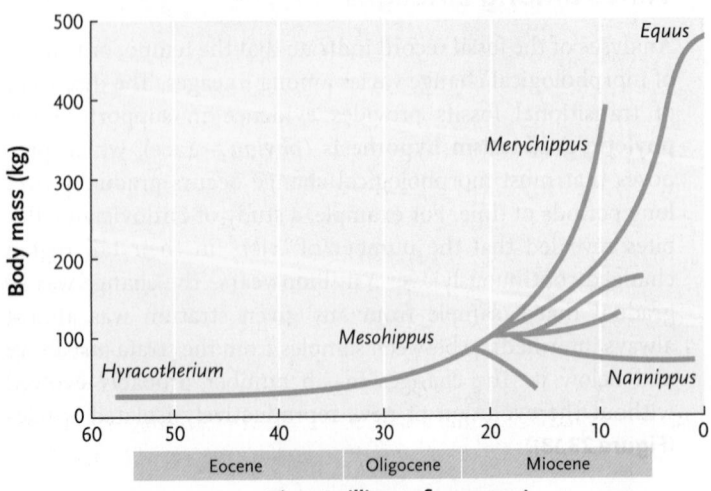

SUMMARY The lineage that includes modern horses includes many branches that did not leave any living descendants. If a branch other than *Equus* had survived, Marsh's description of trends in horse evolution would have been very different. All evolutionary lineages have extinct branches, and any attempt to trace a linear evolutionary path—as Marsh did for horses and many people do for humans—imposes artificial order on an inherently disorderly history.

think like a scientist Modern horse breeders have used artificial selection to produce miniature horses that resemble their full-sized ancestors in all respects except body size. Does artificial selection for miniature horses constitute a reversal of the evolutionary trends observed in the horse lineage?

FIGURE 23.18 **Observational Research**

Evidence of Phyletic Gradualism

Hypothesis: The phyletic gradualism hypothesis states that most morphological change within evolutionary lineages results from the accumulation of small, incremental changes over long periods of time.

Prediction: The morphology of fossils from a given stratum will be intermediate between those of fossils from the strata immediately below and above it.

Method: Peter R. Sheldon of Trinity College, Dublin, Ireland, counted the number of "ribs" in the tail region of the exoskeletons of approximately 15,000 trilobite fossils from central Wales, United Kingdom. The fossils had formed over a span of about 3 million years during the Ordovician period. Shelton plotted the mean number of ribs found in successive samples of eight lineages.

Results: Sheldon's data reveal gradual changes in the mean number of "ribs" in these animals with no evidence of speciation.

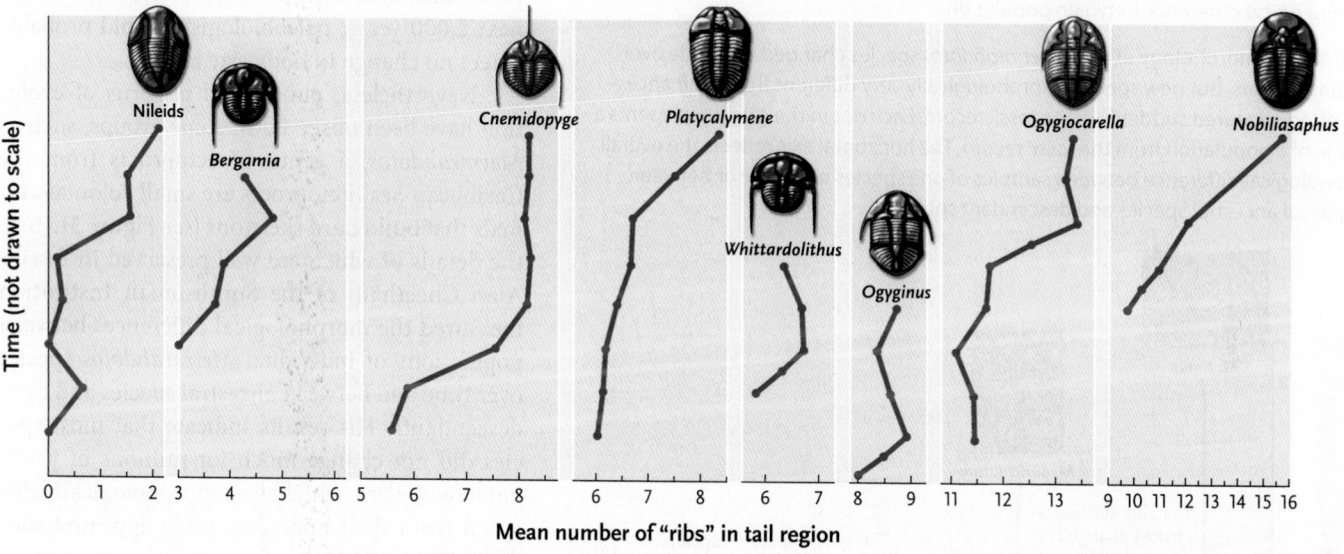

Conclusion: Morphological changes in Ordovician trilobites from central Wales are consistent with the predictions of the phyletic gradualism hypothesis.

Observe
Hypothesize
Predict
Experiment
Interpret

Do you think that the changes in rib number over time in *Bergamia* and *Ogyginus* are the products of balancing selection, directional selection, or some other agent of evolution?

Source: Adapted from P. R. Sheldon. 1987. Parallel gradualistic evolution of Ordovician trilobites. *Nature* 330:561–563.

But for most groups of organisms, the fossil record is not nearly as good as it is for horses or trilobites. In fact, the discovery of intermediate or transitional fossils is fairly rare. Instead, most species appear suddenly in a particular layer, persist for some time with little change, and then disappear from the fossil record. Then another species with different traits suddenly appears in the next higher stratum.

In the early 1970s, Niles Eldredge of the American Museum of Natural History and Stephen Jay Gould of Harvard University published an explanation for the absence of transitional forms, or "missing links." Their **punctuated equilibrium hypothesis** suggested that speciation usually occurs in isolated populations at the edge of a species' geographical distribution. Such populations experience substantial genetic drift and distinctive patterns of natural selection (as described in Section 21.3). According to this hypothesis, morphological variations appear rapidly as new species arise; then, most species exhibit long periods of morphological equilibrium (that is, little change in form), punctuated by brief periods of speciation and rapid morphological evolution. Eldridge and Gould further argued that transitional forms live only for short periods of geological time in small, localized populations—the very conditions that discourage broad representation in the fossil record. Darwin himself used this line of reasoning to explain puzzling gaps in the fossil record: new species appear as fossils only after they become abundant and widespread and begin a period of morphological stasis.

Some evolutionists challenged the hypothesis' definition of rapid morphological change, particularly given our inability to

 FIGURE 23.19 **Observational Research**

Evidence of a Punctuated Pattern of Morphological Change

Hypothesis: The punctuated equilibrium hypothesis states that most morphological change within evolutionary lineages appears during speciation.

Prediction: The fossil record will reveal that most species experienced relatively little morphological change for long periods of time, but that new, morphologically distinctive species arose suddenly.

Method: Cheetham examined numerous fossilized samples of populations of a small marine invertebrate, the ectoproct *Metrarabdotos,* from the Dominican Republic. He measured 46 morphological characters in populations representing 18 species and then used a complex statistical analysis to summarize the extent of morphological difference between populations.

Results: The morphology of most *Metrarabdotos* species changed very little over millions of years, but new species, morphologically very different from their ancestors, often appeared suddenly in the fossil record. Each dot in the graph represents a sample of a population from the fossil record. The horizontal axis reflects the overall morphological difference between samples of one species over time or between samples of ancestral species and descendant species.

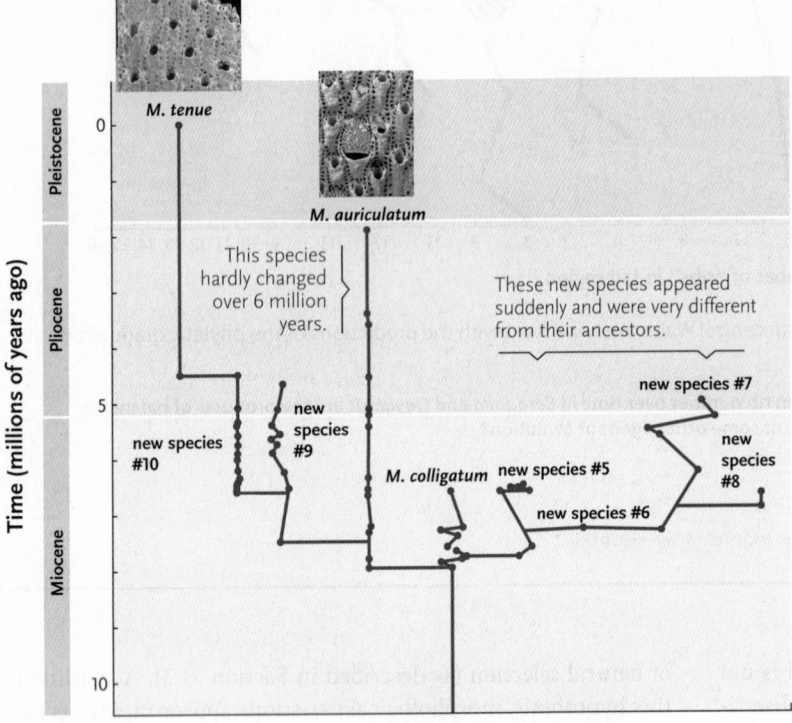

Overall morphological difference

Conclusion: The *Metrarabdotos* lineage exhibits a pattern of morphological evolution that is consistent with the predictions of the punctuated equilibrium hypothesis.

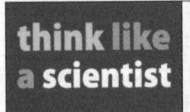 **think like a scientist**

Observe
Hypothesize
Predict
Experiment
Interpret

Which of the *Metrarabdotos* species in the illustration shows some evidence of gradual morphological change?

Source: Adapted from A. H. Cheetham. 1987. Tempo of evolution in a Neogene bryozoan: Are trends in single morphologic characters misleading? *Paleobiology* 13:286–296.

© Cengage Learning 2017

resolve time precisely in the fossil record. To a paleobiologist with a geological perspective, "instantaneous" events occur over tens or hundreds of thousands of years. But to a population geneticist, those time scales may encompass thousands of generations, ample time for gradual microevolutionary change.

Moreover, examples of evolutionary stasis may not be as static as they appear. Alternating periods of directional selection that favor opposite patterns of change could produce the appearance of stasis. For example, if natural selection favored slight increases in body size for 2,000 years and then favored slight decreases for the next 2,000 years, paleobiologists would probably detect no change in body size at all.

Nevertheless, punctuated patterns of evolution have been observed in some groups, such as *Metrarabdotos,* a genus of ectoprocts from the Caribbean Sea. Ectoprocts are small colonial animals that build hard skeletons (see Figure 31.15A), the details of which are well preserved in fossils. Alan Cheetham of the Smithsonian Institution measured the morphological differences between populations of individual *Metrarabdotos* species over time and between ancestral species and their descendants. His results indicate that most species did not change much for millions of years, but new species, which were morphologically different from their ancestors, often appeared suddenly **(Figure 23.19).**

Researchers have found evidence of both gradual and punctuated morphological changes, as well as some intermediate patterns. The punctuated equilibrium hypothesis caused a stir when it was first proposed because it challenged long-held assumptions about the tempo of evolution. But its publication rekindled interest in paleobiology and macroevolution, and like any good idea in science, it inspired much new research. Some of the most interesting results have focused on the evolution of morphological novelties, which biologists now analyze with studies in paleontology, embryology, and genetics.

STUDY BREAK 23.5

1. Did the horse lineage undergo a steady increase in body size over its evolutionary history? Explain your answer.
2. What is the difference between the morphological change predicted by the phyletic gradualism hypothesis and the punctuated equilibrium hypothesis?

A. Allometric growth in humans

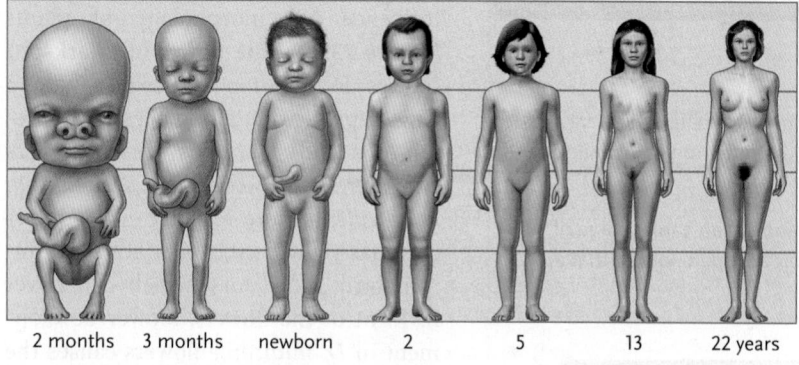

| 2 months | 3 months | newborn | 2 | 5 | 13 | 22 years |

Humans exhibit allometric growth from prenatal development until adulthood. Our heads grow more slowly than other body parts; our legs grow faster.

B. Differential growth in the skulls of chimpanzees and humans

Changes in chimpanzee skull

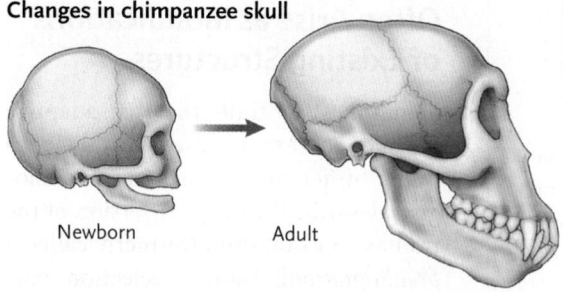

Newborn   Adult

Changes in human skull

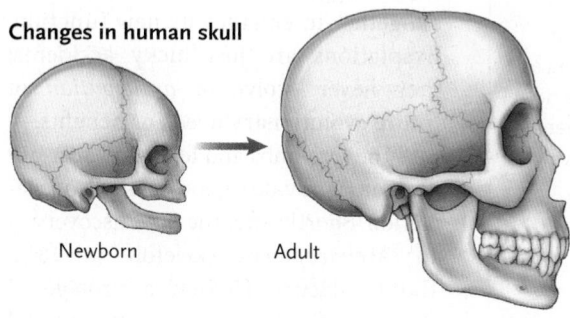

Newborn   Adult

Although the skulls of newborn humans and chimpanzees are remarkably similar, differential patterns of growth make them diverge during development. The jaw of a chimpanzee grows much faster than other regions of the skull. In humans, the different parts of the skull grow at more similar rates.

FIGURE 23.20 Examples of allometric growth.
© Cengage Learning 2017

23.6 The Evolution of Morphological Novelties

The fossil record documents the appearance of distinctive morphological novelties, such as the limbs of four-legged vertebrates and the appearance of crests and horns on some dinosaurs. In this section we consider explanations for how such novelties may arise.

Differential Growth of Body Parts and Changes in the Timing of Developmental Events Produce Distinctive Structures

The morphology of individuals sometimes changes over time because of **allometric growth** (*alios* = other; *metron* = measure), the differential growth of body parts. In humans, for example, the relative sizes of different body parts change because human heads, torsos, and limbs grow at different rates **(Figure 23.20A)**.

Allometric growth can also establish morphological differences in closely related species. For example, the skull proportions of chimpanzees and humans are similar in newborns, but markedly different in adults **(Figure 23.20B)**. The jaw region of the chimp skull grows much faster than other regions, while the different parts of the human skull all grow at more similar rates. Differences in the adult skulls may therefore simply reflect changes in one or a few genes that regulate the pattern of growth.

Changes in the timing of developmental events, called **heterochrony** (*heteros* = different; *khronos* = time), also cause the morphology of closely related species to differ. **Paedomorphosis** (*paedo-* = child; *morphos* = form or shape), the development of reproductive capability in an organism with juvenile characteristics, is a common form of heterochrony.

Many salamanders, for example, undergo metamorphosis from an aquatic juvenile into a morphologically distinct terrestrial adult. However, populations of several species are paedomorphic—they grow to adult size and become reproductively mature without changing to the adult form **(Figure 23.21)**. The evolutionary change causing these differences may be surprisingly simple. In amphibians, including salamanders, thyroid hormone induces metamorphosis (see Chapter 42). Paedomorphosis results from mutations that reduce thyroid hormone production, limit the responsiveness of some developmental processes to thyroid hormone concentration, or both.

FIGURE 23.21 Paedomorphosis in salamanders. Some small-mouthed salamanders (*Ambystoma talpoideum*) undergo metamorphosis, losing their gills and developing lungs (left). Others are paedomorphic: they retain juvenile morphological characteristics, such as gills, after attaining sexual maturity (right).
© Cengage Learning 2017

 **FIGURE 23.22** **Observational Research**

Paedomorphosis in *Delphinium* Flowers

Hypothesis: The narrow tubular shape of the flowers of *Delphinium nudicaule,* which are pollinated by hummingbirds, is the product of paedomorphosis, the retention of juvenile characteristics in a reproductive adult.

Prediction: The flowers of *D. nudicaule* grow more slowly and mature at an earlier stage of development than those of *Delphinium decorum,* a species with broad, open flowers that are pollinated by bees.

D. decorum

D. nudicaule

Gary Head

Gary Head

Method: Edward O. Guerrant of the University of California, Berkeley, measured 42 bud and flower characteristics in *D. nudicaule* and *D. decorum* as their flowers developed and used the number of days since the completion of meiosis in pollen grains as a measure of flower maturity. He then used a complex statistical analysis to compare the characteristics of the buds and flowers of both species.

Results: The mature flowers of *D. nudicaule* resemble the buds of both species more closely than they resemble the flowers of *D. decorum.* Although the time required for maturation of the reproductive structures is similar in the two species, the rate of petal growth (measured as petal blade length) is slower in *D. nudicaule.* As a result, the mature flowers of *D. nudicaule* do not open as widely as those of *D. decorum.* Because of these morphological differences, bees can pollinate flowers of *D. decorum,* but they cannot land on the flowers of *D. nudicaule,* which are pollinated by hummingbirds.

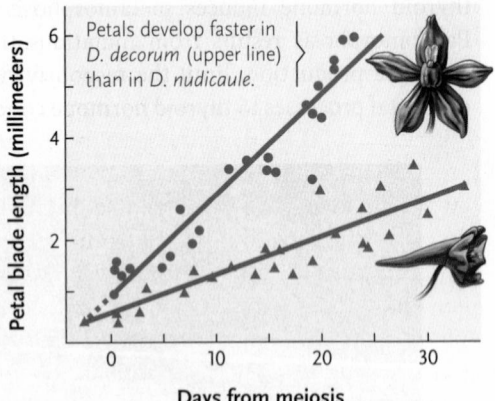

Conclusion: The narrower and more tubular shape of *D. nudicaule* flowers, which mature at an earlier stage of development than *D. decorum* flowers, is the product of paedomorphosis.

 **think like a scientist**

Observe
Hypothesize
Predict
Experiment
Interpret

How might the structural difference in the flowers of *D. decorum* and *D. nudicaule* maintain reproductive isolation between the two species?

Source: Adapted from E. O. Guerrant. 1982. Neotenic evolution of *Delphinium nudicaule* (Ranunculaceae): A hummingbird-pollinated larkspur. *Evolution* 36:699–712.

Changes in developmental rates also influence the morphology of plants **(Figure 23.22)**. The flower of a larkspur species, *Delphinium decorum,* includes a ring of petals that guide bees to its nectar tube and structures on which bees can perch. By contrast, *Delphinium nudicaule,* a more recently evolved species, has tight flowers that attract hummingbird pollinators, which can hover in front of the flowers. Slower development in *D. nudicaule* flowers causes the structural difference: a mature flower in the descendant species resembles an unopened (juvenile) flower of the ancestral species.

Morphological Novelties Often Arise as Modifications of Existing Structures

Sometimes a trait that is adaptive in one context is also advantageous under different circumstances. Biologists describe the original version of the trait as an **exaptation** (formerly called a *preadaptation*). Natural selection may then exaggerate the trait or modify it altogether to enhance its new function. Exaptations are just lucky accidents; they never evolve *in anticipation* of future evolutionary needs or benefits.

The forelimbs and feathers of certain dinosaurs provide a good example of exaptation. Shortly after the first discovery of an *Archaeopteryx* skeleton in 1861, Thomas Henry Huxley, a protégé of Darwin, developed the controversial hypothesis that birds are the living representatives of the theropods, a lineage of bipedal predatory dinosaurs (see Figure 20.13). More than a century passed before John Ostrom of Yale University provided support for Huxley's hypothesis; in 1964, he described a previously unknown theropod dinosaur, *Deinonychus antirrhopus,* with a very birdlike skeleton. But *Archaeopteryx* remained a puzzling creature because of its mix of dinosaurian characteristics—teeth, a flat breastbone, a long tail with many vertebrae, and three clawed fingers on each hand—and avian (that is, birdlike) characteristics—a furcula (wishbone), winglike forelimbs, and feathers. Although *Archaeopteryx* was the

A. Fossil of *Microraptor gui*

Feathers on forelimb

Feathers on hindlimb

Spencer Platt/Getty Images News/Getty Images

B. Hindlimb with feathers

Journal Nature/Xing Xu/Getty Images

C. Reconstruction of *Microraptor gui* with limbs extended for gliding

Julius Csotonyi

FIGURE 23.23 Feathered dinosaurs. *Microraptor gui* is one of many fossils of feathered dinosaurs that researchers have found in China since the mid-1990s. The reconstruction shows how this species may have looked with its four limbs extended for gliding flight. The scale bar in part A is 6 cm.

only feathered dinosaur-like fossil known at that time, scientists hypothesized that at least some of its dinosaur ancestors must have been feathered as well. Nevertheless, the known fossil record suggested that the transition from dinosaurs to feathered birds was abrupt: feathers were absent in the dinosaur ancestors of birds, but present in what biologists consider the earliest known bird.

Since the early 1990s, however, paleontologists have discovered fossils of dozens of feathered dinosaur species; most have been found in early Cretaceous deposits, dated at 124 million years ago, in northern China. The fossilized dinosaur feathers range from filamentous or downy structures that covered their bodies to long stiff feathers on their forelimbs and tails. One dinosaur species, *Microraptor gui*, had well-developed feathers on all four of its limbs as well as its long tail **(Figure 23.23)**. A recent analysis suggests that their feathered forelimbs powered flapping flight, their feathered hindlimbs functioned as rudders for steering, and their feathered tail served as a stabilizer (see Figure 23.23C). Research has confirmed that the feathers in one fossil included β-keratin, the protein found in the feathers of modern birds. Based on this anatomical and biochemical evidence, biologists hypothesize that feathers arose as modifications of ancestral scales in the

skin of some early dinosaurs. Thus, feathers are not unique to birds. They appeared in bird ancestors, perhaps as a form of insulation or as structures used for social communication; birds merely retain this ancestral characteristic.

But flight requires wings as well as feathers. How did wings and flapping flight evolve from the forelimbs of bipedal dinosaurs? Many paleobiologists now think that modern birds are the descendants of ground-dwelling, bipedal dinosaurs that used their feathered forelimbs either to swat at insects and other small prey or to provide lift as they scampered up trees to escape their own predators. Natural selection may then have favored a reduction in body size, enlargement and modification of forelimb feathers and strengthening of muscles associated with flapping flight (described in Section 32.9).

In a 2012 publication, Darla K. Zelenitsky of the University of Calgary, Canada, and her colleagues from Canada, Japan, and the United States proposed another explanation for the evolution of feathered wings. They reported that fossils of *Ornithomimus edmontonicus*, a nonflying, ground-dwelling dinosaur, exhibited age-related variations in the structure of its feathers. Individuals of all ages were covered with supple, filamentous feathers that draped over their bodies, but older individuals also had longer

FIGURE 23.24 Age-related plumage changes in *Ornithomimus edmontonicus*. Ornithomimosaur dinosaurs of all ages were covered with soft, supple, filamentous feathers, but adults also grew long stiff feathers on their forelimbs. Because these animals did not fly, the feathers must have served some other purpose, such as facilitating social interactions.

© Cengage Learning 2017

and stiffer feathers on their forelimbs **(Figure 23.24)**. In living birds, stiff feathers develop shortly before young leave the nest and fly, but in *Ornithomimus* they developed only as individuals reached maturity. Because *Ornithomimus* did not fly and did not prey on other animals (it was herbivorous), the authors conclude that the stiff feathers and winglike structures did not serve for aerial locomotion or predatory behaviors. Instead, the authors propose, these structures may have been secondary sexual characteristics, used for courtship behaviors or for brooding young in the nest. Only later, in the lineage that includes *Microraptor*, did natural selection foster the evolution of true wings that facilitated flight.

Evolutionary Developmental Biology May Explain the Sudden Appearance of Some Morphological Novelties

During much of the twentieth century, the evolutionary biologists who studied morphological novelties in extinct and living organisms worked independently of the scientists studying embryonic development. As a result, evolutionary biologists were unable to construct a coherent picture of the specific developmental mechanisms that contributed to morphological innovations. Since the late 1980s, however, advances in molecular genetics have allowed scientists to explore the genomes of organisms in great detail, fostering a new approach to these studies. **Evolutionary developmental biology**—*evo-devo*, for short—asks how evolutionary changes in the genes regulating embryonic development can lead to the morphological changes that foster adaptive radiations, increasing biodiversity over geological time.

In the life cycle of a multicellular organism, the many different body parts of the adult develop in a highly controlled sequence of steps that is specified by genetic instructions in the single cell of a fertilized egg. Developmental biologists study how regulatory genes control the development of phenotypes and their variations. As described in Section 16.4, **homeobox genes** (described further for plants in Chapter 36) code for

transcription factors that bind regulatory sites on DNA, either activating or repressing the expression of other genes that contribute to an organism's form. In the remainder of this chapter, we describe a few intriguing discoveries about the genetic mechanisms that underlie some macroevolutionary changes in animals. You have already encountered one example when you read about the evolution of limblessness in snakes (see Section 20.3).

Most Animals Share the Genetic Tool Kit That Regulates Their Development

Comparisons of genome sequence data reveal that most animals, regardless of their complexity or position in the Tree of Life, share a set of several hundred homeobox genes that control their development. Collectively, these genes have been dubbed the "genetic tool kit," because they govern the basic design of the body plan by controlling the activity of thousands of other genes. Some of the tool-kit genes must be at least 500 million years old, because all living animals inherited them from a common ancestor alive at that time. Not surprisingly, their structure does not differ much among the animals that possess them, and they generally play the same role in development for all species. For example, as you learned in Section 16.4, genes in the *Hox* family of homeobox genes control the body plans of animals along the anterior–posterior axis, including where appendages—wings in flies and legs in mice—will develop on the animal's body. Some of the same tool-kit genes are also present in plants, fungi, and prokaryotes, suggesting that those genes may date back to the earliest forms of life.

Another example of a highly conserved and widely distributed tool-kit gene, the *Pax-6* gene—a member of the *Pax* family of homeobox genes—triggers the formation of light-sensing organs as diverse as the eye spots in flatworms, the compound eyes of insects and other arthropods, and the camera eyes of vertebrates (see Chapter 41). Like the *Hox* genes, *Pax-6* also codes for a protein that either activates or represses gene transcription. The proteins coded by *Pax-6* in different animals are so similar that when researchers genetically engineered fruit fly larvae to express the *Pax-6* gene taken from a squid or a mouse, the flies responded by developing eyes. The induced eyes were, however, fruit fly eyes—not squid eyes or mouse eyes. Thus, *Pax-6* triggers activity in the fruit fly genes that carry the specific instructions for making an eye typical of that species. Apparently, the ancient genetic sequence for *Pax-6*, the master regulatory gene for eye development, has been conserved over the hundreds of millions of years since the common ancestor of squids, fruit flies, and mice lived.

Evolutionary Changes in Developmental Switches May Account for Much Evolutionary Change

If most animals share the same tool-kit genes, how has evolution produced different body plans among species? What makes a squid, a fruit fly, and a mouse different? Researchers in evo-devo have proposed that morphological differences among species arise when mutations alter the effects of developmental regulatory genes. As described in Section 16.4, the developmental programs of animals involve complex networks of many interacting genes. Varying combinations of tool-kit genes may be expressed at different times and in different body regions. According to this hypothesis, the several hundred tool-kit genes encode proteins that work as either activators or repressors in a multitude of possible combinations. Thus, they can generate an unimaginably large number of different gene expression patterns, each with the potential to alter morphology.

Sean B. Carroll of the Howard Hughes Medical Institute has described the regulatory sites that transcription factors can bind as *switches,* like those we use to turn lights on or off. When a combination of transcription factors turns on a regulatory switch, a gene further downstream (that is, one that is expressed later in time) is activated. When transcription factors turn off a regulatory switch, a downstream gene is inactivated.

Although all the cells in an animal contain exactly the same set of genes, the differential expression of genes in different body regions and at different times during embryonic development causes different structures to be made. Allometric growth can result from evolutionary changes in developmental switches that cause certain body parts to grow faster or larger than others. Similarly, heterochrony can be explained as an evolutionary change in the switches that either delays the development of adult characteristics or speeds up the development of reproductive maturity.

If Carroll's hypothesis is correct, morphological novelties arise when evolutionary changes in developmental switches alter the expression patterns of *existing* genes. This view contrasts markedly with the explanation proposed in the modern synthesis (the unified theory of evolution described in Chapter 20), that most morphological novelties arise as mutations that slowly accumulate in the genes that carry blueprints for building particular structures. According to the modern synthesis, the accumulated mutations eventually create *new* genes that specify the creation of new structures.

Although Carroll's hypothesis proposes a different genetic mechanisms as the basis for evolutionary changes in morphology, mutations in developmental regulatory genes and their effects on morphology are subject to the action of the same microevolutionary processes—natural selection, genetic drift, and gene flow—that influence the frequencies of genotypes and phenotypes in populations. Thus, every morphological change induced by a mutation in a homeobox gene or developmental switch is tested by the success or failure of the individual that carries it.

Numerous studies have shown that changes in the expression of homeobox genes can have dramatic effects on morphology. *Molecular Insights* explains how a change in the timing of the expression of *Hox* genes may have changed the fins of fishes into the limbs of four-legged vertebrates.

In another example, researchers have determined how an adaptive morphological change in a small fish, the three-spined stickleback *(Gasterosteus aculeatus),* results from the deactivation of a homeobox gene. The freshwater stickleback populations in North American lakes are the descendants of marine ancestors that colonized the lakes after the retreat of glaciers between 10,000 and 20,000 years ago. Marine sticklebacks have bony armor along their sides and prominent spines; lake-dwelling sticklebacks have greatly reduced armor and, in many populations, lack spines on their pelvic fins **(Figure 23.25)**.

Natural selection has apparently fostered these morphological differences in response to the dominant predators in each habitat. In marine environments, long spines prevent some predatory fishes from swallowing sticklebacks. But long spines are a liability in lakes, where voracious dragonfly larvae

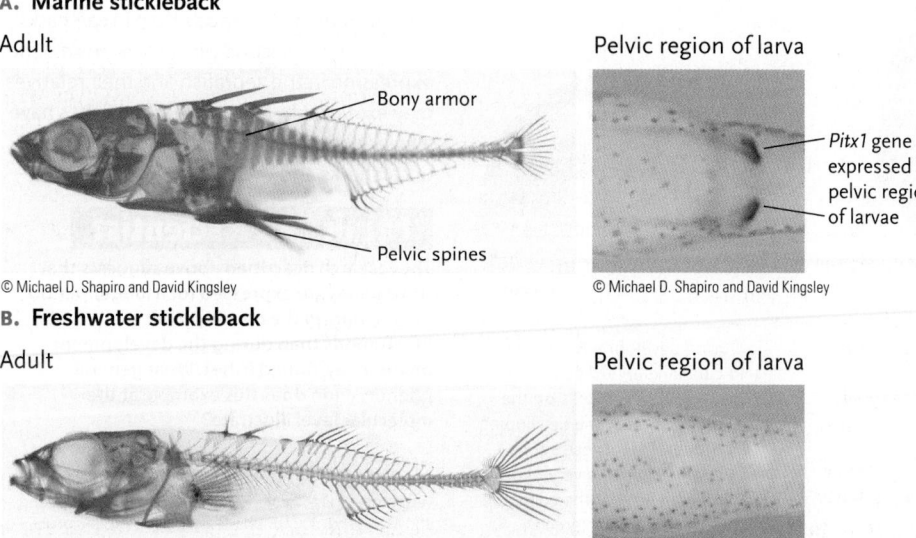

A. Marine stickleback

Adult

Bony armor

Pelvic spines

© Michael D. Shapiro and David Kingsley

B. Freshwater stickleback

Adult

© Michael D. Shapiro and David Kingsley

Pelvic region of larva

Pitx1 gene expressed in pelvic region of larvae

© Michael D. Shapiro and David Kingsley

Pelvic region of larva

© Michael D. Shapiro and David Kingsley

FIGURE 23.25 Sticklebacks with and without pelvic spines. (A) Marine populations of three-spined sticklebacks *(Gasterosteus aculeatus)* have prominent bony plates along their sides and large spines on their dorsal and pelvic fins. The growth of pelvic spines is induced by the expression of the *Pitx1* gene (the purple crescents in the photo on the right) in the pelvic region during embryonic development. **(B)** Many freshwater populations of the same species lack the bony plates and spines. Pelvic spines do not develop in the freshwater sticklebacks because the *Pitx1* gene is not expressed in the pelvic region. Natural selection has apparently fostered these morphological differences in response to the dominant predators in each environment. The skeletons of these specimens, each about 8 cm long, were dyed bright red.

Fancy Footwork: From fins to fingers

Both fishes and tetrapods (four-footed animals) have two pairs of appendages, one anterior and one posterior. In both groups the appendages develop from buds of mesoderm—the middle of the three primary embryonic tissue layers—that thicken by increased cell division. As the buds elongate, cartilage is deposited at localized centers, where bones later form.

In fishes, the bones that support the fin develop along a central axis from the base to the tip of the fin. Like fins, tetrapod limbs have a central axis, but they also have distal structures, the digits (that is, fingers and toes). In tetrapods, centers of cartilage formation generate both the long bones of the limb and the five digits of the foot.

Groups of homeobox (*Hox*) genes control limb development in all animals with paired anterior and posterior appendages. Researchers have studied the expression of the *Hoxd* genes by using a nucleic acid probe that can pair with mRNA products of the genes, resulting in the generation of a color. The color is visualized under the microscope in embryos at different stages of development to determine where and when gene expression occurs. For tetrapods, genes at the 5′ end of the *Hoxd* cluster—designated 5′ *Hoxd* genes—are expressed in two distinct phases (**Figure,** part A). First, gene activity is restricted to the posterior half of the limb;

this period of activity corresponds to development of the long limb bones. Second, the 5′ *Hoxd* genes became active in a band of cells perpendicular to the central axis; the cartilage centers that form the bones of the digits develop in this anterior-posterior band.

In 1995 Paolo Sordino, Frank van der Hoeven, and Denis Duboule at the University of Geneva in Switzerland published a study on *Hoxd* expression in the zebrafish *Danio rerio* (a common aquarium fish and a model research organism). They found that only a single phase of 5′ *Hoxd* gene activity occurs during zebrafish fin development (Figure, part B). Because the second phase of 5′ *Hoxd* gene activity is apparently absent in zebrafishes, they concluded that tetrapod digits must be an entirely novel evolutionary structure fostered by the addition of the second phase of 5′ *Hoxd* gene activity.

Research Question

Is the pattern of 5′ *Hoxd* gene expression seen in zebrafish typical of all fishes? That is, are tetrapods the only vertebrates that have a second phase of 5′ *Hoxd* gene expression?

Experiment

A number of research groups have carried out experiments to answer the question. One study, published in 2007 by Renata Freitas, Guang Jun Zhang, and Martin Cohn at the University of Florida, Gainesville, examined

the pattern of 5′ *Hoxd* gene expression in developing catshark *(Scyliorhinus canicula)* fins. Catsharks are included in an evolutionary lineage (Chondrichthyes, fishes with cartilaginous skeletons) that predates the lineage that includes zebrafishes (Actinopterygii, ray-finned fishes). Thus, data on fin development in catsharks may portray a more ancient developmental pattern than the one observed in zebrafishes.

Results

Their results showed that in the early stages of catshark fin development, the 5′ *Hoxd* genes are expressed in a phase like that seen during fin development in zebrafish and in the first phase of limb development in tetrapods. Unexpectedly, the researchers observed a *second* phase of expression of two of the 5′ *Hoxd* genes along the distal margin of the catshark fin buds (Figure, part C). Strikingly, that second phase of expression is similar to the second phase of 5′ *Hoxd* gene expression in tetrapod limb development.

Conclusion

The second phase of 5′ *Hoxd* gene expression is not uniquely associated with tetrapod limb development, as the earlier zebrafish data had suggested. Indeed, the more recent research suggests that two phases of 5′ *Hoxd* gene expression were present in the common ancestor of sharks, ray-finned fishes, and tetrapods. The authors of the later article hypothesized that digits may have arisen if the *Hoxd* genes in the distal parts of the limb were expressed for a longer period of development in tetrapods than in catsharks. The loss of the second phase of 5′ *Hoxd* gene expression in the zebrafish and their relatives may explain why most ray-finned fishes have a fin skeleton that is much smaller than that of catsharks and their relatives.

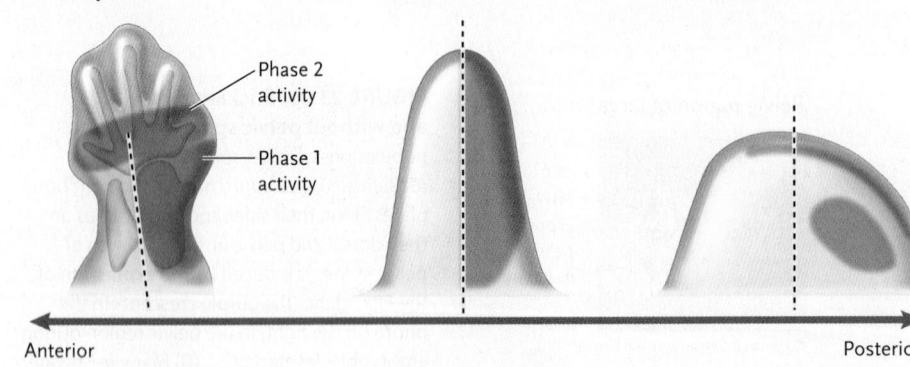

A. Tetrapod

Phase 2 activity

Phase 1 activity

Anterior

B. Zebrafish

C. Catshark

Posterior

During development of the limb and digits in tetrapods, *Hoxd* genes first become active in cells on the posterior side of the limb (blue). Later, these genes are active in a broad band of cells perpendicular to the central axis of the limb (green).

During development of the fin in zebrafish, *Hoxd* genes become active in cells only on the posterior side of the developing fin (blue). Zebrafish exhibit no second phase of *Hoxd* activity in the developing fin.

During development of the fin in catsharks, *Hoxd* genes first become active in cells on the posterior side of the developing fin (blue). Later these genes become active in a very narrow band of cells near the edge of the developing fin (green).

© Cengage Learning 2017

think like a scientist

The research described above suggests that *Hoxd* genes are expressed for a longer period of time during the development of the limbs in tetrapods than during the development of fins in ray-finned fishes. What general phenomenon does this example at the molecular level illustrate?

Sources: P. Sordino, F. van der Hoeven, and D. DuBoule. 1995. Hox gene expression in teleost fins and the origin of vertebrate digits. *Nature* 375:678–681; R. Freitas, G. Zhang, and M. J. Cohn. 2007. Biphasic *Hoxd* gene expression in shark paired fins reveals an ancient origin of the distal limb domain. *PLoS ONE* 2:e754.

grab sticklebacks by their spines and devour them; freshwater sticklebacks that lack spines are more likely to escape from their clutches.

In a paper published in 2004, Michael D. Shapiro of the Stanford University School of Medicine and the Howard Hughes Medical Institute and colleagues at several other institutions described how the presence or absence of spines on the pelvic fins of these fishes is governed by the expression of the gene *Pitx1*. Pelvic spines are part of the pelvic fin skeleton, the fishes' equivalent of a hindlimb. In fact, *Pitx1* also contributes to the development of hindlimbs in four-legged vertebrates and to certain glands and sensory organs in the head.

In long-spined marine sticklebacks, *Pitx1* is expressed in the embryonic buds from which pelvic fins develop, promoting the development of spines. But *Pitx1* is not expressed in the fin buds of freshwater sticklebacks; hence, pelvic spines do not develop. However, freshwater sticklebacks have not *lost* the *Pitx1* gene; it is still expressed elsewhere in the fishes' bodies. In a follow-up paper, published in 2009, the Shapiro team reported that the failure of freshwater sticklebacks to express the *Pitx1* gene in the developing pelvic region results from the deletion of a tissue-specific enhancer of hindlimb expression; without this enhancer, the *Pitx1* gene is not expressed, and the fishes do not grow pelvic spines. Indeed, in 2012, David M. Kingsley of the Stanford University Medical School and 28 colleagues at many institutions published a genomic analysis comparing 21 marine and freshwater stickleback populations to understand better the genetic changes that underlie the fish's colonization of freshwater habitats. Their analysis revealed that only 20% of the genetic differences occurred in coding regions of DNA and that the vast majority of mutations occurred in regulatory regions.

In the next chapter, we examine how biologists explore the evolutionary relationships among species and how they organize that information into a useful framework for researchers in every biological discipline. In the following unit, we will revisit evo-devo and examine some recent discoveries about how changes in homeobox genes have diversified body plans in the major evolutionary groups of organisms.

STUDY BREAK 23.6

1. What is the difference between allometry and heterochrony?
2. What evidence suggests that many developmental control genes have been conserved by evolution?
3. What genetic factor is apparently responsible for the presence or absence of spines in stickleback fish?

 ## Unanswered Questions

Does morphological evolution always proceed gradually or can it occur in great leaps and bounds?

As you read in this chapter, biologists disagree about whether evolutionary changes in morphology can occur very rapidly. Although biologists have proposed various hypotheses to explain the abrupt changes that we sometimes find in the fossil record, evo-devo studies provide insight into one mechanism for how dramatic changes can arise: the spatial redeployment of homeotic genes. *Homeosis* is defined as the complete replacement of one type of organ with another. In one famous example, a *Hox* gene mutation in *Drosophila* replaces the antennae with legs. If such a mutation were to occur in nature, the organism would probably not have a selective advantage. But what if it did? These kinds of mutant phenotypes first inspired Richard Goldschmidt to develop his idea of the "hopeful monster" early in the twentieth century.

Stephen Jay Gould later revised and updated this idea in the context of his punctuated equilibrium hypothesis about the tempo and mode of evolution. The hypothesis suggests that if, on very rare occasions, truly dramatic morphological changes provide a selective advantage, they may lead to the rapid formation of a new species based on only a few genetic differences. What types of organisms are the most likely to exhibit homeotic change in an evolutionary context? The best candidates are those with highly modular bodies made up of repeating units—such as the segments of an insect or the bones in the spine of a vertebrate. Such animals often express different organ identities in different modular units—such as the antennae, claws, and legs of a lobster—and these identities may be redeployed to different positions along the body axis. Plants are among the most modular organisms on Earth. They produce serially repeated structures—a leaf, a bud at the base of the leaf, and a stem—to generate their bodies. Many exciting and promising questions in plant evo-devo relate to how evolutionary homeosis may have generated rapid change in plant morphology.

Has homeosis contributed to the appearance and diversification of the angiosperms?

The sudden appearance of flowering plants, the angiosperms, in the fossil record so puzzled Charles Darwin that he dubbed their evolution an "abominable mystery." How did the gymnosperms, which always bear their male and female reproductive structures separately, give rise to the hermaphroditic (that is, bearing both male and female structures) flower? Our current understanding of the genetics of floral developmental provides a simple solution to this puzzle: the genetic program controlling floral organ identity is homeotic. Thus, it is possible for very simple genetic changes to transform an entirely male set of reproductive organs into a combination of male and female parts. Such models have been outlined by Günter Theissen at the Friedrich-Schiller-Universität Jena in Germany and by David Baum and Lena Hileman at the University of Wisconsin–Madison and the University of Kansas, respectively. In addition to fostering the origin of the angiosperms, homeosis may have played a role in the group's diversification. Commonly observed shifts in the morphology of sepals and petals or in the number of stamens (male reproductive structures) are suggestive of homeotic changes. These examples are more suitable for experimental verification than is the question on the origin of the angiosperms because they are much more recent occurrences. Although these hypotheses are very attractive, they remain to be confirmed through molecular genetic analyses.

think like a scientist One point of disagreement among evolutionary biologists is whether evolutionary changes primarily occur in gene regulatory regions or in the coding regions. Which of these two types of changes do you think is more likely with homeotic evolution?

Elena M. Kramer is an Associate Professor of Organismic and Evolutionary Biology at Harvard University, where she studies the evolution of the genetic mechanisms controlling floral development. To learn more about Dr. Kramer's research, go to http://www.oeb.harvard.edu/faculty/kramer/Site/Home.html.

REVIEW KEY CONCEPTS

For access to MindTap and additional study materials visit www.cengagebrain.com.

23.1 The Fossil Record

- Fossils are the parts of organisms preserved in sedimentary rocks or in oxygen-poor environments (Figures 23.2 and 23.3).
- The fossil record is incomplete because few organisms fossilize completely, because some organisms are more likely to fossilize than others, and because natural processes destroy many fossils.
- Fossils provide a relative dating system, the geological time scale, for the strata in which they occur. Radiometric dating techniques establish the absolute age of rocks and fossils (Figures 23.4, Table 23.1).
- The fossil record provides data on changes in morphology, biogeography, ecology, and the behavior of organisms (Figure 23.5).

23.2 Earth History

- Earth's crust is composed of plates of solid rock that float on a semisolid mantle (Figure 23.6). New crust is constantly generated and old crust is recycled. Currents in the mantle cause the continents to move over geological time (Figure 23.7). The breakup of the continents profoundly influenced biological evolution by separating populations and creating more shallow marine habitats.
- Continental movements caused variations in climate, the extent of glaciation, and sea levels (Figure 23.8). Asteroid impacts and volcanic eruptions have also influenced Earth's environment, sometimes triggering large-scale extinctions.

23.3 Historical Biogeography and Convergent Biotas

- Disjunct species distributions can be produced by dispersal and/or vicariance (Figure 23.9). Dispersal results in a disjunct distribution when a new population is established on the far side of a barrier. Vicariance results in a disjunct distribution when external factors such as continental drift fragment the landscape.
- Continental drift has created 11 major zoogeographical realms, each with a characteristic fauna (Figure 23.10).
- Convergent evolution produces similar adaptations in distantly related species that live in similar environments (Figures 23.11 and 23.12).

23.4 The History of Biodiversity

- Adaptive radiation produces morphologically diverse species within lineages (Figure 23.13), increasing biodiversity.
- Extinction decreases species diversity. Mass extinctions have occurred at least six times in the history of life (Figure 23.14). Tectonic activity, climate change, and asteroid strikes are probable causes of mass extinctions.
- Biodiversity has increased repeatedly since life first evolved, partly in response to increased geographical separation of the continents and partly because complex interactions evolve among existing species (Figures 23.15 and 23.16).

23.5 Interpreting Evolutionary Lineages

- The horse lineage is complex and highly branched. It includes species of various sizes and diverse morphological adaptations (Figure 23.17).
- The tempo of morphological change varies among lineages. Some exhibit a pattern of gradual change through time (Figure 23.18). Other lineages confirm the predictions of the punctuated equilibrium hypothesis (Figure 23.19).

23.6 The Evolution of Morphological Novelties

- Morphological novelties can arise from evolutionary changes in the relative growth rates of body parts (Figure 23.20) or in the timing of developmental events (Figures 23.21 and 23.22).
- Morphological novelties are often modifications of existing structures. An exaptation is a trait that turns out to be useful in a new environmental context even before natural selection refines its form. For example, the feathers and wings of birds evolved from the ancestral feathers and elongated forelimbs of some bipedal dinosaurs (Figures 23.23 and 23.24).
- Evolutionary developmental biology—evo-devo—examines how evolutionary changes in genes that regulate embryonic development can foster changes in body shape and form.
- Most organisms share an ancient tool kit of several hundred genes that regulate the expression of the thousands of genes involved in development. Evolutionary changes in developmental switches may account for many morphological changes. The differential expression of the *Pitx1* gene in sticklebacks determines whether or not a fish grows pelvic spines (Figure 23.25).

TEST YOUR KNOWLEDGE

Remember/Understand

1. The fossil record:
   a. provides direct and indirect evidence about life in the past.
   b. shows that all morphological novelties arise rapidly.
   c. provides abundant data about rare species with local distributions.
   d. is equally good for all organisms that ever lived.
   e. provides no evidence about the physiology or behavior of ancient organisms.

2. The absolute age of a geological stratum is determined by:
   a. the thickness of its rocks.
   b. the particle size in its rocks.
   c. the types of fossils found within it.
   d. its position relative to other layers.
   e. radiometric dating techniques.

3. The evolutionary history of horses demonstrates that:
   a. modern horses are the direct, lineal descendants of the earliest horses.
   b. the leg bones of modern horses are more complex than those of the earliest horses.
   c. horses have always had specialized teeth that allow them to feed on tough grasses.
   d. horses diversified greatly, but only a few types survived to the present.
   e. the first horses lived in open, grassy habitats.

4. The punctuated equilibrium hypothesis:
   a. recognizes that morphological evolution may occur slowly or quickly.
   b. suggests that major morphological novelties arise gradually.
   c. may help explain why there are so many "missing links" in the fossil record.
   d. suggests that the fossil record is usually complete.
   e. links mass extinctions to the impact of asteroids striking Earth.

5. Biologists believe that the overall similarities between marsupial mammals in Australia and placental mammals in North America were caused by:
   a. plate tectonics.
   b. dispersal.
   c. convergent evolution.
   d. homeotic genes.
   e. punctuated equilibria.

6. The differential growth of body parts is called:
   a. allometry.
   b. paedomorphosis.
   c. heterochrony.
   d. phyletic gradualism.
   e. evo-devo.

7. Exaptations are traits that:
   a. prepare some organisms for future environmental changes.
   b. appear in lineages as a result of an adaptive radiation.
   c. evolve in anticipation of a species' future needs.
   d. are useful in new situations before natural selection alters them for a new function.
   e. occur in animals, but not in plants.

8. Adaptive radiations often follow mass extinctions because:
   a. mass extinctions limit the impact of paedomorphosis.
   b. mass extinctions foster allometry and heterochrony.
   c. mass extinctions decimate all forms of life on Earth.
   d. species that form transitional fossils often survive mass extinctions.
   e. extinctions open adaptive zones that had been previously occupied.

9. Homeobox genes are defined as genes that:
   a. bind directly to a regulatory site on DNA.
   b. code for transcription factors activating or repressing genes that influence an organism's form.
   c. determine whether or not a morphological innovation leads to an adaptive radiation.
   d. have been inherited from an ancient ancestor by nearly all forms of life.
   e. help biologists differentiate between plants and animals.

Apply/Analyze

10. The observation that fossils of *Premedosaurus* are found only in Argentina and Northern Europe provides an example of:
    a. a continuous distribution.
    b. a disjunct distribution.
    c. punctuated equilibrium.
    d. heterochrony.
    e. gradualism.

11. **Discuss Concepts** One possible measure of a species' evolutionary success is the number of descendant species it produces. Should our species, *Homo sapiens*, be considered successful under this definition?

Evaluate/Create

12. **Discuss Concepts** Many millions of years from now, continental drift may obliterate the Pacific Ocean, pushing North America into physical contact with Asia. What effects might these events have on the organisms living at that time?

13. **Discuss Concepts** Extinctions are common in the history of life. Why are biologists alarmed by the current wave of extinctions caused by human activity?

14. **Design an Experiment** Design a study to determine whether the wings of birds, bats, and insects and their ability to fly are the products of convergent evolution.

15. **Apply Evolutionary Thinking** The geological evolution of Earth has had an obvious effect on biological evolution. You have read about how the release of oxygen by photosynthetic organisms increased atmospheric oxygen concentration. How are human activities changing the physical environment on Earth? What new selection pressures do these environmental alterations establish?

For selected answers, see Appendix A.

INTERPRET THE DATA

The graphs presented below provide data on the body sizes (above) and brain sizes (below) of modern humans and three extinct species in the evolutionary lineage that includes humans. How would you describe the relative changes of body size and brain size in this lineage through time?

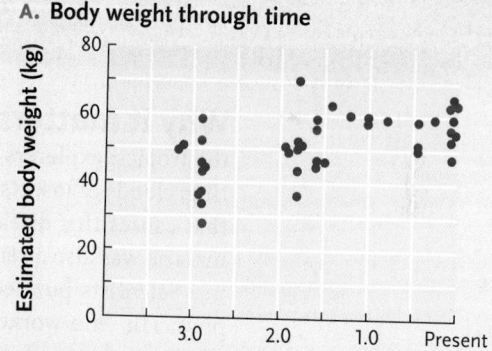

A. Body weight through time

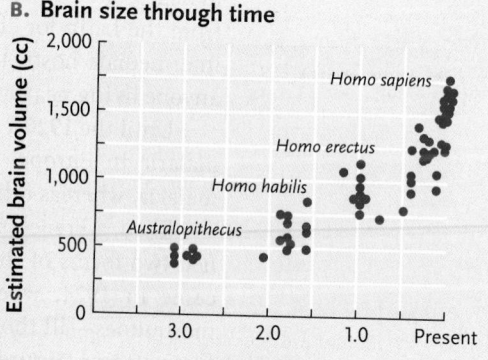

B. Brain size through time

Source: Adapted from D. J. Futuyma. 2009. Evolution, 2e. Sinauer Associates, Sunderland, MA (after S. Jones, R. Martin, and D. Pilbeam, eds. 1992). The Cambridge Encyclopedia of Human Evolution. Cambridge University Press, Cambridge.

24 Systematics and Phylogenetics: Revealing the Tree of Life

Courtesy of Dr. Marc Ducousso

Barbie pagoda fungus *(Podoserpula miranda)* discovered in 2011 in New Caledonia.

Why it matters . . . Mention the word "malaria," and people envision old movies about the tropics: explorers wander through the jungle in pith helmets and sleep under mosquito netting; clouds of insects hover nearby, ready to infect them with *Plasmodium,* the protistan parasite that causes this disease. You may be surprised to learn, however, that less than 100 years ago, malaria was also a serious threat in the southeastern United States and much of Western Europe.

Scientists puzzled over the cause of malaria for thousands of years. Hippocrates, a Greek physician who worked in the fifth century B.C., knew that people who lived near malodorous marshes often suffered from fevers and swollen spleens. Indeed, the name *malaria* is derived from the Latin for "bad air." By 1900, scientists had established that mosquitoes, *Plasmodium*'s intermediate hosts, transmit the parasite to humans. Mosquitoes breed in standing water, and anyone living nearby is likely to suffer their bites.

Until the 1920s, scientists thought that the mosquito species *Anopheles maculipennis* carried malaria in Europe. But some areas with huge populations of these insects had little human malaria, whereas other areas had relatively few mosquitoes and a high incidence of the disease.

Then, a French researcher reported variation in the mosquitoes, and Dutch scientists identified two forms of the "species," only one of which seemed to carry malaria. The breakthrough came in 1924, when a retired public health inspector in Italy discovered that individual mosquitoes—all thought to be the same species—produced eggs with one of six distinctive surface patterns **(Figure 24.1).**

Further research revealed that the name *Anopheles maculipennis* had been applied to six separate mosquito species. Although the adults of these species are almost indistinguishable, their eggs are clearly different. The species are reproductively isolated from each other, and they differ ecologically: some breed in brackish coastal marshes, others in freshwater inland marshes,

A. *Anopheles* mosquito feeding on human blood

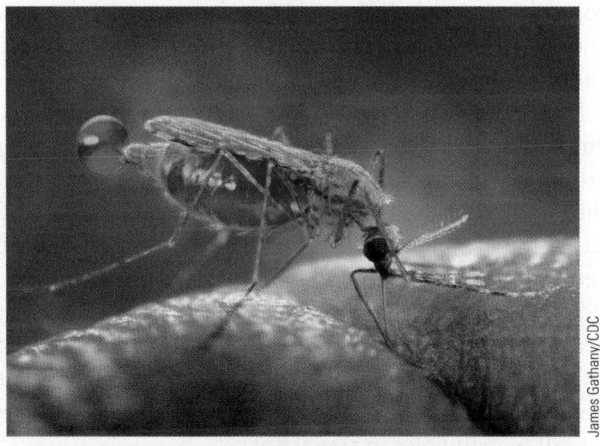

James Gathany/CDC

B. Eggs of six European *Anopheles* mosquito species

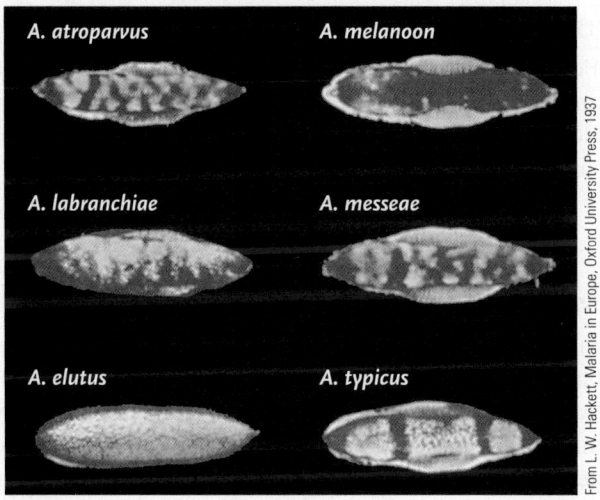

A. atroparvus

A. melanoon

A. labranchiae

A. messeae

A. elutus

A. typicus

From L. W. Hackett, Malaria in Europe, Oxford University Press, 1937

FIGURE 24.1 Carriers of malaria. (A) Like other *Anopheles* mosquitoes, *A. gambiae* frequently take a blood meal from a human host. **(B)** Differences in surface patterns on the eggs of *Anopheles* mosquitoes in Europe helped researchers identify six separate species. The adults of all six species look remarkably alike.

© Cengage Learning 2017

and still others in slow-moving streams. Only some of these species have a preference for human blood, and researchers eventually determined that only three of them routinely transmit malaria to humans.

These discoveries explained why the geographical distributions of mosquitoes and malaria did not always match. And government agencies could finally fight malaria by eradicating the disease-carrying species. Health workers drained marshes to prevent mosquitoes from breeding. They used insecticides to kill mosquito larvae or introduced fish of the genus *Gambusia,* the mosquitofish, which eats them. These targeted control programs were very successful in the early and middle decades of the twentieth century.

Today, with the increased mobility of humans, agricultural products, and other goods, some mosquito species—as well as many other organisms—are invading habitats made more hospitable by global climate change (discussed further in Unit

Seven). Some have expanded their geographical ranges substantially: introduced mosquito species have been discovered on all continents except Antarctica. Mosquitoes carry numerous agents of disease, and biologists are now devising new ways to recognize—and eradicate—the species that pose the greatest threats to human welfare.

The historic eradication of malaria in Europe owed a debt to **systematics,** the branch of biology that studies the diversity of life and its evolutionary relationships. Systematic biologists—*systematists* for short—identify, describe, name, and classify organisms, organizing their observations within a framework that reflects the organisms' evolutionary relationships. In this chapter we briefly describe the traditional approach to classification. We then focus attention on how systematists working today develop hypotheses about the evolutionary relationships of all the branches, twigs, and leaves on the Tree of Life.

24.1 Nomenclature and Classification

The Swedish naturalist Carl von Linné (1707–1778), better known by his Latinized name, Carolus Linnaeus, was the first modern practitioner of **taxonomy,** the science that identifies, names, and classifies new species. A professor at the University of Uppsala, Linnaeus sent ill-prepared students around the world to gather specimens, losing perhaps a third of his followers to the rigors of their expeditions. Although he may not have been a commendable student adviser, Linnaeus developed the basic system of naming and classifying organisms that biologists embraced for two centuries.

Linnaeus Developed the System of Binomial Nomenclature

Linnaeus invented the system of **binomial nomenclature,** in which species are assigned a Latinized two-part name, or **binomial.** The first part of the name identifies a **genus** (plural, *genera*), a group of species with similar characteristics. The second part is the **specific epithet,** or species name. When identifying and naming a new species, Linnaeus used the morphological species concept (described in Section 22.1), assigning the same scientific name to individuals that shared anatomical characteristics.

A combination of the generic name and the specific epithet provides a unique name for every species. For example, *Ursus maritimus* is the polar bear and *Ursus arctos* is the brown bear. By convention, the first letter of a generic name is always capitalized; the specific epithet is never capitalized; and the entire binomial is italicized. In addition, the specific epithet is never used without the full or abbreviated generic name preceding it because the same specific epithet is often given to species in different genera. For instance, *Ursus americanus* is the American black bear, *Homarus americanus* is the Atlantic lobster, and *Bufo americanus* is the American toad. If you were to order just *"americanus"* for dinner, you might be dismayed when your plate arrived—unless you have an adventurous palate!

Nonscientists often use different common names to identify a species. For example, *Bothrops asper,* a poisonous snake native to Central and South America, is called *barba amarilla* (meaning "yellow beard") in some places and *cola blanca* (meaning "white tail") in others; biologists have recorded about 50 local names for this species. Adding to the confusion, the same common name is sometimes used for several different species. Binomials, however, allow people everywhere to discuss organisms unambiguously.

Many binomials are descriptive of the organism or its habitat. *Asparagus horridus,* for example, is a spiny plant. Other species, such as the South American bird *Rhea darwinii,* are named for notable biologists. The naming of newly discovered species follows a formal process of publishing a description of the species in a scientific journal. International commissions meet periodically to settle disputes about scientific names.

Linnaeus Devised the Taxonomic Hierarchy to Organize Information about Species

Linnaeus described and named thousands of species on the basis of their morphological similarities and differences. Keeping track of so many species was no easy task, so he devised a **classification,** a conceptual filing system that arranges organisms into ever more inclusive categories. Linnaeus' classification, called the **taxonomic hierarchy,** includes a nested series of formal categories (from most inclusive to least): domain, kingdom, phylum, class, order, family, genus, species, and subspecies **(Table 24.1);** you may also find it useful to review the traditional taxonomic categories, which were described in Section 1.3 and illustrated in Figure 1.10.) The organisms included

TABLE 24.1	The Taxonomic Hierarchy

The taxonomic hierarchy as developed by Linnaeus and applied to the American black bear (*Ursus americanus*).

Taxonomic category	Taxon
Domain	Eukarya
Kingdom	Animalia
Phylum	Chordata
Class	Mammalia
Order	Carnivora
Family	Ursidae
Genus	*Ursus*
Species	*Ursus americanus*

within any category of the taxonomic hierarchy compose a **taxon** (plural, *taxa*). Woodpeckers, for example, are a taxon (Picidae) at the family level, and pine trees are a taxon *(Pinus)* at the genus level.

Species that are included in the same taxon at the bottom of the hierarchy (that is, in the same genus or family) generally share many characteristics. By contrast, species that are included in the same taxon only near the top of the hierarchy (that is, the same kingdom or phylum) generally share many fewer traits. The hierarchy has been a great convenience for biologists because every taxon is defined by a set of shared characteristics. Thus, when a biologist refers to a member of the family Picidae, all of his or her colleagues understand that the biologist is talking about a medium-sized bird that uses its stout bill to drill holes in tree trunks.

24.2 Phylogenetic Trees

Linnaeus devised the taxonomic hierarchy long before Darwin published his theory of evolution. His goals were to illuminate the details of God's creation and to devise a practical way for naturalists to keep track of their discoveries. But the science of systematics changed in response to Darwin's idea that all organisms in the Tree of Life are descended from a distant common ancestor: systematists began to focus on discovering the evolutionary relationships between groups of organisms.

Systematists Adapted Linnaeus' Approach to a Darwinian Worldview

The taxonomic hierarchy that Linnaeus defined was easily adapted to Darwin's concept of branching evolution, which is itself a hierarchical phenomenon (see Figure 20.11); as we discussed in Chapters 22 and 23, ancestral species give rise to descendant species through repeated branching of a lineage. Organisms in the same genus generally share a fairly recent common ancestor, whereas those assigned to the same higher taxonomic category, such as a class or phylum, share a common ancestor that lived in the more distant past.

In the second half of the nineteenth century, systematists began to reconstruct the **phylogeny** (that is, the evolutionary history) of organisms. Phylogenies are illustrated in **phylogenetic trees,** which are formal hypotheses that identify likely relationships among species and higher taxonomic groups. And like all hypotheses, they are constantly revised as scientists gather new data.

A Phylogenetic Tree Depicts the Evolutionary History of a Group of Organisms

Contemporary evolutionary biologists construct phylogenetic trees to illustrate the hypothesized evolutionary history of organisms. Researchers tailor the breadth of their analyses to match specific research questions. Thus, some trees might include the evolutionary history of all known organisms, others a small cluster of closely related populations within a species, and still others a group somewhere between those extremes. Regardless of how wide a range of organisms is included, all phylogenetic trees share a specific structure and depict key relationships in similar ways **(Figure 24.2).**

For example, phylogenetic trees are usually drawn along an implicit or explicit timeline. In this book, phylogenetic trees are generally depicted vertically; the most ancient organisms and evolutionary events are at the bottom of the tree (often labeled "long ago"), and the most recent are at the top (often labeled "present"). The common ancestor of all species included in the tree is described as the **root** of the tree. In a few cases in this book, trees are presented with the root on the left, with time passing from left to right.

As you know from the discussion in Section 23.5, the tempo of evolution varies within and among lineages. In some cases, evolutionary changes may accumulate slowly in a lineage as the environment shifts over time. This pattern of gradual phyletic change is often described as *anagenesis*. If the changes through time are substantial and the fossil record is incomplete, paleontologists who discover morphologically distinct fossils in different strata may assign them different species names and say that "the ancestral species A evolved into the descendant species B." But the production of such "new" species by anagenesis does not increase biodiversity; rather, it is simply the gradual transformation of one "species" into another as its characteristics shifted over time. Anagenesis is often illustrated by a straight line in a phylogenetic tree.

In other circumstances, an ancestral species undergoes speciation, producing two descendant species, both of which are distinct from their common ancestor. This pattern of evolution is described as *cladogenesis,* a process that *does* increase biodiversity. Cladogenesis is depicted in a phylogenetic tree by a branching pattern, with two descendants arising from their common ancestor. When they first emerge, the two branches may represent new species. But as cladogenesis continues repeatedly through evolutionary time—with branches giving rise to branchlets and branchlets to twigs—each of those new species may become the common ancestor of its own many descendants. Thus, each new species produced by cladogenesis has the potential to become the "root" of its own evolutionary lineage.

When reading a phylogenetic tree, each branching point is called a **node,** and each evolutionary lineage—a node with all of the branches, branchlets, and twigs that emerge from it—is called a **clade** (*klados* = branch). You can identify a clade on any phylogenetic tree by following a lineage from a node to the tips of all of its twigs at the top of the tree. Some clades, such as Aves (birds), include thousands of species, whereas others, such as *Geospiza* (a genus of ground and cactus finches from the Galápagos Islands), include just a few (see Figure 23.13). Like the taxonomic hierarchy, phylogenetic trees have a nested structure: younger and smaller clades, such as the genus *Geospiza*, are nested within larger and older clades, such as Aves. Two clades that emerge from the same node are called **sister clades** (or sister taxa) because they are each other's closest relatives; similarly, two species that emerge from the same node near the very top of the tree are described as **sister species.**

Phylogenetic Trees Allow Biologists to Define Evolutionary Classifications

Evolutionary biologists working today want a classification that mirrors the branching patterns of a group's phylogenetic history. When converting a phylogenetic tree into a classification, they try to identify only **monophyletic taxa** or lineages. A monophyletic taxon comprises one clade—an ancestral species (represented by a node) and *all* of its descendants, but no other species **(Figure 24.3)**. Monophyletic taxa are defined at every level of the taxonomic hierarchy. For example, biologists consider the Felidae (the traditional family-level taxon that includes all cat species) to be monophyletic: all cat species living on Earth today, from house cats to tigers, are the descendants—and the only living descendants—of a common ancestor that lived about 25 million years ago. Thus, the Felidae is one small, but complete, branch on the Tree of Life. At a much broader scale, the Animalia is a monophyletic taxon (at the traditional kingdom level) that comprises all animals, which are the descendants of one common ancestor. Even if biologists have not yet identified the very first animal in the fossil record, they can infer its existence at the root of the phylogenetic tree for Animalia.

Biologists have not always defined strictly monophyletic taxa. Because of missing data, or as a matter of convenience, they sometimes named taxa that included species from different clades or taxa that included some, but not all, of an ancestral species' descendants. A **polyphyletic taxon** is one that includes organisms from different clades, but not their common ancestor. For example, a taxon that included only birds and bats, two clades of vertebrates that are capable of flight, would be considered polyphyletic because it would not include their last common ancestor, a four-legged creature that lived many millions of years before birds and bats first appeared. A **paraphyletic taxon** is one that includes an ancestor and some, *but not all,* of its descendants. For example, people commonly used to define terrestrial dinosaurs and birds as distinct groups. But "terrestrial dinosaurs" was a paraphyletic taxon: as you know from Section 23.6, birds are the descendants of one group of terrestrial dinosaurs. Thus, the monophyletic taxon Dinosauria must include birds, as well as their nonflying relatives.

HOW TO READ PHYLOGENETIC TREES

A. Phylogenetic tree for Anthropoidea

The phylogenetic tree for Anthropoidea illustrates properties that many phylogenetic trees share. The relative positions of the nodes (branching points) define how recently sister clades diverged. Clades that emerge from a node closer to the top of the tree are more closely related than clades that emerged from a node closer to the root of the tree.

Clade of African apes and hominins

Bipedal locomotion

HOMININI

Common ancestor of chimpanzees and humans

Common ancestor of chimpanzees and gorillas

Chimpanzees are more closely related to hominins than to gorillas because they share a more recent common ancestor with hominins than they do with gorillas.

HOMINIDAE

The clade that includes gorillas and the clade that includes both chimpanzees and humans are sister taxa because they emerged from a common node.

HOMINOIDEA

Nodes represent common ancestors that underwent cladogenesis and produced two descendant clades.

ANTHROPOIDEA

Common ancestor of all anthropoids at root of the tree

Time (millions of years ago)

New world monkeys
Old world monkeys
Gibbons
Orangutans
Gorillas
Chimpanzees
Humans

B. Phylogenetic tree with Hominidae rotated

This phylogenetic tree contains exactly the same information about anthropoid relationships as the tree in part A. Clades can be rotated around a node without changing their relationship to other clades. Compare the position of the Hominidae—which includes gorillas, chimpanzees, and humans—in parts A and B.

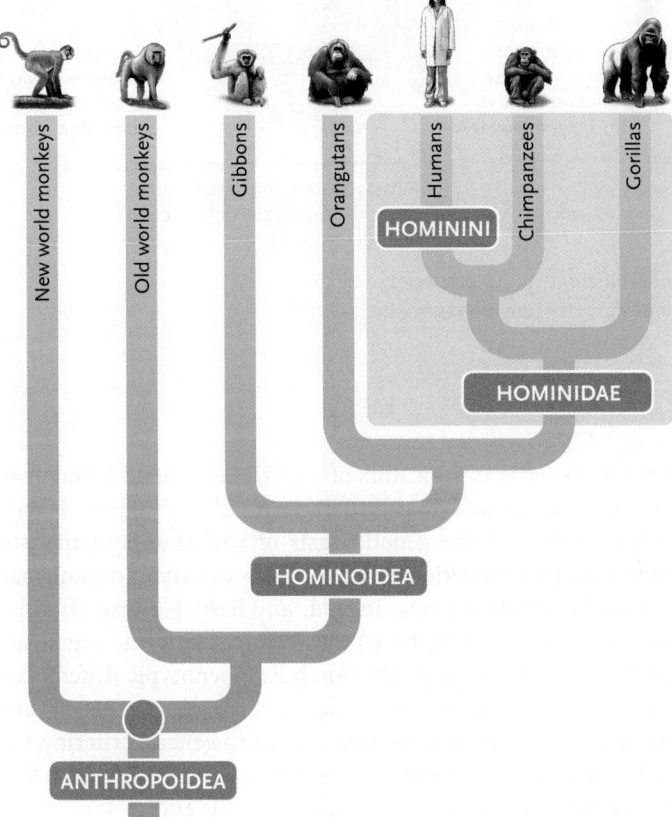

Depiction of time. If a phylogenetic tree includes an explicit time axis, the positions of the nodes reveal when on the geological time scale a clade originated; the length of the vertical branch between nodes indexes how long an ancestral group persisted before it diversified. In most phylogenetic trees in this book, the time scale is not precise; often it is not even specified. The sequence of nodes indicates the order in which clades appeared; the lengths of the branches contain no specific information about the time since two clades diverged.

Depiction of relatedness. Horizontal spacing between clades in most of this book's phylogenetic trees does not indicate their degree of difference or their degree of relatedness. However, when the horizontal distance between species or clades is meaningful (as in, for example, Figure 23.19), how the distances should be interpreted is explained in the horizontal axis label, the figure legend, or the text.

Number of descendants. Most nodes in phylogenetic trees have two branches emerging from them, reflecting the evolution of two descendants from one ancestor. When biologists have not yet discovered the detailed pattern of branching that produced the diversity of clades in the tree, you may see three or more branches emerging from a node or from a horizontal branch. These nodes are currently "unresolved"; future research will allow the portrayal of these evolutionary relationships more precisely.

Relative ages of clades. In most phylogenetic trees in this book, the clades have been arranged from oldest on the left to youngest on the right. But any clade can be rotated around a node without changing the meaning of the phylogenetic tree. When reading a phylogenetic tree, focus on which clades share more recent common ancestors, indicated by the relative positions of the nodes from which they emerge.

SUMMARY Phylogenetic trees provide hypotheses about the evolutionary histories of the organisms included in the analysis. The common ancestor of sister clades is depicted on the node from which the two clades emerge. An implied or explicit timeline identifies the sequence in which new clades arose from their ancestors. Clades with a common ancestor closer to the top of the tree are more closely related than those with a common ancestor closer to the root of the tree.

think like a scientist The phylogenetic tree illustrated in part B positions humans next to orangutans. Does this arrangement suggest that humans are more closely related to orangutans than they are to gorillas?

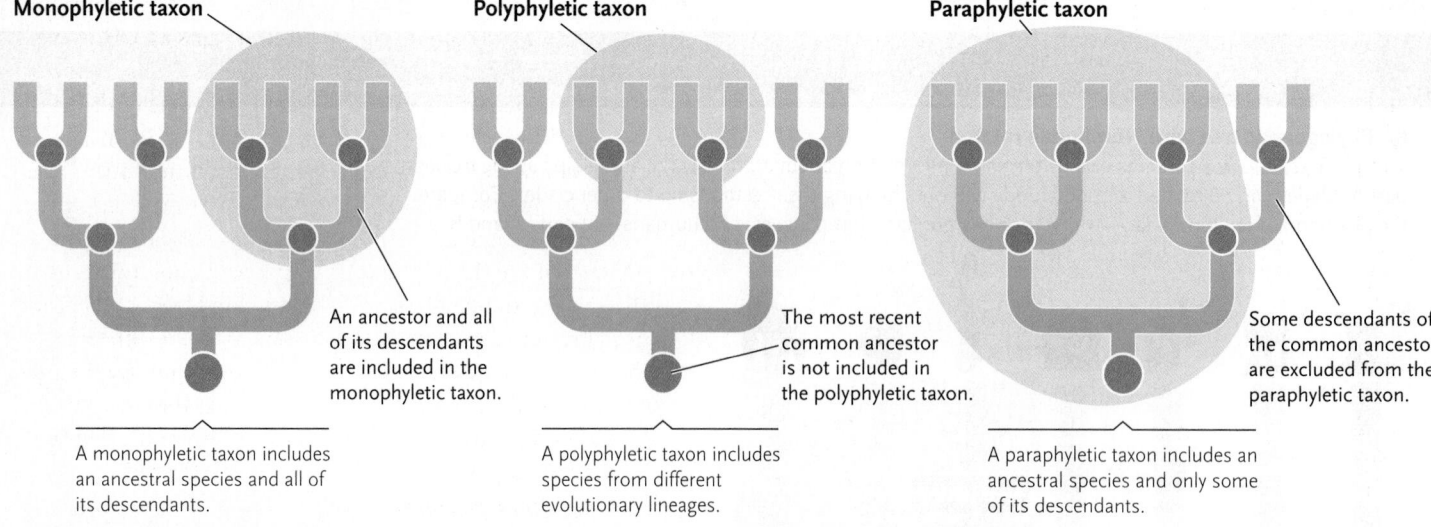

Monophyletic taxon

An ancestor and all of its descendants are included in the monophyletic taxon.

A monophyletic taxon includes an ancestral species and all of its descendants.

Polyphyletic taxon

The most recent common ancestor is not included in the polyphyletic taxon.

A polyphyletic taxon includes species from different evolutionary lineages.

Paraphyletic taxon

Some descendants of the common ancestor are excluded from the paraphyletic taxon.

A paraphyletic taxon includes an ancestral species and only some of its descendants.

FIGURE 24.3 Defining taxa in a classification. Although systematists can identify different groups of species in a phylogenetic tree as a taxon, a major goal of contemporary systematics is to identify taxa that are monophyletic groups.

© Cengage Learning 2017

Many of the phylogenetic trees included in this book identify monophyletic taxa with labels on their branches (see Figure 24.2). The names of these clades are identified in capital letters in a bronze-colored plaque; the major characteristics that define large clades are identified in lowercase letters in a tan-colored plaque. As a convenience, some commonly used names for groups that biologists now recognize as paraphyletic are presented in quotation marks in bronze plaques with dashed, rather than solid, borders.

STUDY BREAK 24.2

1. What is the difference between a phylogenetic tree and a classification?
2. What are the differences between a monophyletic taxon, a polyphyletic taxon, and a paraphyletic taxon?

24.3 Sources of Data for Phylogenetic Analyses

Linnaeus classified organisms on the basis of their morphological similarities and differences—even though he did not understand how those characteristics arose. For example, he defined birds as a class of oviparous ("egg-laying") animals with feathered bodies, two wings, two feet, and a bony beak. No other animals possess all these characteristics, which distinguish birds from "quadrupeds" (his term for mammals), "amphibians" (among which he included reptiles), fishes, insects, and "worms."

Mendel's subsequent work on the heritable basis of morphological variations provided the scientific rationale for this endeavor: modern systematists infer that morphological differences serve as indicators of underlying genetic differences between species and lineages. Today, with our much deeper understanding of the genetic basis of variation, systematists undertake phylogenetic analyses using a variety of organismal and molecular characters. Indeed, any heritable trait (that is, any trait with a genetic basis) that is intrinsic to the organism can be used in a phylogenetic analysis; phenotypic differences caused by environmental variation (see Section 20.1) are excluded. In this section we first consider a general criterion for evaluating characters and then examine examples of how a few specific types of characters are useful in this effort.

The Analysis of Homologous Characters Sheds Light on Evolutionary Relationships

A basic premise of phylogenetic analyses is that phenotypic similarities between organisms reflect their underlying genetic similarities. As you may recall from Figure 20.3, species that are morphologically similar have often inherited the genetic basis of their resemblance from a common ancestor. Similarity that results from shared ancestry, such as the four limbs of all tetrapod vertebrates, is called **homology,** and biologists frequently describe such traits in two or more species as *homologies* or *homologous characters.* Any trait, from genetic sequences to anatomical structures to mating behaviors, can be described as homologous in two or more species as long as they inherited the trait from their common ancestor.

Even though characters are homologous, they may differ greatly among species, especially if their function has changed over time. For example, the stapes, a bone in the middle ear of tetrapod vertebrates, evolved from—and is therefore homologous to—the hyomandibula, a bone that supported the jaw joint of early fishes. The ancestral function of the bone is

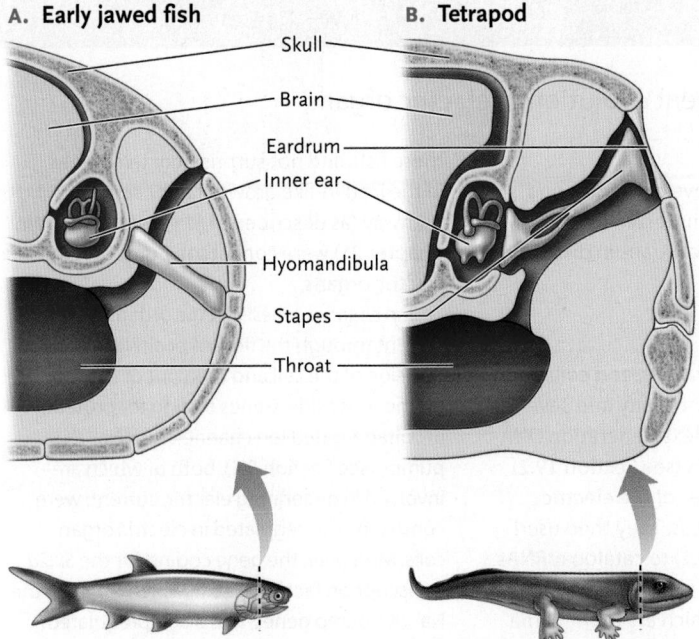

A. Early jawed fish

- Skull
- Brain
- Eardrum
- Inner ear
- Hyomandibula
- Stapes
- Throat

B. Tetrapod

FIGURE 24.4 Homologous bones, different structure and function. The hyomandibula, which braced the jaw joint against the skull in early jawed fishes **(A),** is homologous to the stapes, which transmits sound from the eardrum to the inner ear in the four-legged vertebrates, exemplified here by an early tetrapod **(B).** Both large illustrations show a cross section through the head just behind the jaw joint, as depicted in the small illustrations.

© Cengage Learning 2017

retained in some modern fishes, but its structure, position, and function are different in tetrapods **(Figure 24.4).**

As you know from the discussion in Section 23.3, organisms that are not closely related sometimes bear a striking resemblance to one another, especially when they are exposed to similar patterns of natural selection. Phenotypic similarity that evolved independently in different lineages is called **homoplasy,** which is often the product of convergent evolution (see Section 23.3); biologists describe such similarities as *homoplasies* or *homoplastic characters.* Some biologists use the terms *analogies* or *analogous characters* for homoplastic characters that serve a similar function in different species.

When systematists encounter similar morphological traits, how can they determine whether they are homologous or homoplastic? First, homologous structures are similar in anatomical detail and in their relationship to surrounding structures. For example, the bones in the wings of birds and bats are considered to be homologous **(Figure 24.5).** They include the same structural elements and have similar connections to the rest of the skeleton. Moreover, the fossil record documents that birds and bats inherited the basic skeletal structure of the

forelimb from their most recent common ancestor, a tetrapod vertebrate that lived more than 300 million years ago. However, the large flat surfaces of their wings, as well as flying behavior itself, are homoplastic. The wing surfaces are made of different materials—feathers in birds and membranous skin in bats—and their common ancestor, lacking any hint of such structures on its forelimbs, was confined to life on the ground. Thus, for birds and bats, flight and some of the anatomical structures that produce it are the products of convergent evolution.

Second, in multicellular organisms, homologous characters grow from the same embryonic tissues and in similar ways during development. Systematists have always put great stock in embryological indications of homology on the assumption that evolution has conserved the pattern of embryonic development in related organisms. Indeed, recent discoveries in evolutionary developmental biology (described in Sections 16.4 and 23.6) have revealed that the genetic controls of developmental pathways are very similar across a wide variety of organisms. And, as described in *Molecular Insights,* genomic techniques are revealing remarkable shared similarities in the underlying genetic and cellular mechanisms that have contributed to the evolution of convergent characters in species that are not closely related.

Finally, as you will see in Section 24.4, the structure of a phylogenetic tree can reveal whether two or more species inherited specific similarities from a common ancestor. If they did, the structures are homologous; if not, they are homoplastic.

Morphological Characters Provide Abundant Clues to Evolutionary Relationships

Morphological structures often provide useful information for phylogenetic analyses. Structural differences between organisms, which often reflect underlying genetic differences (see

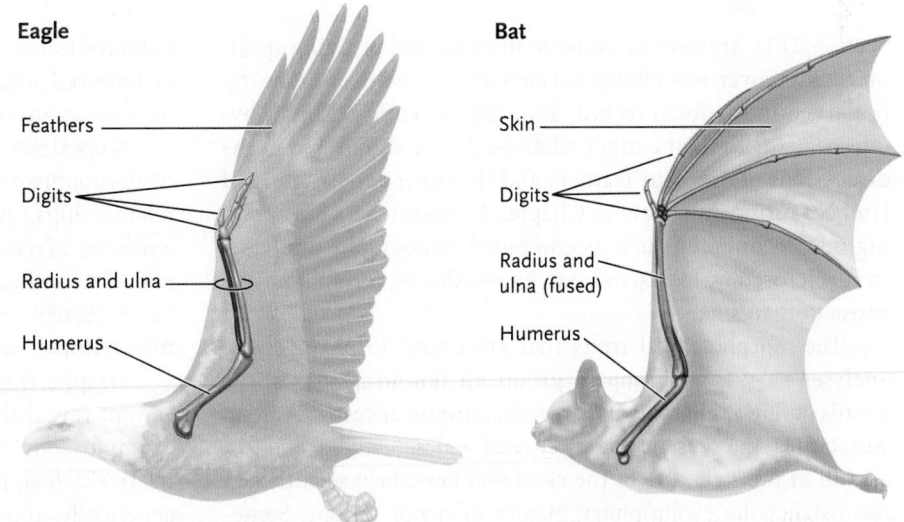

Eagle

- Feathers
- Digits
- Radius and ulna
- Humerus

Bat

- Skin
- Digits
- Radius and ulna (fused)
- Humerus

FIGURE 24.5 Assessing homology. The wing skeletons of birds and bats are homologous structures with the same basic elements. However, similarities in the flat wing surfaces are homoplastic because the surfaces are composed of different tissues.

© Cengage Learning 2017

Electric Organs in Fishes: Deep homology fosters convergent evolution of electric organs

Many distantly related groups of fish have electric organs. In most cases, these organs generate weak electrical impulses that are used for communication (as described in Section 56.5) or for detecting prey, but some are strong enough to stun other organisms. These electrical signals are produced in electric organs by specialized cells called *electrocytes,* which in most cases are derived from cells that would otherwise form skeletal muscle (see Figure 38.5A). Cells that become electrocytes produce specialized structures to conduct unusually large electrical currents. Like other specialized cells, electrocytes differentiate by turning on the expression of genes coding for proteins that make their specialized structures (see Section 16.4) and by turning off genes that code for proteins typical of skeletal muscle cells.

Electric organs in fish are a classic example of homoplastic characters, because they have evolved independently in at least five different lineages that are only distantly related to one another **(Figure).** In fact, Darwin used this example to illustrate the idea of convergent evolution. However, evolutionary developmental biologists have discovered that similar structures produced by convergent evolution are often produced by the same developmental genetic toolkit (described in Section 23.6).

Research Question

Did common genetic, developmental, and physiological mechanisms foster convergent evolution of electric organs in distantly related fishes?

Experiment

Jason Gallant, Lindsay Traeger, and collaborators at Michigan State University and several other institutions used next-generation DNA-sequencing technologies (see Section 19.2) to sequence the genomes of the electric eel, *Electrophorus electricus.* They then used RNA-seq (see Section 19.2) to catalog mRNAs in *E. electricus,* as well as four other species of electric fish, two of which are in the same order as *E. electricus* (Gymnotiformes) and two of which are more distantly related. In all five species, they compared the mRNAs produced in electric organs to those produced in skeletal muscle. In total, hundreds of billions of nucleotides of DNA and RNA sequence were generated.

Results

The researchers identified a number of genes that were consistently upregulated or downregulated in electrocytes in both gymnotiformes and the more distantly related fishes:

Cell growth regulating genes: Electrocytes are much larger than skeletal muscle cells in

these fish, and not surprisingly two genes in the insulin-like growth factor signaling pathway (as described in *Molecular Insights* in Chapter 21) were consistently upregulated in electric organs.

Ion channel genes: Electric organs generate current through the flow of positive ions into one side of the cell and then out of the cell on the other side. Genes coding for proteins in voltage-gated ion channels and Na^+/K^+ pumps (see Section 5.4), both of which are involved in generating electric current, were consistently upregulated in electric organ cells. Moreover, the gene coding for the *Six2a* transcription factor protein, which turns on the Na^+/K^+ pump genes, was also upregulated.

Protein contraction genes: Another transcription factor gene upregulated in electric organs is *Hey1,* which inhibits the *myoD* transcription factor gene, a key mechanism causing cells to develop into skeletal muscle (see Section 16.4 and Figure 16.14). Because *myoD* was inhibited in electrocytes, the genes coding for actin, myosin, and other proteins involved in skeletal muscle cells contractions were all downregulated.

Conclusion

These findings show that electrocytes in these different groups of fishes are produced by many of the same underlying genetic

Section 21.1), are easy to measure in preserved or living specimens. Moreover, morphological characteristics are often clearly preserved in the fossil record, allowing the comparison of living species with their extinct relatives. As you discovered in the discussions of dinosaur feathers and the inner ear structure of lambeosaurine dinosaurs in Chapter 23, researchers are applying new techniques, such as computed tomography and electron microscopy, to expand our knowledge of the anatomy of extinct organisms.

The morphological traits that are useful in phylogenetic analyses vary from group to group. In flowering plants, the details of flower anatomy may reveal common ancestry. Among vertebrates, the presence or absence of scales, feathers, and fur, as well as the structure of the skull and jaws, help scientists to reconstruct the evolutionary history of major groups. Sometimes systematists use obscure characters of unknown function. But differences in the number of scales on the backs of lizards or in the curvature of a vein in the wings of bees may be

good indicators of the genetic differentiation that accompanied or followed speciation—even if we do not know *why* these differences evolved.

Sometimes characteristics found only in the earliest stages of an organism's life cycle can provide evidence of evolutionary relationships. As described in Chapter 31, analyses of the embryos of vertebrates revealed that they are rather closely related to sea cucumbers, sea stars, and sea urchins and even more closely related to a group of nearly shapeless marine invertebrates called sea squirts or tunicates.

Despite their usefulness, morphological characters alone cannot reveal the details of all evolutionary relationships. For example, some salamander species in North America differ in relatively few morphological features, even when they are genetically, physiologically, and behaviorally distinct. Moreover, researchers cannot easily compare the structures of organisms, such as flatworms and dogs, that share very few morphological traits.

and cellular mechanisms, even though they evolved independently. These mechanisms were present in the ancestors of these fishes, long before they evolved electric organs. To produce electrically conducting cells, existing molecular mechanisms simply had to be regulated differently. For example, electrocytes make use of the same ion channels and pumps that are produced by many other cells in vertebrates, but they produce these proteins in much greater quantities, to increase vastly the flow of ions across their membranes. This is an example of a "deep homology," in which adaptations that have evolved independently nevertheless make use of the same preexisting biological mechanisms.

think like a scientist

Evolutionary biologists consider the electric organs of distantly related fishes to be homoplastic characters—the products of convergent evolution—because they are very different structurally. However, as described above, they are all produced by genetic and cellular mechanisms that these fishes inherited from an ancestral vertebrate. Given the shared genetic basis of the development of these organs, are they really homoplastic, or should we consider them homologous?

Source: Jason R. Gallant et al. 2014. Genomic basis for the convergent evolution of electric organs. *Science* 344:1522–1525.

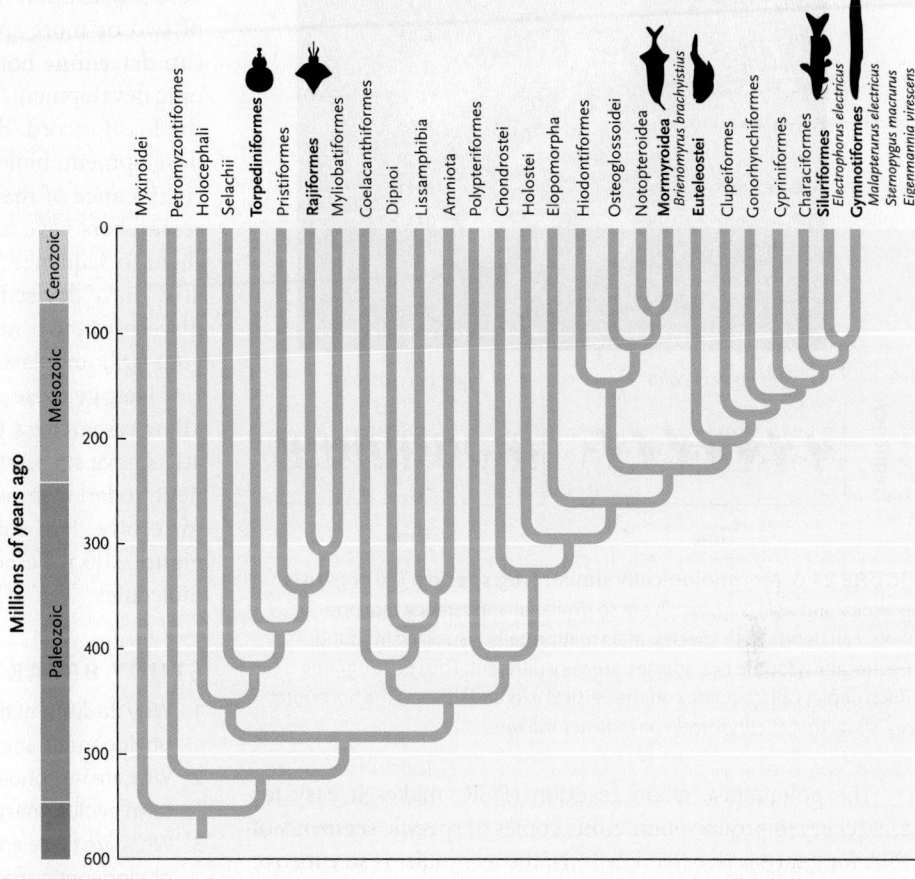

FIGURE **Phylogenetic tree of vertebrate clades; icons identify lineages that have electric organs.** The names of the species included in this study are listed below the clade to which they belong.

Behavioral Characters Are Useful When Animal Species Are Not Morphologically Distinct

When external morphology cannot be used to differentiate animal species, systematists often examine their behaviors for clues about their relationships. For example, two species of tree frog (*Hyla versicolor* and *Hyla chrysoscelis*) commonly occur together in forests of the central and eastern United States. Both species have bumpy skin and adhesive pads on their toes that enable them to climb vegetation. They also have gray backs, white bellies, yellowish-orange coloration on their thighs, and large white spots below their eyes. The frogs are so similar that even experts cannot easily tell them apart.

How do we know that these frogs represent two species? During the breeding season, males of each species use a distinctive mating call to attract females **(Figure 24.6)**. The difference in calls is a prezygotic reproductive isolating mechanism that prevents females from mating with males of a different species (see Section 22.2). Prezygotic isolating mechanisms are excellent systematic characters because they are often the traits that animals themselves use to recognize members of their own species. The two frog species also differ in chromosome number—*H. chrysoscelis* is diploid and *H. versicolor* is tetraploid—which is a postzygotic isolating mechanism.

Molecular Sequences Are Now a Commonly Used Source of Phylogenetic Data

For roughly 200 years, systematists building on Linnaeus' work relied on a variety of organismal traits to analyze evolutionary relationships and classify organisms. Today, most systematists conduct phylogenetic analyses using molecular characters, such as the nucleotide base sequences of DNA and RNA. Because DNA is inherited, shared changes in molecular sequences—insertions, deletions, or substitutions—provide clues to the evolutionary relationships of organisms.

Hyla versicolor   Hyla chrysoscelis

Frequency

Time   Time

FIGURE 24.6 Morphologically similar frog species. The frogs *Hyla versicolor* and *Hyla chrysoscelis* are so similar in appearance that one photo can depict both species. Male mating calls, visualized in sound spectrograms for the two species, are very different. The spectrograms, which depict call frequency on the vertical axis and time on the horizontal axis, show that *H. chrysoscelis* has a faster trill rate.

The polymerase chain reaction (PCR) makes it easy for researchers to produce numerous copies of specific segments of DNA for analysis (see Section 18.1); the technique is so effective that it allows scientists to sequence minute quantities of DNA taken from dried or preserved specimens in museums and even from some fossils. Technological advances have automated the sequencing process, and researchers use analytical software to compare new data to known sequences filed in online data banks. Nuclear DNA is frequently used in phylogenetic analyses, and the publication of complete genome sequences for an ever-expanding list of organisms allows researchers to undertake broad comparative studies.

Molecular sequences have certain practical advantages over organismal characters. First, they provide abundant data: every base in a nucleic acid can serve as a separate, independent character for analysis. Moreover, because many genes have been conserved by evolution, molecular sequences can be compared between distantly related organisms that share no organismal characteristics. They can also be used to study closely related species with only minor morphological differences. Finally, nucleic acids are not directly affected by the developmental or environmental factors that cause the nongenetic morphological variations described in Section 21.1.

Molecular characters have certain drawbacks, however. For example, only four alternative character states (the four nucleotide bases) exist at each position in a DNA or RNA sequence and only 20 alternative character states (the 20 amino acids) at each position in a protein. (You may want to review Sections 14.2 and 15.1 on the structures of these molecules.)

Because of the limited number of character states, researchers may find it difficult to assess the homology of a nucleotide base substitution that appears at the same position in the DNA of two or more species. For organismal characters, biologists can determine homology by analyzing the characters' embryonic development, details of their function, or their presence in the fossil record. But molecular characters have no embryonic development; biologists still do not understand the functional significance of many molecular differences they discover; and researchers have only recently improved techniques that allow them to sequence DNA found in fossils. Nevertheless, systematists have devised complex statistical tools that allow them to discern whether molecular similarities are likely to be homologous or homoplastic.

Despite these potential disadvantages, molecular sequences allow researchers to sample the genome directly, and systematists have successfully used sequence data to analyze phylogenetic relationships that organismal characters were unable to resolve. For example, the phylogenetic tree for animals in Figure 31.6 is based on data from several different nucleic acid molecules.

STUDY BREAK 24.3

1. Why do systematists use homologous characters in their phylogenetic analyses?
2. Why are morphological traits often helpful in tracing the long-term evolutionary relationships within a group of organisms?
3. What are three advantages of using molecular characters in phylogenetic analyses?

24.4 Traditional Classification and Paraphyletic Groups

For a century after Darwin published his theory of evolution, systematists followed an approach called **traditional systematics.** Researchers constructed phylogenetic trees and classified organisms by assessing the amount of phenotypic divergence between lineages, as well as the patterns of branching evolution that had produced them. In other words, they focused on the products of anagenesis (that is, evolutionary change through the accumulation of new or modified characteristics), as well as the products of cladogenesis (that is, the new species and lineages produced through branching evolution). Thus, their classifications did not always strictly reflect the patterns of branching evolution **(Figure 24.7).**

For example, the fossil record for tetrapod (four-legged) vertebrates reveals that the amphibian and mammalian lineages each diverged early. The remaining lineages, collectively called Reptilia, diverged into the Lepidosauromorpha (including living lizards and snakes) and the Archelosauromorpha (including living turtles, crocodilians, and birds).

A. Traditional classification

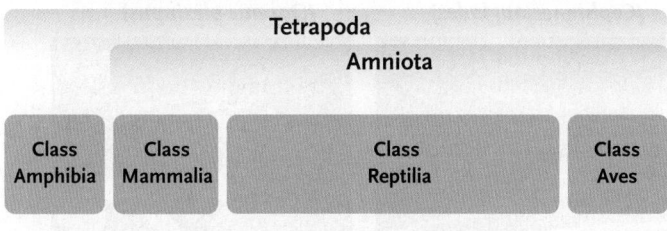

B. Cladistic classification

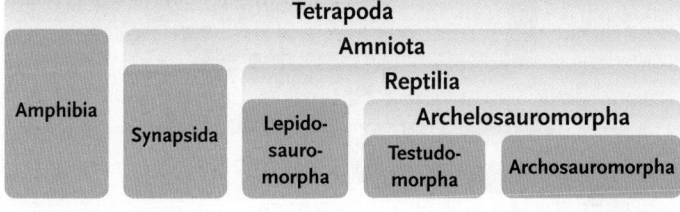

C. Phylogenetic tree

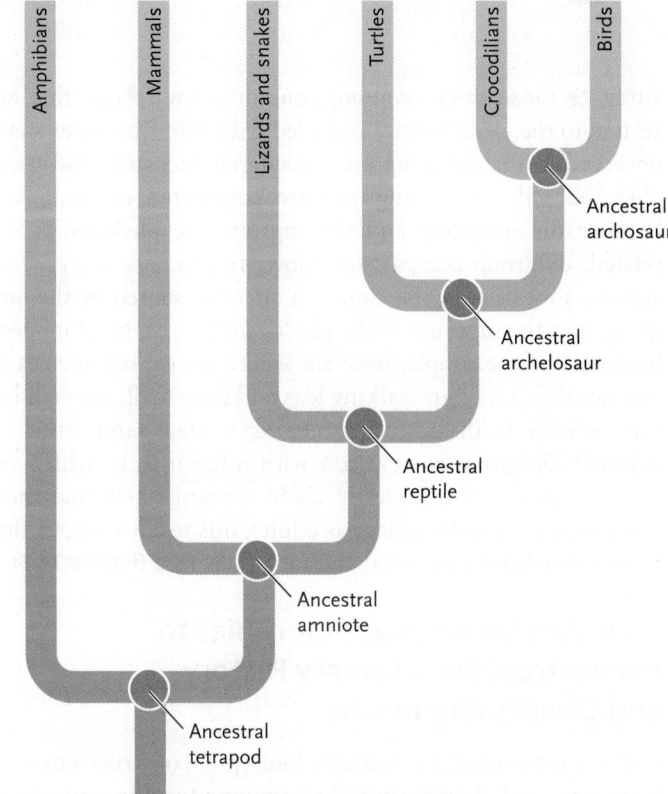

FIGURE 24.7 Phylogenetic trees and classifications for tetrapod vertebrates. (A) The traditional classification and **(B)** the cladistic classifications for tetrapod vertebrates are very different. **(C)** The phylogenetic tree for these animals illustrates why the cladistic classification reflects their evolutionary history better than the traditional classification does.

© Cengage Learning 2017

Thus, although crocodilians, with their scaly skin and sprawling posture outwardly resemble lizards, evolutionary biologists have long recognized that crocodilians share a more recent common ancestor with birds.

Even though the phylogenetic tree shows six living clades, the traditional classification recognizes only four classes of tetrapod vertebrates: Amphibia, Mammalia, Reptilia, and Aves. These groups are given equal ranking because each represents a distinctive body plan and way of life. However, the traditionally defined taxon Reptilia is clearly paraphyletic because, even though crocodilians share a common ancestor with birds, Reptilia includes the former taxon, but not the latter. Traditional systematists justified this definition of Reptilia because it included morphologically similar animals with close evolutionary relationships. Crocodilians were classified with lizards, snakes, and turtles because they share a distant common ancestry and are covered with dry, scaly skin. Traditional systematists also argued that the key innovations initiating the adaptive radiation of birds—a high metabolic rate, wings, and flight—represent such extreme divergence from the ancestral morphology that birds merited recognition as a separate class. As you will learn in Section 24.5, a different approach produces classifications that do not suffer from such inconsistencies.

STUDY BREAK 24.4

Why does a classification produced by traditional systematics sometimes include paraphyletic groups?

24.5 The Cladistic Revolution

In the 1950s and 1960s, some researchers criticized the traditional classifications based on two distinct phenomena—branching evolution and morphological divergence—as inherently unclear. How can we tell *why* two groups are classified in the same higher taxon? They may have shared a recent common ancestor, as did lizards and snakes. Alternatively, they may have retained some ancestral characteristics after being separated on different branches of a phylogenetic tree, as is the case for lizards and crocodilians.

To avoid such confusion, many systematists followed the philosophical and analytical lead of Willi Hennig, a German entomologist (that is, a scientist who studies insects), who published an influential book, *Phylogenetic Systematics,* in 1950; its English translation appeared in 1966. Hennig and his followers argued that classifications should be based solely on evolutionary relationships. This approach, which is called **cladistics,** produces phylogenetic hypotheses and classifications that reflect only the branching pattern of evolution; it ignores morphological divergence altogether.

Cladistic Analyses Focus on Recently Evolved Character States

Like traditional systematics, cladistics analyzes the evolutionary relationships among organisms by comparing their organismal and, more recently, genetic characteristics. Each **character** can exist in two forms, described as **character states.** Evolutionary processes change characters over time from an original, **ancestral character state** to a newer, **derived**

FIGURE 24.8 Outgroup comparison. Most adult insects, such as **(A)** the caddis fly and **(B)** the orange palm dart butterfly, have six walking legs. This comparison of butterflies with other insects suggests that the four walking legs of **(C)** the monarch butterfly represents the derived character state.

A. Caddis fly (Limnephilidae)

Nature's Images/Science Source

B. Orange palm dart butterfly (Cephrenes auglades)

Neil Bowman/FLPA

C. Monarch butterfly (Danaus plexippus)

Millard H. Sharp/Science Source

Outgroup

Monarch butterflies have four walking legs.

Most butterfly species have six walking legs.

Most insects have six walking legs.

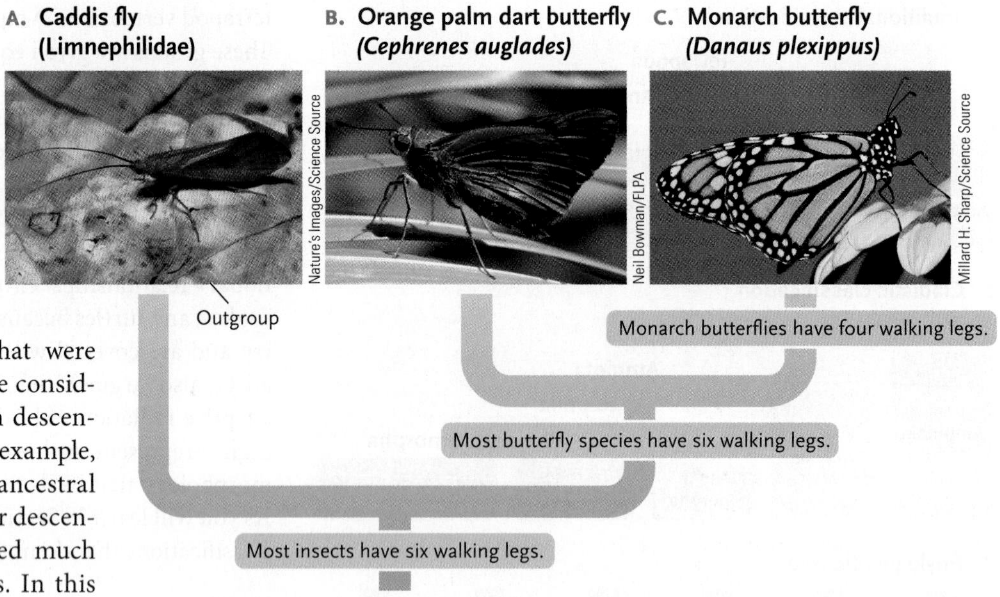

character state. Character states that were present in the ancestors of a clade are considered ancestral; those that are new in descendants are considered derived. For example, ancient fishes, which represent the ancestral vertebrates, had fins; but some of their descendants, the tetrapods, which appeared much later in the fossil record, have limbs. In this example, fins are the ancestral character state, and limbs are the derived character state.

In the jargon of cladistics, a derived character state is called an **apomorphy** (*apo* = away from; *morphe* = form), and a derived character state found in two or more species is called a **synapomorphy** (*syn* = together). The presence of a synapomorphy (that is, a *shared derived character state*) among species provides a clue that they may be members of the same clade. As you learned in Section 23.4, morphological innovations often trigger an adaptive radiation. And once a derived character state becomes established in a species, it is likely to be present in that species' descendants. Thus, unless they are lost or replaced by newer characters over evolutionary time, *synapomorphies can serve as markers for monophyletic lineages.*

Systematists define synapomorphies only when comparing character states among species. Thus, any particular character state is derived *only in relation to* an ancestral character state observed in other organisms—either an older version of the character or its absence. For example, most species of animals lack a vertebral column. However, one animal clade—the vertebrates, including fishes, amphibians, reptiles, birds, and mammals—has that structure. Thus, when systematists compare vertebrates to all other animals, the absence of a vertebral column is the ancestral character state and the presence of a vertebral column is derived.

How can systematists distinguish between ancestral and derived character states? In other words, how can they determine the direction in which a character has evolved? The fossil record, if it is detailed enough, can provide unambiguous information. For example, biologists are confident that the presence of a vertebral column is a derived character state because fossils of animals that lived before vertebrates lack that structure.

In the absence of evidence from fossils, systematists frequently use a technique called **outgroup comparison** to identify ancestral and derived character states. Using this approach, systematists compare characters in the *ingroup*, the clade under study, to those in an *outgroup*, one or a few species that are related to the clade but not included within it. Character states observed in the outgroup are considered ancestral, and those observed *only* in the ingroup are considered derived. And because the outgroup and the ingroup are phylogenetically related, outgroup comparison allows researchers to hypothesize the root (that is, the common ancestor shared by the outgroup and the ingroup) of the phylogenetic tree. Most modern butterflies, for example, have six walking legs, but species in two families have four walking legs and two small, nonwalking legs. Which is the ancestral character state, and which is derived? Outgroup comparison with other insects, which are not included in the butterfly clade, demonstrates that most insects have six walking legs as adults; this result suggests that six walking legs is ancestral and four is derived **(Figure 24.8)**.

Cladistics Uses Synapomorphies to Reconstruct Evolutionary History and Classify Organisms

Following the cladistic method, biologists construct phylogenetic trees and classifications by grouping together only those species that *share derived character states*. Ancestral character states, because they are shared by the ingroup and the outgroup, do not help to define the ingroup. For example, mammals are a clade—a monophyletic lineage because they possess a unique set of synapomorphies: hair, mammary glands, and a reduced number of bones in the lower jaw. The ancestral character states found in mammals, such as a vertebral column and four legs, do not distinguish them from other tetrapod vertebrates. Thus, these shared ancestral character states are not useful in defining the mammal clade.

The results of a cladistic analysis are presented in a phylogenetic tree that illustrates the hypothesized sequence of evolutionary branchings that produced the organisms under study (see Figure 24.7C): a common ancestor is hypothesized at each node, and every branch portrays a strictly monophyletic group. Once a researcher identifies a suitable outgroup and the ancestral and derived character states, a cladistic analysis is straightforward **(Figure 24.9)**. The synapomorphies that define each clade are sometimes listed on the branches.

The use of molecular sequence data in phylogenetic analyses relies on the same logic that underlies analyses based on organismal characters, like those considered above: species included within a clade are expected to exhibit more molecular synapomorphies than do species from different clades. The comparison of sequences in the ingroup to those in the outgroup may allow a researcher to define ancestral and derived character states.

The classifications produced by cladistic analysis often differ radically from those of traditional systematics (compare Figure 24.7B with Figure 24.7A). In a cladistics classification, pairs of nested taxa are defined directly from the two-way branching pattern of the phylogenetic tree. Thus, the clade Tetrapoda (the traditional amphibians, reptiles, birds, and mammals) is divided into two taxa, the Amphibia (tetrapods that do not have an amnion, as discussed in Section 32.3) and the Amniota (tetrapods that have an amnion). The Amniota is subdivided into two taxa on the basis of skull morphology and other characteristics: Synapsida (mammals) and Reptilia (turtles, lizards, snakes, crocodilians, and birds). The Reptilia is further divided into the Lepidosauromorpha (lizards and snakes) and the Archelosauromorpha. The latter taxon is divided into Testudomorpha (turtles) and the Archosauromorpha (crocodilians and birds). Thus, a strictly cladistic classification exactly parallels the pattern of branching evolution that produced the organisms included in the classification. These parallels are the essence and strength of the cladistic method.

Most biologists value the evolutionary focus, clear goals, and precise methods of the cladistic approach. In fact, some systematists advocate abandoning the Linnaean hierarchy for classifying and naming organisms. They propose using a strictly phylogenetic system, called **PhyloCode,** that identifies and names clades instead of pigeonholing organisms into traditional taxonomic categories. This approach is further described in *Unanswered Questions* at the end of this chapter. Although traditional systematics has guided many people's understanding of biological diversity, we use cladistic analyses to describe evolutionary lineages and taxa in Unit Four (Biodiversity), which follows this chapter.

Systematists Use Several Techniques to Identify an Optimal Phylogenetic Tree

In practice, most phylogenetic studies are far more complicated than the examples discussed above and in Figure 24.9. Researchers may collect data on hundreds of characters in dozens of species. After scoring each character state as ancestral or derived in every species, a systematist uses one or more computer programs to generate a set of alternative phylogenetic trees. The output of these analyses is often substantial: an analysis of five species can produce 15 possible phylogenetic trees; an analysis of 50 species can produce 3×10^{76}.

Faced with such an unimaginably large number of alternative hypotheses, how does a systematist decide which phylogenetic tree is the "best" representation of a clade's evolutionary history? This problem is complex, because, when evaluating large data sets, we expect to see some similarities that arise when convergent evolution causes distantly related organisms to evolve similar traits independently; because such traits are not synapomorphies, they are false indicators of relatedness that confound the analysis. We also expect to find some differences between closely related organisms if natural selection or some other microevolutionary process caused a derived character state to be reversed or lost. How can we tell which of the many possible phylogenetic hypotheses is the most likely to represent the evolutionary history of the group? Researchers use several approaches to sort through the alternatives, two of which we describe below.

PARSIMONY APPROACH Many systematists adopt a philosophical concept, the **principle of parsimony,** to identify the optimal phylogenetic tree. This principle states that the simplest plausible explanation of any phenomenon is the best. If we assume that any complex evolutionary change is an unlikely event, then it is extremely unlikely that the same complex change evolved twice in one lineage. Thus, when the principle is applied to phylogenetic analyses, it suggests that the "best" phylogenetic tree is the one that hypothesizes the smallest number of evolutionary changes needed to account for the distribution of character states within a clade; in effect, this approach minimizes the number of homoplasies (that is, the independent evolution of similar traits) in the tree **(Figure 24.10).** To apply the principle, computer programs evaluate the number of evolutionary changes hypothesized by each phylogenetic tree they generate, and the researcher identifies the one with the fewest hypothesized changes as the most plausible.

The principle of parsimony also allows researchers to identify homologous characters and infer their ancestral and derived states. Once the most parsimonious phylogenetic tree is identified, a researcher can visualize the distribution of derived character states and pinpoint when each evolved.

STATISTICAL APPROACHES When comparing two genome sequences, each base in a strand of DNA can be treated as a character with four possible states (A, G, T, and C). One could perform a parsimony analysis on molecular sequence data, like the one illustrated for morphological data in Figure 24.12, to identify the phylogenetic tree that assumes the fewest mutations. But the application of the parsimony approach to molecular data is complicated by several factors. First, given that there are only four possible character states at each position in a nucleic acid, identical changes in nucleotides often arise independently. Second, segments of DNA that do not code for

 FIGURE 24.9 | Research Method

Using Cladistics to Construct a Phylogenetic Tree

Purpose: Systematists construct phylogenetic trees to visualize hypothesized evolutionary relationships among organisms. The cladistic method requires a researcher to group together organisms that share derived characters states. The derived character states identified in the tan plaques are the synapomorphies that define each clade.

Protocol:

1. *Select the organisms to study.* To demonstrate the method, we develop a phylogenetic tree for the nine groups of living vertebrates: lampreys, sharks (and their close relatives), bony fishes, amphibians (frogs and salamanders), turtles, lizards (including snakes), crocodilians (including alligators), birds, and mammals. We also include marine animals called lancelets (Chordata, Cephalochordata) as the outgroup. Lancelets are closely related to, but not included within, the vertebrates (see Chapter 32). The inclusion of an outgroup allows biologists to identify ancestral versus derived character states and root the tree.

2. *Choose the characters on which the phylogenetic tree will be based.* Our simplified example is based on the presence or absence of nine characters: (1) vertebral column, (2) jaws, (3) swim bladder or lungs, (4) paired limbs (with one bone connecting each limb to the body), (5) extraembryonic membranes (such as the amnion), (6) mammary glands, (7) dry, scaly skin somewhere on the body, (8) one opening on each side of the skull in front of the eye, and (9) feathers.

3. *Score the character states in each group.* Because lancelets serve as the outgroup in this analysis, we consider character states observed in lancelets as ancestral; any deviation from the lancelet pattern is considered derived. Because lancelets lack all of the characters in our analysis, the presence of each character is the derived condition. We tabulate data on the distribution of ancestral (−) and derived (+) characters in all species included in the analysis.

4. *Construct the phylogenetic tree from information in the table, grouping organisms that share derived character states.* All groups except lancelets have vertebrae. Thus, we group organisms that share this derived character state on the right-hand branch, identifying them as a monophyletic lineage. Lancelets are on their own branch to the left, indicating that they lack vertebrae.

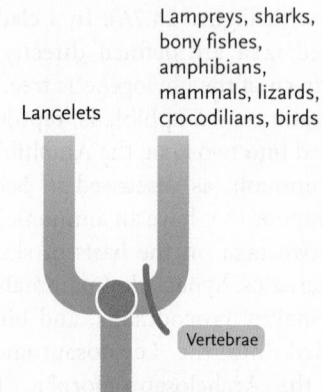

Lancelets

Lampreys, sharks, bony fishes, amphibians, mammals, lizards, crocodilians, birds

Vertebrae

All of the remaining organisms except lampreys have jaws. (Lancelets also lack jaws, but we have already separated them out, and do not consider them further.) Place all groups with jaws, a derived character state, on the right-hand branch. Lampreys are separated out to the left, because they lack jaws. Again, the branch on the right represents a monophyletic lineage.

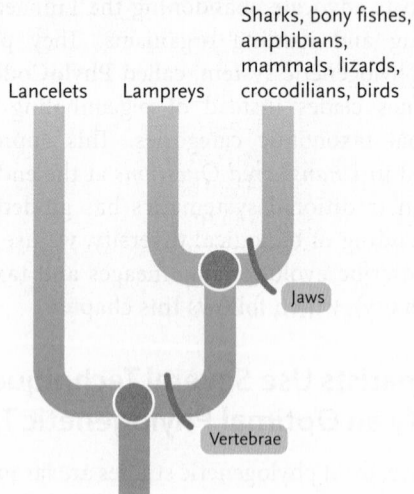

Lancelets Lampreys

Sharks, bony fishes, amphibians, mammals, lizards, crocodilians, birds

Jaws

Vertebrae

	Vertebrae	Jaws	Swim bladder or lungs	Paired limbs	Extraembryonic membranes	Mammary glands	Dry, scaly skin	One opening in front of eye	Feathers
Lancelets	−	−	−	−	−	−	−	−	−
Lampreys	+	−	−	−	−	−	−	−	−
Sharks	+	+	−	−	−	−	−	−	−
Bony fishes	+	+	+	−	−	−	−	−	−
Amphibians	+	+	+	+	−	−	−	−	−
Mammals	+	+	+	+	+	+	−	−	−
Lizards	+	+	+	+	+	−	+	−	−
Crocodilians	+	+	+	+	+	−	+	+	−
Birds	+	+	+	+	+	−	+	+	+

5. *Construct the rest of the phylogenetic tree using the same step-by-step procedure to separate the remaining groups.* In our completed tree, six groups share a swim bladder or lungs; five share paired limbs; and four have extraembryonic membranes during development. Some groups are distinguished by the unique presence of a derived character state, such as feathers in birds.

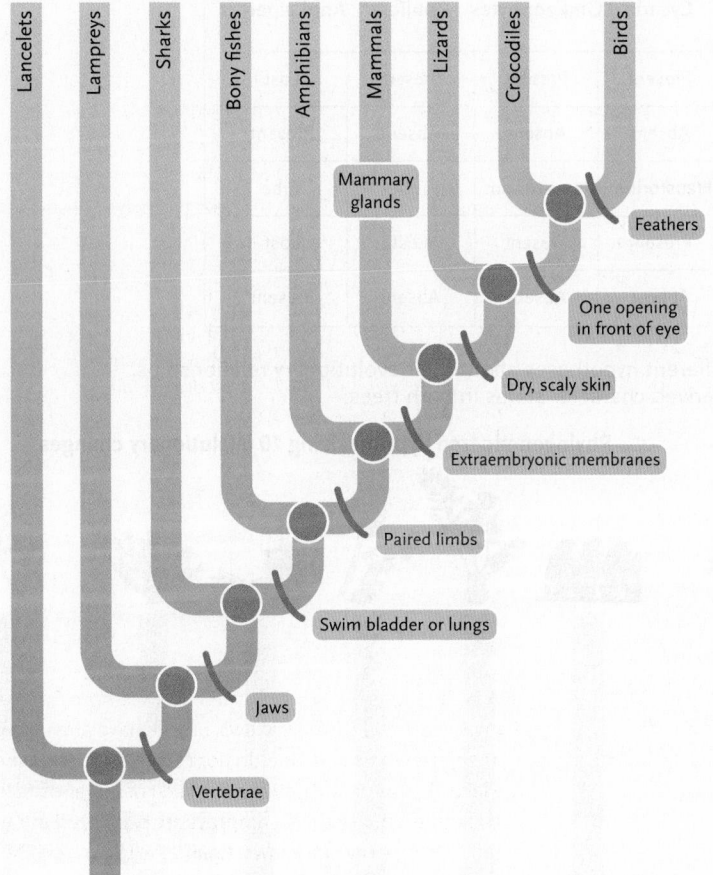

Interpreting the Results: Although phylogenetic trees provide information about evolutionary relationships, the common ancestors represented by the branch points are often hypothetical. You can tell from the tree, however, that birds are more closely related to lizards than they are to mammals. Follow the branches of the tree from birds and lizards back to their node. Next, trace the branches of birds and mammals to their node. You can see that the bird–mammal node is closer to the root of the tree than the bird–lizard node is. Nodes that are closer to the bottom of the tree indicate a more distant common ancestry than those closer to the top. Note also that this simplified example produces a phylogenetic tree that is easy to interpret because the data set includes no homoplastic similarities and no conflicting evidence. Most phylogenetic analyses include many such complications.

proteins are less likely than coding regions to be affected by natural selection. As a result, mutations accumulate faster in noncoding regions, causing them to evolve rapidly. Third, because of the degeneracy of the genetic code (described in Section 15.1), mutations in the third codon position do not often influence the amino acid composition of the protein for which a gene codes. As a result, third codon mutations are often selectively neutral (see Section 21.4), and they accumulate more rapidly than do mutations in the first or second positions. Finally, certain nucleotide substitutions are more common than others: transitions (the substitution of a purine for another purine or a pyrimidine for another pyrimidine) occur more frequently than transversions (substitutions between purines and pyrimidines).

To avoid this problem, systematists develop statistical models of evolutionary change that take into account variations in the evolutionary rates at different nucleotide positions, in different genes, or in different species, as well as changes in evolutionary rates over time. In one statistical approach, the **maximum likelihood method,** systematists compare alternative trees with specific models about the rates of evolutionary change in different regions of DNA. The tree that is most likely to have produced the observed distribution of molecular character states is identified as the best hypothesis.

To illustrate how phylogenetic trees are constructed from DNA sequence data, we cite an example using the **genetic distance method,** which calculates the overall proportion of bases that differ between two species **(Figure 24.11).** The genetic distance between closely related species is smaller than the genetic distance between distantly related species, because the gene pools of closely related species have accumulated distinctive mutations for a shorter period of time. Systematists can construct a phylogenetic tree from these data by making multiple comparisons of genetic distance between pairs of species and then between groups of species; branch lengths in these trees are proportional to the amount of genetic change that has occurred since two species or clades diverged from their common ancestor. Although not as powerful as the maximum likelihood method, the genetic distance method does not depend on assumptions about the evolutionary likelihood of different types of mutations. It also requires much less computing power, which is useful when comparing billions of bases of homologous DNA sequences.

THE PRINCIPLE OF PARSIMONY

A. Distribution of character states in six clades of vascular plants

Ferns represent the outgroup: all of its character states are considered ancestral.

Characters (possible states)	Ferns (outgroup)	Gnetophytes	Cycads	Ginkgophytes	Conifers	Angiosperms
Archegonium (present or lost)	Present	Lost	Present	Present	Present	Lost
Double fertilization (absent or present)	Absent	Present	Absent	Absent	Absent	Present
Pollen tube growth (haustorium or tube)	Absent	Tube	Haustorium	Haustorium	Tube	Tube
Sperm flagella (present or lost)	Present	Lost	Present	Present	Lost	Lost
Vessels (absent or present)	Absent	Present	Absent	Absent	Absent	Present

Two alternative phylogenetic trees for the six clades illustrate different hypotheses about their evolutionary relationships. Bars across the branches mark the hypothesized evolution of derived character states in both trees.

B. Phylogenetic tree hypothesizing five evolutionary changes

C. Phylogenetic tree hypothesizing 10 evolutionary changes

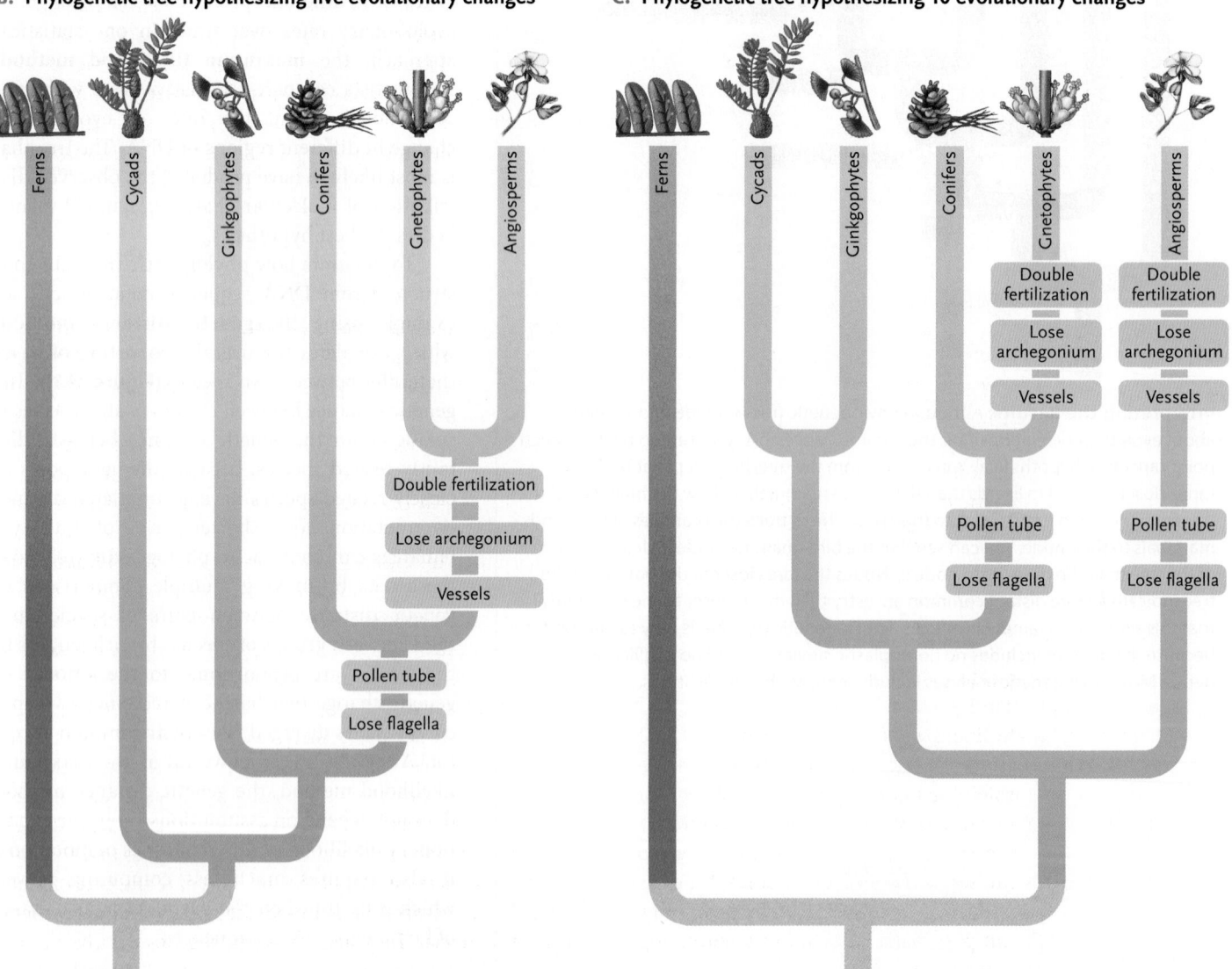

SUMMARY Based on this set of five characters, the phylogenetic tree with five hypothesized evolutionary changes (**B**) is more parsimonious than the tree that hypothesizes 10 evolutionary changes (**C**). In the absence of additional data, a systematist would accept the more parsimonious tree as the best working hypothesis. (The characters and clades of vascular plants are discussed in detail in Chapter 28.)

think like a scientist Do the five characters in the table provide sufficient information to develop a hypothesis about the evolutionary relationship of cycads and ginkgophytes?

STUDY BREAK 24.5

1. How does outgroup comparison facilitate the identification of ancestral and derived character states?
2. What characteristics are used to group organisms in a cladistic analysis?
3. How is the principle of parsimony applied in phylogenetic analyses?

THINK OUTSIDE THE BOOK

Access the web page for the Tree of Life project at http://www.tolweb.org/tree/. Select a group of animals or plants that is of interest to you, and study the structure of its phylogenetic tree. How many major clades does it include? On the basis of what shared derived characters are those clades defined?

24.6 Phylogenetic Trees as Research Tools

In addition to providing a wealth of information about the patterns of branching evolution across the entire spectrum of living organisms, phylogenetic trees are useful tools that facilitate research in all areas of biology.

Molecular Clocks Estimate the Time of Evolutionary Divergences

Although many biological molecules have been conserved by evolution, different adaptive changes and neutral mutations accumulate in separate lineages from the moment they first diverge. Because mutations that arise in noncoding regions of DNA do not affect protein structure, they are probably not often eliminated by natural selection. If mutations accumulate in these segments at a reasonably constant rate, differences in their DNA sequences can serve as a **molecular clock,** indexing the time at which two species diverged. Large differences imply divergence in the distant past, whereas small differences suggest a more recent common ancestor.

Because different molecules exhibit individual rates of evolutionary change, every molecule is an independent clock, ticking at its own rate. Researchers study different molecules to track evolutionary divergences that occurred at different times in the past. For example, mitochondrial DNA (mtDNA) evolves relatively quickly; it is useful for dating evolutionary divergences that occurred within the last few million years. Studies of mtDNA have illuminated aspects of the evolutionary history of humans, as described in Section 32.12. By contrast, chloroplast DNA (cpDNA) and genes that encode ribosomal RNA evolve much more slowly, providing information about divergences that date back hundreds of millions of years.

To calibrate molecular clocks, researchers examine the degree of genetic difference between species in relation to their time of divergence estimated from the fossil record. Alternatively, the clock can be calibrated biogeographically with independent data on when volcanic islands first emerged from the sea or when landmasses separated **(Figure 24.12).**

The reliability of molecular clocks depends on the constancy of evolutionary change in the DNA segment analyzed. Some researchers have noted that even DNA segments that are thought to be selectively neutral may show variable rates of evolution. Many factors can influence the rates at which mutations accumulate, and researchers must be cautious when evaluating divergence times estimated with this technique, especially if there are no independent data to corroborate the estimates.

Phylogenetic Trees Allow Biologists to Propose and Test Hypotheses

Accurate phylogenetic trees are essential tools for analyses that biologists describe as the "comparative method." With this approach, researchers compare the characteristics of different species to assess the homology of their similarities and infer where on the phylogenetic tree a particular trait appeared. The comparative method is used to study almost any sort of organismal trait, but in this section we will focus on parental care behavior.

As noted in Figure 24.7B, birds and crocodilians are included within the Archosauria, a clade that also includes nonavian dinosaurs (that is, extinct terrestrial dinosaurs that are not included within the bird clade), pterosaurs (an extinct group of flying vertebrates, not closely related to bats, but with wing surfaces formed by skin), and a number of other groups that became extinct in the early Mesozoic era. Crocodilians and birds share certain anatomical characteristics, such as a four-chambered heart and the one-way flow of air through their lungs. They also share behavioral characteristics, including the production of mating calls (songs in birds and roars in crocodilians), nest-building behavior, and parental care of their young. Female crocodilians guard their nests and keep them moist with urine. They also excavate the nest as the young hatch, and then carry them to standing water. Young stay with

 **FIGURE 24.11** | **Research Method**

Using Genetic Distances to Construct a Phylogenetic Tree

Purpose: Systematists use data on genetic distances (the overall differences in DNA sequences) among species to reconstruct their phylogenetic tree.

Protocol:

1. Calculate the genetic distance between each pair of species. The genetic distances between humans and three species of great apes are shown in the first table.

	Chimpanzee	Gorilla	Orangutan
Human	1.37	1.75	3.40
Chimpanzee		1.81	3.44
Gorilla			3.50

2. Identify the pair of species with the smallest genetic distance in the first table; in this example, the smallest distance is between chimpanzee and human (genetic distance = 1.37). These two species therefore form a cluster of two closely related species.

3. Calculate the average genetic distances between the chimpanzee–human cluster and each of the other species in the analysis, gorilla and orangutan. For example, the genetic distance between the chimpanzee–human cluster and gorilla is the average of the human–gorilla genetic distance and the chimpanzee–gorilla genetic distance [(1.75 + 1.81)/2 = 1.78]. The newly calculated genetic distances are shown in the second table.

	Gorilla	Orangutan
Chimpanzee–human cluster	1.78	3.42
Gorilla		3.50

4. Identify the two groups (individual species or clusters) with the smallest genetic distance in the second table. In this example, the next smallest genetic distance is between the chimpanzee–human cluster and gorilla (genetic distance = 1.78). Thus, chimpanzee–human–gorilla forms the next cluster, leaving orangutan as the outgroup.

5. Because there are only four species in our example, these genetic distance calculations define the phylogenetic tree shown below. If the analysis included additional species, we would repeat the calculation described in steps 2 to 4 as many times as necessary to complete the phylogenetic tree.

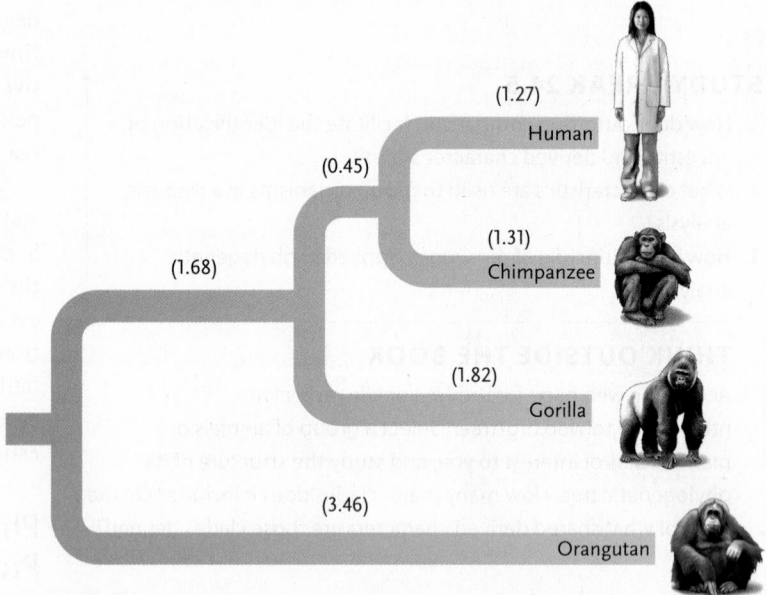

Interpreting the Results: In our phylogenetic tree, the length of each branch is proportional to the amount of genetic change that has occurred in that branch. A longer branch length indicates that the genome of that species has evolved at a correspondingly higher rate (that is, undergone more evolutionary change since the two sister branches emerged from their common ancestor).

Source: A. Scally et al. 2012. Insights into hominid evolution from the gorilla genome sequence. *Nature* 483:169–175.

their mother for about a year, feeding on scraps of food that fall from her mouth.

Did similar parental care behavior evolve independently in birds and crocodilians, or is it truly a synapomorphy? Did most Mesozoic archosaurs care for their young as birds and crocodilians do today? The comparative method seeks answers to these questions by examining the phylogenetic tree for archosaurs **(Figure 24.13).** As you can readily see, crocodilians and birds lie on widely separated branches of the archosaur tree, with pterosaurs and nonavian dinosaurs positioned between them. The most parsimonious inference about the evolution of parental care behavior is that it evolved once in the common ancestor of crocodilians and birds. If that inference is correct,

then nonavian dinosaurs and pterosaurs probably also cared for their young in the nest. Indeed, that prediction was confirmed in 1995, when Mark A. Norell of George Washington University and his colleagues discovered a fossil of a nonavian dinosaur *(Oviraptor)* sitting on a nest full of eggs.

STUDY BREAK 24.6

1. What assumption underlies the use of genetic sequence differences between species as a molecular clock?
2. Are birds more closely related to nonavian dinosaurs or to crocodilians?

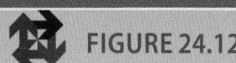 **FIGURE 24.12** **Observational Research**

Do Molecular Clocks Tick at a Constant Rate?

Hypothesis: Some DNA segments accumulate mutations at a constant rate, allowing researchers to use genetic differences between species as molecular clocks.

Prediction: The genetic differences observed between pairs of species will be proportional to the time since they diverged from their common ancestor.

Method: Robert C. Fleischer of the Smithsonian Institution and colleagues at the Smithsonian and Pennsylvania State University compared gene sequences among pairs of related bird species (honeycreepers) and fruit fly species *(Drosophila)* on different Hawaiian Islands. They used radiometric dating of volcanic rocks to estimate the age of each island and used the estimates of island age to measure the divergence time between pairs of species. This approach assumes that birds and flies founded populations on each new island shortly after the island's emergence from the Pacific Ocean and that newly founded populations began to diverge from ancestral populations immediately. Thus, if a population of *Drosophila* now lives on Oahu, the researchers estimate that it diverged from its ancestor 3.7 million years ago, which is the estimated time of Oahu's origin. The researchers plotted a summary statistic of genetic differences between species pairs versus estimated island age to test their hypothesis about constant rates of evolution in the DNA segments they studied.

Results: The extent of genetic difference in the cytochrome *b* sequences of Hawaiian honeycreepers and in the *Yp1* gene sequences of Hawaiian fruit flies is strongly correlated with the time since the members of each species pair were separated. The results indicate that these DNA sequences have evolved at constant rates.

A. Hawaiian Islands (with time of origin)

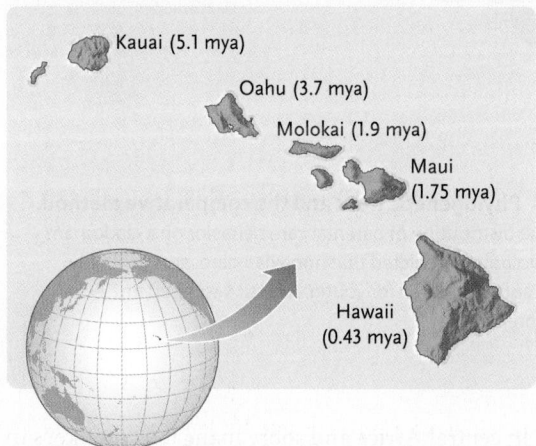

B. Hawaiian honeycreepers

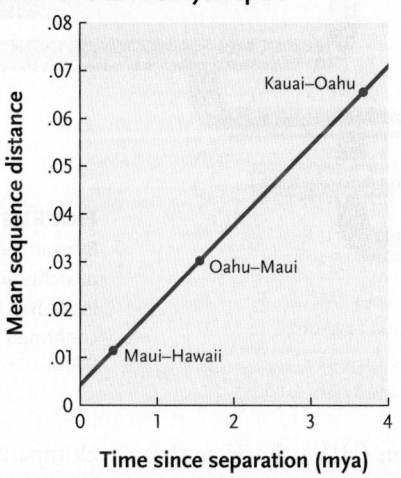

C. Hawaiian *Drosophila*

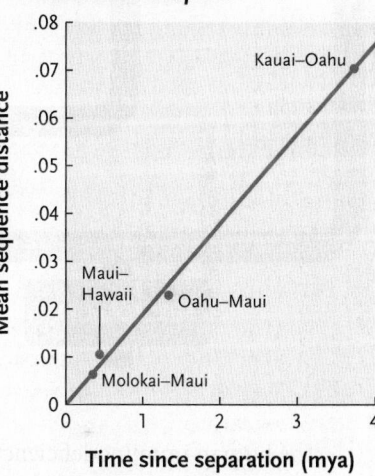

Conclusion: Because the gene sequences evolve at a constant rate, they can be used as molecular clocks. Each 0.016 units of genetic difference in the cytochrome *b* sequence in Hawaiian honeycreepers and each 0.019 units of genetic difference in the *Yp1* sequence in Hawaiian *Drosophila* indexes 1 million years of independent evolution.

Observe
Hypothesize
Predict
Experiment
Interpret

Which of the two gene sequences surveyed—cytochrome *b* honeycreepers or *Yp1* sequence in *Drosophila*—appears to have evolved more quickly?

Source: R. C. Fleischer et al. 1998. Evolution on a volcanic conveyor belt: Using phylogeographic reconstructions and K–Ar-based ages of the Hawaiian Islands to estimate molecular evolutionary rates. *Molecular Ecology* 7:533–545.

24.7 Molecular Phylogenetic Analyses

The application of molecular techniques to phylogenetic analyses has allowed systematic biologists to resolve some evolutionary puzzles that could not be addressed with older techniques. In this section we describe how molecular analyses have enabled public health researchers to identify sources of HIV infection in humans and elucidate the relationships among the three domains of the Tree of Life.

Molecular Phylogenetic Analyses Pinpoint the Origins of Infectious Diseases

Phylogenetic analyses also allow physicians and public health workers to identify the origin of infectious agents and follow their spread through a population. Many pathogenic organisms and viruses mutate as they proliferate, establishing derived character states that are ripe for phylogenetic analysis.

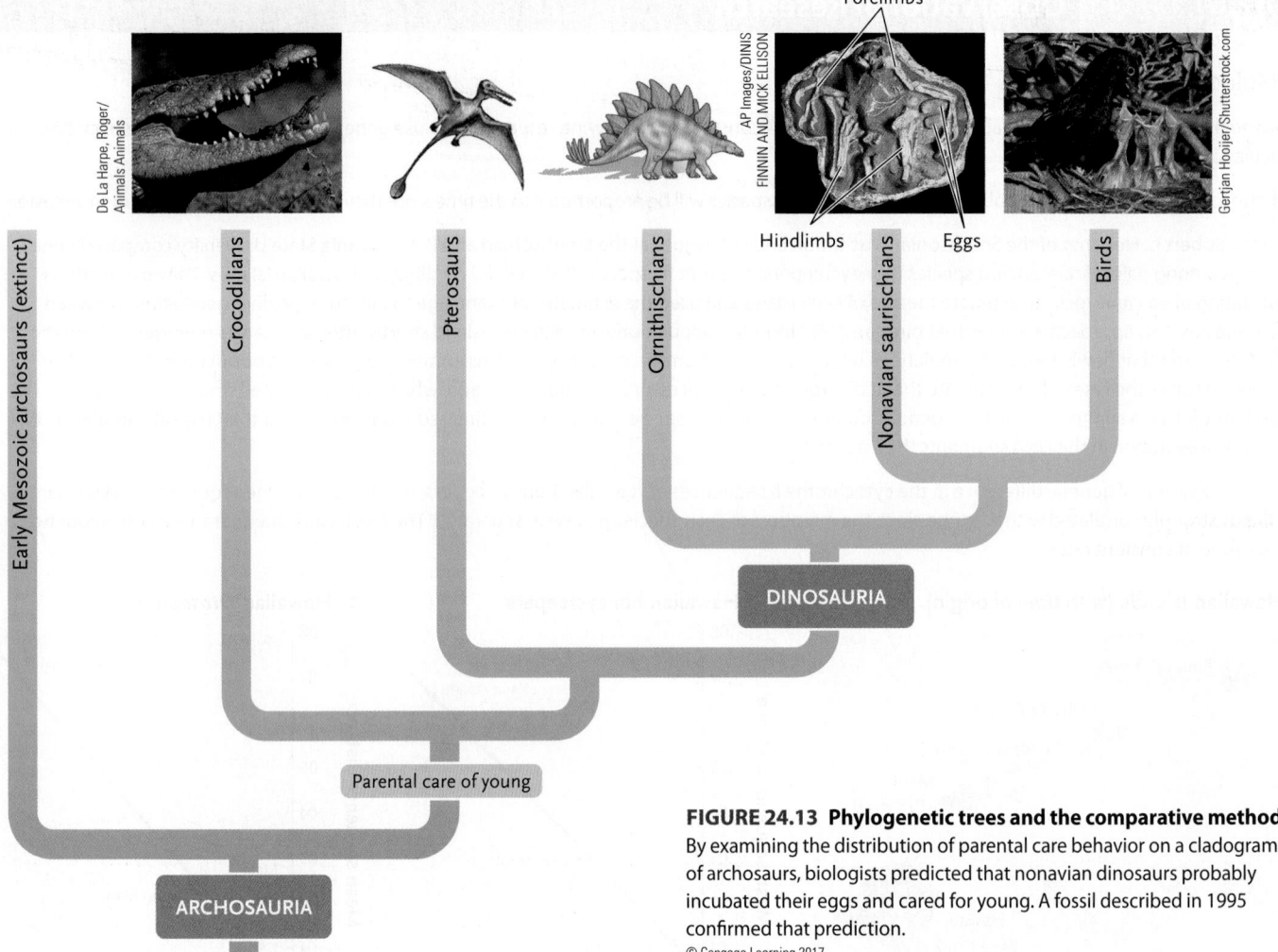

FIGURE 24.13 Phylogenetic trees and the comparative method. By examining the distribution of parental care behavior on a cladogram of archosaurs, biologists predicted that nonavian dinosaurs probably incubated their eggs and cared for young. A fossil described in 1995 confirmed that prediction.

© Cengage Learning 2017

The human immunodeficiency virus (HIV), the agent that causes acquired immunodeficiency syndrome (AIDS) in humans, began to infect large numbers of people in the 1980s. As its devastating effects on humans became apparent, scientists scrambled to discover its origin. Genetic analyses linked it to the lentiviruses, specifically simian immunodeficiency virus (SIV), which infects dozens of monkey species, as well as chimpanzees, in Africa. Surprisingly, SIV does not cause illness in those animals, perhaps because their populations developed immunity to it after a long period of exposure.

Two distinct strains of HIV infect humans: HIV-1 is common in central Africa, and HIV-2 is common in West Africa. Did these strains evolve within human hosts, or did they exist before the virus was first transmitted to humans? An analysis by Beatrice H. Hahn of the University of Alabama at Birmingham and the Howard Hughes Medical Institute and colleagues at other institutions identified three major clades of SIV. The clade that infects chimpanzees includes HIV-1, and one of the clades that infect monkeys includes HIV-2 **(Figure 24.14).** Thus, the two strains of HIV apparently originated in nonhuman hosts. Scientists suspect that the transmission to humans occurred multiple times when hunters who were butchering bush meat—

chimpanzees in central Africa and sooty mangabey monkeys in West Africa—acquired the virus through cuts on their hands.

Analyses of Gene Sequences Have Revealed the Branching Pattern of the Entire Tree of Life

On a very grand scale, molecular phylogenetics has revolutionized our view of the entire Tree of Life. The first efforts to create a phylogenetic tree for all forms of life were based on morphological analyses. However, these analyses did not resolve the branches of the Tree containing prokaryotes, which lack significant structural variability for analysis, or the relationships of those branches to eukaryotes.

In the 1960s and early 1970s, biologists organized living systems into five kingdoms. All prokaryotes were grouped into the kingdom "Monera". The eukaryotic organisms were grouped into four kingdoms: Fungi, Plantae, Animalia, and "Protista." The kingdom "Protista" was always recognized as a polyphyletic "grab bag" of unicellular or acellular organisms (discussed further in Chapter 27). Unfortunately, phylogenetic analyses based on morphology were unable to sort them into distinct evolutionary lineages.

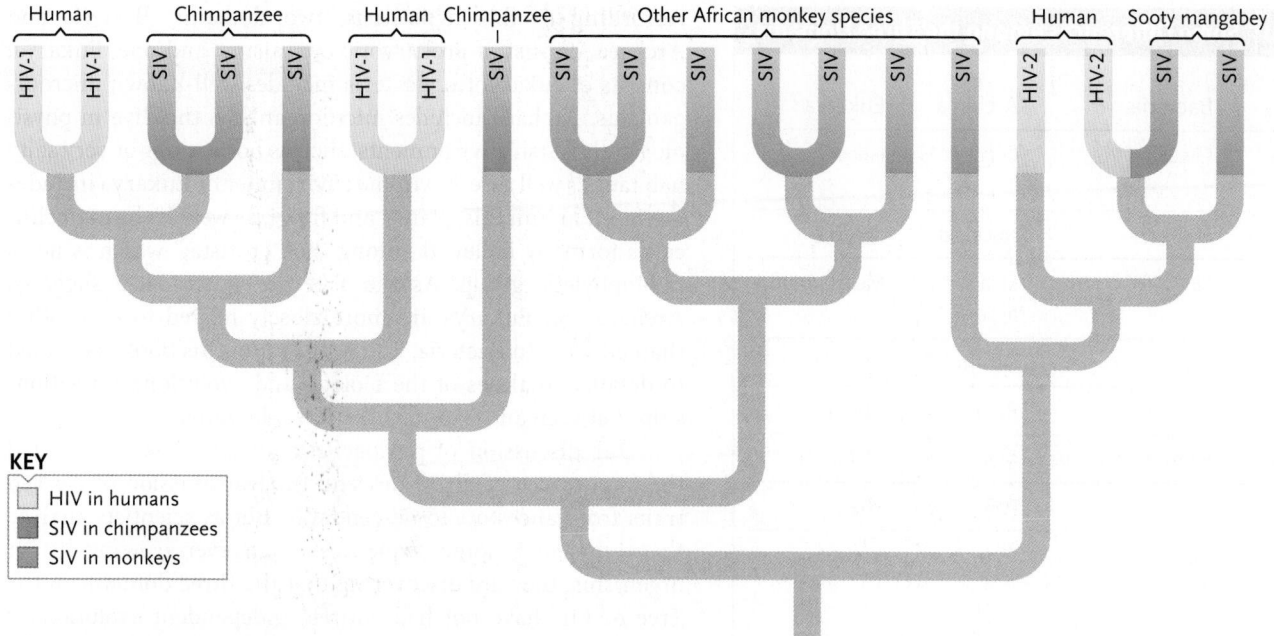

FIGURE 24.14 Phylogenetic trees and public health. A phylogenetic tree for strains of simian immunodeficiency virus (SIV) and human immunodeficiency virus (HIV) suggests that the virus was transmitted to humans independently from chimpanzees and sooty mangabey monkeys.

KEY
- HIV in humans
- SIV in chimpanzees
- SIV in monkeys

In the 1970s, biologists realized that molecular phylogenetics provides an alternative approach. They simply needed to identify and analyze molecules that have been conserved by evolution over billions of years. Carl R. Woese, a microbiologist at the University of Illinois at Urbana-Champaign, identified the ribosomal RNA (rRNA) molecule of the small subunit as a suitable molecule for analysis. Ribosomes, the structures that translate messenger RNA molecules into proteins (see Section 15.4), are remarkably similar in all forms of life. They are apparently so essential to cellular processes that the genes specifying ribosomal structure exhibit similarities in their nucleotide sequences in organisms as different as bacteria and humans. Thus, it is possible to compare the sequences of these genes in a phylogenetic analysis.

The phylogenetic tree based on rRNA sequences divides living organisms into three primary lineages called domains—Bacteria, Archaea, and Eukarya—which differ in many molecular and cellular characters (**Figure 24.15** and **Table 24.2**).

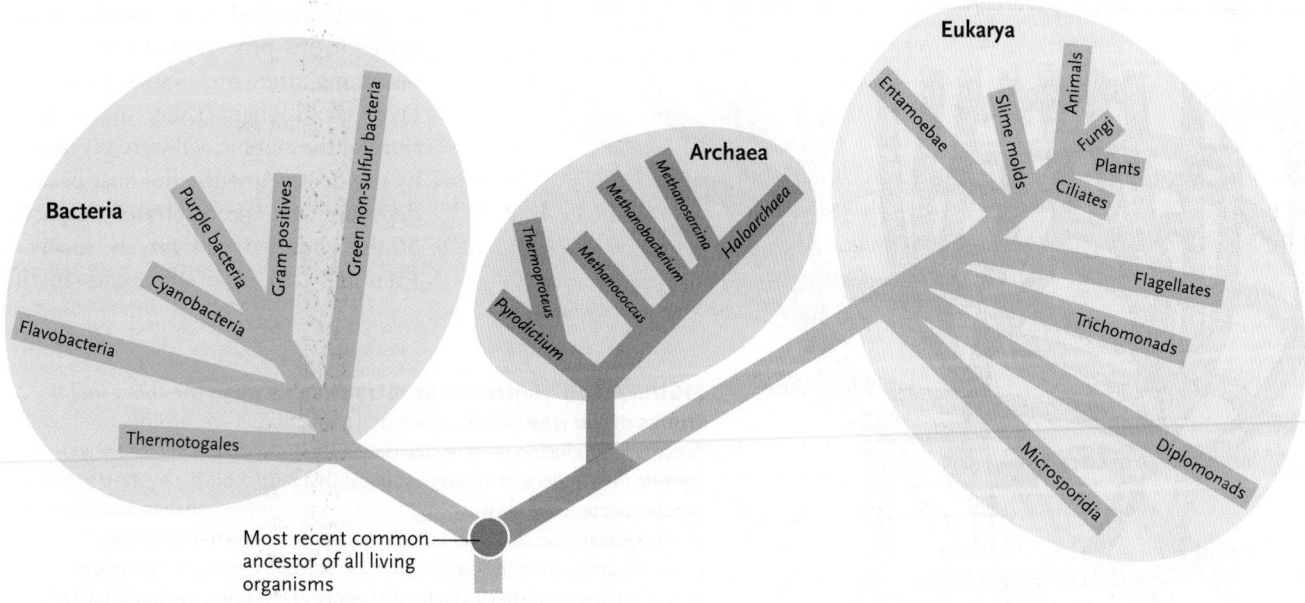

FIGURE 24.15 Three domains in the Tree of Life. Carl R. Woese's analysis of rRNA sequences suggests that all living organisms can be classified into one of three domains: Bacteria, Archaea, and Eukarya.

© Cengage Learning 2017

TABLE 24.2 Some Differences among the Three Domains

Character	Bacteria	Archaea	Eukarya
Chromosome structure	Circular	Circular	Linear
DNA location	Nucleoid	Nucleoid	Nucleus
Chromosome segregation	Binary fission	Binary fission	Meiosis/mitosis
Introns in genes	Rare	Common	Common
Operons	Present	Present	Absent
Initiator tRNA	Formylmethionine	Methionine	Methionine
Ribosomes	70S	70S	80S
Membrane-enclosed organelles	Absent	Absent	Present
Membrane lipids	Ester-linked	Ether-linked	Ester-linked
Peptidoglycan in cell wall	Present	Absent	Absent
Methanogenesis	Absent	Present	Absent
Temperature tolerance	Up to 90°C	Up to 120°C	Up to 70°C

According to this hypothesis, two domains, Bacteria and Archaea, consist of prokaryotic organisms, and one, Eukarya, consists of eukaryotes. Bacteria includes well-known microorganisms. Archaea includes microorganisms that live in physiologically harsh environments, such as hot springs or very salty habitats, as well as less extreme environments. Eukarya includes the familiar animals, plants, and fungi, as well as the many lineages formerly included among the "Protista," which is not a monophyletic group. As the Tree in Figure 24.15 suggests, Archaea and Eukarya are more closely related to each other than either is to Bacteria. The next unit of this book is devoted to detailed analyses of the biology and evolutionary relationships between and within these three domains.

Our discussion of phylogenetic analysis has emphasized the importance of direct descent: the transmission of derived traits from ancestors to descendants. But as scientists analyze the complete genome sequences of an ever-growing list of organisms, they are discovering that the three domains in the Tree of Life have not had entirely independent evolutionary histories. Although inheritance from one generation to the next has produced the clades we recognize today, **horizontal gene transfer,** the movement of genetic material between organisms by any mechanism other than descent, has also been important (discussed in Section 17.1 and Chapter 26). Horizontal gene transfer, which is most common in prokaryotes, can occur between different species, introducing genes from one species into another. In bacteria, three major mechanisms facilitate horizontal gene transfer: conjugation, transformation, and transduction (discussed in Section 17.1).

Horizontal gene transfer can also occur through viral infection, and by the incorporation of one organism into another. For example, biologists hypothesize that mitochondria evolved from an aerobic bacterium that was engulfed by an anaerobic archean. The archean acquired some unique genes through this process (see Chapter 25), becoming the first eukaryotic cell. Over evolutionary time, many functions of the aerobic cell were taken over by the host cell while the host became dependent on the respiratory capacity of the engulfed cell for its survival. Eventually, many of the genes in the

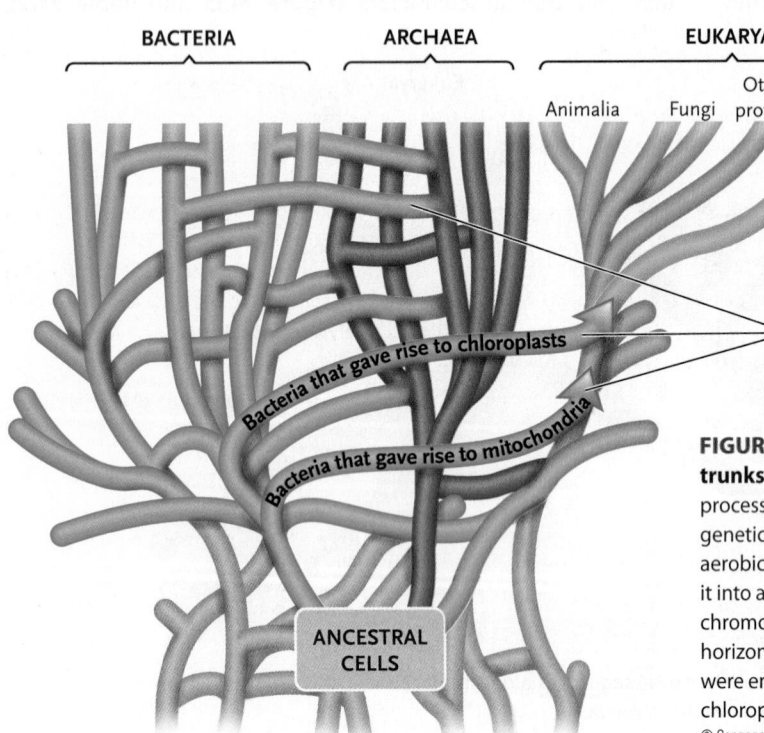

FIGURE 24.16 Horizontal gene transfer between the three major trunks of the Tree of Life. Horizontal gene transfer was an important process in producing present-day clades. As shown in this stylized phylogenetic tree, molecular analyses suggest that mitochondria evolved from an aerobic bacterium that was engulfed by an anaerobic archaean, transforming it into a eukaryotic cell. Over time, many genes on the mitochondrial chromosome were transferred to nuclear chromosomes, an example of horizontal gene transfer. Later in the history of life, photosynthetic bacteria were engulfed by eukaryotic cells; the engulfed bacteria evolved into chloroplasts, which are responsible for photosynthesis in algae and plants.
© Cengage Learning 2017

mitochondrial chromosome were transferred to nuclear chromosomes, thus causing those nuclear chromosomes to contain bacterial genes that had been transferred horizontally. Later in evolutionary time a photosynthetic bacterium was engulfed by a eukaryotic cell, adding more unique genes to the organism. The engulfed bacterium evolved into a chloroplast, and the cells containing it founded lines that gave rise to eukaryotic algae and plants. In fact, genomic analysis has shown that a more accurate portrait of the evolutionary history of life on Earth involves complex horizontal gene transfers between the three major trunks on the Tree of Life **(Figure 24.16)**. Thus, a true

portrait of the relationships among living systems looks more like a web than the tree Darwin envisioned (see Figure 20.11).

STUDY BREAK 24.7

1. Did HIV originate in humans, or was the infection acquired from other animals?
2. Why was a phylogenetic analysis of prokaryotes based on molecular sequence data more successful than the analysis based on morphological data?

 ## Unanswered Questions

Should we abandon the traditional Linnaean hierarchy in favor of a more evolutionary classification?

Diligent Kindly Professors Cannot Often Fail Good Students—or some equally silly mnemonic device for remembering the Linnaean taxonomic hierarchy—is all that many students recall about systematics. Even if they remember the underlying rank names—is *G* for "group" or "genus"?—they often forget that Linnaeus conceived his system of classification more than a century before Darwin articulated his theory of evolution, which revolutionized our understanding of biological diversity. In the more than 150 years since Darwin published *On the Origin of Species,* systematists have sought not only to categorize life's diversity but, more important, to understand its origins. The broad relevance of studies in systematics has become increasingly clear as biologists have discovered that systematic principles are as important in tracing the emergence and spread of HIV as they are in distinguishing a duck from a dove.

More than a century and a half after it was first published, Darwin's theory has an increasingly revolutionary impact. Perhaps the most striking recent example is a call for the complete abandonment of the Linnaean taxonomic hierarchy. Although biologists thought they had reconciled the perspectives of Darwin and Linnaeus, a growing minority of systematists now believe that any effort to catalog and categorize life's diversity must be explicitly phylogenetic and free of the arbitrary ranks that Linnaeus invented. This movement, which has been codified in the PhyloCode initiative, is fueled largely by newly available molecular data, vastly improved phylogenetic methods, and increasingly fast computers. These advances offer the potential to reconstruct accurate and fully resolved phylogenetic trees at a scale never before possible. For the first time, biologists see real progress in accurately reconstructing the entire Tree of Life. Although we are still far from achieving this goal, every day millions of new, phylogenetically informative DNA fragments are being sequenced and analyzed by thousands of computers running around the clock.

Although PhyloCode's synthesis of taxonomy and evolutionary systematics may be long overdue, this attempted coup is not without controversy. For example, some systematists contend that such a radical revision of our taxonomic system will introduce confusion and instability in the naming of species. Even the revolution's adherents recognize that we still face many challenging limitations to the synthesis between

taxonomic practice and Darwinian principles. Nowhere is this more evident than in the definition of species.

During Linnaeus' time, species were viewed as immutable natural types created by God. Darwin, however, formulated his theory on the principle that species change over time. Although the truth of this basic hypothesis is no longer a subject of debate, its practical implications for delimiting species boundaries and understanding how new species form are among the most exciting areas of study in modern systematics. Most practicing systematists view species as real (that is, biologically meaningful) categories, but the criteria for recognizing species vary dramatically among systematists working on different types of organisms (plants versus animals, or organisms that reproduce asexually versus those that reproduce sexually). Using new molecular tools and sophisticated genetic experiments, evolutionary biologists are beginning to probe the precise genetic basis of species. Over the past decade, a small number of "speciation genes" have been identified; more such discoveries are sure to follow in the coming years. Although many of these studies have been restricted to model research organisms, such as fruit flies, the new tools offered by the fields of genomics and bioinformatics offer the potential to address similar questions in an increasingly broad array of organisms.

Simply put, the systematics of today is not that of your grandparents. Given the enormous challenge involved in categorizing and understanding the origin and evolutionary relationships of millions of species, many additional changes are on the horizon. For the next generation of systematists, however, a better mnemonic to remember may be "Do Keep Probing Charles' *Origin* For Good Systematics."

think like a scientist How easily does the Linnaean system of classification accommodate the phenomenon of horizontal gene transfer? Is a completely phylogenetic view of biodiversity, such as PhyloCode, more consistent with the idea that genes have moved horizontally between clades?

Richard Glor is an associate professor in the Department of Ecology and Evolutionary Biology at the University of Kansas and curator of the Herpetology Division of the KU Biodiversity Institute. To learn more about Dr. Glor's research, go to http://biodiversity.ku.edu/herpetology/people/rich-glor.

REVIEW KEY CONCEPTS

For access to MindTap and additional study materials visit www.cengagebrain.com.

24.1 Nomenclature and Classification

- Linnaeus invented a system of binomial nomenclature in which each species receives a unique two-part name.
- Species are organized into a taxonomic hierarchy, which largely reflects the pattern of branching evolution (Table 24.1). Species classified in the same lower-level taxon have a more recent common ancestor than species classified in the same higher-level taxon.

24.2 Phylogenetic Trees

- Phylogenetic trees are hypotheses that portray the branching pattern of evolution (Figure 24.2). Most phylogenetic trees have an implicit or explicit timeline that indicates the relative times for cladogenesis. Branch points are described as nodes, and monophyletic lineages are called clades. Two lineages that share a node are called sister clades. Clades can be rotated at nodes without changing the meaning of the tree.
- Evolutionary biologists define monophyletic taxa directly from the structure of a phylogenetic tree (Figure 24.3). Biologists never intentionally define paraphyletic or polyphyletic taxa.

24.3 Sources of Data for Phylogenetic Analyses

- Homologies are characters that two or more descendant species inherited from their common ancestor (Figure 24.4). Homologous structures are similar in anatomical detail, and they often show a similar pattern of embryonic development. Characters that are similar for any other reason are described as homoplasies (Figure 24.5).
- Systematists often use morphological characters in phylogenetic analyses.
- If morphological characters do not distinguish species, behavioral characters may be useful (Figures 24.6).
- Systematists today use genetic sequence data to reconstruct phylogenetic history.

24.4 Traditional Classification and Paraphyletic Groups

- In traditional systematics, biologists used the similarities and differences between organisms to construct a phylogenetic tree and a classification.
- Traditional classifications sometimes included paraphyletic groups (Figure 24.7A).

24.5 The Cladistic Revolution

- Derived character states can serve as markers of clades.
- Systematists use evidence from the fossil record, as well as outgroup comparison, to identify which character states are derived and which are ancestral (Figure 24.8).
- Cladistic analyses use synapomorphies (derived character states) to construct phylogenetic hypotheses (Figure 24.7B).
- The construction of a simple phylogenetic tree using cladistic methods is straightforward (Figure 24.9).
- Systematists use the principle of parsimony (Figure 24.10), maximum likelihood techniques, and genetic distance data (Figure 24.11) to construct and evaluate alternate phylogenetic hypotheses.

24.6 Phylogenetic Trees as Research Tools

- If a nucleic acid accumulates mutations at a reasonably constant rate, it can be used as a molecular clock (Figure 24.12).
- Phylogenetic trees allow biologists to propose and test hypotheses about when certain characters evolved (Figure 24.13).

24.7 Molecular Phylogenetic Analyses

- Molecular techniques allow systematists to address questions that they could not have posed in the mid-twentieth century. Phylogenetic analyses are useful tools in medicine because they allow researchers to track the origin and spread of infectious diseases (Figure 24.14).
- Sequence data from the small subunit ribosomal RNA allowed researchers to recognize the three domains of life (Figure 24.15). Horizontal gene transfer complicates our understanding of the history of biodiversity (Figure 24.16).

TEST YOUR KNOWLEDGE

Remember/Understand

1. The evolutionary history of a group of organisms is called its:
   a. classification.
   b. taxonomy.
   c. phylogeny.
   d. domain.
   e. outgroup.

2. In the Linnaean hierarchy, the organisms classified within the same taxonomic category are called:
   a. a phylum.
   b. a taxon.
   c. a genus.
   d. a binomial.
   e. an epithet.

3. Which of the following does *not* help systematists determine whether a morphological character state is ancestral or derived?
   a. outgroup comparison
   b. patterns of embryonic development
   c. studies of the fossil record
   d. studies of the character in more related species
   e. dating of the character by molecular clocks

4. In a cladistic analysis, a systematist groups together organisms that share:
   a. derived homologous traits.
   b. derived homoplastic traits.
   c. ancestral homologous traits.
   d. ancestral homoplastic traits.
   e. all of the above.

5. A monophyletic taxon is one that includes:
   a. an ancestor and all of its descendants.
   b. an ancestor and some of its descendants.
   c. organisms from different evolutionary lineages.
   d. an ancestor and those descendants that still resemble it.
   e. organisms that resemble each other because they live in similar environments.

6. Which of the following is *not* an advantage of using molecular characters in a systematic analysis?
   a. Molecular characters provide abundant data.
   b. Systematists can compare molecules among species that are morphologically very similar.
   c. Systematists can compare molecules among species that are morphologically very different.
   d. Nucleotide sequences in DNA are generally not influenced by environmental factors.
   e. Systematists can easily determine whether base substitutions in the DNA of two species are synapomorphies.

7. Which of the following underlying assumptions allows differences in a particular molecular sequence to be used as a molecular clock?
   a. The sequence never experiences any mutations.
   b. The sequence codes for a protein.
   c. The sequence accumulates mutations at a reasonably constant rate.
   d. The sequence is part of a mitochondrial gene.
   e. The sequence codes for small subunit ribosomal RNA.

Apply/Analyze

8. When systematists study morphological or behavioral traits to reconstruct the evolutionary history of a group of animals, they assume that:
   a. similarities and differences in phenotypic characters reflect underlying genetic similarities and differences.
   b. the animals use exactly the same traits to identify appropriate mates.
   c. differences in these traits caused speciation in the past.
   d. the adaptive value of these traits can be explained.
   e. variations in these traits are produced by environmental effects during development.

9. Which of the following pairs of structures are homoplastic?
   a. the wing skeleton of a bird and the wing skeleton of a bat
   b. the wing of a bird and the wing of a fly
   c. the eye of a fish and the eye of a human
   d. the bones in the foot of a duck and the bones in the foot of a chicken
   e. the toes on the foot of a lizard and the toes on the foot of a human

10. To construct a phylogenetic tree by applying the principles of parsimony to molecular sequence data, one would:
    a. start by making assumptions about variations in the rates at which different DNA segments evolve.
    b. group together organisms that share the largest number of ancestral sequences.
    c. group together organisms that share derived sequences, matching the groups to those defined by morphological characters.
    d. group together organisms that share derived sequences, minimizing the number of hypothesized evolutionary changes.
    e. identify derived sequences by studying the embryology of the organisms.

11. **Discuss Concepts** In the past, systematists commonly used the amino acid sequences of proteins and DNA sequences in phylogenetic analyses. Think about the genetic code (see Section 15.1), and explain why phylogenetic hypotheses based on DNA sequences may be more accurate than those based on amino acid sequences.

12. The following table provides information about the distribution of ancestral and derived character states for six systematic characters (labeled 1 through 6) in five species (labeled A through E). A "d" means that the species has the derived state, and an "a" means that it has the ancestral state. Construct a phylogenetic tree for the five species using the principle of parsimony; in other words, assume that each derived character state evolved only once in this group of organisms. Mark the branches of the phylogenetic tree to show where each character changed from the ancestral to the derived state.

Species	Character					
	1	2	3	4	5	6
A	a	a	a	a	a	a
B	d	a	a	a	a	d
C	d	d	d	a	a	a
D	d	d	d	a	d	a
E	d	d	a	d	a	a

13. Create an imaginary phylogenetic tree for an ancestral species and its 10 descendants. Circle a monophyletic group, a polyphyletic group, and a paraphyletic group on the tree. Explain why the groups you identify match the definitions of the three types of groups.

Evaluate/Create

14. Traditional evolutionary systematists identified Reptilia, not including birds, as one class of vertebrates, even though we know that this taxon is paraphyletic. Describe disadvantages of defining paraphyletic taxa in a classification.

15. Imagine that you are a systematist studying a group of little-known flowering plants. You discover that the phylogenetic tree based on flower morphology differs dramatically from the phylogenetic tree based on DNA sequences. How would you try to resolve the discrepancy? Which tree would you believe is more accurate?

16. **Design an Experiment** Imagine that you are trying to determine the evolutionary relationships among six groups of animals that look very much alike because they have few measurable morphological characters. What data would you collect to reconstruct their phylogenetic history?

17. **Apply Evolutionary Thinking** Is the construction of parsimonious phylogenetic trees more consistent with the predictions of the phyletic gradualism hypothesis or the punctuated equilibrium hypothesis of evolutionary change? You may want to review material in Section 23.5 before answering this question.

For selected answers, see Appendix A.

INTERPRET THE DATA

The phylogenetic tree for 12 cat species (Felidae) reproduced at right was assembled from molecular sequence data. Which species is the domestic cat's closest relative? Which clade is the sister taxon to tigers? Are bobcats more closely related to cougars or to ocelots?

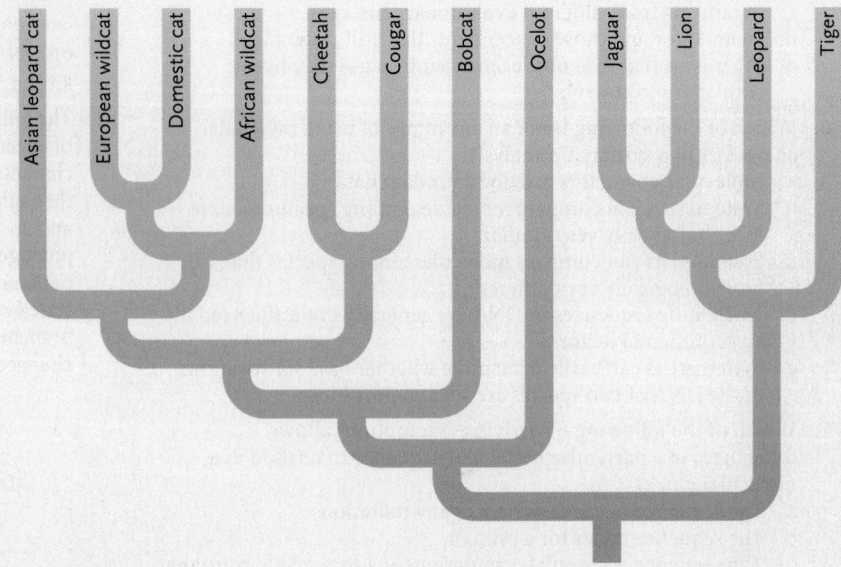

© Cengage Learning 2017

Source: From Warren E. Johnson et al. 2006. The late Miocene radiation of modern Felidae: A genetic assessment. *Science* 311:73–77.

The Origin of Life

Calcium carbonate alkaline vent chimneys in the Lost City Hydrothermal Field in the mid-Atlantic Ocean.

National Science Foundation (University of Washington/Woods Hole Oceanographic Institution)

Why it matters . . . In 1927, Belgian priest and astronomer George Lemaître proposed the Big Bang Theory, which is now the dominant scientific theory about the origin of the universe. According to this theory, an incomprehensibly vast explosion about 14 billion years ago produced the matter and energy of our universe. As the universe expanded, slight variations in temperature and density led to an uneven distribution of matter. Gravitational attraction caused gas to form clouds and then collapse into more concentrated collections of matter. Stars, in which nuclear reactions generated the major elements we find in present-day biological molecules, were formed. Stars that were aging before the birth of our solar system expanded to produce red supergiants, which eventually exploded as supernovae. Such an explosion ejects the contents of a star at near light speed, in effect distributing all the elements of the periodic table across the universe. When our Sun formed, some of the elements coalesced with it and also formed the planets of our solar system. In other words, stardust, the remnants of long-ago stars, is the elemental origin of life on Earth. The origin of cellular life required two other fundamental conditions: a continual source of energy, and a temperature range in which liquid water can form. The continual energy source is, of course, the Sun. The relatively narrow temperature range suitable for life relates to the nature of the Sun, the distance of Earth from the Sun, and the properties of the atmosphere that cause trapping of heat.

Earth is estimated to have formed approximately 4.6 billion years ago. When and how did living cells form from inert materials? Answering these questions is difficult because we must rely on indirect geological evidence and on experimental data that are open to debate and multiple interpretations.

Figure 25.1 outlines the key events in the origin and early evolution of life, which we will examine in this chapter. The earliest events are uncertain, but probably include the development

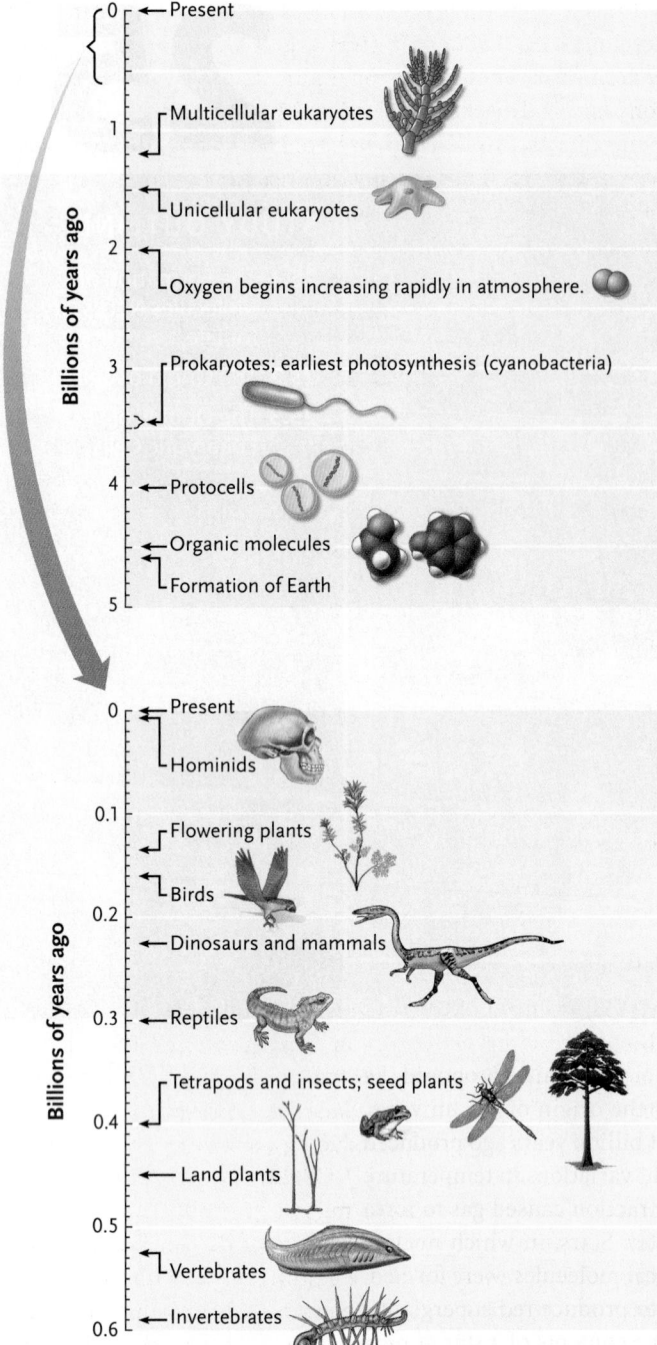

FIGURE 25.1 A timeline for the evolution of cellular life, in the context of the evolution of major present-day organisms. The dates for the origin of cellular life derive from geological evidence. Each date is the subject of debate, as discussed in the text. A more complete presentation of the timeline for the evolution of organismal life is in Table 23.1.

© Cengage Learning 2017

development of the first eukaryotic cells. Fossils of cells that resemble modern eukaryotic cells appeared about 1.5 billion years ago.

25.1 The Formation of Molecules Necessary for Life

All present-day living cells have the following characteristics: (1) a boundary membrane separating the cell interior from the exterior; (2) one or more nucleic acid molecules whose nucleotide sequences make up the genetic information that determines the properties of the cell; (3) a system for reading that genetic information to make RNA molecules and proteins and, through them, other biological molecules; and (4) a metabolic system to provide energy for these activities.

To understand the origin of life, we need to describe possible ways that these first cells could have come into existence through a series of evolutionary events. Evolution of structures as complex as even the simplest living cells could not have occurred without natural selection in favor of variants that reproduce more efficiently. In other words, some form of natural selection would have to have taken place before the first cells existed, when reproduction happened in single molecules or complexes of molecules that were much simpler than even the simplest cells today.

The scientific study of the origin of life thus asks the following questions: What were the first self-replicating molecules, and how did they store genetic information? When and how were these molecules surrounded by a boundary membrane? What energy sources could have supported the metabolism of these first self-replicating molecules and protocells? And what additional evolutionary events would have produced cells as we know them today, with genetic information systems based on DNA, RNA, and proteins? But before considering any of those questions, we must first explore what organic molecules are likely to have been present in the oceans and atmosphere of the early Earth.

Most scientists study the origin of life by assuming that it originated from nonliving matter on Earth, through chemical and physical processes no different from those operating today. Hypotheses made under these assumptions are testable to the extent that the chemical and physical processes can be duplicated in the laboratory.

Scientists have considered a possible extraterrestrial origin of life. Analysis of meteorites has shown that they contain some organic molecules that are characteristic of living organisms. Could a living cell or organism, or perhaps a spore, have arrived in such a way? Most scientists believe it is unlikely that a cell or an organism could have survived a long journey in space, even if protected from radiation, or that it could have survived intense heating while traveling through Earth's atmosphere and the actual impact with Earth. At this point the hypothesis that life arrived on Earth by interplanetary transport is not testable, so it cannot be ruled out. Nonetheless, even if a living

of **protocells,** primitive cell-like structures that have some of the properties of life and that might have been the precursors of cells. Some geological evidence shows that prokaryotic cells were present in the Archaean eon by 3.5 billion years ago (see Table 23.1 for the geological time scale). The activity of photosynthetic prokaryotes generated an oxygen-enriched atmosphere, a condition that may have been necessary for the

organism arrived from space and spawned a population on this planet, life would still have had to arise from nonliving matter in a similar way on the organism's home planet.

Conditions on Primordial Earth Led to the Formation of Organic Molecules

The formation of Earth led to a planet with three layers. In its center is the core, which is surrounded by the mantle. Both the core and the mantle are rich in iron. Surrounding the mantle is the relatively thin crust, which consists mainly of silicates. Volcanic activity on early Earth released gases from the crust and mantle such as CO_2, H_2, and N_2, which helped form the first atmosphere. Water vapor emitted from volcanoes formed the first oceans, lakes, and rivers.

Primordial Earth met several basic conditions necessary for life to begin. Although Earth's gravitational pull was strong enough to retain an atmosphere, it was not strong enough to compress the atmospheric gases into liquid form. Earth's distance from the Sun was such that, on average, sunlight warmed the surface enough to keep much of the liquid water from freezing, but not enough to boil the water. *Liquid water is essential for the chemistry of biological systems* (see Section 2.4).

In the primordial atmosphere, any molecular oxygen would have reacted with elements of the crust and atmosphere to form oxides. Spontaneous reactions of hydrogen, nitrogen, and carbon would have produced ammonia (NH_3) and methane (CH_4). As Earth's surface cooled, natural sources of energy caused chemical bonds to break and reform, creating a variety of molecules. In addition to sunlight and electrical discharges during storms and volcanic activity, radioactivity from atomic decay and heat from volcanoes, geysers, and hydrothermal (hot water) vents in the seafloor all acted on the primordial atmosphere and crust—as they still do today. One or more of these sources of energy enabled inorganic molecules such as ammonia, methane, molecular hydrogen (H_2), and water to react to form the simple organic molecules that are the building blocks for life. Wherever a steady release of energy and the right mix of inorganic molecules caused organic molecules to be synthesized more rapidly than they were broken down, organic molecules would have accumulated slowly over time.

The Oparin–Haldane Hypothesis Initiated Scientific Investigations into the Origin of Life

Scientific efforts to explain the origin of life began with a major hypothesis proposed independently in the 1920s by two investigators, Aleksandr I. Oparin, a Russian plant biochemist at Moscow State University in Russia, and J. B. S. Haldane, a British geneticist and evolutionary biologist at Cambridge University in England. Their hypothesis rested on the critical assumption that Earth's primordial atmosphere was radically different from today's atmosphere. They proposed that, rather than being an oxygen-rich (oxidizing) atmosphere as it is now, the early atmosphere was a strongly reducing atmosphere, due

to the concentration of substances such as hydrogen (H_2), methane (CH_4), ammonia (NH_3), and water, which are *fully reduced*—they contain the maximum possible number of electrons and hydrogens (see Section 7.1). They reasoned that the primordial atmosphere would have had an abundance of electrons and hydrogens for reduction reactions, which could create organic molecules from inorganic elements and compounds. Energy to drive the reductions, according to the hypothesis, came from solar energy and other natural sources such as the electrical energy of lightning in atmospheric storms and from volcanic activity.

Oparin and Haldane proposed that reduction reactions occurring on the primordial Earth produced great quantities of organic molecules. The molecules accumulated because the two main routes by which such substances break down today, chemical attack by oxygen and decay by microorganisms, could not take place. According to Oparin and Haldane's hypothesis, the organic substances would have become so concentrated that the oceans and other bodies of water resembled a prebiotic soup.

The presence of a reducing atmosphere is essential to the Oparin–Haldane hypothesis. Molecular oxygen (O_2) or substantial quantities of less reduced molecules such as CO_2 and N_2 can reverse reductions by removing electrons and hydrogens from organic molecules (see Section 7.1), causing newly formed organic molecules to be broken down quickly.

Chemistry Simulation Experiments Support the Oparin–Haldane Hypothesis

In 1953, Stanley L. Miller, a graduate student in Harold Urey's laboratory at the University of Chicago, tested the Oparin–Haldane hypothesis. He created a laboratory simulation of conditions Oparin and Haldane believed existed on early Earth; namely, involving a reducing atmosphere consisting of H_2, CH_4, and NH_3, as well as H_2O (**Figure 25.2**). After one week of running the apparatus, as much as 15% of the carbon was now in the form of organic compounds, including amino acids. The significance of the finding at the time was enormous; amino acids, which are essential to cellular life, could be made under the conditions scientists believed existed on early Earth. The experiment is now known as the *Miller–Urey experiment*.

The synthesis of complex biological molecules in a reducing atmosphere in the Miller–Urey experiment supported the Oparin–Haldane hypothesis. However, many scientists believe that early Earth did not have a reducing atmosphere at the time key organic molecules were formed. Rather, the atmosphere at that time is thought to have been neutral (neither reducing nor oxidizing), consisting mostly of gases released by volcanic activity, notably CO_2, N_2, CO, and H_2S. Indeed, experiments using these gases in the Miller–Urey apparatus, in addition to the original gases used, resulted in the formation of many other molecules of biological significance. Produced, for example, were all the building blocks of complex biological molecules,

FIGURE 25.2 | Experimental Research

The Miller–Urey Apparatus Demonstrating That Organic Molecules Can Be Synthesized Spontaneously under Conditions Simulating Primordial Earth

Question: Do chemistry simulation experiments support the Oparin–Haldane hypothesis that reduction reactions on primordial Earth produced organic molecules?

Experiment: Stanley Miller set up an experiment to simulate chemical conditions of early Earth **(Figure)**. He placed components of a reducing atmosphere—hydrogen, methane, and ammonia—in a closed apparatus. Water vapor was added to the "atmosphere" by boiling water in one part of the apparatus, and it was removed by cooling and condensation in another part. This simulated the cycle of release of water vapor into the atmosphere, and condensation into water. Miller exposed the gases to an energy source in the form of continuously sparking electrodes to simulate lightning.

~4.5 billion years ago

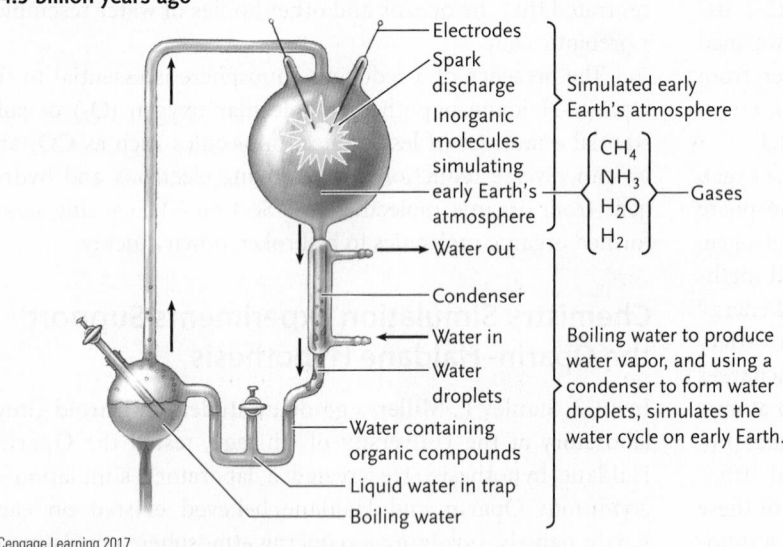

Electrodes
Spark discharge
Inorganic molecules simulating early Earth's atmosphere
Water out
Condenser
Water in
Water droplets
Water containing organic compounds
Liquid water in trap
Boiling water

Simulated early Earth's atmosphere
CH_4
NH_3
H_2O
H_2
Gases

Boiling water to produce water vapor, and using a condenser to form water droplets, simulates the water cycle on early Earth.

© Cengage Learning 2017

Results: After running the apparatus for only a week, Miller found a large assortment of organic compounds in the water, including urea, amino acids, and lactic, formic, and acetic acids—as much as 15% of the carbon was now in the form of organic compounds. Two percent of the carbon was in the form of amino acids, which form easily under sufficiently reducing conditions.

Conclusion: The synthesis of organic molecules, including biological molecules, in a reducing atmosphere supported the Oparin–Haldane hypothesis.

think like a scientist

Observe
Hypothesize
Predict
Experiment
Interpret

What aspect of the Miller–Urey experimental design ruled out the possibility that amino acids were synthesized by living organisms such as bacteria?

Source: S. Miller. 1953. A production of amino acids under possible primitive Earth conditions. *Science* 117:528–529. (Figure redrawn from an original courtesy of S. L. Miller, Copyright 1955 by the American Chemical Society.)

© Cengage Learning 2017

including amino acids; fatty acids; the purine and pyrimidine building blocks of nucleic acids; sugars such as glyceraldehyde, ribose, deoxyribose, glucose, and fructose; and phospholipids, which form the lipid bilayers of biological membranes.

Scientists Have New Theories about the Sites for the Origin of Organic Molecules

The discovery of "black smoker" hydrothermal vents in 1977 led Günter Wächtershäuser to propose that the production of energy-rich iron–sulfur compounds in those environments provided the chemical conditions necessary for life to originate. Many such vents exist in today's oceans, emitting bursts of mineral-rich water superheated to up to 400°C by submarine volcanoes. Scientists exploring hydrothermal vents find complex ecosystems associated with them.

Life might have originated near oceanic hydrothermal vents because reducing conditions existed there along with an abundance of the chemicals that are essential for life. Even now, there are high levels of hydrogen gas, methane, and ammonia around the vents. Based on simulation experiments, scientists hypothesize that hydrothermal vents could have produced a lot more organic material than that generated in the Miller–Urey experiment. Critics of the hydrothermal-vent origin of life theory argue that the temperature at the vents is too high to permit the origin of life because the organic molecules are too unstable and would be destroyed by the high temperature as soon as they form. Supporters of the theory counter that the necessary organic molecules for life do not form at the vent itself, but somewhere in the gradient between the hot water at the vent and the much colder water surrounding the vent.

An even more promising candidate for the site of life's origin is alkaline hydrothermal vents, which were discovered as recently as 2000 (see chapter-opening photo). Unlike "black smoker" hydrothermal vents, which are extremely hot and acidic, alkaline hydrothermal vents produce only moderately high temperatures (30–90°C) and have moderately alkaline pH levels, two properties that are compatible with biological metabolism. They are also thought to produce a steady release of energy lasting tens of thousands of years, much longer than "black smoker" hydrothermal events, which appear and disappear with a much shorter time scale. We will see the importance of this in the next section, when we discuss chemical evolutionary processes that are thought to have produced the first cells.

STUDY BREAK 25.1

1. Why is the reducing nature of early Earth's atmosphere key to the Oparin–Haldane theory of the origin of molecules necessary for life?
2. How do the theories about the sites for the origin of life differ?

25.2 The Origin of Cells

No matter where organic molecules originated, they do not qualify as life. In this section, we discuss the key stage in the origin of life—the formation of the first cells.

The Evolution of Molecular Replicators, Molecules That Store and Reproduce Genetic Information

To determine how organic building blocks such as amino acids and nucleotides assemble into macromolecules (proteins and nucleic acids, respectively), researchers have proposed and tested several processes. One process is the concentration of subunits by the evaporation of water. Another is *dehydration synthesis (condensation),* in which subunits assemble into larger molecules through removal of the elements of a molecule of water (see Section 3.1). Experiments with these processes under simulated conditions showed that both evaporation and condensation reactions can produce polypeptide chains from amino acids, polysaccharides from glucose and other monosaccharides, and nucleotides and nucleic acids from nitrogenous bases, ribose, and phosphates.

But while organic polymers are a necessary prerequisite for life as we know it, the essence of life is the *reproduction of genetic information.* In cells, genetic information is stored in sequences of nucleotides in DNA molecules. The information is "read" to determine the properties of the cell as genes are transcribed and translated to produce RNA and protein molecules. Those RNA and protein molecules in turn produce and make up the structures of the cell. No cell lives forever, so the information in a particular cell continues to exist on Earth only if that cell is able to reproduce, making more cells with the same genetic information.

Genetic information as we know it today occurs only in cellular life, and in viruses that require some of the machinery of cellular life to help them reproduce. The smallest bacterial genomes contain hundreds of genes and produce hundreds of different proteins and RNA molecules. Those bacterial cells contain hundreds of other different small-molecule metabolites. The earliest forerunners of life on Earth must have been much simpler, but they would still have had to contain some form of genetic information that was able to reproduce itself. That genetic information was probably not in the form of DNA molecules. It may have been in the form of another polymer (RNA is a leading candidate), or some other physical structure that can occur in many different possible forms, such as the different sequences of nucleotides in a DNA polymer. Molecules that are able to store and reproduce genetic information in this way are called **molecular replicators.**

Just as living things evolve by natural selection, molecular replicators can evolve by natural selection if certain forms of those replicators are able to reproduce themselves more quickly than other forms. For example, isolated RNA molecules placed in the right chemical environment can replicate to produce new RNA molecules with complementary nucleotide sequences. When those complementary RNA molecules replicate, more RNA molecules with the original sequence are produced. If RNA molecules with certain sequences replicate more quickly than RNA molecules with other sequences, over time there will be a greater number of RNA molecules with the more efficiently replicating sequences. The natural environment is selecting for RNA molecules of a particular sequence in the same way that natural selection favors cellular organisms that are able to reproduce more effectively. If the replication of RNA molecules is imperfect, producing occasional mutations, the RNA sequence will evolve by natural selection whenever a new mutation produces a sequence that can replicate even faster than the earlier pre-mutation sequence could. Thus, a limited form of natural selection would have appeared in the early history of life on Earth as soon as there were molecular replicators.

Why would some RNA molecules have replicated faster than others? Remember that RNA molecules can fold into three-dimensional shapes like proteins, and some RNAs fold to form **ribozymes** that can catalyze chemical reactions just like protein enzymes (see Section 6.6). Polymerase ribozymes have been produced experimentally that speed up the synthesis of new complementary RNA polymers from RNA templates. If a polymerase ribozyme is more likely to facilitate the replication of other RNA molecules with the same sequence, over time more and more copies of that RNA molecule will exist. In other words, that sequence is favored by natural selection acting on a molecular level. The sequence of the polymerase ribozyme will evolve as mutations to that sequence occur that improve the functioning of the ribozyme. Just such an evolutionary process has been demonstrated using artificial selection in the laboratory.

The hypothesis that RNA molecules once functioned both as the store of genetic information in molecular replicators and as the main enzymes is the **RNA world model.** It is a very appealing explanation for a pre-cellular stage in the origin of life because RNA is the only molecule that combines these two key functions of living systems. By serving as both the genetic information and the replication mechanism of a molecular replicator, self-replicating RNA molecules provide two of the key characteristics of life in a system that is much simpler than the system that exists in contemporary organisms, in which information flows from DNA to RNA to proteins.

In the RNA world, DNA would have developed in a later step. At first, DNA nucleotides may have been produced by random removal of an oxygen atom from the ribose subunits of the RNA nucleotides. At some point, the DNA nucleotides paired with the RNA informational molecules and were assembled into complementary copies of the RNA sequences. Some modern-day viruses carry out this RNA-to-DNA reaction using the enzyme reverse transcriptase (see Sections 17.2 and 18.2). Once the DNA copies were made, selection may have favored DNA as the informational storage molecule because it has greater chemical stability and can be assembled into much longer coding sequences than can RNA. RNA was left to function at intermediate steps between the stored information in DNA and protein synthesis, as it does today.

As the RNA-based information system evolved, some RNAs may have acted as tRNA-like molecules, linking to amino acids and pairing with the RNA informational molecules. These associations could have led to the assembly of polypeptides with an ordered sequence of amino acids—the development of an RNA genetic code. When DNA took over information storage from RNA, the DNA would have picked up the code information that could then be transferred back to RNA by transcription.

Modern analysis of the ribosome, the organelle responsible for translation of mRNA (see Section 15.4), has shown that the enzymatic activity that catalyzes the formation of a peptide bond between amino acids is a function of one of the RNA molecules of the ribosome. This finding supports the proposal that, in addition to replicating themselves, RNA molecules also generated the first proteins.

The Evolution of Cellular Membranes

All life today is made up of cells surrounded by a lipid membrane that in most organisms comprises two layers of molecules each with a hydrophobic end and a hydrophilic end like phospholipids. The hydrophilic ends of these molecules point out of both sides of the membrane, toward the aqueous solution inside the cell and surrounding the cell. Cell membranes are useful adaptations for living things because: (1) they enable cells to hold on to the special and useful molecules that they synthesize, including the molecules that make up their genetic information; (2) they enable cells to maintain a chemically distinct intracellular environment with, for example, different concentrations of key ions than the extracellular fluid; and (3) they protect cells from parasitic genetic information and other unwanted molecules.

We know that membranes must have evolved at some point as containers for genetic information, but it is not entirely clear when this happened. It was probably very early in the process, as natural selection acting on molecular replicators relies on more effective ribozymes preferentially replicating other copies of the same RNA sequence. That is most likely to happen if a population of similar or identical RNA molecules is clustered within a distinct compartment, surrounded by some kind of membrane. Also, the evolution of increased genetic complexity would likely have involved several ribozymes specialized for different functions working together and dependent on each other. Only later would the genetic information coding for these molecules have been combined into longer multigenic genomes. Before then, membrane enclosures would have helped to keep this community of RNA molecules together in one functional organism, forming a protocell. *Protocells* are primitive cell-like structures that might have been the precursors of cells. The appearance of protocells is a key event in the origin of life (**Figure 25.3**). *Molecular Insights* describes the experimental construction of protocell-like vesicles containing active ribozymes.

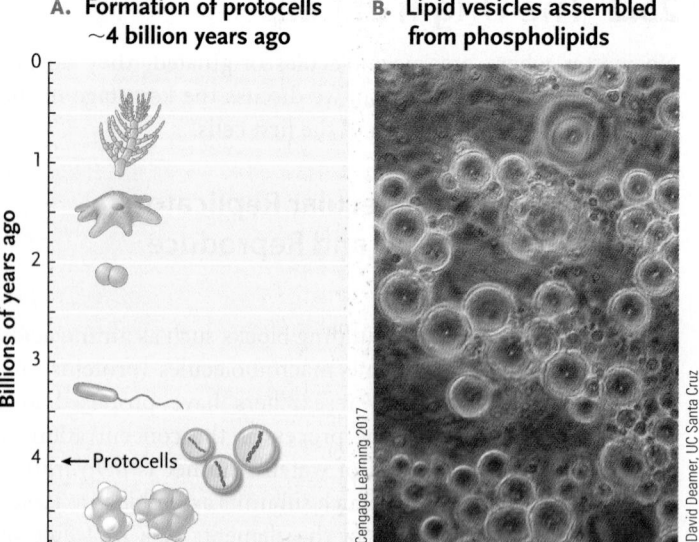

A. Formation of protocells ~4 billion years ago

B. Lipid vesicles assembled from phospholipids

← Protocells

© Cengage Learning 2017

Dr. David Deamer, UC Santa Cruz

FIGURE 25.3 Protocells. (A) As a key event in the origin of life, protocells are thought to have formed approximately 4 billion years ago. **(B)** Phase micrograph of lipid vesicles assembled from phospholipids. Structurally, primitive protocells were vesicle-like, with a lipid membrane surrounding aggregations of major biological molecules.

Experiments have demonstrated that when fatty acids are highly concentrated in solutions, they will often form bilayer membranes spontaneously. The carboxyl groups of fatty acids are not as hydrophilic as the heads of phospholipids, but fatty acids nevertheless can form bilayers organized with the carboxyl groups pointing out towards the aqueous solution on either side of the membrane. The fact that carboxyl groups are only weakly hydrophilic is actually an advantage for abiotic formation and growth of fatty-acid bilayer membranes, because it causes fatty acids to be more likely to flip from one layer to the other of a bilayer. This would have been critically important in the early evolution of biological membranes. Before intracellular metabolic pathways had evolved for lipid synthesis, membranes could only have grown through addition of lipids from the outer surface, in which case there would have to have been flux of fatty acids from the outer to the inner layer of the bilayer.

Early membranes would likely have been heterogeneous mixtures of fatty acids together with whatever other lipids had resulted from abiotic synthesis reactions, and in fact more heterogeneous membranes have been shown to be more chemically stable than those composed purely of fatty acids. Early membranes would also have been considerably more permeable to other molecules, including molecules as large as amino acids and nucleotides that cannot diffuse across phospholipid bilayer membranes without the assistance of transport proteins. This greater permeability would actually have been an advantage for the replication of genetic information, because nucleotide monomers would have been able to diffuse into protocells.

Membranes are thought to have grown primarily by the incorporation of micelles and vesicle-sized aggregates of membrane lipids, a process that has been shown to occur abiotically.

Toward the Evolution of Life in the Lab: The construction of protocell-like vesicles containing an active ribozyme

Life is characterized by membrane-bound cells that grow and divide, and in which enzymatic reactions take place in a controlled and sequestered environment. We have learned in this chapter so far that living cells may have developed from protocells. And the discovery of ribozymes led to the proposal that an RNA world was the first step in the development of a molecular information system that could store, replicate, and translate the information required for protein synthesis. Constructing a simple protocell in the lab that contains an active ribozyme would provide a system with which to explore the possible steps for the origin and early evolution of life. That is, the simplest cellular system capable of evolution requires a self-replicating informational molecule, and a means to compartmentalize that molecule.

Research Goal
Construct stable protocell-like vesicles that are capable of growth and that contain an active ribozyme.

Experiments
Jack Szostak and his research team at the Howard Hughes Medical Institute and Harvard Medical School set out to construct model protocell vesicles. Many chemicals are capable of forming simple vesicles, but the key requirement here is that the vesicle membrane must be stable and capable of growth under conditions that favor ribozyme activity. For example, fatty acids form vesicles that can grow and divide, but ribozymes cannot function in them because they need magnesium ions (Mg^{2+}) for activity, and divalent cations such as Mg^{2+} cause the fatty acids to precipitate.

Results
The researchers found that a mixture of particular, chemically simple *amphiphiles* (molecules with both hydrophilic and lipophilic properties) form vesicles with the desired protocell-like properties. That is, these vesicles are stable and grow, but, importantly, remain intact in the presence of Mg^{2+}. The vesicles are also permeable to Mg^{2+}, as shown by an experiment in which a ribozyme included in the vesicles became active and self-cleaved when Mg^{2+} was added to the solution.

Conclusion
The researchers successfully constructed simple vesicles containing active ribozymes. They concluded that the stability of the vesicles, and their ability to grow and allow the diffusion into them of Mg^{2+}, are "critical for building model protocells in the laboratory and may have been important for early cellular evolution."

think like a scientist
Why was the observation that a ribozyme could become activated within the protocell-like vesicles important?

Source: I. A. Chen, K. Salehi-Ashtiani, and J. W. Szostak. 2005. RNA catalysis in model protocell vesicles. *Journal of the American Chemical Society* 127:13213–13219.

According to one hypothesis, the presence of replicating RNA molecules within protocells would have produced osmotic stress that would energetically favor the incorporation of more lipids. This process could enable membrane-enclosed volumes containing more RNA molecules to compete more effectively for additional membrane lipids. In this way, faster replication of RNA molecules could have been linked to faster growth in volume of the protocells.

To reproduce in a manner analogous to present-day cells, protocells would have to have spontaneously divided their membrane and interior volume into two or more smaller compartments. How this would have happened has been something of a puzzle, as a roughly spherical structure should be energetically the most favorable, and yet the membrane would have to pinch in somehow to divide into two protocells. Moreover, pinching in would increase the surface area required to surround a given volume of interior contents, which could only happen with wholesale incorporation of more lipids from outside. However, a curious phenomenon has been observed in experimental protocells. As more lipids are incorporated into larger protocells, they protrude a tube-shaped structure, and over time, more and more of the membrane surface area is incorporated into the tube, until it largely replaces the earlier spheroid. This tube then spontaneously disintegrates into a number of smaller spheroids. If this were to happen in a protocell containing a large enough number of self-replicating RNA polymers, a portion of that genetic information could be included in each of the newly replicated protocells.

The Evolution of Biological Energy Sources

Life as we know it today is critically dependent on energy to produce and maintain its organized structures, and some sources of energy to drive endergonic reactions (see Section 6.2) would have been required very early in the development of increasing biological complexity. Life on Earth today derives almost all of its energy from photons coming from the Sun, but photosynthesis is not thought to have evolved until hundreds of years after the origin of molecular replicators and cellular life. Instead, the earliest living things are most likely to have evolved in places where energy was released through geochemical activity.

Some geochemical activity persists on Earth today, giving us the opportunity to study environmental conditions that may be similar to those in which life first evolved, but geochemical activity was even more abundant at the time life originated. One very promising candidate for the site of life's origin is the alkaline hydrothermal vents described at the end of Section 25.1. Alkaline hydrothermal vents are sources of free energy because the chemical reactions that take place within them cause the water they release to be chemically distinct from the surrounding ocean water. The vent water is rich in H_2 and CH_4 and has very low concentrations of free H^+, whereas the ocean water has higher concentrations of H^+, CO_2, and bicarbonate (HCO_3^-). At the time life originated on Earth, the oceans were much richer in CO_2 than they are today. By reacting with water, CO_2 produces H^+ and HCO_3^-, so that all three of these molecules would have been even more concentrated in ancient ocean water than they are in oceans today.

Chemical reactions that take place where the vent water encounters ocean water cause calcium carbonate to precipitate out of the water, forming chimneys made of calcium carbonate that are honeycombed with microscopic pores that the water percolates through (see chapter-opening photo). Within these porous chimneys, the two different chemical environments of the vent water and the ocean water encounter each other. Because a fresh supply of alkaline water is constantly released from the vent, the interface between vent water and ocean water is constantly out of equilibrium. You learned in Section 6.2 that free energy is at a minimum when any system reaches equilibrium, so the nonequilibrium conditions produced by the alkaline vents represent a continually renewing source of energy.

One very intriguing source of energy in the vent environment is the H^+ concentration gradient between the ocean water and vent water, because this is still a prime energy source for living things today. Although some ATP is made in living things today directly by substrate-level phosphorylation (see Section 7.1) during the metabolism of organic molecules, most ATP molecules are produced by ATP synthase enzymes in mitochondria and chloroplasts, using energy released from the diffusion of H^+ down a concentration gradient (see Sections 7.4 and 8.2). Scientists studying the alkaline vent environment have hypothesized that the importance of H^+ gradients as sources of energy for all living things may be a relic of the release of energy stored in H^+ gradients in calcium carbonate chimneys. The central role of H^+ diffusion through the ATP synthase protein in present-day ATP synthesis is consistent with the hypothesis that H^+-gradients were key energy sources for the *Last Universal Common Ancestor (LUCA)* of cellular life.

A second source of energy in the vent environment is the difference in electron affinity between the reduced H_2 and CH_4 in the vent water and the more oxidized CO_2 in the ocean water. In Chapter 7, you learned how energy is released as electrons are transferred between reduced molecules such as NADH and the oxidized carrier complexes of the mitochondrial electron transfer system. This is just one example of the many electron transfer systems that release energy in a wide variety of living things today. The interface between the reduced H_2 and CH_4 in vent water and oxidized CO_2 in ocean water would have provided a perfect opportunity for the evolution of energy-releasing electron transfer reactions in protocells. The fact that electron transfer systems are such common and indispensable mechanisms for releasing energy in living things today may be a relic of the evolution of life in an alkaline vent chemical environment. Some of the microscopic compartments in alkaline vent chimneys are only micrometers in diameter—in other words, about the size of a cell. These structures provide a promising model of how protocells could have evolved to make use of the nonequilibrium chemical conditions of alkaline vents. According to one hypothesis, molecular replicators may have taken up residence in microscopic compartments in the porous chimneys, and lipid membranes may have formed on the inner surfaces of the chamber walls of these compartments and across the porous openings **(Figure 25.4)**. By growing within chimneys, protocells would have been ideally placed to take advantage of the enduring H^+ concentration gradients and oxidation–reduction differences between the vent water and surrounding ocean water. The energy provided by these gradients could have supported the synthesis of organic molecules and the maintenance of molecular organization in protocells.

Prokaryotic Cells Were the First Living Cells

The evolution of increasingly complex populations of molecular replicators, enclosed within protocells and making use of some geochemical energy source, culminated in the evolution of the LUCA of life on Earth. By identifying molecular features that are common to all present-day living things, we can construct reasonable hypotheses to describe the LUCA. Those features include: (1) a DNA-based genome, mechanisms for replicating the DNA and transcribing it into RNA; and (2) mechanisms by which the sequence of nucleotides in DNA could code for the amino acid sequences of polypeptides. That coding mechanism included ribosomes, transfer RNAs, and aminoacyl–tRNA synthetase enzymes—all of which are found in every species of cellular organism known today. The cytoplasm of the LUCA must also have contained an oxidative system supplying chemical energy for protein synthesis and assembly of other required molecules, as well as a mechanism of cell division, allowing replicated DNA to be distributed equally between daughter cells. All of these systems were enclosed by a membrane controlling the flow of molecules and ions in and out of the cell.

The stages leading to the LUCA may have taken hundreds of millions of years, beginning after the solidification of the Earth's crust around 4.3 billion years ago. For clues to how early the first living things evolved, we have to look to geological evidence. This includes:

- **Isotope ratios determined by radiometric dating.** Radiometric dating is a technique used to date a rock and is based on the decay of isotopes in the rock (see Section 23.1). Each isotope of an element has a specific decay rate. However, if rock is subjected to biological activity that uses one isotope preferentially over another, the ratio of the two isotopes will differ from that in inert rock from the same time. Thus, an altered isotope ratio in rock or mineral samples can be interpreted as evidence of biological activity. The earliest

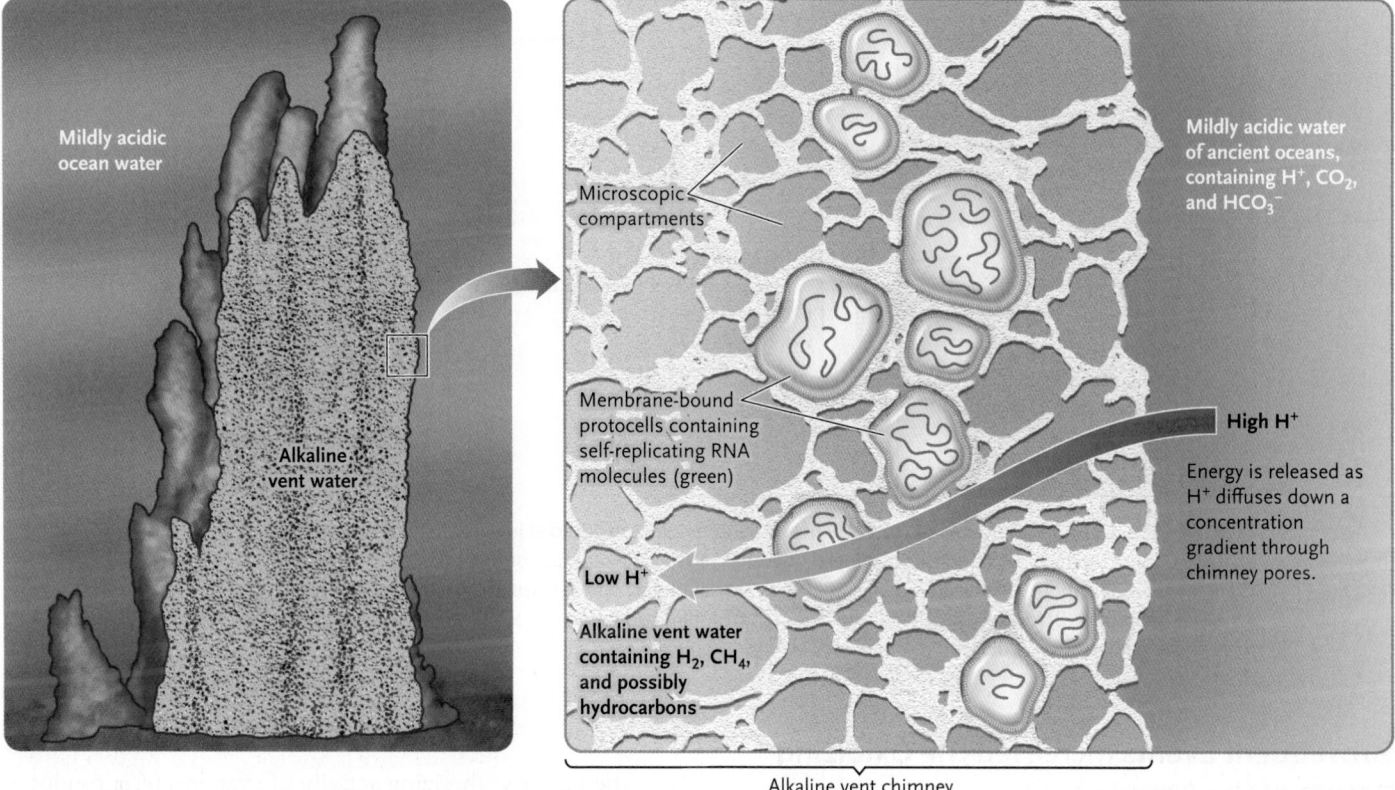

A. Cross-sectional view through an alkaline vent chimney

Mildly acidic ocean water

Alkaline vent water

B. Magnified cross section through part of an alkaline vent chimney

Microscopic compartments

Membrane-bound protocells containing self-replicating RNA molecules (green)

Low H⁺

Alkaline vent water containing H_2, CH_4, and possibly hydrocarbons

Mildly acidic water of ancient oceans, containing H^+, CO_2, and HCO_3^-

High H⁺

Energy is released as H^+ diffuses down a concentration gradient through chimney pores.

Alkaline vent chimney

FIGURE 25.4 A hypothesis for the origin of cells in alkaline vent chimneys. (A) A cross-sectional view through an alkaline vent chimney. Within the porous material of the chimney, warm (30–90°C) alkaline water issuing from the vent mixes with cooler and mildly acidic ocean water. **(B)** A magnified cross-sectional view through a part of the chimney. Membrane-bound protocells containing populations of self-replicating RNA molecules are shown, which are hypothesized to have formed within the microcompartments of the porous chimneys.
© Cengage Learning 2017

rock samples showing a significant alteration of an isotope ratio, interpreted to involve preferential use of $^{12}CO_2$ rather than $^{13}CO_2$ (as is seen in present-day bacteria), date to about 3.7 billion years ago. This may mark the earliest evidence of biochemical reactions by living cells, which, by any present theory, would be prokaryotes.

- **Stromatolites.** Fossil stromatolites are layers of carbonate or silicate rock that resemble present-day stromatolites **(Figure 25.5A).** Geologists think that the fossils formed as layers of phototrophic prokaryotes grew and died, and their remaining forms were filled in by calcium carbonate or silica. ("Phototrophic" refers to an organism that obtains energy from light.) The earliest fossil stromatolites that are accepted by geologists date to about 3.4 billion years ago in the oldest rocks of the Archaean eon. However, the fossil rock is too deformed to allow details of cell structure to be analyzed, and some researchers question whether they are of biological origin.

- **Biosignatures. Biosignatures** are particular organic molecules in sedimentary rocks that could only have been formed by cellular activity. There are limited examples of biosignature molecules. One is a steroidlike molecule that

is found only in the membrane lipids of cyanobacteria, and in no other organism known today. That particular biosignature molecule has been found in rock dated to 2.5 billion years ago, providing evidence for the presence of cyanobacteria by the end of the Archaean eon. Cyanobacteria are photosynthetic organisms. Since photosynthesis is a complex metabolic process, more primitive prokaryotes must have developed at an earlier time.

- **Microfossils.** A **microfossil** is the remains of a cell that has decayed and been filled in by calcium carbonate or silica **(Figure 25.5B).** (Section 23.1 discusses the fossil record and its use in providing information about life in the past.) The earliest microfossils are of filamentous and single-celled prokaryotes, and they date to about 2.0 billion years ago. Those microfossils are considered the most convincing geological evidence of early life. Microfossils from around 1.5 billion years ago include some with larger cells that resemble cells of present-day eukaryotic protists (see Chapter 27).

In sum, prokaryotes probably appeared first about 3.5 billion years ago and eukaryotes first appeared about 1.5 billion years ago, or perhaps earlier.

A. Stromatolites date to
~3.4 billion years ago

B. Microfossils date to
~2.0 billion years ago

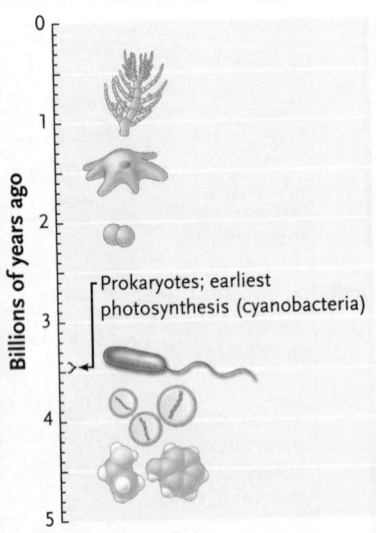

Prokaryotes; earliest
photosynthesis (cyanobacteria)

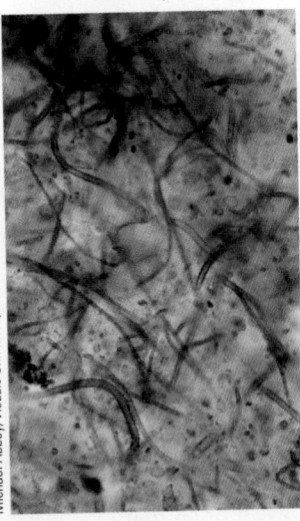

FIGURE 25.5 Geological evidence for early life. (A) Stromatolites exposed at low tide in Western Australia's Shark Bay. These mounds, which consist of mineral deposits made by prokaryotic communities (predominantly cyanobacteria), are about 2,000 years old; they are highly similar in structure to fossil stromatolites that formed 3.4 billion years ago. **(B)** Microfossils—fossil algae from the Gunflint Formation, Ontario, Canada. The earliest microfossils date to about 2.0 billion years ago.

© Cengage Learning 2017

Subsequent Events Increased the Oxidizing Nature of the Atmosphere

According to Richard E. Dickerson of University of California, Los Angeles, and others, the earliest form of photosynthesis evolved about 3.5 billion years ago in the early prokaryotes. This form of photosynthesis probably used electron donors such as hydrogen sulfide (H_2S) that do not release oxygen. However, at some point, an enzymatic system evolved that could use the most abundant molecule of the environment, water (H_2O), as the electron donor for photosynthesis. This reaction split water into protons, electrons, and oxygen, which was released into the atmosphere.

The oxygen released by the water-splitting reaction accumulated in the atmosphere and set the stage for the development of electron transfer systems using oxygen as the final electron acceptor. These transfer systems arose when some cells developed cytochromes that could deliver low-energy electrons to oxygen (see Section 7.4). These cells were able to tap the greatest possible amount of energy from the electrons before releasing them from electron transfer, making the cells highly successful in their environment.

The first water-splitting photosynthesizers confirmed by fossil evidence are the cyanobacteria. Convincing evidence from biosignatures indicates cyanobacteria were present 2.5 billion years ago (see earlier discussion). Present-day stromatolites predominantly contain cyanobacteria but, as stated earlier, the types of prokaryotes in fossil stromatolites cannot be identified with any certainty. If fossil stromatolites do contain cyanobacteria, that would date early water-splitting photosynthesis to about 3.4 billion years ago.

The photosynthesizing activity of cyanobacteria produced a gradual rise in the amount of oxygen in early Earth's atmosphere. That rising oxygen concentration likely killed off many prokaryotic groups to which oxygen is lethal. Around 2 billion years ago, the concentration of oxygen began to increase more rapidly. This abrupt change may correlate with the evolution of eukaryotic cells containing chloroplasts.

These major events established the preconditions for the evolution of eukaryotic cells. The next section traces this evolution, which was pivotal to the later evolution of large-scale multicellularity and the plants, animals, and other organisms of the domain Eukarya.

STUDY BREAK 25.2

1. What properties of alkaline vents suggest that protocells may have evolved in similar environments?
2. What evidence suggests that RNA molecules may have functioned as molecular replicators in early protocells?
3. What properties are the earliest protocells hypothesized to have in common with modern-day cellular organisms, and what properties of modern-day cellular organisms are hypothesized to have been absent in the earliest protocells?

THINK OUTSIDE THE BOOK

Individually or collaboratively, investigate banded iron formations and outline the following in your investigation: (1) what they are; and (2) how researchers consider that they provide evidence for the presence of particular oxygen-producing organisms on early Earth.

25.3 The Origins of Eukaryotic Cells

Present-day eukaryotic cells have several interrelated characteristics that distinguish them from prokaryotes: (1) the separation of DNA and cytoplasm by a nuclear envelope; (2) the presence in the cytoplasm of membrane-bound compartments with specialized metabolic and synthetic functions—mitochondria, chloroplasts, the endoplasmic reticulum (ER), and the Golgi complex, among others; and (3) highly specialized motor (contractile) proteins that move cells and internal cell parts. In this section we discuss how eukaryotes most probably evolved from associations of prokaryotes. As you read, note that there are more good theories for the origin of eukaryotes than there are good data.

The Endosymbiotic Theory Proposes That Mitochondria and Chloroplasts Evolved from Ingested Prokaryotes

The **endosymbiotic theory,** put forward by Lynn Margulis at the University of Massachusetts, Amherst, proposes that the membranous organelles of eukaryotic cells, the mitochondria and chloroplasts, may each have originated from mutualistic (mutually advantageous) relationships between two prokaryotic cells **(Figure 25.6)**.

The theory proposes that the following events occurred:

1. *Generation of endosymbionts.* Mitochondria began to develop when photosynthetic and nonphotosynthetic prokaryotes coexisted in an oxygen-rich atmosphere. The nonphotosynthetic prokaryotes fed themselves by ingesting organic molecules from their environment. These prokaryotes included both anaerobes, which are unable to use oxygen as the final acceptor for electron transfer, and aerobes, which are fully capable of using oxygen. Only the aerobes could exploit fully the energy stored in organic molecules, but predatory anaerobes could capture that energy by eating aerobic cells. These anaerobic prokaryotes had become efficient predators, and lived by ingesting other cells. Among the ingested cells were some aerobic prokaryotes; instead of being digested, some of them persisted in the cytoplasm of the predators and continued to respire aerobically in

their new location. They had become **endosymbionts,** organisms that live symbiotically within a host cell. The cytoplasm of the host anaerobe, formerly limited to the use of organic molecules as final electron acceptors, was now home to an aerobe capable of carrying out the much more efficient transfer of electrons to oxygen. From an evolutionary perspective, cells that had engulfed respiring bacterial cells capable of cellular respiration would have been at a considerable advantage compared to other cells, so natural selection would have favored their persistence. They could ingest simple carbohydrates from their environment and metabolize them as a source of energy.

2. *Formation of the nucleus and other membranous structures.* As a part of the transition to a true eukaryotic cell, the cell also evolved to acquire other membranous structures, the major ones being the nuclear envelope, the ER, and the Golgi complex. Endocytosis, the process of

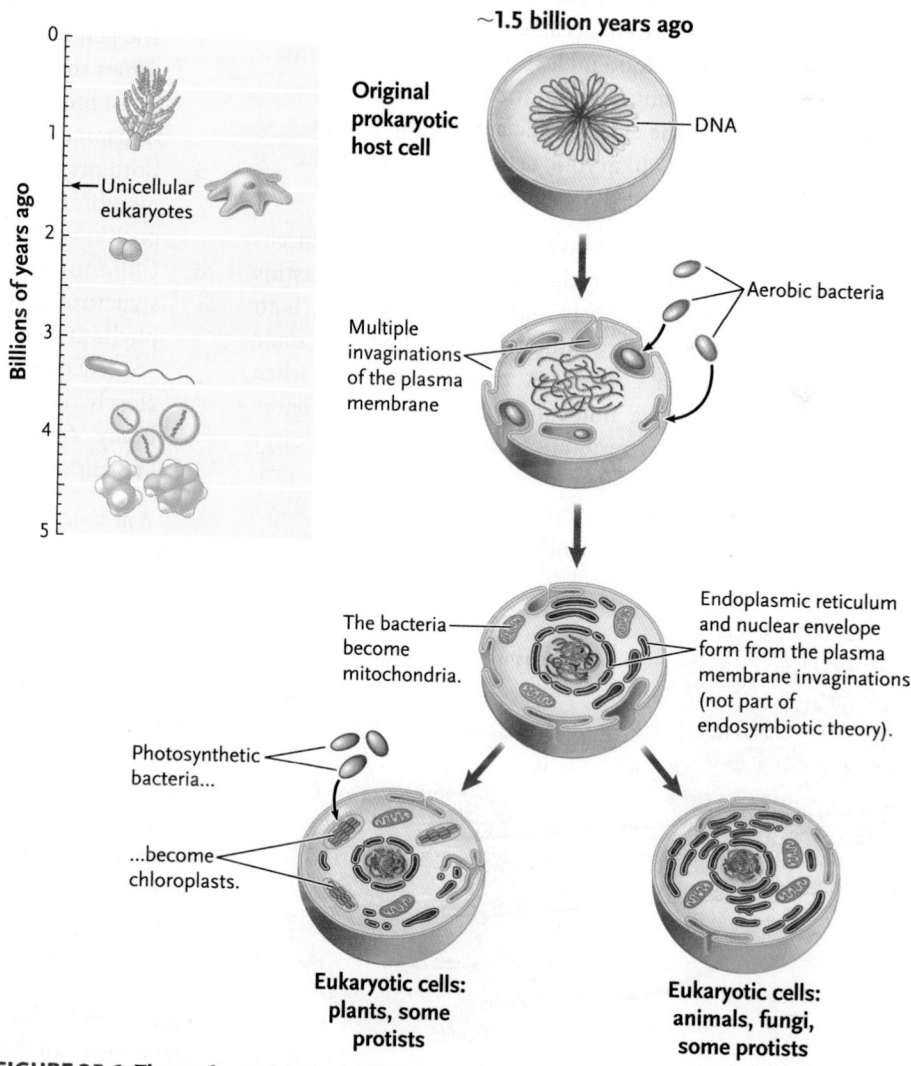

FIGURE 25.6 The endosymbiotic theory. The mitochondrion is thought to have originated from an aerobic prokaryote that lived as an endosymbiont within an anaerobic prokaryote. The chloroplast is thought to have originated from a photosynthetic prokaryote that became an endosymbiont within an aerobic cell that had mitochondria.

© Cengage Learning 2017

infolding of the plasma membrane (see Figure 4.15 and Section 5.5), is thought to be responsible for the evolution of these structures. Researchers think that, in cell lines leading from prokaryotes to eukaryotes, pockets of the plasma membrane formed during endocytosis may have extended inward and surrounded the nuclear region. Some of these membranes fused around the DNA, forming the nuclear envelope and, hence, the nucleus. The remaining membranes formed vesicles in the cytoplasm that gave rise to the ER and the Golgi complex **(Figure 25.7)**.

3. *Transfer of many functions from endosymbiont to the host cell.* Next, many functions duplicated in the aerobic endosymbiont were taken over by the host cell. As part of this transfer of function, most of the genes of the aerobe moved to the cell nucleus and became integrated into the host cell's DNA. At the same time, the host anaerobe became dependent for its survival on the respiratory capacity of the symbiotic aerobe. The ingested aerobe presumably benefited as well, because the host cell brought in large quantities of food molecules to be oxidized. This gradual process of mutual adaptation culminated in transformation of the cytoplasmic aerobe into a mitochondrion. The first eukaryotic cells had appeared, the ancestors of all modern-day eukaryotes.

The endosymbiotic theory also proposes that a similar mechanism led to the appearance of the membrane-bound **plastids** (the general term for chloroplasts and related organelles, both photosynthetic and nonphotosynthetic; see Section 4.4) some time after mitochondria evolved. A plastid originated when an aerobic cell that had a mitochondrion and was therefore capable of efficient cellular respiration, but was unable to carry out photosynthesis, ingested one or more photosynthetic prokaryotes resembling present-day cyanobacteria (see Figure 25.6). Cells that had engulfed photosynthetic prokaryotes were now capable of photosynthesis and no longer needed to ingest simple carbohydrates for energy. That is, the engulfed bacteria within them produced the carbohydrates from carbon dioxide and water, with energy provided by sunlight. A photosynthetic prokaryote gradually changed into a plastid by evolutionary processes similar to those that produced a mitochondrion. A cell with both a plastid and a mitochondrion founded the cell lines that gave rise to the modern eukaryotic algae and plants.

Several Lines of Evidence Support the Endosymbiotic Theory

A lot of evidence supports the endosymbiotic theory, including the following:

1. Mitochondria and plastids are similar in size to bacteria.
2. Both organelles divide by a process similar to binary fission (see Section 10.5).
3. Both organelles typically contain circular DNA molecules that closely resemble bacterial chromosomes.
4. The genomes of both organelles contain codes for ribosomes that resemble those found in bacteria rather than those found in eukaryotes. The rRNA components of the ribosomes are encoded by organelle genes.
5. Both organelles are surrounded by two or more membranes, the innermost of which has a chemical composition similar to that of a bacterial plasma membrane.
6. Chloroplasts resemble cyanobacteria in their internal structure, including the presence of particular chlorophylls and the existence of thylakoids. Moreover, DNA sequence analysis indicates that chloroplast DNA most closely resembles the DNA in modern cyanobacteria, with many of the same genes, supporting the conclusion that an ancient cyanobacterium was the ancestor of chloroplasts.

Another line of evidence supports a key assumption of the endosymbiont hypothesis by showing that engulfed cells or organelles can survive in the cytoplasm of the ingesting cell. Among animals, no less than 150 living genera, distributed among 11 phyla, include species that contain eukaryotic algae or cyanobacteria as residents in the cytoplasm of their cells. For example, larvae of the marine slug *Elysia* initially contain no chloroplasts, but after they begin feeding on algae, chloroplasts from the algal cells are taken up into the cells lining the gut. When the larvae develop into adult snails, the chloroplasts continue to carry out photosynthesis in their new location and produce carbohydrates that are used by the snails. Some of the chloroplast proteins synthesized during this time are encoded by genes in the slug's nuclear genome. The uptake of functional chloroplasts has also been observed among the protists; **Figure 25.8** shows a protist with chloroplasts that closely resemble cyanobacteria. (Protists are discussed in Chapter 27.)

How long did it take for evolutionary mechanisms to produce fully eukaryotic cells? The oldest known convincing microfossil eukaryotes are 1.5 billion years old. If prokaryotic cells first evolved some 3.5 billion years ago, it took up to 2 billion years for

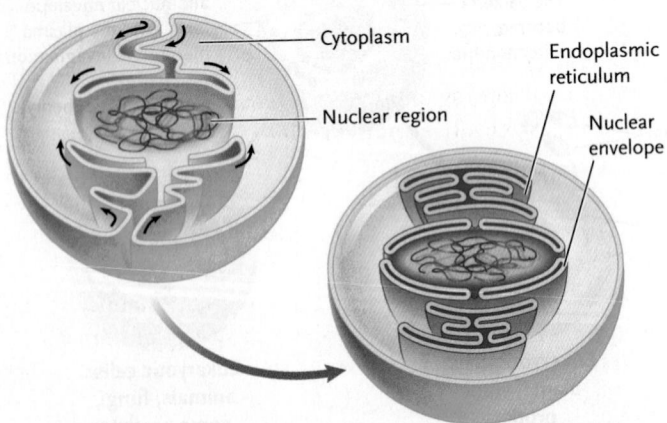

Cytoplasm

Endoplasmic reticulum

Nuclear region

Nuclear envelope

FIGURE 25.7 A hypothetical route for formation of the nuclear envelope and endoplasmic reticulum, through segments of the plasma membrane that were brought into the cytoplasm by endocytosis.

© Cengage Learning 2017

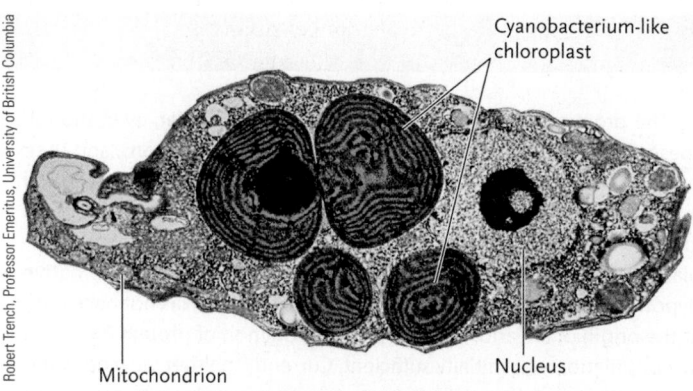

Cyanobacterium-like chloroplast

Mitochondrion

Nucleus

FIGURE 25.8 *Cyanophora paradoxa,* **a protist with chloroplasts that closely resemble cyanobacteria without cell walls.**

Robert Trench, Professor Emeritus, University of British Columbia

eukaryotic cells to evolve from prokaryotes (see Figure 25.1). If so, this long interval probably reflects the complexity of the adaptations leading from prokaryotic to eukaryotic cells. Of course, it is possible that eukaryotic cells evolved more quickly, and we have yet to find the supporting evidence.

Eukaryotic Cells May Have Evolved from a Common Ancestral Line Shared with Archaeans

The system of classification that has gained acceptance among biologists, and the one used in this book, groups all living organisms into three domains. One domain, Eukarya, contains the eukaryotes. The second domain, Bacteria, includes one of two groups of prokaryotes, the bacteria, which consists of both photosynthesizing and nonphotosynthesizing species. The third domain, Archaea, contains the other group of prokaryotes, the archaeans, many of which inhabit extreme environments, including highly saline environments and hot springs.

There is little question that the three domains originated from a common ancestral cell line. However, the events leading from this common ancestry to the three domains of life remain unclear. The most difficult questions surround the role of the archaeans in both bacterial and eukaryotic evolution.

Archaeans have some features that are typical of bacteria, including a circular DNA genome in a nucleoid region without a surrounding nuclear envelope, and no membrane-bound organelles in the cytoplasm. However, the archaeans also have some features that are typically eukaryotic, including the presence of introns (see Section 15.3) in some of their genes. The archaeans also have some characteristics that are unique to their domain, including features of gene and rRNA sequences, and features of cell wall and plasma membrane structure that are found nowhere else among living organisms. The characteristics shared by archaeans and eukaryotes suggest that their roots may lie in a common ancestral line that split off from the line leading to bacteria (see Figure 24.15). At some point, this ancestral line split into the lines leading to Domain Archaea and Domain Eukarya.

Multicellular Eukaryotes Probably Evolved in Colonies of Cells

The first eukaryotes were unicellular. They are the ancestors of the present-day diversity of unicellular eukaryotes. Multicellular eukaryotes evolved from unicellular eukaryotes and then diverged to produce multiple eukaryotic lineages. Molecular clock analysis indicates the first multicellular eukaryote likely arose between 800 and 1,000 million years ago, while the first fossil records (of small algae) are from 600 to 800 million years ago.

According to the prevalent theory, multicellular eukaryotes arose by the aggregation of cells of the same species into a colony. The ability to act in a coordinated way probably increased the capacity of colonies to adapt to changes in the environment. Subsequently, differentiation of cells into various specialized cell types with distinct functions produced organisms with a wider range of capabilities and adaptability. Cell differentiation in a colony would have required cell signals that affected gene expression. That is, because each cell in the colony has the same genome, the development of specific functions (phenotypes) would require intracellular signals that would change the program of gene regulation. (Gene regulation in eukaryotes is discussed in Section 16.4.) Over time, as genomes evolved, the division of function among cells led to the evolution of the tissues and organ systems of complex eukaryotes.

Multicellularity evolved several times in early eukaryotes, producing a number of lineages of algae as well as the ancestors of present day fungi, plants, and animals.

Life May Have Been the Inevitable Consequence of the Physical Conditions of the Primitive Earth

Some researchers maintain that the evolution of life on our planet was an inevitable outcome of the initial physical and chemical conditions established by Earth's origin, among them a reducing atmosphere (at least in some locations), a size that generates moderate gravitational forces, and a distance from the Sun that results in average surface temperatures between the freezing and boiling points of water. Given the same conditions and sufficient time, according to these scientists, it is inevitable that life has evolved or is evolving now on other planets in the universe.

The chapters that follow in this unit trace the course of evolution and its products after eukaryotic cells were added to the prokaryotes already living on Earth. Among prokaryotes, evolution established two major groups, the bacteria and archaea; among eukaryotes, further evolution established the protists, fungi, plants, and animals. The survey of life's diversity begins in the next chapter with a description of present-day bacteria and archaeans.

STUDY BREAK 25.3

What are the key points of the theory of endosymbiont origins for mitochondria and chloroplasts?

What was the first polymer of life?

As discussed in this chapter, many researchers favor RNA for this role. This is because it both self-replicates and can catalyze chemical reactions, and is neatly connected with the contemporary life based on nucleic acids and proteins. There are, however, several problems with this hypothesis. One of them is that synthesis of RNA and its building blocks, nucleotides, is quite difficult under primordial conditions. To circumvent this difficulty, several researchers proposed that other genetic polymers that are simpler to synthesize might have preceded RNA. For example, Albert Eschenmoser from the Swiss Federal Institute of Technology (ETH) in Zurich replaced the difficult-to-synthesize ribose portion of nucleotides by the six-membered sugar pyranose. These polymers are stable and form a double helix. Most recently, Philipp Holliger and his group at the MRC Laboratory of Molecular Biology, Cambridge, United Kingdom, showed that genetic information can be transferred between DNA and its analogs containing different sugars, xeno-nucleic acids. Another proposal is that the initial complement of nucleic acid bases was different than A, U, G, and C. This proposal is motivated by the poor stability of cytosine in water. Although we have no evidence that polymers alternative to contemporary nucleic acids were present on the early Earth, it is important to realize that such a possibility exists and might be used by life elsewhere.

The protein-first hypothesis is currently less popular, even though these polymers are excellent catalysts of chemical reactions, and their building blocks, amino acids, existed on prebiotic Earth and are found in relatively large quantities in meteorites. This is because there is no known mechanism for proteins to self-replicate. Some researchers speculate that a limited replication of proteins is possible. An alternative hypothesis is that replication of individual polymers was not necessary at the origin of life and, instead, the reproduction of protein functions in a population was initially sufficient. Currently, neither view has sufficient experimental support, but as we learn more about the structure and functions of small proteins major surprises might be in store.

think like a scientist If RNA molecules were the first molecules capable of catalyzing biochemical reactions, why were they replaced by proteins?

Courtesy of Andrew Pohorille

Andrew Pohorille heads the NASA Center for Computational Astrobiology and Fundamental Biology at NASA's Ames Research Center. He is also professor of Chemistry and Pharmaceutical Chemistry at the University of California, San Francisco. For his work on the origin of life he was awarded the 2002 NASA Exceptional Scientific Achievement Medal.

REVIEW KEY CONCEPTS

For access to MindTap and additional study materials visit www.cengagebrain.com.

25.1 The Formation of Molecules Necessary for Life

- Living cells are characterized by a boundary membrane, one or more nucleic acid coding molecules in a nuclear region, a system for using the coded information to make proteins, and a metabolic system providing energy for those activities.

- Oparin and Haldane independently hypothesized that life arose *de novo* under the conditions they thought prevailed on the primitive Earth, including a reducing atmosphere that lacked oxygen. Reduction reactions, fueled by natural energy acting on the primitive atmosphere, produced organic molecules. Chemistry simulation experiments support the hypothesis that organic molecules would form under these conditions (Figure 25.2).

- Present thinking is that early Earth's atmosphere in fact was neutral (neither reducing nor oxidizing), which challenges Oparin and Haldane's hypothesis. One new theory proposes that life developed near vents in the seafloor that release geochemical energy.

25.2 The Origin of Cells

- Small organic molecules produced in the early Earth environment combined to form polymers, and some of these polymers were self-replicating. The leading candidate for these early molecular replicators is RNA, because it can serve as a template to direct the synthesis of new RNA molecules with complementary nucleotide sequences.

- Protocells are thought to have formed as lipid membranes assembled spontaneously around aggregates of self-replicating RNA molecules (Figure 25.3).

- The growth and division of protocells containing self-replicating RNA molecules required a continuous release of energy and a supply of simple organic molecules, both of which could have been provided by alkaline vents in the seafloor (Figure 25.4).

- Eventually, a ribosome-based mechanism evolved that synthesized polypeptides using the information coded in RNA sequences, and those polypeptides began to catalyze chemical reactions that had previously been catalyzed by RNA ribozymes.

- Once DNA replaced RNA as the long-term store of genetic information, the complete genetic information system found in all cellular life today had evolved.

- The first living cells were prokaryotes. Eventually, some early cells developed the capacity to carry out photosynthesis using water as an electron donor; the oxygen produced as a by-product accumulated and the oxidizing character of Earth's atmosphere increased. From this time on organic molecules produced in the environment were quickly broken down by oxidation, and life could arise only from preexisting life, as in today's world.

25.3 The Origins of Eukaryotic Cells

- According to the endosymbiotic theory, mitochondria evolved from ingested prokaryotes that were capable of using oxygen as final electron acceptors; chloroplasts evolved from ingested cyanobacteria (Figure 25.6).

- Eukaryotic structures such as the ER, Golgi complex, and nuclear envelope appeared through infoldings of the plasma membrane as a part of endocytosis (Figure 25.7).

- Multicellular eukaryotes probably evolved by differentiation of cells of the same species that had aggregated into colonies. Multicellularity evolved several times, producing lineages of several algae and ancestors of fungi, plants, and animals.

TEST YOUR KNOWLEDGE

Remember/Understand

1. Earth was formed ____ years ago, whereas the oldest known living cell formed about ____ years ago.
   a. 400×10^3; 3.6×10^6
   b. 4.6×10^9; 1.0×10^9
   c. 3.8×10^9; 4.6×10^7
   d. 4.6×10^9; 3.5×10^9
   e. 2.0×10^9; 600×10^6

2. Which of the following is *not* a characteristic of all living organisms?
   a. They replicate genetic information and convert the information into proteins.
   b. They pass genetic information between generations.
   c. They have an internal metabolic system that provides energy for biological activity.
   d. They use external energy to drive internal reactions that require energy.
   e. They use mitochondria for synthesis of ATP.

3. Evidence for the appearance of the following is separated by the greatest amount of time:
   a. nonlife: prokaryotes
   b. photosynthetic prokaryotes: first eukaryotes
   c. nonphotosynthetic bacteria: photosynthetic bacteria
   d. unicellular eukaryotes: multicellular eukaryotes
   e. first prokaryotic cells: first eukaryotic cells

4. According to the Oparin–Haldane hypothesis, the atmosphere when life began was believed to be composed primarily of the following substances that contain the maximum number of electrons:
   a. H_2O, N_2, and CO_2.
   b. H_2, H_2O, NH_3, and CH_4.
   c. H_2O, N_2, O_2, and CO_2.
   d. O_2 and no H_2.
   e. H_2 only.

5. The Miller–Urey experiment:
   a. was based on the understanding that the atmosphere was oxidizing.
   b. was able to synthesize amino acids and macromolecules from reduced gases.
   c. did not require much energy or a continuous energy source to keep synthesizing.
   d. did not require water to produce organic molecules.
   e. used free oxygen as a reactant.

6. An unknown organism was found in a park. It was unicellular, had no nuclear membrane around its DNA, and contained no mitochondria and no chloroplasts. It is most likely a:
   a. eukaryote.
   b. vertebrate or plant.
   c. bacteria or archaean.
   d. plant or fungi.
   e. virus.

7. In a population of self-replicating RNA molecules, natural selection would favor those molecules that:
   a. have longer nucleotide sequences.
   b. have shorter nucleotide sequences.
   c. replicate more rapidly.
   d. replicate with a higher mutation frequency.
   e. are transported across lipid membranes.

8. Which of the following properties of living cells was most likely *not* present in early membrane-bound protocells?
   a. Their membranes were permeable to some molecules but impermeable to others.
   b. They contained one or more self-replicating molecules containing genetic information.
   c. They performed metabolic reactions using energy provided by molecules with relatively high chemical potential energy.
   d. Their membranes were composed of phospholipid molecules.
   e. They were filled with an aqueous solution.

9. As part of the evolution of eukaryotic cell, endocytosis, the process of infolding of the plasma membrane, led to the formation of:
   a. chromosomes.
   b. the cell wall.
   c. ribosomes.
   d. the nuclear envelope.
   e. microtubules.

10. Which of the following is *not* part of the evidence supporting the endosymbiotic theory? Both mitochondria and plastids:
    a. are each the size of many bacterial cells.
    b. have inner membranes with a chemical composition similar to that of a bacterial plasma membrane.
    c. have ribosomes that are similar to bacterial ribosomes.
    d. contain circular DNA.
    e. have DNA similar to nuclear DNA.

11. **Discuss Concepts** What evidence supports the idea that life originated through inanimate chemical processes?

Apply/Analyze

12. **Discuss Concepts** Explain, in terms of hydrophilic and hydrophobic interactions, how protocells might have formed in water from aggregations of lipids, proteins, and nucleic acids.

13. **Discuss Concepts** What conditions would likely be necessary on a planet located elsewhere in the universe for life to originate in a way similar to that which occurred on Earth?

14. **Apply Evolutionary Thinking** In the evolution unit, you learned how changes in the environment can foster evolutionary changes in biological systems. How have changing biological systems influenced the evolution of changes in Earth's physical environment?

Evaluate/Create

15. **Discuss Concepts** Most scientists agree that life on Earth can arise only from preexisting life, but also that life could have originated spontaneously on the primordial Earth. Can you reconcile these seemingly contradictory statements?

16. **Design an Experiment** Suppose you discover a hot spring–fed pool on a remote mountain that has never before been explored by humans. In the pool you find a cellular life form that appears to be prokaryotic. What experiments would you do to distinguish between the alternative hypotheses that this organism evolved on Earth from ancestral prokaryotes, or is descended from a life form that arrived at that location in a meteorite?

For selected answers, see Appendix A.

INTERPRET THE DATA

Studies of ancient rocks and fossils can provide insight into many of the events in Earth's history, including changes in its atmosphere that have taken place over geological time. The **Figure** shows how asteroid impacts and the composition of the atmosphere have changed over time. Use this figure and information in the chapter to answer the following questions.

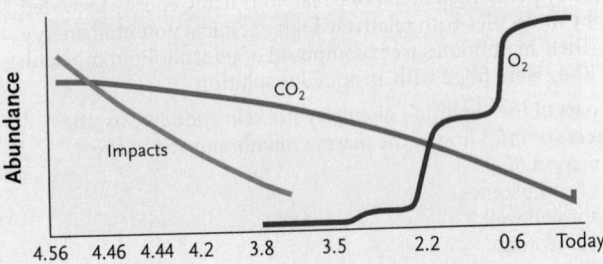

FIGURE Changes over geological time in asteroid impacts (green), atmospheric carbon dioxide concentration (pink), and oxygen concentration (blue).

© Cengage Learning 2017

1. Which occurred first, a decline in asteroid impacts or a rise in the atmospheric level of oxygen?

2. How do modern levels of carbon dioxide and oxygen compare to those at the time when the first cells arose?

3. Which is now more abundant, oxygen or carbon dioxide?

Prokaryotes: Bacteria and Archaea

PTP/Phototake

The bacterium *Clostridium butyricum,* one of the *Clostridium* species that produces the toxin botulin (colorized TEM).

Why it matters . . . You wait in line at a fast-food restaurant, eagerly anticipating your hamburger. In the back of your mind you may worry because, not too many years ago, people were sickened, and a few even died, because their fast-food hamburgers were contaminated by a pathogenic strain of the bacterium *Escherichia coli,* the normally harmless bacteria that inhabit our intestinal tract. Since then, fast-food restaurants have cooked their hamburgers well beyond the point required to kill any lurking pathogenic *E. coli* or other pathogenic species of bacteria.

The bacterium *E. coli* is a prokaryote, an organism lacking a true nucleus. Prokaryotes, the topic of this chapter, are the smallest organisms in the world **(Figure 26.1).** Few species are more than 1 to 2 μm long; from 500 to 1,000 of them would fit side by side across the dot above this letter "i."

Prokaryotes are small, but their total collective mass (their *biomass*) on Earth may be greater than that of all plant life. They colonize every niche on Earth that supports life, meaning essentially everywhere. Huge numbers of bacteria inhabit surfaces and cavities of the human body, including the skin, the mouth and nasal passages, the large intestine, and the vagina. Collectively, bacteria in and on the human body outnumber the body's own cells by about threefold.

On the basis of molecular phylogenetics analyses, biologists classify prokaryotes into two of the three domains of life, the **Bacteria** and the **Archaea** (see Section 24.7 and Figure 24.15). The third domain, the **Eukarya,** includes all eukaryotes. **Table 26.1** compares bacteria, archaea, and eukaryotes. Bacteria are the prokaryotic organisms most familiar to us, including many types responsible for diseases of humans and other animals and many other types found in a wide variety of ecosystems. Many members of the domain Archaea (*arkhaios* = ancient) live under conditions so extreme, including high salinity, acidity, or temperature, that their environments cannot be tolerated by other organisms, including bacteria. In general, archaeans are less well studied, and therefore less well understood, than bacteria.

Table 26.1	Characteristics of the Bacteria, Archaea, and Eukarya		
Characteristic	Bacteria	Archaea	Eukarya
DNA arrangement	Single, circular in most, but some linear and/or multiple	Single, circular	Multiple linear molecules
Chromosomal proteins	Prokaryotic histonelike proteins	Five eukaryotic histones	Five eukaryotic histones
Genes arranged in operons	Yes	Yes	No
Nuclear envelope	No	No	Yes
Mitochondria	No	No	Yes
Chloroplasts	No	No	Yes
Peptidoglycans in cell wall	Present	Present but modified, or absent	Absent
Membrane lipids	Unbranched; linked by ester linkages	Branched; linked by ether linkages	Unbranched; linked by ester linkages
RNA polymerase	One type	Multiple types	Multiple types
Ribosomal proteins	Prokaryotic	Some prokaryotic, some eukaryotic	Eukaryotic
First amino acid placed in proteins	Formyl-methionine	Methionine	Methionine
Aminoacyl–tRNA synthetases	Prokaryotic	Eukaryotic	Eukaryotic
Cell division proteins	Prokaryotic	Prokaryotic	Eukaryotic
Proteins of energy metabolism	Prokaryotic	Prokaryotic	Eukaryotic
Sensitivity to chloramphenicol and streptomycin	Yes	No	No

As a group, prokaryotes have a wide range of metabolic capabilities. Their metabolic activities are crucial for maintenance of the biosphere. In particular, prokaryotes are the key players in the life-sustaining recycling of the elements carbon,

A. 70× magnification **B. 14,000× magnification**

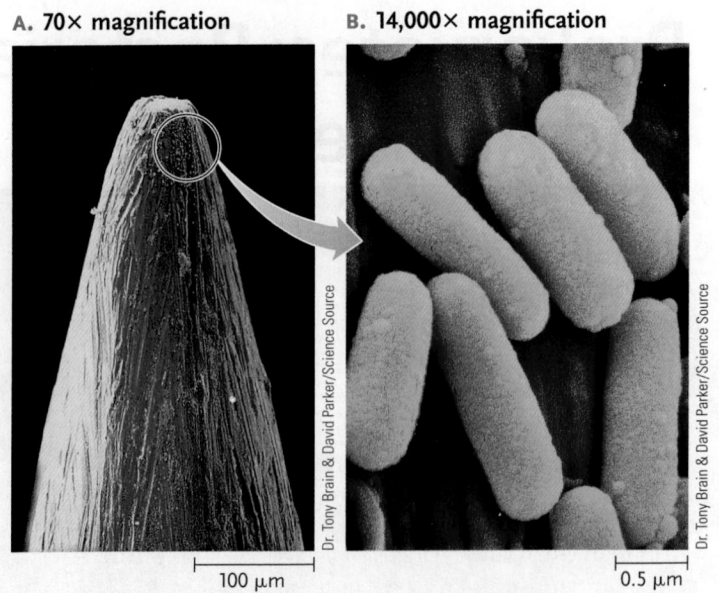

100 μm   0.5 μm

FIGURE 26.1 *Bacillus* bacteria (color-enhanced) on the point of a pin.

nitrogen, and oxygen. For example, prokaryotes (along with other organisms) break down organic material in dead plants and animals, releasing carbon dioxide that is used for plant growth. In addition, prokaryotes are necessary to make nitrogen available to most forms of life. And a significant amount of the oxygen in the atmosphere originates from bacterial photosynthesis. Overall, prokaryotes have an essential role in enabling life of all forms to exist.

Prokaryotes also have specific effects on the lives of humans, both beneficial and harmful. Among other beneficial effects, they are important for the production of certain foods; they carry out chemical reactions that are of importance in industry; they are used for the production of pharmaceutical products; and they are used for bioremediation of polluted sites. Harmful effects include causing certain diseases, many of which are discussed throughout the chapter.

In this chapter you will learn about the structure and function of prokaryotes, and about the diverse organisms in the evolutionary branches of Bacteria and Archaea.

26.1 Prokaryotic Structure and Function

Prokaryotes show great diversity in their ability to colonize areas that can sustain life. Their cells are small and, although prokaryotes show a greater structural diversity than eukaryotes, prokaryotic cells are not as complex in organization as eukaryotic cells. For instance, although prokaryotes do not have a membrane-bound nucleus or organelles, their DNA and some proteins are localized in particular places. They vary in how their cell membrane is protected, and some species have specialized surface structures that protect them from their environment or that enable them to move actively. Prokaryotes also show great diversity in the ways they obtain energy and in their metabolic activities.

The diversity of prokaryotes has arisen through rapid adaptation to their environments as a result of evolution by natural selection. Genetic variability in prokaryotic populations, the basis for the rapid adaptation, derives largely from mutation and to a lesser degree from the transfer of genes between organisms by horizontal gene transfer, that is, by transformation, transduction, and conjugation (see Chapter 17). Prokaryotes have much shorter generation times than eukaryotes, resulting in much more rapid reproduction. Prokaryotes also have smaller genomes (roughly 1,000 times smaller than an average eukaryote), and they have roughly 1,000 times more mutations per gene, per unit time, per individual, than do eukaryotes. Furthermore, prokaryotes typically have much larger population sizes than eukaryotes, which contributes to their greater genetic variability. In short, prokaryotes have an enormous capacity to adapt, and this has been the key to their evolutionary success.

Prokaryotes Are Different in Structure and Organization from Eukaryotic Cells

Like eukaryotic cells, prokaryotic cells are bounded by a plasma membrane. Prokaryotic cells also have a cell wall, which provides reinforcement and protection. Within the cell, the genetic material of prokaryotes is localized to a non-membrane-enclosed area of the cell, rather than being in a membrane-bound nucleus as in eukaryotes. Many prokaryotic species contain few if any internal membranes. They lack, therefore, the membrane-bound organelles characteristic of eukaryotic cells, including mitochondria, endoplasmic reticulum, Golgi complex, and chloroplasts. With few exceptions, the reactions carried out by these organelles in eukaryotes are distributed in

prokaryotes between the cytoplasmic solution and the plasma membrane.

As you learned in Chapter 4 both prokaryotic and eukaryotic cells have cytoskeletons. Prokaryotic cytoskeletons play important roles in creating and maintaining the proper shape of cells, in cell division, and, in certain prokaryotes, in determining polarity of the cells.

Although prokaryotic cells appear relatively simple according to this description, their simplicity is deceptive. Most can use a variety of substances as energy and carbon sources, and they can synthesize almost all of their required organic molecules from simple inorganic raw materials. In many respects, prokaryotes are more biochemically versatile than eukaryotes.

The following specifics of prokaryotic structure illustrate some of the diversity seen in this group of organisms.

SHAPE Three shapes are common among prokaryotes: spheres, rods, and spirals **(Figure 26.2).** The spherical prokaryotes are **cocci** (singular, *coccus*) (*kokkos* = berry). Cylindrical or rod-shaped prokaryotes are **bacilli** (singular, *bacillus*) (*bacillus* = small staff or rod). The spiral prokaryotes are **spirilla** (singular, *spirillum*) (*speira* = coil, twist, wreath), which are twisted helically like a corkscrew, and **vibrios** (*vibrare* = to move quickly to and fro), which are curved and commalike. Among the prokaryotes of all structural types are some that live singly and others that link into chains or aggregates of cells.

INTERNAL STRUCTURES The genome of most prokaryotes consists of a single, circular DNA molecule called the *prokaryotic chromosome*. A few bacterial species, for example the causative agent of Lyme disease (*Borrelia burgdorferi*), have a linear chromosome. Genome sequencing projects have shown that the genomes of bacteria and archaeans vary widely in size (see Figure 19.5). The smallest genome of any organism that can be

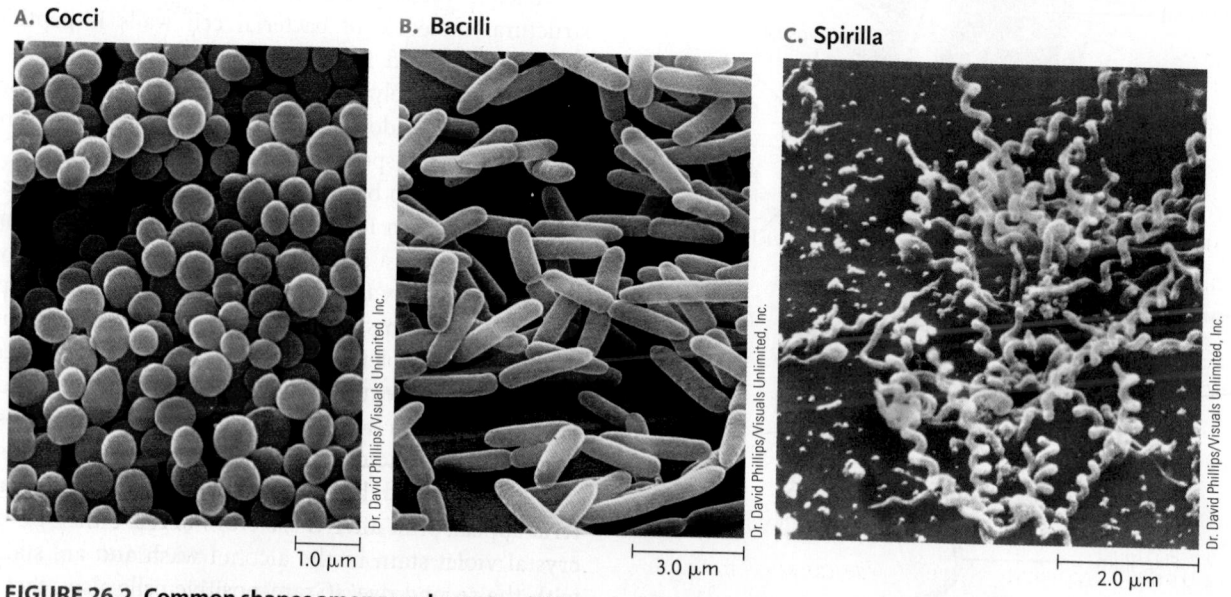

A. Cocci   **B. Bacilli**   **C. Spirilla**

1.0 µm   3.0 µm   2.0 µm

Dr. David Phillips/Visuals Unlimited, Inc.

FIGURE 26.2 Common shapes among prokaryotes. (A) Scanning electron microscope (SEM) image of cocci of *Micrococcus* bacteria (color-enhanced); **(B)** SEM image of bacilli of *Salmonella* bacteria (color-enhanced); **(C)** SEM image of spirilla of *Spiroplasma* bacteria.

cultured in the laboratory, that of the endosymbiotic bacterium found in the leafhopper insect, *Nasuia deltocephalinicola*, is about 112,000 bp and codes for only 127 proteins. The first prokaryote whose genome was sequenced completely, the parasitic bacterium *Mycoplasma genitalium*, has a genome of about 580,000 bp.

In all prokaryotes, the chromosome is packed into a central area of the cell called the **nucleoid.** The nucleoid has no nucleolus, and it has no boundary membranes equivalent to the nuclear envelope of eukaryotes (**Figure 26.3;** and see Sections 4.2 and 4.3). The arrangement of the genetic material in prokaryotes permits the translation of mRNA transcripts of protein-coding genes to begin while transcription is in progress. This is possible because the mRNA is immediately accessible to ribosomes, the cellular structures on which translation takes place. Simultaneous transcription and translation is not possible in eukaryotes because transcription and mRNA processing must be completed in the nucleus, and the mRNA must enter the cytosol where the ribosomes are before translation can begin (see Section 15.3).

Besides the DNA of the nucleoid, many prokaryotes also contain small circles of DNA called **plasmids,** which are distributed in the cytoplasm. The plasmids, which often contain genes with functions that supplement those in the nucleoid,

contain a replication origin that allows them to replicate along with the nucleoid DNA and be passed on during cell division (see Section 14.5).

Prokaryotic ribosomes are smaller than eukaryotic ribosomes and contain fewer proteins and RNA molecules. Archaeal ribosomes resemble those of bacteria in size, but differ in structure. Scientists have demonstrated that, with some differences in detail, bacterial ribosomes carry out protein synthesis by the same mechanisms as those of eukaryotes (see Section 15.4). Protein synthesis in archaeans is a combination of bacterial and eukaryotic processes, with some unique archaeal features. As a result, antibiotics that stop bacterial infections by targeting ribosome activity do not stop protein synthesis of archaeans.

Some prokaryotes are capable of photosynthesis (see Chapter 8). Photosynthetic bacteria and archaeans have complex layers of intercellular membranes formed by invaginations of the plasma membrane on which photosynthesis takes place. Those membranous structures correspond to those that carry out photosynthesis in plants, but they are organized differently.

The cytoplasm of many prokaryotes also contains storage granules holding glycogen, lipids, phosphates, or other materials. The stored material is used as an energy reserve or a source of building blocks for synthetic reactions.

PROKARYOTIC CELL WALLS The plasma membrane of prokaryotes must withstand both high intracellular osmotic pressures and the action of natural chemicals in the environment that have detergent properties. Most prokaryotes have one or more layers of materials coating the plasma membrane to form a cell wall that provides the necessary protection, although the structure of the wall is different in bacteria and archaeans.

Bacteria typically are surrounded by a cell wall. The primary structural molecule of bacterial cell walls is **peptidoglycan,** a polymer formed from a polysaccharide backbone tied together by short polypeptides. (Polysaccharides are described in Section 3.2.) Peptidoglycan varies in chemical structure among different bacterial species.

Differences in bacterial cell wall composition are of clinical importance. In 1882, Hans Christian Gram, a Danish physician, developed a staining method to distinguish two types of bacteria in body fluids, each of which could cause pneumonia. In this **Gram stain technique,** bacteria are treated with the dye crystal violet and then with iodine, which fixes the dye to the cell wall. Next the bacteria are washed with alcohol, and then treated with a second stain, either fuchsin or safranin. **Gram-positive** bacteria appear purple after these steps because they have retained the crystal violet stain. **Gram-negative** bacteria appear pink after these steps because they have lost the crystal violet stain in the alcohol wash and are stained pink with the second dye. (Gram-positive cells also react with the second dye, but the stain does not affect the color imparted by the crystal violet.)

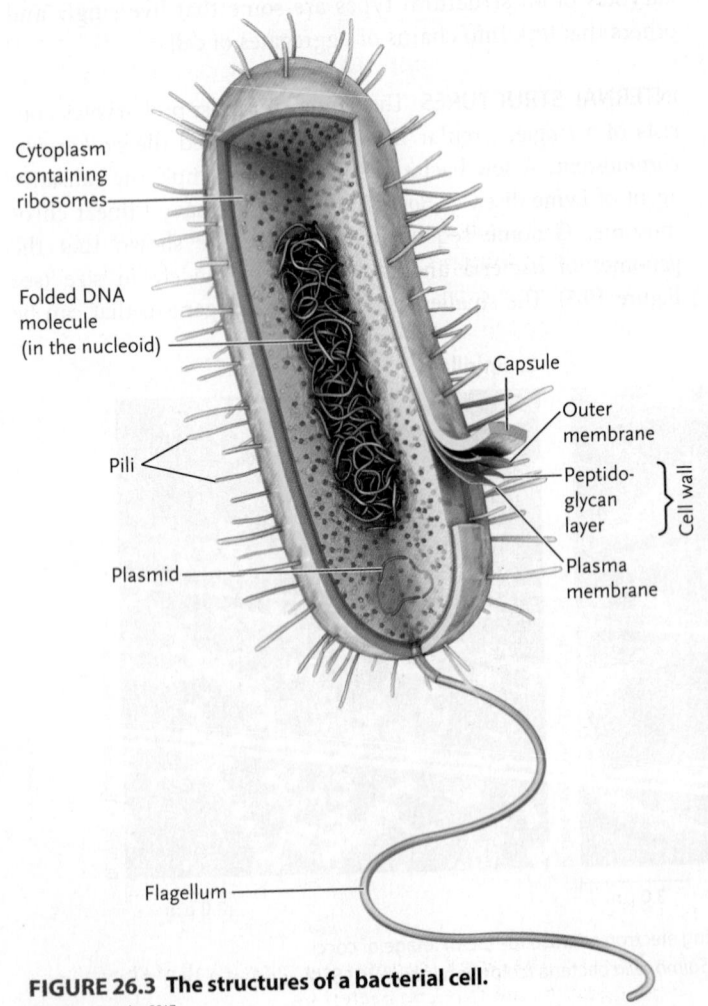

Cytoplasm containing ribosomes

Folded DNA molecule (in the nucleoid)

Pili

Plasmid

Flagellum

Capsule

Outer membrane

Peptido-glycan layer

Cell wall

Plasma membrane

FIGURE 26.3 The structures of a bacterial cell.
© Cengage Learning 2017

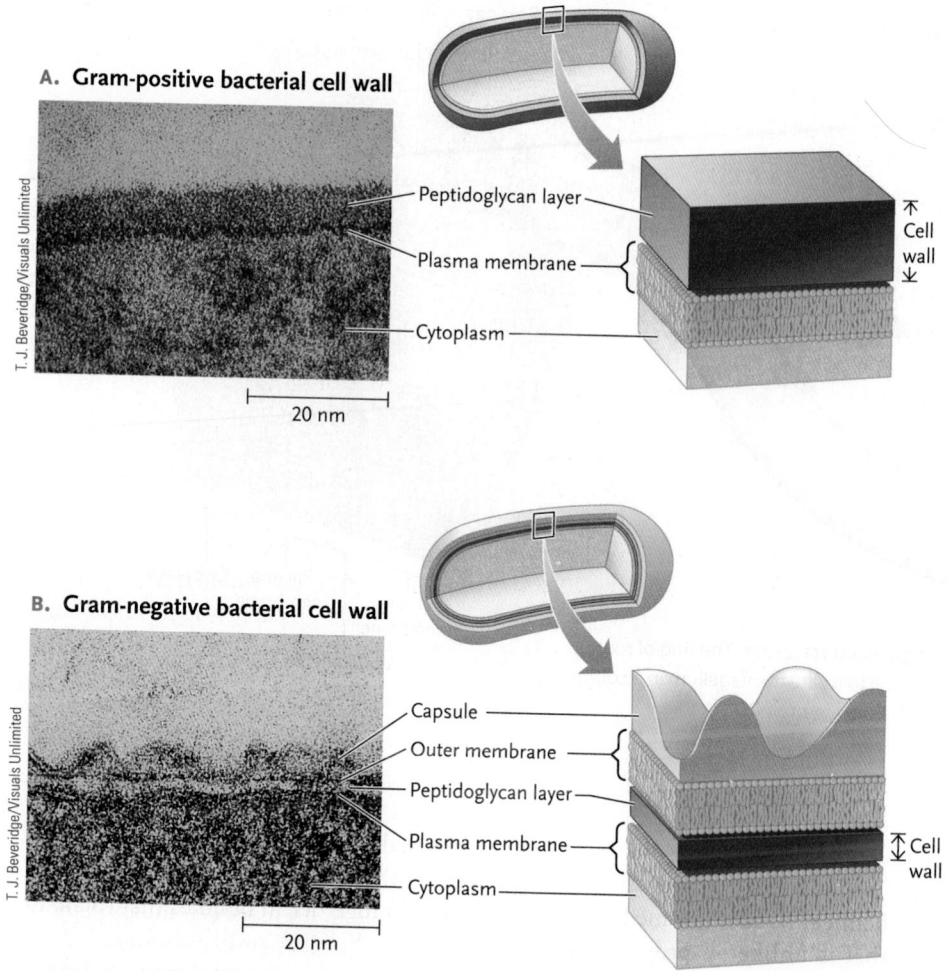

A. Gram-positive bacterial cell wall

T. J. Beveridge/Visuals Unlimited

Peptidoglycan layer

Plasma membrane

Cytoplasm

Cell wall

20 nm

B. Gram-negative bacterial cell wall

T. J. Beveridge/Visuals Unlimited

Capsule

Outer membrane

Peptidoglycan layer

Plasma membrane

Cytoplasm

Cell wall

20 nm

FIGURE 26.4 Cell wall structure in Gram-positive and Gram-negative bacteria. (Micrographs color-enhanced; drawings not to scale.)
© Cengage Learning 2017

The staining difference reflects differences in the cell walls of the bacteria **(Figure 26.4).** The cell wall of typical Gram-positive bacteria consists of a thick peptidoglycan layer (see Figure 26.4A). In contrast, the cell wall of typical Gram-negative bacteria consists of a thin layer of peptidoglycan (see Figure 26.4B). Covering the peptidoglycan layer is the **outer membrane,** an additional boundary membrane that contains **lipopolysaccharides** assembled from lipid and polysaccharide subunits found nowhere else in nature. The outer membrane protects Gram-negative bacteria from potentially damaging substances in their environment. For example, the outer membrane of *E. coli* protects it from the detergent effects of bile released into the intestinal tract (see Chapter 47), which otherwise would lyse ("break open") the bacterium and kill it.

Distinguishing rapidly between Gram-positive and Gram-negative bacteria is important for determining the first line of treatment for bacterial-caused human diseases. Most pathogenic bacteria are Gram-negative species; their outer membrane protects them against the body's defense systems and blocks the entry of drugs such as antibiotics. For example, the antibiotic penicillin is effective against Gram-positive pathogens, blocking new bacterial cell wall formation by inhibiting peptidoglycan

crosslinking. The weakened cell wall soon leads to the death of the bacterium. Penicillin is less effective against Gram-negative pathogens because their outer membrane inhibits entry of the antibiotic.

The walls of many Gram-positive and Gram-negative bacteria are coated with an external layer of polysaccharides called the **glycocalyx,** a slime coat typically composed of polysaccharides. When the glycocalyx is diffuse and loosely associated with the cells, it is a **slime layer;** when it is gelatinous and more firmly attached to the cells, it is a **capsule (Figure 26.5).** Depending on the species, the capsule ranges from a layer that is thinner than the cell wall to many times thicker than the entire cell. The glycocalyx helps protect cells from physical damage and desiccation and from the effects of some antibiotics, and may enable a cell to attach to a surface.

In many bacteria, the capsule prevents bacterial viruses and molecules such as enzymes, antibiotics, and antibodies from reaching the cell surface. In many pathogenic bacteria, the presence or absence of the protective capsule differentiates virulent ("infective") from nonvirulent ("noninfective") forms. For example, normal *Streptococcus pneumoniae* bacteria have capsules and are virulent, causing severe pneumonia in humans and other mammals. Mutant *S. pneumoniae* without capsules are nonvirulent and can easily be eliminated by the body's immune system if they are injected into mice or other animals (see Section 14.1).

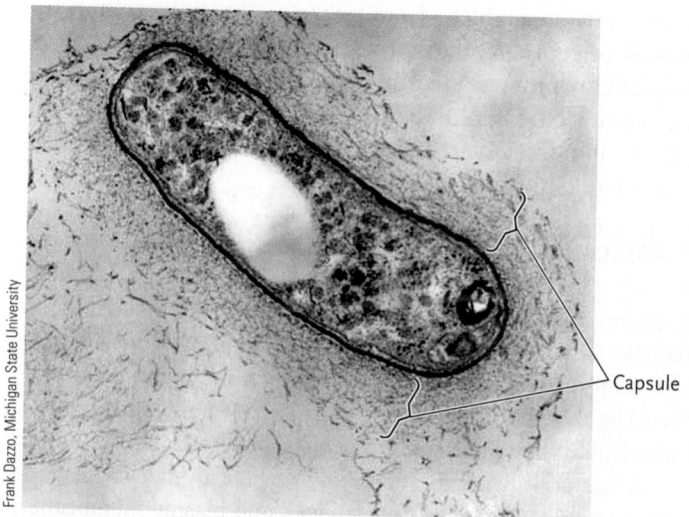

Frank Dazzo, Michigan State University

Capsule

FIGURE 26.5 The capsule surrounding the cell wall of *Rhizobium,* a Gram-negative, soil-dwelling bacterium.

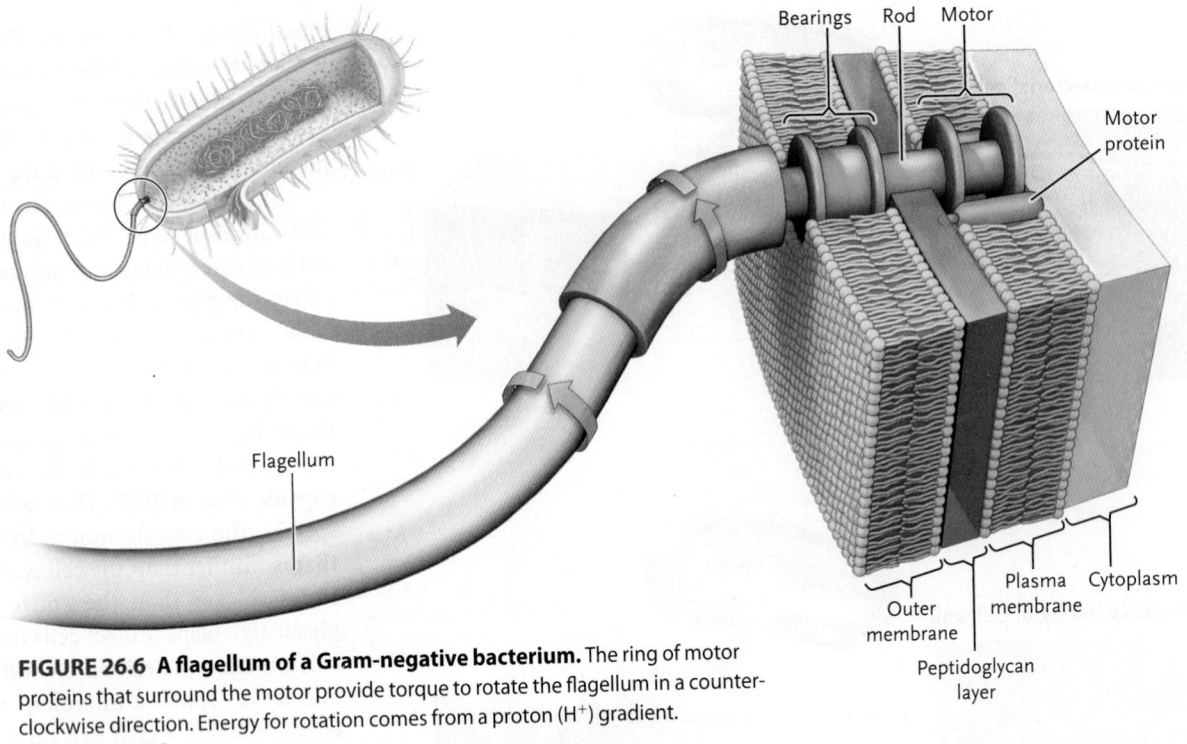

Bearings Rod Motor

Motor protein

Flagellum

Outer membrane

Peptidoglycan layer

Plasma membrane

Cytoplasm

FIGURE 26.6 A flagellum of a Gram-negative bacterium. The ring of motor proteins that surround the motor provide torque to rotate the flagellum in a counter-clockwise direction. Energy for rotation comes from a proton (H⁺) gradient.
© Cengage Learning 2017

Archaeans differ from bacteria by having some unique plasma membrane and cell wall characteristics. The lipid molecules in archaean plasma membranes have a chemical bond between the hydrocarbon chains and glycerol that is unlike that found in the plasma membranes of all other organisms. The exceptional linkage is more resistant to disruption, making the plasma membranes of archaeans more tolerant of the extreme environmental conditions in which many of these organisms live.

The cell walls of some archaeans are assembled from molecules related to the peptidoglycans, but that have different molecular components and bonding structure. Others have walls assembled from proteins or polysaccharides instead of peptidoglycans. The cell walls of archaeans are as resistant to physical disruption as the plasma membranes are; some archaeans can be boiled in strong detergents without disruption. Different archaeans stain as either Gram-positive or Gram-negative.

FLAGELLA AND PILI Many bacteria and archaeans can move actively through liquids and across wet surfaces. The most common mechanism for movement involves the action of **flagella** (singular, *flagellum* = whip) extending from the cell wall **(Figure 26.6)**. These flagella are much smaller and simpler than the flagella of eukaryotic cells and contain no microtubules (eukaryotic flagella are discussed in Section 4.3).

A bacterial flagellum consists of a helical fiber of protein that rotates in a socket in the cell wall and plasma membrane, much like the propeller of a boat. The rotation, produced by what is essentially a tiny electric motor, pushes the cell through liquid. The motor is powered by a proton gradient, which creates an electrical repulsion that makes the flagellum rotate. In some bacteria, flagella group together in bundles that rotate to propel the cell.

Archaeal flagella are analogous, not homologous, to bacterial flagella. That is, they carry out the same function, but they differ in their structure and mechanisms of action. The genes for the two types of flagellar systems are also different.

Some bacteria and archaeans have hairlike shafts of protein called **pili** (singular, *pilus*) extending from their cell walls **(Figure 26.7)**. There are two types of pili. **Common pili** (also called *fimbriae*; singular, *fimbria*), the common form among bacteria and archaeans, are relatively short, there are many per cell, and there are many subvarieties. Functionally, common pili are important for attaching a cell to surfaces, attaching a cell to another prokaryotic cell or to an animal cell, biofilm formation (discussed later in this chapter), cell motility, and the transport of protein and DNA across membranes. Among bacteria, common pili are characteristic primarily of Gram-negative bacteria; relatively few Gram-positive bacteria produce these structures. For example, *Neisseria gonorrhoeae*, the Gram-negative bacterium that causes gonorrhea, has pili that allow it to attach to cells of the throat, eye, urogenital tract, or rectum in humans. **Sex pili** are specialized structures which attach that cell to another bacterium lacking sex pili during conjugation, a joining of two cells that leads to transfer of genetic material from one to the other (see Section 17.1). Sex pili are long structures, there are only two or three of them per cell, and only particular bacterial species can produce them.

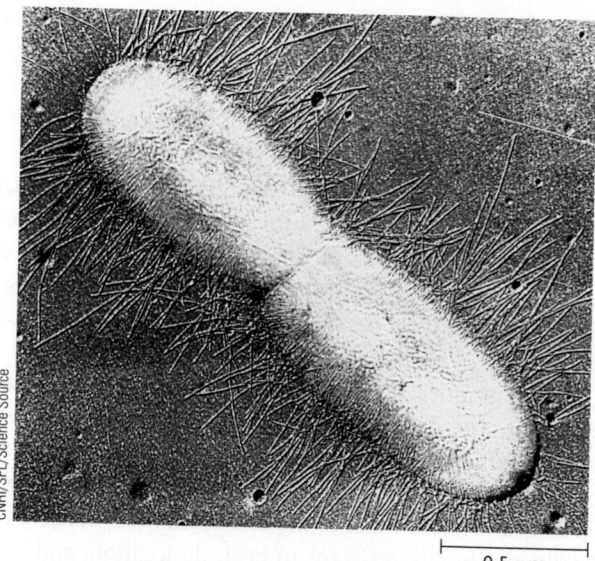

FIGURE 26.7 Common pili extending from the surface of a dividing *E. coli* bacterium (color-enhanced).

	Energy source	
	Oxidation of molecules*	**Light**
CO₂	**CHEMOAUTOTROPH** Found in some bacteria and archaeans; not found in eukaryotes	**PHOTOAUTOTROPH** Found in some photosynthetic bacteria, in some protists, and in plants
Organic molecules	**CHEMOHETEROTROPH** Includes some bacteria and archaeans, and also in protists, fungi, animals, and plants	**PHOTOHETEROTROPH** Found in some photosynthetic bacteria

(Left axis label: **Carbon source**; Top axis label spans "Oxidation of molecules*" and "Light")

*Inorganic molecules for chemoautotrophs and organic molecules for chemoheterotrophs.

FIGURE 26.8 Modes of nutrition among Bacteria and Archaea. All four modes of nutrition occur in the Bacteria, with chemoheterotrophs as the most common type. Among the Archaea, chemoautotrophs are most common, while other archaeans are chemoheterotrophs.

© Cengage Learning 2017

Prokaryotes Have the Greatest Metabolic Diversity of All Living Organisms

All organisms take in carbon and energy in some form, but prokaryotes show the greatest diversity in their modes of securing these resources **(Figure 26.8)**. Prokaryotes are classified on the basis of the source of carbon atoms they use. Some prokaryotes are **autotrophs** (*auto* = self; *trophe* = nourishment), meaning that they, like plants, obtain carbon from CO_2 (which is considered to be an inorganic molecule). Others are **heterotrophs,** meaning that they, like humans and other animals, obtain carbon from organic molecules. Prokaryotic heterotrophs obtain carbon from the organic molecules of living hosts, or from organic molecules in the products, wastes, or remains of dead organisms.

Prokaryotes are also classified according to the source of the energy they use to drive biological activities. **Chemotrophs** (*chemo* = chemical; *trophe* = nourishment) obtain energy by oxidizing inorganic or organic substances, whereas **phototrophs** obtain energy from light. Combining carbon source and energy gives us the following four types (see Figure 26.8):

1. **Chemoautotrophs:** Prokaryotic chemoautotrophs obtain energy by oxidizing inorganic substances such as hydrogen, iron, sulfur, ammonia, nitrites, and nitrates and use CO_2 as their carbon source. They use the electrons they remove in the oxidations to make organic molecules by reducing CO_2 or to provide the energy for ATP synthesis (using an electron transfer system embedded in the plasma membrane). Chemoautotrophy occurs widely among prokaryotes, including many bacteria and most archaeans, but is not found among eukaryotes.

2. **Chemoheterotrophs:** Prokaryotic chemoheterotrophs oxidize organic molecules as their energy source and obtain carbon in organic form. They include most of the bacteria that cause disease in humans, domestic animals, and plants and many bacteria that are responsible for decomposing matter. They are the largest prokaryotic group in terms of numbers of known species.

3. **Photoautotrophs:** Photoautotrophs are photosynthetic organisms that use light as their energy source and CO_2 as their carbon source. They include several groups of bacteria, for example, the *cyanobacteria,* as well as plants and many protists. The cyanobacteria use water (H_2O) as their source of electrons for reducing CO_2. Some other types of phototrophic bacteria use sulfur or sulfur compounds.

4. **Photoheterotrophs:** Photoheterotrophs use light as their ultimate energy source and preexisting organic carbon molecules as reducing agents, but they cannot use CO_2. Photoheterotrophs are limited to three subgroups of proteobacteria.

Prokaryotes Differ in Whether Oxygen Can Be Used in Their Metabolism

Prokaryotes also differ in how their metabolic systems function with respect to oxygen (see Chapter 7). **Aerobes** require oxygen for cellular respiration (in other words, oxygen is the final electron acceptor for that process). **Obligate aerobes** cannot grow without oxygen. **Anaerobes** do not require oxygen to live. **Obligate anaerobes** are poisoned by oxygen, and survive either by fermentation, in which organic molecules are the final electron acceptors (see Section 7.5), or by a form of respiration in which inorganic molecules such as nitrate ions (NO_3^-) or sulfate ions (SO_4^{2-}) are used as final electron acceptors. **Facultative anaerobes** use O_2 when it is present, but under anaerobic conditions, they live by fermentation.

Prokaryotes Fix and Metabolize Nitrogen

Nitrogen is a component of amino acids and nucleotides and, hence, is of vital importance for the cell. Prokaryotes are able to metabolize nitrogen in many forms. For example, a number of bacteria and archaeans are able to reduce atmospheric nitrogen (N_2, the major component of Earth's atmosphere) to ammonia (NH_3), a process called **nitrogen fixation.** The ammonia is ionized quickly to ammonium (NH_4^+), which the cell then uses in biosynthetic pathways to produce nitrogen-containing molecules such as amino acids and nucleic acids. Nitrogen fixation is an exclusively prokaryotic process and is the only means of replenishing the nitrogen sources used by most microorganisms and by all plants and animals. In other words, all organisms use nitrogen that is fixed by bacteria. Examples of nitrogen-fixing bacteria are some of the cyanobacteria and *Azotobacter* species, which are free-living, and *Rhizobium* species, which are in symbiotic relationships with plants (see Chapter 35).

Not all bacteria convert fixed nitrogen directly into organic molecules. Some bacteria carry out **nitrification,** the conversion of ammonium (NH_4^+) to nitrate (NO_3^-). This is carried out in two steps by two types of *nitrifying bacteria.* One type of nitrifying bacterium (for example, *Nitrosomonas* species) converts ammonium to nitrite (NO_2^-), whereas the other (for example, *Nitrobacter* species) converts nitrite to nitrate. Because of this specialization, both types of nitrifying bacteria are usually present in soils and water, with some converting ammonium to nitrite and others using that nitrite to produce nitrate. The nitrate can be used by plants and fungi to incorporate nitrogen into organic molecules. Animals obtain nitrogen in organic form by eating other organisms.

In sum, nitrification makes nitrogen available to many other organisms, including plants and animals and bacteria that cannot metabolize ammonia. You will learn more about nitrogen metabolism in connection with the nitrogen cycle (see Chapter 54). The metabolic versatility of the prokaryotes is one factor that accounts for their abundance and persistence on the planet; another factor is their impressive reproductive capacity.

Prokaryotes Pass Genes to Progeny by Binary Fission or, Rarely, from Cell to Cell by Horizontal Gene Transfer

In prokaryotes, reproduction occurs by binary fission in which a parent cell divides into two daughter cells that are exact genetic copies of the parent (see Figure 10.17). In this process, genetic material moves from generation to generation, that is, by descent.

Conjugation, in which two parent cells join or "mate," occurs in some bacterial and archaeal species. Conjugation depends on genes carried by a plasmid that replicates separately from the prokaryotic chromosome. Usually only the plasmid is passed on during conjugation, but in some bacteria, the plasmid integrates into the chromosome of the host so that host genes transfer from one parent (donor) to the other (recipient). Genetic recombination then occurs. The recombinant cell divides to produce daughter cells that differ in genetic information from either parent. The movement of genetic material between organisms by means other than descent, as is the case with conjugation, is **horizontal gene transfer.** Horizontal gene transfer by conjugation between bacterial cells is described in Section 17.1.

A small number of bacteria can produce an **endospore,** so called because it develops *within* the cell **(Figure 26.9).** The endospore, which typically develops when environmental conditions become unfavorable, is a dormant cell containing the bacterial genome and some of the cytoplasm. It is inactive metabolically and highly resistant to heat, desiccation, and attack by enzymes or other chemical agents. In the first step of endospore formation, binary fission cuts the parent cell into two parts of unequal size. The larger cell then envelops the smaller one (which contains the genome) and surrounds it with a tough, chemically resistant protein coat; the smaller cell develops into the endospore. Rupture of the larger cell releases the endospore to the environment. If environmental conditions become favorable for growth, the spore germinates: it becomes permeable, water enters the cell, its surface coat breaks, and the cell is released in a metabolically active form.

No one is certain how long endospores can survive. There are claims that endospores survive for thousands to millions of years, but the data are controversial. They certainly can survive a very long time.

STUDY BREAK 26.1

1. What distinguishes a prokaryotic cell from a eukaryotic cell?
2. What is the difference between a chemoheterotroph and a photoautotroph?
3. What is the difference between an obligate anaerobe and a facultative anaerobe?
4. What is the difference between nitrogen fixation and nitrification? Why are nitrogen-fixing prokaryotes important?

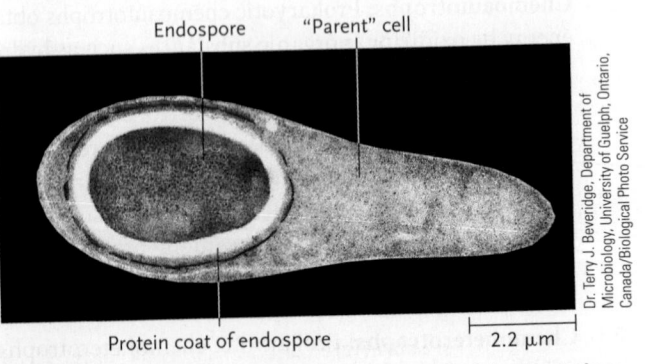

Endospore   "Parent" cell

Protein coat of endospore   2.2 μm

Dr. Terry J. Beveridge, Department of Microbiology, University of Guelph, Ontario, Canada/Biological Photo Service

FIGURE 26.9 A developing endospore within a cell of the bacterium *Clostridium tetani,* **a dangerous pathogen that causes tetanus.**

THINK OUTSIDE THE BOOK

On your own or collaboratively, research a study in which genomic experiments have shown marked changes in gene expression pattern when a bacterium forms a biofilm. Outline how the experiment was done, and summarize the results of the experiments.

26.2 The Domain Bacteria

Figure 26.10 presents a simplified phylogenetic tree of the domains Bacteria and Archaea (the Eukarya is the third domain of life). It shows five major evolutionary groups of the domain Bacteria, which are discussed in this section. Several other bacterial phyla are not included in these groups. The groups of the domain Archaea are discussed in the next section. Table 26.1 compares Bacteria, Archaea, and Eukarya.

Molecular Phylogenetics Analysis Determined the Evolutionary Groups of the Prokaryotes

The phylogenetic tree in Figure 26.10 is a model based on molecular phylogenetics analysis of DNA sequences of a number of prokaryotes, including hundreds of complete genome sequences. As Table 19.1 shows, more than 4,400 bacterial genomes have been sequenced. (Molecular phylogenetics is discussed in Chapter 24.) Phylogenetic tree construction involves using molecular sequence comparisons to hypothesize how observed variations in genes may be explained in terms of evolutionary lineages. An important assumption in such analysis is vertical inheritance, namely that genes are passed from parents to offspring. However, much more so for prokaryotes than for eukaryotes,

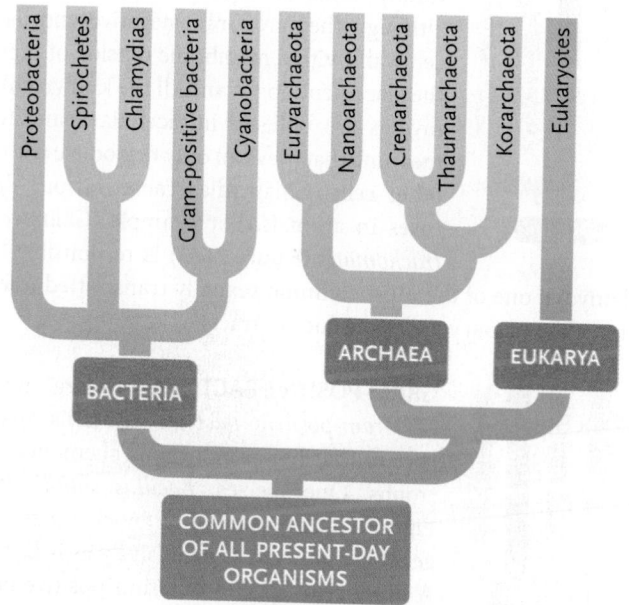

FIGURE 26.10 **An abbreviated and simplified phylogenetic tree of prokaryotes.**
© Cengage Learning 2017

evolution of genes also involves horizontal gene transfer (see Sections 17.1 and 26.1), which makes phylogenetic tree construction more difficult. That is, suppose a gene moved from ancient bacterium A to unrelated ancient bacterium B on a plasmid, and then subsequently moved to bacterium B's chromosome. Based on studying the sequences of the gene in present-day descendants, an investigator might conclude that the two bacterial species are closely related, whereas genome-level studies could show them to be distantly related. The bottom line is that creating one Tree of Life that depicts the evolution of all life forms accurately and completely likely is not possible and, especially relevant to the prokaryotes, we may never understand completely how the earliest prokaryotes and eukaryotes evolved.

The Five Major Evolutionary Groups of Bacteria Have Diverse Properties

The five major evolutionary groups of bacteria are the proteobacteria, spirochetes, chlamydias, and cyanobacteria, all of which are Gram-negative, and the Gram-positive bacteria (see Figure 26.11).

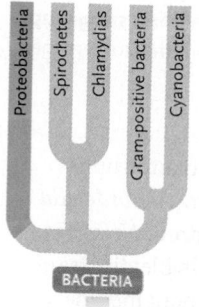

PROTEOBACTERIA The *proteobacteria* are the largest and most metabolically diverse group. These Gram-negative bacteria consist of more than one third of all bacterial species, and include most of the bacteria with medical and agricultural significance. They exhibit a range of metabolic capabilities, including chemoautotrophy, chemoheterotrophy, and photoautotrophy. Some are aerobes, and others are anaerobes. Based on comparisons of rRNA sequences, the proteobacteria are divided into five subgroups:

1. **Alphaproteobacteria.** The alphaproteobacteria consist of over 1,000 known species, and show considerable metabolic diversity. This subgroup includes many species that form symbiotic relationships with plants and animals, as well as some pathogens. For example, *Rhizobium* (see Figure 26.5) and *Agrobacterium* species infect the roots of leguminous plants to form nodules in which nitrogen fixation occurs. *Agrobacterium* infection has been developed as a method for producing transgenic plants (see Section 18.2). *Acetobacter* species convert ethanol into acetic acid, the acid component of vinegar. *Rickettsia* species are obligate intracellular parasites of mammalian cells. Rocky Mountain Spotted Fever is produced as a result of a tick-borne infection involving *Rickettsia*. Mitochondria are thought to have evolved from endosymbiotic aerobic alphaproteobacteria. (The endosymbiotic theory for mitochondrion and chloroplast evolution is described in Section 25.3.)

2. **Betaproteobacteria.** Betaproteobacteria consist of nearly 500 known species. They also are metabolically diverse and include a number of important pathogens. For

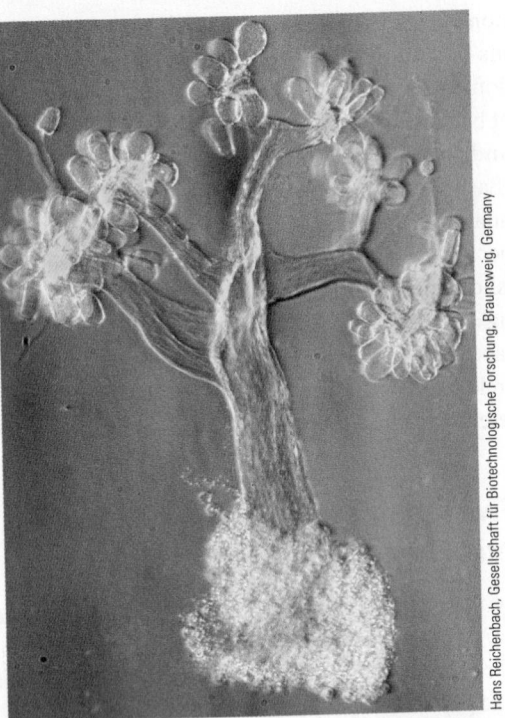

FIGURE 26.11 The fruiting body of *Chondromyces crocatus,* a myxobacterium, a member of the Deltaproteobacteria subgroup of the Proteobacteria. Cells of this species collect together to form the fruiting body.

Hans Reichenbach, Gesellschaft für Biotechnologische Forschung, Braunschweig, Germany

example, *Nitrosomonas* species are chemoautotrophs that carry out nitrification (see Section 26.1). *Bordetella pertussis* causes whooping cough in children, *Neisseria gonorrhoeae* causes the sexually transmissible disease gonorrhea, and *Neisseria meningitidis* causes meningococcal meningitis, an infection of the meninges (membranes that cover the brain and spinal cord).

3. **Gammaproteobacteria.** The gammaproteobacteria are the largest subgroup of the proteobacteria, also metabolically diverse, and consisting of more than 1,500 known species. This subgroup is well known for facultative anaerobes that inhabit mammalian colons, such as *E. coli* (see Figure 26.7). The subgroup also contains some important pathogens such as *Salmonella enterica typhi* (causes typhoid fever) (see Figure 26.2B), *Vibrio cholerae* (causes cholera), *Legionella* (causes legionellosis, one form of which is Legionnaire's disease), *Shigella dysenteriae* (causes dysentery), and *Haemophilus influenzae* (causes ear infections and meningitis, but not influenza). Some gammaproteobacteria are photosynthetic; typically they are anaerobic bacteria that use photosynthesis to produce carbohydrates using chlorophyll, but do not produce oxygen.

4. **Deltaproteobacteria.** The deltaproteobacteria consist of only a few species. They include a group of anaerobes that are involved in the cyclic interchanges of sulfur-containing compounds in the biosphere. Other species have complex life cycles that include multicellular developmental forms. The best-studied examples are the slime-secreting myxobacteria ("slime bacteria"). Myxobacteria glide actively over the slime, typically traveling in swarms of cells that stay together using an intercellular signaling system. Enzymes secreted by the swarms digest "prey"—other bacteria, primarily—that become stuck in the slime. When environmental conditions become unfavorable, as when soil nutrients or water are depleted, myxobacteria form a *fruiting body* **(Figure 26.11),** which contains clusters of spores. When the fruiting body bursts, the spores disperse and form new colonies.

5. **Epsilonproteobacteria.** This subgroup consists of a small group of bacteria that inhabit the animal digestive tract. Examples are *Campylobacter* and *Helicobacter* species, some of which are pathogenic. *C. jejuni* is a major cause of food-borne intestinal disease, and *H. pyloris* is the most common form of peptic ulcers **(Figure 26.12).**

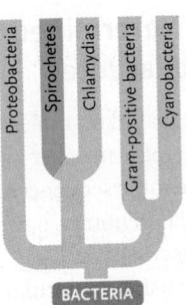

SPIROCHETES The *spirochetes* are Gram-negative bacteria with helically spiraled bodies and an unusual form of movement in which their flagella, which are embedded in the region between the outer membrane and the cell membrane, cause the entire cell to twist in a corkscrew pattern. Some spirochetes are pathogenic to humans. For example, *Treponema pallidum* causes syphilis, and *Borrelia burgdorferi* causes Lyme disease. Beneficial spirochetes in termite intestines aid in the digestion of plant fiber.

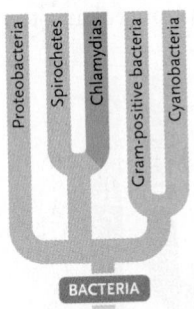

CHLAMYDIAS The *chlamydias* are structurally unusual among bacteria because, although they are Gram-negative and have cell walls with a membrane outside of them, they lack peptidoglycan. All the known chlamydias are obligate intracellular parasites, meaning that they can only reproduce within other cells. Chlamydias cause various diseases in animals. For example, *Chlamydia trachomatis* **(Figure 26.13)** is responsible for chlamydia, one of the most common sexually transmitted infections of the urinary and reproductive tracts of humans.

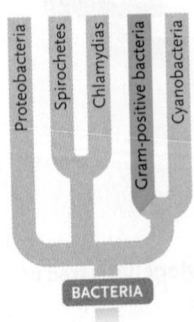

GRAM-POSITIVE BACTERIA The large group of *Gram-positive bacteria* contains many species that live primarily as chemoheterotrophs. One species, *Bacillus subtilis* (see Figure 26.1), is studied by biochemists and geneticists almost as extensively as is *E. coli*. A number of species of Gram-positive bacteria are pathogenic to humans, including *Bacillus anthracis,* which causes anthrax; *Staphylococcus aureus,* which causes a range

FIGURE 26.12 | Experimental Research

Demonstration of the Bacterial Cause of Peptic Ulcers

Question: Do bacteria cause gastric and duodenal ulcers?

Experiment: A gastric ulcer (also called a peptic ulcer) is an erosion of the mucosal layer of the stomach that can potentially penetrate into the deeper layers of the stomach wall. Pepsin and HCl in the stomach irritate the exposed stomach wall, causing extreme pain. Similar ulcers, called duodenal ulcers, can form in the duodenum. In 1982, physicians "knew" that gastric ulcers were caused by excessive amounts of stomach acid that was caused by stress, smoking, personality issues, or susceptibility genes. They treated an ulcer with various kinds of antacids or drugs to inhibit acid production in the stomach. Typically the ulcer symptoms returned after treatment was stopped. Before that time, various researchers had observed the presence of spiral or curved bacteria in biopsies taken from the majority of patients with gastric and duodenal ulcers. The bacteria were considered either to be contaminants or not related to ulcer formation.

In 1982, Barry Marshall and J. Robin Warren of Royal Perth Hospital, Perth, Western Australia, tested the hypothesis that bacteria caused peptic ulcers. They took biopsies of peptic ulcers and duodenal ulcers and assayed for the presence of bacteria by Gram staining and culturing on Petri plates.

Results: The results of the biopsies, published in 1984, are shown in the **Table.**

Association of Bacteria with Biopsy Samples

Biopsy Appearance	Total Samples	Number (%) Associated with Bacteria
Gastric ulcer	22	18 (77%)
Duodenal ulcer	13	13 (100%)
Total	35	31 (89%)

The researchers' results showed that almost all cases of gastric or duodenal ulcers were associated with bacteria. The bacteria had a curved appearance, but the species was unknown at the time. It is now known to be the epsilonproteobacterium *Helicobacter pylori*.

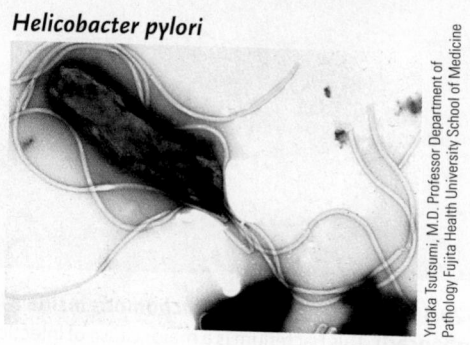

Helicobacter pylori

Yutaka Tsutsumi, M.D. Professor Department of Pathology Fujita Health University School of Medicine

Marshall and Warren's results did not persuade physicians that gastric and duodenal ulcers were bacterial diseases. How could it be shown that *H. pylori* can cause gastric and duodenal ulcers? The accepted way to prove that a bacterium is a pathogen is to fulfill a set of four conditions proposed in the nineteenth century by Robert Koch known as **Koch's postulates:**

1. The bacteria must be present in every case of the disease.

2. The bacteria must be isolated from the infected host and grown in pure culture.

3. The specific disease must be induced when a pure culture of the bacteria is introduced into a healthy host.

4. The bacteria must be isolated from the now experimentally infected host.

Postulates 3 and 4 had not been fulfilled in the study. In a bold and highly unusual move, Marshall decided to try to fulfill postulates 3 and 4 himself. He drank 200 mL of an *H. pylori* suspension, then fasted the rest of the day. Five to eight days later, Marshall began feeling very nauseated to the point of vomiting. On the tenth day, a colleague performed a biopsy, the analysis of which showed the presence of *H. pylori* cells. *Helicobacter pylori* was proven to be a pathogen.

Conclusion: Marshall and Warren's experiments had shown that *Helicobacter pylori* can cause gastric and duodenal ulcers. This bacterium probably accounts for more than 80% of cases of such ulcers in humans in the United States. In *H. pylori*-caused ulcers, the effective treatment is an antimicrobial agent such as an antibiotic. Marshall and Warren received a Nobel Prize in 2005 "for their discovery of the bacterium *Helicobacter pylori* and its role in gastritis and peptic ulcer disease."

think like a scientist

Observe
Hypothesize
Predict
Experiment
Interpret

Marshall developed a bacterial infection after drinking a solution of *Helicobacter*. Speculate on the origin of *Helicobacter* in patients who developed gastric or duodenal ulcers and were positive for the bacterium.

Sources: B. J. Marshall, J. A. Armstrong, D. B. McGechie, and R. J. Glancy. 1985. Attempt to fulfill Koch's postulates for pyloric *Campylobacter. Medical Journal of Australia* 142:436–439; B. J. Marshall and J. R. Warren. 1984. Unidentified curved bacilli in the stomach of patients with gastritis and peptic ulceration. *The Lancet* 1(8390):1311–1315.

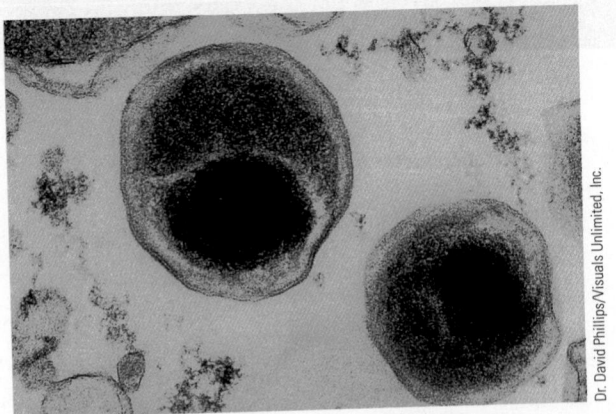

FIGURE 26.13 Cells of *Chlamydia trachomatis* inside a human cell (color-enhanced). This bacterium is a major cause of infectious eye and genital diseases in humans.

of illnesses including pimples, abscesses, toxic shock syndrome, pneumonia, meningitis and sepsis (an infection state involving a whole-body inflammation; commonly referred to as blood poisoning); and *Streptococcus pyogenes* **(Figure 26.14),** which causes of streptococcal pharyngitis ("strep throat"), scarlet fever, and rheumatic fever. Not all species of *Bacillus, Staphylococcus,* and *Streptococcus* are pathogenic, however. Some Gram-positive bacteria are beneficial. *Lactobacillus* species, for example, carry out the lactic acid fermentation used in the production of pickles, sauerkraut, yogurt, and some other fermented foods. Also, bacteria of the Firmicutes and Actinobacteria phyla are major components of bacterial communities in the mammalian large intestine, contributing, along with a variety of Gram-negative bacterial species, to digestion of gut contents (discussed later in this section and in Chapter 47). One unusual group of bacteria, the *mycoplasmas*, is classified in the group Gram-positive bacteria by molecular studies even though they lack a cell wall and, therefore, they react negatively to the Gram stain. Some species

are pathogenic to humans. *Mycoplasma pneumonia*, for instance, causes a form of pneumonia.

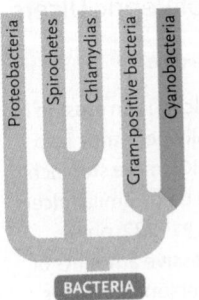

CYANOBACTERIA The *cyanobacteria* **(Figure 26.15)** are Gram-negative photoautotrophs that have a blue-green color and carry out photosynthesis using the same pathways as eukaryotic algae and plants, and use the same chlorophyll that plants have as their primary photosynthetic pigment. They release oxygen as a by-product of photosynthesis.

The direct ancestors of present-day cyanobacteria were the first organisms to use the water-splitting reactions of photosynthesis. As such, they were critical to the appearance of abundant oxygen in the atmosphere, which allowed the evolutionary development of aerobic organisms (see Section 25.2). Chloroplasts probably evolved from early cyanobacteria that were incorporated into the cytoplasm of primitive eukaryotes, which eventually gave rise to the algae and higher plants (see Section 25.3). Besides releasing oxygen, present-day cyanobacteria help fix nitrogen into organic compounds in aquatic habitats and also in lichens that involve a cyanobacterium and a filamentous fungus living together symbiotically (see Chapter 30). (Most lichens involve a green alga and a filamentous fungus.)

A. A population of cyanobacteria covering the surface of a pond

B. Chains of cyanobacterial cells
Some cells in the chains form spores. The heterocyst is a specialized cell that fixes nitrogen.

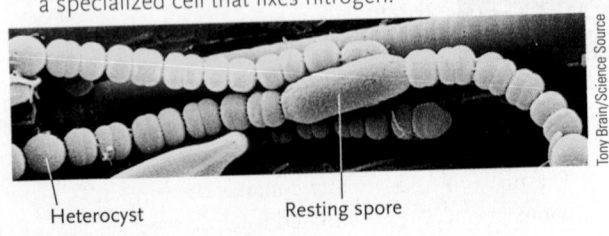

Heterocyst   Resting spore

FIGURE 26.15 Cyanobacteria.

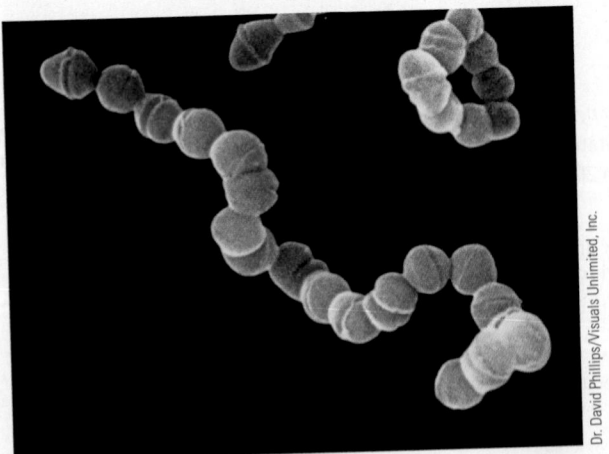

FIGURE 26.14 *Streptococcus* bacteria, a genus of Gram-positive bacteria, forming a long chain of cells, which is typical of the many species in this genus (color-enhanced).

In Nature, Bacteria Typically Are Found in Communities

Researchers grow prokaryotes as individuals in liquid cultures or as isolated colonies on solid media. The results from studies using pure cultures have been crucial in developing an understanding of, among many other things, the nature of the genetic material, DNA replication, gene expression, and gene regulation. But, in nature, bacteria typically are not in pure cultures. Instead, there are mixed populations of bacterial species, perhaps also with archaeal and eukaryotic microbial species. Thus, information from pure cultures in the laboratory may not apply to populations of the same species in nature. We focus our discussion on two types of communities involving bacteria: *microbiomes* and *biofilms*. For humans, you will see that microbiomes and biofilms can be beneficial or harmful to health.

MICROBIOMES A **microbiome** is the complete collection of microorganisms (microbes) associated with a particular organism. The microorganisms generally consist mostly of bacteria, but also archaea, protists, and yeasts and other fungi may be present, as well as viruses. Viruses—mostly bacteriophages—are estimated to outnumber bacterial cells by perhaps a factor of five. Identification of microbes in a microbiome involve both culturing them if they can be cultured, and DNA sequencing analysis (small ribosomal subunit RNA gene sequencing: see Section 24.7), or genome sequencing analysis. To illustrate the properties of microbiomes, here we summarize some important features of the human microbiome:

- The total number of microorganisms in the human microbiome includes about 100 trillion (10^{14}) bacterial cells, about three times the most recent estimates of 37 trillion total human cells. Note, though, that the microbiome is a dynamic community so the numbers can change under different conditions. Collectively the human microbiome weighs about 1.1 kg (2.5 lb).
- The human microbiome is distributed over all sites of the body, including skin, mouth, nasal cavities, throat, stomach, intestines, and urogenital tracts. The population density of microbes varies with location. The large intestines are the most heavily populated, whereas blood and lymph contain very few microbes. The functions of the microbial communities are a subject of ongoing research, including that supported by the U.S.-based Human Microbiome Project (see http://commonfund.nih.gov/hmp/index). For instance, there is evidence that in at least the mouth and intestines, normal constituents of the microbiome provide some protection against pathogens both directly and by being in molecular communication with the immune system. Some bacteria on the skin also are known to protect against certain pathogens.
- The normal microbiome consists of about 1,000 species of bacteria out of an estimated total number of a million bacterial species worldwide. The species diversity of viral and fungal species is less than that of the bacteria. The genetic diversity of the microbiome is much greater than that of the human host. That is, there are approximately 19,900 protein-coding genes in the human genome compared to perhaps as many as 8 million genes collectively among the bacteria, fungi, and viruses in the microbiome. This means that the microbiome has genetic capabilities that extend beyond that of the human genome. For instance, microbes of the gut microbiome synthesize a large number of enzymes for digesting complex dietary carbohydrates, compounds for which we do not synthesize digestive enzymes. (The role of the human microbiome in digestion is discussed more in Chapter 47.)
- The microbiome carries out similar functions in every human but does not necessarily involve the same species of microbes in each person. Certainly there is variability among human populations as to specific functions of their microbiomes, such as the ability to digest a particular dietary compound, or the ability to metabolize a particular drug.
- The human microbiome changes over time. The most dramatic change is between birth and two years of age when the microbiome is being established in the gut. At age two, the gut microbiome looks like a typical adult microbiome. Once established, the gut microbiome can change over time and can be changed intentionally, for example when a person changes their diet or takes particular drugs. The change involves alterations in the proportions of particular species.
- Particular combinations of microbes in the microbiome correlate with certain disease states. For example, a number of research studies show a correlation between microbiome composition and obesity (see Chapter 47), as well as with inflammatory bowel disease, celiac disease, colorectal cancer, and asthma. However, a causal relationship between microbiome composition and disease state has not been established. Moreover, some changes in the microbiome do not affect different people in the same way. For example, alteration of the intestinal microbiome by antibiotics correlates with the development of persistent, and potentially deadly, infections of the Gram-positive bacterium, *Clostridium difficile*, in some people but not others. The symptoms of such infections include watery diarrhea along with abdominal pain. More than 500,000 Americans are sickened and 14,000 killed by *C. difficile* infections each year. A treatment that has a high cure rate (100% in one research study) is *fecal transplantation* in which the intestinal microbiome altered by antibiotic treatment is put back into balance by swallowing coated pills containing concentrated fecal bacteria that are released into the large intestine.

Much remains to be learned about the human microbiome and the microbiomes of other organisms. The research to date has illuminated some surprising facts about the relationship between the microbiome and human health. Many questions remain, including whether particular differences in the microbiome correlate with differences in human health. In the future, identified differences in the microbiome will prove valuable as *biomarkers* (measurable indicators showing a disease state, infection, etc.) for predicting predisposition to specific diseases, and possibly for personalized drug therapies.

BIOFILMS A **biofilm** is a complex aggregation of microorganisms (either one or multiple species) attached, in most cases, to a surface. For instance, biofilms may be found on lake surfaces, on rocks in freshwater or marine environments (making the rocks slippery), surrounding plant roots and root hairs, and on animal tissues such as the intestinal mucosa and teeth (human dental plaque is a biofilm). Biofilms on the surfaces of organisms involve microbes that are part of that organism's microbiome.

Biofilms may be beneficial or detrimental to humans. On the beneficial side, for example, biofilms on solid supports are used in sewage treatment plants for processing organic matter before the water is discharged. On the detrimental side, for example, biofilms adhere to many kinds of surgical equipment and supplies, including catheters and synthetic implants such as pacemakers and artificial joints. When pathogenic bacteria are involved, biofilms can cause infections that are difficult to treat, because pathogenic bacteria in a biofilm are up to 1,000 times more resistant to antibiotics than are the same bacteria in pure liquid cultures. Other examples of medical conditions resulting from the activities of biofilms include middle-ear infections, bacterial endocarditis (an infection of the heart's inner lining or the heart valves), and Legionnaire's disease (an acute respiratory infection caused by breathing in pieces of biofilms containing the pathogenic gammaproteobacterium *Legionella*).

Biofilms can form on a variety of organic and inorganic surfaces, and they can assume many different forms and carry out many different functions for different species. Biofilm formation may be induced by environmental signals that vary depending on the species, including temperature, pH, oxygen level, and the presence of particular amino acids.

Figure 26.16 illustrates a common pattern in the formation of many types of biofilms. The numbers in the following list correspond to the steps in the figure.

1. The species-specific environmental signal induces marked changes in the pattern of gene expression compared to that of free bacteria. In effect, the prokaryote becomes a significantly different organism. This observation is especially important in the case of pathogenic bacteria, since most research on the control of those bacteria is done with liquid cultures. (*Molecular Insights* presents the results of a study showing the gene expression changes in a gum-disease-causing bacterial pathogen.) The challenge is to devise new treatment strategies for biofilm-caused diseases.

2. The genetically reprogrammed bacteria attach to a surface, for example by pili or lipopolysaccharides, and secrete molecules that cause other free bacteria to attach and form a monolayer.

3. The bacteria attaching to the surface begin communicating with one another by sending and receiving chemical signals in a process called **quorum sensing.** The chemical signaling molecules are synthesized and secreted continuously by individual cells, but it is only when the population of cells reaches a certain level (a "quorum") that the concentration of the signal is high enough to be sensed. Recognition of the signal triggers more genetic changes that cause the bacteria to bind more strongly to the surface and to each other, producing small colonies.

4. The bacteria produce an extracellular matrix of polysaccharide polymers and entrapped organic and inorganic materials. The matrix enables the biofilm to mature,

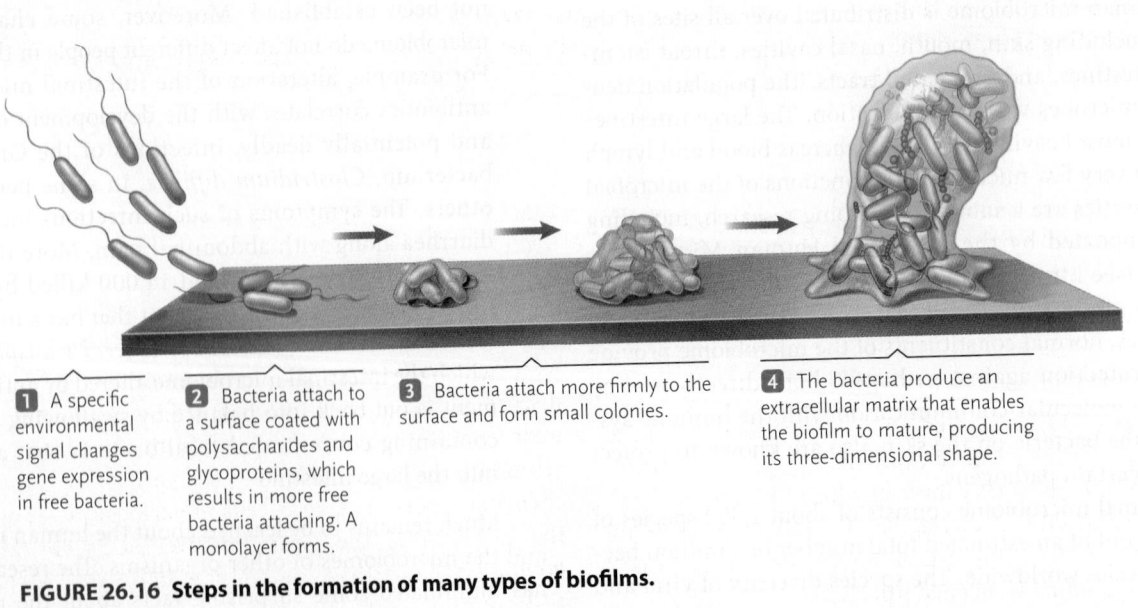

1 A specific environmental signal changes gene expression in free bacteria.

2 Bacteria attach to a surface coated with polysaccharides and glycoproteins, which results in more free bacteria attaching. A monolayer forms.

3 Bacteria attach more firmly to the surface and form small colonies.

4 The bacteria produce an extracellular matrix that enables the biofilm to mature, producing its three-dimensional shape.

FIGURE 26.16 Steps in the formation of many types of biofilms.
© Cengage Learning 2017

Gums the Word: Genome-wide analysis of gene expression changes in an oral cavity pathogen associated with transition from free-living bacteria in the oral microbiome to biofilm

Gingivitis, a nondestructive periodontal disease, is inflammation of the gums. The gums are part of the periodontal tissue, the tissue that supports the teeth. Commonly gingivitis develops due to plaque—bacterial biofilms—on the surfaces of teeth. Untreated, gingivitis can progress to periodontitis, which is a destructive periodontal disease affecting tooth-supporting tissues **(Figure).** Without treatment, periodontitis can result in loosening and then loss of teeth.

Periodontitis is the leading cause of tooth loss in industrial nations, and one of the most frequently occurring infectious diseases in humans. *Porphyromonas gingivalis* is a Gram-negative pathogenic bacterium

Healthy   **Periodontal disease**

Healthy gums
Healthy bone level
Plaque
Tartar
Pocket
Reduced bone level

©Highforge Solutions/Shutterstock.com

FIGURE Healthy gum tissue (left) and gum tissue with periodontal disease (right).

© Cengage Learning 2017

that is present in dental plaque. It is strongly implicated in the onset of periodontitis and the destructive progression of that disease. However, little is known about the physiology of *P. gingivalis* or about gene regulatory changes that occur in the biofilm that could relate to pathogenicity.

Research Question
How does gene expression differ in *P. gingivalis* in free-living cells versus cells in a biofilm?

Experiment
Eric Reynolds and his colleagues at Melbourne Dental School, the University of Melbourne, and Monash University in Australia answered this question using DNA microarrays to study the transcriptome of *P. gingivalis*. (The transcriptome is the complete set of RNA transcripts in a cell. Transcriptomics and DNA microarrays are discussed in Section 19.3 and Figure 19.8.) The researchers grew *P. gingivalis* in culture in fermenters to maintain anaerobic conditions. In one culture the bacterium was grown in a planktonic state (a culture of single cells), and in a second culture it was grown as a biofilm. They harvested cells from each culture, extracted RNA from the cells, and quantified the mRNAs present by DNA microarray analysis to compare genome-wide protein-coding gene expression levels for the two growing states.

Results
The researchers found that 377 genes of the genome (18% of the total number of genes)

were expressed differentially when the bacterium is grown as a biofilm. Of those genes, 191 were up-regulated, meaning they were expressed more in the biofilm compared with the planktonic cells, and the other 186 were down-regulated. Among those up-regulated were genes involved in transport (including signal transduction pathways), and among those down-regulated were genes involved in cell membrane synthesis, DNA replication, and energy production.

Conclusion
The results suggest that the bacterium transitions to a decreased rate of cell replication and, hence, of rate of growth, when it becomes a biofilm. The researchers argue that the changes in gene expression are involved in the adaptive response of the bacterium to growth in the biofilm state. How the changes in gene expression relate to the pathogenicity of *P. gingivalis* when it is growing in a biofilm is presently unknown. One caveat to consider here is that the experiments were done in pure cultures of the bacterium, whereas in the mouth exhibiting gingivitis, *P. gingivalis* is present in a biofilm with other microorganisms.

think like a scientist

Outline in a general way the molecular events that must occur for *P. gingivalis* to make the transition from planktonic to biofilm growth.

Source: A. W. Lo et al. 2009. Comparative transcriptomic analysis of *Porphyromonas gingivalis* biofilm and planktonic cells. *BMC Microbiology* 9:18.

leading to complex three-dimensional forms in which channels supply the bacteria with nutrients for growth. Maturation of the biofilm may also involve trapping other microorganisms, including other bacterial species, algae, fungi, and/or protists, thereby producing diverse microbial communities.

Much remains to be learned about how bacteria form a biofilm, how the change in gene expression during the transition is regulated, and how they interact within the biofilm. Particularly for pathogenic bacteria, having the knowledge of genetic changes will be important for informing scientists about treating a biofilm-associated infection.

Bacteria Cause Human Diseases by Several Mechanisms

As you have learned, some bacteria are pathogenic. Pathogenic bacteria are known for all types of eukaryotic organisms. Let us consider bacteria pathogenic to humans and how they cause diseases. A number of bacterial lineages produce **exotoxins,** toxic proteins that are secreted from the bacterium or are released from the cell when it lyses, and interfere with the biochemical processes of body cells in various ways. For example, the exotoxin of the Gram-positive bacterium *Clostridium botulinum* is found as a contaminant in poorly preserved foods, and causes *botulism*. The botulism exotoxin is one of the most poisonous substances known: a few nanograms of the toxin can cause illness, and a few hundred grams could kill every human on Earth. It acts by interfering with the transmission of nerve impulses. The muscle paralysis produced by the exotoxin can be fatal if the muscles that control breathing are affected. The botulinum exotoxin, under the brand name Botox, is used in low doses for the cosmetic removal of wrinkles, in the treatment of migraine headaches, and of some other medical conditions. Exotoxins are also responsible for the intestinal problems caused by the related *C. difficile* that were mentioned in the discussion of microbiomes. The exotoxins damage the lining of the intestine causing inflammation and leading to the watery diarrhea and abdominal pain.

Some other bacteria cause disease through **endotoxins.** Endotoxins are lipopolysaccharides released from the outer membrane surrounding the cell walls when the bacterium dies and lyses. Endotoxins are natural components of the outer membrane of all Gram-negative bacteria, which include the gammaproteobacteria *E. coli, Salmonella* species, and *Shigella* species. The lipopolysaccharides cause disease by overstimulating the host's immune system, often triggering inflammation (see Chapter 45). Endotoxin release has different effects, depending on the bacterial species and the site of infection, and include typhoid or other fevers, diarrhea, and, in severe cases, organ failure and death. For example, *Salmonella enterica typhi*, the cause of typhoid fever, enter the human intestines and penetrate the intestinal wall, eventually ending up in the lymph nodes. There they multiply, and some of the cells die and lyse, releasing endotoxins into the bloodstream. This both triggers the host's immune response and causes sepsis. Without treatment, the condition can progress to multiple organ system failure and, eventually, death.

Some bacteria release **exoenzymes,** enzymatic proteins that digest plasma membranes and cause cells of the infected host to rupture and die. Exoenzymes may also digest extracellular materials such as collagen, causing connective tissue diseases. Some exoenzymes attack red or white blood cells, leading to anemias, impairment of the immune response, or interference with blood clotting. Among the bacteria that release exoenzymes are particular species of the Gram-positive genera, *Streptococcus* (see Figure 26.14) and *Staphylococcus.* Necrotizing fasciitis (flesh-eating disease), the spectacularly destructive and rapid degeneration of subcutaneous tissues in the skin, is caused by an exoenzyme released by *Streptococcus pyogenes, Staphylococcus aureus,* and some other bacteria.

Some of the ill effects of bacteria have little to do with exotoxins, endotoxins, or exoenzymes, but are caused purely by the body's responses to infection. The severe pneumonia caused by *Streptococcus pneumoniae,* for example, results from massive accumulation of fluid and white blood cells in the lungs in response to the infection. The white blood cells have little effect on the bacteria, however, because of the bacterial cell's protective capsule. As the fluid, white blood cells, and bacteria continue to accumulate, they block air passages in the lungs and impair breathing severely.

Pathogenic Bacteria Commonly Develop Resistance to Antibiotics

Antibiotics are routinely used to treat bacterial infections. These substances, produced as defensive molecules by some bacteria and fungi, or by chemical synthesis, kill or inhibit the growth of other microbial species. For example, streptomycins, produced by soil bacteria, block protein synthesis in their target cells. Penicillins, which are produced by fungi, prevent formation of covalent bonds that hold bacterial cell walls together, weakening the wall and causing the cells to rupture.

Many pathogenic bacteria develop resistance to antibiotics through mutations that allow them to break down the drugs or otherwise counteract their effects (see *Why it matters . . .* in Chapter 21). Resistance is also acquired by horizontal gene transfer when resistance genes on plasmids transfer between bacteria by conjugation, or when DNA is brought into pathogens by other pathways such as transformation and transduction (see Section 17.1). Taking antibiotics routinely in mild doses, or failing to complete a prescribed dosage, contribute to the development of resistance by selecting strains that can survive in the presence of the drug. Antibiotics prescribed for colds and other virus-caused diseases do not affect the virus, but can result in resistance among bacteria present in the body. Antibacterial agents that may result in resistance are also commonly included in such commercial products as soaps, detergents, and deodorants. Resistance is a form of evolutionary adaptation. Antibiotics alter the bacterium's environment, conferring a reproductive advantage on those strains best adapted to the altered conditions.

The development of resistant strains has made tuberculosis, cholera, typhoid fever, gonorrhea, "staph," and other diseases caused by bacteria difficult to treat with antibiotics. For example, as recently as 1988, drug-resistant strains of the Gram-positive bacterium *Streptococcus pneumoniae,* which causes pneumonia, meningitis, and middle-ear infections, were practically unheard of in the United States. Now, resistant strains of *S. pneumoniae* are common and increasingly difficult to treat with antibiotics.

In this section, you have seen that bacteria thrive in nearly every habitat on Earth, including the human body. However, some members of the second prokaryotic domain, the Archaea, the subject of the next section, live in habitats that are too forbidding even for the Bacteria.

STUDY BREAK 26.2

1. What is a biofilm? Give an example of a biofilm that is beneficial to humans and one that is harmful.
2. What is an exotoxin, an endotoxin, and an exoenzyme, and how do they differ with respect to how they cause disease?
3. By what mechanisms can pathogenic bacteria develop resistance to antibiotics?

26.3 The Domain Archaea

The first-studied archaeans were found in extreme environments, such as hot springs, hydrothermal vents on the ocean floor, and salt lakes **(Figure 26.17)**. For that reason, these prokaryotes were called *extremophiles* ("extreme lovers"). Subsequently archaeans have also been found living in normal environments; like bacteria, these are *mesophiles*. The genomes of almost 300 archaean species have been sequenced to date (see Table 19.1).

Many archaeans are chemoautotrophs that obtain energy by oxidizing inorganic substances, whereas others are chemoheterotrophs that oxidize organic molecules. Recent research has identified some archaeans as possible human pathogens, for example a species associated with tooth infections.

Research has shown that archaeans have some eukaryotic features, some bacterial features, and some features that are unique to the group (see Table 26.1; also discussed in Section 25.3).

DNA Sequencing and Genome Sequencing Analyses Reveal Evolutionary Branches in the Archaea

The simplified phylogenetic tree in Figure 26.10 shows five evolutionary groups of the Archaea as determined by DNA sequencing analysis (rRNA coding sequences) and genome sequencing analysis. Two major groups, the *Euryarchaeota* and the *Crenarchaeota*, contain many species of archaeans that have been cultured and examined in the laboratory. Each of the other three groups, the *Nanoarchaeota*, the *Thaumarchaeota*, and the *Korarchaeota*, is represented by one to a few living species, with others in the group recognized by related rRNA coding sequences in DNA from environmental samples.

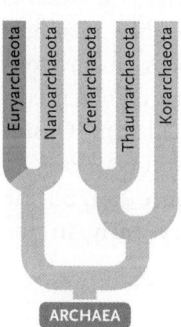

EURYARCHAEOTA The **Euryarchaeota** are found in different extreme environments. They include methanogens, which produce methane; extreme halophiles, which live in high concentrations of salt; and some extreme thermophiles, which live under high-temperature conditions.

Methanogens ("methane generators") live in reducing environments **(Figure 26.18)**, and represent about one-half of all known species of archaeans. All known methanogens, for example *Methanococcus* and *Methanobacterium* species, belong to the Euryarchaeota. Methanogens are obligate

A. Red-purple color produced by archaeans in highly saline water

B. Bright colors produced in hot, sulfur-rich water by the oxidative activity of archaeans that convert H₂S to elemental sulfur

FIGURE 26.17 Typically extreme archaean habitats.

anaerobes, meaning they are killed by oxygen. They are found in the anoxic ("oxygen-lacking") sediments of, for instance, swamps and marshes, as well as in more moderate environments, such as in the microbiomes of rumen in cattle and sheep (see Chapter 47); of the large intestine of dogs and humans; and of the hind-guts of insects such as termites and cockroaches. Methanogens generate energy by converting at least ten different substrates such as carbon dioxide and hydrogen gas, methanol, or acetate into methane gas (CH_4), which is released into the atmosphere.

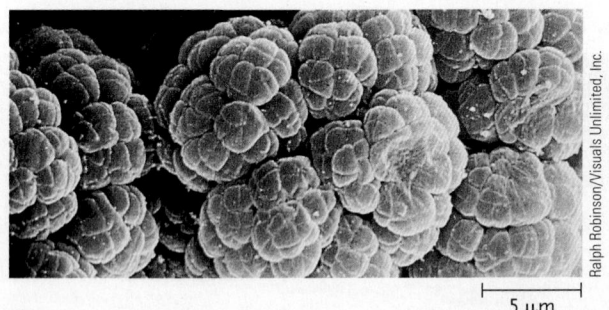

FIGURE 26.18 A colony of cells of the methanogenic archaean *Methanosarcina,* which live in the sulfurous, waterlogged soils of marshes and swamps (color-enhanced).

Halophiles are salt-loving organisms. Extreme halophilic Archaea live in highly saline (salty) environments such as the Great Salt Lake or the Dead Sea, and on foods preserved by salting. Moreover, they *require* a minimum NaCl concentration of about 1.5 M (about 9% solution) to live, and they can live in a fully saturated solution (5.5 M, or 32%). All known extreme halophilic Archaea belong to the Euryarchaeota. Most are aerobic chemoheterotrophs; they obtain energy from sugars, alcohols, and amino acids using pathways similar to those of bacteria. Examples are *Halobacterium* and *Natrosobacterium* species.

Extreme thermophiles live in extremely hot environments. Archaeans of this type live in thermal areas such as ocean-floor hydrothermal vents and hot springs such as those in Yellowstone National Park. Their optimal temperature range for growth is 70–95°C, approaching the boiling point of water. By comparison, no eukaryotic organism is known to live at a temperature higher than 60°C. Some extreme thermophiles, such as *Pyrophilus* species and *Thermoplasma* species, are members of the Euryarchaeota. Some of them, such as *Pyrophilus* species, are obligate anaerobes, whereas others, such as *Thermoplasma* species, are facultative anaerobes that grow on a variety of organic compounds.

Two extreme thermophiles in this group have proved to be valuable for research. Recall that the polymerase chain reaction (PCR) uses a thermostable DNA polymerase for amplifying a segment of DNA (see Section 18.1). Two commercially available thermostable DNA polymerases for PCR are Vent DNA Polymerase, derived from *Thermococcus litoralis,* and Pfu DNA polymerase, derived from *Pyrococcus furiosus.*

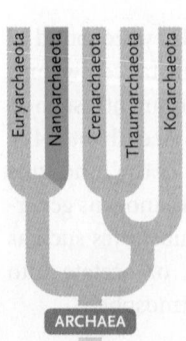

NANOARCHAEOTA The **Nanoarchaeota** group consists of only one known living species, *Nanoarchaeum equitans;* other members are represented only by sequences detected in environmental samples. This extreme thermophile was discovered in a hydrothermal vent off the coast of Iceland. Analysis of its genome shows that it lacks genes for several metabolic pathways, including those for most nucleotides, amino acids,

and lipids. It survives by being a symbiont of a thermophilic member of the Crenarchaeota.

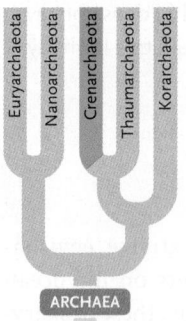

CRENARCHAEOTA The **Crenarchaeota** group contains most of the extreme thermophiles. Their optimal temperature range of 75–105°C is higher than that of the Euryarchaeota. Most are unable to grow at temperatures below 70°C. The most thermophilic member of the group, *Pyrolobus,* grows optimally at 106°C, but dies below 90°C. *Pyrolobus* species lives in ocean-floor hydrothermal vents where the pressure makes it possible to have temperatures above 100°C, the boiling point of water on Earth's surface.

Also within this group are **psychrophiles** ("cold loving"), organisms that grow optimally in cold temperatures in the range of −10 to −20°C. These organisms are found mostly in the Antarctic and Arctic oceans, which are frozen most of the year, and in the intense cold of ocean depths.

Mesophilic members of the Crenarchaeota comprise a large part of plankton in cool, marine waters where they are food sources for other marine organisms. As yet, no individual species of these archaeans has been isolated and characterized.

Crenarchaeota archaeans exhibit a wide range of metabolism with regard to oxygen, and include obligate anaerobes, facultative anaerobes, and aerobes.

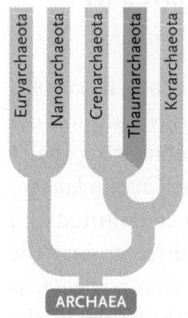

THAUMARCHAEOTA The **Thaumarchaeota** are a newly designated group of the Archaea consisting, at present, of four living species. Other members are represented by sequences found in environmental samples. All living representatives are mesophilic, chemoautotrophic ammonia oxidizers. The establishing archaean for this group is *Cenarchaeum symbiosum.*

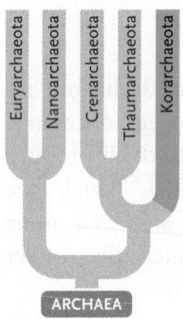

KORARCHAEOTA The **Korarchaeota** group consists of one living species, *Candidatus Korarcheum cryptofilum.* Other members are represented in sequences in environmental samples. The living member is a thermophile that was isolated from Obsidian Pool at Yellowstone National Park (a similar pool is shown in Figure 26.17B). The Korarchaeota diverged very early in the evolution of the Archaea.

Archaea and Eukarya Are Closely Related

As discussed in Section 24.7 and illustrated in Figure 26.10, Archaea and Eukarya are more closely related to each other than either Domain is to Bacteria. Molecular phylogenetics studies of archaea, particularly those involving genome sequencing, are being carried out currently to advance our understanding of the evolution of archaea and of the origin of the eukaryotic cell and eukaryotic species. In 2015, researchers reported that genomic sequencing of uncultured archaea from deep ocean sediment identified a new candidate archaeal group, the **Lokiarchaeota.** Lokiarchaeota contain more eukaryotic-like genes than any of the other known archaeal species. For example, Lokiarchaeota have actins like eukaryotes, whereas other archaea have distantly related actins, and they have the most eukaryotic-like ribosome identified in archaea so far. The researchers concluded that the results support the hypothesis that the eukaryotic cell evolved from an archaean that already had some of the features that define present-day eukaryotic cells. This study has clearly closed the evolutionary gap between Archaea and Eukarya.

From the highly varied prokaryotes, we now turn to eukaryotes, a group which does not equal the metabolic diversity or the conquest of extreme environments we have seen in the Bacteria and Archaea, but which does display new levels of physiological and behavioral complexity. The next six chapters discuss the eukaryotic kingdoms of protists, plants, fungi, and animals.

STUDY BREAK 26.3

1. What distinguishes members of the Archaea from members of the Bacteria and Eukarya?
2. How does a methanogen obtain its energy? In which group or groups of Archaea are methanogens found?
3. Where do extreme halophilic archaeans live? How do they obtain energy? In which group or groups of Archaea are the extreme halophiles found?
4. What are extreme thermophiles and psychrophiles?

 Unanswered Questions

As you learned in this chapter, archaea are morphologically similar to bacteria, yet they have a number of features that are more related to eukaryotes than to bacteria. More specifically, the DNA replication and transcription machineries of archaea are related to those of eukaryotes and distinct from those found in bacteria. Also, recent work has revealed that, in some archaea, aspects of cell division have close parallels in human cells but not in bacteria.

Why do some archaea have multiple replication start sites per chromosome?

Analysis of the available genome sequences of archaea show that the machinery that archaea use to replicate their DNA is, like the transcription machinery, related to that of eukaryotes and distinct from the bacterial apparatus. Recall from Section 14.3 that bacteria have a single replication origin per chromosome, whereas eukaryotes have many replication origins per chromosome. Some archaea replicate their chromosomes from a single origin, but a number of species have multiple replication origins per chromosome. For example, species in the *Sulfolobus* genus (hyperthermophiles of the Crenarchaeota group of the domain Archaea) have three replication origins in their single chromosome. Recent work has demonstrated that in *Sulfolobus islandicus*, each individual origin is specified by a unique initiator protein, thus revealing the *Sulfolobus* chromosome to be a mosaic of three distinct replicons. (A replicon is a region of the chromosome that replicates from a particular origin of replication.) How did this arise? It seems likely that the chromosome has been sculpted by integration of extrachromosomal elements, each bringing in its own origin and initiator protein. Indeed, in *Haloferax volcanii,* a salt-loving member of the Euryarchaeota, a plasmid found in the wild isolate of the species has been observed to be integrated into the main chromosome of a laboratory strain of the organism. Strikingly, the plasmid replication origin remained active when embedded in the main chromosome. Presumably, the acquisition of extra replication origins means that it takes less time to replicate a given chromosome and thus would confer a growth advantage. This strategy of multi-origin chromosomes appears widespread in archaea and is ubiquitous in eukarya. Why then has this situation never arisen in bacteria? Perhaps

one answer may lie in how replication is terminated in bacteria versus archaea and eukarya. All bacteria characterized thus far use dedicated termination systems that arrest replication forks at defined sites; in contrast, archaea and eukarya appear to terminate replication by a more passive mechanism with termination occurring wherever replication forks collide. Perhaps the positional constraints imposed by the bacterial termination systems have rendered their chromosomes nonpermissive to the acquisition of extra origins of replication. Another issue may lie in the mechanisms whereby initiation of replication is controlled. With an increasing number of origins there is clearly an increased regulatory burden on the cell to ensure that replication is controlled appropriately in time and space. It is possible, indeed likely, that the cell cycle machineries of archaea are fundamentally distinct from those of bacteria.

think like a scientist The ability to replicate DNA is fundamental to the propagation of all life on the planet. Yet it is apparent that bacteria use a different mode of replication—single start sites per chromosome as opposed to the multi-origin chromosomes of many archaea and all eukarya. Furthermore, bacteria employ an entirely distinct machinery to replicate their DNA from that employed by archaea and eukaryotes. How could such distinct modes and machineries for such a fundamentally important process have evolved?

Stephen D. Bell and **Rachel Y. Samson** work at the Department of Molecular and Cellular Biochemistry, Indiana University (IU). Steve is also a Professor in the Biology Department at IU. His research group is focused on investigating transcription, DNA replication, and cell division in the archaeal domain of life. Rachel joined Steve's group in 2007 and is researching the molecular and genetic basis of archaeal DNA replication. To learn more about their research, go to http://www.indiana.edu/~sbelllab/.

REVIEW KEY CONCEPTS

For access to MindTap and additional study materials visit www.cengagebrain.com.

26.1 Prokaryotic Structure and Function

- Three shapes are common in prokaryotes: spherical, rodlike, and spiral (Figure 26.2).
- Prokaryotic genomes typically consist of a single, circular DNA molecule packaged into the nucleoid. Many prokaryotic species also contain plasmids, which replicate independently of the main DNA (Figure 26.3).
- Gram-positive bacteria have a cell wall consisting of a thick peptidoglycan layer. Gram-negative bacteria have a thin peptidoglycan cell wall surrounded by an outer membrane (Figure 26.4).
- A polysaccharide capsule, or slime layer, surrounds many bacteria. This sticky, slimy layer both protects the bacteria and helps them adhere to surfaces (Figure 26.5).
- Some prokaryotes have flagella, corkscrew-shaped protein fibers that rotate like propellers, and pili, protein shafts that help bacterial cells adhere to each other or to eukaryotic cells (Figure 26.7).
- Prokaryotes show great diversity in their modes of obtaining energy and carbon. Chemoautotrophs obtain energy by oxidizing inorganic substrates and use carbon dioxide as their carbon source. Chemoheterotrophs obtain both energy and carbon from organic molecules. Photoautotrophs are photosynthetic organisms that use light as a source of energy and carbon dioxide as their carbon source. Photoheterotrophs use light as a source of energy and obtain their carbon from organic molecules (Figure 26.8).
- Some prokaryotes are capable of nitrogen fixation, the conversion of atmospheric nitrogen to ammonia; others are responsible for nitrification, the two-step conversion of ammonium to nitrate.
- Prokaryotes normally reproduce by binary fission. Some prokaryotes are capable of conjugation, in which part of the DNA of one cell is transferred to another cell.

26.2 The Domain Bacteria

- Bacteria are divided into several evolutionary branches (Figure 26.10).
- The proteobacteria are a highly diverse group of Gram-negative bacteria that exhibit a range of metabolic capabilities, including chemoautotrophy, chemoheterotropy, and photoautotrophy. Some are aerobic, others are anaerobes. Five subgroups of proteobacteria are recognized. Free-living proteobacteria include *E. coli* and species that fix nitrogen (Figure 26.11).

- The chlamydias are Gram-negative, intracellular parasites that cause various diseases in animals. They have cell walls with an outer membrane, but they lack peptidoglycans (Figure 26.13).
- The spirochetes are spiral-shaped Gram-negative bacteria that are propelled by twisting movements produced by the rotation of flagella.
- The Gram-positive bacteria are primarily chemoheterotrophs and include many pathogenic species (Figure 26.14).
- The cyanobacteria are Gram-negative photoautotrophs that carry out photosynthesis and release oxygen as a by-product (Figure 26.15).
- In nature, bacteria typically are found in communities such as microbiomes and biofilms. A microbiome is the complete collection of microorganisms associated with a particular organism. The microbiome is distributed over all sites of the body and the microbes involved carry out many functions, including aiding in food digestion and protecting against pathogens. Biofilms are complex aggregations of microorganisms on a surface (Figure 26.16). Microbiomes and biofilms can have harmful or beneficial impacts.
- Bacteria cause disease through exotoxins, endotoxins, and exoenzymes.
- Pathogenic bacteria may develop resistance to antibiotics through mutation of their own genes, or by acquiring resistance genes or plasmids from other bacteria.

26.3 The Domain Archaea

- Members of the domain Archaea have some features that are like those of bacteria, other features that are like those of eukaryotes, and some other features that are uniquely archaean (Table 26.1).
- The archaean plasma membrane contains unusual lipid molecules. The cell walls of archaeans consist of distinct molecules similar to peptidoglycans, or of protein or polysaccharide molecules.
- Five evolutionary groups of Archaea are: the Euryarchaeota, which include the methanogens, the extreme halophiles, and some extreme thermophiles; the Nanoarchaeota, the only living member of which is an extreme thermophile; the Crenarchaeota, which contain most of the archaean extreme thermophiles, as well as psychrophiles and mesophiles; the Thaumarchaeota, all of which are mesophilic chemoautotrophs; and the Korarchaeota, the only living member of which is a thermophile (Figure 26.10).

TEST YOUR KNOWLEDGE

Remember/Understand

1. A urologist identifies cells in a man's urethra as bacterial. Which of the following descriptions applies to the cells?
   a. They have sex pili, which give them motility.
   b. They have flagella, which allow them to remain in one position in the urethral tube.
   c. They are covered by a capsule, which enables them to multiply quickly.
   d. They are covered by pili, which keep them attached to the urethral walls.
   e. They contain a peptidoglycan cell wall, which gives them buoyancy to float in the fluids of the urethra.

2. A bacterium that uses nitrites as its only energy source was found in a deep salt mine. It is a:
   a. chemoautotroph.
   b. parasite.
   c. photoautotroph.
   d. heterotroph.
   e. photoheterotroph.

3. The ___ are all oxygen-producing photoautotrophs.
   a. spirochetes
   b. chlamydias
   c. cyanobacteria
   d. Gram-positive bacteria
   e. proteobacteria

4. At the health center, a fecal sample was taken from a feverish student. Organisms with corkscrewlike flagella and no endomembranes but with cell walls that lack peptidoglycan were isolated as the cause for the illness. These organisms probably belong to the group:
   a. chlamydias.
   b. spirochetes.
   c. Euryarchaeota.
   d. Cyanobacteria.
   e. Archaea.

5. Which of the following is *not* a property of an endospore?
   a. resistant to boiling—must be autoclaved to be killed
   b. metabolically inactive
   c. can potentially survive millions of years
   d. provides a method to preserve bacterial DNA under harsh conditions
   e. is the portion of the bacterial cell that undergoes conjugation

6. Bacterial cells are generally thought to act independently. An exception to this is bacterial cells involved in:
   a. biofilms.
   b. photosynthesis.
   c. peptidoglycan layering.
   d. toxin release.
   e. facultative anaerobic metabolism.

7. Penicillin, an antibiotic, inhibits the formation of cross-links between sugar groups in peptidoglycan. Bacteria treated with penicillin should be:
   a. aerobic.
   b. anaerobic.
   c. Gram-negative.
   d. Gram-positive.
   e. flagellated.

8. The best choice when using/prescribing antibiotics is to:
   a. increase the dosage when the original amount does not work.
   b. determine the kind of bacterium causing the problem.
   c. stop taking the antibiotic when you feel better but the prescription has not run out.
   d. ask the doctor to prescribe a drug as a precaution for an infection you do not have.
   e. choose soaps that are labeled "antibacterial."

9. Archaeans have:
   a. proteins of energy metabolism like bacteria, cell division proteins like bacteria, and mitochondria like eukaryotes.
   b. multiple types of RNA polymerases like eukaryotes, chromosomes like bacteria, and formylmethionine as the first amino acid placed in proteins.
   c. operons like bacteria, multiple types of RNA polymerases like eukaryotes, and cell division proteins like bacteria.
   d. no sensitivity to streptomycin, histones like eukaryotes, and no operons.
   e. histones like eukaryotes, a single type of RNA polymerase like bacteria, and operons like bacteria.

10. Methanogens, obligate anaerobes that generate methane, are:
    a. members of the Proteobacteria.
    b. members of the Cyanobacteria.
    c. members of the Spirochetes.
    d. members of the Euryarchaeota.
    e. members of the Korarchaeota.

Apply/Analyze

11. **Discuss Concepts** The digestive tract of newborn chicks is free of bacteria until, for example, they eat food that has been exposed to the feces of adult chickens. The ingested bacteria establishes a population in the digestive tract that is beneficial for the digestion of food. However, if *Salmonella* are present in the adult feces, this bacterium, which can be pathogenic for humans who ingest it, may become established in the digestive tracts of the chicks. To eliminate the possibility that *Salmonella* might become established, should farmers feed newborn chicks a mixture of harmless known bacteria from a lab culture, or a mixture of unknown fecal bacteria from healthy adult chickens? Design an experiment to answer this question.

Evaluate/Create

12. **Discuss Concepts** Investigators in Australia found that mats of pond scum formed by the bacterium *Botyrococcus braunii* decayed into a substance resembling crude oil when the ponds dried up. Formulate a hypothesis explaining how this process may have contributed to Earth's oil deposits.

13. **Design an Experiment** You receive a sample of an unidentified prokaryote. Design an experiment to see if it is a bacterium or an archaean.

14. **Apply Evolutionary Thinking** Does our understanding of horizontal gene transfer affect our application of the biological species concept to bacteria?

For selected answers, see Appendix A.

INTERPRET THE DATA

The proteobacteria in the genus *Pseudomonas* are important spoilers of food. The rate at which food spoils is dependent on temperature. The data in the **Figure** show the relationship between the square root of the growth rate (how fast the population is growing per day; unitless) and the temperature of the culture.

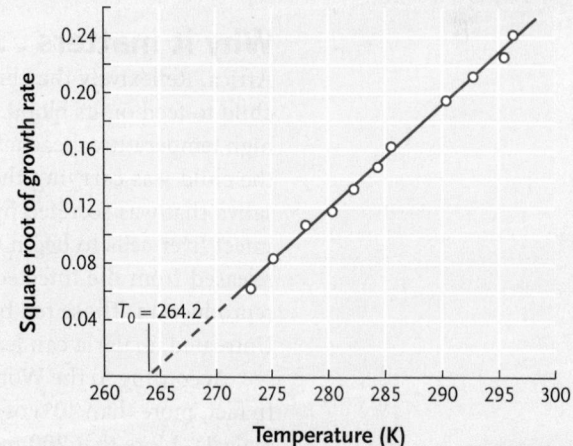

1. Convert the temperature of the data point at the lowest temperature from K (kelvin) to °C. Why are there no data below this temperature?

2. Does typical refrigeration (temperature lowered to 4°C) appear sufficient to limit growth rate compared to room temperature (20°C)?

Source: D. A. Ratkowsky et al. 1982. Relationship between temperature and growth rate of bacterial cultures. *Journal of Bacteriology* 149:1–5. *Journal of Bacteriology* by American Society for Microbiology; Society of American Bacteriologists. Reproduced with permission of American Society for Microbiology (etc.) in the format Journal via Copyright Clearance Center.

27 Protists

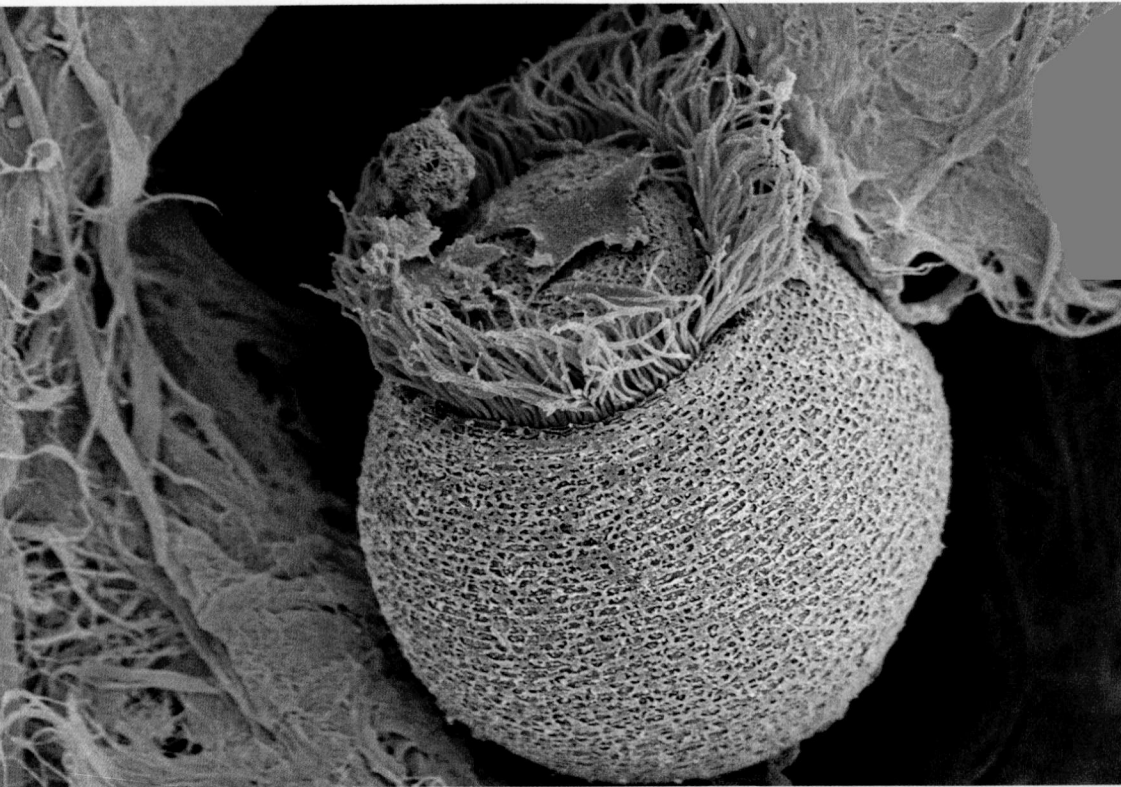

Steve Gschmeissner/Science Source

A ciliated protist (color-enhanced SEM). This protist lives in water, feeding on bacteria and decaying organic matter.

Why it matters . . . A female mosquito lands on the arm of a young sleeping child in Africa. Reflexively the child brushes the mosquito away, but not before the insect has bitten the child to feed on its blood. Ten days later the child experiences flulike symptoms—fever, chills, a high temperature, headaches, nausea, and sweating. The child has *malaria*. The mosquito that bit the child was carrying the parasites that cause malaria. The parasites are carried in mosquito's saliva that was secreted by the salivary glands. Once injected during the bite, the parasites then infect liver cells to begin the reproduction cycle of the parasite. Thousands of new parasites are released from the infected liver cells and then infect red blood cells, where they continue their reproduction. Those red blood cells are killed as parasites are released to infect other blood cells. Untreated, malaria can lead to damage of vital organs and even death.

According to the World Health Organization, in 2012 malaria was endemic in 104 countries. In fact, more than 40% of the world's population lives in areas where there is a risk of contracting malaria. More that 200 million people had the disease in 2010, and approximately 655,000 died, most of them children under age 5.

The parasite that causes malaria is a species of *Plasmodium,* which is a type of protist. (See *Focus on Research: Applied Research* for more on malaria and the *Plasmodium* life cycle.) Most protists are single-celled organisms. Structurally, they are the simplest of all eukaryotes. Protists are highly diverse, and only some of them are parasites or pathogens. In this chapter, we describe several lineages of protists, a diverse array of eukaryotic organisms that are clearly not a monophyletic group. (A *monophyletic group* is a group of organisms that includes a single ancestral species and all of its descendants; see Section 24.2.) Protists are the products of the varied early branching of eukaryotic evolution. They are abundant on Earth and play key ecological, economic, and medical roles in the world's biological communities.

27.1 What Is a Protist?

Protists are easily the most varied of all Earth's organisms. **Figure 27.1** shows a number of protists, illustrating their great diversity. Protists include both microscopic single-celled and large multicellular organisms. They may inhabit aquatic environments, moist soils, or the bodies of animals and other organisms, and they may live as predators, photosynthesizers, parasites, or decomposers. The extreme diversity of the group has made the protists difficult to classify.

Protists Are Evolutionary Groups of Eukaryotes Other Than Fungi, Land Plants, and Animals

The one reasonable certainty about protist classification is that they are not prokaryotes, fungi, plants, or animals. Because protists are eukaryotes, the boundary between them and prokaryotes is clear and obvious. Protists have a true nucleus, with multiple, linear chromosomes and cytoplasmic organelles, including mitochondria (in most but not all species), endoplasmic reticulum, Golgi complex, and chloroplasts (in some species). They reproduce asexually by mitosis or sexually by meiosis and union of gametes, rather than by binary fission as do prokaryotes.

Figure 27.2 shows the phylogenetic relationships between protists and other eukaryotes. This tree represents a consensus based on molecular data, including phylogenomics analysis, which compares sequences of whole genomes, or large segments of genomes. The tree shows the most ancient bifurcation of eukaryotes into the **Unikonta (unikonts),** eukaryotes with a single flagellum, and **Bikonta (bikonts),** eukaryotes with two flagella. Further evolution produced several categories of eukaryotes. As you can see in Figure 27.2, all eukaryotic organisms except animals, land plants, and fungi—three groups that arose from protist ancestors—are protists. Although some protists have features that resemble those of the fungi, land plants, or animals, several characteristics are distinctive. For instance, cell wall components in protists differ from those of the fungi (molds and yeasts). In contrast to land plants, protists lack highly differentiated structures equivalent to true roots, stems, and leaves. They also lack the protective structures that encase developing embryos in plants. Protists are distinguished from animals by their lack of highly differentiated structures such as limbs and a heart, and by the absence of features such as nerve cells, complex developmental stages, and an internal digestive tract. Protists also lack collagen, the characteristic extracellular support protein of animals.

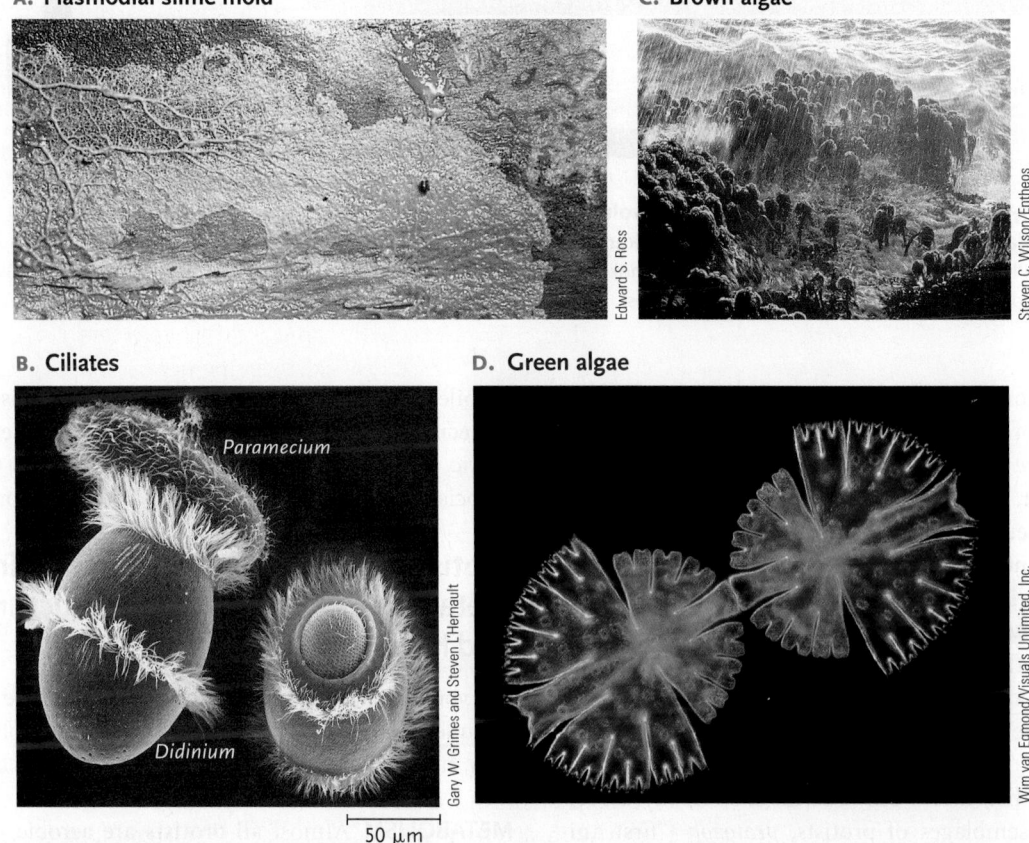

A. Plasmodial slime mold

Edward S. Ross

C. Brown algae

Steven C. Wilson/Entheos

B. Ciliates

Paramecium

Didinium

Gary W. Grimes and Steven L'Hernault

50 µm

D. Green algae

Wim van Egmond/Visuals Unlimited, Inc.

FIGURE 27.1 A sampling of protist diversity. (A) *Physarum,* a plasmodial slime mold (yellow shape, lower part of figure) migrating over a rotting log. **(B)** *Didinium,* a ciliate, consuming another ciliate, *Paramecium.* **(C)** *Postelsia palmaeformis* (the sea palm), a brown alga, thriving in the surf pounding a California coast. **(D)** *Micrasterias,* a single-celled green alga, here shown dividing in two.

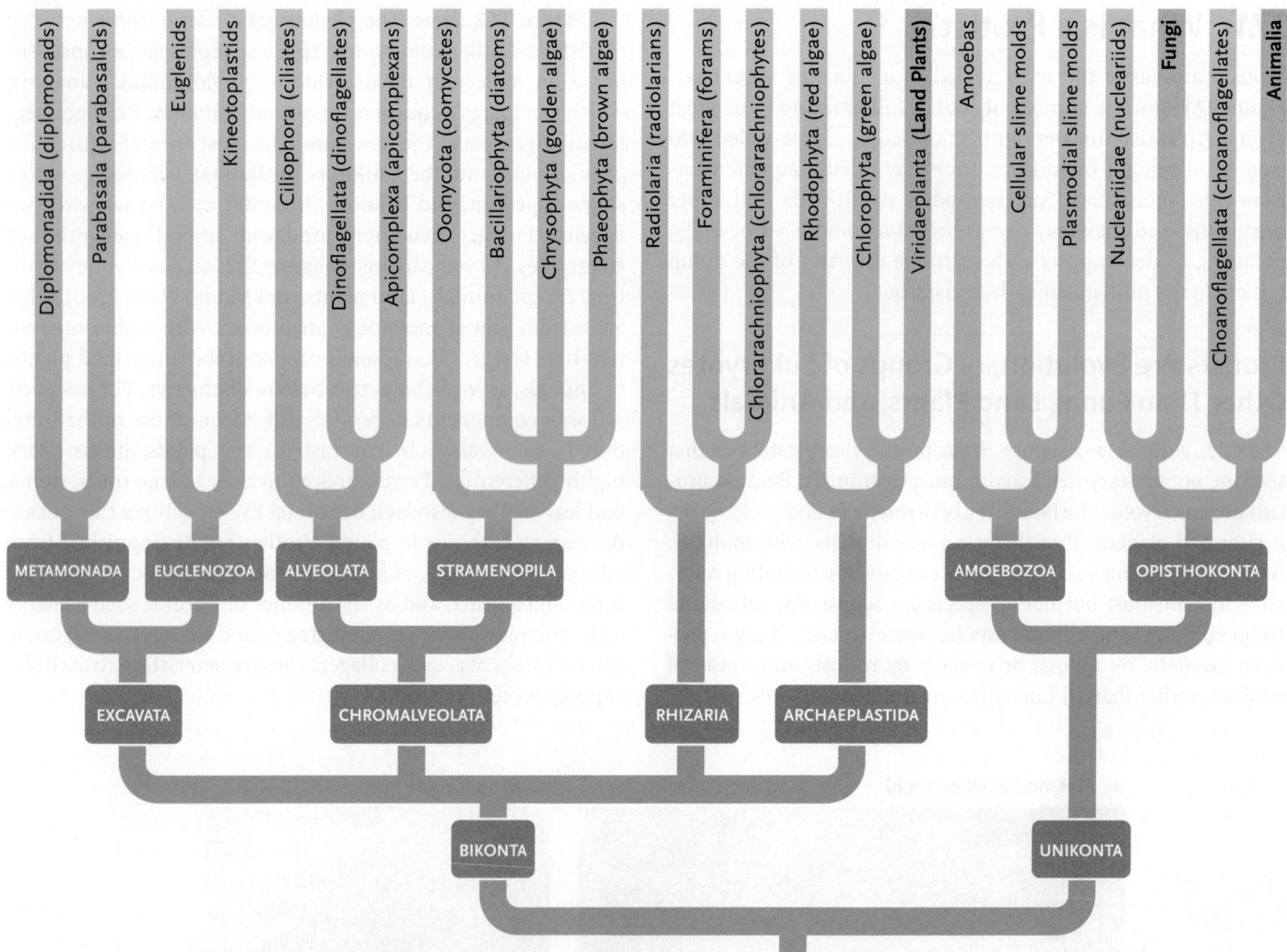

FIGURE 27.2 The phylogenetic relationship between the evolutionary groups of the protists (green branches) and the relationship of the protists with other eukaryotes. The Archaeplastida includes the land plants of the kingdom Plantae (rust-colored branch), and the Opisthokonta includes the animals of the kingdom Animalia and the fungi of the kingdom Fungi (rust-colored branches).
© Cengage Learning 2017

The classification scheme for protists has been challenging to devise because of the diversity of these organisms. Changes and refinements are likely as more sequence data are accumulated. What is clear, though, is that the protists are a polyphyletic assemblage of eukaryotic organisms. As Figure 27.2 shows, many protists are no more closely related to each other than they are to the fungi, plants, or animals.

For simplicity, we will refer to these organisms as protists in this book, with the understanding that it is a collection of distantly unrelated organisms placed together for convenience. We will refer to the major evolutionary clusterings as groups (see Figure 27.2).

You may have heard two broad terms historically associated with large assemblages of protists, *protozoa* ("first animal") (also called *protozoans*) and *algae*. Neither of these assemblages corresponds to a taxonomic group of protists, however. Protozoa is a name for a diverse assemblage of heterotrophic protists, formerly considered to be animals, that are mobile and feed either by ingesting prey or absorbing organic molecules, whereas algae is a name for several sets of photosynthetic protists. The term *algae* will be used in this chapter in association with specific taxonomic groups of protists.

Protist Diversity Is Reflected in Their Metabolism, Reproduction, Structure, and Habitat

As you might expect from the broad range of organisms included, protists are highly diverse in metabolism, reproduction, structure, and habitat.

METABOLISM Almost all protists are aerobic organisms that live either as *heterotrophs*—by obtaining their organic molecules from other organisms—or as *autotrophs*—by producing organic molecules for themselves by photosynthesis. Among the heterotrophs, some protists obtain organic molecules by directly ingesting part or all of other organisms and digesting them internally. Others absorb organic molecules from their environment. A few protists can live as both heterotrophs and autotrophs by supplementing photosynthesis with the active ingestion of food.

REPRODUCTION Reproduction may be asexual by mitosis or sexual by meiotic cell division and formation of gametes. In protists that reproduce by both mitosis and meiosis, the two modes of cell division are combined into a life cycle that is highly distinctive among the different protist groups.

STRUCTURE Many protists live as single cells or as **colonies** in which individual cells show little or no differentiation and are potentially independent. Within colonies, individuals use cell signaling to cooperate on tasks such as feeding or movement. Some protists are large multicellular organisms, in which cells are differentiated and completely interdependent. For example, seaweeds are multicellular marine protists that include the largest and most differentiated organisms of the group. Their structures include a holdfast to secure the organism to rocks, leaflike fronds, and, in some cases, an air bladder for flotation. The giant kelp of coastal waters rival forest trees in size.

Some single-celled and colonial protists have complex intracellular structures, some found nowhere else among living organisms **(Figure 27.3).** For example, many freshwater protists

have a mechanism that maintains water balance in and out of the cell to prevent lysis. Excess water entering cells by osmosis (see Section 5.3) is handled using a specialized cytoplasmic organelle, the **contractile vacuole.** The contractile vacuole gradually fills with water. When it reaches maximum size it moves to the plasma membrane and forcibly contracts, expelling the water to the outside through a pore in the membrane. Many protists also have **food vacuoles** that digest prey or other organic material engulfed by the cells. Enzymes secreted into the food vacuoles digest the organic molecules. Any remaining undigested matter is expelled to the outside by a mechanism similar to the expulsion of water by contractile vacuoles.

The cells of some protists are supported by an external cell wall, or by an internal or external shell built up from organic or mineral matter; in some, the shell takes on highly elaborate forms. Other protists have a **pellicle,** a layer of supportive protein fibers located inside the cell, just under the plasma membrane, which provides strength and flexibility instead of a cell wall (see Figure 27.5).

Almost all protists have structures that provide motility at some time during their life cycle. Some move by amoeboid motion, in which the cell extends one or more lobes of cytoplasm called **pseudopodia** ("false feet"; see Figure 27.19). The rest of the cytoplasm and the nucleus then flow into a pseudopodium, completing the movement. Other protists move by the beating of cilia (many per cell) or flagella (one to a few per cell). (Cilia and flagella are discussed in Section 4.3; remember that protists are eukaryotes so the flagella here are the eukaryotic not the bacterial structures.) In some protists, cilia are arranged in complex patterns, with an equally complex network of microtubules and other cytoskeletal fibers supporting the cilia under the plasma membrane. Among the protists are the most complex single cells known because of the wide variety of cytoplasmic structures they have.

HABITAT Protists live in aqueous habitats, including aquatic or moist terrestrial locations such as oceans, freshwater lakes, ponds, streams, and moist soils, and within host organisms. In bodies of water, small photosynthetic protists collectively make up the **phytoplankton** (*phytos* = plant; *planktos* = drifting), the abundant organisms that capture the energy of sunlight in nearly all aquatic habitats. These photosynthetic protists provide organic substances and oxygen for heterotrophic bacteria and protists and for the small crustaceans and animal larvae that are the primary constituents of **zooplankton** (*zoe* = life, usually meaning animal life); although protists are not animals, biologists often include them among the zooplankton. The phytoplankton and the larger multicellular protists forming seaweeds collectively account for about half of the total organic matter produced by photosynthesis.

In the moist soils of terrestrial environments, protists play important roles among the detritus feeders

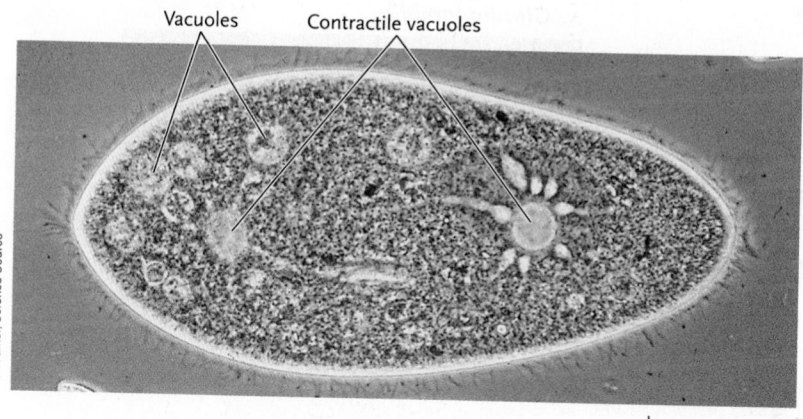

Vacuoles   Contractile vacuoles

MI Walker/Science Source

20 μm

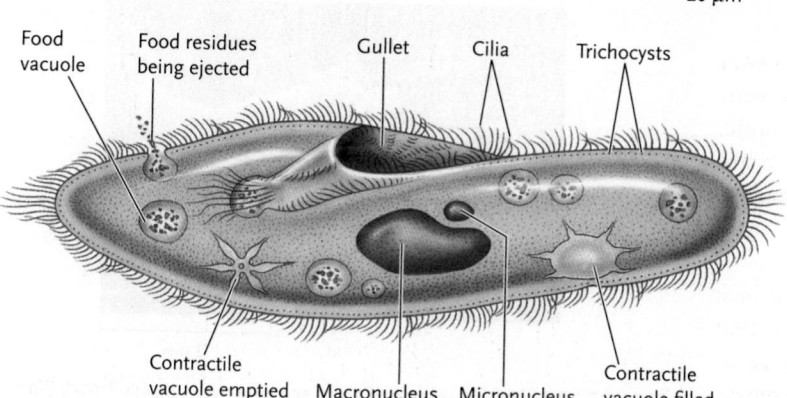

Food vacuole
Food residues being ejected
Gullet
Cilia
Trichocysts
Contractile vacuole emptied
Macronucleus
Micronucleus
Contractile vacuole filled

FIGURE 27.3 A ciliate, *Paramecium,* showing the cytoplasmic structures typical of many protists.
© Cengage Learning 2017

that recycle matter from organic back to inorganic form. In their roles in phytoplankton, zooplankton, and as detritus feeders, protists are enormously important in the world ecosystem.

Protists that live in host organisms are parasites, obtaining nutrients from the host. Indeed, many of the parasites that have significant effects on human health are protists, causing diseases such as malaria, sleeping sickness, and giardiasis.

STUDY BREAK 27.1

What distinguishes protists from prokaryotes? What distinguishes protists from fungi, land plants, and animals?

THINK OUTSIDE THE BOOK

Phylogenomics employs comparisons of genome sequences or the sequences of many genes simultaneously to construct phylogenetic trees. Use the Internet or research literature to outline how recent phylogenomic analysis has provided new insights into the evolutionary relationships among the protists.

27.2 The Protist Groups

This section presents the biological features of each of the groups of protists in Figure 27.2.

The Excavates Include Protists with Unique Flagella and Protists with Modified Mitochondria

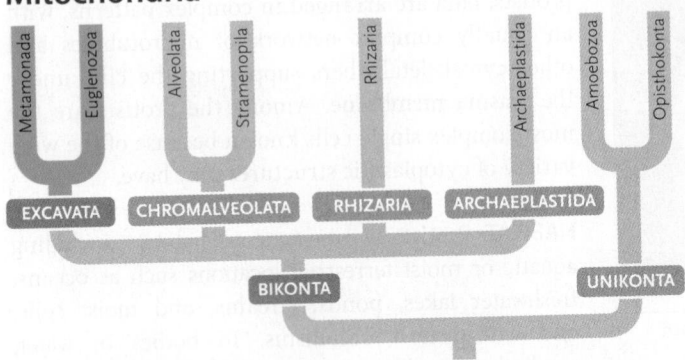

The **Excavata** are a diverse group of protists. Many excavates have a scooped out (excavated) feeding apparatus on the ventral surface of the cell from which the group gets its name. Some excavates have flagella, some are anaerobes, some are photosynthetic, and some are parasites. Two major subgroups of the Excavata are the **Metamonada** and **Euglenozoa.**

METAMONADA **Metamonads** have multiple flagella and lack mitochondria. The lack of functional mitochondria means that metamonads are anaerobic because they are incapable of aerobic respiration and are limited to glycolysis as an ATP source. However, the nuclei of metamonads contain genes derived from mitochondria, meaning that the ancestors of these pro-

tists had mitochondria that were lost or reduced during evolution of the group.

The metamonads consist of the Diplomonadida and Parabasala. *Diplomonad* cells have two nuclei and move by means of multiple freely beating flagella. In addition to lacking mitochondria, they also lack a clearly defined endoplasmic reticulum and Golgi complex. The best-known representative of the group, *Giardia lamblia* **(Figure 27.4A),** infects the mammalian intestinal tract, inducing severe diarrhea and abdominal cramps. *Giardia* is spread by water that is contaminated with feces, in which resistant cysts of the protist can be present in large numbers.

Parabasalids also have freely beating flagella, as well as a sort of fin called an **undulating membrane,** formed by a flagellum buried in a fold of the cytoplasm. The buried flagellum allows parabasalids to move through thick and viscous fluids. Among the Parabasala are the *trichomonads*, including *Trichomonas vaginalis* **(Figure 27.4B),** which is responsible for infections of the urinary and reproductive tracts in both men and women. The infective trichomonad is passed from person to person primarily, but not exclusively, by sexual intercourse. It lives in the vagina in women and in the urethra of both sexes. The infection is usually symptomless in men, but in women *T. vaginalis* can cause severe inflammation and irritation of the vagina and vulva. It is cured easily by drugs.

A. *Giardia lamblia*

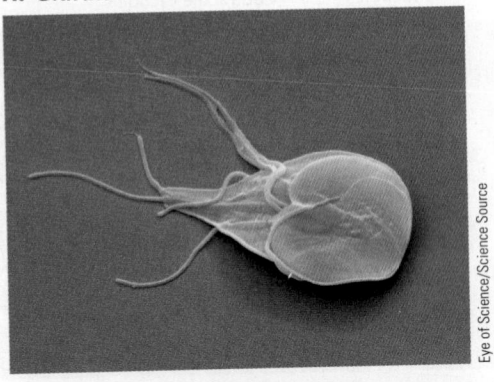

Eye of Science/Science Source

B. *Trichomonas vaginalis*

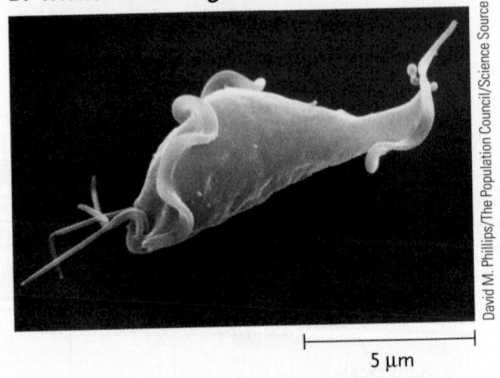

David M. Phillips/The Population Council/Science Source

5 μm

FIGURE 27.4 Metamonads of the Excavata. (A) A diplomonad, *Giardia lamblia,* that causes intestinal disturbances. **(B)** A parabasalid, *Trichomonas vaginalis,* that causes a sexually transmitted disease, trichomoniasis.

EUGLENOZOA Euglenozoans include about 1,800 species, almost all single-celled, highly motile cells that swim by means of flagella. They have functional mitochondria characterized by disc-shaped cristae (inner mitochondrial membranes). While most are photosynthetic, some are facultative heterotrophs. Some can even alternate between photosynthesis and life as a heterotroph. The two major subgroups of the Euglenozoa are the *euglenids* and the *kinetoplastids*.

Euglenids. **Euglenids** are free-living protists with anterior flagella. The best-known species is *Euglena gracilis* **(Figure 27.5)**. With the exception of a few marine species, the euglenids inhabit freshwater ponds, streams, and lakes. Most are autotrophs that carry out photosynthesis by the same mechanisms as plants, using the same photosynthetic pigments, including chlorophylls *a* and *b* and β-carotene. Many of the photosynthetic euglenids, including *E. gracilis,* can also live as heterotrophs by absorbing organic molecules through the plasma membrane. Some euglenids lack chloroplasts and live entirely as heterotrophs.

The cytoplasmic organelles of *E. gracilis* and other euglenids include a contractile vacuole and, in photosynthetic species, chloroplasts (see Figure 27.5). Instead of an external cell wall, the euglenids have a spirally grooved pellicle formed from transparent, protein-rich material. Most of the photosynthetic euglenids, including *E. gracilis,* have an *eyespot* containing carotenoid pigment granules in association with a light-sensitive structure. The eyespot is part of a sensory mechanism that stimulates cells to swim toward moderately bright light or away from intensely bright light so that the organism is positioned in light conditions for optimal photosynthetic activity. The cells swim by whiplike movements of flagella that extend from a pocketlike depression at one end of the cell. Most have two flagella, one rudimentary and short, the other long.

Kinetoplastids. **Kinetoplastids** are a group of nonphotosynthetic, heterotrophic cells that live as animal parasites **(Figure 27.6)**. Their name reflects the structure of the single mitochondrion in their cells, which contains a large DNA–protein deposit called a *kinetoplast.* Most kinetoplastids have two flagella, which are used for movement. In some cases, as in Figure 27.6, one of the flagella is attached to the side of the cell, forming an undulating membrane that is often used to enable the organism to glide along or attach to surfaces.

The kinetoplastids include the *trypanosomes,* responsible for several diseases afflicting millions of humans in tropical regions. *Trypanosoma brucei* (see Figure 27.6) causes African sleeping sickness, transmitted from one host to another by bites of the tsetse fly. Early symptoms include fever, headaches, rashes, and anemia. Untreated, the disease damages the central nervous system, leading to a sleeplike coma and eventual death. The disease has proved difficult to control because the same trypanosome infects wild mammals, providing an inexhaustible reservoir for the parasite. Other trypanosomes, also transmitted by insects, cause leishmaniasis, which can produce disfiguring skin sores and ulcers, and Chagas disease, which can lead to severe brain and heart damage. Chagas disease is common in the Americas and has been reported in the southern United States.

FIGURE 27.5 Body plan and an electron micrograph of *Euglena gracilis.* The plane of section in the electron micrograph has cut off all but the base of the flagellum.
© Cengage Learning 2017

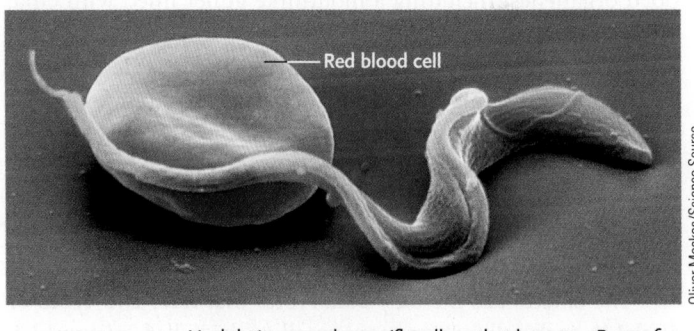

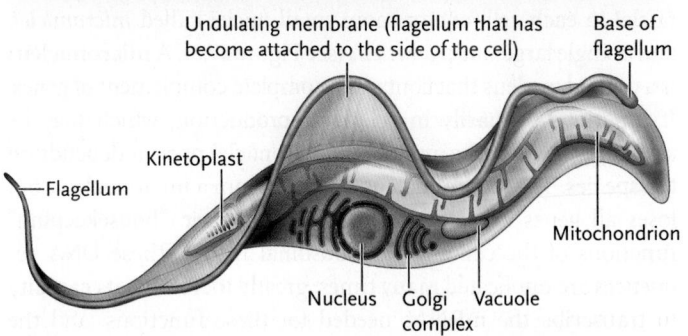

FIGURE 27.6 *Trypanosoma brucei,* the parasitic kinetoplastid that causes African sleeping sickness.
© Cengage Learning 2017

The Chromalveolates Are a Heterogeneous Group of Protists with a Range of Forms and Lifestyles

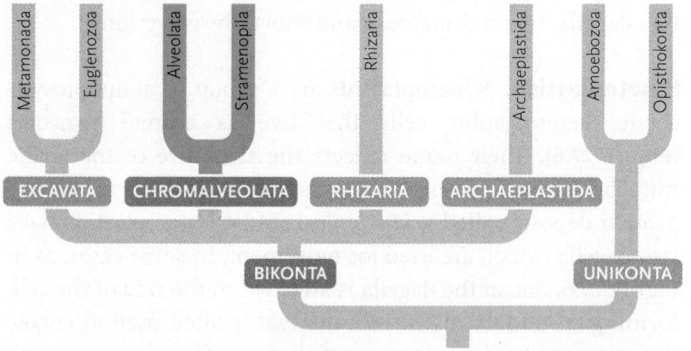

The **Chromalveolata** consists of the **Alveolata** and the **Stramenopila.**

ALVEOLATA The **alveolates** are so called because they have small, flattened, membrane-bound vesicles called *alveoli* (*alvus* = belly) in a layer just under the plasma membrane. The alveolates include two motile, primarily free-living groups, the **Ciliophora** and **Dinoflagellata,** and a nonmotile, parasitic group, the **Apicomplexa.**

Ciliophora. Ciliophora (ciliates) include nearly 10,000 known species of primarily single-celled but highly complex heterotrophic organisms that swim by means of cilia (see Figures 27.1B and 27.3). Essentially any sample of pond water or bottom mud contains a wealth of these creatures.

The organisms in the Ciliophora have many highly developed organelles, including a mouthlike gullet lined with cilia; structures that exude mucins, toxins, or other defensive and offensive materials from the cell surface; contractile vacuoles; and complex systems of food vacuoles. A pellicle reinforces cell shape. A complex cytoskeletal network of microtubules and other fibers anchors the cilia just below the pellicle and coordinates the ciliary beating. The cilia can stop and reverse their beating in synchrony, allowing ciliates to stop, back up, and turn if they encounter negative stimuli.

The ciliates are the only eukaryotes that have two types of nuclei in each cell: one or more small nuclei called *micronuclei*, and a single larger *macronucleus* (see Figure 27.3). A **micronucleus** is a diploid nucleus that contains a complete complement of genes. It functions primarily in cellular reproduction, which may be asexual or sexual. The number of micronuclei present depends on the species. The **macronucleus** develops from a micronucleus, but loses all genes except those required for basic "housekeeping" functions of the cell and for ribosomal RNAs. These DNA sequences are duplicated many times, greatly increasing its capacity to transcribe the mRNAs needed for these functions, and the rRNAs needed to make ribosomes.

In asexual reproduction by mitosis, both types of nuclei replicate their DNA, divide, and are passed on to daughter cells. **Figure 27.7** illustrates sexual reproduction of *Paramecium*, which is initiated when two cells **conjugate** and a cytoplasmic bridge forms between the cells.

Ciliates abound in freshwater and marine habitats, where they feed on bacteria, other protists, and each other. *Paramecium* is a typical member of the group (see Figure 27.3). Its rows of cilia drive it through its watery habitat, rotating the cell on its long axis while it moves forward, or backs and turns. The cilia also sweep water laden with prey and food particles into the gullet, where food vacuoles form. The ciliate digests food in the vacuoles and eliminates indigestible material through an anal pore. Contractile vacuoles with elaborate, raylike extensions remove excess water from the cytoplasm and expel it to the outside. When ciliates are under attack or otherwise stressed, surface organelles called **trichocysts** discharge dartlike protein threads.

Some ciliates live individually whereas others are colonial. Certain ciliates are animal parasites. Others live and reproduce in their hosts as mutually beneficial symbionts. A compartment of the stomach of cattle and other grazing animals contains large numbers of symbiotic ciliates that digest the cellulose in their host's plant diet. The animals then digest the excess ciliates.

One ciliate, *Balantidium coli*, is a human intestinal parasite that causes diarrhea, with stools typically containing blood and pus. It is acquired when humans eat food contaminated by the feces of animals infected by *Balantidium*, particularly pigs.

Dinoflagellata. Dinoflagellata (dinoflagellates) consists of over 4,000 known species, most of which are single-celled organisms in marine phytoplankton. They live as heterotrophs or autotrophs; many can carry out both modes of nutrition. Some contain algae as symbionts. Typically, they have a shell formed from cellulose plates **(Figure 27.8).** The beating of flagella, which fit into grooves in the plates, makes dinoflagellates spin like a top, hence their name (*dinos* = spinning).

The cytoplasmic structures of dinoflagellates include mitochondria, chloroplasts in photosynthetic species, and other internal membrane systems characteristic of eukaryotes. The photosynthetic dinoflagellates contain chlorophylls *a* and *c* along with accessory pigments that make them golden-brown or brown. Algal symbionts give some a green, blue, or red color.

Their abundance in phytoplankton makes dinoflagellates a major primary producer of ocean ecosystems. Some species live as symbionts in the tissues of other marine organisms such as jellyfish, sea anemones, corals, and mollusks. For example, dinoflagellates in coral use the coral's carbon dioxide and nitrogenous waste, while supplying 90% of the coral's nutrition. The vast numbers of dinoflagellates living as photosynthetic symbionts in tropical coral reefs allow the reefs to reach massive size. Without the dinoflagellates, many coral species would die.

Some dinoflagellates are **bioluminescent**—they glow or release a flash of light, particularly when disturbed. The production of light depends on the enzyme *luciferase* and its substrate *luciferin*, which is similar to the system that produces light in fireflies. Dinoflagellate fluorescence can make the sea glow in the wake of a boat at night, and coat nocturnal surfers and swimmers with a ghostly light.

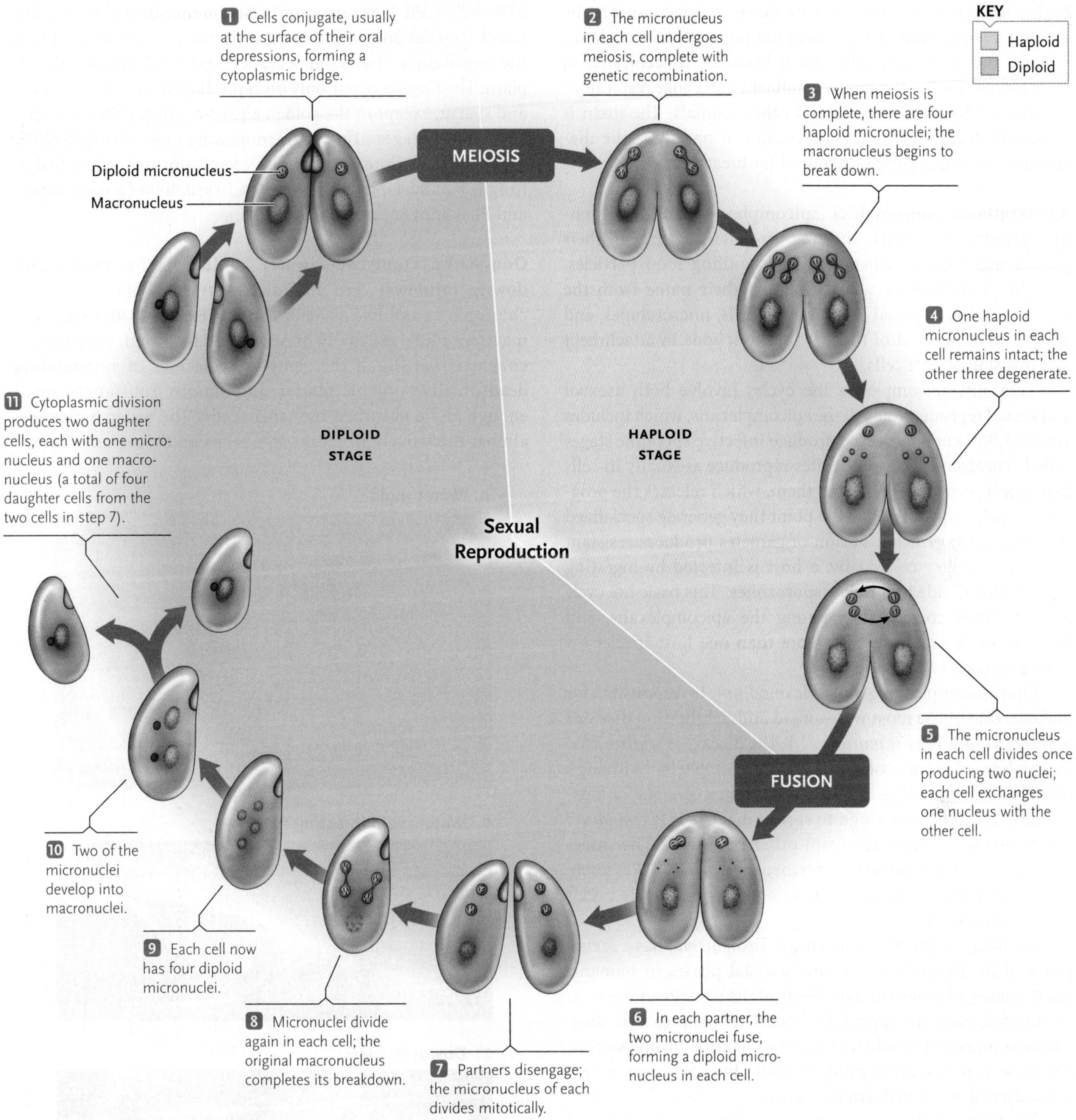

KEY

☐ Haploid
☐ Diploid

1 Cells conjugate, usually at the surface of their oral depressions, forming a cytoplasmic bridge.

2 The micronucleus in each cell undergoes meiosis, complete with genetic recombination.

MEIOSIS

3 When meiosis is complete, there are four haploid micronuclei; the macronucleus begins to break down.

Diploid micronucleus
Macronucleus

4 One haploid micronucleus in each cell remains intact; the other three degenerate.

DIPLOID STAGE

HAPLOID STAGE

11 Cytoplasmic division produces two daughter cells, each with one micronucleus and one macronucleus (a total of four daughter cells from the two cells in step 7).

Sexual Reproduction

5 The micronucleus in each cell divides once producing two nuclei; each cell exchanges one nucleus with the other cell.

FUSION

10 Two of the micronuclei develop into macronuclei.

9 Each cell now has four diploid micronuclei.

8 Micronuclei divide again in each cell; the original macronucleus completes its breakdown.

7 Partners disengage; the micronucleus of each divides mitotically.

6 In each partner, the two micronuclei fuse, forming a diploid micronucleus in each cell.

FIGURE 27.7 Sexual reproduction by conjugation in a ciliate, *Paramecium*.
© Cengage Learning 2017

At times dinoflagellate populations grow to such large numbers that they color the seas red, orange, or brown. The resulting **red tides** are common in spring and summer months along the warmer coasts of the world, including all the United States coasts. Some red-tide dinoflagellates (see Figure 27.8) produce a toxin that interferes with nerve function in the animals that ingest these protists. Fish that feed on plankton, and

FIGURE 27.8 *Karenia brevis*, a toxin-producing dinoflagellate.

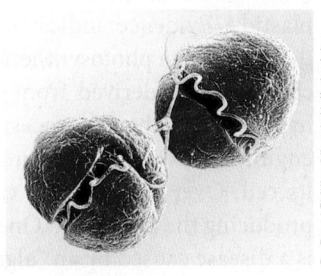

Dr. David Phillips/Visuals Unlimited, Inc.

birds that feed on the fish, may be killed in huge numbers by the toxin. Dinoflagellate toxin does not noticeably affect clams, oysters, and other mollusks, but it becomes concentrated in their tissues. Eating the tainted mollusks can cause respiratory failure and death in humans and other animals. The toxin is especially deadly for mammals because it paralyzes the diaphragm and other muscles required for breathing.

Apicomplexa. Apicomplexa (**apicomplexans**) are all nonmotile parasites of animals. They absorb nutrients through their plasma membranes rather than by engulfing food particles, and they lack food vacuoles. They get their name from the *apical complex,* a special grouping of fibrils, microtubules, and organelles at one end of the cell that functions in attachment and invasion of host cells.

Typically, apicomplexan life cycles involve both asexual and sexual reproduction. All the apicomplexans, which includes almost 4,000 known species, produce infective sporelike stages called *sporozoites.* The sporozoites reproduce asexually in cells they infect, eventually bursting them, which releases the progeny to infect new cells. At some point they generate specialized cells that form gametes. Fusion of gametes produces resistant cells known as *cysts.* Usually, a host is infected by ingesting cysts, which divide to produce sporozoites. This basic life cycle pattern varies considerably among the apicomplexans, and many of these organisms use more than one host species for different stages of their life cycle.

One apicomplexan genus, *Plasmodium,* is responsible for malaria, one of the most widespread and debilitating diseases of humans (see *Why it matters . . .*). The disease is transmitted by the bite of mosquitoes of the genus *Anopheles.* Although the disease is now rare in the United States, *Anopheles* mosquitoes are common enough to spread malaria if *Plasmodium* is introduced by travelers from other countries. The infective cycle of *Plasmodium,* described in *Focus on Research: Applied Research,* is representative of the complex life cycles of apicomplexans.

Another organism in this group, *Toxoplasma,* has a sexual phase of its life cycle in cats and asexual phases in humans, cattle, pigs, and other animals. Cysts of the parasite in the feces of infected cats are spread in household and garden dust. Humans ingesting or inhaling the cysts develop toxoplasmosis, a disease that is usually mild in adults but can cause severe brain damage or even death to a fetus.

Most apicomplexan parasites, including *Plasmodium* and *Toxoplasma,* contain a remnant plastid called an **apicoplast.** (Plastids are discussed in Section 4.4; chloroplasts are a type of plastid.) Evidence indicates that the apicoplast is a plastid derived from a photosynthetic eukaryote. By contrast, the plant chloroplast is derived from a photosynthetic bacterium. It is hypothesized that an ancestor of present-day apicomplexans engulfed a red alga and then incorporated the alga's plastid into its cell. Over evolutionary time, the plastid became reduced, producing the apicoplast. One can muse, therefore, that malaria is a disease caused by an "alga gone bad."

STRAMENOPILA Stramenopila (**stramenopiles**) have two different flagella: one with hollow tripartite projections that give the flagellum a "hairy" appearance and a second one that is plain. The flagella occur only on reproductive cells such as eggs and sperm, except in the golden algae, in which cells are flagellated in all stages. The Stramenopila includes the Oomycota (water molds, white rusts, and mildews—formerly classified as fungi), Bacillariophyta (diatoms), Chrysophyta (golden algae), and Phaeophyta (brown algae).

Oomycota. Oomycota (**water molds, white rusts, and downy mildews**)—are funguslike stramenopiles that lack chloroplasts and live as heterotrophs **(Figure 27.9).** Oomycota may reproduce asexually or sexually. Like fungi, they secrete enzymes that digest the complex molecules of surrounding dead or alive organic matter into simpler substances small enough to be absorbed into their cells. The water molds live almost exclusively in freshwater lakes and streams or moist

A. Water mold

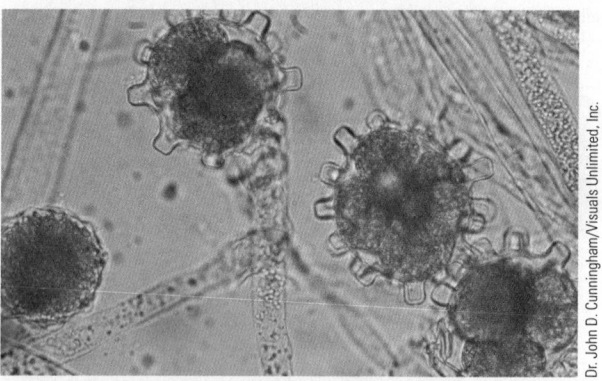

Dr. John D. Cunningham/Visuals Unlimited, Inc.

B. Water mold infecting fish

Heather Angel

C. Downy mildew

W. Merrill

FIGURE 27.9 Oomycota. (A) The water mold *Achlya recurva.* **(B)** The water mold *Saprolegnia parasitica* growing as cottony white fibers on the tail of an aquarium fish. **(C)** A downy mildew, *Plasmopara viticola,* growing on grapes. At times it has nearly destroyed vineyards in Europe and North America.

Focus on Research

Applied Research: Malaria and the *Plasmodium* Life Cycle

Malaria is uncommon in the United States, but it is a major epidemic in many other parts of the world (see *Why it matters...*). Five different species of *Plasmodium* can infect humans and cause malaria. In the life cycle of the parasites **(Figure)**, sporozoites develop in a female *Anopheles* mosquito, which transmits them by its bite to human or bird hosts. The infecting parasites divide repeatedly in their hosts, initially in liver cells and then in red blood cells. Their growth causes red blood cells to rupture in regular cycles every 48 or 72 hours. The ruptured red blood cells clog vessels and release the parasite's metabolic wastes, causing cycles of chills and fever.

The victim's immune system is ineffective because, during most of the infective cycle,

the parasite is inside body cells and thus "hidden" from antibodies. Further, *Plasmodium* regularly changes its surface molecules, continually producing new forms that are not recognized by antibodies developed against a previous form. In this way, the parasite keeps one step ahead of the immune system, often making malarial infections essentially permanent.

Travelers in countries with high rates of malaria are advised to use antimalarial drugs such as chloroquine, quinine, or quinidine as a preventive measure. However, many *Plasmodium* strains in Africa, India, and Southeast Asia have developed resistance to the drugs. Vaccines have proved difficult to develop; because vaccines work

by inducing the production of antibodies that recognize surface groups on the parasites, they are defeated by the same mechanisms the parasite uses inside the body to keep one step ahead of the immune reaction.

While in a malarial region, travelers should avoid exposure to mosquitoes by remaining indoors from dusk until dawn and sleeping inside mosquito nets treated with insect repellent. When out of doors, travelers should wear clothes that expose as little skin as possible and are thick enough to prevent mosquitoes from biting through the cloth. An insect repellent containing DEET should be spread on any skin that is exposed.

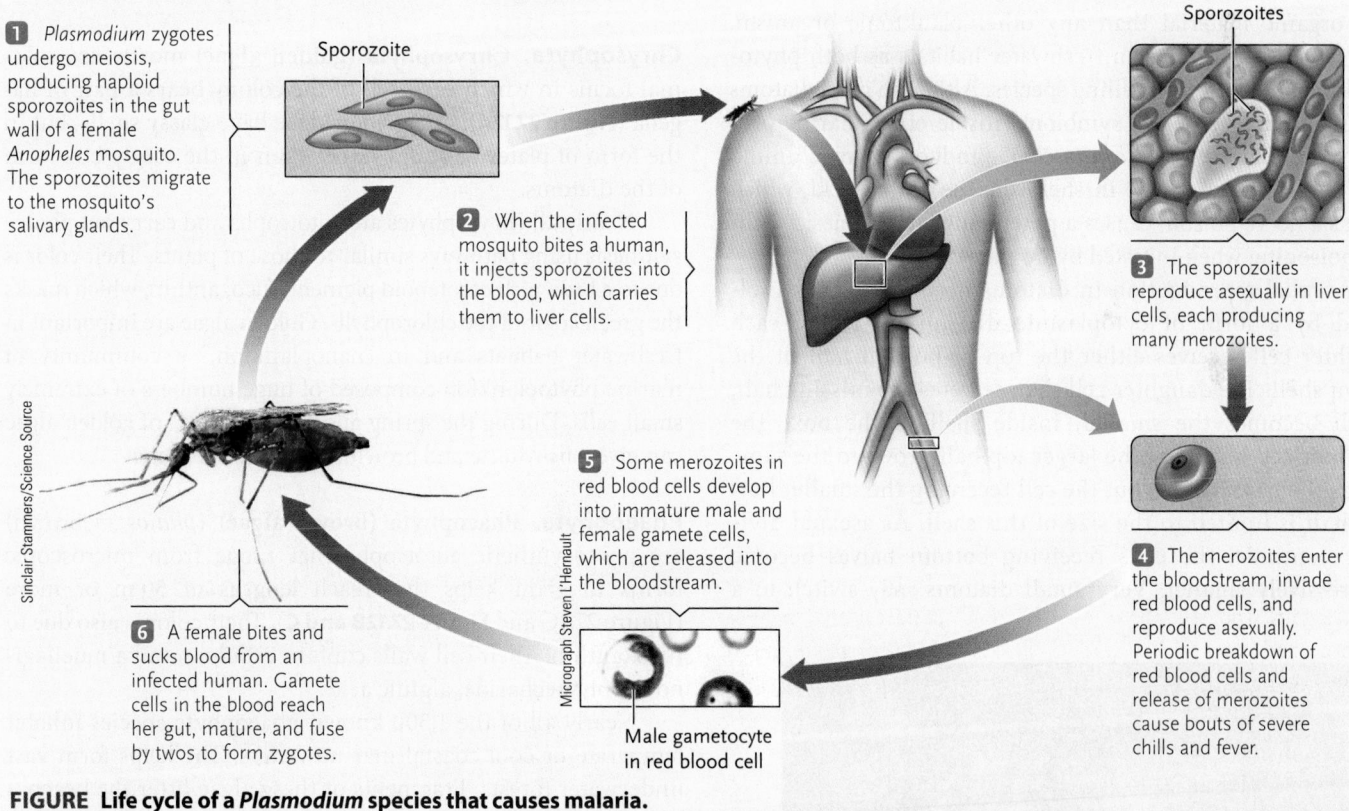

FIGURE Life cycle of a *Plasmodium* species that causes malaria.

1. *Plasmodium* zygotes undergo meiosis, producing haploid sporozoites in the gut wall of a female *Anopheles* mosquito. The sporozoites migrate to the mosquito's salivary glands.

2. When the infected mosquito bites a human, it injects sporozoites into the blood, which carries them to liver cells.

3. The sporozoites reproduce asexually in liver cells, each producing many merozoites.

4. The merozoites enter the bloodstream, invade red blood cells, and reproduce asexually. Periodic breakdown of red blood cells and release of merozoites cause bouts of severe chills and fever.

5. Some merozoites in red blood cells develop into immature male and female gamete cells, which are released into the bloodstream.

6. A female bites and sucks blood from an infected human. Gamete cells in the blood reach her gut, mature, and fuse by twos to form zygotes.

Sporozoite

Sporozoites

Male gametocyte in red blood cell

Sinclair Stammers/Science Source

Micrograph Steven L'Hernault

© Cengage Learning 2017

terrestrial habitats. Most water molds are key decomposers of both aquatic and moist terrestrial habitats. Dead animal or plant material immersed in water commonly becomes coated with cottony water molds. Other water molds parasitize living aquatic animals, such as the mold growing on the fish shown in Figure 27.9B. The white rusts and downy mildews are parasites of land plants. For example, the downy mildew

Phytophthora infestans is a water mold that infects valuable crop plants such as potatoes. Its destruction of potato crops was responsible for the Irish potato famine between 1845 and 1852. Today, related species threaten forests in the United States, Europe, and Australia.

Like fungi, many Oomycota grow as microscopic, nonmotile filaments called **hyphae** (singular, *hypha*), which form a network called a **mycelium** (see Figure 27.9A). However, differences in DNA sequence clearly indicate close evolutionary relationships to the stramenopiles rather than to the fungi. Further, nuclei in hyphae are diploid in the Oomycota, rather than haploid as in the fungi; reproductive cells are flagellated and motile, whereas fungi have no motile stages; and the cell walls of most Oomycota contain cellulose (see Figure 3.8C), whereas fungal cell walls contain a different polysaccharide, chitin (see Figure 3.8D).

Bacillariophyta. **Bacillariophyta (diatoms)** are single-celled organisms that are covered by a glassy silica shell, which is intricately formed and beautiful in many species. The two halves of the shell fit together like the top and bottom of a candy box **(Figure 27.10).** Substances move to and from the plasma membrane through elaborately patterned perforations in the shell. Although flagella are present only in gametes, many diatoms move by an unusual mechanism in which a secretion released through grooves in the shell propels them in a gliding motion.

Diatoms are autotrophs that carry out photosynthesis by pathways similar to those of plants. The primary photosynthetic organisms of marine plankton, diatoms fix more carbon into organic material than any other planktonic organism. They are also abundant in freshwater habitats as both phytoplankton and bottom-dwelling species. Although most diatoms are free living, some are symbionts inside other marine protists. One diatom, *Pseudo-nitzschia,* produces a toxic amino acid that can accumulate in shellfish. The amino acid, which acts as a nerve poison, causes a potentially fatal amnesic shellfish poisoning when ingested by humans.

Asexual reproduction in diatoms occurs by mitosis followed by a form of cytoplasmic division in which each daughter cell receives either the top or bottom half of the parent shell. The daughter cell then secretes the missing half, which becomes the smaller, inside shell of the box. The daughter cell receiving the larger top half grows to the same size as the parent shell, but the cell receiving the smaller bottom half is limited to the size of this shell. As asexual divisions continue, the cells receiving bottom halves become progressively smaller. Very small diatoms may switch to a sexual mode of reproduction in which they enter meiosis and produce flagellated gametes that lose their shells and fuse in pairs to form a zygote. The zygote grows to normal size before secreting a completely new shell with full-size top and bottom halves.

The shells of diatoms are common in fossil deposits. In fact, more diatoms are known as fossils than as living species—some 35,000 extinct species have been described as compared with 7,000 living species. For about 180 million years, the shells of diatoms have been accumulating into thick layers of sediment at the bottom of lakes and seas. Since diatoms store food as oil, fossil diatoms may be a source of oil in many oil deposits. **Figure 27.11** presents the results of genome analysis experiments that provide some insights into the evolution of diatoms.

Grinding the fossilized shells into a fine powder produces *diatomaceous earth,* which is used in abrasives and filters, as an insulating material, and as a pesticide. Diatomaceous earth kills crawling insects by abrading their exoskeleton, causing them to dehydrate and die. Insect larvae are killed in the same way. Insects also die when they eat the powder but larger animals, including humans, are unaffected by it.

Chrysophyta. **Chrysophyta (golden algae)** mostly are colonial forms in which each cell of the colony bears a pair of flagella **(Figure 27.12A).** The golden algae have glassy shells, but in the form of plates or scales rather than in the candy-box form of the diatoms.

Nearly all chrysophytes are autotrophs and carry out photosynthesis using pathways similar to those of plants. Their color is due to a brownish carotenoid pigment, fucoxanthin, which masks the green color of the chlorophylls. Golden algae are important in freshwater habitats and in "nanoplankton," a community of marine phytoplankton composed of huge numbers of extremely small cells. During the spring and fall, "blooms" of golden algae can give a fishy taste and brownish color to the water.

Phaeophyta. **Phaeophyta (brown algae)** (*phaios* = brown) are photosynthetic autotrophs that range from microscopic forms to giant kelps that reach lengths of 50 m or more (Figure 27.1C and **Figure 27.12B and C**). Their color is also due to fucoxanthin. Their cell walls contain cellulose and a mucilaginous polysaccharide, alginic acid.

Nearly all of the 1,500 known phaeophyte species inhabit temperate or cool coastal marine waters. The kelps form vast underwater forests. Fragments of these algae litter the beaches in coastal regions where they grow. Great masses of another brown alga, *Sargassum,* float in an area of the mid-Atlantic Ocean called the Sargasso Sea, which covers millions of square kilometers between the Azores and the Bahamas.

Kelps are the largest and most complex of all protists. Their tissues are differentiated into leaflike *blades,* stalklike *stipes,* and rootlike *holdfasts* that anchor them to the bottom. Hollow, gas-filled bladders give buoyancy to the stipes and blades and help keep them upright. The stalks of some kelps contain tube-like vessels, similar to the vascular elements of plants, which

FIGURE 27.10 Diatom shells. Depending on the species, the shells are either radially or bilaterally symmetrical, as seen in this sample.

Jan Hinsch/Science Source

FIGURE 27.11 **Experimental Research**

The Evolutionary History of Diatoms as Revealed by Genome Analysis

Question: What do diatom genomes reveal about their evolutionary history?

Experiment: Diatoms are members of the chromalveolates. There are two major classes of diatoms, the radially symmetrical centrics and the bilaterally symmetric pennates. The earliest fossil centrics date to 180 million years ago, whereas the earliest fossil pennates date to 90 million years ago. Researchers have long been interested in how the two types of diatoms evolved and diverged.

Researchers compared the genome sequences of two diatoms, the pennate diatom, *Phaeodactylum tricornutum,* and the centric diatom, *Thalassiosira pseudonana,* with the genome sequences of three metazoans, two plants, three green algae, one red alga, two fungi, and ten other chromalveolates.

Results: The table below presents the results of the genome sequence comparisons. In the table, three types of genes are distinguished:

- *Core genes:* Genes found in all of the eukaryotic groups with which the diatom sequences were compared.
- *Diatom-specific genes:* Genes found in both diatoms, but not in the other organisms.
- *Unique genes:* Genes found only in one or other of the two diatoms.

	Diatom species	
	P. tricornutum	*T. pseudonana*
Genome size (Mb[1])	27.4	32.4
Number of genes	10,402	11,776
Core genes	3,523	4,332
Diatom-specific genes	1,328	1,407
Unique genes	4,366	3,912

[1] Mb = 1 megabase = 10^6 base pairs

Notably, the results show that the two diatoms have only about 57% of their genes in common, illustrating a high degree of divergence since the split to the two diatom lineages.

The researchers made some other observations from the genome sequence comparisons:

- 171 of their chloroplast genes are of red algal origin; 108 of these genes are shared by the two species.
- 587 genes are of prokaryotic origin; 56% of these genes are found in both diatoms. These genes are presumed to have been obtained by horizontal gene transfer, the movement of genetic material between organisms other than by descent (see Sections 17.1 and 26.1). These genes represent more than 5% of the total genes of these diatoms, an unusually high percentage for a eukaryote. Most of the genes can be traced to proteobacteria, cyanobacteria, and various archaea. This observation is interpreted to mean that there has been a long-term association between diatoms and bacteria that has led to the transfer of useful genes.
- The diatom-specific genes are evolving faster than other genes in these diatoms or in eukaryotes in general. Although the exact reason for this fact is unknown, it is thought that it may correlate with the extensive diversification of diatoms.

Conclusion: Analysis of the two diatom genome sequences has provided insights into the evolutionary history of diatoms. The analysis provides interesting comparisons with other eukaryotic organisms, and supports the hypothesis that chromalveolates evolved from symbiosis between a photosynthetic red alga and a heterotrophic host (a concept discussed further later in the chapter). The evidence for the extent of horizontal gene transfer, and for the increased rate of evolution of diatom-specific genes, may help explain the high degree of diversity of diatoms and their adaptations for their life in the oceans.

think like a scientist
Observe
Hypothesize
Predict
Experiment
Interpret

What would you do to obtain a better understanding of the evolutionary history of diatoms?

Sources: C. Bowler et al. 2008. The *Phaeodactylum* genome reveals the evolutionary history of diatom genomes. *Nature* 456:239–244; E. V. Armbrust et al. 2004. The genome of the diatom *Thalassiosira pseudonana:* ecology, evolution, and metabolism. *Science* 306:79–86.

FIGURE 27.12 Golden and brown algae. (A) A microscopic, swimming colony of *Synura,* a golden alga. Each cell bears two flagella, which are not visible in this light micrograph. **(B)** A brown alga, *Macrocystis.* Gas bladders keep the blades floating. **(C)** The holdfast, stemlike stipes, and leaflike blades, as seen in another brown alga, the sea palm *Polstelsia palmaeformis.*

A. Golden alga, *Synura*

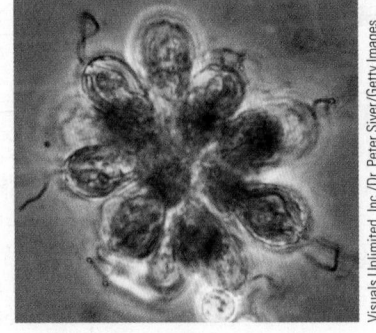

Visuals Unlimited, Inc./Dr. Peter Siver/Getty Images

B. Brown alga, *Macrocystis*

NatalieJean/Shutterstock.com

C. Brown alga, *Polstelsia palmaeformis*

Jeffrey Levinton, State University of New York, Stony Brook

distribute dissolved sugars and other products of photosynthesis rapidly throughout the body of the alga.

Typically, life cycles among the brown algae are complex and in many species consist of alternating haploid and diploid generations (**Figure 27.13**). The large structures recognized as kelps and other brown seaweeds are diploid **sporophytes**, so called because they give rise to haploid spores by meiosis. The spores germinate and divide by mitosis to form a small, independent, haploid **gametophyte** generation, which gives rise to haploid gametes. Variations occur in smaller brown algae, including some life cycles in which the sporophytes and gametophytes are the same size and some in which the gametophyte is larger than the sporophyte.

The alginic acid in brown algal cell walls, called **algin** when extracted, is an essentially tasteless and nontoxic substance used to thicken such diverse products as ice cream, pudding, salad dressing, jellybeans, cosmetics, paper, and floor polish. Brown algae are also harvested as food crops and as fertilizers.

The Rhizarians Consist of Amoebas with Filamentous Pseudopods

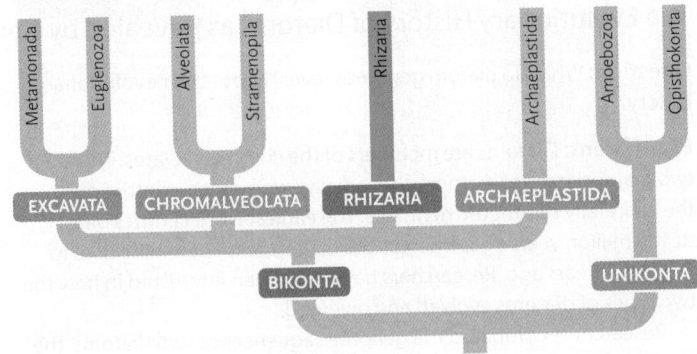

Amoeba is a descriptive term for a single-celled protist that moves by means of temporary cellular projections called pseudopods. Several major groups of protists contain amoebas, which are similar in form but are not all closely related. The

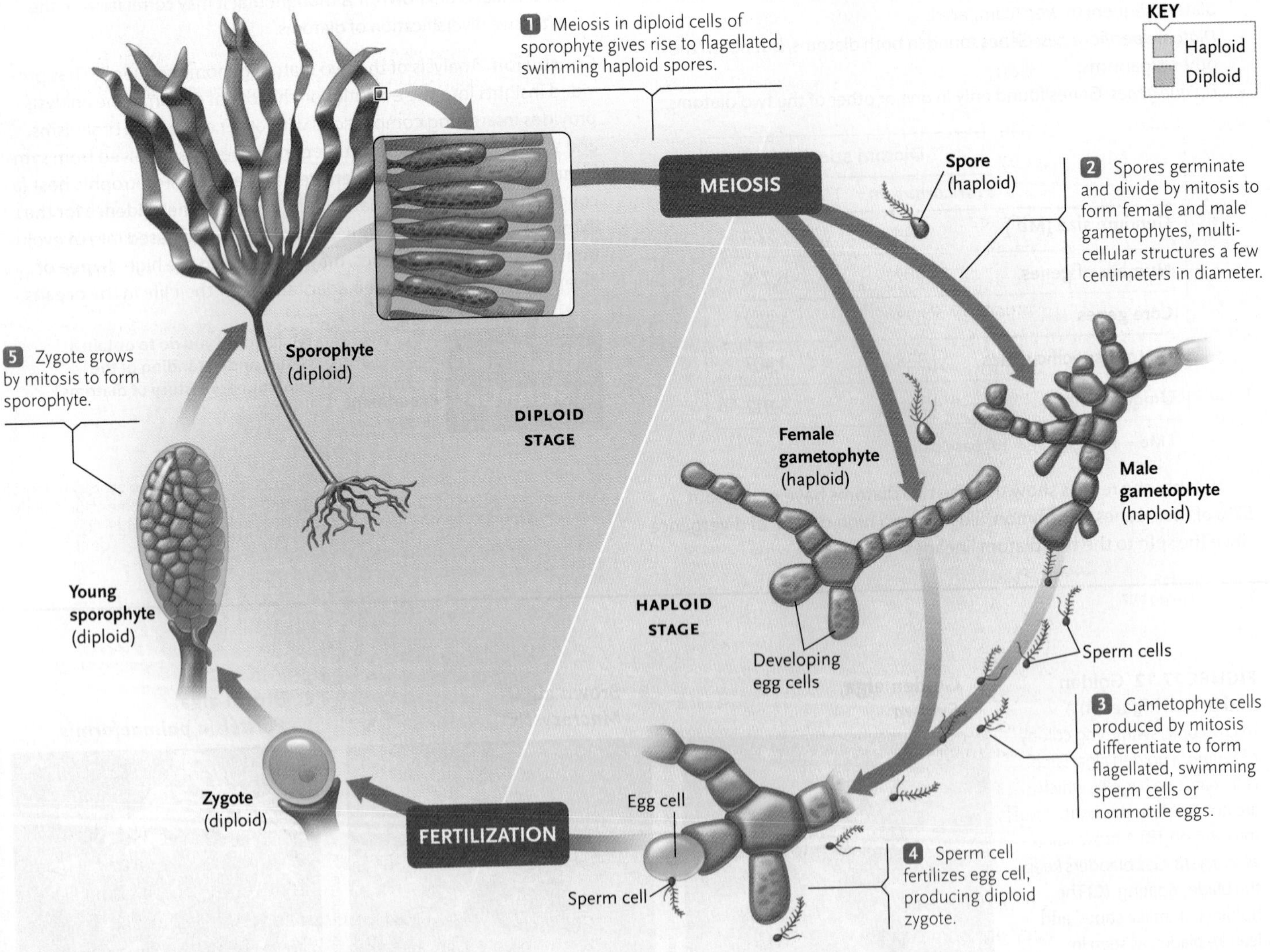

KEY

- Haploid
- Diploid

1 Meiosis in diploid cells of sporophyte gives rise to flagellated, swimming haploid spores.

MEIOSIS

Spore (haploid)

2 Spores germinate and divide by mitosis to form female and male gametophytes, multicellular structures a few centimeters in diameter.

Sporophyte (diploid)

DIPLOID STAGE

5 Zygote grows by mitosis to form sporophyte.

Female gametophyte (haploid)

Male gametophyte (haploid)

Young sporophyte (diploid)

HAPLOID STAGE

Developing egg cells

Sperm cells

3 Gametophyte cells produced by mitosis differentiate to form flagellated, swimming sperm cells or nonmotile eggs.

Zygote (diploid)

Egg cell

FERTILIZATION

Sperm cell

4 Sperm cell fertilizes egg cell, producing diploid zygote.

FIGURE 27.13 The life cycle of the brown alga *Laminaria,* which alternates between a diploid sporophyte stage and a haploid gametophyte stage.

A. Radiolarian skeletons

Gene Shih & Richard Kessel/Visuals Unlimited, Inc.

B. Living foram

Courtesy of Allen W. H. Be and David A. Caron

C. Foram shells

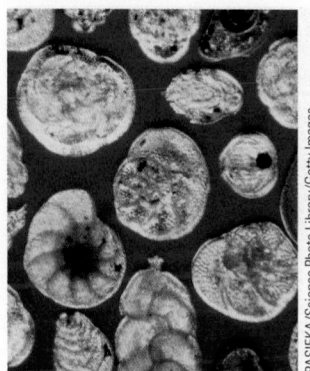

PASIEKA/Science Photo Library/Getty Images

D. Foram body plan

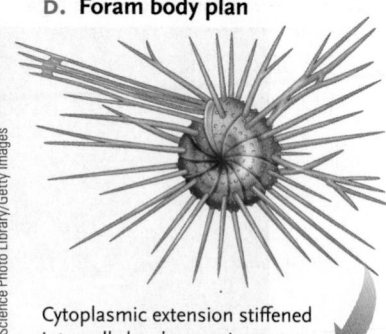

Cytoplasmic extension stiffened internally by glassy spine

FIGURE 27.14 Radiolarians and forams. (A) The internal skeletons of two radiolarian species, possibly *Pterocorys* and *Stylosphaera*. Bundles of microtubules support the cytoplasmic extensions of the radiolarians. **(B)** A living foram, showing the cytoplasmic strands extending from its shell. **(C)** Empty foram shells. **(D)** The body plan of a foram. Needlelike, glassy spines support the cytoplasmic extensions of the forams.

© Cengage Learning 2017

amoebas of the **Rhizaria** produce stiff, filamentous pseudopodia, and many produce hard outer shells, also called *tests*. We consider here two heterotrophic groups of rhizarian amoebas, the Radiolaria and the Foraminifera, and a third, photosynthesizing group, the Chlorarachniophyta.

RADIOLARIA Radiolaria **(radiolarians)** are marine protists that are distinguished by axopods, slender, raylike strands of cytoplasm supported internally by long bundles of microtubules. They engulf prey organisms that stick to the axopods and digest them in food vacuoles.

Radiolarians secrete a glassy internal skeleton from which the axopods project **(Figure 27.14A and B).** Just outside the skeleton, the cytoplasm is crowded with frothy vacuoles and lipid droplets, which provide buoyancy. The skeletons of dead radiolarians sink to the bottom of the ocean and become part of the sediment. Over time, they harden into sedimentary rocks that form an important part of the geological record.

FORAMINIFERA (FORAMS) Foraminifera **(forams)** are also marine protists. Their shells consist of organic matter reinforced by calcium carbonate **(Figure 27.14C and D).** Most foram shells are chambered, spiral structures that, although microscopic, resemble those of mollusks. Forams are identified and classified primarily by the form of the shell; about 250,000 species are known. Some species are planktonic, but they are most abundant on sandy bottoms and attached to rocks along coasts. Their name comes from the perforations in their shells (*foramen* = little hole), through which extend long, slender strands of cytoplasm supported internally by a network of needlelike spines. The forams engulf prey that adhere to the strands and conduct them through the holes in the shell into the central cytoplasm, where they are digested in food vacuoles. Some forams have algal symbionts that carry out photosynthesis, allowing them to live as both heterotrophs and autotrophs.

Marine sediments typically are packed with the shells of dead forams. The sediments may be hundreds of feet thick; the White Cliffs of Dover in England are composed primarily of the shells of ancient forams. Most of the world's deposits of limestone and marble contain foram shells. The great pyramids and many other monuments of ancient Egypt are built from blocks cut from fossil foram deposits. Because distinct species lived during different geological periods, they are widely used to establish the age of sedimentary rocks containing their shells.

CHLORARACHNIOPHYTA Chlorarachniophyta **(chlorarachniophytes)** are green, photosynthetic amoebas that also engulf food. They contain chlorophylls *a* and *b*, but they are phylogenetically distinct from other chlorophyll *b*-containing eukaryotes. Many filamentous pseudopodia extend from the cell surface.

The Archaeplastids Include the Red and Green Algae, and Land Plants

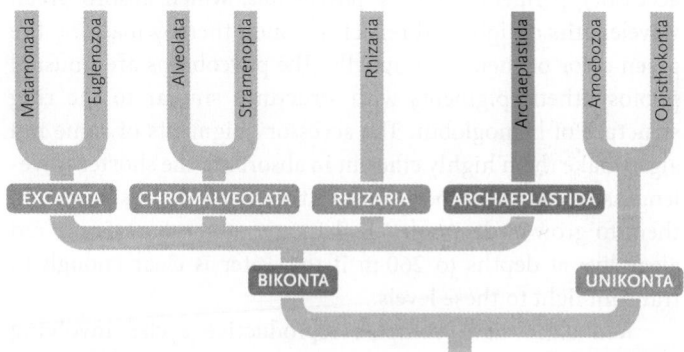

The **Archaeplastida** consist of the Rhodophyta (red algae) and Chlorophyta (green algae), which are protists, and the land plants (the *viridaeplantae*, or "true plants"), which comprise the kingdom Plantae. These three groups are descended from the first eukaryote that acquired a chloroplast from an endosymbiotic cyanobacterium (see Section 25.3 and later in this chapter), and they are all photosynthesizers. Here we describe the red and green algae. Land plants are discussed in Chapters 28 and 29.

A. Filamentous red alga

Wim van Egmond/Visuals Unlimited, Inc.

B. Sheetlike red alga

Andrew J. Martinez/Science Source

FIGURE 27.15 Red algae. (A) *Antithamnion plumula,* showing the filamentous and branched body form most common among red algae. **(B)** A sheetlike red alga growing on a tropical reef.

RHODOPHYTA (RED ALGAE) Rhodophyta (red algae) (*rhodon* = rose) consist of about 4,000 known species, most of which are small marine seaweeds (**Figure 27.15**). Fewer than 200 species are found in freshwater lakes and streams or in soils. Most red algae grow attached to sandy or rocky substrates, but a few occur as plankton. Although most are free-living autotrophs, some are parasites that attach to other algae or plants.

Red algae are typically multicellular organisms, with plantlike bodies that are composed of interwoven filaments. The base of the body is differentiated into a holdfast, which anchors it to the bottom or other solid substrate, and into stalks with leaflike plates. Their cell walls contain cellulose and mucilaginous pectins that give them a slippery texture. In some species, the walls are hardened with stonelike deposits of calcium carbonate. Many of the red algae with stony cell walls resemble corals and occur with corals in reefs and banks.

Although most red algae are reddish in color, some are greenish purple or black. The color differences are produced by accessory pigments, mainly *phycobilins,* which absorb green wavelengths of light and reflect red ones, thereby masking the green color of their chlorophylls. The phycobilins are unusual photosynthetic pigments with structures similar to the ring structure of hemoglobin. The accessory pigments of some red algae make them highly efficient in absorbing the shorter wavelengths of light that penetrate to the ocean depths, allowing them to grow at deeper levels than any other algae. Some red algae live at depths to 260 m if the water is clear enough to transmit light to these levels.

Red algae have complex reproductive cycles involving alternation between diploid sporophytes and haploid gametophytes. No flagellated cells occur in the red algae; instead, gametes are released into the water to be brought together by random collisions in currents.

Extracts containing the mucilaginous pectins of red algal cell walls are widely used in industry and science. Extracted **agar** is used as a moisture-preserving, inert agent in cosmetics and baked goods, as a setting agent for jellies and desserts, and as a solidifying agent for culture medium in the laboratory.

Carrageenan, extracted from the red alga *Eucheuma,* is used to thicken and stabilize paints, dairy products such as pudding and ice cream, and many other creams and emulsions.

Some red algae are harvested as food in Japan and China. *Porphyra,* one of these harvested algae, is used in sushi bars as the *nori* wrapped around fish and rice.

CHLOROPHYTA (GREEN ALGAE) Chlorophyta (green algae) (*chloros* = green) are autotrophs that carry out photosynthesis using the same pigments as plants. They include single-celled, colonial, and multicellular species (**Figure 27.16**; see also

A. Single-celled green alga

Jovana Bila Dubaic/Shutterstock.com

C. Multicellular green alga

Marevision/Age Fotostock/Getty Images

B. Colonial green alga

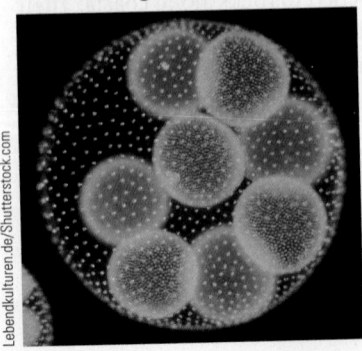

Lebendkulturen.de/Shutterstock.com

FIGURE 27.16 Green algae. (A) A single-celled green alga, *Acetabularia,* which grows in marine environments. Each individual in the cluster is a large, multinucleate cell with a rootlike base, stalk, and cap. **(B)** A colonial green alga, *Volvox.* Each green dot in the spherical wall of the colony is a potentially independent, flagellated cell. Daughter colonies can be seen within the parent colony. **(C)** A multicellular green alga, *Ulva,* common to shallow seas around the world.

Figure 27.1D). Most green algae are microscopic, but some range upward to the size of small seaweeds. Although the multicellular green algae have bodies that are filamentous, tubular, or leaflike, they have relatively little cellular differentiation as compared with the brown algae. However, the most complex green algae, such as the sea lettuce *Ulva* (see Figure 27.16C), have tissues differentiated into a leaflike body and a holdfast.

With at least 16,000 species, green algae show more diversity than any other algal group. Most live in freshwater aquatic habitats, but some are marine, or live on rocks and soil surfaces, on tree bark, or even on snow. The green, slimy mat that grows profusely in stagnant pools and ponds, for example, consists of filaments of a green alga. A few species live as symbionts in other protists or in fungi and animals. Lichens (see Figure 30.15) are the primary example of a symbiotic relationship between green algae and fungi. Many animal phyla, including some marine snails and sea anemones, contain green algal chloroplasts, or entire green algae, as symbionts in their cells.

Life cycles among the green algae are as diverse as their body forms. Many can reproduce either sexually or asexually, and some alternate between haploid and diploid generations. Gametes in different species may be morphologically identical flagellated cells, or cells that have differentiated into a flagellated sperm cell and a nonmotile egg cell. Most common is a life cycle with a multicellular haploid phase and a single-celled diploid phase **(Figure 27.17).**

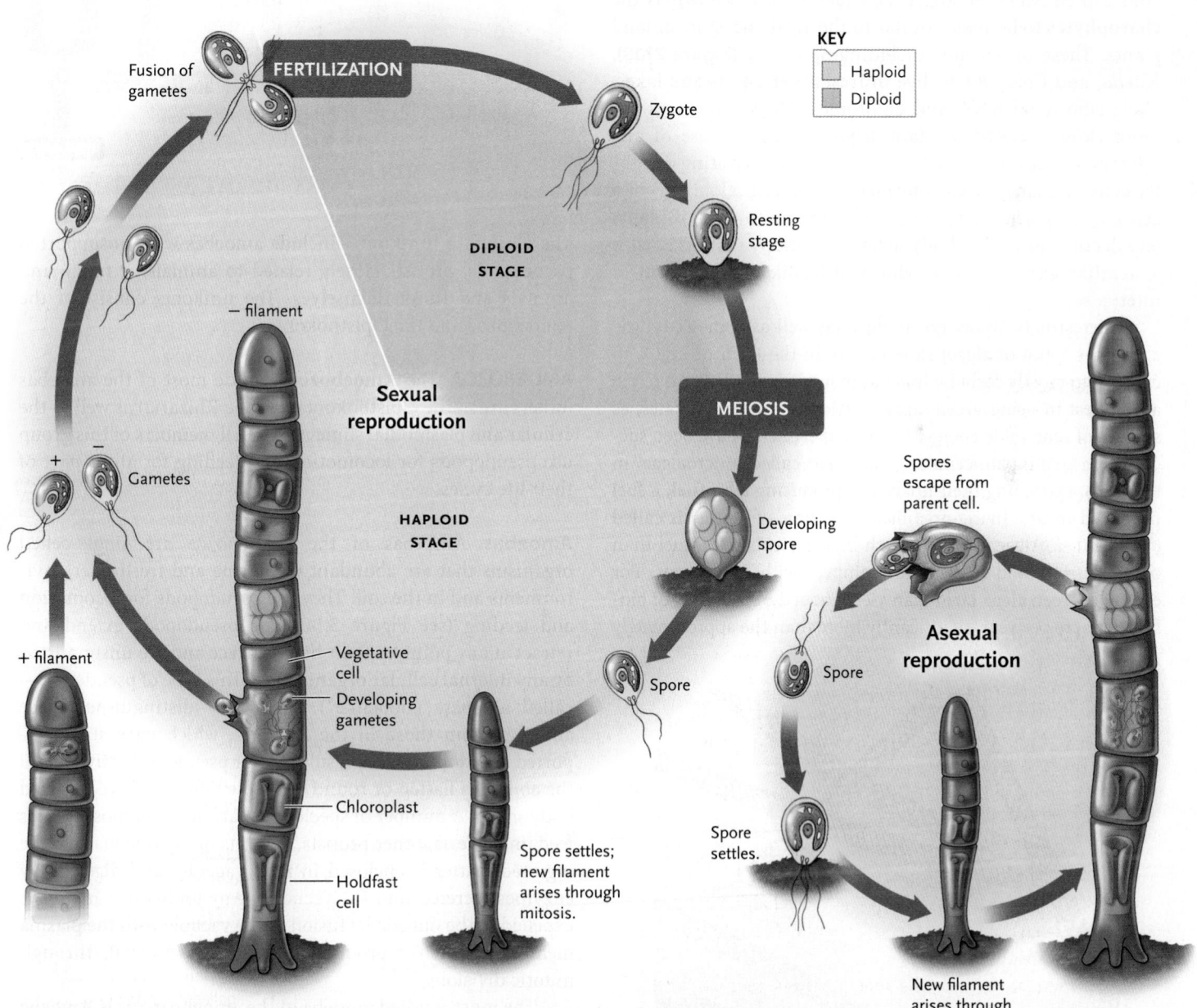

FIGURE 27.17 The life cycle of the green alga *Ulothrix*. In sexual reproduction, the haploid stage is multicellular and the diploid stage is a single cell, the zygote. "+" and "−" are morphologically identical mating types ("sexes") of the alga. In asexual reproduction, all stages are haploid.

© Cengage Learning 2017

Among all the algae, the nucleic acid sequences of green algae are most closely related to those of land plants. In addition, as already noted, green algae use the same photosynthetic pigments as plants, including chlorophylls *a* and *b,* and have the same complement of carotenoid accessory pigments. In some green algae, the thylakoid membranes within chloroplasts are arranged into stacks resembling the grana of plant chloroplasts (see Section 4.4). Green algae contain starches of the same types as plants, and the cell walls of some green algal species contain cellulose, pectins, and other polysaccharides like those of plants. On the basis of these similarities, many biologists propose that some ancient green alga gave rise to the evolutionary ancestors of modern-day plants.

What green alga might have been the ancestor of modern land plants? Many biologists consider a group known as the **charophytes** to be most similar to the algal ancestors of land plants. These organisms, including *Spirogyra* (**Figure 27.18**), *Nitella,* and *Coleochaete,* live in freshwater ponds and lakes. Their ribosomal RNA and chloroplast DNA sequences are more closely related to plant sequences than those of any other green alga. Further, the new cell wall separating daughter cells in charophytes is formed through development of a cell plate, by a mechanism closely similar to that of plants (see Section 10.2). The body form is distinctly plantlike, with a stemlike axis on which whorls of leaflike blades occur at intervals.

Interestingly, many green algae (as well as some red algae and other types of algae) store energy in the form of lipids, in contrast to mostly carbohydrates as in land plants. This has led to an interest in using green algae, particular unicellular ones, as sources of renewable energy. In fact, experiments have been successful in farming unicellular green algae (called "microalgae" in the field), extracting their lipids and producing a **biofuel,** a fuel produced from a living organism. This type of biofuel is called *algal biofuel.* Algae produce much higher yields of biofuel than other sources, and it does not compete with food crops. For example, green algae farms can yield about 2,500 gallons of biofuel per acre per year, significantly more than the approximately

48 gallons for soybeans, and 18 gallons for corn. Moreover, algae can be farmed in areas that are unsuitable for agriculture. As of October 2014, the cost of production of one gallon of algal biofuel was $7.50, far above the present cost of a gallon of gasoline based on fossil fuel. However, despite genomic manipulations to produce a strain of lipid-rich algae to offset culturing and processing costs, the fuel industry considers algae biofuel to be at least 25 years away from becoming mainstream.

The Unikonts Include Protists That Are Closely Related to Animals and Fungi

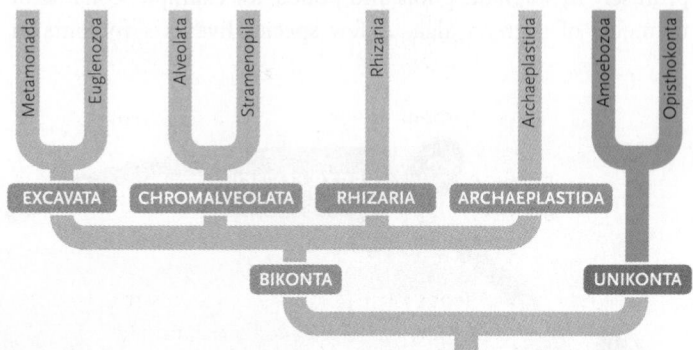

The **Unikonta (unikonts)** include amoebas with unsupported pseudopods, protists closely related to animals or fungi, and animals and fungi themselves. The unikonts consist of the Amoebozoa and the Opisthokonta.

AMOEBOZOA The **Amoebozoa** include most of the amoebas (others are in the Opisthokonta and the Rhizaria) as well as the cellular and plasmodial slime molds. All members of this group use pseudopods for locomotion and feeding for all or part of their life cycles.

Amoebas. Amoebas of the Amoebozoa are single-celled organisms that are abundant in marine and freshwater environments and in the soil. They use pseudopods for locomotion and feeding (see Figure 5.14). The pseudopods extend and retract at any point on their body surface and are unsupported by any internal cellular organization. This type of pseudopod—called a *lobose* ("lobelike") *pseudopod*—distinguishes these amoebas from those in the Rhizaria, which have stiff, supported pseudopods. As a result of their pseudopod activity, and the ability to flatten or round up, these amoebas have no fixed body shape. A number of species are parasites, but most species feed on bacteria, other protists, and bits of organic matter. The ingested matter is enclosed in food vacuoles and digested by enzymes secreted into the vacuoles. Any undigested matter is expelled to the outside by fusion of the vacuole with the plasma membrane. Their reproduction is entirely asexual, through mitotic divisions.

The most-studied amoeba of the amoebozoans is *Amoeba proteus* (**Figure 27.19**). Its natural habitat is in freshwater ponds and streams. Another member, *Acanthamoeba,* which lives in the soil, is widely used as a source of actin and myo-

FIGURE 27.18 The charophyte *Spirogyra,* representative of a group of green algae that may have given rise to the plant kingdom.

Nancy Nehring//iStockphoto.com

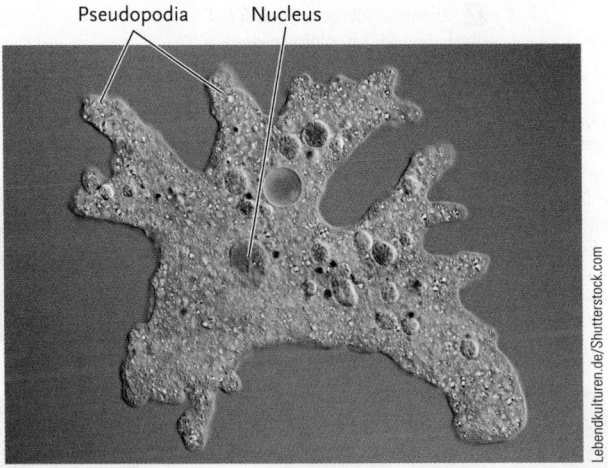

Pseudopodia   Nucleus

FIGURE 27.19 *Amoeba proteus* of the Amoebozoa, perhaps the most familiar protist.

sin for scientific studies of amoeboid motion and cytoplasmic streaming.

The parasitic amoebas include some 45 species that infect the human digestive tract, one in the mouth and the rest in the intestine. One of the intestinal parasites, *Entamoeba histolytica,* causes amoebic dysentery. Cysts of this amoeba contaminate water supplies and soil in regions with inadequate sewage treatment. When ingested, a cyst breaks open to release an amoeba that feeds and divides rapidly in the digestive tract. Enzymes released by the amoebas destroy cells lining the intestine, producing the ulcerations, painful cramps, and debilitating diarrhea characteristic of the disease. Amoebic dysentery afflicts millions of people worldwide. In less-developed countries, it is a leading cause of death among infants and small children. Other parasitic amoebas cause less severe digestive upsets.

Slime Molds. **Slime molds** are heterotrophic protists that, at some stage of their life cycle, exist as individuals that move by amoeboid motion but the remainder of the time exist in more complex forms. They live on moist, rotting plant material such as decaying leaves and bark. The cells engulf particles of dead organic matter, and also bacteria, yeasts, and other microorganisms, and digest them internally. At one stage of their life cycles, they differentiate into a funguslike, stalked structure called a **fruiting body,** which forms spores by either asexual or sexual reproduction. Some species are brightly colored in hues of yellow, green, red, orange, brown, violet, or blue. The two major evolutionary lineages of slime molds, the *cellular slime molds* and the *plasmodial slime molds,* differ in cellular organization.

Cellular slime molds exist primarily as individual cells, either separately or as a coordinated mass. Among the 70 or so species of cellular slime molds, *Dictyostelium discoideum* is best known. **Figure 27.20** shows asexual reproduction (steps 1–5) and sexual reproduction (steps 6–7) in *D. discoideum. (Molecular Insights* discusses bacteria associated symbiotically with *Dictyostelium.)*

Plasmodial slime molds exist primarily as a large composite mass, the **plasmodium,** in which individual nuclei are suspended in a common cytoplasm that is surrounded by a single plasma membrane. (Do not confuse this with *Plasmodium,* the genus of apicomplexans that causes malaria.) There are about 500 known species of plasmodial slime molds. The main phase of the life cycle, the plasmodium (see Figure 27.1A), flows and feeds as a single huge amoeba—a single cell that contains thousands to millions or even billions of diploid nuclei surrounded by a single plasma membrane. Typically, a plasmodium, which may range in size from a few centimeters to more than a meter in diameter, moves in thick, branching strands connected by thin sheets. The movements occur by cytoplasmic streaming, driven by actin microfilaments and myosin (see Section 4.3).

At some point, often in response to unfavorable environmental conditions, fruiting bodies form at sites on the plasmodium. At the tips of the fruiting bodies, nuclei become enclosed in separate cells, each surrounded by its own plasma membrane and cell wall. Depending on the species, either chitin or cellulose may reinforce the walls. These cells undergo meiosis, forming haploid, resistant spores that are released from the fruiting bodies and carried about by water or wind. If they reach a favorable environment, the spores germinate to form flagellated or unflagellated gametes, depending on the species, that fuse to form a diploid zygote. The zygote nucleus then divides repeatedly without an accompanying division of the cytoplasm, forming many diploid nuclei suspended in the common cytoplasm of a new plasmodium.

Both the cellular and plasmodial slime molds, particularly *Dictyostelium* (cellular) and *Physarum* (plasmodial), have been of great interest to scientists because of their ability to differentiate into fruiting bodies with stalks and spore-bearing structures. This differentiation is much simpler than the complex developmental pathways of other eukaryotes, providing a unique opportunity to study cell differentiation at its most fundamental level.

OPISTHOKONTA The **Opisthokonta (opisthokonts)** (*opisthen* = behind or posterior) are a broad group of eukaryotes that includes the fungi, animals, and two protist groups, the choanoflagellates and the nucleariids. If present at some stage of the life cycle, the single flagellum is located at the posterior.

Choanoflagellata (choanoflagellates) (*choanos* = collar) are named for a collar of closely packed microvilli that surrounds the single flagellum by which these protists move and take in food **(Figure 27.21).** The collar resembles an upside-down lampshade. There are about 150 species of choanoflagellates. They live in fresh and marine waters. Some species are mobile, with the flagellum pushing the cells along, as is the case with animal sperm, in contrast to most flagellates, which are pulled by their flagella. Most choanoflagellates, though, are *sessile;* that is, attached via a stalk to a surface. A number of species are colonial with a cluster of cells on a single stalk.

Choanoflagellates have the same basic structure as choanocytes (collar cells) of sponges, and they are similar to collared

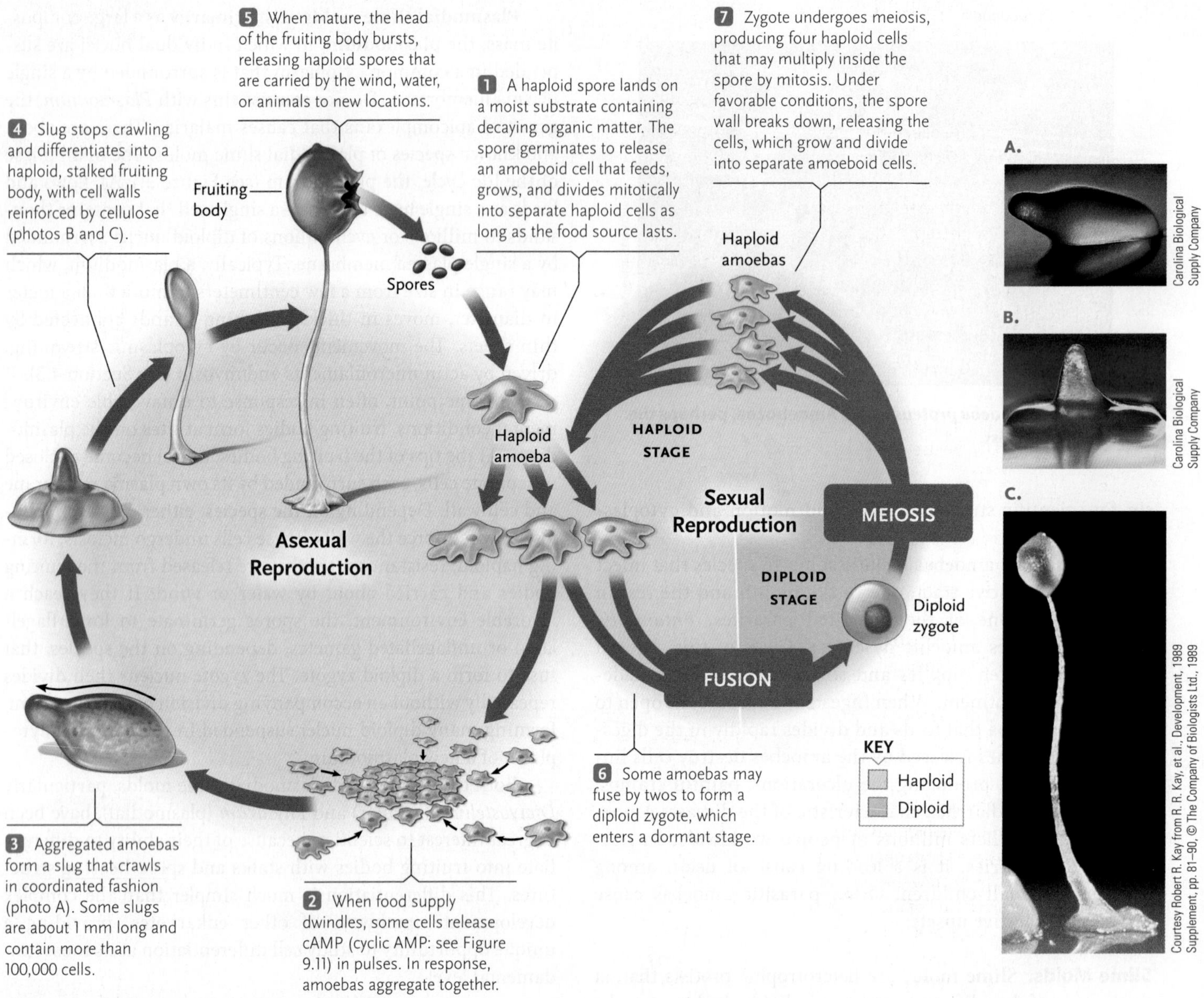

5 When mature, the head of the fruiting body bursts, releasing haploid spores that are carried by the wind, water, or animals to new locations.

1 A haploid spore lands on a moist substrate containing decaying organic matter. The spore germinates to release an amoeboid cell that feeds, grows, and divides mitotically into separate haploid cells as long as the food source lasts.

7 Zygote undergoes meiosis, producing four haploid cells that may multiply inside the spore by mitosis. Under favorable conditions, the spore wall breaks down, releasing the cells, which grow and divide into separate amoeboid cells.

4 Slug stops crawling and differentiates into a haploid, stalked fruiting body, with cell walls reinforced by cellulose (photos B and C).

Fruiting body

Spores

Haploid amoebas

Haploid amoeba

HAPLOID STAGE

Sexual Reproduction

MEIOSIS

Asexual Reproduction

DIPLOID STAGE

Diploid zygote

FUSION

6 Some amoebas may fuse by twos to form a diploid zygote, which enters a dormant stage.

KEY
Haploid
Diploid

3 Aggregated amoebas form a slug that crawls in coordinated fashion (photo A). Some slugs are about 1 mm long and contain more than 100,000 cells.

2 When food supply dwindles, some cells release cAMP (cyclic AMP: see Figure 9.11) in pulses. In response, amoebas aggregate together.

A.

Carolina Biological Supply Company

B.

Carolina Biological Supply Company

C.

Courtesy Robert R. Kay from R. R. Kay, et al., Development, 1989 Supplement, pp. 81–90. © The Company of Biologists Ltd., 1989

FIGURE 27.20 Life cycle of the cellular slime mold *Dictyostelium discoideum*. The light micrographs show **(A)** a migrating slug, **(B)** an early stage in fruiting body formation, and **(C)** a mature fruiting body.
© Cengage Learning 2017

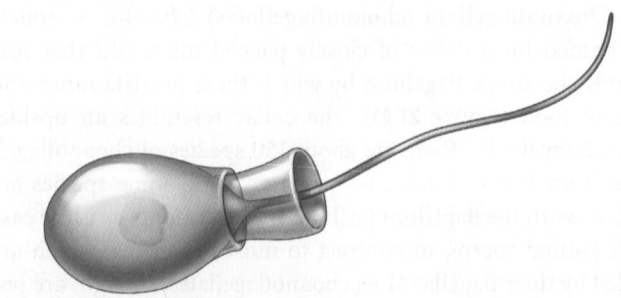

FIGURE 27.21 A choanoflagellate.
© Cengage Learning 2017

cells that act as excretory organs in organisms such as flatworms and rotifers. These morphological similarities, as well as molecular sequence comparison data, indicate that an ancestor of a present-day choanoflagellate is likely to have been the ancestor of animals.

Nucleariidae (nucleariids) are heterotrophic, predominantly spherical amoebae with radiating, fine pseudopods that are not supported by microtubules. Molecular sequence data show that nucleariids are more closely related to fungi than to animals (see Chapter 30).

Dictyostelium the Farmer: Two different bacteria associated with a cellular slime mold

About one-third of wild-collected clones of the cellular slime mold *Dictyostelium discoideum* are associated with bacteria during the sporulation and subsequent mitotic division phase (see Figure 27.20, step 1). These clones are considered to be "farmers" because they transport, seed, and then harvest the bacteria as food. But only about one-half of the bacteria carried by the farmers are used as a food source.

Research Question

Why are only one-half of the bacteria associated with *Dictyostelium* used as a food source?

Experiments

Experiments by researchers at Washington University in St. Louis and Harvard Medical School answered the question. The scientists first identified the food and nonfood bacteria associated with farmer *Disctyostelium*. Then they compared the two types of bacteria to identify chemical compounds that differed between them, and the genetic basis for the difference.

Results

The researchers identified both the food and nonfood bacteria as *Pseudomonas fluorescens* (**Figure**), a gammaproteobacterium (see Section 26.2). They grew the two strains in culture, made an extract of the cells, and looked for differences in metabolites. Chemical analysis of the extracts showed significant amounts of the iron chelator pyochelin in food bacteria, and a previously undescribed molecule chromene and the antifungal pyrrolnitrin in nonfood bacteria. Chromene enhances spore formation (see Figure 27.20, steps 4–5) in farmers, and suppresses spore formation in nonfarmers. Pyrrolnitrin functions similarly and may also inhibit microbial pathogens of other species.

Given the potential toxicity of pyrrolnitrin to *Dictyostelium*, could the farmers have adapted to associating with a potentially lethal bacterium? To answer this question, the researchers tested the effect of various concentrations of pyrolnitrin on farmer and nonfarmer (control) slime molds. The results showed that the nonfarmer was sensitive to the toxin effect of pyrrolnitrin whereas the farmer was not.

Next, the scientists looked for genetic differences between the food and nonfood bacteria. They sequenced the genomes of both bacterial strains and identified a key mutation in the *gacA* gene in the food strain. The gene product, GacA, is part of a system that is part of a response regulation system. When the researchers knocked out the *gacA* gene in the nonfood strain, the metabolite profile of that strain changed to match that of the food strain, and converted the strain into a food source.

Conclusion

The farmer *Dictyostelium* and the associated bacteria appear to have coevolved. A single mutation in a nonfood ancestral bacterial strain that served as a protective role for the slime mold converted it to a food source.

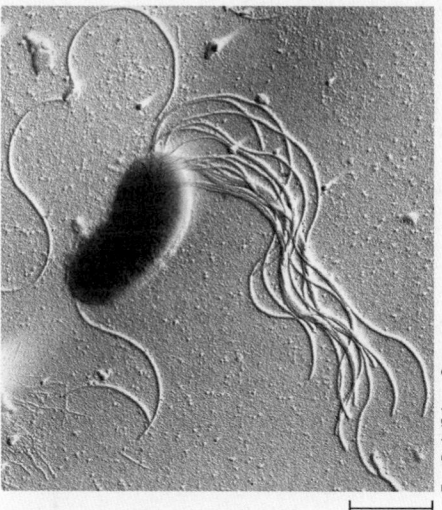

Tony Brain/Science Source

1 μm

FIGURE *Pseudomonas fluorescens* **(transmission electron micrograph).**

Subsequently both food and nonfood bacteria remained associated with the slime mold in a symbiotic relationship.

think like a scientist

Why was it important to knock out the *gacA* experimentally in the research?

Source: P. Stallforth et al. 2013. A bacterial symbiont is converted from an inedible producer of beneficial molecules into food by a single mutation in the *gacA* gene. *Proceedings of the National Academy of Sciences USA* 110:14528–14533.

© Cengage Learning 2017

In Several Protist Groups, Plastids Evolved from Endosymbionts

We have encountered chloroplasts in a number of eukaryotic organisms in this chapter: red algae, green algae, land plants, euglenids, dinoflagellates, stramenopiles, and chlorarachniophytes. How did these chloroplasts evolve?

In Section 25.3 we discussed the endosymbiotic theory for the origin of eukaryotes. In brief, an anaerobic prokaryote ingested an aerobic prokaryote, which survived as an endosymbiont (see Figure 25.6). Over many generations, the endosymbiont became an organelle, the mitochondrion, which was incapable of free living, and the result was a true eukaryotic cell. Cells of animals, fungi, and some protists derive from this ancestral eukaryote. The addition of plastids (the general term for chloroplasts and related organelles) through further endosymbiotic events produced the cells of all photosynthetic eukaryotes, including land plants, algae, and some other protists.

Figure 27.22 presents a model for the origin of plastids in eukaryotes through two major endosymbiosis events. First, in a single **primary endosymbiosis** event perhaps 600 million years ago, a eukaryotic cell engulfed a photosynthetic cyanobacterium (a photosynthetic prokaryote, remember). In some such cells, the

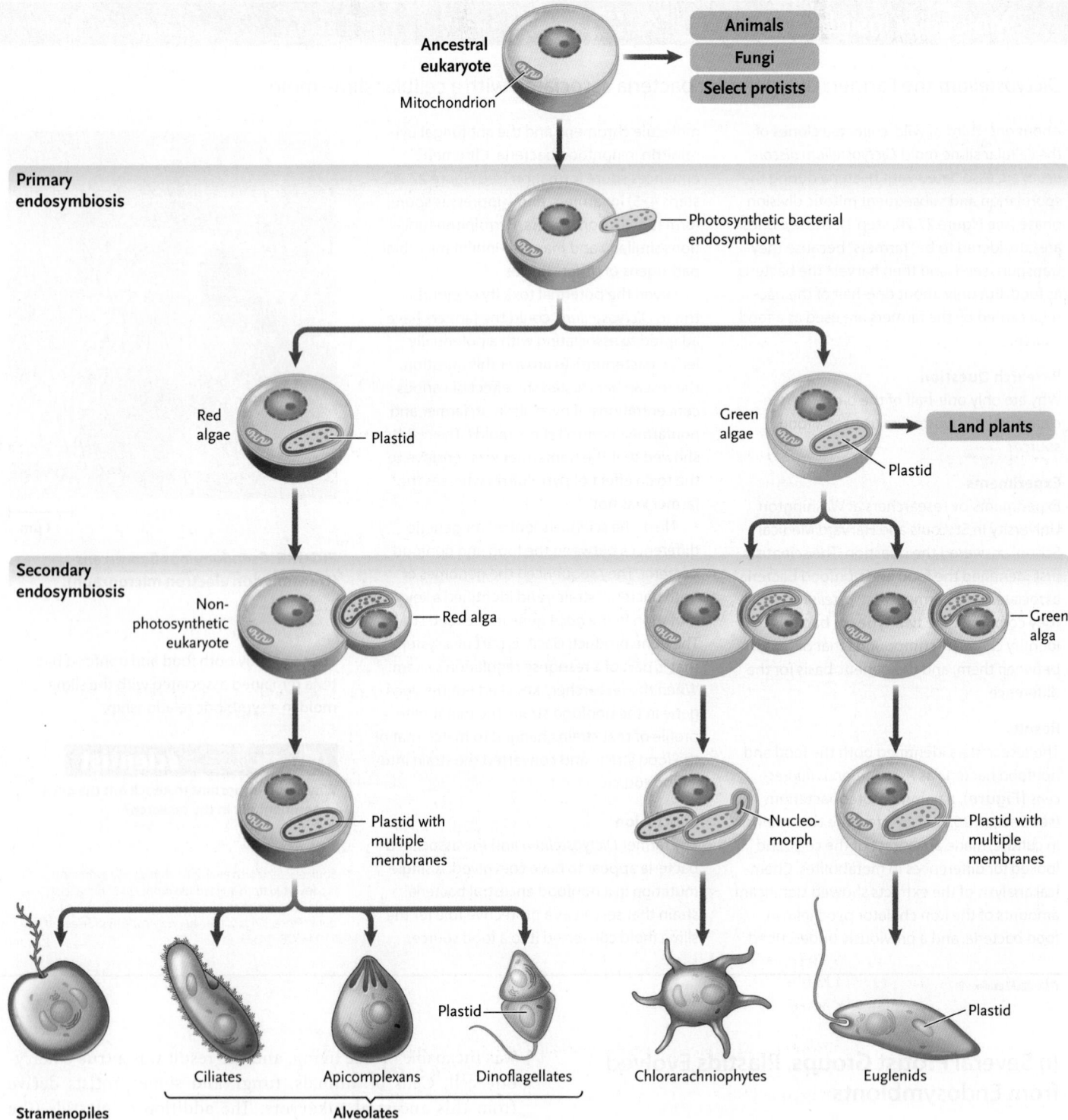

FIGURE 27.22 The origin and distribution of plastids among the eukaryotes by primary and secondary endosymbiosis.
© Cengage Learning 2017

cyanobacterium was not digested, but instead formed a symbiotic relationship with the engulfing host cell—it became an endosymbiont. Over time the symbiont lost genes no longer required for independent existence, and most of the remaining genes migrated from the prokaryotic genome to the host's nuclear genome. The symbiont had become an organelle—a chloroplast. All plastids subsequently evolved from this original chloroplast. Evidence for the single origin of plastids comes from a variety of sequence comparisons, including recent sequencing of the genomes of key protists: a red alga and a diatom.

The first photosynthesizing eukaryote was essentially an ancestral single-celled alga. The chloroplasts of the Archaeplastida—the red algae, green algae, and land plants—result from evolutionary divergence of this organism. Their chloroplasts, which originate from primary endosymbiosis, have two membranes, one from the plasma membrane of the engulfing eukaryote and the other from the plasma membrane of the cyanobacterium.

At least three **secondary endosymbiosis** events led to the plastids in other protists (see Figure 27.22). In each case, a non-photosynthetic eukaryote engulfed a photosynthetic eukaryote, and new evolutionary lineages were produced. In one of these events, a red alga ancestor was engulfed and became an endosymbiont. In models accepted by a number of scientists, the transfer of functions that occurred over evolutionary time led to the chloroplasts of the stramenopiles, ciliates, and dinoflagellates. And from the same photosynthetic ancestor, loss of chloroplast functions occurred in the lineage of the Apicomplexa, which have a remnant plastid (discussed earlier in the chapter). In an independent event, a nonphotosynthetic eukaryote engulfed a green alga ancestor. Subsequent evolution in this case produced the euglenids. In a different event, a similar endosymbiosis involving a green alga led to the chlorarachniophytes. In these protists, the chloroplast is contained still within the remnants of the original symbiont cell, with a vestige of the original nucleus (the nucleomorph) also present.

Note that secondary endosymbiosis has produced plastids with additional membranes acquired from the new host, or series of hosts. For example, euglenids have plastids with three membranes, while chlorarachniophytes have plastids with four membranes (see Figure 27.22). Sequencing the genomes of the chlorarachniophyte's nucleus, chloroplast, and vestigial nucleus is providing interesting information about the early endosymbiosis event that generated these organisms.

In sum, the protists are a highly diverse and ecologically important group of organisms. Their complex evolutionary relationships, which have long been the subject of contention, are now being revised as new information is discovered, including more complete genome sequences. A deeper understanding of protists is also contributing to a better understanding of their recent descendants, the fungi, plants, and animals. We turn to these descendants in the next five chapters, beginning with the fungi.

STUDY BREAK 27.2

1. What is the evidence that the metamonads, which lack mitochondria, derive from ancestors that had mitochondria rather than from ancestors that were in lineages that never contained mitochondria?
2. In primary endosymbiosis, a nonphotosynthetic eukaryotic cell engulfed a photosynthetic cyanobacterium. How many membranes surround the chloroplast that evolved?

 Unanswered Questions

What was the first eukaryote?

Because prokaryotes precede eukaryotes in the fossil record, we assume that eukaryotes arose after prokaryotes. The first eukaryote would have been some sort of protist—a single-celled organism with a nucleus and some rudimentary organelles, perhaps even a half-tamed mitochondrion. One approach to identifying which of the surviving protists is the most ancient has been to infer evolutionary trees from gene sequence data. To determine the earliest branching eukaryote, these trees need to include the prokaryotes. But herein lies the problem—prokaryotes are very distant, evolutionarily speaking, from even the simplest eukaryotes, and the mathematical models used to construct evolutionary trees are not yet up to the job. Initially, these models suggested that some protist parasites, like the excavates *Giardia* and *Trichomonas,* might be the most ancient eukaryotes, and this idea fit nicely with the fact that these protists lacked mitochondria. Indeed, for a time it was thought that the excavates might actually have diverged from the eukaryotic branch of life before the establishment of mitochondria. However, we know now that *Giardia* and *Trichomonas* did have mitochondria initially. The latest research shows that they even have a tiny relic of the mitochondrion, though exactly what it does in these oxygen-shunning parasites remains to be figured out. Thus, trees depicting *Giardia* and *Trichomonas* at the base of the great expansion of eukaryotic life must be viewed with some caution—these protists might be the surviving representatives of the earliest cells with a nucleus, but they might not be. We simply need better methods for identifying just what the first eukaryotes were like.

How many times did plastids arise by endosymbiosis?

For many years researchers thought that the green algae, plants, and red algae were the only organisms to have primary endosymbiosis-derived plastids. However, a second, independent primary endosymbiosis has been recently discovered in which a shelled amoeba has captured and partially domesticated a cyanobacterium. This organism, known as *Paulinella,* is a vital window into the process by which autotrophic eukaryotes first arose some 600 million years ago. *Paulinella* has tamed the cyanobacterium sufficiently to have it divide and segregate in coordination with host cell division, but the endosymbiont is still very much a cyanobacterium and has undergone little of the modification and streamlining we see in the red or green algal plastids.

After a primary endosymbiosis was established, the second chapter in plastid acquisition could take place. Secondary endosymbiosis involves a eukaryotic host engulfing and retaining a eukaryotic alga. Essentially, secondary endosymbiosis can convert a heterotrophic organism into an autotroph by hijacking a photosynthetic cell and putting it to work as a solar-powered food factory. Secondary endosymbiosis results in plastids with three or four membranes, and we know that it occurred at least three times—once for the euglenids, once for the chlorarachniophytes, and once for the chromalveolates (a grouping of stramenopiles and alveolates). We can even tell what kind of endosymbiont was involved by the biochemistry and genetic makeup of the plastid: a green alga for euglenids and chlorarachniophytes, and a red alga for chromalveolates. The number of secondary endosymbioses is hotly debated, largely because not all protistologists support the existence of chromalveolates. Some contend that there were multiple, independent enslavements of different red algae to produce the dinoflagellates, stramenopiles, and apicomplexans. Understanding these events is crucial to confirming or refuting the proposed chromalveolate "supergroup."

A nice example of secondary endosymbiosis-in-action was recently discovered by Japanese scientists who found a flagellate, *Hatena,* with a green algal endosymbiont. *Hatena* has not yet assumed control of endosymbiont division and has to get new symbionts each time it divides, so it appears to be at a very early stage in establishing a relationship. We also want to know how secondary endosymbioses proceed because they have been a major driver in eukaryotic evolution. The stramenopiles, for instance, are the most important ocean phytoplankton and are key to ocean productivity and global carbon cycling. Knowing exactly how they got to be autotrophs in the first place is fundamental to understanding the world we live in.

REVIEW KEY CONCEPTS

For access to MindTap and additional study materials visit www.cengagebrain.com.

27.1 What Is a Protist?

- Protists are eukaryotes that differ from fungi in having motile stages in their life cycles and distinct cell wall molecules. Unlike plants, they lack true roots, stems, and leaves. Unlike animals, protists lack collagen, nerve cells, and an internal digestive tract, and they lack complex developmental stages (Figures 27.1 and 27.2).

- Protists are primarily aerobic organisms that live as autotrophs or heterotrophs, or by a combination of both nutritional modes. Some are parasites or symbionts living in or among the cells of other organisms.

- Protists live in aquatic or moist terrestrial habitats, or as parasites within animals as single-celled, colonial, or multicellular organisms, and range in size from microscopic to some of Earth's largest organisms.

- Reproduction may be asexual by mitotic cell divisions, or sexual by meiosis and union of gametes in fertilization.

- Many protists have specialized cell structures including contractile vacuoles, food vacuoles, eyespots, and a pellicle, cell wall, or shell. Most are able to move by means of flagella, cilia, or pseudopodia (Figure 27.3).

27.2 The Protist Groups

- The most ancient bifurcation separates eukaryotes into the Unikonta and Bikonta. Protist groups are found in both the unikonts and bikonts (Figure 27.2).

- The Excavata are a diverse group of protists, many of which have a scooped out (excavated) feeding apparatus. Some have flagella, some are anaerobes, some are photosynthetic, and some are parasites. The excavates include the Metamonada and the Euglenozoa.

- Metamonads consist of diplomonads and parabasalids, which have flagellated, single cells that lack mitochondria (Figure 27.4).

- Euglenozoans are almost all single-celled, autotrophic or heterotrophic (some are both), motile protists that swim using flagella. The free-living, photosynthetic forms—the euglenids—typically have complex cytoplasmic structures, including eyespots. The parasitic, nonphotosynthetic forms—the kinetoplastids—have characteristic mitochondrial structures (Figures 27.5 and 27.6).

- The Chromalveolata, protists with heterogeneous forms and life styles, includes the Alveolata and the Stramenopila.

- The alveolates include the ciliates, dinoflagellates, and apicomplexans. The ciliates swim using cilia and have complex cytoplasmic structures and two types of nuclei, the micronucleus and macronucleus. The dinoflagellates swim using flagella and are primarily marine organisms; some are photosynthetic. The apicomplexans are nonmotile parasites of animals (Figures 27.7 and 27.8).

- The stramenopiles include the funguslike groups, the Bacillariophyta (diatoms), Chrysophyta (golden algae), and Phaeophyta (brown algae). For most stramenopiles, flagella occur only on reproductive cells. Many Oomycota grow as masses of microscopic hyphal filaments and secrete enzymes that digest organic matter in their surroundings. Diatoms are single-celled organisms covered by a glassy silica shell; golden algae are colonial forms; brown algae are primarily multicellular marine forms that include large seaweeds with extensive cell differentiation (Figures 27.9–27.13).

- Rhizaria, amoebas with filamentous pseudopods supported by internal cellular structures, include the Radiolaria, the Foraminifera, and the Chlorarachniophyta. Many rhizarians produce hard outer shells.

- Radiolaria (radiolarians) are primarily marine organisms that secrete a glassy internal skeleton. They feed by engulfing prey that adhere to their axopods. Foraminifera (forams) are marine, single-celled organisms that form chambered, spiral shells containing calcium. They engulf prey that adhere to the strands of cytoplasm extending from their shells. Chlorarachniophytes engulf food using their pseudopodia (Figure 27.14).

- The Archaeplastida include the Rhodophyta (red algae) and Chlorophyta (green algae), as well as the land plants.

- Red algae are typically multicellular, primarily photosynthetic organisms of marine environments, with plantlike bodies composed of interwoven filaments. They have complex life cycles including alternation of generations, with no flagellated cells at any stage (Figure 27.15).

- Green algae are single-celled, colonial, and multicellular species that live primarily in freshwater habitats and carry out photosynthesis by mechanisms similar to those of plants; all produce flagellated gametes (Figures 27.16–27.18).

- The Unikonta includes the Amoebozoa, protists with unsupported pseudopods, and Opisthokonta, protists closely related to animals and fungi, and animals and fungi themselves.

- Amoebozoans include most amoebas and two heterotrophic slime molds, cellular (which move as individual cells) and plasmodial (which move as large masses of nuclei sharing a common cytoplasm). The amoebas in this group are heterotrophs abundant in marine and freshwater environments and in the soil. They move by extending pseudopodia (Figures 27.19 and 27.20).

- The opisthokonts are a broad group of eukaryotes that includes the fungi, the animals, and two groups of protists, the choanoflagellates and the nucleariids. Choanoflagellates are characterized by a collar of microvilli surrounding a single flagellum. An ancestor of a present-day choanoflagellate is considered likely to have been the ancestor of animals (Figure 27.21). Nucleariids are spherical amoeba with fine pseudopods. They are more closely related to fungi than to animals.

- Several groups of protists, as well as land plants, contain chloroplasts. Present-day chloroplasts and other plastids result from endosymbiosis events that took place millions of years ago: in a primary endosymbiosis event, a eukaryotic cell engulfed a cyanobacterium, which became an endosymbiont. Over time, the symbiont became the chloroplast. This first photosynthesizing organism was a green alga. Evolutionary divergence produced the red algae, green algae, and land plants. By secondary endosymbiosis, in which a nonphotosynthetic eukaryote engulfed a photosynthetic eukaryote, the various photosynthetic protists were produced (Figure 27.22).

TEST YOUR KNOWLEDGE

Remember/Understand

1. Which of the following is a characteristic of protists that is also found in at least one other group?
   a. division by binary fission
   b. multicellular structures
   c. complex developmental stages
   d. peptidoglycan cell walls
   e. organelles and reproduction by meiosis/mitosis

2. Which of the following is *not* found among the protist groups?
   a. life cycles
   b. contractile vacuoles
   c. pellicles
   d. collagen
   e. pseudopodia

3. A member of this group can cause urinary infections and be spread by sexual intercourse. The group is characterized by a flagellum buried in a fold of the cytoplasm that allows the organisms to move through thick and viscous fluids of humans. The group is:
   a. Ciliophora.
   b. Apicomplexa.
   c. Amoebozoa.
   d. Parabasala.
   e. Diplomonadida.

4. When *Paramecium* conjugate:
   a. cytoplasmic division produces four daughter cells, each having two micronuclei and two macronuclei.
   b. one haploid micronucleus in each cell remains intact; the other three degenerate. The micronucleus of each cell divides once, producing two nuclei, and each cell exchanges one nucleus with the other cell. In each partner the two micronuclei fuse, forming a diploid zygote micronucleus in each cell.
   c. the micronucleus of each of the disengaged partners divides meiotically. Macronuclei divide again in each cell, and the original micronucleus breaks down. Each cell has two haploid micronuclei; one of the macronuclei develops into a micronucleus.
   d. the mating cells join together at opposite sites of their oral depression.
   e. the micronucleus in each cell undergoes mitosis. When mitosis is complete, there are four diploid macronuclei; the micronucleus then breaks down.

5. The group Diplomonadida is characterized by:
   a. a mouthlike gullet and hairlike surface. *Paramecium* is an example.
   b. flagella and a lack of mitochondria. *Giardia* is an example.
   c. nonmotility, parasitism, and sporelike infective stages. *Toxoplasma* is an example.
   d. switching between autotrophic and heterotrophic lifestyles. *Euglena* is an example.
   e. large protein deposits and movement by two flagella, which are part of an undulating membrane. *Trypanosoma* is an example.

6. The greatest contributors to protist fossil deposits, and probable source of oil in many oil deposits, are:
   a. Oomycota.
   b. Chrysophyta.
   c. Bacillariophyta.
   d. Sporophyta.
   e. Alveolata.

7. The group with the distinguishing characteristic of gas-filled bladders and a cell wall composed of alginic acid, which is used by humans to thicken such diverse products as ice cream and floor polish, is:
   a. Chrysophyta.
   b. Phaeophyta.
   c. Oomycota.
   d. Bacillariophyta.
   e. none of the preceding.

8. *Plasmodium* is transmitted to humans by the bite of a mosquito (*Anopheles*). Its life cycle is characterized by spores, gametes, and cysts that can "hide" in human cells. This infective protist belongs to the group:
   a. Apicomplexa.
   b. Archaeplastida.
   c. Dinoflagellata.
   d. Oomycota.
   e. Ciliophora.

9. Tripping on a rotten log, a hunter notices a mass resembling mucus that appears to be moving slowly toward what appear to be brightly colored fruiting bodies. The organisms in the mass are:
   a. amoebas in the group Rhizaria.
   b. slime molds.
   c. red algae.
   d. green algae.
   e. charophytes.

10. Endosymbioses that lead to the evolution of euglenoids and, separately, the evolution of chlorarachniophytes were the result of the combining of:
    a. two ancestral nonphotosynthetic prokaryotes.
    b. two ancestral photosynthetic prokaryotes.
    c. a nonphotosynthetic eukaryote with a photosynthetic eukaryote.
    d. a photosynthetic prokaryote with a nonphotosynthetic eukaryote.
    e. mitochondria with an already established plastid.

Apply/Analyze

11. **Discuss Concepts** You decide to vacation in a developing country where sanitation practices and standards of personal hygiene are inadequate. Considering the information about protists covered in this chapter, would you consider it safe to drink water in that country? What treatments could make the water safe to drink? What kinds of foods might be best avoided? What kinds of preparation might make foods safe to eat?

12. **Discuss Concepts** The overreproduction of dinoflagellates, which produces red tides, is sometimes caused by fertilizer runoff into coastal waters. The red tides kill countless aquatic species, birds, and other wildlife. Would you consider drastic cutbacks in the use of fertilizers as a means to lessen the red tides? Why?

Evaluate/Create

13. **Design an Experiment** Design an experiment to demonstrate whether the flagellated protist *Euglena* is phototropic, that is, is attracted to and moves toward light. Also propose a follow-up experiment (on the assumption of a positive result) to determine the wavelength range and light intensity range sufficient to cause phototropic movement.

14. **Apply Evolutionary Thinking** Use the Internet to research why studies of a molecular sensor, receptor tyrosine kinase (see Section 9.3), supports the hypothesis that a choanoflagellate type of protist is the ancestor of animals. Summarize your findings.

For selected answers, see Appendix A.

INTERPRET THE DATA

Marine diatoms are a crucial part of marine productivity because they are food for many organisms. Their productivity is dependent on how well they photosynthesize. The **Figure** shows the response of photosynthesis in the marine diatom *Skeletonema costatum* to light.

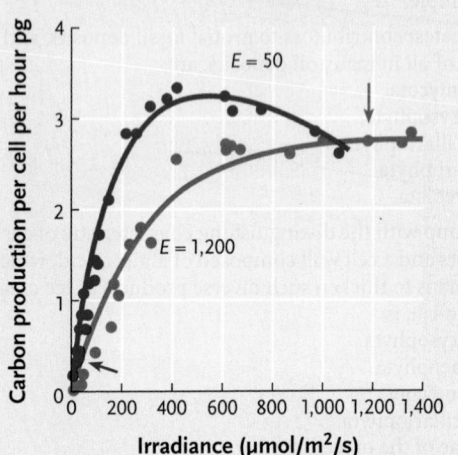

Photosynthesis is expressed on the *y* axis as carbon production in picograms (pg) per cell per hour, and light is expressed on the *x* axis as the number of photons of irradiance that can activate photosynthesis (0 μmol/m^2/s would be complete darkness, and 2,000 μmol/m^2/s would be full sunlight at midday). The blue symbols show a cell culture grown for five days at 50 μmol/m^2/s of irradiance (the treatment irradiance), with an arrow showing the level of photosynthetic production at the end of this period. The red symbols show a different population of cells from the same culture grown for five days at 1,200 μmol/m^2/s of irradiance (the treatment irradiance), with an arrow showing the level of photosynthetic production. Both cultures were then exposed to all of the possible irradiance conditions for 30-minute periods.

1. In each light treatment, at what irradiance was the photosynthetic rate maximized? Was this the same rate as the treatment irradiance? Why or why not?

2. The response of photosynthesis to irradiance at low irradiance is linear. The rate of this linear increase is the same as the organism's photosynthesis efficiency with respect to light. Which treatment has higher photosynthesis efficiency?

Source: T. Anning et al. 2000. Photoacclimation in the marine diatom *Skeletonema costatum. Limnology and Oceanography* 45(8):1807–1817.

Seedless Plants

Ferns, familiar representatives of the seedless plant lineage Monilophyta.

Why it matters . . . Mammals, reptiles, amphibians, insects—hundreds of millions of years before lineages of large land animals emerged on our planet, plants were thriving here. Indeed, because plants are the basis of terrestrial food webs, animals could not evolve until plant communities were present.

There is wide agreement that land plants arose from a lineage of freshwater green algae in the phylum **Charophyta,** a hypothesis supported by molecular and other lines of evidence. Modern descendants of those plant pioneers run the gamut from ground-hugging liverworts to 10-meter-tall tree ferns, sky-scraping redwoods, and a pageant of flowering plant taxa that include sunflowers, cacti, orchids, and grasses, among a host of others. Today there are at least 280,000 living plant species, which collectively make up the **Kingdom Plantae.** Plant scientists who prefer to include green algae in this clade refer to it as the Viridiplantae ("green plants").

All plants are multicellular, and the vast majority are terrestrial autotrophs that carry out photosynthesis using sunlight, CO_2 from the air, water, and dissolved minerals to synthesize the organic compounds plants use in metabolism or store in their tissues. Fossils indicate the first plants had colonized land by at least 475 million years ago (mya) and in part because they were photosynthetic organisms, their arrival would dramatically change conditions on the Earth. For example, as spreading plant populations removed CO_2 from the atmosphere and stored quantities of carbon in their tissues, they altered the global carbon cycle (Chapter 54). The early Earth's sparse soils, which may have been formed originally by ancient microbes, were enriched by the carbon compounds, amino acids, and other substances in decaying plant remains, and the presence and growth of plants also contributed to the weathering of rocky landscapes. Over time, the availability of carbon-rich plant tissues also laid the nutritional foundation for diverse communities of organisms, including lineages of terrestrial animals that eventually came to include *Homo sapiens*.

A. *Mnium punctatum,*
a moss

B. Ferns in a New Zealand
forest

C. Horsetails

D. Liverworts

FIGURE 28.1 Representatives of seedless plants. (A) Close-up of a leafy moss. Mosses evolved relatively soon after plants made the transition to land. **(B)** Ferns in a New Zealand forest. **(C)** Horsetails, also known colloquially as scouring rushes. **(D)** Liverworts growing in a South American rainforest.

The evolutionary transition to life on dry land involved major adaptive challenges. To begin with, surviving on land required avoiding desiccation (drying out) of tissues not bathed in water. It also required means of obtaining water and nutrients from soil, and ways of reproducing sexually in environments where water was not available for the dispersal of eggs and sperm. This chapter begins with an overview of how structural and functional adaptations solved these problems and opened adaptive opportunities for plant life. We will then see how those changes are manifest in the most ancient extant lineages of land plants—seedless plants such as liverworts, mosses, horsetails, ferns, and their relatives **(Figure 28.1).** Over more than 70 million years, seedless plants would dominate many land environments before the emergence of a major new plant lineage—the seed plants we consider in Chapter 29, including gymnosperms such as conifers and angiosperms, the flowering plants.

Fossil finds have contributed immensely to our understanding of the history of plant life on Earth. As you read this chapter, however, bear in mind that early land plants lacked hard parts (such as wood) that fossilize well, and the most common fossil finds are microscopic reproductive cells called **spores** or other bits and pieces. Unfortunately, such spores and fragments seldom occur in conjunction with obvious remnants of other parts, such as leaves, stems, and roots. Even if they do, it can be hard to determine if the fossilized shards all belong to the same individual. Fossils of whole plants are extremely rare. Complicating things further, it is likely that some adaptations to terrestrial life arose in several plant lineages.

We now consider evidence for the emergence of land plants from an algal ancestor and adaptations that were crucial for the survival of early land plants.

28.1 Plant Evolution: Adaptations to Life on Land

Various lines of evidence support the hypothesis that land plants form a monophyletic clade descended from a common green algal ancestor, and a charophyte in particular. To begin with, fossil green algae predate the earliest documented land plant fossils by approximately 270 million years. The groups also share some basic morphological and biochemical traits. For instance, green algae and land plants both have cellulose in their cell walls and both store energy captured during photosynthesis as starch. In their chloroplasts, both have light-absorbing pigments that include chlorophyll *a* and chlorophyll *b*. As noted in the chapter introduction, molecular studies support the hypothesis that land plants arose from a charophyte alga. This evidence, which includes the comparison of 129 gene sequences from 40 plant taxa and the analysis of chloroplast genomes from various charophyte groups, points to the charophyte order Zygnematales as the most likely sister clade (the closest relatives) of land plants. Modern representatives of this clade include filamentous green algae in the genus *Spirogyra* **(Figure 28.2).**

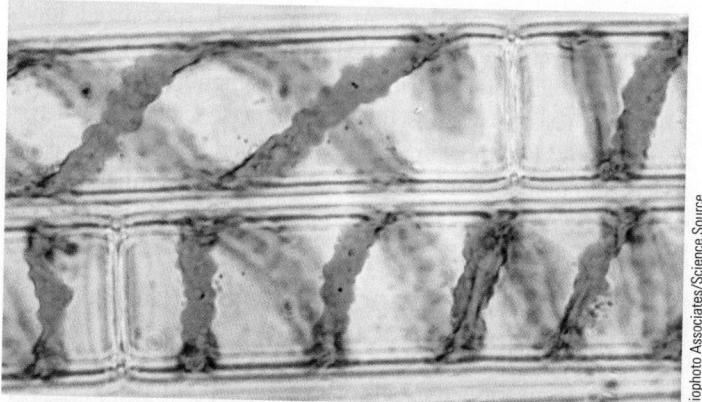

FIGURE 28.2 Cells of a filamentous green alga in the genus *Spirogyra.* The genus was named for the spiral arrangement of chloroplasts in the algal cells.

Biochemical and Structural Adaptations Helped Protect Land Plants against Desiccation and Sheltered Developing Embryos

Desiccation of tissues is not a problem for aquatic algae, but it is a major constraint for plants living on land. Accordingly, natural selection favored the survival of plants having features that protected tissues from drying out. Fossils dating to as early as 450 mya reveal that some of the first land plants had an outer waxy layer called a **cuticle,** which slows water loss **(Figure 28.3A, B).** By about 420 mya, fossils of most land plant lineages show another adaptation, **stomata** (singular, *stoma*; *stoma* = mouth), tiny openings in the cuticle-covered surfaces **(Figure 28.3A, C).** A stoma is formed by a pair of cells that can

change shape and so open up or close a space between them. As you will read in Chapter 34, stomata are the structures plants use to control water loss by evaporation and to take up carbon dioxide for photosynthesis.

Other early adaptations protected plant embryos. One of these was the inherited ability to make **sporopollenin,** a complex polymer that strengthens tissue containing it and helps prevent desiccation. Sporopollenin surrounds the zygotes of some modern charophytes, and in land plants it is a major component of pollen grains and the protective wall of structures that produce reproductive spores (described in detail in Chapter 36). Additional reproductive adaptations that arose in early land plants included multicellular chambers that protect developing gametes, and tissues that shelter the multicellular embryo inside a parent plant. The term **embryophyte** (*phyton* = plant) is a synonym for land plants because all land plants produce a protected embryo during their reproductive cycle.

The Meristem Was an Innovation for Lifelong Growth

Botanists long ago discovered that the body parts of plants generally grow from regions of unspecialized cells called *meristems* ("dividing parts"). This feature evidently arose very early in land plant evolution because meristems exist in all extant plant groups except the liverworts, described in Section 28.3. **Apical meristems** are regions of unspecialized, dividing cells near the tips of branching shoot and root systems that emerged as land plant lineages diversified. Chapter 33 describes in detail how descendants of meristem cells differentiate and form all mature plant tissues, and function in the lifelong growth that is a basic plant characteristic.

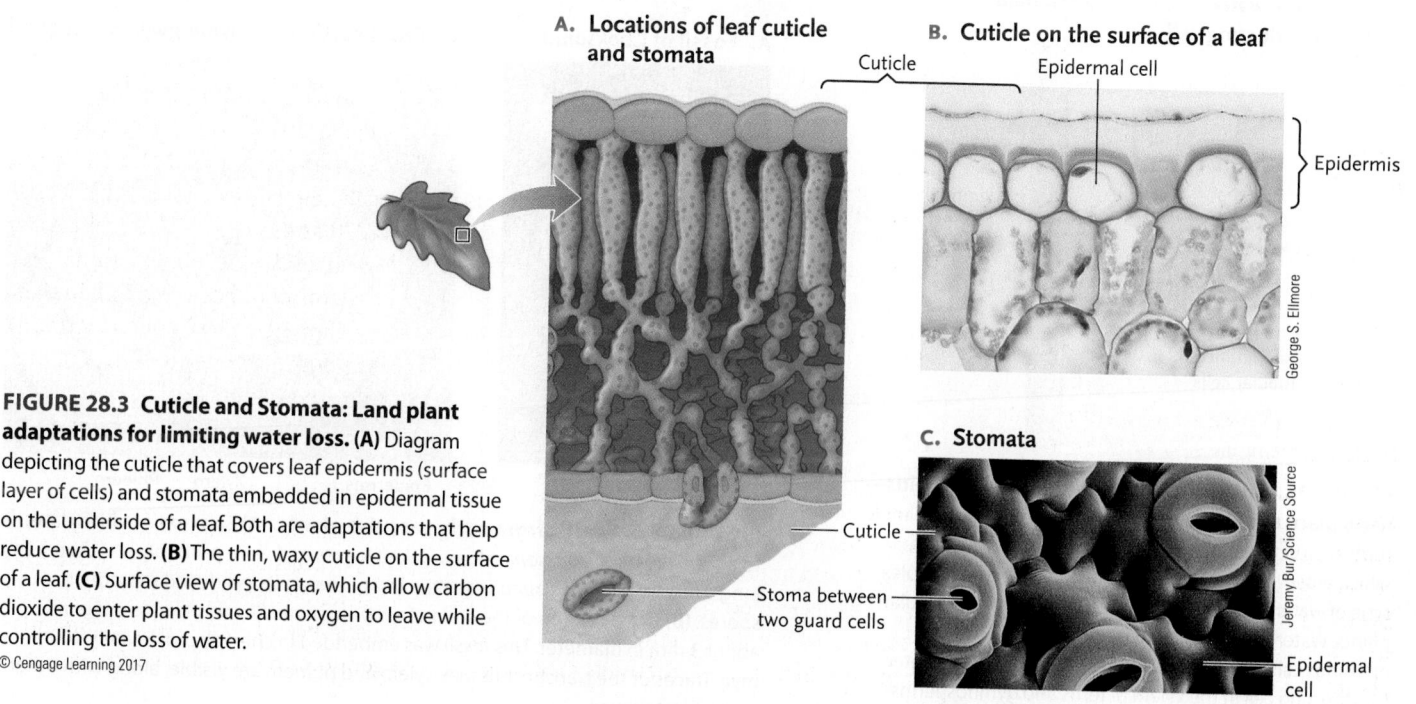

A. Locations of leaf cuticle and stomata

B. Cuticle on the surface of a leaf

Cuticle

Epidermal cell

Epidermis

C. Stomata

Cuticle

Stoma between two guard cells

Epidermal cell

FIGURE 28.3 Cuticle and Stomata: Land plant adaptations for limiting water loss. (A) Diagram depicting the cuticle that covers leaf epidermis (surface layer of cells) and stomata embedded in epidermal tissue on the underside of a leaf. Both are adaptations that help reduce water loss. **(B)** The thin, waxy cuticle on the surface of a leaf. **(C)** Surface view of stomata, which allow carbon dioxide to enter plant tissues and oxygen to leave while controlling the loss of water.
© Cengage Learning 2017

Evolution of Vascular Tissue Enhanced the Transport of Substances within a Large Plant Body and Favored Erect Growth

Multicellular organisms require means of moving substances from cell to cell and between body parts. In the first, minuscule land plants, *plasmodesmata*—the minute channels between plant cells described in Chapter 9—would have been sufficient for moving water and sugars formed in photosynthesis. Yet transporting substances through plasmodesmata would have limited the size of such plants to a few millimeters at most. More efficient means of internal transport arose by about 475 mya; plant fossils dating to that time show evidence of simple tubular cells through which water moved **(Figure 28.4A)**. Lacking the more elaborate transport adaptations of lineages that came later, such plants and their modern descendants are **nonvascular plants,** a clade also called **bryophytes.**

By about 450 mya, fossils provide evidence of the evolution of simple **vascular plants** (Latin *vas* = duct), which have specialized conducting tissues that branch throughout the plant body. Unlike nonvascular plants, these ancestors of modern ferns, gymnosperms, angiosperms, and other groups all could synthesize **lignin,** a tough, rather inert polymer that strengthens secondary cell walls. By about 380 mya, **xylem,** the first vascular tissue built of such lignified cells, had arisen. Xylem distributes water and dissolved mineral ions through the plant body. It contains stacked, lignin-reinforced water-conducting cells called **tracheids (Figure 28.4B),** which is why vascular plants collectively are called **tracheophytes.** (In most flowering plants, which diverged later, more efficient water-conducting cells called *vessel*

elements are also present.) At some point, specialized food-conducting vascular tissue, called **phloem,** also evolved in vascular plants. Phloem distributes sugars manufactured during photosynthesis, as well as other substances. Chapter 34 delves more deeply into how tracheids and other specialized cells of xylem and phloem perform key internal transport functions.

Shoot and Root Systems Evolved as Adaptations for Support and Nutrition in Vascular Plants

The first unquestioned fossils of a vascular plant are specimens in the extinct genus *Cooksonia.* This plant had leafless, "naked" stems with simple, forked branches of equal size **(Figure 28.5A).** Reproductive spores formed in capsules called **sporangia** located at the branch tips and simple xylem wove through the stems. *Rhyniophytes* were other early vascular plants; the name was bestowed because numerous fossil specimens are from rock outcrops near Rhynie, Scotland. *Cooksonia* and rhyniophyte fossils all date to about 425 mya and show similar features, including vascular tissue **(Figure 28.5B).** Most likely none of these species had roots, which are organs specialized to anchor a vascular plant and take up water and minerals. Instead these ancestral forms were supported physically by a **rhizome**—a horizontal, modified stem that can penetrate a substrate and anchor the plant. Cells called *rhizoids,* which can help anchor a plant and take up water, extended into the soil. Stems of Rhynie fossils also reveal evidence of endosymbiotic fungi. As Chapter 35 discusses, most land plants depend in part on fungal symbioses for their nutrition. It is quite likely that such associations also were essential to the early success of plants on land.

A. Structure of an early water-conducting cell

Small pores

Simple nonlignified tubular cells

B. Structure of a tracheid

Pits in tracheid

Lignified secondary cell wall

FIGURE 28.4 Water-conducting cells and tracheids—early plant adaptations for water transport. (A) Simple tubular structure of the earliest water-conducting cells, which arose in bryophytes. **(B)** A tracheid, a more complex type of water-transport cell that arose in seedless vascular plants. Water moves from tracheid to tracheid through openings called pits. Tracheids are the main type of water transporting cell in the xylem of ferns and gymnosperms.

© Cengage Learning 201

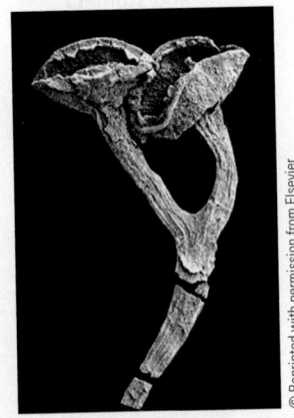

A. Fossil of *Cooksonia*

© Reprinted with permission from Elsevier

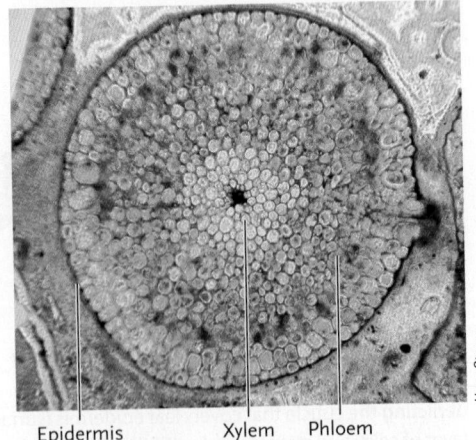

B. Cross section of *Rhynia gwynne-vaughnii*

Epidermis   Xylem   Phloem

Hans Steur

FIGURE 28.5 Fossils from early vascular plants. (A) *Cooksonia,* which dates to about 420 mya. *Cooksonia* stems lacked leaves and were probably less than 3 cm long. The cup-shaped structures at the top of the stems produced reproductive spores. **(B)** Cross section of the stem of a rhyniophyte, *Rhynia gwynne-vaughnii,* about 3 mm in diameter. This fossil was embedded in chert approximately 400 mya. Traces of the transport tissues xylem and phloem are visible, along with other specialized tissues.

LIGNIFIED SHOOT SYSTEMS: MECHANICAL SUPPORT FOR UPRIGHT GROWTH Above ground, the simple stems of early land plants gradually became more specialized, evolving into the **shoot systems** we see in virtually all living vascular plants. Shoot systems have stems and leaves that arise from apical meristems and that absorb light energy from the Sun and carbon dioxide from the air. Lignified cells, including the lignified tracheids of xylem formed a strong, internal scaffold that supports upright, branching stems. Such gravity-defying stems, bearing leaves and other photosynthetic structures, allowed vascular plants to grow taller, increasing the surface area and opportunities for intercepting sunlight. In addition, sporangia or other reproductive structures borne on aerial stems served as platforms for more efficient launching of spores from the parent plant. These adaptations, all related to the presence of lignin in tracheids and some other types of plant cells, collectively were crucial in allowing plants to colonize nearly every land environment.

ENHANCED PHOTOSYNTHESIS: LEAVES WITH STOMATA AND VEINS Structures we think of as "leaves" arose several times during plant evolution. Leaves are modifications of stems and fall into two general categories. **Microphylls** are narrow and have a single strand of vascular tissue, or **vein**. **Megaphylls** are broad leaves with multiple veins. **Figure 28.6** illustrates the basic steps of possible evolutionary pathways for these leaf forms. Among the seedless vascular plants known as lycophytes (see Section 28.3), microphyll-like parts may have evolved as flaplike outgrowths of the plant's main vertical stem (see Figure 28.6A). Club mosses, the most common living lycophytes, still have this type of leaf. In contrast, various lines of evidence suggest that megaphylls arose from modified branches, and by the late Paleozoic era (about 350 mya), they had appeared in several plant groups. Figure 28.6B shows a simplified scenario for this adaptive shift: a slender stem branches into two roughly equal sections in a recurring pattern (called dichotomous branching). As slender branched sections were overtopped by more robust ones, the sections flattened out and a web of tissue having cells with chloroplasts then filled in the gap between branches. Over evolutionary time the tissue gained stomata, which improve the plant's access to carbon dioxide for photosynthesis, and a simple network of veins that transport water.

There is a strong correlation between the density of venation in a plant's leaves and the plant's capacity for photosynthesis—and, ultimately, for producing energy for growth. Timothy Brodribb and Taylor Feild, plant physiologists who compared the leaves of more than 500 living and extinct flowering plants,

A. Development of microphylls as an offshoot of the main vertical axis

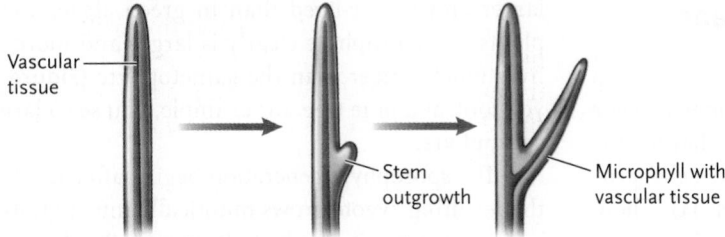

Vascular tissue — Stem outgrowth — Microphyll with vascular tissue

B. Development of megaphylls in a branching pattern

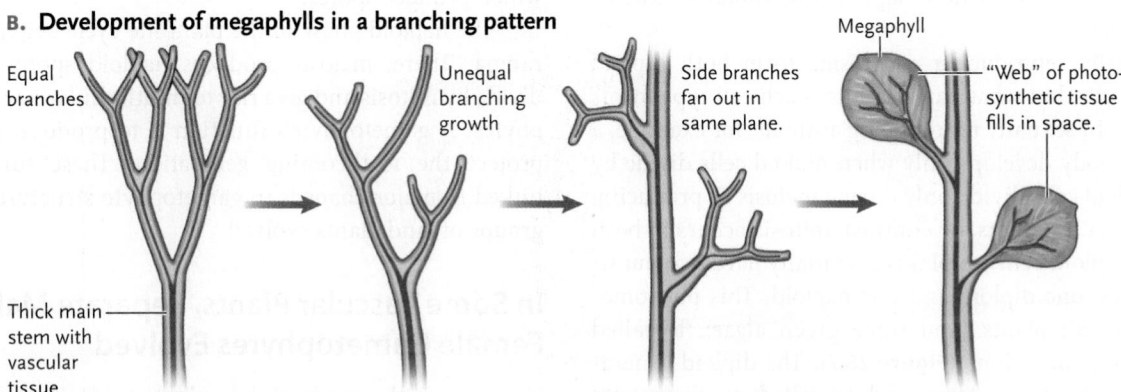

Equal branches — Unequal branching growth — Side branches fan out in same plane. — Megaphyll — "Web" of photosynthetic tissue fills in space.

Thick main stem with vascular tissue

FIGURE 28.6 Evolution of leaves. (A) One type of early leaflike structure, the microphyll, may have evolved as an offshoot of the plant's main vertical axis. Each leaf contained a single vein (vascular strand). The seedless vascular plants known as lycophytes (club mosses) still have this type of leaf. **(B)** In other seedless vascular plants, the form of leaf called a megaphyll could have arisen in a series of steps that began when the main stem evolved a branching growth pattern. In this scenario, small side branches then fanned out and photosynthetic tissue filled the space between them, becoming the leaf blade. With time the small branches became modified into vascular strands. Ferns, horsetails, gymnosperms, and angiosperms all have this type of leaf.

documented a dramatic increase in the number of leaf veins over the course of angiosperm evolution. They discovered that by about 100 mya, leaves of later evolving flowering plant species had up to 10 times more veins than leaves of more ancient species. Brodribb and Feild calculated that even a tripling in the number of veins would have increased the capacity of photosynthesis by 100%. Their findings suggest strongly that the evolution of large leaves with numerous stomata and a dense system of veins was a major factor in the remarkable proliferation of flowering plants that has occurred over the past 150 million years.

SUPPORT AND NUTRITION: ROOTS As upright growth emerged as a basic feature of vascular plants, it required a means of anchoring aerial parts in the soil, as well as effective strategies for obtaining water and mineral nutrients from soil. The evolution of **roots**—anchoring organs that also absorb water and nutrients—fulfilled these requirements. Fossils of early vascular plants do not provide clear evidence of exactly when roots appeared, and it is possible that the feature arose independently in various groups. Ultimately, however, vascular plants developed specialized **root systems,** which generally consist of underground, cylindrical absorptive structures with a large surface area that favors the rapid uptake of soil water and dissolved minerals.

As Plant Lineages Became Adapted to Land, the Diploid Phase Became the Dominant Portion of the Life Cycle

As early plants colonized increasingly drier habitats, major modifications occurred in their life cycles. Recall that in sexually reproducing organisms, meiosis in diploid cells produces haploid (n) reproductive cells (see Chapter 11). These cells may be gametes—sperm or eggs—or they may be spores, which can give rise to a new haploid individual asexually, without mating.

All sexually reproducing organisms form both diploid and haploid cells, but what happens after each cell type forms varies greatly. In sexually reproducing animals, for example, a multicellular body develops only when diploid cells divide by mitosis. Haploid cells divide only during meiosis II, producing haploid gametes. In plants, by contrast, mitosis occurs in both diploid and haploid cells, so plants essentially have two multicellular phases, one diploid and one haploid. This phenomenon, found in all plants (and some green algae), is called **alternation of generations (Figure 28.7).** The diploid generation produces haploid spores and is called a **sporophyte** ("spore producer"). The haploid generation produces gametes and is called a **gametophyte** ("gamete producer").

In the first land plants, the life cycle included long-lived haploid gametophytes and short-lived diploid sporophytes. The bryophytes we consider later in this chapter still show this life cycle pattern. As plant adaptations to terrestrial life accumulated, the haploid gametophyte phase became physically

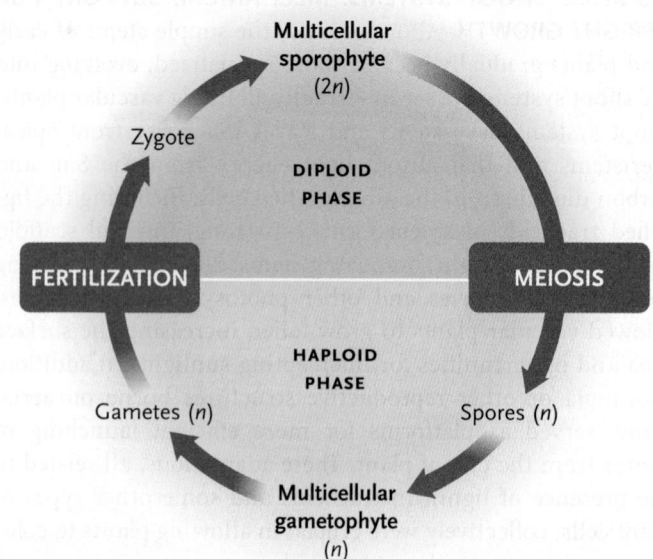

FIGURE 28.7 Overview of alternation of generations, the basic pattern of the plant life cycle. The relative dominance of haploid and diploid phases is different for different plant groups (compare with Figure 28.8).
© Cengage Learning 2017

smaller and less complex and had a shorter lifespan, while just the opposite occurred with the diploid sporophyte phase. Accordingly, in nonvascular plants the sporophyte is a little larger and longer-lived than in green algae, and in vascular plants the sporophyte clearly is larger and more complex and lives much longer than the gametophyte **(Figure 28.8).** When you look at a pine tree, for example, you see a large, long-lived sporophyte.

The sporophyte generation begins after fertilization, when the resulting zygote grows mitotically into a multicellular, diploid organism. Its body will eventually develop sporangia, which produce spores.

The haploid phase of the plant life cycle begins in the sporangia. There, meiosis produces haploid spores. The spores divide by mitosis and give rise to multicellular haploid gametophytes. A gametophyte's function is to produce, nourish, and protect the forthcoming generation. These functions were linked to major changes in gametophyte structure as different groups of land plants evolved.

In Some Vascular Plants, Separate Male and Female Gametophytes Evolved

During sexual reproduction in plants, meiosis produces spores. When a plant makes only one type of spore, it is said to be **homosporous** ("same spore"). A gametophyte that develops from such a spore is bisexual—it can produce both sperm and eggs. The sperm have flagella and swim through liquid water in order to encounter female gametes.

Other vascular plants are **heterosporous.** They produce two types of spores in two different types of sporangia, and those

FIGURE 28.8 Evolutionary trend from dominance of the gametophyte (haploid) generation to dominance of the sporophyte (diploid) generation, represented by existing species ranging from a bryophyte (a moss) to a flowering plant. This trend developed as early plants were colonizing habitats on land. In general, the sporophytes of vascular plants are larger and more complex than those of bryophytes, and their gametophytes are reduced in size and complexity. In this diagram ferns represent seedless vascular plants.
© Cengage Learning 2017

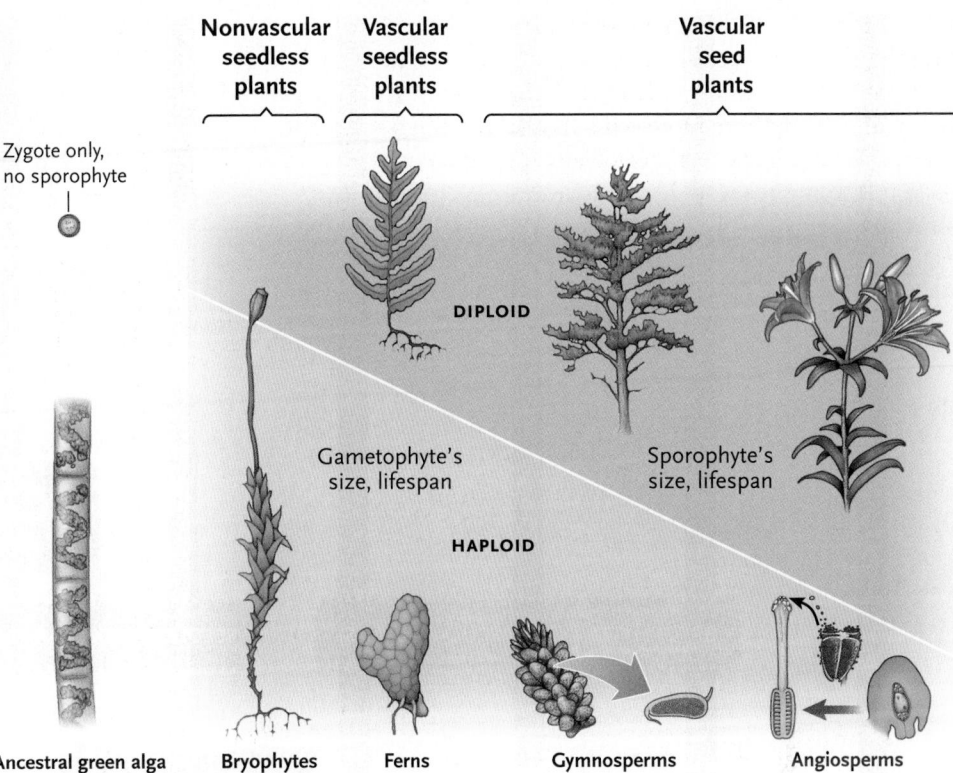

Nonvascular seedless plants

Vascular seedless plants

Vascular seed plants

Zygote only, no sporophyte

DIPLOID

Gametophyte's size, lifespan

Sporophyte's size, lifespan

HAPLOID

Ancestral green alga   Bryophytes   Ferns   Gymnosperms   Angiosperms

---

spores develop into small, sexually different gametophytes. In seed plants—gymnosperms and angiosperms—the smaller spore type develops into a male gametophyte—a *pollen grain*. The larger one develops into a female gametophyte, in which eggs form and fertilization occurs. **Figure 28.9** summarizes these differences.

A. Homospory

Sporophyte

One type of sporangia

One type of spores

Bisexual gametophyte

Male gametes   Female gametes

B. Heterospory

Sporophyte

Male sporangia   Sexually different sporangia   Female sporangia

Sexually different spores

Male gametophyte   Female gametophyte

Male gametes   Female gametes

FIGURE 28.9 Homospory and heterospory compared. In homosporous species, all sporangia make a single type of spore that gives rise to a bisexual gametophyte that produces both male (sperm) and female gametes (eggs). Heterosporous species have two types of sporangia. One type produces spores that develop into male sperm-producing gametophytes and the other type produces spores that develop into female egg-producing gametophytes.
© Cengage Learning 2017

So many new fossils appear in Devonian rocks that paleobotanists (scientists who study fossil plants) have struggled to determine which fossil lineages gave rise to modern plant phyla. Molecular studies are now powerful tools in that effort. What is clear is that as each major lineage came into being, its characteristic adaptations included modifications of existing structures and functions (**Figure 28.10** and **Table 28.1**). The next sections fill out this general picture with respect to the seedless plants. We begin with bryophytes, nonvascular seedless plants that most clearly resemble the plant kingdom's algal ancestors.

STUDY BREAK 28.1

1. How did plant adaptations such as a root system, a shoot system, and a vascular system collectively influence the transition to terrestrial life?
2. Describe the difference between homospory and heterospory, and explain how heterospory paved the way for other reproductive adaptations in land plants.

THINK OUTSIDE THE BOOK

In 2009 researchers studying seaweeds off the coast of California reported discovering a species of red alga with lignin in its tissues. The project's lead scientist was Patrick Martone, now at the University of British Columbia. Use the Internet to learn more about Martone's work. Why was the discovery of lignin in a red alga so surprising, and what are its potential evolutionary implications?

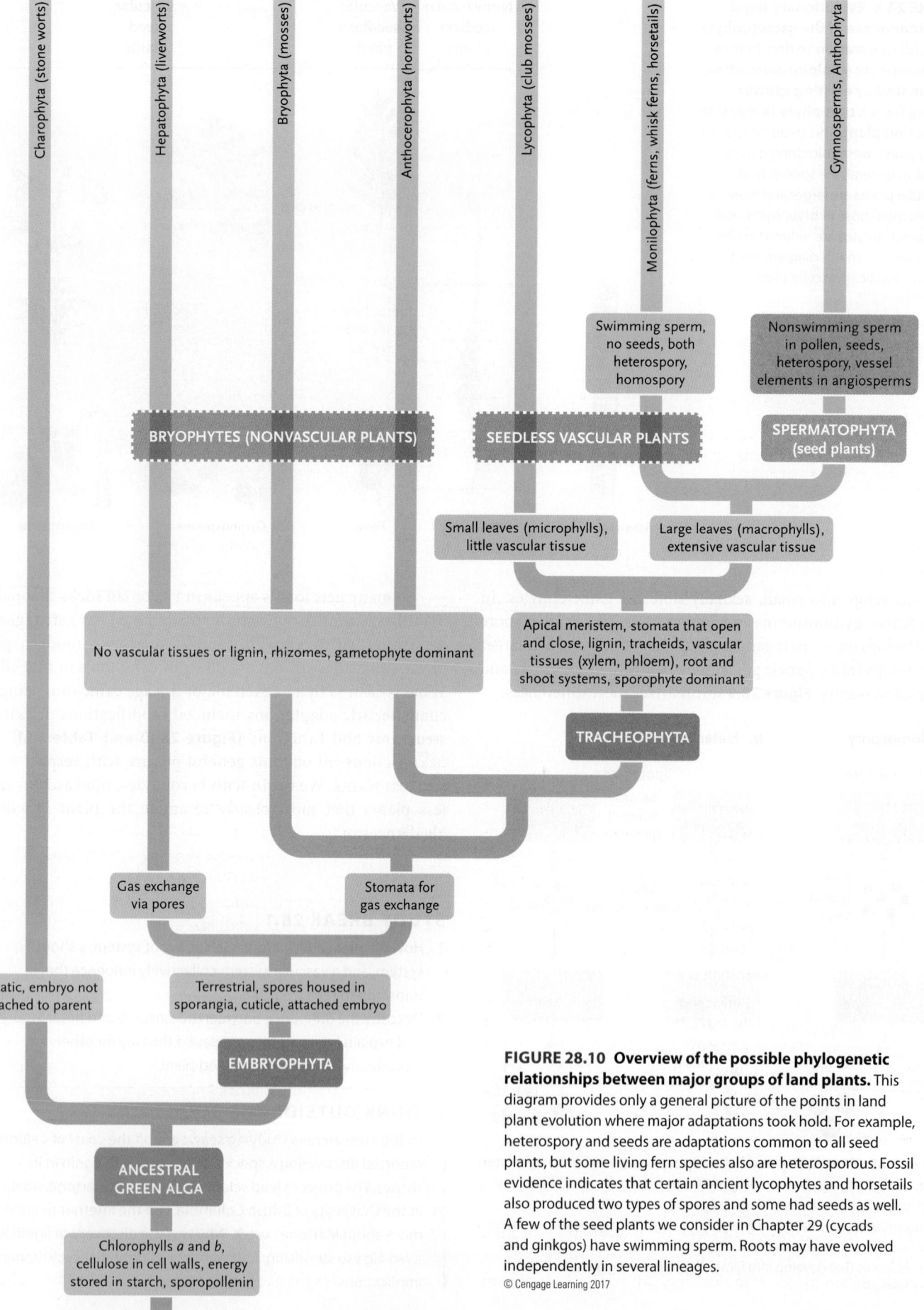

FIGURE 28.10 Overview of the possible phylogenetic relationships between major groups of land plants. This diagram provides only a general picture of the points in land plant evolution where major adaptations took hold. For example, heterospory and seeds are adaptations common to all seed plants, but some living fern species also are heterosporous. Fossil evidence indicates that certain ancient lycophytes and horsetails also produced two types of spores and some had seeds as well. A few of the seed plants we consider in Chapter 29 (cycads and ginkgos) have swimming sperm. Roots may have evolved independently in several lineages.

© Cengage Learning 2017

Labels in figure:

- Charophyta (stone worts)
- Hepatophyta (liverworts)
- Bryophyta (mosses)
- Anthocerophyta (hornworts)
- Lycophyta (club mosses)
- Monilophyta (ferns, whisk ferns, horsetails)
- Gymnosperms, Anthophyta

Swimming sperm, no seeds, both heterospory, homospory

Nonswimming sperm in pollen, seeds, heterospory, vessel elements in angiosperms

BRYOPHYTES (NONVASCULAR PLANTS)

SEEDLESS VASCULAR PLANTS

SPERMATOPHYTA (seed plants)

Small leaves (microphylls), little vascular tissue

Large leaves (macrophylls), extensive vascular tissue

No vascular tissues or lignin, rhizomes, gametophyte dominant

Apical meristem, stomata that open and close, lignin, tracheids, vascular tissues (xylem, phloem), root and shoot systems, sporophyte dominant

TRACHEOPHYTA

Gas exchange via pores

Stomata for gas exchange

Aquatic, embryo not attached to parent

Terrestrial, spores housed in sporangia, cuticle, attached embryo

EMBRYOPHYTA

ANCESTRAL GREEN ALGA

Chlorophylls *a* and *b*, cellulose in cell walls, energy stored in starch, sporopollenin

TABLE 28.1 | Trends in Plant Evolution

Traits derived from algal ancestor: cell walls with cellulose; energy stored in starch; two forms of chlorophyll (*a* and *b*); possibly, sporopollenin in spore wall.

Bryophytes	Tracheophytes				Functions in Land Plants
	Lycophytes, ferns, and their kin	Gymnosperms	Angiosperms		

Cuticle ──→				Protection against water loss, pathogens
Stomata ──→				Regulation of water loss and gas exchange (CO_2 in, O_2 out)
Nonvascular ──→ Vascular ──────────────────────────────────→				Internal tubes that transport water (xylem), nutrients (phloem)
Lignin ──────────────────────────────────────→				Mechanical support for water-conducting tracheids, vertical growth
Apical meristem ──────────────────────────────→				Branching shoot system, vast root system
Roots, stems, leaves (microphylls, macrophylls) ──────────────→				Enhanced uptake, transport of nutrients and enhanced photosynthesis
Haploid phase dominant ──→ Diploid phase dominant ──────────→				Genetic diversity
Homospory (one spore type) ──→ Heterospory (two spore types) ──→				Promotion of genetic diversity
Motile gametes ───────────────────→ Nonmotile gametes ──→				Protection of gametes within parent body
Seedless ───────────────────────→ Seeds ──→				Protection of embryo

28.2 Bryophytes, the Nonvascular Land Plants

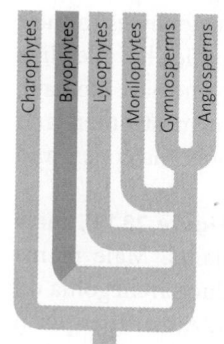

In general, the bryophytes (*bryon* = moss)—liverworts, mosses, and hornworts—are strikingly algalike, with a combination of traits that reflect adaptations to both aquatic and land environments. For example, they produce flagellated sperm that must swim through water to reach eggs, and they do not have a complex vascular system for transporting water. Given these characteristics, it is not surprising that bryophytes almost always grow on wet sites along creek banks or on rocks just above running water; in bogs, swamps, or the dense shade of damp forests; and on moist tree trunks or rooftops. Some species are **epiphytes** (*epi* = upon)—they grow independently (that is, not as a parasite) on another organism or some other moist place, ranging from the splash zone just above high tide on rocky shores to edges of snowbanks.

Despite their reliance on wet or moist surroundings, bryophytes are clearly adapted to land. They have parts that are rootlike, stemlike, and leaflike, although the fibrous "roots" are rhizoids, and bryophyte "stems" and "leaves" did not evolve from the same structures as vascular plant stems and leaves did. (That is, stems and leaves are not homologous in the two groups.) Sporophytes of some bryophyte species also have a water-conserving cuticle and stomata. As mentioned earlier, bryophyte tissues do not contain lignified cells. The absence of this strengthening material and the lack of internal pipelines for efficient nutrient transport partly account for bryophytes' small size—typically less than 20 cm long—and for their tendency to grow sprawled along surfaces instead of upright.

Like most plants, bryophytes also have both sexual and asexual reproductive modes, and as is true of all plants, the life cycle has both gametophyte (*n*) and sporophyte (*2n*) phases— though the bryophyte sporophyte is tiny and lives only a short time. **Figure 28.11** shows the gametophyte of a leafy moss, with miniscule diploid sporophytes that consist of a sporangium and slender stalk. Bryophyte gametophytes produce gametes sheltered within a layer of protective cells called a **gametangium** (plural, *gametangia*). The gametangia in which bryophyte eggs form are flask-shaped structures called **archegonia** (*archi* = first; *gonos* = seed). Flagellated sperm form in rounded gametangia called **antheridia** (*antheros* = flowerlike). The sperm swim through a film of

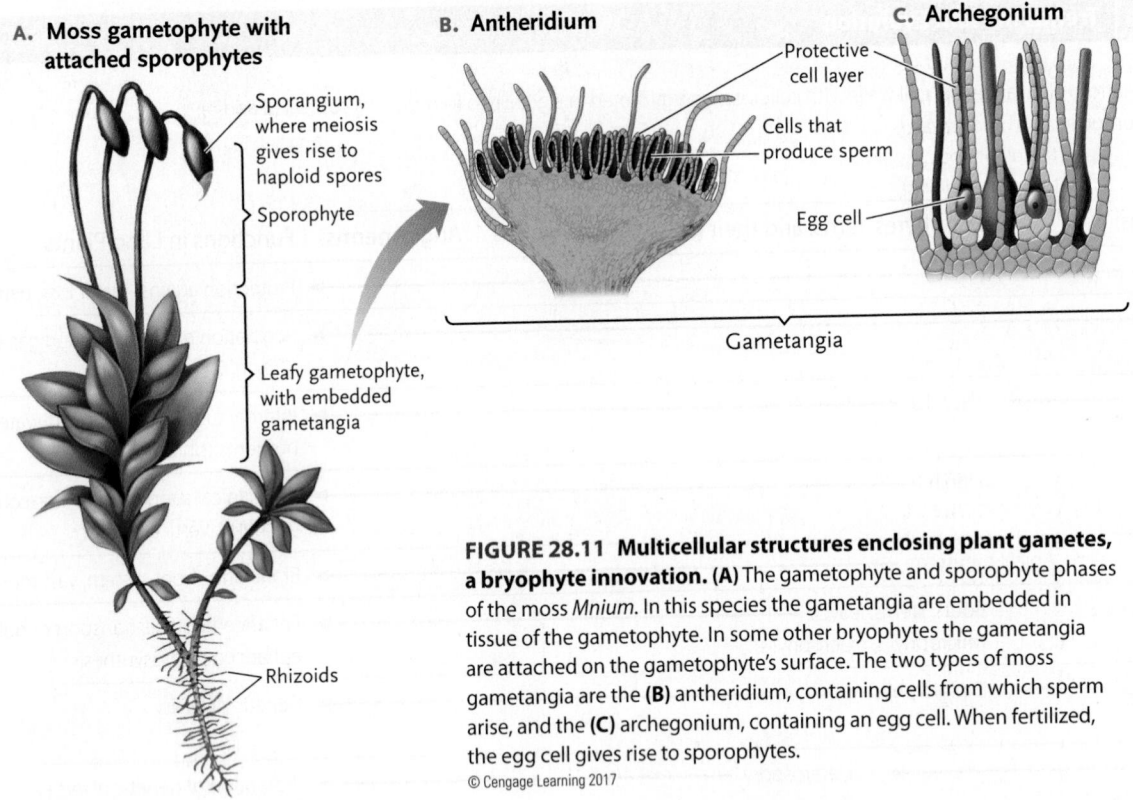

A. Moss gametophyte with attached sporophytes

Sporangium, where meiosis gives rise to haploid spores

Sporophyte

Leafy gametophyte, with embedded gametangia

Rhizoids

B. Antheridium

C. Archegonium

Protective cell layer

Cells that produce sperm

Egg cell

Gametangia

FIGURE 28.11 Multicellular structures enclosing plant gametes, a bryophyte innovation. (A) The gametophyte and sporophyte phases of the moss *Mnium*. In this species the gametangia are embedded in tissue of the gametophyte. In some other bryophytes the gametangia are attached on the gametophyte's surface. The two types of moss gametangia are the **(B)** antheridium, containing cells from which sperm arise, and the **(C)** archegonium, containing an egg cell. When fertilized, the egg cell gives rise to sporophytes.

© Cengage Learning 2017

water on the plant to the archegonia and fertilize eggs. Each fertilized egg gives rise to a diploid embryo sporophyte, which stays attached to the gametophyte, produces spores—and the cycle repeats.

Despite their similarities to more complex plants, bryophytes do differ in important ways. Unlike vascular species, bryophytes have a gametophyte that is much larger than the sporophyte and that obtains its nutrition independently of the sporophyte. The comparatively tiny sporophyte, on the other hand, remains attached to the gametophyte and depends on the gametophyte for much of its nutrition.

Because of their mix of ancient and more "advanced" traits, the position of bryophytes in plant evolution was long an open question. In one view, bryophytes were a side shoot of evolution, completely separate from the path that led to vascular plants. An alternative hypothesis was that the basic bryophyte body plan was similar to the ancestral condition from which higher plants evolved. Most current molecular, biochemical, cellular, and morphological evidence supports the view that bryophytes are not monophyletic—that is, the different groups do not share a recent common ancestor. A widely accepted scenario is that liverworts diverged early on from the lineage that led to all other land plants, with mosses diverging later and hornworts later still. Accordingly, our survey of nonvascular plants begins with the liverworts and mosses and concludes with hornworts—intriguing little plants that molecular studies place as the closest bryophyte relatives (the sister clade) of vascular plants.

Liverworts Were the First Land Plants

Early herbalists thought that liverworts—the small plants that make up the phylum **Hepatophyta**—were shaped like the lobes of the human liver (*hepat* = liver; *wort* = herb). Modern eyes might have more trouble making that connection: many of the 6,000 species of liverworts consist of a flat, branching, ribbon-like plate of tissue closely pressed against damp soil. This simple body, called a *thallus* (plural, *thalli*), is the gametophytic generation. Threadlike rhizoids anchor the gametophytes to their substrate. About two-thirds of liverwort species have leaflike structures, and some have stemlike parts. None have stomata, the openings that regulate gas exchange in most other land plants, although some species do have pores that open and close.

In the liverwort genus *Marchantia* (**Figure 28.12**), male and female gametophytes are separate plants. Male plants produce antheridia and female plants produce archegonia on specialized stalked organs (see Figure 28.12A, B). The sperm released from an antheridium must swim through surface water to encounter an egg inside an archegonium of a female gametophyte. After fertilization, a small, diploid sporophyte develops inside the archegonium, matures there, and produces haploid spores by meiosis. During meiosis, *Marchantia* sex chromosomes segregate, so some spores have the male genotype and others the female genotype. As in other liverworts, the spores develop inside jacketed sporangia that split open to release the spores. A spore that is carried by air currents to a suitable location germinates and gives rise to

FIGURE 28.12 The bryophyte *Marchantia*, the only liverwort to produce (A) male and (B) female gametophytes on separate plants. *Marchantia* also reproduces asexually by way of (C) gemmae, multicellular vegetative bodies that develop in tiny cups on the plant body. Gemmae can grow into new plants when splashing raindrops transport them to suitable sites.

A. Male plant

Male gametophyte

Martin Hutten/National Park Service

B. Female plant

Female gametophyte

Paul Stehr-green/National Park Service

C. Asexual reproduction

Gemmae

© Wayne P. Armstrong, Professor of Biology and Botany, Palomar College, San Marcos, California

a haploid gametophyte, which is either male or female. *Marchantia* also can reproduce asexually by way of **gemmae** (*gem* = bud), small cell masses that form in cuplike growths on a thallus (see Figure 28.12C). Gemmae can grow into new thalli when rainwater splashes them out of the cups and onto an appropriately moist substrate.

Both molecular and fossil evidence support the hypothesis that liverworts were the first land plants. Mitochondrial gene sequence data show that liverworts lack several introns that are present in all other bryophytes and in vascular plants. A recent fossil discovery in Argentina—five morphologically distinct types of liverwort spores in rocks dated to 472 mya—suggests that even then, liverworts had been diversifying for millions of years.

Mosses Share Many Traits with Vascular Plants

Chances are that you have seen, touched, or sat on at least several of the approximately 10,000 species of mosses, and the use of the name **Bryophyta** for this phylum underscores the fact that mosses are the best-known bryophytes. Their spores, produced by the tens of millions in sporangia, give rise to threadlike, haploid gametophytes that grow into the familiar tufts or carpets of moss plants on the surface of rocks, soil, or bark.

Partly because mosses are structurally and functionally most similar to the vascular plants we will consider in following sections, their life cycle is often used to represent bryophytes generally. The cycle, diagrammed in **Figure 28.13**, begins when a haploid (*n*) spore lands on a wet soil surface. After the spore germinates, it elongates and branches into a filamentous web of tissue called a **protonema** ("first thread"), which can become dense enough to color the surface of soil, rocks, or bark visibly green. After several weeks of growth, the budlike cell masses on a protonema develop into leafy, green gametophytes anchored by rhizoids. A single protonema can be extremely prolific, producing bud after bud—and in this way giving rise to a dense clone of genetically identical gametophytes. Leafy mosses also may reproduce asexually by gemmae produced at the surface of rhizoids, as well as on aboveground parts.

When a leafy moss is sexually mature, gametangia develop on its gametophytes and gametes form in them. In some moss genera, plants are unisexual and produce male *or* female gametangia—antheridia at the tips of male gametophytes and archegonia at the tips of female gametophytes. In other genera, plants are bisexual and produce both antheridia and archegonia. Propelled by a pair of flagella, sperm released from antheridia swim through a film of dew or rainwater and down a channel in the neck of the archegonium, attracted by a gradient of a chemical secreted by each egg. Fertilization produces the new sporophyte generation inside the archegonium, in the form of diploid zygotes that develop into small, mature sporophytes, each consisting of a sporangium on a stalk. Moss sporophytes may eventually develop chloroplasts and nourish themselves by photosynthesis, but initially they depend on the gametophytes for food. And even after a moss sporophyte begins photosynthesis, it still must obtain water, carbohydrates, and some other nutrients from the gametophyte.

Certain moss gametophytes are structurally complex, with features somewhat similar to those of higher plants. For example, mosses have stomata that open and close, and some species have a central strand of primitive conducting tissue. One kind of this tissue is made up of elongated, thin-walled, dead and empty cells that conduct water. These specialized cells, called hydroids, have oblique end walls that sometimes are partly dissolved or perforated with pores. Experiments with dyes show that water moves through them, as it does in similar xylemlike arrangements in vascular plants (see Chapters 33 and 34). In a few mosses, the water-conducting cells are surrounded by sugar-conducting tissue resembling the phloem of vascular plants.

A fascinating characteristic of some moss species is their ability to recover quickly from desiccation—a feature that is highly adaptive in moist environments that may occasionally dry out. For instance, when the star moss (*Tortula ruralis*) is experimentally dried until the water content of its tissues is only about 5–10% of normal, its tissues require only about an hour of rehydration before they plump up and begin photosynthesis again.

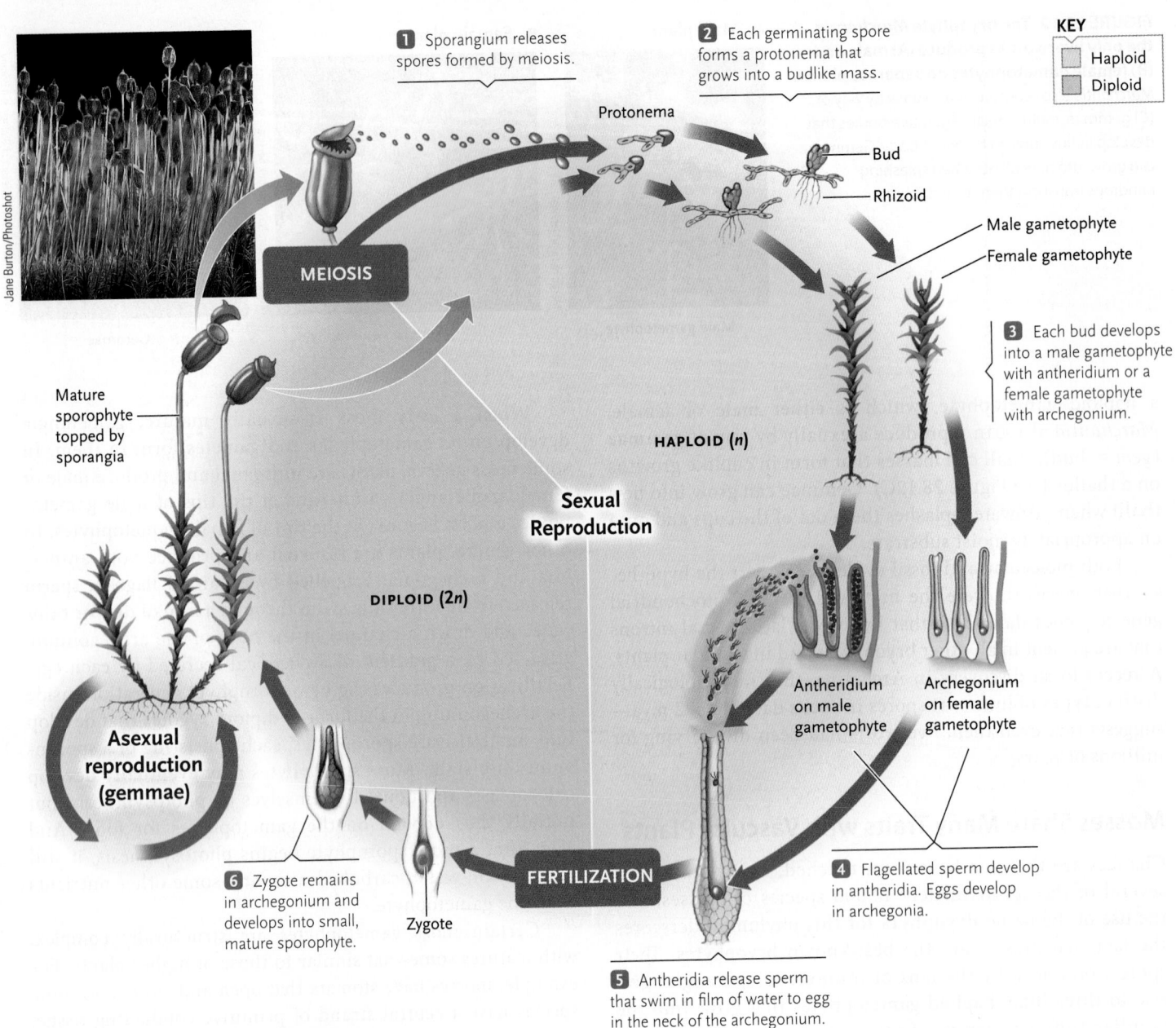

KEY

| | Haploid |
| | Diploid |

1 Sporangium releases spores formed by meiosis.

2 Each germinating spore forms a protonema that grows into a budlike mass.

Protonema

Bud

Rhizoid

Male gametophyte

Female gametophyte

MEIOSIS

3 Each bud develops into a male gametophyte with antheridium or a female gametophyte with archegonium.

HAPLOID (n)

Sexual Reproduction

Mature sporophyte topped by sporangia

DIPLOID (2n)

Antheridium on male gametophyte

Archegonium on female gametophyte

Asexual reproduction (gemmae)

6 Zygote remains in archegonium and develops into small, mature sporophyte.

Zygote

FERTILIZATION

4 Flagellated sperm develop in antheridia. Eggs develop in archegonia.

5 Antheridia release sperm that swim in film of water to egg in the neck of the archegonium.

FIGURE 28.13 Life cycle of a moss, *Polytrichum*.
© Cengage Learning 2017

Hornworts Have Both Plantlike and Algalike Features

The hornwort phylum, **Anthocerophyta**, consists of only about 200 species. Simple, always-open stomata appear in this group. A hornwort gametophyte has a flat thallus, but the sporangium of the sporophyte phase is long and pointed, like a horn (**Figure 28.14**). The sporangia split into ribbonlike sections when they release spores. Sexual reproduction occurs in basically the same way as in liverworts: free-swimming sperm fertilize eggs, from which the sporophytes develop. Hornworts also may reproduce asexually by fragmentation as pieces of a thallus break off, form rhizoids, and develop into new individuals.

Many hornwort species have cell features in common with green algae, including the presence in each cell of a single large chloroplast that contains protein bodies called pyrenoids. Botanists once interpreted the similarities to liverworts and green algae as evidence that hornworts diverged very early in land plant evolution, but more recent molecular and morphological studies support a different scenario. For example, comparisons of introns of the mitochondrial genomes of hornworts and vascular plants suggest strongly that the sequences in vascular plants evolved from ancestral sequences in hornworts. Hornwort spores and other parts also have been found to contain xylan, a polysaccharide that in vascular plants helps strengthen cell walls in transport tissues. Other bryophytes do not synthesize xylan, and its presence in hornworts suggests that

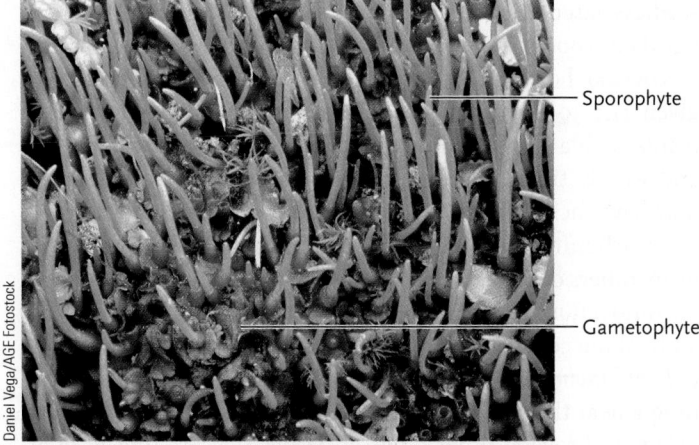

FIGURE 28.14 **The hornwort *Anthoceros*.** The base of each long, slender sporophyte is embedded in the flattened, leafy gametophyte.

STUDY BREAK 28.2
1. Give some examples of bryophyte features that bridge aquatic and terrestrial environments.
2. How do specific aspects of a moss plant's anatomy resemble those of vascular plants?

28.3 Seedless Vascular Plants

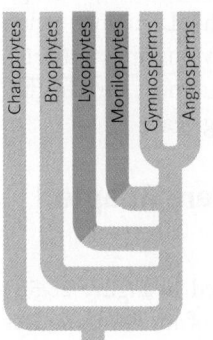

The first vascular plants, which did not "package" their embryos inside protective seeds, were the dominant plants on Earth for almost 200 million years, until seed plants became abundant and diversified across the landscape. The fossil record shows that seedless vascular plants were well established by the late Silurian, some 427 mya, and they flourished until the end of the Carboniferous period, about 250 mya. On one hand, like bryophytes, seedless vascular plants reproduce sexually by releasing spores, and they have swimming sperm that require free water to reach eggs. On the other hand, as in seed plants, the sporophyte of a seedless vascular plant separates from the gametophyte at a certain point in its development and has well-developed vascular tissues (both xylem and phloem). Also, the sporophyte is the larger, longer-lived stage of the life cycle and the gametophytes are very small. Some gametophytes even lack chlorophyll. **Table 28.2** summarizes these characteristics and gives an overview of seedless vascular plant features.

modification of hornwort genetic programs for making and using xylan may have been a crucial step in the evolution of sturdy transport tissues in vascular plants. These discoveries are part of the foundation for the hypothesis that hornworts are the bryophytes most closely related to vascular plants.

In the next section we turn to the seedless vascular plants, which have tissues specialized for transporting water, minerals, and sugars. Without the capacity to move these substances efficiently throughout the plant body, large sporophytes could not have evolved on land. Unlike bryophytes, modern vascular plants are monophyletic—all groups are probably descended from a common ancestor.

TABLE 28.2	Seedless Plant Phyla and Major Characteristics		
Phylum	Common Name	Number of Species*	Common General Characteristics
Bryophytes: Nonvascular plants. Gametophyte dominant, free water required for fertilization, cuticle and stomata present in some.			
Hepatophyta	Liverworts	6,000	Leafy or simple flattened thallus with pores, rhizoids; spores in capsules; cuticle. Moist, humid habitats.
Bryophyta	Mosses	10,000	Feathery or cushiony thallus, some have hydroids; spores in capsules. Moist, humid habitats; colonizes bare rock, soil, or bark.
Anthocerophyta	Hornworts	100	Simple flattened thallus, rhizoids; stomata; hornlike sporangia. Moist, humid habitats.
Seedless vascular plants: Sporophyte dominant, swimming sperm, free water required for fertilization, cuticle and stomata present in all.			
Lycophyta	Club mosses	1,000	Simple leaves, true roots; most species have sporangia on sporophylls. Mostly wet or shady habitats.
Monilophyta	Ferns, whisk ferns, horsetails	13,000	*Ferns:* Finely divided leaves, woody stems in tree ferns; sporangia in sori. Habitats from wet to arid. *Whisk ferns:* Branching stem from rhizomes; sporangia on stem scales. Tropical to subtropical habitats. *Horsetails:* Hollow photosynthetic stem, scalelike leaves with silica in cell walls, sporangia in strobili. Swamps, disturbed habitats.

*Numbers of species are approximate.

Seedless vascular plants once encompassed a huge number of diverse species of trees, shrubs, and herbs. In the late Paleozoic era, they were Earth's dominant vegetation. Some lineages have endured to the present, but collectively these survivors total fewer than 14,000 species. The taxonomic relationships between various lines are still under active investigation, and comparisons of gene sequences from the genomes in plastids, cell nuclei, and mitochondria are revealing previously unsuspected links between some of them. In this book we assign seedless vascular plants to two phyla, the Lycophyta (club mosses and their close relatives) and the Monilophyta (ferns, whisk ferns, and horsetails), although it is almost certain that this arrangement will change in the coming years.

Early Seedless Vascular Plants Were Adapted to Moist Environments

The extinct plant genus *Cooksonia* (depicted in Figure 28.5A) probably was one of the earliest ancestors of modern seedless vascular plants. It and similar plants dominated the mudflats and swamps of the Devonian period (408–360 mya). Like other members of its extinct phylum Rhyniophyta, *Cooksonia* species were small, rootless, and leafless, but their simple stems had a central core of xylem, an arrangement seen in many existing vascular plants. As rhyniophytes and other now-extinct phyla came and went, ancestral forms of both modern phyla of seedless vascular plants appeared. In botanical terms, the earliest

seedless vascular plants were "herbs"—that is, they did not have lignified woody tissue. By the start of the ensuing Carboniferous period, however, the small, herbaceous Devonian plants had given rise to larger shrubby species and to trees—long-lived plants with a single main stem containing both vascular tissue and woody tissue. Other features of some of these early trees may have included bark, roots, leaves, and even seeds.

Carboniferous forests were lush, swampy places dominated by members of the phylum **Lycophyta,** in which roots evolved—an innovation that provided both firm anchoring in the soil and enhanced access to soil water and nutrients. One example is *Lepidodendron*, which had broad, straplike leaves and sporangia near the ends of the branches **(Figure 28.15A)**. It also had xylem and several other types of tissues that are typical of all modern vascular plants, although probably not in the same proportions as seen today.

Carboniferous forests also were home to representatives of the phylum **Monilophyta,** including ferns such as *Medullosa* and giant horsetails such as *Calamites*—huge treelike plants that could have a trunk 30 cm in diameter. Like their modern descendants, these ancient species had a distinctive beadlike arrangement of certain xylem components, hence the group's name (from Latin *moniliformis-*, necklacelike + *phyton,* plant). The strong, upright stems were attached to a system of rhizomes. Ferns populated the forest understory. Some early seed plants were present as well, including now-extinct fernlike plants, called seed ferns, which bore seeds at the tips of leaves **(Figure 28.15B)**.

A. Lycophyte tree (*Lepidodendron*)

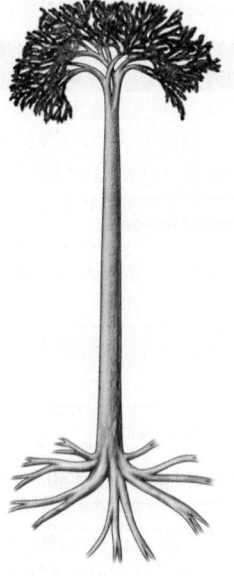

B. Artist's depiction of a Carboniferous forest

Stem of a giant lycophyte (*Lepidodendron*)

Stem of a giant horsetail (*Calamites*)

Seed fern *(Medullosa)*; probably related to the progymnosperms, which may have been among the earliest seed-bearing plants

FIGURE 28.15 Reconstruction of the lycophyte tree *Lepidodendron* and its environment. (A) Fossil evidence suggests that *Lepidodendron* grew to be about 35 m tall with a trunk 1 m in diameter. **(B)** Lush forests of the Carboniferous period were dominated by early seedless vascular plants.

Most modern seedless vascular plants are ferns, and like their ancestors they are confined largely to wet or humid environments because they require external water for reproduction. Except for whisk ferns, their gametophytes do not rely on vascular tissues for water transport, and male gametes swim through water to reach eggs. In the wild, the few vascular seedless plants that are adapted to dry environments such as deserts can reproduce sexually only when seasonal rains make surface water available.

Modern Lycophytes Are Small and Have Simple Vascular Tissues

Unlike their ancient relatives, the most familiar of the 1,000 or so living species of lycophytes are small, ground-hugging species. They include the club mosses shown in **Figure 28.16A**—members of genera such as *Lycopodium* that grow on forest floors—as well as spike mosses and quillworts. Notice that the common terms "spike moss" and "club moss" are misnomers; these plants are not mosses but true vascular species in which the sporophyte is dominant.

A club moss sporophyte has stems and roots that contain a small amount of the vascular tissue xylem. Small, green microphyll-type leaves called **sporophylls** are located near the tips, with a single sporangium tucked at the base of each. A cluster of sporophylls forms a conical **strobilus** (plural, *strobili*). In some species the sporangia release haploid spores produced by meiosis **(Figure 28.16B)**. If a spore germinates (sometimes years after its release), it forms a free-living gametophyte that differs markedly from the sporophyte. Ranging in size from nearly invisible to several centimeters, the gametophyte easily becomes buried under decomposing plant litter. There, rhizoids attach it to a substrate. It cannot photosynthesize, instead obtaining nutrients by way of mycorrhizae, a symbiotic interaction in which a plant's rhizoids or roots associate with a particular fungus. Although

many lycophyte species are homosporous—gametophytes are bisexual and produce both eggs and sperm—some are heterosporous. Regardless, as with ancestral lycophytes, the sperm require water in which they can swim to the eggs. After fertilization, the life cycle comes full circle as the zygote develops into a diploid embryo that grows into a sporophyte.

The Diverse Phylum Monilophyta Includes Ferns, Whisk Ferns, Horsetails, and Their Relatives

Second only to the flowering plants, the monilophytes encompass a large and diverse group of vascular plants—the 13,000 or so species of ferns, whisk ferns, and horsetails. Most ferns are native to tropical and temperate regions. Some floating species are less than 1 cm across, whereas some tropical "tree ferns" grow to 25 m tall. Other species are adapted to life in arctic and alpine tundras, salty mangrove swamps, and semi-arid deserts.

COMPLEX ANATOMICAL FEATURES EVOLVED IN FERNS The sporophyte phase is the familiar body of a fern **(Figure 28.17)**. It produces an aboveground clump of fern leaves, called fronds, that are often finely divided and featherlike. Fronds contain multiple strands of vascular tissue and are an example of large, megaphyll-type leaves—a trait ferns share with seed plants. A typical frond has a well-developed epidermis with chloroplasts in the epidermal cells and stomata on the lower surface. Young fronds are tightly coiled, and as they emerge above the soil these "fiddleheads" (so named because they resemble the scroll of a violin) unroll and expand.

Except for tropical tree ferns, the stems of most ferns are underground rhizomes. The stem's vascular system is organized into a complex, interconnecting network of bundles, each having a core of xylem tissue surrounded by phloem tissue. Roots descend along the length of the rhizomes. Remarkably, a rhizome can live for centuries, growing at its tip and extending outward horizontally through the soil, sometimes over a considerable area. In most ferns, the fronds arise from nodes positioned along the rhizome. A **node** is the point on a stem where one or more leaves are attached.

In most fern species, the sporophyte produces sporangia on the lower surface or margin of some leaves. Often, several sporangia are clustered into a rust-colored **sorus** (plural, *sori*). Sori may be exposed or they may be protected with a flap of tissue. Each sporangium is a delicate case, shaped rather like an old-fashioned pocket watch and covered by a layer of epidermal cells. In the layer, a row of thick-walled cells called the **annulus** ("ring") nearly encircles the sporangium (see inset, Figure 28.17).

Inside the sporangium, haploid spores are produced by meiosis. Meanwhile, the sporangium slowly dries, and in the process the annulus contracts. Eventually the force of the contracting annulus rips open the sporangium, which snaps back on itself, flinging out the mature spores. In this way fern spores can be dispersed up to 2 m away from the parent plant.

A. *Lycopodium* sporophyte

Strobilus

Ed Reschke/Stockbyte/Getty Images

B. Fossilized lycophyte spore

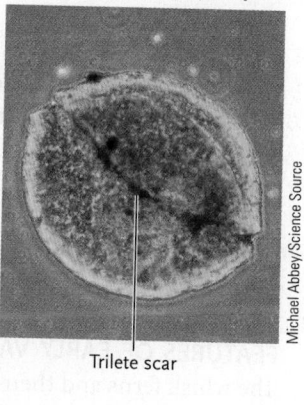

Trilete scar

Michael Abbey/Science Source

FIGURE 28.16 Lycophytes. (A) *Lycopodium* sporophyte, showing the conelike strobili in which spores are produced. **(B)** A fossilized lycophyte spore bearing a characteristic mark called a trilete scar.

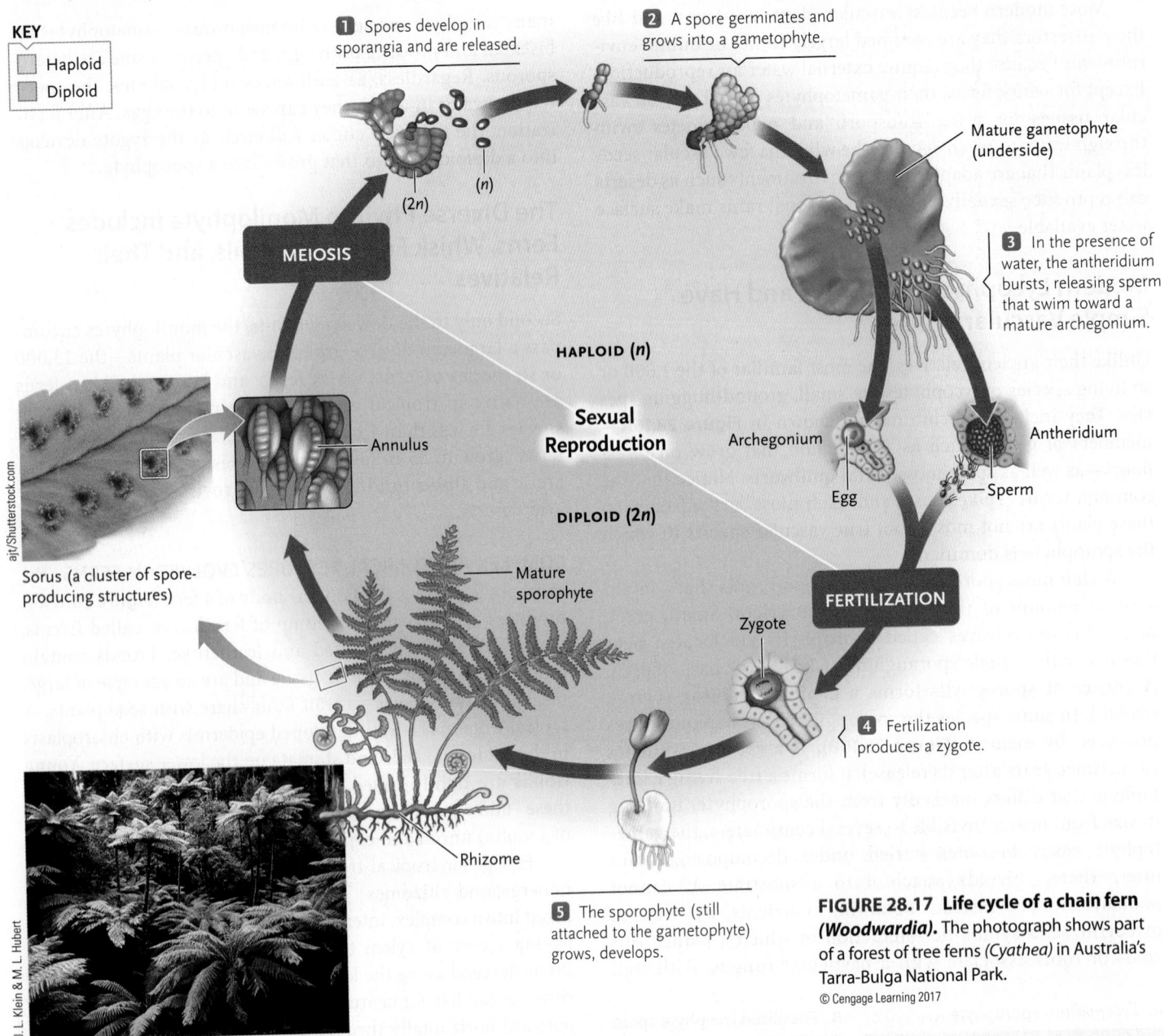

KEY

Haploid
Diploid

1 Spores develop in sporangia and are released.

2 A spore germinates and grows into a gametophyte.

MEIOSIS

(2*n*)

(*n*)

Mature gametophyte (underside)

3 In the presence of water, the antheridium bursts, releasing sperm that swim toward a mature archegonium.

HAPLOID (*n*)

Sexual Reproduction

Annulus

Archegonium

Antheridium

Egg

Sperm

DIPLOID (2*n*)

FERTILIZATION

Sorus (a cluster of spore-producing structures)

Mature sporophyte

Zygote

4 Fertilization produces a zygote.

Rhizome

5 The sporophyte (still attached to the gametophyte) grows, develops.

FIGURE 28.17 Life cycle of a chain fern (*Woodwardia*). The photograph shows part of a forest of tree ferns (*Cyathea*) in Australia's Tarra-Bulga National Park.
© Cengage Learning 2017

Wind may carry them much farther: on board the *Beagle*, Charles Darwin collected fern spores hundreds of miles from shore.

A germinating spore develops into a gametophyte, which is typically a small, heart-shaped plant anchored to the moist soil by rhizoids. Both antheridia and archegonia are present on the lower surface of each gametophyte, where moisture is trapped. Inside an antheridium is a globular packet of haploid cells, each of which develops into a helical sperm cell with many flagella. Eggs develop within archegonia. When water is present, the antheridium bursts, releasing the sperm. If mature archegonia are nearby, the sperm swim toward them, drawn by a chemical attractant that diffuses from the neck of the archegonium, which is open when free water is present.

After a sperm fertilizes an egg, the diploid zygote begins dividing and developing into an embryo, which at this stage obtains nutrients from the gametophyte. In a short time, however, the embryo develops into a young sporophyte that is larger than the gametophyte and has its own green leaf and a root system. The sporophyte now is nutritionally independent, and the parent gametophyte dies.

FEATURES OF EARLY VASCULAR PLANTS IN WHISK FERNS

The whisk ferns and their relatives are represented by two genera, *Psilotum* and *Tmesipteris*, with only about 10 species in all. These rather uncommon plants grow in tropical and subtropical regions, often as epiphytes. In the United States the range for *Psilotum* species **(Figure 28.18)** includes Hawaii, Gulf

FIGURE 28.18 Sporophytes of a whisk fern (*Psilotum*), a seedless vascular plant. Three-lobed sporangia occur at the ends of stubby branchlets; inside the sporangia, meiosis gives rise to haploid spores.

A. Sporophyte stem   B. Sporangia

Strobilus, an aggregation of sporangia at the tip of the horsetail sporophyte

C. Each petal-shaped sporangium of a strobilus contains spores that formed by meiosis.

FIGURE 28.19 A species of *Equisetum*, the horsetails. (A) Vegetative stem. **(B)** Strobili, which bear sporangia. **(C)** Close-up of sporangia and associated structures on a strobilus.

Coast states such as Florida and Louisiana, and parts of the West.

Whisk fern sporophytes are up to 60 cm tall and resemble the extinct *Cooksonia*. Like those early vascular plants, they lack roots and leaves. Instead, small scales adorn an upright branching stem that ascends from a horizontal rhizome system anchored by rhizoids. The rhizoids have mycorrhizal fungi associated with them, which provide enhanced access to key nutrients.

A whisk fern's stem is structurally and functionally multifaceted. Its epidermal cells carry out photosynthesis, whereas its core has the transport tissues xylem and phloem and other anatomical features of more complex vascular plants. Sporangia rest atop some of the stem scales. Inside them, meiotic divisions of specialized cells produce haploid spores.

HORSETAILS, REMNANTS OF AN ANCIENT PAST Only fifteen horsetail species in a single genus, *Equisetum*, have survived to the present **(Figure 28.19).** Unlike towering ancestors such as *Calamites*, most modern horsetails grow less than a meter high, in moist soil along streams and in disturbed habitats, such as roadsides and beds of railroad tracks. Their sporophytes typically have underground rhizomes along with roots that anchor the rhizome to the soil. Whorls of scalelike leaves are arranged around a photosynthetic stem that is stiff and gritty from the silica that horsetails accumulate in their tissues. They came to be known popularly as "scouring rushes" because American pioneers used them to scrub out pots and pans.

Equisetum sporangia are borne in strobili on highly specialized stem structures quite different from the sporophylls of club mosses. Each stalked spore-bearing structure in a strobilus resembles an umbrella and is attached at right angles to a main axis. Haploid spores develop in sporangia attached near the edge of the "umbrella's" underside, and air currents disperse them. They must germinate within a few days to produce gametophytes, which are free-living plants about the size of a small pea.

STUDY BREAK 28.3

1. Compare and contrast the lycophyte and bryophyte life cycles with respect to the sizes and longevity of gametophyte and sporophyte phases.
2. In ferns, whisk ferns, and horsetails, what kinds of structures fulfill the roles of roots and leaves?
3. How does the life cycle of a horsetail differ from that of a fern?

THINK OUTSIDE THE BOOK

Seed ferns, an extinct group mentioned briefly in this section, are assigned to the extinct phylum Pteridospermophyta. They were nearly wiped out by the Permian–Triassic mass extinction event that occurred about 251 mya and became extinct soon thereafter. Seed-bearing conifers living at the time fared better, and conifers were the dominant land plants for at least another 100 million years. Using Internet or library resources, develop a plausible explanation for why the seed fern and conifer lineages had such differing evolutionary fates.

A. Columns of peat cut from an Irish bog

B. Peat processed into a planting medium and soil conditioner

FIGURE 28.20 Peat moss.

28.4 Ecological, Economic, and Research Importance of Seedless Plants

Seedless plants, especially perhaps bryophytes, may be easy to overlook as components of ecological communities and contributors to human economies and well-being. In fact, however, these groups contribute a great deal ecologically, economically, and as model organisms for research.

Bryophytes Have Important Ecological Roles

Bryophytes sometimes are the first colonizers of bare land. When they do so, their small bodies trap particles of organic and inorganic matter, helping to build soil on bare rock and stabilizing soil surfaces with a biological crust in harsh places like embankments created by road construction. Some hornworts harbor mutualistic nitrogen-fixing cyanobacteria, and so increase the amount of nitrogen available to other plants. In arctic tundra, bryophytes constitute as much as half the biomass and are crucial components of the food web that supports animals in that biome. *Sphagnum* and other absorbent "peat" mosses typically grow in bogs **(Figure 28.20)**. People have used them for everything from diapering babies and filtering whiskey to increasing the water-holding capacity of garden soil.

Ancient Seedless Vascular Plants Formed the World's Coal, Oil, and Gas Deposits

In a sense, seedless vascular plants are among the most economically important plants that have ever lived. When leaves, branches, and old trees in swampy Carboniferous landscapes fell to the ground, they became buried in anaerobic sediments. Over geological time, these buried remains were compressed and modified, forming deposits of coal, oil, and natural gas that have been major global energy sources since the Industrial Revolution. Coal, which consists of the fossilized remains of Carboniferous plant life, was the first such fuel to be widely used by human societies and the Carboniferous period is often called the Coal Age. Petroleum (crude oil) and natural gas, often lumped

with coal as "fossil fuels," originated as organic matter buried under ocean or river sediments.

Today, ferns, the most familiar seedless vascular plants, are popular as ornamental plants. Lacy maidenhair ferns (*Adiatum* spp.), sword ferns (*Nephrolepsis exaltata*), and other species are used as houseplants and in private and public gardens. In some areas, the fiddleheads (coiled young fronds) of various edible species are a popular human food item.

Bryophytes and Seedless Vascular Plants Also Have Roles as Model Research Organisms

Several species of bryophytes and seedless vascular plants have played important roles in broadening our understanding of plant biology and evolution. One of these is the lycophyte *Selaginella moellendorffii* discussed in this chapter's *Molecular Insights*; another is the moss *Physcomitrella patens* **(Figure 28.21)**. The recently sequenced genomes of both species have become powerful tools for studying the evolutionary pathways of groups

FIGURE 28.21 Vegetative shoot of *Physcomitrella patens* (false-color SEM, ×20). Leaves are a single layer of cells.

Comparative Genomics Probes Plant Evolution

Plant evolutionary biologists have long hoped to gain a better understanding of how plant genomes changed as seed plants, with their well-developed vascular systems, diverged from simpler vascular seedless plants. Research on this question has been hampered, however, by the lack of a genome sequence for a seedless vascular plant. An international collaboration led by Jo Ann Banks of Purdue University targeted this knowledge gap. The plan was twofold, beginning with sequencing the genome of the spike moss *Selaginella moellendorffii*—a seedless vascular plant in the phylum Lycophyta **(Figure)**. With this lycophyte genome in hand, the team could compare it with genomes available for species representing other major taxa in the lineage of land plants.

pzAxe/Shutterstock.com

FIGURE *Selaginella moellendorffii.*

© Cengage Learning 2017

Research Question

Can comparing the genomes of diverse land plant taxa provide insights into genome changes that accompanied major transitions in the evolution of seed plants?

Experiments

DNA extracted from *S. moellendorffii* leaves, stems, and strobili was sequenced using the whole-genome shotgun sequencing method described in Section 19.2. Annotation of the genome revealed approximately 22,285 protein-coding genes. This information could be compared with similar data available for species representing major branch points in land plant evolution. To represent the starting point for land plant evolution, the researchers used genomic data for the single-celled green alga *Chlamydomonas*. Nonvascular plants (bryophytes) were represented by the genome of a moss in the genus *Physcomitrella* and seedless vascular plants by that of *S. moellendorffii*. Genome data for 15 species of flowering plants represented vascular seed plants.

Comparing the genomes produced fascinating findings—and many avenues for further research. *Chlamydomonas* and all the study's plant species shared 3,814 genes—presumably ones conserved over evolutionary time because they govern features that were strongly favored by selection pressures associated with the transition to terrestrial life. The *Physcomitrella* genome added 3,006 new genes to this "algal" group. In *S. moellendorffii*, a gain of only 516 genes was associated with the evolution of a simple vascular system and the developmental shift to a dominant sporophyte generation. By contrast, the flowering plant genomes contained 1,350 new genes.

Conclusion

The emergence of land plants evidently correlated with the near doubling of the genes (and the set of proteins) that bryophytes inherited from an algal ancestor. Surprisingly, evolution of the first vascular plants—also the first plants to have a dominant sporophyte generation—entailed the gain of relatively few genes. Evidently, many new genes were required during the transition from seedless vascular plants to flowering plants, in which the vascular system, reproductive adaptations, and other structures and functions are the most complex in the plant kingdom.

think like a scientist

In their paper, the Banks team noted that their comparative analysis would have benefited if genome data from a charophyte alga and a gymnosperm, such as a conifer, had been available. How would having such data be useful?

Source: Jo Ann Banks et al. 2011. The *Selaginella* genome identifies genetic changes associated with the evolution of vascular plants. *Science* 332:960–963.

---

throughout the plant kingdom. Ferns and some other seedless vascular plants have long been employed in research on plant growth and development. An especially useful feature is the single, large, centrally located apical cell at the tip of a fern's apical meristems. The cell's size makes it possible to easily track the mitotic divisions that give rise to smaller daughter cells, and ultimately produce growth of shoot and root parts.

STUDY BREAK 28.4

1. What are some important ecological and economic roles of seedless plants?
2. Give some examples of ways seedless plants have been used as model research organisms.

Unanswered Questions

Why is the homosporous lifestyle correlated with unusually large numbers of chromosomes?

My laboratory at the University of Arizona is interested in how plant genomes evolve, and, in particular, how changes in chromosome number and organization occur. In the plant kingdom, and in fact across all eukaryotes, homosporous ferns and lycophytes have the largest numbers of chromosomes: on average $n = 57$. How this strikingly large average chromosome number correlates with reproductive strategies in homosporous lycophytes and ferns has long been a mystery to plant evolutionary biologists. In general, chromosome numbers may increase due to dysploidy—a change in the number of single chromosomes without a change in the overall number of genes—or by polyploidy (duplication of a whole genome). Decades of research have uncovered only rare examples of dysploidy in plants, but polyploidy is much more common.

This observation has fueled hypotheses about why high chromosome numbers are correlated with homospory.

One possible explanation is that the capacity of a single spore to produce both sperm and egg may support higher rates of polyploidy. During polyploidy, the new zygote with duplicated chromosomes is nearly always reproductively isolated from its parental, diploid relatives. This means that for polyploid lineages to survive and establish clades with higher numbers of chromosomes, they must be able to reproduce. If a polyploid individual arises and is the only one of its kind, the reproductive assurance of having homosporous gametophytes will aid in the new line's initial survival. In contrast, a heterosporous plant would need to have a megaspore and a microspore near each other for reproduction, instead of just a single spore.

Alternatively, it may be that having more chromosomes relieves potential deleterious effects of the extreme inbreeding possible in homosporous ferns and lycophytes. Assuming that the overall number of genes remains similar, a greater number of chromosomes will increase independent assortment and thus effectively increase the recombination rate. Whether increased polyploidy, selection for increased independent assortment, or some combination of both is driving the evolution of high chromosome numbers in homosporous lycophytes and ferns remains an open question. New genomic tools are being applied to study the genomes of these groups to give us new insights into how the diversity of plant lifestyles has influenced the diversity of plant genomes.

think like a scientist How would you expect changes in chromosome number to influence recombination and independent assortment? How are changes in chromosome number achieved without creating problems during meiosis? Would you expect all chromosomes to be equally likely to be duplicated or lost?

Courtesy of Michael Barker

Michael S. Barker is an Assistant Professor of Ecology and Evolutionary Biology at the University of Arizona in Tucson. His research focuses on the evolution genomics of plant diversity. To learn more about Barker's work, go to http://barkerlab.net/

REVIEW KEY CONCEPTS

For access to MindTap and additional study materials visit www.cengagebrain.com.

28.1 Plant Evolution: Adaptations to Life on Land

- Plants are thought to have evolved from green algae between 425 and 490 million years ago (Figure 28.2).
- Key adaptations in early land plants include an outer cuticle that helps prevent desiccation, spores protected by a wall containing sporopollenin, tissues containing lignified cells, multicellular chambers that protect developing gametes, and an embryo sheltered inside a parent plant (Figure 28.3).
- Major trends in land plant evolution included the development of lignified tissues including vascular tissues, root systems, and shoot systems, including leaves having stomata; a shift from dominance by a long-lived, larger haploid gametophyte to dominance of a long-lived, larger diploid sporophyte; and a shift from homospory to heterospory with separate male and female gametophytes (Figures 28.4–28.10).

28.2 Bryophytes, the Nonvascular Land Plants

- Existing nonvascular land plants, or bryophytes, include the liverworts (Hepatophyta), hornworts (Anthocerophyta), and mosses (Bryophyta). Liverworts likely were the first land plants.
- Bryophytes release spores and produce flagellated sperm that swim through free water to reach eggs. They lack a vascular system; true roots, stems, and leaves; and lignified tissue. A larger, dominant gametophyte (n) phase alternates with a small, fleeting sporophyte ($2n$) phase (Figures 28.11–28.14).

28.3 Seedless Vascular Plants

- Existing seedless vascular land plants include the lycophytes (club mosses), whisk ferns, horsetails, and ferns. They release spores and have swimming sperm, but have well-developed vascular tissues. The sporophyte is the larger, longer-lived stage of the life cycle and develops independently of the small gametophyte.
- Club mosses (Lycophyta) have sporangia clustered at the bases of specialized leaves called sporophylls. Clusters of sporophylls form a strobilus ("cone"). Haploid spores germinate to form small gametophytes. The life cycle is similar in ferns, whisk ferns, and horsetails (Monilophyta) (Figures 28.15–28.19).
- Ferns are the most diverse group of seedless vascular plants. Most species lack aboveground stems; leaves that arise from nodes along an underground rhizome. Fern leaves typically have well-developed stomata, and the vascular system consists of bundles, each with xylem surrounded by phloem. Sporangia on sporophylls ("fronds") release spores that develop into gametophytes. Sexual reproduction produces a much larger, long-lived sporophyte.

28.4 Ecological, Economic, and Research Importance of Seedless Plants

- Seedless plants are important ecologically, economically, and in scientific research. As colonizers of bare land, bryophytes help build and stabilize loose soil. In the arctic tundra, they constitute as much as half the biomass and are crucial food web components. *Sphagnum* moss species are employed widely as soil amendments and other human uses. Seedless vascular plants of the Carboniferous period formed the world's coal, oil, and natural gas deposits. Ferns are popular ornamental and house plants. Sequencing of several seedless plant genomes is yielding new insights into land plant evolution (Figures 28.20 and 28.21).

TEST YOUR KNOWLEDGE

Remember/Understand

1. Which of the following is *not* an evolutionary trend among plants?
   a. developing vascular tissues
   b. becoming seedless
   c. having a dominant diploid generation
   d. producing nonmotile gametes
   e. producing two types of spores

2. Which of the following were adaptations that occurred as plants made the evolutionary transition to a terrestrial existence?
   a. increased motility of gametes on dry land
   b. a flattened plant body that increased exposure to the Sun
   c. a reduction in the number and distribution of roots
   d. mechanisms for gaining access to nutrients in soil
   e. modifications that allowed stems and leaves to absorb water from the atmosphere

3. Which of the following is *not* generally thought to be accurate with respect to the evolution of leaves?
   a. Leaves arose more than once during land-plant evolution.
   b. Microphylls may have arisen as modifications of the main vertical stem.
   c. Megaphylls may have come about as modifications of stem branches.
   d. The evolution of higher-density leaf venation correlates with increased capacity for photosynthesis.
   e. Only macrophyll-type leaves, which have stomata and a network of veins, are observed in extant plant species.

4. Which is the correct matching of phylum and plant group?
   a. Hepatophyta: liverworts
   b. Anthocerophyta: mosses
   c. Bryophyta: horsetails
   d. Lycophyta: ferns
   e. Monilophyta: hornworts

5. Which feature(s) do ferns share with all other land plants?
   a. sporophyte and gametophyte life cycle stages
   b. gametophytes supported by a thallus
   c. dispersal of spores from a sorus
   d. asexual reproduction by way of gemmae
   e. water uptake by means of rhizoids.

6. In addition to having simple vascular tissue, seedless vascular plants differ from all bryophytes in that:
   a. they do not require free water for fertilization.
   b. the sporophyte generation is dominant.
   c. most do not require a moist, humid, or shaded habitat.
   d. the group is much more diverse, with many more species overall.
   e. most have a feathery or flattened thallus.

7. In comparing bryophytes and seedless vascular plants, which of the following statements is not accurate?
   a. In general, bryophyte traits more closely resemble those of algae.
   b. The leaves, stems, and roots of seedless vascular plants are not homologous with similar bryophyte structures.
   c. Both groups have lignified cells in some body parts.
   d. Bryophytes lack complex conducting tissues (xylem and phloem).
   e. Seedless vascular plants have sexual and asexual reproductive modes, but in bryophytes only asexual reproduction has been observed.

8. Based solely on numbers of species, the most successful seedless plants today are:
   a. horsetails.
   b. ferns.
   c. liverworts.
   d. mosses.
   e. club mosses.

9. Which of the following is/are true with respect to seedless vascular plants?
   a. Ferns are the most numerous species in this group.
   b. Lycophyte species grow the largest of all modern seedless vascular plants.
   c. Seedless vascular plants, including trees, shrubs, and herbaceous ("nonwoody") species dominated Earth's terrestrial environments for nearly 200 million years.
   d. The strobilus, where spore-producing parts sporangia are located, is a reproductive adaptation found in all seedless vascular plants.
   e. In the fern life cycle, the young sporophyte is dependent on the gametophyte, but quickly develops into an independent, larger plant with its own root system and leaves where photosynthesis occurs.

10. A homeowner noticed moss growing between bricks on his patio. Closer examination revealed tiny brown stalks with cuplike tops emerging from green leaflets. These brown structures were:
    a. the sporophyte generation.
    b. the gametophyte generation.
    c. elongated haploid reproductive cells.
    d. archegonia.
    e. antheridia.

Apply/Analyze

11. **Discuss Concepts** Working in the field, you discover a fossil of a previously undescribed plant species. The specimen is small and may not be complete; the bits you have do not include any floral parts. What sorts of observations would you need in order to classify the fossil as a seedless vascular plant with reasonable accuracy? What evidence would you need in order to distinguish between a fossil lycophyte and a fern?

12. **Apply Evolutionary Thinking** As described in Section 28.2, one proposed bryophyte phylogeny considers hornworts and vascular plants (such as lycophytes and ferns) to be sister clades. Among other factors, a shared morphological feature supports this hypothesis: hornwort and vascular plant sporophytes both show indeterminate growth—once a sporophyte develops, it grows for the remainder of the plant's life. Indeterminate growth has not been observed in other bryophyte lineages. Can you think of a different evolutionary explanation for the growth pattern shared by hornworts and vascular plants?

Evaluate/Create

13. **Discuss Concepts** Compare the size, anatomical complexity, and degree of independence of a moss gametophyte and a fern gametophyte. Which one is the most protected from the external environment? Which trends in plant evolution does your work on this question bring to mind?

14. **Design an Experiment** A moss species that you study occurs in lush abundance in the moist, shaded forest floor in a grove of coast redwoods (*Sequoia sempervirens*) in Northern California. When one of the massive old trees abruptly topples over, there is a dramatic increase in the amount of light and air movement in a sizable patch of the moss' habitat. This is a rare opportunity for you to study the effects on the moss of a sudden shift in the environmental conditions in which it grows. Devise an experimental plan to guide your research, including hypotheses about one or more ways the change may affect the species' growth and/or life cycle.

For selected answers, see Appendix A.

INTERPRET THE DATA

Green plants, green algae, red algae, and microscopic freshwater algae called *glaucophytes* are all primary photosynthetic eukaryotes—they have plastids, including chloroplasts where photosynthesis occurs. Biologists have long accepted the hypothesis that plastids arose by way of endosymbiosis of a cyanobacterium during the evolution of lineages leading to primary photosynthetic eukaryotes (see Sections 25.3 and 27.2). Until relatively recently, however, there was no unequivocal molecular evidence to support this assumption. If the assumption is correct, a phylogenetic tree based on molecular data should support the prediction that all primary photosynthetic eukaryotes are descended from a common ancestor. Naiara Rodríguez-Ezpeleta at the University of Montreal led a team that tested this prediction. They already knew that proteins associated with plastids in cyanobacteria and photosynthetic eukaryotes share some amino acid sequences. In the experiment of interest here, they sequenced the amino acids in 143 proteins encoded by nuclear genes of species representing all the primary photosynthetic groups. They used a computerized statistical method called *bootstrapping* to analyze the large number of different data sets from the selected species. Basically, bootstrapping produced a series of phylogenetic trees based on comparisons of the amino acid sequences in the target proteins from each species. It then compared the trees to see where overlaps occurred, and generated a summary tree like the one in the **Figure.**

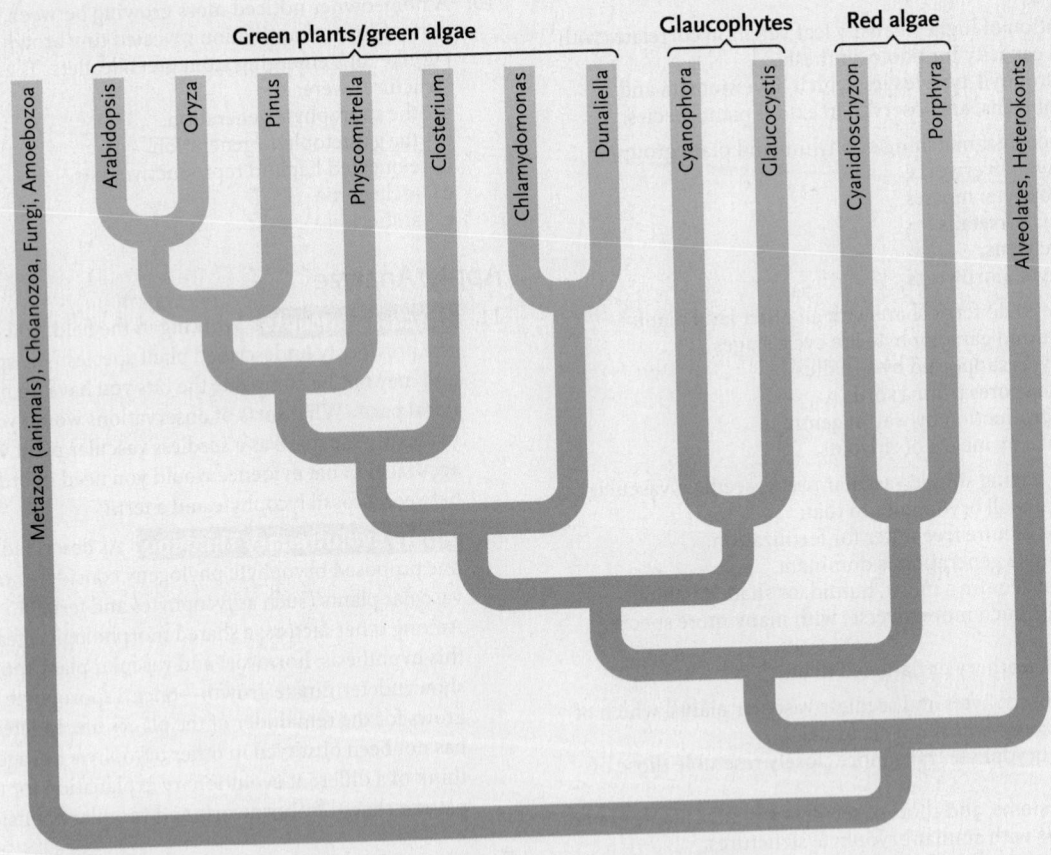

FIGURE **Phylogeny based on a dataset of 143 nuclear encoded proteins.**
© Cengage Learning 2017

1. Does the Figure support or contradict the hypothesis that a single endosymbiosis event correlated with the evolution of organisms containing plastids?

2. Do the data support the prediction that glaucophytes and green plants are sister groups—that is, that they are more closely related to each other than to other groups in the phylogeny?

Source: N. Rodriguez-Ezpeleta et al. 2005. Monophyly of primary photosynthetic eukaryotes: Green plants, red algae, and glaucophytes. *Current Biology* 15:1325–1330.

Seed Plants

A temperate forest with representatives of the two clades of seed plants—conifers (gymnosperms) and rhododendrons (angiosperms).

Why it matters . . . Wildfire blackens a Colorado mountainside, and by the following spring, grass and pine seedlings are beginning to green it again. In a cement cityscape, a dandelion brings a splash of color to a sidewalk crevice. And in Hawaii, successional plants begin to repopulate the seared fringes of an old lava flow **(Figure 29.1).** All these transformations trace to an adaptive journey that began about 360 million years ago (mya) with the evolution of the seed.

A. Small plants encroaching on an old lava flow in Hawaii

B. Dandelions *(Taraxacum)* colonizing a sidewalk crevice

FIGURE 29.1 Seed plants taking root in new habitats.

Together with two other reproductive adaptations—pollen and pollination—the seed was crucial in the radiation of vascular plants into nearly every land environment. By 250 mya, the seed plants called **gymnosperms**—conifers and their relatives—were our planet's dominant flora. Subsequent innovations gave rise to **angiosperms,** the flowering plants, which are the dominant seed plants in most modern habitats.

Seed plants as a whole belong to a nontaxonomic grouping called the Spermatophyta ("seed plants"). This chapter begins with an overview of our current understanding of the evolutionary beginnings of seed plants. We then survey the diversity of extant gymnosperms and angiosperms **(Figure 29.2)** and consider the profound importance of seed plants in all our lives.

29.1 The Rise of Seed Plants

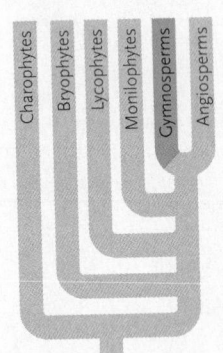

A **seed** consists of the embryo sporophyte, food reserves, and a seed coat that encloses and protects the other parts **(Figure 29.3)**. Each seed develops from a "female" (egg-producing) gametophyte that is enclosed and protected by tissues of the parent sporophyte. A **pollen grain** is an immature male gametophyte. In a sequence described in more detail shortly, sperm develop as the pollen grain matures after pollination. Fertilization unites sperm with an egg that formed in the sheltered female gametophyte, producing a zygote. As an embryo develops from the zygote, other steps transform portions of the female gametophyte into the embryo's food supply and protective seed coat. With these basics in mind, we now consider current hypotheses of how seeds came about.

Multiple Innovations Paved the Way for the Seed

The complex structure we call a seed arose through modifications in reproductive parts and mechanisms that were present in seedless plants (see Chapter 28). Here we focus on four evolutionary milestones: the trend toward reduced gametophytes; the emergence of sexually different spores; protection of female spores and gametophytes within the body of the sporophyte; and the evolution of the *ovule*, the developmental precursor of the seed. Fossils discovered to date do not provide detailed evidence about the sequence of these changes, but each one contributed to evolution of the seed.

REDUCED, PROTECTED GAMETOPHYTES You may recall from Section 28.1 that as lineages of land plants evolved,

A. A gymnosperm

Robert Potts, California Academy of Sciences

B. A flowering plant

© dashingstock/Shutterstock.com

FIGURE 29.2 Some representative seed plants. (A) A ponderosa pine, *Pinus ponderosa,* a conifer. Conifers are gymnosperms, in the phylum Coniferophyta. **(B)** An orchid (*Cattleya rojo*), a showy example of a flowering plant.

A. Seeds of a pine, a gymnosperm

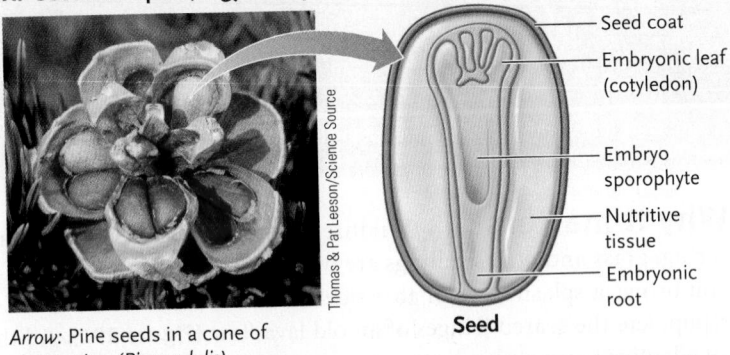

Thomas & Pat Leeson/Science Source

Arrow: Pine seeds in a cone of pinyon pine (*Pinus edulis*)

- Seed coat
- Embryonic leaf (cotyledon)
- Embryo sporophyte
- Nutritive tissue
- Embryonic root

Seed

B. Representative angiosperm seeds, which are contained inside a fruit

Perennou Nuridsany/Science Source

Hairlike tufts attach to the delicate, wind-dispersed fruits of dandelion (*Taraxacum officinale*). Each fruit contains a single seed.

© Denis Tabler/Shutterstock.com

Buttery, edible flesh (botanically, tissue of the ovary) encloses the seed of avocado (*Persea americana*).

FIGURE 29.3 Seeds.
© Cengage Learning 2017

Seedless nonvascular plants	Seedless vascular plants	Progymnosperms	Early gymnosperms (seed ferns)	Later gymnosperms	Angiosperms
← 400 million years ago	~395–345 million years ago	~390 million years ago	~385 million years ago	~360 million years ago	200–135 million years ago
• Large free-living gametophyte • Tiny, short-lived sporophyte • Spores only • Homospory • Swimming sperm	• Larger free-living sporophyte • Smaller gametophyte • Homospory, heterospory • Spores only • Swimming sperm	• Large sporophytes including woody trees • Small, dependent gametophytes • Homospory, heterospory • Spores only	• Sporophytes large, woody • Female gametophytes arise from megaspores partially enclosed in ovules • Male gametophytes in pollen grains • Seeds borne on leaves	• Sporophytes trees or woody shrubs • Gametophytes relatively few cells • Ovules on woody scales of female cones • Pollen on small, short-lived male cones • Most have nonswimming sperm • Seeds	• Diverse growth forms • Ovules enclosed in ovaries (carpels) • Female gametophyte consists of 7 cells • Male gametophyte is a pollen tube containing 2 sperm • Seeds

FIGURE 29.4 Overview of reproductive adaptations in plant evolution. Large, independent sporophytes were the rule by the time seed plants were emerging.
© Cengage Learning 2017

the large, dominant gametophyte generation characteristic of nonvascular plants (bryophytes) became smaller and the sporophyte became the large, dominant generation (see Figure 28.8). A convincing fossil record demonstrates that by the time other adaptations leading to the seed were evolving, seedless vascular plants such as ferns had a much-reduced gametophyte (**Figure 29.4**). This trend continued as seed plants arose; as discussed shortly, gymnosperms have gametophytes consisting of relatively few cells, and angiosperm gametophytes are smaller still. Such small size makes it feasible for a gametophyte to be sheltered by sporophyte's tissues.

HETEROSPORY All bryophytes and most seedless vascular plants are homosporous: sporophytes have sporangia that produce a single type of "bisexual" spore that can give rise to both male and female gametophytes. In some seedless vascular plant groups, however—including some lycophytes and ferns—heterospory arose. Recall that in this reproductive mode, sporophytes have two types of sporangia that produce sexually different spores. **Megaspores** form in **megasporangia**

and give rise to egg-producing female gametophytes. Smaller **microspores** form in **microsporangia** and give rise to male gametophytes that produce sperm.

All seed plants are heterosporous, which suggests that heterospory was established early on in the evolution of seed-bearing lineages. Now-extinct seedless vascular plants called **progymnosperms** are particularly interesting because of two features. Fossilized branches dating to about 385 mya have clusters of exposed sporangia, and differences in the sporangia of certain fossils indicate that some progymnosperms were heterosporous instead of homosporous. The tree *Archaeopteris* depicted in **Figure 29.5** is from a heterosporous line. *Archaeopteris*, or a close relative, has been proposed as a possible ancestor of gymnosperms for another reason as well: microscopic analysis of the sometimes-huge fossilized logs reveals that it formed secondary xylem tissue—wood—much like that of living conifers. Other seedless vascular plants, including tree-sized ferns, do not make true wood. So it is a reasonable hypothesis that the underlying adaptation—a lateral meristem that gives rise to secondary growth (see Chapters 28 and 33)—arose as progymnosperms evolved.

FIGURE 29.5 Fossil-based reconstruction of *Archaeopteris*, a Devonian progymnosperm. Fossil evidence suggests that *Archaeopteris* grew to a height of about 6 meters.
© Cengage Learning 2017

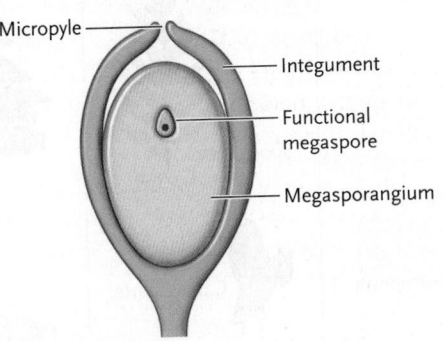

A. General structure of an ovule

Micropyle
Integument
Functional megaspore
Megasporangium

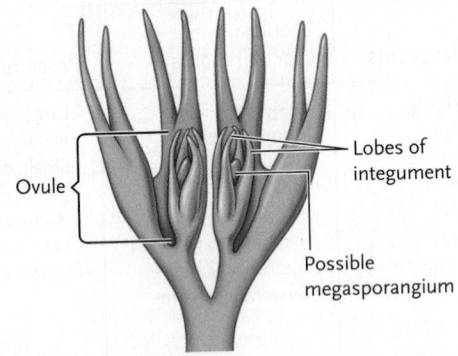

B. Fossil-based reconstruction of a seed fern ovule

Ovule
Lobes of integument
Possible megasporangium

FIGURE 29.6 The ovule, a key reproductive innovation in seed plants. The reconstruction of a seed fern ovule in **(B)** is based on fossils of *Elkinsia polymorpha*, a seed fern species discovered in Elkins, West Virginia. Gradual fusion of its lobed integuments could have produced the structure depicted in part **(A)**.
© Cengage Learning 2017

A SINGLE RETAINED MEGASPORE In seedless plants, meiosis in each megasporangium produces four megaspores that are released to the environment when the mature megasporangium opens. As seed plants arose, however, a new developmental program emerged: three of the four megaspores disintegrated, leaving a single functional megaspore to produce a female gametophyte. Well-preserved progymnosperm fossils show single megaspores alongside shriveled remains of the others, all inside nearly closed megasporangia—that is, instead of releasing the megaspores, the parent plant retained them. As we see next, other structural adaptations helped protect megasporangia, the retained megaspores, and the female gametophyte and egg. Those changes ultimately yielded the first seeds.

THE INTEGUMENT AND OVULE In living seed plants, a tissue called an **integument** ("skin") cloaks each megasporangium. The integument and megasporangium together form an immature **ovule**. In the ovule **(Figure 29.6A)**, the megaspore gives rise to a female gametophyte that produces an egg. After pollination delivers sperm to an opening at one end of the ovule (called the

micropyle), the egg is fertilized. The ovule then develops into a seed, complete with embryo, and the integument becomes the seed coat. Discussions and diagrams of representative gymnosperm and angiosperm life cycles in Sections 29.2 and 29.3 will help you visualize these steps.

Paleobotanists have found evidence of ovules in fossils of an extinct, paraphyletic group of gymnosperms known as **seed ferns**—plants that produced seeds and had leaves resembling fern fronds. Some of these fossils date to about 350 mya and show ovulelike structures, including an integument with fingerlike lobes around each megasporangium **(Figure 29.6B)**.

The Seed Greatly Enhanced Reproductive Success

Compared to a spore released by a seedless plant, a seed dramatically improves the likelihood that a new individual will develop and survive to reproduce. Protected by a seed coat, a seed not only contains a multicellular sporophyte that already has embryonic root and shoot parts, but also contains food to sustain the embryo and seedling until the seedling's shoot and root system can nourish it. In addition, seeds may remain dormant for months, years, or even millennia, sprouting only when external conditions such as temperature and available moisture favor the seedling's growth. Finally, seeds allow plants to disperse the next generation through both space and time, something animals cannot do.

The Evolution of Pollen and Pollination Permitted Sexual Reproduction Free of External Water

While changes involving megasporangia and megaspores were producing the seed, evolutionary changes involving microspores and male gametophytes were giving rise to the pollen grain. The male gametophyte was reduced to a few cells enclosed by a sturdy wall made mostly of the resistant polymer sporopollenin. Other modifications produced the unflagellated, nonswimming sperm that develop within each pollen grain. Over time, many superficially different forms of pollen grains evolved, all microscopic and lightweight **(Figure 29.7)**. These properties are the basis of **pollination**—the transfer of

A. Pine pollen grains

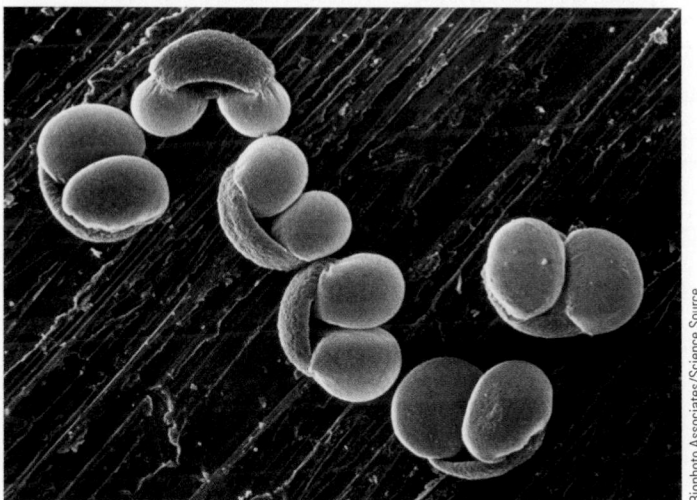

B. Grass pollen grain

C. Chickweed pollen grain

FIGURE 29.7 Examples of pollen grain diversity. The pollen grains of *Pinus* and other conifers have bulging air bladders, an adaptation for wind dispersal.

pollen to female reproductive parts via air currents or on the bodies of animal pollinators.

Pollen and pollination were enormously important adaptations for early seed plants. Unlike bryophytes and seedless plants, which have swimming sperm, plants equipped with nonswimming sperm and a way to deliver them to female gametes no longer required liquid water for reproduction. As a result, they could radiate into new, drier environments. The only extant gymnosperms that have swimming sperm are the cycads and ginkgos we consider in the next section.

STUDY BREAK 29.1

1. What major reproductive adaptations facilitated the evolution of the seed?
2. How did seeds, pollen, and pollination affect the diversification of seed plants?

THINK OUTSIDE THE BOOK

Seed ferns are assigned to the extinct phylum Pteridospermophyta. They were nearly wiped out by the Permian–Triassic mass extinction event that occurred about 251 mya and became extinct soon thereafter. Seed-bearing conifers living at the time fared better, and conifers were the dominant land plants for at least another 100 million years. Using Internet or library resources, develop a plausible explanation for why the seed fern and conifer lineages had such differing evolutionary fates.

29.2 Gymnosperms: The "Naked Seed" Plants

By the Carboniferous period (360–300 mya), when ferns and other seedless plants were dominant, many lines of seed ferns and other gymnosperms had also arisen, and the first true conifers appeared. Gymnosperms radiated extensively during the Permian period (300–250 mya), and much of the Mesozoic era that followed was the age not only of the dinosaurs but of the gymnosperms as well. The habitats of extant species range from tropical forests to deserts, but gymnosperms are most dominant in the cool-temperate zones of the northern and southern hemispheres.

Modern Gymnosperms Include Conifers and a Few Other Groups

Today there are about 800 gymnosperm species **(Table 29.1)**, nearly all of them sizable trees or shrubs. The most widespread and familiar gymnosperms are the conifers (Coniferophyta). Other gymnosperms are the cycads (Cycadophyta), ginkgos (Ginkgophyta), and gnetophytes (Gnetophyta). Our survey of living gymnosperms begins with the cycads, ginkgos, and gnetophytes, remnants of lineages that have all but vanished from modern natural environments.

TABLE 29.1 | **Seed Plants and Major Characteristics**

Gymnosperms: Vascular plants with "naked" seeds borne on exposed ovules of female cones. Nonmotile sperm arise in pollen grains, the immature male gametophytes. Exclusively wind pollinated except for cycads.

Phylum/Group	Common Name	Living Species*	Major Characteristics
Cycadophyta	Cycads	185	Shrubby or treelike; leaves resemble palm fronds; male and female cones (strobili) on separate plants. Pollinated by wind or beetles. Widely distributed in warm climates.
Ginkgophyta	Ginkgo	1	Tree with deciduous fan-shaped leaves. Male, female cones on separate plants.
Gnetophyta	Gnetophytes	70	Woody vines or shrubs. Male and female cones on separate plants. Double fertilization. Limited to deserts, tropics.
Coniferophyta	Conifers	550	Mostly evergreen, long-lived woody trees and shrubs with needlelike or scalelike leaves; pollen and ovulate cones, usually on same plant.

Angiosperms (Anthophyta): Flowering plants; ovules in ovaries that develop from carpels; double fertilization that produces an embryo and endosperm; seeds protected inside fruits; nearly all land habitats, some aquatic.

Major Groups: Magnoliids, Monocots, Eudicots

Magnoliids and other basal angiosperms	Magnolias, laurels, avocado, black peppers, and others	8,000+	Pollen grains have a single groove; seeds of some species have three or more cotyledons.
Monocots	Grasses, palms, lilies, orchids, and others	60,000+	Pollen grains have a single groove; seeds have one cotyledon.
Eudicots	Most fruit trees, roses, cabbages, melons, beans, potatoes, and others	260,000+	Pollen grains have three grooves. Most species have two cotyledons.

*Numbers of species are approximate.

Cycads Are Mesozoic Relicts Adapted to Warm Climates

During the Mesozoic era (251–65 mya), the **Cycadophyta** (*kykas* = palm), or cycads, flourished along with the dinosaurs. The cycad lineage probably arose from seed ferns. Only about 185 species have survived to the present, all confined to the tropics and subtropics.

Superficially, cycads resemble small palms **(Figure 29.8)**. Their cones bear sporangia in which either pollen or ovules develop. Pollination usually occurs when beetles transfer pollen from male plants to the developing gametophyte on female plants. Unlike most other living seed plants, cycads have flagellated sperm that swim to eggs. Various cycad tissues produce poisonous alkaloids that may help deter predatory insects. Sought after by collectors, cycads in some countries are sold in a black-market trade—threatening their survival in the wild.

FIGURE 29.8 A cycad in the genus *Encephalartos.* Note the large female cone and fernlike leaves.

almondd/Shutterstock.com

Ginkgos Are Limited to a Single Living Species

The phylum **Ginkgophyta** is something of a botanical time capsule: the oldest ginkgo fossils go back at least 200 million years, and they closely resemble the leaves and other parts of the phylum's lone living species, the maidenhair tree *(Ginkgo biloba)*. Today, wild ginkgo trees are found only in warm-temperate forests of central China. Although there is no evidence of recent shared ancestry with cycads, as with cycads ovules and pollen are produced on separate trees and sperm are swimming, flagellated cells. Another unusual *Ginkgo*

A. Ginkgo tree

Jimmy Mills/Shutterstock.com

B. Fossil and modern ginkgo leaves

Albert J. Copley/easyFotostock/
AGE Fotostock

© anaken2012/Shutterstock.com

C. Male cones

William Ferguson

D. Ginkgo seeds

Kingsley R. Stern

FIGURE 29.9 *Ginkgo biloba.* **(A)** A ginkgo tree in autumn after its deciduous leaves have turned yellow. **(B)** Fossilized ginkgo leaf compared with a leaf from a living tree. The fossil formed at the Cretaceous–Tertiary boundary. Even though 65 million years have passed, the leaf structure has not changed much. **(C)** Pollen-bearing cones and **(D)** fleshy-coated seeds of the *Ginkgo.*

feature is the species' fan-shaped deciduous leaves that are shed in autumn after turning bright yellow **(Figure 29.9)**. Nursery-propagated male trees often are planted in cities because they are resistant to insects, disease, and air pollution. The female trees are less popular because when their fleshy seeds decompose, the seed coat releases foul-smelling acids.

Gnetophytes Include Simple Seed Plants with Intriguing Features

The phylum **Gnetophyta** contains 70 species in three genera—*Gnetum*, *Ephedra*, and *Welwitschia*. Tropical regions are home to the roughly 30 species of *Gnetum*, which includes both trees and leathery leafed vines called *lianas* **(Figure 29.10A)**. About 35 species of *Ephedra* grow in desert regions of the world **(Figure 29.10B, C)**.

A. *Gnetum* seed pods

Fletcher & Baylis/Science Source

B. *Ephedra* plant

Edward S. Ross

C. *Ephedra* male cones

William Ferguson

D. *Welwitschia* plant with male cones

Mark Boulton/Science Source

FIGURE 29.10 Gnetophytes. (A) Seed pods dangling from a liana (the woody vine) in the genus *Gnetum*. **(B)** Sporophyte of *Ephedra*, with a close-up of **(C)** its pollen-bearing male cones. Female cones develop on separate plants. **(D)** Sporophyte of *Welwitschia mirabilis*, with seed-bearing cones.

The peculiar gnetophyte *Welwitschia* grows in the hot deserts of southern and western Africa. The only exposed part of the plant is a woody disk-shaped stem that bears leaves and cone-shaped strobili; the rest is a deep-reaching taproot. *Welwitschia* produces straplike leaves that split and fray as the plant grows older (**Figure 29.10D**).

Although gnetophytes are structurally and functionally simpler than most other seed plants, recent studies of sexual reproduction mechanisms in *Gnetum* and *Ephedra* species uncovered a two-step process of fertilization—which is a hallmark of angiosperms, the most advanced seed plants. This discovery raised some provocative evolutionary questions, including the possibility that ancient gnetophytes gave rise to flowering plants. Molecular studies have since contradicted this hypothesis. For example, when a research team at the Academia Sinica in Taiwan (China) compared 65 nuclear rRNA sequences from ferns, gymnosperms, and angiosperms, they found no link between the Gnetophyta and angiosperms. Instead, their analysis supported the hypothesis that cycads and ginkgos represent the earliest gymnosperm lineage, with a divergent lineage of gnetophytes and conifers arising later. The cladogram in **Figure 29.11** shows these proposed evolutionary relationships.

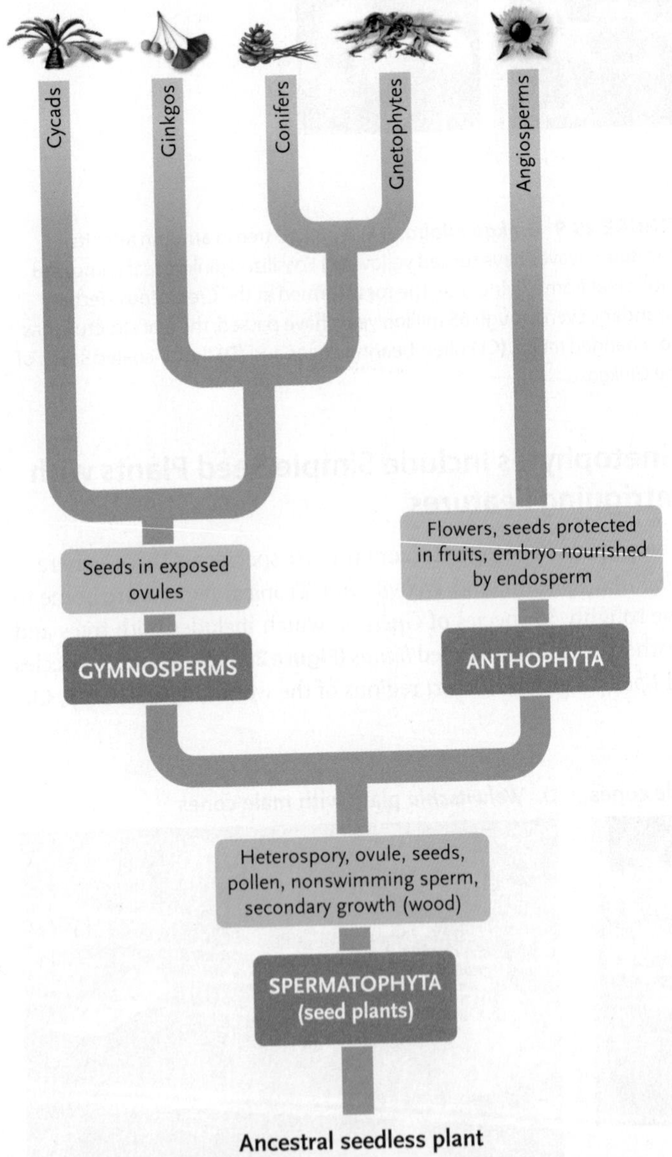

Conifers Are the Most Common Gymnosperms

About 80% of all living gymnosperm species belong to the phylum **Coniferophyta,** or *conifers* ("cone-bearers")—a group that includes the largest and oldest living land plants. For example, the coast redwood *(Sequoia sempervirens)* native to California can grow to more than 112 meters. Tree-ring analysis revealed that one specimen of another California native, the bristlecone pine *(Pinus aristata),* has survived more than 5,000 years, making it the world's oldest known plant that is not part of a clone.

Conifer sporophytes are anatomically and morphologically more complex than any plant sporophyte we have discussed so far. For example, each year most of the roughly 550 conifer species produce a new layer of woody secondary tissue ("growth rings" that are illustrated in Figure 33.25; in some exceptionally long-lived species, such as the sequoias and Douglas fir *(Pseudotsuga menziesii),* the girth of mature trees can easily exceed 6 meters, and some individuals measure more than 15 meters in diameter.

Familiar conifers are the pines, spruces, firs, hemlocks, junipers, cypresses, and redwoods. Most are woody trees or shrubs with needlelike or scalelike leaves—an anatomical adaptation to arid habitats. For instance, needles have a thick cuticle, sunken stomata, and a fibrous epidermis, all traits

FIGURE 29.11 Overview of the possible phylogenetic relationships between the lineages of seed plants. This diagram provides only a general picture of the points in seed plant evolution where major adaptations took hold. For example, heterospory and seeds are shown as adaptations common to all seed plants, but some living fern species also are heterosporous. Fossil evidence indicates that certain ancient lycophytes and horsetails also produce two types of spores and some had seeds as well. Cycads and ginkgos are the only gymnosperms that have swimming sperm.

© Cengage Learning 2017

Cladogram labels: Cycads, Ginkgos, Conifers, Gnetophytes, Angiosperms

Flowers, seeds protected in fruits, embryo nourished by endosperm

Seeds in exposed ovules

GYMNOSPERMS

ANTHOPHYTA

Heterospory, ovule, seeds, pollen, nonswimming sperm, secondary growth (wood)

SPERMATOPHYTA (seed plants)

Ancestral seedless plant

FIGURE 29.12 **Yew seeds.** Yews are coniferous shrubs or trees that produce seeds in fleshy cups instead of in cones. This photograph shows seeds of the common yew *Taxus baccata,* which is native to parts of Europe, Asia, and the Middle East.

Pascal Goetgheluck/Science Source

that reduce the loss of water vapor. A few species, such as junipers and yews, bear seeds in fleshy cones or cuplike structures (**Figure 29.12**). Most conifers produce pollen in clusters of small *pollen cones* and ovules in larger, woody *ovulate cones*. The "naked" seeds develop on the shelflike scales of the ovulate cones.

Most conifers are evergreens. That is, although they shed old leaves (mainly in autumn), enough leaves remain so that the trees still look "green" year-round.

The well-studied pine life cycle (**Figure 29.13**) is a convenient model for gymnosperm life cycles. All but 1 of the 93 pine species are trees (*Pinus mugo,* native to high elevations

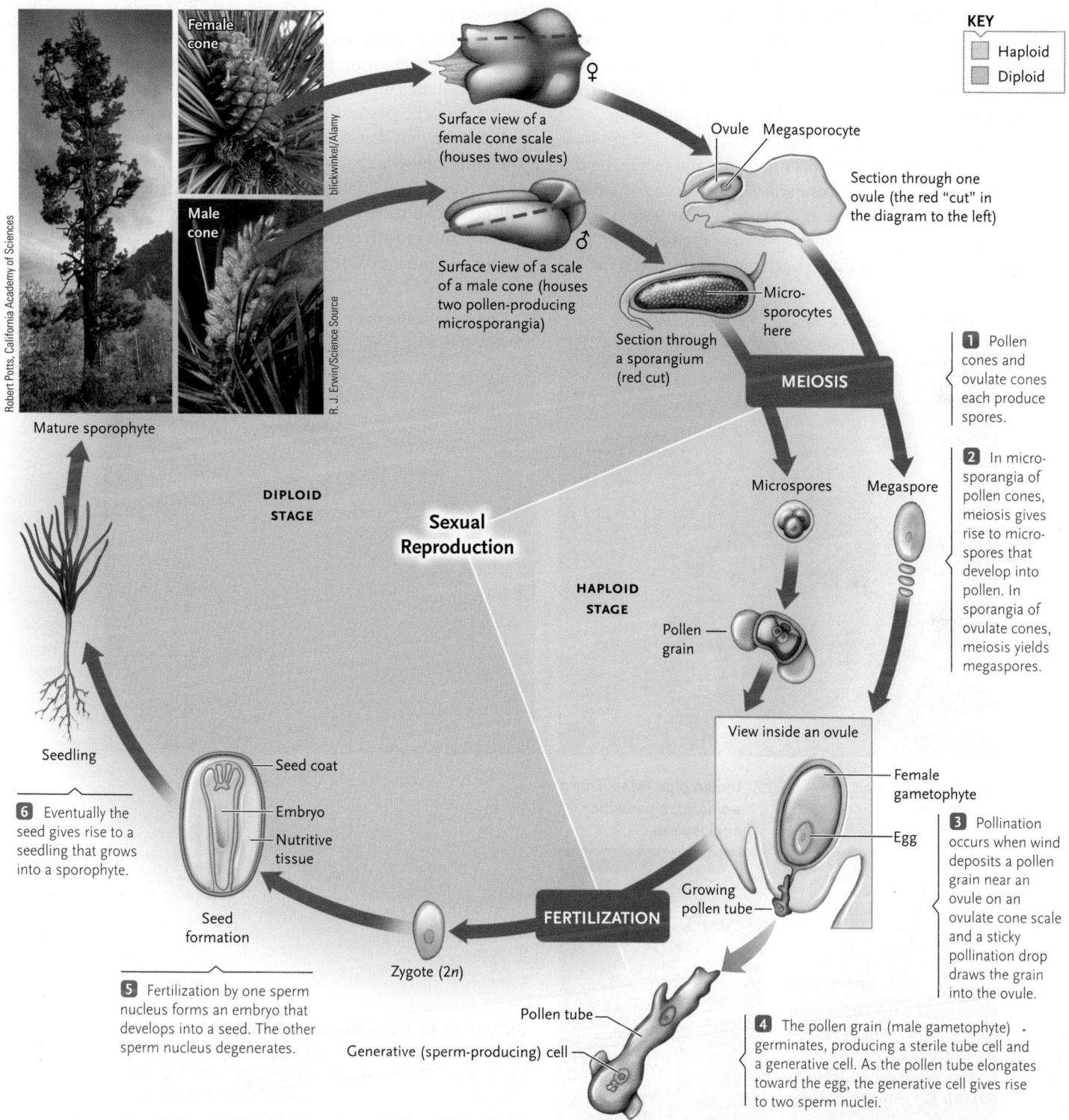

FIGURE 29.13 Life cycle of a representative conifer, a ponderosa pine *(Pinus ponderosa)*. Pines are the dominant conifers in the Northern Hemisphere, and their large sporophytes provide a heavily exploited source of wood.

© Cengage Learning 2017

KEY

Haploid
Diploid

Surface view of a female cone scale (houses two ovules)

Ovule   Megasporocyte

Section through one ovule (the red "cut" in the diagram to the left)

Surface view of a scale of a male cone (houses two pollen-producing microsporangia)

Micro-sporocytes here

Section through a sporangium (red cut)

MEIOSIS

1 Pollen cones and ovulate cones each produce spores.

2 In microsporangia of pollen cones, meiosis gives rise to microspores that develop into pollen. In sporangia of ovulate cones, meiosis yields megaspores.

DIPLOID STAGE

Sexual Reproduction

Microspores   Megaspore

HAPLOID STAGE

Pollen grain

Mature sporophyte

View inside an ovule

Female gametophyte

Egg

3 Pollination occurs when wind deposits a pollen grain near an ovule on an ovulate cone scale and a sticky pollination drop draws the grain into the ovule.

Seedling

Seed coat
Embryo
Nutritive tissue

6 Eventually the seed gives rise to a seedling that grows into a sporophyte.

Seed formation

FERTILIZATION

Growing pollen tube

Zygote (2*n*)

5 Fertilization by one sperm nucleus forms an embryo that develops into a seed. The other sperm nucleus degenerates.

Pollen tube

Generative (sperm-producing) cell

4 The pollen grain (male gametophyte) germinates, producing a sterile tube cell and a generative cell. As the pollen tube elongates toward the egg, the generative cell gives rise to two sperm nuclei.

Robert Potts, California Academy of Sciences

blickwinkel/Alamy

R. J. Erwin/Science Source

Female cone

Male cone

in Europe, is a shrub). Pollen cones are relatively small and delicate, only about 1 cm long. In spring, clusters of pollen cones develop on the lower branches. Each one consists of many small scales, which technically are sporophylls. Two microsporangia develop on the underside of each scale. Inside the microsporangia, spore "mother cells" called *microsporocytes* undergo meiosis and give rise to haploid microspores. Each microspore then undergoes mitosis to develop into a winged pollen grain—an immature male gametophyte. At this stage the pollen grain consists of four cells. Two will degenerate, but the other two—a *generative cell* and a *tube cell*—will function later in reproduction.

Ovulate cones are the familiar woody "pine cones." They typically develop at the tips of upper branches and remain on the tree year-round. Each of their cone scales bears two ovules (megasporangia). Inside each ovule is a spore mother cell called a *megasporocyte*. Meiosis in the megasporocyte produces four haploid megaspores, of which three degenerate, as described earlier.

If and when pollination occurs, the surviving megaspore develops into a mature female gametophyte—a small oval mass of cells with several archegonia at one end, each containing an egg.

In spring, pollen cones release pollen grains—by some estimates, a single pine can release billions of them before its pollen cones shrivel and fall away. The huge numbers ensure that at least some pollen grains will land near the ovules exposed on ovulate cone scales—the "naked seeds" of gymnosperms. Each ovule produces a "pollination drop"—a bit of gooey fluid containing proteins and other substances. As pollen grains stick to the droplet, it sinks inward, transporting pollen into the ovule. Eventually one pollen grain develops a **pollen tube** that digests its way to the female spore mother cell. The pollen tube is the mature male gametophyte, and activity of the tube cell facilitates its growth. Meanwhile, the generative cell gives rise to two unflagellated sperm—essentially, sperm nuclei—and growth of the pollen tube stimulates maturation of the female gametophyte and the production of eggs.

A. Flowering cacti in a desert

B. Alpine angiosperms

C. Corn (*Zea mays*), a grass

D. Indian pipe (*Monotropa uniflora*), a parasitic angiosperm

FIGURE 29.14 Flowering plants. Diverse photosynthetic species are adapted to nearly all environments, ranging from **(A)** deserts to **(B)** snowlines of high mountains. **(C)** Corn is one example of the various grasses humans rely on for food. **(D)** The parasitic flowering plant Indian pipe *(Monotropa uniflora)* lacks chlorophyll for photosynthesis. It obtains food by associating with mycorrhizae that in turn are associated with the roots of photosynthetic plants.

When a pollen tube reaches the female gametophyte, the tube tip bursts, releasing the two sperm nuclei close to the egg. One nucleus fuses with and fertilizes the egg, forming a zygote. The other sperm nucleus degenerates. A seed forms as the zygote develops into an embryo. Finally, the ovulate cone sheds seeds—each of which, recall, includes an embryo, female gametophyte tissue that serves as its food reserve, and a protective seed coat. In pines, the steps just described advance slowly, typically taking six months to two years, depending on the species. Germination of the seed and growth of the embryo into a mature sporophyte completes the life cycle.

STUDY BREAK 29.2

1. Why are gymnosperms called "naked seed" plants?
2. What are distinguishing features of cycads, ginkgos, and gnetophytes?
3. Summarize the steps of the pine life cycle.

29.3 Angiosperms: Flowering Plants

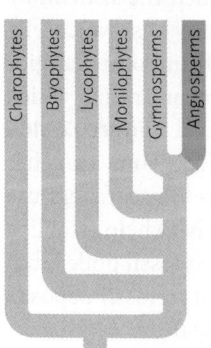

No other plant group even comes close to equaling the sheer numbers of angiosperm species—an estimated 350,000—not to mention the astonishing diversity of the sizes, shapes, habitats, growth habits, and nutrition modes of flowering plants. **Figure 29.14** shows a few examples. In part this evolutionary success reflects the adaptations angiosperms have evolved for sexual reproduction, as well as modes of asexual reproduction we discuss in Chapter 36.

The Fossil Record Is Slowly Revealing Insights about the Origin of Flowering Plants

The evolutionary origin of angiosperms (*angio* = vessel; *sperm* = seed) has stymied plant biologists for well over a hundred years. Charles Darwin famously called it the "abominable mystery," because significant evidence of flowering plants seems to burst into the fossil record, without a fossil sequence that clearly links them to other plant groups. While the most ancient finds of angiosperm pollen date to about 140 mya, the oldest well-documented fossils of whole (or nearly whole) flowering plants date back about 125 million years to the early Cretaceous period. Discovered in China, these remarkable specimens show complex and strikingly modern-looking plants that have leaves, stems, fruits, and seeds **(Figure 29.15)**. Two of these species have been assigned to the genus *Archaefructus,* representing an extinct angiosperm group.

The fossil record has yet to reveal obvious transitional organisms connecting either seedless vascular plants or gymnosperms to flowering plants. And as with gymnosperms, attempts to reconstruct the earliest flowering plant lineages from morphological, developmental, and biochemical characteristics have produced conflicting classifications and family trees. Some researchers hypothesize that flowering plants arose in the Jurassic period (200–145 mya); others propose they arose earlier, in the Triassic (250–200 mya) from now-extinct gymnosperms or from seed ferns. Intriguing support for this view comes from the recent discovery of angiosperm pollen that dates to 240 mya.

As the Mesozoic era ended and the modern Cenozoic era began, great extinctions occurred within both plant and animal kingdoms. Gymnosperms declined, and dinosaurs disappeared. Flowering plants, mammals, and social insects flourished, radiating into new environments. Today we live in what has been called "the age of flowering plants."

Angiosperm Clades Include Monocots, Eudicots, and Various Basal Angiosperms

As you can see in Table 29.1, angiosperms are assigned to the phylum **Anthophyta,** a name that derives from the Greek *anthos,* meaning flower. Angiosperms are the most ecologically diverse plants on Earth, growing on dry land and in wetlands, in freshwater and the seas. They range in size from tiny duckweeds about 1 millimeter long to towering *Eucalyptus* trees more than 100 meters tall. Most angiosperms are free-living photosynthesizers; although as you will read in Chapter 35, some are parasitic on other plants and most also obtain vital nutrients through associations with fungi.

Archaefructus sinensis fossil

© David Dilcher, Biology Department, Indiana University, Bloomington, IN

Sketch of Archaefructus sinensis

FIGURE 29.15 Fossil *of Archaefructus sinensis* (left), thought to have been an early flowering plant. The sketch (right) shows what this small, possibly aquatic plant may have looked like.
© Cengage Learning 2017

TABLE 29.2 **Eudicots and Monocots Compared**

Character	Eudicots	Monocots
Cotyledons	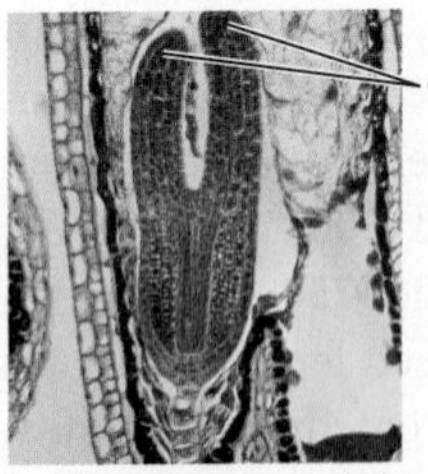 Inside seeds, two cotyledons (seed leaves of embryo)	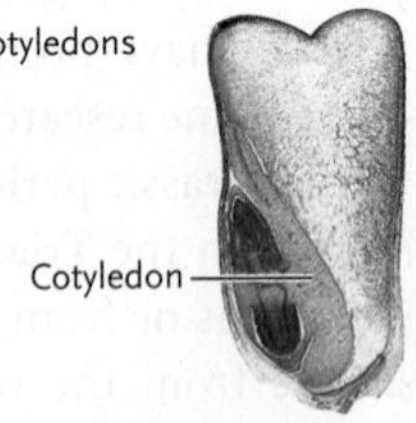 Inside seeds, one cotyledon (seed leaf of embryo)
Floral parts	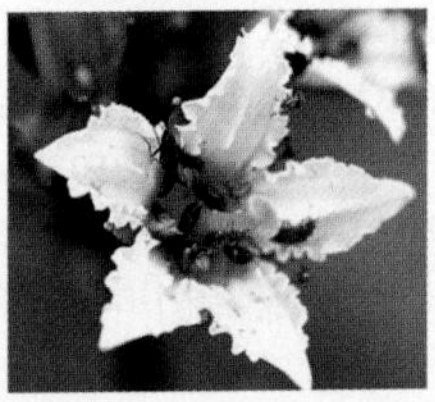 Usually four or five floral parts (or multiples of four or five)	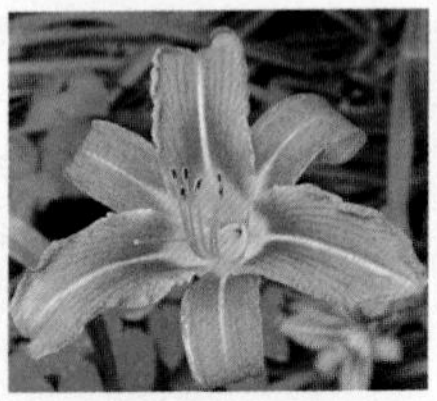 Usually three floral parts (or multiples of three)
Leaf veins	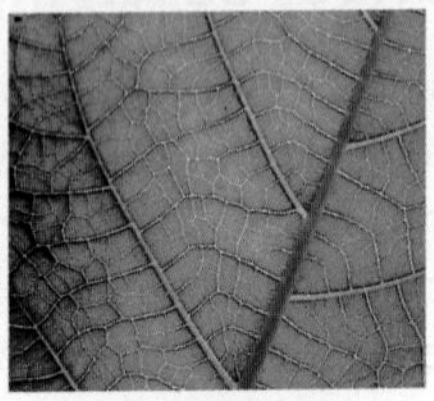 Leaf veins usually in a netlike array	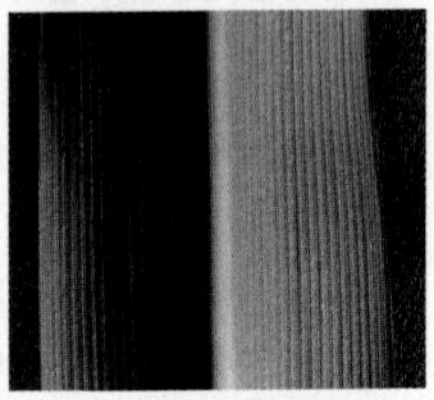 Leaf veins usually running parallel with one another
Pollen pores and furrows	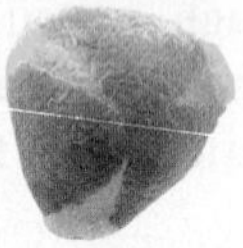 Three pores or furrows (or furrows with pores) in pollen grains	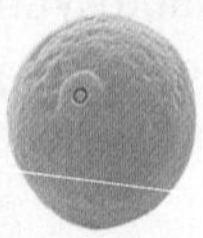 One pore or furrow in the pollen grain surface
Location of vascular bundles	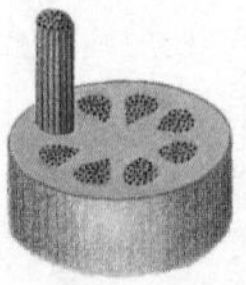 Vascular bundles organized as a ring in ground tissue	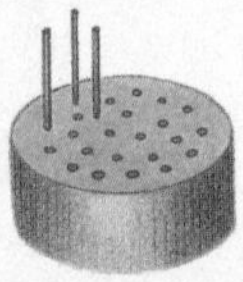 Vascular bundles distributed throughout ground tissue
Root system	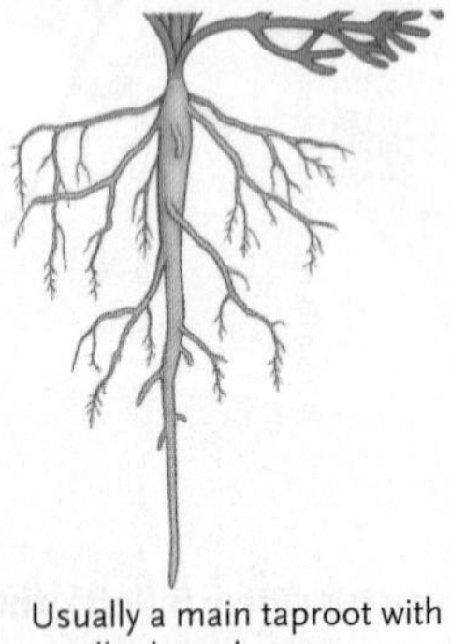 Usually a main taproot with smaller lateral roots	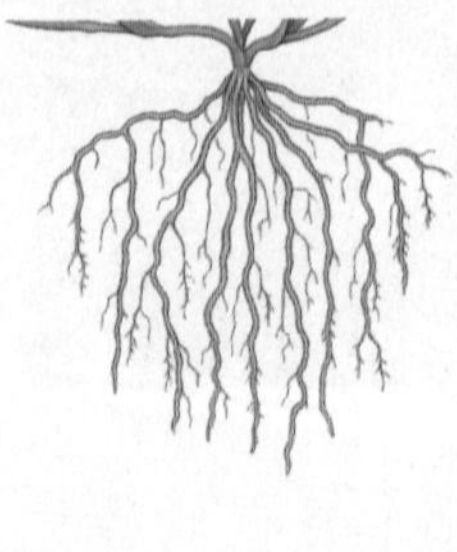 Usually a branching fibrous root system

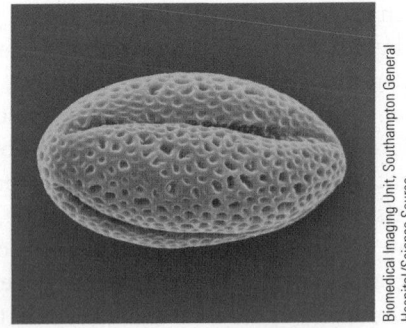

FIGURE 29.16 Eudicot pollen grain. Eudicot pollen grains have three slitlike grooves, only two of which are visible here. Pollen made by all other seed plants, including monocots, have just one groove. See also Table 29.2.

Most angiosperms are either monocots or eudicots. Various features distinguish monocots and eudicots, including the structure of their embryos **(Table 29.2)**. A **monocot** embryo has a single seed leaf called a **cotyledon** ("cuplike hollow"; also called the *scutellum*). By contrast, a **eudicot** embryo generally has two cotyledons. Eudicots also are set apart from other angiosperms by the structure of their pollen grains, which have three grooves (Table 29.2 and **Figure 29.16**), whereas the pollen grains of monocots and all other seed plants have only a single groove. Researchers use this clear structural difference not only to help establish the general type of plant that produced fossil pollen, but also what types of plants were present in fossil deposits of a particular age or geographical location.

Grasses, palms, and orchids are examples of monocots. Eudicots include nearly all angiosperm trees and shrubs, as well as many other types of plants. The structural differences noted in Table 29.2 are widely observed in nature, and we will refer to them often in later chapters as we discuss flowering plant morphology.

Although most angiosperms can fairly easily be categorized as either monocots or eudicots, figuring out the appropriate classification for other angiosperms is an extremely active area of plant research. Genetic and phylogenetic studies are playing a pivotal role in this arena. For example, Robert K. Jansen at the University of Texas in Austin and 15 colleagues in the United States and Germany recently published an ambitious phylogenetic study of 76,583 nucleotide positions in 81 plastid genes from the genomes of 64 plant species. Three species of gymnosperms (a pine, a ginkgo, and a cycad) served as the outgroup. The researchers used their data to generate alternative phylogenetic hypotheses that they evaluated along with six previously published ones, using approaches (such as parsimony) outlined in Chapter 24. Their statistical analysis provided the strongest support for a phylogenetic tree summarized in **Figure 29.17.**

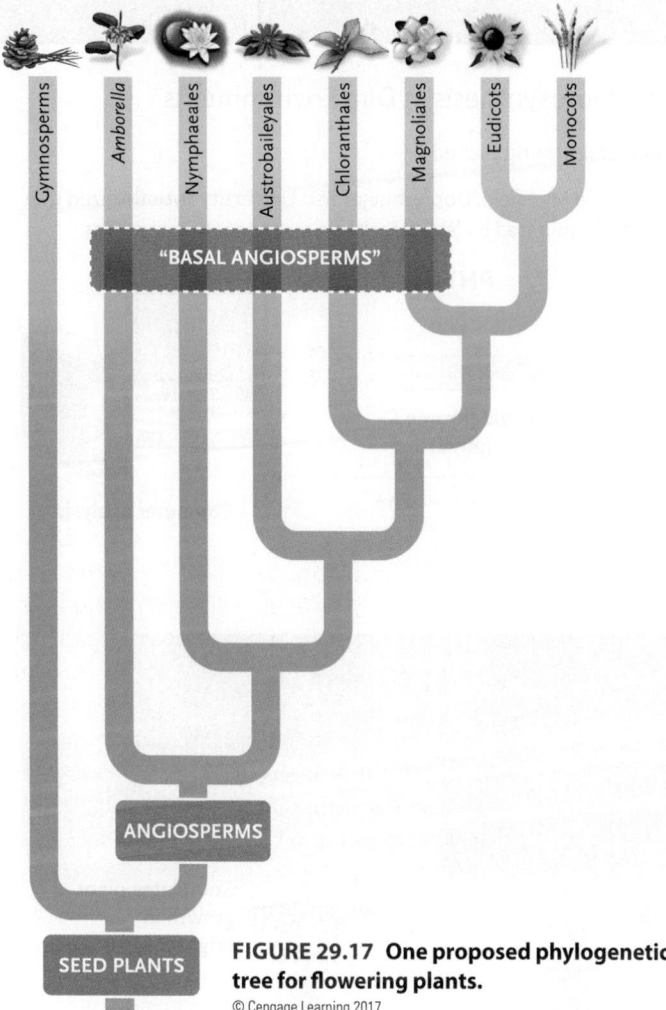

FIGURE 29.17 One proposed phylogenetic tree for flowering plants.
© Cengage Learning 2017

often referred to as *magnoliids*, includes magnolias, laurels, and avocados (see Figure 29.18A), as well as plants that are the sources of spices such as peppercorns and cinnamon. **Austrobaileyales** includes star anise and its close relatives (see Figure 29.18B), whereas water lilies are in the **Nymphaeales** (see Figure 29.18C). The Jansen group's work and several other analyses have identified the shrub *Amborella trichopoda* (see Figure 29.18D) as the closest living relative of the first flowering plants—hence *Amborella*'s divergence near the base of the phylogenetic tree shown in Figure 29.17. *Amborella* occurs only in the dimly lit cloud forests of the South Pacific island of New Caledonia, an environment that may have shaped the evolution of light-sensitive pigments in early angiosperms **(Figure 29.19)**. Morphological differences help make *Amborella* the current leading candidate for "most ancient" living basal angiosperm. *Amborella*'s small white flowers and vascular system are structurally simpler than those of other angiosperms, and its female gametophyte differs as well.

Figure 29.20A gives some idea of the variety of living monocots, which include grasses, palms, lilies, and orchids. There are at least 60,000 species of monocots, including 10,000 grasses and 20,000 orchids. Eudicots are even more diverse, with nearly 200,000 species **(Figure 29.20B)**. They include flowering shrubs and trees, most nonwoody (herbaceous) plants, and cacti.

Evolution of the angiosperm lineage included two powerful reproductive adaptations, the flower and the fruit. Chapter 36 examines the reproductive biology of flowering plants in detail. Here we survey the basic features of flowers and fruits and their adaptive significance.

Flowers Are Specialized Shoots for Sexual Reproduction

Botanically, a **flower** is a reproductive shoot that contains an angiosperm's organs for sexual reproduction. Flower parts arose as natural selection modified sporophylls (leaves or leaflike structures). Depending on the species, up to four types of floral organs develop in flowers: carpels, stamens,

In addition to eudicots and monocots, this proposed tree encompasses five clades of **basal angiosperms** thought to represent the earliest branches of the flowering plant lineage **(Figure 29.18)**. Basic diagnostic features are pollen with a single groove and, in some species, seeds with three or more cotyledons. The largest basal angiosperm group, the **Magnoliales**,

A. Southern magnolia
(*Magnolia grandiflora*)

B. Star anise
(*Illicium floridanum*)

C. Yellow pond lily (*Nuphar polysepala*), a water lily

D. *Amborella trichopoda*

FIGURE 29.18 Representatives of four basal angiosperm clades.

FIGURE 29.19 | **Experimental Research**

Exploring a Possible Early Angiosperm Adaptation for Efficient Photosynthesis in Dim Environments

Question: Did modifications in light-sensitive pigments shape the early evolution of flowering plants?

Experiment: Sarah J. Mathews and J. Gordon Burleigh of the University of Missouri and Michael J. Donoghue of Yale University hypothesized that the first angiosperms may have evolved in the dim understory of moist land habitats dominated by large Mesozoic gymnosperms and ferns.

1. The researchers began by looking at genes designated *PHYA* and *PHYC* that encode phytochromes (pigments) that allow seed plants to detect light of red and far-red wavelengths—wavelengths that predominate in dim light. The nucleotide sequences of *PHYA* and *PHYC* are about 50% identical and evidently are the descendants of a duplicated ancestral *PHY* gene. *PHYC* is sensitive to relatively bright light but apparently does not respond to dim light, and *PHYA* is highly sensitive to dim light and is inactivated by bright light.

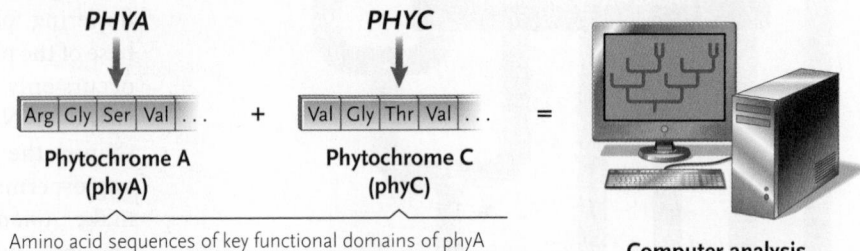

Amino acid sequences of key functional domains of phyA and phyC proteins from conifer, multiple angiosperms

Computer analysis

2. The researchers obtained amino acid sequence data for key functional domains of the phytochrome proteins encoded by *PHYA* and *PHYC* in 45 plant species. Most species were angiosperms; several conifers represented the presumed ancestral gene. Analysis of the data focused both on the number of substitutions in the targeted phytochrome amino acid sequences and on the biochemical effects of the substitutions.

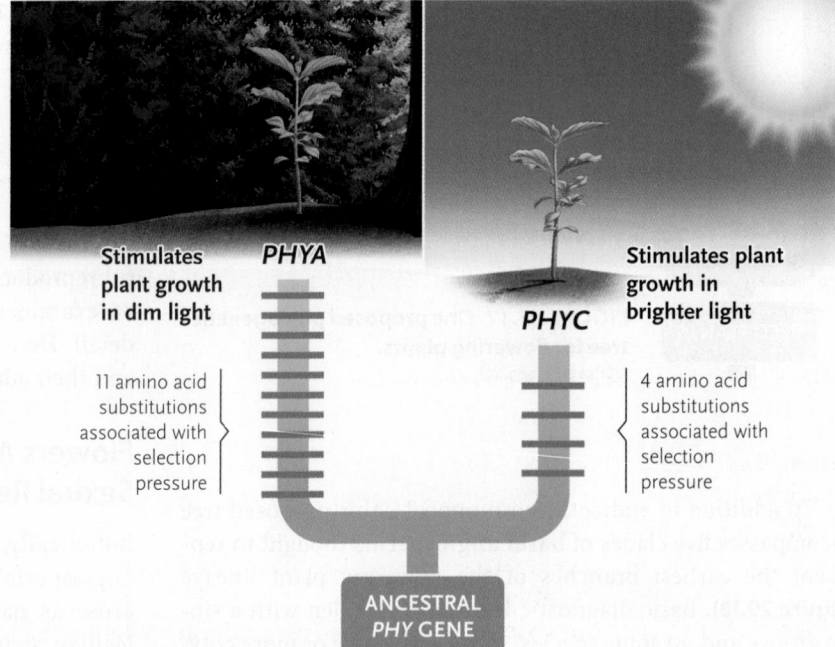

Results: In the tree branch leading from the presumed ancestral *PHY* gene to *PHYA,* 32 amino acid substitutions occurred. Of these, 11 were best interpreted as shifts associated with selection pressure. In the branch leading to *PHYC,* only seven substitutions occurred; four were best interpreted as associated with selection pressure. A phylogenetic tree based on these results displayed genetic divergences as branch points.

Conclusion: Diverging molecular characteristics of *PHYA* and *PHYC* correlated with diverging functions of the phytochromes the genes encode. *PHYA* evidently evolved under strong selection pressure. The availability of phytochrome phyA possibly allowed early angiosperm seedlings to grow in mostly dim light conditions, as in the shade of Mesozoic forests.

think like a scientist

Observe
Hypothesize
Predict
Experiment
Interpret

What might be learned by comparing the *PHY* genes of a moss and a fern with the results of this study?

Source: Sarah Mathews, J. Gordon Burleigh, and Michael J. Donoghue. 2003. Adaptive evolution in the photosensory domain of phytochrome A in early angiosperms. *Molecular Biology and Evolution* 20:1087–1097.

A. Representative monocots

Wheat (*Triticum*)

Tulips (*Tulipa*)

Eastern prairie fringed orchid
(*Platanthera leucophaea*)

B. Representative eudicots

Rose (*Rosa*)

Yellow bush lupine
(*Lupinus arboreus*)

Cherry (*Prunus*)

Claret cup cactus
(*Echinocereus triglochidratus*)

FIGURE 29.20 Examples of monocots and eudicots.

petals, and sepals. The lily (*Lilium*) flower depicted in **Figure 29.21** shows these parts. In the center is the **carpel,** in which female gametophytes form. A carpel includes an **ovary** that shelters one or more ovules against desiccation and attack from herbivores or pathogens. At the top of the carpel, a **stigma** at the tip of the **style** provides a landing platform for pollen grains. (Historically these parts were sometimes collectively called the *pistil*.) Surrounding the carpel are

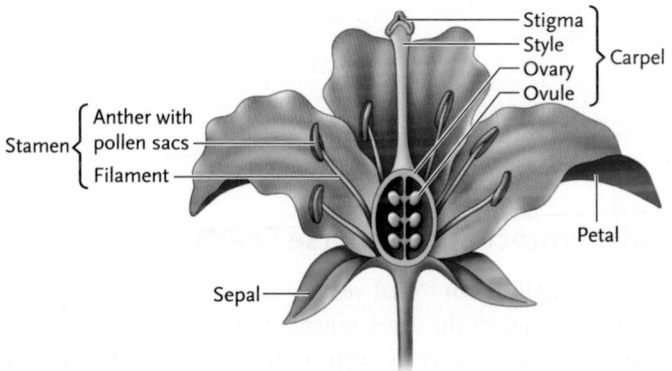

FIGURE 29.21 The parts of a flower.

© Cengage Learning 2017

stamens, in which male gametophytes form. In almost all extant flowering plants, a stamen consists of a stalklike **filament** capped by a bilobed **anther.** Each anther contains four **pollen sacs** where pollen grains develop. Moving outward, **petals** are typically showy leaflike parts that have distinctive colors, patterns, and shapes that help attract animal pollinators. Leaflike **sepals** are usually green and enclose all the other parts in a "bud" that protects the flower before it opens.

In grasses and some other families of angiosperms, natural selection has greatly modified flower structure. Grasses, for example, have tiny flowers that develop in a cluster called an **inflorescence** (see Figure 29.20A). Grasses are wind pollinated, and from an adaptive standpoint do not "need" metabolically expensive parts such as brightly colored petals. Flowers of some grass species have no petals at all.

Double Fertilization Provides Enhanced Nutrition for Embryos

Other angiosperm adaptations made it even more likely that reproduction would succeed. For instance, as noted earlier, the two-step double fertilization process that occurs in

FIGURE 29.22 Life cycle of a flowering plant. The generalized life cycle in this diagram includes double fertilization: the male gametophyte delivers two functional sperm to an ovule. One sperm fertilizes the egg, forming the embryo, and the other fertilizes the endosperm-producing cell, which nourishes the embryo. Figure 36.6 shows the life cycle of a eudicot.
© Cengage Learning 2017

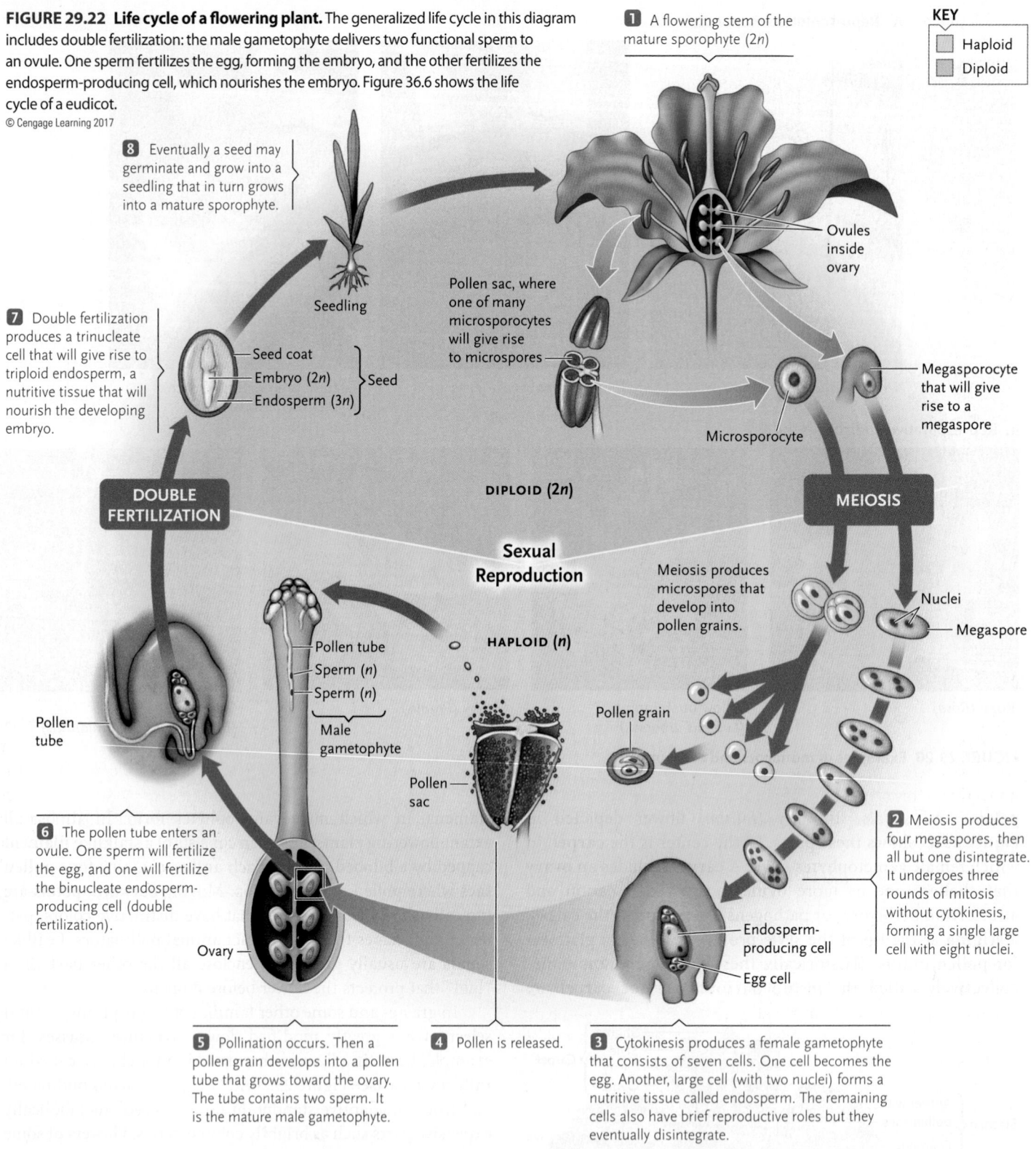

KEY

☐ Haploid
☐ Diploid

1 A flowering stem of the mature sporophyte (2n)

Ovules inside ovary

Pollen sac, where one of many microsporocytes will give rise to microspores

Microsporocyte

Megasporocyte that will give rise to a megaspore

DIPLOID (2n)

MEIOSIS

8 Eventually a seed may germinate and grow into a seedling that in turn grows into a mature sporophyte.

Seedling

7 Double fertilization produces a trinucleate cell that will give rise to triploid endosperm, a nutritive tissue that will nourish the developing embryo.

Seed coat
Embryo (2n) } Seed
Endosperm (3n)

DOUBLE FERTILIZATION

Sexual Reproduction

Meiosis produces microspores that develop into pollen grains.

Nuclei
Megaspore

Pollen tube
Sperm (n)
Sperm (n)

Male gametophyte

HAPLOID (n)

Pollen grain

Pollen sac

Pollen tube

Pollen

6 The pollen tube enters an ovule. One sperm will fertilize the egg, and one will fertilize the binucleate endosperm-producing cell (double fertilization).

Ovary

Endosperm-producing cell

Egg cell

2 Meiosis produces four megaspores, then all but one disintegrate. It undergoes three rounds of mitosis without cytokinesis, forming a single large cell with eight nuclei.

5 Pollination occurs. Then a pollen grain develops into a pollen tube that grows toward the ovary. The tube contains two sperm. It is the mature male gametophyte.

4 Pollen is released.

3 Cytokinesis produces a female gametophyte that consists of seven cells. One cell becomes the egg. Another, large cell (with two nuclei) forms a nutritive tissue called endosperm. The remaining cells also have brief reproductive roles but they eventually disintegrate.

Gnetophytes also occurs in flowering plants. In flowering plants double fertilization gives rise to both an embryo and an enhanced food supply for it, in the form of a nutritive tissue called **endosperm.** Most of the creamy flesh of a corn kernel is endosperm. The angiosperm life cycle diagrammed in **Figure 29.22** illustrates double fertilization.

Fruits Protect and Disperse Seeds

The term **angiosperm** ("seed vessel") refers to the carpel, where fertilization triggers the development of seeds from ovules in the ovary. In turn, an ovary develops into a **fruit** that encloses the angiosperm seed within. A fruit protects a seed and the embryo it contains, and also helps disperse the seed. For example, fleshy

A. Bat pollinating a giant saguaro

B. Hawk moth pollinating an orchid

D. Bee-attracting pattern of a marsh marigold

Visible light   UV light

C. Hummingbird visiting a hibiscus flower

FIGURE 29.23 Coevolution of flowering plants and animal pollinators. The colors and configurations of some flowers, and the production of nectar or odors, have coevolved with specific animal pollinators. **(A)** At night, nectar-feeding bats sip nectar from flowers of the giant saguaro *(Carnegia gigantea),* transferring pollen from flower to flower in the process. **(B)** The hawkmoth *Xanthopan morganipraedicta* has a proboscis long enough to reach nectar at the base of the equally long floral spur of the orchid *Angraecum sesquipedale.* **(C)** A Bahama woodstar hummingbird *(Calliphlox evelynae)* sipping nectar from a hibiscus blossom *(Hibiscus).* The long, narrow bill of hummingbirds coevolved with long, narrow floral tubes. **(D)** Under ultraviolet light, the bee-attracting pattern of a goldpetaled marsh marigold becomes visible to human eyes.

fruits such as blueberries are nutritious food for many animals and their seeds are adapted for surviving digestive enzymes in the animal gut. Fruits also may have hooks, spines, hairs, wings, "parachutes," or sticky surfaces, adaptations that can help move seeds to new locations when they are blown by the wind or adhere to feathers, fur, or clothing of animals that brush against them. Chapter 36 discusses the numerous botanical categories of fruits.

Flowering Plants Coevolved with Animal Pollinators

The evolutionary success of angiosperms correlates not only with adaptations such as flowers and fruits, but also with efficient mechanisms of transferring pollen from male to female reproductive parts. Although nearly all gymnosperms depend on air currents to disperse their pollen, many angiosperms coevolved with pollinators—insects, bats, birds, and other animals that pick up pollen from male floral structures (often while obtaining nectar) and inadvertently transfer it to female reproductive parts.

A flower's reproductive parts typically are positioned so that pollinators will brush against them. In addition, floral features often correlate with specific pollinators. For example, nectar-sipping bats **(Figure 29.23A)** and moths, which forage by night, pollinate intensely sweet-smelling flowers with white or pale petals that are more visible than colored petals in the dark. The Madagascar hawkmoth uncoils a mouthpart the same length—an astonishing 22 cm—as a narrow floral spur of an orchid it pollinates, *Angraecum sesquipedale* **(Figure 29.23B).** Red and

yellow flowers attract birds **(Figure 29.23C),** which have good daytime vision but a poor sense of smell. Hence bird-pollinated plants do not squander metabolic resources to make fragrances. By contrast, flowers of species that are pollinated by beetles or flies may smell like rotten meat, dung, or other decaying matter.

Bees see ultraviolet light and visit flowers with sweet odors and parts that appear to humans as yellow, blue, or purple **(Figure 29.23D).** Produced by pigments that absorb or reflect ultraviolet light, the colors form patterns called "nectar guides" that attract bees—which may pick up or deposit pollen during the visit.

STUDY BREAK 29.3

1. Describe two basic features that distinguish monocots from eudicots, and give some examples of species in each clade.
2. List and briefly describe at least three adaptations that have contributed to the evolutionary success of angiosperms.

29.4 Insights from Plant Genome Research

As more and more whole plant genomes are sequenced and annotated, plant scientists are able to analyze and compare the resulting data in revealing ways. One area of interest is the evolution of fundamental characteristics of major lineages. A case in point is research on the gene *LFY*, which encodes the

Taming a Giant Conifer Genome

Conifers have huge genomes, with estimates for various species ranging from 18 to 35 billion pairs (the human genome has approximately 3.3 billion base pairs). Sequencing such massive genomes is extremely challenging, although having such information can shed light on many unresolved questions about the adaptive heritage and genetic diversity of conifers, other gymnosperms, and flowering plants. A major goal has been to develop high-quality *reference genome sequences* for conifers—complete or nearly complete sequences that allow detailed comparisons among species. Typically, genome researchers today use the whole genome

FIGURE Loblolly pine.

shotgun sequencing approach described in Section 19.2. Even with this powerful method, however, sequencing very large genomes is extremely costly and time-consuming and it can be unusually difficult to sort out functionally important features of the genome. To overcome these problems, a 37-person team led by David B. Neale at the University of California at Davis devised a two-pronged strategy for determining the genome of the loblolly pine, *Pinus taeda* **(Figure).** They began with DNA from haploid (instead of diploid) cells, thereby halving the amount of DNA for initial sequencing. They then modified the sequencing method to make it more efficient.

Research Question

Can a modified sequencing approach produce a high-quality genome sequence for loblolly pine?

Procedure

The researchers extracted and purified DNA from the haploid female gametophyte in a single *P. taeda* seed. After the DNA was broken into fragments, a modified version of the computerized process for reassembly yielded longer-than-usual fragments that were then stitched together into the genome. Subsequent annotation identified functional genes, introns (noncoding regions), repeated sequences, transposable elements, and pseudogenes (see Chapter 19). Finally, the team compared the resulting data to existing genome information for white pine, Norway spruce, and several angiosperms.

Results

The loblolly genome consists of more than 22.18 billion pairs—so far, the largest genome ever sequenced. Initial analysis identified 50,172 functional genes—an estimated 98% of the total—and 20,646 gene families (functionally related genes). A striking finding was that 82% of the *P. taeda* genome consists of repeated sequences inserted into introns that are exceedingly long as a result. The Neale team also identified *P. taeda* genes that appear to help govern secondary growth (wood), disease resistance, and responses to environmental stresses such as drought.

Discussion

Writing in the journal *Genome Biology,* the Neale team concluded that their study "enabled discovery of genes that underlie ecologically and evolutionarily important traits, illuminated larger-scale genomic organization of gene families, and revealed missing genes that evolved in angiosperms and not gymnosperms." In addition, the presence in *P. taeda* of long, repeat-heavy introns supports a hypothesis that the genomes of conifers became so huge because the Coniferophyta lack genetic mechanisms for eliminating repeated sequences and pseudogenes.

think like a scientist

Predict how the *P. taeda* genome might be useful for a researcher employed to help manage forest lands dominated by loblolly pines.

Source: David B. Neale et al. 2014. Decoding the massive genome of loblolly pine using haploid DNA and novel assembly strategies. *Genome Biology* 15:R59.

regulatory protein LEAFY. (Chapter 16 provides a general discussion of regulatory proteins.)

Modification of the *LFY* Gene Demonstrates Evolution of Developmental Events

The LEAFY protein typically controls expression of several genes by binding to the genes' control sequences. All land plants carry the *LFY* gene, but its effects on phenotype vary markedly in different plant groups. In mosses, which arose

almost 400 mya, the LEAFY protein regulates growth throughout the plant. In ferns and gymnosperms, which arose later, LEAFY controls growth in a subset of tissues. In angiosperms, LEAFY regulates gene expression only in the particular type of meristem tissue that gives rise to flowers. Curious about the evolutionary shift from a general to a specific effect, Alexis Maizel and his team at the Max Planck Institute for Developmental Biology in Germany compared *LFY* sequences and their corresponding proteins in 14 species, including a moss, ferns, gymnosperms, and the angiosperms

Arabidopsis (thale cress) and snapdragon. Remarkably, they discovered that the evolutionary honing of the effects of the LEAFY protein correlated with only a handful of changes in the base sequence of the *LFY* gene. Each change affected how—or if—the LEAFY protein regulated the expression of a given gene. Over time, LEAFY took on its highly specific, crucial role in angiosperms, helping to direct the developmental events that produce flowers—another topic we return to in Chapter 36.

Whole Genome Duplication Leads to Polyploidy

You may remember from Chapter 22 that the genetic condition called **polyploidy** results when meiosis or mitosis produces one or more extra copies of the haploid genome in a new individual—an event called *whole genome duplication*. Polyploidy is common in flowering plants, possibly because many flowering plants can self-fertilize or reproduce asexually (see Chapter 36). Many of these angiosperms originated as *allopolyploids*—hybrids of a mating between two closely related species. If such hybrid offspring can interbreed but are reproductively isolated from their parents, a new species has come about within a single generation. Polyploidy likely was a major factor in the rapid diversification of angiosperms. For example, when researchers compared the *Amborella* genome with those of several other flowering plants, they found evidence of a genome-doubling event that apparently occurred about 192 mya, not long before flowering plants and gymnosperms diverged.

Improved Genome Sequencing Has Greatly Advanced Research

Many plant genome-sequencing efforts have a corollary goal of developing improved sequencing methods that will open new opportunities for research. An example is the recent sequencing of the genome of the loblolly pine *(Pinus taeda)* discussed in this chapter's *Molecular Insights*. Sequencing of the genomes of globally important crop plants—including wheat *(Triticum aestivum)*, tomato *(Solanum lycopersicum)*, corn *(Zea mays)*, rice *(Oryza sativa)*, and soybeans *(Glycine max)*—is already providing major new insights into traits such as growth regulation, resistance to diseases and pests, and tolerance to drought.

To date, several dozen plant genomes have been partially or fully sequenced. In addition to the species already noted, these genomes range from cucumber, potato, and cannabis (marijuana) to cultivated apple, orange, and cacao (the source of chocolate). As the genomes of more plant species are sequenced and correlated with evidence from comparative morphology and the fossil record, we can expect a steady stream of new insights about the evolutionary paths by which leaves, flowers, efficient vascular systems and many other plant characteristics have come into being.

STUDY BREAK 29.4

1. How has polyploidy been important in angiosperm evolution?
2. Briefly explain some benefits of sequencing plant genomes.

29.5 Seed Plants and People

Seed plants are both an evolutionary success story in their own right and organisms that shape virtually every aspect of our lives. Like all other animals, humans depend on seed plants for food, clothing, shelter, medicines, and much more.

Seed Plants Nourish All of Humanity

By about 11,000 years ago, agriculture got its start when humans began domesticating seed plants. Today species in the grass family (Poaceae) arguably are the most important seed plants because they produce cereal grains—wheat, corn, rice, oats, sorghum, and barley—that provide the majority of calories humans consume **(Figure 29.24)**. They and other domesticated grasses also nourish people indirectly as feed for livestock and poultry. The grass sugarcane *(Saccharum* spp.) provides 80% of the global sugar supply. Bamboos—more than 1,500 species in at least 10 genera—are fast-growing grasses used for everything from food and building materials to musical instruments.

As sources of human food and other products, members of the pea family (Fabaceae) run a close second to grasses. Familiar foods from this group include soybeans *(Glycine max)* and other beans, as well as peanuts *(Arachis hypogaea)*. The potato family (Solanaceae) encompasses not only potatoes, but also tobacco, tomatoes, peppers, and many other food plants. In tropical regions cassava *(Manihot esculenta,* in the Euphorbiaceae) is an important source of carbohydrates. Species in the rose family (Rosaceae) give us apples, peaches, plums, cherries, and strawberries, among others, as well as a variety of ornamental plants.

A. Selling rice

B. Corn

C. Wheat

FIGURE 29.24 Grasses and the human food supply. Humans use an estimated 50,000 seed plants for food. Nearly two-thirds of the food calories consumed worldwide come from the seeds of three grasses—rice, corn (maize), and wheat.

A. Building with lumber

B. Cotton plants

FIGURE 29.25 Wood and fibers. Most lumber comes from pines and firs. Cotton plants (*Gossypium* spp.) produce the most widely used plant fiber on Earth.

Seed Plants Also Are Essential Sources of Wood, Resins, and Fibers

Trees, especially conifers and angiosperm species such as oaks (*Quercus*), maples (*Acer*), and hickories (*Carya*), are sources of essential building materials such as lumber **(Figure 29.25A)**. Certain pines, firs, and spruces supply much of the raw material for paper, cardboard, and fiberboard. Some conifers produce resins and resin by-products such as turpentine, which are used commercially. In nature such substances deter attacks by microbes and insects.

Angiosperms are indispensable sources of fibers used in clothing and other products. Cotton fibers **(Figure 29.25B)** are raw materials in items ranging from jeans and T-shirts to bedding, rugs, bandages, and upholstery. Linen cloth is made from fibers of the flax plant (*Linum*). Other plant-derived fibers include hemp (used to make rope), sisal (rugs), and ramie (fabric).

Numerous Medicines and Other Drugs Are Derived from Seed Plants

About 25% of medicinal drugs are organic compounds that are derived from seed plants or serve as chemical models for synthetic versions. One well-known example is the cancer-fighting compound taxol originally extracted from yew trees. Bark of willow species (*Salix*) was the original source of salicylic acid, the pain-relieving substance in aspirin.

Some seed plants produce alkaloids, typically as part of the plant's chemical defenses against predation (discussed in Chapter 37). Alkaloids used in medicine include the antimalarial drug quinine, which is derived from the bark of the yellow cinchona tree (*Cinchona*). Other important alkaloids are the anticancer drugs vincristine and vinblastine, both synthesized by the rosy periwinkle (*Catharanthus roseus*) shown in **Figure 29.26A.**

Psychoactive substances have been both a boon and a bane to human societies the world over. Processing of the alkaloid-rich fluid latex synthesized by the opium poppy (*Papaver somniferum*) produces the painkillers morphine and codeine,

as well as the illegal and highly addictive drug heroin. Nicotine from the tobacco plant (*Nicotianum tabacum*) is another powerful alkaloid that in humans acts as an addictive stimulant. Yet another stimulant is the alkaloid we know as caffeine **(Figure 29.26B)**, which occurs in leaves and seeds of a variety of plants, including the coffee plant (*Coffea*) and kola tree (*Cola acuminata*). Cannabinoids in *Cannabis sativa* (marijuana) have hallucinogenic and sedative effects. Taken together, the global economic value of both legal and illegal plant-derived drugs runs to the hundreds of billions of dollars each year.

In Chapter 30 a very different group of organisms, the fungi, takes center stage. Although many fungal species seem superficially plantlike, the two groups are only distantly related. Biologists today are avidly exploring the much closer evolutionary links between fungi and animals.

STUDY BREAK 29.5

1. Name four major ways human societies rely on seed plants.
2. Which seed plant families are most important in terms of the global food supply?

A. Rosy (Madagascar) periwinkle

B. Common legal drugs

FIGURE 29.26 Plant-based pharmaceuticals and drugs. (A) Rosy periwinkle (*Catharanthus roseus*) is the source of the drugs vincristine and vinblastine, which are used to combat various cancers, including certain leukemias, lung cancers, non-Hodgkins lymphoma, and testicular cancer. **(B)** Caffeine comes from coffee trees (*Coffea* spp.). Nicotine is derived from tobacco (*N. tabacum*).

What is the role of gene duplication in the evolution of plant diversity?

What is the genetic basis of the origin of new and complex structures, such as the flower, during the course of evolution? One important factor is gene duplication, thought to be one of the driving forces behind the increase in organismal complexity that we see with evolution. When a plant gene is duplicated, initially there are two identical copies with identical functions (called "redundancy"); often over time one of the copies either is eliminated or becomes nonfunctional, so that the original condition of one gene is restored. But sometimes, through the process of mutation and sequence divergence, the two copies take on different functions. They may divide the functions of the original single gene between them ("subfunctionalization"), or one of the copies may take on entirely new functions, leaving the other copy to perform the original function ("neofunctionalization"). It is this last possibility—the origin of new gene functions after duplication—that is thought to provide the raw material for the origin of new plant structures such as the flower.

Like all gymnosperms, the ancestors of angiosperms produced separate cones with male and female reproductive structures, whereas flowers produce both male and female reproductive organs surrounded by a novel structure unique to flowers, the sterile *perianth* (sepals and petals). What genetic changes occurred in proto-angiosperms that allowed for the development of the complex bisexual flower and the perianth? Of course we cannot look at the genomes of the extinct angiosperm ancestors. However, by comparing the genomes of extant gymnosperms and angiosperms we can identify key flower development genes that are found only in angiosperms. In *Arabidopsis thaliana*, *APETALA1 (AP1)* and *SEPALLATA (SEP)* genes are required for proper flower formation. If *AP1* genes are eliminated, the plants will not form flowers. If *SEP* genes are eliminated, the plants form structures similar to flowers but with all the floral organs replaced by tiny leaves. *AP1* and *SEP* genes are found only in flowering plants; combined with the observation that these genes are

required to form flowers, this suggests that they may have played a role in the evolution of the flower.

AP1 and *SEP* genes appear to have arisen by way of two duplications from a third gene group *(AGL6)* that is found in both gymnosperms and angiosperms. Along with the B- and C-function genes of the ABC model, these genes belong to the MADS-box family of transcription factors, members of which play key roles throughout plant development. Repeated duplication during the course of plant evolution has led to an increase in the number of these regulatory genes from 1 in algae to over 100 in *Arabidopsis;* the proliferation of these key developmental regulators may be one of the driving forces behind the increase in complexity from algae to angiosperms, and is likely to have played a role in the origin of the flower. Thus, duplications in these genes led to the origin in gymnosperms of the B- and C-function genes, which in gymnosperms appear to play a role in specifying reproductive organ identity that is carried into the angiosperms. In contrast, later duplications, coinciding with the origin of the angiosperms, led to the establishment of the *AP1* and *SEP* gene lineages. These angiosperm-specific genes are required for flower formation, and may have been critical, in particular, in the origin of the flower-specific perianth.

think like a scientist If the A function of the ABC model is not found in any species outside the mustard family, explain how the "BCE" model can still account for the formation of four different types of floral organs.

Amy Litt is on the faculty of the Department of Botany and Plant Sciences at the University of California, Riverside. Her main interests lie in the evolution of plant form and how changes in gene function during the course of plant evolution have produced novel plant forms and functions—particularly new flower and fruit morphologies.

Courtesy of Amy Litt

REVIEW KEY CONCEPTS

For access to MindTap and additional study materials visit www.cengagebrain.com.

29.1 The Rise of Seed Plants

- Gymnosperms and angiosperms are the seed-bearing vascular plants. Adaptive trends that facilitated the evolution of the seed included reduction of gametophytes in concert with increasing size of sporophytes; heterospory; and protection of the female gametophyte within an ovule and of the immature male gametophytes within pollen grains. Pollination, and fertilization in the ovule, meant that emerging seed plants did not require liquid water for the union of gametes in sexual reproduction. A seed consists of an embryo and its food reserves enclosed by a seed coat. Now-extinct seed ferns were the first seed plants (Figures 29.1–29.7).

29.2 Gymnosperms: The "Naked Seed" Plants

- During the Mesozoic era (248–65 mya), gymnosperms were the dominant land plants. Gymnosperm phyla include cycads, ginkgos, gnetophytes, and conifers—the dominant living group (Figures 29.8–29.12).

- Conifers produce pollen in small pollen cones and ovules in larger woody ovulate cones. Conifer life cycles may take a year or more from pollination to the shedding of seeds. (Figure 29.13).

29.3 Angiosperms: Flowering Plants

- Angiosperms (Anthophyta) have dominated the land for more than 100 million years and currently are the most diverse plant group. The two main angiosperm clades are monocots and eudicots. Clades of basal angiosperms are represented by magnolias and their relatives (Magnoliales); peppercorns and their relatives (Chloranthales); the star anise group (Austrobaileyales); water lilies (Nymphaeales); and *Amborella*, a single species thought to be the most basal living angiosperm (Figures 29.14–29.20).

- Angiosperm reproductive adaptations include a protective ovary around the ovule, endosperm, fruits that aid seed dispersal, the complex organs called flowers, and the coevolution of flower characteristics with the structural and/or physiological characteristics of animal pollinators (Figures 29.21–29.23).

29.4 Insights from Plant Genome Research

- Sequencing and analysis of plant genomes has begun to yield important insights into plant evolution. Genome analysis also is increasingly being used in studies of the genetic basis of important traits such as disease resistance and drought tolerance.

29.5 Seed Plants and People

- Humans rely on seed plants for much of their food supply. Cereal grains from domesticated grasses, potatoes, and various beans are basic staples for billions of people (Figure 29.24).

- Trees, especially conifers, are essential sources of wood for lumber, paper, packaging and commercially important resins. Plant fibers (cotton, linen, hemp, ramie) are raw materials for fabrics used for clothing, bedding, and many other items (Figure 29.25).

- Seed plant-derived medicines and psychoactive drugs include aspirin, opiates, anticancer compounds, coffee, and marijuana (Figure 29.26).

TEST YOUR KNOWLEDGE

Remember/Understand

1. Which of the following was *not* a reproductive innovation associated with the evolution of the seed?
   a. reduced female gametophytes
   b. heterospory
   c. the ovule
   d. the ovary
   e. the integument

2. Seed plant evolution is also notable for which of the following adaptive feature or features?
   a. pollination by means of water
   b. woody (secondary) growth by way of a lateral meristem
   c. relatively little increase in sporophyte body size over ancestral seedless plants
   d. liberation of free microspores as part of the life cycle
   e. a seasonal loss of all leaves

3. Which modern gymnosperm groups have swimming sperm, an ancient reproductive trait?
   a. ginkgos and cycads
   b. conifers and ginkgos
   c. gnetophytes and cycads
   d. conifers and cycads
   e. gnetophytes and ginkgos

4. The gymnosperm group having species whose sexual reproduction includes double fertilization (normally an angiosperm feature) is:
   a. conifers.
   b. ginkgos.
   c. cycads
   d. gnetophytes.

5. Angiosperms and gymnosperms share which of the following reproductive characteristic(s)?
   a. pollination
   b. ovules protected within an ovary
   c. megaspores and microspores produced by meiosis
   d. seeds
   e. dominant sporophyte generation

6. Which of the following are reproductive features only of angiosperms?
   a. flowers
   b. seeds protected by a seed coat
   c. sperm transfer by way of a growing pollen tube
   d. development of fruit
   e. development of embryo-nourishing tissue within seeds

7. Basal angiosperms include:
   a. monocots.
   b. magnoliids.
   c. the star anise group.
   d. *Amborella*.
   e. eudicots.

8. Which is *not* true of eudicots?
   a. Many species produce grains humans use for food.
   b. The group includes a large number of flowering shrubs.
   c. Most nonwoody plants are eudicots.
   d. There are probably at least three times as many eudicots as monocots.
   e. Eudicots generally have more seed leaves than monocots do.

9. Which of the following is true with respect to seed plant genomes?
   a. To date researchers have only sequenced the genomes of major crop plants.
   b. Improved sequencing methods can provide more detailed genetic information.
   c. Sequencing can provide information relevant to the evolutionary history of seed plant lineages.
   d. The genomes of polyploid species reflect past genome duplication.
   e. Genome duplication is more likely in gymnosperms than in angiosperms.

10. Adaptations that correlate with the adaptive success of angiosperms include:
    a. seeds borne on cones.
    b. protection of the ovule inside an ovary.
    c. seed dispersal by way of fruits.
    d. coevolution with animals that pollinate flowers.
    e. nourishment of developing embryos by tissues of the female gametophyte.

Apply/Analyze

11. **Discuss Concepts** Suggest adjustments in the angiosperm life cycle that would better suit plants to some future world where environments were generally hotter and more arid. Do the same for a colder and wetter environment.

12. **Discuss Concepts** Modern humans emerged about 100,000 years ago. How accurate is it to state that our species has lived in the Age of Wood? Explain.

13. **Discuss Concepts** Compare the size, anatomical complexity, and degree of independence of a Douglas fir female gametophyte and a dogwood female gametophyte. Which one is the most protected from the external environment? Which trend(s) in plant evolution does your work on this question bring to mind?

14. **Apply Evolutionary Thinking** Evolutionary biologist Spencer C. H. Barrett has written that the reproductive organs of angiosperms are more varied than the equivalent structures of any other group of organisms. Explain why you agree or disagree with his view.

Evaluate/Create

15. **Discuss Concepts** Working in the field, you discover a fossil of a previously undescribed plant species. The specimen is small and incomplete; the parts you have do not include any floral organs. What sorts of observations would you need in order to classify the fossil as a vascular seed plant with reasonable accuracy? What evidence would you need in order to distinguish between a fossil monocot and eudicot?

16. **Design an Experiment** You are studying mechanisms that control the development of flowers, and your research to date has focused on eudicots. A colleague has suggested that you broaden your analysis to include representative basal angiosperms. Outline the rationale for this expanded approach and indicate which additional species or group(s) you plan to include. Discuss the type(s) of data you plan to gather and why you feel the information will make your study more complete.

For selected answers, see Appendix A.

INTERPRET THE DATA

In an inversion, a chromosome segment breaks and then reattaches to the chromosome with the order of genes reversed (see Section 13.3). Inversions disrupt recombination during meiosis, and can have evolutionary consequences. Some plant scientists have proposed a "local adaptation" mechanism in which reproductive isolation—the hallmark of speciation—arises after an inversion yields one or more altered phenotypes that confer a selective advantage in the local environment. There was not much real-world support for this hypothesis until a field study conducted by Duke University researchers David B. Lowry and John H. Willis documented local adaptation and reproductive isolation in populations of the yellow monkeyflower, *Mimulus guttatus*. In this species, which is native to a large swath of western North America, there are two distinct ecotypes. Annuals (AN) complete the life cycle in a single growing season and live in inland habitats where soil moisture is scarce in summer. Perennials (PE) survive for at least two growing seasons and live where soil is moist year-round. Lowry and Willis identified a chromosomal inversion that closely tracked this life history difference. They also found evidence that the inversion contributes to several reproductive isolating barriers, including the time frame for flowering (which is when pollination and fertilization occur). In the experiments, AN and PE plants were grown in test plots in both habitats. In addition, to help ensure that findings reflected the influence of the inversion, Lowry and Willis used laboratory breeding techniques to obtain seedlings having the AN or PE arrangement of genes in a variety of genetic backgrounds. The Key indicates colors associated with each set of test plants. **Figure A** graphs the cumulative proportion of *M. guttatus* plants surviving to flower at the inland growing site. **Figure B** graphs the cumulative proportion of plants that survived at the coastal field site. **Figure C** graphs the cumulative proportion of plants that survived at the inland field site.

Key:

- Purple: Seedlings of wild inland annuals (AN)
- Teal: Seedlings of wild coastal perennials (PE)
- Blue: Lab-bred plants having a PE gene arrangement in an AN genetic background
- Orange: Lab-bred plants having the AN arrangement in an AN background
- Green: Lab-bred plants having an AN arrangement in a PE genetic background
- Red: Lab-bred plants having a PE arrangement in a PE background

1. In Figure A, the graphed data suggest that the AN/PE genotypes are associated with reproductive isolation related to flowering time. Explain why this is so.

2. What general conclusion can you draw from the data graphed in Figures B and C? Propose an explanation for the survivorship patterns in Figure C.

Source: D. B. Lowry and J. H. Willis. 2010. A widespread chromosomal inversion polymorphism contributes to a major life-history transition, local adaptation, and reproductive isolation. *PLoS Biol* 8(9):e1000500. doi:10.1371/journal.pbio1000500

A.

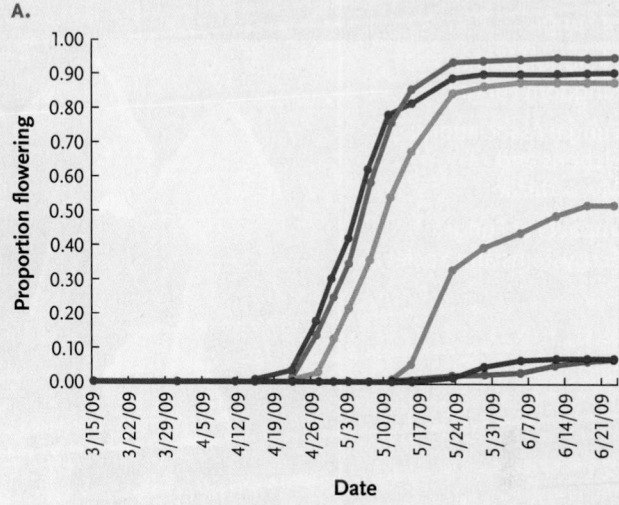

B.

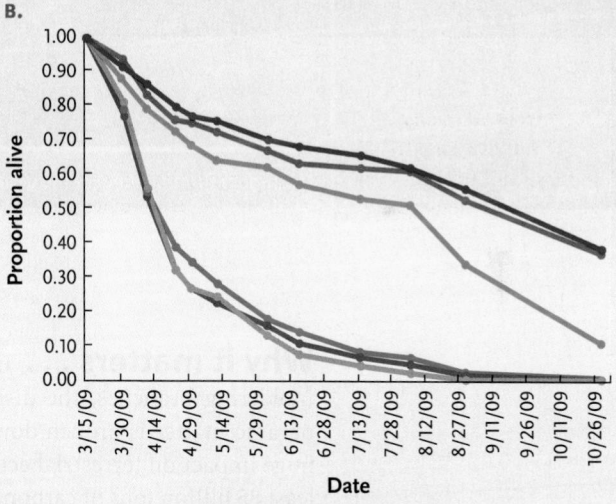

C.

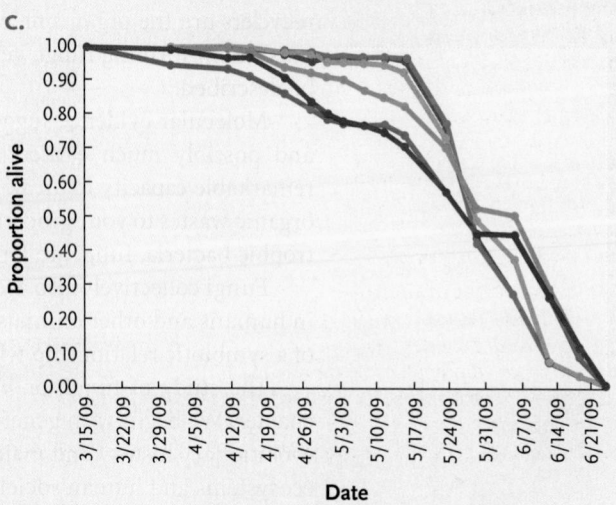

30 Fungi

Mushroom-forming fungi of the genus *Mycena,* which are commonly found on decaying wood.

Why it matters . . . In most natural environments, decay is everywhere—rotting leaves, moldering branches, the disintegrating carcasses of insects or other animals. Gradually, this organic matter is broken down and its elements are recycled in a long-term process that has a huge impact on terrestrial ecosystems. For example, each year the cycling of nutrients returns at least 85 billion tons of carbon, in the form of carbon dioxide, to the atmosphere. Chief among the recyclers are the organisms of the **Kingdom Fungi**—about 100,000 described species of molds, mushroom-forming fungi, yeasts, and their relatives and at least 1.5 million more that are yet to be described.

Molecular evidence suggests that fungi were present on land at least 500 million years ago, and possibly much earlier. In the intervening millennia, evolution equipped fungi with a remarkable capacity to break down organic matter, ranging from living and dead organisms and organic wastes to your groceries, clothing, paper, and even photographic film. Along with heterotrophic bacteria, fungi are Earth's premier decomposers.

Fungi collectively also are a major cause of plant diseases, and a host of species causes disease in humans and other animals. On the other hand, 90% of plants obtain needed minerals by way of a symbiotic relationship with a fungus.

The study of fungi, or *mycology* (*mykes* = mushroom; *logia* = study), is our focus in this chapter. We begin with general characteristics of this kingdom and current understanding of its evolutionary history and major groups. Later sections examine the profound impacts of fungi in ecosystems and human society.

30.1 General Characteristics of Fungi

Fungi are eukaryotes, most are multicellular, and all are heterotrophs, obtaining their nutrients by breaking down organic molecules that other organisms have synthesized. We begin our survey of fungi by examining how fungi differ from other forms of life, how fungi obtain nutrients, and the adaptations for reproduction and growth that enable fungi to spread far and wide through the environment.

As Fungi Evolved, Single-Celled and Multicellular Forms Arose

Two basic body forms, single-celled and multicellular, emerged as the lineages of fungi evolved. A single-celled form is called a **yeast.** A cottony mesh of threadlike filaments makes up multicellular fungi. Some fungal species alternate between yeast and multicellular forms at different stages of the life cycle.

Regardless, a rigid wall usually surrounds the plasma membrane of fungal cells. Generally the polysaccharide **chitin** provides this rigidity, the same function it serves in the external skeletons of insects and other arthropods.

A multicellular fungus exploits food sources by way of its slender filaments, which branch repeatedly as they grow over or into organic matter. Each filament is a **hypha** (*hyphe* = web; plural, *hyphae*), and the combined mass of hyphae is called a **mycelium** (plural, *mycelia*) **(Figure 30.1A).** Hyphae generally are tubular. In most multicellular fungi, cross walls called **septa** (*saeptum* = partition; singular, *septum*) divide hyphae into cell-like compartments that contain organelles. In a relatively few multicellular species, nuclei in growing hyphae divide without cytokinesis, so no septa develop and organelles share a common cytoplasm **(Figure 30.1B).** Such fungi are said to be **coenocytic,** which means "contained in a shared vessel."

Depending on the species, hyphal cells may have more than one nucleus, and septa have pores that permit nuclei and other

A. Multicellular fungus

Honey mushroom
(*Armillaria ostoyae*)

Sporocarp

Mycelium

B. Fungal hyphae

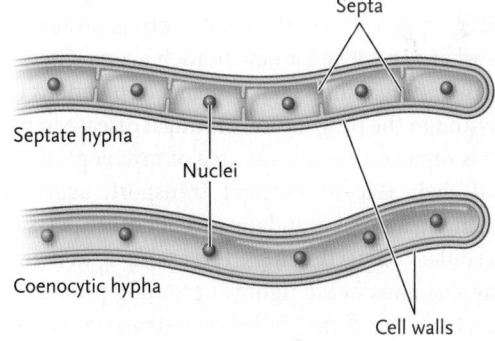

Septa

Septate hypha

Nuclei

Coenocytic hypha

Cell walls

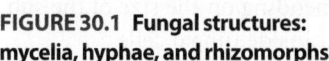

Fungal hyphae

D. Haustoria

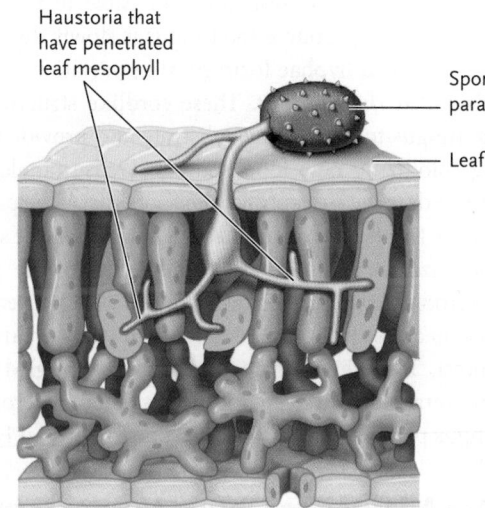

Haustoria that have penetrated leaf mesophyll

Spore of parasitic fungus

Leaf surface

C. Rhizomorphs

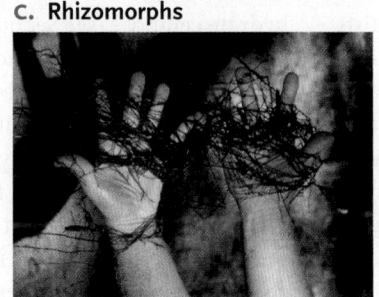

FIGURE 30.1 Fungal structures: mycelia, hyphae, and rhizomorphs.
(A) Sketch of the mycelium of a mushroom-forming fungus, which consists of branching septate hyphae. *Inset:* Micrograph of fungal hyphae. **(B)** The two basic forms of fungal hyphae: septate (top) and coenocytic, or aseptate (bottom). **(C)** Tangled rhizomorphs of *Armillaria ostoyae,* the honey mushroom that was uncovered by biologists working in an Oregon forest. Rhizomorphs are modified hyphae that typically extend over roots or through crevices between a tree's bark and the underlying wood. They are features of some species in the phylum Basidiomycota described in Section 30.2. **(D)** How fungal haustoria parasitize a host.

organelles to move between hyphal cells. These passages also allow cytoplasm to extend from one hyphal cell to the next, throughout the whole mycelium. By a mechanism called **cytoplasmic streaming**, cytoplasm can flow through the hyphae, carrying nutrients from food-absorbing parts of the fungus to reproductive structures or other nonabsorptive parts. Although mycologists commonly consider fungi with cell-like compartments to be "multicellular," they do so with the understanding that the term has a special meaning for these organisms.

A multicellular fungus grows larger as its hyphae elongate and branch. Each hypha elongates at its tip as new wall polymers (delivered by vesicles) are incorporated and additional organelles and cytoplasm are synthesized. As the new hyphae elongate and branch, an extensive mycelium can form quickly. Each forming branch fills with cytoplasm that includes new nuclei produced by mitosis. Although the rapid branching of hyphae is what allows multicellular fungi to grow aggressively—sometimes increasing in mass many times over within a few days—researchers have only recently gained the tools to explore the mechanisms that underlie this phenomenon. Studies spurred by the sequencing of the *Neurospora crassa* genome suggest that multiple steps involving a variety of genes and their interacting protein products determine where and when a new branch arises. Given that the rapid growth of fungal mycelia has such a tremendous impact in ecosystems and in the progression of fungal disease in plants and animals, this topic is a significant area of mycological research.

Beyond their role in nutrient transport, aggregations of hyphae are the structural foundation for all other parts that arise as a multicellular fungus develops. For example, in many fungi a subset of hyphae interweave tightly, becoming prominent reproductive structures sometimes called *sporocarps* (see Figure 30.1A). Grocery store mushrooms are examples. But while a mushroom or some analogous structure may be the most conspicuous part of a given fungus, it usually represents only a small fraction of the organism's total mass. The rest, often invisible to the unaided eye, penetrates the food source the fungus is slowly digesting. In some fungi, modified hyphae form rootlike variations on mycelial threads called *rhizomorphs*. These cordlike structures both anchor the fungus to its existing substrate and provide a means for rapidly colonizing new food sources. Dense tangles of the strands can so damage susceptible tree roots that large patches of forest trees die. In a famous example, U.S. Forest Service scientists found that rhizomorphs of a single individual of the parasitic honey mushroom *Armillaria ostoyae* **(Figure 30.1C)** cover an area of about 880 hectares (roughly 1,665 football fields) in an eastern Oregon forest. By one estimate, the mass of tangled strands extends an average of 1 m deep and nearly 6,000 m across, making the fungus perhaps one of the largest organisms on Earth.

Fungi Are Adapted to Obtain Nutrients by Extracellular Digestion and Absorption

Some major selection pressures have shaped the adaptations by which fungi obtain nutrients. As heterotrophs, fungi must secure nutrients by breaking down organic substances formed by other organisms. Nearly all fungi are terrestrial, but unlike other land-dwelling heterotrophs (such as animals), fungi do not move about in their environment. They also lack mouths or appendages for seizing and handling food items. Instead, fungi have a very different suite of adaptations for obtaining nutrients.

Experiments have shown that most species of fungi can synthesize nearly all their required nutrients from a few raw materials, including water, some minerals and vitamins (especially B vitamins), and a sugar or some other organic carbon source. For many species, carbohydrates in dead organic matter are the carbon sources, and fungi with this mode of nutrition are called **saprobes** (*sapros* = rotten). Other fungi are **parasites** that extract carbohydrates from tissues of a living host, harming it in the process. Most fungi that parasitize living plants produce hyphal branches called **haustoria** (*haustor* = drinker), which penetrate the walls of a host plant's cells and channel nutrients back to the fungal body **(Figure 30.1D)**. Parasitic fungi include those responsible for many devastating plant diseases, such as wheat rust and Dutch elm disease. Still other fungi are nourished by plants with which they have a mutually beneficial symbiotic association, a topic we return to later in this chapter.

Regardless of their nutritional mode, all fungi gain the raw materials to build and maintain their cells by absorption from the environment. Fungi can absorb many small molecules directly, and they gain access to other nutrients through extracellular digestion. In this process, a fungus synthesizes digestive enzymes and packages them in secretory vesicles via the pathway described in Section 15.4. When released to the outside, the enzymes digest nearby organic matter, breaking down larger molecules into absorbable fragments. Fungal species differ in the particular digestive enzymes they synthesize, so a substrate that is a suitable food source for one species may be unavailable to another. Although there are exceptions, in nature fungi typically thrive only in moist environments where they can directly absorb water, dissolved ions, simple sugars, amino acids, and other small molecules. When some of a mycelium's hyphal filaments contact a food source, growth is channeled toward it.

Various parts of a hypha absorb nutrients. Small atoms and molecules pass readily through the tips, and then transport mechanisms move them through the underlying plasma membrane. By contrast, large organic molecules, such as the carbohydrate cellulose (see Section 3.2), cannot directly enter any part of a fungus. To feed on such substances, a fungus must secrete hydrolytic enzymes that break down the large molecules into smaller, absorbable subunits. Depending on the size of the subunit, further digestion may occur inside fungal cells.

As described earlier, fungal adaptations for extracellular digestion make them masters of the decay so vital to terrestrial ecosystems. For instance, in a single autumn, one elm tree can shed 400 pounds of withered leaves; and in a tropical forest, a year's worth of debris may total 60 tons per acre. Without the metabolic activities of saprobic fungi and other decomposers such as bacteria, natural communities would rapidly become buried in their own detritus. As fungi digest dead tissues of other organisms, they also make a major contribution to the recycling of chemical elements from those tissues, including key nutrients such as nitrogen and phosphorus. But the prime example of this recycling virtuosity involves carbon. The respiring cells of fungi and other decomposers give off carbon dioxide, liberating carbon that would otherwise remain locked in the tissues of dead organisms. Each year this activity recycles a vast amount of carbon to plants, the primary producers of nearly all ecosystems on Earth.

All Fungi Reproduce by Way of Spores, but Other Aspects of Reproduction Vary

Biologists have observed striking reproductive variations in fungi, differences that are part of what makes them fascinating to study. As you will learn in the next section, fungi have traditionally been classified on the basis of their reproductive characteristics, although today evidence from molecular analysis also plays a prominent role.

Overall, most fungi can reproduce both sexually and asexually. Although no single diagram can depict all the variations,

Figure 30.2 gives an overview of the life cycle stages that mycologists have observed in several groups of fungi. The figure illustrates three general points. First, as you can see in step 1, genetically different **mating types** exist in many species of fungi (in the figure example, there are two mating types, denoted + and −). Second, the life cycle of multicellular fungi typically involves a haploid (n) stage (step 1 is the end point of this stage), a diploid ($2n$) stage (step 5), and between them a dikaryotic ("two nuclei") stage in which the fungus forms hyphae (and a mycelium) that have two haploid nuclei, making them $n + n$—neither strictly haploid *nor* diploid (step 3). Depending on the type of fungus, this stage may be long lasting or brief, and it is described more fully later in this section. Third, all fungi, whether they are multicellular or in a single-celled (yeast) form, can reproduce via microscopic **fungal spores.** In all but one group the spores are nonmotile (not propelled by flagella). Each spore is a walled single-cell or multicellular structure that is dispersed from the parent body, often via wind or water. The spores of single-celled fungi form inside the parent cell, then escape when the cell wall bursts. In multicellular fungi, spores arise in or on specialized

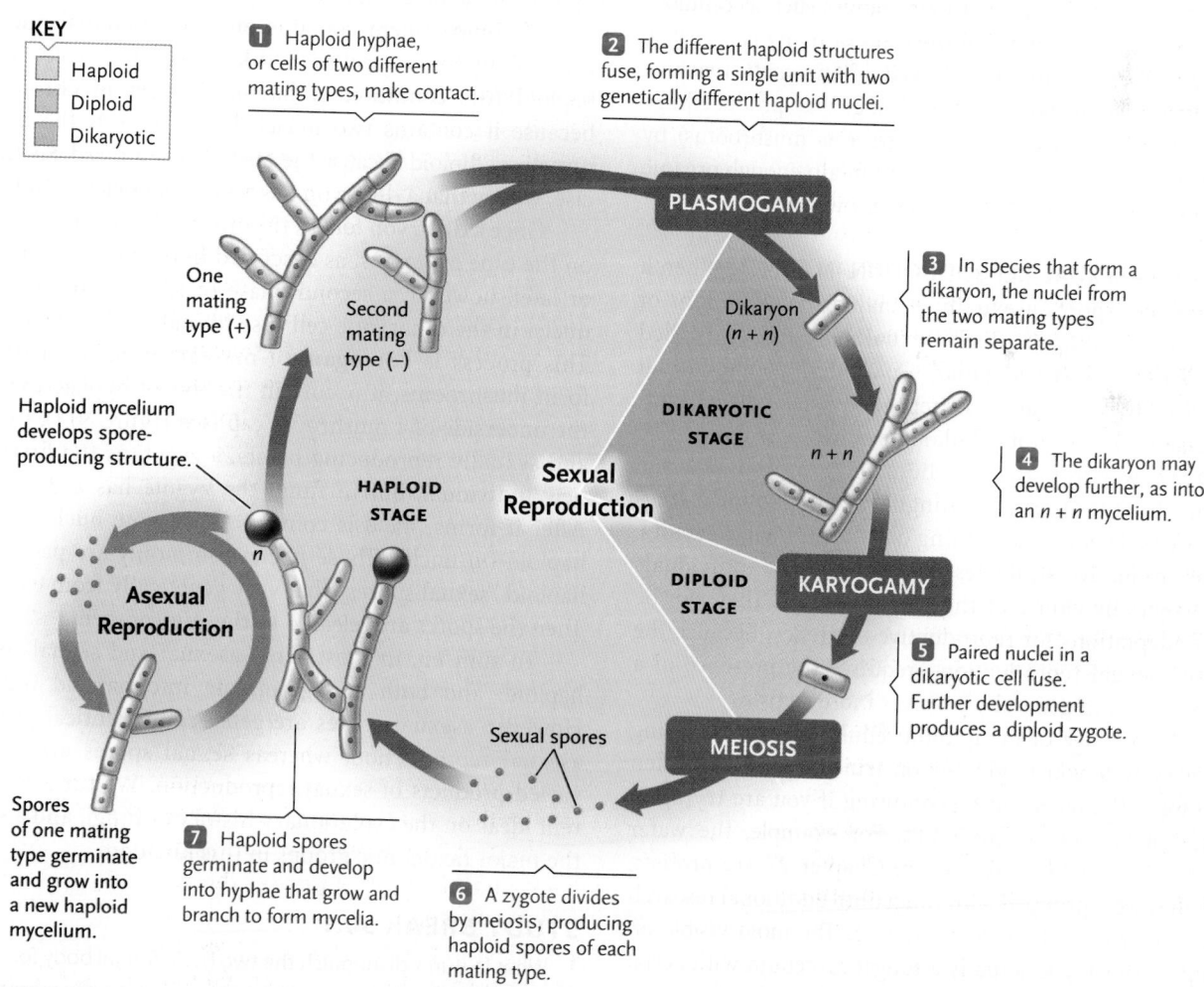

KEY
- Haploid
- Diploid
- Dikaryotic

1 Haploid hyphae, or cells of two different mating types, make contact.

2 The different haploid structures fuse, forming a single unit with two genetically different haploid nuclei.

PLASMOGAMY

One mating type (+)

Second mating type (−)

Haploid mycelium develops spore-producing structure.

HAPLOID STAGE

Sexual Reproduction

Dikaryon ($n + n$)

3 In species that form a dikaryon, the nuclei from the two mating types remain separate.

DIKARYOTIC STAGE

$n + n$

4 The dikaryon may develop further, as into an $n + n$ mycelium.

DIPLOID STAGE

KARYOGAMY

Asexual Reproduction

n

5 Paired nuclei in a dikaryotic cell fuse. Further development produces a diploid zygote.

Sexual spores

MEIOSIS

Spores of one mating type germinate and grow into a new haploid mycelium.

7 Haploid spores germinate and develop into hyphae that grow and branch to form mycelia.

6 A zygote divides by meiosis, producing haploid spores of each mating type.

FIGURE 30.2 A generalized life cycle for many fungi. Most fungi are diploid for only a short time. The duration of the dikaryon stage varies considerably, being lengthy for some species and extremely brief in others. Some types of fungi reproduce only asexually, whereas in others shifts in environmental factors, such as the availability of key nutrients, can trigger a shift from asexual to sexual reproduction or vice versa. For still others sexual reproduction is the norm.

hyphal structures and may develop thick walls that help them withstand cold temperatures or desiccation after they are released.

Reproductive spores represent a crucial fungal adaptation. Most fungi are opportunists, obtaining energy by exploiting unpredictable food sources in the environment. Spores can go dormant for long periods, then germinate and grow when environmental conditions become favorable. Being lightweight, fungal spores are easily disseminated by air or water, increasing the potential for finding suitable substrates and producing a new individual.

In nature generally, opportunistic organisms are adapted to rapidly take advantage of new food sources. Fungi that can quickly degrade simple sugars and starches often are among the first decomposers to exploit a new source of food. Once fungal spores encounter potential food and favorable conditions, they can quickly develop into new individuals that simultaneously feed and rapidly make more spores.

Many opportunistic fungi develop rapidly and reproduce before the food source is depleted. A common trade-off for speed, however, is small, even microscopic, body size. Larger species of fungi are often adapted to exploit food sources such as cellulose and lignin (a complex polymer in the walls of many plant cells), which their predecessors may have lacked the enzymatic machinery to digest efficiently. Some of these fungi may produce huge mycelia (and reproductive sporocarps such as mushrooms) by extracting nutrients from dead trees that contain enough organic material to sustain an extended period of growth.

FEATURES OF ASEXUAL REPRODUCTION IN FUNGI When a fungus produces spores asexually, it may disperse billions of them into the environment. Such asexual spores are also called **conidia** (*konis* = dust; singular, *conidium*). Some fungi (including many yeasts) also can reproduce asexually by budding or fission, or, in multicellular types, when fragments of hyphae break away from the mycelium and grow into separate individuals. In still others, the fungus produces hyphal fragments *or* asexual spores, depending on environmental factors. Asexual reproductive strategies all result in new individuals that are essentially clones of the parent fungus. They can be viewed as adaptations for reproductive efficiency, because the alternative—sexual reproduction—requires the presence of a suitable partner and generally involves more steps.

The asexual stage of many multicellular fungi—including the pale gray fuzz you might see on fruit or bread—is often called a **mold.** The term can be confusing if you are trying to keep track of taxonomic groupings. For example, the water molds and slime molds described in Chapter 27 are protists, although they were grouped with fungi until additional research revealed their true evolutionary standing. The mold visible on an overripe raspberry is actually a fungal mycelium with aerial structures bearing sacs of haploid spores at their tips.

FEATURES OF SEXUAL REPRODUCTION IN FUNGI Although asexual reproduction is the norm, quite a few fungi shift to sexual reproduction when environmental conditions (such as a lack of nitrogen) or other factors dictate. Recall from Chapter 11 that in sexual reproduction two haploid cells unite, and in most species fertilization—the fusion of two gamete nuclei to form a diploid zygote nucleus—soon follows. In fungi, however, the partners in sexual union can be two hyphae, two gametes, or other types of cells; the particular combination depends on the species involved. And in sharp contrast to most other life forms, many days, months, or even years may pass between the time fertilization gets underway and when it is completed.

During the initial sexual stage, called **plasmogamy** (*plasma* = a formed thing; *gamos* = marriage), the cytoplasms of two mating types fuse. Often, genetically distinct hyphae are the "partners" in this fusion, which promotes genetic diversity in new individuals.

The resulting new mycelium contains genetically distinct haploid nuclei that do not necessarily fuse right away and form a zygote. Instead, the differing nuclei may remain separate in the cytoplasm. When two or more genetically different nuclei are present in a mycelium, the mycelium is termed a **heterokaryon** ("different seed"). Usually, a majority of the hyphae in a heterokaryotic mycelium are septate and each hyphal compartment contains only two parental nuclei. This type of mycelium—called a **dikaryon** ("two seeds")—is not haploid (the condition of having one set of chromosomes) because it contains two nuclei. But neither is the dikaryotic mycelium diploid because the nuclei are not fused. So, to be precise, we say that a dikaryon has an $n + n$ nuclear condition.

Once a dikaryon forms, the onset of the next stage depends on the type of fungus, as described in the next section. Sooner or later, however, a second phase of fertilization unfolds: the nuclei in the dikaryotic cell fuse to make a $2n$ zygote nucleus. This process is **karyogamy** ("nuclear union"). In fungi that form mushrooms, it occurs in the tips of hyphae that end on the underside of a mushroom cap (see Figure 30.1). In animals and sexually reproducing plants, a zygote is the first cell of a new individual, but in fungi the zygote has a different fate. After it forms, meiosis converts the zygote nucleus into four haploid (*n*) nuclei. Those nuclei commonly are packaged into haploid "sexual spores" that vary genetically from each parent. Then the spores are released to the environment.

To sum up, in most fungi asexual and sexual spores are haploid, and both can germinate into haploid individuals. However, asexual spores are genetically identical products of asexual reproduction, whereas sexual spores are genetically varied products of sexual reproduction. We turn now to current ideas on the evolutionary history of fungi, and a survey of the major taxonomic groups in this kingdom.

STUDY BREAK 30.1

1. What features distinguish the two basic fungal body forms?
2. What is a fungal spore, and how does it function in reproduction?
3. Fungi reproduce sexually or asexually, but for many species the life cycle includes an unusual stage not seen in other organisms. What is this genetic condition, and what is its role in the life cycle?

reproduction and releases sexual spores. These features can still be useful indicators of the phylogenetic standing of a fungus, although now the powerful tools of molecular analysis are bringing many revisions to our understanding of the evolutionary journey of fungi.

30.2 Evolution of the Kingdom Fungi

For many years fungi were classified as plants, partly because many species seem to have a plantlike appearance. Recent comparisons of RNA and DNA sequences, and the discovery of chitin in nearly all fungal cells, all indicate that fungi and animals are more closely related to each other than they are to most other eukaryotes. Phylogenetic studies, including extensive gene sequence analysis, support the hypothesis that the lineages of both fungi and animals arose from **Opisthokonts**—single-celled, flagellated protists in the clade Opisthokonta, described in Section 27.2.

Fungi Arose from an Amoeba-Like Flagellated Protist

You may remember from Chapter 27 that molecular data show that a single-celled choanoflagellate evidently gave rise to the lineage of animals, whereas fungi are more closely related to **nucleariids**, single-celled amoebas belonging to a different line of opisthokont protists. Among other implications, these findings suggest that multicellular forms of animals and fungi evolved independently.

The first fungi were probably aquatic, like their protist ancestors. Although traces of what may be fossil fungi exist in rock formations nearly 1 billion years old, the oldest fossils that we can confidently assign to the Kingdom Fungi appear in strata laid down about 460 million years ago. When other life forms began to colonize land, fungi may well have been present and even crucial to the survival of early land plants. Supporting this hypothesis are discoveries of what appear to be *mycorrhizae*—symbiotic associations of a fungus and a plant that enhance the plant's uptake of water and minerals—in fossils of ancient bryophytes, the earliest known land plants (see Chapter 28).

Over time, fungi diverged into the strikingly diverse groups that we consider shortly **(Table 30.1).** As the lineages diversified, different adaptations associated with reproduction arose. For this reason, mycologists traditionally assigned fungi to phyla according to the type of structure that houses the final stages of sexual

TABLE 30.1	Summary of Fungal Phyla		
Phylum	Body Type	Key Feature	
Chytridiomycota and other chytrids	One to several cells	Motile spores propelled by flagella; usually asexual	
Zygomycota (zygomycetes)	Hyphal	Sexual stage in which a resistant zygospore forms for later germination	
Glomeromycota (glomeromycetes)	Hyphal	Hyphae associated with plant roots, forming arbuscular mycorrhizae	
Ascomycota (ascomycetes)	Hyphal	Sexual spores produced in sacs called *asci*	
Basidiomycota (basidiomycetes)	Hyphal	Sexual spores (basidiospores) form in basidia of a prominent fruiting body (basidiocarp)	
Cryptomycota (proposed)	Single cell	Sporelike parasites	

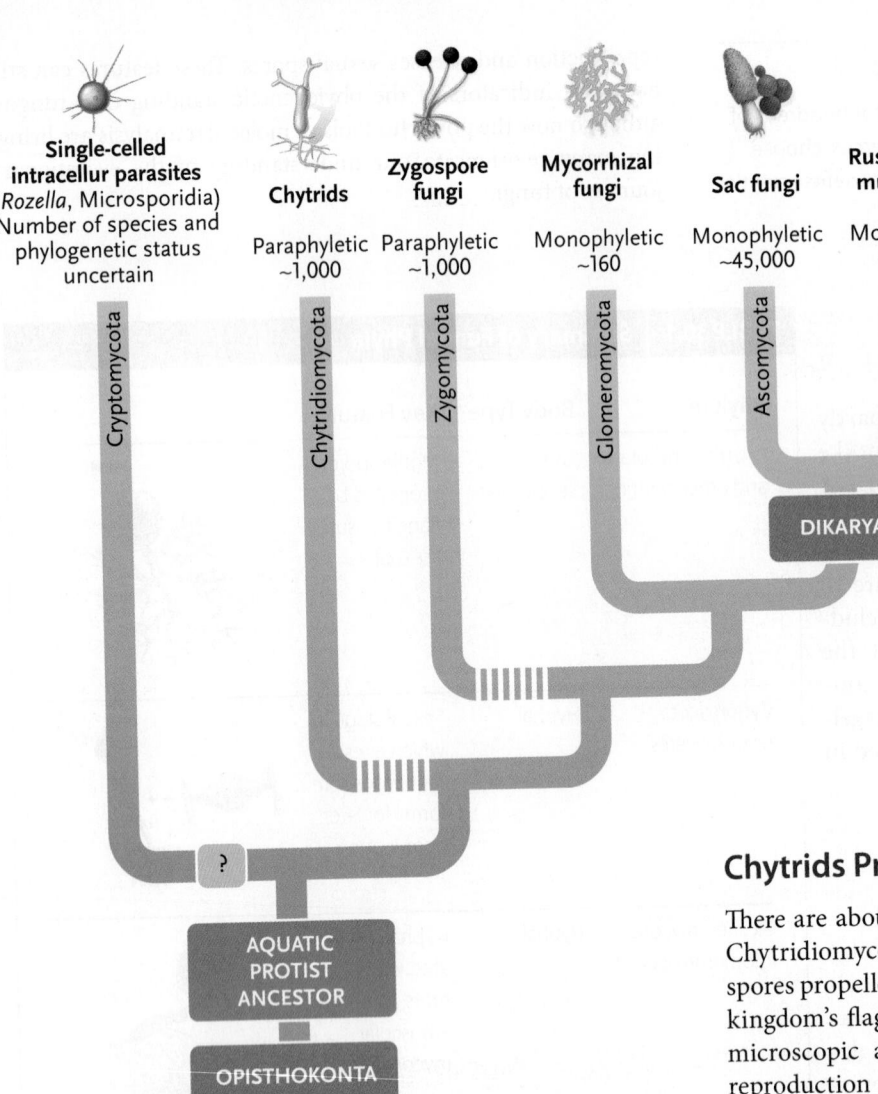

Single-celled intracellur parasites (*Rozella*, Microsporidia)
Number of species and phylogenetic status uncertain

Chytrids
Paraphyletic
~1,000

Zygospore fungi
Paraphyletic
~1,000

Mycorrhizal fungi
Monophyletic
~160

Sac fungi
Monophyletic
~45,000

Rusts, smuts, mushrooms
Monophyletic
~22,000

Cryptomycota

Chytridiomycota

Zygomycota

Glomeromycota

Ascomycota

Basidiomycota

DIKARYA

?

AQUATIC PROTIST ANCESTOR

OPISTHOKONTA

FIGURE 30.3 A phylogeny of fungi.
This scheme represents one view of the general relationships between major groups of fungi and may well be revised as new molecular findings provide more information. The dashed lines indicate that within the chytrids and zygomycetes, lines of descent of some subgroups are not well understood and the phyla are probably paraphyletic. Based on genomic studies, some mycologists place microsporidia and certain fungi formerly classified as chytrids in a new, basal phylum, the Cryptomycota. The question mark indicates ongoing debate about whether the clade meets the technical definition of a phylum.
© Cengage Learning 2017

Fungi Radiated into Several Major Lineages

The evolutionary origins and lineages of fungi have been obscure ever since the first mycologists began exploring the characteristics of this ancient group. Powerful molecular techniques and genomic studies are now advancing scientific understanding of how different groups of fungi arose and may be related. There is general agreement that three phyla, known formally as the Ascomycota, Basidiomycota, and Glomeromycota, are monophyletic. A fourth traditional phylum, a small but diverse array of fungi called the Zygomycota, likely is paraphyletic, and considerable research is underway to sort out the evolutionary relationships among the lineages typically included in it. Recent studies suggest that some members of a fifth traditional phylum, the Chytridiomycota, also should more properly be reclassified into other, separate phyla. An example is the clade called Cryptomycota that we include here at the phylum level **(Figure 30.3).** This clade currently encompasses groups of single-celled organisms that may represent the earliest diverging branches of fungi.

Chytrids Produce Motile, Flagellated Spores

There are about a thousand species of **chytrids** in the phylum Chytridiomycota. Chytrids are the only fungi that produce spores propelled by flagella, a feature they share with the fungal kingdom's flagellated protist ancestors. Nearly all chytrids are microscopic aquatic organisms **(Figure 30.4A),** and chytrid reproduction always requires water through which the spores can swim.

Most chytrids reproduce asexually. They release flagellated spores that swim until coming to rest on a substrate. A resistant cyst then forms around the spore, helping it withstand unfavorable environmental conditions until circumstances improve and the spore germinates—launching the asexual life cycle anew.

A few chytrids reproduce sexually, and of these a subset exhibit alternation of haploid and diploid generations **(Figure 30.4B).** In fact, mycologists have observed a remarkable variety of sexual modes in chytrids, but in all of them spores of different mating types unite. Karyogamy directly follows plasmogamy to produce a 2*n* zygote. This cell may form a mycelium that produces sporangia, or it may directly give rise to either asexual or sexual spores.

Certain chytrids are obligate intracellular parasites. For example, *Batrachochytrium dendrobatidis* infects the skin of amphibians. *B. dendrobatidis* infections have wiped out roughly two-thirds of the species of harlequin frogs (*Atelopus*) of the American tropics **(Figure 30.5).** The epidemic has correlated with global climate change, which has lead to rising average temperature in the frogs' habitats—an environment favoring *B. dendrobatidis* growth.

A. *Chytriomyces hyalinus*

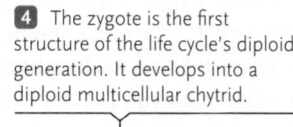

John Taylor/Visuals Unlimited, Inc.

B. Chytrid life cycle

4 The zygote is the first structure of the life cycle's diploid generation. It develops into a diploid multicellular chytrid.

5 Sporangia develop on hyphae of the mature diploid chytrid.

Sporangia

KEY
- Haploid
- Diploid

MITOSIS

Zygote (2n)

Young spore-making chytrid

Mature diploid chytrid

3 The gametangia release flagellated gametes. Female gametes are larger than male gametes. Fertilization occurs when male and female gametes unite (plasmogamy), followed by fusion of their nuclei (karyogamy).

FERTILIZATION

MEIOSIS

1 Meiosis in sporangia of the diploid sporophyte produces flagellated haploid spores that emerge from each sporangium when it bursts. Released spores germinate, giving rise to multicellular haploid chytrids that will develop gametangia as they mature.

DIPLOID STAGE (2n)

Spores (n)

HAPLOID STAGE (n)

Female gamete

Male gamete

Mature haploid chytrid

Young gamete-making chytrid (n)

Female gametangium

Male gametangium

MITOSIS

FIGURE 30.4 Chytrids. (A) *Chytriomyces hyalinus,* one of the few chytrids that reproduces sexually. In some aquatic chytrids, alternation of generations occurs in the life cycle **(B).** The haploid, gamete-producing generation is dominant, as it is in bryophytes (see Chapter 28).
© Cengage Learning 2017

2 Sexually different ("male" and "female") gametangia develop on hyphae of each mature haploid chytrid.

FIGURE 30.5 Chytridiomycosis, a fungal infection.
(A) Chytridiomycosis that has developed in the skin of a frog. The two arrows point to flask-shaped cells of the parasitic chytrid *Batrachochytrium dendrobatidis,* which has devastated populations of harlequin frogs **(B).**

A. Chytridiomycosis in a frog

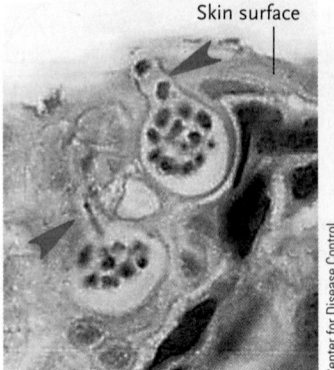

Skin surface

Center for Disease Control

B. Harlequin frog

Courtesy Ken Nemuras

Zygomycetes Form Zygospores for Sexual Reproduction

Approximately a thousand species of fungi are **zygomycetes** assigned to the phylum Zygomycota—fungi that reproduce sexually by way of structures called *zygospores* **(Figure 30.6).** Many zygomycetes are saprobes that feed on plant detritus in soil. In the process their metabolic activities release mineral nutrients that plant roots can take up.

Most zygomycetes have aseptate hyphae, unlike the other multicellular fungi. Mycelia of many zygomycetes may occur in

either a + or − mating type (Figure 30.6, step 1), and the nuclei of the two mating types are equivalent to gametes. Each strain secretes hormones that can stimulate the development of sexual structures in the complementary strain and cause sexual hyphae to grow toward each other. When + and − hyphae come into close proximity, a septum forms behind the tip of each hypha, producing a terminal **gametangium** that contains haploid nuclei. When the gametangia of the two strains touch, cellular enzymes digest the wall between them, yielding a single large, thin-walled cell that contains many nuclei from both parents (steps 2 and 3). In other words, plasmogamy has

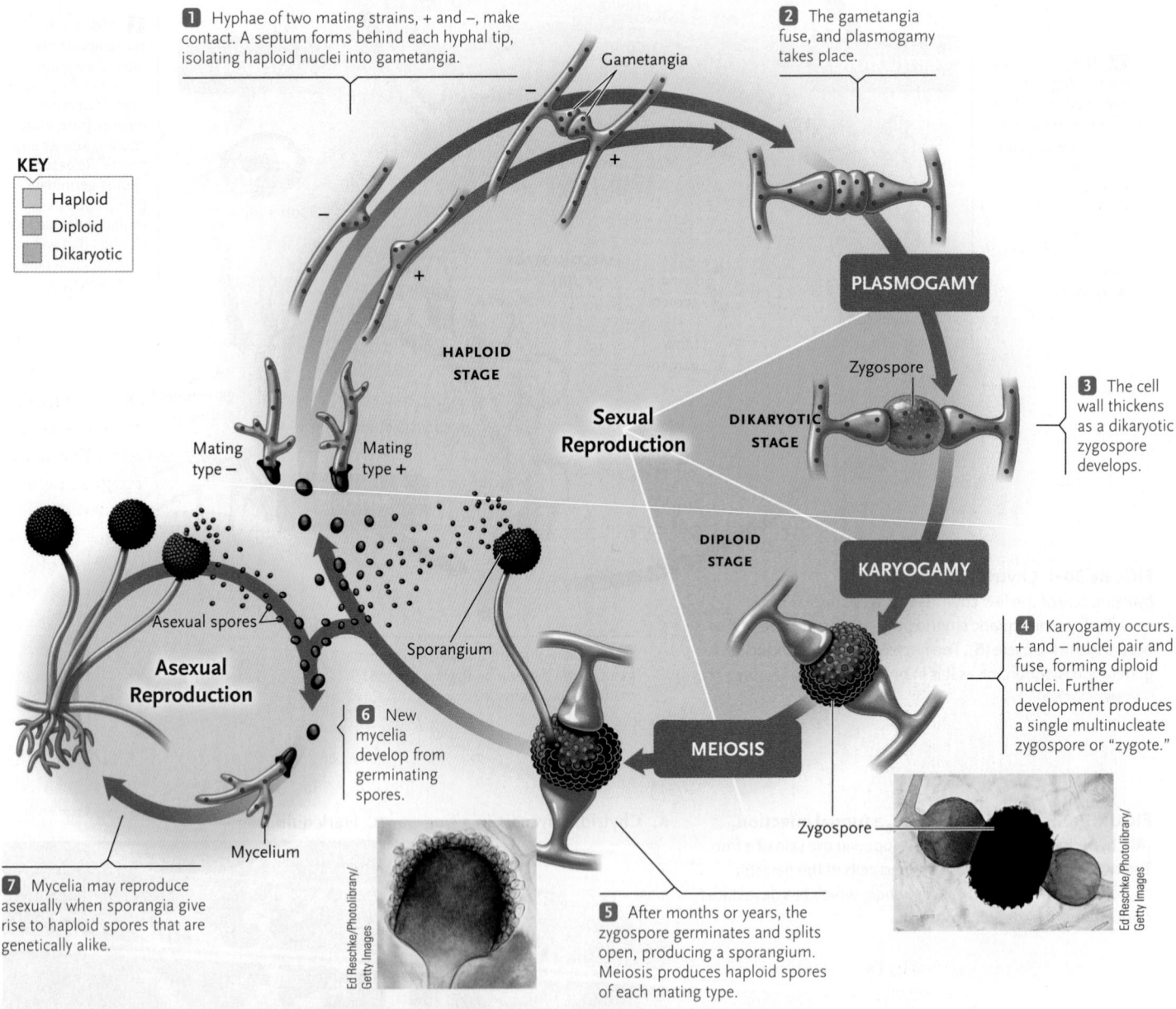

1 Hyphae of two mating strains, + and −, make contact. A septum forms behind each hyphal tip, isolating haploid nuclei into gametangia.

2 The gametangia fuse, and plasmogamy takes place.

Gametangia

KEY
- Haploid
- Diploid
- Dikaryotic

HAPLOID STAGE

Sexual Reproduction

PLASMOGAMY

Zygospore

DIKARYOTIC STAGE

3 The cell wall thickens as a dikaryotic zygospore develops.

Mating type −   Mating type +

DIPLOID STAGE

KARYOGAMY

Asexual spores

Asexual Reproduction

Sporangium

4 Karyogamy occurs. + and − nuclei pair and fuse, forming diploid nuclei. Further development produces a single multinucleate zygospore or "zygote."

MEIOSIS

6 New mycelia develop from germinating spores.

Mycelium

Zygospore

Ed Reschke/Photolibrary/Getty Images

7 Mycelia may reproduce asexually when sporangia give rise to haploid spores that are genetically alike.

5 After months or years, the zygospore germinates and splits open, producing a sporangium. Meiosis produces haploid spores of each mating type.

Ed Reschke/Photolibrary/Getty Images

FIGURE 30.6 Life cycle of the bread mold *Rhizopus stolonifer,* a zygomycete. The black bread mold, *Rhizopus stolonifer,* may produce so many charcoal-colored sporangia that moldy bread looks black. Asexual reproduction is common, but different mating types (+ and −) also reproduce sexually. In both cases, haploid spores form and give rise to new mycelia.

© Cengage Learning 2017

occurred, and this new cell is a dikaryon. Gradually an inner wall forms, thickens, and hardens. This structure, with the multinucleate cell inside it, is a **zygospore**, and gives this fungal group its scientific name. It becomes dormant and can stay dormant for months or years.

Karyogamy follows plasmogamy and eventually the diploid zygospore breaks dormancy. Then, meiosis produces a stalked sporangium (step 5). The sporangium contains haploid spores of each mating type, which are released to the outside **(Figure 30.7)**. When a spore later germinates, it produces either a + or a − mycelium, and the sexual cycle can continue.

Like other fungi, however, zygomycetes usually reproduce asexually, as shown at the lower left in Figure 30.6. When a haploid spore lands on a favorable substrate, it germinates and gives rise to a branching mycelium. Some of the hyphae grow upward, and saclike, thin-walled sporangia form at the tips of these aerial hyphae. Inside the sporangia the asexual cycle comes full circle as new haploid spores arise through mitosis and are released.

Zygomycetes that have aseptate hyphae are structurally simpler than the species in most other fungal groups. Although septa wall off the reproductive structures, in effect the branching mycelium of each fungus is a single, huge, multinucleate cell—the same body structure as found in some algae and certain protists. Because such zygomycetes have numerous nuclei in a common cytoplasm, these fungi are examples of coenocytic species, which means "contained in a shared vessel." By contrast, in other fungal groups septa at least partially divide the hyphae into individual cells, which typically contain two or more nuclei.

Presumably, having hyphae that lack septa confers some selective advantages. One benefit may be that without septa to impede the flow, nutrients can move freely from the absorptive hyphal tips to other hyphae where reproductive parts develop.

FIGURE 30.7 Spore dispersal by the zygomycete *Pilobolus*. *Pilobolus* spores are contained in a sporangium (the dark sac) at the end of a stalked structure. When incoming rays of sunlight strike a light-sensitive portion of the stalk, turgor pressure (pressure against a cell wall due to the movement of water into the cell) inside a vacuole in the swollen portion becomes so great that the entire sporangium may be ejected outward as far as 2 m—a remarkable feat, given that the stalk is only 5 to 10 mm tall.

In zygomycetes, aggregations of "cooperating" hyphae may form body structures specialized for certain functions. However, such structures are more common in the three groups of more complex fungi that we consider next.

Glomeromycetes Form Asexual Spores, at the Ends of Hyphae

All the 160 known **glomeromycetes**—members of the phylum Glomeromycota—are specialized to form the associations called mycorrhizae with plant roots. It would be hard to overestimate their ecological impact, for glomeromycetes collectively make up roughly half of the fungi in soil and form mycorrhizae with an estimated 90% of all land plants. Virtually all glomeromycetes reproduce asexually, by way of spores that form at the tips of hyphae. The hyphae also secrete enzymes that allow them to enter plant roots, where their tips branch into treelike clusters. As described in the next section, the clusters, called arbuscules, nourish the fungus by taking up sugars from the plant and in return supply the plant roots with dissolved minerals from the surrounding soil.

Ascomycetes, the Sac Fungi, Produce Sexual Spores in Saclike Asci

Ascomycetes, fungi in the phylum Ascomycota, include approximately 45,000 species that produce reproductive structures called **asci** (singular, *ascus*; **Figure 30.8**). A few ascomycetes prey on agricultural insect pests and thus have potential for use as biological pesticides. Many more are destructive plant pathogens, including *Venturia inaequalis*, the fungus responsible for apple scab, and *Ophiostoma ulmi*, which causes Dutch elm disease. The orange bread mold *Neurospora crassa* has been important in genetic research, including the elucidation of the one gene–one enzyme hypothesis (see Section 15.1). A few ascomycetes even show trapping behavior, ensnaring small worms that they then digest **(Figure 30.9)**.

Although yeasts and filamentous fungi with a yeast stage in the life cycle occur in all fungal groups except chytrids and glomeromycetes, many of the best-known yeasts are ascomycetes. Yeasts commonly reproduce asexually by fission or budding from the parent cell, but many also can reproduce sexually after the fusion of two cells of different mating types (analogous to the mating types described earlier). Many ascomycete yeasts are found naturally in the nectar of flowers and on fruits and leaves. At least 1,500 species have been described, and mycologists suspect that thousands more are yet to be identified.

Other ascomycetes are multicellular, with tissues built up from septate hyphae. Although septa do slow the flow of nutrients (which, recall, can cross septa through pores), they also confer advantages. For example, septa present barriers to the loss of cytoplasm if a hypha is torn or punctured, whereas in an aseptate zygomycete, fluid pressure may force out a significant

A. Ascocarp

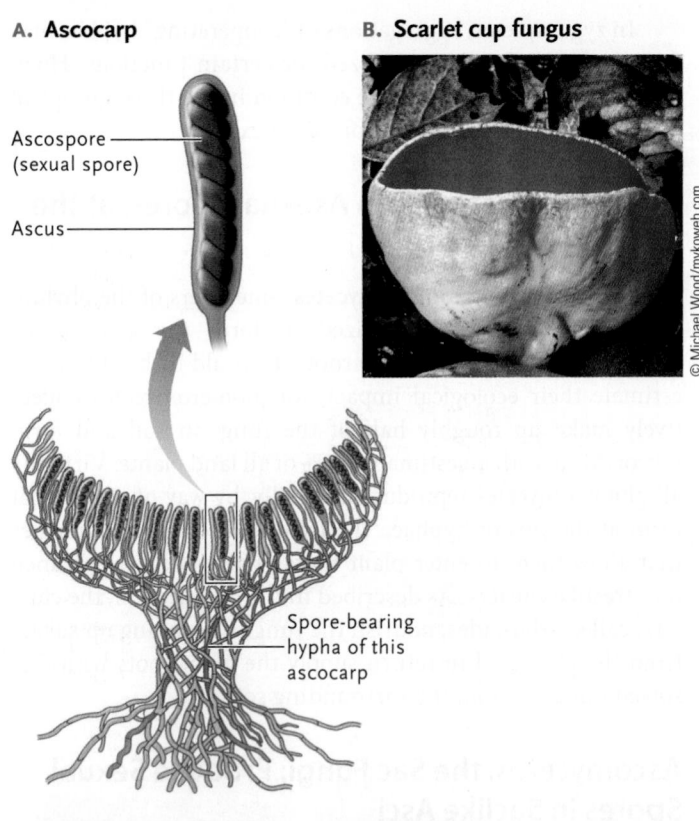

Ascospore (sexual spore)

Ascus

Spore-bearing hypha of this ascocarp

B. Scarlet cup fungus

© Michael Wood/mykoweb.com

C. Morel

© Fred Stevens/mykoweb.com

D. Apple scab

Dave Bevan/Garden World Images/AGE Fotostock

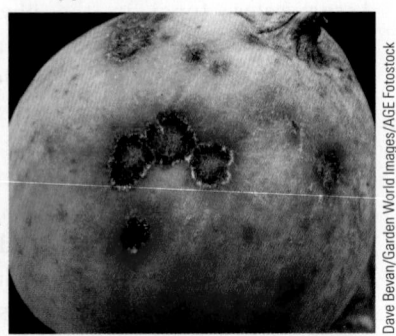

FIGURE 30.8 A few of the ascomycetes, or sac fungi. The examples shown are multicellular species that form ascocarps as reproductive structures. **(A)** A cup-shaped ascocarp, composed of tightly interwoven hyphae. The spore-producing asci occur inside the cup. **(B)** The velvety appearance of the Scarlet cup fungus *(Sarcoscypha)* is due to the thousands of asci in its ascocarp. **(C)** True morels *(Morchella esculenta)*, a prized edible fungus. **(D)** The ascomycete *Venturia inequalis* may grow on apple fruits and leaves, causing apple scab disease. A fungicide can help control the disease, but a severe infection can result in unmarketable fruit.
© Cengage Learning 2017

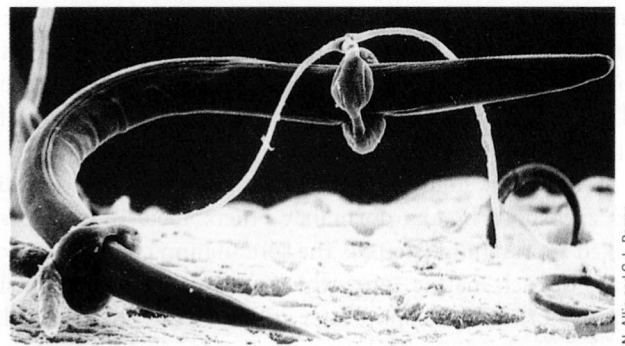

N. Allin and G. L. Barron

FIGURE 30.9 Hyphae of *Arthrobotrys dactyloides*, a trapping ascomycete, which form nooselike rings. When the fungus is stimulated by the presence of a prey organism, rapid changes in ion concentrations draw water into the hypha by osmosis. The increased turgor pressure shrinks the "hole" in the noose and captures this nematode. The hypha then releases digestive enzymes that break down the worm's tissues.

amount of cytoplasm before a breach can be sealed by congealing cytoplasm. In ways that are not well understood, septa can also limit the damage from toxins that are secreted by competing fungi.

As with zygomycetes, certain hyphae in ascomycetes are specialized for asexual reproduction. Many ascomycetes produce the asexual spores called *conidia*. In some of these species, the conidia form in chains that elongate from modified hyphal branches called **conidiophores.** In other ascomycetes, the conidia pinch off from the hyphae in a series of "bubbles," a bit like a string of detachable beads. Either way, an ascomycete can form and release spores much more quickly than a zygomycete can. Each newly formed conidium contains a haploid nucleus and some of the parent hypha's cytoplasm. Conidia and conidiophores of some ascomycete species are visible as the white powdery mildew that attacks grapes, roses, grasses, and the leaves of squash plants.

Ascomycetes can also reproduce sexually and are commonly termed *sac fungi* because most of the events that generate haploid sexual spores occur in saclike cells called *asci* (*askos* = leather bag; singular, *ascus*). In *Neurospora crassa* and other complex ascomycetes, reproductive bodies called **ascocarps** bear or contain the asci. Some ascocarps resemble globes, others flasks or open dishes. In *N. crassa*, an ascocarp begins to develop when a conidium from a hypha of one mating type fuses with a hypha of the opposite mating type (**Figure 30.10,** step 1). Plasmogamy then takes place, with the details differing from species to species. During plasmogamy, the fused sexual structures give rise to dikaryotic hyphae. Asci form at the hyphal tips. Inside them, karyogamy takes place, producing a diploid zygote nucleus. It divides by meiosis, producing four haploid nuclei. In yeasts and some other ascomycetes, cell division stops at this point, but in *N. crassa* and in many other species, a round of mitosis ensues and results in eight nuclei. Regardless, the nuclei, other organelles, and a portion of cytoplasm then are incorporated into ascospores that may germinate on a suitable substrate and continue the life cycle.

Basidiomycetes, the Club Fungi, Form Sexual Spores on Club-Shaped Basidia

The roughly 22,000 species of fungi in the phylum Basidiomycota include the mushroom-forming species, shelf fungi, coral fungi, bird's nest fungi, stinkhorns, smuts, rusts, and

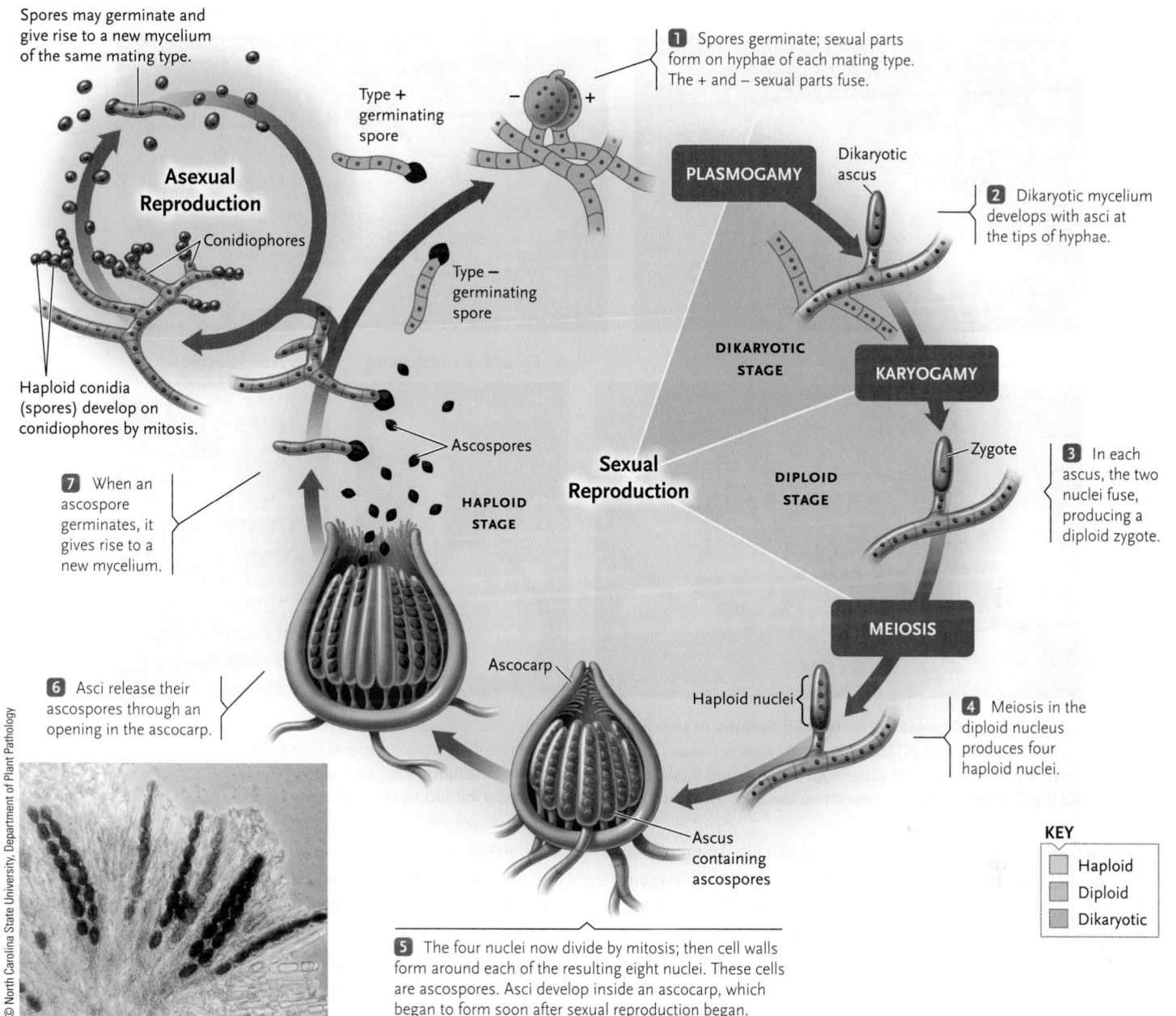

Spores may germinate and give rise to a new mycelium of the same mating type.

Asexual Reproduction

Conidiophores

Haploid conidia (spores) develop on conidiophores by mitosis.

Type + germinating spore

Type − germinating spore

7 When an ascospore germinates, it gives rise to a new mycelium.

Ascospores

HAPLOID STAGE

Sexual Reproduction

6 Asci release their ascospores through an opening in the ascocarp.

Ascocarp

Ascus containing ascospores

1 Spores germinate; sexual parts form on hyphae of each mating type. The + and − sexual parts fuse.

PLASMOGAMY

Dikaryotic ascus

2 Dikaryotic mycelium develops with asci at the tips of hyphae.

DIKARYOTIC STAGE

KARYOGAMY

DIPLOID STAGE

Zygote

3 In each ascus, the two nuclei fuse, producing a diploid zygote.

MEIOSIS

Haploid nuclei

4 Meiosis in the diploid nucleus produces four haploid nuclei.

5 The four nuclei now divide by mitosis; then cell walls form around each of the resulting eight nuclei. These cells are ascospores. Asci develop inside an ascocarp, which began to form soon after sexual reproduction began.

KEY

Haploid
Diploid
Dikaryotic

© North Carolina State University, Department of Plant Pathology

FIGURE 30.10 Life cycle of the ascomycete *Neurospora crassa*.
© Cengage Learning 2017

puffballs **(Figure 30.11)**. **Basidiomycetes** are known commonly as club fungi because their spore-producing cells, called **basidia** (*basis* = base or foundation), usually are club shaped. Some species have enzymes for digesting cellulose and lignin and are important decomposers of woody plant debris. A surprising number of basidiomycetes, including the prized edible oyster mushrooms (*Pleurotus ostreatus;* see Figure 30.11E), also can trap and consume bacteria and small animals such as nematodes by secreting paralyzing toxins or gluey substances that immobilize the prey. This adaptation gives the fungus access to a rich source of molecular nitrogen, an essential nutrient that is often scarce in terrestrial habitats.

Many basidiomycetes take part in vital mutualistic associations with the roots of forest trees, as discussed later in this chapter. Others, the rusts and smuts, are parasites that cause serious diseases in wheat, rice, and other plants.

In a few basidiomycetes, reproduction is usually asexual, by budding or shedding a fragment of a hypha. In general, however, basidiomycetes reproduce sexually, producing large numbers of haploid sexual spores. **Figure 30.12** shows the life cycle of a typical basidiomycete.

Basidia typically develop on a **basidiocarp,** which is the reproductive body of the fungus. A basidiocarp consists of tight clusters of hyphae; the feeding mycelium is buried in the soil or decaying wood. The shelflike bracket fungi on trees are

A. Coral fungus

B. Shelf fungus

C. White-egg bird's nest fungus

D. Fly agaric mushroom

E. Oyster mushrooms

FIGURE 30.11 Representative basidiomycetes, or club fungi. (A) The light red coral fungus *Ramaria*. **(B)** The shelf fungus *Laetiporus*. **(C)** The white-egg bird's nest fungus *Crucibulum laeve*. Each tiny "egg" contains spores. Raindrops splashing into the "nest" can cause "eggs" to be ejected, thereby spreading spores into the surrounding environment. **(D)** The fly agaric mushroom *Amanita muscaria*, which is toxic to humans. **(E)** The oyster mushroom *Pleurotus ostreotus*.

basidiocarps, and about 10,000 species of club fungi produce the basidiocarps we call mushrooms. Each is a short-lived reproductive body consisting of a stalk and a cap. Basidia develop on sheetlike "gills" on the underside of the cap. The basidia undergo meiosis to produce microscopic, haploid **basidiospores** (Figure 30.12, inset) that disperse throughout the environment.

When a basidiospore lands on a suitable food source, it germinates and gives rise to a haploid mycelium. Compatible mating types growing near each other may undergo plasmogamy. The resulting mycelium is dikaryotic, its cells containing one nucleus from each mating type. The dikaryotic stage of a basidiomycete is the feeding mycelium, which can grow for years—a major departure from an ascomycete's short-lived dikaryotic stage. Accordingly, a basidiomycete has many more opportunities for producing sexual spores, and the mycelium can give rise to reproductive bodies many times.

After an extensive mycelium develops, and when environmental conditions such as moisture are favorable, basidiocarps grow from the mycelium and develop basidia. At first, each basidium in the mushroom or other reproductive body is dikaryotic, but then the two nuclei undergo karyogamy, fusing to form a diploid zygote nucleus. The zygote exists only briefly; meiosis soon produces haploid basidiospores, which are wafted away from the basidium by air currents. Basidia can produce huge numbers of spores—for many species, estimates run as high as 100 million spores *per hour* during reproductive periods, day after day. In some species the underlying mycelium can live for many years. As such a mycelium grows, specialized mechanisms during cell division maintain the dikaryotic condition and the paired nuclei in each hyphal cell.

A Grouping Called *Cryptomycota* Includes Single-Celled Parasites That May Represent the Earliest Fungi

As you have read, the phylum Chytridiomycota has long been viewed as the most basal (earliest diverging) fungal clade. Several years ago, scientists studying a single-celled parasitic chytrid genus *Rozella* brought that idea into question when they reported finding no evidence that *Rozella* synthesizes chitin—which, recall, is a basic structural component of fungal cell walls. Although the findings were tentative, the apparent absence of chitin suggested that *Rozella* might represent a new, even more ancient fungal clade that diverged from the protist ancestor of fungi before the genetic machinery underlying chitin synthesis arose. Because this proposed new clade was previously unknown to science, the researchers named it *Cryptomycota*, meaning "hidden fungi."

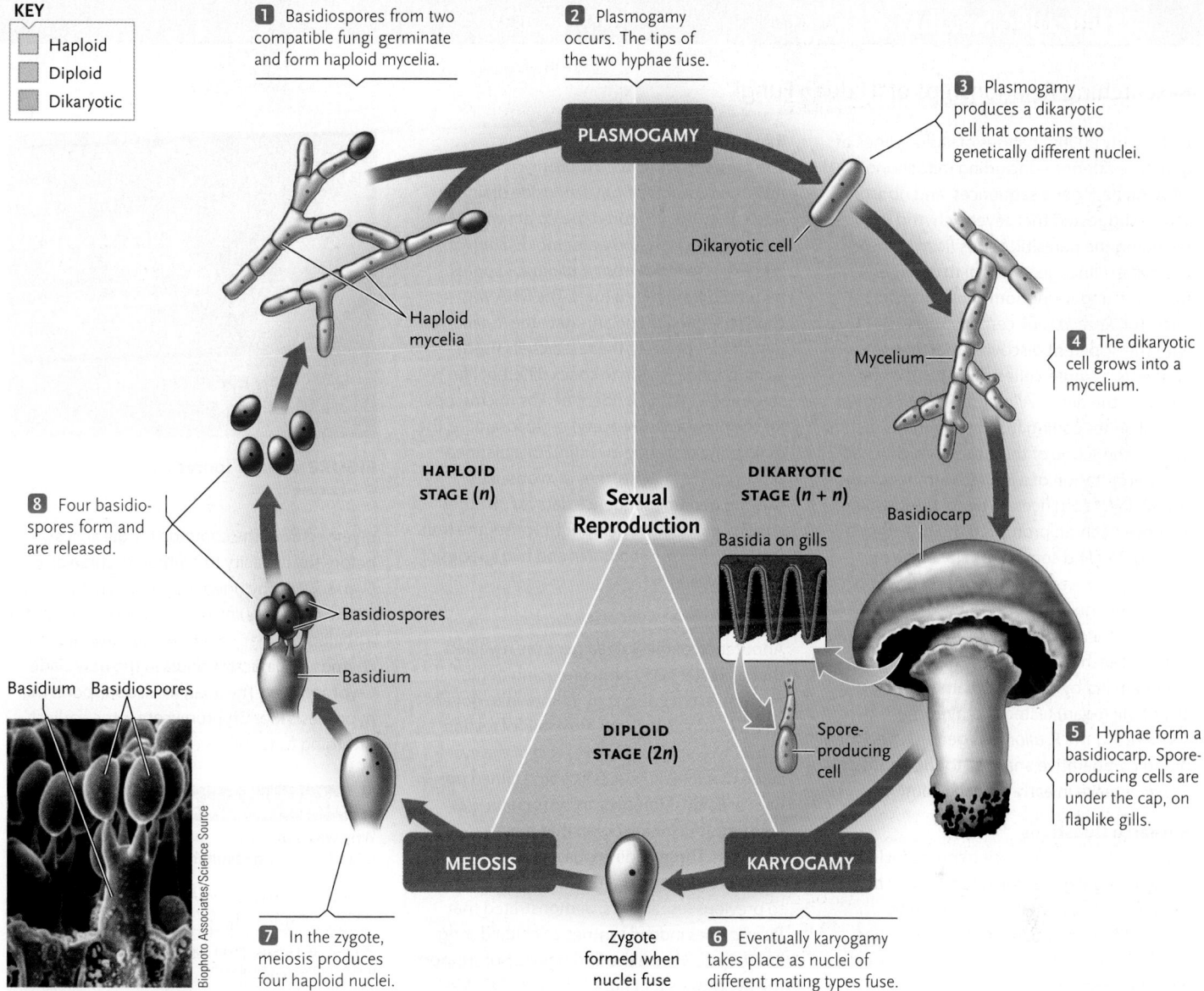

KEY

Haploid
Diploid
Dikaryotic

1 Basidiospores from two compatible fungi germinate and form haploid mycelia.

2 Plasmogamy occurs. The tips of the two hyphae fuse.

PLASMOGAMY

3 Plasmogamy produces a dikaryotic cell that contains two genetically different nuclei.

Dikaryotic cell

4 The dikaryotic cell grows into a mycelium.

Mycelium

Haploid mycelia

HAPLOID STAGE (n)

Sexual Reproduction

DIKARYOTIC STAGE (n + n)

Basidiocarp

Basidia on gills

8 Four basidiospores form and are released.

Basidiospores

Basidium

DIPLOID STAGE (2n)

Spore-producing cell

5 Hyphae form a basidiocarp. Spore-producing cells are under the cap, on flaplike gills.

Basidium   Basidiospores

Biophoto Associates/Science Source

MEIOSIS

KARYOGAMY

7 In the zygote, meiosis produces four haploid nuclei.

Zygote formed when nuclei fuse

6 Eventually karyogamy takes place as nuclei of different mating types fuse.

FIGURE 30.12 Generalized life cycle of the basidiomycete *Agaricus arvensis*, a close relative of the common button mushroom sold in grocery stores. During the dikaryotic stage, *A. arvensis* cells contain two genetically different nuclei, shown here in different colors. *Inset:* Micrograph showing basidia and basidiospores.

Subsequent molecular studies by others have demonstrated that *Rozella* does indeed synthesize chitin as part of the process by which it infects its host (a water mold). That research also produced evidence that a second quirky group of single-celled parasitic fungi, **microsporidia (Figure 30.13),** are related to *Rozella* and therefore also are candidates for inclusion in the Cryptomycota.

There are about 1,500 species of microsporidia. They are known to infect many animals, including humans—especially people who have compromised immune systems. Microsporidia synthesize chitin and physically resemble spores, but they lack mitochondria, stacked Golgi bodies, and some other characteristic features of eukaryotic cells. Genetic studies suggest that the group may have lost many typical fungal features as its highly specialized parasitic lifestyle evolved.

Genome sequences for several microsporidia are now available. As described in this chapter's *Molecular Insights,* comparative studies of microsporidia and *Rozella* genomes support a phylogeny in which the Cryptomycota are the earliest-diverging clade of fungi.

Conidial Fungi Are Species for Which No Sexual Phase Is Known

As noted earlier, fungi generally are classified on the basis of their structures for sexual reproduction. When a sexual phase is absent or has not yet been detected, the fungal species is lumped into a convenience grouping, the conidial fungi (recall that conidia are asexual spores). Other historical names for this grouping are "imperfect fungi" and deuteromycetes.

Researching Relationships of "Hidden Fungi"

In the early 2000s analyses of various lines of genetic evidence—including mitochondrial DNA, nuclear gene sequences, and ribosomal RNA—suggested that several chytrid groups, including the parasitic genus *Rozella,* were the earliest lineages of fungi to diverge from the fungal kingdom's common protist ancestor. *Rozella* took center stage in 2011 with the apparent discovery that *Rozella allomycis,* a single-celled parasite of water molds in the genus *Allomyces,* lacked known fungal genes coding for chitin. Various factors limited the scope of that study, which relied on interpretation of a limited number of ribosomal RNA sequences. Even so, the seeming absence of chitin prompted the hypothesis that *Rozella* and some of its close chytrid relatives represented a previously unknown fungal lineage—the Cryptomycota—that diverged before the evolution of fungi having genetic machinery to synthesize chitin. The team led by Timothy Y. James (University of Michigan) tested this hypothesis by sequencing the *R. allomycis* genome and then doing comparative analyses to shed light on *Rozella*'s place in early fungal evolution.

Research Questions

Does the *Rozella allomycis* genome include known fungal genes for synthesizing chitin? What evolutionary information can be gleaned from comparisons with genomes of other organisms, including fungi, protists, and microsporidia?

Experiments

The James team isolated *R. allomycis* from moist soil collected from a roadside drainage ditch. They then cultured the isolate with mycelia of its host, the water mold *Allomyces,* and extracted DNA from *R. allomycis* spores harvested from the cultures. The DNA was used to sequence and annotate the *R. allomycis* genome. Separately, the researchers used a chitin-binding stain to check different life stages of *R. allomycis* (identified by microscopy) for the presence of *N*-acetyl-D-glucosamine, the main polysaccharide in chitin. For a planned phylogenetic analysis, they also obtained existing genome sequence data for 38 other species, including other fungi, protists, the fruit fly *Drosophila melanogaster,* and two species of microsporidia.

Discussion

Annotation of the *Rozella* genome revealed that of a total 6,350 protein-coding genes, four genes encode enzymes known to govern chitin synthesis in fungi. The walls of resting *R. allomycis* spores **(Figure)** and of the cyst stage that infects host cells both stained positive for chitin. Moreover, in cysts chitin was concentrated in the region that penetrates host cells. These findings, based on much more complete genetic data than was available to previous researchers, demonstrated that *Rozella* does indeed synthesize chitin during its life cycle. Hence the findings do not support the hypothesis that *Rozella* and its close kin

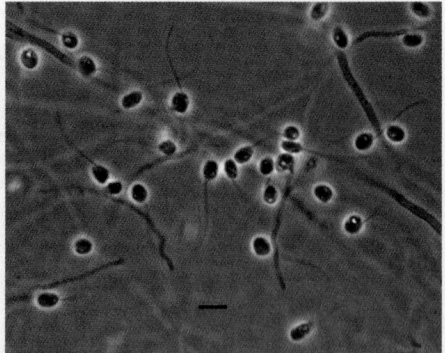

FIGURE *Rozella* **spores.**
© Cengage Learning 2017

diverged from the common fungal ancestor before the capacity to synthesize chitin arose. Even so, the phylogenetic analysis (which compared 200 genes) provided strong evidence for a phylogeny that unites *Rozella,* several other chytrids, and microsporidia in the new clade Cryptomycota. The analysis also supports the hypothesis that Cryptomycota is the earliest diverging fungal clade.

think like a scientist

Why was it important in this research to include staining results?

Source: Timothy Y. James, Adrian Pelin, Linda Bonen, Steven Ahrendt, Divya Sain, Nicholas Corradi, and Jason E. Stajich. 2013. Shared signatures of parasitism and phylogenomics unite cryptomycota and microsporidia. *Current Biology* 23:1548–1553.

© Cengage Learning 2017

FIGURE 30.13 Structure of microsporidia. When a spore germinates, its vacuole expands and forces the coiled "polar tube" outward and into a nearby, soon-to-be host cell. The nucleus and cytoplasm of the parasite enter the host through the tube, launching developmental steps that lead to the development of more microsporidia inside the host. *Molecular Insights* shows spores of the chytridlike *Rozella.* As with the microsporidia, many mycologists now classify *Rozella* in the phylum Cryptomycota.
© Cengage Learning 2017

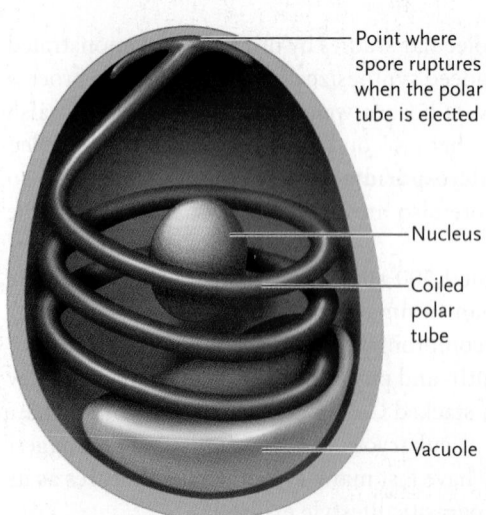

Point where spore ruptures when the polar tube is ejected

Nucleus

Coiled polar tube

Vacuole

Applied Research: Lichens as Monitors of Air Pollution's Biological Damage

Lichens have become reliable pollution-monitoring devices all over the world—in some cases, replacing costly electronic monitoring stations. Different species are vulnerable to specific pollutants. For example, *Ramalina* lichens are damaged by nitrate and fluoride salts. Elevated levels of sulfur dioxide (a major component of acid rain) cause old man's beard (*Usnea trichodea*) to shrivel and die **(Figure),** but strongly promote the growth of a crusty European lichen, *Lecanora conizaeoides*. The sensitivity of yellow *Evernia* lichens to SO$_2$ enabled the scientist who discovered its damage at remote Isle Royale in Michigan to point the finger northward to coal-burning furnaces at Thunder Bay, Canada. Conversely, healthy lichens on damaged trees of Germany's Black Forest lifted suspicion from French coal-burning power plants and allowed investigators to identify the true source of the tree damage: nitrogen oxides from automobile exhausts. The result was Germany's first auto emission standards, which went into effect in the 1990s.

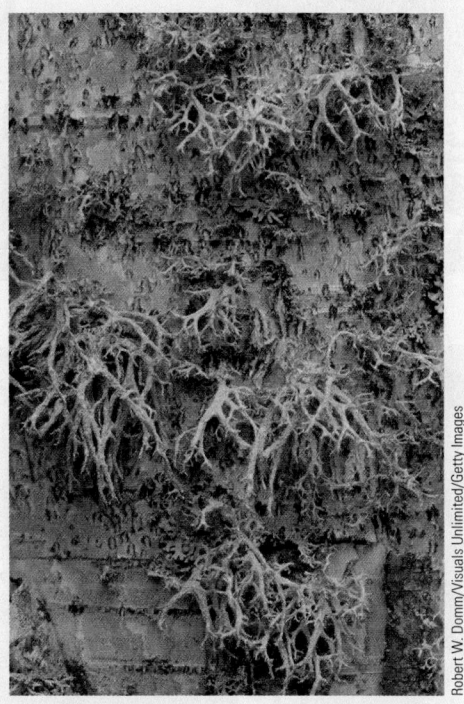

FIGURE *Usnea,* commonly called old man's beard, a pendent (hanging) lichen.

Robert W. Domm/Visuals Unlimited/Getty Images

When researchers discover a sexual phase for a conidial fungus, or when molecular studies establish a clear relationship to a sexual species, the conidial fungus is reassigned to the appropriate phylum. Thus far, some have been classified as basidiomycetes, but most conidial fungi have turned out to be ascomycetes.

STUDY BREAK 30.2

1. Name the major phyla and phylum-level categories of the Kingdom Fungi and describe the reproductive adaptations that distinguish each one.
2. In terms of structure, which are the simplest fungal groups? The most complex?

30.3 Fungal Associations: Lichens and Mycorrhizae

Many fungi are partners in mutually beneficial interactions with plants and other photosynthesizers, with insects, and even with mammals—associations that play major roles in the functioning of ecosystems. A **symbiosis** is a state such as *parasitism* or *mutualism* in which two or more species live together in close

association. Chapter 54 discusses general features of symbiotic associations more fully; here we are interested in some examples of the symbioses fungi form with photosynthetic partners—cyanobacteria, green algae, and plants.

A Lichen Is an Association between a Fungus and a Photosynthetic Partner

Thousands of ascomycetes and a few basidiomycetes form the kind of symbiosis called a *lichen*. Technically, a **lichen** is a single vegetative body that results from an association between a fungus and a photosynthetic partner. The fungal partner in a lichen, called the **mycobiont,** usually makes up about 90% of the whole. The other 10% is the photosynthetic partner, called the **photobiont.** Only about 100 photosynthetic species serve as photobionts. Most frequently, these are green algae of the genus *Trebouxia* or cyanobacteria of the genus *Nostoc.*

Lichens often live in harsh, dry microenvironments, including on bare rock and wind-whipped tree trunks. Yet lichens have vital ecological roles and important human uses. Lichens secrete acid that eats away at rock, breaking it down and facilitating the development of new soil. Some paleobiologists (who study ancient life forms) have suggested that lichens may have been some of the earliest land organisms, covering bare rocks during the Ordovician period (roughly 500 million to 425 million years ago). In this scenario, millennia of decaying lichens would have created the first soils in which the earliest land plants could grow. Today, lichens continue to enhance the survival of other life forms. For instance, in arctic tundra, where plants are scarce, reindeer and musk oxen can survive by eating lichens. Insects and some other invertebrates also consume lichens, and they are nest-building materials for many birds and small mammals. People have derived dyes from lichens; they are even a component of garam masala, an ingredient in Indian cuisine. Some environmental chemists monitor air pollution by observing lichens, most of which cannot grow in heavily polluted air (see *Focus on Research: Applied Research*).

Because lichens are composite organisms, it may seem odd to talk of lichen "species." Biologists do give lichens binomial names, however, based on the characteristics of the mycobiont. More than 13,500 lichens are recognized, each one a unique combination of a particular species of fungus and one or more species of photobiont. The relationship often begins when a fungal mycelium contacts a free-living cyanobacterium, algal cell, or both. The fungus parasitizes the photosynthetic host

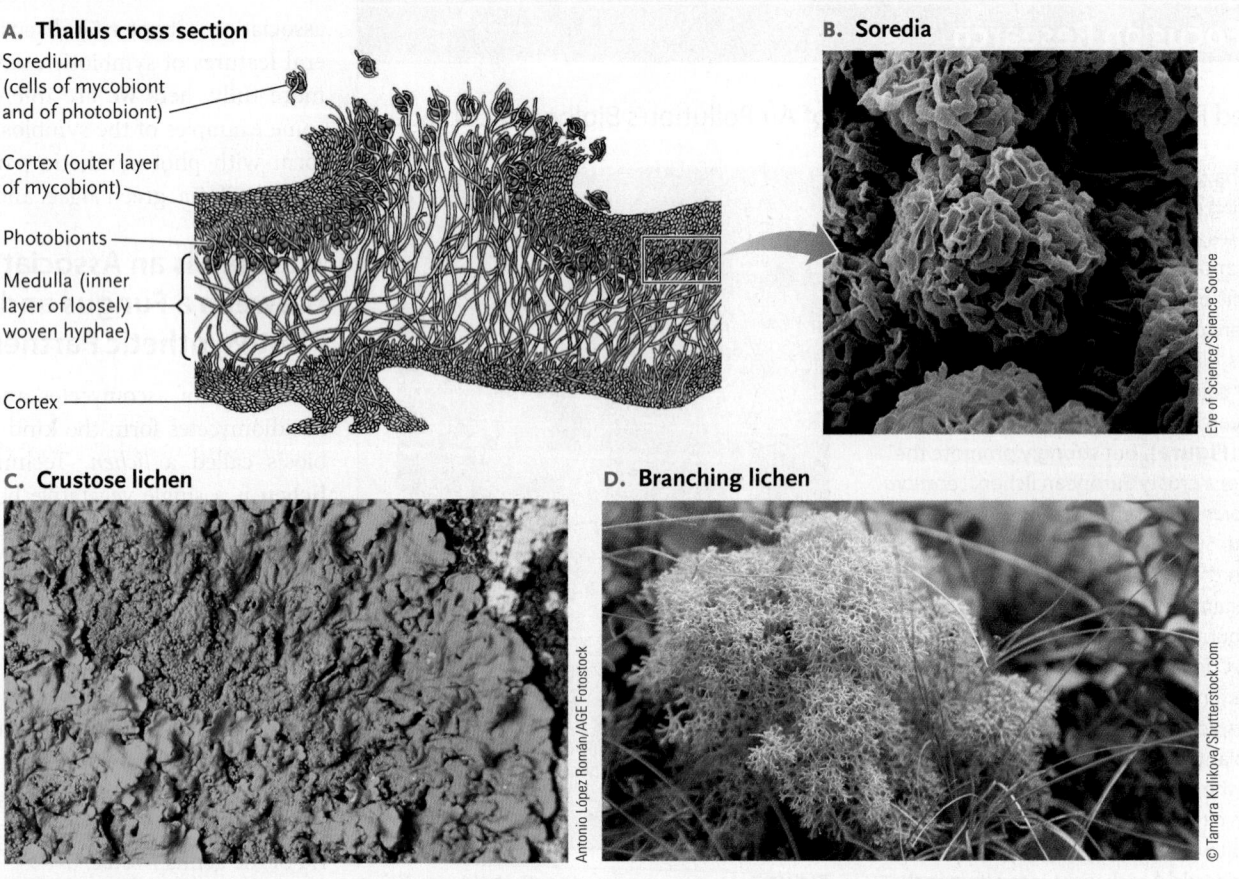

A. Thallus cross section

Soredium (cells of mycobiont and of photobiont)

Cortex (outer layer of mycobiont)

Photobionts

Medulla (inner layer of loosely woven hyphae)

Cortex

B. Soredia

Eye of Science/Science Source

C. Crustose lichen

Antonio López Román/AGE Fotostock

D. Branching lichen

© Tamara Kulikova/Shutterstock.com

FIGURE 30.14 Lichens. (A) Sketch of a cross section through the thallus of the lichen *Lobaria verrucosa*. **(B)** The soredia, which contain both hyphae and algal cells, are a type of dispersal fragment by which lichens reproduce asexually. **(C)** Crustose lichen. **(D)** Erect, branching lichen, *Cladonia rangiferina*.

© Cengage Learning 2017

cell, sometimes killing it. If the host cell can survive, however, it multiplies in association with the fungal hyphae. The result is a tough, pliable body called a **thallus,** which can take a variety of forms **(Figure 30.14A)**. Short, specialized hyphae penetrate photobiont cells of the thallus, which become the fungus's sole source of nutrients. Often, the mycobiont of a lichen absorbs up to 80% of the carbohydrates the photobiont produces.

Benefits for the photobiont are less clear-cut, in part because the drain on nutrients hampers its growth and reproduction. In one view, many and possibly most lichens are parasitic symbioses in which the photobiont does not receive equal benefit. On the other hand, it is relatively rare to find a lichen's photobiont species living independently in the same conditions under which the lichen survives. Studies have also revealed that at least some green algae clearly benefit from the relationship. Such algae are sensitive to desiccation and intense ultraviolet radiation. Sheltered by a lichen's fungal tissues, a green alga can thrive in locales where alone it would die.

Lichen reproduction is quirky. In lichens that involve an ascomycete, the fungus produces ascospores that are dispersed by the wind. The spores germinate to form hyphae that may colonize new photosynthetic cells and so establish new symbioses. A lichen itself can also reproduce in at least two ways. In some types, a section of the thallus detaches and grows into a new lichen. In about one-third of lichens, specialized regions of the thallus give rise asexually to reproductive cell clusters called **soredia** (*soros* = heap; singular, *soredium*). Each cluster includes both algal and hyphal cells **(Figure 30.14B)**. As the lichen grows, the soredia detach and are dispersed by water, wind, or passing animals.

Mycorrhizae Are Symbiotic Associations of Fungi and Plant Roots

A **mycorrhiza** ("fungus root") is a mutualistic symbiosis in which fungal hyphae associate intimately with plant roots. In general, mycorrhizae represent a "win–win" situation for the partners. The fungal hyphae absorb carbohydrates made by the plant, along with some amino acids and perhaps growth factors as well. The growing plant in turn absorbs mineral ions made accessible to it by the fungus. Collectively, the fungal hyphae have a huge surface area for absorbing mineral ions from a large volume of the surrounding soil **(Figure 30.15)**. Dissolved mineral ions accumulate in the hyphae when they are plentiful in the soil and are released to the plant when they are scarce. This service is a survival boon to a great many plants, especially species that cannot readily absorb mineral ions, particularly phosphorus. We will return to mycorrhizal associations in Chapter 35, which focuses on plant nutrition.

FIGURE 30.15 Mycorrhizae of a pine seedling. Appearances can be deceiving: in this photograph, the 4-cm-tall pine seedling still has a small root system. The pale, densely branching underground network below the shoot consists mostly of fungal hyphae that have associated with the roots in mycorrhizae.

STUDY BREAK 30.3

1. What is a lichen, and how does each partner contribute to the whole?
2. What is the biological role of mycorrhizae?

30.4 Impacts of Fungi in Ecosystems and Society

Throughout this chapter we have noted the profound impacts of fungi on ecosystems and other life forms. It is almost impossible to overstate the ecological importance of saprobic species, the fungal decomposers that feed on dead organisms and organic wastes and thereby recycle much of the carbon locked in organic tissues. Mycorrhizal associations of fungi with plant roots (discussed in Section 30.3) make an essential contribution to the nutrition of most terrestrial plants. Here we briefly survey some other ecological roles of fungi, as well as some examples of impacts of fungi in human society.

Endophytic Fungi Are Partners in Mutualisms

As noted in Section 30.3, mycorrhizae are symbiotic associations of certain fungal species with plant roots. In another type of association, species of fungal **endophytes** live within leaves or other plant tissues. Although many of these interactions are not particularly well studied or understood, many, if not most, appear to be mutualistic (mutually beneficial)—that is, the fungus feeds on the host's carbohydrates and other substances, while providing the host with some benefit. For example, many endophytic fungi synthesize compounds thought to deter predation, enhance resistance to abiotic stresses such as drought, or otherwise enhance the plant's survival. Endophytic fungi were originally discovered in grasses and subsequently have been observed in virtually every plant species where researchers have looked for them **(Figure 30.16A).**

A. Endophytic fungus on a grass *(Epichloë typhina)*

B. Mycelium of powdery mildew *(Podosphaera leucotricha)* on an apple leaf *(Malus)*

C. *Candida albicans,* a yeast

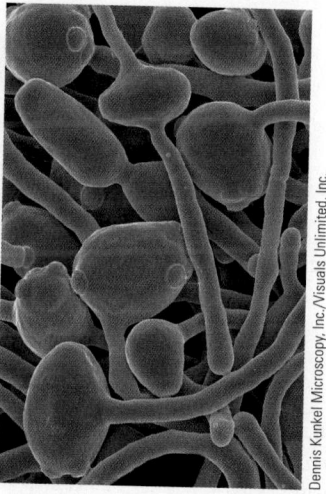

FIGURE 30.16 Examples of endophytic and parasitic fungi.

Some Fungi Parasitize or Cause Disease in Other Organisms

Parasitic fungi feed on a living host without providing a reciprocal benefit. They attack a wide range of hosts, including plants, animals, and other fungi, and are major causes of plant disease. Examples are *V. inequalis*, which causes apple scab disease (see Figure 30.8D), and *Sphaerotheca pannosa*, one of several ascomycetes that cause powdery mildew **(Figure 30.16B)**. The ascomycete *Candida albicans* **(Figure 30.16C)** is a normal resident on human mucous membranes, but produces yeast infections of the mouth and vagina when altered conditions allow it to overgrow. Other ascomycete pathogens on humans range from those that cause nuisance infections like athlete's foot and ringworm, to those that are potentially lethal. *Claviceps purpurea*, a parasite on rye and other grains, causes ergotism, a disease marked by vomiting, hallucinations, convulsions, and, in severe cases, gangrene and even death. Species of *Aspergillus* grow in damp grain or peanuts; over time their metabolic wastes, known as aflatoxins, can cause cancer in people who eat the poisoned food. A basidiomycete, *Cryptococcus neoformans*, causes a form of meningitis in humans.

Fungi Also Are Beneficial Medically, Economically, and As Model Research Organisms

In contrast to fungal parasites and pathogens, other species contribute in vital ways to human welfare and scientific enterprises. Antibiotics and other pharmaceuticals derived from fungi have been crucial in medicine. For example, certain species of *Penicillium* **(Figure 30.17)** are the original source of the penicillin family of antibiotics, and cephalosporins were isolated from fungi in the genus *Cephalosporium*. Cyclosporins are medically important immune suppressants initially discovered in ascomycetes, including *Tricoderma polysporum* and *Cylindrocarpon lucidum*.

Amanita muscaria, the fly agaric mushroom (see Figure 30.11D), has been used as a fly poison. In humans one of its side effects is euphoria, which is why *A. muscaria* was used in religious rituals of ancient societies in Central America, Russia, and India. (Less pleasant side effects can include powerful nausea.) Other species of this genus, including the death cap mushroom *Amanita phalloides*, produce deadly toxins. In eukaryotic cells, the *A. phalloides* toxin, a short peptide called α-amanitin, halts the transcription of protein-coding genes by RNA polymerase II (see Section 15.2), thereby inhibiting protein synthesis. Ingesting as little as 5 mg of the toxin can trigger violent vomiting and diarrhea within 8 to 24 hours. Later, kidney and liver cells start to degenerate; without intensive medical care, death can follow within a few days.

You are probably well aware that humans consume fungi and use them to manufacture food products. Edible mushrooms are nearly all basidiomycetes, although truffles (*Tuber melanosporum*) and a few others (such as the true morel *Morchella esculenta* shown in Figure 30.8C) are ascomycetes. Still other ascomycetes produce the aroma and distinctive flavors of Camembert and Roquefort cheeses.

The yeast *Saccharomyces cerevisiae* produces the ethanol in alcoholic beverages and the carbon dioxide that leavens bread **(Figure 30.18)**. And as you know from previous chapters, *S. cerevisiae* and the mold *Neurospora crassa* both are pivotal

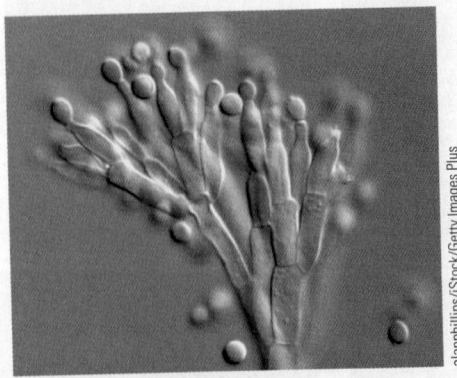

FIGURE 30.17 A *Penicillium* species. Shown here is a species of *Eupenicillium*. Notice the conidia (asexual spores) atop the structures that produce them.

Baker's yeast cells (*Saccharomyces cerevisiae*)

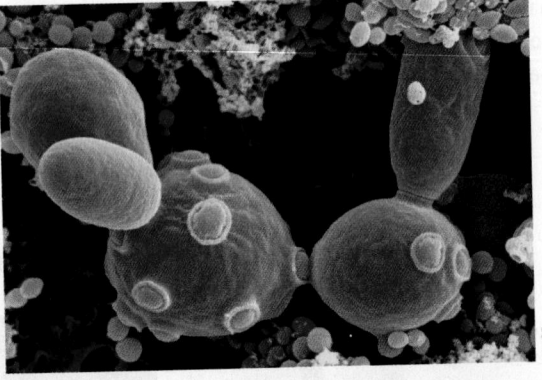

Bread

Beer and wine

FIGURE 30.18 *Saccharomyces cerevisiae* and some products it is used to manufacture.

model organisms in studies of DNA structure and function, and *S. cerevisiae* has also been important in the development of genetic engineering methods. By one estimate, it has been the subject of more genetic experiments than any other eukaryotic microorganism.

STUDY BREAK 30.4

1. In what ways do mycorrhizal, endophytic, and parasitic fungi differ in terms of their interactions with other organisms?
2. Name at least six specific benefits humans obtain from fungi.

 ## Unanswered Questions

Do different fungal species respond differently to environmental changes?

Ecologists Shigeo Yachi and Michel Loreau formulated the "biological insurance" hypothesis, which predicts that ecosystem functions will be maintained more reliably in natural communities when multiple species perform a given ecological function (such as serving as decomposers) but differ in their responses to environmental changes. When environmental conditions shift in such communities, a species that fills the same ecological role but is better suited to the new conditions can then "take over" for a species better adapted to the former conditions. Changes in the composition of fungal communities can result from climate change, loss of host species (for example, due to diseases such as Sudden Oak Death, which is responsible for high mortality rates of dominant tree species on the west coast of the United States), or other ecological perturbations. Nitrogen addition experiments have shown significant changes in community composition of both mycorrhizal and saprobic fungi in response to elevated N levels. Changes in fungal community composition have also been observed in elevated CO_2 treatments, although several longer-term studies indicate that these differences may wane over time. Important avenues for further research include incorporating better representations of environmental and ecological complexity because fungi may respond differently to the interaction of multiple change factors than could be predicted by examining each factor separately, and symbiotic interactions (with other fungi and/or other organisms) may change the type or amount of a species' response to an environmental change. Given the ecological roles and interactions of fungi, changes in fungal communities could have important consequences for ecosystem functions such as the availability of soil nutrients, storage of carbon in organic material, and plant growth. Gaining a better understanding of the mechanisms underlying these patterns and linking changes in community composition to changes in how ecosystems function are important directions for further research.

think like a scientist Ectomycorrhizal fungal communities often exhibit high local diversity, where the root systems of single trees may contain dozens—and a single-host forest stand may contain hundreds—of different fungal symbionts. What factors might allow multiple species of fungi to coexist on a single tree? What factors might favor the dominance of one symbiont over another? How might a better understanding of these phenomena be beneficial to conservation planning?

Courtesy of Todd Osmundson

Todd Osmundson is an Assistant Professor in the Department of Biology at the University of Wisconsin–La Crosse. His research areas include systematics, genetics, biodiversity, ecology, and conservation biology, with an emphasis on fungi. Learn more about Dr. Osmundson's work at http://www.uwlax.edu/profile/tosmundson/.

REVIEW KEY CONCEPTS

For access to MindTap and additional study materials visit www.cengagebrain.com.

30.1 General Characteristics of Fungi

- Fungi are key decomposers contributing to the recycling of carbon and some other nutrients. They occur as single-celled yeasts or multicellular filamentous organisms.

- The fungal mycelium consists of filamentous hyphae that grow throughout the substrate the fungus feeds on (Figure 30.1). A wall containing chitin surrounds the plasma membrane, and in most species septa partition the hyphae into cell-like compartments.

- Pores in septa permit cytoplasm and organelles to move between hyphal cells. Aggregations of hyphae form all other tissues and organs of a multicellular fungus.

- Fungi gain nutrients by extracellular digestion and absorption. Saprobic species feed on nonliving organic matter. Parasitic types obtain nutrients from tissues of living organisms. Many fungi are partners in symbiotic relationships with plants.

- All fungi may reproduce via spores generated either asexually or sexually (Figure 30.2). Some types also may reproduce asexually by budding or fragmentation of the parent body.

- Sexual reproduction usually has two stages. First, in plasmogamy, the cytoplasms of two haploid cells fuse to become a dikaryon containing a haploid nucleus from each parent. Later, in karyogamy, the nuclei fuse and form a diploid zygote. Meiosis then generates haploid spores.

30.2 Evolution of the Kingdom Fungi

- The main phyla of fungi are the Chytridiomycota (which have motile spores), Zygomycota (zygospore-forming fungi), Glomeromycota, Ascomycota (sac fungi), and Basidiomycota (club fungi). A proposed new phylum, the Cryptomycota, includes single-celled parasitic fungi formerly in the Chytridiomycota as well as microsporidia (Figure 30.3).

- Fungal phyla traditionally have been distinguished mainly by the structures of sexual reproduction. When a sexual phase cannot be detected or is absent from the life cycle, the specimen is assigned to an informal grouping, the conidial fungi.

- Chytrids usually are microscopic. They are the only fungi that produce motile, flagellated spores. Many are parasites (Figures 30.4 and 30.5).

- Zygomycetes have aseptate hyphae and are coenocytic, with many nuclei in a common cytoplasm. They sometimes reproduce sexually by way of hyphae that occur in 1 and 2 mating types; haploid nuclei in the hyphae function as gametes. Further development produces the zygospore, which may become dormant.

- The zygospore of a zygomycete breaks dormancy and produces a stalked sporangium containing haploid spores of each mating type, which are released (Figures 30.6 and 30.7).

- Glomeromycetes reproduce asexually, by way of spores that form at the tips of hyphae.

- Most ascomycetes, the cup fungi, are multicellular (Figure 30.8). In asexual reproduction, chains of haploid asexual spores called conidia elongate or pinch off from the tips of conidiophores (modified aerial hyphae; Figure 30.10).

- In sexual reproduction, haploid ascomycete sexual spores called ascospores arise in saclike cells called asci. In the most complex species, reproductive bodies called ascocarps bear or contain the asci. Ascospores can give rise to a new haploid mycelium (Figure 30.10).

- Most basidiomycetes reproduce only sexually. Club-shaped basidia develop on a basidiocarp and bear sexual spores on their surface. When dispersed, these basidiospores may germinate and give rise to a haploid mycelium (Figures 30.11 and 30.12).

- Microsporidia are single-celled sporelike parasites of arthropods, fish, and humans (Figure 30.13).

30.3 Fungal Associations: Lichens and Mycorrhizae

- Many ascomycetes and some basidiomycetes enter into symbioses with cyanobacteria or green algae to produce the communal life form called a lichen, which has a tough, pliable body called a thallus. The algal cells supply the lichen's carbohydrates, most of which are absorbed by the fungus.

- In some lichens a section of the thallus may detach and grow into a new individual. In others, specialized regions of the thallus give rise asexually to reproductive soredia that include both algal and hyphal cells (Figure 30.14).

- In the symbiosis called a mycorrhiza, fungal hyphae make mineral ions and sometimes water available to the roots of a plant partner. The fungus in turn absorbs carbohydrates, amino acids, and possibly other growth-enhancing substances from the plant (Figure 30.15).

30.4 Impacts of Fungi in Ecosystems and Society

- Some species of fungi are endophytes that live within leaves or other plant tissues. The association is a form of mutualism that benefits both partners. Parasitic fungi extract nutrients from a living organism without providing any benefit to the host. Still other fungi cause diseases in plants, animals, and other fungi (Figure 30.16).

- Humans use fungi directly as food or employ various species (such as yeasts) in the preparation of bread, cheeses, and alcoholic beverages. The yeasts *Saccharomyces cerevisiae* and *Neurospora crassa* are widely used in genetic research (Figures 30.17 and 30.18).

TEST YOUR KNOWLEDGE

Remember/Understand

1. Which of the following statements does *not* reflect current understanding of phylogenetic relationships and features among fungi?
   a. Lineages leading to fungi diverged from those leading to plants about 1 billion years ago.
   b. Chytrids are probably a paraphyletic grouping and physically may most closely resemble the earliest fungi.
   c. There are four fungal phyla broadly accepted as monophyletic: the Zygomycota, Glomeromycota, Ascomycota, and Basidiomycota.
   d. Fungal lineages each evolved distinctive reproductive adaptations.
   e. Living members of the Cryptomycota may represent the most basal fungal group.

2. Which of the following events is/are a necessary part of a typical asexual cycle in fungal reproduction?
   a. formation of a dikaryon
   b. hyphae developing into a mycelium
   c. formation of a diploid zygote
   d. plasmogamy, which occurs when hyphae fuse at their tips
   e. production and release of large numbers of spores

3. A trait common to all fungi is:
   a. reproduction via spores.
   b. parasitism.
   c. septate hyphae.
   d. a dikaryotic phase inside a zygospore.
   e. plasmogamy after an antheridium and ascogonium come into contact.

4. The chief characteristic used to classify fungi into the major fungal phyla is:
   a. nutritional dependence on nonliving organic matter.
   b. recycling of nutrients in terrestrial ecosystems.
   c. adaptations for obtaining water.
   d. features of reproduction.
   e. cell wall metabolism.

5. Which of the following fungal reproductive structures are diploid?
   a. basidiocarps
   b. ascospores
   c. conidia
   d. gametangia
   e. zygospores

6. A mushroom is:
   a. the food-absorbing region of an ascomycete.
   b. the food-absorbing region of a basidiomycete.
   c. a reproductive structure formed only by basidiomycetes.
   d. a specialized form of mycelium not constructed of hyphae.
   e. a collection of saclike cells called asci.

7. A zygomycete is characterized by:
   a. usually, aseptate hyphae.
   b. mostly sexual reproduction.
   c. absence of + and − mating types.
   d. the tendency to form mycorrhizal associations with plant roots.
   e. a life cycle in which karyogamy does not occur.

8. Which of the following statements apply/applies to lichens?
   a. It is a fungus that breaks down rock to provide nutrients for an alga.
   b. It colonizes bare rocks and slowly degrades them to small particles.
   c. It spends part of the life cycle as a mycobiont and part as a fungus.
   d. It is an association between a basidiomycete and an ascomycete.
   e. It is an association between a photobiont and a fungus.

Apply/Analyze

9. At lunch George ate a mushroom, some truffles, a little Camembert cheese, and a bit of moldy bread. Which of the following groups was *not* represented in the meal?
   a. Basidiomycota
   b. Ascomycota
   c. conidial fungi
   d. chytrids
   e. Zygomycota

10. In a college greenhouse, a new employee observes fuzzy mycorrhizae in the roots of all the plants. Destroying no part of the plants, she carefully removes the mycorrhizae. The most immediate result of this "cleaning" is that the plants cannot:
    a. carry out photosynthesis.
    b. absorb water through their roots.
    c. transport water up their stems.
    d. extract as much nitrogen from soil.
    e. store carbohydrates in their roots.

11. **Discuss Concepts** A mycologist wants to classify a specimen that appears to be a new species of fungus. To begin the classification process, what kinds of information on body structures and/or functions must the researcher obtain in order to assign the fungus to one of the major fungal groups?

12. **Discuss Concepts** In a natural setting—a pile of horse manure in a field, for example—the sequence in which various fungi appear illustrates ecological succession, which is the replacement of one species by another in a community (see Chapter 52). The first fungi to appear are the most efficient opportunists, for they can form and disperse spores most rapidly. In what order would you expect representatives from each division of fungi to appear on the manure pile? Why?

13. **Discuss Concepts** Humans are fundamentally diploid organisms. Explain how this state of affairs compares with the fungal life cycle, and then compare the two general life cycles in light of the two groups' overall reproductive strategies.

Evaluate/Create

14. **Design an Experiment** Experiments on the orange bread mold *Neurospora crassa,* an ascomycete, were pivotal in elucidating the concept that each gene encodes a single enzyme. As *N. crassa* ascospores arise through meiosis and then mitosis in an ascus, each ascospore occupies a particular position in the final string of eight spores the ascus contains:

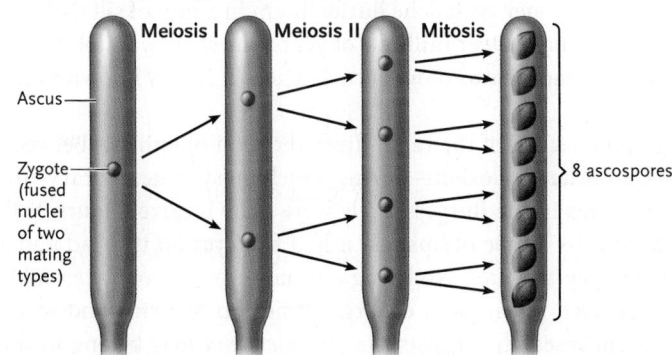

© Cengage Learning 2017

This quirk of ascospore development was extremely useful to early geneticists, because it vastly simplified the task of figuring out which alleles ended up in particular ascospores following meiosis. Recalling genetics topics discussed in Chapter 11, why was the analysis easier?

15. **Apply Evolutionary Thinking** Mycologist John Taylor of the University of California, Berkeley, suggests that a close

biochemical relationship between fungi and animals may explain why fungal infections are typically so resistant to treatment, and why it has proven rather difficult to develop drugs that kill fungi without damaging their human or other animal hosts. Hundreds of fungal genomes have been sequenced, including those of several medically important species. If you are a researcher working to develop new antifungal drugs, how could you make use of this growing genetic understanding? Using Internet resources, can you find examples of antifungal drugs that exploit biochemical differences between animals and fungi?

For selected answers, see Appendix A.

INTERPRET THE DATA

The fungus *Cryptococcus neoformans,* a basidiomycete yeast, is a dangerous pathogen in humans who have a weakened immune system, such as HIV/AIDS patients. The disease it causes, cryptococcosis, produces severe symptoms ranging from a pneumonia-like illness to encephalitis. *C. neoformans* thrives in dark places such as soil, dung, and caves, and light inhibits its sexual reproduction by preventing the development of a dikaryon stage. The recent sequencing of its genome has spurred efforts to identify genes that control this inhibition. Alexander Idnurm and Joseph Heitman of Duke University Medical Center performed a series of experiments using wild-type and mutant *C. neoformans* specimens. As described in Section 30.2, a dikaryon develops when haploid hyphae of two mating types fuse, forming the $n + n$ structure. In the wild *C. neoformans* has two mating types, denoted α and a. In both, the normal gene BWC encodes the light-based inhibition response; there is no inhibition in strains having the mutant *bwc1*. In one series of experiments, Idnurm and Heitman tested how efficiently fusion events occurred in crosses between matings of wild-type parental hyphae (denoted +) and mutant hyphae (denoted by Δ) that were exposed to different light wavelengths. Colored bars in the **Figure** indicate the color (wavelength) of the light stimulus; black denotes darkness, and red denotes dim light at the red/far-red end of the visible spectrum. Green and blue represent wavelengths toward the other end of the visible spectrum. To interpret this graph, you must read the mating combinations vertically.

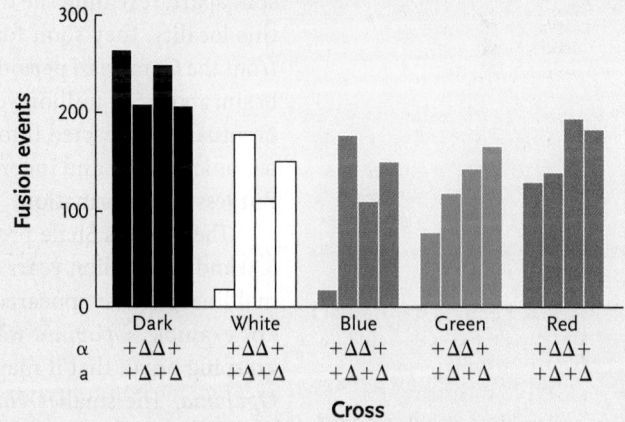

1. In the dark, did matings in which both partners were either wild-type or mutant forms produce a different response to the varying light stimuli than crosses of a wild-type and a mutant?

2. Within the other categories of light stimuli, how did the makeup of the different mating pairs correlate with the response?

Source: A. Idnurm and J. Heitman. 2005. Light controls growth and development via a conserved pathway in the fungal kingdom. *PLoS Biology* 3:e95.

31 Animal Phylogeny, Acoelomates, and Protostomes

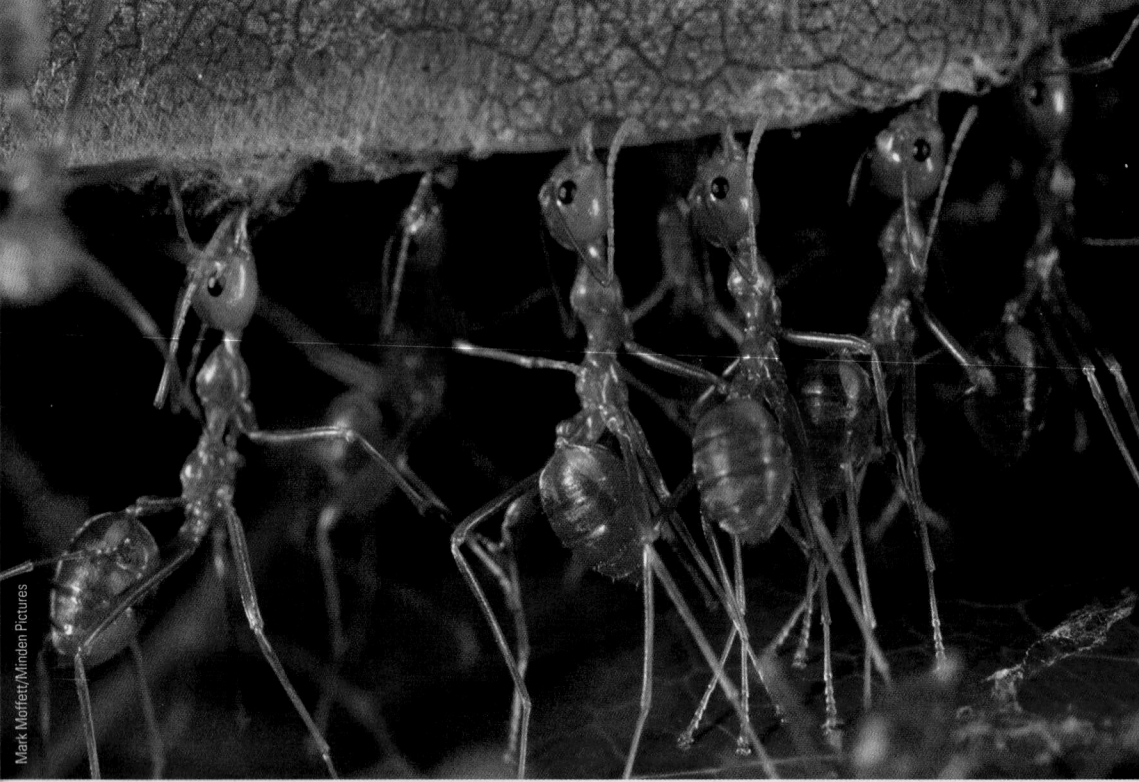

Mark Moffett/Minden Pictures

Weaver ants *(Oecophylla longinoda)* carry a leaf to repair their nest in Papua New Guinea.

Why it matters . . . In 1909, a lucky fossil hunter named Charles Walcott tripped over a rock on a mountain path in British Columbia, Canada. Under the force of his hammer, the rock split apart, revealing the discovery of a lifetime. As Walcott and other workers examined rocks at this locality, they soon found fossils of more than 120 species of previously unknown animals from the Cambrian period. These creatures had lived on the muddy sediments of a shallow ocean basin; about 510 million years ago, an underwater avalanche buried them in a rain of silt that was eventually compacted into finely stratified shale. Over millions of years, the shale was uplifted by tectonic activity and incorporated into the mountains of western Canada. It is now known as the Burgess Shale formation.

The Burgess Shale provides a graphic record of the rapid diversification of animals between 530 and 520 million years ago—the Cambrian explosion—during which most contemporary animal lineages first appeared. Some creatures in the Burgess Shale were truly bizarre **(Figure 31.1)**. For example, *Opabinia* was about as long as a tube of lipstick; it had five eyes on its head and a grasping organ that it may have used to capture prey. No living animals even remotely resemble *Opabinia*. The smaller *Hallucigenia* sported seven pairs of large spines on one side and seven pairs of soft organs on the other. Recent research suggests that *Hallucigenia* may belong in the phylum Onychophora, described in Section 31.7. Nevertheless, most species of the Burgess Shale left no descendants that are still alive today. Thus, this remarkable assemblage of fossils provides a glimpse of some evolutionary novelties that—whether through the action of natural selection or just plain bad luck—were ultimately unsuccessful.

Other animal lineages have shown much greater longevity. Zoologists have described nearly 2 million living species in the kingdom **Animalia.** The familiar **vertebrates,** animals with a backbone, encompass only a small fraction (about 65,000) of that total. The overwhelming majority of

FIGURE 31.1 Animals of the Burgess Shale. *Opabinia* had five eyes and a grasping organ on its head. *Hallucigenia* had seven pairs of spines and soft protuberances.
© Cengage Learning 2017

Opabinia

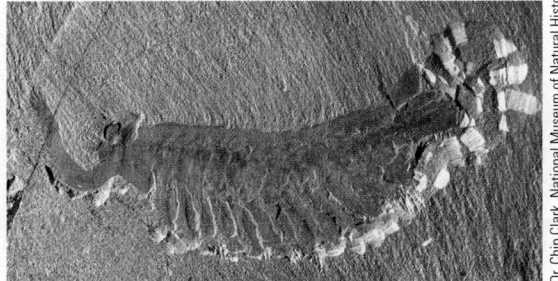

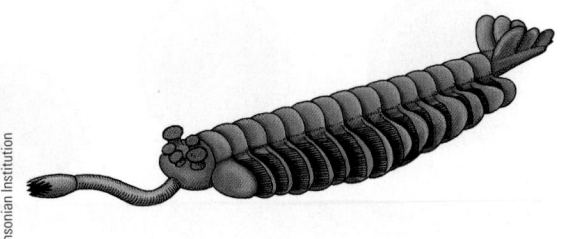

Hallucigenia

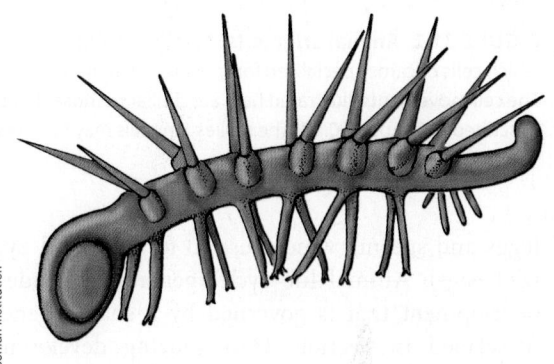

animals fall within the descriptive grouping of **invertebrates,** animals without a backbone.

The remarkable evolutionary diversification of animals resulted from their ability to consume other organisms as food and, for most groups, their ability to move from one place to another. Today, animals are important consumers in nearly every environment on Earth. Their diversification has been accompanied by the evolution of specialized tissues and organ systems, as well as complex behaviors.

In this chapter, we introduce the general characteristics of animals and a phylogenetic hypothesis about their evolutionary history and classification. We also survey some of the major invertebrate phyla; a *phylum* is an ancient monophyletic lineage with a distinctive body plan. In Chapter 32 we examine the *deuterostome* clade, which includes the vertebrates and their nearest invertebrate relatives.

31.1 What Is an Animal?

Biologists recognize the Kingdom Animalia as a monophyletic group that is easily distinguished from the other kingdoms.

All Animals Share Certain Structural and Behavioral Characteristics

Animals are eukaryotic, multicellular organisms. Their cells lack cell walls, a trait that differentiates them from plants and fungi. Because the individual cells of most animals are similar

in size, very large animals, such as elephants, have many more cells than small ones, such as fleas. In large animals, most cells are far from the body surface, but specialized tissues and organ systems deliver nutrients and oxygen to them and carry wastes away.

All animals are **heterotrophs:** they eat other organisms to acquire energy and nutrients. Food is ingested (eaten), then digested (broken down), and its breakdown products are absorbed by specialized tissues. Animals use oxygen to metabolize the food they eat through the biochemical pathways of aerobic respiration, and most store excess energy as glycogen, oil, or fat.

All animals are **motile**—able to move from place to place—at some stage of their life cycles. They travel through the environment to find food or shelter and to interact with other animals. Most familiar animals are motile as adults. However, in some species, such as mussels and barnacles, only the larval stages are motile; they eventually settle down as **sessile** (unable to move from one place to another) adults. The advantages of motility have fostered the evolution of locomotor structures, including fins, legs, and wings that are powered by muscles, which are specialized contractile tissues that move individual body parts. Most animals also have sensory and nervous systems that allow them to receive, process, and respond to information about the environment through which they are traveling.

Animals reproduce either asexually or sexually; in some groups they switch from one mode to the other. Sexually reproducing species produce short-lived, haploid **gametes**

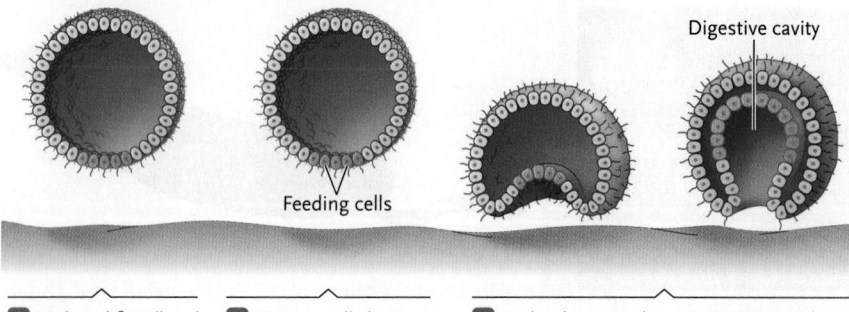

Digestive cavity

Feeding cells

1 Colonial flagellated ancestor with unspecialized cells

2 Certain cells became specialized for feeding and other functions.

3 A developmental reorganization produced a two-layered animal with a sac-within-a-sac body plan.

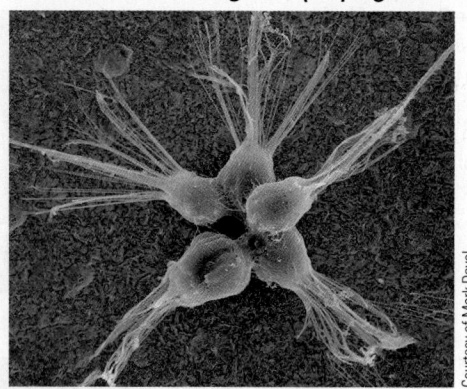

Courtesy of Mark Dayel

FIGURE 31.2 Animal origins. (A) Many biologists agree that animals arose from a colonial, flagellated protist in which cells became specialized for specific functions and a developmental reorganization produced two cell layers. The cell movements illustrated here are similar to those that occur during the early development of many animals, as described in Chapter 50. **(B)** The earliest animals may have resembled living colonial choanoflagellates.

© Cengage Learning 2017

(eggs and sperm), which fuse to form diploid **zygotes** (fertilized eggs). Animal life cycles generally include a period of development that is governed by a shared genetic "toolkit" (described in Section 23.6). During development, mitosis transforms the zygote into a multicelled **embryo;** an embryo develops into a reproductively immature juvenile or a free-living **larva,** which becomes a reproductively mature adult. Larvae often differ markedly from adults and may occupy different habitats and consume different foods.

Animals Probably Arose from a Colonial Choanoflagellate Ancestor

An overwhelming body of morphological and molecular evidence indicates that all animal phyla had a common ancestor. For example, all animals share similarities in the genes that guide their development, their cell-to-cell junctions, and the molecules in their extracellular matrices (see Section 5.5), as well as similarities in the structure of their ribosomal RNAs.

In 1874 the German embryologist Ernst Haeckel proposed that the common ancestor of all animals was a colonial, hollow, ball-shaped organism with unspecialized flagellated cells. According to his hypothesis, its cells became specialized for particular functions, and a developmental reorganization produced a double-layered, sac-within-a-sac body plan **(Figure 31.2A).** As you will see in Chapter 50, the embryonic development of many living animals roughly parallels this hypothetical evolutionary transformation.

Most biologists now agree that the common ancestor of all animals was probably a colonial, flagellated protist that lived at least 700 million years ago, during the Precambrian era. It may have resembled the minute, sessile choanoflagellates that live in both freshwater and marine habitats today **(Figure 31.2B;** see also Figure 27.21). Several lines of morphological and molecular evidence support this hypothesis. First, living choanoflagellates are almost identical in structure to the "collar cells" of sponges, the most ancient living clade of animals; similar cells have also been discovered in other distantly related animal phyla, such as flatworms and echinoderms. Second, DNA sequence data suggest that choanoflagellates and animals are sister clades. Moreover, researchers have identified several genes and proteins for cell adhesion and cell signaling that are present in both choanoflagellates and animals.

STUDY BREAK 31.1

1. What characteristics distinguish animals from plants and fungi?
2. How does the ability of animals to move through the environment relate to their acquisition of nutrients and energy?

31.2 Key Innovations in Animal Evolution

Once established, the animal lineage diversified quickly into an amazing array of body plans. Before the development of molecular sequencing techniques, biologists identified several key morphological innovations that they used to develop hypotheses about the evolutionary relationships of the major animal groups.

Tissues and Tissue Layers Appeared Early in Animal Evolution

The presence or absence of **tissues**—groups of cells that share a common structure and function—divides the animal kingdom into two distinct branches. One branch, the sponges, or **Parazoa** (*para* = alongside; *zoon* = animal), lacks tissues. All other animals, collectively grouped in the **Eumetazoa** (*eu* = true; *meta* = in common with), have tissues.

During the development of eumetazoans, embryonic tissues form as either two or three concentric **primary cell layers** (described further in Chapter 50). The innermost layer, the endoderm (generally illustrated in yellow), eventually develops into the lining of the gut (digestive system) and, in some animals, respiratory organs. The outermost layer, the **ectoderm** (generally illustrated in blue), forms the external body surface and nervous system. Between the two, the **mesoderm** (generally illustrated in red) forms the muscles of the body wall and most other structures between the gut and the body surface. Some animals have a **diploblastic** body plan that includes only two layers, endoderm and ectoderm. However, most animals are **triploblastic,** having all three primary cell layers.

Most Animals Exhibit Either Radial or Bilateral Symmetry

The most obvious feature of an animal's body plan is its shape **(Figure 31.3).** Most animals are **symmetrical;** in other words, their bodies can be divided by a plane into mirror-image halves. The rare exceptions include most sponges, which have irregular shapes and are therefore **asymmetrical.**

Eumetazoans exhibit one of two body symmetry patterns. The **Radiata** includes two phyla, Cnidaria (hydras, jellyfishes, and sea anemones) and Ctenophora (comb jellies), which have **radial symmetry.** Their body parts are arranged regularly around a central axis, like the spokes on a wheel. Thus, any cut down the long axis of a hydra divides it into matching halves. Radially symmetrical animals are usually sessile or slow moving, and their netlike nervous systems receive sensory input from all directions.

All other eumetazoan phyla fall within the **Bilateria,** animals that have **bilateral symmetry.** In other words, only a cut along the midline from head to tail divides them into left and right sides that are essentially mirror images of each other. Bilaterally symmetrical animals also have **anterior** ("front") and **posterior** ("back") ends, as well as **dorsal** ("upper") and **ventral** ("lower") surfaces. As these animals move through the environment, the anterior end encounters food, shelter, or enemies first. Thus, in bilaterally symmetrical animals, natural selection also favored **cephalization,** the development of an anterior head where sensory organs and nervous-system tissue are concentrated.

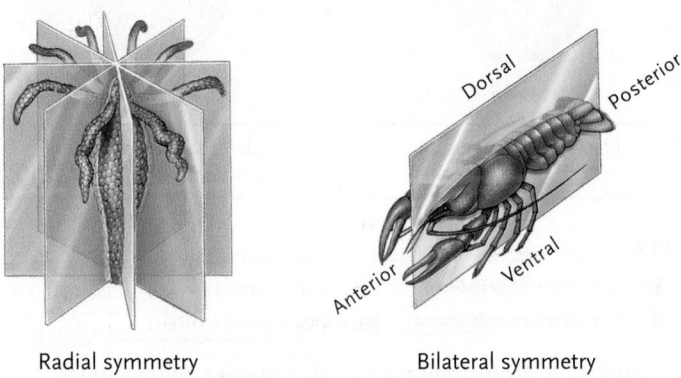

Radial symmetry   Bilateral symmetry

FIGURE 31.3 Patterns of body symmetry. Most animals have either radial or bilateral symmetry.

© Cengage Learning 2017

Many Animals Have Body Cavities That Surround Their Internal Organs

The body plans of many bilaterally symmetrical animals include a body cavity that separates the gut from the muscles of the body wall **(Figure 31.4). Acoelomate** animals (*a* = not; *koilomat* = hollow), such as flatworms (phylum Platyhelminthes), lack such

A. In acoelomate animals, no body cavity separates the gut and body wall.

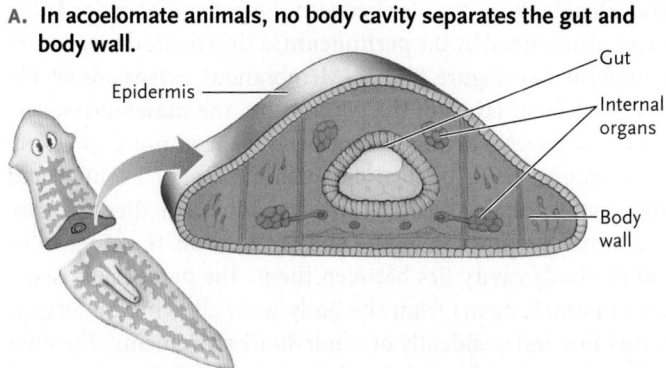

B. In pseudocoelomate animals, the pseudocoelom forms between the gut (a derivative of endoderm) and the body wall (a derivative of mesoderm).

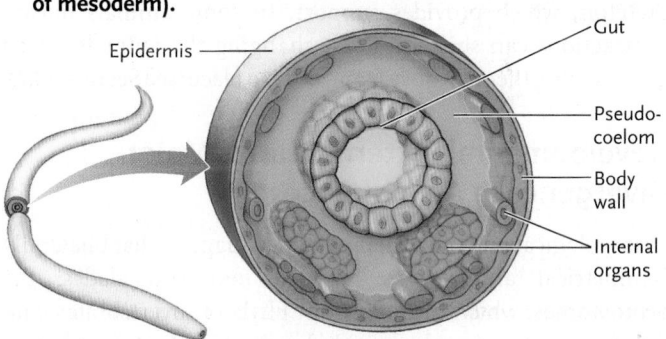

C. In coelomate animals, the coelom is completely lined by peritoneum (a derivative of mesoderm).

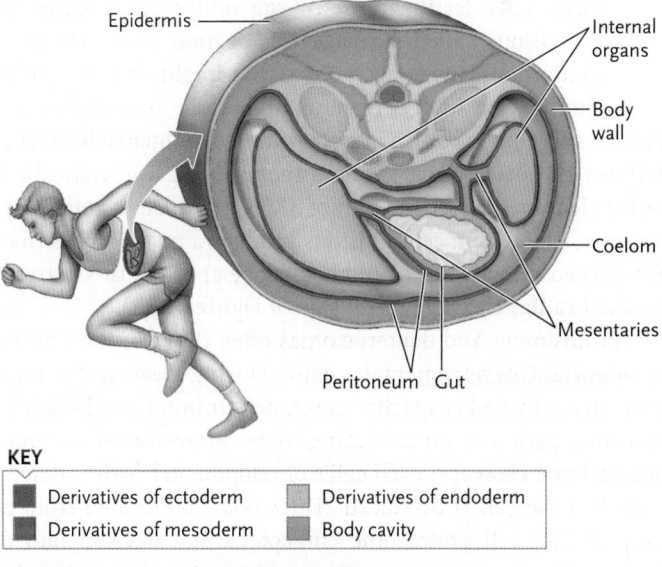

KEY

■ Derivatives of ectoderm	■ Derivatives of endoderm
■ Derivatives of mesoderm	■ Body cavity

FIGURE 31.4 Body plans of triploblastic animals.

© Cengage Learning 2017

a cavity; a continuous mass of tissue, derived largely from mesoderm, packs the region between the gut and the body wall (see Figure 31.4A). **Pseudocoelomate** animals (*pseudo* = false), including the roundworms (phylum Nematoda) and wheel animals (phylum Rotifera), have a **pseudocoelom,** a fluid- or organ-filled space between the gut and the muscles of the body wall, that is partially lined by tissues derived from mesoderm (see Figure 31.4B). Internal organs lie within the pseudocoelom and are bathed by its fluid. **Coelomate** animals have a **coelom,** a fluid-filled body cavity, also between the gut and body wall, that is completely lined by the **peritoneum,** a thin tissue derived from mesoderm (see Figure 31.4C). Membranous extensions of the inner and outer layers of the peritoneum, the **mesenteries,** surround the internal organs and suspend them within the coelom.

Biologists describe the body plan of pseudocoelomate and coelomate animals as a "tube within a tube"; the digestive system forms the inner tube, the body wall forms the outer tube, and the body cavity lies between them. The body cavity separates internal organs from the body wall, allowing the organs to function independently of whole-body movements. The fluid within the cavity also protects delicate organs from mechanical damage. And, because the volume of the body cavity is fixed, the incompressible fluid within it serves as a **hydrostatic skeleton,** which provides support. In some animals muscle contractions can shift the fluid, changing the animals' shape and allowing them to move from place to place (see Section 43.2).

Developmental Patterns Mark a Major Divergence in Animal Ancestry

Embryological and molecular evidence suggests that bilaterally symmetrical animals are divided into two clades: the **protostomes,** which includes most phyla of invertebrates, and the **deuterostomes,** which includes the vertebrates and their nearest invertebrate relatives. Protostomes and deuterostomes differ in several developmental characteristics **(Figure 31.5).**

Shortly after fertilization, an egg undergoes a series of mitotic divisions called **cleavage** (see Section 50.2). The first two cell divisions divide a zygote as you might slice an apple, cutting it into four wedges from top to bottom. In many protostomes, subsequent cell divisions produce daughter cells that lie *between* the pairs of cells below them; this pattern is called **spiral cleavage** (left side of Figure 31.5A). In deuterostomes, by contrast, subsequent cell divisions produce a mass of cells that are stacked directly above and below one another; this pattern is called **radial cleavage** (right side of Figure 31.5A).

Protostomes and deuterostomes often differ in the timing of important developmental events. During cleavage, certain genes are activated at specific times, determining a cell's developmental path and ultimate fate. Many protostomes undergo **determinate cleavage:** each cell's developmental path is determined as the cell is produced. Thus, one cell isolated from a two- or four-cell protostome embryo cannot develop into a functional embryo or larva. By contrast, many deuterostomes have **indeterminate cleavage:** the developmental fates of cells

Protostomes	Deuterostomes

A. Cleavage

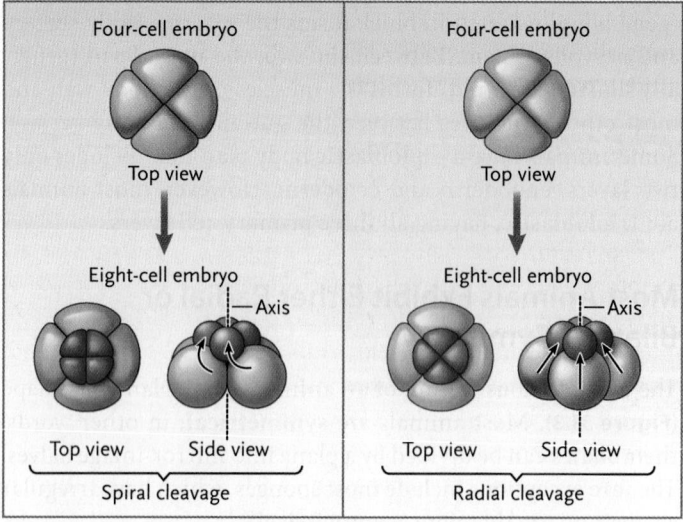

B. Mesoderm and coelom formation

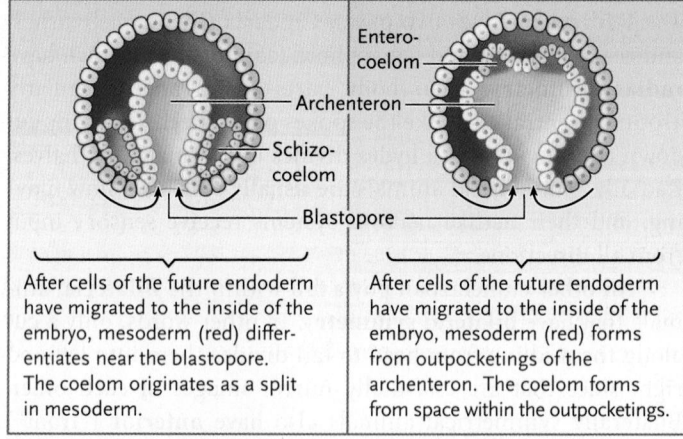

After cells of the future endoderm have migrated to the inside of the embryo, mesoderm (red) differentiates near the blastopore. The coelom originates as a split in mesoderm.

After cells of the future endoderm have migrated to the inside of the embryo, mesoderm (red) forms from outpocketings of the archenteron. The coelom forms from space within the outpocketings.

C. Origin of mouth and anus

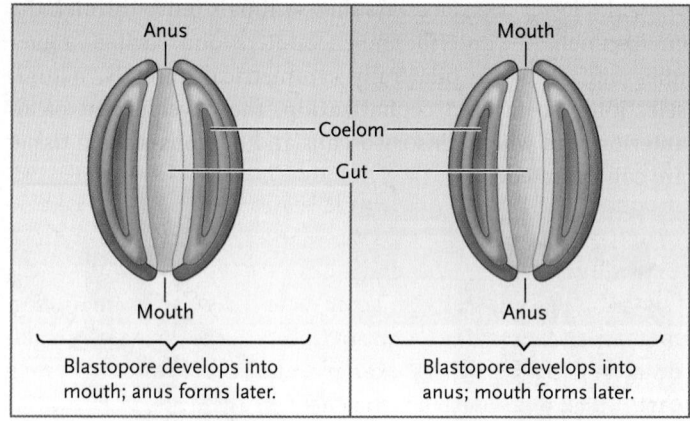

Blastopore develops into mouth; anus forms later.

Blastopore develops into anus; mouth forms later.

KEY

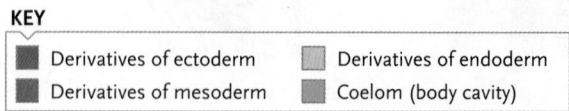

■ Derivatives of ectoderm
■ Derivatives of mesoderm
■ Derivatives of endoderm
■ Coelom (body cavity)

FIGURE 31.5 Protostomes and deuterostomes. The two clades of coelomate animals differ in **(A)** cleavage patterns, **(B)** the origin of mesoderm and the coelom, and **(C)** the origin of the mouth and the anus.
© Cengage Learning 2017

are determined after additional cell divisions. A cell isolated from a four-cell deuterostome embryo will develop into a functional, although smaller than usual, embryo or larva. Like other deuterostomes, humans have indeterminate cleavage; thus, the two cells produced by the first cleavage division sometime separate and develop into identical twins.

As development proceeds after cleavage, some cells migrate to the inside of the embryo through the **blastopore,** an opening on its surface. These cells become the endoderm, which forms the **archenteron,** the developing gut (see Figure 31.5B). Later in development, a second opening at the opposite end of the embryo transforms the pouchlike archenteron into a digestive tube (see Figure 31.5C). In protostomes (*protos* = first; *stoma* = mouth), the blastopore develops into the mouth, and the second opening forms the anus. In some deuterostomes (*deuteros* = second), the blastopore develops into the anus, and the second opening becomes the mouth.

Protostomes and deuterostomes also differ in the origin of mesoderm and the coelom (see Figure 31.5B). In most protostomes, mesoderm originates from a few cells adjacent to the blastopore. As the mesoderm grows and develops, it splits into inner and outer layers. The space between the layers is called a **schizocoelom** (*skhizein* = split). In deuterostomes, mesoderm often forms from outpocketings of the archenteron. The space pinched off from the archenteron by the outpocketing mesoderm is called an **enterocoelom** (*enteron* = intestine).

Several other characteristics also differ in adult protostomes and deuterostomes. For example, the central nervous system of protostomes is generally positioned on the ventral side of the body, and their brain surrounds the anterior section of the digestive tract. By contrast, the nervous system and brain of deuterostomes lie on the dorsal side of the body.

Segmentation Divides the Bodies of Some Animals into Repeating Units

Some phyla in both the protostome and deuterostome clades exhibit varying degrees of **segmentation,** the production of body parts as repeating units. During development, segmentation first arises in the mesoderm, the middle tissue layer that produces most of the body's bulk. In humans and other vertebrates, we see evidence of segmentation in the vertebral column ("backbone"), ribs, and associated muscles, such as the "six-pack abs" that sit-ups accentuate. In some animals, segmentation is also reflected in structures derived from the endoderm and ectoderm. For example, the ringlike pattern on an earthworm or a caterpillar matches the underlying segments.

Segmentation provides several advantages. In markedly segmented animals, such as earthworms and their relatives, each segment may include a complete set of important organs, including respiratory surfaces and parts of the nervous, circulatory, and excretory systems. Thus, a segmented animal may survive damage to the organs in one segment, because those in other segments perform the same functions. Segmentation also improves control over movement, especially in wormlike

animals. Each segment has its own set of muscles, which can act independently of those in other segments. Thus, an earthworm can move its anterior end to the left while it swings its posterior end to the right. Similarly, the segmented backbone and body wall musculature of vertebrates allow greater flexibility of movement than would unsegmented structures.

STUDY BREAK 31.2

1. What is a tissue, and what three primary tissue layers are present in the embryos of most animals?
2. What type of body symmetry do humans have?
3. What is the functional significance of the coelom?
4. What does having a segmented body allow some animals to do?

31.3 An Overview of Animal Phylogeny and Classification

For many years, biologists used the morphological innovations and embryological patterns described earlier to explore the phylogenetic history of animals. These efforts were sometimes hampered by the difficulty of identifying homologous structures in different phyla and by morphological data that led to contradictory interpretations. More recently, researchers have used molecular sequence data to reanalyze animal relationships. Although biologists now recognize nearly 40 animal phyla, in this chapter we focus on 14 phyla, many of which include substantial numbers of species, that represent a diversity of body forms and ways of life **(Table 31.1).** In Chapter 32, we consider three deuterostome phyla, which represent a distinct clade within the animals.

Molecular Phylogenetics Has Refined Our Understanding of Animal Phylogeny

Molecular phylogenies of animal relationships were initially based on the nucleotide sequences in small subunit ribosomal RNA and mitochondrial DNA (see Chapter 15). However, recent research on homeobox gene sequences (described in Sections 16.4 and 23.6) as well as genome-wide studies have largely supported the conclusions of the early molecular analyses. The phylogenetic tree based on molecular and genomic analyses **(Figure 31.6)** includes all of the *major* clades that biologists had defined using morphological innovations and embryological characters. For example, molecular data confirm the distinctions between the Parazoa and the Eumetazoa and between the Radiata and the Bilateria. They also confirm the separation of the deuterostome phyla from all others within the Bilateria.

However, the molecular phylogeny groups many other phyla—including the acoelomate animals, pseudocoelomate animals, protostomes, and a few others—into one taxon, the Protostomia. This group is, in turn, subdivided into two major clades, the Lophotrochozoa and the Ecdysozoa, which were not previously recognized. The name **Lophotrochozoa** (*lophos* = crest; *trochos* = wheel) refers to both the lophophore, a feeding structure found in three phyla (illustrated in Figure 31.15), and the trochophore, a type of larva found in annelids and mollusks (illustrated in Figure 31.22). The name **Ecdysozoa** (*ekdysis* = a stripping or casting off) refers to the cuticle or external skeleton that species within this group secrete and periodically molt (or "cast off") when they experience a growth spurt or begin a different stage of the life cycle (illustrated in Figure 31.33); the molting process is called **ecdysis**.

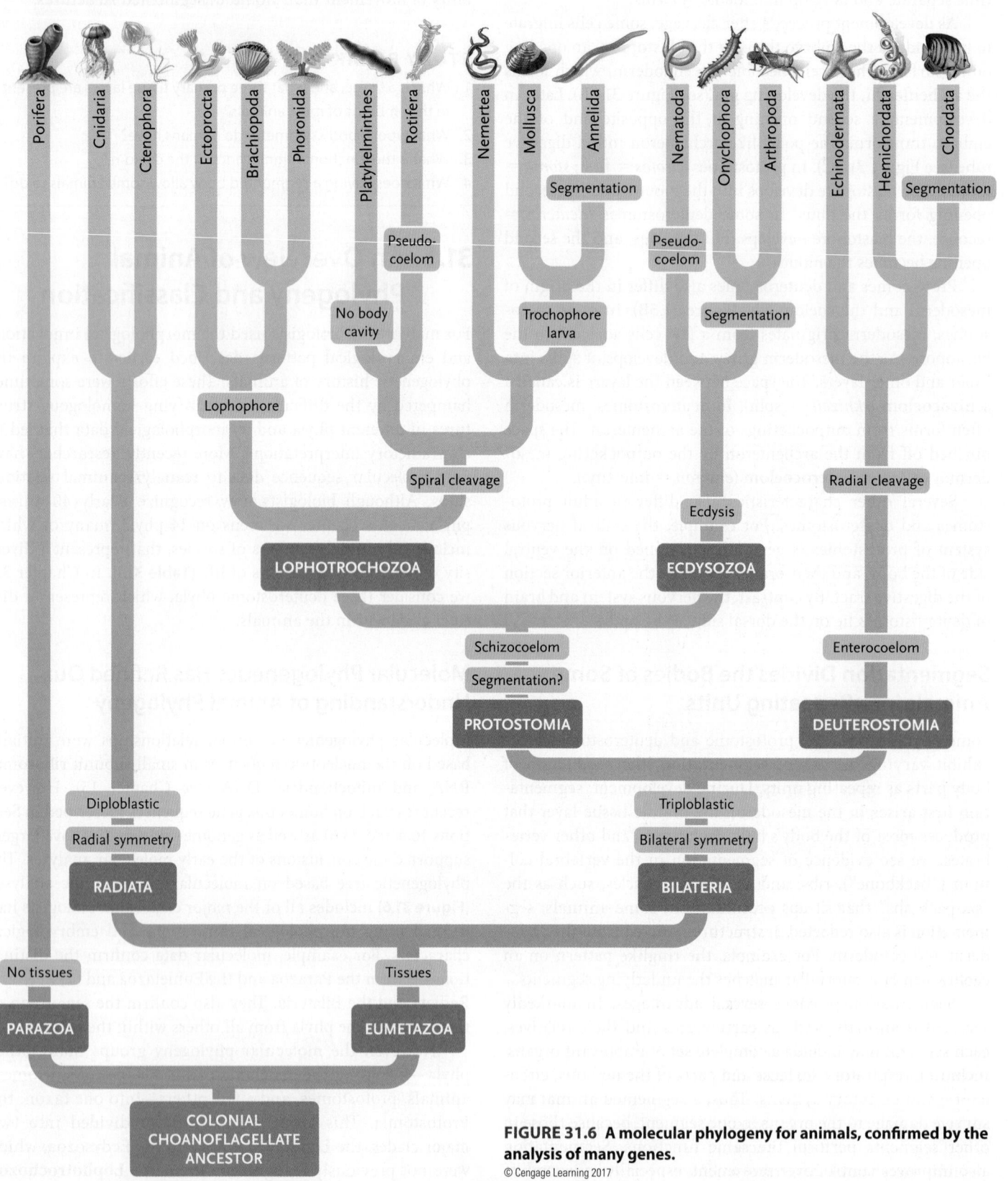

FIGURE 31.6 A molecular phylogeny for animals, confirmed by the analysis of many genes.

© Cengage Learning 2017

TABLE 31.1 | Classification of the Animal Phyla Described in This Book

Major Clade	Phylum	Known Diversity (Approximate Number of Living Species Described)
Parazoa	Porifera	15,000
Eumetazoa		
Radiata	Cnidaria	9,000
	Ctenophora	150
Bilateria		
Protostomia		
Lophotrochozoa	Ectoprocta	5,000
	Brachiopoda	300
	Phoronida	12
	Platyheminthes	20,000
	Rotifera	1,850
	Nemertea	900
	Mollusca	100,000–200,000
	Annelida	9,000
Ecdysozoa	Nematoda	15,000
	Onychophora	180
	Arthropoda	1,150,000
Deuterostomia	Echinodermata	7,000
	Hemichordata	120
	Chordata	65,000

The Molecular Phylogeny Reveals Some Surprising Patterns in the Evolution of Key Morphological Innovations

Phylogenetic trees contain explicit hypotheses about evolutionary change, and the molecular phylogeny has forced biologists to reevaluate the evolution of several important morphological innovations. For example, traditional phylogenies based on morphology and embryology usually inferred that the absence of a body cavity, the acoelomate condition, was an ancestral state and that the presence of a body cavity, the pseudocoelomate or coelomate condition, was derived. But the molecular tree provides a very different view. Because most protostome phyla have a schizocoelom, the molecular phylogeny suggests that this trait evolved in the common ancestor of all protostomes. If that hypothesis is correct, then the acoelomate condition of flatworms (phylum Platyhelminthes) represents the evolutionary *loss* of

the schizocoelom, *not* an ancestral condition. Similarly, the molecular phylogeny hypothesizes that the pseudocoelom evolved independently in rotifers (Lophotrochozoa, phylum Rotifera) and in roundworms (Ecdysozoa, phylum Nematoda). Thus, according to the molecular phylogeny, the pseudocoelomate condition of these organisms is the product of convergent evolution.

Phylogenies based on embryological and morphological characters also suggested that the segmented body plan of several protostome phyla was inherited from a segmented common ancestor and that segmentation arose independently in chordates by convergent evolution. The molecular phylogeny in Figure 31.6, by contrast, suggests that segmentation evolved independently in *three* clades—segmented worms (Lophotrochozoa, phylum Annelida), arthropods and velvet worms (Ecdysozoa, phyla Arthropoda and Onychophora), and chordates (Deuterostomia, phylum Chordata)—rather than in just two. However, in 2010, Nicolas Dray of the Centre Génétique Moléculaire du CNRS, and colleagues from other European research institutes, demonstrated that the development of segmentation in one species of annelid requires expression of the Hedgehog signaling pathway, which also plays a prominent role in the development of segmentation in arthropods. Thus, the genetic basis of segmentation may have originated in the common ancestor of annelids and arthropods. If this conclusion is correct, it is consistent with the traditional view that segmentation was present in the common ancestor of all protostomes. Further research should confirm or refute this hypothesis and reveal more about the evolutionary history of this key morphological innovation in animals.

STUDY BREAK 31.3

1. Which major groupings of animals defined on the basis of morphological characters have been confirmed by molecular sequence studies?
2. What type of body cavity is ancestral within the Deuterostomes?

31.4 Animals without Tissues: Parazoa

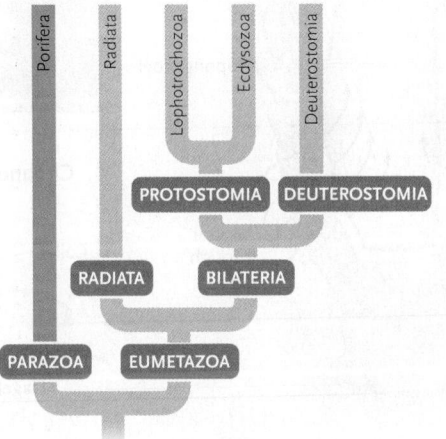

As illustrated in Figure 31.6, Parazoa is the basal animal clade, the sister group to Eumetazoa. It includes the sponges, phylum Porifera (meaning "pore bearers").

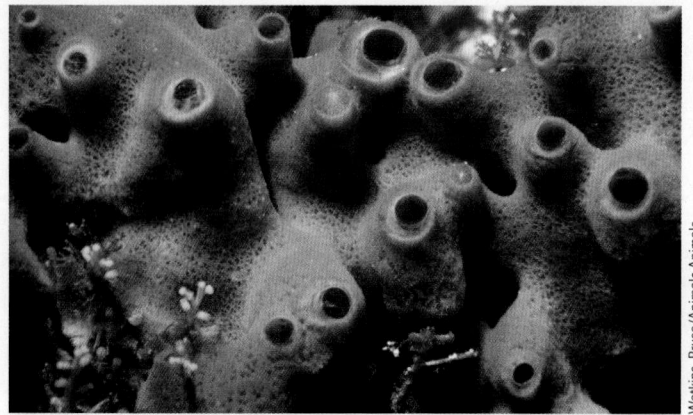

FIGURE 31.7 **Asymmetry in sponges.** Sponges, such as this *Haliclona* species from the Philippines, often have an asymmetrical shape.

Watkins, Bruce/Animals Animals

Sponges lack true tissues: during development, their cells do not form the complex layers typical of other phyla. Mature sponges are sessile, and their shapes are less fixed than those of other animals, because mobile cells allow them to change shape in response to local conditions **(Figure 31.7)**. Sponges have been abundant since the Cambrian, especially in shallow coastal areas. Most of the 15,000 living species are marine and range in size from 1 cm to 2 m at maturity.

Most sponges have simple body plans **(Figure 31.8)**. Flattened cells form an outer layer, the **pinacoderm.** The inner surface of saclike sponges is lined by collar cells, **choanocytes,** each equipped with a beating flagellum and a surrounding

"collar" of modified microvilli. Amoeboid cells wander through the gelatinous **mesohyl** between the two layers; they secrete a supporting skeleton of a fibrous protein and *spicules*, small needlelike structures of calcium carbonate or silica (see Figure 31.8D). The natural sponges that are used for bathing and washing are the fibrous remains of the bath sponge (*Spongia* species), which lacks mineralized parts.

The bodies of most sponges are elaborate filtering systems that capture food particles from the surrounding water. Water flows through pores in the pinacoderm into a central chamber, the **spongocoel,** and then out of the sponge through one or more openings called **oscula** (singular, *osculum*). The beating flagellae of the choanocytes maintain a constant flow of water, and contractile pore cells *(porocytes)* adjust the flow rate. Even a small sponge may filter as much as 20 liters of water per day. The choanocytes capture suspended particles and microorganisms from the water and pass this food to mobile amoeboid cells, which carry nutrients to cells of the pinacoderm. However, not all sponges are filter-feeders. In a surprising discovery in 2012, scientists at the Monterey Bay Research Institute found a predatory sponge in deep water off the California coast. This species, which is shaped like a harp, traps small animals with hooklike spicules on its many vertical branches.

Most sponges are **hermaphroditic:** individuals produce both sperm and eggs. Sperm are released into the spongocoel and then out into the environment through oscula; eggs remain in the mesohyl, where sperm from other sponges, drawn in with water, fertilize them. Zygotes develop into flagellated larvae that are

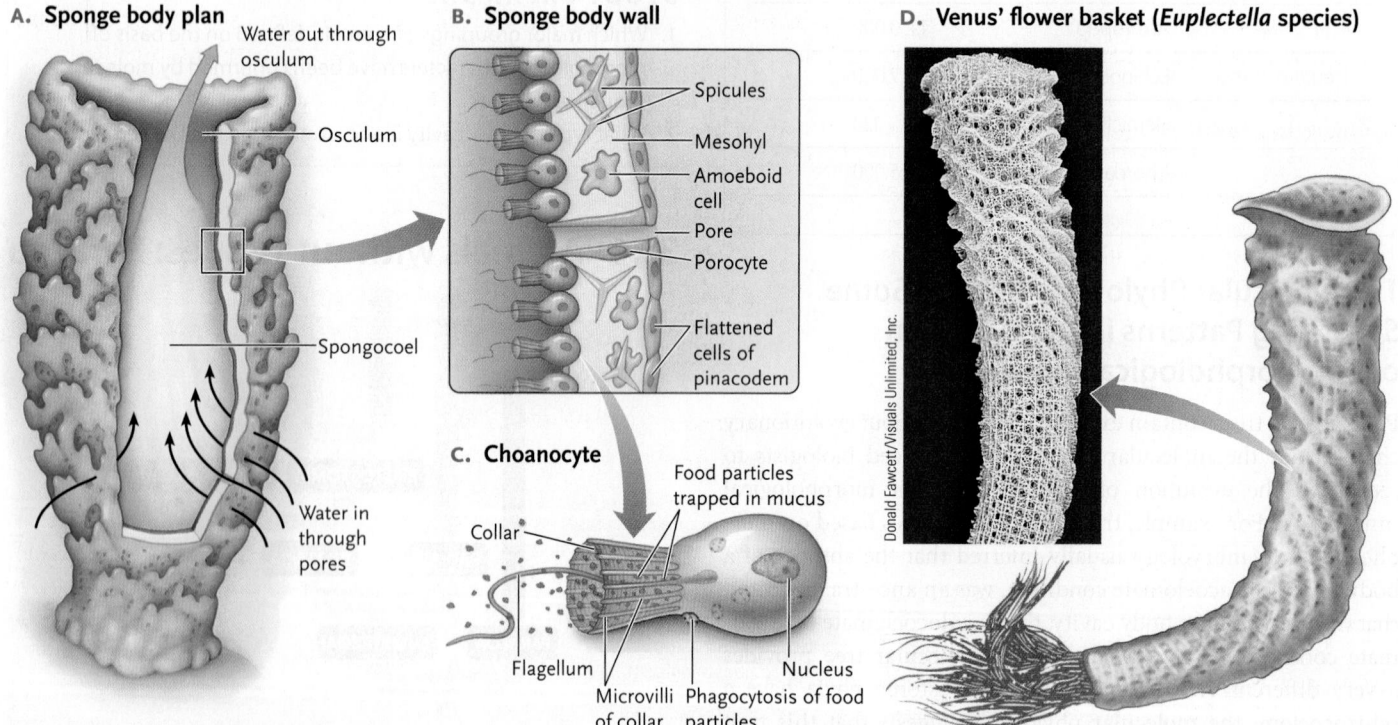

A. **Sponge body plan**

Water out through osculum

Osculum

Spongocoel

Water in through pores

B. **Sponge body wall**

Spicules

Mesohyl

Amoeboid cell

Pore

Porocyte

Flattened cells of pinacodem

C. **Choanocyte**

Food particles trapped in mucus

Collar

Flagellum

Microvilli of collar

Phagocytosis of food particles

Nucleus

D. **Venus' flower basket (*Euplectella* species)**

Donald Fawcett/Visuals Unlimited, Inc.

FIGURE 31.8 **The body plan of sponges.** Most sponges have **(A)** simple body plans and **(B)** relatively few cell types. **(C)** Beating flagella on the choanocytes create a flow of water through incurrent pores, into the spongocoel, and out through the osculum. **(D)** Venus' flower basket, a marine sponge, has spicules of silica fused into a rigid framework.

© Cengage Learning 2017

expelled to fend for themselves. Surviving larvae attach to substrates and undergo **metamorphosis** (a reorganization of form) into sessile adults. Some sponges also reproduce asexually; small fragments break off an adult and grow into new sponges. Many species also produce *gemmules,* clusters of cells with a resistant covering that allows them to survive unfavorable conditions; gemmules germinate into new sponges when conditions improve.

STUDY BREAK 31.4

1. What type of body symmetry do sponges exhibit?
2. How does a sponge gather food from its environment?

31.5 Eumetazoans with Radial Symmetry

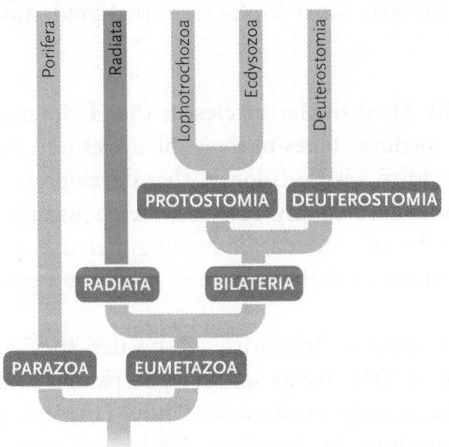

Unlike sponges, eumetazoans have true tissues, which develop from distinct layers in the embryo. Working together, the cells of a tissue perform complex functions beyond the capacity of individual cells. For example, nerve tissue transmits information rapidly through an animal's body, and epithelial tissue forms barriers that surround the body and line body cavities. In this section, we describe eumetazoans with radial symmetry. This symmetry, coupled with a netlike nervous system, enables them to sense stimuli from all directions, an effective adaptation for life in open water.

Two phyla of soft-bodied organisms, Cnidaria and Ctenophora, have radial symmetry and tissues, but they lack organ systems and a coelom. Species in both phyla possess a **gastrovascular cavity** that serves both digestive and circulatory functions. It has a single opening, the mouth. Gas exchange and excretion occur by diffusion because no cell is far from a body surface.

The radiate phyla have a diploblastic body plan with only two tissue layers, the inner *gastrodermis* (an endoderm derivative) and the outer *epidermis* (an ectoderm derivative). Most species also possess a gelatinous *mesoglea* (*mesos* = middle; *glia* = glue) between the two layers. The mesoglea contains widely dispersed amoeboid cells.

Cnidarians Use Nematocysts to Stun or Kill Prey

Nearly all of the 9,000 species in the phylum Cnidaria (*knide* = stinging nettle, a plant with irritating surface hairs) live in the sea. Their body plan is organized around a saclike gastrovascular cavity; the mouth is ringed with tentacles, which push food into it **(Figure 31.9)**. Cnidarians may be vase-shaped, upward-pointing **polyps** or bell-shaped, downward-pointing **medusae** (see Figure 31.9A). Most polyps attach to a substrate at the

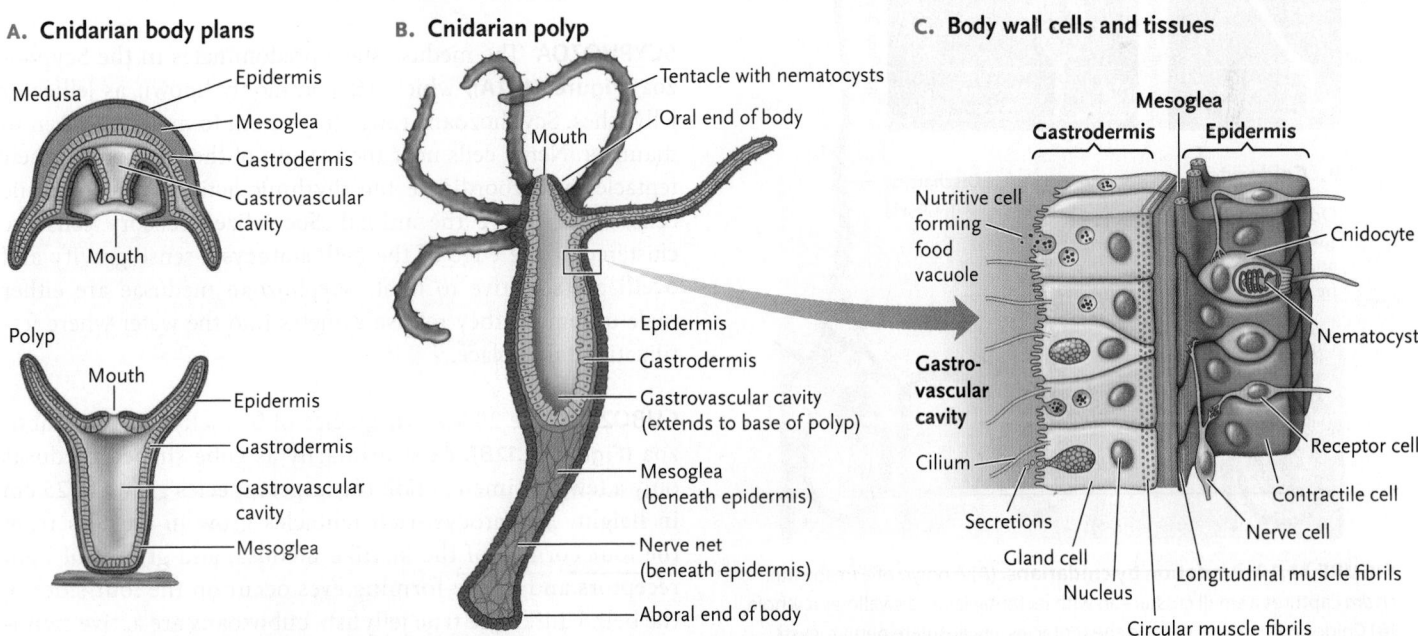

FIGURE 31.9 The cnidarian body plan. (A) Cnidarians exist as either polyps or medusae. **(B)** The body of both forms is organized around a gastrovascular cavity, which extends all the way to the aboral end of the animal. **(C)** The two tissue layers in the body wall, the gastrodermis and the epidermis, include a variety of cell types.

© Cengage Learning 2017

aboral (opposite the mouth) end; medusae are unattached and float.

Cnidarians are the simplest animals that exhibit a division of labor among their specialized tissues (see Figure 31.9B, C). (Sponges have specialized cells, but no tissues.) The gastrodermis includes gland cells and phagocytic nutritive cells, both of which contain circular muscle fibrils. Gland cells secrete enzymes into the gastrovascular cavity for the extracellular digestion of food, which is then engulfed by nutritive cells and exposed to intracellular digestion. The epidermis includes nerve cells, sensory cells, contractile cells, and cells specialized for prey capture. A layer of acellular mesoglea separates the gastrodermis from the epidermis.

Cnidarians prey on crustaceans, fishes, and other animals. The epidermis includes unique cells, **cnidocytes,** each armed with a stinging **nematocyst (Figure 31.10).** The nematocyst is an encapsulated, coiled thread that is fired at prey or predators, sometimes releasing a toxin through its tip. Discharge of nematocysts may be triggered by touch, vibrations, or chemical stimuli. The toxin can paralyze small prey by disrupting nerve cell membranes. The painful stings of some jellyfishes and corals result from the discharge of nematocysts.

Cnidarians engage in directed movements by contracting muscle fibrils in cells of the epidermis and gastrodermis. In medusae, the mesogleal jelly serves as a deformable skeleton against which contractile cells act. Rapid contractions narrow the bell, forcing out jets of water that propel the animal. Polyps use their water-filled gastrovascular cavity as a hydrostatic skeleton. When some cells contract, fluid within the chamber is shunted about, changing the body's shape and moving it in a particular direction.

The **nerve net,** which threads through both tissue layers, coordinates responses to stimuli, but it has no central control organ or brain. Impulses initiated by sensory cells are transmitted in all directions from the site of stimulation.

Many cnidarians exist in only the polyp or the medusa form, but some have a life cycle that alternates between them **(Figure 31.11).** In the latter type, the polyp often produces new individuals asexually from buds that break free of the parent (see Figure 49.2). The medusa is often the sexual stage, producing sperm and eggs, which are released into the water. The four clades of Cnidaria differ in the form that predominates in the life cycle.

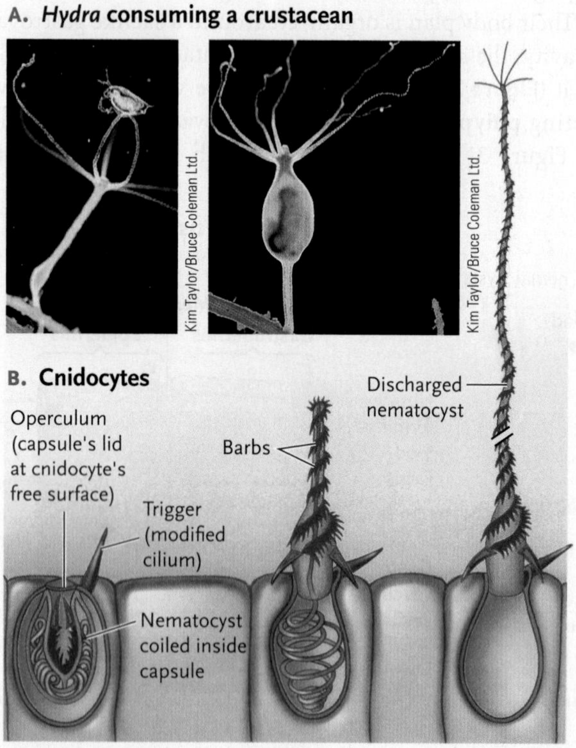

HYDROZOA Most of the species in the Hydrozoa have both polyp and medusa stages in their life cycles (see Figure 31.11). The polyps form sessile colonies that develop asexually from one individual. A colony can include thousands of polyps, which may be specialized for feeding, defense, or reproduction. They share food through their connected gastrovascular cavities.

Unlike most hydrozoans, freshwater species of *Hydra* (see Figure 31.10A) live as solitary polyps that attach temporarily to rocks, twigs, and leaves. Under favorable conditions, hydras reproduce by budding. Under adverse conditions, they produce eggs and sperm; the zygotes, which are encapsulated in a protective coating, develop and grow when conditions improve.

SCYPHOZOA The medusa stage predominates in the Scyphozoa **(Figure 31.12A),** which are commonly known as jellies or jellyfishes. Scyphozoans range from 2 cm to more than 2 m in diameter. Nerve cells near the margin of the bell control their tentacles and coordinate the rhythmic activity of contractile cells, which move the animal. Specialized sensory cells are clustered at the edge of the bell: **statocysts** sense gravity and **ocelli** are sensitive to light. Scyphozoan medusae are either male or female; they release gametes into the water where fertilization takes place.

CUBOZOA The 20 known species of box jellyfish, the Cubozoa **(Figure 31.12B),** exist primarily as cube-shaped medusas only a few centimeters tall; the largest species grows to 25 cm in height. Nematocyst-rich tentacles grow in clusters from the four corners of the boxlike medusa, and groups of light receptors and image-forming eyes occur on the four sides of the bell. Unlike the true jellyfish, cubozoans are active swimmers. They feed on small fishes and invertebrates, immobilizing their prey with one of the deadliest toxins produced

A. *Hydra* consuming a crustacean

Kim Taylor/Bruce Coleman Ltd.

Kim Taylor/Bruce Coleman Ltd.

B. Cnidocytes

Operculum (capsule's lid at cnidocyte's free surface)

Trigger (modified cilium)

Nematocyst coiled inside capsule

Barbs

Discharged nematocyst

FIGURE 31.10 Predation by cnidarians. (A) A polyp of a freshwater *Hydra* captures a small crustacean with its tentacles and swallows it whole. **(B)** Cnidocytes, special cells on the tentacles, encapsulate nematocysts, which are discharged at prey.

© Cengage Learning 2017

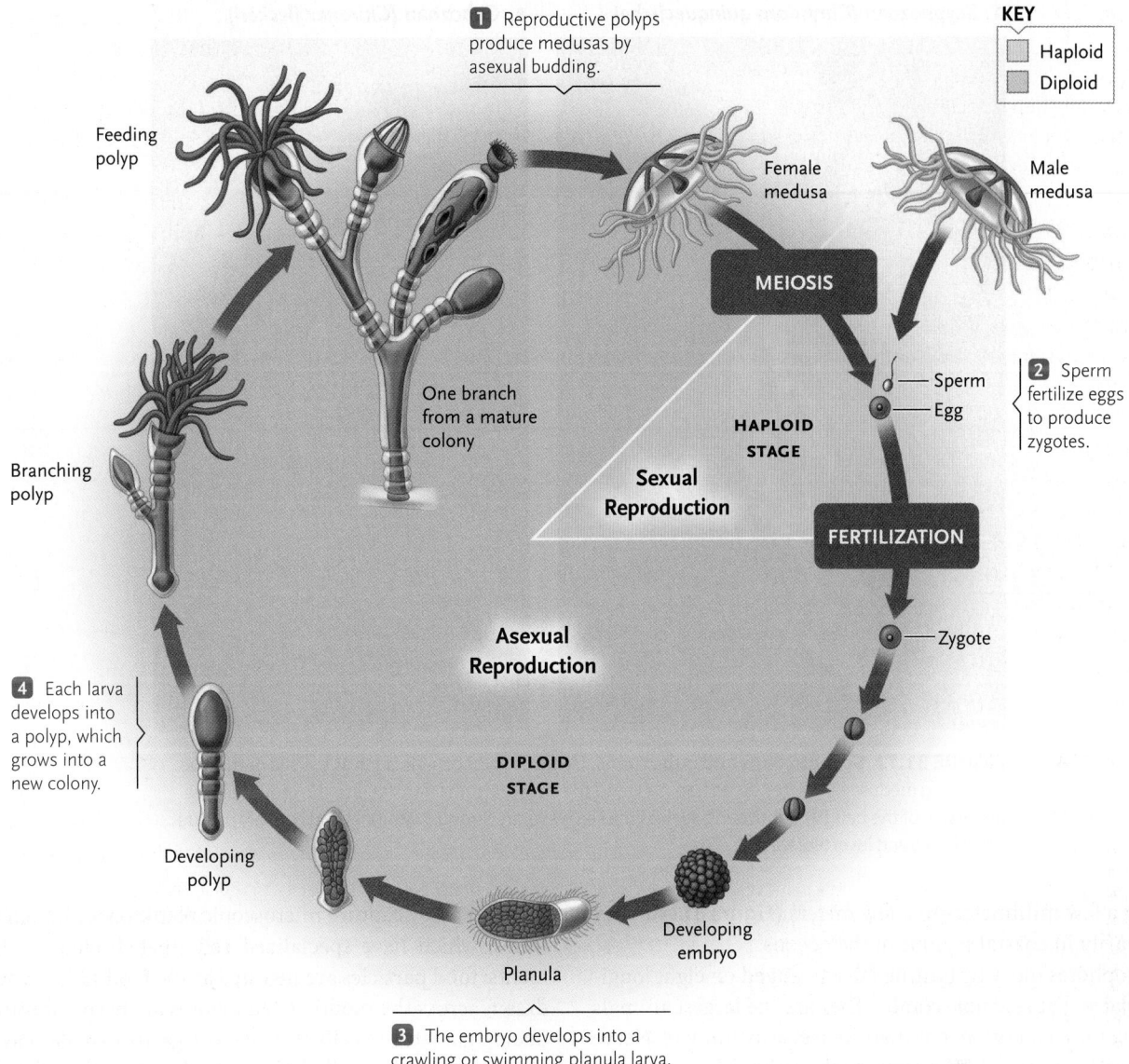

1 Reproductive polyps produce medusas by asexual budding.

Feeding polyp

One branch from a mature colony

Branching polyp

4 Each larva develops into a polyp, which grows into a new colony.

Developing polyp

Planula

3 The embryo develops into a crawling or swimming planula larva.

KEY
Haploid
Diploid

Female medusa

Male medusa

MEIOSIS

Sperm
Egg

2 Sperm fertilize eggs to produce zygotes.

HAPLOID STAGE

Sexual Reproduction

FERTILIZATION

Zygote

Asexual Reproduction

DIPLOID STAGE

Developing embryo

FIGURE 31.11 Life cycle of a *hydrozoan*. The life cycle of *Obelia*, a colonial hydrozoan, includes both polyp and medusa stages.
© Cengage Learning 2017

by animals. Cubozoans live in tropical and subtropical coastal waters, where they sometimes pose a serious threat to swimmers.

ANTHOZOA The Anthozoa includes corals and sea anemones **(Figure 31.13)**. Anthozoans exist only as polyps, and often reproduce by budding or fission; most also reproduce sexually. Corals are always sessile and usually colonial. Most species build calcium carbonate skeletons, which sometimes grow together to form gigantic underwater reefs. The energy needs of many corals are partly fulfilled by the photosynthetic activity of symbiotic protists that live within the corals' cells. For this reason, corals are restricted to shallow water where sunlight can penetrate. Sea anemones are soft-bodied, solitary polyps, ranging from 1 cm to 10 cm in

diameter. Many species occupy shallow coastal waters. Most species are essentially sessile, but some move by crawling slowly or by using their gastrovascular cavity as a hydrostatic skeleton (see Figure 31.13B).

Ctenophores Use Tentacles to Feed on Microscopic Plankton

Like the cnidarians, the 150 species of comb jellies in the phylum Ctenophora (*ktenos* = comb; *-phoros* = bearing) have radial symmetry, mesoglea, and feeding tentacles. However, they differ from cnidarians in significant ways: they lack nematocysts; they expel some waste through anal pores opposite the mouth; and certain tissues appear to be of mesodermal origin. These transparent, and often luminescent, animals range in

A. Scyphozoan (*Chrysaora quinquecirrha*)

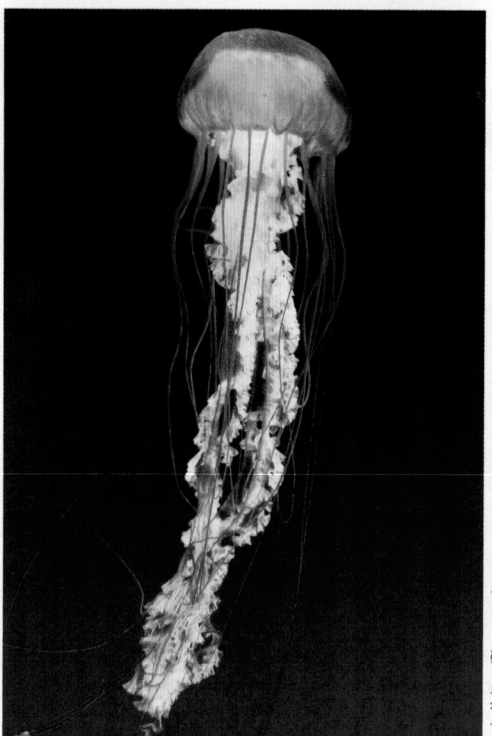

B. Cubozoan (*Chironex fleckeri*)

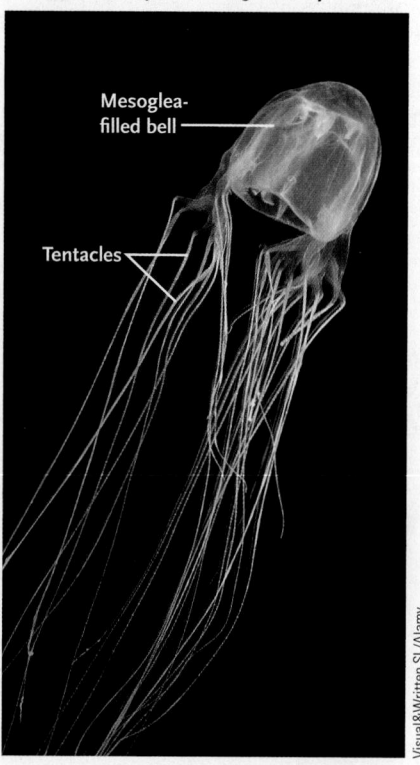

Mesoglea-filled bell

Tentacles

FIGURE 31.12 Scyphozoans and cubozoans. (A) Most scyphozoans, such as the sea nettle, live as floating medusae. Their tentacles trap prey, and the long oral arms transfer it to the mouth on the underside of the bell. **(B)** Cubozoans, such as the sea wasp, are strong swimmers that actively pursue small fishes and invertebrates.

size from a few millimeters to a few meters **(Figure 31.14).** They live primarily in coastal regions of the oceans.

Ctenophores move by beating cilia arranged on eight longitudinal plates that resemble combs. They are the largest animals to use cilia for locomotion, but they are feeble swimmers. Nerve cells connected to the cilia coordinate the animals' movements, and gravity-sensing statocysts help them maintain an upright position. They capture microscopic plankton with their two tentacles, which have specialized cells that discharge sticky filaments; food particles are ingested as the food-laden tentacles are drawn across the mouth. Ctenophores are hermaphroditic, producing gametes in cells that line the gastrovascular cavity. Eggs and sperm are expelled through the mouth or from special pores, and fertilization occurs in the open water.

A. Staghorn coral (*Acropora cervicornis*)

Tentacle of one polyp

Interconnected skeletons of polyps of a colonial coral

B. Sea anemone (*Urticina lofotensis*) escape behavior

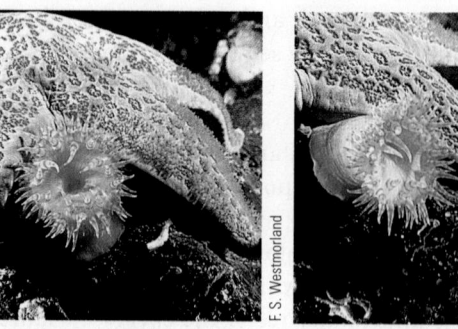

FIGURE 31.13 Anthozoans. (A) Many corals are colonial, and their polyps build a hard skeleton of calcium carbonate. The skeletons accumulate to form coral reefs in shallow tropical waters (see Figure 51.29B). **(B)** A white-spotted sea anemone detaches from its substrate to escape from a predatory sea star.

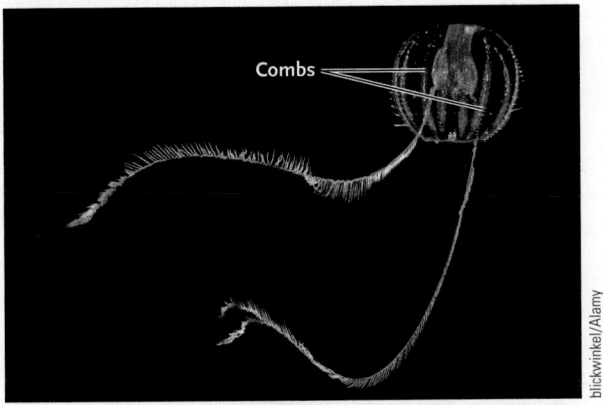

FIGURE 31.14 Ctenophores. Comb jellies, such as this sea gooseberry (*Pleurobrachia pileus*) from the Gulf of Maine, collect microscopic prey on their long, sticky tentacles and then wipe the food-laden tentacles across their mouths.

STUDY BREAK 31.5

1. How do cnidarians capture, consume, and digest their prey?
2. Which group of cnidarians has only a polyp stage in its life cycle?
3. What do ctenophores eat, and how do they collect their food?

31.6 Lophotrochozoan Protostomes

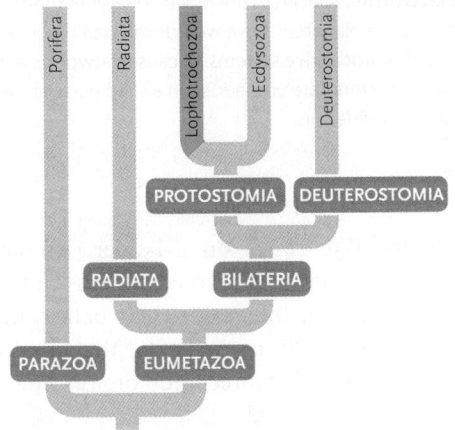

The remaining organisms described in this chapter fall within the group Bilateria: they have bilateral symmetry and triploblastic development (that is, their embryological development includes three tissue layers). Bilaterians also have organ systems, structures that include two or more tissue types that perform specific functions. Most also possess a coelom or pseudocoelom. With bilateral symmetry and sensory organs concentrated at the anterior end of the body, bilaterians generally engage in highly directed, often rapid movements in pursuit of food, mates, or to escape danger. And their complex organ systems accomplish tasks more efficiently than simple tissues. For example, animals that have a tubular digestive

system surrounded by a space (the coelom or pseudocoelom) use muscular contractions of the digestive system to move ingested food past specialized epithelia that break it down and absorb the breakdown products.

Molecular analyses group eight of the most diverse phyla into the monophyletic Lophotrochozoa, one of the two main protostome clades (see Figure 31.6).

The Lophophorate Phyla Share a Distinctive Feeding Structure

Three small groups of aquatic, coelomate animals—the phyla Ectoprocta, Brachiopoda, and Phoronida—possess a **lophophore,** a circular or U-shaped fold with one or two rows of hollow, ciliated tentacles surrounding the mouth **(Figure 31.15).** Molecular sequence data, as well as the lophophore, suggest that these phyla share a common ancestry.

The lophophore, which looks like a crown of tentacles at the anterior end of the animal, serves as a site for gas exchange and waste elimination as well as for food capture. Most lophophorates are sessile suspension-feeders as adults: movement of cilia on the tentacles brings food-laden water toward the lophophore, the tentacles capture small organisms and debris, and the cilia transport them to the mouth. The lophophorates have a complete digestive system, which is U-shaped in most species, with the anus lying outside the ring of tentacles.

A. Ectoprocta (*Plumatella repens*)

B. Brachiopoda (*Terebratulina septentrionalis*)

C. Phoronida (*Phoronis hippocrepia*)

FIGURE 31.15 Lophophorate animals. Although the lophophorate animals differ markedly in appearance, they all use a lophophore—the feathery structures in the photos—to acquire food.

PHYLUM ECTOPROCTA The Ectoprocta (sometimes called Bryozoa) are tiny colonial animals that mainly occupy marine habitats (see Figure 31.15A). They secrete a hard covering over their soft bodies and feed by extending the lophophore through a hole. Each colony, which may include more than a million individuals, is produced asexually by a single animal. Ectoprocta colonies are permanently attached to solid substrates, where they form encrusting mats, bushy upright growths, or jellylike blobs. Nearly 5,000 living species are known.

PHYLUM BRACHIOPODA The Brachiopoda, or lampshells, have two calcified shells that develop on the animal's dorsal and ventral sides (see Figure 31.15B). Most species attach to substrates with a stalk that protrudes through one of the shells. The lophophore is held within the two shells, and the animal feeds by opening its shell and drawing water over its tentacles. Although only 300 species of brachiopods live today, more than 30,000 extinct species are known from the fossil record, mostly from Paleozoic seas.

PHYLUM PHORONIDA The 12 or so species of phoronid worms vary in length from a few millimeters to 25 cm (see Figure 31.15C). They usually build tubes in soft ocean sediments or on hard substrates, and feed by protruding the lophophore from the top of the tube. The animal can withdraw into the tube when disturbed. Phoronida reproduce both sexually and by budding.

Flatworms Have Digestive, Excretory, Nervous, and Reproductive Systems, but Lack a Coelom

The 20,000 flatworm species in the phylum Platyhelminthes (*platys* = flat; *helminth-* = worm) live in aquatic and moist terrestrial habitats. Like cnidarians, flatworms can swim or float in water, but they are also able to crawl over surfaces. They range from less than 1 mm to more than 60 cm in length, but most are just a few millimeters thick. Free-living species eat live prey or decomposing carcasses, whereas parasitic species consume the tissues of living hosts.

Like the radiate phyla, flatworms are acoelomate, but they have a complex structural organization that reflects their triploblastic construction **(Figure 31.16)**. Endoderm derivatives line the gastrovascular (digestive) cavity with cells specialized for the chemical breakdown and absorption of ingested food. Mesoderm, the middle tissue layer, produces muscles and reproductive organs. Ectoderm produces a ciliated epidermis, the nervous system, and the *flame cell system*. Flame cells function in osmoregulation; they regulate the concentrations of salts and water within body fluids, allowing free-living flatworms to live in freshwater habitats. Flatworms do not have circulatory or respiratory systems, but, because all cells of their dorsoventrally (top-to-bottom) flattened bodies are near an interior or exterior surface, diffusion supplies them with nutrients and oxygen.

The flatworm nervous system includes two or more longitudinal ventral nerve cords interconnected by numerous

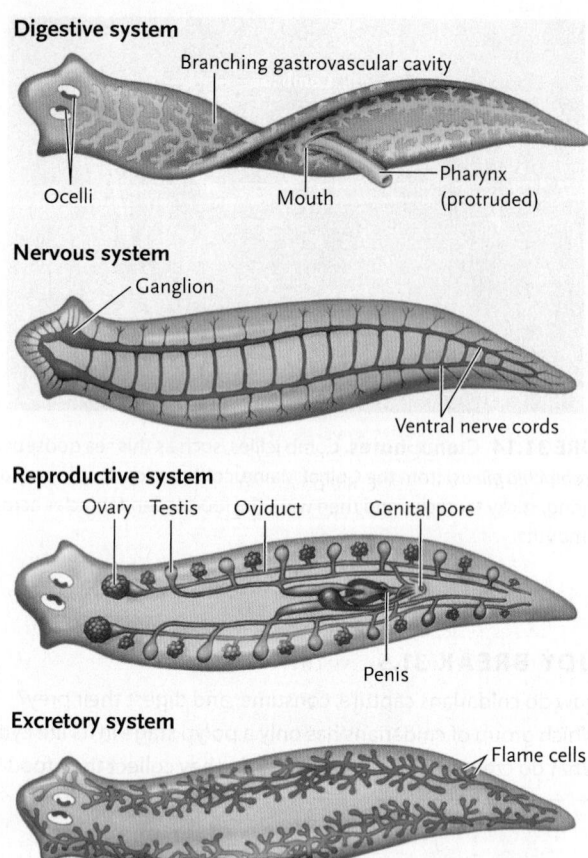

FIGURE 31.16 Flatworms. Species in the phylum Platyhelminthes, exemplified by a freshwater planarian, have well-developed digestive, excretory, nervous, and reproductive systems. Because flatworms are acoelomate, their organ systems are embedded in a solid mass of tissue between the gut and the epidermis.
© Cengage Learning 2017

smaller nerve fibers, like rungs on a ladder. An anterior **ganglion,** a concentration of nervous system tissue that serves as a primitive brain, integrates their behavior. Most free-living species have *ocelli,* or eyespots, that distinguish light from dark and tiny chemoreceptor organs that sense chemical cues.

The phylum Platyhelminthes includes four clades, defined largely by their anatomical adaptations to free-living or parasitic habits.

TURBELLARIA Most free-living flatworms (Turbellaria) live in the sea **(Figure 31.17),** but the familiar planarians and a few others live in freshwater or on land. Turbellarians swim by undulating the body wall musculature, or they crawl across surfaces by using muscles and cilia to glide on mucous trails produced by the ventral epidermis.

The gastrovascular cavity in free-living flatworms is similar to that of cnidarians. Food is ingested and wastes are eliminated through a single opening, the mouth, located on the ventral surface. Food passes from the mouth to a muscular **pharynx** that leads to the digestive cavity (see Figure 31.16).

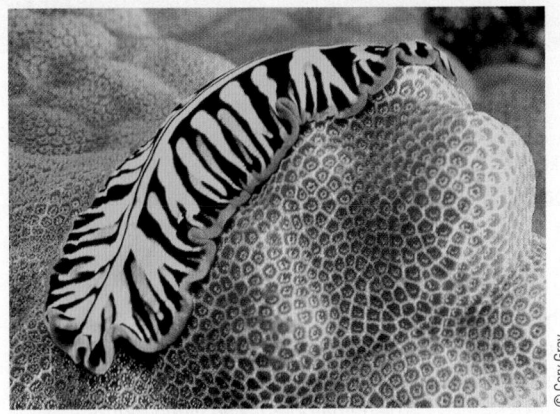

FIGURE 31.17 Turbellaria. Some marine turbellarians, such as *Pseudoceros dimidiatus*, are brightly colored.

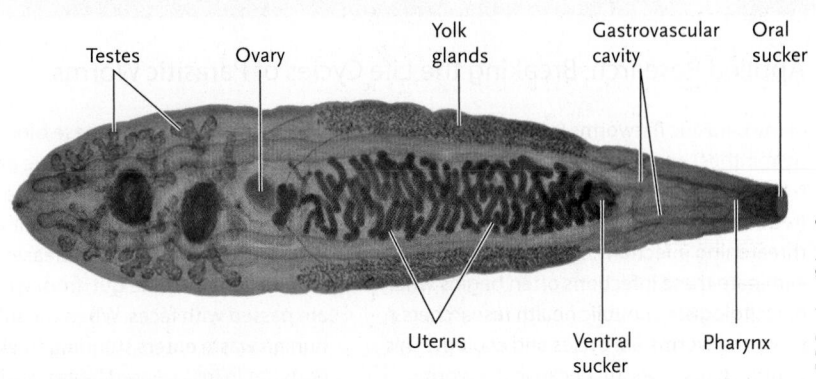

Testes · Ovary · Yolk glands · Gastrovascular cavity · Oral sucker · Uterus · Ventral sucker · Pharynx

E. R. Degginger/Science Source

FIGURE 31.18 Trematoda. The hermaphroditic Chinese liver fluke *(Opisthorchis sinensis)* uses a well-developed reproductive system to produce thousands of eggs.

Chemicals secreted into the saclike cavity digest the food into particles; then cells throughout the gastrovascular surface engulf the particles and subject them to intracellular digestion. In some species, the digestive cavity is highly branched, increasing the surface area for digestion and absorption.

Nearly all turbellarians are hermaphroditic, with complex reproductive systems (see Figure 31.16). When they mate, each partner functions simultaneously as a male and a female. Many free-living species also reproduce asexually by separating their anterior half from their posterior half. Both halves subsequently regenerate the missing parts.

TREMATODA AND MONOGENOIDEA Flukes (Trematoda and Monogenoidea) are parasites that obtain nutrients from tissues of a living host **(Figure 31.18).** Most adult trematodes are **endoparasites,** living in the gut, liver, lungs, bladder, or blood vessels of vertebrates. Monogenes are **ectoparasites,** attaching to the gills or skin of aquatic vertebrates. Flukes are structurally specialized for a parasitic existence. They use suckers or hooks to attach to hosts, and a tough outer covering protects them from chemical attack. They produce numerous eggs that can readily infect new hosts. Monogene flukes usually have simple life cycles with a single host species. Trematodes, by contrast, have complex life cycles and multiple hosts. Humans suffer potentially fatal infections by many flukes, as discussed in *Focus on Research: Applied Research.*

CESTODA Tapeworms (Cestoda) develop, grow, and reproduce within the intestines of vertebrates **(Figure 31.19).** Through evolution, they have lost their mouths and digestive systems and absorb nutrients directly through their body wall. Tapeworms have a specialized structure, the *scolex,* with hooks and suckers that attach to the host's intestine; like the flukes, they also have a protective covering resistant to digestive enzymes.

Most of a tapeworm's body, which may be up to 20 m long, consists of a series of segmental structures, *proglottids,* that contain little more than male and female reproductive systems. New proglottids are generated near the scolex; older proglottids, each carrying as many as 80,000 fertilized eggs, break off from the tapeworm's posterior end, and leave the host's body in feces. Tapeworms are important parasites on humans, pets, and livestock.

Rotifers Are Tiny Pseudocoelomate Animals with a Jawlike Feeding Apparatus

Most of the 1,850 species in the phylum Rotifera *(rota = wheel; ferre = to carry)* live in freshwater **(Figure 31.20).** All are microscopic—about the size of a ciliate protist—but they

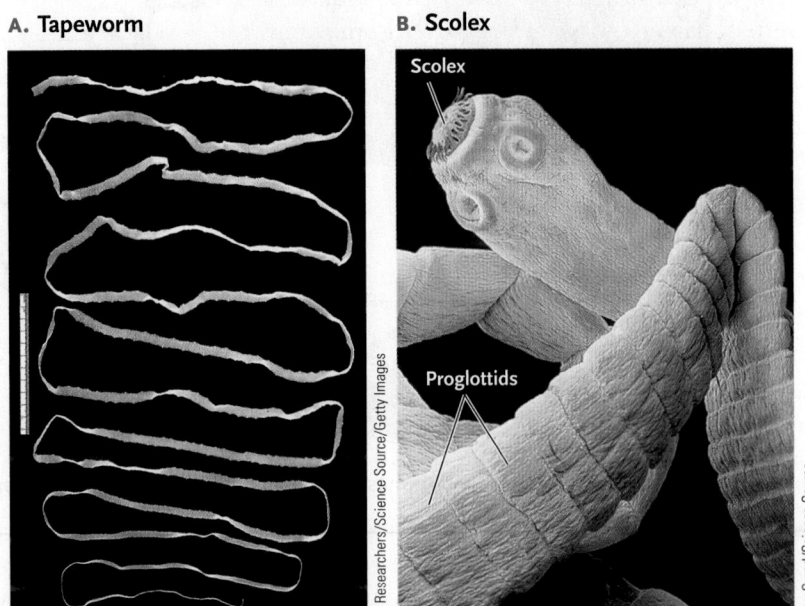

A. Tapeworm · Scolex

B. Scolex · Scolex · Proglottids

Photo Researchers/Science Source/Getty Images

Andrew Syred/Science Source

FIGURE 31.19 Cestoda. (A) Tapeworms have long bodies comprised of a series of proglottids that each produce thousands of fertilized eggs. The scale bar is about 12 cm long. **(B)** The anterior end is a scolex with hooks and suckers that attach to the host's intestinal wall.

Focus on Research

Applied Research: Breaking the Life Cycles of Parasitic Worms

Many parasitic flatworms (phylum Platy-helminthes) and roundworms (phylum Nematoda) call the human body home, frequently causing disfiguring or life-threatening infections. The effort to control or eliminate these infections often begins when parasitologists or public health researchers study the worms' life cycles and ecology. This approach is successful because the worms often have more than one host: humans may be the *primary host,* harboring the sexually mature stage of the parasites' life cycles, but other animals serve as *intermediate hosts* to the larval stages. If researchers can learn the details of a parasite's life cycle, they can identify ways to cut it short, by eliminating intermediate hosts, before the parasite infects a human host.

Roughly 200 million people, most of them in sub-Saharan Africa or in Asia, suffer from *schistosomiasis,* a disease caused by several species of flatworms called blood

flukes (Trematoda). Japanese blood flukes *(Schistosoma japonicum)* mature and mate in blood vessels of the human intestine **(Figure A).** Sharp spines on their eggs rupture the blood vessels, releasing the eggs into the lumen of the gut, from which they are passed with feces. When the infected human waste enters standing freshwater, the eggs hatch into ciliated larvae, which burrow into some aquatic snail species. The larvae feed on the snail's tissues and reproduce asexually. Their offspring leave the snail and, when they contact human skin, bore inward to a blood vessel. They eventually reach the intestine, where they complete their complex life cycle and produce fertilized eggs.

Infected humans mount an immune response against flukes, but it is always a losing battle. Severe infections cause coughs, rashes, pain, and eventually diarrhea, anemia, and permanent damage to the intestines, liver, spleen, bladder, and kidneys. Death

often results. Drug therapy can reduce the symptoms and limit the spread of these para-sites, but schistosomiasis is most common in countries where people have limited access to medical care. Research has demonstrated that the disease can be controlled by proper sanita-tion and the elimination of the snails that serve as intermediate hosts.

The nematodes called *filarial worms* (*Wuchereria* and *Brugia*) cause another debili-tating infection. These large roundworms (up to 10 cm long) live in the human lymphatic system, where they obstruct the normal flow of lymphatic fluid to the bloodstream. Female worms release first-stage larvae, which are acquired by mosquitoes, the intermediate host, when they feed on human blood. The larvae develop into second-stage larvae in mos-quitoes, and when a mosquito bites another human, it may transmit those larvae to a new host. If victims experience severe filarial worm infection, their lymphatic vessels can be so obstructed that surrounding tissues swell gro-tesquely, a condition known as *elephantiasis* **(Figure B).** Public health programs that reduce or eliminate mosquito populations lower the incidence of this disease.

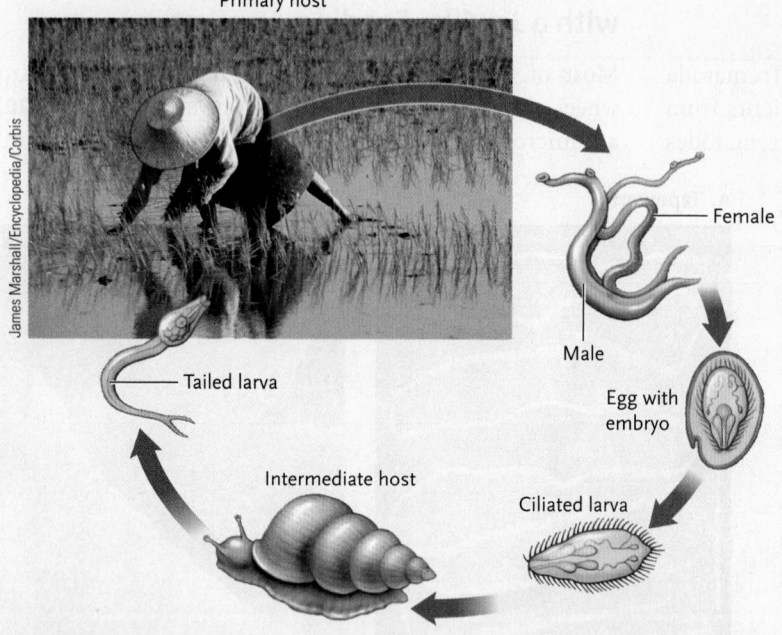

FIGURE A Life cycle of the Japanese blood fluke (*Schistosoma japonicum*).

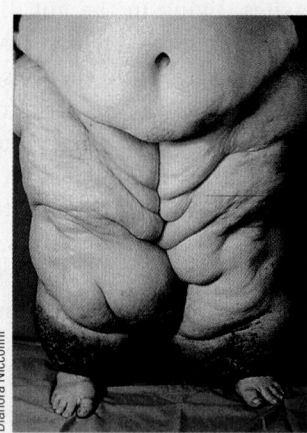

FIGURE B Elephantiasis caused by *Wuchereria bancrofti*.

have well-developed digestive, reproductive, excretory, and nervous systems as well as a pseudocoelom. In some habitats, rotifers make up a large part of the zooplankton, tiny animals that float in open water; some species live in moist terrestrial habitats.

Rotifers use coordinated movements of cilia, arranged in a wheel-like *corona* around the head, to propel themselves in the environment. Cilia also bring food-laden water to their mouths. Ingested microorganisms are conveyed to the *mastax,* a toothed grinding organ, and then passed to the stomach and intestine. Rotifers have a **complete digestive tract:** food enters through the mouth, and undigested waste is voided through the anus.

The life history patterns of some rotifer species are adapted to the ever-changing environments in small bodies of water. During most months, populations include only females that reproduce by **parthenogenesis,** a form of asexual reproduction in which unfertilized eggs develop into diploid females (see Section 49.1). When environmental conditions deteriorate, usually because their habitats become dry, females produce eggs that develop into haploid males. The males fertilize haploid eggs to produce diploid female zygotes. The fertilized eggs have durable shells and food reserves to survive drying or freezing.

However, one group, the bdelloid rotifers, which includes more than 300 species, reproduce *only* by parthenogenesis. Males have never been observed in any bdelloid rotifer species, and genomic studies suggest that the clade has been perpetuated by parthenogenesis for many millions of years. The absence of sexual reproduction—and, thus, genetic recombination—in these species is puzzling because a lack of genetic recombination often leads to the accumulation of deleterious mutations. However, as demonstrated in a study published in 2012, researchers discovered that as much as 10% of the active genes in bdelloid rotifers were acquired through horizontal gene transfer (see Section 24.7) from very distantly related organisms and that 80% of those genes code for enzymes that contribute to bdelloid biochemistry.

A. Rotifer *(Philodina roseola)* body plan

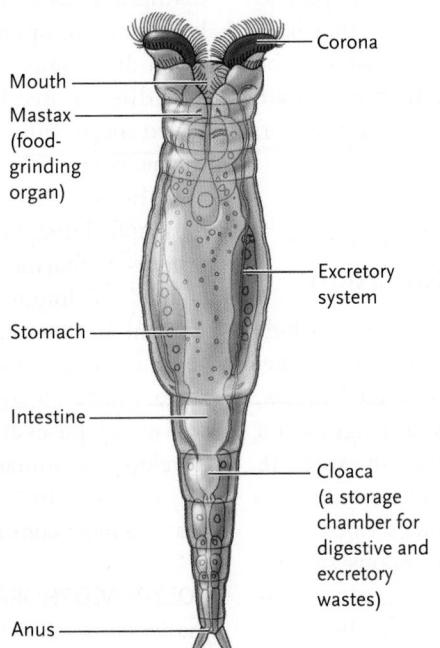

B. Rotifer *(Philodina* species)

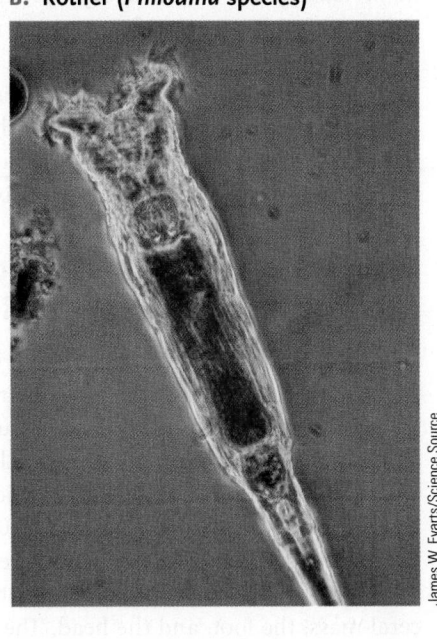

James W. Evarts/Science Source

FIGURE 31.20 Phylum Rotifera. Despite their small size, rotifers have complex body plans and organ systems.

© Cengage Learning 2017

Ribbon Worms Use a Proboscis to Capture Food

The 900 species of ribbon worms or proboscis worms (phylum Nemertea) vary from less than 1 cm to 30 m in length **(Figure 31.21)**. Most species are marine, but a few occupy moist, terrestrial habitats. Although the often brightly colored ribbon worms superficially resemble free-living flatworms, their body plans are more complex. First, they possess both a mouth and an anus; thus, they have a complete digestive tract. Second,

A. Ribbon worm *(Lineus* species)

Marevision/Age Fotostock

B. Ribbon worm anatomy

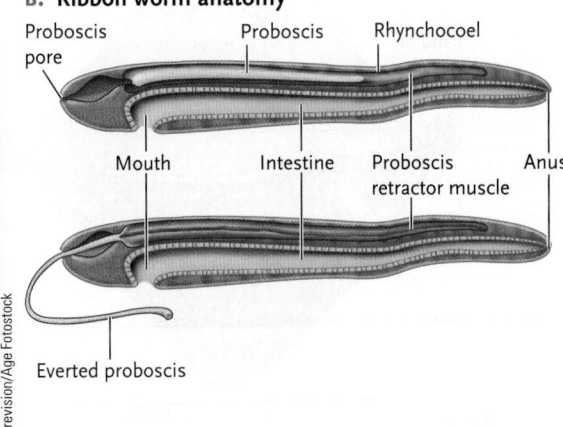

FIGURE 31.21 Phylum Nemertea. (A) The flattened, elongated bodies of ribbon worms are often brightly colored. **(B)** Ribbon worms have a complete digestive system and a specialized cavity, the rhynchocoel, that houses a protrusible proboscis.

© Cengage Learning 2017

nemerteans have a circulatory system in which fluid flows through **circulatory vessels** that carry nutrients and oxygen to tissues and remove wastes (see Section 44.1). Finally, they have a muscular, mucus-covered proboscis, a tube that can be everted (turned inside out) through the mouth or a separate pore to capture prey. The proboscis is housed within a chamber, the *rhynchocoel,* which is unique to this phylum.

Mollusks Have a Muscular Foot and a Mantle That Secretes a Shell or Aids in Locomotion

Most species in the phylum Mollusca (*mollis* = soft)—including clams, snails, octopuses, and their relatives—are marine. However, many clams and snails occupy freshwater habitats, and some snails live on land. Mollusks vary in length from clams less than 1 mm to the giant squids, which can exceed 18 m. Scientists have named roughly 100,000 mollusk species, but experts believe that at least as many are still undescribed.

In mollusks, the body is divided into three regions: the visceral mass, the foot, and the head. The **visceral mass** contains digestive, excretory, and reproductive systems and the heart. The muscular **foot** often provides the major means of locomotion. In more active groups, the **head** is well defined and carries sensory organs and a brain. The mouth often includes a toothed **radula,** which scrapes food into small particles or drills through the shells of prey. Many mollusks are covered by a protective shell of calcium carbonate secreted by the **mantle,** which comprises one or two folds of the body wall that often enclose the visceral mass. The mantle also defines a space, the *mantle cavity*. In aquatic species, the mantle cavity houses the *gills,* delicate respiratory structures with enormous surface area. In most mollusks, cilia on the mantle and gills generate a steady flow of water into the mantle cavity.

The large size of mollusks requires a circulatory system to maintain cells that are far from the body surface. Most mollusks have an **open circulatory system** in which **hemolymph,** a bloodlike fluid, leaves the circulatory vessels and bathes tissues directly (see Figure 44.3A). Hemolymph pools in spaces called *sinuses,* and then drains into vessels that carry it back to the heart.

The sexes are usually separate, although many snails are hermaphroditic. Fertilization may be internal or external. The zygotes of marine species often develop into free-swimming, ciliated **trochophore** larvae **(Figure 31.22),** typical of both this phylum and the phylum Annelida, which we describe next. In some mollusks, the trochophore develops into a second larval stage before metamorphosing into an adult. Some snails, as well as octopuses and squids, have direct development: embryos develop into miniature replicas of the adults.

Mollusca includes eight clades. Here we examine the four that are most commonly encountered.

POLYPLACOPHORA The chitons (Polyplacophora: *poly* = many; *plax* = flat surface) are sedentary mollusks that graze on algae along rocky marine coasts. The oval, bilaterally symmetrical body has a dorsal shell divided into eight plates that allow it to conform to irregularly shaped surfaces **(Figure 31.23).** When a chiton is disturbed or exposed to strong wave action, the muscles of its broad foot maintain a tenacious grip, and the mantle's edge functions like a suction cup to hold fast to the substrate.

GASTROPODA Snails and slugs (Gastropoda: *gaster* = belly; *podos* = foot) are the largest molluscan group, numbering 40,000 species **(Figure 31.24).** Aquatic species use gills to acquire oxygen, but in terrestrial species a modified mantle cavity functions as an air-breathing lung. Gastropods feed on algae, vascular plants, or animal prey. Some are scavengers, and a few are parasites.

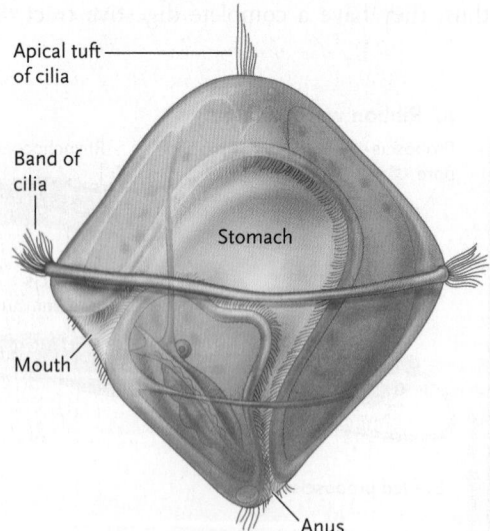

FIGURE 31.22 Trochophore larva. At the conclusion of their embryological development, many mollusks and annelids pass through a trochophore stage. The top-shaped trochophore larva has a band of cilia just anterior to its mouth.
© Cengage Learning 2017

Labels: Apical tuft of cilia; Band of cilia; Stomach; Mouth; Anus

FIGURE 31.23 Polyplacophora. Chitons live on rocky shores, where they use their foot and mantle to grip rocks and other hard substrates. This chiton (*Tonicella lineata*) lives in California.

George Wood/Dreamstime.com

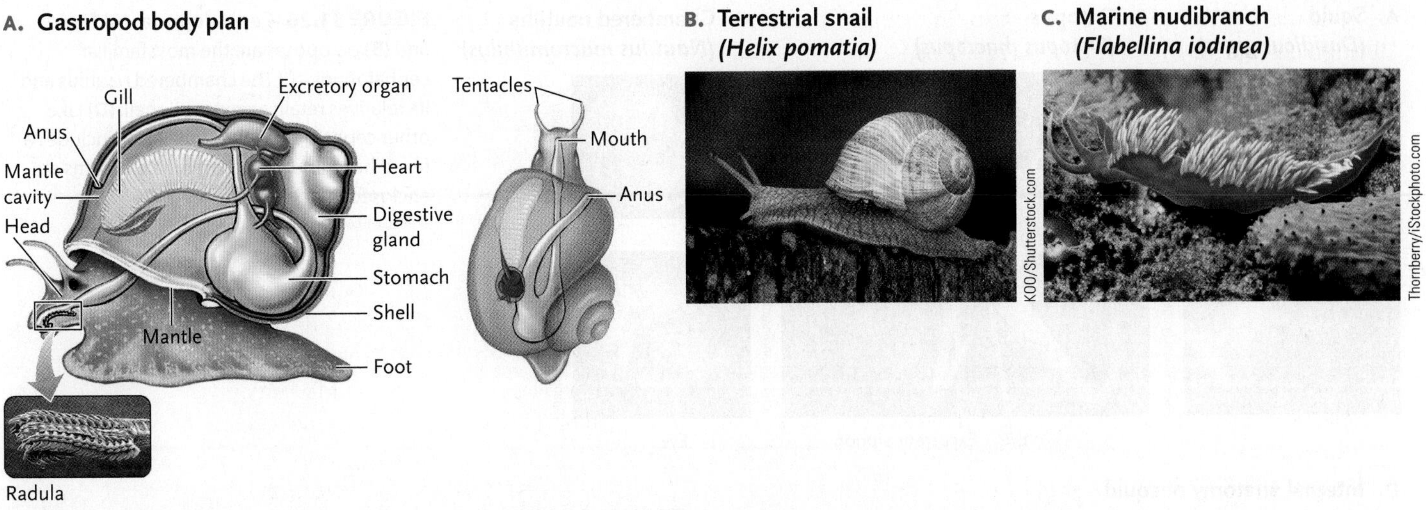

A. Gastropod body plan

Gill
Anus
Mantle cavity
Head
Mantle
Radula

Excretory organ
Heart
Digestive gland
Stomach
Shell
Foot

Tentacles
Mouth
Anus

B. Terrestrial snail (*Helix pomatia*)

K00/Shutterstock.com

C. Marine nudibranch (*Flabellina iodinea*)

Thornberry/iStockphoto.com

FIGURE 31.24 Gastropoda. (A) Most gastropods have a coiled shell that houses the visceral mass. A developmental process called *torsion* causes the digestive and excretory systems to eliminate wastes into the mantle cavity, near the animal's head. **(B)** The edible snail is a terrestrial gastropod. **(C)** Nudibranchs, like this Spanish shawl nudibranch, are shell-less marine snails.

© Cengage Learning 2017

The visceral mass of most snails is housed in a coiled or cone-shaped shell that is balanced above the rest of the body, much as you balance a backpack full of books (see Figure 31.24A, B). Most shelled snails undergo **torsion,** a curious realignment of body parts that is independent of shell coiling. Muscle contractions and differential growth twist the visceral mass and mantle 180° relative to the head-foot. This rearrangement moves the mantle cavity forward so that the head can be withdrawn into the shell in times of danger. It also brings the gills, anus, and excretory openings above the head—a potentially messy configuration, were it not for cilia that sweep away wastes.

Some gastropods, including terrestrial slugs and colorful nudibranchs (sea slugs), are shell-less, a condition that leaves them vulnerable to predators (see Figure 31.24C). Some nudibranchs consume cnidarians and then transfer undischarged

nematocysts to projections on their dorsal surface, where these "borrowed" stinging capsules provide protection.

The nervous and sensory systems of gastropods are well developed. Tentacles on the head include chemical and touch receptors; the eyes detect changes in light intensity but don't form images.

BIVALVIA The clams, scallops, oysters, and mussels (Bivalvia: *bi* = two; *valva* = folding door) are restricted to aquatic habitats. They are enclosed within a pair of shells, hinged together dorsally by an elastic ligament **(Figure 31.25).** Contraction of one or two **adductor muscles** closes the shell by pulling the two sides together; when the muscles relax, the elastic ligament opens the shell by pulling them apart (see Figure 31.25A). Although some bivalves are tiny, giant clams of the South

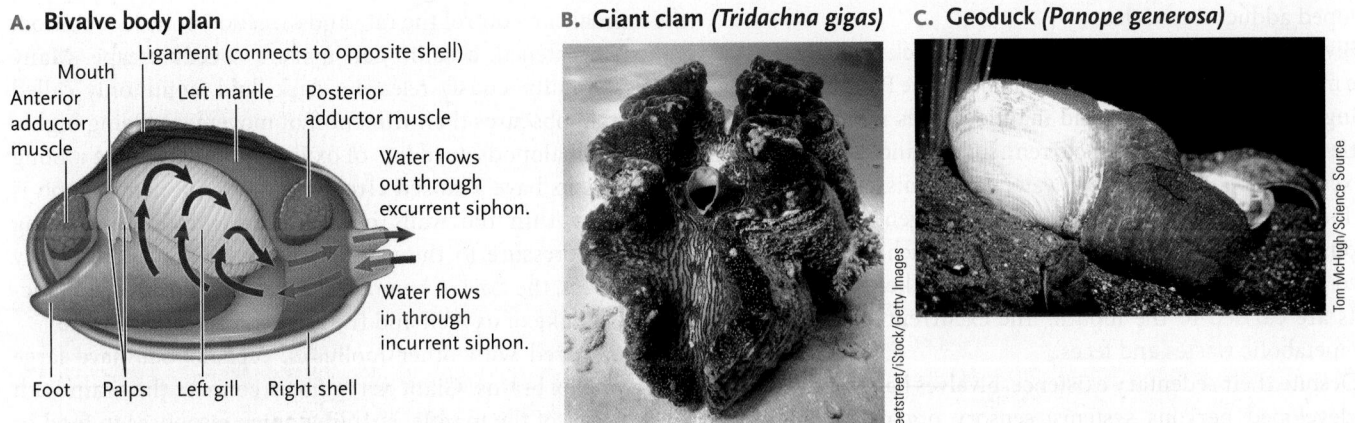

A. Bivalve body plan

Mouth
Anterior adductor muscle
Ligament (connects to opposite shell)
Left mantle
Posterior adductor muscle
Water flows out through excurrent siphon.
Water flows in through incurrent siphon.
Foot
Palps
Left gill
Right shell

B. Giant clam (*Tridachna gigas*)

treetstreet/iStock/Getty Images

C. Geoduck (*Panope generosa*)

Tom McHugh/Science Source

FIGURE 31.25 Bivalvia. (A) Bivalves are enclosed in a hinged two-part shell. Part of the mantle forms a pair of water-transporting siphons. **(B)** Giant clams in the South Pacific sometimes weigh hundreds of kilograms. **(C)** The geoduck is a clam with enormous muscular siphons.

© Cengage Learning 2017

A. Squid
(Dosidicus gigas)

B. Octopus
(Octopus macropus)

C. Chambered nautilus
(Nautilus macromphilus)

Steve Bloom Images/Alamy

Jeff Rotman/Photolibrary/Getty Images

© Christian Slanec/Shutterstock.com

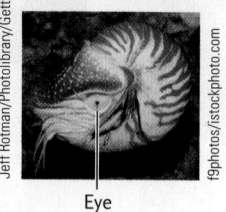

f9photos/istockphoto.com

Eye   Excurrent siphon

Eye

FIGURE 31.26 Cephalopoda. (A) Squids and **(B)** octopuses are the most familiar cephalopods. **(C)** The chambered nautilus and its relatives retain an external shell. **(D)** Like other cephalopods, the squid body includes a fused head and foot; most organ systems are enclosed by the mantle.
© Cengage Learning 2017

D. Internal anatomy of squid

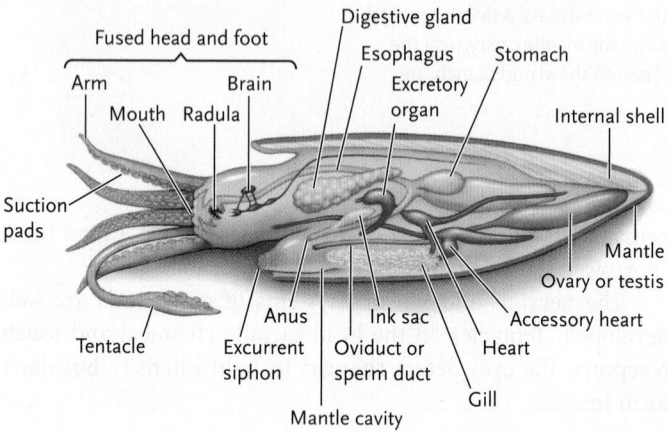

Fused head and foot — Arm, Brain, Mouth, Radula
Digestive gland
Esophagus
Excretory organ
Stomach
Internal shell
Suction pads
Mantle
Ovary or testis
Accessory heart
Heart
Tentacle
Anus
Excurrent siphon
Ink sac
Oviduct or sperm duct
Gill
Mantle cavity

Pacific can be more than 1 m across and weigh 225 kg (see Figure 31.25B).

Adult mussels and oysters are sessile and permanently attached to hard substrates, but many clams are mobile and use their muscular foot to burrow in sand or mud. Some bivalves, such as young scallops, swim by rhythmically clapping their valves ("shells") together, forcing a current of water out of the mantle cavity. The "scallops" that we eat are the scallops' well-developed adductor muscles.

Bivalves have a reduced head, and they lack a radula. Part of the mantle forms two tubes called *siphons* (see Figure 31.25C). Beating of cilia on the gills and mantle carries water into the mantle cavity through the incurrent siphon and out through the excurrent siphon. Incurrent water carries dissolved oxygen and food particles to the gills, where oxygen is absorbed. Mucous strands on the gills trap the food, which is then transported by cilia to *palps*, where final sorting takes place; edible tidbits are carried to the mouth. The excurrent water carries away metabolic wastes and feces.

Despite their sedentary existence, bivalves have moderately well-developed nervous systems; sensory organs that detect chemicals, touch, and light; and statocysts to sense their orientation. When they encounter pollutants, many bivalves stop pumping water and close their shells. When confronted by a predator, some burrow into sediments or swim away.

CEPHALOPODA The octopuses, squids, and nautiluses (Cephalopoda: *kephale* = head) are active marine predators, including the fastest and most intelligent invertebrates **(Figure 31.26)**. They vary in length from a few centimeters to 18 m. Giant squids, the largest invertebrates known, may be the source of "sea monster" stories.

The cephalopod body has a fused head and foot (see Figure 31.26D). The head comprises the mouth and eyes. The ancestral "foot" forms a set of arms and tentacles, equipped with suction pads, adhesive structures, or hooks. Cephalopods use these structures to capture prey and a pair of beaklike jaws to bite or crush it. Venomous secretions often speed the captive's death. Some species use their radula to drill through the shells of other mollusks.

Cephalopods have a highly modified shell. Octopuses have no remnant of a shell at all. In squids and cuttlefishes, it is reduced to a stiff internal support. Only the chambered nautilus and its relatives retain an external shell.

Squids are the most mobile cephalopods, moving rapidly by a kind of jet propulsion. When muscles in the mantle relax, water is drawn into the mantle cavity. When they contract, a jet of water is squeezed out through a funnel-shaped excurrent siphon. By manipulating the position of the mantle and siphon, the animal can control the rate and direction of its locomotion. When threatened, a squid can make a speedy escape. Many species simultaneously release a dark fluid, commonly called "ink," that obscures their direction of movement. Being highly active, cephalopods need lots of oxygen, and they alone among the mollusks have a **closed circulatory system:** hemolymph is confined within the walls of hearts and vessels, providing increased pressure to the vascular fluid. Moreover, accessory hearts speed the flow of hemolymph through the gills, enhancing the uptake of oxygen and the release of carbon dioxide.

Compared with other mollusks, cephalopods have large and complex brains. Giant nerve fibers connect the brain with the muscles of the mantle, enabling quick responses to food or danger. Their image-forming eyes are similar to those of vertebrates. Cephalopods are also highly intelligent. Octopuses, for example, learn to recognize objects with distinctive shapes or colors, and they can be trained to approach or avoid them.

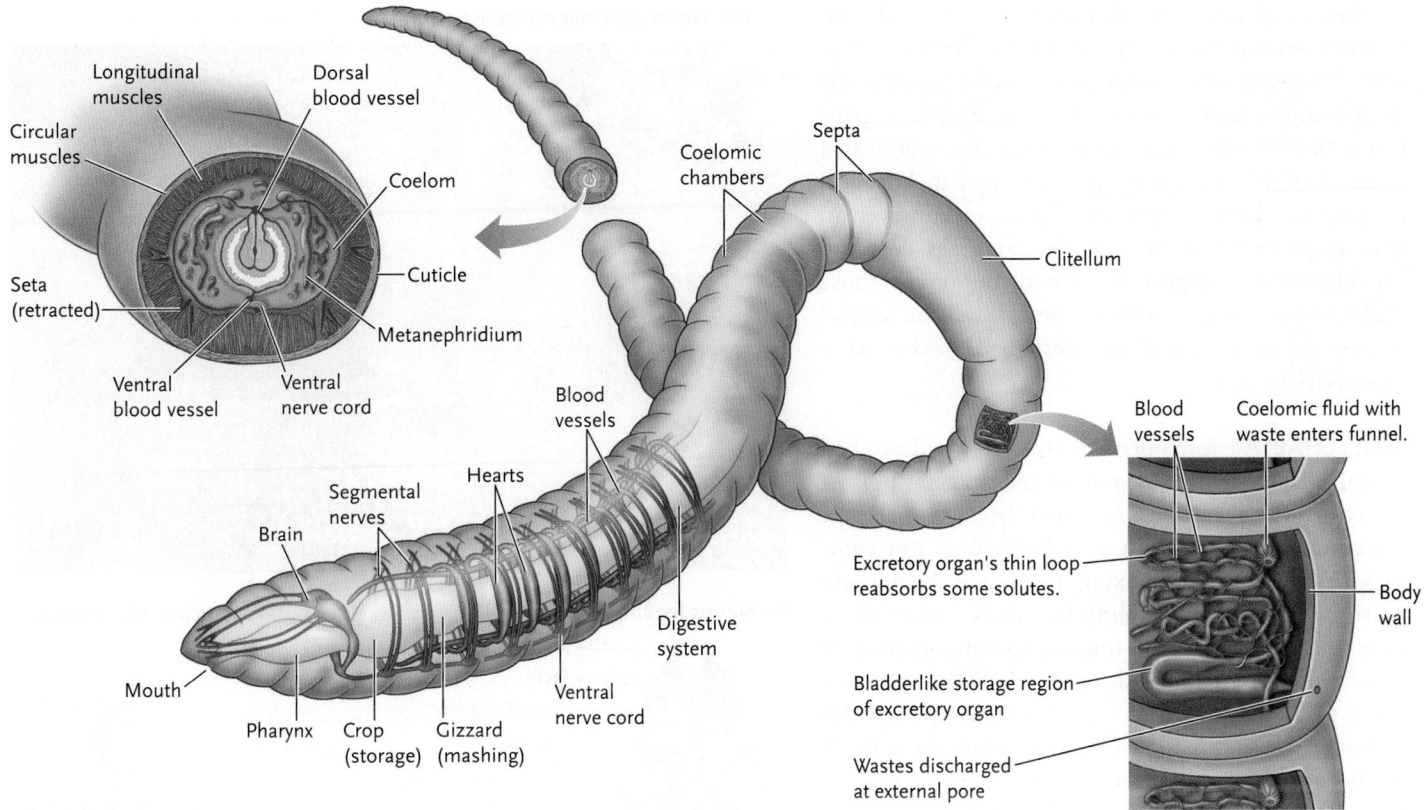

FIGURE 31.27 Segmentation in the phylum Annelida. Although the digestive system, the longitudinal blood vessels, and the ventral nerve cord are not segmented, the coelom, blood vessels, nerves, and excretory organs appear as repeating structures in most segments. The body musculature includes both circular and longitudinal layers that allow these animals to use the coelomic chambers as a hydrostatic skeleton.
© Cengage Learning 2017

Cephalopods have separate sexes and elaborate courtship rituals. Males store sperm within the mantle cavity and use a specialized tentacle to transfer packets of sperm into the female's mantle cavity, where fertilization occurs. Fertilized eggs, wrapped in a protective jelly, are attached to objects in the environment. The young hatch with an adult body form.

Annelids Exhibit a Serial Division of the Body Wall and Some Organ Systems

The 9,000 species of segmented worms (phylum Annelida: *anellus* = little ring) occupy marine, freshwater, and moist terrestrial habitats. Terrestrial annelids eat organic debris; aquatic species consume algae, microscopic organisms, detritus, or other animals. They range from a few millimeters to several meters in length.

The bodies of most annelids are highly segmented: body wall muscles and some organs—including respiratory surfaces, parts of the nervous, circulatory, and excretory systems, and the coelom itself—are divided into similar repeating units **(Figure 31.27).** Body segments are separated by transverse partitions called **septa.** The digestive system and major blood vessels are not segmented and run the length of the animal.

The body wall muscles of annelids have both circular and longitudinal layers (see Figure 31.27). Alternate contractions of these muscle groups allow annelids to make directed movements, using the pressure of the fluid in the coelom as a hydrostatic skeleton. All annelids except leeches also have chitin-reinforced bristles, called **setae** (sometimes written *chaetae*), which protrude outward from the body wall. Setae anchor the worm against the substrate, providing traction.

Annelids have a complete digestive system and a closed circulatory system. However, they lack a discrete respiratory system; oxygen and carbon dioxide diffuse through the skin. The excretory system is composed of pairs of **metanephridia** that usually occur in all body segments posterior to the head. The nervous system is highly developed, with ganglia (local control centers) in every segment, a simple brain in the head, and sensory organs that detect chemicals, moisture, light, and touch.

Most freshwater and terrestrial annelids are hermaphroditic, and worms exchange sperm when they mate. Newly hatched worms have an adult morphology. Some terrestrial annelids also reproduce asexually by fragmenting and regenerating missing parts. Marine annelids usually have separate sexes, and release gametes into the sea for fertilization. The zygotes develop into trochophore larvae that add segments, gradually assuming an adult form.

Recent phylogenetic studies have produced conflicting conclusions about evolutionary relationships within the Annelida. A large genomic analysis of nearly 48,000 amino acid positions in 34 annelid taxa, published in 2011, divided most annelids into one of two clades: Errantia, which are generally very mobile, and Sedentaria, which are generally less active. However, an even more recent analysis of nuclear genes does not provide strong support for the definition of those clades. Nevertheless, both studies as well as several others recognize Clitellata (described below) as a monophyletic group.

"POLYCHAETA" A majority of annelids are described as polychaetes or bristle worms (polychaeta: poly = many; *chaite* = bristles or hair) because of their abundant setae, although polychaeta does not comprise a monophyletic taxon; most polychaetes are marine **(Figure 31.28).** Many live under rocks or in tubes constructed from mucus, calcium carbonate secretions, grains of sand, or small shell fragments. Their setae project from well-developed **parapodia** (singular, *parapodium* = closely resembling a foot), fleshy lateral extensions of the body wall used for locomotion and gas exchange. In more mobile species, sense organs are concentrated on a well-developed head.

Crawling or swimming polychaetes are often predatory; they use sharp jaws in a protrusible muscular pharynx to grab small invertebrate prey. Other species graze on algae or scavenge organic matter. A few tube dwellers draw food-laden water into the tube by beating their parapodia; most others collect food by extending feathery, ciliated, mucus-coated tentacles.

CLITELLATA The Clitellata (*clitella* = a packsaddle) includes annelid worms that had previously been classified as either Oligochaeta (Oligochaeta: *oligos* = few) or Hirudinea (Hirudinea: *hirudo* = leech). Many oligochaetes are terrestrial **(Figure 31.29),** but they are restricted to moist habitats because they quickly dehydrate in dry air or soil. Known as earthworms, terrestrial oligochaetes are generally nocturnal, spending their days in burrows that they excavate. They range in length from a few millimeters to more than 3 m. Aquatic oligochaetes live in mud or detritus at the bottom of lakes, rivers, and estuaries.

Earthworms are scavengers on decomposing organic matter. In his book *The Formation of Vegetable Mould through the Action of Worms*, Darwin noted that earthworms can ingest their own weight in soil every day. He calculated that a typical population of 16,000 worms per hectare (10,000 m²) consumes more than 20 tons of soil in a year. This impressive activity aerates soil and makes nutrients available to plants by mixing the subsoil with the topsoil. Earthworms have

A. Fan worm (*Sabella melanostigma*)

Marco Lijoi/Dreamstime.com

B. *Nereis* feeding structures

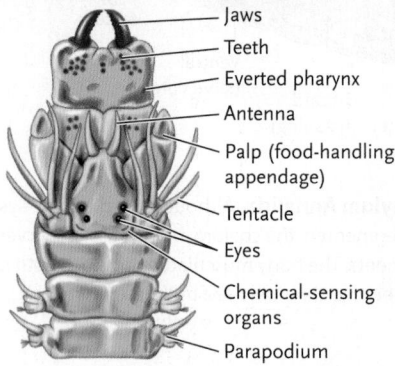

- Jaws
- Teeth
- Everted pharynx
- Antenna
- Palp (food-handling appendage)
- Tentacle
- Eyes
- Chemical-sensing organs
- Parapodium

C. *Proceraea cornuta* setae

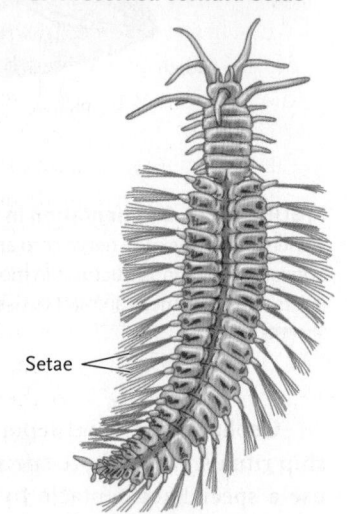

Setae

FIGURE 31.28 Polychaetes. (A) The tube-dwelling fan worm has mucus-covered tentacles that trap small food particles. **(B)** Some polychaetes actively seek food; when they encounter a suitable tidbit, they evert their pharynx, exposing sharp jaws that grab the prey and pull it into the digestive system. **(C)** Many marine polychaetes have numerous setae, which they use for locomotion.
© Cengage Learning 2017

© mashe/Shutterstock.com

FIGURE 31.29 Oligochaetes. Earthworms (*Lumbricus terrestris*) generally move across the ground surface at night.

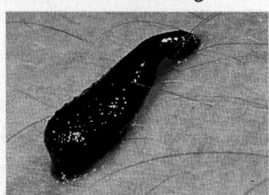
Leech before feeding

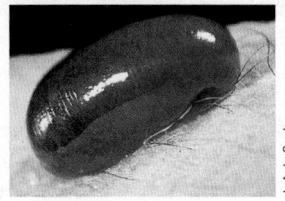
Leech after feeding

J. A. L. Cooke

FIGURE 31.30 Hirudinea. Parasitic leeches consume huge blood meals, as shown by these before and after photos of a medicinal leech *(Hirudo medicinalis)*. Because suitable hosts are often hard to locate, gorging allows a leech to take advantage of any host it finds.

complex organ systems (see Figure 31.28), and they sense light and touch at both ends of the body. In addition, they have moisture receptors, an important adaptation in organisms that must stay damp to allow gas exchange across the skin.

Leeches are mostly freshwater parasites. They have dorsoventrally flattened, tapered bodies with a sucker at each end. Although the body wall is segmented, the coelom is reduced and not partitioned by septa. Many leeches are ectoparasites of vertebrates, but some attack small invertebrate prey.

Parasitic leeches feed on the blood of their hosts. Most attach to the host with the posterior sucker, and then use their sharp jaws to make a small, often painless, triangular incision. A sucking apparatus draws blood from the prey, while a special secretion, *hirudin,* maintains the flow by preventing the host's blood from coagulating. Leeches have a highly branched gut that allows them to consume huge blood meals **(Figure 31.30).** For centuries, doctors used medicinal leeches *(Hirudo medicinalis)* to "bleed" patients; today, surgeons use them to drain excess fluid from tissues after reconstructive surgery, reducing swelling until the patient's blood vessels regenerate and resume this function.

STUDY BREAK 31.6

1. What organ systems are present in free-living flatworms (Turbellaria)? Which of these organ systems is absent in tapeworms (Cestoda)?
2. What anatomical characteristic reveals the close evolutionary relationship of ectoprocts, brachiopods, and phoronid worms?
3. What anatomical structures and physiological systems allow squids and other cephalopods to be much more active than other types of mollusks?
4. Which organ systems exhibit segmentation in most annelid worms?

THINK OUTSIDE THE BOOK

Search the records of your local health department, either online or in person, to discover whether people in your town or city routinely suffered from parasitic worm infections in the past. What public health measures stopped the outbreak?

31.7 Ecdysozoan Protostomes

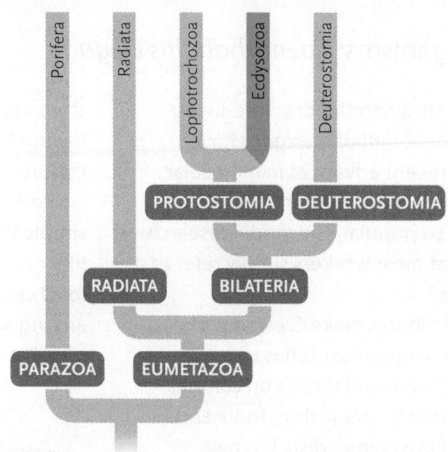

The three phyla in the protostome group Ecdysozoa all have an external covering that they shed periodically. The outer covering protects these animals from harsh environmental conditions, and it helps parasitic species resist host defenses. Although many of these animals live in either aquatic or moist terrestrial habitats, a tough exoskeleton allows one group, the insects, to thrive on dry land and in the air.

Nematodes Are Unsegmented Worms Covered by a Flexible Cuticle

Roundworms (phylum Nematoda: *nema* = thread) are perhaps the most abundant animals on Earth **(Figure 31.31).** A cupful of rich soil, a dead earthworm, or a rotting fruit may contain thousands of them. Although 15,000 species have been described, experts estimate that more than half a million exist. Many nematodes are almost microscopically small, but some species are a meter or more long. They occupy nearly every freshwater, marine, and terrestrial habitat on Earth, consuming detritus, microorganisms, plants, or animals.

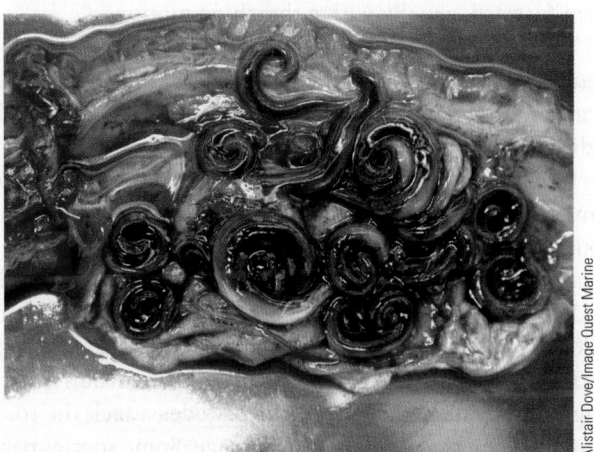

Alistair Dove/Image Quest Marine

FIGURE 31.31 Phylum Nematoda. Some roundworms, such as these *Anguillicola crassus* in the swim bladder of an eel, are parasites of plants or animals. Others are important consumers of dead organisms in most ecosystems.

Model Organisms: *Caenorhabditis Elegans*

Researchers studying the tiny, free-living nematode *Caenorhabditis elegans* have made many recent advances in molecular genetics, animal development, and neurobiology. It is so popular as a model research organism that most workers simply refer to it as "the worm."

Several attributes make *C. elegans* a model research organism. It has an adult size of about 1 mm and thrives on cultures of *E. coli* or other bacteria; thus, thousands can be raised in a culture dish. It is hermaphroditic and often self-fertilizing, which allows researchers to maintain pure genetic strains. It completes its life cycle from egg to reproductive adult within 3 days at room temperature. Furthermore, stock cultures can be kept alive indefinitely by freezing them in liquid nitrogen or in an ultracold freezer set to minus 80°C. Researchers can therefore

store new mutants for later research without having to clean, feed, and maintain active cultures.

Best of all, the worm is anatomically simple **(Figure)**; an adult contains just 959 cells (excluding the gonads). Having a fixed cell number is relatively uncommon among animals, and developmental biologists have made good use of this trait. The

eggs, juveniles, and adults of the worm are completely transparent, and researchers can observe cell divisions and cell movements in living animals with straightforward microscopy techniques. There is no need to kill, fix, and stain specimens for study. And virtually every cell in the worm's body is accessible for manipulation by laser microsurgery, microinjection, and similar approaches.

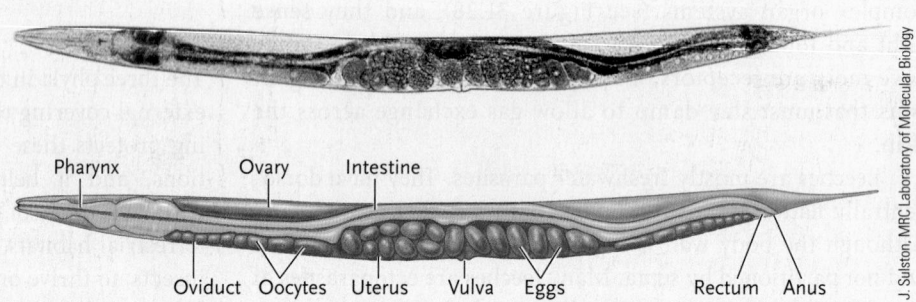

Pharynx   Ovary   Intestine

Oviduct   Oocytes   Uterus   Vulva   Eggs   Rectum   Anus

J. Sulston, MRC Laboratory of Molecular Biology

© Cengage Learning 2017

The roundworm body is cylindrical and usually tapered at both ends. None of the cells have cilia or flagella. Roundworms are covered in a tough but flexible, water-resistant **cuticle,** which must be molted periodically; the cuticle is replaced by the underlying epidermis as the worm grows. The cuticle prevents the animal from dehydrating in dry environments, and, in parasitic species, it resists attack by acids and enzymes in a host's digestive system. Beneath the cuticle and epidermis, a layer of longitudinal muscles extends the length of the body. Alternating contractions of these muscles on a nematode's dorsal and ventral sides push against the substrate and propel the worm forward, usually with a thrashing motion.

The adults of one soil-dwelling species, *Caenorhabditis elegans,* are transparent and contain fewer than 1,000 cells. As *Focus on Research: Model Organisms* explains, biologists have studied this worm inside and out.

Nematodes reproduce sexually, and the sexes are separate in most species. In some, internal fertilization produces many thousands of fertile eggs per day. The eggs of many species can remain dormant if environmental conditions are unsuitable.

Because of their great numbers, nematodes have enormous ecological, agricultural, and medical significance. Free-living species are responsible for decomposition and nutrient recycling in many habitats. Parasitic nematodes attack the roots of plants, causing tremendous crop damage. Some species parasitize animals, including humans. Although some, like pinworms *(Enterobius),* are more of a nuisance than a danger, others, like trichinas or filarial worms, can cause serious disease, disfigurement, or even death (see *Focus on Research: Applied Research*

on page 722). More than 1 billion people worldwide suffer from debilitating and life-threatening nematode infections.

Velvet Worms Have Segmented Bodies and Numerous Unjointed Legs

The 180 or so living species of velvet worms (phylum Onychophora: *onyx* = claw) live under stones, logs, and forest litter in the tropics and in moist temperate habitats in the southern hemisphere. They range in size from 15 mm to 15 cm and feed on small invertebrates.

Onychophorans have a flexible cuticle, segmented bodies, and numerous pairs of unjointed legs **(Figure 31.32)**. Like the annelids, they have pairs of excretory organs in most segments. But unlike annelids, they have an open circulatory system, a specialized respiratory system, relatively large brains, jaws, and tiny claws on their feet. Many produce live young, which, in some species, are nourished within a uterus. Fossil evidence indicates that onychophorans originated during the Cambrian explosion (see *Hallucigenia* in Figure 31.1) and that their body plan has not changed much over the last 500 million years.

Arthropods Are Segmented Animals with a Hard Exoskeleton and Jointed Appendages

The 1.15 million known species of **arthropods** (phylum Arthropoda: *arthron* = joint) include more than half the animal species on Earth—and only a fraction of the living arthropods have been described. This huge clade includes insects,

FIGURE 31.32 Phylum Onychophora. Species in the small phylum Onychophora, such as this *Peripatus* species from a lowland rainforest in Ecuador, have segmented bodies and unjointed appendages.

spiders, crustaceans, millipedes, centipedes, the extinct trilobites, and their relatives.

Arthropods have a segmented body encased in a rigid **exoskeleton.** This external covering is made of chitin, a mix of polysaccharide fibers glued together with glycoproteins, as well as waxes and lipids that block the passage of water. In some marine groups, such as crabs and lobsters, it is hardened with calcium carbonate. The exoskeleton probably first evolved in marine species, providing protection against predators. In terrestrial habitats, it provides support against gravity and protection from dehydration, which contributes to the success of insects in even the driest places on Earth. The exoskeleton is especially thin and flexible at the joints between body segments and in the appendages. Contractions of muscles attached to the inside of the exoskeleton move individual body parts like levers, allowing highly coordinated movements and patterns of locomotion that are more precise than those in soft-bodied animals with hydrostatic skeletons.

Although the exoskeleton has obvious advantages, it is nonexpandable and therefore could limit growth of the animal. But, like other Ecdysozoa, arthropods grow and periodically develop a soft, new exoskeleton beneath the old one, which they shed in the complex process of ecdysis **(Figure 31.33).** After shedding the old exoskeleton, aquatic species swell with water and terrestrial species swell with air before the new one hardens. They are especially vulnerable to predators at these times. "Soft-shelled" crabs, prized as food in many countries, are crabs that have recently molted.

As arthropods evolved, body segments became fused in various ways, reducing the overall number of segments. Each region of the body, along with its highly modified paired appendages, is specialized, but the structure and function vary greatly among groups. In insects (see Figure 31.43), which have three body regions, the **head** includes a brain, sensory structures, and some sort of feeding apparatus. The segments of the **thorax** bear walking legs and, in some insects, wings. The **abdomen** includes much of the digestive system and sometimes part of the reproductive system.

FIGURE 31.33 Ecdysis in insects. Like all other arthropods, this cicada *(Graptopsaltria nigrofuscata)* sheds its old exoskeleton as it grows and when it undergoes metamorphosis into a winged adult.

The coelom of arthropods is greatly reduced, but another cavity, the *hemocoel*, is filled with bloodlike hemolymph. The heart pumps the hemolymph through an open circulatory system, bathing tissues directly.

Arthropods are active animals and require substantial quantities of oxygen. Different groups have distinctive mechanisms for gas exchange, because oxygen cannot cross the impermeable exoskeleton. Marine and freshwater species, such as crabs and lobsters, rely on diffusion across gills. The terrestrial groups—insects and spiders—have developed unique and specialized respiratory systems (described further in Section 46.2).

High levels of activity also require intricate sensory structures. Many arthropods are equipped with a highly organized central nervous system, touch receptors, chemical sensors, **compound eyes** that include multiple image-forming units, and in some, hearing organs.

Arthropod systematics is an active area of research, and scientists are currently using molecular, morphological, and developmental data to reexamine relationships within this immense group. As *Molecular Insights* explains, recent analyses have confirmed that Arthropoda includes two major clades, **Chelicerata** and **Mandibulata,** and that the Mandibulata includes **Myriapoda** and **Pancrustacea** (see the 2010 tree in *Molecular Insights*). In turn, Pancrustacea includes both the animals that are commonly described as "crustaceans" and the **Hexapoda** (insects and their closest relatives). In our discussion of arthropod diversity, we describe arthropod

Arthropod Relationships: Is Crustacea a monophyletic lineage?

For more than 100 years, biologists disagreed about the evolutionary relationships among the major clades of arthropods. Some researchers grouped hexapods and myriapods together (**Figure,** *Traditional tree*) because they share certain morphological characteristics, which may indicate a common ancestry: unbranched appendages, a tracheal system for gas exchange, Malpighian tubules for excretion, and one pair of antennae. However, other researchers grouped hexapods (including insects) and crustaceans together because they share a different set of important morphological characters: jawlike mandibles on the fourth head segment, similar compound eyes, comparable development of the nervous system, and similarities in the structure of thoracic appendages. In 1995, an analysis of mitochondrial DNA (mtDNA) in various arthropods, using an annelid and a mollusk as outgroups, revealed that hexapods and crustaceans shared distinctive similarities in the locations of two genes coding for transfer RNAs, suggesting their close evolutionary relationship (**Figure,** *1995 tree*).

Research Question
Are Crustacea and Hexapoda sister groups, or is one included within the other?

Experiment
In a paper published in 2010, Jerome C. Regier of the University of Maryland Biotechnology Institute and colleagues from the University of Maryland, Duke University, and the Natural History Museum of Los Angeles reported the results of a large-scale phylogenomic analysis of arthropod nuclear protein-coding sequences. They compared DNA sequences in 62 protein-coding genes from 75 arthropod species, including representatives of every major arthropod clade. Five nonarthropods served as outgroups in the analysis.

The researchers restricted their analysis to genes for which exact equivalents (also called *orthologs*, described in Section 19.4) could be identified in each of the 75 arthropod species, as well as the 5 outgroup species. For example, one of the genes included in the analysis codes for the aminoacyl–tRNA–synthetase enzyme that attaches leucine amino acids to selected transfer RNAs. One and only one equivalent gene is present in the genomes of all 75 species.

The researchers obtained sequence data from the 62 genes using polymerase chain reaction (PCR) primers that were designed to amplify unduplicated gene sequences that are homologous in groups that diverged from a common ancestor. They aligned the genes on the basis of translated amino acid sequences, which are highly conserved among species, to ensure that the genes being compared were indeed homologous.

The sequence data were analyzed with four maximum-likelihood strategies (see Section 24.5), each of which used a different set of assumptions about molecular evolution. The different maximum-likelihood analyses produced the same results, providing strong support for the resulting phylogenetic tree. The researchers also confirmed their phylogenetic hypothesis by constructing additional trees using Bayesian and parsimony approaches (also described in Section 24.5).

Results
The phylogenetic tree for Arthropoda (**Figure,** *2010 tree*) reveals that the Chelicerata (including spiders, scorpions, and horseshoe crabs) is the sister group to the Mandibulata (including all other arthropods). Mandibulata includes two major clades: Myriapoda (centipedes and millipedes) and Pancrustacea. The latter clade comprises four smaller clades, including Hexapoda. Several additional studies, using mitochondrial protein-coding genes, nuclear ribosomal genes, and other molecular characters, have largely confirmed these results.

Conclusion
The 2010 analysis resurrected the clade Mandibulata, which had been defined decades before on the basis of the shared mouthparts of these animals. The analysis further revealed that Hexapoda and Crustacea are not sister clades. Instead, Hexapoda is part of the larger clade Pancrustacea. Indeed, as traditionally defined, Crustacea is a paraphyletic group (that is, an ancestor and only some of its descendants) because it does not include the Hexapoda, the most species-rich group within it.

think like a scientist

If we were able to sequence the DNA of a trilobite, where do you think it would fit on the 2010 phylogenetic tree?

Source: J. C. Regier et al. 2010. Arthropod relationship revealed by phylogenomic analysis of nuclear protein-coding sequences. *Nature* 463:1079–1084.

© Cengage Learning 2017

FIGURE 31.34 Trilobita. Trilobites, such as *Olenellus gilberti*, bore many pairs of relatively undifferentiated appendages.

Dr. Chip Clark

body plans and ways of life in terms of the most familiar groupings: Trilobita, Chelicerata, Myriapoda, "crustaceans," and Hexapoda.

TRILOBITA The trilobites (**Trilobita:** *tri* = three; *lobos* = lobe), now extinct, were among the most numerous animals in shallow Paleozoic seas. They disappeared in the Permian mass extinction, but the cause of their demise is unknown. Most trilobites were ovoid, dorsoventrally flattened, and heavily armored, with two deep longitudinal grooves that divided the body into the one median and two lateral lobes for which the group is named (**Figure 31.34**). Their segmented bodies were organized into three body regions: the head, which included a

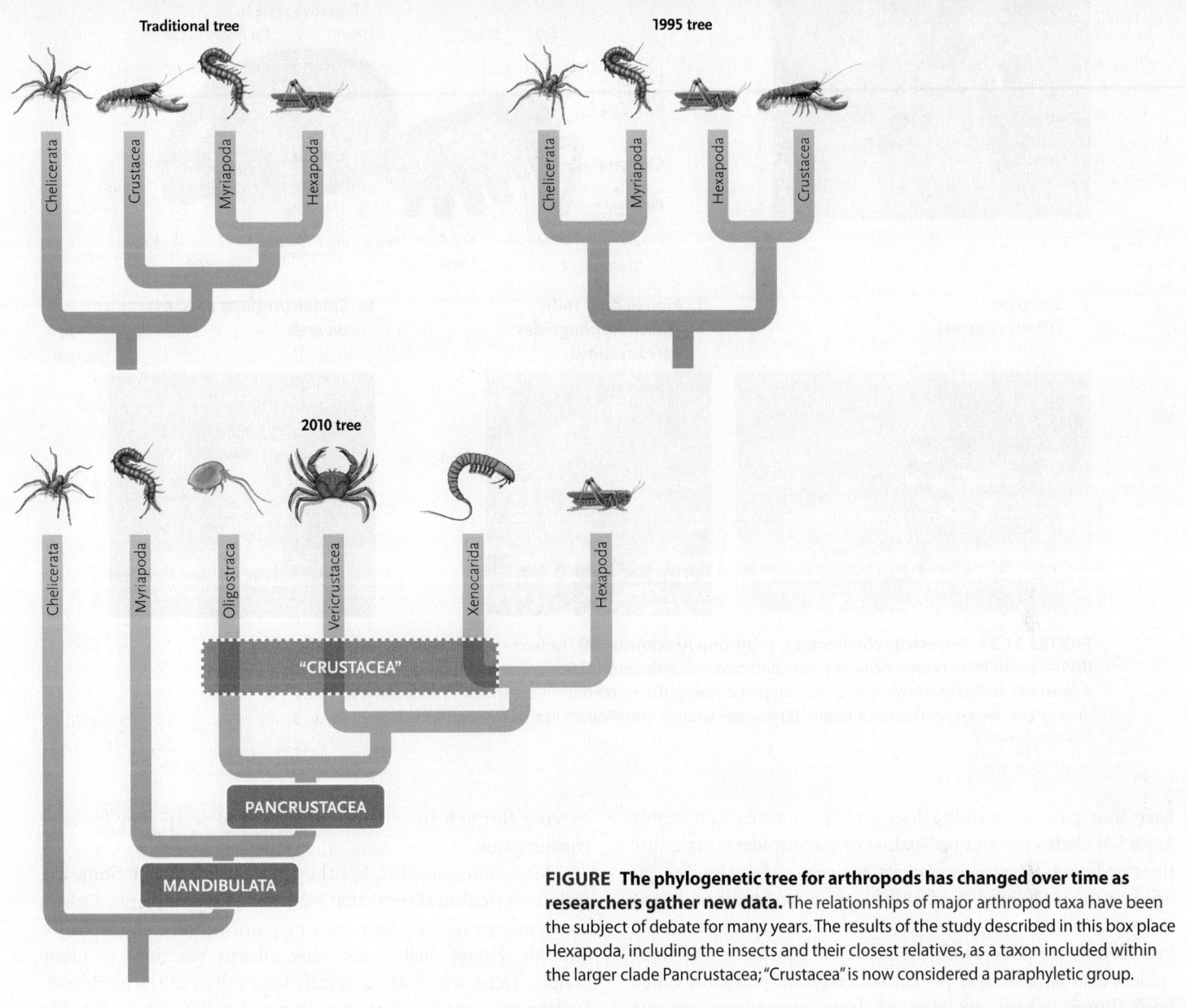

Traditional tree

Chelicerata | Crustacea | Myriapoda | Hexapoda

1995 tree

Chelicerata | Myriapoda | Hexapoda | Crustacea

2010 tree

Chelicerata | Myriapoda | Oligostraca | Vericrustacea | Xenocarida | Hexapoda

"CRUSTACEA"

PANCRUSTACEA

MANDIBULATA

FIGURE The phylogenetic tree for arthropods has changed over time as researchers gather new data. The relationships of major arthropod taxa have been the subject of debate for many years. The results of the study described in this box place Hexapoda, including the insects and their closest relatives, as a taxon included within the larger clade Pancrustacea; "Crustacea" is now considered a paraphyletic group.

pair of sensory **antennae** (chemosensory organs) and two compound eyes, and a thorax and an abdomen, both of which bore pairs of jointed walking legs.

The position of trilobites in the fossil record indicates that they were among the earliest arthropods. Thus, biologists are confident that their three body regions and numerous unspecialized appendages—one pair per segment—represent ancestral traits within the phylum. As you will learn as you read about the other four clades, the subsequent evolution of the different arthropod groups included dramatic remodeling of the major body regions, as well as modifications of the ancestral, unspecialized paired appendages into highly specialized structures that perform different functions.

CHELICERATA In spiders, ticks, mites, scorpions, and horseshoe crabs (subphylum Chelicerata: *chela* = claw; *keras* = horn), the first pair of appendages, the **chelicerae,** are fanglike structures used for biting prey. The second pair of appendages, the *pedipalps,* serve as grasping organs, sensory organs, or walking legs. All chelicerates have two major body regions, the **prosoma** (a fused head and thorax) and the **opisthosoma** (derived from the ancestral abdomen). The group originated in shallow Paleozoic seas, but most living species are terrestrial. They vary in length from less than a millimeter to 20 cm; all are predators or parasites.

The spiders, scorpions, mites, and ticks (Arachnida) represent the vast majority of chelicerates **(Figure 31.35).** Arachnids

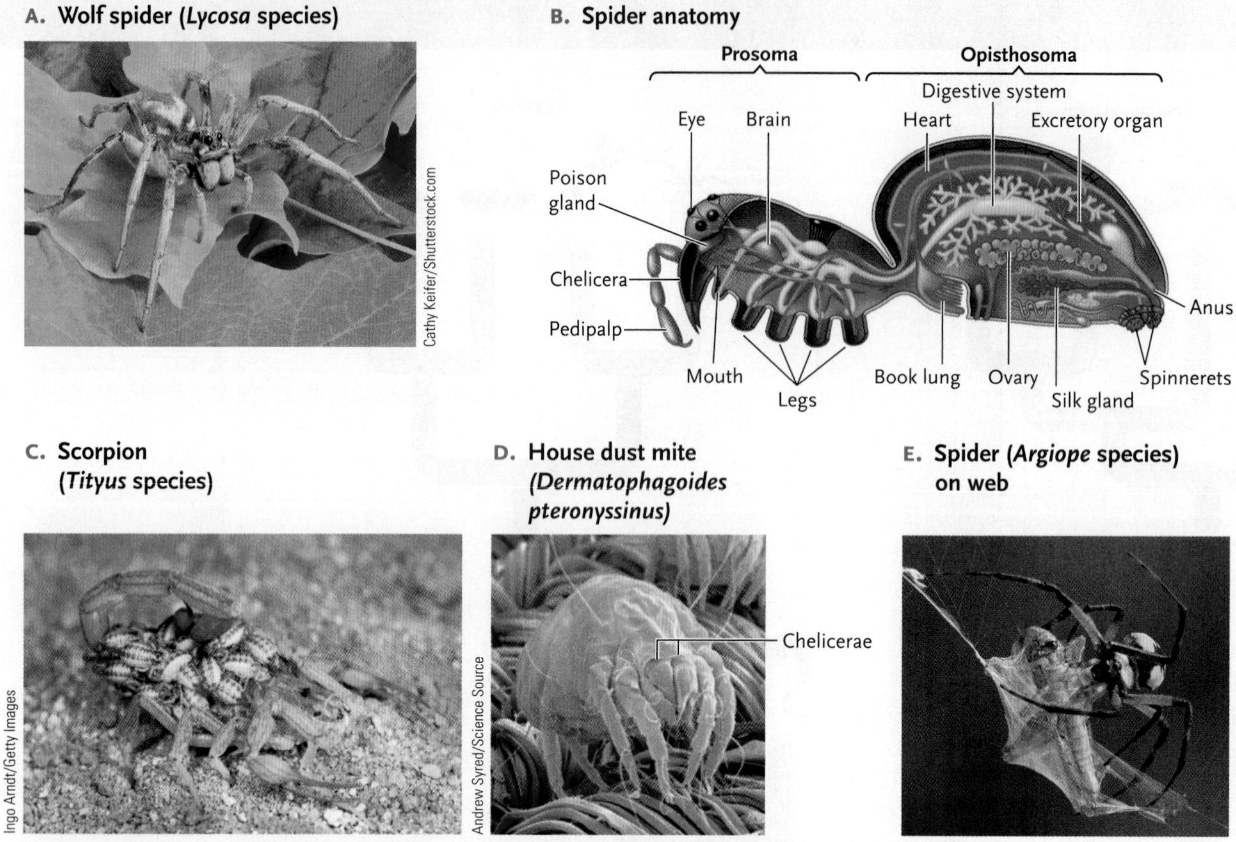

A. Wolf spider (*Lycosa* species)

B. Spider anatomy

Prosoma | Opisthosoma

Eye | Brain | Heart | Digestive system | Excretory organ
Poison gland
Chelicera
Pedipalp
Mouth | Legs | Book lung | Ovary | Silk gland | Spinnerets | Anus

C. Scorpion (*Tityus* species)

D. House dust mite (*Dermatophagoides pteronyssinus*)

Chelicerae

E. Spider (*Argiope* species) on web

FIGURE 31.35 Terrestrial chelicerates, subgroup Arachnida. (A) The wolf spider is harmless to humans. **(B)** The arachnid body plan includes a cephalothorax and abdomen. **(C)** Scorpions have a stinger at the tip of the segmented abdomen. Many protect their eggs and young. **(D)** House dust mites, shown in a scanning electron micrograph, feed on microscopic debris. **(E)** A spider wraps a grasshopper in silk after capturing it on its web.
© Cengage Learning 2017

have four pairs of walking legs on the prosoma and highly modified chelicerae and pedipalps. In some spiders, males use their pedipalps to transfer packets of sperm to females. Scorpions use them to shred food and to grasp one another during courtship. Many predatory arachnids have excellent vision, provided by simple eyes on the prosoma. Scorpions and some spiders also have unique pocketlike respiratory organs, called **book lungs,** which are derived from appendages on the opisthosoma.

Spiders, like most other arachnids, subsist on a liquid diet. They use their chelicerae to inject paralyzing poisons and digestive enzymes into prey and then suck up the partly digested tissues. Many spiders are economically important predators, helping to control insect pests. Only a few are a threat to humans. The toxin of a black widow (*Latrodectus mactans*) causes paralysis, and the toxin of the brown recluse (*Loxosceles reclusa*) destroys tissues around the site of the bite.

Although many spiders hunt actively, others capture prey on silken threads secreted by **spinnerets,** which are modified opisthosomal appendages. Some species weave the threads into complex, netlike webs. The silk is secreted as a liquid, but quickly hardens on contact with air. Spiders also use silk to make nests, to protect their egg masses, as a safety line when moving through the environment, and to wrap prey for later consumption.

Most mites are tiny, but they have a big impact. Some are serious agricultural pests that feed on the sap of plants. Others cause mange (patchy hair loss) or painful and itchy welts on animals. House dust mites cause allergic reactions in many people. Ticks, which are generally larger than mites, are blood-feeding ectoparasites that often transmit pathogens, such as the bacteria that cause Rocky Mountain spotted fever and Lyme disease.

Chelicerata also includes five species of horseshoe crabs (Merostomata), an ancient lineage with a morphology that has not changed much over its 350-million-year history **(Figure 31.36).** Horseshoe crabs are carnivorous bottom feeders in shallow coastal waters. Beneath their characteristic shell, they have one pair of chelicerae, a pair of pedipalps, four pairs of walking legs, and a set of paperlike gills that are derived from ancestral walking legs.

MYRIAPODA The centipedes (Chilopoda) and millipedes (Diplopoda) are classified together in the Myriapoda (*myrias* = ten thousand; *pod-* = feet). Myriapods have two body regions, a head and a segmented trunk **(Figure 31.37).** The head bears one

FIGURE 31.36 **Marine chelicerates.** Horseshoe crabs, such as *Limulus polyphemus,* are included in the Merostomata. In this photo three males—with barnacles growing on their carapaces—attempt to mate with a female in shallow water.

A. Millipede (*Spirobolus* species)

B. Centipede (*Scolopendra* species)

FIGURE 31.37 **Myriapods. (A)** Millipedes feed on living and decaying vegetation. They have two pairs of walking legs on most segments. **(B)** All centipedes are voracious predators, feeding on invertebrates and small vertebrates. Centipedes have one pair of walking legs per segment.

pair of antennae, and the trunk bears one (centipedes) or two (millipedes) pairs of walking legs on most of its many segments. Myriapods are terrestrial, and many species live under rocks or dead leaves. Centipedes are fast and voracious predators, using powerful toxins to kill their prey; they generally feed on invertebrates, but some eat small vertebrates. The bite of some species is harmful to humans. Although most species are less than 10 cm long, some grow to 25 cm. The millipedes are slow but powerful herbivores or scavengers. The largest species attain a length of nearly 30 cm. Although they lack a poisonous bite, they curl into a ball and exude noxious liquids when disturbed.

"CRUSTACEANS" The shrimps, lobsters, crabs, and their relatives (crustaceans: *crusta* = crust or hard shell) represent a lineage that emerged more than 500 million years ago. They are abundant in marine and freshwater habitats. A few species, such as sowbugs and pillbugs, live in moist, sheltered terrestrial environments. In many crustaceans two, or even all three, of the arthropod body regions—head, thorax, and abdomen—are fused; a fused cephalothorax and a separate abdomen is an especially common pattern. The edible "tail" of a lobster, shrimp, or crayfish is actually a highly muscularized abdomen. In some, the exoskeleton includes a **carapace,** a protective covering that extends backward from the head. Crustaceans vary in size from water fleas less than 1 mm long to lobsters that can grow to 60 cm in length and weigh as much as 20 kg.

Crustaceans generally have five pairs of appendages on the head **(Figure 31.38).** Most have two pairs of sensory antennae and three pairs of mouthparts. The latter include one pair of *mandibles,* which move laterally to bite and chew, and two pairs of *maxillae,* which hold and manipulate food. Numerous

A. Crab (*Ocypode* species)

B. Lobster *(Homarus americanus)*

FIGURE 31.38 **Decapod crustaceans. (A)** Crabs, such as this ghost crab, and **(B)** lobsters are typical decapod crustaceans. The abdomen of a crab is shortened and wrapped under the cephalothorax, producing a compressed body. **(C)** Lobsters bear 19 pairs of distinctive appendages; only appendages on the left side of the animal are illustrated in this lateral view, and one pair of mandibles and two pairs of maxillae are not shown. The animal illustrated is a male.
© Cengage Learning 2017

C. Lobster external anatomy

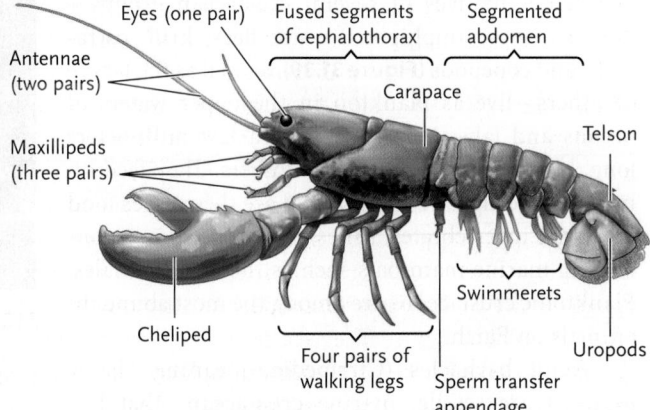

paired appendages posterior to the mouthparts vary among groups.

Most crustaceans are active animals that exhibit complex patterns of movement. Their activities are coordinated by elaborate sensory and nervous systems, including chemical and touch receptors in the antennae, compound eyes, statocysts on the head, and sensory hairs embedded in the exoskeleton throughout the body. The nervous system is similar to that of annelids, but the ganglia are larger and more complex, allowing a finer level of motor control. High levels of activity require substantial amounts of oxygen, and larger species have complex, feathery gills tucked beneath the carapace. Activity also produces abundant metabolic wastes that are excreted by diffusion across the gills or, in larger species, by glands located in the head.

The sexes are typically separate, and courtship rituals are often complex. Eggs are usually brooded on the surface of the female's body or beneath the carapace. Many have free-swimming larvae that, after undergoing a series of molts, gradually assume an adult form.

Crustaceans have so many different body plans that they are divided into many groups and subgroups. The crabs, lobsters, and shrimps (Decapoda, meaning "10 feet") number more than 10,000 species. The vast majority of decapods are marine, but a few shrimps, crabs, and crayfishes occupy freshwater habitats. Some crabs also live in moist terrestrial habitats, where they scavenge dead vegetation, clearing the forest floor of debris.

All decapods exhibit extreme specialization of their appendages. In the American lobster, for example, each of the 19 pairs of appendages is different (see Figure 31.38C). Behind the antennae, mandibles, and maxillae, the thoracic segments have three pairs of *maxillipeds,* which shred food and pass it up to the mouth, a pair of large *chelipeds* (pinching claws), and four pairs of walking legs. The abdominal appendages include a pair specialized for sperm transfer (in males only), *swimmerets* for locomotion and for brooding eggs, and *uropods,* a pair of appendages that, combined with the *telson,* the tip of the abdomen, form a fan-shaped tail. If any appendage is damaged, the animal can autotomize ("drop") it and begin growing a new one before its next molt.

Representatives of several crustacean groups—fairy shrimps, amphipods, water fleas, krill, ostracods, and copepods **(Figure 31.39),** as well as the larvae of others—live as plankton in the upper waters of oceans and lakes. Most are only a few millimeters long, but are present in huge numbers. They feed on microscopic algae or detritus and are themselves food for larger invertebrates, fishes, and some suspension-feeding marine mammals such as the baleen whales. Planktonic crustaceans are among the most abundant animals on Earth.

Adult barnacles (Cirripedia, meaning "hairy footed") are sessile, marine crustaceans that live within a strong, calcified cup-shaped shell **(Figure 31.40).** Their free-swimming larvae attach permanently to substrates—rocks, wooden pilings, the hulls of ships, the shells of mollusks or of other arthropods, even the skin of whales—and secrete the shell, which is actually a modified exoskeleton. To feed, barnacles open the shell and extend six pairs of feathery legs. The beating legs capture microscopic plankton and transfer it the mouth. Unlike most crustaceans, barnacles are hermaphroditic.

HEXAPODA In terms of sheer numbers, diversity, and the range of habitats they occupy, the 1,000,000 or more species of insects and their closest relatives (Hexapoda, *hexapoda* = six feet) are the most successful animals on Earth. They were among the first animals to colonize terrestrial habitats, where most species still live. The oldest insect fossils date from the Devonian, 380 million years ago. Insects are generally small, ranging from 0.1 mm to 30 cm in length. The lineage is divided

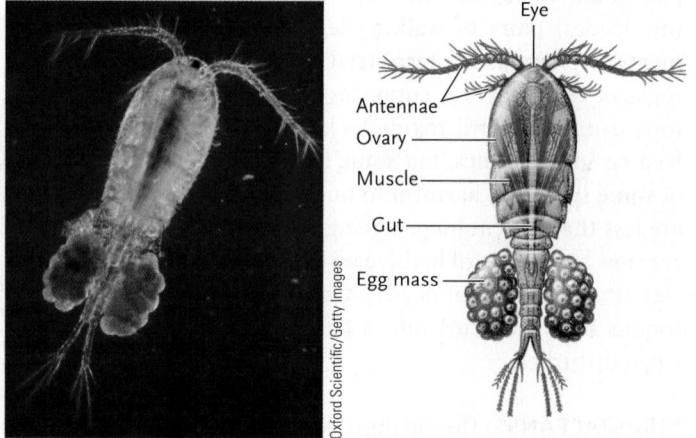

FIGURE 31.39 Copepods. Tiny crustaceans, such as these copepods (*Cyclops* species), occur by the billions in freshwater and marine plankton.
© Cengage Learning 2017

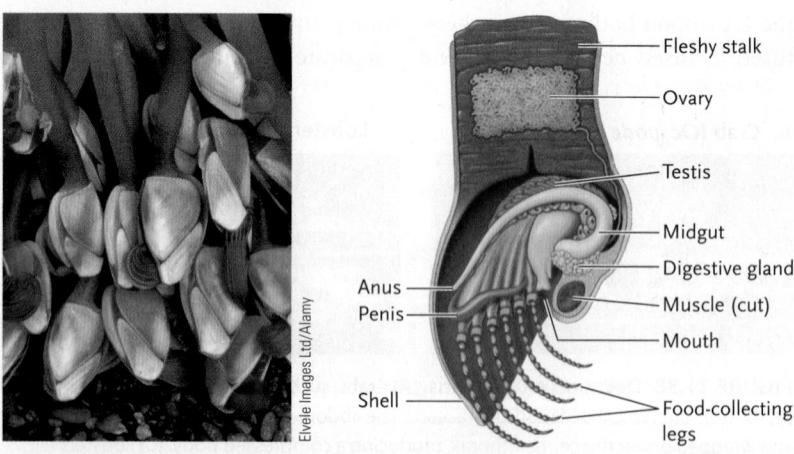

FIGURE 31.40 Barnacles. Gooseneck barnacles (*Lepas anatifera*) attach to the underside of floating debris. Like other barnacles, they open their shells and extend their feathery legs to collect particulate food from seawater.
© Cengage Learning 2017

into about 30 subgroups (**Figure 31.41**). The insect body plan always includes a head, thorax, and abdomen (**Figure 31.42**). The head is equipped with one pair of sensory antennae, multiple mouthparts, and a pair of compound eyes. The thorax has three pairs of walking legs and often one or two pairs of wings. Insects are the only invertebrates capable of flight. Their wings, which are made of lightweight but durable sheets of chitin and sclerotin, arise embryonically from the body wall; unlike the wings of birds and bats, insect wings are not derived from ancestral locomotor appendages.

Studies in evolutionary developmental biology (discussed in Section 23.6) have begun to unravel the genetic changes that fostered certain aspects of the insect body plan. For example, insect wings were a key innovation in their evolutionary

A. Silverfish (Thysanura, *Lepisma saccharina*) are wingless, an ancestral trait within insects.

B. Dragonflies (Odonata, *Epitheca cynosura*) have aquatic larvae that are active predators; adults capture other insects in mid-air.

C. Male praying mantids (Mantodea, *Mantis religiosa*) are often eaten by the larger females during or immediately after mating.

D. This rhinoceros beetle (Coleoptera, *Lucanus cervus*) is one of more than 250,000 beetle species that have been described.

E. Fleas (Siphonoptera, *Ctenophalides canis*) have strong legs with an elastic ligament that allows these parasites to jump on and off their animal hosts.

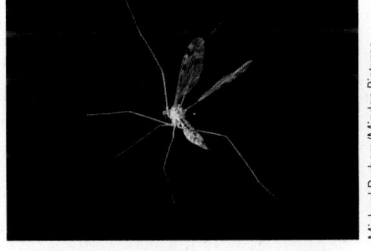

F. Crane flies (Diptera, *Tipula* species) look like giant mosquitoes, but their mouthparts are not useful for biting other animals; the adults of most species live only a few days and do not feed at all.

G. The luna moth (Lepidoptera, *Actias luna*), like other butterflies and moths, has wings that are covered with colorful microscopic scales.

H. Like many other ant species, fire ants (Hymenoptera, *Solenopsis invicta*) live in large cooperative colonies. Fire ants—named for their painful sting—were introduced into southeastern North America, where they are now serious pests.

FIGURE 31.41 Insect diversity. Insects are grouped into about 30 clades, 8 of which are illustrated here.

External anatomy of a grasshopper

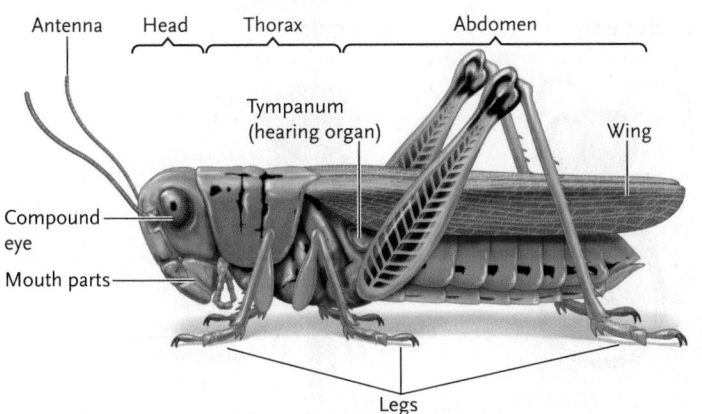

Internal anatomy of a female grasshopper

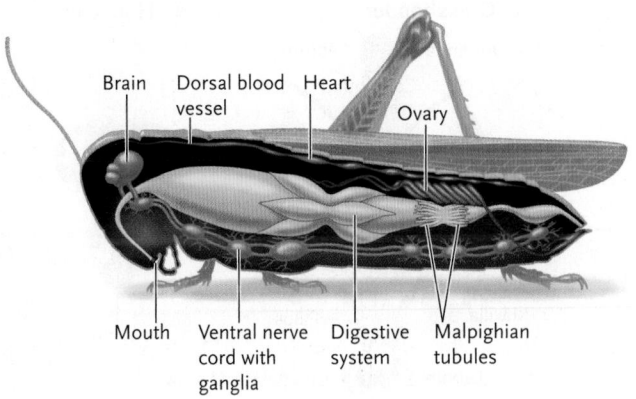

FIGURE 31.42 The insect body plan. Insects have distinct head, thorax, and abdomen. Of all the internal organ systems, only the dorsal blood vessel, ventral nerve cord, and some muscles are strongly segmented.

success. Researchers hypothesize that insect wings were derived from gill-like appendages responsible for gas exchange in wingless insect ancestors. The appendages of aquatic crustaceans, which are closely related to insects, have two branches, one of which is used for gas exchange. Researchers have discovered several developmental genes (including *nubbin, apterous,* and the *Hox* gene *pdm*) that are expressed during the development of both crustacean respiratory appendages and insect wings.

As another example of remarkable discoveries about the origin of the insect body plan, researchers in developmental genetics have discovered the basis for the loss of appendages on the insect abdomen. The *Distal-less* gene (*Dll* for short) is a highly conserved toolkit gene (see Section 23.6) that triggers the development of appendages in all sorts of animals—the legs of chickens, the fins of fishes, the parapodia of polychaete worms, and the diverse appendages of arthropods. All arthropods also have a gene called *Ultrabithorax* (*Ubx* for short). It is one of the *Hox* genes that control the development of structures along the anterior–posterior axis of animals. In insects, the *Ubx* gene contains a unique mutation, not found in other arthropods, that causes the protein for which it codes to repress *Dll,* thereby preventing the formation of appendages wherever *Ubx* is expressed. And because insects express *Ubx* in their abdomen, they do not grow abdominal appendages. All other arthropods, which have the ancestral, nonrepressing form of the *Ubx* gene, have appendages in the posterior region of their body. Thus, one mutation in a *Hox*-family gene has fostered the evolution of a highly distinctive morphological trait in insects—having legs on the thorax, but not on the abdomen.

Insects exchange gases through a specialized **tracheal system,** a branching network of tubes that carry oxygen from small openings in the exoskeleton to tissues throughout the body (see Figure 46.5). Insects excrete nitrogenous wastes through specialized **Malpighian tubules** that transport wastes to the digestive system for disposal with feces (see Figure 48.6). Both of these organ systems are unique among animals. Insect sensory systems are diverse and complex. Besides image-forming compound eyes (see Figure 41.10), many insects have

light-sensing ocelli on their heads. Many also have hairs, sensitive to touch, on their antennae, legs, and other regions of the body. Chemical receptors are particularly common on the legs, allowing the identification of food. And many groups of insects have hearing organs to detect predators and potential mates. The familiar chirping of crickets, for example, is a mating call emitted by males that may repel other males and attract females.

As a group, insects feed in every conceivable way and on most other organisms. Species that eat plants, such as grasshoppers, have a pair of rigid mandibles, which chew food before it is ingested. Behind the mandibles is a pair of maxillae, which may also aid in food acquisition. Insects also have inflexible upper and lower lips, the *labrum* and *labium,* respectively. A tonguelike structure, just dorsal to the labium in chewing insects, houses the openings of the salivary glands. But evolution has modified this ancestral mandibulate pattern in numerous ways **(Figure 31.43).** In houseflies, the mouthparts are adapted for sopping up liquid food. In butterflies and moths, the mouthparts include a long proboscis to drink nectar. And in some biting flies, like mosquitoes and blackflies, the mouthparts have evolved into piercing structures.

After it hatches from an egg, an insect passes through a series of developmental stages called *instars.* Several hormones control development and ecdysis, which marks the passage from one instar to the next. Insects exhibit one of three basic patterns of postembryonic development **(Figure 31.44).** Wingless species simply grow and shed their exoskeleton without undergoing major changes in morphology. Other species undergo **incomplete metamorphosis;** they hatch from the egg as a *nymph,* which lacks functional wings. In many species, such as grasshoppers (Orthoptera), the nymphs resemble the adults. In other insects, such as dragonflies (Odonata), the aquatic nymphs are morphologically very different from the adults.

Most insects undergo **complete metamorphosis:** the larva that hatches from the egg differs greatly from the adult. Larvae and adults often occupy different habitats and consume

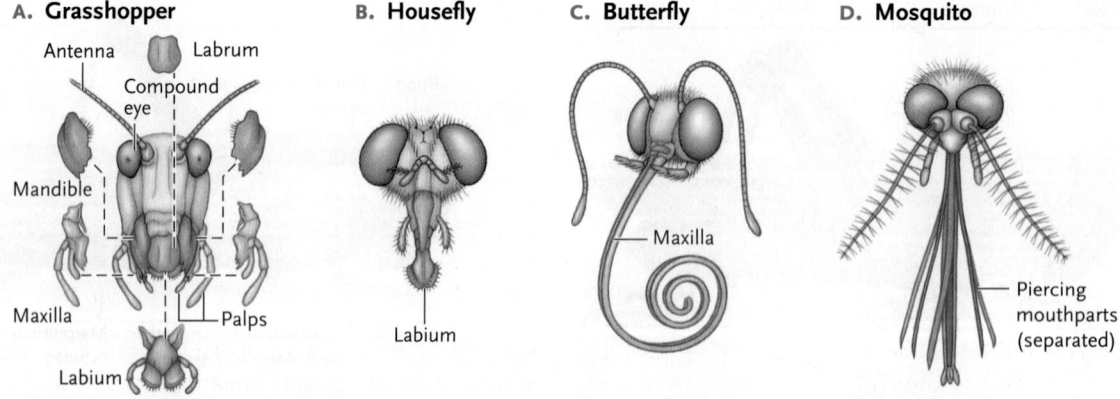

A. Grasshopper

Antenna
Labrum
Compound eye
Mandible
Maxilla
Palps
Labium

B. Housefly

Labium

C. Butterfly

Maxilla

D. Mosquito

Piercing mouthparts (separated)

FIGURE 31.43 Specialized insect mouthparts. The **(A)** ancestral chewing mouthparts have been modified by evolution, allowing different insects to **(B)** sponge up food, **(C)** drink nectar, and **(D)** pierce skin to drink blood.

© Cengage Learning 2017

A. No metamorphosis

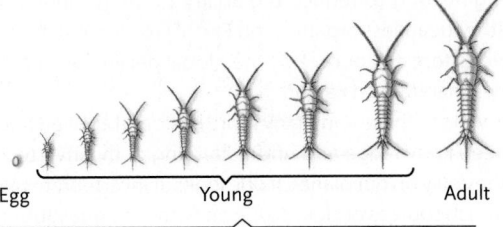

Egg   Young   Adult

Some wingless insects, like silverfish (Thysanura), do not undergo a dramatic change in form as they grow.

B. Incomplete metamorphosis

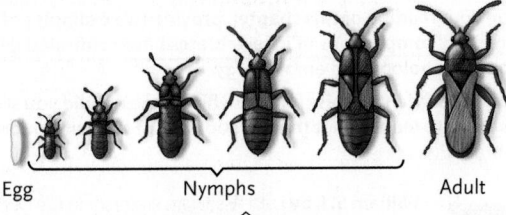

Egg   Nymphs   Adult

Other insects, such as true bugs (Hemiptera), have incomplete metamorphosis; they develop from nymphs into adults with relatively minor changes in form.

C. Complete metamorphosis

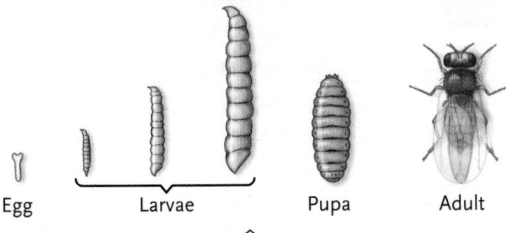

Egg   Larvae   Pupa   Adult

Fruit flies (Diptera) and many other insects have complete metamorphosis; they undergo a total reorganization of their internal and external anatomy when they pass through the pupal stage of the life cycle.

FIGURE 31.44 Patterns of postembryonic development in insects.
© Cengage Learning 2017

different food. The larvae (caterpillars, grubs, or maggots) are often worm-shaped, with chewing mouthparts. They grow and molt several times, retaining their larval morphology. Before they transform into sexually mature adults, they spend a period

of time as a sessile **pupa.** During this stage, the larval tissues are drastically reorganized. The adult that emerges is so different from the larva that it is often hard to believe that they are of the same species. Moths, butterflies, beetles, and flies are examples of insects with complete metamorphosis. Their larval stages specialize in feeding and growth, whereas the adults are adapted for dispersal and reproduction. In some species, the adults never feed, relying on the energy stores accumulated during the larval stage.

The 240-million-year history of insects has been characterized by innovations in morphology, life cycle patterns, locomotion, feeding, and habitat use. Their well-developed nervous systems govern exceptionally complex patterns of behavior, including parental care, a habit that reaches its zenith in the colonial eusocial insects, the ants, bees, and wasps (see Chapter 56). The factors that contribute to the insects' success also make them our most aggressive competitors. They destroy vegetable crops, stored food, wool, paper, and timber. They feed on blood from humans and domesticated animals, sometimes transmitting disease-causing pathogens as they do so. Nevertheless, insects are essential members of terrestrial ecological communities. Many species pollinate flowering plants, including important crops. Many others attack or parasitize species that are harmful to human activities. And most insects are a primary source of food for other animals. Some even make useful products such as honey, beeswax, and silk.

In the next chapter we consider the lineage of deuterostomes, which includes the vertebrates and their closest invertebrate relatives.

STUDY BREAK 31.7

1. What part of a parasitic nematode's anatomy protects it from the digestive enzymes of its host?
2. If an arthropod's rigid exoskeleton cannot be expanded, how does the animal grow?
3. How do the number of body regions and the appendages on the head differ among the four groups of living arthropods?
4. How do the life stages differ between insects that have incomplete metamorphosis and those that have complete metamorphosis?

Unanswered Questions

What are the evolutionary relationships among the invertebrate clades?

If we step back from our vertebrate and terrestrial biases, invertebrates are the dominant form of life on Earth. Many students of natural history are overwhelmed by the sheer numbers of organisms that one encounters, particularly in the aquatic environment, and the challenge of categorizing them taxonomically. Sorting them out in terms of their ecological roles is equally daunting. The problem we face is that invertebrates do it all—they are predators, herbivores, parasites, detritivores, and the primary symbiotic organisms on Earth. This adaptability of form

and function is a fascinating hallmark of the animal way of life, but it poses a legion of questions, many still unanswered. Recent advances in genetic technology have advanced our understanding, but there is much left to do.

In the past two decades, our ability to compare genetic information from various invertebrate groups has led to a remarkable reshuffling of the long-established categories used to classify these organisms. The first categories to be eliminated were groups based on superficial phenotypic resemblances, such as the "pseudocoelomates," which had plagued student understanding of diversity. Today, we have a much deeper knowledge of the evolutionary relationships of these organisms. Perhaps the most exciting discovery is that much of the diversity we see is not the product of slow changes in protein-coding gene sequences, but rather the result of variations in the timing and location of the expression of genes that affect development. Evo-devo, the melding of evolutionary and developmental biology—mostly made possible by intensive studies of two model invertebrates, *Caenorhabditis elegans* and *Drosophila melanogaster*—has revealed that changes in the expression of relatively simple sets of genes have brought about the myriad forms of life we see among the invertebrates. As systematists incorporate these new discoveries in their analyses, our understanding of the evolutionary relationships among the invertebrates will surely change.

What is the genetic basis of the diversity of form and function observed among invertebrates?

Because of advances in genetics research, we are on the cusp of being able to answer some fundamental questions. How does an animal's body develop either radial or bilateral symmetry? How does an organism develop a head with a concentration of nervous system tissue, and how do all the exquisite sensory systems associated with a big brain develop? From where do the respiratory pigments, which increase the capacity of the hemolymph or blood to carry oxygen, increasing an organism's capacity for activity, arise? How does the immune system develop, and what genetic change fosters the quantum leap from a nonspecific defense system to one that responds specifically to foreign invaders? More practically, are there genetic switches that we can manipulate? Can we make blood-feeding invertebrates, such as mosquitoes, or parasitic species, like tapeworms or filarial worms, innocuous? Is it possible to use genetic engineering to reduce the ability of the mosquito *Anopheles gambiae* (the "deadliest organism on Earth") to transmit malaria? Will it be possible to forestall or reverse the global decline of coral reefs, one of the richest habitats on Earth?

The answers to these and many more basic and applied questions lie within a deep knowledge and understanding of the invertebrates and the roles they play on our planet. If one looks at invertebrates as dynamic systems, as rich sources of clues to life on Earth, the questions they pose easily provide a lifetime of investigation and reward.

think like a scientist

1. Based on your reading of this chapter, provide two examples of how recent genetic comparisons of invertebrates have confirmed groupings based on morphology and embryology.

2. Based on the content of this chapter, which genes would you study if you wanted to learn about the genetic basis of diversity among invertebrates?

William S. Irby is an Associate Professor in the Department of Biology at Georgia Southern University in Statesboro. His research focuses on the ecology and evolution of blood-feeding behavior in mosquitoes. To learn more about Irby's work, go to http://cosm.georgiasouthern.edu/biology/people/faculty/william-s-irby/.

REVIEW KEY CONCEPTS

For access to MindTap and additional study materials visit www.cengagebrain.com.

31.1 What Is an Animal?

- Animals are eukaryotic, multicellular, heterotrophs that are motile at some time in their lives.

- Animals probably arose from a colonial flagellated ancestor before the Cambrian period (Figures 31.1 and 31.2).

31.2 Key Innovations in Animal Evolution

- All animals except sponges have tissues, which are organized into either two or three tissue layers.

- Although some animals exhibit radial symmetry, most exhibit bilateral symmetry (Figure 31.3).

- Acoelomate animals have no body cavity. Pseudocoelomate animals have a body cavity between the derivatives of endoderm and mesoderm. Coelomate animals have a body cavity that is entirely lined by derivatives of mesoderm (Figure 31.4).

- Two lineages of bilaterally symmetrical animals differ in developmental patterns (Figure 31.5). Most protostomes exhibit spiral, determinate cleavage, and their coelom forms from a split in a solid mass of mesoderm. Most deuterostomes have radial, indeterminate cleavage, and the coelom usually forms within outpocketings of the primitive gut.

- Four animal phyla exhibit segmentation.

31.3 An Overview of Animal Phylogeny and Classification

- Molecular and genomic analyses have refined our view of animal evolutionary history (Figure 31.6). The molecular phylogeny recognizes some major clades that had been identified on the basis of morphological and embryological characters. Sponges are grouped in the Parazoa. All other clades are grouped in the Eumetazoa. Among the Eumetazoa, the Radiata includes animals with two tissue layers and radial symmetry, and the Bilateria includes animals with three tissue layers and bilateral symmetry.

- Bilateria is further subdivided into Protostomia and Deuterostomia, and the Protostomia includes the Lophotrochozoa and the Ecdysozoa.

- The molecular phylogeny suggests that ancestral protostomes had a coelom and that segmentation arose more than once.

31.4 Animals without Tissues: Parazoa

- Sponges (phylum Porifera) are asymmetrical animals with limited integration of cells in their bodies (Figure 31.7).

- The body of many sponges is a water-filtering system (Figure 31.8). Flagellated choanocytes draw water into the body and capture particulate food.

31.5 Eumetazoans with Radial Symmetry

- The two major radiate phyla, Cnidaria and Ctenophora, have two well-developed tissue layers with a gelatinous mesoglea between them (Figure 31.9). Members of both phyla lack organ systems. All are aquatic.

- Cnidarians capture prey with tentacles and stinging nematocysts (Figures 31.10, 31.12, and 31.13). Their life cycles may include polyps, medusae, or both (Figure 31.11).
- Ctenophores use long tentacles to capture particulate food and use rows of cilia for locomotion (Figure 31.14).

31.6 Lophotrochozoan Protostomes

- This book describes eight phyla within the Lophotrochozoa.
- Three small phyla (Ectoprocta, Brachiopoda, and Phoronida) use a lophophore to feed on particulate matter (Figure 31.15).
- Free-living flatworm species (phylum Platyhelminthes) have well-developed digestive, excretory, reproductive, and nervous systems (Figures 31.16 and 31.17). Parasitic species attach to their animal hosts with suckers or hooks (Figures 31.18 and 31.19).
- The rotifers (phylum Rotifera) are tiny and abundant inhabitants of freshwater and marine ecosystems (Figure 31.20). Movements of cilia in the corona control their locomotion and bring food to their mouths. One group of rotifers includes more than 300 all-female species.
- The ribbon worms (phylum Nemertea) are elongate and often colorful animals with a proboscis housed in a rhynchocoel (Figure 31.21).
- Mollusks (phylum Mollusca) and segmented worms (phylum Annelida) often hatch from their eggs as trochophore larvae (Figure 31.22).
- Mollusks have fleshy bodies that are often enclosed in a hard shell. The molluscan body plan includes a head, a foot, the visceral mass, and a mantle (Figures 31.23–31.26).
- Annelids generally exhibit segmentation of the coelom and of the muscular, circulatory, excretory, respiratory, and nervous systems. They use the coelom as a hydrostatic skeleton for locomotion (Figures 31.27–31.30).

31.7 Ecdysozoan Protostomes

- This book describes three phyla in the taxon Ecdysozoa, which includes animals that periodically shed their cuticle or exoskeleton.
- Roundworms (phylum Nematoda) feed on decaying organic matter or parasitize plants or animals (Figure 31.31). They move by alternately contracting longitudinal muscles on the dorsal and ventral sides of the body wall.
- The velvet worms (phylum Onychophora) have segmented bodies and unjointed legs (Figure 31.32). Some species bear live young, which develop in a uterus.
- The segmented bodies of the arthropods (phylum Arthropoda) have specialized appendages for feeding, locomotion, or reproduction. Arthropods shed their exoskeleton as they grow or enter a new stage of the life cycle (Figure 31.33). They have an open circulatory system, a complex nervous system, and, in some groups, highly specialized respiratory and excretory systems.
- This book describes five groups of Arthropods. The extinct trilobites (subphylum Trilobita), with three-lobed bodies and relatively undifferentiated appendages, were abundant in Paleozoic seas (Figure 31.34). Chelicerate body plans include a prosoma (fused head and thorax) and an opisthosoma (abdomen); appendages on the prosoma include pincers or fangs and pedipalps (Figures 31.35 and 31.36). The remaining groups of arthropods are included in the Mandibulata, which includes two large clades. Myriapods are animals that have a head and an elongate, segmented trunk (Figure 31.37). Pancrustacea includes the paraphyletic "crustaceans" and the Hexapoda. Crustaceans have a carapace that covers the cephalothorax as well as highly modified appendages, including antennae and mandibles (Figures 31.38–31.40). Hexapods have three body regions, three pairs of walking legs on the thorax, and three pairs of feeding appendages on the head (Figures 31.41–31.43). Insects exhibit three patterns of postembryonic development (Figure 31.44).

TEST YOUR KNOWLEDGE

Remember/Understand

1. Which of the following characteristics is *not* typical of most animals?
   a. heterotrophic
   b. sessile
   c. bilaterally symmetrical
   d. multicellular
   e. motile at some stage of life cycle

2. A body cavity that separates the digestive system from the body wall but is *not* completely lined with mesoderm is called a:
   a. schizocoelom.
   b. mesentery.
   c. peritoneum.
   d. pseudocoelom.
   e. hydrostatic skeleton.

3. Protostomes and deuterostomes typically differ in:
   a. their patterns of body symmetry.
   b. the number of germ layers during development.
   c. their cleavage patterns.
   d. the size of their sperm.
   e. the size of their digestive systems.

4. The nematocysts of cnidarians are used primarily for:
   a. capturing prey.
   b. detecting light and dark.
   c. courtship.
   d. sensing chemicals.
   e. gas exchange.

5. Which organ system is absent in flatworms (phylum Platyhelminthes)?
   a. nervous system
   b. reproductive system
   c. circulatory system
   d. digestive system
   e. excretory system

6. Which part of a mollusk secretes the shell?
   a. visceral mass
   b. radula
   c. trochophore
   d. head–foot
   e. mantle

7. Which phylum includes the most abundant animals in soil?
   a. Nematoda
   b. Rotifera
   c. Mollusca
   d. Annelida
   e. Brachiopoda

8. Which body region of an insect bears the walking legs?
   a. head
   b. carapace
   c. abdomen
   d. thorax
   e. trunk

9. Ecdysis refers to a process in which:
   a. bivalves use siphons to pass water across their gills.
   b. arthropods shed their old exoskeletons.
   c. cnidarians build skeletons of calcium carbonate.
   d. rotifers produce unfertilized eggs.
   e. squids escape from predators in a cloud of ink.

Apply/Analyze

10. What is the major morphological innovation seen in annelid worms?
    a. a complete digestive system
    b. image-forming eyes
    c. a respiratory system
    d. an open circulatory system
    e. body segmentation

11. **Discuss Concepts** Many invertebrate species are hermaphroditic. What selective advantages might this characteristic offer? In what kinds of environments might it be most useful?

12. **Discuss Concepts** On a voyage to the ocean bottom, a biologist discovers a worm that appears to be new to science. What characteristics of this animal should the biologist examine to determine whether or not she has discovered a previously undescribed phylum?

13. **Discuss Concepts** What are the relative advantages and disadvantages of radially symmetrical and bilaterally symmetrical body plans?

Evaluate/Create

14. **Discuss Concepts** People who eat raw clams and oysters harvested from sewage-polluted waters often develop mild to severe gastrointestinal infections. These mollusks are suspension feeders. Develop a hypothesis about why people who eat them raw may be at risk.

15. **Design an Experiment** Design an experiment to test the hypothesis that the cuticle of parasitic nematodes protects them from the acids and enzymes present in the digestive systems of their hosts. Your design must include both experimental and control treatments.

16. **Apply Evolutionary Thinking** Many insects have a larval stage that is morphologically different from the adult and that feeds on different foods. What selection pressures may have fostered the evolution of a life cycle with such distinctive life stages? Your answer should address the different biological activities that characterize each life cycle stage.

For selected answers, see Appendix A.

INTERPRET THE DATA

A research team led by Andreas Michalsen of the Kliniken Essen, a hospital in Germany, compared the relative effectiveness of "leech therapy" and the topical application of diclofenac, an anti-inflammatory drug, for the treatment of pain caused by osteoarthritis in the thumb. Leech saliva contains more than 30 biologically active compounds, including some with anti-inflammatory effects. For the group receiving drug therapy, the patients applied the drug to their thumbs at least twice a day for up to 30 days. For the group receiving leech therapy, researchers applied two or three leeches to the affected joint and allowed them to feed until they were satiated; the average feeding time was 50 minutes. Patients in the leech therapy group received just one treatment. The researchers asked patients to score their pain (0 = no pain; 100 = severe pain) before treatment (day 0 in the **Figure**) and again one week, one month, and two months after treatment. The results are presented in the accompanying bar chart. Which treatment had the greater effect in reducing the pain of arthritis patients in this study? Did leech therapy have a prolonged effect?

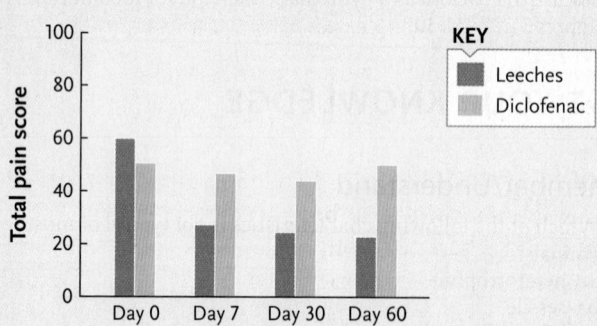

Source: A. Michalsen et al. 2008. Effectiveness of leech therapy in women with symptomatic arthrosis of the first carpometacarpal joint: A randomized controlled trial. *Pain* 137:452–459.

Deuterostomes: Vertebrates and Their Closest Relatives

32

Snow monkeys *(Macaca fuscata)*. These snow monkeys, which have the northernmost distribution of any nonhuman primate, are soaking in a hot spring in Japan.

Why it matters . . . In 1798, naturalists at the British Museum skeptically probed a curious specimen that had been sent from Australia. The furry creature—about the size of a house cat—had webbed front feet, a ducklike bill, and a flat, paddlelike tail **(Figure 32.1).** The scientists eagerly searched for evidence that a prankster had stitched together parts from wildly different animals, but they found no signs of trickery and soon accepted the duck-billed platypus *(Ornithorhynchus anatinus)* as a genuine zoological novelty.

Further study has revealed that the platypus is even stranger than those scientists could have imagined. Like other mammals, the platypus is covered with hair, and females produce milk that the offspring lick off the fur on their mother's belly. But like turtles and birds, a platypus has no teeth, and it reproduces by laying eggs instead of giving birth to its offspring. And like turtles, birds, lizards, snakes, and crocodilians, it has a *cloaca*, a multipurpose chamber through which it releases feces, urine, and eggs. Scientists had never before seen such a weird combination of traits, and they did not quite know what to make of them.

Studies of the platypus under natural conditions have helped biologists make sense of its characteristics. The platypus inhabits streams and lagoons in Australia and Tasmania. It rests in streamside burrows during the day, but at night it slips into the water to hunt for invertebrates. Its dense fur keeps its body warm and dry under water, and its tail serves both as a rudder and as a storehouse for energy-rich fat. It uses its bill to scoop up food and the horny pads that line its jaws to grind up prey. While underwater, the platypus clamps shut its eyes, ears, and nostrils, relying on roughly 800,000 sensory receptors in its bill to detect the movements and weak electrical discharges of nearby prey.

The platypus, with its strange combination of characteristics, illustrates the remarkable diversity of adaptations that enable **vertebrates**—animals with backbones—to occupy nearly

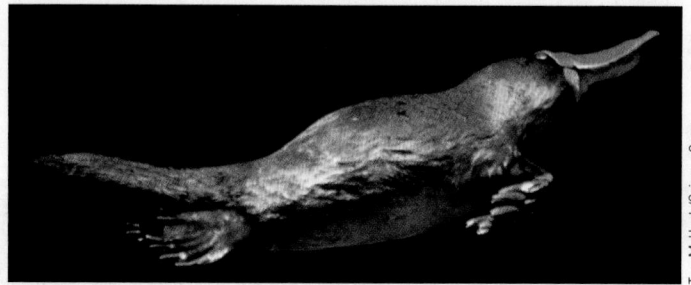

FIGURE 32.1 **A puzzling animal.** Because of its strange mixture of traits, the duck-billed platypus *(Ornithorhynchus anatinus)* amazed the first European zoologists who saw it.

In this chapter, we survey the Deuterostomia, a monophyletic lineage of animals that dates to the Paleozoic. The deuterostomes are defined by features of early embryological development and molecular sequence data (see Chapter 31). There are three living phyla of deuterostomes; we briefly consider two phyla of invertebrate deuterostomes before focusing on the Phylum Chordata, which includes a few thousand species of invertebrates, as well as many living species of vertebrates.

every habitat on Earth. Despite the platypus's mixed characteristics, biologists eventually classified it as a member of the mammal lineage because, like all other mammals, it has hair on its body and produces milk to nourish its offspring. Today biologists know that it is one of just a few remaining survivors of an early lineage of egg-laying mammals.

32.1 Invertebrate Deuterostomes

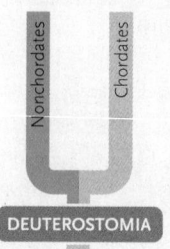

Deuterostome body plans have been so modified by evolution that a casual observer would not readily group the two phyla of invertebrate deuterostomes—Echinodermata and Hemichordata—together with the Phylum Chordata. However, embryological and molecular analyses agree that all three are indeed closely related.

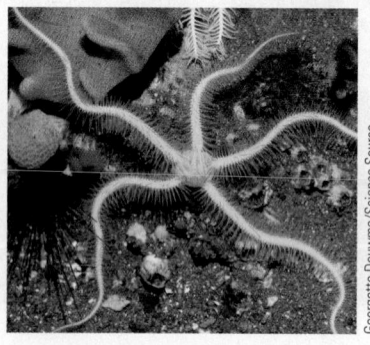

A. Asteroidea: This sea star *(Fromia milleporella)* lives in the intertidal zone.

B. Ophiuroidea: A brittle star *(Ophiothrix* species) lives on coral reefs.

C. Echinoidea: A sea urchin *(Strongylocentrotus purpuratus)* grazes on algae.

D. Holothuroidea: A sea cucumber *(Cucumaria miniata)* extends its tentacles, which are modified tube feet, to trap particulate food.

E. Crinoidea: A feather star *(Oxycomanthus bennetti)* feeds by catching small particles with its numerous arms.

FIGURE 32.2 Echinoderm diversity. Echinoderms exhibit secondary radial symmetry, often organized as five rays around an oral–aboral axis.

Echinoderms Have Secondary Radial Symmetry and an Internal Skeleton

The phylum **Echinodermata** (*echinos* = spiny; *derma* = skin) includes roughly 7,000 species of sea stars, sea urchins, sea cucumbers, brittle stars, and sea lilies. These slow-moving or sessile, bottom-dwelling animals are important herbivores and predators in shallow coastal waters and the ocean depths. Although the phylum was diverse in the Paleozoic, only a remnant of that fauna remains. Living species vary in size from less than 1 cm to more than 50 cm in diameter.

Echinoderms develop from a bilaterally symmetrical, free-swimming larva. But as a larva develops, it assumes a secondary radial symmetry, often organized around five rays or "arms" **(Figure 32.2)**. Many echinoderms have an *oral surface*, with the mouth facing the substrate, and an *aboral surface* facing in the opposite direction. Virtually all echinoderms have an internal skeleton made of calcium-stiffened *ossicles* that develop from mesoderm. In some groups, fused ossicles form a rigid container called a *test*. In most, spines or bumps project from the ossicles.

The internal anatomy of echinoderms is unique among animals **(Figure 32.3)**. They have a well-defined coelom and a complete digestive system (see Figure 32.3A), but no excretory or respiratory systems, and most have only a minimal circulatory system. In many, gases are exchanged and metabolic wastes eliminated through projections of the epidermis and peritoneum near the base of the spines. Given their radial symmetry, there is no head or central brain; the nervous system is organized around nerve cords that encircle the mouth and branch into the rays. Sensory cells are abundant in the skin.

Echinoderms have a unique locomotor system, the **water vascular system,** which consists of fluid-filled canals (see Figure 32.3B). In a sea star, for example, water enters the system through the *madreporite*, a sievelike plate on the aboral surface. A short tube connects it to the *ring canal*, which surrounds the esophagus. The ring canal branches into five *radial canals* that extend into the arms. Each radial canal is connected to numerous **tube feet** that protrude through holes in the ossicles (see Figure 32.3C). Each tube foot has a mucus-covered, suckerlike tip and a small muscular bulb, the *ampulla*, which lies inside the body. When an ampulla contracts, fluid is forced into the tube foot, causing it to lengthen and attach to the substrate. The tube foot then contracts, pulling the animal along. As the tube foot shortens, water flows back into the ampulla, and the tube foot releases its grip on the substrate. The tube foot can then take another step forward, reattaching to the substrate. Although each tube foot has limited strength, the coordinated action of hundreds or thousands of them is so strong that they can hold an echinoderm to a substrate even against strong wave action.

Echinoderms have separate sexes, and most reproduce by releasing gametes into the water. Radial cleavage is so clearly apparent in the transparent eggs of some sea urchins that they are commonly used for demonstrations of cleavage in introductory biology laboratories. A few echinoderms also reproduce asexually by splitting in half and regenerating the missing parts; some can regenerate body parts lost to predators.

Echinodermata includes six clades, one of which, the sea daisies (Concentricycloidea), was discovered only in 1986. These small, medusa-shaped animals occupy sunken, waterlogged wood in the deep sea. Here we describe the five other groups, which are more diverse and better known.

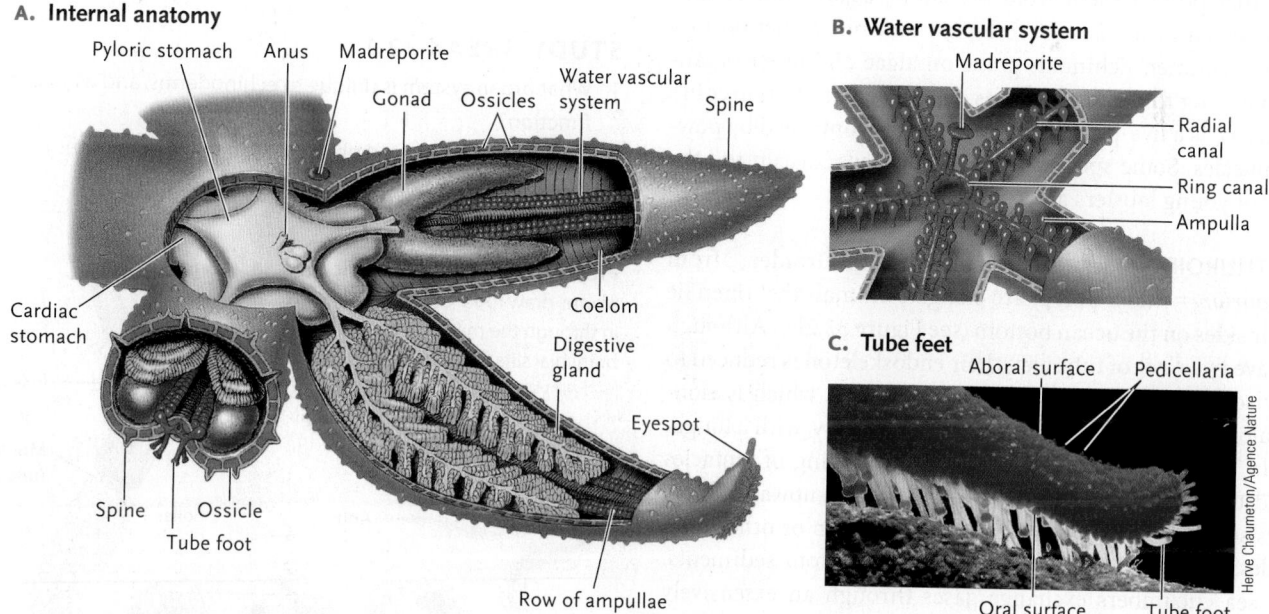

FIGURE 32.3 Internal anatomy of a sea star. (A) The coelom is well developed in echinoderms, as illustrated by this cutaway diagram of a sea star. **(B)** The water vascular system, unique in the animal kingdom, operates the tube feet **(C),** which are responsible for locomotion. Note the pedicellaria on the upper surface of the sea star's arm **(C).**
© Cengage Learning 2017

ASTEROIDEA Sea stars (Asteroidea, from *asteroeides* = starlike) live from rocky shorelines to depths of 10,000 m. The asteroid body plan consists of a central disk surrounded by 5 to 20 radiating "arms" (see Figure 32.2A), with the mouth centered on the oral surface. The ossicles of the endoskeleton are not fused, permitting flexibility of the arms and disk. Small pincers, **pedicellariae,** at the base of short spines remove debris that falls onto the animal's surface (see Figure 32.3C). Many sea stars feed on soft-bodied invertebrates and small fishes. Other species consume bivalve mollusks: after using their tube feet to pry apart the two shells, they evert their stomachs and secrete digestive enzymes onto the mollusk's flesh. Some sea stars are destructive predators of corals, endangering many reefs.

OPHIUROIDEA Brittle stars and basket stars (Ophiuroidea, from *ophioneos* = snakelike) occupy roughly the same range of habitats as sea stars. Their bodies have a well-defined central disk and slender, elongate arms that are sometimes branched (see Figure 32.2B). Ophiuroids can crawl fairly swiftly across substrates by moving their arms in coordinated fashion. As their common name implies, the arms are delicate and easily broken, an adaptation that allows them to escape from predators with only minor losses. Brittle stars feed on small prey, suspended plankton, or detritus that they extract from muddy deposits.

ECHINOIDEA Sea urchins and sand dollars (Echinoidea, *ekhinos* = porcupine) lack arms altogether (see Figure 32.2C). Their ossicles are fused into solid tests, which provide excellent protection but restrict flexibility. The test is spherical in sea urchins and flattened in sand dollars. Five rows of tube feet, used primarily for locomotion, emerge through pores in the test. Most echinoids have movable spines, some with poison glands that protect them from predators; a jab from certain tropical species can cause severe pain and inflammation to a careless swimmer. Echinoids graze on algae and other organisms that cling to marine surfaces. In the center of an urchin's oral surface is a five-part nipping jaw that is controlled by powerful muscles. Some species damage kelp beds, disrupting the habitat of young lobsters and other crustaceans.

HOLOTHUROIDEA Sea cucumbers (Holothuroidea, from *holothourion* = water polyp) are elongate animals that often lie on their sides on the ocean bottom (see Figure 32.2D). Although they have five rows of tube feet, their endoskeleton is reduced to widely separated microscopic plates. The body, which is elongated along the oral–aboral axis, is soft and fleshy, with a tough, leathery covering. Modified tube feet form a ring of tentacles around the mouth, which points to the side or upward. Some species secrete a mucous net that traps plankton or other food particles. Other species extract food from bottom sediments. Many sea cucumbers exchange gases through an extensively branched *respiratory tree* that arises from the rectum, the part of the digestive system just inside the anus at the aboral end of the animal. A well-developed circulatory system distributes oxygen and nutrients to tissues throughout the body.

CRINOIDEA Sea lilies and feather stars (Crinoidea, from *krinon* = lily) are the surviving remnants of a fauna that was diverse and abundant 500 million years ago (see Figure 32.2E). Most species occupy marine waters of medium depth. The central disk and mouth point upward rather than toward the substrate. Between five and several hundred branched arms surround the disk; new arms are added as a crinoid grows larger. The branches of the arms are covered with tiny mucus-coated tube feet, which trap suspended microscopic organisms. The sessile sea lilies have the central disk attached to a flexible stalk that can reach a meter in length. Adult feather stars can swim or crawl weakly, attaching temporarily to substrates.

Acorn Worms Use a Pharynx with Branchial Slits to Acquire Food and Oxygen

The approximately 70 species of acorn worms (phylum Hemichordata, *hemi* = half, *chorda*, referring to the phylum Chordata) are sedentary marine animals that live in U-shaped tubes or burrows in coastal sand or mud. Their soft bodies, which range from 2 cm to 2 m in length, are organized into an anterior proboscis, a tentacled collar, and an elongate trunk **(Figure 32.4).** They use their muscular, mucus-coated proboscis to construct burrows and trap food particles. Acorn worms also have pairs of **branchial slits** (sometimes called *pharyngeal slits* or *gill slits*) in the pharynx, the part of the digestive system just posterior to the mouth. Beating cilia create a flow of water, which enters the pharynx through the mouth and exits through these openings. As water passes through, suspended food is trapped and shunted into the digestive system, and gases are exchanged across the partitions between the slits. This coupling of feeding and respiration reflects the evolutionary relationship between hemichordates and chordates, the phylum that we consider next.

STUDY BREAK 32.1

1. What organ system is unique to echinoderms, and what is its function?
2. How does a perforated pharynx enable hemichordates to acquire food and oxygen from seawater?

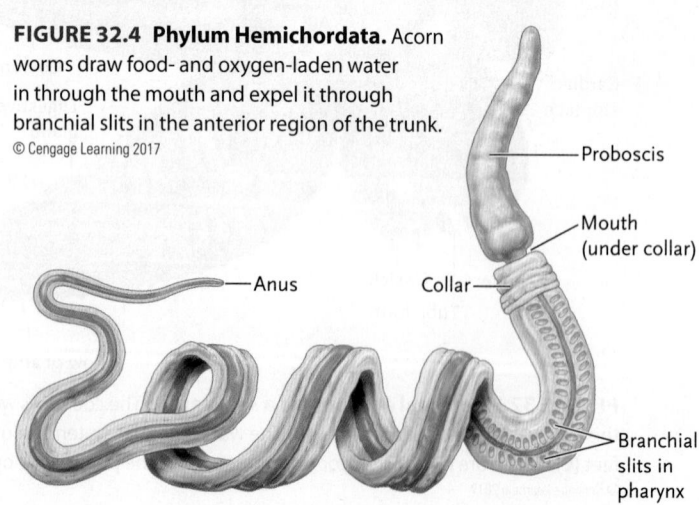

FIGURE 32.4 Phylum Hemichordata. Acorn worms draw food- and oxygen-laden water in through the mouth and expel it through branchial slits in the anterior region of the trunk.
© Cengage Learning 2017

Proboscis

Mouth (under collar)

Anus

Collar

Branchial slits in pharynx

32.2 Overview of the Phylum Chordata

The phylum Chordata contains three subphyla: two lineages of invertebrates, Cephalochordata and Urochordata, and the diverse lineage of vertebrates, Vertebrata.

Key Morphological Innovations Distinguish Chordates from Other Deuterostome Phyla

Species in the phylum **Chordata** are distinguished from other deuterostomes by a set of key morphological innovations: a *notochord, segmental muscles in the body wall and tail, a dorsal hollow nerve chord,* and a *perforated pharynx* **(Figure 32.5).** These structures foster higher levels of activity, unique modes of aquatic locomotion, and more efficient feeding and oxygen acquisition.

NOTOCHORD Early in chordate development, mesoderm that is dorsal to the developing digestive system forms a **notochord** (*noton* = the back; *chorda* = string). This flexible rod, constructed of fluid-filled cells surrounded by tough connective tissue, supports the embryo from head to tail. The notochord forms the skeleton of adult invertebrate chordates. Body wall muscles are anchored to the notochord, and when these muscles contract, the notochord bends, but it does not shorten. As a result, the chordate body swings left and right during locomotion, propelling the animal forward; unlike annelids and other nonchordate invertebrates, the chordate body does not shorten when the animal is moving. Remnants of the notochord persist as gelatinous disks in the backbones of adult vertebrates.

SEGMENTAL BODY WALL AND TAIL MUSCLES Chordates evolved in water, and they swim by contracting segmentally arranged blocks of muscles in the body wall and tail. The chordate tail, which is posterior to the anus, provides much of the propulsion in aquatic species. Segmentation allows each muscle block to contract independently; waves of contractions pass down one side of the animal and then down the other, sweep-ing the body and tail back and forth in a smooth and continuous movement.

DORSAL HOLLOW NERVE CORD The central nervous system of chordates is a hollow nerve cord on the dorsal side of the animal (see Chapter 40). By contrast, most nonchordate invertebrates have solid nerve cords on the ventral side. In vertebrates, an anterior enlargement of the nerve cord forms the brain; in invertebrates, the anterior concentration of nervous system tissue is described as a *ganglion.*

PERFORATED PHARYNX Like the hemichordates described above, most chordates have outpocketings, perforations, or slits in the pharynx during some stage of the animal's life cycle. These paired openings originated as exit holes for water that carried particulate food and oxygenated water into the mouth. Invertebrate chordates exchange gases with water as it passes by the walls of the pharynx, acquiring oxygen and eliminating carbon dioxide. Invertebrate chordates and fishes retain a perforated pharynx throughout their lives. In most air-breathing terrestrial vertebrates, the outpocketings or slits are present only during embryonic development and in larvae.

Invertebrate Chordates Are Small, Marine Suspension Feeders

Two subphyla of invertebrate chordates exhibit the basic chordate body plan in its simplest form.

SUBPHYLUM CEPHALOCHORDATA The 28 species in the subphylum **Cephalochordata** (from *kephale* = head), often called lancelets, occupy warm, shallow marine habitats where they lie mostly buried in sand **(Figure 32.6).** Although generally sedentary, they have well-developed body wall muscles and a prominent notochord. Most species are included in the genus *Branchiostoma* (formerly *Amphioxus*). Lancelet bodies, which are 5 to 10 cm long, are pointed at both ends like the double-edged surgical tools for which they are named. Adults have light receptors on the head, as well as chemical sense organs on

FIGURE 32.5 Diagnostic chordate characters. Chordates have a notochord; a muscular postanal tail; a segmental body wall and tail muscles; a dorsal hollow nerve cord; and a perforated pharynx.
© Cengage Learning 2017

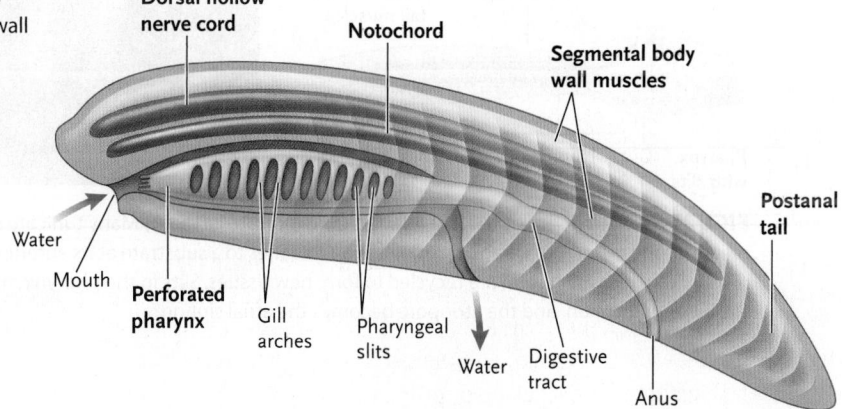

A. Lancelet

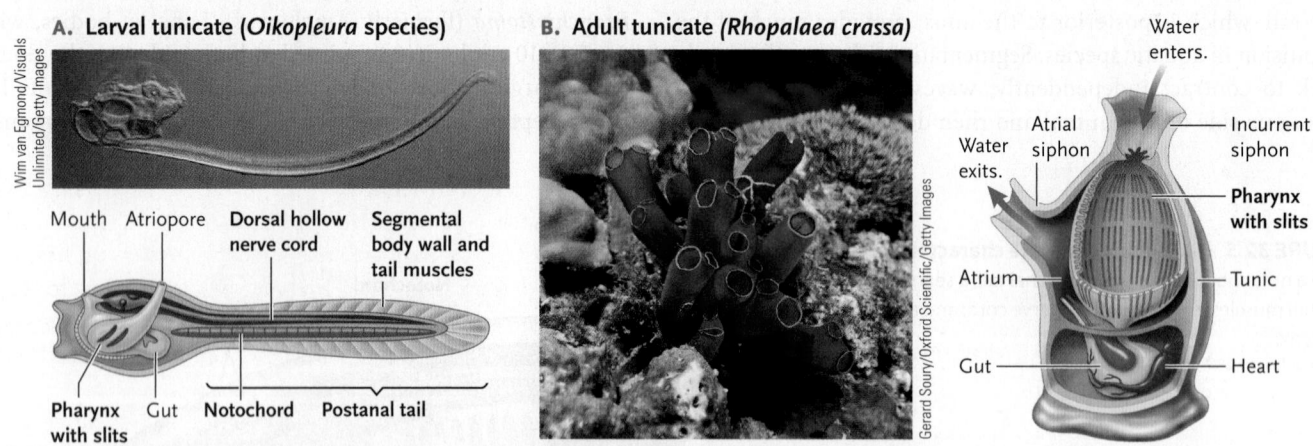

Mouth

Segmental muscles

Heather Angel/Natural Visions/Alamy

B. Lancelet anatomy

Dorsal hollow nerve cord · Notochord · Pharynx with slits · Gut · Postanal tail

Light receptor

Mouth

Oral hood with tentacles · Atrium · Segmental muscles · Atriopore · Anus

FIGURE 32.6 Cephalochordates. (A) The unpigmented skin of adult lancelets reveals their segmental body wall muscles. A cutaway view **(B)** illustrates their internal anatomy.
© Cengage Learning 2017

tentacles that grow from the **oral hood.** Lancelets use cilia to draw food-laden water through hundreds of pharyngeal slits; water flows into a chamber called the **atrium** and is expelled through the **atriopore.** Most gas exchange occurs across the skin.

SUBPHYLUM UROCHORDATA The 1,600 species in the subphylum **Urochordata** (from *oura* = tail) float in surface waters or attach to substrates in shallow marine habitats. The sessile adults of many species, described as tunicates or sea squirts, secrete a gelatinous or leathery "tunic" around their bodies and squirt water through a siphon when disturbed; adults grow to several centimeters **(Figure 32.7).** In the most common group of sea squirts (Ascidiacea), the swimming larvae possess the defining chordate features. Larvae eventually attach to substrates and transform into sessile adults. During metamorphosis, they lose most traces of the notochord, dorsal nerve cord, and tail, and their basketlike pharynx enlarges. In adults, beating cilia pull

water into the pharynx through an **incurrent siphon.** A mucous net traps particulate food, which is carried, with the mucus, to the gut. Water passes through the pharyngeal slits, enters the atrium, and is expelled—along with digestive wastes and carbon dioxide—through the **atrial siphon.** Oxygen is absorbed across the walls of the pharynx.

Vertebrates Possess Several Unique Tissues, Including Bone and Neural Crest

The most distinctive anatomical characteristic of the subphylum Vertebrata is an internal skeleton that provides structural support for muscles and protection for the nervous system and other organs. The skeleton—and the muscles attached to it—also enable most vertebrates to move rapidly through the environment. A vertebrate's skeleton is composed of many separate, bony elements. Indeed, vertebrates are the only animals that

A. Larval tunicate (*Oikopleura* species)

Wim van Egmond/Visuals Unlimited/Getty Images

Mouth · Atriopore · Dorsal hollow nerve cord · Segmental body wall and tail muscles

Pharynx with slits · Gut · Notochord · Postanal tail

B. Adult tunicate (*Rhopalaea crassa*)

Gerard Soury/Oxford Scientific/Getty Images

Water enters.

Atrial siphon · Incurrent siphon

Water exits.

Atrium · Pharynx with slits · Tunic

Gut · Heart

FIGURE 32.7 Urochordates. (A) A tadpolelike tunicate larva. **(B)** Many tunicate species have larva that metamorphose into sessile adults. After a larva attaches to a substrate at its anterior end, the tail, notochord, and most of the nervous system are recycled to form new tissues. Slits in the pharynx multiply, the mouth becomes the incurrent siphon, and the atriopore becomes the atrial siphon.
© Cengage Learning 2017

have **bone,** a connective tissue in which living cells secrete the mineralized matrix that surrounds them (see Figure 38.4D). The **vertebral column,** made up of individual **vertebrae,** surrounds and protects the dorsal nerve cord, and a bony **cranium** surrounds the brain. The cranium, vertebral column, ribs, and sternum (breastbone) make up the **axial skeleton.** Most vertebrates also have a **pectoral girdle** anteriorly and a **pelvic girdle** posteriorly that attach bones in the fins or limbs to the axial skeleton. Bones of the two girdles and the appendages constitute the **appendicular skeleton.** One vertebrate lineage, Chondrichthyes, has lost its bone over evolutionary time; its skeleton is made of cartilage, a dense but flexible connective tissue that is often a developmental precursor of bone (see Section 38.2).

Vertebrates also possess a unique cell type, **neural crest,** which is distinct from endoderm, mesoderm, and ectoderm. Neural crest cells arise next to the developing nervous system, but later migrate throughout a vertebrate's body. They ultimately contribute to many uniquely vertebrate structures, including parts of the cranium, teeth, sensory organs, cranial nerves, and the medulla (that is, the interior part) of the adrenal glands.

Finally, the brains of vertebrates are much larger and more complex than those of invertebrate chordates. Moreover, the vertebrate brain is divided into three regions—the forebrain, midbrain, and hindbrain—each of which governs distinct nervous system functions (see Section 40.1).

STUDY BREAK 32.2

1. On a field trip to a lake, a college student captures a worm-shaped animal with segmented body wall muscles. While examining the specimen in the laboratory the following day, she determines that the main nerve cord runs along the ventral side of the animal. Is this animal a chordate?
2. What structures distinguish vertebrates from invertebrate chordates?

32.3 The Origin and Diversification of Vertebrates

Biologists use molecular, embryological, and fossil evidence to trace the origin of vertebrates and to chronicle their evolutionary diversification.

Vertebrates Probably Arose from an Invertebrate Chordate Ancestor through the Duplication of Genes That Regulate Development

Recent genetic sequence studies suggest that vertebrates are more closely related to urochordates than to cephalochordates. Nevertheless, the fishlike form of adult lancelets probably better represents the anatomy of the earliest vertebrates than does the highly specialized morphology of adult tunicates. The evolution of vertebrates from an invertebrate chordate ancestor was marked by the emergence of neural crest, bone, and other typically vertebrate traits. What genetic changes were responsible for these remarkable developments? Biologists now hypothesize that an increase in the number of homeobox—structure determining—genes may have made the development of more complex anatomy possible. (Homeobox genes are described in Sections 16.4, 20.3, 23.6, and 36.5.)

In animals, one group of homeobox genes—those in the *Hox* gene family—influence the development and location of important structures—such as legs and wings—along the anterior to posterior (that is, head-to-tail) axis of the body. *Hox* genes are arranged on the chromosomes in a particular order, forming what biologists call the *Hox* gene complex. Each gene in the complex governs the development of particular structures. Animal groups with the simplest structure, such as cnidarians, have two *Hox* genes. Those with more complex anatomy, such as insects, have 10. Chordates have as many as 13 or 14. Thus, lineages with many *Hox* genes generally have more complex anatomy than do those with fewer *Hox* genes.

Molecular analyses also reveal that the entire *Hox* gene complex was duplicated several times in the evolutionary history of vertebrates, producing multiple copies of all its genes **(Figure 32.8).** The cephalochordate *Branchiostoma* has just one *Hox* gene complex, but the most primitive living vertebrates, the jawless hagfishes described later, have two. All vertebrates that possess jaws, a derived characteristic, have at least four sets, and some fishes have as many as seven. Evolutionary developmental biologists hypothesize that after the wholesale duplication of *Hox* genes and other toolkit genes, the original copies of these genes maintained their ancestral functions, but the duplicate copies assumed *new* functions, directing the development of novel structures, such as the vertebral column and jaws.

Early Vertebrates Diversified into Numerous Lineages with Distinctive Adaptations

The oldest known vertebrate fossils were discovered in the late 1990s, when scientists in China described several species from the early Cambrian period, about 550 million years ago. Both *Myllokunmingia* and *Haikouichthys* were fish-shaped animals about 3 cm long **(Figure 32.9).** In both species the brain was surrounded by a cranium, which, in these cases, was formed of fibrous connective tissue or cartilage. They also had segmented body wall muscles and fairly well-developed fins, but neither shows any evidence of bone.

The early vertebrates gave rise to numerous descendants, which varied greatly in anatomy, physiology, and ecology. New feeding mechanisms and locomotor structures were often crucial to their success. Today, vertebrates occupy nearly every habitat and feed on virtually all other organisms. Here we briefly introduce the major vertebrate lineages **(Figure 32.10).**

Hox GENES AND THE EVOLUTION OF VERTEBRATES

The *Hox* genes in different animal groups appear to be homologous, indicated here by their color and position in the *Hox* gene complex.

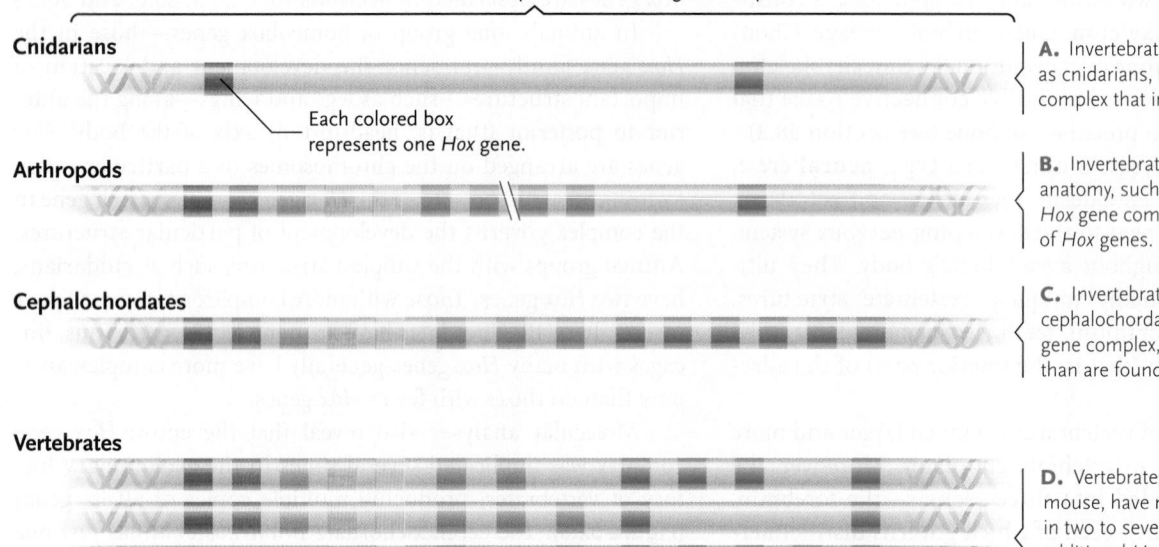

Each row of colored boxes represents one *Hox* gene complex.

Cnidarians

Each colored box represents one *Hox* gene.

A. Invertebrates with simple anatomy, such as cnidarians, have a single *Hox* gene complex that includes just a few *Hox* genes.

Arthropods

B. Invertebrates with more complicated anatomy, such as arthropods, have a single *Hox* gene complex, but with a larger number of *Hox* genes.

Cephalochordates

C. Invertebrate chordates, such as cephalochordates, also have a single *Hox* gene complex, but with even more *Hox* genes than are found in nonchordate invertebrates.

Vertebrates

D. Vertebrates, such as the laboratory mouse, have numerous *Hox* genes, arranged in two to seven *Hox* gene complexes. The additional *Hox* gene complexes are products of wholesale duplications of the ancestral *Hox* gene complex.

SUMMARY Vertebrates have many more individual *Hox* genes than most invertebrates do because the entire *Hox* gene complex was duplicated in the vertebrate lineage. The additional copies of *Hox* genes probably evolved to specify the development of uniquely vertebrate characteristics, such as the cranium, vertebral column, and neural crest cells.

think like a scientist Given the information available to you in this figure, what change in overall body plan accompanied the evolution of an increased number of *Hox* genes *within* the *Hox* gene complex?

© Cengage Learning 2017

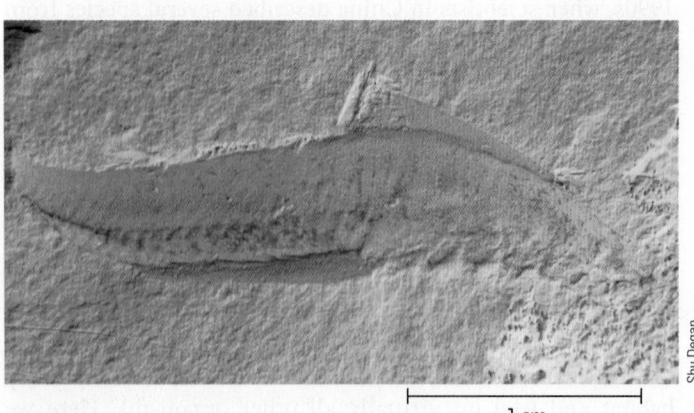

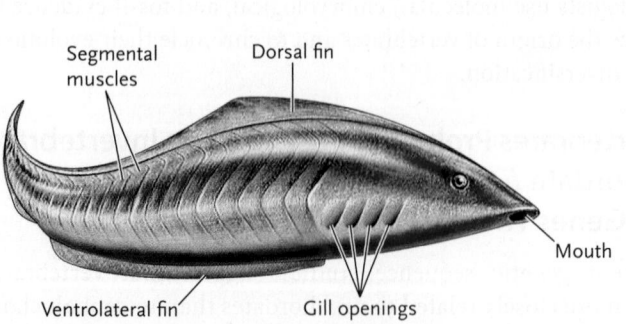

Segmental muscles

Dorsal fin

Ventrolateral fin

Gill openings

Mouth

Shu Degan

1 cm

FIGURE 32.9 An early vertebrate. *Myllokunmingia,* one of the earliest vertebrates yet discovered, had no bones; it was about 3 cm long.

© Cengage Learning 2017

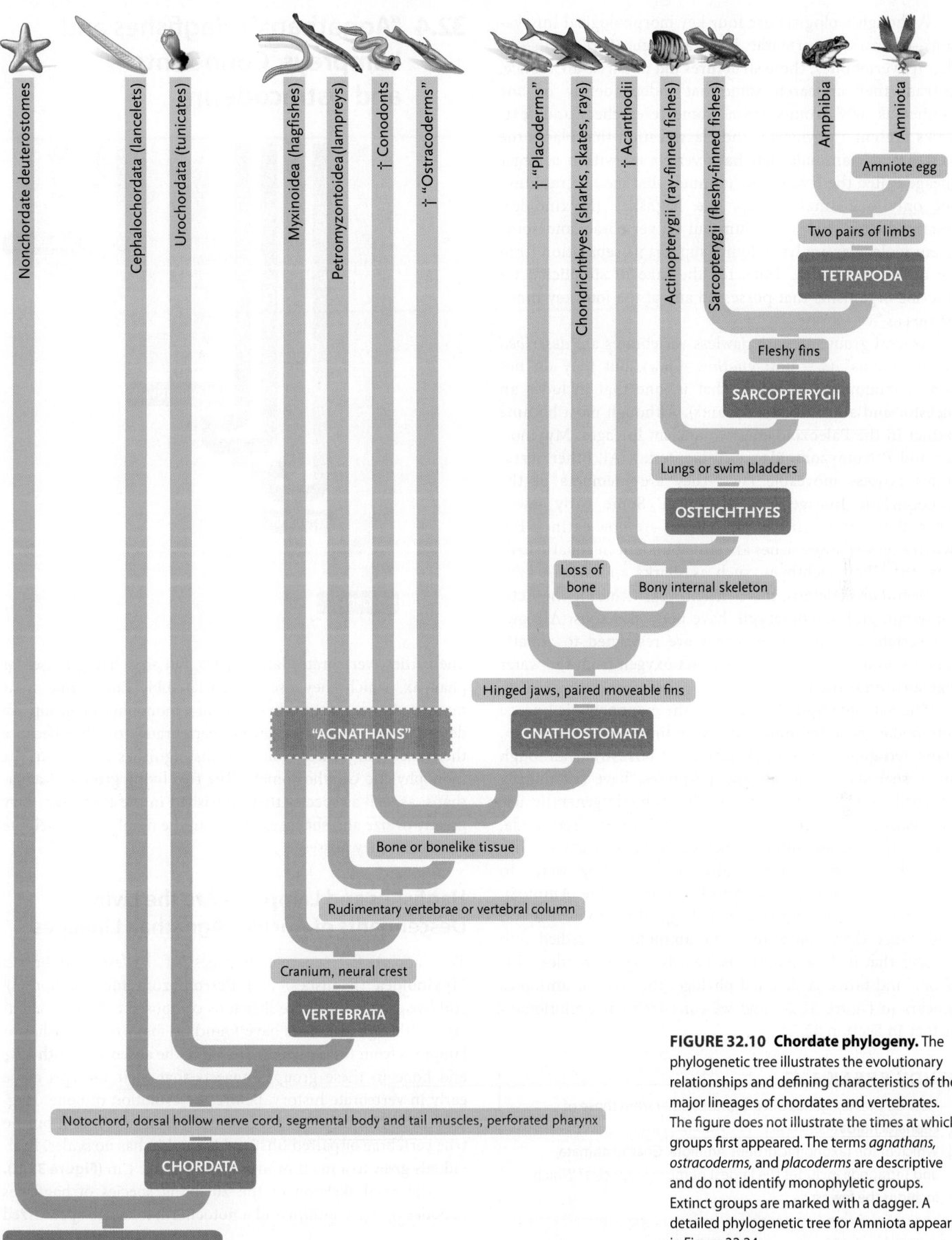

FIGURE 32.10 Chordate phylogeny. The phylogenetic tree illustrates the evolutionary relationships and defining characteristics of the major lineages of chordates and vertebrates. The figure does not illustrate the times at which groups first appeared. The terms *agnathans, ostracoderms,* and *placoderms* are descriptive and do not identify monophyletic groups. Extinct groups are marked with a dagger. A detailed phylogenetic tree for Amniota appears in Figure 32.24.

© Cengage Learning 2017

In the figure, from left to right and bottom to top, the following labels appear:

Nonchordate deuterostomes
Cephalochordata (lancelets)
Urochordata (tunicates)
Myxinoidea (hagfishes)
Petromyzontoidea (lampreys)
† Conodonts
† "Ostracoderms"
† "Placoderms"
Chondrichthyes (sharks, skates, rays)
† Acanthodii
Actinopterygii (ray-finned fishes)
Sarcopterygii (fleshy-finned fishes)
Amphibia
Amniota

Amniote egg
Two pairs of limbs
TETRAPODA
Fleshy fins
SARCOPTERYGII
Lungs or swim bladders
OSTEICHTHYES
Loss of bone
Bony internal skeleton
Hinged jaws, paired moveable fins
"AGNATHANS"
GNATHOSTOMATA
Bone or bonelike tissue
Rudimentary vertebrae or vertebral column
Cranium, neural crest
VERTEBRATA
Notochord, dorsal hollow nerve cord, segmental body and tail muscles, perforated pharynx
CHORDATA
ANCESTRAL DEUTEROSTOME

Although biologists use four key morphological innovations—a cranium, vertebrae, bone, and neural crest cells—to identify vertebrates, these structures did not arise all at once. Instead, they appeared somewhat independently of one another as new groups arose. Some researchers and textbooks present a phylogeny and classification that places the "vertebrates" (animals that have vertebrae) within a larger lineage called the "craniates" (animals that have a cranium). But only one small group, the hagfishes (Myxinoidea, described later), has a cranium but no vertebrae, and some recent molecular analyses do not support its separation from the other vertebrates. Thus, for the sake of simplicity, we describe organisms that possessed any of the four key innovations as "vertebrates."

Several groups of early jawless vertebrates are *described* as "agnathans" (*a* = not; *gnathos* = jaw), but they do not form a monophyletic group (that is, one that includes an ancestor and all of its descendants). Although most became extinct in the Paleozoic era, two ancient lineages, Myxinoidea and Petromyzontoidea, still live today. All other vertebrates possess moveable jaws; they are members of the monophyletic lineage Gnathostomata. Some early jawed fishes, the Acanthodii and placoderms are now extinct. But two lineages of jawed fishes are still abundant in aquatic habitats: the Chondrichthyes, such as sharks and skates, have cartilaginous skeletons; the Osteichthyes, including the Actinopterygii and Sarcopterygii, have bony skeletons. All jawless vertebrates and jawed fishes are restricted to aquatic habitats, and they use gills to extract oxygen from the water that surrounds them.

The Sarcopterygii also includes the monophyletic lineage Tetrapoda; most tetrapods use four limbs for locomotion. Many tetrapods are semiterrestrial or terrestrial, although some, such as sea turtles and porpoises, have secondarily returned to aquatic habitats. Adult tetrapods generally use air-breathing lungs for gas exchange. Within the Tetrapoda, one lineage, the amphibians, includes animals, such as frogs and salamanders, that typically need standing water to complete their life cycles. Another lineage, the Amniota, comprises animals with specialized eggs that can develop on land. Since their appearance, the amniotes diversified into lineages that include mammals, lizards, snakes, turtles, alligators, and birds. A detailed phylogenetic tree for amniotes appears in Figure 32.24, and we consider their evolutionary history in Section 32.7.

STUDY BREAK 32.3

1. How do the *Hox* genes of vertebrates differ from those of cephalochordates?
2. Which of the taxonomic groups Amniota, Gnathostomata, and Tetrapoda includes the largest number of species? Which includes the fewest?

32.4 "Agnathans": Hagfishes and Lampreys, Conodonts and Ostracoderms

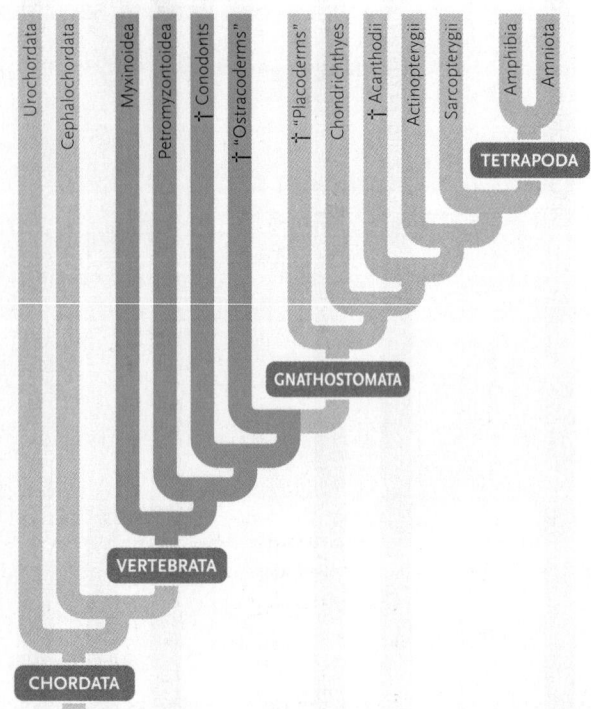

The earliest vertebrates lacked jaws, but they had a muscular pharynx, which they used to suck edible tidbits into their mouths. Although they do not form a monophyletic group, we describe four groups of jawless vertebrates together because their ancestral jawless condition distinguishes them from the monophyletic Gnathostomata. The two living groups of agnathans, as well as species that flourished in the Paleozoic, vary greatly in size and shape, as well as in the number of vertebrate characters they possess.

Hagfishes and Lampreys Are the Living Descendants of Ancient Agnathan Lineages

Two apparently separate lineages of jawless vertebrates, **Myxinoidea** (hagfishes) and **Petromyzontoidea** (lampreys), still live today. Both have skeletons composed entirely of cartilage. Although scientists have found no fossilized hagfishes or lampreys from the early Paleozoic era, the absence of both jaws and bone in these groups suggests that their lineages arose early in vertebrate history, before the evolution of bone. Hagfishes and lampreys have a well-developed notochord, but no true vertebrae or paired fins, and their skin has no scales. Individuals grow to a maximum length of about 1 m **(Figure 32.11)**.

The axial skeleton of the 20 living species of hagfishes includes only a cranium and a notochord; it has no specialized

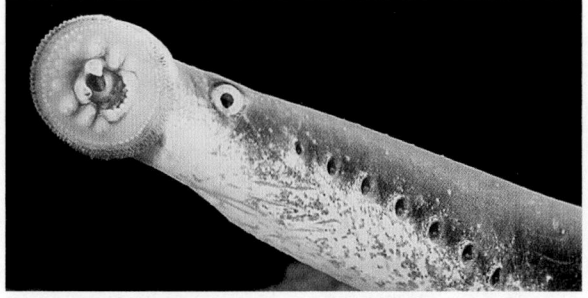

A. **Living jawless fishes**

Hagfish

Tentacles   Gill slits   Slime glands

Lamprey

Oral disk   Gill slits

B. **Mouth of a lamprey**

Heather Angel

FIGURE 32.11 Living agnathans. (A) Two groups of jawless fishes, hagfishes and lampreys, survive today. **(B)** Lampreys use a toothed oral disk to attach to a host and feed on its blood and soft tissues.

© Cengage Learning 2017

structures surrounding the dorsal nerve cord. Some biologists do not even include hagfishes among the Vertebrata, because they lack any sign of vertebrae. Hagfishes are marine scavengers that burrow in sediments on continental shelves. They feed on invertebrate prey and on dead or dying fishes. In response to predators, they secrete an immense quantity of sticky, noxious slime; when no longer threatened, a hagfish ties itself into a knot and wipes the slime from its body. Hagfish life cycles are simple and lack a larval stage.

The 40 or so living species of lampreys have traces of an axial skeleton. Their notochord is surrounded by dorsally pointing cartilages that partially cover the nerve cord; many biologists hypothesize that this arrangement reflects an early

stage in the evolution of the vertebral column. Most lamprey species are parasitic as adults. They have a circular mouth surrounded by a sucking disk with which they attach to a fish or other vertebrate host; they feed on a host's body fluids after rasping through its skin. In most species, sexually mature adults migrate from the ocean or a lake to the headwaters of a stream, where they lay eggs and then die. Their suspension-feeding larvae, which resemble adult cephalochordates, burrow into mud and develop for as long as seven years before undergoing metamorphosis and migrating to the sea or a lake to live as parasitic adults.

Conodonts and Ostracoderms Were Early Jawless Vertebrates with Bony Structures

Mysterious bonelike fossils, most less than 1 mm long, have long been known in oceanic rocks dating from the early Paleozoic era through the early Mesozoic era. Called *conodont* ("cone tooth") elements, these abundant fossils were once described as the support structures of marine algae or the feeding structures of ancient invertebrates. However, in 1983 a research team discovered fossils of intact conodont animals with these toothlike structures in their mouths and pharynxes.

Conodonts were elongate, soft-bodied animals; most were 3 to 10 cm long. They had a notochord, cranium, segmental body wall muscles, and large, moveable eyes **(Figure 32.12)**. The conodont elements at the front of the mouth were forward pointing, hook-shaped structures that apparently functioned to collect food; those in the pharynx were stouter, suitable for crushing small items that had been consumed. Paleontologists now classify conodonts as vertebrates—the earliest vertebrates with bonelike structures.

Because mineralized conodont elements superficially resemble the teeth, as well as the toothlike scales in the skins of later vertebrates, scientists had interpreted conodont elements as the evolutionary precursors of those structures. However, in 2013, Philip C. J. Donoghue of the University of Bristol, working with colleagues at universities in China, Switzerland, the United Kingdom, and the United States, used

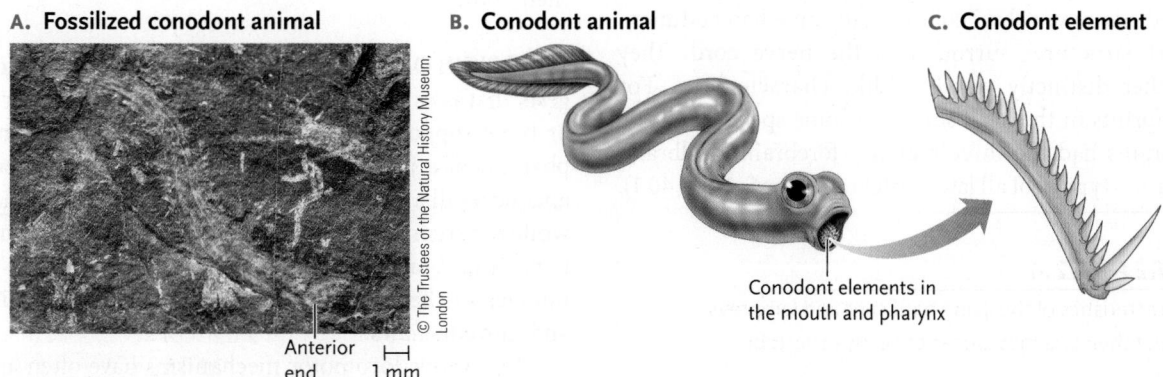

A. **Fossilized conodont animal**

© The Trustees of the Natural History Museum, London

Anterior end   1 mm

B. **Conodont animal**

Conodont elements in the mouth and pharynx

C. **Conodont element**

FIGURE 32.12 Conodonts. (A) The first intact conodont fossils were discovered in 1983. **(B)** Conodont animals were elongate and soft-bodied. **(C)** They had bonelike feeding structures in the mouth and pharynx.

© Cengage Learning 2017

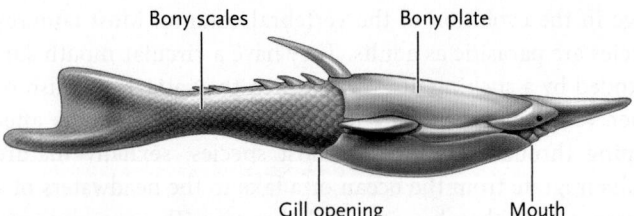

FIGURE 32.13 Ostracoderms. Some ostracoderms, such as *Pteraspis,* had large bony armor plates on the head and small bony scales on the rest of the body; this species was about 6 cm long.
© Cengage Learning 2017

Labels: Bony scales, Bony plate, Gill opening, Mouth

synchrotron radiation X-ray tomographic microscopy to analyze the microstructure of a broad range of conodont elements. Their work revealed that the toothlike elements of later conodont animals had evolved step-by-step from simple structures in their conodont ancestors. Moreover, the early conodont elements are very different from the scales and teeth of later vertebrates. Thus, Donoghue and his colleagues concluded that conodont elements are not the evolutionary precursors of vertebrate scales or teeth. Instead, their superficial similarities are the products of convergent evolution (see Section 23.3).

An assortment of jawless fishes, representing several evolutionary lineages and collectively described as **ostracoderms** (*ostrakon* = shell), were abundant from the Ordovician through the Devonian periods **(Figure 32.13)**. Like their invertebrate chordate ancestors, ostracoderms used their pharynx to extract small food particles and oxygen from mud and water. However, the ostracoderms' muscular pharynx enabled them to *suck* mud and water into their mouths, providing a much stronger flow than the cilia-driven currents of invertebrate chordates. The greater flow rate allowed ostracoderms to collect food and oxygen more rapidly. It also supported a larger body size: although most species were small, some ostracoderms reached a length of 2 m.

The skin of ostracoderms was heavily armored with bony plates and scales. Although some ostracoderms had paired lateral extensions of their armor, most species could not move them the way living fishes move their paired fins. Ostracoderms lacked a true vertebral column, but some had rudimentary support structures surrounding the nerve cord. They also had other distinctly vertebrate-like characteristics. For example, imprints in the head shields of some species indicate that their brains had the three regions—forebrain, midbrain, and hindbrain—typical of all later vertebrates (see Section 40.1).

STUDY BREAK 32.4

1. What characteristics of the living hagfishes and lampreys suggest that their lineages arose very early in vertebrate evolution?
2. What traits in conodonts and ostracoderms are derived relative to those in hagfishes and lampreys?

32.5 Gnathostomata: The Evolution of Jaws

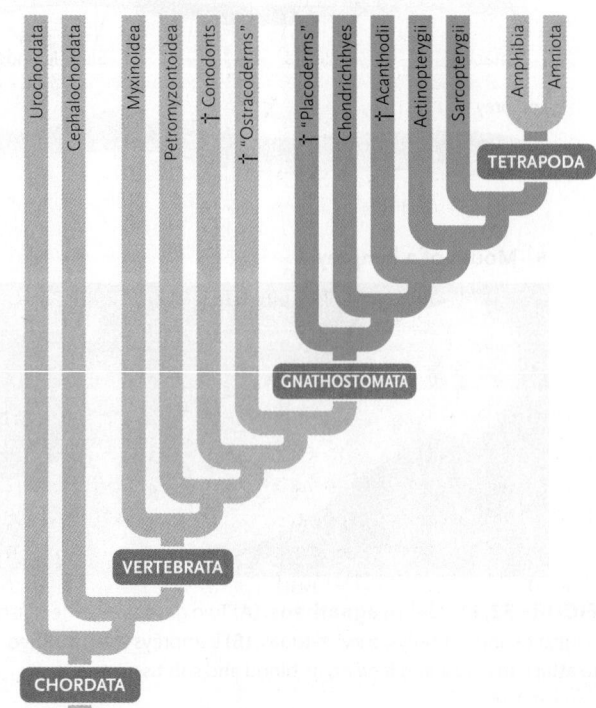

Labels: Urochordata, Cephalochordata, Myxinoidea, Petromyzontoidea, †Conodonts, †"Ostracoderms", †"Placoderms", Chondrichthyes, †Acanthodii, Actinopterygii, Sarcopterygii, Amphibia, Amniota, TETRAPODA, GNATHOSTOMATA, VERTEBRATA, CHORDATA

All vertebrates except the jawless groups described above are classified within **Gnathostomata** ("jawed mouth"). Key derived traits made their feeding and locomotion more efficient than those of their ancestors.

Jawed Fishes First Appeared in the Paleozoic Era

The renowned anatomist and paleontologist Alfred Sherwood Romer of Harvard University described the evolution of jaws as "perhaps the greatest of all advances in vertebrate history." Hinged jaws allow vertebrates to grasp, kill, shred, and crush large food items. Some living species also use their jaws for defense, for grooming, to construct nests, and to transport their young.

THE ORIGIN OF JAWS AND FINS Embryological evidence suggests that jaws evolved from paired **gill arches**—cartilaginous or bony supporting structures between the **gill slits**—in the pharynx of a jawless ancestor **(Figure 32.14)**. The first pair of ancestral gill arches formed bones in the upper and lower jaws, while the second pair was transformed into the hyomandibular bones that braced the jaws against the cranium. Nerves and muscles of the ancestral suspension-feeding pharynx control and move the jaws.

Innovative locomotor mechanisms have often appeared at roughly the same time as innovative feeding mechanisms in the vertebrate lineage; selection probably favored individuals that had the ability to pursue and capture a large range of

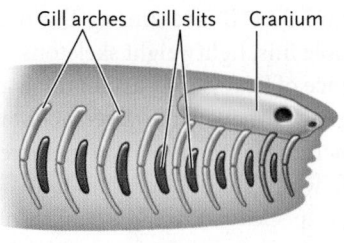

Gill arches Gill slits Cranium

A. Jaws evolved from the anterior pair of gill arches in the pharynx of jawless fishes.

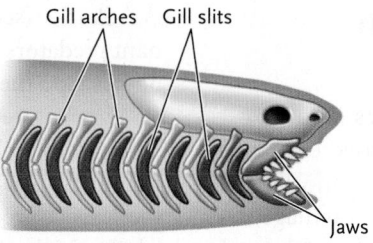

Gill arches Gill slits

Jaws

B. In early jawed fishes, the upper jaw was firmly attached to the cranium.

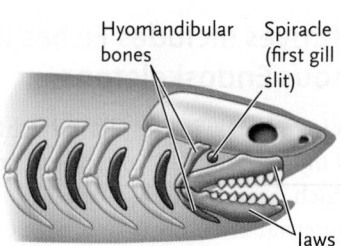

Hyomandibular bones Spiracle (first gill slit)

Jaws

C. In later jawed fishes, the jaws were supported by the hyomandibular bones, which were derived from a second pair of gill arches.

FIGURE 32.14 The evolution of jaws.
© Cengage Learning 2017

animal prey, as well as feed on them. Thus, many early jawed fishes also had fins. The earliest fins were folds of skin and moveable spines that stabilized locomotion and deterred predators. Moveable fins appeared independently in several lineages, and by the Devonian period, most fishes had unpaired (dorsal, anal, and caudal) and paired (pectoral and pelvic) fins **(Figure 32.15)**.

EARLY JAWED FISHES In two early groups of jawed fishes, "placoderms" and spiny sharks, the upper jaw was firmly attached to the cranium (see Figure 32.14B); their inflexible mouths simply snapped open and shut **(Figure 32.16).** Both groups also show evidence of an internal skeleton.

"Placoderms" (placoderm, *plax* = flat surface), now thought to be a paraphyletic group, appeared in the Silurian and diversified in the Devonian and Carboniferous periods, but they left no direct descendants. Some, such as *Dunkleosteus,* reached a length of 10 m. The anterior part of the placoderm body was covered by heavy bony plates and the posterior part by smaller, bony scales. Their jaws had sharp permanent teeth, which could not be replaced as they became worn, and their paired fins had internal skeletons and powerful muscles.

Spiny sharks (Acanthodii, *akantha* = thorn), which persisted from the late Ordovician through the Permian periods, were generally less than 20 cm long. Their small, light scales, streamlined bodies, well-developed eyes, large jaws, and

numerous teeth suggest that they were fast swimmers and efficient predators. The fins of acanthodians had internal skeletal support and were equipped with stiff spines that may have protected these fishes from predators. Acanthodian anatomy suggests that they are the sister group to the Osteichthyes, bony fishes living today.

A. Placoderm (*Dunkleosteus*)

B. *Dunkleosteus* skull

Field Museum Library/Getty Images

C. Spiny shark (*Climatius*)

FIGURE 32.16 Early gnathostomes. (A) Some placoderms were gigantic, up to 10 m in length. **(B)** Sharp teeth that lined the jaws of placoderms could not be replaced as they wore. **(C)** Most acanthodians were small, reaching a total length of no more than 20 cm.
© Cengage Learning 2017

Dorsal fin

Caudal fin

Pectoral fins

Pelvic fins

Anal fin

Andrey Burmakin/Shutterstock.com

FIGURE 32.15 Fish fins. Most fishes have both paired and unpaired fins.
© Cengage Learning 2017

Chondrichthyes Includes Fishes with Cartilaginous Endoskeletons

The 1,050 living species in the **Chondrichthyes** (*khondros* = cartilage; *ikhthys* = fish) have skeletons composed entirely of cartilage, which is much lighter than bone. In most vertebrates, the internal skeleton is first formed of cartilage, but the cartilage is gradually replaced by bone during embryonic development (see Section 38.2). Apparently, some mutation or mutations eliminated the second, bone-forming step in this process when this lineage arose. Because all fishes that lived before the appearance of Chondrichthyes had either bony armor or bony endoskeletons, some biologists suggest that the absence of bone in Chondrichthyes is a derived, not an ancestral, trait.

Most living chondrichthyans are grouped in the Elasmobranchii, which includes the skates, rays, and sharks; nearly all are marine predators **(Figure 32.17)**. Skates and rays are dorsoventrally flattened (see Figure 32.17A). They swim by undulating their enlarged pectoral fins. Most are bottom dwellers that often lie partly buried in sand. They feed on hard-shelled invertebrates, which they crush with massive, flattened teeth. The largest species, the manta ray *(Manta birostris)*, which measures 6 m across, feeds on plankton in the open ocean. Some rays have electric organs that stun prey with as much as 200 volts.

Sharks (see Figure 32.17B) are among the ocean's dominant predators. Flexible fins, lightweight skeletons, streamlined bodies, and the absence of heavy body armor allow most sharks to pursue prey rapidly. Their large livers contain copious amounts of oil, which is lighter than water, increasing their buoyancy. The great white shark *(Carcharodon carcharias)* is the largest predatory species, attaining a length of 10 m. The whale shark *(Rhincodon typus)*, which grows to 18 m, is the largest fish known; it feeds on plankton.

Elasmobranchs—including sharks, skates, and rays—exhibit remarkable adaptations for acquiring and processing food. Their teeth develop in whorls under the fleshy parts of the mouth (Figure 32.17D). New teeth migrate forward as old, worn teeth break free. In many sharks, the upper jaw is loosely attached to the cranium, and it swings down during feeding. As the jaws open, the mouth spreads widely, sucking in large, hard-to-digest chunks of prey, which are swallowed intact. Although the elasmobranch digestive system is short, it includes a corkscrew-shaped **spiral valve,** which slows the passage of material and increases the surface area available for digestion and absorption.

Elasmobranchs also have well-developed sensory systems. In addition to vision and olfaction, they use **electroreceptors** to detect weak electric currents produced by other animals. And their **lateral-line system,** a row of tiny sensors in canals along both sides of the body, detects vibrations in water (see Figure 41.5).

A. Manta ray *(Manta birostris)*

C. Swell shark (Cephaloscylium ventricosum) egg case with developing embryo

D. Sand tiger shark (Eugomphodus taurus) tooth whorls

B. Grey reef shark *(Carcharhinus amblyrhynchos)*

FIGURE 32.17 Chondrichthyes. (A) Skates and rays, as well as **(B)** sharks, are grouped in the Elasmobranchii. **(C)** Many shark egg cases include a large yolk that nourishes the developing embryo. **(D)** The teeth of elasmobranchs develop in whorls, with new teeth ready to migrate forward to replace older, worn teeth.

Chondrichthyans exhibit numerous reproductive specializations. Males have a pair of organs, the **claspers,** on the pelvic fins, which help transfer sperm into the female's reproductive tract. Fertilization occurs internally. In many species, females produce large, yolky eggs with tough, leathery shells (see Figure 32.17C). Others retain the eggs within the oviduct until the young hatch. A few species nourish young within a uterus.

The Actinopterygii and Sarcopterygii Are Fishes with Bony Endoskeletons

In terms of diversity and sheer numbers, the **Osteichthyes** (*oste* = bone; *ichthyes* = fishes)—fishes with bony endoskeletons: a cranium, vertebral column with ribs, and bones supporting their moveable fins—are the most successful of all vertebrates. The endoskeleton provides lightweight support, particularly compared with the heavy bony armor of ostracoderms and placoderms, and enhances their locomotor efficiency. Osteichthyes first appeared in the Silurian period and rapidly diversified into two lineages. **Actinopterygii** (*aktis* = ray; *pteron* = wing), the ray-finned fishes, have fins that are supported by thin and flexible bony rays. **Sarcopterygii** (*sarco* = flesh), the fleshy-finned fishes, have fins that are supported by muscles and a stout internal bony skeleton. Ray-finned fishes have always been more diverse, and they vastly outnumber the fleshy-finned fishes today. The 28,000 living species of bony fishes occupy nearly every aquatic habitat and represent more than 95% of living fish species. Adults range from 1 cm to more than 6 m in length.

Bony fishes have numerous adaptations that increase their swimming efficiency. In many modern ray-finned fishes, a gas-filled **swim bladder** serves as a hydrostatic organ that increases buoyancy (see Figure 32.19A). The swim bladder is derived from an ancestral air-breathing lung that allowed early actinopterygians to gulp air, supplementing their gill respiration in aquatic habitats where dissolved oxygen concentration was low. The scales of most bony fishes are small, smooth, and lightweight. And their bodies are covered with a protective coat of mucus, which retards bacterial growth and smoothes the flow of water.

ACTINOPTERYGII The most primitive living actinopterygians, sturgeons and paddlefishes, have mostly cartilaginous skeletons **(Figure 32.18A).** These large fishes live in rivers and lakes of the northern hemisphere. Sturgeons feed on detritus and invertebrates; paddlefish consume plankton. Gars and bowfins are remnants of a more recent radiation **(Figure 32.18B).** They occur only in the eastern half of North America, where they feed on fishes and other prey. Gars are protected from predators by a heavy coat of bony scales.

Teleosts, the latest radiation of Actinopterygii, are the most diverse, successful, and familiar bony fishes. Evolution has produced a wide range of body forms **(Figure 32.19).** Teleosts have an internal skeleton made almost entirely of bone. On either side of the head, a bony flap of the body wall, the **operculum** (plural, *opercula*), covers a chamber that houses the gills; movements of the opercula help to bring fresh, oxygenated water across the gills. Sensory systems generally include large eyes, a lateral-line system, sound receptors, chemoreceptive nostrils, and taste buds. Variations in jaw structure allow different teleosts to consume plankton, seaweed, invertebrates, or other vertebrates.

Teleosts exhibit remarkable feeding and locomotor adaptations. When some teleosts open their mouths, bones at the front of the jaws swing forward to create a circular opening. Folds of skin extend backward, forming a tube through which they suck food (see Figure 32.19F). Many also have symmetrical caudal fins, posterior to the vertebral column, which provide power for locomotion. And their pectoral fins lie high on the sides of the body, providing fine control over swimming. Some species use their pectoral fins for acquiring food, for courtship, and for care of eggs and young. Some teleosts even use them for crawling on land or gliding in air.

Most marine species produce small eggs that hatch into planktonic larvae. Eggs of freshwater teleosts are generally larger and hatch into tiny versions of the adults. In freshwater species, parents often care for their eggs and young, fanning oxygen-rich water over them, removing fungal growths, and protecting them from predators. Some freshwater species, such as guppies, give birth to live young.

SARCOPTERYGII Two groups of fleshy-finned fishes (Sarcopterygii), the lobe-finned fishes and lungfishes, are now represented by

A. Sevruga sturgeon (*Accipenser stellatus*)

blickwinkel/Alamy

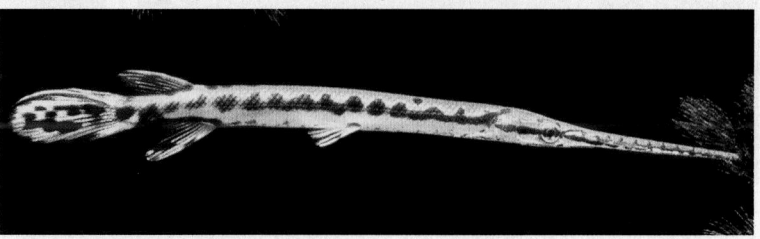

B. Long-nosed gar (*Lepisosteus osseus*)

© Shedd Aquarium/Patrice Ceisel

FIGURE 32.18 Primitive actinopterygians. (A) Sturgeons and **(B)** gars are living representatives of early actinopterygian radiations.

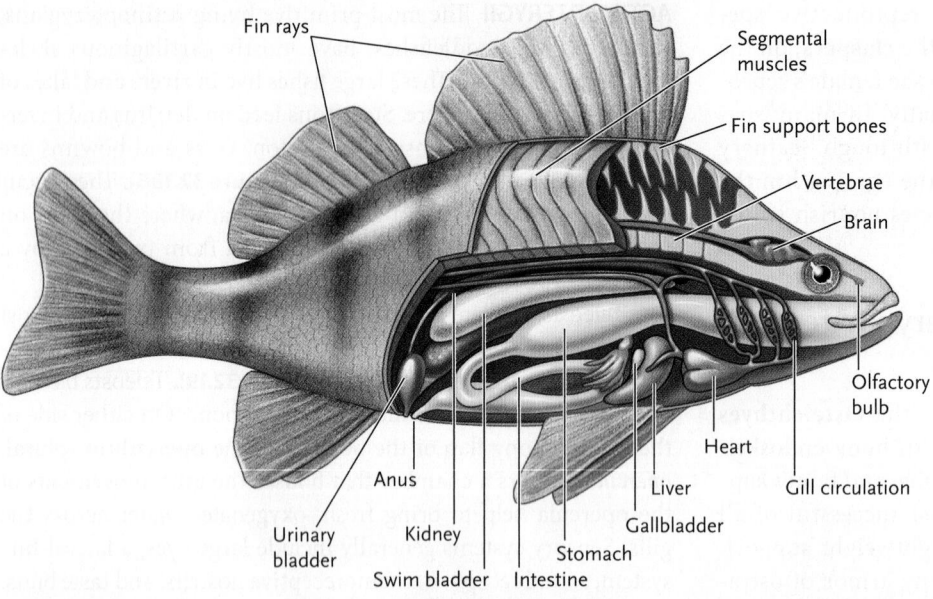

A. Teleost internal anatomy

Fin rays
Segmental muscles
Fin support bones
Vertebrae
Brain
Olfactory bulb
Heart
Gill circulation
Liver
Gallbladder
Stomach
Intestine
Swim bladder
Kidney
Urinary bladder
Anus

B. Sea horses, like the thorny sea horse *(Hippocampus histrix),* use a prehensile tail to hold on to substrates; they are weak swimmers.

C. The long, flexible body of a spotted moray eel *(Gymnothorax moringa)* can wiggle through the nooks and crannies of a reef.

D. Flatfishes, like this panther flounder *(Bothus pantherinus),* lie on one side and leap at passing prey.

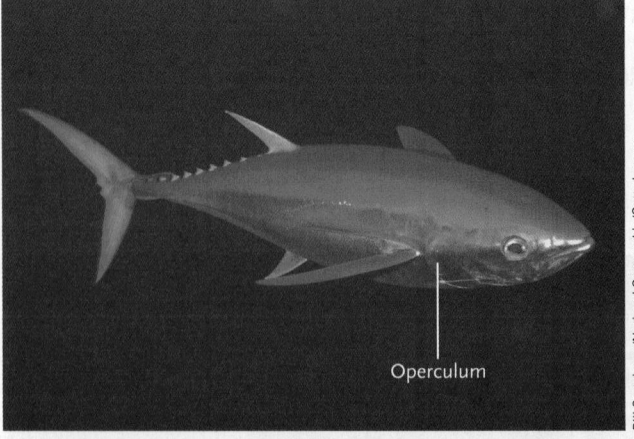

Operculum

E. Open ocean predators, like the yellowfin tuna *(Thunnus albacares),* have strong, torpedo-shaped bodies and powerful caudal fins.

F. Kissing gouramis *(Helostoma temmincki)* extend their jaws into a tube that sucks food into the mouth.

FIGURE 32.19 Teleost diversity. Although all teleosts share similar internal features, their diverse shapes adapt them to different diets and types of swimming.

© Cengage Learning 2017

A. Coelacanth *(Latimeria chalumnae)*

Peter Scoones/The Image Bank/Getty Images

B. African lungfish *(Protopterus annectens)*

Tom McHugh/Science Source

FIGURE 32.20 Sarcopterygians. (A) Two species of lobe-finned fishes, coelacanths, still live in the oceans today. **(B)** The African lungfish is one of only six living lungfish species.

rains begin, water fills the burrow and the fishes awaken from dormancy.

STUDY BREAK 32.5

1. What characteristics of sharks and rays make them more efficient predators than the acanthodians or placoderms?
2. How do the air bladder and fins of ray-finned bony fishes increase their locomotor abilities?
3. How do the lungs of lungfishes allow them to survive in stressful environments?

32.6 Tetrapoda: The Evolution of Limbs

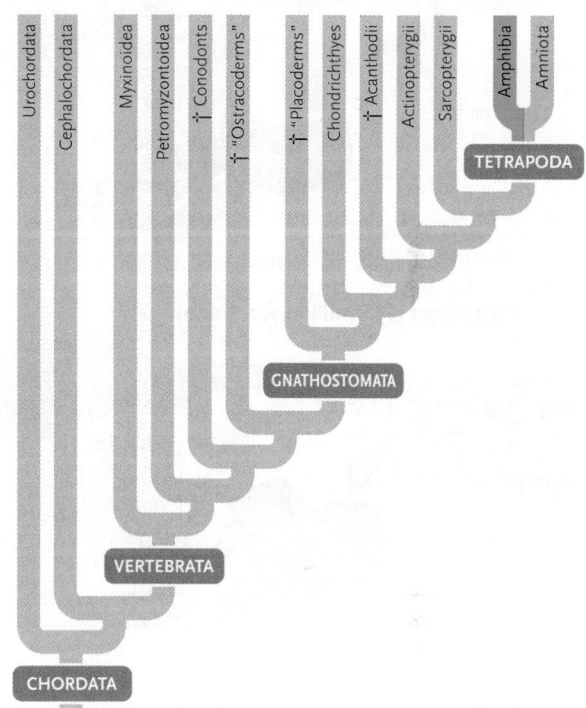

only eight living species **(Figure 32.20).** In the heyday of the Sarcopterygii, during the Devonian period, their sturdy fins probably allowed them to crawl or walk on the muddy bottom in the shallow waters they inhabited. Most species probably also acquired oxygen through primitive lungs to supplement their gill respiration.

Although lobe-finned fishes were once thought to have been extinct for 65 million years, a living coelacanth *(Latimeria chalumnae)* was discovered in 1938 near the Comoros Islands, off the southeastern coast of Africa. We now know that a population of this meter-long fish lives at depths of 70 to 600 m, feeding on fishes and squid. Remarkably, a second population of coelacanths was discovered in 1998, when a specimen was found in an Indonesian fish market, 10,000 km east of the Comoros population. Based on analyses of its DNA, it is a distinct species *(Latimeria menadoensis).*

Lungfishes have changed relatively little over the last 200 million years. Six living species are distributed on southern continents. The Australian lungfishes, which live in rivers and pools, use their lungs to supplement gill respiration when dissolved oxygen concentration is low. The South American and African species, which live in swamps, use their lungs to collect oxygen during the annual dry season, which they spend encased in a mucus-lined burrow in the dry mud. When the

The fossil record suggests that **Tetrapoda** *(tetra = four; pod = foot),* the four-limbed vertebrates and their descendants, arose from a group of fleshy-finned fishes, the *osteolepiforms,* in the late Devonian period, about 380 million years ago. Osteolepiforms and early tetrapods shared several derived characteristics: both had curious infoldings of their tooth surfaces, a trait with unknown function, and the shapes and positions of bones of the skulls and appendages were similar **(Figure 32.21).**

Key Adaptations Facilitated the Transition to Land

Fishes are not adapted to live on land, and early tetrapods faced serious environmental challenges as they ventured out of the water. First, because air is less dense than water, it provides less support against gravity for an animal's body. Second, animals

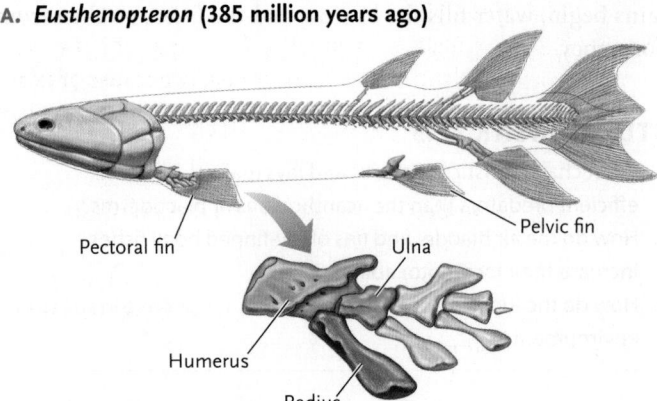

A. *Eusthenopteron* **(385 million years ago)**

Pectoral fin

Pelvic fin

Humerus

Ulna

Radius

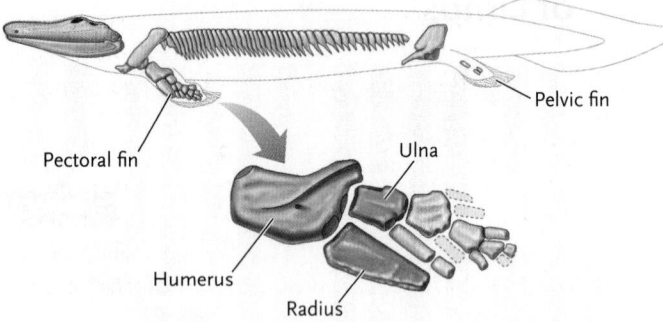

B. *Tiktaalik* **(375 million years ago)**

Pectoral fin

Pelvic fin

Humerus

Ulna

Radius

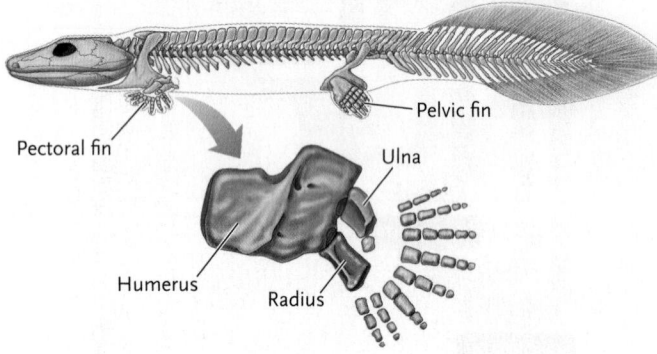

C. *Acanthostega* **(365 million years ago)**

Pectoral fin

Pelvic fin

Humerus

Ulna

Radius

FIGURE 32.21 Evolution of tetrapod limbs. The limb skeleton of *Eusthenopteron*, an osteolepiform fish, exhibits homology with those of *Tiktaalik* and *Acanthostega*, which appear later in the fossil record of sarcopterygian fishes and early tetrapods.
© Cengage Learning 2017

exposed to air inevitably lose body water by evaporation. Third, the sensory systems of fishes, which work well under water, do not function well in air. However, in swampy, late-Devonian habitats dry land also offered distinct advantages. Terrestrial plants, soft-bodied invertebrates, and arthropods provided abundant food, oxygen was more readily available in air than in water, and predators did not yet live in these new habitats.

In some ways, Devonian osteolepiforms had characteristics that may have facilitated the transition to land (see *Eusthenopteron* in Figure 32.21A). Most had strong, stout fins that enabled them to crawl on the muddy bottom of shallow pools, and their vertebral column included crescent-shaped

bones that provided good support. They had nostrils leading to sensory pits that housed olfactory ("smell") receptors. And they almost certainly had lungs to augment gill respiration in the swampy, oxygen-poor waters where they lived.

Over the past few decades, the discovery of new fossils has greatly increased our understanding of the vertebrates' transition to land. In 2006, a team of paleontologists working in the Canadian arctic discovered the find of a lifetime: *Tiktaalik*, which in the local Inuktitut language means "large, freshwater fish." *Tiktaalik*, which lived about 375 million years ago, had characters that identify it as transitional between the osteolepiform fishes and the tetrapod vertebrates that later colonized the land (Figure 32.21B). Its fishlike characters included bony scales on its body, fins, and gills, as well as air-breathing lungs. Its tetrapod characters included a neck; well-developed ribs; a forelimb skeleton that included a humerus, radius, and ulna; and a flattened skull with upward-pointing eyes. Although scientists conclude that *Tiktaalik* was aquatic, it probably lived close to the shore, where it may have captured prey that lived at the water's edge. Although the pelvis and pelvic fins were not described at the time of *Tiktaalik*'s discovery, a subsequent analysis, published in 2014, revealed that they were larger and more robust than those of the osteolepiforms.

Somewhat later in the Devonian, about 365 million years ago, the earliest tetrapods emerged. *Acanthostega* and its close relatives were still somewhat fishlike in the overall form, retaining a large caudal fin, as well as an operculum; the operculum, which covers the gill chamber in living fishes, suggests that this animal still used gill respiration (Figure 32.21C). However, *Acanthostega* had well-defined digits (that is, fingers and toes)—at least eight on both the hands and feet. It also had a moderately sturdy vertebral column with small ribs attached and a well-anchored pelvis. Nevertheless, the structure of its forelimbs suggests that it could not carry its own weight on land. Thus, paleontologists deduce that, like *Tiktaalik*, it occupied shallow-water habitats and captured prey on the shoreline.

In addition to changes in limbs and the vertebral column, life on land also required adaptations in sensory systems. In fishes, for example, the body wall picks up sound vibrations and transfers them to sensory receptors directly. But sound waves are harder to detect in air. Early tetrapods developed a **tympanum,** a specialized membrane on either side of the head that is vibrated by airborne sounds. The tympanum connects to the **stapes,** a bone that is homologous to the hyomandibula, which had supported the jaws of fishes (see Figure 24.4). The stapes, in turn, transfers vibrations to the sensory cells of an inner ear.

Modern Amphibians Are Very Different from Their Paleozoic Ancestors

Most of the more than 6,130 species of living **amphibians**—including frogs, salamanders, and caecilians—are small, and their skeletons contain less bone than those of Paleozoic tetrapods such as *Acanthostega*. All living amphibians are carnivorous as adults, but the aquatic larvae of some species are herbivores.

Most living amphibians have a thin, scaleless skin, well supplied with blood vessels, that is a major site of gas exchange. Because gases must enter the body across a thin layer of water, the skin of most amphibians must remain moist, restricting them to aquatic or wet terrestrial habitats. Adults of some species also acquire oxygen through saclike lungs. The evolution of lungs was accompanied by modifications of the heart and circulatory system that increase the efficiency with which oxygen is delivered to body tissues (see Section 44.1).

The life cycles of many amphibians (*amphi* = of both kinds; *bios* = life) include both larval and adult stages (see Figure 50.3). Eggs are laid and fertilized in water, where they hatch into larvae, such as the tadpoles of frogs, which eventually metamorphose into adults (see Figure 42.8). Although the larvae of many species are aquatic, adults may be aquatic, amphibious, or terrestrial. Some salamanders are paedomorphic; the larval stage attains sexual maturity without changing its form or moving to land. By contrast, some frogs and salamanders reproduce on land and skip the larval stage altogether. But even though they are terrestrial breeders, their eggs dry out quickly unless they are laid in moist places.

Modern amphibians are represented by three clades **(Figure 32.22)**. Populations of practically all amphibian species have declined rapidly in recent years, probably because of exposure to environmental pollutants and fungal infections (see Chapter 55).

ANURA The 5,400 species of frogs and toads have short, compact bodies, and adults lack tails. Their elongate hind legs and webbed feet allow them to hop on land or swim. A few species are adapted to dry habitats, withstanding periods of drought by encasing themselves in mucous cocoons.

CAUDATA Most of the 560 species of salamanders and newts have an elongate, tailed body and four legs. They walk by alternately contracting muscles on either side of the body much the way fishes swim. Species in the most diverse group, the lungless salamanders, are fully terrestrial throughout their lives, using their skin and the lining of the throat for gas exchange.

GYMNOPHIONA The 170 species of caecelians (Gymnophiona, from *gymnos* = naked; *ophioneos* = snakelike) are legless burrowing amphibians with wormlike bodies. They occupy tropical habitats throughout the world. Unlike other modern

A. Northern leopard frog *(Rana pipiens)*

B. Red-spotted newt *(Notophthalmus viridescens)*

C. A caecelian *(Caecelia nigricans)*

FIGURE 32.22 Living amphibians. (A) Frogs and toads have compact bodies and long hind legs; their larvae (inset), called tadpoles, are legless and rely on their tails for locomotion. **(B)** Salamanders and newts have an elongate body and four legs. **(C)** Caecelians are legless, burrowing amphibians.

amphibians, caecelians have small bony scales embedded in their skin. Fertilization is internal, and females give birth to live young.

STUDY BREAK 32.6

1. For early tetrapods, what were the advantages and disadvantages of moving onto the land?
2. What parts of the life cycle in most modern amphibians are dependent on water or very moist habitats?

32.7 Amniota: The Evolution of Fully Terrestrial Vertebrates

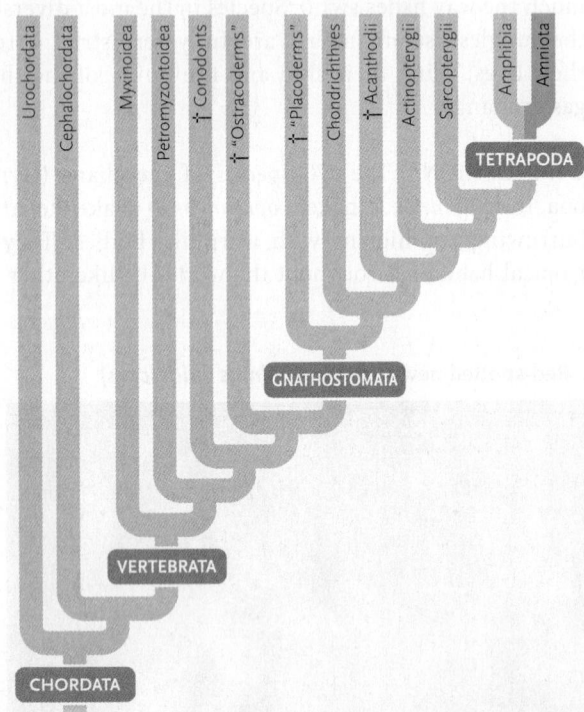

Amniota, vertebrates that can reproduce on land, arose about 320 million years ago, during the Carboniferous period, at a time when seed plants and insects, which served as excellent food resources, began to occupy higher ground. The lineage is named for the amnion, a fluid-filled sac that surrounds the embryo during development.

Key Adaptations Allow Amniotes to Live a Fully Terrestrial Life

Based on the abundance and diversity of their fossils, amniotes were extremely successful, quickly replacing many nonamniote species in terrestrial habitats. Although the fossil record includes abundant skeletons of early amniotes, it provides little direct information about their soft body parts and physiology. For amniotes living today, three key adaptations allow them to

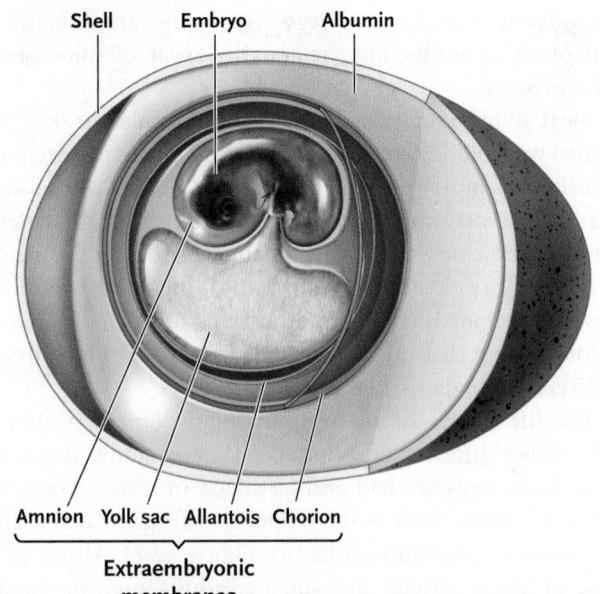

FIGURE 32.23 The amniote egg. A water-retaining egg with four specialized membranes (the amnion, allantois, chorion, and yolk sac) and a hard or leathery shell allowed amniotes and their descendants to reproduce in dry terrestrial environments.
© Cengage Learning 2017

live in dry habitats, freeing them from a dependency on moist surroundings and standing water. First, they have a tough, dry skin. The skin cells are filled with keratin and lipids, substances that are relatively impermeable to water. Thus, amniotes do not dehydrate in air as quickly as amphibians do.

Second, many amniotes produce an **amniote egg,** which can survive and develop on dry land. The eggs of modern reptiles and birds have four specialized membranes and a hard or leathery shell **(Figure 32.23).** The membranes protect the developing embryo and facilitate gas exchange and excretion; the shell, which is perforated by microscopic pores, mediates the exchange of air and water between the egg and its environment. The egg also includes generous supplies of **yolk,** the embryo's main energy source, and **albumin,** a source of nutrients and water. Compared with those of amphibians, amniote eggs are large; and, as amniotes lack a larval stage, the young hatch as miniature versions of the adult. In contrast to reptiles and birds, the eggs of virtually all mammals lack a shell. Instead, mammalian embryos, which have the same four membranes associated with them, implant in the wall of the mother's uterus and receive nutrients and oxygen directly from her.

Third, many amniotes produce uric acid as a waste product of nitrogen metabolism (see Chapter 48). By contrast, fishes and amphibians produce ammonium ions or urea, toxic materials that require lots of water to flush them from body tissues. Because uric acid is less toxic than these other compounds, it can be excreted as a semisolid paste, which conserves body water.

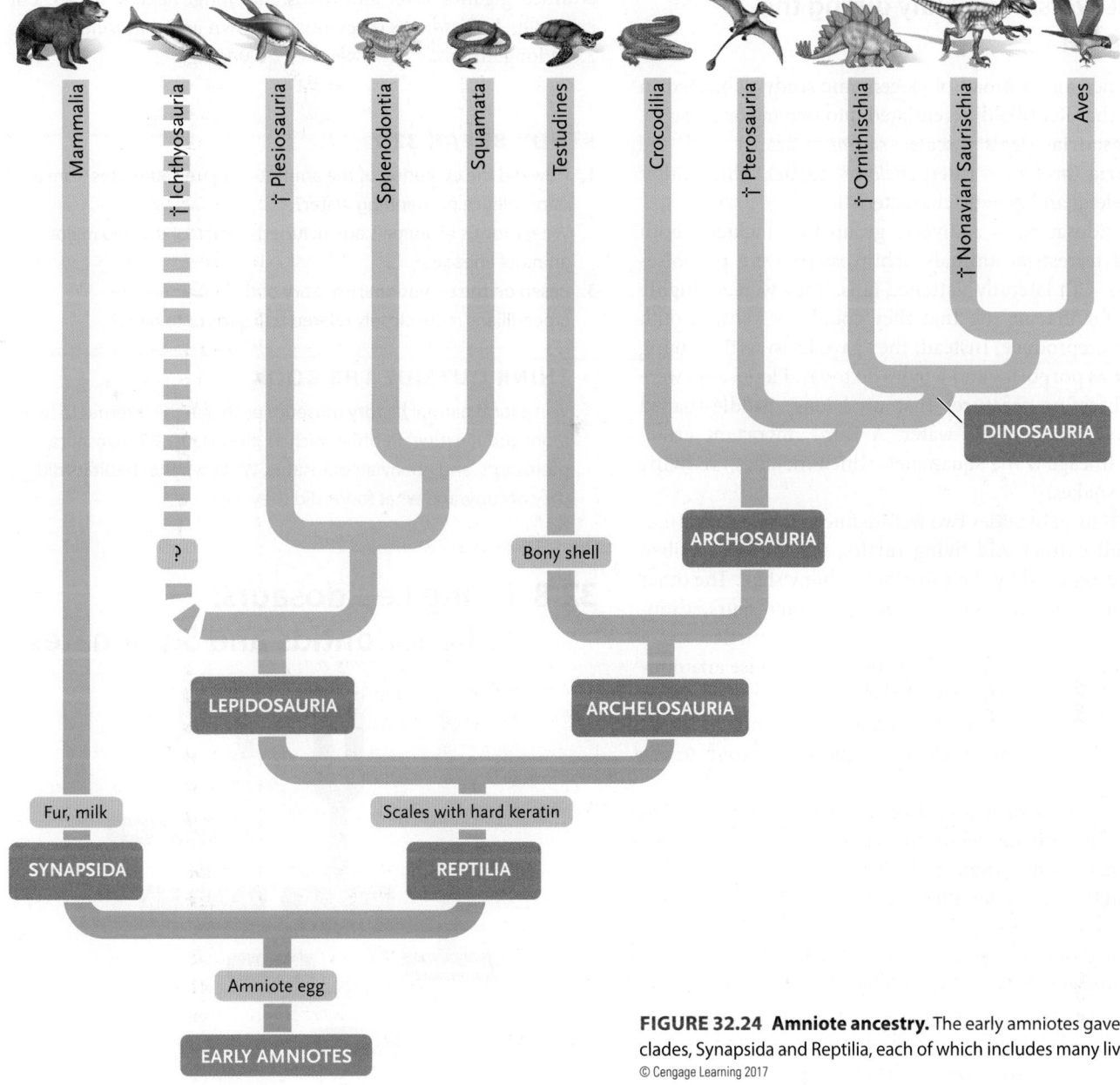

Mammalia

† Ichthyosauria

† Plesiosauria

Sphenodontia

Squamata

Testudines

Crocodilia

† Pterosauria

† Ornithischia

† Nonavian Saurischia

Aves

?

DINOSAURIA

Bony shell

ARCHOSAURIA

LEPIDOSAURIA

ARCHELOSAURIA

Fur, milk

Scales with hard keratin

SYNAPSIDA

REPTILIA

Amniote egg

EARLY AMNIOTES

FIGURE 32.24 Amniote ancestry. The early amniotes gave rise to two clades, Synapsida and Reptilia, each of which includes many living species.
© Cengage Learning 2017

Amniotes Diversified into Two Major Lineages

Based on the morphology of extinct species, the anatomy and physiology of living species, and molecular sequence and genomic data, biologists hypothesize that amniotes differentiated into two lineages during the Carboniferous and Permian periods: Synapsida and Reptilia **(Figure 32.24).** These clades are distinguished by differences in their anatomy and physiology.

Synapsida, the clade that includes all living mammals, comprises animals that have a glandular skin and bodies covered by either fur or hair. Most also give birth to live young, which are nourished by milk produced by their mothers. In addition, all living synapsids are endothermic: they use heat generated by their metabolism to maintain a reasonably high and constant body temperature (see Section 48.8). Researchers agree that Synapsida was the earliest lineage to emerge from the ancestral amniotes in the late Permian period. Early synapsids were small terrestrial predators, but the group has diversified to include mammals with a wide range of sizes and diets (as described in Section 32.10).

Reptilia, the clade that includes all amniotes except the mammals, comprises animals with dry, scaly, nonglandular skin; the bodies of one lineage, the birds, are covered with feathers. Most reptiles lay eggs that can only survive on land, and none produce milk. The Mesozoic era was a time when Reptilia proliferated into numerous and diverse groups. Their living descendants include lizards and snakes, turtles, crocodilians, and birds.

Reptiles Diversified Wildly during the Mesozoic Era

Fossil evidence and a broad phylogenomic study published in 2014 reveal that Reptilia differentiated into two major lineages, the **Lepidosauria** (*lepis* = scale; *sauros* = lizard) and the **Archelosauria** (*arkhon* = ruler; *chelon* = turtle), which differ in many skeletal and genetic characteristics.

The Lepidosauria is a diverse group that included both marine and terrestrial animals. Ichthyosaurs were porpoise-like animals with laterally flattened tails. They were so highly specialized for marine life that they could not venture onto land, even to reproduce. Instead, they gave birth to live young under water as porpoises and whales do today. Plesiosaurs were marine, fish-eating creatures that used long, paddle-shaped limbs to row through the water. A third important group within this lineage is the squamates, which includes the living lizards and snakes.

Archelosauria includes two well-defined clades. Testudines comprises all extinct and living turtles, which, as described below, are recognized by their distinctive bony shell. The other clade, Archosauria, includes crocodilians, pterosaurs, dinosaurs, and birds. The phylogenetic position of turtles was a topic of fierce debate for more than 100 years because anatomical studies did not clearly reveal their relationships to other amniote groups. However, the recent phylogenomic study noted above identifies them clearly as the sister group to the Archosauria.

The Archosauria includes a diverse group of extinct and living reptiles. Crocodilians, which first appeared during the Triassic period, have bony armor and a laterally flattened tail that propels them through water. Pterosaurs, now extinct, were flying predators of the Jurassic and Cretaceous periods. Their wings were composed of thin sheets of skin attached to the sides of the body and supported by an elongate finger. Small pterosaurs may have been active fliers, but large ones, with wingspans as wide as 13 m, probably soared on air currents as vultures do today.

Two lineages of **dinosaurs,** "lizard-hipped" saurischians and "bird-hipped" ornithischians, proliferated in the Triassic and Jurassic periods. As their names imply, they differed in the anatomy of their pelvic girdles.

The largely herbivorous ornithischian dinosaurs had enormous, stocky bodies. This lineage included the armored or plated dinosaurs (*Ankylosaurus* and *Stegosaurus*), the duck-billed dinosaurs (*Hadrosaurus*), horned dinosaurs (*Styracosaurus*), and some with remarkably thick skulls (*Pachycephalosaurus*). The ornithischians were most abundant in the Jurassic and Cretaceous periods.

The saurischian lineage included bipedal carnivores and quadrupedal herbivores. Most carnivorous saurischians were swift runners. Their forelimbs, however, were often ridiculously short. *Tyrannosaurus,* which was 15 m long and stood 6 m high, is the most familiar carnivorous saurischian, but most species were much smaller. One group of small carnivorous saurischians was ancestral to birds (see Section 23.6). By the Cretaceous period, some herbivorous saurischians had also attained gigantic size, and many had long, flexible necks. For example, *Apatosaurus* (previously known as *Brontosaurus*) was 25 m long and may have weighed 50,000 kg.

STUDY BREAK 32.7

1. How did the evolution of the amniote egg free amniotes from a dependence on standing water?
2. What groups of animals are included in each of the two major amniote lineages?
3. Based on the evolutionary history of the amniotes, are crocodilians more closely related to lizards or to birds?

THINK OUTSIDE THE BOOK

Visit a local natural history museum or search the Internet to learn more about extinct Reptilia, such as plesiosaurs, ichthyosaurs, pterosaurs, and nonavian dinosaurs. What types of habitats did they occupy, and what foods did they eat?

32.8 Living Lepidosaurs: Sphenodontids and Squamates

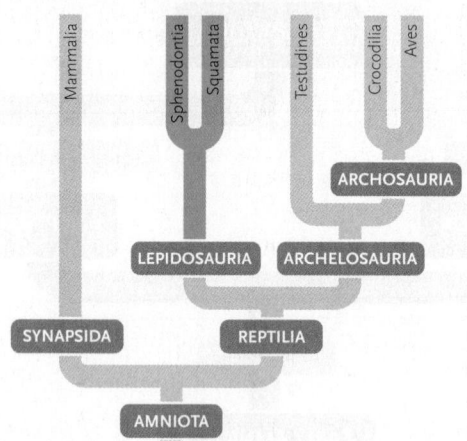

Two groups of lepidosaurs—Sphenodontida and Squamata—still live today. In addition to sharing certain skeletal characteristics, their bodies are encased in a dry, scaly skin.

Living Sphenodontids Are Remnants of a Diverse Mesozoic Lineage

The tuatara (*Sphenodon punctatus*) is one of two living representatives of the sphenodontids (*sphen* = wedge; *odont* = tooth), a diverse Mesozoic lineage **(Figure 32.25A)**. These lizard-shaped animals survive on a few islands off the coast of New Zealand. Adults are about 60 cm long. They live in dense colonies, where males and females defend small territories using vocal and visual displays. They often share underground burrows with seabirds, feeding mainly on invertebrates and small vertebrates. They are primarily nocturnal and maintain low body temperatures during periods of activity.

A. Sphenodontia includes the tuatara *(Sphenodon punctatus)* and one other species.

B. Unlike most geckos, the Madagascar giant day gecko *(Phelsuma madagascariensis)* is active during daylight hours.

C. A western diamondback rattlesnake *(Crotalus atrox)* of the American southwest bares its fangs with which it injects a powerful toxin into prey.

FIGURE 32.25 Living lepidosaurs.

Squamates—Lizards and Snakes—Are Covered by Overlapping, Keratinized Scales

The skin of lizards and snakes (Squamata, *squama* = scale) is composed of overlapping, keratinized scales that protect against dehydration. Squamates periodically shed their skin as they grow, much the way arthropods shed their exoskeletons (see Section 31.7). Most squamates regulate their body temperature behaviorally; they are active only when weather conditions are favorable, and they shuttle between sunny and shady places when they need to warm up or cool down (see Section 48.7 and Figure 51.9).

Most of the 6,000 lizard species are less than 15 cm long **(Figure 32.25B)**. However, the Komodo dragon *(Varanus komodoensis)* grows to nearly 3 m. Lizards occupy a wide range of habitats, but they are especially common in deserts and the tropics; one species *(Zootoca vivipara)* occurs within the Arctic Circle. Most lizards feed on insects, although some eat leaves or meat. The diverse tropical genus *Anolis* has become a frequent subject of research, as described in *Focus on Research: Model Organisms.*

The 3,400 species of snakes evolved from a lineage of lizards that lost their legs over evolutionary time (see Section 20.3). Streamlined bodies make snakes efficient burrowers or climbers **(Figure 32.25C)**. Many subterranean species are only 10 or 15 cm long, but the giant constrictors may grow to 10 m. Unlike lizards, all snakes are predators that swallow prey whole. Snakes have thinner skull bones than their lizard ancestors did, and the bones are connected to each other by elastic ligaments that stretch remarkably, allowing some snakes to swallow food that is larger than their heads. Snakes also have well-developed sensory systems for detecting prey. The flicking tongue carries airborne molecules to sensory receptors in the roof of the mouth. Most snakes can detect vibrations on the ground, and some, such as rattlesnakes, have heat-sensing organs (see Figure 41.22). Many snakes kill by constriction, which suffocates prey, and several groups produce toxins that immobilize, kill, and partially digest the prey before it is swallowed.

STUDY BREAK 32.8

1. In addition to losing their legs over evolutionary time, how do snakes differ from their lizard ancestors?
2. How do the diets of lizards and snakes differ?

Model Organisms: *Anolis* Lizards of the Caribbean

The lizard genus *Anolis* has been a model system for studies in ecology and evolutionary biology since the 1960s, when Ernest E. Williams of Harvard University's Museum of Comparative Zoology first began studying it. With roughly 400 known species—and new ones being described all the time—*Anolis* is one of the most diverse vertebrate genera. Most anoles are less than 10 cm long, not including the tail, and many occur at high densities, making it easy to collect lots of data in a relatively short time. Male anoles defend territories, and their displays make them conspicuous even in dense forests.

Anolis species are widely distributed in South America and Central America, but nearly 40% occupy Caribbean islands. The number of species on an island is generally proportional to the island's size. Cuba, the largest island, has more than 50 species, whereas small islands have just one or two.

Studies by Williams and others suggest that the anoles on some large islands are the products of independent adaptive radiations. Eight of the 10 *Anolis* species now found on Puerto Rico probably evolved on that island from a common ancestor. Similarly, the seven *Anolis* species on Jamaica shared a common ancestor, which was different from the ancestor of the Puerto Rican species. The anole faunas on Cuba and Hispaniola are the products of several independent radiations on each island.

Williams discovered that these independent radiations had produced similar-looking species on different islands. He developed the concept of the *ecomorph*, a group of species that have similar morphological, behavioral, and ecological characteristics even though they are not closely related within the genus. Williams named the ecomorphs after the vegetation that they commonly used **(Figure)**. For example, grass anoles are small, slender species that usually perch on low, thin vegetation. Trunk-ground anoles have chunky bodies and large heads, and they perch low on tree trunks, frequently jumping to the ground to feed. Although the grass anoles or the trunk-ground anoles on different islands are similar in many ways, they are not closely related to each other. Their resemblances are the products of convergent evolution.

Ecomorphs exist because evolutionary processes have accentuated the morphological differences among species that occupy different types of vegetation. Jon Losos of Harvard University has demonstrated that trunk-ground anoles, which have relatively long legs and tails, can run faster on wide surfaces and jump farther than species with relatively short legs. And in nature the trunk-ground anoles run and jump more frequently than the other ecomorphs do.

Different ecomorphs on an island use different parts of their habitats by choosing different perch sites (grass, tree trunks, rocks). When two or more species of the same ecomorph inhabit the same island, they occupy habitats with different temperature and shade conditions (see Figure). For example,

in Puerto Rico, one species of trunk-ground anole (*Anolis gundlachi*) occupies cool, shady uplands; another (*Anolis cristatellus*) lives in warm, fairly open lowland habitats; and a third species (*Anolis cooki*) lives in desert habitats. Other species in Puerto Rico exhibit similar differences in their distributions. These differences in geographical distribution and habitat use presumably allow the different species to avoid competition with each other and gain access to the resources they need to survive and reproduce.

Evolutionary processes have also fostered physiological differences that reinforce the ecological separation established by the lizards' use of different habitats. For example, *A. cristatellus* maintains higher body temperatures than *A. gundlachi,* and neither is physiologically adapted to the environment of the other: *A. cristatellus* dies in the high-altitude forests where *A. gundlachi* thrives, whereas *A. gundlachi* suffers heat stress at body temperatures that are typical for *A. cristatellus.*

Researchers throughout the Americas continue to explore the ecology and evolution of anoles. Some unravel their biogeography and systematic relationships; others focus on the ecology of populations and communities; still others study their social behavior or sensory physiology. With so many species distributed across hundreds of Caribbean islands, the lizard genus *Anolis* provides fertile ground for testing hypotheses about nearly every aspect of vertebrate biology.

A. cooki

Manuel Leal, University of Missouri

A. poncensis

Manuel Leal, University of Missouri

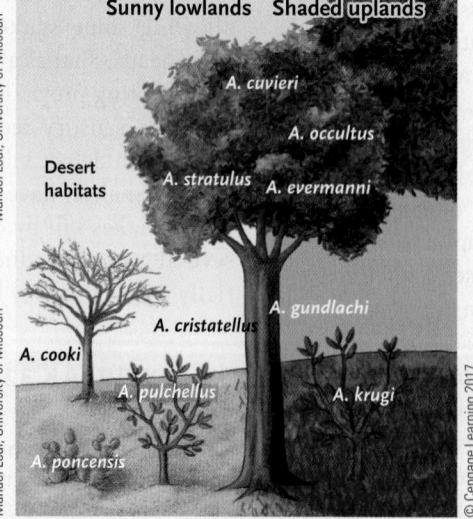

Sunny lowlands   Shaded uplands

Desert habitats

A. cuvieri
A. occultus
A. stratulus   *A. evermanni*
A. gundlachi
A. cristatellus
A. cooki
A. pulchellus
A. krugi
A. poncensis

© Cengage Learning 2017

A. gundlachi

Manuel Leal, University of Missouri

A. krugi

Manuel Leal, University of Missouri

32.9 Living Archelosaurs: Turtles, Crocodilians, and Birds

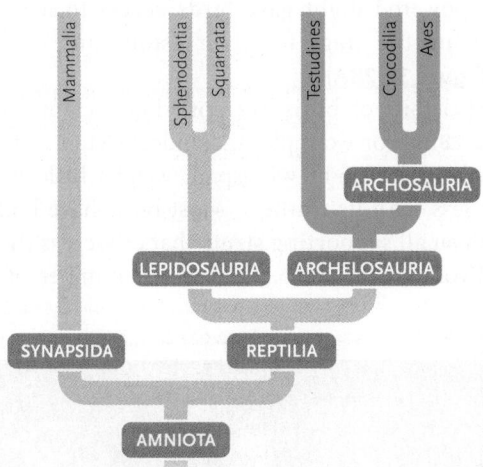

Archelosauria includes two major clades, Testudines and Archosauria, which differ dramatically in their skeletal anatomy. Testudines (turtles) are an ancient group; most species are encased in a bony shell. Only two groups of living archosaurs—Crocodilia and Aves—have survived since the Mesozoic era. Although they bear little resemblance to each other in their external morphology, similarities in their heart and respiratory structures, their

A. The turtle skeleton

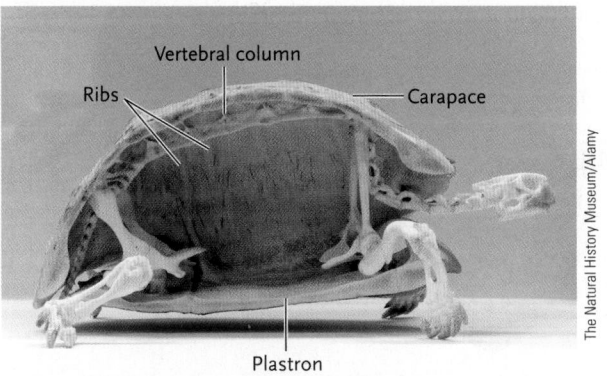

B. Eastern painted turtle (Chrysemys picta)

FIGURE 32.26 Testudines. (A) Most turtles can withdraw their heads and legs into a bony shell. **(B)** Aquatic turtles often bask in sunlight to warm up. The sunlight may also help eliminate parasites that cling to the turtle's skin.

maternal behavior, and their molecular sequences confirm their close relationship.

Turtles Have Bodies Encased in a Bony Shell

The turtle body plan, largely defined by a bony, boxlike shell, has changed little since the group first appeared during the Triassic period, nearly 250 million years ago **(Figure 32.26).** The shell includes a dorsal **carapace** and a ventral **plastron.** A turtle's ribs are fused to the inside of the carapace, and, in contrast to other tetrapods, the pectoral and pelvic girdles lie *within* the ribcage. Large, keratinized scales cover the bony plates that form the shell.

The 330 living species of turtles occupy terrestrial, freshwater, and marine habitats. They range from 8 cm to 2 m in length. All species lack teeth, but they use a keratinized beak and powerful jaw muscles to feed on plants or animal prey. When threatened, most species retract into their shells. Many species are now highly endangered because adults are hunted for meat, their eggs are consumed by humans and other predators, and their young are collected for the pet trade.

Crocodilians Are Semiaquatic, Predatory Archosaurs

In addition to birds, the 21 living species of alligators and crocodiles (Crocodilia, from *crocodilus* = crocodile) are the remnants of a once-diverse archosaur lineage **(Figure 32.27).** The largest species, the Australian saltwater crocodile *(Crocodylus porosus),* grows to 7 m. Crocodilians are aquatic predators that consume other vertebrates. Striking anatomical adaptations distinguish them from living lepidosaurs, including a four-chambered heart and a one-way flow of air through their lungs, characters that are homologous to those of birds.

American alligators *(Alligator mississippiensis)* exhibit strong maternal behavior, which also reflects their relationship to birds. Females guard their nests ferociously and help their young break out of their eggshells. The young stay close to the mother for about a year, feeding on scraps that fall from her mouth and living under her watchful protection.

FIGURE 32.27 Living nonfeathered archosaurs. Crocodilia includes semiaquatic predators, such as this American alligator *(Alligator mississippiensis)* resting in the Okefenokee Swamp, Georgia.

Most alligator and crocodile species are highly endangered. Their habitats have been disrupted by human activities, and they have been hunted for meat and leather. Protection efforts have been extremely successful, however. American alligators, for example, have recovered from the brink of extinction.

Birds Have Key Adaptations That Reduce Body Weight and Power Flight

Birds (Aves) appeared in the Jurassic period as descendants of carnivorous, bipedal dinosaurs. Thus, they are full-fledged members of the archosaur lineage. Their evolutionary relationship to dinosaurs is evident in their skeletal anatomy, the scales on their legs and feet, and their posture when walking. However, powered flight gave birds access to new adaptive zones, setting the stage for their astounding evolutionary success **(Figure 32.28A)**.

The skeletons of birds are both lightweight and strong **(Figure 32.28B)**. For example, the endoskeleton of the frigate bird, which has a 1.5 m wingspan, weighs little more than 100 g, far less than its feathers. Most birds have hollow limb bones with small supporting struts that crisscross the internal cavities. Evolution has also reduced the number of separate

A. Wing movements of an owl during flight

Stephen Dalton/Science Source

B. Skeletal system of birds

- Skull
- Radius
- Ulna
- Pectoral girdle
- Humerus
- Furcula (wishbone)
- Scapula
- Pelvic girdle
- Coracoid
- Keeled sternum

C. Pectoral girdle and flight muscles of bird in frontal view

- Humerus
- Tendon
- Humerus
- Scapula
- Coracoid
- Sternum
- Pectoralis major (lowers wings)
- Supracoracoideus (raises wings)
- Keel of sternum
- Internal structure of bird limb bones

D. Feather structure

- Barbule
- Barb
- Shaft

FIGURE 32.28 Adaptations for flight in birds. (A) The flapping movements of a bird's wing provide *thrust* for forward movement and *lift* to counteract gravity. **(B)** The bird skeleton includes a boxlike trunk, short tail, long neck, lightweight skull and beak, and well-developed limbs. In large birds, limb bones are hollow. **(C)** Two sets of flight muscles attach to a keeled sternum; one set raises the wings, and the other lowers it. **(D)** Flexible feathers form an airfoil on the wing surface.

© Cengage Learning 2017

bony elements in the wings, skull, and vertebral column (especially the tail), making the skeleton light and rigid. And all modern birds lack teeth, which are dense and heavy; they acquire food with a lightweight, keratinized bill. Many species have a long, flexible neck, which allows them to use their bills for feeding, grooming, nest-building, and social interactions.

The bones associated with flight are generally large. The forelimb and forefoot are elongate, forming the structural support for the wing. And most modern birds possess a **keeled sternum** (breastbone) to which massive flight muscles are attached **(Figure 32.28C).** Not all birds are strong fliers, however; ostriches and other bipedal runners have strong, muscular legs but small wings and flight muscles (see Figure 20.2).

Like the skeleton, soft internal organs are modified in ways that reduce weight. Most birds lack a urinary bladder; uric acid paste is eliminated with digestive wastes. Females have only one ovary and never carry more than one mature egg; eggs are laid as soon as they are shelled.

All birds also possess **feathers (Figure 32.28D),** sturdy, lightweight structures derived from scales in the skin of their ancestors. Each feather has numerous barbs and barbules with tiny hooks and grooves that maintain the feathers' structure, even during vigorous activity. Flight feathers on the wings provide lift, contour feathers streamline the surface of the body, and down feathers form an insulating cover close to the skin. Worn feathers are molted and replaced once or twice each year.

Other adaptations for flight allow birds to harness the energy needed to power their flight muscles. Their metabolic rates are eight to ten times higher than those of other comparably sized Reptilia, and they process energy-rich food rapidly. A complex and efficient respiratory system (see Figure 46.7) and four-chambered heart (see Figure 44.5D) enable them to consume and distribute oxygen efficiently. As a consequence of high rates of metabolic heat production, birds maintain a high and constant body temperature (see Section 48.8).

Flying Birds Were Abundant by the Cretaceous Period

Although the earliest known bird, the pigeon-sized *Archaeopteryx*, had feathers, its skeleton was essentially that of a small dinosaur (see Figure 20.13). It had digits and claws on the forelimbs, teeth on its jaws, many bones in its wings and vertebral column, and only a poorly developed sternum. How could flight evolve in so unbirdlike an animal? Some biologists hypothesize that *Archaeopteryx* ran after prey, using its feathered wings like fly swatters. Larger wings would have provided extra lift when it jumped at prey, and gradual evolutionary modifications of the wing bones and muscles could have led to powered flight.

Crow-sized birds with full flight capability appeared by the early Cretaceous period **(Figure 32.29).** They had a keeled sternum and other modern skeletal features. The modern groups of wading birds and seabirds first appear in late Cretaceous rocks; fossils of other modern groups are found in slightly later deposits. Woodpeckers, perching birds, birds of prey, pigeons, swifts,

John Cancalosi/Photolibrary/Getty Images

FIGURE 32.29 Cretaceous birds. Fossils of *Confusciusornis sanctus* and other extinct birds demonstrate that modern birds diversified in the Cretaceous period.

the flightless ratites, penguins, and some other groups were all present by the end of the Oligocene; birds continued to diversify through the Miocene (see Table 23.1).

Modern Birds Vary in Morphology, Diet, Habits, and Patterns of Flight

The 10,000 living bird species show extraordinary ecological specializations, but they share the same overall body plan. Living birds are traditionally classified into nearly 30 groups **(Figure 32.30).** They vary in size from the bee hummingbird *(Mellisuga helenae)* of Cuba, which weighs little more than 1 g, to the ostrich *(Struthio camelus)*, which can weigh as much as 150 kg.

The structure of the bill usually reflects a bird's diet. Seed and nut eaters, such as finches and parrots, have deep, stout bills that crack hard shells. Carnivorous hawks and carrion-eating vultures have sharp bills to rip flesh, and nectar-feeding hummingbirds have slender bills to reach into flowers. The bills of ducks are modified to extract particulate matter from water, and many perching birds have slender bills to feed on insects.

Birds also differ in the structure of their feet and wings. Predators have large, strong talons ("claws"), whereas ducks and other swimming birds have webbed feet that serve as paddles. Long-distance fliers such as albatrosses have narrow wings; those that hover at flowers, such as hummingbirds, have wide ones. The wings of some species, such as penguins, are so specialized for swimming that they are incapable of aerial flight.

All birds have well-developed sensory and nervous systems, and their brains are proportionately larger than those of other reptilian species of comparable size. Large eyes provide sharp vision, and most species also have good hearing, which nocturnal hunters such as owls use to locate prey. Some vultures and other species have a good sense of smell, which they use to find food. Migrating birds use polarized light, changes in air pressure, and Earth's magnetic field for orientation.

A. The waved albatross (Procellariiformes, *Phoebastria irrorata*) has the long, thin wings typical of birds that fly great distances.

B. The saddle-billed stork (Ciconiiformes, *Ephippiorhynchus senegalensis*) searches for aquatic prey, which it captures with its long bill.

C. The bald eagle (Falconiformes, *Haliaeetus leucocephalus*) uses its sharp talons and bill to capture and tear apart prey.

D. A tawny frogmouth (Caprimulgiformes, *Podargus strigoides*) uses its wide mouth to capture flying insects.

E. A rufous hummingbird (Apodiformes, *Selasphorus rufus*) uses its long, thin bill to delve into flowers in search of nectar.

F. The red avadavat (Passeriformes, *Amandava amandava*) uses its stout bill to feed on grass seeds and insects.

FIGURE 32.30 Bird diversity. Birds, the living feathered archosaurs, have diversified into lineages that are adapted to live in different habitats and feed on different foods.

Most birds exhibit complex social behavior, including courtship, territoriality, and parental care. Many species communicate with vocalizations and visual displays to challenge other individuals or attract mates. Most raise their young in a nest, using body heat to incubate eggs. The nest may be a simple depression on a gravelly beach, a cup woven from twigs and grasses, or a feather-lined hole in a tree.

Many bird species embark on a semiannual, long-distance migration (see Section 56.3). The golden plover *(Pluvialis dominica),* for example, migrates 20,000 km twice each year. Migrations are a response to seasonal changes in climate. Birds travel toward the tropics as winter approaches; in spring, they return to high latitudes using seasonally available food sources.

STUDY BREAK 32.9

1. How does the overall structure of turtles distinguish them from other amniotes?
2. What anatomical and behavioral characteristics of crocodilians demonstrate their relatively close relationship to birds?
3. What specific adaptations allow birds to fly?
4. How do the structures of a bird's bill, wings, and feet reflect its dietary and habitat specializations?

32.10 Mammalia: Monotremes, Marsupials, and Placentals

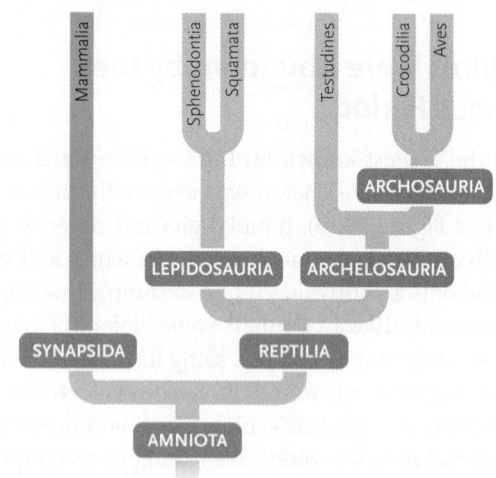

The synapsid lineage, which includes the living **mammals,** was the first group of amniotes to diversify broadly on land. Indeed, during the late Paleozoic era, medium- to large-sized synapsids were the most abundant predators in terrestrial habitats. One particularly successful and persistent branch of the synapsid lineage, the *therapsids*, exhibited many mammal-like characteristics in their legs, skulls, jaws, and teeth. And by the end of the Triassic period, the earliest mammals—most of them no bigger than a rat—had emerged from therapsid ancestors. Several mammalian lineages coexisted with dinosaurs and other Reptilia throughout the Mesozoic, but paleontologists hypothesize that most Mesozoic mammals were active only at night to avoid predatory dinosaurs, which were active during the day. Two mammalian lineages, the egg-laying Prototheria (or monotremes) and the live-bearing Theria (marsupials and placentals), survived the mass extinction that eliminated most dinosaurs at the end of the Mesozoic. The Theria diversified into the mammalian groups that are most familiar today.

Mammals Exhibit Key Adaptations in Anatomy, Physiology, and Behavior

Four sets of key adaptations fostered the success of mammals.

HIGH METABOLIC RATE AND BODY TEMPERATURE Like birds, mammals have high metabolic rates that liberate enough energy from food to maintain high activity levels and enough heat to maintain high body temperatures (see Section 48.8). An outer covering of fur and a layer of subcutaneous fat help retain body heat. Using metabolic heat to stay warm requires lots of oxygen, and mammals have a muscular organ, the diaphragm, that fills their lungs with air (see Figure 46.9). Four-chambered hearts and complex circulatory systems deliver oxygen to active tissues (see Figure 44.5D).

SPECIALIZATIONS OF THE TEETH AND JAWS Mammals also have anatomical features that allow them to feed efficiently. Ancestrally, mammals have four types of teeth (see Figure 47.18): flattened **incisors** nip and cut food; pointed **canines** pierce and kill prey; and two sets of cheek teeth, **premolars** and **molars,** grind and crush food. Moreover, teeth in the upper and lower jaws occlude (that is, fit together) tightly as the mouth is closed; thus, mammals can use their large jaw muscles to chew food thoroughly. The occlusion of premolars and molars allows some mammals to feed on exceptionally tough plant material, such as grasses and twigs.

PARENTAL CARE Mammals provide substantial parental care to their young. In most species, young complete development within a female's uterus, deriving nourishment through the **placenta,** a specialized organ that mediates the delivery of oxygen and nutrients (see Figure 50.13). Females also have **mammary glands,** specialized structures that produce energy-rich milk, a watery mixture of fats, sugars, proteins, vitamins, and minerals. This perfectly balanced diet is the sole source of nutrients for newborn offspring.

COMPLEX BRAINS Finally, mammals have larger brains than other tetrapods of equivalent body size; the difference lies primarily in the **cerebral cortex,** the part of the forebrain responsible for information processing and learning (see Figure 40.6). Most mammals have a keen sense of olfaction; as *Molecular Insights* explains, a huge variety of olfactory receptor proteins send information to the parts of the cerebral cortex that allow mammals to sense and interpret a wide variety of odors. Extensive postnatal care provides opportunities for offspring to learn from older individuals. Thus, mammalian behavior is strongly influenced by past experience and learning, as well as by genetically programmed instincts.

The Major Groups of Modern Mammals Differ in Their Reproductive Adaptations

Biologists recognize a primary distinction between two lineages of modern mammals: the egg-laying Prototheria, or monotremes, and the live-bearing Theria. The Theria, in turn, diversified into two sublineages: the Metatheria, or marsupials, and the Eutheria, or placentals, which also differ in their reproductive adaptations.

MONOTREMES The five living species in **Prototheria** (*protos* = first; *therion* = wild beast), the monotremes, which are limited to the Australian region, reproduce with a leathery-shelled egg **(Figure 32.31).** Newly hatched young lap up milk secreted by modified sweat glands on the mother's belly. The four species of echidnas, or spiny anteaters, feed on ants or termites. The duck-billed platypus (*Ornithorhynchus anatinus*) lives in burrows along riverbanks and feeds on aquatic invertebrates.

MARSUPIALS The 270 species of **Metatheria** (*meta* = between), the marsupials, have short gestation; the young are nourished through a placenta very briefly—sometimes only for 8 to 10 days—before birth. Newborns use their forelimbs to drag themselves across the mother's belly fur and enter her abdominal pouch, the **marsupium,** where they complete development attached to a teat. Marsupials are the dominant native mammals of Australia and a minor component of the South American fauna **(Figure 32.32);** only one marsupial species, the opossum (*Didelphis virginiana*), occurs in North America.

PLACENTALS The 4,000 species of **Eutheria** (*eu* = true), the placental mammals, are the dominant mammals today. They complete embryonic development in the mother's uterus, nourished through a placenta until they reach a fairly advanced stage of development. Some species, such as humans, are helpless at birth, but others, such as horses, are quickly mobile.

Biologists divide eutherians into about 18 groups, traditionally identified as orders, only eight of which include more

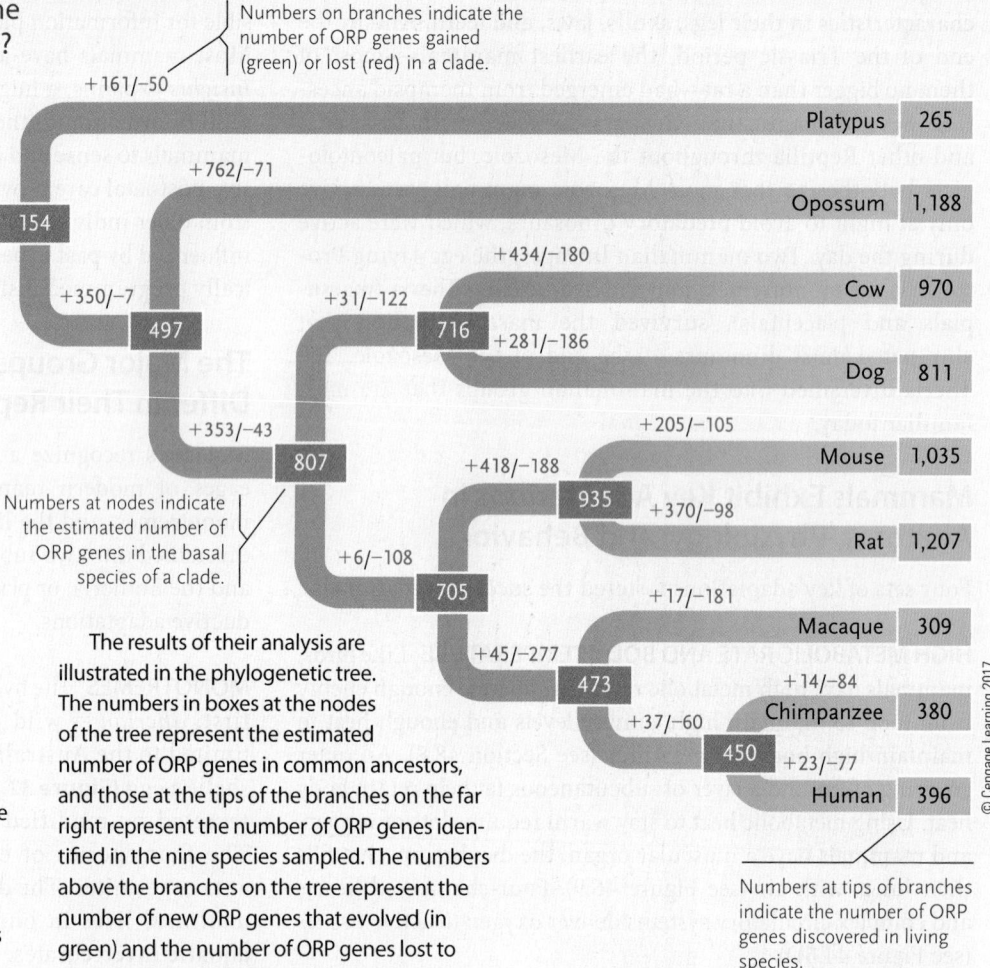

Molecular Insights

Building a Better Nose: How did the sense of smell evolve in mammals?

The sense of olfaction (that is, smell) in vertebrates is made possible by olfactory receptor proteins (ORPs) in specialized cells in their nasal sinuses (see Section 41.5). Humans express hundreds of different ORPs, each of which binds specific airborne molecules, thereby allowing us to detect the presence of many different molecules. Hundreds of different genes—all of which are members of one large multigene family of G-protein–coupled receptors (see Chapters 9 and 19)—code for hundreds of different ORPs. The vertebrate sense of smell evolves when new genes arise through gene duplication and when old genes are lost as mutations turn them into pseudogenes (see Section 19.4).

Research Question

What new olfactory receptor genes evolved in different groups of mammals?

Experiment

Mammals rely more on their sense of smell than do most other vertebrates. It is therefore reasonable to suppose that new olfactory receptor genes evolved in early mammals, producing their acute olfaction. To test this hypothesis, Yoshihito Niimura and coworkers searched a range of mammalian genomes for representatives of the multigene family coding for ORPs. Their work examined the genomes of one monotreme (platypus), one marsupial (opossum), and seven placentals (dog, cow, mouse, rat, macaque monkey, chimpanzee, and human).

The researchers made multiple pair-wise comparisons of the ORP gene sequences in the nine mammal species to construct phylogenetic trees of ORP genes; these phylo-genetic trees illustrate which genes are more closely related to each other. By comparing the ORP gene phylogenetic trees with a phylogenetic tree for the nine mammal species, they were able to estimate how many new ORP genes had evolved in each mammalian lineage and how many ORP genes had been lost through mutation. For example, the opossum genome contains one family of 59 ORP genes that are more closely related to one another than they are to ORP genes in any of the other species. The most parsimonious explanation for this discovery is that a series of gene duplications led to the evolution of 58 *new* ORP genes within the opossum lineage.

The results of their analysis are illustrated in the phylogenetic tree. The numbers in boxes at the nodes of the tree represent the estimated number of ORP genes in common ancestors, and those at the tips of the branches on the far right represent the number of ORP genes identified in the nine species sampled. The numbers above the branches on the tree represent the number of new ORP genes that evolved (in green) and the number of ORP genes lost to mutation (in red) within each lineage.

Many new ORP genes evolved in the lineage that was ancestral to both placental and marsupial mammals. After those clades diverged, new ORP genes continued to evolve in many lineages. The rodents (represented in the study by mouse and rat) experienced the evolution of a strikingly large number of new ORP genes. By contrast, the primates (repre-sented in the study by macaque, chimpanzee, and human) experienced a net loss of ORP genes through mutation.

Conclusion

These findings reflect the relative importance of olfaction in different groups of mammals. The platypus is thought to rely less on olfaction than other mammals do, perhaps because platypuses use electroreceptors in their bill to find food (see Section 41.7); this observation is consistent with the relatively few new ORP genes that evolved in that lineage. The evolution of nocturnal activity in early mammals is thought to have increased the need for olfactory information, an obser-vation that is consistent with the evolution

of many new ORP genes in the marsupial and placental lineages. In fact, of the 1,035 ORP genes in mice, at least 783 appeared after the emergence of mammals, suggesting that most of a mouse's ability to smell has been acquired during the past 200 million years. The numbers of ORP genes lost outpaced the numbers of ORP genes gained only among the primates, as members of that lineage began to rely increasingly on their sense of vision. A decreased reliance on olfaction is typical of all primates, including humans.

think like a scientist

Early primates were almost certainly arboreal, leaping or climbing from one branch to another in trees. How would this behavior favor the evolution of their increased reliance on vision and decreased reliance on olfaction?

Sources: Y. Niimura and M. Nei. 2007. Extensive gains and losses of olfactory receptor genes in mammalian evolution. *PLoS One* 2(8):e708. A. Matsui et al. 2010. Degeneration of olfactory receptor gene repertoires in primates: No direct link to full trichromatic vision. *Molecular Biology and Evolution* 27:1192–1200. The figure represents a new analysis of data from both the 2007 and 2010 publications, kindly provided by Dr. Niimura.

A. Short-nosed echidna *(Tachyglossus aculeatus)*

FLPA/Alamy

B. Duck-billed platypus *(Ornithorhynchus anatinus)*

Jean-Philippe Varin/Science Source

FIGURE 32.31 Monotremes. (A) The short-nosed echidna is terrestrial. **(B)** The duck-billed platypus raises its young in a streamside burrow.

© Kjuuurs/Shutterstock.com

FIGURE 32.32 Marsupials. An Eastern gray kangaroo *(Macropus giganteus)* carries her "joey" in her pouch.

Although early mammals appear to have been insectivorous, the diets of modern eutherians are diverse. Odd-toed ungulates *(ungula* = hoof) such as horses and rhinoceroses (Perissodactyla), even-toed ungulates such as cows and camels (Artiodactyla), and rabbits and hares (Lagomorpha) feed on vegetation. Lions and wolves (Carnivora) consume other animals. Many hedgehogs, moles, and shrews (Lipotyphla) and bats eat insects, but some feed on flowers, fruit, and nectar. Many whales and dolphins prey on fishes and other animals, but some eat plankton. And some groups, including rodents and primates, feed opportunistically on both plant and animal matter.

STUDY BREAK 32.10

1. During the Mesozoic era, why were most mammals active only at night?
2. Which key adaptations in mammals allow them to be active under many types of environmental conditions?
3. On what basis are the three major groups of living mammals distinguished?

32.11 Nonhuman Primates

We now focus our attention on Primates, the mammalian lineage that includes humans, apes, monkeys, and their close relatives. The first Primates appeared early in the Eocene epoch, about 55 million years ago, in forested habitats in North America, Europe, Asia, and North Africa.

Key Derived Traits Enabled Primates to Become Arboreal, Diurnal, and Highly Social

Several derived traits allow primates to be arboreal (to live in trees rather than on the ground). For example, most primates have a more erect posture than other mammals, and they

than 50 living species **(Figure 32.33).** Rodents (Rodentia) make up about 45% of eutherian species, and bats (Chiroptera) comprise another 22%. Our own group, **Primates,** is represented by fewer than 250 living species (less than 5% of all mammalian species), many of which are highly endangered.

Some eutherians have highly specialized locomotor structures. For example, whales and dolphins (Cetacea) are descended from terrestrial ancestors, but their appendages do not function on land; they are now restricted to aquatic habitats. By contrast, seals and walruses (Carnivora) feed underwater but rest and breed on land. Bats (Chiroptera) use wings for powered flight.

A. Capybaras (Rodentia, *Hydrochoerus hydrochaeris*), the largest rodents, feed on vegetation in South American wetlands.

B. Most bats, like the Yuma Myotis (Chiroptera, *Myotis yumanensis*), are nocturnal predators of insects.

C. Walruses (Carnivora, *Obodenus rosmarus*) feed primarily on marine invertebrates in frigid arctic waters.

D. Narwhals (Cetacea, *Monodon monoceros*), which have one incisor that grows into a spiral tusk, may be a source of unicorn legends.

E. Arabian camels (Artiodactyla, *Camelus dromedarius*) use enlarged foot pads to cross hot desert sands.

F. European hares (Lagomorpha, *Lepus europaeus*) are herbivores that feed in open fields and grasslands.

G. Common moles (Lipotyphla, *Talpa europaea*) feed largely on insects that they find underground.

FIGURE 32.33 Eutherian diversity. Among the traditionally defined orders of mammals, only eight include more than 50 living species. Seven of these clades are pictured here. The eighth clade, Primates, is described in Section 32.11.

have flexible hip and shoulder joints, which allow a variety of locomotor activities. They can grasp objects with their hands and feet, because they have nails, not claws, on their fingers and toes; their fingertips are well endowed with sensory nerves that enhance the sense of touch. Unlike other mammals, most primates have an opposable big toe, which can touch the tips of other digits and the sole of the foot; many species also have an opposable thumb.

Most primates are diurnal (active during daylight hours), and, unlike most mammals, they rely more on vision than on their sense of smell (see *Molecular Insights*). Thus, they generally have short snouts and small olfactory lobes of the brain. Most species have forward-facing eyes with overlapping fields of vision, providing excellent depth perception,

which comes in handy when moving through trees. Many species have color vision.

Primate brains—especially the regions that integrate information—are large and complex. As a result, they have an exceptional capacity to learn. Most species live in social groups; thus, young primates, which mature slowly, can interact with and learn from their elders and peers during an extended period of parental care. Females give birth to only one or two young at a time, allowing them to devote substantial attention to each offspring.

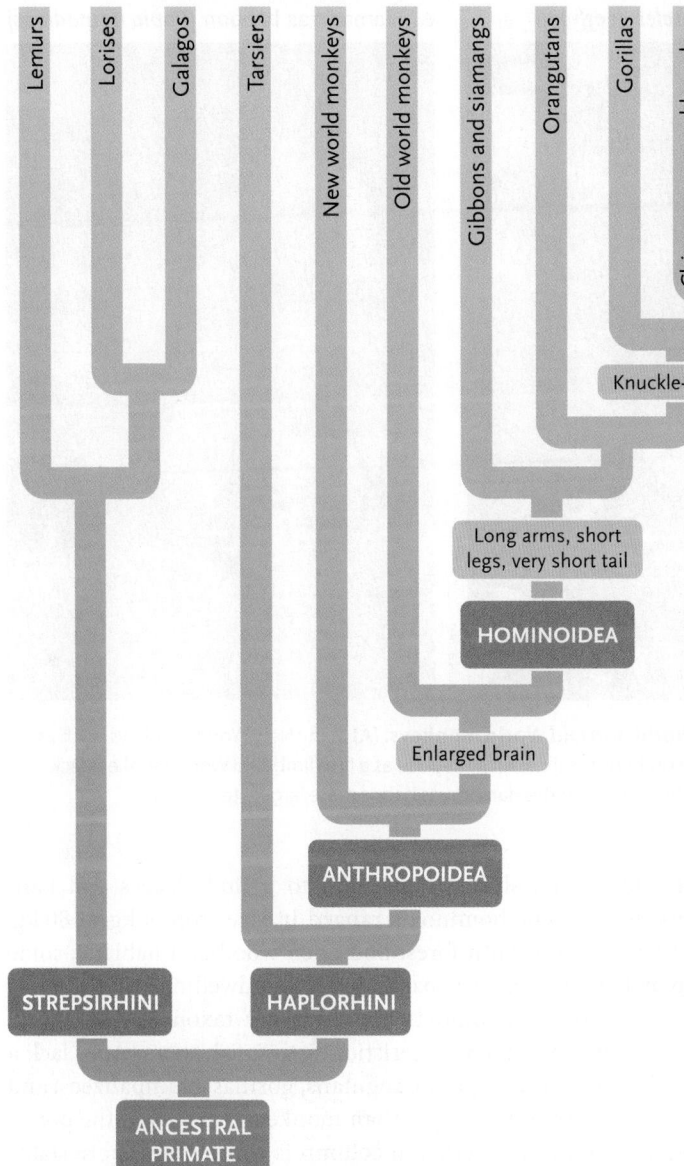

FIGURE 32.34 Primate phylogeny. A phylogenetic tree for the Primates illustrates the two main lineages: Strepsirhini and Haplorhini. Note that chimpanzees and bonobos are the closest living relatives of humans.

© Cengage Learning 2017

Living Primates Include Two Major Lineages

Primatologists recognize two lineages within the Primates **(Figure 32.34)**, the Strepsirhini and the Haplorhini.

STREPSIRHINI The 36 living species of **Strepsirhini** (*streptos* = twisted or turned; *rhinos* = nose)—lemurs, lorises, and galagos— possess many ancestral morphological traits, including moist, fleshy noses and eyes that are positioned somewhat laterally on their heads **(Figure 32.35)**. Strepsirhines generally have short gestation periods and rapid maturation. Today, lemurs survive on Madagascar, a large island off the east coast of Africa; they are ecologically diverse and range in size from 40 g to 7 kg. Some lemurs are arboreal, whereas others spend substantial

FIGURE 32.35 Strepsirhines. The ring-tailed lemur (*Lemur catta*) of Madagascar has ancestral primate characteristics, such as a long snout and wet nose.

Cagan Sekercioglu/Visuals Unlimited, Inc.

time on the ground. Lorises and galagos occupy tropical forest and subtropical woodlands in Africa, India, and Southeast Asia; they are all arboreal and nocturnal.

HAPLORHINI Most species in the **Haplorhini** (*haplos* = single or simple)—the familiar monkeys and apes—have many derived primate characteristics, including compact, dry noses, and forward-facing eyes.

However, tarsiers, which are restricted to tropical forests on the islands of Southeast Asia, exhibit several ancestral traits: small body size (about 100 g), large eyes and ears, and two grooming claws on each foot **(Figure 32.36)**. But they share the derived characteristics of dry noses and forward-facing eyes with the other haplorhines; DNA sequence data link them to the monkeys and apes and not to the strepsirhines described earlier.

The 130 or so species of monkeys, 13 species of apes, and humans constitute the monophyletic haplorhine lineage **Anthropoidea,** which probably arose in Africa. Fossils of a diverse and abundant radiation of forest-dwelling anthropoids, dating from the late Eocene epoch, have been discovered in northern Egypt. Continental drift then established long-term geographical and evolutionary separation of anthropoids in the New World and Old World **(Figure 32.37)**.

By the middle of the Oligocene epoch, about 30 million years ago, the ancestors of the New World monkeys had arrived in South America and begun to diversify there. They probably rafted across the Atlantic, which was narrower at that time, on

A. Spider monkey *(Ateles geoffroyi)*

B. Hamadryas baboon *(Papio hamadryas)*

FIGURE 32.36 Tarsiers. Tarsiers *(Tarsius syrichta)* are haplorhines, but they retain many ancestral characteristics.

FIGURE 32.37 New World and Old World monkeys. (A) Many New World monkeys, such as the spider monkey, have prehensile tails, which they use as a fifth limb. Old World monkeys lack prehensile tails, and many, such as **(B)** the Hamadryas baboon, are largely terrestrial.

trees or other storm debris. New World monkeys now live in Central and South America (see Figure 32.37A). They range in size from tiny marmosets and tamarins (350 g) to hefty howler monkeys (10 kg). Most are exclusively arboreal and diurnal. The larger species may hang below branches by their arms, and some use a prehensile (grasping) tail as a fifth limb.

Anthropoids diversified most spectacularly in the Old World, however, eventually giving rise to two lineages—one ancestral to Old World monkeys and the other to apes and humans. Although many people assume that the apes are descended from Old World monkeys, the fossil record contradicts that impression. The earliest hominoid ("ape") fossils date to the early Miocene, roughly 23 million years ago, but the oldest known Old World monkeys appeared several million years later.

Old World monkeys, which occupy habitats ranging from tropical rainforests to deserts in Africa and Asia, may grow as large as 35 kg (see Figure 32.37B). Many species are sexually dimorphic; in other words, males and females attain different adult sizes (see Section 21.3). Arboreal species use all four limbs for locomotion, but none has a prehensile tail. Some species, such as baboons, often walk or run on the ground.

Within the anthropoid lineage, the **Hominoidea** ("human-like") is a monophyletic group that includes apes and humans. The climate of the early Miocene was wetter than it is today, and eastern Africa, where many early hominoid fossils are found, was covered with extensive forests. A shift to a cooler and drier climate in the middle Miocene, between 17 and 14 million years ago, converted dense forests into more open woodlands. Hominoids probably adopted a more terrestrial

existence and shifted their diets to include leaves and hard foods. Miocene hominoids ranged in size from 4 kg to 80 kg. They occupied both forest and open woodland habitats; some may have been at least partially ground dwelling.

Although hominoids are the sister taxon to Old World monkeys, several characteristics distinguish them. Apes lack a tail, and the great apes (orangutans, gorillas, chimpanzees, and bonobos) are much larger than monkeys. Moreover, the posterior region of the vertebral column is shorter and more stable in apes. Apes also show more complex behavior.

The gibbons and siamangs, which live in tropical forests in Southeast Asia, are the smallest of the apes, ranging in weight from 6 to 11 kg. With extremely long arms and strong shoulders, they hang below branches by their arms and swing themselves forward, a pattern of locomotion called **brachiation (Figure 32.38A).** The much larger orangutan *(Pongo pygmaeus),* now restricted to forested areas on the islands of Borneo and Sumatra, can grow to 90 kg. Orangutans use both hands and feet to climb trees; they sometimes venture onto the ground on all fours.

Gorillas *(Gorilla gorilla),* which are currently restricted to two large central African forests, are the largest of the living primates. Males can weigh 180 kg; females are about half that size. Because of their size, gorillas spend most of their time on the ground. They often use "knuckle-walking" locomotion, leaning forward and supporting part of their weight on the backs of their hands. Gorillas are almost exclusively vegetarian.

Chimpanzees *(Pan troglodytes)* are also forest dwellers, weighing up to 45 kg **(Figure 32.38B).** Like gorillas, they spend

A. Black-handed gibbon
(Hylobates agilis)

B. Chimpanzee
(Pan troglodytes)

FIGURE 32.38 Apes. (A) Small-bodied apes, such as the black-handed gibbon are agile brachiators that swing through the trees with ease. **(B)** Among the large-bodied apes, chimpanzees have opposable thumbs and big toes.

most of their waking hours on the ground; they often knuckle-walk, but sometimes adopt a **bipedal** (two-legged) stance and swagger short distances. Groups of related males form loosely defined communities of up to 50 individuals, which may cooperate in hunts and foraging. Bonobos *(Pan paniscus),* sometimes called *pygmy chimpanzees,* are restricted to a small area in central Africa. Somewhat smaller than chimps, they have longer legs and smaller heads.

The Primates also includes humans *(Homo sapiens),* which occupy virtually all terrestrial habitats. Humans have adaptations that allow an upright posture and bipedal locomotion. They are ground-dwelling animals with extremely broad diets and complex social behavior.

STUDY BREAK 32.11

1. What characteristics of primates allow them to spend a great deal of time in trees?
2. What is the lowest taxonomic group that includes monkeys, apes, and humans? What is the lowest taxonomic group that includes only apes and humans?
3. Which ape species spend the most time on the ground?

32.12 The Evolution of Humans

Genetic analyses of living hominoid species indicate that African hominoids diverged into several lineages between 10 million and 5 million years ago; one lineage, the **hominins,**

includes modern humans and our bipedal ancestors. (Until recently, biologists used the term *hominid* to describe humans and their bipedal ancestors, but that term now defines the group that includes humans, bipedal human ancestors, and the apes.)

Hominins First Walked Upright in East Africa About 6 Million Years Ago

Upright posture and bipedal locomotion are key adaptations that distinguish hominins from apes. Researchers infer these capabilities in early hominin fossils from the anatomy of the skull, spine, pelvis, knees, ankles, and feet. As a consequence of bipedal locomotion, the hands were no longer used for locomotor functions, allowing them to become specialized for other activities, such as tool use. Hominins also eventually developed larger brains.

Since the mid-1990s, paleontologists have uncovered an astounding array of fossils of hominin species that lived in East Africa and South Africa between 7 million and 1 million years ago **(Figure 32.39).** In 2000, researchers found 13 fossils of *Orrorin tugenensis* ("first man" in a local African language), a species that lived about 6 million years ago in East African forests. Shortly thereafter, researchers described the skull of an even older fossil, *Sahelanthropus tchadensis,* but because the pelvis and legs of this species are still unknown, scientists are unsure about its ability to walk bipedally. Indeed, the fossilized remains of most hominins are fragmentary.

In 1994, Tim White of the University of California, Berkeley, and two Ethiopian colleagues made a spectacular discovery of early hominin fossils: they unearthed the remains of 36 individuals of *Ardipithecus ramidus* in the Awash Valley of central Ethiopia. The animals had lived in a closed-canopy woodland about 4.4 million years ago. The fossils were badly deformed and so fragile that they turned to dust in the researchers' hands. After the fossils were carefully excavated and transported to a laboratory at the National Museum of Ethiopia, an international and interdisciplinary team of 47 researchers labored for 15 years to analyze them. They published their findings in a series of 11 papers in *Science* magazine. The work was later hailed by the editors of *Science* as the "breakthrough of the year" in 2009.

Among the fossils that White's team uncovered, the researchers found 125 pieces of one female's skeleton, including bones from her arm and hand, pelvis, leg, ankle, and foot, lower jaw with teeth, and cranium **(Figure 32.40).** Nicknamed "Ardi," she had the body size and brain size of a modern chimpanzee; she stood about 120 cm tall and weighed about 50 kg. Studies of Ardi's anatomy revealed that her face protruded less and her upper canine teeth were shorter and duller than those of modern chimpanzees. The upper parts of her pelvis are shorter and broader than those of living apes, and the anatomy of her hand suggests that she did not knuckle-walk as modern

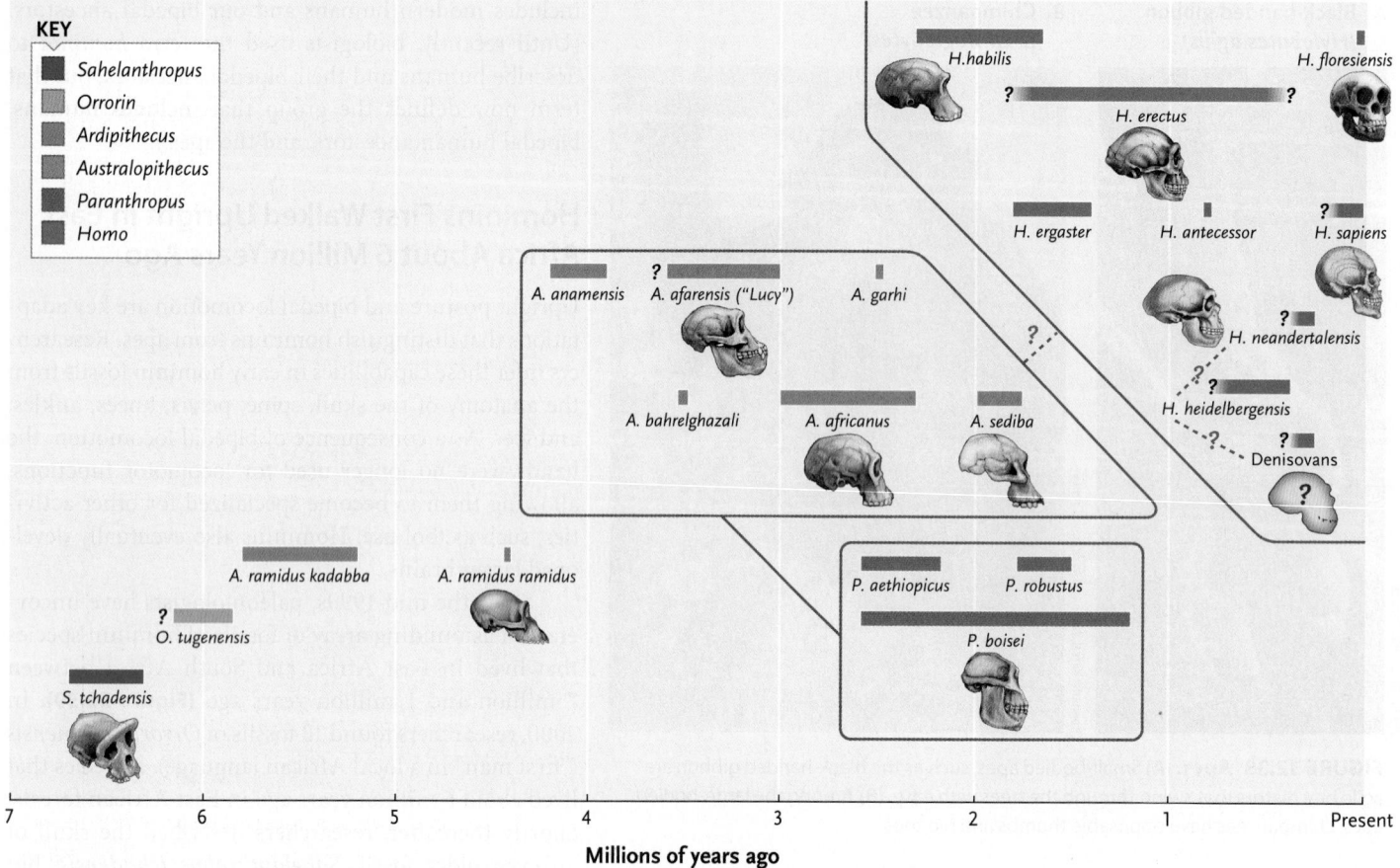

KEY

- Sahelanthropus
- Orrorin
- Ardipithecus
- Australopithecus
- Paranthropus
- Homo

H.habilis

H. floresiensis

H. erectus

H. ergaster   H. antecessor   H. sapiens

A. anamensis   A. afarensis ("Lucy")   A. garhi

H. neandertalensis

A. bahrelghazali   A. africanus   A. sediba

H. heidelbergensis

Denisovans

A. ramidus kadabba   A. ramidus ramidus

P. aethiopicus   P. robustus

O. tugenensis

P. boisei

S. tchadensis

7   6   5   4   3   2   1   Present

Millions of years ago

FIGURE 32.39 Hominin timeline. Several species of hominins lived simultaneously at sites in eastern and southern Africa. The timeline for each species and genus reflects the ages of known fossils. The question marks indicate uncertainty about the precise timing of the origin and extinction of various hominin species as well as the evolutionary relationship between *Australopithecus* and *Homo* and between several species in the latter genus. Some of the skulls pictured are reconstructed from fragmentary fossils.

© Cengage Learning 2017

chimps and gorillas do. These observations suggest that she was able to walk bipedally on the ground, although perhaps not as well as later hominins. Nevertheless, Ardi's foot had an opposable big toe, and the bones of her fingers were long and curved; these characters suggest that she spent considerable time climbing trees, perhaps to escape predators, seek food, or build nests for sleeping. White and his team described Ardi as having an anatomy that was intermediate between that of chimps and later groups of hominins. Some researchers have taken issue with some of the interpretations that White's team published, partly because the fossils were in terrible condition when they were found and partly because, at the time of this writing, few other researchers have yet had the opportunity to study the fossils in detail.

Hominin fossils from 4.2 million to 1.2 million years ago are well known from many sites in East, Central, and South Africa. They are currently assigned to the genera *Australopithecus* (*australis* = southern; *pithekos* = ape) and *Paranthropus* (*para* = beside; *anthropos* = human being). With their large faces, protruding jaws, and small skulls and brains, most of these hominins had an apelike appearance (see Figure 32.39).

Australopithecus anamensis, which lived in East Africa around 4 million years ago, is the oldest known species. It had thick enamel covering on its teeth, a derived hominin characteristic; the structure of a fossilized leg bone suggests that it was bipedal.

Specimens of more than 60 individuals of *Australopithecus afarensis* have been found in northern Ethiopia, including about 40% of a female skeleton, named "Lucy" by its discoverers **(Figure 32.41)**. *A. afarensis* lived 3.5 million to 3 million years ago, but it retained several ancestral characteristics. For example, it still had moderately large and pointed canine teeth, and a relatively small brain. Males and females were 150 cm and 120 cm tall, respectively. Skeletal analyses suggest that *A. afarensis* was fully bipedal, a conclusion supported by fossilized footprints preserved in a layer of volcanic ash.

Other species of *Australopithecus* and *Paranthropus* lived in East Africa or South Africa between 3.7 million and 1 million years ago. Adult males ranged from 40 to 50 kg in weight and from 130 to 150 cm in height; females were smaller. Most species had deep jaws and large molars. Several species had a crest of bone along the midline of the skull, providing a large surface for the attachment of jaw muscles. These anatomical

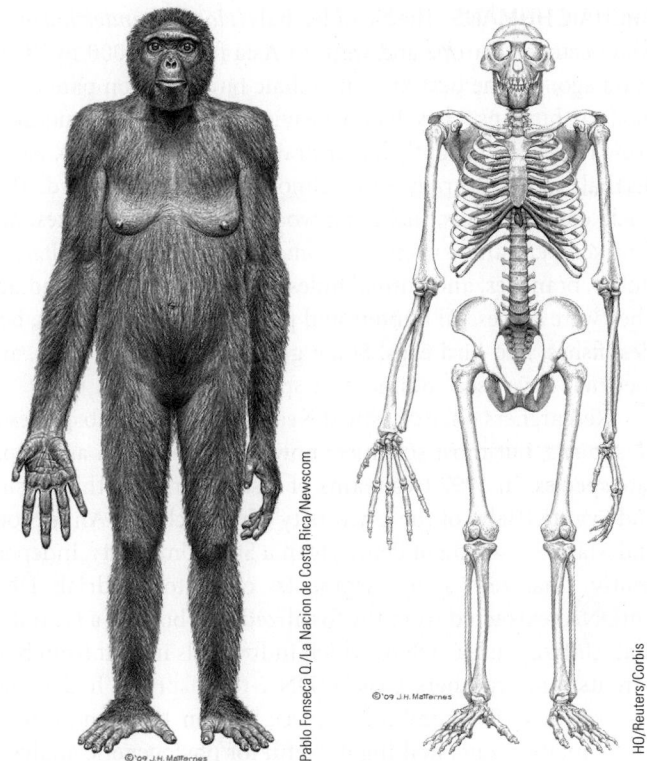

FIGURE 32.40 Ardipithecus. Scientists spent 15 years analyzing and reassembling a fairly complete skeleton of *Ardipithecus ramidus.*

A. "Lucy" (Australopithecus afarensis)

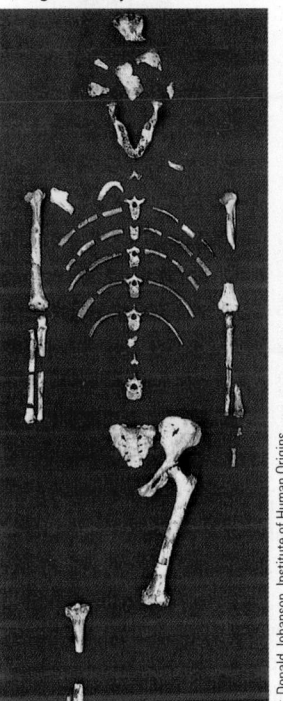

B. Australopithecine footprints

FIGURE 32.41 Australopithecines. (A) Researchers named the most complete fossil of *Australopithecus afarensis* "Lucy." **(B)** Mary Leakey discovered australopithecine footprints, made in soft, damp volcanic ash about 3.7 million years ago. The footprints indicate that australopithecines were fully bipedal.

features suggest that they fed on hard food, such as nuts, seeds, and other vegetable products.

One species, *Australopithecus africanus,* known only from South Africa, had small jaws and teeth, indicating that it probably consumed a softer diet. Yet another species, *Australopithecus sediba,* first described in 2010 from fossils discovered in a South African cave, had even more derived characteristics than *A. africanus:* smaller teeth, longer legs, and a more vertically oriented pelvis. Researchers believe that *A. sediba* may be a descendant of *A. africanus* and an ancestor of the genus that includes modern humans. The phylogenetic relationships of the species classified as *Australopithecus* and *Paranthropus*—and their exact relationships to later hominins—are not yet fully understood. But most scientists agree that *Australopithecus* was ancestral to humans, which are classified in the genus *Homo.*

Homo habilis Was Probably the First Hominin to Manufacture Stone Tools

Pliocene fossils of the earliest humans, which may have included several species, are fragmentary. They are also widely distributed in space and time, complicating analyses of their relationships. For the sake of simplicity, we describe them as belonging to one species, *Homo habilis* ("handy man").

From 2.3 million to 1.7 million years ago, *H. habilis* occupied the woodlands and savannas of eastern and southern Africa, sharing these habitats with various species of *Paranthropus.* The two genera are easy to tell apart because the brains of *H. habilis*

were at least 20% larger, and they had larger incisors and smaller molars than their hominin cousins. Their diet included hard-shelled nuts and seeds, as well as soft fruits, tubers, leaves, and insects. They may also have hunted small prey or scavenged carcasses left by large predators.

Researchers have found numerous tools dating to the time of *H. habilis,* but they are not sure which species used them. Many of the hominin species alive at the time, including *Australopithecus* and *Paranthropus* species, probably cracked marrowbones with rocks or scraped flesh from bones with sharp stones. Paleoanthropologist Louis Leakey was the first to discover evidence of tool *making* at East Africa's Olduvai Gorge, which cuts through a great sequence of sedimentary rock layers. The oldest tools at this site are crudely chipped pebbles, which were probably manufactured by *H. habilis.*

Homo erectus Dispersed from Africa to Other Continents

Early in the Pleistocene epoch, about 1.8 million years ago, new species of humans appeared in East Africa. Most fossils are fragmentary. For convenience, we describe them all as *Homo erectus* ("upright man"), recognizing that they probably represent several species. One nearly complete skeleton suggests that

A. *Homo erectus*   **B. Hand axe**

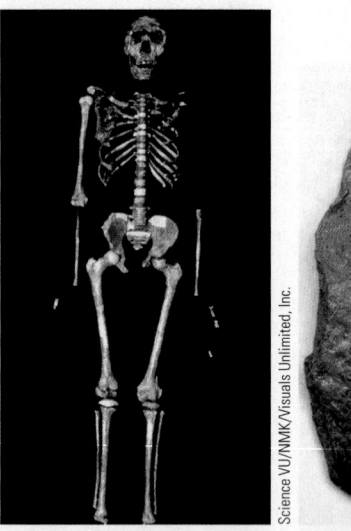

FIGURE 32.42 *Homo erectus.* **(A)** A nearly complete skeleton of *Homo erectus* was discovered in Kenya. **(B)** Hand axes are frequently found at *H. erectus* sites.

H. erectus was taller than its ancestors and had a much larger brain, a thicker skull, and protruding brow ridges **(Figure 32.42).**

   H. erectus made fairly sophisticated tools, including the hand axe (see Figure 32.42B), which they apparently used to cut food and other materials, to scrape meat from bones, and to dig for roots. *H. erectus* probably fed on both plants and animals; they may have hunted and scavenged animal prey. Archaeological data point to their use of fire to process food and to keep themselves warm.

   The pressure of growing populations apparently forced groups of *H. erectus* to move out of Africa about 1.5 million years ago. They dispersed northward from East Africa into both northwestern Africa and Eurasia. Some moved eastward through Asia as far as the island of Java. Recent discoveries in Spain indicate that *H. erectus* also occupied parts of Western Europe.

Modern Humans Are the Only Surviving Descendants of *Homo erectus*

Judging from its geographical distribution, *Homo erectus* was successful in many environments. Fossils from Africa, Asia, and Europe indicate that it gave rise to several species that first appeared more than 400,000 years ago. These early descendants, described as archaic humans, generally had a larger brain, rounder skull, and smaller molars than *H. erectus*. Like *H. erectus* before them, archaic humans left Africa and established populations in the Middle East, Asia, and Europe.

   Some time later, 200,000 to 100,000 years ago, modern humans, *Homo sapiens* ("wise man"), arose in Africa. These modern humans also migrated into Europe and Asia and may have eventually driven the archaic humans to extinction, perhaps through competition for resources rather than hostile interactions.

ARCHAIC HUMANS The Neanderthals (*Homo neanderthalensis*), who occupied Europe and western Asia from 150,000 to 28,000 years ago, are the best-known archaic humans. Compared with modern humans, they had a heavier build, more-pronounced brow ridges, and slightly larger brains (see Figure 32.39). Neanderthals were culturally and technologically sophisticated. They made complex tools, including wooden spears, stone axes, and flint scrapers and knives. At some sites they built shelters of stones, branches, and animal hides, and they routinely used fire. They were successful hunters and probably consumed nuts, berries, fishes, and bird eggs. Some groups buried their dead, and they may have used rudimentary speech.

   Researchers once classified Neanderthals as a subspecies of *H. sapiens,* but most scientists now recognize them as a separate species. In 1997 two teams of researchers, Matthias Kring and Svante Pääbo of the University of Munich and Anne Stone and Mark Stoneking of Pennsylvania State University, independently analyzed short segments of mitochondrial DNA (mtDNA) extracted from the fossilized arm bone of a Neanderthal. Unlike nuclear DNA, which individuals inherit from both parents, only mothers pass mtDNA to offspring. It does not undergo genetic recombination (see Section 11.2), and it has a high mutation rate, making it useful for phylogenetic analyses. Many scientists believe that mutation rates in mtDNA are fairly constant, allowing this molecule to serve as a molecular clock (see Section 24.6). Comparing the Neanderthal sequence with mtDNA from 986 living humans, the researchers discovered three times more differences between the Neanderthals and modern humans than between pairs of modern humans in their sample. These results suggest that Neanderthals and modern humans were different species that diverged from a common ancestor 690,000 to 550,000 years ago—hundreds of thousands of years before modern humans appeared.

   In 2003, researchers discovered eight skeletons of a tiny hominin, tentatively described as *Homo floresiensis,* in a cave on the Indonesian island Flores. Remarkably, these individuals stood just 1 m tall and are estimated to have weighed about 25 kg. The researchers who discovered these remains have hypothesized that they represent a new archaic hominin species, a descendant of *H. erectus,* that lived in Southeast Asia until roughly 12,000 years ago. Some skeptics suggested that these individuals represent a population of modern humans who suffered from some genetic disease or condition that caused their small size. But subsequent analyses of their skeletal remains indicate that *H. floresiensis* branched off from *H. erectus* even earlier than *H. neanderthalensis*. Their diminutive stature may have resulted from selection for a reduction in size, a phenomenon observed in other island-dwelling mammals. Indeed, fossils of miniature elephants have been discovered nearby. The unresolved questions about this archaic human include how and where it first evolved and how it may have interacted with modern humans who lived nearby.

   In an even more startling discovery in 2008, a researcher from the Russian Academy of Science found the 41,000 year-old fossilized finger bone of a previously unidentified hominin in the

Denisova Cave in Southwestern Siberia. Subsequent explorations revealed that the cave had also been occupied by both Neanderthals and *Homo sapiens* for 125,000 years. Svante Pääbo and his colleagues analyzed both mtDNA and nuclear DNA sequences from the Denisovan sample and compared those data to the sequences of Neanderthals and modern humans. Their results suggest that Denisovans and Neanderthals are sister clades, having shared a common ancestor 640,000 years ago. The common ancestor apparently diverged from an African population about 800,000 years ago. Unfortunately, Denisovan fossils are more fragmentary than those of other hominins, and researchers currently have no knowledge of their anatomy or appearance.

MODERN HUMANS Modern humans *(Homo sapiens)* differ from Neanderthals and other archaic humans in having a slighter build, less-protruding brow ridges, and a more prominent chin. The earliest fossils of modern humans found in Africa and Asia are roughly 150,000 years old; those from the Middle East are 100,000 years old. Fossils from about 20,000 years ago are known from Western Europe, the most famous being those of the Cro-Magnon deposits in southern France. The widespread appearance of modern humans roughly coincided with the demise of Neanderthals in Western Europe and the Middle East 40,000 to 28,000 years ago. Although the two species apparently coexisted in some regions for thousands of years, until recently, we have had little concrete evidence about their interactions.

In May 2010, an international team of more than 50 scientists, led by Svante Pääbo, reported a draft sequence of the Neanderthal genome in the journal *Science*. The researchers collected genetic data from fossils of three Neanderthal females who had lived more than 38,000 years ago in what is now Croatia. They compared the Neanderthal DNA sequences to those of five living humans—one each from southern Africa, West Africa, China, Papua New Guinea, and France. Remarkably, the European and the Asians shared between 1% and 4% of their recently derived (that is, newly evolved) nuclear DNA sequences with Neanderthals. Subsequent research, published by another team of researchers in 2011, also found evidence of a small number of archaic human DNA sequences in some modern human populations in central Africa. Moreover, analyses of DNA from the Denisovan fossil suggest that modern humans also interbred with this archaic species: between 4% and 6% of the genome of some populations of contemporary Melanesians is derived from the Denisovan genome. These data collectively suggest that modern humans interbred with archaic humans on several continents as recently as 35,000 years ago. The genes that modern Europeans and Asians share with Neanderthals appear to be involved in cognitive development, skeletal development, and metabolism. Thus, there may well be a little Neanderthal or Denisovan in all of us, no matter where on Earth our recent ancestors lived.

Many researchers believe that all modern human populations are descended from an ancestral population that originated in Africa. DNA sequence data from modern humans confirm this hypothesis. In 1987, Rebecca Cann, Mark Stoneking, and Allan Wilson of the University of California, Berkeley, and their colleagues published an analysis of mtDNA sequences from more than 100 ethnically diverse humans on four continents. They found that contemporary African populations contain the greatest variation in mtDNA. One explanation for this observation is that neutral mutations have been accumulating in African populations longer than in others, marking the African populations as the oldest on Earth. They also found that all human populations contain at least one mtDNA sequence of African origin, suggesting an African ancestry for all modern humans. Cann and her colleagues named the ancestral population, which lived approximately 200,000 years ago, the "mitochondrial Eve."

Other researchers have examined genetic material that males inherit only from their fathers. In 1995, L. Simon Whitfield and his colleagues at Cambridge University published a study on an 18,000-base-pair sequence from the Y chromosome, which does not undergo recombination with the X chromosome. Because the sequence contains no genes, it should not be subject to natural selection. Thus, sequence variations should result only from random mutations, which can serve as a molecular clock. The researchers discovered only three sequence mutations among the five subjects examined—a surprising result, given that the sample included a European, a Melanesian, a South American, and two Africans. By contrast, a chimpanzee exhibited 207 differences from the human version of the sequence. Using a sophisticated statistical analysis, the Whitfield team calculated that the common ancestor of these five diverse humans, dubbed the "African Adam," lived between 37,000 and 49,000 years ago. More recent studies of single nucleotide polymorphisms (see Sections 18.2 and 21.1) in Y chromosomes have estimated that the male common ancestor of all modern humans lived between 115,000 and 156,000 years ago, a range of dates that is much closer to the estimated age of the "mitochondrial Eve."

The limited genetic diversity and relatively recent origin of a common ancestor clearly support the African origin of *H. sapiens*. Follow-up studies on the Y chromosomes of thousands of men from Africa, Europe, Asia, Australia, and the Americas have confirmed that all modern humans are the descendants of a single migration out of Africa.

Ever since Darwin first included humans in his analyses about the diversity of life, the study of human evolution has been a controversial subject. Some disputes arise because the fossil record for humans gives up its secrets slowly and in piecemeal fashion. Other disagreements result from the assumptions that researchers must make about the sizes and geographical ranges of ancient populations, the amount of gene flow they experienced, and how natural selection may have affected them. But intellectual disputes are routine in science, and they challenge researchers to refine their hypotheses and to test them in new ways. Questions about the details of human origins are at the center of one of the liveliest debates in evolutionary biology today; additional research will surely clarify the history of our species.

STUDY BREAK 32.12

1. What trait allows researchers to distinguish between apes and humans?
2. What evidence suggests that Neanderthals and modern humans represent two distinct species?
3. What evidence suggests that modern humans originated in Africa?

THINK OUTSIDE THE BOOK

New results from the comparison of DNA sequences in archaic humans and modern humans appear in the scientific literature almost every month. Search the Internet for the latest discoveries about the complicated history of human ancestry.

Unanswered Questions

What causes the evolution of diversity?

In this chapter, you have read about the extensive diversity—in size, shape, color, structure, lifestyle, and habitats—of the vertebrate animals. Their mechanisms for maintaining themselves—including behaviors such as moving, feeding, reproducing, hiding, fighting, and sleeping—are equally varied. But nearly all vertebrates share some fundamental features. Recent research shows us that this generality is especially true of early development, but it is also apparent in many of the basic homeostatic mechanisms that vertebrates share—features of digestion and metabolism, respiration, and other characteristics regulated by the products of genetic networks and cascades. Biologists had long thought that once we understood the genetics of a variety of species, we would also understand the basis for their evolution and the maintenance of diversity. But now that biologists have sequenced a number of animal genomes, we have actually learned more about the genetic information that is shared by all animal species than we have learned about the genes that promote diversification. One of the great unanswered questions in biology, therefore, is how diversity arises and how it is maintained, given that so much genetic information is shared among even distantly related species. Equally important is why the same kinds of features evolved almost identically time after time in many unrelated lineages.

The ever-increasing body of genetic information is opening the "black box" of why there are so many species, and how they came to be. For example, we have long known that limbs evolved from fins, based on fossil evidence (which continues to accumulate, providing additional supportive evidence). But we have not known what the *mechanisms* for forming a fin or a limb are, and how selection works to modify such structures. Now, with our knowledge of *Hox* and other toolkit genes that control embryonic development, we understand how fins and limbs are formed. We can also experimentally manipulate development and perform selection experiments to test possible pathways of evolution—the "how they came to be."

Why do we see the recurrent, independent evolution of common themes?

A phenomenon that we often see, but do not yet fully understand, is why certain themes—such as body elongation, limblessness, and tooth modification—recur among distantly related vertebrates. The evolution of viviparity, live-bearing reproduction, is one such theme. In the vast majority of animals, reproduction occurs when a female lays her eggs in water and a male sprays sperm over them; typically both parents then abandon the fertilized eggs. But some species in many separate vertebrate lineages have evolved forms of viviparity. Cartilaginous and bony fishes exhibit diverse modes of embryonic nutrition, and in one group, the sea horses, it is the males that become pregnant. Amphibians also exhibit diverse patterns of viviparity: some have pregnant fathers and others have mothers that brood embryos in the skin of their backs, in their stomachs, or in their oviducts. And many squamate reptiles grow placentas similar to those of mammals. All mammals except monotremes are live-bearers with maternal nutrition. In fact, among the living vertebrate groups, birds are the only group that has not evolved the live-bearing habit. Can you think of some reasons why?

In some fishes, amphibians, and squamates, the mother supplies all the nutrition for the developing young through her investment in the egg; but in other species, females resorb their yolks and provide nutrients directly to offspring that are born as fully metamorphosed juveniles. Given that viviparity has evolved independently in approximately 200 vertebrate groups, it is not surprising that we see such a variety of reproductive patterns. Biologists are just beginning to understand the evolution of viviparity and to determine which patterns are shared among different evolutionary lineages and which are not. It appears that the hormonal basis for viviparity is similar in many groups, although the timing, the receptors involved, and the physiological responses vary. Researchers are now identifying candidate genes in the hope of unraveling the genetic networks that have fostered the different modes of viviparity. Ecological studies are revealing the interactions of potential selection regimes that influence reproductive modes. However, we still have much to learn about how genetics, development, physiology, and ecology interact in the evolution of diversity and common recurring themes.

think like a scientist If viviparity has evolved multiple times in different vertebrate lineages, it must have been favored by natural selection. What might some of the benefits of viviparity be?

Marvalee H. Wake is a Professor in the Department of Integrative Biology and the graduate school at the University of California, Berkeley. She studies vertebrate evolutionary morphology, development, and reproductive biology, with the goal of understanding evolutionary patterns and processes. To learn more about Dr. Wake's research, go to http://ib.berkeley.edu/labs/mwake/.

REVIEW KEY CONCEPTS

For access to MindTap and additional study materials visit www.cengagebrain.com.

32.1 Invertebrate Deuterostomes

- Echinoderms have secondary radial symmetry, a five-part body plan, and a unique water vascular system that is used in locomotion and feeding (Figures 32.2 and 32.3).
- Hemichordates collect oxygen and particulate food from seawater that enters the mouth and exits the pharynx through branchial slits (Figure 32.4).

32.2 Overview of the Phylum Chordata

- Chordates share several derived characteristics: a notochord; postanal tail; segmentation of body wall and tail muscles; a dorsal, hollow nerve cord; and a perforated pharynx at some stage of the life cycle (Figure 32.5).
- Two subphyla of invertebrate chordates use their perforated pharynx to collect particulate food (Figures 32.6 and 32.7).
- The subphylum Vertebrata includes animals with a bony endoskeleton, structures derived from neural crest cells, and a brain divided into three regions.

32.3 The Origin and Diversification of Vertebrates

- Vertebrates evolved from an invertebrate chordate ancestor, probably through duplication of the *Hox* gene complex (Figure 32.8).
- The earliest vertebrates were fish-shaped animals with only some vertebrate characteristics (Figure 32.9).
- Vertebrates diversified into numerous lineages (Figure 32.10). Gnathostomata includes all jawed vertebrates. Tetrapoda includes all lineages that ancestrally had four legs. Amniota includes groups descended from animals that produced an amniote egg.

32.4 "Agnathans": Hagfishes and Lampreys, Conodonts and Ostracoderms

- Living agnathans—hagfishes and lampreys—are jawless fishes that lack vertebrae and paired fins (Figure 32.11).
- Conodonts had bonelike elements in the pharynx. Ostracoderms were heavily armored, jawless fishes that sucked particulate food into their mouths (Figures 32.12 and 32.13).

32.5 Gnathostomata: The Evolution of Jaws

- Jaws arose through the evolutionary modification of gill arches, which supported the pharynx of ostracoderms (Figure 32.14). Fins arose at the same time (Figure 32.15).
- The first jawed fishes, placoderms and Acanthodii are now extinct (Figure 32.16). Chondrichthyans have a skeleton composed of cartilage (Figure 32.17). Actinopterygians and Sarcopterygians have a skeleton composed of bone. Actinopterygians have fins supported by bony rays (Figures 32.18 and 32.19). Sarcopterygians have fins supported by muscles and a bony endoskeleton (Figure 32.20).

32.6 Tetrapoda: The Evolution of Limbs

- Tetrapods arose in the late Devonian. Key tetrapod adaptations include a strong vertebral column, girdles, limbs, and modified sensory systems. Paleontologists have discovered several transitional species that reveal the evolution of tetrapods from one group of sarcopterygian fishes (Figure 32.21).

- Modern amphibians are generally restricted to moist habitats. Their life cycles often include larval and adult stages. Caudata (salamanders) are elongate, tailed amphibians. Anura (frogs) have compact bodies, long legs, and no tails. Gymnophiona (caecilians) are legless burrowers (Figure 32.22).

32.7 Amniota: The Evolution of Fully Terrestrial Vertebrates

- Key adaptations in amniotes, the first fully terrestrial vertebrates, included a water-resistant skin, amniote eggs (Figure 32.23), and the excretion of nitrogen wastes as uric acid.
- Amniotes diversified into two lineages (Synapsida and Reptilia) distinguished by differences in their skin, the presence or absence of hair, the presence or absence of milk production, and by differences in their metabolic processes and mechanisms of temperature regulation (Figure 32.24).
- Reptilia split into two lineages: Lepidosaurs and Archelosaurs.

32.8 Living Lepidosaurs: Sphenodontids and Squamates

- Sphenodontids are remnants of a once-diverse lineage (Figure 32.25A).
- Squamates have skin composed of overlapping, keratinized scales (Figure 32.25B, C). Lizards consume a variety of foods, but all snakes are predators on other animals.

32.9 Living Archelosaurs: Turtles, Crocodilians, and Birds

- The turtle body plan includes a bony shell, with a dorsal carapace and ventral plastron (Figure 32.26).
- Crocodilians are semiaquatic predators (Figure 32.27).
- Birds have adaptations that reduce their weight and generate power for flight (Figure 32.28). Modern birds diversified in the Cretaceous period (Figure 32.29) and exhibit adaptations of their bills, feet, wings, and behavior (Figure 32.30).

32.10 Mammalia: Monotremes, Marsupials, and Placentals

- Key adaptations of mammals include endothermy, which allows high levels of activity; modification of the teeth and jaws; extensive parental care of offspring; and large and complex brains.
- There are three major groups of mammals: the Prototheria, to which the monotremes belong, and two lineages of Theria, to which the marsupials and the placentals belong. The three groups differ in their reproductive patterns.
- Monotremes are restricted to Australia and New Guinea (Figure 32.31). Marsupials are abundant in Australia and occur in South America and North America (Figure 32.32). Most living mammals are placentals, occupying nearly all terrestrial and aquatic habitats (Figure 32.33).

32.11 Nonhuman Primates

- Key adaptations allow Primates to be arboreal and diurnal: upright posture and flexible limbs; good depth perception; and a large and complex brain.
- The Primates includes two major lineages (Figure 32.34). The Strepsirhini have many ancestral primate characteristics

(Figure 32.35). The Haplorhini have many derived primate characteristics (Figures 32.36 and 32.37).

- Primates arose in forests about 55 million years ago. The hominoid lineage, which includes apes and humans, arose in Africa about 23 million years ago (Figure 32.38).

32.12 The Evolution of Humans

- Hominins, the lineage that includes humans, arose in Africa between 10 million and 5 million years ago. Hominin anatomy permits bipedal locomotion. Over time, hominins developed larger brains and tool-making behavior. Several genera of hominins occupied sub-Saharan Africa for several million years (Figures 32.39–32.41).

- *Homo habilis* was the first hominin species to make stone tools. *Homo erectus,* which arose in East Africa about 1.8 million years ago, made sophisticated stone tools (Figure 32.42).

- The early descendants of *H. erectus* left Africa in waves, populating Asia and Europe. Neanderthals, the best known of these groups, became extinct about 30,000 years ago.

- Modern humans, *Homo sapiens,* arose approximately 150,000 years ago and migrated out of Africa, eventually replacing archaic humans in Europe and Asia.

TEST YOUR KNOWLEDGE

Remember/Understand

1. Which phylum includes animals that have a water vascular system?
   a. Echinodermata
   b. Hemichordata
   c. Chordata
   d. Tetrapoda
   e. Amniota

2. Which of the following is *not* a characteristic of all chordates?
   a. notochord and postanal tail
   b. segmental body wall and tail muscles
   c. segmented nervous system
   d. dorsal hollow nerve cord
   e. perforated pharynx

3. Which group of vertebrates has adaptations that allow it to reproduce on land?
   a. agnathans
   b. tetrapods
   c. gnathostomes
   d. amniotes
   e. ichthyosaurs

4. Which group of fishes has the most living species today?
   a. sarcopterygians
   b. actinopterygians
   c. chondrichthyans
   d. acanthodians
   e. ostracoderms

5. Modern amphibians:
   a. closely resemble their Paleozoic ancestors.
   b. always occupy terrestrial habitats as adults.
   c. never occupy terrestrial habitats as adults.
   d. are generally larger than their Paleozoic ancestors.
   e. are generally smaller than their Paleozoic ancestors.

6. The Hominoidea is a monophyletic group that includes:
   a. apes and monkeys.
   b. apes only.
   c. humans and human ancestors.
   d. apes and humans.
   e. monkeys, apes, and humans.

7. Which of the following hominin species was the earliest?
   a. *Ardipithecus ramidus*
   b. *Australopithecus afarensis*
   c. *Homo habilis*
   d. *Homo erectus*
   e. *Homo neanderthalensis*

Apply/Analyze

8. Which of the following key adaptations allows amniotes to occupy terrestrial habitats?
   a. the production of carbon dioxide as a metabolic waste product
   b. an unshelled egg that is protected by jellylike material
   c. a dry skin that is largely impermeable to water
   d. a lightweight skeleton with hollow bones
   e. feathers or fur that provide insulation against cold weather

9. Which of the following characteristics does *not* contribute to powered flight in birds?
   a. a lightweight skeleton
   b. efficient respiratory and circulatory systems
   c. enlarged forelimbs and a keeled sternum
   d. a high metabolic rate that releases energy from food rapidly
   e. scaly skin on the legs and feet

10. Which of the following characteristics did *not* contribute to the evolutionary success of mammals?
    a. extended parental care of young
    b. an erect posture and flexible hip and shoulder joints
    c. specializations of the teeth and jaws
    d. enlargement of the brain
    e. high metabolic rate and high body temperature

11. **Discuss Concepts** When tetrapods first ventured onto the land, what new selection pressures did they face? What characteristics might have fostered the success of these animals as they made the transition from aquatic to terrestrial habitats?

12. **Discuss Concepts** Imagine that you unearthed the complete fossilized remains of a mammal. How would you determine the food habits of this now extinct animal?

13. **Discuss Concepts** Many myths about human evolution are embraced by popular culture. Using the information you have learned about human evolution, argue against each of the following myths.
    a. Humans evolved from chimpanzees.
    b. Evolution occurred in a steady linear progression from primitive primate to anatomically modern humans.
    c. All human characteristics, such as bipedal locomotion and an enlarged brain, evolved simultaneously and at the same rate.

14. **Apply Evolutionary Thinking** Birds and crocodiles are both descended from an ancestral archosaur. What shared anatomical and behavioral characteristics reflect this common ancestry? Explain why dinosaurs, which were also members of the archosaur lineage, may have shared these traits as well. Review Figure 32.24 before formulating your answer.

Evaluate/Create

15. **Discuss Concepts** Most sharks and rays are predatory, but the largest species feed on plankton. Construct a hypothesis to explain this observation. How would you test your hypothesis?

16. **Discuss Concepts** Use a pair of binoculars to observe several species of birds that live in different types of environments, such as lakes and forests. How are their beaks and feet adapted to their habitats and food habits?

17. **Design an Experiment** Walking along a rocky coast one day, you discover two small creatures—one lumpy and the other worm-shaped. What anatomical studies would you conduct to determine whether or not they are chordates? What genetic studies might provide supplementary evidence?

For selected answers, see Appendix A.

INTERPRET THE DATA

The phylogenetic tree for vertebrates depicted below was constructed from sequence data for two rRNA mitochondrial genes (12S and 16S). How do the results of this analysis compare with the phylogenetic trees in Figures 32.10 and 32.24? Identify the major clades of vertebrates on the tree depicted below.

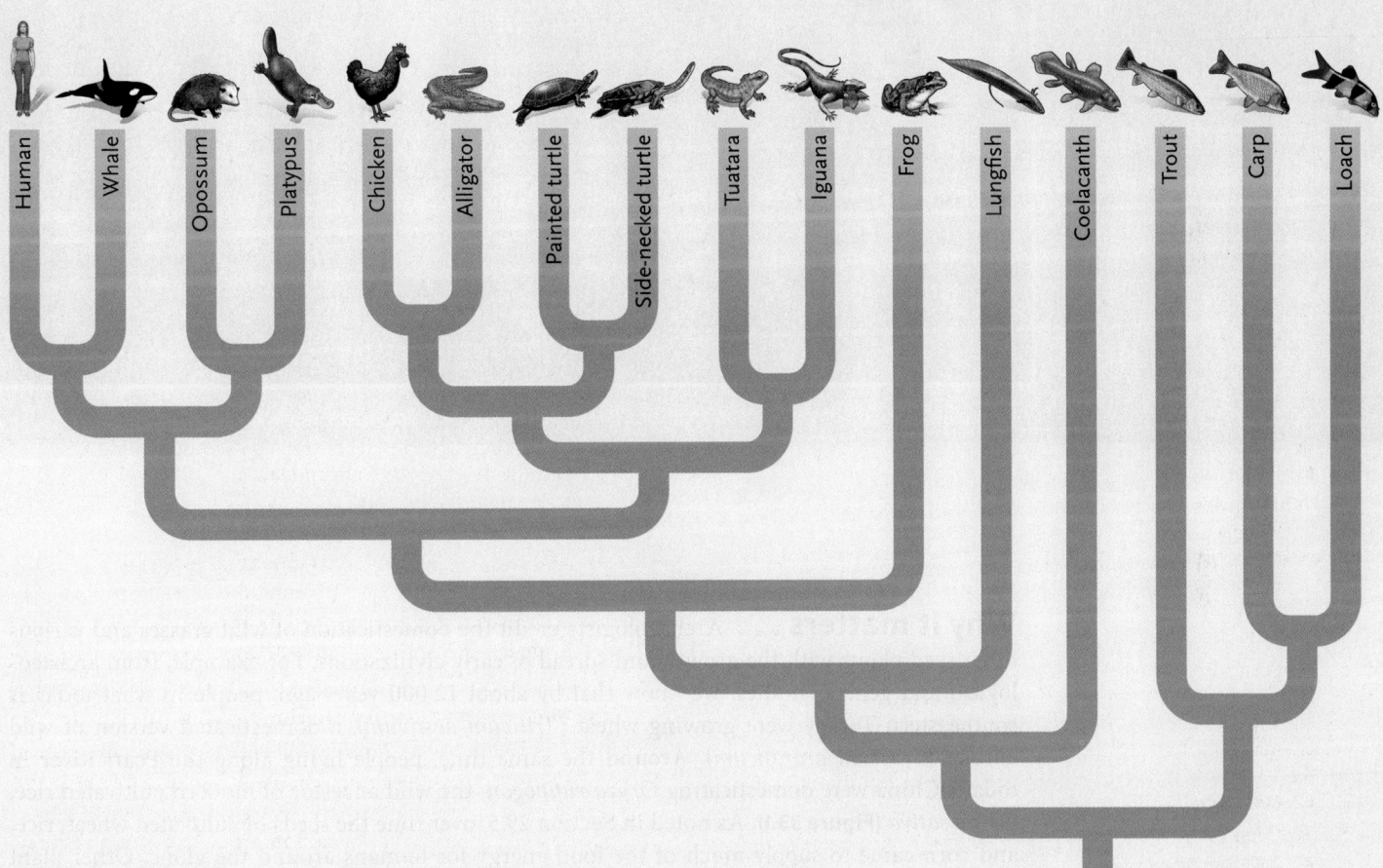

Source: R. Zardoya and A. Meyer. 1998. Complete mitochondrial genome suggests diapsid affinities of turtles. *Proceedings of the National Academy of Sciences, USA* 95:14226–14231. Copyright 1998 National Academy of Sciences, U.S.A.

33

The Plant Body

Basic parts of the shoot system of an apple tree *(Malus domestica)*, including leaves, stems, and vividly colored fruits.

Why it matters . . . Archaeologists credit the domestication of wild grasses and various other seed plants with the growth and spread of early civilizations. For example, from archaeological and genetic studies, we know that by about 12,000 years ago, people in what today is southeastern Turkey were growing wheat *(Triticum aestivum),* a domesticated version of wild emmer *(Triticum araraticum).* Around the same time, people living along the Pearl River in today's China were domesticating *Oryza rufipogon,* the wild ancestor of modern cultivated rice, *Oryza sativa* **(Figure 33.1).** As noted in Section 29.5, over time the seeds of cultivated wheat, rice, and corn came to supply much of the food energy for humans around the globe. Other plant parts—from leaves, flowers, and fruits to stems and roots—are equally as crucial in human societies, providing food, lumber, fibers for paper and clothing, and myriad other materials ranging from inks and pharmaceuticals to rubber and the cork for wine bottles.

This chapter launches our in-depth survey of the structure and functioning of seed plants—their morphology, anatomy, and physiology. *Morphology* refers to the shape and overall appearance of body parts. *Anatomy* is the physical structure of body parts. *Physiology* examines their function, ranging from mechanisms for acquiring nutrients and internal transport of substances to reproduction and defense. In this chapter we begin with an overview of fundamentals of plant structure and functioning. Throughout the unit our focus will be on angiosperms, or flowering plants—in terms of distribution and sheer numbers of species, the most successful plants on Earth.

O. rufipogon

Courtesy of Daderot

Planting O. sativa

Grigvovan/Shutterstock.com

FIGURE 33.1 *O. rufipogon,* **the wild progenitor of modern cultivated rice, being grown at a research facility (left), and planting of domesticated rice (*O. sativa*) in China (right).**

33.1 Basic Concepts of Plant Structure and Growth

Despite the tremendous diversity of seed plants described in Chapter 29, all have the same two-part body plan: A *root system* that in most species is underground, and an aboveground *shoot system*. Like the organ systems of animals (introduced in Chapter 38), both these plant systems consist of **organs**—structures that contain two or more types of tissues and have a definite form and function. Plant organs include roots, leaves, stems, and in angiosperms, flowers and fruits. A **tissue** is a group of cells and intercellular substances that function together in one or more specialized tasks.

Plant organs are built of three tissue systems that Section 33.2 explores in detail. The **ground tissue system,** which makes up most of the primary plant body, includes tissues that function in photosynthesis, storage, and support. The **vascular tissue system** consists of interconnecting cells that form transport channels throughout the plant. This system is subdivided into xylem tissue, which transports water, and phloem tissue, which transports nutrients. The **dermal tissue system** serves as a skinlike protective covering for the plant body. **Figure 33.2** shows the general location of each system in the shoot and root of a tomato plant.

Root and Shoot Systems Perform Different but Integrated Functions

In most plants the **root system** grows below ground. It anchors the plant, and usually provides structural support for aboveground parts. The root system also absorbs water and dissolved minerals from soil and stores carbohydrates.

A plant's **shoot system** is basically a series of repeating components such as stems, leaves, axillary buds and, in angiosperms, flowers. This modular organization correlates with plants' characteristic physiological flexibility: unlike animals, plants have the genetic capacity to modify the development of

organs as environmental conditions shift. In general, during the course of a plant's life cycle, embryonic shoots called *buds* give rise to leaves, flowers, or both, as different gene-guided developmental pathways unfold. Section 33.3 looks more closely at the growth of stems and leaves, and Chapter 36 describes the development of flowers and other reproductive structures.

Some or all of a plant's shoot system is highly adapted for photosynthesis. For example, leaves are organs that greatly increase a plant's surface area and thus its exposure to light. Similarly, aboveground stems position leaves for maximum light exposure—an adaptation for maximizing photosynthesis—and in angiosperms they position flowers for pollination. As described shortly, some shoot system organs also store carbohydrates manufactured during photosynthesis.

Root and Shoot Systems Contain Meristem Tissue from Which New Plant Parts Grow

As you know from experience, in animals growth slows dramatically or stops once the organism reaches a certain size. We also see this pattern, called **determinate growth,** in plant organs such as leaves, flowers, and fruits. Yet most plant species also have parts that can grow throughout the plant's life, a pattern called **indeterminate growth.** Clumps of self-perpetuating embryonic tissue called **meristems** (*merizein* = to divide) are the basis for indeterminate growth, producing new tissues more or less continuously while a plant is alive. (Like the stem cells of animals, meristematic cells are *totipotent,* meaning they have the genetic capacity to give rise to not only new tissues but also a whole new plant—a topic of Chapter 36). The capacity for indeterminate growth gives plants a great deal of structural and functional flexibility—or what biologists often call *plasticity*—in their responses to changes in environmental factors such as light, temperature, water, and the supply of nutrients. This plasticity has major adaptive benefits for an organism that cannot move about. For example, if external factors change the direction of incoming light for photosynthesis, stems can "shift gears" and grow in that direction. Likewise, roots can grow outward toward water.

Meristems Are Responsible for Growth in Both Height and Girth

All vascular plants have a type of meristem that adds length to the plant body, and some also have a different type of meristem for increasing girth **(Figure 33.3).** An increase in length arises from **apical meristems** at the tips of their buds, stems, and roots (see Figure 33.3A). Tissues that develop from apical

A. The root and shoot systems

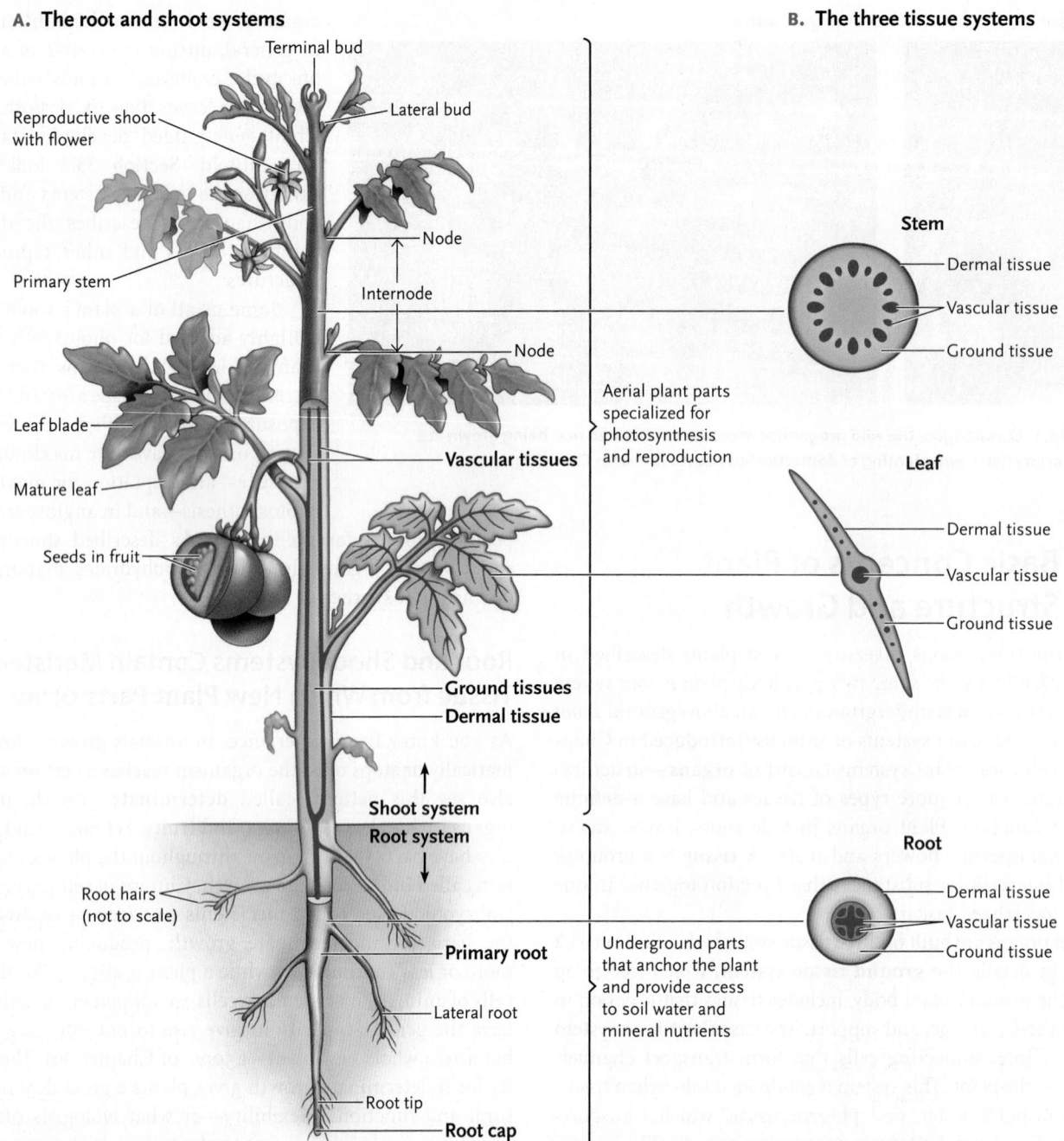

- Terminal bud
- Lateral bud
- Reproductive shoot with flower
- Node
- Primary stem
- Internode
- Node
- Leaf blade
- **Vascular tissues**
- Mature leaf
- Seeds in fruit
- **Ground tissues**
- **Dermal tissue**
- **Shoot system**
- **Root system**
- Root hairs (not to scale)
- **Primary root**
- Lateral root
- Root tip
- **Root cap**

B. The three tissue systems

Aerial plant parts specialized for photosynthesis and reproduction

Stem
- Dermal tissue
- Vascular tissue
- Ground tissue

Leaf
- Dermal tissue
- Vascular tissue
- Ground tissue

Underground parts that anchor the plant and provide access to soil water and mineral nutrients

Root
- Dermal tissue
- Vascular tissue
- Ground tissue

FIGURE 33.2 Basic body plan for the tomato plant *Solanum lycopersicum,* a typical angiosperm.
(A) Vascular tissues (purple) conduct water, dissolved minerals, and organic substances. They thread through ground tissues, which make up most of the plant body. Dermal tissues (epidermis, in this case) cover the surfaces of the root and shoot systems. **(B)** How the three tissue systems are arranged in stems, leaves, and roots.
© Cengage Learning 2017

meristems are called **primary tissues** and make up the **primary plant body.** Growth of the primary plant body is called **primary growth.**

Some species of plants—grasses for example—show only primary growth, which occurs at the tips of roots and shoots. Others, particularly trees and shrubs that have a woody body, show **secondary growth,** which originates at self-perpetuating cylinders of tissue called **lateral meristems.** Secondary growth increases the girth (diameter) of older roots and stems (see Figure 33.3B). The tissues that develop from lateral meristems, called **secondary tissues,** make up the woody **secondary plant body.** Primary and secondary growth can go on simultaneously in a single plant, with primary growth making shoot parts longer and secondary growth adding girth. Plant hormones govern these growth processes and other key events described in Chapter 37.

MERISTEMS IN A VASCULAR PLANT

A. Plants increase in length by cell divisions in apical meristems and by elongation of the daughter cells.

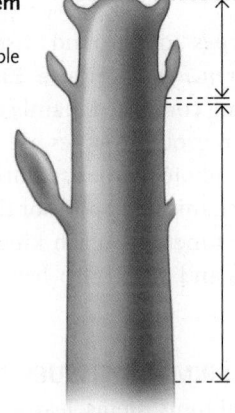

Shoot apical meristem
Dividing cells near all shoot tips are responsible for a shoot's primary tissues and growth.

Cells divide in shoot apical meristem.

New cells elongate and start to differentiate into primary tissues.

Root apical meristem
Dividing cells near root tips are responsible for a root's primary tissues and growth.

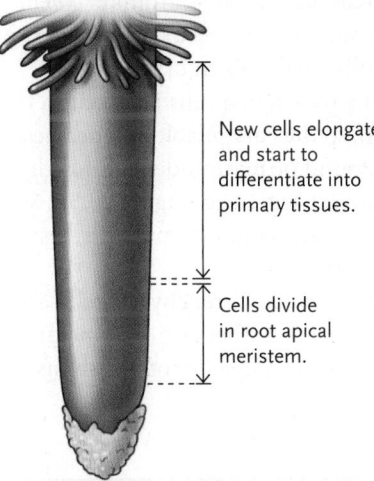

New cells elongate and start to differentiate into primary tissues.

Cells divide in root apical meristem.

B. Some plants increase in girth by way of cell divisions in lateral meristems.

Lateral meristems

Lateral meristems
Dividing cells are responsible for the increase in diameter of shoots and roots.

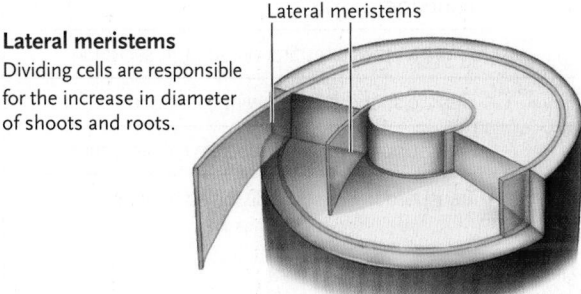

SUMMARY Meristems in different locations are responsible for increases in the length and diameter of the shoots and roots of a vascular plant.

think like a scientist What would happen if part of an apical meristem were removed?

© Cengage Learning 2017

In Angiosperms, Different Overall Patterns of Growth Correlate with Lifespan

Evolution has generated differences in patterns of growth and reproduction we observe in flowering plants. A case in point is secondary growth, which produces woody tissue. Secondary growth demands a substantial investment of energy and other resources. It correlates with a long life and, presumably, repeated opportunities for reproduction. Many eudicots (and all gymnosperms) have secondary growth. By contrast, most monocots (and some eudicots) show little or no secondary growth. They do not have significant woody tissue and are termed *herbaceous* plants. Herbaceous species mature and reproduce quickly, with a relatively small investment of energy and other resources. They are often subdivided into *annuals,* *biennials,* and *perennials,* terms that refer to the pattern of an herbaceous plant's life cycle **(Table 33.1)**.

STUDY BREAK 33.1

1. Compare and contrast the functions of a land plant's root and shoot systems.
2. Explain what meristem tissue is, and name and describe the roles of the basic types of meristems in forming the plant body.

33.2 The Three Plant Tissue Systems

Each of the plant tissue systems introduced in Section 33.1 includes several types of tissue. Each tissue in turn is made up of cells with specializations for different functions **(Table 33.2)**. *Simple* tissues have only one type of cell. *Complex* tissues have organized arrays of two or more types of cells. **Figure 33.4** will help you interpret micrographs of plant tissues.

In all plant tissues, newly forming cells develop a cellulose-rich primary wall (see Section 4.4); this wall forms around the **protoplast,** the botanical term for a plant cell's cytoplasm, organelles, and plasma membrane. In some tissues, the protoplast of various types of cells deposits additional cellulose and lignin inside the primary wall, forming a strong secondary wall. For example, as we will see shortly, a

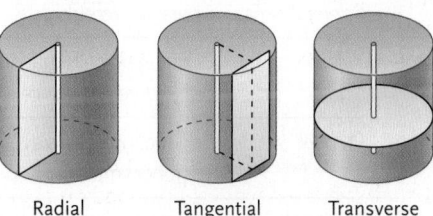

Radial   Tangential   Transverse

FIGURE 33.4 Terms that identify how tissue specimens are cut from a plant. Along the radius of a stem or root, longitudinal cuts give radial sections. Cuts at right angles to a root or stem radius give tangential sections. Cuts perpendicular to the long axis of a stem or root give transverse sections (cross sections).
© Cengage Learning 2017

TABLE 33.1	Distinguishing Features of Annuals, Biennials, and Perennials		
Life Cycle	**Growth Pattern**	**Examples**	
Annual			
Growth and reproduction completed in one growing season	Mainly primary growth	Marigold (eudicot), corn (monocot)	
Biennial			
Completed in two growing seasons; roots, stems, and leaves develop in first season; reproduction (flowering, seed formation) occurs in the second season, followed by death	Limited secondary growth in some species	Carrot, parsley, black-eyed Susan (eudicots)	
Perennial			
Growth and reproduction occur year after year	Many species have secondary growth (woody parts)	Trees, woody shrubs, and vines; irises, tulips (no secondary growth)	

sturdy secondary wall is essential to the functioning of cells that make up the vascular tissue xylem. Our survey of plant tissues begins, however, with ground tissues, which make up the bulk of plant organs.

Ground Tissues Are All Structurally Simple, but Differ in Important Ways

Plants have three types of ground tissue: *parenchyma*, *collenchyma*, and *sclerenchyma* (**Figure 33.5**). Each type is structurally simple, being composed mainly of one kind of cell. Functionally the cells in ground tissues are the "worker bees" of plants, carrying out photosynthesis, storing carbohydrates or water, providing mechanical support for the plant body, and performing other basic functions. Each kind of cell has a distinctive wall structure, and some also have variations in the protoplast.

PARENCHYMA: SOFT PRIMARY TISSUES Most of the soft, moist primary growth of roots, stems, leaves, flowers, and fruits is **parenchyma** (*para* = around; *khein* = fill in, or pour). Parenchyma cells can occur both as part of parenchyma tissue and as individual cells in other tissues.

Most parenchyma cells have only a thin primary wall (and no lignin), so they are pliable and permeable to water. Often the cells are round or many-sided, although they also can be elongated like a sausage, as in Figure 33.5B. Sometimes there are air spaces between parenchyma cells, especially in leaves (see Figure 33.21).

On the whole, parenchyma cells are relatively unspecialized. Yet subsets of them do become differentiated for specialized roles. For example, photosynthesis occurs in parenchyma

TABLE 33.2	Summary of Flowering Plant Tissues and Their Components		
System	**Tissue**	**Cell Types in Tissue**	**Function**
Ground tissue	Parenchyma	Parenchyma cells	Photosynthesis, respiration, storage, secretion
	Collenchyma	Collenchyma cells	Flexible strength for growing plant parts
	Sclerenchyma	Fibers or sclereids	Rigid support, physical/mechanical protection
Vascular tissue	Xylem	Conducting cells (tracheids, vessel elements); parenchyma cells; sclerenchyma cells	Transport of water and dissolved minerals
	Phloem	Conducting cells (sieve-tube elements); parenchyma cells; sclerenchyma cells	Sugar transport
Dermal tissue	Epidermis	Fairly unspecialized epidermal cells; specialized cells such as guard cells	Control of gas exchange; water loss; protection
	Periderm	Cork; cork cambium; secondary cortex	Protection

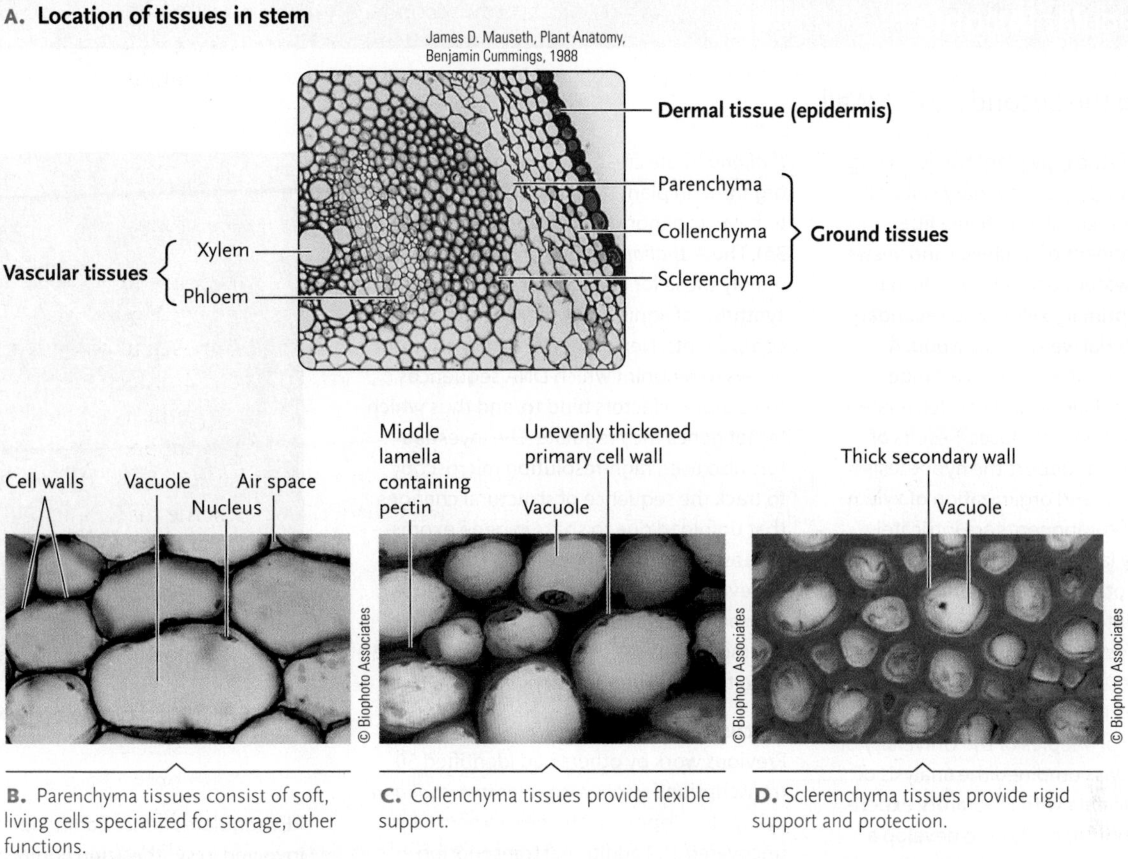

A. Location of tissues in stem

James D. Mauseth, Plant Anatomy,
Benjamin Cummings, 1988

Dermal tissue (epidermis)

Parenchyma
Collenchyma — Ground tissues
Sclerenchyma

Vascular tissues { Xylem
Phloem }

Cell walls   Vacuole   Air space
Nucleus

B. Parenchyma tissues consist of soft, living cells specialized for storage, other functions.

© Biophoto Associates

Middle lamella containing pectin

Unevenly thickened primary cell wall

Vacuole

C. Collenchyma tissues provide flexible support.

© Biophoto Associates

Thick secondary wall

Vacuole

D. Sclerenchyma tissues provide rigid support and protection.

© Biophoto Associates

FIGURE 33.5 Locations and examples of ground, vascular, and dermal tissues in flowering plants.
(A) Ground, vascular, and dermal tissues in a buttercup stem *(Ranunculus)*, transverse section. Ground tissues are simple tissues, while vascular and dermal tissues are complex, containing various types of specialized cells.
(B–D) Examples of ground tissues from the stem of a sunflower plant *(Helianthus annuus)*.
© Cengage Learning 2017

cells in which a large number of chloroplasts develop. The flesh of a potato tuber or an apple is mainly parenchyma specialized for carbohydrate storage. And in many plant species, some parenchyma cells are specialized for short-distance transport of solutes. These cells are common in tissues in which water and solutes must be rapidly moved from cell to cell—for example, in vascular tissues and in tissues that secrete nectar. Parenchyma cells usually remain alive and capable of dividing when they are mature. In fact, their mitotic divisions produce the new cells that often heal wounds in plant parts.

COLLENCHYMA: FLEXIBLE SUPPORT The ground tissue **collenchyma** (see Figure 33.5C) helps support plant parts *(kolla* = glue). Collenchyma cells typically are elongated. Collectively they often form strands or a sheathlike cylinder within growing shoot regions and the stemlike *petiole* at the base of a leaf. The "strings" of celery include collenchyma. Like parenchyma cells, collenchyma cells remain alive and metabolically active as they mature.

Collenchyma is an especially important adaptation in parts such as elongating stems that require structural support

as they grow. Each collenchyma cell develops only a primary wall built of layers of cellulose and pectin, but over time it thickens as the cell synthesizes these components. The result is a strong primary wall that can stretch as the cell elongates. This sturdy but physically flexible structure is an adaptation that ensures that collenchyma is not so rigid as to hamper the growth of plant parts it supports.

SCLERENCHYMA: RIGID SUPPORT AND PROTECTION Mature plant parts gain additional mechanical support and protection from **sclerenchyma** *(skleros* = hard). Most cells of this ground tissue develop thick secondary walls (see Figure 33.5D) that become heavily lignified. Once a sclerenchyma cell is encased in lignin, it dies because its protoplast can no longer take up nutrients or exchange gases with the environment. The walls remain, however, providing protection and support for the life of the plant. This chapter's *Molecular Insights* features current research on the complex genetic events that govern the formation of secondary cell walls.

There are two types of sclerenchyma cells, *sclereids* and *fibers*. **Sclereids** take a variety of shapes. Clumped, roughly

Networking the Secondary Cell Wall

Sclerenchyma is the only plant tissue having cells strengthened by a secondary cell wall. As noted in the text, sclerenchyma fibers are the main component of tracheids and vessel elements, the water-conducting cells that make up both primary xylem and secondary xylem, the material we know as wood. A secondary cell wall is a complex arrangement of lignin, cellulose, and smaller, related polymers called hemicelluloses. Results of many experiments support the hypothesis that the synthesis and organization of xylem secondary wall components are intricately coordinated by large numbers of transcription factors (proteins that regulate gene transcription) interacting with genes whose collective expression produces xylem. Little was known about the extent of this regulatory scheme until a multilaboratory team led by Mallorie Taylor-Teeples of the University of California at Davis combined the analysis of existing genetic data with laboratory experiments and computer analysis to develop a comprehensive map of the xylem transcription factor network.

Research Question

How elaborate is the regulatory network controlling the development of secondary walls of xylem cells?

Experiments

The researchers began with available information on transcription factors and their target genes that function in the formation of xylem in developing roots of *Arabidopsis*

thaliana (thale cress), a major model organism in plant research and the first plant to have its genome sequenced (see Chapter 36). The *A. thaliana* genes of interest were mainly those for enzymes associated with the synthesis of lignin and other secondary wall components. Next, the team used laboratory assays to pinpoint which DNA sequences transcription factors bind to, and thus which target genes they regulate. The investigators also used high-resolution microscopy to track the sequence of structural changes that unfolded due to shifts in gene expression as xylem cells arose and differentiated in developing *A. thaliana* rootlets **(Figure)**. Computerized analysis generated a map of interactions in the regulatory network underlying these sequential changes.

Discussion

Previous work by others had identified 50 transcription factors with roles in *A. thaliana* xylem development. The new experiments uncovered 152 additional transcription factors and an unexpectedly complex regulatory network governing xylem development. Among other findings, computer analysis mapped 617 separate interactions of transcription factors with one another or with regulatory sequences of genes encoding the synthesis of secondary wall components. On average, genes coding for proteins involved in secondary wall development were regulated by five different transcription factors. Using their network analysis, the researchers predicted that a transcription factor called

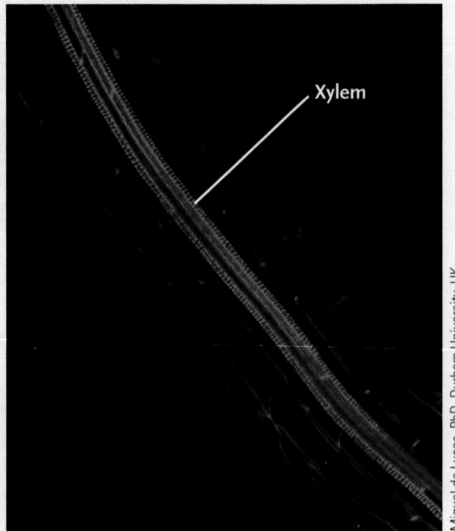

Miguel de Lucas, PhD. Durham University, UK

FIGURE LM of developing root xylem.

REV regulates genes coding for enzymes involved in synthesizing lignin, and they confirmed this prediction in plants with mutations to the gene for REV.

think like a scientist

How might plant scientists shed light on the hypothesis that regulation of secondary wall synthesis in xylem has changed little during the evolution of vascular plants?

Source: M. Taylor-Teeples et al. 2014. An *Arabidopsis* gene regulatory network for secondary cell wall synthesis. *Nature* (December: published online).

cube-shaped sclereids give pears their slightly gritty texture **(Figure 33.6A).** Other sclereids form a thick protective coat around some kinds of seeds, including peach and cherry pits and the hard shells of coconuts, walnuts, and Brazil nuts. **Fibers (Figure 33.6B)** are elongated sclerenchyma cells that are sometimes compared to thick rubber because they are somewhat pliable but resist stretching. Cotton fabric is woven from fibers that develop around the seeds of cotton plants *(Gossypium)*, whereas linen cloth is made from fibers in the stems of flax plants. Tree bark gains a great deal of tensile strength from fibers that are incorporated in it. Fibers also occur in a plant's vascular tissues, the topic we turn to next.

Vascular Tissues Are Specialized for Conducting Fluids

Plant vascular tissues are complex tissues composed of specialized conducting cells, parenchyma cells, and fibers. Xylem and phloem, the two kinds of conducting tissues that arose as vascular plants evolved, are organized into bundles of interconnected cells that extend throughout the plant (see Figure 33.2).

XYLEM: TRANSPORTING WATER AND MINERALS Xylem *(xylon = wood)* conducts water and minerals absorbed from the soil upward from a plant's roots to the shoot. (That

A. Sclereids

Thick secondary wall

B. Fibers

D. E. Atkin and I. L. Rigsby, Richard B. Russel Agricultural Research Service, U.S. Department of Agriculture, Athens, Georgia

Dr. Jack Bostrack/Visuals Unlimited, Inc.

FIGURE 33.6 Examples of sclerenchyma cells. (A) From the flesh of a pear *(Pyrus)*, sclereids called stone cells. Each has a thick, lignified wall. **(B)** Strong fibers from stems of a flax plant *(Linum)*.

process, called *transpiration*, is a central topic of Chapter 34.) As you may recall from Chapter 28, xylem was a key adaptation in early land plant evolution because it not only provided a means for distributing water and minerals but also provided support for upright growth. Xylem contains four types of cells that serve these functions: parenchyma, and fibers and the conducting cells called *tracheids* and *vessel elements* (all schlerenchyma). When both types of conducting cells reach maturity, they lay down secondary cell walls

that are thickened and strengthened by lignin and cellulose. The cell's protoplast then disintegrates and the cell dies—leaving arrays of abutting, empty cells that can serve as pipelines for water and minerals.

Tracheids are elongated cells, with a rather narrow diameter and tapered, overlapping ends **(Figure 33.7A)**. Water can move from cell to cell through openings called *pits*. Usually, a pit in one cell is opposite a pit of an adjacent cell, so water seeps laterally from tracheid to tracheid.

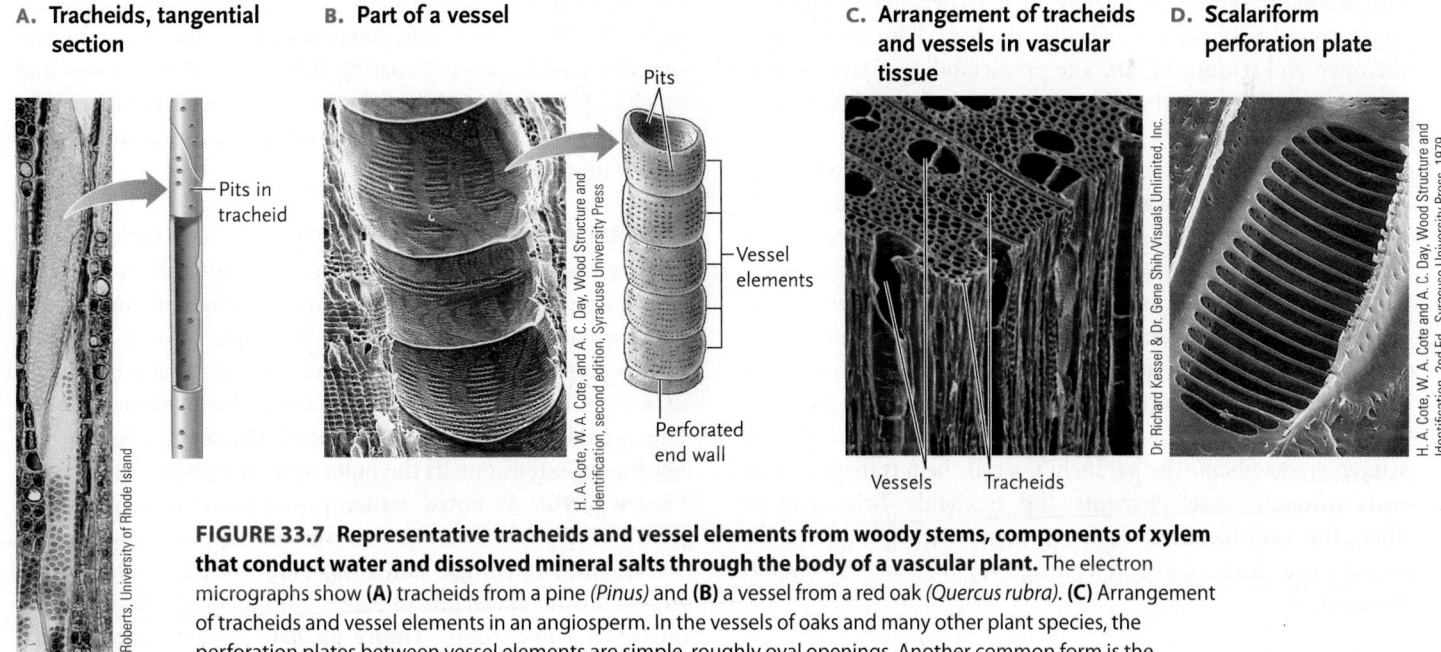

A. Tracheids, tangential section

Pits in tracheid

Alison W. Roberts, University of Rhode Island

B. Part of a vessel

Pits

Vessel elements

Perforated end wall

H. A. Cote, W. A. Cote, and A. C. Day, Wood Structure and Identification, second edition, Syracuse University Press

C. Arrangement of tracheids and vessels in vascular tissue

Vessels   Tracheids

Dr. Richard Kessel & Dr. Gene Shih/Visuals Unlimited, Inc.

D. Scalariform perforation plate

H. A. Cote, W. A. Cote and A. C. Day, Wood Structure and Identification, 2nd Ed., Syracuse University Press, 1979.

FIGURE 33.7 Representative tracheids and vessel elements from woody stems, components of xylem that conduct water and dissolved mineral salts through the body of a vascular plant. The electron micrographs show **(A)** tracheids from a pine *(Pinus)* and **(B)** a vessel from a red oak *(Quercus rubra)*. **(C)** Arrangement of tracheids and vessel elements in an angiosperm. In the vessels of oaks and many other plant species, the perforation plates between vessel elements are simple, roughly oval openings. Another common form is the ladderlike scalariform perforation plate shown in **(D)**.

© Cengage Learning 2017

A. Sieve-tube elements

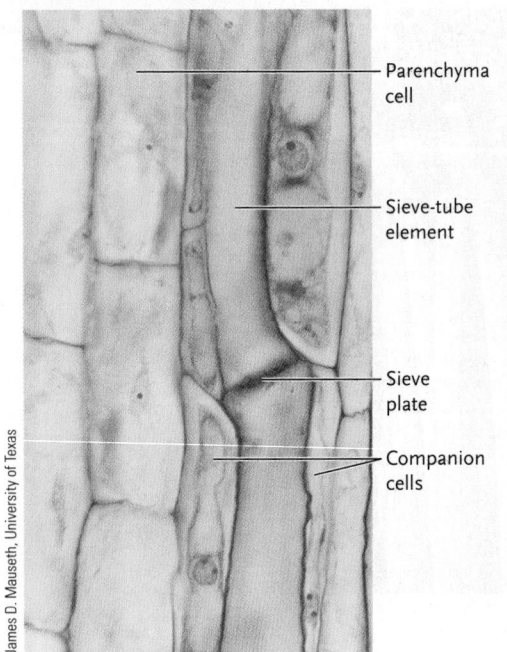

Parenchyma cell

Sieve-tube element

Sieve plate

Companion cells

James D. Mauseth, University of Texas

B. Sieve plate

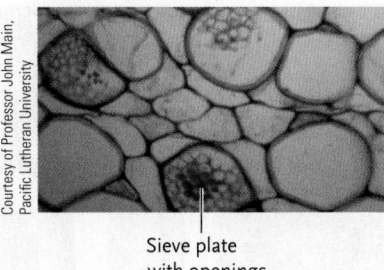

Courtesy of Professor John Main, Pacific Lutheran University

Sieve plate with openings

FIGURE 33.8 Structure of sieve-tube elements. (A) Micrograph showing sieve-tube elements in longitudinal section. Long tubes of sieve-tube elements conduct sugars and other organic compounds. **(B)** SEM of sieve plates of sieve-tube elements in phloem.

Vessel elements are joined end to end in tubelike multicellular columns called **vessels (Figure 33.7B).** Like tracheids, vessel elements have pits through which water can move from cell to cell. However, they also have other structural adaptations that enhance water flow. They have a wider inside diameter than tracheids **(Figure 33.7C),** and as vessel elements mature, enzymes break down portions of their end walls, producing perforations. Some vessel elements have a single, large perforation, so that the end is completely open. Others have a cluster of small, round perforations or ladderlike bars that extend across the open end **(Figure 33.7D).** The predictability of the perforation patterns suggests that this process is under precise genetic control.

Fossils show that the forerunners of modern plant species had only tracheids for water transport. Today, tracheids are still the only conducting cells in most ferns, all gymnosperms, and basal angiosperms such as *Amborella* and some water lilies. Most other angiosperms have both tracheids and vessel elements, however. Presumably, the evolution of vessel elements conferred an adaptive advantage by increasing the efficiency with which water and minerals move throughout the plant.

As already noted, xylem contains parenchyma cells and sclerenchyma fibers. The parenchyma cells help transport minerals through vessel elements and tracheids. Sclerenchyma fibers function like steel cables in concrete, helping keep the xylem tissue fairly rigid and lending overall structural support to the plant.

PHLOEM: A TISSUE THAT TRANSPORTS SUGARS AND OTHER SOLUTES The vascular tissue **phloem** (*phloios* = tree bark) transports solutes, notably the sugars made in photosynthesis, throughout the plant. The main conducting cells of phloem are sieve-tube elements **(Figure 33.8),** which connect end to end, forming a **sieve tube.** As the name implies, their end walls, called sieve plates, are laced with openings. In flowering plants the phloem is strengthened by fibers and sclereids.

Immature sieve-tube elements contain the usual plant organelles. Over time, however, the cell nucleus and internal membranes in plastids break down, mitochondria shrink, and the cytoplasm is reduced to a thin layer lining the interior surface of the cell wall. Even without a nucleus, sieve-tube elements live up to several years in most plants, and much longer in some trees.

In many flowering plants, specialized parenchyma cells called **companion cells** are connected to mature sieve-tube elements by plasmodesmata. Companion cells (which retain their nucleus when mature) assist sieve-tube elements both with the uptake of sugars and with the unloading of sugars in tissues that are growing or storing food. They may also help regulate the metabolism of mature sieve-tube elements. Chapter 35 returns to the functions of phloem cells.

Dermal Tissues Protect Plant Surfaces and Include Various Specialized Cells

A complex tissue called **epidermis** covers the primary plant body in a single continuous layer or sometimes in multiple layers of tightly packed cells. Surface cells of the shoot system secrete a cuticle, a waxy coating that fends off water loss and attacks by microbes. A cuticle coats all external plant parts except the very tips of the shoot and most absorptive parts of roots **(Figure 33.9A).**

EPIDERMAL SPECIALIZATIONS Epidermis also includes cells that are modified in ways that represent major adaptations for plants. An extremely diverse array of **trichomes** are outgrowths of the epidermis. Trichomes may be single cells or elaborate multicellular structures. Some exude sugars that function to attract pollinators, while others produce chemicals designed to deter predators **(Figure 33.9B). Root hairs** are trichomes that develop as extensions in the outer wall of root epidermal cells **(Figure 33.9C).** As noted earlier, root hairs absorb much of a plant's water and minerals from the soil.

Porelike openings called **stomata** (singular, *stoma*) are present in the epidermis of leaves, young stems, flower parts, and even some roots **(Figure 33.9D).** Stomata are situated between specialized epidermal cells called **guard cells** that can open and close. Carbon dioxide for photosynthesis enters plants through open stomata, while water vapor and oxygen exit the plant by the same route. Mechanisms that regulate the

A. Leaf epidermis

Cuticle

Epidermal cell

Parenchyma cell inside leaf

George S. Ellmore

B. Secretory trichomes of *Cannabis sativa*

Antonio Romero/Science Source

C. Root hairs

Root

Root hair

160 µm

Courtesy Mark Holland, Salisbury University

D. Stomata

Cuticle-coated cell
of lower epidermis

Guard cells

One stoma

Dr. Jeremy Burgess/SPL/Science Source

FIGURE 33.9 Structure of epidermal tissue and examples of epidermal specializations.
(A) Cross section of leaf epidermis from a bush lily (*Clivia miniata*). **(B)** Light micrograph of trichomes on a bract (specialized leaf) of marijuana (*Cannabis sativa*). The trichomes secrete various substances including the resinous psychoactive compound THC (tetrahydrocannabinol). In nature, THC deters insect pests. **(C)** Root hairs, trichomes that develop from root epidermis. **(D)** Scanning electron micrograph of a leaf surface showing stomata among cuticle-covered epidermal cells.

opening and closing of stomata (discussed in Chapter 35) are essential to maintaining an appropriate balance between the intake of CO_2 and the loss of water vapor from plant tissues.

DISCOVERING HOW GENES CONTROL STOMATA With their vital role in regulating gas exchange between a plant and its environment, stomata have captured the interest of geneticists probing the molecular underpinnings of plant development. Working with *Arabidopsis thaliana* (thale cress), researchers have identified an enzyme—encoded by the gene *YODA*—that appears to exert overall control over where and how many stomata form. In mutant plants with a defective enzyme, the epidermis is blanketed with stomata packed side by side. The plants often die early in development or are stunted and appear fuzzy—hence the enzyme's name, YODA, recalling the short, hairy Star Wars character. In nonmutated wild-type plants, unequal divisions of precursor cells produce one smaller and one larger daughter cell, and the smaller one gives rise to a stoma's two guard cells. (The larger cell either divides again or becomes an underlying epidermal cell.) When the YODA

enzyme is activated (by phosphorylation), it triggers a cascade of reactions that, by some as-yet-unknown mechanism, either promote or restrict these asymmetric divisions.

Many plants also can apparently modify their stomata in response to changing levels of carbon dioxide in the environment. A flurry of research on this topic began when plant physiologist F. Ian Woodward of Sheffield University hypothesized that a rising concentration of CO_2 in the atmosphere—a hallmark of global climate change—acts as a trigger for the development of fewer stomata in leaves, because under such conditions plants would not require as many stomata to obtain adequate CO_2. When Woodward examined herbarium specimens collected over a recent 200-year period during which atmospheric CO_2 has steadily increased, the number of stomata in leaves of the test specimens had indeed declined. Subsequent research has confirmed that plants can reduce or increase the number of stomata in response to external CO_2 levels. Apparently, a biochemical signal from a plant's mature leaves influences the development of new leaves, adjusting the number of stomata to meet demand. Experiments have identified at least

A. Taproot system

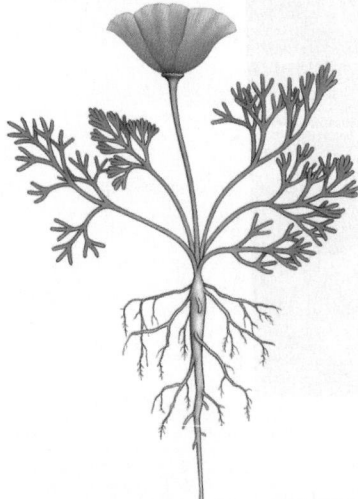

B. Fibrous root system

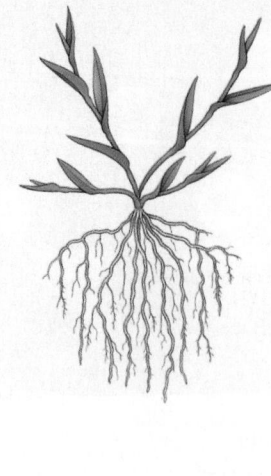

C. Prop roots

Surakit/Shutterstock.com

D. Buttress roots

Hugh Lansdown/Shutterstock.com

E. Adventitious roots

Beth Davidow/Visuals Unlimited, Inc.

F. A storage root

KPG_Payless/Shutterstock.com

FIGURE 33.10 Types of roots. (A) Taproot system of a California poppy (*Eschscholzia californica*). **(B)** Fibrous root system of a grass plant. **(C)** Prop roots of a corn plant. **(D)** A tropical rainforest tree supported by flaring buttress roots. **(E)** The prop roots of mangrove trees *(Rhizophora),* examples of adventitious roots. Sweet potatoes are storage roots **(F).**

© Cengage Learning 2017

one gene that has a central role in these events, and more studies are underway to elucidate the control pathways.

STUDY BREAK 33.2

1. Describe the defining features, cellular components, and functions of the ground tissue system.
2. What are the functions of the vascular tissues xylem and phloem?
3. What are the cellular components and functions of the dermal tissue system?

33.3 Root Systems

Plants must absorb enough water and dissolved minerals to sustain growth and routine cellular maintenance, a task that requires a tremendous root surface. In one study, measurements of the root system of a rye plant *(Secale cereale)* that had been growing for only four months had a surface area of more than 700 m²—about 130 times greater than the surface area of its shoot system. The roots of carrots, sugar beets, and most other plants also store nutrients produced in photosynthesis, some to be used by root cells and some to be transported later to cells of the shoot. As a root system penetrates downward and spreads out, it also anchors the aboveground parts.

Different Types of Root Systems Are Specialized for Particular Functions

Most eudicots begin life with a **taproot system**—a single main root, or taproot, that is adapted for storage and smaller branching roots called **lateral roots (Figure 33.10A).** As the main root grows downward, its diameter increases, and the lateral roots emerge along the length of its older, differentiated regions. The youngest lateral roots are near the root tip. Carrots and dandelions have a taproot system, as do many conifers and some angiosperm trees.

In grasses and many other monocots, the first, embryonic root (called the *radicle*) soon dies and its role is assumed by **adventitious roots** that arose from the young plant's stem. Eventually a monocot's adventitious roots form a **fibrous root system** in which several main roots branch to form a dense mass of smaller ones **(Figure 33.10B).** Fibrous root systems are adapted to absorb water and nutrients from the upper layers of soil, and tend to spread out laterally from the base of the stem. They are important ecologically because dense root networks help hold topsoil in place and prevent erosion.

Adventitious roots can also arise from leaves or other shoot parts. The *prop roots* of a corn plant **(Figure 33.10C)** are adventitious roots that develop from the shoot node nearest the soil surface; they both support the plant and absorb water and nutrients. *Buttress roots* of many rainforest trees are variations on this theme **(Figure 33.10D).**

As this discussion makes clear, roots are extremely diverse. Mangroves and other trees that grow in marshy habitats often have aboveground roots called *pneumatophores* that develop from the main root system and aerate it (**Figure 33.10E**). A sweet potato (*Ipomoea batatas*) (**Figure 33.10F**) is an example of a *storage root* (also called a *root tuber*), a lateral root that stockpiles nutrients that can (in theory) sustain new shoot parts during a subsequent growing season.

Root Structure Is Specialized for Underground Growth

Roots have distinct anatomical parts, each with a specific function. In most plants, primary growth of roots begins when an embryonic root emerges from a germinating seed and its meristems become active. **Figure 33.11** shows the structure of a root tip. Notice that the root apical meristem terminates in a dome-shaped cell mass, the **root cap.** The meristem produces the cap, which in turn surrounds and protects the meristem as the root elongates through the soil. Certain cells in the cap respond to gravity, which guides the root tip downward. Cap cells also secrete a polysaccharide-rich substance that lubricates the tip and eases the growing roots' passage through the soil. Outer root cap cells are continually abraded off and replaced by new cells at the cap's base.

ZONES OF PRIMARY GROWTH IN THE ROOT Primary growth in roots takes place in successive stages, beginning at the root tip and progressing upward. Just inside the root cap some roots have a small clump of apical meristem cells called the **quiescent center.** Unlike other meristematic cells, cells of the quiescent center divide very slowly unless the root cap or the apical meristem is injured, then they become active and can regenerate the damaged part. The quiescent center also may include cells that synthesize plant hormones that control root development.

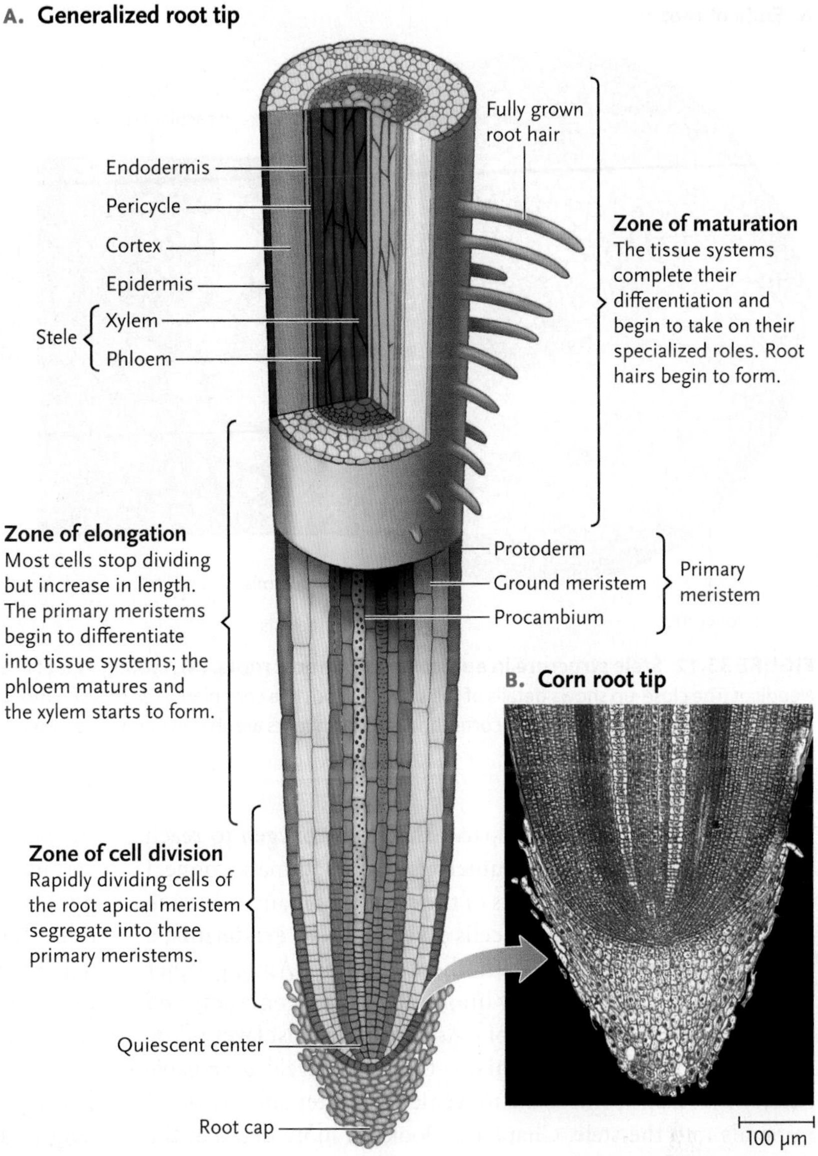

The root apical meristem and the actively dividing cells behind it form the **zone of cell division.** Cells of the root apical meristem segregate into three primary meristems. Cells in the center of the root tip become the procambium; those just outside the procambium become ground meristem; and those on the periphery of the apical meristem become protoderm. As described shortly, primary growth of the stem also includes the formation of procambium, ground meristem, and protoderm.

The zone of cell division merges into the **zone of elongation.** Most of the increase in a root's length comes about here as cells become longer as their vacuoles fill with water. This "hydraulic" elongation pushes the root cap and apical meristem through the soil as much as several centimeters a day.

Above the zone of elongation, cells do not increase in length but they may differentiate further and take on

A. Generalized root tip

Endodermis
Pericycle
Cortex
Epidermis
Stele { Xylem
Phloem

Fully grown root hair

Zone of maturation
The tissue systems complete their differentiation and begin to take on their specialized roles. Root hairs begin to form.

Zone of elongation
Most cells stop dividing but increase in length. The primary meristems begin to differentiate into tissue systems; the phloem matures and the xylem starts to form.

Protoderm
Ground meristem
Procambium
} Primary meristem

B. Corn root tip

Zone of cell division
Rapidly dividing cells of the root apical meristem segregate into three primary meristems.

Quiescent center

100 µm

Biodisc/Visuals Unlimited, Inc.

Root cap

FIGURE 33.11 Tissues and zones of primary growth in a root tip. (A) Generalized root tip, longitudinal section. **(B)** Micrograph of a corn root tip, longitudinal section.
© Cengage Learning 2017

specialized roles in the **zone of maturation.** For instance, epidermal cells in this zone give rise to root hairs, and the procambium, ground meristem, and protoderm complete their differentiation here.

TISSUES OF THE ROOT SYSTEM Coupled with primary growth of the shoot, primary root growth produces a unified system of vascular pipelines extending from root tip to shoot tip. The root procambium produces cells that mature into the root's xylem and phloem (**Figure 33.12**). Ground meristem gives rise to the root's cortex, its ground tissue of starch-storing parenchyma cells that surround the stele. In eudicots, the stele runs through the center of the root (see Figure 33.12A). In corn and some other monocots, the stele forms a ring that divides the ground tissue into cortex and pith (see Figure 33.12B).

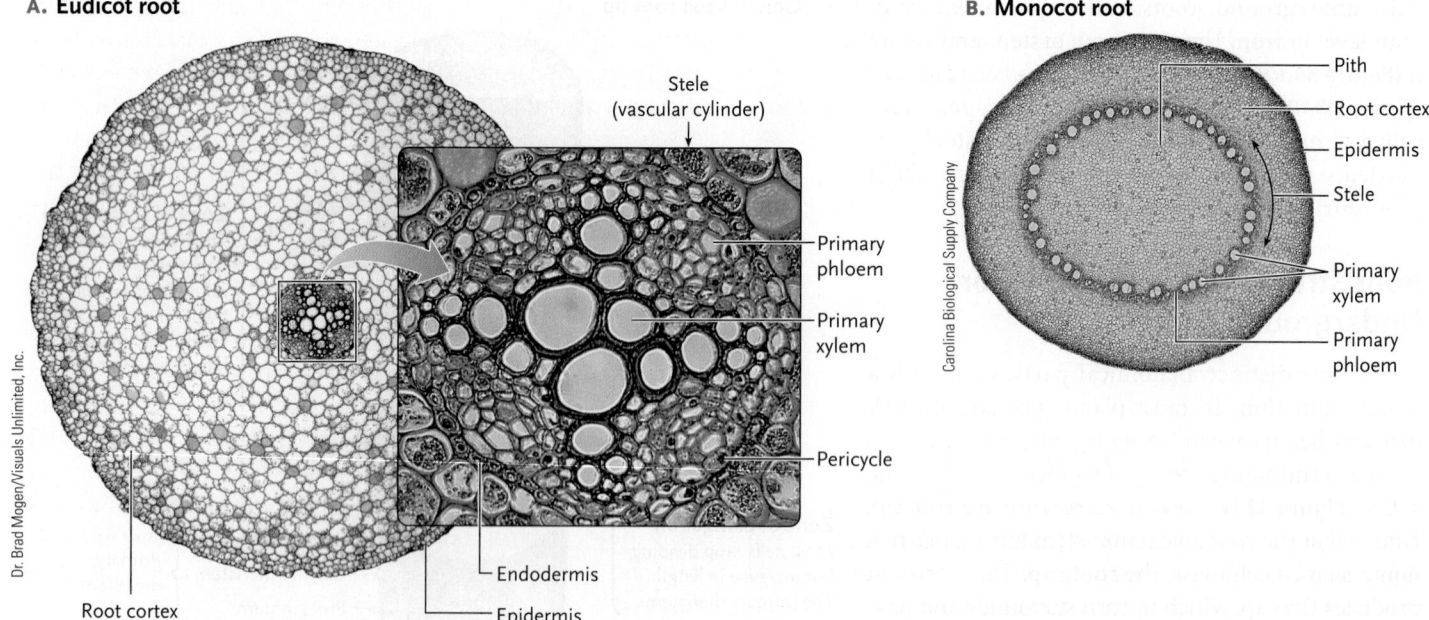

A. Eudicot root

Stele
(vascular cylinder)

Primary
phloem

Primary
xylem

Pericycle

Endodermis

Epidermis

Root cortex

Dr. Brad Mogen/Visuals Unlimited, Inc.

B. Monocot root

Pith

Root cortex

Epidermis

Stele

Primary
xylem

Primary
phloem

Carolina Biological Supply Company

FIGURE 33.12 Stele structure in eudicot and monocot roots. (A) A young root of the buttercup *(Ranunculus),* a eudicot. The close-up shows details of the stele. **(B)** Root of a corn plant *(Zea mays),* a monocot. Notice how the stele divides the ground tissue into cortex and pith. Both roots are shown in transverse section.

© Cengage Learning 2017

The cortex contains air spaces that allow oxygen to reach all of the living root cells. Numerous plasmodesmata connect the cytoplasm of adjacent cells of the cortex. In many flowering plants, the outer root cortex cells give rise to an **exodermis,** a narrow band of cells beneath the epidermis. Among other functions, the exodermis may limit water losses from roots and help regulate the absorption of ions. The innermost layer of the root cortex is the **endodermis,** a thin, selectively permeable barrier that helps control the movement of water and dissolved minerals into the stele. Chapter 34 looks in more detail at the roles of exodermis and endodermis.

Between the stele and the endodermis is the **pericycle,** consisting of one or more layers of parenchyma cells that can still function as meristem. The pericycle gives rise to lateral roots **(Figure 33.13).** In response to chemical growth regulators, rudimentary roots, or **root primordia,** arise at specific sites in

the pericycle. Gradually, the lateral roots emerge and grow out through the cortex and epidermis, aided by enzymes released by the root primordium that help break down the intervening cells. The distribution and frequency of lateral root formation partly control the overall shape of the root system—and the extent of the soil area it can penetrate.

In some cells in the developing root epidermis, the outer surface becomes extended into the trichomes called root hairs (see Figure 33.9C). Root hairs can be more than a centimeter long and can form in less than a day. Collectively, the thousands or millions of them on a plant's roots greatly increase the plant's absorptive surface. Root hair structure supports this essential function. Each hair is a slender tube with thin walls made sticky on their surface by a coating of pectin. Soil particles tend to adhere to the wall, providing an intimate association between the hair and the surrounding earth, thus

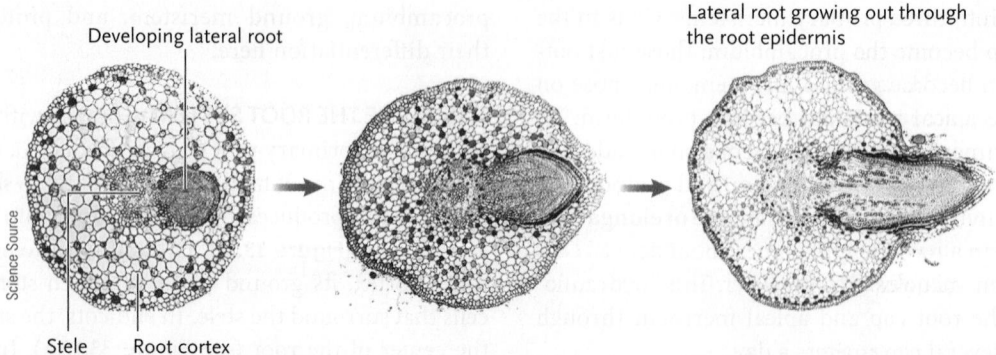

Developing lateral root

Lateral root growing out through
the root epidermis

Stele   Root cortex

Science Source

FIGURE 33.13 Micrographs showing the formation of a lateral root from the pericycle of a willow tree *(Salix).* These micrographs show transverse sections.

facilitating the uptake of water molecules and mineral ions from soil. When plants are transplanted, rough handling can tear off much of the fragile absorptive surface. Unable to take up enough water and minerals, the transplant may die before new root hairs can form.

STUDY BREAK 33.3

1. Compare the two general types of root systems.
2. Describe the zones of primary growth in roots.
3. Describe the various tissues that arise in a root system and their functions.

33.4 Primary Shoot Systems

A flowering plant's primary shoot system consists of the main stem and leaves. In this section we take a closer look at how the constituent parts of the primary shoot system are organized and how they grow and function. Chapter 36 discusses the positional cues that guide key aspects of the plant body's growth.

Stems Are Adapted to Provide Support, Routes for Vascular Tissues, Storage, and New Growth

Stems incorporate evolutionary adaptations that serve four main functions. First, stems provide mechanical support, generally along a vertical axis, for body parts such as leaves and flowers. Second, stems house the vascular tissues (xylem and phloem), which transport water and dissolved minerals, hormones, products of photosynthesis, and other substances throughout the plant. Third, stems often have modifications for storing water and food. And finally, buds and specific stem regions contain meristem tissue that gives rise to new cells of the shoot.

THE MODULAR ORGANIZATION OF A STEM As Section 33.1 noted, we can think of a stem as a set of modules. **Figure 33.14A** shows the two basic elements of each stem module, a *node* and an *internode*. A **node** is where one or more leaves attach to a stem. An **internode** is the area between two nodes. The upper angle between the stem and an attached leaf is an **axil.** New primary growth occurs in the embryonic shoots called buds. A **terminal bud** occurs at the apex of each stem **(Figure 33.14B). Lateral buds** in the leaf axils produce new branches or shoots that give rise to flowers.

PATTERNS OF STEM GROWTH Several general patterns of growth evolved in flowering plants. In eudicots, most growth in a stem's length occurs directly below the apical meristem, as internode cells divide and elongate. Internode cells nearest the apex are most active, so the most visible new growth occurs at the ends of stems. By contrast, in grasses and some other monocots the upper cells of an internode stop dividing as the newly formed internode elongates, and cell division is limited to a meristematic region at the base of the internode. The stems of bamboo and other grasses elongate as the growth of such meristems "pushes up" the internodes. This adaptation allows grasses to rapidly grow back after grazing by herbivores (or being chopped off by a lawnmower), because the meristem is not removed.

Terminal buds release a hormone that inhibits the growth of nearby lateral buds, a phenomenon called **apical dominance.** Gardeners who want a bushier plant can stimulate lateral bud growth by periodically cutting off the terminal bud. The flow of hormone signals from the apical meristem then dwindles to a level low enough that lateral buds begin to grow. In nature, apical dominance is an adaptation that directs the plant's resources into growing up toward the light.

PRIMARY GROWTH AND STRUCTURE OF A STEM Primary growth of a shoot and its parts begins in the shoot apical meristem. In addition to giving rise to stem tissue, the shoot apical meristem also periodically generates shoot primordia, bulges that are the first developmental stages of leaves, additional shoots, and reproductive structures such as flowers.

A shoot apical meristem is a dome-shaped mass of cells. When one cell divides, one of its daughter cells becomes an **initial** that remains as part of the meristem. The other daughter cell, called a **derivative,** will be the source of specialized cells. In this way initials function much like stem cells in animals. They replenish the supply of initials in the meristem and also provide derivatives that result in a plant's new growth.

A. Location of nodes and buds

Terminal bud

Apical meristems

Lateral bud in axil

Node

Internode

Node

B. Leaves at a terminal bud

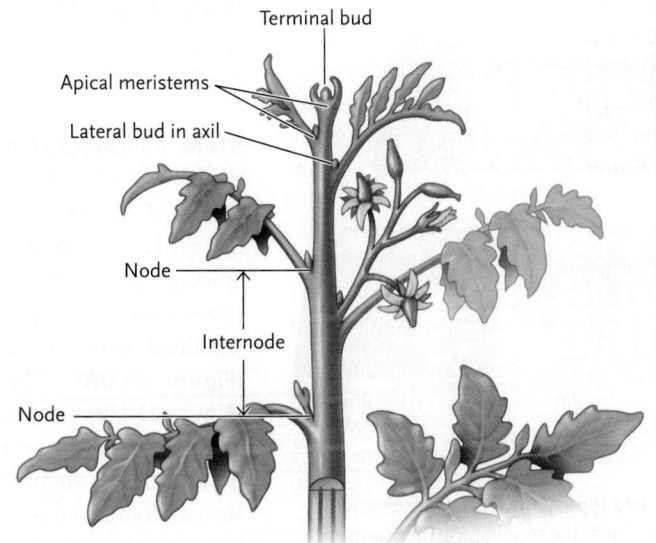

Gary Meszaros/Science Source

FIGURE 33.14 Modular structure of a stem. (A) The arrangement of nodes and buds on a plant stem. **(B)** Formation of leaves at a terminal bud of a dogwood *(Cornus).*
© Cengage Learning 2017

As derivatives differentiate, they give rise to three **primary meristems:** protoderm, procambium, and ground meristem **(Figure 33.15).** These meristems are relatively unspecialized tissues producing cells that in turn differentiate into specialized cells and tissues. In eudicots, the primary meristems are also responsible for elongation of the plant body. Some monocots, such as palms, have a *primary thickening meristem* just under the protoderm that contributes to both stem elongation and its lateral growth.

How do the genetically identical cells of an apical meristem give rise to three types of primary meristem cells, and ultimately to all the specialized cells of the plant? *Focus on*

A. Shoot tip: SEM

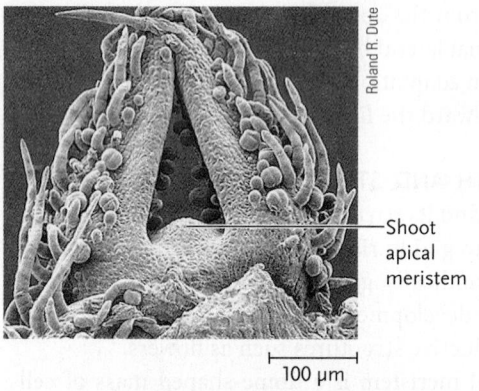

Shoot apical meristem

100 μm

B. Successive stages in primary growth

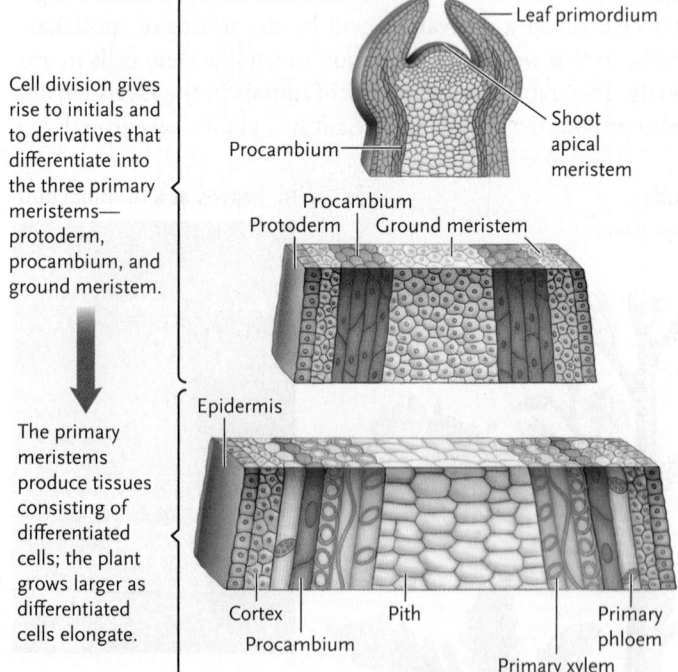

Cell division gives rise to initials and to derivatives that differentiate into the three primary meristems— protoderm, procambium, and ground meristem.

The primary meristems produce tissues consisting of differentiated cells; the plant grows larger as differentiated cells elongate.

Leaf primordium

Procambium

Shoot apical meristem

Procambium

Protoderm   Ground meristem

Epidermis

Cortex   Pith   Primary phloem

Procambium   Primary xylem

FIGURE 33.15 Primary growth in a typical eudicot. (A) Successive stages in primary growth: activity begins at the shoot apical meristem and continues at the primary meristems derived from it. Notice the progressive differentiation of most of the tissue regions. **(B)** Scanning electron micrograph of its surface.

© Cengage Learning 2017

Research: Basic Research describes some experiments that are probing the genetic mechanisms underlying meristem activity.

Each primary meristem occupies a different position in the shoot tip, as shown in Figure 33.15. Outermost is **protoderm,** a meristem that will produce the stem's epidermis. While protoderm cells divide and the resulting derivatives are maturing, the shoot tip continues to grow. Eventually, the protoderm cells differentiate into specific types of epidermal cells, including guard cells and trichomes.

Inward from the protoderm is the **ground meristem,** which will give rise to ground tissue, most of which is parenchyma. **Procambium,** which produces the primary vascular tissues, exists as threadlike strands of cells within the ground meristem. Procambial cells are long and thin, and their spatial orientation foreshadows the future function of the tissues they produce. In most plants, inner procambial cells give rise to xylem and outer procambial cells to phloem. In plants with secondary growth, a thin region of procambium between the primary xylem and phloem remains undifferentiated. As described shortly, later on it will give rise to a lateral meristem.

The developing vascular tissues become organized into **vascular bundles,** multistranded cords of primary xylem and phloem. Wrapped or capped by sclerenchyma, the bundles thread lengthwise through the parenchyma, forming a **stele** (Greek *stele* = pillar; some botanists prefer to call vascular bundles in the shoot the *vascular cylinder*). The stele runs vertically. The ground tissue outside it forms a **cortex,** while tissue inside it is called **pith (Figure 33.16A).** As leaves and buds appear along a stem, some vascular bundles in the stem branch off into these developing tissues.

Both cortex and pith are mainly parenchyma, and in some plants the pith parenchyma stores starch. In the stems of most monocots, vascular bundles are dispersed through the ground tissue **(Figure 33.16B),** so separate cortical and pith regions do not form. In some monocots, including bamboo, the pith breaks down, leaving the stem with a hollow core. The hollow stems of some hard-walled bamboo species are used to make bamboo flutes.

STEM MODIFICATIONS Millennia of adaptation have produced a range of stem specializations, including structures modified for reproduction, storage of food or water, or both. *Stem tubers* are stem regions enlarged by the presence of starch-storing parenchyma cells. The white potato (*Solanum tuberosum*) is a stem tuber; its "eyes" are buds at nodes of the modified stem, and the regions between eyes are internodes **(Figure 33.17A).** The pungent, starchy "root" of ginger **(Figure 33.17B)** is a horizontal modified stem called a *rhizome.* Gladiolas and some other ornamental plants develop fleshy underground stems called *corms* **(Figure 33.17C),** another starch-storage adaptation. Tubers, rhizomes, and corms all have meristematic tissue at nodes from which new plants can be propagated—a vegetative (asexual) reproductive mode. Other plants, including the strawberry, reproduce vegetatively via slender stems called *stolons,* which grow along the soil

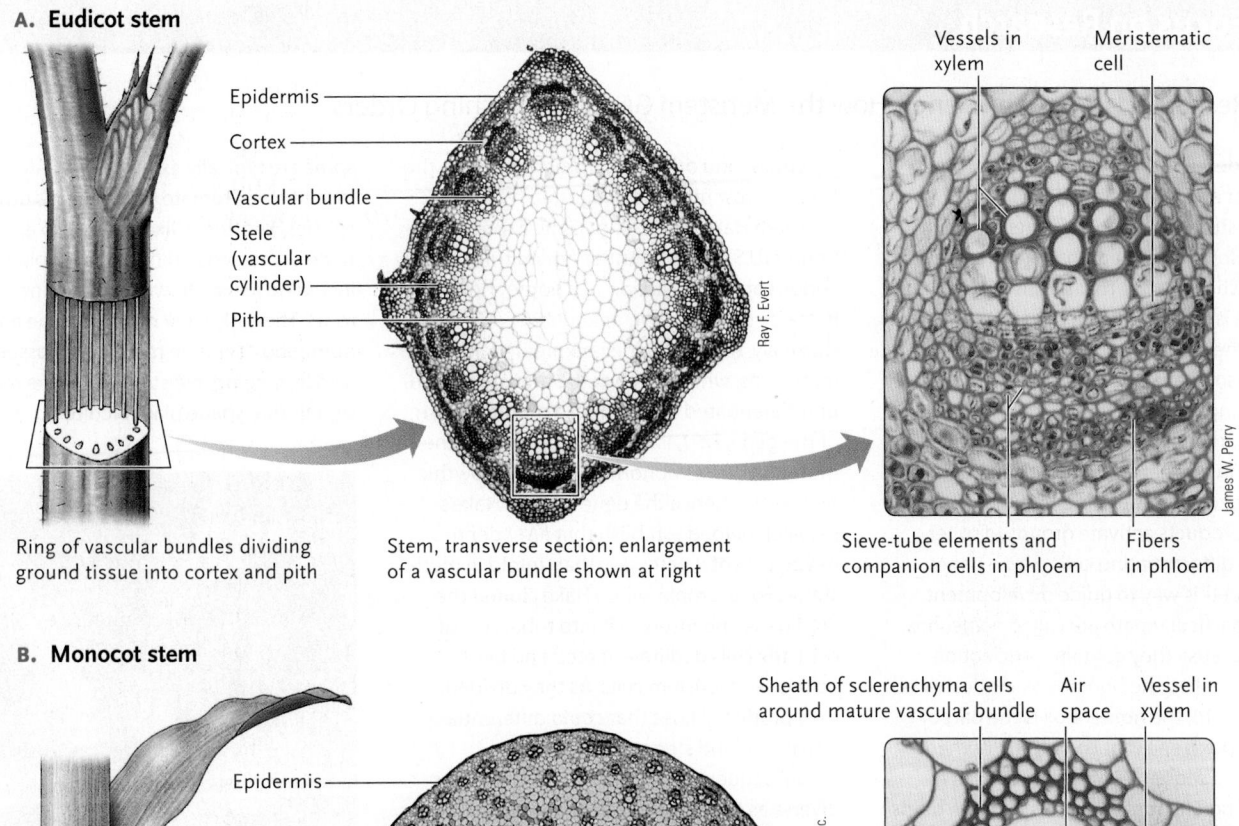

A. Eudicot stem

Epidermis

Cortex

Vascular bundle

Stele (vascular cylinder)

Pith

Ray F. Evert

Ring of vascular bundles dividing ground tissue into cortex and pith

Stem, transverse section; enlargement of a vascular bundle shown at right

Vessels in xylem

Meristematic cell

Sieve-tube elements and companion cells in phloem

Fibers in phloem

James W. Perry

B. Monocot stem

Epidermis

Vascular bundle

Ground tissue

Carolina Biological Supply, Co/Visuals Unlimited, Inc.

Sheath of sclerenchyma cells around mature vascular bundle

Air space

Vessel in xylem

Vascular bundles distributed throughout ground tissue

Stem, transverse section; enlargement of a vascular bundle shown at right

Sieve-tube element in phloem

Companion cell in phloem

James W. Perry

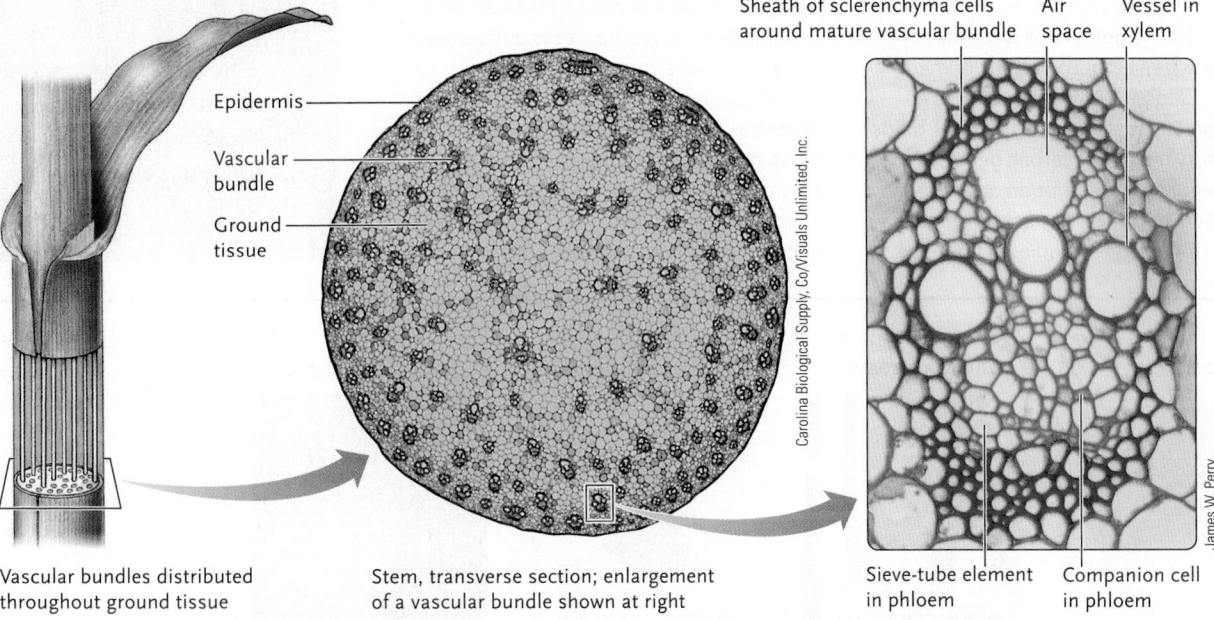

FIGURE 33.16 Organization of cells and tissues inside the stem of a eudicot and a monocot. (A) Part of a stem from alfalfa *(Medicago),* a eudicot. In many species of eudicots and conifers, the vascular bundles develop in a more or less ringlike array in the ground tissue system, as shown here. **(B)** Part of a stem from corn *(Z. mays),* a monocot. In most monocots and some herbaceous eudicots, vascular bundles are scattered through the ground tissue, as shown here.

© Cengage Learning 2017

surface **(Figure 33.17D).** New plants arise at alternating nodes along the stolon.

Other stem modifications are defensive or enhance a plant's capacity to survive environmental stress. Two examples are **thorns (Figure 33.17E),** which are stems modified to deter predation by herbivores, and the fleshy, water-storing stems of cacti **(Figure 33.17F),** which help sustain the plant during extended dry periods in the arid environments in which most cacti live.

Leaves Are Adapted for Photosynthesis, Gas Exchange, and Conserving Water

Each spring a mature maple tree heralds the new season by unfurling roughly 100,000 leaves. Some other tree species produce leaves by the millions. For these and most other plants, leaves are multifaceted organs with evolutionary adaptations that make them the main sites of photosynthesis and permit them to exchange gases and limit the loss of water by evaporation.

Basic Research: Homeobox Genes: How the Meristem Gives Its Marching Orders

How do descendants of some dividing cells in a shoot apical meristem (SAM) "know" to become stem tissues, while others embark on the developmental path that produces leaves or other shoot parts? The full answer to this question is not yet known, but research teams around the world are studying a genetic mechanism that appears to guide the process.

Working with SAM tissue from maize (*Z. mays,* generally known in North America as corn), investigators have identified more than a dozen regulatory genes whose protein products activate groups of other genes in differentiating cells. Some genes that act in this way to guide development along a particular path are called *homeobox genes,* because they contain a nucleotide sequence called the homeobox. As described in Chapter 16, the homeobox is the part of a homeobox gene that codes for a homeo-domain—a sequence of amino acids in a transcription factor. The homeodomain binds to regulatory regions of certain genes; this binding regulates transcription of such genes, turning them on or off. Homeobox genes were first discovered in studies of how legs, antennae, and other structures develop in the fruit fly *Drosophila melanogaster.*

Sarah Hake of the Plant Gene Expression Center (U.S. Department of Agriculture) was curious about the action of a homeobox gene in maize that is known as *knotted-1 (KN-1).* Normally, the *KN-1* gene is expressed in apical meristems, where it keeps the meristem in an undifferentiated state. When a mutated form of the gene, *kn-1,* is expressed, however, the mutation causes abnormal knobby growths on leaves—hence the gene's name. Hake's research helped establish that *KN-1* defines developmental pathways that unfold in meristems. For example, when Hake cloned the *KN-1* gene and inserted it into tobacco leaf cells, the cells dedifferentiated and began acting like meristem cells. As they divided, they produced lines that could differentiate into leaves and stems.

Subsequent studies of *KN-1* in species as diverse as sunflowers and garden peas have led to the identification of the family of what are now called *knotted-1-like* genes, all of which encode regulatory proteins that influence developmental pathways. As in maize, some are typically expressed in SAM tissue. In sunflower, tomato, and perhaps other species, knotted-1-like genes also appear to be expressed in differentiated plant parts including leaves, flowers, stems, and even roots. The early work on SAM tissue and homeobox genes in maize has blossomed into a wide-ranging investigation of the molecular signals that shape plant architecture.

© Cengage Learning 2017

A. Potato tuber

B. Ginger rhizome

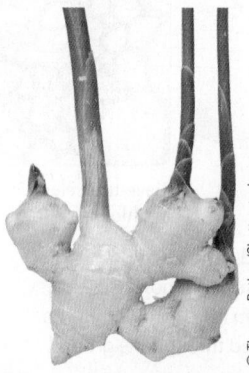

C. Crocus corm

D. Strawberry stolons

E. Honey locust thorns

F. Cactus stems

FIGURE 33.17 A selection of modified stems. (A) A potato (*S. tuberosum*), a tuber. **(B)** Ginger "root," the pungent, starchy rhizome of the ginger plant (*Zingiber officinale*). **(C)** Crocus plants (genus *Crocus*) typically grow from a corm. **(D)** A strawberry plant (*Fragaria ananassa*) and stolon. **(E)** Thorns of honey locust (*Gleditsia triacanthos*). **(F)** Stems of the prickly pear cactus (*Opuntia*).

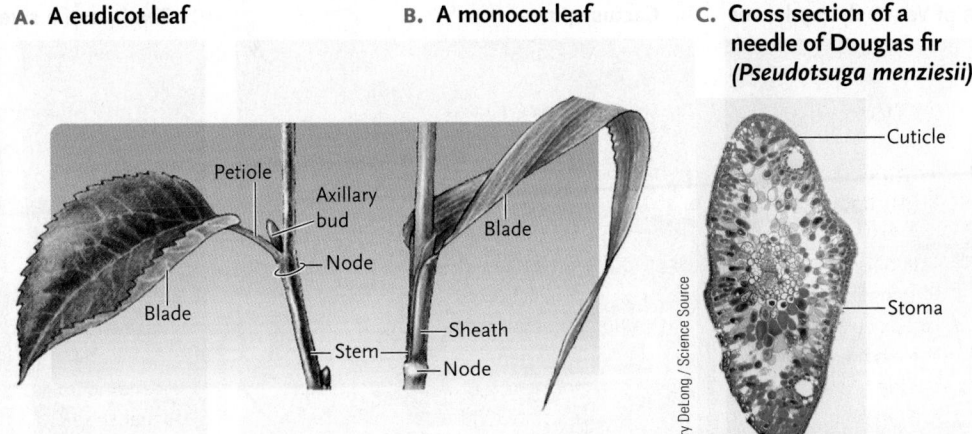

A. A eudicot leaf

Petiole
Axillary bud
Node
Blade
Blade

B. A monocot leaf

Blade
Sheath
Stem
Node

C. Cross section of a needle of Douglas fir (*Pseudotsuga menziesii*)

Cuticle
Stoma

Garry DeLong / Science Source

FIGURE 33.18 Common forms of leaves. Angiosperm leaves typically have thin, flattened blades **(A, B).** Needlelike conifer leaves are thick and slender and have markedly sunken stomata **(C).**
© Cengage Learning 2017

BASIC LEAF ANATOMY Externally, the most conspicuous part of an angiosperm leaf is its thin, flattened **blade,** which provides a large surface area for absorbing sunlight and carbon dioxide. In many eudicot leaves, the blade narrows to a stalklike **petiole,** which attaches the leaf to a stem **(Figure 33.18A).** Petioles can be long, short, or in between, depending on the species. (The stalks of celery and rhubarb are especially long, fleshy petioles.) Unless a leaf's petiole is very short, it holds the leaf away from the stem and helps prevent individual leaves from shading one another. Most monocot leaves lack a petiole **(Figure 33.18B).** Instead, in species such as rye or corn, the blade is long and narrow and its base simply forms a sheath around the stem.

In contrast to the leaves of angiosperms, the leaves of conifers are scalelike or needlelike and have sunken stomata and a thickened cuticle **(Figure 33.18C).** These features are adaptations to the seasonally arid environments where conifers typically grow.

In general, leaves of flowering plants are oriented on the stem axis so that they can capture the maximum amount of sunlight. Chapter 37 describes how the stems and leaves of some plants change position to follow the Sun's movement during the course of a day.

LEAF SHAPES AND MODIFICATIONS Angiosperm leaves are either *simple* or *compound.* Simple leaves have a single blade **(Figure 33.19A),** while in compound leaves the blade is divided into smaller leaflets **(Figure 33.19B).** Leaf edges or margins may be smooth, toothed, or lobed. As with other plant parts, however, the adaptation of land plants to different environments has produced tremendous variety in leaf morphology. For instance, carnivorous plants such as the Venus flytrap (*Dionaea muscipula*) typically live in nutrient-poor soils and have leaves adapted to trap and digest insects **(Figure 33.20A).** In

cacti, photosynthesis occurs in the green, water-storing stems, and leaves are modified as spines that deter predators **(Figure 33.20B).** Leaves or parts of leaves also may be modified into tendrils, like those of the sweet pea **(Figure 33.20C),** or *bracts* that mimic flower petals **(Figure 33.20D).** Epidermal cells on the leaves of the saltbush *Atriplex spongiosa* form balloon-like trichomes **(Figure 33.20E)** that contain concentrated Na^+ and Cl^- taken up from the salty soil. Eventually, the salt-filled epidermal cells burst or fall off the leaf, releasing the salt to the

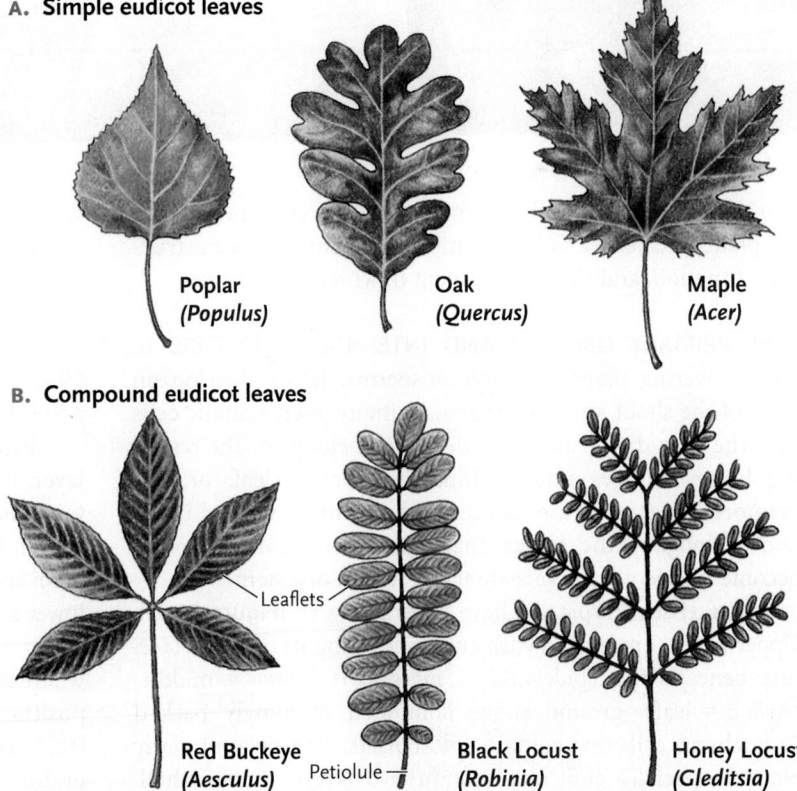

A. Simple eudicot leaves

Poplar (*Populus*)
Oak (*Quercus*)
Maple (*Acer*)

B. Compound eudicot leaves

Leaflets
Red Buckeye (*Aesculus*)
Petiolule
Black Locust (*Robinia*)
Honey Locust (*Gleditsia*)

FIGURE 33.19 Eudicot leaves. (A) Examples of simple eudicot leaves. **(B)** Examples of compound eudicot leaves.
© Cengage Learning 2017

A. Interlocking spines of Venus flytrap leaves

B. Cactus spines

C. Tendrils of a sweet pea

D. *Bougainvillea* **bracts**

E. Salt bladder of a saltbush plant

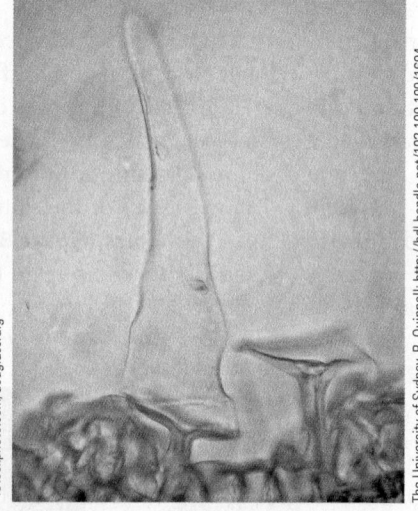

FIGURE 33.20 A few adaptations of leaves.
(A) Margins on the leaves of the Venus flytrap *(D. muscipula)* are modified into interlocking, insect-trapping spines. **(B)** Spines of a Graham dog cactus *(Opuntia)* are leaves modified to deter predation. **(C)** The tendrils of a sweet pea *(Lathyrus odoratus)* help to support the climbing plant's stem. **(D)** Brightly colored bracts of bougainvillea *(Bougainvillea* sp.*)* mimic flower petals, helping attract pollinators to the tiny flower. **(E)** Light micrograph of salt bladders that appear on the leaves of a saltbush plant *(A. spongiosa)*. The "bladders" are trichomes, specialized outgrowths of the leaf epidermis in which excess salt from the plant's tissue fluid accumulates. The salt-laden trichomes eventually burst or slough off.

outside. This adaptation helps control the salt concentration in the plant's tissues—another example of the link between structure, function, and the environment in which a plant lives.

LEAF PRIMARY GROWTH AND INTERNAL STRUCTURE In both flowering plants and gymnosperms, leaves develop on sides of the shoot apical meristem. Initially, meristematic cells near the apex divide and their derivatives elongate. The resulting bulge enlarges into a thin, rudimentary leaf, or **leaf primordium** (see Figure 33.15). As the plant grows and internodes elongate, the leaves that form from leaf primordia become spaced at intervals along the length of a stem.

Leaf tissues typically have several layers **(Figure 33.21)**. Uppermost is epidermis, with cuticle covering its outer surface. Just beneath the epidermis is **mesophyll** (*mesos* = middle; *phyllon* = leaf), ground tissue composed of loosely packed parenchyma cells that contain chloroplasts. The leaves of many plants, especially eudicots, contain two layers of mesophyll. *Palisade mesophyll* cells contain more chloroplasts and are arranged in compact columns with smaller air spaces between them, typically toward the upper leaf surface. *Spongy mesophyll,*

which tends to be located toward the underside of a leaf, consists of irregularly arranged cells with a conspicuous network of air spaces that gives it a spongy appearance. Air spaces between mesophyll cells enhance the uptake of carbon dioxide and release of oxygen during photosynthesis and account for 15% to 50% of a leaf's volume. Mesophyll also contains collenchyma and sclerenchyma cells, which support the leaf blade.

Below the mesophyll is another cuticle-covered epidermal layer. Except in grasses and a few other plants, this layer contains most of the stomata through which water vapor exits the leaf and carbon dioxide enters. For example, the upper surface of an apple leaf has no stomata, while a square centimeter of the lower surface has more than 20,000. A square centimeter of the upper epidermis of a tomato leaf has about 1,200 stomata, whereas the same area of the lower epidermis has 13,000. The positioning of stomata on the side of the leaf that faces away from the Sun may be an adaptation limiting water loss by evaporation through stomatal openings.

Vascular bundles form a lacy network of **veins** throughout the leaf. Eudicot leaves typically have a branching vein pattern; in monocot leaves, veins tend to run in parallel arrays.

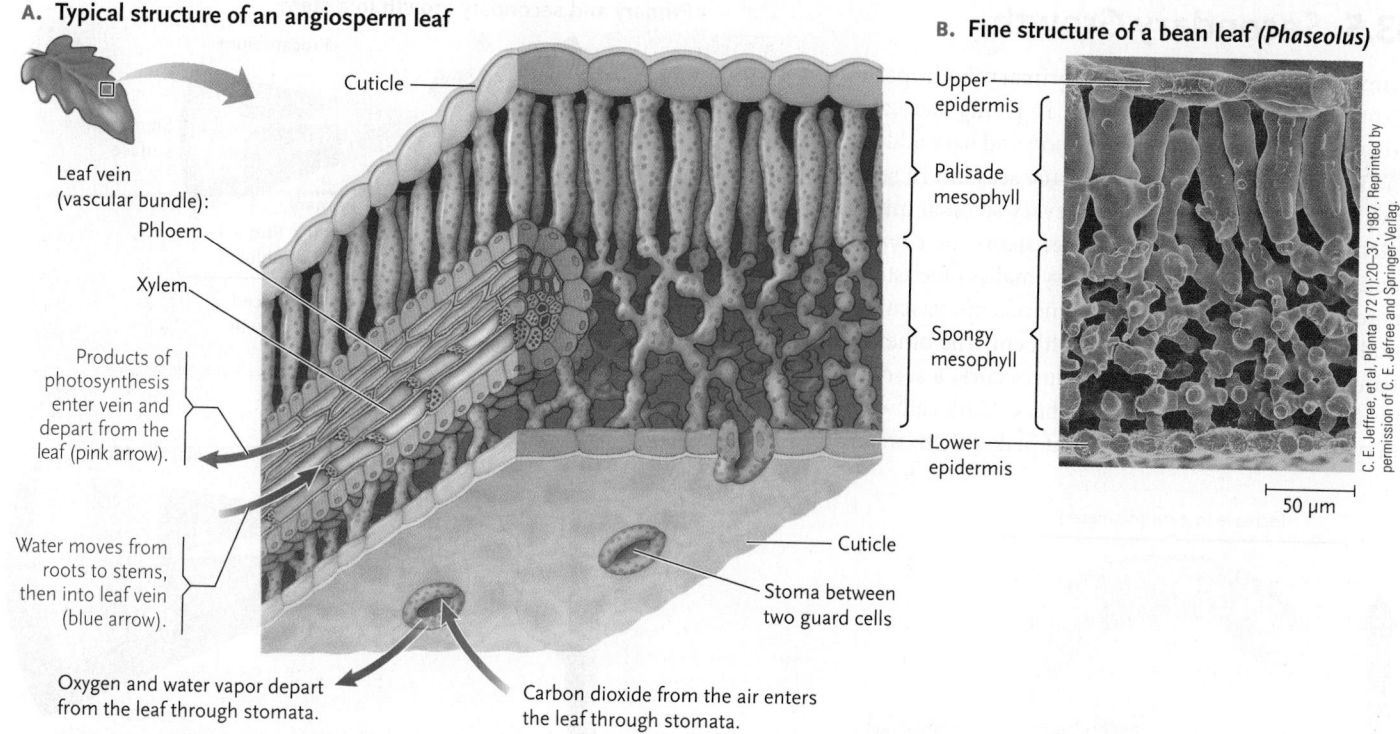

A. Typical structure of an angiosperm leaf

Cuticle

Leaf vein (vascular bundle):

Phloem

Xylem

Products of photosynthesis enter vein and depart from the leaf (pink arrow).

Water moves from roots to stems, then into leaf vein (blue arrow).

Oxygen and water vapor depart from the leaf through stomata.

Carbon dioxide from the air enters the leaf through stomata.

Cuticle

Stoma between two guard cells

B. Fine structure of a bean leaf (Phaseolus)

Upper epidermis

Palisade mesophyll

Spongy mesophyll

Lower epidermis

50 µm

C. E. Jeffree, et al. Planta 172 (1):20–37, 1987. Reprinted by permission of C. E. Jeffree and Springer-Verlag.

FIGURE 33.21 Internal structure of a leaf. (A) Diagram of typical leaf structure for many kinds of flowering plants. See Figure 33.9D for a scanning electron micrograph of stomata. (B) Scanning electron micrograph of tissue from the leaf of a kidney bean plant (Phaseolus), transverse section. Notice the compact organization of epidermal cells.
© Cengage Learning 2017

In temperate regions, most leaves are temporary structures. In deciduous (*deciduus* = that which falls off) species such as birches and maples, hormonal signals cause the leaves to drop from the stem as days shorten in autumn. Other temperate species, such as camellias or hollies, as well as conifers, also drop leaves, but they appear "evergreen" because the leaves may persist for several years and do not all drop at the same time.

Plant Shoots May Have Juvenile and Adult Forms

Leaf shape and other shoot characteristics can mirror the progress of a long-lived plant through its life cycle. Plants that live many years may spend part of their lives in a juvenile phase, then shift to a mature, or adult phase. The differences between juveniles and adults often are reflected in leaf size and shape, in the arrangement of leaves on the stem, or in a change from vegetative growth to a reproductive stage—or sometimes all three. Plants in the genus *Eucalyptus* display a wide range of such variations. For example, in the yellow gum (*Eucalyptus doratoxylon* or *Eucalyptus erythrocorys*) the leaves of a mature tree are blue-green, long and slender with long petioles, and occur in alternating positions on a stem. By contrast, the leaves of a growing seedling are oval with a pale grayish surface, have little or no petiole, and occur opposite one another on the stem. A Southern magnolia tree (*Magnolia grandiflora*) does not flower until its juvenile phase ends, which can be 20 years or more from the time the *M. grandiflora* seed sprouts. In fact, most woody plants must attain a certain size before their meristem

tissue can respond to the hormonal signals that govern flower development, a topic we consider in Chapter 37.

Phase changes provide more examples of the plasticity that characterizes plant development. They almost certainly are associated with changes in the expression of genes that control the development of stem nodes, leaf and flower buds, and other basic aspects of plant growth.

STUDY BREAK 33.4

1. Describe the functions of stems and stem structure, and list the basic steps in primary growth of stems.
2. Explain the general function of leaves and how leaf anatomy supports this role in eudicots and monocots.
3. Describe the steps in primary growth of a leaf and the structures that result from the process.
4. Describe two examples of the life phases of long-lived plant species.

THINK OUTSIDE THE BOOK

On your own or with classmates, make a list of at least five vegetables and five fruits that have not been discussed in this chapter and that you have eaten or seen in the market. Using library resources or the Internet, identify the source plant's genus and species, then describe which anatomical part of the plant each fruit or vegetable is, and identify which tissue or tissues make up the fruit or vegetable. Then describe each item's function in the plant.

33.5 Secondary Growth

Primary growth produces the primary body parts of a seed plant—its root and stem systems. In plants that have secondary growth, the tissues we know as wood and bark add girth to roots and stems over two or more growing seasons. In such plant species, including trees and shrubs, every so often mitosis is reactivated in two types of lateral meristems, or *cambia* (singular, *cambium*). Over time, their activity makes older stems and roots more massive. One of the lateral meristems, **vascular cambium,** produces secondary xylem and phloem. The other lateral meristem, called **cork cambium,** produces **cork,** a secondary epidermis that is a major ingredient in bark. Cork cell walls contain a waxy substance called *suberin* that makes them waterproof.

Primary and secondary growth in a stem

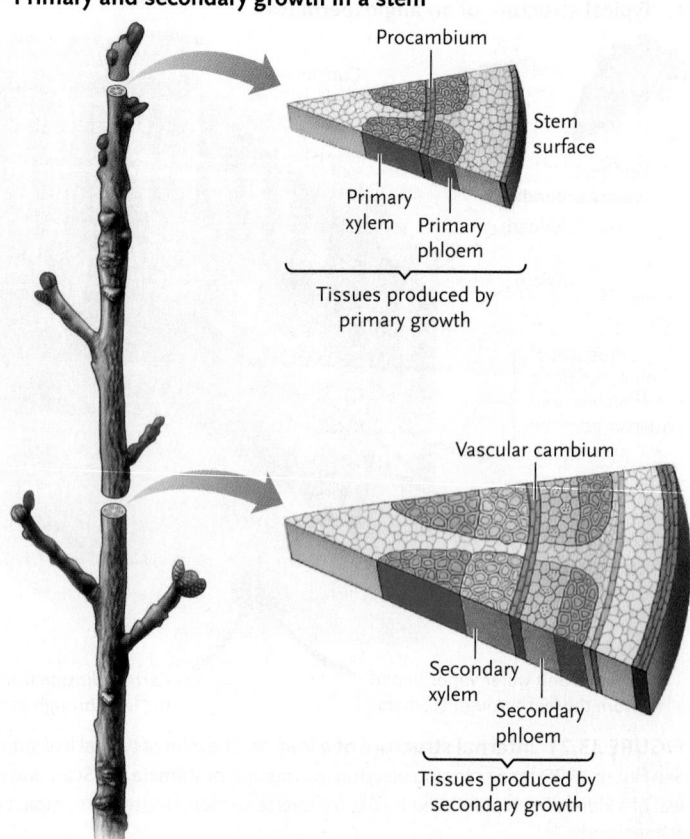

FIGURE 33.22 Secondary and primary growth compared. In a woody plant, primary growth resumes each spring at the terminal and lateral buds. Secondary growth resumes at the vascular cambium inside the stem.
© Cengage Learning 2017

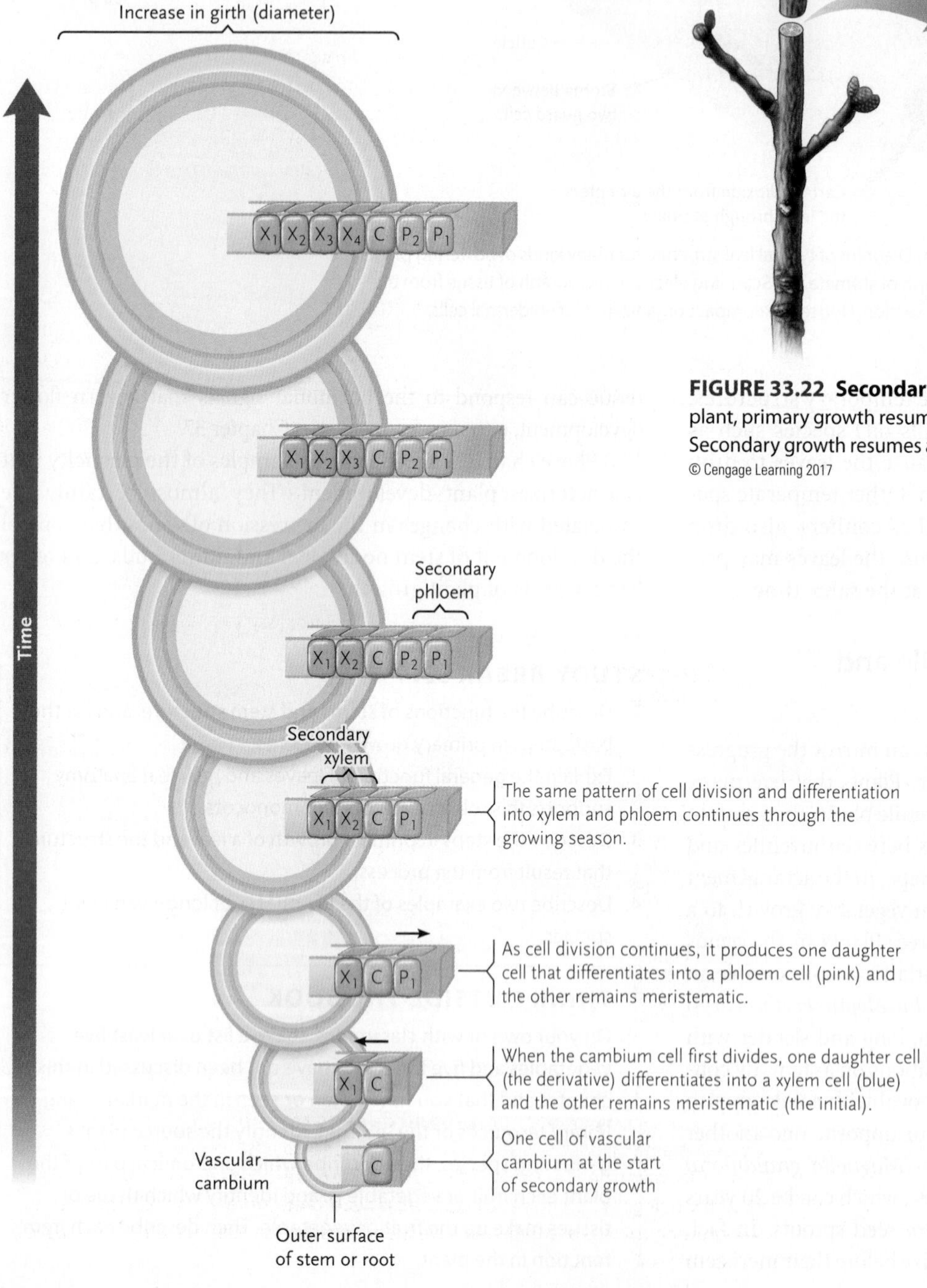

Increase in girth (diameter)

Time

X₁ X₂ X₃ X₄ C P₂ P₁

X₁ X₂ X₃ C P₂ P₁

Secondary phloem

X₁ X₂ C P₂ P₁

Secondary xylem

X₁ X₂ C P₁

The same pattern of cell division and differentiation into xylem and phloem continues through the growing season.

X₁ C P₁

As cell division continues, it produces one daughter cell that differentiates into a phloem cell (pink) and the other remains meristematic.

X₁ C

When the cambium cell first divides, one daughter cell (the derivative) differentiates into a xylem cell (blue) and the other remains meristematic (the initial).

Vascular cambium

C

One cell of vascular cambium at the start of secondary growth

Outer surface of stem or root

FIGURE 33.23 Relationship between the vascular cambium and its derivative cells (secondary xylem and phloem). The drawing shows stem growth through successive seasons. Notice how ongoing divisions displace the cambial cells, moving them steadily outward even as the core of xylem increases the stem or root thickness.
© Cengage Learning 2017

Vascular Cambium Gives Rise to Secondary Growth in Stems

After a woody plant's stem completes its primary growth, each vascular bundle contains a layer of undifferentiated cells between the primary xylem and phloem. These cells, along with parenchyma cells between the bundles, eventually give rise to a cylinder of vascular cambium that encircles the xylem and pith of the stem **(Figure 33.22)**. Vascular cambium consists of two types of initials (which, recall, are the dividing cells in a meristem). Secondary growth takes place mainly as these cells divide in a plane parallel to the vascular cambium, adding girth to the stem. One type of initials gives rise to secondary xylem and secondary phloem cells that mainly conduct fluid vertically through a stem. Secondary xylem forms on the inner face of the vascular cambium and secondary phloem forms on the outer face. Other dividing initials produce *rays*—horizontal columns of parenchyma cells arranged like spokes of a wheel. Ray parenchyma stores carbohydrates or helps move solutes horizontally across the stem. In conifers, rays help transport the sticky resins commonly called *pitch*.

With time, a great deal of secondary xylem forms inside the ring of vascular cambium. This hard secondary xylem is **wood**. The primary phloem cells, which have thin walls, are destroyed as they are pushed outward by secondary growth. Thus, to the outside of the vascular cambium, some secondary phloem also is added each year **(Figure 33.23)**. As a stem's diameter increases, the enlarging mass of new tissue eventually ruptures the cortex. Parts of it split away and carry epidermis with them. Cork cambium replaces the lost epidermis with cork.

Bark consists of the tissues sandwiched between the vascular cambium and the stem surface. It includes the secondary phloem and the **periderm** (*peri* = surrounding; *derma* = skin), which consists of cork, cork cambium, and secondary cortex **(Figure 33.24)**. Tubular openings (called *lenticels*) develop in the periderm. They function a bit like snorkels, permitting exchanges of oxygen and carbon dioxide between the living tissues and the outside air. Natural corks used to seal bottles are manufactured from the especially thick outer bark of the cork oak, *Quercus suber*.

As a tree ages, changes unfold in the appearance and function of the wood itself. In the center of its older stems and roots is **heartwood**, dry tissue that no longer transports water and solutes and is a storage depot for some defensive compounds. In time, these compounds—including resins, oils, gums, and tannins—clog and fill in the oldest xylem pipelines. Typically they darken heartwood, strengthen it, and make it more aromatic and resistant to decay or wood-boring insects. **Sapwood** is located between heartwood and the vascular cambium. Compared with heartwood, it is wet, functions in water transport, and is not as strong.

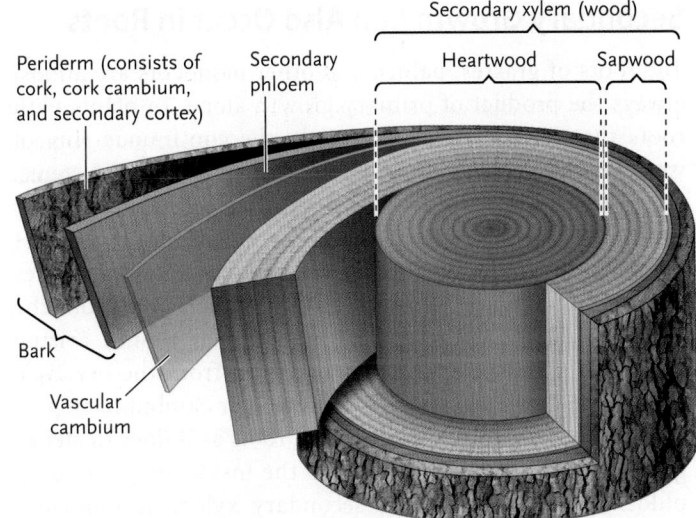

FIGURE 33.24 Structure of a woody stem showing extensive secondary growth. Heartwood, the mature tree's core, has no living cells. Sapwood, the cylindrical zone of xylem between the heartwood and vascular cambium, contains some living parenchyma cells among the nonliving vessels and tracheids. Everything outside the vascular cambium is bark. Everything inside it is wood.
© Cengage Learning 2017

In regions that have large seasonal differences in temperature, trees produce secondary xylem seasonally, with larger-diameter cells produced in spring and smaller-diameter cells in summer. This "spring wood" and "summer wood" reflect light differently, and it is possible to identify them as alternating light and dark bands that represent annual growth layers—the "tree rings" pictured in **Figure 33.25.**

A. A woody stem

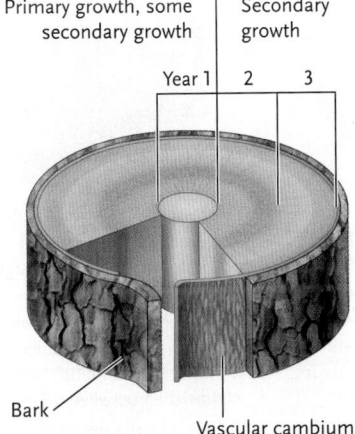

B. Growth rings of American elm (*Ulmus americana*)

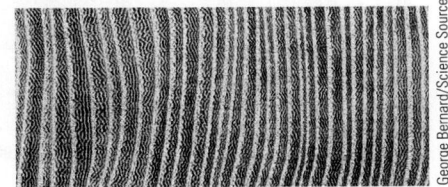

C. *Pinus* growth rings

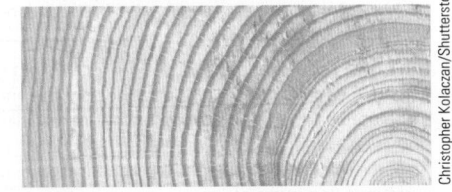

FIGURE 33.25 Secondary growth and tree ring formation. (A) Radial cut through a woody stem that has three annual rings, corresponding to secondary growth in years 2 to 4. **(B)** Growth rings of an elm (*Ulmus*). Each ring corresponds to one growing season. **(C)** Growth rings of a pine, including dark heartwood and lighter sapwood. Pine trees are fast-growing conifers and generally have wider rings than species that grow more slowly.
© Cengage Learning 2017

Secondary Growth Can Also Occur in Roots

The roots of grasses, palms, and other monocots are almost always the product of primary growth alone. In plants with roots that have secondary growth, the continuous ring of vascular cambium develops differently than it does in stems. When their primary growth is done, these roots have a layer of residual procambium between the xylem and phloem of the stele (**Figure 33.26,** step 1). The vascular cambium arises in part from this residual cambium, and in part from the pericycle (step 2). Eventually, the cambial tissues arising from the procambium and those arising from the pericycle merge into a complete cylinder of vascular cambium (step 3). The vascular cambium functions in roots as it does in stems, giving rise to secondary xylem to the inside and secondary phloem to the outside. As secondary xylem accumulates, older roots can become extremely thick and woody. Their ongoing secondary growth is powerful enough to break through concrete sidewalks and even dislodge the foundations of homes.

Cork cambium also forms in roots, where it is produced first by the pericycle and later by initials in secondary phloem. In many woody eudicots and all gymnosperms, most of the root epidermis and cortex fall away, and the surface consists entirely of periderm (step 4).

Secondary Growth Is an Adaptive Response

Plants, like all living organisms, compete for resources, and woody stems and roots confer some advantages. Plants with masses of leaves supported by taller, stronger stems that defy the pull of gravity can intercept more of the light energy from the Sun. With a greater energy supply for photosynthesis, they have the metabolic means to increase their root and shoot systems, and thus are better able to acquire resources—and ultimately to reproduce successfully.

During both primary and secondary growth, the development of new tissues and organs maintains a balance between the shoot system and root system. (This balance is maintained by hormones discussed in Chapter 37.) Leaves and other photosynthetic parts of the shoot must supply root cells with enough sugars to support their metabolism, and roots must provide shoot structures with water and minerals. As long as a plant is growing, this balance is maintained, even as the complexity of the root and shoot systems increases, whether the plant lives for only a few months or—like some bristlecone pines—for 6,000 years.

STUDY BREAK 33.5

1. Explain how and where secondary growth typically occurs in plants.
2. Describe the components of vascular cambium and their roles in secondary growth in stems, including the development of tissues such as bark, cork, and wood.

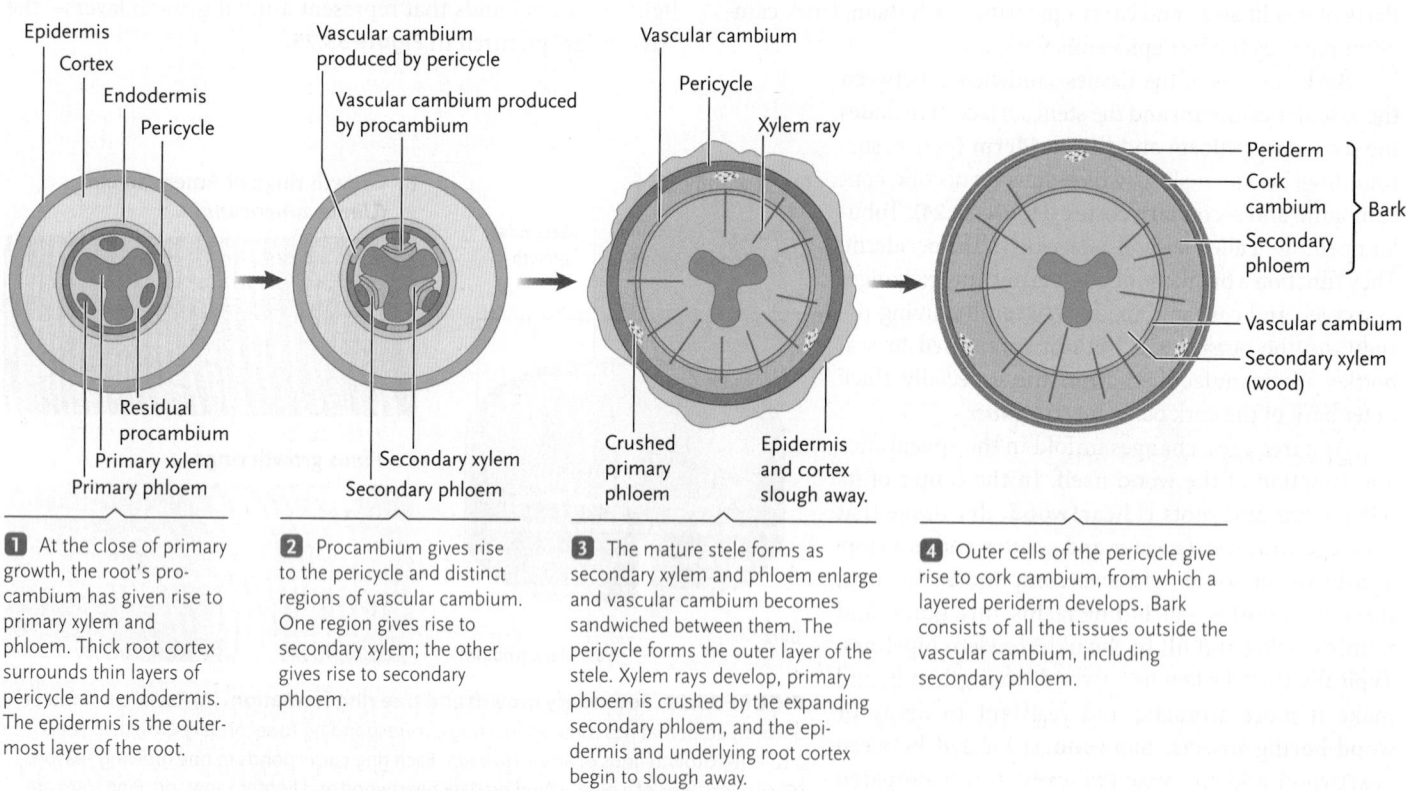

FIGURE 33.26 Secondary growth in the root of one type of woody plant.
© Cengage Learning 2017

1. At the close of primary growth, the root's procambium has given rise to primary xylem and phloem. Thick root cortex surrounds thin layers of pericycle and endodermis. The epidermis is the outermost layer of the root.

2. Procambium gives rise to the pericycle and distinct regions of vascular cambium. One region gives rise to secondary xylem; the other gives rise to secondary phloem.

3. The mature stele forms as secondary xylem and phloem enlarge and vascular cambium becomes sandwiched between them. The pericycle forms the outer layer of the stele. Xylem rays develop, primary phloem is crushed by the expanding secondary phloem, and the epidermis and underlying root cortex begin to slough away.

4. Outer cells of the pericycle give rise to cork cambium, from which a layered periderm develops. Bark consists of all the tissues outside the vascular cambium, including secondary phloem.

How are stem cell reservoirs sustained?

As you have learned in this chapter, a special feature of plants is their ability to continuously grow and generate new organs such as leaves and flowers throughout their life cycle. This unique developmental strategy, which enables plants to flexibly adapt to changing environmental conditions, requires that they continuously produce a supply of new cells from meristems at their growing shoot and root tips. The source of new cells within each meristem is the stem cell reservoir, a small pool of cells that are not committed to a particular organ or tissue fate. During the normal course of plant development, the stem cells slowly divide to replenish the reservoir while simultaneously producing daughter cells that form the different types of tissues and organs.

Because plants such as sequoias can grow and develop for hundreds or even thousands of years, stem cell reservoirs are long lasting and must be highly stable over time. Although it has been demonstrated that mutations that cause excess loss or accumulation of stem cells profoundly affect plant development and architecture, it is not well understood how plant stem cell reservoirs are sustained. Research has shown that the unique identity of shoot stem cells is conferred by a signal from the cells beneath them, which are known as niche cells, because they provide a supportive microenvironment for their stem cell neighbors. In turn the stem cells send a negative signal to the niche cells, via an extracellular signal transduction pathway, to limit the extent of their activity.

This feedback loop maintains the stem cell reservoir at a relatively constant size despite the continuous departure of daughter cells to form tissues and organs. Yet despite its critical importance for plant development, many gaps in our understanding of this regulatory process remain. Key objectives for the future will be to identify all of the components of the stem cell signaling pathways and their outputs at the level of meristem cell division and proliferation, and to determine how additional regulatory elements such as hormones and epigenetic factors modulate the activity of the feedback loop.

think like a scientist Root meristems generate the underground root system, whereas shoot meristems produce the aboveground stems, leaves, and flowers. Considering that they give rise to different tissues, how much overlap might exist between the molecular signatures of stem cells in the root and shoot?

Jennifer Fletcher

Jennifer Fletcher is a research geneticist at the USDA-ARS Plant Gene Expression Center and an adjunct associate professor of plant biology at the University of California, Berkeley. Her research focuses on the molecular genetics of shoot and flower development in *Arabidopsis*. To learn more about Fletcher's work, visit http://pgec.berkeley.edu/jfletcher/.

REVIEW KEY CONCEPTS

For access to MindTap and additional study materials visit www.cengagebrain.com.

33.1 Basic Concepts of Plant Structure and Growth

- The vascular plant body consists of an aboveground shoot system and an underground root system (Figure 33.2).
- Meristems (Figure 33.3) give rise to the plant body and are responsible for a plant's lifelong growth. When a meristem cell divides, one daughter remains an initial that continues to function as meristem. The other daughter is a derivative that differentiates.
- Primary growth of roots and shoots originates at apical meristems. In plants that show secondary growth, lateral meristems increase the diameter of stems and roots.

33.2 The Three Plant Tissue Systems

- Growing plant cells form secondary walls inside the primary walls. Maturing cells become specialized for specific functions, some of which are carried out by walls of dead cells.
- Ground tissues make up most of the plant body, vascular tissues serve in transport, and dermal tissue forms a protective cover. The three types of ground tissues are parenchyma, consisting of parenchyma cells; collenchyma, consisting of collenchyma cells; and sclerenchyma, consisting of fibers or sclereids. Parenchyma is active in photosynthesis, storage, and other tasks. Collenchyma and sclerenchyma provide mechanical support (Figures 33.5 and 33.6).

- Xylem and phloem are vascular tissues. Xylem conducts water and solutes. Its conducting cells are called tracheids and vessel elements (Figure 33.7). In phloem, the food-conducting tissue, living cells called sieve-tube elements join end to end in sieve tubes (Figure 33.8). Both xylem and phloem also contain parenchyma and sclerenchyma cells.
- The dermal tissue, epidermis (Figure 33.9), is coated with a waxy cuticle that restricts water loss. Water vapor and other gases enter and leave the plant through pores called stomata, which are flanked by specialized epidermal cells called guard cells. Epidermal specializations also include trichomes such as root hairs.

33.3 Root Systems

- Roots absorb water and minerals and conduct them to aerial plant parts; they anchor and sometimes support the plant and often store food. Root morphologies include taproot systems, fibrous root systems, adventitious roots, storage roots, and others (Figure 33.10).
- During primary growth of a root, the primary meristem and actively dividing cells make up the zone of cell division, which merges into the zone of elongation. Past the zone of elongation, cells may differentiate in the zone of cell maturation (Figure 33.11).
- A root's xylem and phloem usually are arranged as a central stele (Figure 33.12). Parenchyma outside the stele forms the root cortex. The root endodermis also wraps around the stele. Inside it the pericycle contains parenchyma that can function as meristem. The pericycle gives rise to root primordia from which lateral roots emerge (Figure 33.13). Root hairs greatly increase the absorptive surface of roots.

33.4 Primary Shoot Systems

- The primary shoot system consists of the main stem, leaves, and buds, plus, in angiosperms, any attached flowers and fruits. Stems provide mechanical support, house vascular tissues, and may store food and fluid.

- Stems are organized into modular segments. Nodes are points where leaves and buds are attached; internodes fall between nodes (Figure 33.14A). The terminal bud at a shoot tip consists of shoot apical meristem. Meristem tissue in buds gives rise to leaves, flowers, or both (Figure 33.14B). Apical meristem tissue also occurs just above the root tip.

- Derivatives of the apical meristem produce three primary meristems (Figure 33.15). Protoderm makes the stem's epidermis, procambium gives rise to primary xylem and phloem, and ground meristem gives rise to ground tissue.

- Vascular tissues are organized into vascular bundles, with phloem surrounding xylem in each bundle (Figure 33.16). The bundles form a stele (vascular cylinder). Stem modifications include specializations for defense and storage of food and water (Figure 33.17).

- Monocot and eudicot leaves have blades of different forms, all providing a large surface area for absorbing sunlight and carbon dioxide (Figures 33.18 and 33.19). Leaf modifications such as tendrils, spines and various types of trichomes are adaptive responses to environmental selection pressures (Figure 33.20). Leaf tissues typically have several layers, including a cuticle-covered epidermis and mesophyll (parenchyma cells containing chloroplasts) (Figure 33.21). Leaf characteristics such as shape or arrangement may change over the life cycle of a long-lived plant.

33.5 Secondary Growth

- Secondary growth makes older stems and roots woody and more massive via the activity of vascular cambium and cork cambium.

- Vascular cambium consists of fusiform initials, which generate secondary xylem and phloem, and ray initials, which produce parenchyma cells aligned in horizontal xylem rays (Figures 33.22 and 33.23). Secondary growth takes place as these cells divide.

- Cork cambium produces cork, which replaces epidermis lost when stems increase in diameter. Together, cork cambium and cork make up the periderm (Figure 33.24), the outer portion of bark.

- In root secondary growth, a thin layer of procambium cells between the xylem and phloem differentiates into vascular cambium (Figure 33.26), which gives rise to secondary xylem and phloem. The pericycle produces root cork cambium.

TEST YOUR KNOWLEDGE

Remember/Understand

1. With respect to growth, plants differ from animals in that:
   a. plant growth involves only an increase in the total number of the organism's cells.
   b. plant cells remain roughly the same size after cell division, whereas animal cells increase in size after they form.
   c. all plants form woody tissues during growth.
   d. plants have indeterminate growth; animals have determinate growth.
   e. plants can grow only when young; animals grow throughout their life.

2. Identify the correct pairing of a plant tissue and its function.
   a. epidermis: rigid support
   b. xylem: sugar transport
   c. parenchyma: photosynthesis, respiration
   d. phloem: water and mineral transport
   e. periderm: control of gas exchange

3. Identify the correct pairing of a structure and its component(s).
   a. epidermis: companion cells
   b. phloem: sieve-tube elements
   c. sclerenchyma: nonlignified cell walls
   d. secondary cell wall: cuticle
   e. parenchyma: sclereids

4. Which of the following is *not* part of a stem?
   a. petiole
   b. pith
   c. xylem
   d. procambium
   e. ground meristem

5. Which of the following would be absent in a leaf?
   a. spongy mesophyll
   b. palisade mesophyll
   c. pericycle
   d. vascular bundles
   e. stomata

6. Which of the following is *not* a structure that results from secondary plant growth?
   a. periderm
   b. sapwood
   c. cork
   d. pith
   e. heartwood

7. Which characteristic(s) do eudicots and plants in other clades share?
   a. the position of the vascular bundles
   b. the pattern of leaf veins
   c. the number of grooves in the pollen grains
   d. the number of cotyledons
   e. the formation of flowers

8. The greatest mitotic activity in a root takes place in the:
   a. zone of maturation.
   b. zone of cell division.
   c. zone of elongation.
   d. root cap.
   e. endodermis.

Apply/Analyze

9. A student left a carrot in her refrigerator. Three weeks later she noticed slender white fibers growing from its surface. They were not a fungus. Instead they represented:
   a. lateral roots on a taproot.
   b. adventitious roots.
   c. root hairs on a fibrous root.
   d. root hairs on a lateral root.
   e. young prop roots.

10. A student forgets to water his plant and the leaves start to droop. The structures first affected by water loss and now not functioning are the:
    a. sieve tubes.
    b. sclereids and fibers.
    c. vessel elements and tracheids.
    d. companion cells.
    e. guard cells and stoma.

11. **Discuss Concepts** Baobab trees (*Adansonia* spp.) store water in their trunks (stems) and, although the trees have leaves, considerable photosynthesis also occurs in the trunks. Cacti too have water-storing, photosynthetic stems. In terms of evolutionary adaptation, what does this information suggest about cacti and baobabs?

12. **Discuss Concepts** While camping you notice a "Do Not Litter" sign nailed onto the trunk of a mature oak tree about 7 feet off the ground. When you return five years later, will the sign be at the same height, or will the tree's growth have raised it higher?

13. **Discuss Concepts** Peaches, cherries, and other fruits with pits are produced only on secondary branches that are one year old. To renew the fruiting wood on a peach tree, how often would you prune it? Where on a branch would you make the cut, and why?

14. **Discuss Concepts** African violets and some other flowering plants are propagated commercially using leaf cuttings. Initially, a leaf detached from a parent plant is placed in a growth medium. In time, adventitious shoots and roots develop from the leaf blade, producing a new plant. Which type of cells in the original leaf tissue is the most likely to give rise to the new structures? What property of the cells makes this propagation method possible?

Evaluate/Create

15. **Design an Experiment** The sticky cinquefoil (*Potentilla glandulosa*) is a small, deciduous plant with bright yellow flowers that lives throughout the American West, and its leaf phenotype can vary depending on environmental conditions. Inland, where there are dramatic seasonal temperature swings and unpredictable droughts, plants shed their large "summer leaves" in autumn when the temperature begins to drop. In the spring new leaves are smaller and develop in a compact rosette. This phenotype persists for several months and is thought to be an adaptation that makes the plants drought-resistant (because less water evaporates from reduced leaf surfaces). By contrast, the leaves of *P. glandulosa* plants growing in a coastal climate are always large. In their habitat, seasonal temperature swings are not as great and the annual cycle of winter rain and summer drought is highly predictable. Suppose you decide to explore the hypothesis that the coastal population is genetically capable of exhibiting the same seasonal shift in leaf morphology as the inland plants. Would you need access to a greenhouse where you can control variables, or would it be just as easy to do experiments in the wild? Explain your reasoning and outline your experimental design—including the variable or variables you will test in the first experiment.

16. **Apply Evolutionary Thinking** The first angiosperms may originally have been small, treelike plants in tropical regions, but eventually they began diversifying rapidly into other habitats where early gymnosperms flourished. South African botanist William Bond proposed the "slow seedling" hypothesis to help explain this evolutionary change, and botanists continue to refine it. The hypothesis proposes that angiosperms were able to encroach on many habitats where ancient gymnosperms lived, in part because flowering species increasingly evolved adaptations that made them fast-growing herbaceous plants. Gymnosperms grow more slowly. Based on your reading of this chapter and Chapter 29, what are some structural and biochemical features of gymnosperms (such as conifers) that might result in slower growth, putting them at a competitive disadvantage in this scenario?

For selected answers, see Appendix A.

INTERPRET THE DATA

Rings from long-lived trees may track centuries of annual changes in climatic conditions such as temperature, rainfall, or both. Dendrochronology studies show that Siberian pines *(Pinus sibirica)* growing at high elevations (2,450 m) in the Tarbagatay Pass region of Mongolia may live 500 years, or longer. Growth is faster and rings are wider in warmer years. The reverse is true in cooler years. A research team led by Gordon C. Jacoby of the Tree-Ring Laboratory at the Lamont-Doherty Earth Observatory examined 450 years of *P. sibirica* tree ring data to determine whether the rings provided evidence of a recent climate warming trend in the area. **Figure A** shows the growth fluctuations in a single tree between the years of 1550 and 1995.

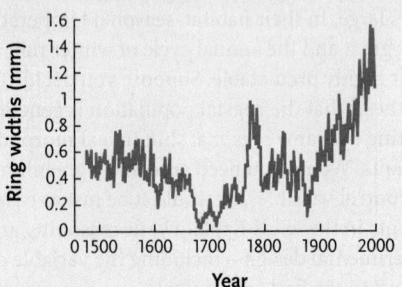

FIGURE A

The upper line in **Figure B** shows the aggregated growth fluctuations of a larger sample of the trees during the same period.

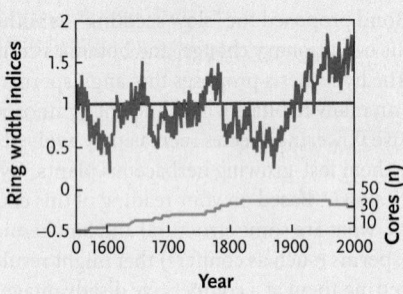

FIGURE B

Figure C correlates the aggregated growth information (solid line) with a reconstruction of annual temperature fluctuations in the region (dotted line).

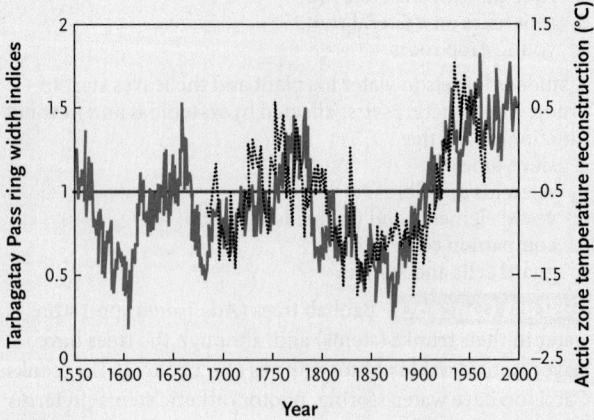

FIGURE C

1. Based on the graphed data, what do you think the research team concluded from this experiment?

2. The researchers included a graph of growth data for a single tree in their research report. Why would such information be relevant?

3. In Figure B, what does the rising trend of the lower line signify? Speculate about why the researchers included that information.

4. In their published report, the researchers included a note that several major volcanic eruptions occurred in the middle of the nineteenth century, spewing massive amounts of smoke and ash into Earth's atmosphere. Why is that fact important for interpreting the graph data?

Source: Adapted from G. Jacoby et al. 1996. Mongolian tree rings and 20th-century warming. *Science* 273:771–777.

Transport in Plants

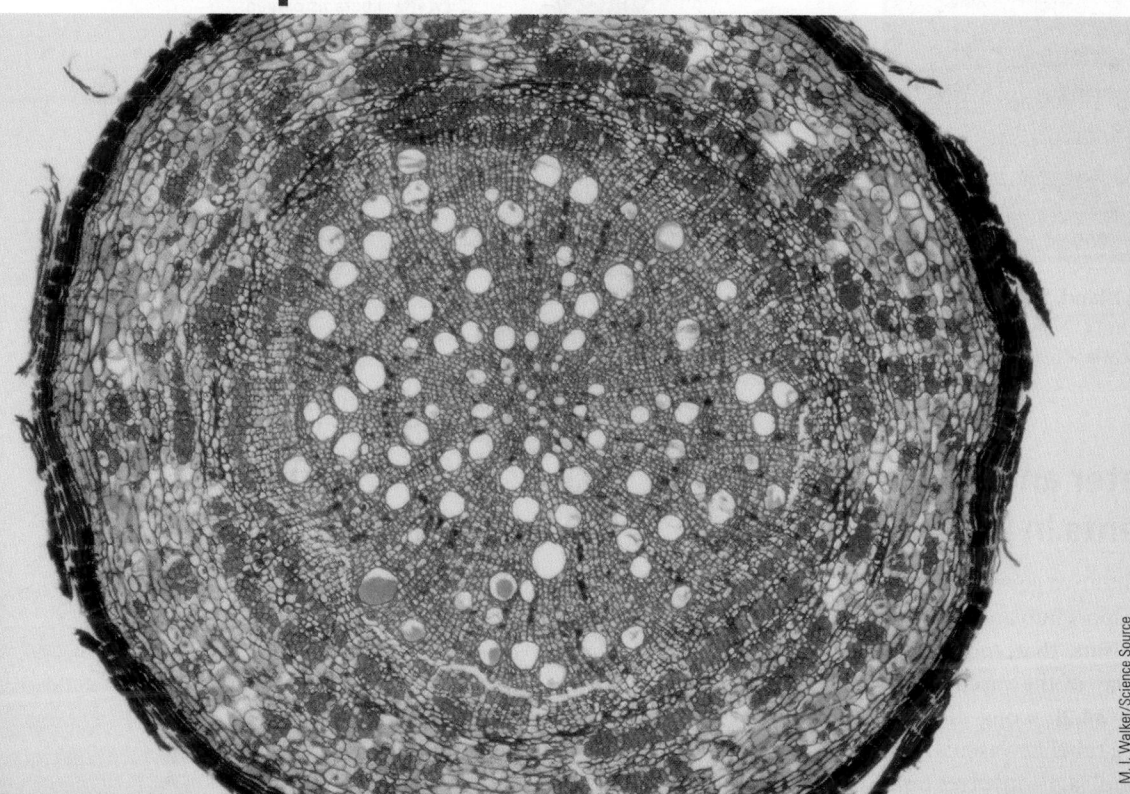

Cross section of the root of an oak (*Quercus* sp.) showing vessels that transport water and nutrients in plants.

M. I. Walker/Science Source

Why it matters . . . In a park on Johns Island near Charleston, South Carolina, sinewy roots of an ancient live oak *(Quercus virginiana)* spread from the tree's base. Some 20 meters above, its leaves catch the sunlight. Dubbed the Angel Oak, the tree is a tourist attraction in part because of its size and strikingly gnarled form **(Figure 34.1).** Its longest limb measures 57 meters; its canopy shades an area equivalent to six regulation tennis courts. In 2012 when developers proposed building an apartment complex near the park where the tree grows, conservation groups successfully opposed the plan out of concern that the development would tap the same pool of groundwater from which the Angel Oak's sprawling roots draw water and mineral nutrients. Estimates of the tree's age range from 400 to 1,400 years, and for all that time its root system has been taking up soil water and minerals, while photosynthesizing cells in the mesophyll of its leaves have been manufacturing sugars that fuel the tree's metabolism. If any part of this natural system breaks down, the tree's long life will quickly come to a close.

In this chapter we look in depth at how vascular plants transport water, minerals, and sugars. Our discussion begins with principles of water and solute movements, including a brief review of osmosis and the operation of transport proteins, topics of Chapter 5. Next we examine how plant roots take up water and minerals and deliver them to the xylem, and how xylem moves those substances from roots to shoot parts. Last we consider how the phloem transports sugar and other organic substances from their sources to the locations where they are used in a plant's diverse metabolic activities. Ultimately, all these materials move by way of the integrated activities of the individual cells, tissues, and organs of a single, smoothly functioning organism—the whole plant.

FIGURE 34.1 **The Angel Oak of Johns Island, South Carolina.** Although the tree's exact age is uncertain, in 1717 it was already imposing enough to be mentioned as a local landmark in a land grant.

34.1 Overview of Water and Solute Movements in Plants

Plant transport mechanisms fall into two general categories—those for short-distance transport into and between cells, and long-distance mechanisms that move substances throughout the plant by way of the vascular tissues xylem and phloem **(Figure 34.2)**. Long-distance transport of substances by a plant's tubelike vascular tissues—the topics of Sections 34.3 and 34.5—involves **bulk flow:** the group movement of molecules in response to a difference in pressure between two locations. In this section our focus is on short-distance mechanisms that move water and solutes into and out of specific cells in roots, leaves, and stems. As you read, keep in mind that the primary plant cell wall does not prevent most solutes from moving into plant cells. Many can diffuse through the wall or, or in the case of mineral ions, enter the cell by active transport. Other solutes—including small molecules and certain large ones, such as proteins, RNAs, and virus particles—cross it by way of the plasmodesmata that connect adjacent cells (see Section 4.4).

Passive and Active Mechanisms Move Substances into and out of Plant Cells

Chapter 5 described two general mechanisms that transport water and solutes across the cell plasma membrane. To briefly recap that discussion, in *passive transport,* substances move down a concentration gradient or, if the substance is an ion, down an electrochemical gradient. *Active transport* requires the cell to expend energy in moving substances *against* a gradient, usually by hydrolysis of ATP. Ions and some larger molecules cross cell membranes assisted by transport proteins embedded in the membrane. In many instances, a hydrogen (H^+) gradient powers the cross-membrane transport of ions or

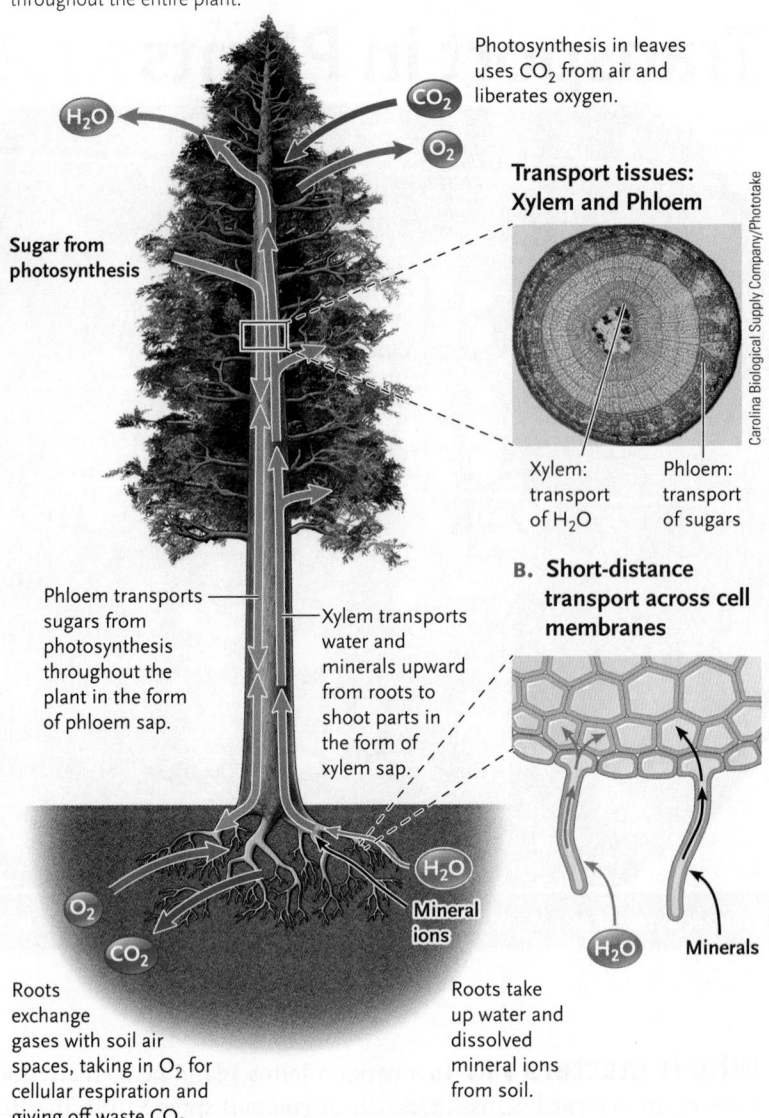

A. Long-distance transport throughout the plant
Vascular tissues—xylem and phloem—distribute substances throughout the entire plant.

Photosynthesis in leaves uses CO_2 from air and liberates oxygen.

H_2O

CO_2

O_2

Sugar from photosynthesis

Transport tissues: Xylem and Phloem

Xylem: transport of H_2O

Phloem: transport of sugars

Phloem transports sugars from photosynthesis throughout the plant in the form of phloem sap.

Xylem transports water and minerals upward from roots to shoot parts in the form of xylem sap.

B. Short-distance transport across cell membranes

O_2

CO_2

H_2O

Mineral ions

H_2O

Minerals

Roots exchange gases with soil air spaces, taking in O_2 for cellular respiration and giving off waste CO_2.

Roots take up water and dissolved mineral ions from soil.

FIGURE 34.2 **Overview of transport routes in plants. (A)** Long-distance transport of water, minerals, and organic substances. **(B)** Short-distance transport of substances between plant cells. Short-distance transport includes movement of water, minerals, sugars, and other materials into, out of, and between cells and into and out of the vascular tissues xylem and phloem. The diagram illustrates the uptake of soil water and solutes by plant root hairs.
© Cengage Learning 2017

molecules. This is the case with substances that enter plant cells by way of the cotransport mechanism called *symport* (see Section 5.4). In symport, the potential energy released as H^+ follows its gradient into the cell is coupled to the uptake of another ion or molecule **(Figure 34.3)**. This is how plant cells take up important mineral ions such as nitrate (NO_3^-) and potassium (K^+). Most organic substances that enter plant cells from the phloem move in by symport as well.

Relative to water, only small amounts of mineral ions and other solutes move into and out of plant cells. As we see next,

A. H⁺ pumped against its electrochemical gradient

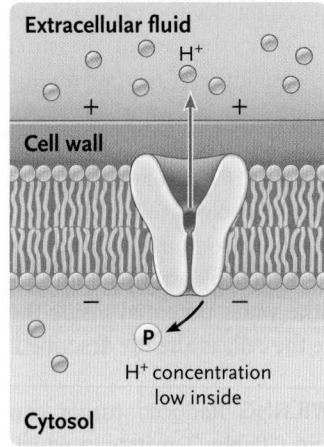

Extracellular fluid

Cell wall

H⁺ concentration low inside

Cytosol

A proton pump moves hydrogen ions (H⁺) out of the cytoplasm, creating an H⁺ gradient that provides energy for transporting other ions and molecules such as sugar into the plant cell. A phosphate group (P) bound earlier to the transporter is released as the ions move to the side of higher concentration.

B. Uptake of cations

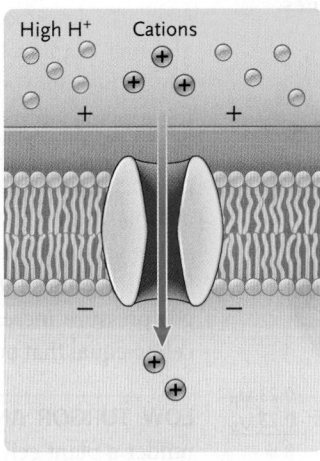

High H⁺   Cations

Cations important in plant nutrition—such as ammonium (NH_4^+)—may enter the cell through selective channel proteins, following the electrochemical gradient created by H⁺ pumping.

C. Cotransport

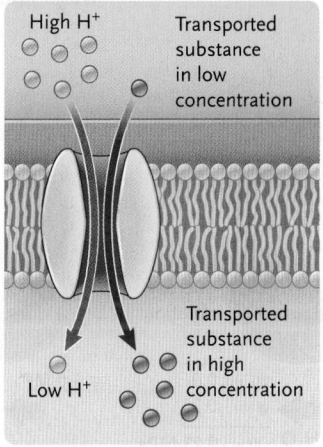

High H⁺   Transported substance in low concentration

Low H⁺   Transported substance in high concentration

This diagram shows the cotransport mechanism called symport, in which the inward diffusion of H⁺ is coupled with the simultaneous active transport of another substance into the cell. Symport moves various substances into plant cells, including the sugar sucrose synthesized during photosynthesis and certain ions taken up by roots.

D. Ion-specific channels

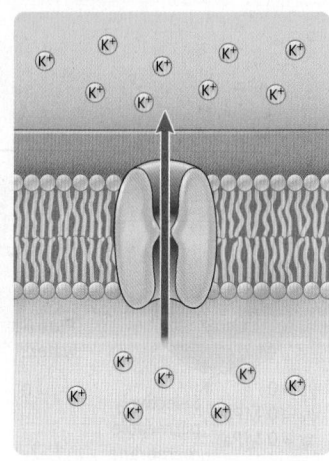

Plant cell membranes have gated channels that open and close in response to stimuli such as stretching and voltage changes. Each channel allows only specific ions, such as K⁺ or Ca²⁺, to pass through.

FIGURE 34.3 Mechanisms that move solutes across the plasma membrane of a plant cell.
© Cengage Learning 2017

large volumes of water enter and exit a plant's cells and tissues by way of osmosis.

In Plants, Water Moves Short Distances by Osmosis

Like other kinds of organisms, plants depend on water for metabolic activities and to maintain the structural integrity of their cells and tissues. Accordingly, the movement of water into and through cells and tissues is one of the most important aspects of a plant's physiology.

Recall from Chapter 5 that **osmosis** is the movement of water across a selectively permeable membrane in response to differences in the concentrations of solutes. The potential energy in water underlies osmosis. Potential energy, remember, is simply stored energy. With respect to water, potential energy is associated with the inherent capacity of water molecules to move from place to place when conditions dictate. The potential energy of water is called **water potential** and is symbolized by the Greek letter Ψ (*psi*, pronounced "sye"). Plant physiologists measure water potential in pressure units called **megapascals** (MPa) and assign a value of 0 MPa to the water potential (Ψ) of pure water in an open container under normal atmospheric pressure and temperature. Shifts away from this baseline are what propel water into or out of cells. Water always moves from regions of higher water potential to regions where the water potential is lower.

Solutes and Pressure Determine Water Potential

Two factors combine to produce water potential. One factor is the effect of dissolved solutes, which is called **solute potential** and is symbolized by Ψ_S. Pure water does not contain solutes, but in water where solutes are present, the solute potential is always a negative value—that is, Ψ_S is always less than 0 MPa. Accordingly, the presence of solutes always reduces water potential. **Figure 34.4** shows this osmotic effect in a laboratory environment, where a solute (in this case, sucrose) is added to pure water on one side of a semipermeable membrane.

Physical pressure also affects water potential. Its influence is called **pressure potential** and is represented by Ψ_P. Unlike solute potential, pressure potential can be either a negative or a positive value. Solute potential and pressure potential jointly produce water potential, a statement represented by the equation $\Psi = \Psi_S + \Psi_P$. In a given situation, knowing solute potential and pressure potential allows you (or a plant scientist) to predict whether water will enter or leave a plant cell, and ultimately the plant as a whole.

INTERACTIONS OF SOLUTE AND PRESSURE POTENTIAL You may recall from Chapter 4 that a typical plant cell has only a thin region of cytoplasm; most of its volume is filled by a large central vacuole where solutes are concentrated (see Figure 4.9B). Because the solute concentration within the cell is usually greater (the Ψ_S

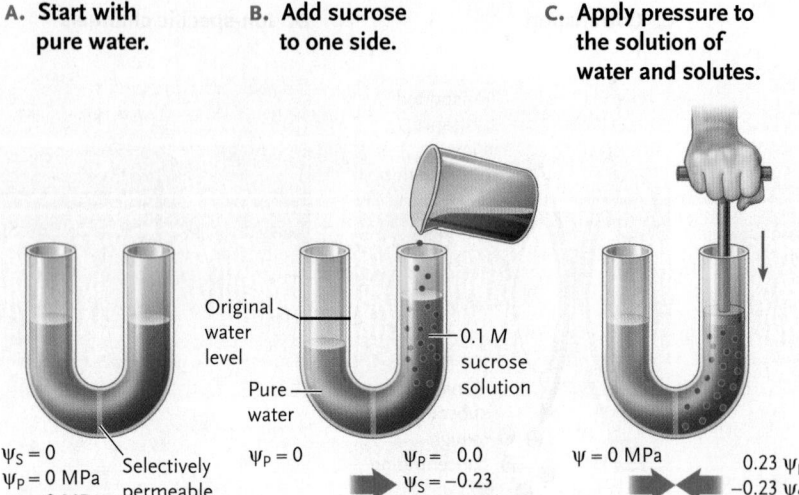

A. Start with pure water.

B. Add sucrose to one side.

C. Apply pressure to the solution of water and solutes.

Original water level

Pure water

0.1 M sucrose solution

$\psi_S = 0$
$\psi_P = 0$ MPa
$\psi = 0$ MPa

Selectively permeable membrane

$\psi_P = 0$

$\psi_P = 0.0$
$\psi_S = -0.23$
$\psi = -0.23$ MPa

H_2O

$\psi = 0$ MPa

0.23 ψ_P
−0.23 ψ_S
0 ψ

Pure water in a curved tube with compartments separated by a selectively permeable membrane

When sucrose is added to the water on one side to form a 0.1 M sucrose solution, the water potential on that side falls. Water moves into the solution by osmosis. ψ_P will increase as osmosis occurs.

By applying enough pressure (ψ_P) to the solution to balance the solute pressure, water potential can be increased to zero, equaling that on the pure-water side of the membrane. Now there is no net movement of water across the membrane. In plant cells, this balancing occurs as the cell takes up enough water to become turgid.

> Plant physiologists assign a value of 0 MPa to the water potential (ψ) of pure water in an open container under normal atmospheric pressure and temperature.

FIGURE 34.4 The relationship between solute potential and water potential. If the water potential is higher on one side of a selectively permeable membrane, water will cross the membrane to the area of lower water potential. This diagram shows pure water on one side of a selectively permeable membrane and a simple sucrose solution on the other side. In an organism, however, the selectively permeable membranes of cells are rarely if ever in contact with pure water.

© Cengage Learning 2017

is more negative) than in the fluid surrounding the cell, the water potential (Ψ) is higher outside plant cells than inside them. Hence water molecules tend to enter the cell by osmosis.

As osmosis drives water into a plant cell, the swelling central vacuole pushes the plasma membrane toward the cell wall. Packed with cellulose microfibrils, the wall has limited flexibility to expand in response to the fluid pressure generated by the incoming water. This pressure, called **turgor pressure** (*turgor* = "to swell"), represents the pressure potential within the cell. When the cell contains so much water that the plasma membrane presses tightly against the cell wall, wall expansion stops and the cell is said to be **turgid**. At this point, backpressure from the stiff wall balances turgor pressure, and Ψ_P in the cell is 0. The water potential inside the cell now equals the potential of the water outside the cell and water no longer enters the cell by osmosis.

MAINTAINING TURGOR PRESSURE It is extremely important for a plant to maintain appropriate turgidity in its cells. Turgid cells are not easily compressed, and this property enables them to support the often-significant weight of large plant organs such as

leaves and flowers. Nonwoody plants in particular depend on turgor pressure to keep aboveground parts erect. As described in Chapter 36, the expansion of growing plant cells also relies on turgor pressure. Given the importance of turgor pressure in maintaining a plant's structural integrity, it is not surprising that when external solutes enter a plant cell's cytoplasm, active transport moves many solutes from the cytoplasm into the central vacuole through channels in the vacuolar membrane (the tonoplast). Water follows the solutes into the vacuole by osmosis until turgor pressure increases the water potential inside the cell to equal that outside the cell, as already described.

LOW TURGOR AND WILTING Reduced turgor can render a plant cell **flaccid** ("flabby"), a state in which its plasma membrane retreats slightly from the cell wall and the wall therefore is less rigid. This situation can arise if osmotic conditions shift (or are manipulated in a laboratory) and the water potential inside the cell equals that outside the cell. Then roughly equal amounts of water enter and leave the cell. The drooping of leaves and stems called **wilting** occurs when a plant loses more water than it gains. Environmental conditions that lead to wilting include drying soil, in which case the water potential in the soil falls below that in the plant. Then turgor pressure inside the cells falls and eventually the protoplast shrinks away from the cell wall. This shrinkage is called **plasmolysis.** The simple experiments shown in **Figure 34.5** demonstrate its effects on a cell. Plasmolysis is reversible if a plant cell is rehydrated in time, but extended plasmolysis generally leads to cell death.

AQUAPORINS: RAPID WATER MOVEMENT ACROSS CELL MEMBRANES A plant cell must compensate fairly quickly for water gains or losses caused by changes in osmotic flow. This means that when osmotic conditions change, a large amount of water must move rapidly across plant cell membranes. Aquaporins, the channel proteins introduced in Chapter 5, allow the bulk flow of water across cell membranes—a mechanism much speedier than normal osmosis. Analysis of the genome of *Arabidopsis thaliana* has revealed 38 genes encoding different aquaporins—evidence that these channel proteins are important in a wide variety of physiological functions in plants. In roots, for example, they facilitate the uptake of water and its movement into the xylem.

Three Pathways Are Available for the Short-Distance Movement of Substances in Plant Tissues

Thus far we have looked at mechanisms by which water and solutes move into and out of plant cells. Substances also travel short distances within and between plant tissues, moving

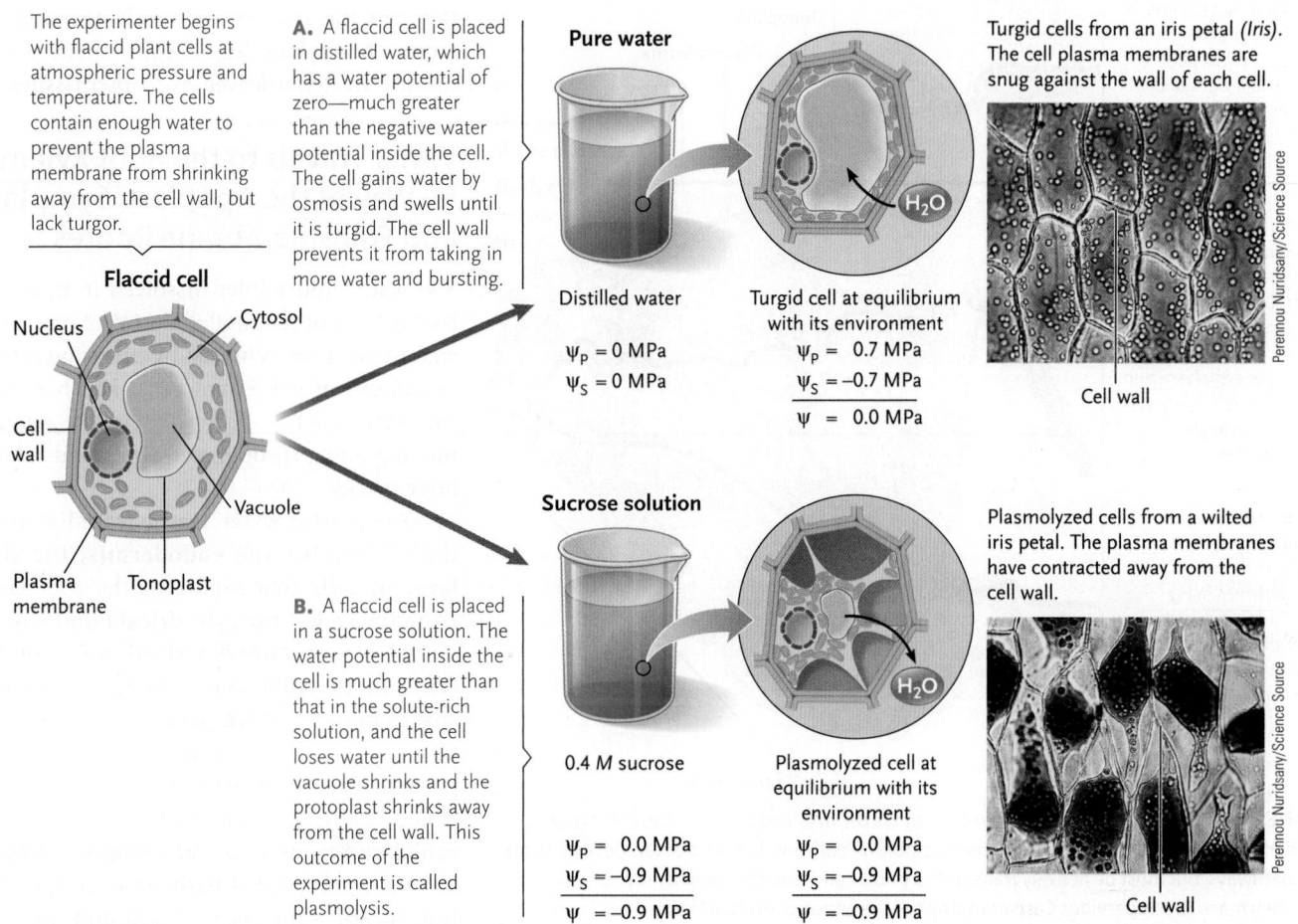

The experimenter begins with flaccid plant cells at atmospheric pressure and temperature. The cells contain enough water to prevent the plasma membrane from shrinking away from the cell wall, but lack turgor.

Flaccid cell

Nucleus
Cytosol
Cell wall
Plasma membrane
Tonoplast
Vacuole

A. A flaccid cell is placed in distilled water, which has a water potential of zero—much greater than the negative water potential inside the cell. The cell gains water by osmosis and swells until it is turgid. The cell wall prevents it from taking in more water and bursting.

Pure water

Distilled water

$\psi_P = 0$ MPa
$\psi_S = 0$ MPa

Turgid cell at equilibrium with its environment

$\psi_P = 0.7$ MPa
$\psi_S = -0.7$ MPa
$\psi = 0.0$ MPa

Turgid cells from an iris petal (*Iris*). The cell plasma membranes are snug against the wall of each cell.

Cell wall

B. A flaccid cell is placed in a sucrose solution. The water potential inside the cell is much greater than that in the solute-rich solution, and the cell loses water until the vacuole shrinks and the protoplast shrinks away from the cell wall. This outcome of the experiment is called plasmolysis.

Sucrose solution

0.4 *M* sucrose

$\psi_P = 0.0$ MPa
$\psi_S = -0.9$ MPa
$\psi = -0.9$ MPa

Plasmolyzed cell at equilibrium with its environment

$\psi_P = 0.0$ MPa
$\psi_S = -0.9$ MPa
$\psi = -0.9$ MPa

Plasmolyzed cells from a wilted iris petal. The plasma membranes have contracted away from the cell wall.

Cell wall

FIGURE 34.5 An experimental demonstration of the effects of different osmotic environments on plant cells. The experimenter begins with flaccid plant cells. The cells contain enough water to prevent the plasma membrane from shrinking away from the cell wall, but lack turgor.
© Cengage Learning 2017

through continuous "compartments" that consist of extracellular spaces and the cytoplasm of living cells **(Figure 34.6).**

PLASMODESMATA AND THE SYMPLAST One such plantwide compartment consists of the protoplasts of living plant cells, which are connected by plasmodesmata. You may remember from Section 4.4 that plasmodesmata are cytosol-filled channels lined by plasma membranes, so connected cells essentially share one continuous surface membrane. Plant physiologists call this continuum the **symplast** ("within cells"). In the **symplastic pathway,** water and its cargo of dissolved minerals moves from cell to cell through plasmodesmata.

Although the symplast was once envisioned as a single, unified entity, research has revealed that the symplast actually is divided into distinct regions called *symplastic domains.* The "borders" of different domains are set by differences in plasmodesmata, including the size of molecules that can pass through the channels from domain to domain. It has long been known that small molecules such as sugars and amino acids pass readily through most plasmodesmata, but more recent research has shown that far from being static passageways, plasmodesmata

are dynamic communication channels. For example, some are capable of active (that is, energy-requiring) transport of macromolecules such as nucleic acids and proteins that are vital to the normal development of cells within a domain (see Chapter 36).

Some plant viruses have become adapted to exploit the dynamic properties of plasmodesmata, employing *viral movement proteins* that alter the structure of plasmodesmata in ways that allow virus particles to use the channels as infective thoroughfares. Chapter 37 looks at plant responses to viruses and other pathogens.

CELL WALLS: THE APOPLAST The porous walls of plant cells and intercellular spaces make up the nonliving **apoplast** (roughly, "not including cells"). Accordingly, the **apoplastic pathway** is an extracellular route that passes through a continuous network of adjoining cell walls and air spaces.

A TRANSMEMBRANE ROUTE A third route is also available for short-distance travel. In this **transmembrane pathway,** water diffuses across plasma membranes or enters cells through membrane aquaporins. As you will see in the next section, the

In the **apoplastic pathway** (red), water moves through nonliving regions—the continuous network of adjoining cell walls and tissue air spaces. However, when it reaches the endodermis, it passes through one layer of living cells.

In the **symplastic pathway** (green), water passes into and through living cells. After being taken up into root hairs, water diffuses through the cytoplasm and passes from one living cell to the next through plasmodesmata.

In the **transmembrane pathway** (black), water that enters the cytoplasm moves between living cells by diffusing across cell membranes, including the plasma membrane and perhaps the tonoplast.

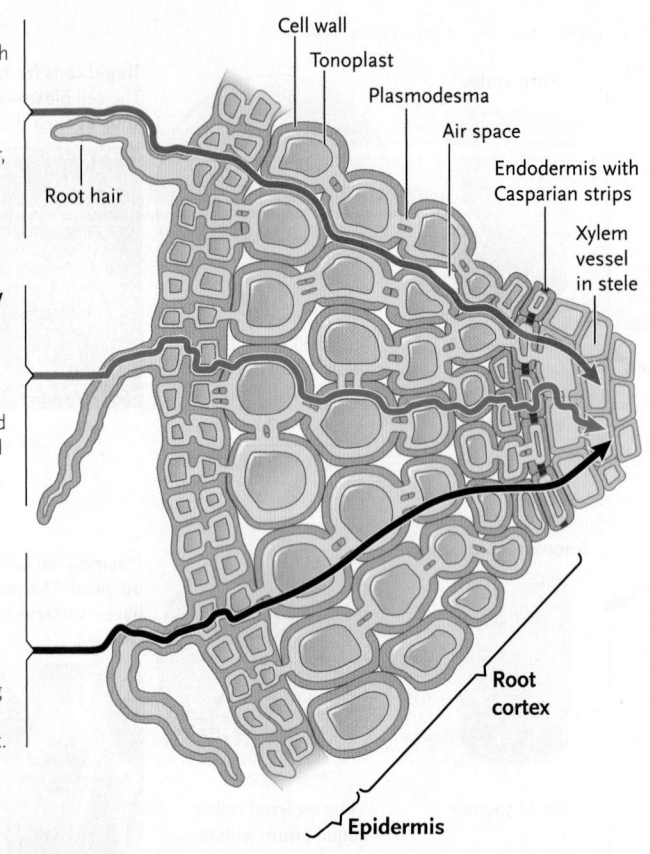

FIGURE 34.6 Pathways for the movement of water into roots. Water can enter roots by way of the apoplast or symplast, or by a transmembrane pathway. Ions also enter roots via these three pathways, but must be actively transported into a cell—and the symplastic pathway—when they reach the impervious Casparian strips of the endodermis (see Figure 34.7). The roots of most flowering plants have both an endodermis surrounding the stele and an exodermis just beneath the epidermis. Both cell layers have a Casparian strip, which helps control the uptake of water and dissolved nutrients.

© Cengage Learning 2017

symplastic, apoplastic, and transmembrane routes all have roles in delivering the water and solutes that enter roots to the xylem, which will transport them throughout the plant.

STUDY BREAK 34.1

1. What is water potential, and why is it important with respect to plant cells?
2. How do solute potential and pressure potential contribute to water potential?
3. Explain how the apoplastic, symplastic, and transmembrane pathways route substances in plant tissues.

34.2 Roots: Moving Water and Minerals into the Plant

Soil around roots provides a plant's water and minerals, but roots do not simply "soak up" these essential substances into the xylem. Instead, water and minerals that enter roots reach the root xylem by first traveling laterally through the root cortex. Only then do they enter the xylem and begin their journey upward to stems, leaves, and other tissues.

Water Travels to the Root Xylem by Way of the Apoplast, Symplast, and Transmembrane Routes

Soil water, and solutes dissolved in it, enters a root by way of the root epidermis. Some of this water enters the symplast by diffusing into the cytoplasm of epidermal cells. Much more of the water roots take up enters the apoplast, moving along through cell walls and intercellular spaces.

Apoplastic water travels rapidly inward until it reaches the **endodermis,** the single layer of cells that separates the root cortex from the **stele**—the cylindrical bundle of vascular tissue that was described in Section 33.4. Cells in the root cortex generally have air spaces between them (which helps aerate the tissue), but endodermal cells are tightly packed. Each endodermal cell also has a lignified, beltlike **Casparian strip** around its radial and transverse walls **(Figure 34.7).** The strip is impregnated with suberin, the waxy, waterproof substance also found in corky parts of roots (see Chapter 33). Being impermeable to water, the Casparian strip blocks the apoplastic pathway at the endodermis, preventing water and minerals in the apoplast from automatically passing on into the stele. Instead, both must enter endodermal cells (and the symplast). Water enters the endodermal cells by osmosis across the cells' plasma membranes, but the semipermeable membrane allows only a subset of the solutes in soil water to cross. Unneeded or potentially toxic solutes may be barred, while desirable ones such as nutritionally important mineral ions move into the cell by facilitated diffusion or active transport. The endodermis also prevents needed substances in the xylem from leaking back into the root cortex. In this way the endodermis provides important control over which substances enter and leave a plant's vascular tissue. The roots of most flowering plants have an additional layer of cells with Casparian strips just inside the root epidermis. This layer, the *exodermis,* functions like the endodermis.

Roots Take Up Ions by Active Transport

Some mineral ions enter the apoplast along with water, but most ions important for plant nutrition tend to be more concentrated in roots than in the surrounding soil, so they

A. Belt-like location of the waxy, water-impervious Casparian strip in the radial and transverse walls of an endodermal cell

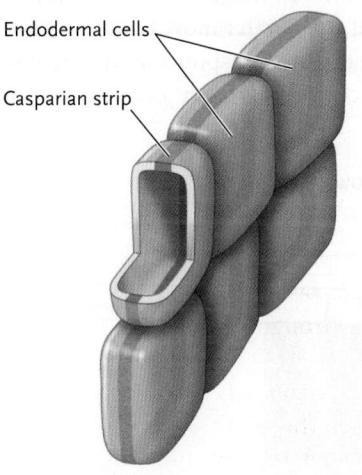

Endodermal cells

Casparian strip

B. How the Casparian strip alters the path of water, forcing it into the endodermal cell —and therefore into the symplast

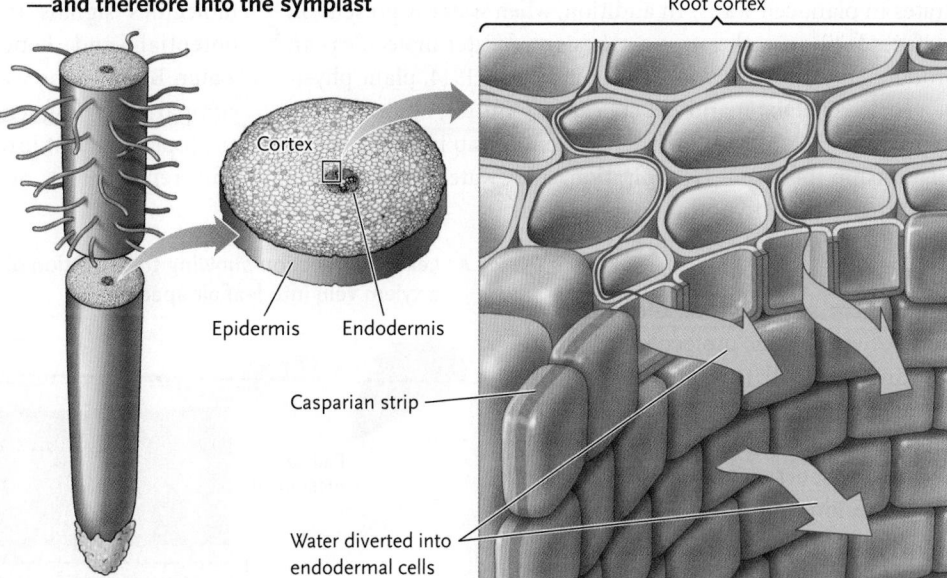

Root cortex

Cortex

Epidermis   Endodermis

Casparian strip

Water diverted into endodermal cells

FIGURE 34.7 How the Casparian strip diverts water into the symplast. The Casparian strip arises as each root endodermal cell develops. Its main constituents are lignin and suberin synthesized by the cell wall.
© Cengage Learning 2017

cannot follow a concentration gradient into root epidermal cells. Instead the epidermal cells actively transport ions inward—that is, ions enter the symplast immediately. They travel to the xylem via the symplastic or transmembrane pathways. Other ions can still move inward following the apoplastic pathway until they reach the Casparian strip of the endodermis. To contribute to the plant's nutrition, however, they must be actively transported into root cortex cells and, as just described, from the endodermis into the stele. In short, mechanisms that control which solutes will be absorbed by root cells ultimately determine which solutes will be distributed through the plant.

Once an ion enters the stele, it diffuses from cell to cell until it is "loaded" into the xylem and becomes part of the **xylem sap,** the dilute solution of water and ions that flows in the xylem. Because the xylem's conducting elements are not living, the water and minerals in xylem sap in effect reenter the apoplast when they reach either tracheids or vessel elements. Once in the xylem, water can move laterally to and from tissues or travel upward in the conducting elements. Minerals distributed to living cells are taken up by active transport. The following section examines how this distribution of water and minerals takes place.

STUDY BREAK 34.2

1. Explain two key differences in how the apoplastic and symplastic pathways route substances laterally in roots.
2. How does an ion enter a root hair and then move to the xylem?

34.3 Transport of Water and Minerals in the Xylem

The height of the Angel Oak described in this chapter's introduction is 20 m, about the same as a typical three-story house. Water-use studies of other large oaks have estimated that in a year, the roots of such a tree take in at least 150,000 L of water, which must travel against the force of gravity from roots to stems and leaves through the tracheids and vessel elements in xylem. Experiments show that of all the water that enters a plant's roots only about 2–5% is used in photosynthesis and in other aspects of metabolism and growth. The rest evaporates from the epidermis of aboveground plant parts. This loss of water vapor from aboveground plant parts is called **transpiration.** A plant's survival depends on replenishing the water it loses through transpiration, yet because mature xylem cells are dead, they cannot expend energy to move water into and through the plant shoot. Instead, as we see next, transpiration generates a water-potential gradient that drives the ascent of xylem sap.

Mechanical Properties of Water Are the Basis for Transporting Xylem Sap from Roots to Shoots

Chapter 2 noted several biologically important mechanical properties of water, three of which underlie its transport in plants. For a brief review here, water molecules are strongly *cohesive:* they tend to form hydrogen bonds with one another.

Water molecules also are *adhesive:* they form hydrogen bonds with molecules of other substances, including the carbohydrates in plant cell walls. In addition, when water is present on surfaces facing air, the strong cohesion of water molecules can produce *surface tension* (see Section 2.4). In 1914, plant physiologist Henry Dixon and his colleague John Joly were the first to propose an explanation for the ascent of sap in terms of the relationship between transpiration and water's mechanical properties. They hypothesized that the transpiration of water from leaves creates tension—that is, negative pressure—that pulls xylem sap upward, and that cohesion between water molecules in the sap sustains this pull throughout the plant. Decades of subsequent experiments have supported and refined this model of xylem transport, which today is called the **cohesion–tension theory of water transport.**

Transpiration Generates the "Pull" That Drives the Ascent of Xylem Sap

Leaf structure is a major factor in transpiration. To begin with, the mesophyll is riddled with air spaces from which mesophyll cells obtain CO_2 for photosynthesis, and open stomata in the leaf epidermis allow the CO_2 to diffuse into those spaces. Also, the cellulose microfibers and other hydrophilic components of mesophyll cell walls attract water, so a film of water on mesophyll cell walls forms an air–water interface. Evaporation of water molecules across the interface helps generate the considerable water vapor that is present in intercellular spaces. These water molecules constitute the water lost via transpiration. The vapor diffuses out through open stomata whenever the air around a leaf is drier than the leaf tissue—that is, as is usually the case, when the water potential of the atmosphere is more negative than the water potential of the leaf.

THE TRANSPIRATION "PULL" The leaf anatomy and properties of water we have been discussing explain how transpiration can drive the rise of xylem sap from roots to shoots. As water evaporates from mesophyll cell walls into leaf air spaces, surface tension in the water film remaining at air–water interfaces pulls the film

inward, producing a curved *meniscus* **(Figure 34.8).** This curved shape produced by surface tension among water molecules signals the development of a negative pressure potential—and, hence, a lower water potential—in a cell's water. Because the water potential in neighboring cells now is comparatively higher, some of their water moves out by osmosis, replacing that lost at the start of transpiration. As the loss and replacement of water continues from cell to cell in

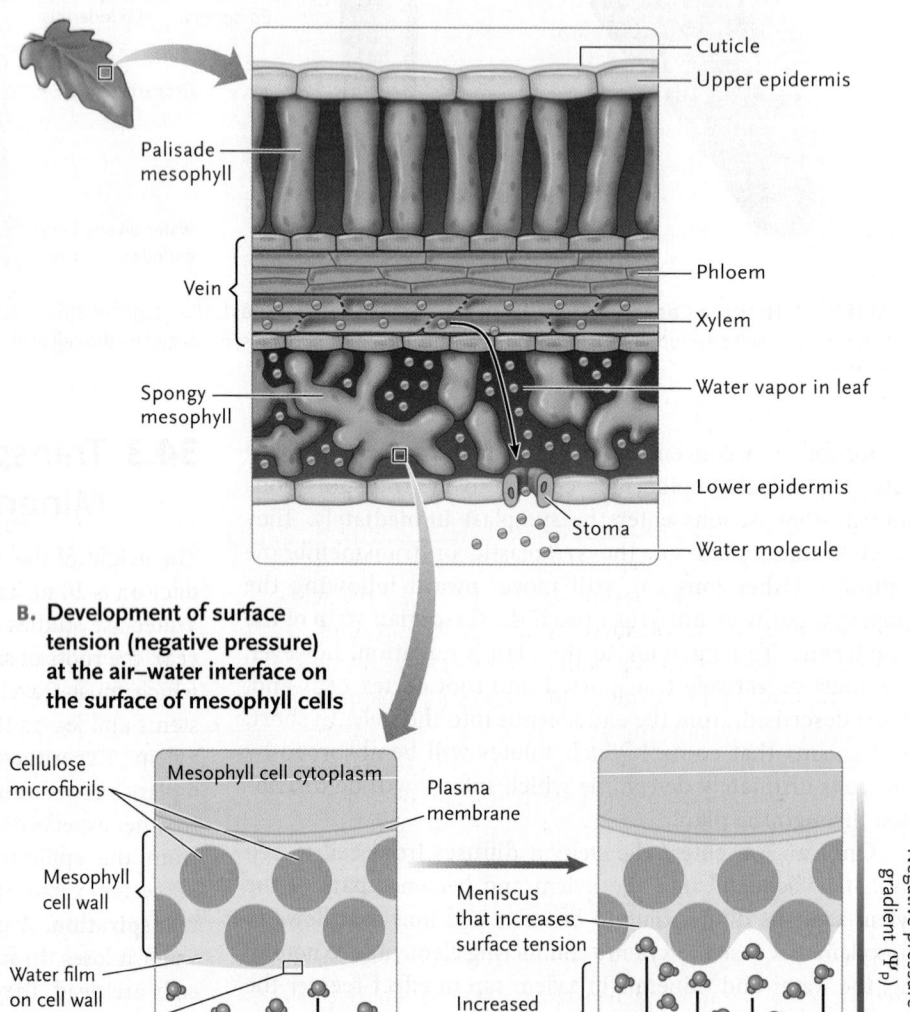

A. Leaf cross section showing the direction of water flow from a xylem vein into leaf air spaces

B. Development of surface tension (negative pressure) at the air–water interface on the surface of mesophyll cells

The water film on the walls of mesophyll cells adheres to hydrophilic wall components such as cellulose. When transpiration begins, evaporation of water from the film is sufficient to replenish the water vapor exiting the leaf through open stomata.

FIGURE 34.8 How transpiration generates the tension to pull xylem sap from roots to the shoot.
© Cengage Learning 2017

With continuing evaporation from the water film, water's adhesiveness pulls the air–water interface into the hydrophilic cell wall. The result is a curved meniscus that increases surface tension. In mesophyll nearest stomata, the rising tension generates a more negative pressure potential with respect to the moist interior of the leaf. Transpiration increases as water molecules follow the tension-induced gradient and move toward and out through stomata. The sustained negative pressure gradient that pulls water from cells and air spaces deeper in the leaf in turn pulls water from the xylem.

mesophyll, eventually replacement water is pulled osmotically out of small xylem veins in the leaf, which in turn withdraw water from larger veins, and so on. The natural cohesion of water maintains the pull on water molecules ever farther from the air–water interface, a pull that extends through the xylem all the way down to a plant's roots.

Thousands of xylem veins lace through every square centimeter of a leaf. In the veins, water molecules are confined in the narrow, tubular vessel elements and tracheids. As we have just described, cohesion of water molecules in veins normally keeps them in a long chain held together by the hydrogen bonds between them. As transpiration occurs, the effects of surface tension and the cohesion of water molecules keeps water moving upward through the xylem and into veins that service mesophyll cells. In addition, water molecules in the water column adhere to the walls of xylem vessels. This adhesion adds to the tension, which is multiplied many times over in all of the leaves and xylem veins of a plant. It increases further as the plant's metabolically active cells take up xylem sap. This is how transpiration moves xylem sap: under continuous tension from above, the entire column of water molecules that initially entered root hairs and root xylem is drawn upward, much as a liquid moves up when you suck on a drinking straw (Figure 34.9). Botanists refer to this root-to-shoot flow as the *transpiration stream*.

Transpiration continues regardless of whether evaporating water is replenished by water rapidly taken up from the soil. If there is too little available soil water, whatever water molecules remain are held ever more tightly by the soil particles. In effect, the action of soil particles reduces the water potential in the soil surrounding plant roots, and as this happens the roots take up water more slowly. Because the water evaporating from the plant's leaves is no longer being fully replaced, turgor pressure drops and the leaves wilt (Figure 34.10). Wilting also occurs when the accumulation of solutes such as NaCl and other salts reduces the water potential in soil. If the water potential finally equals that in leaf cells, a gradient no longer exists. Then movement of soil water into roots and up to the leaves stops entirely.

EFFECTS OF HUMIDITY, TEMPERATURE, AND WIND Three environmental conditions have major effects on the rate of transpiration: relative humidity, air temperature, and air movement. Relative humidity is a measure of the amount of water vapor in air. The less water vapor in the air, the more evaporates from leaves because the water potential is higher in the leaves than in the dry air. Rising air temperature at the leaf surface also speeds evaporation and hence transpiration. Although evaporation cools the leaf a little, the amount of water lost can double for each 10°C rise in air temperature. Air moving across the leaf surface carries water vapor away from the surface and so makes a steeper gradient. Together these factors explain why on extremely hot, dry, breezy days, the leaves of some plants must completely replace their water each hour.

In the Tallest Trees, the Cohesion–Tension Mechanism May Reach Its Physical Limit

Many experiments have tested the premises of the cohesion–tension theory, and thus far the data strongly support it. For example, the theory predicts that xylem sap will begin to move upward at the top of a tree early in the day when water begins to evaporate from leaves. Experiments with various tree species have confirmed that this is the case. The experiments also showed that sap transport peaks at midday when evaporation is greatest, then tapers off in the evening as evaporative water loss slows.

Other experiments have probed the relationship between xylem transport and tree height. George W. Koch at Northern Arizona University and his colleagues studied several of the tallest living redwoods (Figure 34.11), including one that towers nearly 113 m above the forest floor. When the scientists measured the maximum tension exerted in the xylem sap in twigs at the treetops, they discovered that it approached the known physical limit at which the bonds between water molecules in a column of water in a conifer's xylem will rupture. Based on this finding and other evidence, the team has predicted that the maximum height for a healthy redwood tree is 122 to 130 m— so the tallest living redwoods may grow taller still.

Root Pressure Contributes to Upward Water Movement in Some Plants

The cohesion–tension mechanism accounts for upward water movement in tall trees. In some plant species, however—lawn grasses, for instance—a positive pressure can develop in roots and force xylem sap upward. This **root pressure** operates under conditions that reduce transpiration, such as high humidity or low light. In fact, the mechanism that produces root pressure often operates at night, when transpiration slows or stops. Then, active transport of ions into the stele sets up a water potential gradient across the endodermis. Because the Casparian strip of the endodermis tends to prevent ions from moving back into the root cortex, the water potential difference becomes quite large. It can move enough water and dissolved solutes into the xylem to produce a relatively high positive pressure. Although not sufficient to force water to the top of a very tall plant, in some smaller plant species root pressure is strong enough to force water out of leaf openings, in a process called **guttation** (Figure 34.12). Pushed up and out of vein endings by root pressure, tiny droplets of water that look like dew in the early morning emerge from modified stomata at the margins of leaves.

STUDY BREAK 34.3

1. Explain the key steps in the cohesion–tension mechanism of water transport in a plant.
2. What role do properties of water play in the transport of xylem sap?
3. How do environmental conditions affect transpiration?

COHESION–TENSION MECHANISM OF WATER TRANSPORT

A. **The driving force of water movement from roots to aboveground plant parts and into dry air**
Water potential ψ is generally lower in the atmosphere than in leaves. This gradient in water potential drives transpiration (evaporation) of water molecules from stomata in leaves and other aboveground plant parts. The process puts the water in the xylem sap in a state of tension that extends from roots to leaves, driving the upward bulk flow of xylem sap.

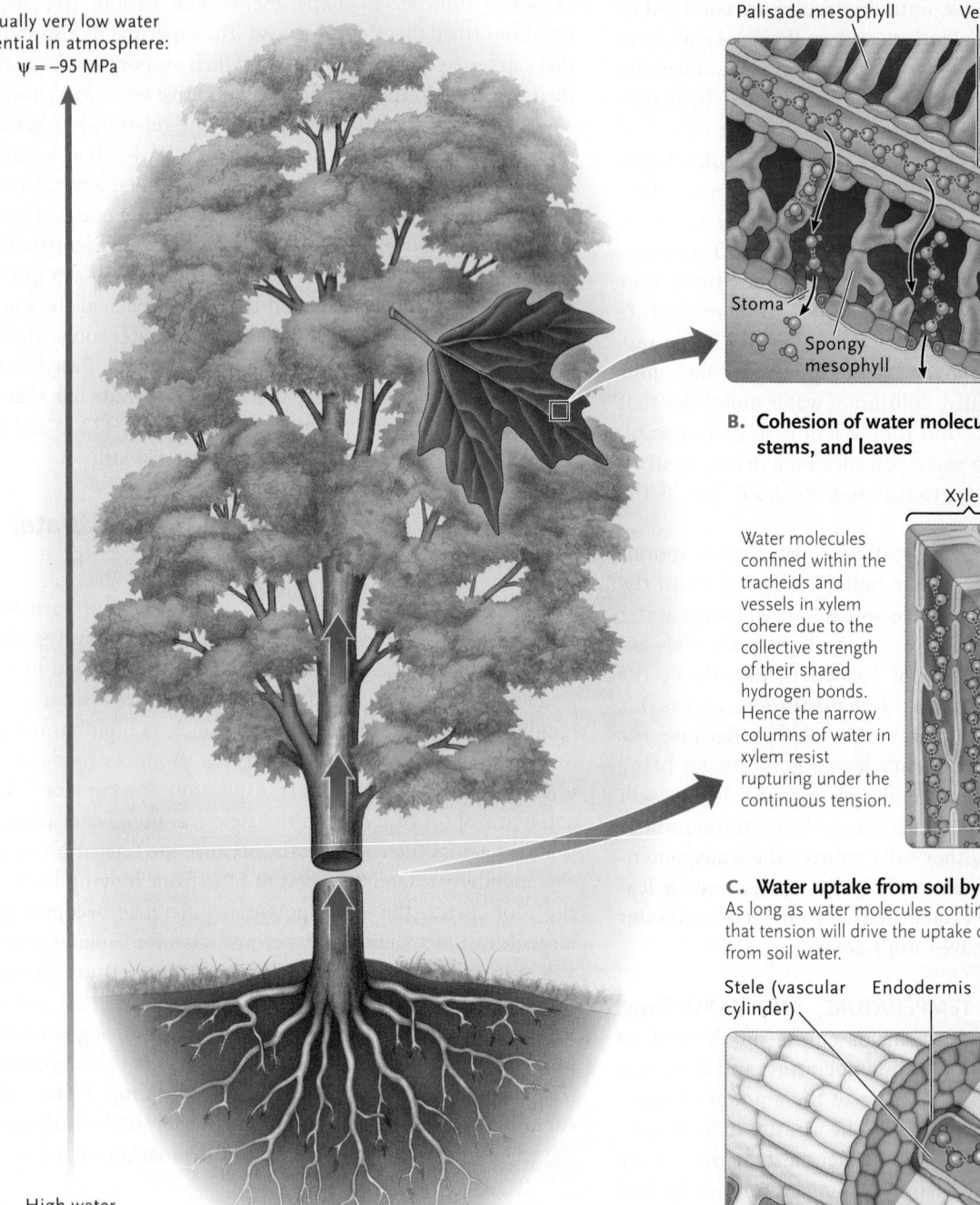

Usually very low water potential in atmosphere:
ψ = −95 MPa

Palisade mesophyll   Vein   Upper epidermis

Stoma
Spongy mesophyll

B. **Cohesion of water molecules in the xylem of roots, stems, and leaves**

Xylem   Vascular cambium   Phloem

Water molecules confined within the tracheids and vessels in xylem cohere due to the collective strength of their shared hydrogen bonds. Hence the narrow columns of water in xylem resist rupturing under the continuous tension.

C. **Water uptake from soil by roots**
As long as water molecules continue to escape by transpiration, that tension will drive the uptake of replacement water molecules from soil water.

Stele (vascular cylinder)   Endodermis   Cortex   Water molecule   Root hair

High water potential in moist soil:
ψ = <0 MPa

SUMMARY In the cohesion–tension mechanism of water transport, transpiration—the evaporation of water from shoot parts—creates tension on the water in xylem sap. This tension, which extends from leaf to root, pulls columns of hydrogen-bonded water molecules upward. Water molecules are not shown to scale.

think like a scientist   On an otherwise hot day, how would a sudden thunderstorm affect transpiration in plants?

FIGURE 34.10 A New Guinea impatiens plant *(Impatiens hawkeri)* wilted and recovered after watering.

FIGURE 34.11 Redwoods *(Sequoia sempervirens)* growing in coastal California have reached recorded heights of over 100 m during lifespans of more than 2,000 years. Such extremely tall trees exemplify the ability of plants to move water and solutes from roots to shoots against the force of gravity over amazingly long distances.

FIGURE 34.12 Guttation, caused by root pressure. The drops of water appear at the endings of xylem veins along the leaf edges of a strawberry plant *(Fragaria)*.

34.4 Stomata: Regulating the Loss of Water by Transpiration

Each day, more than 90% of the water entering a leaf can be lost through transpiration. About 2% of the water remaining in the leaf is used in photosynthesis and other metabolic activities. These physiological statistics emphasize the need for controls over transpiration, for if water loss exceeds water uptake by roots, the resulting dehydration of plant tissues interferes with normal functioning, and the plant may wilt and die.

The cuticle covering the epidermis of leaves and stems reduces the rate of water loss from aboveground plant parts, but it also limits the rate at which CO_2 for photosynthesis can diffuse into the leaf. When stomata are open, carbon dioxide can be absorbed, but unless the relative humidity of external air is 100%, water vapor always moves out through them. However, plants have evolved adaptations that balance water loss with CO_2 uptake. This "transpiration–photosynthesis compromise" involves the regulation of transpiration and gas exchange by opening and closing stomata as environmental conditions change.

OPENING AND CLOSING OF STOMATA Two guard cells flank each stomatal opening **(Figure 34.13)**. Their elastic walls are reinforced by parallel strands of cellulose microfibrils that wrap around the walls in a radial pattern, an arrangement that often is likened to the steel belts in a radial tire. The inner walls are thicker and less elastic than the outer walls.

Stomata open when solutes accumulate in the cells and water follows by osmosis. In a preliminary step, when light strikes guard cells, starch in their chloroplasts is converted to malate. Next, an active transport pump in the plasma membrane begins pumping H^+ ions out of the guard cells. H^+ pumped out of the cells can then follow its concentration gradient back into them. This inward flow of H^+ powers the

A. Open stoma

B. Closed stoma

Guard cells

Stoma

Chloroplast

Stoma

20 μm

FIGURE 34.13 Guard cells and stomatal action. (A) Light micrograph of an open stoma. Water entered collapsed guard cells, which swelled under turgor pressure and moved apart, thus forming the stoma in the leaf of a spiderwort *(Tradescantia)*. Note the chloroplasts in the guard cells. (B) A closed stoma. Water exited the swollen guard cells, which collapsed against each other and closed the stoma.

active transport of K^+ into the guard cells—an example of symport (see Figure 34.3). The flow of H^+ also triggers the inward movement of anions including chloride (Cl^-) and malate through a separate channel. The accumulation of all these solutes in guard cells brings in water by osmosis. As turgor pressure builds in the guard cells, the supportive belts of cellulose microfibrils in the outer cell walls prevent the cells from expanding laterally. Instead each cell elongates as the microfibrils move farther apart. As a result, the two elongating cells bow away from each other and create a stoma ("mouth") between them. Stomata close when the H^+ active transport protein stops pumping. K^+ (and anions) flow passively out of the guard cells, and water follows by osmosis. When the water content of the guard cells dwindles, turgor pressure drops. The guard cells collapse against each other, closing the stomata.

In most plants, stomata open at first light, stay open during daylight, and close at night. Experiments have shown that guard cells respond to a number of environmental and chemical signals, any of which can induce the ion flows that open and close stomata. These signals include light, CO_2 concentration in the air spaces inside leaves, and the amount of water available to the plant.

LIGHT AND CO₂ CONCENTRATION Light induces stomata to open through stimulation of blue-light receptors associated with the plasma membrane of guard cells. When stimulated, the receptors start the signal transduction pathway leading to stomatal opening by triggering activity of the H^+ pumps. Also, as photosynthesis begins in response to light, CO_2 concentration drops in the leaf air spaces as chloroplasts use the gas in carbohydrate production. This drop in CO_2 concentration sets off the series of events increasing the flow of K^+ into guard cells and furthers stomatal opening. The effects of reduced CO_2 concentration have been tested by placing plants in the dark in air containing no CO_2. Even in the absence of light, as the CO_2 concentration falls in leaves, guard cells swell and the stomata open.

Normally, when the Sun goes down, a plant's demand for CO_2 drops as photosynthesis comes to a halt. Yet aerobic respiration continues to produce CO_2, which accumulates in leaves. As CO_2 concentration rises, and the blue-light wavelengths that activated the H^+ pumps wane, K^+ is lost from the guard cells and they collapse, closing the stomata. Thus, at night transpiration is reduced and water is conserved.

WATER STRESS As long as plenty of water is available to a plant's roots, the stomata stay open during daylight. However, if water loss stresses a plant, the stomata close or open only slightly, regardless of light intensity or CO_2 concentration. Stomata arose relatively early in land-plant evolution, probably about 400 million years ago, and the first ones, in nonvascular plants, were simple open pores. For a variety of reasons stomata have proven tricky to study, and only recently did Tim J. Brodribb and his colleague Scott A. M. McAdam at the University of Tasmania show that in seedless vascular plants such as ferns,

stomata operate by basic hydraulics—opening when the water content of their leaves reaches a certain level and closing when it falls below that level. The mechanism in seed plants is more complex. Experiments on these plants have shown that the stress-related closing of stomata depends on a hormone, abscisic acid (ABA), which is released by roots when water is scarce. This mechanism seems especially important in herbaceous (nonwoody) plants. In some elegant studies, test plants were suspended in containers so that only one-half the root system received water. Even though the roots with access to water could absorb enough water to satisfy the needs of all the plants' leaves, the stomata still closed. Tissue analysis revealed that water-stressed roots rapidly synthesize ABA. Transported through the xylem, this hormone stimulates K^+ loss by guard cells, and water moves out of the cell by osmosis (enhanced by aquaporins) so the stomata close. Mesophyll cells also take up ABA from the xylem and release it, with the same effects on stomata, when their turgor pressure falls because of excessive water loss. ABA can also cause stomata to close when the hormone is added experimentally to leaves.

A Biological Clock Helps Govern the Opening and Closing of Stomata

Besides responding to light, CO_2 concentration, and water stress, stomata apparently open and close on a regular daily schedule imposed by a biological clock. Even when plants are placed in continuous darkness, their stomata open and close (for a time) in a cycle that roughly matches the day/night cycle of Earth. Such *circadian rhythms* (*circa* = around; *dies* = day) are also common in animals, and several, including wake/sleep cycles in mammals, are controlled by hormones—a topic pursued in Chapter 40. We consider plant circadian rhythms in Chapter 37.

In Dry Climates, Plants Exhibit Various Adaptations for Conserving Water

Many plants have other evolutionary adaptations that conserve water, including modifications in structure or physiology **(Figure 34.14)**. Oleanders, for example, have stomata at the bottom of pitlike areas of the leaf epidermis lined by hairlike trichomes (Figure 34.14A). Sunken stomata are less exposed to drying breezes, and trichomes help retain water vapor at the pore opening, so that water evaporates from the leaf much more slowly.

The leaves of *xerophytes*—plants adapted to hot, dry environments in which water stress can be severe—have a thickened cuticle that gives them a leathery feel and enhances protection against evaporative water loss. An example is jojoba (*Jimmondsia chinensis*, Figure 34.14C). In still other plants that live in arid landscapes, such as cacti and the baobab tree (*Adansonia*), stems are thick, water-storing cylinders (Figure 34.14D) or pads (see Figure 33.17F). In some plants, coping with water stress includes a capacity to resist *cavitation*,

A. Oleander

B. Oleander leaf

Cuticle

Multilayer epidermis

Trichomes

Recessed stoma

C. Jojoba leaves

D. Baobab tree

E. CAM plant

FIGURE 34.14 Some adaptations that enable plants to survive water stress. (A) Oleanders *(Nerium oleander)* are adapted to arid conditions. **(B)** As shown in the micrograph, oleander leaves have recessed stomata on their lower surface and a multilayer epidermis covered by a thick cuticle on the upper surface. Trichomes are associated with the stomata. **(C)** Leathery leaves of jojoba *(J. chinensis)*. **(D)** A baobab tree *(Adansonia* spp.) native to Madagascar. Depending on the species, a baobab's swollen trunk may store more than 100,000 liters of water. **(E)** *Sedum,* a CAM plant, in which the stomata open only at night.

the formation of physiologically dangerous air bubbles in xylem mentioned in Section 33.2.

A variation on water-conservation mechanisms occurs in CAM plants, including more than half of the world's orchid species, bromeliads such as pineapple, and cacti and other succulents (Figure 34.14E). As discussed in Section 8.4, **crassulacean acid metabolism** (CAM) is a biochemical variation of photosynthesis that was discovered in a member of the family Crassulaceae. CAM plants generally have fewer stomata than other types of plants, and their stomata typically follow a reversed schedule. They are closed during the day when temperatures are higher and the relative humidity is lower, and open at night. At night, the plant temporarily fixes carbon dioxide by converting it to malate, an organic acid. In the daytime, the CO_2 is released from malate and diffuses into chloroplasts, so photosynthesis takes place even though a

CAM plant's stomata are closed. This adaptation prevents heavy evaporative water losses during the heat of the day.

STUDY BREAK 34.4

1. How and when do stomata open and close?
2. In what ways is stomatal functioning important to a plant's ability to manage water loss?

THINK OUTSIDE THE BOOK

Some bromeliad species are *epiphytes*—a nonparasitic plant that grows on another plant. In nature, epiphytes typically grow on the trunks and branches of trees, attaching to their substrate via reduced roots that take up little if any water. Using the Internet or other sources, research a few adaptations that provide epiphytes with access to water and nutrients.

Going with the Phloem

During the development of a vascular plant, the nuclei of sieve-tube elements disintegrate and other cell components are reorganized. This remodeling produces hollow tubes through which phloem sap can move. It is one of numerous developmental events orchestrated by members of a large family of transcription factors collectively called NAC proteins, so named for the initial letters of the first three regulated genes to be identified. Annotation of the *Arabidopsis thaliana* genome (sequenced in 2000) indicated that two NAC proteins—NAC45 and NAC86—function in the differentiation of sieve-tube elements, but until recently their exact role was fuzzy. To clarify the picture, a team of 19 scientists led by Kaori Miyashima Furuta at the University of Helsinki (Finland) blended molecular studies and sophisticated scanning electron microscopy.

Research Question

What role do NAC proteins play in the development of sieve-tube elements?

Experiments

Like the molecular research on xylem formation described in Chapter 33, this study used root tissue from wild-type and mutant *A. thaliana* plants. Mutants lacked normal versions of NAC45 and NAC86. Both groups of plants were grown from seed in the laboratory and root tissue was obtained during the first days of development. In an early step, the team observed phloem development in wild-type rootlets using serial block-face scanning electron microscopy, or SBEM—a technique that allows scientists to assemble a sequence of slicelike micrographs into high-resolution three-dimensional images. The 3-D images captured the normal hollowing-out of sieve-tube elements from start to finish. Members of the research group then grew and analyzed mutant plants lacking functional NAC-regulated enzymes that degrade cell nuclei.

Discussion

Study findings showed that the differentiation of sieve-tube elements is launched by a mechanism in which enzymes break down phloem cell nuclei. Subsequent study of mutants lacking normal NAC proteins showed that either of two NAC transcription factors—NAC45 and NAC86—could drive the remodeling that produces normal, hollow sieve-tube elements **(Figure)**. The NAC targets included genes encoding enzymes that break down the cell nucleus. Thus, sieve elements differentiate through a specialized mechanism of autolysis—that is, destruction of certain cell structures produced by the cell itself.

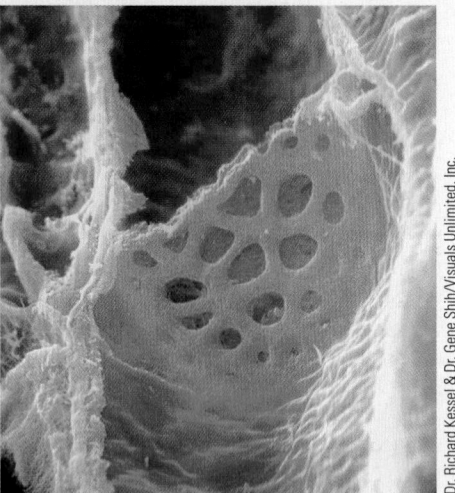

FIGURE Color-enhanced micrograph, showing hollowed out sieve-tube elements. Only a portion of the sieve plate between them remained after the photographer opened up the wall of the sieve tube.

think like a scientist

Why wasn't it surprising that sieve-tube elements are hollowed out by autolysis?

Source: Kaori Miyashima Furuta et al. 2014. *Arabidopsis* NAC45/86 direct sieve element morphogenesis culminating in enucleation. *Science* 345:933–937.

34.5 Transport of Organic Substances in the Phloem

A plant's phloem is another major long-distance transport system, and a superhighway at that: it carries huge amounts of carbohydrates, lesser but vital amounts of amino acids, fatty acids, and other organic compounds, and still other essential substances such as hormones. And unlike the xylem's unidirectional upward flow, the phloem transports substances throughout the plant to wherever they are used or stored. Organic compounds and water in the sieve tubes of phloem are under pressure and driven by concentration gradients. This chapter's *Molecular Insights* discusses genetic controls on phloem development.

Organic Compounds Are Stored and Transported in Different Forms

Plants synthesize tens of thousands of organic compounds, including large amounts of carbohydrates that are stored mainly as starch. Plant cells must often export these compounds for use in distant cells. Yet, with a few exceptions, macromolecules such as starches, proteins, and fat molecules are too large to cross cell membranes and leave the cells where they are made. They also may be too insoluble in water to be transported to other regions of the plant body. Consequently, in leaves and other plant parts, specific reactions convert organic compounds to transportable forms. For example, hydrolysis of starch liberates glucose units, which combine with fructose to form sucrose—the main form in which sugars are transported through the phloem of most plants. Proteins are broken down into amino acids, and lipids converted into fatty acids. These forms are also better able to cross cell membranes by passive or active mechanisms.

Organic Solutes Move by Translocation

In plants, the long-distance transport of substances is called **translocation.** Botanists most often use this term to refer to the

FIGURE 34.15 **Experimental Research**

Translocation Pressure

Hypothesis: High pressure forces phloem sap to flow through sieve tubes from a source to a sink.

Experiment: In the late 1970s, John Wright and Donald Fisher at the University of Georgia devised an experiment to directly measure the turgor pressure in sieve tubes of weeping willow saplings *(Salix babylonica)* under nondestructive conditions, using aphids that feed on *S. babylonica* in the wild. Weeping willow saplings were grown in a greenhouse under natural conditions of light and moisture. Aphids were placed on the trees and allowed to begin feeding by inserting their stylets into sieve tubes in the normal fashion. After being anesthetized by exposure to high concentrations of carbon dioxide, the aphids' bodies were cut away and only their stylets were left embedded in the sieve tubes. A tiny pressure-measuring device called a micromanometer then was glued over the end of each stylet. The micromanometer registered the volume and pressure of phloem sap as it was exuded from the stylet over time periods ranging from 30 to 90 minutes.

A. Aphid releasing honeydew

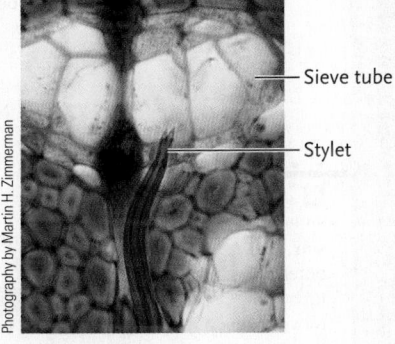

B. Micrograph of aphid stylet in sieve tube

— Sieve tube

— Stylet

Results: In nearly all cases, a high volume of pressurized sap flowed through the severed stylets into the micromanometer during the test periods.

Conclusion: The evidence supports pressure flow as the mechanism that moves phloem sap through sieve tubes.

Other experiments have confirmed that both turgor pressure and the concentration of sucrose are highest in sieve tubes closest to the sap source. Phloem sap also moves most rapidly closest to the source, where pressure is highest.

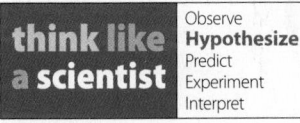

think like a scientist

Observe
Hypothesize
Predict
Experiment
Interpret

In most plants, leaves are the richest source of sucrose in phloem sap. *S. babylonica* is a deciduous tree that drops its leaves in late autumn and then leafs out again starting in the early spring. Might there be a correlation between the concentration of sucrose in its phloem and the age of its leaves?

Source: J. P. Wright and D. B. Fisher. 1980. The direct measurement of sieve tube turgor pressure using severed aphid stylets. *Plant Physiology* 65:1133–1135.

© Cengage Learning 2017

distribution of sucrose and other organic compounds by phloem, and they understand the mechanism best in flowering plants. The phloem of flowering plants contains interconnecting sieve tubes formed by living sieve-tube elements (see Figure 33.8). Sieve tubes lie end to end within vascular bundles, and they extend through all parts of the plant. Water and

organic compounds, collectively called **phloem sap,** flow rapidly through large pores on the sieve tubes' end walls—another example of a structural adaptation that suits a particular function.

Phloem Sap Moves from Source to Sink under Pressure

Over the decades, plant physiologists have proposed several mechanisms of translocation, but it was the tiny aphid, an insect pest, that helped demonstrate that organic compounds flow under pressure in the phloem. An aphid attacks plant leaves and stems, forcing its needlelike stylet (a mouthpart) into sieve tubes to obtain the dissolved sugars and other nutrients inside. Numerous experiments with aphids have shown that in most plant species, sucrose is the main carbohydrate being translocated through the phloem. Studies also verify that the contents of sieve tubes are under high pressure, often five times as much as in an automobile tire. **Figure 34.15** explains a simple and innovative experiment that provided direct confirmation that phloem sap flows under pressure. When a live aphid feeds on phloem sap, this pressure forces the fluid through the aphid's gut and (minus nutrients absorbed) out its anus as a sticky liquid waste called "honeydew." A car parked under an aphid-infested tree might get spattered with honeydew droplets, thanks to the high fluid pressure in the tree's phloem.

Much of what plant physiologists know about the transport of phloem sap has come from studies of sucrose transport in flowering plants. A fundamental discovery is that in flowering plants sucrose-laden phloem sap flows from a starting location, called the *source,* to cells in another site, called the *sink,* along gradients of decreasing solute concentration and pressure. A **source** is any region of the plant where organic substances are produced or released from storage and loaded into the phloem's sieve-tube system. A **sink** is any region where cells use or store organic substances unloaded from the sieve-tube system. What causes the sugar sucrose and other solutes produced in leaf mesophyll to flow from a source to a sink? In flowering plants, the **pressure flow mechanism** builds up at the source end of a sieve-tube system and pushes those solutes by bulk flow toward a sink, where cells remove them. **Figure 34.16** summarizes this mechanism, using sucrose as an example.

Mature photosynthesizing leaves are sources. Another example is a tulip bulb. In spring, stored food is mobilized for transport upward to growing plant parts, but after the plants bloom, the bulb becomes a sink as sugars manufactured in the

CLOSER LOOK FIGURE 34.16

THE PRESSURE FLOW MECHANISM IN THE PHLOEM OF FLOWERING PLANTS

A. Loading at a source

Photosynthetic cells in leaves are a common source of carbohydrates that must be distributed through a plant. Small, soluble forms of these compounds move from the cells into phloem (in a leaf vein).

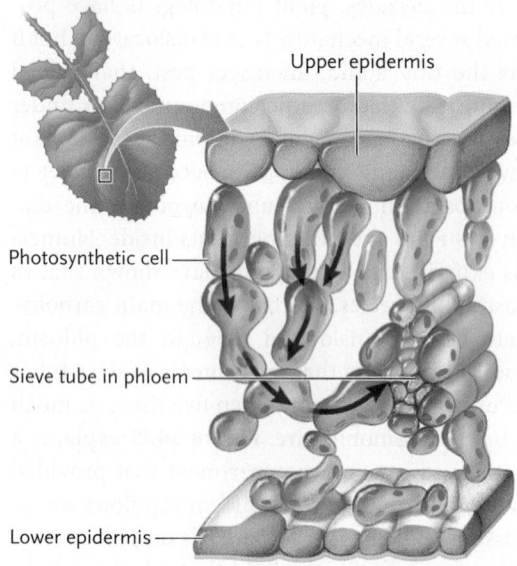

B. The cotransport mechanism that loads sucrose into phloem cells

Pumping of hydrogen ions out of the cell creates a gradient that drives the transport of sucrose into the cell.

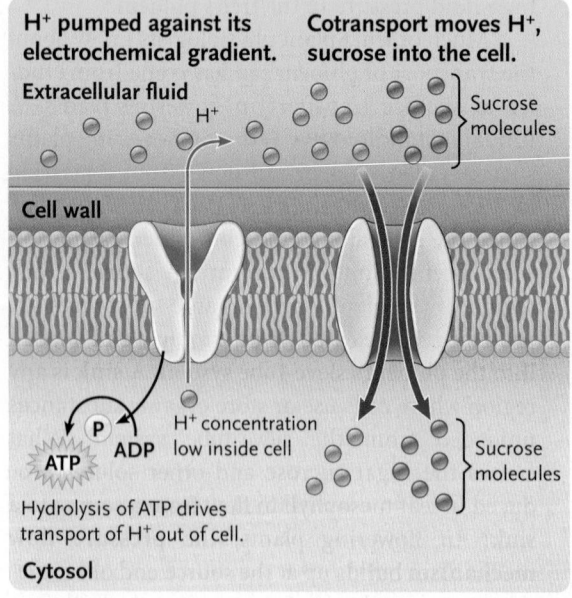

C. Bulk flow from source to sink

Organic solutes are loaded into sieve tubes at a source, such as a leaf, and move by bulk flow toward a sink, such as roots or rapidly growing stem parts.

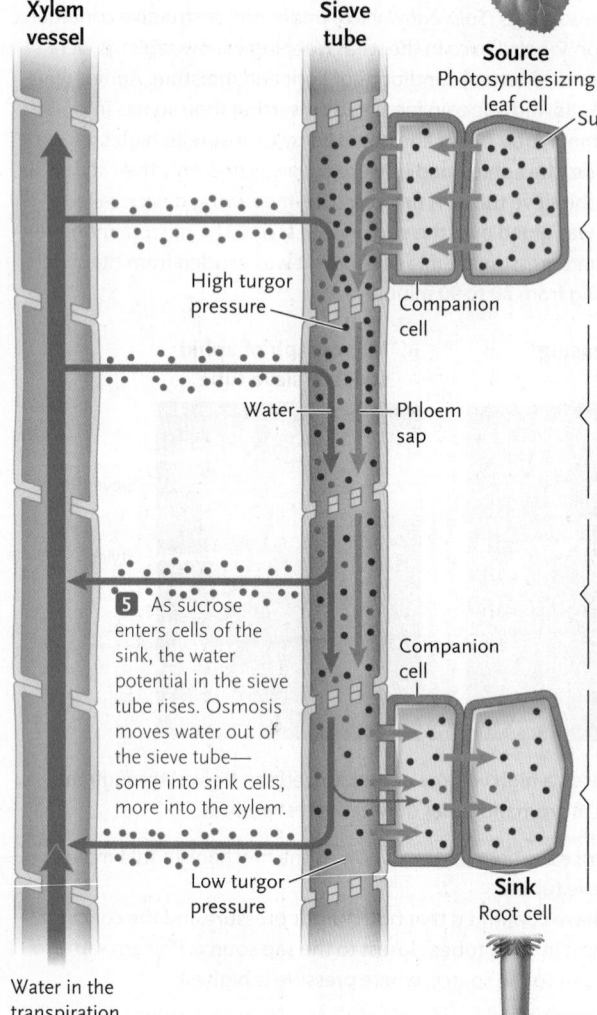

1 Phloem sap forms as active transport loads sucrose into companion cells and then into sieve elements, against concentration gradients.

2 As sucrose becomes more concentrated in the sieve tube, the water potential in the sieve tube falls, so water from xylem enters the tube by osmosis, increasing turgor pressure.

3 Under high pressure, phloem sap moves by bulk flow between a source and a sink. Water moves into and out of the system all along the way.

4 Pressure and sucrose concentration gradually decreases as the sink takes up sucrose from phloem, by active transport from sieve-tube elements into companion cells and then into sink cells.

SUMMARY In the pressure flow mechanism, organic solutes are loaded into sieve tubes at a source, such as a leaf, and move by bulk flow toward a sink, such as roots or rapidly growing stem parts.

think like a scientist At a sink, the unloading of sucrose occurs by active transport. What sort of channel(s) is/are involved?

© Cengage Learning 2017

tulip plant's leaves are translocated into it for storage. Sinks include young leaves, as well as roots and developing fruits. In general, sinks receive organic compounds from sources closest to them. Hence, the lower leaves on a rose bush may supply sucrose to roots, whereas leaves farther up the shoot supply the shoot tip.

Substances carried in phloem must be loaded into sieve-tube elements at sources and unloaded from them at sinks, and both processes have prompted a great deal of research. We know, for example, that in some plants the solutes produced in leaf mesophyll move between the symplast (from cell to cell via plasmodesmata) and the apoplast (adjoining cell walls and air spaces). In addition, active transport mechanisms have a major role in the movement of solutes into and out of cells. Consider sucrose. After this sugar forms inside leaf mesophyll cells, it is exported and travels to sieve tubes and their adjacent companion cells, which interact so closely they often are considered a functional unit. Experiments with tobacco and some other species show that exported sucrose soon leaves the symplast and enters the apoplast next to a small phloem vein (Figure 34.16A). Next, it is actively pumped into companion cells by symport, in which H^+ ions move into the cell through the same carrier that takes up the sugar molecules (Figure 34.16B). Companion cells shuttle the sugar into sieve-tube elements by the same mechanism.

When sucrose is loaded into sieve tubes, its concentration rises inside the tubes. Thus the water potential falls, and water flows into the sieve tubes by osmosis. In fact, the phloem typically carries a great deal of water. As water enters sieve tubes, turgor pressure in the tubes increases, and the sucrose-rich fluid moves by bulk flow into the increasingly larger sieve tubes of larger veins. Eventually, the fluid is pushed out of the leaf into the stem and toward a sink. When sucrose is unloaded at the sink, the water potential in sieve tubes rises, so water tends to flow out of the tubes by osmosis into the surrounding cells, where the water potential is lower. Some of this water "follows the solutes" into sink cells, whereas much more enters the xylem and is recirculated (Figure 34.16C).

Sieve tubes are mostly passive conduits for translocation. The system works because companion cells supply much of the energy that loads sucrose and other solutes at the source, and because solutes are removed at their sinks. As sucrose (and water) enters a sink, for example, its concentration in sieve tubes falls, as does the pressure potential. Thus for sucrose and other solutes transported in the phloem, there is always a gradient of concentration from source to sink—and a pressure gradient that keeps the solute moving along.

As noted previously, phloem sap moving through a plant carries a wide variety of substances, including hormones, amino acids, organic acids, and agricultural chemicals. The phloem also transports organic nitrogen compounds and mineral ions that are withdrawn from dying leaves and stored for reuse in root tissue.

The transport functions of xylem and phloem are closely integrated with phenomena discussed later in this unit—reproduction and embryonic development, and the hormonal regulation of plant growth.

STUDY BREAK 34.5

1. Compare and contrast translocation and transpiration.
2. Using sucrose as your example, summarize how a substance moves from a source into sieve tubes and then is unloaded at a sink. What is this mechanism called, and why?

 ## Unanswered Questions

How can metabolomics advance plant science?

Plants collectively synthesize an estimated 200,000 to 1 million different chemicals. Harnessing this remarkable biosynthetic power will become increasingly important given burgeoning human population growth and the associated need for increased production of nutritious and safe food and for replacing petroleum-based fuels, plastics, and other materials with renewable alternatives. Still, many of the substances plants make are not yet chemically characterized. Both basic science and practical "metabolic engineering" applications will be served by a more complete understanding of the metabolic capabilities of plant tissues.

A key goal is to identify and quantify the compounds that are part of target biosynthetic pathways. Innovations in instrumentation and bioinformatics (use of computational and statistical methods in the study of biological molecules) are now enabling scientists to study the chemical makeup of hundreds, even thousands of metabolites in plant tissues. These advances have launched a discipline called *metabolomics,* which essentially is the study of the small molecule composition of a biological system. There may be thousands of different compounds in an extract from a single tissue type, and the process of characterizing them involves two basic steps—separating them, and then detecting the separated molecules. Although there are a growing number of innovations, gas and liquid chromatography (for separation) and mass spectrometry (for detection) are the most commonly used approaches. A single chromatography/spectrometry run can provide a read-out of hundreds of compounds and their relative levels or absolute concentrations.

The potential applications of metabolomics are vast, and plant scientists have taken a leading role in developing the field. Considerable research is focused on plant storage products such as the starch and protein found in wheat grain. These compounds make up 20% of human caloric intake, and increasing their production is an important step for enhancing global food security. Plant starch is also the major source of biofuels in the United States; therefore, massive investments into research on manipulating starch production have been made. In my laboratory at the University of California at Davis, my colleagues and I use metabolomics to pinpoint compounds that may influence higher levels of storage product synthesis in wheat and other crops. For example, metabolite profiles of plant genotypes that accumulate higher levels of the desired storage product are compared with those that make less. Analysis of metabolites that vary synchronously with the low or high product phenotype may indicate important control points in the pathways. Enzymes catalyzing these steps would be good targets for genetic manipulation to alter accumulation.

Researchers also are tapping metabolomics to address an array of fundamental biological questions in diverse fields such as medicine, environmental science, human nutrition, and food safety. Still, many challenges lie ahead, most of which reflect the complexity of plant metabolism. For instance, because the compounds plants produce vary enormously in their physical and chemical properties, no single extraction and separation technology exists that can measure them all. We must also improve our tools for spatial resolution of metabolites—for example, to see whether a molecule of interest is sequestered within a tissue type or an organelle. Because such data are crucial both for accurate analysis and effective metabolic engineering, researchers are now developing cell-type and even organellar metabolomics to link metabolites with cell-specific processes. A third major need involves analytical refinements that improve the accuracy with which chemical structures revealed by mass spectrometry are classified. At present, more than 60% of the analytes identified from mass spectrometry are classified as "unknown." Among other benefits, advances will help researchers distinguish signatures of endogenous molecules from artifacts of extraction and processing.

think like a scientist The metabolite signature of a tissue is a result of genetic and environmental factors. To identify metabolites that may enhance survival under environmental stress, researchers may compare the metabolite profiles of a stress-sensitive crop plant to stress-tolerant species that may be their wild, nondomesticated relatives. How would you design an experiment to identify the subset of metabolites that may underscore tolerance? What factors must you take into account?

Courtesy of Diane Beckles

Diane M. Beckles is an Associate Professor in the Department of Plant Sciences at the University of California at Davis. Her current research focuses on carbohydrate biosynthesis in crop plants with an emphasis on identifying the factors that dictate its accumulation in seeds and fruits for biotechnological manipulation. To learn more about Dr. Beckles' research, please go to http://www.plantsciences.ucdavis.edu/plantsciences _faculty/beckles/.

REVIEW KEY CONCEPTS

For access to MindTap and additional study materials visit www.cengagebrain.com.

34.1 Overview of Water and Solute Movements in Plants

- Plants have mechanisms for moving water and solutes long distances from the root to shoot or vice versa, and over short distances from cell to cell (Figure 34.2).

- Both passive and active mechanisms move substances into and out of plant cells. Solutes generally are transported by carriers, either passively down a concentration or electrochemical gradient (in the case of ions), or actively against a gradient, which requires ATP energy. An H^+ gradient creates the membrane potential that drives the cross-membrane transport of many ions or molecules (Figure 34.3).

- Most organic substances enter plant cells by symport, a form of cotransport in which the energy of the H^+ gradient is coupled with uptake of a different solute. Some substances cross the plant cell membrane by antiport, in which energy of the H^+ gradient powers movement of a second solute out of cells.

- Driven by water potential (Ψ), water crosses plant cell membranes by osmosis, moving from regions where water potential is higher to regions where it is lower.

- Water potential is the sum of turgor pressure and solute potential. Water potential is measured in megapascals (MPa) (Figures 34.4 and 34.5).

- Water and solutes also move by osmosis into and out of the cell's central vacuole, transported from the cytoplasm across the tonoplast. Aquaporins enhance osmosis across the tonoplast and plasma membrane. Water in the central vacuole is vital for maintaining turgor pressure inside a plant cell.

- Bulk flow of fluid occurs when pressure at one point in a system changes with respect to another point in the system.

34.2 Roots: Moving Water and Minerals into the Plant

- Water and mineral ions entering roots travel laterally through the root cortex to the root xylem, following one or more of three major routes: the apoplastic pathway, the symplastic pathway, and the transmembrane pathway (Figure 34.6A).

- In the apoplastic pathway, water entering roots diffuses between root epidermal cells. Absorbed water and solutes then enter either the symplastic or transmembrane pathway, both of which pass through cells.

- Casparian strips form a barrier that forces water and ions in the apoplast to pass through cells to enter the stele. Ions diffuse from cell to cell to reach the xylem (Figure 34.6B).

34.3 Transport of Water and Minerals in the Xylem

- In xylem, tension generated by transpiration extends down from leaves to roots. By the cohesion–tension mechanism of water transport, water molecules are pulled upward by tension created as water exits a plant's leaves (Figures 34.7 and 34.8).

- In tall trees, negative pressure generated in the shoot drives bulk flow of xylem sap. In some herbaceous species, positive pressure may develop in roots and force xylem sap upward (Figures 34.9 and 34.11).

- Transpiration and CO_2 uptake occur mostly through stomata. Relative humidity, air temperature, and air movement at the leaf surface affect the transpiration rate (Figure 34.12).

34.4 Stomata: Regulating the Loss of Water by Transpiration

- Most plants lose water and take up CO_2 during the day, when stomata are open. The closing of stomata at night conserves water but significantly reduces the inward movement of CO_2.

- Stomata open in response to falling levels of CO_2 in leaves and also to incoming light wavelengths that activate photoreceptors in guard cells.

- Activation of photoreceptors triggers active transport of K^+ into guard cells. Simultaneous entry of anions such as Cl^- and synthesis of negatively charged organic acids increase the solute concentration, lowering the water potential so that water enters by osmosis. As turgor pressure builds, guard cells swell and draw apart, producing the stomatal opening (Figure 34.13).

- Guard cells close when light wavelengths used for photosynthesis wane. The stomata of water-stressed plants close regardless of light or CO_2 needs, possibly under the influence of the plant hormone ABA. The leaves of species native to arid environments typically have water-conserving adaptations (Figure 34.14).

34.5 Transport of Organic Substances in the Phloem

- In flowering plants, phloem sap is translocated in sieve-tube elements. Differences in pressure between sources and sink regions drive the flow (Figure 34.15).

- In leaves, the sugar sucrose is actively transported into companion cells next to sieve-tube elements, then loaded into the sieve tubes through plasmodesmata.

- As the sucrose concentration increases in the sieve tubes, water potential decreases. The resulting influx of water increases pressure inside the sieve tubes, so phloem sap flows in bulk toward the sink, where sucrose and water are unloaded (Figure 34.16).

TEST YOUR KNOWLEDGE

Remember/Understand

1. Short-distance transport mechanisms in plants:
   a. move water and dissolved materials by osmosis.
   b. include the bulk flow of water and solutes.
   c. are not affected by the size of molecules to be transported.
   d. are not affected by the charge of molecules to be transported.

2. Which of the following does *not* have a role in transporting materials between plant cells?
   a. the stele
   b. symport
   c. the cell membrane
   d. stomata
   e. transport proteins

3. Turgor pressure is best expressed as the:
   a. movement of water into a cell by osmosis.
   b. driving force for osmotic movement of water (Ψ).
   c. group movement of large numbers of molecules because of a difference in pressure between two locations.
   d. equivalent of water potential.
   e. pressure exerted by fluid inside a plant cell against the cell wall.

4. Water potential is:
   a. the driving force for the osmotic movement of water into plant cells.
   b. higher in a solution that has more solute molecules relative to water molecules.
   c. a measure of the physical pressure required to halt osmotic water movement across a membrane.
   d. a measure of the combined effects of a solution's pressure potential and its solute potential.
   e. the functional equivalent of turgor pressure.

5. To regulate the flow of water and minerals into roots, the:
   a. Casparian strip of endodermal cells blocks the apoplastic pathway, forcing water and solutes to cross cell plasma membranes in order to pass into the stele.
   b. apoplastic pathway is expanded, allowing a greater variety of substances to move into the stele.
   c. symplastic pathway is modified in ways that make plasma membranes of root cortex cells more permeable to water and solutes.
   d. symplastic pathway shuts down entirely so that substances can move only through the apoplast.
   e. transmembrane pathway augments transport via the apoplast, shunting substances around cells.

6. In seed plants, stomata open when:
   a. water has moved out of the leaf by osmosis.
   b. K^+ and anions flow out of guard cells.
   c. turgor pressure in the guard cells lessens.
   d. the H^+ active transport protein stops pumping.
   e. outward flow of H^+ sets up a concentration gradient that moves K^+ in via symport, while anions enter through other channels.

7. A factor that contributes to the movement of water up a plant stem is:
   a. active transport of water into the root hairs.
   b. an increase in the water potential in the leaf's mesophyll layer.
   c. cohesion of water molecules in stem and leaf xylem.
   d. evaporation of water molecules from the walls of cells in root epidermis and cortex and in the stele.
   e. absorption of raindrops on a leaf's epidermis.

8. In translocation of sucrose-rich phloem sap:
   a. the sap flows toward a source as pressure builds up at a sink.
   b. crassulacean acid metabolism reduces the rate of photosynthesis.
   c. companion cells use energy to load solutes at a source and the solutes then follow their concentration gradients to sinks.
   d. sucrose diffuses into companion cells whereas H^+ simultaneously leaves the cells by a different route.
   e. companion cells pump sucrose into sieve-tube elements.

Apply/Analyze

9. An indoor gardener leaving for vacation completely wraps a potted plant with clear plastic. Temperature and light are left at low intensities. The effect of this strategy is to:
   a. halt photosynthesis.
   b. reduce transpiration.
   c. cause guard cells to shrink and stomata to open.
   d. destroy cohesion of water molecules in the xylem.
   e. increase evaporation from leaf mesophyll cells.

10. **Discuss Concepts** Many popular houseplants are native to tropical rainforests. Among other characteristics, many nonwoody species have extraordinarily broad-bladed leaves, some so ample that indigenous people use them as umbrellas. What environmental conditions might make a broad leaf adaptive in tropical regions, and why?

11. **Discuss Concepts** Insects such as aphids that prey on plants by feeding on phloem sap generally attack only young shoot parts. Other than the relative ease of piercing less mature tissues, suggest a reason why it may be more adaptive for these animals to focus their feeding effort on younger leaves and stems.

12. **Discuss Concepts** So-called systemic insecticides often are mixed with water and applied to the soil in which a plant grows. The chemicals are effective against sucking insects no matter which plant tissue the insects attack, but often do not work as well against chewing insects. Propose a reason for this difference.

13. **Discuss Concepts** Concerns about global climate change and the greenhouse effect center on rising levels of greenhouse gases, including atmospheric carbon dioxide. Plants use CO_2 for photosynthesis, and laboratory studies suggest that increased CO_2 levels could cause a rise in photosynthetic activity. However, as one environmentalist noted, "What plants do in environmental chambers may not happen in nature, where

there are many other interacting variables." Strictly from the standpoint of physiological effects, what are some possible ramifications of a rapid doubling of atmospheric CO_2 on plants in temperate environments? In arid environments?

Evaluate/Create

14. **Design an Experiment** CAM plants typically open their stomata at night instead of during the daytime, in part to limit transpiration on hot days. They include some, but not all, species of succulents (such as cacti and members of the genus *Sedum*). One mild spring afternoon while working in a mountain desert of eastern Oregon you discover a new species of succulent, and although you immediately assume that it is a CAM plant, a quick look with your field microscope reveals that its stomata are wide open. Then you remember that when CAM plants have access to plenty of moisture and are exposed to mild nighttime temperatures, they may shift temporarily to a more common mode of photosynthesis (see Section 8.4). During this period their stomata open during the day and close at night. Your collecting permit allows you to gather a few specimens, which you take back to your lab for testing. Design a simple experiment to determine the basic photosynthetic strategy of the new species, and explain how it will provide the information you seek.

15. **Apply Evolutionary Thinking** A variety of structural features of land plants reflect the conflicting demands for conserving water and taking in carbon dioxide for photosynthesis. Identify at least four fundamental structural adaptations that help resolve this dilemma and explain how each one contributes to a land plant's survival.

For selected answers, see Appendix A.

INTERPRET THE DATA

As you already know, photosynthesizing plant cells require water delivered in the xylem. Recent experiments have revealed that the concentration of ions, especially potassium (K^+), in xylem water influences the velocity of water flow through the xylem, possibly by affecting the pit membranes between xylem vessels. Roots take up K^+ in soil water, but phloem sap also contains K^+ that "recycles" back to the xylem. M. A. Zwieniecki and his colleagues hypothesized that changes in the flow of ions from phloem to xylem can alter the velocity (flow rate) at which water moves through the xylem. Working with red maples *(Acer rubrum)* and sugar maples *(Acer saccharum)*, they devised a girdling experiment that would prevent the recycling of phloem K^+ to xylem without disrupting the movement of xylem water. (Recall that girdling a tree or branch stops the movement of phloem sap beyond the cut.) The experimental design included an apparatus for maintaining normal pressure in the xylem and for adding either deionized water or water containing potassium chloride (KCl), a source of ions, to it. After experimental testing on 43 tree branches, they obtained results for xylem sap flow through the branches as shown in the graphs below.

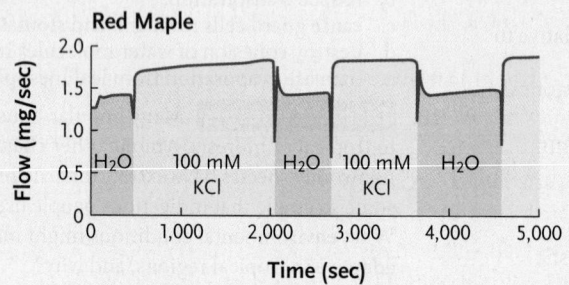

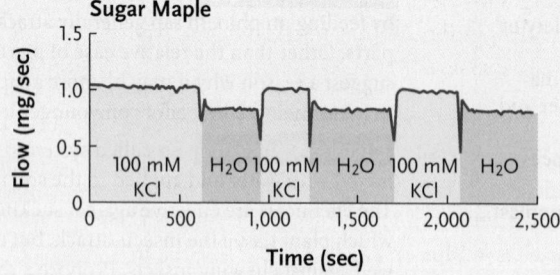

1. What do the different-colored blocks represent?

2. Do the results support or not support a hypothesis that the flow rate of xylem sap slows in response to a decline in ion concentration (KCl)?

3. Were results substantially similar or substantially different for the two species used in the experiment?

Source: M. A. Zwieniecki et al. 2004. A potential role for xylem-phloem interactions in the hydraulic architecture of trees. *Tree Physiology* 24:911–917.

Plant Nutrition

First root of a germinating cabbage seed (*Brassica oleracea,* var. capitata)—the beginning of a root system that will provide the plant with the water and minerals it needs to survive.

Why it matters . . . When botanists talk informally about plant nutrition, the conversation will sometimes include the remark that while animals survive by eating other organisms, plants survive on sunshine and dirt. The comment might generate a laugh, but the underlying point is that, in contrast to animals, plants can generate all the substances required for their metabolism using energy captured via photosynthesis and the water and minerals roots take up from soil. Essentially all the carbon for organic compounds comes from the CO_2 in air, and with enough available water, plant roots gain access to needed hydrogen and oxygen. Roots also provide all the other nutrients vital for normal plant growth and development—elements such as nitrogen, calcium, and phosphorus. In response to the challenge of obtaining soil nutrients, plants have evolved the range of structural and physiological adaptations that we consider in this chapter.

35.1 Plant Nutritional Requirements

No organism grows normally when deprived of a chemical element essential for its metabolism. In the latter half of the nineteenth century, plant physiologists exploited rapid advances in chemistry to probe both the chemical composition of plants and the essential nutrients plants need to survive. In recent times researchers have brought to bear sophisticated methods to expand our understanding of the range of plant nutrients, including those required only in trace amounts.

Plants Require Macronutrients and Micronutrients for Their Metabolism

By weight, the tissues of most plants are more than 90% water. Early researchers could obtain a rough idea of the composition of a plant's dry weight by burning the plant and then analyzing the

FIGURE 35.1 | Research Method

Hydroponic Culture

Purpose: In studies of plant nutritional requirements, using hydroponic culture allows a researcher to manipulate and precisely define the types and amounts of specific nutrients that are available to test plants.

Protocol: In a typical hydroponic apparatus, many plants are grown in a single solution containing pure water and a defined mix of mineral nutrients. The solution is replaced or refreshed as needed and aerated with a bubbling system.

A. Basic components of a hydroponic apparatus

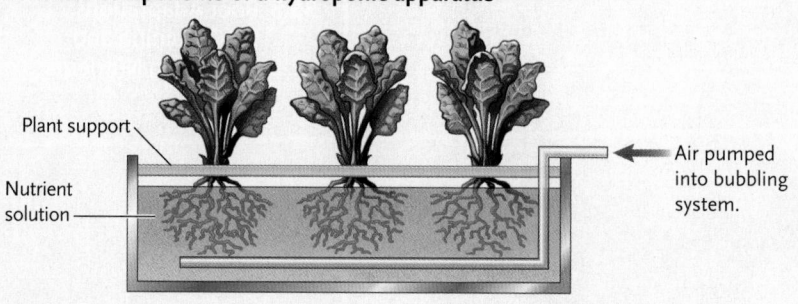

Plant support

Nutrient solution

Air pumped into bubbling system.

B. Procedure for identifying elements essential for proper plant nutrition

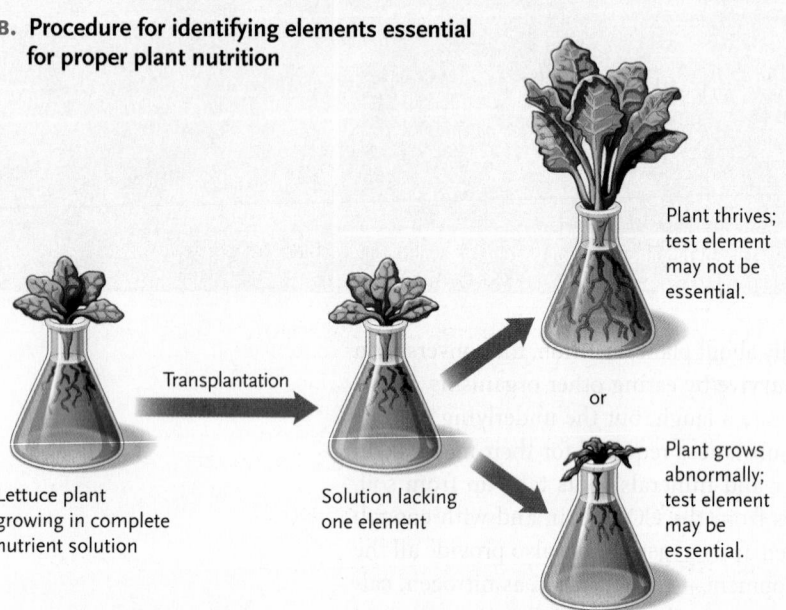

Transplantation

or

Plant thrives; test element may not be essential.

Plant grows abnormally; test element may be essential.

Lettuce plant growing in complete nutrient solution

Solution lacking one element

A "complete" solution contains all the known and suspected essential plant nutrients. An "incomplete" solution contains all but one of the same nutrients, in the same amounts. For experiments, researchers first grow plants in a complete solution, then transplant some of the plants to an incomplete solution.

Interpreting the Results: Normal growth of test plants suggests that the missing nutrient is not essential, whereas abnormal growth is evidence that the missing nutrient may be essential.

© Cengage Learning 2017

depending on the minerals present in the soil where a plant grows, a plant's tissues can contain nonnutritive elements such as gold, lead, arsenic, or uranium.

USING HYDROPONICS IN THE STUDY OF PLANT NUTRITION In 1860, German plant physiologist Julius von Sachs pioneered an experimental method for identifying the minerals absorbed into plant tissues that are essential for plant growth. Sachs carefully measured amounts of compounds containing specific minerals and mixed them in different combinations with pure water. He then grew plants in the solutions, a method now called **hydroponic culture** (*hydro* = water; *ponos* = work). By eliminating one element at a time and observing the results, Sachs deduced a list of six essential plant nutrients, in descending order of the amount required: nitrogen, potassium, calcium, magnesium, phosphorus, and sulfur.

Sachs' innovative research paved the way for decades of increasingly sophisticated studies of plant nutrition, and the eventual identification of many more essential plant nutrients. In the spirit of his work, one basic experimental method involves growing a plant in a solution containing a complete spectrum of known and possible essential nutrients **(Figure 35.1A).** The healthy plant is then transferred to a solution that is identical, except that it lacks one element having an unknown nutritional role **(Figure 35.1B).** Abnormal growth of the plant in this solution is evidence that the missing element is essential. If the plant grows normally, the missing element may not be essential; however, only further experimentation can confirm this hypothesis.

In a typical, modern hydroponic apparatus, the nutrient solution is refreshed regularly, and air is bubbled into it to supply oxygen to the roots. Without sufficient oxygen for respiration, the plants' roots do not absorb nutrients efficiently. (The same effect occurs in poorly aerated or flooded soil.) Variations of this technique are used on a commercial scale to grow some vegetables, such as lettuce and tomatoes.

ESSENTIAL MACRONUTRIENTS AND MICRONUTRIENTS Hydroponics research has revealed that plants generally require 17 essential elements **(Table 35.1).** By definition, an **essential element** is necessary for normal growth and reproduction, cannot be functionally replaced by a different element, and has one or more roles in plant metabolism. With enough sunlight, water, and the 17 essential elements, plants can synthesize all the compounds they need.

ash. This method typically yielded a long list of elements, but the results were flawed. Chemical reactions during burning can dissipate quantities of some important elements, such as nitrogen. Also, plants take up a variety of ions that they do not use;

TABLE 35.1   Essential Plant Nutrients and Their Functions

Element	Commonly Absorbed Forms	Some Known Functions	Some Deficiency Symptoms
Macronutrients			
Carbon*	CO_2	Raw materials for photosynthesis	Rarely deficient
Hydrogen*	H_2O		No symptoms; available from water
Oxygen*	O_2, H_2O, CO_2		No symptoms; available from water and CO_2
Nitrogen	NO_3^-, NH_4^+	Component of proteins, nucleic acids, coenzymes, chlorophylls	Stunted growth; light-green newer leaves; older leaves yellow and die (chlorosis)
Phosphorus	$H_2PO_4^-$, HPO_4^{2-}	Component of nucleic acids, phospholipids, ATP, several coenzymes	Purplish veins; stunted growth; fewer seeds, fruits
Potassium	K^+	Activation of enzymes; key role in maintaining water–solute balance and so influences osmosis	Reduced growth; curled, mottled, or spotted older leaves; burned leaf edges; weakened plant
Calcium	Ca^{2+}	Roles in formation and maintenance of cell walls and in membrane permeability; enzyme cofactor	Leaves deformed; terminal buds die; poor root growth
Sulfur	SO_4^{2-}	Component of most proteins, coenzyme A	Light-green or yellowed leaves; reduced growth
Magnesium	Mg^{2+}	Component of chlorophyll; activation of enzymes	Chlorosis; drooping leaves
Micronutrients			
Chlorine	Cl^-	Role in root and shoot growth, and in photosynthesis	Wilting; chlorosis; some leaves die (deficiency not seen in nature)
Iron	Fe^{2+}, Fe^{3+}	Roles in chlorophyll synthesis, electron transport; component of cytochrome	Chlorosis; yellow and green striping in grasses
Boron	H_3BO_3	Roles in germination, flowering, fruiting, cell division, nitrogen metabolism	Terminal buds, lateral branches die; leaves thicken, curl, and become brittle
Manganese	Mn^{2+}	Role in chlorophyll synthesis; coenzyme action	Dark veins, but leaves whiten and fall off
Zinc	Zn^{2+}	Role in formation of auxin, chloroplasts, and starch; enzyme component	Chlorosis; mottled or bronzed leaves; abnormal roots
Copper	Cu^+, Cu^{2+}	Component of several enzymes	Chlorosis; dead spots in leaves; stunted growth
Molybdenum	MoO_4^{2-}	Component of enzyme used in nitrogen metabolism	Pale green, rolled or cupped leaves
Nickel	Ni^{2+}	Component of enzyme required to break down urea generated during nitrogen metabolism	Dead spots on leaf tips (deficiency not seen in nature)

*Carbon, hydrogen, and oxygen are the nonmineral plant nutrients. All others are minerals.

Nine of the essential elements are **macronutrients,** meaning that plants incorporate relatively large amounts of them into their tissues. Three of these elements—carbon, hydrogen, and oxygen—account for about 96% of a plant's dry mass. Together, these three elements are the key components of lipids and of carbohydrates such as cellulose; with the addition of nitrogen, they form the basic building blocks of proteins and nucleic acids. Plants also use phosphorus in constructing nucleic acids, ATP, and phospholipids, and they use potassium for functions ranging from enzyme activation to mechanisms that control the opening and closing of stomata. Rounding out

the list of macronutrients are calcium, sulfur, and magnesium. All macronutrients except carbon, hydrogen, and oxygen are classified as minerals, which chemists usually define as elements or compounds formed by geological processes and that have a crystalline structure. Minerals are available to plants through the soil as ions dissolved in water.

The other elements essential to plants are also minerals, and are classed as **micronutrients** because plants require them only in trace amounts. Nevertheless, they are just as vital as macronutrients to a plant's health and survival. For example, 5 metric tons of potatoes contain roughly the amount of copper in a single copper-plated penny—yet without it, potato plants are sickly and do not produce normal tubers.

Chlorine, generally present in soil in its anionic form Cl^- (chloride), was identified as a micronutrient nearly a century after Sachs' experiments. Chloride functions in some reactions of photosynthesis and (along with K^+) in the opening and closing of stomata, among other roles. The researchers who discovered its importance in plant nutrition performed hydroponic culture experiments in a California laboratory near the Pacific Ocean, where the air, like coastal air everywhere, contains sodium chloride. The investigators found that their test plants could obtain tiny but sufficient quantities of chloride from the air, as well as from sweat (which also contains NaCl) on the researchers' own hands. Great care had to be taken to exclude chlorine from the test plants' growing environment to prove that it was essential.

In some cases, plant seeds contain enough of certain trace minerals to sustain the adult plant. For example, nickel (Ni^{2+}) is a component of urease, the enzyme required to hydrolyze urea. Urea is a toxic by-product of the breakdown of nitrogenous compounds, and it will kill cells if it accumulates. In the late 1980s, investigators found that barley seeds contain enough nickel to sustain two complete generations of barley plants. Plants grown in the absence of nickel did not begin to show signs of nickel deficiency until the third generation.

Besides the 17 essential elements, some species of plants may require additional micronutrients. Experiments suggest that many, perhaps most, plants adapted to hot, dry conditions require sodium; many plants that photosynthesize by the C_4 pathway (see Section 8.4) appear to be in this group. A few plant species require selenium, which is also an essential micronutrient for animals. Horsetails (*Equisetum*) require silicon; wheat and some other grasses may also need it. Scientists continue to discover additional micronutrients for specific plant groups.

Both micronutrients and macronutrients play vital roles in plant metabolism. Many function as cofactors or coenzymes in protein synthesis, starch synthesis, photosynthesis, and aerobic respiration. As discussed in Section 34.1, some also have a role in creating solute concentration gradients across plasma membranes, which are responsible for the osmotic movement of water. Plants are accomplished hydraulic engineers, and the control of ion concentrations in various compartments is central to their ability to drive water movement.

Nutrient Deficiencies Cause Abnormalities in Plant Structure and Function

Plants differ in the quantity of each nutrient they require—the amount of an essential element that is adequate for one plant species may be insufficient for another. For instance, lettuce and other leafy plants require more nitrogen and magnesium than do other plant types, and alfalfa requires significantly more potassium than does a lawn grass. An adequate amount of an essential element for one plant may even be harmful to another. For example, the amount of boron cotton plants require for normal growth is toxic for lemon trees. For these reasons, the nutrient content of soils is an important factor in determining which plants grow well in a given location.

NUTRIENT DEFICIENCY SYMPTOMS Plants deficient in one or more of the essential elements develop characteristic symptoms (see Table 35.1) that provide clues about the metabolic roles the missing elements play. Deficiency symptoms typically include stunted growth, abnormal leaf color or abnormally formed flowers or stems, or dead spots on leaves or in fruits **(Figure 35.2)**. For instance, iron is a component of the cytochromes on which the cellular electron transfer system depends, and it plays a role in reactions that synthesize chlorophyll. Iron deficiency causes **chlorosis,** a yellowing of plant tissues that results from a lack of chlorophyll (see Figure 35.2A). Because ionic iron (Fe^{3+}) is

A. Iron deficiency

Leaf of a hydrangea plant (*Anomala*) with iron deficiency (left) compared to foliage of a healthy plant (right).

B. Nitrogen deficiency

Leaves of grape vines (*Vitis vinifera*) grown in nitrogen-deficient soil (left) and in soil containing adequate nitrogen (right).

C. Calcium deficiency

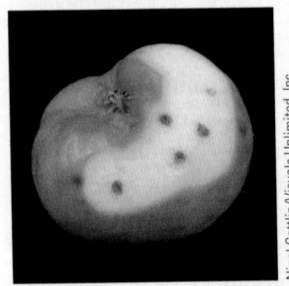

Fruit of an apple tree (*Malus domestica*) lacking sufficient calcium (left) and a normal fruit (right).

FIGURE 35.2 A few examples of mineral deficiency symptoms in plants.

relatively insoluble in water, gardeners often fertilize plants with a soluble iron compound called chelated iron to stave off or cure chlorosis. Similarly, because magnesium is a necessary component of chlorophyll, a plant deficient in this element has fewer chloroplasts than normal in its leaves and other photosynthetic parts. It appears paler green than normal, and its growth is stunted because of reduced photosynthesis.

Plants that lack adequate nitrogen may also become chlorotic (see Figure 35.2B), with older leaves yellowing first because the nitrogen is preferentially shunted to younger, actively growing plant parts. This adaptation is not surprising, given nitrogen's central role in the synthesis of amino acids, chlorophylls, and other compounds vital to plant metabolism. Nutrient deficiencies also affect shoot parts other than leaves. For example, apples from trees grown in calcium-deficient soils develop abnormal gray or brown spots on the skin, and the underlying flesh is brown and dry (see Figure 35.2C). With some other mineral deficiencies, young leaves are the first to show symptoms. These kinds of observations underscore the point that plants use different nutrients in specific, often metabolically complex ways.

FERTILIZERS Soils are more likely to be deficient in nitrogen, phosphorus, potassium, or some other essential mineral than to contain too much. For thousands of years, farmers and gardeners have added nutrients in the form of **fertilizers** to suit the types of plants they wish to cultivate. They may observe the deficiency symptoms of plants grown in their locale or have soil tested in a laboratory, then choose a fertilizer with the appropriate balance of nutrients to compensate for the deficiencies. Packages of commercial fertilizers use a numerical shorthand (for example, 15-30-15) to indicate the percentages of nitrogen, phosphorus, and potassium they contain.

STUDY BREAK 35.1

1. What are the two main categories of the essential elements plants need? Give several examples of each.
2. Do all plants require the same basic nutrients in the same amounts? Explain.

THINK OUTSIDE THE BOOK

Symptoms of a plant nutrient deficiency rarely show up in wild plants. Instead they generally are limited to species cultivated as food crops or common house and garden plants. On your own or with fellow students, develop a hypothesis to explain why this difference exists and describe an experiment that would test your hypothesis.

35.2 Soil

Soil anchors plant roots and is the main source of the water and inorganic nutrients plants require. Soils develop due to the **weathering** of rock and inorganic particles in Earth's crust **(Figure 35.3)**. Wind, rain, and ice are common agents of *physical*

FIGURE 35.3 The weathering of rock. A combination of physical, chemical, and biological weathering is slowly generating soil from these limestone formations in Cappadocia, Turkey.

weathering, which breaks rocks apart into large chunks and smaller particles of various sizes. Minerals in rock are broken down by *chemical weathering* due to agents such as acid rain formed by the interaction of atmospheric water with gaseous acids. *Biological weathering* occurs as lichens and plants secrete organic acids that degrade rocky material, liberating mineral ions into the surrounding soil. The decay of once-living plants and fungi also releases acids that can break down rocky material.

In addition to providing minerals, soil of course is the source of water for most plants, and of oxygen for respiration in root cells. The physical texture of soil is a factor in whether root systems have access to sufficient water and dissolved oxygen. Together, physical and chemical properties of soils have a major impact on the ability of plants to grow, survive, and reproduce in particular habitats.

The Components of a Soil and the Size of the Particles Determine Its Properties

Having been formed by weathering processes, soil is a complex mix of mineral particles, chemical compounds, ions, decomposing organic matter, air, water, and assorted living organisms. The different kinds of soil particles range in size from coarse sand (2.0–0.2 mm) to fine sand (0.2–0.02 mm) to silt (0.02–0.002 mm) and clay (diameter less than 0.002 mm).

INORGANIC SOIL COMPONENTS: THE KEY TO SOIL TEXTURE

The relative proportions of the different sizes of mineral particles give soil its basic texture—gritty if the soil is largely sand, smooth if silt predominates, and dense and heavy if clay is the major component. A soil's texture in turn helps determine the number and volume of pores—air spaces—that it contains. The relative amounts of sand, silt, and clay determine whether a soil is sticky when wet, with few air spaces (mostly clay), or dries quickly and may wash or blow away (mostly sand). Clay soils are more than 30% clay, whereas sandy soils contain less than 20% clay or silt.

ORGANIC SOIL COMPONENTS: IMPROVED ACCESS TO WATER AND NUTRIENTS The mineral particles in soil usually are mixed with various organic components, including **humus**—decomposing parts of plants and animals, animal droppings, and other organic matter. Dry humus is loose and crumbly. It can absorb a great deal of water and thus contributes to the capacity of soil to hold water. Organic molecules in humus are reservoirs of nutrients, including nitrogen, phosphorus, and sulfur, that are vital to living plants.

The piles of bagged humus for sale at garden centers each spring reflect the fact that the amount of humus in a soil also affects plant growth. Its plentiful organic material feeds decomposers whose metabolic activities in turn release minerals that plant roots can take up, but that is not its only value in soil. Humus helps retain soil water and, with its loose texture, helps aerate soil as well. Well-aerated soils containing roughly equal proportions of humus, sand, silt, and clay are called **loams,** and they are the soils in which most plants—including most agricultural crops—do best.

Plants and Other Organisms Influence Soil Features

A square meter of fertile soil contains trillions of bacteria, hundreds of millions of fungi, and several million nematodes, plus an array of other worms and insects. It also contains dead plant roots, leaves, and other parts. Bacteria and fungi decompose this and other organic matter on and in the soil and burrowing creatures such as earthworms aerate the soil. On the other hand, the roots and other tissues of plants may play a key role in shaping the characteristics and composition of soil, including the abundance of soil-dwelling organisms.

Experiments document these soil-shaping activities. For example, Edward Ayres and his colleagues at Colorado State University's Natural Resource Ecology Laboratory studied soil properties in Colorado's San Juan Mountains, where stands of trembling aspen *(Populus tremuloides),* lodgepole pine *(Pinus contorta),* or Engelmann spruce *(Picea engelmannii)* live in close proximity. *P. tremuloides* trees have a more open growth form than pines and spruce trees do, all their leaves drop each year in autumn, and previous research had shown that *P. tremuloides* leaf litter has about twice the nitrogen content of the other two species. With these facts in mind, the Ayres team hypothesized that in their four study areas, such species-specific characteristics would influence the physical, chemical, and biological properties of the soil. The data they gathered supported parts of their hypothesis and also raised questions. For example, they found that in all study areas, the soil in which the aspens grew was significantly warmer—a difference that the team attributed to increased sunlight reaching the ground through the relatively open aspen canopies. The soil littered with aspen leaves also contained more nitrate—a form of nitrogen that plant roots can readily take up—than the nearby soil where the lignin-rich needlelike leaves of pines and spruces accumulated. The study was not designed to attempt a comprehensive analysis of the diversity of the soil's bacterial, fungal, and microscopic animal communities, but the researchers did document markedly different arrays of soil-dwelling organisms associated with the aspens, pines, and spruces in each study area. Clearly, we have a lot more to learn about the intricate interactions between plants, soils, and communities of soil organisms.

The natural development of soils begins as weathering produces *parent material* from rocks. Over time, soils tend to take on a characteristic vertical profile, with a series of layers or **horizons (Figure 35.4).** Each horizon has a distinct texture and composition that varies with soil type. **Topsoil,** the most fertile layer, occurs just below the surface and forms the *A horizon.* This fairly loose layer may be less than a centimeter deep on steep slopes to more than a meter deep in grasslands. It consists of humus mixed with mineral particles and is where the roots of most herbaceous plants are located. Below the topsoil is the **subsoil** or *B horizon,* a layer of larger soil particles containing relatively little organic matter. Mineral ions, including those that serve as plant nutrients, tend to accumulate in the B horizon, and mature tree roots generally extend down into this layer. Under it is the *C horizon,* the parent material consisting of mineral particles and rock fragments that extend down to **bedrock.**

Kenneth W. Fink/Science Source

A horizon: Topsoil
Topsoil contains some percentage of decomposed organic material. It is typically dark in color and its depth usually ranges from 10–30 cm below the surface.

B horizon: Subsoil
The subsoil contains larger soil particles than the A horizon, considerably less organic material, and a greater accumulation of minerals. The absence of much organic material results in lighter color than topsoil. The roots of mature plants often extend into this region, due to its relatively rich mineral content.

C horizon: Parent material
There is usually little if any organic material in the C horizon, which instead consists mostly of the partially weathered fragments and grains of rock from which soil forms. This horizon extends to underlying bedrock.

FIGURE 35.4 Soil horizons in a grassland.

The Characteristics of Soil Affect Root–Soil Interactions

Roots are superbly adapted to penetrate soil and extract needed nutrients from it, but they also are quite sensitive to variations in the properties of soil. In the following section, we consider some adaptations plants have evolved in many otherwise inhospitable soil environments. First, however, we consider the general ways in which soil composition influences the ability of plant roots to obtain water and minerals.

WATER AVAILABILITY As water flows into and through soil, gravity pulls much of the water down through the spaces between soil particles into deeper soil layers. This available water is part of the **soil solution,** a combination of water and dissolved substances that coats soil particles and partially fills pore spaces. The solution develops through ionic interactions between water molecules and soil particles. Clay particles and the organic components in soil (especially proteins) often bear negatively charged ions on their surfaces. The negative charges attract the polar water molecules, which form hydrogen bonds with the soil particles (see Section 2.4).

Unless a soil is irrigated, the amount of water in the soil solution depends largely on the amount and pattern of precipitation (rain or snow) in a region. How much of this water is actually available to plants depends on the soil's composition—the size of the air spaces in which water can accumulate and the proportions of water-attracting particles of clay and organic matter.

The type and size of the particles in a given soil has a major effect on how much water is available to plant roots. Sand particles are small and sandy soil has relatively large air spaces, so water drains rapidly below the top two soil horizons, where most plant roots are located. Soils rich in clay or humus often hold quite a bit of water, although in the case of clay, ample water isn't necessarily an advantage for plants. Whereas a humus-rich soil contains lots of air spaces, the closely layered particles in clay allow few air spaces—and what spaces there are tend to hold tightly the water that enters them. The lack of air spaces in clay soils also severely limits supplies of oxygen available to roots for cellular respiration, and the plant's metabolic activity suffers. Thus, few plants can flourish in clay soils, even when water content is high. (Overwatered houseplants die because their roots are similarly "smothered" by water.) These factors explain why good agricultural soils tend to be sandy or silty loams, which contain a mix of humus and coarse and fine particles.

Chapter 34 described how differences in water potential drive the osmotic movement of water into plant roots. The soil solution is usually quite dilute, with fewer solutes than are dissolved in the water in the cells of plant roots. Accordingly, water tends to move from wet soil, where water potential is higher, into roots, where the water potential is lower. Plants that survive in deserts or salty soils have adaptations that permit their roots to absorb water even when osmotic conditions in soil do not favor water movement into the plant (see Section 34.2).

MINERAL AVAILABILITY Some mineral nutrients enter plant roots as cations (positively charged ions) and some as anions (negatively charged ions). Although both cations and anions may be present in soil solutions, they are not equally available to plants.

Cations such as magnesium (Mg^{2+}), calcium (Ca^{2+}), and potassium (K^+) cannot easily enter roots because they are attracted by the net negative charges on the surfaces of soil particles—mainly clay and humus. To varying degrees, the cations are reversibly bound to negative ions on the surfaces. Attraction in this form is called *adsorption*. Roots do acquire cations, however, through **cation exchange.** In this mechanism one cation, often H^+, replaces a soil cation **(Figure 35.5).** The protons (H^+) come from two main sources. Respiring root cells release carbon dioxide, which dissolves in the soil solution, yielding carbonic acid (H_2CO_3). Subsequent reactions ionize H_2CO_3 to produce bicarbonate (HCO_3^-) and H^+. Roots also excrete H^+ produced by reactions involving organic acids. As H^+ enters the soil solution, it displaces adsorbed mineral cations attached to clay and humus, freeing them to move into roots. Other types of cations may also participate in this type of exchange.

In contrast to cations, anions in the soil solution, such as nitrate (NO_3^-), sulfate (SO_4^{2-}), and phosphate (PO_4^{3-}), are only weakly bound to soil particles. They generally dissolve in the soil solution and so move fairly freely into root hairs. However, because they are so weakly bound compared with cations, anions are more subject to loss from soil by **leaching**—the washing of minerals into deeper soil levels by rain or irrigation.

IMPACT OF SOIL pH The pH of soil affects the availability of many mineral ions. Soil pH is a function of the balance between cation exchange and other processes that raise or lower the concentration of H^+ in soil. In places that receive heavy rainfall—such as both tropical and temperate rainforests—soils tend to become acidic (that is, they have a pH of less than 7). This

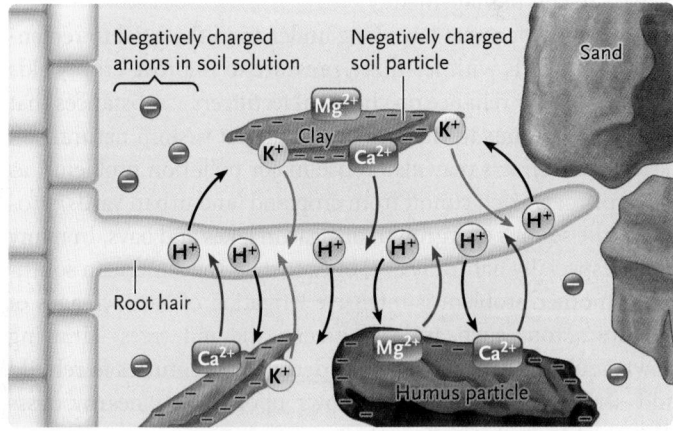

FIGURE 35.5 Cation exchange on the surface of a negatively charged soil particle. When cations come into contact with the negatively charged surface of a soil particle (such as clay), they are adsorbed. As one type of cation, such as H^+, becomes adsorbed, other ions are liberated and can be taken up by plant roots.
© Cengage Learning 2017

acidification occurs in part because moisture promotes the rapid decay of organic material in humus; as the material decomposes, it releases its organic acids. Acid precipitation, which results from the presence of CO_2 and oxides of sulfur and nitrogen oxides in the air (the latter two in large measure from the burning of fossil fuels and industrial emissions), also contributes to soil acidification. By contrast, the soil in arid regions, where precipitation is low, often is alkaline (the pH is greater than 7).

Although most plants are not directly sensitive to soil pH, chemical reactions in very acid (pH less than 5.5) or very alkaline (pH greater than 9.5) soils can have a major impact on whether plant roots take up various mineral cations. For example, experiments have shown that in the presence of OH^- in alkaline soil, calcium and phosphate ions react to form insoluble calcium phosphates. The phosphate captured in these compounds is as unavailable to roots as if it were completely absent from the soil.

For a soil to sustain plant life over long periods, the mineral ions that plants take up must be replenished naturally or artificially. Over the long run, some mineral nutrients enter the soil from the ongoing weathering of rocks and smaller bits of minerals. In the short run, minerals, carbon, and some other nutrients are returned to the soil by the decomposition of organisms and their parts or wastes. Airborne compounds, such as sulfur in volcanic and industrial emissions, may enter soil when they dissolve in rain and fall to earth. Minerals, including compounds of nitrogen and phosphorus, may also enter soil in fertilizers.

Sustainable Agriculture Depends on Proper Soil Management

Directly or indirectly, for more than 10,000 years agriculture has supported complex human societies by providing a reliable supply of the food people consume and feed to livestock, as well as fibers used for clothing and other needs. But agriculture on the scale needed to feed today's billions of people has brought with it serious problems and environmental challenges with respect to soil management.

Although farmers have long understood the need to replenish soil minerals with fertilizer, pressure to improve crop yields has led to heavy reliance on chemical fertilizers—substances that do not add humus to the soil and so do not sustain natural soil fertility. Fertilizers may also cause major pollution problems, as when nitrogen-rich runoff from cropland (and urban yards) promotes the serious overgrowth of algae in lakes and bays. In many areas, especially naturally arid ones such as the American southwest, another problem is intensive irrigation of vast acreages of "thirsty" crops such as leafy greens and nut trees—drawing down aquifers that typically take decades or centuries to rebuild and seriously reducing river flows upon which nearby, less-disturbed ecosystems depend.

Worldwide, an even more pressing issue is soil erosion due to improper tilling, deforestation, or a combination of those or other factors (**Figure 35.6**). *Tilling* refers to plowing to remove plant material remaining from a previous crop and create rows for new seeding. Among other negative impacts, it effectively removes a significant amount of the organic material that forms nutrient-rich humus and leaves the soil more vulnerable to erosion by wind and water. Mismanagement of tilling, irrigation, or fertilizer use also may lead to soil *salinization*—the accumulation of excess salts in soils due to inadequate drainage that often is linked to overirrigation or deforestation (see Chapter 55).

The search for ways to remedy these ills while maintaining necessary food crop production has sparked a movement toward **sustainable agriculture**—methods that include **no-till farming,** in which fields are replanted using techniques that are less physically destructive to the soil and retain the nutrient-dense remains of a previous season's crop. In developed countries, some farmers also are using *precision agriculture technologies* to manage crop production. For example, drone-mounted instruments that monitor chlorophyll production in a swath of cropland—providing an indirect measure of crop growth—allow growers to analyze shifting plant needs for nitrogen and fine-tune applications of nitrogen fertilizers.

In many parts of the world, potentially toxic heavy metals such as arsenic, cadmium, lead, and mercury are present in soil naturally or as industrial contaminants. The use of plants to remove such materials from soil, called *phytoremediation,* is the topic of this chapter's *Focus on Research: Applied Research.*

Science Source

FIGURE 35.6 Soil erosion. Heavy rainfall on an improperly plowed slope in Washington State has eroded topsoil in a portion of a cornfield.

Applied Research: Engineering Solutions to Heavy Metal Contamination

Phytoremediation is the use of plants to remove pollutants from the environment. Major targets are heavy metals such as cadmium, mercury, and arsenic that are naturally present in agricultural soils or enter via industrial pollution. When the roots of crop plants take up heavy metals, those substances may harm the plants' growth; they may also pose a threat to the health of animals, including humans. Rice (*Oryza sativa*) is an example of the latter. Arsenic—a potent human carcinogen—can enter *O. sativa* roots and eventually become incorporated into the rice grains that are a basic foodstuff for billions of people worldwide.

Today much of the scientific search for solutions to heavy metal contamination of soil focuses on the capacity of some plants to accumulate and sequester (permanently retain) large amounts of heavy metals. These "hyperaccumulators" express genes

for enzymes that catalyze the synthesis of *phytochelatins*—peptides that bind with (chelate) metals. The resulting molecular complexes subsequently move into cell vacuoles, where they stay. The enzyme required for phytochelatin synthesis is called PCS (phytochelatin synthase). Genomic studies have identified PCS gene sequences in a range of other plant species, including *Ceratophyllum demersum* **(Figure),** an aquatic plant that naturally accumulates and sequesters arsenic.

Armed with PCS gene sequences, plant scientists in laboratories around the world are attempting to engineer plants having an enhanced capacity to synthesize phytochelatins. For example, a research team at the National Botanical Research Institute in India is using a PCS-coding gene from *C. demersum* in efforts to develop transgenic rice plants in which phytochelatins keep more arsenic sequestered in roots, rather than being

transported to developing grains. Other plant scientists are focusing on creating transgenic plants that can accumulate large quantities of a metal of interest in aboveground tissues that can be harvested (and destroyed), leaving the living plants to continue their "work" of detoxifying a contaminated soil.

FIGURE *Ceratophyllum demersum.*

35.3 Root Adaptations for Obtaining and Absorbing Nutrients

Soil managed for agriculture can be plowed, precisely irrigated, and chemically adjusted to provide air, water, and nutrients in optimal quantities for a particular crop. By contrast, in natural habitats, wide variations in soil minerals, humus, pH, the presence of other organisms, and other factors influence the availability of essential elements. Although adequate carbon, hydrogen, and oxygen are typically available from the air and soil water, other essential elements that must be obtained from soil may not be as abundant. In particular, nitrogen, phosphorus, and potassium are often relatively scarce. The evolutionary solutions to these challenges include an array of adaptations in the structure and functioning of plant roots.

Root Systems Allow Plants to Locate and Absorb Essential Nutrients

Immobile organisms such as plants must locate nutrients in their immediate environment, and for plants the adaptive solution to this problem is an extensive root system. Roots make up 20% to 50% of the dry weight of many plants, and even more in species growing where water or nutrients are especially scarce, such as arctic tundra. As long as a plant lives, its root system continues to grow, taking up water and nutrients from an ever-larger area as it spreads through the surrounding soil. Therefore, although

plants cannot seek water in the same way thirsty animals do, their root systems are fully capable of sensing regions of high nutrient or water availability and accessing those resources as needed. Roots do not necessarily grow *deeper* as a root system branches out, however. In arid regions, a shallow-but-broad root system may be better positioned to take up water from occasional rains that may never penetrate below the first few inches of soil.

You may recall from Section 34.2 that roots take up ions just behind the root tips, where root hairs are present. Root hairs, shown in Figure 33.9C, are a major adaptation for the uptake of mineral ions and water. Over successive growing seasons, long-lived plants such as trees can develop millions, even billions, of root tips, each one a potential absorption site. In a plant such as a mature red oak (*Quercus rubra*), which has a vast root system, the total number of root hairs is astronomical. Even in young plants, root hairs greatly increase the root surface area available for absorbing water and ions. Several years ago a team led by Chinese researcher Keke Yi uncovered a genetic "master switch" that regulates the growth of root hairs. Experiments with *Arabidopsis thaliana* plants revealed that the activity of a transcription factor called RSL4 (for root-hair defective 6-like 4) activates downstream genes that promote the growth of long root hairs, while the absence of RSL4 stunts root hair growth. Apparently, expression of the *RSL4* gene is modulated by external cues, including soil phosphate levels and signals from the plant hormone auxin you will read more about in Chapter 37.

Getting to the Roots of Plant Nutrition

One way that mycorrhizal fungi benefit their host plants is by increasing the amount of phosphate plants take up from the soil.

Research Question

Does a mycorrhizal fungus have a transport protein that can move phosphate into plant roots?

Experiments

To explore this question, Maria J. Harrison and Marianne L. van Buuren at the Samuel Roberts Noble Foundation in Ardmore, Oklahoma, used a cDNA library from a plant whose roots had been colonized by a mycorrhizal fungus in the genus *Glomus* **(Figure).** As described in Section 18.1, the gene sequences in a cDNA library are derived from mRNA, so they represent only the protein-encoding sequences of a genome.

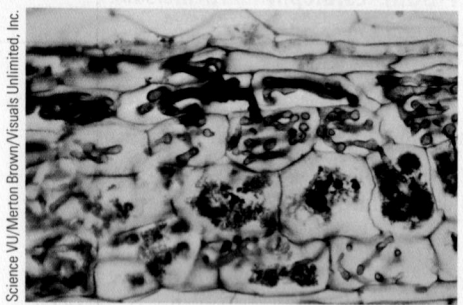

FIGURE Mycorrhizae in root cells of yellow poplar *(Liriodendron tulipifera)* formed by a fungus in the genus *Glomus*.

The cDNA library Harrison and van Buuren used was derived from root tissue in which mycorrhizae were present, so the library contained cDNAs from both the host plant and the mycorrhizal fungus. In an initial step, the researchers probed the cDNA with a radiolabeled yeast gene known to encode a phosphate transporter in the plasma membrane of yeast cells. By way of the probe, the researchers hoped to determine whether any cDNAs in the mixture were similar to the yeast gene.

When mixed with the cDNA, the yeast probe did indeed pair with one of the genes. Subsequent sequencing of the gene revealed that the cDNA coded for a protein with a structure typical of many eukaryotic and prokaryotic membrane transport proteins.

To eliminate the possibility that the yeast probe was identifying a plant cDNA in the library rather than one from the mycorrhizal fungus, the investigators next used the identified cDNA to probe a preparation containing all the DNA of a plant that had *not* been colonized by *Glomus*. No pairing occurred with any of the plant DNA fragments, confirming that the cDNA represented a gene from the fungus. Additional experiments supported this finding.

Harrison and van Buuren carried their investigation further to see whether the fungal gene actually encoded a phosphate transport protein. For this set of experiments, the investigators used a yeast mutant with a nonfunctional phosphate transporter. Because these mutant yeast cells cannot readily take in phosphate, they grow very slowly, even in a phosphate-rich culture medium.

The researchers added the *Glomus* gene to the mutants under conditions that increased the likelihood yeast cells would take up and incorporate the DNA. The yeast cells then began to grow normally, indicating that they could now synthesize a functional phosphate transporter. When radioactive phosphate ions were added to the culture, the cells rapidly became labeled, confirming that they were taking up phosphate ions at a much greater rate than untreated mutants.

Conclusion

This study was the first to reveal the molecular basis of phosphate transport by mycorrhizal fungi. More recent work with potato plants *(Solanum tuberosum)* has identified a gene encoding a phosphate transporter protein that is expressed in parts of potato roots where mycorrhizae form.

think like a scientist

How might having a greater understanding of phosphate transport by mycorrhizal fungi be useful in agronomy (the scientific management of soil properties and crop production)?

Source: M. J. Harrison and M. L. van Buuren. 1995. A phosphate transporter from the mycorrhizal fungus *Glomus versiforme*. *Nature* 378:626–629.

© Cengage Learning 2017

Chapter 34 also mentioned another plant adaptation for gaining access to mineral ions—ion-specific transport proteins in plant cell membranes by which the cells selectively absorb ions from soil. For example, transport channels for potassium ions (K^+) are embedded in the cell membranes of root cortical cells. Such ion transporters absorb more or less of a particular ion depending on chemical conditions in the surrounding soil.

The Discovery That Roots Also Secrete Substances into Soil Expands Our Understanding of Plant Adaptations for Obtaining Nutrients

In addition to acquiring substances from the surrounding soil, roots of various plant species also release into soil a long list of organic compounds, including carbohydrates, amino acids, and various organic and fatty acids, as well as enzymes and other proteins. Today experiments are revealing the details of how such "root exudates" may improve a plant's access to particular nutrients. For example, a study headed by Corey D. Broeckling of Colorado State University showed that the roots of *A. thaliana* and *Medicago truncatula* (barrel medic, a legume) secrete organic compounds that help determine which species of soil fungi can thrive near the roots. As described shortly, such fungi are partners in symbiotic relationships that help nourish many plant species. Similarly, Eric Paterson and his coworkers at the University of Aberdeen in Scotland found that roots of barley plants *(Hordeum vulgare)* exposed to above-normal nitrogen increased their release of organic substances that promote the growth of soil organisms that convert nitrogen into a chemical form plant roots can absorb. Other compounds released by roots enhance the uptake of phosphate. Chapter 37 looks at the role of plant hormones in allowing plants to respond to and even influence the environment their root systems encounter.

Nutrients Move into and through the Plant Body by Several Routes

As discussed earlier in this chapter and in Chapter 34, plants obtain carbon, hydrogen, and oxygen from the air but most mineral ions enter plant roots along with the water in which they are dissolved. Some enter root cells immediately. Others travel in solution *between* cells—in the apoplast—until they meet the endodermis sheathing the root's stele (see Figure 34.6). At the endodermis, the ions are actively transported into the endodermal cells and then into the xylem for transport throughout the plant. Inside cells, most mineral ions enter vacuoles or remain in the cytoplasm, where they are immediately available for metabolic reactions.

Some nutrients, such as nitrogen-containing ions, move in phloem from site to site in the plant, as dictated by growth and seasonal needs. In plants that shed their leaves in autumn, before the leaves age and fall significant amounts of nitrogen, phosphorus, potassium, and magnesium move out of them and into twigs and branches. This evolutionary adaptation conserves the nutrients, which will be used in new growth the next season. Likewise, in late summer, mineral ions move to the roots and lower stem tissues of perennial range grasses that typically die back during the winter. These activities are regulated by hormonal signals, the topic of Chapter 37.

The course of plant evolution has resulted in vitally important symbioses that enhance the capacity of roots to take up nutrients from soil. In the next two subsections we take a closer look at two of these symbioses—mycorrhizae, the associations with fungi introduced in Chapter 30, and interactions with nitrogen-fixing microorganisms.

Mycorrhizae Are Associations with Fungi That Increase Plant Access to Water, Phosphate, and Some Other Nutrients

Chapter 30 noted that **mycorrhizae** are symbiotic associations between a fungus and the roots of a plant. Mycorrhizae promote the uptake of water and nutritionally vital ions—especially phosphate—in at least 80% of plants. This chapter's *Molecular Insights* examines research that revealed how fungi acquire the phosphate they supply to plants.

There are two main kinds of mycorrhizae: **ectomycorrhizae,** in which fungal hyphae remain outside cells in the root cortex, and **endomycorrhizae,** in which hyphae pass through the walls of cells in the root cortex. Both types greatly increase the surface area available to the plant for absorbing water and nutrients.

ECTOMYCORRHIZAE In temperate regions especially, ectomycorrhizae provide many trees and other woody plants with enhanced access to mineral nutrients. The fungal hyphae typically grow as a *mantle*—a dense network of filaments—that cloaks the plant's roots. The hyphae also enter the root cortex, where they weave through the apoplast (the space between cortical cells), as shown in **Figure 35.7A.** The close proximity of the hyphae and cortical cells makes for efficient exchanges of nutrients.

ENDOMYCORRHIZAE Close to 90% of plants, including nearly all angiosperms, benefit from endomycorrhizae. As you might gather from the name (*endo* = "inside"), in these mycorrhizae, the hyphae of the fungal partner do not form a dense sheath on the outside of a root. Instead the hyphae grow through the root epidermis into the root cortex and then move inward through the walls of root cortical cells. In the most common type of endomycorrhiza, when a hypha reaches a cell's plasma membrane, it branches repeatedly, forming treelike **arbuscules (Figure 35.7B).** The arbuscules do not penetrate the plasma membrane but associate closely with it, in a manner akin to a person's fingers pressing into a soft pillow. At this interface, some of the plant's sugars and nitrogenous compounds pass to and nourish the fungus, while some of the minerals the fungus has secured from the soil pass to the plant.

Mycorrhizae may so enhance a plant's nutrition that root-tip epidermal cells do not go to the metabolic expense of producing root hairs. As we see next, another nutritional boon for plants is the suite of soil bacteria whose metabolic activities enhance plants' access to nitrogen.

Plants Depend on Bacteria for an Adequate Supply of Usable Nitrogen

It might seem that plants live surrounded by nitrogen, but in fact very little of that nitrogen is readily available to plant roots. For example, nitrogen steadily enters the soil in organic compounds released when dead organisms and animal wastes decompose, but it is bound up in complex organic molecules such as amino acids and proteins. Air also contains plenty of gaseous nitrogen—at sea level, air is almost 80% N_2 by volume—but plants cannot extract it because they lack the enzymatic machinery necessary to break apart the three covalent bonds in each N_2 molecule ($N\equiv N$). Plants *can* absorb atmospheric nitrogen that reaches the soil in the form of nitrate (NO_3^-) and ammonium ion (NH_4^+), and experiments show that roots of at least some plant species directly take up amino acids. Even so, lack of nitrogen is the single most common limit to plant growth because there usually is not nearly enough nitrogen available in these forms to meet plants' ongoing needs.

Instead, the main natural processes that replenish soil nitrogen and convert it to a form plant roots can absorb are carried out by bacteria. These processes, which we'll now consider, are part of the *nitrogen cycle,* the global movement of nitrogen in its various chemical forms from the environment to organisms and back to the environment, which is described in Chapter 54.

A. Ectomycorrhizae

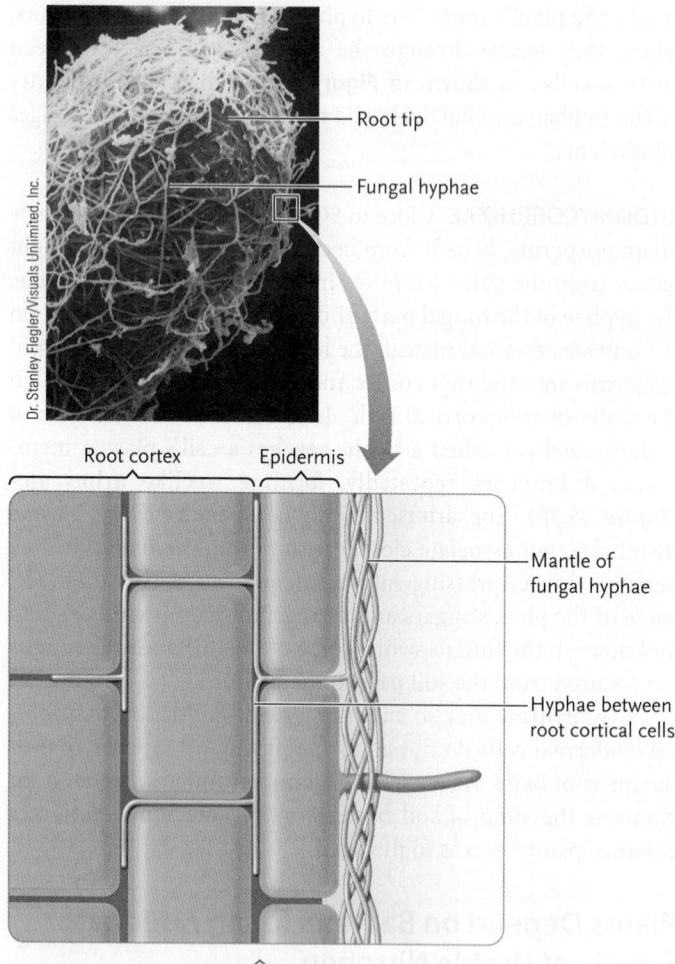

Root tip

Fungal hyphae

Dr. Stanley Flegler/Visuals Unlimited, Inc.

Root cortex   Epidermis

Mantle of fungal hyphae

Hyphae between root cortical cells

In ectomycorrhizae, fungal hyphae form a sheathlike mantle around a plant root. The hyphae cross the root epidermis and infiltrate the spaces between root cortical cells (the apoplast). Carbohydrates from the plant's photosynthesis pass to the fungus, while water, phosphate, and other soil nutrients pass to the plant.

FIGURE 35.7 Mycorrhizae.

B. Endomycorrhizae

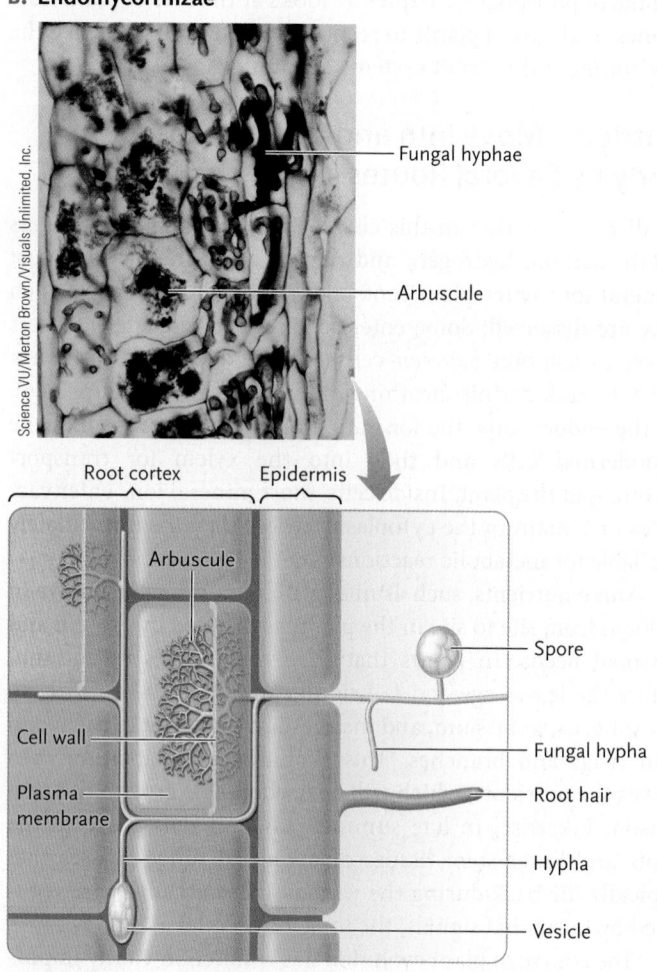

Fungal hyphae

Arbuscule

Science VU/Merton Brown/Visuals Unlimited, Inc.

Root cortex   Epidermis

Arbuscule

Spore

Cell wall

Fungal hypha

Plasma membrane

Root hair

Hypha

Vesicle

In endomycorrhizae, fungal hyphae penetrate root cortex cells. In the arbuscular type of endomycorrhizae shown here, hyphae are sandwiched between the root cell wall and the plasma membrane, where they branch repeatedly into treelike structures. Nutrient exchanges occur across the interface of an arbuscule and the root cell's plasma membrane.

PRODUCTION AND ASSIMILATION OF AMMONIUM AND NITRATE The incorporation of atmospheric nitrogen into compounds that plants can take up is called **nitrogen fixation.** **Figure 35.8** summarizes the basic steps. Metabolic pathways of nitrogen-fixing bacteria in soil or living in mutualistic association with plant roots add hydrogen to atmospheric N_2, producing two molecules of NH_3 (ammonia) and one H_2 for each N_2 molecule. The process requires a substantial input of ATP and is catalyzed by the enzyme **nitrogenase.** In a final step, H_2O and NH_3 react, forming NH_4^+ (ammonium) and OH^-.

Another bacterial process, called **ammonification,** also produces NH_4^+ when *ammonifying bacteria* in soil break down decaying organic matter. Ammonification is an ecologically important process that recycles nitrogen already incorporated into plants and other organisms.

Although plants use NH_4^+ to synthesize organic compounds, most plants absorb nitrogen in the form of nitrate,

NO_3^-. Nitrate is produced in soil by **nitrification,** in which NH_4^+ is oxidized to NO_3^-. Soils generally teem with nitrifying bacteria, which carry out this process. Because of ongoing nitrification, nitrate is far more abundant than ammonium in most soils.

NITROGEN ASSIMILATION Once inside root cells, absorbed NO_3^- is converted by a multistep process back to NH_4^+. In this form, nitrogen is rapidly used to synthesize organic molecules, mainly amino acids. These molecules pass into the xylem, which transports them throughout the plant. In some plants, the nitrogen-rich precursors for needed substances travel in xylem to leaves, where different organic molecules are synthesized. Those molecules then travel to other plant cells in the phloem.

DENITRIFICATION Some soils also contain bacteria that carry out **denitrification**—the conversion of nitrites or nitrates first

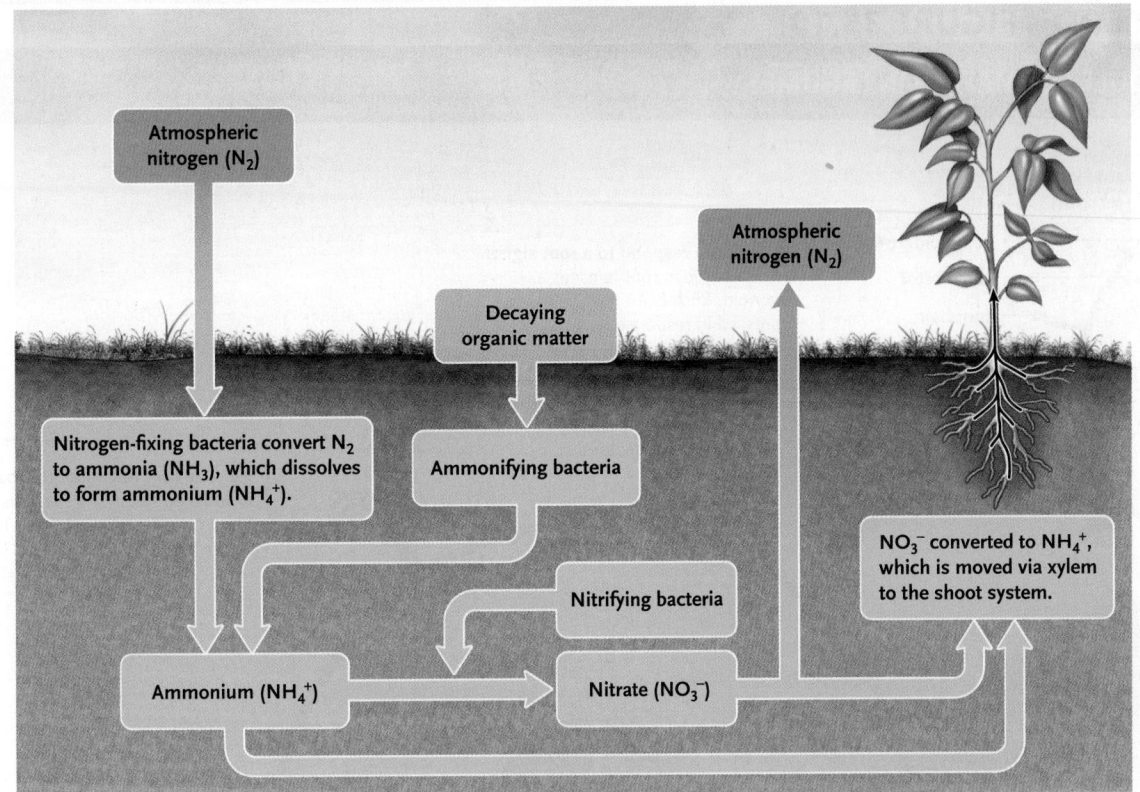

FIGURE 35.8 How plants gain access to nitrogen in the soil. Nitrogen-fixing and ammonifying bacteria in soil convert nitrogen (N_2) from the atmosphere or from decomposing organic material to ammonium (NH_4^+), which plant roots can take up. Other, nitrifying bacteria convert NH_4^+ to nitrate, which also is readily available to roots.
© Cengage Learning 2017

into nitrous oxide (N_2O) and then into molecular nitrogen (N_2) that escapes to the atmosphere. Because denitrification removes nitrates from the soil, it reduces the amount of nitrogen available to plant roots. That said, denitrification is an important part of the global nitrogen cycle discussed in Chapter 54.

Associations with Bacteria Are Vital Sources of Nitrogen for Legumes

Although some nitrogen-fixing bacteria live free in the soil, by far the largest percentage of nitrogen is fixed by species of *Rhizobium* and *Bradyrhizobium,* which form mutualistic associations with the roots of plants in the legume family. The host plant supplies organic molecules that the bacteria use for cellular respiration, and the bacteria supply NH_4^+ that the plant uses to produce proteins and other nitrogenous molecules. In legumes— a large family that includes peas, beans, clover, alfalfa, and alders, among others—the nitrogen-fixing bacteria reside in **root nodules,** localized swellings on roots

(Figure 35.9). Farmers may exploit nitrogen-fixing crops to increase soil nitrogen by rotating crops (for example, planting soybeans and corn in alternating years). When the legume crop is harvested, the root nodules and other tissues remaining in the soil enrich its nitrogen content.

For a plant, an association with nitrogen-fixing bacteria offers the selective advantage of a steady source of absorbable

A. Root nodules on soybean roots

Root nodule

Hugh Spencer/Photo Researchers, Inc.

B. Field experiment with soybeans (*Glycine max*) and *Rhizobium*

NifTAL Project, University of Hawaii, Maui

FIGURE 35.9 The beneficial effect of root nodules. (A) Root nodules on a pea plant *(Pisum sativa).* **(B)** Soybean plants *(Glycine max)* on the left are growing in nitrogen-poor soil. The plants on the right were inoculated with *Rhizobium* cells and developed root nodules.

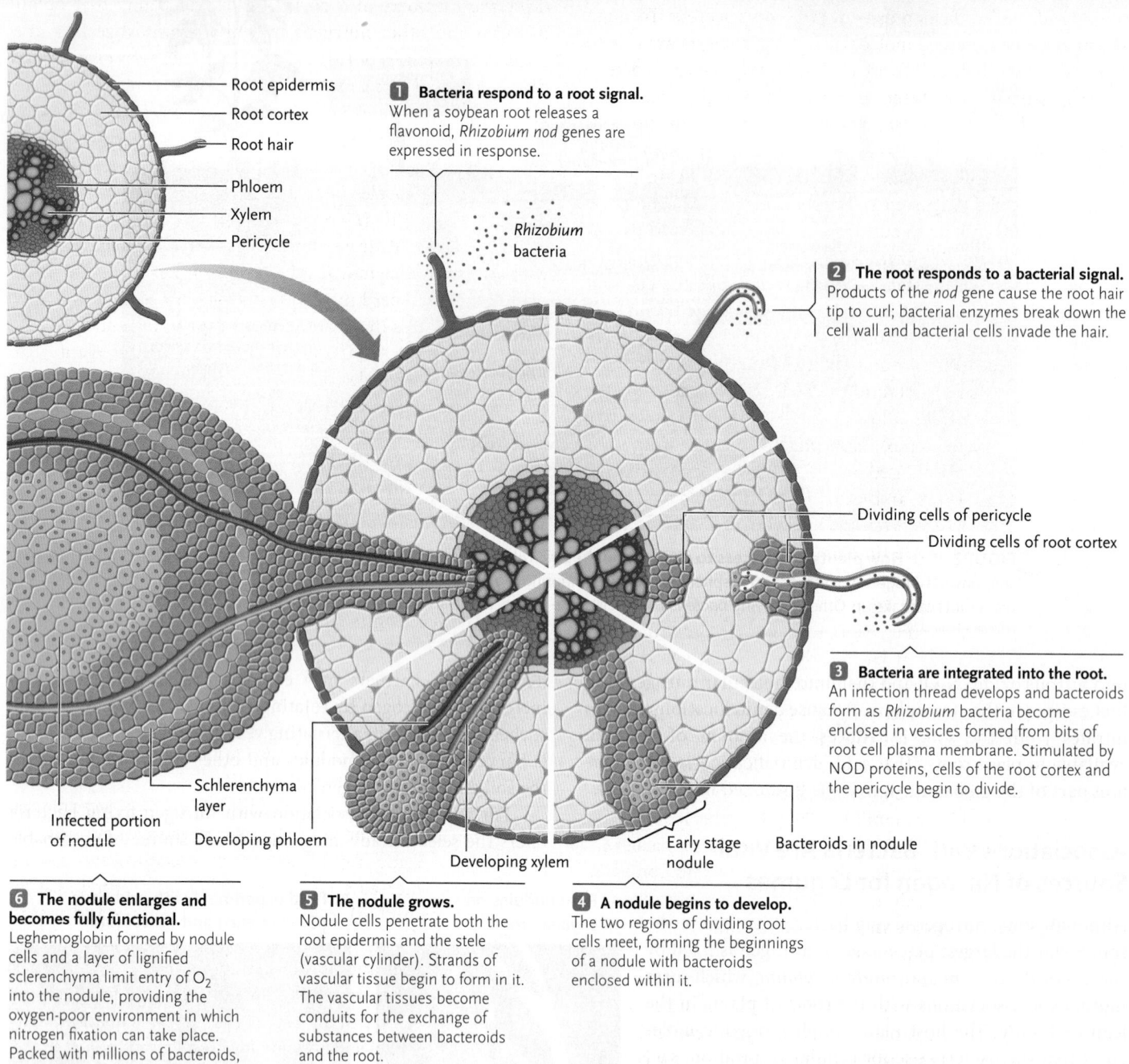

Root epidermis
Root cortex
Root hair
Phloem
Xylem
Pericycle

1 **Bacteria respond to a root signal.** When a soybean root releases a flavonoid, *Rhizobium nod* genes are expressed in response.

Rhizobium bacteria

2 **The root responds to a bacterial signal.** Products of the *nod* gene cause the root hair tip to curl; bacterial enzymes break down the cell wall and bacterial cells invade the hair.

Dividing cells of pericycle
Dividing cells of root cortex

3 **Bacteria are integrated into the root.** An infection thread develops and bacteroids form as *Rhizobium* bacteria become enclosed in vesicles formed from bits of root cell plasma membrane. Stimulated by NOD proteins, cells of the root cortex and the pericycle begin to divide.

Infected portion of nodule
Schlerenchyma layer
Developing phloem
Developing xylem
Early stage nodule
Bacteroids in nodule

6 **The nodule enlarges and becomes fully functional.** Leghemoglobin formed by nodule cells and a layer of lignified sclerenchyma limit entry of O_2 into the nodule, providing the oxygen-poor environment in which nitrogen fixation can take place. Packed with millions of bacteroids, the nodule's girth is much greater than that of the root.

5 **The nodule grows.** Nodule cells penetrate both the root epidermis and the stele (vascular cylinder). Strands of vascular tissue begin to form in it. The vascular tissues become conduits for the exchange of substances between bacteroids and the root.

4 **A nodule begins to develop.** The two regions of dividing root cells meet, forming the beginnings of a nodule with bacteroids enclosed within it.

SUMMARY Legume root nodules typically form as a result of a mutualistic association with the nitrogen-fixing bacteria *Rhizobium* and *Bradyrhizobium*. The association develops as the products of bacterial *nod* genes allow bacteria to infect root cortex cells and proliferate there.

think like a scientist Do all nodule-forming bacteria rely on *nod* gene products for gaining access to root cortex cells?

© Cengage Learning 2017

nitrogen. Decades of research have revealed the details of how this remarkable relationship unfolds. Usually, a single species of nitrogen-fixing bacteria colonizes a single legume species, drawn to the plant's roots by chemical attractants—primarily compounds called flavonoids—that the roots secrete. Through a sequence of exchanged molecular signals, bacteria are able to penetrate a root hair and form a colony inside the root cortex.

An association between a soybean plant *(G. max)* and a bacterium *(Bradyrhizobium japonicum)* illustrates the process. In response to a specific flavonoid released by soybean roots, bacterial genes called *nod* genes (for *nodule*) begin to be expressed (**Figure 35.10,** step 1). Products of the *nod* gene cause the tip of the root hair to curl toward the bacteria and trigger the release of bacterial enzymes that break down the root hair cell wall (Figure 35.10, step 2). As bacteria enter the cell and multiply, they become encased in vesicles formed from the plasma membrane of the host plant cell and are now called **bacteroids.** Meanwhile, invagination of the plasma membrane forms a tube called an **infection thread** that extends into the root cortex and branches, allowing the bacteroids to invade the cortex (Figure 35.10, step 3). The bacteroids now enlarge and become immobile. Stimulated by still other *nod* gene products, cells of the root cortex and pericycle begin to divide. These regions of proliferating cortex cells eventually meet and form the root nodule (Figure 35.10, step 4). As the nodule develops and some of its cells penetrate the root's stele (vascular cylinder), strands of vascular tissue develop within the nodule (Figure 35.10, steps 5 and 6). Typically, each cell in a root nodule contains thousands of bacteroids by way of the nodule's vascular link with the root, the plant takes up some of the nitrogen fixed by the bacteroids, and the bacteroids use some compounds produced by the plant in their metabolism.

Inside bacteroids, N_2 is reduced to NH_4^+ (ammonium) using ATP produced by cellular respiration. Nitrogenase catalyzes the reactions. Ammonium is highly toxic to cells if it accumulates, however. Thus, NH_4^+ is moved out of bacteroids into the surrounding nodule cells immediately and converted to other compounds, such as the amino acids glutamine and asparagine. Small plastids in the nodules, called proplastids, play a crucial role in the process.

Root nodules formed with *Rhizobium* also have adaptations that strictly limit entry of oxygen into the nodule—a crucial function because nitrogenase, the enzyme responsible for nitrogen fixation, is irreversibly inhibited by even low concentrations of O_2. One of these oxygen-limiting features is a layer of lignified sclerenchyma cells positioned close to the nodule's surface (see Figure 35.10, step 5). Another is a factor encoded by bacterial *nod* genes that stimulates nodule cells to produce a protein called **leghemoglobin** ("legume hemoglobin"). Like the hemoglobin of animal red blood cells, leghemoglobin contains a reddish, iron-containing heme group that binds oxygen. Its color gives root nodules a pinkish cast (see Figure 35.9). Leghemoglobin strongly binds oxygen at the nodule surface and offloads just enough oxygen to maintain bacteroid respiration without shutting down the action of nitrogenase.

Some Plants Have Adaptations for Obtaining Nutrients in Unusual Ways

The Venus flytrap, the cobra lily, and various species of sundews are members of a curious group of plants that obtain nitrogen and other nutrients by trapping and digesting animals. A few species of tropical pitcher plants even capture and digest mice and small rats. Such "carnivorous" ("meat eating") plants have elaborate mechanisms for extracellular digestion and absorption that help them to survive in nutrient-deficient, and especially nitrogen-deficient, environments such as sandy areas or boggy locales (where denitrifying bacteria are abundant). The cobra lily *(Darlingtonia californica;* **Figure 35.11A**) is a good example. Its leaves form a "pitcher" that is partly filled with digestive enzymes. Insects lured in by attractive odors often wander deeper into the pitcher, encountering downward-pointing leaf hairs that have a slick, waxy coating and speed the insect's descent into the pool of enzymes. The plant then absorbs monomers released as the animal tissues are digested.

Dodders (**Figure 35.11B**) and thousands of other species of flowering plants are parasites that obtain some or all of their nutrients from the tissues of other plants. Parasitic species develop *haustorial roots* similar to the haustoria of fungi described in Chapter 30. The haustorial roots penetrate deep into the host plant and tap into its vascular tissues. Although some parasitic plants, like mistletoe, contain chlorophyll and thus can photosynthesize (tapping only the host's xylem), dodders and other nonphotosynthesizers rob the host of sugars from the phloem, as well as water and minerals from the xylem.

The snow plant *(Sarcodes sanguinea)* shows a variation on this theme. As its deep red color suggests (**Figure 35.11C),** it lacks chlorophyll, but it doesn't have haustorial roots. Instead, the snow plant's roots take up nutrients from mycorrhizae they "share" with the roots of nearby conifers.

Epiphytes, such as the tropical orchid pictured in **Figure 35.11D,** are not parasitic even though they grow on other plants. Some trap falling debris and rainwater among their leaves, whereas their roots (including mycorrhizae, in the case of the orchid) invade the moist leaf litter and absorb nutrients from it as the litter decomposes. In temperate forests, many mosses and lichens are epiphytes.

These and other strategies plants have evolved for obtaining nutrients and water are only part of the survival equation, however. Plants use nutrients not only for growth and maintenance, but also, of course, for building structures such as pollen, flowers, and seeds used in reproduction—our topic in Chapter 36.

STUDY BREAK 35.3

1. What is a mycorrhiza, and why are mycorrhizal associations so vital to many plants?
2. Distinguish between nitrogen fixation, ammonification, nitrification, and denitrification.
3. Summarize the mechanism by which associations with bacteria supply nitrogen to plants such as legumes.

A. Cobra lily *(D. californica)*

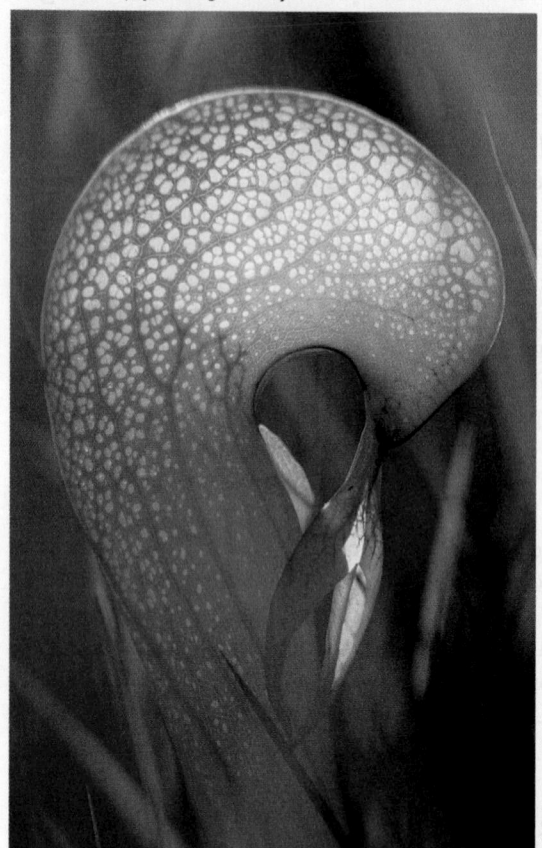

B. Dodder *(Cuscuta)*

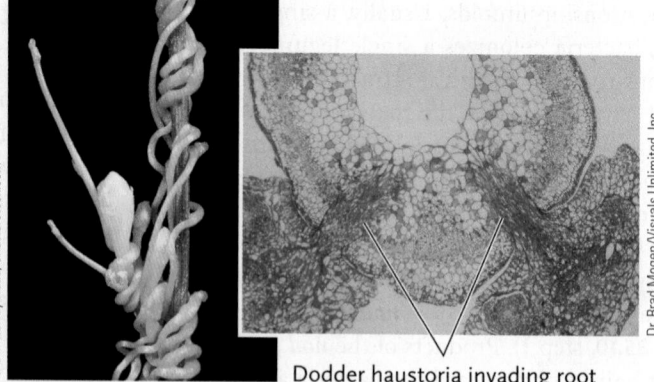

Dodder haustoria invading root

C. Snow plant *(S. sanguinea)*

D. Lady-of-the-night orchid *(Brassavola flagellaris)*

FIGURE 35.11 Some plants with unusual adaptations for obtaining nutrients.
(A) Cobra lily *(D. californica)*, a carnivorous plant. The patterns formed by light shining through the plant's pitcherlike leaves are thought to confuse insects that have entered the pitcher, making an exit more difficult. **(B)** A parasitic dodder, one of the more than 150 *Cuscuta* species. Dodders have slender yellow-orange stems that twine around the host plant before producing haustorial roots *(inset)* that penetrate the host plant and absorb nutrients and water from the host's xylem and phloem. **(C)** Snow plant *(S. sanguinea)*, which pops up in the deep humus of shady conifer forests after snow has melted in spring. This species lacks chlorophyll and does not photosynthesize. Instead its roots intertwine with hyphae of soil fungi that also form associations with the roots of nearby conifers. Radiocarbon studies have shown that the fungi take up sugars and other nutrients from the trees and pass a portion of this food to the snow plant. **(D)** The lady-of-the-night orchid *(Brassavola flagellaris)*, a tropical epiphyte.

THINK OUTSIDE THE BOOK

Botanists have identified a large number of plants that trap and digest small animals. Working with fellow students, do some research online or in the library to develop a list of at least five different variations in the trapping mechanism. Which plant structure or structures have been modified to form each type of trap?

 Unanswered Questions

What is the role of organic nitrogen in plant nutrition and nitrogen cycling?

The classical model of nitrogen cycling in plant ecology has been that plants take up nitrogen in inorganic forms—nitrate and ammonium—made available by microbial conversion of organic nitrogen to mineral nitrogen. This process of nitrogen mineralization involves the decomposition of organic nitrogen-containing substances released into the soil from dead organisms. Such substances include proteins, which comprise a significant proportion of all cells, chitin and peptidoglycan (components of fungal and bacterial cell walls), and DNA and RNA. A fundamental assumption of this classical model is that plants cannot compete with soil microorganisms for mineral nitrogen, and only have

access to what remains after microbial decomposers satisfy their own demand for nitrogen and release excess nitrogen extracted from the organic matter they consume. In recent years, however, studies on nitrogen mineralization in low-nitrogen soils, such as those in many arctic tundra ecosystems, have shown that mineralization rates can be too low to account for the amount of nitrogen taken up by plants in nitrogen-poor environments. This finding raised a fundamental question about nitrogen cycling in arctic tundra and other low-nitrogen environments: How do plants acquire nitrogen from soils where nitrogen mineralization rates are too low to account for plant uptake?

Since the mid 1990s evidence has mounted that plants can bypass the microbial nitrogen mineralization step of the classical nitrogen cycling model. Instead of relying solely on inorganic forms of nitrogen such as nitrate, plants may also acquire nitrogen from organic molecules such as amino acids released into the soil solution as decomposer microorganisms degrade proteinaceous organic matter. The finding that many plants indeed can take up amino acids has raised the possibility that the uptake of organic nitrogen may be an important component of plant nitrogen budgets in low-nitrogen environments, such as arctic tundra. First discovered in arctic tundra plants, amino acid uptake has since been reported for plants in a variety of habitats. Further, we now know that roots possess specific amino acid transporters that can take up a variety of amino acids, and recent studies suggest that root amino acid transport is a trait that can be induced by low nitrogen availability.

think like a scientist Soil microorganisms consume nitrogen-containing organic compounds, and mineralize nitrogen in the process, even after their own nitrogen demands are satisfied. Why don't these microorganisms stop breaking down nitrogenous organic compounds once their demand for nitrogen is met?

Michael Weintraub

Michael Weintraub is an Associate Professor of Soil Ecology in the University of Toledo's Department of Environmental Sciences. His research focuses on developing a mechanistic understanding of key soil processes, to gain insight into how terrestrial ecosystems function, and to predict how they will respond to disturbances such as climate change and nitrogen deposition. To learn more about his work, go to http://www.eeescience.utoledo.edu/Faculty/weintraub/ESELab.htm.

REVIEW KEY CONCEPTS

For access to MindTap and additional study materials visit www.cengagebrain.com.

35.1 Plant Nutritional Requirements

- Plants require 17 essential nutrients (Table 35.1). With enough sunlight and these nutrients, plants can synthesize all the compounds they require to survive.

- Nine essential elements are macronutrients, required in relatively large amounts. Of these, carbon, hydrogen, oxygen, and nitrogen are the main building blocks in the synthesis of carbohydrates, lipids, proteins, and nucleic acids. Macronutrients dissolved in the soil solution are nitrogen, potassium, calcium, magnesium, phosphorus, and sulfur.

- Plants require essential micronutrients in much smaller amounts. Known micronutrients are chlorine, iron, boron, manganese, zinc, copper, molybdenum, and nickel.

- Each plant species requires specific amounts of specific nutrients. Typical deficiency symptoms are stunted growth, yellowing or other abnormal changes in leaf color, and dead spots on leaves and fruit (Figure 35.2).

- Most mineral ions enter plant roots dissolved in water. Inside cells, most mineral ions enter vacuoles or the cell cytoplasm, where they are available for metabolism. Some elements, such as nitrogen and potassium, are mobile—they can move from site to site in phloem as the plant grows.

35.2 Soil

- Soil consists of sand, silt, and clay particles, usually held together by humus and other organic components (Figure 35.3). Humus absorbs considerable water and contributes to the water-holding capacity of soil.

- The relative proportions of various soil mineral particles and humus give soil its basic texture and structure. The best agricultural soils are loams that contain clay, sand, silt, and humus in roughly equal proportions. Topsoil is the most fertile soil layer (Figure 35.4). Plants influence the physical, chemical, and biological features of soil, as when dead plant parts decompose and substances they contain are released to the soil.

- Soil particles are thinly coated by the soil solution, a mixture of water and solutes. Root hairs and other root epidermal cells absorb water and solutes from this solution.

- The amount of water available to plant roots depends mainly on the relative proportions of different soil components. Water moves quickly through sandy soils, whereas soils rich in clay and humus tend to hold water.

- Cations are adsorbed on the negatively charged surfaces of soil particles, potentially limiting their uptake by roots. Cation exchange, in which mineral cations are replaced by H^+, helps make these nutrients available to plants (Figure 35.5). Anions are more weakly bound to soil particles; they move more readily into root hairs but also are more apt to leach out of topsoil. In nature, the soil solution surrounding plant roots generally contains only tiny amounts of essential mineral ions.

- Appropriate soil management is essential for maintaining adequate global food production and avoiding environmental harm. The movement toward sustainable agriculture includes practices such as no-till cultivation and the use of advanced technologies to inform irrigation and the use of fertilizers (Figure 35.6).

35.3 Root Adaptations for Obtaining and Absorbing Nutrients

- Numerous adaptations help plants solve the problems of obtaining and absorbing nutrients. Roots penetrate the soil towards nutrients and water; huge numbers of root hairs greatly increase the root's absorptive surface. Ion-specific transporters in root cortical cells adjust the plant's uptake of particular ions. Mycorrhizal associations between fungi and plant roots enhance the absorption of nutrients, notably phosphorus (Figure 35.7).

- Nitrogen usually is the scarcest nutrient in soil, and nitrogen-fixing bacteria produce much of the usable soil nitrogen. Nitrogen fixation reduces atmospheric N_2 to NH_4^+ (ammonium) in a reaction that requires nitrogenase as a catalyst. Nitrifying bacteria in the soil rapidly convert NH_4^+ to nitrate, the form in which the roots of most plants absorb nitrogen (Figure 35.8).

- In legumes and a few other species, nitrogen-fixing bacteria reside in root nodules in a mutualistic association. Bacteria (bacteroids) enclosed in a root nodule reduce N_2 to NH_4^+. The toxic NH_4^+ is moved out of the bacteroids and converted by nodular proplastids to nitrogen-rich, nontoxic compounds such as amino acids. In plants that do not form root nodules, nitrate absorbed by roots is reduced to ammonium, which then is converted to non-toxic forms (Figures 35.9 and 35.10).

- In many plant species, root cells synthesize amino acids and other organic nitrogenous compounds, and these molecules are transported in phloem throughout the plant. In some plants, the nitrogen-rich precursors travel in xylem to leaves, where different organic molecules are synthesized. Those molecules move to other cells in phloem.

- A few plant species have evolved alternative mechanisms for obtaining some or all of their nutrients (Figure 35.11). Carnivorous plants have structures that physically trap insects or other small animals, and produce solutions of enzymes that digest the animal tissues, releasing absorbable nutrients.

- Some plant species parasitize other plants. The parasite may or may not contain chlorophyll and carry out photosynthesis; species that do not photosynthesize obtain all of their nutrition from the host. Epiphytes grow on other plants but obtain nutrients independently.

TEST YOUR KNOWLEDGE

Remember/Understand

1. Which statement best applies to plant micronutrients?
   a. They typically are not available in loams.
   b. They cannot be replaced by the use of fertilizers.
   c. They appear early on the periodic chart compared with macronutrients.
   d. They are required in large amounts for normal metabolism.
   e. They are essential for normal growth and development.

2. Nutrient runoff from fertilizing lush lawns often causes "algal blooms" in nearby lakes, making swimming impossible. The fertilizer components most likely to have caused the blooms are:
   a. iron, magnesium, and nitrogen.
   b. nitrogen, phosphorus, and sulfur.
   c. nitrogen, potassium, and phosphorus.
   d. selenium, magnesium, and potassium.
   e. nitrogen, magnesium, and nickel.

3. Which of the following are *not* among the ideal soil conditions for growing crops?
   a. extremely large air spaces
   b. sandy or silty loam
   c. blend of sand and clay
   d. less than 5% humus
   e. thick topsoil

4. Which of the following processes contributes to the uptake of mineral ions by plant roots?
   a. chlorosis
   b. osmosis
   c. cation exchange
   d. anion leaching
   e. growth of root hairs

5. Which of the following does *not* influence soil pH?
   a. rainfall
   b. the outward expansion of root systems
   c. release of sulfur and nitrogen oxides into the air
   d. decomposition of organisms
   e. weathering of rock

6. Which of the following is a common process that makes usable nitrogen available to plants?
   a. nitrogen-fixing bacteria synthesizing nitrate
   b. ammonifying bacteria using ammonium to produce nitrate
   c. nitrifying bacteria converting NH_4^+ to NO_3^-
   d. the direct absorption of NH_4^+ by root hairs
   e. the absorption of atmospheric N_2 into the xylem

7. The *nod* genes in the bacteria in soybean nodules allow the bacteria to fix nitrogen. Which of the following, if any, is *not* a step in this process?
   a. The products of *nod* genes cause cells of the root cortex to divide and become the root nodule in which bacteroids fix nitrogen for the plant.
   b. In the cortex cells bacteria enlarge and become immobile, forming bacteroids.
   c. Bacteria enter the root hair cell and multiply, causing the cell plasma membrane to form an infection thread that extends into the root cortex.

   d. Roots release flavonoid, which turns on the expression of bacterial *nod* genes. Products of *nod* genes cause the tip of the root hair to curl toward the bacteria.
   e. Root hairs trigger release of bacterial enzymes that break down root hair cell walls.

8. Being "carnivorous" is a plant adaptation mainly to obtain:
   a. oxygen.
   b. phosphorus.
   c. potassium.
   d. nitrogen.
   e. carbon.

9. Haustorial roots are characteristic of plants that are:
   a. parasites.
   b. epiphytes.
   c. nitrate fixers.
   d. leghemoglobin users.
   e. carnivorous.

10. Identify the correct match of a nutrient with its function.
   a. chlorine: component of several enzymes
   b. potassium: component of nucleic acids
   c. phosphorus: component of most proteins
   d. manganese: role in shoot and root growth
   e. calcium: maintenance of cell walls and membrane permeability

Apply/Analyze

11. **Discuss Concepts** If you want to study factors that affect plant nutrition in nature, what would be the advantages and disadvantages of using a hydroponic culture method?

12. **Discuss Concepts** Gardeners often add a humus-rich "soil conditioner" to garden plots before they plant. Adding the conditioner helps aerate the soil, and the decomposing organic materials in humus provide nutrients. If the plot is for annual plants, it often must be reconditioned year after year, even though the gardener faithfully pulls weeds, fertilizes seedlings, applies chemicals to curtail disease-causing soil microbes, and immediately tosses out the mature plants (along with any plant debris) when they have finished bearing. Suggest some reasons why reconditioning is necessary in this scenario, and some strategies that could help limit the need for it.

13. **Discuss Concepts** One effect of acid rain is to dissolve rock, liberating minerals into soil. Accordingly, can a case be made that acid rain confers environmental benefits, as well as doing harm? What are some other factors, especially with regard to plant adaptations for gaining nutrients, that bear on this question?

14. **Discuss Concepts** Using Table 35.1 as a guide, describe some of the known roles of nitrogen, phosphorus, and potassium in plant function. What are some of the signs that a plant suffers a deficiency in those elements?

Evaluate/Create

15. **Design an Experiment** A plant in your garden is undersized and develops chlorotic leaves even though you fertilize it with a mixture that contains nitrogen, potassium, and phosphorus. After determining that the plant receives enough sunlight for photosynthesis, you next decide to test whether some other aspect of its mineral nutrition is adequate. What specific hypothesis will your experiment test? How will your experimental design test the hypothesis?

16. **Apply Evolutionary Thinking** This chapter's *Focus on Research: Applied Research* discusses phytoremediation, the use of plants to remove environmental pollutants such as heavy metals. As noted, some plant species are hyperaccumulators that take up arsenic and other metallic contaminants and sequester such toxins in shoot parts. How might this activity confer a selective advantage?

For selected answers, see Appendix A.

INTERPRET THE DATA

Numerous studies have shown that the particular plant species growing in a soil influence the soil's chemical properties. With this in mind, in the study of Colorado mountain soil mentioned in Section 35.2, the researchers expected to find differences in the study area's soil chemistry depending on which of the three main tree species—trembling aspen *(Populus tremuloides)*, lodgepole pine *(Pinus contorta)*, or Engelmann spruce *(Picea engelmannii)*—was dominant at a particular site. One factor that affects soil chemistry is the chemical composition of the leaf litter that falls to the ground and decomposes. The **Table** that follows shows values for leaf litter composition as a function of the litter's nitrogen, carbon, cellulose, and lignin content. For each species it also shows the biomass of fine roots (less than 1 mm diameter) and coarse roots (more than 1 mm diameter) in the upper 10 cm of soil.

1. Based on these values, summarize the findings for aspen trees, compared with the pines and spruce trees in the sample.

2. Overall, how does the ratio of lignin to nitrogen in leaf litter seem to correlate with total litter nitrogen content?

3. Did the root biomass among the three tree species differ significantly?

TABLE	Leaf Litter Quality and Root Biomass of Tree Species in a High-Elevation Forest in Colorado							
	Litter N (%)	Litter C (%)	Litter C:N	Litter Cellulose (%)	Litter Lignin (%)	Litter Lignin:N	Fine Root Biomass (g/m^2)	Coarse Root Biomass (g/m^2)
Aspen	0.84 ± 0.08	49.4 ± 0.4	60.4 ± 5.2	42.1 ± 3.0	20.2 ± 2.2	24.1 ± 1.2	321 ± 47	422 ± 134
Pine	0.47 ± 0.03	52.0 ± 0.3	112.1 ± 8.3	59.0 ± 1.0	33.8 ± 0.5	73.2 ± 5.7	215 ± 25	232 ± 87
Spruce	0.41 ± 0.01	48.7 ± 0.1	118.3 ± 4.2	41.1 ± 1.1	19.2 ± 0.7	46.6 ± 1.7	249 ± 28	319 ± 62

Source: Adapted from E. Ayres et al. 2009. Tree species traits influence soil physical, chemical, and biological properties in high elevation forests. *PLoS ONE* 4(6):e5964.

36 Reproduction and Development in Flowering Plants

Ted Kinsman/Science Source

The reproductive structures of an ornamental poppy *(Papaver rhoeas)*. Bright yellow male parts, where pollen grains develop, surround female parts that produce eggs. Within them fertilization occurs and seeds develop.

Why it matters . . . Chocolate gets its start within the fruits of *Theobroma cacao*, a small flowering tree. *T. cacao* evolved in the undergrowth of tropical rainforests in Central America, where the Maya and Aztec peoples domesticated it and incorporated it into their religious and cultural life. Today cacao trees flourish on vast plantations in the tropical lowlands of Central and South America, the West Indies, West Africa, and New Guinea. Unlike most angiosperms, which produce flowers at the tips of floral shoots, *T. cacao* flowers grow directly from buds on the tree trunk. Small flying insects called *midges* pollinate the flowers. Pollination, in turn, is the first step toward fertilization of the eggs, and within about six months, large, podlike fruits develop from them **(Figure 36.1).** Each fruit contains from 20 to 60 seeds—the cacao "beans" that manufacturers process into cocoa, eating chocolate, and other products.

Under natural conditions, *T. cacao* trees reach their maximum pod production by about 10 years of age, but the yield may vary considerably from tree to tree. As the worldwide demand for chocolate has skyrocketed, cacao growers have added cloning to their arsenal of strategies for increasing yields. Combining old-fashioned plant breeding methods with biotechnology, growers create plants that are clones of the best-performing trees. Today much of the chocolate the world consumes comes from orchards of *T. cacao* clones.

Thinking about options for growing cacao trees brings us this chapter's central topic, the reproduction and development of flowering plants. Cacao beans and the seeds of other flowering plants result from sexual reproduction—the union of sperm and eggs that are produced by meiosis. As noted in Chapter 11, this reproductive mode is the source of genetic diversity in offspring, such as differences in the number of seed pods individual cacao trees produce. Genetic diversity, in turn, provides an adaptive advantage in a changing environment, because it increases the chance that as change occurs—perhaps a climatic shift, or the appearance of a new predator or pathogen—at least some offspring will survive and reproduce successfully.

As we will also see in this chapter, many flowering plants are reproductively flexible. They can reproduce either sexually *or* asexually, giving rise to a new generation that is genetically identical to the parents—resulting in phenotypes that are selectively advantageous under certain circumstances. The final section explores techniques humans have harnessed to modify the traits of plants that are sources of food and other products.

36.1 Overview of Flowering Plant Reproduction

In plants as in animals, sexual reproduction occurs when male and female haploid gametes unite to create a diploid fertilized egg. This fertilized egg—the zygote—then may grow and develop into a mature plant.

Once an angiosperm zygote has formed, early development generates an embryo enclosed within a seed. This embryo already contains early versions of the basic plant tissue systems, so the embryo technically is already a **sporophyte**—the diploid, spore-producing body of a plant (see Section 28.1). What most people think of as "a plant" is the sporophyte—for instance, the cherry tree pictured at the top of **Figure 36.2.**

At some point during an angiosperm sporophyte's growth and development, one or more of its vegetative shoots undergo

FIGURE 36.1 Flowers and fruits growing from the trunk of a cacao tree (*Theobroma cacao*) in Central America. Each fruit is the mature ovary of a *T. cacao* flower.

changes in structure and function and become reproductive **floral shoots.** Depending on the species, a floral shoot will give rise to a single flower or a group of flowers—a type of floral shoot called an **inflorescence.** Certain cells in the flowers

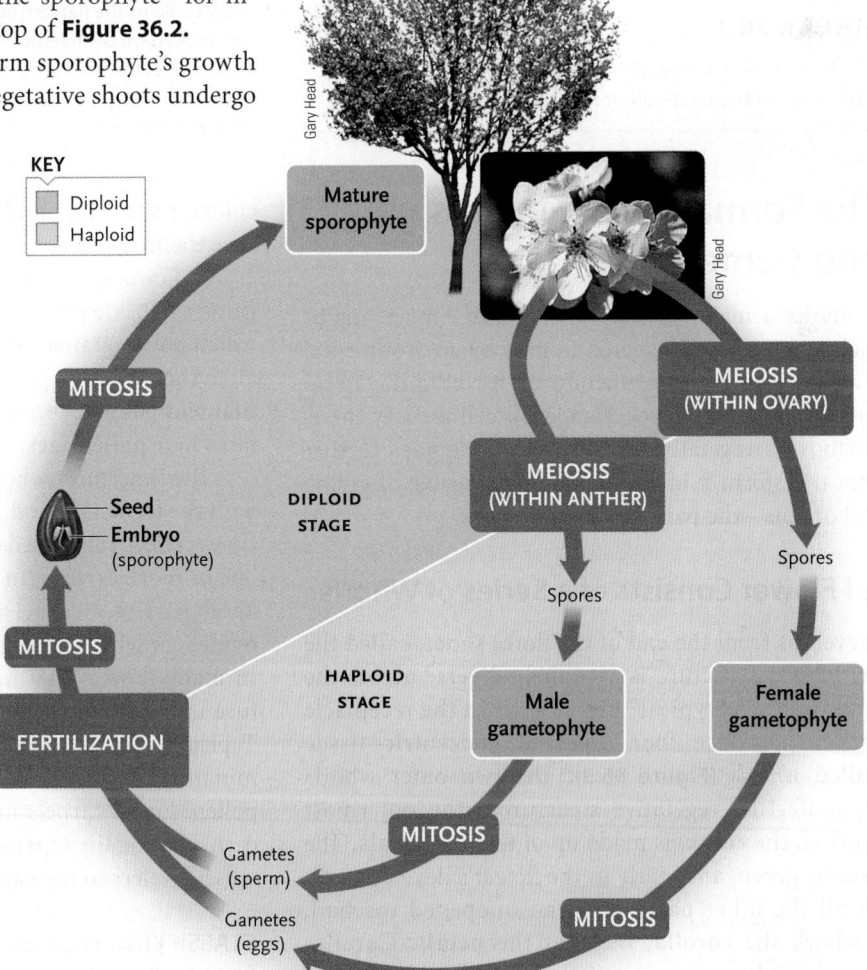

FIGURE 36.2 Overview of the flowering-plant life cycle, using the cherry (*Prunus*) as an example. This type of reproductive cycle, alternation of generations, has a haploid phase in which multicellular gametophytes produce gametes and a diploid phase that begins when two gametes fuse to form a zygote. This zygote develops into a multicellular embryo within a seed and then into a mature sporophyte. Meiotic divisions in the flower of the sporophyte produce spores, which give rise to new gametophytes.

© Cengage Learning 2017

will then divide by meiosis. Unlike meiosis in animals, however, meiosis in plants does not yield haploid gametes. Instead, it gives rise to haploid **spores,** walled cells that subsequently divide by mitosis. By way of these mitotic divisions, each spore grows into a multicellular—but still haploid—gametophyte. The term **gametophyte** means "gamete-producing plant," and you may recall from Chapter 28 that the gametophyte of a vascular plant is a tiny structure that produces haploid sex cells, the gametes, also by mitosis. Male gametophytes produce sperm cells, the male gametes of flowering plants; female gametophytes produce eggs. This division of a life cycle into a diploid, spore-producing generation and a haploid, gamete-producing one is the **alternation of generations** described in Section 28.1.

As you will read shortly, the female gametophyte of a flowering plant usually consists of only seven cells that are embedded in floral tissues. Male gametophytes consist of only three cells and are released into the environment as pollen grains so small that they are measured in micrometers. The pollen grain matures when it reaches a compatible ovule. If fertilization occurs, an embryo develops within a seed.

We turn now to a more detailed discussion of sexual reproduction in angiosperms, beginning with the crucial step in which flowers develop.

STUDY BREAK 36.1

1. What are the two "alternating generations" of plants?
2. How do these two life phases differ in structure and function?

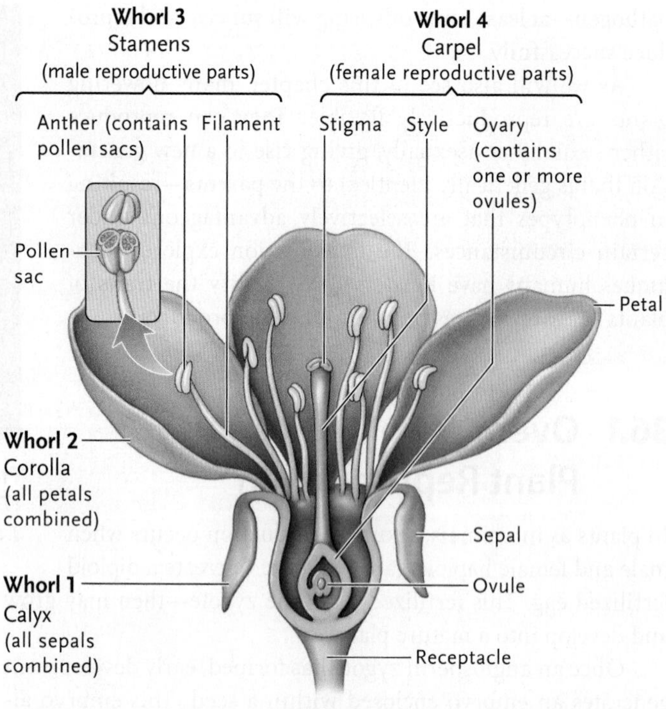

FIGURE 36.3 Structure of a cherry *(Prunus)* flower, with the four whorls indicated. Like the flowers of many angiosperms, it has a single carpel and several stamens. The anthers of each stamen produce haploid pollen. The stigma of the carpel receives the pollen, and the ovule inside the ovary contains the haploid egg. In this diagram, two of the cherry flower's five petals have been removed. For clarity, this diagram shows a single ovule, but as the text notes, flowers of many angiosperm species contain multiple ovules.
© Cengage Learning 2017

36.2 The Formation of Flowers and Gametes

Flowering marks a major developmental shift for an angiosperm. Chemical signals—triggered in part by environmental cues such as day length and temperature—travel to the apical meristem of a shoot and trigger changes in cell activity there. The shoot stops its vegetative growth and undergoes further changes that transform it into a floral shoot capable of giving rise to floral organs—the parts we consider next.

A Typical Flower Consists of a Series of Whorls

A flower develops from the end of the floral shoot, called the **receptacle.** Flower structure is extremely diverse across the angiosperms, but in a "typical" flower, cells in the receptacle differentiate to produce four types of concentric tissue regions called *whorls* **(Figure 36.3).** The two outer whorls consist of nonfertile, vegetative structures. The outermost whorl (whorl 1), the **calyx,** is made up of leaflike **sepals.** The calyx is usually green, and, early in the flower's development, it encloses all the other parts, as in an unopened rosebud. The next whorl, the **corolla,** includes the **petals.** Corollas usually are the "showy" parts of flowers with distinctive

colors, patterning, and shapes. These features often function in attracting bees and other animal pollinators.

A flower's two inner whorls are specialized for making gametes. Inside the corolla is the whorl of **stamens** (whorl 3), in which pollen grains—the male gametophytes—form. In almost all living flowering plant species, a stamen consists of a slender **filament** (stalk) capped by a bilobed **anther.** Each anther contains four **pollen sacs,** in which pollen develops.

The innermost whorl (whorl 4) consists of parts associated with the formation and fertilization of eggs. These parts, sometimes collectively called the *pistil* (now an archaic term), include one or more **carpels,** in which female gametophytes form. The lower part of a carpel is the **ovary.** Inside it is one or more **ovules,** in which an egg develops and fertilization takes place. In many flowers that have more than one carpel, the carpels fuse into a single, common ovary containing multiple ovules. Typically, the carpel's slender **style** widens at its upper end, terminating in the **stigma,** which serves as a landing platform for pollen. Fused carpels may share a single stigma and style, or each may retain separate ones. The name angiosperm ("seed vessel") refers to the carpel.

CLASSIFYING FLOWERS BY WHORL FEATURES The arrangement and number of whorl types varies in different species.

A. Swamp rose-mallow

Dick Poe/Visuals Unlimited, Inc.

B. Squash flower

sigur/Shutterstock.com

C. Tulip

Steve Cordory/Shutterstock.com

D. Corn

David Sieren/Visuals Unlimited, Inc.

Silks of female flowers

Male flowers

E. Goat willow

Vaughan Fleming/Science Source

iStockphoto.com/Ulrich Knaupe

Female catkins

Male catkins

FIGURE 36.4 Examples of complete and incomplete flowers. (A) Complete flower of swamp rose-mallow *(Hibiscus pallustris),* which has all four floral whorls. **(B)** Incomplete flower of a squash plant *(Cucurbita)*—in this case, a "female" plant in which the stamen whorl is missing and the flowers contain only carpels. **(C)** A tulip flower *(Tulipa)* is perfect (and complete)—it has both stamens and carpels. **(D)** Corn *(Zea mays)* is monoecious: each plant produces flowers of both sexual types. In female flowers, long, sticky strands called silks develop from each ovary. Pollination occurs after pollen released by male flowers (called tassels) adheres to the silks. **(E)** The goat willow *(Salix caprea)* is dioecious—individual plants are either male or female and produce either male or female flowers called catkins. Like the flowers of corn, the catkins of a willow plant are examples of incomplete (and imperfect) flowers because they lack the parts of the opposite sexual type.

Flowers with all four whorls are called **complete flowers (Figure 36.4A).** In some species flowers lack one or more of the whorls, and thus botanists describe them as **incomplete flowers (Figure 36.4B).** Botanists also distinguish flowers on the basis of the sexual parts they contain. Most angiosperms produce **perfect flowers,** which have both kinds of sexual parts—that is, both stamens and carpels **(Figure 36.4C). Imperfect flowers** have stamens *or* carpels, but not both. (Notice that all imperfect flowers are also incomplete because they lack one of the whorls.) Species with imperfect flowers are further divided according to whether individual plants produce both sexual types of flowers, or only one. In **monoecious** *(mono-* = one; *oikia* = house) species, such as oaks and corn, each plant has some "male" flowers with only stamens and some "female" flowers with only carpels **(Figure 36.4D).** In **dioecious** ("two houses") species, such as willows, a given plant produces flowers having only stamens or only carpels **(Figure 36.4E).**

EVOLUTIONARY TRENDS IN FLOWER STRUCTURE Chapter 29 discussed the burst of adaptive radiation that occurred as flowering plants diversified into terrestrial environments. Fossils show that this adaptive history has been marked by a variety of changes in flower structure that may reflect selection pressures related to interactions between flowering plants and their animal pollinators (see Section 29.3). **Figure 36.5** provides a closer look at four overall trends in these adaptive changes: a reduction in the number of floral parts, the fusion of floral parts, embedding of the ovary deep in the receptacle, and a shift from radial to bilateral flower symmetry. With this varied angiosperm reproductive anatomy in mind, we turn next to the processes by which male and female gametes come into being.

Pollen Grains Arise from Microspores in Anthers

Most of a flowering plant's reproductive life cycle, from production of sperm and eggs to production of a mature seed, takes place within its flowers. **Figure 36.6** summarizes this cycle as it unfolds in a perfect flower. The microspores that give

Angiosperms and their flowers are extremely diverse. Flower diversity largely reflects adaptations of flower structure that enhance the likelihood of pollination. Evidence from fossils and the morphology of living angiosperms shows four major trends in flower evolution.

Trend 1: Fewer floral parts. In lineages that arose early in angiosperm evolution, flowers typically have many petals and other floral parts, and the number of these parts may vary from blossom to blossom. The flower of the grassland plant called tidy tips is an example. More recently evolved lineages, represented here by a whitemouth dayflower, have fewer parts and specific numbers of them. This shift correlates with increasing specialization of flowers and their pollinators.

A. The many floral parts of the coastal tidy tips *(Layia platyglossa)*

B. The whitemouth dayflower *(Commelina erecta)*, which has three petals (two large and one very small), six stamens, only three of which are fertile, and a single stigma

Trend 2: Fused floral parts. Having separate, often brightly colored petals and numerous distinct anthers and other parts is an ancient floral characteristic geared to attracting a range of pollinators. Other lineages evolved flowers in which petals fuse into tubular or conical shapes that make stamens and stigmas accessible to only specific pollinators.

A. A coneflower *(Rudbeckia)*, which is pollinated by a variety of insects, including bees and butterflies

B. The funnel-shaped flower of devil's trumpet *(Datura metel)*, which opens at night and is pollinated mainly by one species of moth

© Cengage Learning 2017

rise to male gametophytes are produced in a flower bud's anthers (see Figure 36.6, left). The pollen sacs inside each anther hold diploid microsporocytes (or *microspore mother cells*); each microsporocyte undergoes meiosis and eventually produces four small haploid **microspores.** Like most plant cells, the microspores are walled, and inside its wall each microspore divides again, this time by mitosis, producing a *tube cell* and a *generative cell.* The resulting two-celled structure is a **pollen grain:** an immature, haploid male gametophyte.

The generative cell divides by mitosis, forming two sperm cells. Now the immature male gametophyte consists of these two sperm cells plus the tube cell, which will control the development of an elongating cellular extension called a **pollen tube.** When a pollen grain lands on a stigma, this tube grows through the tissues of a carpel and carries the sperm to the ovary (Figure 36.6, step 5). A mature male gametophyte consists of the pollen tube and sperm—the male gametes. Once the pollen tube and sperm have formed, the stage is set for fertilization, which we discuss more thoroughly in Section 36.3.

The walls of pollen grains are hardened by the decay-resistant polymer sporopollenin (see Chapter 29) and are tough enough to protect the male gametophyte during the journey from anther to stigma. The walls are so distinctive that the family to which a plant belongs often can be identified from

Trend 3: Ovaries protected within receptacles.
Older angiosperm lineages have ovaries *superior* to—that is, above—the receptacle, where developing seeds are relatively exposed. The evolution of *inferior* ovaries enclosed within the receptacle offered seeds more protection.

A. Longitudinal section of a tulip flower (*Tulipa*), showing the superior ovary

B. Longitudinal section of a daffodil (*Narcissus*), showing the inferior ovary protected by the flower's fused petals and sepals

Trend 4: Bilateral instead of radial symmetry. Floral organs likely arose through the modification of leaves, and some flowers still have petals and other parts arranged in a spiral or radial pattern commonly seen in the placement of leaves on stems. Radially symmetrical flowers can be sliced into halves anywhere across the center. More specialized flowers show bilateral symmetry—that is, a down-the-middle slice produces halves that are mirror images of each other. Bilaterally symmetrical flowers funnel pollinators toward anthers and stigmas.

A. Radial symmetry of a water lily flower (*Nymphaea*)

B. Bilateral symmetry of a moth orchid (*Phalaenopsis*)

SUMMARY Four general types of changes in flower morphology have occurred over the millennia of angiosperm evolution. These trends, probably driven by selection pressures related to enhanced pollination, are (1) a reduction in the number of floral parts, (2) fusion of floral parts, (3) protection of the ovary within the receptacle, and (4) flowers displaying bilateral symmetry.

think like a scientist Some angiosperm species self-fertilize most or all of the time. Would you expect such species to have large flowers with brightly colored petals?

pollen alone—based on the size and wall sculpturing of the grains, as well as the number of pores in the wall. As noted in Section 29.1, because pollen grains withstand decay, they fossilize well and can provide clues about the evolution of seed plants and the ecological communities that lived in the past.

Eggs and Other Cells of Female Gametophytes Arise from Megaspores

Meanwhile, in the ovary of a flower, one or more dome-shaped masses form on the inner wall. Each mass becomes an ovule (see Figure 36.6, right), which, if all goes well, develops into a seed. Only one ovule forms in the carpel of some flowers, such as the cherry. Dozens, hundreds, or thousands may form in the carpels of other flowers, such as those of a bell pepper plant (*Capsicum annuum*). At one end, the ovule has a small opening, called the **micropyle.**

Inside the cell mass, a diploid megasporocyte (or *megaspore mother cell*) divides by meiosis, forming four haploid **megaspores.** In most species, three of these megaspores disintegrate. The remaining megaspore enlarges and develops into the female gametophyte in a sequence of steps tracked on the right-hand side of Figure 36.6.

First, three rounds of mitosis occur *without* cytoplasmic division; the result is a single cell with eight nuclei arranged in

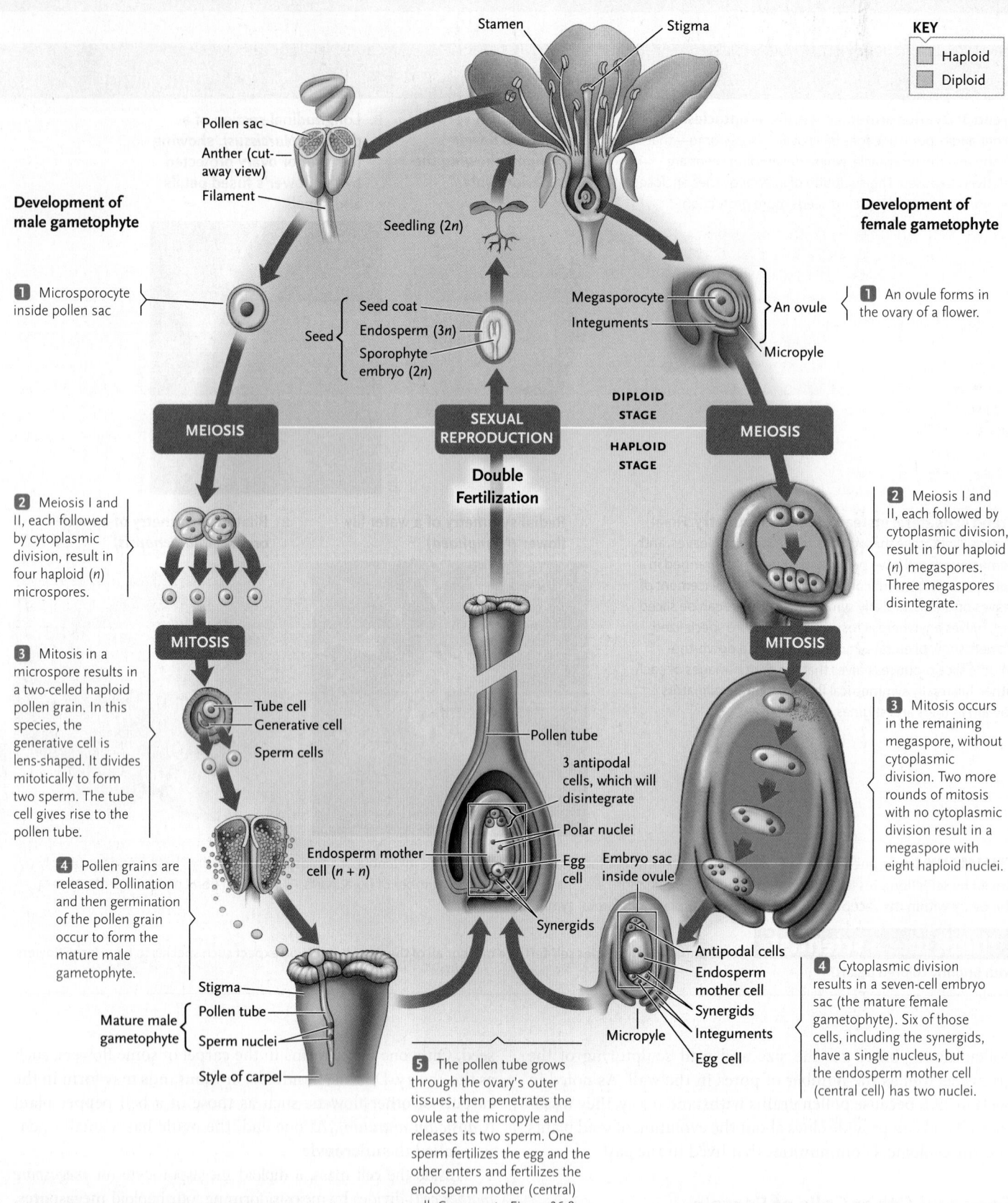

Development of male gametophyte

Stamen

Stigma

KEY
- ☐ Haploid
- ☐ Diploid

Pollen sac
Anther (cut-away view)
Filament

Seedling (2n)

Development of female gametophyte

1 Microsporocyte inside pollen sac

Seed {
Seed coat
Endosperm (3n)
Sporophyte embryo (2n)
}

Megasporocyte
Integuments

An ovule }

1 An ovule forms in the ovary of a flower.

Micropyle

MEIOSIS

SEXUAL REPRODUCTION

DIPLOID STAGE

HAPLOID STAGE

MEIOSIS

Double Fertilization

2 Meiosis I and II, each followed by cytoplasmic division, result in four haploid (n) microspores.

2 Meiosis I and II, each followed by cytoplasmic division, result in four haploid (n) megaspores. Three megaspores disintegrate.

MITOSIS

MITOSIS

3 Mitosis in a microspore results in a two-celled haploid pollen grain. In this species, the generative cell is lens-shaped. It divides mitotically to form two sperm. The tube cell gives rise to the pollen tube.

Tube cell
Generative cell

Sperm cells

Pollen tube

3 antipodal cells, which will disintegrate

3 Mitosis occurs in the remaining megaspore, without cytoplasmic division. Two more rounds of mitosis with no cytoplasmic division result in a megaspore with eight haploid nuclei.

Polar nuclei

Endosperm mother cell (n + n)

Egg cell

Embryo sac inside ovule

4 Pollen grains are released. Pollination and then germination of the pollen grain occur to form the mature male gametophyte.

Synergids

Antipodal cells
Endosperm mother cell
Synergids

4 Cytoplasmic division results in a seven-cell embryo sac (the mature female gametophyte). Six of those cells, including the synergids, have a single nucleus, but the endosperm mother cell (central cell) has two nuclei.

Stigma

Mature male gametophyte {
Pollen tube
Sperm nuclei
}

Style of carpel

Micropyle
Integuments
Egg cell

5 The pollen tube grows through the ovary's outer tissues, then penetrates the ovule at the micropyle and releases its two sperm. One sperm fertilizes the egg and the other enters and fertilizes the endosperm mother (central) cell. Compare to Figure 36.8.

FIGURE 36.6 Life cycle of cherry (Prunus), a eudicot. Pollen grains develop in pollen sacs in the anthers (left, steps 1–3). An embryo sac develops in the single ovule in the cherry flower's ovary, and a haploid egg forms within the embryo sac (right, steps 1–4). When a pollen grain is released and contacts the stigma (left, step 4), it germinates and produces a pollen tube containing two haploid sperm. The tube penetrates the stigma, then grows to the ovary and penetrates the ovule (step 5). Eventually double fertilization occurs, giving rise to an embryonic sporophyte and endosperm that become encased together in a seed coat.

© Cengage Learning 2017

two groups of four. Next, one nucleus in each group migrates to the center of the cell; these two **polar nuclei** ("polar" because they migrate from opposite ends of the cell) may fuse or remain separate. The cytoplasm then divides, and a cell wall forms around the two polar nuclei, forming a single large endosperm mother cell (also called the *central cell*). A wall also forms around each of the other nuclei. Three of these walled nuclei become *antipodal cells,* which eventually disintegrate. Three others form a cluster (called the "egg apparatus") near the micropyle; one of them is an **egg cell** that may eventually be fertilized. The other two, called *synergids,* will have a role in fertilization. The eventual result of all these events is an **embryo sac** containing seven cells and eight nuclei. This embryo sac is the female gametophyte.

Research Has Revealed How Chemical Cues Guide Embryo Sac Development

How do the four different types of embryo sac cells become differentiated, allowing them to fulfill their respective reproductive functions? A study led by Gabriela C. Pagnussat at the University of California, Davis, has provided some insights into this mechanism. Working with *Arabidopsis thaliana* plants, Pagnussat and her colleagues demonstrated that the plant hormone auxin, already known for other growth effects in plants (described in Chapter 37), is a major factor in determining the identities and roles of cells and nuclei within the embryo sac. The team determined that when mitosis begins in a megaspore, an auxin gradient is established along the future axis of the developing embryo sac. The highest auxin concentration occurs near the micropyle, where the pollen tube will enter the ovule. As mitotic cell divisions continue, each of the eight products develops as

an egg, synergid, polar nucleus, or antipodal cell depending on its position within the auxin gradient. Thus the egg cell and synergids develop where auxin is most concentrated, the antipodal cells develop where auxin is lowest, and the polar nuclei develop where there is an intermediate auxin concentration (**Figure 36.7**). Auxin can be said to provide *positional information* for embryo sac development, a topic we return to in Section 36.4.

The formation of gametophytes sets the stage for events by which a seed plant may complete its life cycle. As we see next, pollination is the next crucial step.

STUDY BREAK 36.2

1. What is the biological role of flowers, and what fundamental developmental shift must occur before an angiosperm can produce a flower?
2. Explain the steps leading to the formation of a mature male gametophyte, beginning with microsporocytes in a flower's anthers. Which structures are diploid and which haploid?
3. Trace the development of a female gametophyte, beginning with the megasporocyte in an ovule of a flower's ovary. Which structures are diploid and which haploid?

36.3 Pollination, Fertilization, and Germination

The process by which plants produce seeds—which have the potential to give rise to new individuals—begins with **pollination,** when pollen grains make contact with the stigma of a flower. As discussed in Chapter 29, air or water currents, birds, bats, insects, or other agents make the transfer.

Pollination is the first in a series of events leading to *fertilization,* the fusion of the haploid nuclei of an egg and sperm inside the flower's ovary. The resulting embryo and its ovule mature into a seed housing a young diploid sporophyte, and when the seed sprouts, or *germinates,* the sporophyte begins to grow. We will return to these events shortly. They cannot unfold, however, unless certain "compatibility" requirements are met.

Pollination Requires Compatible Pollen and Female Tissues

Even after pollen reaches a stigma, in most cases pollination and fertilization can take place only if the pollen and stigma are compatible. For example, if pollen from one species lands on a stigma from another, chemical incompatibilities usually prevent pollen tubes from developing.

Flowering plant evolution has resulted in various mechanisms that limit or prevent *inbreeding,* a mating between gametes that are closely related genetically. For example, in monoecious species—in which, recall, individual plants produce flowers having either male *or* female reproductive parts—the stigmas of

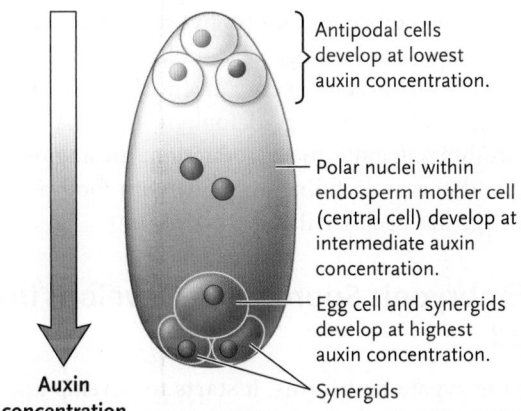

Antipodal cells develop at lowest auxin concentration.

Polar nuclei within endosperm mother cell (central cell) develop at intermediate auxin concentration.

Egg cell and synergids develop at highest auxin concentration.

Auxin concentration

Synergids

FIGURE 36.7 Model for patterning of the female gametophyte by an auxin gradient. A study by Gabriela Pagnussat and coworkers showed that the concentration of auxin (indicated in red) determined the fate of cells in the developing gametophyte. Where auxin was most concentrated, cells became specialized as synergids. At the lowest concentration, they became antipodal cells. Egg cells developed in between these two extremes.
© Cengage Learning 2017

"female" flowers may be receptive to pollen a day or two before or after stamens of "male" flowers release pollen. In another mechanism, even when the sperm-bearing pollen and a stigma are "ready" at the same time, pollination may not lead to fertilization unless the pollen and stigma belong to genetically distinct individuals. In one such scenario, when pollen from a given plant lands on that plant's own stigma, a pollen tube may begin to develop, but stop before reaching the embryo sac. This **self-incompatibility** is a biochemical recognition and rejection process that prevents self-fertilization, and it apparently results from interactions between proteins encoded by *S* (self) genes.

Research has shown that *S* genes usually have multiple alleles—in some species there may be hundreds—and a common type of incompatibility occurs when pollen and stigma carry an identical *S* allele. The result is a biochemical signal that prevents proper formation of the pollen tube **(Figure 36.8)**. For example, studies on plants of the mustard family have revealed that pollen contacting an incompatible stigma produces a protein that prevents the stigma from hydrating the relatively dry pollen grain, an essential step if the pollen tube is to grow. A wide range of self-incompatibility responses has been discovered, however. In cacao, for instance, when incompatible pollen contacts a stigma, a pollen tube grows normally but a hormonal response soon causes the flower to drop off the plant, preventing fertilization.

Self-incompatibility prevents inbreeding and promotes genetic variation, which is the raw material for natural selection and adaptation. Even so, many flowering plants do self-pollinate, either partly or exclusively, because that mode, too, has benefits in some circumstances. For instance, "selfing" may help preserve adaptive traits in a population. It also reduces or eliminates a plant's reliance on wind, water, or animals for pollination, and thus ensures that seeds will form when conditions for cross-pollination are unfavorable, such as when pollinators or potential mates are scarce. Sunflowers, garden peas, and many orchids are self-pollinating.

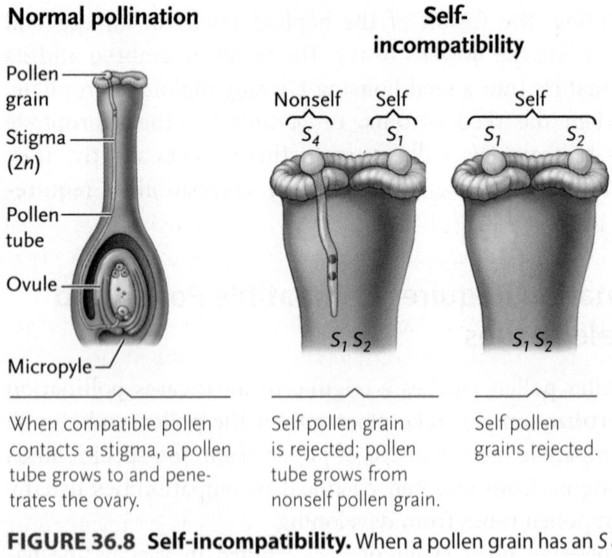

Double Fertilization Occurs in Flowering Plants

If a pollen grain lands on a compatible stigma, it absorbs moisture and germinates a pollen tube, an elongating cell that burrows through the stigma and style toward an ovule (see Figure 36.6 and **Figure 36.9A**). Chemical cues from the two synergid cells lying close to the egg cell help guide the pollen tube toward its destination. Before or during these events, the pollen grain's haploid sperm-producing cell divides by mitosis, forming two haploid sperm. When the pollen tube reaches the ovule, it enters through the micropyle and an opening forms in its tip. By this time one synergid has begun to die (an example of programmed cell death), and the two sperm are released into the disintegrating cell's cytoplasm. Experiments suggest that elements of the synergid's cytoskeleton guide the sperm cells onward, one to the egg cell and the other to the endosperm mother cell.

Next there occurs a remarkable sequence of events called **double fertilization (Figure 36.9B),** which has been observed only in flowering plants and (in a somewhat different version) in the gnetophyte *Ephedra* (see Section 29.2). Typically, one sperm nucleus fuses with the egg to form a diploid ($2n$) zygote. The other sperm nucleus fuses with the endosperm mother cell, forming a cell with a triploid ($3n$) nucleus. Tissues derived from that $3n$ cell are called **endosperm** (*endon* = within; *sperma* = seed). The endosperm nourishes the embryo until its leaves form and photosynthesis has begun.

The evolution of embryo-nourishing endosperm in angiosperms coincided with reduced size of the female gametophyte (see Figure 28.8). In other land plants, such as gymnosperms and ferns, the gametophyte itself contains enough stored food to nourish the embryonic sporophytes. This observation suggests an adaptive advantage of double fertilization. Whereas other seed plants expend energy on the development of a relatively large, nutrient-rich gametophyte that may or may not successfully contribute to reproduction, an angiosperm does not expend energy in filling an ovule with food reserves until after an egg has been fertilized.

The Embryonic Sporophyte Develops inside a Seed

When the zygote first forms, it starts to develop and elongate even before mitosis begins. **Figure 36.10** shows an example of these changes in the eudicot shepherd's purse (*Capsella*). In this example, the nucleus and most other organelles in the zygote become housed in the top half of the cell, and a vacuole takes up most of the lower half (see Figure 36.10, step 1). The first round of mitosis divides the zygote into an upper *apical cell*

Normal pollination

Pollen grain

Stigma (2*n*)

Pollen tube

Ovule

Micropyle

Self-incompatibility

Nonself | Self
S_4 | S_1

Self
S_1 | S_2

S_1S_2

S_1S_2

When compatible pollen contacts a stigma, a pollen tube grows to and penetrates the ovary.

Self pollen grain is rejected; pollen tube grows from nonself pollen grain.

Self pollen grains rejected.

FIGURE 36.8 Self-incompatibility. When a pollen grain has an *S* allele that matches one in the stigma (which is diploid), the result is a biochemical response that prevents fertilization—in this illustration, by preventing the growth of a pollen tube.

© Cengage Learning 2017

A. Pollination

B. Double fertilization

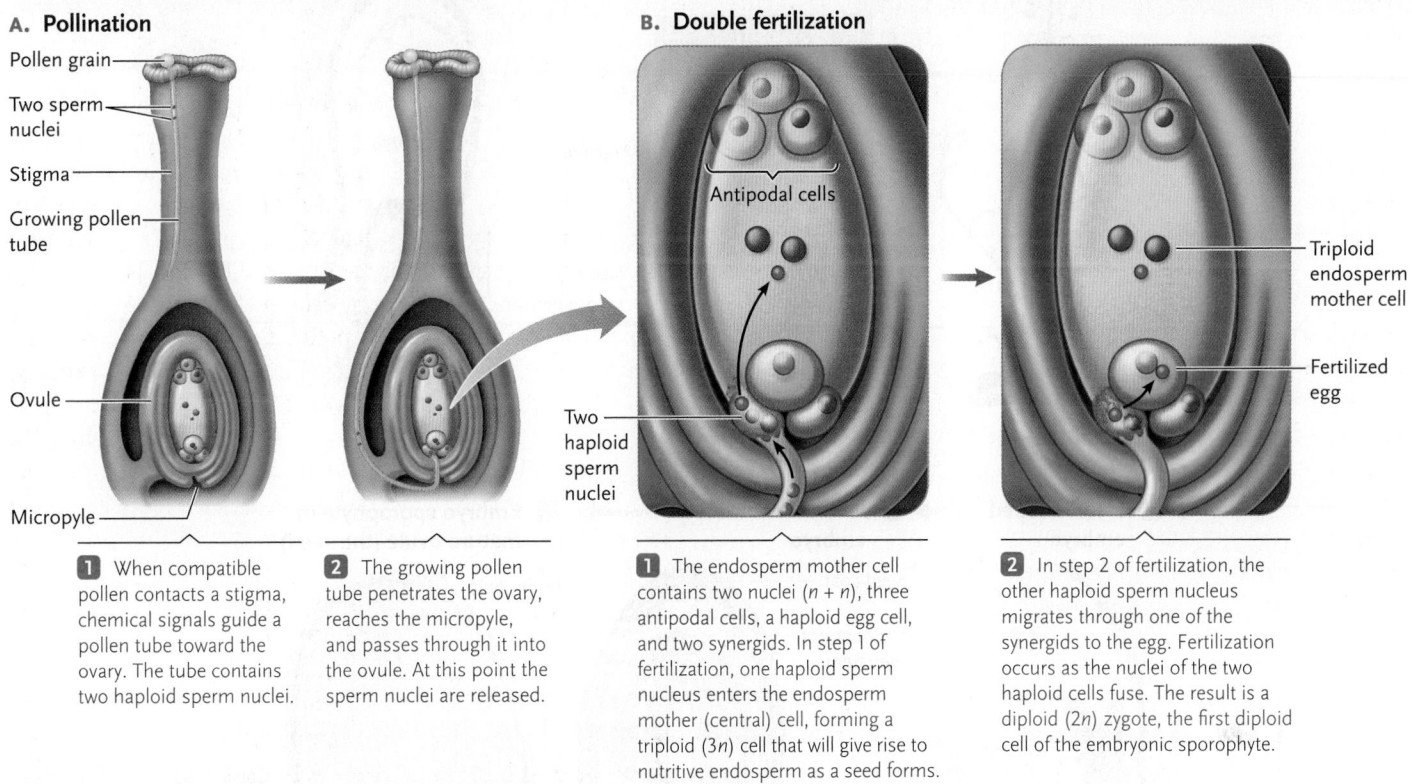

1 When compatible pollen contacts a stigma, chemical signals guide a pollen tube toward the ovary. The tube contains two haploid sperm nuclei.

2 The growing pollen tube penetrates the ovary, reaches the micropyle, and passes through it into the ovule. At this point the sperm nuclei are released.

1 The endosperm mother cell contains two nuclei (n + n), three antipodal cells, a haploid egg cell, and two synergids. In step 1 of fertilization, one haploid sperm nucleus enters the endosperm mother (central) cell, forming a triploid (3n) cell that will give rise to nutritive endosperm as a seed forms.

2 In step 2 of fertilization, the other haploid sperm nucleus migrates through one of the synergids to the egg. Fertilization occurs as the nuclei of the two haploid cells fuse. The result is a diploid (2n) zygote, the first diploid cell of the embryonic sporophyte.

SUMMARY Pollination of a compatible stigma triggers the growth of a pollen tube that delivers two haploid sperm nuclei to the ovule within an ovary. There one sperm nucleus fertilizes the egg and the other nucleus "fertilizes" the endosperm mother cell, triggering the development of endosperm. This process of double fertilization yields an embryonic sporophyte along with endosperm to nourish it.

think like a scientist What sort of experiment would reveal more about chemical signals that guide the migration of a pollen tube toward the ovule?

and a lower *basal cell.* The apical cell then gives rise to the multicellular embryo, while most descendants of the basal cell form a simple row of cells, the **suspensor,** which transfers nutrients from the parent plant to the embryo (step 2).

The first apical cell divisions produce a globe-shaped structure attached to the suspensor. As they continue to grow, eudicot embryos become heart-shaped (step 3). Each lobe of the "heart" is a developing cotyledon, which will provide nutrients for growing tissues in a germinating seedling. Typically, the two cotyledons absorb much of the nutrient-storing endosperm and become plump and fleshy. A familiar example is a mature seed of a sunflower *(Helianthus annuus),* in which no endosperm remains. In some eudicots, however, the cotyledons remain as slender structures; they produce enzymes that digest the seed's ample endosperm and transfer the liberated nutrients to the seedling. Monocots have one large cotyledon; in many monocot species, especially grasses such as corn and rice, the cotyledon absorbs the endosperm after *germination,* the

process described later in this section in which the embryo inside the seed begins to grow.

A **seed** is a mature ovule. By the time the ovule is mature, its integument has developed into a protective **seed coat.** The embryo inside the seed has a lengthwise axis with a root apical meristem at one end and a shoot apical meristem at the other (steps 4 and 5).

Figure 36.11A and **Figure 36.11B** show the structural organization of seeds of two eudicots, the kidney bean *(Phaseolus vulgaris)* and the castor bean plant *(Ricinus communis).* The embryo makes up nearly all of both types of seeds. A mature kidney bean embryo has broad, fleshy cotyledons, and none of the seed's endosperm remains to nourish it. A castor bean embryo has much thinner cotyledons that are surrounded by endosperm, but in other ways the embryos are quite similar. The **radicle,** or embryonic root, is located near the micropyle, where the pollen tube entered the ovule prior to fertilization. The radicle attaches to the cotyledon at a region of cells called

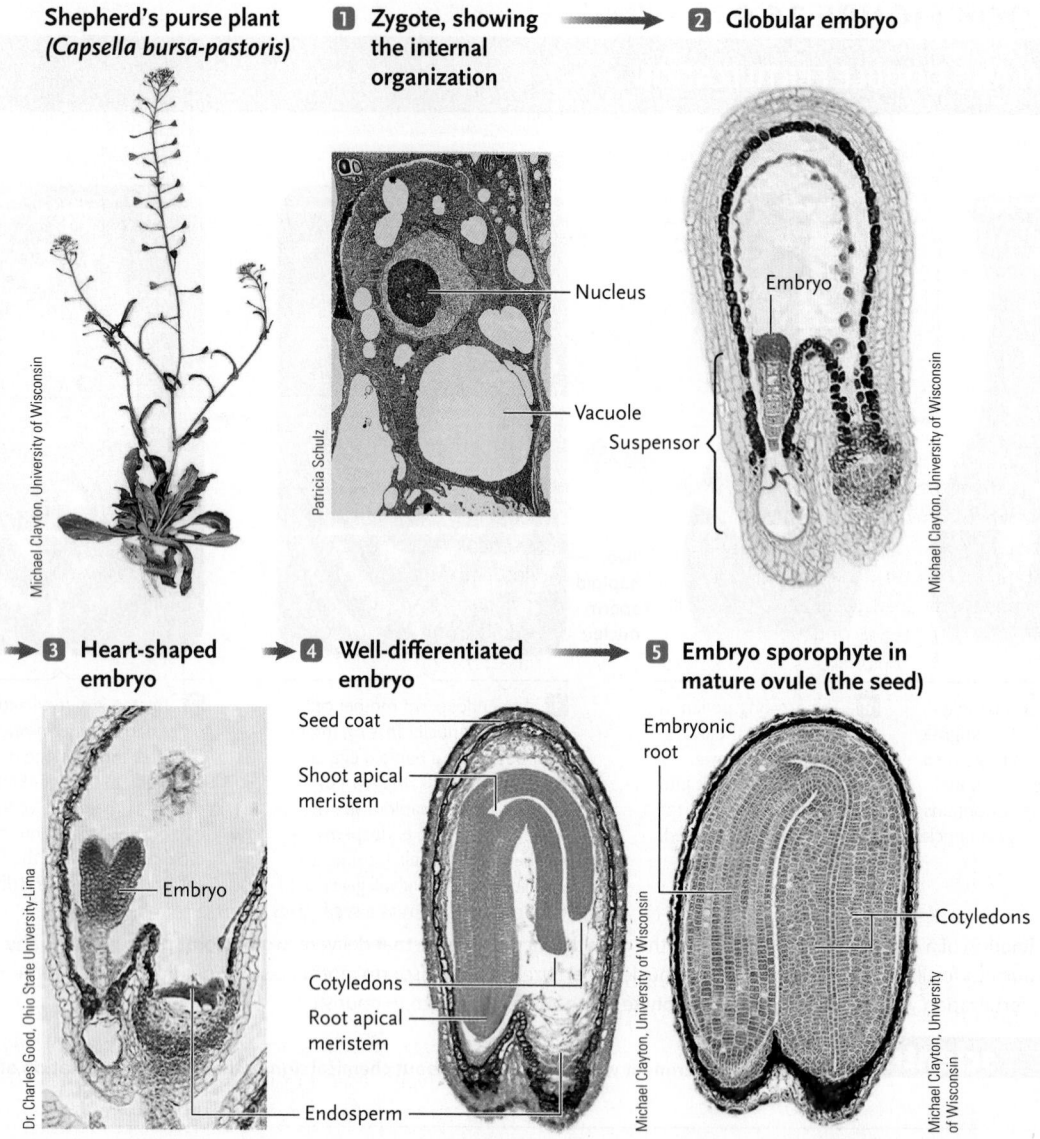

Shepherd's purse plant
(Capsella bursa-pastoris)

1 Zygote, showing the internal organization

2 Globular embryo

Michael Clayton, University of Wisconsin

Nucleus

Vacuole

Patricia Schulz

Embryo

Suspensor

Michael Clayton, University of Wisconsin

3 Heart-shaped embryo

4 Well-differentiated embryo

5 Embryo sporophyte in mature ovule (the seed)

Seed coat

Shoot apical meristem

Embryo

Cotyledons

Root apical meristem

Endosperm

Dr. Charles Good, Ohio State University–Lima

Embryonic root

Cotyledons

Michael Clayton, University of Wisconsin

Michael Clayton, University of Wisconsin

FIGURE 36.10 Stages in the embryonic development of shepherd's purse (Capsella bursa-pastoris), a eudicot. The micrographs are not to the same scale. Figure 36.21 shows more detail of the development of early plant embryos.

the **hypocotyl** ("below the cotyledon"). Beyond the node where the cotyledons attach is the **epicotyl,** which has the shoot apical meristem at its tip and which often bears a cluster of tiny foliage leaves, the **plumule.** At germination, when the root and shoot first elongate and emerge from the seed, the cotyledons are positioned at the first stem node with the epicotyl above them and the hypocotyl below them.

Deconstructing the seed of a monocot can be a bit more complex. To begin with, we see a great deal of variation in the cotyledons of monocots—some are large and plump with stored endosperm, whereas others (such as seeds of orchids and onions) are extremely small and contain little or no food stores for the embryo. **Figure 36.11C** shows the specialized seed of a grass—corn—which includes several structures not seen in the seeds of other angiosperm taxa. Housed inside a kernel (a type of fruit called a *grain*) a corn seed contains a large store of

endosperm. The embryo has a single large cotyledon, a shield-shaped mass called a **scutellum** that absorbs nutrients from the endosperm. In monocot embryos generally, protective tissues blanket the root and shoot apical meristems, and in grasses the shoot apical meristem and plumule are covered by a **coleoptile,** a sheath of cells that protects them during upward growth through the soil. A similar covering, the **coleorhiza,** sheathes the radicle until it breaks out of the seed coat and enters the soil as the young plant's primary root.

Fruits Protect Seeds and Aid Seed Dispersal

Like the corn seed in its kernel, most angiosperm seeds are housed inside fruits, which provide protection and often aid seed dispersal. A **fruit** is a matured ovary. Usually, a fruit begins to develop after pollination and the ensuing fertilization

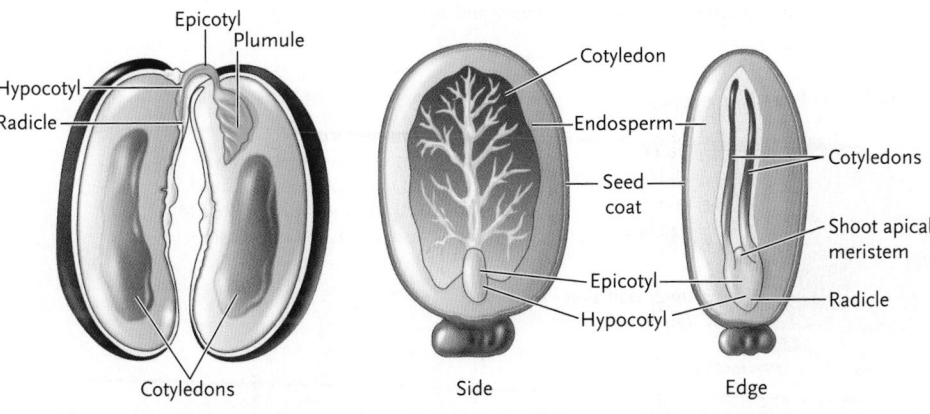

A. Kidney bean *(Phaseolus vulgaris)*

Epicotyl
Plumule
Hypocotyl
Radicle
Cotyledons

B. Castor bean *(Ricinus communis)*

Cotyledon
Endosperm
Seed coat
Cotyledons
Shoot apical meristem
Epicotyl
Hypocotyl
Radicle
Side
Edge

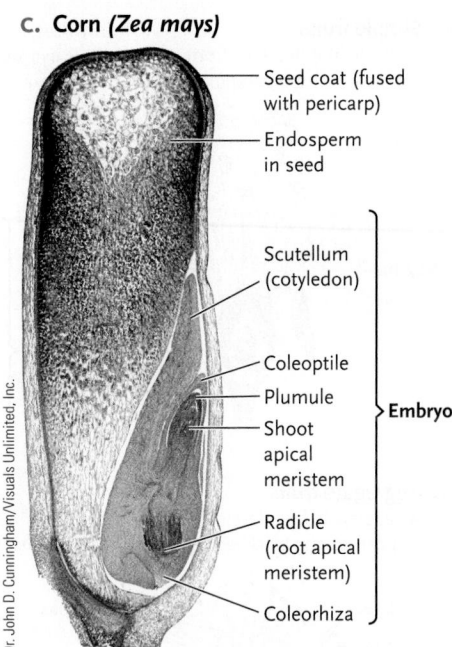

C. Corn *(Zea mays)*

Seed coat (fused with pericarp)
Endosperm in seed
Scutellum (cotyledon)
Coleoptile
Plumule
Shoot apical meristem
Radicle (root apical meristem)
Coleorhiza
Embryo

Dr. John D. Cunningham/Visuals Unlimited, Inc.

FIGURE 36.11 The structure of eudicot and monocot seeds. Eudicot seeds have two cotyledons, which store food absorbed from the endosperm, but the timing of this function varies in different species. **(A)** The cotyledons of a kidney bean *(P. vulgaris)* take up nutrients from endosperm while the seed develops, becoming fleshy. **(B)** In the castor bean *(R. communis)*, the endosperm is thick and the cotyledons are thin until the seed germinates, when the cotyledons begin to take up endosperm nutrients. **(C)** A kernel of corn *(Z. mays)* contains a representative monocot seed, which has been isolated here and is shown in longitudinal section. Monocot seeds have a single cotyledon, which develops into a shield-shaped scutellum that absorbs nutrients from endosperm.

© Cengage Learning 2017

of an egg in a flower's ovule or ovules. The start of ovule growth after fertilization called "fruit set," marks the beginning of the end for a flower **(Figure 36.12)**. The ovary wall gives rise to the **pericarp,** the often-multilayered wall of a fruit. Hormones in pollen grains provide the initial stimulus that turns on the genetic machinery leading to fruit development; additional signals come from hormones produced by the developing seeds.

Geoff Kidd/Science Source

FIGURE 36.12 The developmental shift from flower to fruit. A cluster of rose hips, the fruits of a rose plant *(Rosa)*, with remnants of sepals still attached.

THE FOUR BASIC TYPES OF FRUITS Although fruits are extremely diverse, botanically they are classified into four general types based on combinations of structural features **(Figure 36.13)**. A major defining feature is the number of ovaries or flowers from which a fruit develops. For example, **simple fruits** develop from a single ovary. In many of them, such as peaches, tomatoes, and the cacao fruits pictured in Figure 36.1, at least one layer of the fruit wall—the pericarp—is fleshy and juicy (see Figure 36.13A). Other simple fruits, including grains and nuts, have a thin, dry pericarp, which may be fused to the seed coat. Legumes such as peas and bean plants also produce simple fruits, the peas and beans being the seeds and the shell-like pod the pericarp.

Aggregate fruits are formed from several ovaries in a single flower. Examples are blackberries and raspberries, which develop from clusters of individual ovaries (see Figure 36.13B). A **multiple fruit** develops from the enlarged ovaries of several flowers clustered together in an inflorescence. A pineapple is a prime example of this fruit type (see Figure 36.13C). Pears, apples, and strawberries are examples of **accessory fruits,** in which floral parts in addition to each ovary become incorporated as the fruit develops. For instance, anatomically, the fleshy part of a pear is an expanded receptacle (see Figure 36.13D).

SEED DISPERSAL BY FRUITS AND OTHER MEANS The dispersal of seeds is an essential step in plant sexual reproduction. Seeds that germinate some distance from a parent plant may be more likely to avoid competing with the parent for soil nutrients, sunlight, and other resources. For some species efficient seed dispersal also is a means of colonizing new suitable

A. Simple fruits

A simple fruit develops from ovarian tissue in a single carpel or a set of fused, multiple carpels. Simple fruits are by far the most common and may be fleshy or dry; examples include peaches, tomatoes, nuts, grains, peppers, and legumes.

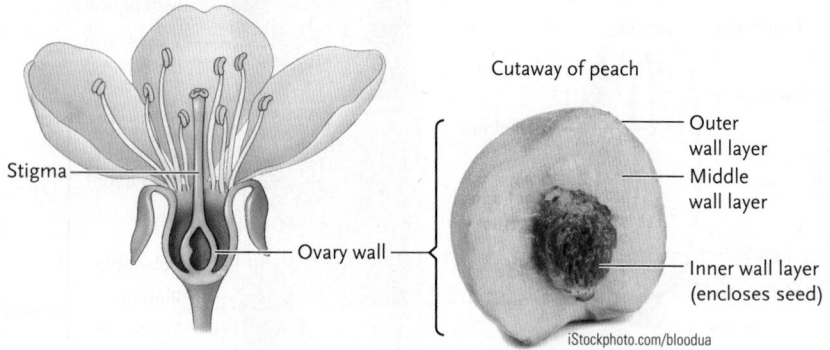

Cutaway of peach

Stigma

Ovary wall

Outer wall layer
Middle wall layer
Inner wall layer (encloses seed)

iStockphoto.com/bloodua

Peach (Prunus persica)
A peach develops from the ovary wall (pericarp) of the flower's single carpel. Sequential layers of the wall give rise to the skin, flesh, and hard pit enclosing the single seed.

B. Aggregate fruits

An aggregate fruit develops from ovarian tissue in separate carpels of a single flower. This group includes raspberries and blackberries, in which small fruits are aggregated into the thimblelike "berry."

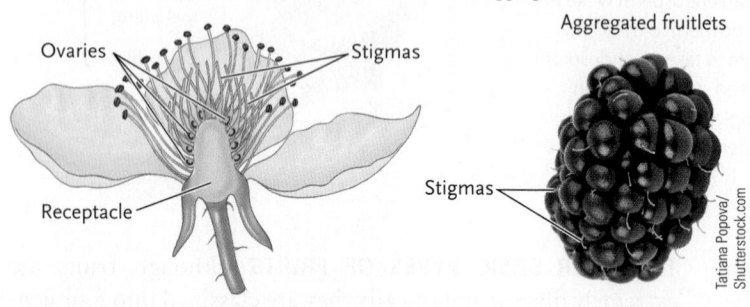

Aggregated fruitlets

Ovaries
Stigmas
Receptacle

Stigmas

Tatiana Popova/Shutterstock.com

Blackberry (Rubus)
Each tiny fruit making up a blackberry contains a seed. Remnants of stigmas look like hairs.

C. Multiple fruits

A multiple fruit such as a pineapple develops from carpels of flowers that are part of an inflorescence.

Flowers in inflorescence

Drew Horne/Shutterstock.com

Serhiy Shullye/Shutterstock.com

Carpel
Remains of flower

Fruits

Pineapple (Ananas comosus)
Inflorescence of a pineapple plant (left) and carpels clustered in a pineapple fruit (right)

D. Accessory fruits

An accessory fruit incorporates tissues from plant parts other than the ovary.

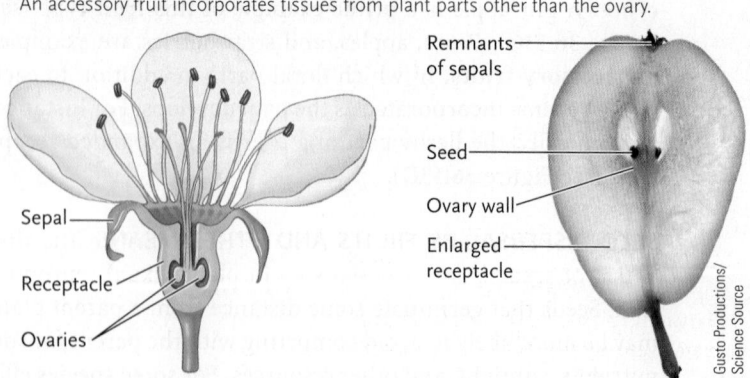

Remnants of sepals
Seed
Ovary wall
Enlarged receptacle

Sepal
Receptacle
Ovaries

Gusto Productions/Science Source

Pear (Pyrus communis)
Pear flowers have multiple (compound) ovaries. The flesh of a pear fruit develops from the receptacle that surrounds the ovaries.

FIGURE 36.13 The four general categories of fruits.
Each category is defined mainly by the floral part or parts from which a plant's fruits form. Fruits of some species contain a single seed but others have more.
© Cengage Learning 2017

habitats. Flowering plants have evolved a fascinating array of adaptations for dispersing seeds, either directly or in fruits. The most common dispersal "agents" are wind, water, and animals.

Figure 36.14 shows a representative sample of these seed dispersal mechanisms. For example, nuts and fleshy fruits are nutritious food for many animals, and the seeds they contain are adapted for surviving digestive enzymes in the animal gut. The enzymes remove just enough of the hard seed coats to increase the chance of successful germination when the seeds are expelled from the animal's body in feces (see Figure 36.14A). Fruits of some species have hooks, spines, hairs, or sticky surfaces, and they are ferried to new locations when they adhere to the fur, feathers, or clothing of a passing animal (see Figure 36.14B). Other fruits and seeds, such as those of maples, dandelions, and milkweeds *(Asclepias),* are wind-dispersed

(see Figure 36.14C, D). Even insects may disperse seeds (see Figure 34.16E). Aquatic plants release seeds or fruits that float away from the parent, and water also may transport buoyant fruits of terrestrial plants, such as the coconut palm (see Figure 36.14F).

Seed Germination Continues the Life Cycle

A mature seed is essentially dehydrated. On average, water accounts for only about 10% of its weight—too little for cell expansion or metabolism. **Germination,** the onset of the growth of a plant embryo, gets underway when the seed begins to imbibe (soak up) water. Ideally, a seed germinates when external conditions favor the survival of the embryo and growth of the new sporophyte. This timing is important, because once germination is underway the embryo loses the

A. Remnants of fleshy fruits in bear scat

Joe McDonald/Visuals Unlimited, Inc.

B. Hook-bearing fruit ("burr") of great burdock *(Arctium lappa)* in a close-up (left) and on an English setter that passed by (right)

Dr. Keith Wheeler/Science Source

Scott Camazine/Science Source

C. Fruit of milkweed *(Asclepias)* releasing airborne seeds

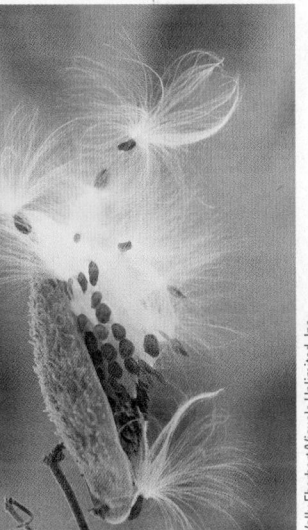

Wally Eberhart/Visuals Unlimited, Inc.

D. Leafcutter ant transporting the seed-containing remains of an apple

Eric Isselee/Shutterstock.com

E. Fruit of a coconut palm *(Cocos nucifera)*

Inga Spence/Science Source

FIGURE 36.14 A gallery of adaptations for dispersal of fruits and seeds. (A) Remnants of red fruits (possibly the "hips" of wild roses) in bear scat demonstrate the adaptive value of fleshy edible fruits in seed dispersal. **(B)** Hooked or spiny seeds of grasses and plants that cling to a passing animal also may move seeds far from the parent plant. **(C)** Seeds dispersed by air currents typically are dry, lightweight, and equipped with structures that resemble wings or parachutes. **(D)** Ants disperse the fruits of many plants, consuming carbohydrate- or protein-rich endosperm but not the seeds. **(E)** Water can transport seeds of plants that live in or near it.

protection of the seed coat and other structures that surround it. Overall, the amount of soil moisture and oxygen, the temperature, day length, and other environmental factors influence when germination takes place.

EVENTS IN GERMINATION The first step in germination is **imbibition,** in which water moves into the seed, attracted to hydrophilic groups of stored carbohydrates and proteins. As water enters, the seed swells and the coat ruptures. As the hydrated embryo resumes growth, the radicle begins its downward growth into the soil. Once the seed coat splits, water and oxygen move more easily into the seed. Metabolism switches into high gear as cells divide and elongate to produce the seedling. Stable enzymes that were synthesized before dormancy become active; other enzymes are produced as the genes encoding them begin to be expressed. Among other roles, the increased gene activity and enzyme production mobilize the seed's food reserves in cotyledons or endosperm. Nutrients released by the enzymes sustain the rapidly developing seedling until its root and shoot systems are established.

We know the most about the events of seed germination in grains, and **Figure 36.15** illustrates them in barley. The seed's endosperm is separated from the pericarp by a thin layer of cells called the **aleurone.** As a hydrating seed imbibes water, the embryo produces a *gibberellin,* a hormone that stimulates aleurone cells to manufacture and secrete hydrolytic enzymes. Some of these enzymes digest components of endosperm cell walls; others digest proteins and starch of the endosperm, releasing nutrient molecules for use by cells of the young root and shoot.

Inside a germinating seed, embryonic root cells are generally the first to divide and elongate, giving rise to the radicle. When the radicle emerges from the seed coat as the primary root, germination is complete. **Figure 36.16** and **Figure 36.17** depict the stages of early development in a kidney bean, a eudicot, and in corn, a monocot. As the young plant grows, its development continues to be influenced by interactions of hormones and environmental factors, as described in Chapter 37.

Most plants give rise to large numbers of seeds because, in the wild, only a tiny fraction of seeds survive, germinate, and eventually grow into another mature plant. Also, flowers, seeds, and fruits represent major investments of plant resources. Asexual reproduction, discussed next, is a more "economical" means by which many plants can propagate themselves.

DORMANCY In some species, the life cycle includes a period of seed **dormancy** (*dormire* = to sleep), in which the seed's metabolic activity is suspended. The conditions required for dormant seeds to germinate vary. Seeds of some species require minimum periods of daylight or darkness, repeated soaking, mechanical abrasion, or exposure to certain enzymes, the high heat of a fire, or a freeze–thaw cycle before they finally break dormancy. A control system based on the hormone abscisic acid, or ABA, underlies the shift from active metabolism to dormancy, and then reactivation and germination. When environmental

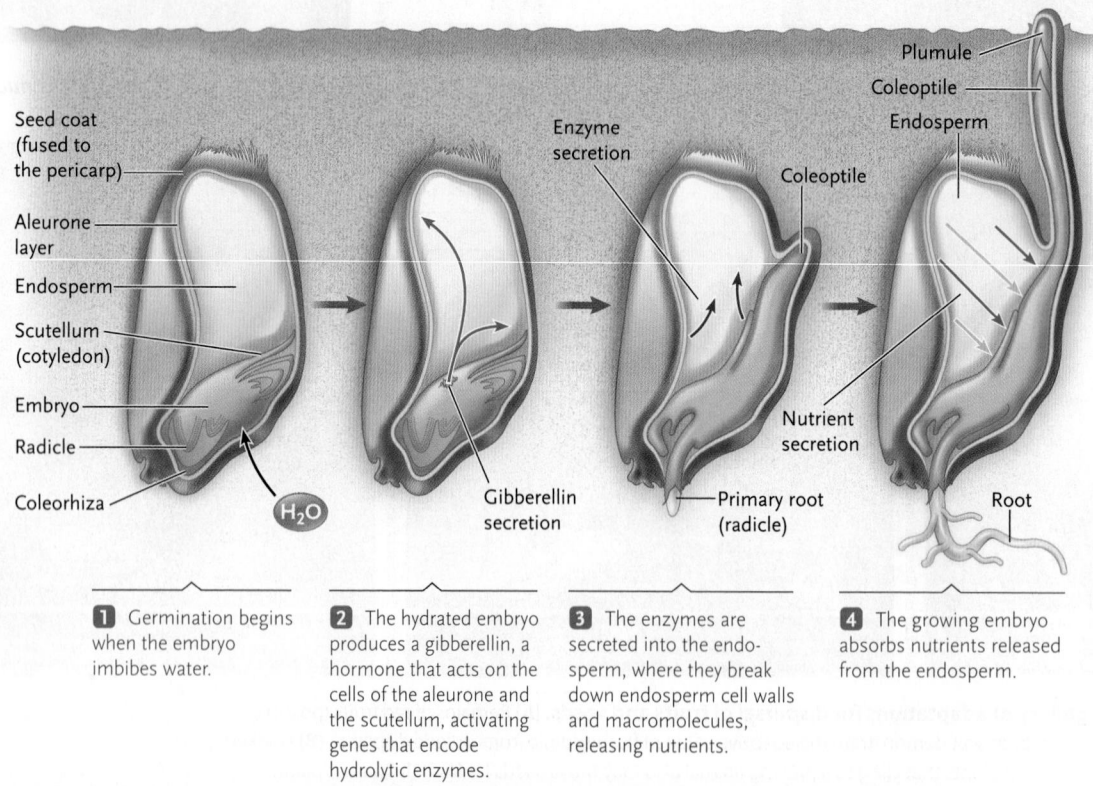

1 Germination begins when the embryo imbibes water.

2 The hydrated embryo produces a gibberellin, a hormone that acts on the cells of the aleurone and the scutellum, activating genes that encode hydrolytic enzymes.

3 The enzymes are secreted into the endosperm, where they break down endosperm cell walls and macromolecules, releasing nutrients.

4 The growing embryo absorbs nutrients released from the endosperm.

FIGURE 36.15 How food reserves are mobilized in a germinated seed within a grain of barley (*Hordeum vulgare*), a monocot. Each grain is a fruit containing a single large seed.

© Cengage Learning 2017

FIGURE 36.16 Stages in the development of a representative eudicot, the kidney bean (P. vulgaris).
© Cengage Learning 2017

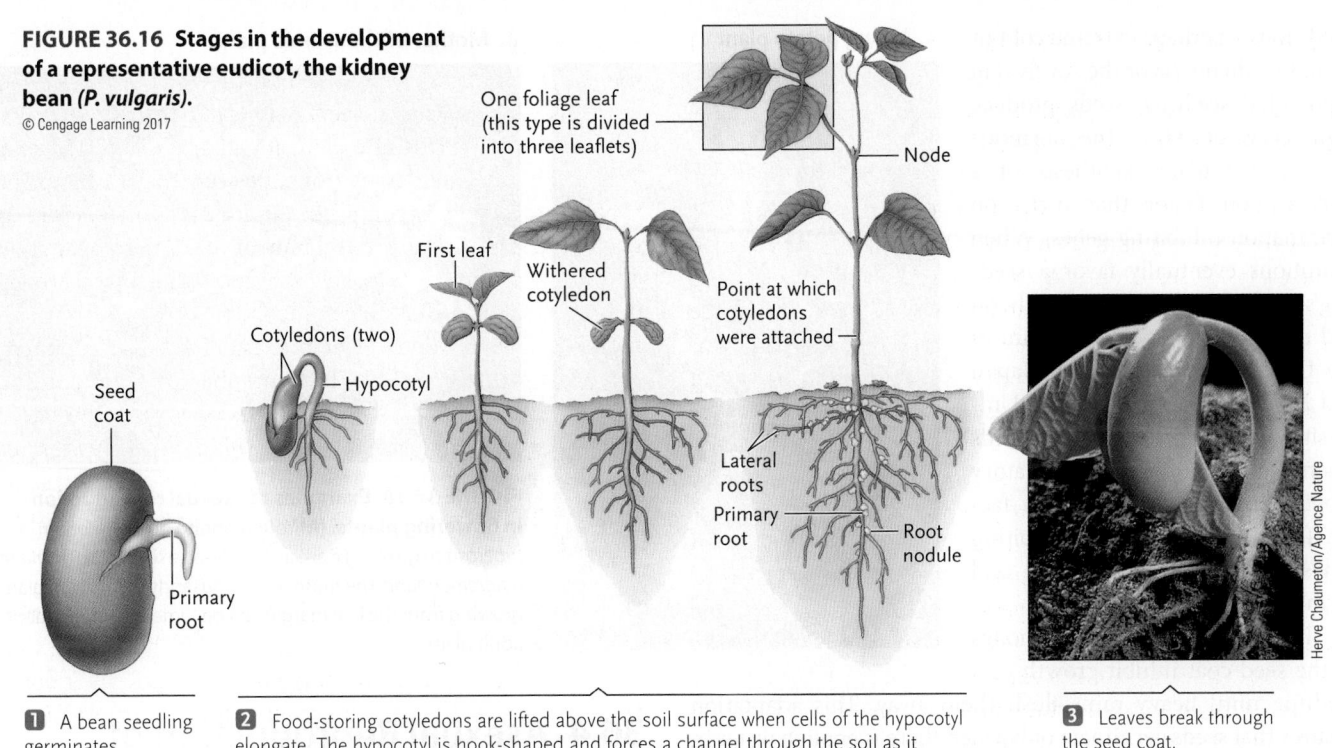

One foliage leaf (this type is divided into three leaflets)

Node

First leaf

Withered cotyledon

Point at which cotyledons were attached

Cotyledons (two)

Hypocotyl

Seed coat

Lateral roots

Primary root

Root nodule

Primary root

Herve Chaumeton/Agence Nature

1 A bean seedling germinates.

2 Food-storing cotyledons are lifted above the soil surface when cells of the hypocotyl elongate. The hypocotyl is hook-shaped and forces a channel through the soil as it grows. At the soil surface, the hook straightens in response to light. For several days, cells of the cotyledons carry out photosynthesis; then the cotyledons wither and drop off. Photosynthesis is taken over by the first leaves that develop along the stem and later by foliage leaves.

3 Leaves break through the seed coat.

FIGURE 36.17 Stages in the development of a monocot, the corn plant (Z. mays). The presence of a coleoptile over developing leaves is typical of grasses but not other monocots.
© Cengage Learning 2017

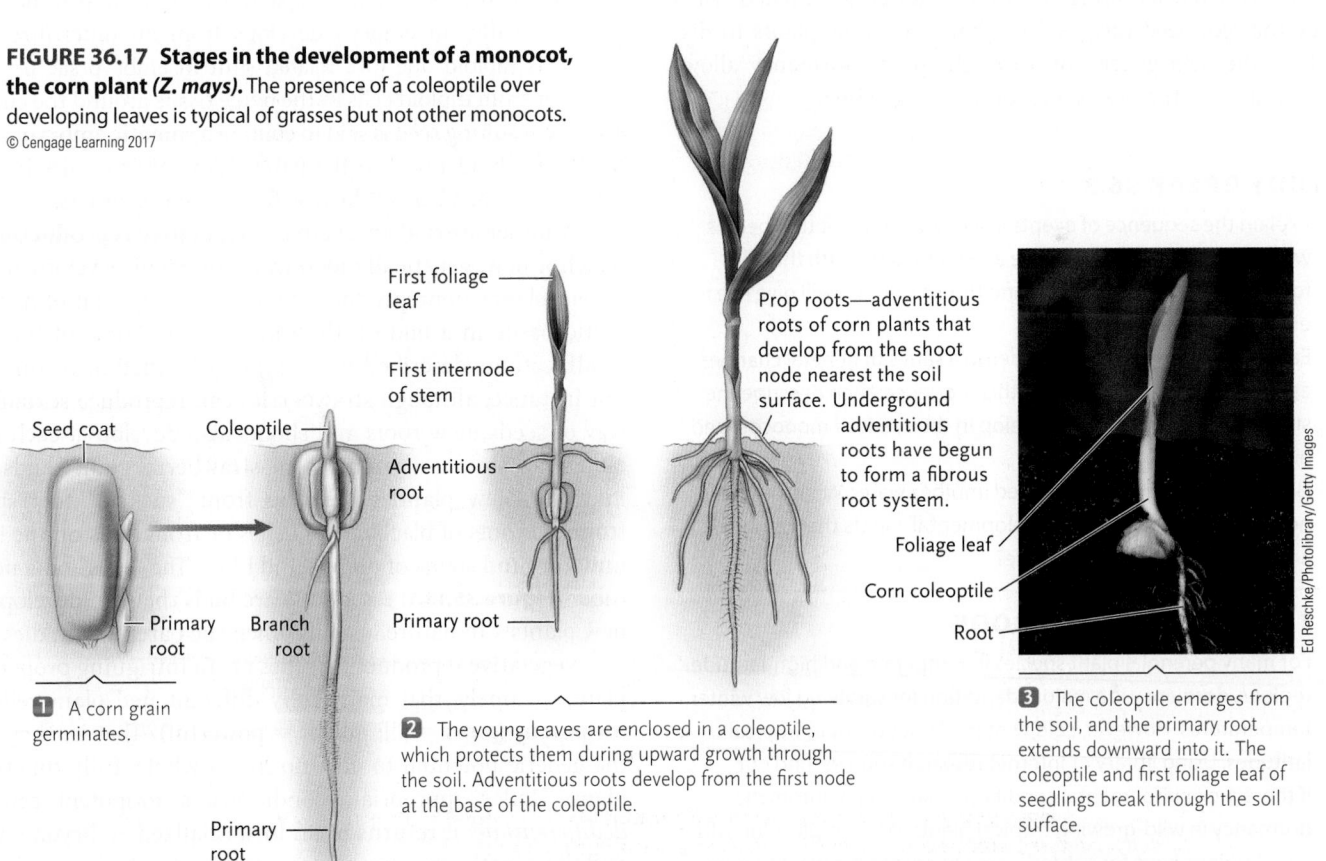

First foliage leaf

First internode of stem

Prop roots—adventitious roots of corn plants that develop from the shoot node nearest the soil surface. Underground adventitious roots have begun to form a fibrous root system.

Seed coat

Coleoptile

Adventitious root

Foliage leaf

Corn coleoptile

Primary root

Branch root

Primary root

Root

Primary root

Ed Reschke/Photolibrary/Getty Images

1 A corn grain germinates.

2 The young leaves are enclosed in a coleoptile, which protects them during upward growth through the soil. Adventitious roots develop from the first node at the base of the coleoptile.

3 The coleoptile emerges from the soil, and the primary root extends downward into it. The coleoptile and first foliage leaf of seedlings break through the soil surface.

conditions—perhaps extreme cold or drought—do not favor the survival of a potential seedling, seeds produce high levels of ABA. The hormone in turn stimulates synthesis of a transcription factor that turns on germination-inhibiting genes. When conditions eventually favor a seedling's survival, the genes switch off and dormancy ends. Kenji Miura of the University of Tsukuba in Japan and Mike Hasegawa at Purdue University recently discovered that this reversal occurs when a regulatory protein binds the transcription factor. Then, the germination-inhibiting genes are no longer expressed and germination can begin.

In some desert plants, hormones in the seed coat inhibit growth of a seedling until heavy rains flush them away. This adaptation ensures that seeds germinate only when there is enough water in the soil to support growth of the plant through the flowering and seed production stages before the soil dries again.

Seeds of some species appear to remain viable for amazing lengths of time. For example, frozen seeds of arctic lupine (*Lupinus arcticus*)—discovered along with fecal material in 10,000-year-old lemming burrows—readily germinated when they thawed. Seed dispersal mechanisms allow plants to distribute the next generation through space. Dormancy allows distribution of the next generation through time.

STUDY BREAK 36.3

1. Explain the sequence of events in a flowering plant that begins with formation of a pollen tube and culminates with the formation of a diploid zygote and the 3n cell that will give rise to endosperm in a seed.
2. Early angiosperm embryos undergo a series of general changes as a seed matures. Summarize this sequence; then describe the structural differences that develop in the seeds of monocots and eudicots.
3. Germination begins when a seed imbibes water. What are the next key biochemical and developmental events that bring an angiosperm's life cycle full circle?

THINK OUTSIDE THE BOOK

For many perennial plant species in temperate and high-latitude regions, dormancy is a major adaptation for surviving low winter temperatures. However, cold winters do not occur at tropical latitudes. Using library or Internet research sources, find out if there are environmental conditions that trigger automatic dormancy in wild-growing tropical plants. What implications do your findings have for gardeners in your area who want to grow tropical species year round?

A. Young potato plant

B. Mother-of-thousands plant

FIGURE 36.18 Examples of asexual reproduction in flowering plants. (A) Young potato plant *(Solanum tuberosum)* growing from an "eye"—an axial bud. **(B)** *Kalanchoe daigremontiana,* the mother-of-thousands plant. Each plantlet growing from the leaf margin can become an independent adult plant.

36.4 Asexual Reproduction of Flowering Plants

As noted in Section 36.1, many flowering plants have the capacity for both sexual and asexual reproduction. Some flowering plants can reproduce asexually through a mechanism called **apomixis,** in which there is no union of haploid maternal and paternal gametes. Typically, an embryo develops from an unfertilized egg (which is diploid due to a related shift in embryo sac development) or from diploid cells in the ovule tissue around the embryo sac. The resulting seed is said to contain a **somatic embryo,** which is genetically identical to the parent. Apomixis occurs in some grasses and members of the rose family, among others.

Another asexual alternative is **vegetative reproduction,** in which a new, genetically identical individual develops from a parent plant's nonreproductive tissue, usually a bit of meristematic tissue in a bud on the root or stem. Some of the stem modifications described in Section 33.2 function in this way. For instance, although strawberries can reproduce sexually by way of seeds, new roots and shoots also develop at each node along the stolons ("runners") a strawberry plant sends out. Likewise, new plantlets develop from "suckers" that sprout from the roots of blackberry bushes or from buds on the short underground stems of onions and lilies. The "eyes" on a potato tuber **(Figure 36.18A)** essentially are buds that can develop into new plants—in nature, usually when the parent plant dies.

Vegetative reproduction relies on an intriguing property of plants—namely, that many fully differentiated plant cells are **totipotent** (*totus* = all; *potens* = powerful). That is, they have the genetic potential to develop into a whole, fully functional plant. Under appropriate conditions, a totipotent cell can *dedifferentiate;* it returns to an unspecialized embryonic state, and the genetic program that guides the development of a new individual switches on.

Vegetative Reproduction Is Common in Nature

Various plant species have developed different mechanisms for vegetative reproduction. In **fragmentation,** cells in a piece of the parent plant dedifferentiate and then can regenerate missing plant parts. For example, when a leaf falls or is torn away from a jade plant (*Crassula* sp.), a new plant can develop from meristematic tissue adjacent to the wound surface in the detached leaf. The "mother of thousands" plant, *Kalanchoe daigremontiana,* produces a wealth of seeds, while more or less simultaneously meristematic tissue in notches along the leaf margin gives rise to tiny plantlets **(Figure 36.18B).** These plantlets eventually fall to the ground, where they can sprout roots and grow to maturity.

In wild plants, most types of asexual reproduction result in offspring located near the parent. As already noted, these clonal populations lack the variability provided by sexual reproduction, which enhances the odds for survival when environmental conditions change. Yet in some situations asexual reproduction offers an advantage. It usually requires less energy than producing complex reproductive structures such as seeds and showy flowers to attract pollinators, and clones are likely to be well suited to the environment in which the parent grows. In addition, it may allow the rapid colonization of disturbed habitats, where seeds might be blown or washed away.

Commercial Growers and Gardeners Often Use Asexual Vegetative Propagation

Gardeners and farmers have long used asexual plant propagation to grow particular crops and trees and some ornamental plants. One strategy is to use *cuttings,* pieces of stems or leaves, to generate new plants. Placed in water or moist soil, a cutting may sprout roots within days or a few weeks. Shoots and leaves follow later.

Fruit trees and wine grapes often are propagated by grafting a bud or branch from a plant with desirable fruit traits—the **scion**—and joining it to a root or stem from a plant with useful root traits—the **stock.** A grafted plant usually produces flowers and fruit identical to those of the scion's parent plant. The scion of a grafted wine grape or walnut variety may be chosen for the quality of its fruit, and the stock for its hardy, disease-resistant root system **(Figure 36.19).**

Vegetative propagation can also be used to grow plants from single cells, using methods of tissue culture that were pioneered in the 1950s by Frederick C. Steward and his coworkers at Cornell University. Those researchers propagated whole carrot plants in the laboratory by culturing carrot root phloem. Their experiments were based on the hypothesis—later confirmed by others—that almost any plant cell that has a nucleus and lacks a secondary cell wall may be totipotent and therefore can develop into a healthy whole plant **(Figure 36.20).** Today, rose bushes and fruit trees from nurseries and commercially important fruits and vegetables such as Bartlett pears, McIntosh apples, Thompson seedless grapes, and asparagus come from plants produced via tissue culture.

FIGURE 36.19 An orchard of grafted English walnut trees (*Juglans regia*) in California's Sacramento Valley. Even when the trees are several decades old, it is easy to distinguish their grafted scion and much darker stock portions.

The method of plant tissue culture Steward pioneered paved the way for research on *somatic embryogenesis* in plants. By way of this procedure, laboratory-cultured totipotent cells give rise to diploid somatic embryos that can be packaged with nutrients and hormones in artificial "seeds." This technique is one of several used to generate the cloned cacao trees described in this chapter's introduction.

Mutations often occur in the DNA of somatic embryos derived from culturing a multicellular mass of cloned cells called a callus. Laboratory screening can identify such *somaclonal* mutants with desirable traits—for example, resistance to a disease that attacks wild-type plants of the same species. Tissue culture propagation then can produce hundreds or thousands of identical plants from a specimen of interest. This technique, called **somaclonal selection,** is now a staple tool in efforts to improve major food crops, such as corn, wheat, rice, and soybeans. The yellow and orange tomatoes common in produce markets are the fruits of plants that were developed by somaclonal selection.

Regardless of how it comes into being, an embryonic sporophyte changes significantly as it begins the developmental journey toward maturity, when it will be capable of reproducing. Next we explore what researchers are learning about these developmental changes.

STUDY BREAK 36.4

1. Describe three modes of asexual reproduction that occur in flowering plants.
2. What is totipotency, and how does tissue culture exploit this property of plant cells?

36.5 Early Plant Development

As you know, unlike animals, plants have specialized body parts such as leaves and flowers that may arise from meristems throughout an individual's life. Accordingly, in plants the biological role of embryonic development is not to generate the tissues and organs of the adult, but to establish a basic body plan—the

FIGURE 36.20 | **Research Method**

Plant Cell Culture

Purpose: To determine if differentiated plant cells can be returned to a totipotent state.

Protocol: Fragments of tissue are excised from a plant, sterilized, and grown in a medium that contains all required nutrients. The procedure disrupts normal interactions among the excised cells. They dedifferentiate, forming an unorganized callus. When cells removed from the callus are grown in a medium containing nutrients and hormones, some develop into plantlets.

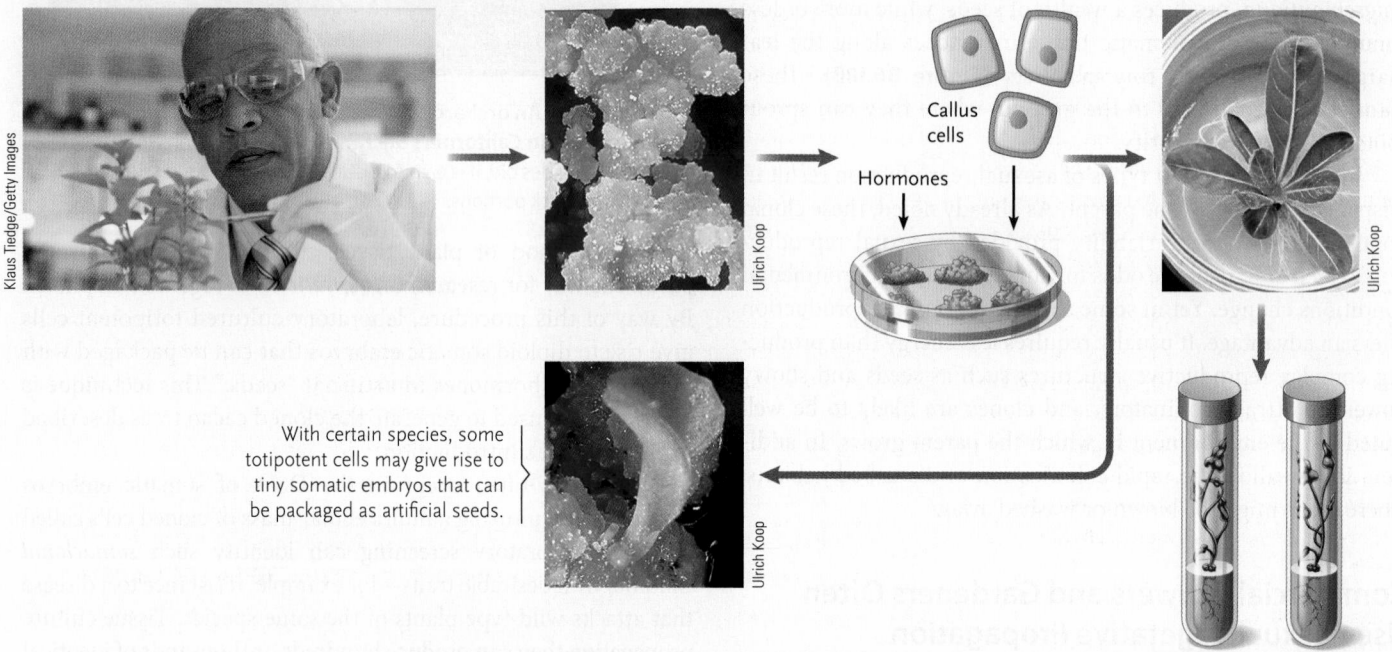

Callus cells

Hormones

With certain species, some totipotent cells may give rise to tiny somatic embryos that can be packaged as artificial seeds.

Klaus Tiedge/Getty Images

Ulrich Koop

Ulrich Koop

Ulrich Koop

Interpreting the Results: Because some cells derived from the callus can give rise to plantlets, they must have regained totipotency.

root–shoot axis and the radial, "outside-to-inside" organization of epidermal, ground, and vascular tissues (see Section 33.1)—and the precursors of the primary meristems. These fundamentals and the stages beyond them all depend on an intricately orchestrated sequence of molecular events that plant scientists are exploring through sophisticated experimentation.

One of the most fruitful approaches has been the study of plants with natural or induced gene mutations that block or otherwise affect steps in development—and accordingly give researchers insight into the developmental roles of the normal, wild-type versions of those abnormal genes. Some of these genes are **homeobox genes** (also called *homeotic genes*), genes that control the formation of structures in development. Many homeobox genes encode transcription factors. Transcription factors are proteins that control the expression of other genes, which in turn direct events in development (see Chapter 16 for a general introduction). Although researchers work with various species to probe the genetic underpinnings of early plant development, the thale cress *(Arabidopsis thaliana)* has become a favorite model organism for plant genetic research (see *Focus on Research: Model Organisms*).

Within Hours, an Early Plant Embryo's Basic Body Plan Is Established

The *A. thaliana* genome sequence has provided a powerful molecular "database" for determining how various genes contribute to shaping the plant body. Experimenters' ability to trace the expression of specific genes has shed considerable light on how the root–shoot axis is set and how the three basic plant tissue systems arise.

FORMATION OF THE ROOT–SHOOT AXIS Soon after fertilization gives rise to an *A. thaliana* zygote, the single cell divides. As with the *Capsella* zygote described earlier, this first round of mitosis produces a small apical cell and a larger basal cell **(Figure 36.21A).** The apical cell receives the lion's share of the cytoplasm, and the basal cell receives the zygote's large vacuole and less cytoplasm. Researchers have confirmed that in this asymmetrical division of the zygote, the daughter cells receive different mixes of mRNAs—the gene transcripts that will be translated into proteins.

Model Organisms: *Arabidopsis thaliana*

For plant geneticists, the little white-flowered thale cress, *A. thaliana,* has attributes that make it a prime subject for genetic research. A tiny member of the mustard family Brassicaceae, *A. thaliana* is revealing answers to some of the biggest questions in plant development and physiology.

Each plant grows only a few centimeters tall, so little laboratory space is required to house a large population. As long as *A. thaliana* is provided with damp soil containing basic nutrients, it grows easily and rapidly in artificial light. Like Mendel's peas, *A. thaliana* is self-compatible and self-fertilizing, and the flowers of a single plant can yield thousands of seeds per mating. Seeds grow to mature plants in just over a month and then flower and reproduce themselves in another 3 to 4 weeks. This permits investigators to perform desired genetic crosses and obtain large numbers of offspring having known, desired genotypes with relative ease. Individual *A. thaliana* cells also grow well in culture.

The *A. thaliana* genome was the first complete plant genome to be sequenced; at this writing researchers have identified approximately 28,000 genes arranged on five pairs of chromosomes. The genome contains relatively little repetitive DNA, so it is fairly easy to isolate *A. thaliana* genes, which can then be cloned using genetic engineering techniques. Cloned genes are inserted into bacterial plasmids and the recombinant plasmids transferred to the bacterial species *Agrobacterium tumefaciens,* which readily infects *A. thaliana* cells. Amplified by the bacteria, the genes and their protein products can be sequenced or studied in other ways.

Typically, researchers use chemical mutagens or recombinant bacteria to introduce changes in the *A. thaliana* genome. These mutants have become powerful tools for exploring molecular and cellular mechanisms that operate in plant development—for example, elucidation of the homeobox genes responsible for flower development described in this chapter. *A. thaliana* mutants are also being used to probe fundamental questions such as how plant cells respond to gravity, and what roles the pigments called phytochromes play in plant responses to light.

Several years ago an ambitious, multinational research effort was launched to determine the functions of all *A. thaliana* genes. This work is providing a comprehensive genetic portrait of a flowering plant—how each gene affects the functioning of not only individual cells but of the plant as a whole. The Arabidopsis Information Resource (TAIR) tallies the percentages of *A. thaliana* genes in different functional categories **(Figure).**

Arabidopsis thaliana

Wally Eberhart/Visuals Unlimited, Inc.

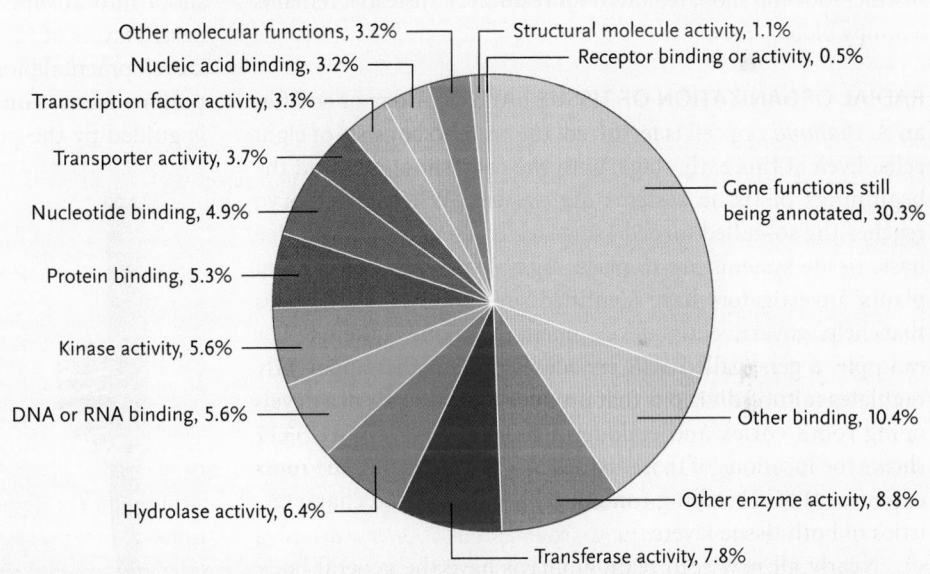

FIGURE The percentages of *A. thaliana* genes that influence different functional categories.
Courtesy of the Arabidopsis Information Resource, 2009.

Translation of differing mRNAs produces proteins that include several transcription factors. As these factors trigger the expression of specific genes, distinct biochemical pathways unfold in the two cells in ways that establish the organization of the plant body. For example, a basal cell initially exports a signaling molecule (a hormone of the auxin family, discussed in Chapter 37) to the apical cell, and this sets in motion steps leading to the development of the various embryonic shoot features **(Figure 36.21B).** Later, gene expression and the flow of chemical signals shift in ways that promote the development of specific structures in the basal cell, including the suspensor.

Several of the genes that influence root–shoot polarity have been identified, and when any of them is disrupted, the result can be a serious defect. For example, when an embryo receives two

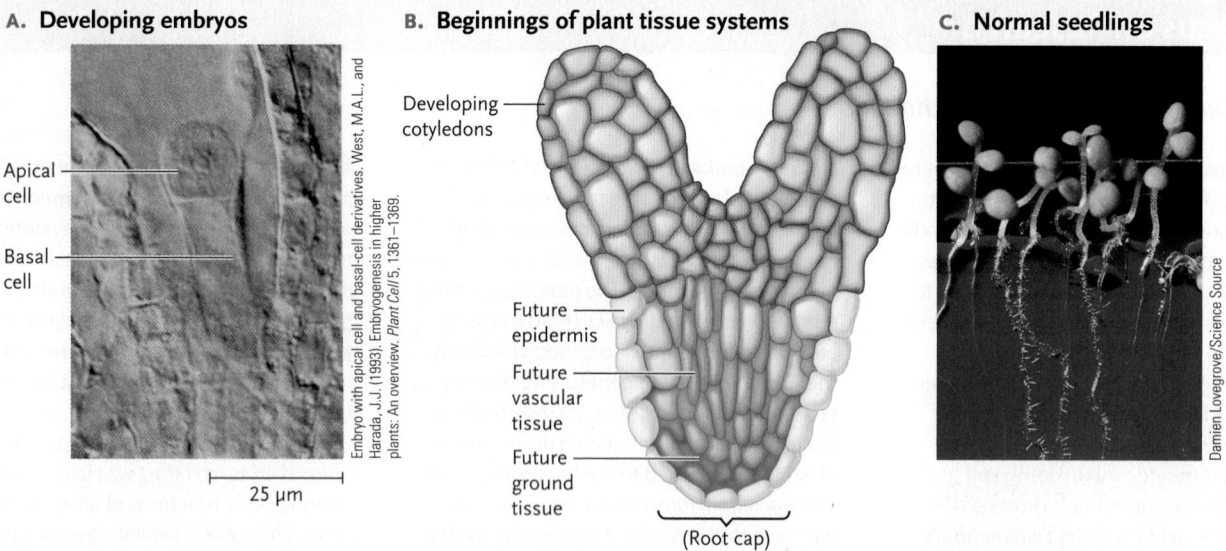

A. Developing embryos

Apical cell
Basal cell

25 μm

Embryo with apical cell and basal-cell derivatives. West, M.A.L., and Harada, J.J. (1993). Embryogenesis in higher plants: An overview. Plant Cell *5, 1361–1369.*

B. Beginnings of plant tissue systems

Developing cotyledons

Future epidermis

Future vascular tissue

Future ground tissue

(Root cap)

C. Normal seedlings

Damien Lovegrove/Science Source

FIGURE 36.21 Stages in the development of the basic body plan of a plant embryo. (A) After a zygote forms, the first round of cell division produces an embryo with an apical cell that contains much of the zygote's cytoplasm, and a larger basal cell that receives the vacuole and less cytoplasm. The division allots different transcription factors to each cell and establishes the plant's root–shoot axis. **(B)** The approximate locations of early embryonic cells that are the forerunners of epidermal, ground, and vascular tissue systems, respectively. **(C)** *A. thaliana* seedlings growing in a culture medium (blue).
© Cengage Learning 2017

copies of a mutant gene called *gnom,* the embryo does not develop distinct root and shoot regions **(Figure 36.21C).** Instead it remains a lumpy blob.

RADIAL ORGANIZATION OF TISSUE LAYERS A day or so after an *A. thaliana* egg cell is fertilized, the embryo consists of eight cells. Even at this early stage, both the root–shoot axis and the beginnings of tissue systems are present. When an embryo reaches the so-called torpedo stage, cells representing all three basic tissue systems are in place. Again working with mutant plants, investigators have identified several *A. thaliana* genes that help govern early development of tissue systems. For example, a gene called *SCR* encodes a protein that apparently regulates mitotic divisions that produce the first cells of a developing root's cortex and endoderm tissue layers. (Figure 33.11 shows the locations of these tissues in a mature root.) The roots of a mutant *SCR* seedling contain cells with jumbled characteristics of both tissue layers.

Nearly all new seed plant embryos have the general body plan we have been discussing. As development proceeds, cells at different sites become specialized in prescribed ways as a particular set of genes is expressed in each type of cell—a process known as *differentiation.* Differentiated cells in turn are the foundation of specialized tissues and organs, which come about through processes we consider next.

A Cell's Position Provides Key Developmental Cues

In both plants and animals, normal development produces ordered spatial arrangements of differentiated tissues. Examples in plants include root and shoot apical meristems at opposite

ends of the root–shoot axis, the cotyledons that divide the shoot into an upper epicotyl and a lower hypocotyl, and the nested layers of vascular, ground, and epidermal tissue systems. Developmental biologists call this progressive ordering of parts **pattern formation,** and a wealth of research has shown that it is guided by the position of cells relative to one another. Like

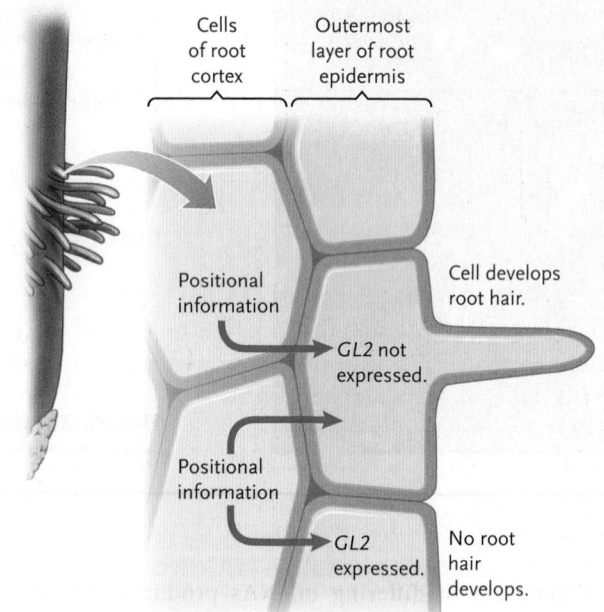

Cells of root cortex

Outermost layer of root epidermis

Positional information

GL2 not expressed.

Cell develops root hair.

Positional information

GL2 expressed.

No root hair develops.

FIGURE 36.22 One model of how positional information influences the development of root hairs. In this model, the only epidermal cells that develop root hairs are those whose inner wall is in contact with two root cortex cells. Such positioning gives rise to signals that block the expression of the *GL2 (GLABRA2)* gene. When *GL2* is expressed, a root epidermal cell will not develop a root hair.
© Cengage Learning 2017

Molecular Insights

Trichomes: Window on development in a single plant cell

The delicate plant cell extensions called *trichomes* are helping to illuminate developmental processes that go on in a single plant cell as it differentiates—that is, as it acquires its ultimate specialized structure and function. In *A. thaliana,* each of these minute protuberances consists of a single cell with a branching tripartite pattern **(Figure A).**

As a trichome differentiates, increases in size, and extends branches in different directions, its chromosomes—and the cell's DNA—are duplicated several times over without mitosis. As a result of the process, called *endoduplication,* the cell has multiple copies of its chromosomes. Experiments that isolate the effects of different mutants have helped confirm that the overall amount of DNA in the cell strongly influences the cell's structure, and also that several genes interact to determine the structure. One of these genes is called *TRY* (for *TRIPTYCHON*).

When it is mutated, the affected plant's trichomes have a double complement of DNA and develop extra branches **(Figure B).** But genes that influence endoduplication are only part of the story. Experiments with other mutants show that numerous other genes also help produce the characteristic three-pronged trichome branching.

These examples underscore how complex molecular interactions that affect multiple aspects of a cell's functioning ultimately determine a cell's form and function. Because the genes that operate in trichomes are also involved in the development of other types of plant cells, studying them promises to shed light on processes that generate differentiated cells throughout the plant body.

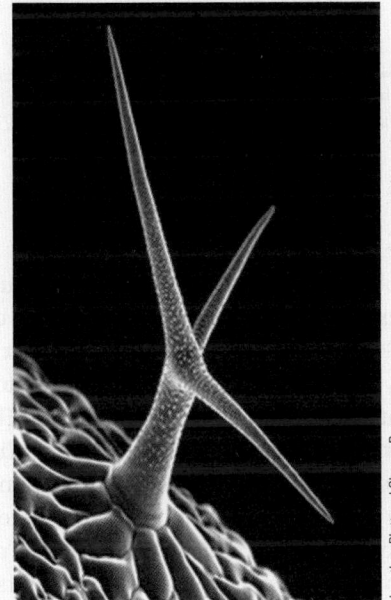

FIGURE A Normal trichome from the epidermis of a leaf *of A. thaliana.*

Jonathan Plett and Sharon Regan

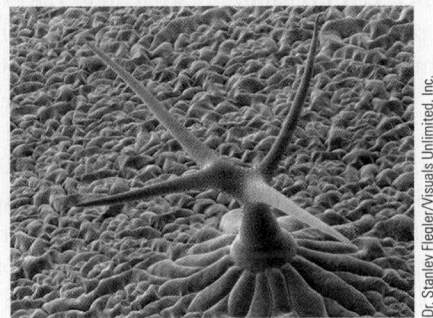

FIGURE B Abnormal trichome from a *try* mutant.

Dr. Stanley Flegler/Visuals Unlimited, Inc.

think like a scientist

In addition to being situated at the surface of leaf and stem epidermis, trichomes are transparent. Given these attributes, what other sorts of cell processes might mutant trichomes be used to study?

Source: Sven Schellman and Martin Hülskamp. 2005. Epidermal differentiation: trichomes in *Arabidopsis* as a model system. *International Journal of Developmental Biology* 49:579–584.

© Cengage Learning 2017

---

the information provided by a hormone gradient (see Section 36.2), this positional information provides cues that "tell" cells where they are in the developing embryo and thus lay the groundwork for an appropriate genetic response.

Many studies have explored how cells in a developing plant or plant part receive and respond to positional information. Experiments have demonstrated, for example, that only certain cells in the epidermis of an embryonic root will give rise to root hairs, the outgrowths of root epidermal cells that take up water and minerals from soil. These specialized root epidermal cells all share the same position with respect to the underlying root cortex—each abuts two cortical cells. By contrast, no root hair extension will develop from an epidermal cell that lines up against only one cortical cell. **Figure 36.22** diagrams one model of what happens next. In this scenario, one or more chemical signals may cross from cortical to epidermal cells by way of plasmodesmata. When an epidermal cell receives signals from a single cortex cell, a series of genes are expressed in a cascade of effects that culminate in the expression of a gene called *GL2* (or *GLABRA2*). The product of *GL2* blocks the formation of root

hairs. If, on the other hand, an epidermal cell aligns with two cortex cells, it receives signals from both and the cascade of gene effects blocks expression of *GL2*—and a root hair develops.

Trichomes are specializations of leaves and stems. *Molecular Insights* gives examples of ways that trichomes have become popular experimental models for studying the differentiation of plant cells.

Morphogenesis Shapes the Plant Body

As tissues of differentiated cells form in a plant embryo, the stage is set for different body regions to develop characteristic shapes and structures that correlate with their function. This process, called **morphogenesis,** shapes the new shoot and root parts produced by dividing cells in meristems. In animals, morphogenesis involves localized cell division and growth, as well as migration of cells and entire tissues from one site to another (see Chapter 50). Plant cells, however, are enclosed within thick walls and usually cannot move. Thus morphogenesis in plants relies on mechanisms that do not require

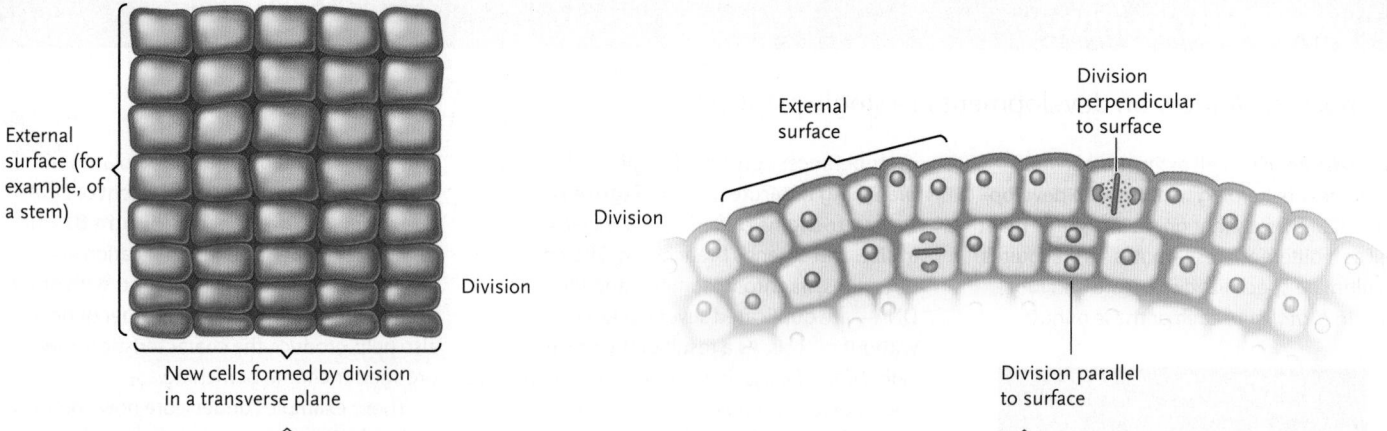

A. Adding length: Division in a transverse plane parallel to the root–shoot axis produces daughter cells stacked like bricks.

B. Adding girth: Cells may divide so that the daughter cell forms parallel to the plant's nearest surface or so that the new cell wall forms perpendicular to the nearest surface.

FIGURE 36.23 Plant cell division in different planes. The external surface nearest the dividing cell is the reference point for establishing division planes.
© Cengage Learning 2017

mobility. One of these mechanisms is **oriented cell division,** which establishes the overall shape of a plant organ, and another is **cell expansion,** which enlarges the cells in specific directions in a developing organ.

ORIENTED CELL DIVISION As described in Chapter 33, roots and stems grow lengthwise as the division and expansion of cells in apical meristems produce columns of cells parallel to the root–shoot axis. The cell divisions occur in a transverse plane—that is, a new cell plate, and then new cell walls, form so that the cells become stacked one atop the other like wooden blocks **(Figure 36.23A).** A plant adds girth—increases in circumference—by way of cell divisions in other planes. For instance, new cell walls may form parallel to the nearest plant surface, or perpendicular both to the nearest surface and to the transverse plane **(Figure 36.23B).**

You may remember from Chapter 10 that the cell plate forms during the cytokinesis phase of mitosis (see Figure 10.10). This step establishes the plane of the middle lamella that will eventually separate the parent and two daughter cells. The capacity of dividing plant cells to synthesize a new cell plate in a different plane from the old one underlies morphogenesis in nearly all plant groups. In meristematic tissue, changes in the plane of cell division establish the direction in which structures such as lateral roots, branches, and leaf and flower buds will grow, and so gives the plant body its overall form.

CELL EXPANSION Once a cell has divided, the daughter cells expand to mature size. Yet plant cells are encased in a primary wall of nonliving material. Botanists are beginning to learn how the cell wall expands to accommodate the enlarging cell within.

Primary cell walls consist of a loose mesh of cellulose microfibrils embedded in a gel-like matrix. As plant cells mature, they may elongate to as much as 100 times their embryonic lengths. During this elongation, the cellulose meshwork is first loosened and then stretched. Turgor pressure supplies the force for stretching. The exact mechanism that loosens the wall structure is not known, although experiments indicate that it depends on a dramatic drop in pH. Some researchers suggest that an auxin in the cell cytoplasm may stimulate a plasma membrane proton pump that moves H^+ into the cell wall (see Section 5.4). The acidic wall conditions may activate hydrolytic enzymes that break bonds between wall components, or they may promote loosening in some other way. We add this important caveat because studies of cell expansion mechanisms have focused on only a few plants, and those models may not apply to all plants.

During expansion, enzyme complexes in the cell's plasma membrane synthesize new cellulose microfibrils from sucrose in the cytoplasm. When each microfibril is fully formed, it is bound in place in the growing wall by pectins and other wall components.

The direction of cell expansion depends on the orientation of the newly formed cellulose microfibrils **(Figure 36.24).** If the microfibrils are randomly oriented, the cell expands equally in all directions. If they are oriented at right angles to the cell's long axis, the cell expands lengthwise. And if new fibrils are deposited parallel to the long axis of the cell, the cell expands laterally.

PATTERNS OF CELL DIVISION DURING EARLY GROWTH Like the first mitotic division that establishes the root–shoot axis in an *A. thaliana* zygote (see Figure 36.21A), cell divisions in a growing plant often are asymmetrical, so that one daughter cell ends up with more cytoplasm than the other. The unequal distribution of cytoplasm means that the daughter cells differ in their composition and structure, and the differences affect how they interact with their neighbors during growth, even though all cells carry the same genes. Their cytoplasmic differences and interactions with one another trigger selective gene expression. Such events seal the developmental fate of particular cell lineages. Their descendant cells divide in prescribed planes and expand in set directions, producing plant parts with diverse shapes and functions.

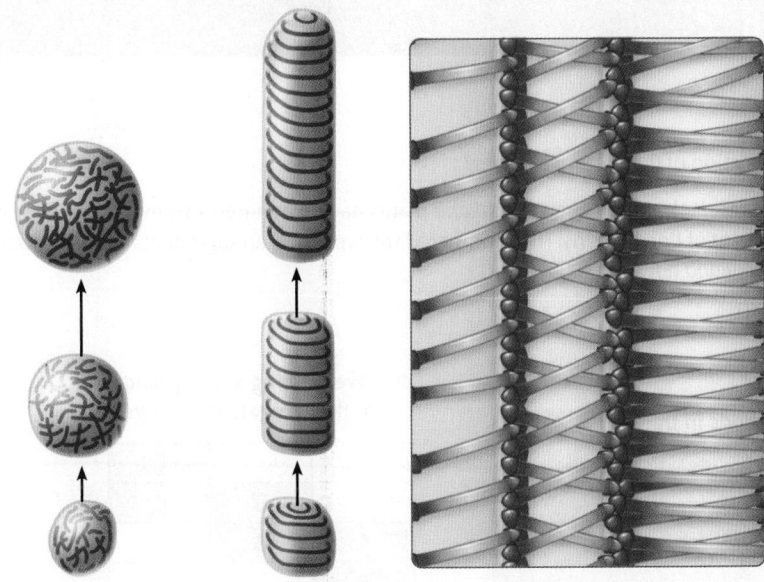

A. When microfibrils are oriented at random, the primary wall is elastic all over, so the cell can grow in all directions.

B. When microfibrils are oriented transversely, the cell can grow only longitudinally.

C. When microfibrils are oriented longitudinally, the cell can grow only laterally.

FIGURE 36.24 Cell expansion and the orientation of cellulose microfibrils. In each cell, microtubules inherited from the parent cell are already oriented in prescribed patterns that govern how cellulose microfibrils will be oriented in the cell wall. Their orientation in turn governs the direction in which a cell can expand.

© Cengage Learning 2017

Regulatory Genes Guide the Development of Floral Organs

Research with several plant species has shed light on the genetic mechanisms that govern the formation of the parts of a flower. For example, classic experiments with *A. thaliana* carried out by Elliot Meyerowitz and his colleagues at the California Institute of Technology showed that *floral organ homeobox* genes regulate the development of the sepals, petals, stamens, and carpels in flowers.

The Meyerowitz team studied plants with various mutations in floral organs. By observing the effects of specific mutations on the structure of *A. thaliana* flowers, the investigators identified three classes of homeobox gene activity, which they named A, B, and C. The different classes appeared to regulate different aspects of normal flower development in eudicots. The gene products are transcription factors. Subsequent studies by other scientists identified an E class of gene activity that apparently is an essential partner in the functioning of the other floral organ homeobox genes. Today, there is widespread acceptance of this **ABCE model**

of floral organ patterning **(Figure 36.25).** Expression of A-, B-, and C-class genes overlaps, and the products of E-class genes function in all four whorls.

Abnormal floral patterns such as those in Figure 36.25C–F show how deactivation of floral organ homeobox genes can affect the development of flower parts. For example, a mutation that deactivates the A-class gene *APETALA2* produces a flower having both carpels and stamens in whorl 1. An intriguing finding is that the A and C activity classes normally oppose each other. When no A gene is expressed, C activity spreads into whorls where the A usually occurs, and vice versa. Also, the E-class genes are "redundant": the transcription factors they encode have such similar effects that an abnormal phenotype develops only when three or more of the genes are deactivated. Subsequent studies have examined many other floral homeobox genes, as well as the genes that control the various gene classes.

As the genes governing flower development are isolated, they can be cloned and their nucleotide sequences defined and manipulated. Such cloned genes already are of keen interest in plant genetic engineering because food grains such as wheat and many other vital agricultural commodities come directly or indirectly from flowers.

Leaves Arise from Leaf Primordia in a Closely Regulated Sequence

A mature leaf may have many millions of differentiated cells organized into tissues such as epidermis, vascular tissues, and mesophyll. As described in Chapter 33, leaves develop from leaf primordia that arise just behind the tips of shoot apical meristems **(Figure 36.26).**

Clonal analysis has opened a window on many aspects of plant development, including how leaf primordia originate and give rise to leaves. In this method, the investigator cultures meristematic tissue that contains a mutated embryonic cell having a readily observable trait, such as the absence of normal pigment. (In the laboratory, this kind of mutation can be induced by chemicals or radiation.) The unusual trait then serves as a marker that identifies the mutant cell's clonal descendants, making it possible to map the growing structure. Researchers have used clonal analysis to study leaf development in garden peas, tomatoes, grasses, and tobacco, among others.

Like flowers, leaves arise through a developmental program that begins with gene-regulated activity in meristematic tissue. Hormones or other signals may arrive at target cells via the stem's vascular tissue, activating genes that regulate development. Studies show that small phloem sieve tubes penetrate a young leaf primordium almost immediately after it begins to bulge out from the underlying meristematic tissue, and xylem soon follows. The early phloem connections are especially vital to the leaf's survival because the leaf does not begin photosynthesis until it attains one-third of its mature size.

FIGURE 36.25 | **Experimental Research**

Probing the Roles of Floral Organ Identity Genes

Question: What are the genetic mechanisms that govern the formation of the parts of a flower?

Experiments: Starting in the 1990s, teams of scientists at various research centers grew *A. thaliana* plants having mutated, inactivated versions of the genes that were suspected of controlling the proper development of floral organs. They compared the types and arrangements of floral organs in the test plants with the organs present in normal, wild-type *A. thaliana* flowers.

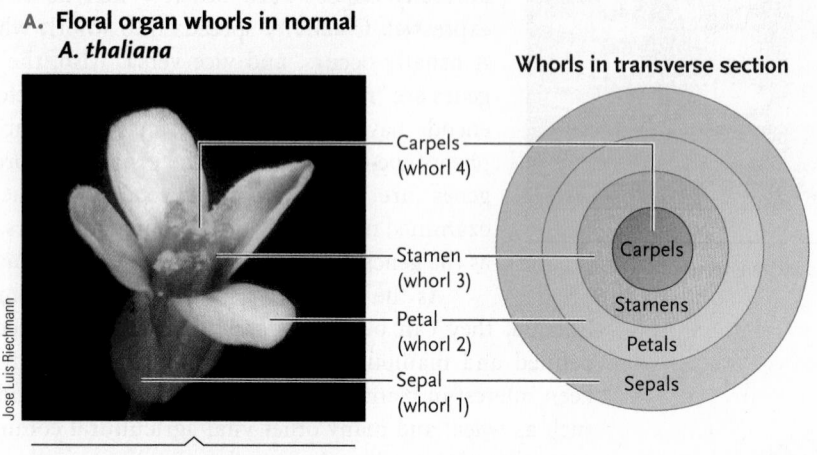

A. Floral organ whorls in normal *A. thaliana*

Carpels (whorl 4)
Stamen (whorl 3)
Petal (whorl 2)
Sepal (whorl 1)

Jose Luis Riechmann

Normal arrangement of organs: carpels in whorl 4, stamens in whorl 3, petals in whorl 2, and sepals in whorl 1

Whorls in transverse section

Carpels
Stamens
Petals
Sepals

B. Overlapping activity fields of floral organ identity genes

		C
	B	C
A	B	
A		
E	E	E

Results: Early experiments showed that at least three classes of homeobox genes—A, B, and C—encoded transcription factors that regulate different aspects of normal floral organ development. Later experiments by others revealed an E class of floral homeobox genes. Various combinations of the different classes of transcription factors produce a normal *A. thaliana* flower's whorls (parts A and B). Particular combinations and dosages of E-class factors, collectively called *SEPALLATA* or *SEP* genes, are required for the normal development of all four whorls. Mutant plants are a window onto this complex system for establishing floral organ identity (parts C–F).

C. When mutation inactivates the *APETALA2* gene, class A genes are not expressed.

Sepals → Carpels
Petals → Stamens

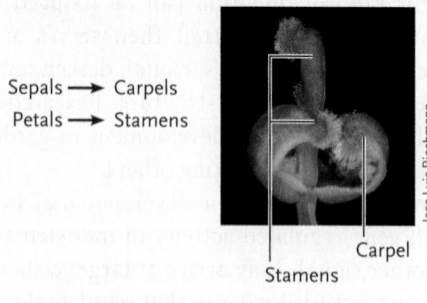

Carpel
Stamens

Jose Luis Riechmann

Stamens replace petals and carpels replace sepals. Organ identity in the other whorls does not change.

D. When mutation inactivates the *APETALA3* or *PISTILLATA* gene, class B genes are not expressed.

Petals → Sepals
Stamens → Carpels

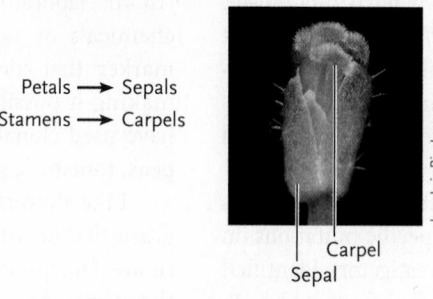

Carpel
Sepal

Jose Luis Riechmann

Carpels replace stamens and sepals replace petals. Organ identity in the other whorls does not change.

E. When mutation inactivates the *AGAMOUS* gene, class C genes are not expressed.

Stamens → Petals
Carpels → Sepals

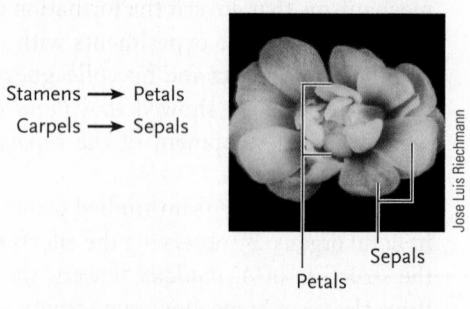

Sepals
Petals

Jose Luis Riechmann

No carpels; instead petals develop in whorl 3 and a version of a floral meristem develops where whorl 4 would normally be. It gives rise to extra petals and sepals. Organ identity in other whorls does not change.

F. When mutation inactivates all four known *SEP* genes, class E genes are not expressed.

blickwinkel/Alamy

Leaflike structures develop but flower parts are poorly defined.

G. SEP4 mutant *A. thaliana* plant

Courtesy of Dr. Martin F. Yanofsky, Distinguished Professor Section of Cell and Developmental Biology, Yanofsky Lab, UCSD

Conclusions: In *A. thaliana,* ABCE activity genes underlie the development of a normal pattern of floral organs. The fields of activity overlap. Further studies have shown that A and C activity apparently counteract each other, and if one is absent the other can spread beyond the whorls where it normally appears. Dosage-dependent activity of E-class *SEP* genes is required for proper expression of all other organ identity genes. The *SEP* genes are redundant; that is, to some degree their products perform the same or similar functions. As a result, their role in establishing floral organ identity only became apparent when researchers developed mutants lacking functional copies of at least three of the genes.

think like a scientist

Observe
Hypothesize
Predict
Experiment
Interpret

Do other flowering plants have a comparable system of floral organ identity genes?

Source: This box summarizes experimental studies by several research teams: D. Weigel and E. M. Meyerowitz. 1993. Activation of floral homeotic genes in *Arabidopsis. Science* 261:1723–1726. J. L. Bowman, J. Alvarez, D. Weigel, E. M. Meyerowitz, and D. R. Smyth. 1993. Control of flower development in *Arabidopsis thaliana* by *APETALA1* and interacting genes. *Development* 119:721–743. S. Pelaz, G. Ditta, S. Kohlami, E. Baumann, and M. Yanofsky. 2000. B and C floral organ identity functions require *SEPALLATA* MADS-box genes. *Nature* 405:200–203.

© Cengage Learning 2017

A growing primordium becomes cone-shaped, wider at its base than at its tip. At a certain point, mitosis speeds up in cells along the flanks of the lengthening cone. In eudicots, the rapid cell divisions occur perpendicular to the surface and produce the leaf blade. In monocot grasses, which have long, narrow leaves, vertical "files" of cells develop as cells in the meristem at the base of the cone divide in a plane parallel to the surface. The ultimate shape of a leaf depends in part on variations in the plane and rate of these cell divisions.

Cells at the leaf tip, and those of xylem and phloem that service it, are the oldest in the leaf, and it is here that photosynthesis begins. Commonly, leaf tip cells also are the first to stop dividing. By the time a leaf has expanded to its mature size, all mitosis has ended and the leaf is a fully functional photosynthetic organ.

In nature, genes that govern leaf and flower development switch on or off in response to changing environmental conditions. In many perennials, new leaves begin to develop inside

A. Leaf primordia of *Coleus*

Leaf primordium   Shoot apical meristem

Triarch/Visuals Unlimited, Inc.

0.25 mm

B. Early stages of leaf development

Next leaf primordium
Leaf primordium
Procambium

Young leaf

C. Leaves of a *Coleus* plant

Nainong/Shutterstock.com

FIGURE 36.26 Early stages in leaf development. (A) Leaf primordia at the shoot tip of a member of *Coleus,* a eudicot genus that includes several popular house plants. **(B)** Diagram showing leaf primordia in different stages of development. **(C)** Fully formed *Coleus* leaf. See also Figure 36.16, which shows the progression of leaves that form during the early growth and development of a eudicot.
© Cengage Learning 2017

buds in autumn, then become dormant until the following spring, when external conditions favor further growth. Environmental cues stimulate the gene-guided production of hormones that travel through the plant in xylem and phloem, triggering renewed leaf growth and expansion. Leaves and other shoot parts also age, wither, and fall away from the plant as hormonal signals change. The far-reaching effects of plant hormones on growth and development are the subjects of Chapter 37.

STUDY BREAK 36.5

1. What is a homeobox gene? Give at least two examples of plant tissues such genes might govern in a species such as *A. thaliana*.

2. What are the two basic mechanisms of morphogenesis in plants? Describe the patterns of cell division by which a plant part (a) grows longer and (b) adds girth.

3. Summarize the gene-guided developmental program that produces a leaf.

Unanswered Questions

What are the chemical cues from female tissues that facilitate compatible pollination?

As you learned in this chapter, if a pollen tube lands on a compatible stigma, chemical cues produced by the female tissue then guide the pollen tube from the stigma to the embryo sac of an ovule. Several lines of evidence point to a role for both sporophytic and gametophytic tissues in proper guidance of pollen tubes to ovules. Recently, it was shown that cysteine-rich polypeptides (LUREs) secreted from synergid cells facilitate pollen tube entry into the ovules of *Torenia fournieri* (wishbone flower). Despite these advances, the identities of the cues released by other pistil tissues remain unknown. What hurdles have hampered the efforts to uncover pollen-tube navigation cues of even the known signaling events described above? First, the pistil tissue within which the pollen tube elongates to the ovule is comprised of several types of tissue, including stigma, style, and transmitting tract, and these tissues are not readily accessible. Second, analyzing the dynamic responses of pollen tubes is difficult given that pollen-tube navigation occurs well within opaque pistils. Third, it appears that multiple, stage-specific, short-range, and readily labile signals produced in minute quantities mediate pollen-tube guidance. However, recent development of highly sensitive global approaches and assays that directly monitor pollen-tube elongation offer hope that guidance cues will be uncovered sooner rather than later.

think like a scientist

1. Do you think pollen-tube guidance signals from different plant species will be similar?

2. What are some of the practical applications of deciphering signals that mediate pollen-tube guidance?

Courtesy of Ravi Palanivelu

Ravi Palanivelu is an Assistant Professor in the Department of Plant Sciences at the University of Arizona. His current research focuses on the isolation and characterization of pollen-tube guidance signals during *Arabidopsis* reproduction, with the long-term goal of understanding the molecular basis of how cells communicate with each other. To learn more about Palanivelu's research, go to http://ag.arizona.edu /research/ravilab.

REVIEW KEY CONCEPTS

For access to MindTap and additional study materials visit www.cengagebrain.com.

36.1 Overview of Flowering Plant Reproduction

- In most flowering plant species, a multicellular diploid sporophyte (spore-producing plant) stage alternates with a multicellular haploid gametophyte (gamete-producing plant) stage. The sporophyte develops roots, stems, leaves, and, at some point, flowers. This life cycle pattern is called alternation of generations (Figure 36.2).

36.2 The Formation of Flowers and Gametes

- A flower develops at the tip of a floral shoot and has up to four whorls, which are supported by the receptacle. The calyx and corolla consist of the sepals and petals, respectively. The third whorl consists of stamens, and carpels make up the innermost whorl (Figure 36.3).

- The anther of a stamen contains sacs where pollen grains develop. Fertilization occurs if pollen lands on a compatible stigma, which is attached to an ovary in which eggs develop.

- Flowers may be classified according to whorls features. A complete flower has all four whorls, whereas incomplete flowers lack at least one whorl. Both male and female parts are present in perfect flowers, but one or the other is absent in imperfect flowers. Monoecious species bear both types of flowers on each plant; in dioecious species the "male" and "female" flowers are on different plants. Major trends in the evolution of flowers include reduced numbers of floral parts, fusion of parts such as petals, positioning ovaries deep within the receptacle, and a shift from radial to bilateral symmetry (Figures 36.4 and 36.5).

- In pollen sacs, pollen grains (immature male gametophytes) develop inside microspores. In a pollen grain, one cell develops into two sperm cells (male gametes). Another cell produces the pollen tube (Figures 36.6 and 36.7).

- An ovule forms inside a carpel. In the ovule, meiosis produces megaspores, of which one survives and develops further, producing the female gametophyte: the seven-celled embryo sac, one cell of which is the haploid egg. A central cell containing polar nuclei helps give rise to endosperm (Figure 36.8).

36.3 Pollination, Fertilization, and Germination

- Upon pollination of a compatible stigma, the pollen grain resumes growth, a pollen tube develops, and mitosis of the male gametophyte's sperm-producing cell produces two haploid sperm nuclei. In a double fertilization process, one sperm nucleus fuses with one egg nucleus to form a diploid ($2n$) zygote. The other sperm nucleus and the two polar nuclei formed in the embryo sac also fuse, forming a $3n$ cell that will give rise to triploid endosperm in the seed (Figure 36.9).

- After the endosperm forms, the ovule expands, and the embryonic sporophyte develops. This mature ovule is a seed. Inside it, the embryo has a lengthwise axis with a root apical meristem at one end and a shoot apical meristem at the other (Figure 36.10).

- Eudicot embryos have two cotyledons. The embryonic shoot consists of an upper epicotyl and a lower hypocotyl; also present is an embryonic root, the radicle. The single cotyledon of a monocot forms a scutellum that absorbs nutrients from endosperm (Figure 36.11).

- A fruit is a matured or ripened ovary. Fruits may be simple, aggregate, or multiple, depending on the number of flowers or ovaries from which they develop. Accessory fruits have tissues that are not derived from an ovary. Fruits also vary in the characteristics of the pericarp (fleshy or dry) that surrounds the seed (Figures 36.12 and 36.13). Common mechanisms for the dispersal of fruits and seeds rely on wind, water, and animals (Figure 36.14).

- Seed germination is a multistep process that begins when the seed becomes hydrated and metabolic activity begins. The emergence of the first embryonic root (the radicle) marks the end of germination (Figures 36.15 and 36.16).

- The seeds of many plants remain dormant, in some cases for decades or longer, until external conditions such as moisture, temperature, and day length favor germination, the survival of the embryo, and the development of a new sporophyte.

36.4 Asexual Reproduction of Flowering Plants

- Many flowering plants can reproduce asexually, with new plants arising by mitosis at nodes or buds along modified stems, by vegetative propagation (Figures 36.18 and 36.19), or by tissue culture from a parent plant's somatic (nonreproductive) cells (Figure 36.20).

36.5 Early Plant Development

- Early on, a new embryo acquires its root–shoot axis, and cells in different regions begin to become specialized for particular functions (Figure 36.21). The position of cells relative to one another guides the formation of the overall pattern of the plant body (Figure 36.22). In morphogenesis, body regions develop characteristic shapes and structures that correlate with their function. (Figure 36.23).

- Dividing plant cells can synthesize a new cell plate in a different plane from the old one. Such changes establish the direction in which structures such as lateral roots, branches, and leaf and flower buds grow (Figure 36.24).

- Chemical signals that help guide morphogenesis appear to act on certain cells in meristematic tissue, activating homeobox genes that ultimately regulate cell division and differentiation (Figures 36.25 and 36.26).

TEST YOUR KNOWLEDGE

Remember/Understand

1. In an angiosperm life cycle, sexual reproduction includes:
   a. meiosis within the male gametophyte to produce sperm.
   b. meiosis within the female gametophyte to produce eggs.
   c. meiosis within the ovary to produce megaspores.
   d. fertilization leading to development of microspores.
   e. fertilization leading to development of megaspores.

2. In a flower that has all four floral whorls, which whorls contain male and female reproductive parts?
   a. whorls 3 and 4 (stamens and carpels)
   b. whorls 1 and 4 (calyx and carpel)
   c. whorls 2 and 3 (corolla and stamens)
   d. whorls 1, 3 and 4 (calyx, stamens, and carpels)
   e. only whorl 4 (carpel)

3. In flowering plants, the term double fertilization refers to:
   a. six sperm fertilizing two groups of three eggs each.
   b. one sperm fertilizing the egg; a second sperm fertilizing the endosperm mother cell.
   c. one microspore becoming a pollen grain; the other microspore becoming a sperm-producing cell.
   d. one pollen grain giving rise to sperm nuclei and a pollen tube.
   e. one sperm fertilizing two endosperm mother cells.

4. From a developmental standpoint, a seed is best described as a(an):
   a. epicotyl.
   b. endosperm.
   c. ovary.
   d. mature spore.
   e. mature ovule.

5. The primary root develops from the embryonic:
   a. epicotyl.
   b. hypocotyl.
   c. coleoptile.
   d. radicle.
   e. plumule.

6. Which of the following is *not* a step in the germination of a monocot seed?
   a. Enzymes secreted into the endosperm digest the endosperm cell walls and macromolecules.
   b. The embryo imbibes water and then produces gibberellin.
   c. The embryo absorbs nutrients released from the endosperm.
   d. Endosperm develops as a food reserve.
   e. Gibberellin acts on the cells of the aleurone and scutellum to encode hydrolytic enzymes.

7. A student cuts off a leaflet from a plant and places it in a glass of water. Within a week roots appear on the base of the cutting. A month later she places the growing cutting into soil and it grows to the full size of the "parent" plant. This is an example of:
   a. cell culture.
   b. fragmentation.
   c. grafting.
   d. vegetative reproduction.
   e. tissue culture propagation.

8. Which of the following is *not* an example of pattern formation in developing plants?
   a. an epidermal cell receiving developmental signals from a cortical cell
   b. the loosening of the cell wall to allow the elongation of selected cells to reach mature size
   c. regulation by homeobox genes of the position of different flower parts
   d. oriented cell division that establishes the shape of an organ
   e. cell expansion that directs specific cells to undergo mitosis at a given time and place

9. During the development of a leaf:
   a. mitotic cell divisions occur on planes specific to different plant groups.
   b. xylem vessels are the first to penetrate the leaf primordium.
   c. the growing leaf primordium becomes wider at its base than at its tip.
   d. the leaf primordium bulges from the region behind the shoot apical meristem.
   e. all of the above occur during leaf development.

Apply/Analyze

10. In spring a lone walnut tree in your backyard develops attractive white flowers, and by the end of summer roughly half the flowers have given rise to the shelled fruits we know as walnuts. Walnut trees are self-pollinating. Assuming that pollination was 100% efficient in the case of your tree, which of the following statements best describes your tree's reproductive parts?
    a. Its flowers are in the botanical category of "imperfect" flowers.
    b. The tree is monoecious.
    c. The tree is dioecious.
    d. The tree has perfect, monoecious flowers.
    e. Answers a and b together provide the best description of the tree's flowers.

11. **Discuss Concepts** A plant physiologist has succeeded in cloning a gene for pest resistance into petunia cells. How can she use tissue culture to propagate a large number of petunia plants that have the gene?

12. **Discuss Concepts** Grocery stores separate displays of fruits and vegetables according to typical uses for these plant foods. For instance, bell peppers, cucumbers, tomatoes, and eggplants are in the vegetable section, while apples, pears, and peaches are displayed with other fruits. How does this practice relate to the biological definition of a fruit?

Evaluate/Create

13. **Design an Experiment** The developmental genetics of flowers are of keen interest in plant biotechnology, especially with regard to food plants such as wheat and rice. Outline a research program for a crop species that would exploit the genetics of flower development, including the effects of homeobox genes, to engineer a more productive variety.

14. **Apply Evolutionary Thinking** Botanists estimate that half or more of angiosperm species may be polyploids that arose initially through hybridization. *Polyploidy*—having more than a diploid set of the parental chromosomes—can result from nondisjunction of homologous chromosomes during meiosis, or when cytokinesis fails to occur in a dividing cell. *Hybridization* is the successful mating of individuals from two different species. Such an interspecific hybrid is likely to be sterile because it has uneven numbers of parental chromosomes, or because the chromosomes are too different to pair during meiosis. A sterile hybrid may reproduce asexually, however, and if by chance its offspring should become polyploid, that plant will be fertile because the original set of chromosomes will have homologs that can pair normally during meiosis. Explain why both the hybrid parent and fertile polyploid offspring may be considered a new species.

For selected answers, see Appendix A.

INTERPRET THE DATA

As Chapter 37 discusses, the plant hormone ABA (abscisic acid) inhibits seed germination, while any of several GAs (gibberellins) stimulate it. For hundreds of plant species, exposure to a substance in wood smoke—one of a family of plant growth regulators called karrikins, or KARs—markedly improves the germination rate. When flames destroy the existing vegetation at a fire scene, more sunlight reaches the ground. In the presence of ample sunlight, KARs apparently promote germination by enhancing the expression of genes that encode GA.

The **Figure** below summarizes experiments by Australian researchers studying KAR activity in mutant *A. thaliana* plants (called gal-3), which do not synthesize GA. The graph tracks germination activity after two groups of identical test seeds began imbibing water in a medium that contained the gibberellin GA_3. All seeds were exposed to identical levels of a karrikin called KAR_1. The red symbols designate seeds of mutant ABA-deficient plants that were supplied with supplementary ABA. The white symbols are ABA-deficient seeds that did not receive the supplement.

Based on this graph:

1. What were the experimental variables in the germination experiment?

2. Did the presence or absence of ABA have a significant effect on the experiment's results?

3. Comparing the upper and lower data sets, what are the *two* basic observations you can make about the effects of the seed exposure to differing GA concentrations?

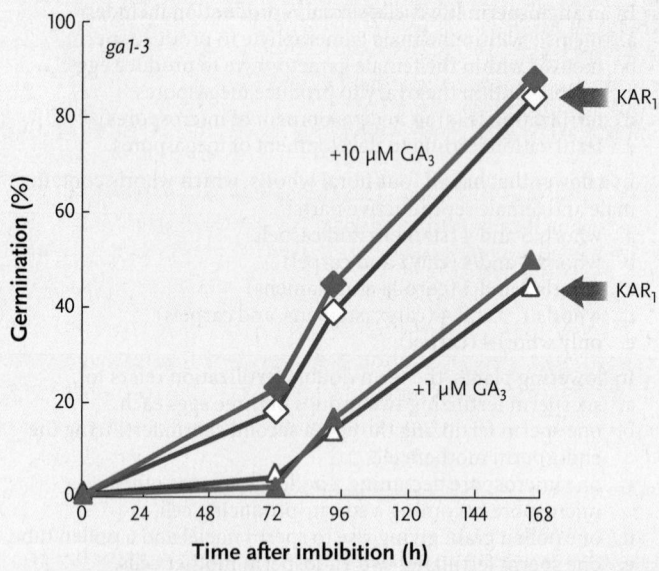

FIGURE Response of phytohormone mutants to KAR_1: Germination of *A. thaliana* ga1-3 mutants in the presence of 1 or 10 μM GA_3 6 1 μM KAR_1.

Source: David C. Nelson, Julie-Anne Riseborough, Gavin R. Flematti, Jason Stevens, Emilio L. Ghisalberti, Kingsley W. Dixon, and Steven M. Smith. 2009. Karrikins discovered in smoke trigger *Arabidopsis* seed germination by a mechanism requiring gibberellic acid synthesis and light. *Plant Physiology* 149:863–873. Plant physiology by AMERICAN SOCIETY OF PLANT PHYSIOLOGISTS. Copyright 2009. Reproduced with permission of AMERICAN SOCIETY OF PLANT BIOLOGISTS in the format Textbook via Copyright Clearance Center.

Plant Signals and Responses to the Environment

Sunflower plants *(Helianthus)* with flower heads that orient toward the Sun's rays—an example of a plant response to shifting light levels in the environment.

Why it matters . . . The creosote bush *Larrea tridentata* won't win many botanical beauty contests, but it is one tough plant. Native to desert areas of the southwestern United States, this shrubby eudicot can withstand droughts of two years or more **(Figure 37.1)**. The species may well be the most drought-tolerant perennial in North America, and individual plants may survive hundreds of years, physiologically "hunkering down" during multiyear dry spells as their fleshy leaves shrivel and reduce photosynthesis to minimal levels. A thick, waxy cuticle helps slow water loss, but the leaves also tend to retain water due to their high internal water potential—a feature related to the plant's adaptations for defense against predation. The leaf tissue is infused with resinous compounds and other chemicals that deter nearly all herbivores, as well as pathogenic fungi and most insect predators. Wet leaves exude a tarry odor that gives the creosote bush its common name.

With adequate rainfall, *L. tridentata* produces small yellow flowers and feathery white fruits packed with seeds (Figure 37.1, inset). However, for a variety of reasons, relatively few *L. tridentata* seeds germinate successfully. Selection pressure related to this high seed mortality may explain another striking *L. tridentata* trait—the formation of large clonal populations by vegetative reproduction. After eight or nine decades, a plant's interior branches may begin to die, eventually leaving what appear to be clusters of separate, widely spaced plants. Carbon dating indicates that some clusters of these genetically identical *L. tridentata* plants are descended from single seeds that germinated 9,000 to 11,000 years ago—qualifying them as some of the oldest living organisms on Earth.

The drought tolerance, chemical defenses, and reproductive flexibility of the creosote bush are prime examples of adaptations plants have evolved for responding to both biotic (living) and abiotic (nonliving) shifts and stresses in their environment. Such adaptations are the focus of this

FIGURE 37.1 Creosote bush, *Larrea tridentata,* **growing in the Mojave Desert.** *Inset:* Flower, fruit, and the resin-impregnated leaves of *L. tridentata.*

chemical environments in which plants live. Collectively these adaptations are the evolutionary solution to a key plant "problem"—the need to respond to changes in the environment despite being rooted in one place.

chapter, starting with the chemical signaling of hormones that regulate many, if not most, aspects of plant growth and development. Next we survey responses that help protect plants against predation or allow them to make adjustments to abiotic factors that are powerful forces in shaping the physical and

37.1 Introduction to Plant Hormones

A **hormone** (*hormon* = to stimulate) is a signaling molecule that regulates or helps coordinate some aspect of growth, metabolism, or development. Plant hormones are called *phytohormones,* especially in the scientific literature. At a fundamental level, hormones control the plant life cycle. They serve as triggers for seed germination and govern the development of a plant's body form, the shift from a vegetative growth phase to a reproductive phase or vice versa, and the timed death of flowers, leaves, and other parts. Beyond those basic roles, hormones mediate changes in the structure and functioning of plant parts in response to external biotic factors such as predation, and abiotic factors such as the availability of light, moisture, and soil nutrients and effects of air currents, gravity, and physical contact with other objects.

TABLE 37.1	Major Plant Hormones and Signaling Molecules		
Hormone/ Signaling Compound	Where Synthesized	Tissues Affected	Effects
Auxins	Apical meristems, developing leaves and embryos	Growing tissues, buds, roots, leaves, fruits, vascular tissues	Promote growth and elongation of stems; promote formation of lateral roots and dormancy in lateral buds; promote fruit development; inhibit leaf abscission; orient plants with respect to light, gravity
Cytokinins	Mainly in root tips	Shoot apical meristems, leaves, buds	Promote cell division; inhibit senescence of leaves; coordinate growth of roots and shoots (with auxin)
Strigolactones	Root tips	Roots, stems, leaves	Regulate branching of roots and stems; promote formation of arbuscular mycorrhizae; promote leaf senescence in response to phosphate deficiency
Gibberellins	Root and shoot tips, young leaves, developing embryos	Stems, developing seeds	Promote cell divisions and growth and elongation of stems; promote seed germination and bolting
Ethylene	Shoot tips, roots, leaf nodes, flowers, fruits	Seeds, buds, seedlings, mature leaves, flowers, fruits	Regulates elongation and division of cells in seedling stems, roots; in mature plants regulates senescence and abscission of leaves, flowers, and fruits
Brassinosteroids	Young seeds; shoots and leaves	Mainly shoot tips, developing embryos	Stimulate cell division and elongation, differentiation of vascular tissue
Abscisic acid	Leaves, chloroplasts, possibly roots in drying soils	Mainly shoot tips, developing embryos, leaves (stomata)	Stimulates cell division and elongation, differentiation of vascular tissue, opening/closing of stomata
Jasmonates	Roots, seeds, probably other tissues	Various tissues, including damaged ones	In defense responses, promote transcription of genes encoding protease inhibitors; possible role in plant responses to nutrient deficiencies
Systemin	Damaged tissues	Damaged tissues	To date known only in tomato; roles in defense, including triggering jasmonate-induced chemical defenses
Salicylic acid	Damaged tissues	Many plant parts	Triggers synthesis of pathogenesis-related (PR) proteins, other general defenses

Some plant hormones are transported from the tissue that produces them to another plant part, whereas others exert their effects in the tissue where they are synthesized. Often, hormonal effects involve changes in gene expression, although sometimes other mechanisms are at work.

All plant hormones are small organic molecules that are active in extremely low concentrations. Hormones that have effects outside the tissue where they are produced typically diffuse to their target site(s) or travel to the site via vascular tissues. Plant hormones vary greatly in their effects, although each one affects a given tissue in a particular way. For instance, some stimulate one or more facets of growth or development, while others have an inhibiting influence. A given hormone also can have different effects in different plant tissues, and the effects can differ depending on a target tissue's stage of development. Adding to the complexity, many physiological responses result from the interaction of two or more hormones. Biologists recognize eight major classes of plant hormones: gibberellins, auxins, cytokinins, strigolactones, ethylene, brassinosteroids, abscisic acid (ABA), and jasmonates **(Table 37.1).** Recent discoveries have added other hormonelike signaling agents to this list.

Plant Hormones Exert Their Effects Via Signal Transduction Pathways

In general, the target cells for a particular hormone have receptors that can bind the hormone and cellular pathways that are activated in response. This mechanism unfolds in three basic steps introduced in Section 9.1. We will briefly review those steps here, and consider current scientific understanding of the signal transduction pathways by which specific hormones exert their effects.

Recall from Chapter 9 that in the first step of a signal transduction pathway, a target cell receptor receives the signal, which may be a molecule or an environmental cue such as sunlight. Next, the signal is transduced—that is, its message is changed into a form that can trigger the cellular response. In the third step, the transduced signal causes the cellular response **(Figure 37.2).** Some transduced signals activate or turn off genes and so alter protein synthesis; others set in motion events that modify existing cell proteins. Various plant hormones and growth factors bind to receptors at the target cell's plasma membrane. Others cross the plasma membrane and bind to

receptors inside the cell. These receptors may be located on the endoplasmic reticulum (ER), in the cytoplasm, or in the nucleus. In many cases, hormone binding causes the receptor to change shape. Regardless, binding of a hormone or growth factor triggers a complex pathway that leads to the cell response—the opening of ion channels, activation of transport proteins, or some other event. Only cells with the appropriate receptor can respond to a particular signaling molecule. For example, certain cells in developing seeds and maturing fruits have receptors for the "ripening hormone" ethylene, but cells in stems generally do not.

We can think of plant hormones and other signaling molecules as external "first messengers" that deliver the initial physiological signal to a target cell. Often, binding of the signal molecule triggers the synthesis of internal second messengers (introduced in Section 9.3). These go-between molecules diffuse through the cytoplasm and provide the main chemical signal that alters cell functioning.

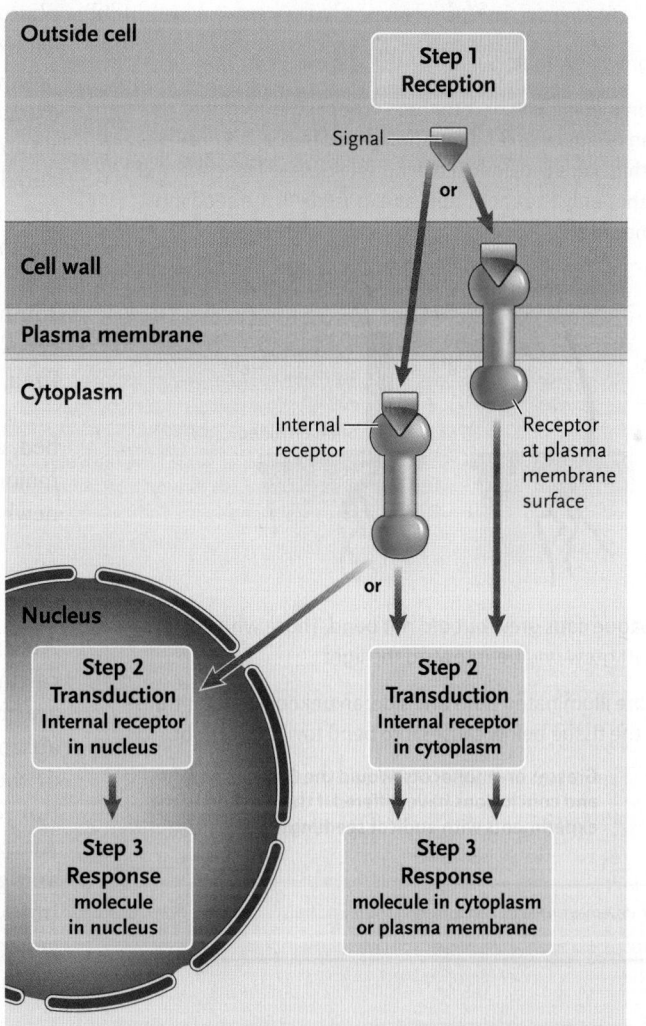

1 An arriving signal binds and activates a receptor in the plasma membrane, cytoplasm, ER, or nucleus. In most cases, the receptor changes shape, which triggers the transduction pathway inside the cell.

2 As the transduction pathway unfolds, receptor activation leads to activation of one or more proteins. This activation step may set in motion a cascade of protein phosphorylation, or it may mobilize second messenger molecules.

3 Phosphorylated proteins or second messenger molecules trigger the cell response, such as a change in ion flow into or out of the cell, a shift in translation of mRNA, or altering gene transcription in the nucleus.

FIGURE 37.2 Overview of signal transduction pathways in plant cells. The three stages of a signal response pathway are reception of the signal, transduction into a form the cell can recognize, and the cell's response. Proteins or second messenger molecules may be the intermediaries in the transduction stage of a signal response pathway (compare Figure 37.8).
© Cengage Learning 2017

FIGURE 37.3 | Experimental Research

The Darwins' Experiments on Phototropism

Question: Why does a plant stem bend toward the light?

Experiment 1: Charles Darwin and his son Francis observed that the first shoot of an emerging grass seedling, which is sheathed by a coleoptile, bends toward sunlight shining through a window. In their first experiment, they removed the shoot tip from a grass seedling and illuminated one side of the seedling.

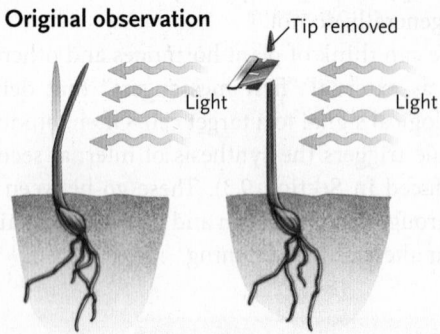

Original observation · Tip removed · Light · Light

Result: The seedling neither grew nor bent.

Experiment 2: The Darwins divided seedlings into three groups. They covered the shoot tips of one group with an opaque cap and the shoot tips of a second group with a translucent cap. In the third group, a light-blocking shield was placed around the shaft of the coleoptile, so that only the shoot tip was exposed. All the seedlings were illuminated from the same side.

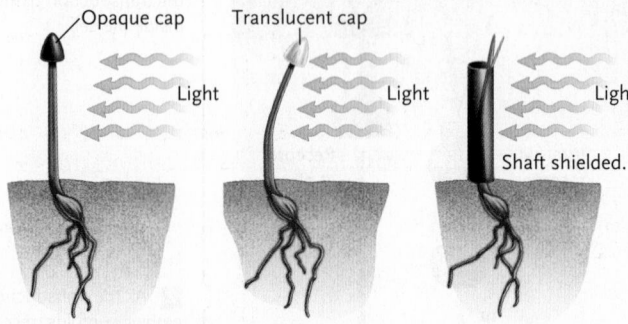

Opaque cap · Translucent cap · Light · Light · Light · Shaft shielded.

Result: The seedlings with opaque caps grew but did not bend. Those with translucent caps and the shielded shaft grew *and* bent toward the light.

Conclusion: When seedlings are illuminated from one side, an unknown factor transmitted from a seedling's tip to the tissue below causes it to bend toward the light.

think like a scientist · Observe / Hypothesize / Predict / Experiment / **Interpret**

Grasses are monocots. Would the Darwins' results and conclusions have differed if they had done the experiments with eudicot seedlings?

Source: C. R. Darwin. 1880. *The Power of Movement in Plants.* London: John Murray.

© Cengage Learning 2017

Second-Messenger Systems Enhance the Plant Cell's Response to a Hormone Signal

Second messengers usually are synthesized in the sequence of chemical reactions that convert an external signal into internal cell activity. For many years the details of plant second-messenger systems were sketchy and hotly debated. Fairly early on, calcium ions were found to play a second messenger role in some hormonal responses. Recent experimental evidence indicates that the hormonal signal from auxins often is conveyed by cAMP (cyclic adenosine monophosphate), a major second messenger in cells of some organisms. Inositol triphosphate (IP_3), a second messenger in plants, fungi, and animals, is involved in the reactions that close plant stomata in response to a signal from abscisic acid, ABA. Other research indicates that a molecule called cGMP (cyclic guanosine monophosphate) serves as a second messenger for auxins, ABA, and a few other plant hormones.

In addition to the basic signal transduction pathways described here, other routes exist that are unique to plant cells. Light is the driving force for photosynthesis, and it is possible that plants have evolved other unique light-related biochemical pathways as well. For instance, experiments are extending our knowledge of how plant cells respond to blue light, which, as noted in Section 34.4 and discussed in more detail shortly, triggers some photoperiod responses such as the opening and closing of stomata. Other exciting research suggests that signal response pathways in plant cells commonly include many steps in which different types of proteins and other molecules are mobilized, not unlike the steps in animal cells.

Auxins were the first plant hormones identified, and we start with them as we consider each major class of plant hormones and discuss some newly discovered signaling molecules as well.

Auxins Promote Growth

Auxins are synthesized mainly in the shoot apical meristem and young stems and leaves. They are crucial to plant growth. Among other effects, auxins are essential for the normal progression of the cell cycle, and they stimulate the elongation of cells in growing stems and coleoptiles. Auxins also mediate growth responses to light and gravity. Indoleacetic acid (IAA) is the most important natural auxin. Botanists often use the general term "auxin" to refer to IAA, a practice we follow here.

EXPERIMENTS THAT LED TO THE DISCOVERY OF AUXINS The path to the discovery of auxins began in the library of Charles Darwin's home in the English countryside (see *Focus on Research: Basic Research* in Chapter 20). Among his interests, Darwin was fascinated by plant **tropisms**—movements such as the bending of a houseplant toward light. This growth response, triggered by exposure to a directional light source, is an example of a **phototropism.**

Working with his son Francis, Darwin explored phototropisms by germinating seeds of two species of grasses, oats *(Avena sativa)* and canary grass *(Phalaris canariensis),* in pots on the sill of a sunny window. Recall from Chapter 36 that the shoot apical meristem and plumule of grass seedlings are sheathed by a coleoptile—a modified leaf that is extremely sensitive to light. Darwin did not know this detail, but he observed that as the emerging coleoptiles grew, within a few days they bent toward the light. He hypothesized that the tip of the coleoptile detected light and communicated that information to the portion of the coleoptile that bent toward the light. Darwin tested this idea in several ways **(Figure 37.3)** and concluded that when seedlings are illuminated from the side, "some influence is transmitted from the upper to the lower part, causing them to bend." In effect, Darwin was stating a key concept with respect to plant hormone activity: the site where a signal is detected and transduced is some distance away from the tissue where the actual response occurs.

The Darwins' observations spawned decades of studies that illustrate how scientific understanding typically advances step-by-step, as one set of experimental findings stimulates new research. First, scientists in Denmark and Poland showed that the bending of a coleoptile toward a light source was caused by something that could move through agar (a jellylike culture material derived from certain red algae) but not through a sheet of the mineral mica, which is impermeable. This finding prompted experiments establishing that indeed the stimulus was a chemical produced in the coleoptile tip. Soon afterward, in 1926, experiments by the Dutch plant physiologist Frits Went confirmed that the growth-promoting chemical diffuses downward from the coleoptile tip **(Figure 37.4).** Using oat seeds, Went first sliced the tips from young coleoptiles that had been grown under normal light conditions. He then placed the tips on agar blocks and left them there long enough for diffusible substances to move into the agar. Meanwhile, the decapitated coleoptiles stopped growing, but growth quickly resumed in seedlings that Went "capped" with the agar blocks (see Figure 37.4A). Clearly, a growth-promoting substance in the excised coleoptile tips had diffused into the agar, and from there into the rest of the coleoptile. Went also attached an agar block to one side of a decapitated coleoptile tip; when the coleoptile began growing again, it bent away from the agar (see Figure 37.4B).

Importantly, Went performed his experiments in total darkness, to avoid any contamination of his results by the possible effects of light. When a flowering plant grows in darkness, metabolic resources are channeled into elongation of the stem rather than into leaf expansion, root growth, and the synthesis of chlorophyll. This phenomenon, called **etiolation (Figure 37.5),** in effect focuses plant growth on finding light—a crucial adaptation for, say, a seedling that has not yet emerged from the soil. As Went knew, exposure of his test plants to even a small amount of light would have led to *de-etiolation,* the shunting of resources into leaf expansion, photosynthesis, and root development.

Went did not determine the mechanism—differential elongation of cells on the shaded side of a coleoptile—by which the growth promoter controlled phototropism. However, he did

A. The procedure showing that IAA promotes elongation of cells below the coleoptile tip

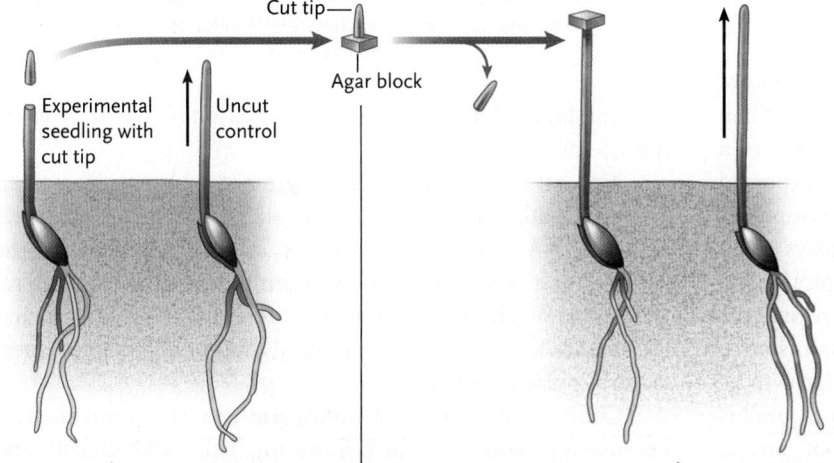

B. The procedure showing that cells in contact with IAA grow faster than those farther away

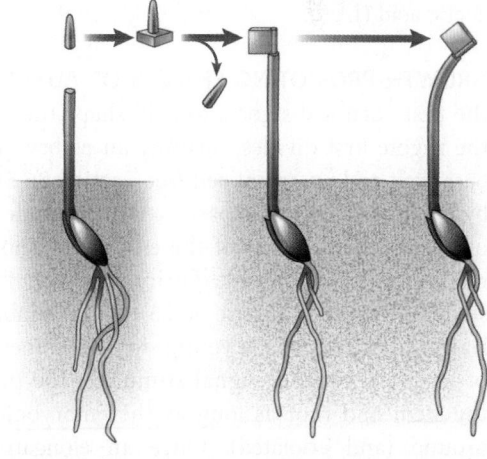

1 After Went cut off the tip of an oat seedling, the coleoptile stopped elongating, while a control seedling with an intact tip continued to grow.

2 He placed the excised tip on an agar block for 1–4 hours. During that time, IAA diffused into the agar block from the cut tip.

3 Went then placed the agar block containing auxin on another detipped oat coleoptile, and the coleoptile resumed elongation, growing about as rapidly as in a control seedling with an intact coleoptile tip.

1 Went removed the tip of a seedling and placed it on an agar block.

2 He placed the agar block containing auxin on one side of the coleoptile tip. Auxin moved into the tip on that side, causing it to bend away from the hormone.

FIGURE 37.4 Two experiments by Frits Went demonstrating the effect of IAA on an oat coleoptile. Went carried out the experiments in darkness to prevent effects of light from skewing the results.

© Cengage Learning 2017

FIGURE 37.5 Effects of the absence of light on young bean plants *(Phaseolus).* The seedling on the left was grown in darkness for several days and has become etiolated. Its leaves are small and pale because it could form carotenoids but not chlorophyll in darkness. It also has a longer stem than the seedling on the right, which was grown in the light. When an etiolated seedling is exposed to light, it greens as chlorophyll synthesis gets underway.

develop a test that correlated specific amounts of the substance, later named auxin (*auxein* = to increase), with particular growth effects. This careful groundwork culminated several years later when other researchers identified auxin as indole-acetic acid (IAA).

GROWTH-PROMOTING EFFECTS OF AUXIN Auxin is one of the first chemical signals to help shape the plant body. When the zygote first divides, forming an embryo that consists of a basal cell and an apical cell (see Section 36.5), auxin exported by the apical cell to the basal cell helps guide the development of the various features of the embryonic shoot. It plays a key role in when and where leaf primordia form in the apical meristem (as shown in Figure 36.26). As the embryo develops, the leaf primordium of the young shoot becomes the main source of IAA; a secondary signal stimulates the primary growth of the stem and root as long as the embryonic plant is underground (and etiolated). Once an elongating shoot breaks through the soil surface, its tip is exposed to sunlight, and the first leaves unfurl and begin photosynthesis. Shortly thereafter the leaf tip stops producing IAA and that task is assumed first by cells at the leaf margin, then by cells at base of the young leaf. Even so, as described in Section 37.3, IAA continues to influence a plant's responses to light and plays a role in its growth responses to gravity as well. IAA also stimulates cell division in the vascular cambium and promotes the formation

FIGURE 37.6 How auxin influences apical dominance. In the runner bean *(Phaseolus coccineus)* plant on the left, auxin moving downward from the apical meristem of the shoot tip has suppressed the development of branches from axillary buds. In the plant on the right, the apical meristem was removed, halting the top-down flow of the auxin signal. As a result, the plant is shorter and numerous branches have developed.

of secondary xylem, as well as the formation of new root apical meristems.

Not all of auxin's effects promote growth, however. Together with the cytokinins and strigolactones we consider shortly, IAA also helps maintain **apical dominance**—the capacity of a plant's apical bud to inhibit the growth of lateral meristems on shoots and to restrict the formation of branches **(Figure 37.6).** Hence, auxin is a signal that the shoot apical meristem is present and active.

Commercial orchardists spray synthetic IAA on fruit trees because it promotes uniform flowering, helps set the fruit, and also helps prevent fruit from dropping off the plant prematurely. Various synthetic auxins are used as **herbicides** (generally, weed killers), essentially stimulating a target plant to "grow itself to death." The most widely used herbicide in the world is the synthetic auxin 2,4-D (2,4-dichlorophenoxyacetic acid). This chemical kills broadleaf eudicots but spares monocots such as grasses. It is used extensively to prevent broadleaf weeds from growing in fields of cereal crops such as corn (which are monocots).

AUXIN TRANSPORT To exert its far-reaching effects on plant tissues, auxin must travel away from its main synthesis sites in shoot meristems and young leaves. Although IAA moves through plant tissues slowly—roughly 1 cm/hr—this rate is 10 times faster than could be explained by simple diffusion. How, then, is auxin transported?

Researchers adapted Fritz Went's agar block method to trace the direction and rate of auxin movements in different kinds of tissues. A team led by Winslow Briggs at Stanford University determined that the shaded side of a coleoptile tip contains more IAA than the illuminated side. Hypothesizing that light causes IAA to move laterally from the illuminated to the shaded region, the team then inserted a vertical, nonporous barrier (a thin slice of mica) between the shaded and illuminated sides of a coleoptile tip. IAA could not cross the barrier, and when the coleoptile tip was illuminated, it did not bend. In addition, the concentrations of IAA in the two sides of the coleoptile tip remained about the same. When the barrier was trimmed so that the separated sides of the tip again touched, the IAA concentration in the shaded area increased significantly, and the tip *did* bend. The study confirmed that IAA initially moves laterally in the coleoptile tip, from the illuminated side to the shaded side, where it triggers the elongation of cells and curving of the tip toward light. Subsequent research showed that IAA then moves downward in a shoot by way of a top-to-bottom mechanism called **polar transport**. That is, IAA in a coleoptile or shoot tip travels from the apex of the tissue to its base, such as from the tip of a developing leaf to the stem. When IAA reaches roots, it moves toward the root tip.

In a stem, parenchyma cells next to vascular bundles apparently transport IAA. The hormone moves through and between cells, traveling by polar transport. As you can see in **Figure 37.7,** the IAA enters at one end by diffusing passively through cell walls, driven by concentration and electrochemical gradients produced by H^+ pumps in the plasma membrane. The hormone exits at the opposite end by active transport across the plasma membrane.

Auxin typically is not found in xylem sap. Instead, it moves into the phloem from parenchyma cells in shoot meristems and young leaves and travels rapidly to plant parts below.

INSIGHTS INTO HOW AUXIN SIGNALS ARE TRANSDUCED

Only in recent years have researchers begun to understand how target cells detect auxins and how subsequent transduction steps unfold. Experiments suggest that auxin binds different receptors, depending on the nature of the ensuing response. Responses that require a change in a gene expression rely on a family of proteins often called simply TIR1, after the original protein discovered. Research in several laboratories confirmed that when IAA in a cell enters the nucleus and binds to TIR1, binding removes an existing inhibition of the transcription of certain genes. The previously repressed genes then are turned on and the cell's activity changes in some way. In responses that do not require a change in gene expression (such as swelling of the protoplast during cell expansion) auxin binds to a

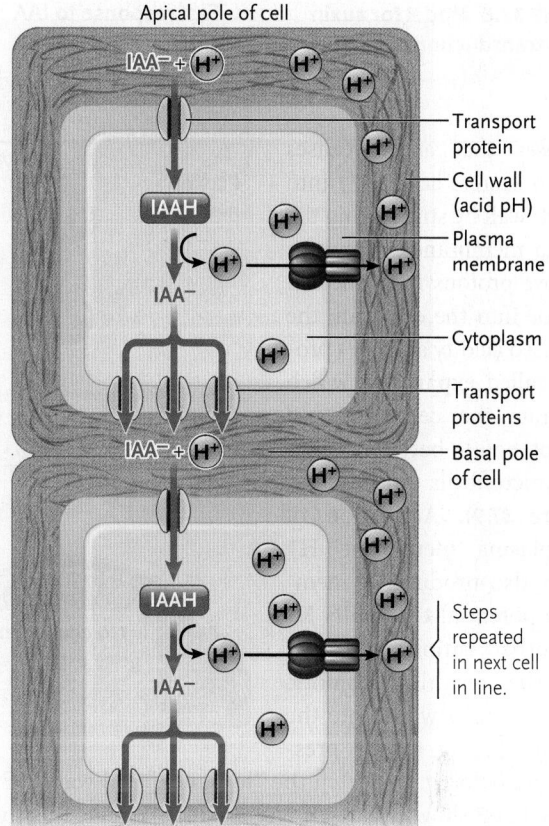

FIGURE 37.7 The polar transport of auxin in plant shoots. Studies of IAA transport in plants have demonstrated that the hormone moves only in one direction, from the shoot tip downward to plant parts below. This diagram shows one model for this polar auxin transport. In this model, auxin enters the cell apex as IAAH (with a hydrogen attached), which forms when hydrogen is attached to IAA^- in the cell wall. In the cytoplasm, H^+ is removed from the auxin molecule, forming IAA^- and H^+. A membrane H^+ pump moves H^+ out of the cell using energy from ATP hydrolysis. Following the gradients, at the basal pole of a cell IAA^- diffuses through the transport proteins into the cell wall, H^+ is added to it again, and then—as IAAH—it moves into the next cell in line.
© Cengage Learning 2017

receptor called ABP1 (for auxin binding protein 1) at the outer surface of the plasma membrane. ABP1 binding triggers additional transduction steps within the cell, although the steps are not well understood. **Figure 37.8** sketches these pathways as researchers currently envision them.

HYPOTHESES ABOUT THE ROLE OF IAA IN PLANT CELL ELONGATION Ever since auxins were discovered, researchers have sought to understand how IAA stimulates plant cells to elongate. As you may recall from Section 36.5, in an elongating plant cell, the cellulose meshwork of the cell wall is first loosened and then stretched by turgor pressure. Several hormones, and auxin especially, increase the stretching of the cell wall. Two major hypotheses have sought to explain this effect, and both may be correct.

Plant cell walls grow much faster in an acidic environment—that is, when the pH is less than 7. The **acid-growth hypothesis**

FIGURE 37.8 Model for auxin signal transduction pathways.
© Cengage Learning 2017

proposes that auxin causes cells to secrete acid (H⁺) into the cell wall by stimulating the plasma membrane H⁺ pumps to move protons from the cell interior into the cell wall; the increased acidity activates proteins called **expansins,** which penetrate the cell wall and disrupt bonds between cellulose microfibrils in the wall **(Figure 37.9).** Activation of the plasma membrane H⁺ pump also produces a membrane potential that pulls K⁺ and other cations into the cell; the resulting osmotic gradient draws water into the cell, increasing turgor pressure and helping to stretch the "loosened" cell walls. (Experiments have shown that all these effects occur shortly after ABP1 binds auxin.)

It is also possible that IAA triggers the expression of genes encoding enzymes that play roles in the synthesis of new wall components. Plant cells exposed to IAA do not show increased growth if they are treated with a chemical that inhibits protein synthesis. On the other hand, certain mRNAs rapidly increase in concentration within 10 to 20 minutes after stem sections are treated with auxin. Expression of mRNAs in response to auxin likely is regulated by the activity of micro RNAs, as described in Chapter 16. It is not yet known exactly which proteins these mRNAs encode.

Cytokinins Enhance Growth and Retard Aging

Cytokinins ("cell movers") play a major role in stimulating cell division and many of their effects come about as they interact with auxins. They were discovered during experiments designed to define the nutrient media required for plant tissue culture. Researchers found that in addition to a carbon source such as sucrose or glucose, minerals, and certain vitamins, cells in culture also required two other substances. One was auxin, which promoted the elongation of plant cells but did not stimulate the cells to divide. The other substance could be coconut milk, which is actually liquid endosperm, or it could be DNA that had been degraded into smaller molecules by boiling. When either was added to a culture medium along with an auxin, the cultured cells would begin dividing and grow normally.

We now know that the active ingredients in both boiled DNA and endosperm are cytokinins, which have a chemical structure similar to that of the nucleic acid base adenine. The

A. Fast response to IAA

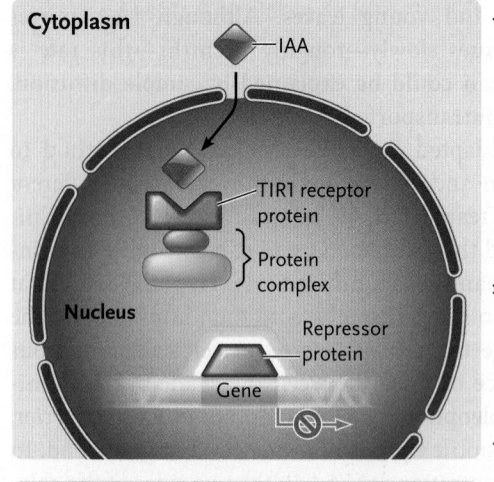

B. IAA responses requiring change in gene expression

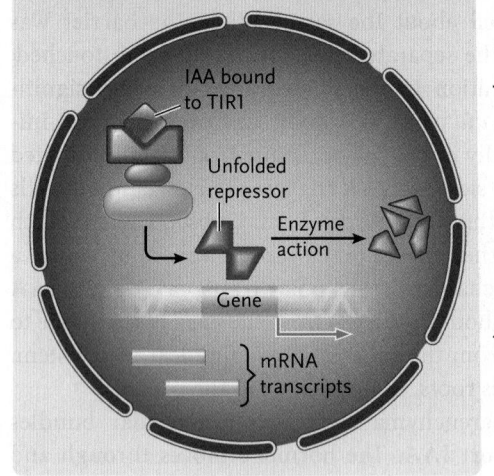

most abundant natural cytokinin is zeatin, so called because it was first isolated from the endosperm of young corn seeds (*Zea mays*). In endosperm, zeatin promotes the burst of cell division that takes place as a fruit matures. As you might expect, cytokinins also are abundant in the rapidly dividing meristems of root and shoot tips. Cytokinins occur not only in flowering plants but also in many conifers, mosses, and ferns. They are also synthesized by many soil-dwelling bacteria and fungi and may be crucial to the growth of mycorrhizae, which help nourish thousands of plant species (see Section 35.3). Conversely, *Agrobacterium* and other microbes that cause plant tumors carry genes that regulate the production of cytokinins.

Cytokinins are synthesized mainly in root tips and travel through the plant in xylem sap. Besides promoting cell division generally, they stimulate the growth of lateral buds and have other developmental and metabolic effects. For example, cytokinins promote the unfurling of young leaves (as leaf cells expand), cause chloroplasts to mature, and retard the *senescence,* or aging, of plant parts. Interactions of cytokinins with auxin and the strigolactones described shortly coordinate the growth of roots and shoots. Investigators culturing tobacco tissues found that the relative amounts of

FIGURE 37.9 The acid-growth hypothesis of how auxin regulates expansion of plant cells. According to the acid-growth hypothesis, plant cells secrete acid (H$^+$) when auxin stimulates the plasma membrane H$^+$ pumps to move protons into the cell wall; the increased acidity activates enzymes called *expansins,* which disrupt bonds between cellulose microfibrils in the wall. As a result, the wall becomes extensible and the cell can expand.

© Cengage Learning 2017

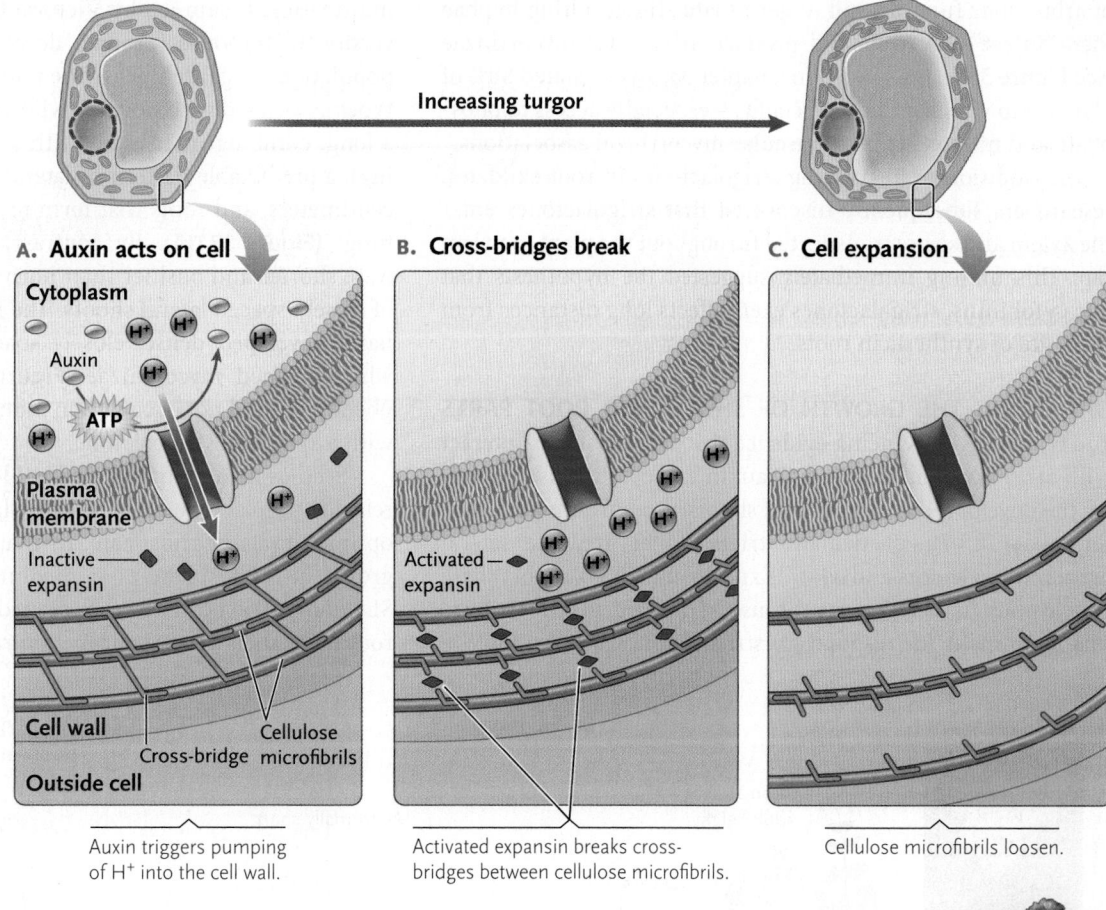

Increasing turgor

A. Auxin acts on cell

Cytoplasm

Auxin

ATP

Plasma membrane

Inactive expansin

Cell wall

Cross-bridge   Cellulose microfibrils

Outside cell

Auxin triggers pumping of H$^+$ into the cell wall.

B. Cross-bridges break

Activated expansin

Activated expansin breaks cross-bridges between cellulose microfibrils.

C. Cell expansion

Cellulose microfibrils loosen.

FIGURE 37.10 Interaction of auxin and cytokinin in the development of plant roots and shoots. Using callus tissue cultured from tobacco pith (spongy parenchyma from a stem), Folke Skoog and Carlos Miller demonstrated how normal development of tobacco plants requires the proportional interaction of auxin and cytokinin.

© Cengage Learning 2017

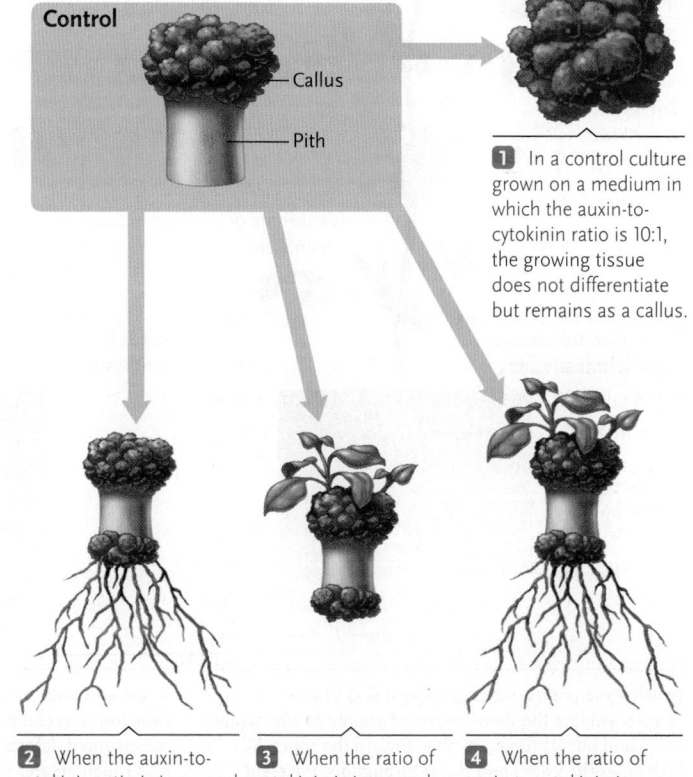

Control

Callus

Pith

1 In a control culture grown on a medium in which the auxin-to-cytokinin ratio is 10:1, the growing tissue does not differentiate but remains as a callus.

2 When the auxin-to-cytokinin ratio is increased to greater than 10:1, the culture produces roots but no differentiated shoot.

3 When the ratio of cytokinin is increased, only shoots develop.

4 When the ratio of auxin to cytokinin is intermediate between the high and low values, both roots and shoots develop.

auxin and a cytokinin strongly influenced not only growth, but also development, as illustrated in **Figure 37.10.**

Natural cytokinins can prolong the life of stored vegetables. Similar synthetic compounds are already widely used to prolong the shelf life of lettuces and mushrooms and to keep cut flowers fresh.

Strigolactones Help Enhance Access to Nutrients and Optimize Plant Growth

Like cytokinins, the hormones called **strigolactones** (SLs) are synthesized mainly in root cells. SLs were identified in the early 2000s after being isolated from root secretions that promote germination of seeds of a parasitic weed in the genus *Striga*. The discovery also stimulated research to identify the adaptive role of this SL-laced root exudate for the plants that synthesize it.

ENHANCING THE DEVELOPMENT OF MYCORRHIZAE Studies on garden peas *(Pisum sativum)* and other plants eventually showed that strigolactones exuded from roots stimulate spores of arbuscular fungi in soil to germinate. The resulting hyphae then "infect" the roots and produce arbuscular mycorrhizae (see Figure 35.7B). As noted in Chapter 35, an estimated 80% of plants reap vital nutritional benefits—especially access to phosphate and nitrate—from arbuscular mycorrhizal associations.

In addition to identifying strigolactones in root exudates, researchers subsequently discovered that strigolactones enter the xylem and so are transported throughout the plant in xylem sap. This finding immediately suggested the hypothesis that like cytokinins, strigolactones exert effects long distances from their site of synthesis in roots.

OPTIMIZING THE GROWTH OF SHOOT AND ROOT PARTS

Today there is convincing evidence that strigolactones interact with other hormones, including auxin and cytokinins, to modify the development of roots and shoots as environmental conditions shift. In particular, strigolactones regulate lateral branching in shoots and roots. Experiments that support these conclusions have relied on the use of mutant plants that are unable to make (or respond to) strigolactones. For example, a

multinational team led by Victoria Gomez-Roldan at the University of Toulouse (France) developed mutant, SL-deficient populations of garden peas. The mutants lacked several phenotypes associated with normal, wild-type pea plants, which have a long, climbing main stem with secondary growth (thickening), a predictable pattern of lateral stem branching every few centimeters, and roots that form mycorrhizae with arbuscular fungi **(Figure 37.11A)**. By contrast, the SL-deficient mutants were shorter and bushier than normal due to the development of closely-spaced lateral shoots. The mutants' root systems similarly developed dense, closely spaced lateral roots, none of which formed mycorrhizae **(Figure 37.11B)**. These abnormal phenotypes did not occur when mutant specimens were treated with a synthetic strigolactone.

Studies using mutants and wild-type plants also have correlated strigolactone activity with plant responses to less-than-optimal levels of phosphate. In that work, the roots of plants grown in a phosphate-poor medium boosted their output of SLs above the normal baseline. Additional lateral roots and root hairs then developed. By contrast, in the shoot fewer lat-

A. Wild-type plant

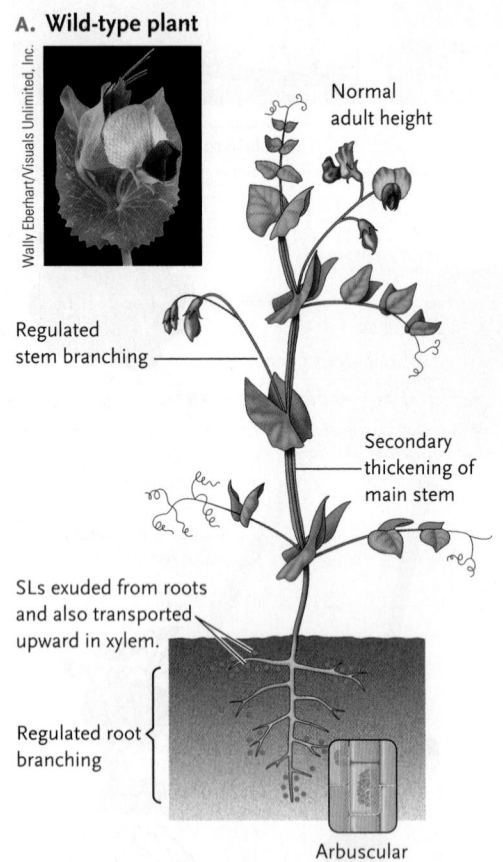

Wally Eberhart/Visuals Unlimited, Inc.

Normal adult height

Regulated stem branching

Secondary thickening of main stem

SLs exuded from roots and also transported upward in xylem.

Regulated root branching

Arbuscular mycorrhizae

In wild-type plants, strigolactones (SLs) exuded by roots stimulate the germination of spores of arbuscular fungi and attract fungal hyphae toward the roots. SLs also enter the xylem and travel throughout the plant, regulating the lateral branching of roots and stems.

B. SL mutant

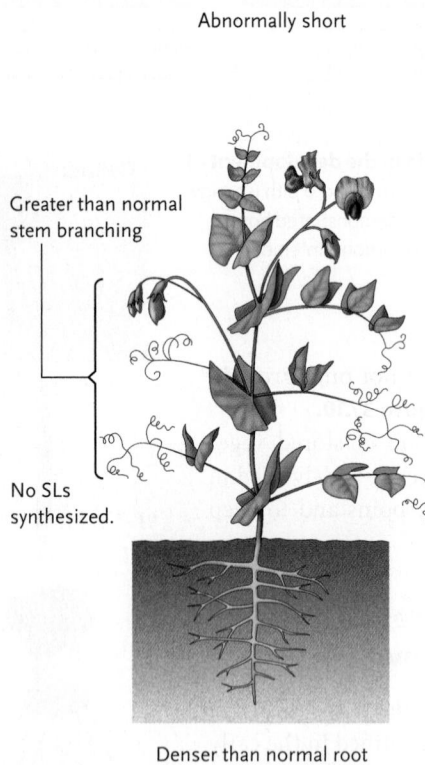

Abnormally short

Greater than normal stem branching

No SLs synthesized.

Denser than normal root system with no mycorrhizae

In experiments with mutant pea plants *(Pisum sativum)* that could not make SLs, vertical growth was stunted, lateral branching of stems and roots was much greater than normal, and no arbuscular mycorrhizae formed.

C. Wild-type plant in phosphate-poor environment

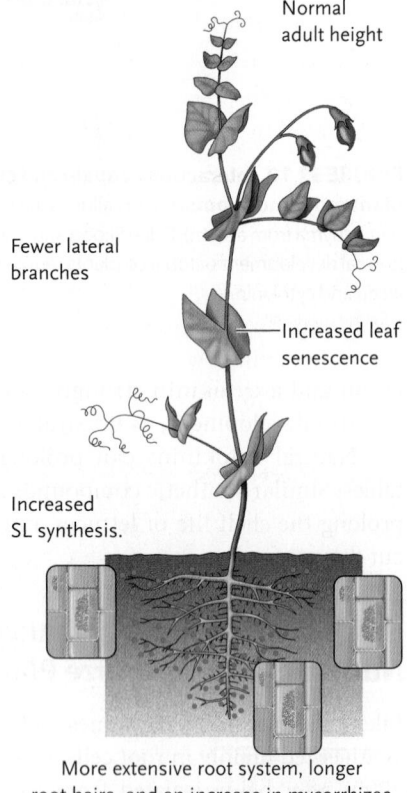

Normal adult height

Fewer lateral branches

Increased leaf senescence

Increased SL synthesis.

More extensive root system, longer root hairs, and an increase in mycorrhizae

In wild-type plants growing in a phosphate-poor environment, SL synthesis increased. Plants grew to normal height, but developed fewer branches and showed early leaf senescence. Branching of roots increased and proportionately more arbuscular mycorrhizae formed.

FIGURE 37.11 A model for the activity of strigolactones in plant growth.

© Cengage Learning 2017

eral buds grew out into branches, and numerous leaves underwent early senescence and died **(Figure 37.11C)**.

Based on these and other experimental findings, there is strong scientific consensus that strigolactones adjust the growth and development of shoot and root systems in ways that foster adequate nutritional support for the whole plant. Inhibiting the development of too many lateral branches, attracting mycorrhizae-forming fungi to roots, and promoting the expansion of root systems when phosphate is scarce all contribute to this outcome. Interactions with cytokinins and IAA undoubtedly are involved in most or all of these activities.

STRIGOLACTONES AND THE EVOLUTION OF LAND PLANTS
Strigolactones and other phytohormones have undoubtedly had roles in the evolution of land plants. Strigolactones have been identified in charophyte algae, the ancestors of land plants, and in the oldest land plant lineages, the bryophytes: hornworts, liverworts, and mosses (see Chapter 28). In bryophytes, strigolactones promote the growth of those plants' rootlike rhizoids; in mosses they stimulate the development of arbuscular mycorrhizae, among other events. While there are still many blanks to fill in, for now it seems reasonable to speculate that strigolactones have been the basis for nutritional and growth-related adaptations throughout the 450-million-year course of land plant evolution. This chapter's *Molecular Insights* examines one proposed model for the emergence of hormone signaling in land plants.

Gibberellins Stimulate Growth in Stems, Fruit, and Seeds

Gibberellins stimulate various aspects of plant growth. Collectively they make up the largest class of plant hormones, with more than 130 recognized chemical variations. Gibberellins have been isolated from fungi, conifers, and from flowering plants, including eudicots and some monocots. They also exist in bryophyte groups as well. In flowering plants, their effects include stimulating the growth of fruits **(Figure 37.12A)**, triggering stem elongation, and prompting seeds and buds to break dormancy. You may recall from Chapter 36 that in barley embryos a gibberellin provides signals during germination that lead to the enzymatic breakdown of endosperm, releasing nutrients that nourish the developing seedling (see Figure 36.15).

Perhaps most apparent to humans is the ability of gibberellins to promote the lengthening of plant stems by stimulating both cell division and cell elongation. Synthesized in shoot and root tips and young leaves, gibberellins (like auxins) modify the properties of plant cell walls in ways that promote cell expansion (although the gibberellin mechanism does not involve acidification of the cell wall). It may be that the two hormones both affect expansins, or are functionally linked in some other way. Experiments show that the general signal transduction pathway for gibberellins is essentially identical to that for auxins—the target cell's activity is altered when a gene repressor in the cell nucleus is inactivated and the gene begins to be expressed (see Figure 37.8).

In most plant species analyzed to date, the main controller of stem elongation is the gibberellin called GA_1. Normally, GA_1 is synthesized in small amounts in young leaves and transported throughout the plant in the phloem. When GA_1 synthesis goes awry, the plant's stature changes dramatically. For example, experiments with a dwarf variety of the garden pea show that these plants and their taller relatives differ at a single gene locus. Normal plants make an enzyme required for gibberellin synthesis; dwarf plants of the same species lack the enzyme, and their internodes barely elongate at all.

Another stark demonstration of the effect gibberellins can have on internode growth is **bolting**—growth of a floral stalk in plants that form vegetative rosettes, such as cabbages *(Brassica oleracea)*, iceberg lettuce *(Lactuca sativa)*, and tarweeds such as the silversword *(Argyroxiphium sandwicense)*. In a rosette plant, stem internodes are so short that the leaves appear to arise from a single node **(Figure 37.12B)**. When these plants flower, however, the stem elongates rapidly and flowers

A. Effect of gibberellin on fruit growth   **B. Silversword rosette**   **C. Bolting silversword**

FIGURE 37.12 Plant responses to gibberellin. (A) Seedless grapes exposed to gibberellin are markedly larger (left) than those from vines that are not treated with the hormone (right). **(B)** Haleakala silversword, one of 28 endangered tarweed species that evolved in the isolation of the Hawaiian Islands, grows only along the upper fringe of Maui's volcanic Haleakala Crater. The plants may grow in a compact rosette form for more than half a century before several weeks of bolting produce a tall floral shoot **(C)**. After flowering, the plant dies.

Investigating the Evolution of Hormone Signaling in Plants

The increasing array of plant genome sequences is proving to be a rich source of information about possible steps in the evolution of hormone signaling in plants. Recently, Chunyang Wang, Yang Liu, and Si-Shen Li, all based at research institutions in China, and Guan-Zhu Han of the University of Arizona used genomic and phylogenetic data from a spectrum of plants and algae to develop hypotheses about the evolutionary origins of the major classes of hormones discussed in this chapter.

Research Question

When did the major classes of plant hormones arise?

Experimental Approach

The team obtained genome sequences of five plant species, including *Arabidopsis thaliana*, representing eudicots; rice (*Oryza sativa*), representing monocots; *Amborella trichopoda*, representing basal angiosperms; the spike moss *Selaginella moellendorffii*, a lycophyte representing early vascular plants; and the moss *Physcomitrella patens*, representing bryophytes. These sequences were compared to those of several green and red algae, and to available genome data for the liverwort

Marchantia polymorpha and four species of charophyte algae—the extant ancestors of land plants described in Chapter 28.

For one part of their analysis, the researchers used the BLAST (Basic Local Alignment Search Tool; see Chapter 19) to search the databases of the selected genome sequences and identify similarities. They then used computerized tools that compare amino acid sequences of proteins to identify potential *orthologs*—genes of different species that arose from a common ancestral gene. In combination with the genome analysis, this step allowed the investigators to develop a phylogenetic tree showing possible evolutionary origins of the nine hormone classes **(Figure)**.

Discussion

The accompanying phylogenetic tree summarizes findings that emerged from this study. The findings suggest—but do not prove—that the emergence of different hormone signaling mechanisms roughly correlate with fundamental events in plant evolution. The tree indicates, for example, that auxins, cytokinins, and strigolactones, all basic growth regulators, arose in charophytes. If the hypothesis is correct, they are the plant kingdom's most ancient hormones. Abscisic

acid, which is crucial in plant responses to water stress, and jasmonates and salicylic acid—both defensive hormones—arose as plants were making the transition to terrestrial life. The model shows gibberellins, which promote shoot elongation and seed germination, originating during the divergence of non-bryophyte lineages from the ground-hugging bryophytes. Brassinosteroid signaling, which helps regulate numerous physiological functions, shows up in the tree at the point just before the emergence of flowering plants. Last, ethylene, which acts in senescence, fruit ripening, and stress responses, appears in the phylogenetic tree after flowering plants were on the scene, but before eudicots and monocots emerged.

think like a scientist

In this study, the only seed plants included were angiosperms. Propose a logical next step for testing whether the authors' evolutionary scenario, particularly regarding the emergence of ethylene, holds up when other seed plants are taken into account.

Source: Chunyang Wang et al. 2015. *Plant Physiology* 167:872–886.

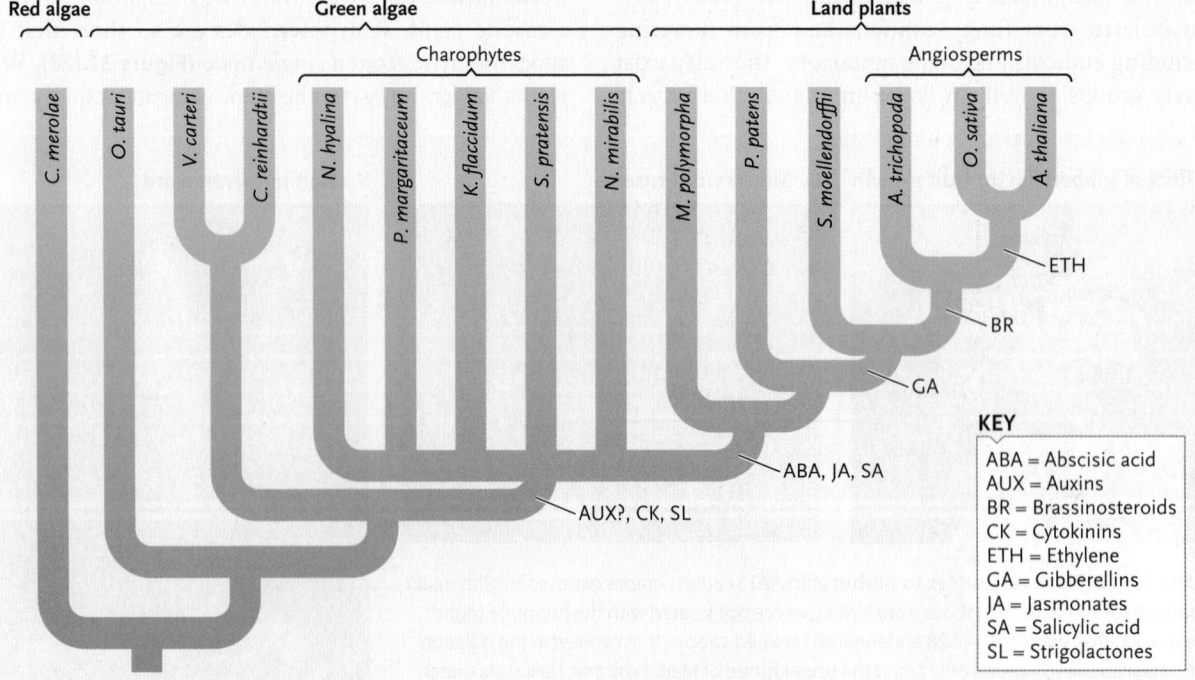

FIGURE A phylogenetic model for the evolution of hormone signaling in plants.
© Cengage Learning 2017

develop on the new stem parts **(Figure 37.12C).** In nature, external cues such as increasing day length or warming after a cold snap stimulate gibberellin synthesis, and bolting occurs soon afterward. This observation supports the hypothesis that in rosette plants and possibly some others, gibberellins switch on internode lengthening when environmental conditions favor a shift from vegetative growth to reproductive growth.

Beyond the effects just mentioned, in monoecious species (having flowers of both sexual types on the same plant), application of a gibberellin encourages proportionately more "male" flowers to develop. As a result, there may be more pollen available to pollinate "female" flowers and, eventually, more fruit produced.

The Gaseous Hormone Ethylene Influences a Range of Events in Plant Growth and Development

Virtually every part of a flowering plant can produce the hormone **ethylene,** the only plant hormone that is a gas. Like auxins, gibberellins, and several other phytohormones, ethylene is a small molecule that easily diffuses from cell to cell. Ethylene helps regulate many aspects of plant physiology including senescence, the abscission or shedding of leaves, the ripening of fleshy fruits, and responses to biotic and abiotic stress.

SENESCENCE AND LEAF ABSCISSION Biologically, **senescence** is a genetically programmed process that leads to the death of specific cells, organs, or the entire organism. As senescence unfolds in plants, affected parts stop growing and begin to deteriorate in an organized way. Changes in gene expression produce enzymes that break down various structural elements of cells and degrade or otherwise alter the cell's chemical constituents. But unlike aging in animals, in plants senescence includes processes that recycle valuable nutrients such as nitrogen, moving them into the phloem for transport to growing roots and reproductive parts such as fruits.

A familiar example of senescence is the seasonal color change of the leaves of deciduous trees. In response to environmental changes such as the shorter days and cooler nights of autumn, ethylene triggers a suite of shifts in gene expression. In leaves, genes that synthesize chlorophyll-degrading chlorophyllases are turned on, as are genes that trigger synthesis of purple anthocyanins and brown tannins. Simultaneously, red-orange carotenoids in leaf cells become more visible. In species-specific ways, these pigments can turn senescing leaves orange, red, yellow, purple, or some combination thereof. Meanwhile, proteases, lipases, and other enzymes break down macromolecules and structures in leaf cells (such as the thylakoids in chloroplasts), liberating their components for recycling via the phloem.

In many plants, senescence is associated with **abscission,** the shedding of leaves, flowers, and fruits. In senescing leaves, ethylene stimulates the activity of enzymes that digest cell walls in an abscission zone at the base of the petiole. The petiole detaches from the stem at that point **(Figure 37.13).**

Various types of cues have an impact on senescence. For instance, in some species the funneling of nutrients into reproductive parts can be stopped by removing newly emerging flowers or seed pods, so a plant's leaves and stems stay green and vigorous much longer **(Figure 37.14).** On the other hand, seasonal shortening of day-length (typical of winter days) appears to be a major environmental trigger for senescence. As an example, when a cocklebur is experimentally induced to flower under winterlike conditions, its leaves turn yellow regardless of whether the nutrient-demanding young flowers are left on or pinched off. These observations underscore the general theme that many plant responses to the environment involve the interaction of multiple molecular signals.

FRUIT RIPENING: A FORM OF SENESCENCE Ripening begins when a fruit starts to synthesize ethylene. Ripening involves

FIGURE 37.13 Abscission zone in a maple (Acer). This longitudinal section is through the base of the petiole of a leaf.

Abscission zone at base of leaf where it joins the stem

N.R. Lersten

Seed pod

Larry D. Nooden

Control plant (pods not removed)

Experimental plant (pods removed)

FIGURE 37.14 Experimental results showing that the removal of seedpods from a soybean plant (Glycine max) delays its senescence.

the conversion of starch or organic acids to sugars and enzyme-driven softening of cell walls. The same kinds of events occur in wounded plant tissues, which also synthesize ethylene.

Ethylene from an outside source can stimulate senescence responses, including ripening, when it binds to specific protein receptors on plant cells. Not all fleshy fruits respond to external ethylene; for example, blackberries, grapes, watermelons, and squashes will not ripen further after picking. By contrast, commercial suppliers routinely use ethylene gas to ripen tomatoes, pineapples, bananas, honeydew melons, mangoes, papayas, and other fruit that typically are picked and shipped while still green. Some ripening fruits, such as bananas, themselves give off ethylene, which is why placing a ripe banana in a closed sack of unripe peaches can cause the peaches to ripen.

Conversely, limiting fruit exposure to ethylene can delay ripening. Apples will keep for months without rotting if they are exposed to a chemical that inhibits ethylene production or if they are stored in an environment that inhibits the hormone's effects—including low atmospheric pressure and a high concentration of CO_2, which binds to ethylene receptors.

ETHYLENE AND SEEDLING GROWTH: THE TRIPLE RESPONSE

In 1901 the Russian botanist Dimitry Neljubov was experimenting with the effects of darkness on pea seedlings when his specimens began exhibiting unusual growth features: their stems were shorter and thicker than normal, and they curved downward and grew horizontally instead of upward **(Figure 37.15)**. At the time, coal gas—which contains ethylene—was widely used in street lamps, and Neljubov soon made the connection between ethylene in the air of his laboratory and

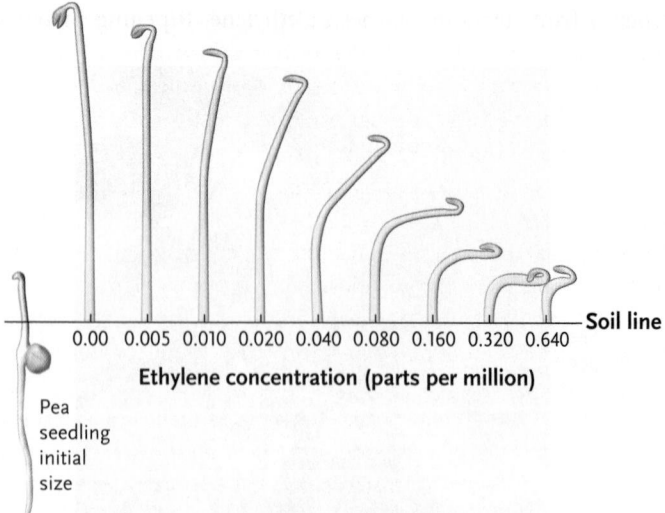

FIGURE 37.15 The triple response of seedlings to ethylene. If a seedling is exposed to mechanical stress that prevents normal elongation of the stem, it synthesizes ethylene. Rising concentration of the hormone triggers a set of three altered growth responses: slowing of normal stem elongation, thickening of the stem, and a shift from vertical to horizontal growth. Together these responses are adaptive because they strengthen the stem and can result in growth away from (or out from under) the stressor. This diagram shows these events in a garden pea seedling.

© Cengage Learning 2017

the seedling growth characteristics he observed. The three effects Neljubov noted—slowed elongation of the stem, stem thickening, and the shift to horizontal growth—became known as the **triple response** to ethylene. In nature the triple response is an adaptation that helps a plant cope with mechanical stress, such as that imposed by a rock that obstructs the normal upward growth of an elongating stem. Mechanical stress triggers the synthesis of ethylene in the plant; the resulting short, thick stem is stronger than a normally developing one, and its horizontal growth may allow the seedling to emerge from under the obstruction and resume normal growth.

Brassinosteroids Help Regulate Plant Growth and Stress Responses

The steroid hormones classed as **brassinosteroids** were first isolated from pollen of canola *(Brassica napus),* a plant in the mustard family. The dozens of brassinosteroids discovered since all appear to be vital for normal plant growth, for they stimulate cell division and elongation in a wide range of plant cell types. Confirmed as plant hormones in the 1980s, brassinosteroids now are the subject of intense research on their sources and effects, many of which involve interactions with auxins. The highest concentrations of brassinosteroids are found in shoot tips and in developing seeds and embryos—all examples of young, actively developing parts. In laboratory studies, the hormones have different effects depending on the tissue where they are active. They promote cell elongation, differentiation of vascular tissue, and elongation of a pollen tube after a flower is pollinated. In roots, on the other hand, brassinosteroids inhibit elongation. Brassinosteroids also regulate the expression of genes associated with a plant's growth responses to light. This role was underscored by experiments using mutant *Arabidopsis* plants that were homozygous for a defective gene called *bri1* (for *brassinosteroid-insensitive receptor 1)*. The results provided convincing evidence that brassinosteroids mediate growth responses to light **(Figure 37.16)**. Recent research has focused on possible roles of brassinosteroids in plant responses to environmental stresses such as drought and high soil salinity.

Abscisic Acid Suppresses Growth and Influences Responses to Environmental Stress

The hormone **abscisic acid** (ABA) has a variety of effects, many of which represent evolutionary adaptations to environmental challenges. Plants synthesize ABA from carotenoid pigments inside plastids in leaves and possibly other plant parts. Several ABA receptors have been identified, and in general, we can group ABA effects into changes in gene expression that inhibit growth over a long time span, and rapid, short-term physiological changes that are responses to immediate stresses, such as drought, in a plant's surroundings. As its name suggests, at one time ABA was thought to play a central role in abscission, although we now know that ethylene is the major abscission trigger.

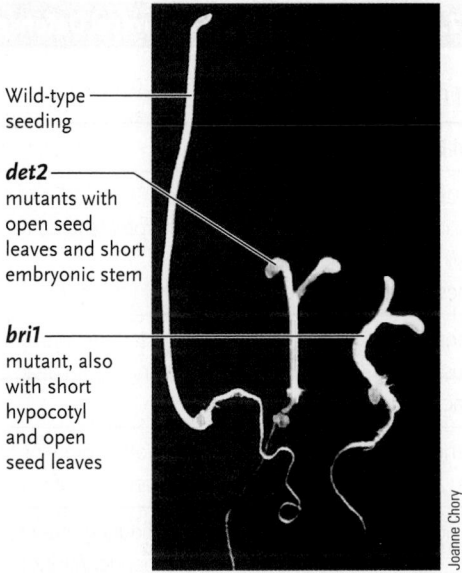

Wild-type seeding

det2 mutants with open seed leaves and short embryonic stem

bri1 mutant, also with short hypocotyl and open seed leaves

Joanne Chory

FIGURE 37.16 Experimental evidence that brassinosteroids can mediate a plant's responses to light by regulating gene expression. In *A. thaliana*, wild-type seedlings synthesize a protein (encoded by the *DET2* gene) that prevents leaves from developing (seedling at left) until photosynthesis is possible, after the seedling breaks out of the dark environment of soil. When the gene is defective, a mutant *det2* plant (center) will develop a short hypocotyl (embryonic stem) and open seed leaves (cotyledons) even when there is no light for photosynthesis. Experiments with *bri1* mutants, which lack functioning receptors for a brassinosteroid, resulted in a similar phenotype (right). These findings supported the hypothesis that a brassinosteroid is necessary for normal expression of the *DET2* gene.

SUPPRESSING GROWTH IN BUDS AND SEEDS

Operating as a counterpoint to growth-stimulating hormones like gibberellins, ABA inhibits growth in response to environmental cues, such as seasonal changes in temperature and light. This growth suppression can last for months or even years. For example, one of ABA's major growth-inhibiting effects is apparent in perennial plants, in which the hormone promotes dormancy in leaf buds—an important adaptive advantage in places where winter cold can damage young leaves.

In some plants that produce fleshy fruits, such as apples and corn, abscisic acid is associated with seed dormancy. As the seed develops, ABA accumulates in the seed coat, and the embryo does not germinate even if it becomes hydrated (see Section 36.3). Experiments with mutant corn plants have demonstrated that this ABA-mediated resistance to premature germination is under genetic control. In plants lacking the gene that makes seeds sensitive to ABA, the seeds (within kernels) germinate while kernels are still on the cob **(Figure 37.17).**

The buildup of ABA in developing seeds does more than just inhibit development, however. As early development draws to a close, ABA stimulates the transcription of certain genes, and large amounts of their protein products are synthesized. These proteins will be sources of nitrogen and other nutrients that the embryo will use when it eventually does germinate— usually after a long period of cool, wet conditions, which

Courtesy of William J. Wiebold, University of Missouri

FIGURE 37.17 Seeds of corn germinating while kernels are still on the cob. When this phenomenon occurs in ABA-resistant corn plants, cobs show masses of prematurely germinating seedlings.

stimulate the breakdown of ABA. Commercial growers often apply ABA and related growth inhibitors to plants slated to be shipped to plant nurseries. Dormant plants suffer less shipping damage, and the effects of the inhibitors can be reversed by applying a gibberellin.

RESPONSES TO ENVIRONMENTAL STRESS

ABA also triggers plant responses to various abiotic environmental stresses, including cold snaps and drought. A great deal of research has focused on how ABA influences plant responses to a lack of water. As you know, stomata regulate the loss of water during transpiration (described in Section 34.4). When a plant is water-stressed, ABA helps prevent excessive water loss by stimulating stomata to close when they might otherwise be open. When a lack of water leads to wilting, mesophyll cells in wilted leaves rapidly synthesize and secrete ABA. The hormone diffuses to guard cells, where an ABA receptor binds it. Binding stimulates the release of K^+ and water from the guard cells, and within minutes the stomata close.

ABA's role in stomatal closure—triggered by water stress or some other environmental cue—begins when the hormone activates a receptor in the plant cell plasma membrane. Experiments have shown that this binding launches a complex sequence of events that transduce the hormone signal. In a first step, binding activates G proteins that in turn activate phospholipase C (see Figure 9.12). This enzyme then stimulates the synthesis of second messengers such as inositol triphosphate (IP_3). The second messenger diffuses through the cytoplasm and binds with calcium channels in structures such as the endoplasmic reticulum and tonoplast. The channels then open, releasing calcium ions that activate protein kinases in the cytoplasm. In turn, these enzymes activate their target proteins by phosphorylating them. An *Arabidopsis* mutant unable to respond to ABA lacks an enzyme that removes phosphate groups from certain proteins—indicating that the cellular response to ABA involves cleaving phosphates needed for the phosphorylation step. As described in Section 9.2, the original hormone signal is greatly amplified by a cascade of activated protein kinases, each one of which can activate a large number of target proteins.

Jasmonates Regulate Growth and Function in Defense

Jasmonates (JAs) are a family of about 20 phytohormones derived from fatty acids. Experiments with *Arabidopsis* and other plants have revealed numerous genes that respond to JAs, including genes that help regulate root growth and seed germination. The JA family is best known, however, as part of the plant arsenal to limit damage by pathogens and predators, the topic of the following section.

37.2 Plant Chemical Defenses

Plants do not have immune systems like those that have evolved in animals (the subject of Chapter 47), but higher plants have evolved an array of means for coping with biotic stressors in the environment. Over the millennia, virtually constant exposure to predation by herbivores and the onslaught of pathogens have resulted in a striking array of chemical defenses that ward off or reduce damage to plant tissues from infectious bacteria, fungi, worms, or plant-eating insects **(Table 37.2).** You will discover in this section that plant defenses include both constitutive and inducible defenses. **Constitutive defenses** are "built-in" barriers to threats, such as bark, the waxy cuticle covering a plant's epidermis, and herbivore-deterring spines, hairs, and chemicals. **Inducible defenses** are responses that arise when an attack occurs. They are mediated by hormones and other biologically active substances. Inducible defenses include general responses to any type of attack and specific responses to particular threats. Some get underway almost as soon as an attack begins, whereas others occur over an extended period. More

TABLE 37.2	Summary of Plant Chemical Defenses
Type of Defense	Effects
General Defenses	
Jasmonate (JA) responses to wounds/injury by pathogens; pathways often include other hormones such as ethylene	Synthesis of defensive chemicals such as protease inhibitors
Hypersensitive response to infectious pathogens (e.g., fungi, bacteria)	Physically isolates infection site by surrounding it with dead cells
Pathogenesis-related (PR) proteins	Enzymes, other proteins that degrade cell walls of pathogens
Salicylic acid (SA)	Mobilized during other responses and independently; induces the synthesis of PR proteins, operates in systemic acquired resistance
Systemin (in tomato)	Triggers JA response
Phytoalexins	Antibiotic
Systemic acquired resistance (SAR)	Long-lasting protection against some pathogens; components include SA and PR proteins that accumulate in healthy tissues
Specific Defenses	
Recognition of PAMPs (chemical features of specific pathogens)	Launches defensive response (e.g., hypersensitive response, PR proteins) against pathogens
Effector-triggered immunity	Resistance proteins bind bacterial virulence effectors, triggering immediate defense responses
Other	
Heat-shock responses (encoded by heat-shock genes)	Synthesis of chaperone proteins that reversibly bind other plant proteins and prevent denaturing caused by heat stress
"Antifreeze" proteins	In some species, stabilize cell proteins under freezing conditions

often than not, multiple chemicals interact as the response unfolds.

Jasmonates Interact with Salicylic Acid and Systemin in a General Response to Wounds

When an insect begins feeding on a leaf or some other plant part, the plant may respond to the resulting wound by launching what in effect is a cascade of chemical responses.

These complex signaling pathways often rely on interactions among jasmonates, ethylene, or some other plant hormone. As the pathway unfolds, it triggers expression of genes leading to chemical and physical defenses at the wound site. For example, in some plants a jasmonate induces a response leading to the synthesis of protease inhibitors, which disrupt an insect's capacity to digest proteins in the plant tissue. The protein deficiency in turn hampers the insect's growth and functioning.

A plant's capacity to recognize and respond to the physical damage of a wound apparently has been subject to strong selection pressure during plant evolution. When a plant is wounded experimentally, numerous defensive chemicals can soon be detected in its tissues. One of these, **salicylic acid,** or SA (a compound similar to aspirin, which is acetylsalicylic acid), has multiple roles in plant defenses, including interacting with jasmonates.

Researchers are regularly discovering new variations of hormone-induced wound responses in plants. For example, experiments have elucidated some of the steps in an unusual pathway that thus far is known only in tomato (*Solanum lycopersicum*) and a few other plant species. As diagrammed in **Figure 37.18,** the wounded plant rapidly synthesizes **systemin,** the first peptide hormone to be discovered in plants. (Several peptide hormones occur in animals, a topic of Chapter 44.) Systemin enters the phloem and is transported throughout the plant. Although details of the signaling pathway have yet to be worked out, when receptive cells bind systemin, their plasma membranes release a lipid that is the chemical precursor of jasmonate. Next jasmonate is synthesized, and it in turn sets in motion the expression of genes that encode protease inhibitors, which protect the plant against attack, even in parts far from the original wound.

The Hypersensitive Response and PR Proteins Are Other General Defenses

Often, a plant that becomes infected by pathogenic bacteria or fungi counters the attack by way of a **hypersensitive response**—a defense that physically cordons off an infection site by surrounding it with dead cells. Initially, cells near the site respond by producing a burst of highly reactive oxygen-containing compounds (such as hydrogen peroxide, H_2O_2) that can break down nucleic acids, inactivate enzymes, or have other toxic effects on cells. Enzymes in the plant cell's plasma membrane catalyze the burst. It begins the process of killing cells close to the attack site, and as the response advances, programmed cell death may also come into play. In short order, the sacrificed dead cells wall off the infected area from the rest of the plant. Thus denied an ongoing supply of nutrients, the invading pathogen dies. A common sign of a successful hypersensitive response is a dead spot surrounded by healthy tissue **(Figure 37.19).**

While the hypersensitive response is underway, salicylic acid triggers other defensive responses by an infected plant.

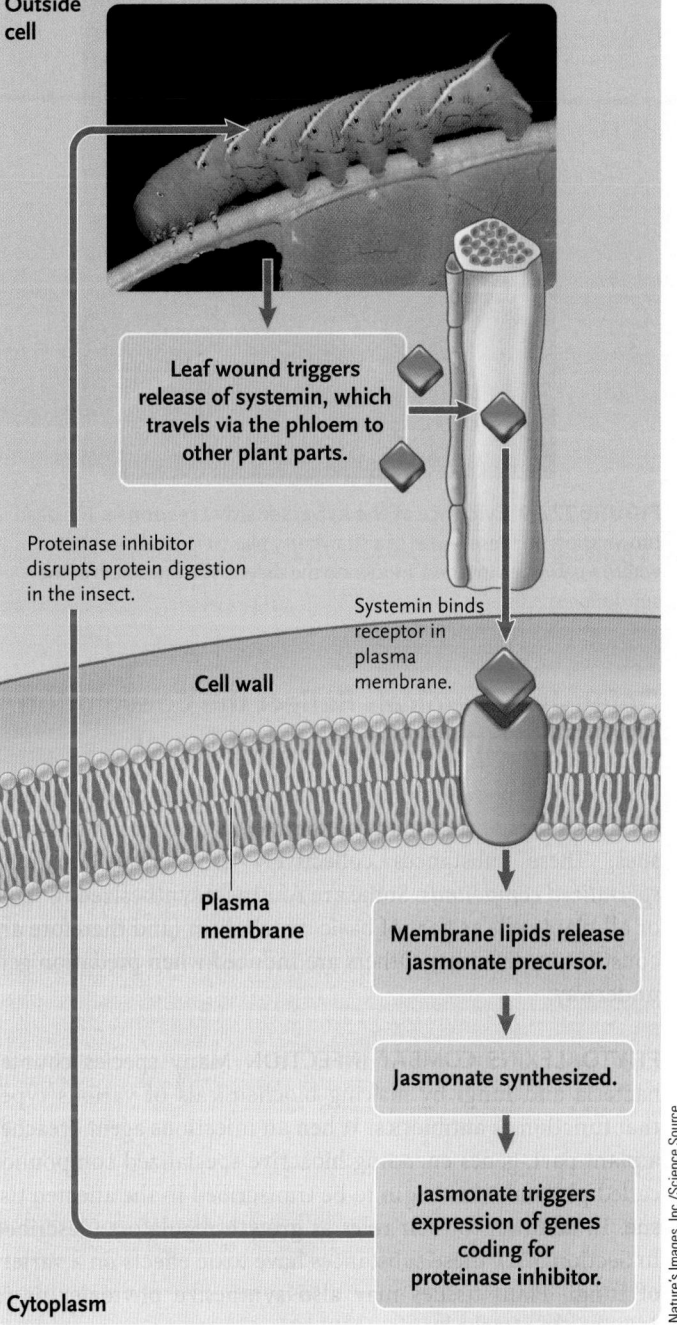

FIGURE 37.18 The systemin response to wounding. When a plant is wounded, it responds by releasing the protein hormone systemin. Transported through the phloem to other plant parts, in receptive cells systemin sets in motion a sequence of reactions that lead to the expression of genes encoding protease inhibitors—substances that can seriously disrupt an insect predator's capacity to digest protein.

One of its effects is to induce the synthesis of **pathogenesis-related (PR) proteins.** Some PR proteins are hydrolytic enzymes that break down components of a pathogen's cell wall. Examples are chitinases that dismantle the chitin in the cell walls of fungi and so kill the cells. In some cases, plant cell receptors also detect the presence of fragments of the disintegrating wall and set in motion additional defense responses.

FIGURE 37.19 Evidence of the hypersensitive response. The dead brown spots on these leaves of a strawberry plant *(Fragaria)* are sites where a pathogen invaded, triggering the defensive destruction of the surrounding cells.

Defensive Chemicals Reflect the Coevolution of Plants, Pathogens, and Herbivores

Millennia of coevolution of plants with their predators have produced tens of thousands of chemical deterrents in plant tissues. These substances collectively are termed *bioactive specialized compounds*. Some are routinely synthesized in most or all plant cells as part of basic metabolism (and therefore are constitutive defenses). Others are induced when predation gets underway.

PHYTOALEXINS COMBAT INFECTION Many species counter bacteria and fungi by making biochemicals of various types that function as antibiotics. When an infectious agent breaches a plant part, genes encoding bioactive specialized compounds called **phytoalexins** begin to be transcribed in the affected tissue. In addition to their roles as growth regulators (described in Section 37.1), these substances have toxic effects on a variety of fungi. Plant tissues may also synthesize phytoalexins in response to attacks by viruses.

DEFENSES AGAINST HERBIVORY Chapter 33 noted some plant structural defenses against herbivory, such as spines, leathery leaves, and irritating hairs (trichomes). A wide range of plant species also deploy bioactive specialized compounds as defenses against feeding herbivores. An estimated 15,000 of these compounds are **alkaloids** such as caffeine, cocaine, and the poison strychnine (in seeds of the nux-vomica tree, *Strychnos nuxvomica*), and various terpenes. The terpene family includes insect-repelling substances in cotton and the resins of conifers and the creosote bush, and essential oils produced by sage and basil plants. Because these terpenes are volatile—they easily diffuse out of the plant into the surrounding air—they also can provide indirect defense to a

FIGURE 37.20 Coevolution of milkweed *(Asclepias syriaca)* and an herbivore, the monarch caterpillar. An alkaloid in *A. syriaca* tissues is toxic to most herbivores (including vertebrates), but not to the monarch butterfly's resistant larval stage.

plant. Released from the wounds created by a munching insect, they attract other insects that prey on the herbivore.

Nearly 10,000 plant defensive chemicals, including tannins such as those in oak leaves and acorns, are compounds classified chemically as **phenolics.** To the human palate phenolics have a bitter taste; it may be some version of this chemical quality that repels many insects, birds, and other herbivorous animals. The leaves of strawberry plants counter herbivory of the two-spotted spider mite by way of phenolics that bind to and disable the insect's digestive enzymes.

Some herbivores have evolved tolerance to the toxic effects of defensive plant chemicals. For example, milkweeds *(Asclepias* spp.) produce a creamy sap that contains alkaloids and terpenes known as cardiac glycosides that are lethal to most insects and also can kill vertebrates (by interfering with heart function). Among the few insects unaffected by these defensive chemicals are the larvae of the monarch butterfly **(Figure 37.20).** The discussion of predator–prey relationships in Chapter 54 looks in detail at the interactions between plants and herbivores.

Plants Respond to Specific Threats by Way of Pattern Recognition Receptors and Effector-Triggered Immunity

One of the most interesting questions with respect to inducible plant defenses is how plants first sense that an attack is underway. Although as noted earlier plants do not have an immune system like that of animals, they do have adaptations for detecting attacks by specific groups of pathogenic bacteria and fungi. One of these adaptations is the capacity to recognize molecular features of invaders—often, proteins associated with the cell wall or flagellum (in the case of bacteria). Such features are called **PAMPs,** an acronym for pathogen-associated molecular patterns. The plasma membranes of plant cells bear specialized *pattern recognition receptors* that detect PAMPs. Detection launches a complex web of signals that culminate in the cell's response, such as changes in gene expression that lead to the synthesis of phytoalexins.

Some plant immune responses are triggered by bacterial *virulence effectors*—proteins produced by pathogenic bacteria that can bypass a plant cell's pattern recognition receptors and so evade the cell's detection of PAMPs. Like the monarch butterfly caterpillar's resistance to milkweed toxins, bacterial virulence effectors reflect the ongoing coevolutionary tit-for-tat between predators and prey. Plants in turn may combat bacterial virulence effectors by way of **effector-triggered immunity.** In this inducible mechanism, **resistance proteins** encoded by so-called *R genes* (for "resistance") trigger an immediate defense response in the plant. Thousands of *R* genes have been identified in a wide range of plant species and confer enhanced resistance to plant pathogens including bacteria, fungi, and nematode worms that attack roots.

Although much remains to be learned about how resistance proteins disable bacterial effectors, experiments have demonstrated a rapid-fire sequence of early biochemical changes that follow binding of the particular resistance protein; these include changes in concentrations of calcium ions inside and outside plant cells and the production of biologically active oxygen compounds that heralds the hypersensitive response. In fact, of the instances of effector triggered immunity plant scientists have observed thus far, most induce the hypersensitive response and the ensuing synthesis of PR proteins, with their antibiotic effects.

Systemic Acquired Resistance Is an Adaptation for Long-Term Protection

The defensive response to a microbial invasion may spread throughout a plant, so that the plant's healthy tissues become less vulnerable to infection. This phenomenon is called **systemic acquired resistance,** and experiments using *Arabidopsis* plants have shed light on how it comes about **(Figure 37.21).** In a key early step, salicylic acid builds up in the affected tissues. By some route, probably through the phloem, the SA passes from the infected organ to newly forming organs such as leaves, which begin to synthesize PR proteins—again, providing the plant with a "home-grown" antimicrobial arsenal. How does the SA exert this effect? It seems that when enough SA accumulates in a plant cell's cytoplasm, a regulatory protein called NPR-1 (for *n*onexpressor of *p*athogenesis-*r*elated genes) moves from the cytoplasm into the cell nucleus. There it interacts with factors that promote the transcription of genes encoding PR proteins.

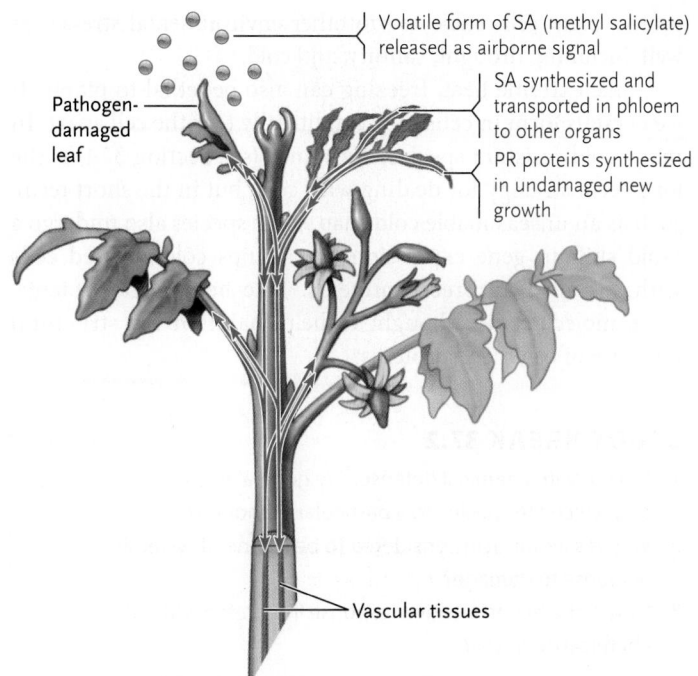

FIGURE 37.21 A proposed mechanism for systemic acquired resistance. When a plant successfully fends off a pathogen, the defensive chemical salicylic acid (SA) is transported in the phloem to other plant parts, where it may help protect against another attack by stimulating the synthesis of PR proteins. In addition, the plant synthesizes and releases a slightly different, more volatile form of SA called methyl salicylate. It may serve as an airborne signal to other parts of the plant as well as to neighboring plants.
© Cengage Learning 2017

In addition to synthesizing SA that will be transported to other tissues by a plant's vascular system, the damaged leaf also synthesizes a chemically similar compound, methyl salicylate. This substance is volatile, and researchers speculate that it may serve as an airborne "harm" signal, promoting defense responses in the plant that synthesized it and possibly in nearby plants as well.

Extremes of Heat and Cold Also Elicit Protective Chemical Responses

Plant cells also contain **heat-shock proteins (HSPs),** a type of chaperone protein (see Section 3.4) found in cells of many species. In general, HSPs bind and stabilize other proteins, including enzymes, which might otherwise stop functioning if they were to become denatured by rising temperature. Plant cells may rapidly synthesize HSPs in response to various stimuli, including a sudden temperature rise. For example, experiments with cells and seedlings of soybean (*Glycine max*) showed that when the temperature rose 10°C to 15°C, in less than 5 min smRNA transcripts coding for as many as 50 different HSPs were present in cells. When the temperature returns to a normal range, HSPs release bound proteins, which can then resume their usual functions. Further studies have revealed that heat-shock proteins help protect plant cells subjected to other environmental stresses as well, including drought, salinity, and cold.

Like extreme heat, freezing can also be lethal to plants. If ice crystals form in cells they can literally tear the cell apart. In many cold-resistant species, dormancy (see Section 37.4) is the long-term strategy for dealing with cold, but in the short term, such as an unseasonable cold snap, some species also undergo a rapid shift in gene expression that equips cold-stressed cells with so-called antifreeze proteins. Like heat-shock proteins, these molecules are thought to help maintain the structural integrity of other cell proteins.

37.3 Plant Movements

Although a plant cannot move from place to place as external conditions change, plants do alter the orientation of their body parts in response to environmental stimuli. As noted earlier in the chapter, growth toward or away from a unidirectional stimulus, such as light or gravity, is called a *tropism*. Tropic movement involves permanent changes in the plant body because cells in particular areas or organs grow differentially in response to the stimulus. Tropisms are fascinating examples of the complex abilities of plants to adjust to their environment. This section will also touch on two other kinds of movements—developmental responses to physical contact, and changes in the position of plant parts that are not related to the location of the stimulus.

Phototropisms Are Responses to Light Mediated by Blue-Light Receptors

Light is a key abiotic stimulus for many kinds of organisms. In general, developmental changes regulated by light are examples of **photomorphogenesis.** Phototropisms, the auxin-dependent growth responses to a directional light source discussed in Figure 37.3, are an example. As the Darwins discovered, if light is more intense on one side of a stem, the stem may curve toward the light **(Figure 37.22A).** Phototropic movements are extremely adaptive for photosynthesizing organisms because they help maximize the exposure of photosynthetic tissues to sunlight.

How do auxins influence phototropic movements? In a coleoptile that is illuminated from one side, IAA moves by polar transport into the cells on the shaded side **(Figure 37.22B–D).** Phototropic bending occurs because cells on the shaded side elongate more rapidly than do cells on the illuminated side.

The main stimulus for phototropism is light of blue wavelengths, which stimulates **blue-light receptors.** Experiments on corn coleoptiles have shown that some blue-light receptors are a family of large, yellow pigment molecules called *phototropins,* which play a role in stimulating the initial lateral transport of IAA to the dark side of a shoot tip. Blue-light-absorbing proteins called **cryptochromes** also mediate the various light-based growth responses. As described later, cryptochromes appear to have a role in other plant responses to light as well.

Gravitropism Orients Plant Parts to the Pull of Gravity

Plants show growth responses to Earth's gravitational pull, a phenomenon called **gravitropism.** After a seed germinates, the primary root curves down, toward the "pull" (positive gravitropism), and the shoot curves up (negative gravitropism).

Several hypotheses seek to explain how plants respond to gravity. The most widely accepted hypothesis proposes that plants detect gravity much as animals do—that is, particles

 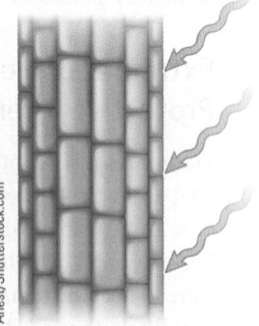 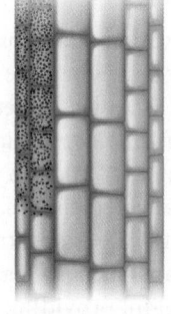 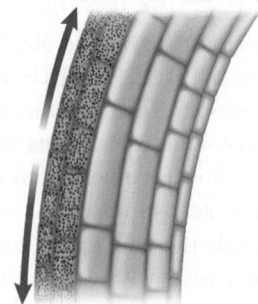

Anest/Shutterstock.com

A. Seedlings bend toward light.

B. Rays from the Sun strike one side of a shoot tip.

C. Auxin (red) diffuses down from the shoot tip to cells on its shaded side.

D. The auxin-stimulated cells elongate more quickly, causing the seedling to bend.

FIGURE 37.22 Phototropism in seedlings. (A) Tomato seedling grown in darkness; its right side was illuminated for a few hours before it was photographed. **(B–D)** Hormone-mediated differences in the rates of cell elongation bring about the bending toward light. (Auxin is shown in red.)
© Cengage Learning 2017

called **statoliths** in certain cells move in the direction gravity pulls them. In the semicircular canals of human ears, calcium carbonate crystals serve as statoliths; in most plants the statoliths are amyloplasts, modified plastids that contain starch grains (see Section 4.4). In eudicot angiosperm stems, amyloplasts often are present in one or two layers of cells just outside the vascular bundles. In monocots such as cereal grasses, amyloplasts are located in tissue near the base of the leaf sheath. In roots, amyloplasts occur in the root cap. If the spatial orientation of a plant cell is shifted experimentally, its amyloplasts sink through the cytoplasm until they come to rest at the bottom of the cell **(Figure 37.23)**.

How do amyloplast movements translate into an altered growth response? The full explanation is probably complex, and there is evidence that somewhat different mechanisms operate in stems and in roots. In stems, the sinking of amyloplasts may provide a mechanical stimulus that triggers a gene-guided redistribution of IAA. For example, when a potted sunflower seedling is turned on its side in a dark room, within 15 to 20 minutes cell elongation decreases markedly on the upper side of the growing horizontal stem, but increases on the lower side. With the adjusted growth pattern, the stem curves upward, even in the absence of light. Using different types of tests, researchers have been able to document the shifting of IAA from the top to the bottom side of the stem. The changing auxin gradient correlates with the altered pattern of cell elongation.

In roots, a high concentration of auxin has the opposite effect—it inhibits cell elongation. If a root is placed on its side, amyloplasts in the root cap accumulate near the side wall that now is the bottom side of the cap. This shift stimulates cell elongation in the opposite wall, and within a few hours the root once again curves downward. In root tips of many plants, however, especially eudicots, researchers could not detect a change in IAA concentration that correlates with the changing position of amyloplasts. Eventually experiments on gravitropism in soybean root tips suggested that the IAA signal is transduced by way of a signal cascade. The hormone induces the accumulation of nitric oxide (NO) at the downward side of the root tip, where NO in turn appears to induce activity cGMP. This second messenger then delivers the original IAA signal. The sequence inhibits the elongation of cells on the tip's downward side, so the root curves downward.

Along with IAA, calcium ions (Ca^{2+}) play a major role in gravitropism. For example, if Ca^{2+} is added to an otherwise untreated agar block that is then placed on one side of a root cap, the root will bend toward the block. In this way, experimenters have been able to manipulate the direction of growth so that the elongating root forms a loop. Similarly, if an actively bending root is deprived of Ca^{2+}, the gravitropic response abruptly stops. By contrast, the negative gravitropic response of a shoot tip is inhibited when the tissue is exposed to excess calcium.

A. Root oriented vertically

B. Root oriented horizontally

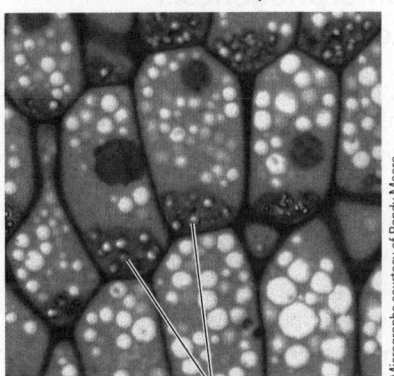
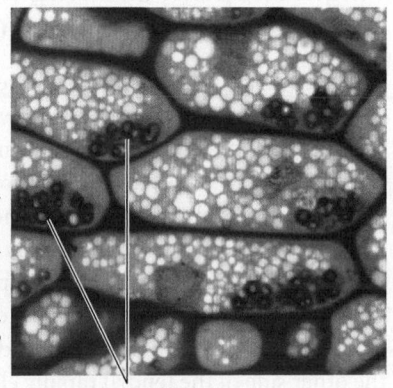

Statoliths

Statoliths

FIGURE 37.23 Evidence that supports the statolith hypothesis. When a corn root was laid on its side, amyloplasts—statoliths—in cells from the root cap settled to the bottom of the cells within 5 to 10 minutes. Statoliths may be part of a gravity-sensing mechanism that redistributes auxin through a root tip.

Just how Ca^{2+} interacts with IAA in gravitropic responses is unknown. One hypothesis posits that calcium functions as an activator. Calcium binds to a small protein called *calmodulin*, activating it in the process. Activated calmodulin in turn can activate a variety of key cell enzymes in many organisms, both plants and animals. It is possible that calcium-activated calmodulin stimulates cell membrane pumps that enhance the flow of both IAA and calcium through a gravity-stimulated plant tissue.

Some of the most active research in plant biology focuses on the intricate mechanisms of gravitropism **(Figure 37.24)**. For example, mounting evidence indicates that in many plants, cells in different regions of stem tissue differ in their sensitivity to IAA, and that gravitropism is linked in some fundamental way to these differences. In a few plants, including some cultivated varieties of corn and radish, the direction of the gravitropic response by a seedling's primary root is influenced by light. Clearly there is much more to be learned.

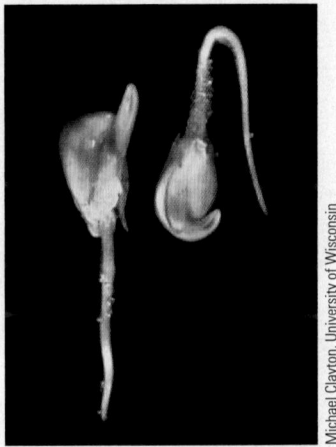

FIGURE 37.24 Gravitropism in corn seedlings. The two corn grains shown here were oriented in opposite directions with respect to gravity. Regardless, the primary root of each one grew downward through the soil and each stem grew upward.

Thigmotropism and Thigmomorphogenesis Are Responses to Physical Contact

Varieties of peas, grapes, and some other plants demonstrate **thigmotropism** (*thigma* = touch), which is growth in response to contact with a solid object. Thigmotropic plants typically are climbers—they have long, slender stems and cannot grow upright without physical support. They often have *tendrils,* modified stems or leaves that can rapidly curl around a fence post or the sturdier stem of a neighboring plant. If one side of a grape vine stem grows against a trellis, for example, specialized epidermal cells on that side of the stem tendril shorten whereas cells on the other side of the tendril rapidly elongate. Within minutes the tendril starts to curl around the trellis, forming tight coils that provide strong support for the vine stem. **Figure 37.25** shows thigmotropic twisting of a cucumber tendril *(Cucumis sativus).* Auxin and ethylene are probably involved in thigmotropism, but most details of the mechanism remain elusive.

The rubbing and bending of stems caused by frequent strong winds, rainstorms, grazing animals, and even farm machinery can inhibit the overall growth of plants and can alter their growth patterns. In this phenomenon, called **thigmomorphogenesis,** a stem stops elongating and instead adds girth when it is regularly subjected to mechanical stress. Simply shaking some plants daily for a brief period will inhibit their upward growth, but although such plants may be shorter, as noted in the earlier discussion of the triple response, their thickened stems will be stronger. Thigmomorphogenesis helps explain why plants growing outdoors are often shorter, have thicker stems, and are not as easily blown over as plants of the same species grown indoors. Trees growing near the snowline of windswept mountains show an altered growth pattern that reflects this response to wind stress.

Research on the cellular mechanisms of thigmomorphogenesis has begun to yield tantalizing clues. In one study, investigators repeatedly sprayed *Arabidopsis* plants with water and imposed other mechanical stresses, then sampled tissues from the stressed plants. The samples contained as much as double the usual amount of mRNA for at least four genes, which had been activated by the stress. The mRNAs encoded calmodulin and several other proteins involved in altering *A. thaliana* growth responses. The test plants were also short, generally reaching only half the height of unstressed controls.

Nastic Movements Are Nondirectional

Tropisms are responses to directional stimuli, such as light striking one side of a shoot tip, but many plants also exhibit **nastic movements** (*nastos* = pressed close together)—reversible responses to nondirectional stimuli, such as mechanical pressure or humidity. We see nastic movements in leaves, leaflets, and even flowers. For instance, certain plants exhibit nastic sleep movements, holding their leaves (or flower petals) in roughly horizontal positions during the day but folding them closer to the stem at night **(Figure 37.26).** At twilight tulip flowers "go to sleep" in this way.

Many nastic movements are temporary and result from changes in cell turgor. For example, the daily opening and closing of stomata in response to changing light levels are nastic movements, as is the traplike closing of the lobed leaves of the Venus flytrap when an insect brushes against hairlike sensory structures on the leaves. The leaves of *Mimosa pudica,* the sensitive plant, also close in a nastic response to mechanical pressure. Each *M. pudica* leaf is divided into pairs of leaflets **(Figure 37.27A).** Touching even one leaflet at the leaf tip triggers a chain reaction in which each pair of leaflets closes up within seconds **(Figure 37.27B).**

In many turgor-driven nastic movements, water moves into and out of the cells in **pulvini** (*pulvinus* = cushion), thickened pads of tissue at the base of a leaf or petiole. Stomatal movements depend on changing concentrations of ions within guard cells, and pulvinar cells drive nastic leaf movements in *M. pudica* and numerous other plants by the same mechanism **(Figure 37.27C).**

FIGURE 37.25 Thigmotropism in a cucumber *(C. sativus)* tendril, which is twisted around a support.

FIGURE 37.26 Nastic sleep movements in a bloodroot *(Sanguinaria canadensis).*

How is the original stimulus transferred from cells in one part of a leaf to cells elsewhere? The answer lies in the polarity of charge across cell plasma membranes (see Section 5.4). Touching an *M. pudica* leaflet triggers an **action potential**—a brief reversal in the polarity of the membrane charge. When an action potential occurs at the plasma membrane of a pulvinar cell, the change in polarity causes potassium ion (K^+) channels to open, and ions flow out of the cell, setting up an osmotic gradient that draws water out as well. As water leaves by osmosis, turgor pressure falls, pulvinar cells become flaccid, and the leaflets move together. Later, when the process is reversed, the pulvinar cells regain turgor and the leaflets spread apart. Action potentials travel between parenchyma cells in the pulvini via plasmodesmata at the rate of about 2 centimeters per second. Chapter 40 describes how animal nerves conduct similar changes in membrane polarity along their plasma membranes.

Stimuli other than touch also can trigger nastic movements. Cotton, soybean, sunflower, and some other plants display *solar tracking,* nastic movements in which leaf blades are oriented toward the east in the morning, then steadily change their position during the day, following the Sun across the sky. Solar tracking movements maximize the amount of time that leaf blades are perpendicular to the Sun, which is the angle at which photosynthesis is most efficient.

A. Undisturbed plant

B. Plant response to touch

C. Leaf folding mechanism

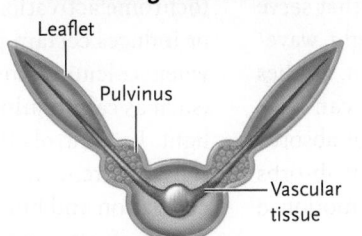

Leaflet

Pulvinus

Vascular tissue

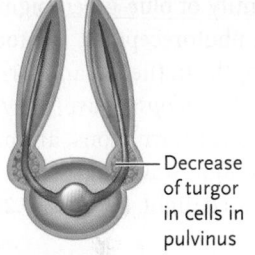

Decrease of turgor in cells in pulvinus

FIGURE 37.27 Nastic movements in leaflets of *Mimosa pudica,* the sensitive plant. (A) In an undisturbed plant the leaflets are open. If a leaflet near the leaf tip is touched, changes in turgor pressure in pulvini at the base cause the leaf to fold closed. **(B, C).** The diagram sketches this folding movement in cross section. Other leaflets close in sequence as action potentials transmit the stimulus along the leaf.
© Cengage Learning 2017

STUDY BREAK 37.3

1. What is the direct stimulus for phototropism? For gravitropism?
2. Explain how nastic movements differ from tropic movements.

37.4 Plant Biological Clocks

Like all eukaryotic organisms, plants have internal time-measuring mechanisms called **biological clocks** that adapt the organism to recurring environmental changes. In plants biological clocks help adjust both daily and seasonal activities.

Circadian Rhythms Are Based on 24-Hour Cycles

Some plant activities occur regularly in cycles of about 24 hours, even when environmental conditions remain constant. These are **circadian rhythms** (*circa* = around; *dies* = day). Chapter 34 noted that stomata open and close on a daily cycle, even when plants are kept in total darkness. The nastic sleep movements described earlier also are examples of a circadian rhythm. Even when a plant that exhibits such movements is kept in constant light or darkness for a few days, its leaves fold into the "sleep" position at roughly 24-hour intervals. In some way, the plant measures time without sunrise (light) and sunset (darkness). Such experiments demonstrate that internal controls, rather than external cues, largely govern circadian rhythms.

Circadian rhythms and other activities regulated by a biological clock help ensure that plants of a single species do the same thing, such as flowering, at the same time. For instance, flowers of the four-o'clock plant (*Mirabilis jalapa*) open predictably every 24 hours—in nature, in the late afternoon. This pattern reflects the coevolution of *M. jalapa* with its main pollinator, a species of nocturnal moth. Although some circadian rhythms can proceed without direct stimulus from light, many biological clock mechanisms are influenced by the relative lengths of day and night.

Photoperiodism Involves Seasonal Changes in the Relative Length of Night and Day

Obviously, environmental conditions in a 24-hour period are not the same in summer as they are in winter. In North America, for instance, winter temperatures are cooler and winter day length is shorter. Experimenting with tobacco and soybean plants in the early 1900s, two American botanists, Wightman Garner and Henry Allard, elucidated a phenomenon they called **photoperiodism,** in which plants respond to changes in the relative lengths of light and dark periods in their environment during each 24-hour period. Through photoperiodism, the biological clocks of plants (and animals) make seasonal adjustments in their patterns of growth, development, and reproduction.

PLANT RESPONSES TO CHANGING PHOTOPERIOD: THE PHYTOCHROME SWITCH

A plant's evolutionary fitness depends heavily on its ability to synchronize its overall growth, and particularly the development of reproductive parts, with favorable light and temperature conditions. More than any other environmental cue, changes in photoperiod indicate seasonal shifts in growing conditions.

Plant responses to changing photoperiod are governed by a family of blue-green pigments called **phytochromes** that serve as photoreceptors. Phytochromes are sensitive to light wavelengths in the red and far-red portions of the spectrum. Studies of *Arabidopsis* have shown that most phytochromes can have two conformations, an inactive one denoted as P_r that absorbs red light wavelengths and another active form, P_{fr}, that absorbs far-red light **(Figure 37.28A).** The P_{fr} form sets in motion a variety of important developmental events in plants, such as seed germination and flowering. These phytochromes are **photoreversible**—that is, the absorption of light triggers the shifts from one conformation to the other. Those changes in turn alter gene expression in ways that stimulate or inhibit plant responses.

Plant cells synthesize phytochrome in its P_r form. Sunlight contains relatively more red light than far-red light, and during daylight hours when red wavelengths dominate, P_r absorbs red light. Absorption of red light triggers the rapid conversion of phytochrome to its active P_{fr} form, and the buildup of P_{fr} in turn triggers growth responses. At sunset, at night, or in shade, where far-red wavelengths predominate, P_{fr} slowly reverts to P_r **(Figure 37.28B).**

THE PHYTOCHROME SIGNAL

Experiments indicate that phytochrome activation stimulates plant cells to take up Ca^{2+} ions, or induces certain plant organelles to release them. Either way, when calcium ions combine with calcium-binding proteins (such as calmodulin), they initiate various cellular responses to light. P_{fr} controls the types of enzymes produced in particular cells; different enzymes are required for seed germination, stem elongation and branching, leaf expansion, and the formation of flowers, fruits, and seeds.

SEED GERMINATION: EVIDENCE OF PHYTOCHROMES IN ACTION

Early evidence for the existence of phytochromes came from experiments on seed germination. The research was prompted by observations that seeds of certain plants, such as the lettuce *L. sativa,* germinated poorly in dim light. In the late 1950s, scientists at the U.S. Department of Agriculture placed groups of lettuce seeds on a moist substrate and then exposed them to flashes of red and far-red light in a regimen summarized in **Figure 37.29.** When the final flash was far-red light—triggering photoreversion to P_r—many seeds failed to germinate. Conversely, when the final flash was red light—triggering the accumulation of P_{fr}—all or nearly all of the seeds germinated.

SHADE AVOIDANCE

In nature, the ongoing conversion of P_r to P_{fr} and back again produces a characteristic ratio of P_r to P_{fr} in leaf cells. For plants adapted to thrive in full sunlight, this balance supplies information about the plant's ongoing access to the red light wavelengths it relies on for photosynthesis (see Section 8.2). If the leaves of other, nearby plants begin to intercept so much sunlight that the ratio of P_r to P_{fr} shifts in favor of P_r, some plants demonstrate an adaptive response called **shade avoidance.** Signal transduction pathways are activated that direct the plant's metabolic resources into more-rapid-than-usual vertical growth. A shade-avoiding plant also may flower earlier than usual, before the physiological emphasis on stem elongation depletes metabolic resources that fuel flower development.

Cryptochromes—which, recall, are sensitive to blue light and influence light-related growth responses—also interact

A. The absorption spectra associated with the interconversion of P_r and P_{fr}

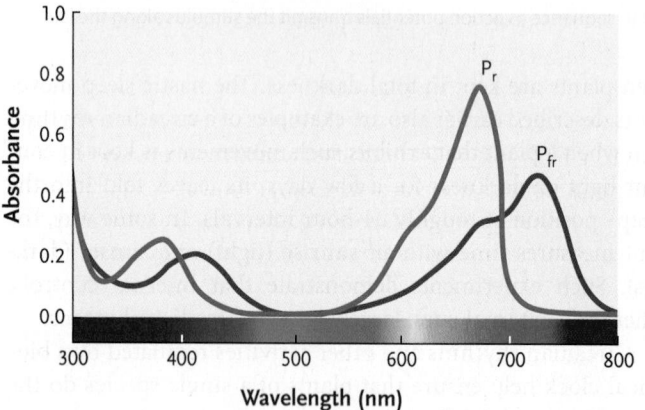

B. Interconversion of phytochrome from the inactive form (P_r) to the active form (P_{fr})

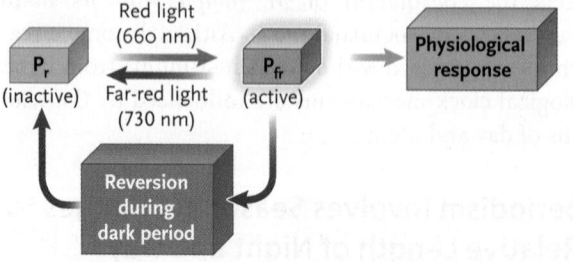

FIGURE 37.28 The phytochrome switching mechanism, which can promote or inhibit growth of different plant parts. The P_r form of phytochrome is sensitive to red light (660 nm), and the P_{fr} form is sensitive to far-red light (730 nm).

© Cengage Learning 2017

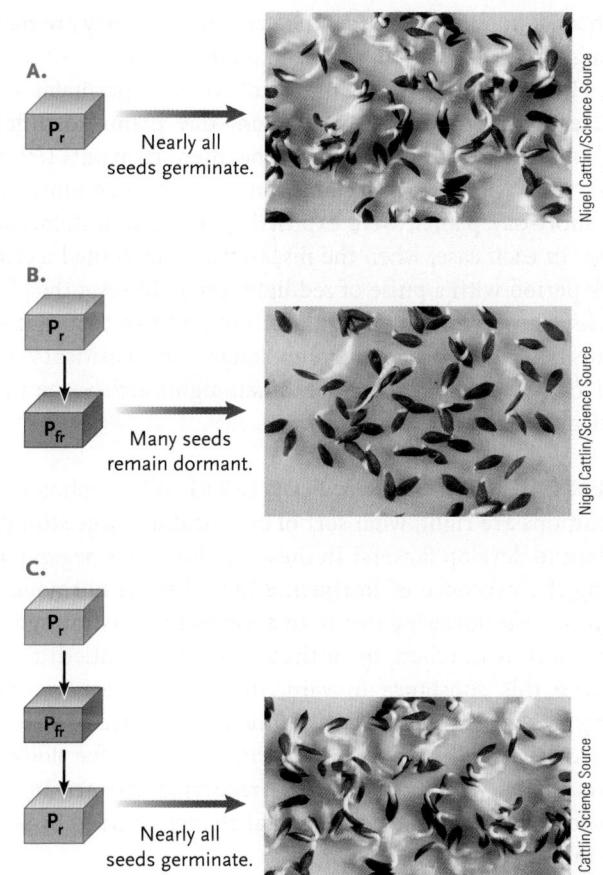

A.

P_r → Nearly all seeds germinate.

B.

P_r
↓
P_fr → Many seeds remain dormant.

C.

P_r
↓
P_fr
↓
P_r → Nearly all seeds germinate.

FIGURE 37.29 Experiment demonstrating how phytochrome switching governs the germination of lettuce seeds. As described in the text and diagrammed in Figure 37.28, the accumulation of P_{fr} serves as signal that a plant is exposed to light of red wavelengths—the main wavelengths in sunlight. Lettuces and some weeds produce tiny seeds that do not germinate in the dark. In this experiment, groups of lettuce seeds were placed on a moist substrate from which they could imbibe water—a precondition for germination—and then subjected to different light conditions. When seeds were exposed only to red light (P_r), more than 90% germinated **(A)**. By contrast, germination was suppressed in seeds exposed to far-red wavelengths *after* the red light **(B)**. When seeds exposed to the red/far-red sequence were again exposed to red light, nearly all of them germinated **(C)**. Based on these results, the researchers concluded that for lettuce seeds, red light is the ultimate trigger for germination.

© Cengage Learning 2017

FIGURE 37.30 Effect of day length on poinsettia (*Euphorbia pulcherrima*). The poinsettia is a short-day plant that in nature blooms in autumn as nights become longer. Commercial growers manipulate the lighting in greenhouses to delay flowering (and the development of the red, white, pink, or multicolored bracts) until the Christmas holiday season when poinsettias are in most demand. The plant on the right has been maintained in long-day conditions and so has not developed flowers or colored bracts.

with phytochrome in producing circadian responses. An intriguing discovery is that cryptochromes occur not only in plants but also in animals such as fruit flies and mice. Only further study will determine if cryptochromes act as circadian photoreceptors in both kingdoms.

Cycles of Light and Dark Often Influence Flowering

Photoperiodism is especially apparent in the flowering process. Like other plant responses, flowering is often keyed to changes in day length through the year and to the resulting changes in environmental conditions. Corn, soybeans, peas, and other annual plants begin flowering after only a few months of growth. Roses and other perennials typically flower every year or after several years of vegetative growth. Carrots, cabbages, and other biennials typically produce roots, stems, and leaves the first growing season, die back to soil level in autumn, then grow a new flower-forming stem the second season.

In the late 1930s, Karl Hamner and James Bonner grew cocklebur plants (*Xanthium strumarium*) in chambers in which the researchers could control environmental conditions, including photoperiod. And they made an unexpected discovery: test plants flowered only when they were exposed to a single night of 8.5 hours of uninterrupted darkness. The length of the "day" in the growth chamber did not matter, but if light interrupted the dark period for even a minute or two, the plant would not flower at all. Subsequent research confirmed that for most angiosperms, it is the length of darkness, not light, that controls flowering.

TYPES OF FLOWERING RESPONSES The photoperiodic responses of flowering plants are so predictable that they can be used to categorize plants. The categories, which refer to day length, reflect the fact that scientists recognized the phenomenon of photoperiodic flowering responses long before they understood that darkness, not light, was the cue. **Long-day plants,** such as irises, daffodils, and spinach, usually flower in spring when nights become shorter and day length becomes longer than some critical value—usually 9 to 16 hours. **Short-day plants,** including cockleburs, chrysanthemums, and the poinsettias shown in **Figure 37.30,** flower in late summer or early autumn when nights become longer and day length becomes shorter than some critical value. **Intermediate-day plants,** such as sugarcane, flower only when day length falls in between the values for long-day and short-day plants. **Day-neutral plants,**

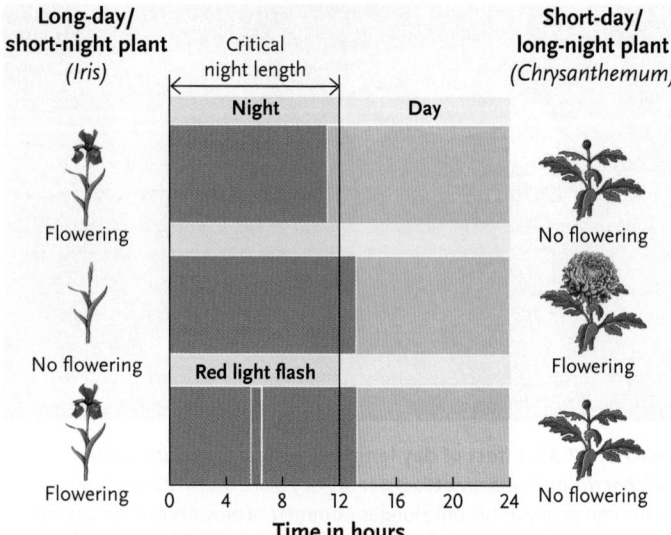

Long-day/ short-night plant *(Iris)*

Critical night length

Short-day/ long-night plant *(Chrysanthemum)*

Night — Day

Flowering — No flowering

No flowering — Flowering

Red light flash

Flowering — No flowering

Time in hours

FIGURE 37.31 Experiments showing that short-day and long-day plants flower in response to night length. Each horizontal bar represents 24 hours. Blue sections indicate night, and yellow sections day. **(A)** Long-day plants such as bearded irises flower when the night is shorter than a critical length, whereas **(B)** short-day plants such as chrysanthemums flower when the night is longer than a critical value. **(C)** When an intense pulse of red light interrupts a long night, both kinds of plants respond as if it were a short night; the irises flower but the chrysanthemums do not.

such as dandelions and roses, flower whenever they are mature enough to do so, without regard to photoperiod.

Figure 37.31 illustrates the results of an experiment to test the responses of short-day and long-day plants to differing night lengths. In this experiment, bearded iris plants (*Iris* species), which are long-day plants, and chrysanthemums, which are short-day plants, were exposed to a range of light conditions. In each case, when the researchers interrupted a critical dark period with a pulse of red light, the light reset the plants' clocks. The experiment provided clear evidence that short-day plants flower only when nights are longer than a critical value—and long-day plants flower only when nights are shorter than a critical value.

CHEMICAL SIGNALS FOR FLOWERING When photoperiod conditions are right, what sort of chemical message stimulates a plant to develop flowers? In the 1930s botanists began postulating the existence of **florigen,** a hypothetical hormone that served as the flowering signal. In a somewhat frustrating scientific quest, researchers spent the rest of the twentieth century seeking this substance in vain. Recently, molecular studies using *Arabidopsis* plants have defined a sequence of steps that may collectively provide the internal stimulus for flowering. Here again, we see one of the recurring themes in plant development—major developmental changes guided by several interacting genes.

Figure 37.32 traces the steps of the proposed flowering signal. To begin with, a gene called *CONSTANS* is expressed in a plant's leaves in tune with the daily light/dark cycle, with expression peaking at dusk (step 1). The gene encodes a regulatory protein called CO (not to be confused with carbon monoxide). As days lengthen in spring, the concentration of CO rises in leaves,

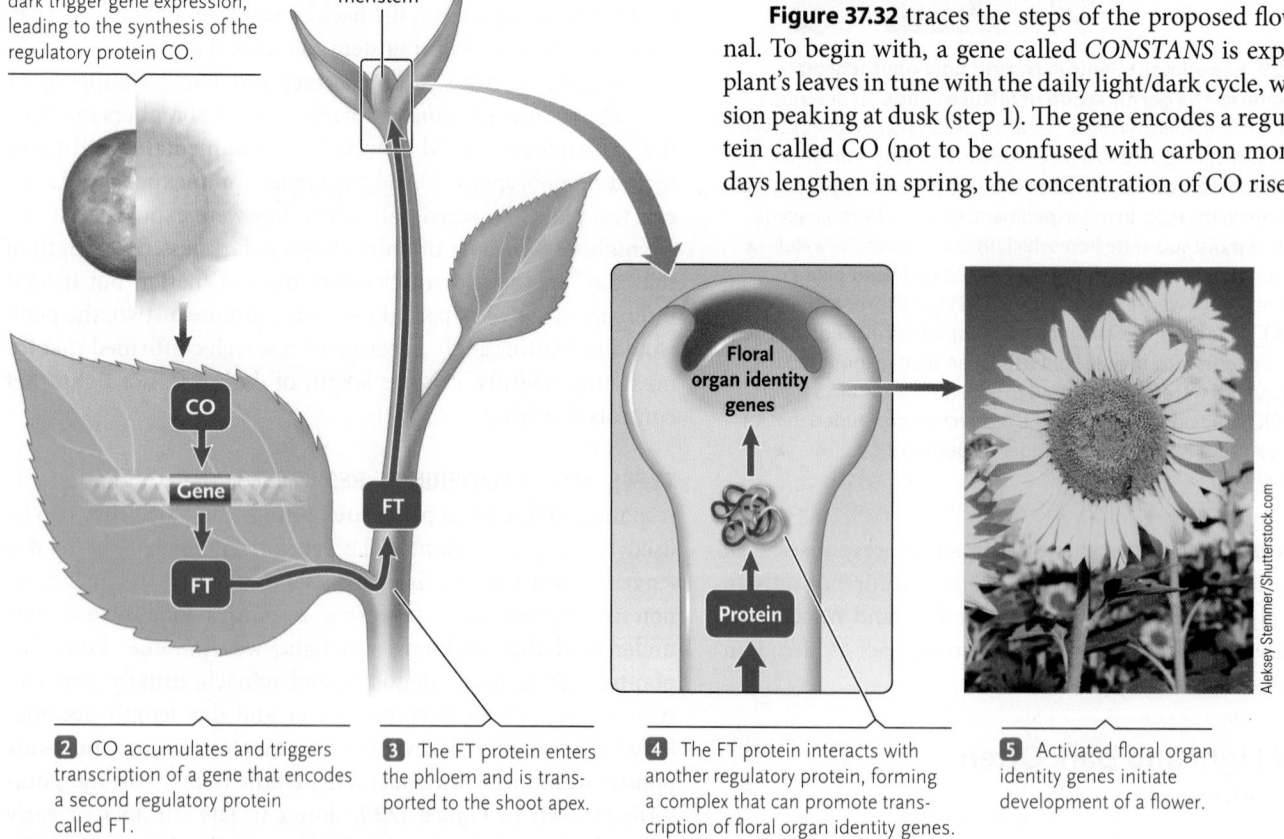

1 Natural cycles of light and dark trigger gene expression, leading to the synthesis of the regulatory protein CO.

Shoot apical meristem

Floral organ identity genes

Protein

2 CO accumulates and triggers transcription of a gene that encodes a second regulatory protein called FT.

3 The FT protein enters the phloem and is transported to the shoot apex.

4 The FT protein interacts with another regulatory protein, forming a complex that can promote transcription of floral organ identity genes.

5 Activated floral organ identity genes initiate development of a flower.

Aleksey Stemmer/Shutterstock.com

FIGURE 37.32 Proposed pathway for the flowering signal. The pathway starts as shifting cycles of light and dark trigger expression of the *CONSTANS* gene. As described in the text, this step is the first in a sequence that leads to the activation of floral organ identity genes in the shoot apical meristem. When these genes are expressed, a flower develops.

Basic Research: Using DNA Microarray Analysis to Track Down Florigen

The more plant scientists learn about plant genomes, the more they are relying on DNA microarray assays to elucidate the activity of plant genes.

Recall from Section 19.3 that a DNA microarray, also called a DNA chip, allows an investigator to explore questions such as how the expression of a particular gene differs in different types of cells. This procedure can be manipulated to reveal the relative amounts of expression of more than one of a cell's genes.

Philip A. Wigge and his colleagues used this method to learn more about the signaling pathway that causes a plant's apical meristem to give rise to flowers. Previous research had established that in leaves, lengthening spring days coincided with rising concentrations of CO, a regulatory protein encoded by the *CONSTANS* gene. But what did CO regulate? Working with *Arabidopsis thaliana*, Wigge's group was able to narrow down the field to four genes, and using microarray analysis of DNA from leaf cells they pinpointed one called FT (for flowering locus T). The researchers found that in leaves, CO causes strong expression of FT: when enough CO is present, FT mRNA is rapidly transcribed, then enters the phloem. (The transport of mRNA in phloem is not unusual.) By contrast, when they tested CO's effects in shoot apex cells, they found that it triggers far less gene expression there. Clearly, CO was not directly triggering

the development of flowers. However, FT mRNA moves in the phloem to the shoot apex, where it is translated into protein. Was that protein the direct flowering signal? Other studies had implicated a regulatory protein called FD, which microarray analysis had shown was expressed *only*—but very strongly—in the shoot apex.

To sort out this final piece of the puzzle, the Wigge team examined flowering responses in normal *A. thaliana* plants, as well as in mutants having a normal FT protein but a defective *fd,* and vice versa. Flowering was abnormal in both types of mutants, possibly because the mutated "partner" suppressed some aspect of the

functioning of the normal protein. On the other hand, in wild-type plants, which had a functioning FD protein, expression of FT triggered a marked increase in the expression of the floral organ gene *APETALA1* **(Figure).** These results have two major implications. First, they support the hypothesis that FT and FD interact in a normal flowering response. Second, the study suggests that FT, the CO-induced signal from leaves, conveys the environmental signal that it is time for a plant to flower. In that sense, FT may be the long-sought florigen. However, only by interacting with FD does FT "know" where to deliver its flowering signal—in the apical meristems of shoots.

FIGURE   Effect of the FT protein on expression of the *APETALA1* (*AP1*) floral organ identity gene. In nature, *Arabidopsis thaliana* is a long-day plant, and the experiment was carried out under long-day (that is, short-night) conditions. Three groups of replicates shown here in yellow, orange, and red respectively, were monitored for both AP1 and FT. After a brief delay, the expression of AP1 closely tracked the appearance of the FT regulatory protein, which had been activated by its interaction with the FD protein.

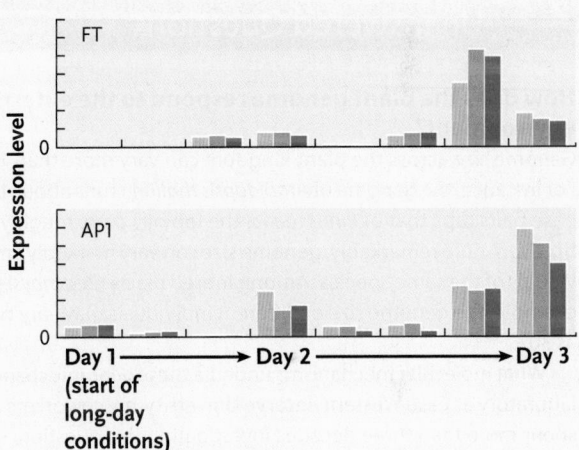

activating a second gene (step 2). The product of this gene, a regulatory protein called FT (for Flowering locus T), travels in the phloem to shoot tips (step 3). There, FT interacts with a second regulatory protein (step 4) that is synthesized only in shoot apical meristems (step 5). The interaction sparks the development of a flower by promoting the expression of floral organ identity genes in the meristem tissue (see Section 36.5). Key experiments that uncovered this pathway all relied on analysis of DNA microarrays, a technique introduced in Section 19.3 and featured in this chapter's *Focus on Research: Basic Research.*

VERNALIZATION AND FLOWERING Flowering is more than a response to changing night length. Temperatures also change with the seasons in most parts of the world, and they too influence flowering. For instance, unless buds of some biennials and perennials are exposed to low winter temperatures, flowers do

not form on stems in spring. Low-temperature stimulation of flowering is called **vernalization** ("making springlike").

In 1915 the plant physiologist Gustav Gassner demonstrated that it was possible to influence the flowering of cereal plants by controlling the temperature of seeds while they were germinating. In one experiment, he maintained germinating seeds of winter rye (*Secale cereale*) at just above freezing (1°C) before planting them. In nature, winter rye seeds in soil germinate during the winter, giving rise to a plant that flowers months later, in summer. Plants grown from Gassner's test seeds, however, flowered the same summer even when the seeds were planted in the late spring. Home gardeners can induce flowering of daffodils and tulips by putting the bulbs in a freezer for several weeks before early spring planting. Commercial growers use vernalization to induce plants such as Easter lilies to flower just in time for seasonal sales.

Dormancy Is an Adaptation to Seasonal Changes or Stress

As autumn approaches and days grow shorter, growth slows or stops in many plants even if temperatures are still moderate, the sky is bright, and water is plentiful. When a perennial or biennial plant stops growing under conditions that seem (to us) quite suitable for growth, it has entered a state of **dormancy.** Ordinarily, its buds will not resume growth until early spring.

Short days and long nights—conditions typical of winter—are strong cues for dormancy. In nature, buds may enter dormancy because less P_{fr} can form when day length shortens in late summer. Other environmental cues are at work also. Cold nights, dry soil, and a deficiency of nitrogen apparently also promote dormancy.

The requirement for multiple dormancy cues has adaptive value. For example, if temperature were the only cue, plants might flower and seeds might germinate in warm autumn weather—only to be killed by winter frost.

A dormancy-breaking process is at work between fall and spring. Depending on the species, breaking dormancy probably involves gibberellins and abscisic acid, and it requires exposure to low winter temperatures for specific periods. The temperature needed to break dormancy varies greatly among species. For example, the Delicious variety of apples grown in Utah requires 1,230 hours near 43°F (6°C); apricots grown there require only 720 hours at that temperature. Generally, trees growing in the southern United States or in Italy require less cold exposure than those growing in Canada or in Sweden.

STUDY BREAK 37.4

1. Summarize the switching mechanism that operates in plant responses to changes in photoperiod.
2. Give some examples of how relative lengths of dark and light can influence flowering.
3. Explain why dormancy is an adaptive response to a plant's environment.

Unanswered Questions

How does the plant genome respond to the external environment?

Genome size across the plant kingdom can vary more than 1,000-fold. For instance, the genome of *Arabidopsis thaliana* runs about 125 million base pairs (bp), that of *Pinus taeda*, the loblolly pine, roughly 21 billion bp. Even more remarkably, genome size can vary markedly among individuals of the same species. Among inbred maize *(Zea mays)* plants, for example, the genome size of different individuals may vary by as much as 30%.

What molecular mechanisms underlie these genomic changes? In my laboratory at Case Western Reserve University, my coworkers and I have spent more than three decades investigating this question, examining environmentally induced heritable changes in flax *(Linum).* Certain flax varieties respond to specific environmental stresses by physically restructuring specific portions of their genome, with the variant genomes being stably transmitted over many generations. Fortunately, the changes occur within a single generation and therefore can be readily tracked. We now know, for example, that the genomic changes occur in the vegetative apical meristem. This tissue is where all new growth occurs, and it also gives rise to the cells (in flowers) that ultimately produce flax gametes. This means that the environment can act as a selective agent to establish advantageous variants that may then be transmitted to the next generation.

We have identified one DNA fragment of interest—*Linum* insertion sequence-1 (LIS-1)-and followed it as responsive flax varieties grow in different environments. LIS-1 is assembled while the plant is growing under inducing conditions and is then (or simultaneously) inserted into a specific genome site with very high frequency. Under the inducing growth conditions, this event occurs in all the test plants, and all their progeny inherit the altered genomic site. Under different conditions, although the insertion occasionally occurs, none of the progeny inherit the altered site. Some other regions of the genome also frequently alter, but in our experiments only insertion of LIS-1 is strongly associated with a specific change in the plant's environment.

The high frequency with which LIS-1 arises and is transmitted to progeny in one environment but not others accords with a hypothesis that the insertion is directly adaptive or is closely linked to an adaptive change. Not all stresses result in the same genomic rearrangements, and the particular genome regions that are restructured under a particular stress are related to the physiological effects of that stress. We are currently investigating the complete genome sequences of the original and induced lines to identify the variable loci and will be using the genome reorganization to point us toward a molecular mechanism. We are also looking at intermediates that may be formed during the induction of the changes. Elucidating these and related mechanisms in flax can help clarify some basic questions—including whether genetic adaptations to stress are an integrated genome-wide reconfiguration rather than single gene mutations. We will also better understand the precise circumstances under which a stressful environment can trigger rapid, limited restructuring of a plant genome. And finally, such studies may also reveal how a stress can then act as the selective force determining which genome variations are transmitted into the next generation.

think like a scientist

1. How does genome reconfiguration differ from mutation systems that normally operate in plants?
2. Given data on environmentally induced heritable changes in flax, how might plant scientists need to modify their thinking about the role of the environment as a mutagen, rather than simply as a selective agent?

Christopher A. Cullis is the Frances Hobart Herrick Professor of Biology at Case Western Reserve University. His research focuses on the mechanisms by which DNA within the cell can change rapidly, particularly in response to external stimuli. He uses flax as a model organism and is interested in developing a flax genome project. Learn more about his research at www.case.edu/artsci/biol/people/cullis.html.

REVIEW KEY CONCEPTS

For access to MindTap and additional study materials visit www.cengagebrain.com.

37.1 Introduction to Plant Hormones

- Hormones and environmental stimuli alter the behavior of target cells, which have receptors to which signal molecules can bind. By means of a response pathway that transduces the hormone signal, a signal can induce changes in the cell's shape or internal structure or influence its metabolism or the transport of substances across the plasma membrane (Figures 37.1 and 37.2).

- Some plant hormones and growth factors may bind to receptors at the target cell's plasma membrane, changing the receptor's shape. This binding often triggers the release of internal second messengers that diffuse through the cytoplasm and provide a chemical signal that alters gene expression.

- Second messengers usually act by way of a reaction sequence that amplifies the cell's response to a signal. The sequence activates a series of proteins, including G proteins and enzymes that stimulate the synthesis of second messengers (such as IP_3) that bind ion channels on endoplasmic reticulum. Binding releases calcium ions, which enter the cytoplasm and activate protein kinases, enzymes that activate specific proteins that produce the cell response.

- At least eight classes of hormones govern flowering plant development, including germination, growth, flowering, fruit set, and senescence.

- Auxins, mainly IAA, promote elongation of cells in the coleoptile and stem and govern apical dominance, among other effects (Figures 37.3–37.9).

- Cytokinins stimulate cell division, promote leaf expansion, and retard leaf senescence (Figure 37.10).

- Strigolactones exuded from roots promote the development of arbuscular mycorrhizae. Those transported in xylem sap regulate lateral branching in shoots and roots (Figure 37.11).

- Gibberellins promote stem elongation and help seeds and buds break dormancy (Figure 37.12).

- Ethylene promotes the triple response in seedlings, senescence, and fruit ripening and abscission (Figures 37.13–37.15).

- Brassinosteroids stimulate cell division and elongation (Figure 37.16).

- Abscisic acid (ABA) promotes stomatal closure (including in response to drought stress) and may trigger seed and bud dormancy (Figure 37.17).

- Jasmonates help regulate growth and have important roles in defense.

37.2 Plant Chemical Defenses

- Plants have diverse chemical defenses that limit damage from bacteria, fungi, worms, or plant-eating insects. The hypersensitive response isolates an infection site by surrounding it with dead cells. Phytoalexins are bioactive compounds that combat pathogenic bacteria and viruses (Figures 37.18 and 37.19).

- Many plants deploy bioactive specialized compounds, including alkaloids and phenolics, as defenses against feeding herbivores (Figure 37.20).

- Recognition of pathogen-activated molecular patterns (PAMPs) enables a plant to chemically detect a specific pathogen and synthesize recognition proteins that mount a defense.

- Systemic acquired resistance provides long-term protection against some pathogens. Salicylic acid has a central role in this response (Figure 37.21).

- Heat-shock proteins can reversibly bind enzymes and other proteins in plant cells and prevent them from denaturing when the plant is under certain types of stress.

- Some plants can synthesize "antifreeze" proteins that stabilize cell proteins when cells are threatened with freezing.

37.3 Plant Movements

- Plants adjust their growth patterns in response to environmental rhythms and unique environmental circumstances. These responses include tropisms.

- In photomorphogenesis, light induces growth responses. Phototropisms, mainly stimulated by blue light, are growth responses to a directional light source (Figure 37.22).

- Gravitropism is a growth response to Earth's gravitational pull. Stems exhibit negative gravitropism (growing upward) whereas roots show positive gravitropism (Figures 37.23 and 37.24).

- Some plants or plant parts demonstrate thigmotropism, growth in response to contact with a solid object (Figure 37.25). Mechanical stress can cause thigmomorphogenesis, which causes the stem to add girth.

- Some plant species show nastic leaf movements in response to certain environmental cues. Changes in fluid pressure in cells of a pulvinus, a pad of tissue at the base of a leaf or petiole, cause the movements (Figures 37.26 and 37.27).

37.4 Plant Biological Clocks

- Plants have biological clocks, internal time-measuring mechanisms with a biochemical basis. Environmental cues can "reset" the clocks, enabling plants to make seasonal adjustments in growth, development, and reproduction.

- In photoperiodism, plants respond to a change in the relative length of daylight and darkness in a 24-hour period. A switching mechanism involving the pigment phytochrome promotes or inhibits germination, growth, and flowering and fruiting (Figures 37.28 and 37.29).

- Long-day plants flower when day length is long relative to night. Short-day plants flower when day length is relatively short, and intermediate-day plants flower when day length falls in between the values for long-day and short-day plants. Flowering of day-neutral plants is not regulated by light. In vernalization, a period of low temperature stimulates flowering (Figures 37.30 and 37.31).

- The direct trigger for flowering may begin in leaves, when the regulatory protein CO triggers the expression of the FT gene. The resulting mRNA transcripts move in phloem to apical meristems where translation of the mRNAs yields a second regulatory protein, which in turn interacts with a third. This final interaction activates genes that encode the development of flower parts (Figure 37.32).

- Dormancy is a state in which a perennial or biennial stops growing even though conditions appear to be suitable for continued growth.

Remember/Understand

1. Which of the following plant hormones does *not* stimulate cell division?
   a. auxins
   b. cytokinins
   c. ethylene
   d. gibberellins
   e. All stimulate cell division.

2. Which is the correct pairing of a plant hormone and its function?
   a. salicylic acid: triggers synthesis of general defense proteins
   b. brassinosteroids: promote responses to environmental stress
   c. cytokinins: stimulate stomata to close in water-stressed plants
   d. gibberellins: slow seed germination
   e. ethylene: promotes formation of lateral roots

3. A characteristic of auxin (IAA) transport is:
   a. IAA moves by polar transport from the base of a tissue to its apex.
   b. IAA moves laterally from a shaded to an illuminated side of a plant.
   c. IAA enters a plant cell in the form of IAAH, an uncharged molecule that can diffuse across cell membranes.
   d. IAA exits one cell and enters the next by means of transporter proteins clustered at both the apical and basal ends of the cells.
   e. All of the above are characteristics of auxin transport in different types of cells.

4. Strigolactones:
   a. mainly interact with IAA to regulate apical dominance.
   b. function only in roots, where they limit lateral branching.
   c. counteract the effects of gibberellin in stimulating stem elongation.
   d. attract fungal hyphae to roots and regulate lateral branching.
   e. are major hormones in plant defenses.

5. Which of the following is *not* an example of a plant chemical defense?
   a. ABA inhibits leaves from budding if conditions favor attacks by sap-sucking insects.
   b. Jasmonate activates plant genes encoding protease inhibitors that prevent insects from digesting plant proteins.
   c. Acting against fungal infections, the hypersensitive response allows plants to produce highly reactive oxygen compounds that kill selected tissue, thus forming a dead tissue barrier that walls off the infected area from healthy tissues.
   d. Chitinase, a PR hydrolytic protein produced by plants, breaks down chitin in the cell walls of fungi and thus halts the fungal infection.
   e. Attack by fungi or viruses stimulates the production of phytoalexins having antibiotic properties.

6. Which of the following statements about plant responses to the environment is *true*?
   a. The heat-shock response induces a sudden halt to cellular metabolism when an insect begins feeding on plant tissue.
   b. In gravitropism, amyloplasts sink to the bottom of cells in a plant stem, causing the redistribution of IAA.
   c. The curling of tendrils around a twig is an example of thigmotropism.
   d. Phototropism results when IAA moves first laterally, then downward in a shoot tip when one side of the tip is exposed to light.
   e. Nastic movements, such as the sudden closing of the leaves of a Venus flytrap, are examples of a plant's ability to respond to specific directional stimuli.

7. Which of the following steps is *not* part of the sequence that triggers flowering?
   a. Cycles of light and dark stimulate the expression of the CONSTANS gene in a plant's leaves.
   b. CO proteins accumulate in the leaves and trigger expression of a second regulatory gene.
   c. mRNA transcribed during expression of a second regulatory gene moves via the phloem to the shoot apical meristem.
   d. Interactions among regulatory proteins promote the expression of floral organ identity genes in meristem tissue.
   e. CO proteins in the floral meristem interact with florigen, a so-called flowering hormone, which provides the final stimulus for expression of floral organ identity genes.

8. Which of the following are means by which contact with a specific infectious bacterium or fungus triggers a plant defensive response?
   a. A plasma membrane receptor binds to a PAMP.
   b. PR proteins begin to break down components of the pathogen cell wall.
   c. Systemin binding leads to the synthesis of jasmonate.
   d. Resistance proteins bind to bacterial virulence effectors that have entered a plant cell.
   e. Salicylic acid molecules are released from the besieged plant cell.

9. In the sequence that unfolds after molecules of a hormone such as ABA bind to receptors at the surface of a target plant cell:
   a. first messenger molecules in the cytoplasm are mobilized, then G proteins carry the signal to second messengers such as protein kinases, which alter the activity of cell proteins such as IP3.
   b. binding activates G proteins, which in turn activate second messengers such as IP3; subsequent steps are thought to involve activation of genes that encode protein kinases.
   c. binding activates phospholipase C, which in turn activates G proteins, which then activate molecules of IP3, a step that leads to the synthesis of protein kinases.
   d. binding stimulates G proteins to activate protein kinases, which then bind calcium channels in ER; the flux of calcium ions activates second messenger molecules that alter the activity of cell proteins or enter the cell nucleus and alter the expression of target genes.
   e. binding activates G proteins, which in turn activate phospholipase C; this substance then stimulates the synthesis of second messenger molecules, the second messengers bind calcium channels in the cell's ER, and finally protein kinases alter the activity of proteins by phosphorylating them.

Apply/Analyze

10. Hanging wire fruit baskets have many holes or open spaces. The major advantage of these spaces is that they:
    a. prevent gibberellins from causing bolting or the formation of rosettes on the fruit.
    b. allow the evaporation of ethylene and thus slow ripening of the fruit.
    c. allow oxygen in the air to stimulate the production of ethylene, which hastens the abscission of fruits.
    d. allow oxygen to stimulate brassinosteroids, which hasten the maturation of seeds in/on the fruits.
    e. allow carbon dioxide in the air to stimulate the production of cytokinins, which promotes mitosis in the fruit tissue and hastens ripening.

11. In nature the poinsettia, a plant native to Mexico, blooms only in or around December. This pattern suggests that:
    a. the long daily period of darkness (short day) in December stimulates the flowering.
    b. vernalization stimulates the flowering.
    c. the plant is dormant for the rest of the year.
    d. phytochrome is not affecting the poinsettia flowering cycle.
    e. a circadian rhythm is in effect.

12. **Discuss Concepts** Synthetic auxins such as 2,4-D can be weed killers because they cause an abnormal growth burst that kills the plant within a few days. Suggest reasons why such rapid growth might be lethal to a plant.

13. **Discuss Concepts** In some plant species, an endodermis is present in both stems and roots. In experiments, the shoots of mutant plants lacking differentiated endodermis in their root and shoot tissue do not respond normally to gravity, but roots of such plants do respond normally. Explain this finding, based on your reading in this chapter.

14. **Discuss Concepts** In *A. thaliana* plants carrying a mutation called pickle (pkl), the primary root meristem retains characteristics of embryonic tissue—it spontaneously regenerates new embryos that can grow into mature plants. However, when the mutant root tissue is exposed to a gibberellin (GA), this abnormal developmental condition is suppressed. Explain why this finding suggests that additional research is needed on the fundamental biological role of GA.

Evaluate/Create

15. **Discuss Concepts** You work for a plant nursery and are asked to design a special horticultural regimen for a particular flowering plant. The plant is native to northern Spain, and in the wild it grows a few long, slender stems that produce flowers each July. Your boss wants the nursery plants to be shorter, with thicker stems and more branches, and she wants them to bloom in early December in time for holiday sales. Outline your detailed plan for altering the plant's growth and reproductive characteristics to meet these specifications.

16. **Design an Experiment** Tiny, thornlike trichomes on leaves are a common plant adaptation to ward off insects. Those trichomes develop very early on, as outgrowths of a seedling's epidermal cells. Biologists have observed, however, that many mature plants develop more leaf trichomes after the fact, as a response to insect damage. Researchers at the University of Chicago decided to study this phenomenon, and specifically wanted to determine the effects, if any, of jasmonate, salicylic acid, and gibberellin in stimulating trichome development. Keeping in mind that plant hormones often interact, how many separate experiments, at a minimum, would the research team have had to carry out to obtain useful initial data? Do you suppose they used mutant plants for some or all of the tests? Why or why not?

17. **Apply Evolutionary Thinking** Cryptochrome occurs in plants and animals. If it was inherited from their shared ancestor, what other major groups of organisms might also have it?

For selected answers, see Appendix A.

INTERPRET THE DATA

Jennifer Nemhauser, Todd Mockler, and Joanne Chory at the Plant Biology Laboratory of the Salk Institute for Biological Studies performed experiments to learn more about the interaction of auxin and brassinosteroids (BR) in regulating the growth of *Arabidopsis* seedlings. In a 2004 research report on this work, graphics for displaying experimental data included the Venn diagram shown here **(Figure)**. Venn diagrams are useful for giving a snapshot of the interrelationships between sets of data, often using overlapping circles. In this diagram, "up" and "down" refer to hormone effects that stimulate or inhibit gene expression. The numerals in each circle represent numbers of seedlings.

1. Considering the two upper overlapping circles, what general finding do the data support?

2. Describe at least two other general findings summarized in the diagram.

3. Why isn't there any overlap between the "up" and "down" bubbles for each hormone?

Source: J. Nemhauser, T. Mockler, and J. Chory. 2004. Interdependency of brassinosteroids and auxin in *Arabidopsis* seedling growth. *PLoS Biology* 2:1460–1471. doi: 10.1371/journal.pbio.0020258

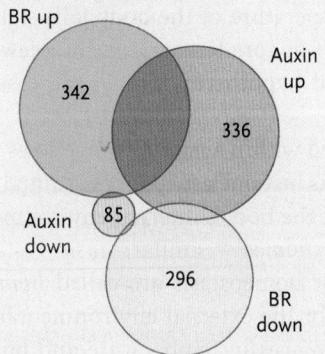

FIGURE Venn diagram showing the overlap between *Yucca* genes responsive to auxin and BR.

38 Introduction to Animal Organization and Physiology

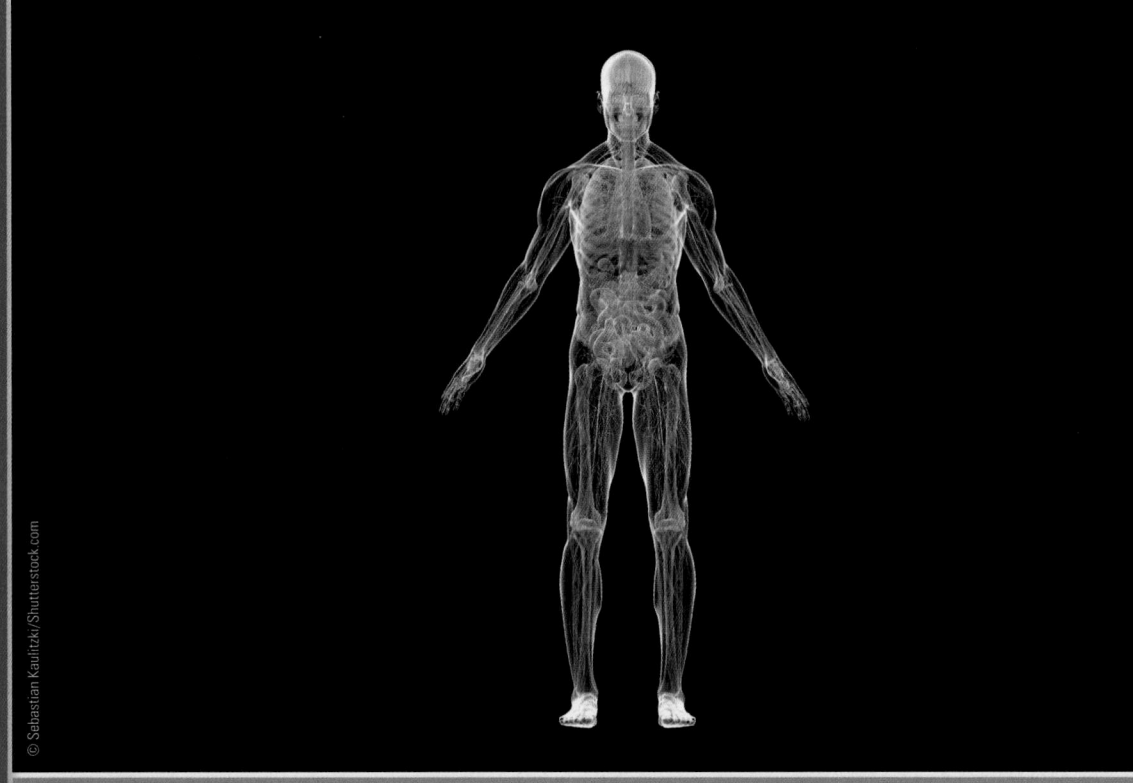

© Sebastian Kaulitzki/Shutterstock.com

See-through image of a human male showing the major organs and the skeleton (computer rendering).

Why it matters . . . On April 10, 1912, the RMS *Titanic* left Southampton, England, on her maiden voyage, bound for New York. A total of 2,223 people were on board, including 899 crew. Four days out, near midnight, the ship hit an iceberg. Some parts of the hull buckled, and rivets popped out below the waterline. As a result, the *Titanic* began to take on water. The captain ordered passengers and crew to the lifeboats. The ship's lifeboats could hold a maximum of 1,178 people, far fewer than the number of people on board. Two hours and forty minutes after striking the iceberg, the *Titanic* sank. Of the people on board, 1,517 died, most of them succumbing to hypothermia in the −2°C water. At that temperature, death likely occurred within 15 minutes. *Hypothermia* is a condition in which the core temperature of the body falls significantly below normal for a prolonged period. In humans, a drop in core temperature of a few degrees affects brain function and leads to confusion. Continued hypothermia, as in the case of the *Titanic* disaster, can lead to death.

Human body temperature is normally regulated within a narrow range so as to provide optimal conditions for cellular life. Body temperature is just one aspect of an animal's internal environment that must be regulated. The regulation of the body's internal environment to maintain it in a relatively stable state is called *homeostasis* (*homeo* = similar; *stasis* = standing or stopping). The processes and activities responsible for homeostasis are called *homeostatic control systems*. These control systems compensate both for the external environmental changes that a human or other animal encounters, such as differences in temperature and humidity, and for changes in their own body systems, such as the availability of nutrients for cells and tissues. As you saw for the unfortunate *Titanic* passengers and crew, there are limits to the ability of the homeostatic control systems for temperature regulation to maintain the stable state, beyond which they fail. Not only can animals die from hypothermia after prolonged exposure to the cold, they can also die from hyperthermia after prolonged exposure to excessive heat. Similarly, an imbalance between energy input (in the form of nutrients) and energy output can be lethal.

All animals have body systems for acquiring and digesting nutrients to provide energy for life, growth, reproduction, and movement. Other body systems seen in animals include the nervous system, endocrine system, muscular system, integumentary system, circulatory system, lymphatic system, respiratory system, excretory system, and reproductive system. Biologists are interested in the *structures* and *functions* of all of these systems. **Anatomy** is the study of the *structures* of organisms, and **physiology** is the study of their *functions*—the physicochemical processes of organisms.

In this unit, we discuss the structures and functions of the body systems of animals. This chapter introduces animal organization and physiology, describing the organization of individual cells into tissues, organs, and organ systems, the major body structures that carry out animal activities. Our discussion continues with a look at how the functions of organ systems coordinate to accomplish homeostasis. The other chapters in this unit proceed from cellular physiology to organ systems. Although the unit emphasizes vertebrates, with particular reference to human physiology, comparisons are made with invertebrates to keep the structural and functional diversity of the animal kingdom in perspective and to illustrate the evolution of the structures and functions involved.

38.1 Organization of the Animal Body

In this section, we discuss the significance of multicellularity and the organization of cells in an animal body.

Multicellularity Led to Important Evolutionary Advances in Animals

The individual cells of animals must be surrounded by an aqueous solution that contains ions and molecules in particular concentrations to maintain osmotic balance and to supply energy and maintenance needs. Most animal cells also require oxygen to serve as the final acceptor for electrons removed in oxidative reactions. Also, waste molecules and other byproducts of their metabolic activities, such as carbon dioxide, need to be released from the cells.

The evolution of multicellularity (see Section 25.3) made it possible for organisms to create an **internal environment** of fluid that cells use for the exchange of materials. By contrast, single-celled organisms exchange materials with fluid in their **external environment.** The internal fluid enables multicellular organisms to occupy a wider range of external environments than is possible for single-celled organisms. For instance, dry terrestrial habitats likely would be lethal to single-celled organisms. Multicellular organisms can also become relatively large because their individual cells remain small enough to exchange ions and molecules with the internal fluid.

The evolution of multicellularity also allowed major life functions to be subdivided among specialized groups of cells. In animals, some groups of cells became specialized for movement, others for functions such as food capture, digestion, internal circulation of nutrients, excretion of wastes, and reproduction. Specialization greatly increases the efficiency with which animals carry out these functions.

The Animal Body Is Typically Organized into Tissues, Organs, and Organ Systems

In most animals, specialized groups of cells are organized into tissues, the tissues into organs, and the organs into organ systems **(Figure 38.1)**. A **tissue** is a group of cells with similar structure and specialized function. The four primary types of tissues are *epithelial, connective, muscle,* and *nervous.* An **organ** integrates two or more different tissues into a structure that performs a particular function or functions. The eye, liver, and stomach are examples of organs. An **organ system** (also called a **body system**) is a collection of organs with related functions that interact to carry out a major body function such as digestion, respiration, or reproduction. The nervous system, for example, is the main regulatory system of the body and consists of the brain, spinal cord, peripheral nerves, and sensory organs.

STUDY BREAK 38.1

1. What are some advantages of an organism being multicellular?
2. What is the difference between a tissue, an organ, and an organ system?

38.2 Animal Tissues

Although the most complex animals may contain hundreds of distinct cell types, all can be classified into one of four basic tissue groups: *epithelial, connective, muscle,* and *nervous* (see Figure 38.1). The properties of the individual cells in each tissue type determine the structure and, therefore, the function of the tissue. More specifically, the structure and integrity of a tissue depend on the structure and organization of the cytoskeleton within the cell, the type and organization of the extracellular matrix (ECM) surrounding the cell, and the junctions holding cells together (see Section 4.5).

Junctions of various kinds link cells into tissues (see Figure 4.27). *Anchoring junctions* form buttonlike spots or belts that weld cells together. Anchoring junctions attach cells to each other or to the extracellular matrix. They are most abundant in tissues subject to stretching, such as skin and heart muscle. In *tight junctions,* plasma membrane proteins of adjacent cells interact to fuse the two cells partly together and thus create a barrier between the cells. For instance, tight junctions in the tissue lining the urinary bladder prevent waste molecules and ions from leaking out of the

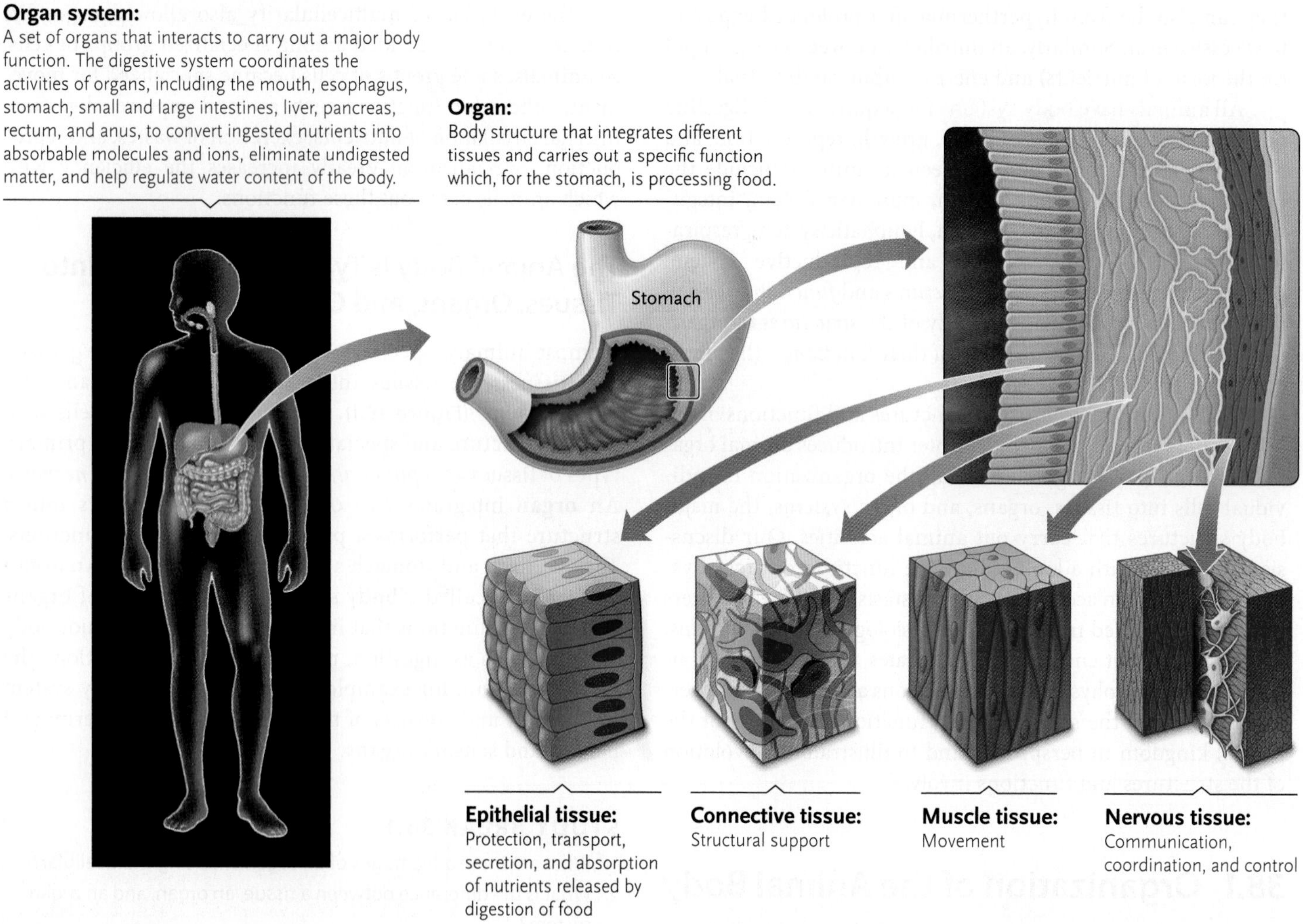

Organ system:
A set of organs that interacts to carry out a major body function. The digestive system coordinates the activities of organs, including the mouth, esophagus, stomach, small and large intestines, liver, pancreas, rectum, and anus, to convert ingested nutrients into absorbable molecules and ions, eliminate undigested matter, and help regulate water content of the body.

Organ:
Body structure that integrates different tissues and carries out a specific function which, for the stomach, is processing food.

Stomach

Epithelial tissue:
Protection, transport, secretion, and absorption of nutrients released by digestion of food

Connective tissue:
Structural support

Muscle tissue:
Movement

Nervous tissue:
Communication, coordination, and control

FIGURE 38.1 Organization of animal cells into tissues, organs, and organ systems, exemplified here by the digestive system.
© Cengage Learning 2017

bladder into other body tissues, and tight junctions form the blood–brain barrier that blocks many potentially harmful substances in the blood from reaching the brain's extracellular fluid. *Gap junctions* open direct channels between cells in the same tissue, allowing ions, small molecules, and electrical signals to flow rapidly from one to another. For example, gap junctions between muscle cells help muscle tissue function as a unit.

Epithelial Tissue Forms Protective, Secretory, and Absorptive Coverings and Linings of Body Structures

Epithelial tissue (*epi* = over; *thele* = covering) consists of sheetlike layers of cells with little ECM material between them **(Figure 38.2)**. Also called **epithelia** (singular, *epithelium*), these tissues protect the internal environment and regulate the exchange of materials between the internal environment and the external environment. Every substance that enters or leaves the internal environment must cross an epithelium.

Some epithelia have tight junctions between cells that create a barrier to prevent substances from passing between adjacent cells. The epithelial cells lining the intestine are organized in this way so that digestion products must pass *through* the cells to enter the bloodstream. Other epithelia have anchoring junctions between cells. These junctions make the epithelium somewhat leaky so that molecules can cross the epithelium by passing between adjacent cells. The epithelium of capillaries (small blood vessels) is a leaky epithelium of this kind, acting as a filter to allow ions and small molecules, but not blood cells and large proteins, to pass out of the blood into the surrounding extracellular fluid.

Epithelia cover exposed body surfaces and the surfaces of internal organs, and line cavities and ducts within the body. They protect body surfaces from abrasion or from invasion by bacteria and viruses and secrete or absorb substances. For example, the epithelium covering a fish's gill structures serves as a barrier to bacteria and viruses and exchanges oxygen, carbon dioxide, and ions with the aqueous environment. In the epidermis of vertebrates, some epithelial cells contain a

A. Simple squamous epithelium

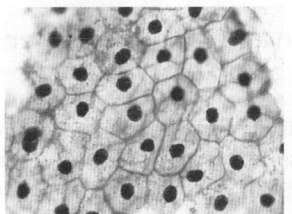

Ray Simmons/Science Source

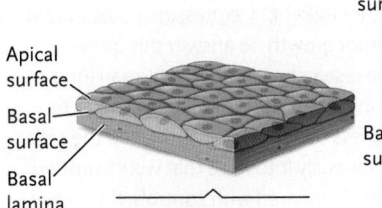

Apical surface
Basal surface
Basal lamina

Description: Layer of flattened cells

Common locations: Blood vessel inner lining (called endothelium); air sacs of lungs

Function: Diffusion

B. Stratified squamous epithelium

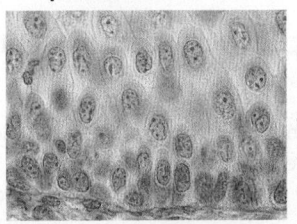

Jose Luis Calvo/Shutterstock.com

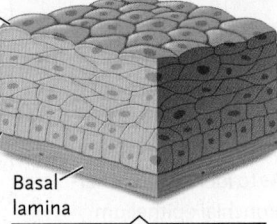

Apical surface
Basal surface
Basal lamina

Description: Several layers of flattened cells

Common locations: Skin and other surfaces subject to abrasion, such as the mouth, esophagus, and vagina

Function: Protection against abrasion; typically not involved in secretion or absorption

C. Cuboidal epithelium

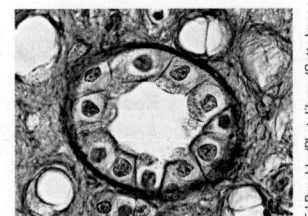

Ed Reschke/Photolibrary/Getty Images

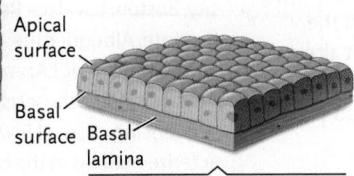

Apical surface
Basal surface
Basal lamina

Description: Layer of cubelike cells; free surface may have microvilli

Common locations: Glands and tubular parts of nephrons in kidneys

Function: Secretion, absorption

FIGURE 38.2 Structures and functions of the principal types of epithelia.
© Cengage Learning 2017

D. Simple columnar epithelium

Dr. Donald Fawcett/Visuals Unlimited, Inc.

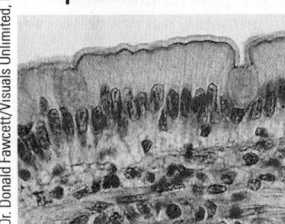

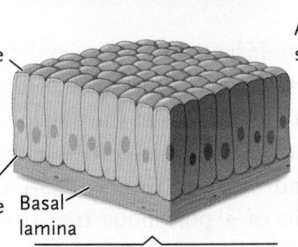

Apical surface
Basal surface
Basal lamina

Description: Layer of tall, slender cells with nuclei near base; free surface may have microvilli or cilia; may contain secretory vesicles

Common locations: Lining of gut, cervical canal, and gallbladder

Function: Secretion, absorption, such as secreting digestive enzymes and absorbing nutrients in the gut; protection; secreting mucus

E. Simple pseudostratified columnar epithelium

Jose Luis Calvo/Shutterstock.com

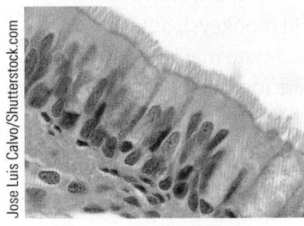

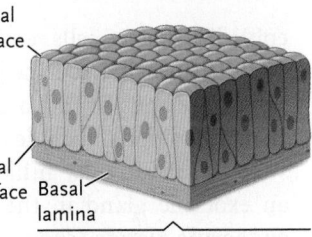

Apical surface
Basal surface
Basal lamina

Description: Single layer of columnar cells of differing heights; some cells do not reach the apical surface. Due to the cell organization, the nuclei are staggered, giving the epithelium the false appearance of stratification. May be ciliated.

Common locations: Nasal cavities, trachea, and upper digestive tract; some parts of male reproductive system

Function: Protection; secretes mucus and moves it across surface

network of keratin, a family of fibrous proteins. Keratin forms protective structures: the fingernails, hair, claws, hooves, and horns of mammals, the feathers of birds, and the scales of reptiles and fish. In arthropods, the surface epithelium secretes a tough cuticle that forms a barrier to the environment and serves as the animal's skeleton.

Because epithelia form coverings and linings, they have an outer surface and an inner surface. The outer, **apical surface** may be exposed to water, air, or fluids within the body. The inner, **basal surface** adheres to a layer of ECM secreted by the epithelial cells called the **basal lamina** (also called *basement membrane*), which fixes the epithelium to underlying tissues, often connective tissues.

In internal cavities and ducts, the apical surface is often covered with *cilia* (see Section 4.3), which beat like oars to move fluids through the cavity or duct. The epithelium lining the oviducts in mammals, for example, is covered with cilia that generate fluid currents to move eggs from the ovaries to the uterus. In some epithelia, including the lining of the small intestine, the free surface is crowded with *microvilli,* fingerlike extensions of the plasma membrane that increase the area available for secretion or absorption.

TYPES OF EPITHELIA Epithelia are classified as *simple*—formed by a single layer of cells—or *stratified*—formed by multiple cell layers (see Figure 38.2). The shapes of cells within an epithelium may be *squamous* (mosaic, flattened, and spread out), *cuboidal* (shaped roughly like dice or cubes), or *columnar* (elongated, with the long axis perpendicular to the epithelial layer). Five principal types of epithelia are

A Primate-Specific miRNA Regulator of Respiratory Tract Epithelial Cell Differentiation and Lung Carcinogenesis

MicroRNAs (miRNAs) are small, noncoding RNAs that regulate gene expression posttranscriptionally (see Chapter 16). Carcinogenesis (the development of cancer) often is associated with decreased or loss of expression of tissue-specific miRNAs. For lung cancer, smoking is a major risk factor. Smoking brings about molecular changes throughout the respiratory tract, including significant alterations of the transcriptome of the bronchial epithelium. The **Figure** shows isolated bronchial epithelial cells.

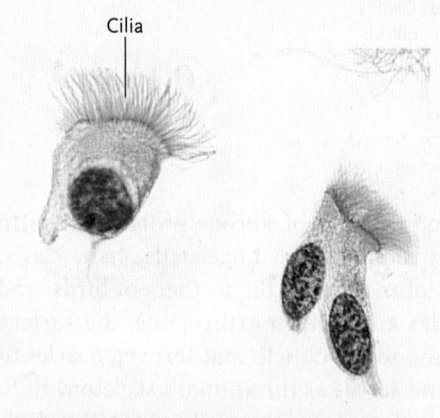

Cilia

Jose Luis Calvo/Shutterstock.com

FIGURE Isolated bronchial epithelial cells.

Research Question
Are miRNAs involved in cancer-associated gene expression differences in the respiratory tract?

Experiments
Collaborating researchers at Boston University, Boston; Lovelace Respiratory Research Institute, Albuquerque; David Geffen School of Medicine at UCLA; and MatTek Corporation, Ashland, Massachusetts, used next-generation sequencing of small RNAs to identify and characterize miRNAs in the bronchial epithelium that could be related to lung cancer. Epithelial miRNAs were compared for individuals who were: (1) healthy and never smoked; (2) healthy current smokers; (3) current or former smokers with lung cancer; and (4) current or former smokers with benign lung disease.

Results
The sequencing results identified an miRNA—miR-4423—with a high expression level in the bronchial epithelium. The gene for miR-4423 is highly conserved in primates (humans and other great apes, Old World monkeys, and New World monkeys), but it is not conserved in nonprimate mammals. Therefore, the researchers focused on this miRNA to determine its role in lung cancer development.

An analysis of miR-4423 in 24 human tissues showed that this miRNA is primarily expressed in the respiratory tract epithelium. Furthermore, miR-4423 expression is down-regulated in most lung tumors and in premalignant lesions of the lungs. This led to the question of whether miR-4423 expression could inhibit lung tumor growth. To answer this question, the researchers modified human lung cells to include a clone of miR-4423 so as to overexpress the miRNA, and injected the cells subcutaneously into mice that were immunodeficient. Compared with controls, the tumors that formed were significantly smaller in size.

Conclusion
The miRNA miR-4423 is a primate-specific regulator of respiratory tract epithelium differentiation. A decrease in miR-4423 expression, and, therefore, a decrease in its function, contributes to lung carcinogenesis.

think like a scientist
From an evolutionary perspective, interpret the fact that the gene for miR-4423 is highly conserved in primates, but not conserved in nonprimate mammals.

Source: C. Perdomo et al. 2013. MicroRNA 4423 is a primate-specific regulator of airway epithelial cell differentiation and lung carcinogenesis. *Proceedings of the National Academy of Sciences USA* 19:18946–18951.

found in the body (see Figure 38.2). (*Molecular Insights* discusses miRNA regulation of epithelial cell differentiation in the lungs and its relationship to lung cancer.)

The cells of some epithelia, such as those forming the skin and the lining of the intestine, divide constantly to replace worn and dying cells. New cells are produced through division of stem cells in the basal layer of the skin. *Stem cells* are undifferentiated cells that divide to produce either more stem cells or differentiating cells that become specialized into one of the many cell types of the body (see Section 18.2).

GLANDS FORMED BY EPITHELIA Epithelia may contain or give rise to cells that are specialized for secretion. Some secretory cells are scattered among nonsecretory cells within the epithelium. Epithelial tissue-derived structures called **glands** are specialized for secreting. Glands are formed during embryonic development by pockets of epithelial tissue that dip inward from the surface and develop the ability to secrete.

In **exocrine glands** (*exo* = external; *crine* = secretion), connecting cells between the secretory gland cells and the epithelial surface cells remain to form a duct between the gland and the surface. Exocrine glands secrete substances to the outside of the body or into a cavity that opens to the outside. Exocrine secretions include mucus, saliva, digestive enzymes, sweat, earwax, oils, milk, and venom (**Figure 38.3A** shows an exocrine gland in the skin of a poisonous tree frog). In **endocrine glands** (*endo* = inside), the connecting cells disappear during development, leaving the secretory gland ductless and, therefore, isolated from the surface. Endocrine glands, such as the pituitary gland, adrenal gland, and thyroid gland (**Figure 38.3B**), release their products—hormones (see Chapters 9 and 42)—directly into the interstitial fluid, to be picked up and distributed by the circulatory system.

Some glands act as both exocrine glands and endocrine glands. The pancreas, for instance, has an exocrine function of secreting pancreatic juice through a duct into the small intestine, where it plays an important role in food digestion, and an endocrine function of secreting the hormones insulin and glucagon into the bloodstream to help regulate glucose levels in the blood.

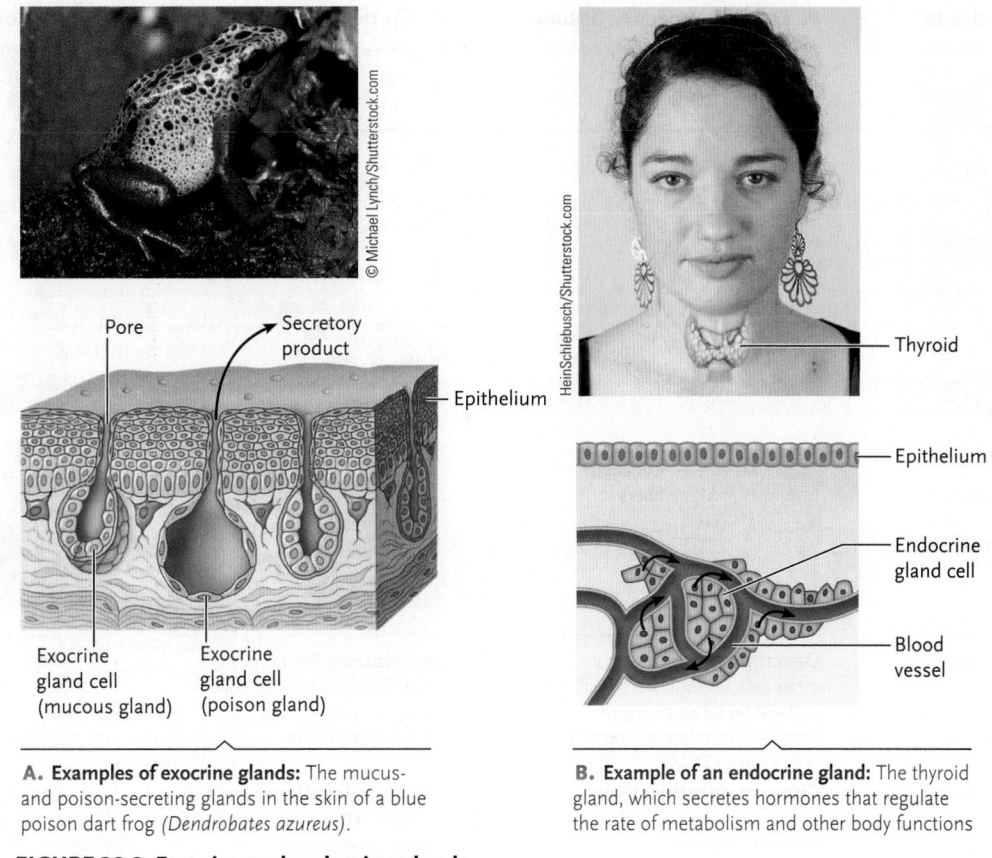

A. Examples of exocrine glands: The mucus- and poison-secreting glands in the skin of a blue poison dart frog (*Dendrobates azureus*).

B. Example of an endocrine gland: The thyroid gland, which secretes hormones that regulate the rate of metabolism and other body functions

FIGURE 38.3 Exocrine and endocrine glands.
© Cengage Learning 2017

Connective Tissue Supports Other Body Tissues

Most animal body structures contain one or more types of **connective tissues.** Connective tissues support other body tissues, transmit mechanical and other forces, and in some cases act as filters. They consist of cells that form networks or layers in and around body structures and that are separated by non-living material, specifically the ECM secreted by the cells of the tissue (see Section 4.5). Many forms of connective tissue have more ECM material (both by weight and by volume) than cellular material.

The mechanical properties of a connective tissue depend on the type and quantity of its ECM. The consistency of the ECM ranges from fluid (as in blood and lymph), through soft and firm gels (as in tendons), to hard and crystalline (as in bone).

In most connective tissues, the ECM consists primarily of the fibrous glycoprotein **collagen** embedded in a network of *proteoglycans*—glycoproteins that are very rich in carbohydrates. Collagen is the most abundant family of proteins in animals and is found in all eumetazoans.

In bone, the glycoprotein network surrounding collagen is impregnated with mineral deposits that produce a hard, yet still somewhat elastic, structure. Another family of glycoproteins, **fibronectin,** aids in the attachment of cells to the ECM and helps hold the cells in position.

In some connective tissues another rubbery protein, **elastin,** adds elasticity to the ECM—it is able to return to its original shape after being stretched, bent, or compressed. Elastin fibers, for example, help the skin return to its original shape when pulled or stretched, and give the lungs the elasticity required for their alternating inflation and deflation. Related to elastin, the protein **resilin** is found only in insects and some crustaceans and is the most elastic material known. Resilin is the basis for the jumping of fleas, crickets, and locusts.

Vertebrates have six major types of connective tissues: *loose connective tissue, dense connective tissue, cartilage, bone, adipose tissue,* and *blood.* Each type has a characteristic function correlated with its structure **(Figure 38.4).**

LOOSE CONNECTIVE TISSUE Loose connective tissue consists of sparsely distributed **fibroblast** cells surrounded by a network of collagen and other glycoprotein fibers **(Figure 38.4A);** the fibroblasts secrete most of these proteins. Loose connective tissues support epithelia and form a corsetlike band around blood vessels, nerves, and some internal organs. They also reinforce deeper layers of the skin. Sheets of loose connective tissue, covered on both surfaces with epithelial cells, form the **mesenteries,** which hold the abdominal organs in place and provide lubricated, smooth surfaces that prevent abrasion to adjacent structures as the body moves.

A. Loose connective tissue

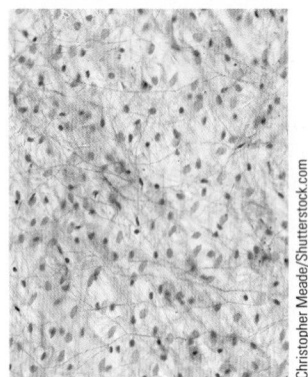

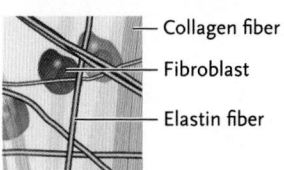

— Collagen fiber
— Fibroblast
— Elastin fiber

Description: Fibroblasts and other cells surrounded by collagen and elastin fibers forming a glycoprotein matrix

Common locations: Under the skin and most epithelia; around blood vessels, nerves, and some internal organs

Function: Support, elasticity, diffusion

B. Dense connective tissue

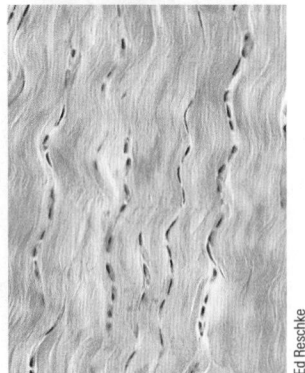

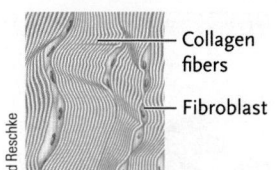

— Collagen fibers
— Fibroblast

Description: Fibroblasts and other cells surrounded by collagen and elastin fibers arranged irregularly or in parallel bundles with a dense ECM

Common locations: Skin, muscle covering, digestive tract, tendons, ligaments

Function: Strength, elasticity

C. Cartilage

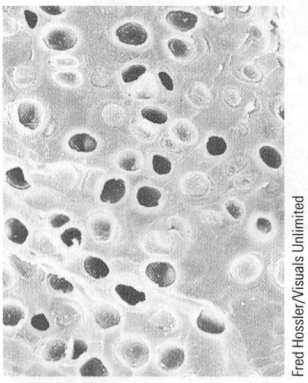

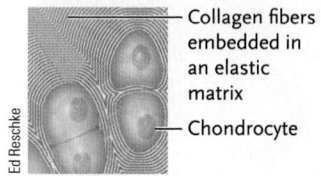

— Collagen fibers embedded in an elastic matrix
— Chondrocyte

Description: Chondrocytes embedded in a pliable, solid matrix of collagen and chondroitin sulfate

Common locations: Ends of long bones, ears, nose, parts of airways, skeleton of vertebrate embryos

Function: Support, flexibility, low-friction surface for joint movement

D. Bone tissue

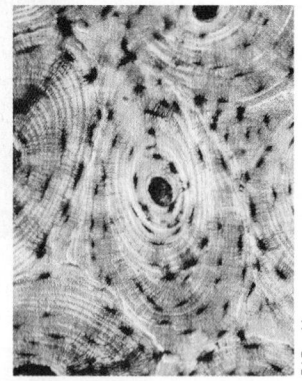

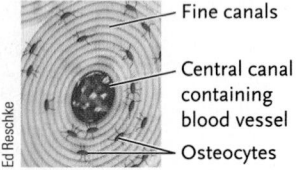

— Fine canals
— Central canal containing blood vessel
— Osteocytes

Description: Osteocytes in a matrix of collagen and glycoproteins hardened with hydroxyapatite

Common locations: Bones of vertebrate skeleton

Function: Movement, support, protection

E. Adipose tissue

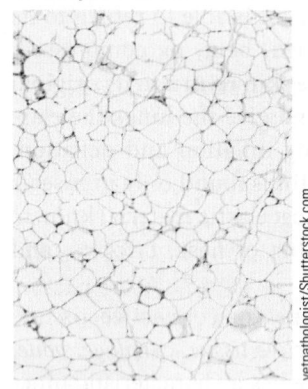

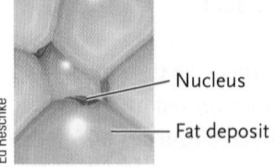

— Nucleus
— Fat deposit

Description: Large, tightly packed adipocytes with little ECM

Common locations: Under skin; around heart, kidneys

Function: Energy reserves, insulation, padding

F. Blood

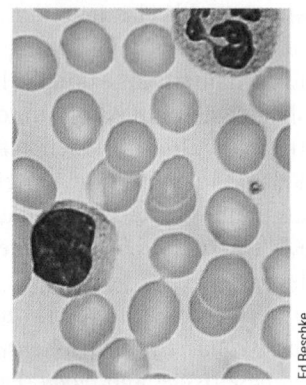

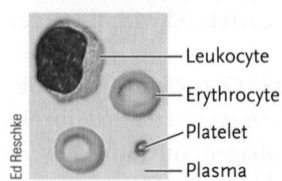

— Leukocyte
— Erythrocyte
— Platelet
— Plasma

Description: Leukocytes, erythrocytes, and platelets suspended in plasma, a fluid ECM

Common locations: Circulatory system

Function: Transport of substances

FIGURE 38.4 Structures and functions of the six major types of connective tissues in vertebrates.

DENSE CONNECTIVE TISSUE In **dense connective tissue,** fibroblasts are sparsely distributed among dense masses of collagen and elastin fibers that are arranged to resist stretch and provide strength **(Figure 38.4B).** In one type of dense connective tissue, the fibers are arranged irregularly so as to resist stretch in specific directions. Examples of this type are found in the dermis of the skin, in the covering of muscles, and in the walls of the digestive tract. In a second type of dense connective tissue, the fibers are lined up in highly ordered, parallel bundles, giving the tissue a fibrous appearance under the microscope (photo in Figure 38.4B). The parallel arrangement produces maximum tensile strength and elasticity. Examples include **tendons,** which attach muscles to bones, and **ligaments,** which connect bones to each other at a joint.

CARTILAGE **Cartilage** consists of sparsely distributed cartilage-producing cells called **chondrocytes,** surrounded by networks of collagen fibers embedded in a tough but elastic matrix of the glycoprotein *chondroitin sulfate* **(Figure 38.4C).** Elastin is also present in some forms of cartilage.

The elasticity of cartilage allows it to resist compression and stay resilient, like a piece of rubber. Bending your ear or pushing the tip of your nose, which are supported by a core of cartilage, shows the flexible nature of this tissue. In humans, cartilage also supports the larynx, trachea, and smaller air passages in the lungs. It forms the disks cushioning the vertebrae in the spinal column and the smooth, slippery capsules around the ends of bones in joints such as the hip and knee. Cartilage also serves as a precursor to bone during embryonic development; in sharks and rays and their relatives (the Chondrichthyes; see Section 32.5), almost the entire skeleton remains as cartilage in adults.

BONE The densest form of connective tissue, **bone,** forms the skeleton, which supports the body, protects softer body structures such as the brain, and contributes to body movements (see Chapter 43).

Mature bone consists primarily of cells called **osteocytes** (*osteon* = bone) embedded in an ECM containing collagen fibers and glycoproteins impregnated with *hydroxyapatite,* a calcium–phosphate mineral **(Figure 38.4D).** The collagen gives bone tensile strength and elasticity and the hydroxyapatite resists compression and allows bones to support body weight. Cells called **osteoblasts** (*blast* = bud or sprout) produce the collagen and mineral of bone—as much as 85% of the weight of bone is mineral deposits. Osteocytes, in fact, are osteoblasts that have become trapped and surrounded by the bone materials they themselves produce. **Osteoclasts** (*clast* = break) are cells that remove the minerals and recycle them through the bloodstream. Bone is reshaped continuously by the bone-building osteoblasts and the bone-degrading osteoclasts.

Bones are not solid. They are porous structures, with microscopic spaces and canals. The structural unit of bone, the **osteon,** consists of a minute central canal surrounded by osteocytes embedded in concentric layers of mineral matter (see Figure 38.4D). A blood vessel and extensions of nerve cells run through the central canal, which is connected to the spaces containing cells by very fine, radiating canals filled with interstitial fluid. The blood vessels supply nutrients to the cells with which the bone is built, and the nerve cells hook up the bone cells to the nervous system.

ADIPOSE TISSUE The connective tissue called **adipose tissue** mostly contains large, densely clustered cells called **adipocytes** that are specialized for fat storage **(Figure 38.4E).** It has little ECM. Adipose tissue also cushions the body and, in mammals, forms an especially important insulating layer under the skin.

The animal body stores limited amounts of carbohydrates, primarily in muscle and liver cells. Excess carbohydrates are converted into the fats stored in adipocytes. The storage of chemical energy as fats offers animals a weight advantage. For example, the average human would weigh about 45 kg (100 pounds) more if the same amount of chemical energy was stored as carbohydrates instead of fats. Adipose tissue is richly supplied with blood vessels, which move fats or their components to and from adipocytes.

BLOOD Blood **(Figure 38.4F)** is considered a connective tissue because its cells are suspended in a fluid ECM, plasma. The straw-colored **plasma** is a solution of proteins, nutrient molecules, ions, and gases.

Blood contains two primary cell types, **erythrocytes** (red blood cells; *erythros* = red) and **leukocytes** (white blood cells; *leukos* = white). Erythrocytes are packed with hemoglobin, a protein that can bind and transport oxygen. Several types of leukocytes all help to protect the body against invading viruses, bacteria, and other disease-causing agents. The blood plasma also contains **platelets,** membrane-bound fragments of specialized blood cells, which take part in the reactions that seal wounds with blood clots.

Blood is the major transport vehicle of the body. It carries oxygen and nutrients to body cells, removes wastes and by-products such as carbon dioxide, and maintains the internal fluid environment, including the osmotic balance between cells and the interstitial fluid. Blood also transports hormones and other signal molecules that coordinate body responses. (The components and roles of blood are described in Chapter 44.)

Muscle Tissue Produces the Force for Body Movements

Muscle tissue consists of cells that have the ability to contract ("shorten"). The contractions, which depend on the interaction of the two proteins **actin** and **myosin,** move body limbs and other structures, pump the blood, and produce a squeezing pressure in organs such as the intestine and uterus. Three types of muscle tissues, *skeletal, cardiac,* and *smooth,* produce body movements in vertebrates **(Figure 38.5).**

SKELETAL MUSCLE Typically, **skeletal muscle** is attached by tendons to the skeleton. Skeletal muscle cells are also called **muscle fibers (Figure 38.5A).** These cells are multinucleate (contain many nuclei in the same cytoplasm) and are packed with actin and myosin molecules arranged in highly ordered, parallel units that give the tissue a banded or striated appearance when viewed under a microscope. Muscle fibers packed side-by-side into parallel bundles surrounded by sheaths of connective tissue constitute muscles, such as the biceps.

Skeletal muscle contracts in response to signals carried by the nervous system. The contractions of skeletal muscles, characteristically rapid and powerful, move body parts and maintain posture. In some cases disorganized muscle contractions (shivering) are used to release heat as a by-product of cellular metabolism. This heat helps mammals, birds, and some other vertebrates maintain their body temperatures when environmental temperatures fall. (Skeletal muscle is discussed further in Chapter 43.)

CARDIAC MUSCLE Cardiac muscle is the contractile tissue of the heart **(Figure 38.5B).** Actin and myosin molecules arranged like those in skeletal muscle give cardiac muscle a striated appearance. However, cardiac muscle cells are short and

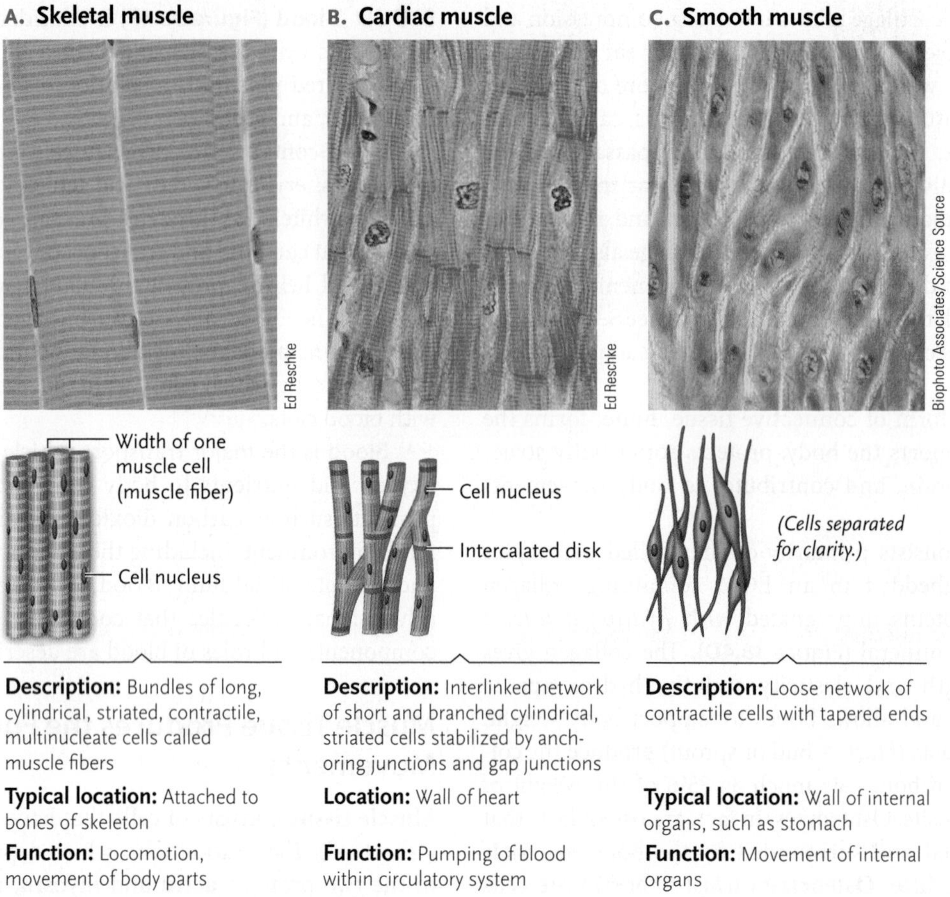

A. Skeletal muscle

Ed Reschke

Width of one muscle cell (muscle fiber)

Cell nucleus

Description: Bundles of long, cylindrical, striated, contractile, multinucleate cells called muscle fibers

Typical location: Attached to bones of skeleton

Function: Locomotion, movement of body parts

B. Cardiac muscle

Ed Reschke

Cell nucleus

Intercalated disk

Description: Interlinked network of short and branched cylindrical, striated cells stabilized by anchoring junctions and gap junctions

Location: Wall of heart

Function: Pumping of blood within circulatory system

C. Smooth muscle

Biophoto Associates/Science Source

(Cells separated for clarity.)

Description: Loose network of contractile cells with tapered ends

Typical location: Wall of internal organs, such as stomach

Function: Movement of internal organs

FIGURE 38.5 Structures and functions of skeletal, cardiac, and smooth muscle.
© Cengage Learning 2017

branched, with each cell connecting to several neighboring cells; the joining point between two such cells is called an **intercalated disk.** Cardiac muscle cells thus form an interlinked network, which is stabilized by anchoring junctions and gap junctions. This network enables heart muscle to contract in all directions, producing a squeezing or pumping action rather than the lengthwise, unidirectional contraction characteristic of skeletal muscle.

SMOOTH MUSCLE Smooth muscle is found in the walls of tubes and cavities in the body, including blood vessels, the stomach and intestine, the bladder, and the uterus. Smooth muscle cells are relatively small and spindle-shaped (pointed at both ends), and their actin and myosin molecules are arranged in a loose network rather than in bundles **(Figure 38.5C).** This loose network makes the cells appear smooth rather than striated when viewed under a microscope. Smooth muscle cells are enclosed by a mesh of connective tissue, and some of them are connected by gap junctions. The gap junctions allow the passage of current that makes smooth muscles contract as a unit, typically producing a squeezing motion. Although smooth muscle contracts more slowly than skeletal and cardiac muscles do, its

contractions can be maintained at steady levels for a much longer time. These contractions move and mix the stomach and intestinal contents, constrict blood vessels, and push an infant out of the uterus during childbirth.

Nervous Tissue Receives, Integrates, and Transmits Information

Nervous tissue contains cells called **neurons** (also called *nerve cells*) that serve as lines of communication and control between body parts. Billions of neurons are packed into the human brain; others extend throughout the body. Nervous tissue also contains **glial cells** (*glia* = glue), which physically support and provide nutrients to neurons, provide electrical insulation between them, and scavenge cellular debris and foreign matter.

A neuron consists of a *cell body*, containing the nucleus and organelles, and two types of cell extensions, dendrites and axons **(Figure 38.6).** *Dendrites* are usually highly branched, whereas *axons* are usually unbranched except at their terminals. The shape, number, and length of dendrites and axons vary from neuron to neuron. Depending on the type of neuron and its location in the body, its axon may extend from about

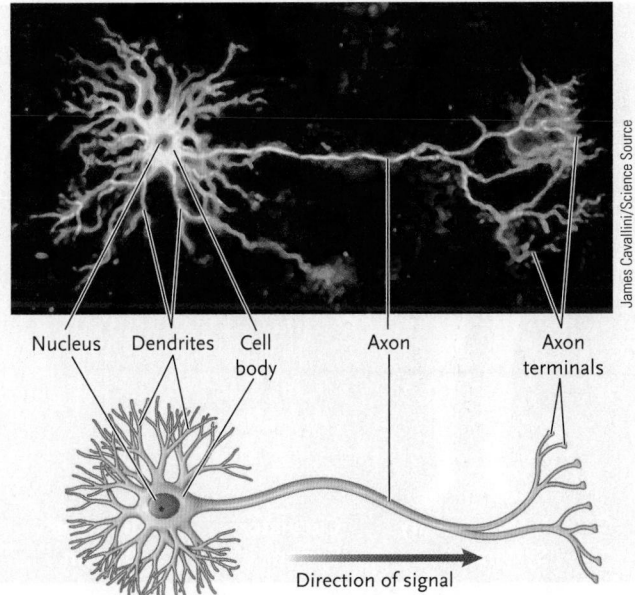

FIGURE 38.6 Neurons and their structure. The micrograph shows a network of motor neurons, which relay signals from the brain or spinal cord to muscles and glands.

© Cengage Learning 2017

Nucleus   Dendrites   Cell body   Axon   Axon terminals

Direction of signal

1 mm or less to more than 1 m. Examples of the latter are axons that run the length of a giraffe's neck.

The cell body or the dendrites of a neuron receive chemical or electrical signals from other neurons. The signals are converted into an electrical wave that is transmitted along the axons to the axon terminals. Depending on the type of neuron, axons either convert the electrical signal at their terminals to a chemical signal that stimulates a response in nearby muscle cells, gland cells, or other neurons, or transfer the electrical signal directly to another neuron through gap junctions. (Neurons and their organization in body structures are discussed further in Chapters 39, 40, and 41.)

The four primary tissue types—epithelial, connective, muscle, and nervous—combine to form the organs and organ systems of animals. The next section depicts the major organs and organ systems of vertebrates, and outlines their main functions.

STUDY BREAK 38.2

1. Distinguish between exocrine and endocrine glands. What is the tissue type of each of these glands?
2. What are the six major types of connective tissue in vertebrates?
3. What three types of muscle tissue produce movements in the body?

THINK OUTSIDE THE BOOK

Individually or collaboratively, determine the type or types of muscle tissues found in invertebrates and give examples of each type.

38.3 Coordination of Tissues in Organs and Organ Systems

In the tissues, organs, and organ systems of an animal, each cell engages in the basic metabolic activities that ensure its own survival, and performs one or more functions of the system to which it belongs. All vertebrates (and most invertebrates) have eleven major organ systems. **Figure 38.7** summarizes structures and functions of these organ systems, which are discussed in detail in the other chapters within this unit.

The functions of the organ systems are coordinated and integrated to accomplish collectively tasks that are vital to all animals, whether a flatworm, a fly, a salmon, a platypus, or a human. These functions include:

1. Acquiring nutrients and other required substances such as oxygen, coordinating their processing, distributing them throughout the body, and disposing of wastes.
2. Synthesizing the protein, carbohydrate, lipid, and nucleic acid molecules required for body structure and function.
3. Sensing and responding to changes in the environment, such as varying temperature, pH, and ion concentrations.
4. Protecting the body against injury or attack from other animals and from viruses, bacteria, and other disease-causing agents.
5. Reproducing and, in many instances, nourishing and protecting offspring through their early growth and development.

Together these tasks maintain homeostasis, preserving the internal environment required for survival of the body.

STUDY BREAK 38.3

What are the major functions of each of the eleven organ systems of the vertebrate body?

38.4 Homeostasis

To live, cells of all organisms must take in nutrients and O_2 from the external environment and eliminate wastes such as CO_2 and particular chemicals to that external environment. A single-celled organism such as an amoeba is in direct contact with the external environment. Although most cells of a multicellular animal are isolated from direct contact with the external environment, those cells have the same needs for nutrient and O_2 input and waste elimination. Those needs are met by the internal environment, namely the aqueous **extracellular fluid (ECF) (Figure 38.8)**.

The ECF has two components:

1. **Plasma,** the fluid portion of blood.
2. **Interstitial fluid** (*inter* = between; *stitial* = that which stands), the fluid that surrounds the cells.

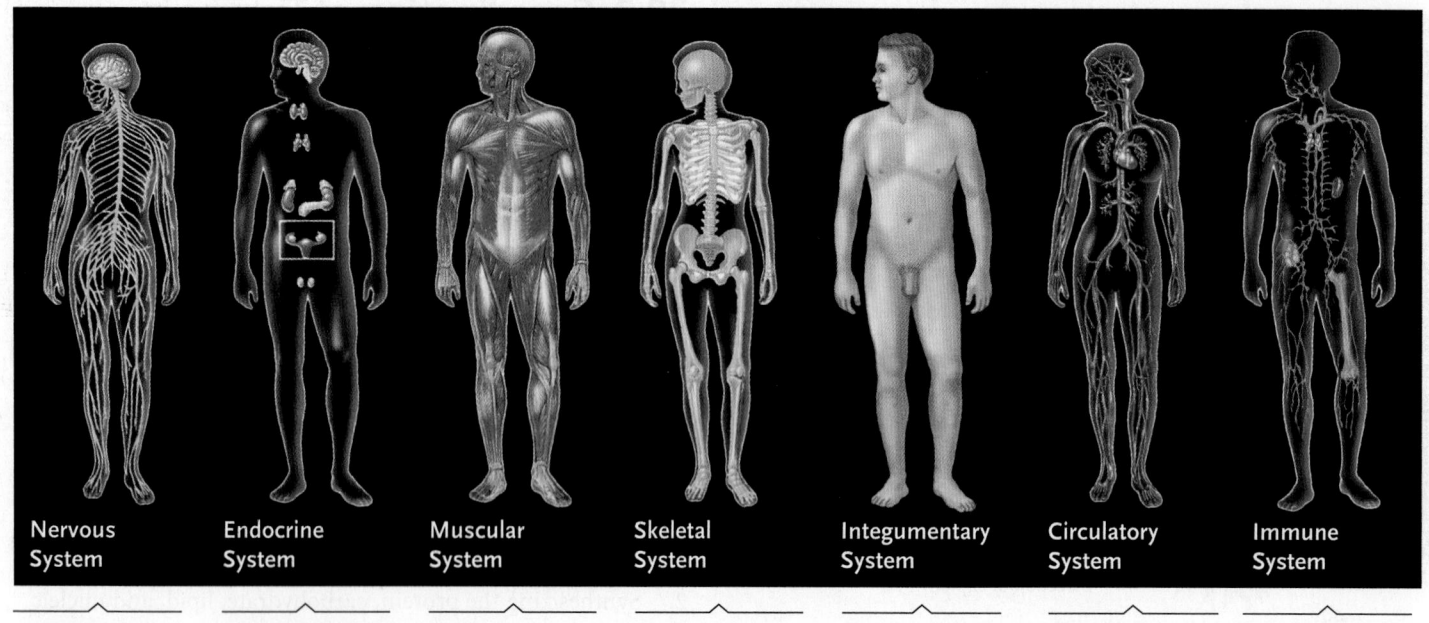

Nervous System	Endocrine System	Muscular System	Skeletal System	Integumentary System	Circulatory System	Immune System
Main structures: Brain, spinal cord, peripheral nerves, sensory organs	**Main structures:** Pituitary, hypothalamus, thyroid, adrenal, pancreas, and other hormone-secreting glands	**Main structures:** Skeletal, cardiac, and smooth muscle	**Main structures:** Bones, tendons, ligaments, cartilage	**Main structures:** Skin, sweat glands, hair, nails	**Main structures:** Heart, blood vessels, blood	**Main structures:** Lymph nodes, lymph ducts, spleen, thymus, bone marrow, and white blood cells
Main functions: Principal regulatory system; monitors changes in internal and external environments and formulates compensatory responses; coordinates body activities	**Main functions:** Regulates and coordinates body activities through secretion of hormones	**Main functions:** Moves body parts; helps run bodily functions; generates heat; moves intestinal lumen contents	**Main functions:** Supports and protects body parts; provides leverage for body movements; stores minerals	**Main functions:** Covers external body surfaces and protects against injury and infection; helps regulate water content and body temperature	**Main functions:** Distributes water, nutrients, oxygen, hormones, and other substances throughout body and carries away carbon dioxide and other metabolic wastes; helps stabilize internal temperature and pH	**Main functions:** Defends against disease-causing microorganisms and viruses (pathogens). *(Cellular components of the immune system are not shown in the figure.)*

FIGURE 38.7 Structures and functions of the organ systems of the human body.
© Cengage Learning 2017

The ECF is the transitional zone connecting the **intracellular fluid (ICF)**—the fluid inside cells—to the external environment. Thus, no matter where a cell is within the body, it can make the exchanges essential to its life with the interstitial fluid. Particular organ systems enable these exchanges between the external and internal environments. The digestive system processes incoming food and transfers the resulting nutrients into the plasma. The nutrients reach all parts of the body by the action of the circulatory system, along with O_2, which enters the blood by the action of the respiratory system. The nutrients and O_2 in the plasma reach the interstitial fluid through the capillaries and, from there, they enter the cells as needed. Waste moves in the opposite direction: from the cells into the interstitial fluid, and then to the plasma. The respiratory system handles removal of CO_2, and the excretory system handles the metabolic wastes.

For optimal function of these systems, the composition and state of the ECF must be maintained within a relatively narrow range so that cells have available the necessary nutrients and O_2 and wastes can be eliminated. Further, other aspects of the internal environment that are important for cellular (and therefore organismal) life, such as temperature, must also be regulated within a narrow range because most cells cannot tolerate much change. When the ECF composition moves outside the normal range of values, regulatory

Extracellular fluid

Cell   Interstitial fluid   Plasma

Blood vessel

FIGURE 38.8 Components of the extracellular fluid (ECF).
© Cengage Learning 2017

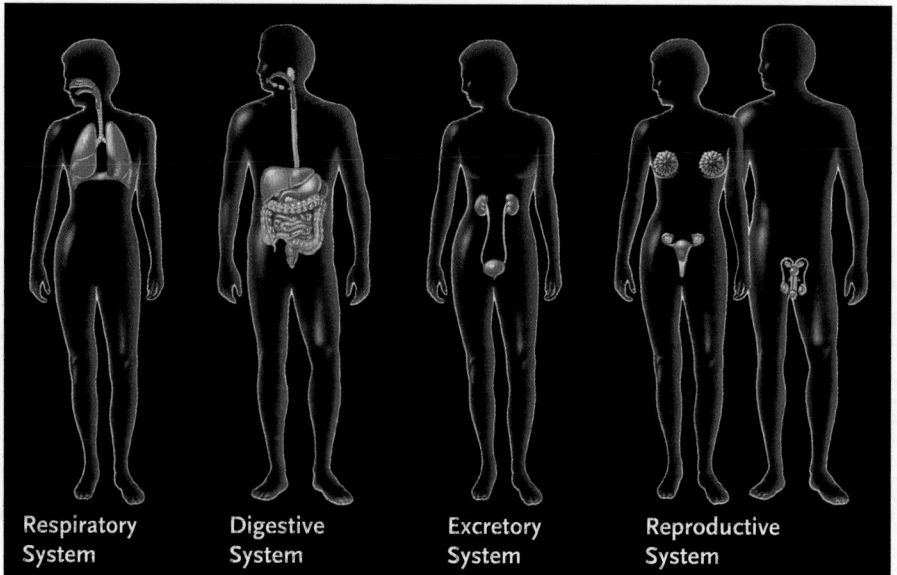

Respiratory System	**Digestive System**	**Excretory System**	**Reproductive System**
Main structures: Lungs, diaphragm, trachea, and other airways	**Main structures:** Oral cavity, pharynx, esophagus, stomach, intestines, liver, pancreas, rectum, anus	**Main structures:** Kidneys, bladder, ureter, urethra	**Main structures:** *Female*: ovaries, oviducts, uterus, vagina, mammary glands *Male*: testes, sperm ducts, accessory glands, penis
Main functions: Exchanges gases with the environment, including uptake of oxygen and release of carbon dioxide	**Main functions:** Converts ingested matter into molecules and ions that can be absorbed into body; eliminates undigested matter; helps regulate water content	**Main functions:** Removes and eliminates excess water, ions, and metabolic wastes from body; helps regulate internal osmotic balance and pH; helps regulate blood pressure	**Main functions:** Maintains the sexual characteristics and passes on genes to the next generation

fluids.) Similarly, internal adjustments are needed for homeostasis during exercise or hibernation. The factors controlled by homeostatic mechanisms all require energy that must be acquired continually from the external environment.

The processes and activities responsible for homeostasis are called **homeostatic control systems;** they are found in all organisms. Humans and other mammals maintain a consistent internal environment, but many kinds of multicellular animals show much more variability in their internal environment than mammals do. Animals fall into two major categories in this regard: **regulators** maintain factors of the internal environment in a relatively constant state, whereas **conformers** have internal environments that match the external environment. For instance, mammals and birds are thermoregulators, meaning that they maintain their body temperature at a relatively constant value regardless of the temperature of the external environment. However, most animals—for example, fishes, reptiles, and insects—are thermoconformers, meaning that their body temperature matches that of the external environment.

Many Factors of the Internal Environment Are Regulated Homeostatically

Factors of the internal environment that are regulated homeostatically include:

1. *Nutrient concentration.* Energy production by cells requires a constant supply of nutrient molecules. The energy generated by catabolizing the nutrients is used for basic cellular processes and any specialized activities of the cell.

2. *Concentration of O_2.* Cellular respiration (see Chapter 7), the process that generates energy from catabolic reactions, requires a constant supply of O_2 for optimal productivity.

3. *Concentration of CO_2.* The CO_2 produced by the catabolic reactions of cellular respiration must be removed as waste or the ECF would become increasingly acidic.

4. *Concentration of waste chemicals.* Particular biochemical reactions in the cell generate products that would be toxic to the cell if not removed as waste.

5. *Concentration of water and NaCl.* The relative concentrations of NaCl and water in the ECF affect how much water enters or leaves a cell's ICF and hence determine the cell's volume (see Chapter 5). These concentrations must be regulated to maintain a cell volume that is optimal for function; swollen or shrunken cells are functionally impaired.

6. *pH.* Changes in pH of the ECF can adversely affect enzymatic activities within cells, as well as nerve cell functions.

mechanisms kick in to attempt to return the ECF composition to within the normal range.

As introduced in *Why it matters . . .* the regulation of the internal environment to maintain it in a relatively stable state is called **homeostasis.** Although the *stasis* part of homeostasis suggests a static, unchanging process, homeostasis is a *dynamic* process in which internal adjustments are made continuously to compensate for external changes. For example, in a state of homeostasis, the compositions of both the ICF and the ECF are in a dynamic **steady state.** *Dynamic* means that a homeostatically regulated factor is changing continuously— materials are moving in both directions between the ICF and the ECF. *Steady state* means that a homeostatically regulated factor does not vary much from a steady level—the compositions of the two fluids are relatively stable. Thus, in a dynamic steady state, there is no *net* movement of materials between the ICF and the ECF. (Note that steady state is not the same as *equilibrium,* a term that, in this context, would mean that the compositions of the ICF and the ECF are identical. In fact, the concentrations of many substances are different in the two

7. *Volume and pressure of plasma.* Both the volume and pressure of the plasma component of the ECF within the vessels must be maintained at levels adequate to distribute the fluid throughout the body. This circulation is vitally important for supplying cells with their needs and removing their wastes.

8. *Temperature.* Body cells (of warm-blooded animals) function optimally within a fairly narrow temperature range. Outside of that range, chemical reactions change their rates and may be inhibited completely. If cells become too cold, the rates of enzymatic reactions decrease too much. If cells become too hot, structural and enzymatic proteins can be denatured and, therefore, become inactive.

Most Homeostatic Control Systems Operate throughout the Body

Local homeostatic controls operate only within an organ where a change in the internal environment needs to be addressed. For example, when you exercise, a skeletal muscle increases its use of O_2, which reduces the concentration of O_2 in the muscle cells and sets up a requirement for increased O_2 uptake from the interstitial fluid. The local homeostatic control in this instance responds to the decreased O_2 concentration by triggering the relaxation of the smooth muscle in the walls of the blood vessels supplying the exercising muscle. As a result, the blood vessels dilate (expand in diameter), increasing the blood flow to the exercising muscle, which brings more O_2. The end result is maintenance of an optimal O_2 concentration for the exercising skeletal muscle.

Systemic homeostatic controls are initiated outside of an organ or organ system to control that organ's or organ system's activity. The body's two major regulatory systems, the nervous system and the endocrine system, are responsible for most systemic homeostatic control. This type of control enables the regulation of several organs or organ systems to be coordinated toward a common goal. For example, if blood pressure drops too low, the nervous system acts on the heart to increase contraction strength and on blood vessels to constrict them so as to increase pressure in the system (see Chapter 44). Blood pH is controlled by both the nervous and endocrine systems (see Chapter 46), blood glucose by the endocrine system (see Chapter 42), internal temperature by the nervous and endocrine systems (see Chapter 48), and oxygen and carbon dioxide concentrations by the nervous system (see Chapter 46).

Homeostasis Is Accomplished by Negative Feedback Control Systems

Homeostatic control systems function primarily by using *negative feedback* to resist change **(Figure 38.9)**. In **negative feedback,** a change in a factor triggers a response to oppose the change and restore the factor to normal by moving the factor in the opposite direction of its initial change. The following list presents the components of a negative feedback control system maintaining homeostasis (see Figure 38.9). In parentheses are descriptions of the specific homeostatic control system used by mammals and birds—thermoregulators—to maintain body temperature within a relatively narrow range around a set point:

- **Stimulus**—an environmental change that triggers a response. (A change in body temperature beyond the normally controlled range around the normal temperature level.)

- **Sensor**—the body component that detects the environmental change. (Temperature-monitoring nerve cells throughout the body.)

- **Integrator**—a control center that receives information from the sensor and compares it with the **set point,** the normal level at which the condition being controlled is to be maintained. The integrator, typically part of the brain or endocrine system, sends out commands to correct any

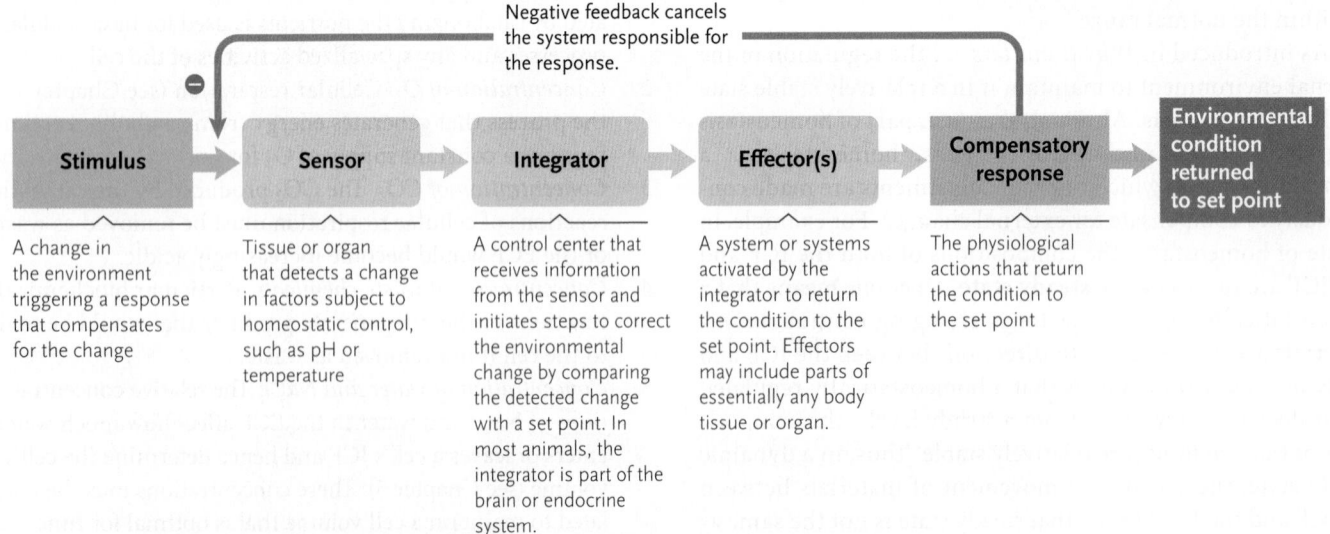

FIGURE 38.9 Components of a negative feedback control system maintaining homeostasis. The sensor, integrator, and effector(s) are physical components in the body, such as a body structure or chemical.

© Cengage Learning 2017

significant change from the set point detected. (The temperature control center in a region of the brain called the *hypothalamus* [see Chapter 40] detects changes in the temperature of the brain and the rest of the body, and compares it with a set point. For humans, the set point has a relatively narrow range centered at about 37°C.)

- **Effector**—a system that is activated by the integrator to bring the homeostatic factor under control back to the set point. Effectors may include parts of essentially any body tissue or organ. They trigger compensatory responses to return the factor to the set point. Once that is achieved, the compensatory responses are turned off.

Let us consider the effectors for body temperature homeostasis in thermoregulating vertebrates in a little more detail. If the temperature falls below the lower limit, as was the case for the passengers and crew of the *Titanic* who ended up in the water (see *Why it matters . . .*), the hypothalamus activates effectors that constrict the blood vessels in the skin. The reduction in blood flow means that less heat is conducted from the blood through the skin to the environment, reducing heat loss from the skin. Other effectors may induce shivering, a physical mechanism to generate body heat. Also, integrating neurons in the brain, stimulated by signals from the hypothalamus, make us consciously sense a chill, which we may counteract behaviorally by, if possible, putting on more clothes or moving to a warmer area.

Conversely, if blood temperature rises above the set point by exposure to high external temperature or by vigorous exercise, the hypothalamus triggers effectors that dilate the blood vessels in the skin, increasing blood flow and heat loss from the skin. Other effectors induce sweating, which cools the skin and the blood flowing through it as the sweat evaporates. And again, through integrating neurons in the brain, we may consciously sense being overheated, which we may counteract behaviorally by shedding clothes, moving to a cooler location, stopping exercise, or taking a dip in a pool or a cool shower.

All mammals have similar homeostatic control systems that maintain or adjust body temperature. **Figure 38.10** illustrates how a dog responds to high environmental temperatures by the mechanisms we have discussed and, in addition, by panting. Birds also pant to reduce their body temperatures. Many terrestrial animals enter or splash water over their bodies to cool off. (See Chapter 48 for a detailed discussion of thermoregulation in animals.)

The homeostatic control system for temperature can also be overwhelmed by prolonged exposure to a particularly high or low temperature. In either case, the end result will be death unless the external temperature returns to a more moderate value. As the *Titanic* story describes, only 15 minutes in near-freezing water can kill a person. Likewise, homeostatic control systems for other regulated factors can also be overwhelmed or be disrupted. For example, the inability to regulate blood glucose concentration homeostatically leads to the disease state of diabetes.

Whereas mammals and birds regulate their internal body temperature within a narrow range around a set point, certain other vertebrates regulate over a broader range. These vertebrates use other, less precise negative feedback mechanisms for their temperature regulation. Snakes and lizards, for example, respond behaviorally to compensate for variations in environmental temperatures. They may absorb heat by basking on sunny rocks in the cool early morning and move to cooler, shaded spots in the heat of the afternoon.

Some invertebrates, such as dragonflies, moths, and butterflies, use muscular contractions equivalent to shivering when their body temperature falls below the level required for

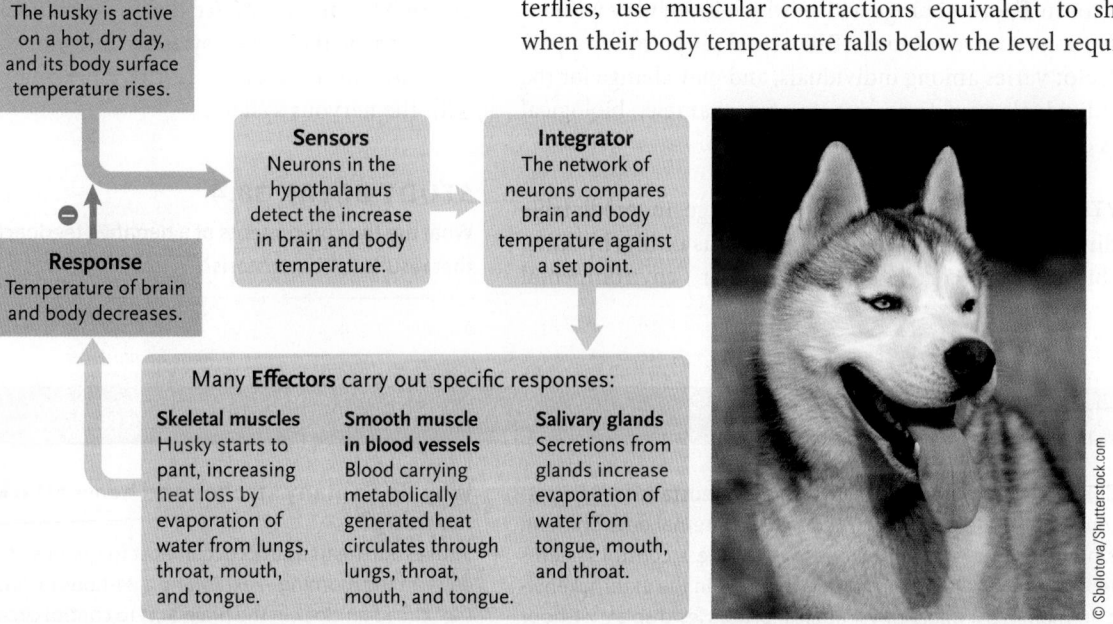

Stimulus
The husky is active on a hot, dry day, and its body surface temperature rises.

Sensors
Neurons in the hypothalamus detect the increase in brain and body temperature.

Integrator
The network of neurons compares brain and body temperature against a set point.

Response
Temperature of brain and body decreases.

Many **Effectors** carry out specific responses:

Skeletal muscles
Husky starts to pant, increasing heat loss by evaporation of water from lungs, throat, mouth, and tongue.

Smooth muscle in blood vessels
Blood carrying metabolically generated heat circulates through lungs, throat, mouth, and tongue.

Salivary glands
Secretions from glands increase evaporation of water from tongue, mouth, and throat.

© Sbolotova/Shutterstock.com

FIGURE 38.10 Homeostatic negative feedback control system maintaining the body temperature of a dog when environmental temperatures are high.
© Cengage Learning 2017

flight. The shivering contractions warm the muscles to flying temperature. All of these physiological and behavioral responses depend on negative feedback mechanisms involving sensors, integrators, and effectors.

Animals Also Have Positive Feedback Mechanisms That Are Not Homeostatic

Under certain circumstances, animals respond to a change in internal or external environmental conditions by a **positive feedback** mechanism that intensifies or adds to the change. Such mechanisms, with some exceptions, are not homeostatic. Positive feedback mechanisms are less common than negative feedback mechanisms.

The birth process in mammals provides an example of positive feedback. During human childbirth, initial contractions of the uterus push the head of the fetus against the cervix, the opening of the uterus into the vagina, causing the cervix to stretch. Sensors that detect the stretching signal the hypothalamus to stimulate the release of the hormone oxytocin from the pituitary gland (see Chapter 42). Oxytocin increases the uterine contractions, intensifying the squeezing pressure on the fetus and further stretching the cervix. The stretching results in more oxytocin release and stronger uterine contraction, repeating the positive feedback circuit and increasing the squeezing pressure until the fetus is pushed out of the uterus entirely. (Human childbirth is described in more detail in Chapter 50.)

Set Points Can Change Due to Biological Rhythms or Altered Environmental Conditions

Homeostatic control systems operate to maintain a homeostatic factor in a normal range within which it can vary without triggering a corrective response. The actual set point for a particular factor varies among individuals, and may change for the same individual over time due to, for example, biological rhythms or altered environmental conditions.

BIORHYTHMS Some regulated factors change in predictable and cycling patterns called **biological rhythms** or **biorhythms.** Many biorhythms correlate with regular environmental changes, such as daily light–dark cycles or the seasons of the year. Biorhythms are generated by changes in the set point of the regulated factor involved. For instance, in mammals, many physiological processes have a 24-hour rhythm called a **circadian rhythm** (*circa* = approximately; *dies* = day), which is generated by a gene-based circadian clock. Humans, for example, have circadian rhythms for several regulated factors in the body, including body temperature and blood pressure. Body temperature is lowest in the early hours of the morning and is highest in late afternoon. That is why you may feel cold when you study late at night—your temperature set point has been lowered. (This chapter's *Unanswered Questions* also discusses circadian rhythms and malfunctions of the circadian clock.)

ALTERED ENVIRONMENTAL CONDITIONS When a set point changes naturally because of an alteration in environmental conditions, it is called **acclimatization.** When a set point changes artificially in a laboratory setting, it is called **acclimation.** Acclimatization examples include: (1) adapting to low oxygen levels when traveling from low to high elevations by increasing the concentration of red blood cells in the blood so as to facilitate adequate oxygen delivery to the tissues (described more in Chapter 46); and (2) adapting to a cold climate or winter cold, or to a hot climate or summer heat. In all cases of acclimatization, the set point changes when the environmental condition is reversed, for example, by returning to lower altitudes.

It is important to distinguish *acclimatization* and *evolutionary adaptation.* Acclimatization is a temporary change in a physiological process that occurs during the life of an animal, whereas evolutionary adaptation is a change that occurs at the genetic level over many generations. For example, native Tibetans are naturally adapted to living at high altitudes as a result of a particular gene allele inherited from an extinct species of humans (see *Molecular Insights* in Chapter 46).

Each of the organ systems introduced in this chapter is described in more detail in the following chapters, beginning with the nervous system.

STUDY BREAK 38.4

What are the components of a negative feedback control system that results in homeostasis?

 Unanswered Questions

As described in the chapter, all animals have homeostatic control systems through which they maintain a consistent internal environment. The physiological processes underlying many of these control systems are shaped by evolution to adapt to daily changes in the external environment. In mammals, many physiological processes display 24-hour rhythms that are generated by an internal gene-based clock called a *circadian clock*. Some malfunctions of systems that are normally under homeostatic control also seem to exhibit a circadian cycle.

Why do so many strokes and heart attacks occur in the morning?

Strokes and heart attacks occur most frequently at a particular time of the day—in the morning—exhibiting a 24-hour, or circadian, rhythm. A "central" circadian clock in the brain acts to control circadian rhythm in sleep–wake cycles. These same proteins are also found throughout the organs and cells of the body, including blood vessels (vascular clock). Recent work has shown that a broken clock also breaks down important physiological signals, such as nitric oxide, to worsen heart attack, stroke, and vascular disease in animal models. (The role of nitric oxide on neuron function is discussed in Chapter 39.) Current research is addressing how circadian clock malfunctions contribute to both acute and chronic disease processes. In our laboratory, we are using mice with targeted disruption of circadian clock genes (knockout or mutant mice) and studying their response to cardiovascular stresses, including hypertension and vascular disease.

Why do we measure blood pressure in the daytime?
High blood pressure is still a major health problem that is integral to the process of cardiovascular disease. In fact, blood pressure is not the same at all times of the day. At night our blood pressure drops, whereas in the day it increases, exhibiting a circadian rhythm. People who do not exhibit this up-and-down pattern in blood pressure, called *nondippers,* have worse cardiovascular disease. Despite this, medical guidelines for healthy blood pressure ranges are based on the "daytime" pressure, which really only relates to the times of day when we have the opportunity to visit the doctor's office. More research needs be done to understand what "bad" nighttime blood pressure is, how we can control nighttime blood pressure, and whether the circadian clock controls the body signals that regulate blood pressure. Bad sleep, shift work, and even jet lag may have unexpected effects on blood pressure of which we need a greater understanding.

Location, location, location: Where in the body does the circadian clock malfunction to cause vascular disease?
Though heart attacks and strokes are the culmination of ongoing vascular disease, their onset is acute, often occurring in the morning around 8 A.M.

In contrast, the progression to vascular disease is chronic, occurring in a time span of years. Our laboratory and others have now demonstrated that the genes comprising the circadian clock can contribute to vascular disease. This finding suggests that, in addition to daily cycles, the circadian clock controls aspects of biology that are also important in the long term. We think of circadian disorders as overt behavioral disorders that include poor sleep, night or shift work, and jet lag, but there may be more to the picture. Recently, we have demonstrated that dysfunction of the circadian clock within local tissue (blood vessels) influences disease risk, independent of the central clock. Thus, circadian malfunction in tissues outside the brain may contribute to diseases such as atherosclerosis, hypertension, diabetes, and even cancer. These basic science observations of circadian function in animals may lay the foundation for clinical studies in humans, which ultimately may change the way we understand and treat arteriosclerosis, hypertension, and heart attack.

think like a scientist How might circadian rhythms have been important for survival in early organisms?

R. Daniel Rudic is an Associate Professor in the Department of Pharmacology and Toxicology at the Medical College of Georgia. To learn more about his research on circadian rhythms and vascular biology, go to http://www.gru.edu /mcg/phmtox/rudiclab/index.php.

R. Daniel Rudic

REVIEW KEY CONCEPTS

For access to MindTap and additional study materials visit www.cengagebrain.com.

38.1 Organization of the Animal Body

- Multicellular organisms have an internal environment of fluid that cells use for the exchange of materials, a distinction from single-celled organisms.

- In most animals, cells are specialized and organized into tissues, tissues into organs, and organs into organ systems. A tissue is a group of cells with similar structure and specialized function. An organ integrates two or more tissues into a structure that performs a particular function or functions. An organ system is a collection of organs with related functions that interact to carry out a major body function (Figure 38.1).

38.2 Animal Tissues

- Primary tissues in animals are epithelial, connective, muscle, or nervous (Figure 38.1). The properties of the cells of these tissues determine the structures and functions of the tissues.

- Various kinds of junctions link cells in a tissue. Anchoring junctions "weld" cells together. In tight junctions, plasma membrane proteins of adjacent cells interact to fuse the cells partly together to form a barrier between the cells. Gap junctions form direct channels for ions, small molecules, and electrical signals between cells in the same tissue.

- Epithelial tissue consists of sheetlike layers of cells that cover body surfaces and the surfaces of internal organs and line cavities and ducts within the body. Epithelial tissues protect the internal environment and regulate the exchange of materials between the internal and external environments (Figure 38.2).

- Glands are secretory structures derived from epithelia. They may be exocrine (connected to an epithelium by a duct that empties on the epithelial surface) or endocrine (ductless, with no direct connection to an epithelium) (Figure 38.3).

- Connective tissue consists of cell networks or layers and an extracellular matrix (ECM). It supports other body tissues, transmits mechanical and other forces, and in some cases acts as a filter (Figure 38.4).

- Loose connective tissue consists of sparsely distributed fibroblasts surrounded by an open network of collagen and other glycoproteins. It supports epithelia and organs of the body and forms a covering around blood vessels, nerves, and some internal organs.

- Dense connective tissue contains sparsely distributed fibroblasts in a matrix of dense masses of collagen and elastin fibers that are arranged to resist stretch and provide strength. The fibers may be irregularly arranged, as in dense connective tissue in the walls of the digestive tract, or lined up in highly ordered parallel bundles, as in tendons and ligaments.

- Cartilage consists of sparsely distributed chondrocytes surrounded by a network of collagen fibers embedded in a tough but highly elastic matrix of branched glycoproteins. Cartilage provides support, flexibility, and a low-friction surface for joint movement (Figure 38.4).

- In bone, osteocytes are embedded in a collagen matrix hardened by mineral deposits. Osteoblasts secrete collagen and minerals for the ECM; osteoclasts remove the minerals and recycle them into the bloodstream (Figure 38.4).

- Adipose tissue consists of cells specialized for fat storage (Figure 38.4).

- Blood consists of a fluid matrix, the plasma, in which erythrocytes and leukocytes are suspended. The erythrocytes carry oxygen to body cells; the leukocytes produce antibodies and initiate the immune response against disease-causing agents (Figure 38.4).
- Muscle tissue contains cells that have the ability to contract forcibly (Figure 38.5). Skeletal muscle, containing long cells called muscle fibers, moves body parts and maintains posture. Cardiac muscle, which contains short contractile cells with a branched structure, forms the heart. Smooth muscle consists of spindle-shaped contractile cells that form layers surrounding body cavities and ducts.
- Nervous tissue contains neurons and glial cells. Neurons communicate information between body parts in the form of electrical and chemical signals (Figure 38.6). Glial cells support the neurons or provide electrical insulation between them.

38.3 Coordination of Tissues in Organs and Organ Systems

- Organs and organ systems are coordinated to carry out vital tasks, including maintenance of internal body conditions; nutrient acquisition, processing, and distribution; waste disposal; molecular synthesis; environmental sensing and response; protection against injury and disease; and reproduction.
- In vertebrates and most invertebrates, the eleven major organ systems are the nervous, endocrine, muscular, skeletal, integumentary, circulatory, immune, respiratory, digestive, excretory, and reproductive systems (Figure 38.7).

38.4 Homeostasis

- Homeostasis is the regulation of an animal body's internal environment—the extracellular fluid (Figure 38.8)—to maintain it in a relatively stable steady state. It is a dynamic state, in which internal adjustments are made continuously to compensate for environmental changes.
- With respect to homeostatic regulation of the internal environment, multicellular organisms may be regulators, which maintain factors of the internal environment in a relatively constant state, or conformers, which have internal environments that match the external environment.
- Homeostatically regulated factors of the internal environment include nutrient concentration; O_2 and CO_2 concentrations; concentrations of waste chemicals, water, and NaCl; pH; volume and pressure of plasma; and temperature.
- Most homeostatic control systems operate systemically, whereas some operate locally.
- Homeostasis is accomplished by negative feedback mechanisms that include a sensor, which detects a change in an external or internal condition; an integrator, which compares the detected change with a set point; and an effector or effectors, which return the condition to the set point if it has varied. Once the factor is back to the set point, the compensatory response is turned off (Figure 38.9).
- Animals also have positive feedback mechanisms, in which a change in an internal or external condition triggers a response that intensifies the change and typically does not result in homeostasis.
- Set points can change as a result of biological rhythms (for example, circadian rhythms of body temperature) or by altered environmental conditions (for example, acclimatization to higher altitudes).

TEST YOUR KNOWLEDGE

Remember/Understand

1. Which tissue type consisting of sheetlike layers of cells can both exchange oxygen and act as a barrier to bacteria?
   a. nervous
   b. epithelial
   c. connective
   d. muscle
   e. heart

2. Which of the following is a constant source of adult stem cells in a mammal?
   a. bone marrow
   b. pancreas
   c. basal lamina
   d. heart muscle
   e. kidney

3. A flexible, rubbery protein in connective tissue is called ___, whereas a more fibrous, less flexible glycoprotein is called ___.
   a. adipose; cartilage
   b. endocrine; exocrine
   c. sweat; hormones
   d. chondroitin sulfate; hydroxyapatite
   e. elastin; collagen

4. Adipose tissue:
   a. provides elasticity under the epithelium.
   b. gives strength to tendons.
   c. insulates and is an energy reserve.
   d. provides movement, support, and protection.
   e. supports the nose and airways.

5. The bones of an elderly woman break more easily than those of a younger person. You would surmise that with aging, the cell type that diminishes in activity is the:
   a. osteocyte.
   b. osteoblast.
   c. osteoclast.
   d. chondrocyte.
   e. fibroblast.

6. Lifting weights will most increase the size of:
   a. skeletal muscle.
   b. smooth muscle.
   c. cardiac muscle.
   d. involuntary muscle.
   e. interlinked, branched muscle.

7. Which muscle types appear striated under a microscope?
   a. skeletal muscles only
   b. cardiac muscles only
   c. skeletal muscles and cardiac muscles
   d. smooth muscles only
   e. skeletal muscles and smooth muscles

8. Which of the following is not a homeostatic response?
   a. In a contest, a student eats an entire chocolate cake in 10 minutes, and his blood glucose level does not change dramatically.
   b. A jogger is sweating after a two-mile jog.
   c. The pupils in humans' eyes constrict when looking at a light.
   d. Physical activity increases carbon dioxide blood levels, which lowers blood pH and increases breathing.
   e. Oxytocin is released by the hypothalamus during human childbirth.

9. When you exercise, a skeletal muscle increases its use of O_2, which reduces the concentration of O_2 in the muscle cells and sets up a requirement for increased O_2 uptake from the interstitial fluid. This is an example of:
   a. osmolarity.
   b. environmental sensing.
   c. integration.
   d. systemic homeostatic control.
   e. local homeostatic control.

10. The system that coordinates other organ systems is the:
   a. skeletal system.
   b. reproductive system.
   c. muscular system.
   d. nervous system.
   e. integumentary system.

Apply/Analyze

11. **Discuss Concepts** Blood is often described as an atypical connective tissue. If you had to argue that blood is a connective tissue, what reasons would you include? What reasons would you include if you had to argue that blood is not a connective tissue?

12. **Discuss Concepts** Positive feedback mechanisms are rare in animals compared with negative feedback mechanisms. Why do you think this is so?

13. **Discuss Concepts** Near the time of childbirth, collagen fibers in the connective tissue of the cervix break down, and gap junctions between the smooth muscle cells of the uterus increase in number. What do you think is the significance of these tissue changes?

14. **Discuss Concepts** Explain how, when driving, you control the car's speed by a typical negative feedback mechanism.

Evaluate/Create

15. **Discuss Concepts** What effect do you think a program of lifting weights would have on the bones of the skeleton? How would you design an experiment to test your prediction?

16. **Design an Experiment** The regulation of temperature in mammals and birds is an example of homeostasis. Design an experiment to observe and measure processes involved in temperature homeostasis in sedentary versus athletic humans during exercise.

17. **Apply Evolutionary Thinking** Large animals (such as many of the vertebrates described in this chapter) often have more types of specialized tissues and organ systems than do small animals (such as flatworms). Why would natural selection favor the evolution of additional tissue types and organ systems in larger organisms?

For selected answers, see Appendix A.

INTERPRET THE DATA

Vasopressin is an important regulator of blood pressure. Taurine has been shown to inhibit vasopressin release. An experiment was done to test the hypothesis that taurine acts as part of a negative feedback system that regulates vasopressin. The experiment measured the release of taurine from rat pituitary cells in the presence of vasopressin (10 nM) **(Figure).** Are these data consistent with the hypothesis? Why?

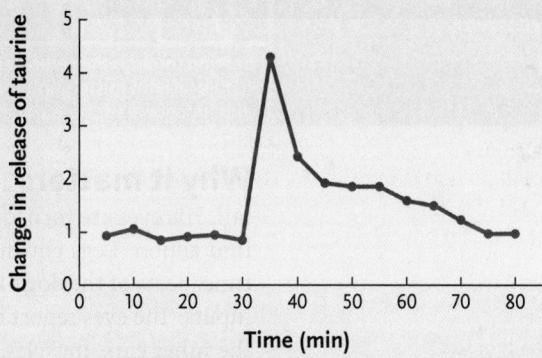

Source: L. Rosso et al. 2004. Vasopressin-induced taurine efflux from rat pituicytes: a potential negative feedback for hormone secretion. *The Journal of Physiology* 554:731–742.

39 Information Flow and the Neuron

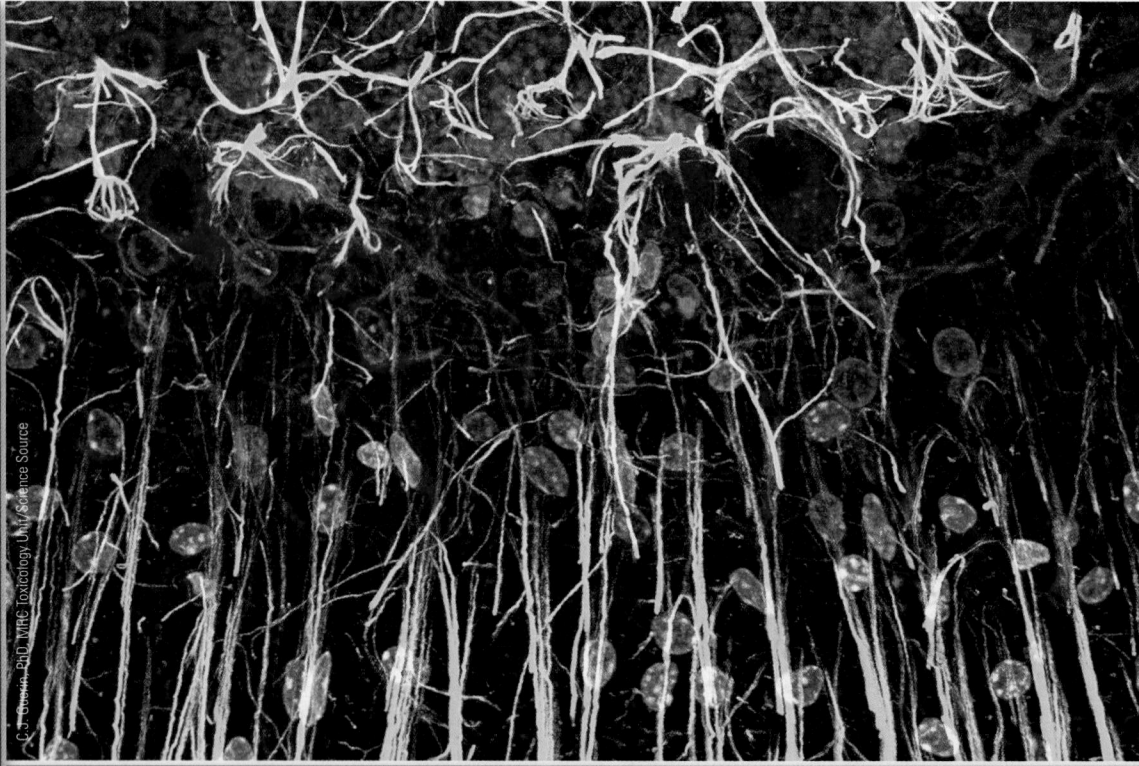

Section through the cerebellum, a part of the brain that integrates signals coming from particular regions of the body (confocal light micrograph). Neurons, the cells that send and receive signals, are red; glial cells, which provide structural and functional support for neurons, are yellow; and nuclei are purple.

FIGURE 39.1 A dog leaps to catch a Frisbee. The coordinated leap involves processing and integration of information by the dog's nervous system.

Why it matters . . . The dog stands alert, muscles tense, motionless except for a wagging tail. His eyes are turned toward his owner, who throws a Frisbee for him to catch. The dog springs into action. Legs churning, eyes following the Frisbee, the dog runs beneath its track. All this time, parts of the dog's brain have been processing information received through various sensory inputs. The eyes report his travel over the ground and the speed and arc of the Frisbee. Sensors in the inner ears, muscles, and joints detect the position of the dog's body, and his brain sends out signals that keep his movements on track and in balance. Other parts of the brain register inputs from homeostatic sensors monitoring body temperature and carbon dioxide levels in the blood, and send signals that adjust heart and breathing rates accordingly.

At just the right instant, signals from the dog's brain cause trunk and leg muscles to contract in a coordinated pattern, and the dog leaps to intercept the Frisbee in midair with a snap of his jaws **(Figure 39.1).** Now the animal turns his head and eyes toward the ground as his brain calculates the motions required to land on his feet. The dog makes a perfect landing and trots back to his owner.

The functions of the dog's nervous system in the chase and capture are astounding in the amount and variety of sensory inputs, the rate and complexity of the brain's analysis and integration of incoming information, and the flurry of signals the brain sends to make compensating adjustments in body activities. Yet they are ordinary in the sense that the same activities take place countless times each day in the nervous system of all but the simplest animals.

All these activities, no matter how complex, depend on the functions of two major cell types: *neurons* and *glial cells*. In most animals, these cells are organized into complex networks called *nervous systems*. This chapter discusses neuron structure and how neurons send and receive signals with the aid and support of glial cells. Chapter 40 considers the neural networks of the brain and its associated structures. Chapter 41 discusses the sensory receptors that detect environmental changes and convert that information into signals for integration by the nervous system.

39.1 Neurons and Their Organization in Nervous Systems

An animal receives stimuli constantly from both internal and external sources. **Neural signaling,** communication by neurons, is the process by which an animal responds appropriately to a stimulus **(Figure 39.2).** In most animals, there are four components of neural signaling:

1. **Reception,** the detection of a stimulus, is performed by specialized sensory receptors such as those in the eye and skin and directly by **neurons.**

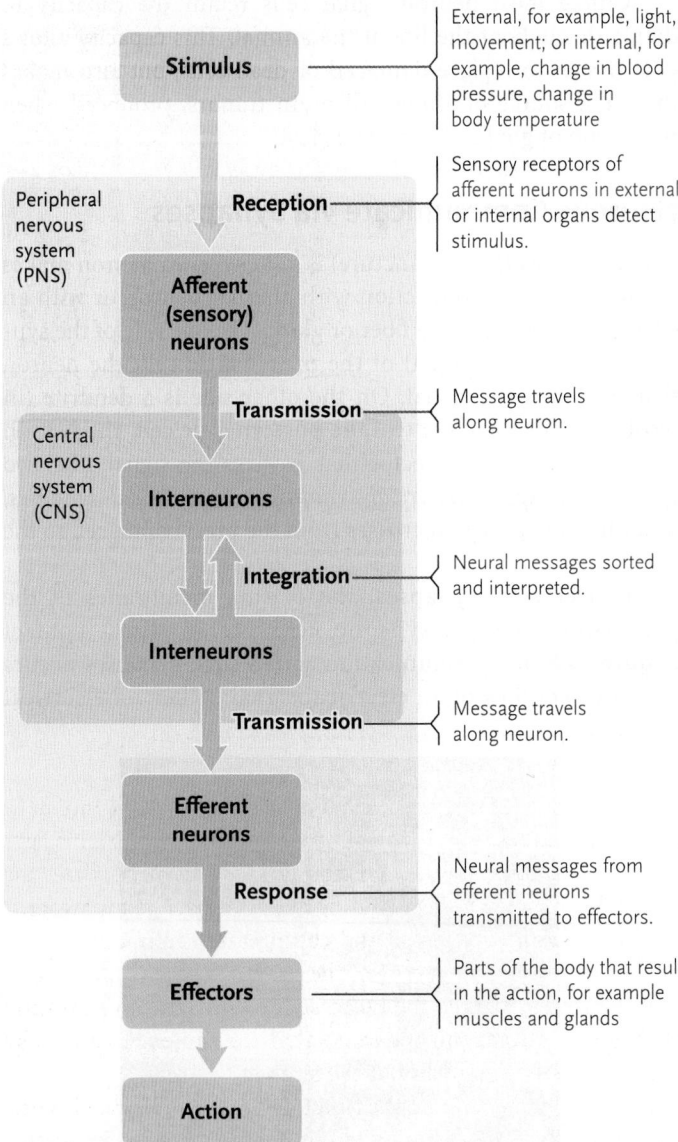

FIGURE 39.2 Neural signaling: the information-processing steps in the nervous system.
© Cengage Learning 2017

2. **Transmission** is the sending of a message along a neuron, and then to another neuron or to a muscle or gland.
3. **Integration** is the sorting and interpretation of neural messages and the determination of the appropriate response(s).
4. **Response** is the "output" or action resulting from the integration of neural messages.

For a dog catching a Frisbee, for instance, sensors in the eye receive light stimuli from the environment, and internal sensors receive stimuli from all the animal's organ systems. The neural messages generated are transmitted through the nervous system and integrated to determine the appropriate response, in this case stimulating the muscles so the dog jumps into the air and catches the Frisbee.

Neurons Are Cells Specialized for the Reception and Transmission of Informational Signals

Neural signaling involves three functional classes of neurons (the blue boxes in Figure 39.2). **Afferent neurons** (*afferre* = to bring to) (also called **sensory neurons**) transmit stimuli collected by their sensory receptors to **interneurons,** which integrate the information to generate an appropriate response. In humans and some other primates, 99% of neurons are interneurons. **Efferent neurons** (*efferre* = to carry away) carry the signals indicating a response away from the interneuron networks to the **effectors,** the muscles and glands. Efferent neurons that carry signals to skeletal muscle are called **motor neurons.** The information-processing steps in the nervous system, therefore, are:

1. Sensory receptors on afferent neurons receive a stimulus.
2. Afferent neurons transmit the information to interneurons.
3. Interneurons integrate the neural messages.
4. Efferent neurons transmit the neural messages to effectors, which act in a way appropriate to the stimulus.

The model neuron that is used commonly to discuss neuron function has an enlarged cell body and two types of extensions or processes, called *dendrites* and *axons* **(Figure 39.3).** The **cell body,** which contains the nucleus and the majority of cell organelles, is the site of synthesis of most of the proteins, carbohydrates, and lipids of the neuron. Dendrites and axons conduct electrical signals, which are produced by ions flowing down concentration gradients through channels in the plasma membrane of the neuron. **Dendrites** (*dendron* = tree) receive the signals and transmit them toward the cell body. **Axons** (also called *nerve fibers*) conduct signals away from the cell body to another neuron or an effector. An axon arises from a junction with the cell body called an **axon hillock,** and ends with a small, buttonlike swelling called an **axon terminal,** which connects the neuron functionally with an adjacent neuron or effector. The shape, number, and length of axons and dendrites vary from neuron to neuron.

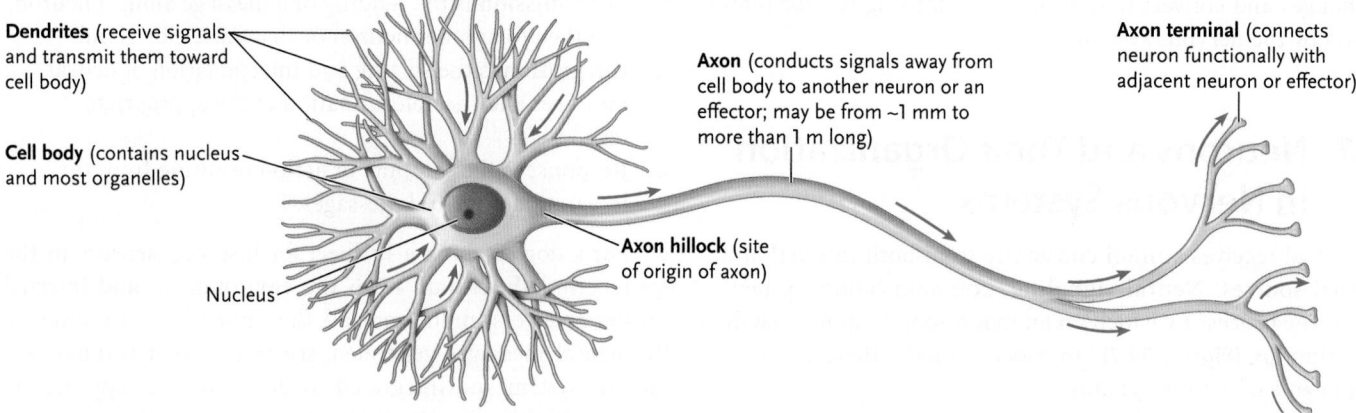

Dendrites (receive signals and transmit them toward cell body)

Cell body (contains nucleus and most organelles)

Nucleus

Axon hillock (site of origin of axon)

Axon (conducts signals away from cell body to another neuron or an effector; may be from ~1 mm to more than 1 m long)

Axon terminal (connects neuron functionally with adjacent neuron or effector)

FIGURE 39.3 Structure of a neuron (nerve cell). The arrows show the direction electrical impulses travel along the neurons.

© Cengage Learning 2017

Connections between axon terminals of one neuron and the dendrites or cell body of a second neuron contribute to the formation of **neural circuits.** A typical neural circuit contains an afferent (sensory) neuron, one or more interneurons, and an efferent neuron. The circuits combine into networks that interconnect the parts of the nervous system. In vertebrates, the afferent neurons and efferent neurons collectively form the **peripheral nervous system (PNS).** The interneurons form the brain and spinal cord, called the **central nervous system (CNS).** The anatomical terms *ganglion* (plural, *ganglia*) and *nucleus* (plural, *nuclei*) are used to refer to clusters of nerve cells typically having related functions. Generally, a **ganglion** is a group of neurons lying outside the CNS, and a **nucleus** refers to a comparable concentration of cells within the CNS. As depicted in Figure 39.2, afferent information is ultimately transmitted to the CNS, where efferent information is initiated. The nervous systems of most invertebrates are also divided into central and peripheral divisions.

Neurons Are Supported Structurally and Functionally by Glial Cells

Glial cells are nonneuronal cells that provide nutrition and support to neurons. One type, called **astrocytes** because they are star-shaped **(Figure 39.4),** occur only in the vertebrate CNS, where they closely cover the surfaces of blood vessels. Astrocytes provide physical support to neurons and help maintain the concentrations of ions in the interstitial fluid surrounding them. Two other types of glial cells—**oligodendrocytes** in the CNS and **Schwann cells** in the PNS—wrap around axons in a jellyroll fashion to form **myelin sheaths (Figure 39.5).** Myelin sheaths act as electrical insulators because of the high lipid content of the many layers of plasma membranes of the myelin-forming cells. The gaps between Schwann cells, called **nodes of Ranvier,** expose the axon membrane directly to extracellular fluids. This structure speeds the rate at which electrical impulses move along the axons covered by glial cells (explained later in the chapter).

Unlike most neurons, glial cells retain the capacity to divide throughout the life of the animal. This capacity allows glial tissues to replace damaged or dead cells, but also makes them the source of almost all brain tumors, produced when regulation of glial cell division is lost.

Neurons Communicate via Synapses

A **synapse** (*synapsis* = juncture) is a site where a neuron makes a communicating connection with another neuron or with an effector such as a muscle fiber or gland. On one side of the synapse is an axon terminal of the **presynaptic cell,** the neuron that transmits the signal. On the other side is a dendrite (in most cases) or a cell body of the **postsynaptic cell,** the neuron or the surface of an effector that receives the signal. The two types of synapses, *electrical synapses* and *chemical synapses,* differ in how the signal crosses from the presynaptic cell to the postsynaptic cell.

In **electrical synapses,** the plasma membranes of the presynaptic and postsynaptic cells are in direct contact **(Figure 39.6A).** Communication across such synapses occurs by the direct flow of an electrical signal. When an electrical

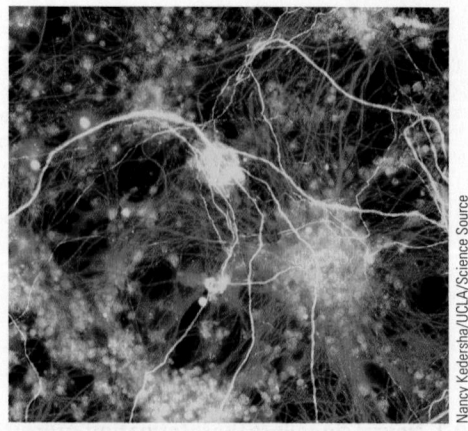

Nancy Kedersha/UCLA/Science Source

FIGURE 39.4 Astrocytes (orange), a type of glial cell, and a neuron (yellow) in brain tissue.

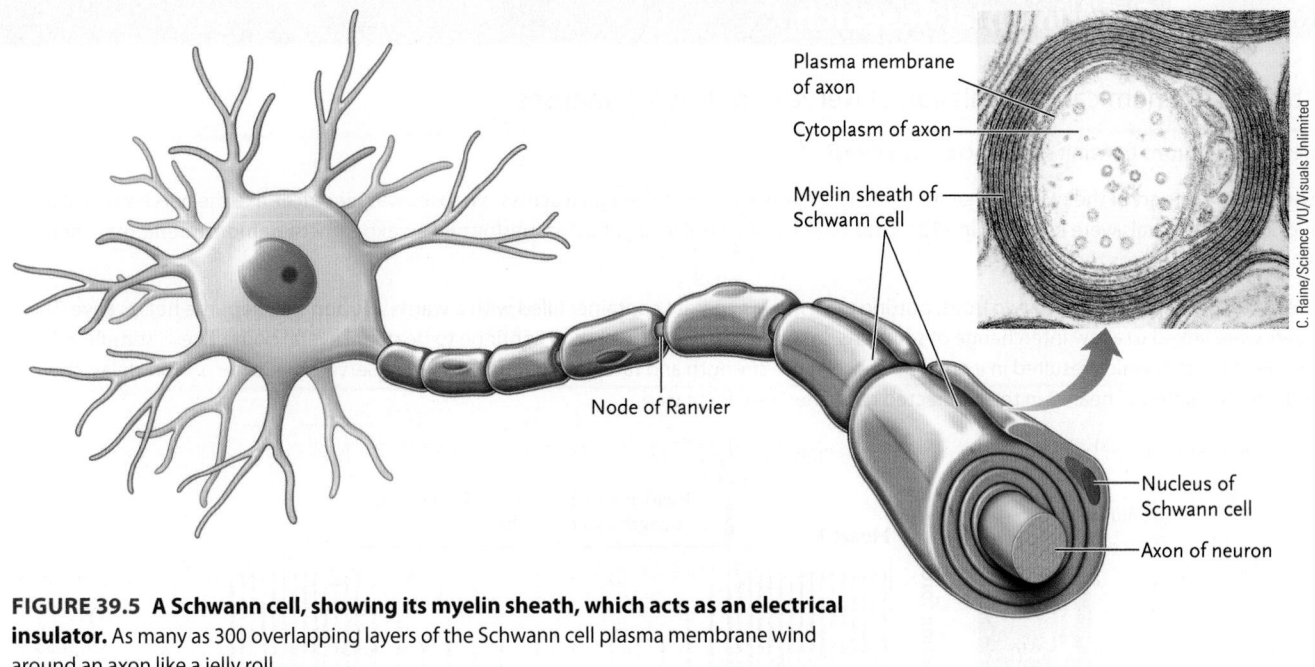

FIGURE 39.5 A Schwann cell, showing its myelin sheath, which acts as an electrical insulator. As many as 300 overlapping layers of the Schwann cell plasma membrane wind around an axon like a jelly roll.
© Cengage Learning 2017

impulse arrives at the axon terminal, gap junctions (see Section 4.5) allow ions to flow directly between the two cells, leading to unbroken transmission of the electrical signal. Although electrical synapses allow the most rapid conduction of signals, this type of connection is essentially "on" or "off" and unregulated. In humans and other animals, electrical synapses typically occur as neural circuit elements where synchrony of firing and speed of response are paramount. Electrical synapses are found, for instance, in cardiac muscle, in the retina of the eye, and in the pulp of a tooth.

A. Electrical synapse

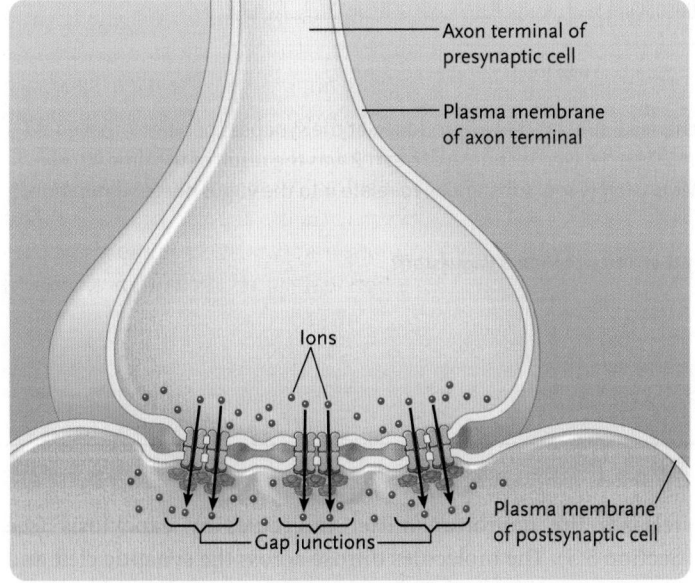

In an electrical synapse, channel proteins in the plasma membranes of the presynaptic and postsynaptic cells form gap junctions. Ions flow through the gap junctions, allowing impulses to pass directly to the postsynaptic cell.

B. Chemical synapse

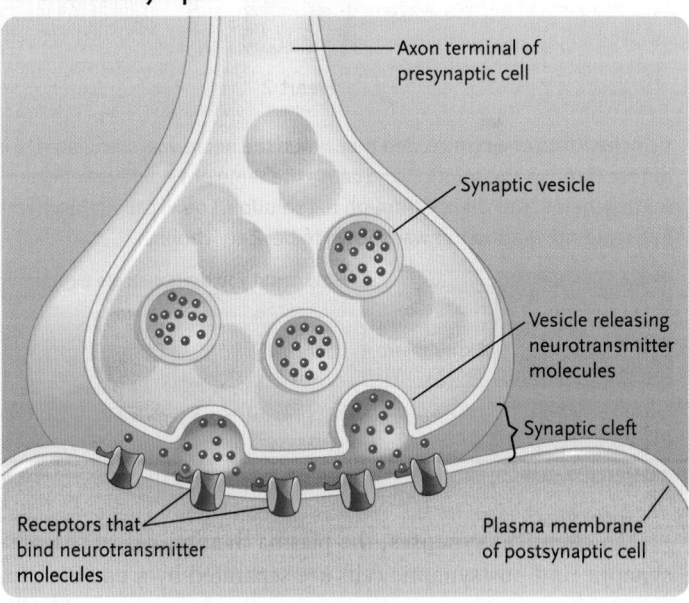

In a chemical synapse, the plasma membranes of the presynaptic and postsynaptic cells are separated by a narrow synaptic cleft. Neurotransmitter molecules released from synaptic vesicles diffuse across the cleft and bind to receptors in the plasma membrane of the postsynaptic cell. The binding opens channels to ion flow that may generate an impulse in the postsynaptic cell.

FIGURE 39.6 The two types of synapses by which neurons communicate with other neurons or effectors, electrical synapses (A) and chemical synapses (B).
© Cengage Learning 2017

Demonstration of Chemical Transmission of Nerve Impulses at Synapses

Question: How do neurons transmit signals across synapses?

Experiment: In the first part of the twentieth century, many scientists thought the signal across synapses was electrical, like the nerve signal itself; others thought that chemicals were involved. In 1921, Otto Loewi of Graz University, Austria, performed an experiment demonstrating that chemicals can transmit a signal across the synapse.

Results: Loewi isolated the hearts from two frogs, putting each into a separate container filled with a warm solution that kept the hearts alive. The two containers were linked to allow interchange of solutions. In this setup, the hearts could continue to beat for several hours. Loewi stimulated the vagus nerve of heart 1, which resulted in a rapid decrease in the strength and rate of its beating. Loewi observed that, after a short delay, the strength and rate of beating of heart 2 in the connected container also decreased.

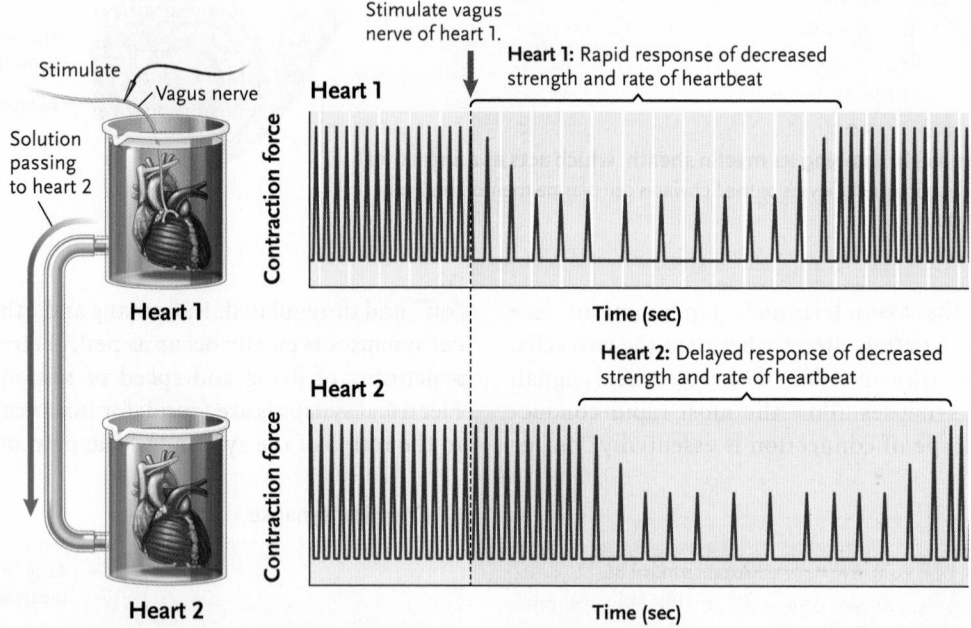

Conclusion: Loewi concluded that, when the nerve was stimulated, some chemical substance was produced at the synapses of heart 1 that could diffuse through the solution and cause the same physiological effect in heart 2. The delayed response of heart 2 was the result of the time it took for the chemical to diffuse through the solution. Loewi called the chemical *Vagusstoff* (vagal substance), to relate it to the vagus nerve stimulation. Subsequently, Vagusstoff was shown to be acetylcholine.

Observe
Hypothesize
Predict
Experiment
Interpret

How could you demonstrate experimentally that acetylcholine is Vagusstoff?

Source: O. Loewi. 1921. Über humorale Übertragbarkeit der Herznervenwirkung. I. *Pflügers Archiv* 189:239–242.

In **chemical synapses,** the plasma membranes of the presynaptic and postsynaptic cells are separated by a narrow gap, about 25 nm wide, called the **synaptic cleft (Figure 39.6B).** Communication across such synapses occurs by means of chemicals called **neurotransmitters.** Neurotransmitter molecules are located near the axon terminal in **synaptic vesicles.** When an electrical impulse arrives at an axon terminal, neurotransmitter-filled synaptic vesicles dock with the presynaptic cell's plasma membrane where it borders the synaptic cleft. Fusion of the synaptic vesicles with the plasma membrane releases the neurotransmitter molecules by exocytosis (see Section 5.5). The molecules diffuse across the synaptic cleft and bind to receptors in the plasma membrane of the postsynaptic cell. If enough neurotransmitter molecules bind to these receptors, the postsynaptic cell generates a new electrical impulse, which travels along its axon to reach a synapse with the next neuron or effector in the circuit. The vast majority of vertebrate synapses are chemical synapses. **Figure 39.7** describes the first experiment to demonstrate the chemical transmission of a nerve impulse across a synapse.

A chemical synapse is more than a simple on–off switch because many factors can influence the generation of a new electrical impulse in the postsynaptic cell, including neurotransmitters that inhibit that cell rather than stimulating it. The balance of stimulatory and inhibitory effects in chemical synapses contributes to the integration of incoming information in a receiving neuron.

Proteomic analysis is providing insights into the complexity of chemical synapses. Subcellular fractionation of neurons of rodent brains has shown large numbers of proteins in different fractions, for example, over 400 proteins in synaptic vesicles and more than 3,000 in *synaptosomes* (isolated synaptic terminals). The functional significance of the proteins of the chemical synapse proteome is the subject of current research, as is the three-dimensional organization of the proteins in synaptosomes. Regarding the latter, a 2014 research paper showed that the average rat brain synaptosome contains 384 synaptic vesicles, each containing proteins ranging in number from hundreds to thousands.

STUDY BREAK 39.1

1. Distinguish between a dendrite and an axon.
2. Distinguish between the functions and locations of afferent neurons, efferent neurons, and interneurons.
3. What is the difference between an electrical synapse and a chemical synapse?

39.2 Signaling by Neurons

All cells of an animal have a **membrane potential,** a separation of positive and negative charges across the plasma membrane. Outside the cell the net charge is positive, and inside the cell it is negative. This charge separation produces **voltage**—an electrical potential difference—across the plasma membrane.

Selective Permeability of the Plasma Membrane to Ions and Other Charged Molecules Produces the Membrane Potential

Membrane potential is the result of the selective permeability of the plasma membrane to charged atoms (ions) and molecules. Because of the action of the **Na^+/K^+ pump** in the membrane, Na^+ is removed from the cytoplasm and K^+ is brought into the cytoplasm (see Section 5.4; the ATPase of the pump hydrolyzes ATP to provide the energy for the pump's activity). This results in a concentration gradient of Na^+ strongly favoring its diffusion into the cell and the opposite for K^+.

Two other aspects of membrane permeability are crucial to the establishment of the membrane potential:

1. The membrane is about 25 to 30 times more permeable to passive diffusion of K^+ than of Na^+. This is because it is easier for K^+ than for Na^+ to cross the membrane due to the fact that the membrane typically has many more protein channels always open for passive K^+ movement than channels open for passive Na^+ movement. These types of ion-specific protein channels are called *leak channels* because they always allow passage of their selected ion.
2. The cytoplasm is rich in large anionic molecules—amino acids, proteins, nucleic acids—that cannot cross the cell membrane.

Because of its concentration gradient, K^+ attempts to diffuse from the cell. But, when it does, large anion molecules are left behind. These anions are not effectively neutralized by cations because the cell membrane does not permit significant Na^+ movement. At steady state, the cell has a negative internal potential produced by the unpaired organic anions that is just strong enough to prevent further K^+ exit. Thus, this steady state forms a balanced disequilibrium between the concentration gradient for K^+ and the electrical gradient counteracting further K^+ movement.

In most cells, the membrane potential does not change. However, neurons and muscle cells use the membrane potential in a specialized way. That is, in response to electrical, chemical, mechanical, and certain other types of stimuli, their membrane potential changes rapidly and transiently. Cells with this property are said to be *excitable cells*. Excitability, produced by a sudden ion flow across the plasma membrane, is the basis for nerve impulse generation.

Resting Membrane Potential Is the Unchanging Membrane Potential of an Unstimulated Neuron

The membrane of a neuron (or a muscle cell, another excitable cell) that is not conducting an impulse exhibits a steady negative membrane potential, called the **resting membrane potential.** The word *resting* in the name means that the membrane potential has reached a steady state and is unchanging. Resting membrane potentials are measured on a relative scale **(Figure 39.8)** and typically fall in the −40 to −90 millivolt (mV) range in isolated neurons and muscle cells. This value indicates that the intracellular fluid is negative relative to the extracellular fluid, which is set to 0 mV. A neuron exhibiting the resting membrane potential is said to be *polarized.*

The distribution of ions inside and outside an axon that produces the resting membrane potential is shown in **Figure 39.9.** As described earlier in this section, the Na^+/K^+ pump is responsible for creating both the imbalance of Na^+ and K^+ inside and outside of the cell and the unpaired charge state between intracellular anions and extracellular positive charges. Voltage-gated ion channels for Na^+ and K^+ open and close when a neuron is stimulated, causing the membrane potential changes.

FIGURE 39.8 | **Research Method**

Measuring Membrane Potential

Purpose: To determine the membrane potentials of unstimulated and stimulated neurons and muscle cells.

Protocol: Prepare a microelectrode by drawing out a glass capillary tube to a tip with a diameter much smaller than that of a cell and filling it with a salt solution that can conduct an electric current. Under a microscope, use a micromanipulator (mechanical positioning device) to insert the tip of the microelectrode into an axon. Place a reference electrode in the solution outside the cell. Use an oscilloscope or voltmeter to measure the voltage between the microelectrode tip in the axon and the reference electrode outside the cell.

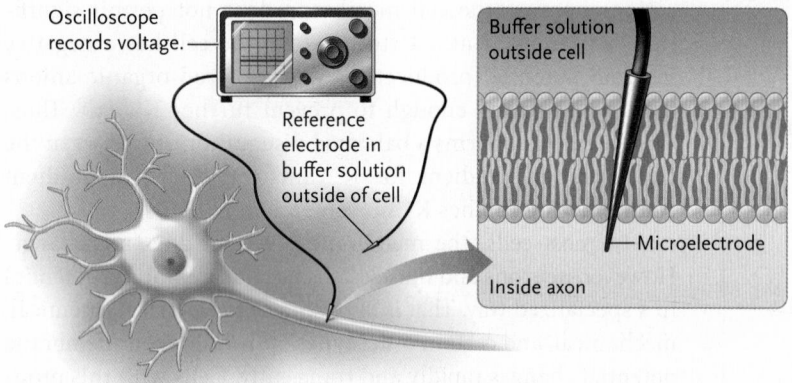

Oscilloscope records voltage.

Reference electrode in buffer solution outside of cell

Buffer solution outside cell

Microelectrode

Inside axon

Interpreting the Results: The oscilloscope or voltmeter measurement is in volts. The extracellular solution is given a value of 0 volts and the voltage measured inside the cell is relative to the extracellular value. Changes in membrane potential caused by stimuli or chemical treatments can be measured and recorded. For an isolated, unstimulated neuron (shown above), the membrane potential is typically about −70 mV.

© Cengage Learning 2017

As noted previously, the tendency for K^+ to diffuse from the cell and leave behind unpaired negative charges is largely responsible for the negative resting membrane potential. The electrical potential necessary to balance this diffusional potential of K^+ is called the **equilibrium potential** for K^+ (E_K).

The **Nernst equation** can be used to calculate the equilibrium potential for a single ion with differing concentrations across a membrane:

$$E_{ion} = \frac{61}{z} \log \frac{C_o}{C_i}$$

where

- E_{ion} = equilibrium potential for the ion in mV
- 61 = a constant
- z = the valence of the ion, which is 1 for K^+ and Na^+, the two ions that contribute to membrane potential
- C_o = concentration of ion in extracellular fluid (mM)
- C_i = concentration of ion in intracellular fluid (mM)

For example, the extracellular concentration of K^+ is 5 mM and its intracellular concentration is 150 mM (see Figure 39.9). Therefore:

$$E_K = 61 \log \frac{5 \text{ mM}}{150 \text{ mM}}$$
$$= 61 \log \frac{1}{30}$$
$$= 61(-1.477)$$
$$= -90 \text{ mV}$$

If K^+ were the only ion moving across the membrane, the E_K would be equal to the resting membrane potential. Typically, the resting membrane potential is a little more positive than E_K, indicating that some positive ion may be leaking into the cell. The obvious candidate is Na^+ because of the strong concentration gradient for this ion between the outside and the inside of the cell. The reason that the positive charge of Na^+ does not swamp the negative charge within the cell is the extremely low permeability of the membrane to Na^+. Indeed, the equilibrium potential for Na^+ (E_{Na}) calculated from the Nernst equation is 61 mV; in other words, if Na^+ diffused freely across the membrane, the charge inside of the cell would have to be raised to +61 mV to maintain the disequilibrium of Na^+ concentration between the inside of the cell and the outside.

Equilibrium potentials provide a convenient way to predict ion movement across the cell membrane. If an equilibrium potential is close to the resting membrane potential, there will be little ion movement even if the membrane is freely permeable to the ion. For example, if a K^+ channel opens in a membrane with a resting membrane potential in the −70 mV range, some small amount of K^+ will leave the cell to bring the membrane potential closer to the E_K, which is generally somewhat more negative than −70 mV. By contrast, if a Na^+ channel opens, both the concentration gradient for Na^+ and the attraction of the intracellular anions will act together to promote significant Na^+ entry into the cell. The effect will be to **depolarize** the cell—make it less negative.

The Membrane Potential Changes Rapidly from Negative to Positive during an Action Potential

When a neuron transmits an electrical impulse, a rapid and transient change in membrane potential called the **action potential** occurs. An action potential begins as a stimulus that causes positive charges from outside the neuron to flow inward, making the cytoplasmic side of the membrane less negative **(Figure 39.10)**. As the membrane potential becomes less negative, the membrane (which was polarized at rest) becomes depolarized. Depolarization proceeds relatively slowly until it reaches

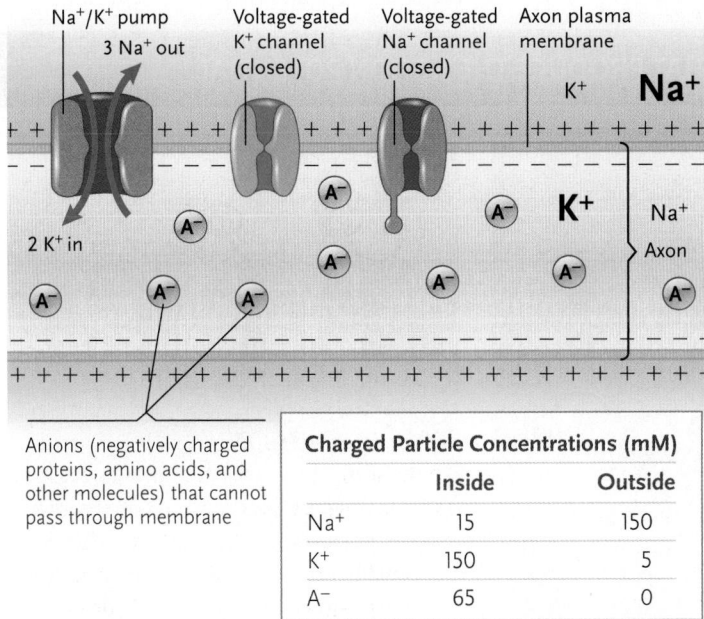

FIGURE 39.9 The distribution of ions inside and outside an axon that produces the resting membrane potential, −70 mV. The distribution of ions that do not directly affect the resting membrane potential, such as Cl⁻, is not shown. The voltage-gated ion channels open and close when the membrane potential changes.

© Cengage Learning 2017

Labels in figure:
- Na⁺/K⁺ pump
- 3 Na⁺ out
- Voltage-gated K⁺ channel (closed)
- Voltage-gated Na⁺ channel (closed)
- Axon plasma membrane
- K⁺
- Na⁺
- 2 K⁺ in
- K⁺
- Na⁺
- Axon
- A⁻
- Anions (negatively charged proteins, amino acids, and other molecules) that cannot pass through membrane

Charged Particle Concentrations (mM)		
	Inside	Outside
Na⁺	15	150
K⁺	150	5
A⁻	65	0

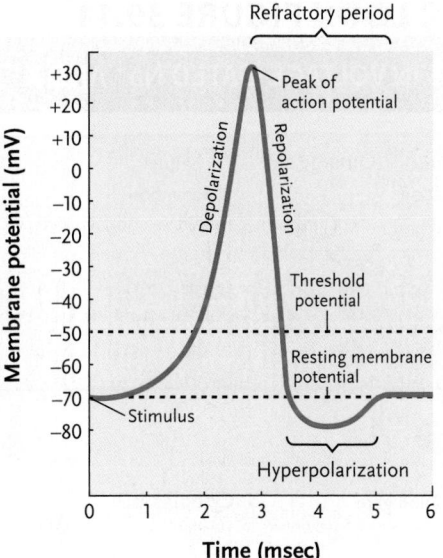

FIGURE 39.10 Changes in membrane potential during an action potential.

© Cengage Learning 2017

Labels in figure:
- Refractory period
- Peak of action potential
- Depolarization
- Repolarization
- Threshold potential
- Resting membrane potential
- Stimulus
- Hyperpolarization
- Membrane potential (mV)
- Time (msec)

a level known as the **threshold potential,** typically 10 to 20 mV more positive than the resting membrane potential. Once the threshold is reached, the action potential fires—and the membrane potential suddenly increases. In less than 1 millisecond (1 msec = one-thousandth of a second), it rises so high that the inside of the plasma membrane becomes positive due to an influx of positive ions across the cell membrane, momentarily reaching a value of +30 mV or more. The potential then falls again rapidly, in many cases dropping to about −80 mV before rising again to the resting membrane potential. When the potential is below the resting value, the membrane is said to be **hyperpolarized** (also called the *undershoot*). The entire change, from initiation of the action potential to the return to the resting membrane potential, may take as little as 1 msec in the fastest neurons. Action potentials take the same basic form in neurons of all types, with differences in the values of the resting membrane potential and the peak of the action potential, and in the time required to return to the resting membrane potential.

An action potential is produced only if a stimulus that causes positive charges from outside the neuron to flow inward is strong enough to cause depolarization that reaches the threshold. This is referred to as the **all-or-nothing principle;** once triggered, the changes in membrane potential take place independently of the strength of the stimulus.

Beginning at the peak of an action potential, the membrane enters a **refractory period** that consists of two phases. During the initial phase, the cell cannot be restimulated. This is followed by a phase during which the threshold required for generation of an action potential is much higher than normal. The refractory period lasts until the membrane has stabilized at the resting membrane potential. As we shall see, the refractory period keeps impulses traveling in a one-way direction in neurons.

The Action Potential Is Produced by Ion Movements through the Plasma Membrane

The action potential is produced by movements of Na⁺ and K⁺ through the plasma membrane. The movements are controlled by specific **voltage-gated ion channels,** membrane-embedded proteins that open and close as the membrane potential changes (see Figure 39.9). Voltage-gated Na⁺ channels have two gates, an *activation gate* and an *inactivation gate,* whereas voltage-gated K⁺ channels have one gate, an *activation gate.*

Figure 39.11 shows how the two voltage-gated ion channels operate to generate an action potential. At the end of an action potential, the membrane potential has returned to its resting state, but the ion distribution has changed slightly. That is, some Na⁺ ions have entered the cell, and some K⁺ ions have left the cell—but not many, relative to the total number of ions, and the distribution is not altered enough to prevent other action potentials from occurring. In the long term, the Na⁺/K⁺ pumps restore the Na⁺ and K⁺ ions to their original locations.

The original theory for the ion movements involved in generating an action potential came from electrophysiology experiments involving electrodes inserted into giant squid axons (~0.5 mm diameter) performed by Alan Hodgkin and his student Andrew Huxley of the University of Cambridge, UK. Hodgkin and Huxley received a Nobel Prize in 1963 "for their discoveries concerning the ionic mechanisms involved in excitation and inhibition in the peripheral and central portions of the nerve cell membrane."

Information about how ions flowing through channels can change membrane potential has also come from experiments

CHANGES IN VOLTAGE-GATED Na⁺ AND K⁺ CHANNELS THAT PRODUCE THE ACTION POTENTIAL

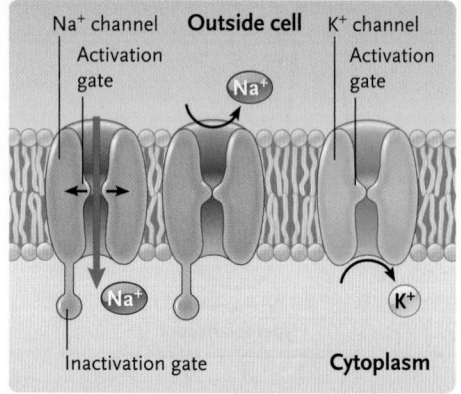

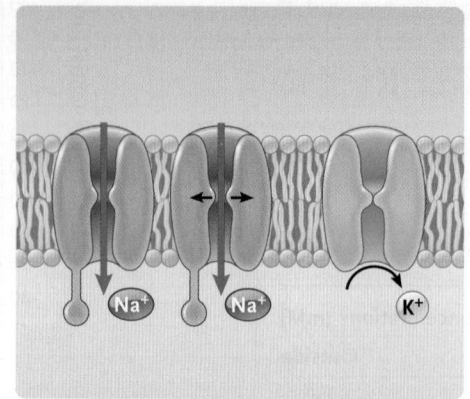

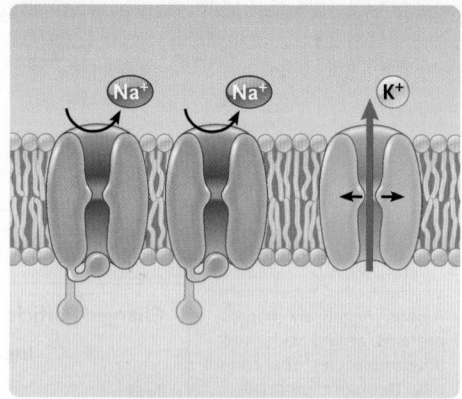

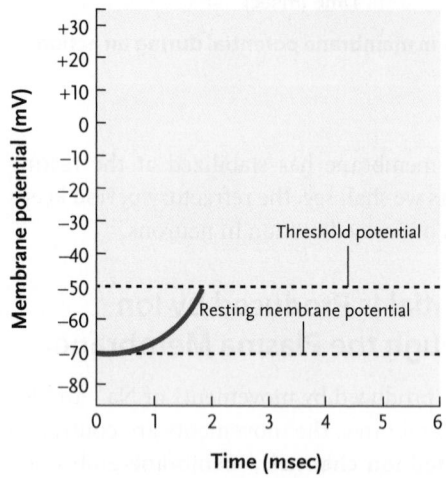

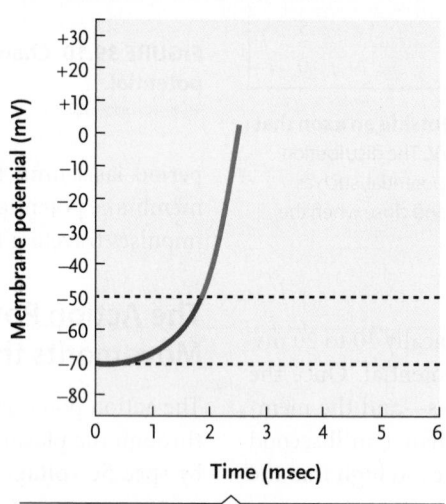

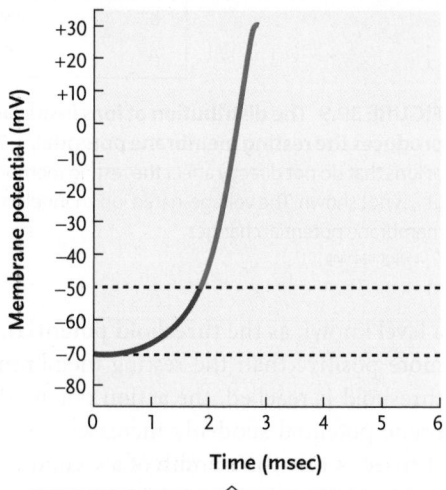

1 When the membrane is at the resting potential, the activation gates of both the Na⁺ and K⁺ channels are closed. As a depolarizing stimulus raises the membrane potential to threshold, the activation gates of the Na⁺ channels open, causing a burst of Na⁺ ions to flow into the axon along their concentration gradient.

2 Once above the threshold, more Na⁺ channels open (as a result of positive feedback), causing a rapid inward flow of positive charges that raises the membrane potential toward the peak of the action potential.

3 The onset of Na⁺ channel inactivation (resembling putting a stopper into a sink) stops the inward flow of Na⁺ and causes the action potential to peak. At the same time, activated K⁺ channels allow K⁺ to flow outward in response to their concentration gradient. The refractory period now begins.

using the **patch-clamp technique.** In the *patch* part of the technique, a micropipette with a tip 1 to 3 μm in diameter is touched to the plasma membrane of a neuron (or other cell type). The contact seals the membrane to the micropipette and, when the micropipette is pulled away, a patch of membrane with one or a few ion channels comes with it. The *clamp* part of the technique refers to a voltage clamp, in which an electronic device holds the membrane potential of the patch at a steady value chosen by the investigator. The investigator can add a stimulus that is expected to open or close ion channels. The amount of current the clamping device needs to keep the voltage constant is directly related to the number and charge of the ions moving through the channels and, hence, measures channel activity.

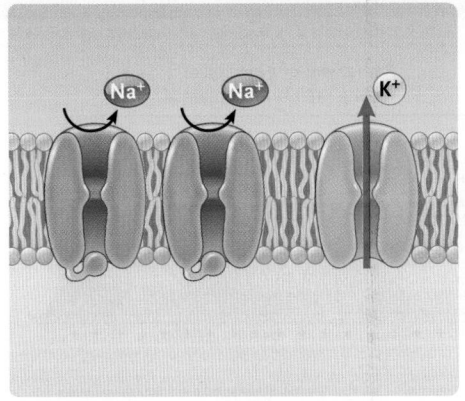

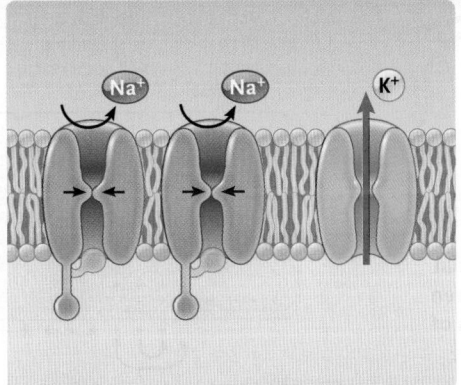

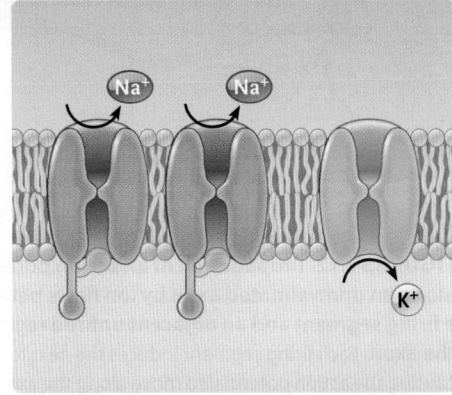

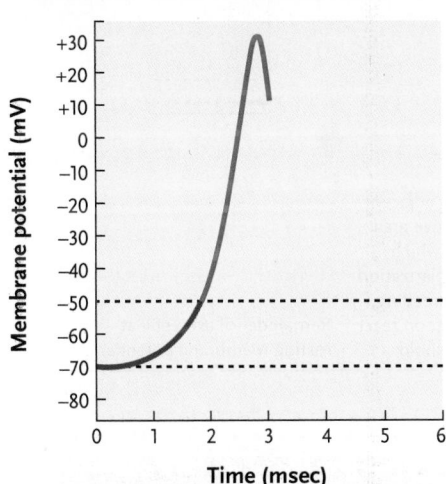

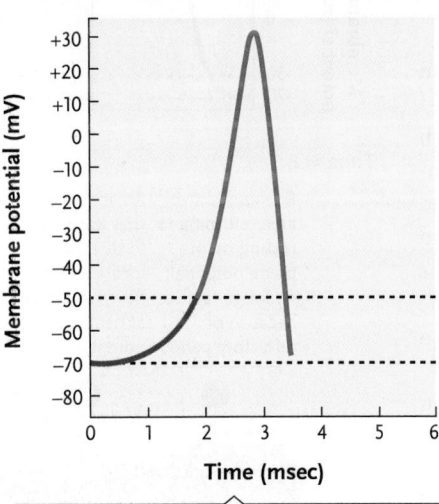

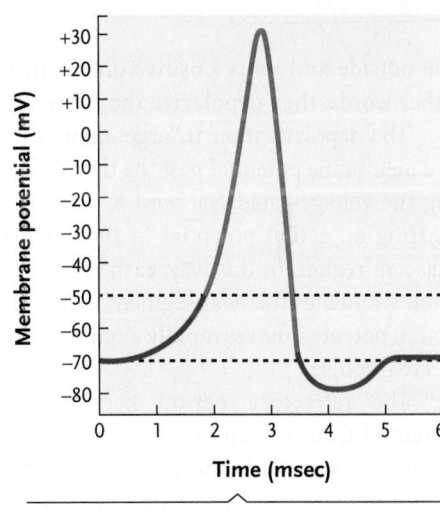

4 The outward flow of K⁺ along its concentration gradient compensates for the inward movement of Na⁺ ions and causes the membrane potential to begin to fall.

5 As the membrane potential reaches the resting value, the activation gate of the Na⁺ channel closes and the inactivation gate opens. In most neurons, the K⁺ activation gate remains open and K⁺ continues to flow outward for a brief time after the membrane reaches the resting membrane potential. This excess outward flow causes hyperpolarization, in which the membrane potential dips briefly below the resting membrane potential (the dip is seen in step 6).

6 Closure of the K⁺ activation gate stabilizes the membrane potential at the resting value. The refractory period has now ended, and the membrane is ready for another action potential.

SUMMARY An action potential is generated when a depolarizing stimulus initiates a specific sequence of opening and closing voltage-gated Na⁺ and K⁺ channels in the axon's plasma membrane. Na⁺ flows into the axon raising the membrane potential until the peak of the action potential, at which point Na⁺ channels are blocked and K⁺ begins to flow outward through K⁺ channels. The outflow of K⁺ lowers the membrane potential. When the resting membrane potential is reached, K⁺ flow stops.

think like a scientist   How might the action potential curve change if you inhibited the voltage-gated K⁺ channel significantly?

Neural Impulses Move by Propagation of Action Potentials

Once an action potential is initiated in a neuron, it is propagated along the surface of the axon as an automatic wave of depolarization traveling away from the stimulation point **(Figure 39.12)**. In a segment of an axon that is generating an

action potential, the outside of the membrane becomes temporarily negative and the inside positive. Because opposites attract, as the region outside becomes negative, local current flow occurs between the area undergoing an action potential and the adjacent downstream inactive area both inside and outside the membrane (*arrows*, Figure 39.12). This current flow makes nearby regions of the axon membrane less positive on

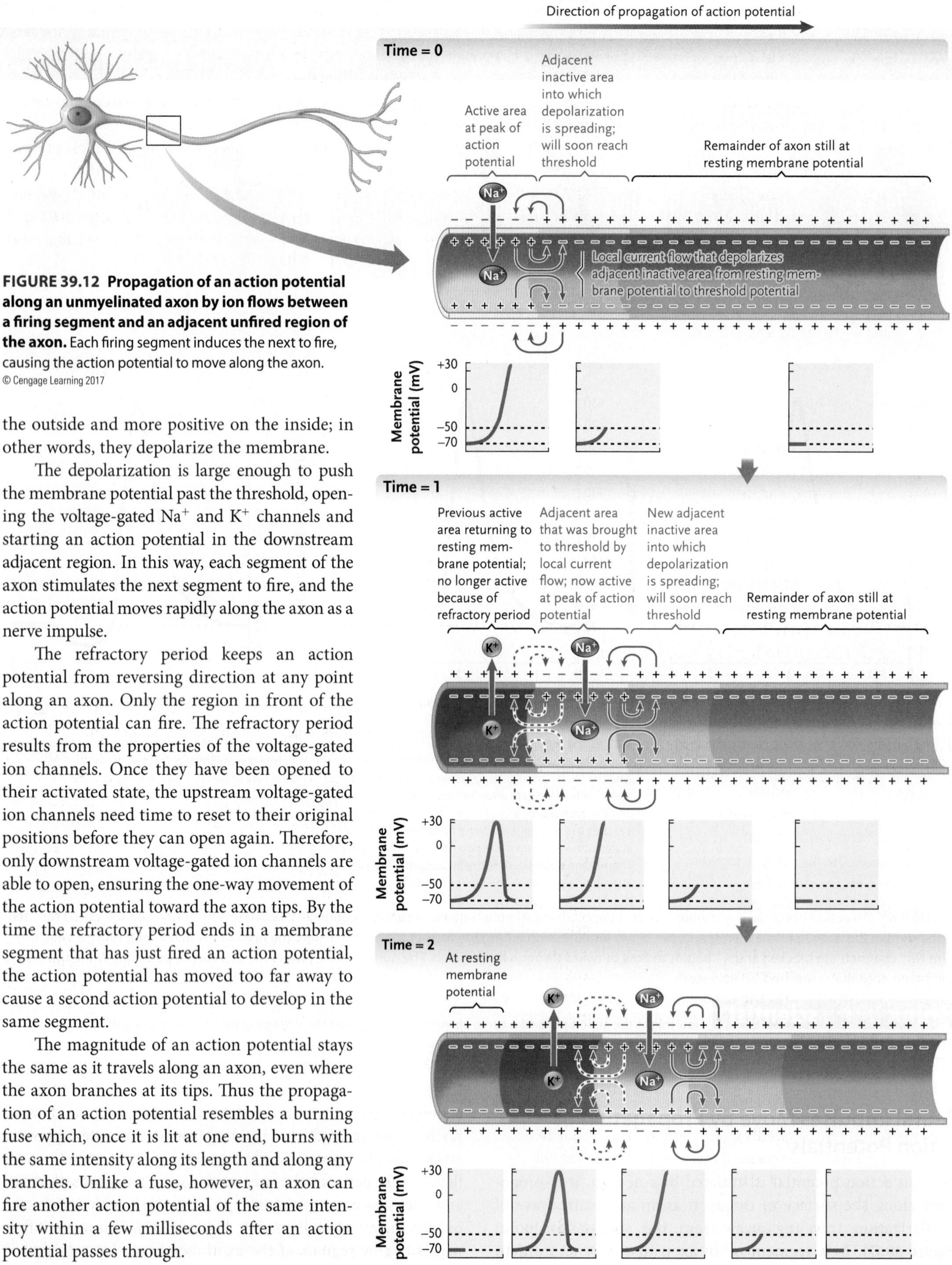

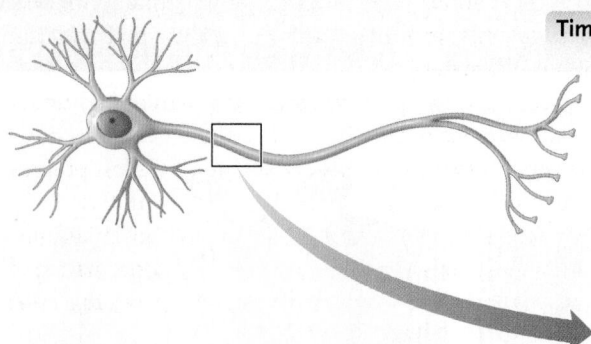

Direction of propagation of action potential →

Time = 0

Active area at peak of action potential

Adjacent inactive area into which depolarization is spreading; will soon reach threshold

Remainder of axon still at resting membrane potential

Na⁺

Na⁺

Local current flow that depolarizes adjacent inactive area from resting membrane potential to threshold potential

Membrane potential (mV): +30, 0, −50, −70

Time = 1

Previous active area returning to resting membrane potential; no longer active because of refractory period

Adjacent area that was brought to threshold by local current flow; now active at peak of action potential

New adjacent inactive area into which depolarization is spreading; will soon reach threshold

Remainder of axon still at resting membrane potential

K⁺

K⁺

Na⁺

Na⁺

Membrane potential (mV): +30, 0, −50, −70

Time = 2

At resting membrane potential

K⁺

K⁺

Na⁺

Na⁺

Membrane potential (mV): +30, 0, −50, −70

FIGURE 39.12 Propagation of an action potential along an unmyelinated axon by ion flows between a firing segment and an adjacent unfired region of the axon. Each firing segment induces the next to fire, causing the action potential to move along the axon.
© Cengage Learning 2017

the outside and more positive on the inside; in other words, they depolarize the membrane.

The depolarization is large enough to push the membrane potential past the threshold, opening the voltage-gated Na⁺ and K⁺ channels and starting an action potential in the downstream adjacent region. In this way, each segment of the axon stimulates the next segment to fire, and the action potential moves rapidly along the axon as a nerve impulse.

The refractory period keeps an action potential from reversing direction at any point along an axon. Only the region in front of the action potential can fire. The refractory period results from the properties of the voltage-gated ion channels. Once they have been opened to their activated state, the upstream voltage-gated ion channels need time to reset to their original positions before they can open again. Therefore, only downstream voltage-gated ion channels are able to open, ensuring the one-way movement of the action potential toward the axon tips. By the time the refractory period ends in a membrane segment that has just fired an action potential, the action potential has moved too far away to cause a second action potential to develop in the same segment.

The magnitude of an action potential stays the same as it travels along an axon, even where the axon branches at its tips. Thus the propagation of an action potential resembles a burning fuse which, once it is lit at one end, burns with the same intensity along its length and along any branches. Unlike a fuse, however, an axon can fire another action potential of the same intensity within a few milliseconds after an action potential passes through.

Due to the all-or-nothing principle of action potential generation, the intensity of a stimulus is reflected in the *frequency* of action potentials—the greater the stimulus, the more action potentials per second, up to a limit depending on the axon type—rather than by the change in membrane potential. For most neuron types, the limit is about 100 action potentials per second.

Both natural and synthetic substances target specific parts of the mechanism generating action potentials. Local anesthetics, such as procaine and lidocaine, bind to voltage-gated Na^+ channels and block their ability to permit ion flow. As a result, sensory nerves in the anesthetized region cannot transmit pain signals. The potent poison of the pufferfish, tetrodotoxin, also blocks voltage-gated Na^+ channels in neurons, potentially causing muscle paralysis and death.

Saltatory Conduction Increases Propagation Rate in Small-Diameter Axons

In the propagation pattern shown in Figure 39.12, an action potential spreads along every segment of the membrane along the length of the axon. For this type of action potential propagation, the rate of conduction increases with the diameter of the axon because a larger axon offers less resistance to the signal. Axons with a very large diameter have evolved in invertebrates such as lobsters, earthworms, and squids, as well as a few marine fishes. Giant axons typically carry signals that produce an escape or prey–capture response, such as the sudden flexing of the tail (abdomen) in lobsters that propels the animal backward.

Although large-diameter axons can propagate impulses as rapidly as 25 m/sec (over twice the speed of the world record 100-meter dash), they take up a great deal of space. In vertebrate central nervous systems with many more neurons, natural selection has led to a mechanism that allows small-diameter axons to conduct impulses rapidly. The mechanism, called **saltatory conduction** (*saltare* = to hop, leap), allows action potentials to "hop" rapidly along axons instead of burning smoothly like a fuse.

Saltatory conduction depends on the insulating myelin sheath that forms around some axons and on the periodic interruption of the sheath at the nodes of Ranvier, which exposes the axon membrane to extracellular fluids. Voltage-gated Na^+ and K^+ channels crowded into the neuronal membrane at the nodes allow action potentials to develop at these positions **(Figure 39.13)**. The inward movement of Na^+ ions produces depolarization. Positive ions spread rapidly with little loss, due to the myelin insulation, to the next node where they cause depolarization, inducing an action potential at that node. As this mechanism repeats, the action potential jumps rapidly along the axon from node to node. Saltatory conduction proceeds at rates up to 130 m/sec, whereas an unmyelinated axon of the same diameter conducts action potentials at about 1 m/sec.

In humans, the optic nerve leading from the eye to the brain is only 3 mm in diameter but is packed with more than a million axons. If those axons were unmyelinated, each would have to be about 100 times thicker to conduct impulses at the same velocity, producing an optic nerve about 300 mm (12 inches) in diameter.

The disease *multiple sclerosis* (*sklerosis* = hardening) underscores the importance of myelin sheaths to the operation of the vertebrate nervous system. In this disease, myelin is lost progressively from axons and replaced by hardened scar tissue. The changes block or slow the transmission of action potentials, producing numbness, muscular weakness, faulty coordination of movements, and paralysis that worsens as the disease progresses.

STUDY BREAK 39.2

1. What mechanism ensures that an electrical impulse in a neuron is conducted in only one direction down the axon?
2. How does having a myelin sheath affect the conduction of impulses in neurons?

THINK OUTSIDE THE BOOK

Individually or collaboratively, explore the research literature to find two examples of neurophysiological research that used tetrodotoxin (TTX) in the experiments. For the two projects, outline the hypothesis being tested or question being asked, how TTX was used, and the key results and conclusions.

39.3 Transmission across Chemical Synapses

Although action potentials are transmitted directly across electrical synapses, they cannot jump across the cleft in a chemical synapse. Instead, the arrival of an action potential causes neurotransmitter molecules to be released by exocytosis from the plasma membrane of the axon terminal, called the *presynaptic membrane* **(Figure 39.14)**. The neurotransmitter diffuses across the cleft and alters ion conduction by activating receptors in the plasma membrane of the postsynaptic cell.

Neurotransmitter communication from presynaptic to postsynaptic cells is a specialized case of cell-to-cell communication and signal transduction (see Chapter 9). Neurobiologists study this phenomenon to understand the function of neurons, but some of the details are similar in many other types of cells.

Two types of receptors bind neurotransmitters, with different consequences:

- For **ionotropic neurotransmitter receptors,** a neurotransmitter binds to the receptor in the postsynaptic membrane, causing an associated ion channel gate to open or close, thereby altering the transmembrane flow of a specific ion or ions in the postsynaptic cell. Figure 39.14 illustrates this type of receptor. (This is an example of a **ligand-gated ion channel,** a channel that opens or closes when a specific ligand molecule [*ligandum* = binding], here a neurotransmitter, binds to the channel.) The time between arrival of an

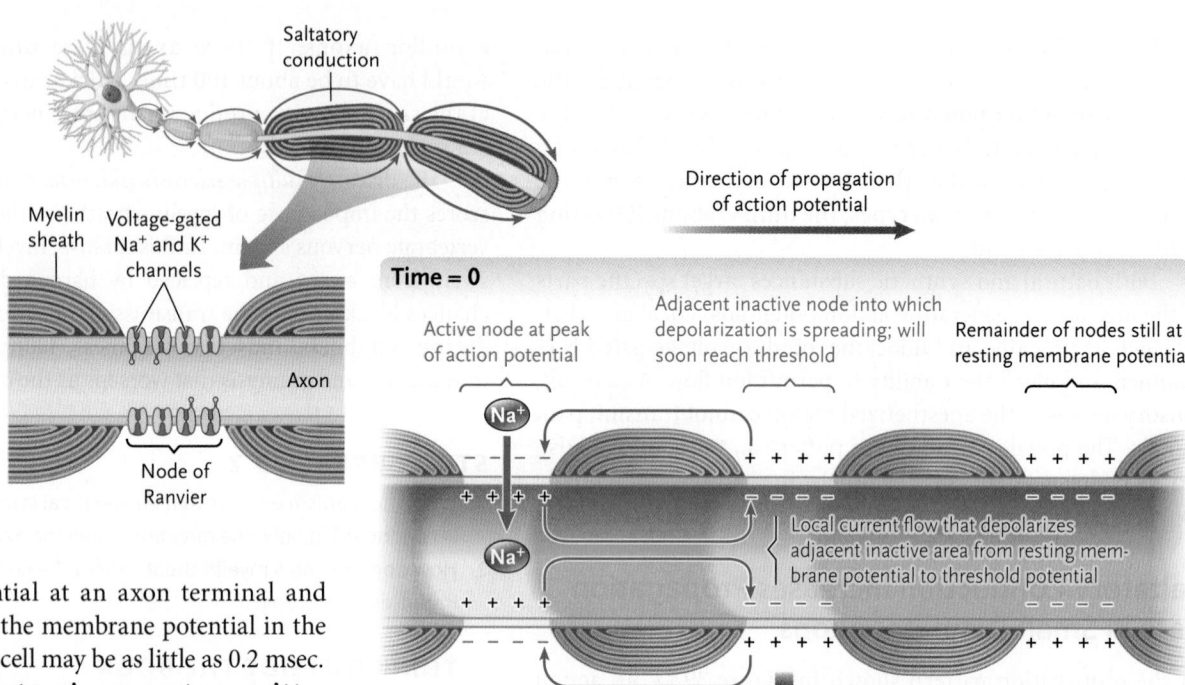

Saltatory conduction

Myelin sheath

Voltage-gated Na⁺ and K⁺ channels

Na^+ and K^+ channels

Axon

Node of Ranvier

action potential at an axon terminal and alteration of the membrane potential in the postsynaptic cell may be as little as 0.2 msec.

- For **metabotropic neurotransmitter receptors,** a neurotransmitter works indirectly and, hence, more slowly (on the order of hundreds of milliseconds). For this type of receptor, which is typically a G-protein–coupled receptor, the neurotransmitters act as *first messengers.* That is, the neurotransmitters bind to the receptor, which activates it and triggers generation of a second messenger such as cyclic AMP or other processes (see Section 9.3). The cascade of *second-messenger* reactions opens or closes ion-conducting channels in the postsynaptic membrane. Neurotransmitters that bind to metabotropic neurotransmitter receptors typically have effects that may last for minutes or hours.

Some neurotransmitters bind only to one or the other of the receptor types, whereas other neurotransmitters bind to both receptor types, bringing about different responses depending on which receptor is involved.

The time required for the release, diffusion, and binding of neurotransmitters across chemical synapses delays transmission as compared with the almost instantaneous transmission of impulses across electrical synapses. However, communication through chemical synapses allows neurons to receive inputs from hundreds to thousands of axon terminals at the same time. Some neurotransmitters have stimulatory effects and others have inhibitory effects, with the information received at a postsynaptic membrane integrated to produce a response. Thus, by analogy, communication by electrical synapses

Direction of propagation of action potential

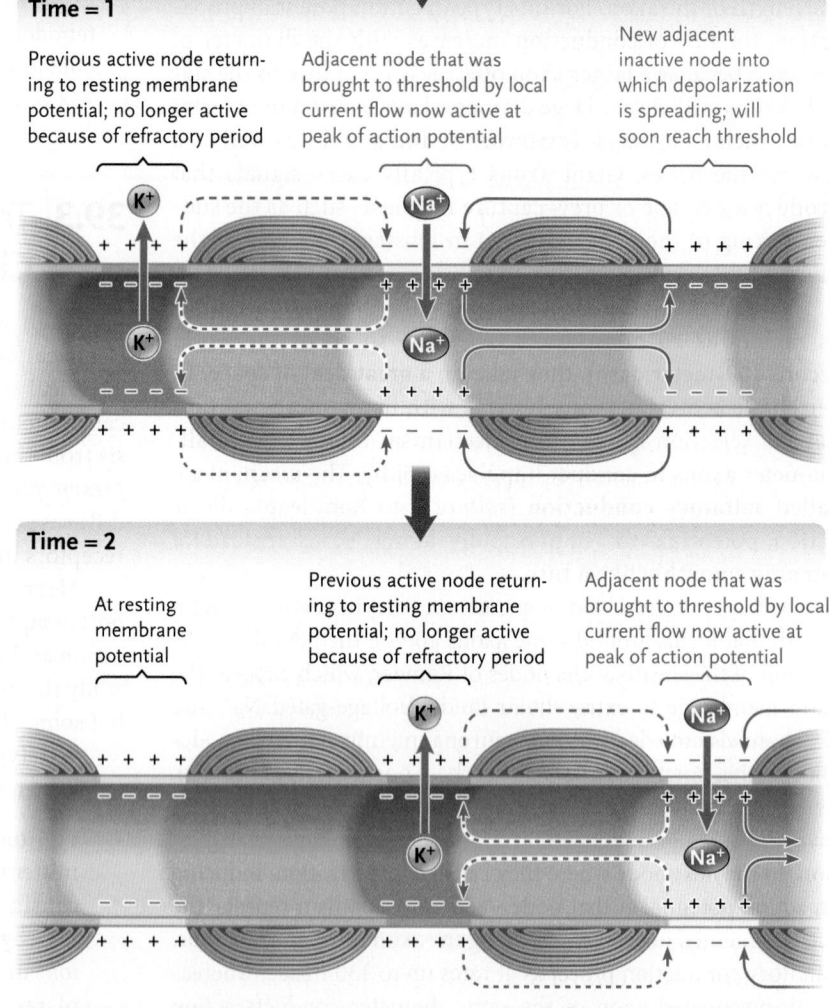

Time = 0

Active node at peak of action potential

Adjacent inactive node into which depolarization is spreading; will soon reach threshold

Remainder of nodes still at resting membrane potential

Local current flow that depolarizes adjacent inactive area from resting membrane potential to threshold potential

Time = 1

Previous active node returning to resting membrane potential; no longer active because of refractory period

Adjacent node that was brought to threshold by local current flow now active at peak of action potential

New adjacent inactive node into which depolarization is spreading; will soon reach threshold

Time = 2

At resting membrane potential

Previous active node returning to resting membrane potential; no longer active because of refractory period

Adjacent node that was brought to threshold by local current flow now active at peak of action potential

FIGURE 39.13 Saltatory conduction of the action potential by a myelinated axon. The action potential jumps from node to node, greatly increasing the speed at which it travels along the axon.

© Cengage Learning 2017

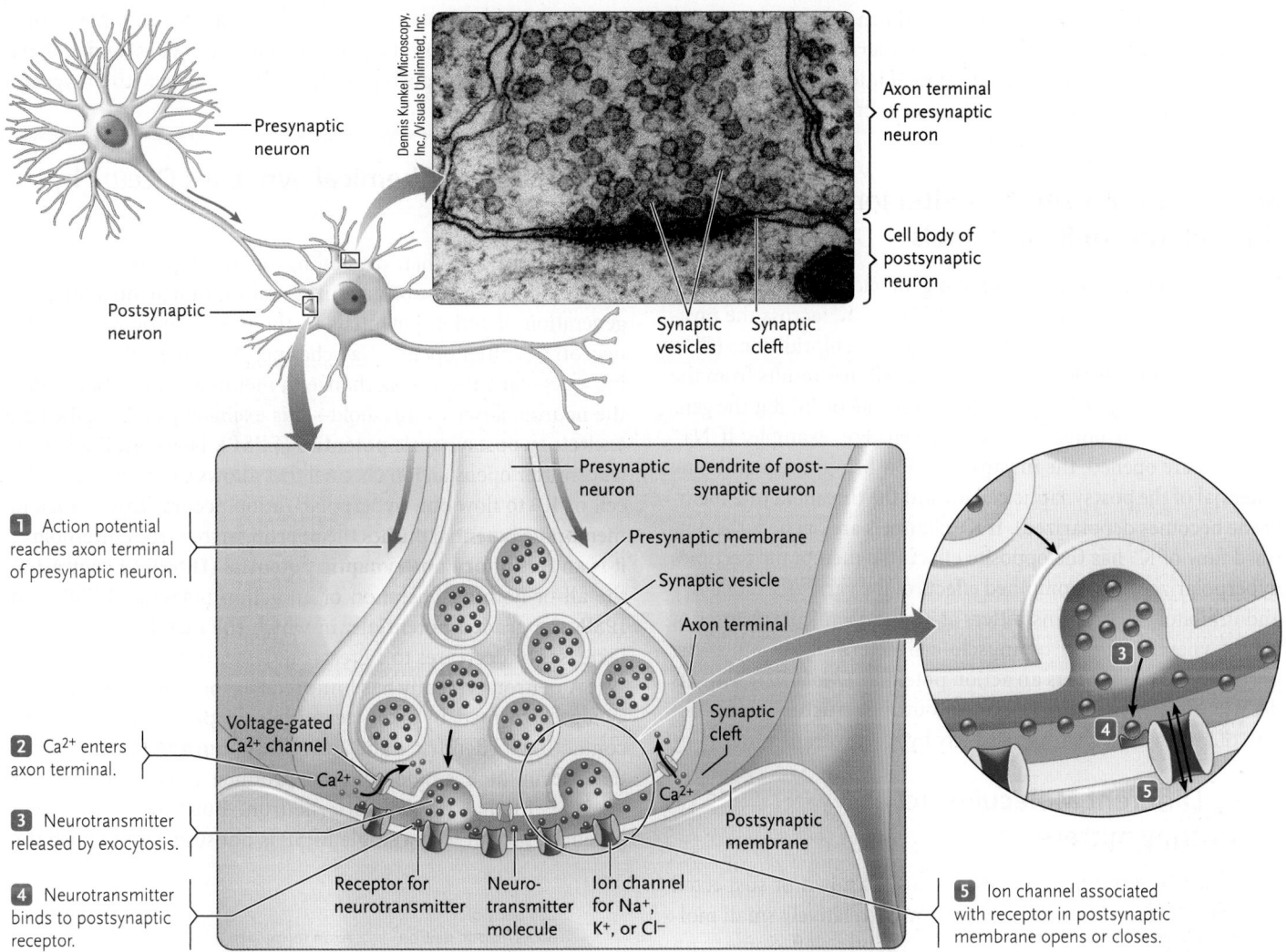

FIGURE 39.14 Structure and function of chemical synapses.
© Cengage Learning 2017

Labels in figure:

Presynaptic neuron

Postsynaptic neuron

Dennis Kunkel Microscopy, Inc./Visuals Unlimited, Inc.

Axon terminal of presynaptic neuron

Cell body of postsynaptic neuron

Synaptic vesicles

Synaptic cleft

1 Action potential reaches axon terminal of presynaptic neuron.

2 Ca^{2+} enters axon terminal.

3 Neurotransmitter released by exocytosis.

4 Neurotransmitter binds to postsynaptic receptor.

Presynaptic neuron

Dendrite of post-synaptic neuron

Presynaptic membrane

Synaptic vesicle

Axon terminal

Voltage-gated Ca^{2+} channel

Ca^{2+}

Synaptic cleft

Ca^{2+}

Postsynaptic membrane

Receptor for neurotransmitter

Neuro-transmitter molecule

Ion channel for Na^+, K^+, or Cl^-

5 Ion channel associated with receptor in postsynaptic membrane opens or closes.

---

resembles the effect of simply touching one wire to another, whereas communication by neurotransmitters through both direct and indirect receptor responses resembles the integration of multiple inputs by a computer chip.

Neurotransmitters Are Released by Exocytosis

Neurotransmitters are stored in synaptic vesicles in the cytoplasm of an axon terminal. The arrival of an action potential at the terminal releases the neurotransmitters by *exocytosis*: the vesicles fuse with the presynaptic membrane and release the neurotransmitter molecules into the synaptic cleft.

The release of neurotransmitters from synaptic vesicles depends on voltage-gated Ca^{2+} channels in the plasma membrane of an axon terminal (see Figure 39.14). Ca^{2+} ions are constantly pumped out of all animal cells by an active transport protein in the plasma membrane, keeping their concentration higher outside than inside. (That is, the equilibrium potential for Ca^{2+}, E_{Ca}, is strongly positive.) As an action potential arrives, the change in membrane potential opens the Ca^{2+} channel gates in the axon terminal, allowing Ca^{2+} to

flow back into the cytoplasm. The rise in Ca^{2+} concentration triggers a protein in the membrane of the synaptic vesicle that allows the vesicle to fuse with the plasma membrane, releasing neurotransmitter molecules into the synaptic cleft.

Each action potential arriving at a synapse typically causes approximately the same number of synaptic vesicles to release their neurotransmitter molecules. For example, arrival of an action potential at one type of synapse causes about 300 synaptic vesicles to release a neurotransmitter called **acetylcholine.** Each vesicle contains about 10,000 molecules of the neurotransmitter, giving a total of some 3 million acetylcholine molecules released into the synaptic cleft by each arriving action potential.

When a stimulus is no longer present, action potentials are no longer generated and a response is no longer needed. In this case, a series of events prevents continued transmission of the signal. When action potentials stop arriving at the axon terminal, the voltage-gated Ca^{2+} channels close and the Ca^{2+} in the axon cytoplasm is quickly pumped to the outside. The drop in cytoplasmic Ca^{2+} stops vesicles from fusing with the presynaptic membrane, and no further neurotransmitter molecules are released. Any free neurotransmitter molecules remaining in the cleft quickly diffuse away, are broken down by enzymes in the cleft, or are pumped back into the axon terminals or into glial cells by active transport. Transmission of information across the synaptic cleft ceases within milliseconds after action potentials stop arriving at the axon terminal.

Most Neurotransmitters Alter Ion Flow through Na$^+$ or K$^+$ Channels

Neurotransmitters work by opening or closing ion channels. Most of these ion channels conduct Na$^+$ or K$^+$ across the post-synaptic membrane, although some regulate chloride ions (Cl$^-$). The altered ion flow in the postsynaptic cell that results from the opening or closing of the gates may stimulate or inhibit the generation of action potentials by that cell. For example, if Na$^+$ channels are opened, the inward Na$^+$ flow brings the membrane potential of the postsynaptic cell toward the threshold (the membrane becomes depolarized). If K$^+$ channels are opened, the outward flow of K$^+$ has the opposite effect (the membrane becomes hyperpolarized). The combined effects of the various stimulatory and inhibitory neurotransmitters at all the chemical synapses of a postsynaptic neuron or muscle cell determine whether the postsynaptic cell triggers an action potential. (*Molecular Insights* describes experiments that worked out the structure and function of an ion channel gated directly by a neurotransmitter.)

Many Different Molecules Act as Neurotransmitters

In all, nearly 100 different substances are known or suspected to be neurotransmitters. Most of them are relatively small molecules that diffuse rapidly across the synaptic cleft. Some axon terminals release only one type of neurotransmitter; others release several types. Depending on the type of receptor to which it binds, the same neurotransmitter may depolarize or hyperpolarize potentials in the postsynaptic cell. **Table 39.1** lists a number of neurotransmitters, the types of molecules they represent, their sites and type of action, and some drugs and other molecules that affect neurotransmission.

STUDY BREAK 39.3

1. What features characterize a substance as a neurotransmitter?
2. Describe how a neurotransmitter that binds to an ionotropic neurotransmitter receptor in a presynaptic neuron controls action potentials in a postsynaptic neuron.

39.4 Integration of Incoming Signals by Neurons

Most neurons receive a multitude of stimulatory and inhibitory signals carried by both direct and indirect neurotransmitters. These signals are integrated by the postsynaptic neuron into a response that reflects their combined effects. The integration depends primarily on the patterns, number, types, and activity of the synapses the postsynaptic neuron makes with presynaptic neurons.

Integration at Chemical Synapses Occurs by Summation

You have learned that, depending on the type of receptor to which it binds, a neurotransmitter may stimulate or inhibit the generation of action potentials in the postsynaptic neuron. If a neurotransmitter opens a Na$^+$ channel, Na$^+$ enters the cell, causing a depolarization. This change in membrane potential pushes the neuron closer to threshold—it is excitatory and is called an **excitatory postsynaptic potential (EPSP)**. However, if a neurotransmitter opens an ion channel that allows Cl$^-$ to flow into the cell or K$^+$ to flow out, hyperpolarization occurs. This change in membrane potential pushes the neuron farther from threshold—it is an **inhibitory postsynaptic potential (IPSP)**. In contrast to the all-or-nothing operation of an action potential, EPSPs and IPSPs are **graded potentials,** in which the membrane potential increases or decreases to a greater or lesser degree.

A neuron typically has hundreds to thousands of chemical synapses formed by axon terminals of presynaptic neurons contacting its dendrites and cell body **(Figure 39.15)**. The events that occur at a single synapse produce either an EPSP or an IPSP in that postsynaptic neuron. But how is an action potential produced if a single EPSP is not sufficient to push the

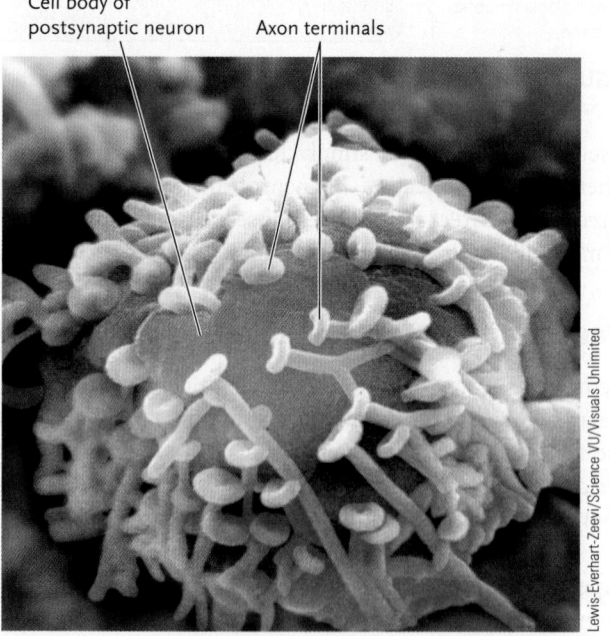

Cell body of postsynaptic neuron   Axon terminals

FIGURE 39.15 The multiple chemical synapses relaying signals to a neuron. The drying process used to prepare the neuron for electron microscopy has toppled the axon terminals and pulled them away from the neuron's surface.

Lewis-Everhart-Zeevi/Science VU/Visuals Unlimited

TABLE 39.1 | Examples of Neurotransmitters and Drugs and Chemicals That Affect Neurotransmission

Neurotransmitter	Site(s) of Action	Action	Examples of Drugs and Other Molecules That Affect Neurotransmission
Acetylcholine			
Acetylcholine	Between some neurons of CNS and at neuromuscular junctions in PNS	Mostly excitatory; inhibitory at some sites	• Curare—blocks release • Atropine—blocks receptors • Nicotine—activates receptors
Monoamines (Biogenic Amines)			
Norepinephrine	CNS interneurons involved in diverse brain and body functions, such as memory, mood, sensory perception, muscle movements, maintenance of blood pressure, and sleep	Excitatory or inhibitory	• Amphetamines—stimulate release of norepinephrine and dopamine and block their reuptake • Methylphenidate (Ritalin)—increases release • Certain antidepressants—prevent reuptake
Dopamine	CNS interneurons involved in many pathways similar to norepinephrine	Mostly excitatory	• Amphetamines—stimulate release of norepinephrine and dopamine and block their reuptake • Cocaine—stimulates release of norepinephrine and dopamine and blocks dopamine reuptake
Serotonin	CNS interneurons in a number of pathways, including those regulating appetite, reproductive behavior, muscular movement, sleep, and some emotional states	Inhibitory or modulatory	• Fluoxetine (Prozac), sertraline (Zoloft), paroxetine (Paxil)—block serotonin reuptake
Amino Acids			
Glutamate	Many CNS pathways, including those involved in vital brain functions such as memory and learning	Excitatory	• Phencyclidine (PCP or angel dust)—blocks receptor
Gamma-aminobutyric acid (GABA)	Many CNS pathways; often acts in same circuits as glutamate	Inhibitory (main inhibitory neurotransmitter in mammalian CNS)	• Alcohol (ethanol)—stimulates GABA neurotransmission, increases dopamine neurotransmission, and inhibits glutamate neurotransmission • Some antianxiety/sedative drugs such as diazepam (Valium) and flunitrazepam (Rohypnol, the "date rape" drug) increase GABA neurotransmission • Tetanus toxin blocks GABA release in synapses that control muscle contraction
Neuropeptides			
Endorphins ("endogenous morphines")	Most act on CNS and PNS, as well as on effectors such as muscle, reducing pain and, in some cases, also inducing euphoria	Inhibitory—modulate pain response	
Enkephalins (subclass of endorphins)	CNS	Inhibitory—modulate pain response	
Substance P	Released by special, unmyelinated sensory neurons in spinal cord	Excitatory—pain perception	
Gaseous Neurotransmitters			
Nitrous oxide (NO)	Diffuses across cell membranes in PNS rather than being released at synapses	Modulatory—relaxes smooth muscles in walls of blood vessels	• Sildenafil citrate (Viagra, an impotency drug)—aids erection by inhibiting enzyme that normally reduces effects of NO in vascular beds of penis

Dissecting Neurotransmitter Receptor Functions

Many ionotropic neurotransmitter receptors are ion channels that are opened or closed by the binding of a neurotransmitter molecule. Each of these receptors has two domains: a hydrophilic portion on the outside surface of the plasma membrane that binds the neurotransmitter, and a hydrophobic transmembrane portion that anchors the receptor in the plasma membrane and forms the ion-conducting channel.

Research Question
Are the two primary activities of these receptors—binding neurotransmitters and conducting ions—dependent on parts of the protein that work independently, or do they each require the entire protein structure?

Experiment
To answer the question, Jean-Luc Eiselé and his coworkers at Institut Pasteur, Paris, and colleagues at Centre Médical Universitaire, Geneva, Switzerland, constructed artificial receptors using regions of the ionotropic

receptors for two different neurotransmitters. The two receptors are not closely related in amino acid sequence, bind different neurotransmitters, and react differently to calcium ions. One receptor, the nicotinic acetylcholine receptor (nAChR), is activated by acetylcholine and by nicotine, and ion conduction by the receptor is enhanced by Ca^{2+}. The other receptor, the serotonin receptor ($5HT_3$), is activated by serotonin, and Ca^{2+} ions block its channel and stop ion conduction.

To create the artificial receptors, the investigators used molecular techniques to break each of the genes for the nicotinic acetylcholine and serotonin receptors into two sequences, one encoding the hydrophilic domain and the other encoding the hydrophobic domain. They then reassembled the parts so that in the protein encoded by the chimeric gene, the part of the acetylcholine receptor located on the membrane surface was joined to the

transmembrane channel of the serotonin receptor (**Figure**). Five versions of the artificial gene, encoding proteins in which the two parts were joined at different positions in the amino acid sequence, were then cloned to increase their quantity and injected into oocytes of the African clawed frog, *Xenopus laevis*. Once in the oocytes, the genes were expressed to produce the chimeric receptor proteins, which became inserted into the oocyte plasma membranes. The properties of the chimeric receptors were then examined.

Results
Of the five artificial receptors, all bound acetylcholine, but only two conducted ions in response to binding the neurotransmitter. Agents that inhibit the normal acetylcholine receptor, such as curare, also inhibited the artificial receptors. Serotonin, in contrast, did not bind nor did it open the receptor channels, and agents that inhibit the normal serotonin receptor had no effect on the artificial receptors. However, elevated Ca^{2+} concentrations blocked the channel, as in the normal serotonin receptor.

Conclusion
The research by Eiselé and his coworkers showed that the parts of a receptor binding a neurotransmitter and conducting ions function independently. Their work also demonstrated the feasibility of constructing chimeric receptors as a means for dissecting the functions of subregions of the receptors.

think like a scientist

As stated in the Results, two of the five artificial receptors conducted ions whereas the other three did not. What is the likely explanation for this result?

Source: J.-L. Eiselé et al. 1993. Chimaeric nicotinic–serotonergic receptor combines distinct ligand binding and channel specificities. *Nature* 366:479–483.

FIGURE Schematic for construction of chimeric receptor genes from parts of the nAChR and serotonin receptor genes.
© Cengage Learning 2017

© Cengage Learning 2017

postsynaptic neuron to threshold? The answer involves summation of the inputs received through those many chemical synapses formed by presynaptic neurons. At any given time, some or many of the presynaptic neurons may be firing, producing EPSPs and/or IPSPs in the postsynaptic neuron. The sum of all the EPSPs and IPSPs at a given time determines the total potential in the postsynaptic neuron and, therefore, how that neuron responds. **Figure 39.16** shows, in a greatly simplified way, the effects of EPSPs and IPSPs on membrane potential, and how summation of inputs over space and time brings a postsynaptic neuron to threshold. The postsynaptic neuron in the figure has three presynaptic neurons forming synapses

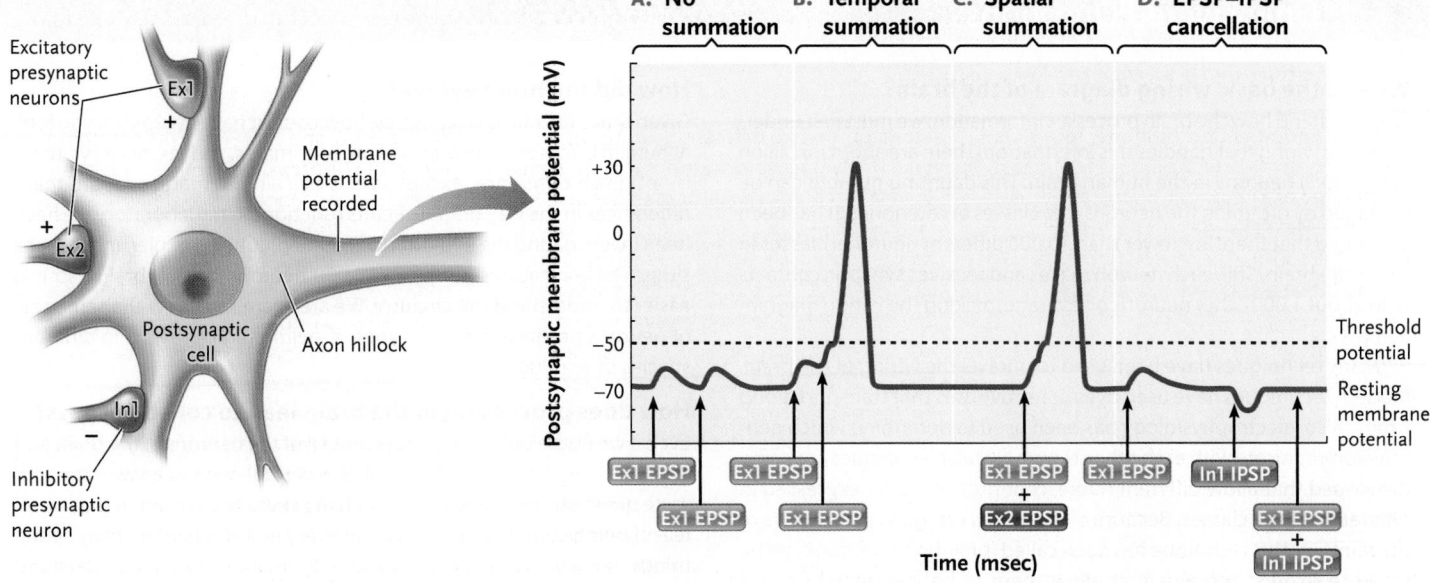

FIGURE 39.16 The effects of EPSPs and IPSPs on membrane potential, and how summation of inputs over space and time brings a postsynaptic neuron to threshold.

© Cengage Learning 2017

A. No summation: The axon of Ex1 releases a neurotransmitter, which produces an EPSP in the postsynaptic cell. The membrane depolarizes, but not enough to reach threshold. If Ex1 input causes a new EPSP after the first EPSP has died down, it will be of the same magnitude as the first EPSP and no progression toward threshold has taken place—no summation has occurred.

B. Temporal summation: If instead, Ex1 input causes a new EPSP before the first EPSP has died down, the second EPSP will sum with the first and a greater depolarization will have taken place. This summation of two (or more) EPSPs produced by successive firing of a single presynaptic neuron over a short period of time is **temporal summation**. If the total depolarization achieved in this way reaches threshold, an action potential is produced in the postsynaptic neuron.

C. Spatial summation: The postsynaptic neuron may be brought to threshold by **spatial summation**, the summation of EPSPs produced by the simultaneous firing of two different excitatory presynaptic neurons, such as Ex1 and Ex2.

D. Summation resulting in cancellation: EPSPs and IPSPs can sum to cancel each other out. In the example, firing of the excitatory presynaptic neuron Ex1 alone produces an EPSP, firing of presynaptic inhibitory neuron In1 alone produces an IPSP, while firing of Ex1 and In1 simultaneously produces no change in the membrane potential.

with its dendrites and cell body; Ex1 and Ex2 are excitatory neurons, and In1 is an inhibitory neuron. The summation point for EPSPs and IPSPs is the axon hillock of the postsynaptic neuron. A high density of voltage-gated Na⁺ channels occurs in that region, resulting in the lowest threshold potential in the neuron.

The Patterns of Synaptic Connections Contribute to Integration

The total number of connections made by a neuron may be very large. Some single interneurons in the human brain, for example, form as many as 100,000 synapses with other neurons. The number of synapses can change through modification, addition, or removal as animals mature and experience changes in their environments. The combined activities of all the neurons in the nervous system, constantly integrating information from sensory receptors and triggering responses by effectors, control the internal activities of animals and regulate their behavior. This behavior ranges from the simple reflexes of a flatworm to the complex behavior of mammals, including consciousness, emotions, reasoning, and creativity in humans.

Although researchers do not yet understand how processes such as ion flow, synaptic connections, and neural networks produce complex mental activities, they continue to find correspondences between them and the types of neuronal communication described in this chapter. In the next chapter we learn about how nervous systems of animals are organized, and how higher functions such as memory, learning, and consciousness are produced.

STUDY BREAK 39.4

How does a postsynaptic neuron integrate signals carried by direct and indirect neurotransmitters?

What is the basic wiring diagram of the brain?

To understand how the brain processes information, we must first understand the wiring that handles this information. There are about 86 billion (8.6 × 10^10) neurons in the human brain. This daunting number can be managed by grouping the neurons into classes or categories. It has been estimated that there are fewer than 10,000 different neuronal classes in the entire brain. Still, each neuron makes and receives synaptic contacts with about 1,000 other neurons on average, making the wiring diagram complex.

Many techniques have been used to unravel the wiring of the brain. Modern techniques have used dyes or retroviruses that transport along axons. Also, electrophysiology has been used to determine which neurons communicate with each other. New molecular techniques are being developed that allow different fluorescent proteins to be expressed in different neuron classes. Because the neurons can glow in all colors of the rainbow, this technique has been called "brainbow." Neurons can be made to express proteins that allow them to be stimulated or suppressed with light. These amazing new techniques will provide valuable information about how the brain works.

Despite all of the advances for gathering information about the brain's wiring diagram, we do not have an adequate means to represent this complexity, nor do we have a means of understanding how information flows through such a complex network. Just as modern DNA sequencing technology allowed a revolution in the field of genomics, similar breakthroughs will need to occur in the categorization of all neurons and their interactions. These breakthroughs will be in data management, computational simulations, and multisite recording techniques.

How did the brain evolve?

Given that the brain is so complex, how could it have evolved? If you look at human brains and the brains of other primates, such as monkeys, there is not much difference, except size. Is size all that matters, or are there differences in the way neural circuits function? In my laboratory we have been investigating the evolution of neural circuits in simpler animals: sea slugs. These creatures have just 10,000 neurons in their brains, so it is easier to understand the circuitry. We are examining how the same sets of neurons produce different types of swimming behaviors in different species of sea slug.

How does processing in the brain lead to consciousness?

Even if we figure out how to represent all of the neurons in the brain and all of the ways that they communicate, we still need to answer the ultimate question, "How does neural activity cause the sensation that we all feel of being conscious and alive?" In every age, it is hard to imagine the things that we do not know. For example, imagine trying to understand how the brain worked before electricity was discovered; it would have seemed like a magical force.

think like a scientist What would it mean to understand the biological nature of conscious thought?

Paul S. Katz is a Professor in the Neuroscience Institute at Georgia State University. His research interests include neuromodulation and the evolution of neuronal circuits underlying behavior. Learn more about Dr. Katz's work at http://tinyurl.com/katzlab.

Paul S. Katz

REVIEW KEY CONCEPTS

For access to MindTap and additional study materials visit www.cengagebrain.com.

39.1 Neurons and Their Organization in Nervous Systems

- The nervous system of an animal: (1) receives information about conditions in the internal and external environment; (2) transmits the message along neurons; (3) integrates the information to formulate an appropriate response; and (4) sends out signals to muscles or glands that accomplish the response (Figure 39.2).

- Neurons have dendrites, which receive information and conduct signals toward the cell body, and axons, which conduct signals away from the cell body to another neuron or an effector (Figure 39.3).

- Afferent neurons conduct information from sensory receptors to interneurons, which integrate the information into a response. The response signals are passed to efferent neurons, which activate the effectors carrying out the response (Figure 39.2).

- The combination of an afferent neuron, an interneuron, and an efferent neuron makes up a basic neuronal circuit. The circuits combine into networks that interconnect the peripheral and central nervous systems.

- Glial cells help maintain the balance of ions surrounding neurons and form insulating layers around the axons (Figure 39.5).

- Neurons make connections by two types of synapses. In an electrical synapse, impulses pass directly from the sending to the receiving cell. In a chemical synapse, neurotransmitter molecules released by the presynaptic cell diffuse across a narrow synaptic cleft and bind to receptors in the plasma membrane of the postsynaptic cell (Figure 39.6).

39.2 Signaling by Neurons

- The membrane potential of a cell depends on the unequal distribution of positive and negative charges across the plasma membrane, which establishes a potential difference across the membrane.

- Three primary conditions contribute to the resting membrane potential of neurons: (1) an Na^+/K^+ active transport pump that sets up concentration gradients of Na^+ ions (higher outside) and K^+ ions (higher inside); (2) greater membrane permeability for K^+ compared with Na^+ for passive diffusion through leak channels; and (3) negatively charged proteins and other molecules inside the cell that cannot pass through the membrane (Figure 39.9).

- An action potential is generated when a stimulus pushes the resting membrane potential to the threshold value at which voltage-gated Na^+ and K^+ channels open in the plasma membrane. The inward flow of Na^+ changes membrane potential rapidly from negative to a positive peak. The potential falls rapidly to the resting value again as the gated K^+ channels allow this ion to flow out (Figure 39.11).

- Action potentials move along an axon as the ion flows generated in one segment depolarize the potential in the next segment (Figure 39.12).
- Action potentials are prevented from reversing direction by a brief refractory period, during which a segment of membrane that has just generated an action potential cannot be stimulated to produce another for a few milliseconds.
- In myelinated axons, ions can flow across the plasma membrane only at nodes where the myelin sheath is interrupted. As a result, action potentials skip rapidly from node to node by saltatory conduction (Figure 39.13).

39.3 Transmission across Chemical Synapses

- Neurotransmitters released into the synaptic cleft bind to receptors in the plasma membrane of the postsynaptic cell, altering the flow of ions across the plasma membrane of the postsynaptic cell and pushing its membrane potential toward or away from the threshold potential (Figure 39.14).
- Neurotransmitters that bind to ionotropic neurotransmitter receptors act directly. Binding to a receptor in the postsynaptic membrane activates it, leading to opening or closing an associated ion channel gate.
- Neurotransmitters that bind to metabotropic neurotransmitter receptors act indirectly. Binding to a receptor in the postsynaptic

membrane triggers generation of a second messenger, which causes an ion-conducting channel to open or close.

- Neurotransmitters are released from synaptic vesicles into the synaptic cleft by exocytosis, which is triggered by entry of Ca^{2+} ions into the cytoplasm of the axon terminal through voltage-gated Ca^{2+} channels opened by the arrival of an action potential.
- Neurotransmitter release stops when action potentials cease arriving at the axon terminal. Neurotransmitters remaining in the synaptic cleft are broken down by enzymes or taken up by the axon terminal or glial cells.
- Types of neurotransmitters include acetylcholine, amino acids, biogenic amines, neuropeptides, and gases such as NO (Table 39.1). Many of the biogenic amines and neuropeptides are also released into the general body circulation as hormones.

39.4 Integration of Incoming Signals by Neurons

- Neurons carry out integration by summing excitatory postsynaptic potentials (EPSPs) and inhibitory postsynaptic potentials (IPSPs); the summation may push the membrane potential of the postsynaptic cell toward or away from the threshold for an action potential (Figure 39.16).
- The combined effects of summation in all the neurons in the nervous system control behavior in animals and underlie complex mental processes in mammals.

TEST YOUR KNOWLEDGE

Remember/Understand

1. Nerve signals travel in the following manner:
   a. A dendrite of a sensory neuron receives the signal; its cell body transmits the signal to a motor neuron's axon, and the signal is sent to the target.
   b. An axon of a motor neuron receives the signal; its cell body transmits the signal to a sensory neuron's dendrite, and the signal is sent to the target.
   c. Efferent neurons conduct nerve impulses toward the cell body of sensory neurons, which send them on to interneurons and, ultimately, to afferent motor neurons.
   d. A dendrite of a sensory neuron receives the signal; the cell's axon transmits the signal to an interneuron, and the signal is transmitted by efferent neurons to effectors.
   e. The axons of oligodendrocytes transmit nerve impulses to the dendrites of astrocytes.

2. Glial cells:
   a. are unable to divide after an animal is born.
   b. called Schwann cells form the insulating myelin sheath around axons.
   c. called astrocytes form the nodes of Ranvier in the brain.
   d. called oligodendrocytes cover the surfaces of blood vessels in the PNS.
   e. are neuronal cells that provide nutrition and support to nonneuronal cells.

3. An example of a synapse could be the site where:
   a. neurotransmitters released by an axon travel across a gap and are picked up by receptors on a muscle cell.
   b. an electrical impulse arrives at the end of a dendrite causing ions to flow onto axons of presynaptic neurons.
   c. postsynaptic neurons transmit a signal across a cleft to a presynaptic neuron.
   d. oligodendrocytes contact the dendrites of an afferent neuron directly.
   e. an on–off switch stimulates an electrical impulse in a presynaptic cell to stimulate other presynaptic cells.

4. The resting membrane potential in neurons requires:
   a. membrane transport channels to be constantly open for Na^+ and K^+ flow.
   b. the inside of neurons to be positive relative to the outside.

   c. the diffusion of K^+ out of the cell and a charge difference between the inside and outside of the axon set up by this movement of K^+.
   d. an active Na^+/K^+ pump, which pumps Na^+ and K^+ into the neuron.
   e. three Na^+ ions to be pumped through three Na^+ gates and two K^+ ions to be pumped through two K^+ gates.

5. The major role of the Na^+/K^+ pump is to:
   a. cause a rapid firing of the action potential so the inside of the membrane becomes momentarily positive.
   b. decrease the resting membrane potential to zero.
   c. hyperpolarize the membrane above resting value.
   d. cause an action potential to enter a refractory period.
   e. maintain the resting membrane potential at a constant negative value.

6. In the propagation of a nerve impulse:
   a. the refractory period begins as the K^+ channel opens, allowing K^+ ions to flow outward along their concentration gradient.
   b. Na^+ ions flow out of the axon with their concentration gradient.
   c. positive charges lower the membrane potential to its lowest action potential.
   d. gated K^+ channels open at the same time as the activation gate of Na^+ channels closes.
   e. the depolarizing stimulus lowers the membrane potential to open the Na^+ gates.

7. Which of the following does *not* contribute to propagation of action potentials?
   a. As the area outside the membrane becomes negative, it attracts ions from adjacent regions; as the inside of the membrane becomes positive, it attracts negative ions from nearby in the cytoplasm. These events depolarize nearby regions of the axon membrane.
   b. The refractory period allows the impulse to travel in only one direction.
   c. Each segment of the axon prevents the adjacent segments from firing.
   d. The magnitude of the action potential stays the same as it travels down the axon.
   e. Up to a limit, increasing the intensity of the stimulus increases the number of action potentials.

8. Which of the following statements best describes saltatory conduction?
   a. It inhibits direct neurotransmitter release.
   b. It transmits the action potential at the nodes of Ranvier and thus speeds up impulses on myelinated axons.
   c. It increases neurotransmitter release at the presynaptic membrane.
   d. It decreases neurotransmitter uptake at chemically gated postsynaptic channels.
   e. It removes neurotransmitters from the synaptic cleft.

9. Transmission of a nerve impulse to its target cell requires:
   a. endocytosis of neurotransmitters by excitatory presynaptic vesicles.
   b. the release of thousands of molecules of neurotransmitter stored in the postsynaptic cell into the synaptic cleft.
   c. Ca^{2+} ions to diffuse through voltage-gated Ca^{2+} channels.
   d. a fall in Ca^{2+} in the cytoplasm to trigger a protein that causes the presynaptic vesicle to fuse with the plasma membrane.
   e. an action potential to open the Ca^{2+} gates so that Ca^{2+} ions, in higher concentration outside the axon, can flow back into the cytoplasm of the neuron.

10. Which of the following matches between neurotransmitter and site(s) of action is correct?
    a. acetylcholine: neuromuscular junctions in PNS
    b. norepinephrine: mostly acts on CNS and PNS, as well as on effectors such as muscle
    c. serotonin: many CNS pathways, including those involved in vital brain functions such as memory and learning
    d. endorphins: CNS interneurons in a number of pathways, including those regulating appetite, reproductive behavior, muscular movement, sleep, and emotional states such as anxiety
    e. nitrous oxide: released by special, unmyelinated sensory neurons in spinal cord

Apply/Analyze

11. **Discuss Concepts** In some cases of attention-deficit/hyperactivity disorder (ADHD), the impulsive, erratic behavior typical of affected people can be calmed with drugs that stimulate certain brain neurons. Based on what you have learned about neurotransmitter activity in this chapter, can you suggest a neural basis for this effect?

12. **Discuss Concepts** Most sensory neurons form synapses either on interneurons in the spinal cord or on motor neurons. However, in many vertebrates, certain sensory neurons in the nasal epithelium synapse directly on brain neurons that activate behavioral responses to odors. Suggest at least one reason why natural selection might favor such an arrangement.

13. **Discuss Concepts** How did evolution of chemical synapses make higher brain functions possible?

Evaluate/Create

14. **Discuss Concepts** Search for Internet resources using the phrase "pediatric neurotransmitter disease" and, for one such disease, explain how the symptoms relate to neurotransmitter function.

15. **Design an Experiment** Design an experiment to test whether neurons are connected via electrical or chemical synapses.

16. **Apply Evolutionary Thinking** A biologist hypothesized that the mechanism for the propagation of action potentials down a neuron evolved only once. What evidence would you collect from animals living today to support or refute that hypothesis?

For selected answers, see Appendix A.

INTERPRET THE DATA

You learned in this chapter that Na^+/K^+ active transport pumps in the plasma membrane of the axons are responsible for creating the imbalance between Na^+ and K^+ inside and outside of the neuron that produces the resting membrane potential. In early research studying the role of ions and the involvement of active transport of ions in neural signaling, investigators used the giant axon of a squid as a model. The diameter of a giant axon is far greater than that of a mammalian axon, which enabled researchers to isolate it easily and use it in *in vitro* experiments. In one early experiment, researchers investigated the active transport of Na^+ out of the axon in response to the presence of cyanide. Experimentally they hooked up a section of axon to a syringe, immersed the axon in artificial seawater, introduced radioactive ^{22}Na (as $^{22}NaCl$) into the axon, and then quantified the transport of ^{22}Na out through the axon's plasma membrane. The rate of ^{22}Na transport out of the axon was determined by measuring the radioactivity released into the fluid surrounding the axon over a period of time. The **Figure** shows the results of the experiment.

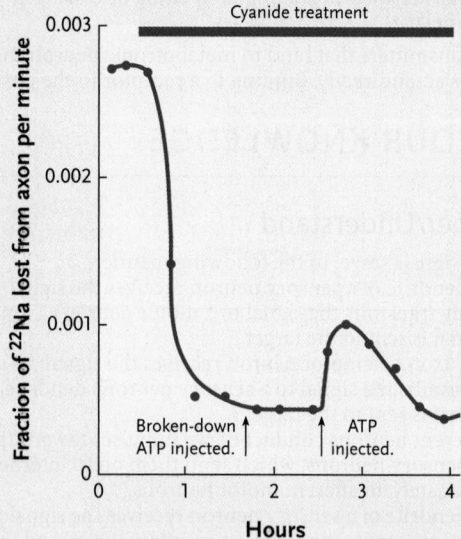

1. What is the effect of cyanide on Na^+ transport out of the squid axon? How do the data show the effect?

2. What is the effect of injecting broken-down ATP (ATP hydrolyzed to break it down into its component parts) on the transport of Na^+ out of the cyanide-treated axon?

3. What is the effect of injecting intact ATP on the transport of Na^+ out of the cyanide-treated axon? What does this result mean with respect to Na^+ active transport?

Source: P. C. Caldwell et al. 1960. The effects of injecting "energy-rich" phosphate compounds on the active transport of ions in the giant axons of *Loligo*. *The Journal of Physiology* 152:561–590.

© Cengage Learning 2017

Nervous Systems

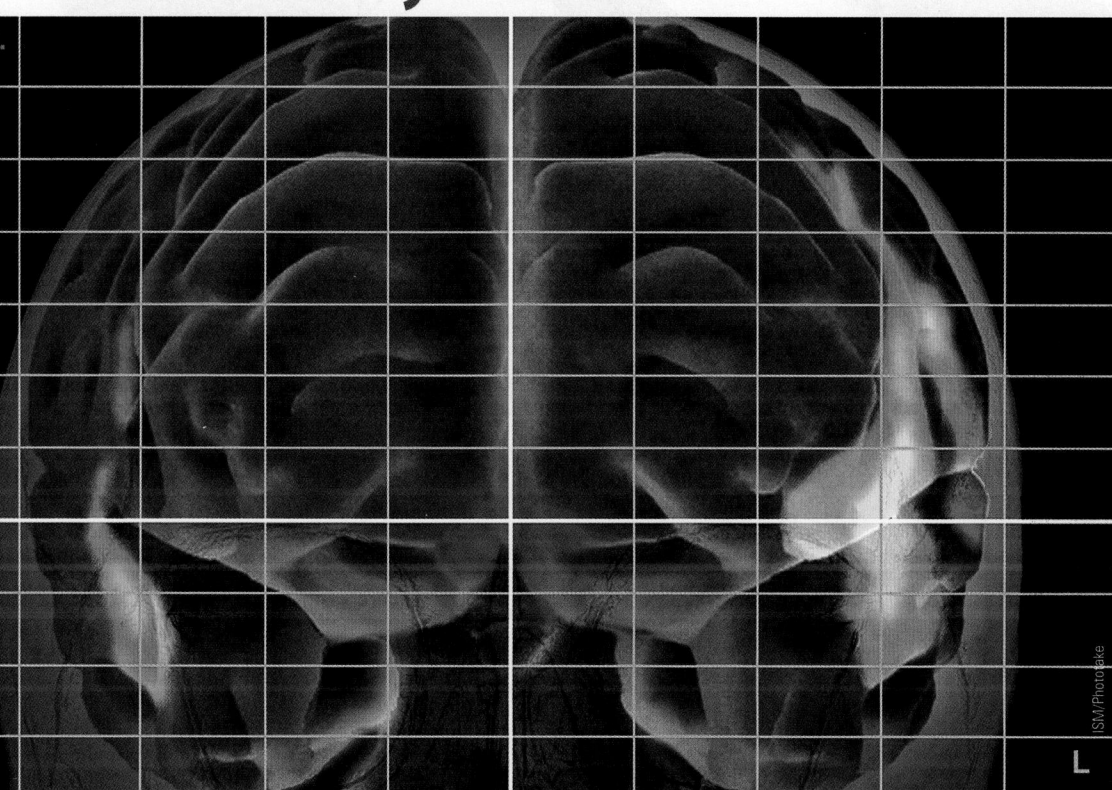

Activity in the human brain while reading aloud. The image combines an fMRI of a male brain with a PET scan, which shows that blood circulation increases in the language, hearing, and vision areas of the brain, especially in the left hemisphere.

40

STUDY OUTLINE

40.1 Invertebrate and Vertebrate Nervous Systems Compared

40.2 The Peripheral Nervous System

40.3 The Central Nervous System and Its Functions

40.4 Memory, Learning, and Consciousness

Why it matters . . . The conductor's baton falls and the orchestra plays the first notes of a Mozart symphony. Unaware of the complex interactions of their nervous systems, the musicians translate printed musical notation into melodious sounds played on their instruments. Although their fingers and arms move to produce precise harmonies, the musicians are only vaguely conscious of these movements, learned through years of practice. Their only conscious endeavor is to interpret the music in line with the conductor's directions.

Near the stage, a housefly moves in random twists and turns. Although far less complex than that of a human, the fly's nervous system contains networks of neurons that work in the same way, in patterns adapted to its lifestyle. The fly does not register the sounds reverberating through the hall as a significant sensory input. However, some of its receptors are exquisitely sensitive to the presence of potential food molecules, including those in the sweat on the conductor's face. The fly swoops and turns and alights on the tip of his nose. When sensory receptors in the fly's footpads detect organic matter on the surface of the nose, they trigger an automated feeding response: the fly's proboscis lowers and its gut begins contractions that suck up the nutrients.

The conductor's eyes notice the insect's approach, and sensory receptors in his skin pinpoint the spot where it lands. Without missing a beat, the conductor's hand flicks toward that exact spot. But his nervous system and effectors, although highly sophisticated, are no match for the escape reflexes of the fly. The fly's sensory receptors detect the motion of the fingers, sending impulses to the fly's leg and wing muscles that launch it into flight long before the fingers reach the nose.

The fly wanders into the orchestra, attracted to potential nutrients on various musicians, and lands on the left hand of the timpanist, who is listening with pleasure to the music while he awaits his entrance late in the first movement. His right hand holds a mallet. With a skill born of long

A. Cnidarian (sea anemone)

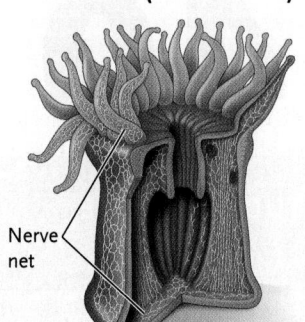

Nerve net

B. Echinoderm (sea star)

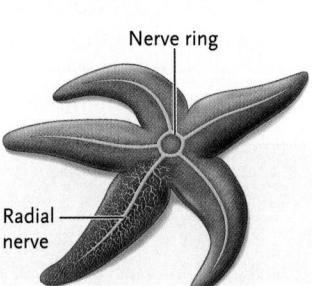

Nerve ring

Radial nerve

C. Planarian (flatworm)

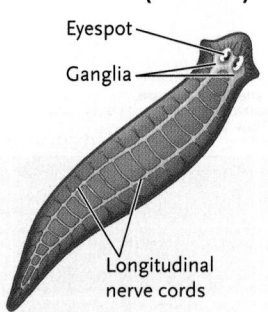

Eyespot

Ganglia

Longitudinal nerve cords

D. Arthropod (grasshopper)

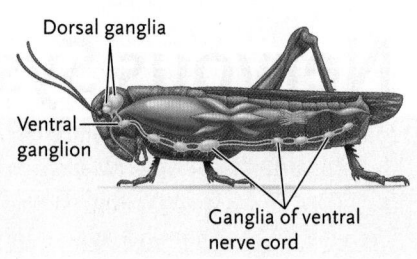

Dorsal ganglia

Ventral ganglion

Ganglia of ventral nerve cord

E. Mollusk (octopus)

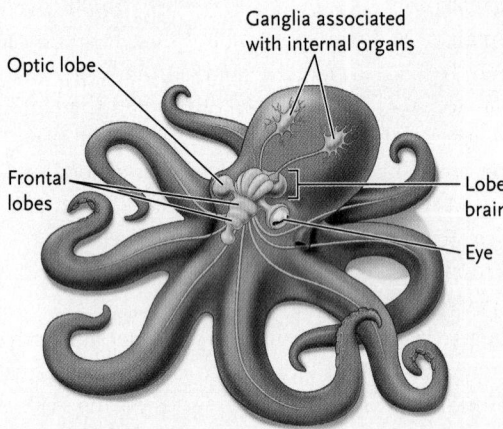

Optic lobe

Ganglia associated with internal organs

Frontal lobes

Lobed brain

Eye

F. Vertebrate (salamander)

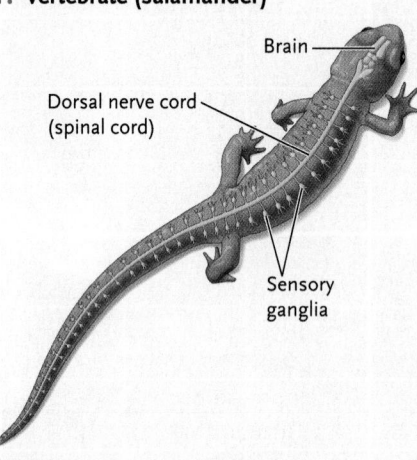

Brain

Dorsal nerve cord (spinal cord)

Sensory ganglia

FIGURE 40.1 Invertebrate and vertebrate nervous systems compared, showing increasing cephalization. The diagrams are not drawn to the same scale.

© Cengage Learning 2017

practice in hitting drums, gongs, and bells with speed and precision, the timpanist deftly swings his mallet and dispatches the fly, ending the latest contest between mammalian and arthropod nervous systems.

The nervous systems underlying these behaviors are one of the features that set animals apart from other organisms. As animals evolved, the need to find food, living space, and mates, and to escape predators and other dangers, provided a powerful selection pressure for increasingly complex and capable nervous systems. Neurons, described in the previous chapter, provide the structural and functional basis for all these systems. We can trace some of the developments along this extended evolutionary pathway by examining the nervous systems of living animals, from invertebrates to mammals, and especially humans.

40.1 Invertebrate and Vertebrate Nervous Systems Compared

The nervous systems of most invertebrates are relatively simple, typically containing fewer neurons, arranged in less complex networks, than vertebrate systems. Our comparative survey of nervous systems begins with the simplest invertebrates.

Cnidarians and Echinoderms Have Nerve Nets

Cnidarians and echinoderms are radially symmetrical animals with body parts arranged regularly around a central axis, like the spokes of a wheel. Their nervous systems, called **nerve nets,** are loose meshes of neurons organized within that radial symmetry.

The nerve nets of cnidarians such as sea anemones extend into each "spoke" of the body **(Figure 40.1A).** Their neurons lack clearly differentiated dendrites and axons. When part of the animal is stimulated, impulses are conducted through the nerve net in all directions from the point of stimulation. Although there is no cluster of neurons that plays the coordinating role of a brain, nerve cells may be more concentrated in some regions. For example, in scyphozoan jellyfish, which swim by rhythmic contractions of their bells, neurons are denser in a ring around the margin of the bell, in the same area as the contractile cells that produce the swimming movements.

In echinoderms, especially sea stars, the nervous system is a modified nerve net, with some neurons organized into **nerves,** bundles of axons enclosed in connective tissue and following the same pathway. A *nerve ring* surrounds the centrally located mouth, and a *radial nerve* branches to connect to nerve nets throughout each arm **(Figure 40.1B).** If the radial nerve serving an arm is cut, the arm can still move, but not in coordination with the other arms.

More Complex Invertebrates Have Cephalized Nervous Systems

More complex invertebrates have neurons with clearly defined axons and dendrites, and more specialized functions. Some neurons are concentrated into a functional cluster called a **ganglion** (plural, *ganglia*). A key evolutionary development in invertebrates is a trend toward *cephalization,* the formation of a distinct head region containing ganglia that constitute a **brain,** the control center of the nervous system that integrates major sensory input with motor output. **Nerve cords**—bundles of nerves—extend from the brain to the rest of the body. Another evolutionary trend is toward bilateral symmetry of the body and the nervous system, in which body parts are mirror images on left and right sides. These trends toward cephalization and bilateral symmetry are seen in flatworms, arthropods, and mollusks:

- *Flatworms.* In flatworms, a brain consisting of a pair of ganglia at the anterior end is connected to nerve nets in the rest of the body by two or more longitudinal nerve cords **(Figure 40.1C).** The brain integrates inputs from sensory receptors, including a pair of anterior eyespots that respond to light. The brain and longitudinal nerve cords constitute the flatworm's **central nervous system (CNS),** the simplest one known, and the nerves from the CNS to the rest of the body constitute the **peripheral nervous system (PNS).**
- *Arthropods.* Arthropods such as insects have a head region that contains a brain consisting of dorsal and ventral pairs of ganglia, and major sensory structures, usually eyes and antennae **(Figure 40.1D).** A ventral nerve cord enlarges into a pair of ganglia in each body segment. In arthropods with fused body segments, as in the thorax of insects, the ganglia are also fused into larger masses forming secondary control centers.
- *Mollusks.* Although different in basic plan from the arthropod system, the nervous systems of mollusks (such as clams, snails, and octopuses) also rely on neurons clustered into paired ganglia and connected by major nerves. Different mollusks have varying degrees of cephalization, with cephalopods having the most pronounced cephalization of any invertebrate group. In the head of an octopus, for example, a cluster of ganglia is fused into a complex, lobed brain with clearly defined sensory and motor regions. Paired nerves link different lobes with muscles and sensory receptors, including prominent optic lobes linked by nerves to large, complex eyes **(Figure 40.1E).** Octopuses are capable of rapid movement to hunt prey and to escape from predators, behaviors that rely on rapid, sophisticated processing of sensory information.

Vertebrates Have Highly Specialized Nervous Systems

In vertebrates, the CNS consists of the brain and spinal cord, and the PNS consists of all the nerves and ganglia that connect the brain and spinal cord to the rest of the body **(Figure 40.1F).**

The head contains specialized sensory organs that are connected directly to the brain by nerves.

The structure of the vertebrate nervous system reflects its pattern of development. The nervous system of a vertebrate embryo begins as the hollow **neural tube** (discussed more in Chapter 50), the anterior end of which develops into the brain and the rest into the **spinal cord.** The cavity of the neural tube becomes the fluid-filled **ventricles** of the brain and the **central canal** through the spinal cord. Adjacent tissues give rise to nerves that connect the brain and spinal cord with all body regions.

Early in embryonic development, the anterior part of the neural tube enlarges into three distinct regions: the **forebrain, midbrain,** and **hindbrain (Figure 40.2).** The hindbrain subdivides and eventually gives rise to the *cerebellum,* which integrates sensory signals from the eyes, ears, and muscle spindles with motor signals from the *telencephalon,* and the *pons,* a major traffic center for information passing between the cerebellum and the higher integrating centers of the adult telencephalon. (**Muscle spindles** are bundles of small, specialized muscle cells wrapped with the dendrites of afferent neurons.) The hindbrain also gives rise to the *medulla oblongata* (commonly shortened to medulla), which controls many vital involuntary tasks such as respiration and blood circulation.

The midbrain gives rise to the adult midbrain, which, with the pons and the medulla, constitutes the brain stem (see Figure 40.2). The midbrain has centers for coordinating reflex responses (involuntary reactions) to visual and auditory (hearing) input and relays signals to the telencephalon.

The forebrain of the embryo eventually gives rise to the *cerebrum* (or adult telencephalon) and the thalamus and hypothalamus of the adult (see Figure 40.2). The cerebrum is the largest part of the brain. It controls higher functions such as thought, memory, language, and emotions, as well as voluntary movements. The *thalamus* is a center that receives sensory input and relays it to the regions of the cerebral cortex concerned with pertinent motor responses, and the *hypothalamus* is the primary center for homeostatic control over the internal environment.

Anatomical analysis and experimental physiological studies have shown the changes that have occurred in the brain regions and brain size of vertebrates during evolution:

- *Brain regions:* Most vertebrates have the same number of brain divisions, suggesting that much of the organization of the brain arose at or soon after the origin of vertebrates. The exception is the absence of a cerebellum in agnathans (jawless vertebrates; see Section 32.4).
- *Brain size:* Looking across the vertebrate phylogenetic tree, brain size varies approximately 30-fold for a given body size. There are also some generalizations that can be made about brain size and the various vertebrate lineages. For example, most bony fishes have larger brains for the same body size than agnathans, ditto for frogs versus salamanders; reptile brains are two to three times larger than the brains of most amphibians of the same body size; and the brains of both

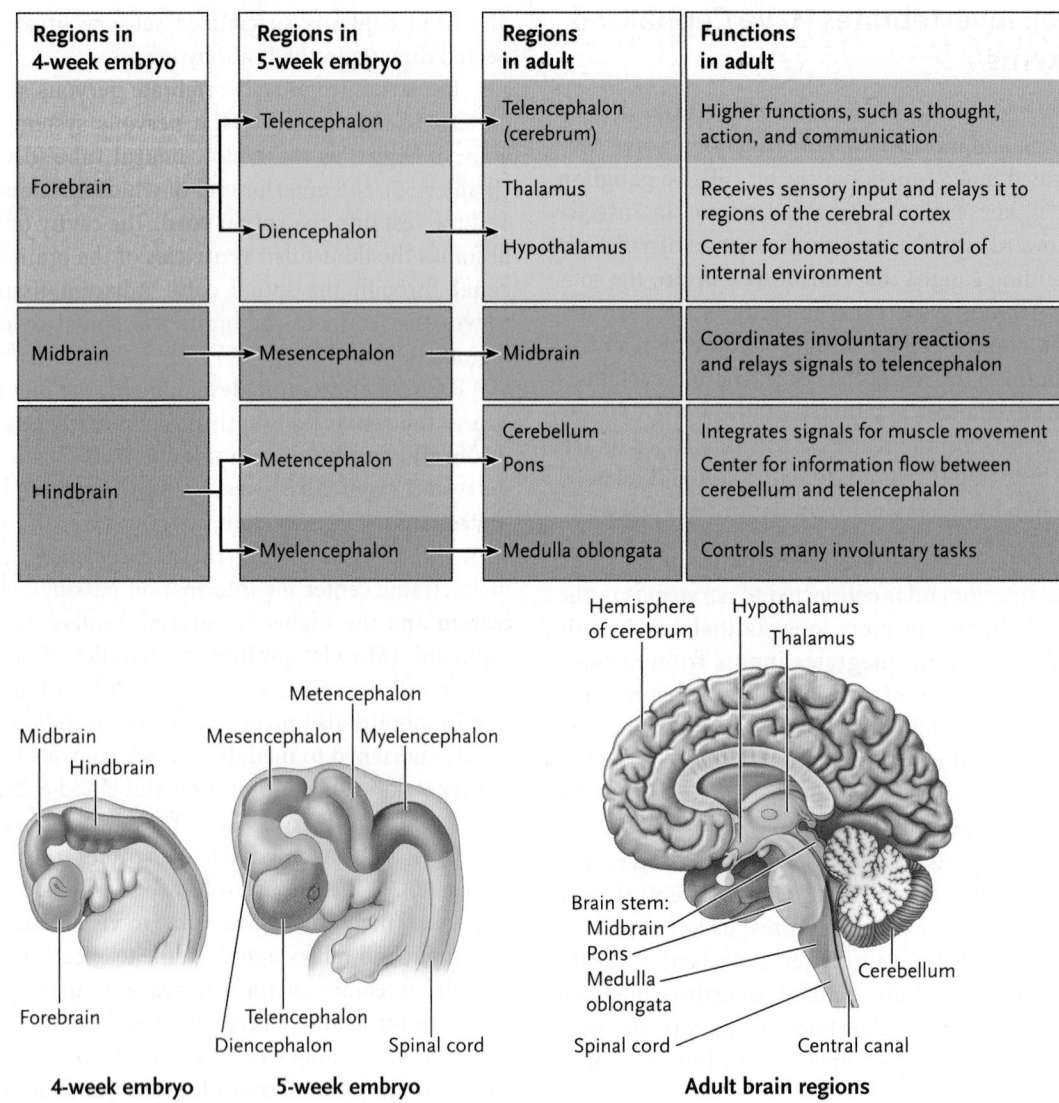

Regions in 4-week embryo	Regions in 5-week embryo	Regions in adult	Functions in adult
Forebrain	Telencephalon	Telencephalon (cerebrum)	Higher functions, such as thought, action, and communication
	Diencephalon	Thalamus	Receives sensory input and relays it to regions of the cerebral cortex
		Hypothalamus	Center for homeostatic control of internal environment
Midbrain	Mesencephalon	Midbrain	Coordinates involuntary reactions and relays signals to telencephalon
Hindbrain	Metencephalon	Cerebellum	Integrates signals for muscle movement
		Pons	Center for information flow between cerebellum and telencephalon
	Myelencephalon	Medulla oblongata	Controls many involuntary tasks

4-week embryo

5-week embryo

Adult brain regions

FIGURE 40.2 Development of the human brain from the anterior end of an embryo's neural tube.
© Cengage Learning 2017

birds and mammals are 6–10 times larger than the brains of reptiles of the same body size. Among mammals, primates and cetaceans have the largest brains for their body size, whereas marsupials, nonplacental mammals, insectivores, and rodents have the smallest brains for their body size.

The relative sizes of some brain regions also have changed during vertebrate evolution, reflecting the functions of particular brain regions in particular vertebrate groups. Generally speaking, the larger a brain region, the more neurons it contains. As a result, more complex processing of incoming neural information can be done. For example, in fishes, the cerebrum is little more than a relay station for olfactory (sense of smell) information. In amphibians, reptiles, and birds, it becomes progressively larger and contains greater concentrations of integrative functions. In mammals, the cerebrum is the major integrative structure of the brain. By contrast, the medulla oblongata has changed little in relative size.

In the following sections, we examine vertebrate nervous systems, and the human nervous system in particular. **Figure 40.3** shows the organization of the vertebrate nervous into: (1) the central nervous system (CNS), consisting of the brain and spinal cord; (2) the peripheral nervous system (PNS), consisting of neurons connecting the CNS and the other parts of the body; and (3) the **enteric nervous system,** a network of nerves in the wall of the digestive tract. Digestion in the digestive tract is controlled by the autonomic nervous system, by the enteric nervous system, and by hormones (see Chapter 47). We begin our discussion of the vertebrate nervous system with the peripheral nervous system.

STUDY BREAK 40.1

1. Distinguish between a nerve net, nerves, and nerve cords.
2. What is cephalization?

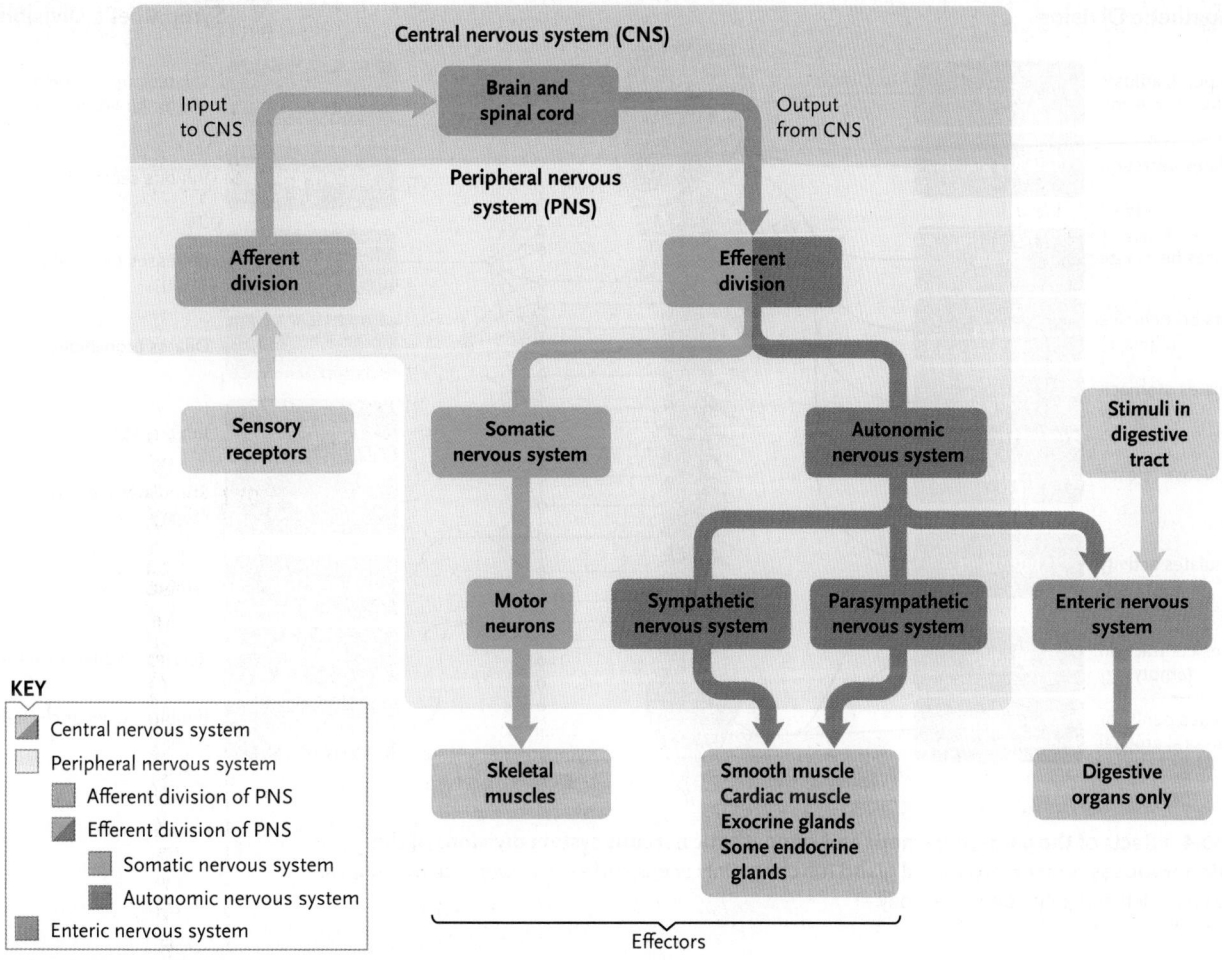

FIGURE 40.3 Organization of the vertebrate nervous system.
© Cengage Learning 2017

40.2 The Peripheral Nervous System

The peripheral nervous system is subdivided into the *afferent division* and the *efferent division* (see Figure 40.3). The **afferent division** of the PNS carries signals to the CNS and includes all the neurons that transmit sensory information from their receptors. The **efferent division** of the PNS carries signals from the CNS to the muscles and glands, which act as effectors to bring about the desired response. In mammals, 31 pairs of **spinal nerves** carry signals between the spinal cord and the body trunk and limbs, and 12 pairs of **cranial nerves** connect the brain directly to the head, neck, and body trunk. The efferent division of the PNS is further subdivided into the *somatic nervous system* and the *autonomic nervous system* (see Figure 40.3).

The Somatic Nervous System Controls the Contraction of Skeletal Muscles, Producing Body Movements

The **somatic nervous system** controls body movements that are primarily conscious and voluntary. Its neurons, called **motor neurons,** carry efferent signals from the CNS to the skeletal muscles. The dendrites and cell bodies of motor neurons are located in the spinal cord, and their axons extend from the spinal cord to the skeletal muscle cells they control. As a result, the somatic portions of the cranial and spinal nerves consist only of axons.

Although the somatic nervous system is primarily under conscious, voluntary control, some contractions of skeletal muscles are unconscious and involuntary. These include the reflexes, shivering, and the constant muscle contractions that maintain body posture and balance.

The Autonomic Nervous System Is Subdivided into Sympathetic and Parasympathetic Nervous Systems

The **autonomic nervous system** consists of nerves to smooth muscle, cardiac muscle, exocrine glands, and some endocrine glands (see Figure 40.3). It controls largely involuntary processes such as blood circulation, secretion by sweat glands, many functions of the reproductive and excretory systems, and contraction of smooth muscles in all parts of the body. The autonomic nervous system has two divisions, the *sympathetic nervous system* and the *parasympathetic nervous system,* which are always active

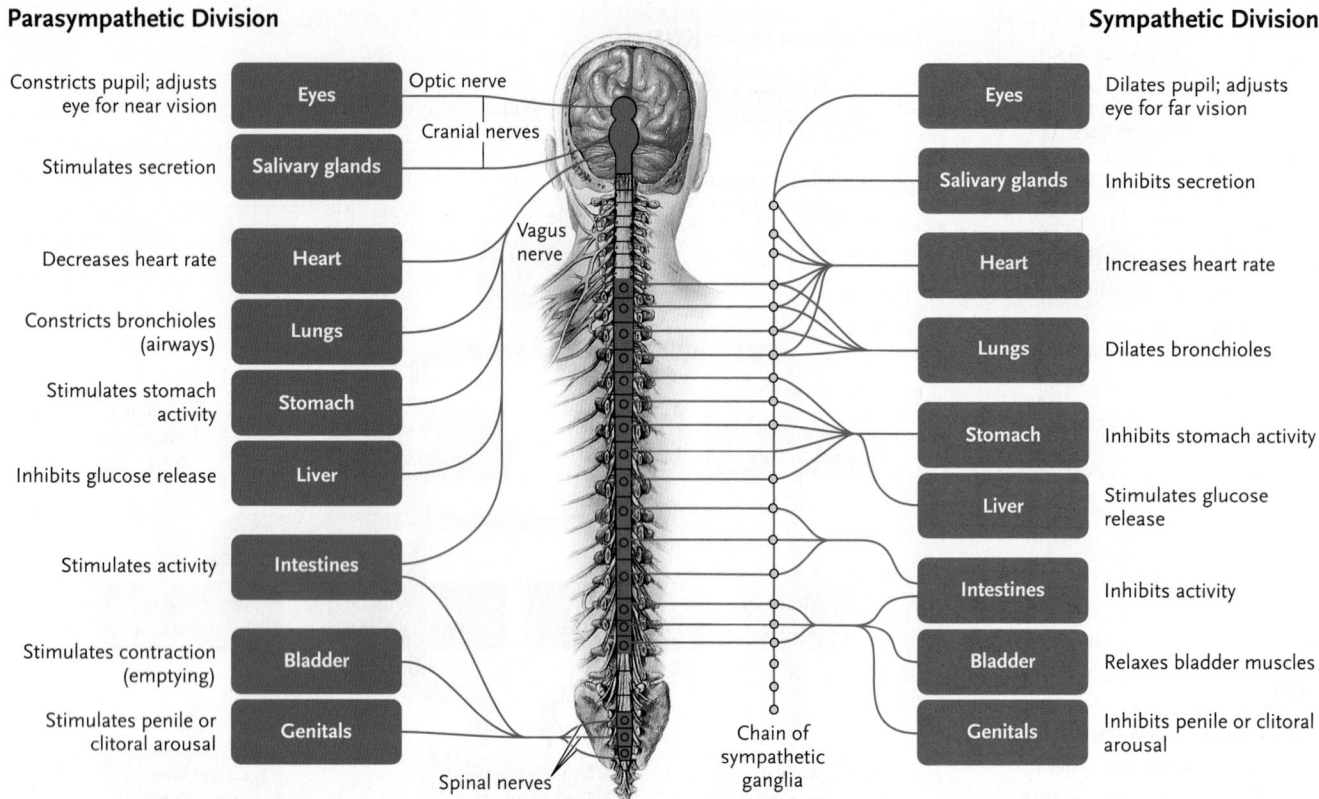

Parasympathetic Division

Constricts pupil; adjusts eye for near vision — Eyes

Stimulates secretion — Salivary glands

Decreases heart rate — Heart

Constricts bronchioles (airways) — Lungs

Stimulates stomach activity — Stomach

Inhibits glucose release — Liver

Stimulates activity — Intestines

Stimulates contraction (emptying) — Bladder

Stimulates penile or clitoral arousal — Genitals

Optic nerve

Cranial nerves

Vagus nerve

Spinal nerves

Sympathetic Division

Eyes — Dilates pupil; adjusts eye for far vision

Salivary glands — Inhibits secretion

Heart — Increases heart rate

Lungs — Dilates bronchioles

Stomach — Inhibits stomach activity

Liver — Stimulates glucose release

Intestines — Inhibits activity

Bladder — Relaxes bladder muscles

Genitals — Inhibits penile or clitoral arousal

Chain of sympathetic ganglia

FIGURE 40.4 Effects of the parasympathetic and sympathetic nervous system divisions of the autonomic nervous system on organ and gland function. Only one side of each division is shown; both are duplicated on the left and right sides of the body.

© Cengage Learning 2017

and have opposing effects on the organs they affect, thereby enabling precise control **(Figure 40.4)**. For example, in the circulatory system, sympathetic neurons stimulate the force and rate of the heartbeat, and parasympathetic neurons inhibit these activities. These opposing effects control involuntary body functions precisely. The autonomic nervous system helps control digestive activities as well; however, those activities are also controlled by the enteric nervous system, which can act independently of the autonomic system, and by hormones.

The pathways of the autonomic nervous system have two neurons in series **(Figure 40.5)**. The first neuron, the **preganglionic**

neuron, has its dendrites and cell body in the CNS, and its axon extends to a ganglion in the PNS. There, it synapses with the dendrites and cell body of the second neuron in the pathway, the **postganglionic neuron.** The axon of the second neuron extends from the ganglion to the effector carrying out the response. The sympathetic ganglia are arrayed as an orderly chain along the spinal cord (see Figure 40.4), whereas the parasympathetic ganglia are irregularly placed and often next to, or within, the effector structures.

The **sympathetic nervous system** predominates in situations involving stress, danger, excitement, or strenuous

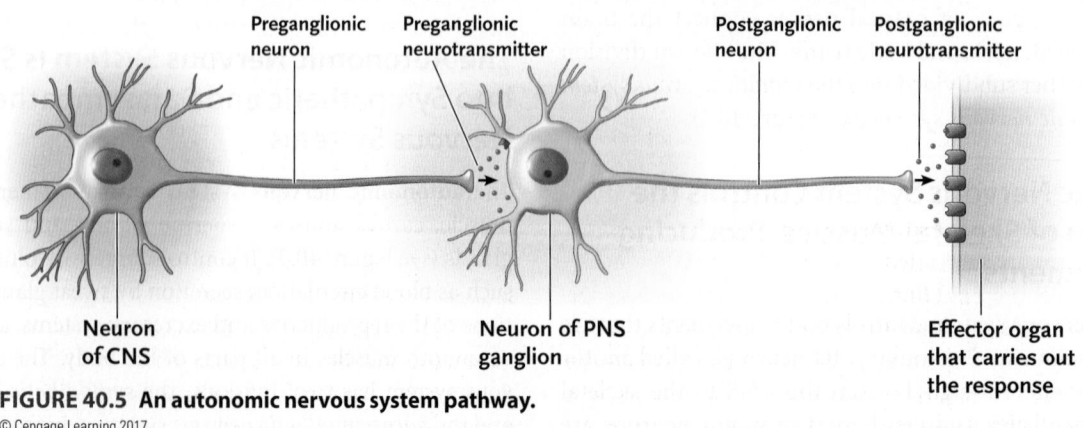

Preganglionic neuron

Preganglionic neurotransmitter

Postganglionic neuron

Postganglionic neurotransmitter

Neuron of CNS

Neuron of PNS ganglion

Effector organ that carries out the response

FIGURE 40.5 An autonomic nervous system pathway.

© Cengage Learning 2017

physical activity. Signals from the sympathetic division increase the force and rate of the heartbeat, raise the blood pressure by constricting selected blood vessels, dilate air passages in the lungs, induce sweating, and open the pupils wide. Activities that are less important in an emergency, such as digestion, are suppressed by the sympathetic system. The **parasympathetic nervous system,** in contrast, predominates during quiet, low-stress situations, such as while relaxing. During relaxation the influence of the parasympathetic division reduces the effects of the sympathetic division, such as rapid heartbeat and elevated blood pressure, and activities such as digestion predominate. The chemical signals used by the divisions of the autonomic nervous system are as follows:

- Sympathetic and parasympathetic preganglionic neurons secrete the neurotransmitter acetylcholine, which binds to receptors on the postganglionic neuron.
- Most sympathetic postganglionic neurons secrete the neurotransmitter norepinephrine, which binds to receptors on cells of the effector organ.
- Most parasympathetic postganglionic neurons secrete acetylcholine, which binds to receptors on cells of the effector organ.

STUDY BREAK 40.2

Which of the two autonomic nervous system divisions predominates in the following scenarios? (a) You are hiking on a trail and suddenly a bear appears in your path. (b) It is a hot, sunny day. You find a shady tree and sit down. Leaning against its trunk, you feel your eyes becoming heavy.

40.3 The Central Nervous System and Its Functions

The central nervous system (CNS) consists of the brain and the spinal cord. The estimated 86 billion neurons in your brain are joined by 10^{14}–10^{15} synapses and organized into complex, linked networks that integrate incoming sensory information from the PNS into compensating responses for managing bodily activities. Those activities include controlling voluntary movements, experiencing emotions, subconsciously regulating your internal environment, perceiving your surroundings and yourself, thought, and memory.

The CNS Is Protected by the Meninges and by Cerebrospinal Fluid

The brain and spinal cord are surrounded and protected by three layers of connective tissue called **meninges** (*meninga* = membrane), and by **cerebrospinal fluid,** which circulates through the ventricles of the brain, through the central canal of the spinal cord, and between two of the meninges. The fluid cushions the brain and spinal cord from jarring movements and impacts, and it both nourishes the CNS and protects it from toxic substances.

The Blood–Brain Barrier Regulates Exchanges between Blood and Brain

The brain is shielded from many changes in the blood by a highly selective **blood–brain barrier.** That is, unlike the epithelial cells forming capillary walls elsewhere in the body, which allow small molecules and ions to pass freely from the blood to surrounding fluids, those forming capillaries in the brain are sealed together by tight junctions (see Section 4.5 and Figure 4.27). The tight junctions set up a blood–brain barrier that prevents most substances dissolved in the blood from entering the cerebrospinal fluid and thus protects the brain and spinal cord from viruses, bacteria, and toxic substances that may be circulating in the blood.

Molecules that are lipid soluble such as steroids, alcohol, and anesthetics can move directly across the epithelial cell membranes by diffusion. Aside from relatively free movement of salts and water, a few other substances—most significantly glucose and ketones, the only molecules that brain and spinal cord cells can oxidize for energy—are moved across the plasma membrane by highly selective transport proteins.

The Brain Integrates Sensory Information and Formulates Responses; the Spinal Cord Relays Signals between the PNS and the Brain and Controls Reflexes

The *brain* is the major center that receives, integrates, stores, and retrieves information. Its interneuron networks generate responses that provide the basis for our voluntary movements, consciousness, emotions, learning, reasoning, language, and memory, among many other complex activities.

Each of the structures of the adult brain (see Figure 40.2) contains both **gray matter,** consisting of nerve cell bodies and dendrites, and **white matter,** consisting of axons, many of them surrounded by myelin sheaths **(Figure 40.6).** (The myelination of the axons gives white matter its color.)

The pons and medulla, along with the midbrain, form a stalklike structure known as the **brain stem,** which connects the "higher" centers with the spinal cord. All but two of the twelve pairs of cranial nerves also originate from the brain stem. These higher centers, which make up most of the mass of the brain in humans, are generally referred to as the *telencephalon (cerebrum)*. The cerebrum, the largest part of the brain in humans, is organized into the left and right *cerebral hemispheres*, which have many fissures and folds (see Figure 40.6). Each hemisphere consists of **cerebral cortex,** a thin outer shell of gray matter covering a thick core of white matter. The *basal nuclei*, consisting of several regions of gray matter, are located deep within the white matter.

The *spinal cord,* which extends dorsally from the brain stem, carries impulses between the brain and the PNS and contains the interneuron circuits that control motor reflexes. In cross section, the spinal cord has a butterfly-shaped core of gray matter surrounded by white matter. Pairs of spinal nerves emerge from the spinal cord at spaces between the vertebrae.

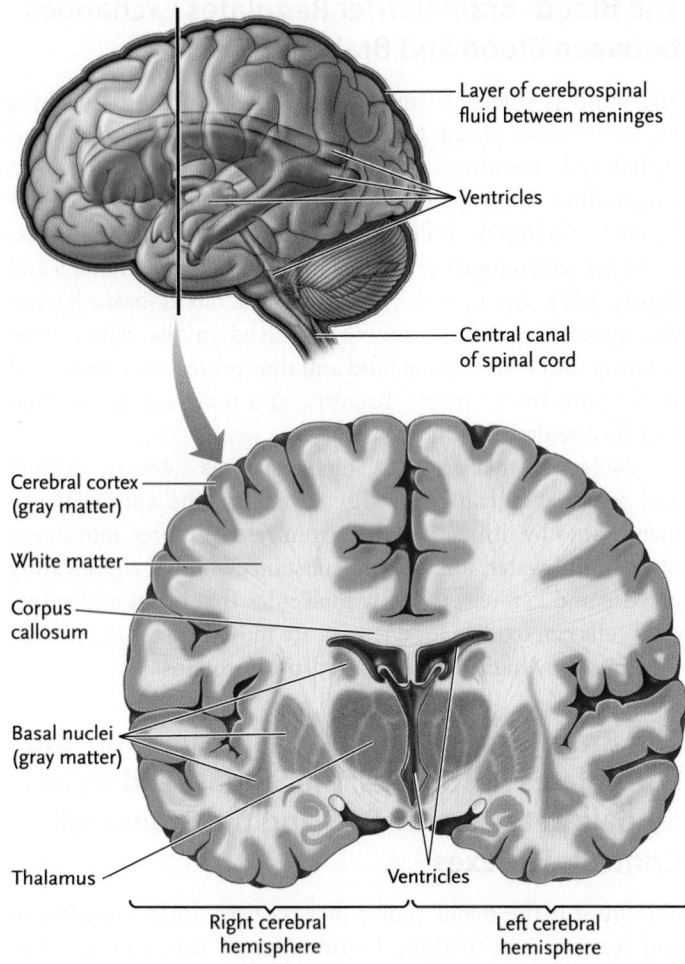

FIGURE 40.6 The human brain, illustrating the distribution of gray matter and white matter, and the locations of the four ventricles (in blue) with their connection to the central canal of the spinal cord.

© Cengage Learning 2017

Layer of cerebrospinal fluid between meninges

Ventricles

Central canal of spinal cord

Cerebral cortex (gray matter)

White matter

Corpus callosum

Basal nuclei (gray matter)

Thalamus

Ventricles

Right cerebral hemisphere

Left cerebral hemisphere

The afferent axons entering the spinal cord make synapses with interneurons in the gray matter, which send axons upward through the white matter of the spinal cord to adjacent segments of the cord and, in some cases, through to the brain. Conversely, axons from interneurons of the brain pass downward through the white matter of the cord and make synapses with the dendrites and cell bodies of efferent neurons in the gray matter of the cord. The axons of these efferent neurons exit the spinal cord through the spinal nerves.

The gray matter of the spinal cord also contains pathways involved in **reflexes,** programmed movements that take place without conscious effort; an example is the **patellar tendon reflex** (commonly called the **knee-jerk reflex**), in which a tap to the tendon just below the knee cap (patella) causes the leg to kick **(Figure 40.7).** The existence of only a single synapse between the efferent and afferent neurons in the reflex facilitates a rapid response—about 50 msec between tap and the start of the kick. The patellar tendon reflex is the classic example of the **stretch reflex.** The main purpose of the patellar tendon reflex is to react to added loads that stretch the muscles of the leg. During normal activity, it serves to maintain balance

should anything cause an unexpected load on the quadriceps muscle, such as a stumble or a trip. Another example is a stretch reflex in your biceps muscle when textbooks are piled onto your arms in the bookstore, or when you catch an object. In general, stretch reflexes are local negative feedback mechanisms that sense and resist change in muscle length when a load is added.

Similarly, a reflex pathway is involved for the rapid withdrawal of your hand if you touch a hot surface. You may recall from experience that when a reflex movement withdraws your hand from a hot surface or other damaging stimulus, you feel the pain shortly *after* the hand is withdrawn. This is the extra time required for impulses to travel from the neurons of the reflex via interneurons to the brain (see substance P in Table 39.1).

The Brain Stem Regulates Many Vital Housekeeping Functions of the Body

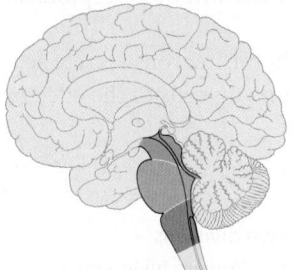

Gray-matter centers in the brain stem control many vital body functions without conscious involvement or control by the cerebrum. Among these functions are the heart and respiration rates, blood pressure, constriction and dilation of blood vessels, coughing, and reflex activities of the digestive system such as vomiting. Damage to the brain stem has serious and sometimes lethal consequences.

A complex network of interconnected neurons known as the **reticular formation** (*reticulum* = little net) runs through the length of the brain stem, connecting to the thalamus at the anterior end and to the spinal cord at the posterior end **(Figure 40.8).** Most incoming sensory input goes to the reticular formation, which integrates the information and then sends signals to other parts of the CNS. The reticular formation has two parts. The **ascending reticular formation,** also called the *reticular activating system,* contains neurons that convey stimulatory signals via the thalamus to arouse and activate the cerebral cortex. It is responsible for the sleep–wake cycle; depending on the level of stimulation of the cortex, various levels of alertness and consciousness are produced. Lesions in this part of the brain stem result in coma. The other part, the **descending reticular formation,** receives information from a variety of sources including the hypothalamus and connects with interneurons in the spinal cord that control skeletal muscle contraction, thereby controlling muscle movement and posture. The reticular formation filters incoming signals, helping to discriminate between important and unimportant ones. Such filtering is necessary because the brain is unable to process every one of the signals from millions of sensory receptors. For example, the action of the reticular formation enables you to sleep through many sounds but waken to specific ones, such as a cat meowing to be let out or a baby crying.

Only a few neurons and a single interneuron are shown; many neurons and interneurons actually participate in the reflex.

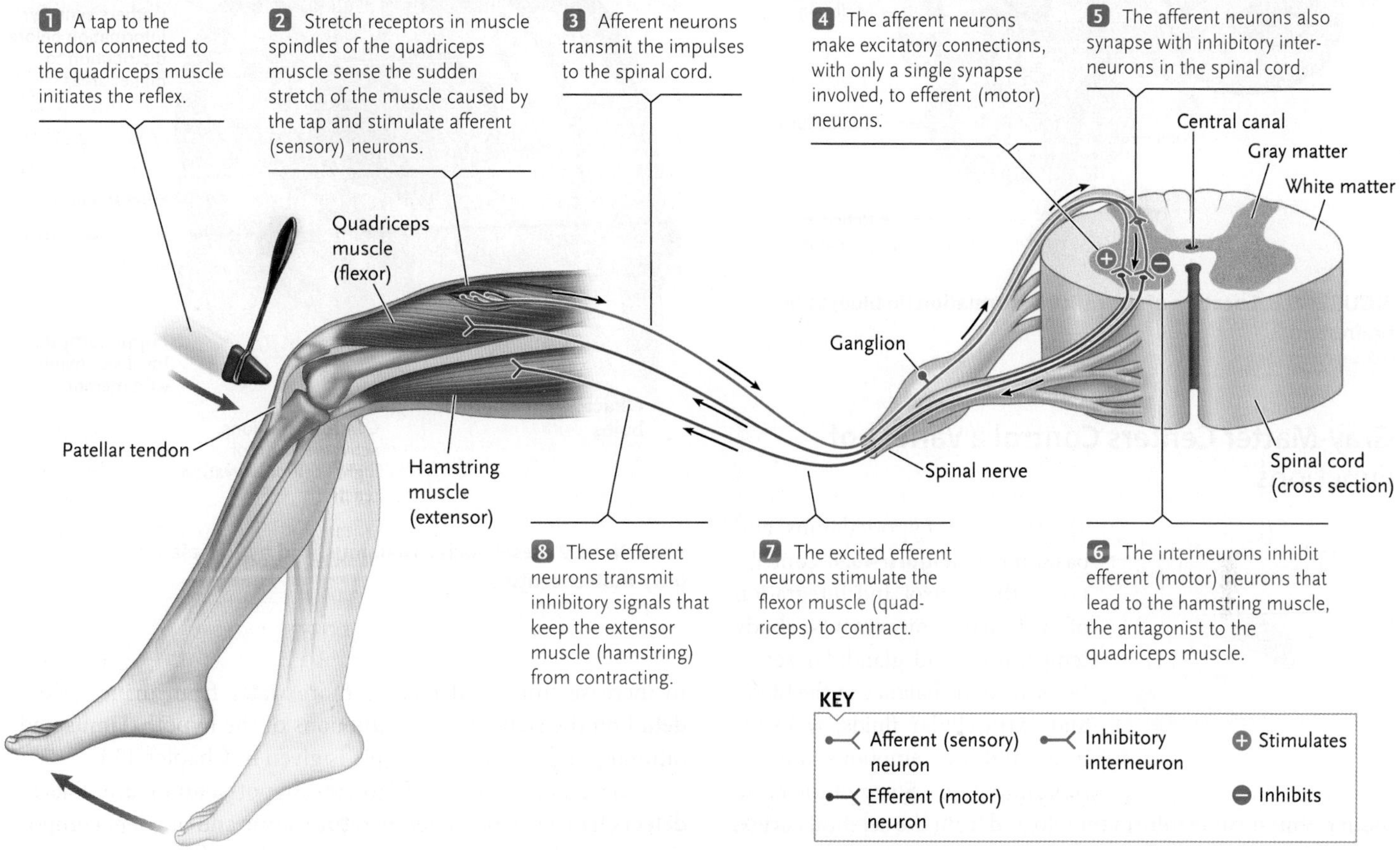

1 A tap to the tendon connected to the quadriceps muscle initiates the reflex.

2 Stretch receptors in muscle spindles of the quadriceps muscle sense the sudden stretch of the muscle caused by the tap and stimulate afferent (sensory) neurons.

3 Afferent neurons transmit the impulses to the spinal cord.

4 The afferent neurons make excitatory connections, with only a single synapse involved, to efferent (motor) neurons.

5 The afferent neurons also synapse with inhibitory interneurons in the spinal cord.

Quadriceps muscle (flexor)

Central canal

Gray matter

White matter

Patellar tendon

Hamstring muscle (extensor)

Ganglion

Spinal nerve

Spinal cord (cross section)

8 These efferent neurons transmit inhibitory signals that keep the extensor muscle (hamstring) from contracting.

7 The excited efferent neurons stimulate the flexor muscle (quadriceps) to contract.

6 The interneurons inhibit efferent (motor) neurons that lead to the hamstring muscle, the antagonist to the quadriceps muscle.

KEY

⤙ Afferent (sensory) neuron

⤙ Inhibitory interneuron

⊕ Stimulates

⤙ Efferent (motor) neuron

⊖ Inhibits

SUMMARY In a patellar tendon reflex (knee-jerk reflex), a tap on the tendon just below the knee causes a sequence of events that causes the leg to kick. The tap stimulates afferent neurons that make excitatory connections to efferent neurons with a single synapse, and to inhibitory interneurons. Excited efferent neurons stimulate the flexor muscle of the leg to contract. The inhibitory interneurons connect to efferent neurons that lead to the extensor muscle of the leg; those neurons inhibit that muscle from contracting.

think like a scientist Physicians sometimes test knee-jerk reflexes in patients. If the response to the test is significantly slower than normal, what might the physician conclude?

© Cengage Learning 2017

The Cerebellum Integrates Sensory Inputs to Coordinate Body Movements

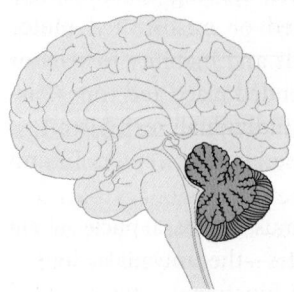

Although the **cerebellum** is an outgrowth of the pons (see Figure 40.8), it is separate in structure and function from the brain stem. Through its extensive connections with other parts of the brain, the cerebellum receives sensory input from receptors in muscles and joints, from balance receptors in the inner ear, and from the receptors of touch, vision, and hearing. These signals convey information about how the body trunk and limbs are positioned, the degree to which different muscles are contracted or relaxed, and the direction in which the body or limbs are moving. The cerebellum integrates these sensory signals and compares them with signals from the cerebrum that control voluntary body movements. Outputs from the cerebellum to the cerebrum, brain stem, and spinal cord modify and fine-tune the movements to keep the body in balance and directed toward targeted positions in space. The human cerebellum also contributes to the learning and memory of complex motor skills such as playing the guitar.

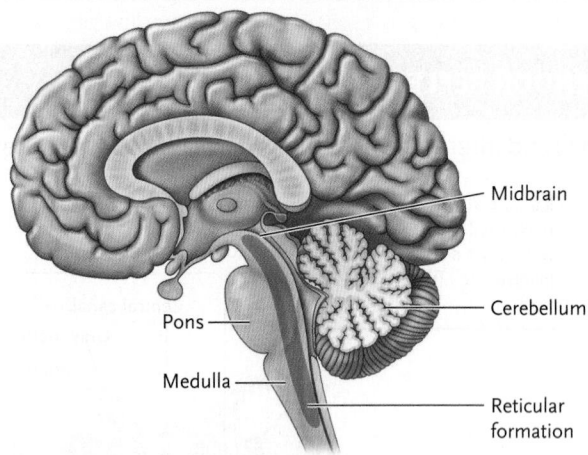

FIGURE 40.8 Location of the reticular formation (in blue) in the brain stem.
© Cengage Learning 2017

Gray-Matter Centers Control a Variety of Functions

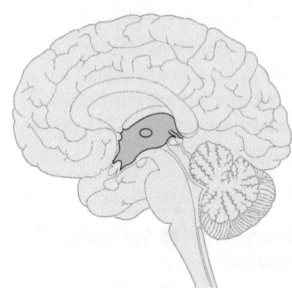

The thalamus, hypothalamus, and basal nuclei (**Figure 40.9**) contribute to the control and integration of voluntary movements, body temperature and glandular secretions, osmotic balance of the blood and extracellular fluids, wakefulness, and the emotions, among other functions. Some of the gray-matter centers route information to and from the cerebral cortex, brain stem, and cerebellum.

The **thalamus** (see Figure 40.9) forms a major switchboard that receives sensory information and relays it to the regions of the cerebral cortex concerned with motor responses to sensory information. Part of the thalamus near the brain stem cooperates with the reticular formation in alerting the cerebral cortex to full wakefulness, or in inducing drowsiness or sleep.

The **hypothalamus** is a region of the brain located in the floor of the cerebrum (see Figure 40.9) that contains clusters of neurons known as *nuclei*. Suspended just below it and connected to it by a stalk of tissue is the *pituitary gland,* consisting of two fused lobes, the *anterior pituitary* and the *posterior pituitary,* which release hormones. The hypothalamus regulates basic homeostatic functions of the body both directly and through the release of hormones.

Some nuclei in the hypothalamus set and maintain body temperature, most notably by regulating basal metabolic rate. This is an interesting example of an endocrine circuit, in which a hypothalamic hormone causes the release of a hormone from the anterior pituitary that in turn causes the release of a thyroid hormone. Other nuclei monitor the osmotic balance of the blood plasma. If osmotic pressure rises above normal levels, these neurons secrete a hormone from their terminals in adjacent areas of the hypothalamus and from the posterior pituitary

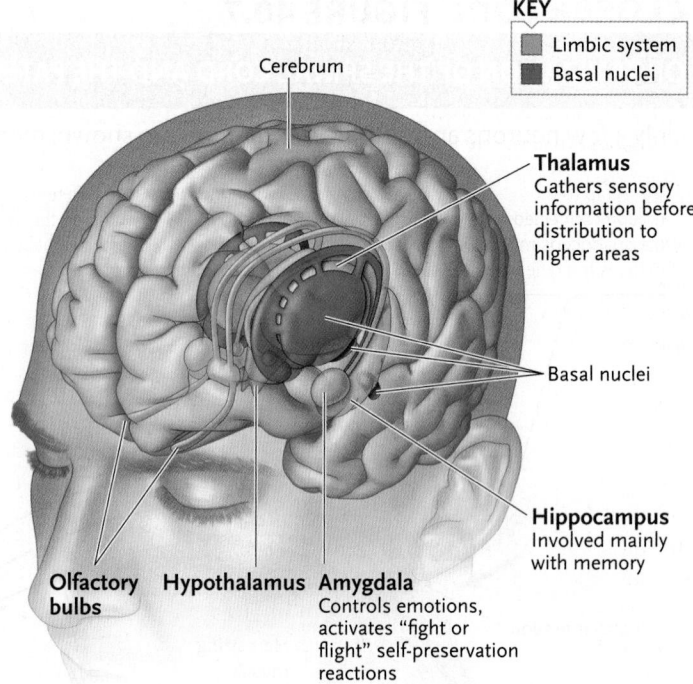

KEY
- Limbic system
- Basal nuclei

Thalamus Gathers sensory information before distribution to higher areas

Basal nuclei

Hippocampus Involved mainly with memory

Olfactory bulbs   Hypothalamus   Amygdala Controls emotions, activates "fight or flight" self-preservation reactions

Cerebrum

FIGURE 40.9 Basal nuclei, thalamus, and hypothalamus gray-matter centers.
© Cengage Learning 2017

to increase thirst and recover more water from urine. (More detail on the structure and functions of the hypothalamus and pituitary and endocrine circuits is given in Chapter 42.)

These and other nuclei of the hypothalamus are able to detect changes in blood temperature, ionic and nutrient composition, and the presence of hormones released in other areas of the body because they are *not* protected by the blood–brain barrier. The wide-ranging influence of the hypothalamic nuclei can be seen in their coordination of autonomic system functions, such as the control of the heartbeat, contraction of smooth muscle cells in the digestive system, and glandular secretions. A particular hypothalamic structure, the *suprachiasmatic nucleus,* is the main generator of the biological clock that times our myriad daily behavioral and metabolic rhythms.

The **basal nuclei** are gray-matter centers that surround the thalamus on both sides of the brain (see Figure 40.9). They moderate voluntary movements directed by motor centers in the cerebrum. Damage to the basal nuclei can affect the planning and fine-tuning of movements, leading to stiff, rigid motions of the limbs and unwanted or misdirected motor activity, such as tremors of the hands and inability to start or stop intended movements at the intended place and time. Parkinson disease, in which affected individuals exhibit all of these symptoms, results from degeneration of centers in and near the basal nuclei.

Parts of the thalamus, hypothalamus, and basal nuclei, along with other nearby gray-matter centers—the amygdala, hippocampus, and olfactory bulbs—form a functional network called the **limbic system** (*limbus* = edge, border), sometimes called our "emotional brain" (see Figure 40.9). The **amygdala** routes information about experiences that have an emotional component through the limbic system. The **hippocampus** is involved in sending information to the frontal lobes, and the **olfactory bulbs** relay inputs from odor receptors to both the cerebral cortex and the limbic system. The olfactory connection to the limbic system may explain why certain odors can evoke particular, sometimes startlingly powerful emotional responses.

The limbic system controls emotional behavior and influences the basic body functions regulated by the hypothalamus and brain stem. Stimulation of different parts of the limbic system produces anger, anxiety, fear, satisfaction, pleasure, or sexual arousal. Connections between the limbic system and other brain regions bring about emotional responses such as smiling, blushing, or laughing.

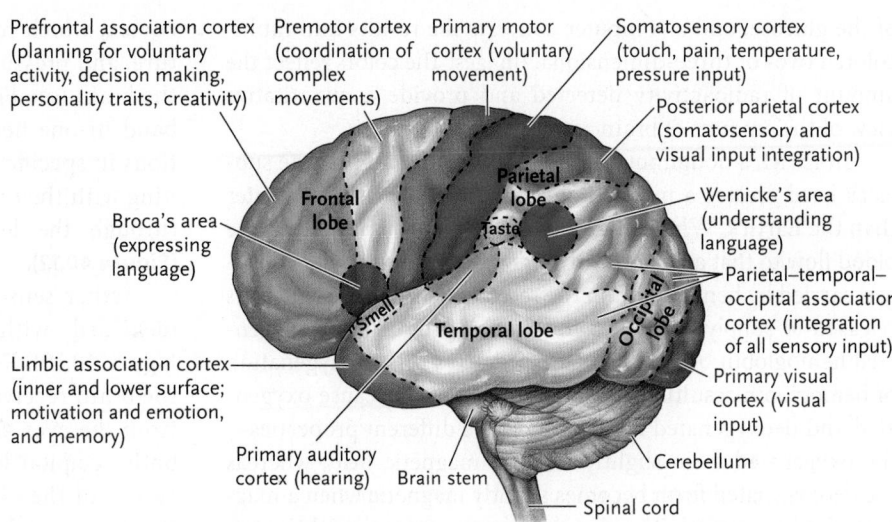

FIGURE 40.10 Functional regions of the cerebral cortex.
© Cengage Learning 2017

(Labels in figure:)
Prefrontal association cortex (planning for voluntary activity, decision making, personality traits, creativity)
Premotor cortex (coordination of complex movements)
Primary motor cortex (voluntary movement)
Somatosensory cortex (touch, pain, temperature, pressure input)
Posterior parietal cortex (somatosensory and visual input integration)
Frontal lobe
Parietal lobe
Taste
Wernicke's area (understanding language)
Broca's area (expressing language)
Smell
Temporal lobe
Occipital lobe
Parietal–temporal–occipital association cortex (integration of all sensory input)
Primary visual cortex (visual input)
Limbic association cortex (inner and lower surface; motivation and emotion, and memory)
Primary auditory cortex (hearing)
Brain stem
Cerebellum
Spinal cord

The Cerebral Cortex Carries Out All Higher Brain Functions in Humans

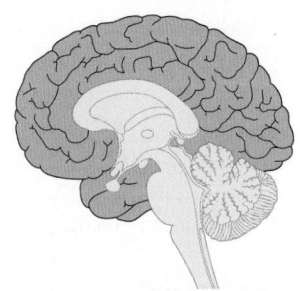

The cerebral cortices of each hemisphere contain the processing centers for the integration of neural input and the initiation of neural output. The white matter of the cerebral hemispheres, by contrast, contains the neural routes for signal transmission between parts of the cerebral cortex, or from the cerebral cortex to other parts of the CNS. No information processing occurs in the white matter. Uniquely in mammals, the cerebral cortex of the cerebral hemispheres is organized into six layers of neurons. These layers are the newest part of the cerebral cortex in an evolutionary sense.

Over the course of vertebrate evolution, the surface area of the cerebral cortex increased by continually folding in on itself, thereby expanding the structure into sophisticated information encoding and processing centers. Primates have cerebral cortices with the largest number of convolutions. In humans, each cerebral hemisphere is divided by surface folds into *frontal, parietal, temporal,* and *occipital lobes* **(Figure 40.10)**. The **frontal lobes** are responsible for voluntary motor activity, expressing language, and elaboration of thought. The **parietal lobes** are mainly responsible for receiving and processing sensory input. The **temporal lobes** receive auditory input. The **occipital lobes** perform initial processing of visual input.

The two cerebral hemispheres can function separately, and each has its own communication lines internally and with the rest of the CNS and the body. The left cerebral hemisphere responds primarily to sensory signals from, and controls movements in, the right side of the body. The right hemisphere has

the same relationships to the left side of the body. Thick axon bundles, forming a structure called the **corpus callosum** (see Figure 40.6), connect the two cerebral hemispheres and coordinate their functions.

STUDYING THE FUNCTIONS OF THE CEREBRAL CORTEX Physicians and researchers have learned much about the functions of various regions of the cerebral cortex by studying normal subjects and patients with brain damage from stroke, infection, tumors, or mechanical disturbance. Scanning techniques such as **positron emission tomography (PET)** and **functional magnetic resonance imaging (fMRI)** allow researchers to identify the functions of specific brain regions in noninvasive ways. For a PET scan of the brain **(Figure 40.11),** an individual is given a dose of radioactive glucose. When an area of the brain becomes active, more glucose is needed there to be metabolized to provide energy. Therefore, the amount of radioisotope that becomes localized to a particular area correlates with the amount of brain activity in that area. The scanner detects the radioactivity

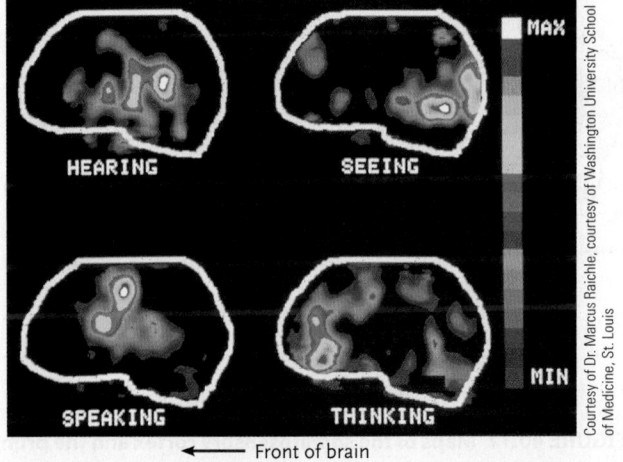

HEARING   SEEING
SPEAKING   THINKING
MAX   MIN
← Front of brain

Courtesy of Dr. Marcus Raichle, courtesy of Washington University School of Medicine, St. Louis

FIGURE 40.11 PET scans showing regions of the brain active when a person performs specific mental tasks. The colors show the relative activity of the sections, with white being the most active.

of the glucose, and a computer converts the results into multi-colored two- or three-dimensional images. The colors reflect the amount of radioactivity detected and provide a quantitative view of the amount of brain activity.

In fMRI, a doughnut-shaped magnet surrounding the subject's head creates a magnetic field about 10,000 times greater than the Earth's. When an area of the brain becomes active, the blood flow to that area suddenly increases, bringing with it oxygen carried by hemoglobin in the red blood cells. This oxygen is released to supply oxygen to the tissues, producing deoxygenated hemoglobin. An fMRI scanner detects this deoxygenation of hemoglobin resulting from neuron activity because oxygenated and deoxygenated hemoglobin have different properties— the oxygenated form slightly repels a magnetic field, whereas the deoxygenated form becomes slightly magnetic when a magnetic field is applied. The chapter-opening image combines an fMRI result with a PET scan to show the brain activity associated with reading aloud.

SENSORY REGIONS OF THE CEREBRAL CORTEX Areas that receive and integrate sensory information are distributed over the cerebral cortex. In each hemisphere, the **somatosensory**

cortex, which registers information on touch, pain, temperature, and pressure, runs in a band across the parietal lobes of the brain (see Figure 40.10). Experimental stimulation of this band in one hemisphere causes prickling or tingling sensations in specific parts on the opposite side of the body, beginning with the toes at the top of each hemisphere and running through the legs, trunk, arms, and hands, to the head **(Figure 40.12).**

Other sensory regions of the cerebral cortex have been identified with hearing, vision, smell, and taste (see Figure 40.10). Regions of the temporal lobes on both sides of the brain receive auditory inputs from the ears, while inputs from the eyes are processed in the primary visual cortex in both occipital lobes. Olfactory input from the nose is processed in the olfactory bulbs, located on the ventral side of the temporal lobes. Regions in the parietal lobes receive inputs from taste receptors on the tongue and other locations in the mouth.

PRIMARY MOTOR CORTEX OF THE CEREBRAL CORTEX The **primary motor cortex** of the cerebral cortex runs in a band just in front of the somatosensory cortex (see Figure 40.10), and

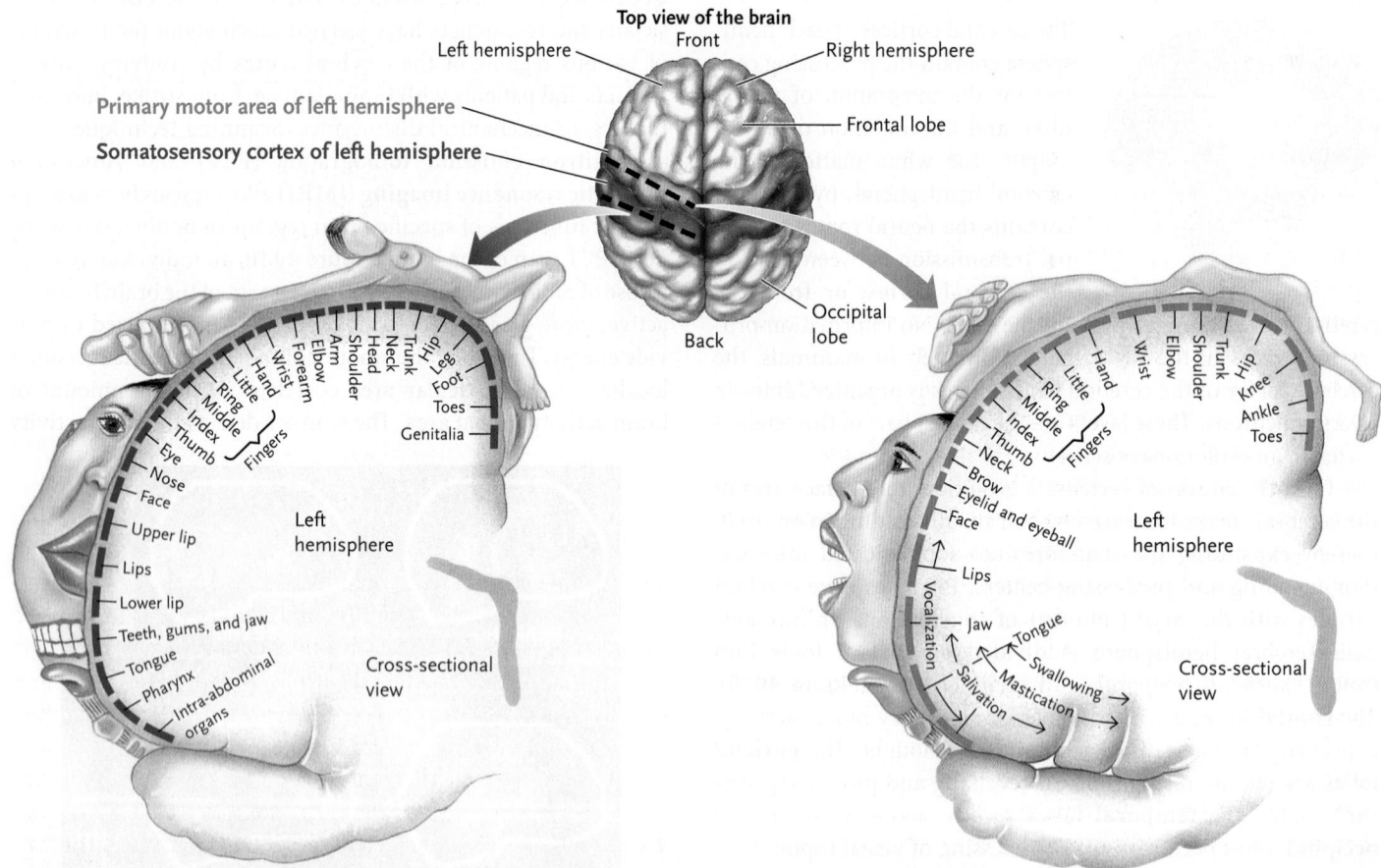

FIGURE 40.12 Maps of the somatosensory cortex and the primary motor cortex. The distorted representations of the human body show the relative distribution of sensory input to the somatosensory cortex from different parts of the body (left), and the relative distribution of motor output from the primary motor cortex to different parts of the body (right).

© Cengage Learning 2017

is involved in coordinating skeletal muscles for voluntary movements. Experimental stimulation of points along this band in one hemisphere causes movement of specific body parts on the opposite side of the body, corresponding generally to the parts registering in the somatosensory cortex at the same level (see Figure 40.11). Other areas that integrate and refine motor control are located nearby.

In both the somatosensory and primary motor cortices, some body parts, such as the lips and fingers, are represented by large regions, and others, such as the arms and legs, are represented by relatively small regions. As shown in Figure 40.12, the relative sizes produce a distorted image of the human body that is quite different from the actual body proportions. The differences are reflected in the precision of touch and movement in structures such as the lips, tongue, and fingers. Such differences are also seen in many other animals, reflecting their adaptations to sensing the environment.

OTHER MOTOR CONTROL AREAS OF THE CEREBRAL CORTEX
The **premotor cortex** of the frontal lobe (see Figure 40.10) controls skeletal muscles in coordinating complex movements. This area of the cerebral cortex is guided by sensory input received and processed by the **posterior parietal cortex** of the parietal lobe. Damage to either of these two motor areas results in an inability to process complex sensory information to carry out movements in space, such as working with tools.

LANGUAGE AREAS OF THE CEREBRAL CORTEX
Language is a written or spoken form of communication used to convey ideas. Language involves integrating *speaking ability* (expression) and *comprehension*. The two main areas of the cerebral cortex specialized for language are **Wernicke's area** in the parietal lobe and **Broca's area** in the frontal lobe (see Figure 40.10), which function in spoken and written language. They are usually present on only one side of the brain—in the left hemisphere in 97% of the human population. Comprehension of spoken and written language depends on coordination of inputs from the visual, auditory, and general sensory association areas by Wernicke's area. Interneuron connections lead from Wernicke's area to Broca's area, where the motor programs for coordination of the lips, tongue, jaws, and other structures producing the sounds of speech are generated before being passed to the primary motor cortex. The brain-scan images in Figure 40.11 illustrate dramatically how these brain regions participate as a person performs different linguistic tasks.

People with damage to Wernicke's area have difficulty comprehending spoken and written words, even though their hearing and vision are unimpaired. Although they can speak, their words usually make no sense. People with damage to Broca's area have normal comprehension of written and spoken language, and know what they want to say, but are unable to speak except for a few slow and poorly pronounced words. Often, such people are also unable to write. Other areas of the brain are involved in language functions as well.

ASSOCIATION AREAS OF THE CEREBRAL CORTEX
The sensory, motor, and language areas comprise about one half of the cerebral cortex. The other areas, called **association areas,** are involved in higher functions:

- The **prefrontal association cortex** of the frontal lobe is the key part of the brain involved in thinking, such as planning for voluntary activity, decision making, and creativity, as well as for personality traits.
- The **parietal–temporal–occipital association cortex** integrates all sensory input, such as locating objects by sight, touch, and/or sound, and relating the parts of the body to the external environment.
- The **limbic association cortex** of the temporal lobe is important for motivation, emotion, and memory.

PLASTICITY
In the context of the nervous system, **plasticity** is experience-dependent change in structure and function. In the brain, plasticity occurs more during development than in adulthood. An example of plasticity is a right-handed person learning to write with the left hand when the right one is broken. In this case, the writing skill now becomes developed in the other side of the cerebral cortex. The basis of plasticity, in part, is the formation of new synaptic connections between neurons (not the addition of new neurons) in response to experience. In this process, called **synaptic plasticity,** a neuron's dendrites can change in shape and extent, thereby establishing new synaptic connections with other neurons. The neuron now can receive and integrate more signals through the new connections. In this way, new neural pathways form.

Synaptic plasticity in the brain is limited by genetic and developmental constraints. For instance, the cerebral cortex regions for learning and adding new memories continue to exhibit synaptic plasticity throughout a person's lifetime, whereas other regions can be modified only for a limited time after birth.

Some Higher Functions Are Distributed in Both Cerebral Hemispheres; Others Are Concentrated in One Hemisphere

Most of the higher functions of the human brain—such as abstract thought and reasoning; spatial recognition; mathematical, musical, and artistic ability; and the associations forming the basis of personality—involve the coordinated participation of many regions of the cerebral cortex. Some of these regions are equally distributed in both cerebral hemispheres, and some are more concentrated in one hemisphere.

Among the functions more or less equally distributed between the two hemispheres is the ability to recognize faces. This function is concentrated along the bottom margins of the occipital and temporal lobes (see Figure 40.10). People with damage to these lobes are often unable to recognize even close relatives by sight but can recognize voices immediately.

Typically some brain functions are more localized in one of the two hemispheres, a phenomenon called **lateralization.**

Sex Differences in the Neural Connections of the Human Brain

Brain function is the result of activity of a complex network of neurons that exists within and between parts of the brain **(Figure).** The entire network of neural connections in the brain is called the **connectome** (see the Human Connectome Project at www .humanconnectomeproject.org).

Research Question

Is there a sex difference in the human connectome?

Experiments

Ragine Verma and his colleagues at the School of Medicine, University of

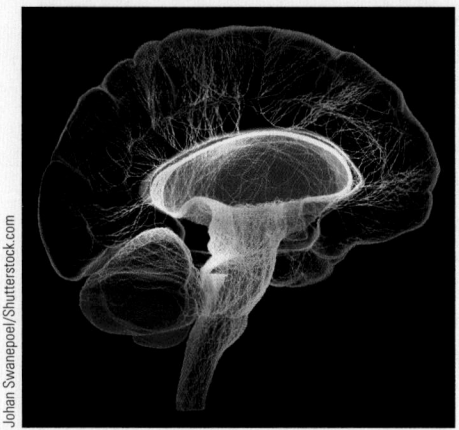

Johan Swanepoel/Shutterstock.com

FIGURE Human brain showing some of the neural connections.

Pennsylvania, mapped and analyzed the human connectome of 949 healthy young individuals (428 males and 531 females). For analysis, the researchers divided the subjects into three groups based on age: 8–13.3 years old, 13.4–17 years old, and 17.1–22 years old, corresponding approximately to the developmental states of childhood, adolescence, and young adulthood, respectively. The scientists mapped the connectomes using a type of MRI technique called diffusion tensor imaging (DTI) that tracks the paths of neural fibers in white matter.

Results

Analysis of the connectomes revealed significant differences in neural connectivity patterns between the sexes. In the region of the brain containing the cerebrum (the largest part of the brain), most connections that were stronger in males than in females were *within* each hemisphere, whereas most connections that were stronger in females than in males were *between* the two hemispheres. In the cerebellum, the part of the brain that plays a major role in motor control, however, the opposite was the case: males showed stronger connections between the left cerebellar hemisphere and the cerebral cortex of the right side.

Analysis of the three groups of subjects provided an insight into the development

of the sex differences of the connectome. In the childhood group the scientists observed only a few gender differences in connectivity, suggesting the beginning of a divergence between the two sexes. The sex differences were greater in both the adolescents and the young adults, but with the adolescents clearly showing an intermediary state between the children and the young adults.

Conclusion

The study revealed significant sex differences in the human connectome indicating different neural connectivity patterns in males and females. Based on the known functions of the brain, the researchers concluded that the results "suggest that male brains are structured to facilitate perception and coordinated action, whereas female brains are designed to facilitate communication between analytical and intuitive processing modes." The sex differences in the connectome develop over time, the divergence beginning at a relatively young age.

think like a scientist

Suggest some other experiments concerning the human connectome.

Source: M. Ingalhalikar et al. 2014. Sex differences in the structural connectome of the human brain. *Proceedings of the National Academy of Sciences USA* 111:823–828.

© Cengage Learning 2017

The unequal distribution of these functions was originally worked out in the 1960s by Roger Sperry and Michael S. Gazzaniga of the California Institute of Technology in subjects who had had their corpus callosum (see Figure 40.6) cut surgically **(Figure 40.13).** Sperry received a Nobel Prize in 1981 "for his discoveries concerning the functional specialization of the cerebral hemispheres."

Studies of people with split hemispheres, as well as surveys of brain activity by PET and fMRI, have confirmed that, for the vast majority of people, the left hemisphere specializes in spoken and written language, abstract reasoning, and precise mathematical calculations. The right hemisphere specializes in nonverbal conceptualizing, intuitive thinking, musical and artistic abilities, and spatial recognition functions such as fitting pieces into a puzzle. The right hemisphere also handles mathematical estimates and approximations that can be made by visual or spatial representations of numbers. Thus the left hemisphere in most people is verbal and mathematical, and the

right hemisphere is intuitive, spatial, artistic, and musical. (*Molecular Insights* discusses sex differences in the neural connections of the human brain.)

Genes Control the Structure and Function of the Brain

More than one third of the protein-coding genes in the human genome are active in the brain, the highest proportion in any part of the human body. Understanding the functions of those genes is crucial to understanding how the brain develops and operates. To this end, scientists at the Allen Institute for Brain Science, located in Seattle, Washington, are studying how gene expression and neural connections in the brain are integrated. The motivating goal is to understand how the human brain works in health and disease. In projects combining genomics approaches with neuroanatomy studies, the researchers are creating Brain Atlases, three-dimensional

FIGURE 40.13 **Experimental Research**

Investigating the Functions of the Cerebral Hemispheres

Question: Do the two cerebral hemispheres have different functions?

Experiment: Roger Sperry and Michael Gazzaniga studied split-brain individuals, in whom the corpus callosum connecting the two cerebral hemispheres had been surgically severed to relieve otherwise uncontrollable epileptic convulsions. In one experiment, they tested how subjects perceived words that were projected onto a screen in front of them.

The retinas of the eyes gather visual information and send signals via the optic nerves to the cerebral hemispheres **(Figure A)**. Light from the *left* half of the visual field reaches light receptors on the *right* sides of the retinas, and parts of the two optic nerves carry signals to the *right* cerebral hemisphere. Light from the *right* half of the visual field reaches light receptors on the *left* sides of the retinas, and signals are sent to the *left* cerebral hemisphere.

The researchers projected words such as COWBOY in such a way that the subjects could see only the left half of the word (COW) with the left eye, and the right half of the word (BOY) with the right eye **(Figure B)**. Sperry asked the subjects to say what word they saw, and he asked them to write the perceived word with the left hand—a hand that was deliberately blocked from the subject's view.

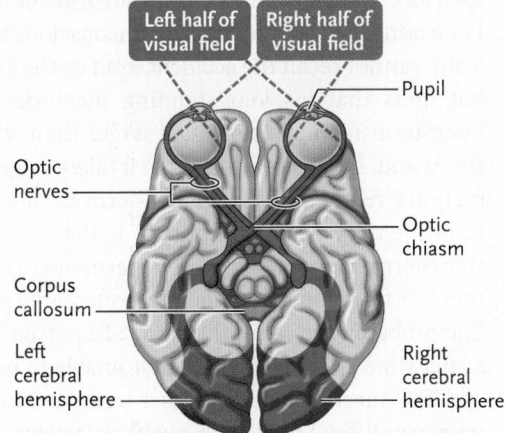

FIGURE A Pathway of visual information from eyes to cerebral hemisphere.

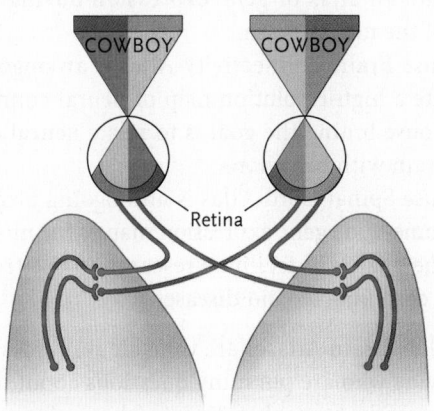

FIGURE B Experimental set up—COW seen by right sides of retinas and BOY by left sides.

Results: The split-brain subjects said the word in the right half of the visual field (BOY), but wrote the word in the left half of the visual field (COW) **(Figure C)**.

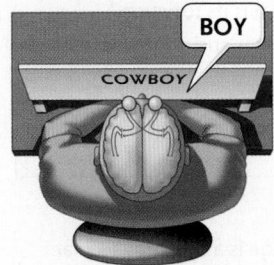

FIGURE C Results given by split-brain subjects.

Conclusions: The studies showed that the left and right hemispheres are specialized in different tasks. The left hemisphere processes language and was able to recognize BOY but received no information about COW. The right hemisphere directs motor activity on the left side of the body and was able to direct the left hand to write COW. However, the subjects could not say what word they wrote. That is, cutting the corpus callosum interrupted communication between the two halves of the cerebrum. In effect, one cerebral hemisphere did not know what the other was doing, and information stored in the memory on one side was not available to the other. In normal individuals, information is shared across the corpus callosum; they would see COWBOY and be able to speak and write the entire word.

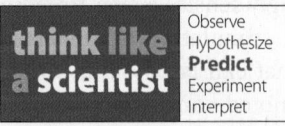

Observe
Hypothesize
Predict
Experiment
Interpret

Consider this study performed on split-brain patients. The researcher presents an image to only one side of the visual field. The patient is asked to describe the object. If, for instance, the researcher presented a flower vase to the right visual field, would the patient be able to describe it?

Source: M. S. Gazzaniga and R. W. Sperry. 1967. Language after section of the central commissures. *Brain* 90:131–148.

maps of gene expression in the brain. Five Brain Atlases have been generated or are in process:

- The Allen Mouse Brain Atlas is a genome-wide, high-resolution atlas of gene expression patterns in the normal adult mouse brain. This was the first Brain Atlas made public (in 2006). It was created as a resource to aid research into brain diseases on the grounds that a number of brain disease models are available in the mouse.
- The Allen Human Brain Atlas is an atlas of the normal human brain integrating genomic and anatomical information. This atlas was created as a resource to inform research into brain injuries and mental health disorders.
- The Allen Developing Mouse Brain Atlas is an ongoing project to create an atlas of gene expression during the development of the mouse brain.
- The Allen Mouse Brain Connectivity Atlas is an ongoing project to create a high-resolution map of neural connections in the mouse brain. The goal is to relate neural circuitry in the brain with behavior.
- The Allen Mouse Spinal Cord Atlas is an ongoing project to create a genome-wide gene expression map of the mouse spinal cord. The goal is to facilitate research in the treatment of spinal cord injuries and diseases.

Overall, the Allen Brain Atlases are valuable resources for researchers worldwide who are pursuing questions about normal brain function and about alterations of brain function caused by disease or injury. Each Brain Atlas is a dynamic database that is updated continually with new information as research continues.

STUDY BREAK 40.3

1. What is the blood–brain barrier, and what is its function?
2. What is the difference between gray matter and white matter?
3. What is the function of the brain stem?
4. Distinguish the structure and functions of the cerebellum from those of the cerebral cortex.

THINK OUTSIDE THE BOOK

The most common cause of brain damage is a cerebrovascular accident (CVA), commonly called a stroke. Use the Internet or research literature to outline the biological cause of a stroke and why a stroke produces different symptoms in different stroke victims.

40.4 Memory, Learning, and Consciousness

We set memory, learning, and consciousness apart from the other functions because they appear to involve coordination of structures from the brain stem to the cerebral cortex. **Memory** is the storage and retrieval of a acquired knowledge for later recall. **Learning** is the acquisition of knowledge or skills as a result of experience, instruction, or both. **Consciousness** may be defined as awareness of ourselves, our identity, and our surroundings, and an understanding of the significance and likely consequences of events that we experience.

Memory Takes Two Forms, Short Term and Long Term

Psychology research and our everyday experiences indicate that humans store acquired knowledge in two stages: (1) **short-term memory,** which lasts for seconds, minutes, or at most an hour or so; and (2) **long-term memory,** which stores information from days to years. Short-term memory, but not long-term memory, is usually erased if a person experiences a disruption such as a sudden fright, a blow, a surprise, or an electrical shock. For example, a person knocked unconscious by an accident typically cannot recall the accident itself or the events just before it, but finds that his long-standing memories are undisturbed. Long-term memory stores are larger than short-term memory stores and, because of this, often it takes longer to retrieve information ("remember") from long-term memory than from short-term memory. Another contrast is that information lost from short-term memory is forgotten permanently, whereas information lost from long-term memory often is forgotten temporarily. You probably can recall the name of a person "coming to you" at a later time after you were at first unable to remember it.

The mechanisms for short-term memory and long-term memory differ. Short-term memory involves transient changes in the functions of synapses, for example, in the amount of a neurotransmitter released in response to a stimulus, or increased responsiveness of a postsynaptic cell to a neurotransmitter. By contrast, storage of long-term memory involves more or less permanent structural and functional changes between existing neurons.

Memories register initially in short-term form. They are then either erased and lost, or committed to long-term form. The intensity or vividness of an experience, the attention focused on an event, emotional involvement, or the degree of repetition may all contribute to the conversion from short-term to long-term memory. People with injuries to the hippocampus cannot remember information for more than a few minutes. Their long-term memory is limited to information stored before the injury occurred.

How are neurons and neuron pathways altered permanently to create long-term memory? One change that has been much studied is **long-term potentiation:** a long-lasting increase in the strength of synaptic connections in activated neural pathways following brief periods of repeated stimulation. The synapses become increasingly sensitive over time, so that a constant level of presynaptic stimulation is converted into a larger postsynaptic output that can last hours, weeks, months, or years. Another change noted consistently as part of long-term memory is synaptic plasticity; that is, more or less permanent alterations occur in the number and the area of synaptic connections between neurons, and in the number and branches of dendrites. Gene transcription and protein synthesis have also been shown to change in interneurons.

Experiments have shown that protein synthesis is critical to long-term memory storage in animals as varied as mollusks, *Drosophila,* and rats. For example, goldfish were trained to avoid an electrical shock by swimming to one end of an aquarium when a light was turned on. The fish could remember the training for about a month under normal conditions; if exposed to a protein synthesis inhibitor while being trained, they forgot the training within a day.

Learning Involves Combining Past and Present Experiences to Modify Responses

All animals appear to be capable of learning to some degree. Learning involves three sequential mechanisms: (1) storing memories; (2) scanning memories when a stimulus is encountered; and (3) modifying the response to the stimulus in accordance with the information stored as memory.

One of the simplest forms of memory is **sensitization**—increased responsiveness to mild stimuli after experiencing a strong stimulus. The process was nicely illustrated by Eric Kandel of Columbia University and his associates in experiments with a shell-less marine snail known as the Pacific sea hare, *Aplysia californica,* which is frequently used in research involving reflex behavior, memory, and learning. Many of its neuron circuits have been worked out completely, allowing investigators to follow the reactions of each neuron active in pathways such as learning. The first time the researchers administered a single sharp tap to the siphon (which admits water to the gills), the slug retracted its gills by a reflex movement. However, at the next touch, whether hard or gentle, the siphon retracted much more quickly and vigorously. Sensitization in *A. californica* has been shown to involve changes in synapses, which become more reactive when more serotonin is released by action potentials. Kandel received a Nobel Prize in 2000 "for discoveries concerning signal transduction in the nervous system." (Learning and its relationship to animal behavior are considered further in Chapter 56.)

Consciousness Involves Different States of Awareness

The spectrum of human consciousness ranges from alert wakefulness to daydreaming, dozing, and sleep. Even during sleep there is some degree of awareness, because sleepers can respond to stimuli and waken, unlike someone who is unconscious. Moving between the states of consciousness has been found to involve changes in neural activity over the entire surface of the telencephalon.

When an individual is fully awake, an *electroencephalogram* (EEG; a chart of the brain electrical activity detectable by electrodes placed on the scalp) records a pattern of rapid, irregular *beta waves* **(Figure 40.14).** With thought processes at rest and eyes closed, the person's EEG pattern changes to slower and more regular *alpha waves.* As drowsiness and light sleep come

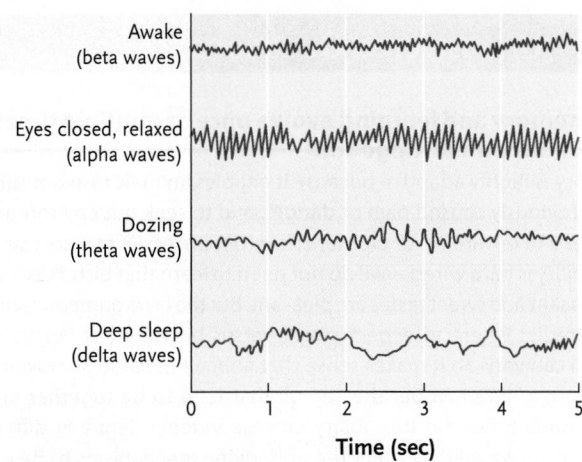

FIGURE 40.14 Brain waves characteristic of various states of consciousness.
© Cengage Learning 2017

on, the array of waves gradually become larger, slower, but again less regular. These slower pulsations are called *theta waves.* During the transition from drowsiness to deep sleep, the EEG pattern shifts to even slower *delta waves.* The heart and breathing rates become slower and the skeletal muscles increasingly relax, although the sleeper may still change position.

Periodically during deep sleep, the delta wave pattern is replaced by the rapid, irregular beta waves characteristic of the waking state. The person's heartbeat and breathing rate increase, the limbs twitch, and the eyes move rapidly behind the closed eyelids, giving this phase its name of **rapid-eye-movement (REM) sleep.** The REM sleep phase occurs about every 1.5 hours while a healthy adult is sleeping, and lasts for 10 to 15 minutes. Sleepers do most of their dreaming during REM sleep, and most research subjects awakened from REM sleep report they were experiencing vivid dreams.

As mentioned earlier, the reticular activating system controls the sleep–wake cycle. It sends signals to the spinal cord, cerebellum, and cerebral cortex, and receives signals from the same locations. The flow of signals along these circuits determines whether we are awake or asleep.

Many other animals also alternate periods of wakefulness and sleep or inactivity. Although sleep obviously has restorative effects on mental and physical functions, the physiological basis of these effects remains unknown.

In the previous chapter we learned about neurons, and in this chapter we have discussed the organization of neurons into nervous systems, as well as the structures of the brain and their functions. In the next chapter, we consider the sensory systems that provide input for the brain to process.

STUDY BREAK 40.4

An aging person often experiences a progressive decline in cognitive function. This typically begins with short-term memory loss and the inability to learn new information. What parts of the brain might be changing?

Did memory and learning evolve once or multiple times within the animal kingdom?

Memory is highly adaptive because it enables animals to avoid stimuli that previously caused pain or danger, and to seek out environments that lead to rewarding or pleasurable experiences. Of course, some of this ability is hard wired—we do not need to learn that bitter tastes are unpleasant and sweet tastes are pleasant. But the environmental stimuli that predict where to expect sweet versus bitter or hot versus cold stimuli can vary. So it makes sense that animals capable of movement can learn and remember the stimuli that tend to be together in the environment. But did this ability emerge independently in different phyla? If so, we might expect the underlying mechanisms to be quite different in the different phylogenetic groups, even if evolutionary adaptations have converged onto similar memory abilities (see Section 23.3 on convergent evolution). Or did memory evolve one time in a common ancestor before the various animal phyla diverged? In this case, we might expect to find shared underlying mechanisms no matter how distantly related the organisms. We do not know the answer to these questions. But given what we do know about memory mechanisms, a common origin seems likely even for animals as distantly related as invertebrates and vertebrates.

Genetic studies have provided many key insights into mechanisms of memory formation. One of the most remarkable findings from genetic investigation is that many of the same genes and biochemical signaling pathways are utilized for memory formation and storage no matter what species one examines. In several cases, genetic pathways relevant to human cognition also appear to underlie memory in animals as distant from humans as fruit flies. Cyclic AMP (cAMP) signaling is one of the best examples; this signaling pathway was first connected to memory in invertebrates of the genera *Aplysia* (sea hares) and *Drosophila* (fruit flies), then in rodents, and more recently in humans. In fact, the conversion from short-term to long-term memory in fruit flies, sea hares, bees, mice, rats, and probably humans involves activation of a gene called the cAMP responsive enhancer binding protein (CREB). The discovery of shared genetic mechanisms of memory supports the idea that memory mechanisms in vertebrate and invertebrate animals share an evolutionary origin. What about similarities in the neural circuits that support memory formation?

Similarities at the neural circuit level between humans and mice or rats are relatively easy to detect because the overall organization of rodent and human brains is similar. Thus, we can compare the requirements for neural circuits that are likely to be related through evolution. The similarities are harder to establish when we compare invertebrates and vertebrates because their brains are so different in size and structure. But even here, there are some hints that the way neural circuits store memories is at least analogous. A good example comes from studies of the neural circuits involved in conversion of short-term to long-term memory. An emerging theme is that different neural circuits may be required for short-term and long-term memory. Initially, this "dissection" of neural circuit function came from precise surgical lesion of brain regions in rodents and chickens. In certain cases, these lesions caused disruptions of either short-term or long-term memory, but not both. Similar effects are seen in people with localized brain damage.

More modern genetic approaches now permit disruptions that are specific not only to brain regions, but to individual neuronal cell types within a region. In our laboratory, we used this approach recently to determine which neurons are involved in short-term and long-term memory formation in fruit flies. We used a Pavlovian learning test in which the flies learn that a specific odor predicts that they will receive an electric shock. We then used precise genetic manipulations to test which neuron cell types require cAMP signaling for either short-term or long-term memory. We focused on a brain region called the *mushroom bodies* because previous work already had demonstrated that this is an olfactory learning center in insects. To our surprise, we found that two different neuronal cell types are required for short-term versus long-term memory.

We do not know why brains have evolved specialized circuits for short-term and long-term storage of the same memory. Perhaps it reflects an important feature for storing information within a neural network. The question is whether this feature emerged independently in different phyla or once in a common ancestor.

think like a **scientist**

You come across two related species of beetles and you observe their behaviors. One beetle species seems always to feed on a single food source from a single plant species; we call it a "specialist." The second beetle species forages widely across many different food sources; it is a "generalist." You decide to study their learning abilities in the lab, and you design a fairly artificial learning task. You try to teach these animals that a pure chemical odor predicts an electric shock. If you had to guess (and you do), which beetle species will exhibit better learning in your experiment? Why?

Josh Dubnau is an Associate Professor at Cold Spring Harbor Laboratory where his research focuses on memory in fruit flies. One theme in Dr. Dubnau's work is an attempt to use both reductionist methods and more holistic observations at the organismal level. You can learn more about Dubnau's research by visiting http://dubnaulab.cshl.edu/.

Josh Dubnau

REVIEW KEY CONCEPTS

For access to MindTap and additional study materials visit www.cengagebrain.com.

40.1 Invertebrate and Vertebrate Nervous Systems Compared

- The simplest nervous systems are the nerve nets of cnidarians. Echinoderms have modified nerve nets, with some neurons grouped into nerves (Figure 40.1A, B).

- Flatworms, arthropods, and mollusks have a simple central nervous system (CNS), consisting of ganglia in the head region (a brain), and a peripheral nervous system (PNS), consisting of nerves from the CNS to the rest of the body (Figure 40.1C–E).

- In vertebrates, the CNS consists of a large brain located in the head and a hollow spinal cord, and the PNS consists of all the nerves and ganglia connecting the CNS to the rest of the body (Figure 40.1F).

- In the vertebrate embryo, the anterior end of the hollow neural tube develops into the brain, and the rest develops into the spinal cord. The embryonic brain enlarges into the forebrain, midbrain, and hindbrain, which develop into the adult structures (Figure 40.2).

40.2 The Peripheral Nervous System

- Afferent neurons in the PNS conduct signals to the CNS, and signals from the CNS travel via efferent neurons to the effectors—muscles and glands—that carry out responses (Figure 40.3).

- The somatic nervous system of the PNS controls the skeletal muscles, producing voluntary body movements, as well as involuntary muscle contractions that maintain balance, posture, and muscle tone (Figure 40.3).

- The autonomic nervous system of the PNS, which controls involuntary functions, is organized into the sympathetic nervous system and the parasympathetic nervous system. The autonomic nervous system controls digestive activities along with the nerves in the digestive tract (the enteric nervous system) and hormones (Figure 40.3). The pathways of the autonomic nervous system include two neurons (Figures 40.4 and 40.5).

40.3 The Central Nervous System and Its Functions

- The brain and the spinal cord are surrounded and protected by the meninges. Cerebrospinal fluid provides nutrients and cushions the CNS. A blood–brain barrier allows only selected substances to enter the cerebrospinal fluid.

- Each adult brain structure contains gray matter and white matter. Functionally, the brain integrates sensory information and generates responses. The cerebrum is divided into right and left cerebral hemispheres, which are connected by a thick band of nerve fibers, the corpus callosum. Each hemisphere consists of the cerebral cortex, a thin layer of gray matter covering a thick core of white matter. Other collections of gray matter, the basal nuclei, are deep in the telencephalon (Figure 40.6).

- The spinal cord carries signals between the brain and the PNS. Its neuron circuits control reflex muscular movements and some autonomic reflexes (Figure 40.7).

- The medulla, pons, and midbrain form the brain stem, which connects the cerebrum, thalamus, and hypothalamus with the spinal cord.

- Gray-matter centers in the pons and medulla control involuntary functions. Centers in the midbrain coordinate responses to visual and auditory sensory inputs.

- The reticular formation receives sensory inputs from all parts of the body and sends outputs to the cerebral cortex that help maintain balance, posture, and muscle tone. It also regulates states of wakefulness and sleep (Figure 40.8).

- The cerebellum integrates sensory inputs on the positions of muscles and joints, along with visual and auditory information, to coordinate body movements.

- The telencephalon's subcortical gray-matter centers control many functions. The thalamus receives, filters, and relays sensory and motor information to and from regions of the cerebral cortex. The hypothalamus regulates basic homeostatic functions of the body and contributes to the endocrine control of body functions. The basal nuclei affect the planning and fine-tuning of body movements (Figure 40.9).

- The limbic system includes parts of the thalamus, hypothalamus, and basal nuclei, as well as the amygdala and hippocampus. It controls emotions and influences the basic body functions controlled by the hypothalamus and brain stem (Figure 40.9).

- The somatosensory cortex of the cerebral cortex registers incoming information on touch, pain, temperature, and pressure from all parts of the body. In general, the right cerebral hemisphere receives sensory information from the left side of the body and vice versa (Figures 40.10 and 40.12).

- The primary motor cortex of the cerebral cortex coordinates skeletal muscles for voluntary movements. The premotor cortex controls skeletal muscles in coordinating complex movements guided by sensory input from the posterior parietal cortex (Figures 40.10 and 40.12).

- Wernicke's area in the parietal lobe and Broca's area in the frontal lobe are the two main areas of the cerebral cortex specialized for language. Wernicke's area integrates visual, auditory, and other sensory information into the comprehension of language; Broca's area coordinates movements of the lips, tongue, jaws, and other structures to produce the sounds of speech (Figure 40.10).

- Association areas of the cerebral cortex are involved in higher functions: the prefrontal association cortex for thinking, decision making, creativity, and personality traits, the parietal–temporal–occipital association cortex for integrating all sensory input, and the limbic association cortex for motivation, emotion, and memory (Figure 40.10).

- The brain shows plasticity, experience-dependent changes in structure and function. Plasticity involves changes in synaptic connections between neurons in response to experience.

- Long-term memory and consciousness are equally distributed between the two cerebral hemispheres. Spoken and written language, abstract reasoning, and precise mathematical calculations are left hemisphere functions; nonverbal conceptualizing, mathematical estimation, intuitive thinking, spatial recognition, and artistic and musical abilities are right hemisphere functions (Figure 40.13).

- Genes control the structure and function of the brain. Brain Atlases—three-dimensional maps of gene expression in the brain—have been or are being generated with the goal of understanding how gene expression and neural connections in the brain are integrated.

40.4 Memory, Learning, and Consciousness

- Memory is the storage and retrieval of a sensory or motor experience or a thought. Short-term memory involves temporary storage of information, whereas long-term memory is essentially permanent, involving structural and functional changes in neuronal connections as a result of synaptic plasticity.

- Learning involves modification of a response through comparisons made with information or experiences that are stored in memory.

- Consciousness is the awareness of ourselves, our identity, and our surroundings. It varies from full alertness to sleep and is controlled by the reticular activating system (Figure 40.14).

TEST YOUR KNOWLEDGE

Remember/Understand

1. Ganglia first became enlarged and fused into a lobed brain in the evolution of:
   a. vertebrates.
   b. annelids.
   c. flatworms.
   d. cephalopods.
   e. mammals.

2. The hindbrain subdivides and eventually gives rise to the:
   a. spinal cord.
   b. cerebellum.
   c. mesencephalon.
   d. thalamus.
   e. cerebrum.

3. The autonomic nervous system is subdivided into:
   a. afferent and efferent systems.
   b. sympathetic and parasympathetic divisions.
   c. skeletal and smooth muscle innervations.
   d. voluntary and involuntary controls.
   e. peripheral and central systems.

4. People with severe insect-sting allergies carry an EpiPen, an autoinjector containing medication that they can use in an emergency. The medication causes smooth muscles in the lung passages to relax so they can breathe, but causes their hearts to pound rapidly. This is an example of stimulation of the:
   a. parasympathetic system.
   b. sympathetic system.
   c. somatic nervous system.
   d. limbic system.
   e. voluntary system.

5. Which of the following statements about the blood–brain barrier is *incorrect*?
   a. It is formed of capillary walls that are composed of tight junctions.
   b. It transports glucose to brain cells by means of transport proteins.
   c. It allows alcohol to pass through its lipid bilayer.
   d. Oxygen can move through the lipid bilayer.
   e. It reduces blood supply to brain cells compared with other body cells.

6. Which one of the following structures participates in a reflex?
   a. the gray matter of the brain
   b. the white matter of the brain
   c. the gray matter of the spinal cord
   d. an interneuron that stimulates an afferent neuron
   e. an interneuron that inhibits an afferent neuron

7. A segment of the brain stem that coordinates spinal reflexes with higher brain centers and regulates breathing and wakefulness is the:
   a. reticular formation.
   b. white matter of the pons.
   c. white matter of the medulla.
   d. hypothalamus.
   e. cerebellum.

8. Cushioning and nourishing the brain and spinal cord and filling the ventricles of the brain is (are):
   a. meninges.
   b. myelin.
   c. cerebrospinal fluid.
   d. ganglia.
   e. glucose.

9. Which structure and function are correctly paired below?
   a. thalamus: relays emotion signals through the limbic system
   b. basal nuclei: relay inputs from odor receptors to the cerebrum
   c. hypothalamus: releases hormones; sets up daily rhythms
   d. amygdala: relays sensory information to the cerebrum
   e. olfactory bulbs: moderate motor centers in the cerebrum

10. A patient had a tumor in Wernicke's area. It was initially diagnosed when the patient could not:
    a. understand the morning newspaper.
    b. hear a child crying.
    c. see the traffic light turn red.
    d. speak.
    e. feel if the car heater was on.

Apply/Analyze

11. **Discuss Concepts** Meningitis is an inflammation of the meninges, the membranes that cover the brain and spinal cord. Diagnosis involves using a needle to obtain a sample of cerebro-spinal fluid to analyze for signs of infection. Why analyze this fluid and not blood?

12. **Discuss Concepts** An accident victim arrives at the emergency department with severe damage to the reticular formation. Based on information in this chapter, describe some of the symptoms that the examining physician might discover.

13. **Discuss Concepts** In the 1930s and 1940s, prefrontal lobotomy, in which neural connections in the frontal lobes of both cerebral hemispheres were severed, was used to treat behavioral conditions such as extreme anxiety and rebellious-ness. Although the procedure calmed patients, it had side effects such as apathy and a seriously disrupted personality. In view of the information presented in this chapter, why do you think the operation had these effects?

Evaluate/Create

14. **Design an Experiment** How would you demonstrate, using mice, that gene activity in the brain is altered by aging?

15. **Apply Evolutionary Thinking** How do paleontologists con-tribute to our understanding of the evolution of the brain?

For selected answers, see Appendix A.

INTERPRET THE DATA

Animal studies are often used to assess the effects of prenatal exposure to drugs. For example, Jack Lipton used rats to study the behavioral effect of prenatal exposure to MDMA (3,4-methylenedioxymeth-amphetamine), the active ingredient in the illicit drug ecstasy. He injected female rats with either MDMA or saline solution when they were 14–20 days pregnant and their offspring's brains were forming. When those offspring were 21 days old, Lipton tested their response to a new environment. He placed each young rat in a new cage and used a photobeam system to record how much each rat moved around before settling down. The **Figure** shows his results.

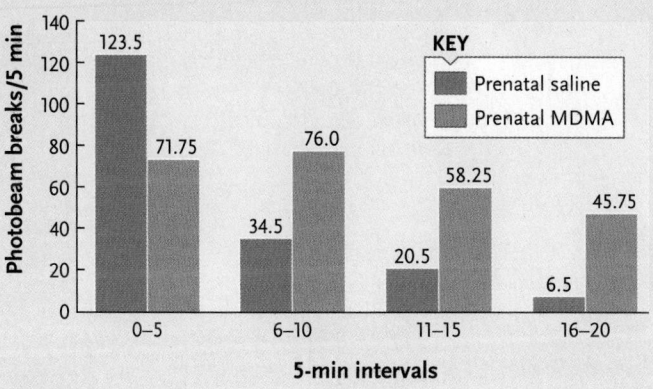

1. Which rats moved most (caused the most photobeam breaks) during the first 5 minutes in a new cage: those prenatally exposed to MDMA or the controls?

2. How many photobeam breaks did the MDMA-exposed rats make during their second 5 minutes in the new cage?

3. Which rats moved most during the last 5 minutes of the study?

4. Does this study support the hypothesis that exposure to MDMA affects a developing rat's brain?

Source: Based on J. B. Koprich et al. 2003. Prenatal 3,4-methylenedioxmeth-amphetamine (ecstasy) alters exploratory behavior, reduces monoamine metabolism, and increases forebrain tyrosine hydroxylase fiber density of juvenile rats. *Neurotoxicology and Teratology* 25:509–517.

41 Sensory Systems

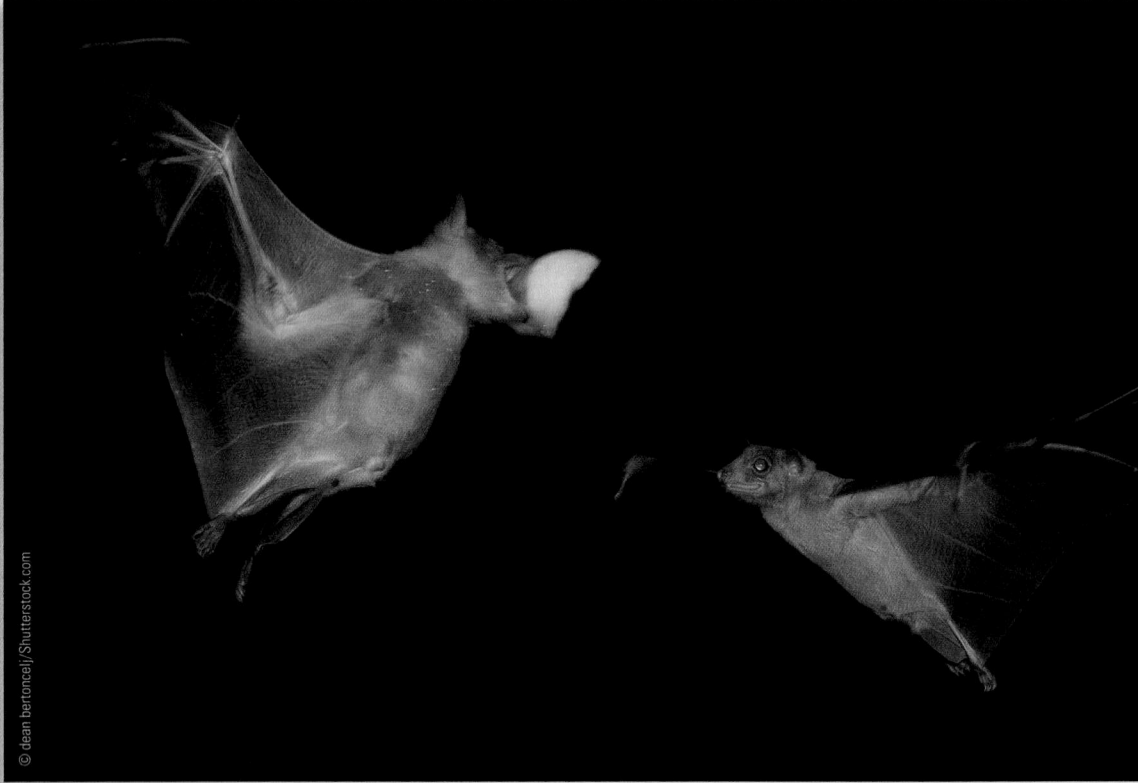

© dean bertoncelj/Shutterstock.com

Two fruit bats fighting for food.

Why it matters . . . An insectivorous bat searches for food. As it flies, the bat emits ultrasonic clicking noises. Receptors in the bat's ears detect echoes of the clicks bouncing off of objects and send signals to the brain, where they are integrated into a sound map that the animal uses to navigate through the environment. This ability, called **echolocation,** is so well developed that a bat can detect and avoid a thin wire in the dark.

Besides recognizing obstacles, the bat's auditory system is keenly tuned to the distinctive pattern of echoes from the fluttering wings of moths. Although the slow-flying moth would seem doomed to become a meal for the foraging bat, some species of moths also have a highly sensitive auditory sense. On each side of its abdomen is an "ear," a thin membrane that resonates at the frequencies of the bat's clicks. The moth's ears register the clicks while the bat is still about 30 m away and initiate a response that turns its flight path directly away from the source of the clicks.

In spite of the moth's evasive turns, if the bat approaches within about 6 m of it, echoes from the moth begin to register in the bat's auditory system, and the bat increases the frequency of its clicks, enabling it to pinpoint the moth's position. But, as the bat closes in, the increased frequency of the clicks sets off another programmed response that alters the moth's flight into sudden loops and turns, ending with a closed-wing, vertical fall toward the ground. After dropping a few feet, the moth resumes its flight and may again be detected by the bat, and so it goes.

Echolocation is not confined to bats. For instance, porpoises use echolocation to locate food fishes in murky waters, and some bird species use echolocation to avoid obstacles and find their nests in dark caves.

Natural selection has produced highly adaptive sensory receptors in all animals. These systems, the subject of this chapter, provide animals with a steady stream of information about their internal and external environments. After integrating the information in the central nervous system (CNS), animals respond in ways that enable them to survive and reproduce.

We begin this chapter with a survey of animal sensory systems and the ways in which they work. Then we examine several individual receptor types and their characteristics.

41.1 Overview of Sensory Receptors and Pathways

A **stimulus** is a change detected by the body. A variety of energy forms are stimuli, including heat, light, sound, and pressure. Stimuli are detected by **sensory receptors.** Sensory receptors associated with eyes, ears, skin, and other surface organs detect stimuli from the external environment. Sensory receptors associated with internal organs detect stimuli arising in the body interior. The receptors occur in three structural forms **(Figure 41.1).** Two of the forms involve peripheral endings of an afferent neuron, and the third form is a specialized cell that synapses with an afferent neuron.

Each sensory receptor has a defined **receptive field,** a region surrounding the receptor within which the receptor

responds to a stimulus. Sensory receptors respond to stimuli in their receptive fields by undergoing a change in membrane potential. The change, called a **receptor potential,** varies in magnitude with the magnitude of the stimulus; thus, it is a graded potential (see Section 39.4). In most receptors, the change is caused by changes in the rate at which channels conduct positive ions such as Na^+, K^+, or Ca^{2+} across the plasma membrane. The conversion of a stimulus into a receptor potential is called **sensory transduction.** If the receptor potential is large enough, it will trigger an action potential in the afferent neuron that travels along the axon into the interneuron networks of the CNS. These interneurons integrate the sensory stimuli, and the brain formulates a compensating response (see Section 40.3).

Five Basic Types of Receptors Are Common to Almost All Animals

Many sensory receptors are positioned individually in body tissues. Others are part of complex sensory organs, such as the eyes or ears, which are specialized for reception of physical or chemical stimuli. Commonly, sensory receptors are classified

A. Sensory receptor consisting of free nerve endings—dendrites of an afferent neuron

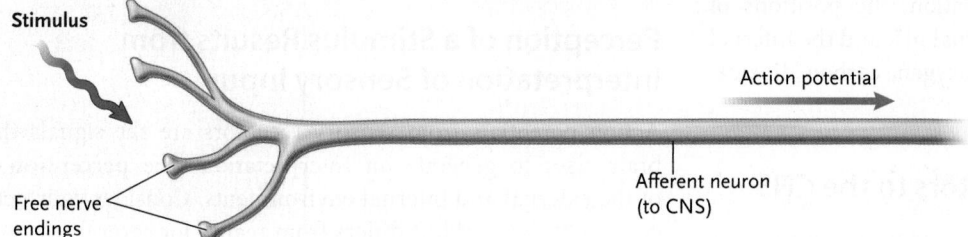

In sensory receptors consisting of free dendrites of afferent neurons, a stimulus causes a change in membrane potential that generates action potentials in the axon of the neuron. Examples are pain receptors and some mechanoreceptors.

B. Sense organ—sensory receptor involving nerve endings of an afferent neuron enclosed in a specialized structure

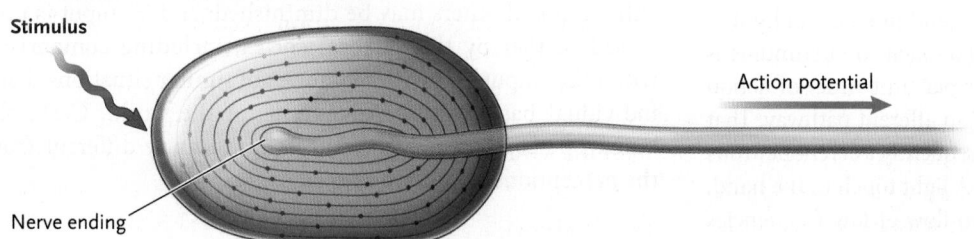

In sensory receptors involving endings of an afferent neuron enclosed in a specialized structure, a stimulus affecting the structure triggers an action potential in the neuron. Some mechanoreceptors are of this type.

C. Sensory receptor formed by a cell that synapses with an afferent neuron

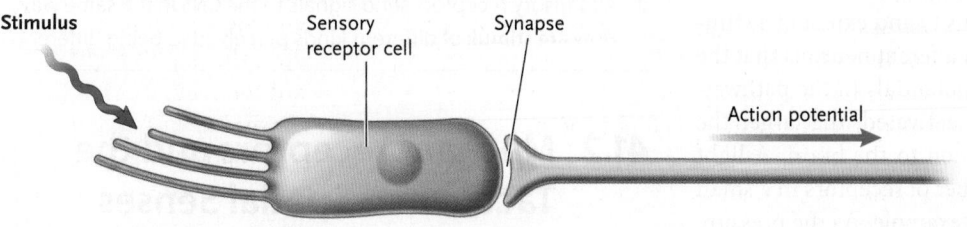

In sensory receptors consisting of separate cells, a stimulus causes a change in membrane potential that releases a neurotransmitter from the cell. The neurotransmitter triggers an action potential in the axon of an afferent neuron to which the sensory receptor cell is synapsed. Examples are photoreceptors, chemoreceptors, and some mechanoreceptors.

FIGURE 41.1 Three forms of sensory receptors.
© Cengage Learning 2017

into five major types, based on the type of stimulus that each detects:

1. **Mechanoreceptors** detect mechanical energy when it deforms membranes. Changes in pressure, body position, or acceleration are detected by mechanoreceptors, for instance. The auditory receptors in the ears are examples of mechanoreceptors.
2. **Photoreceptors** detect the energy of light. In vertebrates, photoreceptors are mostly located in the retina of the eye.
3. **Chemoreceptors** detect specific molecules, or chemical conditions such as acidity. The taste buds on the tongue are examples of chemoreceptors.
4. **Thermoreceptors** detect the flow of heat energy. Receptors of this type are located in the skin.
5. **Nociceptors** detect tissue damage or noxious chemicals; their activity registers as pain. Pain receptors are located in the skin, and also in some internal organs.

In addition to these major types, some animals have receptors that can detect electrical or magnetic fields.

Although humans are traditionally said to have five senses—vision, hearing, taste, smell, and touch—our sensory receptors actually detect more than four times as many kinds of environmental stimuli. Among these are external heat, internal temperature, gravity, acceleration, the positions of muscles and joints, body balance, internal pH, and the internal concentration of substances such as oxygen, carbon dioxide, salts, and glucose.

Afferent Neurons Link Receptors to the CNS

Sensory pathways begin at a sensory receptor and proceed by afferent neurons to a particular location in the CNS. For example, action potentials arising in the retina of the eye travel along the optic nerve to the visual cortex, where they are interpreted by the brain as differences in the pattern, color, and intensity of light.

One way in which the intensity and extent of a stimulus is registered is by the frequency (number per unit time) of action potentials traveling along each axon of an afferent pathway. That is, the stronger the stimulus, the more frequently afferent neurons fire action potentials (see Section 39.2). A light touch to the hand, for example, causes action potentials to flow at low frequencies along the axons of afferent neurons to the somatosensory cortex of the cerebral cortex (see Section 40.3). As the pressure increases, the frequency of action potentials rises in proportion.

The second way in which the intensity and extent of a stimulus are registered is by the number of afferent neurons that the stimulus activates to generate action potentials in the pathway. The more sensory receptors that are activated, the larger the number of axons that carry information to the brain. A light touch activates a relatively small number of receptors in a small area near the surface of the finger, for example. As the pressure increases, the resulting indentation of the finger's surface increases in area and depth, activating more and different types of receptors.

Many Receptor Systems Reduce Their Response When Stimuli Remain Constant

In many systems, the effect of a stimulus is lessened if it continues at a constant level. This **sensory adaptation** reduces the frequency of action potentials generated in afferent neurons.

For example, when you go to bed, you are initially aware of the touch and pressure of the covers on your skin. Within a few minutes, the sensations lessen or are lost even though your position remains the same. The loss reflects adaptation of mechanoreceptors in your skin. By contrast, nociceptors do not adapt to painful stimuli; the frequency of action potentials remains elevated as long as the stimulus is detected.

Sensory adaptation also increases the sensitivity of receptor systems to *changes* in environmental stimuli, which may be more important to survival than keeping track of environmental factors that remain constant. Consider a cat sitting motionless, focused on its prey, a mouse. As long as the environmental stimuli are constant, the cat's position remains fixed. However, if the mouse moves, the cat will respond rapidly and pounce.

Nonadapting receptors, such as those detecting pain, are also essential for survival. Pain signals a potential danger to some part of the body, and the signals are maintained until a response by the animal compensates for the stimulus causing the pain.

Perception of a Stimulus Results from Interpretation of Sensory Input

Action potentials from sensory receptors are the signals the brain uses to generate an interpretation—the **perception**—of the external and internal environments. Consider your perception of the world. It differs from reality for several reasons: (1) you lack receptors for particular types of energy, such as X-rays; (2) sensory input begins to be processed before it reaches the cerebral cortex, so some features of the stimuli may be enhanced and others may be diminished; and (3) input is processed further by the cerebral cortex, including comparison with other input and with memories of similar situations. Each individual has different perceptions of the world. Certainly, human perception of the world is significantly different from the perceptions of other organisms.

STUDY BREAK 41.1

1. Define sensory transduction.
2. All sensory receptors send signals to the CNS in the same way. How are stimuli of different kinds perceived as being different?

41.2 Mechanoreceptors and the Tactile and Spatial Senses

Mechanical stimuli such as touch and movement are detected by mechanoreceptors. The mechanical forces of a stimulus create tension in the plasma membrane of a receptor, causing ion channels to open and producing a receptor potential. If the receptor potential is large enough, an action potential is triggered in the associated afferent neuron leading to the CNS. Sensory information from mechanoreceptors informs the brain of the body's contact with objects in the environment; provides information on the movement, position, and balance of body parts; and underlies the sense of hearing.

Receptors for Touch and Pressure Occur throughout the Body

In vertebrates, mechanoreceptors detecting touch and pressure are embedded in the skin and other surface tissues, in skeletal muscles, in the walls of blood vessels, and in internal organs. In humans, receptive fields for touch in the skin are smallest in the fingertips, lips, and tip of the tongue, giving these regions the greatest sensitivity to mechanical stimuli. In other areas, such as the skin of the back, arms, and legs, the receptive fields are much larger.

You can compare the receptive fields of touch receptors by pressing two toothpicks lightly against a fingertip and then against the skin of your arm or leg. On your fingertip, the points can be quite close together—separated by only a millimeter or so—and still be discerned as separate. On your arm or leg, they must be nearly 5 cm (almost 2 inches) apart to be distinguished.

Human skin contains several types of touch and pressure receptors **(Figure 41.2)**. Some are free nerve endings, the dendrites of afferent neurons with no specialized structures surrounding them. Free nerve endings wrapped around hair follicles respond when the hair is bent, making you instantly aware, for example, of a breath of air, or a spider on your arm. Other mechanoreceptors, such as Pacinian corpuscles, have structures surrounding the nerve endings that contribute to reception of stimuli.

Proprioceptors Provide Information about Movements and Position of the Body

Mechanoreceptors called **proprioceptors** (*proprius* = one's own) detect stimuli that are used in the CNS to maintain body balance and equilibrium and to monitor the position of the head and limbs. The activity of these receptors allows you to touch the tip of your nose with your eyes closed, for example, or reach and brush away that spider on your arm.

TENSION RECEPTORS IN VERTEBRATES Proprioceptors in the muscles and tendons of vertebrates detect the position and movement of the limbs, thereby allowing the CNS to monitor the body's position and help keep the body in balance. They also allow muscles to apply constant force under a constant load, and to adjust almost instantly if the load changes. The sensory receptors that detect the length of a muscle and its contraction are

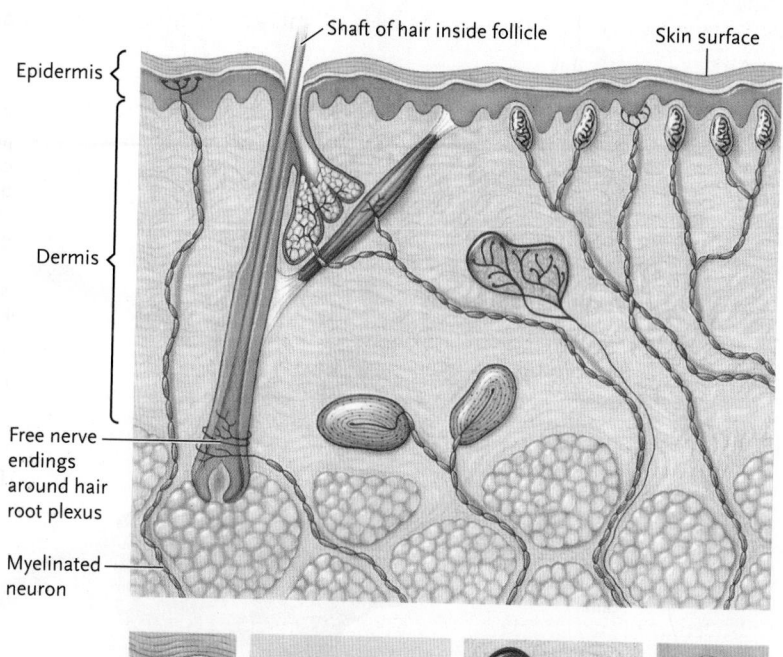

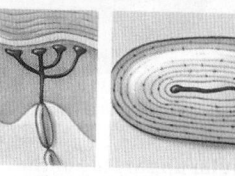 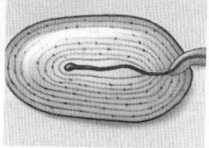 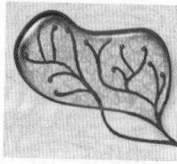 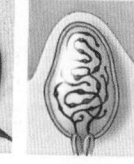

Free nerve endings: light touch

Pacinian corpuscle: deep pressure and vibrations

Ruffini endings: deep pressure

Meissner's corpuscle: light touch, surface vibrations

FIGURE 41.2 Types of mechanoreceptors that detect tactile stimuli in human skin.
© Cengage Learning 2017

muscle spindles, bundles of small, specialized muscle cells wrapped with the dendrites of afferent neurons and enclosed in connective tissue. When you hold a cup while someone fills it with coffee, for example, the muscle spindles in your biceps muscle detect the additional stretch as the cup becomes heavier. Signals from the muscle spindles allow you to compensate for the additional weight by increasing the contraction of the muscle, keeping the cup level with no conscious effort on your part. Proprioceptors are typically slow to adapt, so that the body's position and balance are constantly monitored.

THE VESTIBULAR APPARATUS IN VERTEBRATES The inner ear of most terrestrial vertebrates has two specialized sensory structures, the *vestibular apparatus* and the *cochlea*. The **vestibular apparatus (Figure 41.3)** detects rotational motions of the head and provides information about the up–down positioning of the head, as well as changes in the rate of linear movement of the head. The vestibular apparatus consists of three *semicircular canals*, the *utricle*, and the *saccule*, which are all filled with a fluid called *endolymph*. The **semicircular canals,** which are positioned at angles corresponding to the three planes of space, detect rotational ("spinning") motions. Each canal has a swelling at its base called an *ampulla*, which is

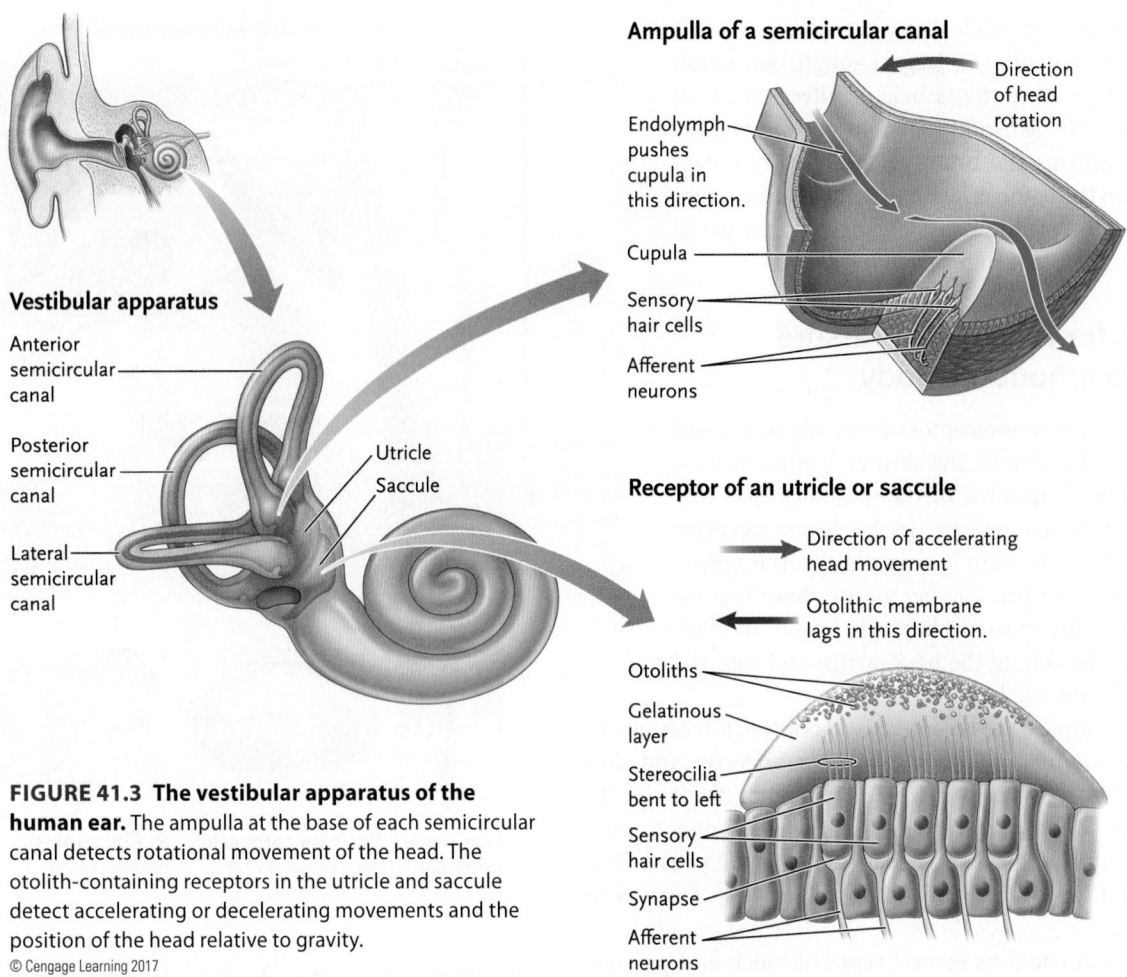

Vestibular apparatus

Anterior semicircular canal

Posterior semicircular canal

Lateral semicircular canal

Utricle
Saccule

Ampulla of a semicircular canal

Direction of head rotation

Endolymph pushes cupula in this direction.

Cupula

Sensory hair cells

Afferent neurons

Receptor of an utricle or saccule

Direction of accelerating head movement

Otolithic membrane lags in this direction.

Otoliths

Gelatinous layer

Stereocilia bent to left

Sensory hair cells

Synapse

Afferent neurons

FIGURE 41.3 The vestibular apparatus of the human ear. The ampulla at the base of each semicircular canal detects rotational movement of the head. The otolith-containing receptors in the utricle and saccule detect accelerating or decelerating movements and the position of the head relative to gravity.

© Cengage Learning 2017

topped with sensory hair cells that synapse with afferent neurons. The surface of a hair cell at the other end from the synapse is covered with **stereocilia** (singular, *stereocilium*), which are actually microvilli (cell processes reinforced by bundles of microfilaments). The stereocilia extend into a gelatinous structure, the **cupula** (*cupula* = little cup), which protrudes into the endolymph of the canals.

When the head rotates horizontally, vertically, or diagonally, the endolymph in the semicircular canal corresponding to that direction lags behind, pulling the cupula with it. The displacement of the cupula bends the sensory hair cells. Channels at the tips of the stereocilia open and K^+ enters, causing the sensory hair cell to depolarize. The depolarization opens Ca^{2+} channels in the main part of the hair cell, and entry of Ca^{2+} causes the release of neurotransmitter from the hair cell into the synapse. The neurotransmitter triggers action potentials in the afferent neuron.

The **utricle** and **saccule** provide information about the position of the head with respect to gravity (up versus down), as well as changes in the rate of linear movement of the head. They are saclike structures in a bony chamber located between the semicircular canals and the cochlea. Oriented approximately 30° to each other, both utricle and saccule contain sensory hair cells with stereocilia. The hair cells are covered with a

gelatinous *otolithic membrane* (which is similar to a cupula) in which **otoliths** (*oto* = ear; *lithos* = stone), small crystals of calcium carbonate, are embedded (see Figure 41.3).

When an animal is upright, the sensory hairs in the utricle are oriented vertically, and those in the saccule are oriented horizontally. When the head is tilted in any direction other than straight up and down, or when there is a change in linear motion of the body, the otolithic membrane of the utricle moves and bends the sensory hairs. Depending on the direction of movement, the hair cells release more or less neurotransmitter, which changes the rate of action potentials in the afferent neurons. The utricle and saccule adapt quickly to the head's motion, decreasing their response when there is no change in the rate and direction of movement. For instance, when you move your head to the left, that new position becomes the "norm." Then, if you move your head again, signals from the utricle and saccule tell your brain that your head is moving to a new position.

THE LATERAL LINE SYSTEM IN AMPHIBIANS AND FISHES

Fishes and some aquatic amphibians detect vibrations and currents in the water through a series of mechanoreceptors along the length of the body called the **lateral line system (Figure 41.4).** In fish, the mechanoreceptors, known as *neuromasts,* also provide information about the fish's orientation with respect to

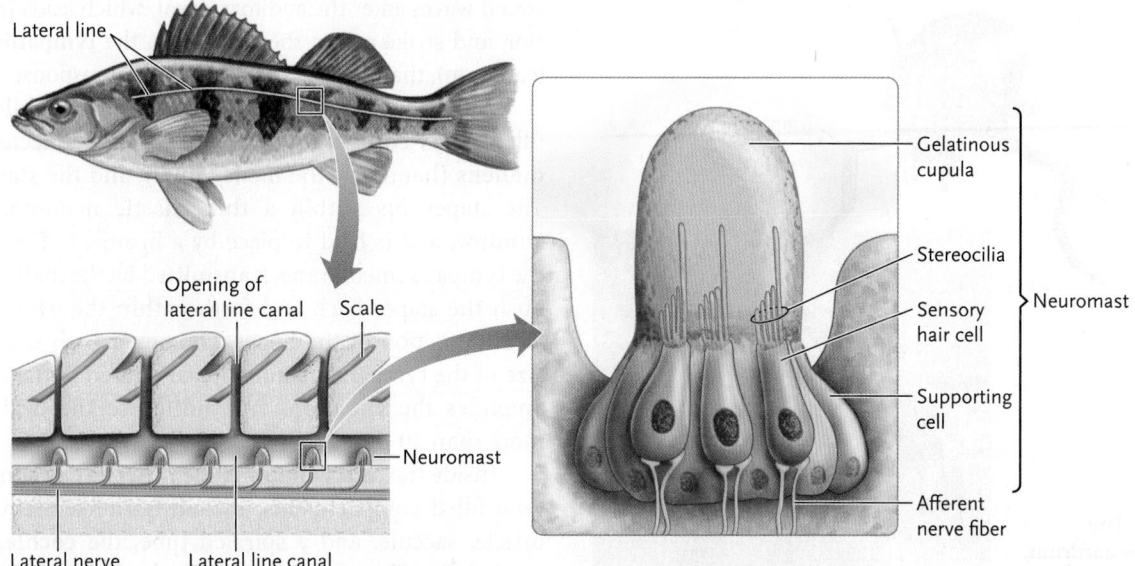

FIGURE 41.4 The lateral line system of fishes. The sensory receptor of the lateral line, the neuromast, has a gelatinous cupula that is pushed and pulled by vibrations and currents transmitted through the lateral line canal. As the cupula moves, the stereocilia of the sensory hair cells are bent, generating action potentials in afferent neurons that lead to the brain.
© Cengage Learning 2017

gravity and its swimming velocity. Each dome-shaped neuromast has sensory hair cells clustered in its base. Stereocilia on the hair cells extend into a cupula similar to that of the vertebrate vestibular apparatus, which moves with pressure changes in the surrounding water. Movement of the cupula bends the stereocilia, which leads to the triggering of action potentials in associated afferent neurons.

Vibrations detected by the lateral line enable fishes to avoid obstacles, orient in a current, and monitor the presence of other moving objects in the water. The system is also responsible for the ability of schools of fish to move in unison, turning and diving in what appears to be a perfectly synchronized aquatic ballet. In actuality, the movement of each fish creates a pressure wave in the water that is detected by the lateral line systems of other fishes in the school. Schooling fish can still swim in unison even if blinded, but if the nerves leading from the lateral line system to the brain are severed, the ability to school is lost.

STATOCYSTS Many aquatic invertebrates, including jellyfish, some gastropods, and some arthropods, have organs of equilibrium called **statocysts** (*statos* = standing; *kystis* = bladder, pouch). Most statocysts are fluid-filled chambers with walls that contain sensory hair cells enclosing one or more movable stonelike bodies called **statoliths (Figure 41.5),** which are similar to vertebrate otoliths. When the animal moves, the statoliths lag behind the movement, bending the sensory hairs and triggering action potentials in afferent neurons. In this way, the statocysts signal the brain about the body's position and orientation with respect to gravity.

STUDY BREAK 41.2

1. What is the function of proprioceptors?
2. What properties qualify proprioceptors as mechanoreceptors?

41.3 Mechanoreceptors and Hearing

Sounds are vibrations that travel as waves produced by the alternating compression and rarefaction of the air. The loudness, or *intensity,* of a sound depends on the amplitude ("height") of the wave. The *pitch* of a sound—a high note or a low note—depends on the frequency of the waves, measured in hertz (cycles per second). The more cycles per second, the higher the pitch. Some animals, such as the bat in the *Why it matters . . .* for this chapter, can hear sounds well above 100,000 hertz. Humans can hear

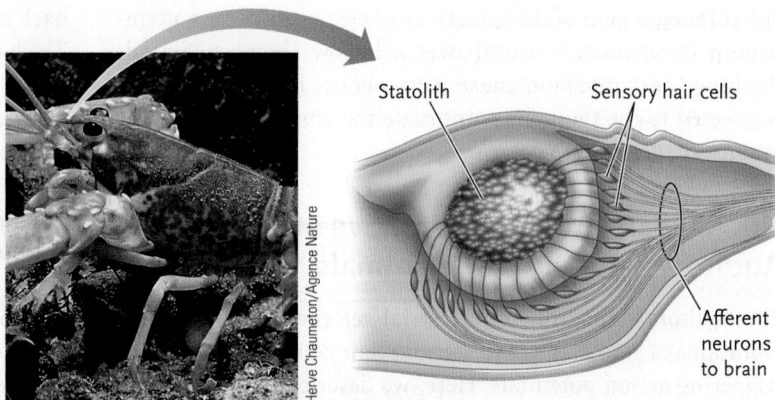

Herve Chaumeton/Agence Nature

FIGURE 41.5 A statocyst, an invertebrate organ of equilibrium, and its location at the base of an antenna in a lobster. The statoliths inside are usually formed from fused grains of sand, as they are in the lobster, or from calcium carbonate.
© Cengage Learning 2017

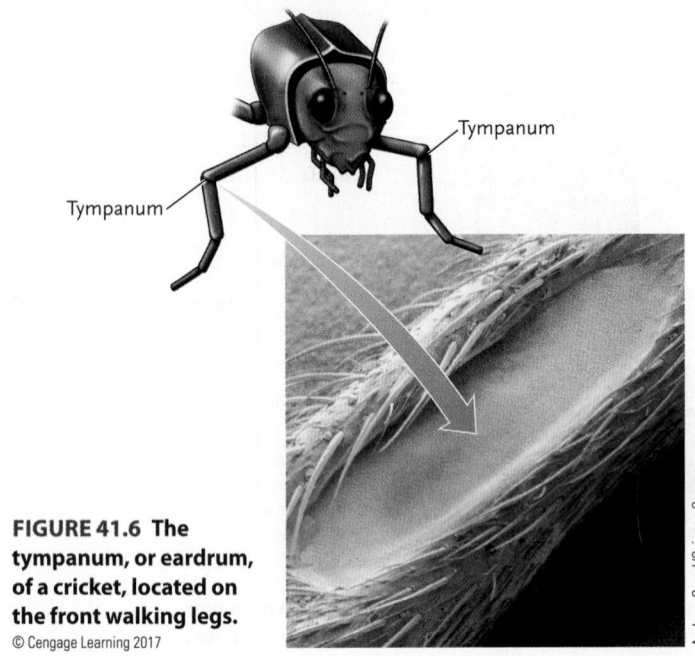

FIGURE 41.6 The tympanum, or eardrum, of a cricket, located on the front walking legs.
© Cengage Learning 2017

Andrew Syred/Science Source

sounds between about 20 and 20,000 hertz, so we are deaf to the bat's sonar clicks.

Invertebrates Have Varied Vibration-Detecting Systems

Most invertebrates detect sound and other vibrations through mechanoreceptors in their skin or other surface structures. An earthworm, for example, quickly retracts into its burrow at the smallest vibration of the surrounding earth, even though it has no specialized structures serving as ears. Squids and octopuses have a system of mechanoreceptors on their head and tentacles, similar to the lateral line of fishes. Many insects have sensory receptors in the form of hairs or bristles that vibrate in response to sound waves.

Some insects, such as moths, grasshoppers, and crickets, have complex auditory organs on either side of the abdomen or on the first pair of walking legs **(Figure 41.6)**. These "ears" consist of a thinned region of the insect's exoskeleton that forms a tympanum (*tympanum* = drum) over a hollow chamber. Sounds reaching the tympanum cause it to vibrate; mechanoreceptors connected to the tympanum translate the vibrations into nerve impulses.

Human Ears Are Representative of the Auditory Structures of Mammals

The auditory structures of terrestrial vertebrates transmit the vibrations of sound waves to sensory hair cells that respond by triggering action potentials. Here, we describe the human ear, which is representative of mammalian auditory structures **(Figure 41.7)**.

The **outer ear** has an external structure, the **pinna** (*pinna* = feather, wing), which concentrates and focuses sound waves. The sound waves enter the auditory canal, which leads from the exterior, and strike a thin sheet of tissue, the **tympanic membrane** (eardrum), that vibrates back and forth in response.

Behind the tympanic membrane is the **middle ear,** an air-filled cavity containing three small, interconnected bones: the **malleus** (hammer), the **incus** (anvil), and the **stapes** (stirrup). The stapes fits within a thin, elastic membrane, the **oval window,** and is held in place by a ligament. The vibrations of the tympanic membrane, transmitted by the malleus and incus, push the stapes back and forth within the oval window. The levering action of the bones, combined with the much larger size of the tympanic membrane compared to the oval window, amplifies the vibrations transmitted to the oval window by more than 20 times.

Inside the oval window is the **inner ear.** It contains several fluid-filled compartments, including the semicircular canals, utricle, saccule, and a spiraled tube, the **cochlea** (*cochlea* = snail shell). The cochlea twists through about two and a half turns; if stretched out flat, it would be about 3.5 cm long. Thin membranes divide the cochlea into three longitudinal chambers, the *vestibular canal* at the top, the *cochlear duct* in the middle, and the *tympanic canal* at the bottom (see Figure 41.7). The vestibular canal and the tympanic canal join at the outer tip of the cochlea, so that the fluid within them is continuous. Within the cochlear duct is the **organ of Corti** (also called the **spiral organ**), which contains the sensory hair cells that detect sound vibrations transmitted to the inner ear (see Figure 41.7).

The vibrations of the oval window pass through the fluid in the vestibular canal, make the turn at the end, and travel back through the fluid in the tympanic canal. At the end of the tympanic canal, they are transmitted to the **round window,** a thin membrane that faces the middle ear.

The vibrations traveling through the inner ear cause the basilar membrane to vibrate in response. The **basilar membrane,** which forms part of the floor of the cochlear duct, anchors the sensory hair cells in the organ of Corti. The stereocilia of these cells are embedded in the **tectorial membrane,** which extends the length of the cochlear canal. When the basilar membrane vibrates, the stereocilia of the hair cells are bent back and forth in relation to the stationary tectorial membrane. The back and forth bending of the sensory hair cells alternately opens and closes ion channels in the stereocilia of the cells, causing alternating depolarization and hyperpolarization. Neurotransmitter release from the sensory hair cells triggers action potentials in the associated afferent neurons.

Pitch discrimination, the ability to distinguish between various frequencies of incoming sound stimuli, depends on the shape and properties of the basilar membrane. The basilar membrane is narrow at the oval window end and wide at its outer end. Due to its shape, different sound frequencies cause different regions of the basilar membrane to vibrate maximally: the narrow end vibrates most with high-frequency pitches, and the wide end vibrates most with low-frequency pitches. Wherever the basilar membrane vibrates, the sensory hair cells in that area move back and forth and trigger action potentials. In

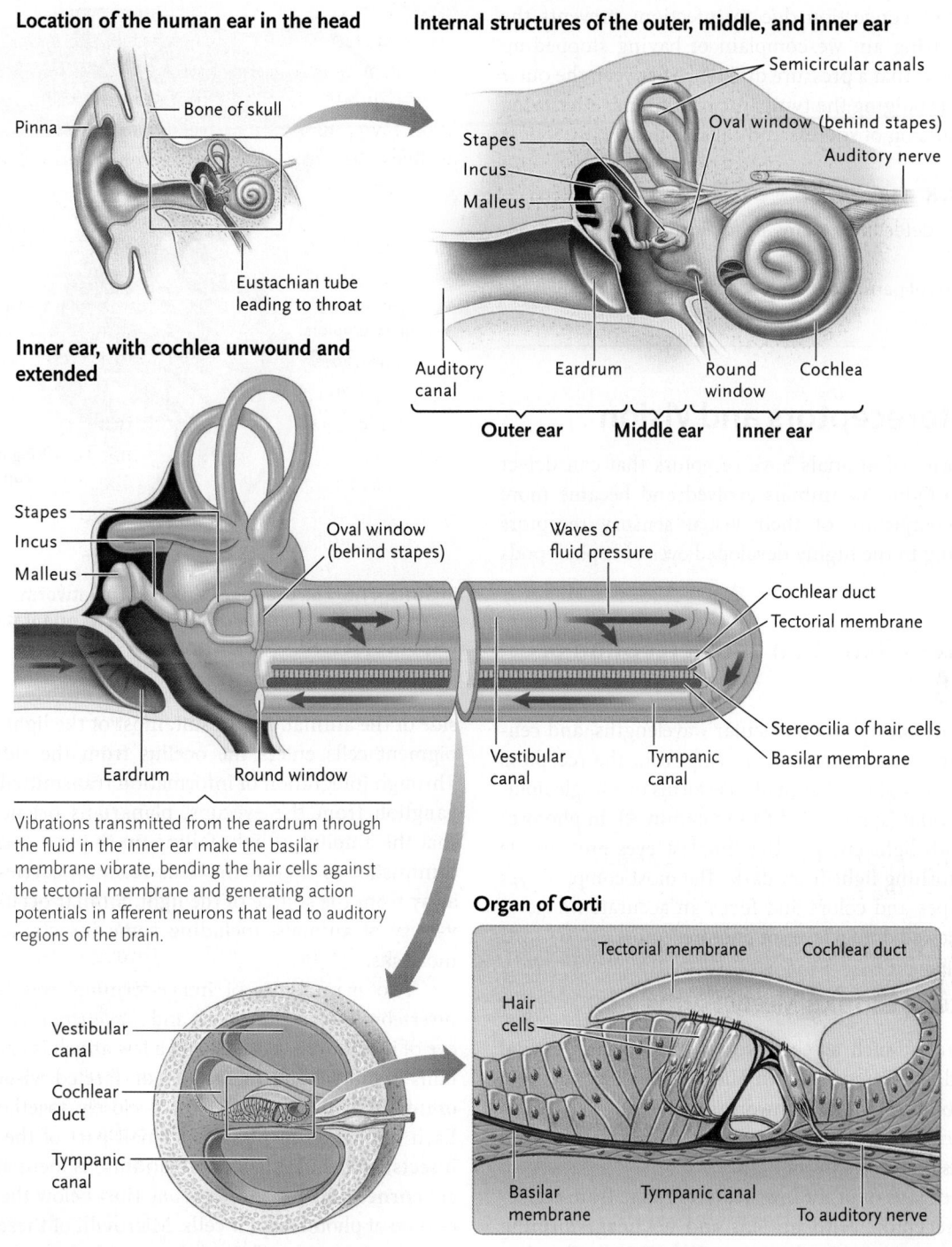

Location of the human ear in the head

Pinna

Bone of skull

Eustachian tube
leading to throat

Internal structures of the outer, middle, and inner ear

Semicircular canals

Oval window (behind stapes)

Auditory nerve

Stapes

Incus

Malleus

Auditory canal

Eardrum

Round window

Cochlea

Outer ear Middle ear Inner ear

Inner ear, with cochlea unwound and extended

Stapes

Incus

Malleus

Oval window (behind stapes)

Waves of fluid pressure

Cochlear duct

Tectorial membrane

Stereocilia of hair cells

Basilar membrane

Eardrum Round window

Vestibular canal

Tympanic canal

Vibrations transmitted from the eardrum through the fluid in the inner ear make the basilar membrane vibrate, bending the hair cells against the tectorial membrane and generating action potentials in afferent neurons that lead to auditory regions of the brain.

Vestibular canal

Cochlear duct

Tympanic canal

Organ of Corti

Tectorial membrane Cochlear duct

Hair cells

Basilar membrane

Tympanic canal

To auditory nerve

FIGURE 41.7 Structures of the human ear.
© Cengage Learning 2017

essence, the sensory hair cells in a particular location in the basilar membrane are tuned to a particular sound frequency.

More than 15,000 hair cells are distributed in small groups along the basilar membrane, enabling a high degree of pitch discrimination. Each group of hairs is connected by synapses to afferent neurons, which in turn are bundled together in the *cochlear nerve,* a cranial nerve that carries information through intermediate regions and into the thalamus. From there, the signals are routed to specific regions in the temporal lobe,

which integrates the information into the perception of sound at a corresponding pitch and loudness.

Another system protects the tympanic membrane from damage by changes in environmental atmospheric pressure. The system depends on the *Eustachian tube* (also called the *auditory tube*), a duct that leads from the air-filled middle ear to the throat (see Figure 41.7). As we swallow or yawn, the tube opens, allowing air to flow into or out of the middle ear to equalize the pressure on both sides of the tympanic membrane.

When swelling or congestion due to infections prevents the tube from admitting air, we complain of having stopped-up ears—we can sense that a pressure difference between the outer and middle ear is bulging the tympanic membrane, interfering with the transmission of sounds and causing pain.

STUDY BREAK 41.3

1. What vibration-detecting systems are found in octopuses and insects?
2. How are sounds of particular frequencies distinguished and "heard" by humans?

41.4 Photoreceptors and Vision

The great majority of animals have receptors that can detect and respond to light. As animals evolved and became more complex, the complexity of their visual sensory receptors increased, leading to the highly developed eyes of cephalopods and vertebrates.

Vision Involves Detection and Perception of Radiant Energy

Photoreceptors detect light at particular wavelengths, and centers in the brain integrate signals arriving from the receptors into a perception of light. All animals use forms of a single lipid-like pigment, *retinal* (synthesized from vitamin A), in photoreceptors to absorb light energy. The simplest eyes are capable only of distinguishing light from dark. The most complex eyes distinguish shapes and colors and focus an accurate image of objects being viewed onto a layer of photoreceptors.

Invertebrate Eyes Take Many Forms

Some invertebrates, such as earthworms, do not have visual organs. Instead, photoreceptors in their skin allow them to sense and respond to light. Earthworms respond negatively to light, as you can discover easily by shining a flashlight on an earthworm outside its burrow at night.

The visual organs of other invertebrates range from collections of photoreceptors with no lens and no image-forming capability to eyes remarkably like those of vertebrates. The photoreceptors of invertebrates are depolarized when they absorb light, and generate action potentials or increase their release of neurotransmitter molecules when they are stimulated.

The simplest eye is the **ocellus** (plural, *ocelli;* also called an *eyespot* or *eyecup*). An ocellus, which detects light but does not form an image, consists of fewer than 100 photoreceptor cells lining a cup or pit. In planarians, for example, photoreceptor cells in a cuplike depression below the epidermis are connected to the dendrites of afferent neurons that are bundled into nerves that travel from the ocelli to the cerebral ganglion **(Figure 41.8)**. Each ocellus is covered on one side by a layer of pigment cells that blocks most of the light rays arriving from the opposite

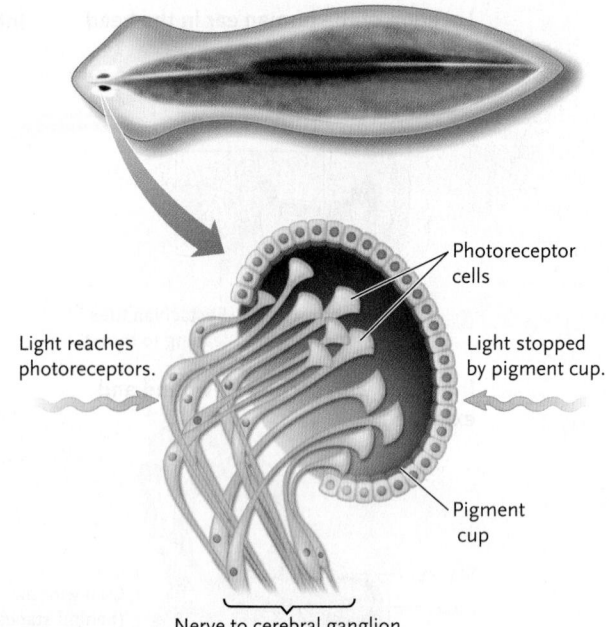

FIGURE 41.8 The ocellus of a planarian flatworm, and the arrangement of pigment cells on which its orientation response is based.
© Cengage Learning 2017

side of the animal. As a result, most of the light received by the pigment cells enters the ocellus from the side that it faces. Through integration of information transmitted to the cerebral ganglion from the eyecups, planarians orient themselves so that the amount of light falling on the two ocelli is equal and diminishes as they swim. This reaction carries them directly away from the source of the light. Similar ocelli are found in a variety of animals, including some insects, arthropods, and mollusks.

Two main types of image-forming eyes have evolved in invertebrates: *compound eyes* and *single-lens eyes*. The **compound eye** of insects, crustaceans, and a few annelids and mollusks contains hundreds to thousands of faceted visual units called **ommatidia** (*omma* = eye) fitted closely together **(Figure 41.9)**. Each ommatidium samples a small part of the visual field. In insects, light entering an ommatidium is focused by a transparent **cornea** and a *crystalline cone* (just below the cornea) onto a bundle of photoreceptor cells. Microvilli of these cells interdigitate like the fingers of clasped hands, forming a central axis rich in **rhodopsin,** a **photopigment** (light-absorbing pigment) consisting of retinal bound covalently to an **opsin** protein. Absorption of light by rhodopsin causes action potentials to be generated in afferent neurons connected to the base of the ommatidium. From these signals, the brain receives a mosaic image of the world. Because even the slightest motion is detected simultaneously by many ommatidia, compound eyes are extraordinarily adept at detecting movement—a lesson soon learned by fly-swatting humans.

The **single-lens eye** of cephalopods **(Figure 41.10)** resembles a vertebrate eye in that both types operate like a camera. In the cephalopod eye, light enters through the transparent cornea, a

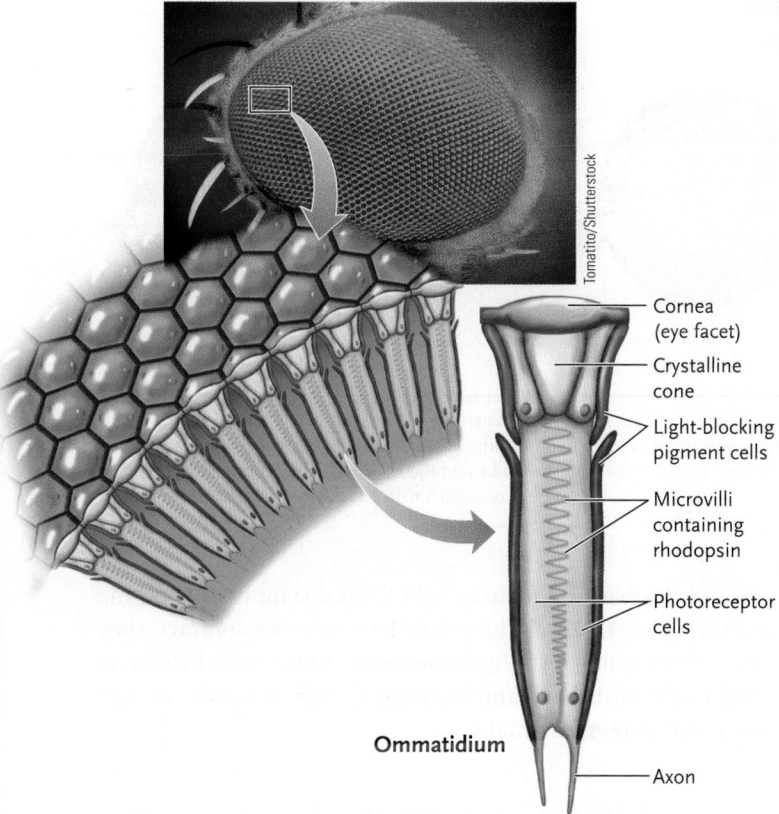

FIGURE 41.9 **The compound eye of a fly.** Each ommatidium has a cornea that directs light into the crystalline cone; in turn, the cone focuses light on the photoreceptor cells. A light-blocking pigment layer at the sides of the ommatidium prevents light from scattering laterally in the compound eye.
© Cengage Learning 2017

Labels for Figure 41.9:
- Cornea (eye facet)
- Crystalline cone
- Light-blocking pigment cells
- Microvilli containing rhodopsin
- Photoreceptor cells
- Ommatidium
- Axon

lens concentrates the light, and a layer of photoreceptors at the back of the eye, the **retina,** records the image. Behind the cornea is the **iris,** which surrounds the **pupil,** the opening through which light enters the eye. Muscles in the iris adjust the size of the pupil to vary the amount of light entering the eye. When the light is bright, circular muscles in the iris contract, shrinking the size of the pupil and reducing the amount of light that enters. In dim light, radial muscles contract and enlarge the pupil, increasing the amount of light that enters the eye.

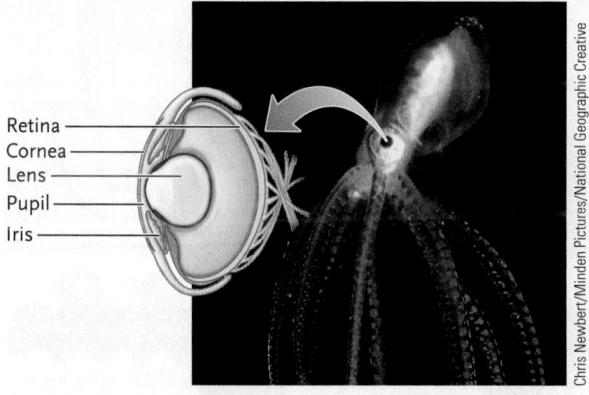

FIGURE 41.10 **The eye of an octopus, a cephalopod mollusk.**
© Cengage Learning 2017

Labels for Figure 41.10:
- Retina
- Cornea
- Lens
- Pupil
- Iris

Muscles move the lens forward and back with respect to the retina to focus the image. This is an example of **accommodation,** a process by which the lens changes to enable the eye to focus on objects at different distances.

A neural network lies under the retina, meaning that light rays do not have to pass through the neurons to reach the photoreceptors. The vertebrate eye has the opposite arrangement. This and other differences in structure and function indicate that cephalopod and vertebrate eyes evolved independently.

The Structure of Vertebrate Eyes

The human eye **(Figure 41.11)** has similar structures—cornea, iris, pupil, lens, and retina—to those of the cephalopod eye just described. Light entering the eye through the cornea passes through the iris and then the lens. The lens focuses an image on the retina, and the axons of afferent neurons originating in the retina converge to form the optic nerve leading from the eye to the brain.

A clear fluid called the **aqueous humor** fills the space between the cornea and lens. This fluid carries nutrients to the lens and cornea, which do not contain any blood vessels. The main chamber of the eye, located between the lens and the retina, is filled with the jellylike **vitreous humor** (*vitrum* = glass). The outer wall of the eye contains a tough layer of connective tissue (the *sclera*). Inside it is a darkly pigmented layer (the *choroid*) that prevents light from entering except through the pupil. It also contains the blood vessels nourishing the retina.

Human photoreceptors, rods and cones, occur in the retina along with layers of neurons that carry out an initial integration of visual information before it is sent to the brain; rods and cones function differently from invertebrate photoreceptors. **Rods** are specialized for detection of light at low intensities. **Cones** are specialized for detection of different wavelengths (colors).

Accommodation does not occur by forward and back movement of the lens, as seen in cephalopods. Rather, the lens

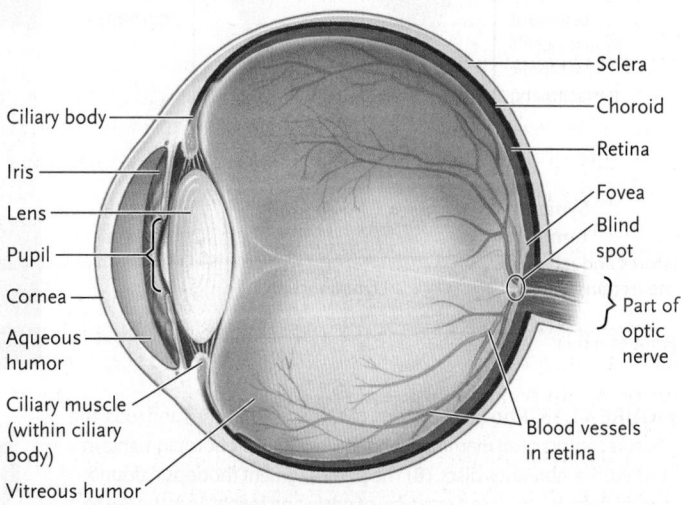

FIGURE 41.11 **Structures of the human eye.**
© Cengage Learning 2017

Labels for Figure 41.11:
- Ciliary body
- Iris
- Lens
- Pupil
- Cornea
- Aqueous humor
- Ciliary muscle (within ciliary body)
- Vitreous humor
- Sclera
- Choroid
- Retina
- Fovea
- Blind spot
- Part of optic nerve
- Blood vessels in retina

FIGURE 41.12 Accommodation in terrestrial vertebrates: the lens changes shape rather than moving forward and back to focus on (A) distant and (B) near objects.
© Cengage Learning 2017

A. Focusing on distant object

Ciliary muscle relaxed

Distant object

Taut ligaments

Lens is flattened.

When the eye focuses on a distant object, the ciliary muscles relax, allowing the ligaments that support the lens to tighten. The tightened ligaments flatten the lens, bringing the distant object into focus on the retina.

B. Focusing on near object

Ciliary muscle contracted

Near object

Slack ligaments

Lens is rounded.

When the eye focuses on a near object, the ciliary muscles contract, loosening the ligaments and allowing the lens to become rounder. The rounded lens focuses a near object on the retina.

of most terrestrial vertebrates is focused by changing its shape. The lens is held in place by fine ligaments that anchor it to a surrounding layer of connective tissue and muscle, the **ciliary body.** These ligaments keep the lens under tension when the ciliary muscle is relaxed. The tension flattens the lens, which is soft and flexible, and focuses light from distant objects on the retina **(Figure 41.12A).** When the ciliary muscles contract, they relieve the tension of the ligaments, allowing the lens to assume a more spherical shape and focusing light from nearby objects on the retina **(Figure 41.12B).**

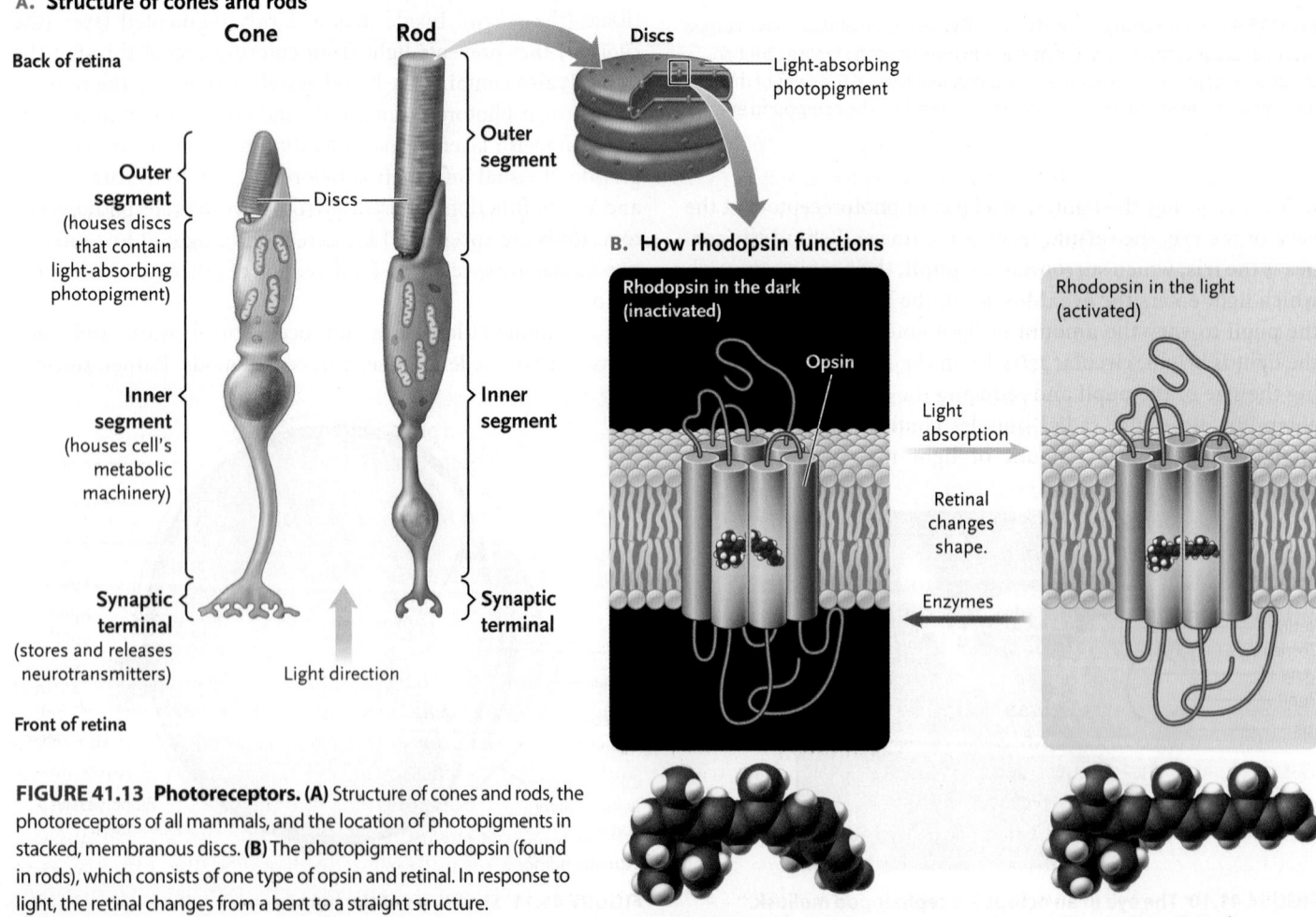

A. Structure of cones and rods

Cone   Rod

Back of retina

Discs

Light-absorbing photopigment

Outer segment (houses discs that contain light-absorbing photopigment)

Discs

Outer segment

B. How rhodopsin functions

Rhodopsin in the dark (inactivated)

Opsin

Light absorption

Retinal changes shape.

Enzymes

Rhodopsin in the light (activated)

Inner segment (houses cell's metabolic machinery)

Inner segment

Synaptic terminal (stores and releases neurotransmitters)

Synaptic terminal

Light direction

Front of retina

FIGURE 41.13 Photoreceptors. (A) Structure of cones and rods, the photoreceptors of all mammals, and the location of photopigments in stacked, membranous discs. **(B)** The photopigment rhodopsin (found in rods), which consists of one type of opsin and retinal. In response to light, the retinal changes from a bent to a straight structure.
© Cengage Learning 2017

cis-Retinal

trans-Retinal

The Retina of Mammals and Birds Contains Rods and Cones and a Complex Network of Neurons

The retina of a human eye contains about 120 million rods and 6 million cones organized into a densely packed, single layer. Neural networks of the retina are layered on top of the photoreceptor cells, so that light rays focused by the lens on the retina must pass through the neurons before reaching the photoreceptors. The light must also pass through a layer of fine blood vessels that covers the surface of the retina.

In mammals and birds with eyes specialized for daytime vision, cones are concentrated in and around a small region of the retina, the **fovea** (see Figure 41.11). The image focused by the lens is centered on the fovea, which is circular and less than a millimeter in diameter in humans. The rods are spread over the remainder of the retina. We can see distinctly only the image focused on the fovea; the surrounding image is what we term *peripheral vision*. Mammals and birds with eyes specialized for night vision have retinas containing mostly rods, and lacking a defined fovea. Some fishes and many reptiles have cones generally distributed throughout their retina and very few rods.

The rods of mammals are much more sensitive than the cones to low-intensity light. In fact, rods can respond to a single photon of light. This is why, in dim light, we can detect objects better by looking slightly to the side of the object. This action directs the image away from the cones in the fovea to the highly light-sensitive rods in surrounding regions of the retina.

SENSORY TRANSDUCTION BY RODS AND CONES Photoreceptors have three parts: (1) an outer segment consisting of stacked, flattened, membranous discs; (2) an inner segment where the cell's metabolic activities occur; and (3) the synaptic terminal, where neurotransmitter molecules are stored and released **(Figure 41.13A)**. The photopigments of rods and cones consist of the light-absorbing pigment, retinal, bonded covalently to an opsin. A different opsin is found in rods and in each type of cone. The photopigments are embedded in the membranous discs of the photoreceptors' outer segments **(Figure 41.13B)**.

Just as we saw in invertebrates, the retinal–opsin photopigment in rods is rhodopsin. In the dark, the retinal of rhodopsin is in an inactive form known as *cis*-retinal (see Figure 41.13B), and the rods steadily release the neurotransmitter glutamate. When rhodopsin absorbs a photon of light, retinal converts to its active form, *trans*-retinal (see Figure 41.13B), and the rods *decrease* the amount of glutamate they release.

Rhodopsin is a membrane-embedded G-protein–coupled receptor (see Section 9.3). An extracellular signal received by a G-protein–coupled receptor activates the receptor, which triggers a signal transduction pathway within the cell, leading to a cellular response. Here, activated rhodopsin triggers a pathway that leads to the closure of Na$^+$ channels in the plasma membrane **(Figure 41.14)**. Closure of the channels hyperpolarizes the photoreceptor's membrane, thereby decreasing neurotransmitter

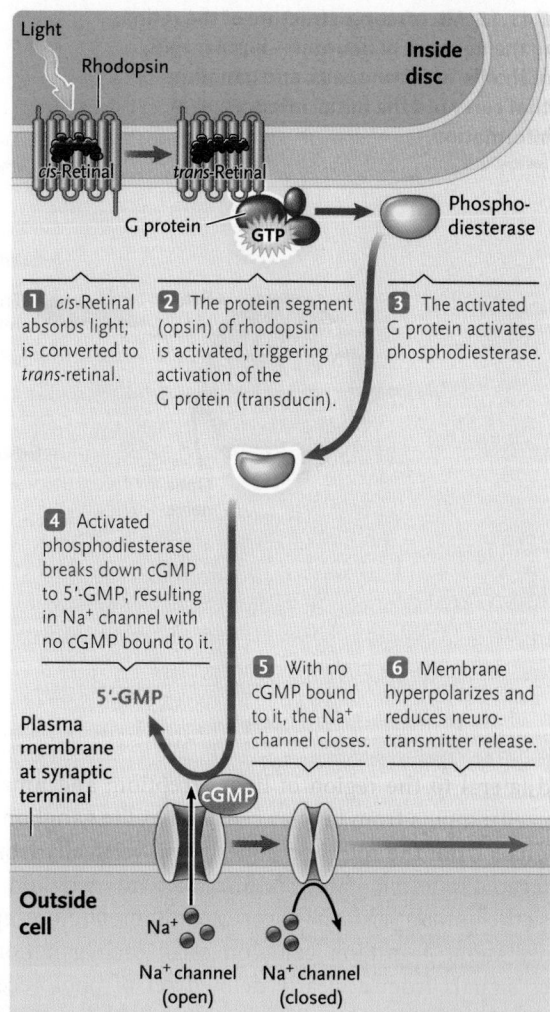

FIGURE 41.14 The signal transduction pathway that closes Na$^+$ channels in photoreceptor plasma membranes when rhodopsin absorbs light.

© Cengage Learning 2017

release. The response is graded: as light absorption increases, the amount of neurotransmitter released is reduced proportionately; and, if light absorption decreases, neurotransmitter release increases proportionately.

VISUAL PROCESSING IN THE RETINA In the retina of all vertebrates, the two types of photoreceptors are linked to a network of neurons that carries out initial processing of visual information. The retina of mammals contains four types of neurons **(Figure 41.15)**. The rods and cones form synaptic connections with **bipolar cells,** where the processing of visual information begins. **Ganglion cells** receive synaptic signals from the bipolar cells. The axons of the ganglion cells extend over the retina and collect at the back of the eyeball to form the optic nerve, which carries action potentials into the brain. The point where the optic nerve exits the eye lacks photoreceptors, resulting in a *blind spot* several millimeters in diameter. Two other types of neurons form lateral connections in the retina: **horizontal cells** receive and integrate inputs from multiple photoreceptors

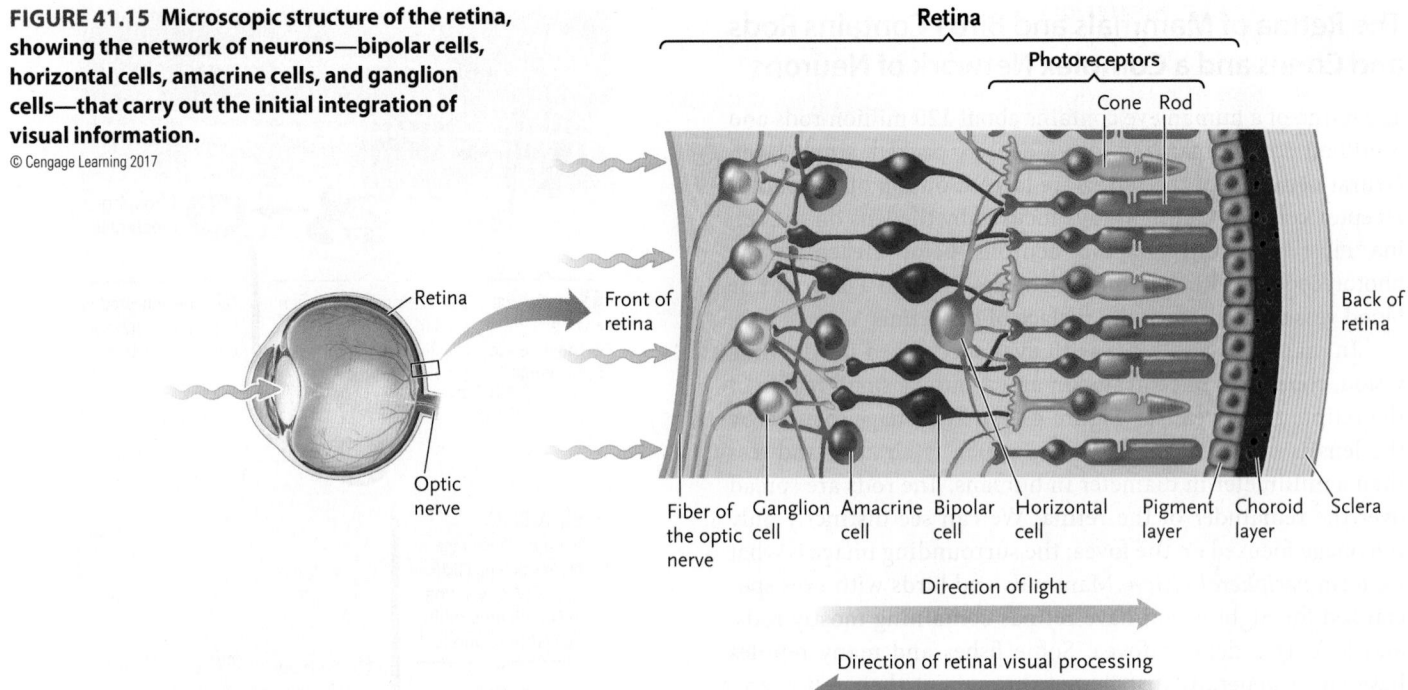

FIGURE 41.15 Microscopic structure of the retina, showing the network of neurons—bipolar cells, horizontal cells, amacrine cells, and ganglion cells—that carry out the initial integration of visual information.
© Cengage Learning 2017

located lateral to the region of light reception; and **amacrine cells** receive inputs from bipolar cells and excite ganglion cells.

Signals from the rods and cones move vertically from the photoreceptors to bipolar cells and then to ganglion cells. Because the human retina has over 120 million photoreceptors, but only about 1 million ganglion cells, each ganglion cell receives signals from a clearly defined set of photoreceptors that constitute the receptive field for that cell. Therefore, stimulating numerous photoreceptors in a ganglion cell's receptive field results in only a single message to the brain from that cell. Receptive fields are typically circular and are of different sizes. Smaller receptive fields result in sharper images because they send more precise information to the brain regarding the location in the retina where the light was received.

Signals may also move laterally from a rod or cone through a horizontal cell and continue to bipolar cells through the horizontal cell's inhibitory connections. To understand this, consider a spot of light falling on the retina. Photoreceptors detect the light and send a signal to bipolar cells and horizontal cells. The horizontal cells inhibit more distant bipolar cells that are outside the spot of light, causing the light spot to appear lighter and its surrounding dark area to appear darker. This type of visual processing is called **lateral inhibition** and serves to both sharpen the edges of objects and enhance contrast in an image.

Three Kinds of Opsin-Containing Photopigments in Cones Underlie Color Vision

Many invertebrates and some species in each class of vertebrates have color vision. Color vision depends on the cone photoreceptors. Most mammals have only two types of cones, but humans and other primates have three types, the *blue, green,*

and *red cones.* The photopigment in each of the human or primate cone types has the same retinal but a different opsin. The three cone opsins are different from the opsin of the rod rhodopsin. The three cone opsins absorb light over different, but overlapping, wavelength ranges, with peak absorptions at 445 nm (blue light) for blue cones, 535 nm (green light) for green cones, and 570 nm (red light) for red cones. The farther a wavelength is from the peak color absorbed, the less strongly the cone responds.

Having overlapping wavelength ranges for the three photoreceptors means that light at any visible wavelength will stimulate at least two of the three types of cones. However, because the maximal absorption of each type of cone is a different wavelength, it is stimulated to a different extent by light at a given wavelength. The differences, relayed to the visual centers of the brain, are integrated into the perception of a color corresponding to the particular wavelength absorbed. Light stimulating all three receptor types equally is seen as white.

Defects in genes encoding the opsin proteins of cone photopigments can result in abnormal color vision. Red–green colorblindness in humans, for example, is caused by a mutation in either the gene for the red cone opsin or the gene for the green cone opsin. Red–green colorblindness is inherited as an X-linked recessive trait (see Chapter 13) because both of those genes are on the X chromosome.

The Visual Cortex Processes Visual Information

Just behind the eyes, the optic nerves converge before entering the base of the brain. A portion of each optic nerve crosses over to the opposite side, forming the **optic chiasm** (*chiasma* = crossing place). Most of the axons enter the **lateral geniculate**

FIGURE 41.16 Neural pathways for vision. Because half of the axons carried by the optic nerves cross over in the optic chiasm, the left half of the field seen by both eyes is transmitted to the visual cortex in the right cerebral hemisphere. The right half of the field seen by both eyes is transmitted to the visual cortex in the left cerebral hemisphere. As a result, the right hemisphere of the brain sees objects to the left of the center of vision, and the left hemisphere sees objects to the right of the center of vision.

© Cengage Learning 2017

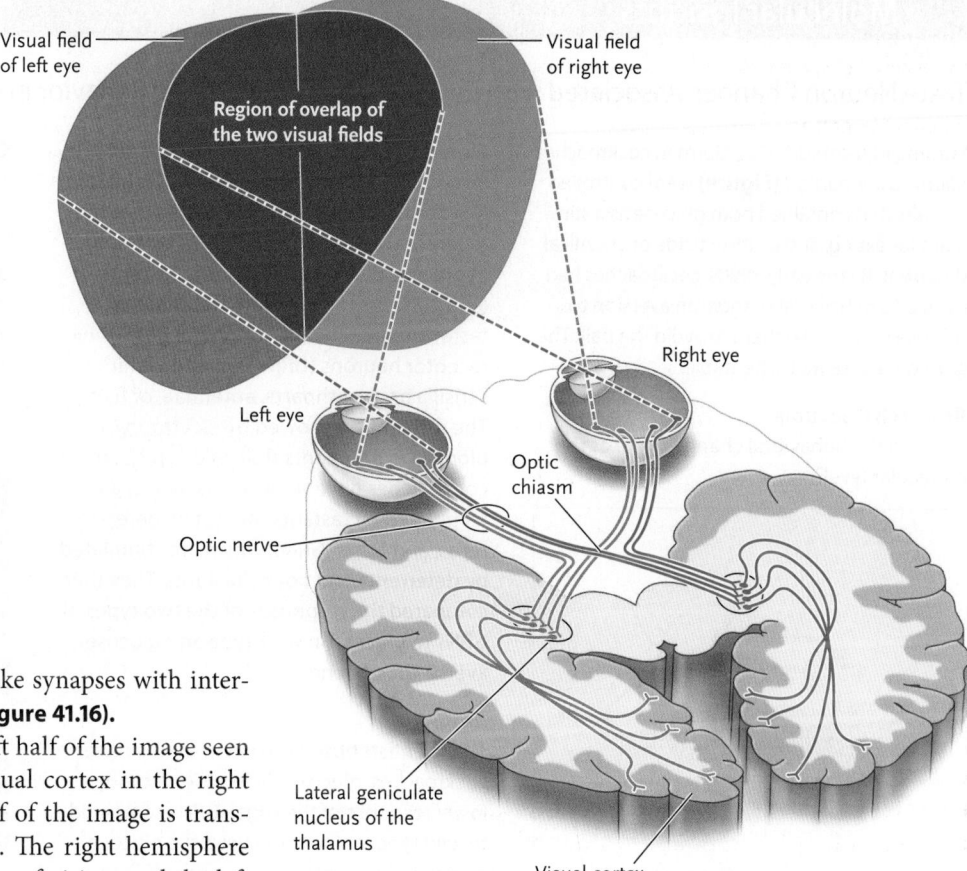

nuclei in the thalamus, where they make synapses with interneurons leading to the visual cortex **(Figure 41.16).**

Because of the optic chiasm, the left half of the image seen by both eyes is transmitted to the visual cortex in the right cerebral hemisphere, and the right half of the image is transmitted to the left cerebral hemisphere. The right hemisphere thus sees objects to the left of the center of vision, and the left hemisphere sees objects to the right of the center of vision. Communication between the right and left hemispheres integrates this information into a perception of the entire visual field seen by the two eyes.

If you look at a nearby object with one eye and then the other, you will notice that the point of view is slightly different. Integration of the visual field by the brain creates a single picture with a sense of distance and depth. The greater the difference between the images seen by the two eyes, the closer the object appears to the viewer.

The two optic nerves together contain more than a million axons, more than all other afferent neurons of the body put together. Almost one third of the cerebral cortex in humans is devoted to visual information. These numbers give some idea of the complexity of the information integrated into the visual image formed by the brain.

STUDY BREAK 41.4

For vertebrate photoreception, define: (a) photopigment; (b) photoreceptor; and (c) receptive field.

41.5 Chemoreceptors

Chemoreceptors form the basis of taste (gustation) and smell (olfaction), and measure the levels of internal body molecules such as oxygen, carbon dioxide, and hydrogen ions.

Invertebrate Receptors for Taste and Smell

In many invertebrates, the same receptors serve for the senses of smell and taste. These receptors may be confined to certain locations or distributed over the body surface. For example, the cnidarian *Hydra* has chemoreceptor cells around its mouth that respond to glutathione, a chemical released from prey organisms ensnared in its tentacles. Stimulation of these chemoreceptors causes the tentacles to retract, resulting in ingestion of the prey. By contrast, earthworms have taste/smell receptors distributed over their entire body surface.

Some terrestrial invertebrates, particularly insects, have clearly differentiated taste and smell receptors. In insects, taste receptors occur inside hollow sensory bristles called **sensilla** (singular, *sensillum*), which may be located on the antennae, mouthparts, or feet **(Figure 41.17).** Pores in the sensilla admit molecules from potential food to the chemoreceptors, which are specialized to detect sugars, salts, amino acids, or other chemicals. Many female insects have chemoreceptors on their ovipositors, which allow them to lay their eggs on food appropriate for the hatching larvae. *Molecular Insights* discusses taste neuron changes that occurred with the emergence of an adaptive behavior in cockroaches.

Insect olfactory receptors detect airborne molecules. Some insects use odor as a means of communication, such as the pheromones released into the air as sexual attractants by female

Taste Neuron Changes Associated with Emergence of an Adaptive Behavior in Cockroaches

Starting in the mid-1980s, German cockroaches (*Blattella germanica*) **(Figure)** were controlled by baits that contained both glucose as a stimulant for eating, and an insecticide or chemical deterrent. By the early 1990s, cockroaches had evolved a behavioral change: an aversion to glucose that caused them to avoid the bait. The glucose-averse trait is heritable.

Research Question

How did the behavioral change occur at the molecular level?

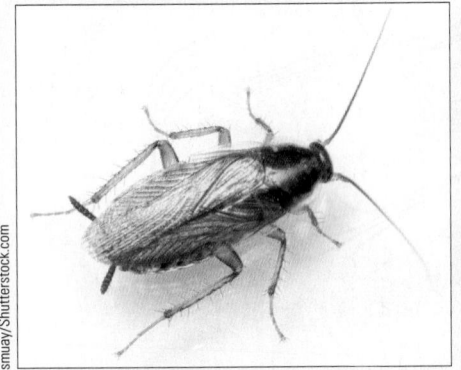

FIGURE The German cockroach, *Blattella germanica*.

smuay/Shutterstock.com

Experiments

Coby Schal and his coworkers at North Carolina State University hypothesized that the glucose-averse trait was caused by changes in glucose detection. In insects, *tastants* (substances that stimulate the sense of taste) are detected in peripheral gustatory receptor neurons (GRNs) located within sensilla on mouthparts, antennae, or feet. The researchers showed by electrophysiological experiments that wild-type German cockroaches have sugar-GRNs, which are stimulated by tastants but not by deterrents, and bitter-GRNs, which are stimulated by deterrents and not by tastants. They then compared the responses of the two types of GRNs to glucose in wild-type and glucose-averse cockroaches.

Results

The scientists observed that in glucose-averse cockroaches, glucose elicits a significantly lower response of the sugar-GRNs compared to wild type. At the same time, the glucose stimulated the bitter-GRNs in the glucose-averse cockroaches, a response that was not seen in the wild type.

Conclusion

The results support the hypothesis that the emergence of glucose-averse cockroaches has occurred as a result of mutational changes that altered the response parameters of the bitter-GRNs so that they now recognize glucose as a deterrent. The authors state: "The rapid emergence of this highly adaptive behavior underscores the plasticity of the sensory system to adapt to rapid environmental change."

think like a scientist

Loss-of-function mutations lead to absence or decreased biological activity of the gene product. Gain-of-function mutations confer a new property on the gene product, causing a new phenotype. For the emergence of glucose-averse cockroaches, was the mutation or mutations of the loss-of-function or gain-of-function type(s)?

From what you learned about genetic changes that affect protein structure and function in Section 15.5, what kinds of base-pair changes could underlie loss-of-function mutations? Gain-of-function mutations?

Source: A. Wada-Katsumato, J. Silverman, and C. Schal. 2013. Changes in taste neurons support the emergence of an adaptive behavior in cockroaches. *Science* 340:972–975.

© Cengage Learning 2017

FIGURE 41.17 Taste receptors on the foot of a fruit fly, *Drosophila melanogaster*.
© Cengage Learning 2017

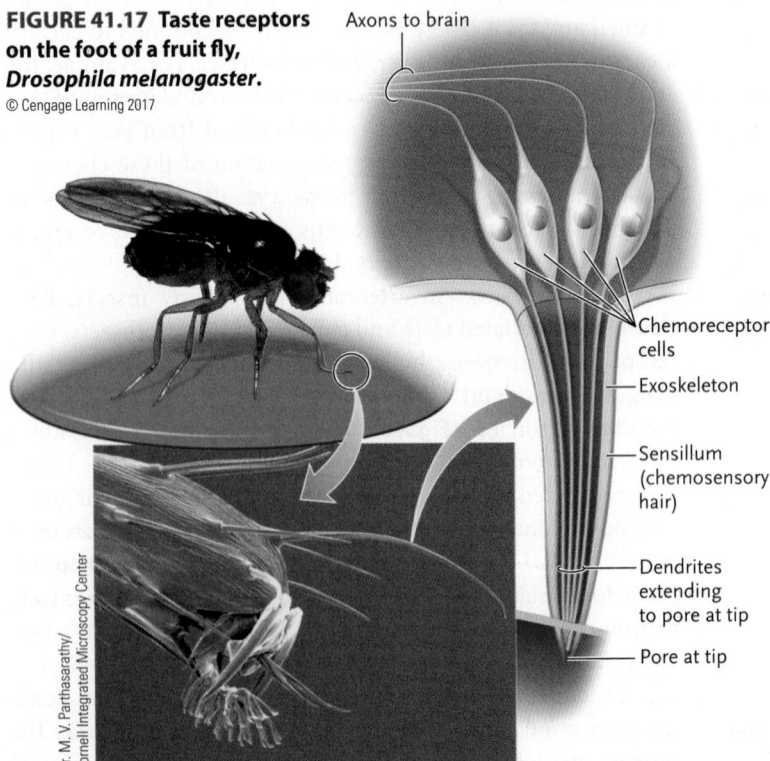

Axons to brain

Chemoreceptor cells

Exoskeleton

Sensillum (chemosensory hair)

Dendrites extending to pore at tip

Pore at tip

Dr. M. V. Parthasarathy/ Cornell Integrated Microscopy Center

moths. Olfactory receptors in the bristles of male silkworm moth antennae **(Figure 41.18)** have been shown experimentally to be able to detect pheromones released by a female of the same species in concentrations as low as one attractant molecule per 10^{17} air molecules. When as few as 40 of the 20,000 receptor cells on its antennae have been stimulated by pheromone molecules, the male moth responds to attract the female's attention. Ants, bees, and wasps identify members of the same hive or nest and communicate by means of odor molecules.

Taste and Smell Receptors Are Differentiated in Terrestrial Vertebrates

In terrestrial animals, taste involves the detection of potential food molecules in objects that are touched, whereas smell involves the detection of airborne molecules. Although both taste and smell receptors have hairlike extensions containing the proteins that bind environmental molecules, the hairs of taste receptors are derived from microvilli and contain microfilaments, whereas the hairs of smell receptors are derived from cilia and contain microtubules. Another significant difference between taste and smell in vertebrates is that information from taste receptors is typically processed in the parietal lobes, whereas information from smell receptors is processed in the olfactory bulbs and the temporal lobes.

Taste Receptors Are Located in Taste Buds

The taste receptors of most vertebrates form part of a structure called a **taste bud,** a small, pear-shaped capsule with a pore at the top that opens to the exterior **(Figure 41.19).** The sensory hairs of the taste receptors pass through the pore of a taste bud and project to the exterior. The opposite ends of the receptor cells form synapses with dendrites of an afferent neuron.

The taste receptors of terrestrial vertebrates are concentrated in the mouth. Humans have about 10,000 taste buds scattered over the tongue, roof of the mouth, and throat. Those on the tongue are embedded in outgrowths called **papillae** (singular, papilla) (*papula* = pimple), which give the surface of the tongue its rough texture.

Taste receptors on the human tongue are thought to respond to five basic tastes: sweet, sour, salty, bitter, and umami ("savory"). Some of the receptors for umami respond to the amino acid glutamate (familiar as monosodium glutamate, or MSG). Recent research indicates that the classes of receptors may all have many subtypes, each binding a specific molecule within that class.

Signals from the taste receptors are relayed to the thalamus. From there, some signals lead to gustatory centers in the cerebral cortex, which integrate them into the perception of taste. Other signals lead to the brain stem and limbic system, which links tastes to involuntary visceral and emotional responses. Through these connections, a pleasant taste may lead to salivation, secretion of digestive juices, sensations of pleasure, and even sexual arousal, whereas an unpleasant taste may produce revulsion, nausea, and even vomiting.

Olfactory Receptors Are Concentrated in the Nasal Cavities in Terrestrial Vertebrates

Receptors that detect odors are located in the nasal cavities. Bloodhounds have more than 200 million receptors in patches of olfactory epithelium in the upper nasal passages; humans have about 5 million olfactory receptors.

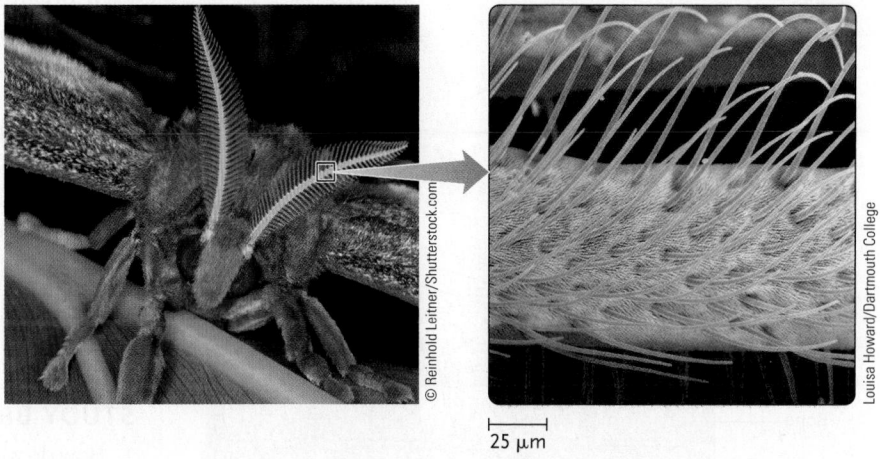

FIGURE 41.18 The brushlike antennae of a male silkworm moth, *Bombyx mori*. Fine sensory bristles containing olfactory receptor cells cover the filaments of the antennae.
© Cengage Learning 2017

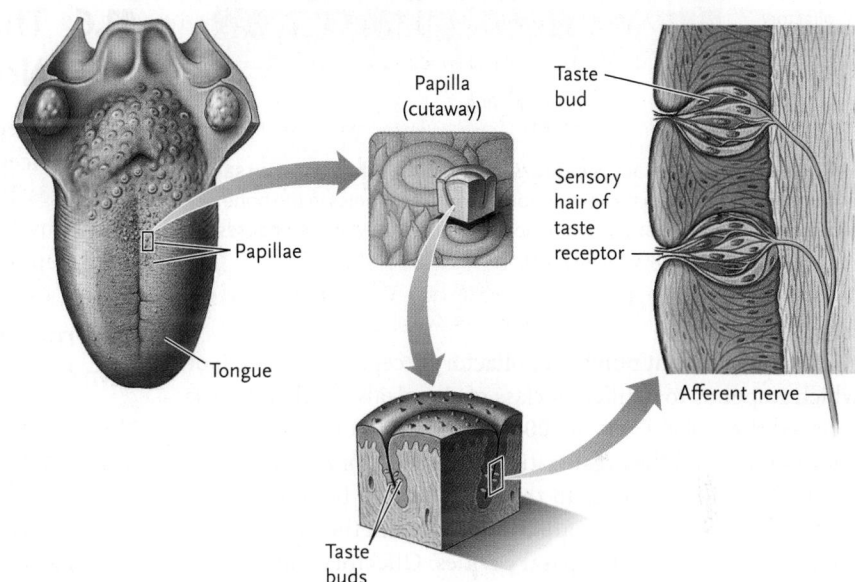

FIGURE 41.19 Taste receptors in the human tongue. The receptors occur in microscopic taste buds that line the sides of the papillae.
© Cengage Learning 2017

On one end, each olfactory receptor cell has 10 to 20 sensory hairs that project into a layer of mucus covering the olfactory area in the nose **(Figure 41.20).** To be detected, airborne molecules must dissolve in the watery mucus solution. On the other end, the olfactory receptors make synapses with interneurons in the olfactory bulbs. Olfactory receptors are the only receptor cells that make direct connections with brain interneurons, rather than via afferent neurons.

From the olfactory bulbs, nerves conduct signals to the olfactory centers of the cerebral cortex, where they are integrated into the perception of tantalizing or unpleasant odors. Most odor perceptions arise from a combination of different olfactory receptors. In the early 1990s, Richard Axel and Linda Buck discovered that about 1,000 different human genes give

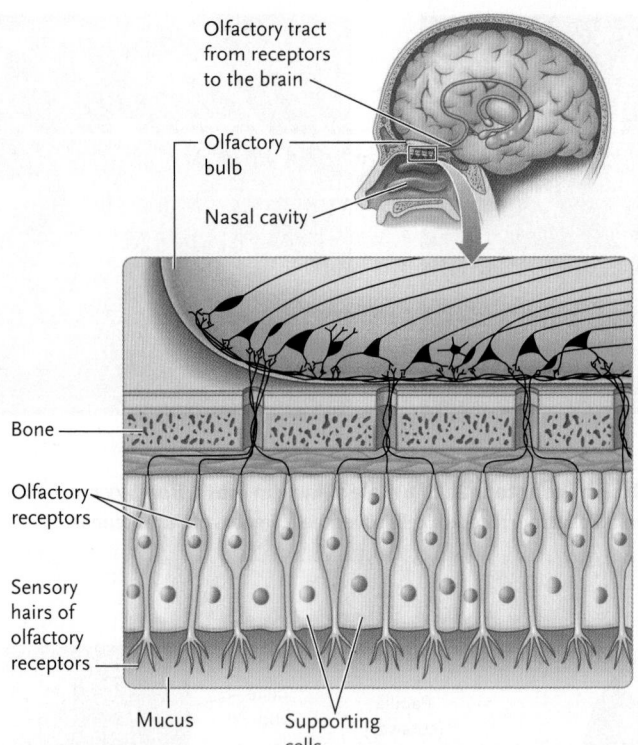

FIGURE 41.20 **Olfactory receptors in the roof of the nasal passages in humans.** Axons from these receptors pass through holes in the bone separating the nasal passages from the brain, where they make synapses with interneurons in the olfactory bulbs.
© Cengage Learning 2017

Labels for figure 41.20:
Olfactory tract from receptors to the brain
Olfactory bulb
Nasal cavity
Bone
Olfactory receptors
Sensory hairs of olfactory receptors
Mucus
Supporting cells

rise to an equivalent number of olfactory receptor types, each of which responds to a different class of chemicals. Axel and Buck received the Nobel Prize in 2004 "for their discoveries of odorant receptors and the organization of the olfactory system."

Olfaction contributes to the sense of taste because vaporized molecules from foods are conducted from the throat to the olfactory receptors in the nasal cavities. Olfactory input is the reason why anything that dulls your sense of smell—such as a head cold, or holding your nose—diminishes the apparent flavor of food.

Many mammals use odors as a means of communication. Individuals of the same family or colony are identified by their odor. Odors are also used to attract mates and to mark territories and trails. Dogs, for example, use their urine to mark home territories with identifying odors.

Bioinformatics Analysis of Whole-Genome Sequences Shows Evolutionary History of Olfactory Receptor Genes

Olfactory receptor genes form the largest multigene family in vertebrates. Bioinformatics analysis of whole-genome sequences has shown that the numbers of olfactory receptor genes vary greatly among vertebrates. At the phylogenetic level, the arrays of olfactory receptor genes in vertebrate groups appear to have changed rapidly; the changes relate to each organism's environment. For example, higher primates with well-developed visual systems have lost a large number of olfactory receptor genes. Also, after tetrapods adapted to their new terrestrial environment, two groups of olfactory receptor genes for detecting airborne odors expanded greatly. Based on molecular phylogenomic analysis, the evolutionary origin of vertebrate olfactory receptor genes can be traced to the common ancestor of all chordate species. By contrast, distinct families of chemoreceptor genes are found in insects, echinoderms, and nematodes, indicating that chemoreceptor genes evolved several times independently.

STUDY BREAK 41.5

1. How do we distinguish different kinds of smells?
2. For terrestrial vertebrates, describe the pathway by which a signal generated by taste receptors leads to a response.

41.6 Thermoreceptors and Nociceptors

Thermoreceptors detect changes in the surrounding temperature. **Nociceptors** respond to stimuli that may potentially damage tissues. Both types of receptors consist of free nerve endings formed by the dendrites of afferent neurons, with no specialized receptor structures surrounding them.

Thermoreceptors Can Detect Warm and Cold Temperatures and Temperature Changes

Most animals have thermoreceptors. Some invertebrates, such as mosquitoes and ticks, use thermoreceptors to locate their warm-blooded prey. Some snakes, including rattlesnakes and pythons, use thermoreceptors called *pit organs* to detect the body heat of warm-blooded prey animals **(Figure 41.21).**

Pit organs

FIGURE 41.21 **The pit organs of an albino Western diamondback rattlesnake *(Crotalus atrox)*, located in depressions on both sides of the head below the eyes.** These thermoreceptors detect infrared radiation emitted by warm-bodied prey animals such as mice.

In mammals, current evidence suggests the existence of five thermoreceptor ranges, beginning at less than 8°C and extending to above 52°C. At the extremes, the neural signals are perceived as painful.

Some neurons in the hypothalamus of mammals also function as thermoreceptors, an ability that has only recently been investigated. Not only do these neurons sense changes in brain temperature, but they also receive afferent thermal information. These neurons are highly sensitive to shifts from the normal body temperature, and trigger involuntary responses such as sweating, panting, or shivering, which restore normal body temperature.

Nociceptors Protect Animals from Potentially Damaging Stimuli

The signals from nociceptors—receptors in vertebrates and invertebrates that detect damaging stimuli—are interpreted by the brain as pain. Pain is a protective mechanism. In humans, pain prompts us to do something immediately to remove or decrease the damaging stimulus. Often pain elicits a reflex response—such as withdrawing the hand from a hot surface—that proceeds before we are even consciously aware of the sensation.

Various types of stimuli cause pain, including mechanical damage such as a cut or a blow to the body, and temperature extremes. Some nociceptors are specific for a particular type of damaging stimulus, whereas others respond to all kinds.

The axons that transmit pain in vertebrates are part of the somatic component of the PNS (see Section 40.2). They synapse with interneurons in the gray matter of the spinal cord, and activate neural pathways in the CNS by releasing the neurotransmitters glutamate or substance P (see Table 39.1). Glutamate-releasing axons produce sharp, prickling sensations that can be localized to a specific body part—the pain of stepping on a sharp stone, for example. Substance P-releasing axons produce dull, burning, or aching sensations, the location of which may not be easily identified—the pain of tissue damage such as stubbing your toe.

As part of their protective function, pain receptors adapt very little, if at all. Some pain receptors, in fact, gradually intensify the rate at which they send out action potentials if the stimulus continues at a constant level.

The CNS also has a pain-suppressing, or gating, system. In response to stimuli such as exercise, sex, and stress, the brain releases *endorphins,* natural painkillers that bind to membrane receptors on substance P neurons, reducing the amount of neurotransmitter released (see Table 39.1).

Nociceptors contribute to the taste of some spicy foods, particularly those that contain hot peppers. In fact, researchers who study pain often use *capsaicin,* the organic compound that gives jalapeños and other peppers their hot taste, to identify nociceptors. To some, the burning sensation from capsaicin is addictive. Nociceptors in the mouth, nose, and throat immediately transmit pain messages to the brain when they detect capsaicin. The brain responds by releasing endorphins, which act as a painkiller and create temporary euphoria.

STUDY BREAK 41.6

What distinguishes thermoreceptors and nociceptors from the other types of sensory receptors discussed previously?

THINK OUTSIDE THE BOOK

You have now learned about various types of sensory receptors. On your own or collaboratively, investigate the role of breakdowns in sensory receptor pathways in causing human disease. Pick two diseases caused in this way and outline how the change in the sensory receptor pathway brings about disease symptoms.

41.7 Magnetoreceptors and Electroreceptors

Some animals can sense magnetic or electrical fields. In so doing, they sense stimuli that humans can detect only with scientific instruments.

Magnetoreceptors Are Used for Navigation

Some animals that navigate long distances, including migrating butterflies, beluga whales, sea turtles, homing pigeons, and foraging honeybees, have **magnetoreceptors** that allow them to detect and use Earth's magnetic field as a source of directional information.

The pattern of Earth's magnetic field differs from region to region yet remains almost constant over time, largely unaffected by changing weather and day/night cycles. As a result, animals with magnetoreceptors are able to monitor their location reliably. Although relatively little is known about this type of receptor, they may depend on the fact that moving a conductor, such as an electroreceptor cell, through a magnetic field generates an electric current.

In one interesting laboratory experiment, Kenneth Lohmann of the University of North Carolina tested a hypothesis that magnetoreception plays a central role in the migration of loggerhead sea turtles **(Figure 41.22).** Lohmann used an experimental system in the laboratory in which the direction loggerhead sea turtle hatchlings swam was analyzed in

FIGURE 41.22 A loggerhead sea turtle.

different magnetic fields. In the normal magnetic field of the Earth, the hatchlings swam in a direction that mimicked the direction they follow normally when migrating at sea. In a magnetic field 180° reversed from that of Earth's magnetic field, hatchlings swam in a direction 180° opposite of the direction they follow when migrating at sea. Lohmann concluded that the loggerhead sea turtle hatchlings have the ability to detect Earth's magnetic field and use it as a way to orient their migration. Lohmann believes that the magnetoreceptor system in the turtles involves magnetite.

Electroreceptors Are Used for Location of Prey or for Communication

Many sharks and bony fishes, some amphibians, and even some mammals (such as the star-nosed mole and duckbilled platypus) have specialized **electroreceptors** that detect electrical fields. The plasma membrane of an electroreceptor cell is depolarized by an electrical field, leading to the generation of action potentials. The electrical stimuli detected by the receptors are used to locate prey or navigate around obstacles in muddy water, or, by some fishes, to communicate. Some electroreception systems are passive—they detect electric fields in the environment, not the animal's own electric currents. Passive systems are used mainly to find prey. For example, the electroreceptors of sharks and rays can locate fish buried under the sand from the electrical currents generated by their prey's heartbeat or by the muscle contractions that move water over their gills.

Other electroreception systems are active—the animal emits and receives low-voltage electrical signals, either to locate prey or to communicate with members of the same species. The electrical signals are generated by special electric organs. A few species, such as the electric eel and the electric catfish, produce discharges on the order of several hundred volts. These discharges are used to stun or kill prey.

STUDY BREAK 41.7
What are three ways electroreceptors are used in aquatic vertebrates?

 ## Unanswered Questions

What happens when the senses get scrambled—when listening to music causes you to "see" colors, or when you "taste" certain words?

Synesthesia ("joined senses") occurs when two senses, normally separate, are perceived together. For the most part, people with synesthesia—synesthetes—are born with it, and it tends to run in families. A recent study by Michael Esterman and his colleagues at the University of California, Berkeley, showed that the posterior parietal cortex, a region of the brain thought to be involved in sensory integration, appears to be crucial to sensory commingling. Some researchers think that this commingling is how the senses function early in development, when the nervous system is still immature. They believe that the senses normally separate from one another around four months after birth. In synesthetes, however, this separation is incomplete and two of their senses remain mingled.

London's Science Museum collaborated with Jamie Ward of University College London in an experiment that paired visual images and music. They wanted to determine if volunteers visiting the museum would prefer combinations of music and images designed by synesthetes over combinations that were designed by nonsynesthetes. Interestingly, people found the synesthetic combinations more pleasing than the nonsynesthetic ones. Thus, it is possible that everyone may have a built-in understanding of what sounds and colors go together.

In an evolutionary context, which "sense" developed first?

The descriptions of the senses in this chapter focus primarily on vertebrate sensory systems, but the nervous systems of many invertebrates can be quite complex. Indeed, squids, sea hares, leeches, horseshoe crabs, lobsters, and cockroaches have been instrumental in helping scientists understand the nervous system. Clearly, vertebrates are not the only multicellular organisms to develop senses.

However, is a nervous system necessary for organisms to have senses? Can single-celled organisms (which, of course, do not have nervous systems) respond to stimuli? The answer is yes. Consider *Paramecium tetraurelia*, a single-celled, ciliated protist that lives in water. It can detect substances in its environment and swim toward certain chemicals while avoiding others. It also responds to solid objects by turning when it runs into one. Thus, it has a "chemical sense" similar to taste or smell, and it responds to a type of "touch." We will probably never know which sense developed first, but some type of touch or chemical sense seems the most likely.

think like a scientist

1. Your college roommate is very fond of scrambled eggs. You serve her scrambled eggs colored with green food coloring. You video your roommate eating the eggs, and show the video in your biology class. What sensory receptors are you hoping to stimulate most noticeably? If you were to suggest to your professor an imaging test for synesthetes, what imaging method might you choose (see Chapter 40)?

2. Does the fact that some prokaryotes contain a microbial rhodopsin suggest to you that the sense of vision has early evolutionary origins?

 **Rona Delay** is an Associate Professor in the Department of Biology at the University of Vermont. Her research centers on understanding how sensory receptors change or transduce information about the external world into a language the brain can understand. The focus of her research is the sense of smell. To learn more about Dr. Delay's work, go to http://www.uvm.edu/~biology/?Page=faculty/ronadelay.php&SM=facultysubmenu.html.

REVIEW KEY CONCEPTS

For access to MindTap and additional study materials visit www.cengagebrain.com.

41.1 Overview of Sensory Receptors and Pathways

- Sensory receptors are formed by the endings of afferent neurons, endings of afferent neurons enclosed in specialized structures, or specialized cells adjacent to the neurons. They detect stimuli such as mechanical pressure, sound waves, light, or specific chemicals. Action potentials generated by the receptors are carried by the axons of afferent neurons to pathways leading to specific parts of the brain, where signals are processed into sensory sensations (Figure 41.1).

- Receptors are specialized as mechanoreceptors, photoreceptors, chemoreceptors, thermoreceptors, and nociceptors. Some animals have receptors that detect electrical or magnetic fields.

- The routing of information from sensory receptors to particular regions of the brain identifies a specific stimulus as a sensation. The intensity of a stimulus is determined by the frequency of action potentials traveling along the neural pathways and the number of afferent neurons carrying action potentials.

- Many sensory systems show sensory adaptation, in which the frequency of action potentials decreases while a stimulus remains constant. Some sensory receptors, such as those related to pain, show no sensory adaptation.

- The brain uses signals from sensory receptors to generate an interpretation of the external and internal environments. That interpretation is our perception of the world. Each type of organism has a unique perception of the world.

41.2 Mechanoreceptors and the Tactile and Spatial Senses

- Mechanoreceptors detect touch, pressure, acceleration, or vibration. Touch and pressure receptors are free nerve endings or encapsulated nerve endings of sensory neurons (Figure 41.2).

- Mechanoreceptors called proprioceptors detect stimuli used by the CNS to monitor and maintain body and limb positions.

- Receptors in muscles, tendons, and joints of vertebrates detect changes in stretch and tension of body parts.

- Proprioceptors based on sensory hair cells generate action potentials when the hairs are moved (Figures 41.3–41.5).

41.3 Mechanoreceptors and Hearing

- Many invertebrates have mechanoreceptors in their skin or other surface structures that detect sound and other vibrations.

- Hearing relies on sensory hair cells in organs that respond to the vibrations of sound waves.

- In terrestrial vertebrates, the ear consists of three parts. The outer ear directs sound to the tympanic membrane (eardrum). Vibrations of the tympanic membrane are transmitted through one or more bones in the middle ear to the fluid-filled inner ear. In the inner ear, the vibrations are transmitted through membranes that bend the stereocilia of hair cells, leading to bursts of action potentials that are reflected in the frequency of the sound waves (Figure 41.7).

41.4 Photoreceptors and Vision

- Invertebrates possess many forms of eyes, from the simplest, an ocellus, to single-lens eyes that are similar to vertebrate eyes (Figures 41.8–41.10).

- The photoreceptors of all animal eyes contain the pigment retinal, which absorbs the energy of light and uses it to generate changes in membrane potential.

- The transparent cornea admits light into the vertebrate eye. Behind the cornea, the iris controls the diameter of the pupil, regulating the amount of light that strikes the lens. The lens focuses an image on the retina lining the back of the eye, where photoreceptors and neurons carry out the initial integration of information detected by the photoreceptors (Figure 41.11).

- In terrestrial vertebrates, the lens is focused by adjusting its shape (Figure 41.12). The retina contains two types of photoreceptors: rods are specialized for detecting light of low intensity; cones are specialized for detecting light of different wavelengths, which are perceived as colors.

- The light-absorbing pigment in photoreceptor cells consists of retinal combined with an opsin protein. When it absorbs light, retinal changes form, initiating reactions that alter the amount of neurotransmitter released by the photoreceptor cells (Figures 41.13 and 41.14).

- Rods and cones are linked to neurons in the retina that perform the initial processing of visual information. The processed signal is sent via the optic nerve through the lateral geniculate nuclei to the visual cortex (Figures 41.15 and 41.16).

41.5 Chemoreceptors

- Chemoreceptors respond to the presence of specific molecules in the environment. In vertebrates, they form parts of receptor organs for taste (gustation) and smell (olfaction).

- Taste receptors detect molecules from food or other objects that come into direct contact with the receptor and are used primarily to identify foods (Figures 41.17 and 41.19).

- Olfactory receptors detect molecules from distant sources; besides identifying food, they are used to detect predators and prey, identify family and group members, locate trails and territories, and communicate (Figures 41.18 and 41.20).

- Olfactory receptor genes vary greatly in number among vertebrates. The evolutionary origin of these genes can be traced to the common ancestor of all chordate species.

41.6 Thermoreceptors and Nociceptors

- Thermoreceptors, which consist of free nerve endings located at the body surface and in limited numbers in the body interior, detect changes in body temperature.

- Nociceptors, located on both the body surface and interior, detect stimuli that can damage body tissues. Information from these receptors is integrated in the brain into the sensation of pain.

41.7 Magnetoreceptors and Electroreceptors

- Some vertebrates have magnetoreceptors that detect magnetic fields (Figure 41.22) or electroreceptors that detect electrical currents and fields.

TEST YOUR KNOWLEDGE

Remember/Understand

1. An ambulance siren in close proximity to a dog can cause the dog to howl in pain. Which receptors are responsible for this response?
   a. thermoreceptors and chemoreceptors
   b. photoreceptors and nociceptors
   c. mechanoreceptors and nociceptors
   d. chemoreceptors and mechanoreceptors
   e. photoreceptors and chemoreceptors

2. The sensation of sheets lessens if you lie still in bed. This response is due to:
   a. sensory adaptation of mechanoreceptors.
   b. sensory adaptation of nociceptors.
   c. pH change receptors associated with sleep.
   d. the vestibular apparatus.
   e. vibration-detecting systems.

3. Which of the following situations is associated with movement and position in the human body?
   a. Statoliths in statocysts bend sensory hairs and trigger action potentials.
   b. If sensory hairs in the utricle are oriented horizontally and those in the saccule are oriented vertically, the person is lying down.
   c. When the head rotates, the endolymph in the semicircular canal pulls the cupula with it to activate sensory hair cells.
   d. Displacement of the utricle and saccule generates action potentials.
   e. If the body is spinning at a constant rate and direction, the cupula is displaced and action potentials are initiated.

4. Neuromast function is best described as:
   a. nonadapting pain receptors.
   b. stereocilia that bend and generate action potentials.
   c. statoliths that detect motion.
   d. motor axons that activate motion.
   e. cupulas that detect vibrations.

5. Structures in the vertebrate ear are activated by sound waves in the following order:
   a. oval window, tympanic membrane, semicircular canals, Golgi tendon organ, incus, malleus, stapes.
   b. organ of Corti, malleus, incus, stapes, auditory nerve, tympanic membrane.
   c. eustachian tube, round window, vestibular canal, tympanic canal, cochlear canal, oval window, pinna.
   d. basilar membrane, tectorial membrane, otoliths, utricle, saccule, malleus, cochlea.
   e. pinna, tympanic membrane, malleus, incus, stapes, oval window, cochlear duct.

6. The eyes of vertebrates and cephalopods are similar in structure and function. A difference between the vertebrate eye and the cephalopod eye is that the vertebrate eye has:
   a. an iris surrounding the pupil, whereas in cephalopods the pupil surrounds the iris.
   b. a lens that changes shape when focusing, whereas in cephalopods the lens moves back and forth to focus.
   c. a retina that moves in the socket when recording the image, whereas in cephalopods the retina changes shape when stimulated.
   d. a pupil that shrinks in size in bright light, whereas cephalopods have a pupil that enlarges in bright light.
   e. retinal synthesized from vitamin A, whereas cephalopods lack retinal.

7. Which of the following events does not occur during light absorption in the vertebrate eye?
   a. The retinal component of rhodopsin changes from *cis* to *trans* form.
   b. Rhodopsin, a G-protein–coupled receptor, triggers a signal transduction pathway to close Na^+ channels in the plasma membrane.
   c. The light stimulus passes from rods and cones to bipolar cells and horizontal cells and then to ganglion cells, whose axons compose the optic nerve.
   d. As light absorption increases, the rhodopsin response causes an increase in the release of neurotransmitters.
   e. When integrating information across the retina, horizontal cells connect the rods and cones, and amacrine cells join with the bipolar cells and ganglion cells.

8. The variety of color seen by humans is directly dependent on the:
   a. activation of three different opsins in cones.
   b. transmission of an image to separate brain hemispheres by the optic chiasm.
   c. transmission of impulses from rods across the lateral geniculate nuclei.
   d. lateral inhibition by amacrine cells.
   e. light stimulation of all photoreceptor types equally.

9. In terrestrial animals:
   a. the hairs of taste receptors are derived from cilia and contain microtubules.
   b. the hairs of smell receptors are derived from microvilli and contain microfilaments.
   c. signals from taste receptors are relayed to the cerebellum.
   d. olfactory receptors are located primarily in the mouth.
   e. nerve connections from the olfactory bulbs lead to the cerebral cortex.

10. In the human response to temperature or pain:
    a. all Ca^{2+} channels act as pain receptors.
    b. cold receptors are activated between 27°C and 37°C.
    c. pain receptors decrease the rate at which they send out action potentials if the pain is constant.
    d. nociceptors activated by capsaicin transmit pain messages to the brain.
    e. the CNS releases glutamate or substance P to dull the pain sensation.

Apply/Analyze

11. **Discuss Concepts** In owls and many other birds of prey, the fovea is located toward the top of the retina rather than at the center as in humans. This arrangement correlates with the birds' hunting behavior, in which they look down when they fly, scanning the ground for a meal. With this arrangement in mind, why do you think a standing owl would turn its head upside down?

12. **Discuss Concepts** A patient made an appointment with her doctor because she was experiencing recurrent episodes of dizziness. Her doctor asked questions to distinguish whether she had sensations of light-headedness (as if she were going to faint), or vertigo (as if she or objects near her were spinning around). Why was this clarification important in the evaluation of her condition?

Evaluate/Create

13. **Discuss Concepts** Humans have about 200 million photo-receptors in two eyes, and about 32,000 sensory hair cells in two ears. About 3% of the somatosensory cortex is devoted to hearing, whereas roughly 30% of it is devoted to visual processing. Suggest an explanation for these differences from the perspective of natural selection and adaptation.

14. **Design an Experiment** The fruit fly *Drosophila melanogaster* can distinguish a large repertoire of odors in the environment. Their response may be to move toward food or away from danger. Moreover, particular odors play an important role in their mating behavior. The olfactory organs of a fruit fly are the antennae and an elongated bulge on the head called the maxillary pulp. Because of the ease with which fruit fly genes can be manipulated, identifying and studying their olfactory receptors likely would contribute significantly to our understanding of neural pathways of odor recognition more generally. How could you identify candidate fruit fly genes that encode components of olfactory receptors?

15. **Apply Evolutionary Thinking** In 2005, researchers took saliva and blood samples from six cats, including domestic cats, a tiger, and a cheetah, and found that all have a defective gene for one of the two chemoreceptor proteins needed to identify food as sweet. (The scientists conjecture that the lack of a sweet tooth may explain why cats are finicky eaters.) What are the evolutionary implications of the finding?

For selected answers, see Appendix A.

INTERPRET THE DATA

The graph in the **Figure** plots relative olfactory bulb size for pairs of closely related birds (a through m), with each pair having a species that is more active at night (nocturnal) and one that is more active during the day (diurnal). The researchers hypothesized that nocturnal birds would have larger olfactory bulbs because they need to depend on senses other than sight.

1. From these data, which, if any, of the bird pairs supports the hypothesis of the researchers?
2. Do any of the pairs provide evidence against the hypothesis?
3. Looking at these data as a whole, can you draw any overall conclusions?

Explain each of your answers.

Source: From S. Healy and T. Guilford. 1990. Olfactory-bulb size and nocturnality in birds. *Evolution* 44:339–346. Copyright © 1990 Wiley Publishing.

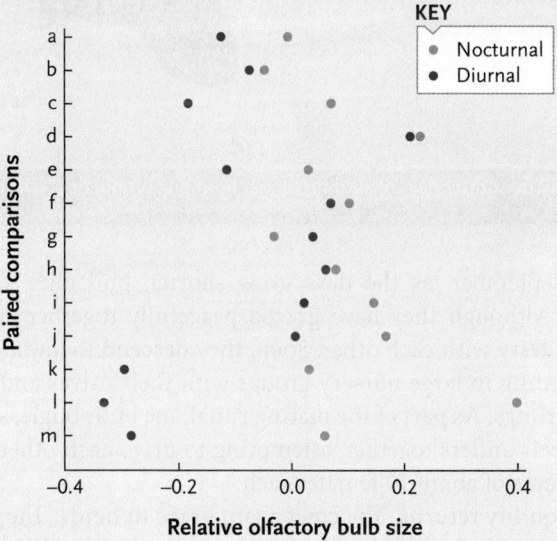

42 The Endocrine System

Beth Davidow/Getty Images

Two North American bull elks contesting for cows. The shorter days of autumn trigger hormone production, battling, and reproductive behavior.

Why it matters . . . Every September, as the days grow shorter, bull elks *(Cervus canadensis)* begin to strut their stuff. Although they have grazed peacefully together at high mountain elevations, they now become testy with each other. Soon, they descend to lower elevations, where the cow elks have been feeding in large nursery groups with their calves and yearlings. The bulls chase away the male yearlings. As part of the mating ritual, the bulls bugle, square off, strut, and circle. Then they clash their antlers together, attempting to drive each other from the cows. The winning males claim harems of about 10 females each.

After the mating season ends, tranquility returns. The cows again graze in herds. The males form now-friendly bachelor groups that also feed quietly in the meadows. The young will be born eight to nine months later, when summer returns.

The transition to mating behavior is triggered by the shortening days of later summer and fall. Detected by the eyes and registered in the brain, reduced day length initiates changes in the secretion of long-distance signaling molecules called **hormones** (*horme* = to excite). Hormones released from one group of cells are transported through the circulatory system to other cells, their target cells, whose activities they change. Among the changes in the elks will be a rise in the concentration of hormones responsible for mating behavior.

We too are driven by our hormones. They control our day-to-day sexual behavior, as well as a host of other functions, including the concentration of salt in our blood, body growth, and the secretion of digestive juices. Along with the central nervous system, hormones coordinate the activities of multicellular life.

The best-known hormones are secreted by cells of the **endocrine system** (*endo* = within; *krinein* = to separate), although hormones actually are produced by many organ systems in the body. The endocrine system, like the nervous system, regulates and coordinates distant organs.

The two systems are structurally, chemically, and functionally related, but they control different types of activities. The nervous system, through its high-speed electrical signals, enables an organism to interact rapidly with the external environment, whereas the endocrine system mainly controls activities that involve slower, longer-acting responses. Typical responses to hormones may persist for hours, weeks, months, or even years.

The mechanisms and functions of the endocrine system are the subjects of this chapter. As in other chapters of this unit, we pay particular attention to the endocrine system of humans and other mammals.

42.1 Hormones and Their Secretion

Cells signal other cells using neurotransmitters, hormones, and local regulators. Recall from Chapters 39, 40, and 41 that a neurotransmitter is a chemical released by an axon terminal at a synapse that affects the activity of a postsynaptic cell. Our focus in this chapter is hormones and local regulators.

The Endocrine System Includes Four Major Types of Cell Signaling

Four types of cell signaling occur in the endocrine system: classical endocrine signaling, neuroendocrine signaling, paracrine regulation, and autocrine regulation.

In **classical endocrine signaling,** hormones are secreted into the extracellular fluid (ECF) by the cells of ductless secretory organs called **endocrine glands (Figure 42.1A).** (In contrast, *exocrine glands,* such as the sweat and salivary glands, release their secretions into ducts that lead outside the body or into the cavities of the digestive tract—see Section 38.2.) The hormones circulate throughout the body in the blood and, as a result, most body cells are exposed to a wide variety of hormones. (The cells of the central nervous system are sequestered from the general circulatory system by the blood–brain barrier—see Section 40.3.) Only *target cells* of a hormone, those with *receptor proteins* recognizing and binding that hormone (see Chapter 9), respond to it. Through these responses, hormones control such vital functions as digestion, osmotic balance, metabolism, cell division, reproduction, and development. The action of hormones may either speed up or inhibit these cellular processes. For example, growth hormone stimulates cell division, whereas glucocorticoids inhibit glucose uptake by most cells in the body. Hormones are cleared from the body by enzymatic breakdown, mainly in the liver and kidneys, but also in target cells themselves. Breakdown products are excreted in urine and feces. Depending on the hormone, the breakdown takes minutes to days.

In **neuroendocrine signaling,** specialized neurons called **neurosecretory neurons** release hormones called **neurohormones** into the circulatory system when stimulated appropriately **(Figure 42.1B).** A neurohormone is distributed by the circulatory system and elicits a response in target cells that have receptors for it. Although both neurohormones and neurotransmitters are secreted by neurons, they act differently: neurohormones are carried to target cells by the blood, whereas neurotransmitters act across a synaptic cleft (see Figure 39.6B).

A. Classical endocrine signaling

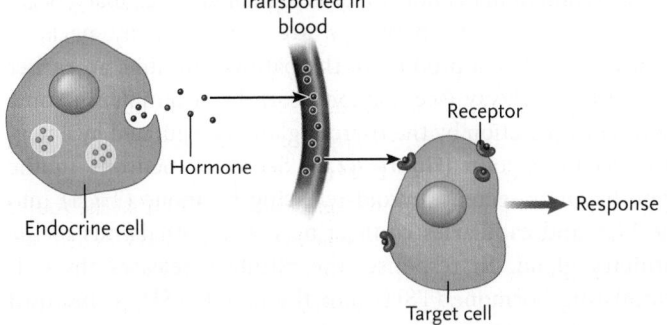

C. Paracrine regulation

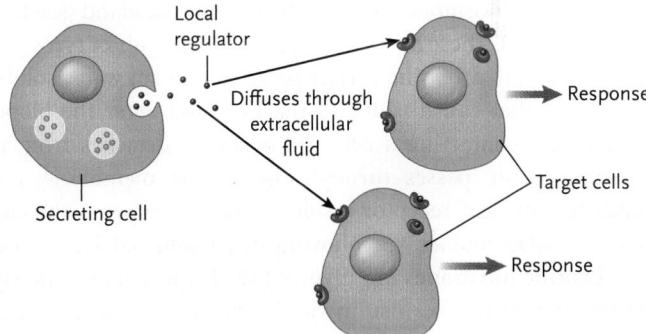

B. Neuroendocrine signaling

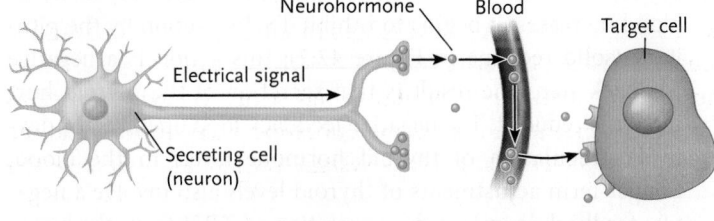

D. Autocrine regulation

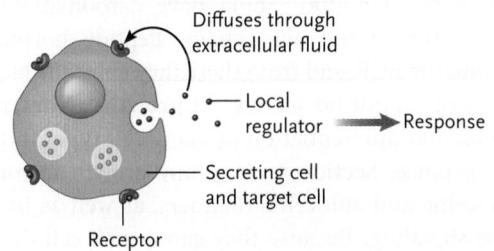

FIGURE 42.1 The four major types of cell signaling in the endocrine system.
© Cengage Learning 2017

However, both neurohormones and neurotransmitters function in the same way—they cause cellular responses by interacting with specific receptors on target cells.

In **paracrine regulation,** a cell releases a signaling molecule that diffuses through the ECF and acts on nearby cells—regulation is *local* rather than at a distance, as is the case with hormones and neurohormones **(Figure 42.1C).**

In **autocrine regulation,** a local regulator acts on the same cells that produced it **(Figure 42.1D).** Many of the growth factors that regulate cell division and differentiation act in both a paracrine and autocrine fashion.

Hormones and Local Regulators Can Be Grouped into Four Classes Based on Their Chemical Structure

More than 60 hormones and local regulators have been identified in humans. Many human hormones are either identical or very similar in structure and function to those in other animals, but other vertebrates, as well as invertebrates, have hormones not found in humans. Most hormones and local regulators fall into four molecular classes: *amine* (see Table 3.1), *peptide* (see Figure 3.16), *steroid* (see Figure 3.12), and *fatty acid-derived* molecules (see Figure 3.10).

Amine hormones are involved in classical endocrine and neuroendocrine signaling. Most amine hormones are based on tyrosine (see Figure 3.14). With one major exception, they are hydrophilic molecules, which dissolve readily into the blood and ECF. On reaching a target cell, they bind to receptors at the cell surface (see Chapter 9). The amine hormones include dopamine, epinephrine, and norepinephrine, already familiar as neurotransmitters released by some neurons (see Section 39.3). The exception is thyroxine and triiodothyronine, two hydrophobic amine hormones secreted by the thyroid gland (see later in the chapter). These hormones are based on a pair of tyrosines. They bind to hydrophilic carrier proteins to form water-soluble complexes that can dissolve in the ECF and enter the bloodstream. On contacting a cell, the hormone detaches from its carrier protein, passes through the plasma membrane, and binds to a nuclear receptor inside the target cell, as is the case with steroid hormones (see following discussion and Section 9.4).

Peptide hormones are involved in classical endocrine signaling (for example, insulin) and neuroendocrine signaling (for example, gonadotropin-releasing hormone). They consist of amino acid chains, ranging in length from as few as 3 amino acids to more than 200. Some have carbohydrate groups attached to the amino acid chain. Peptide hormones are released into the ECF, and from there they enter the blood. One large group of peptide hormones, the **growth factors,** regulates the division and differentiation of many cell types in the body (see, for instance, Section 16.5). Many growth factors act in both paracrine and autocrine manners, as well as in classical endocrine signaling. Because they can switch cell division on or off, growth factors are an important focus of cancer research.

Steroid hormones are involved in classical endocrine signaling. All are hydrophobic molecules derived from cholesterol. As with the thyroid hormones just mentioned, they form complexes with hydrophilic carrier proteins, enabling them to dissolve in the ECF and enter the bloodstream. On contacting a cell, the hormone is released from its carrier protein, passes through the plasma membrane of the target cell, and binds to internal receptors in the nucleus or cytoplasm (see Section 9.4). Steroid hormones include aldosterone, cortisol, and the sex hormones. Steroid hormones may vary little in structure, but produce very different effects. For example, testosterone and estradiol, two major sex hormones responsible for the development of mammalian male and female characteristics, respectively, differ in little more than the presence or absence of a methyl group (see Figure 3.13).

Fatty acid-derived molecules are involved in paracrine and autocrine regulation. **Prostaglandins,** for example, are important as local regulators. Virtually every cell can secrete prostaglandins, and they are present at essentially all times. In semen, they enhance the transport of sperm through the female reproductive tract by increasing the contractions of smooth muscle cells, particularly in the uterus. During childbirth, prostaglandins secreted by the placenta work with the peptide hormone oxytocin to stimulate labor contractions (see Sections 38.4 and 50.4). Other prostaglandins induce contraction or relaxation of smooth muscle cells in many parts of the body, including blood vessels and air passages in the lungs. When released as a product of membrane breakdown in injured cells, prostaglandins may also intensify pain and inflammation.

Many Hormones Are Regulated by Feedback Pathways

The secretion of many hormones is regulated by feedback pathways. Most of these pathways are controlled by negative feedback—that is, a product of the pathway inhibits an earlier step in the pathway (see Section 38.4). For example, in some mammals, secretion by the thyroid gland is regulated by a negative feedback loop **(Figure 42.2).** Secretory neurons in the hypothalamus secrete thyroid-releasing hormone (TRH) into the ECF and capillaries connecting the hypothalamus to the pituitary gland. In response, the pituitary releases thyroid-stimulating hormone (TSH) into the blood. TSH is the most important regulator of thyroid hormone secretion in that it stimulates the thyroid gland to synthesize and release thyroid hormones. When the thyroid hormone concentration in the blood increases, it begins to inhibit TSH secretion by the pituitary (solid red line in Figure 42.2); this action is a negative feedback step. The result is that secretion of the thyroid hormones is reduced. The negative feedback loop operates for day-to-day regulation of thyroid hormone levels in the blood. Longer-term adjustments of thyroid levels also involve a negative feedback loop to reduce secretion of TRH from the hypothalamus (dotted red line in Figure 42.2).

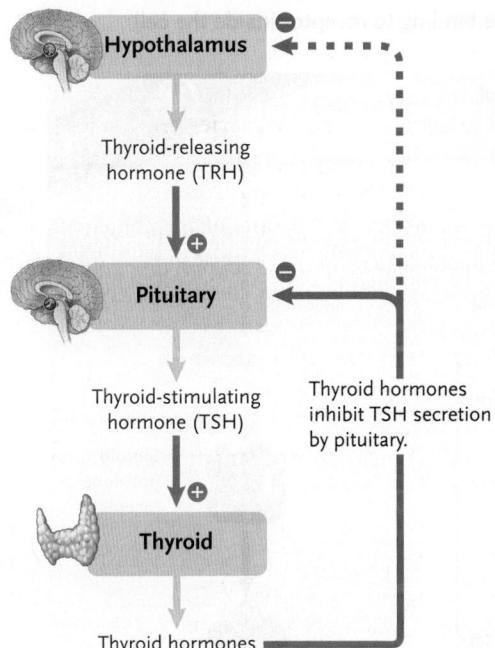

FIGURE 42.2 Negative feedback regulation of the secretion of the thyroid hormones. For day-to-day regulation, increased thyroid hormone levels in the blood trigger a negative feedback loop that inhibits TSH secretion by the pituitary (solid red line, negative sign). For long-term adjustment of thyroid levels, increased thyroid hormone levels trigger a negative feedback loop that inhibits TRH secretion by the hypothalamus (dotted red line, negative sign).
© Cengage Learning 2017

Body Processes Are Regulated by Coordinated Hormone Secretion

Although we will talk mostly about individual hormones in the remainder of the chapter, body processes are affected by more than one hormone. For example, the blood concentrations of glucose, fatty acids, and ions such as Ca^{2+}, K^+, and Na^+ are regulated by the coordinated activities of several hormones secreted by different glands. Similarly, body processes such as oxidative metabolism, digestion, growth, sexual development, and reactions to stress are all controlled by multiple hormones.

In many of these systems, negative feedback loops adjust the levels of secretion of hormones that act in antagonistic ("opposing") ways, creating a balance that maintains homeostasis. For example, consider the regulation of fuel molecules such as glucose, fatty acids, and amino acids in the blood. We usually eat three meals a day and fast to some extent between meals. During these periods of eating and fasting, four hormone systems act in coordinated fashion to keep the fuel levels in balance: (1) insulin and glucagon, secreted by the pancreas; (2) growth hormone, secreted by the anterior pituitary; (3) epinephrine and norepinephrine, released by the sympathetic nervous system and the adrenal medulla; and (4) glucocorticoid hormones, released by the adrenal cortex.

The entire system of hormones regulating fuel metabolism resembles the fail-safe mechanisms designed by human engineers, in which redundancy, overlapping controls, feedback loops, and multiple safety valves ensure that vital functions are maintained at constant levels in the face of changing and even extreme circumstances.

STUDY BREAK 42.1

1. What is the difference between a hormone and an exocrine secretion?
2. How is endocrine signaling different from autocrine signaling?

42.2 Mechanisms of Hormone Action

Hormones control cell functions by binding to receptor molecules on or in their target cells. Small quantities of hormones can typically produce profound effects in cells and body functions as a result of **amplification** (see Figure 9.6). In amplification, binding of a hormone to a receptor triggers activation of many proteins, each of which then activates an even larger number of proteins for the next step in the cellular reaction pathway, and so on, increasing in magnitude for each step in the pathway. It has been estimated that, by amplification, a single molecule of epinephrine acting on a liver cell will liberate 10^6 molecules of glucose from stored glycogen.

Hydrophilic Hormones Bind to Surface Receptors, Activating Protein Kinases Inside Cells

Hormones that bind to receptor molecules in the plasma membrane—primarily hydrophilic amine and peptide hormones—produce their responses through signal transduction pathways (see Section 9.2). In brief, when a surface receptor binds a hormone, the receptor is activated and transmits the signal through the plasma membrane. Within the cell, the signal is *transduced,* changed into a form that causes the cellular response **(Figure 42.3A).** Typically, the reactions of signal transduction pathways involve protein kinases, enzymes that add phosphate groups to proteins. Adding a phosphate group to a protein may activate it or inhibit it, depending on the protein and the reaction. The particular response produced by a hormone depends on the kinds of protein kinases activated, the type of cell that can respond, and the types of target proteins they phosphorylate.

Receptor tyrosine kinases and G-protein–coupled receptors are the two major types of surface receptors that bind hydrophilic hormones. Receptor tyrosine kinases have a built-in protein kinase on the cytoplasmic side of the receptor (Figure 42.3A illustrates this type of receptor). Binding of a signal molecule to this type of receptor turns on the receptor's protein kinase, which activates the receptor. The activated

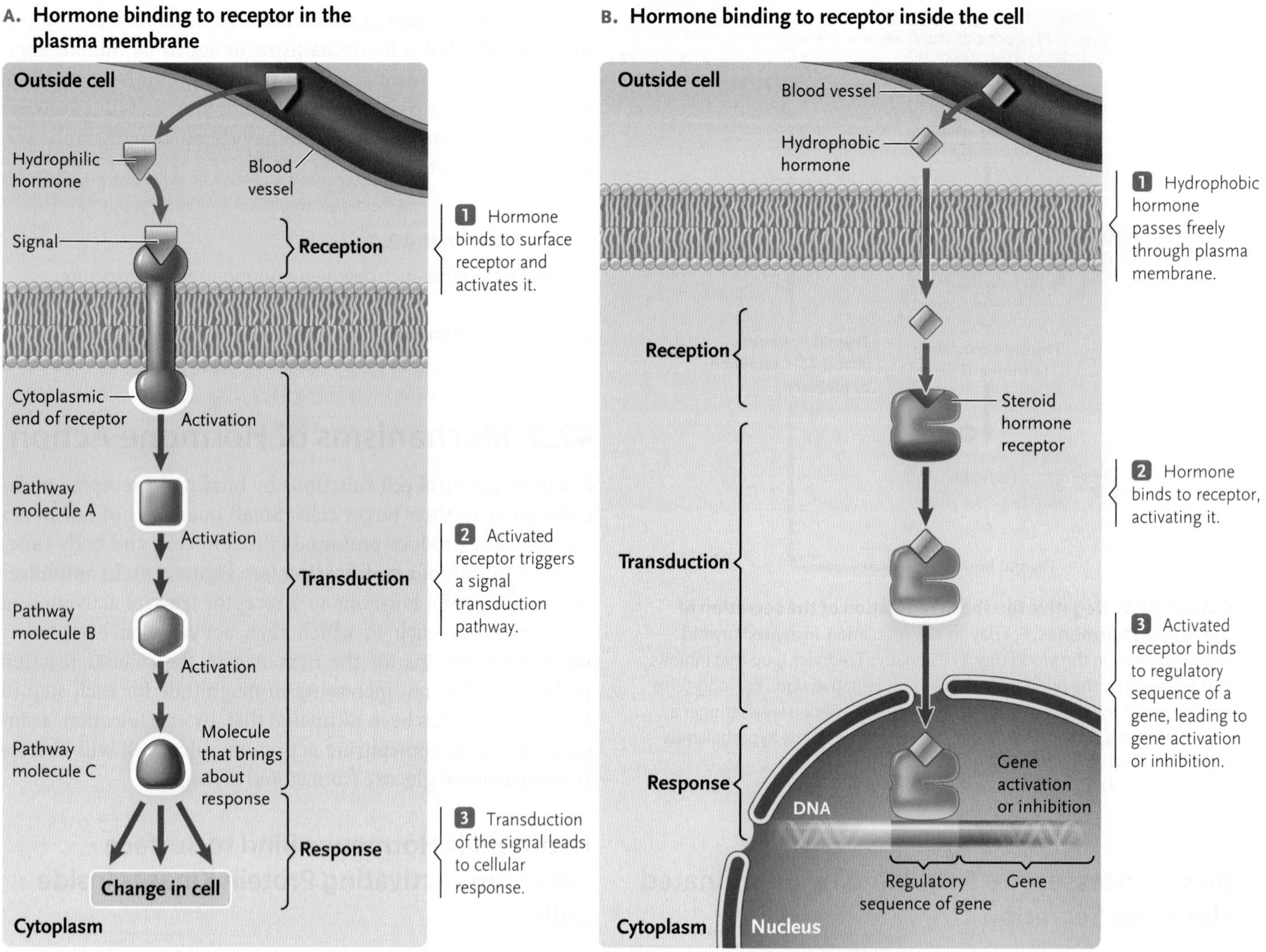

A. Hormone binding to receptor in the plasma membrane

Outside cell

Hydrophilic hormone

Blood vessel

Signal

Reception — ① Hormone binds to surface receptor and activates it.

Cytoplasmic end of receptor — Activation

Pathway molecule A — Activation

Transduction — ② Activated receptor triggers a signal transduction pathway.

Pathway molecule B — Activation

Pathway molecule C — Molecule that brings about response

Response — ③ Transduction of the signal leads to cellular response.

Change in cell

Cytoplasm

B. Hormone binding to receptor inside the cell

Outside cell

Blood vessel

Hydrophobic hormone

① Hydrophobic hormone passes freely through plasma membrane.

Reception

Steroid hormone receptor

② Hormone binds to receptor, activating it.

Transduction

③ Activated receptor binds to regulatory sequence of a gene, leading to gene activation or inhibition.

Response

DNA

Gene activation or inhibition

Regulatory sequence of gene — Gene

Cytoplasm — **Nucleus**

FIGURE 42.3 Outline of the reaction pathways activated by hormones that bind to receptor proteins in the plasma membrane (A) or inside cells (B). In both mechanisms, the signal—the binding of the hormone to its receptor—is transduced to produce the cellular response.
© Cengage Learning 2017

receptor then initiates a signaling cascade within the cell, typically involving protein kinases, which phosphorylate target proteins, resulting in a cellular response. The cytoplasmic reactions that receptor tyrosine kinases control when they are activated are described in detail in Section 9.3. Insulin, a peptide hormone that lowers glucose concentration in the blood, elicits a cellular response using a signal transduction pathway initiated by activating a receptor tyrosine kinase. Insulin acts mainly by binding to receptors on liver cells, adipose tissue (fat), and skeletal muscle cells to stimulate glucose transport into cells, conversion of glucose to glycogen, and other metabolic activities.

G-protein–coupled receptors lack a built-in protein kinase. This type of receptor responds to the binding of a hormone by activating a G protein associated with the cytoplasmic end of the receptor. The activated G protein then activates an effector molecule, which then generates a second messenger molecule. Second messengers activate protein kinases in the

cell, which elicit the cellular response by phosphorylating target proteins. The cytoplasmic reactions that G-protein–coupled receptors control when activated are described in detail in Section 9.3. Glucagon, a peptide hormone that raises glucose concentration in the blood, elicits a cellular response using a signal transduction pathway initiated by activating a G-protein–coupled receptor. When glucagon binds to surface receptors on liver cells, it triggers the breakdown of glycogen stored in those cells into glucose. The glucose then is released into the circulatory system.

Hydrophobic Hormones Bind to Receptors Inside Cells, Activating or Inhibiting Genetic Regulatory Proteins

After passing through the plasma membrane, the hydrophobic steroid and thyroid hormones bind to internal receptors in the nucleus or cytoplasm (**Figure 42.3B,** and described in detail in

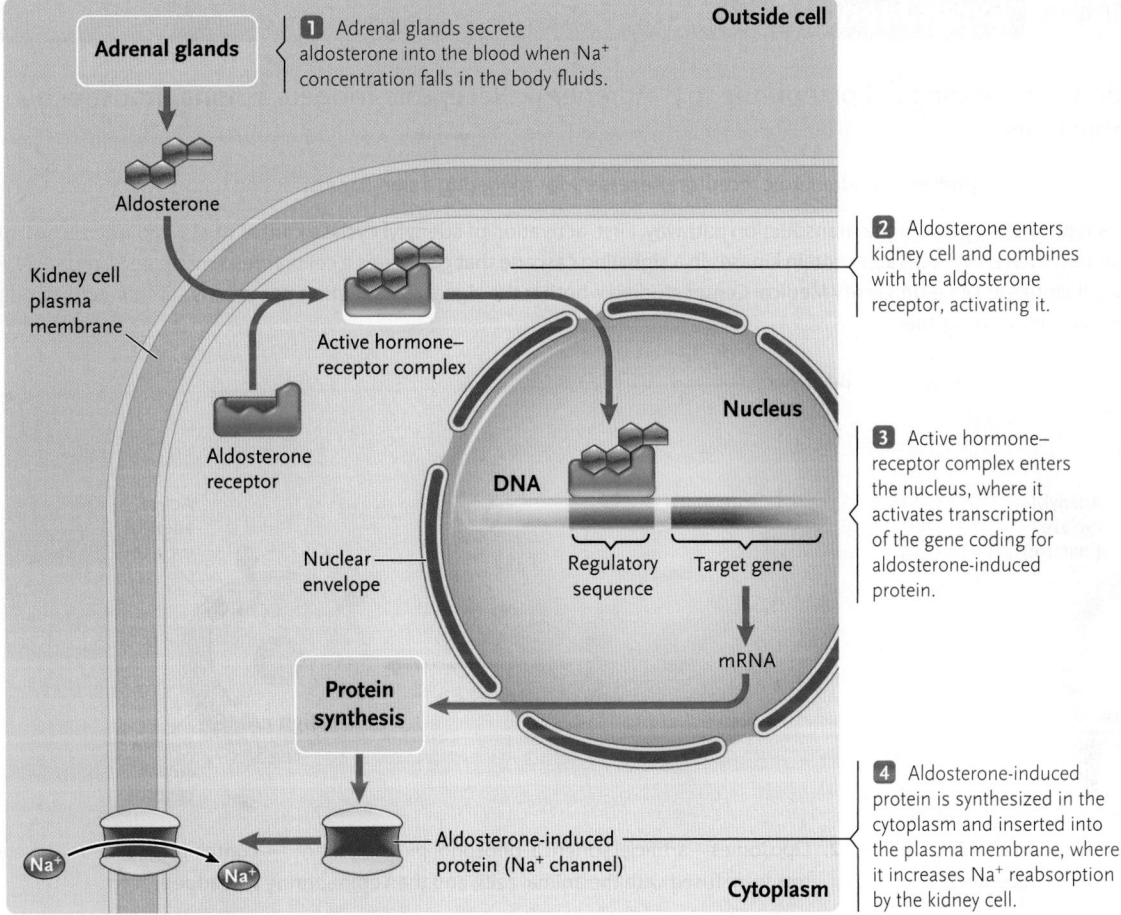

FIGURE 42.4 The action of aldosterone in increasing Na⁺ reabsorption in the kidneys when concentration of the ion falls in the blood.
© Cengage Learning 2017

Within the figure:

- **Adrenal glands**
- **1** Adrenal glands secrete aldosterone into the blood when Na⁺ concentration falls in the body fluids.
- **Outside cell**
- Aldosterone
- Kidney cell plasma membrane
- Active hormone–receptor complex
- Aldosterone receptor
- **2** Aldosterone enters kidney cell and combines with the aldosterone receptor, activating it.
- **Nucleus**
- DNA
- Nuclear envelope
- Regulatory sequence
- Target gene
- **3** Active hormone–receptor complex enters the nucleus, where it activates transcription of the gene coding for aldosterone-induced protein.
- mRNA
- Protein synthesis
- Na⁺ Na⁺
- Aldosterone-induced protein (Na⁺ channel)
- **Cytoplasm**
- **4** Aldosterone-induced protein is synthesized in the cytoplasm and inserted into the plasma membrane, where it increases Na⁺ reabsorption by the kidney cell.

Section 9.4). Binding of the hormone activates the receptor, which then binds to a regulatory sequence of specific genes to elicit a genomic response rather than triggering a signal transduction pathway. Depending on the gene, binding the regulatory sequence either activates or inhibits its transcription, leading to changes in protein synthesis that accomplish the cellular response. The characteristics of the response depend on the specific genes controlled by the activated receptors, and on the presence of other proteins that modify the activity of the receptor.

One of the actions of the steroid hormone aldosterone illustrates the mechanisms triggered by internal receptors **(Figure 42.4)**. If blood pressure falls below optimal levels, aldosterone is secreted by the adrenal glands. The hormone can enter all cells, but affects only kidney cells that contain the aldosterone receptor in their cytoplasm. When activated by aldosterone, the receptor binds to the regulatory sequence of a gene, leading to the synthesis of proteins that increase reabsorption of Na⁺ by the kidney cells. The resulting increase in Na⁺ concentration in body fluids increases water retention and, with it, blood volume and pressure. These changes then feed back negatively on the processes that initiated aldosterone secretion.

(Recent research has shown that aldosterone, and some other steroid hormones, have specific cell surface receptors that trigger signal transduction pathways when the steroid hormone binds. In this way, those steroid hormones can initiate rapid nongenomic responses in addition to the relatively slower genomic responses.)

Target Cells May Respond to More Than One Hormone, and Different Target Cells May Respond Differently to the Same Hormone

A single target cell may have receptors for several hormones and respond differently to each hormone. For example, vertebrate liver cells have receptors for the pancreatic hormones insulin and glucagon (described earlier). Insulin increases glucose uptake and conversion to glycogen, which decreases blood glucose levels, while glucagon stimulates the breakdown of glycogen into glucose, which increases blood glucose levels.

Conversely, particular hydrophilic and hydrophobic hormones interact with different types of receptors on or in a range of target cells. Different responses are then triggered in each target cell type because the receptors trigger different signal

FIGURE 42.5 | **Experimental Research**

Demonstration That Binding of Epinephrine to β-Adrenergic Receptors Triggers a Signal Transduction Pathway within Cells

Question: Is binding of epinephrine to β-adrenergic receptors necessary for triggering a signal transduction pathway within cells?

Experiment: Epinephrine triggers a signal transduction pathway. First, activation of adenylyl cyclase causes the level of the second messenger cAMP to increase, and then cAMP activates protein kinases in a signaling cascade that generates a cellular response (see Section 9.3). Richard Cerione and his colleagues at Duke University Medical Center studied whether the signal transduction pathway is stimulated by binding of epinephrine to β-adrenergic receptors.

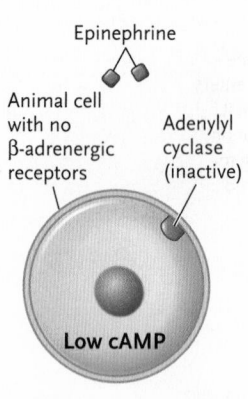

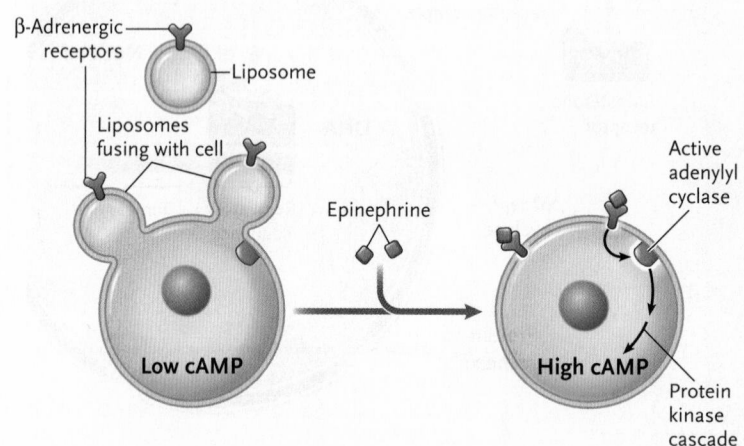

1. Epinephrine was added to animal cells lacking β-adrenergic receptors.

   Result: No change occurred to the low level of cAMP in those cells. This result demonstrated that epinephrine alone was not able to trigger an increase in cAMP.

2. Liposomes—artificial spherical phospholipid membranes—containing purified β-adrenergic receptors were fused with the animal cells, and then epinephrine was added.

   Result: When the liposomes fused with the animal cells, β-adrenergic receptors became part of the fused cell's plasma membrane. Then, adding epinephrine triggered synthesis of cAMP, resulting in high levels of cAMP in the cells. This result demonstrated that β-adrenergic receptors must be present in the membrane for epinephrine to trigger an increase in cAMP in the cell. The simplest interpretation was that epinephrine bound to the β-adrenergic receptors, activating adenylyl cyclase within the cell.

Conclusion: The cellular response depended on binding of the hydrophilic hormone to a specific plasma membrane-embedded receptor.

 Observe Hypothesize Predict **Experiment** Interpret | How could this system be used to study what structural features of the receptor molecules are important for controlling the associated signal transduction pathway?

Source: R. A. Cerione et al. 1983. Reconstitution of the β-adrenergic receptors in lipid vesicles: Affinity chromatography-purified receptors confer catecholamine responsiveness on a heterologous adenylate cyclase system. *Proceedings of the National Academy of Sciences USA* 80:4899–4903.

transduction pathways. For example, the amine hormone epinephrine secreted by the adrenal medulla prepares the body for handling stress (including dangerous situations) and physical activity. (Epinephrine is discussed in more detail in Section 42.4.) In mammals, epinephrine can bind to several different G-protein–coupled receptors (see Section 9.3) known as **adrenergic receptors** because of the adrenal origin of the hormones that bind to them. These receptors are categorized into two types: α-adrenergic receptors and β-adrenergic receptors. (The experimental demonstration that binding of epinephrine to a specific adrenergic receptor triggers a cellular response is described in **Figure 42.5**.) When epinephrine binds

to an α-adrenergic receptor on a smooth muscle cell, such as that of a blood vessel, it triggers a response pathway that causes the cells to constrict, reducing circulation to peripheral organs. When epinephrine binds to β_1-adrenergic receptors on heart muscle cells, the contraction rate of the cells increases, which in turn enhances blood supply. When epinephrine binds to β_2-adrenergic receptors on liver cells, it stimulates the breakdown of glycogen to glucose, which is released from the cell. The overall effect of these, and a number of other, responses to epinephrine secretion is to supply energy to the major muscles responsible for locomotion—the body is now prepared for handling stress or for physical activity.

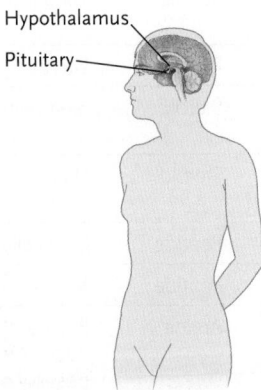

Hypothalamus
Produces and secretes releasing and inhibiting hormones that regulate secretions by the anterior pituitary

Produces ADH and oxytocin, which are stored and released by the posterior pituitary

Anterior pituitary
Secretes ACTH, TSH, FSH, and LH, which stimulate other glands, as well as prolactin, GH, MSH, and endorphins

Posterior pituitary
Stores and releases ADH and oxytocin

Adrenal cortex
Secretes cortisol and aldosterone and small amounts of androgens

Adrenal medulla
Secretes epinephrine and norepinephrine

Islets of Langerhans (in pancreas)
Secrete insulin and glucagon

Ovaries
Secrete estrogens and progestins

Testes
Secrete androgens

Pineal gland
Secretes melatonin

Thyroid gland
Secretes thyroxine and triiodothyronine, calcitonin

Parathyroid glands
Secrete parathyroid hormone

FIGURE 42.6 Major endocrine glands of the human body. Among the other organs that contain hormone-producing cells are the kidneys (see Section 44.2), the stomach and small intestine (see Section 47.4), and the heart (see Section 48.4).
© Cengage Learning 2017

STUDY BREAK 42.2

1. Compare and contrast the mechanisms by which glucagon and aldosterone cause their specific responses.
2. Explain how one type of target cell could respond to different hormones, and how the same hormone could produce different effects in different cells.

In summary, the mechanisms by which hormones work have four major features:

1. Only the cells that contain surface or internal receptors for a particular hormone respond to that hormone.
2. Once bound by their receptors, hormones may produce a response that involves stimulation or inhibition of cellular processes through the specific types of internal molecules triggered by the hormone action.
3. Because of the amplification that occurs in both the surface and internal receptor mechanisms, hormones are effective in very small concentrations.
4. The response to a hormone differs among target organs.

In the next two sections, we discuss the major endocrine cells and glands of vertebrates. The locations of these cells and glands in the human body and their functions are summarized in **Figure 42.6** and **Table 42.1.** Peptide hormones secreted by other body regions, including the stomach and small intestine, the thymus gland, the kidneys, and the heart, are described in the chapters in which these tissues and organs are discussed.

42.3 The Hypothalamus and Pituitary

Hypothalamus
Pituitary

The hormones of vertebrates work in coordination with the nervous system. The action of several hormones is closely coordinated by the hypothalamus and the connected pituitary.

The hypothalamus is a region of the brain located in the floor of the cerebrum (see Section 40.3). The **pituitary gland,** consisting mostly of two fused lobes, is suspended just below it by a slender stalk of tissue that contains both neurons and blood vessels **(Figure 42.7).** The **posterior pituitary** contains axons and nerve endings of neurosecretory neurons that originate in the hypothalamus. The **anterior pituitary** contains nonneuronal endocrine cells that form a distinct gland. The two lobes are separate in structure and embryonic origins.

TABLE 42.1 | The Major Human Endocrine Glands and Hormones

Secretory Tissue or Gland	Hormones	Molecular Class	Target Tissue	Principal Actions
Hypothalamus	Releasing and inhibiting hormones	Peptide	Anterior pituitary	Regulate secretion of anterior pituitary hormones
Anterior pituitary	Thyroid-stimulating hormone (TSH)	Peptide	Thyroid gland	Stimulates secretion of thyroid hormones
	Adrenocorticotropic hormone (ACTH)	Peptide	Adrenal cortex	Stimulates secretion of glucocorticoids by adrenal cortex
	Follicle-stimulating hormone (FSH)	Peptide	Ovaries in females, testes in males	Stimulates egg growth, secretion of female sex hormones, sperm production
	Luteinizing hormone (LH)	Peptide	Ovaries in females, testes in males	Regulates ovulation and secretion of male sex hormones
	Prolactin (PRL)	Peptide	Mammary glands	Stimulates breast development and milk secretion
	Growth hormone (GH)	Peptide	Bone, soft tissue	Stimulates bone growth; helps control metabolism
	Melanocyte-stimulating hormone (MSH)	Peptide	Melanocytes in skin of some vertebrates	Promotes darkening of the skin
	Endorphins	Peptide	Pain pathways of PNS	Inhibit perception of pain
Posterior pituitary	Antidiuretic hormone (ADH)	Peptide	Kidneys	Promotes water reabsorption in kidneys
	Oxytocin	Peptide	Uterus, mammary glands	Promotes uterine contractions; stimulates milk production and secretion
Thyroid gland	Calcitonin	Peptide	Bone	Lowers calcium concentration in blood
	Thyroxine and triiodothyronine	Amine	Most cells	Increase metabolic rate; essential for normal body growth
Parathyroid glands	Parathyroid hormone (PTH)	Peptide	Bone, kidneys, intestine	Raises calcium concentration in blood
Adrenal medulla	Epinephrine and norepinephrine	Amine	Sympathetic receptor sites throughout body	Reinforce sympathetic nervous system; contribute to responses to stress
Adrenal cortex	Aldosterone (mineralocorticoid)	Steroid	Kidney tubules	Increases Na$^+$ reabsorption and K$^+$ excretion in kidneys
	Cortisol (glucocorticoid)	Steroid	Most body cells, particularly muscle, liver, and adipose cells	Increases blood glucose levels
Testes	Androgens, such as testosterone[a]	Steroid	Various tissues	Control male reproductive system development and maintenance
	Oxytocin	Peptide	Uterus	Promotes uterine contractions
Ovaries	Estrogens, such as estradiol[b]	Steroid	Breast, uterus, other tissues	Stimulate maturation of sex organs at puberty, and development of secondary sexual characteristics
	Prolactins, such as progesterone[b]	Steroid	Uterus	Prepare and maintain uterine lining
Pancreas (islets of Langerhans)	Glucagon (alpha cells)	Peptide	Liver cells	Raises glucose concentration in blood
	Insulin (beta cells)	Peptide	Most cells	Lowers glucose concentration in blood
Pineal gland	Melatonin	Amine	Brain, anterior pituitary, reproductive organs, immune system, possibly others	Helps synchronize body's biological clock with day length
Many cell types	Growth factors	Peptide	Most cells	Regulate cell division and differentiation
	Prostaglandins	Fatty acid	Various tissues	Have many diverse roles

[a]Small amounts secreted by ovaries and adrenal cortex.
[b]Small amounts secreted by testes.

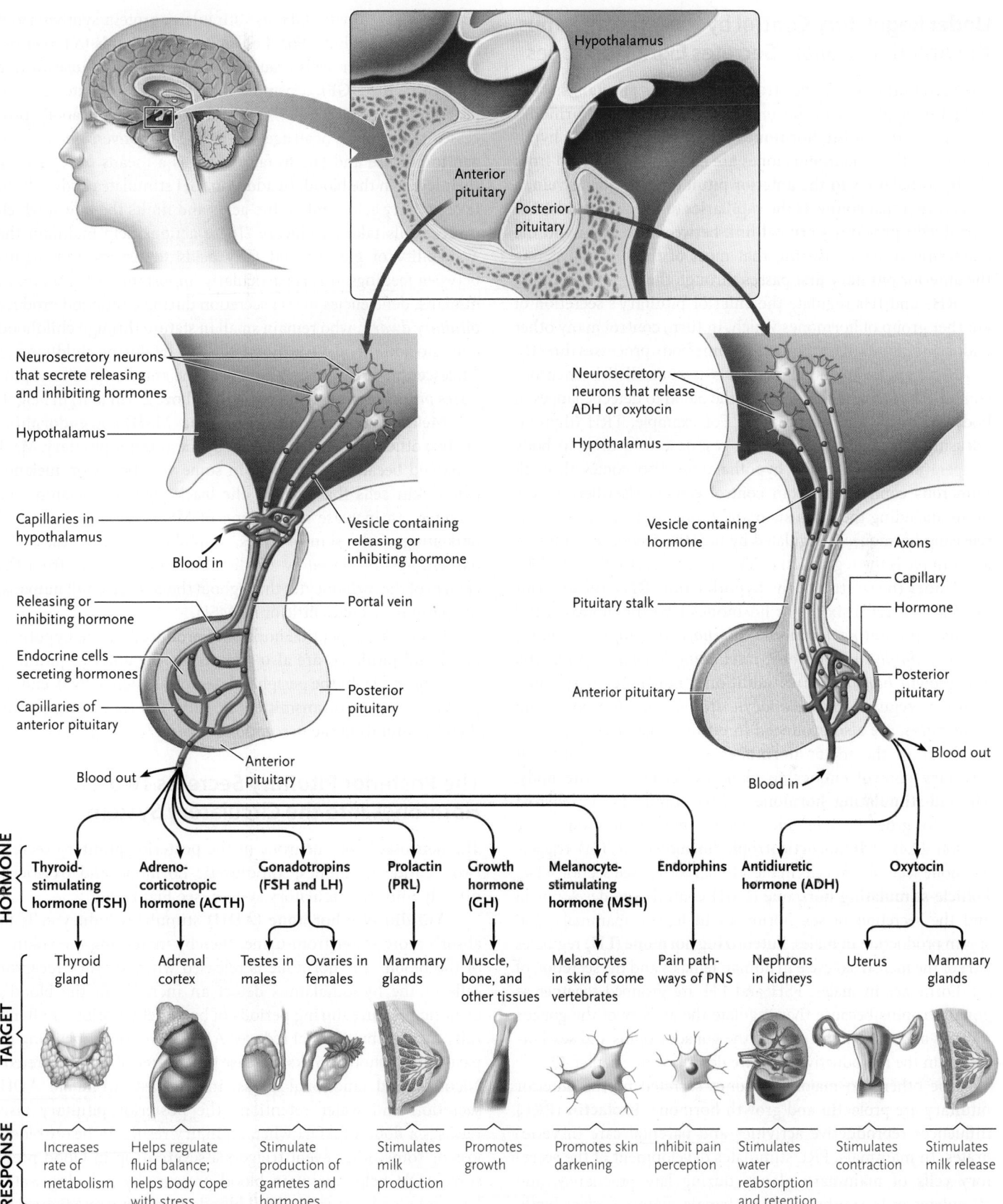

FIGURE 42.7 The hypothalamus and pituitary. Hormones secreted by the anterior and posterior pituitary are controlled by neurohormones released in the hypothalamus.

© Cengage Learning 2017

Under Regulatory Control by the Hypothalamus, the Anterior Pituitary Secretes Eight Hormones

The secretion of hormones from the anterior pituitary is controlled by peptide neurohormones called **releasing hormones (RHs)** and **inhibiting hormones (IHs),** released by the hypothalamus. These neurohormones are carried in the blood from the hypothalamus to the anterior pituitary in a **portal vein,** a special vein that connects the capillaries of the two glands. The portal vein provides a critical link between the brain and the endocrine system, ensuring that most of the blood reaching the anterior pituitary first passes through the hypothalamus.

RHs and IHs regulate the anterior pituitary's secretion of another group of hormones which, in turn, control many other endocrine glands of the body, and some body processes directly.

Secretion of hypothalamic RHs is controlled by neurons containing receptors that monitor the blood to detect changes in body chemistry and temperature. For example, TRH (thyroid-releasing hormone) is secreted in response to a drop in body temperature. Input to the hypothalamus also comes through numerous connections from control centers elsewhere in the brain, including the brain stem and limbic system. Secretion of releasing hormones is regulated by negative feedback pathways; an example is the regulation of TRH secretion (see Figure 42.2).

Under the control of the hypothalamic RHs, the anterior pituitary secretes six major hormones into the bloodstream: *thyroid-stimulating hormone, adrenocorticotropic hormone, follicle-stimulating hormone, luteinizing hormone, prolactin,* and *growth hormone.* Two additional hormones with more complex regulation, *melanocyte-stimulating hormone* and *endorphins,* are also produced in pituitary tissue.

Four of the major hormones secreted by the anterior pituitary control endocrine glands elsewhere in the body. **Thyroid-stimulating hormone (TSH)** stimulates the thyroid gland to grow in size and secrete thyroid hormones (see Section 42.4). **Adrenocorticotropic hormone (ACTH)** triggers hormone secretion by cells in the adrenal cortex (see Section 42.4). **Follicle-stimulating hormone (FSH)** controls egg development and the secretion of sex hormones in female mammals, and sperm production in males. **Luteinizing hormone (LH)** regulates part of the menstrual cycle in human females and the secretion of sex hormones in males. FSH and LH are grouped together as **gonadotropins** because they regulate the activity of the gonads (ovaries and testes). The roles of the gonadotropins and sex hormones in the reproductive cycle are described in Chapter 49.

The other two major hormones secreted by the anterior pituitary are prolactin and growth hormone. **Prolactin (PRL)** influences reproductive activities and parental care in vertebrates. In mammals, PRL stimulates development of the secretory cells of mammary glands during late pregnancy, and stimulates milk synthesis after a female mammal gives birth. Stimulation of the mammary glands and the nipples, as occurs during suckling, leads to PRL release.

Growth hormone (GH) stimulates cell division, protein synthesis, and bone growth in children and adolescents, thereby causing body growth. GH also stimulates protein synthesis and cell division in adults. For these actions, GH binds to target tissues, mostly liver cells, causing them to release **insulin-like growth factor (IGF),** a peptide that directly stimulates growth processes. GH also controls a number of major metabolic processes in mammals of all ages, including the conversion of glycogen to glucose and fats to fatty acids as a means of regulating their levels in the blood. In addition, GH stimulates body cells to take up fatty acids and amino acids and limits the rate at which muscle cells take up glucose. These actions help maintain the availability of glucose and fatty acids to tissues and organs between feedings; this is particularly important for the brain. In humans, deficiencies in GH secretion during childhood produce *pituitary dwarfs,* who remain small in stature through childhood and into adulthood. Overproduction of GH during childhood or adolescence, often due to a tumor of the anterior pituitary, produces *pituitary giants,* who may grow above 2.4 m (8 ft) in height.

Melanocyte-stimulating hormone (MSH) and **endorphins** are two other hormones produced by the anterior pituitary. MSH is named because of its effect in some vertebrates on melanocytes, skin cells that contain the black pigment melanin. For example, an increase in secretion of MSH produces a marked darkening of the skin of fishes, amphibians, and reptiles. The darkening is produced by a redistribution of melanin from the centers of the melanocytes throughout the cells. In adult humans, the pituitary secretes little or no MSH.

Endorphins, peptide hormones produced by the hypothalamus and pituitary, are also released by the anterior pituitary (see Table 42.1). In the peripheral nervous system (PNS), endorphins act as neurotransmitters in pathways that control pain, thereby inhibiting the perception of pain.

The Posterior Pituitary Secretes Two Hormones into the Circulatory System

The neurosecretory neurons in the posterior pituitary secrete two peptide hormones, antidiuretic hormone and oxytocin, directly into the circulatory system (see Figure 42.7).

Antidiuretic hormone (ADH) stimulates kidney cells to absorb more water from urine, thereby increasing the volume of the blood. The hormone is released when sensory receptor cells of the hypothalamus detect an increase in the blood's osmotic pressure during periods of body dehydration or after a salty meal. Ethyl alcohol reduces ADH secretion, explaining in part why alcoholic drinks increase the volume of urine excreted. Nicotine and emotional stress, in contrast, stimulate ADH secretion and water retention. The posterior pituitary also releases a flood of ADH when an injury results in heavy blood loss or some other event triggers a severe drop in blood pressure. ADH helps maintain blood pressure by reducing water loss and also by causing small blood vessels in some tissues to constrict. (The role of ADH in the regulation of mammalian kidney function is discussed more in Chapter 48.)

Hormones with structure and action similar to ADH are also secreted in fishes, amphibians, reptiles, and birds. In amphibians, these ADH-like hormones increase the amount of water entering the body through the skin and from the urinary bladder.

Oxytocin stimulates the ejection of milk from the mammary glands of a nursing mother (see Table 42.1). Stimulation of the nipples in suckling sends neuronal signals to the hypothalamus, and leads to release of oxytocin from the posterior pituitary. The released oxytocin stimulates more oxytocin secretion by a positive feedback mechanism. Oxytocin causes the smooth muscle cells surrounding the individual alveoli of the mammary glands to contract, forcibly expelling the milk through the nipples. The entire cycle, from the onset of suckling to milk ejection, takes less than a minute in mammals. Oxytocin also plays a key role in childbirth, as discussed in Sections 38.4 and 50.4.

In males, oxytocin is secreted into the seminal fluid by the testes. Like prostaglandins, when the seminal fluid is ejaculated into the vagina during sexual intercourse, oxytocin stimulates contractions of the uterus that aid movement of sperm through the female reproductive tract.

In both sexes, oxytocin is known to produce feelings of well-being and calm. *Molecular Insights* discusses experiments showing that the hormone can also increase anxiety and fear.

STUDY BREAK 42.3

1. Summarize the functional relationship between the hypothalamus and the anterior pituitary gland.
2. What, in addition to the hormones they secrete, distinguishes the anterior and posterior pituitary glands?

42.4 Other Major Endocrine Glands of Vertebrates

Besides the hypothalamus and pituitary, the body has seven major endocrine glands or tissues, many of them regulated by the hypothalamus–pituitary connection. These glands are the *thyroid gland, parathyroid glands, adrenal medulla, adrenal cortex, gonads, pancreas,* and *pineal gland* (shown in Figure 42.6 and summarized in Table 42.1).

The Thyroid Hormones Stimulate Metabolism, Development, and Maturation

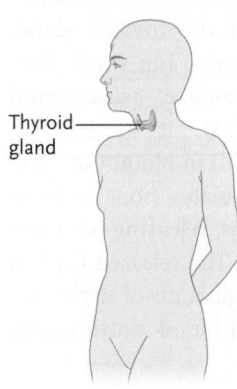

Thyroid gland

The **thyroid gland,** which is located in the front of the throat in humans, has a shape similar to that of a bowtie (see Figure 38.3B). It secretes the same hormones in all vertebrates. The primary circulating thyroid hormone, **thyroxine,** is known as T_4 because it contains four iodine atoms. The thyroid also secretes smaller amounts of a closely related hormone, **triiodothyronine** or T_3, which contains three iodine atoms. A supply

of iodine in the diet is necessary for production of these hormones. Normally, their concentrations are kept at finely balanced levels in the blood by negative feedback loops such as that described in Figure 42.2.

T_4 and T_3 are hydrophobic. Each is transported in the bloodstream bound with hydrophilic carrier proteins to form water-soluble complexes. Both T_4 and T_3 enter cells. Once inside the cell, most of the T_4 is converted to T_3, the form that combines with internal receptors in the nucleus. Binding of T_3 to receptors alters gene expression, which brings about the hormone's effects.

The thyroid hormones are vital to growth, development, maturation, and metabolism in all vertebrates. They interact with GH for their effects on growth and development. Thyroid hormones also increase the sensitivity of many body cells to the effects of epinephrine and norepinephrine, hormones released by the adrenal medulla as part of the "fight-or-flight response" (discussed further later in this chapter).

In amphibians such as frogs, thyroid hormones trigger **metamorphosis,** or change in body form from tadpole to adult **(Figure 42.8).** Thyroid hormones also contribute to seasonal changes in the plumage of birds and coat color in mammals.

In human adults, low thyroid output, *hypothyroidism,* causes affected individuals to be sluggish mentally and physically; they have a slow heart rate and weak pulse, and often feel confused and depressed. Hypothyroidism in infants and children leads to cretinism, that is, stunted growth and diminished intelligence. Overproduction of thyroid hormones in human adults, *hyperthyroidism,* produces nervousness and emotional instability, irritability, insomnia, weight loss, and a rapid, often irregular heartbeat. The most common

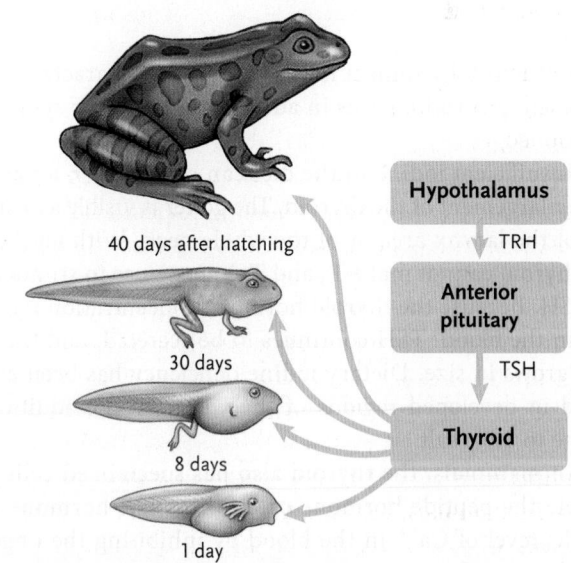

40 days after hatching

30 days

8 days

1 day

Hypothalamus

TRH

Anterior pituitary

TSH

Thyroid

FIGURE 42.8 Metamorphosis of a tadpole into an adult frog, under the control of thyroid hormones. As a part of the metamorphosis, changes in gene activity lead to a change from an aquatic to a terrestrial habitat. TRH, thyroid-releasing hormone; TSH, thyroid-stimulating hormone.
© Cengage Learning 2017

Molecular Insights

Fear-Enhancing Effects of Some Oxytocin Receptors in Mice

Oxytocin is a peptide hormone secreted directly into the circulatory system by neurosecretory neurons in the posterior pituitary. In addition to its major physiological effects described in the chapter, oxytocin also influences behavior. In fact, it has become known as the "love hormone" because it promotes feelings of love, well-being, and social bonding. In other words, oxytocin can reduce anxiety, fear, and stress. However, some observations in humans indicate that oxytocin can cause increased recollection of stressful events.

Research Question

What is the neurobiological basis for fear regulation by oxytocin?

Experiments

To answer the question, Jelena Radulovic and coworkers at Northwestern University, Chicago, Illinois, USA; Tohoku University, Miyagi, Japan; and Jichi Medical University, Tochigi-ken, Japan, used mice in a study that focused on a region of the brain known to be involved significantly with stress and anxiety and that has high levels of oxytocin receptors (Oxtrs). Their experiments used acute social defeat, an event known to increase oxytocin release in the brain area of interest in normal mice, as a stressor.

That is, mice are given a stressful experience by placing them individually in cages with aggressive mice. If the experience produces a fear/social defeat response, the introduced mouse shows a freezing behavior. The researchers tested individuals from three groups of mice: (A) mice manipulated genetically to lack Oxtrs in the brain area of interest so oxytocin could not enter brain cells; (B) mice manipulated genetically to produce increased numbers of Oxtrs so brain cells would be flooded with oxytocin; and (C) control mice with a normal number of receptors.

Results

Individuals of all three mice groups were first exposed to social defeat. Then, 6 hours later the mice were put back in the cages with aggressive mice and their fear response was observed. Group A mice showed no fear, Group B showed an intense fear reaction, and Group C showed a normal fear reaction. In other words, Group A had not remembered the previous stressful experience, whereas Group B had remembered it intensively. These results showed that oxytocin is essential for strengthening the memory of stressful experiences such as these social interactions with aggressive mice.

In another experiment, the three groups of mice experienced social defeat by aggressive mice and then, 6 hours later, they were placed in a box and given a brief electric shock to startle them. Twenty-four hours later, the mice were returned to the box but not given a shock. Group A showed no enhanced fear when returned to the box, whereas Group B showed much greater fear. Group C controls showed an average fear response. Thus, oxytocin can increase fear in future stressful situations.

Conclusion

This is the first study to demonstrate that oxytocin can be associated with social stress, and that the hormone can increase anxiety and fear in response to future stress.

think like a scientist

Is it likely that an identical or similar fear-enhancing effect mechanism of oxytocin occurs in humans?

Source: Y. F. Guzmán et al. 2013. Fear-enhancing effects of septal oxytocin receptors. *Nature Neuroscience* 16:1185–1187.

---

form of hyperthyroidism is *Graves' disease,* characterized by inflamed, protruding eyes in addition to the other symptoms mentioned.

Insufficient iodine in the diet can cause *goiter,* an abnormal enlargement of the thyroid. The goiter is visible as a swelling of the larynx area or of the whole neck. Without iodine, the thyroid cannot make T_3 and T_4 in response to stimulation by TSH. Because the thyroid hormone concentration remains low in the blood, TSH continues to be secreted, and the thyroid grows in size. Dietary iodine deficiency has been eliminated in developed regions of the world by the addition of iodine to table salt.

In mammals, the thyroid also has specialized cells that secrete the peptide hormone **calcitonin.** The hormone lowers the level of Ca^{2+} in the blood by inhibiting the ongoing dissolution of calcium from bone. Calcitonin secretion is stimulated when Ca^{2+} levels in blood rise above the normal range and inhibited when Ca^{2+} levels fall below the normal range.

The Parathyroid Glands Regulate Ca^{2+} Level in the Blood

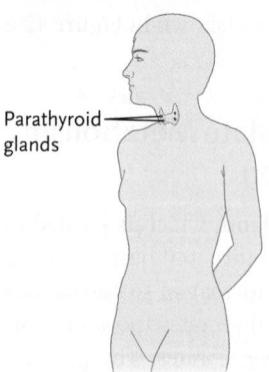

Parathyroid glands

The **parathyroid glands** occur only in tetrapod vertebrates—amphibians, reptiles, birds, and mammals. Each is a spherical structure about the size of a pea. Mammals have four parathyroids located on the posterior surface of the thyroid gland, two on each side. The single hormone they produce, **parathyroid hormone (PTH),** is secreted in response to a fall in blood Ca^{2+} levels. PTH stimulates bone cells to dissolve the mineral matter of bone tissues, releasing both calcium and phosphate ions into the blood. The released Ca^{2+} is then available for enzyme activation, conduction of nerve signals across synapses, muscle contraction, blood clotting, and

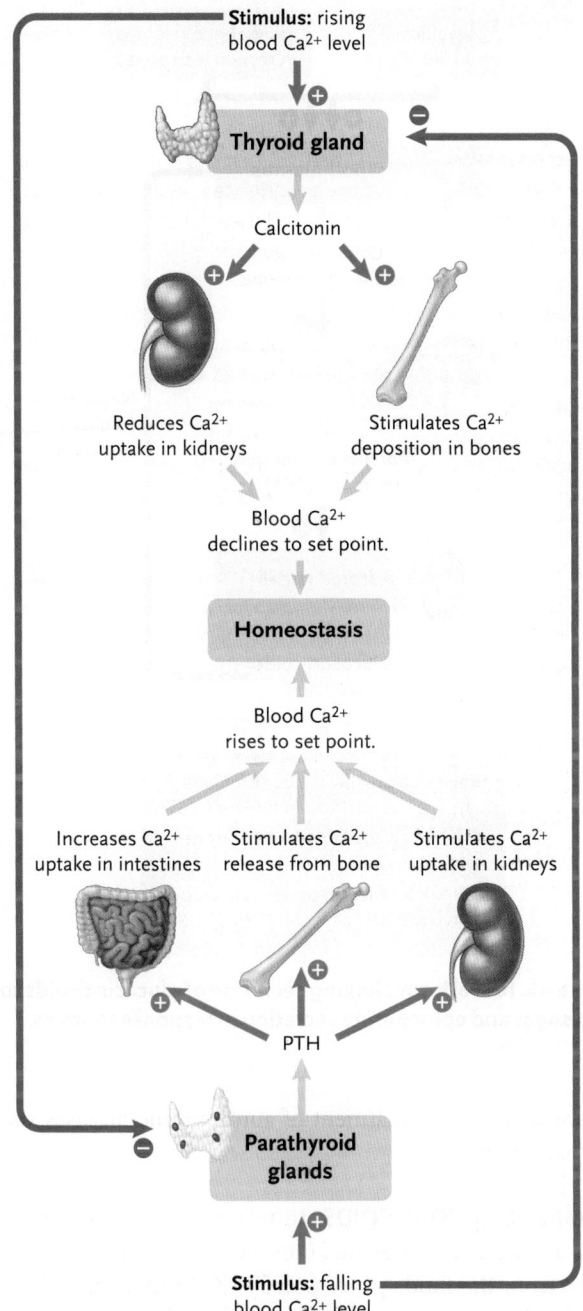

Stimulus: rising
blood Ca²⁺ level

Thyroid gland

Calcitonin

Reduces Ca²⁺
uptake in kidneys

Stimulates Ca²⁺
deposition in bones

Blood Ca²⁺
declines to set point.

Homeostasis

Blood Ca²⁺
rises to set point.

Increases Ca²⁺
uptake in intestines

Stimulates Ca²⁺
release from bone

Stimulates Ca²⁺
uptake in kidneys

PTH

**Parathyroid
glands**

Stimulus: falling
blood Ca²⁺ level

FIGURE 42.9 Negative feedback control of PTH and calcitonin secretion by blood Ca²⁺ levels.
© Cengage Learning 2017

other uses. How blood Ca²⁺ levels control PTH and calcitonin secretion is shown in **Figure 42.9.**

PTH also stimulates enzymes in the kidneys that convert **vitamin D,** a steroidlike molecule, into its fully active form in the body. The activated vitamin D increases the absorption of Ca²⁺ and phosphates from ingested food by promoting the synthesis of a calcium-binding protein in the intestine; it also increases the release of Ca²⁺ from bone in response to PTH.

PTH underproduction causes Ca²⁺ concentration to fall steadily in the blood, disturbing nerve and muscle function—the

muscles twitch and contract uncontrollably, and convulsions and cramps occur. Without treatment, the condition is usually fatal, because the severe muscular contractions interfere with breathing. Overproduction of PTH results in the loss of so much calcium from the bones that they become thin and fragile. At the same time, the elevated Ca²⁺ concentration in the blood causes calcium deposits to form in soft tissues, especially in the lungs, arteries, and kidneys (where the deposits form kidney stones).

The Adrenal Medulla Releases Two "Fight-or-Flight" Hormones

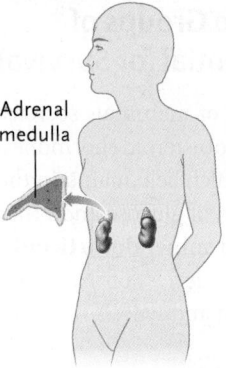

Adrenal
medulla

The adrenal glands (*ad* = to; *renes* = kidneys) of mammals consist of two distinct regions. The central region, the **adrenal medulla,** contains secretory cells of neural crest origin (see Chapter 50); the tissue surrounding it, the **adrenal cortex,** contains endocrine cells with a different developmental origin. The two regions secrete hormones with entirely different functions. Nonmammalian vertebrates have glands equivalent to the adrenal medulla and adrenal cortex of mammals, but they are separate. Most of the hormones produced by these glands have essentially the same functions in all vertebrates. The only major exception is aldosterone, which is secreted by the adrenal cortex or its equivalent only in tetrapod vertebrates.

In most species, the adrenal medulla secretes two amine hormones, **epinephrine** and **norepinephrine,** which are **catecholamines,** chemical compounds derived from the amino acid tyrosine that circulate in the bloodstream. They bind to receptors in the plasma membranes of their target cells. (Epinephrine and norepinephrine are also secreted by some cells of the CNS and neurons of the sympathetic nervous system. In these cases, epinephrine and norepinephrine function as neurotransmitters in a broad array of brain and body functions; see Section 39.3.)

Epinephrine and norepinephrine are secreted from sympathetic neurons and the adrenal medulla when the body encounters stresses such as emotional excitement, danger (fight-or-flight situations), anger, fear, infections, injury, even midterm and final exams. Epinephrine in particular prepares the body for handling stress or physical activity. The heart rate increases. Glycogen and fats break down, releasing glucose and fatty acids into the blood as fuel molecules. In the heart, skeletal muscles, and lungs, the blood vessels dilate to increase blood flow. Elsewhere in the body, the blood vessels constrict, raising blood pressure, reducing blood flow to the intestines and kidneys, and inhibiting smooth muscle contractions, thereby reducing water loss and slowing digestive processes. Airways in the lungs also dilate, helping to increase the flow of air.

The effects of norepinephrine on heart rate, blood pressure, and blood flow to the heart muscle are similar to those of epinephrine. However, in contrast to epinephrine, norepinephrine causes blood vessels in skeletal muscles to constrict. This antagonistic effect is largely canceled out because epinephrine is secreted in much greater quantities.

No known human diseases are caused by underproduction of the hormones of the adrenal medulla, as long as the sympathetic nervous system is intact. Overproduction of epinephrine and norepinephrine, which can occur if there is a tumor in the adrenal medulla, leads to symptoms duplicating a stress response.

The Adrenal Cortex Secretes Two Groups of Steroid Hormones That Are Essential for Survival

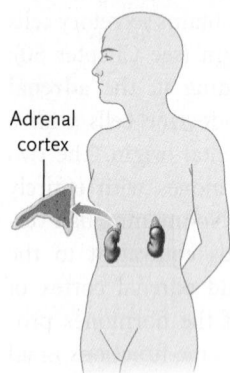

Adrenal cortex

The adrenal cortex of mammals secretes two major types of steroid hormones: **glucocorticoids,** which help maintain the blood concentration of glucose and other fuel molecules, and **mineralocorticoids,** which regulate the levels of Na$^+$ and K$^+$ ions in the blood and ECF.

THE GLUCOCORTICOIDS The glucocorticoids help maintain glucose levels in the blood by three major mechanisms: (1) stimulating the synthesis of glucose from noncarbohydrate sources such as fats and proteins; (2) reducing glucose uptake by body cells except those in the central nervous system; and (3) promoting the breakdown of fats and proteins, which releases fatty acids and amino acids into the blood as alternative fuels when glucose supplies are low. The absence of down-regulation of glucose uptake to the CNS keeps the brain well supplied with glucose between meals and during periods of extended fasting. **Cortisol** is the major glucocorticoid secreted by the adrenal cortex.

Secretion of glucocorticoids is ultimately under control of the hypothalamus **(Figure 42.10).** Low glucose concentrations in the blood or elevated levels of epinephrine secreted by the adrenal medulla in response to stress are detected in the hypothalamus, leading to secretion of ACTH by the anterior pituitary. ACTH promotes the secretion of glucocorticoids by the adrenal cortex.

Overproduction of glucocorticoids makes blood glucose rise and increases fat deposition in adipose tissue and protein breakdown in muscles and bones. The loss of proteins from muscles causes weakness and fatigue. Loss of proteins from bone, particularly collagens, makes the bones fragile and susceptible to breakage. Underproduction of glucocorticoids causes blood glucose concentration to fall below normal levels and diminishes tolerance to stress.

Glucocorticoids have anti-inflammatory properties and, consequently, they are used clinically to treat conditions such as arthritis or dermatitis. They also suppress the immune system

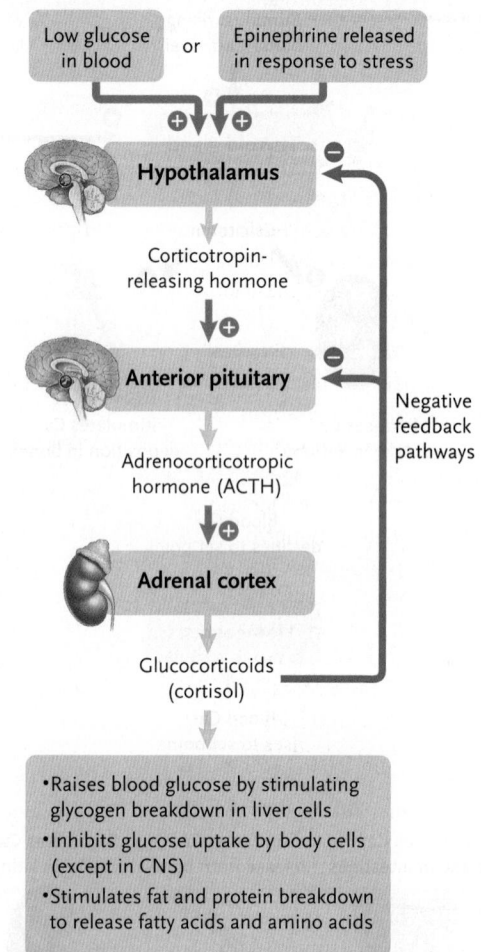

FIGURE 42.10 Pathways linking secretion of glucocorticoids to low blood sugar and epinephrine secretion in response to stress.
© Cengage Learning 2017

and are used in the treatment of autoimmune diseases such as rheumatoid arthritis.

THE MINERALOCORTICOIDS In tetrapods, the mineralocorticoids, primarily **aldosterone,** increase the amount of Na$^+$ reabsorbed from the fluids processed by the kidneys and absorbed from foods in the intestine. They also reduce the amount of Na$^+$ secreted by salivary and sweat glands and increase the rate of K$^+$ excretion by the kidneys. The net effect is to keep Na$^+$ and K$^+$ balanced at the levels required for normal cellular functions, including those of the nervous system. Secretion of aldosterone is tightly linked to blood volume and indirectly linked to blood pressure (see Section 42.2 and Chapter 48).

Moderate overproduction of aldosterone causes excessive water retention in the body, so that tissues swell and blood pressure rises. Conversely, moderate underproduction can lead to excessive water loss and dehydration. Severe underproduction is rapidly fatal unless mineralocorticoids are supplied by injection or other means.

The adrenal cortex also secretes small amounts of androgens, steroid sex hormones responsible for maintenance of male characteristics, which are synthesized primarily by the

Basic Research: Neuroendocrine and Behavioral Effects of Anabolic-Androgenic Steroids in Humans

Anabolic-androgenic steroids (AASs) are synthetic derivatives of the natural steroid hormone testosterone. They were designed to have potent anabolic (tissue building) activity and low androgenic (masculinizing) activity in therapeutic doses.

AASs are used to treat conditions such as delayed puberty and subnormal growth in children, as well as for therapy in chronic conditions such as cancer, AIDS, severe burns, liver and kidney failure, and anemias. Because of their anabolic effects, which include an increase in muscle mass, strength, and endurance, as well as acceleration of recovery from injuries, AASs are used by some athletes such as bodybuilders, weightlifters, baseball players, and football players. In reality, this use is abuse, because the doses typically administered for these purposes are far higher than therapeutic doses. AAS abuse is significant: a recent survey estimated that 2.9–4.0 million Americans ages 13 to 50 have used AASs.

Are AASs harmful at high doses? When researchers gave rodents doses of AASs comparable to those associated with human AAS abuse, they observed significant increases in aggression, anxiety, and sexual behaviors. These changes occur as a result of alterations in the neurotransmitters and other signaling molecules associated with those behaviors. All of these changes have been hypothesized to occur in human AAS abusers.

To study the effect of high doses of AAS on the human endocrine system, R. C. Daly and colleagues at the National Institute of Mental Health in Bethesda, Maryland, administered the AAS methyltestosterone (MT) to normal (medication-free) human volunteers over a period of time in an in-patient clinic. The subjects were examined for the effects of MT on pituitary–gonadal, pituitary–thyroid, and pituitary–adrenal hormones, and the researchers attempted to correlate endocrine changes with psychological symptoms caused by the MT.

The researchers found, for instance, that high doses of MT caused a significant decrease in the levels of gonadotropins and gonadal steroid hormones in the blood. At the same time, thyroxine and TSH levels increased. No significant increases were seen in pituitary–adrenal hormones.

The decrease in testosterone levels correlated significantly with cognitive problems, such as increased distractibility and forgetfulness. The increase in thyroxine correlated significantly with a rise in aggressive behavior, notably anger, irritability, and violent feelings. There were no changes in activities associated with pituitary–adrenal hormones—energy, disturbed sleep, and sexual arousal—as was expected by the lack of change in those hormones.

In sum, behavioral changes associated with high doses of an AAS suggest that AAS-induced hormonal changes may well contribute to the adverse behavioral and mood changes that occur during AAS abuse.

---

gonads. These hormones have significant effects only if they are overproduced, as can occur with some tumors in the adrenal cortex. The result is altered development of the primary or secondary sex characteristics.

The Gonadal Sex Hormones Regulate the Development of Reproductive Systems, Sexual Characteristics, and Mating Behavior

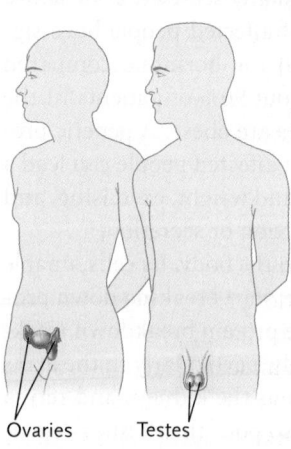

Ovaries   Testes

The **gonads,** the testes and ovaries, are the primary source of sex hormones in vertebrates. The steroid hormones they produce, the **androgens, estrogens,** and **progestins,** have similar functions in regulating the development of male and female reproductive systems, sexual characteristics, and mating behavior. Both males and females produce all three types of hormones, but in different proportions. Androgen production is predominant in males, whereas estrogen and progestin production is predominant in females. An outline of the actions of these hormones is presented here, and a more complete picture is given in Chapter 49.

The **testes** of male vertebrates secrete androgens, steroid hormones that stimulate and control the development and maintenance of male reproductive systems and masculine characteristics. The principal androgen is **testosterone,** the male sex hormone. In young adult males, a jump in testosterone levels stimulates puberty and the development of secondary sexual characteristics, including the growth of facial and body hair, muscle development, changes in vocal cord morphology, and development of normal sex drive. The synthesis and secretion of testosterone by cells in the testes is controlled by the release of luteinizing hormone (LH) from the anterior pituitary, which in turn is controlled by **gonadotropin-releasing hormone (GnRH)** secreted by the hypothalamus.

A large number of synthetic derivatives of androgens, known as **anabolic-androgenic steroids (AASs)** (also referred to as **anabolic steroids**), mimic the effects of androgens. They have been in the news over the years because of their use by bodybuilders and other athletes in sports in which muscular strength is important. *Focus on Research: Basic Research* discusses the potential adverse effects of anabolic-androgenic steroids.

The **ovaries,** under the stimulatory influence of follicle-stimulating hormone (FSH), produce estrogens, steroid hormones that stimulate and control the development and maintenance of female reproductive systems. The principal estrogen is **estradiol,** which stimulates maturation of sex organs at puberty and the development of secondary sexual characteristics. Ovaries also produce progestins, principally **progesterone,** the steroid hormone that prepares and maintains the uterus for implantation of a fertilized egg and the subsequent growth and development of an embryo. The synthesis and secretion of progesterone by cells in the ovaries is controlled by the release of LH from the anterior pituitary, which in turn is controlled by the same GnRH as in males.

The Pancreatic Islet of Langerhans Hormones Regulate Glucose Metabolism

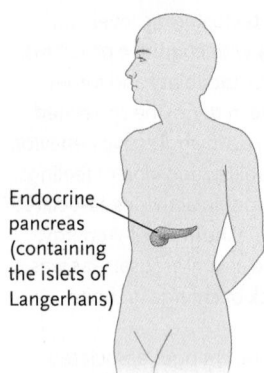

Endocrine pancreas (containing the islets of Langerhans)

Most of the pancreas, a relatively large gland located just behind the stomach, forms an exocrine gland that secretes digestive enzymes into the small intestine (see Chapter 47). However, about 2% of the cells in the pancreas are endocrine cells that form the **islets of Langerhans**. Found in all vertebrates, the islets secrete the peptide hormones insulin and glucagon into the bloodstream.

Insulin and glucagon regulate the metabolism of fuel substances in the body. **Insulin,** secreted by *β (beta) cells* in the islets, acts mainly on cells of skeletal muscles, liver cells, and adipose tissue (fat). (Brain cells do not require insulin for glucose uptake.) Insulin lowers circulating blood glucose, fatty acid, and amino acid levels and promotes their storage. That is, the actions of insulin include stimulation of glucose transport into cells, glycogen synthesis from glucose, uptake of fatty acids by adipose tissue cells, fat synthesis from fatty acids, and protein synthesis from amino acids. Insulin also inhibits glycogen degradation to glucose, fat degradation to fatty acids, and protein degradation to amino acids.

Glucagon, secreted by *α (alpha) cells* in the islets, has effects opposite to those of insulin: it stimulates glycogen, fat, and protein degradation. Glucagon also promotes the use of amino acids and other noncarbohydrates as the input for glucose synthesis; this aspect of glucagon function operates during fasting. Negative feedback mechanisms that are keyed to the concentration of glucose in the blood control secretion of both insulin and glucagon to maintain glucose homeostasis **(Figure 42.11).**

Diabetes mellitus, a disease that afflicts more than 14 million people in the United States and at least 200 million people worldwide, results from problems with insulin production or action. The three classic diabetes symptoms are frequent urination, increased thirst (and consequently increased fluid intake), and increased glucose in the blood (hyperglycemia). Frequent urination occurs because body cells are not stimulated to take up glucose, leading to abnormally high glucose concentration in the blood. Excretion of the excess glucose in the urine requires water to carry it, which causes increased fluid loss from the blood. Food intake is necessary to offset the negative

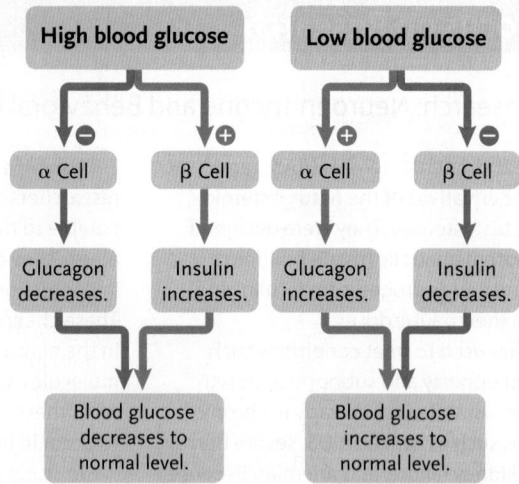

FIGURE 42.11 The action of insulin and glucagon in maintaining the concentration of blood glucose at an optimal level.
© Cengage Learning 2017

energy balance resulting from the decreased glucose uptake by cells or else weight loss will occur. Two of these classic symptoms gave the disease its name: *diabetes* is derived from a Greek word meaning "siphon," referring to the frequent urination; and *mellitus,* a Latin word meaning "sweetened with honey," refers to the sweet taste of a diabetic's urine. (Before modern blood or urine tests were developed, physicians tasted a patient's urine to detect the disease.)

The disease occurs in two major forms called *type 1* and *type 2.* Type 1 diabetes, which occurs in about 10% of diabetics, results from insufficient insulin secretion by the pancreas. This type of diabetes is usually caused by an autoimmune reaction that destroys pancreatic beta cells. To survive, type 1 diabetics must receive regular insulin injections (typically, a genetically engineered human insulin called Humulin); careful dieting and exercise also have beneficial effects because active skeletal muscles do not require insulin to take up and utilize glucose.

In type 2 diabetes, insulin is usually secreted at or above normal levels, but the target cells of affected people have significantly reduced responsiveness to the hormone compared with the cells of normal people. About 90% of patients in the developed world with type 2 diabetes are obese. A genetic predisposition can also be a factor. Most affected people can lead a normal life by controlling their diet and weight, exercising, and taking drugs that enhance insulin action or secretion.

Diabetes has long-term effects on the body. Its cells, unable to utilize glucose as an energy source, start breaking down proteins and fats to generate energy. The protein breakdown weakens blood vessels throughout the body, particularly in the arms and legs and in critical regions such as the kidneys and retina of the eye. The circulation becomes so poor that tissues degenerate in the arms, legs, and feet. At advanced stages of the disease, bleeding in the retina causes blindness. The breakdown of circulation in the kidneys can lead to kidney failure. In addition, in type 1 diabetes, acidic products of fat breakdown (ketones) are produced in abnormally high quantities and accumulate in the blood. The resulting lowering of blood pH can disrupt heart and brain function, leading to coma and death if the disease is untreated.

The Pineal Gland Regulates Some Biological Rhythms

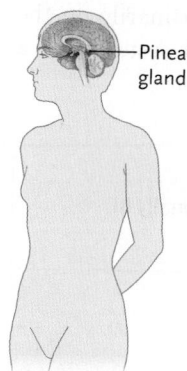
—Pineal gland

The **pineal gland** is found at different locations in the brains of vertebrates. In mammals, for example, the pineal gland is at roughly the center of the brain, whereas in birds and reptiles, it is on the surface of the brain just under the skull. The pineal gland regulates some biological rhythms.

The earliest vertebrates had a third, light-sensitive eye at the top of the head, and some species, such as lizards and tuataras (New Zealand reptiles), still have an eyelike structure in this location. In most vertebrates, the third eye became modified into a pineal gland, which in many groups retains some degree of photosensitivity. In mammals, it is too deeply buried in the brain to be affected directly by light. Nonetheless, specialized photoreceptors in the eyes make connections to the pineal gland.

In mammals, the pineal gland secretes a peptide hormone, **melatonin,** that helps to maintain daily biorhythms. Secretion of melatonin is regulated by an inhibitory pathway. Light hitting the eyes generates signals that inhibit melatonin secretion. Consequently, the hormone is secreted most actively during periods of darkness. Melatonin targets a part of the hypothalamus called the *suprachiasmatic nucleus,* which is the primary biological clock coordinating body activity to a daily cycle. The nightly release of melatonin may help synchronize the biological clock with daily cycles of light and darkness. The physical and mental discomfort associated with jet lag may reflect the time required for melatonin secretion to reset a traveler's daily biological clock to match the period of daylight in a new time zone.

Melatonin also plays a role in other vertebrates. In some fishes, amphibians, and reptiles, melatonin and other hormones produce changes in skin color through their effects on *melanophores,* the pigment-containing cells of the skin. Skin color may vary with the season, the animal's breeding status, or the color of the background.

STUDY BREAK 42.4

1. What effect does parathyroid hormone have on the body?
2. What hormones are secreted by the adrenal medulla, and what are their functions?
3. What are the two types of hormones secreted by the adrenal cortex, and what are their functions?
4. To what molecular class of hormones do estradiol and progesterone belong, and where do they act in the body?

42.5 Endocrine Systems in Invertebrates

In even the simplest animals, such as the cnidarian *Hydra,* hormones produced by neurosecretory neurons control reproduction, growth, and development of some body features. In annelids, arthropods, and mollusks, endocrine cells and glands produce hormones that regulate reproduction, water balance, heart rate, and sugar levels.

Some hormones occur in related forms in invertebrates and vertebrates. For example, fruit flies, mollusks, and humans have insulin-like hormones and receptors, even though molecular studies suggest that their last common ancestor existed more than 800 million years ago. Both invertebrates and vertebrates secrete peptide and steroid hormones, but most of the hormones' structures differ between the two groups, and therefore most have no effect when injected into members of the other group. However, the reaction pathways stimulated by the hormones are the same in both groups, suggesting that these regulatory mechanisms appeared very early in animal evolution.

Hormones have been studied in detail in only a few invertebrate groups, with the most extensive studies focusing on regulation of metamorphosis in insects. Butterflies, moths, and flies undergo the most dramatic changes as they mature into adults. They hatch from the egg as a caterpillar-like *larva.* During the larval stage, growth is accompanied by one or more *molts,* in which an old exoskeleton is shed and a new one forms. The insect then enters a typically nonmotile stage, the *pupa,* in which the body forms a thick, resistant coating, and finally it transforms into an adult.

Three major hormones regulate molting and metamorphosis in insects: (1) **prothoracicotropic hormone (PTTH),** a peptide hormone secreted by neurosecretory neurons in the brain; (2) **ecdysone** (*ekdysis* = getting out), a steroid hormone secreted by the *prothoracic glands;* and (3) **juvenile hormone (JH),** a terpenoid (a type of unsaturated hydrocarbon) secreted by the *corpora allata,* a pair of glands just behind the brain **(Figure 42.12).** The outcome of the molt depends on the level of JH. If it is high, the molt produces a larger larva; if it is low, the molt leads to pupation and the emergence of the adult.

Hormones that control molting have also been detected in crustaceans, including lobsters, crabs, and crayfish. Before

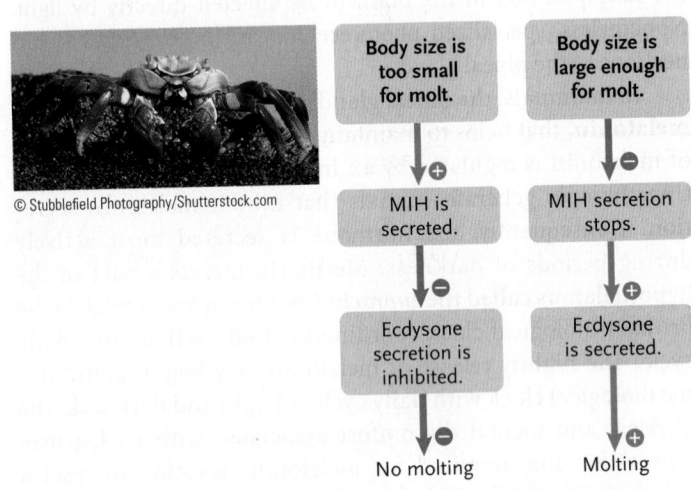

growth reaches the stage at which the exoskeleton is shed, **molt-inhibiting hormone (MIH)**, a peptide neurohormone secreted by a gland in the eye stalks, inhibits ecdysone secretion **(Figure 42.13)**. As body size increases to the point requiring a molt, MIH secretion is inhibited, ecdysone secretion increases, and the molt is initiated.

In the next chapter, we discuss the structure and functions of muscles, and their interactions with the skeletal system to cause movement. Muscle function depends primarily on the action of the nervous system, but the endocrine system plays a role in the control of smooth muscle contraction.

STUDY BREAK 42.5

How do hormones compare structurally and functionally in invertebrates and vertebrates?

FIGURE 42.12 The roles of prothoracicotropic hormone, ecdysone, and juvenile hormone in the development of a silkworm moth.
© Cengage Learning 2017

1 Prothoracicotropic hormone (PTTH) stimulates the prothoracic glands to release ecdysone.

2 Ecdysone promotes growth of a new exoskeleton under the old one and stimulates the release of other hormones that lead to the molt.

3 If the concentration of juvenile hormone (JH) is high, the molt produces a larger larva.

4 If the JH concentration is low, the molt leads to pupation.

5 After the adult emerges from the pupa, JH levels rise again and help trigger full mature sexual behavior.

FIGURE 42.13 Control of molting by molt-inhibiting hormone (MIH), which is secreted by a gland in the eye stalks of crustaceans such as this crab.
© Cengage Learning 2017

© Stubblefield Photography/Shutterstock.com

 ## Unanswered Questions

What is the role of estrogens in breast cancer?

Breast cancer is the most common cancer in Western women, with over 230,000 new cases estimated to be diagnosed in the United States in 2015. The etiology of breast cancer remains largely unknown. A number of risk factors have been identified, including family history and reproductive factors; however, ~80% of newly diagnosed women have no known risk factors. For example, women with an inherited mutation in the *BRCA1* gene have a 70–80% likelihood of developing breast cancer during their lifetime, yet only one third of women with a strong family history of breast cancer carry a *BRCA1* mutation or have some other known mutation that would predispose them to cancer. The failure in identifying strong risk factors has led to an absence of specific targets for breast cancer prevention.

Currently, the only gene that has been successfully targeted for breast cancer prevention is the estrogen receptor (ER) gene. Estrogen induces cell proliferation in the breast and exerts its effects through binding to the ER, leading to activation or inhibition of several downstream signaling targets. ER activity can be inhibited by tamoxifen and other antiestrogens, or by bilateral ovariectomy (removal of both ovaries), which lowers circulating estrogen levels. The link between high lifetime exposure to estrogens and increased breast cancer risk has been established clearly through observational studies in women.

Paradoxically, estrogens also can prevent breast cancer or inhibit the growth of antiestrogen-resistant tumors, likely through their ability to induce differentiation and apoptosis. In fact, the synthetic estrogen diethylstilbestrol (DES) was prescribed for the treatment of breast cancer until it was replaced with antiestrogens in the early 1980s. Further, an exposure to estradiol, as single agent hormone replacement therapy, prevents breast cancer and its metastasis. The seemingly dual effects of estrogen may reflect the critical role that timing of exposure appears to play in affecting breast cancer risk.

How does the timing of estrogenic exposures affect breast cancer risk?

Millions of pregnant women in the 1940s to 1970s took DES in the United States and Europe, initially to prevent miscarriage but also to aid various pregnancy-related conditions, such as morning sickness. It was subsequently proven ineffective and even dangerous. Most notable, young daughters exposed to DES *in utero* exhibited an unusually high incidence of clear cell vaginal adenocarcinoma (a malignant tumor), and the practice of treating pregnant women with DES stopped. The DES daughters have now reached the age when breast cancer is commonly detected (older than 40 years old), and the findings indicate that these daughters exhibit two times higher risk of developing breast cancer as well. Risk also is increased in mothers themselves.

Research from my laboratory at Georgetown University, using preclinical models of breast cancer, shows that maternal exposure to estrogenic dietary compounds during pregnancy, such as high-fat diet, increases mammary cancer risk among female offspring. This increase is not limited to daughters but also is seen in granddaughters and great granddaughters, and mothers, without any further intervening exposures. Our studies suggest that these multigenerational effects are inherited via permanent changes in the epigenome (the set of epigenetic modifications in the genome), rather than through genetic changes, such as mutations. Genes that are epigenetically silenced in the mammary tissue from generation to generation following maternal dietary exposures during pregnancy include those that regulate proliferation of progenitor cells (they are daughters of stem cells) and their differentiation. The increased methylation of these genes, in turn, is proposed to be a hallmark of increased breast cancer susceptibility. Importantly, epigenetic changes are in principle reversible by compounds that demethylate silenced genes, and many such compounds are currently available and used to treat some cancers. Our preclinical studies suggest that these same compounds may also prevent *in utero* estrogen-exposed animals from developing mammary cancer.

The period when estrogens appear to be protective is around puberty, when the breasts begin to grow. Pregnancy is characterized by a 50- to 100-fold increase in circulating estrogen levels, and women who had their first child before age 20 have half the risk of developing breast cancer than women who were over 30 when their first child was born. Other estrogenic exposures at young adulthood also reduce lifetime breast cancer risk. Women who consumed high levels of soy foods during childhood, which contain the estrogenic isoflavone genistein, have a reduced risk of developing breast cancer. Similar findings have been obtained in preclinical animal models. Work done in my laboratory is investigating whether the protective effect of childhood genistein/soy intake is mediated by epigenetic changes that involve genes that may protect mammary epithelial cells from malignant transformation.

think like a scientist What are the benefits and weaknesses of using *in vitro* (tissue-culture based) and *in vivo* (animal model) approaches versus performing human studies to understand the role of estrogen in breast cancer etiology?

Leena Hilakivi-Clarke is a Professor of Oncology at Georgetown University in Washington, D.C. Her research involves understanding why timing of estrogenic exposures, including those originating from diet, determines whether they increase, reduce, or have no effect on breast cancer risk and risk of recurrence of this disease. Learn more about her research at http://clarkelabs.georgetown.edu/index.

Courtesy of Leena Hilakivi-Clarke

REVIEW KEY CONCEPTS

For access to MindTap and additional study materials visit www.cengagebrain.com.

42.1 Hormones and Their Secretion

- Hormones are substances secreted by cells that control the activities of cells elsewhere in the body. The cells that respond to a hormone are its target cells. The best-known hormones are secreted by the endocrine system.

- The endocrine system includes four major types of cell signaling: classical endocrine signaling, in which endocrine glands secrete hormones; neuroendocrine signaling, in which neurosecretory neurons release neurohormones into the circulatory system; paracrine regulation, in which cells release local regulators that diffuse through the ECF to regulate nearby cells; and autocrine regulation, in which cells release local regulators that regulate the same cells that produced them (Figure 42.1).

- Most hormones and local regulators fall into one of four molecular classes: amines, peptides, steroids, and fatty acids.

- Many hormones are controlled by negative feedback mechanisms (Figure 42.2).

42.2 Mechanisms of Hormone Action

- Hormones typically are effective in very low concentrations in the body fluids because of amplification.

- Hydrophilic hormones bind to receptor proteins embedded in the plasma membrane, activating them. The activated receptors transmit a signal through the plasma membrane, triggering signal transduction pathways that cause a cellular response. Hydrophobic hormones bind to receptors in the cytoplasm or nucleus, activating them. The activated receptors control the expression of specific genes, the products of which cause the cellular response (Figure 42.3).

- As a result of the types of receptors they have, target cells may respond to more than one hormone, or they may respond differently to the same hormone.

- The major endocrine cells and glands of vertebrates are the hypothalamus, pituitary gland, thyroid gland, parathyroid glands, adrenal medulla, adrenal cortex, testes, ovaries, islets of Langerhans of the endocrine pancreas, and pineal gland. Hormones are also secreted by endocrine cells in the stomach and intestine, thymus gland, kidneys, and heart. Most body cells are capable of releasing prostaglandins (Figure 42.6).

42.3 The Hypothalamus and Pituitary

- The hypothalamus and pituitary together regulate many other endocrine cells and glands in the body (Figure 42.7).

- The hypothalamus produces hormones (releasing hormones and inhibiting hormones) that control the secretion of six major hormones by the anterior pituitary: thyroid-stimulating hormone (TSH), adrenocorticotropic hormone (ACTH), follicle-stimulating hormone (FSH), luteinizing hormone (LH), prolactin (PRL), and growth hormone (GH). Pituitary tissue also produces melanocyte-stimulating hormone (MSH) and endorphins with more complex regulation.

- The posterior pituitary secretes antidiuretic hormone (ADH), which regulates body water balance, and oxytocin, which stimulates the contraction of smooth muscle in the uterus as a part of childbirth and triggers milk release from the mammary glands during suckling of the young.

42.4 Other Major Endocrine Glands of Vertebrates

- The thyroid gland secretes the thyroid hormones and, in mammals, calcitonin. The thyroid hormones stimulate the oxidation of carbohydrates and lipids, and coordinate with growth hormone to stimulate body growth and development. Calcitonin lowers the Ca^{2+} level in the blood by inhibiting the release of Ca^{2+} from bone. In amphibians, such as frogs, thyroid hormones trigger metamorphosis (Figure 42.8).

- The parathyroid glands secrete parathyroid hormone, which stimulates bone cells to release Ca^{2+} into the blood. PTH also stimulates the activation of vitamin D, which promotes Ca^{2+} absorption into the blood from the small intestine (Figure 42.9).

- The adrenal medulla secretes epinephrine and norepinephrine, which reinforce the sympathetic nervous system in responding to stress. The adrenal cortex secretes glucocorticoids, which help maintain glucose at normal levels in the blood, and mineralocorticoids, which regulate Na^+ balance and ECF volume. The adrenal cortex also secretes small amounts of androgens (Figure 42.10).

- The gonadal sex hormones—androgens, estrogen, and progestins—play a major role in regulating the development of reproductive systems, sexual characteristics, and mating behavior.

- The islet of Langerhans cells of the endocrine pancreas secrete insulin and glucagon, which together regulate the concentration of fuel substances in the blood. Insulin lowers the concentration of glucose in the blood and inhibits the conversion of non-carbohydrate molecules into glucose. Glucagon raises blood glucose by stimulating glycogen, fat, and protein degradation (Figure 42.11).

- The pineal gland secretes melatonin, which interacts with the hypothalamus to set the body's daily rhythms.

42.5 Endocrine Systems in Invertebrates

- Hormones control development and function of the gonads, manage salt and water balance in the body fluids, and control molting in insects and crustaceans.

- Three major hormones—prothoracicotropic hormone (PTTH), ecdysone, and juvenile hormone (JH)—control molting and metamorphosis in insects. Hormones that control molting are also present in crustaceans (Figures 42.12 and 42.13).

TEST YOUR KNOWLEDGE

Remember/Understand

1. Amine hormones are usually:
   a. hydrophilic when secreted by the thyroid gland.
   b. based on tyrosine.
   c. paracrine but not autocrine.
   d. not transported by the blood.
   e. repelled by the plasma membrane.

2. Prostaglandins would be best described as inducers of:
   a. male and female characteristics.
   b. cell division.
   c. nerve transmission.
   d. smooth muscle contractions.
   e. cell differentiation.

3. When the concentration of thyroid hormone in the blood increases, it:
   a. inhibits TSH secretion by the pituitary.
   b. stimulates TRH secretion by the hypothalamus.
   c. stimulates the pituitary to secrete TRH.
   d. stimulates the pituitary to secrete TSH.
   e. activates a positive feedback loop.

4. Which of the following statements about endocrine targeting and reception is *correct*?
   a. The idea that one hormone affects one type of tissue is illustrated when epinephrine binds to smooth muscle cells in blood vessels, as well as to beta cells in heart muscle.
   b. The idea that one hormone affects one type of tissue is illustrated when epinephrine cannot activate both the receptors on liver cells and the beta receptors of heart muscle.

   c. The idea that a target cell can respond to more than one hormone is illustrated by a vertebrate liver cell responding to insulin and glucagon.
   d. The idea that a minute concentration of hormone can cause widespread effects demonstrates the specificity of cells for certain hormones.
   e. The idea that the response to a hormone is the same among different target cells is shown when different liver cells are activated by insulin.

5. The posterior pituitary secretes:
   a. hormones that control the hypothalamus.
   b. IGF, which simulates cell division and protein synthesis.
   c. ADH, which increases water absorption in the kidneys.
   d. oxytocin, which controls egg and sperm development.
   e. prolactin, which stimulates milk synthesis.

6. Blood levels of calcium are regulated directly by:
   a. insulin synthesized by the alpha cells of the pancreas.
   b. PTH made by the pituitary.
   c. vitamin D activated in the liver.
   d. prolactin synthesized by the anterior pituitary.
   e. calcitonin secreted by specialized thyroid cells.

7. If the human body is stressed, glucocorticoids:
   a. promote the breakdown of proteins in the muscles and bones.
   b. increase the amount of sodium reabsorbed from urine in the kidneys.
   c. decrease potassium secretion from the kidneys.
   d. decrease glucose uptake by cells in the nervous system.
   e. inhibit the synthesis of glucose from noncarbohydrate sources.

8. When blood glucose rises:
   a. alpha cells increase glucagon secretion.
   b. beta cells increase insulin secretion.
   c. urination decreases in a person with type 1 diabetes who has not recently received an insulin injection.
   d. glucagon stimulates the breakdown of amino acids into glycogen.
   e. target cells decrease their insulin receptors.

9. In mammals:
   a. the suprachiasmatic nucleus of the pineal gland controls both male and female reproductive systems.
   b. estradiol is produced by the hypothalamus to control ovulation.
   c. melatonin controls anabolic steroid production.
   d. GnRH stimulates LH to control testosterone production.
   e. progesterone increases the secretion of LH from the posterior pituitary.

10. Insect development is regulated by:
   a. ecdysone, a peptide secreted by the brain.
   b. juvenile hormone, a terpenoid secreted by the corpora allata near the brain.
   c. molt-inhibiting hormone, a steroid secreted by the prothoracic glands.
   d. prothoracicotropic hormone, a steroid secreted by the hypothalamus.
   e. melatonin, a peptide secreted by the brain in the larval stage.

Apply/Analyze

11. **Discuss Concepts** A physician sees a patient whose symptoms include sluggishness, depression, and intolerance to cold. What disorder do these symptoms suggest?

12. **Discuss Concepts** A 20-year-old woman with a malignant brain tumor has her pineal gland removed. What kinds of side effects might this loss have?

13. **Discuss Concepts** In integrated pest management, a farmer uses a variety of tools to combat unwanted insects. These include applications of either hormones or hormone-inhibiting compounds to prevent insects from reproducing successfully. How might each of these hormone-based approaches disrupt reproduction?

Evaluate/Create

14. **Design an Experiment** The Environmental Protection Agency (EPA) defines *endocrine disruptors* as chemical substances that can "interfere with the synthesis, secretion, transport, binding, action, or elimination of natural hormones in the body that are responsible for the maintenance of homeostasis (normal cell metabolism), reproduction, development, and/or behavior." The chemicals, sometimes called *environmental estrogens,* come from both natural and human-made sources. A simple hypothesis is that endocrine disruptors act by mimicking hormones in the body. Many endocrine disruptors affect sex hormone function and, therefore, reproduction.

Examples of endocrine disruptors are the synthetic chemicals DDT (a pesticide), dioxins, and natural chemicals such as phytoestrogens (estrogen-like molecules in plants), which are found in high levels in soybeans, carrots, oats, onions, beer, and coffee.

Design an experiment to investigate whether a new synthetic chemical (pick your own interesting scenario) is an endocrine disruptor. (*Hint:* You probably want to work with a model organism.)

15. **Apply Evolutionary Thinking** Which endocrine system evolved earlier, endocrine glands or neurosecretory neurons? Support your conclusion with information obtained from online research.

For selected answers, see Appendix A.

INTERPRET THE DATA

Researchers measured prolactin and growth hormone in the blood plasma of six subjects at regular intervals over a 24-hour period. The investigators were interested in determining if the levels of these two hormones cycle over a 24-hour period. The **Figure** presents the results of their experiments, with the *y* axis values showing the amount of each hormone expressed as a percentage of the 24-hour mean value, and each point showing the average of the six subjects' levels.

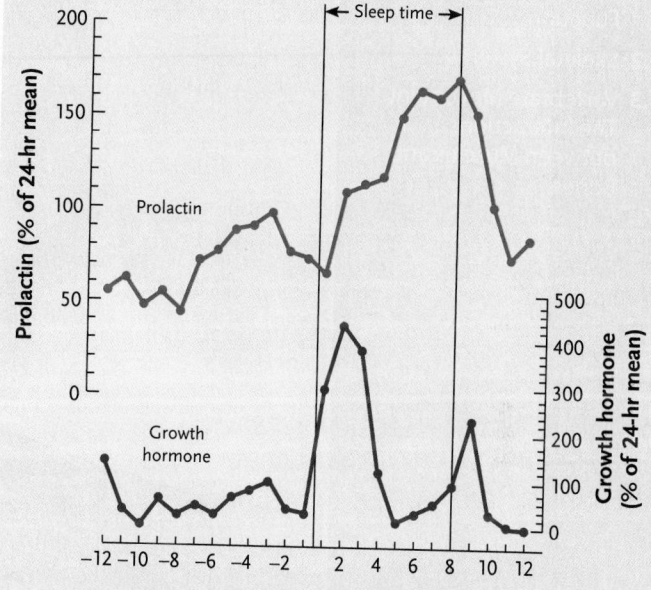

FIGURE Mean concentrations of prolactin and growth hormone in plasma, averaged for six subjects and expressed as a percentage of the average concentrations measured for the 24-hour period.

1. How did the level of growth hormone change over the 24-hour period and, in particular, how did it relate to the sleep period?

2. How did the level of prolactin change over the 24-hour period and, in particular, how did it relate to the sleep period?

3. Is there any consistent relationship between the patterns of prolactin and growth hormone in the plasma in the 24-hour period?

4. Do the results support the hypothesis that the changes in prolactin level over a 24-hour period are associated with the sleep–wake cycle?

Source: From J. F. Sassin et al. 1972. Human prolactin: 24-hour pattern with increased release during sleep. *Science* 177:1205–1207.

43 Muscles, Bones, and Body Movements

Movement of vertebrates such as these dolphins occurs as a result of contractions and relaxations of skeletal muscles. When stimulated by the nervous system, actin filaments in the muscles slide over myosin filaments to cause muscle contractions.

Why it matters . . . A Mexican leaf frog *(Pachymedusa dacnicolor)* detects an approaching cricket. In a total of 260 msec—about one quarter of a second—the frog lunges forward at just the right moment, thrusts out its sticky tongue, and captures the prey. How does the frog move so swiftly, and so surely? As its prey draws near, neuronal signals travel from the frog's brain to the muscles that extend the frog's hind legs, causing those muscles to contract and propel the frog forward on its forelimbs toward the cricket. Within 50 msec after the jump begins, other signals contract the muscles of the lower jaw, opening the mouth. Then, a muscle on the upper surface of the tongue contracts, which raises the tongue and flips it out of the mouth **(Figure 43.1).** Within 80 msec after the lunge begins, the tip of the frog's tongue contacts the cricket. The tongue, with the cricket, folds back into the frog's mouth, aided by contraction of a muscle on the bottom of the tongue.

In Section 38.2 you learned about the three types of muscle tissue: skeletal, cardiac, and smooth. *Skeletal muscle* is so named because most muscles of this type are attached by tendons to

FIGURE 43.1 A Mexican leaf frog *(Pachymedusa dacnicolor)* capturing a grasshopper. The frog's movements were captured using a high-speed video camera linked to a millisecond timer, with a grid in the background that allowed precise measurement of the distances body parts traveled during the capture.

the skeleton of vertebrates. *Cardiac muscle* is the contractile muscle of the heart, and *smooth muscle* is found in the walls of tubes and cavities in the body, including blood vessels and intestines. In this chapter we describe the structure and function of skeletal muscles, the skeletal systems of invertebrates and vertebrates, and how muscles bring about movement.

43.1 Vertebrate Skeletal Muscle: Structure and Function

Vertebrate **skeletal muscles** connect to bones of the skeleton. The cells forming skeletal muscles are typically long and cylindrical and contain many nuclei (shown in Figure 38.5A). Skeletal muscle is controlled by the somatic nervous system.

Most skeletal muscles in humans and other vertebrates are attached at both ends to bones of the skeleton and across a joint. Some, such as those that move the lips, are attached to other muscles or connective tissues under skin. Depending on its points of attachment, contraction of a single skeletal muscle may extend or flex body parts, or may rotate one body part with respect to another. The human body has more than 600 skeletal muscles, ranging in size from the small muscles that move the eyeballs to the large muscles that move the legs.

Skeletal muscles are attached to bones by cords of connective tissue called *tendons* (see Section 38.2). Tendons vary in length from a few millimeters to some, such as those that connect the muscles of the forearm to the bones of the fingers, that are 20 to 30 cm long.

The Striated Appearance of Skeletal Muscle Fibers Results from a Highly Organized Internal Structure

A skeletal muscle consists of bundles of elongated, cylindrical cells called **muscle fibers** that are 10–100 μm in diameter and run the entire length of the muscle **(Figure 43.2)**. Muscle fibers are multinucleate—they contain many nuclei—reflecting their development by fusion of smaller cells. Some very small muscles, such as some of the muscles of the face, contain only a few hundred muscle fibers; others, such as the larger leg muscles, contain hundreds of thousands. In both cases, the muscle fibers are held in parallel bundles by sheaths of connective tissue that surround them in the muscle and merge with tendons. (Connective tissue is described in Section 38.2.) Muscle fibers are supplied with nutrients and oxygen by an extensive network of blood vessels that penetrates the muscle tissue.

Muscle fibers are packed with **myofibrils,** cylindrical contractile elements about 1 μm in diameter that run lengthwise inside the cells. Each myofibril consists of a regular arrangement of **thick filaments** (13–18 nm in diameter) and **thin filaments** (5–8 nm in diameter) (see Figure 43.2). The thick and thin filaments alternate with one another in a packed set.

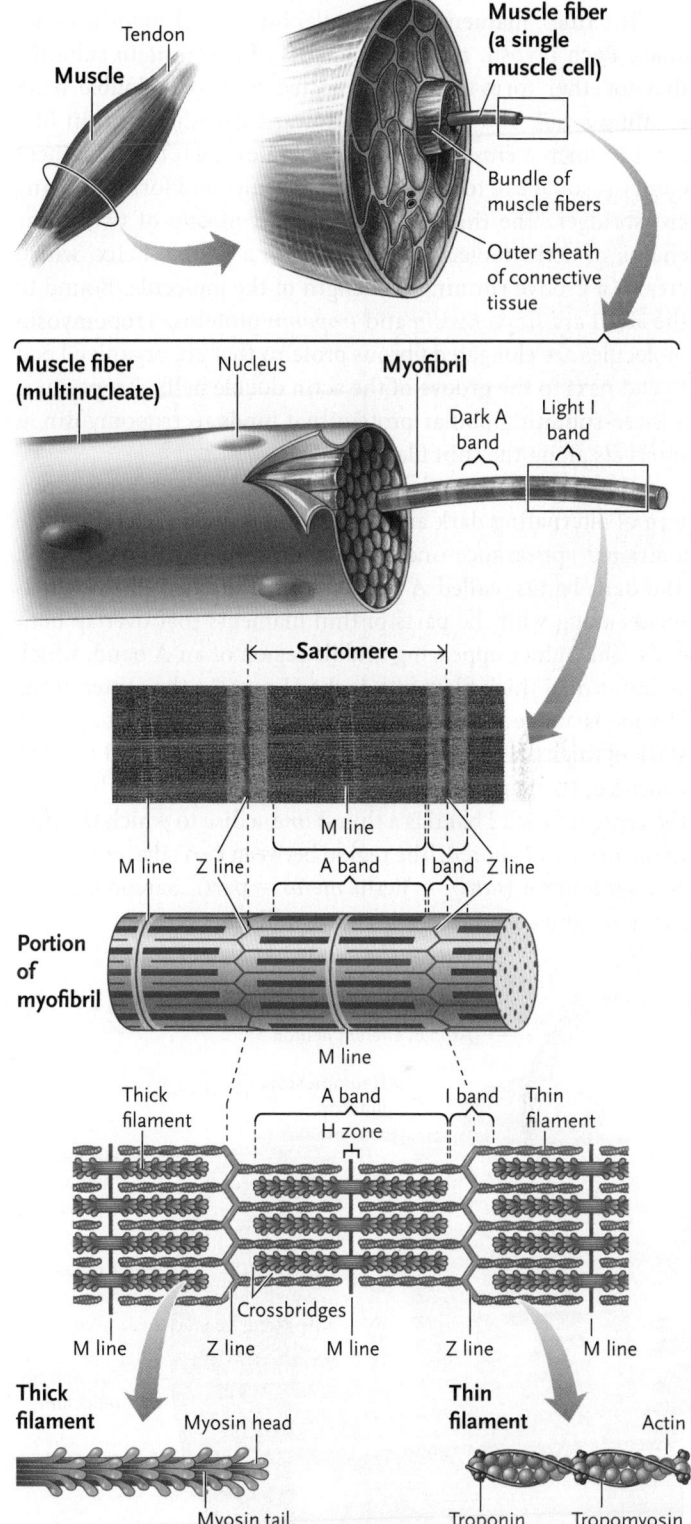

FIGURE 43.2 Skeletal muscle structure. Muscles are composed of bundles of multinucleate cells called muscle fibers; within each muscle fiber are longitudinal bundles of myofibrils. The unit of contraction within a myofibril, the sarcomere, consists of overlapping myosin thick filaments and actin thin filaments. The myosin molecules in the thick filaments each consist of two subunits organized into a head and a double-helical tail. The actin subunits in the thin filaments form twisted, double helices, with tropomyosin molecules arranged head-to-tail in the groove of the helix and troponin bound to the tropomyosin at intervals along the thin filaments.
© Cengage Learning 2017

The thick filaments are parallel bundles of myosin molecules. Each myosin molecule consists of two protein subunits that together form a *head* connected to a long double helix forming a *tail*. The head is bent toward the adjacent thin filament to form a *crossbridge*. In vertebrates, each thick filament contains some 200 to 300 myosin molecules and forms as many crossbridges. The thin filaments consist mostly of two linear chains of actin molecules twisted into a double helix, which creates a groove running the length of the molecule. Bound to the actin are *tropomyosin* and *troponin* proteins. Tropomyosin molecules are elongated fibrous proteins that are organized end to end next to the groove of the actin double helix. Troponin is a three-subunit globular protein that binds to tropomyosin at intervals along the thin filaments.

The arrangement of thick and thin filaments forms a pattern of alternating dark and light bands, giving skeletal muscle a striated appearance under the microscope (see Figure 43.2). The dark bands, called *A bands,* consist of stacked thick filaments along with the parts of thin filaments that overlap both ends. The lighter-appearing middle region of an A band, which contains only thick filaments, is the *H zone*. In the center of the H zone is a disc of proteins called the *M line,* which holds the stack of thick filaments together. The light bands, called *I bands,* consist of the parts of the thin filaments not in the A band. In the center of each I band is a thin *Z line,* a disc to which the thin filaments are anchored. The region between two adjacent Z lines is a **sarcomere** (*sarco-* = flesh; *meros* = part). Sarcomeres are the basic units of contraction in a myofibril.

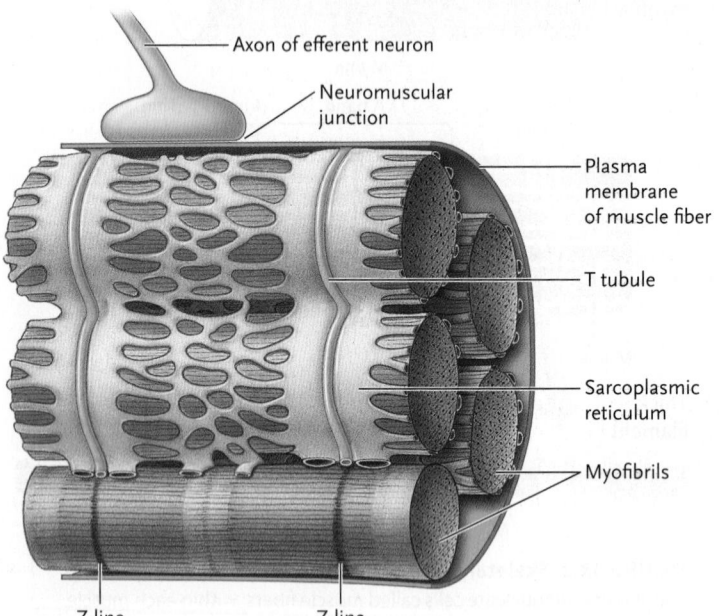

Axon of efferent neuron

Neuromuscular junction

Plasma membrane of muscle fiber

T tubule

Sarcoplasmic reticulum

Myofibrils

Z line   Z line

FIGURE 43.3 Components in the pathway for the stimulation of skeletal muscle contraction by neural signals. T (transverse) tubules are infoldings of the plasma membrane into the muscle fiber originating at each A band–I band junction in a sarcomere. The sarcoplasmic reticulum encircles the sarcomeres (it is cut away from the bottom myofibril in the figure), and segments of it end in close proximity to the T tubules.

© Cengage Learning 2017

At each junction of an A band and an I band, the plasma membrane folds into the muscle fiber to form a **T (transverse) tubule (Figure 43.3).** Encircling the sarcomeres is the **sarcoplasmic reticulum,** a complex system of vesicles modified from the smooth endoplasmic reticulum. Segments of the sarcoplasmic reticulum are wrapped around each A band and I band and are separated from the T tubules in those regions by small gaps.

An axon of an efferent neuron leads to each muscle fiber. The axon terminal makes a single synapse with a muscle fiber called a **neuromuscular junction** (see Figure 43.3). The neuromuscular junction, T tubules, and sarcoplasmic reticulum are key components in the pathway for stimulating skeletal muscle contraction.

During Muscle Contraction, Thin Filaments on Each Side of a Sarcomere Slide over Thick Filaments

The precise control of body motions depends on an equally precise control of muscle contraction by a signaling pathway that starts with action potentials traveling down an efferent neuron. These neural signals carry information from nerves to muscle fibers and trigger contraction.

MECHANISM OF MUSCLE FIBER CONTRACTION Contraction of a muscle fiber involves a process in which the thin filaments on each side of a sarcomere slide over the thick filaments toward the center of the A band, which brings the Z lines closer together, shortening the sarcomeres and contracting the muscle **(Figure 43.4).** This **sliding filament model** of muscle contraction depends on dynamic interactions between actin and myosin proteins in the two filament types. **Figure 43.5** presents a molecular model showing how an action potential arriving at the neuromuscular junction leads to the cyclic actin–myosin interactions that are the basis for muscle fiber contraction. That is, like neurons, skeletal muscle fibers are *excitable,* meaning that the electrical potential of their plasma membrane can change in response to a stimulus. The action potential leads to an increase in the concentration of Ca^{2+} in the cytosol of the muscle fiber (steps 1–3). The increase in Ca^{2+} triggers the crossbridge cycle, in which repeated power strokes pull the actin thin filament over the myosin thick filament (steps 4–6). When action potentials stop arriving at the neuromuscular junction, the muscle fiber stops contracting and relaxes (step 7).

Crossbridge cycles based on actin and myosin power movements in all living organisms, from cytoplasmic streaming in plant cells and amoebae to muscle contractions in animals.

Although the force produced by a single myosin crossbridge is comparatively small, it is multiplied by the hundreds of crossbridges acting in a single thick filament, and by the billions of thin filaments sliding in a contracting sarcomere. The force, multiplied further by the many sarcomeres and myofibrils in a muscle fiber, is transmitted to the plasma membrane of a muscle fiber by the attachment of myofibrils to elements of

 **FIGURE 43.4** **Experimental Research**

The Sliding Filament Model of Muscle Contraction

Question: What is the mechanism of muscle contraction?

Experiment: In 1954, Andrew Huxley and Ralph Niedergerke of the University of Cambridge, UK, and Hugh Huxley and Jean Hanson of the Massachusetts Institute of Technology used high-resolution light microscopy techniques to study how the actin and myosin arrangements changed during muscle contraction.

Results: Their micrographs provided important evidence for the sliding of filaments as the muscle went from being relaxed to fully contracted **(Figure)**. The key observations were that: (1) the I band and the H zone each decrease in length in proportion to the shortening of the sarcomere; and (2) the A band remains constant in length.

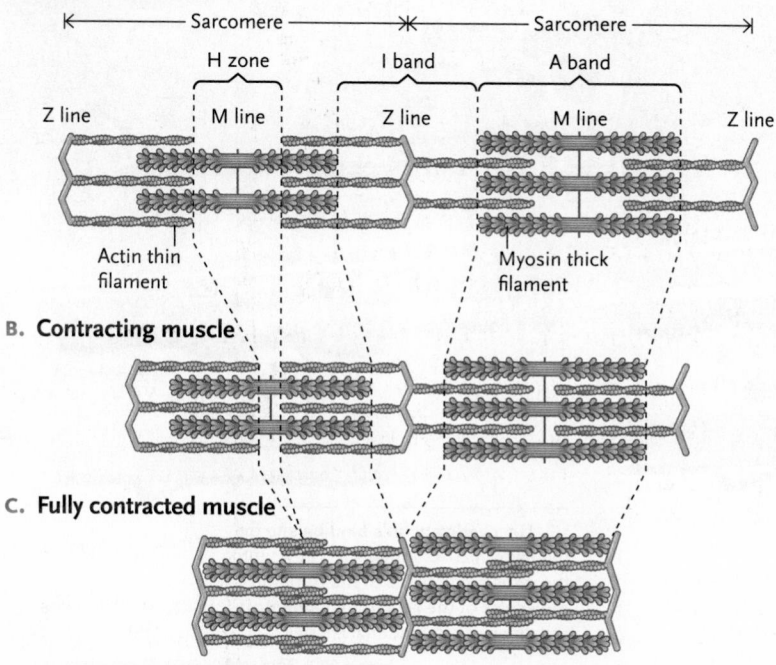

A. **Relaxed muscle**

B. **Contracting muscle**

C. **Fully contracted muscle**

Conclusion: This classic piece of research supported the sliding filament model in which muscle shortening (sarcomere shortening) results from increased overlap of thin and thick filaments, not from any change in length of those filaments. It set the path for research into the molecular basis of muscle contraction.

 **think like a scientist**

Observe
Hypothesize
Predict
Experiment
Interpret

Earlier research by other scientists had reported that myosin has ATPase activity. Interpret that result in terms of the sliding filament model.

Sources: A. F. Huxley and R. Niedergerke. 1954. Structural changes in muscle during contraction. *Nature* 173:971–973; H. Huxley and J. Hanson. 1954. Changes in the cross-striations of muscle during contraction and stretch and their structural interpretation. *Nature* 173:973–976.

the cytoskeleton. From the plasma membrane, it is transmitted to bones and other body parts by the connective tissue sheaths surrounding the muscle fibers and by the tendons.

Several gene mutations affecting muscle and nerve tissues interrupt the transmission of force and cause severe disabilities. Duchenne muscular dystrophy (DMD), for example, is caused by a mutation that weakens the cytoskeleton of the muscle fiber, causing the cells to rupture when contractile forces are generated. The mutation is in the *DMD* gene, which encodes a protein called dystrophin that is found primarily in skeletal muscles. The *DMD* gene is located on the X chromosome and the disease DMD is inherited as an X-linked recessive trait (see Section 13.4).

DEADLY INTERRUPTIONS OF THE CROSSBRIDGE CYCLE The mechanism controlling vertebrate muscle contraction can be blocked by several toxins and poisons. For example, the bacterium *Clostridium botulinum*, which grows in improperly preserved food, produces botulinum toxin. The toxin blocks acetylcholine release in neuromuscular junctions, and affected body muscles are unable to contract. When the diaphragm, the muscle essential for inflating the lungs, becomes paralyzed, the victim dies from respiratory failure. The toxin is so poisonous that 0.0000001 g is enough to kill a human; 600 g could wipe out the entire human population. A very low concentration of the same toxin, under the brand name Botox, is injected under the skin as a cosmetic treatment to remove or reduce wrinkles—if muscles cannot contract, wrinkles cannot form.

In a natural process, within a few hours after an animal dies, Ca^{2+} diffuses into the cytoplasm of muscle cells and initiates the crossbridge cycle, producing *rigor mortis,* a strong tension of essentially all the skeletal muscles that stiffens the entire body. The crossbridges become locked to the thin filaments because ATP production stops (remember that ATP is required to release the crossbridges from actin). The stiffness reverses as actin and myosin are degraded.

The Response of a Muscle Fiber to Action Potentials Ranges from Twitches to Tetanus

A single action potential arriving at a neuromuscular junction usually causes a single, weak contraction of a muscle fiber called a **muscle twitch (Figure 43.6A).** After a muscle twitch begins, the tension of the muscle fiber increases in magnitude for about 30–40 msec, and then peaks as the action potential runs its course through the T tubules and the Ca^{2+} channels begin to close. Tension then decreases as the Ca^{2+} ions are pumped back into the sarcoplasmic reticulum, falling to zero in about 50 msec after the peak.

If a muscle fiber is restimulated after it has relaxed completely, a new twitch identical to the first is generated (see

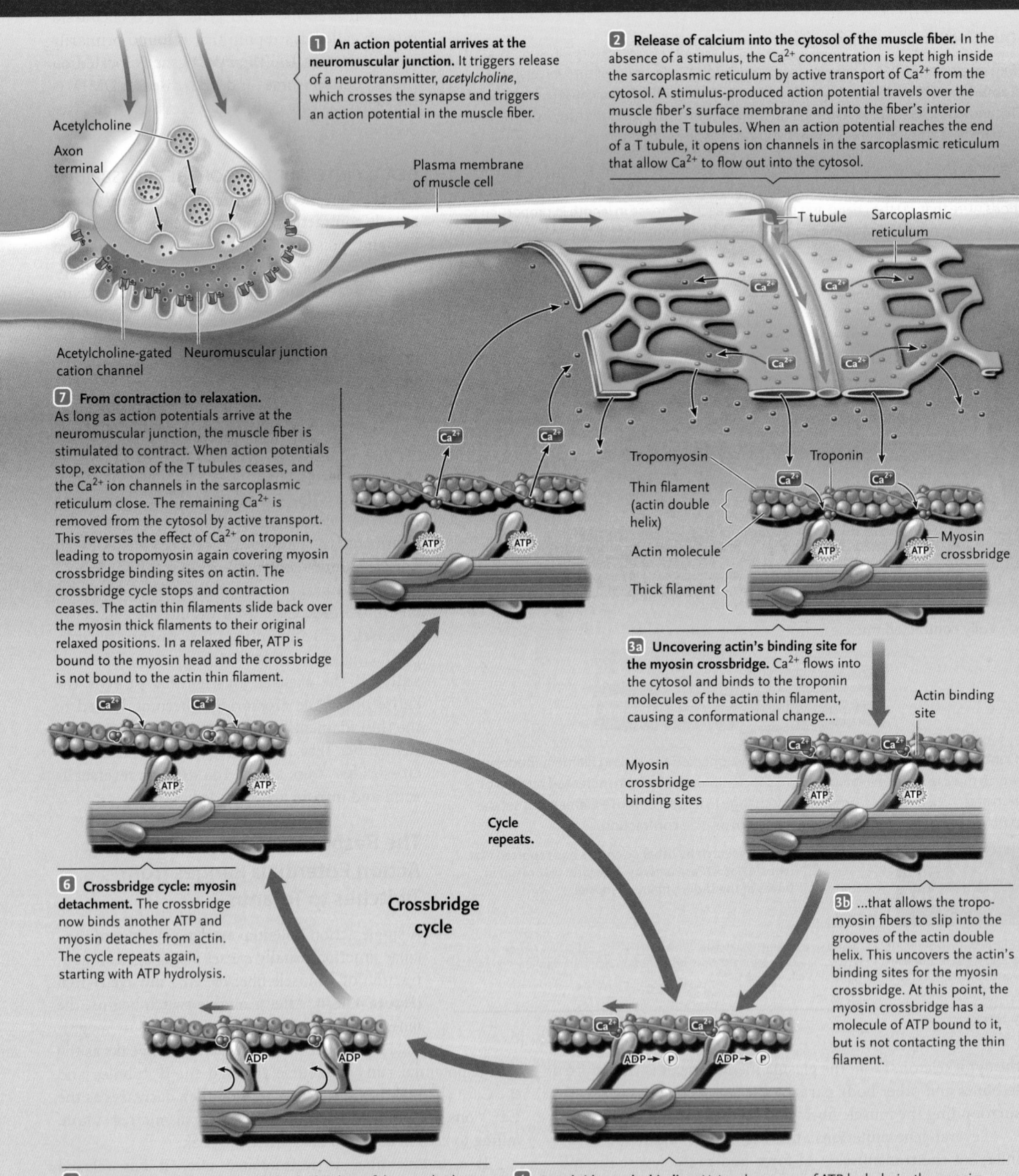

1 **An action potential arrives at the neuromuscular junction.** It triggers release of a neurotransmitter, *acetylcholine*, which crosses the synapse and triggers an action potential in the muscle fiber.

2 **Release of calcium into the cytosol of the muscle fiber.** In the absence of a stimulus, the Ca^{2+} concentration is kept high inside the sarcoplasmic reticulum by active transport of Ca^{2+} from the cytosol. A stimulus-produced action potential travels over the muscle fiber's surface membrane and into the fiber's interior through the T tubules. When an action potential reaches the end of a T tubule, it opens ion channels in the sarcoplasmic reticulum that allow Ca^{2+} to flow out into the cytosol.

Acetylcholine

Axon terminal

Plasma membrane of muscle cell

Acetylcholine-gated cation channel

Neuromuscular junction

T tubule

Sarcoplasmic reticulum

Tropomyosin

Troponin

Thin filament (actin double helix)

Actin molecule

Myosin crossbridge

Thick filament

7 **From contraction to relaxation.** As long as action potentials arrive at the neuromuscular junction, the muscle fiber is stimulated to contract. When action potentials stop, excitation of the T tubules ceases, and the Ca^{2+} ion channels in the sarcoplasmic reticulum close. The remaining Ca^{2+} is removed from the cytosol by active transport. This reverses the effect of Ca^{2+} on troponin, leading to tropomyosin again covering myosin crossbridge binding sites on actin. The crossbridge cycle stops and contraction ceases. The actin thin filaments slide back over the myosin thick filaments to their original relaxed positions. In a relaxed fiber, ATP is bound to the myosin head and the crossbridge is not bound to the actin thin filament.

3a **Uncovering actin's binding site for the myosin crossbridge.** Ca^{2+} flows into the cytosol and binds to the troponin molecules of the actin thin filament, causing a conformational change...

Actin binding site

Myosin crossbridge binding sites

Cycle repeats.

Crossbridge cycle

6 **Crossbridge cycle: myosin detachment.** The crossbridge now binds another ATP and myosin detaches from actin. The cycle repeats again, starting with ATP hydrolysis.

3b ...that allows the tropomyosin fibers to slip into the grooves of the actin double helix. This uncovers the actin's binding sites for the myosin crossbridge. At this point, the myosin crossbridge has a molecule of ATP bound to it, but is not contacting the thin filament.

5 **Crossbridge cycle: snapping back.** The binding of the crossbridge to actin triggers release of the molecular spring in the crossbridge, which snaps back toward the tail producing the power stroke (motor) that pulls the thin filament over the thick filament. ADP is released.

4 **Crossbridge cycle: binding.** Using the energy of ATP hydrolysis, the myosin crossbridge bends away from the tail and binds to an exposed myosin crossbridge binding site on an actin molecule. This bending compresses a molecular spring in the myosin head.

SUMMARY An action potential arriving at the neuromuscular junction results in a release of acetylcholine, which triggers an action potential in the muscle fiber. This action potential leads to an increase in the concentration of Ca^{2+} in the fiber's cytosol, which triggers the crossbridge cycle. Repeated power strokes of the crossbridge cycle pull the actin thin filaments over the myosin thick filament.

think like a scientist Muscles can be studied outside of the body to determine how muscle contraction is controlled. If you added an inhibitor of myosin ATPase to a muscle preparation, what would you expect the result to be on muscle contraction?

© Cengage Learning 2017

Figure 43.6A). However, if a muscle fiber is restimulated before it has relaxed completely, the second twitch is added to the first, resulting in *twitch summation,* a summed, stronger contraction **(Figure 43.6B)**. If action potentials arrive so rapidly (about 25 msec apart) that the fiber cannot relax at all between stimuli, the Ca^{2+} channels remain open continuously and twitch summation produces a peak level of continuous contraction called **tetanus (Figure 43.6C)**. (Do not confuse this with the disease of the same name, in which a bacterial toxin causes uncontrolled and continuous muscle contraction.) Contractile activity decreases if either the stimuli cease or the muscle fatigues.

Tetanus (or tetanic contraction) is an essential part of muscle fiber function. If you lift a moderately heavy weight, for example, many of the muscle fibers in your arms enter tetanus and remain in that state until the weight is released. Even body movements that require relatively little effort, such as standing still but in balance, involve tetanic contractions of some muscle fibers.

Muscle Fibers Differ in Their Rate of Contraction and Resistance to Fatigue

Muscle fibers differ in their rate of contraction and resistance to fatigue, and thus can be classified as slow fibers, fast aerobic fibers, and fast anaerobic fibers. Their properties are summarized in **Table 43.1**. The proportions of the three types of muscle fibers tailor the contractile characteristics of each muscle to suit its function within the body.

Slow muscle fibers contract relatively slowly, and the intensity of contraction is low because their myosin crossbridges hydrolyze ATP relatively slowly. They can remain contracted for relatively long periods without fatiguing. Slow muscle fibers typically contain many mitochondria and make most of their ATP by oxidative phosphorylation (cellular respiration; see Chapter 7). They have a low capacity to make ATP by anaerobic glycolysis. They also contain high concentrations of the oxygen-storing protein **myoglobin,** which greatly enhances their oxygen supplies. Myoglobin is closely related to

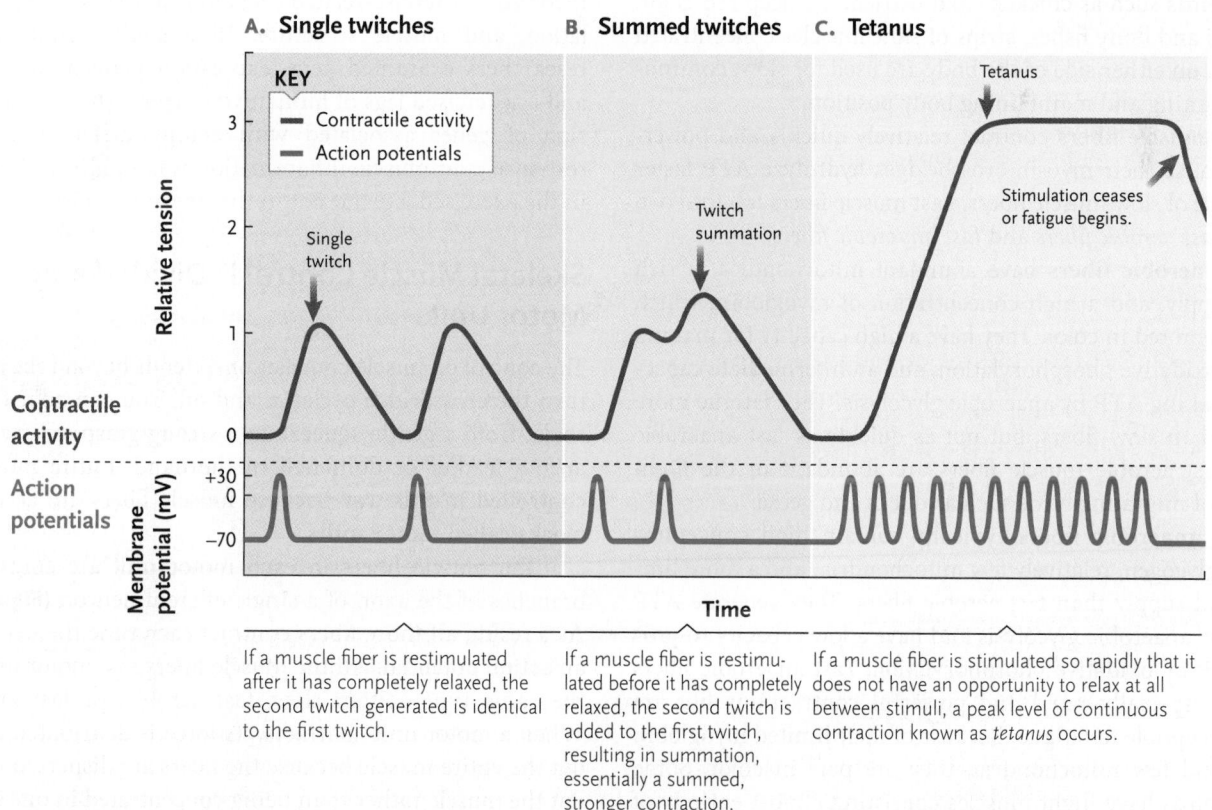

FIGURE 43.6 The relationship of the tension produced in a muscle fiber to the frequency of action potentials.
© Cengage Learning 2017

	Fiber Type		
Property	Slow	Fast Aerobic	Fast Anaerobic
Contraction speed	Slow	Fast	Fast
Contraction intensity	Low	Intermediate	High
Fatigue resistance	High	Intermediate	Low
Myosin–ATPase activity	Low	High	High
Oxidative phosphorylation capacity	High	High	Low
Enzymes for anaerobic glycolysis	Low	Intermediate	High
Mitochondria	Many	Many	Few
Myoglobin content	High	High	Low
Fiber color	Red	Red	White
Glycogen content	Low	Intermediate	High

TABLE 43.1 Characteristics of Slow and Fast Muscle Fibers in Skeletal Muscle

hemoglobin, the oxygen-carrying protein of red blood cells. Myoglobin gives slow muscle fibers, such as those in the legs of ground birds such as chickens and ostriches, a deep red color. In sharks and bony fishes, strips of slow muscles concentrated in a band on either side of the body are used for slow, continuous swimming and maintaining body position.

Fast muscle fibers contract relatively quickly and powerfully because their myosin crossbridges hydrolyze ATP faster than those of slow muscle fibers. Fast muscle fibers fall into two groups: *fast aerobic fibers* and *fast anaerobic fibers*.

Fast aerobic fibers have abundant mitochondria, a rich blood supply, and a high concentration of myoglobin, which makes them red in color. They have a high capacity for making ATP by oxidative phosphorylation, and an intermediate capacity for making ATP by anaerobic glycolysis. They fatigue more quickly than slow fibers, but not as quickly as fast anaerobic fibers. Fast aerobic muscle fibers are abundant in the flight muscles of migrating birds such as ducks and geese.

Fast anaerobic fibers typically contain high concentrations of glycogen, relatively few mitochondria, and a more limited blood supply than fast aerobic fibers. They generate ATP mostly by anaerobic glycolysis and have a low capacity to produce ATP by oxidative phosphorylation. Fast anaerobic fibers produce especially rapid and powerful contractions but are more susceptible to fatigue. Because of their limited myoglobin supply and few mitochondria, they are pale in color. Some ground birds have flight muscles consisting almost entirely of fast anaerobic muscle fibers. These muscles can produce a short burst of intensive contractions allowing the bird to escape a

predator, but they cannot produce sustained flight. Most muscles of lampreys, sharks, fishes, amphibians, and reptiles also contain fast anaerobic muscle fibers, allowing the animals to move quickly to capture prey and avoid danger.

The muscles of humans and other mammals are mixed, and contain different proportions of slow and fast muscle fibers, depending on their functions. Muscles specialized for prolonged, slow contractions, such as the postural muscles of the back, have a high proportion of slow fibers and are a deep red color. The muscles of the forearm that move the fingers have a higher proportion of fast fibers and are a paler red than the back muscles. These muscles can contract rapidly and powerfully, but they fatigue much more rapidly than the back muscles.

The number and proportions of slow and fast muscle fibers in individuals are inherited characteristics. However, particular types of exercises can convert some fast muscle fibers between aerobic and anaerobic types. Endurance training, such as long-distance running, converts fast muscle fibers from the anaerobic to the aerobic type, and regimens such as weight lifting induce the reverse conversion. If the training regimens stop, most of the fast muscle fibers revert to their original types.

Genomics studies are showing that gene expression patterns change during training. For example, in 2010 researchers analyzed the transcriptomes (see Chapter 19) of skeletal muscle biopsy samples from racehorses trained for 10 months using an endurance plus high-intensity exercise program. They found that, compared with untrained racehorses, the training regimen resulted in significant increases in expression of genes involved in exercise-related metabolism, oxidative phosphorylation, and muscle structure. In a similar study in 2014, researchers examined gene expression patterns in exercised and unexercised legs of human volunteers. They found expression of genes associated with energy metabolism, insulin response, and muscle inflammation was significantly changed in the exercised leg, but not in the unexercised leg.

Skeletal Muscle Control Is Divided among Motor Units

The control of muscle contraction extends beyond the ability to turn the crossbridge cycle on and off. You can adjust a handshake from a gentle squeeze to a strong grasp, or exactly balance a feather or dumbbell in the hand. Entire muscles are controlled in this way because muscle fibers are activated in blocks called **motor units.**

The muscle fibers in each motor unit are controlled by branches of the axon of a single efferent neuron **(Figure 43.7).** As a result, all those fibers contract each time the neuron fires an action potential. All the muscle fibers in a motor unit are of the same type—either slow, fast aerobic, or fast anaerobic. When a motor unit contracts, its force is distributed throughout the entire muscle because the fibers are dispersed throughout the muscle rather than being concentrated in one segment.

For a delicate movement, only a few efferent neurons carry action potentials to a muscle, and only a few motor units

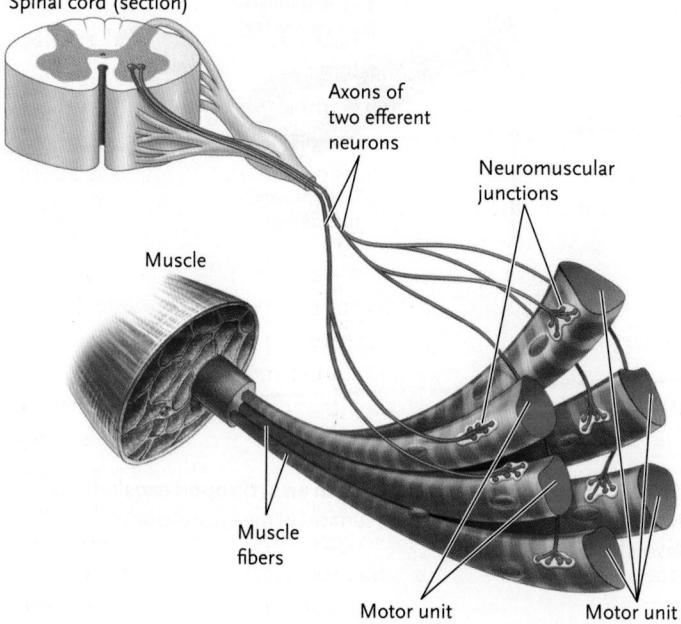

Spinal cord (section)

Axons of
two efferent
neurons

Neuromuscular
junctions

Muscle

Muscle
fibers

Motor unit   Motor unit

FIGURE 43.7 Motor units in vertebrate skeletal muscles. Each motor unit consists of groups of muscle fibers activated by branches of a single efferent (motor) neuron.

© Cengage Learning 2017

contract. For more powerful movements, more efferent neurons carry action potentials, and more motor units contract.

Muscles that can be controlled precisely and delicately, such as those moving the fingers in humans, have many motor units in a small area, with only a few muscle fibers—about 10 or so—in each unit. Muscles that produce grosser body movements, such as those moving the legs, have fewer motor units in the same volume of muscle but thousands of muscle fibers in each unit. In the calf muscle that raises the heel, for example, most motor units contain nearly 2,000 muscle fibers. Other skeletal muscles fall between these extremes, with an average of about 200 muscle fibers per motor unit.

Invertebrates Move Using a Variety of Striated Muscles

Invertebrates also have muscle cells in which actin-based thin filaments and myosin-based thick filaments produce movements by the sliding filament model. Muscles that are striated, which occur in virtually all invertebrates except sponges, have thick and thin filaments arranged in sarcomeres entirely similar to those of vertebrates, except for variations in sarcomere length and the ratio of thin to thick filaments.

In invertebrates, an entire muscle is typically controlled by one or a few motor neurons. Nevertheless, invertebrate muscles are capable of finely graded contractions because individual neurons make large numbers of synapses with the muscle cells. As action potentials arrive more frequently at the synapses, more Ca^{2+} is released into the cells, and the muscles contract more strongly.

43.2 Skeletal Systems

Animal skeletal systems provide physical support for the body and protection for the soft tissues. They also act as a framework against which muscles work to move parts of the body or the entire organism. The three main types of skeletons found in both invertebrates and vertebrates are *hydrostatic skeletons*, *exoskeletons*, and *endoskeletons*.

A Hydrostatic Skeleton Consists of Muscles and Fluid

A **hydrostatic skeleton** (*hydro-* = water; *statikos* = causing to stand) is a structure consisting of muscles and fluid that provides support for the animal or part of the animal; no rigid support, such as a bone, is involved. A hydrostatic skeleton consists of a body compartment or compartments filled with water or body fluids, which are incompressible liquids. When the muscular walls of the compartment contract, they pressurize the contained fluid. If muscles in one part of the compartment are contracted while muscles in another part are relaxed, the pressurized fluid will move to the relaxed part of the compartment, distending it. In short, the contractions and relaxations of the muscles surrounding the compartments change the shape of the animal. Hydrostatic skeletons are the primary support systems of cnidarians, flatworms, roundworms, and annelids (see Chapter 31). In all these animals, compartments containing fluids under pressure make the body semirigid and provide a mechanical support on which muscles act.

In flatworms, roundworms, and annelids, striated muscles in the body wall act on the hydrostatic skeleton to produce creeping, burrowing, or swimming movements. Among these animals, annelids have the most highly developed musculoskeletal systems, with an outer layer of circular muscles surrounding the body, and an inner layer of longitudinal muscles running its length **(Figure 43.8).** Contractions of the circular muscles reduce the diameter of the body and increase the length of the worm. Contractions of the longitudinal muscles shorten the body and increase its diameter. Annelids move along a surface or burrow by means of alternating waves of contraction of the two muscle layers that pass along the body, working against the fluid-filled body compartments of the hydrostatic skeleton.

Many arthropods (see Chapter 31) have hydrostatic skeletal elements. In the larvae of flying insects, for example, internal fluids held under pressure by the muscular body wall provide some body support.

Some structures of echinoderms are supported by hydrostatic skeletons. The tube feet of sea stars and sea urchins, for example, have muscular walls enclosing the fluid of the water vascular system (see Figure 32.2).

In vertebrates, the erectile tissue of the penis is a fluid-filled hydrostatic skeletal structure.

An Exoskeleton Is a Rigid, External Body Covering

An **exoskeleton** (*exo* = outside) is a rigid, external body covering, such as a shell, that provides support. In an exoskeleton, the force of muscle contraction is applied against that covering.

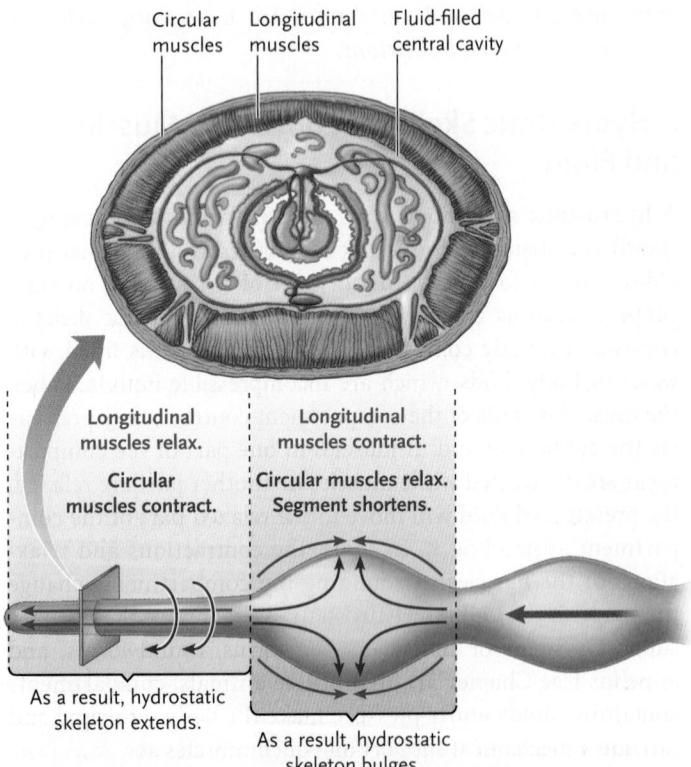

FIGURE 43.8 Movement of an earthworm, showing how muscles in the body wall act on its hydrostatic skeleton. Contraction of the circular muscles reduces body diameter and increases body length, and contraction of the longitudinal muscles decreases body length and increases body diameter.
© Cengage Learning 2017

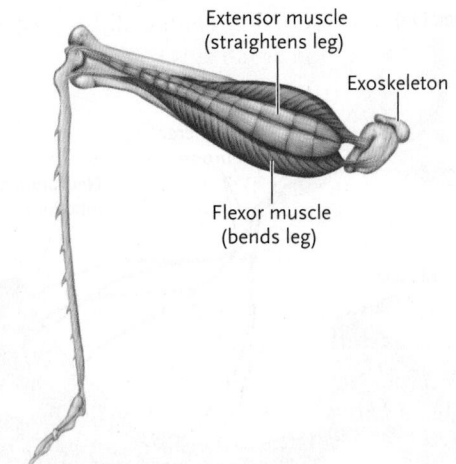

FIGURE 43.9 Muscle attachments in an arthropod exoskeleton. Muscles are attached to the inside surfaces of the exoskeleton in a typical insect leg, such as this one.
© Cengage Learning 2017

An exoskeleton also protects delicate internal tissues such as the brain and respiratory organs.

Many mollusks (see Chapter 31), such as clams and oysters, have an exoskeleton consisting of a hard calcium carbonate shell secreted by glands in the mantle. Arthropods, such as insects, spiders, and crustaceans, have an external skeleton in the form of a chitinous cuticle, secreted by underlying tissue, that covers the outside surfaces of the animals. Like a suit of armor, the arthropod exoskeleton has movable joints, flexed and extended by muscles that extend across the inside surfaces of the joints **(Figure 43.9).** The exoskeleton protects against dehydration, serves as armor against predators, and provides the levers against which muscles work. In many flying insects, elastic flexing of the exoskeleton contributes to the movements of the wings.

In vertebrates, the shell of a turtle or tortoise is an exoskeletal structure, as are the bony plates, abdominal ribs, collar bones, and most of the skull of the alligator.

An Endoskeleton Consists of Supportive Internal Body Structures Such as Bones

An **endoskeleton** (*endon* = within) consists of internal body structures, such as bones, that provide support. In an endoskeleton, the force of contraction is applied against those structures. Like exoskeletons, endoskeletons protect delicate internal tissues such as the brain and internal organs.

Echinoderms have an endoskeleton consisting of *ossicles* (*ossiculum* = little bone), formed from calcium carbonate crystals. The shells of sand dollars and sea urchins are the endoskeletons of these animals.

The endoskeleton is the primary skeletal system of vertebrates. An adult human, for example, has an endoskeleton consisting of 206 bones **(Figure 43.10).** The skull, vertebral column, sternum, and rib cage form the central part of the endoskeleton, called the **axial skeleton** (red in Figure 43.10). The shoulder, hip, leg, and arm bones form the **appendicular skeleton** (green in Figure 43.10).

Bones of the Vertebrate Endoskeleton Are Organs with Several Functions

The vertebrate endoskeleton supports and maintains the overall shape of the body and protects key internal organs. In addition, the endoskeleton is a storehouse for calcium and phosphate ions, releasing them as required to maintain optimal levels of these ions in body fluids. Bones are also sites where new blood cells form.

Bones are complex organs built up from multiple tissues of several kinds, blood vessels, nerves, and, in some cases, stores of adipose tissue. Bone tissue is distributed between dense, compact bone regions, which have essentially no spaces other than the microscopic canals of the osteons (see Figure 38.4D), and spongy bone regions, which are opened by larger spaces (see Figure 43.10). Compact bone tissue generally forms the outer surfaces of bones, and spongy bone tissue the interior. The interior of some flat bones, such as the hip bones and the ribs, are filled with *red marrow,* a tissue that is the primary source of new red blood cells in mammals and birds. The shaft of long bones such as the femur has a large central canal filled with adipose tissue called *yellow marrow,* a source of some white blood cells.

Throughout the life of a vertebrate, calcium and phosphate ions are constantly deposited and withdrawn from bones. Hormonal controls maintain the concentration of Ca^{2+} ions at optimal levels in the blood and extracellular fluids (see Figure 42.9), ensuring that calcium is available for proper functioning of the nervous system, muscular system, and other physiological processes.

Molecular Insights discusses an analysis of a genome of a cartilaginous fish that has provided valuable information about the genetics of bone formation.

STUDY BREAK 43.2

1. How do hydrostatic skeletons, exoskeletons, and endoskeletons provide support to the body? Give an example of each of these types in echinoderms and vertebrates.
2. What are the functions of the bones of the vertebrate endoskeleton?

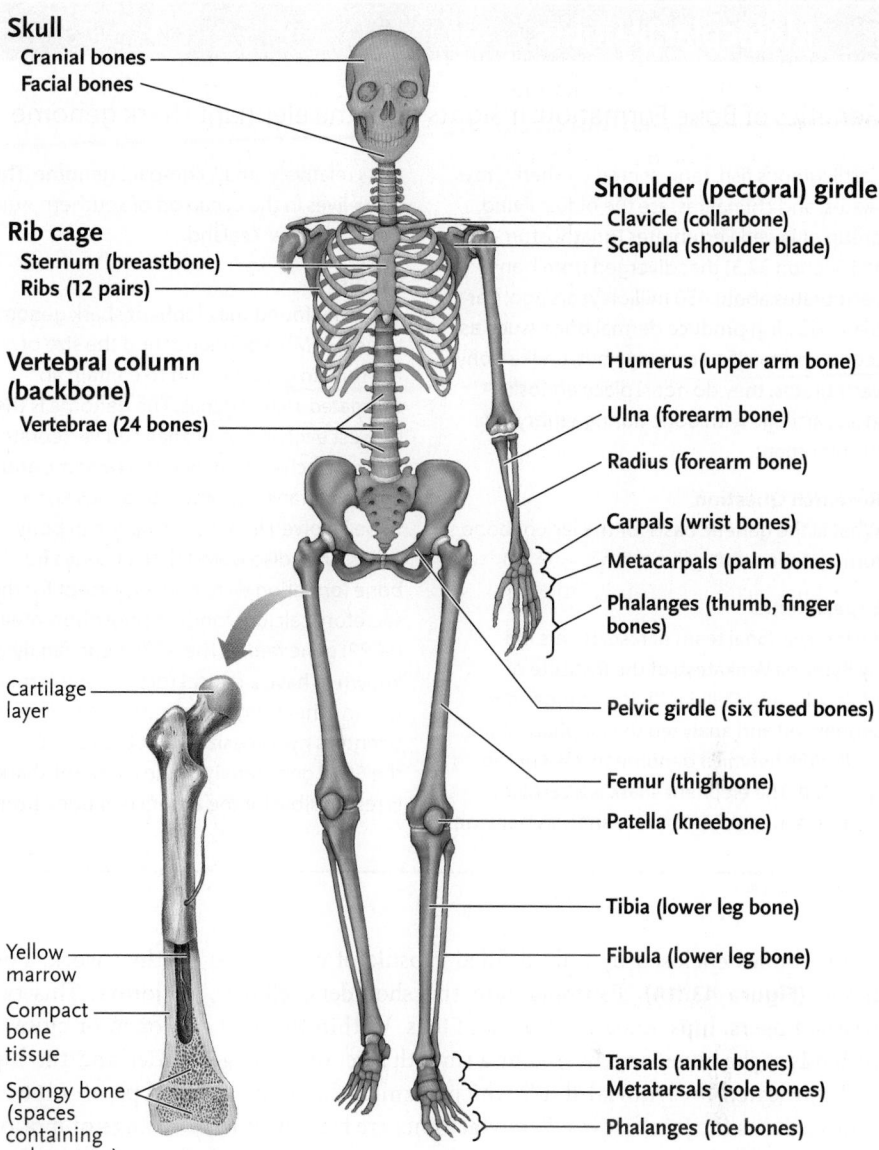

FIGURE 43.10 Major bones of the human body. The axial skeleton is shaded red, and the appendicular skeleton is shaded green. The inset shows the structure of a limb bone, with the location of red and yellow marrow. The internal spaces lighten the bone's structure. The cartilage layer forms a smooth, slippery cushion between bones in a joint.
© Cengage Learning 2017

43.3 Vertebrate Movement: The Interactions between Muscles and Bones

The skeletal systems of all animals act as a framework against which muscles work to move parts of the body or the entire organism. In this section, the muscle–bone interactions that are responsible for the movement of vertebrates are described.

Joints of the Vertebrate Endoskeleton Allow Bones to Move and Rotate

The bones of the vertebrate skeleton are connected by joints. **Synovial joints,** the most-movable joints, consist of the ends

Genetics of Bone Formation: Insights from the elephant shark genome

Cartilaginous fish, represented by sharks, rays, skates, and chimaeras, are the oldest living group of jawed vertebrates (gnathostomes; see Section 32.5) that diverged from bony vertebrates about 450 million years ago. Cartilaginous fish produce dermal bone (such as teeth) and calcified cartilage but, unlike bony vertebrates, they do not replace endoskeleton cartilage with bone during embryonic development.

Research Question

What is the genetic basis for the lack of bone formation in cartilaginous fish?

Experiments

An international team of researchers led by Byrappa Venkatesh of the Institute of Molecular and Cellular Biology, Singapore, sequenced and analyzed the elephant shark *(Callorhinchus milii)* genome to answer the question. The elephant shark is a cartilaginous fish model for genome analysis because

of its relatively small, compact genome. This shark lives in the ocean off of southern Australia and New Zealand.

Results

The team found the elephant shark genome to be 937 Mb, about one third the size of the human genome, and to contain an estimated 18,872 genes. The genome is the slowest evolving of all analyzed vertebrates. The researchers searched the genome and analyzed transcriptomes for genes known to be involved in bone formation in bony fishes. They discovered that all genes for bone formation were present except for the secretory calcium-binding phosphoprotein (SCPP) gene family. The SCPP gene family is known to have a critical role in bone formation in other bony vertebrates. Therefore, the scientists hypothesized that the absence of the SCPP gene family in the elephant shark is responsible for the absence of bone from

its endoskeleton. To test the hypothesis, the researchers knocked down the expression of a member of the gene family in zebrafish, a bony fish. (Knocking down gene expression is described in Section 19.3.) These zebrafish developed with significantly reduced endoskeletal bone formation. This result supported their hypothesis.

Conclusion

The study showed that the lack of SCPP genes in the elephant shark is responsible for the absence of bone in its endoskeleton.

think like a scientist

What was the key result in the study that demonstrated SCPP genes are needed for bone formation?

Source: B. Venkatesh et al. 2014. Elephant shark genome provides unique insights into gnathostome evolution. *Nature* 505:174–179.

---

of two bones enclosed by a fluid-filled capsule of connective tissue **(Figure 43.11A).** Examples are the shoulders, elbows, wrists, fingers, hips, knees, ankles, and toes. Within the joint, the ends of the bones are covered by a smooth layer of cartilage and lubricated by synovial fluid, which permits the bones to slide easily as the joint moves. Synovial joints are held together by straps of connective tissue called **ligaments** that extend across the joints outside the capsule **(Figure 43.11B).** Synovial joints are classified into several different types based on how

they work. For example, the elbow and the knee are **hinge joints.** This type of joint can move only in one direction to open or close the angle between the bones flanking it. Shoulder and the hip joints exemplify **ball-and-socket joints.** This type of joint can rotate about its axis, providing the greatest range of motion of any joint in the body.

In **cartilaginous joints,** which are less movable, the ends of bones are covered with layers of cartilage, but have no fluid-filled capsule surrounding them. Dense connective tissue (see

A. Synovial joint cross section

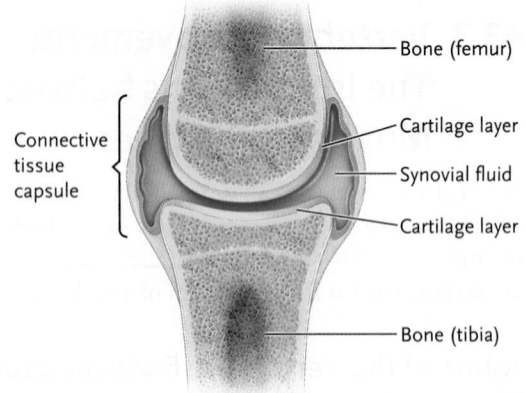

- Bone (femur)
- Cartilage layer
- Synovial fluid
- Cartilage layer
- Bone (tibia)

Connective tissue capsule

B. Ligaments reinforcing the knee joint

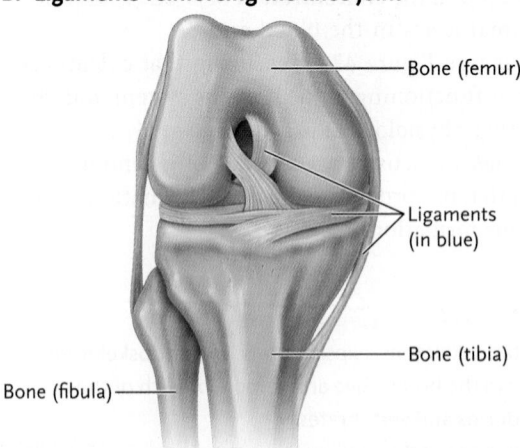

- Bone (femur)
- Ligaments (in blue)
- Bone (tibia)
- Bone (fibula)

FIGURE 43.11 The synovial joint, the human knee. This type of synovial joint is a hinge joint.
© Cengage Learning 2017

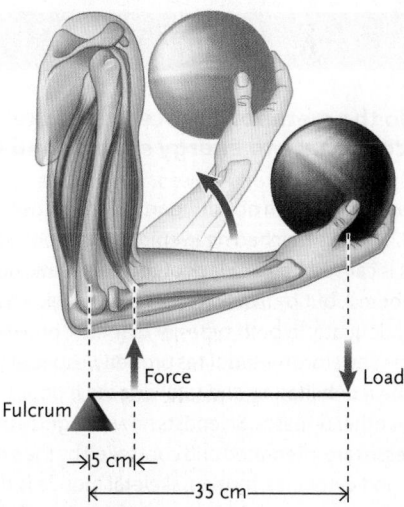

Force   Load

Fulcrum

→|5 cm|←

|←——35 cm——→|

FIGURE 43.12 The body lever: the lever formed by the bones of the forearm. The fulcrum (the hinge of the joint) is at one end of the lever, the load is placed on the opposite end, and then force is exerted at a point on the lever between the fulcrum and the load.
© Cengage Learning 2017

Section 38.2) covers and connects the bones of these joints. They occur between the vertebrae and some rib bones.

The bones connected by movable joints work like levers. A lever is a rigid structure that can move around a pivot point known as a *fulcrum*. Levers differ with respect to where the fulcrum is positioned along the lever and where the force is applied. The most common type of lever system in the body—exemplified by the elbow joint—has the fulcrum at one end, the load at the opposite end, and the force applied at a point between the ends **(Figure 43.12)**. For this lever, the force applied must be much greater than the load, but it increases the distance the load moves as compared with the distance over which the force is applied. This allows small muscle movements to produce large body movements, and also allows

movements such as running or throwing to be carried out at high speed.

At a joint, a muscle that causes movement in the joint when it contracts is called an **agonist**. Most of the bones of vertebrate skeletons are moved by muscles arranged in **antagonistic pairs: extensor muscles** extend the joint, meaning increasing the angle between the two bones, and **flexor muscles** do the opposite. (Antagonistic muscles are also used in invertebrates for movement of body parts—for example, the limbs of insects and other arthropods; see Figure 43.9.) In humans, one such pair is formed by the biceps brachii muscle at the front of the upper arm (the flexor muscle) and the triceps brachii muscle at the back of the upper arm (the extensor muscle) **(Figure 43.13)**.

Vertebrates Have Muscle–Bone Interactions Optimized for Specific Movements

Vertebrates differ widely in the patterns by which muscles connect to bones, and in the length and mechanical advantage of the levers produced by these connections. These differences produce limbs and other body parts that are adapted for either power or speed, or the most advantageous compromise between these characteristics. Among burrowing mammals such as the mole, for example, the limb bones are short, thick, and heavy, and the points at which muscles attach produce levers that are slow to move but that need to apply smaller forces to move a load when compared with a human biceps. In contrast, a mammal such as the deer has relatively light and thin bones with muscle attachments producing levers that can generate rapid movement, moving the body easily over the ground.

STUDY BREAK 43.3

1. What are the features of synovial joints and cartilaginous joints?
2. What are antagonistic muscle pairs?

A. Flexion: When the biceps muscle (the flexor muscle) contracts and raises the forearm, its antagonistic partner, the triceps muscle (the extensor muscle), relaxes.

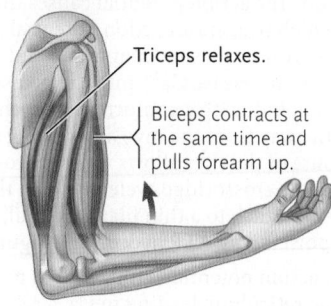

Triceps relaxes.

Biceps contracts at the same time and pulls forearm up.

B. Extension: When the triceps muscle contracts and extends the forearm, the biceps muscle relaxes.

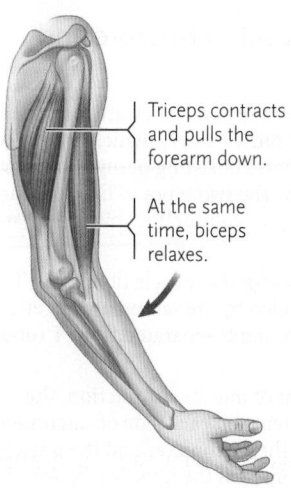

Triceps contracts and pulls the forearm down.

At the same time, biceps relaxes.

FIGURE 43.13 The arrangement and function of skeletal muscles in an antagonistic pair.
© Cengage Learning 2017

How can muscle growth processes be controlled to improve the clinical treatment of muscular dystrophy and related disorders?

Muscle growth is regulated by many different hormones and growth factors, including myostatin, a protein secreted by muscle itself that inhibits this process. Myostatin is therefore a potential therapeutic target for treating muscle wasting, including that which occurs with normal aging, in late stages of cancer, or with Duchenne muscular dystrophy, a degenerative and fatal disease associated with the progressive loss of skeletal muscle mass and cardiac muscle dysfunction. Animals with nonfunctioning and mutant forms of the myostatin gene (for example, the Belgian Blue and Piedmontese cattle breeds) or myostatin knockout mice, in which the gene has been removed experimentally (see Section 18.2 and Figure 18.8, and *Focus on Research: Model Organisms* in Chapter 45), have significantly enhanced musculature that is commonly referred to as *double muscling*. Such mutations and enhanced skeletal muscle mass have also been described in a racing dog breed, the whippet, and recently in people.

Our laboratory develops novel technologies to inhibit myostatin activity—in essence, inhibiting the inhibitor—and thereby stimulate skeletal muscle growth in both clinical and agricultural settings. These include gene therapeutics that use nonpathological or "safe" viruses to infect muscle with myostatin-disrupting genes. We have recently determined that myostatin can also negatively regulate cardiac muscle growth. Thus, disrupting myostatin production, availability, or action may help heart attack patients. Replacing damaged skeletal and cardiac muscle using adult or embryonic stem cells engineered to match either tissue type is another highly promising technique for treating these disorders, especially if combined with other "antimyostatin" technologies that enhance growth of the transplanted cells.

How much do the metabolic processes of skeletal muscle specifically contribute to energy storage and whole body form?

Complications associated with obesity, particularly diabetes mellitus type 2 (see Section 42.4), have reached near-epidemic proportions worldwide. Type 2 diabetes is caused not by a lack of the pancreatic hormone insulin (as in type 1 diabetes), but by insulin resistance, in which tissue responses to insulin are inadequate. In both types of diabetes, however, the body is unable to process and store metabolites properly, especially glucose. Type 2 diabetes can be a debilitating and fatal disease if poorly managed and often aggravates other diseases. Scientists now recognize that growth and metabolic processes are integrated and controlled by the same hormones, growth factors, and cytokines. Indeed, skeletal muscle is the largest consumer of metabolites and has the greatest potential to affect their circulating levels. Recent studies suggest that even small increases in muscle mass can significantly reduce fat mass and improve insulin responsiveness. Enhancing skeletal muscle growth in obese patients with type 2 diabetes could therefore improve treatments for both. The same antimyostatin technologies used to treat muscle growth disorders could therefore be used to treat severe obesity and type 2 diabetes with the goals of increasing muscle mass, decreasing fat mass, and improving insulin sensitivity.

think like a scientist Why do you think changes in skeletal muscle mass ultimately influence fat mass?

Buel (Dan) Rodgers is an Associate Professor in the Department of Animal Sciences at Washington State University. He is also the Director of the Washington Center for Muscle Biology (http://WCMB.wsu.edu) and studies how the endocrine system controls the growth, development, and repair of striated muscle. To learn more about Dr. Rodgers' research, visit http://www.ansci.wsu.edu/people/faculty/dan-rodgers/ or the Rodgers Laboratory Facebook site.

Courtesy of Dan Rodgers

REVIEW KEY CONCEPTS

For access to MindTap and additional study materials visit www.cengagebrain.com.

43.1 Vertebrate Skeletal Muscle: Structure and Function

- Skeletal muscles move the joints of the body. They are formed from long, cylindrical cells called muscle fibers, which are packed with myofibrils, contractile elements consisting of myosin thick filaments and actin thin filaments. The two types of filaments are arranged in an overlapping pattern of contractile units called sarcomeres (Figure 43.2).

- Infoldings of the plasma membrane of the muscle fiber form T tubules. The sarcomeres are encircled by the sarcoplasmic reticulum, a system of vesicles with segments separated from T tubules by small gaps (Figure 43.3).

- In the sliding filament mechanism of muscle contraction, the simultaneous sliding of thin filaments on each side of sarcomeres over the thick filaments shortens the sarcomeres and the muscle fibers, producing the force that contracts the muscle (Figure 43.4).

- The sliding motion of thin and thick filaments is produced in response to an action potential arriving at the neuromuscular junction. The action potential causes the release of acetylcholine, which triggers an action potential in the muscle fiber that spreads over its plasma membrane and stimulates the sarcoplasmic reticulum to release Ca^{2+} into the cytosol. The Ca^{2+} combines with troponin, inducing a conformational change that moves tropomyosin away from the myosin-binding sites on thin filaments. Exposure of the sites allows myosin crossbridges to bind and initiate the crossbridge cycle in which the myosin heads of thick filaments attach to a thin filament, pull, and release in cyclic reactions powered by ATP hydrolysis (Figure 43.5).

- When action potentials stop, Ca^{2+} is pumped back into the sarcoplasmic reticulum, leading to Ca^{2+} release from troponin, which allows tropomyosin to cover the myosin-binding sites in the thin filaments, thereby stopping the crossbridge cycle (Figure 43.5).

- A single action potential arriving at a neuromuscular junction causes a muscle twitch. Restimulation of a muscle fiber before it has relaxed completely causes a second twitch, which is added to the first, causing a summed, stronger contraction. Rapid arrival of action potentials causes the twitches to sum to a peak level of contraction called tetanus. Normally, muscles contract in a tetanic mode (Figure 43.6).

- Muscle fibers occur in three types. Slow muscle fibers contract relatively slowly, but do not fatigue rapidly. Fast aerobic fibers contract relatively quickly and powerfully, and fatigue more quickly than slow fibers. Fast anaerobic fibers can contract more rapidly and powerfully than fast aerobic fibers, but fatigue more rapidly. The fibers differ in their number of mitochondria and capacity to produce ATP (Table 43.1).

- Skeletal muscles are divided into motor units, consisting of a group of muscle fibers activated by branches of a single motor neuron. The total force produced by a skeletal muscle is determined by the number of motor units that are activated in the muscle (Figure 43.7).

- Invertebrate muscles contain thin and thick filaments arranged in sarcomeres, and contract by the same sliding filament mechanism that operates in vertebrates.

43.2 Skeletal Systems

- A hydrostatic skeleton is a structure consisting of a muscle-surrounded compartment or compartments filled with fluid under pressure. Contraction and relaxation of the muscles change the shape of the animal (Figure 43.8).

- In an exoskeleton, a rigid external covering provides support for the body. The force of muscle contraction is applied against the covering. An exoskeleton can also protect delicate internal tissues (Figure 43.9).

- In an endoskeleton, the body is supported by rigid structures within the body, such as bones. The force of muscle contraction is applied against those structures. Endoskeletons also protect delicate internal tissues. In vertebrates, the endoskeleton is the primary skeletal system (Figure 43.10).

- Bone tissue is distributed between compact bone, with no spaces except the microscopic canals of the osteons, and spongy bone tissue, which has spaces filled by red or yellow marrow (Figure 43.10).

- Calcium and phosphate ions are constantly exchanged between the blood and bone tissues. The turnover keeps the Ca^{2+} concentration balanced at optimal levels in body fluids.

43.3 Vertebrate Movement: The Interactions between Muscles and Bones

- The bones of a skeleton are connected by joints. A synovial joint, the most movable type, consists of a fluid-filled capsule surrounding the ends of the bones forming the joint. A cartilaginous joint, which is less movable, has smooth layers of cartilage between the bones with no surrounding capsule (Figure 43.11).

- The bones moved by skeletal muscles act as levers, with a joint at one end forming the fulcrum of the lever, the load at the opposite end, and the force applied by attachment of a muscle at a point between the ends (Figure 43.12).

- At a joint, an agonist muscle, perhaps assisted by other muscles, causes movement. Most skeletal muscles are arranged in antagonistic pairs, in which the members of a pair pull a bone in opposite directions. When one member of the pair contracts, the other member relaxes and is stretched (Figure 43.13).

- In vertebrates, muscles connect to bones in a variety of patterns, giving different properties to the levers produced. Those properties are specialized for the activities of the animal.

TEST YOUR KNOWLEDGE

Remember/Understand

1. Vertebrate skeletal muscle:
   a. is attached to bone by means of ligaments.
   b. may bend but not extend body parts.
   c. may rotate one body part with respect to another.
   d. is found in the walls of blood vessels and intestines.
   e. consists of cells that contain no nuclei.

2. In muscle fiber in a completely relaxed state:
   a. sarcomeres are regions between two H zones.
   b. discs of M line proteins called the A band separate the thick filaments.
   c. I bands are composed of the same thick filaments seen in the A bands.
   d. Z lines are adjacent to H zones, which attach thick filaments.
   e. dark A bands contain overlapping thick and thin filaments with a central thin H zone composed only of thick filaments.

3. The sliding filament contractile mechanism:
   a. causes thin filaments on each side of a sarcomere to slide over thick filaments toward the center of the A band, bringing the Z lines closer together.
   b. is inhibited by the influx of Ca^{2+} into the muscle fiber cytosol.
   c. lengthens the sarcomere to separate the I regions.
   d. depends on the isolation of actin and myosin until a contraction is completed.
   e. uses myosin crossbridges to stimulate delivery of Ca^{2+} to the muscle fiber.

4. During contraction of skeletal muscle:
   a. ATP stimulates Ca^{2+} to move out of the cytosol, which allows tropomyosin to bind myosin causing contraction of the thin filament.
   b. myosin crossbridges use ATP to relax the molecular spring in the myosin head, which pulls the thick filaments away from the thin actin filaments.
   c. actin binds ATP, allowing troponin in the thick filaments to form the myosin crossbridge.
   d. action potentials cause the release of Ca^{2+} allowing tropomyosin fibers to uncover the actin binding sites needed for the myosin crossbridge.
   e. botulinum toxin could increase the release of acetylcholine at the contracting muscle site.

5. When marathoners are running a race, most likely their:
   a. muscles have low concentrations of myoglobin.
   b. slow muscle fibers will do most of the work for the run.
   c. slow muscle fibers will remain in constant tetanus over the length of the run.
   d. fast muscle fibers will be employed in the middle of the run.
   e. slow muscle fibers are using ATP obtained primarily by anaerobic respiration.

6. Which description is characteristic of a motor unit?
   a. A single motor unit's muscle fibers vary among the slow/fast aerobic and slow/fast anaerobic forms.
   b. When receiving an action potential, a motor unit is controlled by a single efferent axon that causes all its fibers to contract.
   c. When a motor unit contracts, certain sections of the muscle as a whole remain relaxed.
   d. If a motor unit controls walking, it is found in large numbers in the same volume of muscle.
   e. If a motor unit controls finger movement, it contains a large number of muscle fibers that are stimulated over a large area.

7. Which of the following is *not* an example of a hydrostatic skeletal structure?
   a. the tube feet of sea urchins
   b. the body wall of annelids
   c. the mantle of squids
   d. the body wall of cnidarians
   e. the penis of mammals

8. Endoskeletons:
   a. protect internal organs and provide structures against which the force of muscle contraction can work.
   b. differ from exoskeletons in that endoskeletons do not support the external body.
   c. cannot be found in echinoderms.
   d. are composed of cartilaginous structures that form the skull.
   e. compose the arms and legs.

9. Connecting the bones of the vertebrate skeleton are:
   a. nonmovable synovial joints.
   b. ligaments holding together connective tissue of fibrous joints.
   c. cartilaginous joints held together by fluid-filled capsules.
   d. synovial joints lubricated by synovial fluid.
   e. fibrous joints that move around a fulcrum allowing small muscle movements to produce large body movements.

10. The movement of vertebrate muscles is:
    a. controlled primarily by cartilaginous joints with agonist muscles.
    b. antagonistic when it causes movement in the joint.
    c. caused by extensor muscles that flex the joint.
    d. caused by flexor muscles that extend the joint.
    e. exemplified by the biceps and triceps contracting or relaxing simultaneously.

Apply/Analyze

11. **Discuss Concepts** A coach must train young athletes for the 100-meter sprint. They need muscles specialized for speed and strength, rather than for endurance. What kinds of muscle characteristics would the training regimen aim to develop? How would it be altered to train marathoners?

12. **Discuss Concepts** What kind of exercise program might the coach in question 1 recommend to an older person developing osteoporosis? Why?

13. **Discuss Concepts** Based on material in this chapter and in Chapter 42 on endocrine controls, outline some possible causes and physiological effects of calcium deficiency in an active adult.

14. **Apply Evolutionary Thinking** What characteristics of vertebrate muscle suggest that the genes for muscle structure were inherited from invertebrate ancestors?

Evaluate/Create

15. **Design an Experiment** Design an experiment using rats to determine whether endurance training alters the proportion of slow, fast aerobic, and fast anaerobic muscle fibers.

For selected answers, see Appendix A.

INTERPRET THE DATA

Osteogenesis imperfecta (OI) is a genetic disorder caused by a mutation in a gene for collagen. As bones develop, collagen forms a scaffold for deposition of mineralized bone tissue. In children with OI, the scaffold forms improperly, causing them to be born with multiple fractures in their fragile bones and to require many corrective surgeries. **Table 1** and **Table 2** show the results of an experimental test of a new drug. Treated children with OI, all less than 2 years old, were compared to similarly affected children of the same age who were not treated with the drug.

TABLE 1	Effect of Drug on Bone Properties in Children with OI		
	Vertebral Area, in cm^2		
Treated Child	(Before Treatment)	(After Treatment)	Fractures Per Year
1	14.7	16.7	1
2	15.5	16.9	1
3	16.7	16.5	6
4	17.3	11.8	0
5	13.6	14.6	6
6	19.3	15.6	1
7	15.3	15.9	0
8	19.9	13.0	4
9	10.5	13.4	4
Mean	11.4	14.9	2.6

TABLE 2	Bone Properties in Children with OI Who Received No Drug Treatment		
	Vertebral Area, in cm^2		
Control Child	(Initial)	(Final)	Fractures Per Year
1	18.2	13.7	4
2	16.5	12.9	7
3	16.4	11.3	8
4	13.5	17.7	5
5	16.2	16.1	8
6	18.9	17.0	6
Mean	16.6	13.1	6.3

1. An increase in the vertebral area during the 12-month period of the study indicates bone growth. How many of the treated children showed an increase?

2. How many of the untreated children showed an increase?

3. How did the rate of fractures in the two groups compare?

4. Do the results shown support the hypothesis that giving young children who have OI this drug, which slows bone breakdown, can increase bone growth and reduce fractures?

The Circulatory System

44

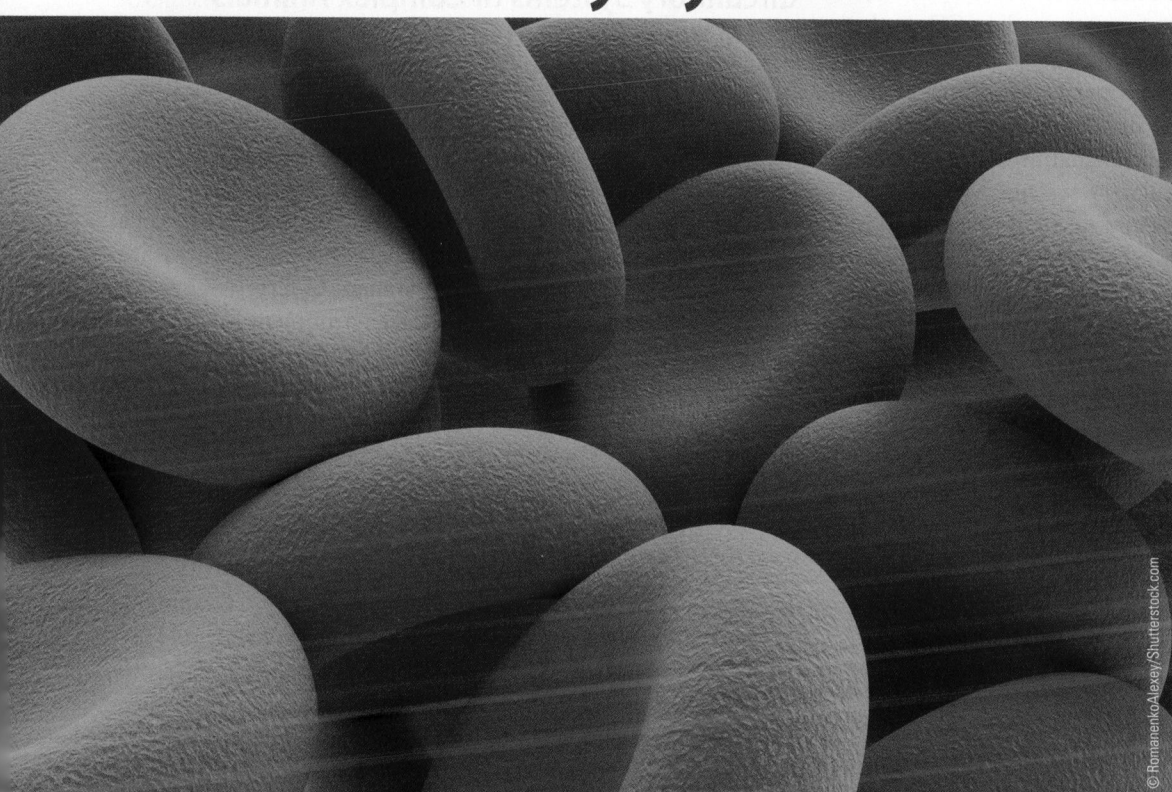

Red blood cells, the oxygen-carrying cells of the blood.

Why it matters . . . Dr. Augustus D. Waller, a physician at St. Mary's Hospital, London, performed experiments in the emerging field of *electrophysiology*. Using himself as his test subject, Waller made an apparatus that detected the electrical currents produced each time the heart beats. He already knew that the heart creates an electrical current as it beats; other scientists had found this out by attaching electrodes directly to the heart of experimental animals. Waller set up two metal pans containing saltwater and connected wires from the pans to a galvanometer. He put his bare left foot in one of the pans, and his right hand in the other one. The indicator of the galvanometer jumped each time his heart beat.

Waller also invented a method for recording the changes in current, which became the first electrocardiogram (ECG). He constructed a galvanometer by placing a column of mercury in a fine glass tube, with a conducting salt solution layered above the mercury. Changes in the current passing through the tube changed the surface tension of the mercury, producing movements that could be detected by reflecting a beam of light from the mercury surface. Waller could record the movements of the reflected light on a photographic plate (**Figure 44.1** shows one of his records).

The beating of the heart is part of the actions of the **circulatory system,** an organ system consisting of a fluid, a heart, and vessels for moving important molecules, and often cells, from one tissue to another. Examples of transported molecules are oxygen (O_2), nutrients, hormones, carbon dioxide (CO_2), and other wastes.

We study the circulatory system in this chapter, with emphasis on the circulation in humans and other mammals. We also discuss the **lymphatic system,** an accessory system of vessels and organs that helps balance the fluid content of the blood and surrounding tissues and participates in the body's defenses against invading disease organisms.

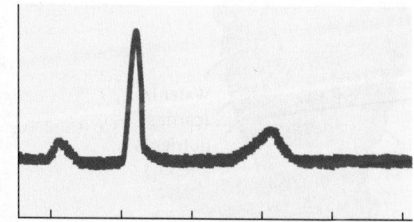

FIGURE 44.1 One of Augustus Waller's early electrocardiograms.
© Cengage Learning 2017

44.1 Animal Circulatory Systems: An Introduction

As you learned in the discussion of homeostasis in Section 38.4, the cells of all organisms must take in nutrients and O_2 from the external environment, and eliminate wastes such as CO_2 to that same environment. For very small or very thin animals, movement of such molecules occurs primarily by diffusion (see Section 5.2). Diffusion is slow over distances of even a few millimeters, however, making this mechanism unsuitable for larger and more complex animals. Circulatory systems overcome the problem by bringing fluids for gas and nutrient exchange close to the cells, where diffusion and other transport mechanisms can be effective. You will learn more about gas exchange in the discussion of the respiratory system in Chapter 46.

Simple Animals Have No Distinct Circulatory System

The least complex animals, including sponges (see Section 31.4), cnidarians (see Section 31.5), and flatworms (see Section 31.6), have no distinct circulatory system. These animals are aquatic or, like parasitic flatworms, live surrounded by the body fluids of a host animal. Their bodies are structured as thin sheets of cells that lie close to the fluids of the surrounding external environment. Substances diffuse between the cells and the external environment through the animal's external surface, or through the surfaces of internal channels and cavities that are open to the environment.

In sponges, water carrying nutrients and O_2 is pumped by surface cells with beating flagella through hundreds of pores in the body wall surrounding a central cavity; it passes through the cavity and leaves through a large exit pore, now carrying CO_2 and wastes **(Figure 44.2A).** Hydras, jellyfish, sea anemones, and other cnidarians have a central **gastrovascular cavity** with a mouth that opens to the outside and extensions that radiate into the tentacles and all other body regions **(Figure 44.2B).** Water enters and leaves through the mouth, serving both digestion and circulation.

Circulatory Systems of Complex Animals Share Basic Elements

In larger and more complex animals, most cells lie in cell layers too deep within the body to exchange substances directly with the external environment by diffusion. Instead, the animals have a set of tissues and organs—a circulatory system—that conducts O_2, CO_2, nutrients, and wastes between body cells and specialized regions of the animal where substances are exchanged with the external environment. In terrestrial vertebrates, for example, oxygen from the environment is absorbed in the lungs and is carried by the blood to all parts of the body; CO_2 released from body cells is carried by the blood to the lungs, where it is released to the environment. Wastes are conducted from body cells to the kidneys, which remove the wastes from the circulation and excrete them into the environment.

The animal circulatory systems carrying out these roles have three basic components:

1. **Fluid:** A specialized fluid medium carries O_2, CO_2, nutrients, and wastes, as well as cells, and plays a major role in homeostasis (see Section 38.4).
2. **Heart:** A muscular heart pumps the fluid through the circulatory system.
3. **Vessels:** Tubular vessels serve as conduits to distribute the fluid pumped by the heart. Vessels conducting blood away from the heart are called **arteries** (singular, *artery*), and vessels conducting blood to the heart are called **veins.**

Animal circulatory systems are either *open* or *closed* **(Figure 44.3).** In an **open circulatory system,** when the heart contracts, arteries leaving the heart release a bloodlike fluid called **hemolymph** directly into body spaces called **sinuses** that surround organs. Thus, there is no distinction between hemolymph and *interstitial fluid,* the fluid immediately surrounding body cells (see Section 38.4). After flowing through the sinuses, the hemolymph reenters the heart when the heart relaxes. In this step, either the hemolymph enters veins and moves into the heart through valves in the heart wall (shown in Figure 44.3A), or it enters the heart directly through pores called *ostia.* In either case, valves prevent hemolymph exiting the heart into the veins or through the pores when the heart contracts. In a **closed circulatory system,** the fluid—blood—is confined to vessels and is distinct from the interstitial fluid. Substances are exchanged between the blood and the interstitial fluid and then between the interstitial fluid and cells.

Most Invertebrates Have Open Circulatory Systems

Arthropods (see Section 31.7) and most mollusks (see Section 31.6) have open circulatory systems with one or more muscular

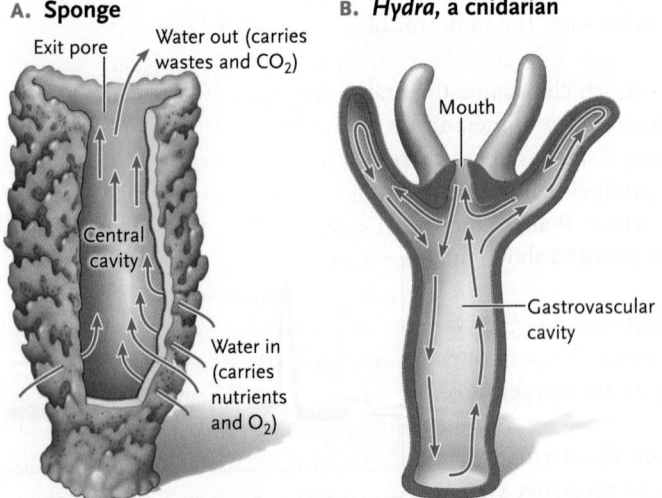

A. Sponge

Exit pore
Water out (carries wastes and CO_2)
Central cavity
Water in (carries nutrients and O_2)

B. *Hydra,* a cnidarian

Mouth
Gastrovascular cavity

FIGURE 44.2 Invertebrates with no circulatory system.
© Cengage Learning 2017

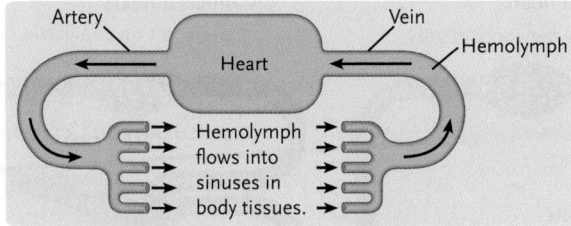

A. Open circulatory system: no distinction between hemolymph and interstitial fluid

Artery
Vein
Heart
Hemolymph
Hemolymph flows into sinuses in body tissues.

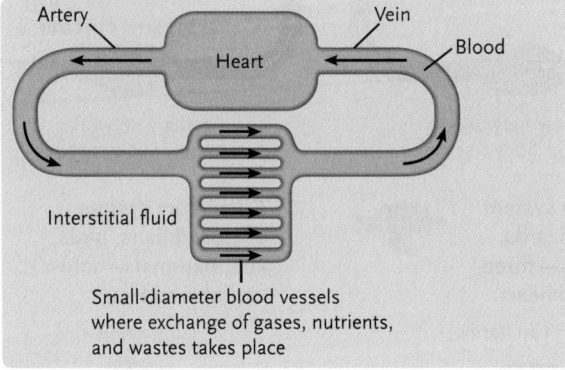

B. Closed circulatory system: blood separated from interstitial fluid

Artery
Vein
Heart
Blood
Interstitial fluid
Small-diameter blood vessels where exchange of gases, nutrients, and wastes takes place

FIGURE 44.3 Open and closed circulatory systems. The open circulatory system **(A)** shows the type with veins carrying the hemolymph back to the heart.
© Cengage Learning 2017

hearts. The open circulatory system of a grasshopper is shown in **Figure 44.4;** in this animal, the hemolymph returns to the heart through pores. In an open system, most of the fluid pressure generated by the heart dissipates when the hemolymph is released into the sinuses. As a result, hemolymph flows relatively slowly. Open systems operate efficiently in these animals because they lead relatively sedentary lives. As a consequence, their tissues do not require O_2 and nutrients at the rate and quantities required by more active species. Among highly mobile and active species with open systems, such as insects and crustaceans, other adaptations compensate for the relatively slow distribution of hemolymph. In insects, for example, O_2 and CO_2 are exchanged efficiently with the environment by

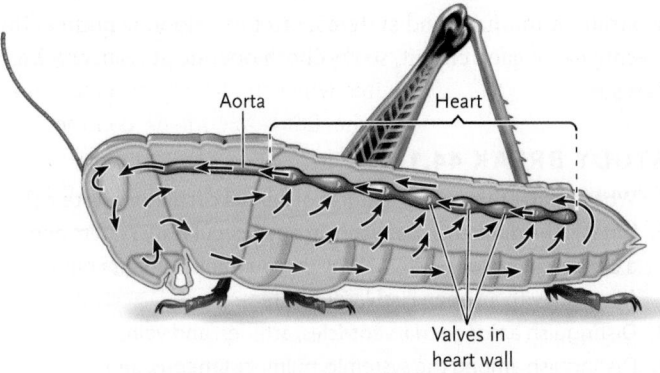

Aorta
Heart
Valves in heart wall

FIGURE 44.4 Open circulatory system of a grasshopper.
© Cengage Learning 2017

specialized air passages called *tracheae* that branch throughout the body rather than by the hemolymph (discussed further in Chapter 46).

Some Invertebrates and All Vertebrates Have Closed Circulatory Systems

Annelids (see Section 31.6), cephalopod mollusks such as squids and octopuses (see Section 31.6), and all vertebrates (see Chapter 32) have closed circulatory systems. In these systems, arteries conduct blood away from the heart at relatively high pressure. From the arteries, the blood enters highly branched networks of microscopic, thin-walled vessels called **capillaries** that are well adapted for diffusion. Nutrients, O_2, and wastes are exchanged between the blood and body tissues as the blood moves through the capillaries. The blood then flows at relatively low pressure from the capillaries to larger vessels, the veins, which carry it back to the heart.

In many animals, but particularly the vertebrates, closed systems allow precise control of the distribution and rate of blood flow to different body regions by means of muscles that contract or relax to adjust the diameter of the blood vessels.

Vertebrate Circulatory Systems Have Evolved from Single to Double Blood Circuits

Several evolutionary trends accompanied the invasion of terrestrial habitats by the different vertebrate groups. Among the most significant is a trend from an effectively single circuit of blood pumped by the heart in sharks and bony fishes to the two separate circuits of birds and mammals. As part of this development, the heart evolved from one series of chambers to a double pump acting in two parallel series. In other words, depending on the vertebrate, the heart may consist of one or two **atria** (singular, *atrium*), the chambers that receive blood returning to the heart, and one or two **ventricles,** the chambers that pump blood from the heart.

THE SINGLE BLOOD CIRCUIT OF FISHES In fishes, the heart consists of two chambers and pumps blood into one circuit **(Figure 44.5A).** The ventricle of the heart pumps blood into arteries leading to capillary networks in the gills, where the blood releases CO_2 and picks up O_2. The oxygenated blood flows through arteries to capillary networks in other body tissues where it delivers O_2 and picks up CO_2. The deoxygenated blood enters veins that carry it back to the atrium of the heart, and the single circuit is repeated. Compared with double blood circuits, single blood circuits provide less O_2. The reason is that blood has to pass through two capillary networks before it is pumped again, and blood pressure drops significantly when blood passes through a capillary network. The drop in blood pressure at the gills has the effect of limiting blood flow to the rest of the body.

The circulatory system of fishes is highly suited to the environments in which these animals live. Their bodies, supported by water and adapted for swimming, do not use as much energy in locomotion as do animals of an equivalent size moving on the land or in the air. Hence, fishes require less O_2 for their activities than terrestrial vertebrates do, and their relatively simple circulatory systems reflect this reduced O_2 requirement.

DOUBLE BLOOD CIRCUITS Vertebrate hearts changed significantly with the evolution of the first air-breathing fishes, such as the lungfish (see Section 32.5). In these fishes, the lung evolved as a respiratory organ in addition to gills. The lung necessitated a separate circuit because it is an additional organ for oxygenating the blood.

In amphibians, the separation into two circuits was accomplished by division of the atrium into two parallel chambers, the left and right atria, to produce a three-chambered heart, and by adaptations that keep oxygenated and deoxygenated blood partially separate as they are pumped into two circuits by the single ventricle **(Figure 44.5B).** The wall between the two atrial chambers is a **septum** (from Latin *saeptum* = partition or wall), in this case the *atrial septum*. In water, amphibians obtain O_2 primarily through the skin; the lungs are used only when the animal is in air. Oxygenated blood from lungs and skin enters veins that lead to the left atrium, whereas deoxygenated blood from the rest of the body enters the right atrium. Simultaneous contraction of the atria pumps the two types of blood into the single ventricle. The two types of blood remain largely (90%) separate because of a smooth pattern of flow and a small tissue flap. Contraction of the ventricle pumps most of the oxygenated blood into the **systemic circuit** that delivers blood to most tissues and cells of the body. Most of the deoxygenated blood enters the **pulmocutaneous circuit** that leads to the lungs and skin, where CO_2 is released and O_2 is picked up. Because the blood flows through separate circuits, the blood leaving the heart flows through only one capillary network in each circuit before returning to the heart. This separation greatly increases the blood pressure and flow in the systemic circuit as compared with that of fishes.

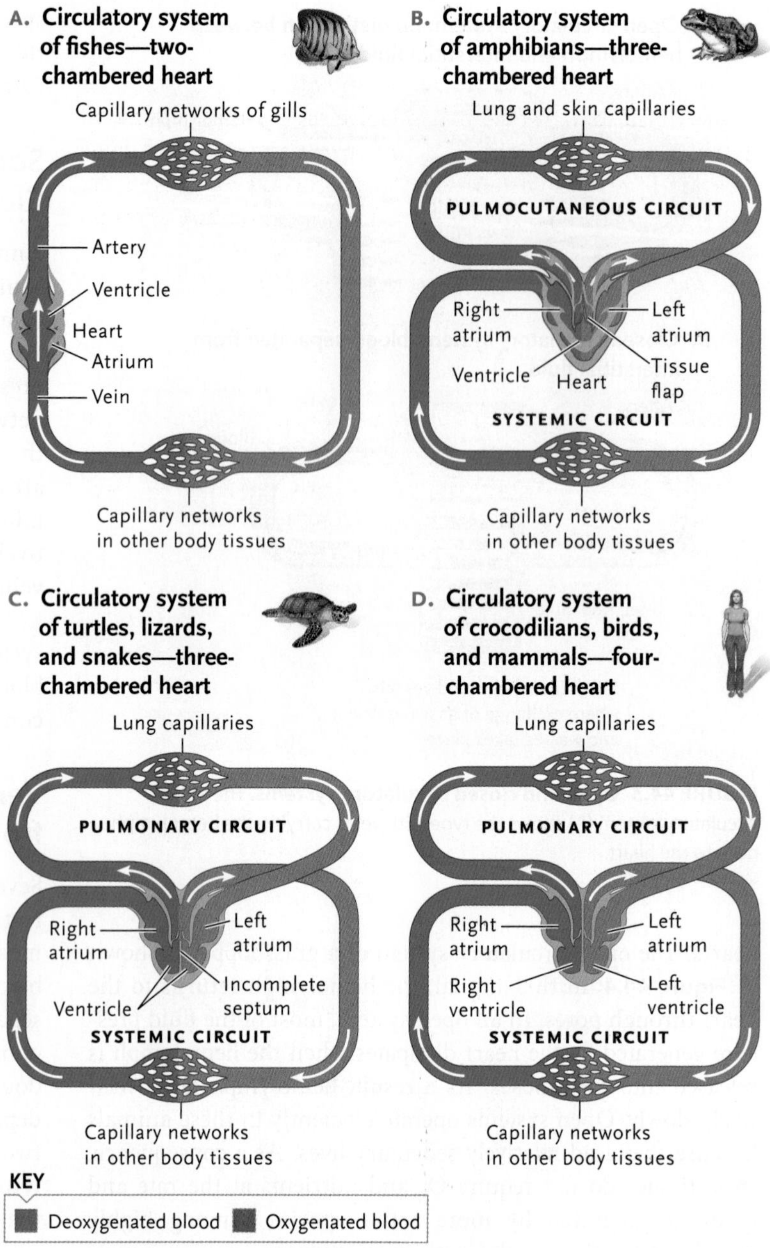

Turtles, lizards, and snakes also have two atria and a single ventricle **(Figure 44.5C).** The ventricle is divided by an incomplete septum (a flap of connective tissue) into right and left halves. The incomplete septum functions to keep the flow of oxygenated and deoxygenated blood almost completely separate in a systemic and a **pulmonary circuit,** but allows some mixing.

Crocodilians (crocodiles and alligators), birds (which share ancestry with crocodilians), and mammals have a septum dividing the heart completely into two, producing a four-chambered heart consisting of two atria and two ventricles **(Figure 44.5D).** In effect, each half of the heart operates as a separate pump, restricting the blood circulation to completely

A. Circulatory system of fishes—two-chambered heart

Capillary networks of gills

Artery
Ventricle
Heart
Atrium
Vein

Capillary networks in other body tissues

B. Circulatory system of amphibians—three-chambered heart

Lung and skin capillaries

PULMOCUTANEOUS CIRCUIT

Right atrium
Left atrium
Ventricle   Heart
Tissue flap

SYSTEMIC CIRCUIT

Capillary networks in other body tissues

C. Circulatory system of turtles, lizards, and snakes—three-chambered heart

Lung capillaries

PULMONARY CIRCUIT

Right atrium
Left atrium
Ventricle
Incomplete septum

SYSTEMIC CIRCUIT

Capillary networks in other body tissues

D. Circulatory system of crocodilians, birds, and mammals—four-chambered heart

Lung capillaries

PULMONARY CIRCUIT

Right atrium
Left atrium
Right ventricle
Left ventricle

SYSTEMIC CIRCUIT

Capillary networks in other body tissues

KEY

■ Deoxygenated blood   ■ Oxygenated blood

FIGURE 44.5 Evolutionary developments in the heart and circulatory system of major vertebrate groups.
© Cengage Learning 2017

separate pulmonary and systemic circuits. Blood is pumped by a ventricle in each circuit, so that both operate at relatively high pressure.

STUDY BREAK 44.1

1. What are the three basic features of animal circulatory systems?
2. What is the difference between an open circulatory system and a closed circulatory system? Why do you think humans could not function with an open circulatory system?
3. Distinguish among atria, ventricles, arteries, and veins.
4. Distinguish among the systemic, pulmocutaneous, and pulmonary circuits.

44.2 Blood and Its Components

In vertebrates, blood is a complex tissue containing a variety of cells suspended in a liquid matrix called the **plasma.** In addition to transporting molecules, blood stabilizes the pH and salt composition of body fluids and serves as a conduit for cells of the immune system. It also helps regulate body temperature by transferring heat between warmer and cooler body regions, and between the body and the external environment (see Chapter 48).

For an average-sized adult human, the total blood volume is 4 to 5 L—more than a gallon—and makes up about 8% of body weight. The plasma, a clear, straw-colored fluid, makes up about 55% of the volume of blood in human males and 58% in human females on average. Suspended in the plasma are three main types of blood cells—*erythrocytes, leukocytes,* and *platelets*—which constitute the hematocrit, the remainder of the blood volume. The typical components of human blood are shown in **Figure 44.6.**

In humans, blood cells develop in red bone marrow primarily in the vertebrae, sternum (breastbone), ribs, and pelvis.

Blood cells originate from cells called *pluripotent* (*plura* = multiple; *potens* = power) *stem cells* that retain the embryonic capacity to divide **(Figure 44.7).** Pluripotent stem cells differentiate into two other types of cells, myeloid stem cells and lymphoid stem cells. Myeloid stem cells give rise to erythrocytes, platelets, and four types of leukocytes: neutrophils, basophils, eosinophils, and monocytes/macrophages. Lymphoid stem cells give rise to two other types of leukocytes that function in the immune system: the B lymphocytes and T lymphocytes (see Chapter 45).

Plasma Is an Aqueous Solution of Proteins, Ions, Nutrient Molecules, and Gases

Plasma is so complex that its complete composition is unknown. Among its known components are water (91–92% of its volume), glucose and other sugars, amino acids, plasma proteins, dissolved gases (mostly O_2, CO_2, and nitrogen), ions, lipids, vitamins, hormones and other signal molecules, and metabolic wastes, including urea and uric acid.

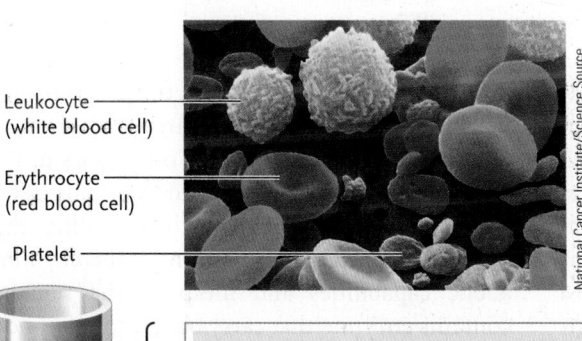

Leukocyte (white blood cell)
Erythrocyte (red blood cell)
Platelet

National Cancer Institute/Science Source

FIGURE 44.6 Typical components of human blood. The colorized scanning electron micrograph shows the three major cellular components. The sketch of the test tube shows what happens when you centrifuge a blood sample. The blood separates into three layers: a thick layer of straw-colored plasma on top, a thin layer containing leukocytes and platelets, and a thick layer of erythrocytes at the bottom. The table shows the relative amounts and functions of the various components of blood.
© Cengage Learning 2017

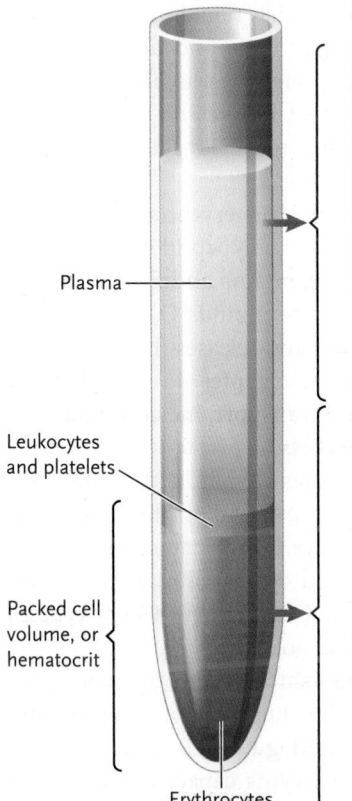

Plasma
Leukocytes and platelets
Packed cell volume, or hematocrit
Erythrocytes

Plasma Portion: 55% (males)–58% (females) of total volume

Component	Percentage of Plasma Volume	Functions
1. Water	91–92	Solvent
2. Plasma proteins (albumin, globulins, fibrinogen, etc.)	7–8	Defense, clotting, lipid transport, roles in extracellular fluid volume, etc.
3. Ions, sugars, lipids, amino acids, hormones, vitamins, dissolved gases	1–2	Roles in extracellular fluid volume, pH, etc.

Cellular Portion (Hematocrit): 45% (males)–42% (females) of total volume

Component	Cells per Microliter	Functions
1. Erythrocytes (red blood cells)	4,800,000–5,400,000	Oxygen, carbon dioxide transport
2. Leukocytes (white blood cells)		
Neutrophils	3,000–6,750	Phagocytosis during inflammation
Lymphocytes	1,000–2,700	Immune response
Monocytes/macrophages	150–720	Phagocytosis in all defense responses
Eosinophils	100–360	Defense against parasitic worms
Basophils	25–90	Secrete substances for inflammatory response and for fat removal from blood
3. Platelets	250,000–300,000	Roles in clotting

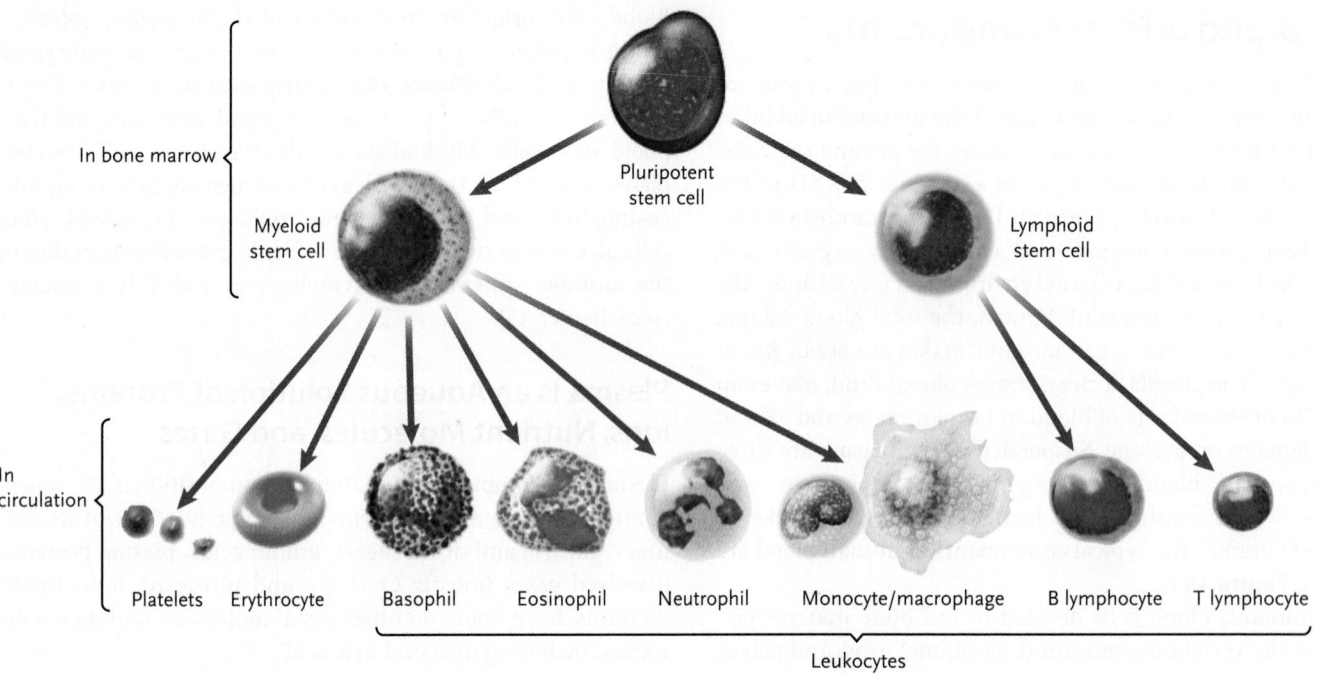

FIGURE 44.7 Major cellular components of mammalian blood and their origins from stem cells.
© Cengage Learning 2017

The plasma proteins fall into three classes, the *albumins,* the *globulins,* and *fibrinogen.* The **albumins,** the most abundant proteins of the plasma, are important for osmotic balance and pH buffering. They also transport a wide variety of substances through the circulatory system, including hormones, therapeutic drugs, and metabolic wastes. The **globulins** transport lipids (including cholesterol) and fat-soluble vitamins. A specialized subgroup of globulins, the *immunoglobulins,* constitute antibodies and other molecules contributing to the immune response. Some globulins are also enzymes. **Fibrinogen** plays a central role in the blood-clotting mechanism. All plasma proteins except immunoglobulins are synthesized in the liver. The immunoglobulins are synthesized by B and T lymphocytes (see Chapter 45).

The ions of the plasma include Na^+, K^+, Ca^{2+}, Cl^-, and HCO_3^- (bicarbonate). The Na^+ and Cl^- ions—the components of common table salt—are the most abundant. Some ions, particularly the bicarbonate ion, help maintain arterial blood at its characteristic pH, which in humans is about pH 7.4 (the bicarbonate ion and its role in pH balance are discussed further in Chapter 46).

Erythrocytes Are the Oxygen Carriers of the Blood

Erythrocytes—the red blood cells—carry O_2 from the lungs to body tissues. Each microliter of human blood contains about 5 million small, flattened, and disclike erythrocytes. Indentations on each flattened surface make them *biconcave*—thinner in the middle than at the edges (see Figure 44.6). They are about

8 μm in diameter and 2 μm thick. Erythrocytes are highly flexible cells, able to squeeze through narrow capillaries.

Erythrocytes derive from precursors in the bone marrow that are generated from myeloid stem cells (see Figure 44.7). As they mature, mammalian erythrocytes lose their nuclei, cytoplasmic organelles, and ribosomes, thereby limiting their metabolic capabilities and lifespan. The remaining cytoplasm contains enzymes that carry out glycolysis, and large quantities of *hemoglobin,* the O_2-carrying protein of the blood. Most of the ATP produced by glycolysis (see Chapter 7) is used to power active transport mechanisms that move ions in and out of erythrocytes.

Hemoglobin, the protein that gives erythrocytes and the blood their red color, consists of four polypeptides, each linked to a nonprotein *heme* group that contains an iron atom in its center (see Figure 3.17). The iron atom binds O_2 molecules as blood circulates through the lungs and releases the O_2 as blood flows through other body tissues. Chapter 46 describes the mechanisms of gas exchange and transport, including the process by which hemoglobin transports O_2.

Some 2 to 3 million erythrocytes are produced in the average human each *second.* The lifespan of an erythrocyte in the circulatory system is about 120 days. At the end of their useful life, erythrocytes are engulfed and destroyed by *macrophages* (*macros* = big; *phagein* = to eat) (see Figure 44.7), a type of leukocyte, in the spleen, liver, and bone marrow.

A negative feedback mechanism keyed to the blood's O_2 carrying capacity stabilizes the number of erythrocytes **(Figure 44.8).** (See Section 38.4 and Figure 38.9 for a discussion of negative feedback.) If the O_2 carrying capacity in the blood

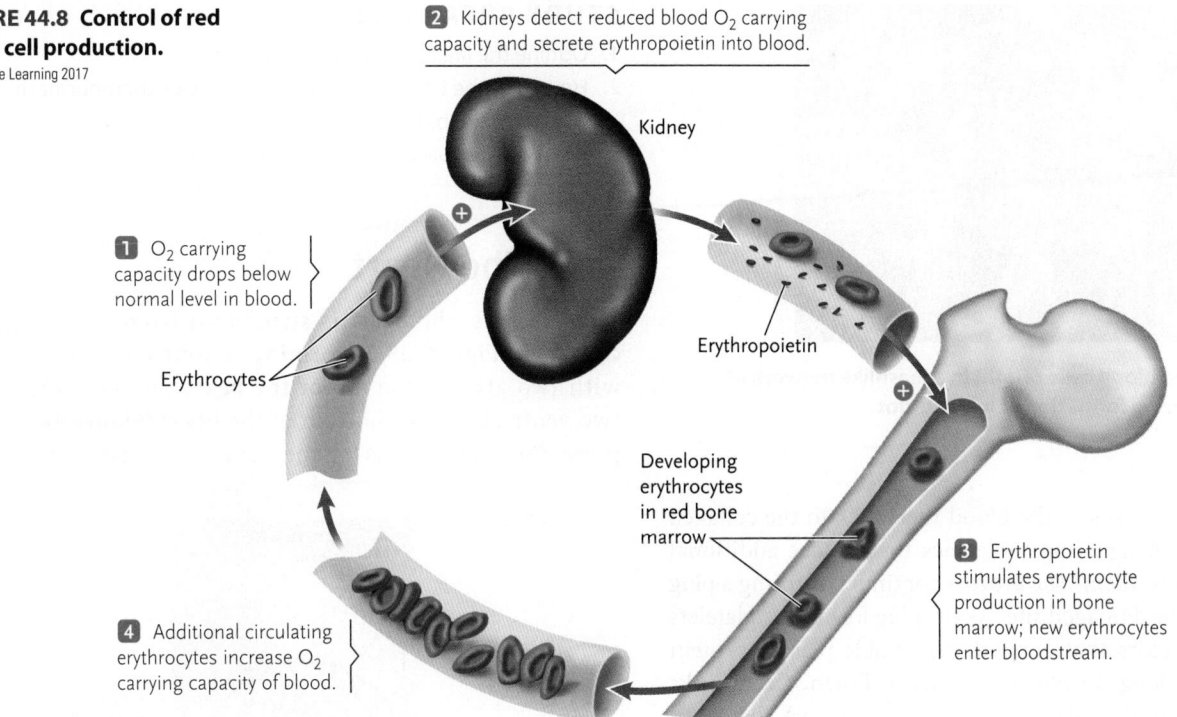

FIGURE 44.8 Control of red blood cell production.
© Cengage Learning 2017

2 Kidneys detect reduced blood O_2 carrying capacity and secrete erythropoietin into blood.

Kidney

1 O_2 carrying capacity drops below normal level in blood.

Erythropoietin

Erythrocytes

Developing erythrocytes in red bone marrow

3 Erythropoietin stimulates erythrocyte production in bone marrow; new erythrocytes enter bloodstream.

4 Additional circulating erythrocytes increase O_2 carrying capacity of blood.

drops below the normal level (step 1), the kidneys detect it and they secrete the hormone **erythropoietin** (EPO) into the blood (step 2). EPO stimulates stem cells in bone marrow to increase erythrocyte production (step 3). EPO is also secreted after blood loss and when mammals move to higher altitudes. As new red blood cells enter the bloodstream, the O_2 carrying capacity of the blood rises (step 4). If the O_2 carrying capacity of the blood rises above normal levels, EPO production falls in the kidneys and red blood cell production drops. The gene encoding human EPO has been cloned, allowing researchers to produce this protein in large quantities. It can then be injected into the body to stimulate erythrocyte production; for example, in patients with anemia (lower-than-normal hemoglobin levels) caused by kidney failure or chemotherapy. It can also supplement or even replace blood transfusions. Some endurance athletes such as triathletes, bicycle racers, marathon runners, and cross-country skiers have used EPO to increase their erythrocyte levels to enhance performance. The use of EPO (a type of blood doping) is deemed illegal by the governing organizations of most endurance sports and, as a result, many athletes have been sanctioned or banned in recent years.

Human blood groups are determined by antigens, the carbohydrate portions of particular glycoproteins on the surfaces of erythrocytes. Section 12.2 described the antigens of the human ABO blood group and how they are important in transfusions.

Disorders of erythrocytes are responsible for a number of human disabilities and diseases. The *anemias,* which result from too few, or malfunctioning, erythrocytes, prevent O_2 from reaching body tissues in sufficient amounts. Shortness of breath, fatigue, and chills are common symptoms of anemia.

Anemia can be produced, for example, by blood loss from a wound, by certain infections, by dietary insufficiencies, and by certain genetic disorders.

Leukocytes Provide the Body's Front Line of Defense against Disease

Leukocytes eliminate dead and dying cells from the body, remove cellular debris, and defend against invading organisms. They are called white cells because they are colorless, in contrast to the strongly pigmented red blood cells. Also unlike red blood cells, leukocytes retain their nuclei, cytoplasmic organelles, and ribosomes as they mature, and hence are fully functional cells.

Some types of leukocytes are capable of continued division in the blood and body tissues. The specific types of leukocytes and their functions in the immune system are discussed in Chapter 45. Some leukocytes are produced from precursor cells in the bone marrow derived from myeloid stem cells, and others are produced from lymphoid stem cells in the bone marrow (see Figure 44.7).

Platelets Induce Blood Clots That Seal Breaks in the Circulatory System

Blood **platelets** are oval or rounded small cell fragments about 2–4 μm in diameter that contain enzymes and other factors that take part in blood clotting. Platelets are shed from the surface of very large cells in the bone marrow that are derived from myeloid stem cells. When blood vessels are damaged, collagen fibers in the extracellular matrix are exposed to the

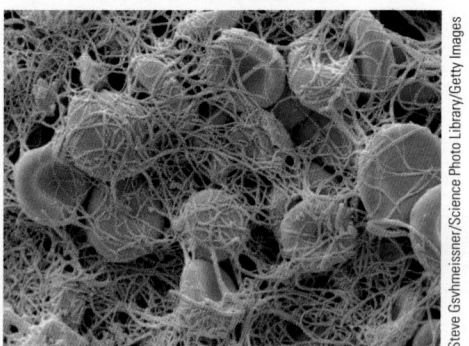

FIGURE 44.9 Red blood cells caught in a meshlike network of fibrin threads during formation of a blood clot.

Steve Gsvhmeissner/Science Photo Library/Getty Images

leaking blood. Platelets in the blood then stick to the collagen fibers and release signaling molecules that induce additional platelets to stick to them. The process continues, forming a plug that helps seal the damaged site. As the plug forms, the platelets release other factors that convert the soluble plasma protein fibrinogen into long, insoluble threads of **fibrin**. Cross-links between the fibrin threads form a meshlike network that traps blood cells and platelets and further seals the damaged area **(Figure 44.9).** The entire mass is a blood clot.

Mutations or diseases that interfere with the enzymes and factors taking part in the clotting mechanism can have serious effects and lead to uncontrolled bleeding. In the most common form of hemophilia, for example, a mutation in a single protein (called *clotting factor VIII*) interferes with the clotting reaction (see Section 13.2). Bleeding is uncontrolled in afflicted individuals; even small cuts and bruises can cause life-threatening blood loss.

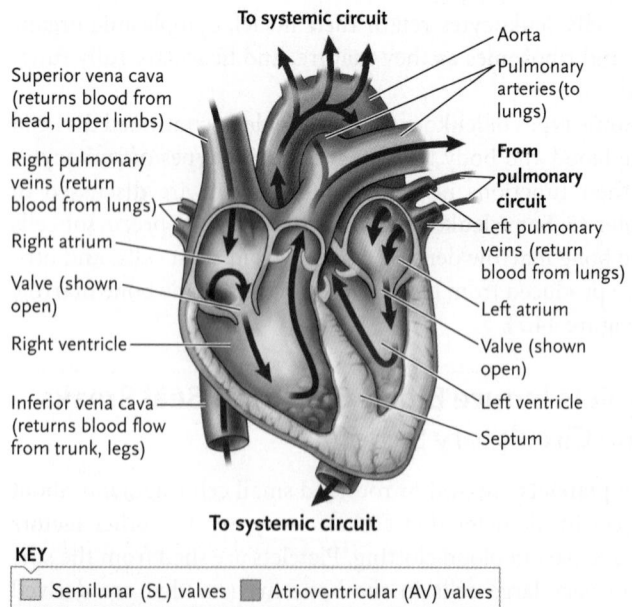

FIGURE 44.10 Cutaway view of the human heart showing its internal organization.

© Cengage Learning 2017

1. Outline the life cycle of an erythrocyte.
2. How does the body compensate for a lower-than-normal level of oxygen in the blood?
3. What are the roles of leukocytes and platelets?

44.3 The Heart

In mammals, the heart is structured from cardiac muscle cells (see Figure 38.5) forming a four-chambered pump, with two atria at the top of the heart pumping blood into two ventricles at the bottom of the heart **(Figure 44.10).** The powerful contractions of the ventricles push the blood at

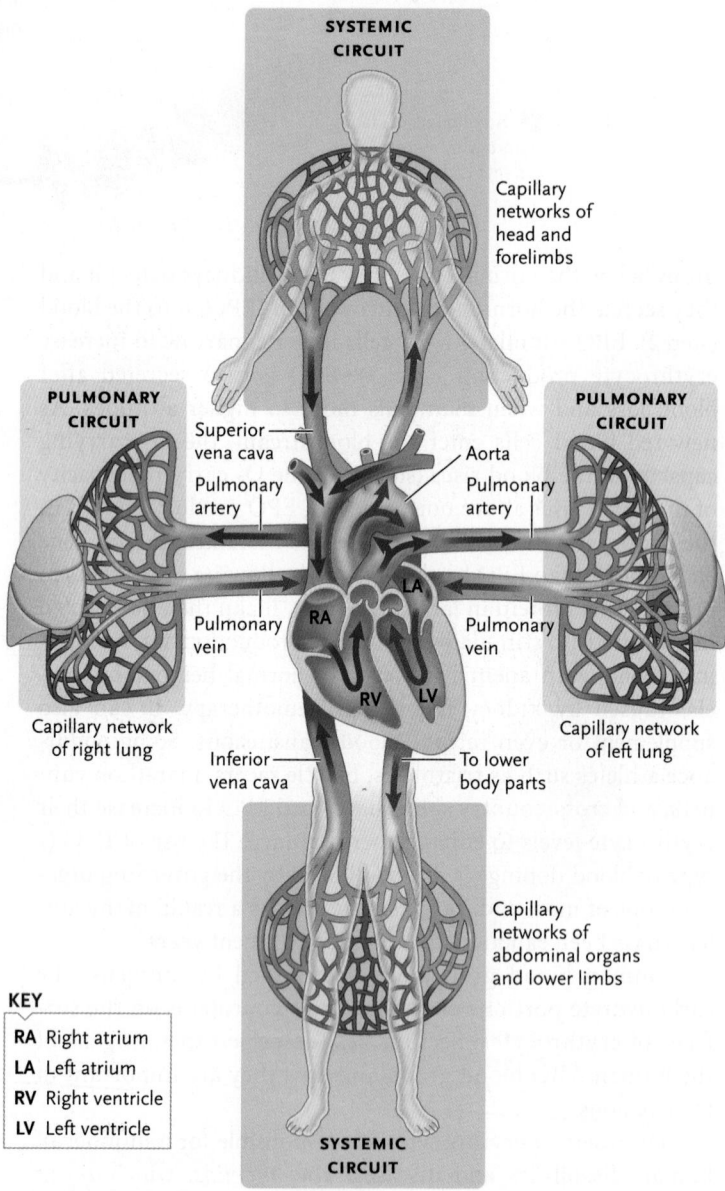

FIGURE 44.11 The pulmonary and systemic circuits of mammals. The right half of the heart pumps blood into the pulmonary circuit, and the left half of the heart pumps blood into the systemic circuit.

© Cengage Learning 2017

relatively high pressure into arteries leaving the heart. This arterial pressure is responsible for the blood circulation. Valves in the heart keep the blood from moving backward. Between each atrium and ventricle is an **atrioventricular valve (AV valve),** and between the ventricle and the arteries leaving the heart (the aorta and pulmonary arteries) are **semilunar valves (SL valves).**

The mammalian heart pumps blood through the completely separate systemic circuit and pulmonary circuits of blood vessels **(Figure 44.11):**

1. The right atrium (toward the right side of the body) receives blood returning to the heart in vessels coming from the entire body, except for the lungs: the *superior vena cava* conveys blood from the head and forelimbs, and the *inferior vena cava* conveys blood from the abdominal organs and the hind limbs. This blood is depleted of O_2 and has a high CO_2 content.

2. The right atrium pumps blood into the right ventricle, which contracts to push the blood into the **pulmonary arteries** leading to the lungs. In the capillaries of the lungs, the blood releases CO_2 and picks up O_2. The oxygenated blood completes this pulmonary circuit by returning in **pulmonary veins** to the heart.

3. Blood returning from the pulmonary circuit enters the left atrium, which pumps the blood into the left ventricle.

4. The left ventricle, the most thick-walled and powerful of the heart's chambers, contracts to send the oxygenated blood coursing into a large artery, the **aorta,** which branches into arteries leading to all body regions except the lungs. In the capillary networks of these body regions, the blood releases

O_2 and picks up CO_2. The O_2-depleted blood collects in veins, which complete the systemic circuit. The blood from the veins enters the right atrium (step 1, above).

The amount of blood pumped by the two halves of the heart is normally balanced so that neither side pumps more than the other. More details of gas exchange in the pulmonary circuit are presented in Chapter 46.

The heart also has its own circulation, called the *coronary circulation*. The aorta gives off two **coronary arteries** that course over the heart. The coronary arteries branch extensively, leading to dense capillary beds that serve the cardiac muscle cells. The blood from the capillary networks collects into veins that empty into the right atrium.

The Heartbeat Is Produced by a Cycle of Contraction and Relaxation of the Atria and Ventricles

Average heart rates vary among vertebrates, depending on body size and the overall level of metabolic activity. A human heart beats 72 times per minute, on average. The resting heart rate of a trained endurance athlete may be only 60% of this rate. The heart of a flying bat may beat 1,200 times a minute, whereas that of an elephant beats only 30 times a minute. The period of ventricular contraction and emptying of the heart is called the **systole** and the period of relaxation and filling of the heart between contractions is called the **diastole.** The systole–diastole sequence of the heart is called the **cardiac cycle (Figure 44.12).** At an average heart rate, the cardiac cycle takes a little over 0.8 sec.

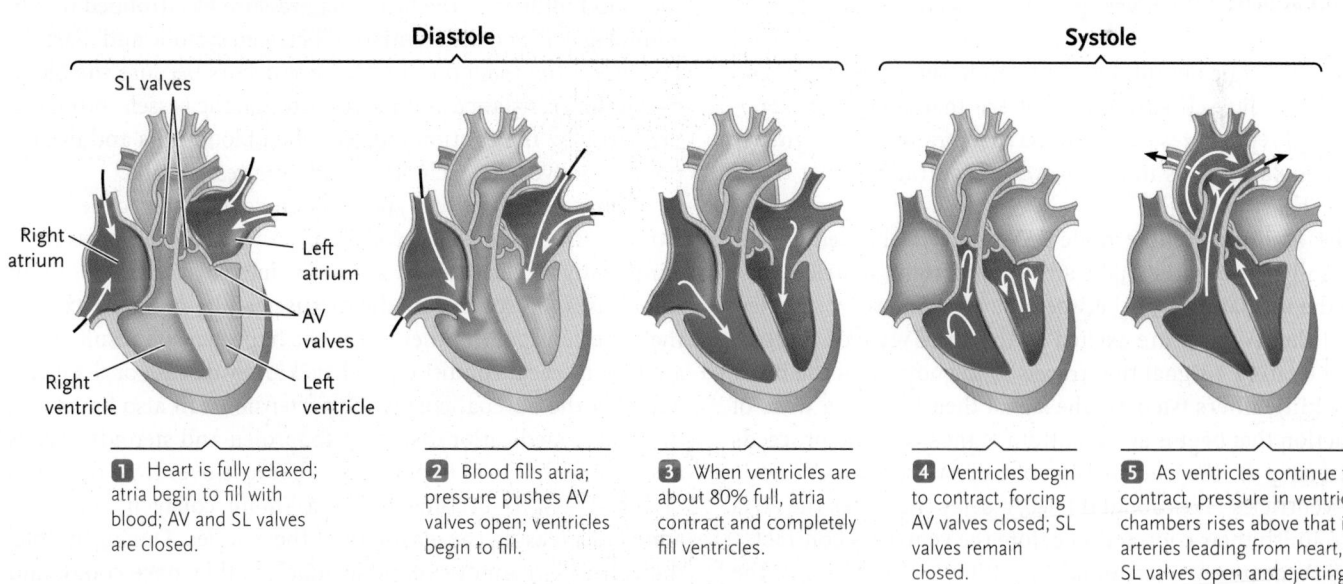

FIGURE 44.12 The cardiac cycle. The figure shows one cardiac cycle beginning at the start of diastole.
© Cengage Learning 2017

Diastole

SL valves

Right atrium

Left atrium

AV valves

Right ventricle

Left ventricle

Systole

1 Heart is fully relaxed; atria begin to fill with blood; AV and SL valves are closed.

2 Blood fills atria; pressure pushes AV valves open; ventricles begin to fill.

3 When ventricles are about 80% full, atria contract and completely fill ventricles.

4 Ventricles begin to contract, forcing AV valves closed; SL valves remain closed.

5 As ventricles continue to contract, pressure in ventricle chambers rises above that in arteries leading from heart, forcing SL valves open and ejecting blood into aorta and pulmonary arteries. About two-thirds of blood in ventricles enters arteries. Ventricles now relax, lowering pressure in ventricle chambers below that in arteries. Blood pressure in ventricles now closes SL valves. The cycle begins again.

The heart valves make a "lub-dub" sound when the heart beats, which you can hear by listening to the heart through a stethoscope. The "lub" sound occurs when the AV valves are pushed shut by the contraction of the ventricles; the "dub" sound is made when the SL valves are forced shut as the ventricles relax. *Heart murmurs* are abnormal sounds produced by turbulence created in the blood when one or more of the valves fails to open or close completely and blood flows backward.

In an adult human at rest, each ventricle pumps roughly 5 L of blood per minute—an amount roughly equivalent to the entire volume of blood in the body. At maximum rate and strength, the human heart pumps about five times the resting amount.

The Cardiac Cycle Is Initiated within the Heart

Contraction of cardiac muscle cells is triggered by action potentials that spread along the muscle cell membranes. Some crustaceans, such as crabs and lobsters, have **neurogenic hearts,** that is, hearts that beat under the control of signals from the nervous system. Other animals, including all insects and all vertebrates, have **myogenic hearts,** that is, hearts that maintain their contraction rhythm autonomously. It is the ability to sustain contractile rhythmicity in the mammalian heart that makes heart transplants possible.

Figure 44.13 illustrates the electrical control of the cardiac cycle. Two nodes, the *sinoatrial node* and *atrioventricular node,* generate the electrical signals. The **sinoatrial node (SA node)** controls the rate and timing of cardiac contraction in mammalian myogenic hearts by coordinating the contractions of individual cardiac muscle cells. The SA node consists of **pacemaker cells,** which are specialized cardiac muscle cells in the upper wall of the right atrium near the entry point for blood from the systemic circuit. Ion channels in these cells open and close in a cyclic, self-sustaining pattern that alternately depolarizes and repolarizes their plasma membranes. The regularly timed depolarizations initiate waves of contraction that travel over the heart (step 1).

The **atrioventricular node (AV node)** is located in the heart wall between the right atrium and right ventricle, just above the insulating layer between the atria and the ventricles. Cells of the AV node are excited by the atrial wave of contraction, generating a signal that travels to the bottom of the heart via **Purkinje fibers** (step 3). The signal then induces a wave of contraction that begins at the bottom of the heart and proceeds upward (step 4). The transmission of a signal from the AV node to the ventricles takes about 0.1 sec, a delay that gives the atria time to finish their contraction before the ventricles contract.

As Augustus Waller found (see *Why it matters . . .*), the electrical signals passing through the heart can be detected by attaching electrodes to different points on the surface of the body. The signals change in a regular pattern corresponding to the electrical signals that trigger the cardiac cycle, producing what is known as an **electrocardiogram** (**ECG;** also **EKG,** from German *elektrokardiogramm*). Many malfunctions of the heart alter the ECG pattern in characteristic ways, providing clues to the location and type of heart disease.

Arterial Blood Pressure Cycles between a High Systolic and a Low Diastolic Pressure

The pressure that a fluid in a confined space exerts is called *hydrostatic pressure.* That is, fluid in a container exerts some pressure on the wall of the container. Blood vessels are essentially tubular containers that are part of a closed system filled with fluid. Hence, the blood in vessels exerts hydrostatic pressure against the walls of the vessels. **Blood pressure** is the measurement of that hydrostatic pressure on the walls of the arteries as the heart pumps blood through the body. Blood pressure is determined by the force and amount of blood pumped by the heart and the size and flexibility of the arteries. In any person, blood pressure changes continually in response to activity, temperature, body position, emotional state, diet, and medications being taken.

For most healthy adults at rest, the systolic pressure measured at the upper arm using a blood pressure cuff is between 90 and 120 mm Hg, and the diastolic pressure is between 60 and 80 mm Hg. The numbers, written in the form 120/80 mm Hg and stated verbally as "120 over 80," refer to the height of a column of mercury in millimeters that would be required to balance the pressure exactly. The systolic and diastolic blood pressures in the pulmonary arteries are typically much lower, about 24/8 mm Hg.

The blood pressure in the systemic and pulmonary circuits is highest in the arteries leaving the heart, and drops as the blood passes from the arteries into the capillaries. By the time the blood returns to the heart, its pressure has dropped to 2 to 5 mm Hg, with no differentiation between systolic and diastolic pressures. The reduction in pressure occurs because the blood encounters resistance as it moves through the vessels, produced primarily by the friction created when blood cells and plasma proteins move over each other and over vessel walls.

Some people have **hypertension,** commonly called *high blood pressure,* a medical condition in which blood pressure is chronically elevated above normal values to at least 140/90. In some cases, no specific medical cause can be found to explain the hypertension. In other cases, the hypertension results from another medical condition, such as kidney disease or disorders affecting the adrenal cortex. Hypertension can also be caused by certain medications, such as ibuprofen and steroids. Age is also a contributor to hypertension because over time the walls of blood vessels become stiffer as more collagen fibers are added, decreasing the elasticity of the arteries. During systole, these arteries cannot expand as much as they once could, and this results in a higher arterial blood pressure.

Hypertension is rarely severe enough to cause symptoms. However, in the long term, the increased pressure in the arteries can cause damage to organs. Hence, hypertension is treated because of the correlation with an increased risk for a number of medical conditions, including myocardial infarction (heart

THE ELECTRICAL CONTROL OF THE CARDIAC CYCLE

The top part of the figure shows how a signal originating at the SA node leads to ventricular contraction. The bottom part of the figure shows the electrical activity for each of the stages as seen in an ECG. The colors in the hearts show the location of the signal at each step and correspond to the colors in the ECG.

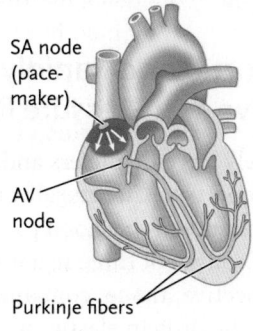
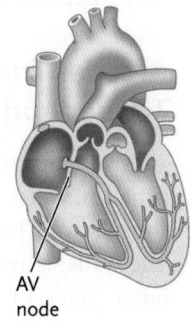
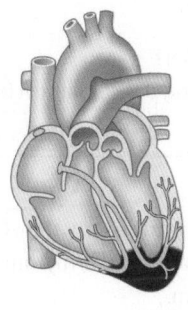
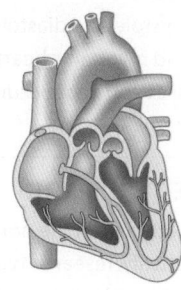

1 Depolarization signal. Pacemaker cells of SA node generate rhythmic depolarization signals, initiating waves of contractions that travel over the heart. The waves cause cells of the atria to contract in unison and fill the ventricles with blood.

2 Contraction signal. A layer of connective tissue that separates and insulates the atria from the ventricles prevents a contraction signal from the SA node from spreading directly to the ventricles.

3 Signal from AV node. AV node cells excited by the atrial wave of contraction produce a signal that travels to the bottom of the heart via Purkinje fibers, which travel through the insulating layer to the bottom of the heart, where they branch into the walls of the ventricles.

4 Wave of contraction. The signal carried by the Purkinje fibers induces a wave of contraction that begins at the bottom of the heart and proceeds upward, squeezing the blood from the ventricles into the aorta and pulmonary arteries.

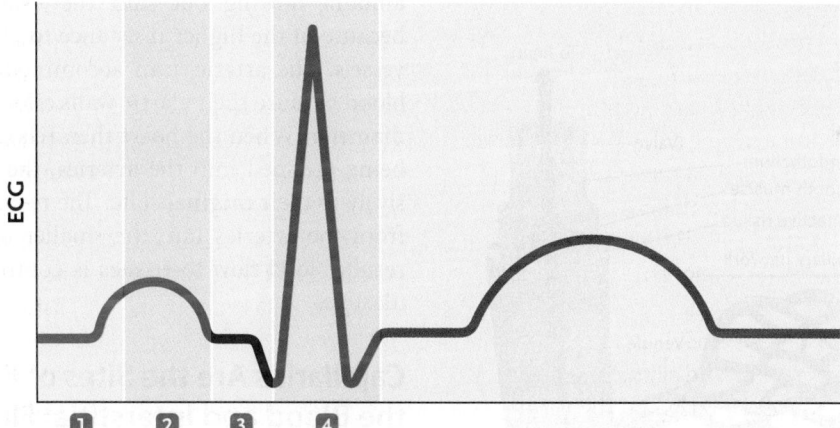

SUMMARY Waves of rhythmic depolarization signals generated by the SA node cause waves of contraction that travel over the heart, starting with the atria. AV node cells excited by atrial contraction generate a signal that travels to the bottom of the heart and then induces a wave of contraction that moves upward, causing the ventricles to contract.

think like a scientist What effect do you think a dysfunction of the SA node would have on the cardiac cycle?

attack), cardiovascular accident (stroke), chronic kidney disease, and retinal damage. Treatments to reduce hypertension typically involve lifestyle changes, such as weight loss and regular exercise. In the case of moderate to severe hypertension, drugs are also prescribed.

What happens to blood pressure during exercise? During static exercise involving a sustained contraction of a muscle group or groups, such as weight lifting, both systolic and diastolic pressures increase. In elite weight lifters, for instance, blood pressure during lifts can reach 300/150 mm Hg. However, during dynamic exercise involving intermittent and rhythmical muscle contractions, such as running, bicycling, and swimming, only the systolic pressure increases.

STUDY BREAK 44.3

1. What is the role of each of the four chambers of the mammalian heart in blood circulation?
2. Distinguish between systole and diastole.
3. How do neurogenic and myogenic hearts differ?
4. Describe the electrical events that occur during the cardiac cycle in a mammalian heart.

THINK OUTSIDE THE BOOK

Arrhythmia is any variation from the normal heartbeat rhythm. On your own or collaboratively, research two examples of arrhythmia and outline the physiological reasons for them.

44.4 Blood Vessels of the Circulatory System

Both the systemic and pulmonary circuits consist of a continuum of different blood vessel types that begin and end at the heart **(Figure 44.14).** From the heart, large arteries carry blood and branch into progressively smaller arteries that deliver the blood to the various parts of the body. When a small artery reaches the organ it supplies, it branches into yet smaller vessels, the **arterioles.** Within the organ, arterioles branch into capillaries, the smallest vessels of the circulatory system. The capillaries form a network in the organ that is used to exchange substances between the blood and the surrounding cells. Capillaries rejoin to form small **venules,** which merge into the small veins that leave the organ. The small veins progressively join to form larger veins that eventually become the large veins that enter the heart.

Arteries Transport Blood Rapidly to the Tissues and Serve as a Pressure Reservoir

Arteries have relatively large diameters and, therefore, provide little resistance to blood flow. Structurally, they are adapted to the relatively high pressure of the blood passing through them. The walls of arteries consist of three major tissue layers: (1) an outer layer of connective tissue containing collagen fibers mixed with fibers of the protein elastin, which gives the vessel recoil ability; (2) a relatively thick middle layer of vascular smooth muscle cells also mixed with elastin fibers; and (3) a one-cell-thick inner layer of flattened cells that forms an **endothelium,** a specialized type of simple squamous epithelial tissue that lines the entire circulatory system (see Figure 44.14).

In addition to being the conduits for blood traveling to the tissues, arteries also act as a pressure reservoir to generate the force for blood movement when the heart is relaxing. When contraction of the ventricles pumps blood into the arteries, the amount of blood flowing into the arteries is greater than the amount flowing out into the smaller vessels downstream because of the higher resistance to blood flow in those smaller vessels. The arteries can accommodate the excess volume of blood because their elastic walls allow the arteries to expand in diameter. When the heart then relaxes and blood is no longer being pumped into the arteries, the arterial walls recoil passively to their original state. The recoil pushes the excess blood from the arteries into the smaller downstream vessels. As a result, blood flow to tissues is continuous during systole and diastole.

Capillaries Are the Sites of Exchange between the Blood and Interstitial Fluid

Capillaries, which thread through nearly every tissue in the body, are arranged in networks that bring them within 10 μm of most body cells. They form an estimated 2,600 km^2 of total surface area for the exchange of gases, nutrients, and wastes with the interstitial fluid. Capillary walls consist of a single layer of endothelial cells.

CONTROL OF BLOOD FLOW THROUGH CAPILLARIES Blood flow through the capillary networks is controlled by contraction of smooth muscle in the arterioles **(Figure 44.15).** The capillaries themselves do not have smooth muscle but, in many cases, a small ring of smooth muscle called a *precapillary sphincter* is present at the junction between an arteriole and a

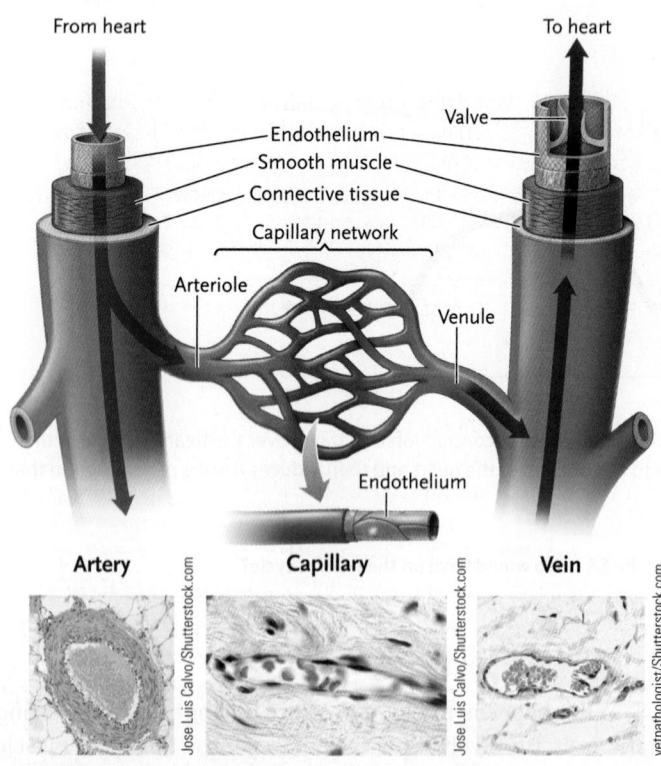

FIGURE 44.14 The structure of arteries, capillaries, and veins and their relationship in blood circuits.
© Cengage Learning 2017

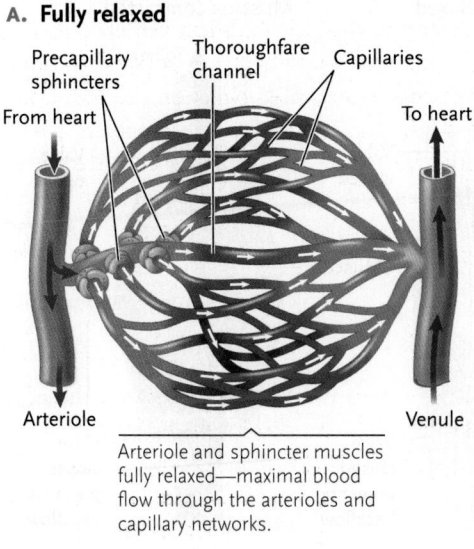

A. **Fully relaxed**

Precapillary sphincters

Thoroughfare channel

Capillaries

From heart

To heart

Arteriole

Venule

Arteriole and sphincter muscles fully relaxed—maximal blood flow through the arterioles and capillary networks.

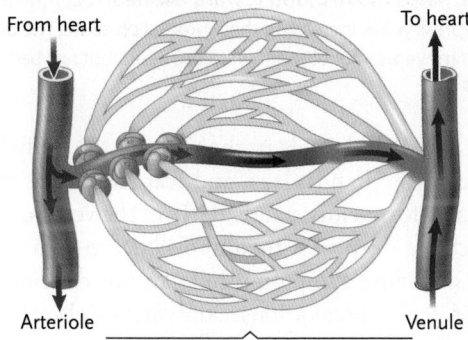

B. **Fully contracted**

From heart

To heart

Arteriole

Venule

Arteriole and sphincter muscles fully contracted—blood flow through arterioles and capillary networks is limited to a minimal amount through the thoroughfare channel.

FIGURE 44.15 **Control of blood flow through arterioles and capillary networks when arteriole and sphincter muscles are (A) fully relaxed and (B) fully contracted.**
© Cengage Learning 2017

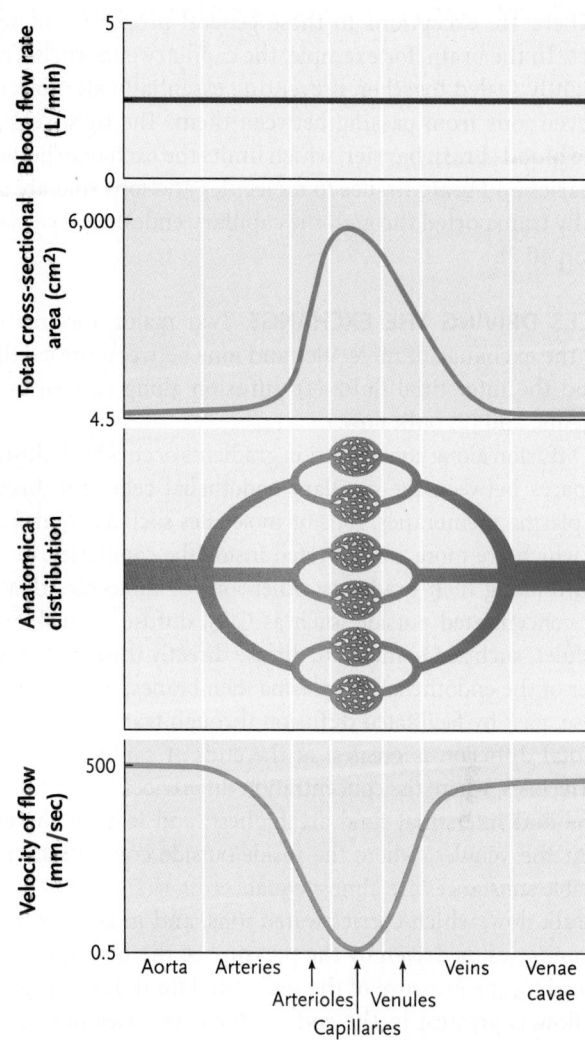

FIGURE 44.16 **Blood flow rate and velocity of flow in relation to total cross-sectional area of the blood vessels.** The blood flow rate is identical throughout the circulatory system and is equal to the cardiac output. The velocity of flow in the different types of blood vessels is inversely related to the total cross-sectional area of all the vessels of a particular type: for example, the velocity is highest in the aorta, which has the smallest cross-sectional area, and lowest in the capillaries, which collectively have the largest total cross-sectional area.
© Cengage Learning 2017

capillary. Variation in the contraction of arteriole and sphincter smooth muscles adjusts the rate of flow through the capillary networks between the two flow limits. For example, during exercise, flow of blood through the capillary networks is increased several-fold over the resting state by relaxation of the precapillary sphincters.

CONTROL OF BLOOD VOLUME TO CAPILLARIES BY ARTERIOLES The volume of blood flowing through an organ is adjusted by regulating the internal diameter of the arterioles of the organ. Although their total surface area is astoundingly large, the diameter of individual capillaries is so small that red blood cells must squeeze through most of them in single file. As a result, each capillary presents a high resistance to blood flow, so blood slows considerably as it moves through the capillaries **(Figure 44.16)**. The slow movement of blood through the capillaries maximizes the time for exchange of substances between blood and tissues. As they leave the tissues, the capillaries

rejoin to form venules and then veins. Veins have a reduced total cross-sectional area compared with capillaries, so the rate of flow increases as blood returns to the heart.

EXCHANGE OF SUBSTANCES ACROSS CAPILLARY WALLS In most body tissues, narrow spaces between the capillary endothelial cells allow water, ions, and small molecules such as glucose to pass freely between the blood and interstitial fluid. Erythrocytes, platelets, and most plasma proteins are too large to pass between the cells and are retained inside the capillaries, except for molecules that are transported through the endothelial cells by specific carriers. Leukocytes, however, are able to squeeze actively between the cells and pass from the blood to the interstitial fluid.

There are exceptions to these general properties in some tissues. In the brain, for example, the capillary endothelial cells are tightly sealed together, preventing essentially all molecules and even ions from passing between them. The tight seals set up the **blood–brain barrier,** which limits the exchange between capillaries and brain tissues to molecules and ions that are specifically transported through the capillary endothelial cells (see Section 40.3).

FORCES DRIVING THE EXCHANGE Two major mechanisms drive the exchange of molecules and ions between the capillaries and the interstitial fluid: (1) diffusion along concentration gradients; and (2) bulk flow.

Diffusion along concentration gradients occurs both through the spaces between the capillary endothelial cells and through their plasma membranes. Ions or molecules such as O_2 and glucose, which are more concentrated inside the capillaries, diffuse outward along their gradients; other ions or molecules that are more concentrated outside, such as CO_2, diffuse inward. Some molecules, such as O_2 and CO_2, diffuse directly through the lipid bilayer of the endothelial cell plasma membranes; others, such as glucose, pass by facilitated diffusion through transport proteins. The total diffusion is greatest at the ends of capillaries nearest the arterioles, where the concentration differences between blood plasma and interstitial fluid are highest, and least at the ends nearest the venules, where the inside/outside concentrations of diffusible substances are almost equal.

Bulk flow, which carries water, ions, and molecules out of the capillaries, is driven by the pressure of the blood, which is higher than the pressure of the interstitial fluid. Like diffusion, bulk flow is greatest in the ends of the capillaries nearest the arterioles, where the pressure difference is highest, and drops off steadily as the blood moves through the capillaries and the pressure difference becomes smaller.

Venules and Veins Serve as Blood Reservoirs in Addition to Conduits to the Heart

The walls of venules and veins are thinner than those of arteries and contain little elastin. Many veins have flaps of connective tissue that extend inward from their walls. These flaps form one-way valves that keep blood flowing toward the heart (see Figure 44.14).

Rather than stretching and contracting elastically, like arteries, the relatively thin walls of venules and veins can expand and contract over a relatively wide range, allowing them to act as blood reservoirs as well as conduits. At times, the venules and veins may contain from 60–80% of the total blood volume of the body. The stored volume is adjusted by skeletal muscle contraction and by the valves, in response to metabolic conditions and signals carried by hormones and neurotransmitters **(Figure 44.17).**

When you sit without moving for long periods of time, as you might during an airline flight, the lack of skeletal muscular activity greatly reduces the return of venous blood to the heart.

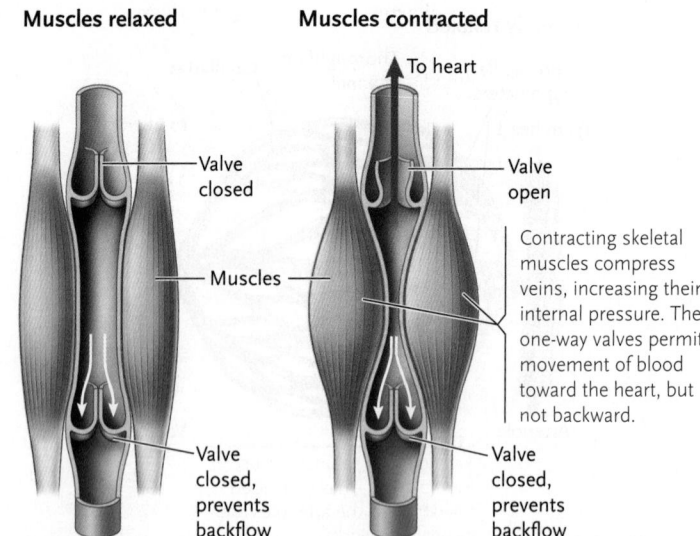

Muscles relaxed   **Muscles contracted**

FIGURE 44.17 How skeletal muscle contraction, and the valves inside veins, helps move blood toward the heart. Contracting skeletal muscles compress veins, increasing their internal pressure. The one-way valves permit movement of blood toward the heart, but not backward.
© Cengage Learning 2017

As a result, the blood pools in the veins of the body below the heart, making the hands, legs, and feet swell. The motionless blood can also form clots, particularly in the veins of the legs, a condition called **deep vein thrombosis.** Deep vein thrombosis often does not cause symptoms, but can cause serious medical problems if a clot breaks loose and moves elsewhere in the body, such as to the lungs. Raising the arms and getting up at intervals to exercise or contracting and relaxing the leg muscles as you sit can relieve this condition.

Disorders of the Circulatory System Are Major Sources of Human Disease

The layer of endothelial cells lining the arteries and veins is normally smooth and does not impede blood flow. However, several conditions, including bacterial and viral infections, chronic hypertension, smoking, and a diet high in fats, can damage the endothelial cells, exposing the underlying smooth muscle tissue, which begins a cycle of injury and repair leading to lesions. (*Focus on Research: Applied Research* in Chapter 3 discusses the relationship between fats and cholesterol and coronary artery disease, and the *Unanswered Question* in Chapter 3 discusses how the enzymes that synthesize fatty acids and cholesterol are regulated in cells.) Thickened deposits of material called *atherosclerotic plaques* may form at the damaged sites **(Figure 44.18).** The plaques, which consist of cholesterol-rich fatty substances, smooth muscle cells, and collagen deposits, reduce the diameter of the blood vessel and impede blood flow. Worse, the damaged endothelial lining may stimulate platelets to adhere and trigger the formation of blood clots. The clots further reduce the vessel diameter and flow and may break loose, along with segments of plaque material, to block finer vessels in other regions of the body.

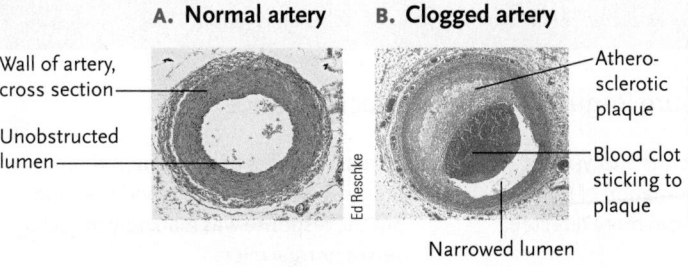

A. Normal artery

Wall of artery, cross section

Unobstructed lumen

Ed Reschke

B. Clogged artery

Athero-sclerotic plaque

Blood clot sticking to plaque

Narrowed lumen

Biophoto Associates/Science Source

FIGURE 44.18 Atherosclerosis. (A) A normal coronary artery. **(B)** A coronary artery that is partially clogged by an atherosclerotic plaque.

Atherosclerosis has its most serious effects in the smaller arteries of the body, particularly in the fine coronary arteries that serve the heart muscle. Here, the plaques and clots reduce or block the flow of blood to the heart muscle cells. Serious blockage can cause a heart attack—the death of cardiac muscle cells deprived of blood flow. The blockage of arteries in the brain by plaque material or blood clots released from atherosclerotic arteries is also a common cause of **stroke (cerebrovascular accident)**—a loss of critical brain functions because of the death of nerve cells in the brain. Heart disease is the leading cause of death in North America; stroke is the fifth leading cause of death in North America.

The risk of heart attacks and stroke can be reduced by avoiding the conditions that damage the blood vessel endothelium. A diet low in saturated fats and cholesterol, avoidance of cigarette smoke, and a program of exercise can reduce epithelial damage and plaque deposition. Medication and exercise, or exercise alone, can also reduce the effects of hypertension. There are good indications that these preventive programs can also reduce the size of existing atherosclerotic plaques.

STUDY BREAK 44.4

1. How is blood flow through capillary networks controlled?
2. Explain how, in contrast to most body tissues, the brain does not allow exchange of molecules and ions with blood.
3. Describe the two major mechanisms that drive the exchange of molecules and ions between the capillaries and the interstitial fluid.

44.5 Maintaining Blood Flow and Pressure

Arterial blood pressure is the principal force moving blood to the tissues. The three main mechanisms for regulating blood pressure are controlling *cardiac output* (the pressure and amount of blood pumped by the left and right ventricles), the degree of constriction of the blood vessels (primarily the arterioles), and the total blood volume. The sympathetic division of the autonomic nervous system and the endocrine system interact to coordinate these mechanisms.

Cardiac Output Is Controlled by Regulating the Rate and Strength of the Heartbeat

Regulation of the rate and strength of the heartbeat starts at stretch receptors called *baroreceptors* (a type of mechanoreceptor; see Section 41.2), located in the walls of blood vessels. By detecting the amount of stretch of the vessel walls, baroreceptors provide information about blood pressure. The baroreceptors in the cardiac muscle, the aorta, and the carotid arteries (which supply blood to the brain) are the most crucial. Signals sent by the baroreceptors go to the medulla within the brain stem. In response, the brain stem sends signals via the autonomic nervous system that adjust the rate and force of the heartbeat. The heart beats more slowly and contracts less forcefully when arterial pressure is above normal levels, and it beats faster and contracts more forcefully when arterial pressure is below normal levels.

The O_2 content of the blood, detected by chemoreceptors in the aorta and carotid arteries, also influences cardiac output. If the O_2 concentration falls below normal levels, the brain stem integrates this information with the signals sent from baroreceptors and issues signals that increase the rate and force of the heartbeat. Too much O_2 in the blood has the opposite effect, reducing the cardiac output.

Hormones Regulate Both Cardiac Output and Arteriole Diameter

Hormones secreted by several glands contribute to the regulation of blood pressure and flow. As part of the stress response, the adrenal medulla reinforces the action of the sympathetic nervous system by secreting epinephrine and norepinephrine into the bloodstream (see Section 42.4). Epinephrine in particular raises the blood pressure by increasing the strength and rate of the heartbeat and stimulating vasoconstriction of arterioles in some parts of the body, including the skin, gut, and kidneys. At the same time, by inducing the vasodilation of arterioles that deliver blood to the heart, skeletal muscles, and lungs, epinephrine increases the blood flow to these structures. *Molecular Insights* shows how gene manipulation experiments illuminated the role of a receptor for these hormones in regulating blood pressure.

The adrenal cortex and the posterior pituitary release hormones that regulate blood pressure. Those hormones and their effects are described in Chapter 48.

Local Controls Regulate Arteriole Diameter

Several automated mechanisms operate locally to increase the flow of blood to body regions engaged in increased metabolic activity, such as the muscles of your legs during an extended uphill bike ride. Low O_2 and high CO_2 concentrations, produced by the increased oxidation of glucose and other fuels, induce vasodilation of the arterioles serving muscles. Vasodilation, which increases the flow of blood and the O_2 supply, occurs when a signaling molecule is released from arteriole endothelial cells in body regions engaged in increased metabolic activity. In

Identifying the Role of a Hormone Receptor in Blood Pressure Regulation Using Knockout Mice

When the body experiences stress, the sympathetic division of the autonomic nervous system stimulates the adrenal medulla to release the hormones epinephrine and norepinephrine. These hormones increase the strength and rate of the heartbeat and change arteriole diameter, causing a temporary increase in blood pressure. The hormones exert their effects by binding to specific membrane-embedded receptors on target cells, activating them and triggering a cellular response via a signal transduction pathway. These G-protein–coupled receptors (see Section 9.3) are known as *adrenergic receptors* because of the adrenal origin of the hormones that bind to them.

Different types of adrenergic receptors are responsible for different responses to epinephrine and norepinephrine. For blood pressure regulation, the α_1 adrenergic receptor is a key receptor. Binding of norepinephrine, and to a lesser extent, epinephrine, to α_1 receptors on arteriolar smooth muscle causes vasoconstriction of arterioles, thereby contributing to an increase in blood pressure.

Researchers have identified three subtypes of α_1 adrenergic receptors: α_{1A}, α_{1B}, and α_{1D}. Some experimental results using drugs suggested that the α_{1D}-adrenergic receptor plays an important role in the control of blood pressure. Because the drugs used could not affect just one subtype of receptor selectively, confirmation of that conclusion was needed from more directed studies.

Research Question

What is the role of the α_{1D}-adrenergic receptor in the regulation of blood pressure by vasoconstriction?

Experiments

Gozoh Tsujimoto and colleagues at the Tokyo University of Pharmacy and Life Sciences, and the National Children's Medical Research Center, Tokyo, Japan, answered the question by making and studying mice with a knockout (deletion) of each of the two copies of the α_{1D} gene encoding the α_{1D} receptor (the technique for making a gene knockout is described in Section 18.2 and Figure 18.8). They compared the cardiovascular functions of the $\alpha_{1D}^-/\alpha_{1D}^-$ knockout mice with normal mice.

Results

The cardiovascular functions of the $\alpha_{1D}^-/\alpha_{1D}^-$ knockout mice were:

1. Modestly hypotensive (slightly lower-than-normal blood pressure) but a normal heart rate.
2. Normal levels of circulating norepinephrine and epinephrine.
3. Increasing doses of norepinephrine progressively increased blood pressure, but the response was markedly reduced versus normal mice.
4. Contraction of blood vessels caused by norepinephrine was reduced considerably. This was shown in experiments in which the researchers measured the effects of norepinephrine on contraction of segments of the aorta (see also Figure 44.19). Far less contraction was seen than in normal mice. This result showed that the α_{1D} receptor is directly involved in vascular smooth muscle contraction.

Conclusion

The researchers concluded that the α_{1D} adrenergic receptor participates directly in sympathetic nervous system-driven regulation of blood pressure by vasoconstriction.

think like a scientist

What was the key element of the experimental design that enabled the researchers to draw the conclusions they did about the role of the α_{1D}-adrenergic receptor in cardiovascular function?

Source: A. Tanoue et al. 2002. The α_{1D}-adrenergic receptor directly regulates arterial blood pressure via vasoconstriction. *Journal of Clinical Investigation* 109:765–775.

© Cengage Learning 2017

1980 Robert Furchgott of the State University of New York, Health Science Center at Brooklyn, New York, reported the first evidence for a vasodilatory signaling molecule produced by arterial endothelial cells **(Figure 44.19)**. Subsequently, Louis Ignarro of the UCLA School of Medicine and Ferid Murad of the University of Texas Medical School at Houston determined that Furchgott's signaling molecule was the gas nitric oxide (NO). The three researchers received the Nobel Prize in 1998 for their "discoveries concerning nitric oxide as a signaling molecule in the cardiovascular system." NO is quickly broken down enzymatically after its release, ensuring that its effects are local.

STUDY BREAK 44.5

1. Why is it important to regulate arterial blood pressure?
2. What are the three main mechanisms for regulating blood pressure?
3. How does epinephrine affect blood pressure?

44.6 The Lymphatic System

Under normal conditions, a little more fluid from the blood plasma in the capillaries enters the tissues than is reabsorbed from the interstitial fluid into the plasma. The **lymphatic system** is an extensive network of vessels that collects the excess interstitial fluid and returns it to the venous blood **(Figure 44.20)**. The interstitial fluid picked up by the lymphatic system is called **lymph.** This system also collects fats that have been absorbed from the small intestine and delivers them to the blood circulation (see Chapter 47). The lymphatic system is a key component of the immune system (see Chapter 45).

Vessels of the Lymphatic System Extend throughout Most of the Body

Vessels of the lymphatic system collect the lymph and transport it to *lymph ducts* that empty into veins of the circulatory system. The *lymph capillaries*, the smallest vessels of the

 FIGURE 44.19 | **Experimental Research**

Demonstration of a Vasodilatory Signaling Molecule

Question: What causes arterial vasodilation?

Experiment: In live mammals, the neurotransmitter acetylcholine (see Chapter 39) is a potent vasodilator. Furchgott conducted *in vitro* experiments with strips of rabbit aorta to determine how acetylcholine-caused vasodilation occurred. He attached the aorta strips to an apparatus that recorded the tension of the strips. Contraction increases tension whereas relaxation (which is equivalent to vasodilation in the intact blood vessel) decreases tension. To the strips he first added norepinephrine to cause them to contract, and then added acetylcholine. He compared two key experimental setups: (1) aorta strips from which the endothelial cells had been removed by rubbing; and (2) gently handled aorta strips with intact endothelium.

Results: Furchgott's results are shown in the **Figure.** Relaxation occurred on acetylcholine treatment only in the strips with intact endothelium. He obtained the same result with a number of other rabbit arteries and with a variety of arteries from different laboratory animals. Furchgott hypothesized that the endothelial cells produced the vasodilatory signaling molecule, which he named *endothelium-derived relaxing factor (EDRF)*.

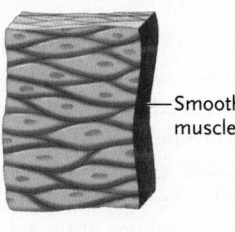 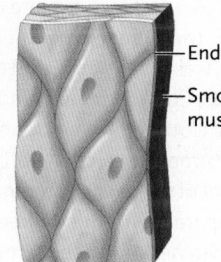

Norepinephrine-treated aorta smooth muscle strip (no endothelium)

Norepinephrine-treated complete aorta strip (endothelium present)

Endothelium
Smooth muscle
Smooth muscle

Acetylcholine: No relaxation | Relaxation

Conclusion: Vasodilation of arteries occurs as a result of release of a signaling molecule from the endothelial cells of the arteries. Subsequently, Furchgott's EDRF signaling molecule was determined to be nitric oxide (NO) (see Section 39.3). NO released from endothelial cells is also the signal molecule responsible for arteriole vasodilation.

 **think like a scientist** — Observe Hypothesize Predict **Experiment** Interpret — **How might you demonstrate experimentally that NO is EDRF?**

Source: R. F. Furchgott and J. V. Zawadzki. 1980. The obligatory role of endothelial cells in the relaxation of arterial smooth muscle by acetylcholine. *Nature* 288:373–376.

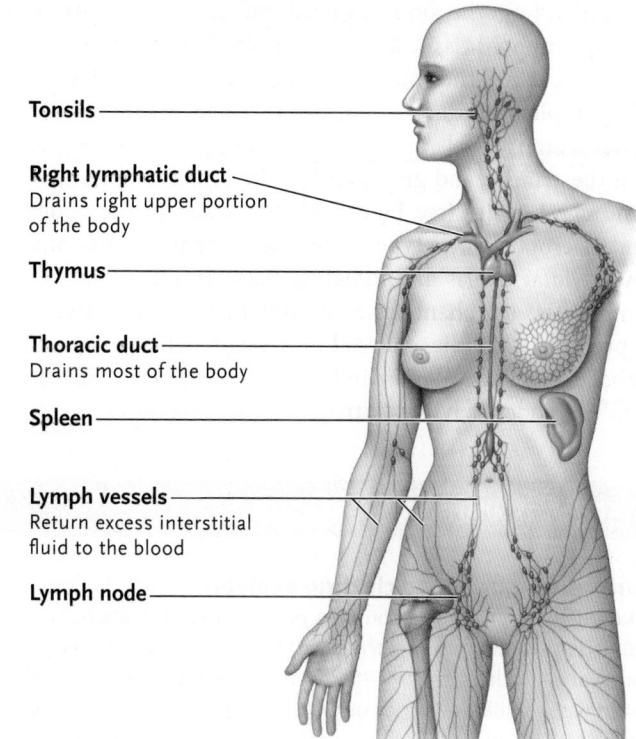

Tonsils

Right lymphatic duct
Drains right upper portion of the body

Thymus

Thoracic duct
Drains most of the body

Spleen

Lymph vessels
Return excess interstitial fluid to the blood

Lymph node

FIGURE 44.20 The human lymphatic system. Patches of lymphoid tissue in the small intestine and in the appendix also are part of the lymphatic system.

consist of a single layer of endothelial cells surrounded by a thin network of collagen fibers. Interstitial fluid enters the lymph capillaries—becoming lymph—at sites in their walls where the endothelial cells overlap, forming a flap that is forced open by the higher pressure of the interstitial fluid. The openings are wide enough to admit all components of the interstitial fluid, including infecting bacteria, damaged cells, cellular debris, and lymphocytes.

The lymph capillaries merge into *lymph vessels* containing one-way valves that prevent the lymph from flowing backward. The lymph vessels lead to the thoracic duct and the right lymphatic duct (see Figure 44.20), which empties the lymph into a vein beneath the clavicles (collarbones), adding it to the plasma in the vein.

Movements of the skeletal muscles adjacent to the lymph vessels and breathing movements help move the lymph through the vessels, just as they help move the blood through veins. Over a day's time, the lymphatic system returns about 3 to 4 L of fluid to the bloodstream.

Lymphoid Tissues and Organs Act as Filters and Participate in the Immune Response

The tissues and organs of the lymphatic system include the *lymph nodes,* the *spleen,* the *thymus,* and the *tonsils.* They play primary roles in filtering viruses, bacteria, damaged cells, and cellular debris from the lymph and bloodstream,

lymphatic system, are distributed throughout the body, intermixed intimately with the capillaries of the circulatory system. Although they are several times larger in diameter than the blood capillaries, the walls of lymph capillaries also

and in defending the body against infection and cancer. Patches of lymphoid tissue are also scattered in other regions of the body, such as the small intestine and the appendix.

The **lymph nodes** are small, bean-shaped organs spaced along the lymph vessels and clustered along the sides of the neck, in the armpits and groin, and in the center of the abdomen and chest cavity (see Figure 44.19). Spaces in the nodes contain macrophages, a type of leukocyte that engulfs and destroys cellular debris and infecting bacteria and viruses in the lymph. The lymph nodes also contain other leukocytes, which produce antibodies that aid in the destruction of invading pathogens (discussed more in Chapter 45). Cancer cells that lodge in the nodes may be destroyed or may remain to grow and divide, forming new tumors within the nodes. Therefore, during surgery to reduce the risk of cancer spread, lymph nodes near a tumor typically are inspected and may be removed.

The lymph nodes may become enlarged and painful if large numbers of bacteria or viruses carried by the lymph become trapped inside them. A physician usually checks for swollen nodes, particularly in the neck, armpits, and groin, as indicators of an infection in the region of the body served by the nodes.

STUDY BREAK 44.6
What is lymph, and how does it enter the lymph capillaries?

Unanswered Questions

How did vertebrate blood clotting evolve?

Vertebrate blood clotting is a complex process involving about two dozen different proteins found in the blood plasma. The system has the properties of a biochemical amplifier in that the exposure of a tiny amount of tissue initiates a series of proteolytic events, one protease activating another successively, the climax being a large amount of localized thrombin that transforms fibrinogen into a fibrin clot.

Many years ago when I was a graduate student working in a laboratory devoted to blood proteins, I asked myself the question, How could blood clotting ever have evolved? The process seemed much too complicated to have been concocted in one fell swoop, I reasoned. Rather, it must have begun in a simpler fashion and gradually become more complicated. Other people had been asking similar questions about complex organs—the evolution of the eye, for example, had been considered by many scientists, including Charles Darwin—but facts about the evolution of individual proteins were just beginning to emerge.

In particular, Vernon Ingram, a scientist working at MIT, had just determined the amino acid sequences of the alpha and beta chains of hemoglobin and found that these two proteins were about 45% identical. They must be the products of a gene duplication, he reported. Given his observation, it seemed to me that the proteases involved in blood clotting also ought to be the products of gene duplication. At the time, none of their amino acid sequences was known, but several had unique properties that distinguished them from other proteases, and it stood to reason that they were related. Some nonproteolytic proteins were also involved, and they must have been added to the process independently.

So where to start? I decided to compare the blood clotting process in a wide range of animals. As it happens, most animals, invertebrate and vertebrate alike, have a kind of blood (see Section 46.2), and in most cases it can be coagulated by various stressful events. But the vertebrate process looked unique and must have evolved independently of the system that occurs in lobsters (Crustacea), for example. Among the vertebrates, the most primitive (early diverging) creature I could get my hands on was the jawless fish called the lamprey (see Section 32.4, where its place in the vertebrate lineage is discussed).

Because I was hoping to find a simple, predecessor scheme, I was somewhat disappointed to find that lampreys have a rather sophisticated coagulation process that involved many of the proteins observed in mammals. Certainly, a small amount of tissue factor provoked a thrombin generation that converted fibrinogen to fibrin, just as occurs in humans, at least in a general way. At the time there was no way to determine whether lampreys use the equivalent of *all* the clotting factors on the way to generating thrombin. Some of the most important proteins in the human system occur in minute amounts, and there was no possibility of ever isolating them from the lamprey, short of collecting several barrels of lamprey blood.

Many years later, after the sequences for many clotting factors had been reported for humans, my colleague Da-Fei Feng and I were able to align the sequences with a computer and make a phylogenetic tree. It fit the notion of a series of gene duplications very well. Moreover, it implied that most of the duplications had occurred a long time ago, at the very dawn of the vertebrates.

In 2003, the complete genomic sequence of a modern bony fish, the pufferfish, was determined, and one could scan through it and see what genes it has. Computer searching revealed that all but a few of the more peripheral clotting proteins found in humans were present; the central theme of thrombin clotting fibrinogen was clearly the same. However, the amount of sequence difference between human and pufferfish proteins compared with the degree of difference observed between the duplicated genes suggested some of the gene duplications had occurred not long before the appearance of bony fish. Why not look at the lamprey genome, which diverged 50 million to 100 million years before the appearance of bony fish, to see if the preduplication genes were there? Unhappily, the lamprey genome had not yet been totally sequenced.

Still, preliminary searches of the available data suggested that the lamprey lacked at least two of the mainline clotting factors (factors VIII and IX), each of which in higher vertebrates is the obvious result of gene duplication. Significantly, these two proteins, factors VIII and IX, interact with each other during the clotting reaction. The conclusion, which has borne the test of time, was that some time between the divergence of jawless fish like the lamprey and jawed ones like the pufferfish, the genes for both proteins duplicated, paving the way for an added interaction in the series of events that gives rise to the fibrin clot.

At about the same time as the 2003 pufferfish report, the complete genome of a urochordate called the sea squirt (see Section 32.2) was also published. It was well known that the bloodlike fluid of sea squirts does not clot, and a preliminary computer search suggested that none of the clotting factors found in vertebrates was present. But a later reexamination of the genome sequences that focused more closely on connections between introns and exons found that the gene for a protein very much like vertebrate fibrinogen was actually present. But what is it doing in sea squirts, whose blood does not clot? It is a genuine mystery waiting to be solved.

REVIEW KEY CONCEPTS

For access to MindTap and additional study materials visit www.cengagebrain.com.

44.1 Animal Circulatory Systems: An Introduction

- Only the simplest invertebrates—the sponges, cnidarians, and flatworms—have no circulatory systems (Figure 44.2).

- Animals with circulatory systems have a muscular heart that pumps a specialized fluid, such as blood, from one body region to another through tubular vessels. The blood carries O_2 and nutrients to body tissues and carries away CO_2 and wastes.

- Most invertebrates have an open circulatory system, in which the heart pumps hemolymph into vessels that empty into body spaces called sinuses before returning to the heart. Some invertebrates and all vertebrates have a closed system, in which the blood is confined in blood vessels throughout the body and does not mix directly with the interstitial fluid (Figure 44.3).

- In invertebrates, open circulatory systems occur in arthropods and most mollusks, whereas closed circulatory systems occur in annelids and in mollusks such as squids and octopuses. In vertebrates, the circulatory system has evolved from a heart with a single series of chambers, pumping blood through a single circuit, to a four-chambered heart that pumps blood through separate pulmonary and systemic circuits (Figures 44.4 and 44.5).

44.2 Blood and Its Components

- Mammalian blood is a fluid connective tissue consisting of erythrocytes, leukocytes, and platelets, suspended in a fluid matrix, the plasma (Figure 44.6).

- Plasma contains water, ions, dissolved gases, glucose, amino acids, lipids, vitamins, hormones, and plasma proteins. The plasma proteins include albumins, globulins, and fibrinogen.

- Erythrocytes contain hemoglobin, which transports O_2 between the lungs and all body regions (Figure 44.7). A negative feedback mechanism keyed to the oxygen carrying capacity of the blood stabilizes the number of erythrocytes in the circulation (Figure 44.8).

- Leukocytes defend the body against infecting pathogens.

- Platelets are functional cell fragments that trigger clotting reactions at sites of damage to the circulatory system.

44.3 The Heart

- The mammalian heart is a four-chambered pump. Two atria at the top of the heart pump the blood into two ventricles at the bottom of the heart, which pump blood into two separate pulmonary and systemic circuits of blood vessels (Figures 44.10 and 44.11).

- In both circuits, the blood leaves the heart in large arteries, which branch into smaller arteries, the arterioles. The arterioles deliver the blood to capillary networks, where substances are exchanged between the blood and the interstitial fluid. Blood is collected from the capillaries in small veins, the venules, which join into larger veins that return the blood to the heart (Figure 44.11).

- Contraction of the ventricles pushes blood into the arteries at a peak (systolic) pressure. Between contractions, the blood pressure in the arteries falls to a minimum (diastolic) pressure. The systole–diastole sequence is the cardiac cycle (Figure 44.12).

- Contraction of the atria and ventricles is initiated by signals from the SA node (pacemaker) of the heart (Figure 44.13).

44.4 Blood Vessels of the Circulatory System

- Blood is carried from the heart to body tissues in arteries; small branches of arteries, the arterioles, deliver blood to the capillaries, where substances are exchanged with the interstitial fluid. The blood is collected from the capillaries in venules and then returned to the heart in veins (Figure 44.14).

- The walls of arteries consist of an inner endothelial layer, a middle layer of smooth muscle, and an outer layer of elastic fibers. The smallest arteries, the arterioles, constrict and dilate to regulate blood flow and pressure into the capillaries.

- Capillary walls consist of a single layer of endothelial cells. Blood flow through capillaries is controlled by variation in contraction of the smooth muscles of arterioles and precapillary sphincters (Figure 44.15).

- In the capillary networks, the rate of blood flow is considerably slower than that in arteries and veins. This maximizes the time for exchange of substances between blood and tissues. Diffusion along concentration gradients and bulk flow drive the exchange of substances (Figure 44.16).

- Venules and veins have thinner walls than arteries, allowing the vessels to expand and contract over a wide range. As a result, they act as blood reservoirs, as well as conduits.

- The return of blood to the heart is aided by pressure exerted on the veins when surrounding skeletal muscles contract and by respiratory movements. One-way valves in the veins prevent the blood from flowing backward (Figure 44.17).

44.5 Maintaining Blood Flow and Pressure

- Blood pressure and flow are regulated by controlling cardiac output, the degree of blood vessel constriction (primarily arterioles), and the total blood volume. The autonomic nervous system and the endocrine system coordinate these mechanisms.

- Regulation of cardiac output starts with baroreceptors, which detect blood pressure changes and send signals to the medulla. In response, the brain stem sends signals via the autonomic nervous system that alter the rate and force of the heartbeat.

- Hormones secreted by several glands contribute to the regulation of blood pressure and flow.

- Local controls respond primarily to O_2 and CO_2 concentrations in tissues. Low O_2 and high CO_2 concentration causes dilation of arteriole walls, increasing the arteriole diameter and blood flow. High O_2 and low CO_2 concentrations have the opposite effects. NO released by arterial endothelial cells acts locally to increase arteriole diameter and blood flow (Figure 44.19).

44.6 The Lymphatic System

- The lymphatic system, a key component of the immune system, is an extensive network of vessels that collect excess interstitial fluid—which becomes lymph—and returns it to the venous blood (Figure 44.20).

- The tissues and organs of the lymphatic system include the lymph nodes, the spleen, the thymus, and the tonsils. They remove viruses, bacteria, damaged cells, and cellular debris from the lymph and bloodstream, and defend the body against infection and cancer.

TEST YOUR KNOWLEDGE

Remember/Understand

1. Compared with vertebrates, most invertebrates:
   a. lead more mobile lives.
   b. require a higher level of oxygen.
   c. have more complex layers of cells.
   d. have blood and interstitial fluid mixing directly in body spaces.
   e. require faster delivery and greater quantities of nutrients.

2. Which circulatory system description best matches the group(s) of animals?
   a. Cephalopod mollusks such as squids and octopuses have open circulatory systems with ventricles that pump blood away from the heart.
   b. Fishes have a single-chambered heart with an atrium that pumps blood through gills for oxygen exchange.
   c. Amphibians have the most oxygenated blood in the pulmocutaneous circuit and the most deoxygenated blood in the systemic circuit.
   d. Amphibians and reptiles use a two-chambered heart to separate oxygenated and deoxygenated blood.
   e. Birds and mammals pump blood to separate pulmonary and systemic systems from two separate ventricles in a four-chambered heart.

3. A healthy student from the coastal city of Boston enrolls at a college in Boulder, Colorado, a mile above sea level. An analysis of her blood in her first months at college would show:
   a. decreased macrophage activity.
   b. increased secretion of erythropoietin by the kidneys.
   c. increased signaling ability of platelets.
   d. anemia caused by malfunctioning erythrocytes.
   e. increased mitosis of leukocytes.

4. A characteristic of blood circulation through or to the heart is that:
   a. the superior vena cava conveys blood to the head.
   b. the inferior vena cava conveys blood to the right atrium.
   c. the pulmonary arteries convey blood from the lungs to the left atrium.
   d. the pulmonary veins convey blood into the left ventricle.
   e. the aorta branches into two coronary arteries that convey blood from heart muscle.

5. The heartbeat includes:
   a. the systole when the heart relaxes and fills.
   b. the diastole when the heart contracts and empties.
   c. pressure that causes the AV valves to open, filling the ventricles.
   d. rising pressure in the ventricles to open the AV valves and close the SL valves.
   e. the "lub" sound when the SL valves open and the "dub" sound when the AV valves close.

6. Keeping the mammalian cardiac cycle balanced is/are:
   a. an AV node between the right atria atrium and right ventricle, which induces a wave of contraction.
   b. pacemaker cells, which compose the AV node and signal the SA node.
   c. an insulating layer that isolates the SA node from the right atrium.
   d. ion channels in pacemaker cells, which close to depolarize their plasma membranes.
   e. neurogenic stimuli from the nervous system.

7. Hydrostatic pressure is best described as:
   a. the uncoordinated contractions that occur during heart attacks.
   b. a premature ventricular contraction that signifies a skipped beat.
   c. a high point of pressure called diastolic blood pressure.
   d. an arterial vasodilation caused by carbon dioxide.
   e. the pressure of blood on the walls of arteries.

8. Characteristics of veins and venules are:
   a. thick walls.
   b. large muscle mass in the walls.
   c. a large quantity of elastin in the walls.
   d. low blood volume compared with arteries.
   e. one-way valves to prevent backflow of blood.

9. When capillaries exchange substances:
   a. red blood cells move through the capillary lumen in double file.
   b. blood flow resistance is lower than it is in arteries and veins.
   c. water, ions, glucose, and erythrocytes pass freely between blood and tissues.
   d. diffusion along a concentration gradient and bulk flow are operating.
   e. diffusion is greatest closest to the venules.

10. How do lymphatic vessels and capillaries differ from other vessels and capillaries in the circulatory system?
    a. Lymph capillaries are larger in diameter than blood capillaries.
    b. Lymph capillary walls contain several layers of endothelial cells, which allows for efficient filtering of lymph.
    c. Lymph vessels are only distributed in specific regions of the body.
    d. Movement of lymph does not require skeletal muscle movement.
    e. The lymphatic system moves up to 8 L of fluid through the bloodstream daily.

Apply/Analyze

11. **Discuss Concepts** *Aplastic anemia* develops when certain drugs or radiation destroy red bone marrow, including the stem cells that give rise to erythrocytes, leukocytes, and platelets. Predict some symptoms a person with aplastic anemia would be likely to develop. Include at least one symptom related to each type of blood cell.

12. **Discuss Concepts** Carbon monoxide (CO), a component of cigarette smoke, binds irreversibly to hemoglobin. What might be the impact of this phenomenon on a smoker's health?

13. **Discuss Concepts** In some people, the pressure of the blood pooling in the legs leads to a condition called *varicose veins*, in which the veins stand out like swollen, purple knots. Explain why this might happen, and why veins closer to the leg surface are more susceptible to the condition than those in deeper leg tissues.

Evaluate/Create

14. **Design an Experiment** Mice in which the apolipoprotein E gene has been knocked out (deleted) by genetic engineering methods have high levels of plasma cholesterol and readily develop atherosclerosis, particularly on diets high in cholesterol. The immunosuppressant drug rapamycin is being touted also to be a drug that can affect atherosclerosis. Design an experiment to determine whether and at what dose rapamycin is effective in reducing atherosclerosis caused by dietary cholesterol.

15. **Apply Evolutionary Thinking** What is the evolutionary advantage of closure of the septum between the two ventricles to create a double circulatory system?

For selected answers, see Appendix A.

INTERPRET THE DATA

Nifedipine is an antihypertension medication that is also used to reduce the workload on the heart. The study below compares the effects of two different types of capsules used to administer the drug to patients (GITS—red line, and Cotracten X—blue line). Each graph shows changes with time following initial administration of the capsules. **Figure A** compares the levels of nifedipine in the blood with each capsule type. The ability of the different capsules to reduce workload on the heart by altering blood pressure or heart rate is shown in **Figures B** and **C,** respectively.

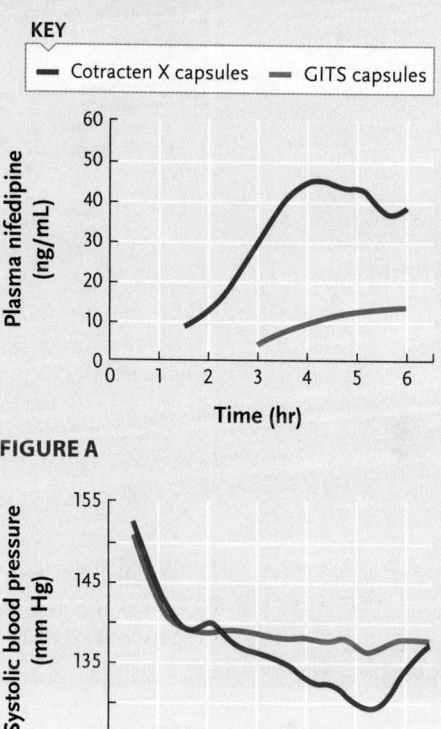

FIGURE A

FIGURE B

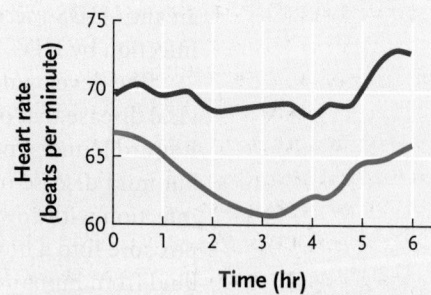

FIGURE C

1. Are elevated plasma levels of nifedipine (above 10 ng/mL) required for the drug to perform its actions?

2. Do both types of capsule appear to cause a reduction in the workload of the heart?

3. Which type of capsule would you expect to have fewer side effects?

Source: M. J. Brown and C. B. Toal. 2008. Formulation of long-acting nifedipine tablets influences the heart rate and sympathetic nervous system response in hypertensive patients. *British Journal of Clinical Pharmacology* 65:646–652.

45

Defenses against Disease

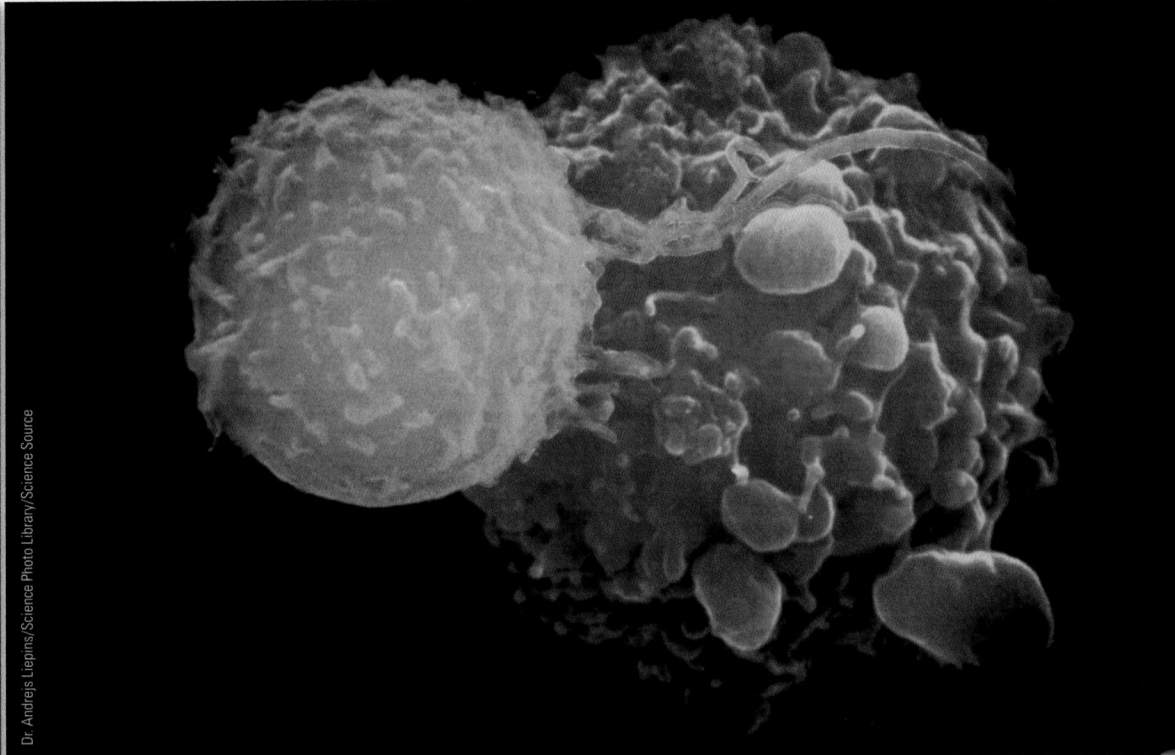

Dr. Andrejs Liepins/Science Photo Library/Science Source

Death of a cancer cell. A cytotoxic T cell (orange) induces a cancer cell (mauve) to undergo apoptosis (a type of programmed cell death). Cytotoxic T cells are part of the body's immune response system programmed to seek out, attach themselves to, and kill cancer cells and pathogen-infected host cells.

Why it matters . . . Acquired immunodeficiency syndrome (AIDS) is a constellation of disorders that follows infection by the **human immunodeficiency virus (HIV)** (see Figure 17.11). AIDS was first identified in the early 1980s, and at the end of 2013 was infecting about 35 million people worldwide. The virus continues to spread. One important aim of researchers is to develop an anti-AIDS *vaccine*—a substance that, when swallowed or injected, provides protection against infection by HIV.

The development of vaccines began with efforts to control smallpox, a dangerous infectious viral disease that once infected millions of people worldwide. In 1796, an English country doctor, Edward Jenner, knew that milkmaids never got smallpox if they had contracted cowpox, a similar but mild disease of cows that can be transmitted to humans. Jenner tested whether a deliberate infection with cowpox would protect humans from smallpox. He scratched material from a cowpox sore into a boy's arm. Six weeks later, after the cowpox infection had subsided, he scratched fluid from human smallpox sores into the boy's skin. (Jenner's use of the boy as an experimental subject would now be considered unethical.) The boy remained free from smallpox. Jenner tested other subjects similarly with the same results. His technique became the basis for worldwide **vaccination** (*vacca* = cow) against smallpox. With improved vaccines, smallpox has now been eradicated from the human population.

Vaccination takes advantage of the **immune system** (*immunis* = exempt), the inherited mechanisms that protect the body from pathogens. The immune system is our main defense against infectious disease. This chapter focuses on the immune system and other defenses against infection, such as the skin. Our description emphasizes human and other mammalian systems, in which most of the scientific discoveries revealing the structure and function of the immune system have been made. At the end of the chapter, we compare briefly the mammalian systems with the protective systems of nonmammalian vertebrates and invertebrates. Natural protection systems are not confined to animals: all organisms have some ability to defend themselves against infectious agents such as viruses. For example, restriction enzymes (see Section 18.1), the molecular biologist's tool for cloning and other DNA manipulations, are a bacterium's natural defense against infecting bacteriophages. Plants also have defense mechanisms against infectious agents (see Chapter 37).

45.1 Three Lines of Defense against Pathogens

Every organism is exposed constantly to *pathogens,* disease-causing viruses or organisms such as infectious bacteria, protists, fungi, and parasitic worms. Humans and other mammals have three lines of defense against these threats:

1. **Physical barriers;** they are not part of the immune system.
2. The **innate immune system,** the inherited mechanisms that protect the body from many kinds of pathogens in a nonspecific way. The innate immune system is found in all animals.
3. The **adaptive immune system,** the inherited mechanisms leading to the synthesis of molecules that target pathogens in a specific way. The adaptive immune system is only found in vertebrates.

Reaction to an infection takes minutes in the case of the innate immune system versus days for the adaptive immune system. Importantly, the three lines of defense are not either/or systems. Rather, they are cumulative actions of the body to defend against pathogens, meaning that physical barriers and the innate immune system still function against a specific pathogen even when the adaptive immune system is operational.

Epithelial Surfaces Are Anatomical Barriers That Help Prevent Infection

The body surface—the skin covering the body exterior and the epithelial surfaces covering internal body cavities and ducts, such as the lungs and intestinal tract—is the first line of defense. The tight junctions between epithelial cells of the body surface (see Section 4.5) keep most pathogens (as well as toxic substances) from entering the body.

Many epithelial surfaces are coated with a mucus layer secreted by the epithelial cells that protects against pathogens, as well as toxins and other chemicals. In the respiratory tract, ciliated cells constantly sweep the mucus, with its trapped bacteria and other foreign matter, into the throat, where it is coughed out or swallowed.

Many of the body cavities lined by mucous membranes have environments that are hostile to pathogens. For example, the strongly acidic environment of the stomach kills most bacteria and destroys many viruses that are carried there. Most of the

pathogens that survive the stomach acid are destroyed by the digestive enzymes and bile secreted into the small intestine. The vagina, too, is acidic, which prevents many pathogens from surviving there. The mucus coating in some locations contains the enzyme lysozyme, which is secreted by specialized epithelial cells. Lysozyme breaks down the walls of some bacteria, causing them to lyse.

The normal array of microorganisms of the human body can also act as a nonspecific defense mechanism by inhibiting colonization by pathogens. For example, the healthy human vaginal microbiome includes bacteria of the genus *Lactobacillus* (a type of Gram-positive bacterium). (Recall from Section 26.2 that a microbiome is the collection of microorganisms found associated with an organism, or with parts of the organism.) Lactobacilli have been shown to inhibit the growth of several species of pathogenic bacteria *in vitro;* this inhibition may be due in part to the production of lactic acid. Lactic acid contributes to low vaginal acidity, which, as you just learned, is unfavorable for the growth of many potential pathogens. Lactobacilli also have been shown to produce antimicrobial compounds and may compete with potential pathogens for adherence to vaginal epithelial cells.

The Immune System Protects the Body from Pathogens That Have Crossed External Barriers

The body's second line of defense is a series of generalized internal chemical, physical, and cellular reactions that attack pathogens that have breached the first line. These defenses include inflammation, which creates internal conditions that inhibit or kill many pathogens, and specialized cells that engulf or kill pathogens or infected body cells.

This second line of defense involves mechanisms of the *innate immune system.* (Innate means inborn/born with.) **Innate immunity** is the initial response by the body to eliminate cellular pathogens, such as bacteria and viruses, and prevent infection. Innate immunity provides an immediate, *nonspecific* response; that is, it targets any invading pathogen and has no memory of prior exposure to the pathogen. ("Memory" here means immunological memory, in which the body has a cellular record that it has been exposed previously.) It provides some protection against invading pathogens while a more powerful, specific response system is mobilized.

The third and most effective line of defense involves mechanisms of the *adaptive immune system,* and **adaptive immunity** (also called **acquired immunity**) is the term for this response. Adaptive immunity is *specific:* it recognizes individual pathogens and mounts an attack that neutralizes or eliminates them directly. It is so named because it is stimulated and shaped by the presence of a specific pathogen or foreign molecule. Adaptive immunity takes several days to become protective. It is triggered by specific molecules on pathogens that are recognized as being foreign to the body. The body retains a cellular memory of its first exposure to a foreign molecule, which enables it to respond more quickly if the pathogen is encountered again in the future.

The innate immune system and the adaptive immune system together constitute the vertebrate immune system. The defensive reactions of the system are termed the **immune response.** Functionally, the two components of the immune system interconnect and communicate at the chemical and cellular levels. The immune system is the product of a long evolutionary history of compensating adaptations by both pathogens and their targets. Over millions of years of vertebrate history, the mechanisms by which pathogens attack and invade have become more efficient, but the defenses of animals against the invaders have kept pace.

White Blood Cells Are Key Participants in the Innate and Adaptive Immune Systems

White blood cells—leukocytes—and their derivatives, along with several types of plasma proteins, are responsible for the activities of the two immune systems. **Table 45.1** lists the major types of leukocytes and their functions, Figure 44.7 presents drawings of the leukocyte types, and **Figure 45.1** shows a photograph of a neutrophil and a lymphocyte. Most types of leukocytes are present in the blood only transiently as they travel to

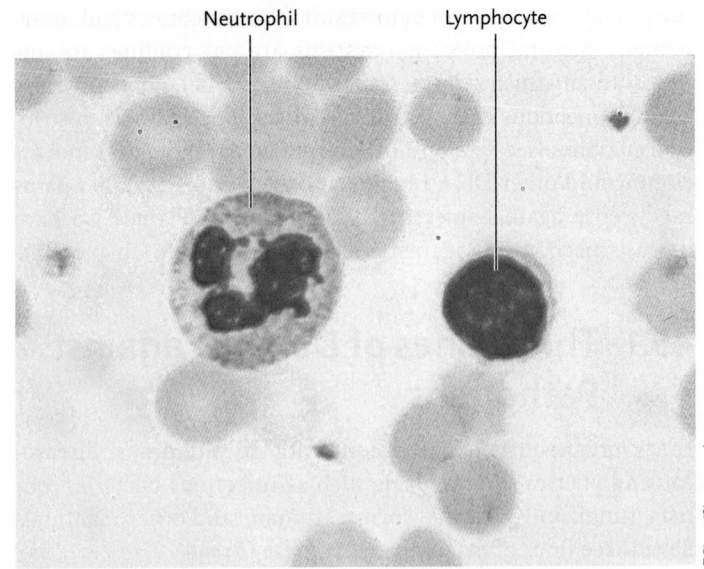

FIGURE 45.1 Neutrophil (left) and lymphocyte (right) (light micrograph).

T-Photo/Shutterstock.com

TABLE 45.1	Major Types of Leukocytes and Their Functions
Type of Leukocyte*	Function
Monocyte	Differentiates into a macrophage when released from blood into damaged tissue
Macrophage	Phagocyte that engulfs infected cells, pathogens, and cellular debris in damaged tissues; helps activate lymphocytes carrying out immune response
Neutrophil	Phagocyte that engulfs pathogens and tissue debris in damaged tissues
Eosinophil	Secretes substances that kill eukaryotic parasites such as worms
Lymphocyte	Main subtypes involved in innate and adaptive immunity are natural killer (NK) cells, B cells, plasma cells, helper T cells, and cytotoxic T cells. NK cells function as part of innate immunity to kill virus-infected cells and some cancerous cells of the host. The other cell types function as part of adaptive immunity: they produce antibodies, destroy infected and cancerous body cells, and stimulate macrophages and other leukocyte types to engulf infected cells, pathogens, and cellular debris.
Basophil	Located in blood, responds to IgE antibodies in an allergic response by secreting histamine, which stimulates inflammation

*Drawings of the types of leukocytes are shown in Figure 44.7.

the tissues for their roles in the immune systems. Note that several of the types of white blood cells are **phagocytes,** cells that engulf bacteria or other cellular debris by the process of **phagocytosis** (see Sections 4.3 and 5.5; Figure 5.14 shows an amoeba engulfing its prey by phagocytosis).

Most leukocytes originate from stem cells in the bone marrow, from which they are released into the blood. The exception is lymphocytes (see Section 44.2), which form in the bone marrow and then migrate to the **thymus** (a type of lymphoid tissue), where they mature. From the thymus, mature T lymphocytes are released into the blood and into the lymphatic system (see Section 44.6 and Figure 44.20). The lymphatic system, which includes an extensive network of vessels, the lymph nodes, the spleen, the thymus, and the tonsils, has a function in immunity. Potential pathogens that gain access to the lymph are filtered through the lymph nodes, where white blood cells act to inactivate them. The spleen, the largest lymphoid tissue, performs a similar function to inactivate pathogens in blood coming through it.

STUDY BREAK 45.1

1. What features of epithelial surfaces protect against pathogens?
2. What are the key differences between innate immunity and adaptive immunity?

45.2 Innate Immunity: Nonspecific Defenses

In most cases, the body needs 7–10 days to develop a fully effective adaptive immune response against a pathogen that is invading the body for the first time. In the meantime, innate immunity holds off invading pathogens, killing or containing them until adaptive immunity comes fully into play. Here we look at the internal mechanisms of innate immunity in vertebrates: secreted molecules and cellular components. As you will see, cellular pathogens (such as bacteria) and viral pathogens elicit different responses.

Innate Immunity Provides an Immediate, General Defense against Invading Cellular Pathogens

Cellular pathogens—typically microorganisms—usually enter the body when injuries break the skin or other epithelial surfaces. The innate immune system is available immediately to try to combat those pathogens. In particular, neutrophils and macrophages, which are both phagocytic cells, play an important role in innate immunity.

RECOGNITION OF A PATHOGENIC MICROORGANISM How does the host body recognize the pathogen as foreign? The answer is that the host has mechanisms to distinguish self from nonself. The innate immune system recognizes **pathogen-associated molecular patterns** that are associated with pathogenic organisms but are absent in the host. An example is the carbohydrates found typically on bacterial cell surfaces. The patterns are recognized by *pattern recognition receptors* of phagocytic cells. Mammals have several classes of pattern recognition receptors, one of which is the **toll-like receptors.** Toll-like receptors are found on the cell surface and within the cell on various membrane-bound compartments, such as the ER, lysosomes, and endocytic vesicles (see Section 4.3). A repertoire of ten toll-like receptors have been identified in humans. Each type recognizes a different, specific set of molecular patterns on pathogens. These receptors recognize distinct or overlapping pathogen-associated molecular patterns such as DNA, RNA, protein, and lipids. Recognition of a pathogen by a toll-like receptor triggers a cellular response by the phagocytic cell. (Receptors and cellular responses triggered by them are discussed more generally in Chapter 9) Several types of responses can occur depending on the receptor. For example, one response is to cause the phagocytic cell to engulf and destroy the pathogenic microorganism. Another is to secrete chemicals with antimicrobial properties, or chemicals that contribute to *inflammation* or that activate the *complement system*.

ANTIMICROBIAL PEPTIDES All of our epithelial surfaces, namely skin, the lining of the gastrointestinal tract, the lining of the nasal passages and lungs, and the lining of the genitourinary tracts, are protected by antimicrobial peptides called **defensins.** Some defensins are secreted by the epithelial cells, whereas others are secreted by neutrophils. And some defensins are secreted continuously, whereas others are secreted only in response to detection of pathogen-associated molecular patterns (such as by toll-like receptors). The defensins attack the plasma membranes of the pathogens, eventually disrupting them, thereby killing the cells.

INFLAMMATION A tissue's rapid response to injury, including infection by most pathogens, involves **inflammation**

(*inflammare* = to set on fire), the heat, pain, redness, and swelling that initially or exclusively occur at the site of an infection. Inflammation at the site of an infection is called **local inflammation.** This inflammatory response is highly similar regardless of the triggering event, be it pathogen, mechanical damage, or chemical injury. Overall, the response operates to isolate the injured area and stop the spread of damage.

Several interconnecting mechanisms initiate local inflammation (**Figure 45.2**). Consider bacteria entering a tissue as a result of a wound. Toll-like receptors on **macrophages** ("big eaters"; see Table 45.1)—a type of phagocyte—in the area of the wound recognize and bind to surface molecules on the pathogen, activating the macrophage to phagocytize (engulf) the pathogen (see Figure 45.2, step 1). Activated macrophages also secrete **cytokines,** molecules that bind to receptors on other host cells and, through signal transduction pathways, trigger a response. Usually, too few macrophages are present in the area of a bacterial infection to remove all of the bacteria.

The tissue damage also activates **mast cells,** which then release **histamine,** an inflammatory signaling molecule (step 2). The histamine, along with cytokines from activated macrophages, dilates local blood vessels around the infection site and increases their permeability, which increases blood flow and leakage of fluid from the vessels into body tissues (step 3). The response initiated by cytokines directly causes the heat, redness, and swelling of inflammation.

Cytokines also make the endothelial cells of the blood vessel wall stickier, causing circulating **neutrophils** and **monocytes** (see Table 45.1) to attach to the wall in massive numbers (step 3). From there, the neutrophils and monocytes are attracted to the infection site by **chemokines,** proteins also secreted by activated macrophages (step 4). Once they are located in the damaged tissue, the monocytes differentiate into macrophages. Both the neutrophils (which, like macrophages, have toll-like receptors that enable them to recognize pathogens) and the new macrophages engulf the pathogens (step 5).

Once a neutrophil or macrophage has engulfed a pathogen, it uses a variety of mechanisms to destroy it. These mechanisms include attacks by enzymes and defensins located in lysosomes and the production of toxic chemicals. The harshness of these attacks usually kills the neutrophils as well, whereas macrophages usually survive to continue their pathogen-scavenging activities. Dead and dying neutrophils, in fact, are a major component of the pus formed at infection sites. The pain of inflammation is caused by the migration of macrophages and neutrophils to the infection site and their activities there.

Some pathogens, such as parasitic worms, are too large to be engulfed by macrophages or neutrophils. In that case, macrophages, neutrophils, and **eosinophils** (see Table 45.1) cluster around the pathogen and secrete lysosomal enzymes and defensins in amounts that are often sufficient to kill the pathogen.

If tissue damage is extensive, or the infection spreads to the blood, a **systemic inflammation**—inflammation throughout the body—likely will occur. In systemic inflammation, chemical signaling molecules released by infected tissues may stimulate the

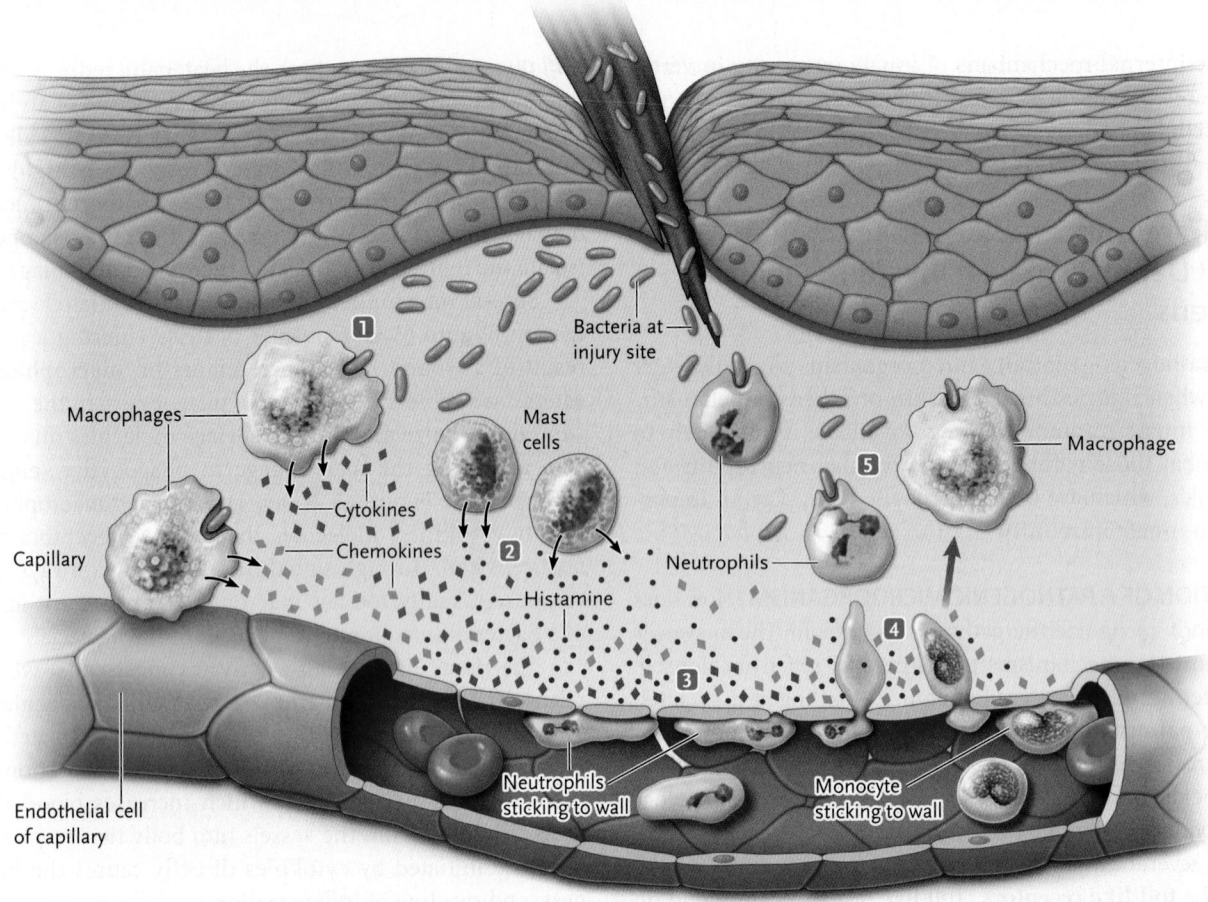

| **1** A break in the skin introduces bacteria, which reproduce at the wound site. Activated resident macrophages engulf the pathogens and secrete cytokines and chemokines. | **2** Mast cells in the area are activated by the tissue damage and release histamine. | **3** Histamine and cytokines dilate local blood vessels and increase their permeability. The cytokines also make the blood vessel wall sticky, causing neutrophils and monocytes to attach. | **4** Chemokines attract neutrophils and mono-cytes, which squeeze out between cells of the blood vessel wall and migrate to the infection site. | **5** The newly arrived monocytes differentiate into macrophages. Neutrophils and macrophages engulf the pathogens and destroy them. |

FIGURE 45.2 The steps producing inflammation.
© Cengage Learning 2017

release of neutrophils from the bone marrow, thereby increasing the number of white blood cells circulating in the blood. Systemic inflammation may also involve the onset of **fever.** That is, several chemicals released by macrophages in response to the infection act as **pyrogens** (*pyro* = fire; *gen* = production). Pyrogens stimulate the hypothalamus to release *prostaglandins,* particular chemical messengers that act locally (see Section 42.1). The prostaglandins act to turn up the hypothalamic thermostat regulating body temperature, which produces the fever. How fever fights infection is uncertain. Some models propose that it enhances phagocytosis or interferes with bacterial propagation.

THE COMPLEMENT SYSTEM The **complement system** is a group of more than 30 interacting plasma proteins that circulate in the blood and interstitial fluid. Complement proteins are activated when they recognize molecules on the surfaces of pathogens. Then, some of the proteins assemble into **membrane attack complexes,** which insert themselves into the plasma membrane of many types of bacterial cells and create pores that allow ions

and small molecules to pass through. As a result, the bacteria can no longer maintain osmotic balance, and they swell and lyse.

Combating Viral Pathogens Requires a Different Innate Immune Response

Recognition of specific nonself pathogen-associated molecular patterns on the surfaces of cellular pathogens (such as bacteria) in the extracellular fluid is key to initiating innate immune responses. However, most viruses are within infected cells and not in the extracellular fluid. Therefore, a host uses other strategies to provide some immediate protection against viral infections until the adaptive immune system, which can discriminate between viral and host proteins, is effective. Two strategies involve *interferon* and *natural killer cells.*

INTERFERON When a virus infects a cell, the cell synthesizes and releases cytokines called **interferons** in response to being exposed to the viral genome. Interferons act both on the infected cell that produces them, an autocrine effect, and on neighboring uninfected cells, a paracrine effect (see Section 42.1). They work by binding to cell surface receptors, triggering a signal transduction pathway that changes the gene expression pattern of the cells. The key changes include activation of a ribonuclease enzyme that degrades most cellular RNA and inactivation of a protein required for protein synthesis, thereby inhibiting most protein synthesis in the cell. These effects on RNA and protein synthesis inhibit replication of the viral genome, while putting the cell in a weakened state from which it may or may not recover.

NATURAL KILLER CELLS Cells that have been infected with a virus must be destroyed, and that is the role of **natural killer (NK) cells.** NK cells are a type of **lymphocyte,** a leukocyte that carries out most of its activities in tissues and organs of the lymphatic system (see Figure 44.20). NK cells circulate in the blood and kill target host cells—not only cells that are infected with a virus, but also some cells that have become cancerous.

NK cells can be activated by cell surface receptors or by interferons secreted by virus-infected cells. NK cells are not phagocytes; instead, they secrete granules containing **perforin,** a protein that creates pores in the target cell's membrane. Unregulated diffusion of ions and molecules through the pores causes osmotic imbalance, swelling, and rupture of the infected cell. NK cells also kill target cells indirectly by secreting *proteases* (protein-degrading enzymes) that pass through the pores. The proteases trigger **apoptosis** (a type of programmed cell death) by activating other enzymes that cause DNA degradation which, in turn, induces pathways leading to the cell's death.

How does an NK cell distinguish a target cell from a normal cell? The surfaces of most vertebrate cells have particular *major histocompatibility complex (MHC) proteins* on them. You will learn about the role of these proteins in adaptive immunity in the next section; for now, just consider them to be tags on the cell surface. NK cells monitor the level of MHC proteins and respond differently depending on their level. An appropriately high level, as on normal cells, inhibits the killing activity of NK cells. Viruses often inhibit the synthesis of MHC proteins in the cells they infect, lowering the levels of those proteins and identifying them to NK cells. Cancer cells also have low or, in some cases, no MHC proteins on their surfaces, which makes them a target for destruction by NK cells as well.

STUDY BREAK 45.2

1. What are the usual characteristics of the inflammatory response?
2. What processes specifically cause each characteristic of the inflammatory response?
3. What is the complement system?
4. Why does combating viral pathogens require a different response by the innate immune system than combating bacterial pathogens? What are the two main strategies a host uses to protect against viral infections?

45.3 Adaptive Immunity: Specific Defenses

Adaptive immunity is a vertebrate-specific defense mechanism that recognizes specific molecules as being foreign and clears those molecules from the body. The foreign molecules recognized may be free, as in the case of toxins, or they may be on the surface of viruses or cells such as pathogenic bacteria, cancer cells, pollen, or cells of transplanted tissues and organs.

Adaptive immunity develops only after the body is exposed to the foreign molecules and, hence, takes several days to become effective. This would be a significant problem in the case of invading pathogens were it not for the innate immune system, which combats the invaders in a nonspecific way within minutes after they enter the body. Adaptive immunity retains a memory of the foreign molecule that triggered the response, enabling a rapid, more powerful response if that pathogen is encountered again, whereas innate immunity retains no memory of exposure to the pathogen.

In Adaptive Immunity, B lymphocytes and T lymphocytes Detect Foreign Molecules and Clear Them from the Body

A foreign molecule that triggers an adaptive immune response is called an **antigen** ("*anti*body *gen*erator"). Antigens are macromolecules; most are large proteins (including glycoproteins and lipoproteins) or polysaccharides (including lipopolysaccharides). Some nucleic acids can also act as antigens, as can various large, artificially synthesized molecules.

Some antigens enter the body from the environment, for example, antigens on pathogens introduced beneath the skin, antigens in vaccinations, and inhaled or ingested macromolecules, such as toxins. Other antigens are produced within the body, for example, proteins encoded by viruses that have infected cells and altered proteins produced by mutated genes, such as those in cancer cells.

Two types of lymphocytes, B cells and T cells, recognize antigens in the body. **B cells** differentiate from stem cells in the bone marrow (see Section 44.2). Differentiated B cells are released into the blood and carried to capillary beds serving the tissues and organs of the lymphatic system. Like B cells, **T cells** originate from stem cells in the bone marrow. In immature form, they are released into the blood and carried to the thymus, where they mature. As you will see, two types of T cells—*helper T cells* and *cytotoxic T cells*—are involved in adaptive immunity.

The role of lymphocytes in adaptive immunity was demonstrated by experiments in which all of the leukocytes in mice were killed by irradiation with X-rays. These mice were then unable to develop adaptive immunity. Injecting lymphocytes from normal mice into the irradiated mice restored the response; other body cells extracted from normal mice could not restore the response. (For more on the use of mice as an experimental organism in biology, see *Focus on Research: Model Organisms.*)

Model Organisms: The Mighty Mouse

© Vasiliy Koval/Shutterstock.com

The mouse (*Mus musculus*) and its cells have been used to great advantage as models for research on mammalian developmental genetics, immunology, and cancer. The availability of the mouse as a research tool enables scientists to carry out experiments with a mammal that would not be practical or ethical with humans.

The small size of the mouse makes it relatively inexpensive and easy to maintain in the laboratory, and its short generation time, compared with most other mammals, allows genetic crosses to be carried out within a reasonable time span. For example, within 18 to 22 days after fertilization, a female gives birth to a litter of about 5 to 10 offspring.

Mice have a long and highly productive history as experimental animals. The first

example of a lethal allele was found in mice, and pioneering experiments on the transplantation of tissues between individuals were conducted with mice. During the 1920s, Fred Griffith laid the groundwork for the research showing that DNA is the hereditary molecule in his work with pneumonia-causing bacteria in mice (see Section 14.1). More recently, genetic experiments with mice have revealed more than 500 mutations that cause hereditary diseases, immunological defects, and cancer in mammals, including humans.

The mouse has also been the mammal of choice for experiments that introduce and modify genes through genetic engineering. Genetic engineering has produced, for example, "knockout" mice, in which a gene of interest is completely nonfunctional (see Section 18.2). The effects of the lack of function of a gene in the knockout mice often allow investigators to determine the role of the normal form of the gene by comparing normal mice with knockout mice. Some knockout mice have been developed to be defective in genes homologous to human

genes that cause serious diseases, such as cystic fibrosis. This allows researchers to study the disease in mice with the goal of developing cures or therapies.

The mouse has also been the main experimental choice for studies of immune responses. The results obtained have provided valuable insights into how the human immune systems work. However, mice and humans are separated by 65 million years of evolution and, as a result, the two organisms have significant differences in immune system mechanisms. These differences lead researchers to be cautious in relating mouse studies directly to humans when using mice in clinical studies involving the immune systems. For instance, the mouse model of multiple sclerosis has some differences from the human disease.

In 2002 the sequence of the mouse genome was reported. The sequence is enabling researchers to refine and expand their use of the mouse as a model organism for studies of mammalian biology and mammalian diseases.

There are two types of adaptive immune responses: *antibody-mediated immunity* and *cell-mediated immunity*.

- In **antibody-mediated immunity** (also called **humoral immunity**), B-cell derivatives called **plasma cells** secrete **antibodies,** highly specific protein molecules that circulate in the blood and lymph recognizing and binding to antigens and clearing them from the body.
- In **cell-mediated immunity**, a particular type of T cell becomes activated and, with other cells of the immune system, attacks infected cells directly and kills them.

The body's reaction to an infection is not an either/or choice between antibody-mediated immunity and cell-mediated immunity. Rather, both responses typically occur and, depending on the infection, the relative strength of the two responses varies. Complicated chemical signaling controls how the two responses develop.

The steps of the adaptive immune response are similar for antibody-mediated immunity and cell-mediated immunity:

1. *Antigen encounter and recognition:* lymphocytes encounter and recognize an antigen.
2. *Lymphocyte activation:* the lymphocytes are activated by binding to the antigen and proliferate by cell division to produce many clones of identical cells.

3. *Antigen clearance:* the many clones of activated lymphocytes clear the antigen from the body.
4. *Development of immunological memory:* some activated lymphocytes differentiate into **memory cells** that circulate in the blood and lymph, ready to initiate a rapid immune response upon a later exposure to the same antigen.

The following discussions of antibody-mediated immunity and cell-mediated immunity expand on these steps.

Antibody-Mediated Immunity Involves the Recognition of Antigens by T Cells, and the Activation of B Cells and Their Differentiation into Antibody-Secreting Cells

Antibody-mediated immunity begins as soon as an antigen is encountered and recognized in the body. Antigens are recognized by T cells that are activated as a result of the interaction and differentiate and proliferate, producing cells that activate B cells. Activated B cells differentiate into antibody-secreting cells. The antibodies are specific to the antigens that elicited the antibody-mediated immunity and they clear those antigens from the body.

ANTIGEN ENCOUNTER AND RECOGNITION BY LYMPHOCYTES

Antigens are encountered by B cells and T cells in the lymphatic system. Each B cell and each T cell is specific for a particular antigen, meaning that the cell can bind to only one particular molecular structure. The binding is specific because the plasma membrane of each B cell and T cell is studded with thousands of identical receptors for the antigen, the **B-cell receptors** on B cells and the **T-cell receptors** on T cells **(Figure 45.3).**

B-cell receptors and T-cell receptors are encoded by different genes and thus have different structures. The B-cell receptor on a B cell **(Figure 45.3A)** corresponds to the antibody secreted by that cell when it is activated and differentiates into a plasma cell (see discussion later in this chapter). Like its corresponding antibody, a B-cell receptor is a protein consisting of four polypeptide chains. At one end, the protein has two identical **antigen-binding sites,** the regions that bind to a specific antigen. At the opposite end from the antigen-binding sites are *transmembrane domains,* which embed in the plasma membrane. T-cell receptors **(Figure 45.3B)** consist of a protein made up of two different polypeptides. T-cell receptors also have an antigen-binding site at one end and transmembrane domains at the other end.

Considering the entire populations of B cells and T cells in the body, there are multiple cells that can recognize each antigen but, most important, the populations (in normal persons) contain cells capable of recognizing any antigen. For example, each of us has about 10 trillion B cells that collectively have about 100 million different kinds of B-cell receptors. Importantly, these cells are present *before* the body has encountered the antigens.

The binding between antigen and receptor is an interaction between two molecules that fit together like an enzyme and its substrate. A given B-cell receptor or T-cell receptor does not bind to the whole antigen molecule, but to small regions of it called **epitopes** or **antigenic determinants.** Therefore, several different B cells and T cells, each with different receptors, may bind to the population of a particular antigen encountered in the lymphatic system.

ANTIBODIES
Antibodies are the core molecules of antibody-mediated immunity. Antibodies are large, complex Y-shaped molecules that belong to a class of proteins known as **immunoglobulins** (immunoglobulin [Ig]). Each antibody molecule consists of four polypeptide chains: two identical **light chains** and two identical **heavy chains** about twice or more the size of the light chain **(Figure 45.4).**

Each polypeptide chain of an antibody molecule has a **constant (C) region** and a **variable (V) region.** The constant region of each antibody type has the same amino acid sequence for that part of the heavy chain, and likewise for that part of the light chain. The variable region of both the heavy and light chains, by contrast, has a different amino acid sequence for each antibody molecule in a population. The variable regions

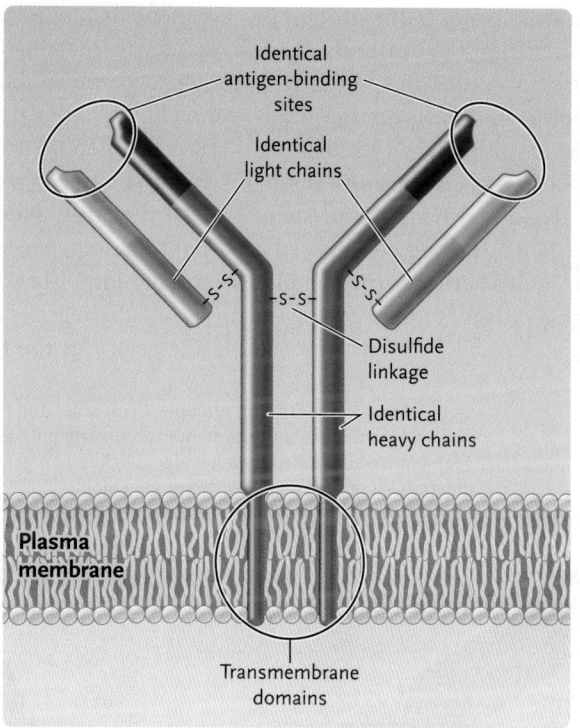
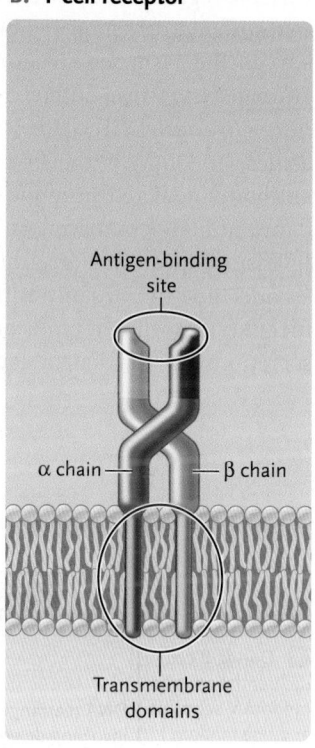

A. B-cell receptor

Identical antigen-binding sites

Identical light chains

Disulfide linkage

Identical heavy chains

Plasma membrane

Transmembrane domains

B. T-cell receptor

Antigen-binding site

α chain — — β chain

Transmembrane domains

FIGURE 45.3 Antigen-binding receptors on B cells and T cells. The B-cell receptor corresponds to the antibody secreted by the cell when it is activated and differentiates into a plasma cell.
© Cengage Learning 2017

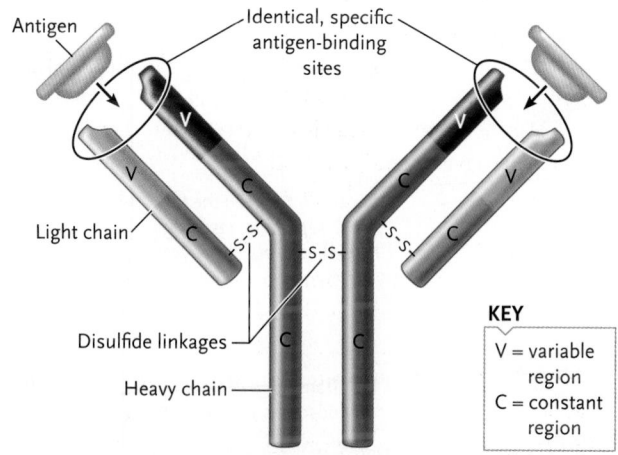

Antigen

Identical, specific antigen-binding sites

Light chain

Disulfide linkages

Heavy chain

KEY

V = variable region
C = constant region

FIGURE 45.4 The arrangement of light and heavy polypeptide chains in an antibody molecule. Disulfide (—S—S—) linkages hold the chains together, and the whole molecule folds into a Y-shape. The bonds between the two arms of the Y form a hinge that allows the arms to flex independently of one another. As shown, two sites, one at the tip of each arm of the Y, bind the same antigen.
© Cengage Learning 2017

are the top halves of the polypeptides in the arms of the Y-shaped molecule; the three-dimensional folding of the variable regions of each arm creates the antigen-binding site. The antigen-binding site is the same for the two arms of an antibody molecule because its two heavy chains are identical, as are its two light chains, as mentioned earlier. However, the antigen-binding sites are different from antibody molecule to antibody molecule because of the amino acid differences in the variable regions of the two chain types.

Different constant regions of the heavy chains in the tail part of the Y-shaped structure determine the **antibody class.** Humans have five different classes of antibodies—*IgM, IgG, IgA, IgE,* and *IgD.* The B-cell receptors on B cells are **IgM** molecules, and IgM is the first type of antibody secreted in an antibody-mediated immune response. **IgG** is the most abundant antibody circulating in the blood and lymphatic system. IgG is produced in large amounts when the body is exposed a second time to the same antigen. It also activates the complement system when it binds an antigen. **IgA** is found mainly in secretions at particular locations in the body where the

antibodies it secretes bind to surface groups on pathogens and block their attachment to body surfaces. **IgE** is secreted by plasma cells of the skin and tissues lining the gastrointestinal tract and respiratory tract. IgE binds to basophils and mast cells, triggering release of histamine, which causes an inflammatory response (see discussion later in this chapter). **IgD** also serves as a B-cell receptor.

THE GENERATION OF ANTIBODY DIVERSITY The human genome has ~19,900 protein-coding genes, far fewer than necessary to encode 100 million different antibodies if two genes encoded one antibody, one gene for the heavy chain and one for the light chain. Instead, antibody diversity is generated during B-cell differentiation by a process involving three rearrangements of DNA segments that encode parts of the light and heavy chains. **Figure 45.5** shows the process that produces light-chain genes for an IgM antibody, the B-cell receptor; the production of IgM heavy-chain genes, and the light-chain and heavy-chain genes for the other classes of antibodies, is similar. The genes for the two different subunits of the T-cell receptor

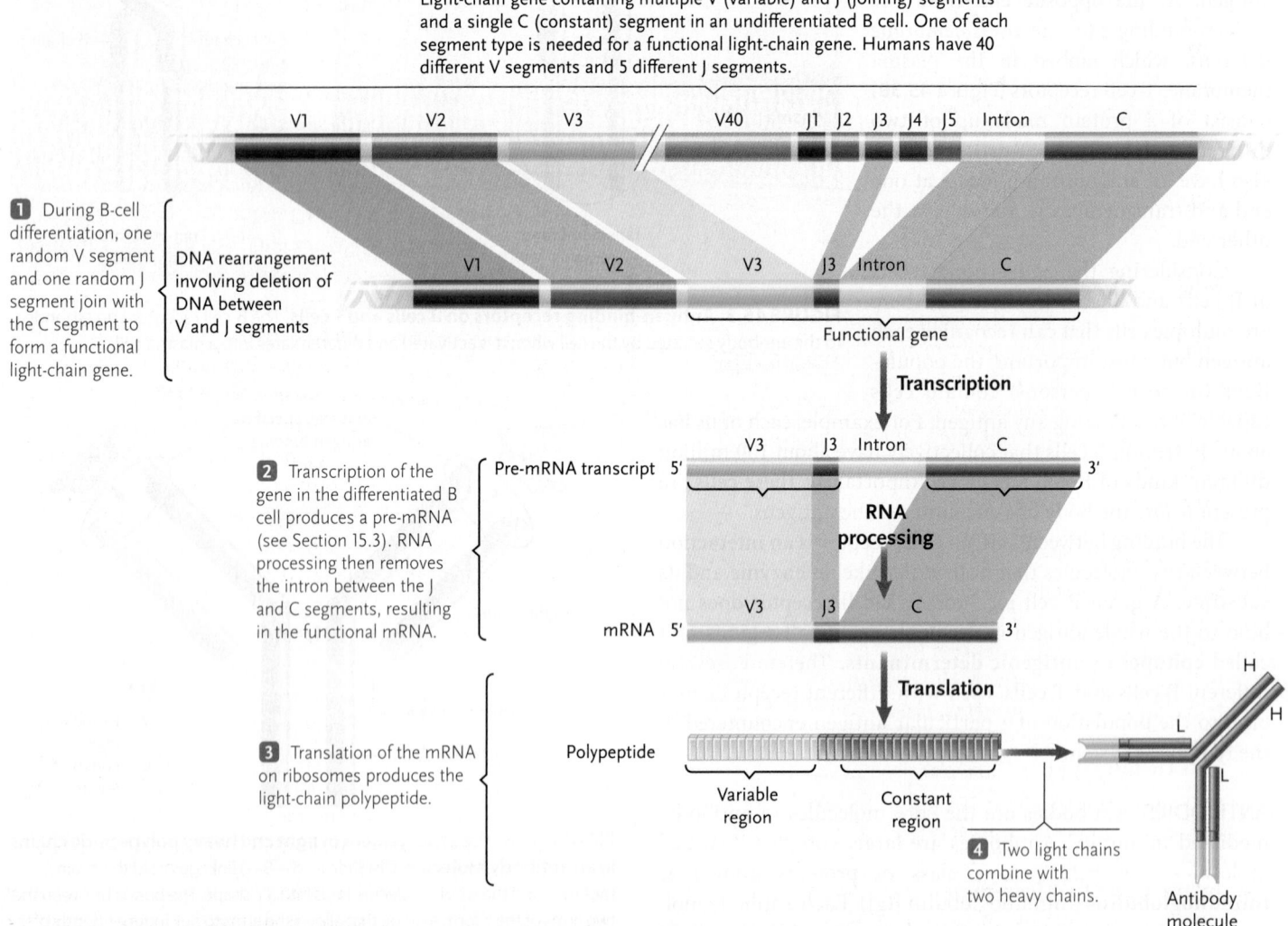

1 During B-cell differentiation, one random V segment and one random J segment join with the C segment to form a functional light-chain gene.

2 Transcription of the gene in the differentiated B cell produces a pre-mRNA (see Section 15.3). RNA processing then removes the intron between the J and C segments, resulting in the functional mRNA.

3 Translation of the mRNA on ribosomes produces the light-chain polypeptide.

4 Two light chains combine with two heavy chains.

FIGURE 45.5 The DNA rearrangements producing a functional light-chain gene, in simplified form.
© Cengage Learning 2017

undergo similar rearrangements to produce the great diversity in antigen-binding capability of those receptors.

Light-chain genes have one C segment, but heavy-chain genes have five types of C segments, each of which encodes one of the constant regions of IgM, IgD, IgG, IgE, and IgA. In the assembly of functional heavy-chain genes, the inclusion of one of the five C segment types therefore specifies the class of antibody that will be made by the cell.

LYMPHOCYTE ACTIVATION Let us now follow the development of an antibody-mediated immune response for a bacterial pathogen in the circulation or in a tissue. Circulating viruses are dealt with in the same way. We follow the steps of recognition of an antigen by lymphocytes, activation of lymphocytes by antigen binding, and production of antibodies **(Figure 45.6).** T cells are key players in the initial steps of this process. Two phases are involved: (1) particular T cells are activated; and (2) those T cells help B cells become activated and secrete antibodies. The numbered steps in the following description correspond to the events shown in Figure 45.6:

1. **Engulfment of bacterium.** A type of phagocyte called a **dendritic cell** engulfs a bacterium. Dendritic cells, which have the same origin as leukocytes, are so named because they have many surface projections resembling dendrites of neurons. They recognize a bacterium as foreign through the same recognition mechanism used by macrophages in the innate immune system.

2. **Degradation of bacterium and release of antigens.** Engulfment of a bacterium activates the dendritic cell; the activated cell migrates to a nearby lymph node. Then the vesicle within the dendritic cell that contains the engulfed bacterium fuses with a lysosome, which leads to degradation of the bacterium's proteins into short peptides. Those peptides function as antigens.

3. **Presentation of antigens on dendritic cell surface.** Within the dendritic cell, the antigens bind intracellularly to **class II major histocompatibility complex (MHC)** proteins. These proteins are named for a large cluster of 128 genes encoding them, called the **major histocompatibility complex (MHC).** Each individual of each vertebrate species has a unique combination of MHC proteins on almost all body cells, meaning that no two individuals of a species except identical siblings are likely to have exactly the same MHC proteins on their cells. The two classes of MHC proteins, class I and class II, have different functions in adaptive immunity, as you will see.

   An antigen bound to a class II MHC protein migrates to the cell surface, where the antigen is displayed. The cell is now an **antigen-presenting cell (APC),** which can present the antigen to T cells.

4. **Interaction of antigen-presenting cell with lymphocyte.** The key function of an antigen-presenting cell is to present the antigen to a lymphocyte. For antibody-mediated immunity, that lymphocyte is a **CD4⁺ T cell,** so called because it has CD4 receptors on its surface in addition to T-cell receptors. A specific $CD4^+$ T cell having a T-cell receptor with an antigen-binding site that recognizes an epitope of a bacterial antigen binds to the antigen on the antigen-presenting cell, the CD4 receptor interacts directly with the class II MHC protein, and the two cells become linked together.

5. **Activation of T cell.** When the antigen-presenting cell binds to the $CD4^+$ T cell, the antigen-presenting cell secretes an **interleukin** (meaning "between leukocytes"), a type of cytokine that activates the associated T cell.

6. **Production of helper T cells.** The activated T cell secretes other interleukins, which act in an autocrine manner (see Section 42.1) to stimulate **clonal expansion,** the proliferation of the activated $CD4^+$ T cell by cell division to produce a clone of cells. These clonal cells differentiate into **helper T cells,** so named because they assist with the activation of other lymphocytes, in this case B cells. A helper T cell is an example of an **effector T cell,** meaning that it is involved in effecting—bringing about—the specific immune response to the antigen.

7. **Presentation of antigens on B cell surface.** Antibodies are produced in and secreted by activated B cells. The activation of a B cell requires the B cell to present the antigen on its surface, and then to link with a helper T cell that has differentiated as a result of encountering and recognizing the *same* antigen. Antigen presentation on a B cell surface begins when B-cell receptors on the B cell bind to antigens on the surface of the bacterium it encounters. The bacterium plus B-cell receptor is then taken into the cell where the antigen is processed in the same way as in dendritic cells (see steps 2 and 3). The result is the presentation of antigen pieces on the B-cell surface in a complex with class II MHC proteins.

8. **Interaction of B cell with helper T cell.** When a B cell encounters a helper T cell (from step 6) displaying the same antigen, usually in a lymph node or in the spleen, the two cells become tightly linked together.

9. **Activation of B cell.** The linkage between cells stimulates the helper T cell to secrete interleukins that activate the B cell and then stimulate the B cell to proliferate, producing a clone of those B cells with identical B-cell receptors.

10. **Production of plasma cells and memory B cells.** Many of the cloned cells differentiate into relatively short-lived *plasma cells,* which now secrete the same antibody that was displayed on the parental B cell's surface to circulate in lymph and blood and attack the pathogen. The other cloned cells differentiate into **memory B cells,** which are long-lived cells that set the stage for a much more rapid response should the same antigen be encountered later.

Remember that there is an enormous diversity of randomly generated lymphocytes in the body, each with a particular receptor that may potentially recognize a particular antigen. **Clonal selection** is the term used for the process by which a

THE ANTIBODY-MEDIATED IMMUNE RESPONSE, ILLUSTRATED FOR A BACTERIAL PATHOGEN

Antigen recognition and T-cell activation and differentiation

1 Engulfment of bacterium. A bacterium is taken up by a dendritic cell by phagocytosis.

2 Degradation of bacterium and release of antigens. The bacterium is degraded within a lysosome. The released bacterial proteins are broken into pieces, which act as antigens.

3 Presentation of antigens on dendritic cell surface. Antigens bind to class II MHC proteins within the cell and the interacting molecules then are displayed on the cell surface. The cell is now an antigen-presenting cell (APC).

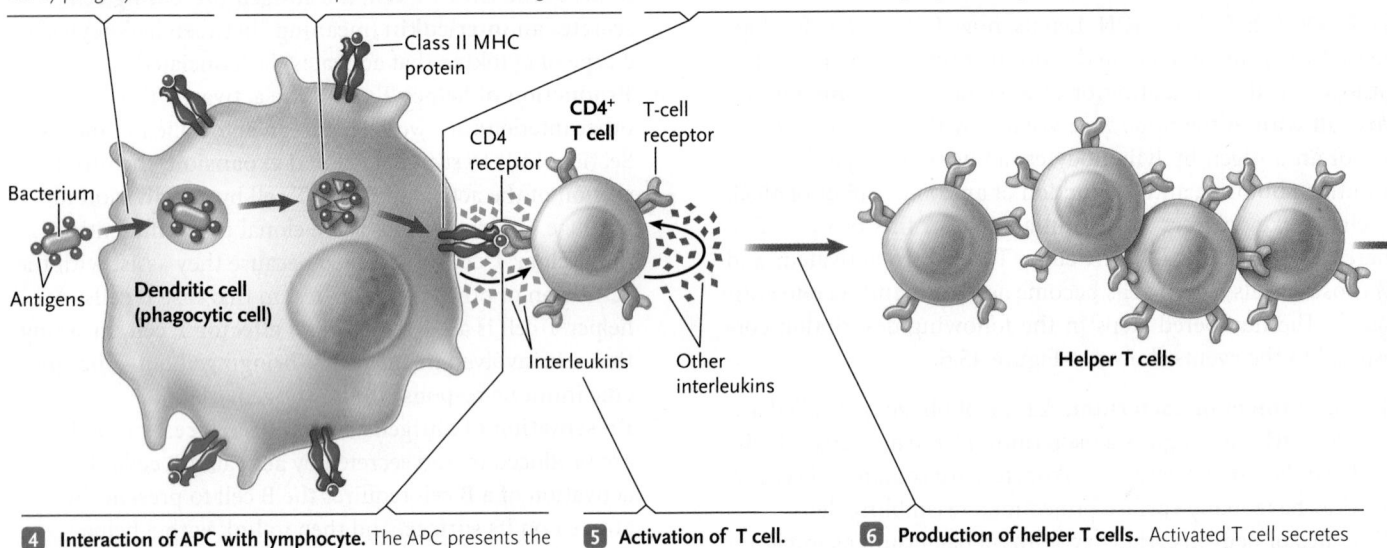

4 Interaction of APC with lymphocyte. The APC presents the antigen to a CD4+ T cell with a T-cell receptor that recognizes the antigen. The CD4 receptor binds directly to the class II MHC protein. The APC and the T cell become linked together.

5 Activation of T cell. The APC secretes interleukins, which activate the T cell.

6 Production of helper T cells. Activated T cell secretes other interleukins, which stimulate the T cell to proliferate to produce a clone of cells. The cloned cells differentiate into helper T cells and memory helper T cells.

B-cell activation by helper T cells and differentiation of B cells into antibody-producing cells

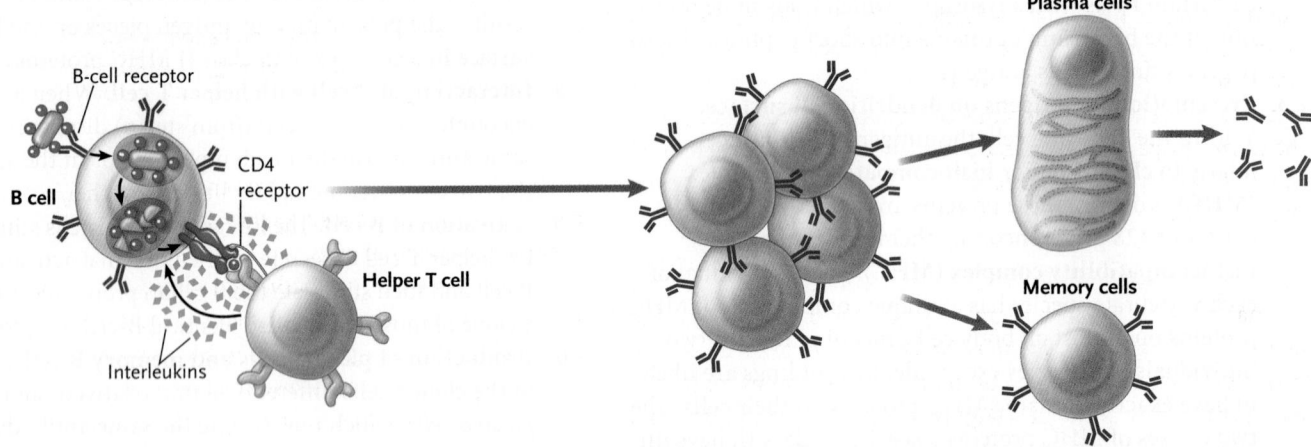

7 Presentation of antigens on B cell surface. The B-cell receptor binds to antigen on the bacterium. Bacterium is engulfed and its macromolecules degraded. The antigens produced are displayed on cell surface bound to class II MHC proteins.

8 Interaction of B cell with helper T cell. The T-cell receptor of a helper T cell recognizes the antigen on the B cell and links the two cells together.

9 Activation of B cell. Interleukins secreted by the helper T cell activate the B cell and then stimulate B-cell proliferation to produce a clone of cells.

10 Production of plasma cells and memory B cells. Some cloned B cells differentiate into plasma cells, which secrete antibodies, while a few differentiate into memory B cells.

SUMMARY A dendritic cell engulfs a pathogen, presents pathogen antigens on its surface, and then interacts with a CD4+ T cell, activating it. The T cell produces helper T cells. Next, a B cell with pathogen antigens on its surface interacts with a helper T cell, triggering activation and proliferation of the B cell, and differentiation into antibody-producing plasma cells and memory B cells.

think like a scientist What is the key cellular event involved in the antibody-mediated immune response responsible for the production of antibodies specific for an antigen?

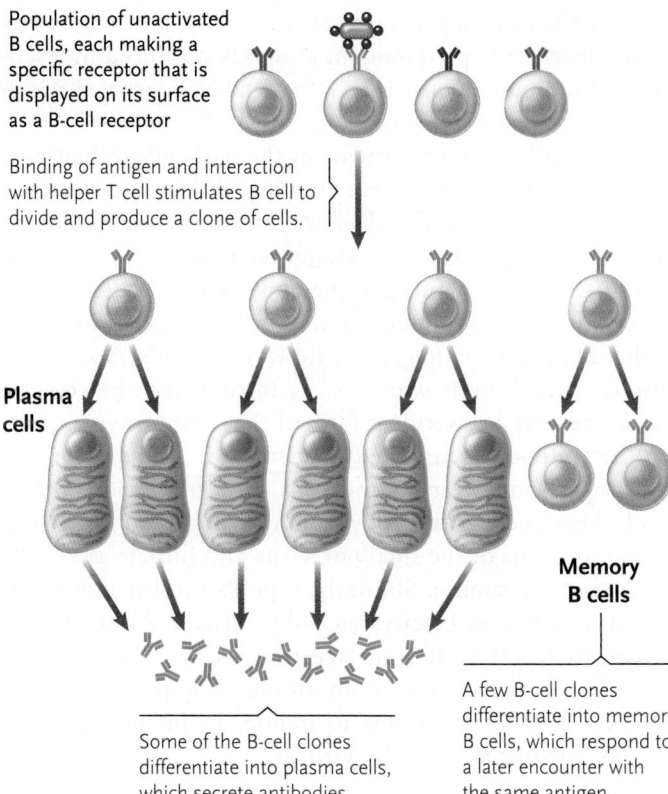

Population of unactivated B cells, each making a specific receptor that is displayed on its surface as a B-cell receptor

Binding of antigen and interaction with helper T cell stimulates B cell to divide and produce a clone of cells.

Plasma cells

Memory B cells

Some of the B-cell clones differentiate into plasma cells, which secrete antibodies.

A few B-cell clones differentiate into memory B cells, which respond to a later encounter with the same antigen.

FIGURE 45.7 Clonal selection. The binding of an antigen to a B cell already displaying a specific antibody to that antigen and interaction with helper T cells stimulates the B cell to divide and differentiate into plasma cells, which secrete the antibody, and memory cells, the cells that remain in the circulation ready to mount a response against the antigen at a later time.
© Cengage Learning 2017

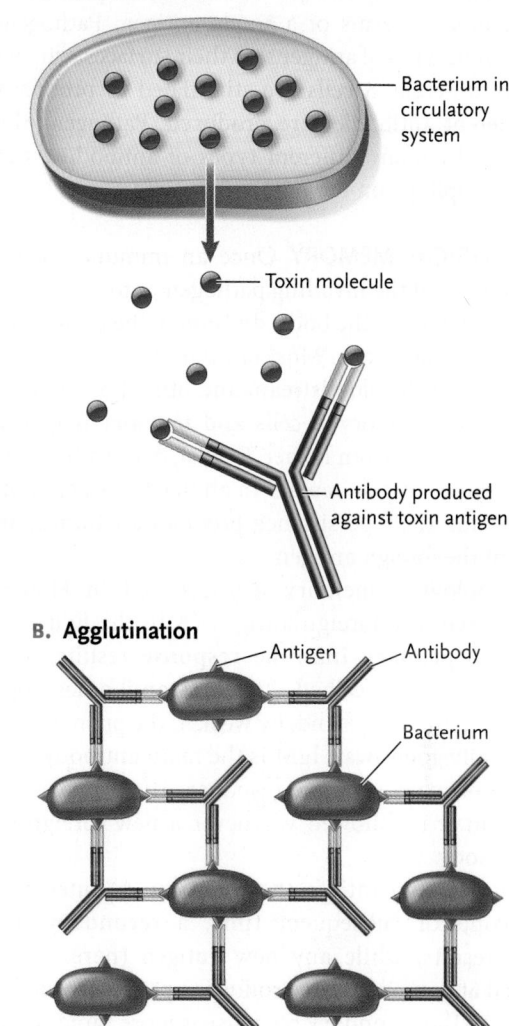

A. Neutralization

Bacterium in circulatory system

Toxin molecule

Antibody produced against toxin antigen

B. Agglutination

Antigen　　Antibody

Bacterium

FIGURE 45.8 Neutralization (A) and agglutination (B), two examples of mechanisms for clearing antigens from the body.
© Cengage Learning 2017

particular preexisting lymphocyte is selected specifically for cloning when it recognizes a foreign antigen (**Figure 45.7;** this figure is an expansion of step 10 of Figure 45.6). The result of clonal selection is a pool of antibody-secreting cells that collectively produce large amounts of antibodies directed against the same antigen. The process of clonal selection was proposed in the 1950s by several scientists, most notably F. Macfarlane Burnet, Niels Jerne, and David Talmage.

CLEARING THE BODY OF FOREIGN ANTIGENS Two important mechanisms to clear foreign antigens from the body are *neutralization* and *agglutination* (**Figure 45.8**). In **neutralization,** toxins produced by invading bacteria, such as tetanus toxin, can be *neutralized* by antibodies (**Figure 45.8A**). The antibodies bind to the toxin molecules, preventing them from carrying out their damaging action. In **agglutination** (clumping), the pathogen is immobilized by the antibodies (**Figure 45.8B**). For instance, antibodies will bind to antigens on the surfaces of intact bacteria at an infection site or in the circulatory system. Because the two arms of an antibody molecule bind to different copies of the antigen molecule, an antibody molecule may bind to two bacteria with the same antigen. A population of antibodies against the bacterium, then, link many bacteria together into a lattice causing *agglutination* of the bacteria. Agglutination

immobilizes the bacteria, preventing them from infecting cells. Antibodies can also agglutinate viruses, also preventing them from infecting cells.

More important, antibodies aid the innate immune response triggered by the pathogens. That is, antibodies bound to antigens stimulate the complement system. Membrane attack complexes are formed and insert themselves into the plasma membranes of the bacteria, leading to their lysis and death. In the case of viral infections, membrane attack complexes can insert themselves into the membranes surrounding enveloped viruses, which disrupts the membrane and prevents the viruses from infecting cells.

Antibodies also enhance phagocytosis of bacteria and viruses. Phagocytic cells have receptors on their surfaces that recognize the heavy-chain end of antibodies (the end of the molecule opposite the antigen-binding sites). Antibodies bound to bacteria or viruses therefore bind to phagocytic cells, which then engulf the pathogens and destroy them.

For simplicity, the adaptive immune response has been described here in terms of a single antigen. Pathogens have many different types of antigens on their surfaces, which means that many different B cells are stimulated to proliferate and many different antibodies are produced. Pathogens therefore are attacked by many different types of antibodies, each targeted to one epitope on the pathogen's surface.

IMMUNOLOGICAL MEMORY Once an immune reaction has run its course and the invading pathogen or toxic molecule has been eliminated from the body, division of the plasma cells and helper T cell clones stops. Most or all of the clones die and are eliminated from the bloodstream and other body fluids. However, long-lived memory B cells and **memory helper T cells** (which differentiate from helper T cells), derived from encountering the same antigen, remain in an inactive state in the lymphatic system. Their persistence provides an **immunological memory** of the foreign antigen.

Immunological memory is illustrated in **Figure 45.9**. When exposed to a foreign antigen (A in the figure) for the first time, a **primary immune response** results, following the steps already described. The first antibodies appear in the blood in 3 to 14 days and, by week 4, the primary response has essentially gone away. IgM is the main antibody type produced in a primary immune response. The primary immune response curve is followed whenever a new foreign antigen enters the body.

When a foreign antigen (here, antigen A) enters the body for a second or subsequent time, a **secondary immune response** results, while any new antigen (here, antigen B) introduced at the same time produces a primary response (see Figure 45.9). The secondary response is more rapid than a primary response because it involves the memory B cells and memory T cells that have been stored in the meantime, rather than having to initiate the clonal selection of a new B cell and T cell. Moreover, less antigen is needed to elicit a secondary

response than a primary response, and many more antibodies are produced. The predominant antibody produced in a secondary immune response is IgG; the switch occurs at the gene level in the memory B cells.

Immunological memory forms the basis of vaccinations, in which antigens in the form of living or dead pathogens or antigenic molecules themselves are introduced into the body. After the primary immune response, memory B cells and memory T cells remaining in the body can mount an immediate and intense immune reaction against similar antigens in the dangerous pathogen. In Edward Jenner's experiment (discussed in *Why it matters . . .*) introducing the cowpox virus, a related, less virulent form of the smallpox virus, into healthy individuals initiated a primary immune response. After the response ran its course, a bank of memory B and T cells remained in the body, able to recognize quickly the similar antigens of the smallpox virus and initiate a secondary immune response. Similarly, a polio vaccine developed by Jonas Salk uses inactivated polio viruses. Although the viruses are inactive, their surface groups can still act as antigens. The antigens trigger an immune response, leaving memory B and T cells able to mount an intense immune response against active polio viruses.

ACTIVE AND PASSIVE IMMUNITY **Active immunity** is the production of antibodies in the body in response to exposure to a foreign antigen—the process that has been described up until now. **Passive immunity** is the acquisition of antibodies as a result of direct transfer from another person or by their injection. This form of immunity provides immediate protection against antigens that the antibodies recognize without the person receiving the antibodies having developed an immune response. Examples of passive immunity include the transfer of IgG antibodies from mother to fetus through the placenta and the transfer of IgA antibodies in the first breast milk fed from the mother to the baby. Compared with active immunity, passive immunity is a short-lived phenomenon in that the antibodies typically break down within a month. However, in that time, the protection plays an important role. For example, a breast-fed baby is protected until it is able to mount an immune response itself, an ability that is not present until about a month after birth. Importantly, in passive immunity, the body does not develop a memory of the pathogen, which means that there is no protection against future infection with the same pathogen.

DRUG EFFECTS ON ANTIBODY-MEDIATED IMMUNITY Several drugs used to reduce the rejection of transplanted organs target helper T cells. Cyclosporin A, used routinely after organ transplants, blocks the activation of helper T cells and, in turn, the activation of B cells. Unfortunately, cyclosporin and other immunosuppressive drugs also leave the treated individual more susceptible to infection by pathogens.

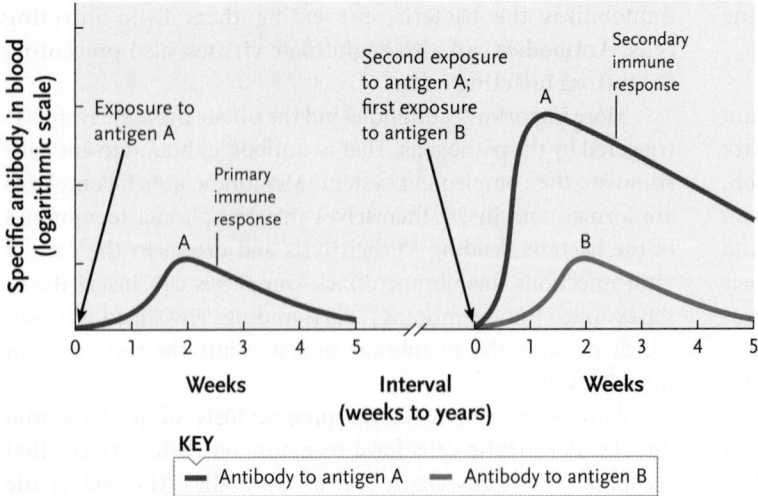

KEY

— Antibody to antigen A   — Antibody to antigen B

FIGURE 45.9 Immunological memory: primary and secondary responses to the same antigen.

© Cengage Learning 2017

FIGURE 45.10 | Research Method

Production of Monoclonal Antibodies

Purpose: Injecting an antigen into an animal produces a collection of different antibodies that react against different parts of the antigen. Monoclonal antibodies are produced to provide antibodies that all react against the same epitope of a single antigen.

Protocol:

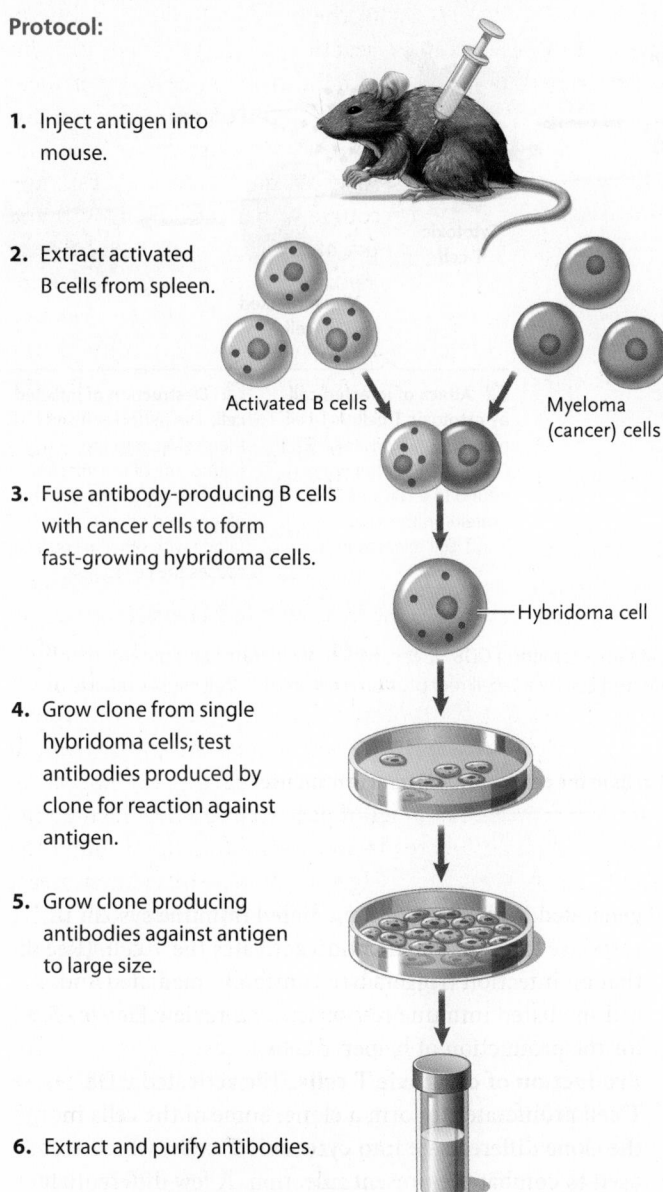

1. Inject antigen into mouse.

2. Extract activated B cells from spleen.

Activated B cells

Myeloma (cancer) cells

3. Fuse antibody-producing B cells with cancer cells to form fast-growing hybridoma cells.

Hybridoma cell

4. Grow clone from single hybridoma cells; test antibodies produced by clone for reaction against antigen.

5. Grow clone producing antibodies against antigen to large size.

6. Extract and purify antibodies.

Outcome: Monoclonal antibodies are highly specific in their ability to bind to the same part of a single antigen. They are widely used in scientific research and have many medical applications.

Source: G. Köhler and C. Milstein. 1975. Continuous cultures of fused cells secreting antibody of predefined specificity. *Nature* 256:495–497.

Antibodies Have Many Uses in Research

The ability of the antibody-mediated immune system to generate antibodies against essentially any antigen provides an invaluable research tool to scientists, who can use antibodies to identify biological molecules and determine their locations and functions in cells. To obtain the antibodies, a molecule of interest is injected into a test animal such as a mouse, rabbit, goat, or sheep. In response, the animal develops antibodies capable of binding to the molecule. The antibodies are then extracted and purified from a blood sample.

To identify the cellular location of a molecule, antibodies made against the molecule are combined with a visible marker such as a dye molecule or heavy metal atom. When added to a tissue sample, the marked antibodies can be seen in the light or electron microscope localized to cellular structures such as membranes, ribosomes, or chromosomes, showing that the molecule forms part of the structure.

Antibodies can also be used to "grab" a molecule of interest from a preparation containing a mixture of different cellular molecules. For an example, see *Molecular Insights* in Chapter 9.

Injecting a molecule of interest into a test animal typically produces a wide spectrum of antibodies that react with different parts of the antigen. Some of the antibodies also cross-react with other, similar antigens, producing false results that can complicate the research. These problems have been solved by producing **monoclonal antibodies,** each of which reacts only against the same segment (epitope) of a single antigen.

Georges Köhler and Cesar Milstein pioneered the production of monoclonal antibodies in 1975. In their technique, a test animal (usually a mouse) is injected with a molecule of interest **(Figure 45.10)**. After the animal has developed an immune response, fully activated B cells are extracted from the spleen and placed in a cell culture medium. Because B cells normally stop dividing and die within a week when cultured, they are induced to fuse with cancerous lymphocytes called *myeloma cells,* forming single, composite cells called **hybridomas.** Hybridomas combine the desired characteristics of the two cell types—they produce antibodies like fully activated B cells, and they divide continuously and rapidly like myeloma cells.

Single hybridoma cells are then separated from the culture and used to start clones. Because all the cells of a clone are descended from a single hybridoma cell, they all make the same highly specific antibody, able to bind the same part of a single antigen. In addition to their use in scientific research, monoclonal antibodies are also widely used in medical applications such as pregnancy tests, screening for prostate cancer, and testing for AIDS and other sexually transmitted diseases.

In Cell-Mediated Immunity, Cytotoxic T Cells Expose "Hidden" Pathogens to Antibodies by Destroying Infected Body Cells

In **cell-mediated immunity,** cytotoxic T cells directly destroy host cells infected by pathogens, particularly those infected

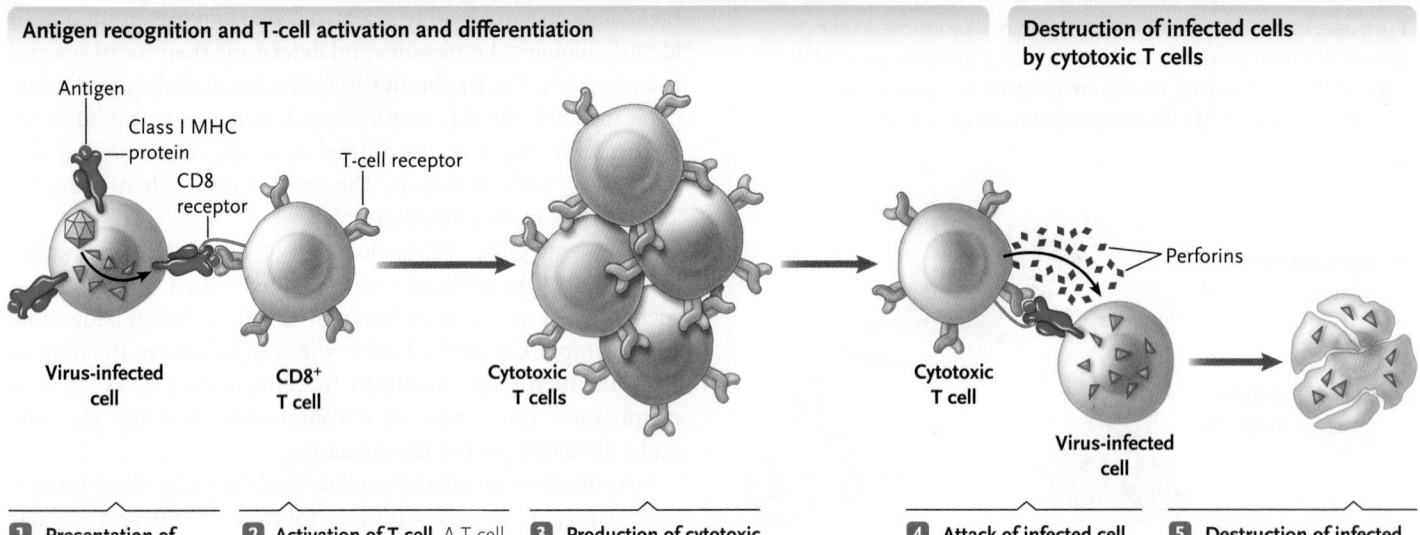

Antigen recognition and T-cell activation and differentiation

Destruction of infected cells by cytotoxic T cells

1 Presentation of antigens on cell surface. Viral proteins are degraded into fragments that act as antigens. The antigens are displayed on the cell surface bound to class I MHC proteins.

2 Activation of T cell. A T-cell receptor on a CD8+ T cell recognizes an antigen bound to a class I MHC protein on an infected cell, the CD8 receptor binds directly to the class I MHC protein, and the two cells link together. The interaction along with cytokines from helper T cells activates the T cell.

3 Production of cytotoxic T cells. The activated CD8+ T cell proliferates and forms a clone. The cloned cells differentiate into cytotoxic T cells and memory cytotoxic T cells.

4 Attack of infected cell by cytotoxic T cell. A T-cell receptor on a cytotoxic T cell recognizes the antigen bound to a class I MHC protein on the infected cell. The T cell releases perforins.

5 Destruction of infected cell. The perforins insert themselves into the membrane of the infected cell, forming pores. Leakage of ions and other molecules (along with other events) causes the cell to lyse.

SUMMARY Presented pathogen antigens on the surface of an infected cell activate an interacting CD8+ T cell, which proliferates and differentiates into cytotoxic T cells and memory T cells. Recognition of surface antigens on the infected cell by a T-cell receptor on a cytotoxic T cell causes release of perforins from the T cell, which leads to destruction of the infected cell.

think like a scientist What is the purpose of memory cytotoxic T cells in the cell-mediated immune response?

© Cengage Learning 2017

by a virus (**Figure 45.11**). For this type of immunity, T cells are the only type of lymphocyte involved. The numbered steps in the following description correspond to the events depicted in the figure:

1. **Presentation of antigens on cell surface.** The killing process begins when some of the pathogens break down inside infected host cells, releasing antigens that are fragmented by enzymes in the cytoplasm. The cell becomes an antigen-presenting cell with antigen fragments bound to class I MHC proteins displayed on the cell surface.

2. **Activation of T cell.** The antigen-presenting cell encounters a **CD8+ T cell,** a cell that has CD8 receptors on its surface in addition to T-cell receptors. A specific CD8+ T cell having a T-cell receptor with an antigen-binding site that recognizes an epitope of the antigen binds to the antigen on the antigen-presenting cell, the CD8 receptor interacts directly with the class I MHC protein, and the two cells become linked together. This linkage event plus the presence of cytokines secreted by helper T cells

generated in the antibody-mediated immune system in response to the same infection activates the T cell. (Recall that an infection triggers both antibody-mediated and cell-mediated immune responses, and review Figure 45.6 for the production of helper T cells.)

3. **Production of cytotoxic T cells.** The activated CD8+ T cell proliferates to form a clone. Some of the cells in the clone differentiate into **cytotoxic T cells,** which are used to combat the present infection. A few differentiate into **memory cytotoxic T cells,** which are long-lived cells that enable a rapid response should the same antigen be encountered on infected cells in the future. Cytotoxic T cells are another type of effector T cell.

4. **Attack of infected cell by cytotoxic T cell.** T-cell receptors on the cytotoxic T cell again recognize the antigen fragment bound to class I MHC proteins on the infected cells (the antigen-presenting cells). The cytotoxic T cell releases perforins.

5. **Destruction of infected cell.** The infected cell undergoes apoptosis brought about by the perforins, as well as

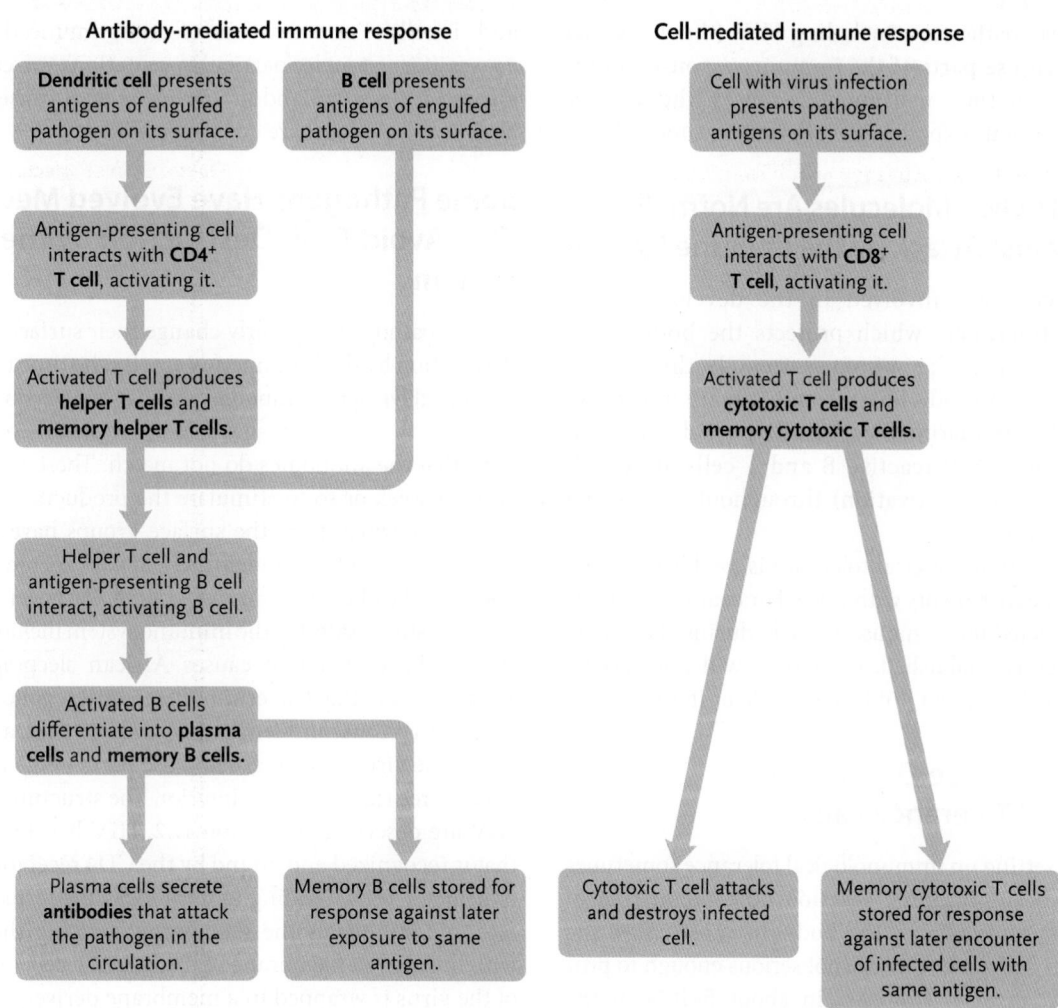

Antibody-mediated immune response

Dendritic cell presents antigens of engulfed pathogen on its surface.

B cell presents antigens of engulfed pathogen on its surface.

Antigen-presenting cell interacts with **CD4⁺ T cell**, activating it.

Activated T cell produces **helper T cells** and **memory helper T cells**.

Helper T cell and antigen-presenting B cell interact, activating B cell.

Activated B cells differentiate into **plasma cells** and **memory B cells**.

Plasma cells secrete **antibodies** that attack the pathogen in the circulation.

Memory B cells stored for response against later exposure to same antigen.

Cell-mediated immune response

Cell with virus infection presents pathogen antigens on its surface.

Antigen-presenting cell interacts with **CD8⁺ T cell**, activating it.

Activated T cell produces **cytotoxic T cells** and **memory cytotoxic T cells**.

Cytotoxic T cell attacks and destroys infected cell.

Memory cytotoxic T cells stored for response against later encounter of infected cells with same antigen.

FIGURE 45.12 Summary of the steps of the antibody-mediated and cell-mediated immune responses.
© Cengage Learning 2017

proteases secreted by the cytotoxic T cell acting in a way similar to that seen for natural killer cells (see discussion earlier in the chapter). Rupture of dead, infected cells releases the pathogens to the interstitial fluid, where they are open to attack by antibodies and phagocytes.

Cytotoxic T cells can also kill cancer cells if their class I MHC molecules display fragments of altered cellular proteins that do not normally occur in the body.

Figure 45.12 summarizes the steps of the antibody-mediated and cell-mediated immune responses.

STUDY BREAK 45.3

1. How, in general, do antibody-mediated and cell-mediated immune responses help clear the body of antigens?
2. Describe the general structure of an antibody molecule.
3. What are the principles of the mechanism used for generating antibody diversity?
4. What is clonal selection?
5. How does immunological memory work?

THINK OUTSIDE THE BOOK

In this section you learned about the DNA rearrangements that occur to produce functional light-chain and heavy-chain genes. Susumu Tonegawa received a Nobel Prize in 1987 for his discovery of this mechanism of antibody diversity. He published the research in: N. Hozumi and S. Tonegawa. 1976. Evidence for somatic rearrangement of immunoglobulin genes coding for variable and constant regions. *Proceedings of the National Academy of Sciences USA* 73:3628–3632. Read the paper and make a flow chart outlining the experiments the investigators used in their research.

45.4 Malfunctions and Failures of the Immune System

The immune system is highly effective, but it is not foolproof. Some malfunctions of the immune system cause the body to react against its own proteins or cells, producing **autoimmune disease.** In addition, some viruses and other pathogens have evolved means to avoid destruction by the immune system.

A number of these pathogens, including HIV (the virus that causes AIDS), even use parts of the immune response to promote infection. Another malfunction causes the *allergic reactions* that most of us experience from time to time.

An Individual's Own Molecules Are Normally Protected against Attack by the Immune System

B cells and T cells are involved in the development of **immunological tolerance,** which protects the body's own molecules from attack by the immune system. In this system, molecules present in an individual from birth are not recognized as foreign by circulating B and T cells, and do not elicit an immune response. Self-reactive B and T cells are eliminated (by apoptosis or inactivation) throughout life so that self-tolerance is maintained.

Evidence that immunological tolerance is established early in life comes from experiments with mice. For example, if a foreign protein is injected into a mouse at birth, during the period in which tolerance is established, the mouse will not develop antibodies against the protein if it is injected later in life.

Autoimmune Disease Occurs When Immunological Tolerance Fails

The mechanisms setting up immunological tolerance sometimes fail, leading to an **autoimmune reaction**—the production of antibodies against molecules of the body. In most cases, the effects of such anti-self antibodies are not serious enough to produce recognizable disease. However, in about 5–10% of the human population, anti-self antibodies cause serious problems.

There are many examples of autoimmune diseases. *Type 1 diabetes* (see Section 42.4) is caused by an autoimmune reaction against the pancreatic beta cells producing insulin. The anti-self antibodies gradually eliminate the beta cells until the individual is incapable of producing insulin. **Systemic lupus erythematosus (lupus)** is caused by production of a wide variety of anti-self antibodies against blood cells, blood platelets, and internal cell structures and molecules such as mitochondria and proteins associated with DNA in the cell nucleus. People with lupus often become anemic and have problems with blood circulation and kidney function because antibodies, combined with body molecules, accumulate and clog capillaries and the microscopic filtering tubules of the kidneys. Lupus patients may also develop anti-self antibodies against the heart and kidneys. **Rheumatoid arthritis** is caused by a self-attack on connective tissues, particularly in the joints, causing pain and inflammation. **Multiple sclerosis** results from an autoimmune attack against a protein of the myelin sheaths insulating the surfaces of neurons. Multiple sclerosis can seriously disrupt nervous function, producing such symptoms as muscle weakness and paralysis, impaired coordination, and pain.

The causes of most autoimmune diseases are unknown. In some cases, an autoimmune reaction can be traced to injuries that expose body cells or proteins that are normally inaccessible to the immune system, such as the lens protein of the eye, to B

and T cells. For a number of autoimmune diseases, genetic causes (for example, particular mutations, or combinations of alleles in the body) and/or environmental causes (for example, chemicals or drugs) are considered likely.

Some Pathogens Have Evolved Mechanisms That Avoid Their Destruction by the Immune System

Several pathogens regularly change their surface groups to avoid destruction by the immune system. By the time the immune system has developed antibodies against one version of the surface proteins, the pathogens have switched to different surface proteins that the antibodies do not match. These new proteins take another week or so to stimulate the production of specific antibodies, by which time the surface groups have changed again. The changes continue indefinitely, always keeping the pathogens one step ahead of the immune system. Pathogens that use these mechanisms to sidestep the immune system include *Trypanosoma brucei*, the protist that causes African sleeping sickness (see Section 27.2); the bacterium that causes gonorrhea; and the viruses that cause influenza, the common cold, and AIDS.

Some viruses, such as HIV, use parts of the immune system to get a free ride to the cell interior. The structure and life cycle of HIV are described in Section 17.2. HIV has a surface molecule that is recognized and bound by the CD4 receptor on the surface of helper T cells. Binding to CD4 locks the virus to the cell surface and stimulates the membrane covering the virus to fuse with the plasma membrane of the helper T cell. (The protein coat of the virus is wrapped in a membrane derived from the plasma membrane of the host cell in which it was produced.) The fusion introduces the virus into the cell, initiating the infection. Within the cell, the viral-encoded reverse transcriptase makes a DNA copy of the viral RNA genome. The DNA copy integrates into the host cell's genome, where it is replicated as the cell divides. In this integrated state, the virus is protected from attack by the immune system. When the helper T cell is stimulated by an antigen, the integrated HIV genome is expressed to produce new viral RNA genomes and mRNAs that direct host cell ribosomes to make viral proteins, and new HIV particles are generated. Once released from the cell (which kills the cell), the viral particles may infect more body cells or another person. HIV also infects and kills macrophages. With time, more and more helper T cells and macrophages are destroyed, eventually crippling the body's immune response. The infected person becomes susceptible to opportunistic infections by viruses, bacteria, and fungi, and to development of otherwise rare forms of cancer. These secondary infections signal the appearance of full-blown AIDS. Steady debilitation and death typically follow within a period of years in untreated persons. Fortunately, the development of AIDS can be greatly slowed by drugs that interfere with reverse transcription of the viral genomic RNA into the DNA copy that integrates into the host's genome.

Some pathogens enter a **latent state** during an infection, meaning that the pathogen cannot be isolated from the infected organism and identified. Pathogens in a latent state are not recognized by the immune system. For example, herpes simplex virus can enter peripheral nerves and remain in a latent state for years. Its sudden emergence from latency causes cold sores. Similarly, the bacterium *Mycobacterium tuberculosis,* the causative agent of tuberculosis, may enter a latent state during which no symptoms of tuberculosis are exhibited and the person is not infectious. Emergence from this latent state causes the development of tuberculosis.

Allergies Are Produced by Overactivity of the Immune System

The substances responsible for allergic reactions form a distinct class of antigens called **allergens,** which induce B cells to secrete an overabundance of IgE antibodies **(Figure 45.13).** The IgE antibodies, in turn, bind to receptors on mast cells in connective tissue and on **basophils,** a type of leukocyte in blood (see Table 45.1), inducing them to secrete histamine, which produces severe inflammation. Most of the inflammation occurs in tissues directly exposed to the allergen, such as the surfaces of the eyes, the lining of the nasal passages, and the air passages of the lungs. Signaling molecules released by the activated mast cells also stimulate mucosal cells to secrete floods of mucus and cause smooth muscle in airways to constrict (histamine also causes airway constriction). The resulting allergic reaction can vary in severity from a mild irritation to serious and even life-threatening debilitation. **Asthma** is a severe response to allergens involving constriction of airways in the lungs. **Antihistamines**—substances that block histamine receptors—are usually effective in countering the effects of the histamine released by mast cells.

An individual is *sensitized* by a first exposure to an allergen, which may produce only mild allergic symptoms or no reaction at all **(Figure 45.13A).** However, the sensitization produces memory B and T cells. At the next and subsequent exposures, the system is poised to produce a greatly intensified allergic response **(Figure 45.13B).**

In some persons, inflammation stimulated by an allergen is so severe that the reaction brings on a life-threatening condition called **anaphylactic shock.** Among other symptoms, extreme constriction of air passages in the lungs interferes with breathing, and massive leakage of fluid from capillaries causes the blood pressure to drop precipitously. Death may result in minutes if the condition is not treated promptly. In persons who have become sensitized to the venom of wasps and bees, for example, a single sting may bring on anaphylactic shock within minutes. Allergies developed against drugs such as penicillin and certain foods can have the same drastic effects. Anaphylactic shock can be controlled by immediate injection of epinephrine (adrenaline), which reverses the condition by constricting blood vessels and dilating air passages in the lungs.

STUDY BREAK 45.4

1. What is immunological tolerance?
2. Explain how a failure in the immune system can result in an allergy.

THINK OUTSIDE THE BOOK

A number of so-called new-generation drugs have been developed to treat patients with rheumatoid arthritis. One of them is abatacept (Orencia). Use the Internet or research literature to develop an outline of how abatacept works with respect to the adaptive immune system described in this section.

A. Initial exposure to allergen

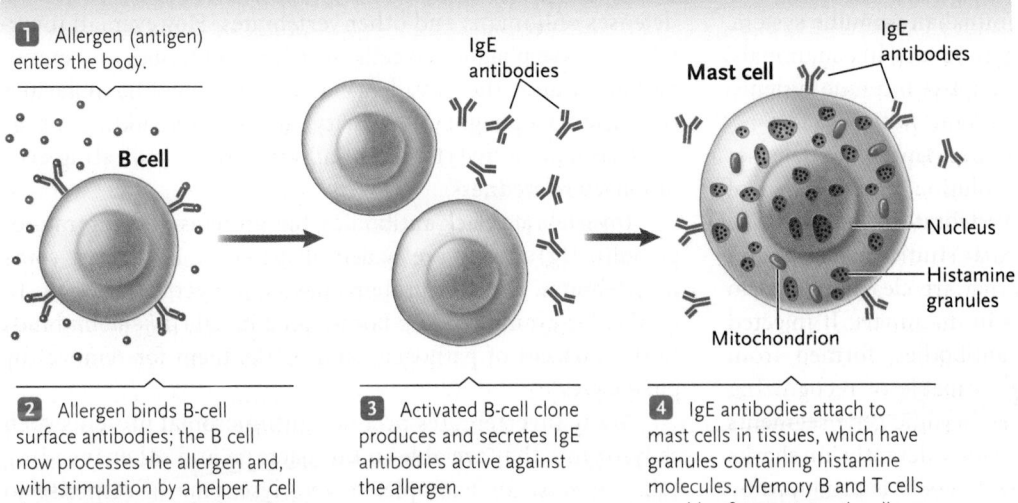

1 Allergen (antigen) enters the body.

B cell

IgE antibodies

Mast cell

IgE antibodies

Nucleus

Histamine granules

Mitochondrion

2 Allergen binds B-cell surface antibodies; the B cell now processes the allergen and, with stimulation by a helper T cell (not shown), proceeds through the steps leading to cell division and antibody production.

3 Activated B-cell clone produces and secretes IgE antibodies active against the allergen.

4 IgE antibodies attach to mast cells in tissues, which have granules containing histamine molecules. Memory B and T cells capable of recognizing the allergen are also produced.

B. Further exposures to allergen

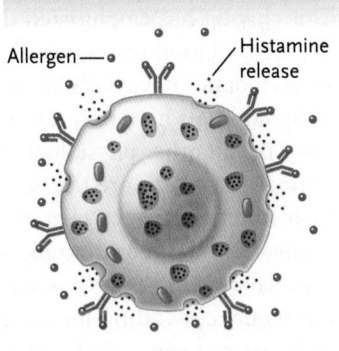

Allergen

Histamine release

5 After the first exposure, when the allergen enters the body, it binds with IgE antibodies on mast cells; binding stimulates the mast cell to release histamine and other substances.

FIGURE 45.13 The response of the body to allergens. (A) The steps in sensitization after initial exposure to an allergen. **(B)** Production of an allergic response by further exposures to the allergen.

© Cengage Learning 2017

Unique Immune System of the Atlantic Cod

The Atlantic cod (*Gadus morhua*) **(Figure)** is a cold-adapted teleost fish (see Section 32.5) that is important for the commercial fishing and fish farming industries. The cod's immune system is unusual compared with other teleosts: high levels of IgM, a low antibody response following exposure to a pathogen, and high levels of neutrophils in the circulating blood (see Figure 45.1).

Research Question

What is the genetic basis for the Atlantic cod's unusual immune system?

FIGURE Atlantic cod.

Gertjan Hooijer/Shutterstock.com

Experiments

Kjetill Jakobsen and his colleagues at the University of Oslo, Norway, sequenced and analyzed the genome of the cod to identify the genes for its immune system.

Results

The Atlantic cod genome was calculated to be 830 Mb, with an estimated 20,095 protein-coding genes. The researchers found that most genes for vertebrate immunity are present, but notably absent are the genes for class II MHC, the CD4 receptor, and a protein that aids in assembly and transport of class II MHC to the cell surface. These three proteins are part of the antibody-mediated immune response of adaptive immunity that deals with pathogens (see Figure 45.5). Without the proteins, the antibody-mediated immune response cannot operate. Indeed, in mammals, class II MHC is highly important as demonstrated by the immune defects shown in mice engineered genetically to lack the molecule.

The scientists discovered that the Atlantic cod appears to have compensated for the lack of class II MHC by having 10 times more class I MHC genes than other vertebrates. Class I MHC is involved in the cell-mediated immune response of adaptive immunity (see Figure 45.10). The interpretation is that, in the Atlantic cod, bacterial pathogens can activate the class I MHC/CD8+ T cell system.

Conclusion

The immune system seen in the Atlantic cod is unique among vertebrates that have been studied. The adaptation of the class I MHC system to respond to bacterial pathogens illustrates the remarkable plasticity of adaptive immunity in jawed vertebrates.

think like a scientist

How would you show if the absence of the three genes is specific to the Atlantic cod or if it is common to all of the cod family?

Source: B. Star et al. 2011. The genome sequence of Atlantic cod reveals a unique immune system. *Nature* 477:207–210.

© Cengage Learning 2017

45.5 Evolved Defenses against Pathogens in Other Animals

This chapter has emphasized the mammalian immune system, the focus of most immunology research. In fact, like mammals, all vertebrates have both innate and adaptive immune systems. Other animals have only an innate immune system.

Like all other physiological systems, mammalian defenses against pathogens are the result of evolution, and evidence of their functions can be seen in other vertebrate groups and also in invertebrates. For example, molecular studies in sharks and rays have revealed DNA sequences that are clearly related to the sequences coding for antibodies in mammals. If injected with an antigen, sharks produce antibodies, formed from light- and heavy-chain polypeptides, capable of recognizing and binding the antigen. Although embryonic gene segments for the two polypeptides are arranged differently in sharks than they are in mammals, antibody diversity is produced by the same kinds of genetic rearrangements in both. Sharks also mount highly efficient nonspecific defenses, including production of a steroid that appears to kill bacteria and neutralize viruses. *Molecular Insights* discusses an unusual immune system in the Atlantic cod.

Invertebrates lack an adaptive immune system, but do have an innate immune system. That is, their reactions to invading pathogens most closely resemble the nonspecific defenses of humans and other vertebrates. However, all invertebrates have phagocytic cells, which patrol tissues and engulf pathogens and other invaders. Some of the signaling molecules that stimulate phagocytic activity, such as interleukins, appear to be similar in invertebrates and vertebrates, indicating evolutionary relatedness.

Invertebrates lack antibodies, but proteins of the immunoglobulin (Ig) family are widely distributed. In at least some invertebrates, these Ig proteins have a protective function. In moths, for example, an Ig-family protein called *hemolin* binds to the surfaces of pathogens and marks them for removal by phagocytes.

Many invertebrates produce antimicrobial proteins such as lysozyme that are able to kill bacteria and other invading cells. Insects, for example, secrete lysozyme in response to bacterial infections.

STUDY BREAK 45.5

Compare invertebrate and mammalian immune defenses.

How does HIV evade the adaptive immune system?

In cell-mediated immunity, a pathogen-infected antigen-presenting cell presents an antigen fragment bound to a class I major histocompatibility complex (MHC) protein to a CD8$^+$ T cell, stimulating the T cell to differentiate into cytotoxic T cells. The mature cytotoxic T cells specifically recognize viral antigen bound to MHC on infected cells and destroy them. Cytotoxic T cells act particularly against host cells infected by viral pathogens. HIV infects host cells but, rather than being eliminated by the host, this virus establishes a chronic infection that leads to the development of AIDS. That is, HIV evades the adaptive immune system.

My research group at the University of Michigan Medical School has investigated the mechanism of this evasion. We have learned that the virus down-regulates the display of class I MHC proteins on the surface of HIV-infected cells, which limits the presentation of viral antigens by those cells. The down-regulation occurs by the action of the HIV Nef (*negative factor*) protein. Nef binds to class I MHC molecules and inhibits them from moving through the Golgi complex to the cell surface. Without class I MHC molecules on the cell surface to present antigens, the immune response is compromised. The action of Nef, therefore, enhances the ability of HIV to cause a persistent infection that ultimately leads to AIDS.

A major research question addressed by my laboratory is: How does Nef disrupt the movement of class I MHC molecules from the ER through the Golgi complex to the cell surface? Recent experiments have shown that the disruption results from an interaction of Nef with two cellular proteins, adaptor protein 1 and β-COP. The normal function of these cellular proteins is to help target other cellular proteins to their correct locations within the cell. By sequentially linking the two cellular proteins to MHC, Nef promotes the trafficking of the class I MHC proteins to

lysosomes (where they are degraded), rather than to the cell surface. Nef links adaptor protein to a tyrosine residue in the MHC-I protein.

Different types of MHC-I proteins are expressed by cells, and these proteins are highly polymorphic. The diversity of MHC-I proteins allows cells to present a wide range of antigens. All cells express six alleles of MHC-I: two alleles of MHC-I HLA-A, two alleles of MHC-I HLA-B, and two alleles of MHC-I HLA-C. MHC-I HLA-A and HLA-B proteins have the tyrosine amino acid that Nef needs to attach adaptor protein 1 and disrupt antigen presentation. However, HLA-C molecules lack this tyrosine and HLA-C is not down-regulated by Nef. HLA-C may be protective in HIV disease. There is growing evidence that individuals who naturally express higher levels of HLA-C have a more favorable prognosis in HIV disease. This may result from the fact that HLA-C is resistant to the effects of Nef.

My research team is now working to develop pharmaceutical reagents aimed at blocking Nef action as a means of reducing HIV's attack on the immune system.

think like a scientist HLA-C is the most recently evolved of the human MHC-I proteins. Can you think of why new MHC-I molecules might evolve in a species?

Courtesy of Kathleen Collins

Kathleen Collins is a Professor of Internal Medicine and Microbiology and Immunology at the University of Michigan at Ann Arbor. She does research on the strategies HIV utilizes to establish a persistent infection and teaches virology to undergraduate and graduate students. Learn more about her research at http://www.med.umich.edu/microbio/bio/collins.htm.

REVIEW KEY CONCEPTS

For access to MindTap and additional study materials visit www.cengagebrain.com.

45.1 Three Lines of Defense against Pathogens

- Humans and other vertebrates have three lines of defense against pathogens: (1) a nonspecific system involving physical barriers set up by epithelial surface (not part of the immune system); (2) innate immunity, a nonspecific inborn system that defends the body against pathogens and toxins penetrating the first line; and (3) adaptive immunity, a specific inborn system that recognizes and eliminates particular pathogens and retains a memory of that exposure so as to respond rapidly if the pathogen is encountered again. The response is carried out by lymphocytes, a specialized group of leukocytes (Table 45.1 and Figure 45.1).

45.2 Innate Immunity: Nonspecific Defenses

- In the innate immune system, molecules on the surfaces of pathogens are recognized as foreign by receptors on host cells. The pathogen is then combated by the inflammation and complement systems.

- Epithelial surfaces secrete defensins, a type of antimicrobial peptide, in response to attack by a microbial pathogen. Defensins disrupt the plasma membranes of pathogens, killing them.

- Inflammation is characterized by heat, pain, redness, and swelling at the infection site. Several interconnecting mechanisms initiate inflammation, including pathogen engulfment, histamine secretion, cytokine release, and local blood vessel dilation and permeability increase (Figure 45.2).

- Large arrays of complement proteins are activated when they recognize molecules on the surfaces of pathogens. Some complement proteins form membrane attack complexes, which insert themselves into the plasma membrane of many types of bacteria and cause their lysis. Fragments of other complement proteins coat pathogens, stimulating phagocytes to engulf them.

- Two nonspecific defenses are used to combat viral pathogens: interferons and natural killer cells.

45.3 Adaptive Immunity: Specific Defenses

- Adaptive immunity is a vertebrate-specific defense mechanism that is carried out by B and T cells, and targets particular pathogens or toxin molecules.

- Antibodies consist of two light and two heavy polypeptide chains, each with variable and constant regions. The variable regions of the chains combine to form the specific antigen-binding site (Figure 45.4).

- Antibodies occur in five different classes: IgM, IgD, IgG, IgA, and IgE. Each class is determined by its constant region.

- Antibody diversity is produced by genetic rearrangements in developing B cells that combine gene segments into intact genes encoding the light and heavy chains. The rearrangements producing heavy-chain genes and T-cell receptor genes are similar. The light- and heavy-chain genes are transcribed into pre-mRNAs, which are processed into finished mRNAs, which are translated on ribosomes into the antibody polypeptides (Figure 45.5).

- The antibody-mediated immune response has two general phases: T-cell activation and B-cell activation and antibody production. T-cell activation begins when a dendritic cell engulfs a pathogen and displays antigens derived from the pathogen on its surface, making the cell an antigen-presenting cell. The antigen-presenting cell secretes interleukins, which activate the T cell. The T cell then secretes other interleukins, which stimulate the T cell to proliferate, producing a clone of cells. The clonal cells differentiate into helper T cells (Figures 45.6 and 45.12).

- B-cell receptors on B cells recognize antigens on a pathogen and engulf it. The B cells then display the antigens. The T-cell receptor on a helper T cell activated by the same antigen binds to the antigen on the B cell. Interleukins from the T cell stimulate the B cell to produce a clone of cells with identical B-cell receptors. The clonal cells differentiate into plasma cells, which secrete antibodies specific for the antigen, and memory B cells, which provide immunological memory of the antigen encounter (Figure 45.6).

- Clonal expansion is the process of selecting a lymphocyte specifically for cloning when it encounters an antigen from among a randomly generated, large population of lymphocytes with receptors that specifically recognize the antigen (Figure 45.7).

- Antibodies clear the body of antigens by neutralizing or agglutinating them, or by aiding the innate immune response (Figure 45.8).

- In immunological memory, the first encounter of an antigen elicits a primary immune response and later exposure to the same antigen elicits a rapid secondary response with a greater production of antibodies (Figure 45.9).

- Active immunity is the production of antibodies in the body in response to an antigen. Passive immunity is the acquisition of antibodies by direct transfer from another person.

- Antibodies are widely used in research to identify, locate, and determine the functions of molecules in biological systems.

- Monoclonal antibodies are made by isolating fully active B cells from a test animal, fusing them with cancer cells to produce hybridomas, and using single hybridomas to start clones of cells, all of which make highly specific antibodies against the same epitope of an antigen (Figure 45.10).

- In cell-mediated immunity, cytotoxic T cells recognize and bind to antigens displayed on the surfaces of infected body cells, or to cancer cells. They then kill the infected body cell (Figures 45.11 and 45.12).

45.4 Malfunctions and Failures of the Immune System

- In immunological tolerance, molecules present in an individual at birth normally do not elicit an immune response.

- In some people, the immune system malfunctions and reacts against the body's own proteins or cells, producing autoimmune disease.

- Some pathogens have evolved mechanisms that avoid their destruction by the immune system.

- The first exposure to an allergen sensitizes an individual by leading to the production of memory B and T cells, which cause a greatly intensified response at the next and subsequent exposures.

- Most allergies result when antigens act as allergens by stimulating B cells to produce IgE antibodies, which leads to the release of histamine. Histamine produces the symptoms characteristic of allergies (Figure 45.13).

45.5 Evolved Defenses against Pathogens in Other Animals

- Vertebrates have both the innate immune system and the adaptive immune system.

- Invertebrates lack an adaptive immune system and rely on innate immune systems for nonspecific defenses, including surface barriers, phagocytes, and antimicrobial molecules.

TEST YOUR KNOWLEDGE

Remember/Understand

1. Which of the following most accurately describes mammal defenses against disease-causing viruses or organisms?
   a. The three lines of defense work independently of each other in defending against a particular pathogen.
   b. Physical barriers are part of the immune system.
   c. The adaptive immune system reacts faster to pathogens than the innate immune system.
   d. Once the adaptive immune system is activated in response to a specific pathogen, the innate immune system stops functioning against that specific pathogen.
   e. White blood cells are key participants in adaptive immunity but not innate immunity.

2. Which of the following is *not* a component of the inflammatory response?
   a. macrophages   d. mast cells
   b. neutrophils   e. eosinophils
   c. B cells

3. When a person's immune system resists infection by a pathogen after being vaccinated against it, this is the result of:
   a. innate immunity.
   b. immunological memory.
   c. a response with defensins.
   d. an autoimmune reaction.
   e. systemic inflammation.

4. One characteristic of a B cell is that it:
   a. has the same structure in both invertebrates and vertebrates.
   b. recognizes antigens held on class I major histocompatibility complex proteins.
   c. binds virus-infected cells and kills them directly.
   d. makes many different B-cell receptors on its surface.
   e. has a B-cell receptor on its surface, which is the IgM molecule.

5. Antibodies:
   a. are each composed of four heavy and four light chains.
   b. display a variable end, which determines the antibody's location in the body.
   c. belonging to the IgE group are the major antibody class in the blood.
   d. found in large numbers in the mucous membranes belong to class IgG.
   e. function primarily to identify and bind antigens free in body fluids.

6. Place the following steps of recognition of an antigen by lymphocytes in the order in which they would occur:
   (1) Bacteria degraded
   (2) Dendritic cell engulfs bacteria
   (3) T cells activated
   (4) Antigens bind to class II MHC proteins
   (5) Plasma cells and memory cells produced
   a. 1, 2, 3, 4, 5.   d. 2, 1, 4, 3, 5.
   b. 3, 2, 1, 4, 5.   e. 5, 4, 3, 2, 1.
   c. 4, 3, 1, 2, 5.

7. An antigen-presenting cell:
   a. can be a CD8$^+$ T cell.
   b. derives from a phagocytic cell and exposes an antigen to a lymphocyte.
   c. secretes antibodies.
   d. cannot be a B cell.
   e. cannot stimulate helper T cells.

8. Antibodies function to:
   a. deactivate the complement system.
   b. neutralize natural killer cells.
   c. clump bacteria and viruses for easy phagocytosis by macrophages.
   d. eliminate the chance for a secondary response.
   e. kill viruses inside of cells.

9. After Sally punctured her hand with a dirty nail, she received both a vaccine and someone else's antibodies against tetanus toxin. The immunity conferred here is:
   a. both active and passive.
   b. active only.
   c. passive only.
   d. first active; later passive.
   e. innate.

10. Drugs are administered to patients to enhance the immune response when treating:
    a. organ transplant recipients.
    b. anaphylactic shock.
    c. rheumatoid arthritis.
    d. HIV infection.
    e. type I diabetes.

Apply/Analyze

11. **Discuss Concepts** HIV wreaks havoc with the immune system by attacking helper T cells and macrophages. Would the impact be altered if the virus attacked only macrophages? Explain.

12. **Discuss Concepts** Given what you know about how foreign invaders trigger immune responses, explain why mutated forms of viruses, which have altered surface proteins, pose a monitoring problem for memory cells.

13. **Discuss Concepts** Cats, dogs, and humans may develop myasthenia gravis, an autoimmune disease in which antibodies develop against acetylcholine receptors in the synapses between neurons and skeletal muscle fibers. Based on what you know of the biochemistry of muscle contraction (see Section 43.1), explain why people with this disease typically experience severe fatigue with even small levels of exertion, drooping of facial muscles, and trouble keeping their eyelids open.

Evaluate/Create

14. **Design an Experiment** Space, the final frontier! Indeed, but being in space causes some problems. Astronauts in space show a decline in their ability to mount an immune response and, consequently, develop a decreased resistance to infection. Two potentially important differences in physiology in space versus on Earth are more fluid flowing to the head and a lack of weight bearing on the lower limbs. Could they be involved somehow in the deleterious effect on the immune system? Design an experiment to be done on Earth to answer this question.

15. **Apply Evolutionary Thinking** Defensins are found in a wide range of organisms, including plants as well as animals. What are the evolutionary implications of this observation?

For selected answers, see Appendix A.

INTERPRET THE DATA

In 2003, Michelle Khan and her coworkers published their findings on a 10-year study in which they followed cervical cancer incidence and human papillomavirus (HPV) status in 20,514 women. All women who participated in the study were free of cervical cancer when the test began. Papanicolaou (Pap) tests were taken at regular intervals, and the researchers used a DNA probe hybridization test to detect the presence of specific types of HPV in the women's cervical cells.

The results are shown as a graph of the incidence rate of cervical cancer by HPV type **(Figure).** Women who are HPV positive are often infected by more than one type, so the data were sorted into groups based on the women's HPV status ranked by type: either positive for HPV16; or negative for HPV16 and positive for HPV18; or negative for HPV16/18 and positive for any other cancer-causing HPV; or negative for all cancer-causing HPV.

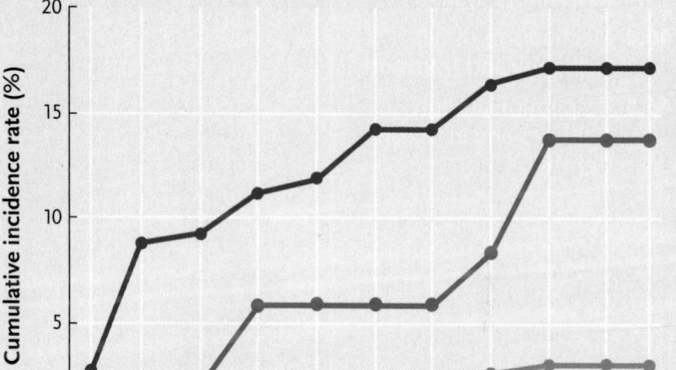

KEY
— HPV16 positive
— HPV16 negative and HPV18 positive
— HPV16/18 negative but positive for other HPV
— Negative for all cancer-causing HPV

FIGURE Cumulative incidence rate of cervical cancer correlated with HPV status in 20,514 women ages 16 years and older.

1. At 110 months into the study, what percentage of women who were not infected with any type of cancer-causing HPV had cervical cancer? What percentage of women who were infected with HPV16 also had cervical cancer?

2. In which group would women infected with both HPV16 and HPV18 fall?

3. Is it possible to estimate from this graph the overall risk of cervical cancer associated with infection of cancer-causing HPV of any type?

4. Do these data support the conclusion that being infected with HPV16 or HPV18 raises the risk of cervical cancer?

Source: Based on M. J. Khan et al. 2005. The elevated 10-year risk of cervical neoplasia in women with human papillomavirus (HPV) type 16 or 18 and the possible utility of type-specific HPV testing in clinical practice. *Journal of the National Cancer Institute* 97:1072–1079.

46 Gas Exchange: The Respiratory System

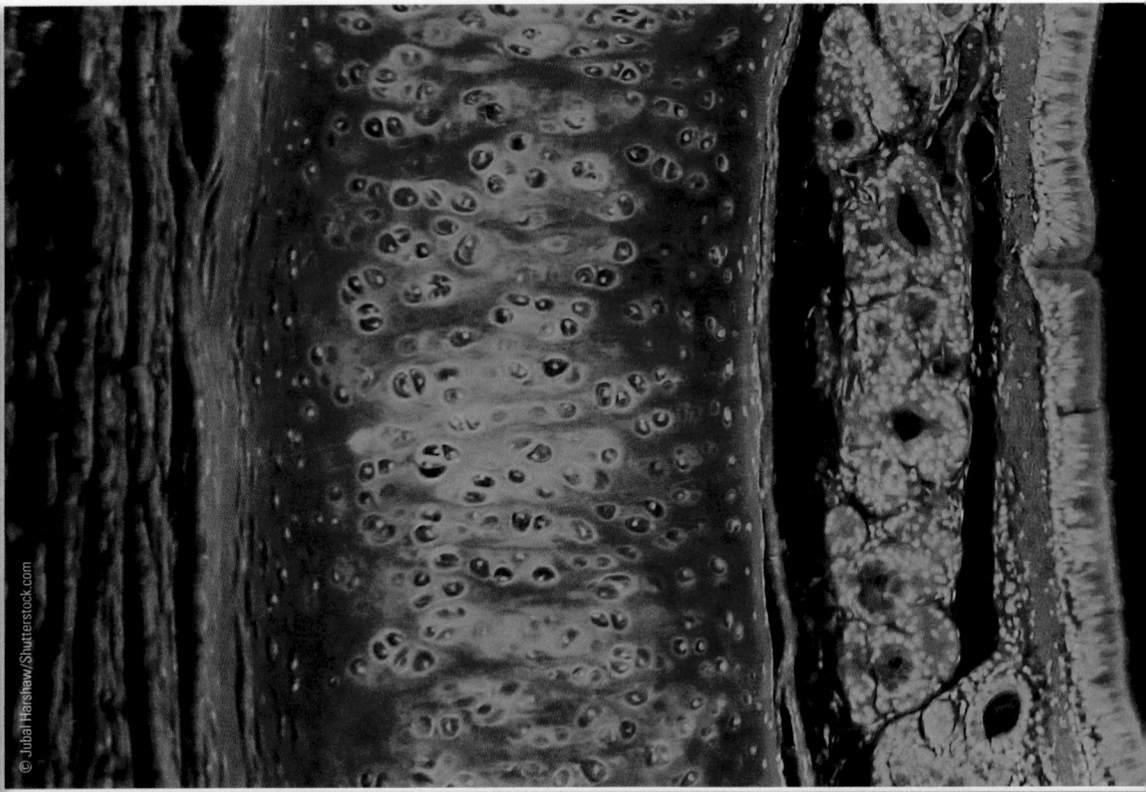

© Jubal Harshaw/Shutterstock.com

Cross section of a trachea (windpipe) with cilia on the surface (right side of photo) (fluorescent micrograph).

Why it matters . . . Late in 1999, a Learjet N47BA took off from Orlando International Airport. As the jet continued its climb toward a cruising altitude of 39,000 feet, the pressure of the outside air dropped steadily and with it the availability of the oxygen (O_2) that all animal life requires, including the pilots and passengers on the jet. Normally, the cabin pressure in aircraft is maintained at a level equivalent to an altitude of 8,000 feet, more than sufficient to keep O_2 available to all on board. But the pressurization system was not functioning normally.

When humans experience increasingly higher altitudes, each breath brings less O_2 into the body. Of all the cells affected by reduced O_2, the ones most sensitive are those of the eyes and brain. Without an O_2 supply at 25,000 feet, most people progress from fully alert to unconscious in about 3 minutes; at 40,000 feet, the progression takes only 15 seconds.

Communication with the plane was lost. The jet continued its climb, eventually reaching an altitude of 46,000 feet. Military pilots sent to investigate saw no movement in the Learjet cabin. Evidently, the aircraft was maintaining its course through the autopilot.

Many hours later, the aircraft crashed into a field near Aberdeen, South Dakota. The subsequent investigation pointed to failure of the cabin pressurization system as the cause of the accident. This tragic loss of life emphasizes the vital importance of O_2 to the survival of humans and other animals. In this chapter we discuss the respiratory system, the system that allows an animal to exchange CO_2 produced in the body for O_2 from the surroundings. The respiratory systems of animals reflect the environmental conditions under which they live, and this general principle has resulted in a truly remarkable array of adaptations.

46.1 The Function of Gas Exchange

Physiological respiration (also called *external respiration*) is the sequence of events by which animals exchange gases with their surroundings—how they take in O_2 from the outside environment and deliver it to body cells, and remove CO_2 from body cells and deliver it to the environment **(Figure 46.1)**. The absorbed O_2 is used as the final electron acceptor for the oxidative reactions that produce ATP in mitochondria (see Section 7.4). The CO_2 released to the environment is a product of those oxidative reactions. Because they use O_2 and release CO_2, these ATP-producing reactions are called *cellular respiration*.

How gas exchange occurs in an animal depends on its respiratory medium—air or water—and the nature of its respiratory surface. The **respiratory medium** is the environmental source of O_2 and the "sink" for released CO_2. For aquatic animals the respiratory medium is water; for terrestrial animals, it is air. Amphibians and some fishes use both water and air as respiratory media.

The **respiratory surface,** formed by a layer of epithelial cells, provides the interface between the body and the respiratory medium. Oxygen is absorbed across the respiratory surface, and CO_2 is released. In all animals, the exchange of gases across the respiratory surface occurs by simple diffusion, movement of molecules from a region of higher concentration to a region of lower concentration (see Section 5.2). The rate of diffusion becomes higher with larger concentration gradients and with increasing temperature.

Generally, the concentration of O_2 is higher in the respiratory medium than on the internal side of the respiratory surface, and thus the net diffusion of O_2 is inward. Carbon dioxide moves in the opposite direction because the CO_2 concentration is higher on the internal side of the respiratory surface than in the respiratory medium.

Respiratory surfaces typically have two structural properties that favor a high rate of diffusion:

1. *Respiratory surfaces are thin.* The rate of diffusion is inversely proportional to the square of the distance over which the diffusion occurs. Diffusion rates are therefore higher through thin surfaces such as the single layer of epithelial cells forming many respiratory surfaces.
2. *Respiratory surfaces have large surface areas.* The rate of diffusion is directly proportional to the surface area across which diffusion occurs, meaning that large surface areas allow for higher rates of gas exchange than small surface areas.

In some relatively small animals, such as sponges, ctenophores, roundworms, flatworms, and some annelids, the entire body surface serves as the respiratory surface. All these animals are invertebrates that live in aquatic or moist environments.

In larger animals, specialized structures, *gills* and *lungs,* form the primary respiratory surface for exchanging gases with water and air, respectively. In insects, a **tracheal system,** an extensive system of branching tubes, channels air from the outside to the internal organs and most individual cells of the animal.

Because gases must dissolve in water to enter and leave epithelial cells, the respiratory surface must be wetted to function in gas exchange, either directly by the respiratory medium or by a thin film of water. For this reason, in water-breathing animals, **gills** are *evaginations* of the body: they extend outward into the respiratory medium. In terrestrial animals, **lungs** are typically pockets or *invaginations* of the body surface, buried deeply in the body interior where they are less susceptible to drying out. Terrestrial animals also have adaptations that moisten dry air before it reaches the respiratory surface. For example, in humans and other mammals, moisture is added to air as it passes through the mouth, nasal passages, throat, and air passages leading to the lungs.

The organ system responsible for gas exchange is termed the **respiratory system.** The respiratory system consists of all the parts of the body involved in exchanging air between the external environment and the blood. In mammals, this includes the airways leading to and into the lungs, the lungs themselves, and the structures of the chest used to move air through the airways into and out of the lungs.

Evolutionary Adaptations That Increase Ventilation and Perfusion of the Respiratory Surface Maximize the Rate of Gas Exchange

Two primary evolutionary adaptations help animals maintain the difference in concentration between gases outside and inside

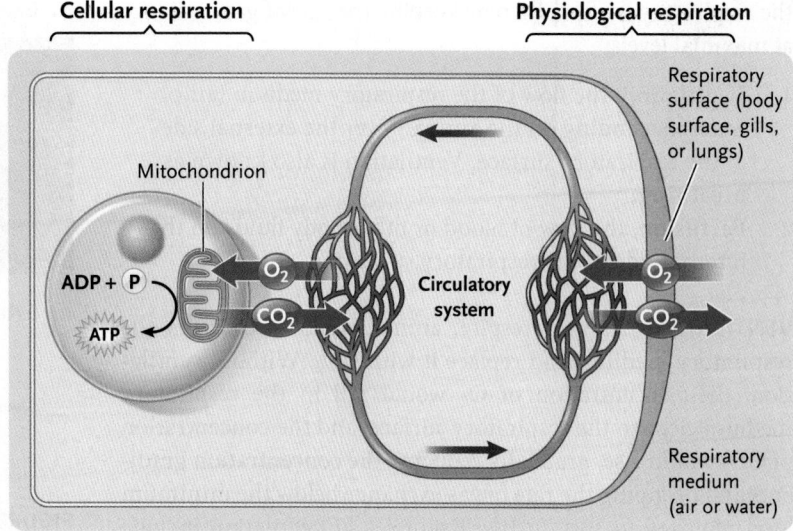

FIGURE 46.1 The relationship between cellular respiration and physiological respiration.
© Cengage Learning 2017

the respiratory surface, thereby keeping the rate of gas exchange at maximal levels:

1. **Ventilation,** the flow of the respiratory medium (air or water, depending on the animal) over the external side of the respiratory surface. Ventilation is also known as **breathing.**
2. **Perfusion,** the flow of blood or other body fluids on the internal side of the respiratory surface.

VENTILATION As they respire, animals remove O_2 from the respiratory medium and replace it with CO_2. Without ventilation, the concentration of O_2 would fall in the respiratory medium close to the respiratory surface, and the concentration of CO_2 would rise, gradually reducing the concentration gradients and dropping the rate of gas exchange below the minimum level required to sustain life. Examples of ventilation include the one-way flow of water over the gills in fish and many other aquatic animals and the in-and-out flow of air in the lungs of most vertebrates and in the tracheal system of insects.

PERFUSION The constant replacement of blood or another fluid on the internal side of the respiratory surface helps to keep the inside/outside concentration differences of O_2 and CO_2 at a maximum. In animals without a circulatory system, such as roundworms and flatworms, body movements help circulate body fluids beneath the skin. Most animals without a circulatory system are small or have thin, greatly flattened bodies, because all body cells must be located close to the respiratory surface to exchange O_2 and CO_2 adequately. In animals with a circulatory system, the circulatory system brings blood to the internal side of the respiratory surface, transporting CO_2 from all cells of the body—no matter how far they are from the respiratory surface—to exchange for O_2, which is then taken to all cells of the body.

Evolutionary Adaptations That Increase the Area of the Respiratory Surface Maximize the Quantity of Gases Exchanged

Most animals have evolutionary adaptations that increase the quantity of gases exchanged by increasing the area of the respiratory surface. In animals whose skin serves as the respiratory surface, an elongated or flattened body form increases the area of the respiratory surface **(Figure 46.2A).**

In animals with gills, the respiratory surface is increased by highly branched structures that include many fingerlike or platelike projections **(Figure 46.2B).** Similarly, in animals with lungs or tracheae, the respiratory surface is increased by a multitude of branched tubes, folds, or pockets **(Figure 46.2C).**

Water and Air Have Advantages and Disadvantages as Respiratory Media

Because their respiratory surfaces are exposed directly to the environment, water breathers easily keep the respiratory

A. **Extended body surface: flatworm**

C. **Lungs: human**

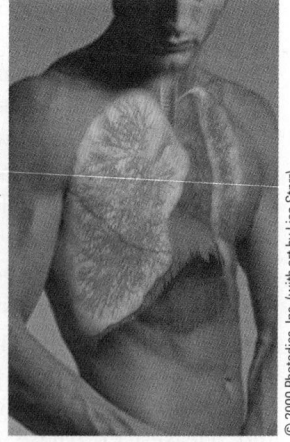

B. **External gills: salamander**

FIGURE 46.2 Evolutionary adaptations increasing the area of the respiratory surface. (A) The flattened and elongated body surface of a flatworm. **(B)** The highly branched, feathery structure of the external gills in an amphibian, the salamander (*Salamandra salamandra*). **(C)** The many branches and pockets expanding the respiratory surface in the human lung.

surface wetted. However, aquatic animals face several significant challenges in obtaining O_2 from water compared with terrestrial animals:

- Water contains approximately one-thirtieth as much O_2 as air does (at 15°C). Therefore, to obtain the same amount of O_2, an aquatic animal must process 30 times more of its respiratory medium than a terrestrial animal with similar energy needs.
- Water is about 1,000 times as dense as air and about 50 times as viscous. Therefore, it takes significantly more energy to move water than air over a respiratory surface.
- The rate of diffusion of gases in water is about 10,000-fold slower than in air.
- Temperature and solutes affect the O_2 content of water. As either the temperature or the amount of solutes increases, the amount of gas that can dissolve in water decreases. Therefore, for obtaining O_2, aquatic animals that live in warm water are at a disadvantage compared with those that live in cold water. And, because levels of solutes (such as sodium chloride) are higher in seawater than in freshwater, aquatic animals living in seawater are at a disadvantage.

For these reasons, ventilation in most aquatic animals takes place in a one-way direction, which, as you will see later, increases the efficiency of gas exchange.

Turning now to air breathers, the relatively high O_2 content, low density, and low viscosity of air greatly reduce the energy required to ventilate the respiratory surface. These advantages allow animals with lungs to breathe in and out, reversing the direction of flow of the respiratory medium, without a large energy penalty.

Living in air does have its disadvantages. One major disadvantage is that water evaporates constantly from the respiratory surface unless the air is saturated with water vapor. Therefore, except in an environment with 100% humidity, animals lose water (and heat) by evaporation during breathing and must replace the water to keep the respiratory surface from drying and causing the death of the surface cells.

We next turn to the adaptations that allow water-breathing and air-breathing animals to obtain O_2 and release CO_2 in aquatic and terrestrial environments. These adaptations allow animals to exploit the advantages and circumvent the disadvantages of water and air as respiratory media.

STUDY BREAK 46.1

1. Distinguish between the roles of the respiratory medium and the respiratory surface in respiratory systems.
2. What is an advantage of water over air as a respiratory medium? What are two key advantages of air over water as a respiratory medium?

46.2 Evolutionary Adaptations for Respiration

Although most animals that live in water exchange gases through the skin or gills, some, such as whales, seals, and dolphins, exchange gases through lungs (which originally evolved in aquatic creatures). And, although most animals that live on land exchange gases through lungs, some, such as sow bugs and land crabs, exchange gases through gills, and others, such as insects, exchange gases using a tracheal system.

Aquatic Gill Breathers Exchange Gases More Efficiently Than Skin Breathers

Gills provide water breathers, and a few air breathers, with more efficient gas exchange than skin breathers have. In combination with the organized circulatory system common to gilled animals, gills also allow animals to live in more diverse habitats, and to achieve greater body mass, than animals that breathe primarily or exclusively through the skin.

Structurally, gills are respiratory surfaces that are branched and folded evaginations of the body surface. **External gills,** which lack protective coverings, extend out from the body and contact the water directly. With no protective coverings, external gills are exposed to mechanical damage and must be immersed in water to keep them from collapsing or drying. For these reasons, animals with external gills, including some annelids and mollusks **(Figure 46.3A),** aquatic insects, the larval forms of some bony fishes, and some amphibians, are limited to relatively protected aquatic environments.

Internal gills, by contrast, are located within chambers of the body that have a cover providing physical protection for the gills and protecting them from drying. Water must be brought to internal gills. Covered internal gills allow animals to live in highly diverse aquatic habitats and even in moist terrestrial

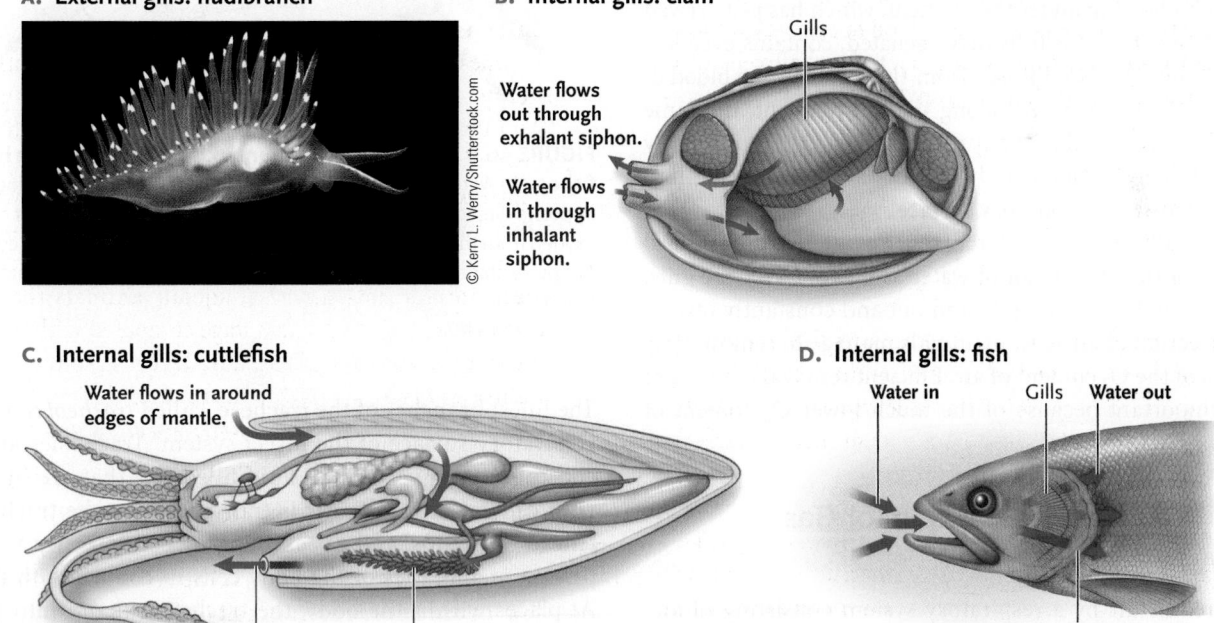

FIGURE 46.3 External and internal gills. (A) The external gills of a nudibranch *(Flabellina trophina).* **(B)** The internal gills in a clam. **(C)** The internal gills in a cuttlefish. **(D)** Internal gills of a bony fish. Water enters through the mouth and passes over the filaments of the gills before exiting through an opening at the edges of the flaplike protective covering, the operculum.

© Cengage Learning 2017

habitats. Most crustaceans, mollusks, sharks, and bony fishes have internal gills. Some invertebrates, such as clams and oysters, use beating cilia to circulate water over their internal gills **(Figure 46.3B)**. Others, such as the cuttlefish, use contractions of the muscular mantle to pump water over their gills **(Figure 46.3C)**. In adult bony fishes, the gills extend into a chamber covered by gill flaps or *opercula* (singular, *operculum*; "little lid") on either side of the head. The operculum also serves as part of the one-way pumping system that ventilates the gills: water enters the mouth, flows over the gills, and exits through the gill covers, all in one direction **(Figure 46.3D)**.

Many Animals with Internal Gills Use Countercurrent Flow to Maximize Gas Exchange

Sharks, fishes, and some crabs take advantage of one-way flow of water over the gills to maximize the amounts of O_2 and CO_2 exchanged with water. In this mechanism, called **countercurrent exchange,** the water flowing over the gills moves in a direction opposite to the flow of blood under the respiratory surface.

Figure 46.4 illustrates countercurrent exchange in the uptake of O_2. At the point where fully oxygenated water first passes over a gill filament in countercurrent flow, the blood flowing beneath it in the opposite direction is also almost fully oxygenated. However, O_2 concentration is still higher in the water than in the blood, and the gas diffuses from the water into the blood, raising the concentration of O_2 in the blood almost to the level of the fully oxygenated water. At the opposite end of the filament, much of the O_2 has been removed from the water, but the blood flowing under the filament, which has just arrived from body tissues and is fully deoxygenated, contains even less O_2. As a result, O_2 also diffuses from the water to the blood at this end of the filament. All along the gill filament, the same relationship exists, so that at any point, the water is more highly oxygenated than the blood, and O_2 diffuses from the water into the blood across the respiratory surface.

The overall effect of countercurrent exchange is the removal of 80–90% of the O_2 content of water as it flows over the gills. In comparison, by breathing in and out and constantly reversing the direction of air flow, mammals manage to remove only about 25% of the O_2 content of air. Efficient removal of O_2 from water is important because of the much lower O_2 content of water compared with air.

Insects Use a Tracheal System for Gas Exchange

Insects breathe air by a respiratory system consisting of air-conducting tubes called **tracheae** (singular, *trachea*). The tracheae are invaginations of the outer epidermis of the animal, reinforced by rings of chitin, the material of the insect exoskeleton. They lead from the body surface and branch so extensively inside the animal that almost every cell is served by a microscopic branch **(Figure 46.5)**. Some branches even penetrate inside larger cells, such as those of insect flight muscles.

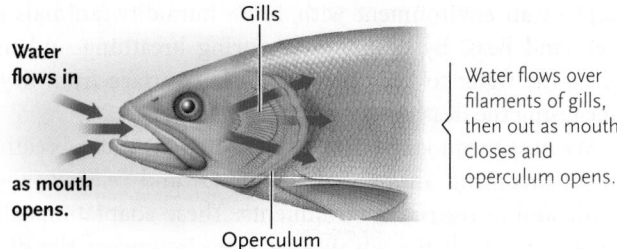

A. The flow of water around the gill filaments

B. Countercurrent flow in fish gills, in which the blood and water move in opposite directions

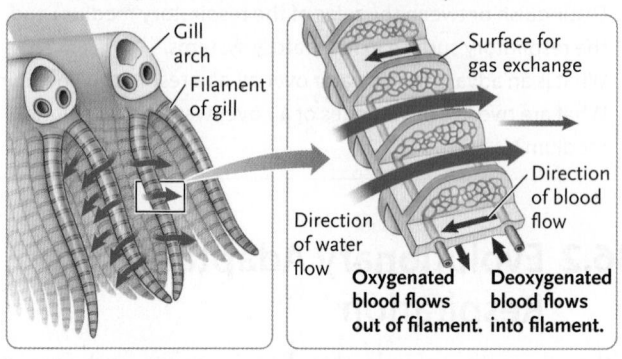

C. In countercurrent exchange, blood leaving the capillaries has the same O_2 content as fully oxygenated water entering the gills.

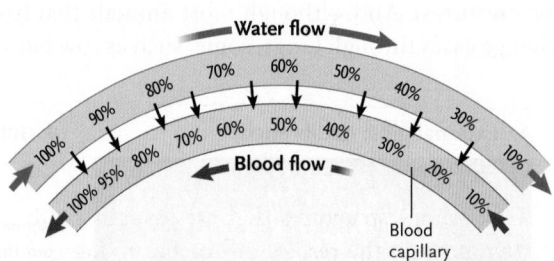

FIGURE 46.4 Ventilation and countercurrent exchange in bony fishes. (A) Water flows around the gill filaments. **(B)** Water and blood flow in opposite directions through the gill filaments. **(C)** Countercurrent exchange: oxygen from the water diffuses into the blood, raising its oxygen content. The percentages indicate the degree of oxygenation of water (blue) and blood (red).

© Cengage Learning 2017

The finest branches of the tracheae, called *tracheoles,* form the respiratory surface of the insect system. Tracheoles are dead-end tubes with very small fluid-filled tips that are in contact with cells of the body. Air is transported by the tracheal system to those tips, and gas exchange occurs directly across the plasma membranes of the body cells in contact with the tips. At places within the body, the tracheae expand into internal air sacs that act as reservoirs to increase the volume of air in the system.

Air enters and leaves the tracheal system at openings in the insect's chitinous exoskeleton called **spiracles** (*spiraculum* = air hole). In adult insects, the spiracles are located in a row on either side of the thorax and abdomen. The spiracles open and close in coordination with body movements to compress and

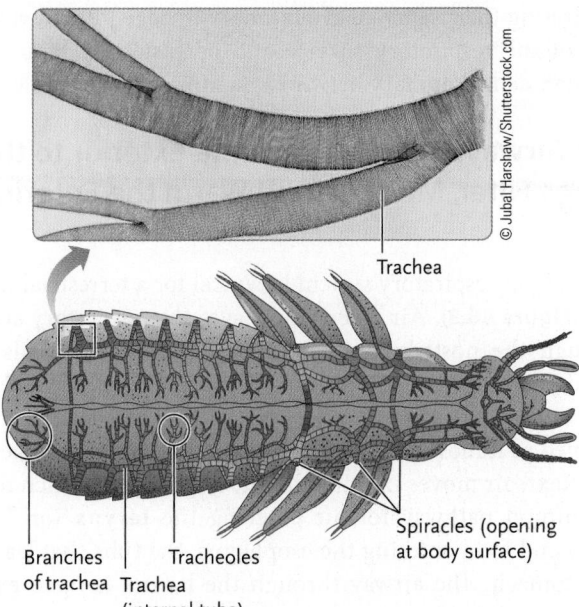

FIGURE 46.5 **The tracheal system of insects.** Chitin rings, visible in the photomicrograph, reinforce many of the tracheae.
© Cengage Learning 2017

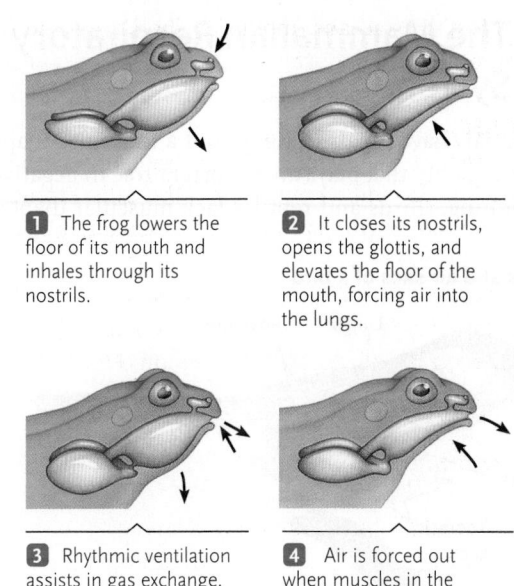

1 The frog lowers the floor of its mouth and inhales through its nostrils.

2 It closes its nostrils, opens the glottis, and elevates the floor of the mouth, forcing air into the lungs.

3 Rhythmic ventilation assists in gas exchange.

4 Air is forced out when muscles in the body wall above the lungs contract and the lungs recoil elastically.

FIGURE 46.6 **Positive pressure breathing in an amphibian (a frog).**
© Cengage Learning 2017

expand the air sacs and pump air in and out of the tracheae. During insect flight, alternating compression and expansion of the thorax by the flight muscles also pumps air through the tracheal system.

Lungs Allow Animals to Live in Completely Terrestrial Environments

Lungs are respiratory structures that are invaginated into the body and that contain the respiratory medium. Lungs are one of the primary adaptations that allowed animals to invade terrestrial environments fully. Some fishes and amphibians have lungs, as do all reptiles, birds, and mammals.

In some fishes, such as lungfishes, lungs and air breathing evolved as adaptations to survive in oxygen-poor water or temporarily in air when the water level dropped and exposed them. The lungs of these fishes consist of thin-walled sacs, which branch off from the mouth, pharynx, or parts of the digestive system; air is obtained by **positive pressure breathing,** a gulping or swallowing motion that forces air into the lungs.

The lungs of mature amphibians such as frogs and salamanders are also thin-walled sacs with relatively little folding or pocketing. Amphibians also fill their lungs by positive pressure breathing, in this case using a rhythmic motion of the floor of the mouth as the pump, in coordination with opening and closing of the nostrils **(Figure 46.6).**

The lungs of reptiles, birds, and mammals have many pockets and folds that increase the area of the respiratory surface, which contains dense, highly branched capillary networks. Mammalian lungs consist of millions of tiny air pockets, the **alveoli,** each surrounded by dense capillary networks. Reptiles and mammals fill their lungs by **negative pressure**

breathing—by muscular contractions that expand the lungs, lowering the pressure of the air in the lungs and causing air to be pulled inward. (Mammalian negative pressure breathing is described in more detail in the next section.)

In birds, a one-way flow system provides the most complex and efficient vertebrate lungs **(Figure 46.7).** Birds have a pair of lungs that are compact, rigid structures. In addition to paired lungs, birds have nine pairs of air sacs that branch off the respiratory tract and are outside of the lungs, occupying a large proportion of the thoracic and abdominal cavities. The air sacs, which collectively contain several times as much air as the lungs, act as a bellows that sends air flowing in one direction through the lungs, rather than in and out as in other vertebrates. Within the lungs, air flows through an array of fine, parallel tubes called **parabronchi.** A capillary network crosses the parabronchi in a perpendicular direction. Thus, the blood flows in a crosscurrent pattern relative to the flow of air through the parabronchi. As a result, breathing in birds is more efficient than breathing in mammals, although less efficient than the countercurrent exchange system of fish gills.

STUDY BREAK 46.2

1. What advantages do gills confer on a water-breathing animal over skin breathing?
2. What is countercurrent exchange, and how is it beneficial for gas exchange?
3. How does the tracheal system of insects facilitate gas exchange with the cells of the body?
4. Distinguish between positive pressure breathing and negative pressure breathing in animals with lungs.

46.3 The Mammalian Respiratory System

All mammals have a pair of lungs and a muscular diaphragm in the chest cavity that plays an important role in negative pressure breathing. (Birds and reptiles lack muscular diaphragms, illustrating their separate evolutionary lineage.) Rapid ventilation of the respiratory surface and perfusion by blood flow through dense capillary networks maximize gas exchange.

A. Lungs and air sacs of a bird

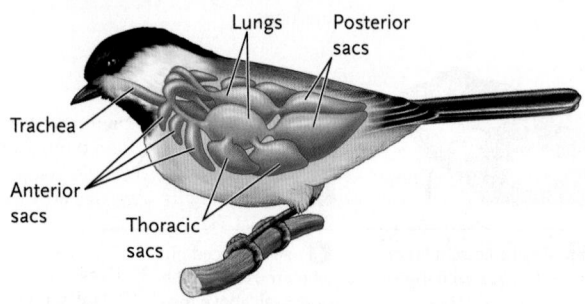

B. One-way air flow

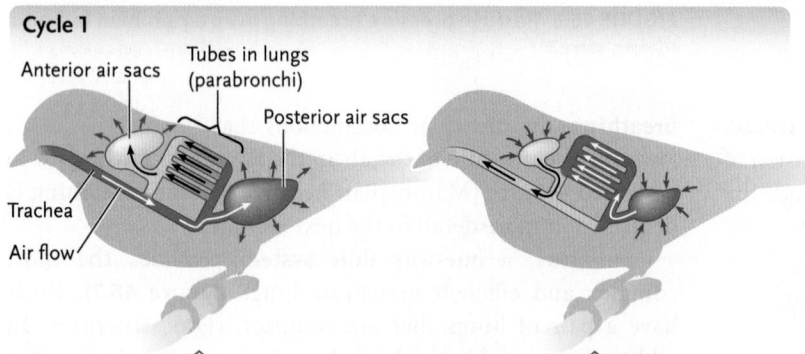

1 During the first inhalation, most of the air flows directly to the posterior air sacs, which expand. The anterior air sacs also expand with air that has already crossed the respiratory surface.

2 During the following exhalation, both anterior and posterior air sacs contract. Air from the posterior sacs flows into the gas-exchanging parabronchi of the lungs. O_2 from the air diffuses into capillaries that cross the parabronchi.

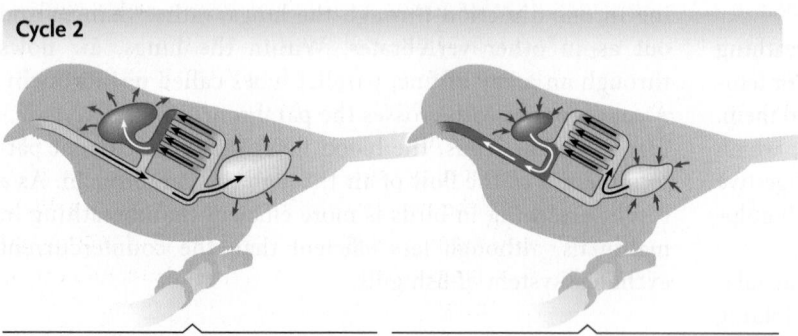

1 During the next inhalation, air from the lung (now deoxygenated and containing CO_2 that diffused from the capillaries) moves into the expanded anterior air sacs.

2 In the second exhalation, air from the anterior sacs is expelled to the outside through the trachea.

FIGURE 46.7 One-way air flow in bird lungs. (A) Unlike mammalian lungs, bird lungs do not expand and contract. Changes in pressure in the expandable anterior and posterior air sacs move air in and out. **(B)** Air flows in one direction through the parabronchi of the lungs; blood flows in a capillary network that runs perpendicular to the parabronchi. O_2 diffuses out of the parabronchi into the capillaries and CO_2 diffuses in the opposite direction. Two cycles of inhalation and exhalation are needed to move a specific volume of air through the bird respiratory system.

© Cengage Learning 2017

The Airways Leading from the Exterior to the Lungs Filter, Moisten, and Warm the Entering Air

The human respiratory system is typical for a terrestrial mammal **(Figure 46.8).** Air enters and leaves the respiratory system through the nostrils and mouth. Hairs in the nostrils and mucus covering the surface of the airways filter out and trap dust and other large particles. Inhaled air is moistened and warmed as it moves through the mouth and nasal passages.

Next, air moves into the throat, or **pharynx,** which forms a common pathway for air entering the **larynx** (or "voice box") and food entering the esophagus, the tube that leads to the stomach. The airway through the larynx is open except during swallowing.

From the larynx, air moves into the trachea (or "windpipe"), which branches into two airways, the **bronchi** (singular, *bronchus*). The bronchi lead to the two elastic, cone-shaped lungs, one on each side of the chest cavity. Inside the lungs, the bronchi narrow and branch repeatedly, becoming progressively narrower and more numerous. The terminal airways, the **bronchioles,** lead into cup-shaped pockets, the *alveoli* (singular, *alveolus*; shown in Figure 46.8 insets).

Each of the 250 million alveoli in each lung is surrounded by a dense network of capillaries. By the time inhaled air reaches the alveoli, it has been moistened to the saturation point and brought to body temperature. The many alveoli provide an enormous area for gas exchange. If the alveoli of an adult human were flattened out in a single layer, they would cover an area approaching 100 square meters—about the size of a tennis court.

The larynx, trachea, and larger bronchi are nonmuscular tubes encircled by rings of cartilage that prevent the tubes from compressing. The largest of the rings, which reinforces the larynx, stands out at the front of the throat as the Adam's apple; smaller supporting rings can be felt at the front of the throat just below the larynx. The walls of the smaller bronchi and the bronchioles contain smooth muscle cells that contract or relax to control the diameter of these passages, and with it, the amount of air flowing to and from the alveoli.

The epithelium lining each bronchus contains cilia and mucus-secreting cells. Bacteria and airborne particles such as dust and pollen are trapped in the mucus and then moved upward and into the throat by the beating of the cilia lining the airways. Infection-fighting macrophages

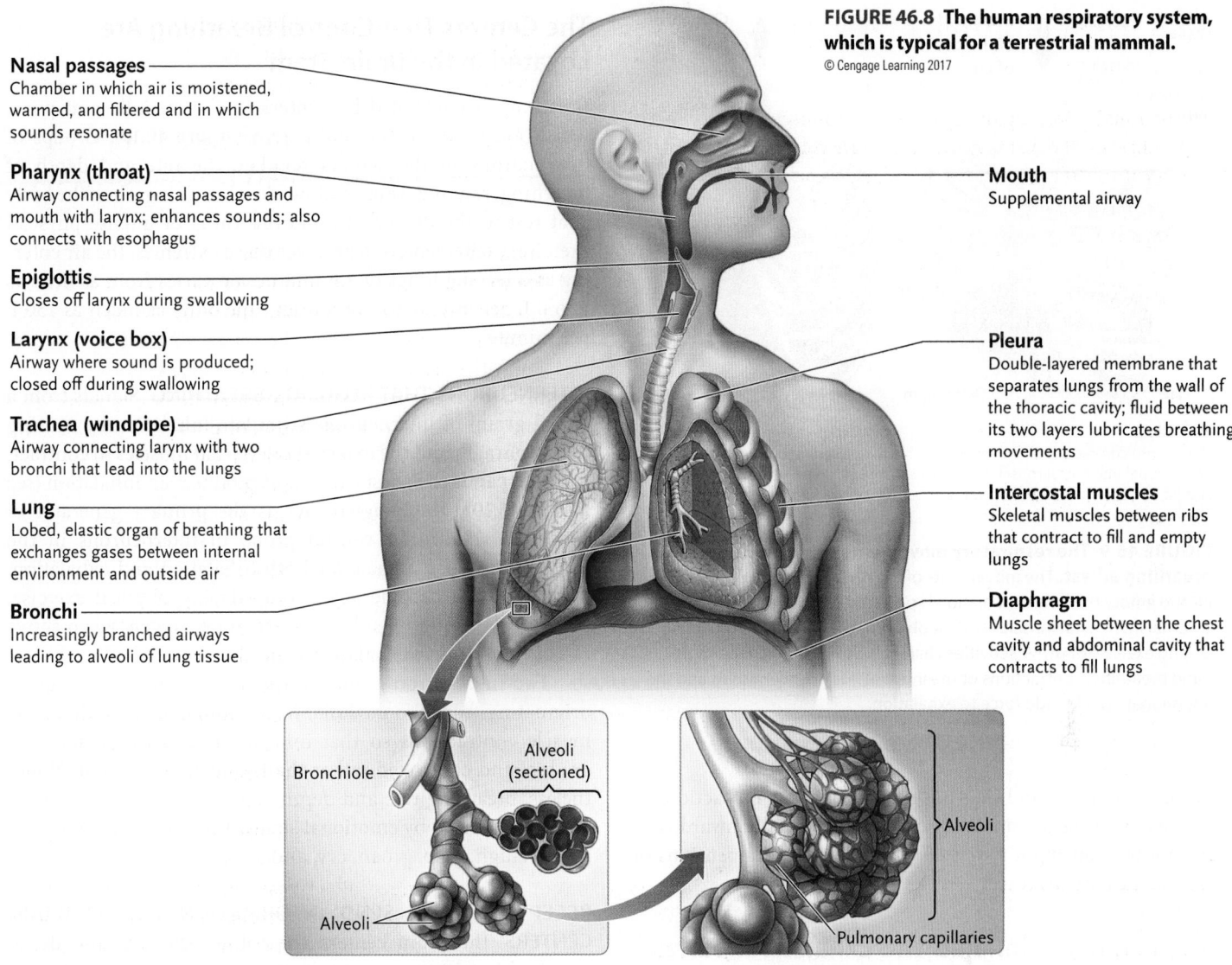

FIGURE 46.8 The human respiratory system, which is typical for a terrestrial mammal.
© Cengage Learning 2017

Nasal passages
Chamber in which air is moistened, warmed, and filtered and in which sounds resonate

Pharynx (throat)
Airway connecting nasal passages and mouth with larynx; enhances sounds; also connects with esophagus

Epiglottis
Closes off larynx during swallowing

Larynx (voice box)
Airway where sound is produced; closed off during swallowing

Trachea (windpipe)
Airway connecting larynx with two bronchi that lead into the lungs

Lung
Lobed, elastic organ of breathing that exchanges gases between internal environment and outside air

Bronchi
Increasingly branched airways leading to alveoli of lung tissue

Mouth
Supplemental airway

Pleura
Double-layered membrane that separates lungs from the wall of the thoracic cavity; fluid between its two layers lubricates breathing movements

Intercostal muscles
Skeletal muscles between ribs that contract to fill and empty lungs

Diaphragm
Muscle sheet between the chest cavity and abdominal cavity that contracts to fill lungs

Bronchiole

Alveoli (sectioned)

Alveoli

Alveoli

Pulmonary capillaries

also patrol the respiratory epithelium. The epithelium of alveoli contains some cells that secrete a phospholipoprotein surfactant that facilitates lung expansion. (A surfactant is a chemical compound that reduces surface tension. Surface tension is discussed in Section 2.4.)

Tobacco smoke, by paralyzing the cilia lining the respiratory tract, interferes with the processes that clear bacteria and airborne particles from the lungs. As a result, the bacteria and foreign matter may persist in the lungs.

Contractions of the Diaphragm and Muscles between the Ribs Ventilate the Lungs

The lungs are located in the rib cage above the *diaphragm,* a dome-shaped sheet of skeletal muscle separating the chest cavity from the abdominal cavity. The lungs are covered by a double layer of epithelial tissue called the **pleura.** The inner pleural layer is attached to the surface of the lungs, and the outer layer is attached to the surface of the chest cavity. A narrow space between the inner and outer layers is filled with slippery fluid,

which allows the lungs to move within the chest cavity without rubbing or abrasion as they expand and contract.

Contraction of muscles between the ribs and the diaphragm brings air into the lungs by a negative pressure mechanism. As an inhalation begins, the diaphragm contracts and flattens, and one set of muscles between the ribs, the *external intercostal* muscles, contracts, pulling the ribs upward and outward **(Figure 46.9).** These movements expand the chest cavity and lungs, lowering the air pressure in the lungs below that of the atmosphere. As a result, air is drawn into the lungs, expanding and filling them.

The expansion of the lungs is much like filling two rubber balloons. Like balloons, the lungs are elastic, and resist stretching as they are filled. Also like balloons, the stretching stores energy, which can be released to expel air from the lungs. When a person at rest exhales, the diaphragm and muscles between the ribs relax, and the elastic recoil of the lungs expels the air.

When physical activity increases the body's demand for O_2, other muscles help expel the air by forcefully reducing the volume of the chest cavity. Contractions of abdominal wall muscles increase abdominal pressure, exerting an upward-directed force

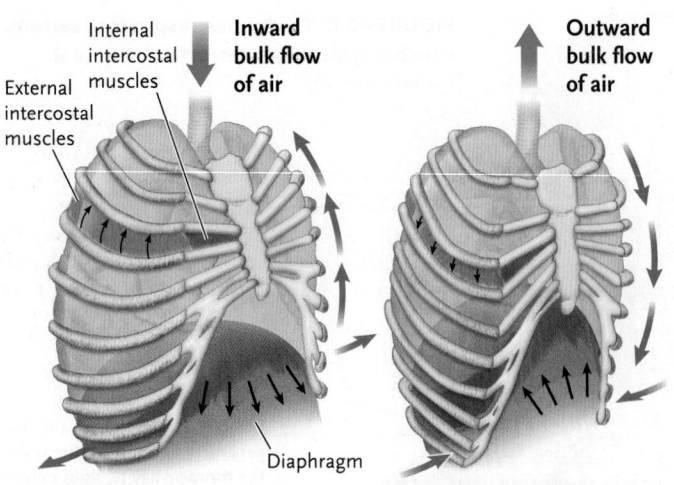

Inhalation.
Diaphragm contracts and moves down. The external intercostal muscles contract and lift rib cage upward and outward. The lung volume expands.

Exhalation during breathing or rest.
Diaphragm and external intercostal muscles return to the resting positions. Rib cage moves down. Lungs recoil passively.

FIGURE 46.9 The respiratory movements of humans during breathing at rest. The movements of the rib cage and diaphragm fill and empty the lungs. Inhalation is powered by contractions of the external intercostal muscles and diaphragm, and exhalation is passive. During exercise or other activities characterized by deeper and more rapid breathing, contractions of the internal intercostal muscles and the abdominal muscles add force to exhalation.

© Cengage Learning 2017

on the diaphragm and thus pushing it upward. Contractions of *internal intercostal* muscles pull the chest wall inward and downward, causing it to flatten. As a result, the dimensions of the chest cavity decrease.

The Volume of Inhaled and Exhaled Air Varies over Wide Limits

The volume of air entering and leaving the lungs during inhalation and exhalation is called the **tidal volume.** In a person at rest, the tidal volume amounts to about 500 mL. As physical activity increases, the tidal volume increases to match the body's needs for O_2; at maximal levels, the tidal volume reaches about 3,400 mL in females and 4,800 mL in males. The maximum tidal volume is called the **vital capacity** of an individual.

Even after the most forceful exhalation, about 1,000 mL of air remains in the lungs in females, and about 1,200 mL in males; this is the **residual volume** of the lungs. In fact, the lungs cannot be deflated completely because small airways collapse during forced exhalation, blocking further outflow of air. Because air cannot be removed from the lungs completely, some gas exchange can always occur between blood flowing through the lungs and the air in the alveoli. A disadvantage of this two-way system of breathing compared with one-way systems is that incoming fresh air always mixes with the residual air, which is mostly deoxygenated.

The Centers That Control Breathing Are Located in the Brain Stem

Breathing is controlled by centers in the medulla and pons, which form part of the brain stem **(Figure 46.10).** Groups of interneurons in the centers regulate the rate and depth of breathing, ranging from shallow, slow breathing when the body is at rest to the deep and rapid breathing of intense physical exercise, excitement, or fear. Over these extremes, the air entering and leaving lungs of a human male varies from as little as 5 to 6 L per minute to (for a brief time only) as much as 150 L per minute.

INTERNEURONS THAT REGULATE BREATHING Signals from a dorsal group of interneurons stimulate inhalation by causing the diaphragm and the external intercostal muscles to contract, which expands the chest cavity and produces an inhalation (see Figure 46.10). These signals act as the primary generator of breathing rhythm. A ventral group of interneurons in the medulla can send signals for both inhalation and exhalation. These neurons become active only during physical exercise, fear, or other situations that require more oxygen, when active rather than passive exhalation is needed.

Two interneuron groups in the pons modulate the signals originating from the medulla, fine-tuning and smoothing the muscle contractions so that inhalations and exhalations are gradual and controlled rather than sudden and abrupt. Nonetheless, breathing rate and depth can be modified consciously and, for example, by emotional states. Thus breathing is altered as you laugh, gasp, groan, cry, and sigh.

RECEPTORS THAT SEND INFORMATION TO THE BRAIN CENTERS The brain centers controlling the rate and depth of breathing integrate sensory information sent by chemoreceptors that monitor O_2 and CO_2 levels in the blood and body fluids. The integration of sensory information serves to match breathing rate to the metabolic demands of the body. The chemoreceptors are located centrally on the surface of the medulla, and peripherally in **carotid bodies** in the carotid arteries leading to the brain and in **aortic bodies** in the large arteries leaving the heart (see Figure 46.10).

Surface receptors of the medulla detect changes in pH in the cerebrospinal fluid; the pH is determined mostly by the CO_2 concentration in the blood. (Remember that pH decreases as CO_2 levels increase.) The receptors in the carotid and aortic bodies detect changes in CO_2 and O_2 concentrations in the blood.

The CO_2 receptors in the medulla have the greatest effects on breathing. If increased body activities cause the CO_2 concentration to rise in the blood, the medulla receptors trigger interneuron groups in the medulla that increase the rate and depth of breathing. If CO_2 concentration falls, the receptors send signals to the medulla that lead to a slowing of the rate and depth of breathing.

The peripheral receptors in the carotid and aortic bodies detect changes in pH or O_2 concentration in arterial blood. When these receptors detect a decrease in blood pH they send signals to the medulla that cause the medulla to increase the rate and depth of breathing. Although the receptors in the carotid and aortic bodies also detect the O_2 level in arterial blood, the receptors do not respond until the blood O_2 level falls below 60% of normal. This reaction makes the O_2 receptors act as a backup system that comes into play only when blood O_2 concentration falls to critically low levels.

Thus, the level of CO_2 in the blood and body fluids is much more closely monitored, and has a much greater effect on breathing, than the O_2 level. This reflects the fact that small fluctuations in blood pH have much greater effects on the ability of hemoglobin to carry oxygen, and on enzyme activity in the blood and interstitial fluid, than fluctuations in the O_2 level.

LOCAL CONTROLS Other, automated controls within the lungs match the rates of ventilation and perfusion by responding to O_2 concentrations in the blood. If air flow lags behind capillary blood flow, so that the O_2 level falls in the blood, the reduced O_2 concentration causes smooth muscles in the walls of arterioles in the lungs to contract. This reduces the flow of blood, thereby giving it more time to pick up O_2. Conversely, if blood flow lags behind, the rising blood O_2 concentration causes the smooth muscle cells in arteriole walls to relax, dilating the arterioles and increasing the rate of blood flow through lung capillaries. These local controls, in combination with the neural controls that regulate rate and depth of breathing, ensure that the respiratory system meets the body's varying need to obtain O_2 and release CO_2.

STUDY BREAK 46.3

1. Explain how inhalation and exhalation occur in a mammal at rest.
2. What is the most important feedback stimulus for breathing?
3. What is the role of the chemoreceptors in the medulla?

THINK OUTSIDE THE BOOK

Individually or collaboratively, use the Internet or the research literature to answer the following questions: What is asthma, and what causes it? How do anti-inflammatory drugs and bronchodilators operate to relieve the symptoms of asthma? How is asthma similar to and different from chronic obstructive pulmonary disease (COPD)?

46.4 Mechanisms of Gas Exchange and Transport

In both the lungs and body tissues, gas exchange occurs when the gas diffuses from an area of higher concentration to an area of lower concentration. In this section, we consider the mechanics of gas exchange between air and the blood in mammals, and

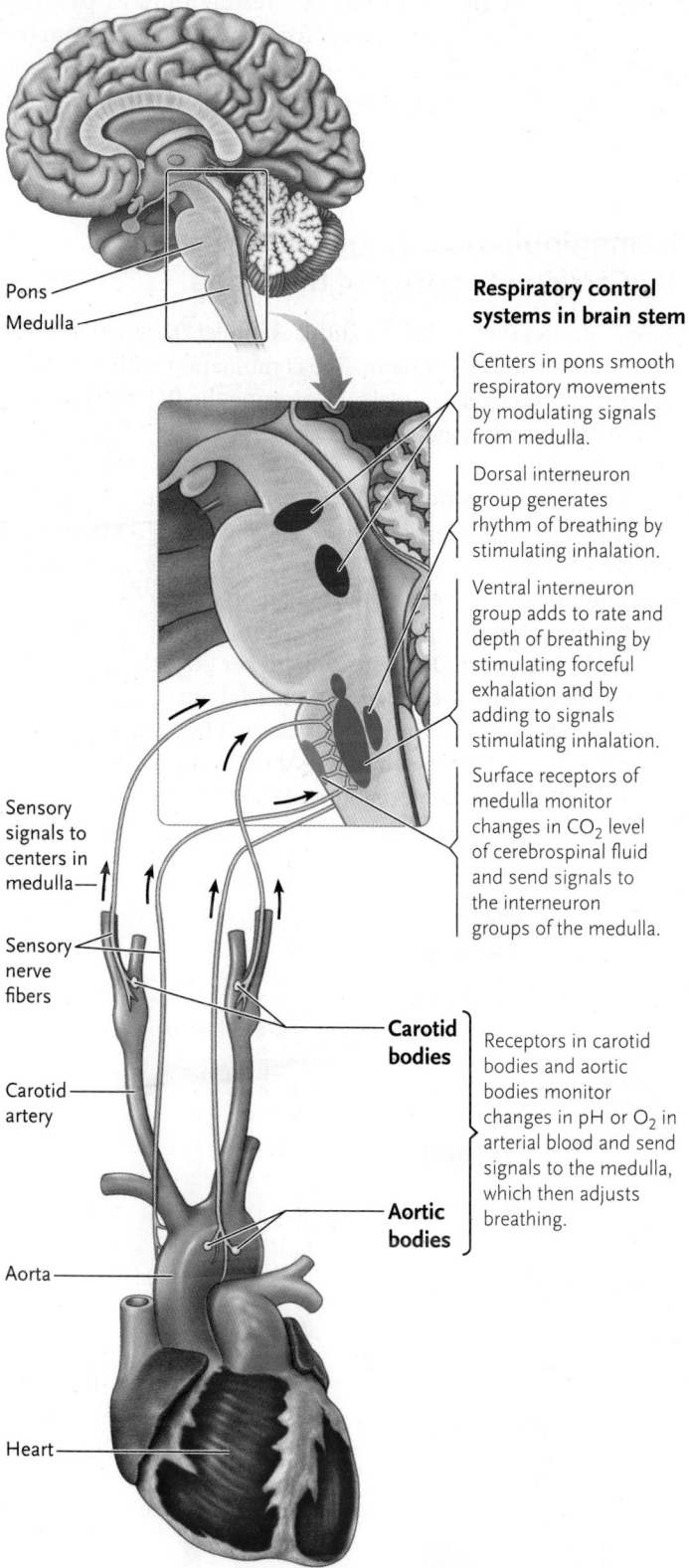

Respiratory control systems in brain stem

Centers in pons smooth respiratory movements by modulating signals from medulla.

Dorsal interneuron group generates rhythm of breathing by stimulating inhalation.

Ventral interneuron group adds to rate and depth of breathing by stimulating forceful exhalation and by adding to signals stimulating inhalation.

Surface receptors of medulla monitor changes in CO_2 level of cerebrospinal fluid and send signals to the interneuron groups of the medulla.

Pons
Medulla
Sensory signals to centers in medulla
Sensory nerve fibers
Carotid bodies
Carotid artery
Aortic bodies
Aorta
Heart

Receptors in carotid bodies and aortic bodies monitor changes in pH or O_2 in arterial blood and send signals to the medulla, which then adjusts breathing.

FIGURE 46.10 Control of breathing. Centers in the pons (red) and centers in the medulla (blue) control the rhythm, rate, and depth of breathing. Receptors in the carotid arteries and aorta detect changes in the levels of O_2 and CO_2 in blood and body fluids. Signals from these receptors are integrated in the respiratory centers of the medulla and pons.
© Cengage Learning 2017

the means by which gases are transported between the lungs and other body tissues. A major part of this story involves hemoglobin (see Figure 3.17D).

The Proportion of a Gas in a Mixture Determines Its Partial Pressure

For gases, concentration differences typically are considered as differences in pressure. When gases are present in a mixture, the pressure of each individual gas, called its **partial pressure,** is determined by its proportion in the mixture. Air, water, and blood all contain mixtures of gases, including oxygen, carbon dioxide, nitrogen, and other gases, so each gas exerts only a part of the total gas pressure. For example, the proportion of O_2 in dry air is about 21%, or 21/100. In dry air at sea level, the total atmospheric pressure under standard conditions is 760 mm Hg. The partial pressure of O_2, written as P_{O_2}, is equivalent to $760 \times 21/100$, or about 160 mm Hg. The proportion of CO_2 in dry air is about 0.04%, so its partial pressure, P_{CO_2}, is equivalent to $760 \times 0.04/100$, or about 0.3 mm Hg. For O_2 to diffuse inward across a respiratory surface, its partial pressure outside the surface must be greater than inside; for CO_2 to diffuse outward, its partial pressure inside must be greater inside than outside.

In the lungs, even though the P_{O_2} is reduced by mixing with the air in the residual volume, it is still much higher than the P_{O_2} in deoxygenated blood entering the network of capillaries in the lungs **(Figure 46.11).** As a result, O_2 readily diffuses passively down its partial pressure gradient from the alveolar air into the plasma solution in the pulmonary capillaries. Conversely, the P_{CO_2} in the lungs is much lower than the P_{CO_2} in the tissues. As a result, CO_2 readily diffuses passively down its partial pressure gradient from the pulmonary capillaries into the alveolar air.

Hemoglobin Greatly Increases the O_2-Carrying Capacity of the Blood

After entering the plasma, O_2 diffuses into erythrocytes, where it combines with hemoglobin. The combination with hemoglobin removes O_2 from the plasma, lowering the P_{O_2} of the plasma and allowing additional O_2 molecules to diffuse from alveolar air to the blood.

Recall from Section 44.2 that a mammalian hemoglobin molecule has four heme groups, each containing an iron atom that can combine reversibly with an O_2 molecule. A hemoglobin molecule can therefore bind a total of four molecules of O_2. The combination of O_2 with hemoglobin allows blood to carry about 60 times more O_2 (about 200 mL per liter) than it could if the O_2 simply dissolved in the plasma (about 3 mL per liter). About 98.5% of the O_2 in blood is carried by hemoglobin and about 1.5% is carried in solution in the blood plasma.

The reversible combination of hemoglobin with O_2 is related to the partial pressure of O_2 in a pattern shown by the

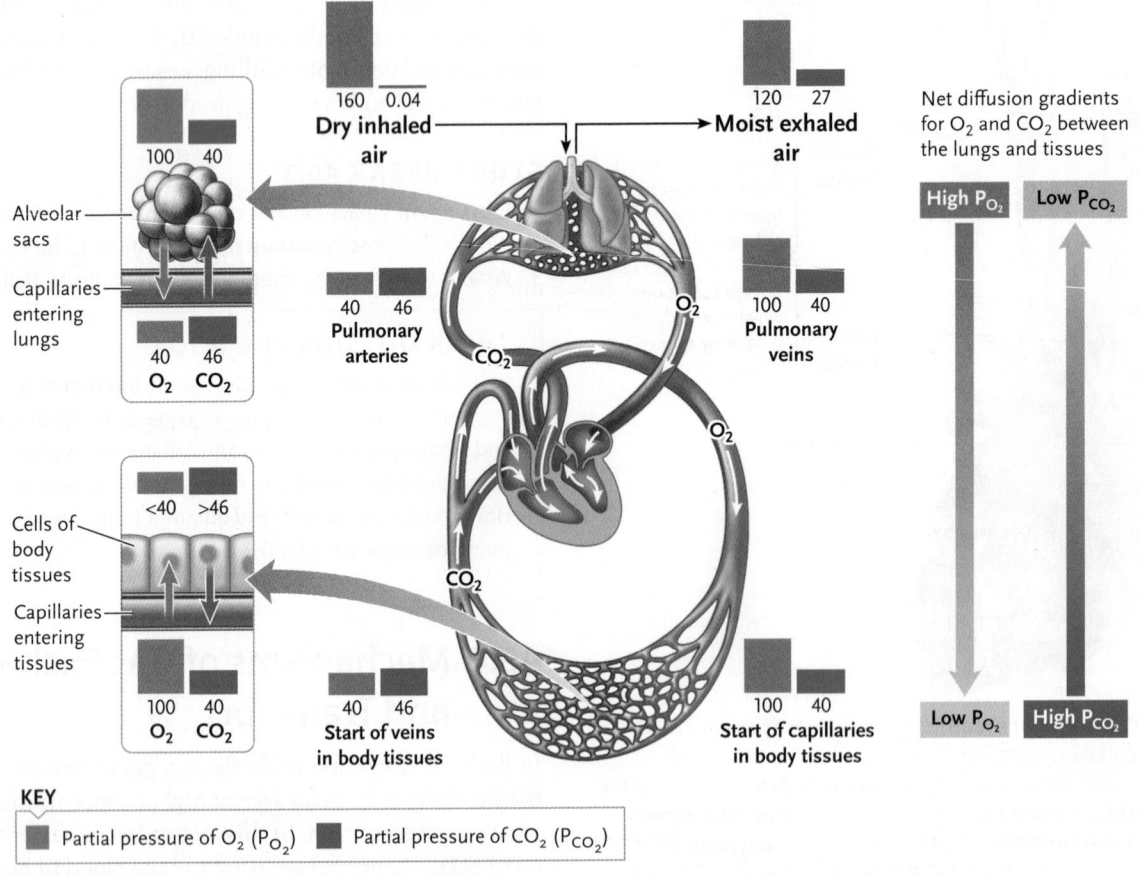

FIGURE 46.11 The partial pressures of O_2 (P_{O_2}) and CO_2 (P_{CO_2}) in various locations in the body.
© Cengage Learning 2017

hemoglobin–O₂ dissociation curve in **Figure 46.12A.** (The curve is generated by measuring the amount of hemoglobin saturated at a given P_{O_2}.) The curve is S-shaped with a plateau region, rather than linear. The top, plateau part of the curve above 60 mm Hg is in the blood P_{O_2} range found in the pulmonary capillaries where O_2 is binding to hemoglobin. For this part of the curve, the blood remains highly saturated with O_2 over a relatively large range of P_{O_2}. Even at P_{O_2} levels much higher than shown on the graph (P_{O_2} theoretically can go up to 760 mm

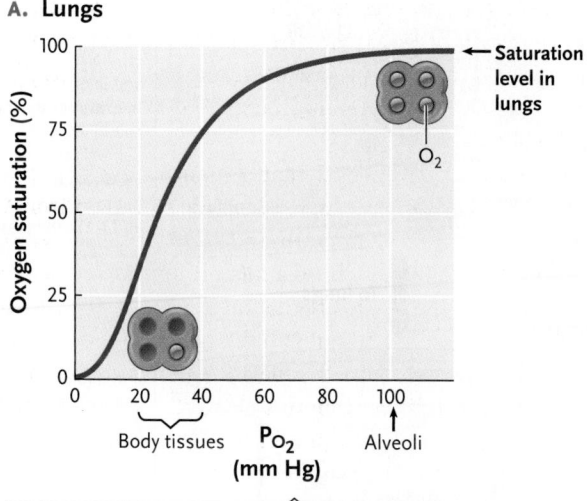

A. Lungs

In the alveoli, in which the P_{O_2} is about 100 mm Hg and the pH is 7.4, most hemoglobin molecules are 100% saturated, meaning that almost all have bound four O_2 molecules.

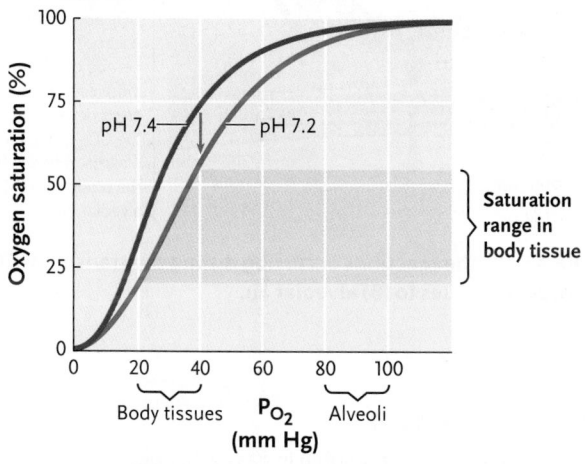

B. Body tissues

In the capillaries of body tissues, where the P_{O_2} varies between about 20 and 40 mm Hg depending on the level of metabolic activity and the pH is about 7.2, hemoglobin can hold less O_2. As a result, most hemoglobin molecules release two or three of their O_2 molecules to become between 25% and 50% saturated. The drop in pH to 7.2 (red line) in active body tissues reduces the amount of O_2 hemoglobin can hold as compared with pH 7.4. The reduction in binding affinity at lower pH increases the amount of O_2 released in active tissues.

FIGURE 46.12 Hemoglobin–O₂ dissociation curves, which show the degree to which hemoglobin is saturated with O₂ at increasing P_{O_2}, in lungs (A) and body tissues (B).

Hg), only a small extra amount of O_2 will bind to hemoglobin. The steep part of the curve between 0 and 60 mm Hg is in the blood P_{O_2} range found in the capillaries in the rest of the body. For this part of the curve, small changes in P_{O_2} result in a large change in the amount of O_2 bound to hemoglobin.

The P_{O_2} of alveolar air is about 100 mg Hg (see Figure 46.11). As a result, most of the hemoglobin molecules are fully saturated in the blood leaving the alveolar networks, meaning that most of the hemoglobin molecules are bound to four O_2 molecules (see Figure 46.12A). The P_{O_2} of blood plasma has risen to approximately the same level as in the alveolar air, about 100 mm Hg. The blood has also changed color, reflecting the bright red color of oxygenated hemoglobin as compared with the darker red color of deoxygenated hemoglobin.

The oxygenated blood exiting from the alveoli collects in venules, which merge into the pulmonary veins leaving the lungs. These veins carry the blood to the heart, which pumps the blood through the systemic circulation to all parts of the body.

As the oxygenated blood enters the capillary networks of body tissues **(Figure 46.12B)**, it encounters regions in which the P_{O_2} in the interstitial fluid and body cells is lower than that in the blood, ranging from about 40 mm Hg downward to 20 mm Hg or less (see Figure 46.11). As a result, O_2 diffuses from the blood plasma into the interstitial fluid, and from the fluid into body cells. As O_2 diffuses from the blood plasma into body tissues, it is replaced by O_2 released from hemoglobin.

Several factors contribute to the release of O_2 from hemoglobin, including increased acidity (lower pH) in active tissues. The acidity increases because oxidative reactions release CO_2, which combines with water to form carbonic acid (H_2CO_3). The lowered pH alters hemoglobin's conformation, reducing its affinity for O_2, a phenomenon called the **Bohr effect.** As a result of the reduced affinity, O_2 is released, diffuses into cells, and is used in cellular respiration.

The net diffusion of O_2 from blood to body cells continues until, by the time the blood leaves the capillary networks in the body tissues, much of the O_2 has been removed from hemoglobin. The blood, now with a P_{O_2} of 40 mm Hg or less (see Figures 46.11 and 46.12B), returns in veins to the heart, which pumps it through the pulmonary arteries to the lungs for another cycle of oxygenation.

Carbon Dioxide Diffuses down Concentration Gradients from Body Tissues into the Blood and Alveolar Air

The CO_2 produced by cellular respiration diffuses from active cells into the interstitial fluid, where it reaches a partial pressure of about 46 mm Hg. Because this P_{CO_2} is higher than the 40 mm Hg P_{CO_2} in the blood entering the capillary networks of body tissues (see Figure 46.11), CO_2 diffuses from the interstitial fluid into the blood plasma **(Figure 46.13A)**.

About 10% of the CO_2 remains in solution as a gas in the plasma. Another 30% of the CO_2 binds to hemoglobin in erythrocytes. The remaining 60% of the CO_2 combines with water to produce carbonic acid (H_2CO_3), which dissociates into bicarbonate (HCO_3^-) and H^+ ions. The reaction takes place slowly in the blood plasma and rapidly inside erythrocytes, where an enzyme, *carbonic anhydrase,* catalyzes the reaction.

Most of the H^+ ions produced by the dissociation of carbonic acid combine with hemoglobin or with proteins in the plasma. The combination, by removing excess H^+ from the blood solution, *buffers* the blood pH, helping to maintain it at its near-neutral set point of 7.4. (Buffers are discussed in Section 2.5. The discussion includes the carbonic acid–bicarbonate buffer system used to buffer blood pH in humans and many other animals.) pH homeostasis is important because many enzymes are sensitive to minor changes in pH. The combined pathways absorbing CO_2 in the blood—solution in the plasma, conversion to bicarbonate, and combination with hemoglobin—help maintain the concentration gradient for gaseous CO_2 and keep its diffusion from the interstitial fluid into the blood at optimal levels.

The blood leaving the capillary networks of body tissues (the systemic circulation) is collected in venules and veins and returned to the heart, which pumps it through the pulmonary arteries into the lungs. As the blood enters the capillary networks surrounding the alveoli, the entire process of CO_2 uptake is reversed **(Figure 46.13B)**. The P_{CO_2} in the blood, now about 46 mm Hg, is higher than the P_{CO_2} in the alveolar air, about 40 mm Hg (see Figure 46.11). As a result, CO_2 diffuses from the blood into the air. The diminishing CO_2 concentrations in the plasma, along with the lower pH encountered in the lungs, promote the release of CO_2 from hemoglobin. As CO_2 diffuses away, bicarbonate ions in the blood combine with H^+ ions, forming carbonic acid molecules that break down into water and additional CO_2. This CO_2 adds to the quantities diffusing from the blood into the alveolar air. By the time the blood leaves the capillary networks in the lungs, its P_{CO_2} has been reduced to the same level as that of the alveolar air, about 40 mm Hg (see Figure 46.11).

Carbon monoxide (CO), a colorless, odorless gas produced when fuels are incompletely burned, as in automobile exhaust or in a faulty furnace, gas appliance, or space heater, and also present in cigarette smoke, binds to hemoglobin if it is inhaled into the lungs. It binds so strongly that it displaces O_2 from hemoglobin and reduces the amount of O_2 carried to body tissues drastically. If CO is inhaled in high quantity for even a few minutes, the reduction in oxygen delivered to the brain can lead to unconsciousness and brain damage. Sustained exposure leads to death by *hypoxia* (lack of oxygen). Because the brain regulates breathing based on CO_2 levels in blood rather than on O_2 levels, victims breathing CO can die from hypoxia without noticing anything amiss up to the point of unconsciousness.

The CO–hemoglobin molecule, *carboxyhemoglobin,* is bright red. This has led to a myth often seen in textbooks that victims of CO poisoning turn a "classic cherry red" in color. This occurs in less than 2% of cases.

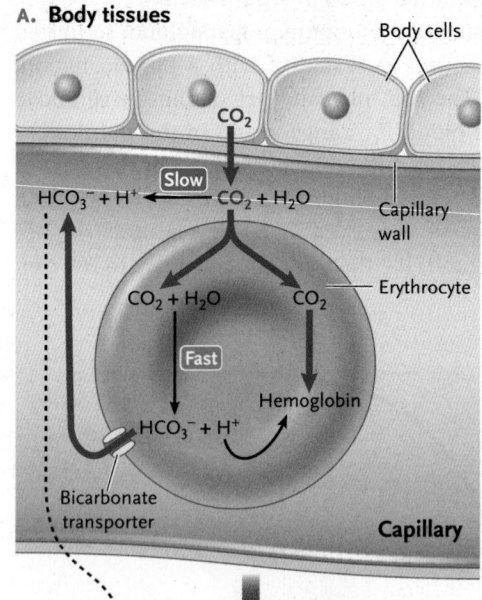

A. Body tissues

In body tissues, some of the CO_2 released into the blood remains in solution in the plasma. Some of the remaining CO_2 diffuses into erythrocytes and combines with hemoglobin. The rest combines with water to form HCO_3^- and H^+, both in the plasma and in erythrocytes. In erythrocytes, the H^+ combines with hemoglobin; the HCO_3^- is transported out to add to the HCO_3^- in the plasma.

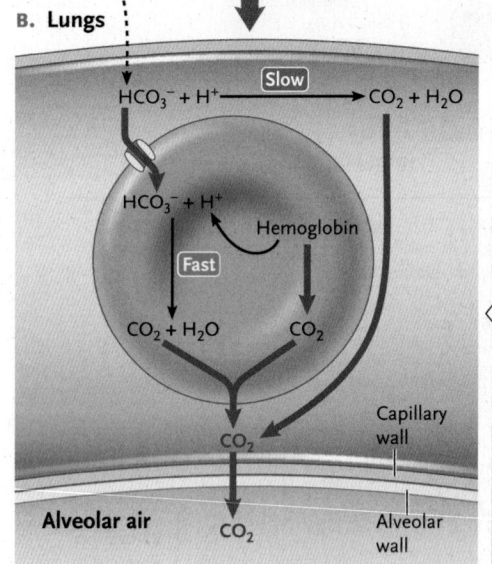

B. Lungs

In the lungs, the reactions are reversed. Some of the HCO_3^- in the blood plasma combines with H^+ to form CO_2 and water. However, most of the HCO_3^- is transported into erythrocytes, where it combines with H^+ released from hemoglobin to form CO_2 and water. CO_2 is released from hemoglobin. The CO_2 diffuses from the erythrocytes and, with the CO_2 in the blood plasma, diffuses from the blood into the alveolar air.

FIGURE 46.13 The reactions occurring during the transfer of CO_2 from (A) body tissues to (B) alveolar air.
© Cengage Learning 2017

STUDY BREAK 46.4

1. Explain the role of hemoglobin in gas exchange.
2. Why is carbon monoxide potentially lethal?

46.5 Respiration at High Altitudes and in Ocean Depths

This chapter's introduction described some challenges to respiration that arise when humans travel to high altitude. In this concluding section we look at the effects of high altitude on respiration, along with the effects of increased pressure when humans and other mammals dive underwater.

Altitude Adaptation in Tibetans Caused by Gene Allele Inherited from an Extinct Species of Humans

The Tibetan plateau **(Figure)** sits more than 4,000 m (13,100 ft) above sea level and has an atmospheric oxygen pressure about 40% lower than at sea level. That, plus cold climate and limited resources, has made it inhospitable for most human inhabitation. The traditional thinking is that adaptation to such high altitudes involves increasing the concentration of oxygen-carrying hemoglobin in the blood; this is seen, for instance, in highlanders in the Andes of South America. However, Tibetans have adapted, surprisingly, by having less hemoglobin in their blood. This condition is interpreted to confer less risk of blood clots and strokes that may occur with blood thickened by high concentrations of red blood cells. Clearly, then, Tibetans are able to use smaller amounts of oxygen efficiently. Research in 2010 showed that Tibetans have single nucleotide polymorphisms (SNPs; see

FIGURE Tibetan plateau and mountains.

Section 18.3) in several genes of possible significance to altitude adaptation. One gene in particular, *EPAS1*, encodes a transcription factor that is induced under hypoxic (low oxygen) conditions and is involved in regulating the production of hemoglobin in the body. Specifically, individuals with the SNP variant in *EPAS1* had lower levels of hemoglobin than individuals homozygous for the ancestral normal allele.

Research Question

What is the origin of the *EPAS1* "Tibetan Altitude gene" variant?

Experiments and Results

1. An international team of researchers sequenced the *EPAS1* gene in the genomes of 40 Tibetans and 40 Han Chinese. These two peoples once were part of the same population, splitting into two groups between 2,750 and 5,500 years ago. Han Chinese do not live at high altitudes. The results showed that all of the Tibetans and only two of the 40 Han Chinese had the *EPAS1* SNP variant pattern associated with low hemoglobin levels. The investigators searched a catalog of genome sequences in the 1000 Genomes Project (see Chapter 19 *Why it matters...*) and found no other living person represented in the database who had the same SNP variant pattern.

2. The scientists analyzed DNA sequences from ancient humans and showed that the *EPAS1* SNP variant pattern closely matched DNA from a girl's finger bone found in Denisova Cave in the Siberian Alta Mountains.

3. The researchers compared the full *EPAS1* gene sequence in worldwide populations and interpreted the data to mean that Tibetans and Han Chinese inherited the complete *EPAS1* variant gene from Denisovans in approximately the past 40,000 years.

Conclusion

The *EPAS1* gene variant in Tibetans and Han Chinese was inherited from Denisovans, a conclusion supported by computer modeling. Likely the ancestors of present-day Tibetans and Han Chinese got the gene by interbreeding with Denisovans when Tibetans and Han Chinese were still part of one group. In the present day, only a low frequency of Han Chinese have the gene variant.

think like a scientist

If both ancestral Tibetans and Han Chinese inherited the *EPAS1* gene variant from Denisovans, why is the variant present at such low frequency in present-day Han Chinese?

Source: E. Huerta-Sánchez et al. 2014. Altitude adaptation in Tibetans caused by introgression of Denisovan-like DNA. *Nature* 512:194–197.

High Altitudes Reduce the P_{O_2} of Air Entering the Lungs

As altitude increases, atmospheric pressure decreases, and with it, the P_{O_2} of alveolar air and the concentration gradient of O_2 across the respiratory surface. At 5,500 m (18,000 ft), the dry air pressure is about 380 mm Hg and the P_{O_2} is only $380 \times 21/100 =$ about 80 mm Hg, half that at sea level.

Humans who travel from sea level to elevations of 1,800 m (about 6,000 ft) or more may experience one or more unpleasant symptoms, including headache, blurred vision, dizziness, nausea, and fatigue. However, a person who stays at higher altitude will adjust to live and function normally through **acclimatization,** a process by which, over a period of days, the body compensates to permit adequate oxygen delivery to the tissues. That is, in response to a drop in blood O_2 detected at the kidneys, the kidneys increase the secretion of the hormone

erythropoietin (EPO), which stimulates erythrocyte production. If the person subsequently returns to live at lower altitudes, the then-higher blood O_2 signals a reduction in erythrocyte production. However, the erythrocyte count remains high for several weeks after high-altitude exposure. For this reason, athletes often train at high altitudes to increase their erythrocyte count, with the idea that it will improve their stamina and endurance at lower altitudes.

People who live at high altitudes from childhood develop more permanent changes, including an increase in the number of alveoli and more extensive capillary networks in the lungs. These anatomical characteristics are retained if they move to lower altitudes. *Molecular Insights* describes a genetic study concerning altitude adaptation in Tibetans.

Some mammals evolutionarily adapted to high altitudes show genetically determined changes that are present throughout life. For example, deer mice that live at above 4,372 m (about

FIGURE 46.14 | **Experimental Research**

Demonstration of a Molecular Basis for High-Altitude Adaptation in Deer Mice

Question: Do mutations in hemoglobin genes underlie high-altitude adaptation in deer mice?

Experiment: Hemoglobin of adult mammals consists of two α-globin and two β-globin polypeptides. Hemoglobin binds oxygen and transports it in the blood. In the low-oxygen environment of high altitudes, mammals can suffer hypoxia, a condition that results when insufficient oxygen is carried to the body tissues in arterial blood. Some mammals, such as deer mice *(Peromyscus maniculatus),* live and thrive in a wide variety of environments, including at high altitude.

Deer mice have two gene loci each for the α-globin and β-globin polypeptides. Jay Storz and his colleagues at the University of Nebraska–Lincoln, and collaborators at other universities, tested the hypothesis that, compared with their lowland relatives, deer mice living at high altitude have mutations in their α-globin and β-globin genes that lead to hemoglobin molecules adapted to carrying oxygen in low-oxygen environments.

The experimenters isolated DNA from deer mice living at high altitude and at low altitude and compared the sequences of the two α-globin and two β-globin genes.

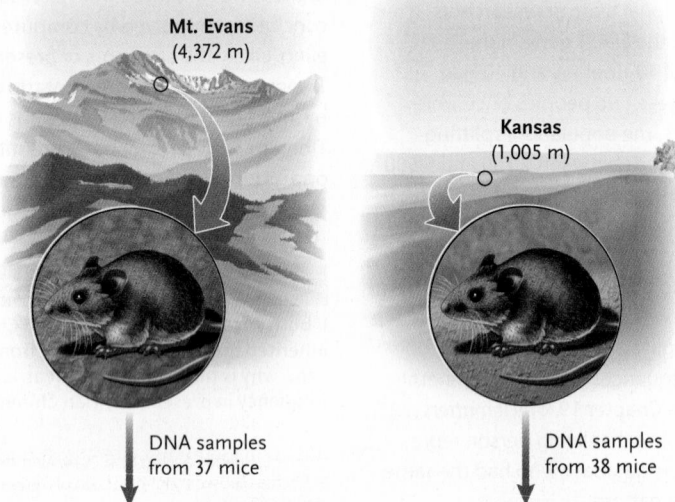

Mt. Evans
(4,372 m)

Kansas
(1,005 m)

DNA samples
from 37 mice

DNA samples
from 38 mice

The two α-globin and two β-globin genes from each mouse
were cloned, sequenced, and compared.

Results: The investigators identified a number of mutations in the four genes of the high-altitude mice. The researchers determined that the mutations increase the oxygen-binding affinity of hemoglobin compared with the low-altitude hemoglobin. This increased binding affinity in turn would increase the O_2 concentration in arterial blood, thereby enabling the mice to tolerate chronic hypoxia.

Conclusion: The research identified specific mutations that correlate with the evolutionary adaptation of deer mice to high altitude.

Interestingly, the hemoglobin of high-altitude deer mice is functionally similar to fetal hemoglobins in humans and other placental mammals. Fetal hemoglobin consists of two α-globin and two γ-globin polypeptides (the γ-globin polypeptide is related to the β-globin polypeptide). The sequence difference from adult hemoglobin enables fetal hemoglobin to carry O_2 efficiently in the low-O_2 intrauterine environment.

think like
a scientist

Observe
Hypothesize
Predict
Experiment
Interpret

The alpine regions in which the high-altitude mice now live were covered with glaciers in the last Ice Age and were not free of ice until about 10,000 years ago. What does that information say about the adaptation of deer mice to high altitude?

Source: J. Storz et al. 2009. Evolutionary and functional insights into the mechanism underlying high-altitude adaptation of deer mouse hemoglobin. *Proceedings of the National Academy of Sciences USA* 106:14450–14455.

14,343 ft) have hemoglobin molecules with greater affinity for O_2 than the hemoglobin molecules of their lower-altitude relatives **(Figure 46.14).** As a result, hemoglobin becomes saturated with O_2 at the lower partial pressures typical of high altitudes. Increased binding affinity of hemoglobin for O_2 is also seen in birds adapted to life at high altitudes, such as the bar-headed goose *(Anser indicus).* These birds have been observed flying over the peaks of the Great Himalayas, which have altitudes greater than 6,000 m.

Diving Mammals Are Evolutionarily Adapted to Survive the High Partial Pressures of Gases at Extreme Depths

As a mammal such as a seal or whale dives from the surface, each additional 10 m of depth increases the partial pressure of dissolved gases by about 1 atmosphere. Below about 25 m or so, the pressure becomes so great that the lungs collapse and cease to function. Evolutionary adaptations of diving mammals such as seals and whales allow these animals to survive the extreme pressure and lack of lung function, in some species for over an hour at ocean depths of more than a mile.

Among these adaptations are more blood per unit of body weight and more red blood cells, which are stored in the spleen and released during a dive. In addition, the muscles of these animals contain much greater quantities of the O_2-binding protein myoglobin than the muscles of land-dwelling mammals do. In all, the adaptations pack about twice as much O_2 per kilogram of body weight into a seal, for example, than into a human.

Other adaptations decrease O_2 consumption during a deep and prolonged dive. The heart rate slows by 80–90% and the circulation of blood to internal organs and muscles is cut by as much as 95%, leaving only the brain with its normal blood supply. Even though most of the blood supply to muscles is cut off, the muscles continue to work by shifting to anaerobic oxidation. The lactic acid produced by anaerobic respiration in the muscles is not released into the blood until the animal returns to the surface.

These combined adaptations give seals and whales an amazing ability to dive to great depths and remain underwater for extended periods. Although average dives are on the order of 10 to 20 minutes, some sperm whales, tracked by sonar, have reached depths of 2,250 m (more than 7,000 ft) and remained underwater for as long as 82 minutes.

STUDY BREAK 46.5

What are the key evolutionary adaptations that diving mammals use to survive at significant ocean depths?

Unanswered Questions

Does developmental nicotine exposure alter development of respiratory neurons in the brain stem?

In this chapter you learned that the muscles of breathing are controlled in the brain stem, by groups of interneurons in the medulla oblongata and pons. The control of respiratory muscles (and thus breathing) by these neurons is called *central ventilatory control*. Neonatal mammals that are exposed to nicotine *in utero* and in early postnatal development show various breathing abnormalities, such as reduced ventilatory output, increased frequency and duration of apneas (suspension of breathing), and delayed arousal in response to hypoxia (reduced blood oxygen levels) during sleep. One or several of these abnormalities may underlie *sudden infant death syndrome* (SIDS), also called *crib death* because victims are typically found dead in their crib. Clinical studies have shown that exposure to tobacco smoke is the number one risk factor for SIDS. Accordingly, laboratories, such as our own, are using animal models to examine how developmental nicotine exposure alters development of central ventilatory control. Although there are many chemicals in tobacco smoke, nicotine is known to alter neuronal development; such neurotoxic compounds are often called *neuroteratogens*. Moreover, pregnant women who smoke are routinely prescribed nicotine patches. The theory behind this practice is that, although addictive, nicotine is otherwise safer than tobacco (interestingly, the same theory is behind the booming use of e-cigarettes). We disagree. In addition to alterations in nervous system development caused by nicotine alone, this practice assumes that the patients really do stop or reduce their smoking behavior, hardly a solid assumption.

Several research methods can be used to study the consequences of developmental nicotine exposure, depending on the question being addressed. In all of our experiments, neonatal rodents are exposed to nicotine *in utero* by implanting a small osmotic pump under the skin of a female rat that is 4 to 5 days pregnant. The pump releases nicotine at a prescribed rate, and the developing neonates are exposed to nicotine as the drug passes from mother to neonate via the placenta. The pups continue to receive nicotine via breast milk after birth, simulating a mother that continues to smoke after delivery, secondhand smoke exposure, or both. When the neonates are born (on the twenty-first day of pregnancy), we study their breathing responses while they are awake or asleep using a device called a *plethysmograph*. This device senses the tiny pressure changes that accompany breathing in these small animals, and by adjusting chamber size, animals can be studied from birth to adulthood.

We can also dissect the brain stem, spinal cord, and rib cage from a neonatal animal, and place the preparation in a chamber for *in vitro* studies. Remarkably, this preparation is able to maintain rhythmic firing of respiratory neurons for up to 6 hours, allowing us to apply drugs and neurotransmitters to brain stem respiratory neurons while recording the electrical activity of neurons and respiratory muscle nerves.

Finally, we can prepare a brain stem slice, containing the most important central respiratory neurons and the hypoglossal nerve (this nerve innervates the tongue muscles, and it contains axons with rhythmic, respiratory-related activity), for detailed electrophysiological studies using the patch clamp technique (see Section 39.2). This preparation allows us to examine how developmental nicotine exposure influences the membrane potential and firing properties of respiratory neurons. During recording, we can also add dyes to the intracellular electrode to fill the neurons, allowing us to visualize them and make detailed analyses of their architecture.

To date, our studies have shown abnormal breathing in awake neonates, as well as an increase in inhibitory neurotransmission in respiratory neurons. Recently, we have shown that the intrinsic biophysical properties of respiratory neurons are altered in the nicotine-exposed pups; this means that when the cells are subjected to natural synaptic input from neighboring neurons (for example, from chemoreceptor neurons), their response is altered because the change in intrinsic properties is due to changes in the cell membrane. Recent anatomical findings show that the size and branching pattern of the dendrites of brain stem respiratory neurons are smaller and less complex in the nicotine-exposed animals. Current and future studies are directed at understanding the detailed cellular mechanisms that lead to these changes in intrinsic membrane properties and cell size. Understanding these mechanisms will, we hope, lead to the development of drugs that can counteract nicotine's impact on the brain, as well as to an increased awareness and acceptance of the link between prenatal nicotine exposure and breathing abnormalities, resulting in more aggressive smoking prevention strategies.

think like a scientist

1. If nicotine affects surface receptors of the medulla oblongata, is an infant's breathing rate being altered by carbon dioxide level or oxygen level?

2. Given your understanding of the peripheral nervous system and control of heart rate (from Section 40.2), is research investigating the effects of nicotine on SIDS and breathing focused on the somatic system or the autonomic system, and if the autonomic system, which division?

Ralph Fregosi is Professor of Physiology and Neuroscience at the University of Arizona at Tucson. He does research on the neural control of breathing and teaches physiology to undergraduate and graduate students. Learn more about his research at http://physiology.arizona.edu/person/ralph-f-fregosi-phd.

REVIEW KEY CONCEPTS

For access to MindTap and additional study materials visit www.cengagebrain.com.

46.1 The Function of Gas Exchange

- Physiological respiration is the process by which animals exchange O_2 and CO_2 with the environment (Figure 46.1).

- The two primary operating features of gas exchange are the respiratory medium, either air or water, and the respiratory surface, a wetted epithelium over which gas exchange takes place.

- In some invertebrates, the skin serves as the respiratory surface. In other invertebrates and all vertebrates, gills or lungs provide the primary respiratory surface (Figure 46.2).

- Simple diffusion of molecules from regions of higher concentration to regions of lower concentration drives the exchange of gases across the respiratory surface. The area of the respiratory surface determines the total quantity of gases exchanged by diffusion.

- The concentration gradients of O_2 and CO_2 across the respiratory surface are kept at optimal levels by ventilation and perfusion.

46.2 Evolutionary Adaptations for Respiration

- Animals breathing water keep the respiratory surface wetted by direct exposure to the environment. The high density and viscosity of water, and its relatively low O_2 content as compared with air, require water-breathing animals to expend significant energy to keep their respiratory surface ventilated.

- Air is high in O_2 content, allowing air-breathing animals to maintain higher metabolic levels than water breathers. The low density and viscosity of air as compared with water allow air breathers to ventilate the respiratory surface with relatively little energy. To accommodate water loss by evaporation, lungs typically are invaginations of the body surface, allowing air to become saturated with water before it reaches the respiratory surface.

- Gills are evaginations of the body surface. Water moves over the gills by the beating of cilia or is pumped over the gills by contractions of body muscles (Figures 46.2B and 46.3A–D).

- Water moves in a one-way direction over the gills of sharks, bony fishes, and some crabs, allowing these animals to use countercurrent exchange to maximize the exchange of gases over the respiratory surface (Figure 46.4).

- Insects breathe by means of tracheae, air-conducting tubes that lead from the body surface and send branches to essentially every cell in the body. Gas exchange takes place in the fluid-filled tips at the ends of the branches (Figure 46.5).

- Lungs consist of an invaginated system of branches, folds, and pockets. They may be filled by positive pressure breathing, in which air is forced into the lungs by muscle contractions, or by negative pressure breathing, in which muscle contractions expand the lungs, lowering the air pressure inside them and allowing air to be pulled into the lungs (Figures 46.6–46.9).

46.3 The Mammalian Respiratory System

- Air enters the respiratory system through the nose and mouth and passes through the pharynx, larynx, and trachea. The trachea divides into two bronchi, which lead to the lungs. Within the lungs, the bronchi branch into bronchioles, which lead into the alveoli, which are surrounded by dense networks of blood capillaries (Figure 46.8).

- Mammals inhale by a negative pressure mechanism. Air is exhaled passively by relaxation of the diaphragm and the external intercostal muscles between the ribs, and elastic recoil of the lungs. During deep and rapid breathing, the expulsion of air is forceful, driven by contraction of the internal intercostal muscles (Figure 46.9).

- The tidal volume of the lungs is the air moved in and out of the lungs during an inhalation and exhalation. The vital capacity is the total volume of air a person can inhale and exhale by breathing as deeply as possible. The air remaining in the lungs after as much air as possible is exhaled is the residual volume of the lungs.

- Breathing is controlled by a combination of local chemical controls and regulation by centers in the brain stem. These controls match the rate of air and blood flow in the lungs, and link the rate and depth of breathing to the body's requirements for O_2 uptake and CO_2 release (Figure 46.10).

- The basic rhythm of breathing is produced by interneurons in the medulla. When more rapid breathing is required, another group of interneurons in the medulla sends signals reinforcing inhalation and producing forceful exhalation. Two interneuron groups in the pons modulate the signals from the medulla, fine-tuning and smoothing muscle contractions for breathing.

- Sensory receptors in the medulla, the carotid bodies, and the aortic bodies detect changes in the levels of O_2 and CO_2 in the blood and body fluids. The control centers in the medulla and pons adjust the rate and depth of breathing to compensate for changes in the blood gases.

46.4 Mechanisms of Gas Exchange and Transport

- The partial pressure of O_2 is higher in the alveolar air than in the blood in the capillary networks surrounding the alveoli, causing O_2 to diffuse from the alveolar air into the blood. Most of the O_2 entering the blood combines with hemoglobin inside erythrocytes (Figure 46.11).

- A hemoglobin molecule can combine with four O_2 molecules. The large quantities of O_2 that combine with hemoglobin maintain a large gradient in partial pressure between O_2 in the alveolar air and in the blood (Figure 46.12).

- In body tissues outside the lungs, the O_2 concentration in the interstitial fluid and body cells is lower than in the blood plasma. As a result, O_2 diffuses from the blood into the interstitial fluid, and from the fluid into body cells.

- The partial pressure of CO_2 is higher in the tissues than in the blood. About 10% of this CO_2 dissolves in the blood plasma, about 30% combines with hemoglobin, and about 60% is converted into H^+ and HCO_3^- (bicarbonate) ions (Figures 46.11 and 46.13A).

- In the lungs, the partial pressure of CO_2 is higher in the blood than in the alveolar air. As a result, the reactions packing CO_2 into the blood are reversed, and the CO_2 is released from the blood into the alveolar air (Figure 46.13C).

46.5 Respiration at High Altitudes and in Ocean Depths

- In mammals that move to high altitudes, the number of red blood cells and the amount of hemoglobin per cell increase. These changes are reversed if the animals return to lower altitudes.

- Humans living at higher altitudes from birth develop more alveoli and capillary networks in the lungs.

- Some mammals and birds adapted to high altitudes have forms of hemoglobin with greater affinity for O_2, allowing saturation at the lower P_{O_2} typical of high altitudes (Figure 46.14).

- Marine mammals adapted to deep diving have a greater blood volume per unit of body weight, and their blood contains more red blood cells, with a higher hemoglobin content, than other mammals. Their muscles also contain more myoglobin than those of land mammals, allowing more O_2 to be stored in muscle tissues. During a dive, the heartbeat slows, and circulation is reduced to all parts of the body except the brain.

TEST YOUR KNOWLEDGE

Remember/Understand

1. Which of the following describes a respiratory medium?
   a. the liver of an amphibian, in which the rate of diffusion is high
   b. neurons in the human brain, where CO_2 moves from the neurons to the blood
   c. the O_2 in the blood of humans
   d. epithelial cells of fish that are in contact with the air
   e. air and water

2. Which of the following describes a respiratory surface?
   a. a surface consisting of multiple layers of epithelial cells
   b. the exoskeleton of an insect
   c. the nasal passages of a mammal
   d. thin surface consisting of a single layer of epithelial cells
   e. the outer membrane of a mitochondrion

3. A geothermal HVAC (heating, ventilating, and air conditioning) system air-conditions your house by passing warm air pulled from the house past a continuous flow of cool water pulled from the ground. How this heat exchanger works is analogous to:
   a. countercurrent exchange of gases in fish gills.
   b. diffusion of O_2 from blood to cells in shark tissues.
   c. diffusion of CO_2 from cells to blood in crabs.
   d. use of O_2 in cells in insects.
   e. excretion of CO_2 from mammalian cells.

4. Tracheal systems are characterized by:
   a. closed circulatory tubes that move gases.
   b. spiracles that move gases between cells and body fluids.
   c. body movements that compress and expand air sacs to pump air.
   d. positive pressure breathing, which swallows air into the body.
   e. negative pressure breathing, which lowers air pressure at the respiratory surfaces.

5. The partial pressure of O_2 in the atmosphere is 160 mm Hg, but when O_2 moves into the blood its partial pressure is one third lower than the atmospheric partial pressure. Which structure is it moving through at this partial pressure?
   a. alveoli
   b. bronchi
   c. bronchioles
   d. tracheae
   e. pharynges

6. As a speed skater finishes the last lap of a race:
   a. his diaphragm and rib muscles contract when he exhales.
   b. positive pressure brings air into his lungs.
   c. his lungs undergo an elastic recoil when he inhales.
   d. his tidal volume is at vital capacity.
   e. his residual volume momentarily reaches zero.

7. A teenager is frightened when she is about to step onto the stage but then remembers to breathe deeply and slowly as she faces the audience. What is occurring here?
   a. Interneurons in the medulla cause the rib muscles to relax, followed later by stimulation and contraction of the intercostal muscles.
   b. Signals from the pons override the initial brain stem stimuli.
   c. The limbic system stabilized her emotional state, so there is no change in the mechanical movement of air.
   d. The brain signals the aortic bodies in the carotid arteries to adjust the breathing rate.
   e. Initial low CO_2 blood levels causing high pH are followed by increased CO_2 levels that lower pH.

8. Oxygen enters the blood in the lungs because relative to alveolar air:
   a. the CO_2 concentration in the blood is high.
   b. the CO_2 concentration in the blood is low.
   c. the O_2 concentration in the blood is high.
   d. the O_2 concentration in the blood is low.
   e. the process is independent of gas concentrations in the blood.

9. A hemoglobin O_2 dissociation curve:
   a. demonstrates that hemoglobin is about 50% saturated in the alveoli.
   b. shifts to the left when pH rises.
   c. demonstrates that hemoglobin holds less O_2 when the pH is higher.
   d. illustrates that oxygen saturation is not dependent on CO_2 levels.
   e. demonstrates why hemoglobin can bind O_2 at high pH in the lungs and release it at lower pH in the tissues.

10. The majority of CO_2 in the blood:
    a. is in the form of carbonic acid and bicarbonate ions.
    b. dissociates to add H^+ to the blood to raise its pH to 7.4.
    c. has a lower P_{CO_2} than the P_{CO_2} in the alveolar air.
    d. increases in the lung capillaries, which have a higher pH than the tissue capillaries.
    e. can be displaced on the hemoglobin molecule by CO if CO is inhaled.

Apply/Analyze

11. **Discuss Concepts** Smoking has traditionally been considered to reduce the ability of athletes to run without becoming exhausted. Why might this be true?

12. **Discuss Concepts** People are occasionally found unconscious from breathing too much CO_2 (as from a charcoal heater placed indoors) or too much CO (as from auto exhaust in a closed garage). Would it be more advantageous to give pure O_2 to a person breathing too much CO_2 than simply moving the person to fresh air? Why? Which—pure O_2 or fresh air—would be best for a person unconscious from breathing CO? Why?

13. **Discuss Concepts** Hyperventilation, or overbreathing, is breathing faster or deeper than necessary to meet the body's needs. Hyperventilation reduces the CO_2 content of blood, but does not significantly increase the amount of O_2 available to tissues. Why might this be so?

Evaluate/Create

14. **Design an Experiment** Propose a hypothesis for the effect of zero gravity on respiration, and design an experiment to test the hypothesis.

15. **Apply Evolutionary Thinking** From what you have learned in this chapter and in Chapter 32, do you think lungs evolved once, or on several occasions? Justify your answer.

For selected answers, see Appendix A.

INTERPRET THE DATA

Lycopene, which is abundant in tomatoes, is thought to have a potential benefit in protecting against many types of cancers. In the study, the results of which are shown in the **Figure,** mice were exposed to tobacco smoke or normal air. The amount of lung tissue damage caused by each exposure was measured in mice that were given tomato juice (1) or water (2). To estimate the destruction of the alveolar wall, a destructive index (DI) was calculated from histological slides of the lung. All lungs will show some level of damage, but a DI above 10% is considered significant destruction.

1. What is the effect of tobacco smoke on the lung tissue in these mice?

2. Do these data suggest that lycopene has a protective effect against tobacco's effects?

Source: Based on S. Kasagi et al. 2006. Tomato juice prevents senescence-accelerated mouse P1 strain from developing emphysema induced by chronic exposure to tobacco smoke. *American Journal of Physiology–Lung Cellular and Molecular Physiology* 290:L396–L404.

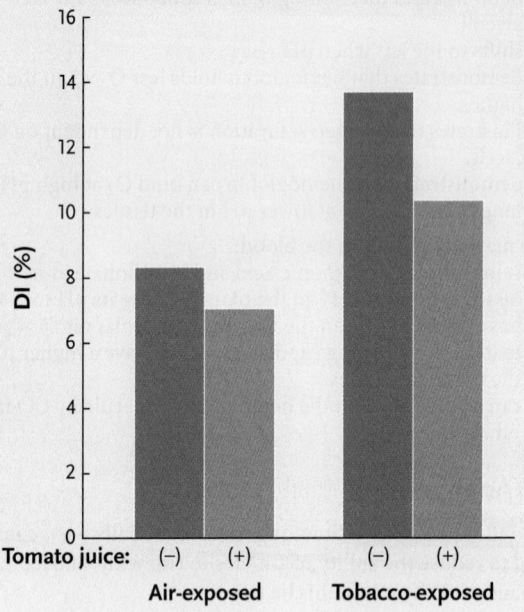

Animal Nutrition

Two bald eagles *(Haliaeetus leucocephalus),* fighting over food.

© Sergey Uryadnikov/Shutterstock.com

Why it matters . . . In the inky darkness, a deep-sea anglerfish (a member of the order Lophiiformes) lies in wait, its gaping mouth lined with sharp teeth. Just above the mouth a glowing lure resembling a tiny fish is suspended from a fishing-rod–like spine that projects from the fish's dorsal fin **(Figure 47.1).**

A hapless fish is attracted to the lure. As it comes within range, the mouth of the anglerfish expands suddenly, creating a powerful suction that whips the prey in. The backward-angling fangs keep the prey from escaping. Contractions of throat muscles send the prey to the stomach of the anglerfish. In the fish's digestive tract, acids and enzymes dissolve the body of the prey, gradually breaking it into molecules small enough to be absorbed into the bloodstream. In this function, the digestive system of the anglerfish is the same as that of any other vertebrate, including humans—it provides nutrients that allow the animal to live. And the adaptations of the anglerfish for feeding are no more remarkable than those of many other animals.

Animal **nutrition**—which includes the processes by which food is ingested, digested, and absorbed into body cells and fluids—is the subject of this chapter. Our discussion begins with the basic categories of animal foods and **ingestion,** the feeding methods used to take food into the digestive cavity. Then we examine the process of **digestion,** which is the splitting of complex molecules in foods, such as carbohydrates, proteins, and fats, into chemical subunits small enough to be absorbed into an animal's body fluids and cells. The chapter also presents the main structural and functional features of *digestive systems,* with special emphasis on humans and other mammals. The adaptations animals use to obtain and digest food are among their most strongly defining anatomical and functional characteristics.

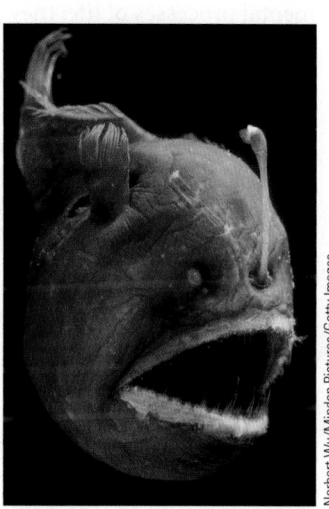

FIGURE 47.1 A deep-sea anglerfish (Melanocetus johnsoni), with its rod and lure lit and ready to attract prey.

Norbert Wu/Minden Pictures/Getty Images

47.1 Feeding and Nutrition

All organisms require sources of matter and energy for metabolism, homeostasis (maintaining their internal environment in a stable state; see Section 38.4), growth, and reproduction. For animals, meeting these nutritional requirements involves **feeding,** the uptake of food from the surroundings. Animals use various feeding methods ranging from the ingestion of molecules in liquid solutions to eating entire organisms in one gulp. Once the food is ingested, digestive processes convert its molecules into absorbable subunits. In this section, we survey animal nutritional requirements and feeding methods as an introduction to animal digestive processes.

Animals Require Both Organic and Inorganic Molecules for Nutrition

Plants and other photosynthesizers need only sunlight as an energy source and a supply of simple inorganic precursors such as water, carbon dioxide, and minerals to make all the organic molecules they require. In contrast, animals require a constant diet of organic molecules as a source of both energy and nutrients that they cannot make for themselves.

Animals are classified according to their sources of organic molecules. **Herbivores,** such as antelopes, horses, bison, giraffes, kangaroos, manatees, and grasshoppers, obtain organic molecules primarily by eating plants. **Carnivores,** such as cats, Tasmanian devils, penguins, sharks, and spiders, primarily eat other animals. We say "primarily" because many herbivores eat animal matter at times, and a number of carnivores occasionally eat plant material. An antelope will eat insects as it grazes, and a grizzly bear, although primarily carnivorous, also eats berries. **Omnivores,** such as crows, cockroaches, and humans, eat both plants and animals and, digestive enzymes allowing, may consume any source of organic matter.

Organic molecules are the basis for two of the most fundamental processes of life: they act as fuels for oxidative reactions that supply energy, and they are the building blocks for making complex biological molecules. Energy in fuels is measured in *calories* (see Section 2.4). In animal nutrition, food energy is presented as *kilocalories* (kcal = 1,000 calories = 1 Calorie [uppercase C] = 4.2 kilojoules). Carbohydrates, proteins, and fats are the three main types of energy-rich foodstuffs consumed by animals. Fat provides about twice as many calories per gram as each of the other two.

Animals whose intake of organic fuels is inadequate, or whose assimilation of such fuels is abnormal, suffer from **undernutrition.** The opposite condition, **overnutrition,** is caused by excessive intake of specific nutrients. Both undernutrition and overnutrition are types of **malnutrition.**

An animal suffering from undernutrition is starving for one or more nutrients, taking in fewer calories than needed for daily activities. Animals with chronic undernutrition lose weight because they have to use energy-providing molecules of their own bodies as fuels. Mammals use stored fats and glycogen (animal starch) first. Once those stores have been used up, proteins are metabolized as fuels. The use of proteins as fuels leads to muscle wastage and, in the long term, to organ and brain damage, which eventually leads to death.

Organic molecules also serve as building blocks for carbohydrates, lipids, proteins, and nucleic acids. Animals can synthesize many of the organic molecules that they do not obtain directly in the diet by converting one type of building block into another. Typically they cannot make certain amino acids and fatty acids from other organic molecules. These required organic building blocks are called **essential amino acids** and **essential fatty acids** because they must be obtained in the diet. If they are not obtained in the diet over a period of time, the animal may have serious health consequences. For instance, protein synthesis cannot continue unless all 20 amino acids are present. In the absence of essential amino acids in the diet, the animal would have to break down its own proteins to provide them for new protein synthesis.

Animals must also take in **vitamins,** organic molecules required in small quantities that the animal cannot synthesize for itself, and **essential minerals,** which are species-specific required inorganic elements such as calcium, iron, and magnesium. Many vitamins are *coenzymes,* nonprotein organic subunits associated with enzymes that assist in enzymatic catalysis (see Section 6.4).

The essential amino acids, fatty acids, vitamins, and minerals are known collectively as an animal's **essential nutrients.** The list of essential nutrients differs from animal to animal. For instance, pet foods for cats and dogs have different compositions that provide the essential nutrients for each species.

Animals Obtain Nutrients in Fluid, Particle, or Bulk Form

Animals fall into one of four groups according to overall feeding methods and the physical state of the organic molecules they consume—*fluid feeders, suspension feeders, deposit feeders,* and *bulk feeders* (**Figure 47.2**). These feeding methods are adaptations that allow animals to obtain the food they need in their particular environments.

Fluid feeders ingest liquids that contain organic molecules in solution. Invertebrate fluid feeders include aphids, mosquitoes, leeches, and spiders. Vertebrate fluid feeders include birds such as hummingbirds (**Figure 47.2A**), which feed on flower nectar; parasitic fishes such as lampreys, which feed on body fluids of their hosts; and some bats, which feed on nectar or blood. Many fluid feeders have mouthparts specialized to reach the nourishing fluid, for example, needlelike mouthparts to pierce body surfaces in mosquitoes, and long bills and tongues to extend deep within flowers in nectar-feeding birds. Some fluid feeders use enzymes or other chemicals to liquefy their food or to keep it liquid during feeding. For example, spiders inject digestive enzymes that liquefy tissues inside their victim and then suck up the liquid, and the saliva of mosquitoes,

A. Fluid feeder

© Steve Byland/Shutterstock.com

B. Suspension feeder

© Jo Crebbin/Shutterstock.com

Baleen

C. Deposit feeder

Caradan/Shutterstock.com

D. Bulk feeder

© Cappi Thompson/Shutterstock.com

FIGURE 47.2 Adaptations for feeding. (A) A juvenile ruby-throated hummingbird *(Archilochus colubris)*, an example of a fluid feeder. **(B)** A gray whale *(Eschrichtius robustus)*, an example of a suspension feeder. **(C)** A fiddler crab *(Uca* species), an example of a deposit feeder. **(D)** A great blue heron *(Ardea herodias)*, an example of a bulk feeder.

leeches, and vampire bats contains an anticoagulant that prevents blood from clotting.

Suspension feeders obtain nutrients by ingesting small organisms that are suspended in water, such as bacteria, archaea, single-celled protists, algae, and small crustaceans, or fragments of these organisms. Among the suspension feeders are aquatic invertebrates such as clams, mussels, barnacles, many fishes, and some birds and whales. These animals strain food particles suspended in water through a body structure covered with sticky mucus or through a filtering network of bristles, hairs, or other body parts. The trapped particles are then funneled into the animal's mouth. For example, bits of organic matter are filtered from water by the sievelike fringes of horny fiber hanging in the mouths of baleen whales such as the gray whale **(Figure 47.2B).**

Deposit feeders pick up or scrape particles of organic matter from solid material they live in or on. For instance, earthworms are deposit feeders that eat their way through soil, taking the soil into their mouth and digesting and absorbing any organic material it contains. The fiddler crab *(Uca* species) has claws of markedly different sizes **(Figure 47.2C).** The small claw picks up sediment and moves it to the mouth where the contents are sifted. The edible parts of the sediment are ingested,

and the rest is put back on the sediment as a small ball. The feeding-related movement of the small claw over the larger claw looks like the crab is playing the large claw like a fiddle and hence gives the crab its name.

Bulk feeders are animals that consume sizeable food items whole or in large chunks. Most mammals, birds, and fishes, and all reptiles and adult amphibians, eat this way. Depending on the animal, adaptations for bulk feeding include teeth for tearing or chewing, and claws and beaks for holding large food items, and jaws that are hinged or otherwise modified to permit a food mass to enter the mouth. **Figure 47.2D** shows a great blue heron eating a fish.

We now take up the processes by which animals, having fed, undertake the mechanical and chemical breakdown of food into absorbable molecular subunits.

STUDY BREAK 47.1

1. What are carnivores, herbivores, and omnivores?
2. What are essential nutrients, and are they the same for all animals?
3. What is the difference between deposit feeders and suspension feeders?

47.2 Digestive Processes

Digestive processes break food molecules into molecular subunits that can be absorbed into body fluids and cells. The breakdown occurs by **enzymatic hydrolysis** (see Figure 3.4). Specific enzymes speed these reactions: **amylases** catalyze the hydrolysis of starches, **lipases** break down fats and other lipids, **proteases** hydrolyze proteins, and **nucleases** digest nucleic acids. Depending on the animal, the enzymatic hydrolysis of food molecules may take place inside or outside the body cells.

Intracellular Digestion Takes Place within Cells; Extracellular Digestion Occurs in an Internal Pouch or Tube

In **intracellular digestion,** cells take in food particles by endocytosis (see Section 5.5). Inside the cell, the endocytic vesicle containing the food particles fuses with a lysosome, a vesicle containing hydrolytic enzymes (see Figure 4.16). The molecular subunits produced by the hydrolysis pass from the vesicle to the cytosol. Any undigested material remaining in the vesicle is released to the outside of the cell by exocytosis (see Section 5.5). Only a few animals, primarily sponges and some cnidarians, break down food exclusively by intracellular digestion. In sponges, water containing particles of organic matter and microorganisms enters the animal's saclike body through pores in the body wall (see Figure 31.8B). In the body cavity, individual *choanocytes* (collar cells) lining the body wall trap the food particles, take them in by endocytosis, and transport them to amoeboid cells, which digest them intracellularly (see Figure 31.8C).

Extracellular digestion takes place outside body cells, in a pouch or tube that is enclosed within the body. Epithelial cells lining the pouch or tube or accessory organs connected to it secrete enzymes that digest the food. Processing food in specialized compartments in this way prevents the animal from digesting its own body tissues.

Most invertebrates and all vertebrates digest food primarily by extracellular digestion. From an adaptive standpoint, extracellular digestion greatly expands the range of available food sources by allowing animals to digest much larger food items than single cells can take in. Extracellular digestion also allows animals to eat large batches of food, which can be stored and digested while the animal continues other activities.

Saclike Extracellular Digestive Systems Have a Single Opening through Which Food Enters and Undigested Matter Exits

Some animals, including flatworms and cnidarians such as hydras, corals, and sea anemones, have a saclike digestive system with a single opening, a mouth, that serves both as the entrance for food and the exit for undigested material. These animals lack a separate vascular system, so water taken into the

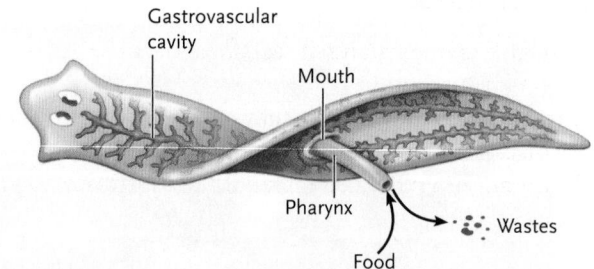

FIGURE 47.3 The digestive system of the flatworm *Dugesia*. The gastrovascular cavity (in blue) is a blind sac, with one opening to the exterior through which food is ingested and wastes are expelled.
© Cengage Learning 2017

digestive chamber also serves to circulate nutrients and other materials through the various tissue layers. Because the digestive chamber contributes both to digestion and to circulation, it is called a **gastrovascular cavity.** In the flatworm *Dugesia,* for example, food enters a **pharynx** that protrudes from the mouth, and then enters the gastrovascular cavity **(Figure 47.3).** Glands in the cavity wall secrete enzymes that begin the digestive process. Cells lining the cavity take up the partially digested material by endocytosis and complete digestion intracellularly. Undigested matter is released to the outside through the protrusible pharynx.

Digestive Tracts Typically Process Nutrients in Five Successive Steps

For most invertebrates and all vertebrates, the extracellular digestive system consists of a *digestive tract* plus accessory digestive organs, the organs that secrete chemicals required for digestion. A **digestive tract** (also called *gastrointestinal [GI] tract, digestive tube, gut,* or *alimentary canal*) is tubelike with two openings that form a separate mouth and anus. In the digestive tract, the digestive contents move in one direction from the mouth to the anus through specialized regions of the tube. Structurally, the inside of the digestive tube—called the **lumen**—is external to all body tissues. In other words, although entirely enclosed, the lumen is, in a functional sense, *outside* of the body.

In most animals with a digestive tract, digestion occurs in five successive steps, with each step taking place in a specialized region of the tube. The tube thus acts as a biological disassembly line, with food entering at one end and passing through as many as five areas in which food processing occurs.

1. **Mechanical processing:** chewing, grinding, and tearing break food chunks into smaller pieces, increasing their mobility and the surface area exposed to digestive enzymes.
2. **Secretion** of enzymes and other digestive aids: enzymes and other substances that aid the process of digestion, such as acids, emulsifiers, and lubricating mucus, are released into the tube.

3. **Enzymatic hydrolysis:** food molecules are broken down through enzyme-catalyzed reactions into absorbable molecular subunits.
4. **Absorption:** the molecular subunits are absorbed from the digestive contents into body fluids and cells.
5. **Elimination:** undigested materials are expelled through the anus.

The material being digested is pushed along by muscular contractions of the wall of the digestive tube. During its progress through the tube, the digestive contents may be stored temporarily at one or more locations. The storage allows animals to take in larger quantities of food than they can process immediately, so that feedings can be spaced in time rather than continuous.

DIGESTION IN AN ANNELID The earthworm (genus *Lumbricus,* **Figure 47.4A**) is a deposit feeder. Food entering the mouth passes to the muscular pharynx. Contraction of the pharynx moves the food through the **esophagus** into the **crop,** where the contents are mixed with lubricating mucus. The **gizzard** carries out mechanical processing of the food, pulverizing it using muscular contractions and the sand particles it contains. The pulverized mixture enters a long **intestine,** where the organic matter is hydrolyzed by secreted enzymes. The molecular subunits are absorbed by cells lining the intestine as the mixture is moved along by muscular contractions of the intestinal wall. Undigested residue is expelled through the anus.

DIGESTION IN AN INSECT Herbivorous insects such as the grasshopper **(Figure 47.4B)** tear leaves and other plant parts into small particles with hard external mouth parts. Food particles enter the mouth and pass through the pharynx, where salivary secretions moisten the mixture before it enters the esophagus and passes into the crop and then into the gizzard, which grinds it into smaller pieces. These food particles enter the **stomach,** in which food is stored and digestion begins. Insect stomachs have saclike outgrowths, the *gastric ceca* (*caecus* = blind), where enzymes hydrolyze the digestive contents; the products of digestion are absorbed through the walls of the ceca. The undigested contents then move into the intestine for further digestion and absorption. At the end of the intestine, water is absorbed from the undigested matter and the remnants are expelled through the anus. The digestive systems of other arthropods are similar to the insect system.

DIGESTION IN A BIRD A pigeon **(Figure 47.4C)** picks up food (such as seeds) with its bill. In the mouth the food is moistened by mucus-filled saliva and swallowed whole (birds have no teeth). The food passes through the pharynx into the esophagus. The anterior end of the esophagus is tubelike; at the posterior end is the pouchlike crop, in which the bird can store large quantities of food. From the crop, the food passes into the anterior glandular portion of the stomach, called the *proventriculus,* which secretes digestive enzymes and acids. The posterior end is the gizzard, in which the food is ground into fine particles, aided by ingested bits of sand and gravel. The food particles are

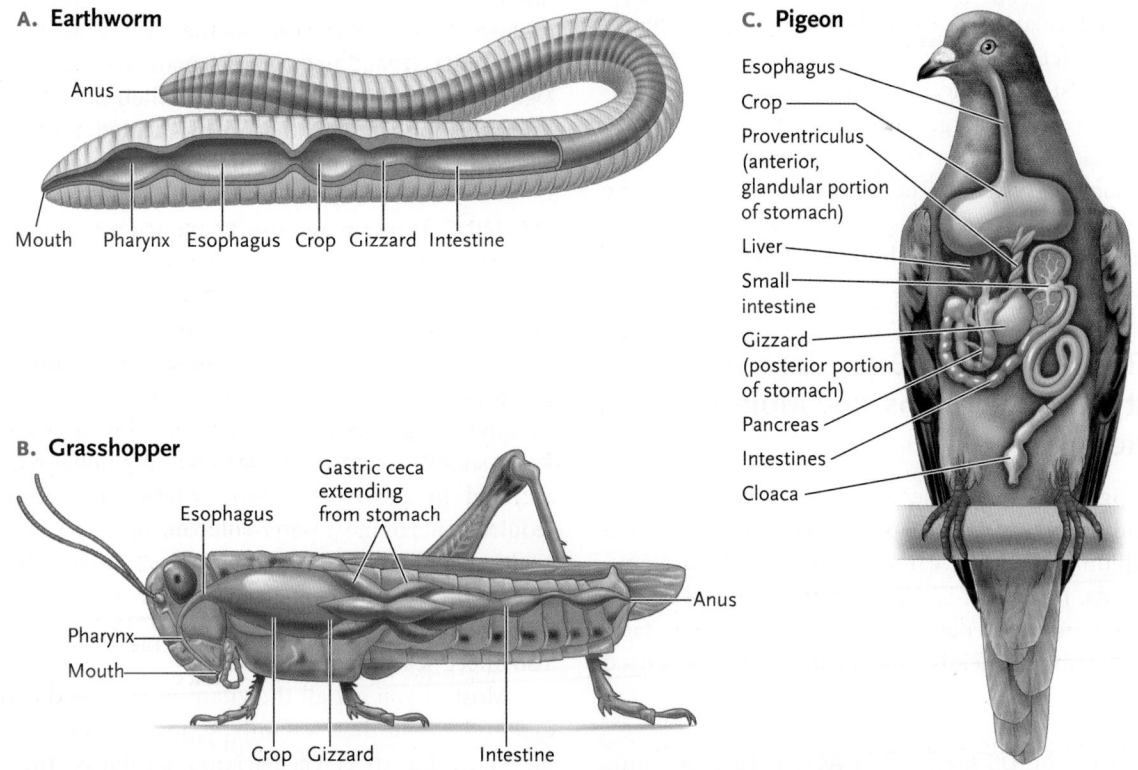

A. Earthworm

Anus

Mouth Pharynx Esophagus Crop Gizzard Intestine

C. Pigeon

Esophagus
Crop
Proventriculus (anterior, glandular portion of stomach)
Liver
Small intestine
Gizzard (posterior portion of stomach)
Pancreas
Intestines
Cloaca

B. Grasshopper

Gastric ceca extending from stomach

Esophagus

Pharynx
Mouth

Anus

Crop Gizzard Intestine

FIGURE 47.4 The digestive systems of an annelid, the earthworm (A), an insect, the grasshopper (B), and a bird, the pigeon (*Columba*) (C).

released into the intestine, where the liver secretes *bile* (aids fat digestion) and the pancreas secretes digestive enzymes. (The liver and pancreas are accessory digestive organs. Bile and pancreatic digestive enzymes are discussed more in Section 47.3.) The molecular subunits produced by enzymatic digestion are absorbed as the mixture passes along the intestine, and the undigested residues are expelled through the anus.

Many of the structures of the pigeon's digestive system, including the mouth, pharynx, esophagus, stomach, intestine, liver, and pancreas, occur in almost all vertebrates.

STUDY BREAK 47.2

1. Distinguish between intracellular digestion and extracellular digestion.
2. What are the five steps of food processing in a digestive tube?

THINK OUTSIDE THE BOOK

Use the Internet or a research paper search tool such as PubMed (http://www.ncbi.nlm.nih.gov/PubMed/) to list the research question being addressed in three projects involving bird digestive systems.

47.3 Digestion in Humans and Other Mammals

Mammals digest foods using the same five steps as other animals with a digestive tract: mechanical processing, secretion of enzymes and other digestive aids, enzymatic hydrolysis, absorption of molecular subunits, and elimination. The mammalian digestive tract is a series of specialized digestive regions that perform these steps, including the mouth, pharynx, esophagus, stomach, small and large intestines, rectum, and anus **(Figure 47.5)**. These regions are under the control of the nervous and endocrine systems. The accessory organs for the mammalian digestive system consist of the *salivary glands,* the *exocrine pancreas,* the *liver,* and the *gallbladder* (all are discussed in this section).

Humans Require Specific Essential Amino Acids, Fatty Acids, Vitamins, and Minerals in Their Diet

The human digestive system meets our basic needs for fuel molecules and for a wide range of nutrients, including the molecular building blocks of carbohydrates, lipids, proteins, and nucleic acids. If the diet is adequate, the digestive system also absorbs the essential nutrients—the amino acids, fatty acids, vitamins, and minerals that cannot be synthesized within our bodies.

ESSENTIAL AMINO ACIDS AND FATTY ACIDS There are nine essential amino acids for adult humans: lysine, histidine, tryptophan, phenylalanine, threonine, valine, methionine, leucine,

and isoleucine. (Histidine originally was thought to be an essential amino acid only for infants, but studies have shown that it is also essential for adults.) The proteins in animal food such as fish, meat, egg whites, milk, and cheese are complete, meaning that they supply all the essential amino acids in appropriate quantities to satisfy nutritional requirements. In contrast, the proteins of most plants used for food are incomplete, meaning that they do not provide sufficient amounts of each essential amino acid. Therefore, vegetarians and especially vegans, who eat a diet with no animal-derived nutrients, must choose their foods carefully to obtain all of the essential amino acids in sufficient quantities to avoid protein malnutrition. Such diets typically include combinations of foods, each of which provides some amino acids, that together contain all of the essential amino acids in sufficient quantities.

If the diet lacks one or more essential amino acids, many enzymes and other proteins cannot be synthesized in sufficient quantities. The resulting protein deficiency is most damaging to the young, who synthesize proteins rapidly for development and growth. Even mild protein starvation during pregnancy or for some months after birth can affect a child's mental and physical development adversely.

Only two fatty acids, linoleic acid and linolenic acid, are essential in the human diet. Both are required for synthesis of phospholipids forming parts of biological membranes and certain hormones. Because almost all foods contain these fatty acids, most people have no problem obtaining them. However, people on a low-fat diet that is deficient in linoleic acid and linolenic acid are at serious risk for developing coronary heart disease. This is illustrated in the case of Hindu vegetarians from India. Their diet consists mainly of low-fat grains and legumes—clearly a low-fat diet—yet their rate of coronary heart disease is higher than that in the United States and Europe, where dietary fat content is higher.

VITAMINS Humans require 13 known vitamins in their diet. Many metabolic reactions depend on vitamins, and the absence of one vitamin can affect the functions of the others. These essential nutrients fall into two classes: **water-soluble vitamins** (hydrophilic vitamins) and **fat-soluble vitamins** (hydrophobic vitamins) (summarized in **Table 47.1**). The body stores excess fat-soluble vitamins in adipose tissues, but any amount of most water-soluble vitamins above daily nutritional requirements is excreted in the urine. Thus, meeting the daily minimum requirements of most water-soluble vitamins is critical for good health. The body can tap its stores of fat-soluble vitamins to meet daily requirements; however, these stores are depleted quickly, so prolonged deficiencies of the fat-soluble vitamins also affect health negatively.

Most of you get all the vitamins you need through a normal and varied diet. Vitamin supplements are usually necessary only for strict vegetarians, newborns, the elderly, and individuals who have medical conditions or take medication that affects the body's uptake of nutrients.

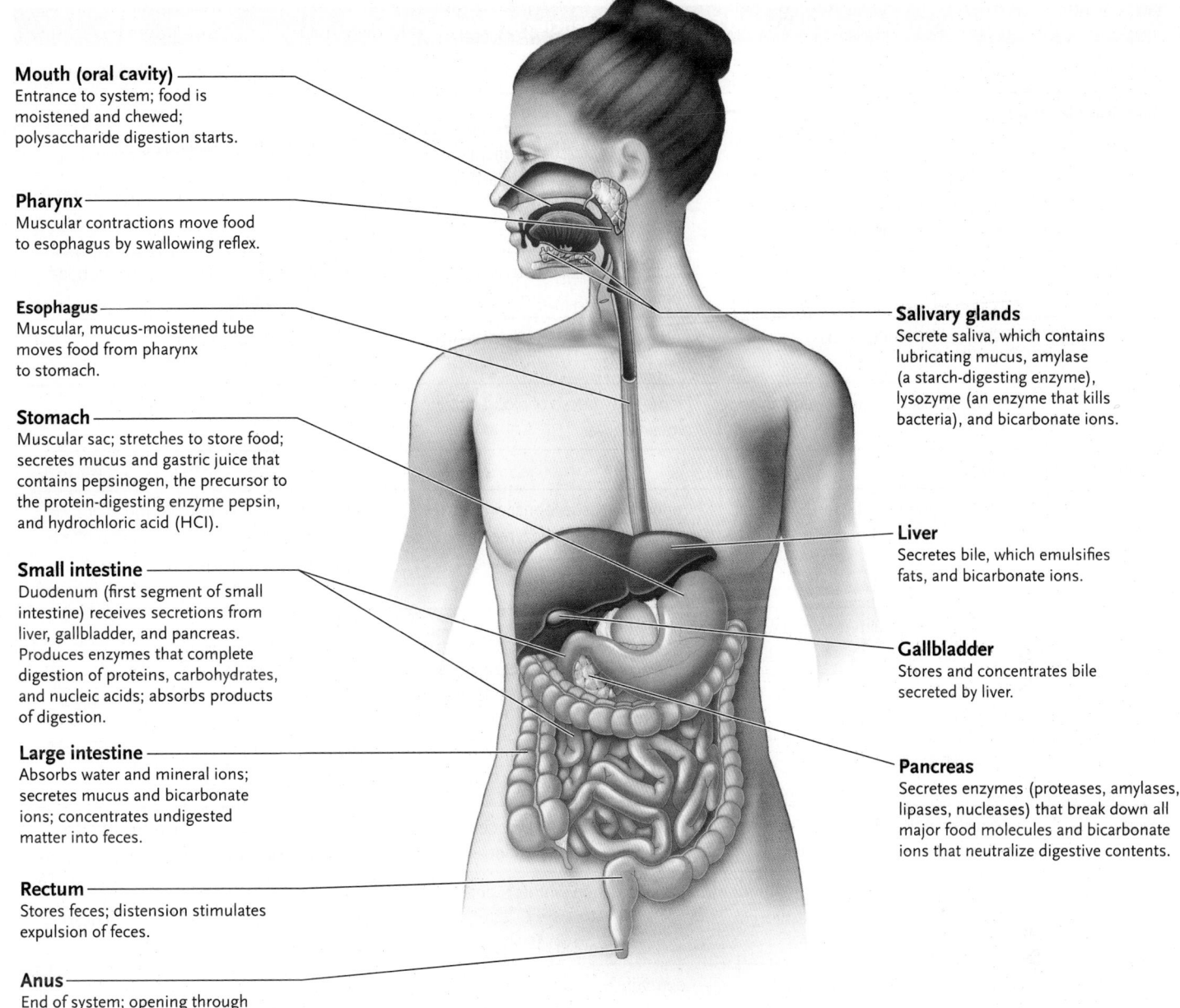

Mouth (oral cavity)
Entrance to system; food is moistened and chewed; polysaccharide digestion starts.

Pharynx
Muscular contractions move food to esophagus by swallowing reflex.

Esophagus
Muscular, mucus-moistened tube moves food from pharynx to stomach.

Stomach
Muscular sac; stretches to store food; secretes mucus and gastric juice that contains pepsinogen, the precursor to the protein-digesting enzyme pepsin, and hydrochloric acid (HCl).

Small intestine
Duodenum (first segment of small intestine) receives secretions from liver, gallbladder, and pancreas. Produces enzymes that complete digestion of proteins, carbohydrates, and nucleic acids; absorbs products of digestion.

Large intestine
Absorbs water and mineral ions; secretes mucus and bicarbonate ions; concentrates undigested matter into feces.

Rectum
Stores feces; distension stimulates expulsion of feces.

Anus
End of system; opening through which feces are expelled.

Salivary glands
Secrete saliva, which contains lubricating mucus, amylase (a starch-digesting enzyme), lysozyme (an enzyme that kills bacteria), and bicarbonate ions.

Liver
Secretes bile, which emulsifies fats, and bicarbonate ions.

Gallbladder
Stores and concentrates bile secreted by liver.

Pancreas
Secretes enzymes (proteases, amylases, lipases, nucleases) that break down all major food molecules and bicarbonate ions that neutralize digestive contents.

FIGURE 47.5 The human digestive system.
© Cengage Learning 2017

Vitamin D (calciferol) differs from other essential vitamins because humans can synthesize it themselves, through the action of ultraviolet light on lipids in the skin. However, many people are not exposed to enough sunlight to make sufficient quantities of the vitamin, and so must rely on dietary sources. And, although humans cannot make vitamin K, much of the requirement for this vitamin is supplied through the metabolic activity of bacteria living in the large intestine. Vitamin K deficiency, therefore, is exceedingly rare in healthy persons. Vitamin K plays a role in blood clotting, so individuals with vitamin K deficiency will bruise easily and show increased blood clotting times. Vitamin K deficiency can be caused by long-term antibiotic therapy because the antibiotics kill intestinal bacteria.

Other mammals have essentially the same vitamin requirements as humans, with some differences. For example, most mammals, with the exception of primates, guinea pigs, and fruit bats, can synthesize vitamin C. No known animal can synthesize B vitamins, but ruminants such as cattle and deer are supplied with those vitamins by microorganisms that live in the digestive tract (see Section 47.5).

MINERALS Some minerals are essential in the human diet **(Table 47.2).** *Macrominerals,* such as potassium, are required in amounts ranging from 50 mg to more than a gram per day. *Trace minerals,* such as iodine, are required only in small amounts, some less than 1 mg per day. All of the minerals, although listed as elements, are ingested as compounds or as ions in solution.

TABLE 47.1	Vitamins: Sources, Functions, and Effects of Deficiencies in Humans		
Vitamin	Common Sources	Main Functions	Selected Effects of Chronic Deficiency
Fat-Soluble Vitamins			
A (retinol)	Liver; fish oils; milk; eggs. (Beta-carotene, which is converted to vitamin A in the body, is in many plant foods such as sweet potato, spinach, and carrots.)	Component of visual pigments; bone metabolism; epithelial tissue maintenance	Night blindness; total blindness; skin disorders; decreased immunity
D	Fish liver oil; egg yolk; fortified milk; produced in skin exposed to sunshine	Calcium and phosphorus absorption from gut	Bone deformities (rickets) in children; bone softening in adults
E (tocopherol)	Nuts; seeds; vegetable oils	Antioxidant; maintenance of cell membranes	Neuromuscular problems
K	Intestinal bacteria; green vegetables	Promotes synthesis of blood-clotting protein by liver	Abnormal blood clotting, bleeding
Water-Soluble Vitamins			
B_1 (thiamin)	Yeast; cereal grains; beans; nuts; meat	Connective tissue formation; needed for folate utilization; coenzyme forming part of enzyme in oxidative reactions	Beriberi (nervous system disorder that includes impaired sensory perception, limb weakness and pain, weight loss, cardiovascular malfunction)
B_2 (riboflavin)	Whole grains, poultry, fish, egg white, milk, lean meat	Coenzyme	Skin lesions
B_6 (pyridoxine)	Spinach, whole grains, tomatoes, potatoes, meat	Coenzyme in amino acid and fatty acid metabolism	Skin, muscle, and nerve damage
B_{12} (cobalamin)	Eggs; meats; dairy products	Coenzyme in nucleic acid metabolism; red blood cell formation	Anemia; brain and nervous system damage
Biotin	Legumes; egg yolk; some synthesized by colon bacteria	Coenzyme in fat and glycogen formation and amino acid metabolism	Scaly skin (dermatitis); sore tongue; brittle hair; depression; weakness
C (ascorbic acid)	Fruits and vegetables, especially citrus, berries, cantaloupe, cabbage, broccoli, green pepper	Vital for collagen synthesis; antioxidant	Scurvy (weakness, anemia, gum disease, and skin problems); delayed wound healing; impaired immunity
Folic acid	Leafy vegetables; legumes; whole grains; yeast; liver; egg yolks	Coenzyme in nucleic acid and amino acid metabolism	Anemia; diarrhea; impaired growth; birth defects
Niacin	Fruits and vegetables; nuts; grains; meats	Coenzyme of oxidative phosphorylation	Pellagra (diarrhea, dermatitis, dementia)
Pantothenic acid	In many foods (meat, yeast, egg yolk especially)	Coenzyme in carbohydrate and fat oxidation; fatty acid and steroid synthesis	Fatigue; tingling in hands; headaches; nausea

A normal and varied diet supplies adequate amounts of the essential minerals. Supplements may be required for those on a strict vegetarian diet, the very young, and the aged. Deficiencies of essential minerals have adverse effects, as Table 47.2 indicates. Excess intake of minerals can also have adverse effects. For instance, ingesting excess iron has been linked to liver, heart, and blood vessel damage, and too much sodium can lead to elevated blood pressure and excess water retention in tissues.

We now turn to the structures that extract nutrients from ingested foods. We begin with a survey of digestive structures common to all vertebrates.

Four Major Layers of the Digestive Tract Each Have Specialized Functions in Digestion

The wall of the digestive tract in mammals and other vertebrates contains four major layers, each with specialized functions. **Figure 47.6** shows these layers for the stomach.

1. The **mucosa,** which contains epithelial and glandular cells, lines the inside of the digestive tract. The epithelial cells absorb digested nutrients, and the glandular cells secrete enzymes, substances such as lubricating mucus

	TABLE 47.2	Minerals: Sources, Functions, and Effects of Deficiencies in Humans[a]		

Mineral	Common Sources	Main Functions	Selected Effects of Deficiency
Macrominerals			
Calcium (Ca)	Leafy green vegetables; legumes; whole grains; nuts; dairy products; eggs	Bone and tooth formation; blood clotting; neural and muscle action	Stunted growth; loss of bone mass
Chlorine (Cl)	Table salt; vegetables; meat; dairy products; eggs	HCl formation in stomach; contributes to body's acid–base balance; necessary for neural function and water balance	Muscle cramps; impaired growth; poor appetite
Magnesium (Mg)	Green leafy vegetables; legumes; nuts; dairy products; meat	Required for many enzymes; in bones and teeth; ATP processing	Weak, sore muscles; nervous system problems
Phosphorus (P)	Whole grains; legumes; nuts; dairy products; eggs; meats	In bones and teeth; component of nucleic acids, ATP, and phospholipids; energy processing	Muscular weakness; loss of minerals from bone
Potassium (K)	Many vegetables and fruits; whole grains; dairy products	Muscle and neural function; water balance; acid–base balance; main positive ion in cell	Muscular weakness; cardiac abnormalities or failure
Sodium (Na)	Table salt; dairy products; eggs; meat	Acid–base balance; water balance; muscle and neural function; main positive ion in extracellular fluid	Muscle cramps
Sulfur (S)	Proteins from food sources, including legumes, nuts, dairy products, eggs, and meat	Component of body proteins	Same as protein deficiencies
Trace Minerals[b]			
Iodine (I)	Seafood; iodized salt	Thyroid hormone formation	Goiter (enlarged thyroid), with metabolic disorders
Iron (Fe)	Green leafy vegetables; legumes; whole grains; nuts; eggs; meats (particularly liver)	Component of hemoglobin, myoglobin, and electron carriers	Iron-deficiency anemia; weakness
Zinc (Zn)	Some vegetables; whole grains; legumes; nuts; fish; meats; many other foods	Component of many enzymes and some transcription factors; protein synthesis; DNA synthesis; cell division; immunity; wound healing	Impaired growth; loss of appetite; impaired immune function

[a]All of the minerals in this table have harmful effects when excess amounts are consumed.
[b]Other trace minerals not listed in the table are chromium (Cr), cobalt (Co), copper (Cu), fluorine (F), manganese (Mn), molybdenum (Mo), and selenium (Se).

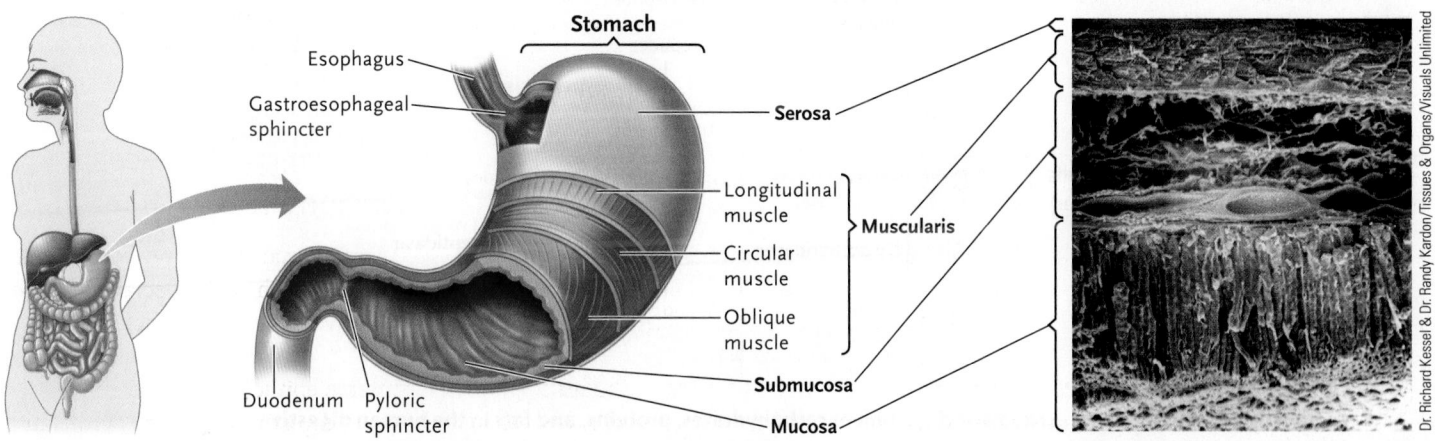

FIGURE 47.6 Layers of the digestive tract wall in vertebrates, as seen in the stomach wall.
© Cengage Learning 2017

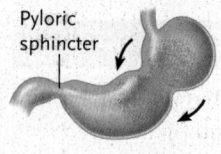
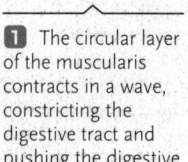
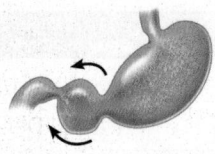
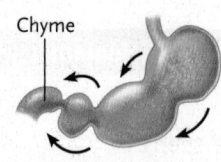

Pyloric sphincter

Chyme

1 The circular layer of the muscularis contracts in a wave, constricting the digestive tract and pushing the digestive contents onward.

2 The longitudinal layer contracts, shortening and expanding the digestive tract and making space for the contents to advance.

3 Partially processed food (chyme) enters the small intestine.

FIGURE 47.7 The waves of peristaltic contractions moving food through the stomach.
© Cengage Learning 2017

that aid digestion, and substances that adjust the pH of the digestive contents.

2. The **submucosa** is a thick layer of connective tissue that contains neuron networks (the enteric nervous system; see Chapter 40) and blood and lymph vessels. It gives the digestive tract its elasticity and its ability to distend. The neuron networks help control digestive activity,

and the blood and lymph vessels have branches that go into the mucosa and into the surrounding muscle layer.

3. In most regions of the digestive tract, the **muscularis** is formed by two smooth muscle layers, a *circular layer* and a *longitudinal layer*. (The stomach uniquely has an additional *oblique layer* running diagonally around its wall.) These two muscle layers coordinate their activities to push the digestive contents through the digestive tract **(Figure 47.7).** In this process, called **peristalsis,** the circular muscle layer contracts in a wave that passes along the digestive tract, constricting it and pushing the digestive contents onward. In a wave just in front of the advancing constriction, the longitudinal layer contracts, shortening and expanding the tube and making space for the contents to advance.

4. The outermost layer of the digestive tract, the **serosa,** consists of connective tissue that secretes an aqueous, slippery fluid. The fluid lubricates the areas between the digestive organs and other organs, reducing friction between them as they move together as a result of muscle movement.

Powerful rings of smooth muscle called **sphincters** form valves between major regions of the digestive tract. By contracting and relaxing, sphincters control the passage of the

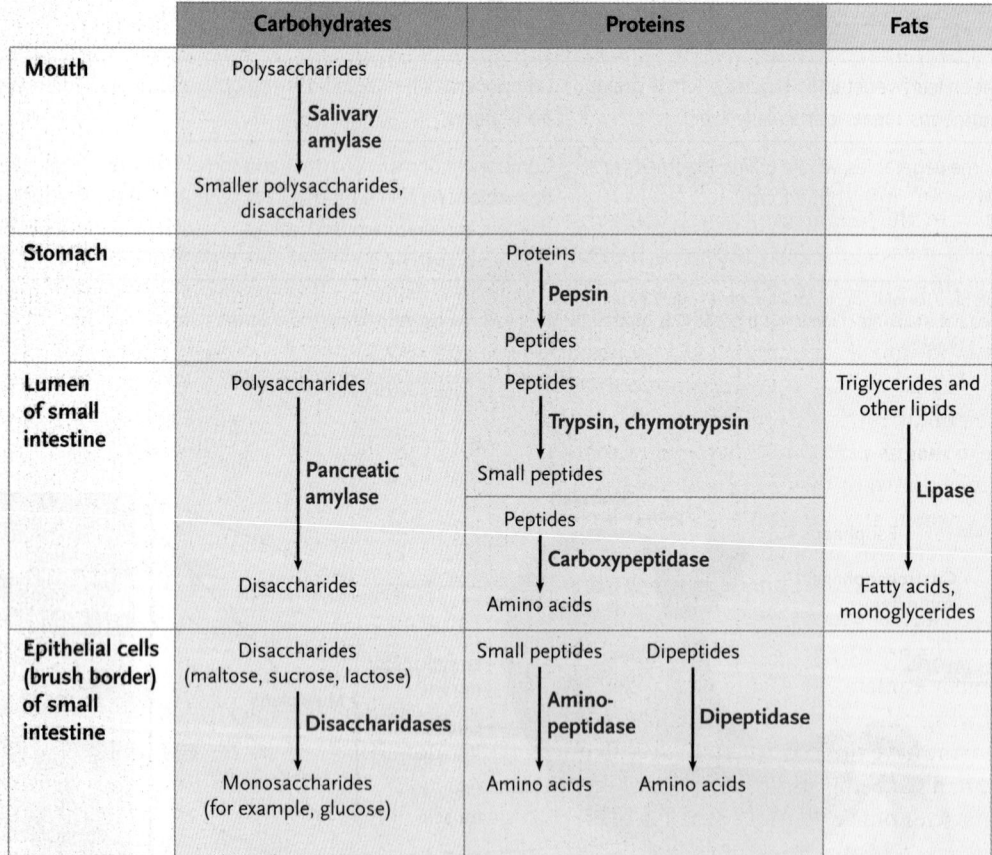

FIGURE 47.8 Enzymatic digestion of carbohydrates, proteins, and fats in the human digestive system.
© Cengage Learning 2017

digestive contents from one region to the next, and ultimately through the anus.

The specialized regions of the digestive tract that perform the sequential processes of digestion in humans allow us to extract nutrients efficiently from the highly varied foods we ingest. **Figure 47.8** summarizes the enzymatic digestion of carbohydrates, proteins, and fat in the regions of the human digestive tract. The rest of this section describes those digestion processes.

Food Begins Its Journey through the Digestive Tract in the Mouth, Pharynx, and Esophagus

The human digestive tract in its normal contracted state in a living adult is about 4.5 m long. Fully relaxed, as in a cadaver, it is about twice as long. The first step in the digestive process is **mastication** or **chewing,** in which ingested food is sliced, torn, and ground into small pieces by the teeth. During chewing, the food is mixed with **saliva,** a secretion of three pairs of salivary glands into the mouth.

Saliva, which is more than 99.5% water, moistens the food. Saliva contains **salivary amylase,** which hydrolyzes starches to the disaccharide maltose (see Figure 47.8). It also contains mucus, which lubricates the food mass, facilitating the formation of a **bolus,** a ball of chewed or liquid food, in preparation for swallowing. Bicarbonate ions (HCO_3^-) in saliva neutralize acids in the food and keep the pH of the mouth between 6.5 and 7.5, which is the optimal range for salivary amylase to function. Saliva also contains *lysozyme,* an enzyme that kills bacteria by breaking open their cell walls. Some 1 to 2 L of saliva are secreted into the mouth each day.

Little digestion of food occurs in the mouth. After chewing, the bolus is pushed by the tongue to the back of the mouth, where touch receptors detect the pressure and trigger the **swallowing reflex (Figure 47.9).** The reflex is an involuntary action produced by contractions of muscles in the walls of the pharynx that direct food into the esophagus. Peristaltic contractions of the esophagus, aided by mucus secreted by the esophagus, propel the bolus toward the stomach. The passage of a bolus down the esophagus stimulates the **gastroesophageal sphincter** at the junction between the esophagus and the stomach (see Figure 47.6) to open and admit the bolus to the stomach. After the bolus enters the stomach, the sphincter closes tightly. If the closure is imperfect, the acidic stomach contents can enter the esophagus and produce the irritation and pain we recognize as *acid reflux* or heartburn.

You can consciously initiate the swallowing reflex. However, once the swallowing reflex has begun, you cannot stop it voluntarily, as you might have noticed when you get that feeling of a piece of food or a pill being stuck in the throat or chest. This is because the muscles of the pharynx and upper esophagus are skeletal muscles, which you can control, while the muscles below are smooth muscles, which you cannot control.

Involuntary movements of the tongue and soft palate at the back of the mouth prevent food from backing into the mouth

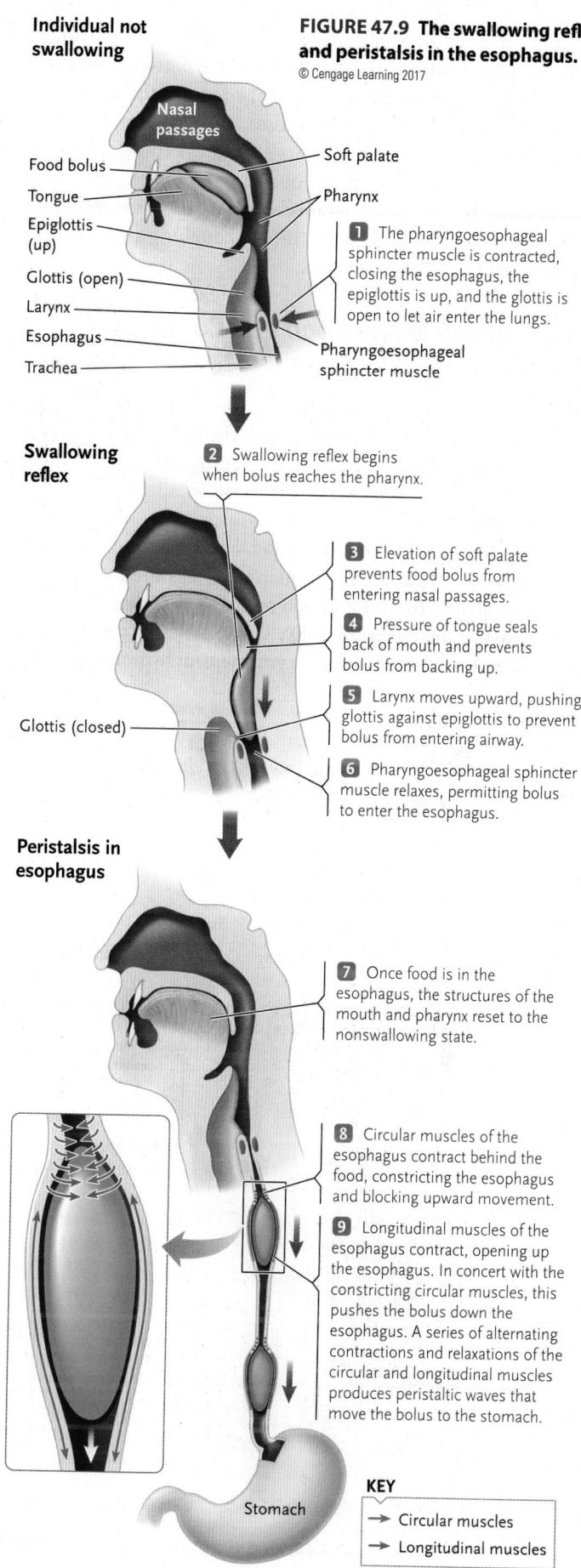

FIGURE 47.9 The swallowing reflex and peristalsis in the esophagus.
© Cengage Learning 2017

Individual not swallowing

Nasal passages
Food bolus
Tongue
Epiglottis (up)
Glottis (open)
Larynx
Esophagus
Trachea
Soft palate
Pharynx
Pharyngoesophageal sphincter muscle

1 The pharyngoesophageal sphincter muscle is contracted, closing the esophagus, the epiglottis is up, and the glottis is open to let air enter the lungs.

Swallowing reflex

2 Swallowing reflex begins when bolus reaches the pharynx.

3 Elevation of soft palate prevents food bolus from entering nasal passages.

4 Pressure of tongue seals back of mouth and prevents bolus from backing up.

Glottis (closed)

5 Larynx moves upward, pushing glottis against epiglottis to prevent bolus from entering airway.

6 Pharyngoesophageal sphincter muscle relaxes, permitting bolus to enter the esophagus.

Peristalsis in esophagus

7 Once food is in the esophagus, the structures of the mouth and pharynx reset to the nonswallowing state.

8 Circular muscles of the esophagus contract behind the food, constricting the esophagus and blocking upward movement.

9 Longitudinal muscles of the esophagus contract, opening up the esophagus. In concert with the constricting circular muscles, this pushes the bolus down the esophagus. A series of alternating contractions and relaxations of the circular and longitudinal muscles produces peristaltic waves that move the bolus to the stomach.

Stomach

KEY
→ Circular muscles
→ Longitudinal muscles

or nasal cavities. Entry into the trachea (the airway to the lungs) is blocked by closure of the *glottis* (the space between the vocal cords) and an upward movement of the *larynx* (the voice box) at the top of the trachea, which closes against a flaplike valve, the **epiglottis.** You can feel the larynx and the front of the epiglottis bob upward if you place your hand on your throat while you swallow. If these blocking mechanisms fail, touch receptors in the nasal passages and larynx trigger coughing and sneezing reflexes that clear these passages.

The Stomach Stores Food and Continues Digestion

The stomach is a muscular, elastic sac that stores food and adds secretions for the process of digestion. The mucosal layer of the stomach is an epithelium covered with tiny **gastric pits** that are entrances to millions of **gastric glands.** These glands extend deep into the stomach wall and contain cells that secrete some of the products needed to digest food.

The entry of food into the stomach activates stretch receptors in its wall. Signals from the stretch receptors stimulate the secretion of **gastric juice (Figure 47.10),** which contains **pepsinogen,** the precursor for the digestive enzyme **pepsin;** hydrochloric acid (HCl); and lubricating mucus. The stomach secretes about 2 L of gastric juice each day.

Pepsinogen is secreted by **chief cells** in the gastric glands and is converted to pepsin by the highly acidic conditions of the stomach (see Figure 47.10). Once produced, pepsin itself catalyzes the reaction that converts more pepsinogen to pepsin. (This is a positive feedback mechanism; see Section 38.4.) Pepsin begins the digestion of proteins by cleaving certain amino acid linkages to produce peptides (short chains of amino acids; see Figure 47.8). The activation of pepsinogen illustrates a common theme in the digestive system: powerful hydrolytic enzymes that would be dangerous to the cells secreting them are synthesized in the form of inactive precursors and are not converted into an active form until they are exposed to the digestive contents.

Parietal cells (see Figure 47.10) secrete H^+ and Cl^-, which combine to form HCl in the lumen of the stomach. The HCl lowers the pH of the digestive contents to pH 2 or lower, the level at which pepsin reaches optimal activity. To put this pH in perspective, lemon juice is pH 2.4, and sulfuric acid or battery acid is approximately pH 1 (recall that pH is a log scale; see Section 2.5). The acidity of the stomach also helps break up food particles and causes proteins in the digestive contents to unfold, which exposes their peptide linkages to hydrolysis by

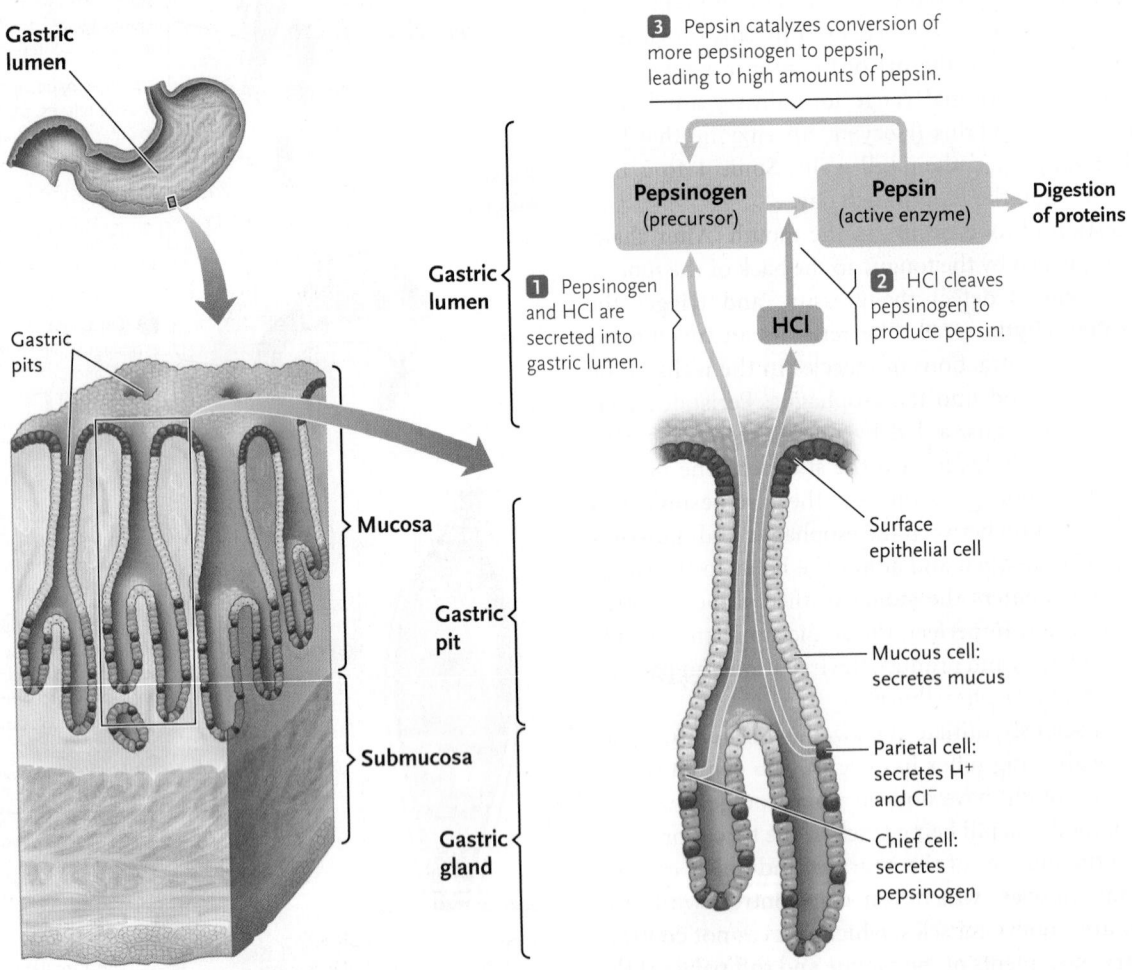

FIGURE 47.10 Cells that secrete mucus, pepsin, and HCl in the stomach lining.

pepsin. The acid also kills most of the bacteria that reach the stomach and stops the action of salivary amylase.

A thick coating of alkaline mucus, secreted by **mucous cells** (see Figure 47.10), protects the mucosal layer of the stomach from attack by pepsin and HCl. (Note that *mucus* is the noun, and *mucous* is the adjective.) Underneath the mucous barrier, tight junctions between cells prevent gastric juice from seeping into the stomach wall. Even so, some breakdown of the mucosal layer does occur. However, the damage is normally repaired quickly by the rapid division of mucosal cells.

Most bacteria cannot survive the highly acidic environment of the stomach, but one, *Helicobacter pylori,* thrives there. In some people, the bacterium breaks down the mucous barrier, exposing the stomach wall to attack by HCl and pepsin. The resulting lesion, known as a **peptic ulcer** or **stomach ulcer,** causes stomach bleeding and pain (see Figure 26.12). Ulcers are treated by taking an antibiotic that kills *H. pylori.* An untreated ulcer may perforate the stomach wall, with potentially fatal consequences.

As part of the digestive process, contractions of the stomach walls continually mix and churn the contents, which can amount to as much as 2 L when the stomach is full. Peristaltic contractions of the stomach wall move the digestive contents toward the **pyloric sphincter** (*pylorus* = gatekeeper) at the junction between the stomach and small intestine. The arrival of a strong stomach contraction relaxes and opens the valve briefly, releasing a pulse of the stomach contents, now called **chyme,** into the small intestine (see Figure 47.7).

Depending on the volume and composition of the stomach contents, it can take from 1 to 6 hours for the stomach to empty after a meal. Feedback controls that regulate the rate of gastric emptying tend to match it to the rate of digestion, so that food is not moved along faster than it can be processed chemically. In particular, when chyme with high fat content and high acidity enters the **duodenum**—the first ~20 cm of the small intestine—it stimulates the mucosal layer to secrete hormones that slow stomach emptying. Fat is digested more slowly than other nutrients, and it is digested in the lumen of the small intestine. Therefore, further emptying of the stomach is prevented until the processing of fat has been completed in the small intestine. This is why a fatty meal, such as a greasy pizza, feels so heavy in the stomach. The highly acidic chyme that empties into the duodenum is neutralized by bicarbonate.

Most Digestion and Absorption of Nutrients Occurs in the Small Intestine

Most digestion and absorption of nutrients occurs in the small intestine. No absorption of nutrients occurs in the mouth, pharynx, or esophagus; with the exception of a few substances, such as alcohol, aspirin, caffeine, and water, little absorption occurs in the stomach.

The "small" in the small intestine refers to its diameter, about 3 cm. It is about 6 m long and complexly coiled within the abdominal cavity. The lining of the small intestine folds into ridges that are densely covered by microscopic, fingerlike extensions, the **intestinal villi** (singular, *villus*). In addition, the epithelial cells covering the villi have a **brush border** consisting of fingerlike projections of the plasma membrane called **microvilli (Figure 47.11).** The intestinal villi

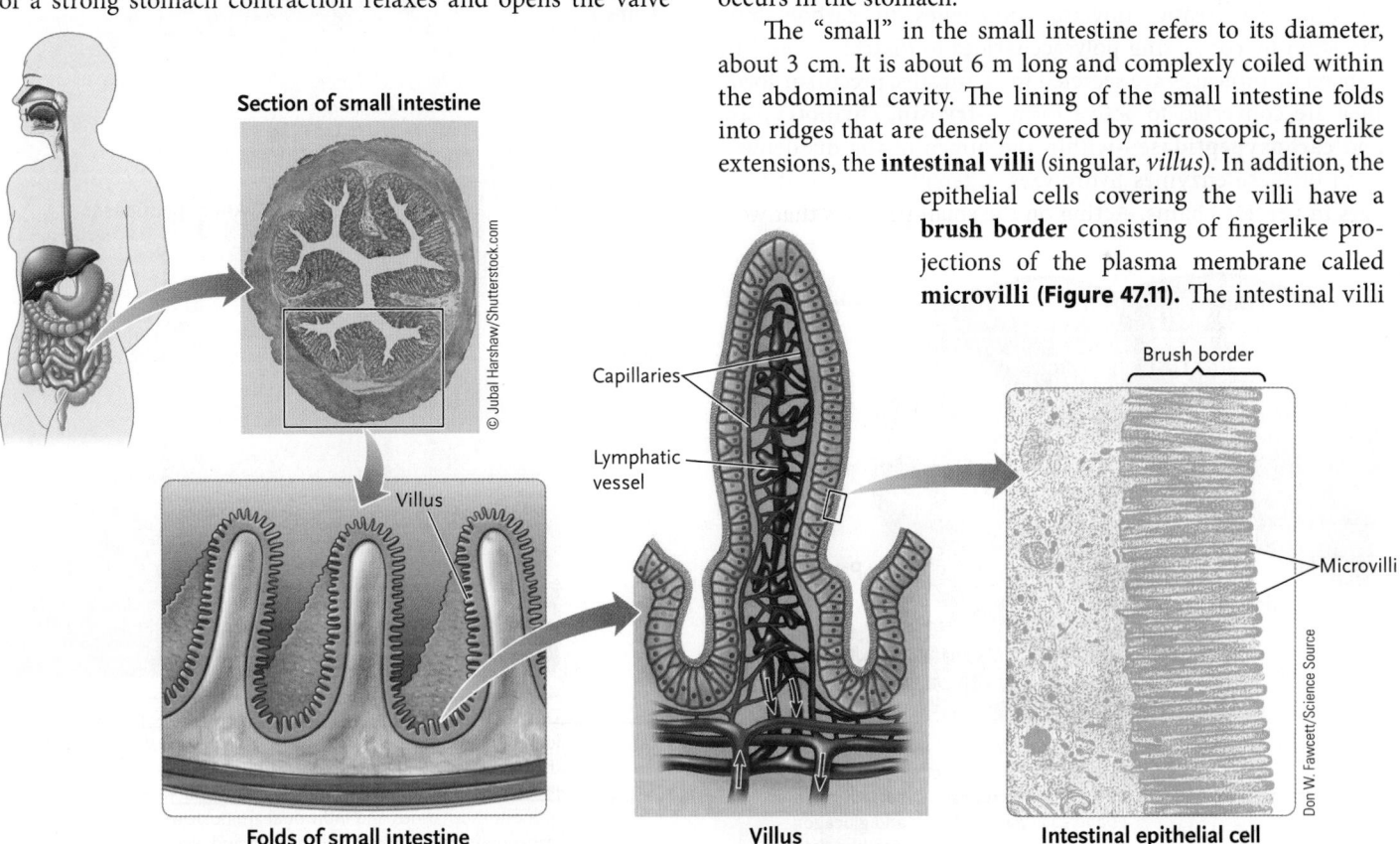

FIGURE 47.11 The structure of villi in the small intestine. The plasma membrane of individual epithelial cells of the villi extends into fingerlike microvilli, which greatly expand the absorptive surface of the small intestine. Collectively, the microvilli form the brush border of an epithelial cell of the intestinal mucosa.

© Cengage Learning 2017

and microvilli are estimated to increase the absorptive surface area of the small intestine to as much as 300 m^2, which is about the size of a doubles tennis court.

SECRETIONS OF THE PANCREAS, LIVER, AND MUCOSA OF THE SMALL INTESTINE

When the stomach contents are emptied into the small intestine, they are mixed with secretions of the pancreas, liver, and the mucosa of the small intestine. About 7 to 9 L of these secretions enter the small intestine daily. About 95% of this amount is reabsorbed as water and nutrients as the digestive contents travel along the small intestine. Movement of the contents from the duodenum to the end of the small intestine takes about 3 to 5 hours.

In humans, the **pancreas** is an elongated, flattened gland located between the stomach and duodenum (see Figures 47.5 and 47.12). Recall from Section 42.4 that most of the pancreas consists of an exocrine gland, with the remaining 2% consisting of endocrine cells that secrete the hormones insulin and glucagon. Exocrine cells in the pancreas secrete bicarbonate ions ($H_2CO_3^-$) and digestive enzymes—**pancreatic enzymes**—via a duct into the lumen of the duodenum **(Figure 47.12)**. The bicarbonate ions neutralize the acid in the chyme, allowing optimal activity of the pancreatic enzymes (see Figure 47.8).

All of the pancreatic enzymes—which include an amylase, proteases, nucleases, and lipase—act in the lumen of the small intestine. Pancreatic amylase contributes to carbohydrate digestion by converting polysaccharides to disaccharides. The pancreatic proteases are secreted in an inactive precursor form; they are converted to active forms—**trypsin, chymotrypsin,** and **carboxypeptidase**—within the lumen of the duodenum. Each of these enzymes hydrolyzes different amino acid linkages in peptide chains. Acting on the small peptides that were produced in the stomach, their action produces small peptides and amino acids (see Figure 47.8). The pancreatic nucleases digest DNA and RNA to nucleotides.

The **liver** secretes bicarbonate ions and **bile,** a mixture of substances including **bile salts,** cholesterol, and **bilirubin.** These substances enter the duodenum via the same duct used by the secreted pancreatic enzymes. Bile salts are derivatives of cholesterol and amino acids that aid fat digestion. Fat is water-insoluble, yet it must be transferred from the aqueous chyme to the aqueous body fluids. In the small intestine, ingested fat aggregates into large, oily globules of triglyceride droplets. Bile salts attach to the surfaces of the globules and, by their detergent action, *emulsify* the fats—break them into tiny lipid droplets—during the churning motions of the small intestine. Mixing of oil and vinegar to make a salad dressing is a household example of emulsification. The emulsion increases the surface area available for hydrolysis of the triglycerides by pancreatic **lipase** to produce monoglycerides and fatty acids (see Figure 47.8).

Bilirubin, a waste product derived from worn-out red blood cells, is the yellow pigment that gives bile its color. Bacterial enzymes in the intestines modify bilirubin, resulting in the characteristic brown color of feces.

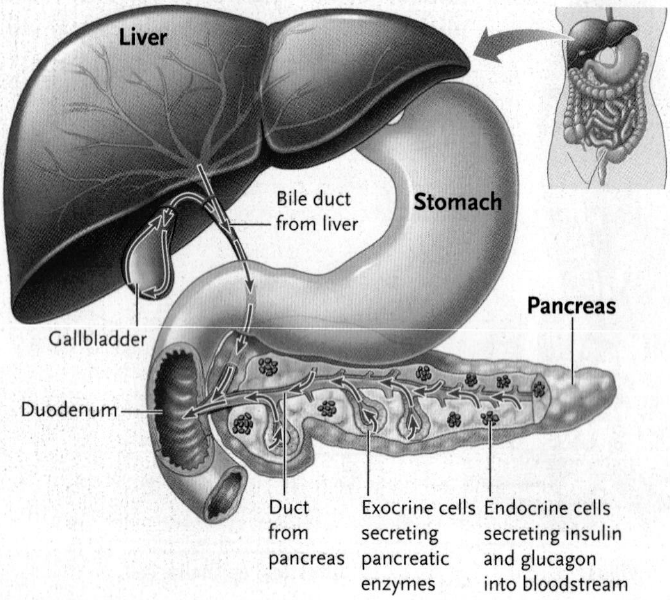

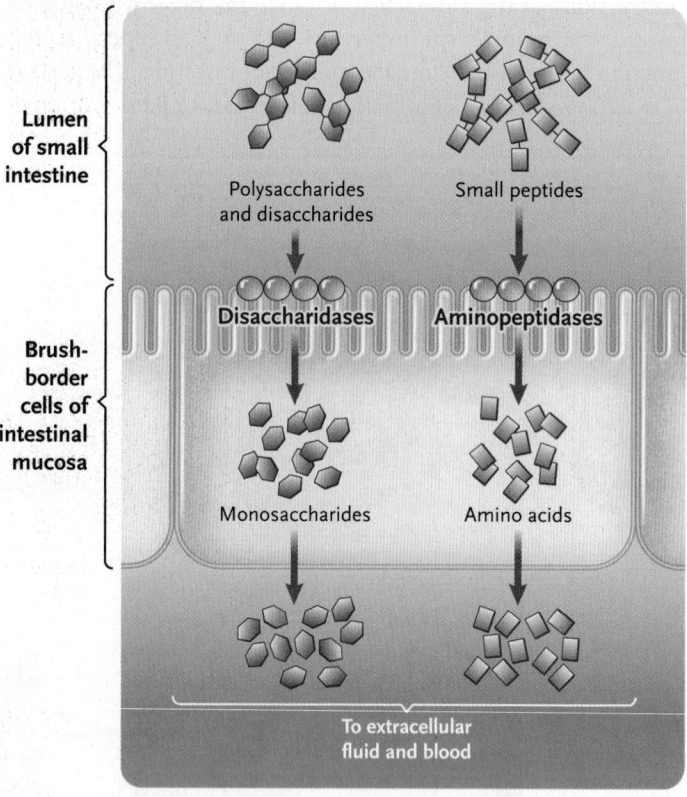

Disaccharides and small peptides in the digestive products in the lumen of the small intestine are absorbed by brush-border cells. As the molecules are transported across the plasma membranes of those cells, embedded enzymes in the membranes hydrolyze them to monosaccharides and individual amino acids, which are then transported into the extracellular fluid and blood.

FIGURE 47.13 Absorption and digestion of disaccharides and small peptides by the brush-border cells of the small intestinal mucosa.
© Cengage Learning 2017

FIGURE 47.12 The ducts that deliver bile and pancreatic juice to the duodenum of the small intestine.
© Cengage Learning 2017

The liver secretes bile continuously. Between meals, when no digestion is occurring, bile is stored in the **gallbladder** (see Figure 47.12), where it is concentrated by the removal of water. After a meal, entry of chyme into the small intestine stimulates the gallbladder to release stored bile into the small intestine.

Brush-border epithelial cells on the villi of the small intestine mucosa secrete an aqueous salt and mucus mixture into the lumen. The mucus serves to protect and lubricate the digestive mixture. The aqueous salt solution provides water for the enzymatic digestion of food, which mostly involves hydrolysis reactions. No digestive enzymes are secreted by the brush-border cells.

DIGESTION BY ENZYMES OF THE BRUSH-BORDER CELLS
Complete digestion of fats, but not of carbohydrates and proteins, takes place in the lumen of the small intestine.

Monoglycerides and fatty acids associate with bile salts to form water-soluble microdroplets called micelles. Micelles carry their contents to the microvillar membranes of the brush-border epithelial cells where the monoglycerides and fatty acids enter the cells by passive diffusion through the plasma membranes. Within the cells, triglycerides are resynthesized from the monoglycerides and fatty acids. The triglycerides aggregate and are coated with lipoproteins from the endoplasmic reticulum to form chylomicrons, which leave the cell by exocytosis and enter the lymphatic system.

FIGURE 47.14 Absorption of monoglycerides and fatty acids, the products of fat digestion in the lumen of the small intestine, by the brush-border cells of the small intestinal mucosa.
© Cengage Learning 2017

Carbohydrate and protein digestion are completed by digestive enzymes that are integral proteins embedded in the plasma membranes of the brush-border cells of the small intestine mucosa **(Figure 47.13).** That is, disaccharides and small peptides are transported across the plasma membranes of the brush-border cells where embedded **disaccharidases** hydrolyze disaccharides to monosaccharides, and embedded **aminopeptidases** hydrolyze small peptides to individual amino acids.

One of the brush-border cell disaccharidases is *lactase,* the enzyme that breaks down the milk sugar lactose into its component glucose and galactose monosaccharides. Many adults lose the capacity to synthesize lactase and, thus, cannot absorb lactose. The lactose remaining in the intestine is broken down by bacteria, producing excess methane and CO_2. The accumulating gases distend the large intestine, producing pain, discomfort, and other symptoms known as **lactose intolerance.** Taking tablets containing lactase before eating milk products may limit or prevent those symptoms.

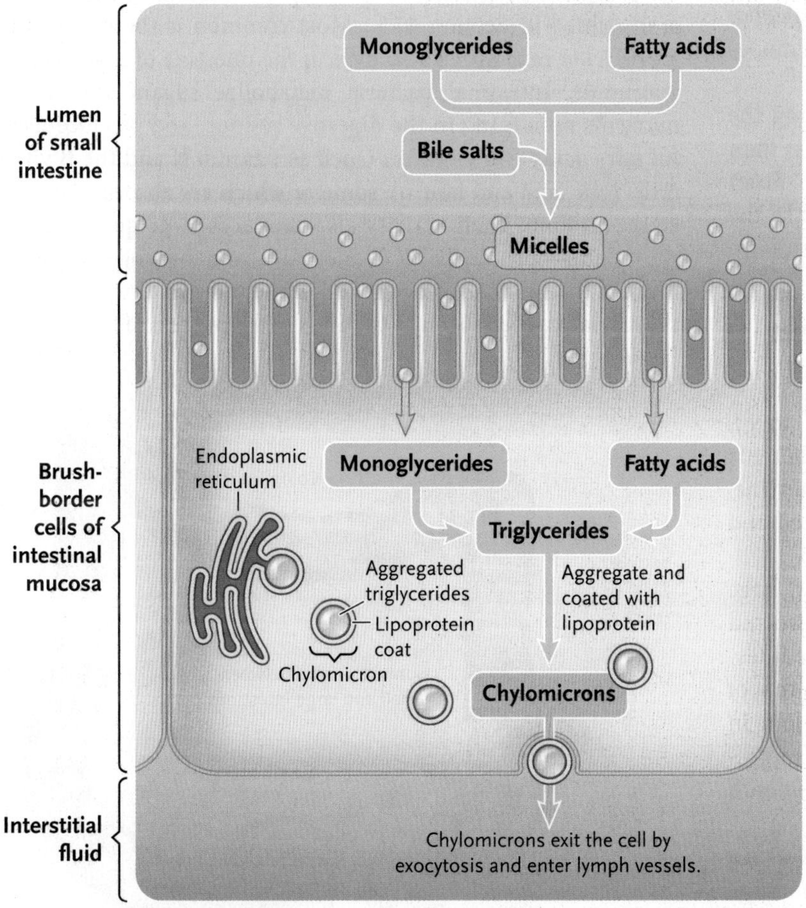

ABSORPTION BY THE BRUSH-BORDER CELLS OF THE SMALL INTESTINAL MUCOSA
All products of carbohydrate and protein digestion are water-soluble. In the small intestine, lumen contents are absorbed by the brush-border cells by active transport or facilitated diffusion (see Chapter 5), depending on the substance. You have just learned about the absorption of disaccharides and small peptides and their breakdown into monosaccharides and amino acids (see Figure 47.13). Amino acids in the lumen contents are absorbed also, but no further digestion is necessary. The final products of digestion are then transported from the mucosal cells into the extracellular fluids, from which they enter the bloodstream in the capillary networks of the submucosa.

Nucleotides produced in the small intestine by the action of pancreatic nucleases are water-soluble. They are absorbed by brush-border cells and their digestion is completed by cytosolic enzymes. **Nucleotidases** break them down into nucleosides, and **nucleosidases** and **phosphatases** convert the nucleosides to nitrogenous bases, five-carbon sugars, and phosphates. These products move to the bloodstream as for carbohydrate and protein digestion products.

The monoglyceride and fatty acid products of fat digestion are fat-soluble but water-insoluble. They enter the cell as shown in **Figure 47.14.** Once in the cell, they are used to resynthesize triglycerides, which aggregate and are coated with a layer of lipoprotein to form **chylomicrons.** Cholesterol absorbed in the small intestine is also packed into the chylomicrons. Chylomicrons leave the epithelial cell by exocytosis and enter the lymphatic system.

Processing in the Liver

The capillaries absorbing nutrient molecules in the small intestine collect into veins that join to form a larger blood vessel, the **hepatic portal vein,** which leads to capillary networks in the liver. There, some of the nutrients leave the bloodstream and enter liver cells for chemical processing. Among the reactions taking place in the liver is the combination of excess glucose units into glycogen, which is stored in liver cells. This reaction reduces the glucose concentration in the blood exiting the liver to about 0.1%. If the glucose concentration in the blood entering the liver falls below 0.1% during a period of fasting between meals, the reaction reverses. The reversal adds glucose to return the blood concentration to the 0.1% level before it exits the liver. The hormonal activities controlling blood glucose levels in the blood are described in Section 42.4.

The liver also synthesizes the lipoproteins that transport cholesterol and fats in the bloodstream, detoxifies ethyl alcohol and other toxic molecules, and inactivates steroid hormones and many types of drugs.

As a result of the liver's activities, the blood leaving the liver has a markedly different concentration of nutrients than the blood carried into the liver by the hepatic portal vein. From the liver, blood is carried to the heart and then pumped by the heart to deliver nutrients to all parts of the body.

The Large Intestine Primarily Absorbs Water and Mineral Ions from Digestive Residues

The small intestine reabsorbs all but about 1 L of the 7 to 9 L of fluid released from the stomach. The remaining contents of the small intestine then move on to the large intestine. By this point, almost all nutrients have been hydrolyzed and absorbed.

A sphincter at the junction between the small and large intestines controls the passage of material between the two and prevents backward movement of the contents. The large intestine has an average diameter of 7.6 cm, more than twice that of the small intestine, but it is relatively short, about 1.5 m long in humans, as compared with the 6-m length of the small intestine. The inner surface of the large intestine is relatively smooth and contains no villi.

The large intestine has several distinct regions **(Figure 47.15)**. At the junction with the small intestine, a part of the large intestine forms a blind pouch called the **cecum.** A fingerlike sac, the **appendix,** extends from the cecum. The appendix is on average 100 mm long and 7 mm in diameter; it is a vestigial structure with no known function in digestion. Since it contains patches of lymphoid tissue, it is presumed to have some role as part of the immune system. The cecum merges with the **colon,** the main part of the large intestine, which forms an inverted U. At its distal end, the colon connects with the final segment of the large intestine, the **rectum.**

The large intestine secretes mucus and bicarbonate ions and absorbs water and other ions, primarily sodium and chloride. The absorption of water condenses and compacts the digestive contents into solid masses, the **feces.** Normally, by the time the fecal matter reaches the rectum, it contains less than 200 mL of the fluid that enters the digestive tract each day. Diarrhea, by contrast, is an abnormal condition in which the fecal matter is highly fluid. The most common cause of diarrhea is a higher-than-normal rate of movement of materials through the small intestine, which does not leave adequate time for the normal amount of water to be absorbed. The higher rate of movement can occur as a result of irritation of the small intestine wall caused by bacterial and viral infections, or because of emotional stress.

About 30–50% of the dry matter of feces in humans and other vertebrates consists of more than 500 species of bacteria that are part of the gut microbiome, the complete collection of microorganisms associated with the gut. (The gut microbiome is discussed in Section 47.5.) Most common is the bacterium *Escherichia coli,* which also lives in the intestine of many other mammals. Intestinal bacteria metabolize sugars and other nutrients remaining in the digestive residue, and produce useful fatty acids and vitamins (such as vitamin K and the B vitamins folic acid and biotin), some of which are absorbed in the large intestine. Their activity also produces large quantities of gas—*flatus*—primarily CO_2, methane, and hydrogen sulfide. Most of the gas is absorbed through the intestinal mucosa, and the rest is expelled through the anus in the process of **flatulence.**

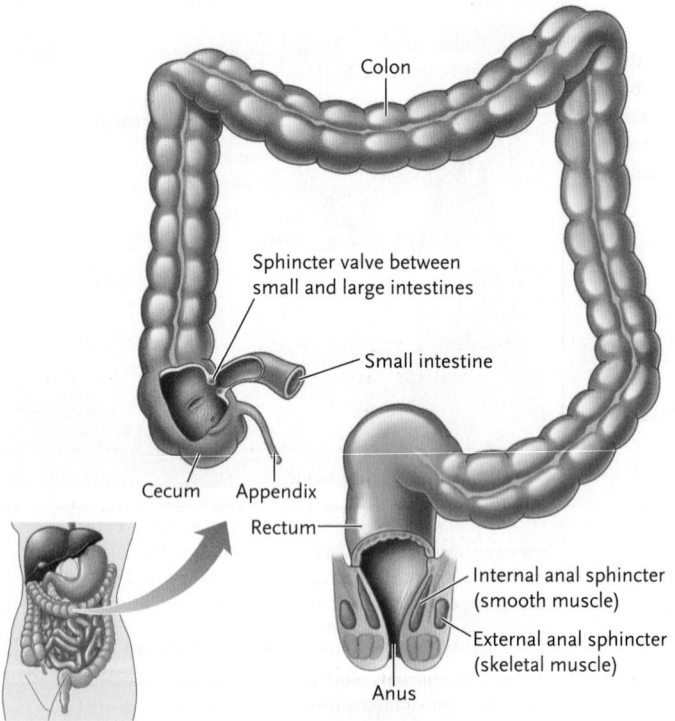

FIGURE 47.15 The human large intestine.
© Cengage Learning 2017

The amount and composition of the gas produced depend on the type of food being ingested and the particular population of bacteria present in the large intestine. Some foods, such as beans, contain carbohydrates that humans cannot digest but that can be metabolized by the gas-producing intestinal bacteria. After eating such foods, humans may produce more gas than usual, and flatulence is more likely to occur.

When feces enter the rectum, they stretch the rectal wall. The stretching triggers a **defecation reflex** that opens the **anal sphincter** and expels the feces through the anus. Because the anal sphincter contains rings of voluntary skeletal muscle as well as involuntary smooth muscle (see Figure 47.15), you can resist the defecation reflex by tightening the striated muscle ring voluntarily.

Having completed the journey of ingested food through the digestive tract, we now turn our attention to the mechanisms that regulate the activities of the digestive system. These mechanisms coordinate one region of the digestive tract with another, and help match the production of nutrients with the body's needs.

STUDY BREAK 47.3

1. What are the two classes of vitamins? Which of the two types is more critical in the diet and why?
2. What are the four layers of the mammalian gut? Which layer is responsible for peristalsis?
3. How is pepsin produced? What is its function?
4. Distinguish between the functions of the small and large intestines in the process of digestion.

THINK OUTSIDE THE BOOK

Use the Internet or research literature to outline the mechanisms involved in vomiting, and to determine how vomiting is distinguished from regurgitation.

47.4 Regulation of the Digestive Process

The digestive process is regulated and coordinated at many steps by controls that are largely automated. The autonomic nervous system, the enteric nervous system (local neuron networks in the digestive tract wall), and endocrine glands interact in these controls, in response to sensory information gathered by receptors in the digestive tract. The integration of the controls speeds up or slows down digestion to produce maximum efficiency in the breakdown of food molecules and the absorption of the nutrients.

Much of the control of the digestive system originates in the neuron networks of the submucosa. Other controls, particularly those regulating appetite and oxidative metabolism, originate in the brain, in control centers forming part of the hypothalamus.

The Digestive Tract Itself Has a Number of Control Systems

The movement of food through the digestive tract is controlled by receptors in and hormones secreted by various parts of the digestive system. Control starts with the mouth. Saliva is secreted constantly into the mouth. The presence of food activates receptors that increase the rate of salivary secretion by as much as 10-fold.

Swallowed food expands the stomach and sets off signals from stretch receptors in the stomach walls that increase the rate and strength of stomach contractions. At the same time, chemoreceptors in the stomach respond to the presence of food molecules, particularly proteins, with secretion of a hormone, **gastrin.** After traveling through the circulatory system, gastrin returns to the stomach, where it stimulates the secretion of HCl and pepsinogen (see Figure 47.10). These molecules are then used in the digestion of the protein that, as part of the swallowed meal, was responsible for their secretion. Gastrin also stimulates stomach and intestinal contractions, activities that keep the digestive contents moving through the digestive system when a new meal arrives.

Three hormones secreted when food is present in the duodenum also participate in regulating the digestive processes:

1. **Secretin.** When chyme is emptied into the duodenum, its acidic nature stimulates the release of the hormone **secretin** from glandular cells in the small intestine; the hormone enters the bloodstream. Secretin inhibits further gastric emptying to prevent more acid from entering the duodenum until the newly arrived chyme is neutralized. It also inhibits gastric secretion to reduce acid production in the stomach, and it stimulates HCO_3^- secretion into the lumen of the duodenum to neutralize the acid. If the acid is not neutralized, the duodenal wall will become damaged, and pancreatic enzymes secreted into the duodenum will be inactivated.
2. **Cholecystokinin (CCK).** Fat, and to a lesser extent protein, in the chyme that enters the duodenum stimulates the release of the hormone **cholecystokinin (CCK).** CCK inhibits gastric activity, thereby allowing time for nutrients in the duodenum to be digested and absorbed. It also stimulates the secretion of pancreatic enzymes, used to digest the macromolecules in the chyme.
3. **Glucose-dependent insulinotropic peptide (GIP).** The hormone **glucose-dependent insulinotropic peptide (GIP)** acts primarily to stimulate insulin release by the pancreas. When a meal is ingested, the body must change its metabolic state to use and store the new nutrients absorbed. Those activities are mostly under the control of insulin. Therefore, when a meal enters the digestive tract, GIP secretion is stimulated to trigger the release of insulin. Insulin is particularly important in stimulating the uptake and storage of glucose and so, not surprisingly,

glucose in the blood draining the duodenum is directly detected as well by pancreatic β-cells, the cells from which insulin is secreted (see Section 42.4).

The Hypothalamus Exerts Overall Controls

The hypothalamus contains two interneuron centers that work in opposition to control appetite and oxidative metabolism. One center stimulates appetite and reduces oxidative metabolism, and the other center stimulates the release of a peptide hormone called **α-melanocyte-stimulating hormone (α-MSH),** which inhibits appetite.

A major link in the control pathways is the peptide hormone *leptin* (*leptos* = thin), discovered in mice by Jeffrey Friedman and his coworkers at the Rockefeller University. Fat-storing cells secrete leptin when the deposition of fat increases in the body. Leptin travels in the bloodstream and binds to receptors in both centers in the hypothalamus. Binding stimulates the center that reduces appetite and inhibits the center that stimulates appetite. At the same time, leptin binds to receptors on body cells, triggering reactions that oxidize fatty acids rather than converting them into fats. When fat storage is reduced, leptin secretion drops off, and signals from other pathways activate the appetite-stimulating center in the hypothalamus and turn off the appetite-inhibiting center.

These controls closely match the activity of the digestive system to the amount and types of foods that are ingested, and they coordinate appetite and oxidative metabolism with the body's needs for stored fats.

STUDY BREAK 47.4

1. What differentiates the consequences of secretion of gastrin and cholecystokinin (CCK)?
2. How does the hypothalamus regulate the digestive process?

47.5 Digestive Specializations in Vertebrates

Natural selection has modified the basic vertebrate digestive system into a multitude of structural and functional variations. The most common modifications are in the form of the mouth, teeth, and jaws; the structure and function of the esophagus, stomach, and cecum; and the length of the digestive tract. In addition, vertebrates show variation in the symbiotic microorganisms in their digestive tracts.

Teeth Are Adapted to Feeding Methods

To anthropologists and paleontologists, dentition (the number, kind, and arrangement of teeth) opens a window to an animal's diet and feeding method—and hence reveals a great deal about its habitat and lifestyle. For example, snakes have sharp, pointed teeth that curve backward into the mouth, which helps to ensure that prey (dead or living) does not slip out of the animal's mouth as muscles contract to swallow. The dentition is combined with specializations in jaw structure; many snakes have jaws with elastic connections that allow them to open wide enough to swallow prey whole.

Tooth specialization is especially evident among mammals **(Figure 47.16).** Typically, mammals have four types of upper and lower teeth. **Incisors,** located at the front of the mouth, are flattened, chisel-shaped teeth used to nip or cut food. Horses use their prominent incisors to clip off blades of grasses. Pointed **canines** at the sides of the incisors are specialized for biting and piercing. Carnivores such as wolves and tigers use their long, sharp canines to pierce and kill prey, but the canines are minimally developed or absent in many herbivores. The blocky teeth at the sides of the mouth, the **premolars** and **molars,** have surface bumps, or *cusps,* that are used in crushing, grinding, and shearing food. Large

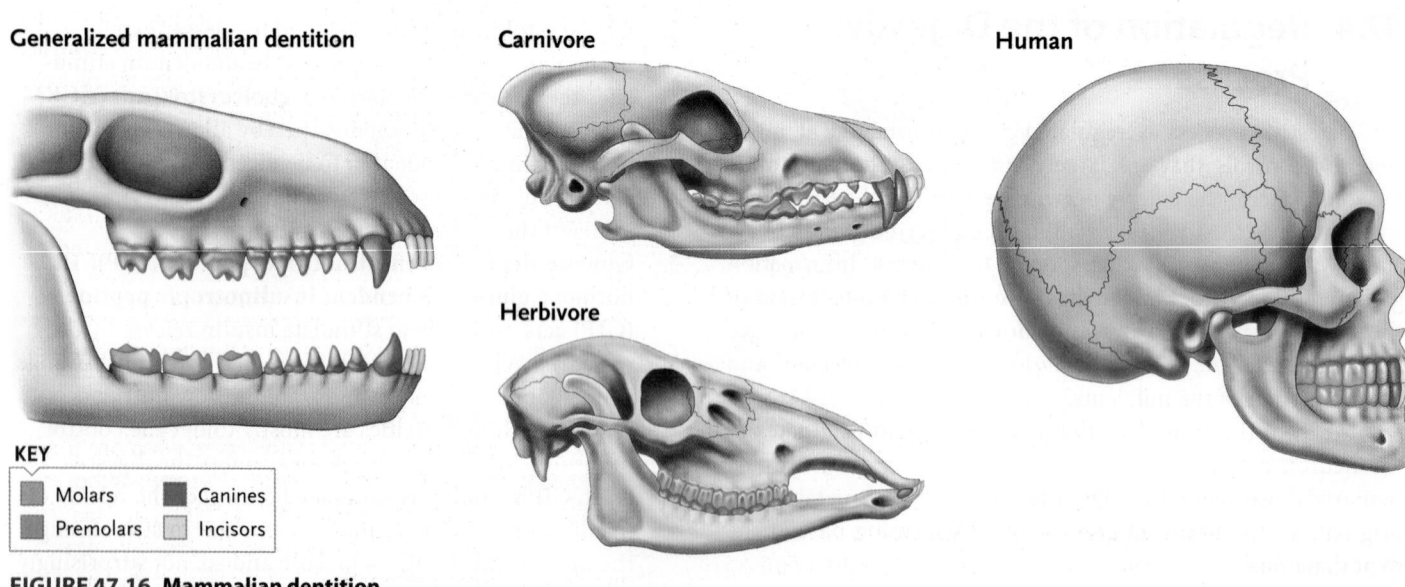

Generalized mammalian dentition

KEY

- Molars
- Premolars
- Canines
- Incisors

Carnivore

Herbivore

Human

FIGURE 47.16 Mammalian dentition.
© Cengage Learning 2017

premolars and molars with a ridged surface are characteristic of animals, such as deer, that consume fibrous plant material. The premolars and molars of some carnivores, such as cats, have sharp shearing surfaces that can slice meat efficiently. All four types of teeth are typically well developed in omnivores, such as humans.

The Length of the Intestine and Specializations of the Digestive Tract Reflect Feeding Patterns

There is a strong correlation between diet and the length of the digestive system **(Figure 47.17).** Vertebrates that feed primarily on nutrient-rich foods such as meat, blood, nectar, or insects, including carnivores (such as the dog), generally have a relatively short intestine. In contrast, herbivores (such as the rabbit) have a long intestinal tract and specializations of the esophagus, stomach, and cecum, or other structures that can store large volumes of plant material. Both the longer intestinal tract and greater storage capacity allow an herbivore to extract more nutrients from plant matter, which is relatively difficult to digest. Both types of intestine appear during the life cycle of frogs: a frog tadpole, which eats algae primarily, has a relatively long, coiled intestine, but after metamorphosis, the adult frog, which eats insects primarily, has a short intestine.

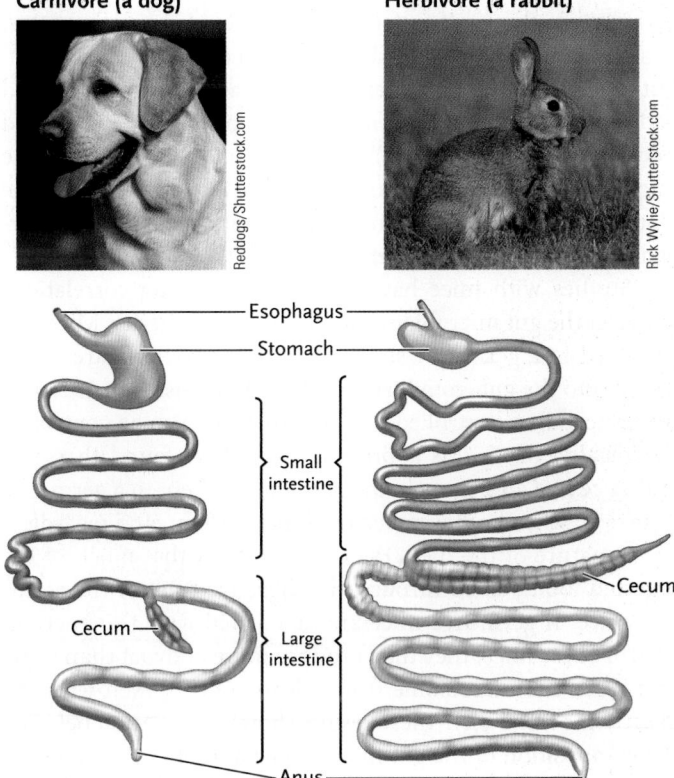

FIGURE 47.17 Comparison of the length of the digestive tract in a carnivore and an herbivore. The carnivore has a relatively short digestive tract, while the herbivore's digestive tract is much longer.
© Cengage Learning 2017

Specializations of the digestive tract in vertebrates include the crop and gizzard (areas of the stomach) in birds that you learned about earlier in the chapter.

Symbiotic Microorganisms in Gut Microbiomes Aid Digestion in a Number of Organisms

You learned in Section 26.2 about **microbiomes,** complete collections of microorganisms associated with a particular organism, and especially about the human microbiome. Microbiomes consist mostly of bacteria, but archaea, protists, yeasts and other fungi, and viruses may also be present. Microbiomes in digestive tracts—**gut microbiomes**—play important roles in digestion in many organisms.

Let us first consider herbivores. Many herbivores use the hydrolytic capabilities of microorganisms in their gut microbiomes such as bacteria, protists, and fungi to aid digestion of plant material, housing them in specialized structures of the esophagus, stomach, or cecum. Unlike vertebrates, the microorganisms can synthesize *cellulase*, the enzyme that hydrolyzes the cellulose of plant cell walls into glucose subunits. This arrangement is a classic example of *mutualism* (a type of symbiosis in which both interacting organisms benefit); the herbivores benefit from the digestive capabilities of the microorganisms, and the microorganisms benefit from an ideal habitat and an abundant supply of nutrients. (Mutualism is discussed more in Chapter 53.)

The most remarkable adaptations for mutualistic digestion of plant matter among vertebrates occur in the **ruminants,** which include cattle, deer, goats, sheep, pronghorns, and antelopes. These animals have a complex, four-chambered stomach **(Figure 47.18).** The first three chambers are derived from the esophagus. After the ruminant's teeth tear, cut, and grind plant matter, boluses are swallowed. Most of them go to the **rumen** (the largest chamber), and the rest go to the **reticulum** (step 1). In these two chambers, symbiotic microorganisms hydrolyze cellulose in the plant matter into fuels for fermentation reactions (the oxygen level in the chambers is too low to support mitochondrial reactions). The fermentations generate various products, including alcohols, amino acids, and fatty acids, which are used as nutrients by the ruminants. Methane, another product, collects in the fermentation chambers. Ruminants belch the gas in huge quantities; a cow potentially can release more than 400 L of methane per day. In fact, cattle are estimated to contribute 20% of the methane polluting our atmosphere.

As part of the digestive process, a ruminant "chews its cud"—it regurgitates material from the rumen and reticulum, rechews it, and swallows it again (step 2). This process crushes the plants into smaller fragments, exposing more surface area to the microbial enzymes, and gives the enzymes more time to act. As a result, the food material is broken down further.

Reswallowed cud, consisting of matter that has been digested and liquefied by the microorganisms, bypasses the rumen and reticulum and instead goes to the **omasum,** where water is absorbed from the mass (step 3).

Chewing, swallowing, regurgitation, rechewing,
and reswallowing of food through esophagus

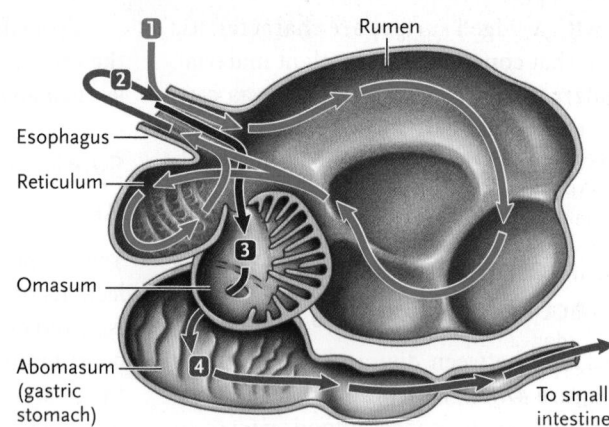

1 Swallowed boluses go to rumen and reticulum, where fermentation reactions by symbiotic microorganisms begin digesting the plant matter.

2 The animal chews its cud by regurgitating material, rechewing it, and swallowing it again.

3 Reswallowed cud goes to the omasum, where water is absorbed.

4 Matter then moves to the abomasum, where typical gastric digestion occurs.

Rumen

Esophagus

Reticulum

Omasum

Abomasum (gastric stomach)

To small intestine

FIGURE 47.18 A ruminant, the pronghorn *(Antilocapra americana),* and its four-chambered system that digests plant matter with the aid of mutualistic microorganisms.
© Cengage Learning 2017

Matter then moves to the *abomasum* (the ruminant's gastric stomach) (step 4). There, the addition of acids and pepsin to the food mass kills the microorganisms and starts the process of typical vertebrate digestion. As the food mass moves to the small intestine, the dead microorganisms, which are a rich source of proteins, vitamins, and other nutrients, are digested and absorbed along with other hydrolyzable molecules in the digestive contents.

Although the ruminant digestive system is uniquely specialized, many other vertebrate species also have esophageal or gastric chambers containing plant-digesting symbiotic microorganisms. These include the camel, sloth, and langur monkey, and marsupials such as kangaroos and wallabies. One bird, the South American hoatzin *(Opisthocomus hoazin),* is known to have a crop in which microorganisms break down plant matter.

Many herbivorous vertebrates house symbiotic, plant-digesting microorganisms in the cecum. Horses, elephants, rhinos, rabbits, koalas, some rodents, and some reptiles and birds, including the iguana and chicken, are all examples. However, because the cecum and the remainder of the large intestine have little capacity to absorb nutrients, microbial digestion in the cecum is not as productive as microbial digestion in the stomach.

Mutualistic microorganisms that synthesize essential amino acids and vitamins, and digest particular components of food that otherwise are indigestible, are found in the gut microbiomes of all mammals. Let us focus on the role of the human gut microbiome in digestion.

The human gut microbiome is the largest and most diverse community of microorganisms in the human body. Microorganisms in that microbiome, mostly the bacteria, aid in extracting energy and nutrients from foodstuffs. Evolutionarily, the activities of the mutualistic gut bacteria have facilitated the use of a wider variety of foodstuffs than otherwise would have been the case. That is, the gut microorganisms encode a large variety of enzymes that human cells do not encode. For example, the human genome encodes fewer than 20 carbohydrate-digesting

enzymes, whereas bacteria in the gut microbiome encode hundreds. For instance, the genome of just one intestinal bacterium, the Gram-negative obligate anaerobe *Bacterioides thetaitomicron,* encodes more than 260 such enzymes. As a result of the bacterial-encoded enzymes, complex carbohydrates in the diet can be broken down to monosaccharides in the gut and the monosaccharides absorbed into the bloodstream.

Gut microorganisms can provide some amino acids for humans. As you learned earlier, there are nine essential amino acids for adult humans. Those amino acids cannot be synthesized by human cells and therefore must be provided in other ways. Some, of course, are obtained by digestion of proteins in the diet. But at least some of the essential amino acids can be produced and secreted by certain gut microorganisms and then absorbed into the bloodstream. Many other chemical compounds the microorganisms synthesize and secrete are also absorbed into the bloodstream (e.g., see Section 47.3), illustrating that the microorganisms have significant interactions with their human host.

Studies with mice have revealed interesting correlations between the gut microbiome and leanness or obesity. Mice can be raised in a germ-free state, meaning no bacteria are introduced into the gut from birth on. Microorganisms then can be introduced in a controlled way, and diet can also be controlled specifically. Using this approach, researchers found that normal mice had 40% more body fat than mice kept in a germ-free state even when both were fed the same amount, and regardless of the nature of the diet. The explanation is that much of the ingested food passed through the digestive tract of the germ-free mice. If germ-free mice are inoculated with the microbiome of obese mice, they then produce more body fat than those that receive a microbiome from a lean control, even when on exactly the same diet. These results show a correlation between the composition of the gut microbiome and obesity. It appears that, in some way, regulatory systems link diet, microbiome composition, and metabolism. Could microbiome composition play a role in obesity in humans? *Molecular Insights* describes research experiments designed to answer that question.

Association of Particular Bacterial Populations in the Gut Microbiome with Obesity in Humans

The human gut microbiome contains an estimated 10^{13}–10^{14} microorganisms. Gut microorganisms consist largely of bacteria, with two groups predominating, the Firmicutes (a phylum of Gram-positive bacteria) and the Bacteroidetes (a phylum of Gram-negative bacteria). Jeffrey Gordon and his group at Washington University School of Medicine, St. Louis, have shown that genetically obese mice (genotype *ob/ob*) have about 50% fewer Bacteroidetes, and correspondingly more Firmicutes, than do normal, lean siblings (genotype *ob⁺/ob⁺*), leading to the conclusion that particular distributions of bacteria in the gut microbiome correlate with obesity.

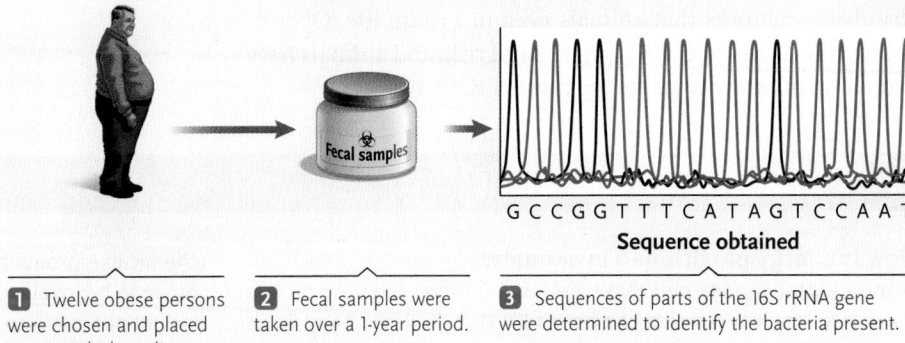

1 Twelve obese persons were chosen and placed on a weight-loss diet.

2 Fecal samples were taken over a 1-year period.

3 Sequences of parts of the 16S rRNA gene were determined to identify the bacteria present.

FIGURE A
© Cengage Learning 2017

Research Question

Are particular populations of bacteria in the gut microbiome associated with obesity in humans?

Experiment

Gordon's group studied the relationship between intestinal populations of the two types of bacteria and body fat in humans. They worked with 12 people classified as being obese according to the standard criterion of having a body mass index (BMI) of ≥30. BMI is calculated from the formula: BMI = weight (kg)/[height (m)]^2. BMI does not measure body fat directly, although it correlates well with the amount of body fat found in nonathletes. **Figure A** shows the experimental design of Gordon's research.

Sequencing specific regions of the 16S rRNA gene (which encodes an rRNA found only in the ribosome of prokaryotes) enabled the researchers to identify the bacteria in the fecal sample. The regions sequenced vary significantly among species, permitting the identification of species present in a mixed sample without having to culture the organisms. Analyzing these sequences is one of the methods used in molecular phylogenetics (see Section 24.7).

Results

Most of the bacteria present in the obese individuals and in two lean individuals (controls) were Firmicutes or Bacteroidetes. As shown in **Figure B:** (1) Before embarking on the diet, obese individuals had more Firmicutes and fewer Bacteroidetes than

lean individuals; and (2) over the course of the diet, as weight loss occurred, the relative abundance of Firmicutes decreased, and the relative abundance of Bacteroidetes increased.

Conclusion

Obese and lean individuals showed significant differences in their intestinal content of Firmicutes and Bacteroidetes. Moreover, the data from the effects of the weight-loss diet indicate a dynamic relationship between the intestinal populations of the two groups of bacteria and body fat content. Taken with the results of the mouse study, this suggests that manipulation of intestinal microbial communities could be a potential way to control or treat obesity. However, we must be cautious. The results of this research must not be interpreted to mean that microbiome composition is the *cause* of obesity. That is, gut microbiome composition is only one contributing factor in obesity; other contributing factors include genetics (as in the mouse study), hormones, neurological effects, the environment, and lifestyle choices.

think like a scientist

How could you use germ-free, genetically normal adult mice (mice without bacteria in their intestines) to test further the relationship between the bacterial population in genetically obese mice and obesity?

Sources: R. E. Ley et al. 2006. Human gut microbes associated with obesity. *Nature* 444:1022–1023; P. J. Turnbaugh et al. 2006. An obesity-associated gut microbiome with increased capacity for energy harvest. *Nature* 444:1027–1031.

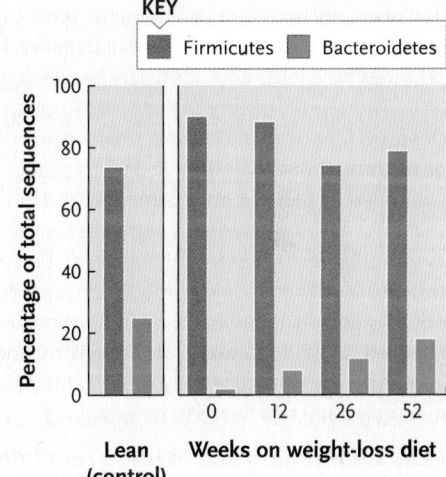

KEY
Firmicutes   Bacteroidetes

Percentage of total sequences

Lean (control)

Weeks on weight-loss diet
0   12   26   52

FIGURE B
© Cengage Learning 2017

© Cengage Learning 2017

We have seen that animals use various strategies to extract the available nutrients from foods. These nutritional strategies involve multiple steps combining both mechanical and chemical processing, which convert complex foodstuffs into the absorbable subunits that animals need to sustain life. Obtaining food is costly in terms of energy and risk, and animals have evolved many ways to make the most of it.

STUDY BREAK 47.5

1. How are the different types of teeth used in feeding?
2. What roles do symbiotic microorganisms play in digestion?

Unanswered Questions

How is energy partitioned in animals?

In this chapter you learned that ingested food molecules are broken down into smaller units that can be readily absorbed into body fluids and cells. The ingested energy is assimilated into the organism and partitioned into four main categories: maintenance metabolism, growth, reproduction, and storage. The top priority for energy allocation is maintenance metabolism—the energy required to seek and digest food and to support essential life processes. Researchers are studying the regulation of energy partitioning among the other categories when available energy exceeds that required for maintenance. The findings indicate that priorities for energy allocation vary between species, as well as during an individual's life history. In the case of growth, for example, some species grow for a short period (*determinant growth*), whereas other species grow throughout their lives (*indeterminant growth*). Animal growth is controlled by genetic, environmental, and nutritional factors, but how such cues are integrated with each other and with other processes, such as reproduction and storage, remains to be determined. For example, the cues used to shut down growth and reproduction during fasting are not fully known. (In humans, malnourished juveniles become growth-retarded and adult females stop menstruating.) The extent to which severe energy restriction may result in metabolic adaptation (reduced metabolic rate) also needs to be examined.

How are foraging and feeding regulated?

In this chapter you also learned about the control of appetite and oxidative metabolism. Although some of the players are known—among them hunger centers in the brain, appetite-stimulating hormones such as neuropeptide Y, and appetite-suppressing or satiety hormones such as leptin—a number of questions remain. For example, how are various sensory inputs, such as smell, integrated to initiate feeding? Similarly, how are other inputs, such as gut contents or increasing nutrient concentration in the blood, integrated to terminate feeding? Also, how is an animal's feeding strategy matched to nutrient quantity and quality in the environment and to the animal's metabolic pattern; for example, sluggish or active (tortoise or hare)? Inevitably, a complex of integrating chemicals—produced by the brain, gut, fat cells, and other tissues—will prove to be involved. Knowledge of this complex system will provide insight not only into obesity but into eating disorders such as anorexia as well.

How are lipids taken up and utilized?

Adipose and other cells take up lipids (mostly triglycerides) that have been hydrolyzed from chylomicrons and very-low-density lipoprotein (VLDL) by the enzyme lipoprotein lipase (LPL). Low-density lipoproteins (LDLs) are VLDL remnants and are the major means of delivering cholesterol to tissues. Normally, LDLs are taken up by receptor-mediated endocytosis. Cholesterol can be scavenged by high-density lipoproteins (HDLs) through the action of the enzyme lecithin cholesterol acyltransferase (LCAT), taken up by the liver, and removed from the body. Researchers have found that defects in LPL, LCAT, and LDL disrupt normal lipid metabolism and lead to severe health risks. For example, genetic defects in the LDL receptor are associated with familial hypercholesterolemia, in which excess cholesterol is left in the blood and causes heart disease and other abnormalities. Continued research on lipid and lipoprotein metabolism will be important for understanding obesity, type 2 diabetes, and cardiovascular disease.

think like a scientist

1. In the most general terms, how is allocation to maintenance metabolism different for tortoises and hares (hint: think temperature regulation)?

2. Given what you learned about lipids in the chapter on biological molecules, what is the importance of triglycerides in the diet of animals?

Mark Sheridan is Vice Provost for Graduate and Postdoctoral Affairs/Dean of the Graduate School and a Professor of Biology at Texas Tech University. His research examines the control of growth, development, and metabolism in vertebrates with current emphasis on elucidating the molecular basis of hormone multifunctionality and endocrine disruption of growth by environmental estrogens. Find out more about Dr. Sheridan's research at http://www.biol.ttu.edu/facultylist.aspx?id=mark.sheridan@ttu.edu.

Courtesy of Mark Sheridan

REVIEW KEY CONCEPTS

For access to MindTap and additional study materials visit www.cengagebrain.com.

47.1 Feeding and Nutrition

- Animals obtain organic molecules by eating other organisms. Herbivores primarily eat plants, carnivores primarily eat other animals, and omnivores eat animals, plants, and other sources of organic nutrients. The organic molecules are used as fuels for oxidative reactions providing energy and as building blocks for making complex biological molecules.

- Animals require essential substances in their diets—amino acids, fatty acids, vitamins, and minerals—that they cannot make for themselves.

- Animals may be classified with respect to feeding methods and the physical state of the organic molecules they eat. Fluid feeders ingest liquids containing organic molecules in solution. Suspension feeders eat small particles of organic matter or small organisms in suspension in fluids. Deposit feeders ingest small organic particles or organisms that are part of solid matter that the feeders live in or on. Bulk feeders consume large pieces of organisms, or entire large organisms (Figure 47.2).

47.2 Digestive Processes

- Digestion is the process of mechanical and chemical breakdown of food into molecular subunits small enough to be absorbed into body fluids and cells.
- Digestion may be intracellular or extracellular. Extracellular digestion allows food to be eaten in large batches, stored, and broken down while the animal carries out other activities.
- In animals with extracellular digestion, the digestive processes take place in an internal body cavity that is either a pouch or sac with one opening that serves as both mouth and anus, or a tube with two openings forming a mouth on one end and an anus on the other end (Figures 47.3 and 47.4).
- In animals with a digestive tract (also called digestive tube or gut), digestion occurs in five stages: (1) mechanical processing, including chewing and grinding of food; (2) secretion of enzymes and other digestive aids into the digestive tract; (3) enzymatic hydrolysis of food molecules into molecular subunits; (4) absorption of the molecular subunits across cell membranes; and (5) elimination of undigested matter.
- Food particles and molecules are pushed through the digestive tract by muscular contractions of its wall. Storage of food at various locations in the tube allows animals to digest food while engaged in other activities.

47.3 Digestion in Humans and Other Mammals

- Adult humans require 9 essential amino acids, 13 vitamins (Table 47.1), and a large number of essential minerals (Table 47.2).
- The mouth, pharynx, esophagus, stomach, intestine, and anus are common to the digestive system of mammals, including humans, and most vertebrates (Figure 47.5).
- The wall of the vertebrate digestive tract is formed from four layers of tissues: the mucosa, the submucosa, the muscularis, and the serosa (Figure 47.6).
- Coordinated contractions of the circular and smooth muscles produce peristaltic waves that move the digestive contents from the mouth to the anus (Figure 47.7).
- The enzymatic digestion of carbohydrates, proteins, and fat take place in particular regions of the human digestive tract (Figure 47.8)
- Digestion begins in the mouth, where the teeth break the food into smaller bits. Salivary amylase, an enzyme that digests starch, is secreted into the food in the mouth. After chewing, the food is swallowed and travels through the pharynx and esophagus to reach the stomach (Figure 47.9).
- In the stomach, hydrochloric acid, the protein-digesting enzyme pepsin, and mucus are added to the food mass. The stomach churns the acid contents into chyme, which is released in pulses into the small intestine (Figure 47.10).
- Absorption of nutrients begins in the small intestine. Specializations of the small intestine to optimize absorption are the intestinal villi and microvilli (Figure 47.11).

- In the small intestine, digestive juices from the pancreas and liver add enzymes and digestive aids to the food mass (Figure 47.12). The pancreatic juice contains digestive enzymes and bicarbonate ions that neutralize the acidity of the digestive contents. The liver secretion, bile, contains bile salts, which emulsify fats, cholesterol, bilirubin, and additional bicarbonate ions.
- Fat is digested to completion in the small intestine. Completion of carbohydrate and protein digestion is done by digestive enzymes that are embedded in the plasma membranes of the brush-border cells of the small intestine mucosa as the incompletely digested molecules are absorbed by those cells. The final products of digestion exit the cells and enter the bloodstream (Figure 47.13).
- The digestion products of fat digestion, the monoglycerides and fatty acids, are coated with bile salts to form microdroplets (micelles) that deliver the products to the brush-border epithelial cells. The monoglycerides and fatty acids enter the cells, where they are used to resynthesize triglycerides. These are coated with a layer of lipoproteins to form chylomicrons, which leave the epithelial cell and enter the lymphatic system (Figure 47.14).
- Absorbed nutrients are delivered to the liver, where excess glucose is converted into glycogen and fats, and some of the amino acids are converted into plasma proteins or sugars. The liver also synthesizes cholesterol from lipids, carbohydrates, and other substances.
- The large intestine absorbs water and mineral ions from the digestive contents. At the end of the large intestine the undigested remnants, the feces, are expelled from the anus (Figure 47.15).

47.4 Regulation of the Digestive Process

- Digestion is regulated by signals from the autonomic nervous system, by the activity of neuron networks in the digestive tube wall, and by hormones secreted by the digestive system. The regulatory mechanisms operate in response to signals from sensory receptors that monitor the volume and composition of the digestive contents.
- One interneuron center of the hypothalamus stimulates appetite and reduces oxidative metabolism, and another inhibits appetite. Together these controls match the activity of the digestive system to the amount and types of ingested foods, and they coordinate appetite and oxidative metabolism with the body's needs for stored fats.

47.5 Digestive Specializations in Vertebrates

- Common variations in vertebrate digestive systems include modifications of the teeth, length of the digestive tract, and structure and function of the stomach.
- Mammals have four basic types of teeth—incisors for cutting, canines for piercing, and premolars and molars for cutting, grinding, and smashing food (Figure 47.16).
- Carnivores and vertebrates that eat other nutrient-rich foods have a relatively short digestive tract. Herbivores, which eat nutrient-poor foods, typically have a relatively long digestive tract that includes extensive storage regions (Figure 47.17).
- Many herbivores have digestive chambers in which symbiotic microorganisms in gut microbiomes digest plant matter into molecules that can be absorbed by the host (Figure 47.18).
- The gut microbiomes of all mammals contain mutualistic microorganisms that synthesize essential amino acids and vitamins, and digest particular components of food that otherwise are indigestible.

TEST YOUR KNOWLEDGE

Remember/Understand

1. Required molecules that animals cannot synthesize are called:
   a. nutrients.
   b. essential nutrients.
   c. enzymes.
   d. proteins.
   e. carbohydrates.

2. Which of the following accurately describes a feeding style?
   a. Deposit feeders obtain nutrients from organic molecules in solution.
   b. Deposit feeders scrape organic matter from solid material on which they live.
   c. Fluid feeders digest organisms suspended in water.
   d. Fluid feeders strain food with networks of mucus or bristles and hairs.
   e. Suspension feeders consume sizable food whole or in chunks.

3. The order of successive steps in digestion is:
   a. absorption follows enzymatic hydrolysis.
   b. secretion of enzymes follows absorption of digestive material.
   c. mechanical processing follows enzyme secretion.
   d. mechanical processing follows enzymatic hydrolysis.
   e. enzymatic hydrolysis precedes secretion of digestive aids.

4. The esophagus, crop, gizzard, and intestine are found in:
   a. birds and mammals.
   b. insects and mammals.
   c. flatworms and birds.
   d. earthworms and birds.
   e. sponges and cnidarians.

5. Which of the following is *not* an essential nutrient in humans?
   a. vitamin B_6
   b. calcium
   c. glycogen
   d. linoleic acid
   e. vitamin K

6. A specialized region of the digestive tract is/are the:
   a. submucosa formed by circular and longitudinal layers.
   b. serosa lining the digestive tract for absorption.
   c. mucosa composed of thick, elastic connective tissue for movement.
   d. muscularis, an outer layer that secretes a slippery material to prevent friction with other organs.
   e. sphincters, which form valves between major digestive organs.

7. If the fat in whole milk is ingested:
   a. the stomach, with its high pH, will stimulate cells of the duodenum to hasten stomach emptying.
   b. parietal cells in the stomach will absorb it.
   c. in the small intestine, bile salts emulsify the fats and then lipase hydrolyzes them.
   d. lactase deficiency in the small intestine would prevent its digestion.
   e. microvilli will absorb the fat in the form of chylomicrons directly into the blood of the hepatic portal vein.

8. The role of the liver in digestion is to:
   a. synthesize aminopeptidase and dipeptidase to digest polypeptides.
   b. synthesize lipase to form free fatty acids.
   c. secrete trypsin to break the bonds in polypeptides.
   d. secrete bicarbonate ions and bile salts to help emulsify fats.
   e. store bile between meals.

9. Which of the following best describes regulation of digestion?
   a. GIP inhibits insulin release from the pancreas.
   b. Gastrin stimulates pancreatic secretion of HCl and pepsinogen.
   c. Secretin stimulates gastric emptying into the duodenum.
   d. CCK stimulates gastric activity to activate the duodenum.
   e. Leptin binds different hypothalamic receptors to inhibit appetite.

10. An example of a digestive specialization is seen in:
    a. the long intestines characteristic of herbivores.
    b. the incisors being the dominant teeth in wolves.
    c. the canine teeth being the dominant teeth in deer.
    d. salivary lipase being made by humans.
    e. cellulose being made by humans.

Apply/Analyze

11. **Discuss Concepts** As a person ages, the number of cells in the body steadily decreases and their energy needs decline. If you were planning a diet for an older person, what kind(s) of nutrients would you emphasize, and why? Which ones would you recommend the person consume less of? Include vitamins and minerals in your answer.

12. **Discuss Concepts** Formulate a healthy diet for a young, actively growing 7-year-old child, and explain why you have included each part of the diet. Refer to Question 11, above, for some issues to consider.

13. **Discuss Concepts** A baby develops symptoms of protein deficiency, and the attending physician suggests the cause is a genetic defect leading to a nonfunctional enzyme associated with digestion. Name at least three enzymes that might be likely suspects, and for each one explain how the defect would result in a protein deficiency.

Evaluate/Create

14. **Design an Experiment** Design experiments to test whether cigarette smoke affects the functioning of the various parts of the digestive system.

15. **Apply Evolutionary Thinking** What is the advantage of a tubelike digestive system over a saclike digestive system?

For selected answers, see Appendix A.

INTERPRET THE DATA

The human *AMY-1* gene encodes salivary amylase, an enzyme that breaks down starch. The number of copies of this gene varies, and people who have more copies generally make more of the enzyme. In addition, the average number of *AMY-1* copies differs among cultural groups.

George Perry and his colleagues hypothesized that duplications of the *AMY-1* gene would confer a selective advantage in cultures in which starch is a large part of the diet. To test this hypothesis, the scientists compared the number of copies of the *AMY-1* gene among members of seven cultural groups that differed in their traditional diets. The **Figure** shows their results.

1. Starchy tubers are a mainstay of Hadza hunter-gatherers in Africa, whereas fishing sustains Siberia's Yakut. Almost 60% of Yakut had fewer than 5 copies of the *AMY-1* gene. What percentage of the Hadza had fewer than 5 copies?

2. None of the Mbuti (rainforest hunter-gatherers) had more than 10 copies of *AMY-1*. Did any European Americans have more than 10 copies of *AMY-1*?

3. Do these data support the hypothesis that a starchy diet favors duplications of the *AMY-1* gene?

Source: G. Perry et al. 2007. Diet and the evolution of human amylase gene copy number variation. *Nature Genetics* 39:1256–1260.

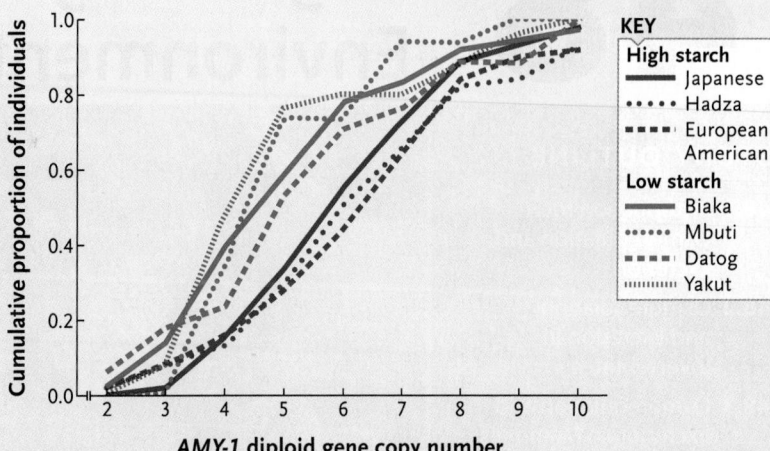

FIGURE Number of copies of the *AMY-1* gene among members of cultures with traditional high-starch or low-starch diets.

48 Regulating the Internal Environment

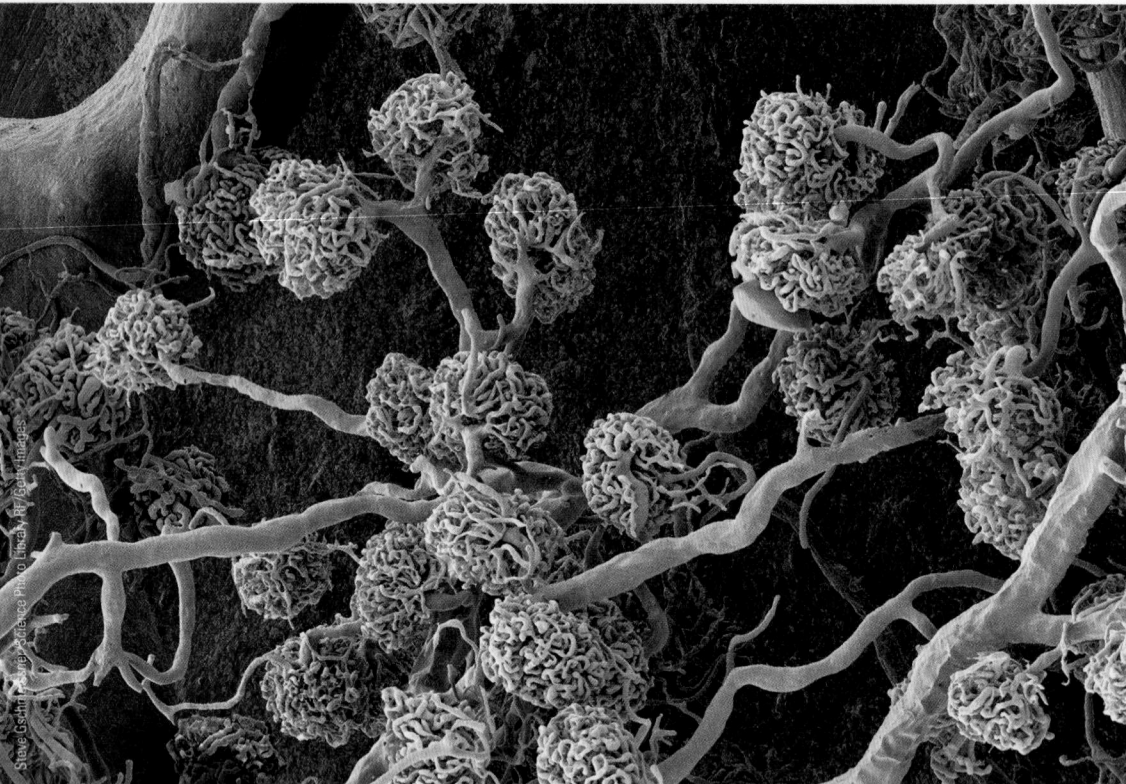

Glomeruli, ball-like tufts of capillaries in the nephrons of a kidney. Glomeruli filter water and solutes from the blood as the first step in urine formation (color-enhanced EM).

Why it matters . . . In August 1943, a combination of circumstances led to the World War II bomber *Lady Be Good* flying over the Sahara Desert by mistake. When the fuel had nearly run out, the crew members parachuted from the aircraft, landing some 440 miles from their base on the North African coast. The bomber later crash-landed many miles away.

The eight survivors began a northward trek with only half a canteen of water among them, in desert heat that reached 130°F during the day. In a testimony to the physiological mechanisms that conserve water and cool the body, they continued onward for eight days. But then, one by one, they died as prolonged dehydration and the heat overwhelmed their homeostatic control systems for *water balance* (the equilibrium in inward and outward flow of water) and temperature regulation, and hyperthermia exceeded their capacity to survive.

This story illustrates the trials of animal life under changing environmental conditions. Water and required nutrients may become more or less abundant. Temperatures may rise or fall. Through their homeostatic control systems, animals have an astounding capacity to compensate for fluctuating external conditions and to maintain the internal environment of their bodies within the relatively narrow limits that cells can tolerate (homeostasis is discussed in Chapter 38).

These limits, and the compensating mechanisms that maintain them, are the subjects of this chapter. First we examine **osmoregulation,** the control of water and ion balance, and the closely related topic of **excretion,** which helps maintain the body's water and ion balance while ridding the body of metabolic wastes. Then we consider **thermoregulation,** the control of body temperature.

48.1 Introduction to Osmoregulation and Excretion

Living cells contain water, are surrounded by water, and constantly exchange water with their environment. For the simplest animals, the water of the external environment surrounds cells directly. For more complex animals, an aqueous extracellular fluid (ECF) surrounds the cells, and is separated from the external environment by a body covering. In animals with a circulatory system, the ECF includes both the interstitial fluid immediately surrounding cells and the plasma, the fluid portion of the blood, or other circulated fluid.

In this section, we review the mechanisms cells use to exchange water and solutes with the surrounding fluid through osmosis and diffusion.

Osmosis Is a Form of Passive Diffusion

In osmosis (see Section 5.3 and Figures 5.9 and 5.10), water molecules move across a selectively permeable membrane from a region where they are more highly concentrated to a region where they are less highly concentrated. The difference in water concentration is produced by differing numbers of solute molecules or ions on the two sides of the membrane. The side of the membrane with a *lower* solute concentration has a *higher* concentration of water molecules, so water will move osmotically to the other side, where water concentration is *lower*. Proteins are among the most important solutes in establishing the conditions that produce osmosis.

The total solute concentration of a solution, called its **osmolarity,** is measured in *osmoles*—the number of solute molecules and ions (in moles)—per liter of solution. Because the total solute concentration in the body fluids of most animals is less than 1 osmole, osmolarity is usually expressed in thousandths of an osmole, or *milliosmoles* (mOsm). The osmolarity of body fluids in humans and other mammals, birds, and reptiles is about 300 mOsm/L. Marine teleosts have similar osmolarities of ~330 mOsm/L, but marine animals such as sharks and rays and marine invertebrates have much higher osmolarities of ~1,000 mOsm/L, and freshwater animals such as freshwater teleosts have lower osmolarities of ~290 mOsm/L. Freshwater invertebrates have even lower osmolarities of ~225 mOsm/L.

Considering solutions on either side of a selectively permeable membrane, a solution of higher osmolarity is said to be *hyperosmotic* to a solution of lower osmolarity, and a solution of lower osmolarity is said to be *hypoosmotic* to a solution of higher osmolarity. If the solutions on either side of a membrane have the same osmolarity, they are said to be *isoosmotic*. Water moves across the membrane between solutions that differ in osmolarity (see Figure 5.9), whereas when two solutions are isoosmotic, no net water movement occurs.

Animals Have Different Evolutionary Adaptations for Keeping Osmosis from Swelling or Shrinking Their Cells

For metabolic stability, animals must keep their cellular fluids and ECFs isoosmotic. In some animals, called **osmoconformers,** the osmolarity of the cellular and extracellular solutions matches the osmolarity of the environment. Most marine invertebrates are osmoconformers. Other animals, called **osmoregulators,** use control mechanisms to keep the osmolarity of cellular and extracellular fluids the same, but at levels that may differ from the osmolarity of the surroundings. Most freshwater and terrestrial invertebrates, and almost all vertebrates, are osmoregulators.

For terrestrial animals, one of the greatest challenges to osmoregulation is the limited supply of water in the environment, as the crew of the *Lady Be Good* experienced.

Excretion Is Closely Tied to Osmoregulation

Control over osmolarity is maintained partly by removing certain molecules and ions from cells and body fluids and releasing them into the environment; thus, excretion is closely related to osmoregulation. Animals excrete H^+ ions to keep the pH of body fluids near the neutral levels required by cells for survival. They also excrete toxic products of metabolism, such as nitrogenous (nitrogen-containing) compounds resulting from the breakdown of proteins and nucleic acids, and breakdown products of poisons and toxins. Excretion of ions and metabolic products is accompanied by water excretion because water serves as a solvent for those substances. Animals that take in large amounts of water may also excrete water to maintain osmolarity.

Microscopic Tubules Form the Basis of Excretion in Most Animals

Except in the simplest animals, minute tubular structures—**excretory tubules**—carry out osmoregulation and excretion **(Figure 48.1)**. The tubules are immersed in body fluids at one end (the *proximal end*), and open directly or indirectly to the body exterior at the other end (the *distal end*). The tubules are formed from a **transport epithelium**—a layer of cells with specialized transport proteins in their plasma membranes. The transport proteins move specific molecules and ions into and out of the tubule by either active or passive transport, depending on the particular substance and its concentration gradient.

Typically, the tubules function in a four-step process:

1. **Filtration.** Filtration is the nonselective movement of water and a number of solutes—ions and small molecules, but not large molecules such as proteins—into the proximal end of the tubules through spaces between cells. In animals with a closed circulatory system, such as humans, the water and

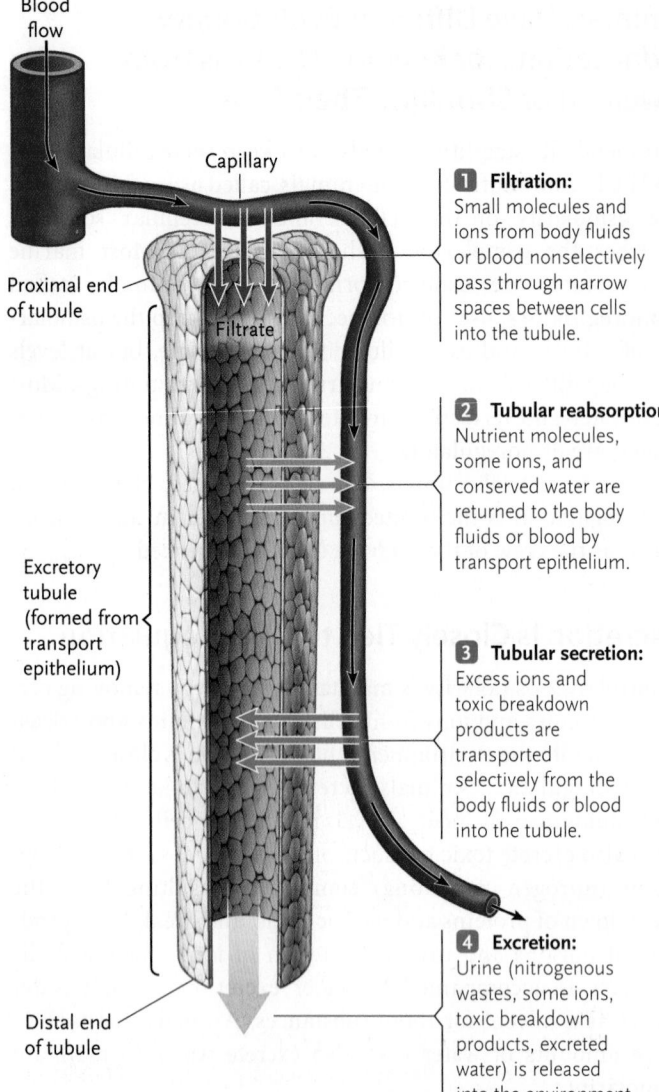

Blood flow

Capillary

Proximal end of tubule

Filtrate

Excretory tubule (formed from transport epithelium)

Distal end of tubule

1 Filtration: Small molecules and ions from body fluids or blood nonselectively pass through narrow spaces between cells into the tubule.

2 Tubular reabsorption: Nutrient molecules, some ions, and conserved water are returned to the body fluids or blood by transport epithelium.

3 Tubular secretion: Excess ions and toxic breakdown products are transported selectively from the body fluids or blood into the tubule.

4 Excretion: Urine (nitrogenous wastes, some ions, toxic breakdown products, excreted water) is released into the environment.

FIGURE 48.1 Common structures and operations of the tubules carrying out osmoregulation and excretion in animals. The tubules are typically formed from a single layer of cells with transport functions.
© Cengage Learning 2017

3. **Tubular secretion.** Tubular secretion is a selective process in which specific small molecules and ions are transported from the ECF and blood into the tubules. Both secretion and filtration play an important role in eliminating substances from the body fluid or blood. The difference between the two processes is that filtration is nonselective whereas tubular secretion is selective for substances transported.

4. **Excretion.** The fluid containing waste materials—urine—is released from the body into the environment from the distal end of the tubule. In some animals, the fluid is stored in a bladder; in others, it is concentrated into a solid or semisolid form. Urine contains no cells or large molecules.

In all vertebrates and many invertebrates, the excretory tubules are concentrated in specialized organs, the *kidneys*, which are discussed in later sections.

Animals Excrete Nitrogen Compounds as Metabolic Wastes

The metabolism of ingested food is a source of both energy and molecules for the biosynthetic activities of an animal. Importantly, metabolism of ingested food produces water—called *metabolic water*—that is used in chemical reactions and is involved in physiological processes such as the excretion of wastes.

The nitrogenous products of the breakdown of proteins, amino acids, and nucleic acids are excreted by most animals as *ammonia, urea,* or *uric acid,* or a combination of these substances **(Figure 48.2).** The particular molecule or combination of molecules produced depends on a balance among toxicity, water conservation, and energy requirements.

AMMONIA Ammonia (NH_3) forms by a series of biochemical steps beginning with the removal of amino groups ($—NH_3^+$) from amino acids as a part of protein breakdown. Ammonia is readily soluble in water, but it is also highly toxic. Therefore, ammonia must either be excreted or converted to a nontoxic derivative. Due to its toxicity, ammonia can be excreted from the body only in dilute solutions, making this path possible only in animals with a plentiful supply of water. Those animals include aquatic invertebrates, teleosts, and larval amphibians.

UREA All mammals, most amphibians, turtles, some marine fishes, and some terrestrial invertebrates convert ammonia in a series of steps to *urea,* a soluble and relatively nontoxic substance. Although producing urea requires more energy than forming ammonia, excreting urea instead of ammonia requires only about 10% as much water.

URIC ACID Water is conserved further in some animals, including terrestrial invertebrates, lizards and snakes, and birds, by the formation of uric acid instead of ammonia or urea.

solutes come from the capillary blood plasma and exchange with the intratubular fluid (shown in Figure 48.1), with the movement into the tubules similarly driven by hydrostatic pressure of the plasma in the capillaries, namely blood pressure. In animals with an open circulatory system, the water and solutes come from body fluids, with movement into the tubules driven by the higher pressure of the body fluids compared with the fluid inside the tubule. (Open and closed circulatory systems are described in Section 44.1.)

2. **Tubular reabsorption.** Tubular reabsorption is a selective process in which some molecules (for example, glucose and amino acids) and ions are transported by the transport epithelium from the lumen of the tubule back into the ECF and eventually into the blood (in animals with closed circulatory systems) as the filtered solution moves through the excretory tubule.

Uric acid is nontoxic, and so insoluble that it precipitates in water as a crystal. (The white substance in bird droppings is uric acid.) The embryos of reptiles and birds also conserve water by forming uric acid, which is stored as a waste product.

Although making uric acid requires even more energy than making urea, molecule for molecule it contains four times as much nitrogen as ammonia. And, because uric acid precipitates from water, it can be excreted as a concentrated paste. These factors conserve about 99% of the water that would be required to excrete an equivalent amount of nitrogen as ammonia.

In the sections that follow, we look at the specifics of osmoregulation and excretion in different animal groups, beginning with the invertebrates.

STUDY BREAK 48.1

Define the terms *osmosis, osmolarity, hypoosmotic, osmoregulator,* and *transport epithelium.*

48.2 Osmoregulation and Excretion in Invertebrates

Except for the simplest groups, most invertebrates, whether osmoconformers or osmoregulators, carry out excretion by specialized excretory tubules.

Most Marine Invertebrates Are Osmoconformers; All Freshwater and Terrestrial Invertebrates Are Osmoregulators

Most marine invertebrates are osmoconformers. They release water, certain ions, and nitrogenous wastes—usually in the form of ammonia—directly from body cells to the surrounding seawater. The cells of these animals do not swell or shrink because the osmolarity of their intracellular and extracellular fluids and the surrounding seawater is the same, about 1,000 mOsm/L. Therefore, they do not have to expend energy to maintain their osmolarity. However, osmoconformers do expend energy to keep some ions, such as Na^+, at lower concentrations inside cells than in the surroundings.

By contrast, all freshwater invertebrates such as flatworms and mussels are osmoregulators because their cells could not survive if their internal ion concentrations were reduced to freshwater levels. These animals must expend energy to keep their internal fluids hyperosmotic to their surroundings. Although osmoregulation is energetically expensive, these invertebrates can live in more varied habitats than can osmoconformers.

Freshwater osmoregulators have cellular and extracellular fluids with an osmolarity higher than that of the external environment, causing water to move constantly from the surroundings into their bodies. This excess water must be excreted, at a considerable cost in energy, to maintain the hyperosmotic state of their body fluids. These animals must also obtain the salts required to keep their body fluids hyperosmotic to fresh water. The salts are obtained from foods, and by actively transporting salt ions from the water into their bodies (even fresh water contains some dissolved salts). This active ion transport occurs through the skin or gills.

Terrestrial annelids (earthworms), arthropods (insects, spiders and mites, millipedes, and centipedes), and mollusks (land snails and slugs) must obtain salts from their surroundings, usually in their foods. Although they do not have to excrete water entering by osmosis, they must constantly replace water lost from their bodies.

In Invertebrate Osmoregulators, Specialized Excretory Tubules Participate in Osmoregulation and Carry Out Excretion

Invertebrate osmoregulators typically use specialized tubules for carrying out excretion. Three common types of these specialized tubules are *protonephridia,* found in flatworms and larval mollusks; *metanephridia,* found in annelids and most adult mollusks; and *Malpighian tubules,* found in insects and other arthropods. Each processes body fluids in a different way.

PROTONEPHRIDIA The flatworm *Dugesia* provides an example of the simplest form of invertebrate excretory tubule, the **protonephridium** (*protos* = first; *nephros* = kidney). In *Dugesia,* two branching networks of protonephridia run the length of the body **(Figure 48.3).** The proximal branches of the tubule network end with a *flame cell* where filtration occurs. Each flame cell

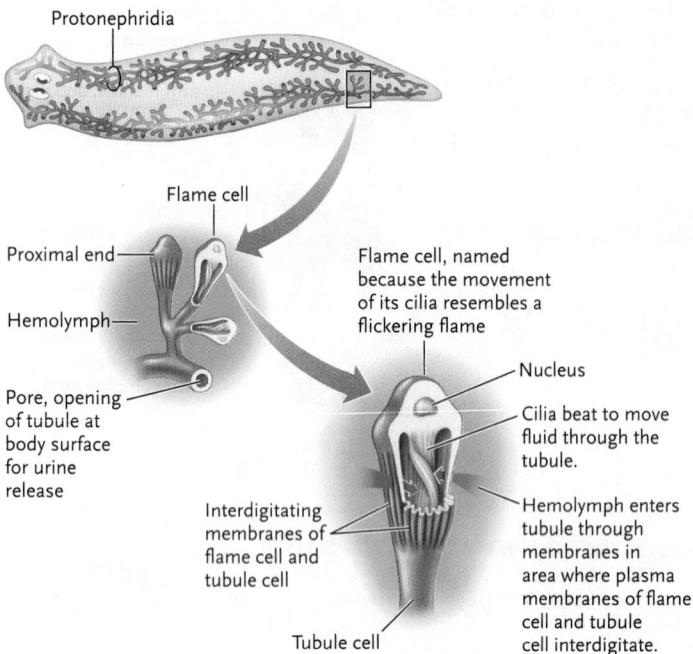

FIGURE 48.3 **The protonephridia of the flatworm** *Dugesia*, **showing a flame cell.**
© Cengage Learning 2017

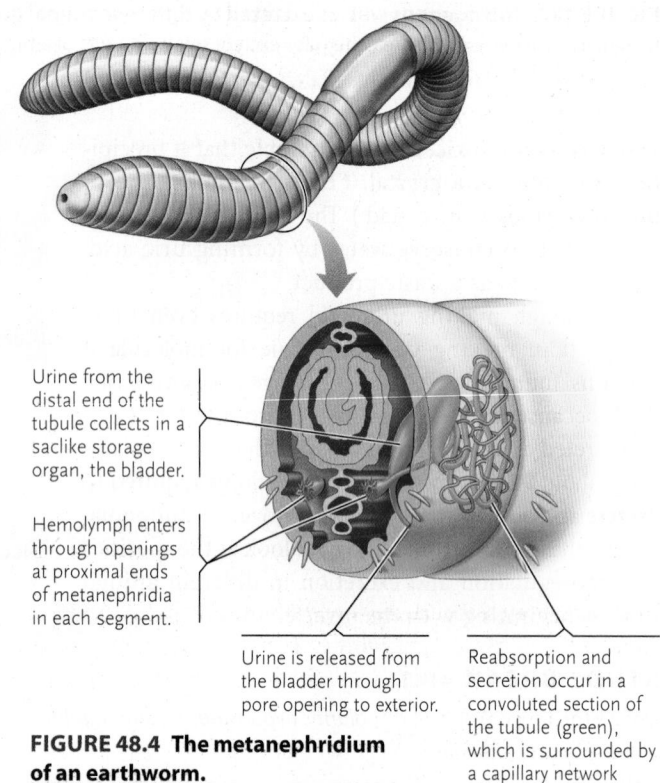

FIGURE 48.4 **The metanephridium of an earthworm.**
© Cengage Learning 2017

contains a bundle of cilia that extend into the tubule and beat to move fluid through the tubule. When the hemolymph, the invertebrate equivalent to blood, passes through the protonephridia, some molecules and ions are reabsorbed and others, including nitrogenous wastes, are secreted into the tubules. The urine produced is excreted through pores at the distal ends of the tubules where they reach the body surface.

METANEPHRIDIA The excretory tubule of most annelids and adult mollusks, the **metanephridium** (*meta* = between), has a funnel-like proximal end surrounded with cilia that admits hemolymph in the filtration step of excretion. As hemolymph moves through the tubule, some molecules and ions are reabsorbed and other ions and nitrogenous wastes are secreted into the tubule and excreted from the distal end at the body surface.

Figure 48.4 shows the arrangement and operation of metanephridia in an earthworm. The proximal ends of a pair of metanephridia are located in each body segment, one on either side of the animal. Each tubule of the pair extends into the following segment, where it bends and folds into a convoluted arrangement surrounded by a network of blood vessels.

MALPIGHIAN TUBULES The excretory tubule of insects, the **Malpighian tubule,** has a closed proximal end that is immersed in the hemolymph **(Figure 48.5).** The distal ends of the tubules empty into the gut. Unlike other excretory systems, Malpighian tubules do not use pressure for the filtration step. Instead, they secrete K^+ into the lumen of the proximal segment, which makes the tubule fluid more positively charged and draws in

Cl^- ions from the hemolymph surrounding the tubule. The accumulation of KCl makes the tubule fluid more concentrated, causing water from the hemolymph to move in by osmosis. Organic wastes, in particular uric acid, are then secreted into

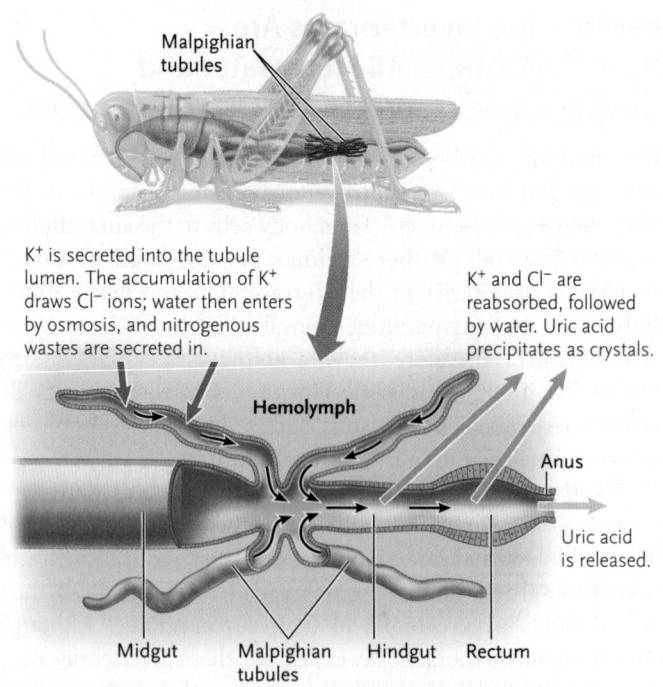

FIGURE 48.5 **Excretion through Malpighian tubules in a grasshopper.**
© Cengage Learning 2017

the tubule. The contents of the tubule empty into the gut. When the fluid reaches the hindgut and rectum, K+ and Cl− are reabsorbed, followed by water. As the water is removed, uric acid precipitates as crystals, which mix with the undigested matter in the rectum and are released with the feces. In this way, nitrogenous waste excretion is accomplished with minimal water loss.

STUDY BREAK 48.2

How are protonephridia, metanephridia, and Malpighian tubules different? In which animal groups are each of these excretory tubules found?

48.3 Osmoregulation and Excretion in Mammals

In all vertebrates, specialized excretory tubules, called **nephrons,** contribute to osmoregulation and carry out excretion. Nephrons are located in a specialized organ, the kidney. We begin our survey of vertebrate osmoregulation and excretion with a description of the structure and function of the mammalian kidney.

The Kidneys and Ureters, the Bladder, and the Urethra Form the Urinary System

Mammals have a pair of kidneys, located on either side of the vertebral column at the back of the abdominal cavity **(Figure 48.6).** Internally, the mammalian kidney is divided into an outer **renal cortex** surrounding a central region, the **renal medulla.**

The **renal artery** carries blood into the kidney, where metabolic wastes and excess ions are moved into urine. The filtered blood is routed away from the kidney by the **renal vein.** The urine leaving individual nephrons is processed further in **collecting ducts** and then drains into a central cavity in the kidney called the **renal pelvis.**

From the renal pelvis, the urine flows through a duct called the **ureter** to the **urinary bladder,** a storage sac located outside the kidneys. Urine leaves the bladder through the **urethra,** a tube that (in most mammals) opens to the outside. In human females, the opening of the urethra is just in front of the vagina; in males, the urethra opens at the tip of the penis. The two kidneys and ureters, the urinary bladder, and the urethra constitute the mammalian urinary system.

Two sphincter muscles control the flow of urine from the bladder to the urethra. In human infants, urination is an autonomic reflex triggered by stretch receptors in the bladder wall. When the bladder becomes full, the sphincters relax, smooth muscles in the bladder wall contract, and the urine is forced to the exterior. At about two years of age, children learn to override the autonomic reflex by consciously contracting their sphincters until urination is convenient.

Mammalian Nephrons Are Differentiated into Regions with Specialized Functions

Each human kidney has more than a million nephrons. Nephrons are differentiated into regions that perform successive steps in excretion. At its proximal end, a human nephron forms the **Bowman's capsule,** an infolded region that cups around a ball of arterial capillaries called the **glomerulus** (see Figure 48.6). The capsule and glomerulus are located in the renal cortex. As blood flows through the glomerulus, plasma filters through the glomerular capillaries into Bowman's capsule; this process, called **glomerular filtration,** is the first step of urine formation.

Following Bowman's capsule, the nephron forms a **proximal convoluted tubule** that descends into the renal medulla in a U-shaped bend called the **loop of Henle** and then ascends again to form a **distal convoluted tubule.** The distal tubule drains the urine into a collecting duct that leads to the renal pelvis. As many as eight nephrons may drain into a single collecting duct. The combined activities of the proximal convoluted tubule, the loop of Henle, the distal convoluted tubule, and the collecting duct convert the filtrate entering the nephron into urine.

Unlike most capillaries in the body, the capillaries in the glomerulus do not lead directly to venules. Instead, they form another arteriole that branches into a second capillary network called the **peritubular capillaries.** These capillaries thread around the proximal and distal convoluted tubules and the loop of Henle. Molecules and ions that are reabsorbed from the filtrate are transferred indirectly between the nephron and the peritubular capillaries. First, the molecules or ions pass through the one-cell-thick wall of the tubule, then they diffuse through the interstitial fluid, and finally they pass into the capillary through its one-cell-thick wall.

Most human kidney diseases attack the nephrons, adversely affecting their filtering capability. One genetic disease of the kidneys, polycystic kidney disease (PKD), is characterized by the development from the tubules of fluid-filled cysts. *Molecular Insights* discusses the role of miRNAs in PKD.

Mammalian Nephrons Interact with Surrounding Kidney Structures to Produce Hyperosmotic Urine

In mammals, urine is hyperosmotic to body fluids. All other vertebrates except for a few aquatic bird species produce urine that is hypoosmotic (or isoosmotic) to body fluids. Production of hyperosmotic urine is a water-conserving adaptation involving an interaction between the activities of the mammalian nephrons and the highly ordered structure of the mammalian kidney. Three features underlie this interaction:

1. The arrangement of the loop of Henle, which descends through the medulla and returns to the cortex again.
2. Differences in the permeability of successive regions of the nephron, established by a specific group of membrane transport proteins in each region.

Involvement of miRNAs with the Development of Polycystic Kidney Disease

Polycystic kidney disease (PKD) is one of the most common life-threatening genetic diseases. PKD leads to kidney failure, which necessitates lifelong dialysis or a kidney transplant. In patients with PKD, fluid-filled cysts develop from kidney tubules and are lined by altered epithelial cells that secrete fluid. The excessive fluid expands the cysts, which compress surrounding normal nephrons, leading to kidney failure. Experiments have shown that miRNA expression is altered in PKD.

Research Question

Do changes in miRNA expression contribute to the development of PKD?

Experiments and Results

Vishal Patel and colleagues at the University of Texas Southwestern Medical Center, the Université Paris-Descartes Institut Cochin, and Yale University School of Medicine

performed a number of experiments to answer the question:

1. First, they used microarray analysis (see Section 19.3) to compare miRNAs in normal control mice and in mice genetically modified to be an animal model of PKD. They observed that 64 miRNAs were expressed differentially in kidneys of the two groups of mice. In particular, four members of the six-member evolutionarily conserved miR-17~92 miRNA cluster were expressed at higher levels in the PKD model. One of the miRNAs in the cluster had been shown previously to be implicated in kidney cyst formation in *Xenopus*.

2. Using transgenic mice, the researchers demonstrated that kidney-specific overexpression of the miR-17~92 cluster miRNAs results in the formation of kidney cysts.

3. Deletion of the genes for the miRNA cluster by gene knockout in the mouse model of PKD slowed kidney cyst growth, improved kidney function, and prolonged survival compared with PKD mice with the miRNA cluster intact.

Conclusion

The experiments "demonstrate a pathogenic role of miRNAs in mouse models of PKD."

think like a scientist

Do the results suggest any research direction for developing a therapeutic treatment for PKD?

Source: V. Patel et al. 2013. miR-17~92 miRNA cluster promotes kidney cyst growth in polycystic kidney disease. *Proceedings of the National Academy of Sciences USA* 110:10765–10770.

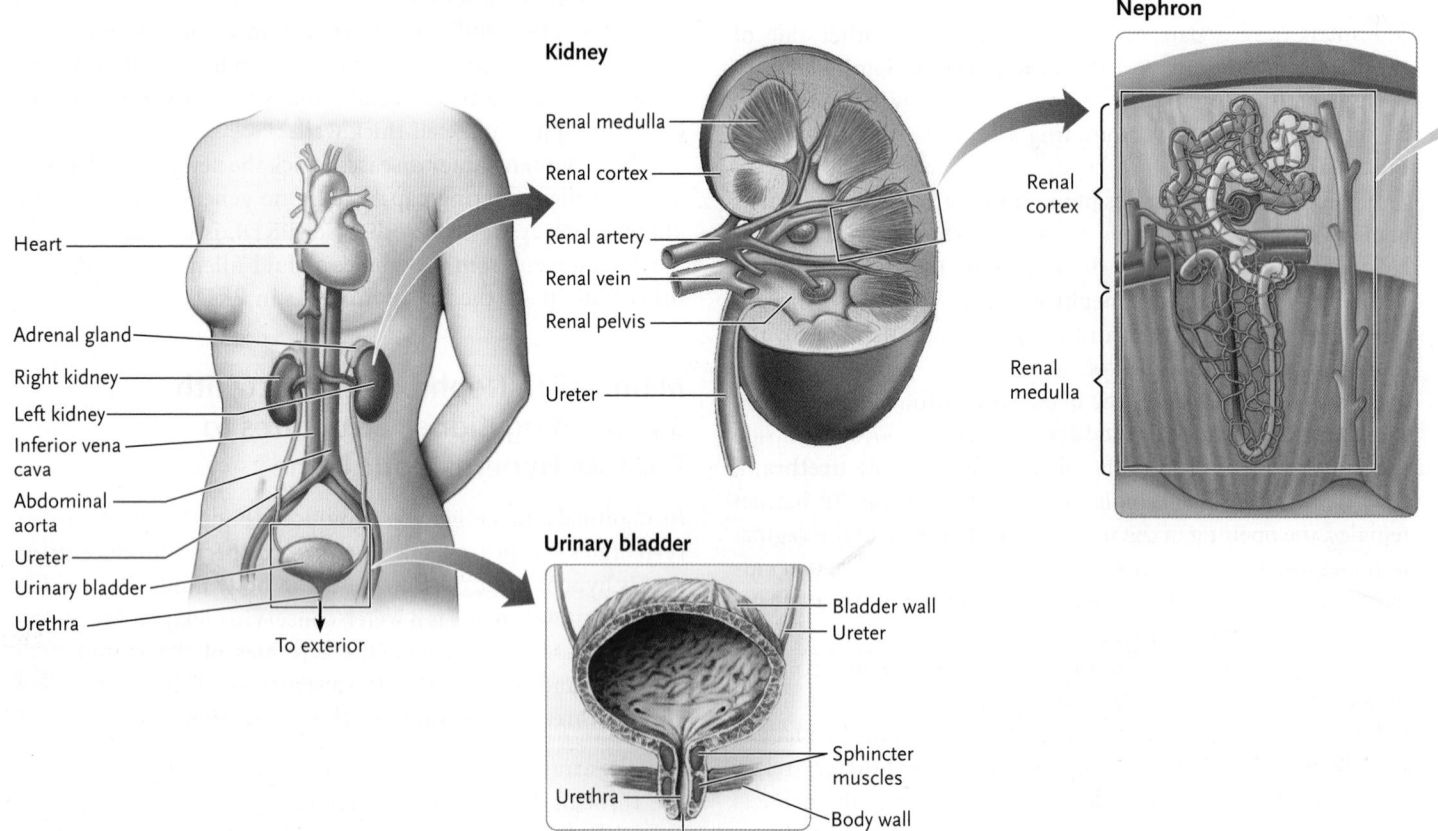

FIGURE 48.6 Human kidneys, urinary system, and nephron.

Segment	Location	Permeability and Movement	Osmolarity of Filtrate and Urine	Result of Passage
Bowman's capsule	Cortex	Water, ions, small nutrients, and nitrogenous wastes move through spaces between epithelia	300 mOsm/L, same as surrounding interstitial fluid	Water and small substances, but not proteins, pass into nephron
Proximal convoluted tubule	Cortex	Na^+ and K^+ actively reabsorbed, Cl^- follows; water leaves through aquaporins; HCO_3^- reabsorbed into plasma of peritubular capillaries; glucose, amino acids, and other nutrients actively reabsorbed	300 mOsm/L	67% of ions, 65% of water, 50% of urea, and all nutrients return to interstitial fluid; pH maintained
Descending segment of loop of Henle	Cortex into medulla	Water leaves through aquaporins; no movement of ions or urea	From 300 mOsm/L at top to 1,200 mOsm/L at bottom of loop	Additional water returned to interstitial fluid
Ascending segment of loop of Henle	Medulla into cortex	Na^+ and Cl^- actively transported out; no entry of water; no movement of urea	From 1,200 mOsm/L at bottom to 150 mOsm/L at top of loop	Additional ions returned to interstitial fluid
Distal convoluted tubule	Cortex	K^+ and Na^- secreted via active transport into urine; Na^+ and Cl^- reabsorbed; water moves into urine through aquaporins; HCO_3^- reabsorbed into plasma of peritubular capillaries	From 150 mOsm/L at beginning to 300 mOsm/L at junction with collecting duct	Ion balance, pH balance
Collecting ducts	Cortex through medulla, empties into renal pelvis	Water moves out via aquaporins; no movement of ions; some urea leaves at bottom of duct	From 300 mOsm/L to 1,200 mOsm/L at junction with renal pelvis	More water and some urea returned to interstitial fluid; some H^+ added to urine

Nephron

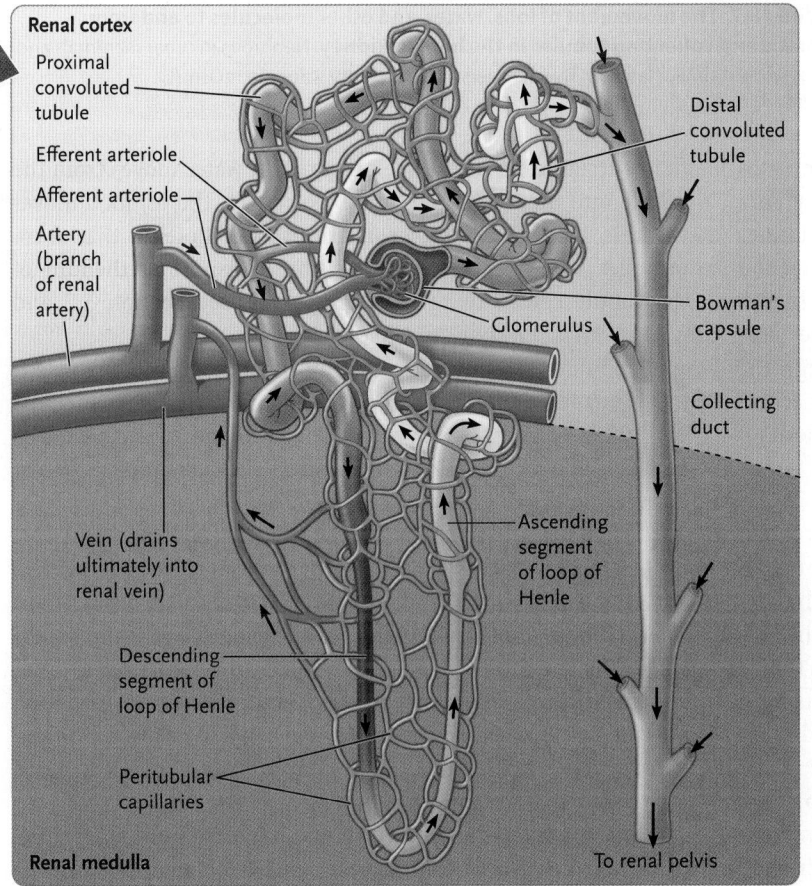

Renal cortex
Proximal convoluted tubule
Efferent arteriole
Afferent arteriole
Artery (branch of renal artery)
Glomerulus
Distal convoluted tubule
Bowman's capsule
Collecting duct
Vein (drains ultimately into renal vein)
Ascending segment of loop of Henle
Descending segment of loop of Henle
Peritubular capillaries
Renal medulla
To renal pelvis

3. A gradient in the concentration of molecules and ions in the interstitial fluid of the kidney, which increases gradually from the renal cortex to the deepest levels of the renal medulla.

These features interact to conserve nutrients and water, balance salts, and concentrate wastes for excretion from the body.

Three processes are involved in forming urine: *glomerular filtration, tubular reabsorption,* and *tubular secretion.* These processes are described in the next two subsections.

Glomerular Filtration Begins the Process of Excretion

Glomerular filtration begins the process of excretion (shown in **Figure 48.7** and summarized in **Table 48.1**). The single layer of epithelial cells forming the glomerular capillaries is perforated by many pores that make the vessels much more permeable to water and solutes than other capillaries in the body. The higher pressure of the blood drives fluid containing ions, small nutrient molecules such as glucose and amino acids, and nitrogenous waste molecules (primarily urea) through the pores of the glomerular capillaries into Bowman's capsule. Blood cells and plasma proteins are too large to pass and are retained inside the capillaries.

The blood pressure driving fluid through the glomerular capillaries into Bowman's capsule is maintained because the diameter of the **afferent arteriole** delivering blood to the glomerulus is larger than that of the **efferent arteriole** that takes blood away from the glomerulus. That is, because more blood can enter the glomerulus through the afferent arteriole than can leave through the efferent arteriole, glomerular capillary pressure is maintained at a high level as a result of blood damming up in the capillaries.

In humans, Bowman's capsules collectively filter about 180 L (47.5 gallons) of fluid each day, from a daily total of 1,400 L (369.5 gallons) of blood that pass through the kidneys. The human body contains only about 2.75 L of blood plasma, meaning that the kidneys filter a fluid volume equivalent to 65 times the volume of the blood plasma each day. On average, more than 99% of the filtrate, mostly water, is reabsorbed in the nephrons, leaving about 1.5 L to be excreted daily as urine.

Reabsorption and Secretion Take Place in the Remainder of the Nephron

The fluid filtered into Bowman's capsule contains water, other small molecules, and ions at essentially the same concentrations as the blood plasma. As the filtrate flows through the tubules, substances of value are returned to the plasma in the peritubular capillaries (see Figure 48.6) through the selective process of tubular reabsorption. In addition, selective movement of substances from the blood of the peritubular capillaries into the tubules occurs by tubular secretion. Tubular secretion is a second avenue for substances to enter the tubules from the blood, the first being glomerular filtration. By the time the fluid reaches the distal end of the tubules and passes through the collecting ducts, reabsorption out of the tubules and secretion into them have markedly altered the concentrations of all components of the filtrate.

THE PROXIMAL CONVOLUTED TUBULE Reabsorption of water, ions, and nutrients into the interstitial fluid is the main function of the proximal convoluted tubule. Na^+/K^+ pumps in the epithelium of the proximal convoluted tubule move Na^+ and K^+ from the filtrate into the interstitial fluid surrounding the tubule (see Figure 48.7). The movement of positive charges sets up a voltage gradient that causes Cl^- ions to be reabsorbed from within the tubule with the positive ions. Specific active transport proteins reabsorb essentially all the glucose, amino acids, and other nutrient molecules from the filtrate into the interstitial fluid, making the filtrate hypoosmotic to the interstitial

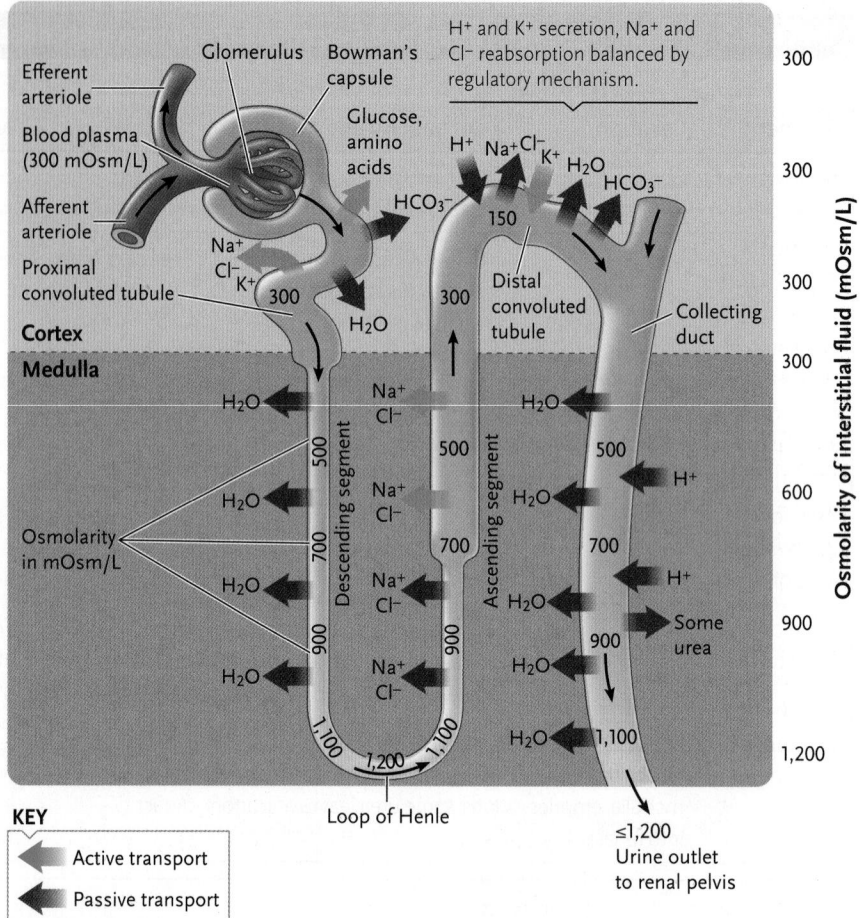

FIGURE 48.7 The movement of ions, water, and other molecules to and from nephrons and collecting tubules in the human kidney. Nephrons in other mammals and in birds work in similar fashion. The numbers are osmolarity values in mOsm/L.
© Cengage Learning 2017

fluid surrounding the tubule. As a result, water moves from the tubule into the interstitial fluid by osmosis. The osmotic movement is aided by *aquaporins,* transport proteins that form passages for water molecules in the transport epithelium of the tubule cells. (To review aquaporins, see Section 5.2 and *Unanswered Questions* in Chapter 5.)

Tubular secretion moves some substances into the proximal convoluted tubule; notably, toxins and drugs that were processed by the liver are moved by active transport into the lumen of the tubule (not shown in the figure). The substances are then excreted in the urine.

In all, the proximal convoluted tubule reabsorbs about 67% of the Na^+, K^+, and Cl^- ions, 65% of the water, 50% of the urea, and essentially all the glucose, amino acids, and other nutrient molecules in the filtrate. The ions, nutrients, and water reabsorbed by the tubule are transported into the interstitial fluid, and then into capillaries of the peritubular network. Although half of the urea is reabsorbed, the constant flow of filtrate through the tubules keeps the concentration of nitrogenous wastes low in body fluids.

The proximal convoluted tubule has structural specializations that fit its function. The epithelial cells that make up its walls are carpeted on their inner surface by a brush border of microvilli. Like the brush border of epithelial cells in the small intestine (see Section 47.3), these microvilli greatly increase the surface area available for reabsorption and secretion.

THE DESCENDING SEGMENT OF THE LOOP OF HENLE
The loop of Henle has dual functions: (1) to recapture water in the descending segment; and (2) to reabsorb salts in the ascending segment.

The filtrate leaving the proximal convoluted tubule enters the descending segment of the loop of Henle, where water is reabsorbed. As this segment descends, it passes through regions of increasingly higher solute concentrations in the interstitial fluid of the medulla (see Figure 48.7). (The generation of this concentration gradient is described later.) As a result, more water moves out of the tubule by osmosis as the fluid travels through the descending segment.

Aquaporins in the descending segment allow the rapid transport of water. The outward movement of water concentrates the molecules and ions inside the tubule, gradually increasing the osmolarity of the fluid to a peak of about 1,200 mOsm/L at the bottom of the loop. This is the same as the osmolarity of the interstitial fluid at the bottom of the medulla.

THE ASCENDING SEGMENT OF THE LOOP OF HENLE
The fluid then moves into the ascending segment of the loop of Henle, where Na^+ and Cl^- are reabsorbed into the interstitial fluid. As this segment ascends, it passes through regions of gradually lessening osmolarity in the medulla (see Figure 48.7). The ascending segment has membrane proteins that transport salt ions, but no aquaporins. Because water is trapped in the ascending segment, the osmolarity of the urine is reduced as the salt ions move out of the tubule.

By the time the fluid reaches the cortex at the top of the ascending loop, its osmolarity has dropped to about 150 mOsm/L. During the travel of fluid around the entire loop of Henle, water, nutrients, and ions have been conserved and returned to body fluids, and the total volume of the filtrate in the nephron has been reduced greatly. Urea and other nitrogenous wastes have been concentrated in the filtrate. Little secretion occurs in either the descending or ascending segments of the loop of Henle.

THE DISTAL CONVOLUTED TUBULE
Additional water is recovered by osmosis from the fluid in the distal convoluted tubule. In response to hormones triggered by changes in the body's salt concentrations (described in Section 48.4), a varying amount of K^+ is secreted into the fluid, and varying amounts of Na^+ and Cl^- ions are reabsorbed. Bicarbonate ions are reabsorbed from the filtrate as in the proximal tubule.

The amounts of urea and other nitrogenous wastes remain the same. By the time the fluid—now urine—enters the collecting ducts at the end of the nephron, it is isoosmotic with blood plasma (about 300 mOsm/L) but very different in composition.

THE COLLECTING DUCTS
The collecting ducts concentrate the urine. These ducts, which are permeable to water but not salt ions, descend from the cortex through the medulla of the kidney. As the ducts descend, they encounter the gradient of increasing solute concentration in the interstitial fluid of the medulla. This increase makes water move osmotically out of the ducts and greatly increases the concentration of the urine, which can become as high as 1,200 mOsm/L at the bottom of the medulla. Near the bottom of the medulla, the walls of the collecting ducts contain passive urea transporters that allow a portion of this nitrogenous waste to pass from the duct into the interstitial fluid. This urea adds significantly to the concentration gradient of solutes in the medulla.

At its maximum value of 1,200 mOsm/L, reached when water conservation is at its maximum, the urine at the bottom of the collecting ducts is about four times more concentrated than body fluids. It can also be as low as 50 to 70 mOsm/L when very dilute urine is produced in response to conditions such as excessive water intake. (The regulatory mechanism for this adjustment of urine concentration is discussed later in Section 48.4.)

The hyperosomotic state of the interstitial fluid toward the bottom of the medulla would damage the medulla cells if they were not protected against osmotic water loss. The protection comes from high concentrations of otherwise inert organic molecules called *osmolytes* in the cytoplasm of these cells. The osmolytes, primarily a sugar alcohol called *sorbitol*, raise the osmolarity of the cells to match that of the surrounding interstitial fluid.

Urine flows from the end of the collecting ducts into the renal pelvis, and then through the ureters into the urinary bladder, where it is stored. From the bladder, urine exits through the urethra to the outside.

Terrestrial Mammals Have Additional Water-Conserving Evolutionary Adaptations

Terrestrial mammals have other evolutionary adaptations that complement the water-conserving activities of the kidneys. One is the location of the lungs deep inside the body, which reduces water loss by evaporation during breathing (see Section 46.1). Another is a body covering of keratinized skin. Skin is nearly impermeable so that it almost eliminates water loss by evaporation, except for the controlled loss through evaporation of sweat in mammals with sweat glands.

Water-conserving adaptations reach their greatest efficiency in desert rodents such as the kangaroo rat (Figure 48.8). The proportion of nephrons with long loops extending deep into the kidney medulla of kangaroo rats is very high, allowing them to excrete urine that is 20 times more concentrated than body fluids. Further, most of the water in the feces is absorbed in the large intestine and rectum. Lacking sweat glands, kangaroo rats lose little water by evaporation from the body surface. Much of the moisture in their breath is condensed and recycled

	Kangaroo rat	Human
Water gain (mL)		
From ingesting food	6.0	850
From drinking liquids	0.0	1,400
By metabolism	54.0	350
	60.0	2,600
Water loss (mL)		
In urine	13.5	1,500
In feces	2.6	200
By evaporation	43.9	900
	60.0	2,600

FIGURE 48.8 A comparison of the sources of water for a human and a kangaroo rat (genus *Dipodomys*). Water conservation in the kangaroo rat is so efficient that the animal never has to drink water.
© Cengage Learning 2017

by specialized passages in the nasal cavities. They stay in burrows during daytime, and come out to feed only at night.

About 90% of the kangaroo rat's daily water supply is generated from oxidative reactions in its cells. (Humans, by contrast, can make up only about 12% of their daily water needs from this source.) The remaining 10% of the kangaroo rat's water comes from its food. These structural and behavioral adaptations are so effective that a kangaroo rat can survive in the desert without ever drinking water.

Marine mammals, including whales, seals, and manatees, eat foods that are high in salt content and never drink fresh water. They are able to survive the high salt intake because they produce urine that is more concentrated than seawater. As a result, they are easily able to excrete all the excess salt they ingest in their diet.

STUDY BREAK 48.3

1. Describe the structure of a human nephron from the proximal end to the distal end.
2. The urine entering the collecting ducts at the end of the nephron has an osmolarity essentially the same as that of fluids in other parts of the body. How is the urine subsequently made more concentrated?

THINK OUTSIDE THE BOOK

On your own or collaboratively, research the symptoms and possible causes of chronic kidney disease. Then outline the hypothesis and research approach for a study of chronic kidney disease that uses the rat as the experimental organism.

48.4 Regulation of Mammalian Kidney Function

Mammalian excretory functions are integrated into overall body functions by three primary control systems, which link kidney functions to blood pressure, to the osmolarity and pH of body fluids, and to the body's water balance:

- An autoregulation system located entirely within the kidney keeps glomerular filtration constant during variations in arterial blood pressure, as when we move from sitting to standing.
- Two hormonal control systems compensate for excessive loss of salt and body fluids and adjust the rate of water uptake in the kidneys to compensate for excessive water intake or loss. These two hormonal systems regulate interactions between the kidneys and the rest of the body.

The three systems work together, but we will discuss them separately.

An Autoregulation System Controls Glomerular Filtration Rate

An autoregulation system responds almost instantly to small variations in blood pressure to keep the glomerular filtration rate constant. Two mechanisms are involved in the autoregulation:

1. *Vascular control of blood pressure.* Increased blood pressure in the afferent arteriole is detected by receptors in the vessel. In response to the stretching, the smooth muscles of the arterial wall contract, which constricts the blood vessel, thereby constraining the blood flow to the glomerulus and reducing the glomerular filtration rate. In the opposite condition of reduced blood pressure in the arteriole, the receptors signal the blood vessel to dilate (expand), thereby increasing blood flow into the glomerulus and, therefore, glomerular filtration rate, even though arterial blood pressure has dropped.

2. *Feedback control from the distal convoluted tubule to the juxtaglomerular apparatus.* The **juxtaglomerular apparatus** (*juxta* = near) is located at a point where the distal convoluted tubule contacts the afferent arteriole carrying blood to the glomerulus (**Figure 48.9,** blow up). The feedback control mechanism is triggered by salt concentration in the distal convoluted tubule. If the glomerular filtration rate increases above normal, more fluid than normal is filtered, and a higher salt concentration results in the distal convoluted tubule. Specialized tubule cells in the juxtaglomerular apparatus monitor the salt level of the fluid flowing past them in the tubule. If the salt level rises, those cells secrete chemical messengers that act locally (by paracrine regulation; see Section 42.1 and Figure 42.1) on the adjacent afferent arteriole, causing it to constrict. As a result, glomerular blood flow is reduced and glomerular filtration rate decreases to

normal. If the glomerular filtration decreases below normal, the specialized tubule cells detect the lower salt level that results, secretion of the chemical messengers from those cells stops and the afferent arteriole dilates, thereby allowing an increase in the glomerular filtration rate.

The Renin–Angiotensin–Aldosterone System (RAAS) Responds to Na$^+$ by Triggering Na$^+$ Reabsorption

Major changes in blood volume and pressure occur when the body loses or gains Na$^+$ in excessive amounts. Excessive Na$^+$ loss may result from prolonged and heavy sweating, repeated vomiting, severe diarrhea, or insufficient Na$^+$ uptake in the diet. The Na$^+$ loss reduces the osmolarity of body fluids, which causes less water to be reabsorbed in the kidneys. The water loss reduces the volume of blood and interstitial fluid and causes the blood pressure to drop. Excessive Na$^+$ intake in salty foods may have the opposite effects. The body must compensate for significant changes in Na$^+$.

The **renin–angiotensin–aldosterone system (RAAS)** is the most important hormonal system involved in regulating Na$^+$ **(Figure 48.9).** At normal body salt concentrations, the RAAS allows about 10 g of salt to be excreted in the urine each day. If excessive Na$^+$ is excreted, blood pressure and body fluid volume drop, and the glomerular filtration rate falls below levels that can be restored by the juxtaglomerular apparatus autoregulation. In response to this condition, the following events take place (see Figure 48.9):

- Cells in the juxtaglomerular apparatus secrete the enzyme **renin** into the bloodstream. (The RAAS also is activated to secrete renin when blood pressure or blood volume decreases independently of Na$^+$ levels, as in the case of a hemorrhage.)
- Renin cleaves the plasma protein *angiotensinogen* to produce **angiotensin I.**
- **Angiotensin-converting enzyme (ACE)** converts angiotensin I to **angiotensin II.**

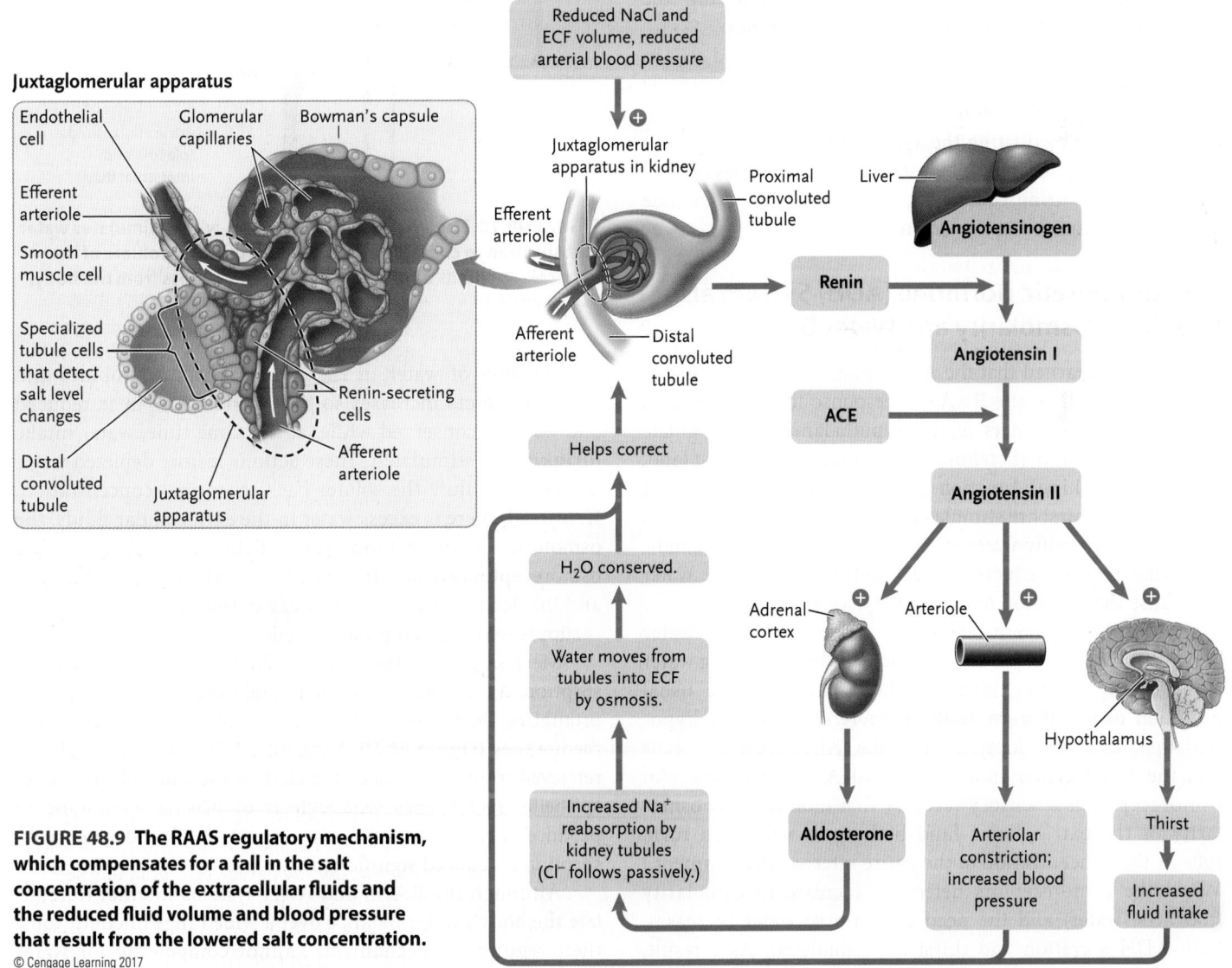

FIGURE 48.9 The RAAS regulatory mechanism, which compensates for a fall in the salt concentration of the extracellular fluids and the reduced fluid volume and blood pressure that result from the lowered salt concentration.

© Cengage Learning 2017

Angiotensin II has three effects: (1) it raises blood pressure quickly by constricting arterioles in most parts of the body; (2) it stimulates synthesis of the steroid hormone **aldosterone** and its secretion from the adrenal cortex; and (3) it stimulates thirst so that more water is brought into the body. The aldosterone increases Na^+ reabsorption in the kidneys, which raises the osmolarity of body fluids. As a result, water moves from the tubules into the interstitial fluid, which conserves water. Angiotensin II may also stimulate secretion of **antidiuretic hormone (ADH)** (also called *vasopressin*) by the posterior pituitary (antidiuretic means "against urine output") (not shown in the figure). ADH increases water absorption in the kidneys. Overall, the combined effects of angiotensin II act to return the blood pressure to normal.

In the opposite situation, when salt intake is too high, both body fluid volume and blood pressure rise above normal. Under these conditions, renin secretion is inhibited and, as a result, angiotensin II production and aldosterone synthesis are not stimulated. The reduction in angiotensin II lowers blood pressure by allowing arterioles to dilate; the reduction in aldosterone increases Na^+ loss in the urine by retarding the reabsorption of Na^+ and Cl^- from the kidney tubules.

Elevated blood pressure also stimulates specialized cells in the heart to release **atrial natriuretic factor (ANF),** a peptide hormone that inhibits renin release. ANF increases the filtration rate as well by dilating the arterioles that deliver blood to glomeruli and by inhibiting aldosterone release. As less Na^+ is reabsorbed and urine volume increases, both plasma volume and blood pressure return to normal.

The Antidiuretic Hormone (ADH) System also Regulates Osmolarity and Water Balance

You have just learned that the ADH system may be stimulated by angiotensin II of the RAAS in response to an increase in Na^+. The control centers of the hypothalamus that regulate ADH secretion (and, therefore, urinary output) and thirst (and, therefore, drinking) function together. That is, both ADH secretion and thirst are stimulated by a bodily water deficit and suppressed by bodily water excess. Logically, the same conditions that require reduced urinary output to conserve water also cause the sensation of thirst to replace water.

The ADH system regulates osmolarity and water balance—and, therefore, urinary output—by increasing water reabsorption in the kidneys without changing the usual excretion of salt **(Figure 48.10).** *Osmoreceptors* in the hypothalamus that are located near the ADH-secreting cells and the thirst center monitor control ADH secretion. (An **osmoreceptor** is a sensory receptor that monitors the osmolarity of the extracellular fluid bathing it, which, in turn, reflects the concentration of the entire internal environment.) When the osmoreceptors detect an increase in osmolarity (too little water) and the need to conserve water increases, both ADH secretion and thirst are stimulated. As a result,

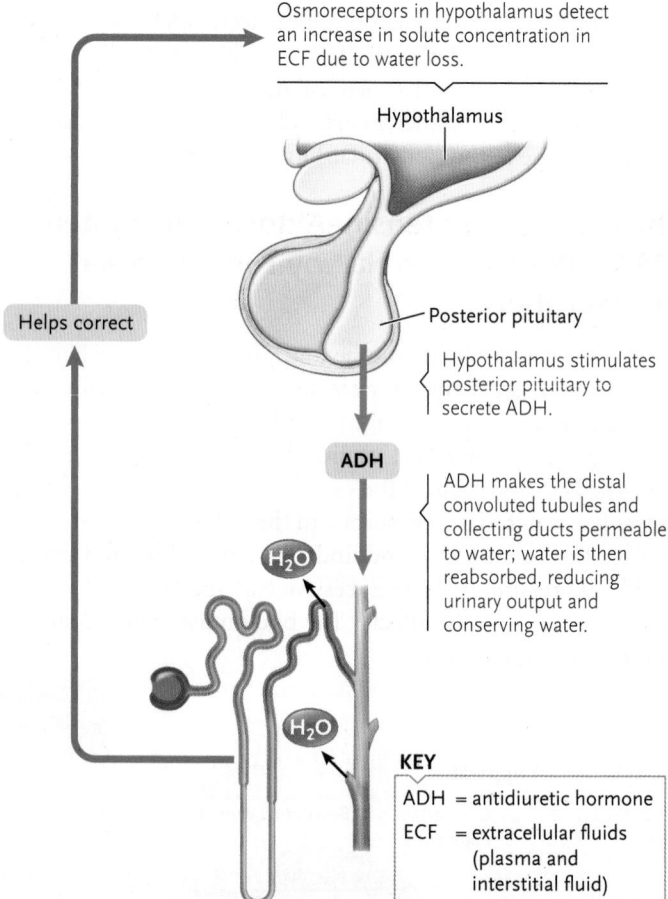

FIGURE 48.10 The ADH regulatory system, which stimulates water reabsorption to compensate for a loss in the fluid volume of the extracellular fluids because of excessive water loss from the body.
© Cengage Learning 2017

reabsorption of water in the distal convoluted tubules and collecting ducts increases so that urinary output is reduced and water is conserved while, at the same time, water intake (drinking) is stimulated. These actions restore depleted water stocks, and dilute the solutes back to normal concentration. If, instead, there is excess water in the extracellular fluids, the osmolarity of those fluids drops below normal levels. The osmoreceptors are not stimulated so ADH secretion is reduced and this leads to increased urinary output. In addition, thirst is suppressed, reducing water intake.

Mechanistically, the action of ADH increases water reabsorption in the distal convoluted tubule and collecting ducts by promoting the insertion of more aquaporins into the epithelial membranes **(Figure 48.11).** Without ADH, water channels are retrieved from the distal convoluted tubule and collecting duct epithelial membranes (endocytosis of plasma membrane to produce intracellular vesicles; see Figure 48.11), and water reabsorption is reduced significantly.

Although the RAAS and ADH systems interact to regulate the body's water balance over a wide range of conditions, their regulatory mechanisms cannot compensate for water losses for more than a few days if water is unavailable. Dehydration becomes fatal when water loss amounts to about 12% of the normal fluid volume of the body.

STUDY BREAK 48.4
Outline the roles of the RAAS and ADH system in regulating mammalian kidney function.

48.5 Kidney Function in Nonmammalian Vertebrates

Most nonmammalian vertebrates secrete hypoosmotic urine; only a few species of aquatic birds produce hyperosmotic urine. The particular evolutionary adaptations that maintain osmolarity and water balance among these animals vary depending on whether retention of water or salts is the major issue.

Marine Fishes Have Evolutionary Adaptations for Conserving Water and Excreting Salts

Marine teleosts live in seawater, which is strongly hyperosmotic to their body fluids. As a result, they lose water continually to their environment by osmosis and must replace it by drinking continually (**Figure 48.12A**). The kidneys of marine teleosts play little role in regulating salt in their body fluids because they cannot produce hyperosmotic urine that would both remove salt and conserve water. Instead, excess Na^+, K^+, and Cl^- ions are eliminated from the body by specialized cells in the gills, called *chloride cells*. Those cells transport Cl^- actively into the surrounding seawater; in addition, the Na^+ and K^+ ions are transported actively to maintain electrical neutrality. Certain other ions in the ingested seawater, such as Ca^{2+} and Mg^{2+}, are removed by the kidneys in an isoosmotic urine. Some salt is taken in in food and swallowed water, and some salt is gained by diffusion into gill tissue. On balance, a marine teleost is able to retain most of the water it drinks and eliminate most of the salt, allowing its body fluids to remain hypoosmotic to the surrounding water with no need to secrete hyperosmotic urine. Nitrogenous wastes are released from the gills, primarily as ammonia, by simple diffusion. The kidneys play little role in nitrogenous-waste removal.

Sharks and rays have a different adaptation to seawater. The osmolarity of their body fluids is maintained close to that of seawater by retaining high levels of urea in body fluids, along with another nitrogenous waste, *trimethylamine oxide*. The match in osmolarity keeps sharks and rays from losing water to the surrounding sea by osmosis, and they do not have to drink seawater continually to maintain their water balance. Excess salts ingested with food are excreted in the kidney and by specialized secretory cells in a *rectal salt gland* located near the anal opening.

Freshwater Fishes and Amphibians Have Evolutionary Adaptations for Excreting Water and Conserving Salts

The body fluids of freshwater fishes and aquatic amphibians (no amphibians live in seawater) are hyperosmotic to the surrounding water, which usually ranges from about 1 to 10 mOsm/L. Water therefore moves osmotically into their tissues (**Figure 48.12B**), as described earlier (see Section 48.2). Such animals rarely drink, and they excrete large volumes of dilute urine to get rid of excess water. In freshwater fishes, salt ions lost with the urine and by diffusion out of the gill tissue are replaced by salt in foods and by active transport of Na^+ and K^+ into the body by the gills; Cl^- follows to maintain electrical neutrality. Aquatic amphibians obtain salt in the diet and by active transport across the skin from the surrounding water. Nitrogenous wastes are excreted from the gills as ammonia in both freshwater fishes and aquatic amphibians.

Terrestrial amphibians must conserve both water and salt, which is obtained primarily in foods. In these animals, the kidneys secrete salt into the urine, causing water to enter the urine by osmosis. In the bladder, the salt is reclaimed by active transport and returned to body fluids. The water remains in the bladder, making the urine very dilute; during times of drought, it is reabsorbed as a water source. Terrestrial amphibians also have behavioral adaptations that help minimize water loss, such as seeking shaded, moist environments and remaining inactive during the day.

Larval amphibians, which are completely aquatic, excrete nitrogenous wastes from their gills as ammonia. Adult amphibians excrete nitrogenous wastes through their kidneys as urea.

Reptiles and Birds Have Evolutionary Adaptations for Excreting Uric Acid to Conserve Water

Terrestrial reptiles (lizards and snakes) conserve water by secreting nitrogenous wastes in the form of an almost water-free paste of uric acid crystals. Further water conservation occurs as the epithelial cells of the cloaca, the common exit for the digestive and excretory systems, absorb water from feces and urine before those wastes are excreted. Most birds conserve water by the same processes—they excrete nitrogenous wastes as uric acid and absorb water from the urine and feces in the cloaca. In reptiles, the scales covering the skin allow almost no water to escape through the body surface.

Reptiles and birds that live in or around seawater, including reptiles such as crocodilians, sea snakes, and sea turtles and birds such as seagulls, penguins, and pelicans, take in large quantities of salt with their food and rarely or never drink fresh water. These animals typically excrete excess salt through specialized *salt glands* located in the head, which remove salts from the blood by active transport. The salts are secreted to the environment as a water solution in which salts are two to three times more concentrated than in body fluids. The secretion exits through the nostrils of birds and lizards, through the mouth of marine snakes, and as salty tears from the eye sockets of sea turtles and crocodilians. Neural and hormonal controls, essentially the same as those regulating osmolarity in mammals, control the rate of fluid secretion and its salt concentration. **Figure 48.13** illustrates the salt glands of a seagull. The glands are located near the eyes and consist of blind-ended tubules lined with active salt-secreting cells. The cells do not carry out filtration or excretion. They are active only when the bird is dehydrated or overloaded with salt.

In sum, the evolutionary adaptations described in this section permit excretion of toxic wastes and allow animals to maintain the salt concentration of body fluids at levels that keep cells from swelling or shrinking.

STUDY BREAK 48.5

1. How do marine and freshwater teleosts differ in water, salt, and nitrogenous-waste regulation?
2. Reptiles and birds excrete nitrogenous wastes in the form of uric acid. Is there an advantage to doing this as opposed to the mammalian process of excreting nitrogenous wastes as urea?

48.6 Introduction to Thermoregulation

Another vital challenge for animals is maintaining their internal environment at temperatures that can be tolerated by body cells. Animal cells can survive only within a temperature range from about 0°C to 45°C (32°F–113°F). Not far below 0°C, the lipid bilayer of a biological membrane changes from a fluid to a frozen gel, which disrupts vital cell functions, and ice crystals destroy the cell's organelles. As temperatures approach 45°C, most proteins and nucleic acids unfold from their functional form. Either condition leads quickly to cell death. Animals therefore usually maintain internal body temperatures somewhere within the 0°C–45°C limits.

Temperature regulation—*thermoregulation*—is based on negative feedback pathways in which temperature receptors (*thermoreceptors*) detect changes from a temperature *set point*. Signals from the receptors trigger physiological and behavioral responses that return the temperature to the set point (thermoreceptors are discussed in Section 41.6; homeostatic control systems involving negative feedback mechanisms and set points are discussed in Section 38.4). All of the responses triggered by negative feedback mechanisms involve adjustments in the rate of heat-generating oxidative reactions within the body, coupled with adjustments in the rate of heat gain or loss at the body surface. The particular adaptations accomplishing these responses vary widely among species. And, although body temperature is regulated closely around a set point in all endotherms, the set point itself may vary over the course of a day and between seasons.

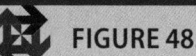 **FIGURE 48.11** | **Experimental Research**

ADH-Stimulated Water Reabsorption in the Kidney Collecting Duct

Question: How does ADH cause water reabsorption in the kidney collecting ducts?

Experiment: The peptide hormone ADH causes the reabsorption of water in the kidney by increasing the number of aquaporin2 (AQP2) water channels in the plasma membrane of the collecting duct cells. AQP2 is the only aquaporin expressed exclusively in the collecting duct. Mark Knepper and his colleagues at the National Institutes of Health, Bethesda, Maryland, and the University of Aarhus, Denmark, investigated the mechanism by which ADH causes AQP2 to appear in the plasma membrane. They tested the "shuttle" hypothesis, which states that ADH induces the movement of AQP2 from intracellular vesicles (IVs) to the plasma membrane of cells on the lumen side of the collecting duct. The researchers first attached sections of collecting duct from rat kidneys to a pipet and passed an ADH solution through the duct **(Figure A).** They then counted the AQP2 in the IVs and in the plasma membrane of the lumen cells. To do this, they cut treated ducts into thin sections and added an antibody that specifically recognizes AQP2. The antibody had gold particles attached to it. Under an electron microscope, the electron-dense gold is easily visualized. Counting the gold particles therefore quantified the AQP2.

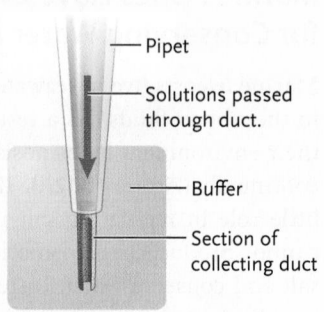

Pipet

Solutions passed through duct.

Buffer

Section of collecting duct

FIGURE A

Results: The researchers measured the distribution of AQP2 in the absence of ADH, in the presence of ADH, and then after ADH had been washed away. Their results are expressed as the ratio of AQP2 in the plasma membrane of lumen cells to AQP2 in IVs:

1. Before ADH treatment: 0.32 ± 0.03

2. In presence of ADH: 0.38 ± 0.14

3. After ADH removal: 0.35 ± 0.04

Conclusion: ADH induces a change in distribution of AQP2, consistent with the shuttle hypothesis of AQP2 channels moving from IVs to the plasma membrane of lumen cells. The redistribution of AQP2 channels is reversible, as shown by the results after ADH removal.

Figure B shows the current molecular model for the ADH-induced redistribution of AQP2 channels in collecting duct epithelial cells.

© Cengage Learning 2017

Thermoregulation Allows Animals to Reach Optimal Physiological Performance

Within the 0°C–45°C range of tolerable temperatures, an animal's *organismal performance*—the rate and efficiency of its

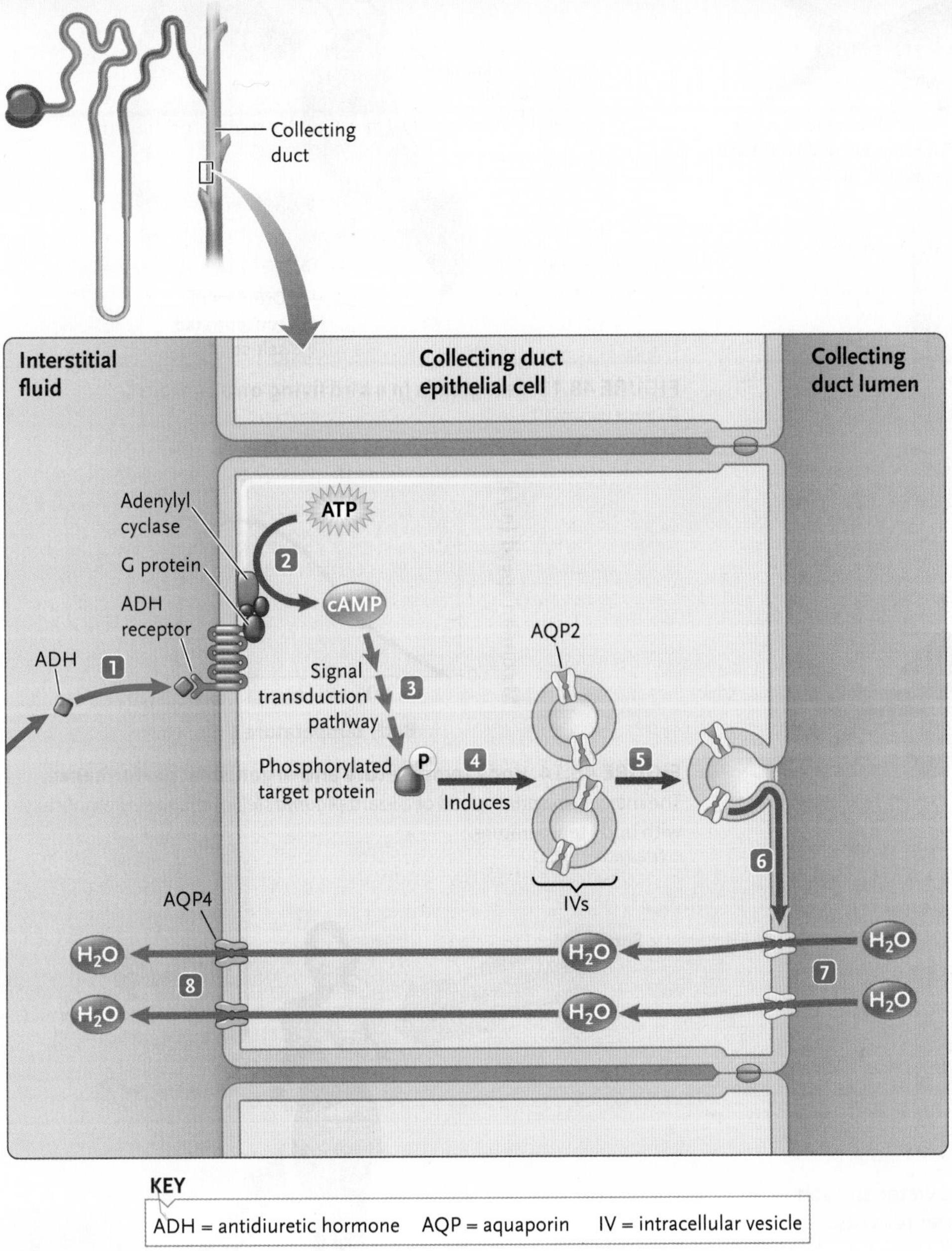

Collecting duct

Interstitial fluid

Collecting duct epithelial cell

Collecting duct lumen

Adenylyl cyclase

G protein

ADH receptor

ADH

ATP

2

cAMP

Signal transduction pathway

3

Phosphorylated target protein

P

4

Induces

AQP2

5

IVs

6

AQP4

H_2O

8

H_2O

H_2O

H_2O

7

H_2O

H_2O

1. ADH in the circulation passes through the capillary wall, diffuses through the interstitial fluid surrounding the collecting duct, and binds to the ADH G-protein–coupled receptor in the epithelial cell plasma membrane, activating it (see Section 9.3).

2. Activated receptor leads to activation of adenylyl cyclase, which produces cAMP from ATP.

3. cAMP triggers a signal transduction pathway (see Section 9.3) that leads to a phosphorylated target protein.

4. Phosphorylated target protein induces IVs with AQP2 channels in their membranes to move to plasma membrane of cells on lumen side of the collecting duct.

5. By exocytosis, the IVs fuse with the plasma membrane.

6. Completion of exocytosis results in AQP2 channels added to the plasma membrane.

7. Water from the collecting duct lumen is reabsorbed through the AQP2 channels into the epithelial cells.

8. Water exits the epithelial cells via AQP4 channels in the plasma membrane on interstitial fluid side of the cell.

KEY

ADH = antidiuretic hormone   AQP = aquaporin   IV = intracellular vesicle

FIGURE B

think like a scientist

Observe
Hypothesize
Predict
Experiment
Interpret

In diabetes insipidus, the kidneys are unable to conserve water. One heritable form of the disease, central diabetes insipidus, can be treated by desmopressin, a synthetic replacement for ADH. Another heritable form of the disease, nephrogenic diabetes insipidus, does not respond to treatment with desmopressin. Hypothesize which components shown in Figure B might have been affected by the mutations in the two forms of the disease.

Source: S. Nielsen et al. 1995. Vasopressin increases water permeability of kidney collecting duct by inducing translocation of aquaporin-CD water channels to plasma membrane. *Proceedings of the National Academy of Sciences USA* 92:1013–1017.

biochemical, physiological, and whole-body processes—varies greatly. For example, the speed at which the Middle Eastern lizard *Agama stellio* can sprint is low when the animal's body temperature is cold, rises smoothly with body temperature until it levels to a fairly broad plateau, and then drops off dramatically with further increases in body temperature **(Figure 48.14)**. The temperature range that provides good organismal performance varies from one species to another.

A. Marine teleosts

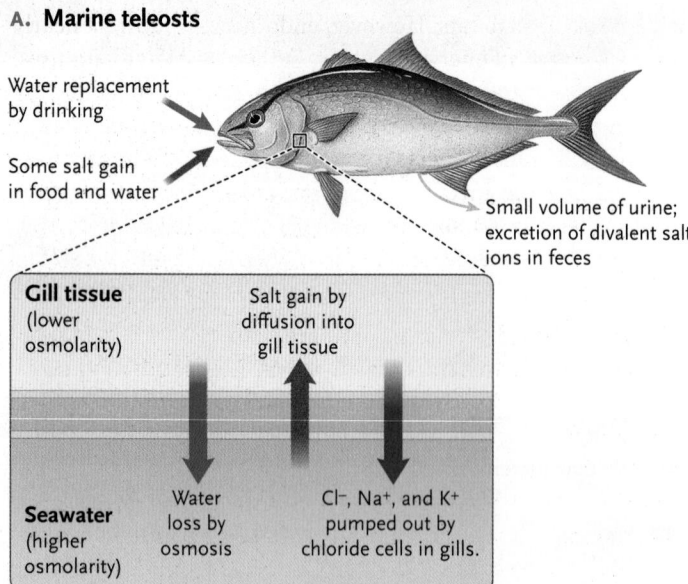

Water replacement by drinking

Some salt gain in food and water

Small volume of urine; excretion of divalent salt ions in feces

Gill tissue (lower osmolarity)

Salt gain by diffusion into gill tissue

Seawater (higher osmolarity)

Water loss by osmosis

Cl⁻, Na⁺, and K⁺ pumped out by chloride cells in gills.

B. Freshwater teleosts

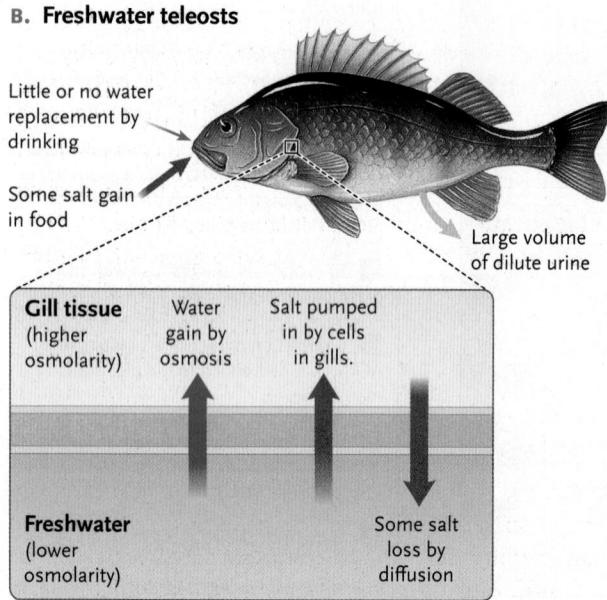

Little or no water replacement by drinking

Some salt gain in food

Large volume of dilute urine

Gill tissue (higher osmolarity)

Water gain by osmosis

Salt pumped in by cells in gills.

Freshwater (lower osmolarity)

Some salt loss by diffusion

FIGURE 48.12 The mechanisms balancing the water and salt content of (A) marine teleosts and (B) freshwater teleosts.
© Cengage Learning 2017

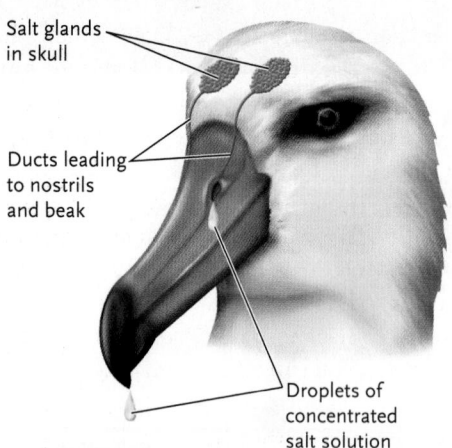

Salt glands in skull

Ducts leading to nostrils and beak

Droplets of concentrated salt solution

FIGURE 48.13 Salt glands in a bird living on a seacoast.
© Cengage Learning 2017

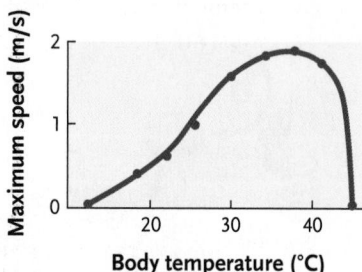

FIGURE 48.14 Body temperature and organismal performance. The maximum sprint speed of a lizard (*Agama stellio*) changes dramatically with body temperature.
© Cengage Learning 2017

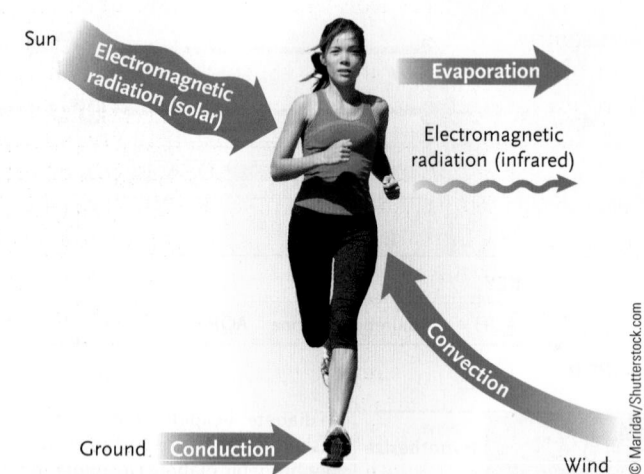

Sun

Electromagnetic radiation (solar)

Evaporation

Electromagnetic radiation (infrared)

Convection

Ground | Conduction

Wind

© Maridav/Shutterstock.com

FIGURE 48.15 Heat flow to (in red) and from (in blue) a runner on a hot, sunny day. Unlike conduction, convection, and evaporation, which take place through the kinetic movement of molecules, electromagnetic radiation is transmitted through space as waves of energy.
© Cengage Learning 2017

Animals Exchange Heat with Their Environments by Conduction, Convection, Radiation, and Evaporation

As part of thermoregulation, animals exchange heat with their environments. As with all physical objects, heat flows into animals if they are cooler than their surroundings and flows outward if they are warmer. This heat exchange occurs by four mechanisms: *conduction, convection, radiation,* and *evaporation* **(Figure 48.15)**.

Conduction is the flow of heat between atoms or molecules in direct contact. An animal loses heat by conduction when it contacts a cooler object and gains heat when it contacts an object that is warmer. **Convection** is the transfer of heat from a body to a fluid, such as air or water, that passes over its surface. The movement maximizes heat transfer by replacing fluid that has absorbed or released heat with fluid at the original temperature. **Radiation** is the transfer of heat energy as electromagnetic radiation. Any object warmer than absolute zero (−273°C) radiates heat; as the object's temperature rises, the amount of heat it loses as radiation increases as well. Animals also gain heat through radiation, particularly by absorbing radiation from the Sun. **Evaporation** is heat transfer through the energy required to change a liquid to a gas. Evaporation of water from a surface is an efficient way to transfer heat; when the water in sweat evaporates from the body surface, the body cools down because heat is being transferred to the evaporated water in the surrounding air.

All animals gain or lose heat by a combination of these four mechanisms. If you run on a hot summer day, for example, you lose heat by the evaporation of sweat from the skin, by convection as air flows over the skin, and by outward infrared radiation. You gain heat from internal biochemical reactions (especially oxidations), by absorbing infrared and solar radiation, and by conduction as the feet contact the hot ground. To maintain a constant body temperature, the heat gained and lost through these pathways must balance.

Ectothermic and Endothermic Animals Rely on Different Heat Sources to Maintain Body Temperature

Different animals use one of two major strategies to balance heat gain and loss. Animals that obtain heat primarily from the external environment are **ectotherms** (*ecto* = outside); those obtaining most of their heat from internal physiological sources are **endotherms** (*endo* = inside). All ectotherms generate at least some heat from internal reactions, however, and endotherms can obtain heat from the environment under some circumstances.

Virtually all invertebrates, fishes, amphibians, and lizards and snakes are ectotherms. These animals are described popularly as cold-blooded, although the body temperature of some, such as an active lizard, may be as high as or higher than ours on a sunny day. Ectotherms regulate body temperature by controlling the rate of heat exchange with the environment. Through behavioral and physiological mechanisms, they adjust body temperature toward a level that allows optimal physiological performance. However, most ectotherms are unable to maintain optimal body temperature when the temperature of their surroundings departs too far from that optimum, particularly when environmental temperatures fall. As a result, the body temperatures of ectotherms fluctuate with environmental temperatures, and they typically are less active when it is cold. Nevertheless, ectotherms are highly successful, particularly in warm environments.

The endotherms—birds, mammals, some fishes, sea turtles, and some invertebrates—keep their bodies at an optimal temperature by regulating two processes: (1) the amount of heat generated by internal oxidative reactions; and (2) the amount of heat exchanged with the environment. Because endotherms use internal heat sources to maintain body temperature at optimal levels, they can remain active over a broader range of environmental temperatures than ectotherms, and thus inhabit a

wider range of habitats. However, endotherms require a nearly constant supply of energy to maintain their body temperatures. Because that energy is provided by food, endotherms typically consume much more food than ectotherms of equivalent size.

The difference between ectotherms and endotherms is reflected in their metabolic responses to environmental temperature **(Figure 48.16)**. For example, the metabolic rate of a resting mouse *increases* steadily as the environmental temperature falls from 25°C to 10°C (77°F to 50°F). This increase reflects the fact that to maintain a constant body temperature in a colder environment, endotherms must process progressively more food and generate more heat to compensate for their increased rate of heat loss. In this respect, an endotherm can be likened to a house in winter. To maintain a constant internal temperature, the homeowner must burn more oil or gas on a cold day than on a warm day.

By contrast, the metabolic rate of a resting lizard typically *decreases* steadily over the same temperature range. Because ectotherms do not maintain a constant body temperature, their biochemical and physiological functions, including oxidative reactions, slow down as environmental and body temperatures decrease. Thus, an ectotherm consumes and uses less energy when it is cold than when it is warm. This difference between ectotherms and endotherms is so fundamental that even samples of living tissue extracted from an ectotherm consume energy more slowly than equivalent samples from an endotherm.

Ectothermy and endothermy represent different strategies for coping with the variations in environmental temperature that all animals encounter; neither strategy is inherently superior to

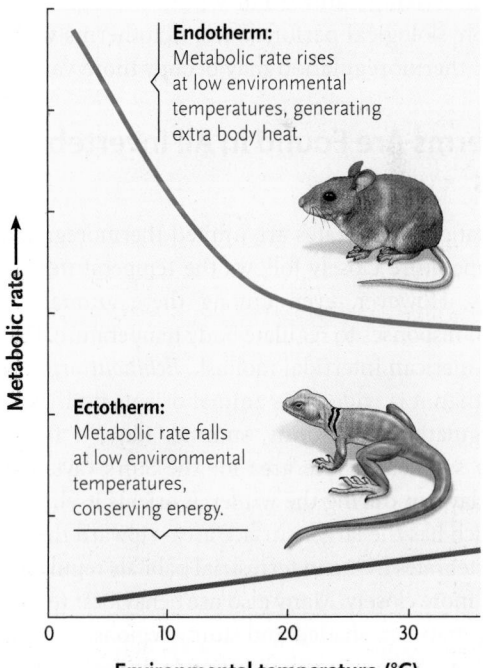

Endotherm: Metabolic rate rises at low environmental temperatures, generating extra body heat.

Ectotherm: Metabolic rate falls at low environmental temperatures, conserving energy.

Metabolic rate →

Environmental temperature (°C)

0   10   20   30

FIGURE 48.16 Metabolic responses of ectotherms and endotherms to cooling environmental temperatures. At any temperature, the metabolic rates of endotherms are always higher than those of ectotherms of comparable size.

© Cengage Learning 2017

the other. Endotherms can remain fully active over a wide temperature range. Cold weather does not prevent them from foraging, mating, or escaping from predators, but it does increase their energy and food needs—and, to satisfy their need for food, they may not have the option of staying curled up safely in a warm burrow. Ectotherms do not have the capacity to be active when environmental temperatures drop too low; they move sluggishly and are unable to capture food or escape from predators. However, because their metabolic rates are lower under such circumstances, so are their food needs, and they do not have to look for food and expose themselves to danger to the extent that endotherms do.

In the next two sections, we examine in more detail how ectothermic and endothermic animals regulate their body temperatures.

STUDY BREAK 48.6
Distinguish between ectothermy and endothermy. Give one advantage and one disadvantage for each form of thermoregulation.

48.7 Ectothermy

Ectotherms vary widely in their ability to regulate internal body temperatures. For example, most aquatic invertebrates have such limited ability for thermoregulation that their body temperatures closely match those of the surrounding environment. These species live in or seek warm or temperate environments, where temperatures fall within a range that produces optimal physiological performance. Ectotherms with a greater ability for thermoregulation may occupy more varied habitats.

Ectotherms Are Found in All Invertebrate Groups

Most aquatic invertebrates are limited thermoregulators whose body temperature closely follows the temperature of their surroundings. However, even among these animals, some use behavioral responses to regulate body temperature. For example, a South American intertidal mollusk, *Echinolittorina peruviana,* is longer than it is wide. This animal orients itself as a means of thermoregulation. On sunny, summer days, it faces the Sun, offering a smaller surface area for the Sun's rays. On overcast summer days, or during the winter, it orients itself with a lateral side—which has the larger surface area—toward the Sun's rays.

Invertebrates living in terrestrial habitats regulate body temperatures more closely. Many also use behavioral responses, such as moving between shaded and sunny regions, to regulate body temperature. Some winged arthropods, including bees, moths, butterflies, and dragonflies, use a combination of behavioral and heat-generating physiological mechanisms for thermoregulation. For example, in cool weather, these animals warm up before taking flight by vibrating the large flight muscles in the thorax, in a mechanism similar to shivering in humans.

Most Fishes, Amphibians, and Reptiles Are Ectotherms

Vertebrate ectotherms—fishes, amphibians, and reptiles (turtles, lizards, and snakes)—also vary widely in their ability for thermoregulation. Most aquatic species have a more limited thermoregulatory capacity than that found among terrestrial species, particularly the reptiles. Some fishes, however, are highly capable thermoregulators.

FISHES The body temperatures of most fishes remain within one or two degrees of their aquatic environment. However, many fishes use behavioral mechanisms to keep body temperatures at levels allowing good physiological performance. Freshwater species, for example, may use opportunities provided by the thermal stratification of lakes and ponds (see Figure 51.24). During hot summer days, they remain in deep, cool water, moving to the shallows to feed only during early morning and late evening when air and water temperatures fall.

AMPHIBIANS AND REPTILES The body temperatures of most amphibians also closely match environmental temperatures. Some terrestrial amphibians bask in the sun to raise their body temperature and seek shade to lower body temperature. However, basking can be dangerous to amphibians because they lose water rapidly through their permeable skin.

Thermoregulation is more pronounced among terrestrial reptiles. Some lizard species can maintain temperatures that are nearly as constant as those of endotherms. For small lizards, the most common behavioral thermoregulatory mechanism is shuttling between sunny (warmer) and shady (cooler) regions; in the deserts, lizards and other reptiles retreat into burrows during the hottest part of summer days. Some, such as the desert iguana *(Dipsosaurus dorsalis),* lose excess heat by *panting*—rapidly moving air in and out of the airways. The air movement increases heat loss by convection and by evaporation of water from the respiratory tract.

Lizards also frequently adjust their posture to foster heat exchange with the environment and control the angle of their body relative to the rays of the Sun. Snakes and lizards can often be found on large rocks and on roads on chilly nights, taking advantage of the heat retained by the stone or concrete.

Ectotherms Can Compensate for Seasonal Variations in Environmental Temperature

Many ectotherms undergo physiological changes, called **thermal acclimatization,** in response to seasonal shifts in environmental temperature. These changes allow the animals to attain good physiological performance at both winter and summer temperatures.

For example, in the summer a bullhead catfish (*Ameiurus* species) can survive water temperatures as high as 36°C (97°F), but it cannot tolerate temperatures below 8°C (46°F). In the winter, however, the bullhead cannot survive water temperatures above 28°C (82°F), but can tolerate temperatures near 0°C (32°F).

Another acclimatizing change involves the phospholipids of biological membranes (see *Focus on Research: Basic Research* in Chapter 5). For example, membrane phospholipids have higher proportions of double bonds in carp living in colder environments than in carp living in warmer environments. The higher proportion of double bonds makes it harder for the membrane to freeze. A higher proportion of cholesterol also protects membranes from freezing.

When seasonal temperatures fall below 0°C, some ectotherms add molecules to their body fluids that act as antifreeze molecules to depress their freezing point and retard ice crystal formation. For example, antifreeze proteins allow fishes such as the winter flounder to remain active in seawater as cold as −18°C (−9°F).

Ectotherms thus control body temperature primarily by regulating heat exchange with the environment. Internal-heat generating mechanisms contribute to the control mechanisms in some species, but are never the primary source of body heat. The opposite conditions occur among endotherms: although these animals also regulate heat exchange with the environment, their primary sources of body heat are internal.

STUDY BREAK 48.7

1. Describe two mechanisms an ectothermic animal can use to regulate its temperature.
2. What is thermal acclimatization?

48.8 Endothermy

Endotherms—mostly birds and mammals—have elaborate and extensive thermoregulatory adaptations. Highly specialized features of body structure interact with both physiological and behavioral mechanisms to keep the body temperature within a narrow range. Typically, the body temperatures of fully active individuals are held constant at levels between about 39°C and 42°C (102°F and 108°F) in birds, and 36°C and 39°C (97°F and 102°F) in mammals. These internal temperatures are maintained in the face of environmental temperatures that may range over much greater extremes, from as low as −42°C to as high as +48°C (−45°F to +120°F). Some highly specialized endotherms can even survive temperatures beyond these limits.

We begin by describing the basic homeostatic control systems that maintain body temperature, with primary emphasis on the human system. Later sections discuss variations in the responses of other mammals and of birds, and daily and seasonal variations in the temperature set point.

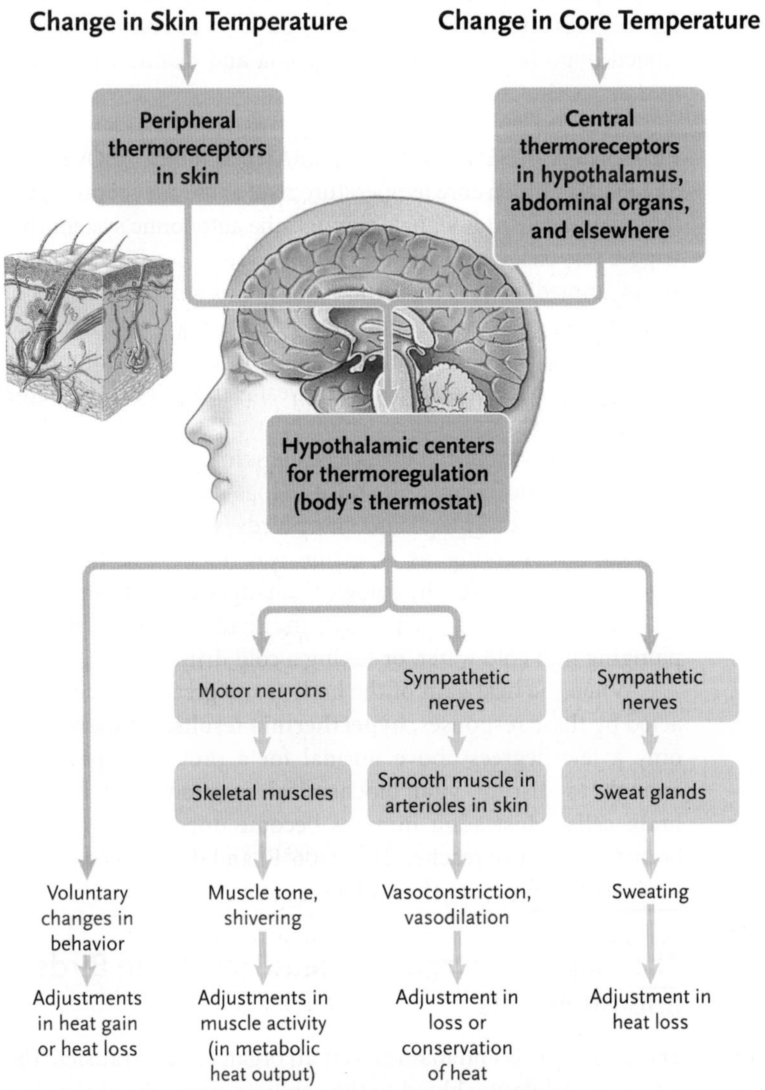

Information from Thermoreceptors Located in the Skin and Internal Structures Is Integrated in the Hypothalamus

Thermoreceptors (discussed in Section 41.6) are found in various locations in the human body, including the **integument** (skin; introduced in Chapter 38), spinal cord, and hypothalamus. Current evidence suggests the existence of five thermoreceptor types, each active in specific temperature ranges, beginning at less than 8°C and extending to above 52°C. Signals from the thermoreceptors are integrated in the hypothalamus and other regions of the brain to bring about compensating physiological and behavioral responses **(Figure 48.17)**. The responses keep body temperature close to the set point. In humans, the set point that is subject to homeostatic regulation is the **core temperature,** the temperature within the central core of the body consisting of the abdominal and thoracic organs, the CNS, and the skeletal muscles. The optimal

FIGURE 48.17 The physiological and behavioral responses of humans and other mammals to changes in skin and core temperature.
© Cengage Learning 2017

temperature for core tissue function is 37.8°C (100°F). Human body temperature is taken typically orally, where the normal temperature traditionally is given as 37°C (98.6°F). However, normal body temperature varies among individuals and by time of day for any given individual, giving a "normal" range of 35.5°C (96.0°F)–37.7°C (99.9°F) for oral measurements. Of course, temperature also varies with location. The outer shell of the body (that part surrounding the central core) is generally cooler than the core and can vary widely in temperature depending on environmental conditions and activity.

The hypothalamus was identified as a major thermoreceptor and response integrator in mammals by experiments in which various regions of the brain were heated or cooled with a temperature probe. Within the brain, only the hypothalamus produced thermoregulatory responses such as shivering or panting. The role of the hypothalamus in temperature regulation is also seen, for instance, when it raises the core temperature in response to an infection so as to produce a fever.

RESPONSES WHEN CORE TEMPERATURE FALLS BELOW THE SET POINT When thermoreceptors signal a fall in core temperature below the set point, the hypothalamus triggers compensating responses by sending signals through the autonomic nervous system. Among the immediate responses is constriction of the arterioles in the skin (vasoconstriction), which reduces the flow of blood to capillary networks in the skin. The reduced flow cuts down the amount of heat delivered to the skin and lost from the body surface. The reduction in flow is most pronounced in the skin covering the extremities, where blood flow may be reduced by as much as 99% when core temperature falls.

Another immediate response is contraction of the smooth muscles erecting the hair shafts in mammals and feather shafts in birds, which traps air in pockets over the skin, reducing convective heat loss. The response is minimally effective in humans because hair is sparse on most parts of the body—it produces the goose bumps you experience when the weather gets chilly. However, in mammals with fur coats or in birds, erection of the hair or feather shafts significantly increases the thickness of the insulating layer that covers the skin.

Immediate behavioral responses triggered by a reduction in skin temperature also help reduce heat loss from the body. Mammals may reduce heat loss by moving to a warmer locale, curling into a ball, or huddling together.

If these immediate responses do not return body temperature to the set point, the hypothalamus triggers further responses, most notably the rhythmic tremors of skeletal muscle we know as shivering. The heat released by the muscle contractions and the oxidative reactions powering them can raise the total heat production of the body substantially. At the same time, the hypothalamus triggers secretion of epinephrine (from the adrenal medulla) and thyroid hormones (see Section 42.4). These hormones increase heat production by stimulating the oxidation of fats and other fuels. The generation of heat by oxidative mechanisms in nonmuscle tissue throughout the body is termed **nonshivering thermogenesis.**

In human newborn babies and many other mammals, the most intense heat generation by nonshivering thermogenesis takes place in a specialized **brown adipose tissue** (also called **brown fat**) that can produce heat rapidly. Heat is generated by a mechanism that uncouples electron transport from ATP production in mitochondria (see Section 7.4); the heat is transferred throughout the body by the blood. Animals that hibernate or are active in cold regions, as well as the young of many others, contain brown adipose tissue. In most mammals, brown adipose tissue is concentrated between the shoulders in the back and around the neck. In human newborn babies, this tissue accounts for about 5% of body weight. Typically the tissue shrinks as humans age, until it is absent or essentially so in most adults. However, if exposure to cold is ongoing, as is the case with male Finlanders who work outside during the year, significant amounts of the tissue remain.

If none of these responses succeeds in raising body temperature to the set point, the result is **hypothermia,** a condition in which the core temperature falls below normal for a prolonged period. In humans, a drop in core temperature of only a few degrees affects brain function and leads to confusion; continued hypothermia can lead to coma and death (see *Why it matters . . .* in Chapter 38).

RESPONSES WHEN CORE TEMPERATURE RISES ABOVE THE SET POINT When core temperature rises above the set point, the hypothalamus sends signals through the autonomic system that trigger responses lowering body temperature (see Figure 38.10). As an immediate response, the signals relax smooth muscles of arterioles in the skin (vasodilation), increasing blood flow and with it, the heat lost from the body surface. In addition, in humans and other mammals with sweat glands, signals from the hypothalamus trigger the secretion of sweat, which absorbs heat as it evaporates from the surface of the skin.

Some endotherms, including dogs (which have sweat glands only on their feet) and many birds (which have no sweat glands), use panting as a major way to release heat (see Figure 38.10). These physiological changes are reinforced by behavioral responses such as seeking shade or a cool burrow, plunging into cold water, or taking a cold drink.

When the heat gain of the body is too great to be counteracted by these responses, **hyperthermia** results. An increase of only a few degrees above normal for a prolonged period is enough to disrupt vital biochemical reactions and damage brain cells. Most adult humans become unconscious if their body temperature reaches 41°C (106°F) and die if it goes above 43°C (110°F) for more than a few minutes.

The Skin Is an Organ of Heat Transfer in Birds and Mammals

The skin of birds and mammals is an organ of heat transfer. The arterioles delivering blood to the capillary networks of the skin constrict or dilate to control blood flow and, with it, the amount of heat transferred from the body core to the surface. The deliberate variation of skin temperature that occurs in this way helps maintain the constancy of the core temperature.

The outermost living tissue of human skin, the **epidermis,** consists of cells that grow and divide rapidly **(Figure 48.18),** becoming packed with fibers of a highly insoluble protein, *keratin* (see Section 4.3). When fully formed, the epidermal cells die and become compacted into a tough, nearly impermeable layer that limits water loss primarily to evaporation of the fluids secreted by the sweat glands.

The sweat glands and hair follicles are embedded in the layer below the epidermis. Called the **dermis,** it is packed with connective tissue fibers such as collagen, which resist compression, tearing, or puncture of the skin. The dermis also contains thermoreceptors and the dense networks of arterioles, capillaries, and venules that transfer heat between the skin and the environment.

The innermost layer of the skin, the **hypodermis,** contains larger blood vessels and additional reinforcing connective tissue. The hypodermis also contains an insulating layer of fatty tissue below the dermal capillary network, which ensures that heat flows between the body core and the surface primarily through the blood. The insulating layer is thickest in mammals that live in cold environments, such as whales, seals, walruses, and polar bears, in which it is known as *blubber.*

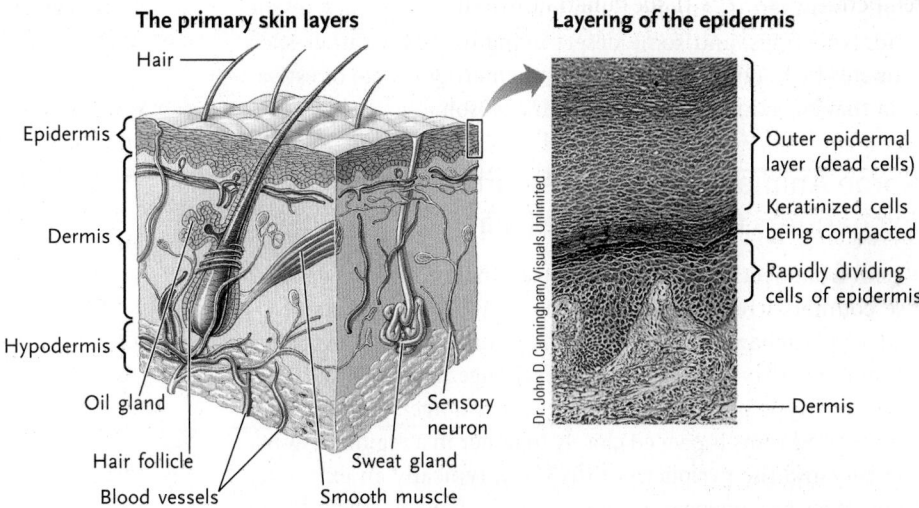

The primary skin layers

Hair
Epidermis
Dermis
Hypodermis
Oil gland
Hair follicle
Blood vessels
Sensory neuron
Sweat gland
Smooth muscle

Layering of the epidermis

Dr. John D. Cunningham/Visuals Unlimited

Outer epidermal layer (dead cells)
Keratinized cells being compacted
Rapidly dividing cells of epidermis
Dermis

FIGURE 48.18 The structure of human skin.
© Cengage Learning 2017

The Set Point Varies in Daily and Seasonal Rhythms in Many Birds and Mammals

The temperature set point in many birds and mammals varies in a regular cycle during the day. In some, the daily variations are relatively small and not obviously keyed to changes in environmental temperature. In others, larger variations are correlated with daily or seasonal temperature changes.

Camels undergo a daily variation of as much as 7°C (13°F) in set-point temperature. During the day, a camel's set point gradually resets upward, an adaptation that allows its body to absorb a large amount of heat. The heat absorption conserves water that would otherwise be lost by evaporation to keep the body at a lower set point. At night, when the desert is cooler, the thermostat resets again, allowing the body temperature to cool several degrees, releasing the excess heat absorbed during the day.

When the environmental temperature is cool, having a lowered temperature set point greatly reduces the energy required to maintain body temperature. In many animals, the lowered set point is accompanied by reductions in metabolic, nervous, and physical activity (including slower respiration and heartbeat), producing a sleeplike state known as **torpor.**

Many small animals and birds enter into **daily torpor,** a period of inactivity keyed to variations in daily temperature.

Typically these animals expend more energy per unit of body weight to keep warm than larger animals, because the ratio of body surface to volume increases as body size decreases. Hummingbirds, for example, feed actively during the daytime, when their set point is close to 40°C (104°F). During the cool of night, however, the set point drops to as low as 13°C (55°F); this allows the birds to conserve enough energy to survive overnight without feeding. Some nocturnal animals, including bats and small rodents such as deer mice, become torpid in cool locations during daylight hours when they do not actively feed. At night, their temperature set point rises and they become fully active.

Many animals enter a prolonged state of torpor tied to the seasons, triggered in most cases by a change in day length that signals the transition between summer and winter. The importance of day length has been shown by laboratory experiments in which animals have been induced to enter seasonal torpor by changing the period of artificial light to match the winter or summer day length.

Extended torpor during winter, called **hibernation** (*hibernus* = relating to winter), greatly reduces metabolic expenditures when food is unobtainable. Typically, hibernators must store large quantities of fats to serve as energy reserves. The drop in body temperature during hibernation varies with the mammal. In some, such as hedgehogs, woodchucks, and squirrels, body temperature may fall by 20°C (36°F) or more. Body temperature even drops to near 0°C in some small hibernating mammals and, in the Arctic ground squirrel, the body *supercools* (goes to a below-freezing, unfrozen state) during hibernation, with body temperature dropping to about −3°C. Some ectotherms, including amphibians and reptiles living in northern latitudes and even some insects, also become torpid during winter.

Some mammals enter seasonal torpor during summer, called **estivation** (*aestivus* = relating to summer), when environmental temperatures are high and water is scarce. Some ground squirrels, for example, remain inactive in the cooler temperatures of their burrows during extreme summer heat.

Many ectotherms, among them land snails, lungfishes, many toads and frogs, and some desert-living lizards, weather such climates by digging into the soil and entering a state of estivation that lasts throughout the hot dry season.

Some Animals Use Countercurrent Heat Exchangers to Retain Core Heat

Recall from Section 46.2 that many animals with internal gills use countercurrent exchange to maximize the amounts of O_2 and CO_2 exchanged with water. A countercurrent exchange system can also be used for heat exchange. **Figure 48.19** illustrates one type of *countercurrent heat exchanger*. Here, a set of arteries and veins is packed closely together in a region between the core and the periphery of the body, typically an extremity (one artery and one vein are shown in the figure). Warm blood from the core flows through arteries toward cold peripheral tissues and organs. In the countercurrent heat exchanger, that warm arterial blood flows next to cold blood from the body periphery that is in the veins. By conduction, heat moves from the artery to the vein and the now-warmer venous blood continues to the core.

Countercurrent heat exchanges of this type are found in many endotherms. In birds they are often in the legs, thereby limiting heat loss from the feet. In mammals, they are also often in the limbs; for example, in a dolphin's fluke and flippers, and in the legs of an arctic wolf.

Another type of countercurrent exchange that helps to limit heat loss takes place in the nasal passages of birds and mammals. The air in the lungs of these animals is at 100% relative humidity because the cells of the lung capillaries must be moist to allow oxygen breathed in to pass into those cells and on into the pulmonary veins. The air expelled from the lungs therefore has the potential to cause significant water loss from the body. And, because the lung air is warm, there is the potential also for significant heat loss. An adaptation to limit water and heat loss is folds in the nasal cavity. These folds can be elaborate and represent a relatively large surface area. In outgoing breaths, some of the water in the air condenses on the nasal folds because the surfaces are cooler than those of the lung interior. Inhaled air is drier than the air leaving the lungs. The

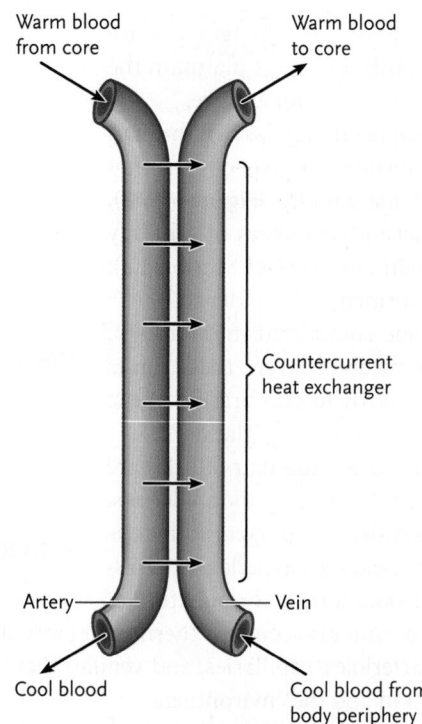

FIGURE 48.19 A countercurrent heat exchanger for conserving core heat.
© Cengage Learning 2017

inhaled air evaporates some of the water on the nasal folds, which cools the folds for the next outgoing breath. This countercurrent exchange system involves opposing fluids separated by time rather than space as in the system discussed previously. This system is important for thermoregulation because it limits the loss of heat in expelled air. By this mechanism, mammals on average reclaim up to 45% of the water in expelled air, along with its heat content. Kangaroo rats reclaim up to 88% of their body water in this fashion.

STUDY BREAK 48.8

Describe how thermoreceptors and negative feedback pathways achieve temperature regulation in endotherms.

 Unanswered Questions

How can we use genetics to understand how the kidney works in humans?

A variety of disorders of kidney function are inherited as Mendelian traits. For example, the disorder of renal growth known as polycystic kidney disease (PKD; see *Molecular Insights* in this chapter) was recognized as a familial phenotype well before the structure of DNA was defined. Many other complex and common renal phenotypes (for example, blood pressure and susceptibility to the development of decreased kidney filtration function) also have genetic components.

The use of molecular cloning methods in the 1980s and 1990s led to enormous growth in our understanding of the specific transport proteins used by the kidney to fine-tune the extracellular environment. These discoveries were quickly followed by the elucidation of a large number of rare inherited disorders caused by mutations in the genes encoding many of these proteins. Careful observation of these human phenotypes, coupled with a biochemical understanding of the effects of these mutations, has in turn led to an improved understanding of basic renal physiology.

In the urine produced by a healthy kidney, protein is normally present only in very small amounts; little protein gets through the glomerular filtration barrier, and most of what does get through is reabsorbed in the renal tubule. By contrast, a large fraction of human kidney diseases are characterized by protein in the urine, a condition called proteinuria. Diabetic kidney disease, for example, the most common form of kidney disease in the Western world, is characterized by increased urine protein and the abnormal accumulation of extracellular protein in the renal glomerulus, leading to a gradual decline in kidney function.

For the past decade, most of my own laboratory's research has focused on inherited diseases of the renal glomerulus, using methods of molecular cloning (see Section 18.1) to identify disease genes. Mutations in the actin-regulatory proteins α-actinin-4 and INF2 both cause late onset proteinuria and slowly progressive kidney dysfunction. Based on our studies and the research of other labs, we believe that fine regulation of the actin cytoskeleton is critical to maintaining the complex architecture of glomerular podocytes, a specialized epithelial cell that forms the final part of the glomerular filter. When this architecture is disturbed, the podocytes become more easily injured.

Large amounts of protein (on the order of 3 g/day) are the central feature of a serious medical condition known as the nephrotic syndrome. By studying very early onset forms of the nephrotic syndrome, genes for proteins that are unique to the podocyte cell–cell junction have been shown to be mutated in these disorders. For example, the cell surface protein nephrin has been shown to play important structural and signaling functions. Despite the growing list of genes known to be important to normal filtering of the blood, much work needs to be done before we

can answer the question of how the products of these genes interact to produce a functioning kidney.

Humans have many disadvantages as experimental organisms. We cannot do invasive experiments on them, we cannot do planned matings, and they take a long time to grow. Nonetheless, the careful study of renal phenotypes in humans followed by genetic analysis has led to a much better understanding of kidney disease and kidney function. It is important both for medicine and for understanding the kidney to use the information gained through genetic studies to inform biology.

think like a scientist People of recent African ancestry develop kidney disease at much greater rates than people of other origins. This is largely the result of two coding sequence variants in the *APOL1* gene. Individuals harboring two of these variant copies (one on the maternal allele and one on the paternal allele) are at higher risk of a variety of types of kidney disease. These variants are not seen in people without recent African ancestry. What are possible explanations for such differences in allele frequencies between populations? Why would a deleterious allele persist in the human population?

Martin Pollak is Chief of Nephrology at the Beth Israel Deaconess Medical Center in Boston and a Professor of Medicine at Harvard Medical School. The focus of his laboratory is to learn more about the causes of kidney disease in patients and families by studying genetics. To learn more, visit https://sites.google.com/site/pollakfsgs/home.

REVIEW KEY CONCEPTS

For access to MindTap and additional study materials visit www.cengagebrain.com.

48.1 Introduction to Osmoregulation and Excretion

- Solute concentration is measured as osmolarity in milliosmoles per liter of solution (mOsm/L). A solution can be comparatively hyperosmotic, hypoosmotic, or isoosmotic to another solution. Movement of water from a region of higher osmolarity to a region of lower osmolarity across a selectively permeable membrane occurs by osmosis.

- Osmoconformers allow the osmolarity of their body fluids to match that of the environment. Osmoregulators keep the osmolarity of body fluids different from that of the environment.

- Molecules and ions must be removed from the body to keep cellular and extracellular fluids isoosmotic. In most animals, ECF is filtered through tubules formed from a transport epithelium and released to the exterior of the animal as urine (Figure 48.1).

- Nitrogenous wastes are excreted as ammonia, urea, or uric acid, or as a combination of these substances (Figure 48.2).

48.2 Osmoregulation and Excretion in Invertebrates

- Most marine invertebrates are osmoconformers; they expend little or no energy on maintaining water balance. Freshwater and terrestrial invertebrates are osmoregulators. They must expend energy to excrete water that moves into their cells by osmosis.

- The cells of the simplest marine invertebrates exchange water and solutes directly with the surrounding seawater. More complex invertebrates have specialized excretory tubules (Figures 48.3–48.5).

48.3 Osmoregulation and Excretion in Mammals

- In mammals and other vertebrates, excretory tubules are concentrated in the kidney.

- The mammalian excretory tubule, the nephron, has a proximal end at which filtration takes place, a middle region in which reabsorption and secretion occur, and a distal end that releases urine. A network of capillaries surrounding the nephron takes up ions and water and other molecules absorbed by the nephron. The urine leaving individual nephrons is processed further in collecting ducts and then pools in the renal pelvis. From there it flows through the ureter to the urinary bladder, and through the urethra to the exterior of the animal (Figure 48.6).

- Glomerular filtration, a process in which plasma of the blood filters through the glomerular capillaries into Bowman's capsule, is the first step of excretion. By the process of tubular reabsorption in the proximal convoluted tubule and loop of Henle, substances of value (Na^+, Cl^-, K^+, water, HCO_3^-, and nutrients) are returned to the plasma as the filtrate flows through the nephron tubules. By the process of tubular secretion, selective movement of substances (such as detoxified poisons) from the blood of the peritubular capillaries takes place. Subsequently, the concentration of salts is balanced between the urine and the interstitial fluid surrounding the nephron, and more water is reabsorbed (Figure 48.7 and Table 48.1).

48.4 Regulation of Mammalian Kidney Function

- The kidney's autoregulation system responds rapidly to small variations in blood pressure so as to keep the glomerular filtration rate constant. The autoregulation system involves: (1) vascular control of blood pressure; and (2) a feedback control mechanism between the tubule and specialized cells in the juxtaglomerular apparatus (Figure 48.9).

- When blood volume and blood pressure drop, the hormones of the renin–angiotensin–aldosterone system (RAAS) raise blood pressure by stimulating arteriole constriction and increasing NaCl reabsorption in the kidneys (Figure 48.9).
- ADH, which increases water reabsorption, is released from the pituitary when osmoreceptors detect an increase in the osmolarity of body fluids (Figures 48.10 and 48.11).

48.5 Kidney Function in Nonmammalian Vertebrates

- Marine teleosts continually drink seawater to replace body water lost by osmosis to their hyperosmotic environment. Excess salts and nitrogenous wastes are excreted by the gills (Figure 48.12A).
- The body fluids of sharks and rays are isoosmotic with seawater. They do not lose water by osmosis, and do not drink seawater. Excess salts are excreted in the kidney and by a rectal salt gland.
- Body fluids of freshwater fishes and amphibians are hyperosmotic to their environment, and these animals must excrete the excess water that enters by osmosis. Body salts are obtained from food and, in fishes, through the gills (Figure 48.12B). Nitrogenous wastes are excreted from the gills of fishes and larval amphibians as ammonia and through the kidneys of adult amphibians as urea.
- Reptiles and birds conserve water by secreting nitrogenous wastes as uric acid and by absorbing water from urine and feces in the cloaca.

48.6 Introduction to Thermoregulation

- Animals must maintain body temperature at a level that provides optimal physiological performance. Heat flows between animals and their environment by conduction, convection, radiation, and evaporation (Figures 48.14 and 48.15).
- Ectothermic animals obtain heat energy primarily from the environment; endothermic animals obtain heat energy primarily from internal reactions (Figure 48.16).

48.7 Ectothermy

- Ectotherms obtain heat energy externally and control body temperature primarily by physiological or behavioral methods of regulating heat exchange with the environment.
- Many animals undergo thermal acclimatization, a structural or metabolic change in the limits of tolerable temperatures as the environment alternates between warm and cool seasons.

48.8 Endothermy

- Endotherms obtain heat energy primarily from internal reactions and maintain body temperature over a narrow range by balancing internal heat production against heat loss from the body surface.
- Internal heat production is controlled by negative feedback pathways triggered by thermoreceptors. When deviations from the temperature set point occur, signals from the receptors bring about compensating responses such as changes in blood flow to the body surface, sweating or panting, and behavioral modifications (Figure 48.17).
- The skin of endotherms is water-impermeable, reducing heat lost by direct evaporation of body fluids. The blood vessels of the skin regulate heat loss by constricting or dilating. A layer of insulating fatty tissue under the vessels limits losses to the heat carried by the blood. The hair of mammals and feathers of birds also insulate the skin. Erection of the hair or feathers reduces heat loss by thickening the insulating layer (Figure 48.18).
- The temperature set point in many birds and mammals varies in daily and seasonal patterns. During cooler conditions, a lowered set point is accompanied by torpor.
- Some animals use countercurrent heat exchangers to retain core heat. One form of exchanger is a set of arteries and veins arrayed close together in a region between the core and the body periphery. The opposing directions of fluid flow in the vessels allows heat to move from the warm blood in the arteries to warm up the cold blood in the veins which is heading toward the core. Another form of exchanger involves the nasal folds in birds and mammals. The folds provide an exchange mechanism to conserve body water and its heat content (Figure 48.19).

TEST YOUR KNOWLEDGE

Remember/Understand

1. Which of the following statements about osmoregulation is true?
   a. In freshwater invertebrates, salts move out of the body into the water because the animal is hypoosmotic to the water.
   b. A marine teleost has to fight gaining water because it is isoosmotic to the sea.
   c. Most land animals are osmoconformers.
   d. Vertebrates are usually osmoregulators.
   e. Terrestrial animals can regulate their osmolarity without expending energy.

2. One role of tubules in excretion is to:
   a. absorb H^+ ions to buffer body fluids.
   b. transport proteins across the transport epithelium.
   c. reabsorb glucose and amino acids.
   d. move toxic substances from the filtrate into the cells composing the transport tubules.
   e. filter by maintaining a lower pressure in the fluid outside the tubule than inside it.

3. Products of metabolism in humans, as in:
   a. terrestrial amphibians, can include urea, which requires more energy to produce than ammonia.
   b. birds and reptiles, can include uric acid, which is nontoxic and excreted as a paste.
   c. sharks, are primarily excreted as ammonia.
   d. hydra, must be isoosmotic with the water ingested.
   e. other mammals, cannot include water as water comes only from what they drink.

4. Filtration and/or excretion can be performed by:
   a. ciliated metanephridia in insects.
   b. protonephridia containing flame cells in flatworms.
   c. a nephron and bladder in insects.
   d. Malpighian tubules on the segments of earthworms.
   e. the hindgut, which reabsorbs Na^+ and K^+ into the hemolymph of earthworms.

5. A mammalian nephron contains the:
   a. Bowman's capsule, which delivers the filtrate to the glomerulus.
   b. Bowman's capsule, which filters fluids, 99% of which will be excreted.
   c. proximal convoluted tubule, which moves Na^+ and K^+ into the filtrate of the interstitial fluids.
   d. proximal convoluted tubule, which reabsorbs K^+, Na^+, Cl^-, and H_2O.
   e. proximal convoluted tubule, which lacks microvilli to ease fluid movement through it.

6. Which of the following correctly describes a part of kidney function?
   a. Collecting ducts dilute urine because they are permeable to salt but not water.
   b. In the ascending loop of Henle, Na^+ and Cl^- move into the tubules because the osmolarity of the filtrate is increased.
   c. The descending loop of Henle receives filtrate from the ascending loop.
   d. The distal convoluted tubule pumps water into the tubule by active transport.
   e. The renal pelvis receives urine from the collecting ducts and carries it to the ureters.

7. Which of the following is an example of autoregulation of kidney function?
   a. The RAAS regulates Na^+ by secreting renin when blood pressure or blood volume decreases.
   b. The ADH system regulates water balance by decreasing water reabsorption and increasing excretion of salt.
   c. Receptors in the juxtaglomerular apparatus detect a higher salt concentration in the distal convoluted tubule and trigger constriction of the afferent arteriole to reduce glomerular filtration rate.
   d. ANF is released by the kidney to increase renin release.
   e. Angiotensin II lowers blood pressure by constricting arterioles.

8. Deficient water levels in humans are prevented by:
   a. osmoreceptors in the hypothalamus that detect decreases in salt concentrations.
   b. the hypothalamus stimulating the posterior pituitary to secrete a hormone that allows the collecting ducts and distal convoluted tubules to be permeable to water.
   c. inhibiting ADH, which causes a rise in osmolarity of the ECF.
   d. producing dilute urine.
   e. drinking alcohol, which stimulates aldosterone to raise the osmolarity of body fluids.

9. Which best exemplifies ectothermy?
   a. The metabolic rate increases as the temperature decreases.
   b. Body temperature remains constant when environmental temperatures change.
   c. Food demand increases when temperatures drop.
   d. Virtually all invertebrate groups are ectotherms.
   e. No vertebrate groups are ectotherms.

10. Unique to endotherms is:
   a. torpor.
   b. thermal acclimatization.
   c. a nonchanging body temperature.
   d. response to seasonal temperature changes.
   e. thermoregulation by a hypothalamus.

Apply/Analyze

11. **Discuss Concepts** A urinalysis reveals glucose, urea, hemoglobin, and sodium. Which of these substances are abnormal in urine, and why?

12. **Discuss Concepts** As a person ages, nephron tubules lose some of their ability to concentrate urine. What is the effect of this change?

13. **Discuss Concepts** Shivering increases air movement over the body surface. What effect does this air movement have on heat conservation in the shivering animal?

14. **Discuss Concepts** What heat transfer processes might account for the change in body temperature when a mammal's body temperature undergoes daily variations?

Evaluate/Create

15. **Design an Experiment** Design experiments to show the role of fluid consumption in thermoregulation during endurance exercise.

16. **Apply Evolutionary Thinking** Humans produce urea as an excretion product, whereas reptiles and birds produce uric acid. Indeed, human kidneys are not as efficient as those of reptiles and birds. What does this mean in an evolutionary sense?

For selected answers, see Appendix A.

INTERPRET THE DATA

Products labeled as "organic" fill an increasing amount of space on supermarket shelves. What does that label mean? A food that carries the USDA's organic label must be produced without pesticides such as malathion and chlorpyrifos that conventional farmers typically use on fruits, vegetables, and grains.

Does eating organic food significantly affect the level of pesticide residues in a child's body? Chensheng Lu of Emory University used urine testing to find out. For 15 days, the urine of 23 children (ages 3–11) was monitored for breakdown products of pesticides. During the first 5 days (phase 1), children ate their standard, nonorganic diet. For the next 5 days (phase 2), they ate organic versions of the same types of foods and drinks. Then, for the final 5 days (phase 3), the children returned to their nonorganic diet. The results are shown in the **Table.**

1. During which phase of the experiment did the children's urine contain the lowest level of the malathion metabolite?

2. During which phase of the experiment was the maximum level of the chlorpyrifos metabolite detected?

3. Did switching to an organic diet lower the amount of pesticide excreted by children?

Source: C. Lu et al. 2006. Organic diets significantly lower children's dietary exposure to organophosphorus pesticides. *Environmental Health Perspectives* 114:260–263.

Levels of Metabolites (Breakdown Products) of Malathion and Chlorpyrifos in the Urine of Children Taking Part in a Study of Effects of an Organic Diet*

Study Phase	No. of Samples	Malathion Metabolite		Chlorpyrifos Metabolite	
		Mean (µg/L)	Maximum (µg/L)	Mean (µg/L)	Maximum (µg/L)
1. Nonorganic	87	2.9	96.5	7.2	31.1
2. Organic	116	0.3	7.4	1.7	17.1
3. Nonorganic	156	4.4	263.1	5.8	25.3

*The difference in the mean level of metabolites in the organic and inorganic phases of the study was statistically significant.

49 Animal Reproduction

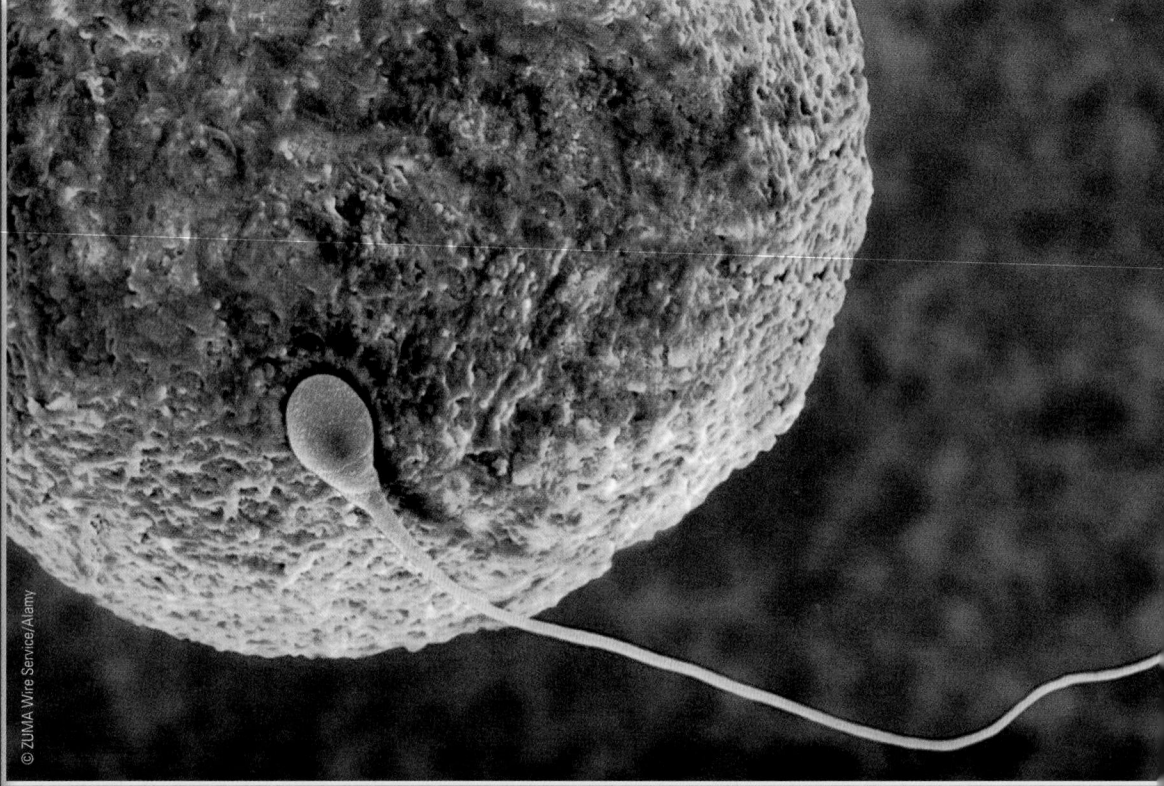

Fertilization, the union of gametes such as the sperm and egg shown here, initiates development of a new individual.

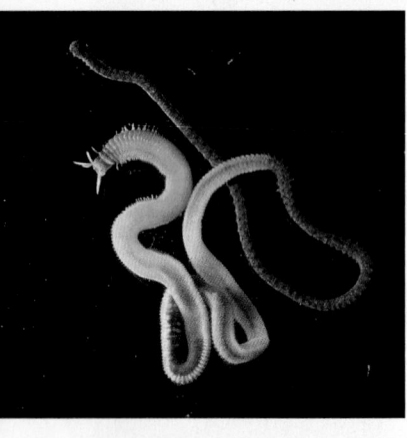

FIGURE 49.1 The palolo worm. Gametes are packed into segments of the tail section (in blue).

Why it matters . . . It is seven days after the October full moon and night is falling. Inhabitants of the Samoan island of Tutuila are in boats on the island's large lagoon. They are awaiting the palolo worm *(Eunice viridis)* **(Figure 49.1),** which has appeared in the water as the moon rises on this same night of the lunar year for as long as the islanders can remember.

The moon rises and there they are, thousands of worms, squirming in the water like animated spaghetti. The boaters scoop up the worms by the netful and dump them into buckets. When the buckets are full, the islanders return to shore where steaming pots are waiting, for palolo worms are a delicacy that are savored only once a year.

The worms that come to the surface are not complete individuals. They are tail sections about 10 to 20 cm long that break from adults after they become filled with eggs or sperm (see Figure 49.1). The adults are polychaete annelids (see Section 31.6) that live in burrows in coral reefs of the Samoan and Fiji islands. These annelids develop tail segments once a year, just after the October full moon. On the seventh night following the full moon, the tails break off and swim to the surface, where—if they are not netted—they disintegrate and release eggs and sperm by the millions. The anterior ends of the worms, safe in their burrows, survive to produce tails the next year.

The story of the palolo worm concerns the *reproductive cycle* of an animal. There is a biological clock in the worms, timed by periods of moonlight, that sets the appearance of the mating swarm precisely. Most animals show cycles in reproductive activity, and those cycles often are linked to the seasons.

The swarm of the palolo worms is only one of many adaptations that accomplish mating in animals. For animals that reproduce by eggs and sperm, the adaptations are as diverse as the number of species on Earth. This diversity allows individuals of the same species to find each other and unite eggs and sperm. Within the diversity, however, are underlying patterns that are shared by all animals.

Both the underlying patterns and the diversity of animal reproduction are the subjects of this chapter. We also discuss the development of eggs and sperm, and the union of egg and sperm that begins the development of a new individual. The next chapter continues with the events of development after eggs and sperm have united.

49.1 Animal Reproductive Modes: Asexual and Sexual Reproduction

All organisms have a life cycle in which they grow, develop, and reproduce; each life cycle is carried out according to instructions encoded in DNA. Rather than survival of the individual, reproduction is the means of passing on the genes of an individual to new generations of the species. As such, it is among the most vital functions of living organisms.

Two basic modes of reproduction operate in the animal kingdom. In **asexual reproduction,** a single individual gives rise to offspring without fusion of **gametes** (egg and sperm); that is, there is no genetic input from another individual. In **sexual reproduction,** male and female parents produce offspring through the union of egg and sperm generated by meiosis (meiosis is discussed in Chapter 11).

Asexual Reproduction Produces Offspring with Genes from Only One Individual

Many aquatic invertebrates and some terrestrial annelids and insects reproduce asexually. Asexual reproduction is rare among vertebrates. In asexual reproduction, one or many cells of a parent develop directly into a new individual. In a few animals that undergo asexual reproduction, the cells taking part are genetically varied products of meiosis, but in most they are the products of mitosis. The offspring therefore are genetically identical to one another and to the parent. In other words, they are genetic clones of the parent. For this reason, asexual reproduction of this kind is also called **clonal reproduction.**

Genetic uniformity of offspring can be advantageous in environments that remain stable and uniform. Asexual reproduction tends to preserve gene combinations, producing individuals that are successful in such environments. Further, individuals do not have to expend energy to produce gametes or find a mate. Asexual reproduction can also bring reproductive advantages to individuals living in sparsely populated areas, or to sessile animals, which cannot move from place to place.

Asexual reproduction involving mitosis occurs in animals by three basic mechanisms: *fission, budding,* and *fragmentation.*

- **Fission.** In **fission,** the parent separates into two or more offspring of approximately equal size. Planarians (flatworms), for instance, reproduce asexually by fission;

depending on the species, they may divide by transverse or longitudinal fission.

- **Budding.** In **budding,** a new individual grows and develops while attached to the parent. Sponges, tunicates, and some cnidarians reproduce asexually by this mechanism. The offspring may break free from the parent, or remain attached to form a *colony.* In the cnidarian *Hydra,* for example, an offspring buds and grows from one side of the parent's body and then detaches to become a separate individual **(Figure 49.2).**

- **Fragmentation.** In **fragmentation,** pieces separate from the body of a parent and develop *(regenerate)* into new individuals. Many species of cnidarians, flatworms, annelids, and some echinoderms can reproduce by fragmentation.

Some animals produce offspring by the growth and development of an egg without fertilization. The offspring may be haploid or diploid depending on the species. This form of asexual reproduction is called **parthenogenesis** (*parthenos* = virgin; *genesis* = origin). Because the egg from which a parthenogenetic offspring is produced derives from meiosis in the female parent, the offspring are not genetically identical to the parent or to each other.

Parthenogenesis occurs in some invertebrates, including certain aphids, bees, and water fleas. In water fleas, for instance, the female can produce two types of eggs. One type of egg develops only if it is fertilized, whereas the other develops by parthenogenesis. The conditions set the switch for which form of reproduction occurs: sexual reproduction occurs when there is environmental stress, whereas asexual reproduction occurs under favorable environmental conditions. The two types of environmental conditions, and, hence, the different reproductive cycles, tend to correlate with the seasons. In bees, haploid male drones are produced parthenogenetically from unfertilized eggs laid by reproductive females (queens) while new queens and sterile workers develop from fertilized eggs.

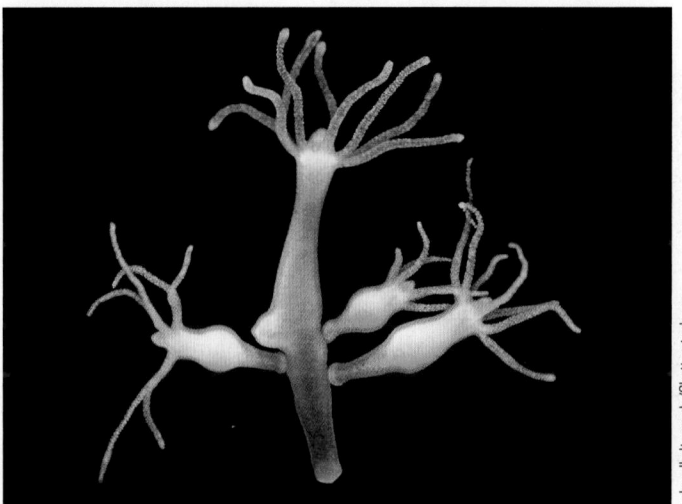

FIGURE 49.2 Asexual reproduction by budding in *Hydra* (stained, visualized by dark field light microscopy).

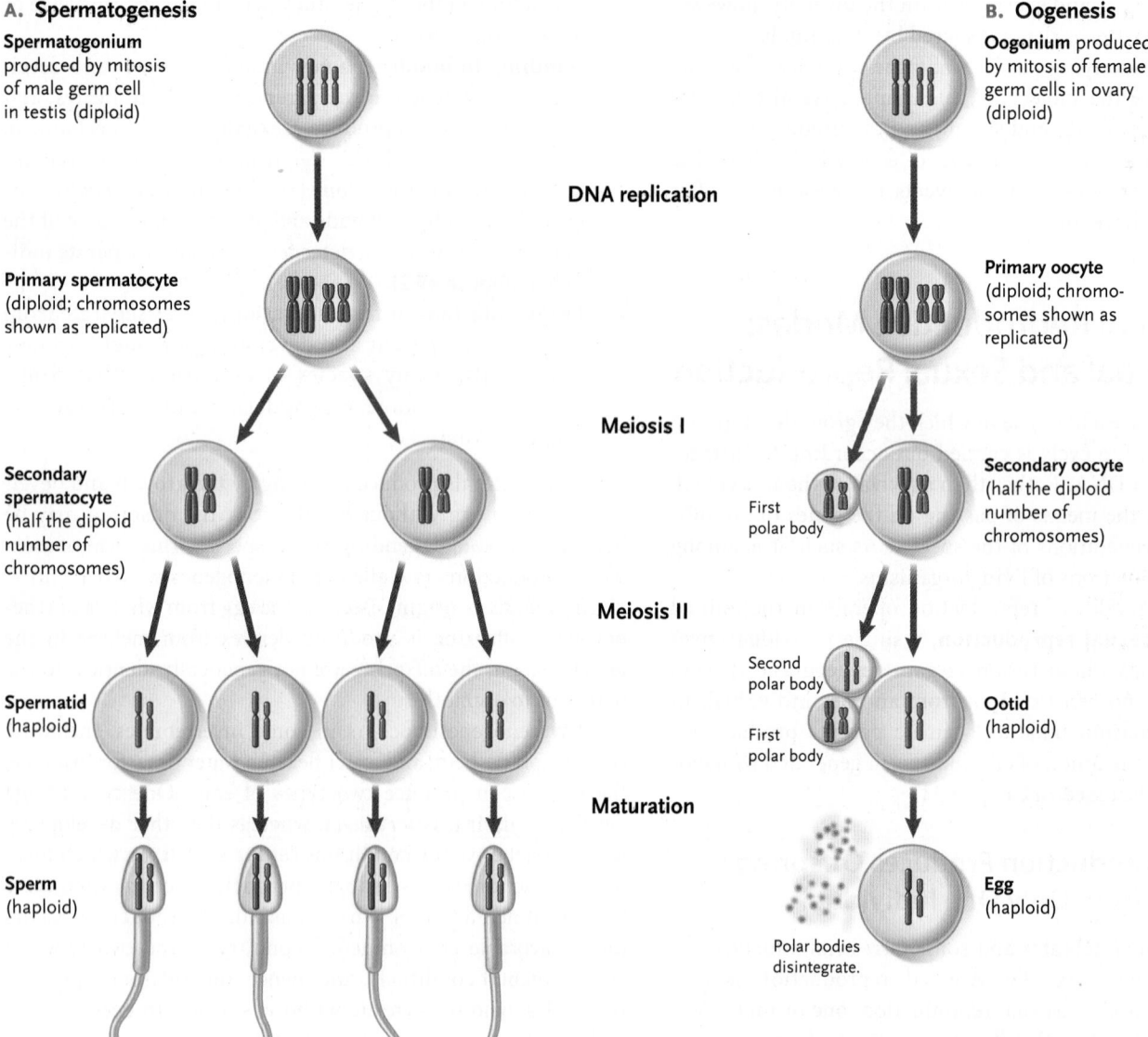

A. Spermatogenesis

Spermatogonium produced by mitosis of male germ cell in testis (diploid)

Primary spermatocyte (diploid; chromosomes shown as replicated)

Secondary spermatocyte (half the diploid number of chromosomes)

Spermatid (haploid)

Sperm (haploid)

DNA replication

Meiosis I

Meiosis II

Maturation

B. Oogenesis

Oogonium produced by mitosis of female germ cells in ovary (diploid)

Primary oocyte (diploid; chromosomes shown as replicated)

First polar body

Secondary oocyte (half the diploid number of chromosomes)

Second polar body

First polar body

Ootid (haploid)

Polar bodies disintegrate.

Egg (haploid)

FIGURE 49.3 The meiotic divisions producing eggs and sperm.
(A) Spermatogenesis. **(B)** Oogenesis. The first polar body may or may not divide, depending on the species, so that either two or three polar bodies may be present at the end of meiosis. Two are shown in this diagram.
© Cengage Learning 2017

Parthenogenesis also occurs in some vertebrates, for example, in certain fish, salamanders, amphibians, lizards, and turkeys. In these animals, an egg, produced by meiosis, typically doubles its chromosomes to produce a diploid cell that begins development. In single-sex species where females have two identical sex chromosomes, the offspring are female, whereas in single-sex species where males have two identical sex chromosomes, the offspring are male. For instance, all whiptail lizards (*Cnemidophorus* species) are females, produced

solely by parthenogenesis. These females go through the motions of mating and copulation with each other.

Sexual Reproduction Generates Diversity among Offspring

Animals reproduce sexually by the union of sperm and eggs produced by meiosis. The overriding advantage of sexual reproduction is the generation of genetic diversity among offspring. This diversity increases the chance that, in a changing environment, at least some offspring will grow and reproduce successfully. Diversity also increases the chance that offspring may be able to live and reproduce in environments previously unoccupied by the species.

Two mechanisms that are part of meiosis give rise to the genetic diversity in eggs and sperm: *genetic recombination* (see Section 11.2) and the *independent assortment* of chromosomes of maternal and paternal origin (see Section 12.1). Genetic recombination mixes the alleles of parents into new combinations within chromosomes. Independent assortment results in random combinations of maternal and paternal chromosomes in gamete nuclei. Additional variability is generated at fertilization when eggs and sperm from genetically different individuals fuse together at random to initiate the development of new individuals. To these sources of variability are added random DNA mutations, which are the ultimate source of variability for both sexual and asexual reproduction.

The disadvantages of sexual reproduction include the expenditure of energy and raw materials in producing gametes and finding mates, and producing fewer offspring per parent. (There is one set of offspring per two parents rather than one set of offspring per parent as in asexual reproduction.) The need to find mates, for instance, can expose animals to predation and takes time from finding food and shelter and caring for existing offspring. Concerning offspring number, consider a female that reproduces asexually. All of her progeny will be females, each of whom will produce more offspring by asexual reproduction. However, in sexual reproduction, one half of the offspring of a mating pair are females, and one half are males (on average). Therefore, only one half of the progeny produced by sexual reproduction of a female produce offspring for the following generation.

We turn now to the mechanisms of sexual reproduction, which include both cellular and whole-organism activities. The next section discusses cellular mechanisms.

STUDY BREAK 49.1

What are the advantages and disadvantages of asexual reproduction? Of sexual reproduction?

49.2 Cellular Mechanisms of Sexual Reproduction

The cellular mechanisms of sexual reproduction are **gametogenesis,** the formation of male and female gametes, and **fertilization,** the union of gametes that initiates development of a new individual. The pairing of a male and a female for the purpose of sexual reproduction is **mating.**

Gametogenesis Involves the Coordinated Events of Meiosis and Sperm and Egg Development

Gametes in most animals form from **germ cells,** a cell line that is set aside early in embryonic development and remains distinct from the other, **somatic cells** of the body. During development, the germ cells collect in specialized gamete-producing organs, the **gonads**—the **testes** (singular, *testis*) in males and **ovaries** in females. Mitotic divisions of the germ cells produce **spermatogonia** in males and **oogonia** in females; these are the cells that enter meiosis (see Chapter 11) to give rise to gametes by **spermatogenesis** in males **(Figure 49.3A)** and **oogenesis** in females **(Figure 49.3B).** In humans, each gamete has only one of 2^{23} possible combinations of parental chromosomes. In some animals, the germ cells also give rise to families of cells that assist gamete development.

Meiosis reduces the number of chromosomes from the diploid level characteristic of somatic cells of the species, in which there are two copies of each chromosome, to the haploid level of gametes, in which there is one copy of each chromosome. The fusion of a haploid sperm and egg during fertilization restores the diploid number of chromosomes and produces a **zygote,** the first cell of a new individual.

SPERMATOGENESIS Spermatogenesis (see Figure 49.3A) produces mature, haploid sperm cells, also called **spermatozoa** (singular, *spermatozoon*) or simply *sperm*. The sperm of most animal species are motile cells, driven through a watery medium by the whiplike beating of a flagellum that extends from the posterior end of the cell **(Figure 49.4).** During maturation from spermatid to sperm, most of the cytoplasm is lost, except for mitochondria, which surround the base of a flagellum. These mitochondria produce the ATP used as the energy source for flagellar beating. At the head of the sperm, a specialized secretory vesicle, the **acrosome,** forms a cap over the nucleus. The acrosome contains enzymes and other proteins that help the sperm attach to and penetrate the surface coatings of an egg of the same species.

OOGENESIS Oogenesis (see Figure 49.3B) produces mature, haploid egg cells, also called **ova** (singular, *ovum*) or simply *eggs*. The eggs of all animals are nonmotile cells, typically much larger than sperm of the same species. Whereas in spermatogenesis, all four products of meiosis develop into functional sperm, in oogenesis, only one of the cell products of meiosis develops into

A. Human sperm

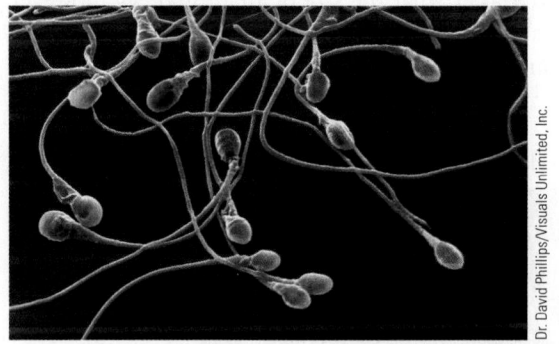

Dr. David Phillips/Visuals Unlimited, Inc.

B. Sperm structure

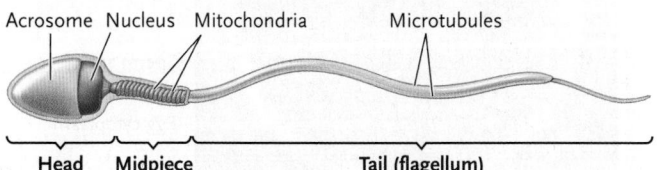

Acrosome Nucleus Mitochondria   Microtubules

Head   Midpiece   Tail (flagellum)

FIGURE 49.4 Spermatozoa. (A) Photomicrograph of human sperm. **(B)** Structure of a sperm.

© Cengage Learning 2017

a functional egg **(Figure 49.5).** That egg retains almost all of the cytoplasm of the parent cell. The other products form nonfunctional cells called **polar bodies** (see Figure 49.3B). The unequal cytoplasmic divisions concentrate nutrients and other molecules required for development in the egg. In most species, the polar bodies disintegrate eventually and do not contribute to fertilization or embryonic development.

The oocytes of most animals do not complete meiosis until fertilization. In mammals, for example, oocytes stop developing at the end of the first meiotic prophase within a few weeks after a female is born. The oocytes remain in the ovary at this stage of development until the female is sexually mature. Then, one to several oocytes advance to the metaphase of the second meiotic division and are released from the ovary at intervals ranging from days to months, or at certain seasons, depending on the species. (In humans, some oocytes may therefore remain in prophase of the first meiotic division for perhaps 50 years until menopause.) As in other animals, meiosis is completed at fertilization to produce the fully mature egg.

An egg typically has specialized features, which include: (1) stored nutrients and cytoplasmic determinants required for at least the early stages of embryonic development (see Section 16.4 for a discussion of cytoplasmic determinants); (2) egg coats of one or more kinds that protect the egg from mechanical injury and infection and, in some species, protect the embryo after fertilization; and (3) mechanisms that prevent the egg from being fertilized by more than one sperm cell.

The amount of stored nutrients in an egg varies with the animal. Mammalian eggs are microscopic, containing few stored nutrients. In mammals, the embryo develops inside the mother and is supplied with nutrients by the mother's body. In contrast, the relatively large eggs of birds and reptiles contain all the nutrients required for complete embryonic development: the "yolk" contains the egg cell, and the "white" contains the nutrients. No matter what the size of an animal egg, however, most of the volume is cytoplasm, and the egg nucleus is microscopic or nearly so in all species.

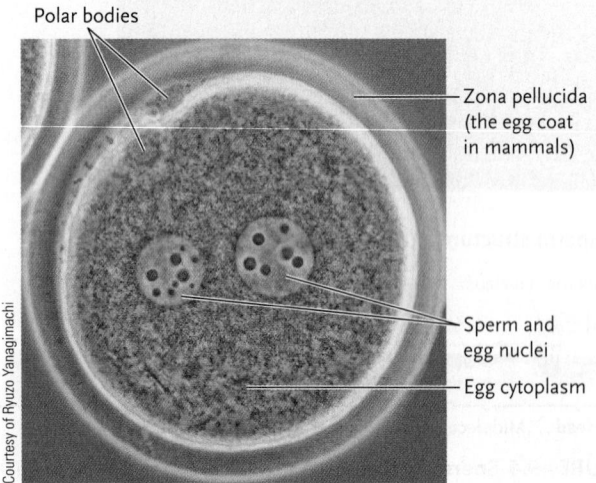

Polar bodies

Zona pellucida (the egg coat in mammals)

Sperm and egg nuclei

Egg cytoplasm

Courtesy of Ryuzo Yanagimachi

FIGURE 49.5 A mature hamster egg that has been fertilized.

Egg coats are surface layers added during oocyte development or fertilization in many species. The **vitelline coat,** called the **zona pellucida** in mammals (see Figure 49.5) is a gel-like matrix of proteins, glycoproteins, or polysaccharides located immediately outside the plasma membrane of the egg cell. Insect eggs have additional outer protein coats that form a hard, water-impermeable layer for preventing desiccation. Amphibians and some echinoderms have an additional outer egg jelly layer instead of a tough protein coat that protects the egg from drying out (see Figure 49.8).

In birds, reptiles, and one group of egg-laying mammals, the **monotremes,** the egg white, a thick solution of proteins, surrounds the vitelline coat. Outside the white is the *shell* of the egg, flexible and leathery in reptiles and mineralized and brittle in birds. Both the egg white and the shell are added while the egg—fertilized or not—is in transit through the **oviduct,** the tube through which the egg moves from the ovary to the outside of the body. In mammals, the egg is surrounded by **follicle cells** during its development. These cells, which grow from ovarian tissue, nourish the developing egg. They also make up part of the zona pellucida while the egg is in the ovary, and remain as a protective layer after it is released.

Fertilization Requires an Internal or External Aquatic Medium

Eggs and sperm are delivered from the ovaries and testes to the site of fertilization by oviducts in females and by sperm ducts in males; in many species, external accessory sex organs participate in the delivery. **Figure 49.6** shows examples of invertebrate and vertebrate reproductive systems. The nonmotile eggs move through the oviducts on currents generated by the beating of cilia lining the oviducts, or by contractions of the oviducts or the body wall.

Depending on the species, fertilization may be *external,* taking place in a watery medium outside the bodies of both parents, or *internal,* taking place in a watery fluid inside the body of the female. In **external fertilization,** which occurs in most aquatic invertebrates, bony fishes, and amphibians, sperm and eggs are shed into the surrounding water. The sperm swim until they collide with an egg of the same species. The process is helped by synchronization of female and male gamete release, and by the enormous quantities of gametes released, as exemplified by the palolo worms in *Why it matters . . . .* In some animals, such as sea urchins and amphibians, the sperm are attracted to the egg by diffusible attractant molecules released by the egg.

Most amphibians, even terrestrial species such as toads, mate in an aquatic environment. Frogs typically mate by a reflex response called *amplexus,* in which the male clasps the female tightly around the body with his forelimbs **(Figure 49.7).** The embrace stimulates the female to shed a mass of eggs into the water through the **cloaca**—the cavity in reptiles, birds, amphibians, and many fishes into which both the intestinal and genital tracts empty. As the eggs are released, they are fertilized by sperm released by the male.

Internal fertilization takes place in invertebrates such as annelids, some arthropods, and some mollusks, and in vertebrates such as reptiles, birds, mammals, some fishes, and some salamanders. In these animals, the sperm are released by the male close to or inside the entrance of the reproductive tract of the female. The sperm swim through fluids in the reproductive tract until they reach and fertilize each egg. In some species, molecules released by the egg attract the sperm to its outer coats. The physical act involving the introduction of the male's accessory sex organ (for example, penis) into a female's accessory sex organ (for example, vagina) to accomplish internal fertilization is known as **copulation.** Internal fertilization makes terrestrial life possible by providing the aquatic medium required for fertilization inside the female's body without the danger of gametes drying by exposure to the air.

Male reptiles, birds, and mammals have accessory sex organs that place sperm directly inside the reproductive tract of females, where fertilization takes place. In reptiles and birds, sperm fertilize eggs as they are released from the ovary and travel through the oviducts, before the shell is added. In mammals, the male's penis delivers sperm into the female's vagina. Unlike the cloaca of the reptiles and birds, which has both sexual and excretory functions, the vagina is specialized for reproduction. The introduced sperm swim into the tubular oviducts containing the eggs, and fertilization takes place.

Fertilization Involves Fusion of a Sperm and an Egg, Which Activates the Egg for Development

Once a sperm touches the outer surface of an egg of the same species, receptor proteins in the sperm plasma membrane bind the sperm to the vitelline coat or zona

A. Insect (the fruit fly, *Drosophila*)

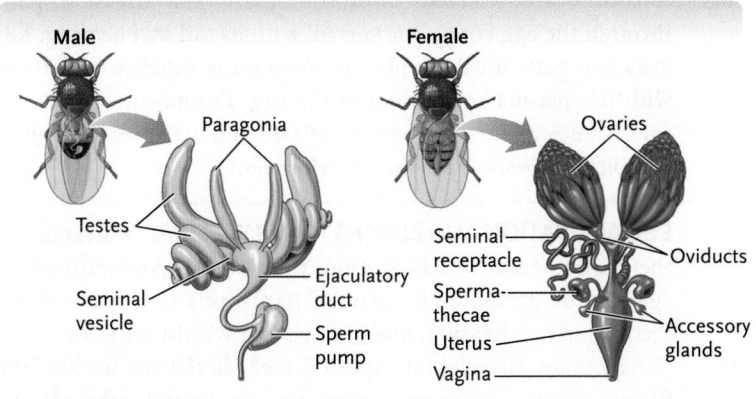

B. Amphibian (frog)

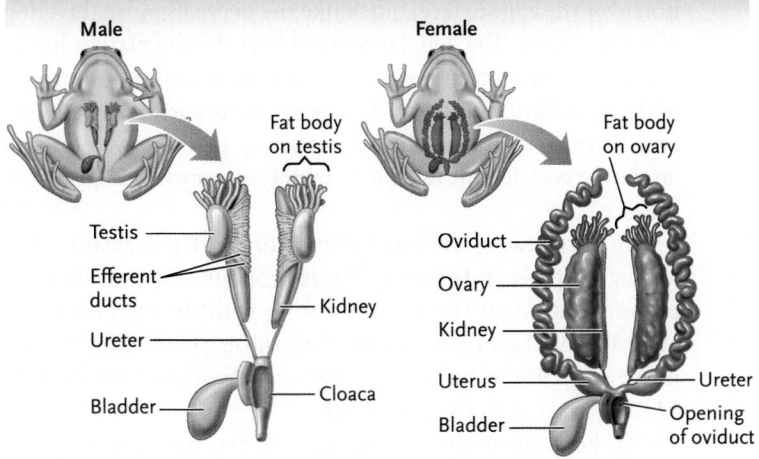

C. Mammal (cat)

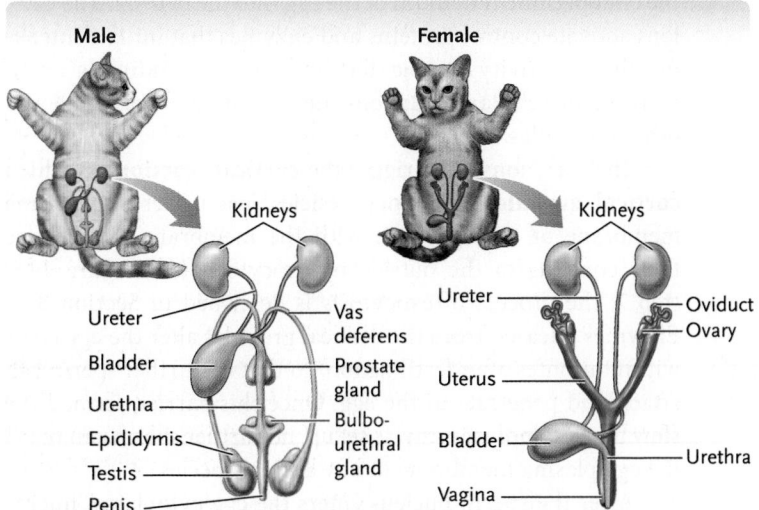

FIGURE 49.6 Some reproductive systems. (A) An insect, *Drosophila* (fruit fly). **(B)** An amphibian, a frog. **(C)** A mammal, a cat. Female systems are shown in blue, and male systems in yellow.

© Cengage Learning 2017

Hans Pfletschinger

FIGURE 49.7 A male leopard frog *(Rana pipiens)* clasping a female in a mating embrace known as amplexus. The tight squeeze by the male frog stimulates the female to release her eggs, which can be seen streaming from her body, embedded in a mass of egg jelly. Sperm released by the male fertilize the eggs as they pass from the female.

pellucida. In most animals, only a sperm from the same species as the egg can recognize and bind to the egg surface.

Species recognition is highly important in animals that carry out external fertilization, because the water surrounding the egg may contain sperm of many different species. It is less important in internal fertilization, because structural adaptations and behavioral patterns of mating usually limit sperm transfer from males to females of the same species.

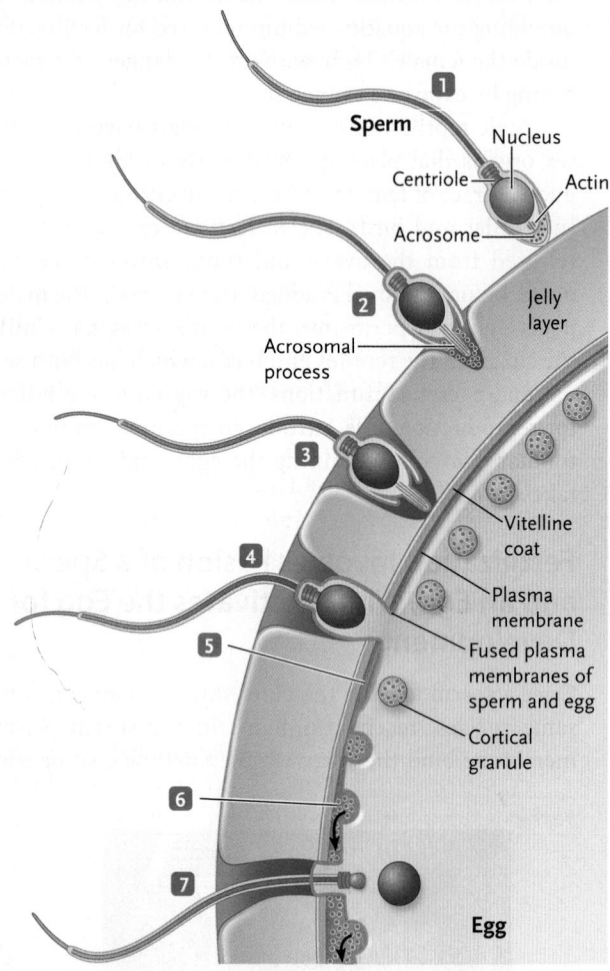

1 A sperm contacts the jelly layer of the egg.

2 The acrosomal reaction begins: enzymes contained in the acrosome are released and dissolve a path through the jelly layer.

3 Proteins in its plasma membrane bind the sperm to the vitelline coat.

4 The sperm lyses a hole in the vitelline coat. The sperm and egg plasma membranes fuse.

5 Membrane depolarization produces the fast block to polyspermy.

6 The fusion of egg and sperm triggers the release of Ca²⁺ ions, which trigger the cortical reaction, the fusion of secretory cortical granules with the egg's plasma membrane. The enzymes of the granules released to the outside alter the egg coats, producing the slow block to polyspermy.

7 The sperm nucleus enters the egg. The sperm nucleus fuses with the egg nucleus (not shown).

FIGURE 49.8 Steps of fertilization in a sea urchin.
© Cengage Learning 2017

FERTILIZATION Figure 49.8 shows the steps of fertilization in the sea urchin, a well-studied organism for invertebrate development and reproduction, and an often-used model for human fertility research. The steps are similar in other organisms. The attachment of sperm to egg triggers a rapid series of events starting with the **acrosomal reaction,** in which enzymes in the acrosome are released from the sperm and digest a path through the egg coats. The sperm, with its tail still beating, follows the path until its plasma membrane touches and fuses with the plasma membrane of the egg. Fusion introduces the sperm nucleus into the egg cytoplasm and activates the egg to complete meiosis and begin development.

EGG ACTIVATION AND BLOCKS TO POLYSPERMY Polyspermy, the phenomenon in which more than one sperm fertilizes an egg, can be prevented by two mechanisms: a *fast block* within seconds of fertilization, and a *slow block* within minutes.

In many invertebrate species, such as the sea urchin, the fusion of egg and sperm opens ion channels in the plasma membrane of the egg, spreading a wave of electrical depolarization over the egg surface, much like the nerve impulse traveling along a neuron. The depolarization alters the egg plasma membrane's potential from negative to positive so that it cannot fuse with any additional sperm. Because it occurs within a few seconds after fertilization, the barrier set up by the wave of depolarization is called the **fast block to polyspermy** (see Figure 49.8, step 5).

In vertebrates, the wave of membrane depolarization following sperm–egg fusion is not as pronounced, and does not prevent additional sperm from fusing with the egg. However, any additional sperm nuclei entering the egg cytoplasm usually break down and disappear, so that only the first sperm nucleus to enter fuses with the egg nucleus.

In both invertebrates and vertebrates, fusion of egg and sperm triggers the release of stored calcium (Ca²⁺) ions from the endoplasmic reticulum of the egg into the cytosol. The Ca²⁺ ions activate control proteins and enzymes that initiate intense metabolic activity in the fertilized egg, including a rapid increase in cellular oxidations and synthesis of proteins and other molecules.

The Ca²⁺ ions also trigger the **cortical reaction,** in which **cortical granules,** secretory vesicles just under the plasma membrane of the egg, fuse with the membrane and release their contents to the outside by exocytosis (see Figure 49.8, step 7; the process of exocytosis is described in Section 5.5). Enzymes released from the cortical granules alter the egg coats within minutes after fertilization, so that no further sperm can attach and penetrate to the egg. Once this barrier, termed the **slow block to polyspermy,** is set up, no further sperm can reach the egg plasma membrane in any animal species.

After the sperm nucleus enters the egg cytoplasm, microtubules move the sperm and egg nuclei together in the egg cytoplasm and they fuse. The chromosomes of the egg and sperm nuclei then assemble together and enter mitosis. The subsequent, highly programmed events of embryonic development that convert the fertilized egg into an individual capable of independent existence are described in the next chapter.

Reproductive Systems May Be Oviparous or Viviparous in Animals with Internal Fertilization

In animals with internal fertilization, three major types of support for embryonic development have evolved: *oviparity,* meaning egg laying; *viviparity,* meaning giving birth to live offspring; and *ovoviparity,* meaning giving birth to live offspring that first hatch internally from eggs.

- **Oviparous** animals (*ovum* = egg; *parere* = to bring forth, to bear) lay eggs that contain the nutrients needed for development of the embryo outside the mother's body. Examples are insects, spiders, most reptiles, and birds. The only oviparous mammals are the *monotremes:* the echidnas (spiny anteaters—found in Australia and New Guinea) and the duck-billed platypus (*Ornithorhynchus anatinus*—found in Australia).
- **Viviparous** animals (*vivus* = alive) retain the embryo within the mother's body and nourish it during at least early embryo development. All mammals except the monotremes are viviparous. Viviparity is seen also in all other vertebrate groups except for the crocodiles, turtles, and birds. In viviparous animals, development of the embryo takes place in a specialized saclike organ, the **uterus** *(womb).* Among mammals, one group, called the *placental mammals* or *eutherians,* has a specialized temporary structure, the **placenta,** which connects the embryo to the uterus. The placenta facilitates the transfer of nutrients from the blood of the mother to the embryo and the movement of wastes in the opposite direction. Humans are an example of placental mammals. The other group of mammals, the *marsupials* or *metatherians,* originally were called *nonplacental mammals* because of a belief that they lacked a placenta. In fact, they do have a placenta, but it derives from a different tissue than that of eutherians and does not connect the embryo and the uterus. Instead it provides nutrients to the embryo from an attached membranous sac containing yolk for only the early stages of its development. In many metatherians, the embryo is then born at an early stage and crawls over the mother's fur to reach the **marsupium,** an abdominal pouch within which it attaches to nipples and continues its development. Kangaroos, koalas, wombats, and opossums are marsupials.
- **Ovoviviparous** animals retain fertilized eggs within the body and the embryo develops using the nutrients provided by the egg. There is no uterus or placenta involved. When development is complete, the eggs hatch inside the mother and the offspring are released to the exterior. Ovoviviparity is seen in some fishes, lizards, and amphibians, many snakes, and many invertebrates.

Hermaphroditism Is a Variation on Sexual Reproduction

Some animals have evolved modified mechanisms that they use as their normal sexual reproduction process. One of these mechanisms is **hermaphroditism** (from *Hermaphroditos,* the son of *Hermes* and *Aphrodite,* a Greek god and goddess), in which both mature egg-producing and mature sperm-producing tissue is present in the same individual. Most flatworms, earthworms, land snails, and numerous other invertebrates are hermaphroditic. In humans and other mammals, there are rare cases of individuals who have both testicular and ovarian tissues. They are not true hermaphrodites because they are not both male and female; rather, they are genetically either male or female, but with ambiguous genitalia. Hence, they are called *pseudohermaphrodites* rather than true hermaphrodites.

Most hermaphroditic animals do not fertilize themselves, because anatomical barriers prevent individuals from introducing sperm into their own body, or the egg and sperm mature at different times. The prevention of self-fertilization maintains the genetic variability of sexual reproduction.

Hermaphroditism takes two forms: **simultaneous hermaphroditism,** in which individuals develop functional ovaries and testes at the same time, and **sequential hermaphroditism,** in which individuals change from one sex to the other. The two earthworms shown in **Figure 49.9** provide a common example of simultaneous hermaphroditism. The only known vertebrate simultaneous hermaphrodites are hamlets (genus *Hypoplectrus*), a group of predatory sea basses. Sequential hermaphroditism is seen among a number of invertebrates (for example, the slipper shell *Crepidula fornicata,* a gastropod), and some ectothermic vertebrates, notably fishes (for example, the clownfish, genus *Amphiprion*). In some species, the initial sex is male (as with the slipper shell and the clownfish), and in others it is female.

FIGURE 49.9 Simultaneous hermaphroditism in the earthworm.
The photograph shows copulation by a mating pair of earthworms, in which each earthworm releases sperm that fertilize eggs in its partner.

1. What are egg coats, and what is their function? What egg coats do mammalian and bird eggs have?

2. How is the slow block to polyspermy brought about?

THINK OUTSIDE THE BOOK

Use the Internet or research literature to outline a model for how hermaphroditism may have evolved.

49.3 Sexual Reproduction in Humans

Except for structural details, human reproduction is typical of that of eutherian (placental) mammals. Internally, these mammals have a pair of gonads, either ovaries or testes. The gonads have a dual function in mammals, as they do in all vertebrates: they both produce gametes and secrete hormones that are responsible for sexual development and mating behavior (see Section 42.4). Males have ducts that carry sperm from the testes to the exterior. Females have an oviduct that leads from each ovary to the uterus, in which fertilized eggs implant and proceed through embryonic development. Nutrients from the mother and wastes from the embryo are exchanged through the placenta. After birth, the newborn offspring is nourished with milk secreted by the mother's mammary glands.

In this section we survey reproductive structures and functions in humans as representative of eutherian mammals. Our story of human development continues in the next chapter, which traces the process from fertilization to birth.

Human Female Sexual Organs Function in Oocyte Production, Fertilization, and Embryonic Development

Human females have a pair of ovaries suspended in the abdominal cavity (**Figure 49.10**). An oviduct leads from each ovary to the uterus, which is a hollow, saclike organ with walls containing smooth muscle. The uterus is lined by the **endometrium,** which is formed by layers of connective tissue with embedded glands and is richly supplied with blood vessels. If an egg is fertilized and begins development, it must implant in the endometrium to continue developing. The lower end of the uterus, the **cervix,** opens into a muscular canal, the **vagina,** which leads to the exterior. Sperm enter the female reproductive tract via the vagina and, at birth, the baby passes from the uterus to the outside through the vagina.

At the birth of a female, each of her ovaries contains about 1 million oocytes, arrested at the end of the first meiotic prophase. Of these oocytes, about 200,000 to 400,000 survive until a female becomes sexually mature; about 400 are **ovulated**—released into the oviducts as immature eggs—during the lifetime of a woman. The egg is released into the abdominal cavity and is pulled into the nearby oviduct by the current produced by the beating of the cilia lining the oviduct. The cilia also propel the egg through the oviduct and into the uterus. Fertilization of the egg occurs in the oviduct.

The external female sex organs, collectively called the **vulva,** surround the opening of the vagina. Two folds of tissue, the **labia minora,** run from front to rear on either side of the opening to the vagina. These folds are partially covered by a pair of fleshy, fat-padded folds, the **labia majora,** which also run from front to rear on either side of the vagina. At the anterior end of the vulva, the labia minora join to partly cover the head of the **clitoris.** The rest of the clitoris is within the body. The clitoris contains erectile tissue and has the same embryonic origins as the penis. A pair of **greater vestibular glands,** with openings near the entrance to the vagina, secretes a mucus-rich fluid that lubricates the vulva. The opening of the urethra, which conducts urine from the bladder, is located between the clitoris and the vaginal opening. Most nerve endings associated with erotic sensations are concentrated in the clitoris, in the labia minora, and around the opening of the vagina. When a human female is born, a thin flap of tissue, the **hymen,** partially covers the opening of the vagina. This membrane, if it has not already been ruptured by physical exercise or other disturbances, is broken by the first sexual intercourse.

Ovulation in Human Females Occurs in a Monthly Cycle

Reproduction in human females is under neuroendocrine control, involving complex interactions between the hypothalamus, pituitary, ovaries, and uterus. Under this control, approximately every 28 days from puberty to menopause, a

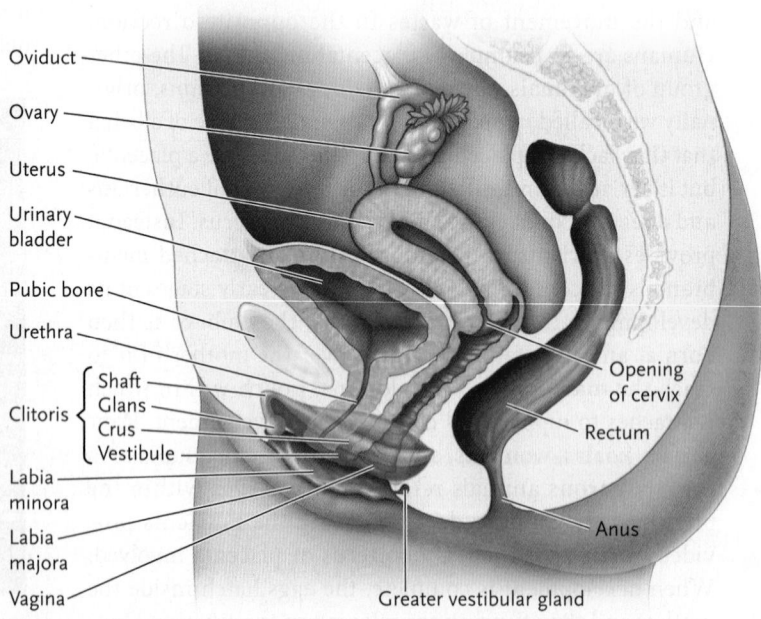

FIGURE 49.10 The reproductive organs of a human female.
© Cengage Learning 2017

female releases an egg from one of her ovaries. The cyclic events in the ovary leading to ovulation are known as the **ovarian cycle.** This cycle is coordinated with the **uterine cycle,** or **menstrual cycle** (*menstruus* = monthly), events in the uterus that prepare it to receive the egg if fertilization occurs. While the cycle is commonly considered a monthly one, there is a lot of variation among human females with a range of about 15–45 days, and an average is about 28–29 days.

THE OVARIAN CYCLE The ovarian cycle produces a mature egg **(Figure 49.11).** The starting point for the cycle is a primary oocyte in prophase of meiosis division I. The beginning of the cycle is triggered by an increase in the release of **gonadotropin-releasing hormone (GnRH)** by the hypothalamus. This hormone stimulates the pituitary to release **follicle-stimulating hormone (FSH)** and **luteinizing hormone (LH)** into the bloodstream **(Figure 49.12A).** FSH stimulates 6 to 20 primary oocytes in the ovaries to be released from prophase of meiosis I and continue through the meiotic divisions. As the primary oocytes develop into secondary oocytes—which arrest in metaphase of meiosis II—they become surrounded by cells that form a **follicle** (day 2 of the cycle; **Figure 49.12B).** During this follicular phase, the follicle grows and develops and, at its largest size, becomes filled with fluid and may reach 12–15 mm in diameter. Usually only one follicle develops to maturity with

release of the egg (secondary oocyte) by ovulation. If two or more follicles develop and their eggs are ovulated, multiple births of nonidentical siblings can result.

As the follicle enlarges, FSH and LH interact to stimulate the follicular cells to secrete **estrogens** (female sex hormones), primarily **estradiol** (see Section 42.4) **(Figure 49.12C).** Initially, the estrogens are secreted in low amounts; at this level, the estrogens have a *negative* feedback effect on the pituitary, inhibiting its secretion of FSH. As a result, FSH secretion declines briefly. However, estrogen secretion increases steadily, and its level peaks at about 12 days after follicle development begins (day 14 of cycle). The high estrogen level now has a *positive* feedback effect on the hypothalamus and pituitary, increasing the release of GnRH and stimulating the pituitary to release a burst of FSH and LH. The increased estrogen levels also convert the mucus secreted by the uterus to a thin and watery consistency, making it easier for sperm to swim through the uterus.

The burst in LH secretion stimulates the follicle cells to release enzymes that digest away the wall of the follicle, causing it to burst and release the egg (see Figure 49.12); this is ovulation. LH also initiates the last phase of the menstrual cycle, the *luteal phase.* That is, LH causes the follicle cells remaining at the surface of the ovary to grow into an enlarged, yellowish structure, the **corpus luteum** (*corpus* = body; *luteum* = yellow) (see Figure 49.11). Acting as an endocrine gland (see

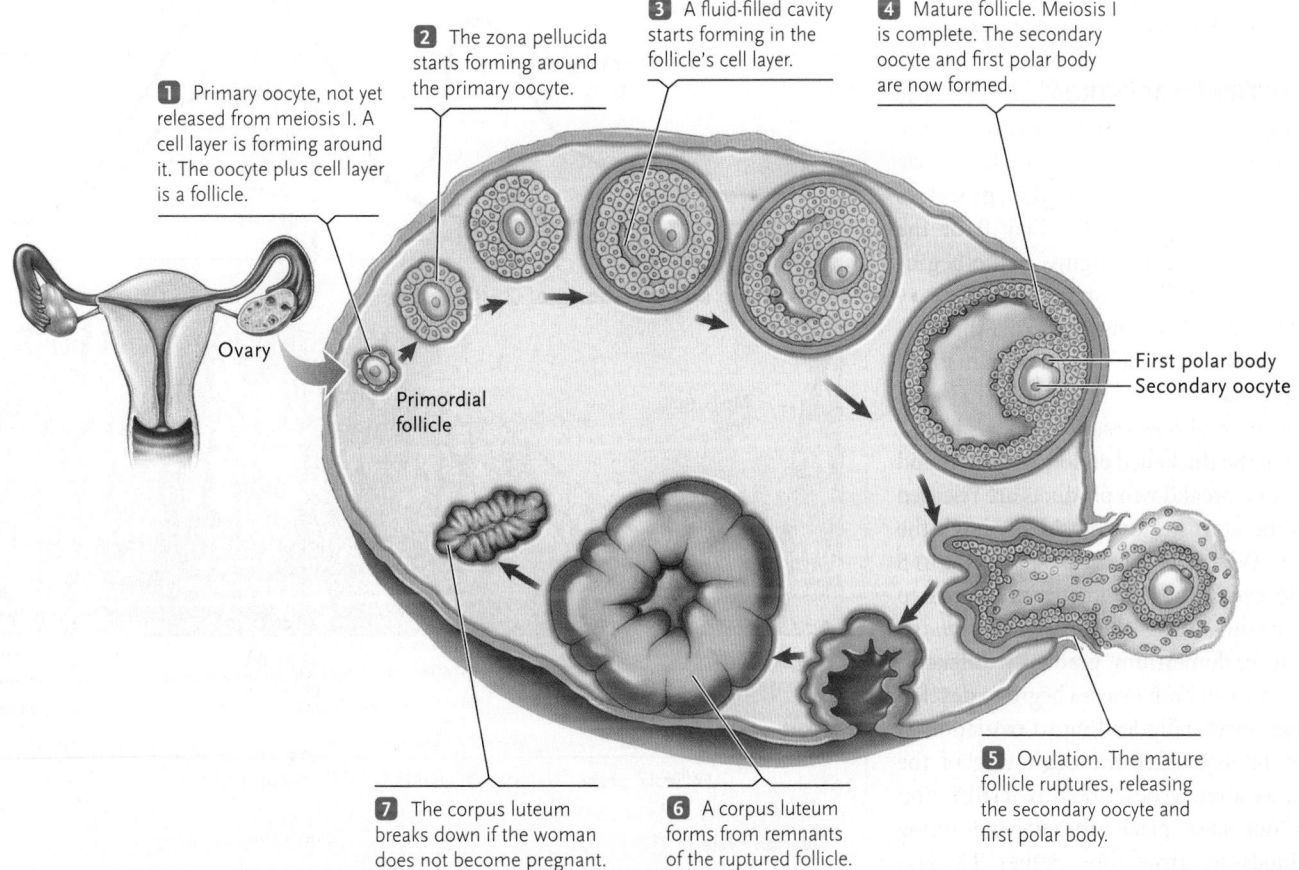

FIGURE 49.11 The growth of a follicle, ovulation, and formation of the corpus luteum in a human ovary.
© Cengage Learning 2017

Section 42.1), the corpus luteum secretes several hormones: estrogens, large quantities of **progesterone,** and **inhibin.** Progesterone, a female sex hormone, stimulates growth of the uterine lining and inhibits contractions of the uterus. Both progesterone and inhibin have a negative feedback effect on the hypothalamus and pituitary. Progesterone inhibits the secretion of GnRH. Without GnRH, the pituitary does not release FSH and LH. FSH secretion from the pituitary is also inhibited directly by inhibin. The fall in FSH and LH levels diminishes the signal for follicular growth, and no new follicles begin to grow in the ovary.

If fertilization does not occur, the corpus luteum gradually degenerates as cells are destroyed by phagocytosis and blood supply is cut off. By about 10 days after ovulation, little tissue remains, meaning that estrogen, progesterone, and inhibin are no longer secreted. In the absence of progesterone, *menstruation* begins (described in the next section). As progesterone and inhibin levels decrease, FSH and LH secretion is no longer inhibited, and a new monthly cycle begins.

THE UTERINE (MENSTRUAL) CYCLE The hormones that control the ovarian cycle also control the uterine (menstrual) cycle **(Figure 49.12D),** keeping the processes connected physiologically. Day 0 of the monthly cycle in the figure is the beginning of follicular development in the ovary (see Figure 49.12B); in the uterus, this correlates with the time at which menstrual flow begins.

Menstrual flow results from the breakdown of the thickened endometrium. Blood and tissue breakdown products are released from the uterus to the outside through the vagina. When the flow ceases, at day 4 to 5 of the cycle, the endometrium begins to grow again—this is the *proliferative phase.* As the endometrium gradually thickens, the oocytes in both ovaries begin to develop further, eventually leading to ovulation at about 14 days after the beginning of the cycle, as already described. If fertilization does not take place, the uterine lining continues to grow for another 14 days after ovulation—this is the *secretory phase.* At the end of that time, the absence of

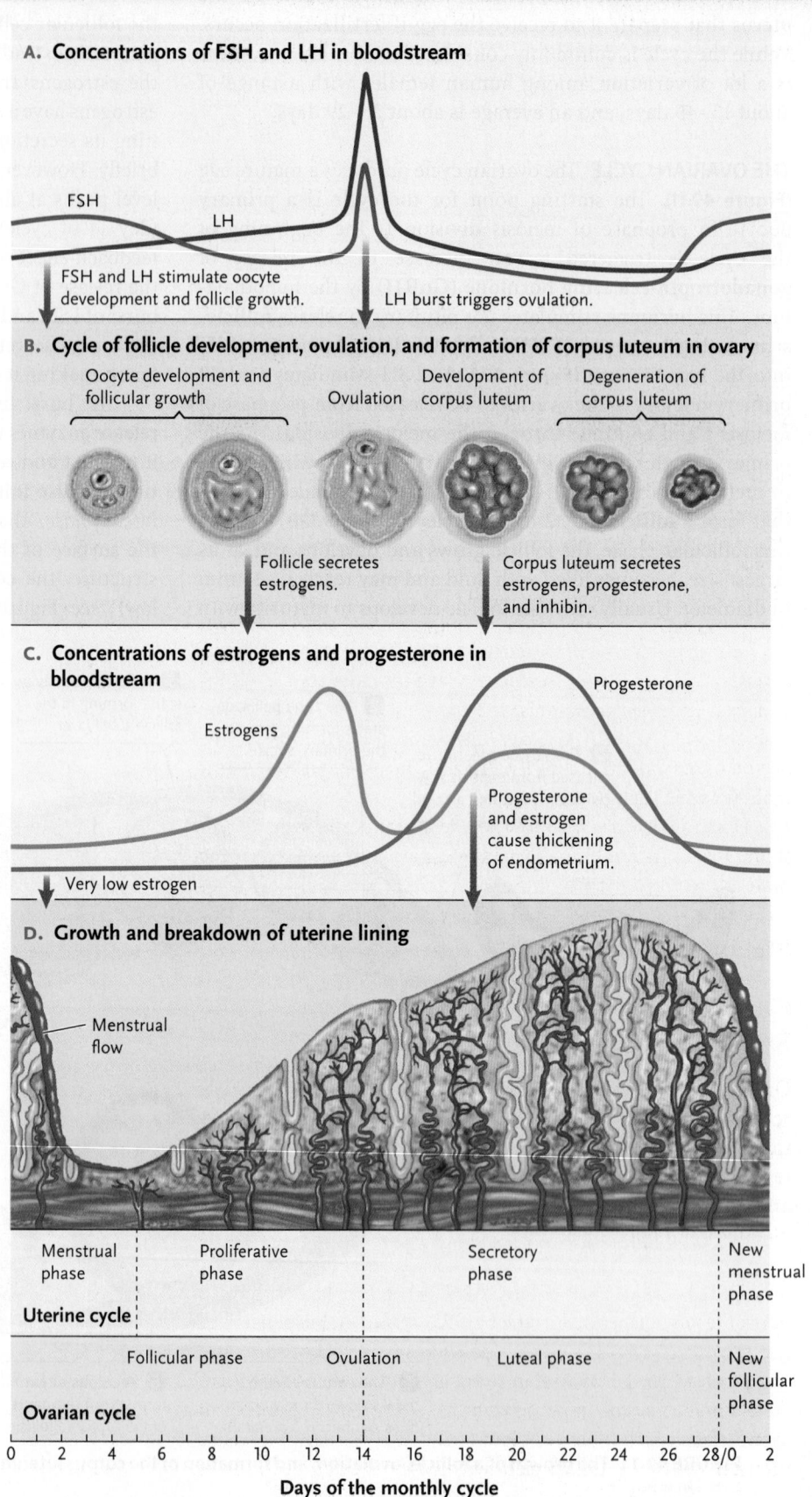

A. **Concentrations of FSH and LH in bloodstream**

FSH

LH

FSH and LH stimulate oocyte development and follicle growth.

LH burst triggers ovulation.

B. **Cycle of follicle development, ovulation, and formation of corpus luteum in ovary**

Oocyte development and follicular growth   Ovulation   Development of corpus luteum   Degeneration of corpus luteum

Follicle secretes estrogens.

Corpus luteum secretes estrogens, progesterone, and inhibin.

C. **Concentrations of estrogens and progesterone in bloodstream**

Estrogens

Progesterone

Progesterone and estrogen cause thickening of endometrium.

Very low estrogen

D. **Growth and breakdown of uterine lining**

Menstrual flow

| Menstrual phase | Proliferative phase | Secretory phase | New menstrual phase |

Uterine cycle

| Follicular phase | Ovulation | Luteal phase | New follicular phase |

Ovarian cycle

0   2   4   6   8   10   12   14   16   18   20   22   24   26   28/0   2

Days of the monthly cycle

think like a scientist If the level of estrogens remains very low and does not increase as shown in the figure, what effect would that have on the menstrual cycle?

© Cengage Learning 2017

progesterone results in the contraction of arteries supplying blood to the uterine lining, shutting down the blood supply and causing the lining to disintegrate. The menstrual flow begins. Contractions of the uterus, no longer inhibited by progesterone, help expel the debris. Prostaglandins released by the degenerating endometrium add to the uterine contractions, making them severe enough to be felt as the pain of "cramps," and also sometimes causing other effects such as nausea, vomiting, and headaches.

Menstruation—the menstrual flow—occurs only in human females and our closest primate relatives, gorillas and chimpanzees. In other mammals, the uterine lining is completely reabsorbed if a fertilized egg does not implant during the period of reproductive activity. The uterine cycle in those mammals is called the *estrous* cycle.

Hormonal tests can show where a woman is in her menstrual cycle. In nonhuman primates, various cues inform the male of the reproductive state of a female. **Figure 49.13** describes a research study demonstrating that vocal cues may signal the fertility state of human females.

Human Male Sexual Organs Function in Sperm Production and Delivery

Organs that produce and deliver sperm make up the male reproductive system **(Figure 49.14).** The testes are located outside the abdominal cavity. Sperm, which are produced by the testes, pass through tubules that enter the abdominal cavity and join with the urethra, the duct that carries urine from the bladder to an opening at the tip of the penis.

MALE REPRODUCTIVE STRUCTURES Human males have a pair of testes, suspended in the baglike **scrotum.** Suspension in the scrotum keeps the testes cooler than the body core, at a temperature that provides an optimal environment for sperm development. Some land mammals such as elephants and monotremes have relatively low body temperatures and have internal testes (testes carried within the body). A testis is

FIGURE 49.13 | **Experimental Research**

Vocal Cues of Ovulation in Human Females

Question: Does the pitch of the female voice change in association with ovulation in human females?

Experiment: Primate females other than humans display well-characterized visual or olfactory cues that signal their reproductive state. Some research has documented particular cues associated with ovulation in humans; those cues relate to femininity and female attractiveness. Based on those observations, Gregory Bryant and Martie Haselton of the University of California, Los Angeles, hypothesized that vocal cues associated with female attractiveness would increase in frequency (pitch) over the menstrual cycle. A higher voice pitch is a signal associated with higher levels of female sex hormones, and correlates with being younger, both of which correlate with high fertility. To test their hypothesis, the researchers recruited 69 women with normal menstrual cycles and collected two sets of vocal samples from them saying "Hi, I'm a student at UCLA." One set was taken during a high-fertility phase of the cycle (at a time near ovulation) and the other set was taken during a low-fertility phase (at the luteal phase). The phase of the menstrual cycle at the time of vocal sampling was confirmed directly by hormonal tests. The samples were analyzed for the pitches of the voices and significant differences were determined using statistical methods.

Results: The average pitch of women's voices when the subjects said a simple sentence was significantly higher during the high-fertility phase of the menstrual cycle compared with during the low-fertility phase **(Figure).** The researchers confirmed in blind studies that the difference in frequency is readily detectable.

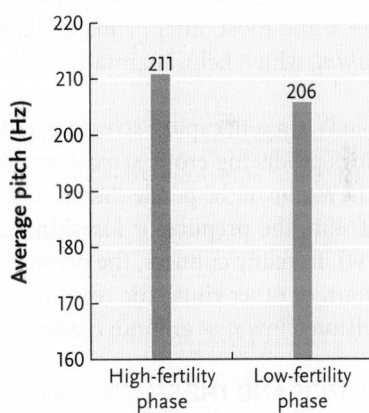

Conclusion: The change in vocal pitch seen for high-fertility (approaching ovulation) versus low-fertility phases of the menstrual cycle is seen as evidence for a cyclic fertility cue in the female human voice. Potentially the cue signals to males that females are in a high-fertility state.

think like a scientist | Observe Hypothesize Predict **Experiment** Interpret | If you wanted to determine whether vocal cues for ovulation is evolutionarily conserved in mammals, what experiment(s) would you do?

Source: G. A. Bryant and M. G. Haselton. 2009. Vocal cues of ovulation in human females. *Biology Letters* 5:12–15.

© Cengage Learning 2017

packed with about 125 m of **seminiferous tubules,** in which sperm proceed through all the stages of spermatogenesis **(Figure 49.15)**. The entire process, from spermatogonium to sperm, takes about 9–10 weeks. The testes produce about 130 million sperm each day.

Supportive cells called **Sertoli cells** completely surround the developing spermatocytes in the seminiferous tubules. They supply nutrients to the spermatocytes and seal them off from the body's blood supply. Other cells located in the tissue surrounding the developing spermatocytes, the **Leydig cells,** produce the male sex hormones, known as **androgens,** particularly **testosterone** (see Figure 49.15 and Section 42.4).

Mature sperm flow from the seminiferous tubules into the **epididymis,** a coiled storage tubule attached to the surface of each testis. Rhythmic muscular contractions of the epididymis move the sperm into a thick-walled, muscular tube, the **vas deferens** (plural, *vasa deferentia*), which leads into the abdominal cavity. Just below the bladder, the vasa deferentia empty into the urethra. During ejaculation, muscular contractions force the sperm into the urethra and out of the penis. The sperm are activated and become motile as they come in contact with alkaline secretions added to the ejaculated fluid by accessory glands.

Most of the interior of the penis is filled with three cylinders of spongelike tissue that become filled with blood and cause erection during sexual arousal. Although the human penis depends solely on engorgement of spongy tissue for erection, the males of many mammalian species, including bats, rodents, walruses, and most other primates, have a bone in the penis, the *baculum*, which helps maintain the penis in an erect state.

The penis ends in a soft, caplike structure, the **glans.** Most of the nerve endings producing erotic sensations are crowded into the glans and the region of the penile shaft just behind the glans. A loose fold of skin, the **prepuce** or **foreskin,** covers the glans (see Figure 49.14). In many cultures, the prepuce is removed for hygienic, religious, or other ritualistic reasons by the procedure called **circumcision** (*circum* = around; *caedere* = cut).

ACCESSORY GLANDS AND THE SEMEN About 150–350 million sperm are released in a single ejaculation. Before they leave the body, these cells are mixed with the secretions of several accessory glands, forming the fluid known as **semen.** In humans, about two-thirds of the volume is produced by a pair of **seminal vesicles,** which secrete a thick, viscous liquid, the **seminal fluid,** into the vasa deferentia near the point where they join with the urethra. The seminal fluid contains prostaglandins that, when ejaculated into the female, trigger contractions of the female reproductive tract that help move the sperm into and through the uterus.

The large **prostate gland,** which surrounds the region where the vasa deferentia empty into the urethra, adds a thin, milky fluid to the semen. The alkaline prostate secretion, which

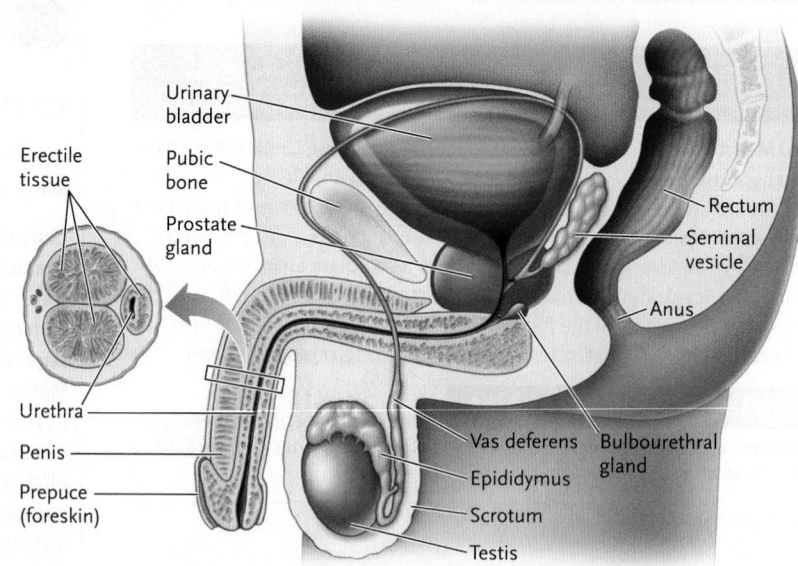

FIGURE 49.14 The reproductive organs of a human male.
© Cengage Learning 2017

makes up about one-third of the volume of the semen, raises the pH of the semen, and of the vagina, to about pH 6, the level of acidity best tolerated by sperm. The raised pH also activates motility of the sperm. As part of the prostate secretion, a fast-acting enzyme converts the semen to a thick gel when it is first ejaculated. The thickened consistency helps keep the semen from draining from the vagina when the penis is withdrawn. A second, slower-acting enzyme in the prostate secretion then gradually breaks down the semen clot and releases the sperm to swim freely in the female reproductive tract.

Finally, a pair of **bulbourethral glands** secretes a clear, mucus-rich fluid into the urethra before and during ejaculation. This fluid lubricates the tip of the penis and neutralizes the acidity of any residual urine in the urethra. In total, the secretions of the accessory glands make up more than 95% of the volume of semen; less than 5% is sperm.

Hormones Also Regulate Male Reproductive Functions

Many of the hormones regulating the menstrual cycle, including GnRH, FSH, LH, and inhibin, also regulate male reproductive functions. Testosterone, secreted by the Leydig cells in the testes, also plays a key role **(Figure 49.16)**.

In sexually mature males, particular neurons in the hypothalamus secrete GnRH in brief pulses every 2 to 3 hours. The GnRH stimulates the anterior pituitary to secrete LH and FSH. LH stimulates the Leydig cells to secrete testosterone, which stimulates spermatogenesis and controls the growth and function of male reproductive structures (masculinizing effects that produce primary and secondary characteristics of the male). FSH stimulates Sertoli cells to secrete a protein and other molecules that are required for spermatogenesis.

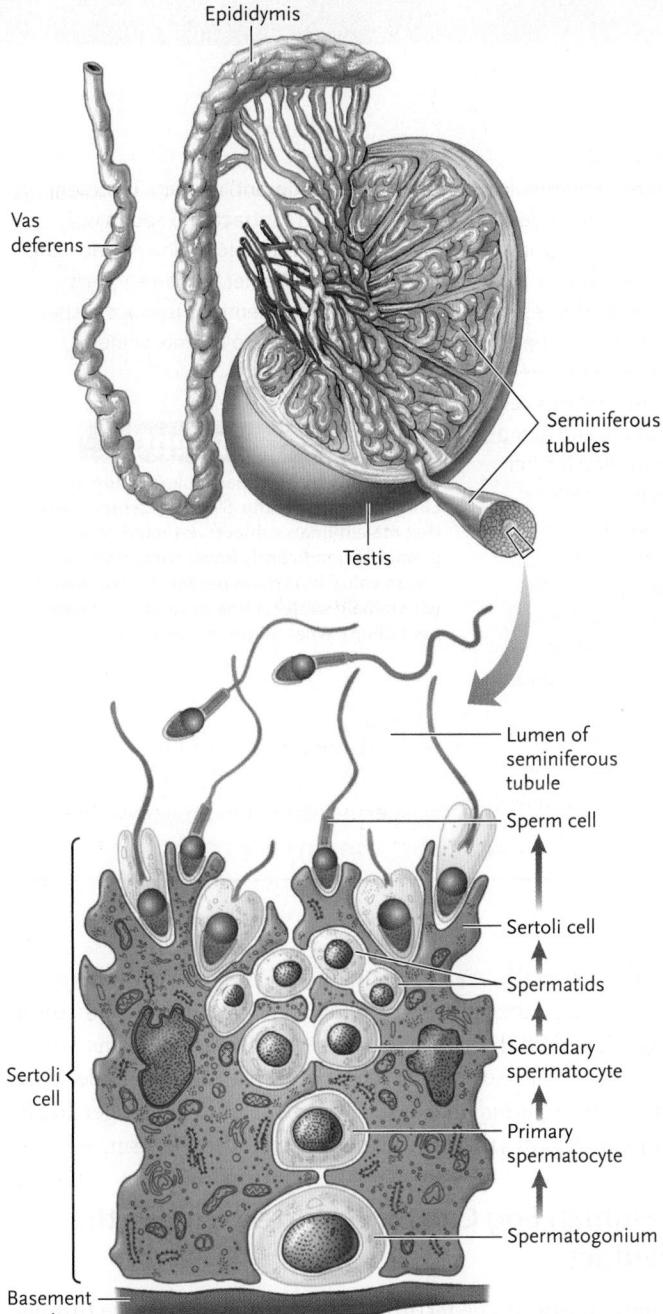

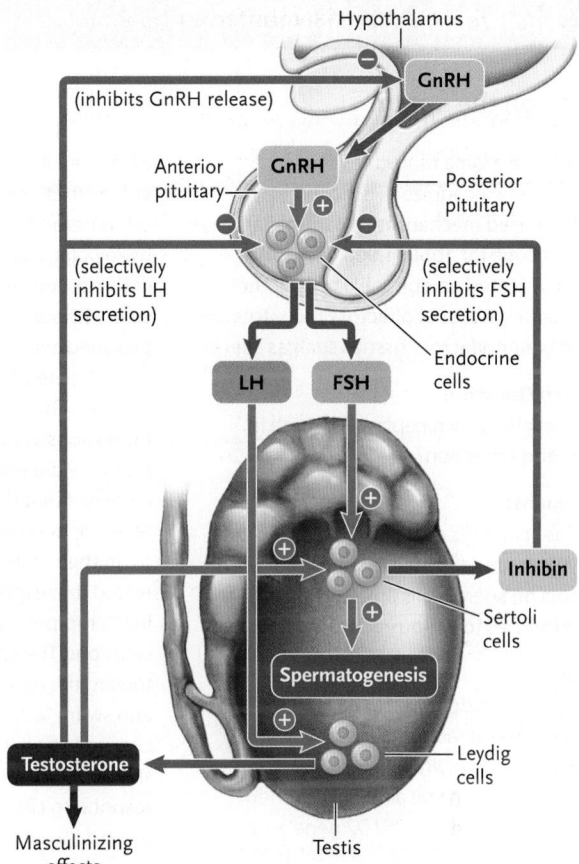

FIGURE 49.16 Hormonal regulation of reproduction in the male, and the negative feedback systems controlling hormone levels.
© Cengage Learning 2017

FIGURE 49.15 Structure of seminiferous tubules and the stages of spermatogenesis. Spermatogonia are located nearest the outer wall and mature sperm cells nearest the tubule lumen. Sertoli cells surround the developing spermatocytes completely and protect them from attack by the immune system.
© Cengage Learning 2017

The secretory rate of LH is controlled by testosterone (see Figure 49.16). Testosterone, the product of stimulation of Leydig cells, inhibits LH secretion by two negative feedback mechanisms. The predominant one acts on the hypothalamus to decrease GnRH release, thereby decreasing both LH and FSH release by the anterior pituitary. The sites of action for this negative feedback mechanism are neurons that release specific neurotransmitters, which stimulate the GnRH-secreting neurons to release the hormone. The other feedback mechanism acts directly on the anterior pituitary in a selective manner to reduce LH secretion.

The secretory rate of FSH is controlled by *inhibin*, a peptide hormone secreted by Sertoli cells (see Figure 49.16). Inhibin acts directly on the anterior pituitary in a selective manner to reduce FSH secretion.

Human Copulation Follows a Typical Mammalian Pattern

When the male is sexually aroused, the sphincter muscles controlling the flow of blood to the spongy erectile tissue of the penis relax, allowing the tissue to become engorged with blood. (The penis is a hydrostatic skeleton structure; see Section 43.2.) As the spongy tissue swells, it maintains the pressure by compressing and almost shutting off the veins draining blood from the penis. The engorgement produces an erection in which the penis lengthens, stiffens, and enlarges. During continued sexual arousal, lubricating fluid secreted by the bulbourethral glands may be released from the tip of the penis.

Female sexual arousal results in enlargement and erection of the clitoris, in a process analogous to erection of the penis in males. The labia minora also become engorged with blood and

Egging on the Sperm

Odorants are aroma molecules that can be specifically recognized ("smelled") by receptor-based mechanisms. Mammalian genomes code for about 1,000 olfactory receptors located on epithelial cells of the nose. A small subset of olfactory receptors are located in nonolfactory tissues such as sperm.

Research Question

Do sperm olfactory receptors function in sperm–egg attraction?

Experiments

Marc Spehr and his colleagues at Ruhr University Bochum set out to identify olfactory receptors on sperm cells and to characterize their responses to various odorants.

Results

The researchers identified in human testicular tissue an active olfactory gene *OR1D2* (previously known as *hOR17-4*), which was not previously known to be active. They then molecularly cloned the *OR1D2* gene and expressed it in cultured human embryonic kidney (HEK) cells. They analyzed a number of odorants and found that the aromatic aldehyde *bourgeonal* triggered a strong response in the genetically altered cells, but not in cells without the introduced gene. This result provided evidence that the product of the *OR1D2* gene confers responsiveness to the odorant. They then tested untreated HEK cells for responsiveness to the odorant, and found that they did not respond, providing further evidence that the product of the olfactory receptor gene was essential for the response.

In their final experiment, Spehr's group tested the response of human sperm cells in micropipettes to gradients of bourgeonal solutions. The sperm swam consistently toward the regions of highest concentration, and swam faster and more directly as the concentration increased. This experiment showed that human sperm can detect and respond to chemical attractants by swimming toward the source of the attractant.

Conclusion

An olfactory (odorant) receptor is present on a sperm cell's surface and sperm will swim towards a chemical known to activate that receptor. Whether human eggs actually release such chemicals that act as sperm attractants remains to be determined.

think like a scientist

A later study by Matthias Laska's group at Linköpking University, Sweden, demonstrated that male human subjects detected bourgeonal at significantly lower concentrations (mean value = 13 parts per billion) compared with female subjects (mean value = 26 parts per billion). What might this result mean?

Sources: M. Spehr et al. 2003. Identification of a testicular odorant receptor mediating human sperm chemotaxis. *Science* 299:2054–2058; P. Olsson and M. Laska. 2010. Human male superiority in olfactory sensitivity to the sperm attractant odorant bourgeonal. *Chemical Senses* 35:427–432.

swell in size, and lubricating fluid is secreted onto the surfaces of the vulva by the greater vestibular glands. In addition to these changes, the nipples become erect by contraction of smooth muscle cells, and the breasts swell due to engorgement with blood.

Insertion of the penis into the vagina and the thrusting movements of copulation lead to the reflex actions of ejaculation, including spasmodic contractions of muscles surrounding the vasa deferentia, accessory glands, and urethra. During ejaculation, the sphincter muscles controlling the exit from the bladder close tightly, preventing urine from being released from the bladder and mixing with the ejaculate. Ejaculation is usually accompanied by **orgasm,** a sensation of intense physical pleasure that is the peak—climax—of excitement for sexual intercourse, followed by feelings of relaxation and gratification.

The motions of copulation stretch the vagina and stimulate the clitoris. The stretching and stimulation can also induce orgasm in females. The vaginal stretching also stimulates the hypothalamus to secrete oxytocin, which induces contractions of the uterus. The contractions keep the sperm in suspension and aid their movement through the reproductive tract. Uterine contractions are also induced by the prostaglandins in the semen.

Sperm reach the site of fertilization in the oviducts within 30 minutes after their ejaculation into the vagina. Of the millions of sperm released in a single ejaculation, only a few hundred actually reach the oviducts. After orgasm, the penis, clitoris, and labia minora gradually return to their unstimulated size. Females can experience additional orgasms within minutes or even seconds of a first orgasm, but most males enter a *refractory period* that lasts for 15 minutes or longer before they can regain an erection and have another orgasm.

A Human Egg Can Be Fertilized Only in the Oviduct

A human egg can be fertilized only during its passage through the third of the oviduct nearest the ovary. If the egg is not fertilized during the 12- to 24-hour period that it is in this location, it disintegrates and dies. However, sperm do not swim randomly for a chance encounter with the egg. Rather, they first swim up the cervical canal to reach the oviduct, and then are propelled up the oviduct by contractions of its smooth muscles. Further, researchers have found evidence that eggs may release chemical attractant molecules that the sperm recognize, causing them to swim directly toward the egg. (*Molecular Insights* describes some of this research.)

To reach the egg, the fertilizing sperm uses enzymes in its plasma membranes to penetrate the layer of follicle cells surrounding the egg, and then by the acrosomal reaction dissolves a path through the zona pellucida coating the egg surface **(Figure 49.17).** The process is similar to that already described

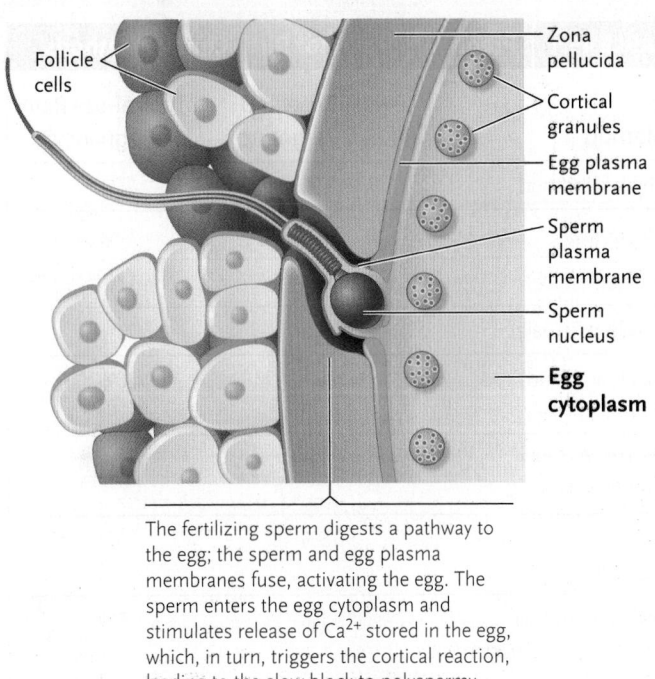

Follicle cells

Zona pellucida

Cortical granules

Egg plasma membrane

Sperm plasma membrane

Sperm nucleus

Egg cytoplasm

The fertilizing sperm digests a pathway to the egg; the sperm and egg plasma membranes fuse, activating the egg. The sperm enters the egg cytoplasm and stimulates release of Ca²⁺ stored in the egg, which, in turn, triggers the cortical reaction, leading to the slow block to polyspermy.

FIGURE 49.17 Fertilization in mammals.
© Cengage Learning 2017

in Figure 49.8 for sea urchin fertilization. As soon as the first sperm cell reaches the egg, the sperm and egg plasma membranes fuse, and the sperm cell enters the cytoplasm of the egg. Although only one sperm fertilizes the egg, the combined release of acrosomal enzymes from many sperm greatly increases the chance that a complete channel will be opened through the zona pellucida. Partly for this reason, a low sperm count is often a source of male infertility. Low sperm count has a number of causes, including infection, heat, frequent intercourse, smoking, and excess alcohol consumption.

The membrane fusion activates the egg. The sperm that has entered the egg releases nitric oxide, which stimulates the release of stored Ca²⁺ in the egg. The Ca²⁺ triggers cortical granule release to the outside of the egg. Enzymes from the cortical granules crosslink molecules in the zona pellucida, hardening it and sealing the channels opened by acrosomal enzymes. The enzymes also destroy the receptors that bind sperm to the surface of the zona pellucida. As a result, no further sperm can bind to the zona or reach the plasma membrane of the egg; this is the slow block to polyspermy. The Ca²⁺ also triggers the completion of meiosis of the egg (recall that, up to that point, it is a secondary oocyte arrested in meiosis II). The sperm and egg nuclei then fuse and the cell is now a zygote. Mitotic divisions of the zygote soon initiate embryonic development.

The first cell divisions of embryonic development take place while the fertilized egg is still in the oviduct. By about 7 days after ovulation, the embryo passes from the oviduct and implants in the uterine lining. During and after implantation, cells associated with the embryo secrete **human chorionic gonadotropin (hCG),** a hormone that keeps the corpus luteum

in the ovary from breaking down. Excess hCG is excreted in the urine; its presence in urine or blood provides the basis of pregnancy tests.

The continued activity of the corpus luteum keeps estrogen and progesterone secretion at high levels, maintaining the uterine lining and preventing menstruation. The high progesterone level also thickens the mucus secreted by the uterus, forming a plug that seals the opening of the cervix from the vagina. The plug keeps bacteria, viruses, and sperm cells from further copulation from entering the uterus.

Later in development, about 10 weeks after implantation, the placenta takes over the secretion of progesterone, hCG secretion drops off, and the corpus luteum regresses. However, the corpus luteum continues to secrete the hormone *relaxin,* which inhibits contraction of the uterus until the time of birth is near.

Infertility Has Many Possible Causes

About 10–15% of couples in the United States are infertile. **Infertility** is defined as the inability for the female of the couple to get pregnant after 12 months of frequent, contraceptive-free intercourse. Infertility may result from a cause in either member of the couple, or a combination of factors involving both members of the couple.

MALE INFERTILITY Male infertility may result from a problem with the sperm or with the delivery of the sperm into the vagina. Problems with the sperm include low sperm concentration or altered sperm shape that affects mobility. For example, testosterone deficiency can lead to a reduced sperm count, and sexually transmitted diseases may temporarily affect sperm motility. In many cases, however, the root cause of low sperm count is not known. Problems with sperm delivery include sexual issues such as erectile dysfunction and blockage of the epididymis or ejaculatory ducts. Environmental factors such as a person's health or lifestyle can also affect male fertility. Such factors include obesity, being underweight, malnutrition, emotional stress, alcohol or drug dependency, exposure to pesticides or other chemicals, and tobacco smoking.

FEMALE INFERTILITY Female infertility may result from any of a variety of causes including physical or hormonal changes. For example, inflammation of the Fallopian tubes (most frequently caused by *Chlamydia* infection) can cause damage, thereby affecting movement of the egg through the tube. Hormonal defects such as low levels of the hypothalamus–pituitary controlled LH and FSH, or certain medications, can block ovulation. Hormone deficiencies such as this can result from various perturbations of normal activity of the hypothalamus and pituitary, including tumors, injury, and excessive exercise. Many of the environmental factors that cause female infertility are the same as those that cause male infertility. In addition, female athletes with vigorous training regimens, such as cycling and running, may experience dysfunctions of their menstrual cycles, a condition called *athletic menstrual cycle irregularities*. The dysfunctions include amenorrhea (no menstrual periods) and cycles of normal length but without ovulation.

STUDY BREAK 49.3

Outline the roles of follicle-stimulating hormone (FSH) and luteinizing hormone (LH) in the ovarian cycle of a human female.

THINK OUTSIDE THE BOOK

The concept of athletic menstrual cycle irregularities was just introduced. Use the Internet or research literature to learn about possible hormonal differences in female athletes and to determine whether they might be involved in athletic menstrual cycle irregularities.

49.4 Methods for Preventing Pregnancy: Contraception

In human society, an unwanted pregnancy can be inconvenient at the least, or at the worst can have serious physical and social repercussions, particularly for the mother. Many methods exist for achieving **contraception**—the prevention of pregnancy—some old and others relatively new.

The oldest method of contraception is total abstinence from sex. Unfortunately, millions of years of animal evolution have stacked the cards against total abstinence by making the sex drive among the most powerful of compulsions. Millions of unwanted children attest to the failures of this method. Other methods of preventing pregnancy include techniques for: (1) preventing the sperm from reaching the site of fertilization; (2) preventing ovulation; or (3) interfering with implantation if fertilization does occur. **Table 49.1** lists the most common contraceptive techniques and their reliability, based on 1 year of use. Note that, while we discuss each method individually, combinations of particular methods may be used to reduce further the chance of pregnancy.

Vasectomy and Tubal Ligation Are the Most Effective Methods for Preventing Fertilization

A natural technique for preventing fertilization is the *rhythm method*, which consists of avoiding intercourse during the time of the month when the egg can be fertilized. Because sperm can survive for as long as 5 days in the female reproductive tract, intercourse should be avoided from 5 days before ovulation and, for safety's sake, for another 4 or 5 days after ovulation. The method is difficult to apply because of the typical unpredictability of the time of ovulation (and the power of the sex drive).

TABLE 49.1 | Pregnancy Rates for Birth Control Methods

Method	Lowest Expected Rate of Pregnancy[a]	Typical-use Rate of Pregnancy[b]
Rhythm method	1–9%	25%
Withdrawal	4%	19%
Condom (male)	3%	14%
Condom (female)	5%	21%
Diaphragm and spermicidal jelly	6%	20%
Vasectomy (male sterilization)	0.1%	0.15%
Tubal ligation (female sterilization)	0.5%	0.5%
Contraceptive pill (combination estrogen/progestin)	0.1%	5%
Contraceptive pill (progestin only)	0.5%	5%
Implant (progestin)	0.09%	0.09%
Intrauterine device (IUD) (copper T)	0.6%	0.8%

[a]Rate of pregnancy when the birth control method was used correctly every time.
[b]Rate of pregnancy when the method was used typically, meaning that it may not have been always used correctly every time.

Source: U.S. Food and Drug Administration, http://www.fda.gov/fdac/features/1997/conceptbl.html. Data reported in 1997 for effectiveness of methods in a 1-year period.

Another natural method to prevent fertilization is *withdrawal*—starting sexual intercourse, but withdrawing the penis before ejaculation. Unfortunately, once ejaculation begins, it proceeds as a series of reflexes that is extremely difficult to interrupt. In addition, some sperm may be present in lubrication produced before ejaculation.

The *condom*, a thin, close-fitting sheath of latex, lambskin, or polyurethane worn over the penis, is one of the traditional methods of preventing ejaculated sperm from entering the vagina. Condoms made from latex may also provide a barrier to the transmission of disease between sexual partners (condoms made from natural skin do not block viruses such as HIV). Pouchlike "female condoms," inserted into the vagina, prevent ejaculated sperm from entering the uterus.

The *diaphragm* is a cuplike rubber device that blocks the cervix in females. (The similar *cervical cap* is smaller and fits more closely over the cervix.) Typically a spermicidal jelly or cream is also used. To be most effective, a diaphragm and the spermicidal jelly must be inserted no more than an hour before intercourse, and left in place for the recommended time afterward.

The *intrauterine device (IUD)* is a small plastic or copper device inserted into the uterus just inside the cervix. The IUD remains in place as a long-term preventive measure. Depending on the type, a single IUD is approved for 5–10 years of use. It is not completely clear how an IUD works; it is thought to cause a mild inflammation of the uterine lining that interferes with sperm function so that fertilization is less likely to occur. IUDs may also interfere with implantation of a fertilized egg but this may be a smaller contribution to their contraceptive effect. The IUD is effective as long as it is not deflected from its correct position in the uterus; unfortunately, this may happen without warning or the user's awareness. A few women also experience unpleasant side effects from the IUD such as cramps, uterine infections, or excessive menstrual bleeding.

Fertilization can also be prevented surgically, by cutting and closing off either the vasa deferentia in males or the oviducts in females. In *vasectomy,* the procedure for males, an incision is made in the scrotum and each vas deferens is severed and tied off. After vasectomy, the seminal fluid is still produced and ejaculated, but it does not contain sperm. In *tubal ligation,* the procedure for females, the oviducts are cut and tied off, or seared with heat (cauterized) to close them. The ligation prevents eggs from being fertilized or reaching the uterus. Neither vasectomy nor tubal ligation interferes with the production of sex hormones by the ovaries or testes, or results in any change in sexual behavior. Both operations are highly effective in preventing pregnancy. Although they can be reversed, the procedures are difficult and not always successful.

The Oral Contraceptive Pill Is the Most Effective Method for Preventing Ovulation

The primary method used to prevent ovulation is the *oral contraceptive pill,* or simply "the pill," containing a combination of estrogen and *progestin* (a synthetic form of progesterone) or progestin alone. In this highly effective method, the pill is taken daily for 20–21 days after the end of the menstrual flow and then stopped (placebo pills are taken for the remaining days of the cycle to maintain the routine of pill taking) to allow menstruation, and then the next month's course is begun. If pregnancy is desired, the pill is simply not taken after the menstrual flow.

The pill works by inhibiting the secretion of FSH and LH by the pituitary; without these hormones, ovulation does not occur. When the pill is stopped after 20–21 days, the resulting drop in progestin concentration causes the uterine lining to break down and initiates the menstrual flow. Since ovulation does not occur, fertilization and pregnancy are not possible.

Most pregnancies among women taking the pill result from failure to take it on schedule—often simply by forgetting to take the pill for a day or two at the wrong time of the month. Some women, about one in four, experience unpleasant side effects, such as nausea, tenderness of the breasts, irritability, nervousness, or changes in skin color or texture. Modern versions of the pill have almost eliminated the more serious side effects, such as increased incidence of breast cancer and formation of blood clots. However, cigarette smoking significantly increases the risk of heart attacks and strokes for women taking the pill. This risk increases with age and with the number of cigarettes smoked per day.

The Morning-After Pill Blocks Ovulation or Fertilization

Whatever the method of birth control, its effectiveness is improved if sex partners are highly motivated to avoid pregnancy, and careful in its use. The effectiveness of condoms, for example, is greatly improved if the penis is withdrawn immediately after ejaculation (before the semen has time to spread under the condom and leak into the vagina).

Another method used to prevent pregnancy is the so-called *emergency contraception pill,* commonly referred to as the "morning-after pill." These pills are administered after intercourse has occurred as a means to prevent pregnancy. A high-dosage synthetic progestin emergency contraception pill called Plan B (levonorgestrel) is available in the United States without prescription to women who are 18 or older. This pill is highly effective if taken within 72 hours after unprotected sexual intercourse. Pregnancy tests do not work until significantly after this time. Research data show that Plan B works by blocking ovulation or fertilization. It may also inhibit implantation of a fertilized egg by altering the endometrium, but it has no effect if the process of implantation has begun.

Another emergency contraception pill is *mifepristone (RU-486),* which contains a molecule that binds to and blocks progesterone receptors in the uterine lining. The blockage prevents the lining from responding to progesterone and causes it to break down (that is, a menstrual period is initiated), taking with it any embryo that may have implanted. Mifepristone is approved in the United States for terminating pregnancies up to 49 days postconception; the time period is longer in some foreign countries. It is available only by prescription.

In this chapter we have focused on animal reproduction up to the point of the fertilized egg. In the next chapter, we address the final stage of reproduction in sexually reproducing organisms, the development of a new individual from the fertilized egg.

STUDY BREAK 49.4
How does the oral contraceptive pill prevent pregnancy?

What molecules are responsible for sperm recognition of egg coats?

Although fertilization is necessary for sexual reproduction, how sperm recognize eggs, particularly in mammals, has been an enigma. Fertilization success or failure may be attributed to how well complementary molecules on each gamete interact with each other. Furthermore, as discussed in Section 22.2, one of the prezygotic species-isolating mechanisms relies on the fact that gamete receptors from one species often do not recognize receptors from other species. However, in artificial environments (such as experimental situations), gametes from different species sometimes do recognize each other and form hybrids. These hybrids are usually explained by structural similarity of the molecules responsible for gamete recognition.

During fertilization in mammals, sperm must penetrate through several layers that surround the egg. Using motility and their own surface hyaluronidase, the sperm first move through the cumulus mass, a remnant of follicle cells (referred to as *cumulus cells*) and the sticky hyaluronate-containing matrix that cumulus cells produce. Next, the sperm bind to the zona pellucida, a tough acellular glycoprotein coat. In mammals, this coat is composed of three to four proteins, often referred to as ZP1, ZP2, ZP3, and ZP4. Mice, the mammal in which fertilization has been studied the most, produce ZP1, ZP2, and ZP3.

Starting in the 1980s, researchers have studied the function of zona pellucida glycoproteins. Early experiments studied the function of individual proteins using a competition assay. For this assay, zona pellucida proteins were dissolved and purified. Then, each was added individually to sperm to determine if it could occupy putative receptors on the plasma membrane of sperm cells and prevent the sperm from binding to cumulus-free eggs. Only ZP3 did this.

ZP3 also induces the acrosome reaction in sperm, which is necessary for sperm to penetrate the zona pellucida (discussed in Section 49.2). Subsequent research found that ZP3 bound to a form of the enzyme β-1,4-galactosyltransferase found on the sperm plasma membrane. Curiously, another ZP3 binding protein, sp56, was also found within the acrosome. These proteins may act sequentially to allow sperm to adhere to the zona pellucida (before and during or after the acrosome reaction).

Recent mouse genetic studies suggest that, in the context of an entire zona pellucida, sperm instead bind to a specific region of ZP2. This work also identified an enzyme, released from the cortical granules at fertilization, that cleaves a sequence within this region of ZP2. The action of this enzyme activates a block to polyspermy. Further work is necessary to resolve the early biochemical data and the more recent results from genetic experiments. It is also important to clarify the role of other zona pellucida receptors on the sperm that have been identified. Finally, recent evidence suggests that sperm begin the acrosome reaction in the cumulus cell mass prior to interaction with the zona pellucida. Regardless, it seems likely that fertilization includes a number or perhaps a sequence of molecular adhesive steps that must function properly for normal fertility and that there may be molecular redundancy in such a fundamental process.

How is membrane fusion regulated during the sperm acrosome reaction?

To release the acrosome, the outer acrosomal membrane fuses at hundreds of points with the plasma membrane overlying it. These fusion points expand, forming vesicles that are released, freeing the acrosomal contents. Acrosomal proteins are released in two phases. First, the more soluble acrosomal proteins are released. Then, acrosomal proteins that initially remained bound to a protein matrix within the acrosome are released. Regulated membrane fusion during secretion has been studied in most detail in neurons. In neurons, each secretory vesicle forms a single fusion point with the plasma membrane. Although membrane fusion during the acrosome reaction involves hundreds of fusion points, and secretory vesicles in somatic cells form just one, we found that proteins promoting membrane fusion in neuronal secretion at the synapse were also involved in the sperm acrosome reaction.

Proteins called SNAREs (from *Soluble N*-ethylmaleimide-sensitive factor *A*ttachment protein *RE*ceptor) promote membrane fusion in neuronal secretion. So-called "neuronal" SNAREs are also found in sperm near the acrosome. We have studied mice deficient in a protein called *complexin* that regulates SNARE function, and have found that sperm from these mice have defective acrosome reactions, as well as markedly reduced fertility. Complexin appears to promote SNARE complex formation and then suspend the SNARE complex in an intermediate state in which membrane fusion can be completed readily by a Ca^{2+} influx into sperm.

Despite some molecular similarity to the neuronal process, membrane fusion in sperm is clearly unique. It is important to identify the important steps that prepare sperm for membrane fusion and to determine how SNARE function is regulated during the acrosome reaction. This should help understand sperm malfunctions that result in defects in the acrosome reaction, a frequent cause of male infertility.

think like a scientist

One of the procedures used in fertility clinics is called *intracytoplasmic sperm injection (ICSI)*. A sperm is aspirated into a pipette, and the pipette is driven through the zona pellucida and egg membrane. The sperm is then injected directly into an egg, and the egg develops into an embryo. Predict the types of fertility problems for which this procedure could be used to produce offspring.

David Miller is a Professor in the Department of Animal Sciences at University of Illinois, Urbana-Champaign. His laboratory studies the molecular underpinnings of sperm maturation and fertilization. You can learn more about Miller's research by visiting http://ansci.illinois.edu/labs/miller-lab.

For access to MindTap and additional study materials visit www.cengagebrain.com.

49.1 Animal Reproductive Modes: Asexual and Sexual Reproduction

- In asexual reproduction, a single parent gives rise to offspring without genetic input from another individual. In sexual reproduction, offspring are produced by the union of gametes—eggs and sperm—from two parents.

- Asexual reproduction involving mitosis occurs in animals by fission, budding, or fragmentation (Figure 49.2). In parthenogenesis, a form of asexual reproduction, females produce eggs that develop without being fertilized.

- In sexual reproduction, genetic variability is produced by the meiotic processes of genetic recombination and independent assortment.

49.2 Cellular Mechanisms of Sexual Reproduction

- Sexual reproduction includes two cellular processes, gametogenesis and fertilization, and a whole-organism process, mating. Gametogenesis is the formation of male and female gametes by meiotic cell division, followed by differentiation of the gametes. Fertilization is the union of gametes that initiates development of new individuals (Figure 49.3).

- Gametogenesis takes place in the testes of males and in the ovaries of females. Sperm and eggs are delivered to the site of fertilization by sperm ducts in males and oviducts in females. External reproductive structures aid the delivery in many species.

- In male gametogenesis—spermatogenesis—each cell entering meiosis produces four haploid motile sperm cells. In female gametogenesis—oogenesis—each cell entering meiosis produces one haploid egg cell; the other division products are nonfunctional polar bodies (Figure 49.3).

- The egg contains stored nutrients and information required for at least the early stages of embryonic development. It is covered by one or more protective coats, and it has a mechanism that blocks additional sperm from entering after fertilization (Figure 49.5).

- Fertilization, which follows mating in most animals, may be external, in which sperm and eggs are shed into the surrounding water, or internal, in which sperm are released close to or inside the female reproductive ducts via copulation (Figure 49.6).

- When a sperm and egg touch during fertilization, their plasma membranes fuse, introducing the sperm nucleus into the egg cytoplasm. The sperm and egg nuclei then fuse to form a diploid zygote nucleus and initiate embryonic development (Figure 49.8).

- Oviparous animals lay eggs in which development of new individuals takes place outside the female's body. In viviparous animals, development takes place inside the female's body. In ovoviviparous animals, fertilized eggs are retained within the body while the embryo develops, the eggs hatch within the mother, and the offspring are then released from the body.

- In hermaphroditism, single individuals produce both mature egg-producing tissue and mature sperm-producing tissue (Figure 49.9).

49.3 Sexual Reproduction in Humans

- In females, eggs released from the ovaries travel through the oviducts to the uterus, which opens into the vagina, the entrance for sperm and the exit for offspring during birth (Figure 49.10).

- The ovarian cycle produces an egg. The cycle begins with the release of GnRH by the hypothalamus, which stimulates the release of FSH and LH from the anterior pituitary. FSH stimulates oocytes in the ovaries to begin meiosis. One oocyte typically develops to maturity and is surrounded by cells that form a follicle (Figures 49.11 and 49.12).

- The enlarging follicle secretes estrogens, causing a burst in FSH and LH blood concentrations; at about day 14 of the cycle, the LH stimulates ovulation, the bursting of the follicle and the release of the egg. The remainder of the follicle forms the corpus luteum, which secretes estrogens, progesterone, and inhibin (Figures 49.11 and 49.12).

- Day 0 of the monthly uterine (menstrual) cycle correlates with the beginning of follicular development in the ovary and the beginning of the menstrual flow. Secretion of estrogen from the developing follicle stimulates the growth of a new endometrium. If fertilization does not occur, progesterone and inhibin maintain the endometrium until day 28 of the cycle, when the corpus luteum regresses. Without progesterone, the endometrium breaks down and is released as the menstrual flow (Figure 49.12).

- In males, sperm develop in seminiferous tubules in the testes and are released into the epididymis. When a male ejaculates, sperm travel from the epididymis to the vas deferens, and then through the urethra and the penis. The seminal vesicles, prostate gland, and bulbourethral glands add fluids to the sperm traveling to the outside (Figures 49.14 and 49.15).

- Sperm production in males is also controlled by LH and FSH. LH stimulates Leydig cells in the testes to secrete testosterone, which stimulates sperm production. FSH stimulates Sertoli cells in the testes to secrete molecules needed for spermatogenesis (Figure 49.16).

- During copulation, sperm are ejaculated into the vagina of the female. The sperm then swim through the female reproductive tract, aided by contractions of the oviduct and guided by molecules released by the egg. On contact with the egg in the oviduct, the acrosomes of sperm release enzymes that digest a path through the coats of the egg. When the fertilizing sperm contacts the egg, the sperm and egg plasma membranes fuse, releasing the sperm nucleus into the egg cytoplasm and activating the egg. The egg completes meiosis, and the sperm and egg nuclei fuse, producing the zygote (Figure 49.17).

- As the embryo implants, the hormone hCG sustains the corpus luteum, which continues to secrete estrogen and progesterone at high levels. These hormones maintain the uterine lining and prevent menstruation.

- Infertility has many causes, including physical problems, hormonal changes, or environmental factors.

49.4 Methods for Preventing Pregnancy: Contraception

- Methods of contraception work by preventing sperm from reaching the site of fertilization, by preventing ovulation, or by interfering with implantation (Table 49.1).

- Methods for preventing fertilization include the rhythm method, the condom, the diaphragm or cervical cap, the IUD, and a vasectomy or tubal ligation.

- The oral contraceptive pill prevents ovulation. It contains a combination of estrogen and the progesterone-like progestin, which inhibiting the secretion of FSH and LH and follicle formation.

- The morning-after pill blocks ovulation or fertilization, and may also inhibit implantation.

TEST YOUR KNOWLEDGE

Remember/Understand

1. Asexual reproduction is most successful in:
   a. changing environments.
   b. sessile animals.
   c. densely settled populations.
   d. land animals.
   e. genetically varied individuals.

2. Which of the following processes does not increase genetic diversity?
   a. parthenogenesis
   b. random DNA mutations
   c. genetic recombination
   d. independent assortment
   e. random combinations of paternal and maternal chromosomes

3. Gametogenesis has parallel stages in egg and sperm formation. The stage in eggs that is equivalent to spermatids is the:
   a. primary oocyte.
   b. oogonium.
   c. ovum.
   d. ootid and polar bodies.
   e. secondary oocyte and polar body.

4. External fertilization provides a male with relative certainty of paternity and is found in:
   a. amphibians.
   b. birds.
   c. sharks.
   d. reptiles.
   e. mammals.

5. The slow block to polyspermy:
   a. is caused by a change in membrane potential from negative to positive.
   b. triggers the movement of Ca^{2+} from the cytosol to the endoplasmic reticulum.
   c. triggers a decrease in egg oxidation and protein synthesis.
   d. describes the fusion of egg and sperm nuclei.
   e. includes the fusion of cortical granules with the egg's plasma membrane.

6. Some placental animals provide nutrients to their embryos from an attached membranous yolk-containing sac. They are called:
   a. oviparous animals.
   b. ovoviviparous animals.
   c. metatherians.
   d. eutherians.
   e. hermaphrodites.

7. Which activity is a step in the ovarian cycle?
   a. FSH stimulates the pituitary to release GnRH.
   b. When FSH and LH levels fall, the corpus luteum shrinks and the uterine lining breaks down.
   c. Luteinizing hormone stimulates the uterus to make progesterone.
   d. Estrogen levels initially have a positive feedback effect on the pituitary, which is followed by higher estrogen levels causing negative feedback.
   e. A fully developed corpus luteum inhibits uterine lining growth.

8. During spermatogenesis in mammals, sperm travels from the:
   a. Sertoli cells past the epididymis and urethra, through the vas deferens to the prepuce.
   b. seminal vesicles past the prostate gland, through the glans and prepuce to the bulbourethral glands.
   c. vestibular glands past the Leydig cells, through the accessory glands and epididymis to the vas deferens.
   d. labia past the bulbourethral glands, through the vas deferens and urethra to the epididymis.
   e. seminiferous tubules past the Sertoli cells, through the epididymis and vas deferens to the urethra.

9. The secondary oocyte in humans is fertilized in the:
   a. uterus.
   b. vagina.
   c. oviduct.
   d. cervical canal.
   e. ovary.

10. The most effective method to prevent fertilization is:
    a. the oral contraceptive.
    b. the IUD.
    c. the morning-after pill.
    d. vasectomy in men and tubal ligation in women.
    e. the rhythm method.

Apply/Analyze

11. **Discuss Concepts** How would an "antipregnancy vaccine" that stimulates a woman's immune system to develop antibodies against human chorionic gonadotropin (hCG) prevent pregnancy?

12. **Discuss Concepts** Men sometimes have reduced fertility because of *testicular varicoceles,* varicose veins in the testes in which blood pools. Based on what you now know of the conditions under which sperm develop properly, how do you think this condition might impair sperm development?

13. **Discuss Concepts** Spermatogenesis produces four sperm for each primary spermatocyte, but oogenesis produces only one egg for each primary oocyte. Why might these different outcomes be adaptive?

14. **Discuss Concepts** Sertoli cells protect spermatocytes from attack by antibodies during their development in the human male. What structures might protect the oocyte and egg from attack by antibodies in the human female?

15. **Discuss Concepts** Compare the advantages and disadvantages of sexual and asexual reproduction for an aphid and a parasitic worm.

Evaluate/Create

16. **Discuss Concepts** It may be possible to develop a birth control drug that would prevent conception by interfering with fertilization. Outline the design for such a drug and explain exactly how it would work. (There may be more than one design that, in theory, would be effective.)

17. **Design an Experiment** Design experiments to determine if, and at what dose, vitamin E can decrease menstrual cramping significantly.

18. **Apply Evolutionary Thinking** The nematodes *Caenorhabditis elegans* and *Caenorhabditis briggsae* are both hermaphroditic. Phylogenetic evidence indicates that the last common ancestor of these two species had a normal male–female mechanism of reproduction. What does this evidence suggest about their hermaphroditism?

For selected answers, see Appendix A.

INTERPRET THE DATA

Contamination of water by agricultural chemicals affects the reproductive function of some animals. Are there effects on humans? Epidemiologist Shanna Swan and her colleagues studied sperm collected from men in four cities in the United States **(Table).** The men were partners of women who had become pregnant and were visiting a prenatal clinic, so all were fertile. Of the four cities, Columbia, Missouri, is located in the county with the most farmlands. New York City represents an area with no agriculture.

Data from a Study of Sperm Collected from Men Who Were Partners of Pregnant Women Who Visited Prenatal Health Clinics in One of Four Cities

	Location of Clinic			
	Columbia, Missouri	Los Angeles, California	Minneapolis, Minnesota	New York, New York
Average age	30.7	29.8	32.2	36.1
Percent nonsmokers	79.5	70.5	85.8	81.6
Percent with history of STD*	11.4	12.9	13.6	15.8
Sperm count (million/mL)	58.7	80.8	98.6	102.9
Percent motile sperm	48.2	54.5	52.1	56.4

*STD stands for sexually transmitted disease.

1. Where did researchers record the highest and lowest sperm counts?

   In which cities did samples show the highest and lowest sperm motility (ability to move)?

2. Aging, smoking, and STDs adversely affect sperm. Could differences in any of these variables explain the regional differences in sperm count?

3. Do these data support the hypothesis that living near farmlands can adversely affect male reproductive function?

Source: S. Swan et al. 2003. Geographic differences in semen quality of fertile US males. *Environmental Health Perspectives* 111:414–420.

50

Animal Development

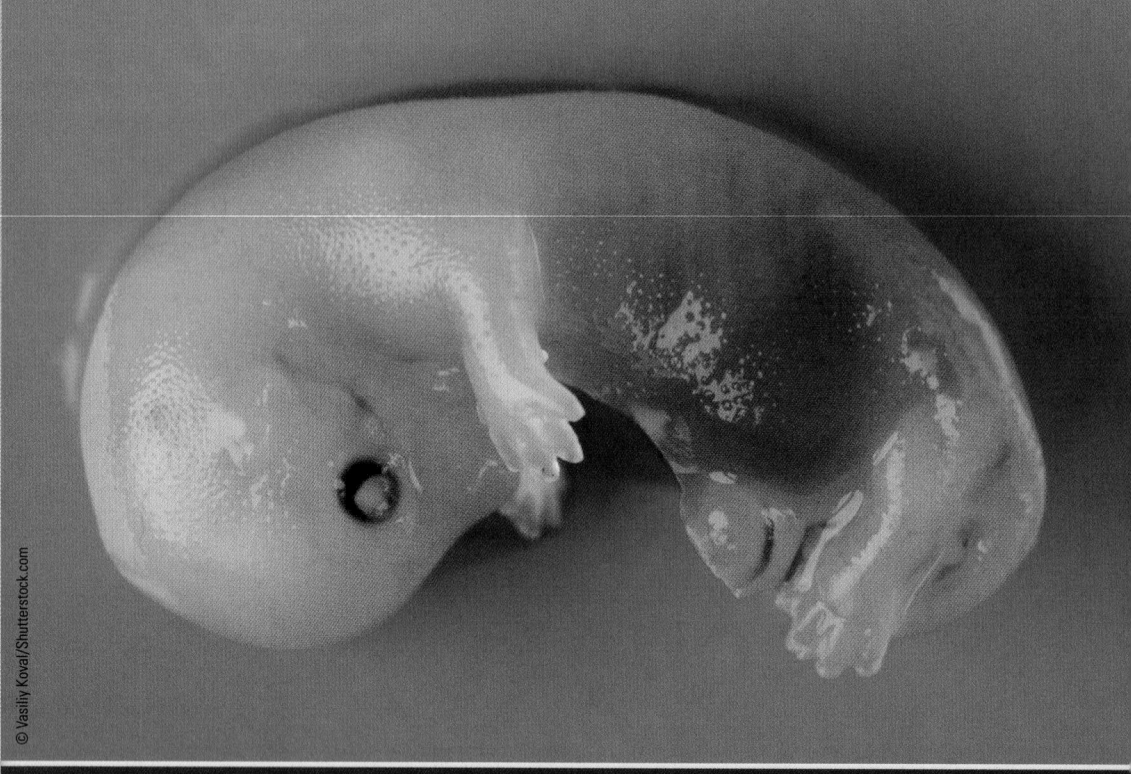

Mink embryo.

© Vasiliy Koval/Shutterstock.com

Why it matters . . . The uterine contractions announcing birth are taking place at shorter intervals and with greater intensity. The mother-to-be endures the discomfort and apprehension with the knowledge that the child that has been growing in her body will soon come into the world. It formed from a fertilized egg, about the size of a period on this page, and grew through a program of cell divisions, complex cell migrations, and molecular interactions.

Over the course of its development, the baby's body formed all the organ systems required for independent existence and, at its birth, they are already working to sustain its life. Most astonishing, perhaps, is the baby's brain. It formed from a tube of nerve tissue that bulged outward and enlarged, continually adding nerve cells and connecting them into complex circuits—all as part of the automated events of development. Still, the human brain is unique only in the degree of its complexity and integrative capacity; the brains of other mammals are basically similar and develop through the same embryonic pathways.

The baby enters the outside world passing head first through the cervix, and then the vagina. It is a girl, who with further luck and good care will continue developing through childhood, puberty, adult life, and old age, all through programs built into her DNA. As part of these passages, she may bring her own child into the world.

People have tried since ancient times to understand how development and birth take place. The scientific quest began with Aristotle, who observed chick development and correctly interpreted the functions of the placenta and umbilical cord in humans. The investigators who followed Aristotle concentrated on describing developmental changes in **morphology,** which is the form or shape of an organism, or of a part of an organism. More recently, investigators began to trace the molecular underpinnings of the morphological events.

In this chapter we survey the results of these investigations. We take up the story of animal development where the previous chapter left off, with the fertilized egg. We continue with the early events leading from the fertilized egg to the primary tissues of the embryo, and then trace the development of organs from these tissues. Next, we describe human development as representative of the process in mammals. Then, we survey the cellular bases of these mechanisms. At the cellular level, the development of an adult animal from a fertilized egg involves cell division, in which more cells are produced by mitosis; **cell differentiation,** in which changes in gene expression establish cells with specialized structure and function; and **morphogenesis** ("form creation"), the generation of the body form of the animal as differentiated cells end up in their appropriate sites.

50.1 Mechanisms of Embryonic Development

Fertilization of an egg by a sperm cell produces a zygote. Embryonic development begins at this point and ultimately produces a free-living organism. All the instructions required for development are packed into the fertilized egg.

Developmental Information and Components for Growth Are Stored in the Fertilized Egg

Once the zygote is formed, mitotic divisions begin the developmental activity (see Section 49.2).

INFORMATION STORAGE IN THE EGG As introduced in Section 16.4, the initiation of development depends primarily on the DNA in the zygote nucleus, and on *cytoplasmic determinants,* that is, mRNA and protein molecules stored in the cytoplasm. Most of the mRNA and protein molecules are maternal in origin because the fertilizing sperm contributes essentially no cytoplasm to the zygote. The mRNAs and proteins direct the first stages of development up until the genes of the zygote become active.

OTHER COMPONENTS OF THE EGG In addition to cytoplasmic determinants, the egg cytoplasm also contains ribosomes and other cytoplasmic components required for protein synthesis and the early cell divisions of embryonic development. For example, the egg cytoplasm contains all the tubulin molecules required to form the spindles for early cell divisions. It also contains mitochondria, nutrients stored in granules in the yolk and in lipid droplets, and, in many animals, pigments that color the egg or regions of it.

Yolk contains nutrients. The eggs of insects, reptiles, and birds contain large amounts of yolk, which supplies all of the nutrients for development of the embryo. The eggs of placental mammals, in contrast, contain very little yolk, which supplies nutrients only for the earliest stages of development.

Depending on the species, the yolk may be concentrated at one end or in the center of the egg, or distributed evenly throughout the cytoplasm. Its distribution influences the rate and location of cell division during early embryonic development. Typically, cell division proceeds more slowly in the region of the egg containing the yolk. In the large, yolky eggs of birds and reptiles, cell division takes place only in a small, yolk-free patch at the surface of the egg.

Unequal distribution of yolk and other components in a mature egg is termed **polarity.** In most species, the egg nucleus is located toward one end of the egg. This end of the egg, called the **animal pole,** typically gives rise to surface structures and the anterior end of the embryo. The opposite end of the egg, the **vegetal pole,** typically gives rise to internal structures such as the gut and the posterior end of the embryo. Yolk, when unequally distributed in the egg cytoplasm, is most commonly concentrated in the vegetal half of the egg.

Fertilization of the egg launches the early stages of development. As you follow these stages, keep in mind that the developmental processes involved work together to produce the body plan of the adult. For a bilaterally symmetrical animal such as a human or a dog, for example, the adult must develop three body axes: the anterior–posterior (head–tail) axis, the dorsal–ventral (back–front) axis, and the left–right axis (see Figure 31.3).

Cleavage, Gastrulation, and Organogenesis Are Early Stages in Development

The three early stages in development are *cleavage, gastrulation,* and *organogenesis.* With modifications, these three stages are common to the early development of most animals. **Figure 50.1** outlines these stages in the life cycle of a frog.

CLEAVAGE Soon after fertilization, the zygote begins **cleavage,** a series of mitotic divisions in which cycles of DNA replication and division occur *without* the production of new cytoplasm. As a result, the cytoplasm of the egg is partitioned into successively smaller cells without increasing the overall size or mass of the embryo **(Figure 50.2).** These cleavage-stage cells are called **blastomeres** (*blastos* = bud or offshoot; *meros* = part or division). In the frog *Xenopus laevis,* for example, twelve cleavage divisions produce an embryo of about 4,000 blastomeres.

The initial cleavage divisions produce a **morula** (*morula* = mulberry), a solid ball or layer of blastomeres. As cleavage divisions continue, the ball or layer hollows out to form the **blastula** (*ula* = small), in which the blastomeres enclose a fluid-filled cavity, the **blastocoel** (*koilos* = hollow) (shown for the frog life cycle in Figure 50.2).

GASTRULATION Once cleavage is complete, the blastomeres undergo extensive cellular rearrangements. This morphogenetic process is called **gastrulation,** and the result is the production of an embryo with three distinct primary cell layers (see Figure 50.1). During the process the embryo is termed a **gastrula**

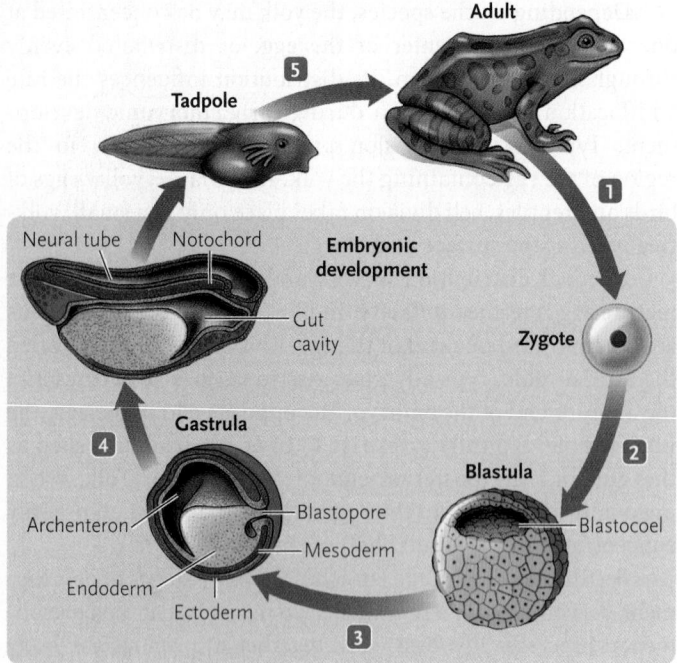

Neural tube   Notochord

Embryonic development

Gut cavity

Zygote

Gastrula

Archenteron
Blastopore
Mesoderm
Endoderm
Ectoderm

Blastula

Blastocoel

Tadpole

Adult

1 **Fertilization:** A sperm penetrates an egg and their nuclei fuse, producing a zygote.

2 **Cleavage:** Mitotic cell divisions produce a morula, a solid ball of blastomeres. Further divisions produce a blastula, a hollowed ball of blastomeres.

3 **Gastrulation:** Cell divisions, cell migrations, and rearrangements produce a gastrula, an early embryo that has primary tissue layers.

4 **Organogenesis:** Cell divisions, cell movements, and other cellular mechanisms produce the major tissues and organ systems, and a body organization characteristic of the species.

5 **Metamorphosis:** The animal develops into the adult, with characteristic adult appearance and all tissues and organs carrying out their specialized functions.

FIGURE 50.1 The first three cleavage divisions of a frog embryo, which convert the fertilized egg into the eight-cell stage. Note that the cleavage divisions cut the volume of the fertilized egg into successively smaller cells, called blastomeres.

© Cengage Learning 2017

TABLE 50.1	Origins of Adult Tissues and Organs in the Three Primary Tissue Layers
Primary Tissue Layer	**Adult Tissues and Organs**
Ectoderm	Skin and its elaborations, including hair, feathers, scales, and nails; nervous system, including brain, spinal cord, and peripheral nerves; lens, retina, and cornea of eye; lining of mouth and anus; sweat glands, mammary glands, adrenal medulla, and tooth enamel
Mesoderm	Muscles; most of skeletal system, including bones and cartilage; circulatory system, including heart, blood vessels, and blood cells; internal reproductive organs; kidneys and outer walls of digestive tract
Endoderm	Lining of digestive tract, liver, pancreas, lining of respiratory tract, thyroid gland, lining of urethra, and urinary bladder

the **mesoderm** (*meso* = middle) between the ectoderm and the endoderm. Gastrulation establishes the body pattern—each tissue and organ of the adult animal originates in one of the three primary cell layers of the gastrula **(Table 50.1).**

As gastrulation proceeds, embryonic cells begin to differentiate: they become recognizably different in biochemistry, structure, and function. The developmental potential of the cells also becomes more limited than that of the fertilized egg from which they originated. In other words, although a fertilized egg is **totipotent,** meaning that it is capable of producing all the various types of cells of the adult, progressively the cells produced become more specialized. Totipotent cells give rise to **pluripotent** cells, which can give rise to most, but not all, adult cell types, and then pluripotent cells give rise to **multipotent** cells, which give rise to cells with particular functions. Thus, for example, a multipotent mesoderm cell may develop into muscle or bone but not normally into skin or brain.

ORGANOGENESIS **Organogenesis** involves rearrangements of the three germ layers to produce tissues and organs. At the end of organogenesis, the embryo has the body organization characteristic of its species.

(*gaster* = gut or belly). The details of gastrulation differ among animal groups, but the result is the same for all. That is, some surface cells of the blastula move to an interior position and three **germ layers** are formed: the outer **ectoderm** (*ecto* = outside; *derma* = skin), the inner **endoderm** (*endo* = inside), and

A. Fertilized egg

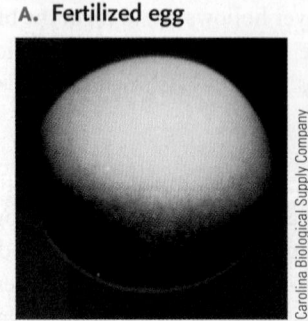

B. Two-cell stage

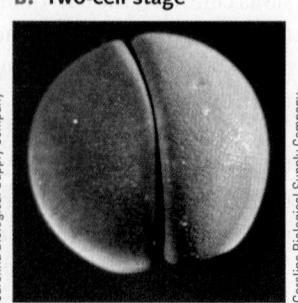

C. Four-cell stage

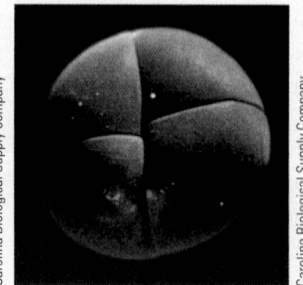

D. Eight-cell stage

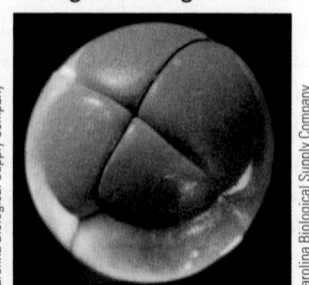

Carolina Biological Supply Company

FIGURE 50.2 Stages of animal development shown in a frog.

Several Cellular Processes Are Responsible for Development

Development in all animals is accomplished by a number of cellular processes that are under genetic control but are influenced to some extent by the environment (for example, temperature affects the rate of cell division):

1. Mitotic cell divisions.
2. Cell migrations.
3. **Selective cell adhesions,** in which cells make and break specific connections to other cells or to the extracellular matrix.
4. **Induction,** in which one group of cells (the inducer cells) causes or influences another nearby group of cells (the responder cells) to follow a particular developmental pathway (see Section 16.4).
5. **Determination,** in which the developmental fate of a cell is set, and it is committed to becoming a particular cell type. Typically, determination is the result of induction.
6. **Differentiation,** which follows determination, establishes a cell-specific developmental program in cells. Differentiation results in cell types with clearly defined structures and functions.
7. **Apoptosis,** programmed cell death, in which tissues no longer required for continued development of the organism are removed.

Examples of these cellular processes in animal development are discussed in the following three sections.

STUDY BREAK 50.1

1. How do cleavage divisions differ from cell division in an adult organism?
2. What are the primary cell layers of the embryo, and what process is responsible for producing them?

50.2 Major Patterns of Cleavage and Gastrulation

With the principles of early embryonic development established, we describe cleavage and gastrulation in three animal groups that have been models in *embryology* (the study of embryos and their development): sea urchins, amphibians, and birds. Later in the chapter, we describe cleavage and gastrulation in humans and other mammals, which resemble the pattern in birds.

Sea Urchin Gastrulation Follows a Symmetrical Pattern That Reflects an Even Distribution of Yolk

The sea urchin is an echinoderm, a type of deuterostome (see Section 32.1). Cleavage divisions first produce a solid mass of blastomeres, the morula (**Figure 50.3,** step 1). Continued divisions produce a blastula containing about a thousand cells (step 2).

Cleavage

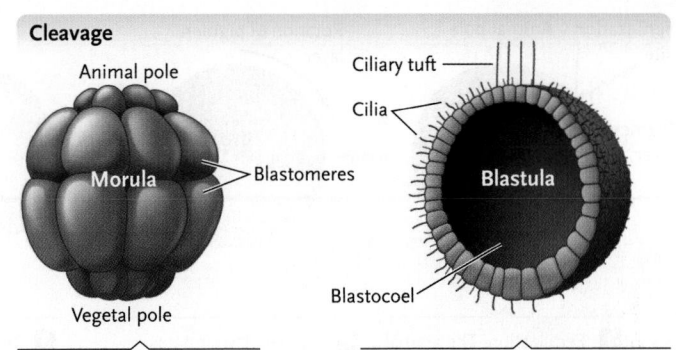

1 Cleavage divisions first produce a solid ball of blastomeres called a morula.

2 Further cleavage divisions produce a blastula, a hollow ball in which blastomeres enclose a fluid-filled cavity.

Gastrulation

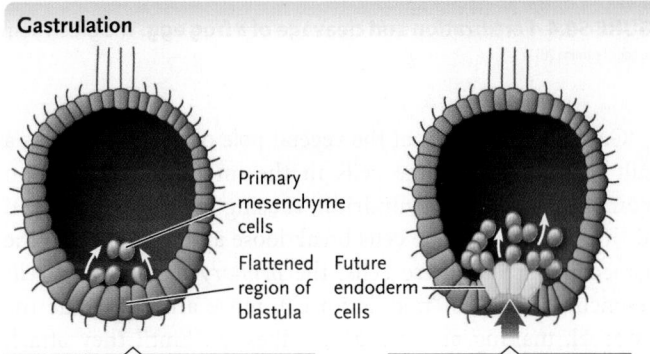

3 Gastrulation begins at the vegetal pole. Some cells are induced to change shape to produce a flattened region. Primary mesenchyme cells (future mesoderm) then break loose and migrate into the blastocoel.

4 The flattened vegetal pole consisting of future endoderm cells begins to invaginate.

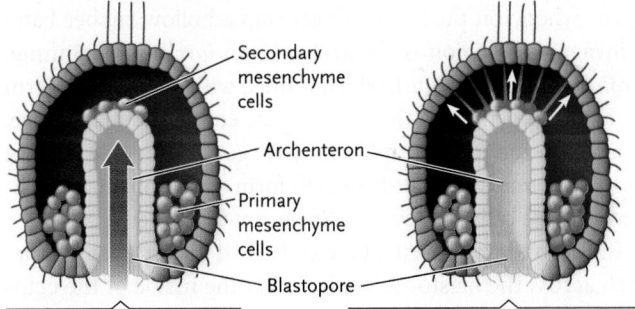

5 Invagination continues. Secondary mesenchyme cells (also future mesoderm) form at the top of the archenteron.

6 Secondary mesenchyme cells send out extensions that stretch across the blastocoel and adhere to the ectoderm.

KEY

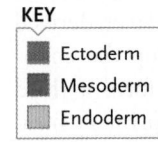

- Ectoderm
- Mesoderm
- Endoderm

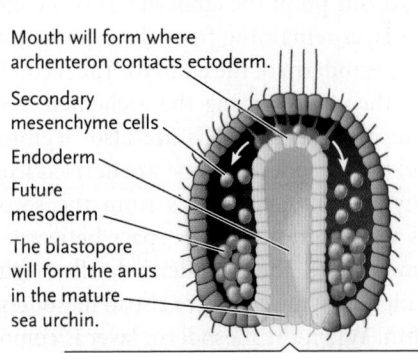

Mouth will form where archenteron contacts ectoderm.

Secondary mesenchyme cells

Endoderm

Future mesoderm

The blastopore will form the anus in the mature sea urchin.

FIGURE 50.3 Cleavage and gastrulation in the sea urchin.
© Cengage Learning 2017

7 Ectoderm and endoderm layers have formed; mesoderm cells are between them, some derived from primary mescenchyme cells and others from secondary mesenchyme cells.

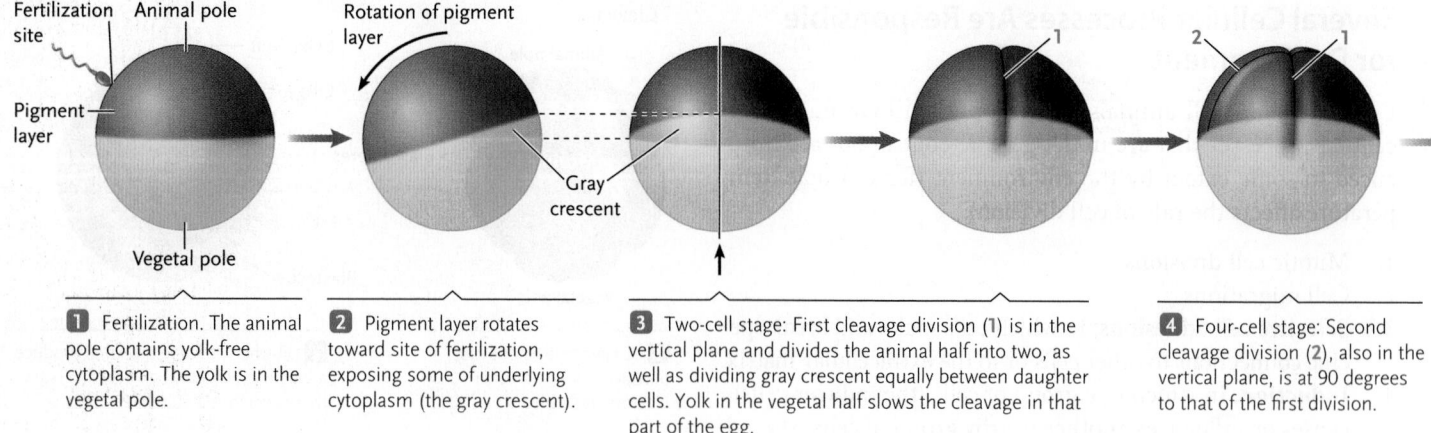

① Fertilization. The animal pole contains yolk-free cytoplasm. The yolk is in the vegetal pole.

② Pigment layer rotates toward site of fertilization, exposing some of underlying cytoplasm (the gray crescent).

③ Two-cell stage: First cleavage division (1) is in the vertical plane and divides the animal half into two, as well as dividing gray crescent equally between daughter cells. Yolk in the vegetal half slows the cleavage in that part of the egg.

④ Four-cell stage: Second cleavage division (2), also in the vertical plane, is at 90 degrees to that of the first division.

FIGURE 50.4 Fertilization and cleavage of a frog egg. The green numbers designate the cleavage divisions.
© Cengage Learning 2017

Gastrulation begins at the vegetal pole of the blastula. As a result of induction, some cells in the middle of that region become elongated and cylindrical, causing the region to flatten and thicken. Then, some cells break loose and migrate into the blastocoel (step 3). These cells, the *primary mesenchyme cells* (mesenchyme means "middle juice"), move around inside the blastocoel, making and breaking adhesions, until they attach along the ventral sides of the blastocoel. These cells eventually become mesoderm cells (see step 7), from which the larval skeleton is produced. Next, the flattened vegetal pole of the blastula folds inward from the surface in a process called **invagination** (steps 4 and 5). (Picture invagination as being similar to what happens when you push your finger into a hollow rubber ball.) The invaginated region is the **archenteron** (*arche* = beginning; *enteron* = intestine or gut), which is lined with future endoderm cells, and its opening at the vegetal pole end is the **blastopore.** The archenteron forms the primitive gut of the sea urchin embryo. *Secondary mesenchyme cells* form at the top of the archenteron. These cells also eventually become mesoderm cells.

The secondary mesenchyme cells send out extensions that stretch across the blastocoel and contact the inside of the ectoderm (step 6). These extensions make tight adhesions and then contract, pulling the invaginated cell layer inward and thereby eliminating most of the blastocoel.

At this point the embryo has two complete cell layers. The outer layer remaining from the original blastula surface makes up the ectoderm of the embryo. The second, inner layer, derived from the cells forming the archenteron, makes up the endoderm. Mesodermal cells are also beginning to form a third layer, the mesoderm. Some are derived from the primary mesenchyme cells and others from the secondary mesenchyme cells that migrate into the space between ectoderm and endoderm (step 7). The mesodermal cells originating from secondary mesenchyme cells give rise to mesodermal organs of the sea urchin. When the mesoderm layer is complete, the embryo has three complete layers: ectoderm, mesoderm, and endoderm. At this point, cells within each layer begin to differentiate, as evidenced by the synthesis of different proteins in each layer.

As the ectoderm, mesoderm, and endoderm develop, the embryo lengthens into an ellipsoidal shape with the blastopore marking its posterior end. From this point on, organ systems differentiate through further cell division, cell migrations, selective cell adhesions, induction, and differentiation.

An archenteron and a blastopore are not present in the embryos of all organisms. For those that do have them, the blastopore gives rise to the anus or mouth of the embryo, depending on the animal group (see Section 31.2). In deuterostomes, which include echinoderms and chordates, the blastopore develops into the anus and the mouth forms at the opposite end of the embryonic gut where the archenteron contacts and fuses with the ectoderm. In the protostomes, which include annelids, arthropods, and mollusks, the blastopore develops into the mouth, and the anus forms at the opposite end of the embryonic gut.

Amphibian Cleavage and Gastrulation Are Influenced by an Unequal Distribution of Yolk

In amphibian eggs, such as those of frogs, yolk is concentrated in the vegetal half, which gives it a pale color. The animal half is darkly colored by a layer of pigment granules just below the surface. **Figure 50.4** shows the steps of fertilization and cleavage of a frog egg. The sperm typically fertilizes the egg in the animal half (step 1). After fertilization, the pigmented layer of cytoplasm rotates toward the site of sperm entry, exposing a crescent-shaped region of the underlying cytoplasm at the side opposite the point of sperm entry (step 2). This region, called the **gray crescent,** establishes the dorsal–ventral axis of the embryo, with the gray crescent marking the future dorsal side. The first cleavage division runs perpendicular to the long axis of the gray crescent and divides the crescent equally between the resulting two cells (step 3).

If one of the first two blastomeres does not receive gray crescent material, and the cells are separated experimentally, the cell without gray crescent divides to produce a disordered

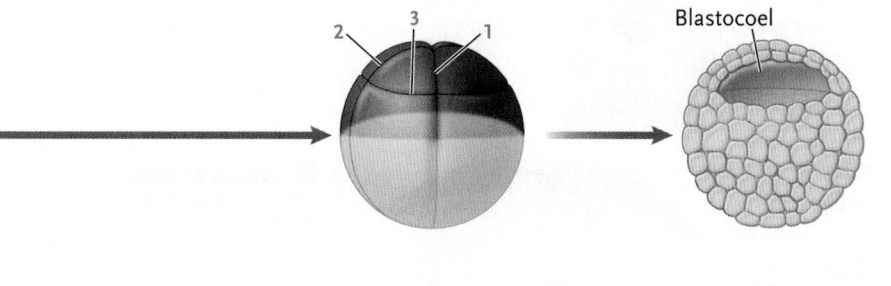

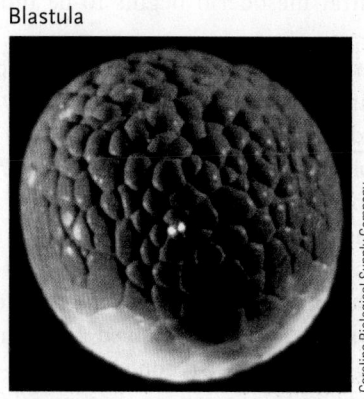

Blastocoel

Blastula

5 Eight-cell stage: Third cleavage division (**3**) is in the horizontal plane within the animal half of the cell. Divisions **1** and **2** have now continued through the vegetal half of the egg.

6 Blastula stage (cross section): More cleavage divisions produce the blastula, which contains at least 128 cells and has a fluid-filled blastocoel.

mass that stops developing. The cell receiving the gray crescent produces a normal embryo. Thus cytoplasmic material localized in the gray crescent is essential to normal development in frog embryos.

The second cleavage division is also in the vertical plane and at 90 degrees to the first cleavage division, producing the four-cell stage (step 4). At this point, the yolk in the vegetal half of the egg has slowed the progression of the first two cleavage furrows from going through that part of the egg. The third cleavage division is in the horizontal plane in the animal half of the egg and produces the eight-cell stage (step 5). By now the cleavage furrows of the first two cleavage divisions have continued all the way through the vegetal half of the egg. Embryo development continues with more divisions, first producing a morula (defined as a 16- to 64-cell stage in amphibians), and then a blastula (defined as at least 128 cells in amphibians) (step 6).

Gastrulation begins when cells from the animal pole move across the embryo surface and reach the region derived from the gray crescent (**Figure 50.5,** steps 1–2). This site is marked by a crescent-shaped depression rotated clockwise 90° called the

dorsal lip of the blastopore. The depression results from invagination of cells from the surface. With continued inward migration of additional cells, the depression eventually forms a complete circle, which is the blastopore.

Cells now migrate into the blastopore by **involution,** a process in which the cells entering from the outer layer of the embryo spread over the internal surfaces of the remaining exterior cells (step 3). A consequence of involution is that the pigmented cell layer of the animal half expands to cover the entire surface of the embryo. The cells of the vegetal half are enclosed by the movement and show on the outside as a yolk plug in the blastopore (step 4). In amphibians, the blastopore gives rise to the anus.

Continued involution moves cells into the interior and upward (see steps 3 and 4), forming two layers that line the inside top half of the embryo. The uppermost of these layers is induced to become the dorsal mesoderm (see step 4). The layer beneath it, which contains cells originating from both the outer surface of the embryo and the yolky interior, becomes the endoderm (shown in yellow). The pigmented cells remaining at the surface of the embryo form the ectoderm (shown in

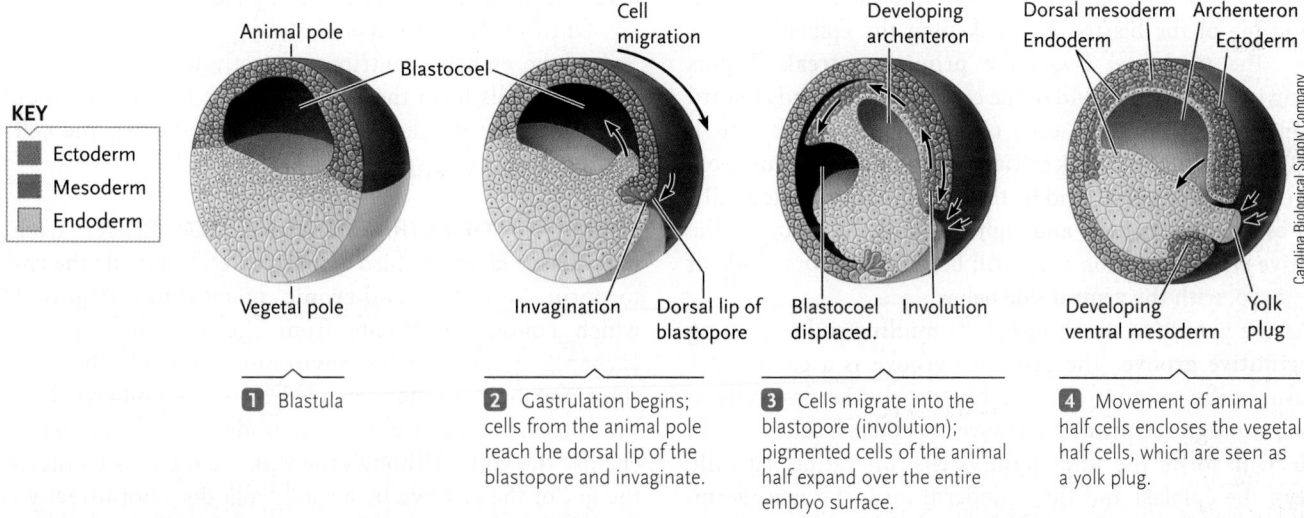

KEY

- Ectoderm
- Mesoderm
- Endoderm

1 Blastula

2 Gastrulation begins; cells from the animal pole reach the dorsal lip of the blastopore and invaginate.

3 Cells migrate into the blastopore (involution); pigmented cells of the animal half expand over the entire embryo surface.

4 Movement of animal half cells encloses the vegetal half cells, which are seen as a yolk plug.

FIGURE 50.5 Gastrulation in a frog embryo.
© Cengage Learning 2017

blue). The ventral mesoderm begins to be induced near the vegetal pole.

As the mesoderm and endoderm form, the depression created by the inward cell migration gradually deepens and extends inward as the archenteron (see steps 3 and 4), which displaces the blastocoel. The cells of the three primary cell layers continue to increase in number by further cell migration and divisions as development proceeds.

During frog gastrulation, cells of the dorsal lip of the blastopore are inducer cells that control blastopore formation; if the cells in the dorsal lip are removed and transplanted elsewhere in the egg, they cause a second blastopore—and a second embryo—to form in this region (see Section 50.5).

The events of gastrulation in frogs thus include the same developmental processes as in sea urchins—cell divisions, cell migrations, selective adhesions, induction, and differentiation.

Gastrulation in Birds Proceeds at One Side of the Yolk

Gastrulation in amniotes (see Section 32.7) such as birds and reptiles is modified by the distribution of yolk, which occupies almost the entire volume of the egg. The portion of the cytoplasm that divides to give rise to the primary tissues of the embryo is confined to a thin layer at the egg surface. Although eggs of mammals (which are also amniotes) have little yolk, gastrulation follows a similar pattern, as discussed in Section 50.4.

CLEAVAGE AND GASTRULATION IN BIRDS The early cleavage divisions in birds produce a disclike layer of cells at the surface of the yolk called the **blastodisc (Figure 50.6,** step 1). When blastodisc formation is complete, the layer contains about 20,000 cells. The cells of the blastodisc then separate into two layers, called the **epiblast** (top layer) and **hypoblast** (bottom layer). The flattened cavity between them is the blastocoel (step 2).

Gastrulation begins as cells in the epiblast stream toward the midline of the blastodisc, thickening the epiblast in this region. The thickened layer—the **primitive streak**—begins forming in the posterior end of the embryo and extends toward the anterior end as more cells of the epiblast move into it (step 3). The primitive streak initially designates the future posterior end of the embryo, and by the time it has elongated fully, it has established the left and right sides of the embryo. The primitive streak forms on what will become the dorsal side of the embryo, with the ventral side below.

As the primitive streak forms, its midline sinks, forming the **primitive groove.** The primitive groove is a conduit for migrating cells to move into the blastocoel. The first cells to migrate through the primitive groove are epiblast cells (step 4), which will form the endoderm. Cells migrating laterally between the epiblast and the endoderm form the mesoderm. The epiblast cells left at the surface of the blastodisc form the ectoderm (see step 4). Thus all three of the primary tissue layers

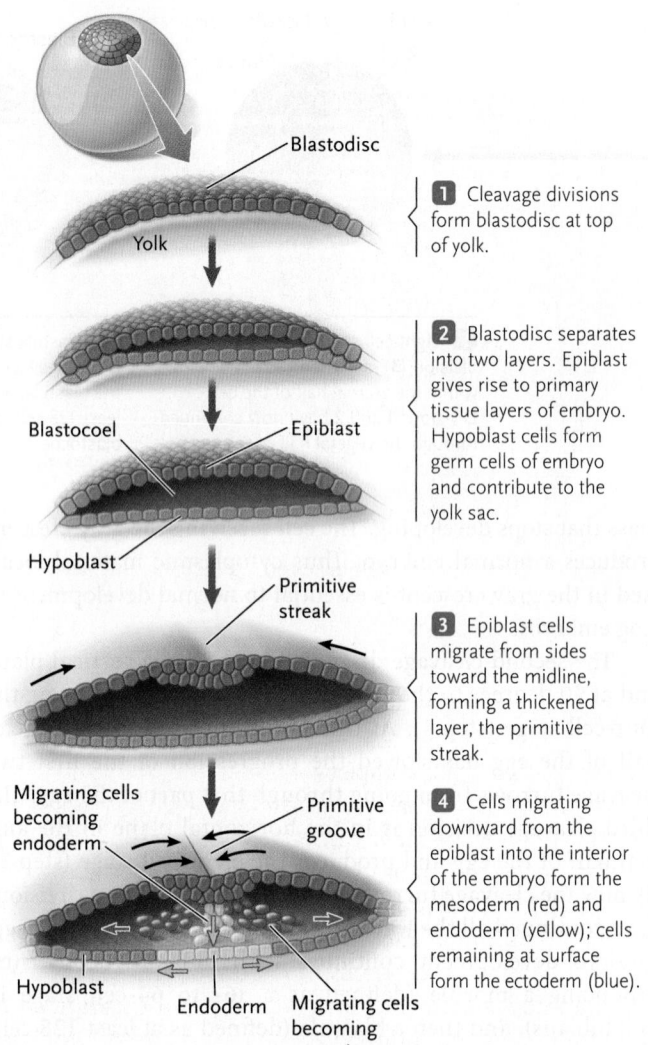

1 Cleavage divisions form blastodisc at top of yolk.

2 Blastodisc separates into two layers. Epiblast gives rise to primary tissue layers of embryo. Hypoblast cells form germ cells of embryo and contribute to the yolk sac.

3 Epiblast cells migrate from sides toward the midline, forming a thickened layer, the primitive streak.

4 Cells migrating downward from the epiblast into the interior of the embryo form the mesoderm (red) and endoderm (yellow); cells remaining at surface form the ectoderm (blue).

FIGURE 50.6 Gastrulation in a bird embryo.
© Cengage Learning 2017

of the chick embryo arise from the epiblast. Once the three layers are formed, gastrulation is complete.

Of the cells in the hypoblast, only a few near the posterior end of the embryo contribute directly to the embryo. These hypoblast cells form the *germ cells* that, later in development, migrate to the developing gonads and establish the cell line leading to eggs and sperm (see Section 49.2).

FORMATION OF EXTRAEMBRYONIC MEMBRANES Each primary tissue layer of a bird embryo extends outside the embryo to form four **extraembryonic membranes (Figure 50.7),** which conduct nutrients from the yolk to the embryo, exchange gases with the environment outside the egg, and store metabolic wastes removed from the embryo. The **yolk sac** consists of extensions of mesoderm and endoderm that enclose the yolk. Although the yolk sac remains connected to the gut of the embryo by a stalk, yolk does not directly enter the embryo by this route. Instead, it is absorbed by blood vessels in the membrane, which transport the nutrients to the

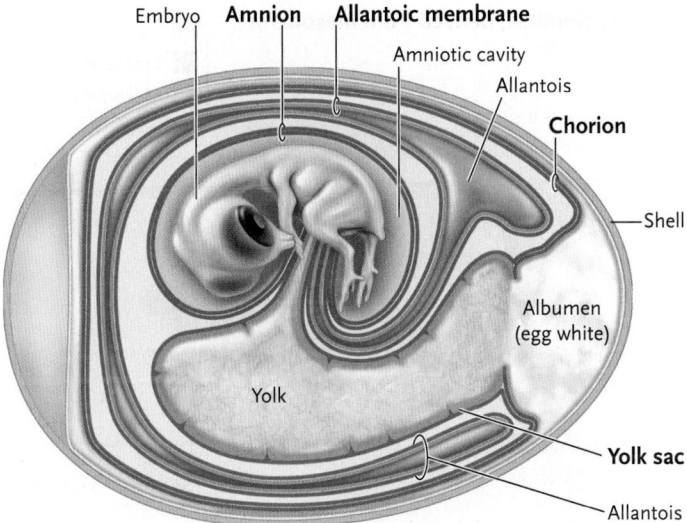

Embryo **Amnion** **Allantoic membrane**

Amniotic cavity

Allantois

Chorion

Shell

Albumen
(egg white)

Yolk

Yolk sac

Allantois

FIGURE 50.7 The four extraembryonic membranes in a bird embryo (in bold).
© Cengage Learning 2017

embryo. The **chorion,** produced from ectoderm and mesoderm, is the outermost membrane, which surrounds the embryo and yolk sac completely, and lines the inside of the egg shell. This membrane exchanges oxygen and carbon dioxide with the environment through the shell of the egg. The **amnion** is the innermost membrane, which closes over the embryo to form the *amniotic cavity.* The cells of the amnion secrete *amniotic fluid* into the cavity, which bathes the embryo and provides an aquatic environment in which it can develop. Reptilian and mammalian embryos are also surrounded by an amnion and amniotic fluid. The adaptation of providing the embryo with an aquatic environment made the development of fully terrestrial vertebrates possible. The evolutionary importance of the amnion to the fully terrestrial vertebrates is recognized by classifying them together as amniotes (see Section 32.7). A membrane derived from mesoderm and endoderm that has bulged outward from the gut forms a sac called the **allantois.** This sac closely lines the chorion and fills much of the space between the chorion and the yolk sac. The allantois stores nitrogenous wastes (primarily uric acid—see Section 48.1) removed from the embryo. In addition, the part of the allantoic membrane that lines the chorion forms a rich bed of blood capillaries that is connected to the embryo by arteries and veins. This circulatory system delivers carbon dioxide to the chorion and picks up the oxygen that is absorbed through the shell and chorion.

STUDY BREAK 50.2

1. What is the role of the gray crescent in amphibian development?
2. What evidence indicates that cells of the dorsal lip of the blastopore act as inducer cells?
3. What are the extraembryonic membranes in birds, and what are their functions?

50.3 From Gastrulation to Adult Body Structures: Organogenesis

Following gastrulation, organogenesis—the process by which the ectoderm, mesoderm, and endoderm develop into organs—gives rise to an individual with the body organization characteristic of its species. Organogenesis involves the same mechanisms used in gastrulation—cell division, cell migrations, selective cell adhesion, induction, and differentiation—plus an additional mechanism, *apoptosis,* in which certain cells are programmed to die (apoptosis is also discussed in Section 45.2 and later in this section). As with other aspects of development, the details of organogenesis differ among animal groups. The frog and the chick are used here as examples.

The Nervous System Develops from Ectoderm

In vertebrates, organogenesis begins with **neurulation,** the development of the nervous system from ectoderm. Preliminary to neurulation, cells of the dorsal mesoderm form the **notochord,** a solid rod of tissue that extends the length of the embryo under the dorsal ectoderm. Dorsal mesoderm cells under the ectoderm then induce the ectoderm cells above them to thicken and flatten into a longitudinal band called the **neural plate** (**Figure 50.8,** step 1).

Next, the neural plate sinks downward along its midline (step 2), creating a deep, longitudinal groove. At the same time, the edges elevate along the sides of the neural plate. The groove becomes deeper and the edges move together (step 3). Next, the edges fuse together and close over the center of the groove, converting the neural plate into a **neural tube** that runs the length of the embryo (step 4). The neural tube then pinches off from the overlying ectoderm, which closes over the tube (step 5). The ectoderm that comprises the neural tube is called *neural ectoderm* (colored light green in figures) to distinguish it from surface ectoderm (colored blue in figures). The central nervous system, including the brain and spinal cord, develops directly from the neural tube.

During formation of the neural tube, cells of the **neural crest**—the region where the neural tube pinches off from the ectoderm—migrate to many locations in the developing embryo and become numerous different types of cells that contribute to a variety of organ systems. Neural crest cells are also ectodermal cells; they are colored dark green in Figure 50.8. (The neural crest is one of the defining features of vertebrates.) Some cells develop into cranial nerves in the head; others contribute to the bones of the inner ear and skull, the cartilage of facial structures, and the teeth. Still others form ganglia of the autonomic nervous system, peripheral nerves leading from the spinal cord to body structures, and nerves of the developing gut. Neural crest cells also move to the skin, where they form pigment cells, and to the adrenal glands, where they form the medulla of the kidney. The migration of neural crest cells contributes to the development of all vertebrates.

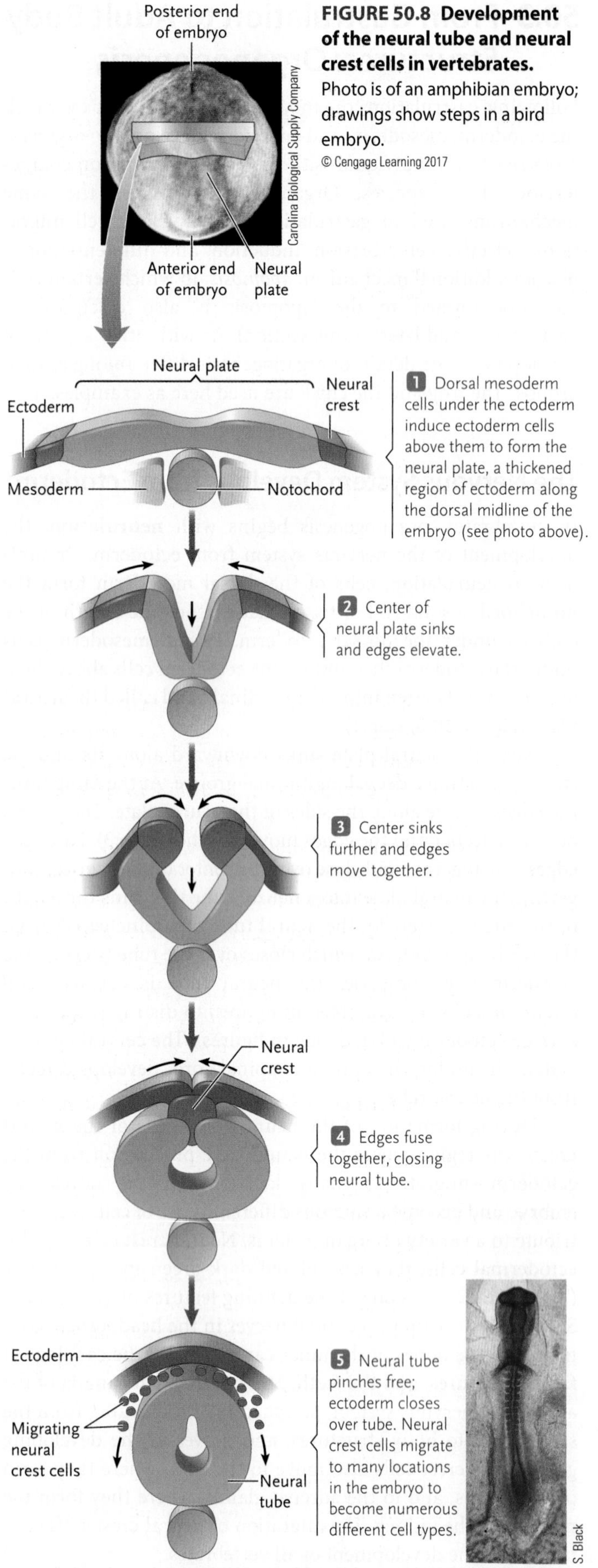

Posterior end of embryo

FIGURE 50.8 Development of the neural tube and neural crest cells in vertebrates.
Photo is of an amphibian embryo; drawings show steps in a bird embryo.
© Cengage Learning 2017

Anterior end of embryo **Neural plate**

Neural plate

Ectoderm

Neural crest

Mesoderm — Notochord

1 Dorsal mesoderm cells under the ectoderm induce ectoderm cells above them to form the neural plate, a thickened region of ectoderm along the dorsal midline of the embryo (see photo above).

2 Center of neural plate sinks and edges elevate.

3 Center sinks further and edges move together.

Neural crest

4 Edges fuse together, closing neural tube.

Ectoderm

Migrating neural crest cells

Neural tube

5 Neural tube pinches free; ectoderm closes over tube. Neural crest cells migrate to many locations in the embryo to become numerous different cell types.

S. Black

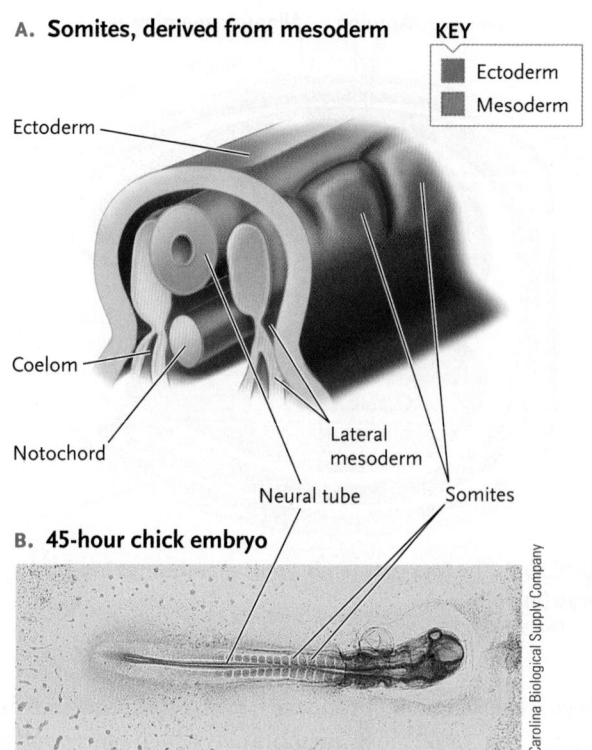

A. Somites, derived from mesoderm

KEY
- Ectoderm
- Mesoderm

Ectoderm

Coelom

Notochord

Lateral mesoderm

Neural tube

Somites

B. 45-hour chick embryo

Carolina Biological Supply Company

FIGURE 50.9 Later development of the mesoderm. (A) The somites develop into segmented structures such as the vertebrae, the ribs, and the musculature between the ribs. The lateral mesoderm gives rise to other structures, such as the heart and blood vessels and the linings of internal body cavities. **(B)** The somites in a 45-hour chick embryo.
© Cengage Learning 2017

Other structures differentiate in the embryo while the neural tube is forming. On either side of the notochord, the mesoderm separates into blocks of cells called **somites,** spaced one after the other along both sides of the notochord **(Figure 50.9).** The somites give rise to the vertebral column, the ribs, the repeating sets of muscles associated with the ribs and vertebral column, and muscles of the limbs. The mesoderm outside the somites, which extends around the primitive gut (lateral mesoderm in Figure 50.9), splits into two layers, one covering the surface of the gut, and the other lining the body wall. The space between the layers is the coelom of the adult (see Section 31.2).

Sequential Inductions and Differentiation Are Central to Eye Development

We now discuss the development of the eye to show how cellular mechanisms interact in organogenesis. Eyes develop by the same pathway in all vertebrates.

The brain forms at the anterior end of the neural tube from a cluster of hollow vesicles that swell outward from the neural tube **(Figure 50.10,** step 1). One paired set of vesicles, the *optic vesicles*, develop into the eyes. The figure depicts the optic vesicles in the brain of a frog embryo; the morphology of the forebrain, midbrain, and hindbrain in embryos differs among vertebrates.

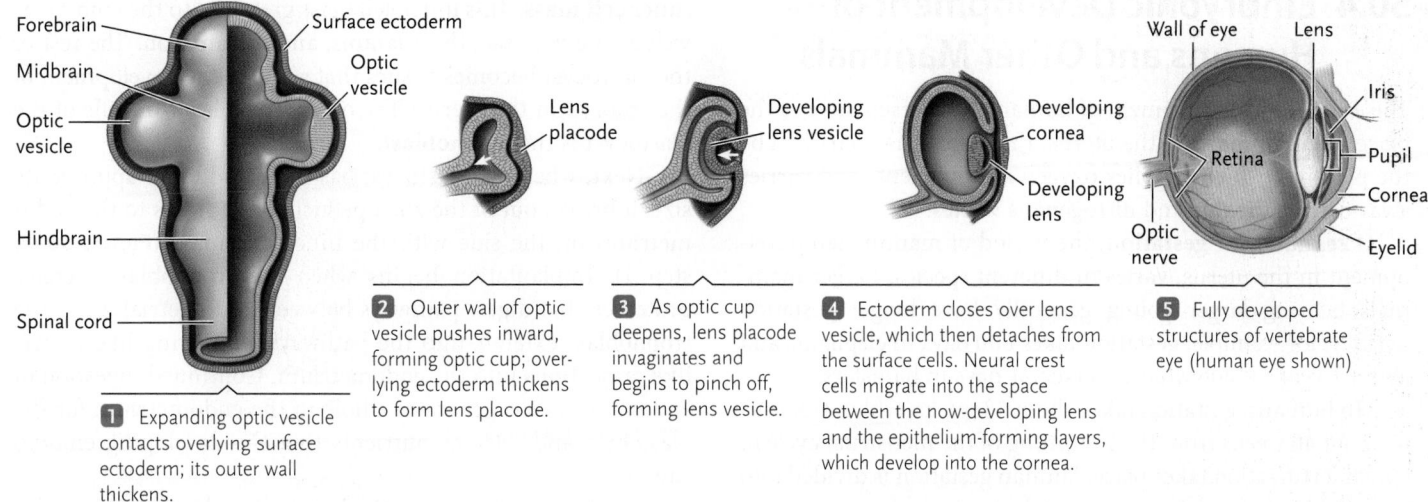

1 Expanding optic vesicle contacts overlying surface ectoderm; its outer wall thickens.

2 Outer wall of optic vesicle pushes inward, forming optic cup; overlying ectoderm thickens to form lens placode.

3 As optic cup deepens, lens placode invaginates and begins to pinch off, forming lens vesicle.

4 Ectoderm closes over lens vesicle, which then detaches from the surface cells. Neural crest cells migrate into the space between the now-developing lens and the epithelium-forming layers, which develop into the cornea.

5 Fully developed structures of vertebrate eye (human eye shown)

FIGURE 50.10 Stages in the formation of the vertebrate eye from the optic vesicle of the brain and the surface ectoderm. (The brain and spinal cord, colored light green, are derived from the neural tube, which consists of neural ectoderm.)

© Cengage Learning 2017

The optic vesicles are derived from a specific region of neural ectoderm in the neural plate. The simultaneous expression of a specific set of transcription factors—Pax6, Six3, and Rx1—in the most anterior tip of the neural plate is responsible for optic vesicle development. The optic vesicles grow outward and contact the overlying surface ectoderm, where they induce a series of developmental responses in both tissues. The outer surface of the optic vesicle thickens and flattens at the region of contact and then pushes inward, transforming the optic vesicle into a double-walled *optic cup*, which eventually becomes the retina. The optic cup induces the surface ectoderm to thicken into a disclike swelling, the *lens placode* (step 2). The center of the lens placode invaginates—sinks inward—toward the optic cup, and its edges eventually fuse together, forming a ball of ectodermal cells, the *lens vesicle* (step 3).

Ectoderm now closes over the lens vesicle, which then detaches from the surface ectoderm (step 4). The lens vesicle becomes the developing lens, the cells of which begin to synthesize *crystallins,* which are transparent, lens-specific proteins that collect into clear, glassy deposits. The lens cells finally lose their nuclei and form the elastic, crystal-clear lens.

As the developing lens invaginates, it induces the overlying surface ectoderm to begin differentiation into the cornea. First, the overlying ectoderm secretes layers of collagen. Neural crest cells then migrate into the collagen layers and form new layers of cells that eventually mature into the clear cornea. Relatively little is known about the development of the cornea, but the transcription factor Pax6 (see earlier discussion) is important for the process. The gene for Pax6 is not expressed in most tissues of adult mammals, but it is expressed in the adult cornea. The gene products that result from the action of Pax6 are important for healing wounds of the cornea.

Other cells contribute to accessory structures of the eye. For example, mesoderm and neural crest cells contribute to the reinforcing tissues in the wall of the eye and the muscles that move the eye. Figure 50.10, step 5, shows a fully developed vertebrate eye.

Many experiments have shown that the initial induction by the optic vesicle is necessary for development of the eye. For example, if an optic vesicle is removed before lens formation, the ectoderm fails to develop a lens placode and vesicle. Moreover, placing a removed optic vesicle under the ectoderm in other regions of the head causes a lens to form in the new location. Or, if the ectoderm over an optic vesicle is removed and ectoderm from elsewhere in the embryo is grafted in its place, a normal lens will develop in the grafted ectoderm, even though in its former location it would not differentiate into lens tissue.

Eye development also demonstrates differentiation. Ectoderm cells that are induced to form the lens of the eye synthesize crystallins; in other locations, ectoderm cells typically synthesize a different protein, *keratin,* as their predominant cell product. Keratin is a component of surface structures such as skin, hair, feathers, scales, and horns. In other words, as a response to induction by the optic vesicle, the genes of the ectoderm cells coding for crystallin are activated, whereas genes coding for keratin are not expressed.

STUDY BREAK 50.3

1. What is the outcome of organogenesis?
2. What tissues or organs develop from the neural tube and neural crest cells?

50.4 Embryonic Development of Humans and Other Mammals

The embryonic development of humans is representative of the placental mammals. In the uterus, the embryo is nourished by the placenta, which supplies oxygen and nutrients and carries away carbon dioxide and nitrogenous wastes.

Pregnancy or **gestation,** the period of mammalian development in the uterus, varies in different species. Larger mammals bearing larger young generally have longer gestation periods; for example, gestation takes 600 days in elephants and about 1 year in blue whales, versus 21 days in hamsters.

In humans, gestation takes about 38 weeks, which amounts to about 40 weeks from the beginning of the menstrual cycle in which fertilization takes place. Human gestation is divided into three **trimesters,** each approximately 3 months long.

The major developmental events in human gestation—cleavage, gastrulation, and organogenesis—take place during the first trimester. By the fourth week, the embryo's heart is beating, and by the end of the eighth week, the major organs and organ systems have formed. From this point until birth, the developing human is called a **fetus.** Only 5 cm long by the end of the first trimester, the fetus grows during the second and third trimesters to an average length of 50 cm and an average mass of 3.5 kg (or about 19.7 inches and 7.7 pounds). The period of gestation ends with birth.

Cleavage and Implantation Occupy the First 2 Weeks of Development

Human fertilization occurs when the egg is in the first third of the oviduct leading from the ovary to the uterus (see Section 49.3). After fertilization, cleavage divisions take place during passage of the developing embryo down the oviduct (also called the fallopian tube in mammals) and while it is still enclosed in the zona pellucida—the original coat of the egg **(Figure 50.11).**

By day 4, the morula, a 16- or 32-cell ball, has been produced. By the time the endometrium (uterine lining) is ready for implantation (about 7 days after ovulation; see Section 49.3), the morula has reached the uterus and has undergone further cell divisions and differentiation into a type of blastula called a **blastocyst,** which is a characteristic of mammalian development. At this time, the blastocyst is a single-cell-layered hollow ball that is made up of about 70–100 cells, and contains a fluid-filled cavity. The blastocoel, the fluid-filled cavity, has a dense mass of cells localized to one side called the

inner cell mass. This inner cell mass gives rise to the embryo as well as the yolk sac, the allantois, and the amnion. The rest of the blastocyst becomes tissues that support the development of the embryo in the uterus. The outer single layer of cells of the blastocyst is the **trophoblast.**

Next, when the blastocyst has increased to an appropriate size, it breaks out of the zona pellucida and sticks to the endometrium on the side with the inner cell mass **(Figure 50.12, step 1).** Implantation begins when the trophoblast secretes proteases that digest pathways between endometrial cells. The trophoblast extends into the pathways, appearing like finger-like projections into the endometrium. Continued digestion of endometrial cells generates a hole in the endometrium for the blastocyst and releases nutrients that the developing embryo can use.

The inner cell mass of the burrowing blastocyst separates into the *embryonic disc*, which consists of two distinct cell layers (see Figure 50.12, step 1). The layer farther from the blastocoel is the *epiblast*, which gives rise to the embryo proper, and the layer nearer the blastocoel is the *hypoblast,* which gives rise to part of the extraembryonic membranes. When implantation is complete, the blastocyst has completely burrowed into the endometrium and is covered by a layer of endometrial cells (step 2).

Mammalian Gastrulation and Neurulation Resemble the Reptilian–Bird Pattern

Gastrulation proceeds as in birds (see Figure 50.6), with the formation of a primitive streak in the epiblast. Some epiblast cells remain in place, becoming the ectoderm, whereas others

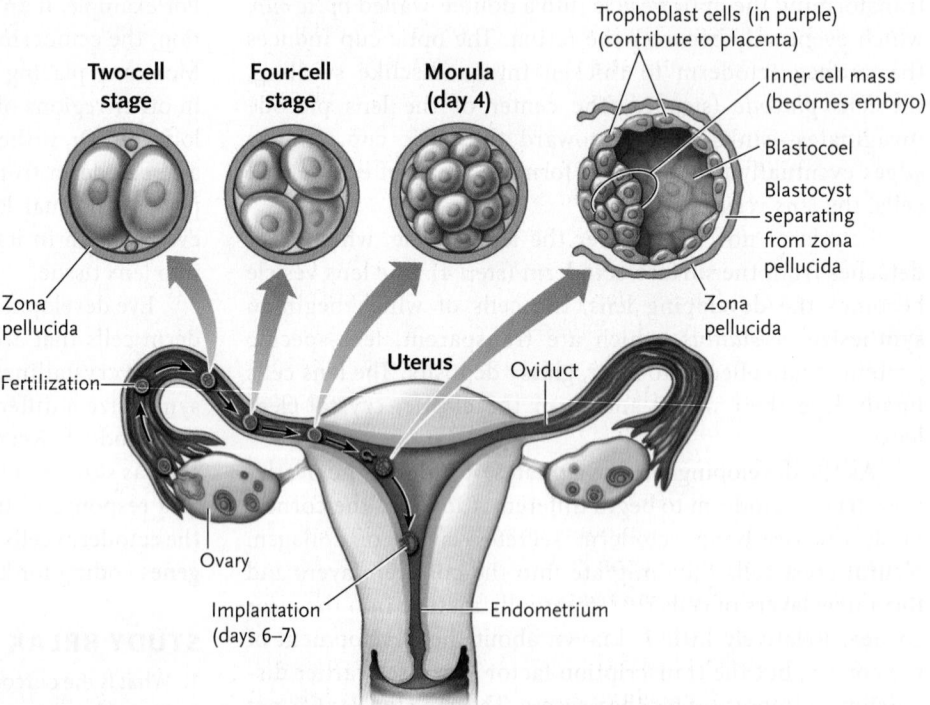

FIGURE 50.11 Early stages in the development of the human embryo.
© Cengage Learning 2017

enter the streak to form the endoderm and mesoderm. The ectoderm, mesoderm, and endoderm are located initially in three layers. From this initial arrangement, the endoderm folds to form the primitive gut, and becomes surrounded with ectoderm and mesoderm. Neurulation in human and other mammalian embryos takes place essentially as in birds (see Figure 50.8).

Extraembryonic Membranes Give Rise to the Amnion and Part of the Placenta

Next, a layer of cells separates from the epiblast along its top margin (see Figure 50.12, step 2). The fluid-filled space created by the separation becomes the amniotic cavity, and the layer of ectodermal cells forming its roof becomes the amnion, the extraembryonic membrane surrounding the cavity. The amnion expands until eventually it surrounds the embryo completely and suspends it in amniotic fluid. As in birds, the hypoblast develops into the yolk sac. However, in mammals, the mesoderm of the yolk sac gives rise to the blood vessels in the embryonic portion of the placenta.

Although the amnion is expanding around the embryonic disc, blood-filled spaces form in maternal tissue, and trophoblast cells grow rapidly around both the embryo and amnion to form the chorion (step 3). Then, a connecting stalk forms between the embryonic disc and the chorion, while the chorion begins to grow into the endometrium as fingerlike extensions called **chorionic villi** (step 4). When complete, the chorionic villi greatly increase the surface area of the chorion. Where these villi grow into the endometrium is the area of the future placenta. As the chorion develops, mesodermal cells of the yolk sac grow into it and form a rich network of blood vessels, the embryonic circulation of the placenta. The expanding chorion stimulates the blood vessels of the endometrium to grow into the maternal circulation of the placenta (step 5).

Within the placenta of humans, apes, monkeys, and rodents, the maternal circulation opens into spaces in which the maternal blood bathes the trophoblast layers of the placenta (step 6). The fetal capillaries are positioned next to the trophoblast so that there can be as few as two cell layers separating the maternal and fetal blood. (Different types of placentas are found in other mammals.) The embryonic circulation remains closed, however, so that the embryonic and maternal blood do not mix directly. This prevents the mother from developing an immune reaction against cells of the embryo, which, because it is genetically different from the mother, may be recognized as foreign by the mother's immune system. Eventually, the placenta and its blood circulation grow to cover about a quarter of the inner surface of the enlarged uterus and reach the size of a dinner plate.

As the embryonic blood circulation develops, this connecting stalk between the embryo and placenta develops into the **umbilical cord,** a long tissue with blood vessels linking the embryo and the placenta. The vessels in the umbilical cord are derived from the extraembryonic membrane, the *allantois.* They conduct blood between the embryo and the placenta (shown in the inset for Figure 50.12, step 6).

Within the placenta, nutrients and oxygen pass from the mother's circulation into the circulation of the embryo. Besides nutrients and oxygen, many other substances taken in by the mother—including alcohol, caffeine, drugs, and toxins in cigarette smoke—can pass from mother to embryo. Carbon dioxide and nitrogenous wastes pass from the embryo to the mother, and are disposed of by the mother's lungs and kidneys.

If the presence of a genetic disease such as cystic fibrosis or Down syndrome is suspected, tests can be carried out on cells removed from the embryonic portion of the placenta or from the amniotic fluid, which contains cells derived from the embryo. The test using cells of the placenta is called *chorionic villus sampling;* the test using cells derived from the amniotic fluid is called *amniocentesis* (see Section 13.4). Chorionic villus sampling can be carried out as early as the eighth week, compared with 14 weeks for amniocentesis. Both tests carry some degree of risk to the embryo.

Further Growth of the Fetus Culminates in Birth

By the end of its fourth week, a human embryo is 3 to 5 mm long, which is 250 to 500 times the size of the zygote. It has a tail and **pharyngeal arches,** which are embryonic features of all vertebrates. The pharyngeal arches contribute to the formation of the face, neck, mouth, larynx, and pharynx. After 6 weeks, the embryo is about 13 mm long and has a more humanlike appearance with, for example, eyes and developing limbs apparent **(Figure 50.13A).** At 16 weeks, the fetus's bones and muscle tissue continue to form, producing a more complete skeleton; its skin begins to form; its limbs with fingers and toes have developed; its organ systems have formed; and its sex is identifiable **(Figure 50.13B).** At birth the baby will have grown to about 3 times the length and 17 times the weight of the fetus at this stage.

Figure 50.14 shows the physical events **(A)** and hormonal events **(B)** of birth. As the period of fetal growth comes to a close, the fetus typically turns so that its head is downward, pressed against the cervix. A steep rise in the levels of estrogen secreted by the placenta at this time causes cells of the uterus to express the gene for the receptor of the hormone *oxytocin.* The receptors become inserted into the plasma membranes of those cells. Oxytocin—which is secreted by the pituitary gland (see Section 42.3)—binds to its receptor, triggering the smooth muscle cells of the uterine wall to contract and begin the rhythmic contractions of labor. These contractions mark the beginning of **parturition** (*parturire* = to be in labor), the three-stage process of giving birth. Typically at the beginning of labor, or sometimes in the first stage, the amniotic membrane bursts, releasing the amniotic fluid. The amniotic fluid helps lubricate the birth canal.

IMPLANTATION OF A HUMAN BLASTOCYST IN THE ENDOMETRIUM OF THE UTERUS AND THE ESTABLISHMENT OF THE PLACENTA

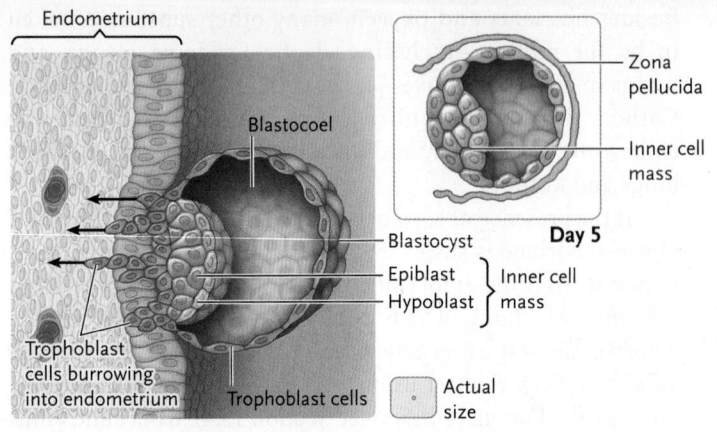

1 **Days 6–7:** Surface cells of the blastocyst attach to the endometrium and start to burrow into it. Implantation is under way.

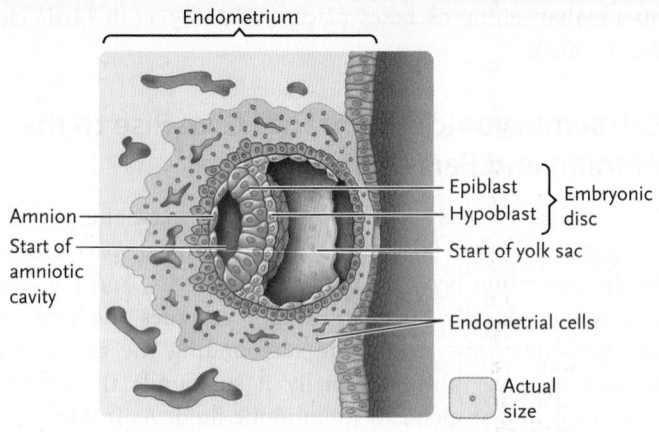

2 **Days 10–11:** A layer of epiblast cells separates, producing the amniotic cavity. The cells above the cavity become the amnion, which eventually surrounds the embryo. The hypoblast begins to form around the yolk sac.

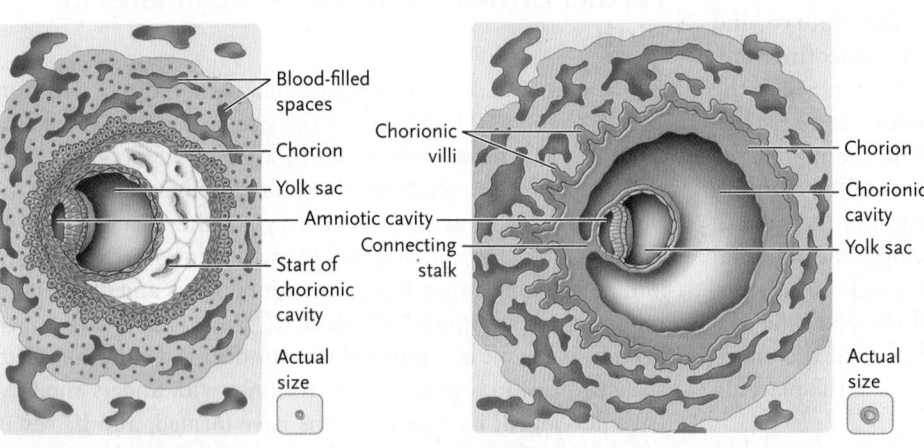

3 **Day 12:** Blood-filled spaces form in maternal tissue. The chorion forms, derived from trophoblast cells, and encloses the chorionic cavity.

4 **Day 14:** A connecting stalk has formed between the embryonic disk and chorion. Chorionic villi, which will be features of a placenta, start to form.

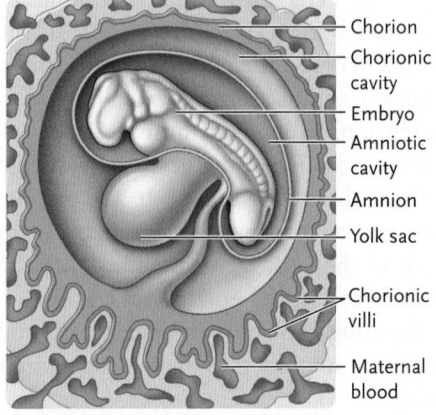

5 **Day 25:** The chorion continues to grow into the endometrium, producing the chorionic villi. The chorion growth stimulates blood vessels of the endometrium to grow into the maternal circulation of the placenta.

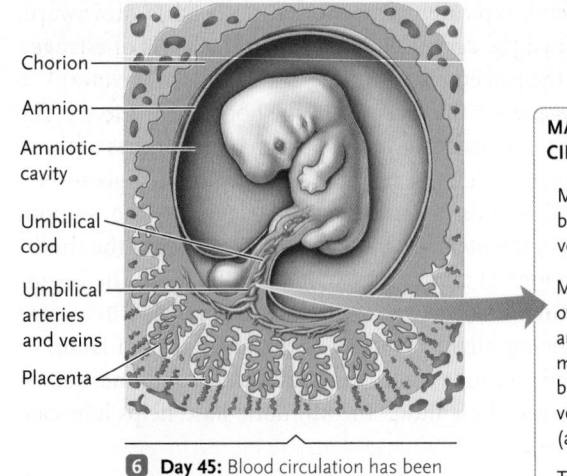

6 **Day 45:** Blood circulation has been established through the umbilical cord to the placenta.

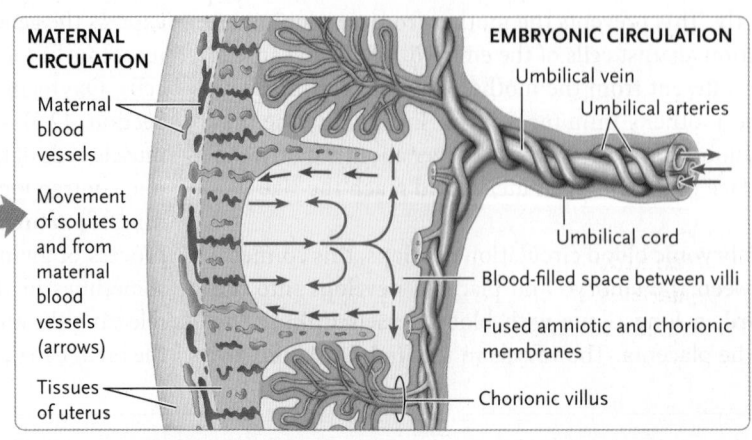

SUMMARY Implantation of a human blastocyst in the endometrium occurs within the first two weeks of development. When implantation is complete, the embryonic disc (epiblast plus hypoblast) is connected to the chorion by a stalk, chorionic villi have started to form, and an amniotic cavity is present. As development continues, the placenta forms and blood circulation is established through the umbilical cord of the embryo and the placenta. The blood circulation of the embryo is kept separate from the maternal circulation.

think like a scientist Why is it important to keep the circulation of the embryo and fetus separate from that of the mother?

© Cengage Learning 2017

A. 6 weeks　　**B. 4 months**

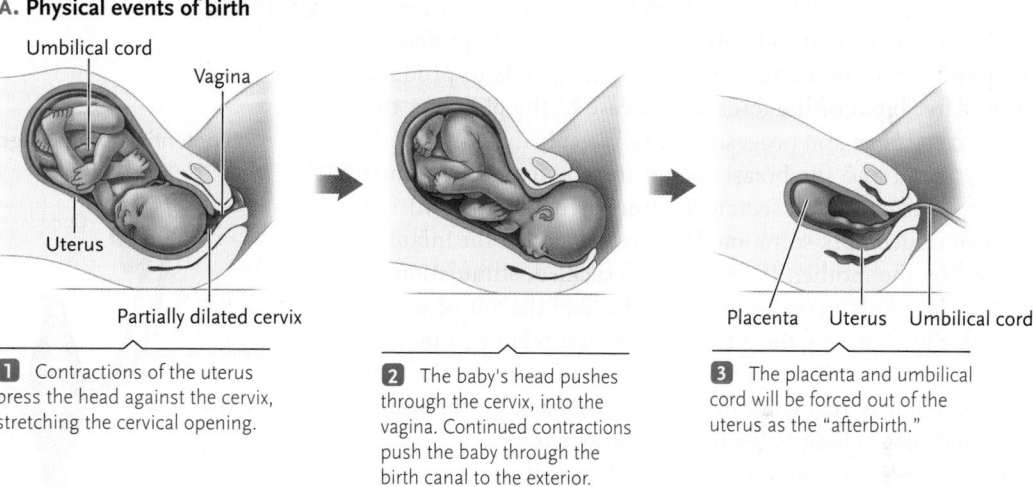

FIGURE 50.13 **The human embryo after 6 weeks (A) and 4 months (B) of development.**

FIGURE 50.14 Birth of a baby.
© Cengage Learning 2017

A. Physical events of birth

Umbilical cord

Vagina

Uterus

Partially dilated cervix

1 Contractions of the uterus press the head against the cervix, stretching the cervical opening.

2 The baby's head pushes through the cervix, into the vagina. Continued contractions push the baby through the birth canal to the exterior.

Placenta　Uterus　Umbilical cord

3 The placenta and umbilical cord will be forced out of the uterus as the "afterbirth."

B. Hormonal events of birth

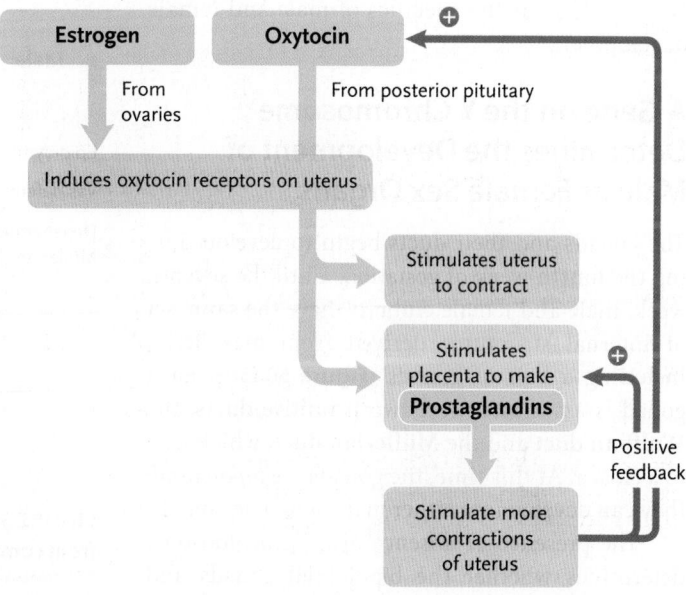

In the first stage of labor, the contractions push the fetus further against the cervix, dilating it up to 10 cm in diameter (Figure 50.14A, step 1). This stage is the longest, typically taking from several hours to as many as 24 hours in a first pregnancy. In response to cervix dilation, stretch receptors in the walls send nerve signals to the hypothalamus, which responds by stimulating the pituitary to secrete more oxytocin. In turn, the oxytocin stimulates more forceful contractions of the uterus, pressing the fetus more strongly against the cervix, and further stretching its walls. The positive feedback cycle continues, steadily increasing the strength of the uterine contractions.

The second stage of labor is the actual birth of the baby (Figure 50.14A, step 2). Continued uterine contractions begin to push the baby through the cervix and the vagina. At this point, stretch receptors in the vagina trigger contractions of the abdominal wall that, in synchrony with the uterine contractions, greatly increase the force pushing the baby through the birth canal. "Pushing" is often aided by voluntary abdominal contractions. Once the head emerges from the birth canal, the rest of the body follows quickly and the entire fetus is forced through the vagina to the exterior, still connected to the placenta by the umbilical cord (Figure 50.14A, step 3).

In the third stage of labor, continuing contractions of the uterus expel the placenta and any remnants of the umbilical cord and embryonic membranes as the **afterbirth.**

After the baby takes its first breath, the umbilical cord is cut and tied off by the birth attendant. The short length of umbilical cord still attached to the infant dries and shrivels within a few days. Eventually it separates entirely and leaves a scar, the **umbilicus** or navel, to mark its former site of attachment during embryonic development.

The Mother's Mammary Glands Become Active after Birth

Before birth, estrogen and progesterone secreted by the placenta stimulate the growth of the mammary glands in the mother's breasts. However, the high levels of these hormones prevent the mammary glands from responding to *prolactin,* the pituitary hormone that stimulates the glands to produce milk. After birth of the fetus and release of the placenta, the levels of estrogen and progesterone fall steeply in the mother's bloodstream, and the breasts begin to produce milk (stimulated by prolactin) and secrete it (stimulated by oxytocin).

Continued milk secretion depends on whether the infant is suckled by the mother. If the infant is suckled, stimulation of the nipples sends nerve impulses to the hypothalamus, which responds by signaling the pituitary to release a burst of prolactin and oxytocin. Hormonal stimulation of milk production and secretion is regulated by positive feedback mechanisms, and continues as long as the infant is breastfed.

So far, we have followed the development of a generic human, but certain aspects of development differ depending on the offspring's sex. Next we look at the specifics of male and female development.

A Gene on the Y Chromosome Determines the Development of Male or Female Sex Organs

The gonads and their ducts begin to develop during the fourth week of gestation. Until the seventh week, male and female embryos have the same set of internal structures derived from mesoderm, including a pair of gonads **(Figure 50.15A).** Each gonad is associated with two primitive ducts, the **Wolffian duct** and the **Müllerian duct,** which lead to a cloaca. At this time, the gonads are *bipotential:* they can develop into either male or female gonads.

The presence or absence of a Y chromosome determines whether the bipotential gonads and

internal ducts develop into male or female gonads and their associated internal ducts. If the fetus has XY sex chromosomes, a single gene on the Y chromosome, *SRY* (*Sex-determining Region of the Y*), becomes active in the seventh week. The protein encoded by the gene sets a molecular switch that causes the primitive gonads to develop into testes. The fetal testes then secrete two hormones, testosterone and the *anti-Müllerian hormone (AMH).* The testosterone stimulates development of the Wolffian ducts into the male reproductive tract, including the epididymis and vas deferens **(Figure 50.15B),** and seminal vesicles (see Figure 49.14). AMH causes the Müllerian ducts to degenerate and disappear. Testosterone additionally stimulates the development of the male genitalia.

If the fetus has XX sex chromosomes, no SRY protein is produced and the primitive gonads, under the influence of the estrogens and progesterone secreted by the placenta, develop into ovaries. The Müllerian ducts develop into the oviducts, uterus, and part of the vagina, and the Wolffian ducts degenerate and disappear **(Figure 50.15C).** The female sex hormones additionally stimulate the development of the female external genitalia.

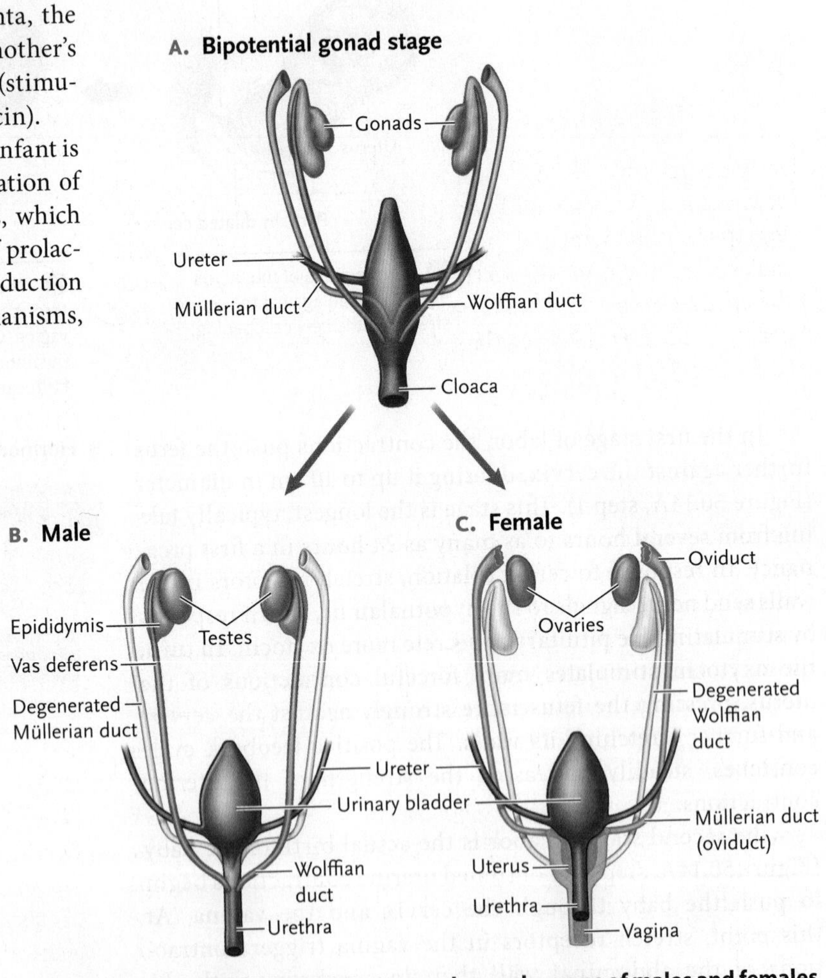

FIGURE 50.15 Development of the internal sexual organs of males and females from common bipotential origins.

© Cengage Learning 2017

Development Continues after Birth

Once fetal development is over, humans and other mammals, as well as most other animals, follow a prescribed course of further growth and development that leads to the adult, the sexually mature form of the species. In humans, the internal and external sexual organs mature and secondary sexual characteristics appear at puberty. Similar changes occur in most mammals. There are, in fact, many examples among different animal groups of developmental changes that take place after hatching or birth. In some cases, offspring hatch that are distinctly different in structure from the adult. Examples among invertebrates include insects such as *Drosophila* and butterflies, in which eggs hatch to produce larva that undergo metamorphosis into the adult. Frogs similarly hatch as tadpoles, which undergo metamorphosis to produce the adult.

STUDY BREAK 50.4

1. Distinguish between the roles of the trophoblast and inner cell mass of the blastocyst in mammalian development.
2. What hormone would you use to induce labor in a pregnant woman?

50.5 The Cellular Basis of Development

In the preceding sections, you learned about the processes of development from a mainly structural point of view. Underlying those developmental processes are specific cellular and molecular events. In this section, you will learn about some of the cellular events that underlie the stages of development. The genetic and molecular regulation of development is discussed in Section 16.4.

Cell Division Varies in Orientation and Rate during Embryonic Development

Recall from earlier in the chapter that *morphogenesis* is the generation of the body form as differentiated cells end up in their appropriate positions. In animals, morphogenesis occurs by changes in cell shape, cell position, and cell adhesion. For embryo morphogenesis, the *orientation* and *rate* of mitotic cell division have special significance. Regulation of these two features of mitotic cell division occurs at all stages of development.

The orientation of cell division refers to the angles at which daughter cells are added to older cells as development proceeds. Orientation is determined by the location of a furrow that separates the cytoplasm after mitotic division of the nucleus (furrowing is discussed in Section 10.2). The furrow forms in alignment with the spindle midpoint. Therefore, when the spindle is centrally positioned in the cell, the furrow leads to symmetrical division of the cell. However, when the spindle is displaced to one end of the cell, the furrow leads to asymmetrical division of the cell into a smaller and a larger cell. Little is known about how spindle positioning is regulated.

The rate of cell division primarily reflects the time spent in the G_1 period of interphase (see Section 10.2); once DNA replication begins, the rest of the cell cycle is usually of uniform length in all cells of the same species. As an embryo develops and its cells differentiate, the time spent in interphase increases and varies in length in different cell types. Therefore, different cell types proliferate at various rates, giving rise to tissues and organs with different cell numbers. Ultimately, the rate of cell division is under genetic control.

Frog egg cleavage provides examples of how both changes in orientation and rate of mitotic division affect development. The first two cleavages start at the animal pole and extend to the vegetal pole, producing four equal blastomeres (see Figure 50.2). The third cleavage occurs equatorially. However, because there is yolk in the vegetal region of the embryo, this cleavage furrow forms not at the equator but up higher toward the animal pole. The result is an eight-cell embryo with four small blastomeres in the animal region of the embryo, and four large blastomeres in the vegetal region. All the blastomeres then divide at the same rate up until the twelfth division. Thereafter, blastomeres in the animal region divide rapidly whereas blastomeres in the vegetal region divide more slowly because division is inhibited by yolk. As a result, the animal region becomes filled with many small cells, whereas the vegetal region contains a relatively small number of yolk-filled blastomeres.

Cell Shape Changes and Cell Migrations Play Important Roles in Development

We have seen that embryonic cells undergo changes in shape that generate cell migrations, such as the infolding of surface layers to produce endoderm or mesoderm. Entire cells also migrate during the embryonic growth of animals, both singly and in groups. Both the shape changes and the whole-cell migrations are produced by microtubules (powered by dyneins and kinesins) and microfilaments (powered by myosins) (see Section 4.3). Cell migrations are also produced by changes in the rate of growth or by the breakdown of microtubules and microfilaments. Generally speaking, changes in both cell shape and cell migration play important roles in cleavage, gastrulation, and organogenesis.

CHANGES IN CELL SHAPE Changes in cell shape typically result from reorganization of the cytoskeleton. For example, during the development of the neural plate in frogs (see Figure 50.8), the ectoderm flattens and thickens. This occurs because microtubules within cells in the ectoderm layer lengthen and slide farther apart, causing the cells to transition from a cubelike to a columnar shape (**Figure 50.16A**).

Once formed, the neural plate sinks downward along its midline as a result of a change in cell shape from columnar to

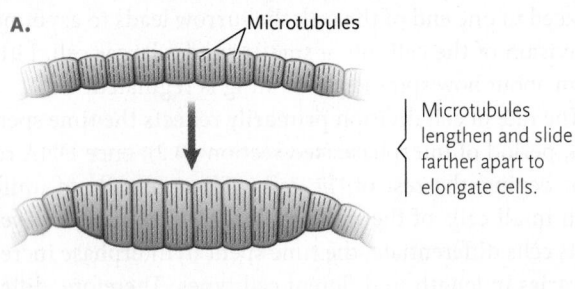

Microtubules lengthen and slide farther apart to elongate cells.

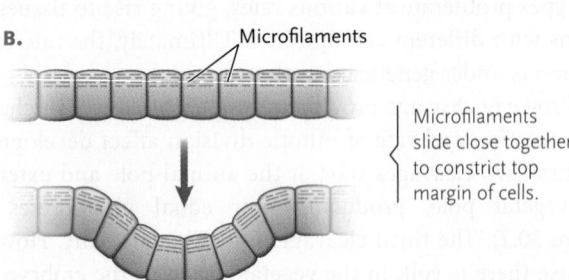

Microfilaments slide close together to constrict top margin of cells.

FIGURE 50.16 The roles of microtubules (A) and microfilaments (B) in the changes in cell shape that produce developmental movements, such as neural plate formation.

© Cengage Learning 2017

wedgelike **(Figure 50.16B)**. As the tops of the cells narrow, the entire cell layer is forced inward—it invaginates. How does this occur? Each wedge-shaped cell contains a group of microfilaments arranged in a circle at its top. The microfilaments slide over each other, tightening the ring like a drawstring and narrowing the top of the cell.

WHOLE-CELL MIGRATIONS Among the most striking examples of whole-cell migrations in embryonic development are the migrations of cells during gastrulation to form the three germ layers—the ectoderm, mesoderm, and endoderm—and the often long-distance migrations of neural crest cells. These whole-cell migrations involve the coordinated activity of microtubules and microfilaments and large-scale reorganization of the cytoskeleton.

Cell migration proceeds in four stages:

1. *Polarization.* The front and back ends of a cell becomes defined by a polarization process. Polarization is triggered either by diffusible signals or by signals from the extracellular matrix (ECM). In response to a polarizing signal, the cell reorganizes its cytoskeleton so that the cell has a front and a back end.
2. *Protrusion of the front end of the cell.* Actin microfilaments assemble into long, parallel bundles, causing the front end of the cell to protrude.
3. *Adhesion.* In order to migrate, a cell must be anchored to a substrate that it can push on. Therefore, the cell at this stage adheres to the surrounding extracellular matrix. (Some molecular aspects of the adhesion process are described later.)

4. *Release of rear adhesion, and forward migration to a new adhesion point.* Adhesion of the cell to the substrate at its rear end is released, and the cell migrates forward and adheres to a new attachment point. Repetition of this stage results in continued forward movement in a fashion somewhat resembling how an amoeba moves (see Section 4.3 and Figure 5.15).

How do the cells know where to go? Typically, cells migrate over the surface of stationary cells in one of the embryo's layers. In many developmental systems, migrating cells follow tracks formed by molecules of the extracellular matrix that is secreted by the cells along the route over which they travel. An important track molecule is *fibronectin,* a fibrous, elongated glycoprotein of the extracellular matrix. Migrating cells can recognize and adhere to the fibronectin. In response, internal changes in the cells trigger movement in a direction based on the alignment of the fibronectin molecules. For instance, paths of fibronectin guide primordial germ cells to the gonads, and cells of the heart to the midline of the embryo. Experiments in which the formation of fibronectin pathways was prevented have shown how crucial those pathways are for embryonic development. For example, gastrulation does not occur in amphibians lacking fibronectin, and in mammals, mesoderm and neural tube development is defective in the absence of fibronectin.

Some migrating cells follow concentration gradients instead of molecular tracks. The gradients are created by the diffusion of molecules (often proteins) released by cells in one part of an embryo. Cells with receptors for the diffusing molecule follow the gradient toward its source, or move away from the source.

Selective Cell Adhesions Underlie Cell Migrations

Selective cell adhesion, the ability of an embryonic cell to make and break specific connections to other cells, is closely related to cell migration. As development proceeds, many cells break their initial adhesions and move, forming new adhesions in different locations. Final cell adhesions hold the embryo in its correct shape and form. Junctions of various kinds, including tight, anchoring, and gap junctions, reinforce the final adhesions (see Section 4.5).

The selective nature of cell adhesions was first demonstrated in a classic experiment by Johannes Holtfreter of the University of Rochester and his student P. L. Townes. The researchers removed pieces of ectoderm, mesoderm, and endoderm from living amphibian embryos in the neurulation stage, separated them into individual cells, and added the cells in various combinations to a culture medium. Initially the cells clumped together at random into a ball. After a few hours, they sorted themselves out and moved into arrangements resembling their normal locations in the gastrula **(Figure 50.17)**.

Cell–cell adhesion is mediated by membrane proteins called **cell adhesion molecules** (CAMs; see Section 4.5). The

FIGURE 50.17 | **Experimental Research**

Demonstrating the Selective Adhesion Properties of Cells

1. Holtfreter and Townes separated ectoderm, mesoderm, and endoderm tissue from amphibian embryos soon after the neural tube had formed. They used embryos from amphibian species that had cells of different colors and sizes, enabling them to follow under the microscope where each cell type ended up. (The colors shown here are for illustrative purposes only.)

2. The researchers placed the tissues individually in alkaline solutions, which caused the tissues to break down into single cells.

3. Holtfreter and Townes combined suspensions of single cells in various ways. Shown here are ectoderm + mesoderm, and ectoderm + mesoderm + endoderm. When the pH was returned to neutrality, the cells formed aggregates. Through a microscope, the researchers followed what happened to the aggregates.

Results: The cell types did not remain mixed. In time, each cell type sorted itself from the others and separated spatially. In the ectoderm + mesoderm mixture, the ectoderm moved to the periphery of the aggregate, surrounding mesoderm cells in the center. In no case did the two cell types remain randomly mixed. In the ectoderm + mesoderm + endoderm mixture, cell sorting in the aggregates generated cell positions reflecting the positions of the cell types in the embryo. That is, the endoderm cells separated from the ectoderm and mesoderm cells and became surrounded by them. In the end, the ectoderm cells were located on the periphery, the endoderm cells were internal, and the mesoderm cells were between the other two cell types.

Question: Do cells make specific connections to other cells?

Experiment: Johannes Holtfreter and P. L. Townes demonstrated that cells make specific connections to other cells, meaning that cells have selective adhesion properties.

Amphibian embryos of different species

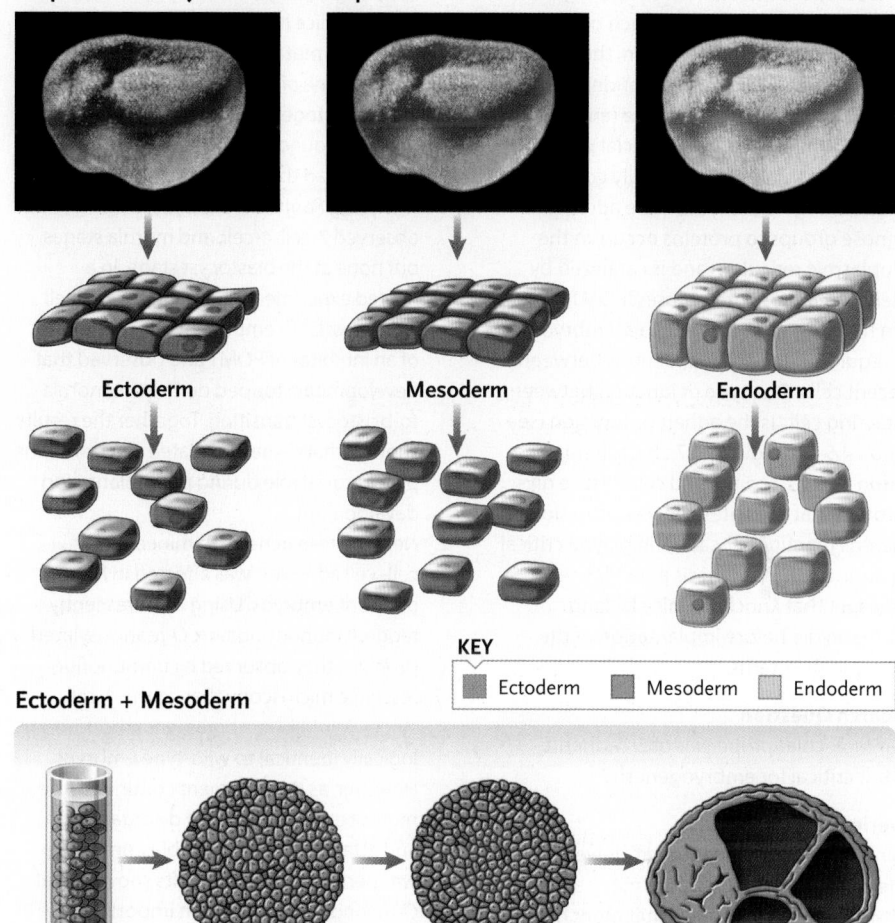

KEY
Ectoderm | Mesoderm | Endoderm

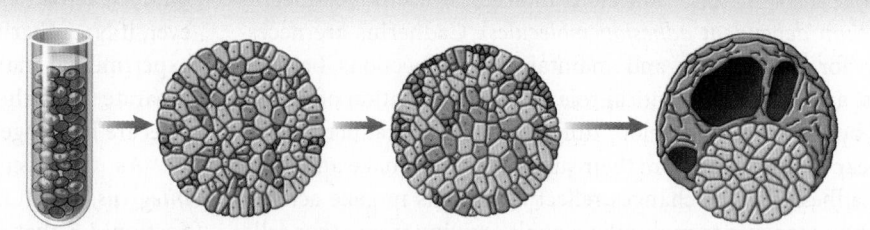

Conclusion: Holtfreter interpreted the results to mean that cells have selective affinity for each other—cells have selective adhesion properties. Specifically, he proposed that ectoderm cells have positive affinity for mesoderm cells but negative affinity for endoderm cells, whereas mesoderm cells have positive affinity for both ectoderm cells and endoderm cells. In modern terms, these properties result from cell surface molecules that give cells specific adhesion properties.

think like a scientist
Observe
Hypothesize
Predict
Experiment
Interpret

As the main text discusses, subsequent research identified cadherins as one type of cell surface proteins responsible for selective cell adhesions. Suppose you have a cloned cDNA for a cadherin, which is proposed to be one type of cell surface protein responsible for selective cell adhesions. How could you use that cloned cDNA to test directly whether or not it has that role?

Source: P. L. Townes and J. Holtfreter. 1955. Directed movements and selective adhesion of embryonic amphibian cells. *Journal of Experimental Zoology* 128:53–120.

An Essential Role of Protein *O*-Mannosylation in Embryonic Development

Most of the known proteins at the cell surface are glycoproteins; that is, proteins modified by the attachment of carbohydrate groups (see Section 3.4). One such protein modification is *O*-mannosylation, the addition of mannose (a monosaccharide; see Figure 3.5) to serine or threonine (amino acids). *O*-mannosylation is a crucial protein modification that is evolutionarily conserved amongst eukaryotes. The addition of mannose groups to proteins occurs in the endoplasmic reticulum and is catalyzed by *protein O-mannosyltransferases* (POMTs).

In humans and other animals, embryogenesis requires adhesive interactions between adjacent cells. One type of junction between contacting cells is the adherens junction (see Section 4.5 and Figure 4.27). E-cadherin (see Section 50.5) is a conserved cell surface glycoprotein that mediates cell–cell adhesion at adherens junctions. E-cadherin plays a critical role during embryogenesis as evidenced by the fact that knockout mice lacking E-cadherin die before implantation of the embryo in the uterus.

Research Question

What molecular properties of E-cadherin make it critical for embryogenesis?

Experiments and Results

Sabine Strahl and colleagues at University of Heidelberg, Heidelberg, Germany and Ludwig-Maximilians-Universität München,

Germany performed the experiments to answer the question.

1. First, they used genetic engineering to produce mice heterozygous for a non-functional mutant allele of *Pomt2*, which encodes one of two POMTs in mammals. They bred together *Pomt2*$^{+/-}$ heterozygous mice and found that none of the living progeny had the genotype *Pomt2*$^{-/-}$. Analyzing *Pomt2*$^{-/-}$ embryos, they observed 2-cell, 4-cell, and morula stages, but none at the blastocyst stage. In a related experiment, they cultivated 2-cell stage *Pomt2*$^{-/-}$ embryos in the presence of an inhibitor of POMT and observed that development stopped during the morula-to-blastocyst transition. Together the results showed that *O*-mannosylated glycoproteins play a crucial role during preimplantation development.

2. Next the researchers examined whether cell–cell adhesion was affected in POMT-deficient embryos. Using a fluorescently tagged antibody against *O*-mannosylated proteins, they observed by immunofluorescence microscopy that, at the 4-cell stage, *Pomt2*$^{-/-}$ embryos were morphologically identical to wild-type embryos. However, as development continued, the mutant embryos became disorganized and attachment between blastomeres was perturbed. These results showed that *O*-mannosylation plays an important role in

cell–cell adhesion during embryonic development, a process that is critical for the morula-to-blastocyst transition. As already indicated, adherens junctions mediated by E-cadherin are involved in cell–cell adhesion. The group observed that mutant embryos were deficient in E-cadherin at the sites of reduced cellular adhesion.

3. The scientists demonstrated that E-cadherin is a target of protein *O*-mannosylation, a fact that was previously unknown.

Conclusion

The study identified E-cadherin as an *O*-mannosylated glycoprotein and established a functional relationship between *O*-mannosylated glycoproteins and E-cadherin-mediated cell–cell adhesion. The interpretation is that *O*-mannosylated E-cadherin is important for the establishment of adherens junctions during the morula-to-blastocysts transition in mammals.

think like a scientist

The researchers bred together *Pomt2*$^{+/-}$ mice. Using molecular techniques, the scientists could determine the genotype of progeny mice. For the living progeny, what genotypic classes did they observe, and in what ratio would you predict they would have found them?

Source: M. Lommel et al. 2013. Protein O-mannosylation is crucial for E-cadherin-mediated cell adhesion. *Proceedings of the National Academy of Sciences USA* 110:21024–21029.

major class of cell adhesion molecules is the **cadherins** (*calcium-dependent adhesion molecules*). Cadherins are necessary for establishing and maintaining connections between cells, and they play a critical role in the organization of cells in the body. As cells develop, different types of cadherins may appear or disappear from their surfaces as they make and break cell adhesions. The changes reflect alterations in gene activity, often in response to molecular signals arriving from other cells. For example, in the neural plate stage of neurulation in the frog (see Figure 50.8), N-cadherin is on neural plate cells, keeping those cells together, whereas E-cadherin is on the adjacent ectodermal cells, keeping those cells together. (*Molecular Insights* describes molecular properties of E-cadherin that make it critical for embryogenesis.) The neural tube is produced when the neural plate cells separate from the ectodermal cells, whereas both cell types retain their respective cadherin type. The neural crest cells have neither cadherin bound to them, so they do not

bind to each other and they disperse (as described earlier). However, if N-cadherin is expressed in the ectodermal cells through experimental manipulation, the forming neural tube does not separate from the flanking ectodermal cells because all of the cells are held together by N-cadherin.

As they begin to disperse, neural crest cells produce *integrins*, which insert into the plasma membrane. Recall from Section 4.5 that integrins are receptor proteins that span the plasma membrane. On the cytoplasmic side, integrins bind to actin microfilaments of the cytoskeleton. Integrins function to integrate changes outside and inside the cell by communicating changes in the extracellular matrix to the cytoskeleton. In frog neurulation, the integrin receptors of the neural crest cells interact with proteins in the extracellular matrix in a way that causes cytoskeletal changes. The cytoskeletal changes move the cells along their migration pathways. More generally, integrins are involved in many key developmental processes.

The development of a new organism involves the cell shape changes, cell migrations, and cell adhesions we have just discussed and, in addition, the programmed emergence of particular types of cells, or cell lineages, as we will see next.

Fate Mapping Correlates Adult Structures with Regions of the Embryos from Which They Developed

From the early days of studying development, embryologists have focused on describing not only how embryos form and develop, but exactly how adult tissues and organs are produced from the cells of the embryo. Thus, an important goal in embryology is to trace cell lineages from embryo to adult. For most organisms, it is not possible to trace lineages at the individual cell level, primarily because of the complexity of the developmental process and the typical opacity of embryos. However, it has been possible to map adult or larval structures onto the region of the embryo from which each structure developed. This type of study is called *fate mapping,* and the result is called a **fate map.** Experimentally, fate mapping is done by following development of living embryos under the microscope, either using species in which the embryo is transparent, or by marking cells so they can be followed. Cells may be marked with vital dyes (dyes that do not kill cells), fluorescent dyes, or radioactive labels. Fate maps have been produced for a number of organisms, including the chick, *Xenopus,* and *Drosophila.*

In most cases a fate map is not detailed enough to relate how particular cells in the embryo gave rise to cells of the adult. The exception is the fate map of the nematode *Caenorhabditis elegans,* an organism that has a fixed, reproducible developmental pattern. This animal has a transparent body, and scientists have been able to map the fate—trace the **cell lineage**—of every somatic and germ-line cell as the zygote divides and the resulting embryo differentiates into the 959-cell adult hermaphrodite or the 1,031-cell adult male **(Figure 50.18).** All somatic cells of the adult can be traced from five somatic *founder cells* produced during early development. Knowing the cell lineages of *C. elegans* has been a valuable tool for research into the genetic and molecular control of development in this organism, because mutations affecting development have an easily visible effect. Among the results of such studies, we have learned that not only are proteins encoded by regulatory genes involved in the control of development, but so are some miRNAs (see Section 16.4). For example, genetic manipulation makes it possible to knock out the functions of specific miRNA-coding genes. Although knockouts of most miRNA-coding genes do not affect development, a few do cause defects in the timing of development.

Induction Depends on Molecular Signals Made by Inducing Cells

Recall from Section 50.1 that induction is the process by which a group of cells (the inducer cells) causes or influences a nearby group of cells (the responder cells) to follow a particular developmental pathway. Recall also that induction is the major process responsible for determination, in which the developmental fate of a cell is set, and that induction occurs when signal molecules interact with surface receptors on the responding cells, typically triggering changes in gene activity.

A German scientist, Hans Spemann of the University of Freiburg, carried out the first experiments identifying induction in embryos in the 1920s. He and his doctoral student, Hilde Mangold, found that, if the dorsal lip of a newt embryo was removed and grafted into a different position on another

A. Founder cells

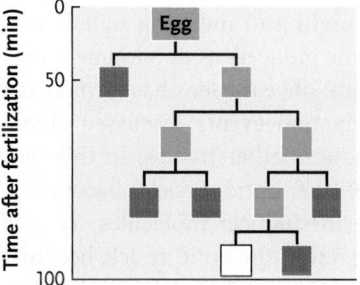

FIGURE 50.18 Fate map showing cell lineages of *C. elegans.* (A) The founder cells (blue) produced in early cell divisions from which all adult somatic cells are produced. The cell in white gives rise to germ-line cells. **(B)** The cell lineage for cells that form the intestine. The detailed lineages for the other parts of the adult are not shown.

© Cengage Learning 2017

B. Cell lineage for intestinal cells

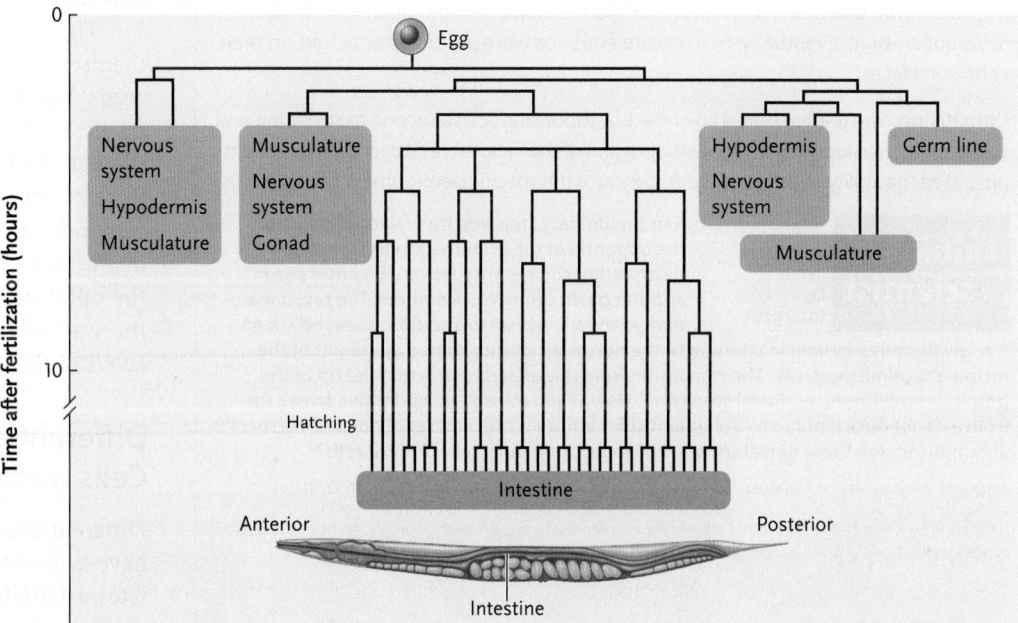

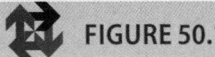 **FIGURE 50.19** | **Experimental Research**

Spemann and Mangold's Experiment Demonstrating Induction in Embryos

Question: Does induction occur in embryonic development?

Experiment: Hans Spemann and Hilde Mangold performed transplantation experiments with newt embryos.

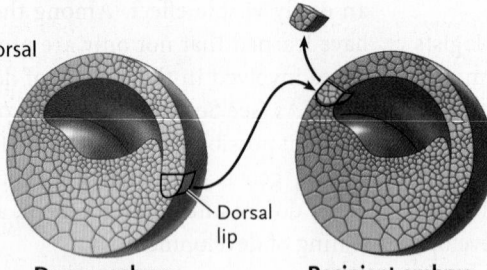

1. The researchers grafted the dorsal lip of the blastopore from one newt embryo onto a different position—the ventral side—of another embryo. The two embryos were from different newt species that differed in pigmentation, allowing the fate of the transplanted tissue to be followed easily during development.

Dorsal lip

Donor embryo   **Recipient embryo**

2. The embryo with the transplant was allowed to develop.

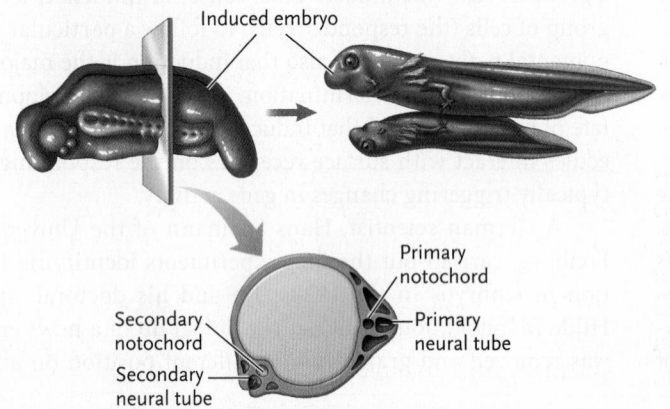

Induced embryo

Primary notochord

Secondary notochord

Primary neural tube

Secondary neural tube

Result: At the ventral location on the recipient embryo where the dorsal lip of the blastopore was grafted, another embryo developed simultaneously with the recipient embryo. Eventually, two mature embryos were produced attached on their ventral surfaces.

Conclusion: The grafted dorsal lip of the blastopore induced a second gastrulation and subsequent development in the ventral region of the recipient embryo. The result demonstrated the ability of particular cells to induce the specific development of other cells.

| Observe | Hypothesize | Predict | Experiment | **Interpret** |

As a preliminary step, you transplanted cells from the upper tip of the primitive streak from one gastrulating chick embryo to an off-center site in another gastrulating chick embryo. The result was that, at the site of transplantation, a second set of dorsal structures formed in addition to the dorsal structures formed as a result of the recipient's primitive streak. The conclusion from this experiment is that the tip of the primitive streak induces dorsal structures. Now, if you transplant a primitive streak tip from a donor duck embryo to a recipient chick embryo, a second set of dorsal structures is also induced. Are these dorsal structures formed from duck cells or chick cells?

Source: H. Spemann and H. Mangold. 1924. Über induktion von Embryonalanlagen durch implantation artfremder Organisatoren. *Roux's Archives of Developmental Biology* 100:599–638. [Reprinted as English translation by V. Hamburger: Induction of embryonic primordia by implantation of organizers from a different species. *International Journal of Developmental Biology* 45:13–38 (2001).]

newt embryo, on the ventral side for instance, cells moving inward from the dorsal lip induced a neural plate, a neural tube, and eventually an entire embryo to form in the new location **(Figure 50.19).** On the basis of his pioneering research, Spemann proposed that the dorsal lip is an *organizer,* acting on other cells to alter their course of development. This action is now known as *induction,* and the cells responsible for induction are known generally as inducer cells. Spemann received a Nobel Prize in 1935 "for his discovery of the organizer effect in embryonic development."

Induction in development typically involves paracrine signaling, meaning that the inducer molecule secreted by the inducer cell changes the cellular behavior of responder cells that are located near the inducer cells (see Section 42.1). For a cell to respond to the inducer molecule, it must have a receptor for that molecule. In addition, the responder cell must be able to respond to the specific inducer, a condition called *competence.* That is, simply having a receptor is not sufficient to generate a response; in addition, the responder cell must be expressing a genetic program that makes it competent to respond. Induction by the optic vesicle, discussed earlier in this chapter (see Figure 50.10), provides an example. You learned in the earlier discussion that transplanting the optic vesicle under ectoderm in other regions of the head induces a lens to form in the new location. However, if the optic vesicle is placed under ectoderm in areas other than the head, lens tissue does not develop. The interpretation is that only head ectoderm is competent to respond to the signals from the optic vesicle and produce a lens.

Tissues responding to inducing signals may themselves become inducers as development proceeds. For instance, once the lens has formed as a result of the induction events discussed above, the lens itself induces other tissues. In this particular example, the optic vesicle becomes a responder tissue for inducer molecules secreted by the lens. As a result, the optic vesicle becomes the optic cup (see Figure 50.11, step 2) and then the wall of the optic cup differentiates into the layers of the retina.

Differentiation Produces Specialized Cells without Loss of Genes

Differentiation is the process by which cells that have committed to a particular developmental fate by the determination process (see Section 50.1) now develop into specialized cell types with distinct structures and functions. As part of differentiation, cells concentrate on the production of molecules characteristic of their specific types. For example, 80–90% of the total protein that lens cells synthesize is crystallin.

Research into differentiation confirmed that as cells specialize, they retain all the genes of the original egg cell; except in rare instances, differentiation does not occur through selective gene loss. Several definitive experiments supporting this conclusion were carried out several decades ago by Robert Briggs and Thomas King of Lankenau Hospital Research Institute in Philadelphia (now Fox Chase Cancer Center), and extended by John B. Gurdon of the University of Cambridge, United Kingdom. In a typical experiment, the nucleus of a fertilized frog egg was destroyed by ultraviolet light. A micropipette was then used to transfer a nucleus from a fully differentiated tissue, intestinal epithelium, to the enucleated egg. Some of the eggs receiving the transplanted nuclei subsequently developed into normal tadpoles and adult frogs. This outcome was possible only if the differentiated intestinal cells still retained their full complement of genes. This conclusion was extended to mammals in 1997 when Ian Wilmut and his colleagues successfully cloned a sheep—Dolly—starting with an adult cell nucleus. (This experiment is described in Section 18.2.)

Apoptosis Is a Normal Part of Development

Induction and differentiation build complex, specialized organs from the three fundamental tissue types. Complementing these processes is *apoptosis*—a type of programmed cell death—a process that removes cells in a regulated way at a specific point in development. Apoptosis is an active process that is a normal part of development in animals, and also has non-developmental roles, such as removing cells that are damaged beyond repair, or that are infected with viruses.

An example of apoptosis is afforded by *C. elegans*. In this worm, division of the fertilized egg eventually generates 1,090 cells, of which exactly 131 die by apoptosis at prescribed times to produce the 959 cells characteristic of an adult hermaphrodite. (*C. elegans* cell fates are summarized earlier in this section; see Figure 50.18.) Apoptosis has numerous functions in animal development, including controlling the number of cells in particular tissues (such as neurons), generating complex organs (such as the heart and the retina), removing structures not needed (such as mammary tissue in male mammals, the tails of frogs, and tissue to create the vaginal opening), and controlling the spacing and orientation of tissues (such as neuron spacing and orientation). In humans, for example, a developing fetus in the uterus produces about three times more neurons than are found at birth; the others are destroyed by apoptosis. Likewise, in the early development of human fetuses, tissue fills in the spaces between what will become the digits (fingers and toes); apoptosis removes that tissue prior to birth (see later discussion for more details about this example).

Apoptosis is characterized by specific morphological changes in cells. These changes include: (1) cell shrinkage, which can also result in a loss of cell-to-cell contact; (2) degradation of chromatin; (3) blebbing (production of small bulges) of the cell membrane; and (4) phagocytosis of the whole cell by macrophages, or fragmentation of the cell, followed by phagocytosis of the fragments.

C. elegans was the model organism for genetic research that first delineated a pathway for apoptosis. Sydney Brenner (then at The Molecular Sciences Institute, Berkeley, CA), H. Robert Hurvitz (MIT, Cambridge, MA), and John Sulson (then at The Wellcome Trust Sanger Institute, Cambridge, UK) received a Nobel Prize in 2002 "for their discoveries concerning genetic regulation of organ development and programmed cell death." Apoptosis is triggered by a *death signal*, which is either a particular extracellular signaling molecule that binds to a receptor in the plasma membrane of a cell targeted for apoptosis, or a molecular signal produced within the cell.

Figure 50.20A illustrates apoptosis involving an extracellular death signal. In the absence of a death signal, the membrane receptor is inactive. This allows a protein associated with the outer mitochondrial membrane, CED-9 (encoded by the *ced-9* cell death gene), to bind to and inactivate CED-4 (encoded by the *ced-4* gene). CED-4 in active form is required to activate the CED-3 protein (encoded by the *ced-3* gene), which is a protease of a class of proteases called *caspases*. Active CED-3 caspase is needed to turn on the cell death program. But here, in the absence of a death signal, CED-9 inactivates CED-4, which then cannot activate CED-3 so no apoptosis occurs. Cells with molecular switches set in this way are those that normally survive to form the adult nematode.

Binding of a death signal to the receptor activates the receptor and the events that follow are typical of signal transduction pathways (**Figure 50.20B**) (see Sections 9.1–9.3). In this case, activation of the receptor leads to inactivation of CED-9. Because CED-9 no longer inhibits CED-4, CED-4 is active and, in turn, it activates the CED-3 protease. Activated CED-3 initiates a cascade of reactions, including the activation of other caspase proteases, and nucleases that degrade cell structures and chromosomes as part of the cell death program.

Homologs of the CED-3, CED-4, and CED-9 proteins are involved in the apoptosis pathway that is common to all animals that have been studied. Genetic studies have demonstrated their importance in development. For instance, worms deficient in CED-4 have 15% more cells than normal worms, yet they are viable. However, mice with deletions of their *ced-4* homologs late in embryonic development exhibit severe head and face anomalies and brain overgrowth. Mice with deletions of their *ced-3* homologs die around birth due to massive cell overgrowth in their nervous systems.

Let us consider the example of the role of apoptosis in the formation of digits in tetrapods. Some tetrapods such as humans, mice, and chickens have separated digits, whereas others, such as ducks and a number of other bird species, frogs, bats, and otters, have webbing between digits. In all cases, early

A. No death signal

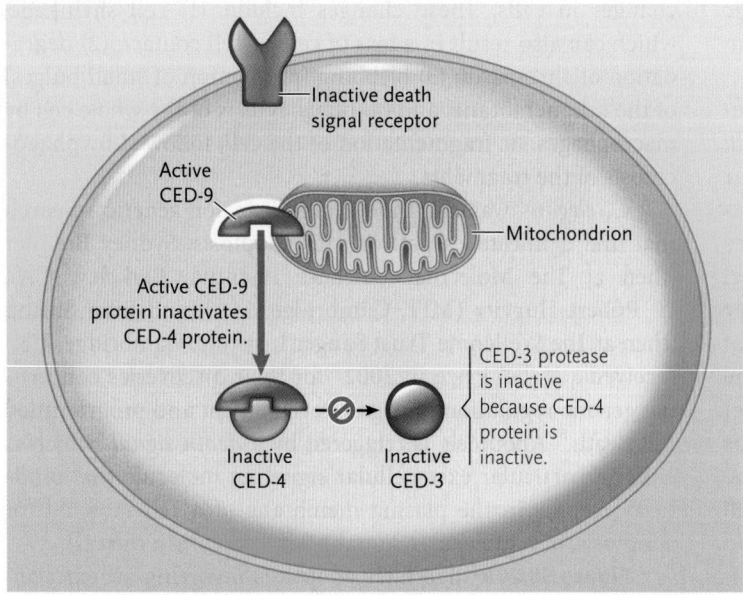

Inactive death signal receptor

Active CED-9

Active CED-9 protein inactivates CED-4 protein.

Mitochondrion

Inactive CED-4

Inactive CED-3

CED-3 protease is inactive because CED-4 protein is inactive.

Apoptosis is inhibited as long as CED-9 protein is active; cell remains alive.

B. Death signal

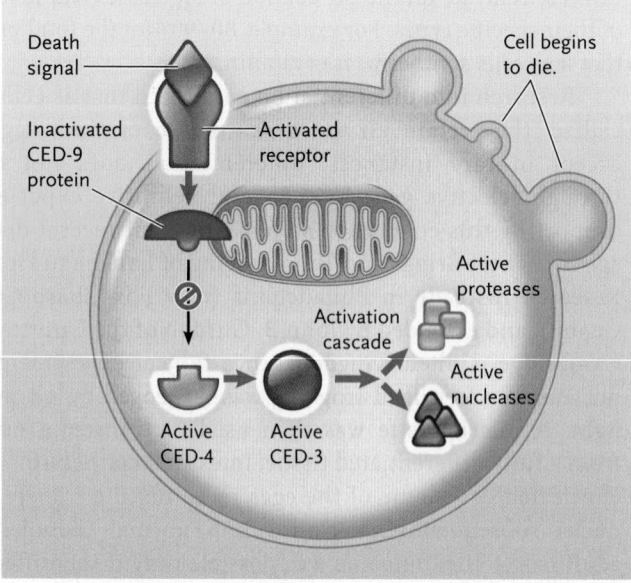

Death signal

Cell begins to die.

Inactivated CED-9 protein

Activated receptor

Active proteases

Activation cascade

Active nucleases

Active CED-4

Active CED-3

When a death signal is present, CED-9 is inactive and unable to inhibit the activation of proteins for apoptosis; the cell dies.

FIGURE 50.20 The molecular basis of apoptosis in *C. elegans*. (A) In the absence of a death signal, no apoptosis occurs. **(B)** When a death signal is present, the cell dies.
© Cengage Learning 2017

development produces a primordial limb with connected digits. The removal or not of the webbing by apoptosis then is under genetic control. Consider the chicken and the duck, which have highly similar feet except for the webbing in the duck. During development of the feet, bone morphogenic proteins (BMPs) are expressed in the tissue between the digits in both birds. BMP is a death signal, triggering the apoptosis events that result in the removal of the tissue between the digits, and those events produce the separated digits we see in the chicken. In the development of the duck foot, however, a protein called Gremlin is expressed in the tissue between the digits that inhibits the action of the BMPs. As a result, the duck has webbed feet. We can interpret these molecular data to mean that the evolution of web-footed birds likely involved the inhibition of BMP-mediated apoptosis in the tissue between the developing digits. The inhibition of BMP-mediated apoptosis by Gremlin is also involved in the development of the wing of the bat; the wing is webbing between the digits of the forefeet.

Recent research is showing that miRNAs are involved in apoptosis and, in fact, may be critical for its regulation in a number of developing tissues. For instance, in the frog,

Xenopus, a particular miRNA is expressed during eye development in the neural layer of the retina and is required for correct formation of the eye structure. Inhibition of that miRNA results in a reduction in eye size, and that reduction occurs because of a significant increase in apoptosis in the retina. Research has shown that the miRNA involved negatively regulates the *ced-4* homolog, confirming its direct role in the regulation of apoptosis during normal development.

STUDY BREAK 50.5

1. What are the key cellular events that contribute to morphogenesis in animals?
2. What is induction? What molecules are involved?
3. How does the activation state of CED-9 play a crucial role in apoptosis?

THINK OUTSIDE THE BOOK

Use the Internet or research literature to outline evidence associating an increase in the level of N-cadherin with breast cancer.

How do cells that comprise animals know what sex an animal should be as it grows up?

This should be an easy question to answer, because we know that in mammals it is the presence of the Y chromosome (containing the gene *SRY* discussed earlier in this chapter) that determines the development of testes in males. Testes then secrete testosterone and anti-Müllerian hormone (AMH), which stimulate Wolffian duct development and Müllerian duct regression. In females, ovaries develop because of the absence of SRY protein and the Müllerian ducts form. Steroid hormones secreted from the gonads then act to differentiate other tissues, including the brain. However, in nature there are many reports of animals that are gynandromorphic; that is, they have both female and male characteristics. These animals can be either bilaterally symmetrical or mosaics (containing a mix of male and female cells). Gynandromorphs have been seen in chickens, song birds, blue crabs, spiders, and insects, including butterflies and carpenter bees, as well as others. Some gynandromorphs that have been well studied include a song bird, the gynandromorphic Australian zebra finch, which had a testis and male plumage on the right side of the body and an ovary and female plumage on the left side of the body. Avian sex determination is chromosomal, and birds have a *ZW* sex-determination system that is reversed compared to the mammalian *XY* system. Female birds have heterogametic chromosomes *(ZW)* and males have homogametic chromosomes *(ZZ)*, and the sex-determining gene in birds is the *Z*-linked *DMRT1* gene. The gynandromorphic zebra finch was studied by Dr. Arthur P. Arnold at the University of California, Los Angeles, where he and a team of researchers determined that the right side of the brain of the bird contained no *W* sex chromosomes while the left side contained both *Z* and *W* chromosomes. The fact that these animals develop naturally indicates that there are still many unanswered questions about sexual differentiation. What is not yet known is if the sex imposed on a developing cell comes entirely from hormonal signals originating from the outside of the cell, or if there are also signals coming from within the cells from the cell's own genome. Also, in vertebrates, how do sex chromosomes, acting on individual cells, play a role in the differences that develop between male and female tissues, including the brain?

What initiates childbirth?

We learned in this chapter that there are a series of hormonal and physical events that are associated with childbirth. There is an increase in the secretion of oxytocin from the posterior pituitary gland and oxytocin receptors are found in the smooth muscle cells of the uterine wall. Oxytocin stimulates the muscle to begin the rhythmic contractions of labor marking the beginning of birth or parturition. This process has been well studied, but the initiation of labor is a complex process that has yet to be fully explained. What fetal-derived signals lead to the initiation of labor? How do these signals play a role in preterm labor? What roles do signals, such as prostaglandins, from the placenta play in regulating the timing of birth? What roles do the fetal hypothalamus and adrenal glands play in initiating childbirth? What roles other fetal organs, such as the lungs, play? What role do hormones, such as estradiol from the mother, or cortisol and corticotropin-releasing factor from the fetal lungs or surfactant from the fetal lungs play in sending the signals to begin the process of parturition? There are several exciting avenues of research in this area. First, Louis J. Muglia M.D., Ph.D. from Cincinnati Children's Hospital Medical Center is researching the timing of birth and the challenges to prevent premature births. His research has demonstrated that prostaglandins are essential for the initiation of parturition and he has determined that the regulation of COX-1 is important for the initiation of parturition in mice. COX-1 (cyclooxygenase-1) is a protein that acts as an enzyme to speed up the production of prostaglandins in the stomach and the expression of the *COX-1* gene increases in the uterus during pregnancy. Identifying regulatory factors, such as COX-1, may provide critical information into timing the onset of parturition. Next, Carole Mendelson, Ph.D. from the University of Texas Southwestern Medical Center has determined that the fetal lungs are also involved in initiating childbirth. Using mice, Dr. Mendelson and her colleagues have demonstrated that surfactant-A (SP-A), a protein synthesized in the developing lung, serves as a hormonal signal that acts on the mother's uterus to indicate when the fetal lungs are mature and prepared for the critical transition from life in amniotic fluid of the womb to life breathing air. In humans the fetal lung starts producing SP-A by gestational week 32. Combined, the research by Drs. Muglia and Mendelson illustrates that combined signals from the fetus contribute to the initiation of childbirth but there is still much to learn.

think like a scientist

Based on the first unanswered question, do you think there are mammals that are gynandromorphs? Does your answer to this question alter the significance of the research?

Courtesy of Laura Carruth

Laura Carruth, an Associate Professor of Neuroscience at Georgia State University, studies the hormonal factors involved in how the brain responds to stress during development and how sexual differentiation of the brain is regulated. Her work focuses on model systems in songbirds (the Australian zebra finch) and mice. To learn more about Dr. Carruth's research, go to http://neuroscience.gsu.edu/profile/laura-carruth/.

REVIEW KEY CONCEPTS

For access to MindTap and additional study materials visit www.cengagebrain.com.

50.1 Mechanisms of Embryonic Development

- Developmental information is stored in both the nucleus and cytoplasm of the fertilized egg. The mRNA and protein molecules that direct the first stages of development are the cytoplasmic determinants.

- The unequal distribution of yolk and other components makes eggs polar. The animal pole typically gives rise to surface structures and the anterior end of the embryo, whereas the vegetal pole typically gives rise to internal structures of the embryo such as the gut.

- Following fertilization, cleavage divisions produce the morula. The morula hollows out to form the blastula, which develops into the gastrula, the stage in which rearrangements of cells produce the ectoderm, mesoderm, and endoderm. Gastrulation establishes the body pattern, in that the organs and other structures of embryo arise from these three tissue layers (Figures 50.1 and 50.2; Table 50.1).

- Development proceeds as a result of cell division, cell movements, selective adhesions, induction, determination, and differentiation.

50.2 Major Patterns of Cleavage and Gastrulation

- In sea urchin eggs, yolk is distributed evenly. As a result, cleavage divisions take place at the same rate in all regions of the embryo, and gastrulation follows a symmetrical pattern (Figure 50.3).

- In amphibian eggs, yolk is distributed unequally, with most in the vegetal pole. As a result, the rate of cell division is more rapid in the animal pole, and gastrulation shows an asymmetrical pattern (Figures 50.4 and 50.5).

- In bird and reptile embryos, the cleavage divisions give rise to a flat disc of cells at the top of the yolk, which divides into the epiblast and the hypoblast. In gastrulation, cells of the epiblast migrate to the interior to form the endoderm and the mesoderm. The epiblast cells left at the surface form the ectoderm (Figure 50.6).

- In birds and reptiles, the yolk sac, chorion, amnion, and allantoic membrane form from extensions of the primary tissue layers. These extraembryonic membranes conduct nutrients from the yolk to the embryo, exchange gases with the environment, and store metabolic wastes (Figure 50.7).

50.3 From Gastrulation to Adult Body Structures: Organogenesis

- In organogenesis, the three primary tissues give rise to the tissues and organs of the embryo. Organogenesis begins with neurulation, the development of the nervous system from ectoderm (Figures 50.8 and 50.9).

- The mesoderm splits into somites, which give rise to the vertebral column and to the muscles of the ribs, vertebral column, and limbs (Figure 50.9).

- Development of the eye from optic vesicles is illustrative of the inductions and differentiations common to organogenesis in vertebrates (Figure 50.10).

50.4 Embryonic Development of Humans and Other Mammals

- In humans, as in other placental mammals, cleavage divisions produce a morula that differentiates into a blastocyst. The blastocyst implants into the endometrium of the uterus, and its inner cell mass separates into the epiblast and hypoblast. The epiblast produces the ectoderm, mesoderm, and endoderm of the embryo (Figures 50.11 and 50.12, steps 1–2).

- Gastrulation, neurulation, differentiation of cell layers, and formation of extraembryonic membranes occur by mechanisms similar to those of bird and reptile embryos. Differentiation of ectoderm, mesoderm, and endoderm into their final tissues and organs is also similar in birds and reptiles.

- Extraembryonic membranes form in mammals by processes that are also similar to the reptilian–bird pattern. However, some of the membranes have altered functions, reflecting the minimal amount of yolk in mammalian embryos, and maintenance of the embryo by the placenta (Figure 50.12, steps 3–5).

- The placenta is connected to the embryo by the umbilical cord, which conducts blood between the embryo and the placenta (Figure 50.12, step 6).

- Fetal growth proceeds until birth, when the baby is forced from the uterine cavity and through the vagina by contractions of the uterus, stimulated by oxytocin (Figures 50.13 and 50.14).

- The mother's mammary glands secrete milk once the offspring is born. Suckling by the offspring stimulates prolactin and oxytocin release from the pituitary, which stimulates milk production and secretion from the glands, respectively.

- Embryos develop internal male or female sex organs from the same primitive structures. The presence or absence of a Y chromosome, which carries the key *SRY* gene, determines whether the internal structures develop into male or female sexual organs (Figure 50.15).

- Most animals continue development after hatching or birth, leading to the adult, the sexually mature form of the species.

50.5 The Cellular Basis of Development

- Development in animals involves the regulation of specific cellular events, including cell division, cell migration, and cell adhesion.

- Cell division in development varies in orientation and rate.

- Cell movements in development occur through changes in cell shape or the migrations of entire cells. Shape changes are produced by microtubules or microfilaments. Cell migrations involve the activity of microtubules and microfilaments and large-scale reorganization of the cytoskeleton. In many developmental systems, migrating cells follow tracks formed by molecules of the extracellular matrix (Figure 50.16).

- Selective cell adhesions, which depend on membrane proteins called CAMs, the major class of which is the cadherins, underlie many cell movements. The final cell adhesions hold the embryo in its correct shape and form (Figure 50.17).

- For some organisms, the origins of adult or larval structures have been mapped to regions of the embryo from which each structure derived (Figure 50.18).

- Induction results from the effects of signaling molecules of the inducing cells on the responder cells that are near the inducer cells. To change its cellular behavior, the responder cell must have a receptor for the inducer molecule and it must be competent to respond (Figure 50.19).

- In differentiation, cells change from embryonic form to specialized types with distinct structures and functions. Differentiation occurs by differential gene activation.

- Apoptosis—a type of programmed cell death—plays an important role in normal development by removing tissues present during development but not in the adult tissue or organ (Figure 50.20).

TEST YOUR KNOWLEDGE

Remember/Understand

1. Major contributors to the cleavage patterns of a zygote are the:
   a. sperm and egg cytoplasm.
   b. sperm and egg chromosomes.
   c. ribosomes and mitochondria.
   d. egg nucleus and yolk.
   e. pigments.

2. The process by which cells undergo mitosis without a corresponding increase in cytoplasm is called:
   a. polarity.   d. organogenesis.
   b. induction.   e. cleavage.
   c. gastrulation.

3. Which of the following mechanisms does *not* contribute to zygote development?
   a. meiosis   d. determination
   b. mitosis   e. induction
   c. selective cell adhesions

4. A major event during gastrulation is:
   a. the movement of surface cells of the blastula into an interior position.
   b. the displacement of the archenteron by the blastocoel.
   c. the formation of the coelom from the endoderm.
   d. the extension of ectoderm and endoderm to form the yolk sac.
   e. the development of ectoderm to form epidermal and neural tissues.

5. The presence of the gray crescent in a frog embryo:
   a. corresponds to the location of the future mouth.
   b. corresponds to the location of sperm entry into the egg.
   c. will cause the formation of a disordered mass of cells.
   d. contains concentrated amounts of darkly colored pigment granules from the animal pole.
   e. indicates that the egg has been fertilized.

6. To contribute to the formation of a nervous system:
   a. the neural crest develops into motor neurons.
   b. the neural tube is converted into a neural plate.
   c. the notochord induces the overlying ectoderm to become a neural plate.
   d. the roof of the archenteron induces the formation of the neural tube.
   e. somites give rise to the autonomic nervous system.

7. In mammalian development:
   a. the morula develops into a trophoblast.
   b. the chorionic villi allow the blastocyst to move down the oviduct.
   c. the allantois takes over the work of the amnion.
   d. the pharyngeal arches transform into the pharynx, larynx, and nasal cavities.
   e. prolactin stimulates parturition.

8. In the development of the female sex organs:
   a. all ducts in the 7-week embryo become Wolffian ducts.
   b. the *SRY* gene is activated.
   c. anti-Müllerian hormone is secreted.
   d. the Müllerian ducts develop into oviducts.
   e. the mother secretes oxytocin.

9. In the embryonic development of the eye:
   a. the optic vesicle cells permanently adhere to each other to prevent movement, whereas the optic cup cells are very motile.
   b. the optic cup induces the surface ectoderm to form the lens placode.
   c. gradients determine that the ectoderm overlying the lens vesicle develops into the optic vesicle.
   d. microtubules powered by myosins and microfilaments powered by dyneins move the eye components around in the head region.
   e. cadherins function in the presence of calcium to allow the lens placode and optic cup to break apart.

10. Which of the following statements about development is incorrect?
    a. As microtubules lengthen and slide apart, the shape of cells within the ectoderm changes.
    b. The extracellular matrix molecule fibronectin guides the movement of cells during development.
    c. The furrow formed during mitosis of cells in the developing embryo is centrally located to ensure all cells divide symmetrically.
    d. Cell surface cadherins regulate the ability of developing cells to adhere to one another.
    e. Integrins are responsible for transmitting information from the extracellular matrix to the cytoskeleton.

Apply/Analyze

11. **Discuss Concepts** Experimentally, it is possible to divide an amphibian egg so that the gray crescent is wholly within one of the two cells formed. If the two cells are separated, only the cell with the gray crescent will form an embryo with a long axis, notochord, nerve cord, and back musculature. The other forms a shapeless mass of immature gut and blood cells. Propose an explanation for these outcomes.

12. **Discuss Concepts** Developmental biologist Lewis Wolpert once observed that "it is not birth, marriage or death, but gastrulation which is truly the most important time in your life." In what sense is he correct?

13. **Discuss Concepts** Arguably, in sexually reproducing animals, development begins when eggs and sperm form in the parents. In a paragraph, explain the rationale for this idea.

14. **Discuss Concepts** Investigators discovered a *Drosophila* protein that triggers development of the nerve cord on the ventral side of the embryos. When an mRNA encoding the protein was injected into cells on the ventral side of *Xenopus* embryos, dorsal structures were formed on the ventral side, including incomplete heads. What do these findings suggest about the evolution of embryonic development?

Evaluate/Create

15. **Design an Experiment** You learned in this chapter about the molecular basis of apoptosis in the nematode worm, *C. elegans*. In humans, the *BCL2* gene is a homolog of the worm *ced-9* gene. What experiment would convince you that the human BCL2 protein is functional homologous to the worm CED-9 protein?

16. **Apply Evolutionary Thinking** In this chapter, you have learned that cleavage and gastrulation differ among vertebrates that have different amounts of yolk in their eggs. What are some features of early development that have been conserved across vertebrate groups during their evolution?

For selected answers, see Appendix A.

INTERPRET THE DATA

People considering fertility treatments should be aware that such treatments raise the risk of multiple births and that multifetal pregnancies are associated with an increased risk of some birth defects. The **Table** shows the results of Yiwei Tang's study of birth defects reported in Florida from 1996 to 2000. Tang compared the incidence of various defects among single and multiple births. She calculated the relative risk for each type of defect based on type of birth, and corrected for other differences that might increase risk: maternal age, income, race, previous adverse pregnancy experience, education, Medicaid participation during pregnancy, and the infant's sex and number of siblings. A relative risk of less than 1 means a defect occurs less often with multiple births than single births. A relative risk of greater than 1 means that multiples are more likely to have a defect.

1. What was the most common type of birth defect in the single-birth group?

2. Was that defect more or less common in the multiple-birth group than among single births?

3. Tang found that multiples have more than twice the risk of single newborns for one type of defect. Which type?

4. Does a multiple pregnancy increase the relative risk of chromosomal defects in offspring?

Source: Y. Tang et al. 2006. The risk of birth defects in multiple births: A population-based study. *Maternal and Child Health Journal* 10:75–81.

Prevalence, per 10,000 Live Births, of Various Types of Birth Defects among Multiple and Single Births

	Prevalence of Defect		Relative Risk
	Multiples	Singles	
Total birth defects	358.50	250.54	1.46
Central nervous system defects	40.75	18.89	2.23
Chromosomal defects	15.15	14.20	0.93
Gastrointestinal defects	28.13	23.44	1.37
Genital/urinary defects	72.85	58.16	1.31
Heart defects	189.71	113.89	1.65
Musculoskeletal defects	20.92	25.87	0.92
Fetal alcohol syndrome	4.33	3.63	1.03
Oral defects	19.84	15.48	1.29

Ecology and the Biosphere

51

National Oceanic and Atmospheric Administration (NOAA)

Stormy weather in the biosphere. Atmospheric disturbances such as Hurricane Katrina, seen here in a satellite photograph just before its arrival in New Orleans on August 28, 2005, often have a dramatic impact on living systems.

Why it matters . . . Large-scale weather patterns that develop over the Pacific Ocean have a major impact on organisms living in Asia, North America, and South America, as well in the seas that lie between them **(Figure 51.1)**. In most years, air flows from a high-pressure system over the eastern Pacific toward a low-pressure system over the western Pacific. These winds move surface water from east to west and bring predictably heavy precipitation to parts of Asia and Australia, providing much of the rainfall on which natural vegetation and agriculture depend. Winds also usually blow from the poles toward the equator along the western sides of continents, and Earth's rotation causes these winds to push ocean surface water westward, away from the coast. The displaced surface water is replaced by cold, deep, nutrient-rich water carried by vertical currents called *upwellings*. The nutrients support complex marine food webs in the shallow water above the continental shelf (see Figure 51.1A). For example, the Peru Current along the west coast of South America once supported a rich anchovy fishery.

Ocean currents vary seasonally, however, and in late December or early January, a warm, nutrient-poor current flows eastward along the equator and then north and south along the coastlines of Central and South America. Peruvian fishermen call this warm current El Niño (Spanish for "the child"), because it reaches their coast around Christmas. It usually persists for only a few weeks.

Sometimes these weather patterns shift, causing widespread changes in the usual rainfall patterns. For example, in the winter of 1997–1998, record rainfall caused mudslides in California and flooding along the normally arid coast of Ecuador and Peru. But the annual rains never arrived in Asia and Australia, and fires consumed tropical rainforests in Indonesia and Malaysia. What caused these major climatic dislocations? Every 3 to 7 years, interactions between the upper layers of the Pacific Ocean and the atmosphere produce El Niño, a climatic event with global

A. Usual pattern of Pacific Ocean currents

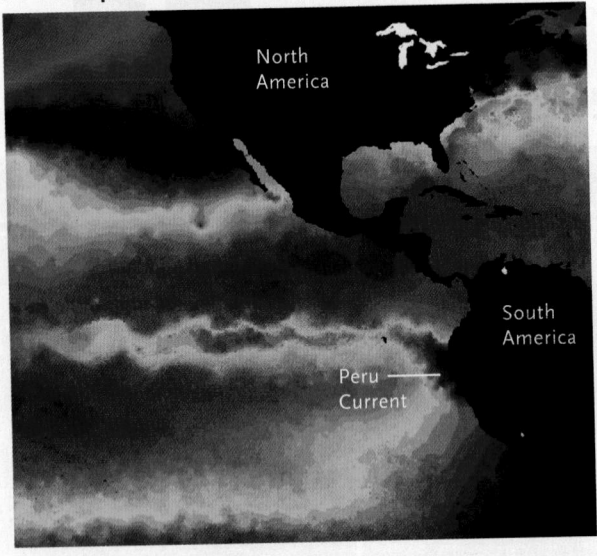

B. Pacific Ocean currents in an El Niño year

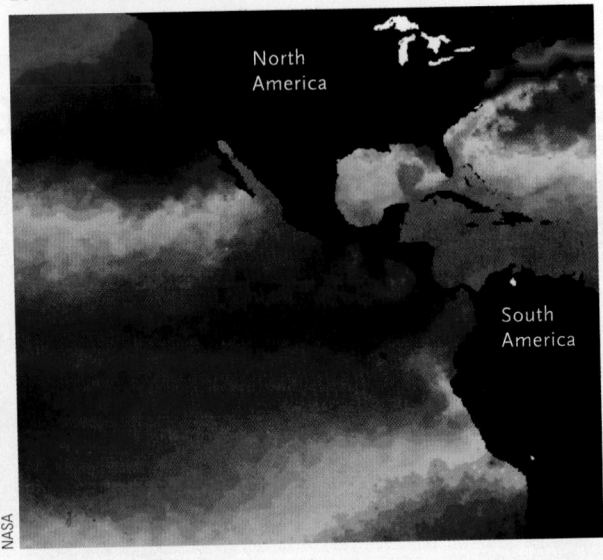

In most years, the powerful Peru Current carries cold water from the ocean bottom to the surface off the west coast of South America. The cold surface water then flows westward along the equator toward a large pool of warm water in the western Pacific Ocean. In this satellite photo taken on May 31, 1988, dark red indicates the warmest water and dark green the coldest upwelled water.

During an El Niño event, equatorial winds reverse directions and warm surface water flows eastward along the equator from the western Pacific Ocean toward South America. In this satellite photo taken on May 13, 1992, the warm water (orange) reached the west coast of South America, suppressing the upwelling of cold water by the Peru Current.

FIGURE 51.1 El Niño and Pacific Ocean currents.

consequences. The 1997–1998 El Niño altered weather patterns worldwide, killing more than 2,000 people and causing at least $30 billion in property damage.

In strong El Niño years, atmospheric pressure systems change markedly over the Pacific, altering the prevailing winds and ocean currents. Equatorial winds weaken; surface currents reverse direction, flowing from west to east; and a huge pool of warm ocean water accumulates in the eastern Pacific. During these shifts, the heavy rain that usually falls on Asia and Australia is instead delivered to the central and eastern Pacific. Thus, in strong El Niño years, Asia and Australia receive less rain than usual, and the west coasts of the Americas receive more. In the United States, winter temperatures are unusually high in the north central states and unusually low in the southern states.

El Niño episodes also alter sea surface temperature. When the warm current flowing from west to east reaches the continental shelf, it displaces the cold water of the Peru Current and prevents the usual upwelling (see Figure 51.1B). These changes in ocean currents have catastrophic effects on marine life. Lacking sufficient nutrients, phytoplankton (that is, tiny floating photosynthetic organisms) die, followed by fishes that eat phytoplankton, and seabirds that eat fishes. In combination with overfishing, the El Niño of 1972 drove the Peruvian anchovy population to the brink of extinction.

Some El Niño years are followed by a weather pattern called La Niña: the low-pressure system over the western Pacific is accentuated, pulling air and ocean surface water from east to west. Low ocean surface temperatures extend from the coast of South America to Samoa. La Niña's effect on winter weather is opposite that of El Niño: parts of Asia and Australia are unusually wet; the northern United States experiences periods of cold, wet weather, whereas the southern region is unusually warm and dry.

El Niño and La Niña are two extremes of a global climate cycle called the El Niño Southern Oscillation, or ENSO (the name refers to fluctuations in air pressure over the tropical Pacific). ENSO, a product of large-scale interactions between the ocean and atmosphere, has a major impact on the **biosphere,** which comprises all organisms on Earth and the places where they live.

The story of an uncommonly severe weather pattern that resulted from these environmental interactions introduces our unit on **ecology,** the study of interactions between organisms and their environments. All environments have both **biotic** ("biological") and **abiotic** ("nonbiological") components. The biotic environment includes all the organisms found in a particular place; the abiotic environment includes temperature, moisture, soil chemistry, and other physical factors. In addition to living organisms, the biosphere includes three abiotic components, which surround Earth's geological bulk like a skin. The **hydrosphere** encompasses all the water, including oceans and polar ice caps. The **lithosphere** includes the rocks, sediments, and soils of the crust. Finally, the **atmosphere** includes gases and airborne particles that envelop the planet.

In this chapter, we survey the biosphere with a wide-angle lens. After first describing how ecologists study the relationships between organisms and their environments, we examine Earth's environmental diversity, as well as some mechanisms—the products of evolution—that allow organisms to cope with environmental variations in space and time. We then consider how variations in the physical environment influence the large-scale distributions of organisms on land, in freshwater, and in the sea.

51.1 The Science of Ecology

The subject matter of ecology is so vast and so diverse that ecological research is often linked to work in genetics, physiology, anatomy, behavior, paleontology, and evolution, as well as geology, geography, and environmental science. The connections between research in ecology and evolution are especially strong, because organisms exhibit evolutionary responses to their ecological relationships, and because ecological relationships change as organisms evolve. Moreover, many ecological phenomena occur over huge areas and long time spans. Thus, ecologists must devise clever ways to determine how environments influence organisms, how organisms respond to environmental factors, and how organisms change the environments in which they live.

Today, the science of ecology encompasses two related disciplines. The major research questions of *basic ecology* relate to the distribution and abundance of species and how they interact with each other and with the physical environment. Using these data as a baseline, workers in *applied ecology* develop conservation plans and amelioration programs to limit, repair, and mitigate ecological damage caused by human activities. Research in applied ecology overlaps to some extent with that in environmental science, but the latter discipline also encompasses perspectives from the social sciences, such as economics and political science.

Ecologists Study Levels of Organization Ranging from Individual Organisms to the Biosphere

Ecology can be divided into five increasingly complex and inclusive levels of organization (see the upper five panels in Figure 1.2). In **organismal ecology,** researchers study the genetic, biochemical, physiological, morphological, and behavioral adaptations of organisms to the abiotic environment. We have described many such adaptations in Units V and VI; we describe the evolution of animal behavior in Chapter 56.

Population ecology, the subject of Chapter 52, focuses on **populations,** groups of individuals of the same species that live together. Population ecologists study how the size and other characteristics of populations change in space and time. Research in **community ecology,** discussed in Chapter 53, examines groups of populations that occur together in one area. Community ecologists study interactions between species, analyzing how predation, competition, and environmental disturbances influence a community's development, organization, and structure. Ecologists studying **ecosystems,** considered in Chapter 54, explore the complex cycling of nutrients and the flow of energy between the biotic components of an ecological community and the abiotic environment. Finally, the largest-scale ecological studies focus on the biosphere, which we consider in this chapter. We discuss biodiversity and conservation biology in Chapter 55.

Ecologists Test Hypotheses with Observational and Experimental Data

Ecology has its roots in descriptive natural-history studies that date back to the ancient Greeks. Modern ecology was born in 1870 when the German biologist Ernst Haeckel coined the term *Oekologie* (*oikos* = house). Contemporary researchers still gather descriptive information about ecological relationships, examples of which are highlighted in *Observational Research* figures; these basic observations often provide the background for more analytical studies, examples of which are described in *Experimental Research* figures. The distinction between these two approaches is discussed in Section 1.4.

Most ecologists create hypotheses about ecological relationships and how they change through time or differ from place to place. Like other scientists, some ecologists formalize these ideas in mathematical models that express clearly defined, but hypothetical, relationships among important variables in a system. Manipulation of a model, usually with the help of a computer, can allow researchers to ask what would happen if some of the variables or their relationships change. Thus, researchers can simulate natural events and large-scale experiments before investing time, energy, and money in field and laboratory work. Bear in mind, however, that mathematical models are no better than the ideas and assumptions they embody, and useful models are constructed only after basic observations define the relevant ecological variables and their relationships.

Ecologists often conduct field or laboratory studies to test the predictions of their hypotheses. In controlled experiments, researchers compare data from an experimental treatment (in which one or more variables are artificially manipulated) with data from a control (in which nothing is changed). Sometimes the distributions of species create "natural experiments," eliminating the need to manipulate variables. Studies of how islands of different sizes harbor different numbers of bird species (described in Section 53.7) provide an example of a natural experiment.

STUDY BREAK 51.1

1. Why are studies of ecosystems more "inclusive" than studies of populations?
2. How do ecologists use mathematical models in their research?

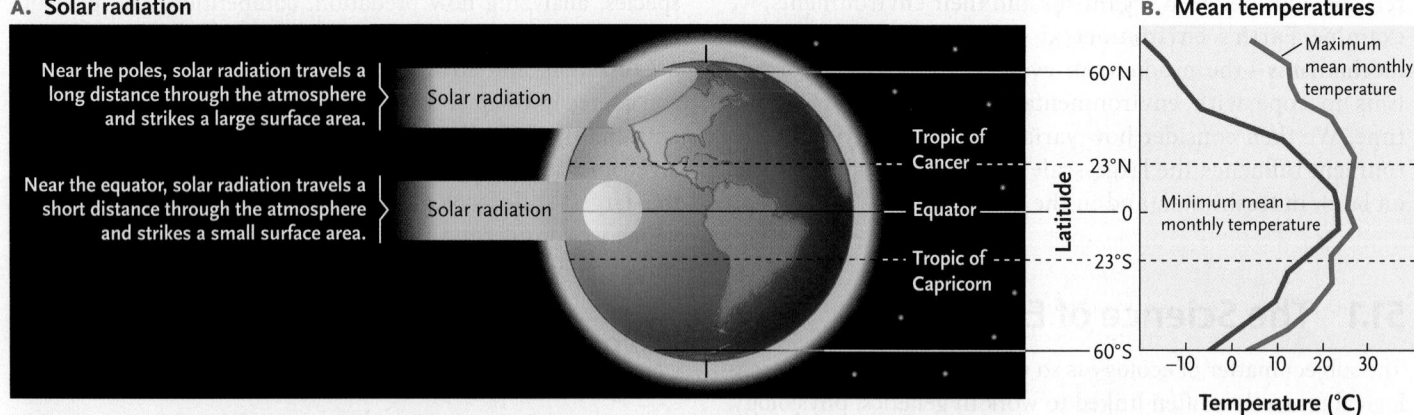

A. Solar radiation

Near the poles, solar radiation travels a long distance through the atmosphere and strikes a large surface area.

Solar radiation

Near the equator, solar radiation travels a short distance through the atmosphere and strikes a small surface area.

Solar radiation

Tropic of Cancer - - - - - 23°N
Equator - - - - 0
Tropic of Capricorn - - - 23°S

60°N

60°S

B. Mean temperatures

Maximum mean monthly temperature

Minimum mean monthly temperature

Latitude

Temperature (°C)
−10   0   10   20   30

FIGURE 51.2 Latitudinal variation in solar radiation and temperature. (A) Solar radiation is more intense near the equator than near the poles. **(B)** Minimum and maximum mean monthly temperatures—as well as the range of mean monthly temperatures—vary with latitude.
© Cengage Learning 2017

51.2 Environmental Diversity of the Biosphere

Numerous abiotic factors—sunlight, temperature, humidity, wind speed, cloud cover, and rainfall—contribute to a region's **climate,** the weather conditions prevailing over an extended period of time. Climates vary on global, regional, and local scales, and they undergo seasonal changes almost everywhere. As you will discover as you read this chapter, variations in climate over space and time influence the lives of all organisms.

Variations in Incoming Solar Radiation Create Global Climate Patterns

A global pattern of environmental diversity results from latitudinal variation in incoming solar radiation, Earth's rotation on its axis, and its orbit around the Sun.

SOLAR RADIATION Earth's spherical shape causes the intensity of incoming solar radiation to vary from the equator to the poles **(Figure 51.2).** When sunlight strikes Earth directly at a 90° angle, as it does near the equator, it travels the shortest possible distance through the radiation-absorbing atmosphere and falls on the smallest possible surface area (see Figure 51.2A). When sunlight arrives at an oblique angle, as it does near the poles, it travels a longer distance through the atmosphere and shines on a larger area. Thus, solar radiation is more concentrated near the equator than it is at higher latitudes, causing latitudinal variation in temperature (see Figure 51.2B).

SEASONALITY Earth is tilted on its axis at a fixed position of 23.5° from the perpendicular to the plane on which it orbits the Sun **(Figure 51.3).** This tilt produces seasonal variation in the duration and intensity of incoming solar radiation. The Northern Hemisphere receives its maximum illumination—and the Southern Hemisphere its minimum—on the June solstice (around June 22), when the Sun shines directly over the Tropic

of Cancer (23.5° N latitude). The reverse is true on the December solstice (around December 22), when the Sun shines directly over the Tropic of Capricorn (23.5° S latitude). Twice each year, on the vernal and autumnal equinoxes (around March 21 and September 23, respectively), the Sun shines directly over the equator.

Earth's tilt is permanent, and only the **tropics**—the latitudes between the Tropics of Cancer and Capricorn—ever receive solar radiation from directly overhead. Tropical regions experience only small seasonal changes in temperature and day length: environmental temperature is high, and days last approximately 12 hours throughout the year. (Tropical seasonality is reflected in the alternation of wet and dry periods rather than warm and cold seasons.) Seasonal variation in temperature and day length increases steadily toward the poles. Polar

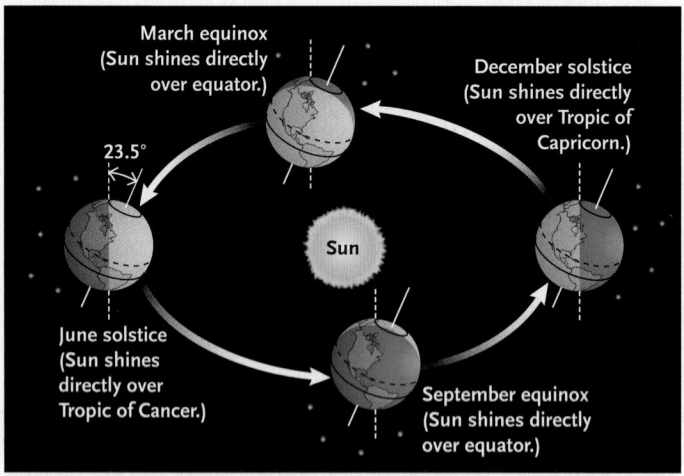

March equinox (Sun shines directly over equator.)

December solstice (Sun shines directly over Tropic of Capricorn.)

23.5°

Sun

June solstice (Sun shines directly over Tropic of Cancer.)

September equinox (Sun shines directly over equator.)

FIGURE 51.3 Seasonal variation in solar radiation. Earth's fixed tilt on its axis causes the Northern Hemisphere to receive more sunlight in June and the Southern Hemisphere to receive more in December. These differences are reflected in seasonal variations in day length and temperature, which are more pronounced at the poles than at the equator.
© Cengage Learning 2017

winters are long and cold with periods of continuous darkness, and polar summers are short with periods of continuous light.

AIR CIRCULATION Sunlight warms air masses, causing them to expand, lose pressure, and rise in the atmosphere. The unequal heating of air at different latitudes initiates global air movements, producing three circulation cells in each hemisphere **(Figure 51.4)**. Warm equatorial air masses rise to high altitude before spreading north and south. They eventually sink back to Earth at about 30° N and S latitude. At low altitude, some air masses flow back toward the equator, completing low-latitude circulation cells. Others flow toward the poles, rise at 60° latitude, and divide at high altitude. Some air flows toward the equator, completing the pair of middle-latitude circulation cells. The rest moves toward the poles, where it descends and flows toward the equator, forming the polar circulation cells.

The flow of air masses at low altitude creates winds near the planet's surface. But the surface rotates beneath the atmosphere, moving rapidly near the equator, where Earth's diameter is greatest, and slowly near the poles. Latitudinal variation in the speed of rotation deflects the movement of the rising and sinking air masses from a strictly north–south path into belts of easterly and westerly winds (see the right side of Figure 51.4); this deflection is called the *Coriolis effect*. Winds near the equator are called the *trade winds;* those further from the equator are the temperate (that is, between roughly 23° and 66° latitude) westerlies and easterlies, named for the direction from which they blow.

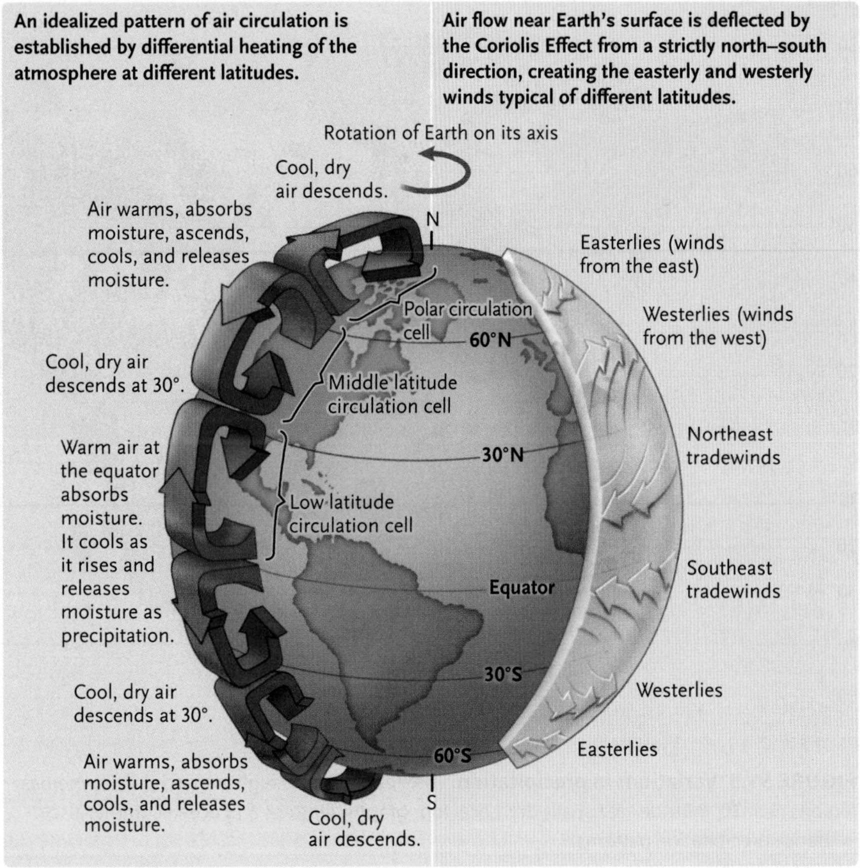

PRECIPITATION Differences in solar radiation and global air circulation create latitudinal variations in rainfall **(Figure 51.5)**. Warm air holds more water vapor than cool air does. As air near the equator heats up, it absorbs water, primarily from the oceans. However, the warm air masses expand as they rise, and their heat energy is distributed over a larger volume, causing their temperature to drop. A decrease in temperature without the actual *loss* of heat energy is called **adiabatic cooling.** After cooling adiabatically, the rising air masses release moisture as rain (see labels on the left side of Figure 51.4). Torrential rainfall is characteristic of warm equatorial regions, where rising, moisture-laden air masses cool as they reach high altitude.

As cool, dry air masses descend at 30° latitude, increased air pressure at low altitude compresses them, concentrating their heat energy, raising their temperature, and increasing their capacity to hold moisture. The descending air masses absorb water from the land, so these latitudes are typically dry. Some air masses continue moving toward the poles in the

An idealized pattern of air circulation is established by differential heating of the atmosphere at different latitudes.

Air flow near Earth's surface is deflected by the Coriolis Effect from a strictly north–south direction, creating the easterly and westerly winds typical of different latitudes.

Rotation of Earth on its axis

Cool, dry air descends.

Air warms, absorbs moisture, ascends, cools, and releases moisture.

Cool, dry air descends at 30°.

Warm air at the equator absorbs moisture. It cools as it rises and releases moisture as precipitation.

Polar circulation cell

Middle latitude circulation cell

Low-latitude circulation cell

Cool, dry air descends at 30°.

Air warms, absorbs moisture, ascends, cools, and releases moisture.

Cool, dry air descends.

60°N

30°N

Equator

30°S

60°S

N

S

Easterlies (winds from the east)

Westerlies (winds from the west)

Northeast tradewinds

Southeast tradewinds

Westerlies

Easterlies

FIGURE 51.4 Global air circulation. Latitudinal variations in the intensity of solar radiation cause equatorial air masses to warm and rise, initiating a global pattern of air movement in three circulation cells in each hemisphere. Air masses moving near Earth's surface are deflected from a strictly north–south flow by the planet's rotation, creating easterly and westerly winds.
© Cengage Learning 2017

lower atmosphere. When they rise at 60° latitude, they cool adiabatically and release precipitation (see Figure 51.4), creating moist habitats in the northern and southern temperate zones.

OCEAN CURRENTS Latitudinal variations in solar radiation also warm the oceans' surface water unevenly. Because the volume of water increases as it warms, sea level is about 8 cm higher at the equator than at the poles. The volume of water associated with this "slope" is enough to cause surface water to move in response to gravity. The trade winds and temperate westerlies also contribute to the mass flow of water at the ocean surface. Thus, surface water flows in the direction of prevailing winds, forming major currents. Earth's rotation, the positions of landmasses, and the shapes of ocean basins also influence their movement.

Oceanic circulation is generally clockwise in the Northern Hemisphere and counterclockwise in the Southern **(Figure 51.6)**. The trade winds push surface water toward the equator and westward until it contacts the eastern edge of a continent. Swift, narrow, and deep currents of warm, nutrient-poor water run toward the poles, parallel to the east coasts of continents. For example, the Gulf Stream flows northward along the east coast

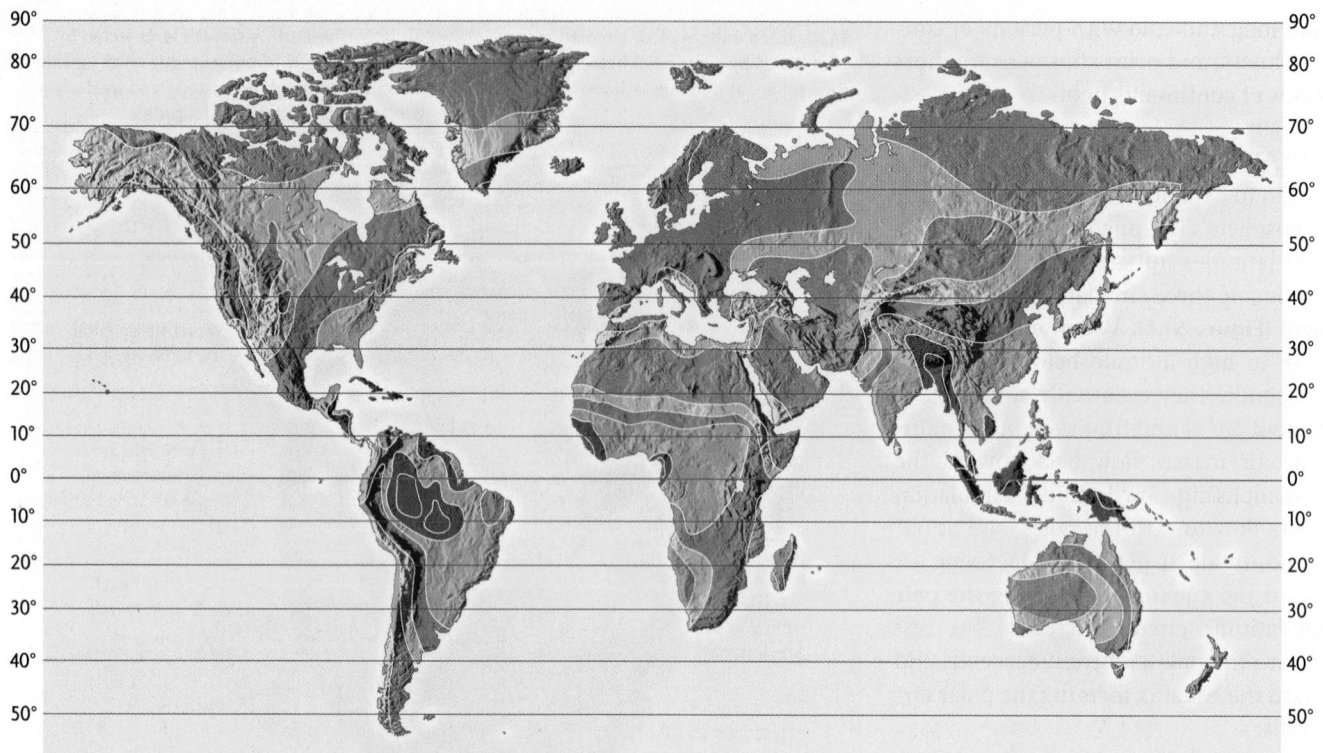

FIGURE 51.5 Variations in precipitation. The tropics receive high annual rainfall, whereas regions near 30° latitude are usually dry. Local topographic features and ocean currents also influence precipitation patterns.

© Cengage Learning 2017

KEY

Precipitation (cm/yr)

- Under 25
- 25 to 50
- 50 to 100
- 100 to 200
- 200 to 250
- Over 250

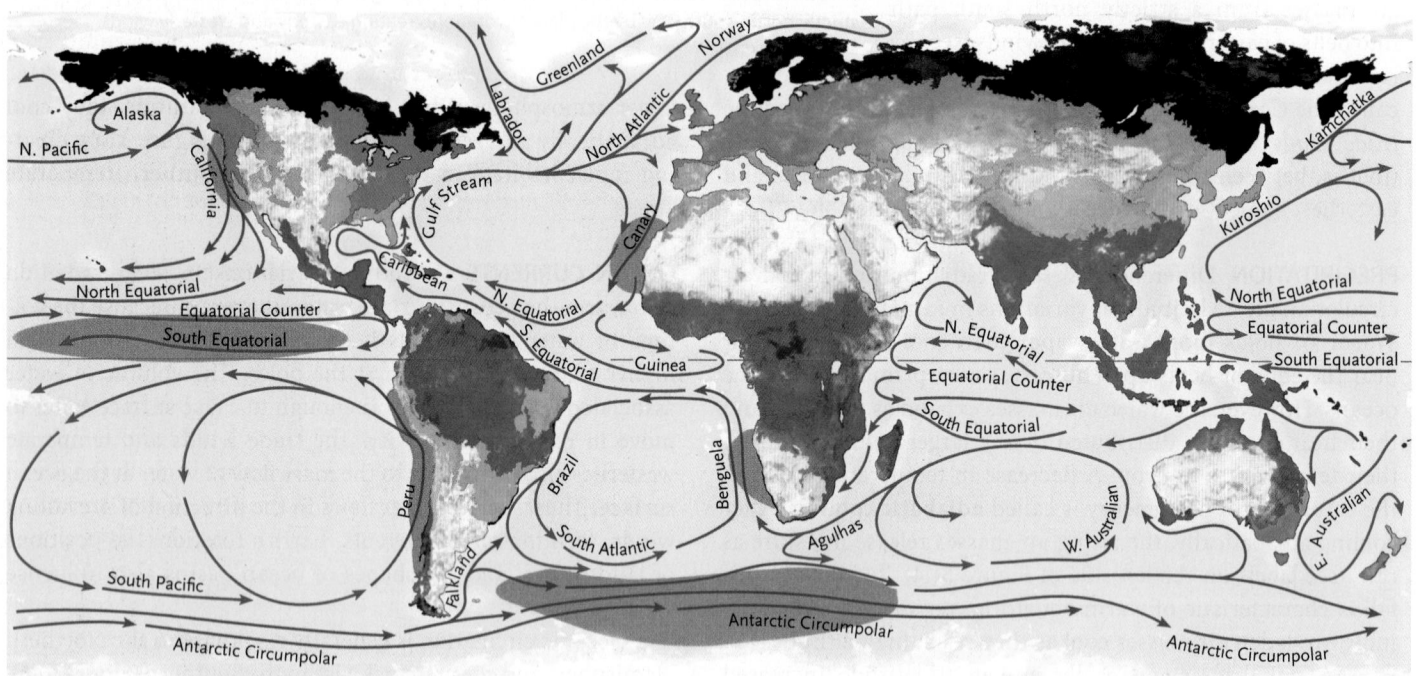

FIGURE 51.6 Ocean currents. Prevailing winds, Earth's rotation, gravity, the shape of ocean basins, and the positions of landmasses establish the direction and intensity of surface currents in the oceans. In general, warm currents flow away from the equator, and cold currents flow toward it.

© Cengage Learning 2017

KEY

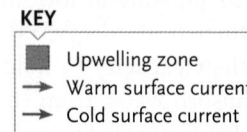

- Upwelling zone
- Warm surface current
- Cold surface current

of North America, carrying warm water toward northwestern Europe. Cold water returns from the poles toward the equator in slow, broad, and shallow currents, such as the Peru Current, that parallel the west coasts of continents.

Regional and Local Effects Overlay Global Climate Patterns

Although global and seasonal patterns determine a site's climate, regional and local effects also influence abiotic conditions.

PROXIMITY TO THE OCEAN Currents running along seacoasts exchange heat with air masses flowing above them, moderating the temperature over nearby land. Breezes often blow from the sea toward the land during the day and in the opposite direction at night **(Figure 51.7).** These local effects sometimes override latitudinal variations in temperature. For example, the climate in London is much milder than that in Minneapolis, even though Minneapolis is slightly further south. Minneapolis has a **continental climate** that is not moderated by the distant ocean, but London has a **maritime climate** that is tempered by winds crossing the nearby North Atlantic Current.

Ocean currents also affect moisture conditions in coastal habitats. For example, air masses absorb water as they move from west to east across the Pacific Ocean. They cool as they cross the cold California Current, and when they reach land in northern California and Oregon during winter, their water vapor condenses into heavy fog and rain. During summer, however, land is warmer than the adjacent ocean. The air masses heat up as they cross the land, and they accumulate water, creating dry conditions.

Some regions experience **monsoon cycles** caused by seasonal reversals of wind direction. In the North American southwest, for example, summer heat causes air masses over land to rise, creating a zone of low pressure. Moist air from the nearby Gulf of California flows inland, where it rises and cools adiabatically, releasing substantial precipitation. Summer monsoon rains deliver one third to one half of the annual rainfall in Arizona and New Mexico. During the winter, when land is cooler than the nearby ocean, low-pressure systems form over the ocean, and winds blow from the land to the sea; thus, winters in the southwest are generally dry. Seasonal monsoon cycles also deliver torrential rainfall to parts of Africa, Asia, and South America.

THE EFFECTS OF TOPOGRAPHY Mountains, valleys, and other topographic features also influence regional climates. In the Northern Hemisphere, south-facing slopes are warmer and drier than north-facing slopes because they receive more solar radiation. In addition, adiabatic cooling causes air temperature to decline 3°–6°C for every 1,000 m increase in altitude.

Mountains also establish regional and local rainfall patterns. For example, after a warm air mass picks up moisture from the Pacific Ocean, it moves inland and reaches the Sierra Nevada, which parallels the California coast. As it rises to cross the mountains, the air cools adiabatically and loses moisture, releasing heavy rainfall on the windward side **(Figure 51.8).** After the now-dry air crosses the peaks, it descends and warms, absorbing moisture and forming a **rain shadow.** Habitats on the leeward side of mountains, such as the Great Basin Desert in western North America, are typically drier than those on the windward side.

MICROCLIMATE Although climate influences the overall distributions of organisms, the abiotic conditions that immediately surround them—the **microclimate**—have the greatest effect on survival and reproduction. For example, a fallen log on the forest floor creates a microclimate in the underlying soil that is shadier, cooler, and moister than surrounding soil exposed to sun and wind. Many animals, including some insects, worms, salamanders, and snakes, occupy these sheltered sites and avoid the effects of prolonged exposure to the elements.

STUDY BREAK 51.2

1. How does Earth's spherical shape influence temperature and air movements at different latitudes?
2. What causes seasonality of the climate in the temperate zone?
3. Why do dry conditions occur at 30° N and S latitude?
4. Briefly describe how mountains influence local precipitation.

A. Daytime sea breeze

2 Cool air descends and replaces air over land through onshore flow.

1 Warm air ascends.

B. Nighttime land breeze

2 Cool air descends and replaces air over the sea through offshore flow.

1 Warm air ascends.

FIGURE 51.7 Sea breezes and land breezes. (A) On a summer afternoon, when the land is warmer than the adjacent ocean, warm air rises over the land, and a cool sea breeze blows inland from the ocean. **(B)** At night, when the ocean is warmer than the land, the pattern is reversed.
© Cengage Learning 2017

FIGURE 51.8 Formation of a rain shadow. White numbers indicate elevation (meters) followed by mean annual precipitation (centimeters) for the Sierra Nevada of California.
© Cengage Learning 2017

51.3 Organismal Responses to Environmental Variation and Climate Change

Daily and seasonal variations in physical factors have profound effects on the biology of individual organisms. Moreover, large-scale variations in environmental conditions, including those caused by global climate change, often influence the distributions of populations.

Organisms Use Homeostatic Responses to Cope with Environmental Variation

Many animals exhibit diverse homeostatic responses—biochemical, behavioral, physiological, and morphological—that enable them to maintain relatively constant conditions within their cells and tissues. Although the ability to use these responses almost certainly has a genetic basis, only some responses to environmental variation are *obligate* (that is, must always be used). *Molecular Insights* describes one such evolutionary response at the molecular level. Many behavioral and physiological responses are *facultative*. In other words, animals may use them or not, as their immediate conditions demand. Here we provide two brief examples of facultative behavioral and physiological responses to variations in environmental temperature through space and time.

Like many ectothermic animals, lizards often use behaviors to regulate body temperature (see Figure 1.16 and Section 48.7). They commonly bask in sunny spots to raise body temperature and seek shaded places to cool off. Many

Anolis lizard species (see *Focus on Research: Model Organisms* in Chapter 32) are distributed over broad elevational ranges, and populations living at high altitude encounter cooler environments than do those at low elevation. While they were graduate students at Harvard University, Paul E. Hertz, now of Barnard College, and Raymond B. Huey, now of the University of Washington, hypothesized that *Anolis* populations living at cool, high elevations would bask more frequently than those living at warm, low elevations. Hertz and Huey tested their hypothesis by observing *Anolis cybotes* and its close relative *Anolis shrevei* along an elevational gradient in the Dominican Republic **(Figure 51.9)**. Their results indicate that basking frequency increases steadily with elevation. Moreover, the body temperatures of the lizards decrease much less with increasing elevation than do air temperatures at the same localities. The researchers therefore concluded that increased basking frequency by lizards at high elevation partially compensates for the lower environmental temperatures they encounter.

The state of extreme physiological sluggishness called *torpor* is a facultative response to daily variations in environmental temperature. Endothermic animals use the heat generated by the metabolic breakdown of food to maintain high body temperature (see Section 48.6). However, small endotherms, such as hummingbirds, have a large relative surface area through which they lose body heat. When environmental temperature is low, they may lose heat faster than they can generate it, risking the total depletion of their energy reserves and death by starvation. The problem is particularly acute at night, when hummingbirds cannot feed. F. Reed Hainsworth and Larry Wolf of Syracuse University discovered that the purple-throated carib (*Eulampis jugularis*), a West Indian hummingbird, often becomes torpid at night, lowering its body temperature from 40° to 20°C. Because torpor reduces the temperature difference between their bodies and the environment, torpid birds lose heat less rapidly. At the nighttime environmental temperatures they usually encounter, the torpid hummingbirds may use 80% less energy than they would if they had not entered a temporarily dormant state.

How Does an Octopus Function at 0°C? RNA editing and cold adaptation

Octopus species occur in all of the world's oceans at an astonishing range of environmental temperatures. Those living in tropical waters may experience temperatures between 25°C and 35°C, whereas those living in the Arctic or Antarctic function at temperatures close to or below 0°C all year long. How does the nervous system of these active predators function at freezing temperatures? Researchers have long known that the transmission of information by neurons, the information conducting cells of the nervous system, depends on the flow of Na^+ ions through voltage-gated channels into the neuron, immediately followed by the flow of K^+ ions, also through voltage-gated channels, to the outside of the cell (see Section 39.2). Once the outflow of K^+ ions returns the voltage difference across the cell membrane to its resting state, the K^+ channel closes, allowing the cell to transmit another message soon thereafter. Scientists have also shown that the voltage-gated K^+ channel is more temperature sensitive than the voltage-gated Na^+ channel. Thus, adaptation of the octopus' nervous system to polar waters may depend on some change in the kinetics of K^+ channel closing, allowing neurons to transmit messages quickly despite the low temperature.

Research Question

What genetic or molecular mechanism is responsible for cold adaptation of the octopus nervous system?

Experiment

Sandra Garrett and Joshua J. C. Rosenthal at the Medical Sciences campus of the University of Puerto Rico, San Juan, hypothesized that differences in the sequences of genes coding for neuronal voltage-gated K^+ channel proteins had evolved among octopus species to maintain the channels' function under varying temperature regimes. They compared sequences in octopuses sampled in the tropics and in Antarctica and found that the sequences were virtually identical, differing at only four positions. None of the sequence differences produced changes in the protein that could explain how the voltage-gated channels function at very low temperatures. However, when Garrett and Rosenthal studied the kinetics of octopus ion channels, expressed in the oocytes of a laboratory frog, they found that channels coded by the genes of the Antarctic octopus opened and closed much more slowly than the channel from the tropical octopus. They hypothesized that some posttranscriptional mechanism (see Section 16.3), rather than the evolution of different gene sequences, was responsible for this aspect of cold adaptation.

Further research revealed that the mRNA products of the gene for this part of the K^+ channel are modified by adenosine deaminases, enzymes that convert adenosine in the RNA to inosine. The inosine is read as guanosine when the mRNA is translated, sometimes resulting in a change in the amino acid sequence of the protein produced. This form of posttranscriptional modification is an example of *RNA editing*. The researchers determined that one of the modified sites on the protein lies just at the edge of the pore through which K^+ ions move. Editing of the RNA causes replacement of an isoleucine amino acid with a valine, and that one change causes the channel to close twice as fast as the protein produced by unedited RNA from the same species. Thus, RNA editing prepares the neuron to carry additional information in half the time that an unedited channel requires.

The researchers subsequently sampled six additional octopus species from a range of latitudes and different ocean temperatures, sequencing both the genomic DNA for the K^+ gene and cDNAs made from transcribed mRNAs for that gene. By comparing genomic and cDNA sequences, they determined how much RNA editing occurs in the ion channel mRNAs of all six species. As they predicted,

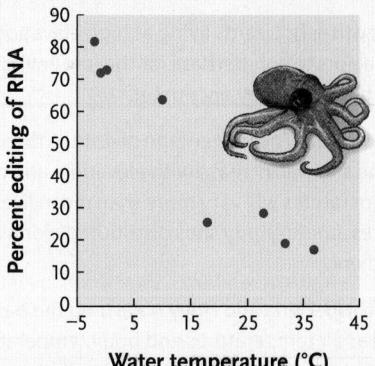

FIGURE **The extent of RNA editing in the voltage-gated K^+ channels in octopus neurons varies with water temperatures where the octopuses live.** Species from cold Arctic and Antarctic waters experience more RNA editing than species from the temperate zone or the tropics.
© Cengage Learning 2017

the amount of RNA editing that replaces isoleucine with valine at the site described above is correlated with the water temperature at which the octopuses live (**Figure**).

Conclusion

Variations in the rate at which the voltage-gated K^+ channel closes in octopus neurons are not caused by the evolution of differences in the genes coding for proteins that make up the channel. Instead, the functional differences—and, hence, cold adaptation in this aspect of octopus biology—result from posttranscriptional editing of the mRNA that directs the construction of those proteins.

think like a scientist

If differences in the voltage-gated K^+ channels in the neurons of different octopus species do not reflect differences in protein-coding DNA sequences, is the structure of the ion channels subject to natural selection?

Source: S. Garrett and J. C. Rosenthal. 2012. RNA editing underlies temperature adaptation in K^+ channels from polar octopuses. *Science* 335:848–851.

© Cengage Learning 2017

Global Climate Change Affects the Ecology of Many Organisms

As described in Section 54.4, virtually all scientists agree that Earth's atmosphere and oceans are getting steadily warmer. What effect will rapid global climate change have on biological systems? Biologists hypothesize that, on the spatial scale of the biosphere, rising temperatures will affect the geographical distributions of populations, species, and communities. Models of climate change predict that the distributions of polar species will contract to even higher latitudes, the ranges of temperate and tropical species will expand or shift toward the poles, and

FIGURE 51.9 | Observational Research

How Do Lizards Compensate for Elevational Variation in Environmental Temperature?

Hypothesis: Lizards living at high elevation can use behaviors to compensate for the low environmental temperatures they encounter.

Prediction: The percentage of lizards observed basking in the sun will increase with elevation, and mean air temperatures will vary more than mean lizard body temperatures among study sites distributed along an elevational gradient.

Method: Hertz and Huey measured the basking behavior as well as air temperatures and body temperatures of two closely related species *of Anolis* lizards distributed along an elevational gradient in the Dominican Republic. They surveyed populations of lizards at sea level, 550 m, 1,100 m, and 2,200 m elevation. They then compared the percentages of lizards basking and the mean air and lizard body temperatures at the four study sites.

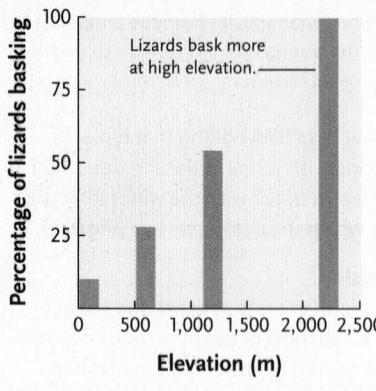
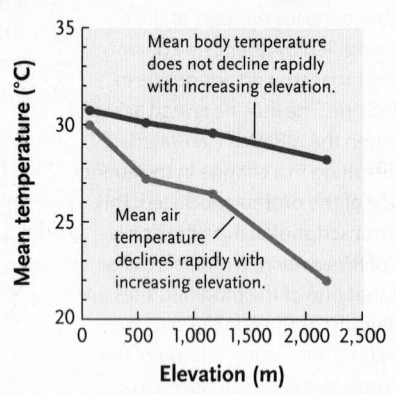

Result: The percentage of lizards basking increased steadily with elevation. Mean air temperature differed by as much as 8°C among study sites, but mean body temperature differed by only 2°C.

Conclusion: Lizards living at high elevation bask in patches of sun more frequently, partially compensating for the low environmental temperatures in their habitats.

Observe
Hypothesize
Predict
Experiment
Interpret

If global climate change caused air temperatures in the habitats occupied by these lizards to increase, how would you expect the lizards' behavior to change?

Source: P. E. Hertz and R. B. Huey. 1981. Compensation for altitudinal variation in the thermal environment by some *Anolis* lizards on Hispaniola. *Ecology* 62:515–521.

© Cengage Learning 2017

changes in the geographical distributions and timing of springtime activities in a wide variety of herbaceous plants, trees, invertebrates, and vertebrates over roughly the past century. Their analysis, published in 2003, suggested that the geographical ranges of 99 species of butterflies, birds, and alpine herbs in the Northern Hemisphere have shifted dramatically into habitats that had previously been too cold for them. Some species have expanded their distributions northward an average of 6.1 km per decade. Other species have shifted their distributions to higher elevation, an average of 6.1 m per decade. Their analysis also indicated that for 172 diverse species (including plants, butterflies, amphibians, and birds) springtime growth and reproduction have occurred on average 2.3 days earlier per decade. If these trends continue at the same rate, spring flowering and animal reproduction will occur one month earlier in the year 2130 than it did in 2000.

Numerous studies published after Parmesan and Yohe's landmark paper have documented similar shifts in the latitudinal and elevational distributions of many species of animals and plants. A parallel analysis of less detailed data on 677 species of plants and animals suggested that 62% of the species surveyed showed trends toward earlier flowering, breeding, or growth. And for 434 species in which researchers documented a change in geographical distribution, 80% of the shifts were in the direction predicted by climate change models.

Climate warming is also changing the combinations of species that occur together within ecological communities. For example, in a paper published in 2008, Craig Moritz of the University of California, Berkeley, and his colleagues from Berkeley and Colorado State University compared the elevational distributions of small mammals in Yosemite National Park to data compiled a century earlier by naturalist Joseph Grinnell. The minimum temperatures recorded at Yosemite have increased about 3°C over the last century, reflecting the worldwide trend of global warming. The distributions of half of the 28 species they monitored had shifted upward an average of 500 m. Some species that had occupied low elevation sites expanded their ranges to include higher elevations. By contrast, some species that had lived only at middle and high elevation sites no longer live at middle elevations; their distributions are now restricted to higher elevations. As a result of these shifts in species distributions, different combinations of small mammals now occupy middle and high elevation sites in the park, and some species that live at high elevation may face extinction as environmental temperatures continue to rise and other species expand upward into their habitats **(Figure 51.10).**

lowland species will move to higher elevations. The models also predict that global warming will change the timing of important biological events. For example, plants whose flowering is triggered by warm springtime temperatures will flower earlier in the season; similarly, migratory animals will return from their wintering grounds and begin reproducing earlier in the year.

Camille Parmesan, then of the University of Texas at Austin, and Gary Yohe of Wesleyan University tested these predictions with a massive literature review. They surveyed studies of

FIGURE 51.10 Endangered by climate change. The elevational range of the alpine chipmunk (*Tamius alpinus*) has shifted more than 600 m upward in the last 100 years, confining its populations to relatively small areas high in the Sierra Nevada of California.

The effects of global climate change are not restricted to terrestrial environments. Ocean temperatures are also rising, and these changes are reflected in the distributions of marine species. For example, among invertebrates and fishes on the California coast, cold-adapted species have become less abundant and warm-adapted species more abundant. Comparable changes have been noted in other communities from Antarctica to the Arctic.

The geographical distributions of species and ecological communities have often changed with climate shifts over evolutionary time, but the rate of global warming has accelerated in your lifetime. As you will discover as you read Chapters 53 and 54, the factors that govern the structures of communities and ecosystems are complex, and scientists are unable to predict all of the consequences of these changes. In the next section, we describe how today's climate affects terrestrial species and community distributions on a biosphere-wide scale. You can be certain that, 50 years from now, biology texts will paint a very different portrait of these large-scale associations.

STUDY BREAK 51.3

1. How does the behavior of *Anolis* lizards in the Dominican Republic change over elevation?
2. What effect is global warming likely to have on the geographical distributions of organisms?

51.4 Terrestrial Biomes

In Section 23.3 we described how convergent evolution produces morphological and physiological similarities in species that occupy similar environments. Early in the twentieth century, two American ecologists, Frederic Clements of the Carnegie Institution in Washington and Victor Shelford of the University of Illinois, generalized this observation by defining a **biome** as a vegetation type plus its associated microorganisms, fungi, and animals. Although vegetation is superficially similar throughout a biome, the particular species within it vary from place to place. For example, in eastern North America, the temperate deciduous forest biome includes beech-maple forests in the north and oak-hickory forests in the south. Before surveying eight major terrestrial biomes—*tropical forests, savannas, deserts, chaparral, temperate grasslands, temperate deciduous forests, evergreen coniferous forests,* and *tundra*—we consider how environmental factors influence their overall distribution.

Environmental Variation Governs the Distribution of Terrestrial Biomes

Because organisms—and the communities they form—are sensitive to abiotic factors, climate is the main determinant of biome distribution. A **climograph** portrays the particular combination of temperature and rainfall conditions where each terrestrial biome occurs **(Figure 51.11)**. For example, some deserts, grasslands, savannas, and tropical forests occur in areas that have comparable mean annual temperatures but vastly different rainfall. Conversely, some biomes, such as boreal forests, temperate deciduous forests, and savannas, are found under similar moisture conditions but different temperature regimes.

Although the climograph provides a general portrait of the temperature and moisture conditions where the different biomes occur, it does not address the details of environmental variation. For example, the climograph includes only mean annual temperature and rainfall, not seasonal variation in these factors. Two regions may have the same mean temperature even though one experiences blazingly hot summers and bitterly cold winters and the other has moderate temperature throughout the year; we would expect them to harbor different organisms. Moreover, the distributions of communities are also influenced by nonclimatic factors, such as regional variations in soil structure and mineral composition (see Section 35.2).

Because temperature and rainfall exhibit latitudinal patterns (displayed in Figures 51.2 and 51.5), the distributions of some terrestrial biomes appear as bands on a world map **(Figure 51.12)**. But regional and local climatic variations influence these broad patterns. For example, chaparral is common in certain coastal habitats, whereas grasslands occur further inland at similar latitudes. Comparable bands of distinct vegetation form on mountainsides because temperature and moisture conditions also change with elevation.

Tropical Forests Include Earth's Most Species-Rich Communities

Three types of **tropical forests**—rainforest, deciduous forest, and montane forest—sweep across the parts of Africa, Asia, Australia, and Central and South America that receive intense solar radiation and heavy rainfall.

DISTRIBUTION OF BIOMES ON A CLIMOGRAPH

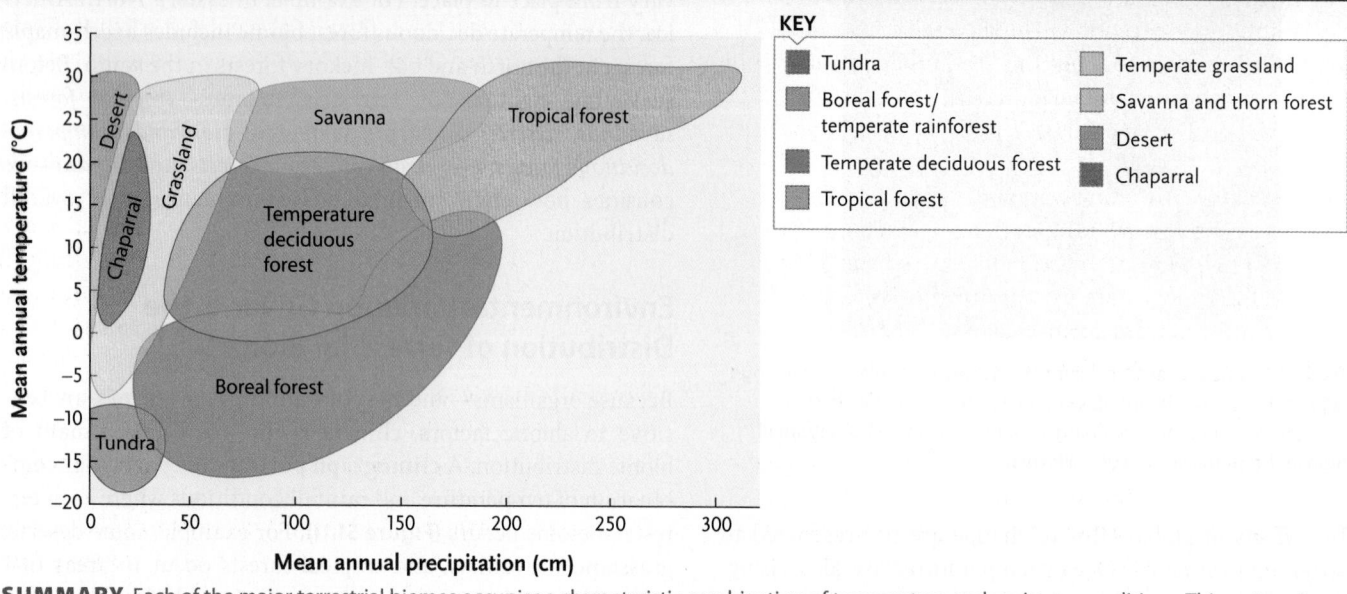

KEY

- Tundra
- Boreal forest/ temperate rainforest
- Temperate deciduous forest
- Tropical forest
- Temperate grassland
- Savanna and thorn forest
- Desert
- Chaparral

SUMMARY Each of the major terrestrial biomes occupies a characteristic combination of temperature and moisture conditions. This pattern is reflected in the global distributions of terrestrial biomes illustrated in Figure 51.12.

think like a scientist Will the effects of global warming change the distribution of biomes on the climograph?

FIGURE 51.12 Terrestrial biomes. Climate governs the distributions of the world's major terrestrial biomes. High mountain ranges are indicated by stripes because different biomes occur at different elevations. See the key in Figure 51.11 for color coding of biomes.

Tropic of Cancer

Equator

Tropic of Capricorn

Tropical rainforests grow where some rain falls every month, mean annual rainfall exceeds 250 cm, mean annual temperature is at least 25°C, and humidity is above 80%. Limited by neither temperature nor water, the productivity of a tropical rainforest is exceptionally high (see Section 54.2). Trees replace their leaves throughout the year, producing a continuous rain of detritus that ants, land crabs, and other scavengers quickly consume. Decomposition is rapid in the hot, moist environment, and little litter accumulates on the ground. Because nutrients released by decomposition are promptly absorbed by vegetation or leached by rain, soil in tropical rainforests is nutrient-poor, with low humus content (see Section 35.2).

Tropical rainforests are usually layered (see Figure 53.20). The crowns of tall trees form a dense, tangled canopy that intercepts most incoming sunlight 40 to 45 m above the ground (**Figure 51.13**). Even the largest trees grow only shallow roots in the thin soil, but many have wide *buttresses,* woody lateral extensions of their trunks, that stabilize them in the ground. Shade-tolerant shrubs and small trees form understory layers below the canopy. The woody stems of lianas climb through both layers, and epiphytes, such as bromeliads and orchids, cover the trunks and branches of trees, especially in sunlit

openings. In mature rainforests, the ground is surprisingly bare of leafy vegetation, because very little sunlight reaches the forest floor.

Tropical rainforests probably harbor more plant and animal species than all other terrestrial biomes combined. Ecologists have proposed numerous hypotheses to explain both the evolution and maintenance of the large number of species living in these communities (see Section 53.7), but no single hypothesis explains the pattern adequately. In fact, we do not even have a complete species list for any rainforest community, largely because most animals live in the highly productive canopy, which ecologists have only recently begun to study in detail (see *Focus on Research: Basic Research*). The most extensive tracts of tropical rainforest occur in South America, central and western Africa, and Southeast Asia. Unfortunately, they are being cleared at an alarming rate (see Chapter 55); some experts predict that, if the current rate of logging continues, this biome will all but disappear before the middle of the twenty-first century.

Habitats centered at 20° north and south of the equator experience a pronounced summer rainy season and winter dry season. **Tropical deciduous forests** occur where winter drought reduces photosynthesis, and most trees drop their leaves. For example, the monsoon forests of Southeast Asia, which harbor teak and other tropical hardwoods, are as lush as tropical rainforests in the rainy season; but many trees are bare in the dry season.

High elevations in the tropics support distinctive **tropical montane forests,** or "cloud forests," which are frequently enveloped in mist. The trees, often no more than 3 m tall, are densely covered with epiphytes, which thrive in the moisture-laden air. Cloud-forest plants grow slowly because productivity is limited by low temperatures, high humidity (making transpiration difficult), and sunlight-blocking clouds.

Savannas Grow Where Moderate Rainfall Is Highly Seasonal

Grasslands with scattered trees, the biome called **savanna,** grow in areas adjacent to tropical deciduous forests **(Figure 51.14).** Seasonality in tropical and subtropical savannas is determined by the availability of water; although annual rainfall averages 90 to 150 cm, droughts typically last for months. Grasses are successful in semiarid conditions because their shallow roots harvest water efficiently. With the onset of seasonal rains, they grow quickly, reaching a height of 2 to 3 m. During the dry season, grasses die back and frequently burn, but their underground parts remain alive and resprout when water again becomes available. Shrubby trees outcompete grasses in moist, low-lying areas or on rocky ground, but periodic fires and grazing mammals eliminate many trees as seedlings.

The largest savannas stretch across eastern and southern Africa; smaller patches occur in India, Australia, and South America. African savannas are home to large herbivorous mammals, including antelopes, zebras, giraffes, and elephants,

FIGURE 51.13 Tropical forest. Many tropical rainforest trees, such as these at Chachagua, Costa Rica, are covered with lianas and epiphytes.
© Cengage Learning 2017

Banana Republic images/Shutterstock.com

Basic Research: Exploring the Rainforest Canopy

Biological diversity in tropical rainforests has fascinated naturalists for centuries. Sadly, most of its organisms live beyond our reach. The forest canopy extends from 9 or 10 m above the ground to heights as great as 45 m. Until recently, the canopy was inaccessible and largely unexplored. Early ecologists were able to study canopy-dwelling species only when they found a fallen tree or followed loggers into the forest. In the 1930s, a clever botanist trained monkeys to retrieve plants from the canopy, but these efforts provided little data about the ecological interactions that govern life in the treetops.

Many ecologists still study canopy-dwelling organisms from the safety of the ground. Binoculars provide a good view of fairly large vertebrates. And a hike along a ridgetop can provide a canopy-level view of trees growing in an adjacent valley or ravine. Some researchers use ropes to hoist nets or traps into the canopy, lowering them periodically to see what they have caught. Others spray a fog of insecticide into the canopy to kill small invertebrates, which then rain down onto plastic sheets spread below the trees. These ground-based techniques have led to the discovery of hundreds—perhaps thousands—of new arthropod species. Ecologists now collect huge samples of arthropods to study the species composition and structure of communities and to monitor changes in these communities over time. But distant observations and mass sampling techniques do not provide detailed data about which insects are feeding on a tree, how often hummingbirds pollinate a flower, or when a tiny lizard hunts its prey.

Today many ecologists routinely risk life and limb to collect detailed ecological data in the rainforest canopy. They climb trees and crawl along stout branches. Many build stable observation decks with walkways, allowing study on either side of the "trail" **(Figure)**.

What does this newfound access to the rainforest canopy add to our knowledge of organisms that live there? Researchers can measure the physical environment of the canopy and observe the physiological and behavioral adaptations of its plants and animals. For example, researchers are gathering data on the feeding habits and behavior of small animals that never venture to the ground, such as fruit-eating bats and birds. When coupled with information about the movement patterns of these animals, the data provide insight into the dispersal of seeds in the fruits. And an understanding of seed dispersal provides information for studies of the population ecology of rainforest trees.

Canopy ecologists have also discovered fascinating relationships between plants and their animal pollinators. For example, Donald Perry, a freelance biologist, discovered that birds are attracted to the sweet nectar of the vine *Norantea sessilis*. Feeding birds step on the vine's sturdy flowers; their feet become covered with the plant's pollen, which is embedded in a gummy substance. When the birds visit another vine of the same species, they transfer the pollen to that plant's flowers, providing cross-pollination, which appears to be necessary for the vine's reproduction.

Nalini M. Nadkarni

FIGURE   A platform in the canopy of a tropical rainforest in Costa Rica provides a comfortable perch for Donald Perry to survey the pollinating activity of birds and bees.

Research in the tropical rainforest canopy promises exciting discoveries about ecological relationships in this unique biome, which is the most threatened on Earth (see Chapter 55). Such research is essential for developing a public appreciation of tropical forests and for creating conservation plans to preserve them.

MattiaATH/Shutterstock.com

FIGURE 51.14 Savanna. The African savanna, a warm grassland with scattered stands of shrubby trees, has an enormous concentration of large ungulates (hoofed, herbivorous mammals), such as these wildebeests (*Connochaetes taurinus*).

© Cengage Learning 2017

some of which fall prey to savanna predators, such as lions, leopards, cheetahs, and wild dogs. Grazing mammals follow the seasonal cycle of grasses, migrating away from dry areas to greener pastures.

Thorn forests grow at the arid borders of true savanna, where large mammals are less abundant. Grasses and other plants that store energy in large underground root systems grow among scrubby trees. Thorn forests are also highly seasonal, growing dramatically in the rainy season and dying back during the annual dry season, which may last for 8 to 9 months.

Deserts Develop Where Little Precipitation Falls

Deserts form where rainfall averages less than 25 cm per year. The hot deserts of the American Southwest, northern Chile, Australia, northern and southern Africa, and Arabia occur near 30° latitude, where descending air masses create very dry conditions. Cool deserts, such as the Gobi and Kyzyl-Kum of Asia and the Great Basin of North America, form in massive rain shadows at higher latitudes.

Desert conditions are often extreme. Rainfall arrives infrequently in heavy, brief pulses; and sudden runoff erodes topsoil, which often has high mineral content but little organic matter. Dry air and scant cloud cover allow most sunlight to reach the ground, raising daytime air and ground temperatures as high as 45°C and 70°C, respectively. At night, the surface loses heat quickly; in some deserts, temperatures drop below freezing in winter.

Desert vegetation is always sparse because arid environments do not favor large, leafy plants. Some deserts, such as the Namib of Africa and the Atacama-Sechura of South America, receive so little rainfall that large areas are practically devoid of vegetation. By contrast, the hot Sonoran Desert in northern Mexico, southeastern California, and southern Arizona harbors a diverse flora, including deep-rooted shrubs and shallow-rooted cacti **(Figure 51.15)**. Mesquite and cottonwood trees grow long taproots into the permanent water supply below streambeds. Perennial plants often protect their tissues from herbivores with spines or toxic chemicals, and many use CAM photosynthesis to conserve water (see Section 9.4). After seasonal rains, annual plants germinate, mature, flower, and produce seeds before brutally dry conditions resume.

Deserts also support abundant animals, most of them fairly small. Ants, birds, and rodents often subsist on seeds. Some seed-eating mammals never drink, surviving on the water they extract from food. Insects, some lizards, and mammals consume the sparse vegetation. Scorpions, lizards, and birds feed primarily on insects; and snakes, owls, and foxes prey on other animals. Most desert animals avoid the midday heat and dehydrating conditions; many retreat into underground burrows, where water vapor from their respiration

FIGURE 51.15 Desert. The warm Sonoran Desert in Saguaro National Park near Tucson, Arizona, is home to columnar saguaro cacti *(Carnegiea gigantea)* and other drought-adapted plants.
© Cengage Learning 2017

cools and moistens the air. Many species are nocturnal or active only in the early morning and late afternoon.

Chaparral Grows Where Winters Are Cool and Wet and Summers Are Hot and Dry

A scrubby mix of short trees and low shrubs called **chaparral** dominates narrow sections of coastal land between 30° and 40° latitude, where winters are cool and wet and summers are hot and dry. Seasonal rainfall averages only 25 to 60 cm per year. Chaparral occurs in central and southern California, central Chile, southwestern Australia, southern Africa, and the Mediterranean region.

Chaparral shrubs are dense, with hard, tough, evergreen leaves **(Figure 51.16)**. They grow woody stems above ground and large root systems in the soil. Many species, such as sages (genus *Salvia*), produce toxic, aromatic compounds that inhibit the germination and growth of other plants. Just after the winter rains, the shrubs are covered with new leaves and flowers, and the vegetation teems with insects and breeding birds. During the hot, dry summers, however, most plants are dormant, and lightning sparks frequent fires. The aromatic oils and resins of many species, such as eucalyptus, make them highly flammable. Their aboveground parts burn swiftly, but they quickly resprout from large root crowns. Other species release seeds from fire-resistant cones or pods, and their seedlings grow in ash-enriched soil.

FIGURE 51.16 **Chaparral.** Chaparral covers broad expanses of hills in coastal areas of central and southern California.
© Cengage Learning 2017

Temperate Grasslands Are Subject to Periodic Disturbance

Temperate grasslands include the prairies of North America, the steppes of central Asia, the pampas of South America, and the veldt of southern Africa. They stretch across the interiors of continents, where winters are cold and snowy and summers are warm and fairly dry. Only 25 to 100 cm of rain falls unevenly through the year. Temperate grasslands are in a near-constant state of flux: seasonal drought, periodic fires, and grazing by mammals destroy the seedlings of shrubs and trees, preventing them from displacing perennial grasses and herbaceous plants (see Section 53.6). Grassland soil is rich in organic matter because the aboveground parts of most plants die and decompose annually.

In North America, shortgrass prairie **(Figure 51.17A)** covers much of the west, where winds are strong, rainfall light and infrequent, and evaporation rapid. Drought-tolerant perennials have deep roots, and their underground rhizomes, which store energy, resprout quickly after a fire. Tallgrass prairie **(Figure 51.17B)** once occupied moister regions to the east of the shortgrass prairie. It boasted an abundance of legumes and sunflowers, often 3 m tall, but most of it was converted to farmland long ago; small patches still exist in nature preserves and in glades within eastern deciduous forests.

North American grasslands are still occupied by large grazing mammals, including pronghorns and bison, which once numbered in the millions. The most familiar burrowing mammal is the prairie dog, a rodent, but pocket gophers, ground squirrels, and jackrabbits are also common. Wolves were the primary large predators until they were hunted nearly to extinction. Coyotes, foxes, ferrets, hawks, and owls still take small prey today.

Temperate Deciduous Forests Experience Seasonal Dormancy

At temperate latitudes, with warm summers, cold winters, and annual precipitation between 75 and 250 cm, **temperate deciduous forests** grow at low to middle elevations. In winter, low temperatures reduce photosynthetic rates, and snow and ice can damage leaves. Thus, most plants shed their leaves and grow new ones in spring **(Figure 51.18)**. The thick layer of leaf litter, which releases mineral nutrients as it decomposes, enriches the soil. Decomposition is slow, however, because the growing season is only about 7 months long.

Temperate deciduous forests harbor fewer species than tropical forests. Trees form a canopy 10 to 35 m high, and woody shrubs form an understory below it. Herbaceous plants and a ground layer of mosses or liverworts grow below the shrubs. Many herbaceous plants, including some terrestrial orchids, flower early in spring, before trees produce

A. Shortgrass prairie

B. Tallgrass prairie

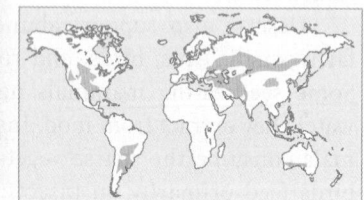

FIGURE 51.17 **Temperate grassland. (A)** The western plains of North America were once covered with shortgrass prairie, as shown in Custer State Park, South Dakota. Bison *(Bison bison)* were the dominant large herbivores. **(B)** Tallgrass prairie, such as this lush patch in eastern Kansas, once covered the eastern plains.
© Cengage Learning 2017

Summer

Winter

Thomas E. Hemmerly

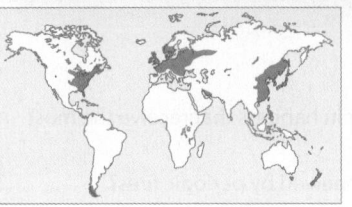

FIGURE 51.18 Temperate deciduous forest. Seasonal variations in temperature and water change the character of this forest south of Nashville, Tennessee.
© Cengage Learning 2017

sunlight-blocking leaves; others flower near the end of the growing season.

Forests of ash, beech, birch, chestnut, elm, and oak stretched unbroken across eastern North America, Europe, and eastern Asia before farmers cleared the land. In North America, introduced diseases and insects have nearly eliminated the once dominant species, such as American chestnut and American elm. Today, beech, birch, and maple predominate in the Northeast; oak–hickory forests dominate farther south and west; and oak woodlands merge into tallgrass prairie to the west. Before the arrival of Europeans, deer, bison, bears, and pumas roamed the forests with many smaller species of animals. Today, small mammals such as voles, mice, chipmunks, squirrels, rabbits, opossums, and raccoons predominate, although deer and bears—species that thrive in habitats that have been disturbed by human activities—have recently surged in abundance.

Evergreen Coniferous Forests Predominate at High Northern Latitudes

The **boreal forest,** or **taiga** (Russian for "swamp forest"), is a circumpolar expanse of evergreen coniferous trees in Europe, Asia, and North America **(Figure 51.19).** Snow blankets the ground during long and extremely cold winters, but most precipitation falls during the short summer. In the northernmost taiga, plants grow quickly during long (18-hour) summer days.

Stands of white spruce and balsam fir dominate North America's boreal forest. Their needle-shaped leaves have a thick cuticle and recessed stomata that conserve water during winter, when groundwater is frozen. Fallen needles acidify the thin soil, which speeds the leaching of most nutrients; few shrubs and herbaceous plants grow beneath the conifers. Lightning-sparked fires are common; some deciduous trees grow in areas opened by fire, but conifers eventually replace them. Cold streams, marshes, ponds, and lakes often dot the landscape; at flat, poorly drained sites, peat mosses, shrubs, and stunted trees dominate acidic bogs, called *muskegs.*

Most taiga is relatively undisturbed by humans, and it still harbors its native animals. Moose, elk, and deer are the dominant large herbivores. Hare, as well as squirrels, porcupines, and other rodents, also feed on plants. Some small animals are active all winter in runways they dig beneath the snow. Wolves, lynx, and wolverines prey on herbivores. Grizzly bears and black bears roam the forest, devouring seeds, berries, fishes, and small animals. Mosquitoes, black flies, and gnats are superabundant near bogs and lakes in summer.

Other types of coniferous forests grow in more southerly coastal lowlands where winters are mild and wet and the summers are cool. For example, a **temperate rainforest,** supported by heavy rain and fog, parallels the North American coast from Alaska into northern California. In western Washington State, the rainforest on the Olympic Peninsula receives 500 cm of rainfall per year, as much as some tropical forests. This temperate rainforest harbors some of the world's tallest trees, including Douglas fir and Sitka spruce to the north and coast redwoods to the south.

FIGURE 51.19 Boreal forest. Single-species stands of tall conifers dominate this boreal forest, the predominant forest at high latitudes in the Northern Hemisphere.
© Cengage Learning 2017

Tundra Comprises a Vast, Treeless Plain in the Northernmost Habitats

The treeless **arctic tundra** stretches from the boreal forests to the polar ice cap in Europe, Asia, and North America. Covering almost 5% of the land on Earth, this biome is windswept and wet. Winter temperatures are consistently below freezing. Historically, the 2-month summer has been so cool that only the topmost layer of soil ever thaws, leaving the ground below perpetually frozen; in some areas, this **permafrost** is more than 500 m thick. However, as global environmental temperatures have increased, researchers have recently observed the thawing of permafrost, part of a positive feedback system between climate and soils. Permafrost is a major repository of carbon in the biosphere. As it thaws, microbial action releases increased quantities of carbon dioxide and methane into the atmosphere; both of these greenhouse gases contribute to global warming, which will, in turn, result in additional permafrost thawing (see Section 54.4).

Although less than 25 cm of precipitation falls each year, evaporation is slow, and permafrost is impermeable; thus, low-lying soil remains permanently waterlogged, forming bogs **(Figure 51.20A)**. Anaerobic conditions and low temperatures retard decomposition, and soggy masses of detritus accumulate.

Plants in the tundra are short because the weak sunlight and minimal growing season provide barely enough energy and warmth for photosynthetic activity; moreover, strong winter winds shred any plants with a high profile. The vegetation consists of low-growing lichens, mosses, grasses, perennial herbs, dwarf shrubs, and a few stunted trees, usually less than 1 m tall. During summer's nearly continuous sunlight, plants flower profusely, and their fruits ripen fast.

Some animals, including herbivorous arctic hares, lemmings, and willow ptarmigans, as well as predatory snowy owls, wolves, foxes, and lynx, are permanent tundra residents. In summer, herds of herbivorous musk oxen, caribou, and reindeer migrate there from boreal forests, and migratory shorebirds and waterfowl arrive to breed. Flying insects abound in summer, especially mosquitoes and black flies, which reproduce in boggy habitats.

A similar biome, called **alpine tundra,** occurs on high mountaintops throughout the world **(Figure 51.20B).** Dominant plants form cushions and mats that withstand the buffeting of strong winds. Winter temperatures are well below freezing, and shaded patches of snow persist even in summer. The thin, fast-draining soil is nutrient-poor, and photosynthetic activity is low.

STUDY BREAK 51.4

1. Which terrestrial biomes occur in habitats that receive the most rainfall?
2. Which terrestrial biomes are renewed by periodic fires?
3. Which terrestrial biomes have the tallest vegetation? Which ones have the shortest?
4. In which terrestrial biomes are the trees usually evergreen?

THINK OUTSIDE THE BOOK

What biome surrounds the community where you live? If you live in a city, how might you define an urban biome? How far would you need to travel to be in a different biome?

A. Arctic tundra

Dr. Peter Kuhry, Arctic Center, University of Lapland-Finland

B. Alpine tundra

Darrell Gulin/Encyclopedia/Corbis

FIGURE 51.20 Tundra. (A) Rain and snowmelt cannot percolate through the arctic tundra's permafrost. In summer, water accumulates in ponds and bogs, as shown in this aerial photograph of the tundra in northern Russia. **(B)** Compact, short plants form the alpine tundra, which occurs on mountaintops at more temperate latitudes, such as those in Alaska's Denali National Park.
© Cengage Learning 2017

51.5 Freshwater Environments

Biomes are traditionally defined for terrestrial environments. Their aquatic counterparts comprise several distinctive habitats in either freshwater or marine environments. In freshwater environments, water with a salt concentration below 0.5% accumulates or moves through a landscape. Ecologists distinguish between *lotic* systems, where water flows through channels, and *lentic* systems, where water stands in an open basin. All freshwater environments interact with the surrounding terrestrial habitats, because runoff from the land carries a nearly constant input of nutrients. Highly productive **wetlands** often occur at the borders between freshwater and terrestrial environments. These marshes and swamps may harbor an astounding array of microorganisms, algae, plants, invertebrates, and vertebrates.

Streams and Rivers Carry Water Downhill to a Lake or Sea

Flowing-water environments start as seeps on high ground. As the water flows downhill, it collects into narrow streams, which merge to form wide rivers **(Figure 51.21)**. Streams and rivers include three habitats. *Riffles* are shallow, fast-moving, turbulent stretches over a rough bottom of pebbles or rocks. *Pools* are deep, slow-moving areas with a smooth sand or mud bottom. *Runs* are deep, fast-moving stretches over smooth bedrock or sand. Streams generally have high flow rate, low volume, and lots of riffles and pools. As they merge into rivers, flow rate declines, but flow volume increases, and runs and pools predominate. Flow rate and volume also vary seasonally with the rate of water input from rainfall and snowmelt and geographically with elevation and topography.

Physical factors change over the length of a flowing-water system. The concentration of suspended particulate material is low in streams, but high in rivers, which are often turbid with silt. Temperature also rises as water flows downstream to warmer lowland habitats. Because oxygen is more soluble in cold water than in warm water, dissolved oxygen is usually higher in streams than in rivers. Erosion of the streambed and surrounding land has always provided the solute content of flowing water. Today, agricultural runoff and industrial and municipal wastes provide major input. In unpolluted streams, organic detritus provides more than 95% of the nutrients and energy entering aquatic food webs. This input is particularly important in streams flowing through dense forests, where vegetation blocks the sunlight necessary for photosynthesis.

The flow of water affects every aspect of life in streams and rivers. In swift-moving riffles, primary producers cling permanently to fixed substrates, because phytoplankton are swept away by the current. Insect larvae and other invertebrates attach to the undersides of rocks; many species have a flattened shape to maintain a low profile in the current. By contrast, large rivers have dense populations of algae and cyanobacteria, which attach to rocks and other substrates (that is, surfaces to which organisms can adhere), and rooted aquatic plants at the river's edge.

Lakes Are Bodies of Standing Water That Accumulates in Basins

Lakes and other standing-water biomes are generally fed by rainfall and by streams and rivers that carry water from surrounding lands **(Figure 51.22)**. Because the availability of light affects photosynthesis by a lake's phytoplankton and plants, ecologists often distinguish between the **photic zone** of a lake, the surface water that sunlight penetrates, and the deeper **aphotic zone,** which is darker.

LAKE ZONATION Every lake includes zones, defined by depth and distance from the shore, that provide distinctive environments **(Figure 51.23)**. In the **littoral zone,** the shallow water near the shore, sunlight penetrates to the bottom. Enriched by nutrients made available by decomposers and runoff, the

A. A stream　　　　　　　　　**B. A river**

FIGURE 51.21 Stream and river habitats. (A) In streams, such as this one in Virginia, water flows quickly through narrow channels, often with a rocky bottom. **(B)** In rivers, such as the Rio Napo in Ecuador, water flows more slowly through broad channels, and suspended sediments often make the water murky.
© Cengage Learning 2017

FIGURE 51.22 Lakes. Lago Di Piani, a lake in the Dolomites region of Italy.

littoral zone has high photosynthetic activity and is occupied by many species. Rooted aquatic plants, such as cattails and water lilies, grow above the surface, and "floating aquatics," such as duckweed, are common. Submerged vegetation harbors a rich community of microorganisms, epiphytes, and invertebrates. Numerous animals—insects, worms, snails, crayfish, fishes, frogs, turtles, and water birds—use the littoral zone to feed and reproduce.

The **limnetic zone,** the sunlit water beyond the littoral, supports communities of plankton: the primary photosynthesizers are phytoplankton—cyanobacteria, diatoms, and green algae; they are eaten by zooplankton—rotifers, copepods, and other tiny heterotrophs. Small fishes, which feed on plankton, are themselves consumed by larger fishes, such as bass.

Photosynthesis is impossible, however, in the **profundal zone,** the perpetually dark water below the limnetic zone. Nevertheless, a constant rain of detritus from the limnetic zone supports a community of bacterial decomposers, as well as the worms, clams, insect larvae, and catfish that feed on them or on dead or dying material.

SEASONAL CHANGES IN TEMPERATE LAKES In temperate areas, seasonal temperature variations induce changes in the vertical zonation of lakes **(Figure 51.24).** Like other liquids, water gets denser as it cools. But water has a unique property: it reaches maximum density at 4°C, with the density declining as it gets colder. Thus, water at 4°C sinks below water that is either warmer or colder, and ice floats because it is less dense than very cold water.

During winter, ice forms on the surface of temperate zone lakes. Water temperature varies from near freezing just

below the ice to 4°C at the bottom. Differences in the density of water at 0° and 4°C maintain this thermal stratification. In spring, as the ice melts, the warmer, denser water sinks; and the surface temperature gradually rises to 4°C. For a brief time, the temperature is uniform at all depths. Winds blowing across the lake create vertical currents that cause a **spring overturn,** mixing surface water with deep water. Oxygen at the surface moves to the bottom, and nutrients from the bottom move to the surface.

By midsummer, sunlight heats the top layer of the limnetic zone, called the **epilimnion,** to temperatures above 4°C. In large lakes, the epilimnion may be more than 10 m deep. In the deep water of the lake's profundal zone, called the **hypolimnion,** the temperature remains near 4°C. However, at the boundary between the epilimnion and the hypolimnion, water temperature changes abruptly over a narrow depth range, called the **thermocline.** The thermocline prevents vertical mixing because warm surface water floats above the thermocline, and cool deep water stays below it. During summer, nutrient-rich detritus sinks to the bottom of the lake, where decomposition depletes the oxygen dissolved in the hypolimnion. In autumn, declining sunlight and winds cause the epilimnion to cool, and as the water becomes denser, it sinks, eliminating the thermocline. Winds then mix the water vertically once again during an **autumn overturn,** and dissolved gases and nutrients are equalized at all depths.

Photosynthetic activity in the limnetic zone varies with the seasonal overturns. In spring, increased sunlight, warm temperatures, and the sudden availability of nutrients induce a bloom of photosynthesis and growth. As the season progresses and the thermocline prevents vertical mixing, nutrient levels dwindle in the epilimnion, and photosynthetic activity declines. By late summer, nutrient shortages limit photosynthesis. After the autumn overturn, nutrient cycling drives a short burst of productivity. But as days get shorter and temperature declines, productivity remains low until spring.

TROPHIC NATURE OF LAKES Ecologists classify lakes by their nutrient content and rates of photosynthetic activity. **Oligotrophic lakes** are poor in nutrients and organic matter, but rich in oxygen. Their low productivity keeps the water crystal clear, making them popular recreational sites. By contrast, **eutrophic lakes** are rich in nutrients and organic matter. The decomposition of organic matter depletes oxygen in the hypolimnion when the lake is stratified, and high productivity in the

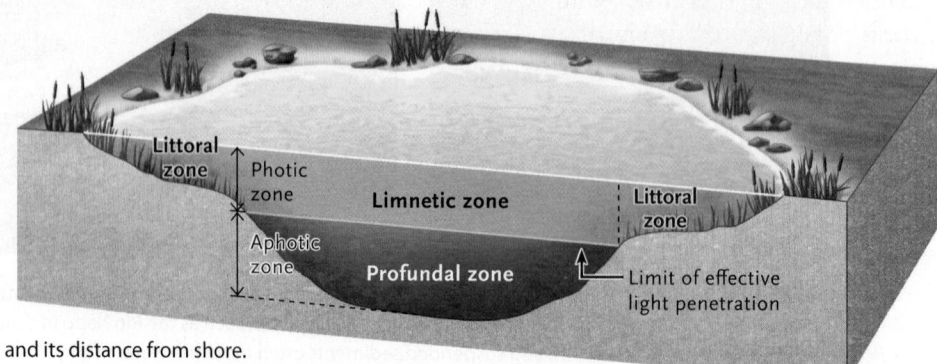

FIGURE 51.23 Lake zonation.
The zonation in a lake is based on the water's depth and its distance from shore.
© Cengage Learning 2017

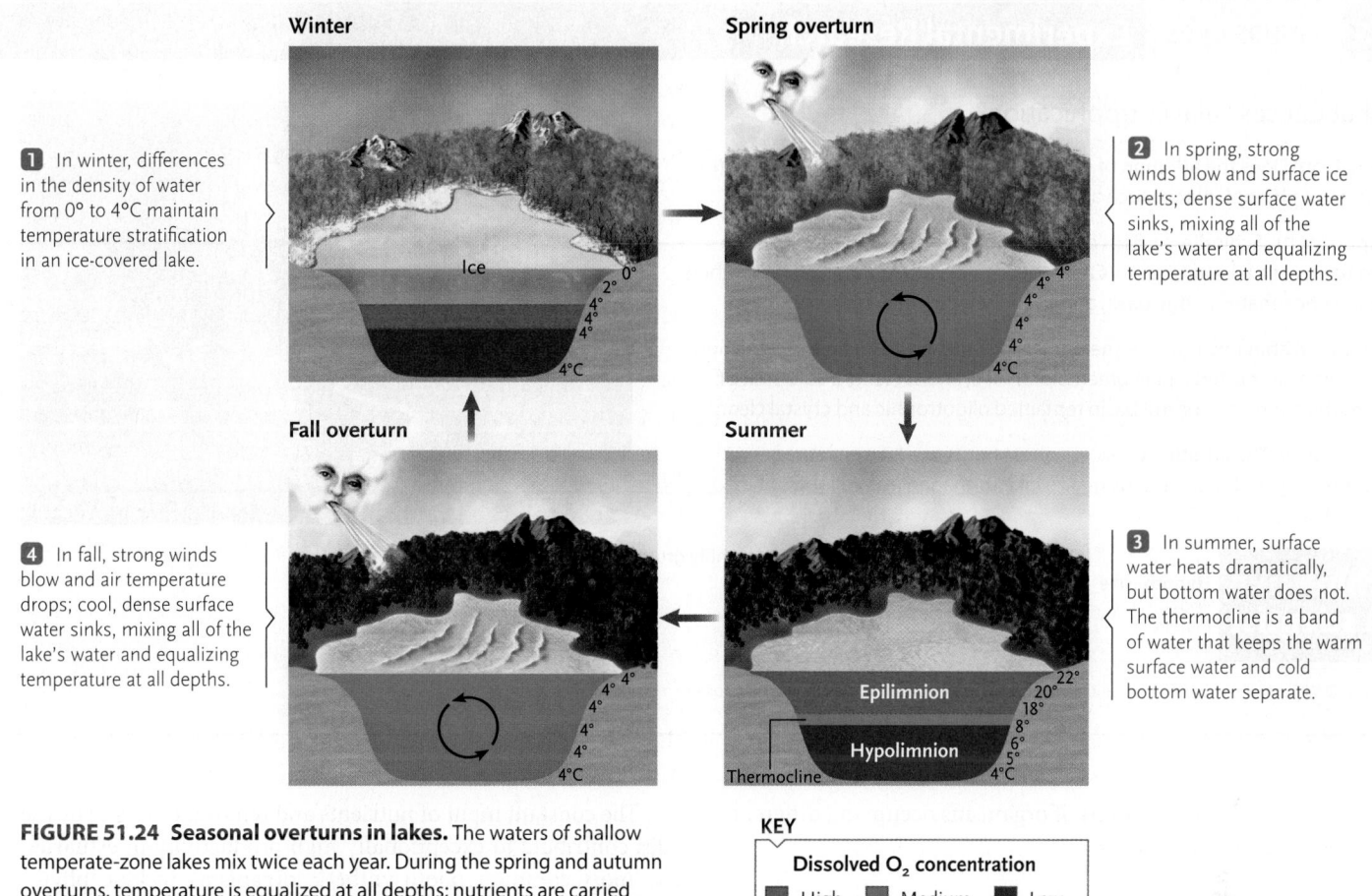

Winter

1 In winter, differences in the density of water from 0° to 4°C maintain temperature stratification in an ice-covered lake.

Ice

0°
2°
4°
4°
4°
4°C

Spring overturn

2 In spring, strong winds blow and surface ice melts; dense surface water sinks, mixing all of the lake's water and equalizing temperature at all depths.

4° 4°
4°
4°
4°
4°C

Fall overturn

4 In fall, strong winds blow and air temperature drops; cool, dense surface water sinks, mixing all of the lake's water and equalizing temperature at all depths.

4° 4°
4°
4°
4°C

Summer

3 In summer, surface water heats dramatically, but bottom water does not. The thermocline is a band of water that keeps the warm surface water and cold bottom water separate.

Epilimnion
22°
20°
18°
8°
6°
5°

Hypolimnion

Thermocline
4°C

FIGURE 51.24 Seasonal overturns in lakes. The waters of shallow temperate-zone lakes mix twice each year. During the spring and autumn overturns, temperature is equalized at all depths; nutrients are carried upward from the bottom; and oxygen is carried downward from the surface.
© Cengage Learning 2017

KEY

Dissolved O$_2$ concentration
■ High   ■ Medium   ■ Low

epilimnion often chokes the water with seasonal blooms of cyanobacteria and filamentous algae. Eutrophic lakes are often thick and "soupy" making them unattractive for recreation. Over long periods of time, as sediments accumulate, lakes naturally change from oligotrophic to eutrophic; their basins eventually fill with sediments, and terrestrial plants invade.

The addition of nutrients to a lake often disrupts its trophic condition (see *Why It Matters . . .* in Chapter 54). In a classic experiment published in 1974, David Schindler and his colleagues at The Experimental Lakes Project in Ontario, Canada, experimentally separated the two basins of a lake with a plastic curtain. The researchers added phosphates to one basin and used the other basin as a control. Within 2 months, the artificially enriched basin sported a bloom of cyanobacteria, a sign of eutrophication; the control basin remained oligotrophic and crystal clear **(Figure 51.25).**

STUDY BREAK 51.5

1. How does the availability of dissolved oxygen vary from the headwaters of a stream to the mouth of a river?
2. What factors cause the seasonal overturns in lakes?
3. Why are oligotrophic lakes better for recreational purposes than eutrophic lakes?

51.6 Marine Environments

Marine environments, in which salinity (salt concentration) averages about 3%, cover nearly three fourths of Earth's surface and account for a large fraction of its photosynthetic activity. They also mediate important global processes: marine phytoplankton process large amounts of carbon dioxide, generating oxygen and moderating a major cause of global climate change (see Chapter 54).

Most of the marine environments described below exist at the edges of open ocean waters. As with standing freshwater environments, depth and distance from shore govern the physical characteristics of marine habitats. Ecologists describe ocean zonation in several ways **(Figure 51.26),** including the distinction between the photic and aphotic zones. Another major distinction is between the **pelagic province,** the water, and the **benthic province,** the bottom sediments. The pelagic province includes the **neritic zone,** the shallow water above the continental shelves, and the **oceanic zone,** the deep water beyond them. The benthic province is divided into the **intertidal zone,** the shoreline that is alternately submerged and exposed by tides, and the **abyssal zone,** the bottom sediments that lie permanently below deeper water. Here we describe five marine environments—*estuaries, rocky and sandy coasts, continental shelves and oceanic banks, open ocean,* and *benthic regions*—that

FIGURE 51.25 | **Experimental Research**

What Causes Lake Eutrophication?

Question: Does the addition of excess phosphorus to a lake encourage the growth of phytoplankton, such as cyanobacteria?

Experiment: Schindler and his colleagues experimentally separated the two basins of a lake in Ontario, Canada, with a plastic curtain. The researchers added phosphates to one basin and used the other basin as a control.

Results: Within two months, the artificially enriched basin (in the upper left of the photo) sported a pale green bloom of cyanobacteria, a sure sign of eutrophication; the control basin remained oligotrophic and crystal clear.

Conclusion: The addition of excess phosphorus to a lake encourages blooms of cyanobacteria, causing the lake to change from oligotrophic to eutrophic.

D. W. Schindler, *Science* 184:897–899.

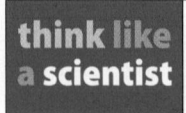

think like a scientist | Observe **Hypothesize** Predict Experiment Interpret

Why do new housing developments on hilly ground often pose a danger to nearby lakes?

Source: D. W. Schindler. 1974. Eutrophication and recovery in experimental lakes: Implications for lake management. *Science* 184:897-899.

© Cengage Learning 2017

---

harbor particular associations of organisms occupying different marine zones and provinces.

Estuaries Form Where Rivers Meet the Sea

Estuaries are coastal regions where seawater mixes with fresh water from rivers, streams, and runoff **(Figure 51.27).** Salinity is low where fresh water enters the estuary and high on the tidal side. After heavy rainfall, fresh water floods into the habitat, reducing salinity and raising water temperature. At high tide, cold, salty water flows in from the sea. All estuarine organisms must tolerate these variable conditions.

Variations in local topography influence an estuary's physical features. Chesapeake Bay in Maryland, Mobile Bay in Alabama, and San Francisco Bay in California are broad, shallow estuaries. The estuaries in Alaska and British Columbia are narrow and deep, as are Norway's fjords. Many estuaries are bordered by **salt marshes,** tidal wetlands dominated by emergent grasses and reeds (Figure 51.27A). In tropical estuaries, the roots of densely packed mangrove trees penetrate the muddy bottom, accumulating sediments and slowly adding land to the shoreline (Figure 51.27B).

The constant input of nutrients and removal of wastes by the tides contribute to exceptionally high productivity in estuaries. The most common photosynthetic organisms include phytoplankton, salt-tolerant grasses and reeds that can withstand submergence at high tide, and algae that grow in mud and on plant surfaces. Roots and stems trap organic matter, which decomposes. The detritus (and bacteria clinging to it) supports nematodes, snails, crabs, and fishes; suspension-feeding mollusks and

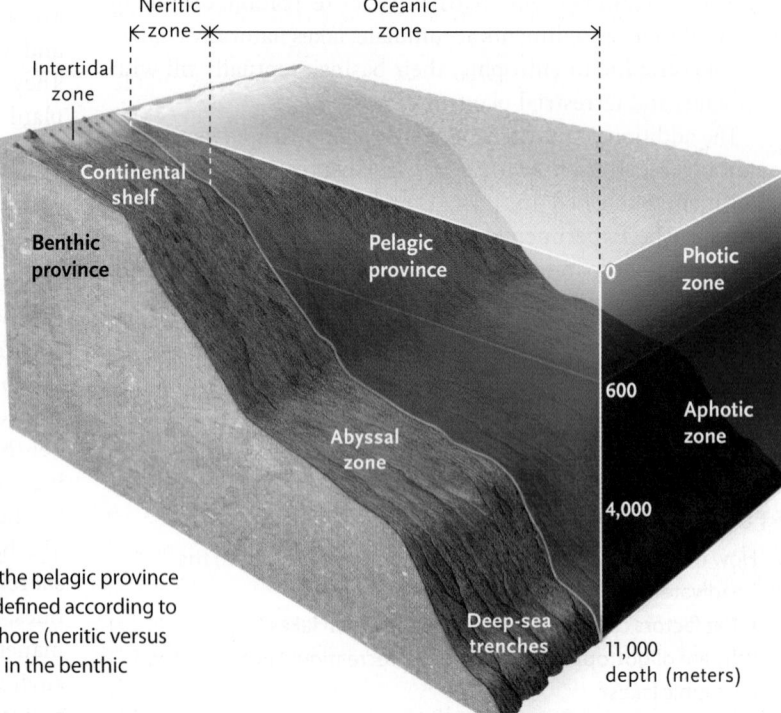

FIGURE 51.26 Ocean zonation. Ecologists divide the ocean into the pelagic province (the water) and the benthic province (the ocean bottom). Zones are defined according to the depth of water (photic versus aphotic zones) and distance from shore (neritic versus oceanic zones in the pelagic province, intertidal versus abyssal zones in the benthic province). The different zones are not drawn to scale.
© Cengage Learning 2017

A. Salt marsh grasses

B. Mangroves

FIGURE 51.27 Estuaries. (A) The salt marsh grass (*Spartina* species) is the most common plant in a South Carolina estuary. **(B)** Red mangroves *(Rhizophora mangle)* are abundant in the Florida Keys.

arthropods capture edible particles in the slowly moving water. Many marine arthropods and fishes breed in calm, shallow estuaries, where their young find abundant food and refuge from predators in the complex vegetation. Migratory birds use estuaries as rest stops, and shore birds and waterfowl use their muddy bottoms as rich feeding grounds, particularly at low tide.

Rocky and Sandy Coasts Experience Cyclic Periods of Exposure and Submergence

The intertidal zone, the area between low and high tide marks, is one of the most stressful habitats on Earth. On rocky shores, residents are battered by waves and floating debris. Sessile species, such as mussels and barnacles, attach to substrates with special structures or cement. Motile species, such as limpets and sea stars, simply hang on to rocks. Organisms that live high on the shore dry out at low tide, freeze in winter, and bake in summer. Exposed animals often seal themselves inside shells, and intertidal algae have thick polysaccharide coats that adsorb water and prevent dehydration.

Biotic interactions also take their toll. Organisms throughout the intertidal zone compete for attachment sites to avoid being washed away (see Figure 53.12). At low tide, predatory birds and mammals attack from above; at high tide, predatory fishes move in from the sea. Because the tides often scour detritus from the rocky intertidal, communities on rocky shores are largely supported by the photosynthetic algae and phytoplankton.

Rocky shores often have three zones **(Figure 51.28)**. The *upper intertidal* is submerged only during the highest tide of the lunar cycle. It is sparsely populated by barnacles, sturdy algae, and grazing and predatory snails. The *middle intertidal* is submerged daily during the highest regular tide and exposed during the lowest. Its tide pools are occupied by red, brown, and green algae, grazing and predatory mollusks, sponges, sea anemones, worms from several phyla, hermit crabs, echinoderms, and small fishes. Biodiversity is greatest in the *lower intertidal,* which is exposed only during the lowest tide of the

lunar cycle. It is occupied by dense beds of algae, tunicates, echinoderms, other invertebrates, and fishes.

Sandy shores are composed of loose sediments that waves and currents constantly rearrange. Large plants and algae cannot grow on such unstable substrates, but organic debris imported from offshore or from nearby land supports animals that feed on detritus. Animals live in burrows, which they must frequently repair as the substrate shifts. Crabs and shorebirds live as scavengers or predators above the high tide mark. At

FIGURE 51.28 Vertical zonation in the intertidal. A rocky shore in the Pacific Northwest clearly exhibits vertical zonation. The distance between low and high tide marks on this rocky shore is about 3 m.

night, beach hoppers and ghost crabs leave their burrows, seeking food. Marine worms, clams, crabs, and other invertebrates live in the sand between the high and low tide marks.

Light Penetrates the Shallow Water over Continental Shelves and Oceanic Banks

The neritic zone includes the shallow water over continental shelves and oceanic banks, underwater landmasses that rise to within 300 m of the surface. Although small in area, the neritic zone is highly productive and species-rich **(Figure 51.29).** Runoff from the land brings a steady inflow of nutrients, and upwelling (that is, vertical currents) and waves circulate nutrients from the bottom to the photic zone.

In temperate regions, giant kelp forests, which are among the most productive ecosystems, occupy some continental shelves and banks (see Figure 51.29A). Kelp are enormous algae that attach to the bottom with giant holdfasts; their stipes ("stems") reach upward with fronds fanning out into the water. Sea anemones, snails, echinoderms, lobsters, and other invertebrates live in the kelp, where fishes and other predators consume them. Even where kelp does not grow, continental shelves and banks teem with life. Most of the important fisheries in the temperate zone occur there.

In the tropics, the warm but nutrient-poor water above continental shelves is often occupied by **coral reefs** (see Figure 51.29B). Sunlight penetrates the clear water all the way to the bottom. Photosynthetic dinoflagellates, living as endosymbionts of the coral animals (see Section 27.2), and coralline algae are largely responsible for the photosynthesis that takes place there. Coral animals also feed on microscopic organisms and suspended particles. The reefs are the remains of corals, algae, and other organisms, and their structural complexity rivals that of tropical rainforests. Tides and currents carve ledges and caverns. Storms frequently disturb the reefs, creating openings in which new coral colonies can grow (see Section 53.5). A reef may be festooned with as many as 750 species of corals and a dizzying variety of algae. The diversity of coral skeletons provides a complex structure that is used by invertebrates from nearly every phylum and by a host of herbivorous and carnivorous fishes.

In the Open Ocean, Photosynthesis Occurs Only in the Sunlit Upper Layers

The oceanic zone lies beyond the continental shelves. Though generally low in nutrients, it is locally enriched by runoff from land and by upwelling bottom waters. The open ocean is typically cold, except in the tropics. The surface water is illuminated by sunlight, which warms it somewhat and allows photosynthesis. Most photosynthesis is undertaken in the top 50 m, however, because seawater filters light. Photosynthetic activity varies seasonally, as it does on land.

In the photic zone, "pastures" of phytoplankton are eaten by zooplankton, including copepods, shrimplike krill, small worms, cnidarians, and the larvae of invertebrates and fishes. Consumers that can actively swim against the currents, such as squids, fishes, marine turtles, and whales, are called **nekton (Figure 51.30).** Some consumers feed on plankton, and some prey on other nekton. Low light levels in water between about 50 and 600 m allow little photosynthesis, but organic matter sinks to these depths, supporting organisms that live there.

A. **Kelp forest**

B. **Coral reef**

FIGURE 51.29 Neritic zone. (A) Kelp forests, such as this one off the coast of California, often grow in the neritic zone along the coast at temperate latitudes. **(B)** This coral reef in Indonesia illustrates the structural complexity and biological diversity found in reef communities.

FIGURE 51.30 Open ocean. A humpback whale *(Megaptera novaeangliae)* breaches (leaps out of the water).

FIGURE 51.31 Deep sea. A deep-sea anglerfish *(Himantolophus groenlandicus)* uses a bioluminescent lure to attract prey to its formidable jaws.

Moreover, many fishes and some mobile invertebrates travel into the sunlit zone to feed on organisms near the surface.

No sunlight ever penetrates the deepest part of the oceanic zone, below 600 m. Some of these abyssal regions, such as the Marianas Trench, are more than 10 km below the surface. Scientists have explored the deepest water in the ocean only during the past few decades, but we know that it is a cold (2°–3°C), dark environment, where organisms live under tremendous pressure from the ocean above. Abyssal communities are surprisingly diverse, although population densities tend to be low. The denizens of the abyssal zone include invertebrates, bony fishes, and sharks. Many fishes and invertebrates are bioluminescent, producing spots of light that may serve for communication, as spotlights to identify potential prey, or as lures to entice prey within reach of their large jaws **(Figure 51.31).**

The Benthic Province Includes the Rocks and Sediments of the Ocean Bottom

The benthic province extends from the intertidal zone to the deep-sea trenches. In the oceanic zone, bottom sediments are composed of soft mud, fine particles of silt, detritus, and the shells of dead microscopic organisms. Species living in and on the bottom are collectively called **benthos.** Sunlight never strikes the benthic province of the open ocean, which is inhabited by bacteria, fungi, and a variety of animals. Sessile invertebrates, such as sponges, sea anemones, and clams, live amidst the sediments, and many motile animals, including worms, mollusks, crustaceans, echinoderms, and fishes, feed on the organic remains that sink from pelagic communities.

In 1977, researchers first found communities thriving near hydrothermal vents at a depth of 2,500 m near the Galápagos Rift, a volcanically active boundary between two crustal plates (see Section 23.2). Near-freezing water seeps into fissures where it is heated to temperatures of 350°C or higher. Pressure forces the heated water upward, and minerals are leached from porous rocks as the water spews out through vents in the seafloor. This hydrothermal outpouring releases hydrogen sulfide, which chemosynthetic bacteria use as an energy source to fix carbon dioxide, just as plants and algae use sunlight for photosynthesis.

Some of these bacteria live as endosymbionts of giant clams and tube-dwelling worms **(Figure 51.32).** Worms in the genus *Riftia* lack digestive systems, subsisting entirely on the organic compounds produced by their chemosynthetic partners, which are so numerous that they account for up to half of the worms' body weight. The worms' blood contains both hemoglobin and sulfide-binding proteins, which deliver oxygen and sulfides to the bacteria. Deep-sea communities also include sea anemones, crustaceans, and fishes. Researchers have located hydrothermal vent ecosystems in the South Pacific, the North Pacific, the Gulf of California, and the Atlantic.

Research in the deep sea has advanced substantially since the first discovery of hydrothermal vent communities. In 1996, a team of Japanese researchers used a remotely operated submersible vehicle to collect specimens at depths of nearly 11,000 m in the Mariana Trench. Their collections included bacteria described as obligate barophiles (that is, organisms that can survive and reproduce *only* under conditions of extreme pressure). Preliminary screening of the bacteria collected by these

FIGURE 51.32 Deep benthos. Giant tube-dwelling worms *(Riftia)* and their chemosynthetic bacterial endosymbionts are common in hydrothermal vent communities on the deep ocean floor.

and other scientists has revealed that they produce previously unknown compounds, some of which appear to have potentially useful tumor-suppressing and antibacterial properties.

In an extraordinary feat of engineering, on March 26, 2012, film director and diver James Cameron piloted the deep submergence vehicle *Deepsea Challenger* to the deepest known site on the ocean floor, the Challenger Deep in the Mariana Trench. Not surprisingly, his vehicle was equipped with strong lights and good cameras, as well as a hydraulic arm to collect samples. Cameron's achievement now allows scientists to explore directly one of the most extreme environments on Earth.

Recent research in the deepest reaches of the ocean reveals that communities also exist in areas far from hydrothermal vents. These "cold seep" communities thrive on broad expanses of the seafloor, where extremely salty water percolates upward from the underlying rocks and sediment, carrying abundant minerals, hydrogen sulfide, and methane to areas that are accessible to organisms. Chemosynthetic bacteria, which grow in large mats, can metabolize these molecules, serving as a food source for animal communities that include sponges, worms, and bivalve mollusks.

STUDY BREAK 51.6

1. What is the difference between the benthic and pelagic provinces of the ocean?
2. Which marine environments experience the largest fluctuations in salinity (salt concentration) over time?
3. Which marine regions receive abundant energy input from sunlight?
4. What is the source of nutrients and energy for the benthos of the oceanic zone?
5. What organisms are responsible for the synthesis of organic compounds in hydrothermal vent and cold seep communities, and how do they differ from those in the photic zone?

Unanswered Questions

How will biomes change in response to anthropogenic (human-induced) global warming?

Will biomes remain largely intact and simply shift their geographical distributions northward or upward (to higher elevations)? Or will the biomes we recognize today become disrupted, and new biomes arise as species associate in different combinations? Parmesan and Yohe estimated that 59% of wild species around the world have already shown some change in their geographical distributions in response to the relatively small level of global warming—a 0.7°C rise in average temperature—over the past 100 years. Documented responses to global warming vary from species to species, however. For example, only 20% of butterfly species in Spain, France, and North Africa have shifted their southern range boundaries northward, but 70% of butterfly species in the United Kingdom and Scandinavia have expanded their northern range borders further northward, sometimes by as much as 300 km over the past 30 years.

Thus, although the distributions of some species appear stable, the ranges of others are showing strong responses to global warming. At least two hypotheses may explain these patterns. First, some species may be stressed by rising temperatures but have not yet shown a measurable response. Second, the geographical distributions of some species may not be governed primarily by climate. Whatever the reason, the fact that we observe large variation in the response of different species suggests that not all species in a community are moving together. Thus, the existing communities of birds, butterflies, and trees are being disrupted—with some species moving and others not. Biologists have also noted differences in the response of different taxonomic groups. Butterflies in Europe and North America seem to be shifting their distributions northward and upward at about the same rate that temperatures are changing, but plants appear to lag behind. Alpine herbs in Switzerland, for example, have shifted their distributions upward at about half the rate that one might expect from the rate of regional warming, and it was not until 30 years after warming began that tree seedlings in Sweden started to colonize alpine habitats at higher elevations, shifting the treeline upward.

Even bigger questions remain. Will the vegetation that currently lives in the tropics expand into what is now the temperate zone and cover more of the planet? Some studies suggest that tropical lowland trees are already at their physiological limit—already showing signs of stress by shutting down photosynthesis on the hottest and driest days. Furthermore, climate model projections from a 2007 report of the Intergovernmental Panel on Climate Change consistently show substantial drying, as well as warming, in middle latitudes. If this projection is correct, many plants and animals now living in the wet tropics will be unable to shift northward into what may become an extreme desert climate. And what will happen to the arctic tundra? Researchers have already collected strong evidence that shrubs and trees are encroaching northward into the tundra of Alaska and Canada. The permafrost is melting, and the soil is drying. How do these observed changes in plant and animal distributions relate to the future? Human activity has already caused Earth's mean annual temperature to rise by 0.7°C in the past 100 years. Climate model projections suggest that further increases between 1.8°C and 4.0°C are likely; some models suggest the rise will be over 6.0°C. Can the tundra biome survive even the lowest projections—more than twice the warming it has already experienced?

How will evolution shape the ways that wild species respond to climate change? Which groups of organisms are likely to adapt, and which are likely to become extinct?

Populations are evolving all the time in response to changing selection regimes. Global warming is one of many human-driven environmental changes that could foster genetic change. Biologists have known for decades that organisms are locally adapted to the climatic conditions under which they routinely live. Scientists have documented local genetic changes toward more warm-adapted genotypes in fruit flies, mosquitoes, and the algal symbionts of corals. Do the observed genetic changes suggest that these species are adapting to anthropogenic global warming? Will other species follow suit? The fossil record suggests that during the Pleistocene glaciations, when Earth's temperature shifted between glacial periods (4°–8°C colder than now) and interglacial periods (today's temperatures), very few species became extinct and few experienced substantial morphological evolution. But, before the Pleistocene, Earth was much hotter than it is today, and the atmosphere had higher levels of CO_2. During the transition from these very warm, high CO_2 conditions to the colder, low CO_2 conditions of the Pleistocene, a large proportion of species became extinct. To how much climate change can organisms adapt? At what point is climate change extreme enough that species come to the limit of their genetic variation, can no longer adapt, and become extinct?

think like a scientist Given what you have learned about the effects of climate change on species' distributions, in which types of habitats or geographical regions should we witness the most dramatic loss of species in the near future?

Camille Parmesan is a Professor in the Marine Institute at Plymouth University, UK, where she holds the National Marine Aquarium Chair in the Public Understanding of Oceans and Human Health. Her research has focused on current impacts of climate change on all wildlife, both on land and in the oceans. To learn more about Dr. Parmesan's research, go to https://www.plymouth.ac.uk/staff/camille-parmesan.

REVIEW KEY CONCEPTS

For access to MindTap and additional study materials visit www.cengagebrain.com.

51.1 The Science of Ecology

- Ecology is the study of the interactions between organisms and their environments. Basic ecology focuses on undisturbed natural systems, whereas applied ecology considers the effects of human disturbance.

- Ecologists conduct research at five levels of organization: organisms, populations, communities, ecosystems, and the biosphere.

- Ecologists test hypotheses about ecological relationships with experimental or observational data. They sometimes frame hypotheses in mathematical models.

51.2 Environmental Diversity of the Biosphere

- The biosphere encompasses all the regions on Earth where organisms live, including the atmosphere, hydrosphere, and lithosphere.

- Latitudinal variations in solar radiation establish global climate patterns (Figure 51.2). Earth's tilt on its axis causes seasonal variation in solar radiation and climate (Figure 51.3). Seasonal variations in day length and temperature increase steadily from tropical latitudes toward the poles.

- Unequal heating of the atmosphere causes air masses to flow in circulation cells that create worldwide wind and precipitation patterns (Figures 51.4 and 51.5). Ocean currents generally flow clockwise in the Northern Hemisphere and counterclockwise in the Southern Hemisphere (Figure 51.6).

- The oceans and local topographical features influence regional and local climates. Proximity to the ocean has a moderating effect on terrestrial climates (Figure 51.7). Habitats are generally wetter on the windward sides of mountains than on the leeward sides (Figure 51.8).

51.3 Organismal Responses to Environmental Variation and Climate Change

- Organisms use homeostatic responses to cope with environmental variation. Animals often use facultative behavioral and physiological mechanisms to respond to environmental temperature (Figure 51.9).

- Global climate change is affecting the ecology of many organisms. Many species are experiencing shifts in their geographical ranges or in the timing of their reproduction (Figure 51.10).

51.4 Terrestrial Biomes

- Biomes are general types of vegetation and other associated organisms. Climate is the major determinant of terrestrial biome distributions (Figures 51.11 and 51.12).

- Tropical forest occurs at low latitudes where seasonality is determined by variations in rainfall rather than by day length and temperature (Figure 51.13). Tropical rainforests are the most species-rich terrestrial biome, but they grow on nutrient-poor soils.

- Savanna is tropical and subtropical grassland with scattered trees (Figure 51.14). Long dry seasons, fires, and grazing by large mammals prevent trees from replacing perennial grasses.

- Deserts form in arid regions where precipitation is low and temperature varies widely on a daily and seasonal basis (Figure 51.15).

- Chaparral is a coastal biome dominated by dense, woody shrubs and trees that resprout after periodic fires (Figure 51.16). Chaparral occurs where winters are mild and wet and summers hot and dry.

- Temperate grassland grows where winters are cold, summers are warm, and rainfall is moderate (Figure 51.17). Tree seedlings are eliminated from grasslands by droughts, periodic fires, and grazing by mammals. Grassland soils are rich and deep.

- Temperate deciduous forest flourishes at middle latitudes with abundant rainfall. The seasonality of the climate is reflected in the annual loss and regrowth of leaves (Figure 51.18).

- The boreal forest, or taiga, includes dense stands of coniferous trees at high latitudes, where winters are long and cold (Figure 51.19).

- Tundra is the northernmost biome, where plants grow in shallow topsoil over a layer of permafrost (Figure 51.20). The brief growing season and winter winds cause tundra plants to be very short.

51.5 Freshwater Environments

- Freshwater environments include both flowing-water and standing-water systems.

- The physical characteristics of flowing-water environments change from the headwaters of a stream to the mouth of a river (Figure 51.21).

- The physical characteristics of standing-water environments change with the depth of water and distance from shore (Figures 51.22 and 51.23). Lakes exhibit marked vertical zonation and, in the temperate zone, undergo a seasonal mixing of their waters (Figure 51.24). Lakes are generally classified by their nutrient status and productivity (Figure 51.25).

51.6 Marine Environments

- The oceans exhibit marked zonation based on water depth and distance from shore (Figure 51.26).
- Estuaries are highly productive tidal environments where rivers provide a constant input of nutrients and freshwater, and the tides carry away wastes (Figure 51.27).
- The intertidal zone is a stressful environment that is alternately submerged and exposed (Figure 51.28).
- Highly productive and diverse shallow-water communities thrive on continental shelves and oceanic banks. Kelp forests predominate at middle to high latitudes; coral reefs occur in the tropics (Figure 51.29).
- The open ocean is highly stratified because photosynthesis is possible only in the uppermost 50 m of water. Plankton are the primary producers in the uppermost layers (Figure 51.30). The deep sea includes many predatory species (Figure 51.31).
- Organisms of the seafloor occupy the benthic province. Falling detritus supports most benthic communities, but chemosynthetic bacteria support communities near deep-sea hydrothermal vents (Figure 51.32) and cold seeps.

TEST YOUR KNOWLEDGE

Remember/Understand

1. The lithosphere includes all:
   a. oceans.
   b. ice caps.
   c. rocks, soils, and sediments.
   d. gases and airborne particles.
   e. places where organisms live.

2. Earth's 23.5° tilt on its axis directly causes:
   a. latitudinal variation in average annual rainfall.
   b. ocean currents to rotate clockwise in the Northern Hemisphere.
   c. microclimates to vary dramatically over short distances.
   d. low rainfall on the leeward side of mountain ranges.
   e. seasonal variation in the amount of solar radiation.

3. Adiabatic cooling causes rising air masses to:
   a. absorb moisture from Earth's surface.
   b. release precipitation.
   c. change the direction of the El Niño current.
   d. flow toward the equator from the poles.
   e. be deflected from a strictly northward or southward flow.

4. The term *rain shadow* describes the:
   a. low rainfall that is typical on the leeward side of mountains.
   b. low rainfall that is typical at 30° latitude.
   c. high rainfall that is typical on the windward side of mountains.
   d. blocking of rain by vegetation in dense tropical forests.
   e. low rainfall that is typical in the interior of continents.

5. The major climatic factors that govern the distributions of terrestrial biomes are:
   a. temperature only.
   b. rainfall only.
   c. wind speed only.
   d. temperature and rainfall.
   e. temperature, rainfall, and wind speed.

6. The major source of nutrients in the headwaters of a small stream is from:
   a. dead leaves and other organic matter from adjacent land.
   b. photosynthesis by phytoplankton.
   c. photosynthesis by floating aquatic plants.
   d. the activity of chemoautotrophic bacteria.
   e. minerals from the underlying bedrock.

Apply/Analyze

7. Which biome experiences the highest annual rainfall?
   a. tropical rainforest
   b. tropical savanna
   c. chaparral
   d. temperature grassland
   e. arctic tundra

8. From which biome are trees excluded by periodic fires and grazing herbivores?
   a. tropical rainforest
   b. thorn forest
   c. chaparral
   d. temperate grassland
   e. arctic tundra

9. During the spring overturn in a temperate zone lake:
   a. oxygen is carried from the surface to the bottom, and nutrients are carried from the bottom to the surface.
   b. nutrients are carried from the surface to the bottom, and oxygen is carried from the bottom to the surface.
   c. nutrients and oxygen are carried from the bottom waters to the surface waters.
   d. nutrients and oxygen are carried from the surface waters to the bottom waters.
   e. oxygen concentration remains constant at all depths, and nutrients sink to the bottom.

10. In which habitat must organisms adjust regularly to changing salinity?
    a. salt marsh
    b. coral reef
    c. benthic province
    d. estuary
    e. riffle

Evaluate/Create

11. **Discuss Concepts** (a) Temperate grassland and chaparral often burn in lightning-induced fires, which stimulate the germination of seeds and regrowth of existing vegetation. Do you think that companies or the government should sell fire insurance to people who build expensive homes in places where periodic fires are virtually inevitable? (b) Boreal forests generally harbor many fewer species of trees than tropical forests do. Develop three hypotheses to explain this pattern. What data would you collect to test your hypotheses? (c) Many regions have been developed for agriculture, industry, and human habitation. Have our activities created new biomes? What physical environments are created by development, and what plants and animals occupy developed areas?

12. **Design an Experiment** Design an experiment to test the hypothesis that streams receive much of their nutrients and energy from material that falls into them from overhanging vegetation.

13. **Apply Evolutionary Thinking** If the geographical ranges of species change in response to global warming, what new selection pressures will organisms face as they move into ecological communities where they have not previously occurred? Your answer should address the effects of novel species interactions, as well as the effects of encountering different physical environments.

For selected answers, see Appendix A.

INTERPRET THE DATA

Climate scientists at NOAA, the National Oceanographic and Atmospheric Administration, collect data about sea surface temperature (SST) and atmospheric conditions to predict El Niño and La Niña events before they occur. Researchers then calculate SST anomalies, the differences between current SSTs and historical averages. *Positive* anomalies of 0.5°C or more indicate that surface waters are much warmer than usual, a sign that El Niño is coming. *Negative* anomalies of 0.5°C or more indicate that surface waters are much cooler than usual, a sign of an upcoming La Niña. Anomalies that are within 0.5°C of average SSTs are identified as neutral conditions. The accompanying **Figure** presents SST anomalies in the eastern Pacific Ocean from 1950 through 2008.

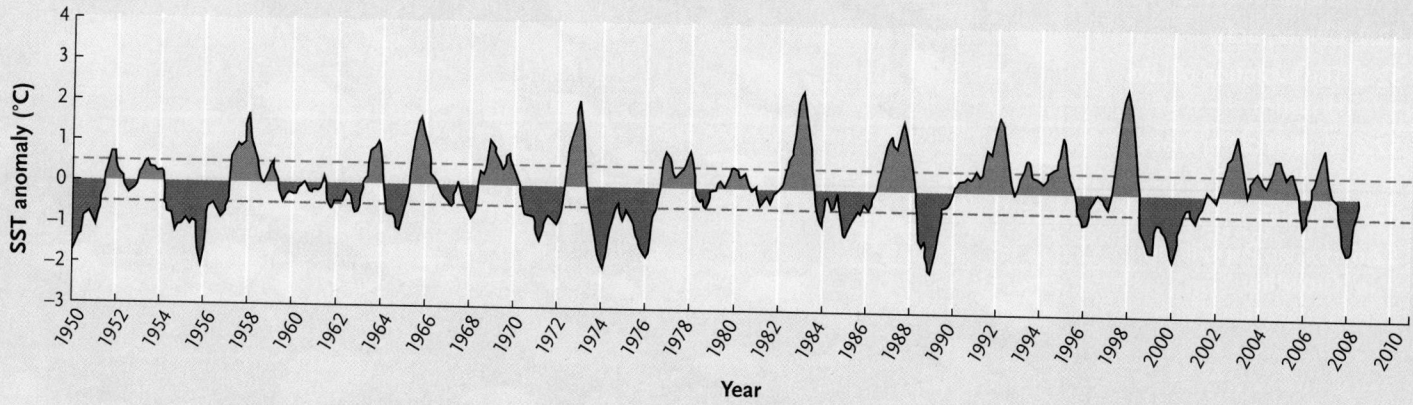

1. How many times have SSTs predicted a likely El Niño event or a likely La Niña event over the time represented in the graph?

2. Are all SST anomalies of equal magnitude and duration?

3. Is there a fixed pattern in the distribution of El Niño and La Niña events? In other words, does one type of event *always* follow the other?

Source: NOAA Center for Weather and Climate Prediction Center, NSW/NOAA.

52

Population Ecology

A population of Australasian gannets *(Morus serrator)* nesting on the coast of New Zealand. Gannets lay their eggs on small mounds of sandy soil. Like other seabirds, they nest in large breeding colonies close to the sea, giving them ready access to the fishes they consume.

Why it matters . . . When humans immigrate to new places, they often transport familiar plants and animals from home, introducing them into their new gardens, fields, and forests. Some organisms fail to survive in the new environments. But other species—such as the European starlings *(Sturnus vulgaris)* and house sparrows *(Passer domesticus)* that are now so common in North America—flourish, and sometimes become pests.

In 1859, an Australian rancher released a few pairs of European rabbits *(Oryctolagus cuniculus)* for sport hunting in the state of Victoria. The rabbits bred rapidly, sometimes producing litters of four or five offspring every month. They had no natural predators in Australia, and by 1900, an estimated 20 million rabbits had overrun much of the continent. Their advance was limited only by extreme climates, clay soil, and lack of food or water. The rabbits destroyed natural vegetation and the pastures that supported a large sheep industry. The government tried in vain to poison the rabbits. Ranchers introduced predators, hoping that they would eat rabbits faster than the rabbits could reproduce. But the rabbits continued to multiply. Eventually, the government built a "rabbit-proof fence" that stretched more than 3,200 km (2,000 miles) to keep the rabbits out of the rich pasture lands in Western Australia **(Figure 52.1).**

In 1950, scientists tackled the devastating problems caused by the introduced rabbits. Biologists collected myxoma virus (a relative of smallpox) from infected rabbits in South America and released it among the European rabbits in Australia. At first, the virus was lethal to European rabbits, which had never evolved resistance to it. The first epidemic of myxomatosis killed more than 99% of infected rabbits. But in the following season, the virus killed only 90% of infected rabbits, and within a few years, the virus was killing only half the rabbits it infected. Clearly, some rabbits were becoming resistant to the virus. Resistant rabbits survived and reproduced, comprising a larger percentage of the population over time (see Section 21.3 to review natural selection). Subsequent research showed that the virus had also become less virulent. Today, wildlife-control agents develop and release more deadly viruses to control the rabbit population.

This brief history of an introduced population identifies several questions about **population dynamics**—how the characteristics of populations change through time and vary from place to place—that we consider in this chapter. For example, why do some populations, such as the rabbits in Australia, grow explosively, whereas others maintain reasonably stable numbers over long periods of time? How do interactions with abiotic and biotic factors in the environment influence the characteristics of populations? Finally, how do populations respond evolutionarily to their interactions with the environment?

FIGURE 52.1 Introduced organisms. European rabbits multiplied so rapidly and destroyed so much vegetation in Australia that the government built a fence across the country to prevent their spread.

John Carnemolla/Corbis

52.1 Population Characteristics

Populations have characteristics that transcend those of the individuals they comprise. For example, every population has a **geographical range,** the overall spatial boundaries within which it lives. Geographical ranges vary enormously. A population of snails might inhabit a small tidepool, whereas a population of marine phytoplankton might occupy an area that is orders of magnitude larger. Every population also occupies a **habitat,** the specific environment in which it lives, as characterized by its biotic and abiotic features. Ecologists also measure other population characteristics, such as size, distribution in space, and age structure.

A Population's Size and Density Determine the Amount of Resources It Uses

Population size is simply the number of individuals in a population at a specified time. **Population density** is the number of individuals per unit area or per unit volume of habitat. Species with large body sizes generally have lower population densities than species with smaller body sizes because each large organism consumes more of the available resources than each small organism does **(Figure 52.2).** Although population size and density are related measures, knowing a population's density provides more information about its relationship to the resources it uses. For example, if a population of 200 oak trees occupies 1 hectare (10,000 m^2), its population density is 200/10,000 m^2 or one tree per 50 m^2. But if a population of 200 oaks is spread over 5 hectares, its density is one tree per 250 m^2. Clearly, the second population is less dense than the first, and its members will have greater access to sunlight, water, and nutrients.

Ecologists measure population size and density to monitor and manage populations of endangered species, economically important species, and agricultural pests. For large-bodied species that live in open habitats, a simple head count provides accurate information. For example, ecologists survey the size and density of African bush elephant (*Loxodonta africana*) populations by flying over herds, either in small aircraft or with unmanned aerial vehicles (that is, "drones"), and counting individuals. Researchers use a variation on that technique to estimate population size in tiny organisms that live at high population densities. To estimate the density of aquatic phytoplankton, for example, you might collect water samples of known volume from representative areas in a lake and use a microscope to count the organisms; you could then extrapolate their population size and density based on the estimated volume of the entire lake. In other cases, researchers use the mark-release-recapture sampling technique **(Figure 52.3).**

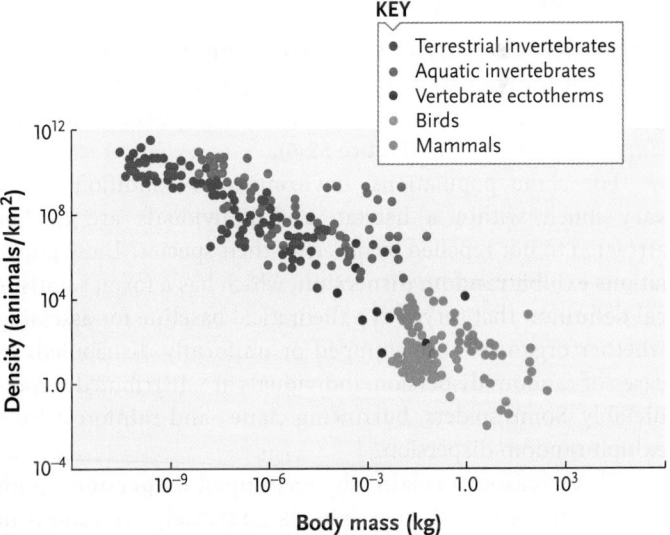

FIGURE 52.2 Population density and body size. Population density generally declines with increasing body size among animal species. Similar trends exist for other types of organisms.

© Cengage Learning 2017

FIGURE 52.3   Research Method

Using Mark-Release-Recapture to Estimate Population Size

Purpose: Ecologists use the mark-release-recapture technique to estimate the population size of mobile animals that live within a restricted geographic range.

Protocol: A sample of organisms is captured, marked in some permanent but harmless way, and released. Insects and reptiles are marked with ink or paint, birds with rings on their legs, and mammals with ear tags or collars. Some time later, a second sample of organisms is captured, and the researcher notes what proportion of the second sample carries the mark. That proportion tells us what percentage of the total population was captured and marked at the first sampling. The total population size is estimated as (number marked) × (number in the second sample/number of marked recaptures).

Michael C. Singer, University of Texas

Interpreting the Results: Imagine that you capture 120 butterflies, mark each with a black spot on its wing, and release them. A week later, you capture a second sample of 150 butterflies, and find that 30 of them have the black mark. Thus, you had marked one out of every five butterflies (30/150) on your first field trip. Because you captured and marked 120 individuals on that first excursion, you would estimate that the total population size is 120 × (150/30) = 600 butterflies.

The technique is based on several assumptions that are critical to its accuracy: (1) that being marked has no effect on survival; (2) that marked and unmarked animals mix randomly in the population; (3) that no migration into or out of the population takes place during the estimating period; and (4) that marked individuals are just as likely to be captured as unmarked individuals. (Sometimes animals become "trap shy" or "trap happy," a violation of the fourth assumption.)

Populations Differ in How They Are Distributed in Space

Populations also vary in their **dispersion,** the spatial distribution of individuals within the geographical range. Ecologists define three theoretical patterns of dispersion: *random, clumped,* and *uniform* (**Figure 52.4**).

For some populations, environmental conditions don't vary much within a habitat, and individuals are neither attracted to nor repelled by others of their species. These populations exhibit **random dispersion,** which has a formal statistical definition that serves as a theoretical baseline for assessing whether organisms are clumped or uniformly distributed. In cases of random dispersion, individuals are distributed unpredictably. Some spiders, burrowing clams, and rainforest trees exhibit random dispersion.

Three reasons explain why a **clumped dispersion**—with individuals grouped together—is extremely common in nature. First, suitable conditions often have a patchy distribution. For example, certain pasture plants may be clumped in small, scattered areas where cowpats fell months before, locally enriching the soil. Second, some animals live in social

groups (see Section 56.7). Mates are easy to locate within groups, and individuals may cooperate in rearing offspring, feeding, or defending themselves from predators. Third, some organisms are clumped because of their reproductive pattern. Plants and animals that produce asexual clones, such as aspen trees and sea anemones, often occur in large aggregations (see Chapters 36 and 49). In other species, seeds, eggs, or larvae lack dispersal mechanisms, and offspring grow near their parents.

When the individuals in a population repel each other because resources are in short supply, they tend to be evenly spaced in their habitat, a pattern called **uniform dispersion.** For example, creosote bushes (*Larrea tridentata*) are uniformly distributed in the dry scrub deserts of the American Southwest. Mature bushes deplete the surrounding soil of water and secrete toxic chemicals, making it impossible for seedlings to grow. Moreover, seed-eating ants and rodents that live at the base of mature bushes consume any seeds that fall nearby. Territorial behavior, the defense of an area and its resources, produces uniform dispersion in animals (see Section 56.4).

Whether the spatial distribution of a population appears to be random, clumped, or uniform depends partly on how large an area an ecologist studies. Oak seedlings may be randomly dispersed on a spatial scale of a few square meters, but over an entire mixed hardwood forest, they are clumped under the parent trees.

In addition, the dispersion of animal populations often varies through time in response to natural environmental rhythms. Few habitats provide a constant supply of resources throughout the year, and many animals move from one habitat to another on a seasonal cycle. For example, tropical birds and mammals are often widely dispersed in deciduous forests during the wet season. But during the dry season, they crowd into narrow "gallery forests" along watercourses where evergreen trees provide food and shelter.

A Population's Age Structure, Generation Time, and Sex Ratio Influence How Quickly It Will Grow

All populations have an **age structure,** a statistical description of the relative numbers of individuals in each age class (see Section 52.6). Individuals can be roughly categorized as prereproductive (younger than the age of sexual maturity), reproductive, or postreproductive (older than the maximum age of reproduction). A population's age structure reflects its recent growth history and predicts its future growth potential. Populations that include many prereproductive individuals grew

Clumped

A clumped dispersion pattern is one in which individuals are grouped more closely to each other than if they are randomly dispersed.

Random

A random dispersion pattern, in which organisms are distributed independently of each other, serves as a statistical yardstick for evaluating other dispersion patterns.

Uniform

A uniform dispersion pattern is one in which individuals are more widely separated from each other than they are if they are randomly dispersed.

FIGURE 52.4 Dispersion patterns. Schooling fishes, such as these Maldives bulleyes *(Priacanthus hamrur)* from the Indian Ocean, exhibit a clumped pattern of dispersion. A random pattern of dispersion, which is fairly rare in nature, occurs in organisms, such as these dandelions *(Taraxacum officianale)*, that are neither attracted to nor repelled by each other. Creosote bushes *(Larrea tridentata)* in Death Valley, California, exhibit uniform dispersion.
© Cengage Learning 2017

rapidly in the recent past and will continue to grow larger as the young individuals mature and reproduce.

Another characteristic that influences a population's growth is its **generation time,** the average time between the birth of an organism and the birth of its offspring. Generation time is usually short in species that reach sexual maturity at a small body size **(Figure 52.5).** Their populations often grow rapidly because of the speedy accumulation of reproductive individuals.

Populations also vary in their **sex ratio,** the relative proportions of males and females. In general, the number of females in a population has a bigger impact on population growth than the number of males because only females actually produce offspring. Moreover, in many species, one male can mate with several females, and the number of males may have little effect on the population's reproductive output. In northern elephant seals *(Mirounga angustirostris),* for example, mature bulls fight for dominance on the beaches where the seals mate, and only a few males may ultimately inseminate a hundred or more females. Thus, the presence of other males in the group has little effect on the size of future generations. However, in animals that form lifelong pair bonds, such as geese and swans, the number of males does influence reproduction in the population.

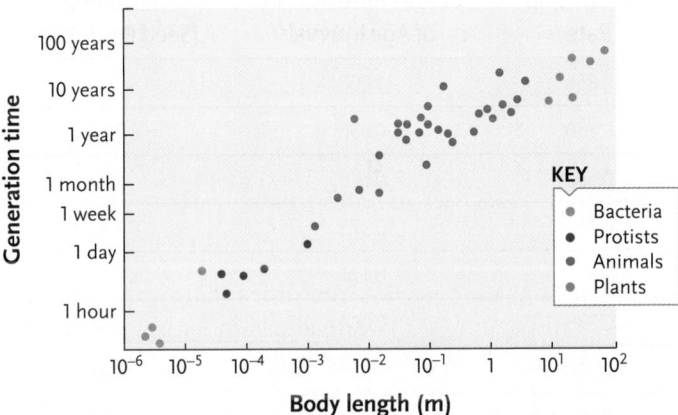

FIGURE 52.5 Generation time and body size. Generation time increases with body size among bacteria, protists, plants, and animals. The logarithmic scale on both axes compresses the data into a straight line.
© Cengage Learning 2017

STUDY BREAK 52.1

1. What is the difference between a population's size and its density?
2. What do the three patterns of dispersion imply about the relationships between individuals in a population?

52.2 Demography

Populations grow larger through the birth of individuals and the **immigration** (movement into the population) of organisms from neighboring populations. Conversely, death and **emigration** (movement out of the population) reduce population size. **Demography** is the statistical study of the processes that change a population's size and density through time.

Ecologists use demographic analysis to predict a population's future growth. For human populations, these data help governments anticipate the need for social services such as schools and hospitals. Demographic data also allow conservation ecologists to develop plans to protect endangered species. For example, demographic data on the northern spotted owl (*Strix occidentalis caurina*) helped convince the courts to restrict logging in the owl's primary habitat, the old growth forests of the Pacific Northwest. *Life tables* and *survivorship curves* are among the tools ecologists use to analyze demographic data.

Life Tables Summarize a Population's Survival and Reproductive Rates

Although every species has a characteristic lifespan, few individuals survive to the maximum age possible. Mortality results from starvation, disease, accidents, predation, or the inability to find a suitable habitat. Life insurance companies first developed techniques for measuring mortality rates, but ecologists adapted these approaches to the study of nonhuman populations.

A **life table** summarizes the demographic characteristics of a population **(Table 52.1).** To collect life-table data for short-lived organisms, demographers typically mark a **cohort,** a group of individuals of similar age, at birth and monitor their survival until all members of the cohort die. For organisms that live more than a few years, a researcher might sample the population for one or two years, recording the ages at which individuals died, and then extrapolate those results over the species' lifespan.

In any life table, the lifespan of the organisms is divided into age intervals of convenient length: days, weeks, or months for short-lived species; years or groups of years for longer-lived species. Mortality can be expressed in two complementary ways. **Age-specific mortality** is the proportion of individuals alive at the start of an age interval that died during that age interval. Its more cheerful reflection, **age-specific survivorship,** is the proportion of individuals alive at the start of an age interval that survived until the start of the next age interval. Thus, in Table 52.1, the age-specific mortality rate during the three-to-six-month age interval is $195/722 = 0.270$, and the age-specific survivorship rate is $527/722 = 0.730$. For any age interval, the sum of age-specific mortality and age-specific survivorship always equals 1. Life tables also summarize the proportion of the cohort that survived to a particular age, a statistic that identifies the probability that any randomly selected newborn will still be alive at that age. For the 3-to-6-month age interval in Table 52.1, this probability is $722/843 = 0.856$.

Life tables also include data on **age-specific fecundity,** the average number of offspring produced by surviving females during each age interval. Table 52.1 shows, for example, that plants in the 3-to-6-month age interval each produced an average of 300 seeds. In some species, including humans, fecundity is highest in individuals of intermediate age. Younger individuals have

TABLE 52.1	Life Table for a Cohort of 843 Individuals of the Grass *Poa annua* (Annual Bluegrass)					
Age Interval (in months)	Number Alive at Start of Age Interval	Number Dying during Age Interval	Age-Specific Mortality Rate	Age-Specific Survivorship Rate	Proportion of Original Cohort Alive at Start of Age Interval	Age-Specific Fecundity (Seed Production)
0–3	843	121	0.144	0.856	1.000	0
3–6	722	195	0.270	0.730	0.856	300
6–9	527	211	0.400	0.600	0.625	620
9–12	316	172	0.544	0.456	0.375	430
12–15	144	90	0.625	0.375	0.171	210
15–18	54	39	0.722	0.278	0.064	60
18–21	15	12	0.800	0.200	0.018	30
21–24	3	3	1.000	0.000	0.004	10
24–	0	—	—	—	—	—

Source: Based on M. Begon and M. Mortimer. *Population Ecology.* Sunderland, MA: Sinauer Associates, 1981. Adapted from R. Law. 1975.

A. Dall sheep (Ovis dalli)

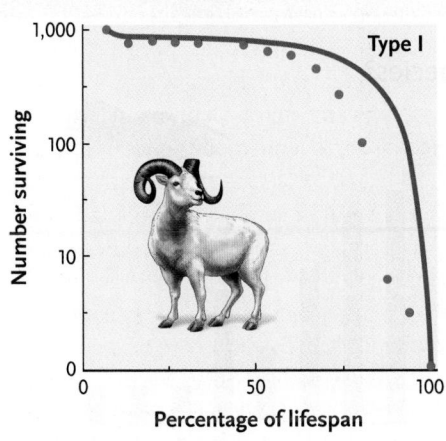

B. Five-lined skink (Eumeces fasciatus)

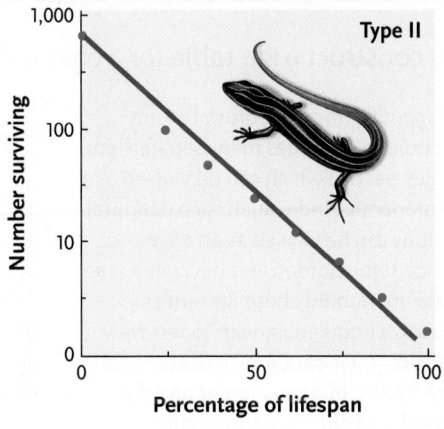

C. Perennial desert shrub (Cleome droserifolia)

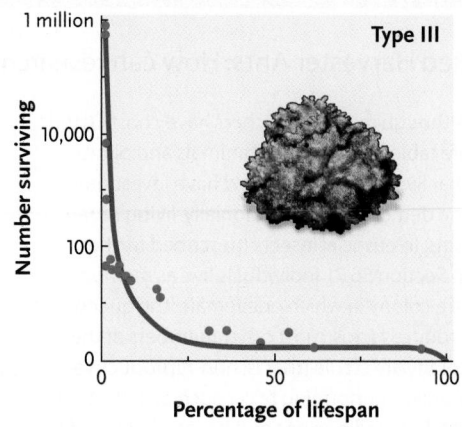

KEY

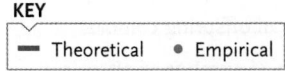

FIGURE 52.6 Survivorship curves. The survivorship curves of many organisms (pink data points) roughly match one of three idealized patterns (blue curves). Note that the number surviving is plotted on a logarithmic scale on the vertical axis to illustrate the proportion of the original cohort surviving to a particular age.

© Cengage Learning 2017

not yet reached sexual maturity, and older individuals are past their reproductive prime. However, in some plants and animals, fecundity increases steadily with age.

Although researchers have developed life tables for many species of organisms that live as individuals, few have done so for colonial organisms, such as ants and bees, that often live in cooperative groups called *colonies. Molecular Insights* describes a recent study about the construction of a life table for a colonial ant species.

Survivorship Curves Depict Changes in Survival Rate over the Lifespan

Survivorship data are depicted graphically in a **survivorship curve,** which displays the rate of survival for individuals over the species' average lifespan. Ecologists have identified three generalized survivorship curves (blue lines in **Figure 52.6**), although most organisms exhibit survivorship patterns that fall between these idealized patterns.

Type I curves reflect high survivorship until late in life, when mortality takes a great toll. Type I curves are typical of large animals that produce few young and provide them with extended care, which reduces juvenile mortality. For example, large mammals, such as the Dall's sheep *(Ovis dalli),* produce only one or two offspring at a time and nurture them through their vulnerable first year (see Figure 52.6A).

Type II curves reflect a relatively constant rate of mortality in all age classes, a pattern that produces steadily declining survivorship. Many lizards, such as the five-lined skink *(Eumeces fasciatus),* as well as songbirds and small mammals, face a constant probability of mortality from predation, disease, and starvation (see Figure 52.6B).

Type III curves reflect high juvenile mortality, followed by a period of low mortality once the offspring reach a critical age and size. For example, *Cleome droserifolia*, a desert shrub from the Middle East, experiences extraordinarily high mortality in its seed and seedling stages. Researchers estimate that for every million seeds produced, fewer than 1,000 germinate, and only about 40 individuals survive their first year. Once a plant becomes established, however, its likelihood of future survival is higher, and the survivorship curve flattens out. Many plants, insects, marine invertebrates, and fishes exhibit type III survivorship (see Figure 52.6C).

STUDY BREAK 52.2

1. What statistics are usually included in a life table?
2. Which type of survivorship curve is characteristic of humans in industrialized countries? Explain your answer.

52.3 The Evolution of Life Histories

The analysis of life tables reveals how natural selection has produced different **life histories**—the lifetime patterns of growth, maturation, and reproduction—that maximize the number of surviving offspring an individual produces.

Organisms Face Trade-Offs in Their Allocation of Resources

Every organism is constrained by a finite **energy budget,** the total amount of energy that it can accumulate and use to fuel its activities. An organism's energy budget is like a savings account. When the individual accumulates more energy than it needs, it makes deposits to this account—energy is stored as starch, glycogen, or fat. When it expends more energy than it harvests, it makes withdrawals from its energy stores. But unlike a bank account, an organism's energy budget cannot be overdrawn, and no loans against future "earnings" are possible.

Red Harvester Ants: How can researchers construct a life table for a colonial species?

Although many researchers have constructed life tables for long-lived animals and plants that live as individuals, few have investigated the demographics of colonially-living organisms. In eusocial insects (described further in Section 56.7), individuals live as members of a colony in which one female, the queen, produces eggs; most other members of the colony are sterile (that is, non-reproductive) workers, performing tasks, such as collecting food or tending eggs and larvae, that benefit the colony as a whole. The queen also produces some fertile male and female offspring, which mate with individuals from the same or another colony; a fertilized female may then leave the nest to found a new daughter colony. Thus, social insects survive and reproduce as a colony of closely related individuals, not as a population of independent organisms.

Research Question

If most individuals in a colony do not reproduce, how can researchers create a life table for colonial organisms?

Experiment

As part of a larger project, Deborah M. Gordon of Stanford University and colleagues at Stanford and Colgate University used molecular techniques to develop a life table for a population of red harvester ants

(Pogonomyrmex barbatus). The ant population includes many separate colonies, each of which can be viewed as a reproductive individual; each daughter colony can be viewed as an offspring. Since 1985, Gordon and her colleagues have monitored about 300 ant colonies in a population near Rodeo, New Mexico. Their data, which documents the "births" of new colonies and the "deaths" of old ones, suggest that a colony begins to produce daughter colonies four to five years after it is founded and that colonies can live for up to 25 years. Thus, a colony can potentially produce daughter colonies annually for about 20 years. Long-term monitoring of the population, colony by colony, provided annual data about the founding of new colonies (equivalent to the "births" of new individuals) and the demise of existing colonies (the "deaths" of individuals).

To determine which parent colonies gave rise to which offspring colonies, the researchers collected samples of 20 workers from each of 265 colonies that ranged in age from one to 28 years. They extracted DNA from each ant and used PCR to amplify five short tandem repeat (STR, also called microsatellite) loci (see Section 18.2) in each

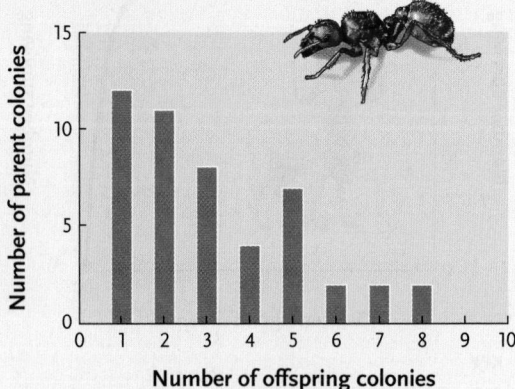

FIGURE A Distribution of number of offspring colonies. Most parent colonies produced three or fewer daughter colonies.

DNA sample. The sizes of the amplified DNA fragments were determined, revealing the genotype at each STR locus for each ant. Because all the sterile workers in a colony are daughters of the same queen (see Section 56.7), the genotype data allowed the researchers to infer the genotype of the queen in each colony. Using a maximum likelihood approach, they then identified which offspring queens were the daughters of which parent queens, and, hence, which offspring colonies were the descendants of which parent colonies.

Organisms use the energy they harvest for three broadly defined functions: maintenance (the preservation of good physiological condition), growth, and reproduction. And when an organism devotes energy to any one of these functions, the balance in its energy budget is reduced, leaving less energy for the other functions.

Life History Patterns Vary Dramatically among Species

A fish, a deciduous tree, and a mammal illustrate the dramatic variations that exist in life history patterns **(Figure 52.7).** Larval coho salmon *(Oncorhynchus kisutch)* (see Figure 52.7A) hatch in the headwaters of a stream, where they feed and grow for about a year before assuming their adult body form. After swimming downstream to the ocean, they remain at sea for a year or two, feeding voraciously and growing rapidly. Eventually, salmon use sun-compass, geomagnetic, and chemical cues to return to the rivers and streams where they hatched. The fishes swim upstream, and each female

lays hundreds or thousands of relatively small eggs. After spending all of their energy reserves on the upstream journey and reproduction, their condition deteriorates and they die.

Most deciduous trees in the temperate zone, such as oaks (genus *Quercus*), begin their lives as seeds in late summer. The seeds remain dormant (that is, metabolically inactive) until the following spring or a later year. After germinating, trees collect nutrients and energy and continue to grow throughout their lives. Once they achieve a critical size, they may produce thousands of seeds annually for many years. Thus, growth and reproduction occur simultaneously through much of the trees' lives (see Figure 52.7B).

European red deer *(Cervus elaphus;* see Figure 52.7C) are born in spring, and young remain with their mothers for an extended period, nursing and growing rapidly. After weaning, they feed on their own. Female red deer begin to breed after reaching adult size in their third year, producing one or two offspring annually until they are about 16 years old, when they reach their maximum lifespan and die.

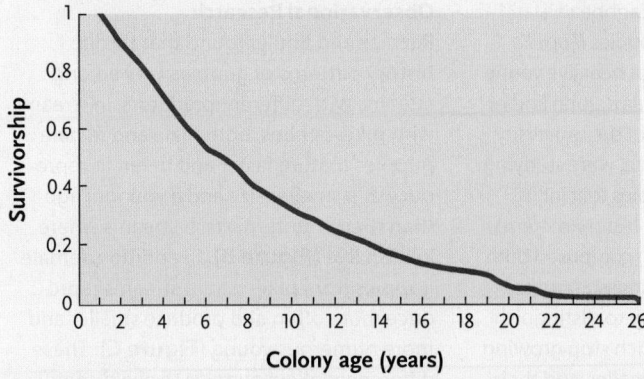
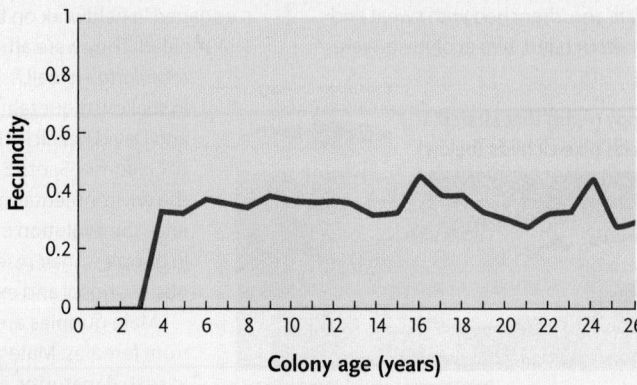

FIGURE B Graphical representation of a life table for red harvester ant colonies. Researchers constructed a survivorship curve (left) and a fecundity curve (right) for colonies of red harvester ants. Note that the vertical axis of the survivorship curve has an arithmetic scale.

Finally, Gordon and her colleagues constructed a life table for the population of red harvester ants, using data on the mortality of colonies to estimate survival probabilities and the founding of new daughter colonies to estimate the fecundity of colonies of different ages.

Results

Their analysis revealed that most parent colonies produce three or fewer daughter colonies **(Figure A).** The life table that the researchers constructed (in graphical format,

see **Figure B**) reveals that only a small proportion of colonies survive for as long as 25 years; about 5% of existing colonies die each year. Moreover, once a colony begins to produce daughter colonies when it is four years old, it continues to produce daughter colonies at a steady rate until the end of its life.

Conclusion

For colonial animals, life tables can be constructed by designating colonies as reproductive individuals and daughter colonies as offspring. In the case of the red harvester ant,

colonies that survive to age four are likely to contribute to the next generation of colonies.

think like a scientist

Given the data on survivorship and fecundity in the graphical life table in Figure B, what is the age of colonies that contribute the most overall to the next generation of offspring colonies?

Source: K. K. Ingram et al. 2013. Colony life history and lifetime reproductive success of red harvester ant colonies. *Journal of Animal Ecology* 82:540–550.

How can we summarize the similarities and differences in the life histories of these organisms? All three species harvest energy throughout their lives. Salmon and deciduous trees continue to grow until old age, whereas deer reach adult size fairly early in life. Salmon produce many offspring in a single reproductive episode, whereas deciduous trees and deer reproduce repeatedly. However, most trees produce thousands of seeds annually, whereas deer produce only one or two young each spring.

What factors have produced these variations in life history patterns? Life history traits—like all population characteristics—are modified by natural selection. Thus, organisms exhibit evolutionary adaptations that increase the fitness of individuals.

A. Coho salmon *(Oncorhynchus kisutch)*

B. English oak *(Quercus robur)*

C. Red deer *(Cervus elaphus)*

FIGURE 52.7 Three organisms with very different life histories.

Basic Research: The Evolution of Life History Traits in Guppies

Some years ago, drenched with sweat and with fishnets in hand, two ecologists were

Male guppy (right) that shared a stream with pike-cichlids (below)

David Reznick/University of California, Riverside

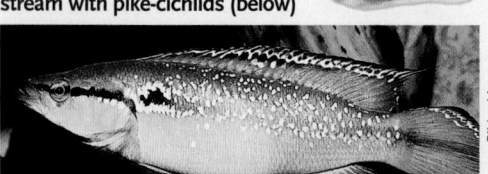

David Reznick/University of California, Riverside

Male guppy (right) that shared a stream with killifish (below)

Hippocampus Bildarchiv

FIGURE A Male guppies from streams where pike-cichlids live (top) are smaller, more streamlined, and have duller colors than those from streams where killifish live (bottom). The pike-cichlid prefers to eat large guppies, and the killifish feeds on small guppies. Guppies are shown approximately life-size; adult pike-cichlids grow to 16 cm in length, and adult killifish grow to 10 cm.

engaged in fieldwork on the Caribbean island of Trinidad. They were after guppies (*Poecilia reticulata*)—small fish that bear live young in shallow mountain streams. John Endler and David Reznick, then of the University of California, Santa Barbara, were studying the environmental variables that influence the evolution of life history patterns in guppies. Their research comprised both observational and experimental studies.

Male guppies are easy to distinguish from females. Males, which stop growing at sexual maturity, are smaller, and their scales have bright colors that serve as visual signals in intricate courtship displays. The drably colored females continue to grow larger throughout their lives.

In the mountains of Trinidad, guppies living in different streams— and even in different parts of the same stream—are eaten by one of two other fish species **(Figure A)**. In some streams, a large pike-cichlid (*Crenicichla alta*) prefers mature guppies and tends not to spend time hunting small, immature ones. In other streams, a small killifish (*Rivulus hartii*) preys on immature guppies but does not have much success with the larger adults.

Observational Research

Reznick and Endler found that the life history patterns of guppies vary among streams with different predators. In streams with pike-cichlids, both male and female guppies mature faster and begin to reproduce at a smaller size and a younger age than their counterparts in streams where killifish live **(Figure B)**. In addition, female guppies from pike-cichlid streams reproduce more often and produce smaller and more numerous young **(Figure C).** These differences allow guppies to avoid some predation. Those in pike-cichlid streams

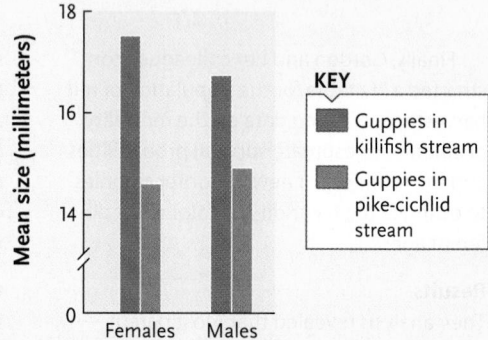

KEY
- Guppies in killifish stream
- Guppies in pike-cichlid stream

FIGURE B Guppies in streams occupied by pike-cichlids are smaller than those in streams occupied by killifish.

Each species' life history is, in fact, a highly integrated "strategy"— not in the human sense of planning ahead, but as a suite of selection-driven adaptations.

Ecologists Analyze the Individual Components of Life Histories

In analyzing life histories, ecologists often compare the number of offspring produced with the amount of care provided to each. They also consider the number of reproductive episodes in the organism's lifetime, and the age at which it first reproduces. Because these characteristics evolve together, a change in one trait is likely to influence the success of the others.

FECUNDITY VERSUS PARENTAL CARE If a female has a fixed amount of energy for reproduction, she can package that energy in various ways. By way of illustration, a female duck with 1,000 units of energy for reproduction might lay 10 eggs that each contain 100 units of energy. A salmon, which has higher fecundity, might lay 1,000 eggs, each endowed with 1 unit of energy. The

amount of energy invested in each offspring *before* it is born represents the **passive parental care** that the female provides. Passive parental care is provided through yolk in an egg, endosperm in a seed, or, in mammals, nutrients that cross the placenta.

Many animals, especially birds and mammals, also provide **active parental care** to offspring *after* their birth. In general, species that produce many offspring in a reproductive episode—such as the coho salmon—provide relatively little active parental care *to each offspring*. In fact, female coho salmon, which produce 2,400 to 4,500 eggs, die before their eggs even hatch. Conversely, species that produce only a few offspring at a time—such as the European red deer—provide a lot of care to each. A red deer doe nurses its single fawn for up to 8 months before weaning it.

ONE REPRODUCTIVE EPISODE VERSUS SEVERAL A second life history characteristic adjusted by natural selection is the number of reproductive episodes in an organism's lifetime. Some organisms, such as the coho salmon, devote all of their stored energy to a single reproductive event, a reproductive

begin to reproduce when they are smaller than the size preferred by that predator. And those from killifish streams grow quickly to a size that is too large to be consumed by killifish. Reznick and Endler hypothesized that the differences in guppy life history traits was fostered by their exposure to different predators.

Experimental Research
Although these life history differences were correlated with the distributions of the two predatory fishes, they might result from some other, unknown differences between the streams. Endler and Reznick investigated this possibility with controlled laboratory experiments. They shipped groups of guppies to California, where they bred guppies from each kind of stream for two generations. Both experimental populations were raised under identical conditions in the absence of predation. Even when predators were absent, the two experimental populations retained their life history differences. These results demonstrated that the observed life history differences had a heritable genetic basis; thus, different patterns of natural selection in the streams occupied by different predators could influence these characteristics in the guppies.

Endler and Reznick also conducted experiments to examine the role of predation in the *evolution* of the guppy size differences directly. They raised guppies for many generations in the laboratory under three experimental conditions—some alone, some with killifish, and some with pike-cichlids. As predicted, the guppy lineage that was subjected to predation by killifish became larger at maturity. Individuals that were small at maturity were frequently eaten, and their reproduction was limited. The lineage that was raised with pike-cichlids showed a trend toward earlier maturity. Individuals that matured at a larger size faced a greater likelihood of being eaten before they had reproduced.

Finally, when they first visited Trinidad, Endler and Reznick had introduced guppies from a pike-cichlid stream into another stream that contained killifish but no pike-cichlids or guppies. Eleven years later, the introduced guppy population had changed. As the researchers predicted, the guppies had become larger in size and reproduced more slowly, characteristics that are typical of natural guppy populations that live and die with killifish.

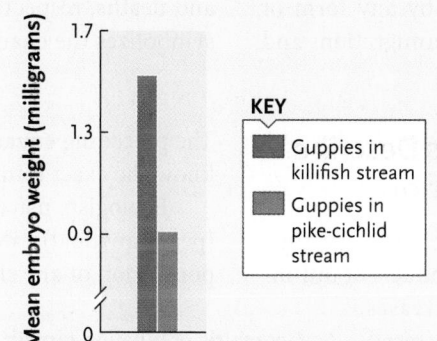

FIGURE C Female guppies from streams occupied by pike-cichlids reproduce more often (shorter time between broods) and produce more young per brood and smaller young (lower embryo weight) than females living in streams occupied by killifish.

KEY
■ Guppies in killifish stream
■ Guppies in pike-cichlid stream

pattern called **semelparity** (*semel* = once; *parus* = giving birth). Any adult that survives the upstream migration is likely to leave some surviving offspring. Other species, such as deciduous trees and red deer, reproduce multiple times, a reproductive pattern called **iteroparity** (*iterum* = again). In contrast to salmon, individuals of these species devote only some of their energy budget to reproduction at any time, with the balance allocated to maintenance and growth. Moreover, in some plants, invertebrates, fishes, and reptiles, larger individuals produce more offspring than small ones do. Thus, one advantage of iteroparity is that by using only part of the energy budget for reproduction, continued growth may result in greater fecundity at a later age. However, if an organism does not survive until the next breeding season, the potential advantage of putting energy into maintenance and growth is lost.

EARLY REPRODUCTION VERSUS LATE REPRODUCTION Individuals that first reproduce at the earliest possible age may stand a good chance of leaving some surviving offspring. But the energy devoted to reproduction is no longer available for maintenance and growth. Thus, early reproducers may be smaller and less healthy than individuals that delay reproduction in favor of these other functions. Conversely, an individual that delays reproduction may increase its chance of survival and its future fecundity by becoming larger or more experienced. But there is always some chance that it will die before the next breeding season, leaving no offspring at all. Thus, a finite energy budget and the risk of mortality establish a trade-off in the timing of first reproduction. Mathematical models suggest that delayed reproduction will be favored by natural selection if a sexually mature individual has a good chance of surviving to an older age, if organisms grow larger as they age, and if larger organisms have higher fecundity. Early reproduction will be favored if adult survival rates are low, if animals don't grow larger as they age, or if larger size does not increase fecundity.

Life history characteristics not only vary from one species to another, but they also vary among populations of a single species. *Focus on Research: Basic Research* describes how predation influences life history characteristics in natural populations of guppies (*Poecilia reticulata*) in Trinidad.

52.4 Models of Population Growth

We now examine mathematical models of population growth that describe very different responses to changes in a population's density. *Geometric* and *exponential* models apply when populations experience unlimited growth. The *logistic* model applies when population growth is limited, often because available resources are finite. These simple models are tools that help ecologists refine their hypotheses, but they do not provide entirely accurate predictions of population growth in nature. In the simplest versions of these models, ecologists define births as the production of offspring by any form of reproduction, and ignore the effects of immigration and emigration.

Geometric and Exponential Models Describe Population Growth without Limitation

Sometimes populations increase in size for a period of time with no apparent limits on their growth. In models of unlimited population growth, population size increases steadily by a constant ratio **(Figure 52.8)**. Bacterial populations provide the most obvious examples, but multicellular organisms also sometimes exhibit exponential population growth.

GEOMETRIC POPULATION GROWTH Bacteria reproduce by binary fission: a parent cell divides in half, producing two daughter cells, which each divide to produce two granddaughter cells. Generation time in a bacterial population is simply the time between successive cell divisions. And if no bacteria in the population die, the population doubles in size each generation. This pattern of population growth is described as "geometric," because the population grows by a constant multiplier every generation; in our example of binary fission in bacteria, the multiplier is 2.

Bacterial populations grow quickly under ideal temperatures and with unlimited space and food. Consider a laboratory population of the human intestinal bacterium *Escherichia coli,* for which the generation time can be as short as 20 minutes. If we start with a population of one bacterium, the population doubles to two cells after one generation, to four cells after two generations, and to eight cells after three generations (Figure 52.8A). After only eight hours, or 24 generations, the population will number more than 16 million. And after a single day, or 72 generations, the population will number nearly 5×10^{21} cells. Although other bacteria grow more slowly than *E. coli,* it is no wonder that pathogenic bacteria, such as those causing cholera or plague, can quickly overtake the defenses of an infected animal.

EXPONENTIAL POPULATION GROWTH IN OTHER ORGANISMS By contrast to bacteria, many plants and animals live side-by-side with their offspring. In these populations, births increase a population's size and deaths decrease it. Over a given time period:

Change in population size =
Number of births − Number of deaths

We express this relationship mathematically by defining N as the population size; ΔN (pronounced "delta N") as the change in population size; Δt as the time period during which the change occurs; and B and D as the absolute numbers of births and deaths, respectively, *during that time period.* Thus, $\Delta N/\Delta t$ symbolizes the change in population size over time, and

$$\Delta N/\Delta t = B - D$$

The preceding equation applies to any population for which we know the exact numbers of births and deaths.

Ecologists usually express births and deaths as *per capita* (per individual) rates, allowing them to apply the model to a population of any size. The per capita birth rate, symbolized b,

A. Geometric population growth

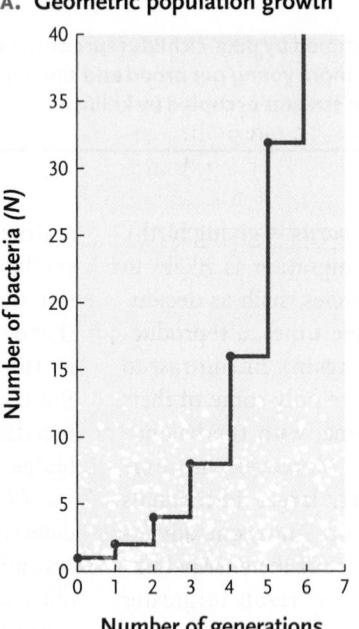

Number of bacteria (N)

Number of generations

B. Exponential population growth

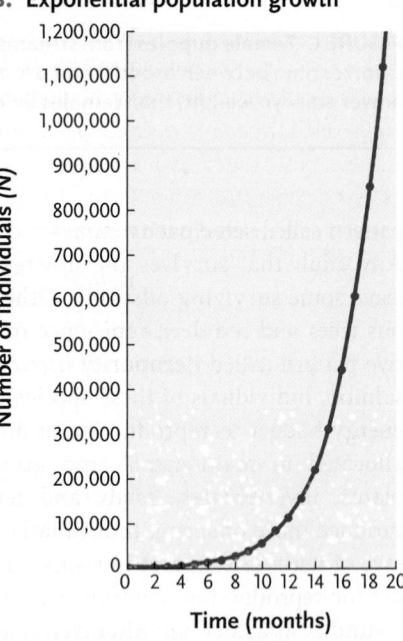

Number of individuals (N)

Time (months)

FIGURE 52.8 Changes in population size predicted by two models of unlimited population growth. (A) If all members of a bacterial population divide simultaneously, a plot of population size over time forms a stair-stepped curve in which the steps get larger as the number of dividing cells increases. **(B)** Exponential population growth produces a J-shaped curve of population size plotted against time. Although the per capita growth rate *(r)* remains constant, the increase in population size gets larger every month because more individuals are reproducing.
© Cengage Learning 2017

is simply the number of births in the population during the specified time period divided by the population size: $b = (B/N)$. Similarly, the per capita death rate, d, is the number of deaths divided by the population size: $d = (D/N)$. If, for example, in a population of 2,000 field mice, 1,000 mice are born and 200 mice die during one month, then $b = 1,000/2,000 = 0.5$ births per individual per month, and $d = 200/2,000 = 0.1$ deaths per individual per month. Of course, no mouse can give birth to half an offspring, and no individual can die one tenth of a death. But these rates tell us the per capita birth and death rates *averaged over all mice in the population*. Per capita birth and death rates are always expressed over a specified time period. For long-lived organisms, such as humans, time is measured in years; for short-lived organisms, such as fruit flies, time is measured in days. We can calculate per capita birth and death rates from data in a life table.

We can now revise the population growth equation to use per capita birth and death rates instead of the actual numbers of births and deaths. The change in a population's size during a given time period ($\Delta N/\Delta t$) depends on the per capita birth and death rates, as well as on the number of individuals in the population. Mathematically, we can write

$$\Delta N/\Delta t = B - D = bN - dN = (b - d)N$$

or, in the notation of calculus,

$$dN/dt = (b - d)N$$

This equation describes the **exponential model of population growth.** (Note that in calculus, dN/dt is the notation for the population growth rate; the "d" in dN/dt is *not* the same "d" that we use to symbolize the per capita death rate.)

The difference between the per capita birth rate and the per capita death rate, $b - d$, is the **per capita growth rate** of the population, symbolized by r. Like b and d, r is always expressed per individual per unit time. Using the per capita growth rate, r, in place of $(b - d)$, the exponential growth equation is written

$$dN/dt = rN$$

If the birth rate exceeds the death rate, r has a positive value ($r > 0$), and the population is growing. In our example with field mice, $r = 0.5 - 0.1 = 0.4$ mice per mouse per month. If the birth rate is lower than the death rate, however, r has a negative value ($r < 0$), and the population is getting smaller. In populations where the birth rate equals the death rate, $r = 0$, and the population's size is not changing—a situation known as **zero population growth,** or ZPG. Even under conditions of ZPG, births and deaths still occur, but the numbers of births and deaths cancel each other out.

As long as a population's per capita growth rate is positive ($r > 0$), the population will increase in size. In our hypothetical population of field mice, we started with $N = 2,000$ mice, and calculated a per capita growth rate of 0.4 mice per individual per month. In the first month, the population grows by $0.4 \times 2,000 = 800$ mice. At the start of the second month, $N = 2,800$ and r still $= 0.4$. Thus, in the second month, the

population grows by $0.4 \times 2,800 = 1,120$ mice. Notice that even though r remains constant, the *increase* in population size gets larger each month simply because more individuals are reproducing. In less than two years, the mouse population will grow to more than 1 million. A graph of exponential population growth has a characteristic J shape, getting steeper through time (Figure 52.8B). The population grows at an ever-increasing pace because the change in a population's size depends on the number of individuals in the population as well as its per capita growth rate.

POPULATION GROWTH UNDER IDEAL CONDITIONS Imagine a hypothetical population living in an ideal environment—one with unlimited food and shelter, no predators, parasites, or disease, and a comfortable abiotic environment. Under such circumstances, which are admittedly unrealistic, the per capita birth rate is very high, the per capita death rate is very low, and the per capita growth rate, r, is as high as it can possibly be. This maximum per capita growth rate, symbolized r_{max}, is the population's **intrinsic rate of increase.** Under these ideal conditions, the exponential growth equation is

$$dN/dt = r_{max}N$$

When populations are growing at their intrinsic rate of increase, population size increases very rapidly. Across a wide variety of bacteria, protists, and animals, r_{max} varies inversely with generation time: species with short generation time (that is, those that mature quickly) have higher intrinsic rates of increase than those with long generation time **(Figure 52.9).**

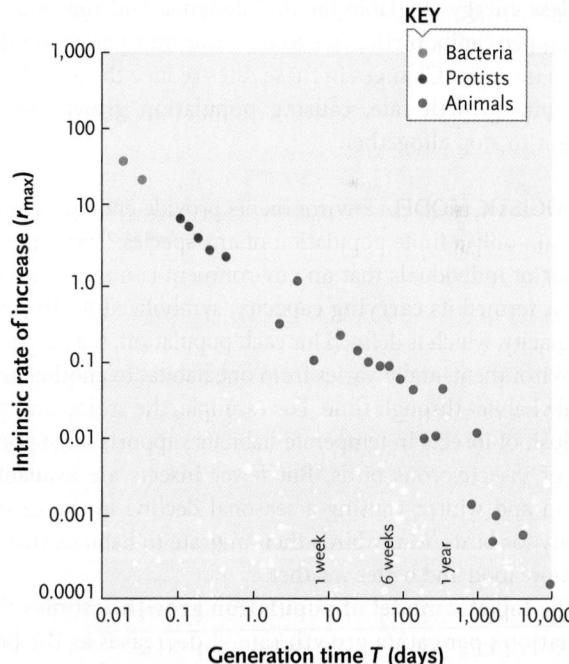

FIGURE 52.9 Generation time and r_{max}. The intrinsic rate of increase (r_{max}) is high for bacteria, protists, and animals with short generation time and low for those with long generation time. The logarithmic scale on both axes compresses the data into a straight line.
© Cengage Learning 2017

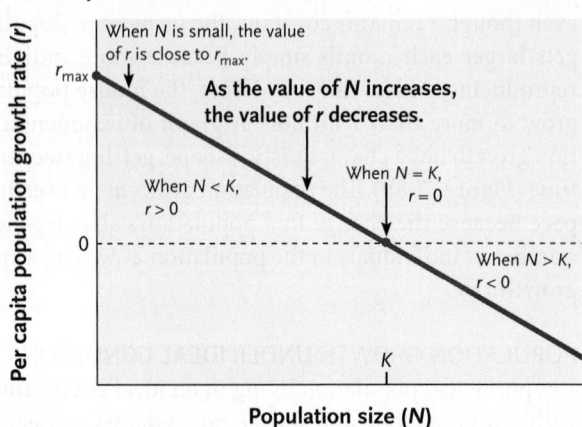

A. The predicted effect of *N* on *r*

Per capita population growth rate (*r*)

r_{max}

When *N* is small, the value of *r* is close to r_{max}.

As the value of *N* increases, the value of *r* decreases.

When *N* < *K*, *r* > 0

When *N* = *K*, *r* = 0

0

When *N* > *K*, *r* < 0

K

Population size (*N*)

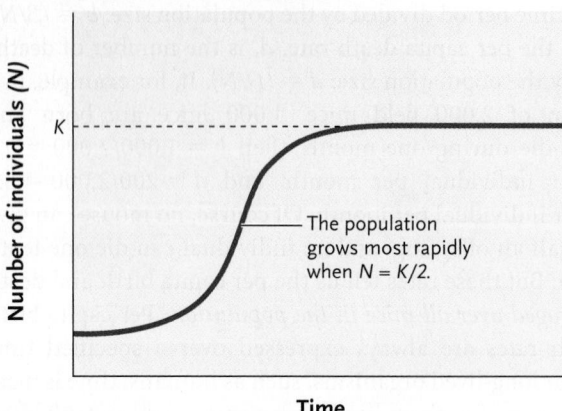

B. Population size through time

Number of individuals (*N*)

K

The population grows most rapidly when *N* = *K*/2.

Time

FIGURE 52.10 The logistic model of population growth. (A) The logistic model assumes that the per capita population growth rate *r* decreases linearly as population size *(N)* increases. **(B)** The logistic model predicts that population size increases quickly at first, but then slowly approaches the carrying capacity *(K)*.

The Logistic Model Describes Population Growth When Resources Are Limited

The geometric and exponential models predict unlimited population growth. But we know from even casual observations that the population sizes of most species are somehow limited—we are not knee-deep in bacteria, rosebushes, or garter snakes. What factors limit the growth of populations? As a population gets larger, it uses more vital resources, and a shortage of resources may eventually develop. As a result, individuals may have less energy available for maintenance and reproduction, causing per capita birth rates to decrease and per capita death rates to increase. Changes in these rates reduce the population's per capita growth rate, causing population growth to slow down or to stop altogether.

THE LOGISTIC MODEL Environments provide enough resources to sustain only a finite population of any species. The maximum number of individuals that an environment can support indefinitely is termed its **carrying capacity,** symbolized *K*. The carrying capacity, which is defined for each population, is a property of the environment, and it varies from one habitat to another and in a single habitat through time. For example, the spring and summer flush of insects in temperate habitats supports large populations of insectivorous birds. But fewer insects are available in autumn and winter, causing a seasonal decline in the carrying capacity for birds. Many birds then migrate to habitats that provide more food and better weather.

The **logistic model of population growth** assumes that a population's per capita growth rate, *r*, decreases as the population gets larger **(Figure 52.10A)**. In other words, population growth slows as the population size approaches the carrying capacity. The mathematical expression (*K* − *N*) tells us how many individuals can be added to a population before it reaches carrying capacity. And the expression (*K* − *N*)/*K*

indicates what *percentage* of the carrying capacity is still available.

To create the logistic model, we factor the impact of carrying capacity into the exponential model by letting $r = r_{max}(K - N)/K$. This calculation reduces the per capita growth rate *r* from its maximum value (r_{max}) as *N* increases:

$$dN/dt = r_{max}N(K - N)/K$$

The calculation of how *r* varies with population size is straightforward **(Table 52.2)**. In a very small population (*N* is much smaller than *K*), plenty of resources are still available; the value of (*K* − *N*)/*K* is close to 1, and the per capita growth rate *r* is therefore close to the maximum possible (r_{max}). Under these conditions, population growth is close to exponential. If a population is large (*N* is close to *K*), few additional resources are available, the value of (*K* − *N*)/*K* is small, and the per capita growth rate *r* is very low. When the size of the population exactly equals the carrying capacity, (*K* − *N*)/*K* becomes zero, and so does the population growth rate—the situation defined as ZPG.

The logistic model of population growth predicts an S-shaped graph of population size over time, with the population slowly approaching its carrying capacity and remaining at that level **(Figure 52.10B)**. According to this model, the population grows slowly when the population size is small, because there are few individuals reproducing. It also grows slowly when the population size is large because the per capita population growth rate is low. The population grows quickly (*dN/dt* is highest) at intermediate population sizes, when a sizable number of individuals are breeding and the per capita population growth rate *r* is still fairly high (see Table 52.2).

INTRASPECIFIC COMPETITION The logistic model assumes that vital resources become increasingly limited as a population grows larger. Thus, the model is a mathematical portrait of

TABLE 52.2	The Effect of N on r and ΔN in a Hypothetical Population Exhibiting Logistic Growth in Which $K = 2{,}000$ and $r_{max} = 0.04$ per Capita per Year			
N (population size)	$(K - N)/K$ (% of K available)	$r = r_{max}(K - N)/K$ (per capita growth rate)	$\Delta N^* = rN$ (change in N)
50	0.990	0.0396	2
100	0.950	0.0380	4
250	0.875	0.0350	9
500	0.750	0.0300	15
750	0.625	0.0250	19
1,000	0.500	0.0200	20
1,250	0.375	0.0150	19
1,500	0.250	0.0100	15
1,750	0.125	0.0050	9
1,900	0.050	0.0020	4
1,950	0.025	0.0010	2
2,000	0.000	0.0000	0

*ΔN rounded to the nearest whole number.

intraspecific (within species) **competition,** the dependence of two or more individuals in a population on the same limiting resource. For animals, limiting resources can be food, water, nesting sites, refuges from predators, and, for sessile species (those permanently attached to a surface), space. For plants, sunlight, water, inorganic nutrients, and growing space can be limiting. The pattern of uniform dispersion described earlier often reflects intraspecific competition for limited resources.

In some very dense populations, the accumulation of poisonous waste products may also reduce survivorship and reproduction. Most natural populations live in open systems where wastes are consumed by other organisms or flushed away. But the buildup of toxic wastes is common in laboratory cultures of microorganisms. For example, yeast cells ferment sugar and produce ethanol as a waste product. Thus, the alcohol content of wine rarely exceeds 14% by volume, the ethanol concentration that poisons winemaking yeasts.

LOGISTIC GROWTH IN THE LABORATORY AND IN NATURE

The logistic model of population growth is a hypothesis about the effects of resource limitation. Under some circumstances, populations conform very well to the predictions of the logistic model **(Figure 52.11)**. In simple laboratory cultures, relatively small organisms, such as *Paramecium,* some crustaceans, and flour beetles, often show an S-shaped pattern of population growth (see Figure 52.11A). Moreover, large animals that have been introduced into new environments sometimes exhibit a pattern of population growth that matches the predictions of the logistic model (see Figure 52.11B).

Nevertheless, some assumptions of the logistic model are unrealistic. For example, the model predicts that survivorship and fecundity respond immediately to changes in a population's density. But many organisms exhibit a delayed response, called a **time lag.** Some time lags occur because fecundity is usually determined by the availability of resources at some time in the past, when individuals were adding yolk to eggs or endosperm to seeds. Moreover, when food resources become scarce, individuals may

A. Rhizopertha dominica

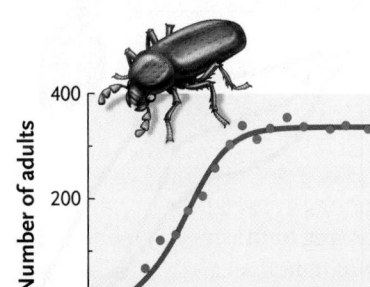

A laboratory population of the grain borer beetle showed logistic growth when its food was replenished weekly.

B. Ovis musiman

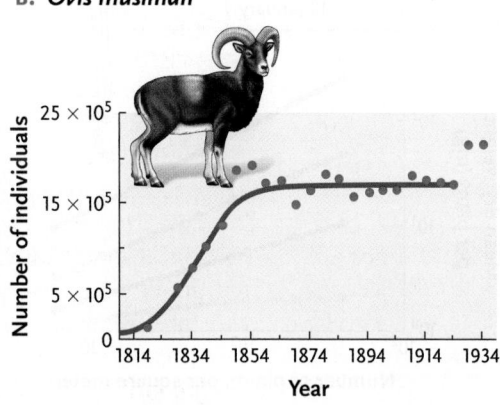

European mouflon sheep introduced into Tasmania exhibited logistic population growth; these data represent 5-year averages, which smooth out annual fluctuations in population size.

C. Daphnia magna

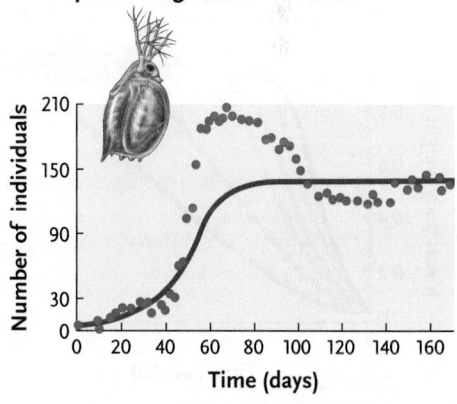

A laboratory population of the water flea overshot its carrying capacity; when population density increased, individuals relied upon stored energy reserves, causing a time lag in the appearance of density-dependent effects.

FIGURE 52.11 Examples of logistic population growth.
© Cengage Learning 2017

KEY

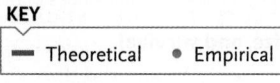

use stored energy reserves to survive and reproduce, and the effects of crowding may not be felt until those reserves are depleted. As a result, the population size may temporarily overshoot its carrying capacity (see Figure 52.11C). Deaths may then outnumber births, causing the population size to drop below the carrying capacity, at least temporarily. Time lags can cause a population to oscillate around its carrying capacity.

Another unrealistic assumption of the logistic model is that the addition of new individuals to a population always decreases survivorship and fecundity, no matter how small the population is. But in small populations, modest population growth probably doesn't have much effect on these processes. In fact, most organisms probably require a minimum population density to survive and reproduce. For example, some plants flourish in small clumps that buffer them from physical stresses, such as strong wind, whereas a single individual living in the open would suffer adverse effects. And in some animal populations, a minimum population density is necessary for individuals to find mates—an important issue in conservation biology (see Chapter 55).

STUDY BREAK 52.4

1. How does the prediction of the exponential model of population growth differ from that of the logistic model?
2. What is carrying capacity? Is it a property of a habitat or of a population?
3. What is a time lag?

52.5 Population Dynamics

Long-term studies on many species have shown that the size, density, age structure, and geographical ranges of most populations change somewhat from season to season and from year to year. As you have seen, some populations experience dramatic variations in size and other characteristics through time, whereas other populations appear to be much more stable. What environmental factors influence these aspects of population dynamics? Why do these characteristics fluctuate more in some populations than in others?

Density-Dependent Factors Often Regulate Population Size

Many factors that affect population dynamics are **density-dependent**: their influence increases or decreases with the density of the population. Examples of density-dependent environmental factors include intraspecific competition and predation. The logistic model includes the effects of density-dependence in its assumption that per capita birth and death rates change with a population's density.

THE EFFECTS OF CROWDING Numerous laboratory and field studies show that crowding (high population density) decreases the individual growth rate, adult size, and survival of plants and animals **(Figure 52.12)**. Organisms living in extremely dense populations are unable to harvest enough resources; they grow

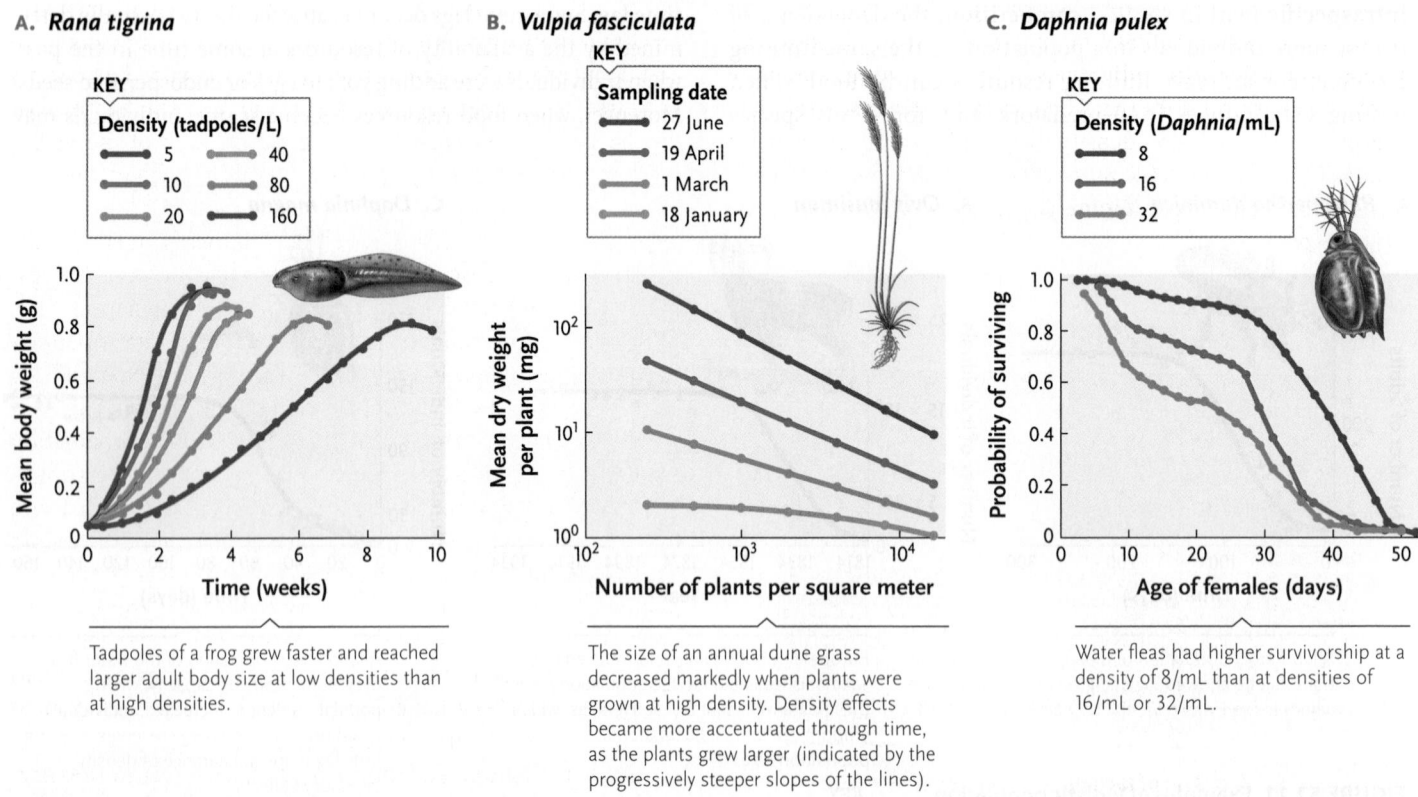

A. *Rana tigrina*

B. *Vulpia fasciculata*

C. *Daphnia pulex*

Tadpoles of a frog grew faster and reached larger adult body size at low densities than at high densities.

The size of an annual dune grass decreased markedly when plants were grown at high density. Density effects became more accentuated through time, as the plants grew larger (indicated by the progressively steeper slopes of the lines).

Water fleas had higher survivorship at a density of 8/mL than at densities of 16/mL or 32/mL.

FIGURE 52.12 Effects of crowding on individual growth, size, and survival.
© Cengage Learning 2017

A. *Capsella bursa-pastoris*

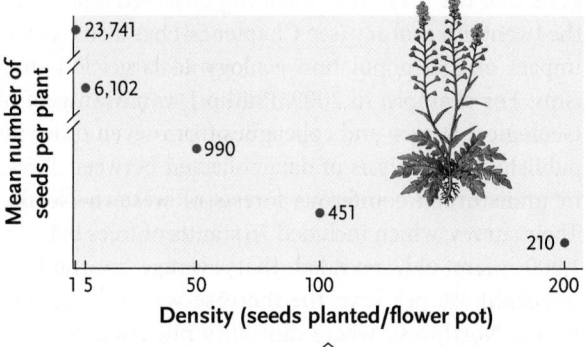

The number of seeds produced by shepherd's purse decreased dramatically with increasing density in experimental pots.

B. *Parus major*

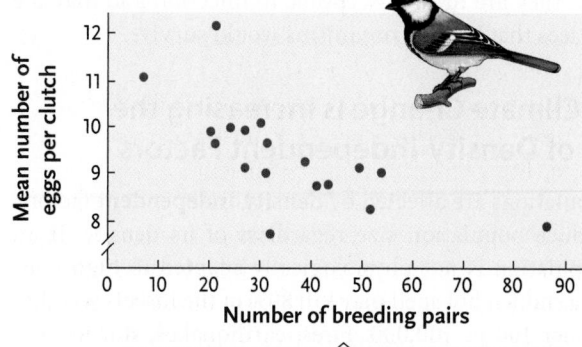

The mean number of eggs produced by great tits, a woodland bird, declined as the number of breeding pairs in Marley Wood increased.

FIGURE 52.13 Effects of crowding on fecundity.
© Cengage Learning 2017

slowly and tend to be small, weak, and less likely to survive. Gardeners understand this relationship, thinning their plants to a density that maximizes the number of vigorous individuals.

Crowding also has a negative effect on reproduction **(Figure 52.13)**. When resources are in short supply, each individual has less energy available for reproduction after meeting its basic maintenance needs. Hence, females in crowded populations produce either fewer offspring or smaller offspring that are less likely to survive.

In some species, crowding stimulates developmental and behavioral changes that may influence the density of a population. For example, migratory locusts *(Locusta migratoria)* can develop into either solitary or migratory forms in the same population. Migratory individuals have longer wings and more body fat, characteristics that allow them to disperse great distances. High population density increases the frequency of the migratory form, and huge numbers of locusts move away from the area of high density **(Figure 52.14)**, reducing the size of the original population.

FIGURE 52.14 A swarm of locusts. Migratory locusts *(Locusta migratoria)*, moving across an African landscape, can devour their own weight in plant material every day.

These studies confirm the assumptions of the logistic equation, but they do not prove that natural populations are regulated by density-dependent factors. A convincing demonstration requires experimental evidence that an increase in population density causes population size to decrease, and that a decrease in density causes it to increase. In one study conducted in the 1960s, Robert Eisenberg of the University of Michigan experimentally increased the numbers of aquatic snails in some ponds, decreased them in others, and maintained natural densities in control ponds. Although adult survivorship did not differ between experimental and control treatments, snails in the high-density ponds produced fewer eggs, and those in the low-density ponds produced more eggs than those living at the control density. In addition, the survival rates of young snails declined as density increased. After four months, the densities in the two experimental groups converged on those in the control, providing strong evidence of density-dependent population regulation.

DENSITY-DEPENDENT INTERACTIONS Our discussion of the logistic equation described *intraspecific competition* as the primary density-dependent factor regulating population size. *Interspecific competition* (that is, competition between populations of different species) also exerts density-dependent effects on population growth, a topic we consider in Section 53.1.

Predation can also cause density-dependent population regulation. As a particular prey species becomes more numerous, predators may consume more of it because it is easier to find and catch. Once a prey species has exceeded a threshold density, predators may consume a larger *percentage* of the prey population, which is a density-dependent effect. For example, on rocky shores in California, sea stars concentrate their feeding on the most abundant of several invertebrate species. When one prey species becomes common, predators feed on it disproportionately, drastically reducing its numbers.

Like predation, *parasitism* and *disease* cause density-dependent regulation of plant and animal populations. Infectious microorganisms spread quickly in a crowded population.

In addition, if individuals crowded together are weak or malnourished, they are more susceptible to infection and may die from diseases that healthy organisms would survive.

Global Climate Change Is Increasing the Impact of Density-Independent Factors

Some populations are affected by **density-independent** factors, which reduce population size regardless of its density. If an insect population is not physiologically adapted to high temperature, a sudden hot spell may kill 80% of the insects whether they number 100 or 100,000. Fires, earthquakes, storms, and other natural disturbances may contribute directly or indirectly to density-independent mortality. But because such factors do not cause a population to fluctuate around its carrying capacity, density-independent factors do not *regulate* population size, although they may reduce it.

For many years, ecologists have recognized the strong effects of density-independent factors on populations of small-bodied species that cannot buffer themselves against environmental change. Their populations grow exponentially for a time, but shifts in climate or random events cause high mortality before populations reach a size at which density-dependent factors regulate their numbers. When conditions improve, populations grow exponentially—at least until another density-independent factor causes them to crash again. For example, a small Australian insect, *Thrips imaginis,* feeds on pollen and flowers of plants in the rose family; they are frequently abundant enough to damage the blooms. *Thrips* populations grow exponentially in spring, when many flowers are available and the weather is warm and moist **(Figure 52.15)**. But populations crash predictably during summer because *Thrips* do not tolerate extremely hot and dry conditions. After the crash, a few individuals survive in remaining flowers, forming the stock from which the population grows exponentially the following spring.

Recent research on the effects of global climate change suggests that the very rapid warming observed since the middle of the twentieth century (see Chapter 54) has also had a significant impact on the population ecology of large, long-lived organisms. For example, in 2009, Phillip J. van Mantgem of the U.S. Geological Survey and colleagues from seven other institutions published an analysis of data collected between 1955 and 2007 in undisturbed coniferous forests of western North America. Their survey, which included 76 stands of trees between 200 and 1,000 years old, revealed that average mortality rates had increased 4% per year; the increase was most apparent in the Pacific Northwest, where mortality rates doubled in as little as 17 years **(Figure 52.16)**. Increased mortality was evident at all elevations and in trees of all species, sizes, and ages. After considering several possible causes of the increased mortality, the researchers concluded that climate warming—about 0.5°C per decade in western North America—and the concomitant lengthening of summer droughts was the major contributing factor. The extended droughts not only prevent tree seedlings from establishing themselves and renewing the forests, but they also limit the photosynthetic activity of established trees and leave them more vulnerable to infestation by insect and fungal pests. These environmental stresses may lead to the large-scale death of western North American forests.

Interacting Environmental Factors Influence Population Dynamics in Complex Ways

Sometimes several density-dependent factors influence a population at the same time. For example, on small islands in the West Indies, the spider *Metepeira datona* is rare wherever lizards (*Ameiva festiva, Anolis carolinensis,* and *Anolis sagrei*)

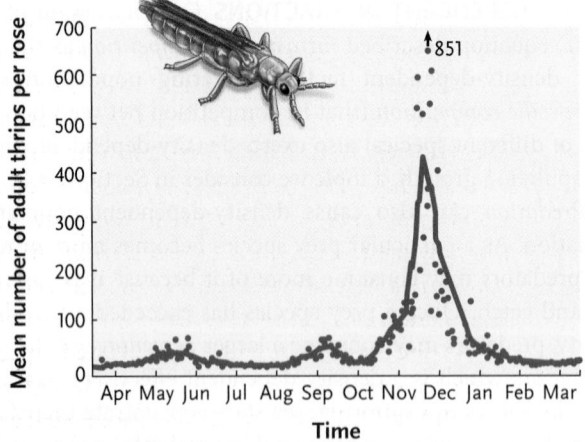

FIGURE 52.15 Booms and busts in a *Thrips* population. Populations of the Australian insect *Thrips imaginis* grow exponentially when conditions are favorable during spring (which begins in September in the southern hemisphere). The populations crash in summer, however, when hot and dry conditions cause high mortality rates.

© Cengage Learning 2017

FIGURE 52.16 Global climate change and mortality in trees. Like other conifers in western North America, red fir trees (*Abies magnifica*) in Sequoia National Park, California, have experienced increased mortality, almost certainly as a result of climate warming.

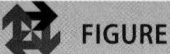

FIGURE 52.17 | Experimental Research

Evaluating Density-Dependent Interactions between Species

Question: Does the population density of lizards on Caribbean islands have any effect on the population density of spiders?

Experiment: Spiller and Schoener built fences to enclose a series of study plots on a small island in the Bahamas. They excluded all individuals of three lizard species from the experimental enclosures, but left resident lizards undisturbed in the control enclosures. They then made monthly measurements of population densities of the web-building spider *Metepeira datona* in both experimental enclosures and control enclosures.

Results: Over the 20-month course of the experiment, spider densities were as much as five times higher in the experimental enclosures than in the control enclosures.

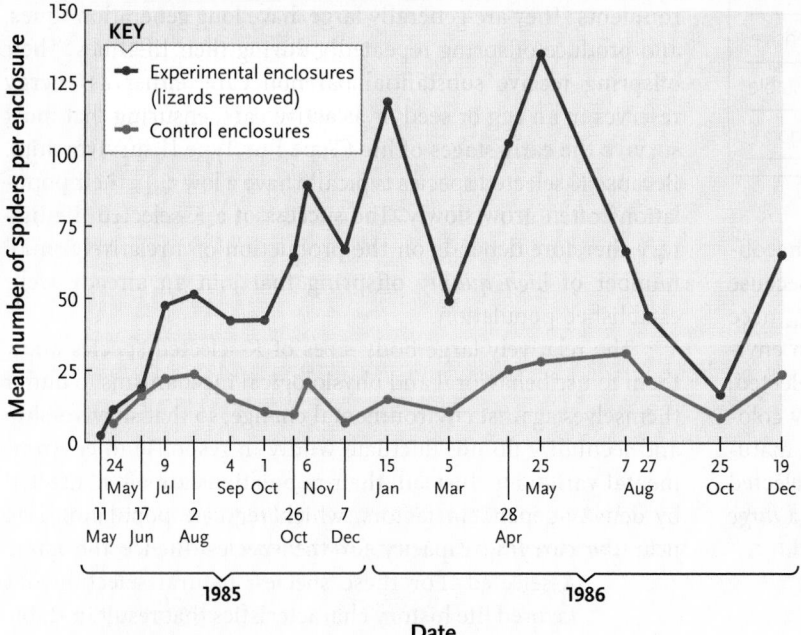

Conclusion: Spiller and Schoener concluded that the presence of lizards has a large impact on spider populations. The lizards not only compete with the spiders for insect food, but they also appear to prey on the spiders.

Observe Hypothesize Predict Experiment Interpret

If you look closely at the graph, you will notice that the numbers of spiders in control enclosures and in enclosures with lizards removed sometimes increase or decrease at roughly the same time. What factors might account for these parallel changes in spider numbers?

are abundant, but common where the lizards are rare or absent. To test whether the presence of lizards limits the abundance of spiders, David Spiller and Tom Schoener of the University of California, Davis, built screened fences around plots on islands where these species occur. They eliminated lizards from experimental plots, but left them in control plots. After two years, spider populations in some experimental

plots were five times denser than those in control plots **(Figure 52.17)**. In this case, lizards had two density-dependent effects on spider populations: they preyed upon spiders, and they competed with them for the insects that both eat.

Density-dependent factors can also interact with density-independent factors, limiting population growth. For example, food shortage caused by high population density (a density-dependent factor) may lead to malnourishment; in turn, malnourished individuals may be more likely to succumb to the stress of extreme weather (a density-independent factor).

Populations can also be affected by density-independent factors in a density-dependent manner. For example, animals often retreat into shelters to escape environmental stresses, such as floods or severe heat. If a population is small, most individuals can fit into a limited number of available refuges. But if a population is large, only a small proportion will find suitable shelter; the larger the population is, the greater the percentage of individuals that will experience the stress. For example, although the density-independent effects of weather limit *Thrips* populations, it is the availability of flowers in summer—clearly a density-dependent factor—that regulates the size of the *Thrips* stock from which the population grows the following spring. Hence, both types of factors influence the size of *Thrips* populations.

Life History Characteristics Govern Fluctuations in Population Size through Time

Even casual observation reveals tremendous variation in how rapidly population size changes in different species. For example, new weeds often appear in a vegetable garden overnight, whereas the number of oak trees in a forest may remain relatively stable for years. Why do some species have the potential for explosive population growth, but others do not? The answer lies in how natural selection has molded life history strategies that are adapted to different ecological conditions. Ecologists describe two divergent life history patterns—***r*-selected** species and ***K*-selected** species—with very different characteristics **(Table 52.3, Figure 52.18)**. These strategies represent extremes on a continuum of possible patterns, and the life histories of most species actually fall somewhere between them.

Species with an *r*-selected life history are adapted to function well in rapidly changing environments. They are generally small, have short generation times, and produce numerous,

TABLE 52.3	Characteristics of *r*-Selected and *K*-Selected Species	
Characteristic	*r*-Selected Species	*K*-Selected Species
Maturation time	Short	Long
Lifespan	Short	Long
Mortality rate	Usually high	Usually low
Reproductive episodes	Usually one	Usually several
Time of first reproduction	Early	Late
Clutch or brood size	Usually large	Usually small
Size of offspring	Small	Large
Active parental care	Little or none	Often extensive
Population size	Fluctuating	Relatively stable
Tolerance of environmental change	Generally poor	Generally good

tiny offspring, often in a single reproductive event. The offspring receive little or no parental care of any kind. Because species with short generation times tend to have high r_{max} (see Figure 52.9), their populations grow exponentially when environmental conditions are favorable—hence the name *r*-selected. Although their numerous offspring disperse and rapidly colonize available habitats, most die before reaching sexual maturity (Type III survivorship). Thus, the success of an r-selected life history depends on flooding the environment with a *large quantity* of young, only a few of which may be successful.

A. An *r*-selected species

B. A *K*-selected species

FIGURE 52.18 Life history differences. (A) An *r*-selected species, such as quinoa (*Chenopodium quinoa*), matures in one growing season and produces many tiny seeds, which were a traditional food staple for the indigenous people of North and South America. **(B)** A *K*-selected species, such as the coconut palm (*Cocos nucifera*), grows slowly and produces a few large seeds repeatedly during its long life.

Because they have small body size, *r*-selected species lack physiological mechanisms to buffer them from environmental variation. Thus, as described earlier for the Australian thrips living in roses, survivorship and fecundity are often greatly influenced by density-independent factors, and population size fluctuates markedly. In good years, survivorship and fecundity may be high, and the population explodes. In bad years, survivorship and fecundity may be low, and the population crashes. Populations of *r*-selected species are often so greatly reduced by changes in abiotic environmental factors, such as temperature or moisture, that they never grow large enough to face a shortage of limiting resources. Thus, the carrying capacity for the species cannot be estimated, and changes in their population size cannot be described by the logistic model of population growth.

By contrast, *K*-selected species thrive in more stable environments. They are generally large, have long generation times, and produce offspring repeatedly during their lifetimes. Their offspring receive substantial parental care, either as energy reserves in an egg or seed or as active care, ensuring that most survive the early stages of life (Type I or Type II survivorship). Because *K*-selected species typically have a low r_{max}, their populations often grow slowly. The success of a *K*-selected life history therefore depends on the production of a relatively small number of *high quality* offspring that join an already well-established population.

The relatively large body sizes of *K*-selected species allow them to use behavioral and physiological mechanisms to buffer themselves against environmental change, so that survivorship and fecundity do not fluctuate wildly in response to environmental variations. Instead, their populations are often affected by density-dependent factors, which regulate population size near the carrying capacity for the species—hence the name *K*-selected. For these species, natural selection has favored life history characteristics that result in stable population sizes: the production of relatively few offspring, extensive parental care, good competitive ability, a long lifespan, and repeated reproductions. Many large terrestrial vertebrates are examples of *K*-selected species.

Metapopulation Structure Allows Local Populations to Exchange Individuals

Although the models of population growth discussed earlier ignore the effects of immigration and emigration, individuals frequently disperse from one local population to another. To describe the dynamics of such movements, ecologists define a **metapopulation** as a group of neighboring populations that exchange individuals. All local populations within a metapopulation are not equal; they often differ in size, population growth rates, the suitability of their habitats, their exposure to predators, and other factors. Moreover, some may decline steadily (even to extinction), while others may increase in size.

FIGURE 52.19 | **Observational Research**

Do Immigrants from Source Populations Prevent Extinction of Sink Populations of the Bay Checkerspot Butterfly?

Paul Ehrlich

Map of serpentine habitat patches near Morgan Hill

KEY
- Occupied habitat patches
- Unoccupied habitat patches

Morgan Hill

10 km

Hypothesis: Populations of the bay checkerspot butterfly (*Euphydryas editha bayensis*) living on small patches of suitable habitat are "sink" populations that frequently become extinct. Populations in large habitat patches can serve as a "source" of individuals to recolonize small habitat patches nearby.

Prediction: Because the bay checkerspot butterfly is a weak flyer, small patches of suitable habitat that are close to a large source population will be recolonized frequently. Patches of suitable habitat that are far from a large source population will be recolonized only rarely.

Method: Susan Harrison, Dennis D. Murphy, and Paul R. Ehrlich of Stanford University surveyed 59 small patches of serpentine grassland near San Jose, California, in 1986 and 1987. They estimated the "quality" of each patch based on the presence or absence of food plants on which bay checkerspots depend and on aspects of the physical environment that are important to these butterflies. They also measured the distance of each patch from Morgan Hill, a very large patch of suitable habitat that had sustained a bay checkerspot population for many years. In patches where they found butterflies, they estimated bay checkerspot population sizes.

Results: A complex statistical analysis revealed that both distance from the Morgan Hill population and habitat patch quality were important factors in determining whether bay checkerspots would be present or absent in small habitat patches. The authors noted that only the nine high-quality habitat patches near Morgan Hill (red on the map) were occupied by bay checkerspots. Of 50 unoccupied habitat patches, 6 were near Morgan Hill but of low quality; 18 were of high quality but far from Morgan Hill; and 26 were too far from Morgan Hill and of too low quality to support a population of bay checkerspots.

Conclusion: Populations of bay checkerspot butterflies that occupy large patches of suitable habitat serve as source populations for individuals that recolonize small patches of suitable habitat where butterfly populations frequently become extinct. However, because the bay checkerspot is a weak flyer, it recolonizes small patches of suitable habitat only if they are close to a source population.

think like a scientist

Observe
Hypothesize
Predict
Experiment
Interpret

What impact would residential and commercial development have on metapopulations of bay checkerspot butterflies?

Source: S. Harrison et al. 1988. Distribution of the bay checkerspot butterfly, *Euphydryas editha bayensis:* Evidence for a metapopulations model. *The American Naturalist* 132:360–382.

Under favorable circumstances, a population may produce numerous offspring, some of which emigrate and join nearby populations, where they breed, providing a genetic connection between local populations (see the discussion of gene flow in Section 21.3). Thus, dispersal and gene flow between local populations maintain the metapopulation.

Populations that are either stable or increasing in size are described as **source populations** because they are a possible source of immigrants to other populations. Those that decline in size are called **sink populations** because they receive available immigrants. Individuals usually move from source populations to sink populations, and sink populations persist because they receive immigrants from source populations in the metapopulation.

The bay checkerspot butterfly, *Euphydryas editha bayensis* **(Figure 52.19),** provides an example of metapopulation dynamics. This species is restricted to serpentine grassland (see Figure 53.19) in the San Francisco Bay area because its larvae eat plants that grow only in that community. Human disturbance has fragmented much of the butterfly's natural habitat into patches of varying size, each of which may support a local butterfly population. The life cycle of these butterflies is always a race against time, because the larvae must feed and mature before dry summer weather kills their food plants. Populations in small patches often become extinct, but those occupying larger patches, where food plants stay alive longer, generally survive the seasonal drought. (Global climate warming has exacerbated the problem in recent years, causing food plants to

POPULATION CYCLES IN PREDATORS AND THEIR PREY

Predator–prey interactions may contribute to density-dependent regulation of both populations.

A. Predictions of a predator–prey model

A mathematical model predicts cycles in the numbers of predators and prey because of time lags in each species' responses to changes in the density of the other. (Predator population size is exaggerated in this graph: predators are usually less common than prey.)

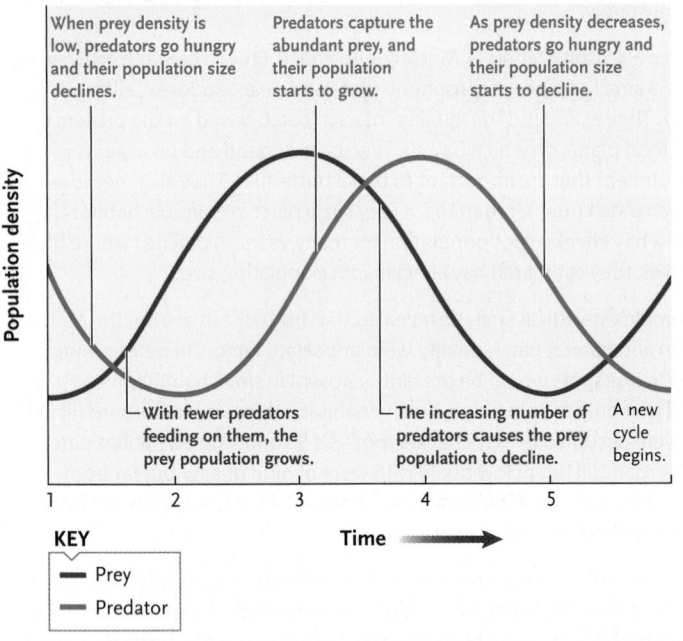

When prey density is low, predators go hungry and their population size declines.

Predators capture the abundant prey, and their population starts to grow.

As prey density decreases, predators go hungry and their population size starts to decline.

With fewer predators feeding on them, the prey population grows.

The increasing number of predators causes the prey population to decline.

A new cycle begins.

Population density

Time

KEY
— Prey
— Predator

B. Lynx and hare population sizes through time

The interaction between the Canada lynx *(Lynx canadensis)* and the snowshoe hare *(Lepus americanus)* was often described as a cyclic predator–prey interaction (see photo). The abundances of lynx and hare are based on counts of pelts that trappers sold to Hudson's Bay Company over a 90-year period.

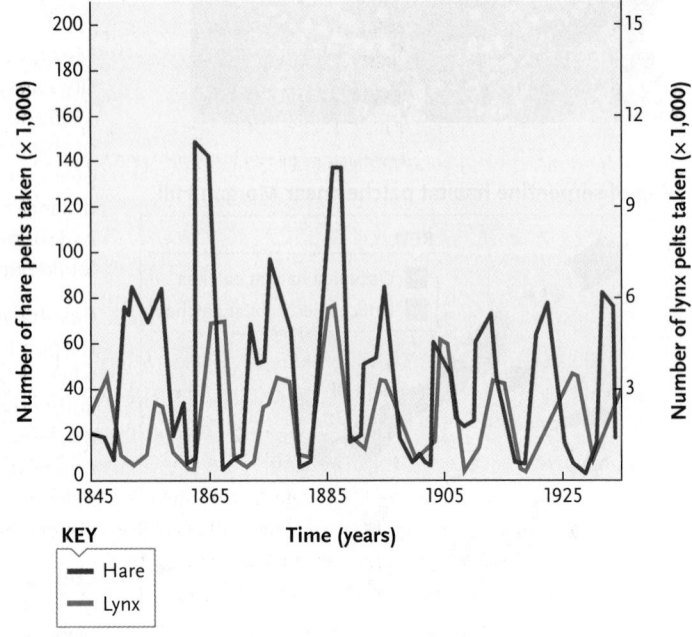

Number of hare pelts taken (× 1,000)

Number of lynx pelts taken (× 1,000)

Time (years)

KEY
— Hare
— Lynx

Tom Brakefield/Getty Images

SUMMARY As the graph in **(B)** illustrates, the number of predators (lynx) often rises and falls with the number of their prey (hare), but the timing of changes in the predator and prey population sizes does not match the predictions of the mathematical model precisely. Recent research has shown that population cycles in snowshoe hares are caused by complex interactions between the hare, its food plants, and its predators.

think like a scientist What is the approximate length of the population cycles of lynx and hare? According to the data in the graph, in which years did the population size of lynx increase almost instantaneously after an increase in the hare population? In which years did the lynx population size increase some time after an increase in the hare population?

© Cengage Learning 2017

die earlier in the season—and pushing the butterfly to the brink of extinction.) Butterfly populations in larger habitat patches still serve as source populations for emigrants that repopulate small habitat patches the following year. But the bay checkerspot is a poor flyer, and it cannot disperse long distances. Thus, small patches of suitable habitat harbor bay checkerspots only if they are close to a larger patch that serves as a source.

Some Species Exhibit Regular Cycles in Population Size

The population densities of many insects, birds, and mammals in the far north fluctuate between species-specific lows and highs in a multiyear cycle. Arctic populations of small rodents vary in size over a four-year cycle, whereas snowshoe hares, ruffed grouse, and lynxes have 10-year cycles. Ecologists documented such cyclic fluctuations more than a century ago, but none of the general hypotheses so far proposed explains the cycles in all species. The availability and quality of food, the abundance of predators, the prevalence of disease-causing microorganisms, and variations in weather may influence population growth. Furthermore, a cycling population's food supply and predators are themselves influenced by the population's size.

Theories of *intrinsic control* suggest that as an animal population grows, individuals undergo hormonal changes that increase aggressiveness, reduce reproduction, and foster dispersal to other areas. The dispersal phase of the cycle may be dramatic. For example, when populations of the Norway lemming (*Lemmus lemmus*), a rodent that lives in the Scandinavian arctic, reach their peak density, aggressive interactions drive younger and weaker individuals away from their place of birth. The exodus of many thousands of lemmings, scrambling over rocks and even cliffs, was sometimes incorrectly portrayed in nature films as a suicidal mass migration. Researchers do not yet know how widespread these hormonal and behavioral changes are among different species or exactly what regulates them.

Other explanations focus on *extrinsic control,* such as the relationship between a cycling species and its food or predators. A dense population may exhaust its food supply, increasing mortality and decreasing reproduction. But experimental food supplementation does not always prevent a decline in mammal populations, indicating that other factors are also at work.

Some mathematical models as well as some laboratory experiments on protists or small arthropods suggest that the cycles of predators and their prey are induced by time lags in each population's response to changes in density of the other **(Figure 52.20)**. In the past, the 10-year cycles of snowshoe hares (*Lepus americanus*) and their feline predators, Canada lynxes (*Lynx canadensis),* were often cited as a classic example of such an interaction. But ecological relationships are usually complex, and recent research has cast doubt on this straightforward explanation. Hare populations exhibit a 10-year fluctuation even on islands where lynxes are absent. Thus, the lynx cannot be solely responsible for the hare's cycle, although cycles in the hare populations may trigger cycles in populations of their predators.

Charles Krebs and his colleagues at the University of British Columbia studied hare and lynx interactions with a large-scale, multiyear experiment in the southern Yukon. They fenced experimental areas where they added food for the hares, excluded mammalian predators, or applied both experimental treatments; unmanipulated plots served as controls. Where mammalian predators were excluded, hare densities approximately doubled relative to the controls. Where food was added, hare densities tripled. But in plots where predators were excluded *and* food was added, the hare densities increased 11-fold. Krebs and his colleagues concluded that neither food availability nor predation alone is solely responsible for arctic hare population cycles; instead, complex interactions between the hares, their food plants, and their predators create the cyclic fluctuations in hare population size.

52.6 Human Population Growth

How do human populations compare with those of other species we have studied? The worldwide human population surpassed 7 billion in 2011, and it is projected to reach 9.7 billion by 2050. Like many other species, most humans live in discrete populations, which vary in their demographic traits and access to resources. Although many of us live comfortably, more than a billion people are malnourished or starving, lack clean drinking water, and live without adequate shelter or health care. Even if it were possible to double the food supply, increased agricultural production would inevitably increase pollution and contribute to spoiled croplands, deforestation, and desertification, which are described in Chapter 55.

Human Populations Have Sidestepped the Usual Density-Dependent Controls

For most of human history, our population grew slowly; but over the past 250 years, the worldwide human population has grown dramatically **(Figure 52.21).** Demographers have identified three ways in which humans have avoided the effects of density-dependent regulating factors.

First, humans have expanded their geographical range into virtually every terrestrial habitat on Earth. Our early ancestors lived in tropical and subtropical grasslands, but by 40,000 years ago, they had dispersed through much of the world (see Section 32.12). Their success resulted from their ability to solve ecological problems by building fires, assembling shelters, making clothing and tools, and planning community hunts. Vital survival skills spread from generation to generation and from one population to another because language allowed the communication of complex ideas and knowledge.

Second, humans have increased the carrying capacities of habitats they occupy. About 11,000 years ago, many populations shifted from hunting and gathering to agriculture. They cultivated wild grasses, diverted water to irrigate crops, and used domesticated animals for food and labor. Such innovations increased the availability of food, raising both the carrying capacity and the population growth rate. In the mid-eighteenth century, people harnessed the energy in fossil fuels, and industrialization began in Western Europe and North America. Food supplies and the carrying capacity increased again, at least in the industrialized countries, through the use of synthetic fertilizers, pesticides, and efficient methods of transportation and food distribution.

FIGURE 52.21 Human population growth. The worldwide human population grew slowly until 250 years ago, when it began to increase explosively. The dip in the mid-fourteenth century represents the death of 60 million Asians and Europeans from the bubonic plague. The table shows the years when the human population reached each additional billion people.

© Cengage Learning 2017

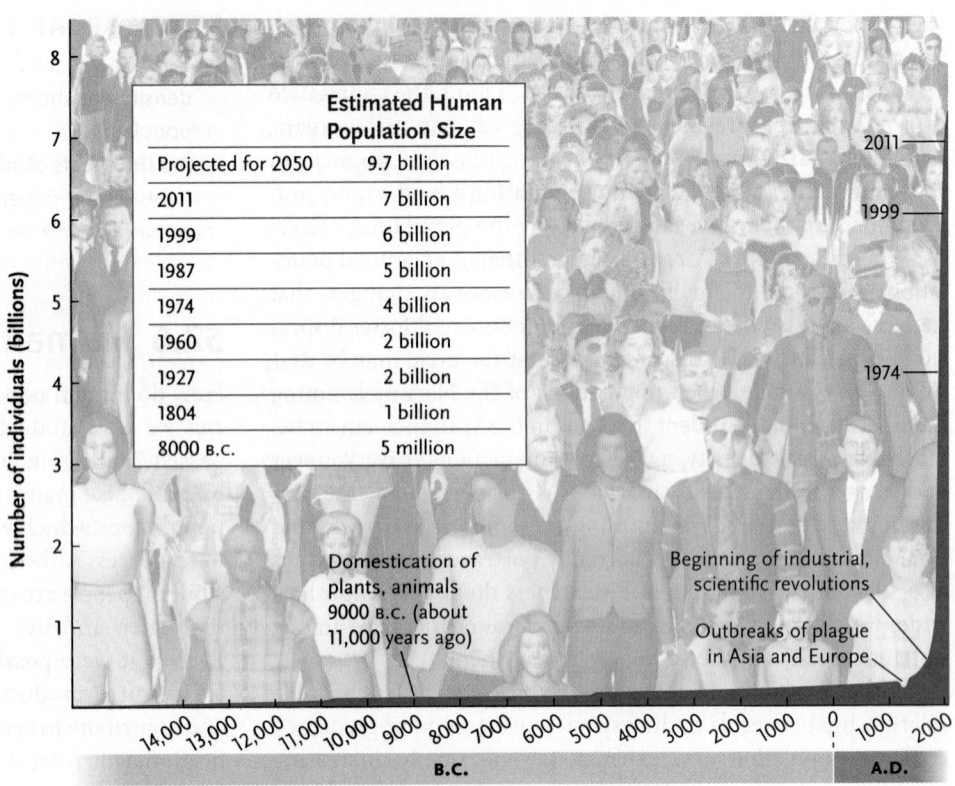

Estimated Human Population Size	
Projected for 2050	9.7 billion
2011	7 billion
1999	6 billion
1987	5 billion
1974	4 billion
1960	2 billion
1927	2 billion
1804	1 billion
8000 B.C.	5 million

Third, advances in public health have reduced the effects of critical population-limiting factors such as malnutrition, contagious diseases, and poor hygiene. Over the past 300 years, modern plumbing and sewage treatment, improvements in food handling and processing, and medical discoveries have reduced death rates sharply. Births now greatly exceed deaths, especially in less industrialized countries, resulting in rapid population growth.

Age Structure and Economic Development May Now Control Our Population Growth

Where have our migrations and technological developments taken us? It took more than 100,000 years for the human population to reach 1 billion, 123 years to reach the second billion, and only 12 years to jump from 6 billion to 7 billion (see the inset table in Figure 52.21). Continued population growth may now be an inevitable consequence of our age structure and economic development.

A. Mean annual population growth rates, 2013

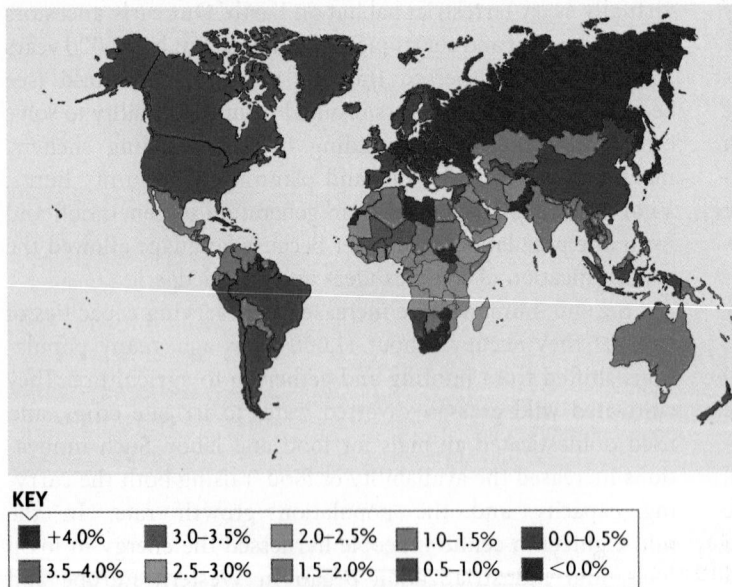

KEY

■ +4.0%	■ 3.0–3.5%	■ 2.0–2.5%	■ 1.0–1.5%	■ 0.0–0.5%
■ 3.5–4.0%	■ 2.5–3.0%	■ 1.5–2.0%	■ 0.5–1.0%	■ <0.0%

B. Actual and projected population sizes for major world regions

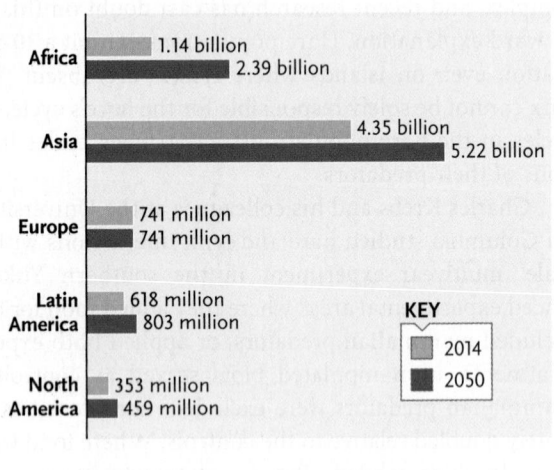

FIGURE 52.22 Local variation in human population growth rates. (A) Average annual population growth rates varied among countries in 2013. **(B)** In some regions, the population is projected to increase greatly by 2050; the population of Europe will likely remain unchanged. Source: Population Reference Bureau, Washington, D.C.

© Cengage Learning 2017

POPULATION GROWTH AND AGE STRUCTURE On a worldwide scale, the annual growth rate for the human population averaged roughly 1.2% (r = 0.012 new individuals per individual per year) between 2001 and 2014. The rate of population growth appears to be declining, but even so, the human population will continue to grow at least until 2050.

The annual population growth rates of individual nations vary widely, however, ranging from slightly below 0% (that is, the population is getting smaller) to roughly 4.0% in 2013 **(Figure 52.22A)**. The industrialized countries of Western Europe as well as Japan have achieved nearly zero population growth, and in some cases negative growth, but other countries—notably those in Africa, Latin America, and Asia—will experience increases in population size over the next 15 to 40 years **(Figure 52.22B)**. Some recent population declines, especially in countries of the Middle East, reflect the recent flight of many refugees from political turmoil and war.

For all long-lived species, differences in age structure are a major determinant of differences in population growth rates **(Figure 52.23)**. The uniform age structure of countries with zero growth—with approximately equal numbers of people of reproductive and prereproductive ages—suggests that individuals have just been replacing themselves and that these populations will not experience a growth spurt when today's children mature. By contrast, the narrow-based age structure of countries with negative growth illustrates a continuing decrease in population size. Reproductives have been producing very few offspring, and the small group of prereproductives may not even replace themselves. Countries with rapid growth have a broad-based age structure, with many youngsters born during

A. Hypothetical age distributions for populations with different growth rates

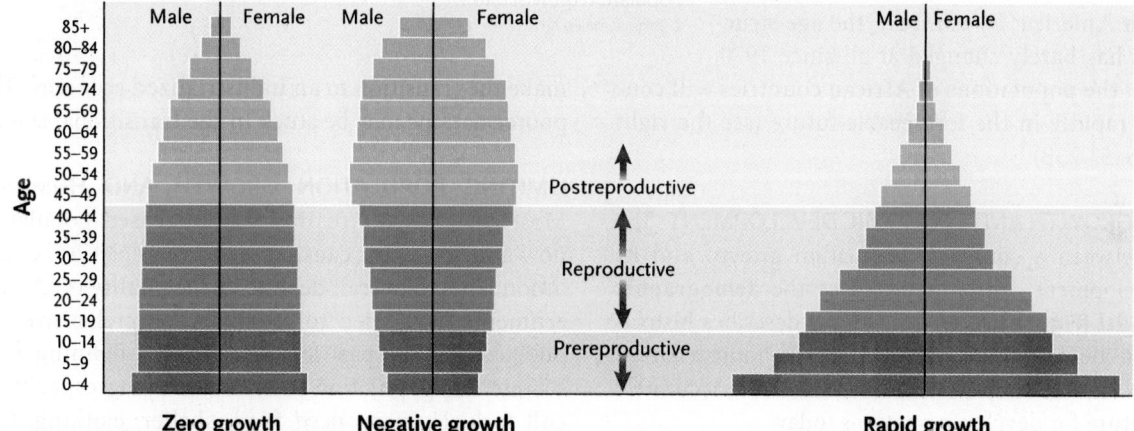

B. Age pyramids for North America and Africa in 2014

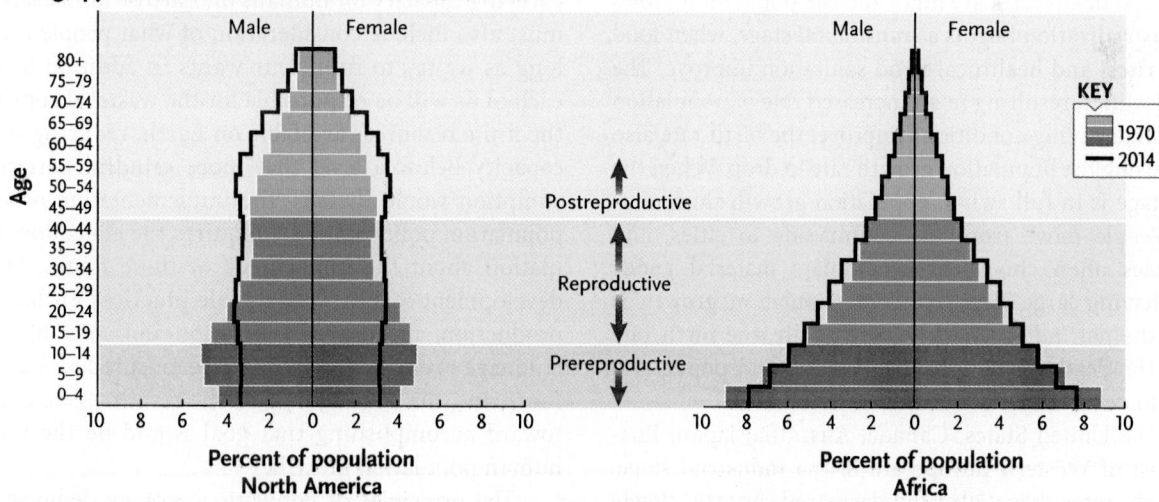

FIGURE 52.23 Age-structure diagrams. (A) Hypothetical age-structure diagrams differ for countries with zero, negative, and rapid population growth rates. The width of each bar represents the proportion of the population in each age class. **(B)** The age structure in North America changed substantially between 1970 and 2014; the bulge of 5- to 24-year olds in 1970 represented the "baby boom" generation born after the end of World War II. By contrast, the age structure of Africa barely changed at all during that time period, indicating that human populations in Africa will continue to grow rapidly.

© Cengage Learning 2017

the previous 15 years. Worldwide, more than one quarter of the human population falls within this pre-reproductive base. This age class will soon reach sexual maturity. Even if each woman produces only two offspring, populations will continue to grow rapidly because so many individuals are reproducing.

The age structure of some human populations has changed over the last half-century. For example, in North America (Canada and the United States), the age structure has shifted from one that generated moderate population growth in 1970 to an age structure that will generate very little population growth in 2014 (see the left panel in Figure 52.23B). The average number of children per family has declined to the two that are necessary to replace their parents in the population. Nevertheless, the U.S. population will continue to grow slowly for the next couple of generations largely because of continued immigration. Similar, but smaller, changes have occurred in Asia and Latin America. By contrast, the age structure in Africa has barely changed at all since 1970, indicating that the populations of African countries will continue to grow rapidly in the foreseeable future (see the right panel in Figure 53.23B).

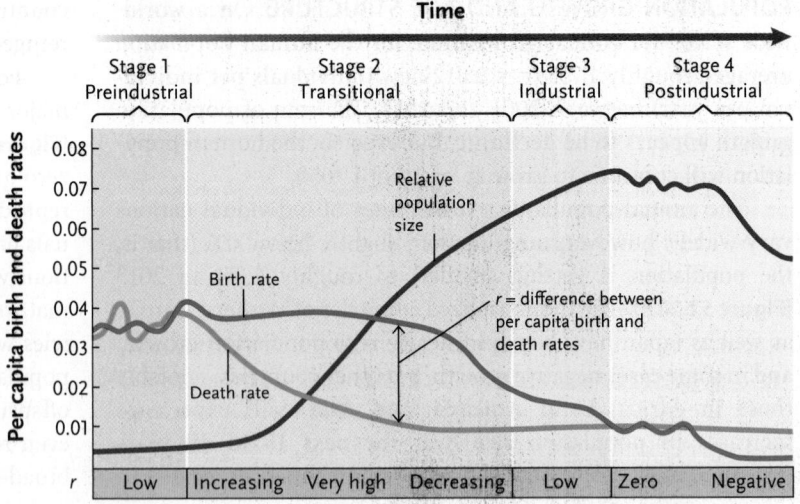

FIGURE 52.24 The demographic transition. The demographic transition model describes changes in the birth and death rates and relative population size as a country passes through four stages of economic development. The bottom bar describes the net population growth rate, r.
© Cengage Learning 2017

POPULATION GROWTH AND ECONOMIC DEVELOPMENT The relationship between a country's population growth and its economic development can be depicted by the **demographic transition model (Figure 52.24)**. This model describes historical changes in demographic patterns in the industrialized countries of Western Europe; we do not know if it accurately predicts the future for developing nations today.

According to this model, during a country's *preindustrial* stage, birth and death rates are high, and the population grows slowly. Industrialization begins a *transitional* stage, when food production rises, and health care and sanitation improve. The death rate declines, resulting in an increased rate of population growth. Later, as living conditions improve, the birth rate also declines, causing the population growth rate to drop. When the *industrial* stage is in full swing, population growth slows dramatically. People move from the countryside to cities, and urban couples often choose to accumulate material goods instead of having large families. Zero population growth is reached in the *postindustrial* stage. Eventually, the birth rate falls below the death rate, r falls below zero, and population size begins to decrease.

Today, the United States, Canada, Australia, Japan, Russia, and most of Western Europe are in the industrial stage. Their growth rates have already decreased or are slowly decreasing. In some Western European countries, birth rates are lower than death rates, and populations are getting smaller, indicating their entry into the postindustrial stage. Many less industrialized countries are in the transitional stage, but they may not have enough skilled workers or enough capital to make the transition to an industrialized economy. Thus, many poorer nations may be stuck in the transitional stage.

LIMITING POPULATION GROWTH AND HUMAN IMPACT Most governments realize that increased population size is now a major factor causing resource depletion, excessive pollution, and an overall decline in the quality of life. Many governments have tried to establish comprehensive population policies over the past few decades, but defining a maximum sustainable population size, the carrying capacity, is a difficult task. Humans *need* food, shelter, clothing, transportation, and a host of other resources. But the definition of a carrying capacity for humans on Earth—or in each country—must also include consideration of what people *want*. And as long as we try to fulfill our wants in addition to our needs, each of us will be responsible for the wasteful consumption of the finite resources available on Earth, reducing the carrying capacity below a level that more mindful patterns of consumption would dictate. The implementation of an effective population policy not only requires the education of the population about the importance of these issues, but also the development of more sustainable practices for land use, food production, and energy generation and consumption. In the language of environmentalists, we must reduce our **ecological footprint,** the sum total of all the resources we use. One step toward accomplishing that goal would be the limitation of human population growth.

The principles of population ecology demonstrate that a slowing of population growth—or an actual decline in population size—can be achieved only by decreasing the birth rate or increasing the death rate. And because increasing mortality is neither a rational nor humane means of population control, most governments are attempting to lower birth rates with family planning programs. These programs educate people about ways to produce an optimal family size on an economically feasible schedule. Programs vary in their details, but all provide information on methods of birth control (see Section 49.4). When thoughtfully developed and carefully administered, family planning programs cause birth rates to decline significantly.

All species face limits to their population growth. We have postponed the action of most factors that limit population growth, but no amount of invention can expand the ultimate limits set by resource depletion and a damaged environment. We now face two options for limiting human population growth: we can make a global effort to limit our population growth, or we can wait until the environment does it for us.

STUDY BREAK 52.6

1. How have humans sidestepped the controls that regulate populations of other organisms?
2. How does the age structure of a population influence its future population growth?

 ## Unanswered Questions

Are there universal governing principles in population ecology, similar to the laws of physical sciences? Or is the natural world so complex that each population must be considered individually, leaving us with just a series of case studies?

These types of broad questions motivated the founders of modern ecological studies, such as G. Evelyn Hutchinson and Robert MacArthur. Many ecologists have attempted to codify aspects of population ecology in terms of specific principles, sometimes imposing artificial dichotomies in the process. For example, this chapter considered whether or not natural populations are subject to either density-dependent or density-independent regulation. Ecologists have also attempted to uncover basic patterns in community ecology, which are described in the next chapter. We do know that some general principles are often important in governing the structure of populations or natural communities but, as yet, we cannot apply any of them to a specific system without also including a detailed study of that system.

What is the importance of scale in ecology?

Although an individual population can be a meaningful object of study, ecologists often collect data from multiple populations of the same species to compare results among the "replicates." However, although the separate populations may appear to be replicates, they are often quite different in appearance, age structure, life history, or other characteristics. Ecologists confronted with such variation might seek explanations in the differences between the populations' environments, including both abiotic and biotic factors. However, such local variation may also be attributable to the larger context, such as the landscape or surrounding communities.

One important manifestation of this question applies to how the populations of a species are distributed in space. In many cases, discrete populations are widely separated from one another. Such separation is easy to imagine in terms of fish that live in lakes or organisms that live on islands, but it also applies in a diversity of other organisms. For example, many plants and animals, such as the bay checkerspot butterflies discussed in this chapter, are found only on chemically distinct patches of soil that are distributed like islands across a terrestrial environment. Many lizards live in rock outcrops that dot the landscape. Other organisms live on cool, wet mountaintops surrounded by desert. Such isolation is often exaggerated by human modification of the landscape, which progressively fragments and isolates habitable environments from one another.

How does such subdivision change the dynamics of the individual populations? How does it change the way they evolve? These are questions that have challenged population and evolutionary biologists for decades. The effects of humans on the environment are making the answers to these questions more than a theoretical concern.

What is the importance of evolution in ecological interactions?

Most research in ecology, ranging from formal models of population growth and regulation to empirical studies, treats populations as if they were unchanging—as if they were not evolving. This implicit perspective does not deny that evolution is happening, but treats it as if it happens on such a long time scale that it need not be considered in contemporary studies. However, many recent studies have shown that populations may evolve quickly, often on a year-to-year basis. If this observation is generally true, then ecological studies that do not include evolutionary change may be compromised. For example, the monitoring and management of commercially exploited fish populations are based entirely on models of population growth, demography, and life histories similar to those considered in this chapter. Commercial fisheries often capture a large proportion of a population every year, focusing on the largest adults. Although research has clearly shown that these practices are likely to select for earlier maturity at a smaller size, these findings have not yet been incorporated into fisheries-management policy. More generally, ecologists have not yet included sufficient emphasis on the interaction between evolution as it occurs on the scale of our day-to-day existence and the modeling and empirical study of ecological processes.

think like a scientist

1. Why do you think the mathematical models describing population growth introduced in this chapter are not considered biological laws?

2. Considering life table data, why do you think population ecologists often "ignore" evolutionary change? How can molecular biology be used to incorporate evolution into population level studies?

Courtesy of David Reznick

David Reznick is a Professor of Biology at the University of California, Riverside. He studies natural selection both from an experimental perspective and by testing evolutionary theory in natural populations. He works primarily with guppies on the island of Trinidad. Learn more about Dr. Reznick's work at http://www.biology.ucr.edu/people/faculty/Reznick.html.

REVIEW KEY CONCEPTS

For access to MindTap and additional study materials visit www.cengagebrain.com.

52.1 Population Characteristics

- A population's size and density can be measured directly or with sampling techniques (Figures 52.2 and 52.3).

- Organisms within a population may be clumped, uniformly distributed, or randomly distributed within their habitat (Figure 52.4). Clumped dispersion is the most common, but animals may change their dispersion pattern seasonally.

- The relative numbers of individuals of different ages determine a population's age structure. Generation time, the average time between an individual's birth and the birth of its offspring, generally increases with body size (Figure 52.5). A population's sex ratio is the relative proportion of males and females.

52.2 Demography

- Demography is the study of the survivorship, reproduction, immigration, and emigration patterns that influence population characteristics.

- Life tables summarize age-specific mortality, survivorship, and age-specific fecundity of surviving individuals (Table 52.1).

- Survivorship curves depict a population's survival pattern over its lifespan. Ecologists define three general patterns of survivorship: high survivorship until late in life, a constant mortality level at all ages, and high juvenile mortality (Figure 52.6).

52.3 The Evolution of Life Histories

- An organism's energy budget mandates trade-offs in the allocation of energy to maintenance, growth, and reproduction (Figure 52.7).

- Natural selection has molded several interacting components of life history variation based upon the allocation of resources to growth, maintenance, and reproduction: the trade-off between fecundity and parental care; whether to reproduce once versus multiple times; and the age of first reproduction.

52.4 Models of Population Growth

- Bacteria reproduce by binary fission, and their populations double in size each generation (Figure 52.8A).

- The exponential growth model, $dN/dt = rN$, describes unlimited population growth. A graph of exponential growth is J-shaped (Figure 52.8B).

- The maximum population growth rates of populations, symbolized r_{max}, vary inversely with their generation times (Figure 52.9).

- The logistic model, $dN/dt = r_{max} N(K - N)/K$, includes the effects of resource limitation. The carrying capacity, K, is the maximum population size that an environment can sustain. The per capita population growth rate, r, decreases as N approaches K. A graph of logistic growth is S-shaped (Figure 52.10, Table 52.2).

- Some populations exhibit logistic growth in the laboratory and in nature, but time lags in response to increased density may cause N to oscillate around K (Figure 52.11).

52.5 Population Dynamics

- Density-dependent factors regulate population size by reducing individual growth rates, adult size, survivorship, and fecundity (Figures 52.12 and 52.13). Competition within populations or between species, predator-prey interactions, parasites, and infectious diseases can cause density-dependent population regulation (Figure 52.14).

- Abiotic environmental factors, which affect a population regardless of its size, cause density-independent limitation of population size (Figure 52.15).

- Global climate change is exaggerating the effects of density-independent factors on populations of large, long-lived organisms (Figure 52.16).

- Interactions between density-dependent and density-independent factors often influence population size (Figure 52.17).

- The life history patterns of most organisms fall between two extremes: r-selected species and K-selected species (Figure 52.18), which differ in many life history characteristics (Table 52.3).

- Within metapopulations, sink populations are often replenished by individuals dispersing from a source population (Figure 52.19).

- Some animal populations exhibit cyclic fluctuations in size (Figure 52.20). No general model has successfully explained all population cycles.

52.6 Human Population Growth

- Human populations have sidestepped density-dependent population regulation by expanding into most terrestrial habitats, increasing carrying capacity, and reducing death rates with improved medical care and sanitation (Figures 52.21 and 52.22).

- Age structure has a large influence on human population growth rates (Figure 52.23). In countries with large numbers of young people, populations will continue to grow rapidly as those individuals reach sexual maturity. The populations of countries with a uniform age structure will not experience much growth in the foreseeable future.

- The demographic transition model describes the influence of economic development on population growth (Figure 52.24).

- Many governments encourage population control through family planning programs.

TEST YOUR KNOWLEDGE

Remember/Understand

1. Ecologists sometimes use mathematical models to:
   a. avoid conducting laboratory studies or field work.
   b. simulate natural events before conducting detailed field studies.
   c. make basic observations about ecological relationships in nature.
   d. collect survivorship and fecundity data to construct life tables.
   e. determine the geographical ranges of populations.

2. The number of individuals per unit area or volume of habitat is called the population's:
   a. geographical range.
   b. dispersion pattern.
   c. density.
   d. size.
   e. age structure.

3. A uniform dispersion pattern implies that members of a population:
   a. cooperate in rearing their offspring.
   b. work together to escape from predators.
   c. use resources that are patchily distributed.
   d. may experience intraspecific competition for vital resources.
   e. have no ecological interactions with each other.

4. The model of exponential population growth predicts that the per capita population growth rate r:
   a. does not change as a population gets larger.
   b. gets larger as a population gets larger.
   c. gets smaller as a population gets larger.
   d. is always at its maximum level (r_{max}).
   e. fluctuates on a regular cycle.

5. According to the logistic model of population growth, the absolute number of individuals by which a population grows during a given time period:
   a. gets steadily larger as the population size increases.
   b. gets steadily smaller as the population size increases.
   c. remains constant as the population size increases.
   d. is highest when the population is at an intermediate size.
   e. fluctuates on a regular cycle.

6. One reason why human populations have been able to sidestep the factors that usually control population growth is that:
   a. the carrying capacity for humans has remained constant since humans first evolved.
   b. agriculture and industrialization have increased the carrying capacity for our species.
   c. the population growth rate (r) for the human population has always been small.
   d. the age structure of human populations has no impact on its population growth.
   e. plagues have killed off large numbers of humans at certain times in the past.

Apply/Analyze

7. One day you caught and marked 90 butterflies in a population. A week later, you returned to the population and caught 80 butterflies, including 16 that had been marked previously. What is the size of the butterfly population?
   a. 170
   b. 450
   c. 154
   d. 186
   e. 106

8. A population of 1,000 individuals experiences 462 births and 380 deaths in 1 year. What is the value of r for this population?
   a. 0.842/individual/year
   b. 0.462/individual/year
   c. 0.380/individual/year
   d. 0.820/individual/year
   e. 0.082/individual/year

9. Which example might reflect density-dependent regulation of population size?
   a. An exterminator uses a pesticide to eliminate carpenter ants from a home.
   b. Mosquitoes disappear from an area after the first frost.
   c. The lawn dies after a month-long drought.
   d. Storms blow over and kill all the willow trees along a lake.
   e. The size of a clam population declines as the number of predatory herring gulls explodes.

10. A K-selected species is likely to exhibit:
    a. a Type I survivorship curve and a short generation time.
    b. a Type II survivorship curve and a short generation time.
    c. a Type III survivorship curve and a short generation time.
    d. a Type I survivorship curve and a long generation time.
    e. a Type II survivorship curve and a long generation time.

11. **Discuss Concepts** How could you define the worldwide carrying capacity for humans? What factors would you have to take into account?

Evaluate/Create

12. **Discuss Concepts** Choose an animal or plant species that lives in your environment and identify the density-dependent and density-independent factors that might influence its population size. How could you demonstrate conclusively that the factors are either density-dependent or density-independent?

13. **Discuss Concepts** Many city-dwellers have noted that the density of cockroaches in apartment kitchens appears to vary with the habits of the occupants: people who wrap food carefully and clean their kitchen frequently tend to have fewer arthropod roommates than those who leave food on kitchen counters and clean less often. Interpret these observations from the viewpoint of a population ecologist.

14. **Design an Experiment** Design an experiment using fruit flies or some other small laboratory animal to test the hypothesis that delaying the age of first reproduction will decrease a population's per capita birth rate. Your experimental design should include experimental and control groups as well as details about your experimental methods and the data you would collect.

15. **Apply Evolutionary Thinking** Many animals, including humans and other primates, live long beyond their reproductive years. Develop an evolutionary hypothesis to explain this observation, and design a study that might test it.

For selected answers, see Appendix A.

INTERPRET THE DATA

Gregory M. Erickson of Florida State University and colleagues from Florida State and the University of Alberta analyzed a fossil assemblage of 22 individuals of *Albertosaurus sarcophagus,* a relative of *Tyrannosaurus rex.* All 22 had died at roughly the same time—but at different ages—perhaps because of a drought or starvation. They used growth lines present in the fossilized leg bones to determine the age at death of each specimen. They had no data for very young individuals, and assumed a 60% mortality rate between birth and age two. The researchers constructed the following **Table** extrapolated to an initial cohort of 1,000 individuals:

Age (years)	Number Surviving	Age (years)	Number Surviving	Age (years)	Number Surviving
0	1,000	10	—	20	73
1	—	11	309	21	56
2	400	12	291	22	—
3	—	13	273	23	36
4	382	14	255	24	—
5	—	15	218	25	—
6	364	16	182	26	—
7	—	17	164	27	—
8	345	18	127	28	18
9	327	19	109		

Did this population of animals exhibit a Type I, Type II, or Type III survivorship pattern—or something intermediate? What does the life table tell you about the age at which *A. sarcophagus* was most vulnerable to the factors causing mortality?

Source: After G. M. Erickson et al. 2006. Tyrannosaur life tables: An example of nonavian dinosaur population biology. *Science* 313:213–217.

Population Interactions and Community Ecology

Three interacting populations. Ladybird beetles *(Coccinella septempunctata)* feed on aphids (order Hemiptera), which consume the sap of plants.

Why it matters . . . In some open woodlands in Central America, flocks of chestnut-headed oropendolas *(Psarocolius wagleri),* members of the blackbird family, build hanging nests in isolated trees **(Figure 53.1).** Female giant cowbirds *(Molothrus oryzivorus)* often bully their way into a colony, laying an egg or two in each oropendola nest. Cowbirds are *brood parasites* on oropendolas, tricking them into caring for cowbird young. The cowbird chicks grow faster than oropendola chicks, and they consume much of the food that the oropendolas bring to their own offspring. Because cowbird chicks take food away from their oropendola nest mates, we might expect adult oropendolas to eject cowbird eggs and chicks from their nests—but often they do not.

Why do some oropendolas care for offspring that are not their own? In an ingenious study conducted in the 1960s, Neal Smith of the Smithsonian Tropical Research Institute determined that cowbird chicks could actually increase the number of offspring that some oropendolas raise. Oropendola chicks are frequently parasitized by botfly larvae, which feed on their flesh. The aggressive cowbird chicks snap at adult botflies and pick fly larvae off their nest mates. Although cowbird chicks eat food meant for oropendola chicks, they also protect them from potentially lethal parasites; twice as many young oropendolas survive in nests with cowbird chicks as in nests without them.

In other areas of Central America, oropendolas build nests near the hives of bees or wasps. These oropendolas chase cowbirds from their colonies, and when a cowbird does manage to sneak an egg into one of their nests, the oropendolas frequently eject it. Why do oropendolas in these colonies reject cowbird eggs? Smith determined that the swarms of bees and wasps keep botflies away from the oropendola colonies. At these sites, twice as many oropendola chicks survive in nests without cowbirds as in those that include them. Thus the oropendolas derive no

FIGURE 53.1 Potential victims of brood parasitism. Chestnut-headed oropendolas *(Psarocolius wagleri)* rear their young in elaborate hanging nests. Some populations of oropendolas are subject to brood parasitism by giant cowbirds *(Molothrus oryzivorus).*

TABLE 53.1		Population Interactions and Their Effects
Interaction		Effects on Interacting Populations
Predation	+/−	Predators gain nutrients and energy; prey are killed or injured.
Herbivory	+/−	Herbivores gain nutrients and energy; plants are killed or injured.
Parasitism	+/−	Parasites gain nutrients and energy; hosts are killed or injured.
Competition	−/−	Both competing populations lose access to some resources.
Commensalism	+/0	One population benefits; the other population is unaffected.
Mutualism	+/+	Both populations benefit.

benefit from having cowbird chicks in their nests, and natural selection has favored discriminating behavior in oropendolas that nest near bees and wasps.

The story of the oropendolas, cowbirds, botflies, bees, and wasps provides an example of the population interactions that characterize life in an **ecological community,** an assemblage of species living in the same place. As this story reveals, the presence or absence of certain species may alter the effects of such interactions in almost unimaginably complex ways. We begin this chapter with a description of some of the many ways that populations in a community interact. We then examine how population interactions and other factors, such as the kinds of species present and the relative numbers of each species, influence a community's characteristics.

53.1 Population Interactions

Interactions between populations usually provide benefits or cause harm to the interacting organisms, as compared with organisms that are not engaged in the interaction **(Table 53.1).** Because interactions with other species often affect the survival and reproduction of individuals, many of the relationships that we witness today are the products of long-term evolutionary modification. Before examining several general types of population interactions, we briefly consider how natural selection has shaped the relationships between interacting species.

Coevolution Produces Reciprocal Adaptations in Species That Interact Ecologically

Population interactions change constantly. New adaptations that evolve in one species exert selection pressure on another, which then evolves adaptations that exert selection pressure on the first. The evolution of genetically based, reciprocal adaptations in two or more interacting species is described as **coevolution.**

Some coevolutionary relationships are straightforward. For example, ecologists describe the coevolutionary interactions between some predators and their prey as a race in which each species evolves adaptations that temporarily allow it to outpace the other. When antelope populations suffer predation by cheetahs, natural selection fosters the evolution of faster speed in the antelopes. Cheetahs then experience selection for increased speed so that they can overtake and capture antelopes. Other coevolved interactions provide benefits to both partners. For example, the flower structures of different monkey-flower species have evolved characteristics that allow them to be visited by either bees or hummingbirds (see Figure 22.7).

Although one can hypothesize a coevolutionary relationship between any two interacting species, documenting the evolution of reciprocal adaptations is difficult. As our introductory story about oropendolas and their parasites illustrated, coevolutionary interactions often involve more than two species. Indeed, most organisms experience complex interactions with numerous other species in their communities, and the simple portrayal of coevolution as taking place between two species rarely does justice to the complexity of these relationships.

Predation and Herbivory Define Many Relationships in Ecological Communities

Because animals acquire nutrients and energy by consuming other organisms, **predation** (the interaction between predatory animals and the animal prey they consume) and **herbivory** (the interaction between herbivorous animals and the plants they eat) are often the most conspicuous relationships in ecological communities.

ADAPTATIONS FOR FEEDING Both predators and herbivores have evolved remarkable characteristics that allow them to feed effectively. Carnivores use sensory systems to locate animal prey and specialized behaviors and anatomical structures to capture and consume it. For example, a rattlesnake (genus *Crotalus*) uses heat sensors on its head (see Figure 41.21) and chemical sensors in the roof of its mouth to find rats or other endothermic prey. Its hollow fangs inject toxins that kill the prey and begin to digest its tissues even before the snake consumes it. And elastic ligaments connecting the bones of its jaws and skull allow a snake to swallow prey that is larger than its head. Herbivores have comparable adaptations for locating and processing their food plants. Insects use chemical sensors on their legs and heads to identify edible plants and sharp mandibles or sucking mouth-parts to consume plant tissues or sap. Herbivorous mammals have specialized teeth to harvest and grind tough vegetation (see Section 47.5).

All animals must select their diets from a variety of potential food items. Some species, described as *specialists*, feed on one or just a few types of food. Among birds, for example, the Everglades kite (*Rostrhamus sociabilis*) consumes just one prey species, the apple snail (*Pomacea paludosa*). Other species, described as *generalists*, have broader tastes. Crows (genus *Corvus*) consume food ranging from grains to insects to carrion.

How does an animal select what food to eat? Some mathematical models, collectively described as **optimal foraging theory,** predict that an animal's diet is a compromise between the costs and benefits associated with different types and sizes of food. Assuming that animals try to maximize their energy intake in a given feeding time, their diets should be determined by the time and energy it takes to pursue, capture, and consume a particular kind of food compared with the energy that food provides. For example, a cougar (*Puma concolor*) will invest more time and energy hunting a mountain goat (*Oreamnos americanus*) than a jackrabbit (*Lepus townsendii*), but the payoff for the cougar is a bigger meal.

Food abundance also affects food choice. When prey are scarce, animals often take what they can get, settling for food that has a low benefit-to-cost ratio. But when food is abundant, they may specialize, selecting types that provide the largest energetic return. Bluegill sunfish (*Lepomis macrochirus*), for example, feed on *Daphnia* and other small crustaceans. When crustacean density is high, the fish hunt mostly large *Daphnia*, which provide more energy for their effort; but when prey density is low, bluegills feed on *Daphnia* of all sizes **(Figure 53.2).**

DEFENSES AGAINST HERBIVORY AND PREDATION Because herbivory and predation have a negative impact on the organisms being consumed, plants and animals have evolved mechanisms to avoid being eaten. Some plants use spines, thorns, and

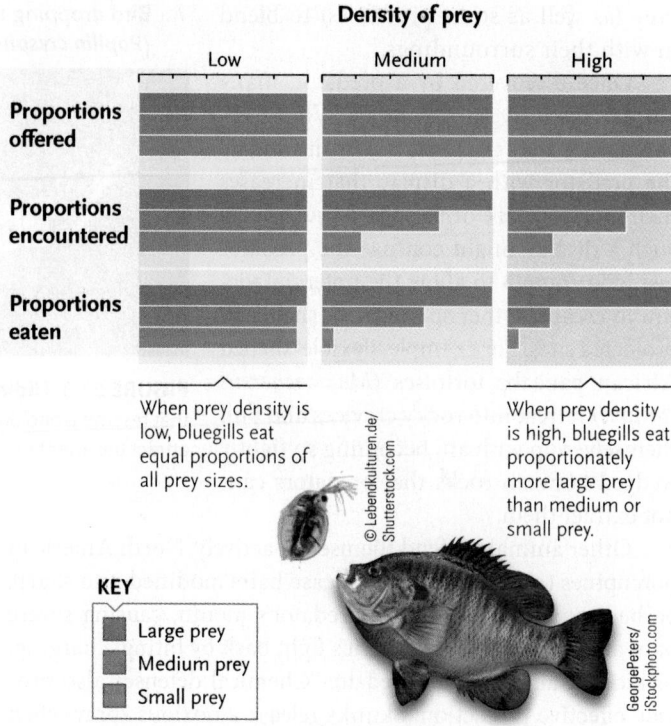

FIGURE 53.2 An experiment demonstrating that prey density affects predator food choice. Researchers tested the food-size preferences of captive bluegill sunfish (*Lepomis macrochirus*) by offering them equal numbers of small, medium, and large-sized prey (*Daphnia magna*) at three different prey densities. Because large prey are the easiest to find, bluegills encountered them more frequently than small or medium-sized prey, especially at the highest prey density. The bluegills' selection of prey varied with prey density; they strongly preferred large prey when prey of all sizes were abundant.
© Cengage Learning 2017

irritating hairs to protect themselves from herbivores. Many plant tissues also contain poisonous chemicals that deter herbivores from feeding. For example, plants in the milkweed family (Asclepiadaceae) exude a milky, irritating sap that contains cardiac glycosides, even small amounts of which are toxic to vertebrate heart muscle. Other compounds mimic the structure of insect hormones, disrupting the development of insects that consume them. Most of these poisonous compounds are volatile, giving plants their typical aromas; some herbivores have coevolved the ability to recognize these odors and avoid the toxic plants. Recent research indicates that some plants increase their production of toxic compounds in response to herbivore feeding. For example, potato and tomato plants that have been damaged by herbivores produce higher levels of protease-inhibiting chemicals; these compounds prevent herbivores from digesting proteins they have just consumed, reducing the food value of these plant tissues.

Many animals have evolved an appearance that provides a passive defense against predation **(Figure 53.3).** Caterpillars that look like bird droppings, for example, may not attract much attention from a hungry predator. And as you learned in Chapter 1 (see Figure 1.9), **cryptic coloration** helps some prey (as well as some predators) to blend in with their surroundings.

Once discovered by a predator, many animals first try to run away. When cornered, they may try to startle or intimidate the predator with a display that increases their apparent size or ferocity **(Figure 53.4)**. Such a display might confuse the predator just long enough to allow the potential victim to escape. Other species seek shelter in protected sites. For example, flexible-shelled African pancake tortoises (*Malacochersus tornieri*) retreat into rocky crevices and puff themselves up with air, becoming so tightly wedged between rocks that predators cannot extract them.

Other animals defend themselves actively. North American porcupines (genus *Erethizon*) release hairs modified into sharp, barbed quills that stick in a predator's mouth, causing severe pain and swelling. Other species fight back by biting, charging, or kicking an attacking predator. Chemical defenses also provide effective protection. Skunks release a noxious spray when threatened, and some frogs and toads produce neurotoxic skin secretions that paralyze and kill mammals. Some insects even protect themselves with poisons acquired from plants. The caterpillars of monarch butterflies (*Danaus plexippus*) are immune to the cardiac glycosides in the milkweed leaves they eat. They store these chemicals at high concentration, even through metamorphosis, making adult monarchs poisonous to vertebrate predators.

Poisonous or repellant species often advertise their unpalatability with bright, contrasting patterns, called **aposematic coloration (Figure 53.5)**. Although a predator might attack a black-and-white skunk, a yellow-banded wasp, or an orange monarch butterfly once, it quickly learns to associate the gaudy color pattern with pain, illness, or severe indigestion—and rarely attacks these easily recognized animals again.

A. Bird dropping mimic
(Papilio cresphontes)

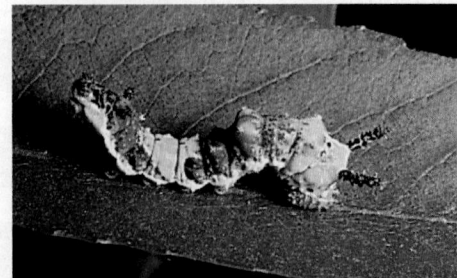

Edward S. Ross

B. Damaged leaf mimic
(Mimetica species)

Dr. Morley Read/Science Source

FIGURE 53.3 Hiding in plain sight. Some animals, such as **(A)** giant swallowtail butterfly larvae that resemble bird droppings and **(B)** some katydids that resemble insect-damaged leaves, do not attract the attention of predators.

Mimicry, in which one species evolves an appearance resembling that of another **(Figure 53.6),** is also a form of defense. In **Batesian mimicry,** named for English naturalist Henry W. Bates, a palatable or harmless species, the **mimic,** resembles an unpalatable or poisonous one, the **model.** Any predator that eats the poisonous model will subsequently avoid other organisms that resemble it. Ecologists have identified many examples of Batesian mimicry between vertebrate mimics and models and between invertebrate mimics and models, but few vertebrates are known to mimic invertebrates. In 2015, Gustavo A. Londoño at the University of California, Riverside, and colleagues at the Universidad Icesi, Cali, Colombia, reported an apparent example of a bird that mimics a poisonous insect. The cinereous mourner (*Laniocera hypopyrra*), a bird that lives in Amazonian forests, lays its eggs in open, cup-shaped nests, and their young are vulnerable to predation during their first few weeks of life. In this species, chicks hatch from their eggs with a luxurious growth of yellow feathers, which look remarkably like the fuzzy hair and spines on the poisonous larvae of flannel moths (Megalopygidae), the model, that live in the same forests (see Figure 53.6A). The cinereous mourner chicks not only resemble the caterpillars in appearance, but, when they are unattended in the nest, they tuck their bills and feet under their feathers and wriggle the way these

G. Ronald Austing/Science Source

FIGURE 53.4 Startle defenses. A short-eared owl (*Asio flammeus*) increases its apparent size when threatened by a predator.

Courtesy of Ken Nemuras

FIGURE 53.5 Aposematic coloration. Poisonous animals, such as the harlequin toad (*Atelopus varius*), often have bright warning coloration.

caterpillars do. Their appearance and behavior presumably protect them from predators that feed on young birds.

In **Müllerian mimicry,** named for German zoologist Fritz Müller, two or more unpalatable species share a similar appearance, which reinforces the lesson learned by a predator that attacks any species in the mimicry complex. Tropical butterflies in the genus *Heliconius* often form Müllerian mimicry complexes (see Figure 53.6B). Although the color patterns of species vary geographically across South America, the appearance of members of a mimicry complex always match.

Despite the effectiveness of many antipredator defenses, coevolution has often molded the responses of predators to overcome them. For example, when threatened by a predator, the pinacate beetle (*Eleodes longicollis*) raises its rear end and sprays a noxious chemical from a gland at the tip of its abdomen. Although this behavior deters many would-be predators, grasshopper mice (genus *Onychomys*) of the American Southwest circumvent this defense: they grab the beetles, shove their abdomens into the ground, rendering the beetle's spray ineffective, and then eat the beetles headfirst (**Figure 53.7**).

Interspecific Competition Occurs When Different Species Depend on the Same Limiting Resources

Populations of different species often use the same limiting resources, causing **interspecific competition** (competition *between* species). The competing populations may experience increased mortality and decreased reproduction, responses that are similar to the effects of intraspecific competition (see Section 52.4). Interspecific competition reduces the size and population growth rate of one or more of the competing populations.

Community ecologists identify two main forms of interspecific competition. In **interference competition,** individuals of one species harm individuals of another species directly. Animals may fight for access to resources, as when lions chase smaller scavengers such as hyenas and jackals from their kills. Similarly, many plant species, including creosote bushes (see Figure 52.4), release toxic chemicals that prevent other plants from growing nearby. In **exploitative competition,** two or more populations use ("exploit") the same limiting resource. The presence of one species reduces resource availability for the others, even in the absence of snout-to-snout or root-to-root confrontations. For example, in the deserts of the American Southwest, many bird and ant species feed largely on seeds. Thus, each seed-eating species may deplete the food supply available to others.

COMPETITIVE EXCLUSION In the 1920s, the Russian mathematician Alfred J. Lotka and the Italian biologist Vito Volterra independently proposed a model of interspecific competition, modifying the logistic equation (see Section 52.4) to describe

A. Batesian mimicry

Cinereus mourner (*Laniocera hypopyrra*) chick, the mimic

Flannel moth caterpillar (Megalopygidae), the model

B. Müllerian mimicry

Heliconius erato

Heliconius melpomene

FIGURE 53.6 Mimicry. (A) Batesian mimics are harmless animals that mimic a dangerous one. The newly hatched chicks of the cinereous mourner (a bird) is a Batesian mimic of the poisonous caterpillars of flannel moths. The chicks even wriggle the way the caterpillars do. **(B)** Müllerian mimics are poisonous species that share a similar appearance. Two distantly related species of butterfly have nearly identical patterns on their wings.

the effects of competition between two species. In their model, an increase in the size of one population reduces the population growth rate (*r*) of the other.

A Russian biologist, G. F. Gause, tested this general model experimentally in the 1930s, using species that might compete with each other for resources in a simple laboratory system. He grew cultures of two *Paramecium* species (ciliate protists)

A. Pinacate beetle (*Eleodes longicollis*)

B. Grasshopper mouse (*Onychomys* species)

FIGURE 53.7 Coevolution of predators and prey. (A) When disturbed by a predator, the pinacate beetle sprays a noxious chemical from its posterior end. **(B)** Grasshopper mice overcome this defense by shoving a beetle's rear end into the soil and dining on it headfirst.

Gause's Experiments on Interspecific Competition in *Paramecium*

Question: Can two species of *Paramecium* coexist in a simple laboratory environment?

Experiment: Gause grew populations of two species, *Paramecium aurelia* and *Paramecium caudatum,* alone (single species cultures) or together (mixed cultures) in small bottles in his laboratory. To determine whether the growth of these populations followed the predictions of the logistic equation, Gause had to maintain a reasonably constant carrying capacity in each culture. Thus, he fed the cultures a broth of bacteria, and he eliminated their waste products (by centrifuging the cultures and removing some of the culture medium) on a regular schedule. He then monitored their population sizes through time.

Results: When grown separately, *P. caudatum* **(A)** and *P. aurelia* **(B)** each exhibited logistic population growth. But when the two species were grown together in a mixed culture **(C),** *P. aurelia* persisted and *P. caudatum* was nearly eliminated from the culture.

A. *P. caudatum* alone

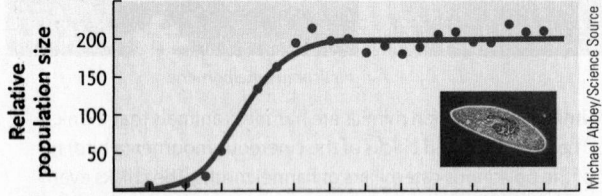

B. *P. aurelia* alone

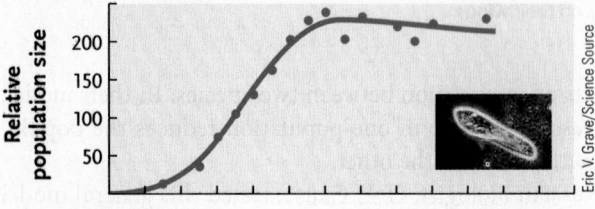

C. Mixed culture

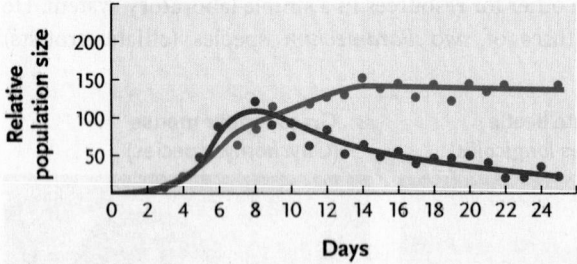

Conclusion: Because one species was almost always eliminated from mixed species cultures, Gause formulated the competitive exclusion principle: populations of two or more species cannot coexist indefinitely if they rely on the same limiting resources and exploit them in the same way.

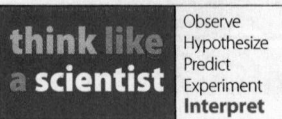

think like a scientist

Observe
Hypothesize
Predict
Experiment
Interpret

The results of this study suggest that when the two species are grown together in mixed culture, *P. aurelia* persists and *P. caudatum* becomes extinct. But does the presence of *P. caudatum* appear to have any effect on *P. aurelia* in the mixed cultures?

Source: G. F. Gause. 1934. *The Struggle for Existence.* Williams & Wilkins Company, London.

under constant laboratory conditions, regularly renewing food and removing wastes. Both species fed on bacteria suspended in the culture medium. When grown alone, each species exhibited logistic growth. When grown together in the same dish, however, *Paramecium aurelia* persisted at high density, but *Paramecium caudatum* was nearly eliminated **(Figure 53.8).** These results inspired Gause to define the **competitive exclusion principle:** populations of two or more species cannot coexist indefinitely if they rely on the same limiting resources and exploit them in the same way. One species is inevitably more successful, harvesting resources more efficiently and producing more offspring than the other.

THE NICHE CONCEPT Ecologists developed the concept of the **ecological niche** as a tool for visualizing resource use and the potential for interspecific competition in nature. We define a population's niche by the resources it uses and the environmental conditions it requires over its lifetime. In this context, the niche includes food, shelter, and nutrients, as well as abiotic conditions, such as light intensity and temperature, which cannot be depleted. In theory, one could identify an almost infinite variety of conditions and resources that contribute to a population's niche. In practice, ecologists usually analyze a few critical resources for which populations might compete. Sunlight, soil moisture, and inorganic nutrients are important resources for plants. Food type, food size, and nesting sites are important for animals.

Ecologists distinguish the **fundamental niche** of a population, the range of conditions and resources that it can possibly tolerate and use, from its **realized niche,** the range of conditions and resources that it actually uses in nature. Realized niches are smaller than fundamental niches, partly because all tolerable conditions are not always present in a habitat, and partly because some resources are used by other species. We can visualize competition between two populations by plotting their fundamental and realized niches with respect to one or more resources **(Figure 53.9).** If the fundamental niches of two populations overlap, they *might* compete in nature.

EVALUATING COMPETITION IN NATURE The observation that several populations use the same resource does not demonstrate that competition is occurring. For example, all terrestrial animals consume oxygen, but they do not compete for oxygen because it is usually plentiful.

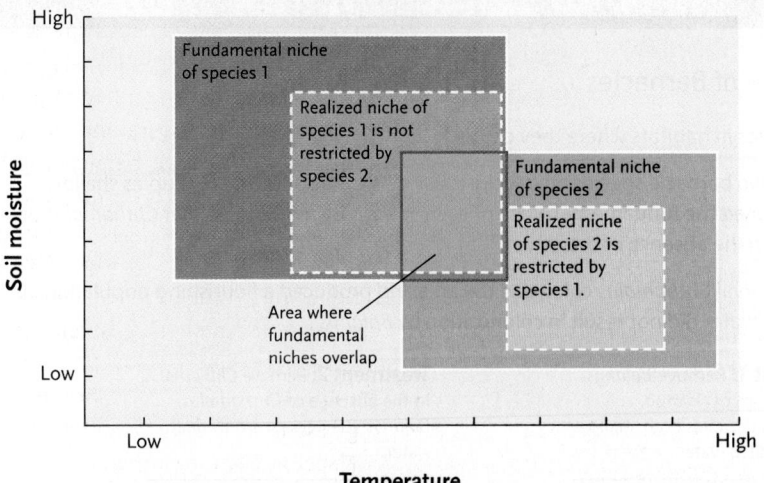

FIGURE 53.9 **Fundamental versus realized niches.** In this hypothetical example, both species 1 and species 2 can survive intermediate temperature and soil moisture conditions, as indicated by the green shading where their fundamental niches overlap. Because species 1 actually occupies virtually all of this overlap zone, its realized niche is not affected by the presence of species 2. By contrast, the realized niche of species 2 is restricted by the presence of species 1; species 2 occupies only the warmer and dryer parts of the habitat.
© Cengage Learning 2017

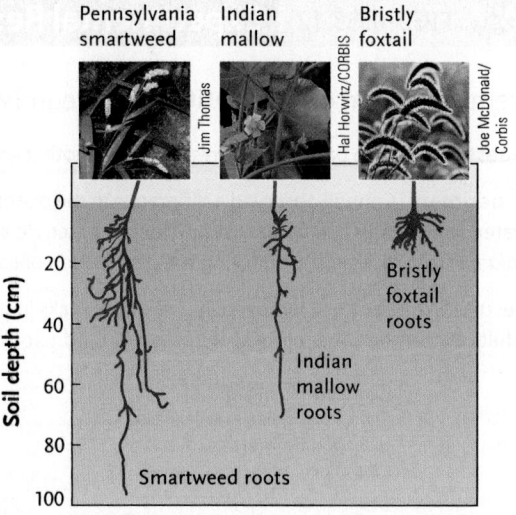

FIGURE 53.10 **Resource partitioning.** The root systems of three plant species that grow in abandoned fields partition water and nutrient resources in soil. Pennsylvania smartweed *(Polygonum pensylvanicum)* has a deep taproot that branches at many depths; Indian mallow *(Abutilon theophrasti)* has a moderately deep taproot; and bristly foxtail grass *(Setaria faberi)* has a shallow root system.
© Cengage Learning 2017

Nevertheless, two general phenomena provide *indirect* evidence that interspecific competition in the past probably fostered the evolution of behaviors, structures, and ecological differences that enable species to avoid interspecific competition in the present. The first is the extremely common observation of **resource partitioning,** the use of different resources or the use of resources in different ways, by species living in the same place. For example, weedy plants might compete for water and dissolved nutrients in abandoned fields. But they reduce the level of competition by partitioning these resources, collecting them from different depths in the soil **(Figure 53.10).**

A second phenomenon that suggests the importance of competition is observed in comparisons of species that are sometimes sympatric (that is, living in the same place) and sometimes allopatric (that is, living in different places); allopatry and sympatry were discussed in Section 22.3 in the context of speciation. In several studies of animals, researchers have documented **character displacement:** allopatric populations are morphologically similar and use similar resources, but sympatric populations are morphologically different and use different resources. The differences between the sympatric populations allow them to coexist without competing. Differences in bill size among sympatric finch species on the Galápagos Islands (see Section 20.2) may be the product of character displacement **(Figure 53.11).**

Data on resource partitioning and character displacement merely suggest the possible importance of interspecific competition in nature. To demonstrate *conclusively* that interspecific competition limits natural populations, one must show that the presence of one population reduces the population size or distribution of its presumed competitor. In a classic field

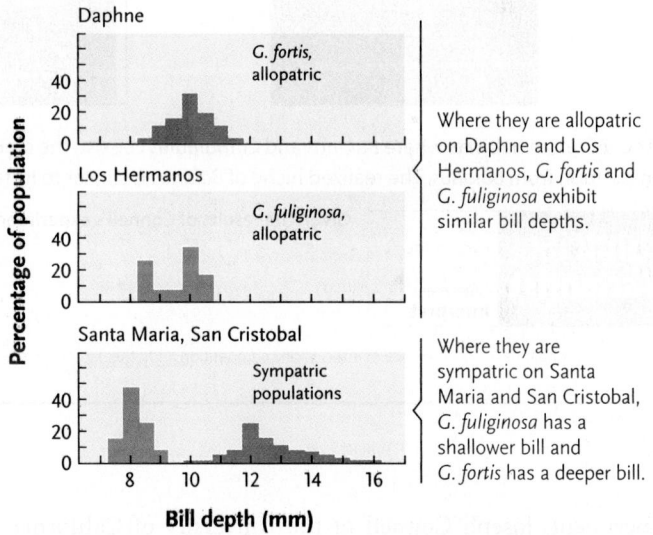

Where they are allopatric on Daphne and Los Hermanos, *G. fortis* and *G. fuliginosa* exhibit similar bill depths.

Where they are sympatric on Santa Maria and San Cristobal, *G. fuliginosa* has a shallower bill and *G. fortis* has a deeper bill.

FIGURE 53.11 **Character displacement.** *Geospiza fortis* and *Geospiza fuliginosa* exhibit character displacement in the depth of their bills, a trait that is correlated with the sizes of seeds they eat.
© Cengage Learning 2017

FIGURE 53.12 | **Experimental Research**

Demonstration of Competition between Two Species of Barnacles

Question: Do two barnacle species limit one another's realized niches in habitats where they coexist?

Experiment: Connell observed a difference in the distributions of two barnacle species on a rocky coast: *Chthamalus stellatus* occupies shallow water, and *Balanus balanoides* lives in deeper water. He then determined the fundamental niche of each species by removing either *Chthamalus* or *Balanus* from rocks and monitoring the distribution of each species in the absence of the other.

Results: When Connell removed *Balanus* from rocks in deep water, larval *Chthamalus* colonized the area and produced a flourishing population of adults. By contrast, the removal of *Chthamalus* from rocks in shallow water did not result in colonization by *Balanus*.

Control: No treatment. *Chthamalus* occupies only shallow water and *Balanus* occupies only deep water.

Treatment 1: Remove *Balanus*. In the absence of *Balanus*, *Chthamalus* occupies both shallow water and deep water.

Treatment 2: Remove *Chthamalus*. In the absence of *Chthamalus*, *Balanus* still occupies only deep water.

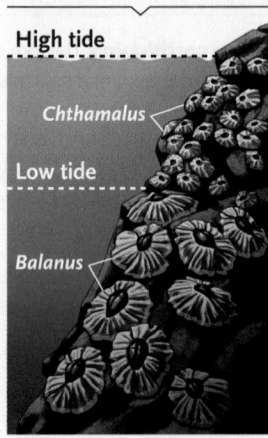
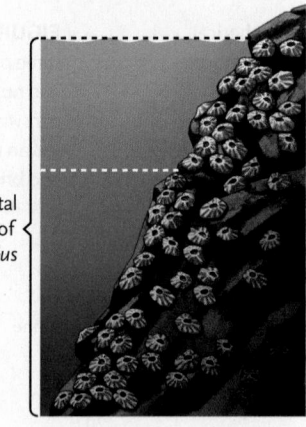

Conclusion: In habitats where *Balanus* and *Chthamalus* coexist, the realized niche of *Chthamalus* is smaller than its fundamental niche because of competition from *Balanus*. The realized niche of *Balanus* is similar to its fundamental niche because it is not affected by the competitive interaction.

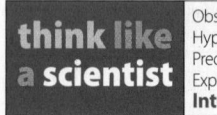

Observe
Hypothesize
Predict
Experiment
Interpret

Given the results of Connell's experiment, which species of barnacle would you describe as the stronger competitor?

Source: J. H. Connell. 1961. The influence of interspecific competition and other factors on the distribution of the barnacle *Chthamalus stellatus*. *Ecology* 42:710–723.

experiment, Joseph Connell of the University of California, Santa Barbara, determined that competition between two barnacle species caused the realized niche of one species to be smaller than its fundamental niche **(Figure 53.12).**

Connell first observed the distributions of barnacles in undisturbed habitats. *Chthamalus stellatus* is generally found in shallow water on rocky coasts, where it is periodically exposed to air. *Balanus balanoides* typically lives in deeper water, where it is usually submerged.

Connell determined the fundamental niche of each species by removing either *Chthamalus* or *Balanus* from rocks and monitoring the distribution of each species in the absence of the other. When Connell removed *Balanus* from rocks in deep water, larval *Chthamalus* colonized the area and produced a

flourishing population of adults. Connell had observed that *Balanus* physically displaced *Chthamalus* from these rocks. Thus, interference competition from *Balanus* prevents *Chthamalus* from occupying areas where it would otherwise live. By contrast, the removal of *Chthamalus* from rocks in shallow water did not result in colonization by *Balanus*. *Balanus* is apparently unable to live in habitats that are frequently exposed to air. Connell therefore concluded that competition from *Chthamalus* does not affect the distribution of *Balanus*. Thus, the competitive interaction between these two species is asymmetrical: *Balanus* has a substantial effect on *Chthamalus*, but *Chthamalus* has virtually no effect on *Balanus*.

Understanding the limits of a species' fundamental and realized niches allows researchers to manage introduced species

FIGURE 53.13 The invasion of the toads. Populations of the Central and South American marine toad *(Rhinella marina)* exploded after it was introduced into Australia as a pest control agent.

(see *Why It Matters . . .* in Chapter 52 and Section 55.2). For example, the cane toad *(Rhinella marina),* which is native to Central and South America, was introduced into Australia in 1935 to control insect pests of sugarcane, an important agricultural crop. It has since increased its numbers to more than 2 million and its geographical range in Australia to more than 1.2 million square kilometers **(Figure 53.13).** It poses a serious threat to native wildlife because it competes for food with some native animal species, feeds on others, and is toxic to predators that consume it. In 2014, Reid Tingley at the University of Melbourne, Australia, and colleagues from Australia and Brazil published an analysis of the fundamental and realized niches of cane toads in South America and Australia. Their results indicate that environmental conditions in Australia match the fundamental niche of cane toads, and, in the absence of competing species or natural predators, it has expanded dramatically its realized niche in Australia. By contrast, in Central and South America, the cane toad does not occupy the entirety of its fundamental niche because of competition from a closely related toad species in cooler and drier habitats at the southern end of its range. Had the biologists who introduced cane toads into Australia understood the fundamental and realized niches of cane toads, they might have sought a different pest-control agent.

In Symbiotic Associations, the Lives of Two or More Species Are Closely Intertwined

Some species have a physically close ecological association called **symbiosis** (*sym* = together; *bio* = life; *sis* = condition). Biologists define three types of symbiotic interactions— *commensalism, mutualism,* and *parasitism*—that differ in their effects.

Commensalism, in which one species benefits and the other experiences neither benefit nor harm, is rare in nature, because few species are unaffected by their interactions with another. One possible example is the relationship between cattle egrets *(Bubulcus ibis),* birds in the heron family, and the large grazing mammals with which they associate **(Figure 53.14).** Cattle egrets feed on insects and other small animals that their commensal partners flush from grass. Feeding rates of egrets are higher when they associate with large grazers than when they do not. The birds clearly benefit from this interaction, but the presence of birds has no apparent positive or negative impact on the mammals.

Mutualism, in which both partners benefit, is extremely common. The coevolved relationships between flowering plants and animal pollinators are largely mutualistic (see Section 29.3). Animals that feed on a plant's nectar or pollen carry its gametes from one flower to another **(Figure 53.15).** Similarly, animals that eat the fruits of flowering plants disperse the seeds, "planting" them in a pile of nutrient-rich feces. These mutualistic relationships between plants and animals do not require active cooperation. Each species simply exploits the other for its own benefit.

Some associations between either bacteria or fungi and plants are also mutualistic. For example, mycorrhizae are fungi that grow alongside the roots of many plant species. These fungi facilitate the plants' uptake of nitrogen and phosphorus from the soil, and the plants provide the fungi with carbohydrates in return. Another important mutualism is the close relationship between the nitrogen-fixing bacterium *Rhizobium* and leguminous plants, such as peas, beans, and clover (see Section 35.3).

Mutualistic relationships between animal species are also common. For example, some small marine fishes feed on parasites that attach to the mouths and gills of large predatory fishes **(Figure 53.16).** Parasitized fishes hover motionless while the "cleaners" scour their tissues. The relationship is mutualistic because the cleaner fishes get a meal, and the larger fishes are relieved of parasites.

The relationship between the bullhorn acacia tree *(Acacia cornigera)* of Central America and a small ant

FIGURE 53.14 Commensalism. Cattle egrets *(Bubulcus ibis)* feed on insects and other small animals flushed by the movements of large grazing mammals, such as this African buffalo *(Syncerus caffer).*

A. Flowering yucca plant

© Douglas Knight/Shutterstock.com

B. Female yucca moth

Harlo H. Hadow

A female yucca moth uses highly modified mouthparts to gather the sticky yucca pollen and roll it into a ball. She carries the pollen to another flower, and after piercing its ovary wall, she lays her eggs. She then places the pollen ball into the opening of the stigma.

C. Yucca moth larva

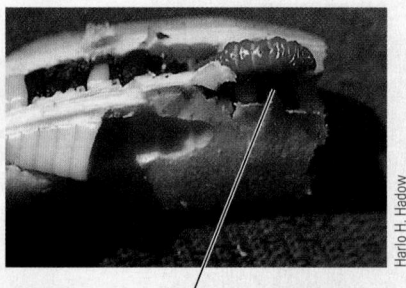

Harlo H. Hadow

When moth larvae hatch from the eggs, they eat some of the yucca seeds and gnaw their way out of the ovary to complete their life cycle. Enough seeds remain undamaged to produce a new generation of yuccas.

FIGURE 53.15 Mutualism between plants and animals. Several species of yucca plants (*Yucca* species) are each pollinated exclusively by one species of yucca moth (*Tegeticula* species). The adult stage of each moth appears at the time of year when its yucca plant flowers. These species are so mutually interdependent that the larvae of each moth species can feed on only one type of yucca, and the flowers of each yucca can be fertilized by only one species of moth. Most plant–pollinator mutualisms are much less specific.

species (*Pseudomyrmex ferruginea*) is one of the most highly coevolved mutualisms known **(Figure 53.17).** Each acacia is inhabited by an ant colony that lives in the tree's swollen thorns. The ants swarm out of the thorns to sting—and sometimes kill—herbivores that touch the tree. The ants also clip any vegetation that grows nearby. Thus, acacia trees that are

© James A Dawson/Shutterstock.com

FIGURE 53.16 Mutualism between animal species. A giant moray (*Gymnothorax javanicus*) remains nearly motionless in the water while bluestreak cleaner wrasses (*Labroides dimidiatus*) carefully remove and eat ectoparasites attached to its mouth. The moray is a predator, and the striped cleaner wrasse is a potential prey—but their mutualistic interaction supersedes a possible predator–prey interaction.

colonized by ants grow in a space free of herbivores and competitors, and occupied trees grow faster and produce more seeds than unoccupied trees. In return, the plants produce sugar-rich nectar consumed by adult ants and protein-rich structures that the ants feed to their larvae. Ecologists describe the coevolved mutualism between these species as *obligatory,* at least for the ants; they cannot subsist on any other food sources.

Parasitism is a type of interaction in which one species, the **parasite,** uses another, the **host,** in a way that is harmful to the host. Parasite–host relationships are like predator–prey relationships: one population of organisms feeds on another. But parasites rarely kill their hosts quickly because a dead host is useless as a continuing source of nourishment.

Tapeworms and other parasites that live *within* a host are **endoparasites.** Many endoparasites acquire their hosts passively when a host accidentally ingests the parasite's eggs or larvae (see *Focus on Research: Applied Research,* Chapter 31). Endoparasites generally complete their life cycle in one or two host individuals. By contrast, leeches (see Figure 31.30), aphids, mosquitoes, and other parasites that feed on the *exterior* of a host are **ectoparasites.** Most animal ectoparasites have elaborate sensory and behavioral mechanisms that allow them to locate specific hosts, and they feed on numerous host individuals during their lifetimes. Some plants, such as mistletoes (genus *Phoradendron*), live as ectoparasites on the trunks and branches of trees; their roots penetrate the host's xylem and extract water and nutrients.

Not all parasites feed directly on a host's tissues. The giant cowbirds described earlier are brood parasites, as are other species of cowbirds and cuckoos. Although oropendolas sometimes benefit from the presence of cowbirds, most brood parasites have negative effects on their hosts. For example,

A. Ants patrolling an acacia

Martin Shields/Getty Images

B. Cleared area around an acacia

Nicholas Smythe/Science Source

FIGURE 53.17 A highly coevolved mutualism. (A) Bullhorn acacia trees *(Acacia cornigera)* provide colonies of small ants *(Pseudomyrmex ferruginea)* with homes in hollow enlarged thorns as well as other resources. Although individual ants are small, they are numerous and aggressive. **(B)** Because the ants attack herbivores and remove vegetation near their tree, acacias occupied by ants grow in a space that is free of herbivores and competitors.

brood parasitism by the brown-headed cowbird *(Molothrus ater)* has played a large role in the near-extinction of Kirtland's warbler *(Dendroica kirtlandii).*

The feeding habits of some insects, called **parasitoids,** fall somewhere between true parasitism and predation. A female parasitoid lays eggs in the larva or pupa of another insect species, and her young consume the tissues of the living host. Some insects respond to the presence of parasitoids in the environment with a "behavioral immune" response. In 2013, Todd A. Schlenke and colleagues at Emory University reported that when female fruit flies *(Drosophila melanogaster)* see female parasitoid wasps in their vicinity, the fruit flies preferentially lay their eggs in decaying fruit with high alcohol content. The alcohol renders the fly larvae less susceptible to parasitoid infection, leading to a fivefold increase in their likelihood of developing into an adult.

53.2 The Nature of Ecological Communities

Ecologists have often debated the nature of ecological communities, asking if they have emergent properties that transcend the interactions among the populations they contain.

Most Ecological Communities Blend into Neighboring Communities

How do complex population interactions affect the organization and functioning of ecological communities? In the 1920s, ecologists in the United States developed two extremely different hypotheses about the nature of ecological communities. Frederic Clements of the University of Minnesota championed an *interactive* view of communities. He described communities as "superorganisms," assemblages of species bound together by complex population interactions. According to this view, each species in a community requires interactions with a set of ecologically different species, just as every cell in an organism requires services that other types of cells provide. Clements believed that once a mature community was established, its **species composition**—the particular combination of species that occupy the site—was at *equilibrium*. If a fire or some other environmental factor disturbed the community, it would return to its predisturbance state.

Henry A. Gleason of the University of Michigan proposed an alternative, *individualistic* view of ecological communities. He believed that population interactions do not always determine species composition. Instead, a community is just an assemblage of species that are individually adapted to similar environmental conditions. According to Gleason's hypothesis, communities do not achieve equilibrium; rather, they constantly change in response to disturbance and environmental variation.

In the 1960s, Robert Whittaker of Cornell University suggested that ecologists could determine which hypothesis was correct by analyzing communities along environmental gradients, such as temperature or moisture **(Figure 53.18)**. According

A. Interactive hypothesis

The interactive hypothesis predicts that species within communities exhibit similar distributions along environmental gradients (indicated by the close alignment of several curves over each section of the gradient) and that boundaries between communities (indicated by arrows) are sharp.

Environmental gradient

B. Individualistic hypothesis

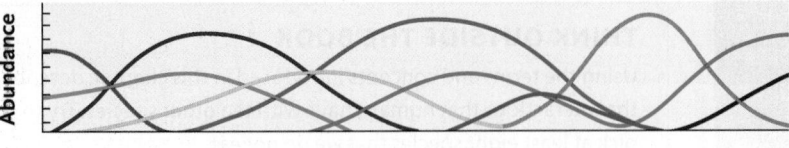

The individualistic hypothesis predicts that species distributions along the gradient are independent (indicated by the lack of alignment of the curves) and that sharp boundaries do not separate communities.

Environmental gradient

C. Siskiyou Mountains

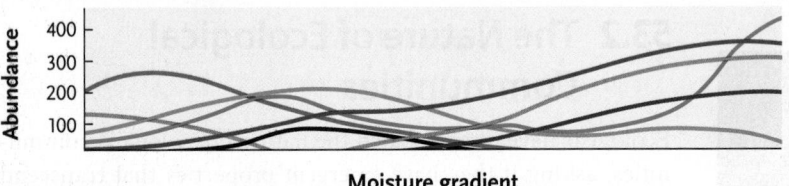

Moisture gradient

Most gradient analyses support the individualistic hypothesis, as illustrated by distributions of tree species along moisture gradients in Oregon's Siskiyou Mountains and Arizona's Santa Catalina Mountains.

D. Santa Catalina Mountains

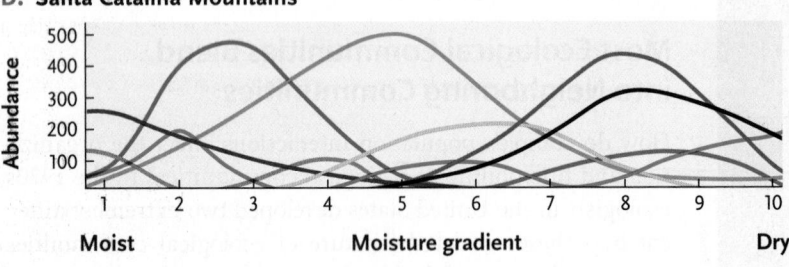

Moist   Moisture gradient   Dry

FIGURE 53.18 Two views of ecological communities. Each colored line represents the abundance of one plant species over the environmental gradient.
© Cengage Learning 2017

to Clements' interactive hypothesis, species that typically occupy the same communities should always occur together. Thus, their distributions along the gradient would be clustered in discrete groups with sharp boundaries between groups (see Figure 53.18A). According to Gleason's individualistic hypothesis, each species is distributed over the section of an environmental gradient to which it is adapted. Different species would have unique distributions, and species composition would change continuously along the gradient. In other words, communities would not be separated by sharp boundaries (see Figure 53.18B).

Most gradient analyses support Gleason's individualistic view of ecological communities. Environmental conditions vary continuously in space, and most plant distributions match these patterns (see Figure 53.18C, D). Species occur together in assemblages because they are adapted to similar conditions, and the species compositions of the assemblages change gradually across environmental gradients.

Nevertheless, the individualistic view does not fully explain all patterns observed in nature. Ecologists recognize certain assemblages of species as distinctive communities—redwood

forests, coral reefs, and the microorganisms that occupy the digestive systems of animals (see Section 47.5) are good examples. But the borders between adjacent communities are often wide transition zones, called **ecotones.** Ecotones are generally rich with species because they include plants and animals from both neighboring communities as well as some species that thrive only under transitional conditions. In some places, however, a discontinuity in a critical resource or some important abiotic factor produces a sharp community boundary. For example, chemical differences between soils derived from serpentine rock and sandstone establish sharp boundaries between communities of native California wildflowers and introduced European grasses **(Figure 53.19).**

STUDY BREAK 53.2

1. Which view of communities suggests that they are just chance assemblages of species that happen to be adapted to similar abiotic environmental conditions?
2. Why would you often find more species living in an ecotone than you would in the communities on either side of it?

FIGURE 53.19 Sharp community boundaries. Soils derived from serpentine rock have high magnesium and heavy metal content, which many plants cannot tolerate. Although native California wildflowers (bright yellow in this photograph) thrive on serpentine soil at the Jasper Ridge Preserve of Stanford University, introduced European grasses (green in this photograph) competitively exclude them from adjacent soils derived from sandstone.

Jasper Ridge Biological Preserve

FIGURE 53.20 Layered forests. Tropical forests include a canopy of tall trees and an understory of short trees and shrubs. Huge vines (lianas) climb through the trees, eventually reaching sunlight in the canopy, and epiphytic plants grow on trunks and branches, increasing the structural complexity of the habitat.
© Cengage Learning 2017

Canopy
Epiphyte
Understory
Liana
Herb layer
Buttress

53.3 Community Characteristics

Although the species composition of any ecological community—whether terrestrial, freshwater, or marine—may vary somewhat over geographical gradients, every community has certain characteristics that define its overall appearance and structure.

The Growth Forms of Plants Establish a Terrestrial Community's Overall Appearance

The growth forms of plants—their sizes and shapes—vary markedly in different environments. Warm, moist environments support complex vegetation with multiple vertical layers. For example, tropical forests include a canopy, formed by the tallest trees; an understory of shorter trees and shrubs; an herb layer under openings in the canopy; vinelike lianas; and epiphytes, which grow on the trunks and branches of trees **(Figure 53.20).** By contrast, physically harsh environments are occupied by low vegetation with simple structure. For example, trees on mountaintops buffeted by cold winds are short, and the plants below them cling to rocks and soil. Other environments support growth forms between these extremes (see Section 51.4).

Foundation Species Moderate the Abiotic Environment within a Community

Within many ecological communities, one common species can function as a **foundation species,** defining the nature of a community by creating locally stable environmental conditions. For example, trees are the foundation species in forested

ecosystems, because their form defines the physical structure of the community, and their leaves and branches moderate short-term fluctuations in abiotic environmental factors such as temperature, runoff from rainfall, and wind speed.

Saltmarsh cordgrass *(Spartina alterniflora)* is a foundation species in the wetlands surrounding Narragansett Bay, Rhode Island, because patches of this meter-high grass slow the velocity of the incoming tide and stabilize the stony beach habitat along the shore. In the absence of *Spartina,* tidal surges move the stones on the beach, disrupting the germination and growth of several small, herbaceous plant species. John E. Bruno of Brown University surveyed the plants growing adjacent to more than 350 *Spartina* patches of varying size. His research revealed that large patches of *Spartina* are more effective than small patches in moderating tidal effects: the percentage of a stony beach occupied by herbaceous plants was directly proportional to the length of the *Spartina* patch that bordered the beach **(Figure 53.21).** Thus, the presence of *Spartina,* a foundation species, allows other species in the community to become established and survive.

Communities Differ in Species Richness and the Relative Abundance of Species

Communities differ greatly in their **species richness,** the total number of species that live within them. For example, the harsh environment on a low desert island may support just a

A. Saltmarsh cordgrass (*Spartina alterniflora*)

B. The effect of cordgrass patch length on the distribution of herbaceous plants

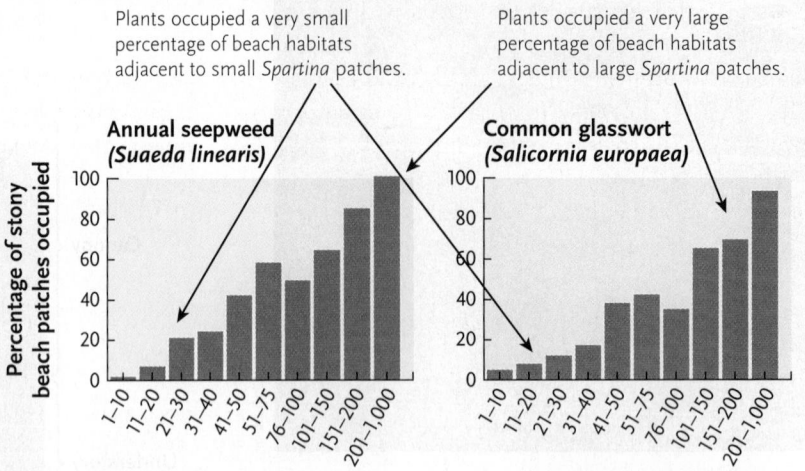

Plants occupied a very small percentage of beach habitats adjacent to small *Spartina* patches.

Plants occupied a very large percentage of beach habitats adjacent to large *Spartina* patches.

Annual seepweed (*Suaeda linearis*)

Common glasswort (*Salicornia europaea*)

Percentage of stony beach patches occupied

Length of adjacent *Spartina* patch (m)

FIGURE 53.21 Foundation species. (A) Saltmarsh cordgrass, the foundation species at the edges of Narragansett Bay, Rhode Island. (B) The presence or absence of herbaceous plants on stony beaches is strongly influenced by the length of the adjacent patch of cordgrass. The plants occupy very few beaches next to small cordgrass patches, but nearly all of the beaches next to large ones.

© Cengage Learning 2017

few species of microorganisms, fungi, algae (photosynthetic protists), plants, and arthropods. By contrast, tropical forests, which grow under milder physical conditions, include many thousands of species. Ecologists have studied global patterns of species richness (described in Section 53.7) for decades. Today, as human disturbance of natural communities has already reached a tipping point, conservation biologists focus on such studies to determine which regions of Earth are most in need of preservation (see Chapter 55).

Within every community, populations differ in their commonness or the **relative abundance** of individuals. Some communities have just one or two **dominant species,** which are either exceptionally numerous or exceptionally large, as well as a number of rare species, each represented by just a few individuals. Foundation species are often dominant in the

communities they occupy. In other communities, species are present in more equal numbers. For example, in a temperate deciduous forest in West Virginia, tulip poplar (*Liriodendron tulipifera*) and sassafras (*Sassafras albidum*) are dominant, together accounting for nearly 85% of the trees (measured as a proportion of the total number of stems and trunks). By contrast, a tropical forest in Costa Rica may include more than 200 tree species, each making up only a small percentage of the total number of trees present.

Species richness and relative abundance together contribute to a community characteristic that ecologists call **species diversity.** To demonstrate the concept of species diversity, we will compare three hypothetical forest communities **(Figure 53.22).** Two of the communities include 50 trees distributed among 10 species. In Forest A, the dominant

Forest A: moderate species diversity

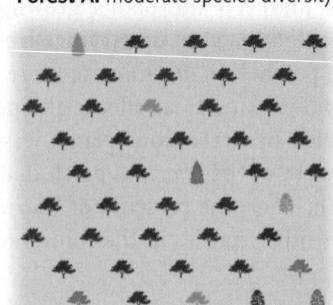

Forest B: high species diversity

Forest C: low species diversity

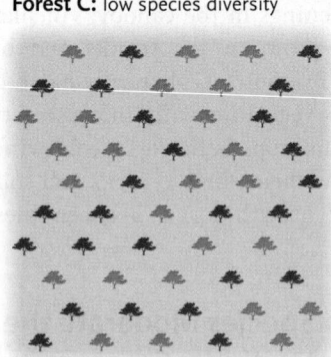

FIGURE 53.22 Species diversity. In this hypothetical example, each of three forests contains 50 trees. Forest A and forest B each include 10 tree species, but forest C includes only two tree species. Because forest A is dominated by one tree species, but forest B is not, ecologists would say that forest B is more diverse. Forest C, with only two tree species, is less diverse than the others.

© Cengage Learning 2017

species is represented by 39 individuals, two species by two individuals each, and seven species by one individual each. In Forest B, each of the 10 species is represented by five individuals. Although both communities have the same species richness (10 species), Forest A is less diverse than Forest B, because more than three-quarters of its trees are of the same species. The third forest has only two tree species (Forest C in Figure 53.22); it is therefore less diverse than either of the others.

Feeding Relationships within a Community Determine Its Trophic Structure

All ecological communities, regardless of their species richness, also have a trophic structure (*trophe* = nourishment) that comprises all of the plant–herbivore, predator–prey, host–parasite, and potential competitive interactions **(Figure 53.23)**.

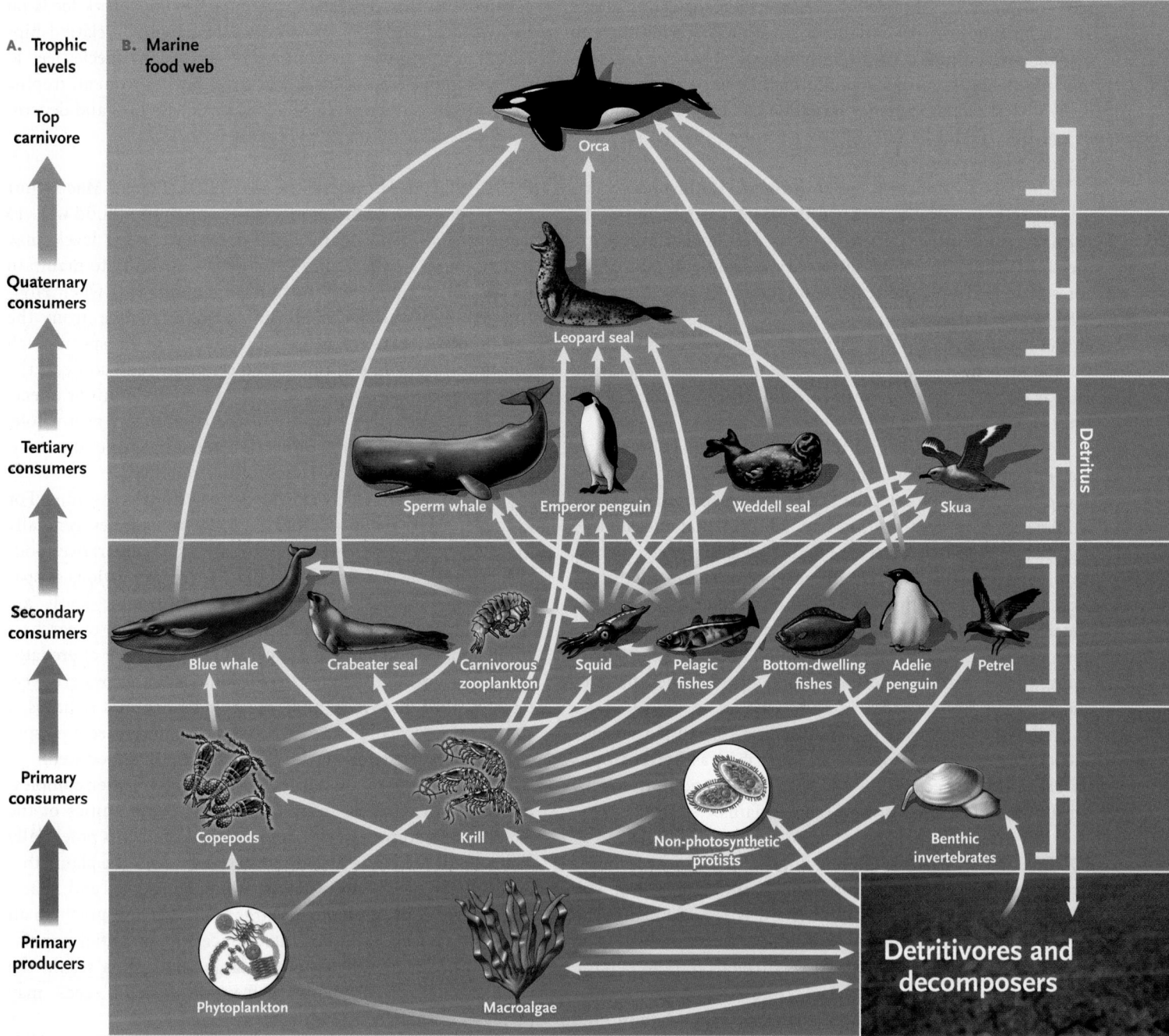

FIGURE 53.23 A simplified depiction of the trophic links that exist among species in the marine food web off the coast of Antarctica. The different trophic levels in this and all other food webs are linked by the feeding relationships within the community. The part of the food web that includes detritivores and decomposers is not detailed in this illustration. A terrestrial food web is illustrated in Figure 54.2.

© Cengage Learning 2017

TROPHIC LEVELS We can visualize the structure of a community as a hierarchy of **trophic levels,** defined by the feeding relationships among its species (see Figure 53.23A). In most terrestrial and aquatic communities, photosynthetic organisms are the **primary producers,** the first trophic level. Primary producers are often described as **autotrophs** (*auto* = self) because they capture sunlight and convert it into chemical energy, using simple *inorganic* molecules acquired from the environment to build larger *organic* molecules that other organisms can use. Plants are the dominant primary producers in terrestrial communities. Multicellular algae (macroalgae) and plants are the major primary producers in shallow freshwater and marine environments. Photosynthetic cyanobacteria and phytoplankton are the primary producers in open water, and chemosynthetic bacteria are the most abundant primary producers in hydrothermal vent and cold seep communities on the deep seafloor (see Section 51.6).

Animals, by contrast, are **consumers.** Herbivores, which feed directly on producers, form the second trophic level, the **primary consumers.** Carnivores that feed on herbivores are the third trophic level, or **secondary consumers,** and carnivores that feed on other carnivores form the fourth trophic level, the **tertiary consumers.** For example, songbirds feeding on herbivorous insects are secondary consumers, and falcons feeding on songbirds are tertiary consumers. Some organisms, such as humans and some bears, are **omnivores,** feeding at several trophic levels simultaneously.

A separate and distinct trophic level includes organisms that extract energy from the organic detritus ("refuse") produced at other trophic levels. Scavengers, or **detritivores,** are animals such as earthworms and vultures that ingest dead organisms, digestive wastes, and cast-off body parts such as leaves and exoskeletons. **Decomposers** are small organisms, such as bacteria and fungi, that feed on dead or dying organic material. As described in Chapter 54, detritivores and decomposers serve a critical ecological function because their activity reduces organic material to small inorganic molecules that producers can assimilate.

All of the consumers in a community—the animals, fungi, and diverse microorganisms—are described as **heterotrophs** (*hetero* = other) because they acquire energy and nutrients by consuming other organisms or their remains.

FOOD CHAINS AND WEBS Ecologists depict the trophic structure of a community in a **food chain,** a portrait of who eats whom. Each link in a food chain is represented by an arrow pointing from food to consumer, thereby illustrating the direction in which energy flows through the food chain. Simple, straight-line food chains are rare in nature because most consumers feed on more than one type of food, and because most organisms are eaten by more than one type of consumer. These complex relationships are portrayed as a **food web,** a set of interconnected food chains with multiple links.

In the food web for the waters off the coast of Antarctica (see Figure 53.23B), most organisms at the bottom of the food web are tiny, and they occur in vast numbers. Huge pastures of phytoplankton (microscopic algae and diatoms) are responsible for most photosynthesis. They are consumed by herbivorous zooplankton (some protists, copepods, and shrimplike krill), which are in turn eaten by larger species, such as carnivorous zooplankton, squids, fishes, and suspension-feeding baleen whales. Some of these secondary consumers are themselves eaten by birds and mammals at higher trophic levels. The top carnivore in this ecosystem, the orca (*Orcinus orca*), feeds on carnivorous birds and mammals. Ecological relationships within any food web are complex because many species feed at more than one trophic level. Moreover, wastes from all trophic levels are consumed and processed by detritivores and decomposers (described further in Section 54.2).

FOOD-WEB ANALYSIS In the late 1950s, Robert MacArthur of Princeton University pioneered the analysis of food webs to determine how the many links between trophic levels may contribute to a community's **stability**—its ability to maintain its species composition and relative abundances when environmental disturbances eliminate some species from the community. MacArthur hypothesized that in species-rich communities, where animals feed on many food sources, the absence of one or two species would have only minor effects on the structure and stability of the community as a whole. He therefore proposed a connection between species diversity, food-web complexity, and community stability.

Recent research has confirmed MacArthur's reasoning. For example, the average number of links per species generally increases with increasing species richness. Comparative food-web analyses have revealed that the relative proportions of species at the highest, middle, and lowest trophic levels are reasonably constant across communities. When researchers compared the number of prey species to the number of predator species in food webs from 92 communities of freshwater invertebrates, they discovered that, regardless of species richness, a community typically includes between two and three prey species for every predator species that is present in a food web.

Interactions among species in a food web are often complex, indirect, and hard to unravel. In desert communities of the American Southwest, for example, rodents and ants potentially compete for seeds, their main food source. And the plants that produce the seeds compete for water, nutrients, and space. Rodents generally prefer to eat large seeds, but ants prefer small seeds. Thus, feeding by rodents reduces the potential population sizes of plants that produce large seeds. As a result, the population sizes of plants that produce small seeds may increase, ultimately providing more food for ants.

Some analyses of food webs focus on interactions in which predators or prey have a significant influence on the growth rates and sizes of other populations in the community; these *strong interactions* can affect overall community structure. In the next section we provide examples of strong interactions when we describe how consumers influence the competitive interactions among populations of their prey.

STUDY BREAK 53.3

1. What plant growth forms are common in tropical forests?
2. What is the difference between species richness and relative abundance?
3. Peregrine falcons are predatory birds that have been introduced into many North American cities, where they feed primarily on pigeons. The pigeons eat mostly vegetable matter. To what two different trophic levels do pigeons and peregrine falcons belong?

53.4 Effects of Population Interactions on Community Characteristics

Numerous studies have shown that interspecific competition and predation can influence a community's species composition.

Interspecific Competition Can Reduce Species Richness within Communities

Interspecific competition can cause the local extinction of species or prevent new species from becoming established in a community, thus reducing its species richness. During the 1960s and early 1970s, ecologists emphasized competition as the primary factor structuring communities. Observations of resource partitioning and character displacement suggested that some process had fostered the evolution of differences in resource use among coexisting species, and competition provided the most straightforward explanation of these patterns.

Seeking to uncover direct evidence of competition, ecologists undertook many field experiments on competition in natural populations. The experiment on barnacles depicted in Figure 53.12 is typical of this approach, in which researchers determine whether adding or removing a species changes the distribution or population size of its presumed competitors. In the early 1980s, two independent reviews of the literature on these field experiments, one by Joseph Connell and the other by Thomas W. Schoener of the University of California, Davis, suggested that competition is sometimes a potent force. Connell's survey, which included 527 published experiments on 215 species, identified competition in roughly 40% of the experiments and more than 50% of the species. Schoener's review, which used different criteria to evaluate 164 experiments on approximately 400 species, found that competition affected more than 75% of the species.

Although these early reviews confirm the importance of competition, the ecological literature on which they were based probably contains several significant biases. First, ecologists who set out to study competition are more likely to study interactions in which they think competition occurs, and they are more likely to publish research that documents its importance. Accordingly, the literature includes more studies of competition in *K*-selected species than in *r*-selected species (review Section 52.5). Recall that populations of *r*-selected species, such as herbivorous insects, rarely reach carrying capacity, and competition may not limit their population sizes. Thus, the Connell and Schoener surveys may *overestimate* the importance of competition. Another bias, which Connell called "the ghost of competition past," *underestimates* the importance of competition. If, as many ecologists believe, the evolution of resource partitioning and character displacement is the result of past competition, we are unlikely to witness much competition today, even though it was once important in structuring those population interactions.

Ecologists have still not reached consensus about whether interspecific competition strongly influences the species composition and structure of most communities. Forest ecologists and vertebrate ecologists, who often study *K*-selected species, generally believe that competition has a profound effect on species distributions and resource use. Insect ecologists and marine ecologists, who often study *r*-selected species, argue that competition is not the major force governing community structure, pointing instead to predation or parasitism and physical disturbance.

Predators Can Boost Species Richness by Stabilizing Competitive Interactions among Their Prey

Predators can influence the species richness and structure of communities by reducing the population sizes of their prey. On the rocky coast of the American Northwest, for example, algae and sessile invertebrates compete for attachment sites on rocks, a requirement for life on a wave-swept shore. California mussels (*Mytilus californianus*) are the strongest competitors for space, eliminating other species from the community. But at some sites, predatory sea stars (*Pisaster ochraceus*) preferentially feed on mussels, reducing their numbers and creating space for other species to grow. Because the interaction between *Pisaster* and *Mytilus* affects other species as well, it qualifies as a strong interaction.

In the 1960s, Robert Paine of the University of Washington conducted species removal experiments to evaluate the effects of *Pisaster* predation **(Figure 53.24).** In predator-free experimental plots, mussels outcompeted barnacles, chitons, limpets, and other invertebrate herbivores, reducing species richness from 18 species to two or three. In control plots that contained predators, however, all 18 species persisted. Ecologists describe predators such as *Pisaster* as **keystone species,** those that have a greater effect on community structure than their numbers might suggest.

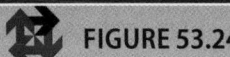 **FIGURE 53.24** | **Experimental Research**

Effect of a Predator on the Species Richness of Its Prey

Question: Does feeding by a predator influence the species richness and relative abundances of the species on which it feeds?

Experiment: The predatory sea star *Pisaster ochraceus* preferentially feeds on California mussels *(Mytilus californianus),* which are the strongest competitors for space in rocky intertidal habitats in Washington State. Paine removed *Pisaster* from caged experimental study plots, but left control study plots undisturbed. He then monitored the species richness of *Pisaster's* invertebrate prey over many years.

Results: Paine documented an increase in California mussel populations in the experimental plots, as well as complex changes in the feeding relationships among species in the intertidal food web. The overall effect of removing *Pisaster,* the top predator in this food web, was a rapid decrease in the species richness of invertebrates and algae. By contrast, control plots maintained their species richness over the entire course of the experiment.

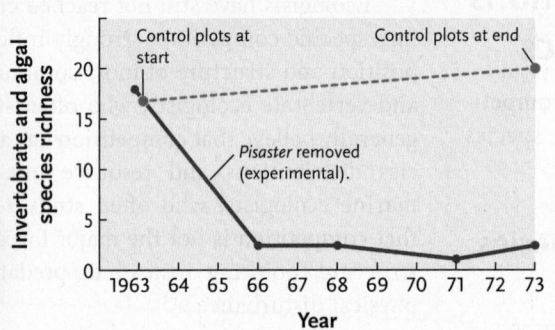

Conclusion: Predation by the sea star *Pisaster ochraceus* maintains the species richness of its prey by preventing mussels from outcompeting other invertebrates on rocky shores.

Suggest one possible explanation for why the full effects of *Pisaster* removal were not evident until three years after the experiment was begun.

Source: R. T. Paine. 1974. Intertidal Community Structure. Experimental studies on the relationship between a dominant competitor and its principal predator. *Oecologia* 15:93–120.

© Cengage Learning 2017

Herbivores May Counteract or Reinforce Competition among Species They Eat

Herbivores also exert complex effects on communities. In the 1970s, Jane Lubchenco, then of Harvard University, studied herbivory by a periwinkle snail *(Littorina littorea),* a keystone species on rocky shores in Massachusetts **(Figure 53.25).** Periwinkles preferentially graze on the tender green alga *Enteromorpha.* In tidepools, which are usually submerged, *Enteromorpha* outcompetes other algae. Moderate feeding by periwinkles, however, eliminates some *Enteromorpha,* allowing less competitive algal species to grow. Moderate herbivory by periwinkles therefore increases algal species richness in tidepools. But on high rocks, which are more frequently exposed to air during low tide, the dehydration-resistant red alga *Chondrus* is competitively dominant. Periwinkles do not

eat the tough *Chondrus,* however, feeding instead on the less abundant and competitively inferior *Enteromorpha.* Thus, on exposed rocks, feeding by the snails reduces algal species richness.

STUDY BREAK 53.4

1. How is the scientific literature on interspecific competition potentially "biased"?
2. What are keystone species, and how do they influence species richness in communities?

53.5 Effects of Disturbance on Community Characteristics

Recent research tends to support the individualistic view that many communities are not in equilibrium and that their species composition changes frequently. Environmental disturbances—storms, landslides, fires, floods, and cold spells—often eliminate some species, providing opportunities for others to become established.

Frequent Disturbances Keep Some Communities in a Constant State of Flux

Physical disturbances are common in some environments. For example, lightning-induced fires commonly sweep through grasslands, powerful hurricanes routinely demolish patches of forest, and waves wash over communities that live at the edge of the sea.

Joseph Connell and his colleagues conducted an ambitious long-term study of the effects of disturbance on coral reefs, shallow tropical marine habitats that are among the most species-rich communities on Earth (see Section 51.6). In some parts of the world, reefs are routinely battered by violent storms, which wash corals off the substrate, creating bare patches in the reef. The scouring action of storms creates opportunities for coral larvae to settle on bare substrates and start new colonies; ecologists use the word *recruitment* to describe the process in which young individuals join a population.

From 1963 to 1992, Connell and his colleagues tracked the fate of the Heron Island Reef at the south end of Australia's Great Barrier Reef **(Figure 53.26).** The inner flat and protected crests of the reef are sheltered from severe wave action during storms, whereas some pools and exposed crests are routinely disturbed by storms. Because corals live in colonies of variable size, the researchers monitored coral abundance by measuring the percentage of the seafloor that colonies covered. They revis-

FIGURE 53.25 | Experimental Research

The Complex Effects of an Herbivorous Snail on Algal Species Richness

Question: How does feeding by periwinkle snails *(Littorina littorea)* influence the species richness of algae in intertidal communities?

Experiment: Lubchenco manipulated the densities of periwinkle snails in tidepools and on exposed rocks in a rocky intertidal habitat by creating enclosures that prevented snails from either entering or leaving her study plots. She then monitored the species composition of algae in the study plots and graphed them against periwinkle density.

Results: The effects of periwinkle density on algal species richness varied dramatically between study plots in tidepools and on exposed rocks.

Periwinkle snails *(Littorina littorea)*

In tidepools

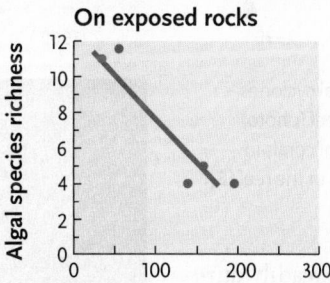

In tidepools, snails at low densities eat little algae, and *Enteromorpha* competitively excludes other algal species, reducing species richness. At high snail densities, heavy feeding on all species reduces algal species richness. At intermediate snail densities, grazing eliminates some *Enteromorpha*, allowing other species to grow.

***Enteromorpha* growing in shallow water**

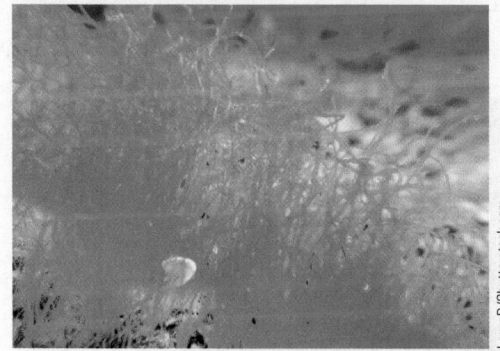

On exposed rocks

On exposed rocks, periwinkles never eat much *Chondrus*, but they consume the tender, less successful competitors. Thus, feeding by periwinkles reinforces the competitive superiority of *Chondrus*: as periwinkle density increases, algal species richness declines.

***Chondrus* growing on exposed rocks**

Conclusion: Crazing by periwinkle snails has complex effects on the species richness of competing algae. In tidepools, where periwinkle snails preferentially feed on *Enteromorpha*, the competitively dominant alga, snails at an intermediate density remove some *Enteromorpha*, which allows weakly competitive algae to grow, increasing species richness. Feeding by snails at either low or high densities reduces algal species richness. On exposed rocks, where periwinkle snails rarely eat the competitively dominant alga *Chondrus*, feeding by snails reduces algal species richness.

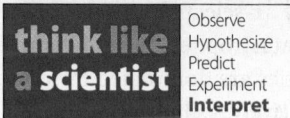

think like a scientist Observe Hypothesize Predict Experiment **Interpret**

After reading all of this section of your text (Section 53.5), interpret Lubchenco's result for snails and algae in tidepools within the context of the intermediate disturbance hypothesis.

Source: J. Lubchenco. 1978. Plant species diversity in a marine intertidal community: Importance of herbivore food preference and algal competitive abilities. *The American Naturalist* 112:23–39.

© Cengage Learning 2017

ited marked study plots at intervals, photographing and identifying individual coral colonies.

Five major cyclones (marked by gray arrows in Figure 53.26) crossed the reef during the 30-year study period. Coral communities in the exposed areas of the reef were in a nearly continual state of flux. In exposed pools, four of the five cyclones reduced the percentage of cover, often drastically. On exposed crests, the cyclone of 1972 eliminated virtually all of the corals, and subsequent storms slowed the recovery of these areas for 20 years. By contrast, corals in sheltered areas suffered much less storm

A. Exposed areas

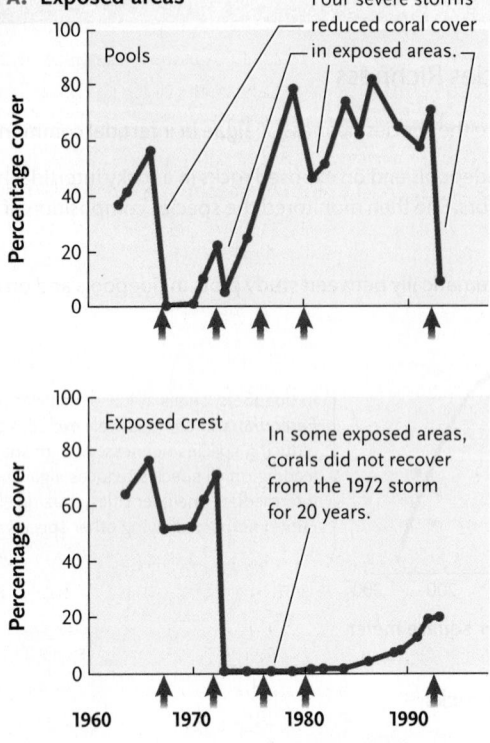

Pools

Four severe storms reduced coral cover in exposed areas.

Exposed crest

In some exposed areas, corals did not recover from the 1972 storm for 20 years.

B. Sheltered areas

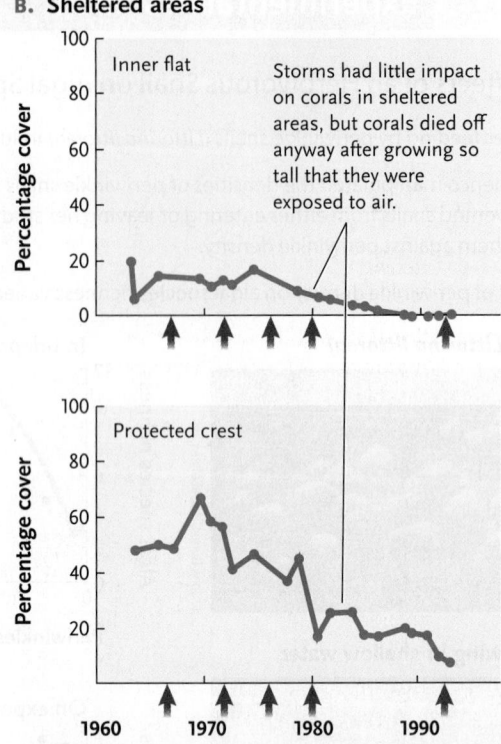

Inner flat

Storms had little impact on corals in sheltered areas, but corals died off anyway after growing so tall that they were exposed to air.

Protected crest

FIGURE 53.26 The effects of storms on corals. Five tropical cyclones (marked by gray arrows) damaged corals on the Heron Island Reef (photo) during a 30-year period. Storms reduced the percentage cover of corals in exposed parts of the reef **(A)** much more than in sheltered parts of the reef **(B)**.
© Cengage Learning 2017

damage. Nevertheless, their coverage also declined steadily during the study as a natural consequence of the corals' growth. As colonies grew taller and closer to the ocean's surface, their increased exposure to air resulted in substantial mortality.

Connell and his colleagues also documented *recruitment,* the growth of new colonies from settling larvae, in their study plots. They discovered that the rate at which new colonies developed was almost always higher in sheltered areas than in exposed areas. However, recruitment rates were extremely variable, depending in part on the amount of space that storms or coral overgrowth had made available.

This long-term study of coral reefs illustrates that frequent disturbances prevent some communities from ever reaching any kind of equilibrium. Changes in the coral reef community at Heron Island result from the combined effects of external disturbances (storms) that remove coral colonies from the reef and internal processes (growth and recruitment) that either eliminate colonies or establish new ones. In this community, growth and recruitment are slow processes, and disturbances are frequent. Thus, the community never attains equilibrium.

Moderate Levels of Disturbance May Foster High Species Richness

According to the **intermediate disturbance hypothesis,** proposed by Connell in 1978, species richness is greatest in communities that experience fairly frequent disturbances of moderate intensity. Moderate disturbances create some openings for *r*-selected species to arrive and join the community, but they allow *K*-selected species to survive. Thus, communities that experience intermediate levels of disturbance contain a rich mixture of species. Where disturbances are severe and frequent, communities include only *r*-selected species that complete their life cycles between catastrophes. Where disturbances are mild and rare, communities are dominated by long-lived *K*-selected species that competitively exclude other species from the community.

Several studies in diverse habitats have confirmed the predictions of the intermediate disturbance hypothesis. For example, Colin R. Townsend and his colleagues at the University of Otago studied the effects of disturbance at 54 stream sites in the Taieri River system in New Zealand. Disturbance occurs in these communities when water flow from heavy rains moves the rocks, soil, and sand in the streambed, disrupting the habitats where animals live. Townsend and his colleagues measured how much of the substrate moved in different

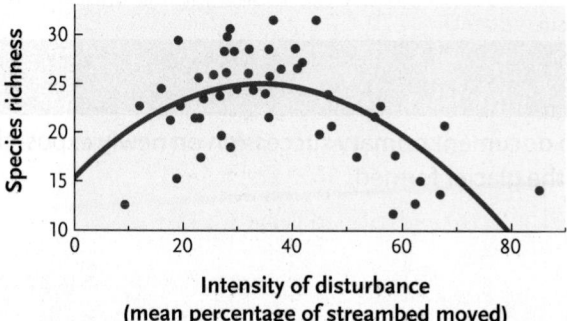

FIGURE 53.27 An observational study that supports the intermediate disturbance hypothesis. In the Taieri River system in New Zealand, species richness was highest in stream communities that experienced an intermediate level of disturbance.

© Cengage Learning 2017

streambeds to index the intensity of the disturbance. Their results indicated that species richness is highest in areas that experience intermediate levels of disturbance **(Figure 53.27).**

Some ecologists have also suggested that species-rich communities recover from disturbances more readily than do less diverse communities. For example, David Tilman and his colleagues at the University of Minnesota conducted large-scale experiments in Midwestern grasslands on the relationship between species number and the ability of communities to recover from disturbance. Their results demonstrate that grassland plots with high species richness recover from drought faster than plots with fewer species.

STUDY BREAK 53.5

1. How might disturbances from storms allow coral reefs to be rejuvenated by the recruitment of young individuals?
2. How do moderately severe and moderately frequent disturbances influence a community's species richness?

53.6 Ecological Succession: Responses to Disturbance

In response to disturbance, communities undergo **ecological succession,** a somewhat predictable series of changes in species composition over time.

Succession Begins after Disturbance Alters a Landscape or Changes the Species Composition of an Existing Community

Primary succession begins when organisms first colonize a habitat after a harsh disturbance, such as a volcanic eruption or the advance of a glacier **(Figure 53.28).** Such sites are effectively sterile because they have either been covered by molten lava or physically stripped of all traces of organisms that had lived there.

In many terrestrial habitats, lichens (see Section 30.3), which derive nutrients from rain and bare rock, are usually the first visible colonizers of such inhospitable habitats. They secrete mild acids that erode rock surfaces, initiating the slow development of soil, which is enriched by the organic material lichens produce. After lichens modify a site, mosses (see Section 28.2) colonize patches of soil and grow quickly.

As soil accumulates, hardy *r*-selected plants—grasses, ferns, and broad-leaved herbs—colonize the site from surrounding areas. Their roots break up rock, and as they die, their decaying remains enrich the soil. Detritivores and decomposers facilitate these processes. As the soil gets deeper and richer, increased moisture and nutrients support bushes and, eventually, trees. Late successional stages are often dominated by *K*-selected species with woody trunks and branches that position leaves in sunlight and large root systems that acquire water and nutrients from soil.

In the classical view of ecological succession, long-lived species eventually dominate a community, and new species join it only rarely. This relatively stable, late successional stage is called a **climax community** because the dominant vegetation replaces itself and persists until an environmental disturbance damages or eliminates it, allowing other species to invade. Local climate and soil conditions, the surrounding communities where colonizing species originate, and chance events determine the species composition of climax communities. However, recent research suggests that even "climax communities" change slowly in response to environmental fluctuations, as described below.

By contrast, **secondary succession** occurs after a much milder disturbance, such as a fire, a storm, or human activity, destroys or disrupts existing vegetation. Such milder disturbances do not sterilize the habitat, and the presence of soil makes the disturbed sites ripe for colonization. Moreover, in many habitats, the soil contains numerous seeds, described as a *seed bank,* that germinate after the disturbance. The early stages of secondary succession proceed more rapidly than the early stages of a primary succession, but later stages resemble those of primary succession.

Secondary succession in the north temperate zone is well studied in abandoned farms, called "old fields," where forests were cleared centuries ago. Because the transformation from old field back to forest takes at least a hundred years, ecologists use historical records to find the age of different stands of vegetation and reconstruct the successional sequence by comparing stands of different ages. In the Piedmont region of southeastern North America, an abandoned field is covered by crabgrass (genus *Digitaria*), an annual plant, during the first growing season. The following year, crabgrass is replaced by horseweed (*Conyza canadensis),* which cannot persist because it secretes substances that inhibit the germination of its own seeds. Ragweed (*Ambrosia artemisiifolia),* another annual, dominates during the third year, but it is gradually replaced by perennial asters (genus *Erigeron*) and broomsedges (genus *Andropogon*), which are, in turn, replaced by shrubs. Ten to

PRIMARY SUCCESSION FOLLOWING GLACIAL RETREAT

The retreat of glaciers at Glacier Bay, Alaska, has allowed ecologists to document primary succession on newly exposed rocks and soil that had been scraped clean of living organisms when the glacier formed.

1 The glacier has retreated about 8 m per year since 1794.

2 The site was covered with ice less than 10 years before this photo was taken. As a glacier retreats, the flow of meltwater leaches minerals, especially nitrogen, from the newly exposed substrate.

3 Once lichens and mosses become abundant, mountain avens (genus *Dryas*) grow on the nutrient-poor soil. Benefitting from associated nitrogen-fixing bacteria, these pioneers spread rapidly over glacial till.

4 Within 20 years, shrubby willows (genus *Salix*), cottonwoods (genus *Populus*), and alders (genus *Alnus*)—all of them symbiotic with nitrogen-fixing microorganisms—grow in drainage channels.

5 In time, young conifers, mostly hemlocks (genus *Tsuga*) and spruce (genus *Picea*), join the community.

6 After 80 to 100 years, forests of Sitka spruce (*Picea sichensis*) and western hemlock (*Tsuga heterophylla*) replace earlier successional species.

SUMMARY As glaciers form, they eliminate all plants and animals from the landscape. When glaciers retreat, primary succession begins, and the site is gradually reoccupied by a somewhat predictable sequence of *r*-selected species that are eventually replaced by *K*-selected species.

think like a scientist Based on your reading about terrestrial biomes in Chapter 51, what would primary succession be like after glacial retreat from an area in the very far north that was occupied by Arctic tundra?

© Cengage Learning 2017

fifteen years after the field was abandoned, pine (genus *Pinus*) seedlings germinate. Growing pines cast substantial shade and their fallen needles acidify the soil, making the site unsuitable for the plants from earlier successional stages. Because pines are intolerant of shade, pine seedlings do not flourish under mature pine trees. Thus, after 50 to 100 years, pines are replaced by a taller mixed hardwood forest of oaks (genus *Quercus*) and hickories (genus *Carya*), the seedlings of which are more shade tolerant than pines. The hardwood forest forms the climax community in the thick, moist soil after more than a century of successional change.

Similar climax communities sometimes arise from alternative successional sequences. For example, hardwood forests also develop in sites that were once ponds. During **aquatic succession,** debris from rivers and runoff accumulates in a body of water, causing it to fill in at its margins. The pond is transformed into a swamp, inhabited by plants adapted to a semisolid substrate. As larger plants get established, their high transpiration rates dry the soil, allowing other plant species to colonize. Given enough time, the site may become a meadow or forest, where an area of moist, low-lying ground is the only remnant of the original pond.

Community Characteristics Change during Succession

Several characteristics undergo directional change as succession proceeds. First, because *r*-selected species are short-lived and *K*-selected species long-lived, species composition changes rapidly in the early stages, but slowly in the late stages of succession. Second, species richness increases rapidly during the early stages because new species join the community faster than resident species become extinct; as succession proceeds, however, species richness stabilizes or may even decline. Third, in terrestrial communities that receive sufficient rainfall, the maximum height and total mass of the vegetation increase steadily as large species replace small ones, creating the complex structure of the climax community.

Because plants influence the physical environment below them, the community itself increasingly moderates the microclimate (described in Section 51.2). The shade cast by a forest canopy retains soil moisture and reduces temperature fluctuations. The trunks and canopy also reduce wind speed. By contrast, the short vegetation in an early successional stage does not effectively shelter the space below it.

Although ecologists usually describe succession in terms of vegetation, animals undergo succession, too. As the vegetation shifts, new resources become available, and animal species replace each other over time. Herbivorous insects, which often have strict food preferences, undergo succession along with their food plants. And as the herbivores change, so do their predators, parasites, and parasitoids. In old-field succession in eastern North America, different successional stages harbor a changing assortment of bird species **(Figure 53.29).**

Several Hypotheses Help to Explain the Processes Underlying Succession

Differences in dispersal abilities, maturation rates, and lifespans among species are at least partly responsible for ecological succession. Early successional stages harbor many *r*-selected species because they produce numerous small seeds that colonize open habitats and grow quickly. Mature successional stages are dominated by *K*-selected species because they are long-lived. Nevertheless, coexisting populations inevitably affect one another. Although the role of population interactions in succession is generally acknowledged, ecologists debate the relative importance of processes that either facilitate or inhibit the turnover of species in a community.

The **facilitation hypothesis** suggests that species modify the local environment in ways that make it less suitable for

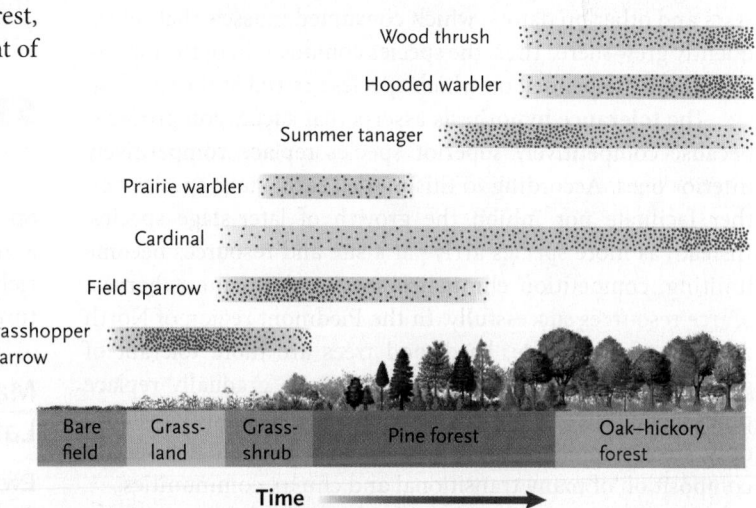

FIGURE 53.29 Succession in animals. Changes in bird species composition during secondary succession in an abandoned agricultural field in eastern North America parallel changes in plant species composition. Residence times of several representative species are illustrated. The density of stippling inside each bar illustrates the relative density of each species through time.
© Cengage Learning 2017

themselves but more suitable for colonization by species typical of the next successional stage. For example, when lichens first colonize bare rock, they produce a small quantity of soil, which is required by mosses and grasses that grow there later. According to this hypothesis, changes in species composition are both orderly and predictable because the presence of each stage facilitates the success of the next. Facilitation is very important in primary succession, but it may not be the best model of interactions that influence secondary succession.

The **inhibition hypothesis** suggests that new species are prevented from occupying a community by whatever species are already present. According to this hypothesis, succession is neither orderly nor predictable because each stage is dominated by whichever species happen to colonize the site first. Species replacements occur only when individuals of the dominant species die of old age or when an environmental disturbance reduces their numbers. Eventually, long-lived species replace short-lived species, but the precise species composition of the climax community is up for grabs.

Inhibition appears to play a role in some secondary successions that follow environmental disturbances. For example, rocky intertidal communities on sheltered shores in the Gulf of Maine include some habitat patches that are dominated by an alga (*Ascophyllum nodosum*) and other patches dominated by a mussel (*Mytilus edulis*). Do these patches represent different alternative states for the climax community in this habitat? In winter, small patches in this habitat are sometimes scoured clean by seaborne ice, after which the patch undergoes succession. In 2009, Peter Petraitis of the University of Pennsylvania and colleagues from several other institutions reported on the fate of habitat patches in which they had experimentally simulated "ice scour." If mussels colonized the cleared site first, they grew faster than the algae and eventually dominated the community. But if the alga colonized first, it provided cover for sea stars and other predators, which consumed mussels that subsequently grew there. Thus, the species composition of the mature community depended on which species arrived at the site first.

The **tolerance hypothesis** asserts that succession proceeds because competitively superior species replace competitively inferior ones. According to this model, early-stage species neither facilitate nor inhibit the growth of later-stage species. Instead, as more species arrive at a site and resources become limiting, competition eliminates species that cannot harvest scarce resources successfully. In the Piedmont region of North America, for example, hardwood trees are more tolerant of shade than pine trees are, and hardwoods gradually replace pines during succession. Thus, the climax community includes only strong competitors. Tolerance may explain the species composition of many transitional and climax communities.

At most sites, succession probably results from a combination of facilitation, inhibition, and tolerance, coupled with interspecific differences in dispersal, growth, and maturation rates. Moreover, within a community, the patchiness of abiotic factors also strongly influences plant distributions and species composition. For example, in the deciduous forests of eastern North America, maples (genus *Acer*) predominate on wet, low-lying ground, but oaks (genus *Quercus*) are more abundant at higher and drier sites. Thus, a deciduous forest climax community is more often a spatial mosaic of species than a uniform stand of trees.

Disturbance and density-independent factors also play important roles, in some cases speeding successional change. In northern forests, for example, moose prefer to feed on deciduous shrubs, accelerating the rate at which conifers replace them. In other cases, disturbance inhibits successional change, establishing a *disturbance climax* or **disclimax community.** In many grassland communities (see Section 51.4), grazing by large mammals and periodic fires kill the seedlings of trees that would otherwise become established. Thus, disturbance prevents the succession from grassland to forest, and grassland persists as a disclimax community.

On a local scale, disturbances often destroy small patches of vegetation, returning them to an earlier successional stage. A hurricane may knock over a few trees in a forest, creating small, sunny patches of open ground. Locally occurring *r*-selected species take advantage of the resources that are suddenly available and quickly colonize the openings. These local patches then undergo succession that is out of synchrony with the immediately surrounding forest. Thus, moderate disturbance, accompanied by succession in local patches, can increase species richness.

STUDY BREAK 53.6

1. What is the difference between primary succession and secondary succession?
2. How does a climax community differ from early successional stages?
3. How do the three hypotheses about the causes of ecological succession view the role of population interactions in the successional process?

53.7 Variations in Species Richness among Communities

Species richness often varies among communities according to a recognizable pattern. Two large-scale patterns of species richness—latitudinal trends and island patterns—have captured the attention of ecologists for more than a century.

Many Types of Organisms Exhibit Latitudinal Gradients in Species Richness

Ever since Darwin and Wallace traveled the globe (see Section 20.2), ecologists have recognized broad latitudinal trends in species richness. For many, but not all, plant and animal groups, species richness follows a latitudinal gradient, with the most species in the tropics and a steady decline in numbers toward the poles **(Figure 53.30).** Several general hypotheses may explain these striking patterns.

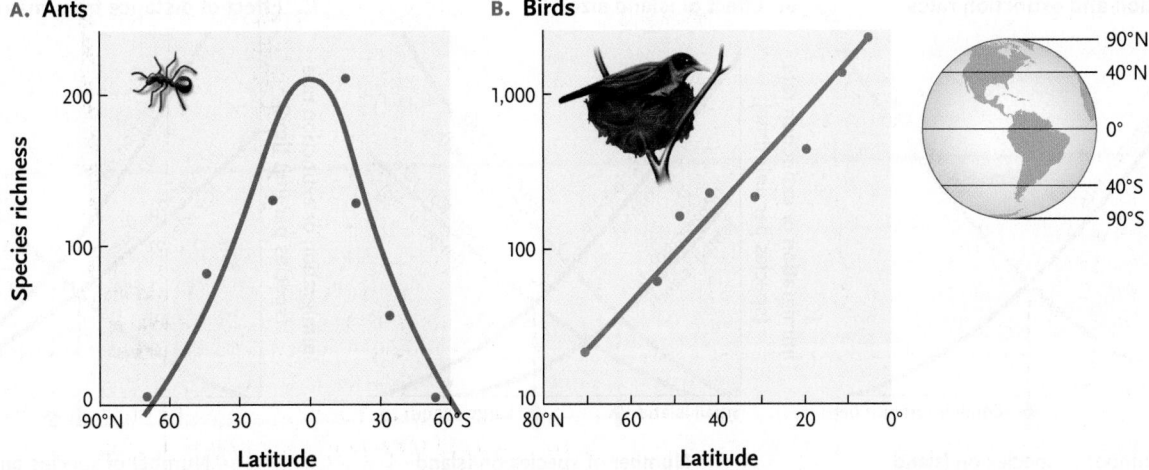

FIGURE 53.30 Latitudinal trends in species richness. The species richness of many animals and plants varies with latitude, as illustrated here **(A)** for ants and **(B)** for birds of North and Central America. The species-richness data for birds are based on records of where the species breed.
© Cengage Learning 2017

Some hypotheses propose historical explanations for the *origins* of high species richness in the tropics. The benign climate in tropical regions allows some tropical organisms to have more generations per year than their temperate counterparts. And, given the small seasonal changes in temperature, tropical species may be less likely than temperate species to migrate seasonally from one habitat to another, thus reducing gene flow between geographically isolated populations (see Section 22.3). These factors may have fostered higher speciation rates in the tropics, accelerating the accumulation of species. Tropical communities may also have experienced severe disturbance less often than communities at higher latitudes, where periodic glaciations have caused repeated extinctions. Thus, new species may have accumulated in the tropics over longer periods of evolutionary time.

Other hypotheses focus on ecological explanations for the *maintenance* of high species richness in the tropics. Some resources are more abundant, predictable, and diverse in tropical communities. Tropical regions experience more intense sunlight, warmer temperatures in most months, and higher annual rainfall than temperate and polar regions (see Chapter 51). These factors provide a long and predictable growing season for the lush tropical vegetation, which supports a rich assemblage of herbivores, and through them many carnivores and parasites. Furthermore, the abundance, predictability, and year-round availability of resources allow some tropical animals to have specialized diets. For example, tropical forests support many species of fruit-eating bats and birds, which could not survive in temperate forests where fruits are not available year-round.

Species richness may therefore be a self-reinforcing phenomenon in tropical communities. Complex webs of population interactions and interdependency have coevolved in relatively stable and predictable tropical climates. Predator–prey, competitive, and symbiotic interactions may prevent individual species from dominating communities and reducing species richness.

The Theory of Island Biogeography Explains Variations in Species Richness

Although the species richness of communities may be stable over time, species composition is often in flux as new species join a community and others drop out. In the 1960s, Robert MacArthur of Princeton University and Edward O. Wilson of Harvard University addressed the question of why communities vary in species richness, using islands as model systems. Islands provide natural laboratories for studying ecological phenomena, just as they do for evolution (see *Focus on Research: Basic Research* in Chapter 22). Island communities are often small, have well-defined boundaries, and are isolated from surrounding communities.

In developing the **equilibrium theory of island biogeography,** MacArthur and Wilson sought to explain variations in species richness on islands of different size and different levels of isolation from other landmasses **(Figure 53.31).** They hypothesized that the number of species on any island was governed by a give and take between two processes: the immigration of new species to the island and the extinction of species already there (see Figure 53.31A).

According to the MacArthur–Wilson model, the mainland harbors a *species pool* from which species immigrate to offshore islands. Seeds and small arthropods are carried by wind or floating debris; some animals, such as birds, arrive under their own power. When few species are already on an island, the rate at which *new* species immigrate to the island is high. But as more species inhabit the island over time, the immigration rate declines because there are fewer species left in the mainland pool that can still arrive on the island as *new* colonizers.

A. Immigration and extinction rates

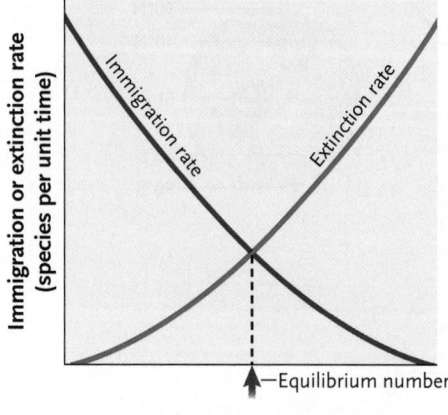

The number of species on an island at equilibrium (indicated by the gray arrow) is determined by the rate at which new species immigrate and the rate at which species already on the island become extinct.

B. Effect of island size

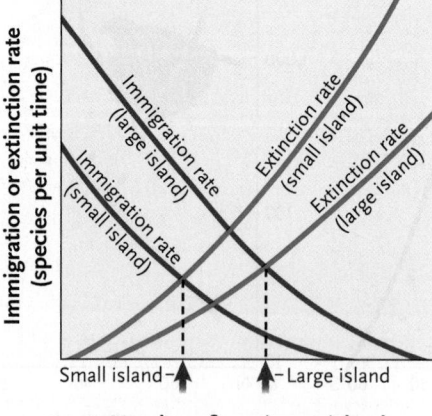

Immigration rates are higher and extinction rates lower on large islands than on small islands. Thus, at equilibrium, large islands have more species than small ones.

C. Effect of distance from mainland

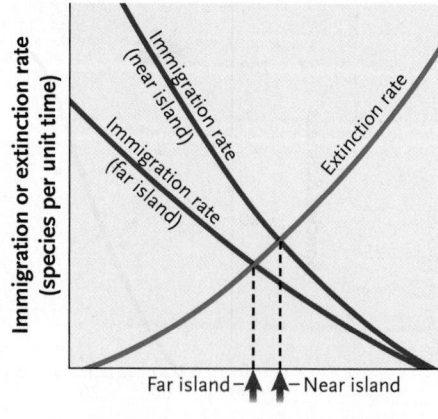

Organisms leaving the mainland locate nearby islands more easily than distant islands, causing higher immigration rates on near islands. Thus, near islands support more species than far ones.

FIGURE 53.31 **Assumptions and predictions of the theory of island biogeography.**
© Cengage Learning 2017

Once a species immigrates to an island, its population grows and persists for some time. But as the number of species on the island increases, the rate at which those species go extinct rises. The extinction rate increases through time partly because there are more species that can go extinct there. In addition, as the number of species on the island increases, competition and predator–prey interactions can reduce the population sizes of some species and drive them to extinction.

According to MacArthur and Wilson's theory, an equilibrium between immigration and extinction determines the number of species that ultimately occupy an island. In other words, once equilibrium is reached, the number of species remains relatively constant because one species already on the island becomes extinct in about the same time it takes a new species to immigrate to the island. The model does not specify which species immigrate to the island or which ones already on the island become extinct. It simply predicts that the number of species on the island is in equilibrium, although species composition is not. The ongoing processes of immigration and extinction establish a constant turnover in the roster of species that live on any island.

The MacArthur–Wilson model explains why some islands harbor more species than others. The model assumes that large islands have higher immigration rates than small islands do, partly because they present a larger target for dispersing organisms, and partly because once immigrants arrive on a large island, they will encounter sufficient resources to establish a population there. Moreover, it assumes that large islands have lower extinction rates because they can support larger populations and provide a greater range of habitats and resources. Thus, the model predicts that, at equilibrium, large islands have more species than small islands (see Figure 53.31B). Similarly, it predicts that islands near the mainland have higher immigration rates than distant

islands do, because dispersing organisms are more likely to land on islands that are close to their point of departure. Distance does not affect extinction rates. Thus, at equilibrium, islands that lie closer to a mainland source have more species than more distant islands (see Figure 53.31C).

The equilibrium theory's predictions about the effects of area and distance are generally supported by observational data on plants and animals (**Figure 53.32**). Daniel Simberloff, one of Wilson's graduate students at Harvard University, was the first person to test the theory's predictions experimentally; he monitored the immigration of arthropods to, and extinction of arthropods on, individual red mangrove trees in the Florida Keys (**Figure 53.33**). The trees, with canopies that spread from 11 to 18 m in diameter, grow in shallow water and are isolated from their neighbors; thus, each tree is an island that harbors an arthropod community. The species pool on the Florida mainland includes about 1,000 arthropod species, but each mangrove island contains no more than 40 species at one time.

After cataloging the species on each island, Simberloff and Wilson hired an extermination company to eliminate all arthropods on them (see Figure 53.33A). Simberloff then monitored both the immigration of arthropods to the islands and the extinction of species that became established on them. He surveyed six islands regularly for two years and at intervals thereafter.

The results of this experiment confirm several predictions of MacArthur and Wilson's theory (see Figure 53.33B). Arthropods recolonized the islands rapidly, and within eight or nine months the number of species living on each island had reached an equilibrium that was near the original species number. In addition, the island nearest to the mainland had more species than the most distant island. However, immigration and extinction were

A. Distance effect

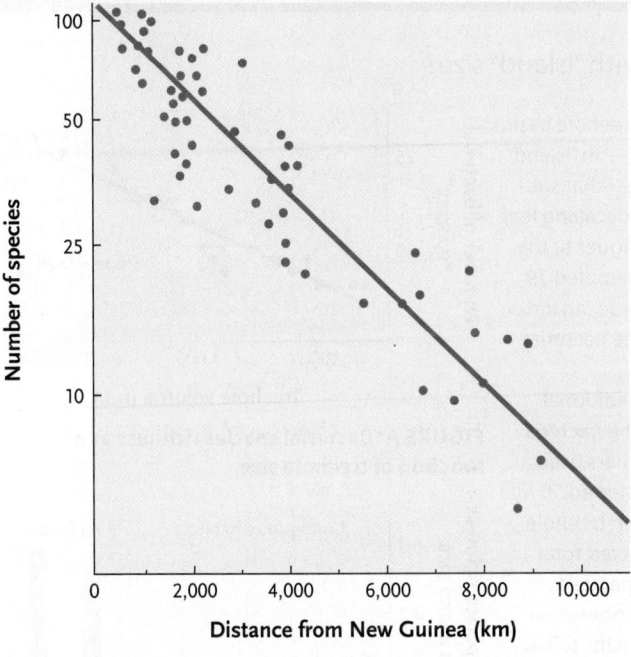

Distance from New Guinea (km)

The number of lowland bird species on islands of the South Pacific declines with the islands' distance from the species source, the large island of New Guinea. Data in this graph were corrected for differences in the sizes of the islands. The number of bird species on each island is expressed as a percentage of the number of bird species on an island of equivalent size close to New Guinea.

B. Area effect

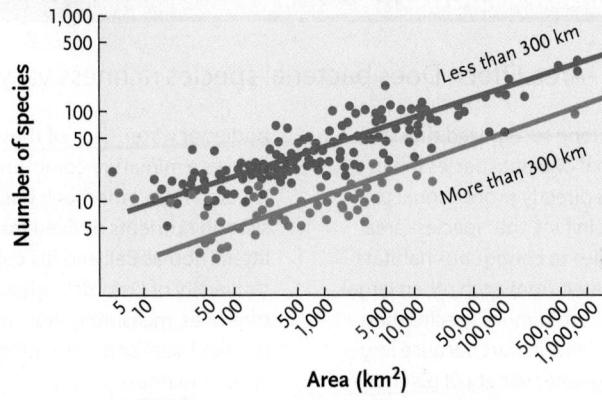

Area (km²)

The number of bird species on tropical and subtropical islands throughout the world increases dramatically with island area. The data for islands near to a mainland source and islands far from a mainland source are presented separately to minimize the effect of distance. Notice that the "distance effect" reduces the number of bird species on islands that are more than 300 km from a mainland source.

FIGURE 53.32 Factors that influence bird species richness on islands. (A) Fewer bird species occupy islands that are distant from the mainland source. **(B)** More bird species occupy large islands than small ones.

© Cengage Learning 2017

A. The process of defaunation

B. The return of species richness over time

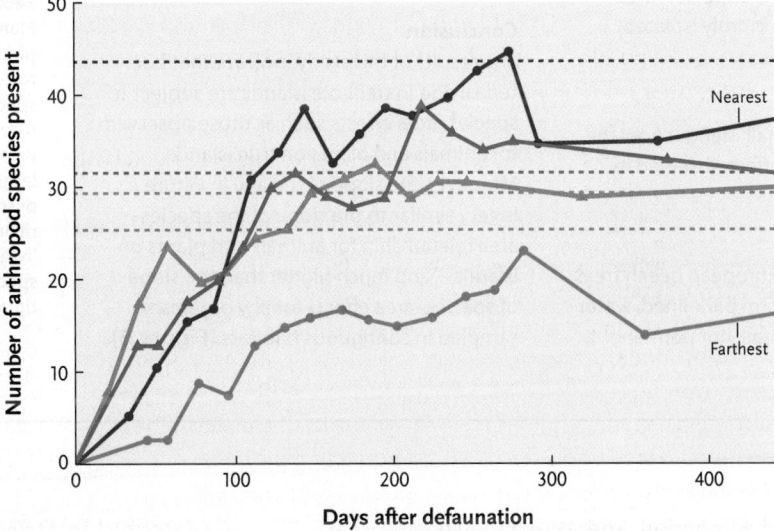

Days after defaunation

FIGURE 53.33 An experimental test of the theory of island biogeography. (A) After cataloguing the arthropods, Simberloff and Wilson hired an exterminating company to erect a tent over each mangrove island. Once the islands were fully covered, exterminators used methyl bromide—a pesticide that does not harm trees and leaves no residue—to eliminate all living arthropods. **(B)** On three of four islands, species richness gradually returned to the predefaunation level (indicated by color-coded dashed lines on the graph). The most distant island had not reached its predefaunation species richness after two years.

© Cengage Learning 2017

The Species–Area Effect: Does bacterial species richness vary with "island" size?

Ecologists have long recognized that the number of animal or plant species inhabiting an island is directly proportional to the island's size. In fact, the "species–area effect" also applies to contiguous habitats that are not isolated from each other: larger areas generally harbor more species than smaller areas, at least in part because larger areas include a greater variety of distinctive resources that facilitate resource partitioning by resident species.

Nevertheless, the species–area effect is less dramatic in contiguous habitats than it is among islands of different sizes. In other words, a graph of species richness versus area for contiguous habitats has a lower slope than a graph of species richness versus area among isolated islands (see Figure 53.31B for graph of island data), probably because colonization rates from one habitat patch to another are higher and extinction rates within habitat patches are lower in contiguous habitats than on islands. Previous research had shown an especially limited species–area effect (that is, a low slope of the graph) for bacteria sampled in contiguous habitats; this result probably reflects the ubiquity of bacteria and the ease with which they colonize areas adjacent to those where they already occur. But researchers had not previously investigated the species–area effect for bacteria living in truly isolated islandlike habitats.

Research Question

Do bacteria living on small islands show the dramatic species–area effect observed in plants and animals?

Experiment

The buttresses of large European beech trees (*Fagus sylvatica*) often form bark-lined, watertight basins that hold small, but permanent,

bodies of water. Each of these treehole basins houses a miniature community—an "island" isolated from other such basins—that subsists on nutrients derived from decaying leaf litter. Thomas Bell and his colleagues at the University of Oxford, England, sampled 29 treeholes, measuring their volume (an index of island size) and estimating the bacterial species richness in each.

Instead of using laborious traditional culture techniques to identify the bacteria in each community, the researchers used a molecular approach. They transferred 50 mL of water and sediment from each treehole into sterile vials, and then extracted total DNA from subsamples. They separated out gene fragments for the 16S ribosomal subunit (16S rRNA), which is specific to bacteria, and then amplified them using PCR. The resulting product was analyzed with gel electrophoresis.

Results

Each band appearing on the gel was scored as a distinct bacterial "species." The number of bands identified in the sample from each community served as an estimate of its species richness. Species richness in these 29 bacterial communities was directly proportional to "island" size, as measured by the volume of each basin (**Figure A**).

Conclusion

The results of this study indicate that bacteria living in treehole islands are subject to species–area effects such as those observed for animals and plants on true islands. Moreover, the slope of the line in Figure A is very similar to the slope of the species–area relationship for animals and plants on islands—and much higher than the slope of species–area effects for any organisms sampled in contiguous habitats (**Figure B**).

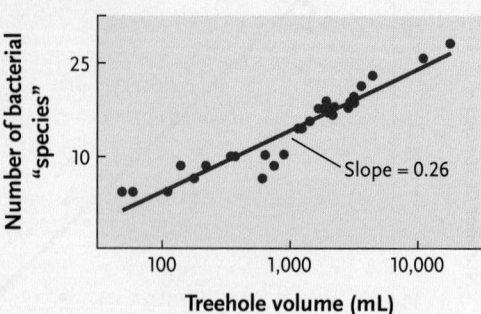

FIGURE A Bacterial species richness as a function of treehole size.

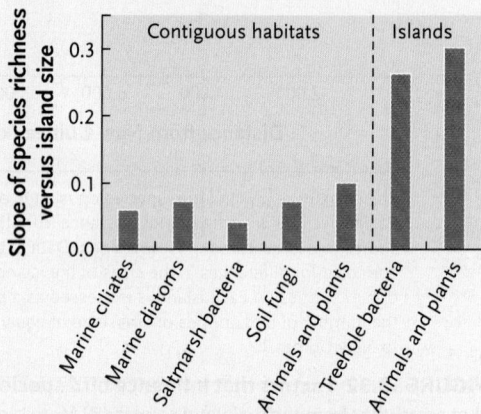

FIGURE B Slope of species–area function for organisms in contiguous habitats and on islands.

think like a scientist

Plants in the clade Bromeliaceae, which includes pineapples, often grow as epiphytes in tropical forest canopies. The leaves of some bromeliad species form a cup that collects water and detritus, much the way British treeholes do. Bromeliads vary greatly in size, and the cups of large bromeliads hold more water than the cups of small ones. Would you expect the species richness of microorganisms and invertebrates living in bromeliad cup communities to exhibit species–area functions more like those of contiguous habitats or like those of islands?

Source: T. Bell et al. 2005. Larger islands house more bacterial taxa. *Science* 308:1884.

incredibly rapid, and Simberloff and Wilson suspected that some species went extinct even before they had noted their presence. The researchers also discovered that three years after the experimental treatments, the species composition on the islands was still changing constantly and did not remotely resemble the species composition in the islands before they were defaunated.

As described in *Molecular Insights*, the equilibrium view of species richness also applies to mainland communities, which exist as islands in a metaphorical sea of dissimilar habitat. Lakes are "islands" in a "sea" of dry land, and mountaintops are habitat "islands" in a "sea" of low terrain. Species richness in these communities is partly governed by the immigration of new species from distant sources and the extinction of species already present. As human activities disrupt environments across the globe, undisturbed sites function as islandlike refuges for threatened and endangered species. Conservation biologists now apply the general lessons of MacArthur and Wilson's theory to the design of nature preserves (see Chapter 55).

In the next chapter we examine ecosystems, which include ecological communities interacting with their abiotic environments, focusing on the movements of energy and nutrients.

STUDY BREAK 53.7

1. What factors may promote the maintenance of high species richness in tropical communities?
2. According to the equilibrium theory of island biogeography, what are the effects of an island's size and its distance from the mainland on the number of species that can occupy it?

Unanswered Questions

Do species interactions change predictably across environments?

As we learned in this chapter, the population interactions that occur between species range from mutualistic to parasitic. Some biologists have suggested that we should expect more competitive interactions between species in some kinds of environments, but more positive interactions in others. Community ecology will become a more quantitative and predictive discipline if researchers focus on how abiotic and biotic environmental factors—such as the presence of particular community members, environmental gradients, or global climate change—influence the strength of the interactions between species. For example, as physical environments become more stressful, the abundance and distribution of species should be determined less by resource limitation and more by the stress itself. Accordingly, plants tend to compete far less with each other in stressful environments than they do under ideal growing conditions. Scientists are now engaged in the intellectual feedback of theory development and experimental testing aimed at generating a predictive framework for particular types of interactions and their consequences for community structure.

What is the relative importance of positive versus negative interactions for community structure?

It was once suggested that ecologists in capitalist societies, such as the United States, tend to more often study competition and predation, but ecologists in socialist societies tend to study mutualism. Although the truth of this anecdote is unclear, it is remarkable that ecologists still do not agree on the relative importance of positive interactions (for example, mutualism or commensalism) versus negative interactions (such as predation or competition) in generating community structure. Advances in this area of study may result from "factorial" experiments, in which two or more types of interactions are manipulated. For example, one might examine the relative effects of excluding pollinators versus excluding herbivores on the success of a plant population. In factorial experiments, the researcher can conclude that one factor has a bigger effect than the other, because all other factors were controlled. These sorts of experiments may eventually lead to an emerging picture of the relative importance of positive versus negative population interactions.

How does the evolutionary history of a species influence its ecology today?

The great evolutionary biologist Theodosius Dobzhansky once noted that "Nothing in biology makes sense except in the light of evolution." Although we know a great deal about both ecology and evolutionary biology, researchers are only beginning to explore the impact of an organism's evolutionary history on its ecology. This very active area of research includes the use of phylogenetic information (see Chapter 24), selection experiments (see Chapter 21), and a knowledge of the genetic basis of particular traits (see Chapter 12). For example, are closely related species more likely to compete with each other than more distantly related species are? Do organisms that are well adapted to particular environments fare poorly in other environments? Why do some organisms specialize in their resource use? Are the population dynamics that species experience shaped by past evolutionary events? These questions are currently being addressed, and the answers uncovered by researchers may unravel many current mysteries about the ecology of populations and communities.

think like a scientist

1. Is mutualism really just reciprocal parasitism? Are the benefits a partner gains in a mutualism just "taken" from the other partner (or are benefits garnered by other means)?
2. Would you expect close relatives to share the same parasites? Why, and what might the consequence of this be for community structure of hosts and parasites?

Anurag Agrawal is a Professor in the Departments of Entomology and Ecology and Environmental Biology at Cornell University. He studies the evolutionary and community ecology of plant–insect interactions. To learn more about Dr. Agrawal's research, go to http://www.herbivory.com.

Anurag Agrawal

REVIEW KEY CONCEPTS

For access to MindTap and additional study materials visit www.cengagebrain.com.

53.1 Population Interactions

- Coevolution is the evolution of reciprocal adaptations in species that interact ecologically (Figure 53.1).

- Predators and herbivores use diverse adaptations to select, locate, capture, and ingest an appropriate diet (Figure 53.2). Plants have both structural and chemical defenses against herbivores. Animal prey may try to hide or escape from predators, defend themselves actively, or advertise their unpalatability (Figures 53.3–53.5). Some animal species mimic the appearance of poisonous species (Figure 53.6). Predators may evolve adaptations to counter prey defenses (Figure 53.7).

- Interspecific competition results if two or more populations use the same limiting resources; competition may lead to the extinction of one competitor (Figure 53.8). Ecologists use the ecological niche concept to visualize a population's resource use (Figure 53.9). Observations of resource partitioning (Figure 53.10) and character displacement (Figure 53.11) suggest that competition may be important, but only field experiments can demonstrate that competition occurs (Figure 53.12). Knowledge of an invasive species' fundamental and realized niches can inform practices to manage its populations (Figure 53.13).

- Symbiosis is a close ecological association between species. In commensal interactions, one species benefits and the other is unaffected (Figure 53.14). In mutualistic interactions, both partners benefit (Figures 53.15–53.17). In parasitic interactions, one species benefits and the other is harmed.

53.2 The Nature of Ecological Communities

- An interactive view suggests that species in a community are bound together in a complex web of necessary biotic interactions; an individualistic view recognizes communities as loose assemblages of organisms that have similar physical requirements (Figure 53.18).

- Ecotones occur where adjacent communities grade into one another; sharp boundaries occur between communities where a critical resource or an important abiotic factor is discontinuous (Figure 53.19).

53.3 Community Characteristics

- In warm, moist environments, vegetation is tall and has a complex physical structure (Figure 53.20). In harsh, cold environments, vegetation is short and has a simple physical structure.

- Foundation species moderate the physical environment within communities (Figure 53.21).

- Communities differ in species richness and the relative abundances of species. Both characteristics contribute to a community's species diversity (Figure 53.22).

- Organisms are classified as producers, consumers, detritivores, or decomposers. Ecologists depict the trophic structure (feeding relationships) of communities in food webs (Figure 53.23). Food-web analyses seek to identify generalities about trophic structure and its relationship to community stability.

53.4 Effects of Population Interactions on Community Characteristics

- Interspecific competition often affects the species composition and structure of communities.

- Predators may increase species richness by reducing the population size of the competitively most successful prey, thus allowing other prey species to occupy the community (Figure 53.24).

- Herbivores sometimes increase species richness of the organisms they eat and sometimes decrease it (Figure 53.25).

53.5 Effects of Disturbance on Community Characteristics

- Environmental disturbances may eliminate populations from a community. Some communities, such as coral reefs, experience such frequent disturbance that their species composition is never at equilibrium (Figure 53.26).

- Disturbances of intermediate intensity and frequency allow both r-selected and K-selected species to occupy a site, increasing species richness (Figure 53.27).

53.6 Ecological Succession: Responses to Disturbance

- Ecological succession is a somewhat predictable change in species composition over time.

- Primary succession occurs in completely new habitats or those that have been scoured of all organisms that had lived at the site (Figure 53.28). Secondary succession occurs where a community existed in the past (Figure 53.29).

- Species composition changes quickly and species richness rises rapidly during early successional stages. Early stages include short-lived r-selected species; later stages include long-lived K-selected species. Some communities eventually achieve a relatively stable climax state.

- Most communities include a mosaic of species that reflect patchiness in underlying local environmental conditions and the mixture of relatively undisturbed and recently disturbed sites.

53.7 Variations in Species Richness among Communities

- Communities near the equator have higher species richness than those near the poles (Figure 53.30). Explanations for this latitudinal gradient focus on either the origin or the maintenance of high species richness in the tropics.

- The equilibrium theory of island biogeography predicts that the number of species on an island represents a balance between the immigration of new species and the extinction of species already present (Figure 53.31). Studies show that large islands harbor more species than small islands, and islands near a mainland source have more species than distant islands (Figures 53.32 and 53.33).

TEST YOUR KNOWLEDGE

Remember/Understand

1. According to optimal foraging theory, predators:
   a. always feed on the largest prey possible.
   b. always feed on the prey that are easiest to catch.
   c. choose prey based on the costs of capturing and consuming it compared to the energy it provides.
   d. feed on plants when animal prey are scarce.
   e. have coevolved mechanisms to overcome prey defenses.

2. The use of the same limiting resource by two species is called:
   a. brood parasitism.
   b. interference competition.
   c. exploitative competition.
   d. mutualism.
   e. optimal foraging.

3. The range of resources that a population can possibly use is called:
   a. its fundamental niche.
   b. its realized niche.
   c. character displacement.
   d. resource partitioning.
   e. its relative abundance.

4. Differences in the bill sizes of finch species living on the same island in the Galápagos may be caused by:
   a. predation.
   b. character displacement.
   c. mimicry.
   d. interference competition.
   e. cryptic coloration.

5. A keystone species:
   a. is usually a primary producer.
   b. has a critically important role in determining the species composition of its community.
   c. is always a predator.
   d. usually reduces the species diversity in a community.
   e. usually exhibits aposematic coloration.

6. Species richness is often highest in communities where disturbances are:
   a. very frequent and severe.
   b. very frequent and of moderate intensity.
   c. very rare and severe.
   d. of intermediate frequency and moderate intensity.
   e. very rare and mild.

7. The change in the species composition of a terrestrial community from bare and lifeless rock to climax vegetation is called:
   a. disturbance.
   b. competition.
   c. secondary succession.
   d. primary succession.
   e. facilitation.

Apply/Analyze

8. Bacteria that live in the human intestine assist digestion and feed on nutrients the human consumed. This relationship might best be described as:
   a. commensalism.
   b. mutualism.
   c. endoparasitism.
   d. ectoparasitism.
   e. predation.

9. The equilibrium theory of island biogeography predicts that the number of species found on an island:
   a. increases steadily until it equals the number in the mainland species pool.
   b. is greater on large islands than on small ones.
   c. is smaller on islands near the mainland than on distant islands.
   d. can never reach an equilibrium number.
   e. is greater for islands near the equator than for islands near the poles.

10. **Apply Evolutionary Thinking** Five processes can foster microevolutionary change: gene flow, genetic drift, mutation, natural selection, and nonrandom mating (see Section 21.3). Which of those processes might contribute to the evolution of Batesian mimicry in two butterfly species? Would the same processes affect both the mimic and the model similarly? Which processes might have contributed to the evolution of the mutualistic relationship between ants and acacia trees, and how would their action on the two mutualists differ?

Evaluate/Create

11. The table below shows how many individuals were recorded for each of five species in five separate communities (a–e). Which community has the highest species diversity?

Community	Species 1	Species 2	Species 3	Species 4	Species 5
a	90	10	0	0	0
b	80	10	10	0	0
c	25	25	25	25	0
d	2	4	6	8	80
e	20	20	20	20	20

12. **Discuss Concepts** After reading about the two potential biases in the scientific literature on competition, describe how future studies of competition might avoid such biases.

13. **Discuss Concepts** How do human activities disrupt the process of succession in terrestrial communities? Would you describe most of our activities as mild disturbances, moderate disturbances, or severe disturbances?

14. **Discuss Concepts** Humans are destroying natural communities at an ever-increasing pace. Using the predictions of the theory of island biogeography, develop hypotheses about what might happen as patches of natural habitats get smaller and smaller. How would you test these hypotheses?

15. **Design an Experiment** Chaparral, a community of woody shrubs that is fairly common in California, often grows adjacent to grassland. The two communities are consistently separated by a "bare zone," usually less than 1 m wide, where no vegetation of either type grows. Ecologists have proposed two possible explanations for this strip of bare soil: (1) that the leaves of chaparral shrubs release harmful, water-soluble chemicals that keep the grass seeds from germinating in the adjacent soil; and (2) that small mammals living in the dense cover provided by chaparral consume the grass seeds before they germinate; the animals do not venture very far from the shrubs because they would be easy targets for predatory hawks. Design a set of field experiments to test the two hypotheses.

For selected answers, see Appendix A.

INTERPRET THE DATA

The Mediterranean shrub *Hormathophylla spinosa* loses as much as 80% of its flowers and fruits to herbivorous mammals each year, and biologists interpret the spines on its flowering stems as an antiherbivore adaptation. Jose M. Gomez and Regino Zamora of the University of Granada, Spain, conducted an exclosure experiment in which they used fences to protect some shrubs from feeding by herbivores and left other shrubs unprotected as controls. The accompanying graph illustrates the density of thorns on the experimental and control groups over a period of two years. How did the protected shrubs respond to the experimental reduction of feeding on the flowers and fruits? How did the unprotected shrubs respond to the control treatment? What benefits would unprotected shrubs derive from their response?

Source: J. M. Gomez and R. Zamora. 2002. Thorns as induced mechanical defense in a long-lived shrub (*Hormathophylla spinosa*, Cruciferae). *Ecology* 83(4):885–890. *Ecology* by Ecological Society of America. Copyright 2002. Reproduced with permission of Ecological Society of America in the format Textbook via Copyright Clearance Center.

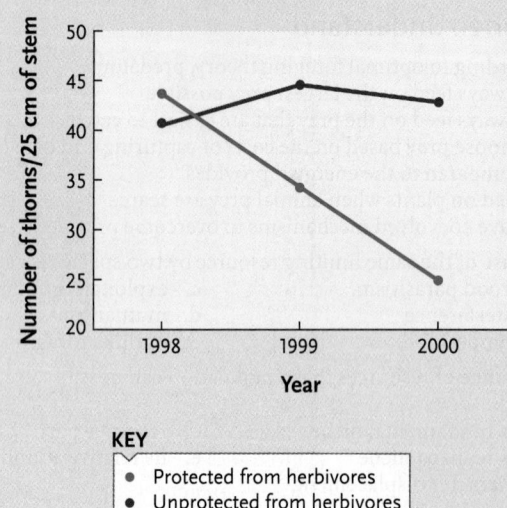

KEY
- Protected from herbivores
- Unprotected from herbivores

Ecosystems and Global Change

Silver Springs, Florida. This small river was the site of one of the earliest comprehensive studies of ecosystem structure and function.

Why it matters . . . Poor Lake Erie, the shallowest of the Great Lakes. Several major industrial cities, including Toledo, Cleveland, Erie, and Buffalo, sprawl along its shoreline. Most of its water comes from the river that flows past Detroit; other rivers that flow into it carry runoff from agricultural fields in Canada and the United States.

When Europeans first settled along its shores roughly 300 years ago, Lake Erie was a wetland paradise. Fishes and waterfowl reproduced in marshes and bays. Even after steel mills and oil refineries were built nearby in the 1860s and 1870s, the lake supported a busy fishing industry and was a famous recreation area.

By 1970, wetlands had been filled for building, bays had been dredged for shipping lanes, and the shoreline had been converted to beaches. Worst of all, household sewage, industrial effluent, and agricultural runoff had so polluted the lake that it no longer supported the activities that had made it famous **(Figure 54.1)**. The water was murky with algae and cyanobacteria; dead fishes washed up on the shore; local health departments closed beaches; and the fishing industry collapsed.

How can a vibrant natural resource become a foul-smelling dump? The answer lies in the human activities that disrupt an **ecosystem,** a biological community and the physical environment with which it interacts. Between the 1930s and the 1970s, Lake Erie's concentration of phosphorus, which had been a

FIGURE 54.1 Pollution of Lake Erie. A steel mill in Lackawanna, New York, discharged industrial wastes into Lake Erie until 1983, when the mill was closed.

1283

limiting nutrient, tripled, largely from household detergents and agricultural fertilizers. High phosphorus concentrations encouraged the growth of photosynthetic algae, changing the phytoplankton community. The density of coliform bacteria, which originate in the human gut and serve as indicators of organic pollution, also skyrocketed as a result of the surge in sewage and nutrients entering the lake.

Increased phytoplankton and bacterial populations depleted oxygen in the lake's waters, contributing to changes elsewhere in the lake. Mayflies (*Hexagenia* species), whose larvae live in well-oxygenated bottom sediments, had once been so abundant that their aerial breeding swarms were a public nuisance. But they became nearly extinct in the polluted lake, replaced by oligo-chaete worms, snails, and other invertebrates. Along with over-fishing, changes in the bottom fauna shifted the composition of the fish community; the catch of desirable food fishes declined to almost zero by the mid-1960s.

In 1972, Canada and the United States began efforts to restore the lake. They spent billions of dollars to reduce the influx of phosphates and limited fishing of the most vulnerable native species. Nonnative salmon (*Onchorhynchus* species) and other predatory fishes were introduced in the hope that they could bring the lake back to its original condition. Even the accidental introduction of zebra mussels (*Dreissina polymorpha*), an aquatic pest, inadvertently helped the effort because they feed on phytoplankton.

But, although somewhat improved, Lake Erie will never return to its former glory. Some native species are now extinct there, and the introduced species that replaced them function differently within the ecosystem. The lake still suffers periods of uncontrolled algal growth, fish kills, and high levels of harm-ful bacteria.

This story of an ecological disaster and partial recovery introduces ecosystem ecology, the branch of ecology that ana-lyzes the flow of energy and the cycling of materials between an

FIGURE 54.2 Terrestrial food webs. Energy and nutrients move through food webs in all ecosystems. Primary producers fix inorganic carbon into organic molecules; herbivores and carnivores feed on living fungi, plants, and animals; and detritivores and decomposers recycle nutrients from material that is no longer alive. Each box in this diagram represents many species. A marine food web is illustrated in Figure 53.23.

© Cengage Learning 2017

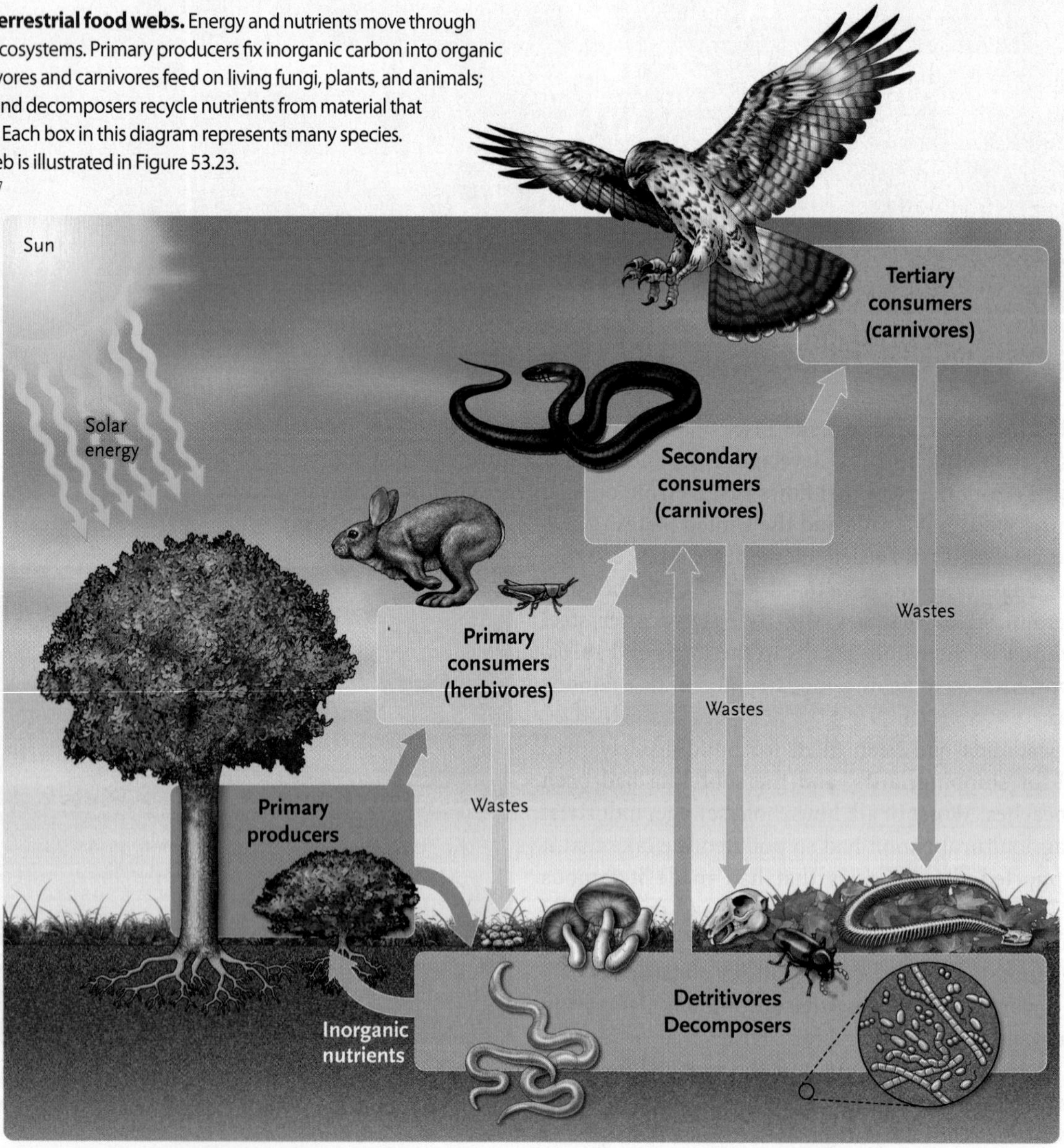

ecosystem's living and nonliving components. These processes make the resident organisms highly dependent on each other and on their physical surroundings. Ultimately, the Lake Erie ecosystem unraveled because human activities disrupted the flow of energy and the cycling of materials on which the organisms depended.

54.1 Modeling Ecosystem Processes

All organisms require steady supplies of energy and nutrients for their maintenance, growth, and reproduction. Studies of ecosystems often focus on the inputs and outputs (that is, the gains and losses) of energy and nutrients to the ecosystem as a whole, as well as the transfer of energy and nutrients within and between the ecosystem's biotic and abiotic components. Although the movements of energy and nutrients through an ecosystem are sometimes coupled, as when you eat a meal that contains both calories and nutrients, the inputs and outputs of energy and nutrients are fundamentally different (see Section 1.1). In virtually all ecosystems, sunlight constantly renews the supply of available energy, but, as dictated by the laws of thermodynamics (see Section 6.1), most of that energy is lost as heat in cellular respiration. By contrast, virtually all the nutrients that will ever be available for biological systems are already present on Earth, and they are constantly recycled between the abiotic and biotic components of ecosystems in what ecologists describe as **biogeochemical cycles.**

Researchers use several types of models to describe ecosystem processes. Food webs define the pathways through which energy and nutrients move within the biotic component of an ecosystem. Compartment models describe how nutrients move between living and nonliving nutrient reservoirs. Simulation models allow ecologists to predict how ecosystems will respond to perturbations of ecosystem processes.

Food Webs Illustrate the Transfer of Energy and Nutrients among Organisms

Food webs define the pathways by which energy and nutrients move through an ecosystem's biotic components. In most ecosystems, they move simultaneously through primary producers, organisms that feed on other living organisms, and those that feed on detritus (that is,

material that is no longer alive) **(Figure 54.2)**. As you learned in Section 53.3, food webs include the producer, herbivore, and carnivore trophic levels, as well as detritivores and decomposers. Because detritivores and decomposers subsist on the remains and waste products of other organisms, they link all of the other trophic levels together.

Compartment Models Track the Movement of Nutrients between Food Webs and Abiotic Reservoirs

Ecologists use a **compartment model** to describe nutrient cycling **(Figure 54.3)**. Two criteria divide ecosystems into four compartments where nutrients accumulate. First, nutrient molecules and ions are described as either *available* or *unavailable*, depending on whether or not they can be assimilated by organisms. Second, nutrients are present either in *organic* material, the living or dead tissues of organisms, or in *inorganic* material, such as rocks and soil. For example, minerals in dead leaves on the forest floor are in the available organic compartment because they are in the remains of organisms that detritivores

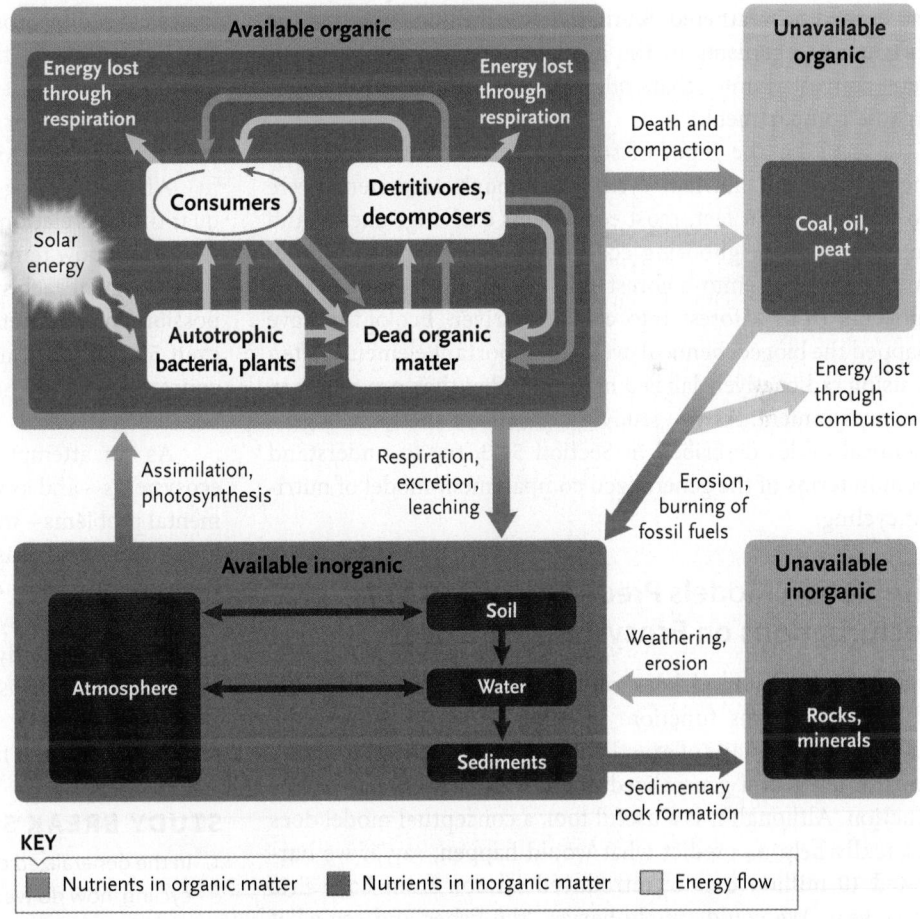

KEY

| Nutrients in organic matter | Nutrients in inorganic matter | Energy flow |

FIGURE 54.3 A compartment model of nutrient cycling. Nutrients are recycled through four major compartments within ecosystems. Processes that move nutrients from one compartment to another are indicated on the arrows. The upper left quadrant includes the living organisms in food webs and their organic remains; the oval arrow represents animal predation on other animals. The pathways of energy flow are also illustrated. Unlike nutrients, energy is not conserved, but is instead lost from the ecosystem.

© Cengage Learning 2017

can eat. But calcium ions in limestone rocks are in the unavailable inorganic compartment because they exist in a nonbiological form that producers cannot assimilate.

Nutrients move rapidly within and between the available compartments. Living organisms are in the available organic compartment, and whenever heterotrophs consume food, they recycle nutrients within that reservoir (indicated by the oval arrow in the upper left of Figure 54.3). Producers acquire nutrients from the air, soil, and water of the available inorganic compartment. Consumers also acquire nutrients from the available inorganic compartment when they drink water or absorb mineral ions through the body surface. Several processes routinely transfer nutrients from organisms to the available inorganic compartment. As one example, respiration releases carbon dioxide, moving both carbon and oxygen from the available organic compartment to the available inorganic compartment.

By contrast, the movement of materials into and out of the unavailable compartments is generally slow. Sedimentation, a long-term geological process, converts ions and particles of the available inorganic compartment into rocks of the unavailable inorganic compartment. Materials are gradually returned to the available inorganic compartment when rocks are uplifted and eroded or weathered. Similarly, over millions of years, the remains of organisms in the available organic compartment were converted into coal, oil, and peat of the unavailable organic compartment.

Except for the input of solar energy, we have described energy flow and nutrient cycling as though ecosystems were closed systems. In fact, most ecosystems exchange energy and nutrients with neighboring ecosystems. For example, rainfall carries nutrients into a forest ecosystem, and runoff carries nutrients from a forest into a lake or river. Ecologists have mapped the biogeochemical cycles of important elements, often by using radioactively labeled molecules that they can follow in the environment. As you study the details of the four biogeochemical cycles described in Section 54.3, try to understand them in terms of the generalized compartment model of nutrient cycling.

Simulation Models Predict the Effects of Perturbations on Ecosystem Processes

The compartment model described above is a *conceptual model* of how ecosystems function. In other words, it ignores the nuts-and-bolts details of exactly how a specific ecosystem functions in favor of a generalized portrait of how all ecosystems function. Although it is a useful tool, a conceptual model does not really help us predict what would happen, say, if we harvested 10 million tons of introduced salmon from Lake Erie every year. We could simply harvest the fishes and see what happens. But ecologists prefer less intrusive approaches to study the potential effects of disturbances.

To understand how an ecosystem will respond to specific changes in physical factors, energy flow, nutrient availability, and climate change, ecologists turn to **simulation modeling.**

Researchers gather detailed information about a specific ecosystem and then create a series of mathematical equations that define its most important relationships. For example, one set of equations might describe how nutrient availability limits photosynthesis by autotrophs. Another might relate population growth of zooplankton to the abundance of phytoplankton. Other equations would relate the population dynamics of primary carnivores to the availability of their food, and still others would describe how the densities of primary carnivores influence reproduction in populations at both lower and higher trophic levels. Thus, a complete simulation model is a set of interlocking equations that collectively predict how changes in one feature of an ecosystem might influence others.

Creating a simulation model is no easy task, because the relationships within every ecosystem are complex. First, you would identify the important species, estimate their population sizes, and measure the average energy and nutrient content of each. Next, you would describe the food webs in which they participate, measure the quantity of food each species consumes, and estimate the growth and reproduction of individuals in each population. For the sake of completeness, you would also determine the ecosystem's energy and nutrient gains and losses caused by erosion, weathering, precipitation, and runoff. You would repeat these measurements seasonally to identify annual variation in the factors. Finally, you might repeat the measurements over several years to determine the effects of year-to-year variation in climate and chance events.

After collecting these data, you would write equations that quantify the relationships in the ecosystem, including information about how temperature and other abiotic factors influence the ecology of each species. At last, you could begin to predict—possibly in great detail—the effects of harvesting 10 million or even 50 million tons of salmon annually from Lake Erie. Of course, you would have to refine the model whenever new data became available.

As we attempt to understand larger and more complex ecosystems—and as we create larger and more complex environmental problems—modeling becomes an increasingly important tool. If a model is based on well-defined ecological relationships and good empirical data, it can allow us to make accurate predictions about ecosystem changes without the need for costly and environmentally damaging experiments. But like all ideas in science, a model is only as good as its assumptions, and models must constantly be adjusted to incorporate new ideas and recently discovered facts.

STUDY BREAK 54.1

1. In the generalized compartment model of biogeochemical cycling, how do we classify the compartments where nutrients accumulate?

2. What are the advantages and disadvantages of relying on conceptual models that describe ecosystem function?

3. What data must ecologists collect before constructing a simulation model of an ecosystem?

54.2 Energy Flow and Ecosystem Energetics

Ecosystems receive a steady input of energy from an external source, which in almost all cases is the Sun. But as energy flows through an ecosystem, much of it is lost as heat without being used by organisms. In this section, we consider the details of energy flow and the efficiency of energy transfer from one trophic level to another.

Sunlight Provides the Energy Input for Practically All Ecosystems

Every minute of every day, Earth's atmosphere intercepts roughly 19 kcal of solar energy per square meter. (Recall from Chapter 2 that 1 kcal = 1,000 calories.) About half that energy is absorbed, scattered, or reflected by gases, dust, water vapor, and clouds without ever reaching the planet's surface (see Chapter 51). Most energy that reaches the surface falls on bodies of water or bare ground, where it is absorbed as heat or reflected back into the atmosphere; reflected energy warms the atmosphere, as we discuss later in this chapter. Only a small percentage contacts primary producers, and most of that energy evaporates water, driving transpiration in plants (see Section 34.3).

Ultimately, photosynthesis converts less than 1% of the solar energy that arrives at Earth's surface into chemical energy. But primary producers capture enough energy to create an average of several kilograms of dry plant material per square meter per year. On a global scale, they produce more than 150 billion metric tons of new biological material annually. Some of the solar energy that producers convert into chemical energy is transferred to consumers at higher trophic levels.

The rate at which producers convert solar energy into chemical energy is an ecosystem's **gross primary productivity.** But like all other organisms, producers use energy for their own maintenance. After deducting the energy devoted to these functions, which are collectively called *cellular respiration* (see Section 7.1), whatever chemical energy remains is the ecosystem's **net primary productivity.** In most ecosystems, net primary productivity is between 50% and 90% of gross primary productivity. In other words, producers use between 10% and 50% of the energy they capture for their own respiration.

Ecologists generally measure primary productivity in units of energy captured (kcal/m^2/yr) or in units of biomass created (g/m^2/yr). *Biomass* is the dry weight of biological material per unit area or volume of habitat. (We measure biomass as the *dry* weight of organisms because their water content, which fluctuates with water uptake or loss, has no energetic or nutritional value.) You should not confuse an ecosystem's productivity with its **standing crop biomass,** the total dry weight of plants present at a given time. Net primary productivity is the *rate* at which the standing crop produces *new* biomass.

The energy captured by plants is stored in biological molecules—mostly carbohydrates, lipids, and proteins. Ecologists can convert units of biomass into units of energy or vice versa as long as they know how much carbohydrate, protein, and lipid a sample of biological material contains (4.2 kcal/g of carbohydrate; nearly 4.1 kcal/g of protein; and 9.5 kcal/g of lipid). Thus, net primary productivity is a measure of the rate at which producers accumulate energy, as well as the rate at which new biomass is added to an ecosystem. Because it is far easier to measure biomass than energy content, ecologists usually measure changes in biomass to estimate productivity. New biomass takes several forms: the growth of existing producers; the creation of new producers by reproduction; and the storage of energy as carbohydrates or lipids. Because herbivores eat all three forms of new biomass, net primary productivity also measures how much new energy is available for primary consumers.

Primary Productivity Varies Greatly on Global and Local Scales

The potential rate of photosynthesis in any ecosystem is proportional to the intensity and the duration of sunlight, which vary geographically and seasonally (see Section 51.2). Sunlight is most intense and day length least variable near the equator. By contrast, light intensity is weakest and day length most variable near the poles. Thus, producers at the equator can photosynthesize nearly 12 hours a day, every day of the year. Near the poles, photosynthesis is virtually impossible during the long, dark winter; in summer, however, plants can photosynthesize around the clock.

Sunlight is not the only factor that influences the rate of primary productivity, however. Temperature and the availability of water and nutrients have significant effects. Across a wide variety of terrestrial ecosystems, net aboveground primary productivity increases with both mean annual temperature and mean annual precipitation **(Figure 54.4).** Thus, tropical rainforests, which are both hot and wet, have exceptionally high productivity. Although many of the world's deserts receive plenty of sunshine and experience high mean annual temperature, they have low rates of productivity because water is in short supply and the soil is nutrient-poor. And although arctic tundra is very wet for much of the year, low temperatures inhibit photosynthesis, reducing productivity **(Table 54.1).** Thus, mean annual net primary productivity varies greatly on a global scale **(Figure 54.5),** reflecting variations in these environmental factors (see Chapter 51).

In systems with sufficient precipitation, a shortage of mineral nutrients may be limiting. All plants need specific ratios of macronutrients and micronutrients for maintenance and photosynthesis (see Section 35.1). But plants withdraw nutrients from soil, and if nutrient concentration drops below a critical level, photosynthesis may decrease or stop altogether. In every ecosystem, one nutrient inevitably runs out before the supplies of other nutrients are exhausted. The element in short supply is called a **limiting nutrient** because its absence limits productivity. Productivity in agricultural fields is subject to the same constraints as productivity in natural ecosystems. Farmers increase

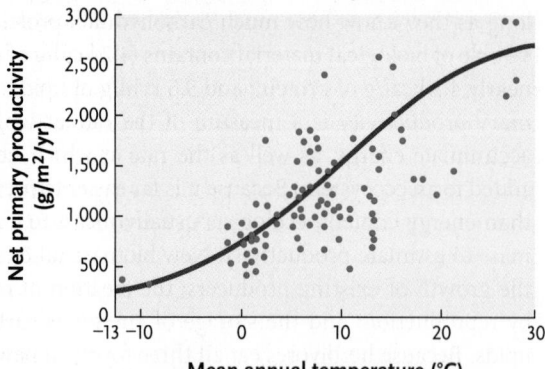

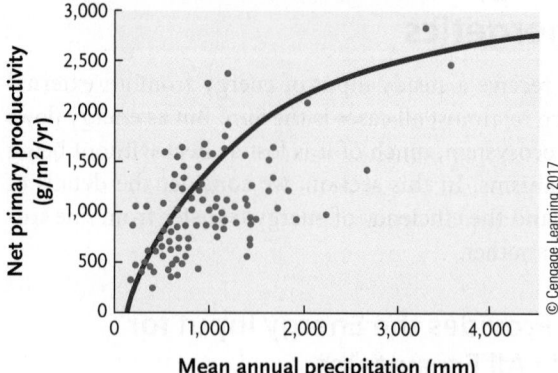

FIGURE 54.4 Temperature and rainfall influence net primary productivity. Mean annual net primary productivity increases with both **(A)** mean annual temperature and **(B)** mean annual precipitation across many terrestrial ecosystems. These data include only aboveground productivity.

TABLE 54.1	Standing Crop Biomass and Net Primary Productivity of Different Ecosystems	
Ecosystem	Mean Standing Crop Biomass (g/m²)	Mean Net Primary Productivity (g/m²/yr)
Terrestrial Ecosystems		
Tropical rainforest	45,000	2,200
Tropical deciduous forest	35,000	1,600
Temperate rainforest	35,000	1,300
Temperate deciduous forest	30,000	1,200
Savanna	4,000	900
Boreal forest (taiga)	20,000	800
Woodland and shrubland	6,000	700
Cultivated land	1,000	650
Temperate grassland	1,600	600
Tundra and alpine tundra	600	140
Desert and thornwoods	700	90
Extreme desert, rock, sand, ice	20	3
Freshwater Ecosystems		
Swamp and marsh	15,000	2,000
Lake and stream	20	250
Marine Ecosystems		
Open ocean	3	125
Upwelling zones	20	500
Continental shelf	10	360
Kelp beds and reefs	2,000	2,500
Estuaries	1,500	1,500
World Average	**3,600**	**333**

From R. H. Whittaker. 1975. *Communities and Ecosystems.* 2nd ed. Macmillan.

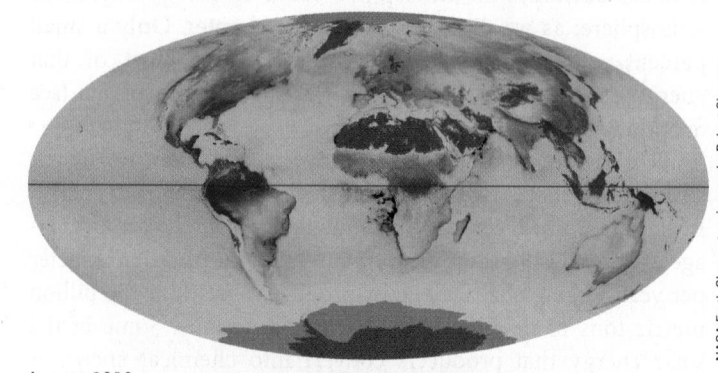

August 2010

December 2010

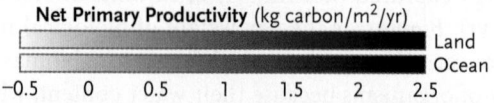

Net Primary Productivity (kg carbon/m²/yr)

Land
Ocean

−0.5   0   0.5   1   1.5   2   2.5

FIGURE 54.5 Seasonal and global variation in net primary productivity. Satellite data from 2010 provide a visual portrait of net primary productivity across Earth's surface in August and December. The key indicates productivity measured as kilograms of carbon fixed by photosynthesis per square meter per year. Note that productivity in the northern hemisphere is high during the northern summer (August), but low during the northern winter (December).
© Cengage Learning 2017

productivity by irrigating (adding water to) and fertilizing (adding nutrients to) their crops.

In freshwater and marine ecosystems, where water is always readily available, the depth of the water and the combined availability of sunlight *and* nutrients govern the rate of primary productivity. Productivity is high in near-shore ecosystems where sunlight penetrates shallow, nutrient-rich waters. Kelp beds and coral reefs, for example, which occur along temperate and tropical coastlines respectively, are among the most productive ecosystems on Earth (see Table 54.1). By contrast, productivity is low in the open waters of a large lake or ocean: sunlight penetrates only the upper layers, and nutrients sink to the bottom. Thus, the two requirements for photosynthesis, sunlight and nutrients, are available in different places.

Although ecosystems vary in their net primary productivity, the differences are not always proportional to variations in their standing crop biomass (see Table 54.1). For example, biomass in temperate deciduous forests and temperate grasslands differs by a factor of 20, but the difference in their rates of net primary productivity is only twofold. Most biomass in trees is present in nonphotosynthetic tissues such as wood. As a result, their ratio of productivity to biomass is low ($1{,}200 \text{ g/m}^2 \div 30{,}000 \text{ g/m}^2 = 0.040$). By contrast, grasslands don't accumulate much biomass because annual mortality, herbivores, and fires remove plant material as it is produced; and their productivity to biomass ratio is much higher ($600 \text{ g/m}^2 \div 1{,}600 \text{ g/m}^2 = 0.375$).

Some ecosystems contribute more than others to overall net primary productivity (**Figure 54.6**). Ecosystems that cover large areas make substantial contributions, even if their productivity is low. Conversely, geographically restricted ecosystems make large contributions if their productivity is high. For example, the open ocean and tropical rainforests contribute about equally to total global productivity, but for different reasons. Open oceans have low productivity, but they cover nearly two thirds of Earth's surface. Tropical rainforests cover only a small area, but they are highly productive.

Some Energy Is Always Lost before It Is Transferred from One Trophic Level to the Next

Net primary productivity ultimately supports all the consumers in food webs. Herbivores and carnivores eat some of the biomass at every trophic level except the highest; uneaten biomass eventually dies and is consumed by detritivores and decomposers. However, consumers at every trophic level assimilate only a portion of the material they ingest, and unassimilated material is passed as feces, which supports detritivores and decomposers.

As energy is transferred from producers to consumers, some is stored in new consumer biomass, called **secondary productivity.** Nevertheless, two factors cause energy to be lost from the ecosystem every time it flows from one trophic level to another. First, animals use much of the energy they assimilate

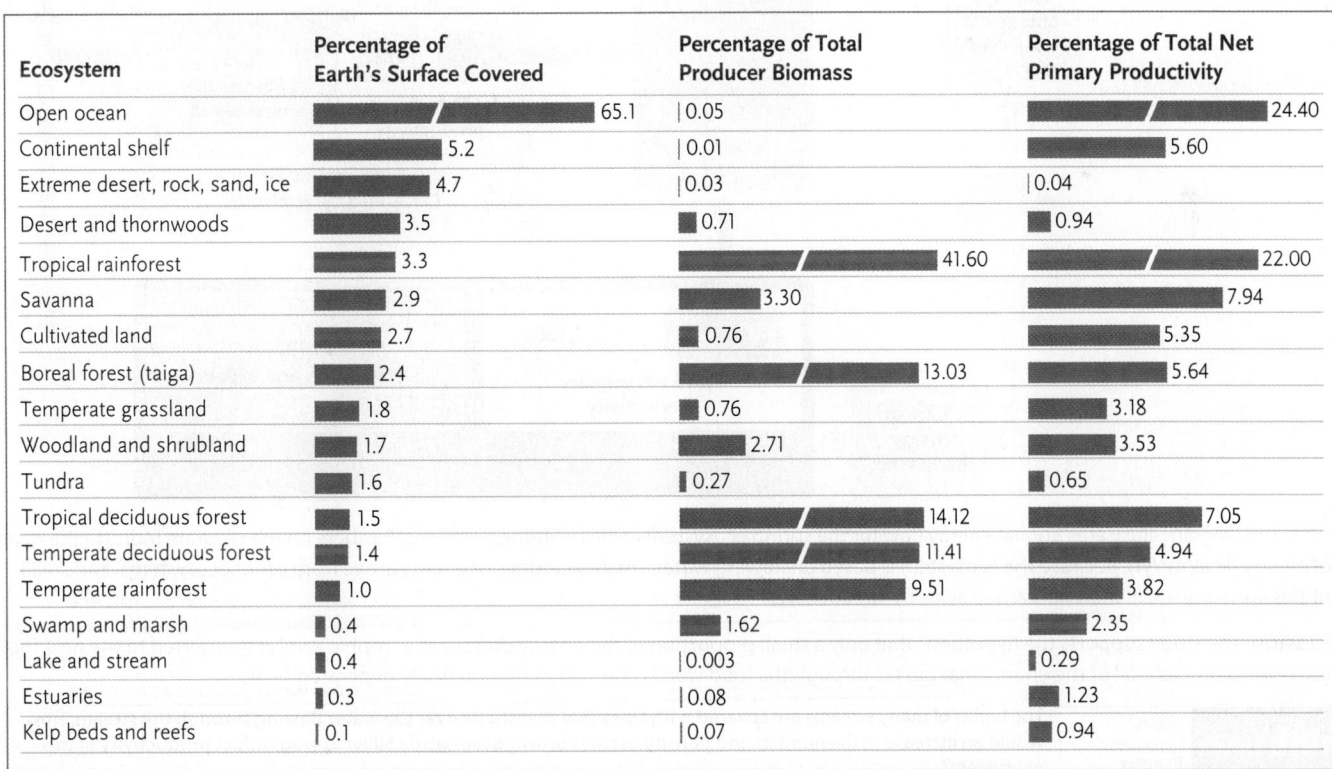

Ecosystem	Percentage of Earth's Surface Covered	Percentage of Total Producer Biomass	Percentage of Total Net Primary Productivity
Open ocean	65.1	0.05	24.40
Continental shelf	5.2	0.01	5.60
Extreme desert, rock, sand, ice	4.7	0.03	0.04
Desert and thornwoods	3.5	0.71	0.94
Tropical rainforest	3.3	41.60	22.00
Savanna	2.9	3.30	7.94
Cultivated land	2.7	0.76	5.35
Boreal forest (taiga)	2.4	13.03	5.64
Temperate grassland	1.8	0.76	3.18
Woodland and shrubland	1.7	2.71	3.53
Tundra	1.6	0.27	0.65
Tropical deciduous forest	1.5	14.12	7.05
Temperate deciduous forest	1.4	11.41	4.94
Temperate rainforest	1.0	9.51	3.82
Swamp and marsh	0.4	1.62	2.35
Lake and stream	0.4	0.003	0.29
Estuaries	0.3	0.08	1.23
Kelp beds and reefs	0.1	0.07	0.94

FIGURE 54.6 Biomass and net primary productivity. The percentage of Earth's surface that an ecosystem covers is not proportional to its contribution to the total biomass of producers or its contribution to the total net primary productivity.

FIGURE 54.7 | **Observational Research**

What Is the Pattern of Energy Flow within the Silver Springs Ecosystem?

Hypothesis: Only a small percentage of the energy present in a trophic level is transferred to the next higher trophic level in the ecosystem.

Prediction: The energy content of the organisms present in each trophic level will decline steadily from the lowest to highest trophic levels.

Method: Howard T. Odum and his team analyzed energy flow in an aquatic ecosystem at Silver Springs, Florida. The producers in this small spring are mostly aquatic plants. The herbivores include snails, shrimp, insects, fishes, and turtles. The carnivores include a variety of invertebrates and fishes. The top carnivores are large fish. Sunlight is available as an energy source all year round. After defining the food web in this ecosystem, the researchers estimated the biomass and energy content (kcal/g) of each trophic level. They then constructed a diagram that illustrates how much energy is present at each trophic level, how much is transferred to the next higher trophic level, and how much energy is lost as it works its way through the food web.

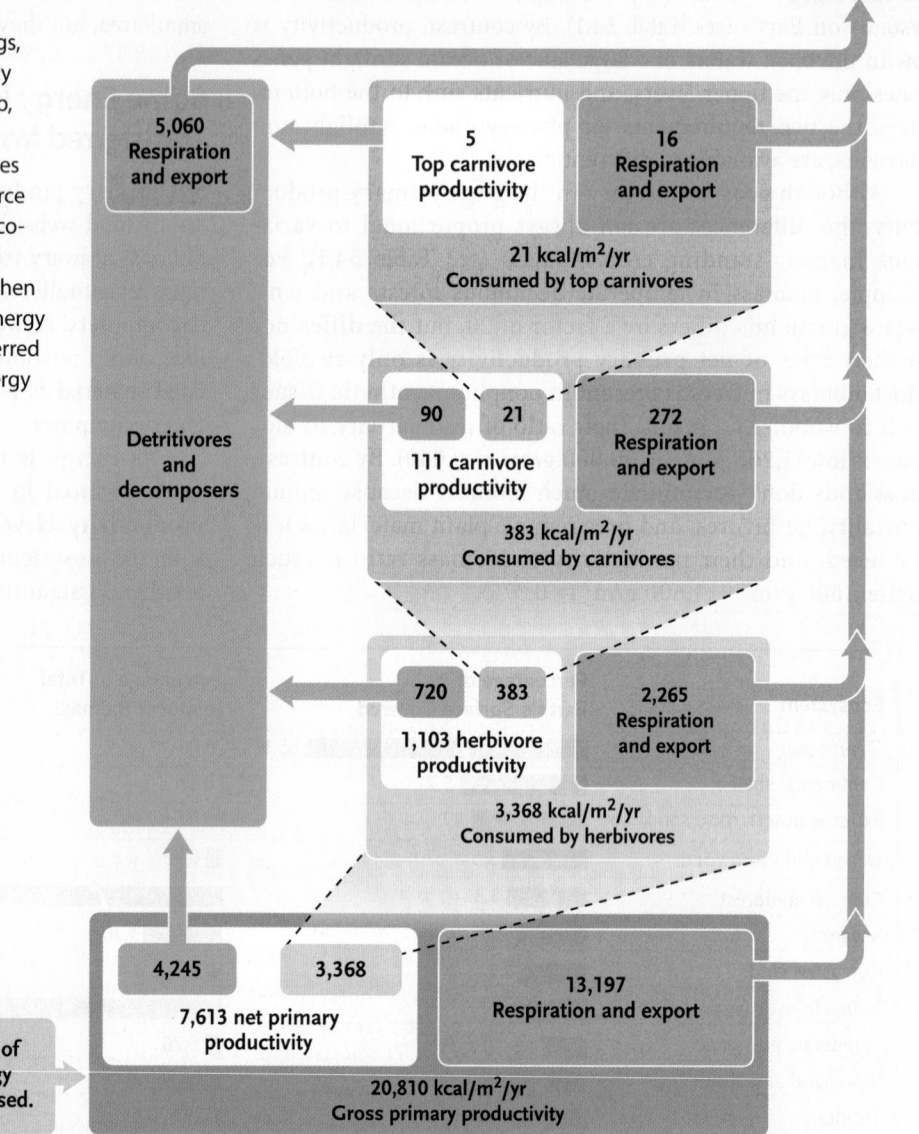

Results: The diagram illustrates annual energy flow for the spring ecosystem at Silver Springs, Florida. Numbers on the diagram indicate the quantity of energy (kcal/m²/yr). Because the ecosystem is based on flowing water, small quantities of energy arrive from other ecosystems, and small quantities are exported in material carried away by stream flow.

Conclusion: The study supports the hypothesis that only a small proportion of the energy present at a trophic level is transferred to the next higher trophic level. Ultimately, all of the energy that passes through the food web is released as metabolically generated heat.

think like a scientist

Observe
Hypothesize
Predict
Experiment
Interpret

The banks of many streams are covered with trees that cast shade over the water flowing through the stream. How would an increase in the number and density of trees growing alongside Silver Springs affect productivity in this ecosystem?

Source: H. T. Odum. 1957. Trophic structure and productivity of Silver Springs, Florida. *Ecological Monographs* 27:55–112.

for maintenance or locomotion rather than the production of new biomass. Second, as dictated by the second law of thermodynamics, no biochemical reaction is 100% efficient; thus, some of the chemical energy liberated by cellular respiration is always converted to heat, which most organisms do not use.

Ecological efficiency is the ratio of net productivity at one trophic level to net productivity at the trophic level below it. For example, if the plants in an ecosystem have a net primary productivity of 100 $g/m^2/yr$ of new tissue and the herbivores that eat those plants produce 10 $g/m^2/yr$, the ecological efficiency of the herbivores is 10%. The efficiencies of three processes—harvesting food, assimilating ingested energy, and producing new biomass—determine the ecological efficiencies of consumers.

Harvesting efficiency is the ratio of the energy content of food consumed to the energy content of food available. Predators harvest food efficiently when prey are abundant and easy to capture (see Section 53.1).

Assimilation efficiency is the ratio of the energy absorbed from consumed food to the food's total energy content. Because animal prey is relatively easy to digest, carnivores absorb between 60% and 90% of the energy in their food; assimilation efficiency is lower for prey with indigestible parts such as bones or exoskeletons. Herbivores assimilate only 15% to 80% of the energy they consume because cellulose is not very digestible.

Production efficiency is the ratio of the energy content of new tissue produced to the energy assimilated from food. Production efficiency varies with maintenance costs. For example, endothermic animals often use less than 10% of their assimilated energy for growth and reproduction because they use energy to maintain body temperature (see Section 48.8). Ectothermic animals, by contrast, channel more than 50% of their assimilated energy into new biomass.

The overall ecological efficiency of most organisms is between 5% and 20%. As a rule of thumb, only about 10% of the energy accumulated at one trophic level is converted into biomass at the next higher trophic level, as illustrated by energy transfers at Silver Springs, Florida **(Figure 54.7)**. Producers in the Silver Springs ecosystem convert 1.2% of the solar energy they intercept into chemical energy (represented by 20,810 kcal of gross primary productivity). However, they use about two thirds of this energy for respiration, leaving only one third to be included in new plant biomass, the net primary productivity. All consumers in the food web (on the right in Figure 54.7) ultimately depend on this energy source, which dwindles with each transfer between trophic levels. Energy is lost to respiration and export (that is, the transport of energy-containing materials out of the ecosystem by flowing water) at each trophic level. In addition, substantial energy flows to the detritivores and decomposers (on the left in Figure 54.7) as organic wastes and uneaten biomass. To determine the ecological efficiency of any trophic level, we divide its productivity by the productivity of the level below it. For example, the ecological efficiency of midlevel carnivores at Silver Springs is 111 kcal/yr ÷ 1,103 kcal/yr = 10.06%.

The low ecological efficiencies that characterize most energy transfers illustrate one advantage of eating "lower on the food chain." Even though humans digest and assimilate meat more efficiently than vegetables, we might be able to feed more people if we all ate more vegetables directly instead of first passing these crops through another trophic level, such as cattle or chickens, to produce meat. The production of animal protein is costly because much of the energy fed to livestock is used for their own maintenance rather than the production of new biomass. But despite the economic—not to mention health-related—logic of a more vegetarian diet, a change in our eating habits alone will not eliminate food shortages or the frequency of malnutrition. Many regions of Africa, Australia, North America, and South America support vegetation that is suitable only for grazing by large herbivores. These areas could not produce significant quantities of edible grains and vegetables.

Ecological Pyramids Illustrate the Cumulative Effects of Energy Losses

All organisms in a trophic level are the same number of energy transfers from the ecosystem's ultimate energy source. Plants are one energy transfer removed from sunlight; herbivores are two transfers away; carnivores feeding on herbivores are three transfers away; and carnivores feeding on other carnivores are four transfers away. As energy works its way up a food web, energy losses are multiplied in successive energy transfers, greatly reducing the energy available to support the highest trophic levels.

Consider a hypothetical example in which ecological efficiency is 10% for all consumers. Assume that the plants in a small field annually produce new tissues containing 100 kcal of energy. Because only 10% of that energy is transferred from one trophic level to the next higher trophic level, the 100 kcal in plants produces only 10 kcal of new herbivorous insects; only 1 kcal of new songbirds, which feed on insects; and only 0.1 kcal of new falcons, which feed on songbirds. Thus, after three energy transfers, only 0.1% of the energy from primary productivity is present in the highest trophic levels.

The inefficiency of energy transfer from one trophic level to the next has profound effects on ecosystem structure. Ecologists illustrate these effects in diagrams called **ecological pyramids (Figure 54.8)**. Trophic levels are drawn as stacked blocks, with the size of each block proportional to the energy, biomass, or numbers of organisms present; primary producers are illustrated at the bottom of the pyramid and higher-level consumers at the top. The energy or biomass in detritivores and decomposers is often illustrated alongside the pyramid for the other trophic levels.

Pyramids of energy typically have wide bases and narrow tops because each trophic level contains on average only about 10% as much energy as the trophic level below it, as illustrated by the pyramid of energy for Silver Springs, Florida (see Figure 54.8A).

The progressive reduction in productivity at higher trophic levels, as illustrated in Figure 54.7, usually establishes a **pyramid**

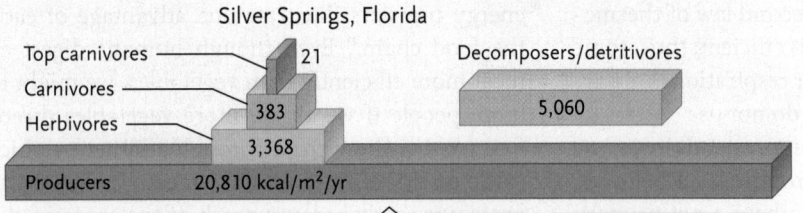

A. Pyramid of energy

Silver Springs, Florida

The amount of energy (kcal/m²/yr) passing through each trophic level of the food web decreases as it moves to higher trophic levels.

B. Pyramids of biomass

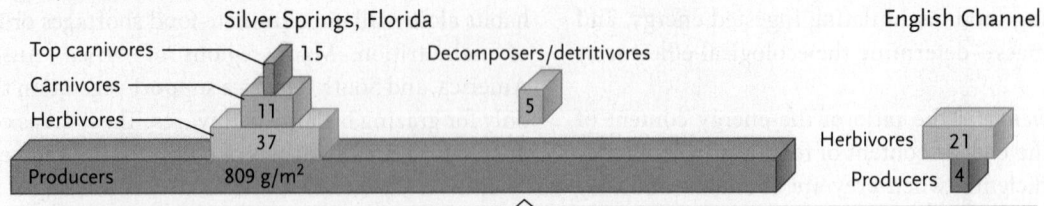

The pyramid of standing crop biomass (g/m²) for the food web at Silver Springs (left) is bottom heavy, as it is for most ecosystems. Some marine ecosystems, such as that in the English Channel (right), have an inverted pyramid of biomass because producers are quickly eaten by primary consumers. Only the producer and herbivore trophic levels are illustrated here.

C. Pyramids of numbers

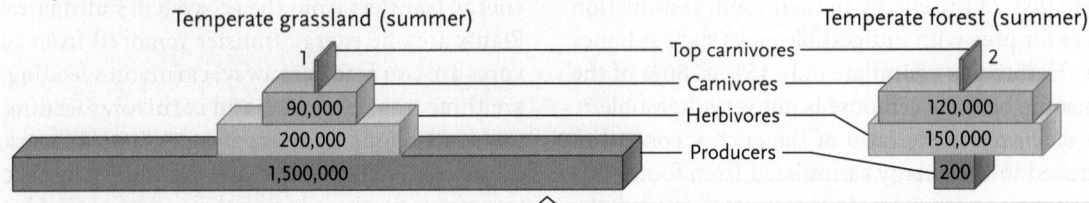

The pyramid of numbers (number of individuals per 1,000 m²) for temperate grasslands (left) is bottom-heavy because individual producers are small and very numerous. The pyramid of numbers for forests (right) may have a narrow base because herbivorous insects vastly outnumber the trees they eat. Detritivores and decomposers (soil animals and microorganisms) are not included because they are difficult to count.

FIGURE 54.8 Ecological pyramids. Pyramids of energy, biomass, and numbers reflect the multiplicative loss of energy from one trophic level to another.
© Cengage Learning 2017

of biomass (see Figure 54.8B). The biomass at each trophic level is proportional to the chemical energy temporarily stored there. Thus, in terrestrial ecosystems, the total biomass of producers is generally greater than the total biomass of herbivores, which is, in turn, greater than the total biomass of predators. Populations of top predators—animals such as mountain lions or alligators—contain too little biomass and energy to support another trophic level; thus, they have no nonhuman predators.

Freshwater and marine ecosystems sometimes exhibit inverted pyramids of biomass (see right side of Figure 54.8B). In the open waters of a lake or ocean, primary consumers (zooplankton) eat the primary producers (phytoplankton) almost as soon as they appear. As a result, the standing crop of primary consumers at any moment in time is actually larger than the standing crop of the primary producers that support them. Food webs in these ecosystems are stable, however, because the producers have exceptionally high **turnover rates.** In other words, the phytoplankton divide and their populations grow so quickly that feeding by zooplankton doesn't endanger their populations or reduce their productivity. And on an annual

basis, the *cumulative total* biomass of primary producers far outweighs that of primary consumers.

The reduction of energy and biomass also affects the population sizes of organisms at the top of a food web. Top predators are often relatively large animals, thus concentrating the limited biomass at the highest trophic levels in relatively few individuals (Figure 54.8C). The extremely narrow top of this **pyramid of numbers** has grave implications for conservation biology. Because each top predator must patrol a large area to find sufficient food, the members of a population are often widely dispersed within their habitats. As a result, they are highly sensitive to hunting, habitat destruction, and random events, which can lead to local extinction (see Chapter 55).

Top Predators Experience Biological Magnification

As described above, top predators generally have small population sizes, partly because of inefficient energy transfer between trophic levels and partly because top predators are usually

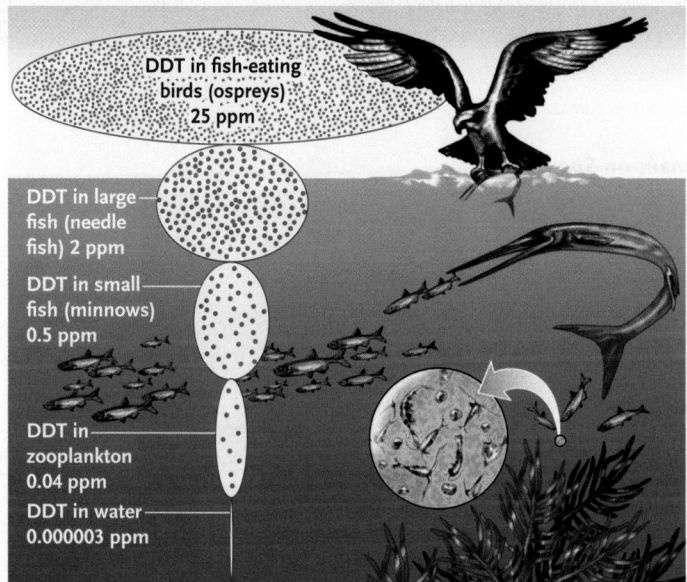

FIGURE 54.9 Biological magnification. In this food web near Long Island Sound, New York, DDT concentration (measured in parts per million, ppm) was magnified nearly 10 million times between zooplankton and ospreys.
© Cengage Learning 2017

larger than their prey. Thus, materials, including both nutrients and toxic compounds, become *concentrated* in individuals at higher trophic levels; the accumulation of harmful compounds is called **biological magnification (Figure 54.9).**

The synthetic organic pesticide DDT (dichloro-diphenyl-trichloroethane) is the best known example of biological magnification. It was first used widely during World War II to kill mosquitoes that transmitted malarial parasites (*Plasmodium* species) and body lice that carried the bacteria causing typhus (*Rickettsia rickettsii*). After the war, people used DDT to kill agricultural pests, disease vectors, and insects in homes and gardens.

Although DDT is nearly insoluble in water, it is more mobile than its users expected. Winds carry it as a vapor, and water transports it as fine particles. DDT is also highly soluble in lipids, and consumers accumulate the DDT from all of the organisms they eat in their lifetimes. Primary consumers, such as herbivorous insects, may ingest relatively small quantities. But a songbird that eats many insects will accumulate a moderate amount, and a predator that feeds on songbirds will accumulate even more. Thus, DDT and other nondegradable poisons become concentrated in organisms at higher trophic levels.

After the war, DDT moved rapidly through ecosystems, affecting organisms in ways that no one had predicted. Songbirds died in cities; salmon succumbed in streams and rivers; and agricultural pests flourished because they evolved resistance to the pesticide (see Figure 20.12 and *Why it matters . . .* in Chapter 21) and because DDT killed the natural predators that had kept their populations in check. Eventually, top carnivores in some food webs were pushed to the brink of extinction. The reproduction of bald eagles, peregrine falcons, ospreys, and brown pelicans was disrupted because one DDT breakdown product interferes with the deposition of calcium in their eggshells. When birds tried to incubate their eggs, the shells cracked beneath the parents' weight. Even today, traces of DDT are found in the tissues of nearly all species, including human fat and breast milk.

Since the 1970s, DDT has been banned in the United States, except for restricted applications to protect public health. Many hard-hit species have partially recovered, but some birds still lay thin-shelled eggs because they pick up DDT at their winter ranges in Latin America. As recently as 1990, the California State Department of Health recommended that a fishery off the coast of California be closed because DDT from industrial waste discharged 20 years earlier was still moving through that ecosystem. Moreover, DDT is still used in other countries, and some enters the United States on imported fruit and vegetables.

Consumers Sometimes Regulate Ecosystem Processes in a Trophic Cascade

As you know from the preceding discussion, numerous abiotic factors—the intensity and duration of sunlight, rainfall, temperature, and the availability of nutrients—have significant effects on primary productivity. Primary productivity, in turn, has profound effects on populations of herbivores and the predators that feed on them. But what effect does feeding by these consumers have on primary productivity?

Recent research suggests that consumers may sometimes influence rates of primary productivity, especially in ecosystems with low species diversity and relatively few trophic levels. For example, food webs in North American salt marshes depend primarily on the productivity of salt marsh cordgrass (*Spartina alterniflora*), a foundation species that defines the nature of that coastal ecosystem (see Section 53.3). Cordgrass is consumed by herbivores, including the marsh periwinkle snail (*Littoraria irrorata*). In the southeastern United States, these herbivores are in turn consumed by blue crabs (*Callinectes sapidus*), mud crabs (*Eurytium limosum*), and terrapins (*Malaclemys terrapin*).

For many years, ecologists believed that the productivity of cordgrass was largely determined by the availability of nutrients in the salt marsh. However, research by Brian R. Silliman and Mark D. Bertness of Brown University showed that the productivity of cordgrass is actually regulated by what ecologists call a **trophic cascade**—predator–prey effects that reverberate through the population interactions at two or more trophic levels **(Figure 54.10)**. Populations of the herbivorous periwinkle snails are controlled by the crabs and turtles that eat them. In the presence of these predators, snail populations are reduced in size, and the cordgrass grows luxuriantly. But when these predators are removed from the system, the snail populations grow rapidly, and feeding by the snails virtually eliminates the cordgrass, converting a highly productive ecosystem into a barren mudflat. Thus, cordgrass productivity is

FIGURE 54.10 | **Experimental Research**

A Trophic Cascade in Salt Marshes

A. Herbivore damage and *Spartina* biomass

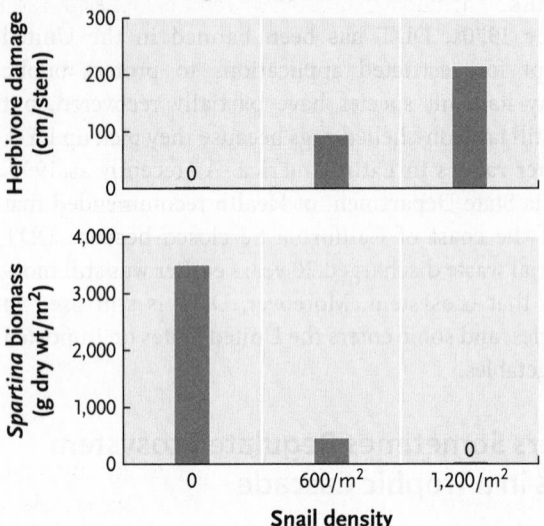

Snail density

C. Trophic cascade

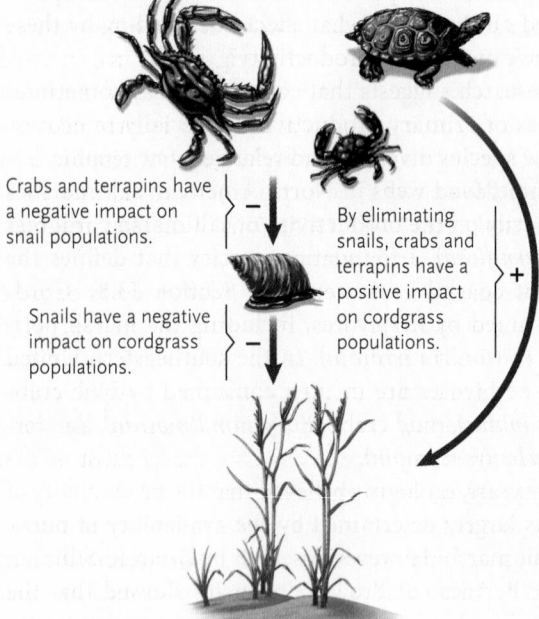

Crabs and terrapins have a negative impact on snail populations.

Snails have a negative impact on cordgrass populations.

By eliminating snails, crabs and terrapins have a positive impact on cordgrass populations.

B. Effect of snails on *Spartina*

Spartina grew luxuriantly in an enclosure with no snails.

Spartina was virtually eliminated from an enclosure with snails at high density.

Question: How is the primary productivity of salt marsh grasses influenced by herbivores and their predators?

Experiment: Silliman and Bertness conducted a set of experiments to measure the effects of herbivores and their predators on the primary productivity of salt marsh cordgrass *(Spartina alterniflora)* along the coast of Sapelo Island, Georgia:

1. The researchers measured the effects on cordgrass of feeding by an herbivorous snail *(Littoraria irrorata)* by controlling snail densities within screened 1-m² enclosures (no snails, 600 snails, or 1,200 snails per enclosure) and measuring herbivore damage to cordgrass stems.

2. They created other enclosures that protected juvenile snails from their predators (crabs and terrapins) and compared the densities of juvenile snails that became established in protected versus unprotected areas.

3. They tethered snails to cordgrass stems at various locations in the marsh and monitored them to determine the rate at which they were consumed by predators.

4. All three experiments were replicated eight times in different parts of the marsh. The results reported here are from parts of the marsh where the cordgrass grew the tallest.

Results: The first experiment showed that herbivore damage increased with snail density, and cordgrass biomass decreased with snail density **(A).** The second experiment demonstrated that snail predators greatly reduce the density of juveniles that become established in the population: snail densities averaged 8.3 individuals/m² in unprotected enclosures versus 305/m² in protected enclosures. The third experiment revealed that 98% of the tethered snails were eaten by predators within 24 hours.

The researchers demonstrated that predators remove most herbivorous snails in this ecosystem, substantially reducing their impact on cordgrass. After predators are experimentally excluded from the system, snail densities increase dramatically, and high snail densities result in the near elimination of cordgrass **(B).**

Conclusion: The three experiments strongly suggest that cordgrass productivity is controlled by a top-down trophic cascade **(C).**

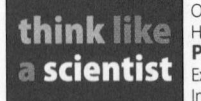

think like a scientist
Observe
Hypothesize
Predict
Experiment
Interpret

Both crabs and turtles are considered delicacies in many cultures. What effects would the harvesting of crabs and turtles have on the cordgrass salt marsh ecosystem?

Source: B. R. Silliman and M. D. Bertness. 2002. A trophic cascade regulates salt marsh primary production. *Proceedings of the National Academy of Sciences USA* 99:10500–10505.

© Cengage Learning 2017

controlled *indirectly* by the abundance of predators that eat the snails that eat the cordgrass.

The direct impact of turtle and crab predation on herbivorous snails—and its indirect impact on cordgrass—is an example of what ecologists call *top-down control* of ecosystem structure: predators at the top of the food web influence both the herbivores they eat and the plants on which herbivores feed. Top-down trophic cascades may be common. In 2014, William J. Ripple of Oregon State University, working with colleagues at many other institutions, published a review of research on the ecological impact of top predators on four continents. Their analysis summarized the diverse effects of top-down control: increasing the abundance of many types of animals, changing the dynamics of infectious diseases, and reducing damage to agricultural crops. Sadly, most of these top carnivores are highly endangered (as described in Chapter 55), and their impact on ecosystems is reduced as their populations decline.

However, top-down trophic cascades are not universal, and some ecologists have argued that many ecosystems are also influenced by the nutritional content of primary producers and their interactions with the herbivores that consume them. For example, the low nitrogen content of most plants may frequently limit the productivity of herbivores. Indeed, experimental nitrogen supplementation has increased the biomass of both plants and herbivores in several ecosystems. Thus, *bottom-up control,* as trophic cascades that start with primary producers are called, is also important. Interactions within ecosystems are complex, and bottom-up and top-down cascades can influence overlapping subsets of species in a food web simultaneously.

STUDY BREAK 54.2

1. What is the difference between gross primary productivity and net primary productivity?
2. What environmental factors influence rates of primary productivity in terrestrial and aquatic ecosystems?
3. Why is energy lost from an ecosystem at every transfer from one trophic level to the trophic level above it?
4. How can the presence of predators influence an ecosystem's productivity?

54.3 Nutrient Cycling in Ecosystems

The availability of nutrients is as important to ecosystem function as the input of energy. Photosynthesis—the conversion of solar energy into chemical energy—requires carbon, hydrogen, and oxygen, which producers acquire from water and air. Producers also need nitrogen, phosphorus, and other minerals (see Table 35.1). A deficiency in any of these minerals can reduce primary productivity.

Earth is essentially a closed system with respect to matter. Thus, unlike energy, for which there is a constant cosmic input, Earth already contains virtually all the nutrients that will ever

be available for biological systems. Biogeochemical cycles constantly circulate nutrient ions or molecules between the abiotic environment and living organisms. Unlike energy, which flows through ecosystems and is gradually lost as heat, matter is conserved in biogeochemical cycles. Although there may be local shortages of specific nutrients, Earth's overall supplies of these chemical elements are never depleted.

Nutrients take various forms as they pass through biogeochemical cycles. Some materials, such as carbon, nitrogen, and oxygen, form gases, which move through global *atmospheric cycles*. Geological processes move other materials, such as phosphorus, through local *sedimentary cycles*, carrying them between dry land and the seafloor. Rocks, soil, water, and air are the reservoirs where mineral nutrients accumulate, sometimes for many years.

The Hydrologic Cycle Recirculates All the Water on Earth

Although it is not a mineral nutrient, water is the universal intracellular solvent for biochemical reactions. Nevertheless, only a fraction of 1% of Earth's total water is present in biological systems at any time.

The cycling of water, called the **hydrologic cycle,** is global, with water molecules moving from the ocean into the atmosphere, to the land, through freshwater ecosystems, and back to the ocean **(Figure 54.11).** Solar energy causes water to evaporate from oceans, lakes, rivers, soil, and living organisms, entering the atmosphere as a vapor and remaining aloft as a gas, as droplets in clouds, or as ice crystals. It falls as precipitation, mostly in the form of rain and snow. When precipitation falls on land, water flows across the surface or percolates to great depth in the soil, eventually reentering the ocean reservoir through the flow of streams and rivers.

The hydrologic cycle maintains its global balance because the total amount of water that enters the atmosphere is equal to the amount that falls as precipitation. Most water enters the atmosphere through evaporation from the ocean, which represents the largest reservoir on the planet. A much smaller fraction evaporates from terrestrial ecosystems, and most of that results from transpiration in green plants.

The constant recirculation provides fresh water to terrestrial organisms and maintains freshwater ecosystems such as lakes and rivers. Water also serves as a transport medium that moves nutrients within and between ecosystems, as demonstrated in a series of classic experiments in the Hubbard Brook Experimental Forest, described in *Focus on Research: Basic Research.*

The Carbon Cycle Includes a Large Atmospheric Reservoir

Carbon atoms provide the backbone of most biological molecules, and carbon compounds store the energy captured by photosynthesis (see Section 8.1). Carbon enters food webs when

Basic Research: Studies of the Hubbard Brook Watershed

Because water always flows downhill, local topography affects the movement of dissolved nutrients in terrestrial ecosystems. A **watershed** is an area of land from which precipitation drains into a stream or river system. Thus, each watershed represents a part of an ecosystem from which nutrients exit through a single outlet, much the way a bathtub empties through a single drain. When several streams join to form a river, the watershed drained by the river encompasses all of the smaller watersheds drained by the streams. For example, the Mississippi River watershed covers roughly one third of the United States, and it includes watersheds drained by the Illinois, Missouri, and Tennessee Rivers, as well as many other watersheds drained by smaller streams and rivers.

Because watersheds are relatively self-contained units, they are ideal for large-scale field experiments about nutrient flow in ecosystems. Herbert Bormann of Yale University and Gene Likens of Cornell University have conducted a classic experiment on this topic since the 1960s. Bormann and Likens manipulated small watersheds of temperate deciduous forest in the Hubbard Brook Experimental Forest in the White Mountain National Forest of New Hampshire. They measured precipitation and nutrient input into the watersheds, the uptake of nutrients by vegetation, and

the amount of nutrients leaving the watershed via streamflow. Nutrients exported in streamflow were monitored in water samples collected from V-shaped concrete weirs built into bedrock below the streams that drained the watersheds **(Figure A).** Impermeable bedrock underlies the soil, preventing water from leaving the system by deep seepage.

After collecting several years of baseline data on six undisturbed watersheds, the researchers cut all the trees in one small watershed in 1965 and 1966. They also applied herbicides to prevent regrowth. After establishing this experimental treatment, they monitored the output of nutrients in streams that drained experimental and control watersheds. They attributed differences in nutrient export between undisturbed watersheds (controls) and the clear-cut watershed (experimental treatment) to the effects of deforestation.

Bormann and Likens determined that vegetation absorbed substantial water and conserved nutrients in undisturbed watersheds. Plants used about 40% of the precipitation for transpiration. The rest contributed to runoff and groundwater. Control watersheds lost only about 8 to 10 kg of calcium per hectare (10,000 square meters) each year, an amount that was replaced by the erosion of bedrock and input from rain.

Moreover, control watersheds actually accumulated about 2 kg of nitrogen per hectare per year and slightly smaller amounts of potassium.

By contrast, the experimentally deforested watershed experienced a 40% annual increase in runoff. During a 4-month period in the summer, runoff increased 300%. Some mineral losses were similarly large. The net loss of calcium was 10 times higher than in the control watersheds **(Figure B)** and the loss of potassium 21 times higher.

Phosphorus losses did not increase; this mineral was apparently retained by the soil. However, the loss of nitrogen was an astronomical 120 kg per hectare per year. So much nitrogen entered the stream draining the experimental watershed that the stream became choked with algae and cyanobacteria. Thus, the results of the Hubbard Brook experiment suggest that deforestation increases flooding, decreases the fertility of ecosystems, and leads to nutrient enrichment of nearby aquatic ecosystems.

The Hubbard Brook Watershed project is an example of long-term ecological research (LTER), one of many in a network of such projects that provide remarkable insights into ecosystem processes. The network itself is funded by the National Science Foundation and described in detail at its web site: http://www.lternet.edu/.

Gene E. Likens from Gene E. Likens et al., Ecological Monographs, 40(1): 23–47, 1970.

FIGURE A   A Weir used to measure the volume and nutrient content of water leaving a watershed by streamflow.

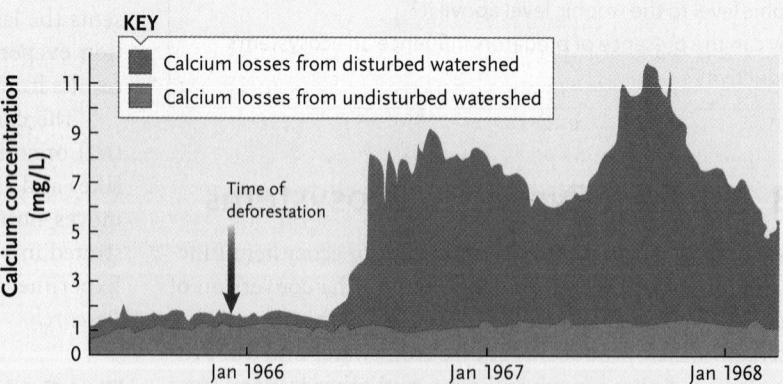

KEY
- Calcium losses from disturbed watershed
- Calcium losses from undisturbed watershed

Time of deforestation

FIGURE B   Calcium losses from a deforested watershed were much greater than those from controls. The arrow indicates the time of deforestation in early winter. Mineral losses did not increase until after the ground thawed the following spring; increased runoff also caused large water losses from the watershed.

© Cengage Learning 2017

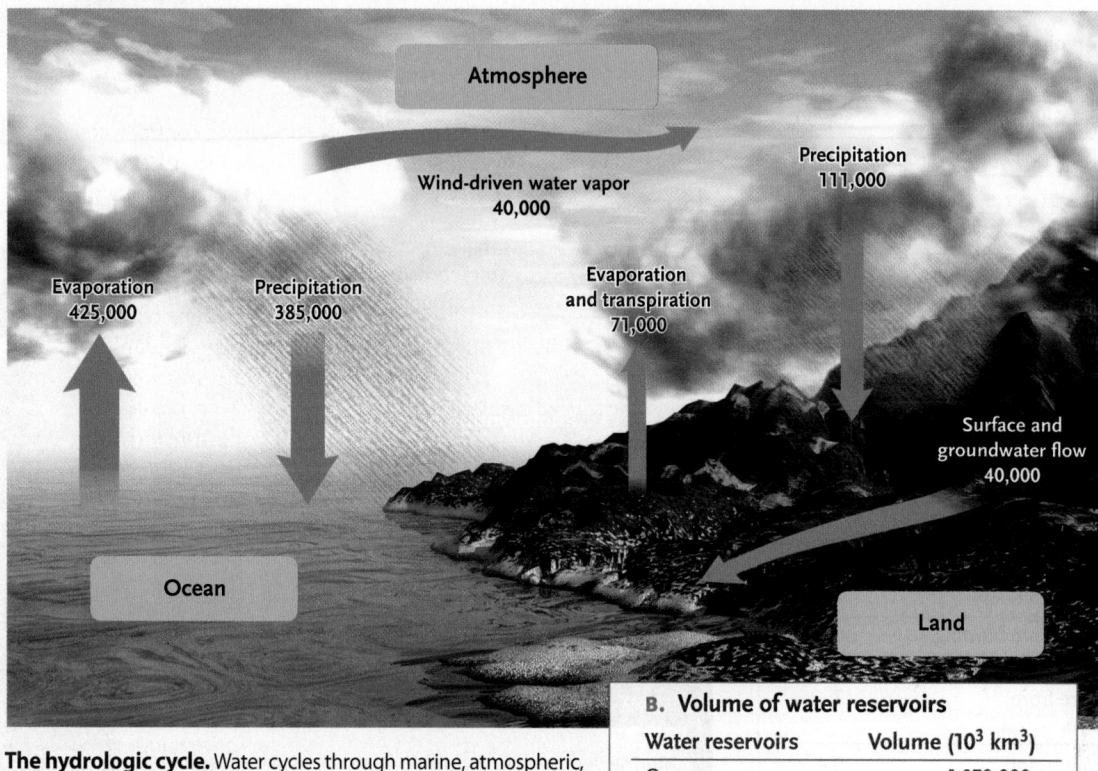

A. The water cycle

Atmosphere

Wind-driven water vapor
40,000

Precipitation
111,000

Evaporation
425,000

Precipitation
385,000

Evaporation
and transpiration
71,000

Surface and
groundwater flow
40,000

Ocean

Land

FIGURE 54.11 The hydrologic cycle. Water cycles through marine, atmospheric, and terrestrial reservoirs. **(A)** Labels on the arrows list the amount of water (km³/yr) moved among reservoirs by various processes. The amount of water entering the atmosphere through evaporation from the ocean plus evaporation and transpiration from land (425,000 + 71,000 = 496,000 km³/yr) is equal to the amount falling as precipitation on the oceans and land (385,00 + 111,000 = 496,000 km³/yr). Similarly, the quantity of water moving from ocean to land as wind-driven water vapor is balanced by the flow of runoff from the land to the ocean. **(B)** The oceans are by far the largest of the six major reservoirs of water on Earth.

© Cengage Learning 2017

B. Volume of water reservoirs

Water reservoirs	Volume (10^3 km³)
Oceans	1,370,000
Polar ice, glaciers	29,000
Groundwater	4,000
Lakes, rivers	230
Soil moisture	67
Atmosphere (water vapor)	14

producers convert atmospheric carbon dioxide (CO_2) into carbohydrates. Heterotrophs acquire carbon by eating other organisms or detritus. Although carbon moves somewhat independently in the sea and on land, a common atmospheric pool of CO_2 creates a global **carbon cycle (Figure 54.12).**

The largest reservoir of carbon is sedimentary rock, such as limestone or marble. Rocks are in the unavailable inorganic compartment, and they exchange carbon with living organisms at an exceedingly slow pace. Most *available* carbon is present as dissolved bicarbonate ions (HCO_3^-) in the ocean. Soil, the atmosphere, and plant biomass form other significant, but much smaller, reservoirs of available carbon. Atmospheric carbon is mostly in the form of molecular CO_2, a product of aerobic respiration. Volcanic eruptions also release CO_2 into the atmosphere.

Sometimes carbon atoms leave the organic compartments for long periods of time. Some organisms in marine food webs build shells and other hard parts by incorporating dissolved carbon into calcium carbonate ($CaCO_3$) and other insoluble salts. When shelled organisms die, they sink to the bottom and

are buried in sediments. The insoluble carbon that accumulates as rock in deep sediments may remain buried for millions of years before tectonic uplifting brings it to the surface, where erosion and weathering dissolve sedimentary rocks and return carbon to an available form.

Over hundreds of millions of years, carbon atoms were also transferred to the unavailable organic compartment when soft-bodied organisms were buried in habitats where low oxygen concentration prevented decomposition. Under suitable geological conditions, these carbon-rich tissues were slowly converted to gas, petroleum, or coal, which humans now use as fossil fuels.

The Nitrogen Cycle Depends on the Activity of Diverse Microorganisms

All organisms require nitrogen to construct nucleic acids, proteins, and other biological molecules. Earth's atmosphere had a high nitrogen concentration long before life originated. Today, a global **nitrogen cycle** moves this element between the huge atmospheric pool of gaseous molecular nitrogen (N_2) and

A. Amount of carbon in major reservoirs

Carbon reservoirs	Mass (10^{15} g)
Sediments and rocks	77,000,000
Ocean (dissolved forms)	39,700
Soil	1,500
Atmosphere	750
Biomass on land	715

B. Annual global carbon movement between reservoirs

Direction of movement	Mass (10^{15} g)
From atmosphere to plants (carbon fixation)	120
From atmosphere to ocean	107
To atmosphere from ocean	105
To atmosphere from plants	60
To atmosphere from soil	60
To atmosphere from burning fossil fuel	6.5
To atmosphere from burning plants	2
To ocean from runoff	0.4
Burial in ocean sediments	0.1

C. The global carbon cycle

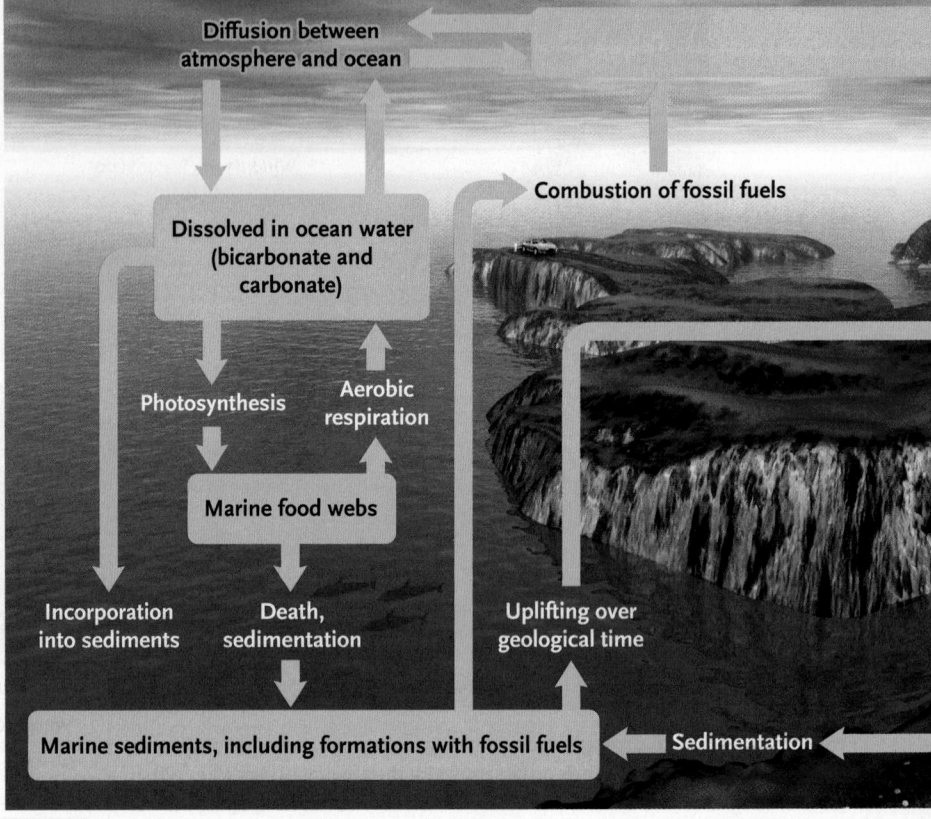

FIGURE 54.12 The carbon cycle. Marine and terrestrial components of the global carbon cycle are linked through an atmospheric reservoir of carbon dioxide. **(A)** By far, the largest amount of Earth's carbon is found in sediments and rocks. **(B)** Earth's atmosphere mediates most movements of carbon. **(C)** In this illustration of the carbon cycle, boxes identify major reservoirs, and labels on the arrows identify the processes that cause carbon to move between reservoirs.
© Cengage Learning 2017

several much smaller pools of nitrogen-containing compounds in soils, marine and freshwater ecosystems, and living organisms **(Figure 54.13)**.

Molecular nitrogen is abundant in the atmosphere, but triple covalent bonds bind its two atoms so tightly that most organisms cannot use it. However, three biochemical processes—nitrogen fixation, ammonification, and nitrification **(Table 54.2)**—convert molecular nitrogen or unusable nitrogen compounds into

nutrients that primary producers can assimilate and use for the production of biological molecules. Secondary consumers obtain their nitrogen by consuming primary producers, thereby initiating the movement of nitrogen through the food webs of an ecosystem.

In **nitrogen fixation** (see Section 35.3), molecular nitrogen (N_2) is converted into ammonia (NH_3) and ammonium ions (NH_4^+). Certain bacteria, including *Azotobacter* and *Rhizobium*,

TABLE 54.2	Biochemical Processes That Influence Nitrogen Cycling in Ecosystems		
Process	Organisms Responsible	Products	Outcome
Nitrogen fixation	Bacteria: *Rhizobium, Azotobacter, Frankia* Cyanobacteria: *Anabaena, Nostoc*	Ammonia (NH_3), ammonium ions (NH_4^+)	Assimilated by primary producers
Ammonification of organic detritus	Soil bacteria and fungi	Ammonia (NH_3), ammonium ions (NH_4^+)	Assimilated by primary producers
Nitrification (1) Oxidation of NH_3 (2) Oxidation of NO_2^-	Bacteria: *Nitrosomonas, Nitrococcus* Bacteria: *Nitrobacter*	Nitrite (NO_2^-) Nitrate (NO_3^-)	Used by nitrifying bacteria Assimilated by primary producers
Denitrification of NO_3^-	Soil bacteria	Nitrous oxide (N_2O), molecular nitrogen (N_2)	Released to atmosphere

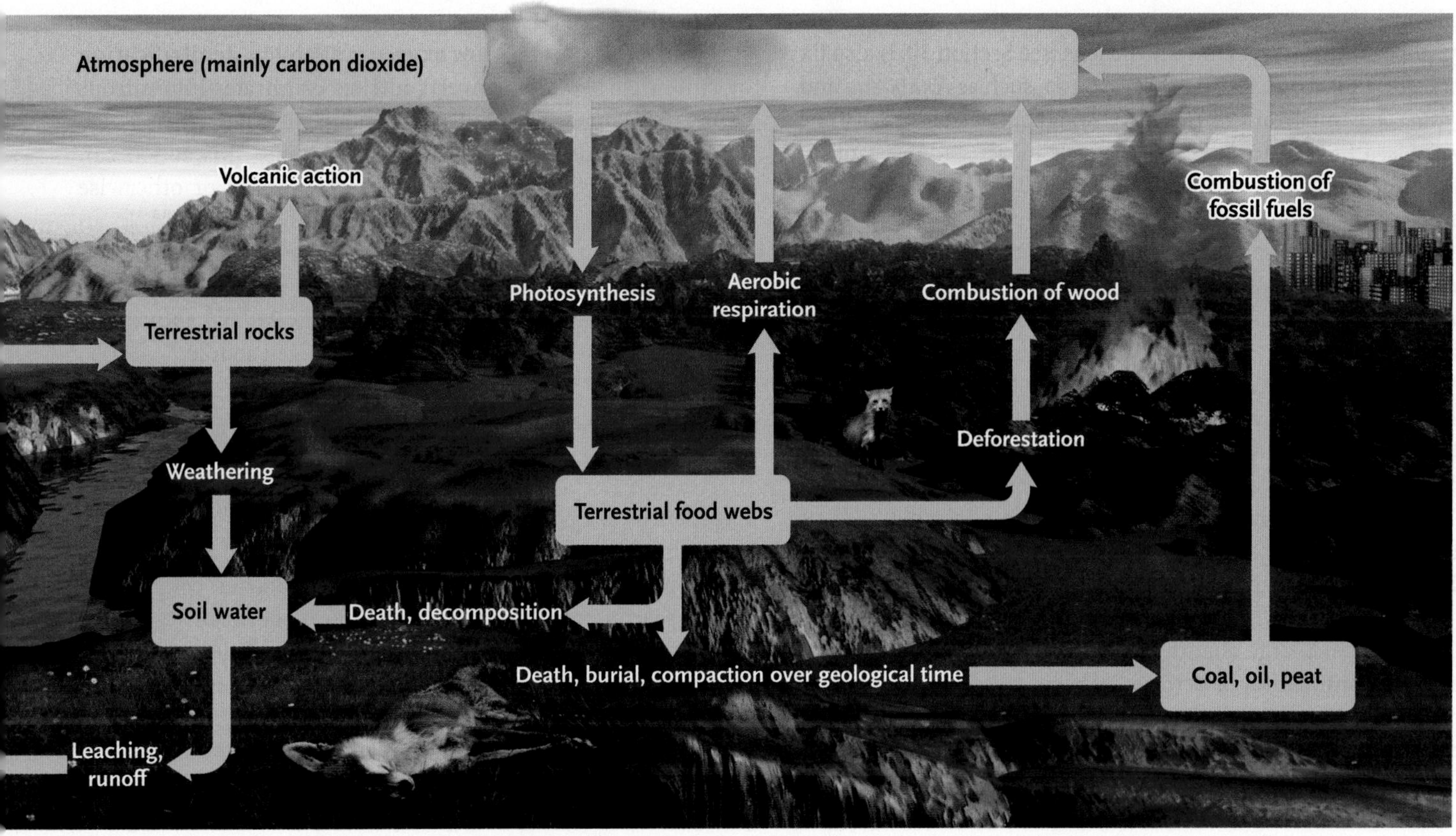

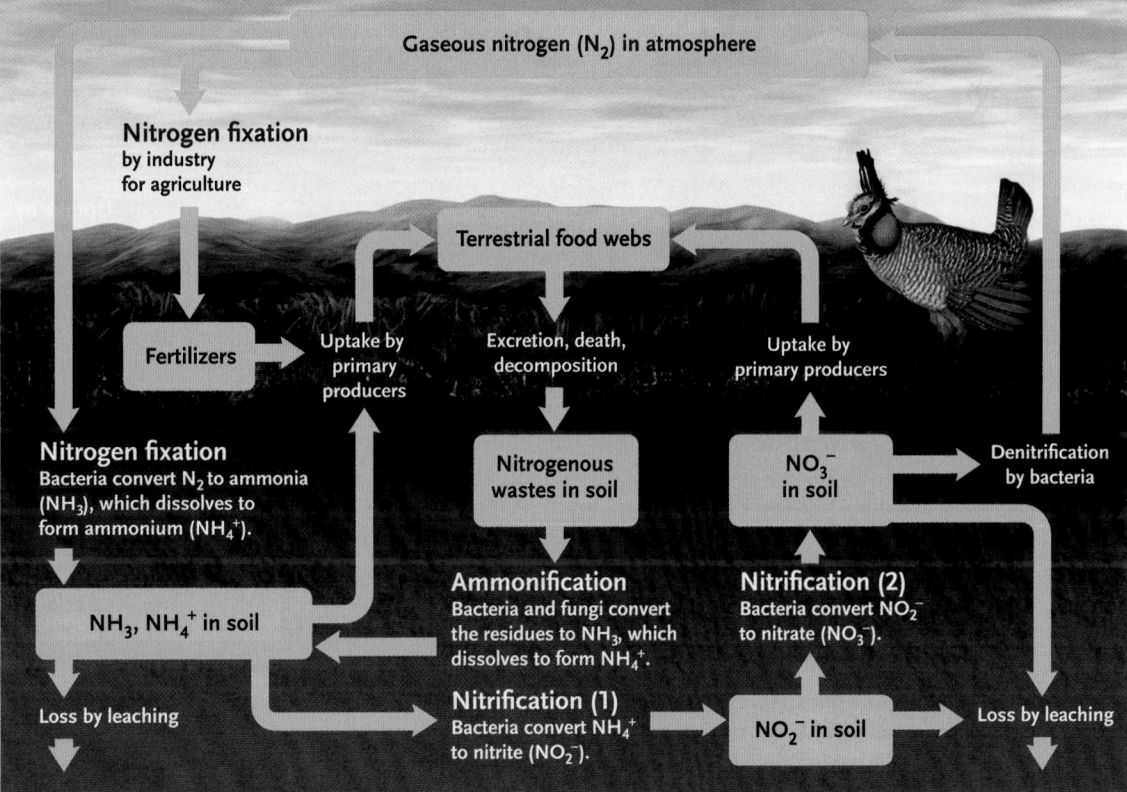

Atmosphere (mainly carbon dioxide)

Volcanic action

Photosynthesis   Aerobic respiration   Combustion of wood   Combustion of fossil fuels

Terrestrial rocks

Weathering

Terrestrial food webs

Deforestation

Soil water ← Death, decomposition

Death, burial, compaction over geological time → Coal, oil, peat

Leaching, runoff

FIGURE 54.13 The nitrogen cycle in terrestrial ecosystems.
Nitrogen cycles through terrestrial ecosystems when unavailable molecular nitrogen is made available through the action of nitrogen-fixing bacteria. Other bacteria recycle nitrogen within the available organic compartment through ammonification and two types of nitrification, converting organic wastes into ammonium ions and nitrates. Denitrification converts nitrate to molecular nitrogen, which returns to the atmosphere. Runoff carries various nitrogen compounds from terrestrial ecosystems into oceans, where it is recycled in marine food webs.
© Cengage Learning 2017

Gaseous nitrogen (N_2) in atmosphere

Nitrogen fixation
by industry for agriculture

Terrestrial food webs

Fertilizers   Uptake by primary producers   Excretion, death, decomposition   Uptake by primary producers

Nitrogen fixation
Bacteria convert N_2 to ammonia (NH_3), which dissolves to form ammonium (NH_4^+).

Nitrogenous wastes in soil

NO_3^- in soil   Denitrification by bacteria

NH_3, NH_4^+ in soil

Loss by leaching

Ammonification
Bacteria and fungi convert the residues to NH_3, which dissolves to form NH_4^+.

Nitrification (2)
Bacteria convert NO_2^- to nitrate (NO_3^-).

Nitrification (1)
Bacteria convert NH_4^+ to nitrite (NO_2^-).

NO_2^- in soil   Loss by leaching

which collect molecular nitrogen from the air between soil particles, are the major nitrogen fixers in terrestrial ecosystems. The cyanobacteria partners in some lichens (see Section 30.3) also fix molecular nitrogen. Other cyanobacteria, such as *Anabaena* and *Nostoc*, are important nitrogen fixers in aquatic ecosystems; the water fern (genus *Azolla*) plays that role in rice paddies. Collectively, these organisms fix an astounding 200 million metric tons of nitrogen each year; nitrogen fixation can also result from lightning and volcanic action. Plants and other primary producers assimilate and use this fixed nitrogen in the biosynthesis of amino acids, proteins, and nucleic acids, which then circulate through food webs.

Some plants, including legumes (such as beans and clover), alders (*Alnus* species), and some members of the rose family (Rosaceae), are mutualists with nitrogen-fixing bacteria. These plants acquire nitrogen from soils much more readily than plants that lack such mutualists. Although these plants have the competitive edge in nitrogen-poor soil, nonmutualistic species often displace them in nitrogen-rich soil.

In addition to nitrogen fixation, several other biochemical processes make large quantities of nitrogen available to producers. **Ammonification** of detritus by bacteria and fungi converts organic nitrogen into ammonia (NH_3), which dissolves into ammonium ions (NH_4^+) that plants can assimilate; some ammonia escapes into the atmosphere as a gas. **Nitrification** by certain bacteria produces nitrites (NO_2^-), which are converted by other bacteria to usable nitrates (NO_3^-). All of these compounds are water-soluble, and water rapidly leaches them from soil into streams, lakes, and oceans.

Under conditions of low oxygen availability, **denitrification** by still other bacteria converts nitrites or nitrates into nitrous oxide (N_2O) and then into molecular nitrogen (N_2), which enters the atmosphere, completing the cycle. This action can deplete supplies of soil nitrogen in waterlogged or otherwise poorly aerated environments, such as bogs and swamps. In an interesting twist on the usual predator–prey relationships, several species of flowering plants that live in nitrogen-poor soils, such as the Venus' fly trap *(Dionaea muscipula)*, capture and digest small insects as their primary nitrogen source.

The Phosphorus Cycle Includes a Large Sedimentary Reservoir

Phosphorus is a crucial element for the energy transfers within the cells of all organisms and for construction of nucleic acids (see Sections 6.3 and 14.2). Because phosphorus compounds lack a gaseous phase, this element moves between terrestrial and marine ecosystems in a sedimentary cycle **(Figure 54.14).** Earth's crust is the main reservoir of phosphorus, as it is for other minerals, such as calcium and potassium that undergo sedimentary cycles.

Phosphorus is present in terrestrial rocks in the form of phosphates (PO_4^{3-}). In the **phosphorus cycle,** weathering and erosion carry phosphate ions from rocks to soil and into

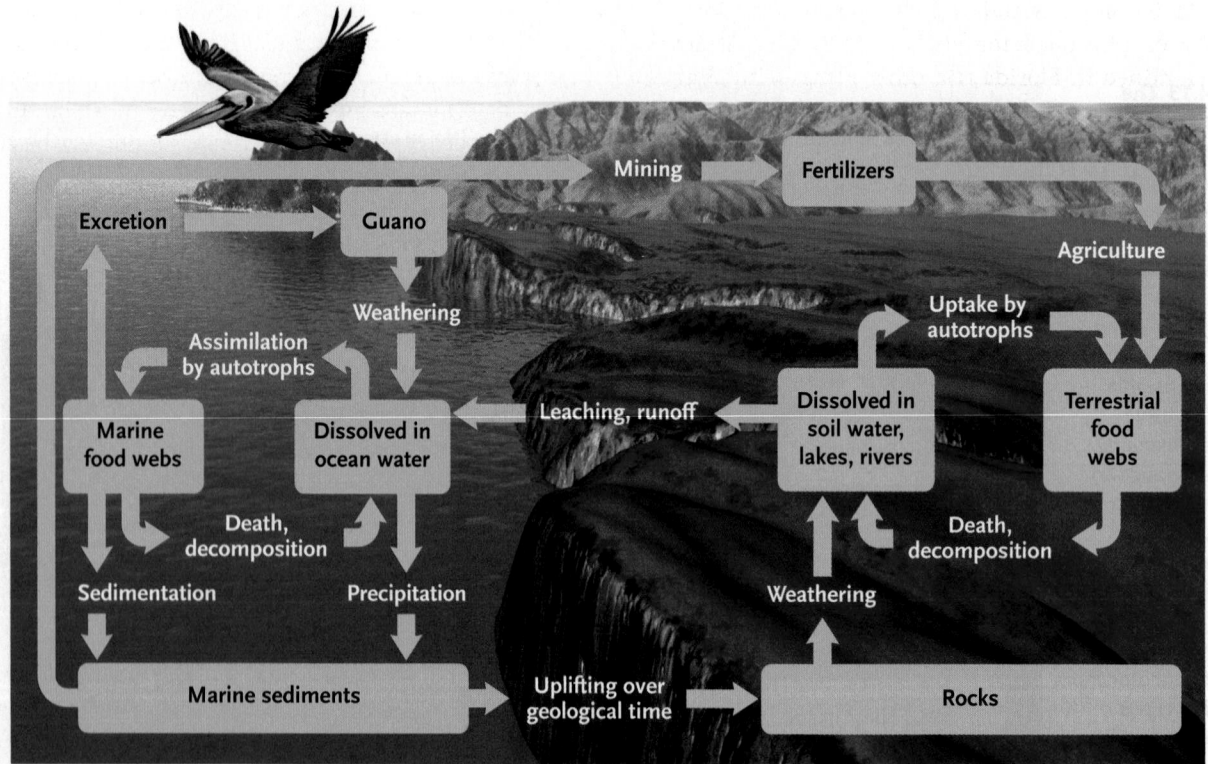

FIGURE 54.14 The phosphorus cycle. Phosphorus becomes available to biological systems when wind and rainfall dissolve phosphates in rocks and carry them into adjacent soil and freshwater ecosystems. Runoff carries dissolved phosphorus into marine ecosystems, where it precipitates out of solution and is incorporated into marine sediments.
© Cengage Learning 2017

streams and rivers, which eventually transport them to the ocean. Once there, some phosphorus enters marine food webs, but most of it precipitates out of solution and accumulates for millions of years as insoluble deposits, mainly on continental shelves. When parts of the seafloor are uplifted and exposed, weathering releases the phosphates.

Plants absorb and assimilate dissolved phosphates directly, and phosphorus moves easily to higher trophic levels. All heterotrophs excrete some phosphorus as a waste product in urine and feces, which are decomposed, and producers readily absorb the phosphate ions that are released. Thus, phosphorus cycles rapidly *within* terrestrial communities.

Supplies of available phosphate are generally limited, however, and plants acquire it so efficiently that they reduce soil phosphate concentration to extremely low levels. Thus, like nitrogen, phosphorus is a common ingredient in agricultural fertilizers, and excess phosphates are pollutants of freshwater ecosystems. For many years, phosphate for fertilizers was obtained from *guano* (the droppings of seabirds that consume phosphorus-rich food), which was mined on small islands off the Pacific coast of South America. Most phosphate for fertilizer now comes from phosphate rock mined in Florida and other places with abundant marine deposits.

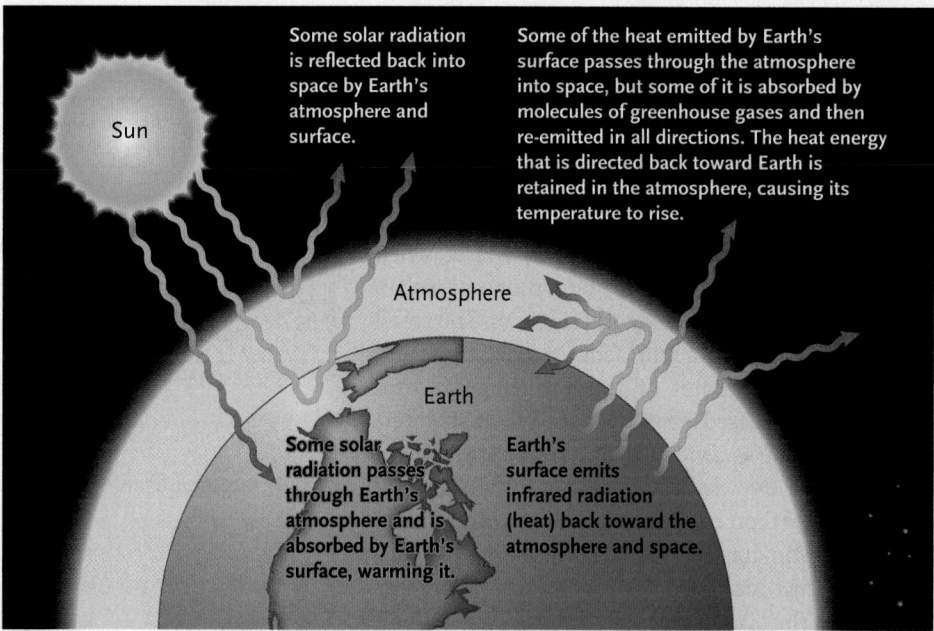

FIGURE 54.15 The greenhouse effect. Most incoming solar radiation passes through Earth's atmosphere and strikes the planet's surface, raising its temperature. The Earth's surface emits infrared radiation (heat energy), some of which is absorbed by molecules of greenhouse gases in the atmosphere. Greenhouse gases then emit heat energy, which is directed back toward Earth, maintaining the temperature of the atmosphere.
© Cengage Learning 2017

STUDY BREAK 54.3

1. How does the global hydrologic cycle maintain its balance?
2. What processes move large quantities of carbon from an organic compartment to an inorganic compartment?
3. What microorganisms drive the global nitrogen cycle, and how do they do it?
4. What is Earth's main reservoir for phosphorus, and why is it recycled at such a slow rate from that reservoir?

54.4 Human Activities and Anthropogenic Global Change

Human activities—industrial processes, agricultural practices, development, and many of our daily behaviors—disrupt biogeochemical cycles by moving materials from one nutrient compartment to another at unnaturally rapid rates. In this section we consider anthropogenic (that is, originating with human activities) global change, which threatens organisms and ecosystems everywhere. We focus on examples from the disruption of the global carbon and nitrogen cycles.

Changes in the Global Carbon Cycle Are Warming the Earth

The combustion of fossil fuels (oil, coal, and peat) and deforestation (described in Section 55.1) are transferring carbon from reservoirs of organic nutrients to atmospheric reservoir of inorganic nutrients at an unprecedented rate. Virtually all scientists agree that the resulting change in the worldwide distribution of carbon is having severe consequences for Earth's climate, as well as terrestrial and aquatic ecosystems.

Concentrations of certain gases in the lower atmosphere have a profound effect on global temperature, which in turn has enormous impact on precipitation and wind patterns. Molecules of carbon dioxide (CO_2), water, ozone, methane, nitrous oxide, and other compounds collectively act like the panes of glass in a greenhouse (hence they are described as *greenhouse gases*). These gases allow the short wavelengths of visible light to reach Earth's surface; but they impede the escape of longer, infrared wavelengths back into space, trapping much of that energy as heat. In short, greenhouse gases foster the accumulation of heat in the lower atmosphere, a warming action known as the **greenhouse effect,** which prevents Earth from being a cold and lifeless planet **(Figure 54.15).**

Since the late 1950s, scientists have measured atmospheric concentrations of CO_2 and other greenhouse gases at several remote sampling sites, which are free of local contamination and reflect the average concentrations of these gases in the atmosphere. These studies, coupled with estimates of historical levels, indicate that concentrations of greenhouse gases have increased steadily for more than 100 years **(Figure 54.16).** The graph of atmospheric CO_2 concentration through time has a regular zigzag pattern that follows an annual cycle in the growth of primary producers in the northern hemisphere (blue line in Figure 54.16A). Photosynthesis withdraws so much CO_2 from the atmosphere during the northern hemisphere summer that its concentration falls. The concentration is higher during the northern hemisphere winter, when aerobic respiration continues, returning carbon to the atmosphere, and photosynthesis slows. The small fluctuations in the data represent seasonal highs and lows, but the midpoint of the annual peaks and troughs has increased steadily. Many scientists interpret these data as evidence of a rapid buildup of atmospheric CO_2, which represents a shift in the distribution of carbon in the major reservoirs on Earth. Scientists estimate that the atmospheric CO_2 concentration has increased 35% in the last 150 years and 16% in the last 30 years.

What has caused the increase in the atmospheric concentration of CO_2? Burning of fossil fuels and wood is the largest contributor, because CO_2 is a combustion product of this process (Figure 54.16B). Today, humans burn more wood and fossil fuels than ever before. In a report published in 2015, the Commonwealth Science and Industrial Research Organization (CSIRO), an agency of the Australian federal government, estimated that CO_2 emissions from the combustion of fossil fuels increased 400% between 1940 and 2015. Although about 60% of CO_2 emissions are absorbed by forests and marine phytoplankton, the remaining 40% has remained in the atmosphere.

Moreover, the capacity of primary producers to withdraw CO_2 from the atmosphere and use it to build other organic molecules appears to be decreasing. Vast tracts of tropical forests are being cleared and burned (see Section 55.1), and in some habitats, tree mortality rates have risen, largely as a result of climate warming. A study published by an international team of dozens of scientists in *Nature* in 2015 estimated that the rate at which forests in the Amazon basin are accumulating biomass—a measure of the ability of these forests to absorb atmospheric CO_2—has been declining since 1983, and it has declined 30% since 2000. Although the Amazon forests still represent one of the largest ecosystem carbon pools on Earth, the decline in biomass accumulation has been caused by steadily increasing tree mortality rates.

Why is an increase in the atmospheric CO_2 concentration so alarming? Virtually all climate researchers agree that increasing atmospheric concentrations of CO_2 and other greenhouse gases have raised the mean annual temperature of Earth's atmosphere and oceans (see red line in Figure 54.16A). According to the Intergovernmental Panel on Climate Change, an agency of the United Nations, the average global surface temperature increased more than 0.7°C during the twentieth century, with

A. Atmospheric CO_2 concentration and global mean temperature

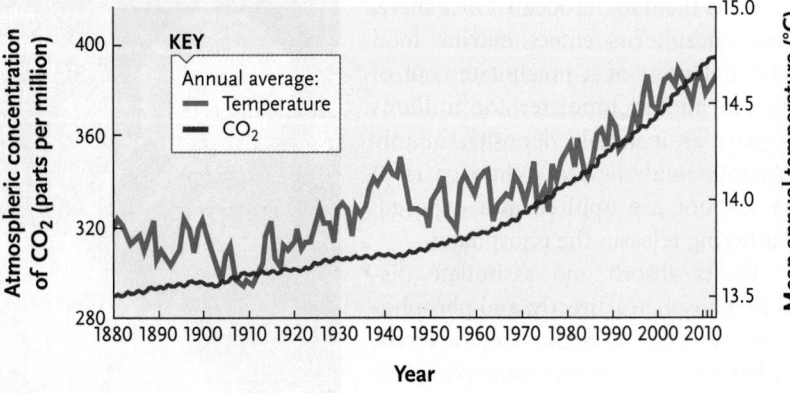

B. CO_2 released from the combustion of fossil fuels

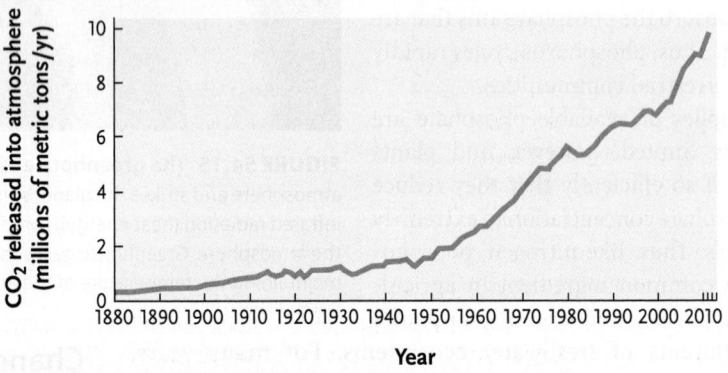

FIGURE 54.16 Atmospheric carbon dioxide and temperature. (A) Data collected at remote monitoring stations reveal that the concentration of CO_2 in the atmosphere and the mean annual global temperature have risen in parallel for the past 100 years. **(B)** Carbon dioxide emissions from the burning of fossil fuels have been a major source of atmospheric CO_2 since 1950.
© Cengage Learning 2017

most of the increase occurring after 1980. Scientists now predict that the mean temperature of the lower atmosphere will rise 2.0° to 6.5°C during the twenty-first century. Ocean temperatures are also increasing, and because water expands when heated, global sea level may rise as much as 0.4 m just from this temperature-induced expansion. In addition, rising atmospheric temperature at the poles has already melted significant portions of the ice sheets in the Arctic and Antarctic, which will raise sea level even more.

Climate scientists worry that Earth has reached a "tipping point" in which further climate warming is inevitable because of positive feedback loops in the climate system: the effects of climate warming cause the rate of warming to increase. For example, warming has caused permafrost in the tundra to thaw and release huge stores of carbon dioxide and methane, another greenhouse gas, into the atmosphere. Similarly, although the polar ice caps reflect most of the solar energy that strikes them, climate warming has reduced their size, thereby exposing more ocean water to the warming rays of the Sun. Finally, climate change has increased the frequency and severity of forest fires in many habitats, releasing additional carbon dioxide into the atmosphere. All of these effects accelerate the rates of warming through positive feedback. Most scientists believe that atmospheric levels of greenhouse gases will continue to increase at least until the middle of the twenty-first century and that global temperature will inevitably rise by several degrees. Possible solutions to counteract the increasing rate of climate change include the reduction of greenhouse gas emissions and the preservation, and planting, of large tracts of forest, which will absorb carbon dioxide from the atmosphere.

Human Activities Have Altered the Nitrogen Cycle

Human activities are also disrupting the nitrogen cycle, primarily through the use of nitrogen-containing fertilizers, the farming and processing of livestock, manufacturing, and the combustion of fossil fuels. In a study published in 2013, researchers from Canada, China, and the United States found that global nitrogen fixation by humans increased more than 200% between 1960 and 2008 **(Figure 54.17)**.

Of all nutrients required for primary production, nitrogen is often the least abundant. Many agricultural crops deplete soil nitrogen, and irrigation fosters soil erosion and leaching, which remove even more. Traditionally, farmers rotated their crops, alternately cultivating nitrogen-fixing legumes and other types of plants in the same fields. In combination with other soil-conservation practices, crop rotation stabilized soils and kept them productive, sometimes for thousands of years.

In traditional agriculture, nearly all the nitrogen in living systems was made available by nitrogen-fixing microorganisms. Today, however, industrialized agriculture relies on the application of synthetic nitrogen-containing fertilizers. The production of synthetic fertilizers is expensive, and it uses fossil fuels both as a raw material and as an energy source. In addition, large-scale livestock production leaves behind mountains of organic wastes, which through the action of the anaerobic microorganisms that decompose it, release nitrous oxide (NO_2), a greenhouse gas. Rain and runoff leach fertilizer from agricultural fields and animal wastes from livestock yards into aquatic ecosystems, where they contribute to eutrophication, artificially enriching the waters and allowing producers to expand their populations (see Section 51.5).

Industrial activities also fix tremendous quantities of nitrogen in the production of drugs, dyes, explosives, fuels, paints, plastics, and synthetic fibers (see Figure 54.17). Moreover, whenever we burn fossil fuels in combustion engines, molecular nitrogen (N_2), which is abundant in the atmosphere, combines with molecular oxygen (O_2) to form nitric oxide (NO). This molecule readily converts to nitrogen dioxide (NO_2), another greenhouse gas, or nitric acid vapors (HNO_3), which are major components of smog and acid precipitation. Sulfur-based impurities in fossil fuels also contribute to the acidification of rainwater (as described in Section 55.2).

According to the Millennium Ecosystem Assessment, a report published in March 2005 with the support of the United

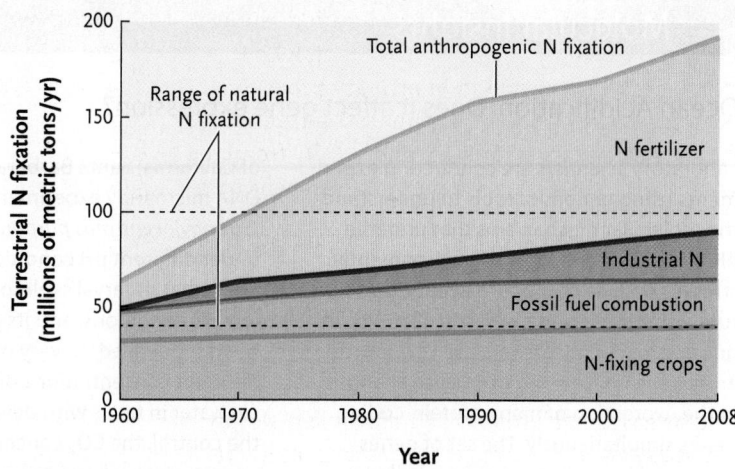

FIGURE 54.17 Anthropogenic nitrogen fixation. The amount of nitrogen fixed by humans has risen sharply since the middle of the twentieth century. The production of fertilizers for agricultural and other industrial uses is responsible for most of the increase.

© Cengage Learning 2017

Nations, human activities more than doubled the amount of nitrogen released from land into other nitrogen reservoirs in the second half of the twentieth century. The report projects that the amount will double again over the next 50 years. These transfers of material contribute significantly to the pollution of aquatic ecosystems, acid precipitation, and global climate change.

The Impact of Global Change on Ecosystems

Global change is happening now, and many studies have revealed its effects on living systems. Global warming has caused shifts in the geographical distributions of many species, the timing of animal migrations, and the timing of reproduction by animals and plants (see Section 51.3). Climate change has also changed the dynamics of population processes, increasing the impact of density-independent factors that regulate population growth of *r*-selected species and the long-term survivorship of *K*-selected species (see Section 52.5). At the community level of organization, global change creates disturbances that alter the species composition of ecological communities and the interactions between the species that still occupy a site (see Chapter 53). Here we provide three brief examples of the impact of climate change on ecosystem functions and properties. In Chapter 55, we assess the impact of global change on biodiversity and the conservation of threatened and endangered species.

ACIDIFICATION OF AQUATIC HABITATS Terrestrial and marine ecosystems absorb roughly equal shares (~30%) of the CO_2 that humans release into the atmosphere each year. In both types of ecosystems, primary producers use CO_2 to create biomass through photosynthesis. In aquatic ecosystems, however, CO_2 reacts with water to form carbonic acid (H_2CO_3), which dissociates into hydrogen ions (H^+) and bicarbonate ions (HCO_3^-). Some marine organisms—principally echinoderms, mollusks,

Ocean Acidification: Does it affect gene expression?

Laboratory scientists are conducting experiments using genomics tools to understand the molecular mechanisms that underlie physiological responses to environmental stresses such as changes in acidity, temperature, osmolarity, and oxygen availability. In one approach, researchers use DNA microarrays (see Chapter 19) to analyze changes in the expression of many protein-coding genes simultaneously. The set of genes expressed in a cell at any time is called a *transcriptome;* the study of transcriptomes is *transcriptomics*. Knowing how ocean acidification affects the transcriptomes of marine organisms would help reveal the physiological mechanism that organisms can use to tolerate a more acidic ocean.

Research Question

Does the pH of seawater affect the transcriptome of sea urchins?

Experiment

To answer the question, Anne Todgham and Gretchen Hofmann of the University of California, Santa Barbara, performed DNA microarray experiments on sea urchin (*Strongylocentrotus purpuratus*) larvae under different pH conditions. They already knew that its larval skeleton is altered by low pH conditions, and its genome had been sequenced. To vary pH, they bubbled different concentrations of CO_2 through seawater in tanks with developing larvae. In the control, the CO_2 concentration matched current atmospheric conditions, producing a pH of 8.01. Two experimental conditions reflected predictions (made by the Intergovernmental Panel on Climate Change in 2007) of atmospheric CO_2 concentrations in the year 2100. The moderate treatment, which assumed a modest increase in atmospheric CO_2 concentration, produced a pH of 7.96. The high treatment, which assumed a large increase in atmospheric CO_2 concentration, produced a pH of 7.88. Todgham and Hofmann's experiment is outlined in the **Figure.**

Results

The researchers found statistically significant changes in transcriptomes of larvae growing in moderate and high CO_2 conditions compared to that of larvae in the control condition:

1. Moderate CO_2—83 of the 1,057 genes analyzed showed decreases in expression; no genes showed increases in expression. Approximately half the genes with decreased expression are involved in biomineralization (that is, the process of building a mineralized skeleton), the cellular stress response, and energy metabolism.

2. High CO_2—178 genes showed changes in expression, with 160 showing decreases and 18 showing increases. Approximately 70% of the genes showing decreased expression are involved in apoptosis (programmed cell death), the cellular stress response, and energy metabolism.

3. The responses to moderate and high CO_2 conditions were largely distinct: only 10 genes showed altered expression under both treatments.

Conclusion

This transcriptomics study demonstrated that acidification affects the expression of numerous genes, many of which are involved in biomineralization, cellular stress response, metabolism, and apoptosis. Surprisingly, ocean acidification mostly decreased gene expression. Moreover, the study showed that pathways other than calcification, a process previously shown to be affected by pH changes, are altered by acidification. Thus, scientists must consider a wide range of physiological processes when assessing the possible effects of ocean acidification on marine organisms.

think like a scientist

Assuming that increasing concentrations of atmospheric CO_2 produce only a moderate decrease in ocean water pH, what effect will that change be likely to have on sea urchins?

Source: A. Todgham and G. E. Hofmann. 2009. Transcriptomic response of sea urchin larvae *Strongylocentrotus purpuratus* to CO_2-driven seawater acidification. *The Journal of Experimental Biology* 212:2579–2594.

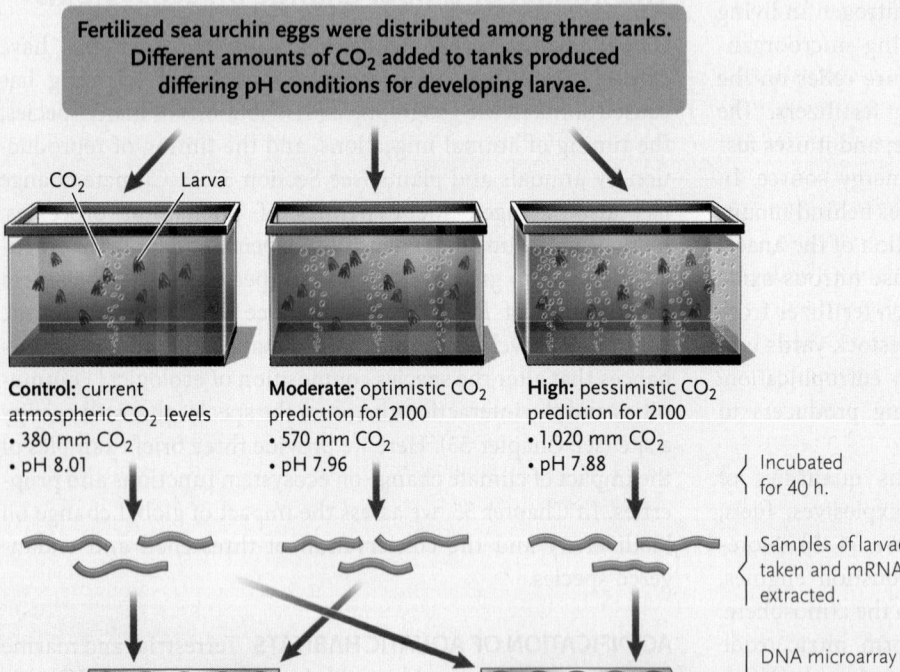

Fertilized sea urchin eggs were distributed among three tanks. Different amounts of CO_2 added to tanks produced differing pH conditions for developing larvae.

CO_2 Larva

Control: current atmospheric CO_2 levels
- 380 mm CO_2
- pH 8.01

Moderate: optimistic CO_2 prediction for 2100
- 570 mm CO_2
- pH 7.96

High: pessimistic CO_2 prediction for 2100
- 1,020 mm CO_2
- pH 7.88

Incubated for 40 h.

Samples of larvae taken and mRNA extracted.

DNA microarray analysis (see Figure 19.8) comparing expression of 1,057 genes in Moderate to Control and in High to Control.

DNA microarray

© Cengage Learning 2017

corals, and microscopic protists called foraminiferans—produce shells of calcium carbonate ($CaCO_3$), which they synthesize from bicarbonate ions and calcium ions. When those organisms die, their shells sink to the ocean bottom, and their carbon content becomes part of the unavailable carbon reservoir.

Because CO_2 is soluble in water, the oceans readily absorb it from the atmosphere—so much so that ecologists describe the oceans as a "sink" for atmospheric CO_2. Some environmental scientists have even suggested that the transfer of CO_2 from the atmosphere to the oceans may mitigate rising atmospheric CO_2 concentration. However, increasing the concentration of dissolved CO_2 fosters the formation of more carbonic acid in marine ecosystems, decreasing oceanic pH; in other words, the oceans are becoming more acidic as they absorb the ever more plentiful supplies of this gas. If the CO_2 concentration of the atmosphere continues to increase at its present rate, scientists predict a decrease in ocean surface pH of 0.2 to 0.4 by the end of the twenty-first century. This increased acidity of ocean water, combined with higher environmental temperatures, is thought to be responsible for the worldwide deaths of corals and their endosymbiotic algae, a phenomenon described as "bleaching" because the dead corals turn a ghostly white. Acidification of ocean waters also disrupts the formation of calcium carbonate shells by marine organisms. As described in *Molecular Insights*, researchers are using genomic tools to study the effects of ocean acidification on marine life.

DECLINING PRIMARY PRODUCTIVITY Experimental research suggests that the local enhancement of atmospheric CO_2 concentration fosters increased photosynthesis in many plant species. Thus, in principle, higher atmospheric concentrations of CO_2 should increase productivity in terrestrial ecosystems. Unfortunately, global climate change has altered rainfall patterns, causing droughts in many habitats that had been lush and wet in the past. As you learned earlier in this chapter, rainfall is a key ingredient for high primary productivity. Thus, droughts have caused productivity in Amazon forests to decline steadily for the past 30 years. Climate warming and drought are the probable causes of similar mass deaths of trees in the Western United States (see Section 52.5), further reducing these forests' ability to absorb carbon from the atmosphere.

Productivity also appears to be declining in some marine habitats, especially in the open ocean. Warming of surface water prevents it from mixing with nutrient-rich deep water, just as it does in temperate lakes in summer (see Figure 51.24). In the absence of the nutrients that mixing brings to the photic zone of the ocean, productivity by phytoplankton inevitably declines.

DEAD ZONES IN SHALLOW MARINE ENVIRONMENTS In the discussion of lake ecology (see Figure 51.25), you have already learned how the addition of excess phosphorus can turn a clear, oligotrophic lake into a soupy, eutrophic one. Nitrogen has a similar effect on aquatic ecosystems, where it is often a limiting nutrient. Runoff, especially from fertilized agricultural land, provides much more nitrogen than aquatic photosynthetic organisms can absorb and use, disrupting the dynamics of food webs and entire ecosystems.

For example, the Mississippi River carries so much excess nitrogen, much of it from fertilizers, into the Gulf of Mexico that it fosters a population explosion, called a "bloom," of phytoplankton in shallow, near-shore environments. When the phytoplankton die at the end of their growing season, they are decomposed by aerobic (that is, oxygen using) bacteria, which, in turn, experience a bloom of their populations. The decomposers deplete the water of oxygen, resulting in the mass deaths of fish and other animals. Like the devastation in Lake Erie described in the opening of this chapter, coastal environments in the Gulf of Mexico, from Texas to the Florida Panhandle, became a "dead zone," where few native organisms—many of them economically important—were able to live. As in Lake Erie, the devastation can be partially reversed if farmers upstream use nitrogen-containing fertilizers more judiciously.

STUDY BREAK 54.4

1. What is the greenhouse effect, and how does an increase in atmospheric CO_2 concentration affect it?
2. What human activities release the most CO_2 into the atmosphere?
3. What agricultural practices contribute to the disruption of the nitrogen cycle?

THINK OUTSIDE THE BOOK

Examine the ingredients listed on the labels of three cleaning products that you use in your household or residence hall. Search the Internet or sources in your library for information about how the disposal of these compounds on land or in sewage might contribute to the disruption of a biogeochemical cycle.

Unanswered Questions

How does the carbon cycle of a forest respond to climate change and urbanization?

As you have read in this chapter, human influences on the environment can have dramatic unforeseen consequences for ecosystems, altering energy flow and nutrient cycling. Given the complexity of ecosystems—the myriad scales of influence and multiple interactions among the organisms, the physical environment, and climate change variables—the

precise response of an ecosystem is difficult to predict, even with advanced ecosystem models. We do know with certainty, however, that the carbon, nitrogen, and water cycles of forested ecosystems in the northeastern United States are changing, and they are likely to continue to do so.

Carbon cycle research in forested ecosystems often entails building an ecosystem model from quantitative data on the various pools and fluxes of carbon in the ecosystem and how these change with time. Scientists then correlate these changes with the environmental conditions and derive a mechanistic understanding that they can use to make predictions about how the ecosystem will respond to future changes. In theory, *gross primary productivity* (GPP) should be predictable from a basic understanding of photosynthesis and a general description of the ambient environmental conditions. In practice, however, the complexity of canopy architecture and leaf positioning, the timing of recurring natural phenomena, and the effects of herbivory and leaf losses from abiotic factors all make accurate predictions more difficult. Furthermore, the problem is dynamic because age-related changes in stand structure, disturbance, invasion, drought, seasonality, and pests or pathogens all add spatial and temporal complexities. Scientists should be able to predict *net primary productivity* (NPP), a key parameter used by ecologists to classify the world's ecosystems, from measurements of the cellular respiration and the relative abundances of representative organisms from the ecosystem.

Quantifying GPP and NPP on a large spatial scale can be challenging, and discovering the underlying mechanisms that control ecosystem responses to changes in environmental conditions is difficult. For example, studies at Black Rock Forest, a deciduous-oak-dominated forest in New York State, revealed that temporal heterogeneity (seasonal variation in leaf and stem respiration) and spatial heterogeneity (variations in canopy and hill-slope position) are important factors that must be included in models of canopy respiration. Nevertheless, some simplifications may be possible. For example, although the basal rate of respiration is quite variable and subject to acclimation, it may be predictable from basic plant properties such as their nitrogen concentrations. Furthermore, the temperature coefficient of respiration is relatively constant, greatly simplifying the construction of an ecosystem model. To consider the impact of tree respiration on ecosystem form and function fully, my research team experimented with models that explicitly consider physiological linkages between photosynthesis and respiration, as mediated by leaf carbohydrate pools. We found that when we included direct linkages to carbon gain in the analysis, the model correctly predicted a large (23%) decrease in the estimated nighttime canopy respiration during the growing season. This result emphasizes the need for a process-based modeling approach when estimating forest productivity.

Our research at Black Rock Forest has also demonstrated that human activities in New York City (60 miles to the south) may be influencing tree growth in both urban and rural areas, with significant changes in seedling size, biomass allocation, herbivory, stomatal densities, nutrient concentrations, efficiency of water use, and rates of key physiological processes such as photosynthesis and respiration. Urbanization has a clear effect on the land area developed, but current research is showing that human activities in urban areas also influence forested ecosystems in the surrounding rural areas. Understanding how human activity, climate change, and forest ecosystems interact is crucial if we are to make prudent and sustainable development decisions, preserving the health of the ecosystems and the services they provide.

think like a scientist Through what mechanisms might the presence of a large city influence primary productivity in a forest 60 miles away?

Kevin Griffin

Kevin Griffin is a Professor at Columbia University's Lamont-Doherty Earth Observatory. His research centers on processes in plant and ecosystem ecology, the goal of which is to increase our understanding of both the role of vegetation in the global carbon cycle and the interactions between the carbon cycle and Earth's climate system. To learn more about Dr. Griffin's research, go to http://www.ldeo.columbia.edu/user/griff.

REVIEW KEY CONCEPTS

For access to MindTap and additional study materials visit www.cengagebrain.com.

54.1 Modeling Ecosystem Processes

- Ecosystems include biological communities and the abiotic environmental factors with which they interact (Figure 54.1).
- Food webs define the pathways along which energy and nutrients move through the biological components of an ecosystem (Figure 54.2).
- Compartment models describe energy flow and nutrient cycling in ecosystems (Figure 54.3).
- Simulation models are interlocking mathematical equations that define the relationships between populations and between populations and the physical environment. They allow researchers to predict the effects of changes in ecosystem structure and function.

54.2 Energy Flow and Ecosystem Energetics

- Photosynthesis converts only a small portion of the solar energy that reaches Earth into chemical energy.
- An ecosystem's gross primary productivity is the rate at which producers convert solar energy into chemical energy. Producers use some energy for respiration; some is converted to heat; and some remains in the ecosystem as net primary productivity.

- Primary productivity is measured in units of energy captured or biomass produced per unit area per unit time. Net primary productivity indexes the energy available to support heterotrophs.
- On land, primary productivity is limited by the availability of sunlight, water, and nutrients; temperature; and the amount of photosynthetic tissue present (Figure 54.4). In marine and aquatic ecosystems, primary productivity is limited when sunlight and nutrients are not available in the same place. Ecosystems vary in productivity and in their contributions to Earth's total productivity (Figures 54.5 and 54.6, Table 54.1).
- Only a fraction of the energy at any trophic level is converted into biomass at higher trophic levels. As energy passes through a food web, an average of 90% is lost at each transfer between trophic levels, limiting the number of trophic levels that a food web can support (Figure 54.7).
- Ecological pyramids portray the effects of energy losses. For terrestrial ecosystems, pyramids of energy, biomass, and numbers generally have broad bases and narrow tops (Figure 54.8).
- Top predators suffer from biological magnification, the accumulation of toxins that have travelled upward through a food web (Figure 54.9).
- Feeding by predators can influence primary productivity through a trophic cascade (Figure 54.10). In some ecosystems, higher trophic levels are strongly influenced by the nutrient content of primary producers.

54.3 Nutrient Cycling in Ecosystems

- Earth is a closed system with respect to matter.
- Nutrients circulate in biogeochemical cycles between living organisms and nonliving reservoirs. Some biogeochemical cycles are atmospheric; others are sedimentary.
- Water circulates through the atmosphere, oceans, and terrestrial and freshwater ecosystems in a global hydrologic cycle. Water evaporates from the oceans and land and falls as precipitation. Runoff and streamflow return excess precipitation from the land to the oceans (Figure 54.11).
- The carbon cycles in terrestrial and aquatic ecosystems are linked through an atmospheric pool of CO_2, which primary producers assimilate. Respiration returns carbon to the atmosphere as CO_2. Earth's largest reservoir of carbon is unavailable in sedimentary rock. Other large reservoirs include coal, oil, and peat, as well as dissolved bicarbonate and carbonate ions in seawater (Figure 54.12).
- Nitrogen is cycled between living organisms and an atmospheric pool of nitrogen gas. Bacteria and cyanobacteria make nitrogen available to the food web through the processes of nitrogen fixation, ammonification, and nitrification. Denitrification converts nitrogen compounds to molecular nitrogen, which enters the atmosphere (Figure 54.13, Table 54.2).
- Phosphorus undergoes a sedimentary cycle. Weathering and erosion of rock make phosphorus available; it is leached from soil and carried to the ocean. Dissolved phosphates precipitate out of seawater, forming insoluble deposits, which are eventually uplifted by tectonic processes (Figure 54.14).

54.4 Human Activities and Anthropogenic Global Change

- Certain gases in the lower atmosphere create a greenhouse effect that traps heat near Earth's surface (Figure 54.15). The combustion of fossil fuels and wood has increased the atmospheric concentration of CO_2 substantially in recent decades (Figure 54.16), enhancing the greenhouse effect and raising the average surface temperature of the planet. Global warming has already had profound effects on the landscape and biological systems, and these effects are likely to increase for the foreseeable future.
- Human activities—particularly nitrogen fixation for the production of fertilizers and other materials, irrigation, livestock farming, and the use of combustible fuels—have disrupted the nitrogen cycle (Figure 54.17). The consequences of shifting nitrogen among its natural reservoirs include the pollution of aquatic ecosystems, acid precipitation, and global climate change.
- Global change has had, and will continue to have, detrimental effects on ecosystems. The acidification of aquatic habitats contributes to the death of corals and interferes with the synthesis of calcium carbonate in hard-shelled animals. Climate change has fostered droughts that result in the death of forests, reducing primary productivity and the capacity to withdraw carbon from the atmosphere. Nitrogen pollution from the Mississippi River has created a dead zone in shallow water ecosystems in the Gulf of Mexico.

TEST YOUR KNOWLEDGE

Remember/Understand

1. Which of the following events would move energy and material from a detritivore into a higher trophic level in a food web?
   a. A beetle eats the leaves of a living plant.
   b. An earthworm eats dead leaves on the forest floor.
   c. A robin catches and eats an earthworm.
   d. A falcon eats a robin.
   e. A bacterium decomposes the feces of an earthworm.

2. The total dry weight of plant material in a forest is a measure of the forest's:
   a. gross primary productivity.
   b. net primary productivity.
   c. cellular respiration.
   d. standing crop biomass.
   e. ecological efficiency.

3. Which of the following ecosystems has the highest rate of net primary productivity?
   a. open ocean
   b. temperate deciduous forest
   c. tropical rainforest
   d. desert and thornwoods
   e. cultivated land

4. Some freshwater and marine ecosystems exhibit an inverted pyramid of:
   a. biomass.
   b. energy.
   c. numbers.
   d. turnover.
   e. ecological efficiency.

5. Which process moves nutrients from the available organic compartment to the available inorganic compartment?
   a. respiration
   b. erosion
   c. assimilation
   d. sedimentation
   e. photosynthesis

6. Which of the following materials has a sedimentary cycle?
   a. water
   b. oxygen
   c. nitrogen
   d. phosphorus
   e. carbon

7. Nitrogen fixation converts:
   a. atmospheric molecular nitrogen to ammonia.
   b. nitrates to nitrites.
   c. ammonia to molecular nitrogen.
   d. ammonia to nitrates.
   e. nitrites to nitrates.

Apply/Analyze

8. Endothermic animals exhibit a lower ecological efficiency than ectothermic animals because:
   a. endotherms are less successful hunters than ectotherms.
   b. endotherms eat more plant material than ectotherms.
   c. endotherms are larger than ectotherms.
   d. endotherms produce fewer offspring than ectotherms.
   e. endotherms use more energy to maintain body temperature than ectotherms.

9. The amount of energy available at the highest trophic level in an ecosystem is determined by:
   a. only the gross primary productivity of the ecosystem.
   b. only the net primary productivity of the ecosystem.
   c. the gross primary productivity and the standing crop biomass.
   d. the net primary productivity and the ecological efficiencies of herbivores.
   e. the net primary productivity and the ecological efficiencies at all lower trophic levels.

10. Which of the following statements is supported by the results of studies at the Hubbard Brook Experimental Forest?
    a. Most of the energy captured by primary producers is lost before it reaches the highest trophic level in an ecosystem.
    b. Deforested watersheds experience significantly less runoff than undisturbed watersheds.
    c. Deforested watersheds lose more calcium and nitrogen in runoff than undisturbed watersheds.
    d. Nutrients generally move through biogeochemical cycles very quickly.
    e. Deforested watersheds generally receive more rainfall than undisturbed watersheds.

11. **Discuss Concepts** If you were growing a vegetable garden, identify the factors that might affect its primary productivity. How would you increase productivity? Identify some possible consequences of your gardening activities to nearby ecosystems.

Evaluate/Create

12. **Discuss Concepts** A lake near your home became overgrown with algae and pond-weeds a few months after a new housing development was built nearby. What data would you collect to determine whether the housing development might be responsible for the changes in the lake?

13. **Discuss Concepts** Some politicians question whether the recent increase in atmospheric temperature results from our release of greenhouse gases into the atmosphere. They argue that atmospheric temperature has fluctuated widely over Earth's history, and the changing temperature is just part of a historical trend. What information would allow you to refute or confirm their hypothesis? In addition, describe the pros and cons of reducing greenhouse gases as soon as possible versus taking a "wait and see" approach to this question.

14. **Discuss Concepts** If you could design the ideal farm animal—one that was grown as food for humans—from scratch, what characteristics would it have?

15. **Design an Experiment** Design an experiment to test the hypothesis that predators in an aquatic ecosystem regulate the ecosystem's primary productivity. Establish as many experimental ponds as you wish, and imagine stocking them with organisms at different trophic levels. If the hypothesis is correct, describe the results you would expect to record from each of your experimental treatments.

16. **Apply Evolutionary Thinking** The dramatic loss of energy as it is transferred upward through a food web produces pyramids of energy, biomass, and numbers that are very narrow at the top. Which agent of evolution (see Section 21.3) would you expect to influence allele frequencies in the relatively small populations that are typical of top predators? What are the long-term *evolutionary* effects of small population size?

For selected answers, see Appendix A.

INTERPRET THE DATA

Amy Rosemond of Vanderbilt University and two colleagues at the Oak Ridge National Laboratory conducted a study of primary productivity in the algal community of Walker Branch, a stream in eastern Tennessee. In their experiment, they compared the productivity of algae that were eaten by snails (grazed) to the productivity of algae that were protected from herbivores (ungrazed) under three experimental treatments: (1) the addition of nitrogen to the stream; (2) the addition of phosphorus to the stream; and (3) the addition of both nitrogen and phosphorus to the stream. They compared the results of these treatments to control treatments of both grazed and ungrazed algae that received no nutrient enrichment. How would you interpret their results, presented in the accompanying graph? Was algal productivity controlled by the presence of herbivores, the addition of nutrients, or a combination of the two factors?

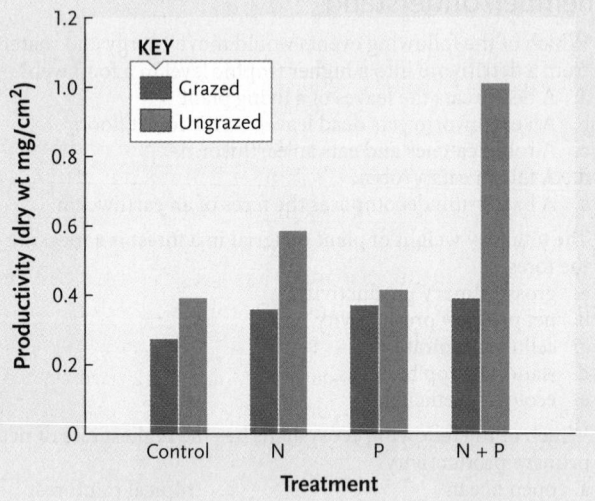

Source: Adapted from A. D. Rosemond et al. 1993. Top-down and bottom-up control of stream periphyton: effects of nutrients and herbivores. *Ecology* 74:1264–1280. *Ecology* by Ecological Society of America.

Biodiversity and Conservation Biology

Florida panther (*Puma concolor coryi*). Between 100 and 160 individuals of this endangered subspecies survive.

Why it matters . . . Someone seems to have disappeared. Investigators thoroughly checked the missing subject's known haunts, but found no trace. They questioned others in the neighborhood, but came up with few leads. The subject was last seen alive in 1978. With so cold a trail to follow, investigators reluctantly marked the case file "Missing and Presumed Extinct."

The subject in this case was Miss Waldron's red colobus monkey, *Procolobus badius waldroni* **(Figure 55.1).** Named for a traveling companion of the taxonomist who first described it in 1933, this distinctively colored subspecies lived in large and noisy social groups in a remote forest on the border between Ivory Coast and Ghana in West Africa.

John Oates of the City University of New York led a research team that tried to locate Miss Waldron's red colobus. They used every imaginable method, including visual and auditory censuses, searching for scat ("dung") in natural habitats, interviewing local people, and looking in marketplaces where monkey meat is commonly traded. In 2000, more than 20 years after the last confirmed sighting, the researchers concluded that this monkey is probably extinct. A later search by a member of the team, William S. McGraw of Ohio State University, found the skin of one monkey that a hunter had shot six months before. But McGraw searched in vain for a living monkey, and he concluded that even if a few are still alive, the population is so small that continued hunting will surely eliminate it.

Procolobus badius waldroni may be the first primate subspecies to become extinct in more than 100 years—and only the second in the last 500 years. Monkeys and other primates are among the most closely monitored and protected species on Earth. Nonetheless, Oates and his colleagues concluded, these monkeys probably became extinct because they

FIGURE 55.1 Miss Waldron's red colobus. *Procolobus badius waldroni*, which weighed about 10 kg, may be the first primate subspecies to become extinct in more than 100 years.

were hunted locally for food by a growing human population and because humans have destroyed their natural habitats.

Miss Waldron's red colobus is just one of many species driven to extinction every year. Current threats to biodiversity—all of which ultimately result from our disruption of natural populations, communities, and ecosystems—are massive. The likely loss of this monkey should warn us that many taxa are at risk, even those that are most rigorously protected.

When ecologists speak of **biodiversity,** they are referring to the richness of living systems at all levels of biological organization (see Figure 1.2). At the most fundamental level, biodiversity encompasses the *genetic variation*—both *within* populations and *between* populations of a species—that is raw material for adaptation, speciation, and evolutionary diversification (see Chapters 21 and 22). At a higher level of organization, biodiversity includes *species richness* within communities (see Section 53.3). The number and variety of species within a community influence its overall characteristics, population interactions, and trophic structure. Finally, biodiversity exists at the *ecosystem level.* Complex networks of interactions bind species in an ecosystem together, and because different ecosystems interact within the biosphere, damage to one ecosystem can reverberate through others (see Chapters 51 and 54).

In this chapter we first describe how human activities threaten biodiversity and reflect on why we should protect it. We then consider theoretical and practical approaches to conservation biology, the scientific discipline that focuses on preserving Earth's biological resources.

55.1 The Biodiversity Crisis on Land, in the Sea, and in River Systems

Biodiversity is declining dramatically, perhaps faster than ever before in Earth's history. In this section, we describe the three broadest threats to biodiversity: the clearing of forests; the commercial overexploitation of marine fish populations; and hydrologic alterations of freshwater ecosystems. Bear in mind that these and other challenges are exacerbated by global change, which we have discussed in previous chapters.

Deforestation Disrupts the Carbon Cycle and May Lead to Desertification

Forests are among the habitats that humans most frequently clear and convert. According to the United Nations Forest Resources Assessment released in 2010, global deforestation occurs at a rate of 5.6 million hectares per year, or 15,000 hectares per day. In other words, an area of forest equivalent to 27,000 football fields is cleared of all trees every day.

Deforestation does not occur uniformly across the globe. Today, more than 90% of deforestation occurs in tropical regions, where many groups of organisms exhibit their highest diversity (described in Section 53.7). Forests are most often cut to clear land for grazing livestock; as a result, a few species of domesticated animals and the grasses they consume replace what had been a species-rich community. Brazil has experienced the most extensive recent damage, accounting for 25% of all deforestation during the late twentieth century **(Figure 55.2)**. This assessment is particularly troubling because Brazil contains approximately 27% of the planet's total aboveground woody biomass.

Compounding the direct environmental damage, most tropical forests are burned as they are cleared, a process that adds CO_2 to the atmosphere, enhancing the greenhouse effect and increasing the rate of global warming (see Section 54.4). According to the Intergovernmental Panel on Climate Change, a Nobel Prize–winning agency of the United Nations, forest burning contributes nearly 20% of all greenhouse gases released into the atmosphere. Ironically, intact forests remove substantial quantities of carbon dioxide from the atmosphere, a capacity that is diminished with every tree felled.

Once a forest is cut, heavy grazing or farming drains nutrients from the soil. To remain productive, even the best agricultural or grazing lands require either the application of fertilizers or long periods during which the land lies fallow, allowing plants

FIGURE 55.2 Deforestation in the Amazon Basin. Satellite photos of Rondonia, in the Brazilian Amazon, show how much of the forest was cut between 1975 and 2009. Each photo illustrates an area approximately 60 by 85 km.

to replenish the soil naturally. Unfortunately, the soil where tropical forests grow is often of marginal value (for reasons described in Section 51.4), and it is rapidly degraded; it becomes hard, even more nutrient-poor, unable to retain water, and likely to wash away.

When large tracts of subtropical forest are cleared and overused, the land often undergoes **desertification:** the ground-water table recedes to deeper levels; less surface water is available for plants; soil accumulates high concentrations of salts (a process called *salinization*); and topsoil is eroded by wind and water. In other words, the habitat is converted to desert.

Desertification speeds the loss of biodiversity locally, sometimes eliminating entire ecosystems. For example, desertification has decimated habitats in the Sahel region of Africa, just south of the Sahara **(Figure 55.3)**. Excessive grazing of cattle and goats by an ever-expanding human population is the main reason for the Sahara's southward expansion at a rate of 5.5 to 8 km per year. Because the sand dunes of the expanding desert shift constantly, agriculture and grazing are nearly impossible, resulting in frequent famines among the inhabitants.

Desertification and salinization have also begun in the Everglades, a unique, shallow "river of grass" that covers much of southern Florida. The amount of fresh water flowing through South Florida to the Everglades has decreased approximately 70% since 1948, when an extensive network of canals and levees was built to reduce flooding. The rapidly growing human population in South Florida contributes directly to desertification, as groundwater is tapped for domestic use and to irrigate lawns, golf courses, and agricultural fields. Salt water from the Gulf of Mexico now intrudes into the water table, causing salinization of the soil. The Comprehensive Everglades Restoration Plan (CERP), approved by the U.S. Congress in 2000, seeks to restore the natural flow of the Everglades by 2030. According to a biennial report published in 2014, the progress of everglades restoration has been slowed by inadequate government funding. If the plan is implemented in a timely manner, this project may halt or reverse the desertification process.

Sadly, deforestation, desertification, and global warming reinforce each other in a positive feedback cycle (see Section 54.4). If scientists' projections are correct, desertification will lead to an increase in the average global temperature, speeding evaporation and the retreat of forests, which, in turn, will increase rates of desertification. If deforestation and desertification continue, we will soon lose a large proportion of Earth's forests and face a decrease in the area of habitable land.

A. **The Sahel region of Africa**

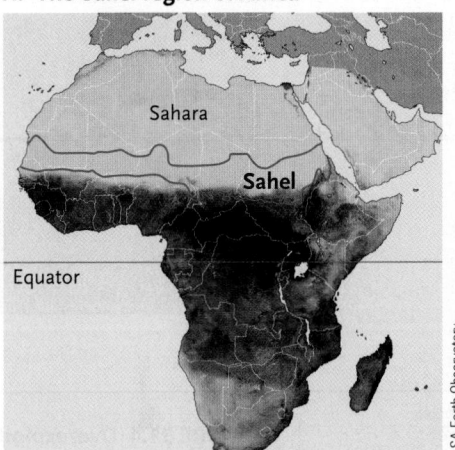

B. **Women preparing millet, a grain, in the Sahel**

FIGURE 55.3 Desertification in the Sahel. (A) A satellite photo taken near the end of the dry season in June 2005 illustrates the severe desertification in parts of the Sahel region of Africa. Dark green areas are densely vegetated; light green areas are sparsely vegetated, and sand-colored areas are barren. **(B)** People who live in this region can barely eke out a living on the land.

Overexploitation Can Drive Species to Extinction

Many local extinctions result from **overexploitation,** the excessive harvesting of an animal or plant species. At a minimum, overexploitation leads to declining population sizes in the harvested species. In the most extreme cases, a species may be wiped out completely. Moreover, overexploitation can foster evolutionary changes in the exploited population, much the way guppies respond to natural predators in the streams of Trinidad (described in *Focus on Research: Basic Research* in Chapter 52).

Humans have overexploited populations in every habitat we occupy. As an example, consider overexploitation in marine ecosystems, the only environment from which we routinely harvest predators (such as tuna) as food. The fishery on the Grand Banks off the coast of Newfoundland, Canada, provides a sad example **(Figure 55.4)**. For hundreds of years, fishers used traditional line and small-net fishing to harvest a large but sustainable catch. During the twentieth century, however, new technology allowed them to locate and exploit schools of fishes more efficiently. As a result, roughly half the fish species harvested there are now overfished. Haddock (*Melanogrammus aeglefinus*) and yellowtail flounder (*Limanda ferruginea*) have been essentially eliminated from the Grand Banks; their populations will probably never recover. And because fishers preferentially harvest the oldest and largest individuals, which fetch a higher market price, Atlantic cod (*Gadus morhua*) now mature at a younger age (three years compared with five or six years) and smaller size.

As a consequence of overfishing, the average yield of the Grand Banks has declined to less than 10% of the highest historic levels. In the mid-1960s, Atlantic cod yielded a minimum of 350,000 tons per year. By the mid-1970s, the catch dropped to 50,000 tons per year. The Canadian government finally closed the fishery in 1993, after the cod catch fell below 20,000 tons

A. The Grand Banks

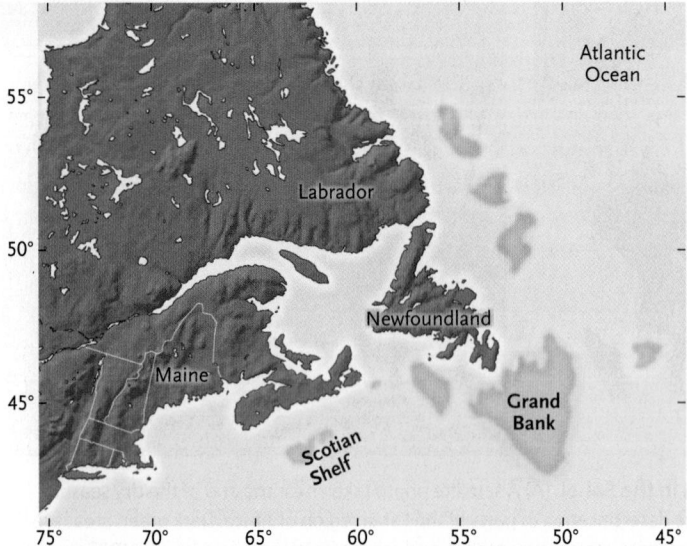

B. Atlantic cod (*Gadus morhua*)

blickwinkel/Alamy

FIGURE 55.4 Overexploitation of North Atlantic fisheries. (A) The Grand Banks (sand-colored shading) were severely overfished in the late twentieth century, leading to the near extinction of many species, including the **(B)** Atlantic cod.
© Cengage Learning 2017

for several consecutive years. But the damage had already been done: the most heavily exploited species are less marketable because of their smaller size, fish populations have decreased to dangerously low levels, and the fishing industry is itself imperiled.

This sequence of events has been replicated in fisheries around the world. Indeed, Ransom A. Myers and Boris Worm of Dalhousie University in Nova Scotia estimated that by 2003 modern fishing techniques had reduced the biomass of large predatory fishes by about 90% in marine ecosystems. In a 2014 publication, based on ecosystem modeling, Villy Christensen of the University of British Columbia and colleagues in France, Italy, and Spain concluded that overfishing had changed the trophic structure of marine ecosystems on a global scale; they predicted that fish communities in the future will be dominated by small prey fish, rather than by the large predators that were abundant in the last century.

Hydrologic Alterations Endanger Freshwater and Wetland Ecosystems

Rivers have always played a key role in the development of human settlements because they provide a source of fresh water, a place to discard wastes, and a means to transport goods. Since ancient times, humans have also dammed rivers to capture reliable supplies of fresh water. When the human population—and the dams they built— were small, these **hydrologic alterations** (that is, changes to the pathways through which water moves in the hydrologic cycle) had primarily local effects. More recently, we have constructed massive dams that

capture vast quantities of fresh water in reservoirs **(Figure 55.5)**. We distribute water from these reservoirs for agricultural, industrial, and domestic uses. The dams also generate hydroelectric power and allow us to control water flow to mitigate flooding of low-lying land. Today, the large scale and ubiquity of these hydrologic alterations have made freshwater ecosystems among the most endangered on Earth.

The damming of rivers and the diversion of their flow wreaks havoc on many interconnected ecosystems. As you learned in Section 51.5, the physical characteristics of rivers change predictably from the head-waters of streams to the estuaries where they empty into the sea. River-dwelling organisms are adapted to specific physical conditions—such as temperature, depth, and flow rate—that are characteristic of each section

NASA

FIGURE 55.5 Three Gorges Dam. In 2008, China completed the Three Gorges Dam, the world's largest hydroelectric dam, across the Yangtze River in Hubei Province. The dam is 2,300 m long and 185 m high. The reservoir behind it measures 660 km by 1.1 km and holds 39.3 km³ of water. More than 1.25 million people had to be relocated from areas inundated by the reservoir. The photo on the right illustrates the size of the reservoir behind the dam two years before construction was completed.
© Cengage Learning 2017

of the river system. And downstream, the flow of water supports distinct communities of organisms, each adapted to the different environments in floodplains, wetlands, and estuaries.

In 2002, Stuart E. Bunn and Angela H. Arthington of Griffith University in Australia identified four ways in which hydrologic alterations threaten freshwater biodiversity. First, the flow rate and volume of rivers are key determinants of their physical habitats, which have a major impact on the organisms that live there. For example, before it was dammed, the River Otra in Norway experienced low winter water flows, but raging summer floods. These conditions established a regular pattern of disturbance that eliminated many rooted plants from the riverbed (see Section 53.5). Now dammed, the river has a more regulated flow regime that allows a huge accumulation of plant biomass.

Second, the life histories of aquatic species, which evolved in response to natural flow patterns, are disrupted by changes in river flows. For example, reproduction by many aquatic invertebrates and fishes is triggered by temperature and day-length cues. Because the water released through dams is often drawn from the depths of the reservoirs behind them, it is colder than the natural flow, changing the cues available to organisms. Researchers working in China discovered that the cold water released by dams delayed spawning by as much as 30 days in some fish species.

Third, dams reduce a river system's "connectivity" (that is, the continuity of flow through a river and its streams and tributaries). Reduced connectivity prevents fishes and other animals from migrating freely through a river system. For example, salmon undertake a spawning migration from the sea, swimming upstream into the tributaries and streams where they reproduce. Dams hamper this already difficult upstream journey. In the Pacific Northwest, more than 400 hydroelectric dams on the Columbia River system prevent many salmon from reaching their spawning grounds; they have reduced the breeding habitat for Chinook salmon (*Onchorhynchus tshawytscha*) by 75%. Similar problems have eliminated migratory fish species from rivers throughout the world.

Finally, dams and reservoirs facilitate the introduction and success of nonnative species that thrive in disturbed habitats (discussed further below). For example, several species of large fishes collectively described as "Asian carp" have become established in the Mississippi River and its tributaries. These fishes feed voraciously on plankton. For years ecologists feared that they would spread into the Great Lakes, where they would outcompete native fish species that are the basis of a large fishing industry. In 2002 and 2004, the U.S. Army Corp of Engineers built two electric barriers in a canal that connected a tributary of the river to Lake Michigan, hoping to block the fishes' advance.

In 2015, environmental DNA analyses found evidence of carp in 24 samples in the canal, one of them just one city block from Lake Michigan. The Army Corps of Engineers has proposed building a $15 billion barrier to separate the Great Lakes from the Chicago Area Water System, which connects Lake Michigan to the Mississippi River. Politicians have balked at the hefty price of the project, and representatives of the transportation industry, which uses the existing waterway to ship 15 million tons of goods each year, have raised objections to the plan. Government officials are now exploring less expensive, short-term solutions to the problem.

Freshwater ecosystems are now under severe pressure. In the worst cases, human-induced hydrologic alterations have practically eliminated them: the Nile and the Colorado River now rarely discharge much water into the sea. Even in less dramatic cases, the effects of hydrologic alterations on biodiversity have been profound. Freshwater fish species have experienced marked declines in the last few decades. One 2006 estimate suggested that 56% of the freshwater fish species endemic to the Mediterranean region, more than 30% of the native species in North America, and 25% of those found in East Africa are now threatened with extinction. The status of freshwater invertebrates, though not as well documented, is probably comparable.

Conservation biologists rank the restoration of natural flow patterns in river systems among their highest priorities. Indeed, government and public support for dam removal has increased markedly in the United States since 1990, and an increasing number of dams are demolished each year. Fisheries biologists, ecologists, and environmental scientists are actively studying the effects of dam removals to determine how long it takes for river systems to recover from anthropogenic hydrologic alterations.

STUDY BREAK 55.1

1. What factors have increased the likelihood of desertification in southern Florida?
2. What are the consequences of the overexploitation of fish populations?
3. How does the construction of a dam disrupt the lives of river-dwelling organisms?

THINK OUTSIDE THE BOOK

Search the Internet for updates about the spread of Asian carp in North America. What is the federal government doing to prevent their spread? Have they successfully invaded the Great Lakes? If so, what impact have they had on the native lake fishes?

55.2 Specific Threats to Biodiversity

Although the clearing of tropical forests, overexploitation of marine fisheries, and damming of rivers endanger entire ecosystems, many other human activities imperil natural populations. In this section, we briefly describe some of these threats.

Habitat Fragmentation Threatens Many Populations

When humans first colonize a pristine habitat, they build roads and then clear isolated areas for specific uses. Although this pattern of development initially affects only local populations,

FIGURE 55.6 **Observational Research**

Near-complete Extinction of Small Mammals in Tropical Forest Fragments

Question: How long do populations of small mammals persist in small fragments of tropical forests?

Hypothesis: Populations of small mammals will eventually become extinct on small islands that were created when a large patch of forest was flooded to establish a reservoir **(A).**

Prediction: Using principles from the theory of island biogeography (see Section 53.7), Gibson and his colleagues predicted that extinctions would be more rapid in small forest fragments than in large fragments.

Method: The researchers conducted on-the-ground surveys on 16 small islands (red) of varying size (0.3–56.3 ha) five times over a period of 20 years, recording all of the small mammal species they encountered. The first surveys took place 5 to 7 years after the forest fragments were formed.

Result: Small mammal populations quickly became extinct in the habitat fragments **(B).** Extinction varied with the size of habitat fragments such that most species disappeared from small islands within the first five to seven years after fragmentation. Twenty-five years after forest fragmentation, nearly all small native mammals became extinct on islands of any size. The Malayan field rat *(Rattus tiomanicus)* was the only mammal that persisted in all fragments; researchers believe that it colonized the fragments after they were separated from each other and from the surrounding forest.

A. Islands in Chiew Larn Reservoir

B. Extinction of small mammals

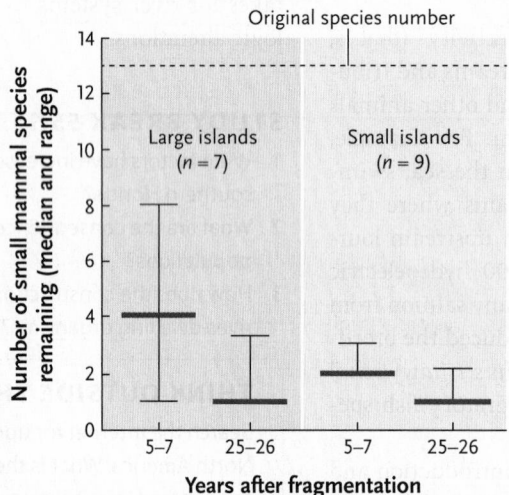

Conclusion: Habitat fragments experienced size-dependent rates of extinction, but all habitat fragments in this study eventually lost all of their native small mammals.

Observe	**The forest fragments that Gibson and his colleagues studied**
Hypothesize	**existed on islands in the newly created reservoir. How do the**
Predict	**results illustrated in B relate to the predictions of the theory of**
Experiment	**island biogeography (described in Section 53.7)?**
Interpret	

Source: L. Gibson et al. 2013. Near-complete extinction of native small mammal fauna 25 years after forest fragmentation. *Science* 341:1508–1510.

Habitat fragmentation is a threat to biodiversity because small habitat patches can sustain only small populations of organisms. As you learned in Section 52.4, a habitat's *carrying capacity,* the maximum population size that it can support, varies with available resources. Populations that occupy small habitat patches inevitably experience low carrying capacities, a problem that is especially acute for species at the higher trophic levels (see Section 54.2). Furthermore, fragmented habitat patches are often separated by unsuitable habitat that organisms may be unable or unwilling to cross. As a result, individuals from one isolated population are unlikely to migrate into another, reducing gene flow between them. The combination of small population size and genetic isolation reduces genetic variability and fosters extinction (see Section 21.3).

In addition to reducing the amount of undisturbed habitat, habitat fragmentation jeopardizes the quality of the habitat that remains. Human activities create noise and pollution that spread into nearby areas. The removal of natural vegetation disrupts the local physical environment, exposing the borders of the remaining habitat to additional sunlight, wind, and rainfall. Increased runoff compacts the soil and makes it waterlogged. These phenomena are collectively described as **edge effects.**

The effects of habitat fragmentation are often profound. Luke Gibson of the National University of Singapore and colleagues from Australia, Canada, China, Thailand, and the United States studied the effects of habitat fragmentation on communities of small tropical mammals in southern Thailand **(Figure 55.6).** They surveyed mammal species composition on 16 small islands that were created in 1986–1987, when a river was dammed to establish a reservoir. The dam flooded 165 km² of continuous forest that had harbored 13 native small mammal species. The many islands within the reservoir are forest fragments that survived in areas that had been hilltops in the

the negative effects spread rapidly to a regional scale. The remaining areas of *intact* habitat are inevitably reduced to small, isolated patches, a phenomenon that ecologists describe as **habitat fragmentation.**

forest before it was flooded (see Figure 55.6A). Islands included in the study ranged in size from 0.3 to 5.6 hectares. (Recall that one hectare = 10,000 m² = 2.471 acres). The researchers first made surveys five to seven years after the fragmentation event and then sampled the islands four more times over 20 years.

Gibson and his colleagues discovered that the species richness of small mammals declined rapidly after the island fragments were formed (see Figure 55.6B). Five to seven years after fragmentation of the forest, when the first surveys were made, between 1 and 6 of the 13 resident mammal species had become extinct on larger islands (those with area greater than 10 ha), and between 10 and 12 had become extinct on the smaller islands (area less than 5 ha). Twenty years later (that is, 25 or 26 years after the islands' creation), small native mammals had virtually disappeared from all of the islands. The only small mammal present on every island was the nonnative Malayan field rat *(Rattus tiomanicus)*, which the researchers believe colonized the islands after the flooding of the forest; this invasive rat may have contributed to the extinction of the other species, either through competition or predation.

Many Forms of Pollution Overwhelm Species and Ecosystems

The release of **pollutants**—materials or energy in forms or quantities that organisms do not usually encounter—poses another major threat to biodiversity.

Although chemical pollutants, the by-products or waste products of agriculture and industry, are released locally, many spread in water or air, sometimes on a continental or global scale. Within North America, for example, winds carry airborne pollutants from coal-burning power plants to the Northeast **(Figure 55.7).** Sulfur dioxide (SO_2), which dissolves in water vapor and forms sulfuric acid, falls as **acid precipitation,** acidifying soil and bodies of water. Many lakes in northeastern North America have experienced a precipitous drop in pH from historical readings near 6 to values that are now well below 5—a 10-fold increase in acidity. Although the lakes once harbored lush aquatic vegetation and teemed with fishes, they are now crystal clear and nearly devoid of life (also see Section 54.4).

As residents of major cities and industrial areas know all too well, wastes produced by the combustion of fossil fuels in factories and automobile engines cause terrible local pollution, increasing rates of asthma and other respiratory ailments. Some airborne pollutants, notably CO_2, also join the general atmospheric circulation, where they contribute to the greenhouse effect and global warming.

Like air pollution, water pollution originates locally but has a much broader impact. Oil spills, for example, disrupt local ecosystems, killing most organisms near the spill. Because oil floats on water, it spreads rapidly to nearby areas.

An explosion and fire on the Deepwater Horizon oil rig in April 2010 allowed more than 200 million gallons of oil to spill into the Gulf of Mexico. The uncapped well continued to spew oil for three months, until it was partially plugged in July 2010. The oil spill had a devastating short-term effect on organisms in the Gulf and adjacent wetlands; scientists are still unable to predict all of its long-term effects on this delicate ecosystem. As of mid-2014, the cleanup effort had cost the operator of the well more than $14 billion.

Pollution can also have serious effects on terrestrial ecosystems. As a recent disaster in India, Nepal, and Pakistan illustrates, the application of synthetic compounds to agricultural fields or livestock can have dire and far reaching consequences (see the discussion of biological magnification in Section 54.2). For thousands of years, enormous populations of vultures (several *Gyps* species)—estimated at more than 40 million birds—consumed the abandoned carcasses of farm animals across South Asia. In the early 1990s, however, farmers began to administer diclofenac, an inexpensive anti-inflammatory drug, to injured livestock. Within a few years, vultures began to disappear; in 2006, scientists estimated that their populations had declined by 99% and that all vulture species in South Asia were on the verge of extinction. Researchers determined that diclofenac, which causes fatal kidney failure in birds, was responsible for the deaths: vultures were ingesting substantial doses of the drug from the livestock carcasses they ate.

Since 2006, the governments of Bangladesh, India, Nepal, and Pakistan have collaborated on efforts to ban diclofenac from veterinary use, and various non-governmental organizations have been educating the public about the dangers of the drug. As a result of these efforts, populations of several vulture species have stabilized, and some have shown signs of recovery. Nevertheless, conservation biologists predict that it will take decades for vulture populations to return to their earlier densities. Despite the apparent reversal of this environmental disaster, the

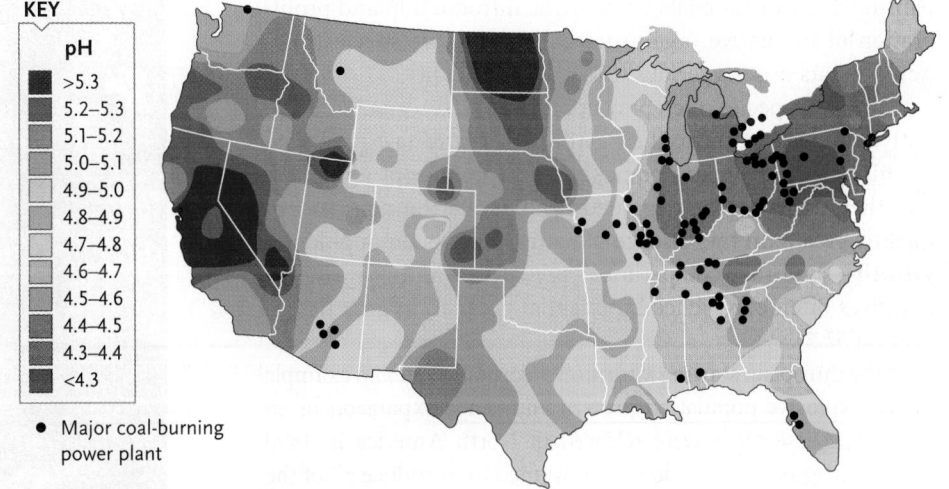

KEY

pH
- >5.3
- 5.2–5.3
- 5.1–5.2
- 5.0–5.1
- 4.9–5.0
- 4.8–4.9
- 4.7–4.8
- 4.6–4.7
- 4.5–4.6
- 4.4–4.5
- 4.3–4.4
- <4.3

● Major coal-burning power plant

FIGURE 55.7 Acid precipitation. Coal-burning power plants (indicated by black dots) release air pollution that is carried northeast, where it falls as acid precipitation. The map shows the average pH of rainfall.
© Cengage Learning 2017

FIGURE 55.8 Starling range expansion. After being introduced in New York City in 1890, European starlings *(Sturnus vulgaris)* increased their numbers and quickly extended their breeding range westward across North America. They reached the west coast by 1960 and Alaska by 1970.
© Cengage Learning 2017

decline in vulture populations has had a tremendous impact on urban and rural communities in South Asia. Livestock carcasses are now consumed by growing populations of feral dogs, many of which carry rabies. India has the world's highest human death toll from rabies—30,000 per year—and two thirds of the cases are caused by dog bites.

The Introduction of Invasive Species Often Eliminates Native Species

As humans travel from one habitat to another, we inevitably carry other species with us. Seeds cling to our legs, insects accompany us in our food and possessions, and some organisms hitch a ride on boats or cars. The introduction and proliferation of nonnative organisms, called **invasive species,** into new habitats poses a serious threat to biodiversity.

Invasive species often prey upon, parasitize, or outcompete native species, leading to their extinction. Many have *r*-selected life histories; they mature quickly and reproduce prodigiously, and they thrive in the degraded habitats that humans so frequently create. In the absence of natural checks on population growth—such as competitors, predators, and parasites—invasives often experience exponential population growth (see Section 52.5).

The European starling *(Sturnus vulgaris)* provides an example of the explosive population growth and range expansion of an invasive. These birds were released in North America in 1890 when a misguided individual, who wanted to introduce all of the bird species mentioned by Shakespeare into North America, imported them into Brooklyn, New York. Within 70 years, they had spread across the continent **(Figure 55.8);** their population size is now estimated at 200 million. Starlings pose a serious

threat to native birds, including several woodpecker species, because they successfully compete with them for nesting sites in natural cavities in trees.

Introduced plants often transform entire ecosystems. One of the best-known examples is kudzu *(Pueraria lobata)*, a fast-growing species from Asia. In the early 1900s, it was widely planted in the southeastern United States as a source of animal feed. Later, a government agency planted it to stabilize soils and decrease erosion on deforested hillsides. But when kudzu has access to abundant nutrients and water, it can grow up to 30 cm *per day*. It spread quickly across the South, literally overgrowing almost all native plants **(Figure 55.9).**

FIGURE 55.9 Kudzu, the vine that ate the South. Kudzu *(Pueraria lobata)*, an introduced vine, grows so quickly that it often covers living trees or even abandoned buildings.

A. Woolly adelgids

B. Hemlocks killed by woolly adelgids

C. Eastern hemlock and woolly adelgid ranges

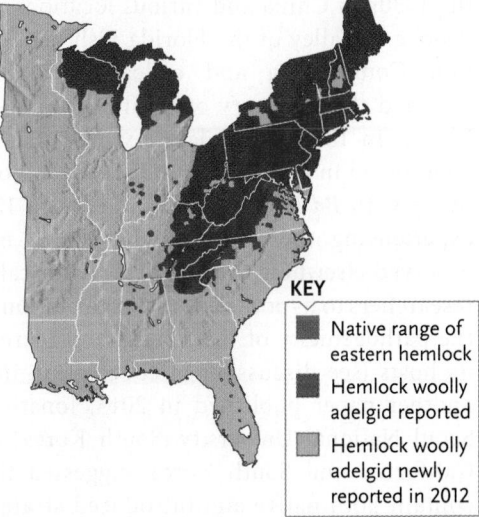

KEY

- Native range of eastern hemlock
- Hemlock woolly adelgid reported
- Hemlock woolly adelgid newly reported in 2012

FIGURE 55.10 Hemlock woolly adelgid. (A) The aphidlike woolly adelgid *(Adelges tsugae)* feeds on the sap of **(B)** eastern hemlocks *(Tsuga canadensis)*, often killing the trees. This insect pest is spreading **(C)** and may someday endanger hemlocks throughout their geographical range.
© Cengage Learning 2017

Invasive insects often become pests of agricultural crops and native plants. The hemlock woolly adelgid *(Adelges tsugae)* was accidentally introduced into North America from Asia. The adelgid kills eastern hemlocks *(Tsuga canadensis)* by feeding on their sap. It now threatens the trees from North Carolina to Massachusetts **(Figure 55.10)**. But adelgids endanger far more than these evergreen trees. Hemlocks buffer the physical conditions below them: hemlock stands are cool in summer and warm in winter, sustaining a unique community of organisms that includes ruffed grouse *(Bonasa umbellus)*, turkey *(Meleagris gallopavo)*, white-tailed deer *(Odocoileus virginianus)*, and snowshoe hare *(Lepus americanus)*. Infested stands rarely survive more than a few years, and the communities established under pure stands of eastern hemlock will likely become extinct because of feeding by the adelgid.

The Spread of Disease-Causing Organisms Endangers Many Species

Pathogenic viruses, bacteria, and fungi frequently decimate populations of the species they infect, especially when they are introduced into new environments. Because native species had no prior exposure to these pathogens, they never evolved resistance to them, leading to devastating outbreaks of disease. For example, amphibians have been in a worldwide decline since 1980. Nearly 40% of the roughly 7,000 described species are now threatened with extinction, and populations of more than half of all amphibian species are declining in numbers. The sudden change in the status of these animals has sparked an enormous research effort aimed at identifying the primary causal factors. Although climate change, habitat destruction, and pollution have undoubtedly taken a great toll, scientists now attribute many recent amphibian declines and extinctions to infection by the chytrid fungus *Batrachochytrium dendrobatidis,* known in the scientific literature as *Bd.* The fungus has been implicated in the extinction of 200 amphibian species, including frogs,

salamanders, and caecilians; more than 1,000 other species are known to suffer from *Bd* infections **(Figure 55.11)**.

This pathogenic fungus was first described in 1998 by researchers investigating skin infections in amphibians from North America, Central America, and Australia. Scientists quickly learned that *Bd* is very strange indeed: it is the only species in its group known to infect vertebrates; it infects only amphibians, feeding on keratin in their skin and in the mouthparts of their tadpoles; and even though it has an aquatic life cycle, it can infect fully terrestrial amphibians that never enter standing water. In 2013, Louise A. Rollins-Smith and her colleagues at Vanderbilt University discovered that *Bd* impairs lymphocyte production in amphibians, thereby disrupting their immune response to the infection. Moreover, Sarah J., Sapsford and her colleagues at James Cook University, Australia, reported, also in 2013, that this pathogen grows well at cool temperatures and that it is most prevalent in aquatic habitats that are interconnected. Thus, the scientists suggest that the infection may travel from high elevation to low elevation habitats via streamflow.

Since 2004, researchers have been screening amphibian specimens in zoological museums for *Bd* DNA. The oldest

FIGURE 55.11 Chytrid fungus infection. These mountain yellow-legged frogs *(Rana muscosa)* from King's Canyon National Park in the southern Sierra Nevada of California succumbed to chytrid fungus infection.

CHAPTER 55 BIODIVERSITY AND CONSERVATION BIOLOGY | **1317**

known infections dated to 1894 in Brazil, 1911 in Korea, and the 1930s in China and various locations in Africa. In 2015, Brooke L. Talley of the Florida Fish and Wildlife Conservation Commission and colleagues at other institutions reported the discovery of *Bd* infection in frogs collected in Illinois in 1888. Their data revealed that the infection was widespread in Illinois by 1900. Thus, Illinois frogs have been living with *Bd* infections for more than 126 years—without experiencing the devastating levels of mortality that are observed elsewhere today. These historical results prompted researchers to hypothesize that coevolution may have reduced the pathogenicity of the fungus and increased resistance of its hosts (see discussion of coevolution in Section 53.1). In another paper published in 2015, Jonathan J. Fong of the Seoul National University, South Korea, and colleagues in California and South Korea suggested that many regions contain both native and introduced strains of *Bd*; amphibians exposed for the first time to a novel strain may succumb to the infection, but those exposed to native strains may be somewhat resistant.

Why did *Bd* start to devastate amphibian populations only in the 1980s? Some researchers suggest that even small increases in temperature and related changes in cloud cover and humidity—all the result of global climate warming—have favored the growth of the pathogenic fungus in some habitats with high amphibian diversity. Other researchers argue that climate warming and pollution stress amphibians, making them more susceptible to infection by many pathogens.

Biologists are working feverishly to learn more about *Bd* and its role in amphibian declines before a majority of amphibian species become extinct. In a broader—and even more frightening—context, ecologists who study the dynamics of disease in natural population are just beginning to grapple with the likely effects of climate change and other consequences of human activity on the spread and success of pathogenic organisms, including those that infect humans.

Human Activities Are Causing a Dramatic Increase in Extinction Rates

As you may remember from Section 23.4, extinction has been common in the history of life; roughly 10% of the species alive at any time in the past became extinct within 1 million years. These *background extinction rates* eliminated perhaps seven or eight species per year. Paleobiologists have also documented six *mass extinctions,* during which extinction rates increased greatly above the background rate for short periods of geological time (see Figure 23.14).

At present, Earth appears to be experiencing the greatest mass extinction of all time. According to Edward O. Wilson of Harvard University, extinction rates today may be 1,000 times the historical background rate, meaning that thousands of species are being driven to extinction each year. The vast majority of extinctions are a direct result of the destructive human activities discussed previously.

If humans are causing the current mass extinction, why didn't it begin long ago? The answer lies in our increased rate of population growth (see Section 52.6). During the nineteenth and twentieth centuries, improvements in food production, sanitation, and health care increased human life expectancy. Our ever-increasing population consumes resources and produces wastes at an escalating rate. And until we change the way we live in relation to the environment that we share with all other species, our negative impact will grow along with our global population.

STUDY BREAK 55.2

1. What environmental factor has caused the demise of vulture populations throughout South Asia?
2. How do extinction rates today compare with the background extinction rate evident in the fossil record?

THINK OUTSIDE THE BOOK

Search the Internet for updated information about *Bd* infections to amphibian populations. Does the infection appear to be spreading, both in terms of the species affected and their geographic locations?

55.3 Ecosystem Services That Biodiversity Provides

Given the many ways in which human activities are fostering biodiversity losses, we should reflect on the present and future value of biodiversity and contemplate why we might want to preserve it. Ecologists and environmentalists value biodiversity for three types of **ecosystem services** they provide: provisioning services, regulating and support services, and cultural services.

Provisioning Services Benefit Humans Directly

Most ecosystems provide *provisioning services*, including the availability of plants, animals, and materials that are useful to humans and, indeed, all other organisms. Even today, scientists search for natural products that might provide us with better medicines, food, and other useful materials. The development of a new medicine often begins when a scientist analyzes a traditional folk remedy or screens naturally occurring compounds for curative properties. Chemists then isolate and purify the active ingredient and devise a way to synthesize it in the laboratory. More than half of the 150 most commonly prescribed drugs were developed from natural products in this manner.

For example, *Taxol,* a drug treatment for breast and ovarian cancer, was isolated from the narrow strip of vascular cambium beneath the bark of the Pacific yew tree, *Taxus brevifolia* **(Figure 55.12).** Unfortunately, a fully grown, 100-year-old tree produces only a tiny amount of Taxol, and six trees must be destroyed to extract enough to treat one patient. Pacific yew trees are not abundant, and they grow slowly. Harvesting them for Taxol extraction could quickly lead to their extinction—and

FIGURE 55.12 The Pacific yew tree. The slow-growing Pacific yew *(Taxus brevifolia)* is the original source of Taxol, a compound that effectively fights several cancers.

an end to the natural source of this life-saving compound. However, after much research, scientists now synthesize this widely used drug in the laboratory.

Although humans have practiced agriculture for more than 10,000 years, wild plants and animals still serve as sources of genetic traits that may improve agricultural crops and domesticated livestock. For example, corn *(Zea mays)* is an *annual* plant (see Section 33.1). Its cultivation requires yearly planting, a laborious activity that leads to the erosion of topsoil. Farmers would rather grow a perennial strain of corn, one that would produce grain for years after a single planting. In 1978, botanists discovered teosinte *(Zea diploperennis)* a perennial plant closely related to corn, in the mountains of western Mexico. Researchers crossed the two species, producing a *perennial* corn (see Section 33.1). If they can increase the yield of this hybrid, it may prove to be an economically valuable crop.

Today, many agricultural researchers use genetic engineering, the transfer of selected genes from one species into another (see Section 18.2), to alter crop plants more precisely than they can using hybridization. The transferred genes may increase resistance to pests or environmental stress, promote faster growth, or increase shelf life after harvesting. However, many scientists and environmentalists fear that genetically modified crops may create environmental hazards that will inadvertently endanger biodiversity. For example, a genetically modified plant or animal that escaped into a natural habitat might compete with naturally occurring species. Or a genetically modified plant might poison harmless animals, as well as insect pests.

Regulating and Support Services Benefit All Forms of Life

Humans and other species derive indirect benefits from a variety of ecosystem processes that depend on biodiversity. In the long run, these *regulating services* and *support services* may be even more valuable to humans than the direct benefits of provisioning services.

Regulating services are critical functions that preserve and recycle the resources upon which ecosystems depend, including soil formation and maintenance, air and water purification, and flood control. Recall, for example, how the presence of vegetation retains water and soil nutrients in forest ecosystems (as described in *Focus on Research: Basic Research* in Chapter 54), maintaining their viability. Photosynthetic organisms can also mitigate the effects of rising atmospheric CO_2 levels. As you learned in Section 54.4, the combustion of fossil fuels releases CO_2 and other gaseous wastes into the atmosphere, increasing the greenhouse effect and fostering global warming. Photosynthetic organisms absorb CO_2 and use it for essential metabolic processes; thus, forests and communities of marine phytoplankton withdraw CO_2 from the atmosphere, transferring it to a different nutrient compartment in the biosphere (see Figure 54.3), a phenomenon called *carbon sequestration*. Recent research indicates that carbon sequestration by these organisms will be increasingly important if we hope to limit the damage caused by fossil fuel combustion.

Support services are ecosystem functions that enable populations of organisms to function within their food webs. For example, fungi and soil microorganisms are critical for the maintenance of biogeochemical cycles because they decompose organic wastes, reducing them to molecules that primary producers can assimilate and use (see Chapter 54). Similarly, insect pollinators are responsible for fruit and seed production by many flowering plant species, including important agricultural crops. In a study published in 2012, Nicholas W. Calderone of Cornell University estimated that, in 2010, honeybees *(Apis mellifera)* pollinated more than $19 billion worth of crops in the United States. Recent declines in bee populations—largely caused by the overuse of chemical pesticides, as well as mite infestations and habitat fragmentation—endanger agricultural ecosystems and productivity. With the extinction of natural honeybee populations, farmers must now rent hives from beekeepers when their crops are flowering; unfortunately, healthy hives of domesticated bees are also in decline because of the stresses noted above.

Cultural Services Reflect Biodiversity's Intrinsic Worth

Some ethicists argue that we should preserve biodiversity because it has intrinsic worth, independent of its direct or indirect value to humans. They note that humans are just one species among millions in the remarkable network of life. Countering this position is the view that our immediate needs should always rank above those of other species and that we should use them to maximize our own welfare. The latter view inevitably leads to the disruption of natural environments and the loss of biodiversity. Framed in this way, the debate lies more within the realms of philosophy and public policy than biology.

Nevertheless, many people feel an emotional or spiritual connection to natural landscapes and the plants and animals they harbor. Moreover, nature provides us with recreational opportunities, such as hiking and snorkeling, that are not available in the highly disturbed environments where many of us live. Thus, biodiversity enhances human existence in intangible ways.

STUDY BREAK 55.3

1. How does biodiversity serve as a storehouse of genetic information that is potentially useful to humans?
2. What ecosystem services do naturally occurring organisms provide to humans?

55.4 Which Species and Ecosystems Are Most Threatened by Human Activities?

To slow the current rate of extinction and loss of biodiversity, conservation biologists must first identify which species are most likely to become extinct. They must also know where

remaining populations of those species live so they can direct conservation efforts to specific geographical locations.

The IUCN Maintains *The Red List of Threatened Species*

The International Union for the Conservation of Nature (IUCN) is a global network of government agencies and non-profit organizations that monitors the status of species and ecosystems worldwide. In addition to sponsoring scientific research, organizing conferences about environmental challenges, and working to find practical solutions to those problems, the organization maintains *The IUCN Red List of Threatened Species,* the catalogue of species that are threatened or **endangered** (that is, in danger of extinction throughout all or a significant portion of their ranges). Government agencies throughout the world rely on data in the Red List to establish policies and enact laws that protect threatened and endangered species. The Red List categorizes higher taxa according to the status of the species they comprise; three categories in the listing represent "threatened" status: vulnerable, endangered, and critically endangered **(Figure 55.13).**

Conservation Biologists Identified Biodiversity Hotspots

If we are to preserve biodiversity generally, we must know how biodiversity is distributed. Although species richness within communities generally increases from the poles to the tropics

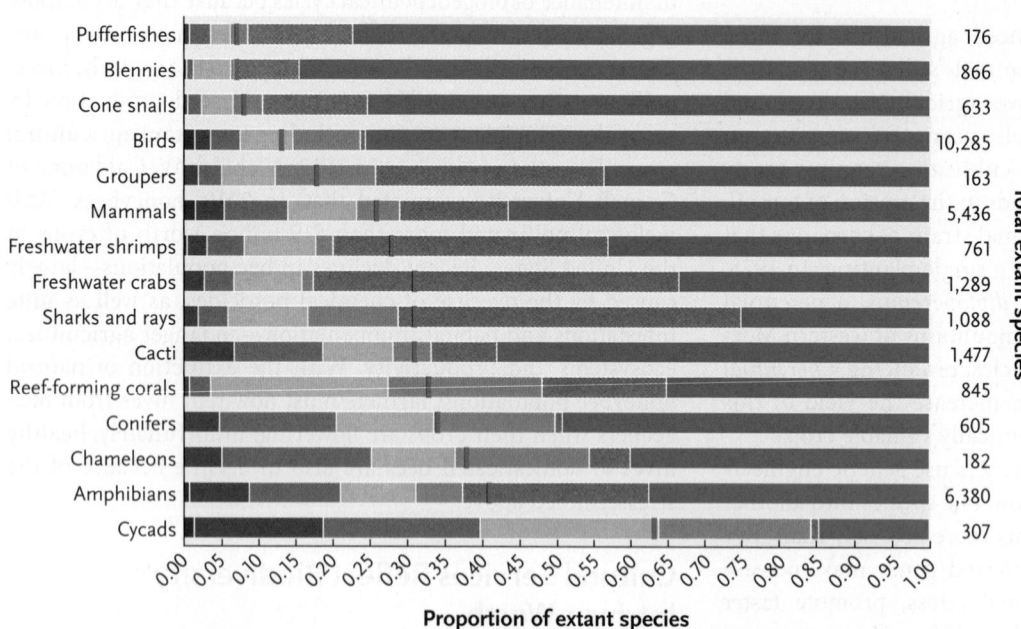

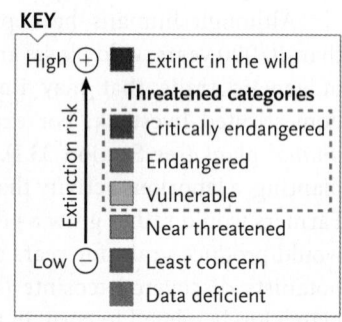

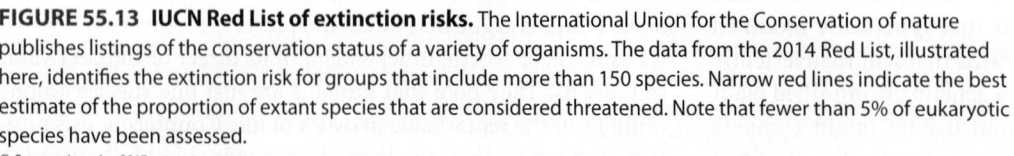

FIGURE 55.13 IUCN Red List of extinction risks. The International Union for the Conservation of nature publishes listings of the conservation status of a variety of organisms. The data from the 2014 Red List, illustrated here, identifies the extinction risk for groups that include more than 150 species. Narrow red lines indicate the best estimate of the proportion of extant species that are considered threatened. Note that fewer than 5% of eukaryotic species have been assessed.

© Cengage Learning 2017

(see Section 53.7), broad latitudinal surveys do not provide enough detail to be useful in this effort.

In a survey published in 2000, Norman Myers of Oxford University and his colleagues in England and the United States pinpointed 25 **biodiversity hotspots,** areas where biodiversity is both concentrated and endangered by human encroachment. To qualify as a biodiversity hotspot under Myers' criteria, an area must harbor at least 1,500 **endemic** plant species (those that are found nowhere else), and it must have already lost at least 70% of its natural vegetation. As human activity in natural habitats has increased, the number of recognized terrestrial biodiversity hotspots has now grown to 34, although some workers identify more than 60.

Myers used the number of endemic species as a criterion for identifying hotspots because endemics tend to have highly specific habitat or dietary requirements, low dispersal ability, and restricted geographical distributions. Indeed, locally distributed species account for much of Earth's biodiversity; and if the local habitats where these species occur are at risk of development, the species are also at risk. Although the 25 original hotspots occupy only 1.4% of Earth's land surface, they include the only remaining habitat for approximately 45% of all terrestrial plant species and 35% of all terrestrial vertebrate species.

Researchers Now Pinpoint Sites Where Extinctions Are Imminent

The identification of biodiversity hotspots tells us where biodiversity is both concentrated and threatened, but most of these areas are large and heavily populated. Conservation biologists need even more detailed information to identify specific localities where their efforts will have the greatest impact.

Building on Myers' pioneering study, a team of 30 collaborators in the United States, Australia, and the United Kingdom, pinpointed sites where extinctions are imminent. In a paper published in 2005, they identified 595 locations in tropical forests, on islands, or in mountainous regions where 794 *trigger species*—highly endangered species of mammals, birds, reptiles, amphibians, and coniferous trees—are each confined to a single site **(Figure 55.14).**

The researchers used strict criteria for including a site in the list. First, it must harbor at least one species that has been officially designated as endangered by the World Conservation Union, an international organization. Second, it must contain at least 95% of the world population of that species. Third, it must have clearly definable boundaries within which habitats are distinct from those outside the boundary; examples of such bounded

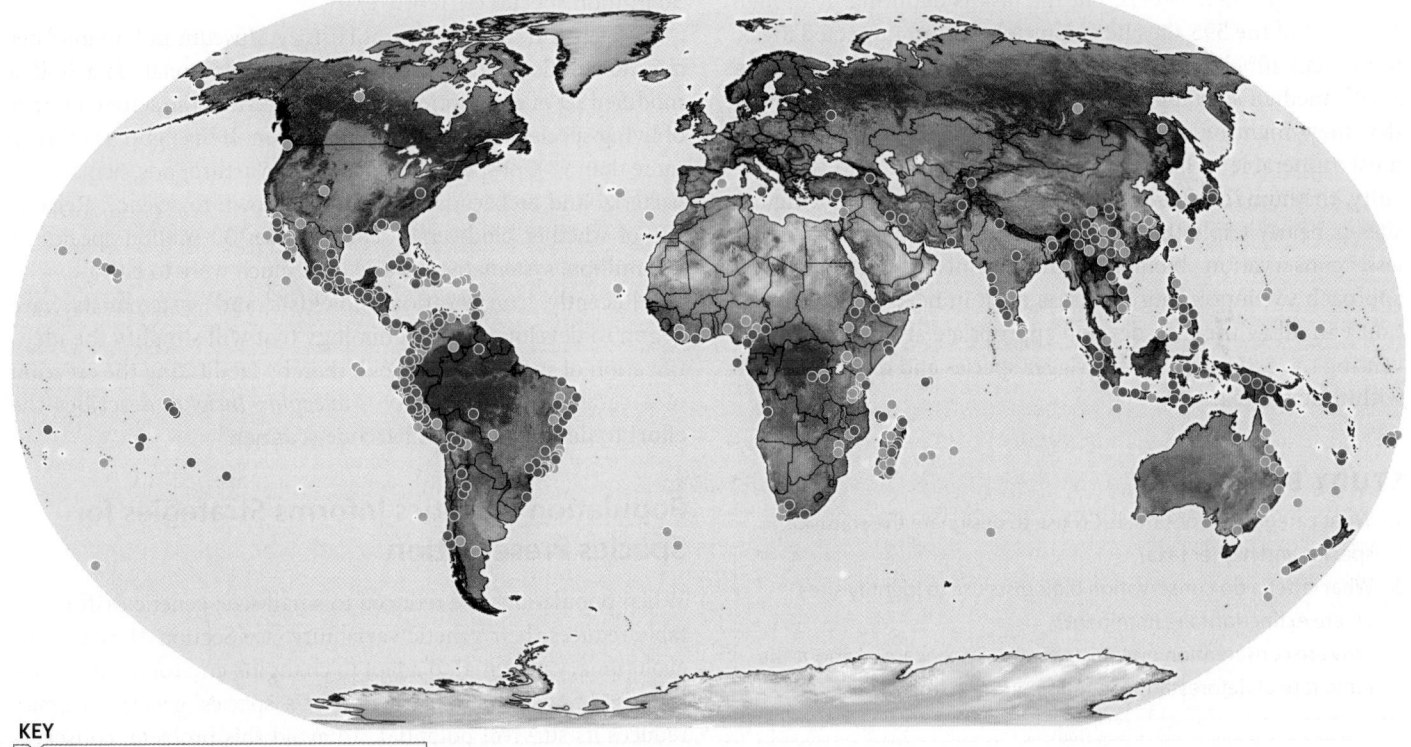

KEY
- Unprotected or protection status unknown
- Completely or partially protected

FIGURE 55.14 Pinpointing imminent extinctions. Taylor Ricketts and his colleagues identified 595 localities worldwide where at least one species of vertebrate or coniferous tree is in imminent danger of extinction. Sites marked in yellow are fully or partially protected. Those marked in red are unprotected or their protection status is unknown. The researchers mapped red localities over yellow ones to highlight areas in need of protection.

© Cengage Learning 2017

habitats include lakes, mountaintops, and forest fragments. The boundaries of the site thus define the area to be conserved. Although the 595 sites of imminent extinction are included within the biodiversity hotspots that Myers identified, the new approach has the practical advantage of pinpointing localized sites where conservation biologists can focus their efforts.

The research team also noted that 794 trigger species are in danger of imminent extinction, compared with 245 species from the same taxonomic groups that are known to have become extinct in the last 500 years. Thus, the rate of extinction in these groups of organisms is accelerating rapidly. Their analysis also reveals that the proportion of extinctions in mainland habitats (as opposed to islands) is also growing: only 20% of historical extinctions occurred in low-lying mainland areas, whereas more than 60% of trigger species live in mainland habitats today. Moreover, their study detected a taxonomic shift in extinction: 53% of historical extinctions were of bird species, but 51% of the trigger species are amphibians. Finally, the data reveal a geographic shift in extinctions: whereas only 21% of historical extinctions were in the New World tropics, 50% of the trigger species live in Central America, South America, and on Caribbean islands. These results indicate that species living in the New World tropics, especially in wet forests, are in the greatest danger.

The data from the new analysis allow conservation biologists to target their efforts, but the task is daunting. Although one third of the 595 sites lie completely within protected areas, more than 40% lack any protection at all. Most of the sites are small (median size approximately 12,000 hectares), suggesting that they might be easy to protect, but small sites are also the most vulnerable to human encroachment. Adding to the difficulty, the human population density in areas surrounding the sites is nearly triple the average density worldwide. Nevertheless, conservation biologists are optimistic that this new approach to pinpointing the areas most in need of their attention will allow them to develop appropriate strategies for preventing the extinction of the trigger species and others that live within these areas.

STUDY BREAK 55.4

1. What categories does the IUCN use to designate the status of species and higher taxa?
2. What criteria do conservation biologists use to identify sites where extinctions are imminent?
3. Why are conservation biologists especially concerned about the rapid rate of deforestation in the New World tropics?

55.5 Conservation Biology: Principles and Theory

Conservation biology is an interdisciplinary science that focuses on the maintenance and preservation of biodiversity. In this section we describe how conservation biologists use theoretical concepts from systematics, population genetics, behavior, and ecology to develop ways to protect habitats and the endangered species that live within them. We introduce practical applications of conservation theory in the next section.

Systematics Organizes Our Knowledge of the Biological World

To develop a conservation plan for any habitat, scientists must start with an inventory of its species. Their primary tool is systematics, the branch of biology that discovers, describes, and organizes our knowledge of biodiversity (see Chapter 24). Cataloguing the diversity of life may be the most daunting task that biologists face. After more than 200 years of work, systematists have described and named approximately 1.6 million species. However, they realize that this number represents only a fraction of the total.

In 1982, Terry Erwin of the Smithsonian Institution studied beetle biodiversity at the Tambopata National Reserve in Southern Peru. He sprayed biodegradable insecticide into the canopy of one large tree and collected 15,869 individual beetles, which he sorted into 3,429 species. More than 90% of the individual beetles he collected belonged to species that had not yet been described. Erwin used this astounding result and a complex mathematical model to predict that approximately 30 million species currently exist.

Nigel Stork of the Natural History Museum in London later questioned Erwin's conclusions. Using additional data and a modified set of assumptions, he estimated that the actual number of living species was closer to 100 million. If his figure is correct, more than 98% of species—most of them arthropods, nematodes, bacteria, and archaeans—are still unknown to science. Regardless of whether biodiversity encompasses 30 million species or 100 million, systematists clearly have much work to do.

Recently, conservation biologists and systematists have begun to develop a new technology that will simplify the identification of species in the field, thereby facilitating the creation of a catalog of biodiversity. *Molecular Insights* describes the effort to develop a "DNA barcode scanner."

Population Genetics Informs Strategies for Species Preservation

When populations are reduced to small size, genetic drift inevitably reduces their genetic variability (see Section 21.3) and the evolutionary potential to adapt to changing environments. Thus, the loss of even a small fraction of a species' genetic diversity reduces its survival potential. To avoid this problem, conservationists strive not only to increase the population sizes of threatened or endangered species but also to maintain or increase their genetic variation, both within and between populations.

For example, the whooping crane (*Grus americana*) was once an abundant bird in wet grassland environments through much of central North America **(Figure 55.15)**. By the early 1940s, excessive hunting and habitat destruction had caused their numbers to decline to just 21 individuals in two isolated populations.

Molecular Insights

Developing a DNA Barcode System

Everyone is familiar with the checkout scanners in stores: the cashier quickly passes an item's barcode over the scanner, and the register identifies it and records its price. The system works because the barcode on every item contains unique identifying information.

Research Question

In 2003, Paul Hebert, a population geneticist at the University of Guelph, Ontario, Canada, and his colleagues proposed an analogous method, called DNA barcoding, for identifying animal species quickly and accurately. The researchers envisioned using a handheld device to rapidly analyze DNA in the field; the resulting data would be sent to a database by cell phone, and minutes later an identification and a description of the species would appear on the instrument's screen.

Experiments

Hebert proposed using the first part of the *COI* (cytochrome oxidase 1) gene—a sequence of about 500 nucleotides—as the DNA barcode for animals because this mitochondrial gene varies greatly between animal species. Moreover, it appears to have no inserted or deleted DNA segments in most animals, making the alignment and comparison of sequences straightforward. Hebert's hope is that any *COI* gene sequence obtained in the field will provide a unique identifier for the species from which the DNA sample was obtained.

Results

Hebert's early tests of his barcode approach were promising and caught the attention of other researchers. For example, he and his collaborators first analyzed the *COI* gene sequence in the skipper butterflies of Costa Rica. Although adult skippers look pretty much alike, their caterpillars vary in appearance and in their food plant preferences, leading researchers to wonder if butterflies that had been assigned to one species (*Astraptes fulgerator*) might actually

FIGURE Visualizations of the *COI* gene sequence used in the barcodes of five fish species.
© Cengage Learning 2017

represent several. Analyses of the *COI* gene sequence sampled from 484 adults allowed Hebert and his colleagues to identify 10 distinctive DNA barcodes, suggesting that there are at least 10 species of skipper in Costa Rica rather than just one.

In early 2007, Hebert and his colleagues reported that they had used the DNA barcode to analyze 2,500 specimens of 643 North American bird species. The results were impressive: barcode differences between species were an order of magnitude greater than the differences within species, allowing the unambiguous identification of species from a short DNA sequence. Interestingly, the barcode analysis identified 15 probable new species that had not been previously identified and revealed that 8 supposed species of gull may be variants of just one species.

Plant researchers quickly noted that the *COI* gene sequence would not work as a DNA barcode for plants because of its much slower rate of evolution in plants. However, in 2009, a large group of plant scientists proposed using a combination of two chloroplast genes in DNA barcoding analyses.

While the handheld analytical device is not yet ready for use in the field, Hebert's idea caught on quickly. In 2004 a consortium of major natural history museums and herbariums started the Barcode of Life Initiative, with the goal of creating a database of

DNA barcodes linked to specimens already identified in their collections. The approach potentially could replace the traditional methods of systematic analysis using organismal and genetic characters to identify species. Researchers have also developed a method for visualizing differences in the *COI* gene sequence among species **(Figure)**.

Conclusion

Taken together, the results to date support using DNA barcodes, and using the *COI* gene sequence specifically for the barcode analysis, as a means of identifying animal species. Many researchers have joined the effort, and the Consortium for the Barcode of Life (http://barcodeoflife.org) now includes hundreds of member organizations around the world. They enter the results of their research into the Barcode of Life Data System (www.boldsystems.org), which, as of early 2015, included nearly four million entries for more than 240,000 described species.

think like a scientist

How might DNA barcoding be useful in the effort to track the spread of Asian carp through North American rivers and lakes?

Source: P. D. N. Hebert et al. 2003. Biological identifications through DNA barcodes. *Proceedings of the Royal Society B* 270:313–321.

© Cengage Learning 2017

This population bottleneck and the resultant loss of genetic variability apparently contributed to developmental deformities of the spine and trachea that had not been seen previously.

During the 1970s, biologists began an aggressive conservation program. In addition to preserving habitats in the crane's summer and winter ranges, they initiated a carefully controlled captive breeding program designed to minimize the effects of inbreeding. Although more than 300 whooping cranes now survive in several wild and captive populations, recent research reveals that they still have a remarkably low level of genetic

FIGURE 55.15 Whooping cranes. Endangered whooping cranes *(Grus americana)* winter in the Aransas National Wildlife Refuge in Corpus Christi, Texas.

variability. As expected, the genetic effects of a severe population bottleneck may persist long after a population begins to increase in size.

Studies of Population Ecology and Behavior Are Essential to Conservation Plans

Conservation programs for animals also require data about target species' ecology and behavior, including their feeding habits, movement patterns, and rates of reproduction.

Sea otters *(Enhydra lutris)* are predatory marine mammals that live along the coastline of the North Pacific Ocean. In the early 1700s, they numbered approximately 300,000 individuals **(Figure 55.16),** but commercial hunting reduced their numbers to about 3,000 individuals by the start of the twentieth century. Sea otters are keystone predators (see Section 53.4), and the destruction of sea otter populations had profound, cascading effects on the communities in which they lived (see Section 54.2). As the numbers of sea otters plummeted, populations of sea urchins, one of their favored prey, exploded; burgeoning sea urchin populations decimated local kelp beds, disrupting the communities of animals that live and breed among these giant algae.

International treaties ended nearly all hunting of sea otters in 1911, and the populations subsequently recovered to about one third of their original levels. Conservation biologists facilitated the recovery by reintroducing otters into southeastern Alaska, British Columbia, Washington, and California. Before deciding where otters should be reintroduced, scientists had to assess the resources available at different sites and determine how far individual otters would move, how rapidly they would reproduce, and how quickly their populations would spread. The reintroduction effort was successful at first. However, populations in California have experienced high mortality since the mid-1990s, and nearly half of those dying have been adults in their reproductive prime. Researchers have identified parasitic infections and heart disease as leading causes of death, suggesting that some coastal environments are so badly degraded that they may no longer support populations of this species.

Using complex mathematical models, conservation biologists often conduct a **population viability analysis** (PVA) to determine how large a population must be to ensure its long-term survival. PVAs evaluate phenomena that may influence

A. Sea otter

B. Geographical range of sea otters

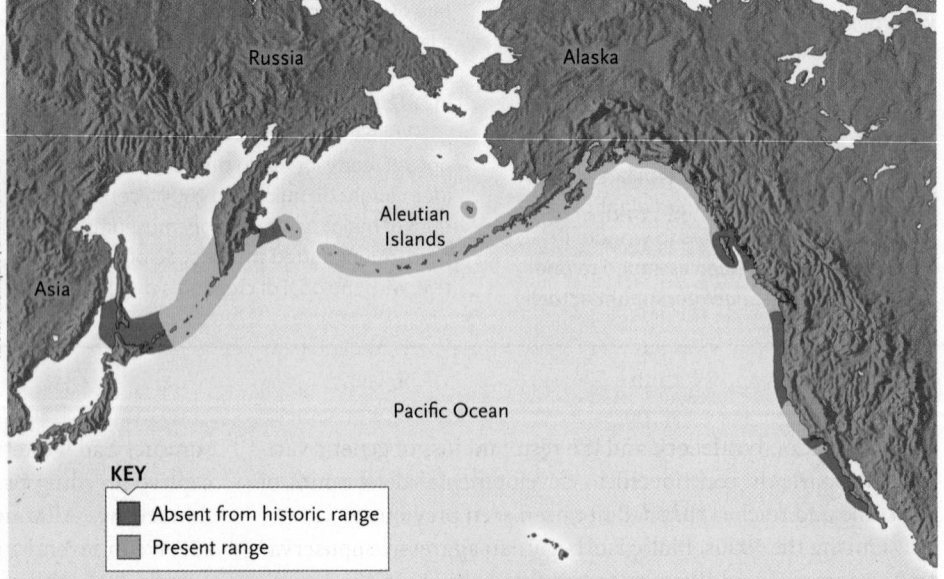

FIGURE 55.16 Sea otters. After being hunted nearly to extinction, **(A)** sea otters *(Enhydra lutris)* have been reintroduced in many parts of their historical range **(B).**

© Cengage Learning 2017

Applied Research: Preserving the Yellow-Bellied Glider

Predicting the future is never easy, especially the future of a threatened species. But population viability analysis (PVA) allows conservation biologists to predict how a species will fare under a range of possible scenarios. An effective PVA for an animal species requires detailed information about its diet, predators, mating habits, habitat preferences, space requirements, demography, geographical distribution, responses to climatic fluctuations and human disturbances, and a host of other aspects of its biology.

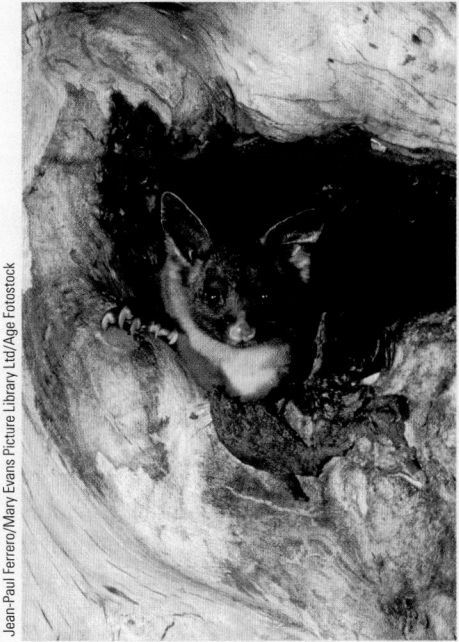

Jean-Paul Ferrero/Mary Evans Picture Library Ltd/Age Fotostock

The Australian yellow-bellied glider (*Petaurus australis*) provides an example of how PVA is essential for a conservation effort. This marsupial, about the size of a squirrel, lives in small family groups in undisturbed *Eucalyptus* forests along Australia's eastern coast. Each glider family maintains a home range (the area it uses for feeding and other activities) of 25 to 85 hectares; the home ranges of neighboring families do not overlap. As a result, the population density of gliders has never been high. But glider populations have declined precipitously as forests have been cleared, and the species is now considered threatened.

Using data from nearly 20 published papers, two Australian conservation biologists, Russ Goldingay of the University of Wollongong and Hugh Possingham of The University of Adelaide, conducted a PVA for this species. They estimated age distributions in glider populations, as well as survival probabilities, litter sizes, sex ratios, lifespan, and home range sizes. They analyzed these data using a mathematical model that predicts the viability for populations of various sizes. In most PVAs, a population is considered viable if it has a 95% probability of surviving for 100 years. Goldingay and Possingham introduced additional complexity to their analysis by assessing the effects of unpredictable environmental events, such as drought, on breeding success. They also conducted sensitivity analyses to examine how changing the values of specific parameters—such as litter size, mortality rates of the different

age classes, or the frequency and severity of droughts—might influence the general predictions of the viability model.

After many thousands of these calculations, the researchers concluded that a viable population of gliders would require at least 150 family groups. They also suggested that a population of that size would need approximately 18,000 hectares (roughly 70 square miles). Currently, only one of the 15 existing conservation reserves is that large.

Regional governments in Australia listed the yellow-bellied glider as a protected species after the PVA was completed, and they have adopted more stringent forest management practices. Glider populations have rebounded in subsequent years, although isolated populations are still at risk. Their recovery has prompted a change in the IUCN Red List listing from "near threatened" to "least concern."

As a result of this PVA, conservation biologists were able to determine which of the remaining forest tracts are large enough to sustain a yellow-bellied glider population. Thus, they now know where to concentrate their limited resources to secure the future survival of this species. Although predicting the future is difficult, PVAs allow conservation biologists to make accurate and reliable recommendations for selective transplants that will contribute to the conservation of threatened species.

Source: R. Goldingay and H. Possingham. 1995. Area requirements for viable populations of the Australian gliding marsupial *Petaurus australis*. *Biological Conservation* 73:161–167.

the longevity of the population or species: habitat suitability, the likelihood of catastrophic events, and other factors that may cause fluctuations in demographics, population size, or genetic variability. When conducting a PVA, researchers must decide what level of risk is acceptable for a given survival time. For example, should a conservation plan attempt to ensure a 95% probability that the species will survive for 100 years, or should it specify a 99% survival probability? An increase in either the survival probability or the survival time requires an increase in the size of the population that must be conserved. The **minimum viable population size** identifies the smallest population that fits the specifications of the conservation plan. *Focus on Research: Applied Research* describes how biologists used PVA in the conservation of an Australian marsupial, the yellow-bellied glider (*Petaurus australis*).

Principles of Community Ecology and Landscape Ecology Guide Large-Scale Preservation Projects

Many conservation efforts focus on the preservation of entire communities or ecosystems. These projects often depend on the work of community and landscape ecologists.

SPECIES–AREA RELATIONSHIPS As you learned in Chapter 53, community composition is dynamic: some species become extinct and others join the community through immigration. If we view patches of intact habitat as islands in a sea of unsuitable terrain, we can apply the predictions of the theory of island biogeography (see Section 53.7) to the design of protected areas. For example, we might expect that the number of

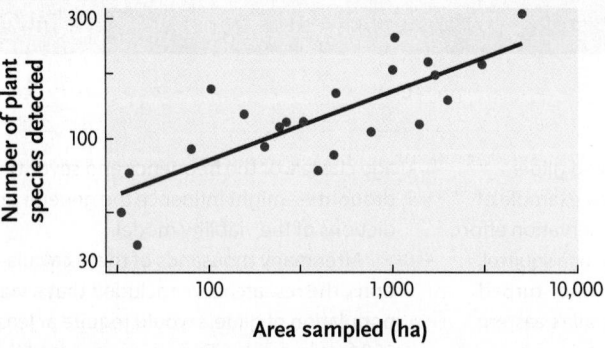

FIGURE 55.17 The species–area relationship. Data on plant distributions in nature preserves in Western Australia illustrate the relationship between the size of the area sampled and the number of species documented as present. Note that both axes are plotted on a logarithmic scale.

© Cengage Learning 2017

species a patch will support depends on its size and proximity to larger patches.

Indeed, ecologists recognized long ago that large habitat patches sustain more species than small patches do **(Figure 55.17)**. When plotted on a logarithmic or semi-logarithmic scale, the species richness of a community increases steadily with the area sampled. In other words, for relatively small habitat patches, even minor increases in area allow a large increase in the number of resident species; nevertheless, as habitat patches get larger, the number of species present must eventually level off (once all species in the area are recorded as present). You encountered an example of the species–area relationship in our discussion of bird species richness on islands of different sizes (see Figure 53.32B).

As habitats become increasingly fragmented, edge effects exaggerate the species–area relationship in mainland habitat patches **(Figure 55.18)**. Consider two hypothetical patches of habitat: one is 100 m on a side, with a total area of 10,000 m²; the other is 200 m on a side, with a total area of 40,000 m².

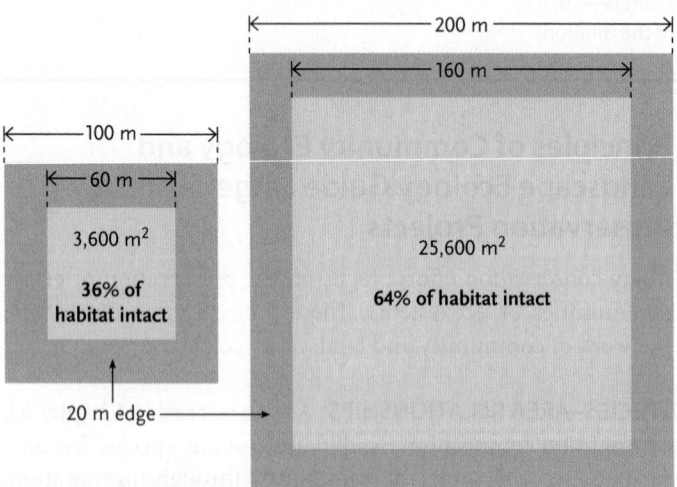

FIGURE 55.18 Edge effects and patch size. This hypothetical example illustrates how a 20-m-wide edge disrupts a larger fraction of a small habitat patch than a large habitat patch.

© Cengage Learning 2017

Now, imagine that edge-effect disturbances penetrate 20 m into each patch from all directions. The small patch contains only 3,600 m² of intact habitat, but the large patch contains 25,600 m² of intact habitat. Although the large patch is only four times larger than the small patch, the large patch contains more than seven times as much *intact* habitat.

LANDSCAPE ECOLOGY Conservation biologists often use **landscape ecology** to design the size and geometry of nature reserves and other protected areas. Landscape ecology analyzes how large-scale ecological factors—such as the distribution of vegetation, topography, and human activity—influence local populations and communities.

When conservation biologists first applied concepts from landscape ecology to the design of protected areas, they debated whether nature preserves should comprise one large habitat patch or several smaller patches. Based on the species–area relationship, large patches should harbor more species than small patches; large patches would also experience proportionately smaller edge effects; and large patches would better support populations of large animals that need substantial resources. Nevertheless, some conservation biologists argued that clusters of physically separate preserves are more effective in maintaining metapopulations of endangered species (see Section 52.5), especially if the patches are interconnected by corridors of intact habitat. Individuals could move between preserves, reviving any local populations that experience a decline.

Ellen I. Damschen of North Carolina State University and several colleagues conducted an ambitious long-term field experiment to test the effects of landscape corridors on plant species richness **(Figure 55.19)**. Their results, published in 2006, suggest that habitat patches connected by corridors retain more native plant species than isolated patches do, and that corridors did not promote the entry of introduced (that is, nonnative) species. Thus, corridors appear to be a useful feature in the design of nature preserves.

Landscape corridors are a key feature of efforts to prevent the extinction of the Florida panther (*Puma concolor coryi*, shown in the chapter-opening photo). This subspecies is critically endangered: according to the U.S. Fish and Wildlife Service only 100 to 160 individuals remain from a much larger population that once ranged throughout the southeastern United States. Other panther subspecies still inhabit the western states. Panthers are large predators, and each female requires nearly 20,000 hectares (more than 75 square miles) for hunting and breeding; males each require more than twice as much space.

Although the state and federal governments have set aside several panther conservation areas in Florida, 52% of the habitat panthers occupy is privately owned, and most of it is highly fragmented. Panthers frequently cross roads, and most panther deaths in Florida are caused by accidents with motor vehicles. Protected landscape corridors might enable panthers to move more safely between conservation areas. A preliminary study

FIGURE 55.19 | **Experimental Research**

Effect of Landscape Corridors on Plant Species Richness in Habitat Fragments

Question: Do landscape corridors connecting habitat patches influence the species richness of native and nonnative plants within the habitat patches?

Experiment: Damschen and her colleagues studied changes in the community composition and species richness of the plants in open habitat patches within a longleaf pine *(Pinus palustris)* forest in South Carolina. Their experimental design included both isolated patches and patches that were connected to one another by a landscape corridor. All patches included the same land area, and their large size (1.375 ha each, including the landscape corridors) allowed the researchers to make a realistic assessment of the effects of landscape corridors. After creating the patches of open habitat within the forest in 2000, the researchers catalogued all plant species occurring in the patches through 2005, although they were unable to collect data in 2004.

Results: Over the course of the study, habitat patches that were connected by landscape corridors harbored increasingly more plant species than did unconnected habitat patches. The researchers also noted that the difference in species richness between the two experimental treatments was caused by a difference in the number of native plant species present. The number of nonnative species in connected and unconnected habitat patches was similar.

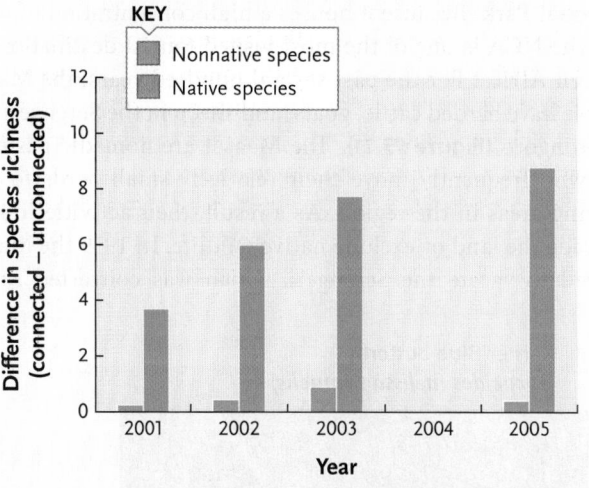

KEY
- Nonnative species
- Native species

Conclusion: Landscape corridors between patches of open habitat in longleaf pine forests increase the species richness of native species in open habitat patches, but they do not foster the entry of nonnative species.

think like a scientist

Observe
Hypothesize
Predict
Experiment
Interpret

Based on the results of this experiment, predict the effect that landscape corridors would have on the species richness of small animals (including insects) in patches of longleaf pine forests.

Source: E. I. Damschen et al. 2006. Corridors increase plant species richness at large scales. *Science* 313:1284–1286.

© Cengage Learning 2017

---

found that panthers already use such corridors, typically along wooded riverbanks, when they are available. The Florida Fish and Wildlife Service proposed the creation of an ambitious 2,500 hectare network of protected corridors alongside the Caloosahatchee River to link several significant habitat fragments in neighboring counties.

BETA-DIVERSITY Conservation biologists now often focus their efforts on preserving assemblages of organisms rather than individual species. As you know from Section 53.2, communities grade into one another as species composition changes across environmental gradients. That discussion focused on diversity *within* well-defined communities, a characteristic that ecologists identify as alpha diversity. But conservation biologists are increasingly interested in diversity *across* communities, which they call *beta-diversity*. Beta-diversity reflects the increasing numbers of species present in an area that includes a wide variety of habitats, vegetation types, and small-scale environments. The concept of beta-diversity is reflected in the slope of the species–area relationship: as the size of an area increases, so does the number of distinct environmental features it includes; and that environmental diversity supports a larger number of species.

By basing the design of nature preserves on the conservation of beta-diversity, conservation biologists can establish reserves that will protect more species. An ideal reserve system might include several large areas, each including a diversity of small-scale environments suitable for species that do not disperse readily, and many small reserves interconnected by landscape corridors. A reserve system distributed over an important environmental gradient would include species that replace each other across it. And as the global climate continues to warm over the coming decades, reserve systems that include protected areas at different elevations or latitudes might allow some species to migrate into cooler environments over time.

STUDY BREAK 55.5

1. How does a population bottleneck increase the likelihood that a species will become extinct?
2. How does a population viability analysis assist in the development of a conservation plan for a species?
3. Would a single large nature preserve or several small preserves experience greater edge effects?

55.6 Conservation Biology: Practical Strategies and Economic Tools

Conservation biology seeks to protect native species, communities, and ecosystems from the effects of human activity. Meeting that goal and reversing some of the existing damage requires the integration of biological research with economic and social realities.

Conservation Efforts Aim to Preserve, Conserve, and Restore Habitats

Conservation groups often highlight efforts to preserve individual animal species, such as the giant panda *(Ailuropoda melanoleuca)* or California condor *(Gymnogyps californianus)*. The preservation of "charismatic megavertebrates," as these large animals are sometimes described, attracts substantial public support. Nonetheless, there is little point in trying to preserve natural populations of individual species if their habitats are in jeopardy. An alternative to species-based conservation focuses on the preservation of intact habitats; individual species are conserved as a consequence of preserving the habitats on which they depend. Conservation biologists approach this goal with a continuum of approaches, which fall into three general categories: *preservation, mixed-use conservation,* and *restoration.*

CONSERVATION THROUGH PRESERVATION In many countries, habitats are preserved when an individual or organization purchases them and enforces strict standards of land use. In sensitive habitats, people may be excluded altogether; in other cases, access is restricted and the exploitation of resources is controlled. This approach works well in countries with efficient law enforcement and a tradition of private land ownership. In the United States, for example, the Nature Conservancy has purchased tracts of land to preserve native species in every state.

The preservation approach has been successful in preserving portions of the Pine Bush habitat near Albany, New York **(Figure 55.20).** This unique ecosystem arose approximately 11,000 years ago at the end of the last glacial period, when a massive deposit of sand was left near the western margin of Albany's current city limits. This sandy region formed an inland pine-barrens habitat in which pitch pine *(Pinus rigida),* scrub oak *(Quercus ilicifolia),* and dwarf chestnut oak *(Quercus*

prinoides) are now the dominant vegetation. The Pine Bush is home to more than 50 plant and animal species that the state and federal government list as threatened or endangered. The habitat itself was once vulnerable because it lies within Albany's city limits; however, since 1988, the Pine Bush has been jointly owned and protected by New York state, local municipalities, and the Nature Conservancy.

Although conservation through preservation is effective, the financial cost of such efforts is high. In 2012, Stuart H. M. Butchart of Birdlife International, an organization in Cambridge, England, working with colleagues from many countries, estimated that lowering the extinction risk category of all threatened birds on the IUCN Red List would cost roughly one billion dollars per year, approximately 10 times the current funding level. Extending protection and effective management to all terrestrial sites for a wider range of taxa would cost in excess of $75 billion per year. Clearly, governments across the globe will have to increase conservation funding dramatically to even begin to achieve these important goals.

MIXED-USE CONSERVATION When outright preservation is impractical, conservation biologists advocate mixed-use conservation, which combines the protection of some land parcels with the controlled development of others.

The Ngorongoro Conservation Area (NCA) in Tanzania provides an example of mixed-use conservation. The NCA covers 829,000 hectares of grassland and borders the Serengeti National Park. Because it houses a high concentration of wildlife, the NCA is one of the most visited tourist destinations in eastern Africa. For the past several hundred years, the Maasai people have herded cattle, goats, and sheep in the Serengeti and Ngorongoro **(Figure 55.21).** The Maasai are nomadic pastoralists who frequently move their relatively small herds to new grazing areas in the region. As a result, their activities do not degrade the land or exclude native wildlife. In 1959 the Maasai agreed to vacate the Serengeti, which was converted into a

A. Albany Pine Bush

B. Karner Blue butterfly
   (Lycaeides melissa samuelis)

FIGURE 55.20 The Albany Pine Bush habitat. (A) The Pine Bush lies entirely within the city limits of Albany, New York. It is home to about 50 threatened or endangered plant and animal species, including **(B)** the Karner Blue butterfly.

FIGURE 55.21 Mixed-use conservation. The Maasai use the Ngorongoro Conservation Area to graze cattle and goats.

national park, in return for retaining the rights to live and herd livestock within the NCA. The government of Tanzania helped create the necessary infrastructure within the NCA, including a constant water supply, as well as social services. Under this agreement, 40,000 indigenous residents, most of them Maasai, live in this large and valuable conservation area.

CONSERVATION THROUGH RESTORATION Conservation biologists sometimes create restoration plans to reestablish the vitality of a previously disrupted community or ecosystem. This effort requires the removal of contaminants, impediments to the natural flow of water, and barriers to animal movement, as well as the restoration of natural processes, such as periodic fires or floods. Most restoration projects also require replanting key plant communities and long-term management once restoration is complete.

Not all degraded habitats can be restored, and not all potential restoration projects are equally feasible. When making project decisions, restoration ecologists consider a number of factors: will the restored habitat be suitable for rare or endangered species, and will its creation increase endemic biodiversity? Would the restoration reunite previously fragmented land parcels? Will the restored habitat experience the periodic disturbances, such as fires or floods that are essential for its continued existence? What are the costs of implementing the plan and maintaining the area? Finally, would the restored land be valued by local residents, and will they support and maintain it?

A successful restoration project is currently underway in the Brazilian Atlantic Forest, sponsored by the Instituto de Pesquisas Ecológicas (IPÊ), a Brazilian nongovernmental organization. In western São Paulo state, near the Morro do Diabo state park, IPÊ is trying to recreate the natural Brazilian Atlantic Forest ecosystem by planting native trees in habitat corridors between remaining forest fragments. These corridors of native tree species should facilitate the preservation of species in those forest patches and supply valuable botanical resources for endemic wildlife and local residents.

Successful Conservation Plans Must Incorporate Economic Factors

Biologists can almost always develop a plan to conserve a species, community, or ecosystem. But to be successful, a plan must be economically feasible, and it must provide direct benefits to local residents whose lives it will affect.

LOCAL INVOLVEMENT Early conservation efforts simply set aside protected areas in which most human activities were banned. Local people were denied access to resources within the preserve—resources that were sometimes essential for their survival. Not surprisingly, these plans generated antipathy towards conservationists and the organisms they were trying to preserve.

For example, the northern spotted owl (*Strix occidentalis caurina*) lives only in old growth coniferous forests of the Pacific Northwest, where many local residents worked in forestry or supporting industries. The suggestion that the owl be listed as an endangered species triggered a bitter political battle between conservationists and local residents because the conservation plan for the owls required closing large tracts of forest to logging. Washington State listed the owl as an endangered species in 1988, but local residents, who lost jobs when logging was reduced, remain hostile to these conservation efforts.

Conservation plans are more successful if they provide local residents with benefits that depend on the existence of a preserve. Chitwan National Park provides an excellent example. For more than 100 years, this area, located in south central Nepal near the northern border of India, was recognized as a hunting ground for local royalty. These activities decimated local populations of large mammals, especially the Bengal tiger (*Panthera tigris*) and one-horned rhinoceros (*Rhinoceros unicornis*). The area was opened for settlement during the late 1950s, and immigrants swarmed into the fragile grassland. As the human population exploded, the populations of tigers and rhinos dwindled by the mid-1960s.

The area was converted into Royal Chitwan National Park in 1973. Today, humans are excluded from the park for most of the year. But each January, after the monsoon rains end and the grasses have dried, local residents are welcomed into the park to harvest the grass, which they use to thatch roofs, make mats, and feed domestic animals **(Figure 55.22)**. The area surrounding the Chitwan National Park has been designated as a buffer zone. Residents of the buffer zone receive 50% of the total revenue earned from park activities, including entrance fees. This income has changed the lifestyle of the people in the buffer zone, and they now value Chitwan and argue for its preservation. Today, more than 500 one-horned rhinos survive in Nepal, most of them in Chitwan National Park. And the Bengal tiger population of Chitwan has increased to more than 120 breeding adults.

FIGURE 55.22 Conservation and the local economy. Local residents support conservation efforts at the Royal Chitwan National Park, Nepal, because officials open the park for a grass harvest each year.

ECOTOURISM In some preserves, governments enlist local residents in park development and operations, providing them with a viable livelihood. The most successful approach has been the development of **ecotourism,** in which visitors, often from wealthier countries, pay a fee to visit a nature preserve. Local people work as guides, cooks, and logistical and support staff.

Not everyone agrees that ecotourism is helpful. Critics note that increased human traffic may degrade habitats, and unregulated ecotourism can eventually lead to overdevelopment. For example, several million people visit national parks in the western United States annually. Traffic jams, automobile accidents, and long lines routinely plague visitors at the most popular sites. Cranky ecotourists call for the construction of more roads and parking lots, which are inconsistent with the purpose of a national park because cars increase local air pollution and occasionally kill wildlife. In 2006, the government began charging a $20 fee for each automobile entering Yosemite National Park in California, hoping to limit the number of visitors arriving in private vehicles and to increase reliance on public transportation.

COUNTRYWIDE ECONOMIC APPROACHES In the mid-1990s, conservation biologists and economists developed the concept of **ecosystem valuation,** in which ecosystem services—such as carbon dioxide processing or water retention and purification,

which are best provided by intact ecosystems—are assigned an economic value. These estimated values are used to negotiate contracts in which a private company or conservation organization pays a community, state, or country to maintain intact ecosystems. By one 2014 estimate, the gross global ecosystem valuation is roughly $125 trillion per year, which is far more than the value of all goods produced by all humans on the planet.

The implementation of ecosystem valuation exchanges is determined on a case-by-case basis, depending on what ecosystem services the paying organization wants to preserve. Costa Rica has led the way in this effort by creating valuation contracts with several corporations. For example, in 1998, the Monteverde Conservation League signed a contract with a local electrical company to ensure the continued flow of water from the Bosque Eterno de los Ninos, a forest preserve. The company had plans to build a hydroelectric dam on the Rio Esperanza, and feared that deforestation upstream would disrupt water flow through the dam. The contract specifies that the electrical company will pay the people who live upstream to preserve their forests rather than cutting them. Thus, both the forests and water flow are preserved, maintaining the forest ecosystem and generating badly needed electricity.

Biodiversity is a precious resource that is disappearing rapidly throughout the world. It can still be conserved through a monumental effort to catalog the diversity of living organisms and develop an understanding of their ecological relationships. Perhaps the major challenges for conservation biologists are to educate the human population about the value of biodiversity and to develop conservation plans that will enlist the support of people who live among the threatened species.

STUDY BREAK 55.6

1. Is the Pine Bush habitat in New York State an example of preservation, mixed-use conservation, or restoration?
2. How has the establishment of the Royal Chitwan National Park in Nepal been a successful conservation effort? How do conservation biologists measure its success?
3. How can the concept of ecosystem services be used to foster conservation of threatened habitats and species?

 Unanswered Questions

Conservation biology is a young science in a rapidly changing world. Our field is brimming with unanswered questions about scarcity, diversity, and extinction. Many of the questions proposed by conservation biologists are interesting, but only a subset may offer answers useful to policy makers, field practitioners, and nongovernmental organizations trying to find cost-effective ways to head off the global biodiversity crisis. I have selected two areas where basic conservation science can inform global policies and local efforts to prevent the current extinction crisis from dooming many species to oblivion.

How can we preserve species that are likely to become extinct?

Some estimates state that up to 50% of all species on Earth could disappear by the end of the century as a result of (1) land conversion in the tropics, where more than 50% of all species occur, and (2) climate change. Professor E. O. Wilson of Harvard University has calculated that more than 100 species go extinct each day. So, ask the cynics, "What are they? Can you name them?"

To save species from imminent extinction, we need to know which species are highly endangered, where they live, what resources they require, and why they are threatened. Led by Taylor Ricketts, biologists from all major conservation organizations used published data to identify the "trigger species" of vertebrates and conifers most likely to go extinct in the next 20 years. We found 794 species in 585 locations for which the entire global population, ranked as endangered or critically endangered, is limited to a single site. Half of these sites contain rare, endemic amphibians, many of which are threatened by the deadly chytrid fungus. Can we find a way to stop the spread of this fungus, or reduce its virulence, to avoid a mass extinction of amphibians? Half of all threatened localities are on tropical mountains. Can we save these habitats as watersheds that provide vital ecosystem services to communities living downstream? What about the 350,000 to 400,000 species of vascular plants, perhaps 10% of which are known only from a single site? What about the millions of species of invertebrates, some so rare that their entire distribution may be limited to the crowns of a few tropical forest tree species? Can we protect all of these areas that are the last refuge for many species? Or is the task so overwhelming that we should try to save only what seems feasible?

Can we find the political will to protect global biodiversity?

Answers to this second question could provide a solution to those posed above. A new global treaty, REDD (Reduced Emissions from Deforestation and Degradation), is currently being negotiated. It offers the greatest conservation opportunity of our lifetimes. Can we seize it? If most of the world's endemic species live in the tropics and a high percentage occur in moist tropical forests, can we forge a new agreement that values these forests for the carbon they sequester, and in so doing, protect much of the world's biodiversity? Deforestation and degradation of moist tropical forests account for 20% of the greenhouse gas emissions recorded annually. These emissions could become much larger if peat swamp forests in Sumatra and other carbon-rich areas are burned, cleared, and converted to plantations. Can we harness the political will to create a carbon financing mechanism that rewards tropical countries that protect their carbon-rich and species-rich forests? Will REDD be a turning point in the race to secure a more stable climate and avoid biological catastrophe?

think like a scientist Write down 20 unanswered questions for which scientists must quickly find answers to save life on Earth. A recently published paper (W. J. Sutherland et al. 2009. One hundred questions of importance to the conservation of global biological diversity. *Conservation Biology* 23:557–567) offers a longer list. Does your list overlap with the most pressing 100 questions identified by biologists, conservationists, and policymakers?

Eric Dinerstein is Director of WildTech and the Biodiversity and Wildlife Solutions Program at RESOLVE, leading a team of biologists who are helping to add biodiversity information to Global Forest Watch. He was the World Wildlife Fund chief scientist and Vice President for Science, and a coarchitect and coauthor of the Global 200 ecoregions, an analysis to identify the most biologically important ecoregions on Earth in the terrestrial, freshwater, and marine realms. To learn more about Dr. Dinerstein's work, go to http://www.wri.org/profile/eric-dinerstein.

REVIEW KEY CONCEPTS

For access to MindTap and additional study materials visit www.cengagebrain.com.

55.1 The Biodiversity Crisis on Land, in the Sea, and in River Systems

- Deforestation is occurring at an alarming rate, especially in tropical regions (Figures 55.1 and 55.2). In addition to adding greenhouse gases to the atmosphere, deforestation may lead to desertification and the loss of entire ecosystems (Figure 55.3). Deforestation, desertification, and global warming reinforce each other in a positive feedback cycle.
- Overexploitation of natural populations reduces their sizes and may induce evolutionary responses in the exploited populations (Figure 55.4).
- Dams and other factors causing hydrologic alterations endanger freshwater ecosystems by changing the structure of riverine habitats, disrupting life histories of aquatic species, decreasing connectivity in river systems, and facilitating the establishment of nonnative species (Figure 55.5).

55.2 Specific Threats to Biodiversity

- Habitat fragmentation reduces the size of intact habitat patches, and edge effects diminish the quality of remaining habitat (Figure 55.6). Only small populations, which are subject to genetic drift and a greatly increased likelihood of extinction, can inhabit small habitat patches.
- Although pollution is released locally, it often spreads regionally and globally, especially in bodies of water and the atmosphere (Figure 55.7). Biological magnification causes pollution to have a detrimental effect on species that feed at higher trophic levels.
- Invasive species often contribute to the extinction of native species through competition, predation, or parasitism (Figures 55.8–55.10). Humans frequently introduce invasives into communities either intentionally or inadvertently.
- The decline of amphibian species is at least partly caused by the spread of a pathogenic chytrid fungus (Figure 55.11). New and introduced disease-causing organisms pose serious threats to biodiversity.
- Although extinction has been common in the history of life, human activities have recently initiated what may be the greatest mass extinction of all time. Some biologists estimate that extinction rates today may be 1,000 times the background extinction rate.

55.3 Ecosystem Services That Biodiversity Provides

- Diverse ecosystems provide provisioning services with direct benefits to humans and other organisms. Natural populations can be sources of useful natural products, as well as genetic resources that can improve domesticated crops and animals (Figure 55.12).
- Diverse ecosystems provide regulating and support services with indirect benefits to humans. Regulating services include the retention of water and nutrients in ecosystems and the purification of air and water resources. Support services include the decomposition of organic wastes and the pollination of flowering plants. Some regulatory and support services help to counteract the harmful effects of human activities.
- Ethicists and environmentalists argue that biodiversity should be preserved simply because of its intrinsic worth.

55.4 Which Species and Ecosystems Are Most Threatened by Human Activities?

- The IUCN monitors the conservation status of species and higher taxa, publishing its findings in *The Red List of Threatened Species* (Figure 55.13).
- Biodiversity hotspots harbor large numbers of endemic species and are threatened by human activities.
- Conservation biologists have pinpointed areas where several groups of vertebrates and coniferous plants are in imminent danger of extinction (Figure 55.14). An analysis of these data suggests that the extinction rate is accelerating and that species in the New World tropics are especially at risk.

55.5 Conservation Biology: Principles and Theory

- Conservation biology draws its theoretical foundation from systematics, population genetics, population ecology, behavior, community ecology, and landscape ecology.
- Systematists provide taxonomic inventories of biodiversity that are helpful for establishing conservation priorities.
- Conservation biologists design breeding programs to maintain or increase the genetic variability of species being preserved (Figure 55.15).
- Conservation biologists study the population ecology and behavior of targeted species (Figure 55.16), and they may use population viability analyses to determine the minimum viable population size necessary to conserve threatened species.

- Studies in community ecology have established the generality of the species–area effect: large habitat patches harbor more species than small habitat patches do (Figure 55.17).
- From the perspective of landscape ecology, biologists have debated the advantages and disadvantages of establishing one large reserve versus several smaller ones that are connected by habitat corridors (Figures 55.18 and 55.19).
- Conservation biologists now include analyses of diversity across communities (called *beta-diversity*) in their designs for nature preserves in an effort to preserve whole assemblages of organisms.

55.6 Conservation Biology: Practical Strategies and Economic Tools

- Efforts to conserve communities or ecosystems follow one of three general strategies. *Preservation* requires the restriction or prohibition of human access to the area (Figure 55.20). The financial cost of preservation is high. *Mixed-use conservation,* an approach that balances the conflicting demands of habitat preservation and development, allows local residents to use the protected area in limited ways (Figure 55.21). *Restoration* attempts to recreate natural communities and ecosystems in places that have already been degraded by human activities.
- Conservation plans must also incorporate economic and social factors to win local support. Most conservation plans now include the involvement of local residents to generate revenue for their communities (Figure 55.22). Ecosystem valuation also encourages the preservation of ecosystems by assigning them a significant economic value.

TEST YOUR KNOWLEDGE

Remember/Understand

1. The greatest extinction in the history of life on Earth:
   a. occurred at the end of the Permian period.
   b. occurred at the end of the Cretaceous period.
   c. occurred at the end of the Ordovician period.
   d. occurred at the end of the Cambrian era.
   e. may be occurring now.

2. Which of the following activities is the most fundamental cause of the worldwide crisis in river ecosystems?
   a. overexploitation of predatory fishes
   b. damming of rivers
   c. pollution from power plants
   d. invasion by nonnative species
   e. deforestation

3. The Red List published by the International Union for the Conservation of Nature lists:
   a. biodiversity hotspots.
   b. the conservation status of species and higher taxonomic groups.
   c. places where a large proportion of species are likely to face extinction.
   d. locations where acid precipitation is likely to damage ecosystems.
   e. the provisioning services provided by a variety of ecosystems.

4. Deforestation:
   a. is a problem only in the tropics.
   b. may speed desertification.
   c. is slowed by grazing and farming.
   d. permanently enriches the soil.
   e. leads to the formation of lush grasslands.

5. Chemical pollutants:
   a. can spread rapidly from the places they are released.
   b. do not appear to influence global climate change.
   c. have contributed to global mass extinctions.
   d. rarely affect natural bodies of water.
   e. rarely influence animals feeding at higher trophic levels.

6. Population viability analyses allow conservation biologists to:
   a. identify the source population from which an individual dispersed to a sink population.
   b. determine how large an area must be preserved for the protection of a threatened species.
   c. identify whether individuals of a threatened species are reproductively mature.
   d. predict the minimum population size of a threatened species that is likely to survive.
   e. predict whether a threatened species will use habitat corridors.

7. Beta-diversity is a measure of:
   a. species diversity across community boundaries.
   b. the number of species within one community.
   c. the number of endemic species found in a particular place.
   d. the likelihood that a particular species will become extinct in the next 20 years.
   e. the minimum population size needed to conserve an endangered species.

8. For which of the following species has the use of habitat corridors been proposed as an important conservation tool?
   a. sea otters
   b. bay checkerspot butterflies
   c. Florida panthers
   d. whooping cranes
   e. Eastern hemlocks

Apply/Analyze

9. Which of the following is most likely to be a biodiversity hotspot?
   a. a patch of forest in the middle of North America that is 500 km from the nearest big city
   b. a series of uninhabitable sand dunes in the Sahara Desert
   c. a botanical garden that houses representatives of 25,000 plant species
   d. a tropical island with many endemic species and a growing human population
   e. a suburban neighborhood where fields have been converted to backyards and playgrounds

10. The main goal of restoration ecology is the reestablishment of:
    a. natural patterns of water flow.
    b. the vitality of a degraded ecosystem.
    c. the historical corridors linking forest fragments.
    d. the natural barriers to animal movement.
    e. ecotourism.

11. **Discuss Concepts** How do concepts from population genetics, metapopulation dynamics, and beta-diversity apply to the design of nature preserves? Do they suggest different ideal designs for nature preserves?

Evaluate/Create

12. **Discuss Concepts** National parks are often established in ecologically sensitive areas. In many places they have become so popular that visitors endanger the ecosystems the parks were originally designed to preserve. How can the goals of conservationists, who work to maintain intact ecosystems, be balanced with those of citizens who wish to visit intact ecosystems? In other words, how would you regulate domestic ecotourism?

13. **Discuss Concepts** Imagine that you are a conservation biologist who has been asked to develop a conservation plan for a species of lizard that lives in the deserts of the American Southwest. What sorts of data would you collect before developing a final plan?

14. **Design an Experiment** Devise a field study to determine whether the species–area relationship applies to aquatic ecosystems, such as ponds and lakes, as it does to terrestrial habitats.

15. **Apply Evolutionary Thinking** Overexploitation of marine fish stocks has depleted natural populations and caused a reduction in the age and size at which many fish species become reproductively mature. What sort of government regulations of fishing might reverse the current trend toward smaller adult size? Explain your answer in terms of the selection pressures that fishing places on targeted species.

For selected answers, see Appendix A.

INTERPRET THE DATA

The accompanying **Table** compares statistics on imminent extinctions versus historical extinctions in five groups of organisms. The data are divided into three geographic categories. In which of the three habitat categories are extinctions accelerating the most? In which group of organisms are extinctions accelerating the most? What do these data suggest about where conservation biologists should focus their efforts if the goal is to preserve as many species as possible?

Distribution of Species Facing Imminent Extinction (i.e., Trigger Species) and Historically Extinct Species among Taxa and Islands, Mountains, and Low Mainland Areas

Taxon	Islands[a]		Mountains[b]		Low mainland[c]		Total	
	Trigger species	Extinct species	Trigger species	Extinct species	Trigger species	Extinct species	Trigger species	Extinct species
Mammals	80	49	35	5	16	19	131	73
Birds	128	121	51	1	38	7	217	129
Reptiles[d]	7	8	0	0	8	1	15	9
Amphibians	88	19	268	11	52	4	408	34
Conifers	9	0	12	0	2	0	23	0
Total	312	197	366	17	116	31	794	245

[a]Islands are landmasses smaller than Greenland and include mountainous sections of islands.
[b]Mountains are mountains on mainland landmasses (not on islands).
[c]Low mainland regions are low-lying regions of continental mainlands.
[d]Reptiles include only turtles and tortoises, crocodilians, and iguanid lizards.

Source: T. H. Ricketts et al. 2005. Pinpointing and preventing imminent extinctions. *Proceedings of the National Academy of Sciences USA* 102:18497–18501.

56 Animal Behavior

PAUL NICKLEN/National Geographic Creative

Musk oxen *(Ovibos moschatus)*. The social behavior of a herd of musk oxen includes encircling their young to protect them from predators.

Why it matters . . . Male white-crowned sparrows *(Zonotrichia leucophrys)* are handsome birds with a song that birdwatchers describe as a "plaintive whistle" followed by a "husky trilled whistle." This distinctive song is a critical feature of a male white-crown's **behavioral ecology,** the set of actions that it can perform to interact with stimuli in its environment. A male sparrow's song is one of the ways he struts his stuff. The song not only announces his presence to rival males, but it also signals to females that he is available as a potential mate. Experienced birders easily recognize this song, which differs from that of song sparrows *(Melospiza melodia)* and swamp sparrows *(Melospiza georgiana),* as sound spectrograms illustrate **(Figure 56.1).** In fact, every songbird species produces vocal signals that are characteristic of its species and its species alone.

Every animal species' *behavioral repertoire* (that is, the range of behaviors it performs) includes a variety of actions and responses. For example, in early spring, male white-crowned sparrows leave their wintering grounds in Mexico and fly thousands of kilometers to their northern breeding range; migrating birds expend immense quantities of energy on the trip, and many die before completing it. On arrival, they occupy habitat patches that contain the resources necessary for breeding—suitable cover, potential nesting sites, and abundant food. Then, they start to sing, repeating their song thousands of times a day. Males also perform elaborate courtship behaviors. And once the young hatch, the chicks communicate with their parents, eliciting the care they need before leaving the nest.

All of these behaviors require the expenditure of substantial time and energy; many also attract the attention of predators. Given the *costs* of these behaviors, what *benefits* do the birds gain from performing them? The ultimate evolutionary benefit is obvious: with luck, individuals that perform these complex behaviors may leave more surviving offspring than those that do not.

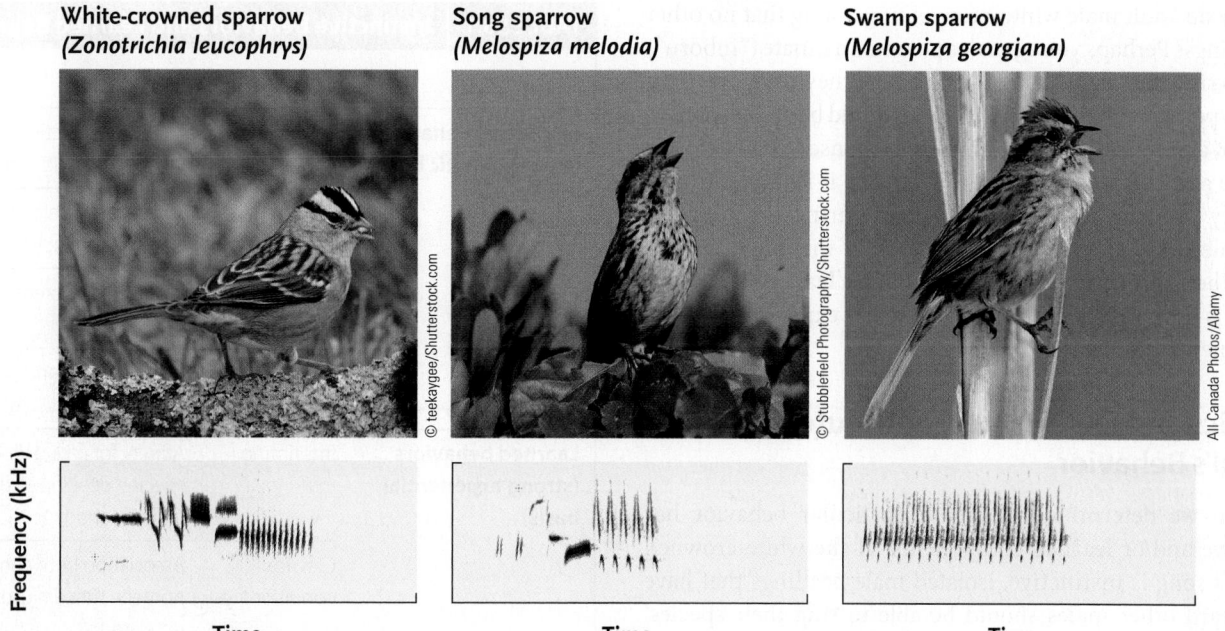

FIGURE 56.1 Songbirds and their songs. Sound spectrograms (visual representations of sound graphed as frequency versus time) illustrate differences in the songs of the white-crowned sparrow, the song sparrow, and the swamp sparrow.

© Cengage Learning 2017

Thus, each species' behavioral repertoire is a set of evolved responses, based on the relative costs and benefits of performing them, that ultimately contribute to the reproductive success and fitness of individuals.

The study of **animal behavior** involves discovering how animals respond to specific stimuli and why they respond in predictable and characteristic ways. A comprehensive approach to animal behavior studies first crystallized in the 1930s, when European researchers—notably Konrad Lorenz, Niko Tinbergen, and Karl von Frisch, who shared a Nobel Prize for their work in 1973—developed the discipline of **ethology,** which focuses on how evolutionary processes shape inherited behaviors and the ways that animals respond to specific stimuli. Tinbergen identified four basic questions that any broad study of animal behavior should address: (1) What mechanisms trigger a specific behavioral response? (2) How does the expression of a behavior develop as an animal matures? (3) What is the behavior's ecological function and how does it increase an animal's chances of surviving and reproducing? (4) How did the behavior evolve?

Advances in **neuroscience**—the integrated study of the structure, function, and development of the nervous system—now allow researchers to explore Tinbergen's first and second questions about the **proximate causes** of behavior—*how* genetic, cellular, physiological, and anatomical mechanisms underlie an animal's ability to detect and respond to stimuli in species-specific ways.

Comparable advances in genetic analysis and evolutionary theory enable scientists to address Tinbergen's third and fourth questions about the **ultimate causes** of behavior—*why* behaviors have adaptive value and *why* the genetic and physiological mechanisms that underlie specific behaviors are subject to microevolutionary change. The behavior of animals is closely tied to their ecological circumstances, and if particular alleles contribute to the development of a behavior that enhances an animal's fitness in the environment it occupies, natural selection will cause the frequency of those alleles to increase in the next generation (see Section 21.3).

In this chapter, after first considering the genetic and experiential bases of animal behavior, we examine the neurophysiological and endocrinological control of specific behaviors. We then turn our attention to the ecology and evolution of several broad categories of animal behavior: orientation, navigation, and migration; habitat selection and territoriality; communication; reproductive behavior and mating systems; and social behavior, including behaviors described as altruistic.

56.1 Instinctive and Learned Behaviors

For many years, animal behaviorists debated whether animals are born with the ability to perform most behaviors completely or whether experience is necessary to shape their actions. Extensive research in neuroscience has demonstrated that no behavior is determined entirely by genetics or entirely by environmental factors. Instead, behaviors develop through complex gene–environment interactions, which we illustrate with a description of how male white-crowned sparrows learn their adult song.

Why do adult male white-crowns sing a song that no other species sings? Perhaps young males possess an innate ("inborn") ability to sing the "right" song the first time they try. According to this hypothesis, their distinctive song would be an **instinctive behavior,** a genetically "programmed" response that appears in complete and functional form the first time it is used. Or perhaps they produce the song only after hearing the songs of adult male white-crowns that live nearby. If that hypothesis is correct, their species-specific song would be a **learned behavior,** one that is dependent on having a particular experience during development.

Genetics and Experience Contribute to an Animal's Behavior

How can we determine whether a particular behavior has instinctive and/or learned components? If the white-crowned sparrow's song is instinctive, isolated male nestlings that have never heard other males should be able to sing their species' song when they mature. But if the learning hypothesis is correct, young birds deprived of certain essential experiences should not sing "properly" as adults.

In a set of pioneering experiments conducted at Rockefeller University, Peter Marler tested these alternative hypotheses. He took newly hatched white-crowns from nests in the wild and reared them individually in soundproof cages. Some chicks listened to recordings of a male white-crowned sparrow's song when they were 10 to 50 days old; others did not. The juvenile males in both groups first started to vocalize when they were about 150 days old, initially producing whistles and twitters that only vaguely resembled the songs of adults. But gradually the young males that had listened to their species' song began to sing better and better approximations of that song. At about 200 days of age, they were right on target, producing a song that was nearly indistinguishable from the one they had heard months before. By contrast, males in the group that had not heard white-crown songs never came close to singing the way wild males do.

These results revealed that learning is essential for a young male white-crowned sparrow to acquire the full song of its species. Although birds isolated as nestlings did sing instinctively, the acoustical experience of listening to their species' song early in life enabled them to reproduce it months later. We can therefore reject the hypothesis that white-crowned sparrows hatch from their eggs with the ability to produce the "right" song. Their species-specific song—and presumably those of other songbirds—has both instinctive and learned components.

Although early researchers generally classified behaviors as *either* instinctive *or* learned, we now know that most behaviors include *both* instinctive *and* learned components. Nevertheless, some behaviors have a strong instinctive basis, whereas others are mostly learned (**Table 56.1**). Researchers use a variety of techniques to unravel the instinctive and learned components of specific behaviors. In some studies, observations and

TABLE 56.1	Instinctive Behaviors and Learned Behaviors	
	Behavior	Description
Instinctive behaviors (strong genetic basis)	Fixed action patterns	Stereotyped actions, often initiated by a sign stimulus
	Feeding behaviors	Innate food preferences and hunting tactics
	Defensive behaviors	Responses to predators
	Reproductive behaviors	Mating habits and parental care activities
Learned behaviors (strong experiential basis)	Imprinting	Affinity for caretaker species, developed during critical period
	Classical conditioning	Association between phenomena that are usually unrelated
	Operant conditioning	Trial-and-error learning
	Cognition	Insight learning in novel situations
	Habituation	Loss of responsiveness

experiments based on simple environmental manipulation are sufficient. In others, researchers conduct more elaborate deprivation experiments (as described above), breeding studies, or experiments based on gene knockouts.

Instinctive Behaviors Are Performed without Prior Experience

Instinctive behaviors—which are often grouped into functional categories, such as feeding behaviors, defensive responses, mating behaviors, and parental care activities—can be performed without the benefit of prior experience. We therefore assume that they have a strong genetic basis and that natural selection has preserved them as adaptive behaviors.

STEREOTYPED BEHAVIORS Many instinctive behaviors are highly stereotyped; in other words, when triggered by a specific cue, they are performed over and over in almost exactly the same way. Such behaviors are called **fixed action patterns,** and the simple cues that trigger them are called **sign stimuli.** For example, sign stimuli and fixed action patterns govern the transfer of food from herring gull (*Larus argentatus*) parents to their offspring: very young chicks secure food from their parents through a begging response (the fixed action pattern), which is triggered by a red spot on the lower bill of an adult (the sign stimulus). This cue "releases" the begging behavior of hungry baby gulls, which peck at the spot on the parent's bill. In

FIGURE 56.2   **Experimental Research**

The Role of Sign Stimuli in Parent–Offspring Interactions

Question: What feature of the parent's head triggers pecking behavior in young herring gulls?

Experiment: Niko Tinbergen and A. C. Perdeck tested the responses of young herring gull *(Larus argentatus)* chicks to cardboard cutouts of an adult herring gull's head and bill. They waved these models in front of the chicks and recorded how often a particular model elicited a pecking response from the chicks. One cutout included an entire gull's head with a red spot near the tip of the bill, another cutout included just the bill with the red spot, and the third cutout included the entire head but lacked the red spot.

Result: Young herring gulls pecked at the model of the bill with a red spot almost as often as they pecked at the model of an entire head with a red spot, but they pecked much less frequently at the model of an entire head that lacked a red spot.

Herring gulls *(Larus argentatus)*

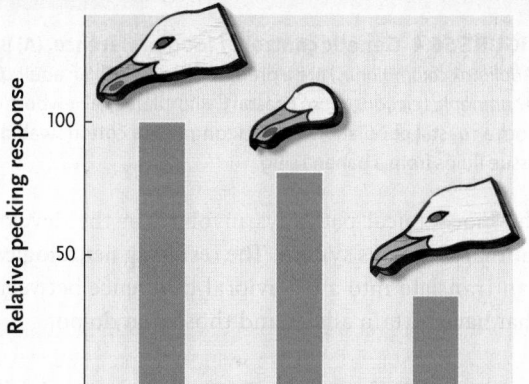

Relative pecking response (y-axis): 50, 100
Model presented (x-axis)

Conclusion: Begging behavior by young herring gulls is triggered by a simple sign stimulus, the red spot on the parent's bill. Experimental tests revealed that herring gull chicks respond more to the presence of the contrasting spot than they do to the outline of an adult's head.

think like a scientist
Observe
Hypothesize
Predict
Experiment
Interpret

Why might natural selection favor the evolution of a sign stimulus–fixed action pattern interaction as the mechanism through which herring gulls provide food to their offspring? What are the advantages of this interaction between offspring and parent relative to one in which parents simply regurgitate food for their offspring as soon as the parents return to the nest?

Source: N. Tinbergen and J. C. Perdeck. 1950. On the stimulus situation releasing the begging response in the newly hatched herring gull chick *(Larus argentatus argentatus* Pont.). *Behaviour* 3:1–39.

© Cengage Learning 2017

turn, the tactile stimulus delivered by the pecking chick serves as a sign stimulus that induces the adult bird to regurgitate food stored in its crop. The baby gulls then feed on the chunks of seafood that lie before them. We know that the spot on the parent's bill releases the begging response of the young gull because the same response is triggered by an artificial bill that looks only vaguely like a herring gull's bill, as long as it has a dark contrasting spot near the tip **(Figure 56.2)**. Thus, even very simple cues can activate fixed action patterns.

Natural selection has molded the behavior of some parasitic species to exploit the relationship between sign stimuli and fixed action patterns for their own benefit. For example, birds that are brood parasites lay their eggs in the nests of other species (see *Why it matters . . .* at the beginning of Chapter 53). When a brood parasite's egg hatches, the alien nestling produces exaggerated versions of sign stimuli that are ordinarily exhibited by its hosts' own chicks: opening its mouth, bobbing its head, and calling vigorously. These behaviors elicit vigorous feeding by the foster parents, and the young brood parasite often receives more food than the hosts' own young **(Figure 56.3).**

Although instinctive behaviors are often performed completely the first time an animal responds to a stimulus, they can be modified by an individual's experiences. For example, the fixed action patterns of a young herring gull change through time. Although the youngster initially begs by pecking at almost anything remotely similar to an adult gull's bill, it eventually learns to

FIGURE 56.3 Exploitation of a releaser. This young European cuckoo *(Cuculus canorus)*, a brood parasite, stimulates feeding behavior by its foster parent, a hedge sparrow *(Prunella modularis)*. It secures food by displaying exaggerated versions of the sign stimuli used by the host offspring to release feeding behavior by the parents.

recognize the distinctive visual and vocal features associated with its parents. The chick uses this information to become increasingly selective about the stimuli that elicit its begging behavior. Thus, even largely instinctive behaviors can be modified in response to particular experiences during their early performances.

INDIVIDUAL DIFFERENCES IN INSTINCTIVE BEHAVIORS

Because the performance of instinctive behaviors does not depend on prior experience, behavioral differences between individuals may reflect genetic differences between them. Stevan Arnold, then at the University of Chicago, tested that hypothesis by studying the innate responses of captive newborn garter snakes (*Thamnophis elegans*) to the olfactory stimuli provided by potential food items that they had never before encountered. Arnold measured the snakes' responses to cotton swabs that had been dipped in a smelly extract of banana slug (*Ariolimax columbianus*), a shell-less mollusk. A snake "smells" by tongue-flicking, which draws volatile chemicals into a special sensory organ in the roof of its mouth. If the young snake had been born to a mother captured in coastal California, where adult garter snakes regularly eat banana slugs, it almost always began tongue-flicking at the slug-scented cotton swab (**Figure 56.4**). By contrast, newborn snakes whose parents came from inland California, where banana slugs do not occur, rarely tongue-flicked at the swabs. Thus, although the coastal and inland snakes belong to the same species, their instinctive responses to the volatile chemicals associated with banana slugs were markedly different.

In another experiment, Arnold tested whether newborn snakes would feed on bite-sized chunks of slug. After a brief flick of the tongue, 85% of the newborn snakes from a coastal population routinely struck at the slug and swallowed it, despite having had no prior experience with this prey. By contrast, only 17% of newborn snakes from the inland population ate slugs consistently, even when no other food was available.

Arnold hypothesized that coastal and inland garter snakes possess different alleles at one or more gene loci controlling their odor-detection mechanisms, leading to differences in their behavior. To test this hypothesis, Arnold crossbred coastal and inland snakes. If genetic differences contribute to the different food preferences of the two snake populations, then hybrid offspring, which receive genetic information from each parent, should behave in an intermediate fashion. Results of the experiment confirmed his prediction: when presented with bite-sized chunks of slug, 29% of the newborn snakes of mixed parentage consumed them every time.

Many additional experiments have confirmed that genetic differences between individuals can translate into behavioral differences between them. *Molecular Insights* describes a striking example of a single gene that influences the grooming behavior of mice. Bear in mind that single genes do not control complex behavior patterns directly. Instead, the alleles present affect the kinds of enzymes that cells can produce, influencing

A. Banana slug

B. Adult coastal garter snake eating a banana slug

C. Newborn coastal garter snake "smelling" slug extract

FIGURE 56.4 Genetic control of food preference. (A) Banana slugs (*Ariolimax columbianus*) are a preferred food of **(B)** an adult garter snake (*Thamnophis elegans*) from coastal California. **(C)** A newborn garter snake from a coastal population flicks its tongue at a cotton swab drenched with tissue fluids from a banana slug.

the biochemical pathways involved in the development of an animal's nervous system. The resulting neurological differences can translate into a behavioral difference between individuals that have certain alleles and those that do not.

Learned Behaviors Depend on an Individual's Prior Experience

Unlike instinctive behaviors, learned behaviors are not performed completely the first time an animal responds to a specific stimulus. Instead, they change in response to the environmental stimuli that an individual experiences. Behavioral scientists generally define **learning** as a process in which experiences change an animal's behavioral responses. Different types of learning occur under different environmental circumstances. In this section we consider *imprinting, classical conditioning, operant conditioning, cognition,* and *habituation.*

Some animals learn the identity of a caretaker or the key features of a suitable mate during a **critical period** (sometimes

Molecular Insights

A Knockout by a Whisker: What is the function of *disheveled* genes in mice?

Almost all eukaryotic organisms share a series of developmental interactions called the *wingless/Wnt* pathway. The pathway was originally discovered in fruit flies, in which mutations in the pathway cause changes in the wings and other segmental structures.

Research Question
The mouse genome includes three genes closely related to *disheveled (dsh)*, one of the genes in the *Drosophila wingless/Wnt* pathway. No function has been identified for the proteins they encode, but they are highly active in both embryos and adults. Their function must be important, but what could it be?

Experiment
Nardos Lijam and his coworkers in several laboratories, including Case Western University, the Universities of Colorado and Maryland, and the National Institutes of Health in Bethesda, Maryland, answered this question by developing a line of mice that lacked *Dvll,* one of the *disheveled* genes. First they constructed an artificial copy of *Dvll* with the central section scrambled so that no functional proteins could be made from its encoded directions (**Figure,** step 1). Next they introduced the artificial gene into embryonic mouse cells and injected the cells into very early mouse embryos (**Figure,** step 2). Some mice grown from these embryos were heterozygotes, with one normal copy of the *Dvll* gene and

one nonfunctional copy. Matings between heterozygotes produced some individuals that carried two copies of the altered *Dvl1* gene and no normal copies. Individuals in which the normal gene is eliminated are called knockout mice for the missing gene (**Figure,** step 3). (The procedure for making knockout mice is described in Section 18.2.)

Surprisingly, the knockout mice grew to maturity with no apparent morphological defects in any tissue examined, including the brain. Their motor skills, sensitivity to pain, cognition, and memory all appeared to be normal. However, their social behavior was a different story. When housed with normal mice, the knockouts failed to take part in common social activities: grooming, tail pulling, mounting, and sniffing. Rather than building nests and sleeping in huddled groups, as normal mice do, the knockouts tended to sleep alone in half-built nests. Heterozygotes for the *Dvll* gene—that is, mice with one normal and one altered copy of the gene—behaved normally in all these social activities.

The knockout mice also jumped around wildly in response to abrupt, startling sounds whereas the response of normal mice was less extreme (**Figure,** step 4). Because a specific neural circuit in the brain inhibits the startle response of normal mice, the reaction of the knockout mice suggested that

this inhibitory circuit was probably altered. Humans with schizophrenia, obsessive-compulsive disorders, Huntington disease, and some other brain dysfunctions also show an intensified startle reflex similar to that of the *Dvll* knockout mice.

Conclusion
The analysis revealed that the *Dvll* gene modifies developmental pathways affecting complex social behavior in mice, and probably in other mammals. It was one of the first genes affecting mammalian behavior to be identified. The similarity in startle-reflex intensity between the knockout mice and humans with neurological or psychiatric disorders also suggests that mutations in the *Dvl* genes and the *wingless* developmental pathway may underlie some of these disorders. If so, further studies of the *Dvl* genes may give us clues to the molecular basis of these ailments, as well as a possible means to their cure.

think like a scientist

If a recessive genetic mutation contributed to the development of a mental disorder in humans, could the genetic basis of the disorder be eliminated from the population? *Hint:* Review Section 21.4.

Source: N. Lijam et al. 1997. Social interaction and sensorimotor gating abnormalities in mice lacking *Dvl1. Cell* 90:895–905.

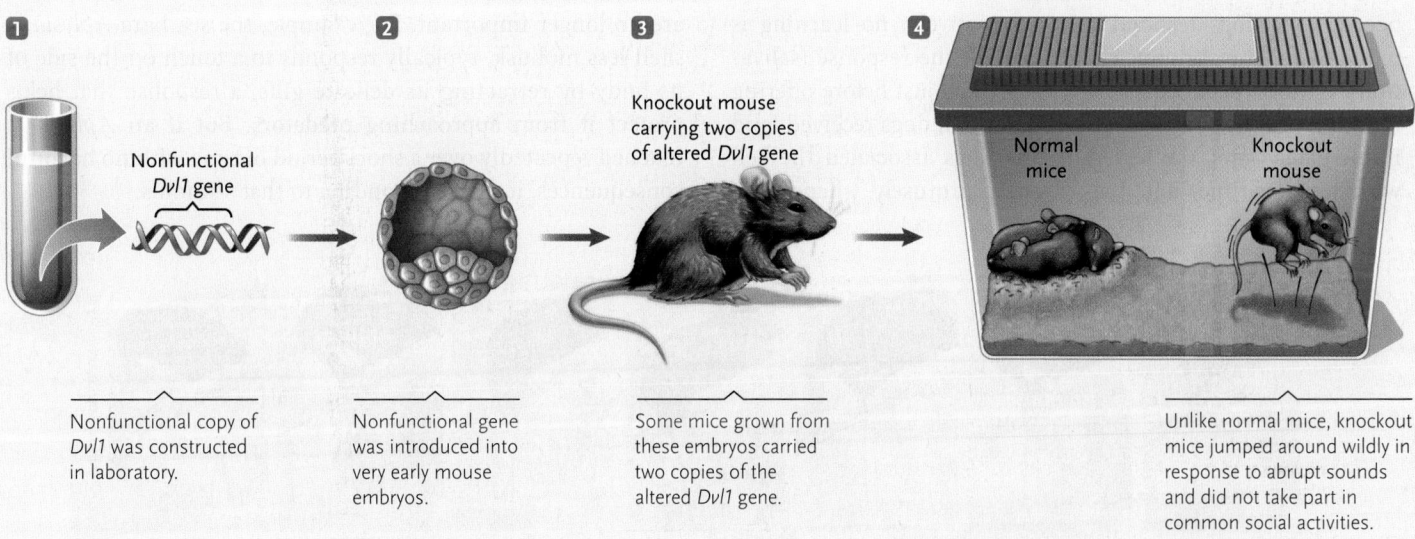

1 Nonfunctional *Dvl1* gene

Nonfunctional copy of *Dvl1* was constructed in laboratory.

2 Nonfunctional gene was introduced into very early mouse embryos.

3 Knockout mouse carrying two copies of altered *Dvl1* gene

Some mice grown from these embryos carried two copies of the altered *Dvl1* gene.

4 Normal mice / Knockout mouse

Unlike normal mice, knockout mice jumped around wildly in response to abrupt sounds and did not take part in common social activities.

© Cengage Learning 2017

FIGURE 56.5 Imprinting. Having imprinted on him shortly after hatching, young greylag geese (*Anser anser*) frequently joined Konrad Lorenz for a swim.

called a *sensitive period*), a time-limited phase of early development. This type of learning is called **imprinting.** For example, newly hatched geese imprint on their mother's appearance and identity, staying near her for months; in this case, the critical period for imprinting is 12 to 16 hours after hatching. When geese reach sexual maturity, they try to mate with other geese, which exhibit the visual and behavioral stimuli on which they had imprinted as youngsters. When Konrad Lorenz tended a group of newly hatched greylag geese (*Anser anser*), they imprinted on him instead of an adult of their own species **(Figure 56.5).** The male geese not only followed Lorenz about, but they also courted humans when they achieved sexual maturity.

Other forms of learning can occur throughout an animal's lifetime. Russian physiologist Ivan Pavlov's classic experiments with dogs explored **classical conditioning,** a type of learning in which animals develop a mental association between two phenomena that are usually unrelated. Dogs generally salivate when they eat. The food is called an *unconditioned stimulus* because the dogs respond to it instinctively; no learning is required for the stimulus (food) to elicit the response (salivation). In his experiment, Pavlov rang a bell just before offering food to dogs. After about 30 trials in which dogs received food immediately after the bell rang, the dogs associated the bell with feeding time, and they drooled profusely whenever it

rang—even when no food was forthcoming. Thus, the bell became a *conditioned stimulus,* one that elicited a particular learned response. In classical conditioning, an animal learns to respond to a conditioned stimulus when it precedes an unconditioned stimulus that normally triggers the response.

In another form of associative learning, called *trial-and-error learning* or **operant conditioning,** animals learn to link a voluntary activity, called an *operant,* with its favorable consequences, called a *reinforcement.* For example, a laboratory rat will explore a new cage randomly. If the cage is equipped with a bar that releases food when it is pressed, the rat will eventually lean on the bar by accident (the operant) and immediately receive a morsel of food (the reinforcement). After just a few such experiences, a hungry rat will learn to press the bar in its cage more frequently—as long as bar-pressing behavior is followed by access to food. Laboratory rats also learn to press bars to turn off disturbing stimuli, such as bright lights.

Many researchers have wondered whether animals can solve a novel problem by using insight, "thinking up" a solution rather than making a series of trial-and-error attempts at resolving it. This process, called **cognition,** implies that an animal is aware of its circumstances, defines a specific goal, and then uses reasoning to achieve the goal. Bernd Heinrich of the University of Vermont demonstrated that common ravens (*Corvus corax*) used cognition to solve a problem that they (and presumably their ancestors) had never before encountered: how to capture a morsel of food that dangled below their perches on a string **(Figure 56.6).** After looking at the string and looking at the food, a raven pulled up a loop of string, secured it against the perch with its foot, and then reached down again to pull up another loop. After repeating these actions six or eight times, the raven gained access to the food.

Animals typically lose their responsiveness to frequent stimuli that are not quickly followed by the usual reinforcement. This learned loss of responsiveness, **habituation,** saves the animal the time and energy of responding to stimuli that are no longer important. For example, the sea hare *Aplysia,* a shell-less mollusk, typically responds to a touch on the side of its body by retracting its delicate gills, a response that helps protect it from approaching predators. But if an *Aplysia* is touched repeatedly over a short period of time with no harmful consequences, it stops responding to that stimulus.

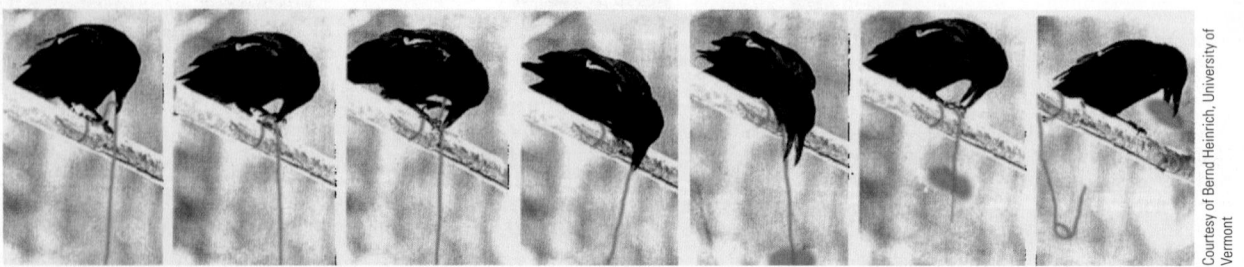

FIGURE 56.6 Cognition in ravens (*Corvus corax*). Ravens quickly figured out that they could retrieve food dangling on a string by repeating a series of actions in which they pulled up a loop of string and then held it in place with their foot. The string and food bundle have been colorized in these photos to make them more visible.

56.2 Neurophysiological and Endocrine Control of Behavior

Research in neuroscience has shown that all behavioral responses, even those that are either mostly instinctive or mostly learned, depend on an elaborate foundation provided by the biochemistry and structure of neurons (nerve cells). The neurons that regulate an innate response, as well as those that make it possible for an animal to learn something, are products of complex developmental processes in which genetic information and environmental contributions are intertwined. Although the anatomical and physiological basis for some behaviors is present at birth, an individual's experiences alter cells of its nervous systems in ways that produce particular patterns of behavior.

Research on many animal species has also revealed that hormones are the chemical signals triggering the performance of specific behaviors. They often accomplish this function by regulating the development of neurons and neural networks or by stimulating the cells within endocrine glands to release chemical signals. In this section we use examples from research on several animal species to explore general principles about how neurophysiology and endocrine function influence behavior.

Discrete Neural Circuits in Specific Brain Regions Control Singing Behavior in Songbirds

Marler's experiments (see Section 56.1) help explain the physiological underpinnings of singing behavior in male white-crowned sparrows. If acoustical experience shapes this behavior, a sparrow chick's brain must be able to acquire and store information present in the songs of other males. Then, months later, when the young male starts to sing, its nervous system must have matched its vocal output to the stored memory of the song that it had heard earlier. Eventually, when it achieves a good match, the sparrow's brain must "lock" on the now complete song and continue to produce it when the bird is singing.

Additional experiments have provided detailed information about the nature of the sparrow's nervous system. Young birds that did not hear recorded song during their critical period, between 10 and 50 days old, never produced the full song of their species, even if they heard it later in life. In addition, young birds that heard recordings of *other* bird species' songs during the critical period never generated replicas of those songs as they matured. These and other findings suggested that certain neurons in the young male's brain are influenced only by the acoustical signals from individuals of its own species, and only during the critical period. Neuroscientists have identified the neuron clusters, called *nuclei* (singular, *nucleus*), that make song learning and song production possible.

Moreover, every behavioral trait appears to have its own neuroanatomical basis. For example, a male zebra finch, *Taeniopygia guttata,* another songbird, can discriminate between the songs of strangers and the songs of established neighbors on adjacent **territories.** (In many bird species, territories are plots of land, defended by individual males or breeding pairs, within which the territory holders have exclusive access to food and other necessary resources. Territories are discussed further in Section 56.4.) The ability to discriminate between the songs of neighbors and those of strangers also involves a nucleus in the forebrain. Cells in this nucleus fire frequently the first time that the song of a new zebra finch is played to a test subject. But as the song is played again and again, these cells cease to respond, indicating that the bird becomes habituated to a now familiar song, although it still reacts to the songs of strangers. The neurophysiological networks that make this selective learning possible enable male zebra finches to behave differently toward familiar neighbors, which they largely ignore, and unfamiliar singers, which they attack and drive away.

The Activation of Specific Genes Fosters the Development of Nuclei That Regulate a Bird's Song

Molecular and cellular techniques reveal that learning depends on the expression of specific genes in neurons. When a bird is exposed to relevant acoustical stimuli, such as the songs of potential rivals of its own species, certain genes are "turned on" within neurons in the song-controlling nuclei of the bird's brain. For example, when a zebra finch hears the elements of its species' song, a gene called *zenk* becomes active in certain regions of the brain, producing an enzyme that changes the structure and function of the neurons **(Figure 56.7)**. In effect, the ZENK enzyme programs the neurons of the bird's brain to "anticipate" key acoustical events of potential biological importance. When these events occur, they trigger additional changes in the bird's brain that affect its actions. As a result, a territory

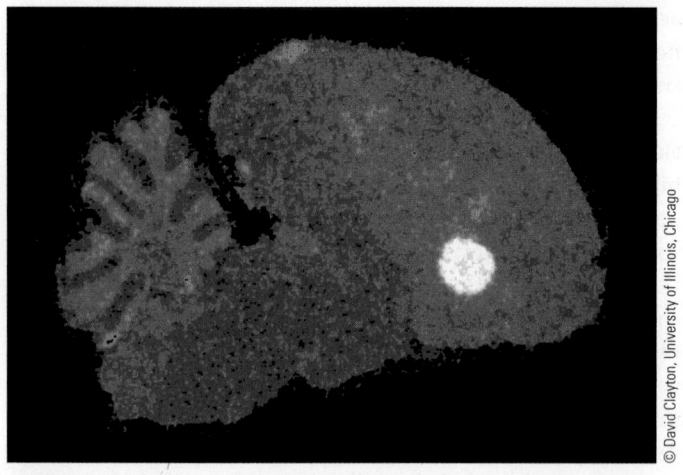

FIGURE 56.7 Gene expression and behavior. A section of zebra finch *(Taeniopygia guttata)* brain was stained in the laboratory to illuminate the localized expression of the *zenk* gene, which helps a male bird reproduce his species' song.

owner habituates to (that is, learns to ignore) a singing neighbor with which it has already adjusted territorial boundaries; but it retains the ability to detect and repel new intruders of its own species, which threaten to its continued control of its territory.

Hormones Regulate the Development of Cells and Networks That Form the Neural Basis of Behavior

How did the neurons in an adult zebra finch acquire the remarkable capacity to change in response to specific stimuli? In zebra finches, only males sing courtship songs. Very early in a male songbird's life, certain cells in its brain produce the hormone estrogen, which affects target neurons in an area of the developing brain called the *higher vocal center*. The presence of this hormone leads to a complex series of biochemical changes that result in the production of more neurons in parts of the brain that regulate singing. By contrast, the brains of developing females do not produce estrogen, and in the absence of this hormone, the number of neurons in the higher vocal center of females *declines* over time **(Figure 56.8)**. Experiments have shown that when young female zebra finches are given estrogen, they produce more neurons in the higher vocal center; but the treated females do not sing later in life unless they are also treated with androgens (male hormones).

Thus, genetically induced hormone production contributes to song learning and singing behavior in male zebra finches by regulating the numbers and types of neurons in the brain centers that produce those behaviors. The development of these neurons primes them for additional changes in response to specific acoustical experiences during the bird's development. For example, specific stimuli, such as the songs of either familiar or unfamiliar males, can trigger the expression of specific genes that influence the behavior of adult birds.

Changing Hormone Concentrations Alter the Behavior of Animals as They Mature

Just as estrogen influences the development of singing ability in zebra finches, other hormones mediate the development of the nervous system in other species. Indeed, a change in the concentration of a certain hormone is often the physiological trigger that induces important changes in an animal's behavior as it matures.

In honeybees *(Apis mellifera)*, worker bees perform different tasks for the colony's welfare as they grow older: bees that are less than 15 days old tend to care for larvae and maintain the hive, whereas those that are more than 15 days old often leave the hive to forage for nectar and pollen. (Food choice and foraging theory are discussed in Section 53.1.) These behavioral changes are induced by rising concentrations of juvenile hormone (see Section 42.5), which is released by a gland near the bee's brain **(Figure 56.9)**. Despite its name, circulating levels of juvenile hormone actually increase as a honeybee gets older.

Juvenile hormone may exert its effect on the bee's behavior by stimulating genes in certain brain cells to produce proteins that affect nervous system function. One such chemical, *octopamine,* stimulates neural transmissions and reinforces memories. It is concentrated in the antennal lobes, a part of the bee's brain that contributes to the analysis of chemical scents in the bee's external environment. Octopamine is present at higher concentrations in the older, foraging bees that have higher levels of juvenile hormone. And when extra juvenile hormone is administered to bees experimentally, their production of octopamine increases. Thus, increased octopamine levels in the antennal lobes may help a foraging bee home in on the odors of flowers where it can collect nectar and pollen.

The honeybee example illustrates how genes and hormones interact in the development of behavior. Genes code for the

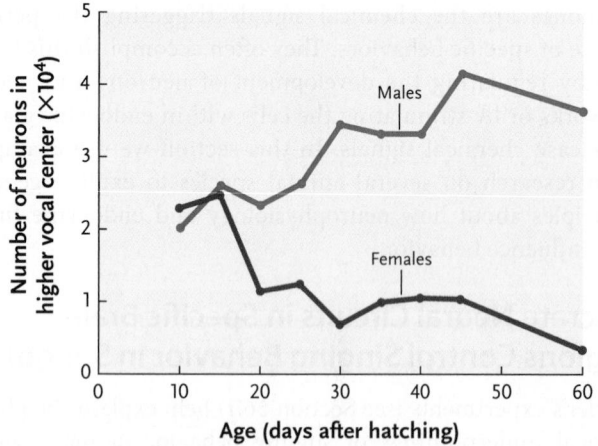

FIGURE 56.8 Hormonally induced changes in brain structure. The brains of young male zebra finches secrete estrogen, a hormone that stimulates the production of neurons in the higher vocal center; these changes contribute to song learning and singing behavior. Lacking this hormone, the brains of young female zebra finches lose neurons in the higher vocal center; females do not sing their species songs.

© Cengage Learning 2017

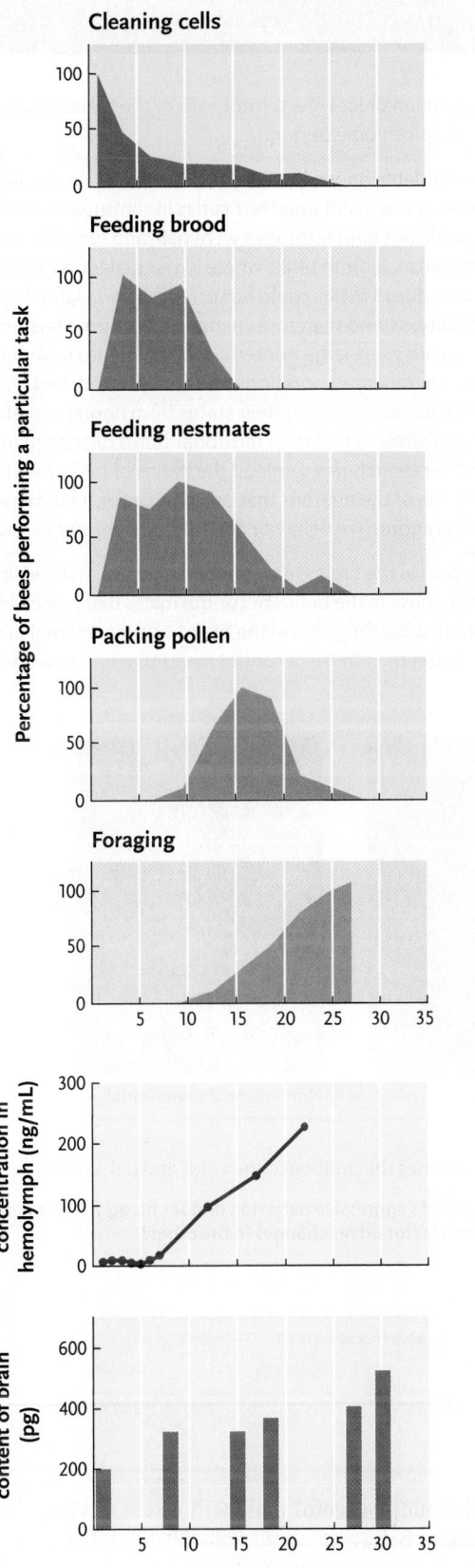

FIGURE 56.9 Age and task specialization in honeybee (*Apis mellifera*) workers. Young bees typically clean cells and feed the brood, and older workers leave the hive to forage for food. These behavioral changes are correlated with rising levels of juvenile hormone in their hemolymph and rising octopamine concentration in their brains.
© Cengage Learning 2017

production of hormones, which change the intracellular environment of assorted target cells. The hormones then directly or indirectly change the genetic activity and enzymatic biochemistry in their targets. If the cells in question are neurons, the changes in their biochemistry translate into changes in the animal's behavior.

Hormone Levels Affect Reproductive Activity in Many Animals

The African cichlid fish (*Haplochromis burtoni*) provides an example of how hormones regulate reproductive behavior. Some adult males maintain nesting territories on the bottom of Lake Tanganyika in East Africa. Territory holders are brightly colored, and they exhibit elaborate behavioral displays that attract egg-laden females to their territories. These males defend their real estate aggressively against neighboring territory holders and against incursions by males that have no territories of their own. By contrast, nonterritorial males are much less colorful and aggressive; they do not control a patch of suitable nesting habitat, and they make no effort to court females.

The behavioral differences between the two types of males are caused by differences in their levels of circulating sex hormones. Recall from Section 49.3 that gonadotropin-releasing hormone (GnRH) stimulates the testes to produce testosterone and sperm. When the circulating testosterone is carried to the brain, it modulates the activity of neurons that regulate sexual and aggressive behavior. In territorial fish the GnRH-producing neurons in the hypothalamus are large and biochemically active, but in nonterritorial fish they are small and inactive. In the absence of GnRH, the testes do not produce testosterone; the testosterone-deficient fish do not court females with sexual displays, nor do they usually attack other males.

What causes the differences in the neuronal and hormonal physiology of the two types of male fish? Russell Fernald and his students at Stanford University conducted laboratory experiments in which they manipulated the territorial status of males **(Figure 56.10)**. Four weeks later, they compared the coloration and behavior, as well as the size of the GnRH-producing cells in the brains of the experimental fishes, with those of the control males that had retained their original status. Males that had held territories in the past, but had then been defeated by another male, quickly lost their bright colors and stopped being combative. Moreover, their GnRH-producing cells were smaller than those of the successful territory-holding controls. Conversely, males that gained a territory in the experiment quickly developed bright colors and displayed aggressive behaviors toward other males. And the GnRH-producing cells in their brains were larger than those of fishes that had maintained their status as non–territory-holding controls.

The neuronal, hormonal, and behavioral differences between the two experimental groups of males are therefore correlated with a key environmental variable: success or failure in the acquisition and maintenance of a territory. The

FIGURE 56.10 | **Experimental Research**

Effects of the Social Environment on Brain Anatomy and Chemistry

Question: How does the acquisition or loss of a territory affect the brain anatomy and chemistry of an African cichlid fish (*Haplochromis burtoni*)?

Experiment: Fernald and his students housed groups of male cichlids in aquariums in their laboratory. Males that established and maintained territories in the aquariums were brightly colored, whereas those that could not hold territories were pale and drab. The researchers then moved some small territorial males into tanks where larger males had already established territories. The newly introduced males could not establish and maintain territories under these experimental conditions, and therefore changed status from territorial to nonterritorial. The researchers also moved some large nonterritorial males into tanks with smaller territorial males. Under these experimental conditions, the newly introduced males quickly established and maintained territories, changing their status from nonterritorial to territorial. Other males, left in their original tanks so that their territorial status did not change, served as controls. Four weeks later, the researchers examined the brains of the experimental and control fish and measured the size of the neurons that produce GnRH, a hormone that stimulates bright coloration as well as aggressive behavior and mating behavior in males.

A. African cichlid fish (*Haplochromis burtoni*)

Nonterritorial male

Territorial male

Russell Fernald, Stanford University

Result: The GnRH-producing cells in the brains of experimental males that had lost their territories were much smaller than those in the brains of control males that had maintained their territories. By contrast, the GnRH-producing cells in the brains of experimental males that had gained territories were much larger than those of control males that had never held territories.

B. GnRH-secreting cells

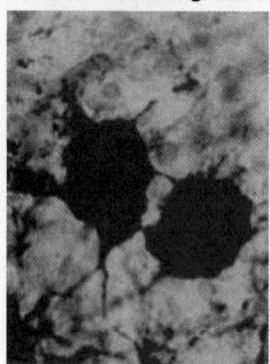

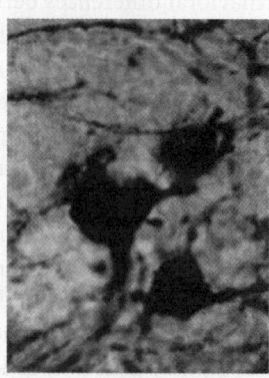

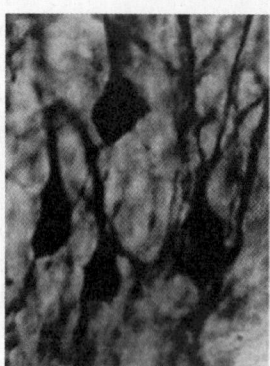

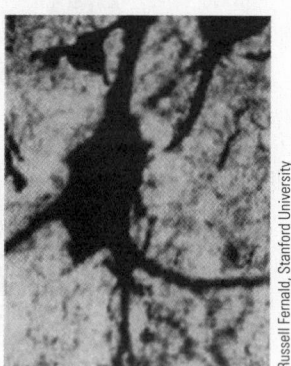

Russell Fernald, Stanford University

Territorial control   10 μm

Territorial to nonterritorial experimentals

Nonterritorial control

Nonterritorial to territorial experimentals

Conclusion: Changes in social status influence the size of brain cells producing hormones that influence the color and behavior of males.

think like a scientist

Observe
Hypothesize
Predict
Experiment
Interpret

Do changes in GnRH production influence a male cichlid's aggressive behavior, or does his aggressive behavior influence his GnRH production? In other words, which factor drives changes in the other?

Source: R. C. Francis et al. 1993. Social regulation of the brain-pituitary-gonadal axis. *Proceedings of the National Academy of Sciences USA* 90:7794–7798.

neurons that process this information transmit their input to the hypothalamus where it affects the size of the GnRH cells, which in turn dictates the hormonal state of the male. A decrease in GnRH production can turn a feisty territorial male into a subdued drifter, biding his time and building his energy reserves for a future attempt at defeating a weaker male and taking over his territory. If successful in regaining territorial status, the male's GnRH levels will increase again,

and the once-peaceful male will revert to vigorous sexual and aggressive behavior.

Note the general similarity of these processes to those described for the white-crowned sparrow's song learning: the fish's brain possesses cells that can change their biochemistry, structure, and function in response to well-defined social stimuli. These physiological changes make it possible for the fish to modify its behavior, depending on its social circumstances. In the following sections, we consider the ecology and evolution of specific categories of behavior.

STUDY BREAK 56.2

1. What research results suggest that certain neurons in a young male bird's brain are influenced only by acoustical signals from members of its own species and only during a critical period?
2. What is the role of the ZENK enzyme in song learning?
3. What is the effect of estrogen on the development of neurons in the higher vocal center of young zebra finches?
4. How does the loss of its territory change the brain chemistry of an African cichlid fish?

56.3 Migration and Wayfinding

Most animals move through their environments at some stage of their life cycles. Although some species move only short distances to find suitable environmental conditions, many others undertake large-scale movements on a seasonal schedule.

Migrating Animals Make Long Round-trips on a Seasonal Cycle

Many animal species undertake a seasonal **migration,** traveling from the area where they were born to a distant and initially unfamiliar destination, and returning to their birth site later. The Arctic tern (*Sterna paradisaea*), a seabird, makes an annual round-trip migration of up to 70,000 km **(Figure 56.11).** Many

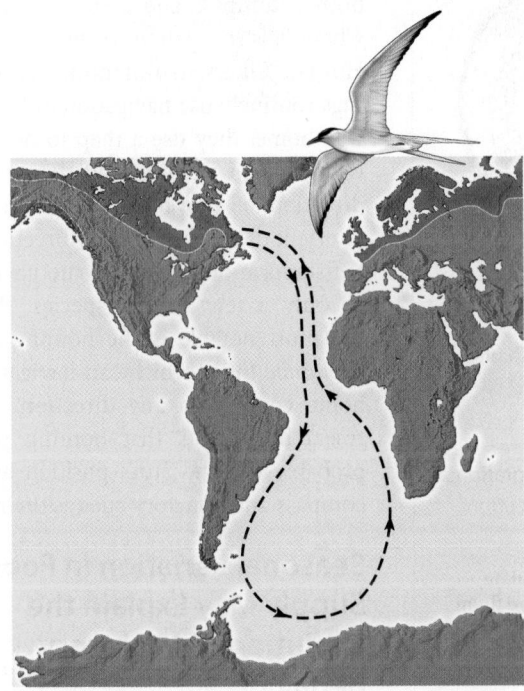

FIGURE 56.11 Long-distance migration. Arctic terns (*Sterna paradisaea*) migrate from the high Arctic to Antarctica each year, a round-trip journey up to 70,000 km. This species' summer breeding range is shaded red on the map.
© Cengage Learning 2017

FIGURE 56.12 Migrating arthropods. Spiny lobsters (*Panulirus argus*) make seasonal migrations between coral reefs and the open ocean floor. As many as 50 individuals march in single file for several days.

other vertebrate species, including gray whales and salmon, also undertake long and predictable journeys. Even some arthropods migrate long distances. For example, spiny lobsters (*Panulirus* species) form long conga lines as they move between coral reefs and the open ocean floor on a seasonal cycle **(Figure 56.12).**

Animals Use Wayfinding Mechanisms to Guide Their Movements

Moving animals use various wayfinding mechanisms to arrive at their destination. Biologists group these mechanisms into three general categories: *piloting, compass orientation,* and *navigation.* Many species probably use a combination of these mechanisms to guide their movements.

The simplest wayfinding mechanism is **piloting,** in which animals use familiar landmarks to guide their journey. For example, gray whales (*Eschrichtius robustus*) migrate from Alaska to Baja California and back using visual cues provided by the Pacific coastline of North America.

Some animals also use specific landmarks to identify their nest site or places where they have stashed food. In a famous experiment published in 1938, Niko Tinbergen showed that female digger wasps (*Philanthus triangulum*), which nest in soil, use visual landmarks to find their nests after flying off in search of food. Tinbergen arranged pinecones in a circle around one nest while the female was still inside. As she left, she flew around the area, apparently noting nearby landmarks. Tinbergen then moved the circle of pinecones a short distance away. Each time the female returned, she searched for her nest within the pinecone circle—but she never found it unless the pinecones were returned to their original position.

In a more sophisticated wayfinding mechanism, **compass orientation,** animals move in a particular direction, often over a specific distance or for a prescribed length of time. Some day-flying migratory birds, for example, orient themselves using the Sun's position in the sky in conjunction with an internal biological clock. The internal clock allows the bird to use the Sun as a compass, compensating for changes in its position through the day; the clock may also allow some birds to estimate how far they have traveled since beginning their journey. Other migratory animals use polarized light or Earth's magnetic field as a compass.

Experimental Analysis of the Indigo Bunting's Star Compass

Question: Do indigo buntings *(Passerina cyanea)* use the positions of stars in the night sky to orient their migrations?

Experiment: Emlen placed individual buntings in cone-shaped test cages. He lined the sides of the cages with blotting paper, placed inkpads on the bottom, and kept the cages in an outdoor enclosure so that the birds had a full view of the sky. Whenever a bird made a directed movement, its inky footprints indicated the direction in which it was trying to fly. Emlen predicted that the footprints would show the buntings' inclination to migrate south in autumn and north in spring.

Results: On clear nights in autumn, the footprints pointed to the south; on clear nights in spring, they pointed north. On cloudy nights, when buntings could not see the stars, their footprints were evenly distributed in all directions.

Indigo bunting

© John L. Absher/Shutterstock.com

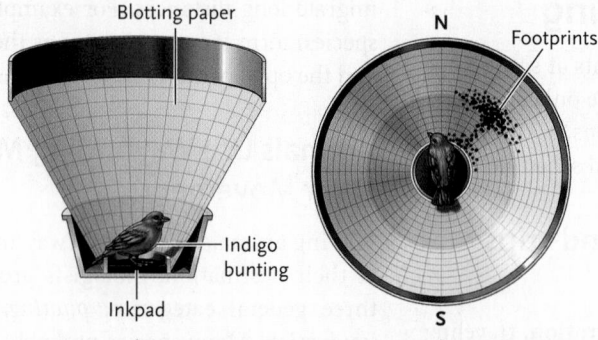

Side (left) and overhead (right) views of the test cage with blotting paper on the sides and an inkpad on the bottom

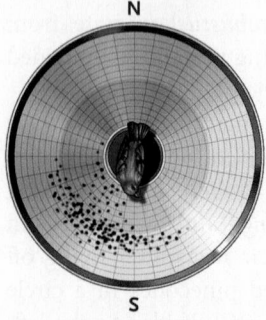
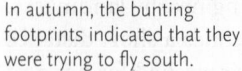

In autumn, the bunting footprints indicated that they were trying to fly south.

In spring, the bunting footprints indicated that they were trying to fly north.

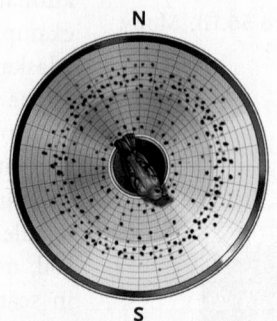

On cloudy nights, when buntings could not see the stars, their footprints indicated a random pattern of movement.

Conclusion: Indigo buntings use the positions of the stars to direct their seasonal migrations. When they could see the stars above their test cages, they moved in the predicted direction; but when clouds obscured their view of the stars, they moved in random directions.

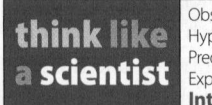

think like a scientist

Observe
Hypothesize
Predict
Experiment
Interpret

In this experiment, does the position of the stars stimulate buntings to make their migration, or does it simply give them cues about the direction in which they should fly?

Source: S. T. Emlen. 1967. Migratory orientation of the indigo bunting, *Passerina cyanea*. Part I: Evidence for use of celestial cues. *The Auk* 84:309–342.

© Cengage Learning 2017

Some birds that migrate at night use the positions of stars to determine their direction. The indigo bunting *(Passerina cyanea)*, for example, flies about 3,500 km from the northeastern United States to the Caribbean or Central America each fall, and makes the return journey each spring. Stephen Emlen of Cornell University demonstrated that these birds use celestial cues to direct their migration **(Figure 56.13)**. Emlen confined individual buntings in cone-shaped test cages. He lined the sides of the cages with blotting paper, placed inkpads on the bottom, and kept the cages in an outdoor enclosure so that the birds had a full view of the sky. Whenever a bird made a directed movement, its inky footprints indicated the direction in which it was trying to fly. Emlen found that on clear nights in fall, the footprints pointed to the south; on clear nights in spring, they pointed north. On cloudy nights, when the buntings could not see the stars, their footprints were evenly distributed in all directions, indicating that their compass required a view of the stars.

The most complex wayfinding mechanism is **navigation,** in which an animal moves toward a specific destination, using both a compass and a "mental map" of where it is in relation to the destination. Human hikers in unfamiliar surroundings routinely use navigation to find their way home: they use a map to determine their current position and the necessary direction of movement and a compass to orient themselves in that direction. Scientists have documented true navigation in only a few animal species. Perhaps the most notable is the homing pigeon *(Columba livia),* which can navigate to its home coop from any direction. Recent research suggests that homing pigeons probably use the Sun's position as their compass and olfactory cues as their map.

Seasonal Variation in Food Supply May Explain the Evolution of Migratory Behavior

Migratory behavior entails obvious costs, such as the time and energy devoted to the journey and the risk of death from exhaustion or predator attack. Why then do some species migrate? What benefits accrue to an individual that undertakes a costly migration?

For migratory birds, the most widely accepted hypothesis focuses on seasonal changes in food supplies. The amount of insect food available in northern forests increases explosively during the warm spring and summer, providing abundant resources to produce eggs and rear offspring. Then, during the late fall and winter, insects all but disappear. A few bird species that forage on seeds and dormant insects do not head south. However, energy supplies are more predictably available in tropical overwintering grounds, and migratory birds may have a better chance of surviving there. The following spring they return north to exploit the food bonanza on their summer breeding grounds.

The two-way migratory journeys may provide other benefits as well. Avoiding the northern winter is probably adaptive because endotherms must increase their metabolic rates just to stay warm in cold climates (see Section 48.8). But in summer the days are longer at high latitudes than they are in the tropics (see Section 51.2), giving adult birds more time to collect enough food to rear a brood.

Seasonal changes in food supply also underlie the migration of monarch butterflies (*Danaus plexippus*), the larvae of which eat only milkweed (*Asclepias* species) leaves. In eastern North America, milkweed plants grow only during spring and summer. Many adult monarchs head south in late summer, when milkweeds are beginning to die, migrating as much as 4,000 km from eastern and central North America to central Mexico, where they cluster in spectacular numbers **(Figure 56.14)**. Unlike migrant birds, these insects do not feed on their overwintering grounds. Instead, their metabolic rate decreases in the cool mountain air, and the butterflies become inactive for months, thereby conserving precious energy reserves. When spring arrives, the butterflies become active again and begin the return migration to northern breeding habitats. The northward migration is slow, however, and many individuals stop along the way to feed and lay eggs. But their offspring, and their offspring's offspring, continue the northward migration through the summer; some descendants eventually reach southern Canada for a final round of breeding. The summer's last generation then returns south to the spot where their ancestors, two to five generations removed, spent the previous winter.

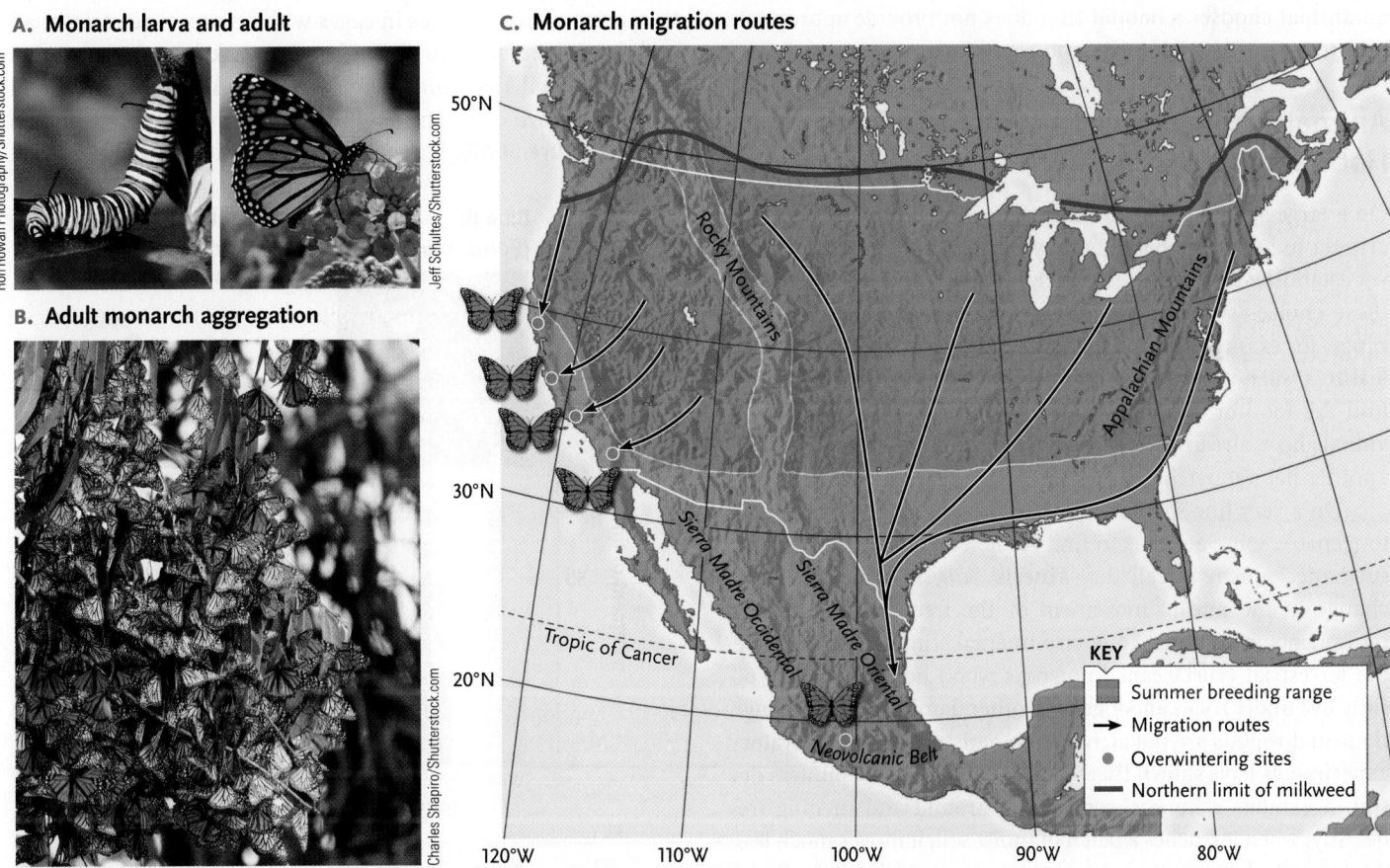

A. Monarch larva and adult

B. Adult monarch aggregation

C. Monarch migration routes

KEY
- Summer breeding range
→ Migration routes
• Overwintering sites
— Northern limit of milkweed

FIGURE 56.14 Monarch butterfly migrations. (A) Monarch butterflies (*Danaus plexippus*) feed primarily on milkweed plants. **(B)** When milkweed plants in their breeding range die back at the end of summer, millions of monarchs migrate south, where they aggregate in forests. **(C)** Butterflies that live and breed east of the Rocky Mountains migrate to Mexico. After passing the winter in a semidormant state, they migrate northward the following spring. Monarchs living west of the Rocky Mountains winter in coastal California.

© Cengage Learning 2017

For other animals, the migration to breeding grounds may provide special conditions necessary for reproduction. For example, gray whales migrate south to breeding grounds in quiet, shallow lagoons where predators are rare and warm water temperatures will not stress their calves.

STUDY BREAK 56.3

1. What is the difference between piloting, compass orientation, and navigation?
2. What is the most probable selection pressure that has fostered seasonal migrations in birds?

56.4 Habitat Selection and Territoriality

The geographical range of nearly every animal species includes a mosaic of habitat types. The breeding range of white-crowned sparrows, for example, encompasses forests, meadows, housing developments, and city dumps. An animal's choice of habitat is critically important because the habitat provides food, shelter, nesting sites, and the other organisms with which it interacts. If an animal chooses a habitat that does not provide appropriate resources, it will not survive and reproduce.

Animals Use Multiple Criteria for Selecting Habitats

On a large spatial scale, animals almost certainly use multiple criteria to select the habitats they occupy, but no research has yet established any general principles about how animals make these choices. When a migrating bird arrives at its breeding range, for example, it probably cues on large-scale geographical features, such as a pond or a patch of large trees. If it does not find the food or nesting resources it needs—or if other individuals have already depleted those resources—it may move to another habitat patch.

On a very fine spatial scale, basic responses to physical factors enable some animals to find suitable habitats. The simplest such mechanism is called a **kinesis** (*kinesis* = movement), a change in the rate of movement or the frequency of turning movements in response to environmental stimuli. For example, the terrestrial crustaceans known as wood lice (Isopoda) typically live under rocks and logs or in other damp places. Although these arthropods are not attracted to moisture *per se*, laboratory experiments have shown that when a wood louse encounters dry soil, it exhibits a kinesis, scrambling around and turning frequently; when it reaches a patch of moist soil, it moves much less. As a result, these animals accumulate in moist habitats. Biologists infer that this behavior is adaptive because wood lice exposed to dry soil quickly dehydrate and die. Other animals may exhibit a **taxis** (*taxis* = arrangement), a response that is directed either toward or away from a specific stimulus. For example, cockroaches (Blattodea) exhibit negative phototaxis:

they actively avoid light and seek darkness, which are behaviors that make them harder for visually oriented predators to detect.

Genetics and Learning Influence Habitat Selection

Biologists generally assume that habitat selection is adaptive and has been shaped by natural selection in most animal species. For example, some animals instinctively select habitats where they are well camouflaged, a means of avoiding detection by predators (see Figure 53.3); predators would discover and eliminate any individual that fails to select a matching background—along with any alleles responsible for the mismatch. Many insects have a genetically determined preference for the plants that they eat during their larval stage. Adults often restrict their mating and egg-laying activities to these food plants, effectively selecting the habitats where their offspring will live and feed, as described in the discussion of sympatric speciation (see Section 22.3).

Vertebrates sometimes exhibit such innate preferences, as demonstrated by two closely related European bird species, blue tits (*Parus caeruleus*) and coal tits (*Parus ater*). Adult blue tits feed mostly in oak trees, whereas coal tits prefer to feed in pines. When Linda Partridge of Edinburgh University reared the young of both species in cages without any vegetation at all and then offered them a choice between oak branches and pine branches, coal tits immediately gravitated toward pines and blue tits toward oaks, strongly suggesting that the preference is innate **(Figure 56.15)**. Further research demonstrated that each

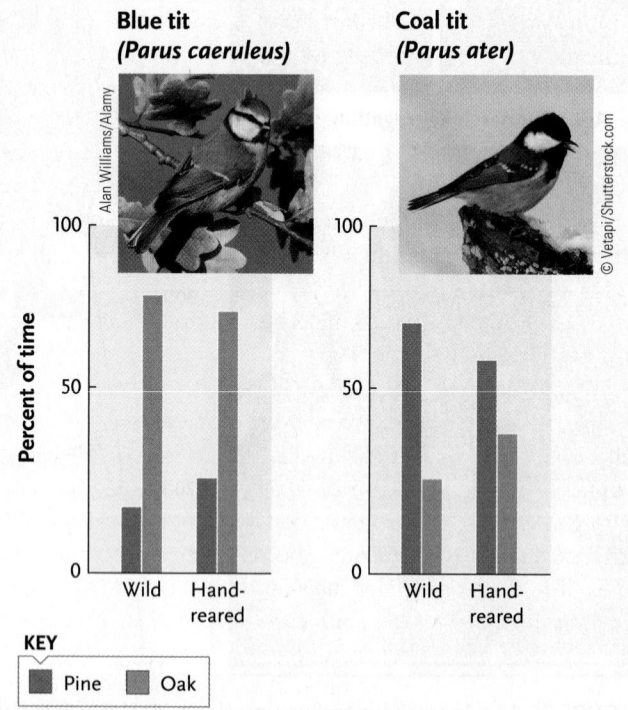

FIGURE 56.15 Habitat selection by birds. Wild blue tits show a strong preference for oak trees, and coal tits show a strong preference for pine trees. Birds that were hand-reared in a vegetation-free environment showed identical, though slightly weaker, preferences.
© Cengage Learning 2017

species feeds most successfully in the tree species it prefers. Thus, natural selection probably fostered these preferences.

Habitat preferences can be molded by experiences early in life, however. For example, the tadpoles of red-legged frogs (*Rana aurora*) usually live in aquatic habitats cluttered with sticks, strands of algae, and plant stems; when given a choice in the laboratory, they prefer striped backgrounds to plain ones. By contrast, tadpoles of the closely related cascade frog (*Rana cascadae*) live over gravel bottoms, and they prefer plain substrates over striped ones. However, when red-legged frogs are reared over plain substrates and cascade frogs over striped substrates, they no longer exhibit preferences for their usual substrates.

Animals Sometimes Defend Patches of Habitat for Their Exclusive Use

Under some circumstances, animals may defend a **territory** from other members of their species, retaining more or less exclusive use of the resources it contains. Territorial behavior occurs in all major groups of vertebrates, many insects, and some other invertebrates, but it is by no means universal. In many organisms, territorial behavior occurs only during the breeding season.

Animals establish and defend territories only if some critical resource is in short supply. Moreover, the resource must be fixed in space so that the area around it can be defended. For example, during the breeding season, most songbirds defend a territory within which they build a nest and collect food for their young. By contrast, most sea birds, such as terns and penguins, do not defend feeding territories. Although they defend a tiny area around the nest, which they build on shore, they never attempt to defend the patches of ocean where they collect food; fishes come and go at will and thus do not constitute a defendable resource.

Territorial defense is always a costly activity. Patrolling territory borders, performing displays hundreds of times per day, and chasing intruders take time and energy. Moreover, territorial displays increase an animal's likelihood of being injured or captured by a predator.

Experiments conducted by Catherine Marler and Michael Moore of Arizona State University illustrate the cost of territorial behavior in Jarrow's spiny lizard (*Sceloporus jarrovi*). Male lizards ordinarily exhibit strong territoriality only during the autumn mating season, when elevated blood levels of testosterone stimulate their aggressive behavior. The researchers implanted small doses of testosterone under the skin of experimental animals in June and July, during the *nonmating* season; controls received a placebo treatment. Testosterone-enhanced males were more active and displayed more often than control males. But even though experimental males spent less time feeding, they used about 30% more energy per day than control males. Over the course of about 7 weeks, a significantly higher percentage of experimental males died—a clear sign that engaging in territorial behavior is costly.

Maintaining a territory has definite benefits, however, such as having access to nesting sites, food supplies, and refuges from predators. For example, the surgeonfish (*Acanthurus lineatus*), which lives in the coral reefs around American Samoa, may engage in as many as 1,900 chases per day to defend a small territory from other algae-eating fish species. But territory holders consume up to five times more food than nonterritorial fish.

STUDY BREAK 56.4

1. Why do wood lice tend to occur in moist parts of their habitat?
2. What are the costs of maintaining a territory, and what are the benefits?

56.5 The Evolution of Communication

Territorial behavior is a specific example of **communication,** the conveyance of information to other individuals. In the formal language of animal behavior studies, all communication systems involve an interaction between a *signaler,* the animal that transmits information, and a *signal receiver,* the animal that intercepts the information and makes a behavioral response. Natural selection has adjusted the ability of signalers to transmit information and the ability of receivers to get the message.

Animal Signals Can Activate Different Sensory Receptors in Receivers

Biologists categorize animal signals according to the sensory receptors, or "channels," through which the signal acts: *acoustical, visual, chemical, tactile,* or *electrical.* Each channel has specific advantages.

Bird songs are examples of **acoustical signals;** a signaler produces a sound that is heard by a signal receiver. Many animals use the acoustical channel, including a host of nocturnal and burrow-dwelling insects and amphibians. These signals reach distant receivers, even at night and in cluttered environments where visual signals are less effective.

Because humans frequently use facial expressions and body language to send messages, **visual signals** are a familiar form of communication. In many animals, visual signals are *ritualized;* in other words, they have become exaggerated and stereotyped over evolutionary time, forming an easily recognized visual display **(Figure 56.16).** Visual displays can even be useful at night or in the darkness of the deep sea; some animals, such as fireflies and certain fishes, send bioluminescent signals to distant receivers.

Many species release **chemical signals,** which carry messages to signal receivers through the olfactory channel. Scent marking (spraying) by male dogs is an example. In particular, mammals and insects often communicate through **pheromones,** distinctive volatile chemicals released in minute amounts to

FIGURE 56.16 **Visual displays.** The courtship display of a male wandering albatross *(Diomedea exulans)* includes ritualized postures and movements of the wings and body.

FIGURE 56.18 **Tactile signals.** Grooming by hyacinth macaws *(Anodorhynchus hyacinthinus)* removes parasites and dirt from feathers. The close physical contact also promotes friendly relations between groomer and groomee.

influence the behavior of members of the same species. For example, a worker ant's body contains a battery of glands, each releasing a different pheromone **(Figure 56.17).** One set of pheromones recruits fellow workers to battle colony invaders; another set stimulates workers to collect food that has been discovered outside the colony. Other animals release pheromones to attract mates. Female silkworm moths *(Bombyx mori)* produce bombykol, a single molecule of which can generate a message in specialized receptors on the antennae of any male silkworm moth that is downwind (see Figure 41.18).

In many species, touch conveys important messages from a signaler to a receiver. **Tactile signals** operate only over very short distances, but for social animals living in close company, they play a significant role in the development of friendly bonds between individuals **(Figure 56.18).**

Some freshwater fish species, especially those that occupy murky tropical rivers where visual signals could not be seen, use weak **electrical signals** to communicate. These fishes have electric organs that can release charges of variable intensity, duration, and frequency, allowing substantial modulation of the

message that a signaler sends. Among the New World knife-fishes (Gymnotiformes), including the electric eel *(Electrophorus electricus),* electrical discharges can signal threats, submission, or a readiness to breed.

Honeybees Use Several Communication Channels to Transmit Complex Messages

When animals need to convey a complex message, they may use several channels of communication simultaneously. For example, as Karl von Frisch demonstrated, the dance of the honeybee *(Apis mellifera)* involves tactile, acoustical, and chemical communication **(Figure 56.19).**

When a foraging honeybee discovers pollen or nectar, it returns to its colony and performs a complex dance on the vertical surface of the honeycomb in the complete darkness of the hive. A bee first dances a half circle in one direction, then dances in a straight line while "waggling" its abdomen back and forth, and finally dances a half circle in the other direction (see Figure 56.19A). As the dancer moves, it also produces a brief buzzing sound with each waggle. The dance attracts a crowd of workers that maintain physical contact with the dancer; the dancer responds to their acoustical signals by regurgitating a sample of the food it found, providing a chemical cue about its discovery. The dance itself directs other workers to the food source. Von Frisch determined that the duration of the waggles and buzzes that the bee makes on the straight run conveys information about the *distance* to the food: the longer the time spent waggling and buzzing, the further the food is from the hive (Figure 56.19B). The angle of the straight run relative to vertical on the honeycomb

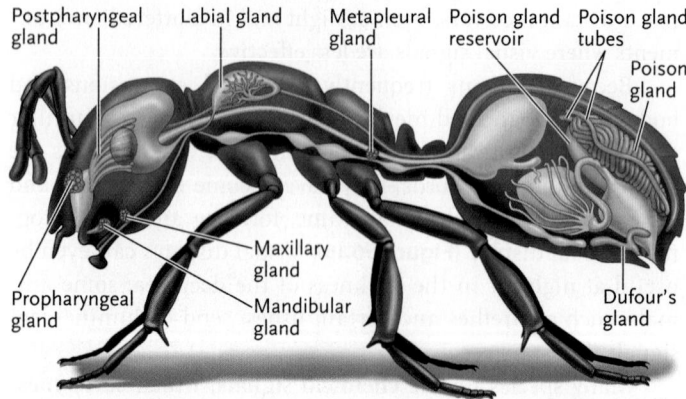

FIGURE 56.17 **Chemical signals.** An ant's body contains a host of pheromone-producing glands, each of which manufactures and releases its own volatile chemical or chemicals.

© Cengage Learning 2017

indicates the *direction* of the food source relative to the position of the Sun (see Figure 56.19C).

According to von Frisch's interpretation of his results, honeybees performed two types of dances: the "round dance" for food sources close to the hive and the "waggle dance" (described above) for food sources at a greater distance. However, in an analysis published in 2008—85 years after von Frisch's pioneering work—Kathryn E. Gardner and her colleagues at Cornell University demonstrated that the two dances represent ends of a continuum, which they called an "adjustable waggle dance." The round dance described by von Frisch is really just a waggle dance with a very short waggle phase.

Biologists Use Evolutionary Hypotheses to Analyze Communication Systems

Signal receivers often respond to communication from signalers in predictable ways. For example, a male white-crowned sparrow generally avoids entering a neighboring territory simply because it hears the song of the resident male. Similarly, young male baboons often retreat without a fight when they see an older male's visual threat display **(Figure 56.20)**, even though they may lose the chance to mate with a female. Why do these receivers behave in ways that appear to be beneficial to their rivals, but not to themselves?

When biologists try to explain behavioral interactions, their hypotheses focus on how an animal's actions may allow it to contribute offspring to the next generation. In our first example, the retreating male sparrow avoids wasting time and energy on a battle he is likely to lose. Moreover, ousting the current resident might be more tiring and risky than finding a suitable unoccupied breeding site. This hypothesis predicts that resident males should almost always win physical contests. In cases when an intruder does win a territory from a resident, it may do so only after a prolonged series of exhausting clashes. Observations of territorial species—whether birds, lizards, frogs, fishes, or insects—generally support these predictions.

Applying a similar argument to competition among male baboons, we can predict that smaller or younger males will concede females to threatening older rivals without fighting. The signal receiver retreats after receiving the threat because he judges that he would be demolished in real combat—after all, a male baboon's canine teeth are not just for show. Evolutionary analyses therefore suggest that both the signaler and the signal receiver benefit from the transfer of information in their communication system.

An evolutionary analysis also helps to explain the strange yell of ravens (*Corvus corax*), which scavenge carcasses of deer, elk, or moose in northern forests during winter. When one of these large birds comes across a food bonanza, it may call loudly, attracting a crowd of hungry ravens. The calling behavior and sharing of resources puzzled Bernd Heinrich of The University of Vermont. Wouldn't a quiet raven eat more, survive longer, and produce more offspring than a noisy bird that attracted competitors to the food? If natural selection favored the raven's calling behavior, we might expect that the cost of calling (in terms of lost food) would be offset by a reproductive benefit for

A. Waggle dance

B. Coding distance to the food source

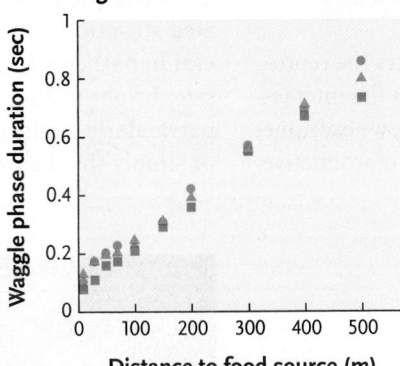

Distance to food source (m)

C. Coding direction to the food source

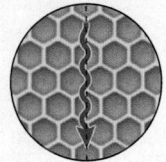 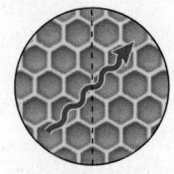 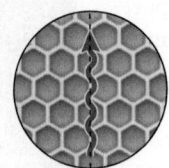

When the bee moves straight down the comb, other bees fly to the source directly away from the Sun.

When the bee moves 45° to the right of vertical, other bees fly at a 45° angle to the right of the Sun.

When the bee moves straight up the comb, other bees fly straight toward the Sun.

FIGURE 56.19 Dance communication by honeybees. Foraging honeybees (*Apis mellifera*) transmit information about the location of a food source by dancing on a vertical honeycomb in the hive. **(A)** The waggle dance conveys information about the location of a newly discovered food source. **(B)** The length of the waggle phase on the straight run of the dance indicates the distance to the food source. The graph shows data from three separate hives, each coded in a different color. **(C)** The dancing bee indicates the direction to a distant food source by the angle of the waggle run.
© Cengage Learning 2017

FIGURE 56.20 Threat displays. The threat display of a dominant male mandrill (*Mandrillus sphinx*), used to drive away rival males, exposes large canine teeth.

the individual caller. Heinrich noticed that paired, territory-owning adults did not yell loudly when they found goat carcasses that he had hauled into the Maine woods; instead, they fed quietly. Only young, wandering ravens that happened upon a carcass in another bird's territory advertised their discovery. The signals of these birds attracted other nonterritorial ravens, which collectively overwhelmed the resident pair's attempts to drive them away. Only then was a wanderer likely to have a chance to feed in an area that would otherwise be off-limits.

56.6 The Evolution of Reproductive Behavior and Mating Systems

In many animal species, communication coordinates the reproductive activities of males and females and governs the interactions between parents and offspring. In this section, we examine how several elements of behavior contribute to the reproductive success of individuals.

Males and Females Use Different Reproductive Strategies

In sexually reproducing species, males and females often differ in their overall **reproductive strategies,** the set of behaviors that lead to reproductive success. This difference arises in part from a fundamental difference in the amount of **parental investment,** the time and energy devoted to the production and rearing of offspring, provided by the two sexes. Because eggs are much larger than sperm, females almost always contribute more energy than males to the production of a gamete.

A male might increase the number of offspring that carry his alleles simply by mating with multiple females, especially if he does not spend time and energy providing parental care to his offspring. Thus, in many animal species, males compete intensely for access to females, and any trait that increases a male's access or attractiveness to females has a big reproductive payoff.

Entirely different selection pressures operate on females. Their reproductive output is generally limited by the number of eggs they can produce, and mating with multiple males will not increase that number. But the success of her offspring may depend on the attributes of their father or the territory he holds. Thus, females of many species choose their mates carefully. In some cases, females mate with males whose territories include abundant resources, ensuring an ample food supply for their young. In other cases, females choose robust males that will contribute "good genes" (that is, alleles that confer a high likelihood of surviving and reproducing) to her offspring, increasing their chances of long-term success.

Male Competition for Females and Female Mate Choice Foster Sexual Selection

Male competition for access to females coupled with the females' choice of mates establishes a form of natural selection called **sexual selection,** that is, selection for mating success (see Section 21.3). As a result of sexual selection, males are larger than females in many species, and males have ornaments and weapons, such as horns and antlers, that are useful for attracting females, as well as for butting, stabbing, or intimidating rival males. Males typically show off these elaborate structures in complex **courtship displays** to attract the attention of females. For example, male peafowl *(Pavo cristatus)* strut in front of females while spreading a gigantic fan of tail feathers, which they shake, rattle, and roll **(Figure 56.21).**

Why should females choose males that display exaggerated structures conspicuously? Biologists have developed several hypotheses to explain the attraction. First, a male's large size, bright feathers, or large horns might indicate that he is particularly healthy, that he can harvest resources efficiently, or simply that he has managed to survive to an advanced age.

Angela Forker/Dreamstime.com

FIGURE 56.21 Sexual selection for ornamentation. The attractiveness of a peacock *(Pavo cristatus)* to females depends in part on the number of eyespots on his extraordinary tail. The offspring of males with elaborate tails are more successful than the offspring of males with plainer tails.

These traits are, in effect, signals of male quality, and if they reflect a male's genetic makeup, he is likely to fertilize a female's eggs with sperm containing successful alleles. In some cases, big, showy males hold large, rich territories, and females that choose them gain access to the resources their territories contain.

The degree to which females *actively* choose genetically superior mates varies among species. In the northern elephant seal *(Mirounga angustirostris),* for example, female choice is passive. Large numbers of females gather on beaches to give birth to their pups before becoming sexually receptive again. Males locate these clusters of females and fight to keep other males away. The winners have exceptional reproductive success, but only after engaging in violent combat with rival males. In this kind of mating system, females are practically guaranteed that the father of their offspring is a large and powerful male in superb physiological condition; these attributes may well be associated with alleles that will increase their offspring's chances of living long enough to reproduce.

In other species, females exercise more active mate choice, copulating only after inspecting a group of potential partners. Among birds, active female mate choice is most apparent at **leks,** display grounds where each male courts attentive females from a small territory. Male sage grouse *(Centrocercus urophasianus),* a lekking bird of western North America, gather in open areas among stands of sagebrush. Each male defends just a few square meters, where it struts in circles while emitting booming calls and showing off its elegant tail feathers and big neck pouches **(Figure 56.22).** Females wander among the displaying males, presumably analyzing the males' visual and acoustical displays. Eventually, each female selects a mate from among the dozens of males present. Females repeatedly favor males that come to the lek daily, defend their small area vigorously, and display more frequently than the average lek participant. In other words, favored males can sustain their territorial defense and high display rate over long periods, an ability that may correlate with useful genetic traits.

Experimental studies of peafowl suggest that the top peacocks at a lek may indeed supply advantageous alleles to their offspring. In nature, peahens prefer males whose tails have many ornamental eyespots (see Figure 56.21). In an experiment on captive birds, some females were mated to males with highly attractive tails, but others were paired with males whose tails were less impressive. The offspring of both groups were reared under uniform conditions for several months and then released into an English woodland. After three months on their own, the offspring of fathers with impressive tails survived better and weighed significantly more than did those whose fathers had less attractive tails. Apparently, a peahen's mate choice does provide her offspring with a survival advantage.

Another hypothesis argues that females select showy males even though their ornate structures and elaborate displays may impede their locomotion or attract the attention of a predator. According to this hypothesis, any male that survives *despite* carrying such a handicap must have a very strong constitution indeed, and he will pass those successful alleles—as well as the alleles responsible for the ornamental handicap—to the female's offspring.

Patterns of Parental Care and Territoriality Influence Mating Systems

In the examples of mate choice just described, successful males inseminate many females, increasing their reproductive success dramatically. But one male mating with many females is only one of several **mating systems,** the ways in which males and females pair up. Some species are **promiscuous:** individuals do not form close pair bonds, and both males and females mate with multiple partners. Other species are **monogamous:** one male and one female form a long-term association. Finally, some species are **polygamous:** *either* males *or* females may have multiple mating partners. If one male mates with many females, the relationship is called **polygyny;** if one female mates with multiple males, it is called **polyandry.**

Mating systems appear to have evolved to maximize reproductive success, partly in response to the amount of parental care that offspring require and partly in response to other aspects of a species' ecology. For example, the young of most songbird species, like the white-crowned sparrow, are helpless upon hatching; all they can do is open their mouths and peep, signaling to their parents that they are ready to be fed. These young require lots of parental care, and they are more likely to flourish if both parents bring food to the nest. As you might expect, nearly all songbirds are monogamous, and males and females team up to provide parental care to their offspring.

In some other bird species, such as red-winged blackbirds *(Agelaius phoeniceus),* males establish large, resource-filled territories, and females select mates largely by the quality of the real estate a male holds. Any male with an exceptionally fine

FIGURE 56.22 Lekking behavior. Male sage grouse *(Centrocercus urophasianus)* use their ornamental feathers in visual courtship displays performed at a lek, where each male has his own small territory. The smaller brown females observe the prancing males before choosing a mate.

territory will be desirable, even if another female has already established herself there. A second female may judge that more resources are available in his territory than in a neighboring one, despite competition with the other female. However, if many females have already settled in a male's territory, intense competition from them may make it less attractive. Given this pattern of habitat and mate choice by females, red-winged blackbirds have a polygynous mating system; males may fertilize the eggs of multiple females and provide little if any direct care to their offspring.

The Comparative Method Reveals How Nest-building Behaviors Evolved in Swallows and Martins

Even before a pair of birds or other animals has mated, they often prepare a nest or den where they rear their young. The locations and structures of nests vary widely, sometimes even within a clade (that is, a monophyletic evolutionary lineage). For example, among the clade of birds that includes swallows and martins—two species of which are pictured in the opening photo

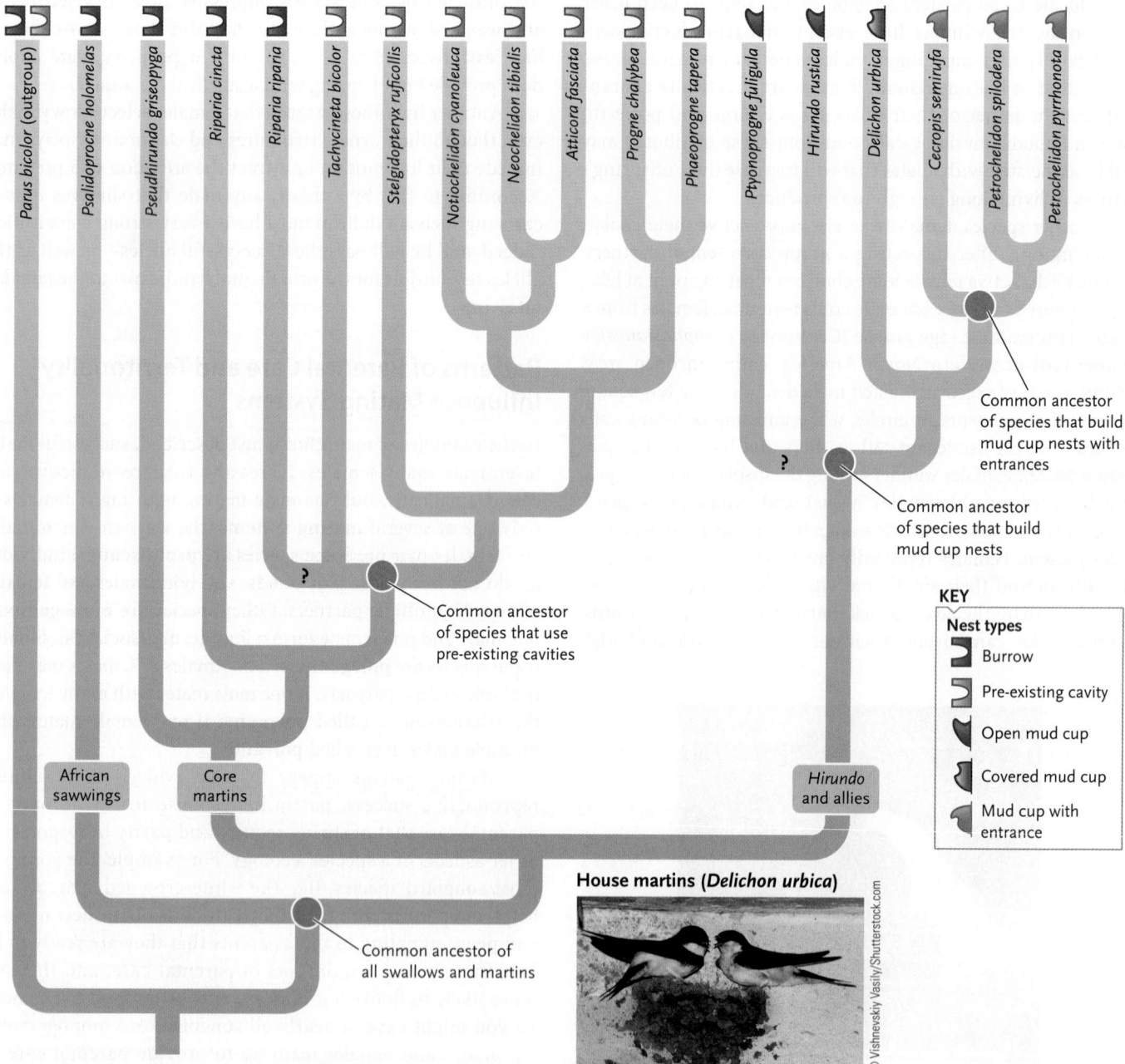

Common ancestor of species that build mud cup nests with entrances

Common ancestor of species that build mud cup nests

Common ancestor of species that use pre-existing cavities

African sawwings

Core martins

Hirundo and allies

KEY
Nest types

Burrow

Pre-existing cavity

Open mud cup

Covered mud cup

Mud cup with entrance

Common ancestor of all swallows and martins

House martins (*Delichon urbica*)

© Vishnevskiy Vasily/Shutterstock.com

FIGURE 56.23 The comparative method, applied to nesting behavior in swallows and martins. Superimposing nest-building behaviors on a phylogenetic tree for 17 species of swallows and martins allowed researchers to determine the pattern in which these diverse behaviors evolved. The question marks indicate phylogenetic relationships that the analysis could not resolve precisely.
© Cengage Learning 2017

of Chapter 22—some species excavate long nesting burrows in soil, others nest in preexisting cavities in rocks or trees, and still others build mud nests of varying complexity on vertical substrates. Given the extraordinary variety of nesting habits in these birds, biologists long wondered about the evolution of their nest-building behavior. Which type of nest is ancestral within the group? Do closely related species share the same nesting habits?

David W. Winkler of Cornell University and Frederick H. Sheldon of the Academy of Natural Sciences in Philadelphia used "the comparative method," combining phylogenetic analysis with observations on behavior, to investigate these questions. Using data from DNA hybridization studies, they constructed a phylogenetic tree for 17 swallow and martin species; a more distantly related bird species, the tufted titmouse (*Parus bicolor*), served as the outgroup **(Figure 56.23)**. (You may wish to review the construction of phylogenetic trees in Section 24.2.) The phylogenetic analysis revealed that both the outgroup and an African species in the clade, the black sawwing (*Psalidoprocne holomelas*), branched off the tree first; thus, their behaviors probably represent the ancestral condition. The remaining species fall into one of two subclades, the "core martins" and "*Hirundo* and its allies."

Winkler and Sheldon then superimposed data about nesting behaviors of the 18 species on the phylogenetic tree. Their results indicate that digging a nest burrow is the ancestral behavior because it is observed not only in the titmouse and African sawwings, but also in some "core martin" species that branched off the tree early in the group's evolution (on the left side of Figure 56.23). The analysis also shows that each of the other two nesting behaviors evolved once, presumably in the common ancestor of species that exhibit the behavior. Within the "core martins" clade, more recently derived species (all descended from one common ancestor) nest in preexisting cavities. Within the "*Hirundo* and its allies" clade, all species build mud nests on vertical substrates. The phylogenetic tree also illustrates how the three different patterns of mud nest construction (open mud cup, covered mud cup, and mud cup with entrance) evolved within the latter clade. Thus, superimposing nesting behaviors on the phylogenetic tree allowed the researchers to identify when each of the observed nest-building behaviors first evolved.

STUDY BREAK 56.6

1. For monogamous species, what characteristics of males should increase their attractiveness to females?
2. What activities do male and female sage grouse perform at a lek?
3. Why might a female red-winged blackbird settle on a male's territory if it was already occupied by another female?

56.7 The Evolution of Social Behavior

Social behavior, the interactions that animals have with other members of their species, has profound effects on an individual's reproductive success. Some animals are solitary, getting together only briefly to mate (rhinoceroses and leopards); others spend most of their lives in small family groups (gorillas); still others live in groups with thousands of relatives (termites and honeybees). Some species, such as some African antelopes and humans, live in large social units composed primarily of nonrelatives.

Group Living Carries Both Benefits and Costs

Ecological factors have a large impact on the reproductive benefits and costs of social living. Groups of cooperating predators frequently capture prey more effectively than they would on their own. For example, white pelicans (*Pelecanus erythrorhynchos*) often encircle a school of fish before moving in for the kill. Conversely, prey that are subject to intense predation often gain safety in numbers. Those living in groups have more watchful eyes to detect an approaching predator. In addition, a predator may be confused when multiple prey scatter in many directions. Finally, few predators have the capacity to capture every individual in a prey cluster, so that some prey escape while the predator pursues others.

Some prey species, such as musk oxen (*Ovibos moschatus*), join forces to defend themselves actively (see the chapter-opening photo). Even some insects, such as Australian sawfly caterpillars (*Perga dorsalis*), exhibit cooperative defensive behavior **(Figure 56.24)**. When predators disturb the caterpillars, all members of the group rear up and writhe about, regurgitating sticky, pungent oils that they have collected from the eucalyptus leaves they eat. Although the caterpillars can store these oils safely, they are toxic and repellent to bird predators.

A group of sawflies regurgitates more repellent eucalyptus oils than a single individual, which may explain why these insects form their simple societies. If this hypothesis is correct, solitary individuals should be at greater risk of being eaten than those that live communally. Birgitta Sillén-Tullberg of the University of Stockholm, Sweden, tested this prediction by offering sawfly caterpillars to young great tits (*Parus major*), a songbird species. Birds that received caterpillars one at a time

FIGURE 56.24 Social defensive behavior. Australian sawfly (*Perga dorsalis*) caterpillars clump together on tree branches. These larvae each regurgitate yellow blobs of sticky, aromatic fluid. The accumulation of fluid from a large group of caterpillars successfully deters some predators.

FIGURE 56.25 Colonial living. Royal penguins (*Eudyptes schlegeli*) on Macquarie Island, between New Zealand and Antarctica, experience both benefits and costs from living together in huge colonies.

FIGURE 56.26 The cost of living alone. A solitary olive baboon (*Papio anubis*) confronts a leopard (*Panthera pardus*) bravely, but without much chance of survival.

consumed an average of 5.6, but those that received them in groups of 20 ate an average of only 4.1 caterpillars. As Sillen-Tullberg had predicted, the caterpillars were somewhat safer in a group than on their own.

In some environments, the costs of social clumping can be significant. These costs may include increased competition for food. For example, when thousands of royal penguins (*Eudyptes schlegeli*) crowd together in huge colonies **(Figure 56.25)**, the pressure on the local food supplies is great, increasing the risk of starvation. Communal living also facilitates the spread of contagious diseases and parasites. Nestlings in large colonies of cliff swallows (*Petrochelidon pyrrhonota*) are often stunted in growth because their nests are swarming with blood-feeding parasites, which move easily from nest to nest under crowded conditions. Such costs are probably why the vast majority of animals do not live in large, complex societies.

Fitness Varies among the Members of a Dominance Hierarchy

Recognizing the costs as well as the benefits of social living, biologists have examined features of social living that appear to reduce the fitness of some individuals. For example, some animal species form **dominance hierarchies,** social systems in which each individual's behavior is governed by its place in a highly structured social ranking. In a typical dominance hierarchy, the dominant or *alpha* individual rules the roost; subordinate individuals typically concede valuable resources to more dominant animals without so much as a peep of protest.

Although dominant individuals gain first access to resources, they also incur costs. Frequent challenges from lower ranking individuals may induce a stress response in dominant animals, which must constantly defend their status. For example, in some primates, wild dogs, and other mammals, dominant males have higher blood levels of cortisol and other stress-related hormones (see Section 42.4) than do subordinates. Elevated cortisol levels may induce high

blood pressure, the disruption of sugar metabolism, and other pathological conditions.

Why does a subordinate remain in the group when dominant companions reduce its chances for reproductive success? A possible explanation is that survival rates and reproductive success may be even lower for animals that live by themselves: a solitary baboon surely quickens the pulse of a passing leopard **(Figure 56.26)**. A subordinate member of a group gains the benefits, such as protection against predators, that come with being part of the group. Low-ranking males may even have the chance to mate with one of the group's females when dominant males are not watching, thus ensuring some representation of their alleles in the next generation. And if a low-ranking individual lives long enough, its social superiors may be toppled by predation, accidents, or old age, and a one-time subordinate may find itself high on the social register with food and mates galore.

Kin Selection May Explain Altruistic Behavior in Some Animal Species

In some species, group members appear to sacrifice their own reproductive success to help individuals that are not their direct descendants; such behaviors are collectively called **altruism.** For example, many birds and mammals that live in groups emit an "alarm call" when they see a predator. The call alerts other members of the group to the imminent danger, but the caller may draw attention to itself and be more likely to be captured. Why would an animal endanger itself to help others? Altruistic behavior, by definition, appears to contradict a basic premise of Darwinian evolutionary theory, namely that natural selection favors traits that increase an *individual's* relative fitness (see Section 21.3).

INCLUSIVE FITNESS AND KIN SELECTION Behavioral ecologist William D. Hamilton of University College, London, provided a solution to this puzzle. He recognized that alleles favoring altruism could be propagated indirectly if altruistic individuals sacrificed personal reproduction to help their relatives reproduce. Helping relatives in this way can propagate the helper's

Calculating Coefficients of Relatedness

Purpose: Kin selection theory suggests that the extent of altruistic behavior exhibited by one individual to another is directly proportional to the percentage of alleles they share. The hypothesis therefore predicts that individuals are more likely to help close relatives because, by increasing a close relative's fitness, the individual is helping to propagate some of its own alleles. Researchers calculate the coefficient of relatedness between individuals to test this prediction.

Protocol: To calculate the coefficient of relatedness between any two individuals, we first draw a family tree that shows all of the genetic links between them. The alleles of a parent are shuffled by recombination and independent assortment in the gametes they produce, so we can calculate only the average percentage of a parent's alleles that offspring are likely to share.

We start by considering *half* siblings, those who share only one genetic parent. Each sibling receives half of its alleles from its mother. Because a parent has only two alleles at each gene locus, the probability of sibling A getting a particular allele from its mother is 0.5 (decimal notation for 50%). Similarly, the probability of sibling B getting the same allele from its mother is also 0.5. Statistically, the probability that two independent events—in this case, the transfer of an allele to sibling A *and* the transfer of the *same* allele to sibling B—will both occur is the product of their separate probabilities. For half-siblings, we multiply the probabilities of the two links in the chain that connect them to calculate the likelihood that both siblings receive the same allele from their mother: $0.5 \times 0.5 = 0.25$; thus, the coefficient of relatedness for half siblings is 0.25.

Now consider two *full* siblings, who share the same genetic mother *and* father. They are connected to each other by two chains of relatedness; by multiplying the probabilities associated with the links in each chain, they share 25% of their alleles through the mother *and* 25% of their alleles through the father. After multiplying the probabilities for the links in each chains, we *add* the probabilities for the two chains. Thus, the coefficient of relatedness for full siblings is $0.25 + 0.25 = 0.50$.

Half siblings

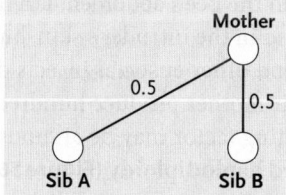

Relatedness = (0.5)(0.5) = 0.25

Full siblings

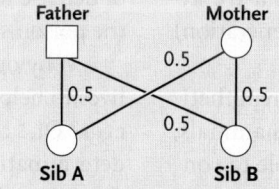

Relatedness
Through mother = (0.5)(0.5) = 0.25
Through father = (0.5)(0.5) = 0.25
Total relatedness = 0.25 + 0.25 = 0.5

First cousins

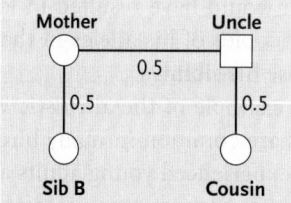

Relatedness = (0.5)(0.5)(0.5) = 0.125

Interpreting the Results: Each link drawn between a parent and an offspring or between full siblings indicates that those two individuals share, on average, 50% of their alleles. We can calculate the coefficient of relatedness between any two individuals by multiplying out the probabilities across all of the links between them. Thus, the degree of relatedness between a niece and an uncle is 0.25, and the degree of relatedness between first cousins is 0.125.

own genes because family members share many of the same alleles. Hamilton defined the term **inclusive fitness** to describe the sum of an individual's classical fitness (that is, the number of successful offspring it produces) *plus* the fitness gained through reproduction by its relatives.

Hamilton framed his idea in a cost–benefit analysis, called **Hamilton's rule.** The analysis includes two actors, the altruist and the beneficiary of the altruistic act. It also includes three critical variables: B is the fitness benefit to the beneficiary (the average number of *additional* offspring that the beneficiary produces); r is the **coefficient of relatedness,** a measure of the percentage of alleles that the altruist and the beneficiary share; and C is the fitness cost to the altruist (the average number of offspring that the altruist will not produce because it acted altruistically). The product of r and B indexes the fitness benefit to the altruist through the representation

of its alleles in the additional offspring produced by the beneficiary. According to Hamilton's Rule, altruism is favored by natural selection when $rB > C$. In other words, altruism is favored when the net fitness benefit to the altruist, measured as the added propagation of its alleles through reproduction by its relatives, is greater than the fitness cost of acting altruistically.

We can quantify the average percentage of alleles that relatives are likely to share by calculating their coefficient of relatedness **(Figure 56.27)**. We start by considering half siblings who, by definition, share only one genetic parent. Half siblings share on average 25% of their alleles by inheritance from their shared parent, making their degree of relatedness 0.25. By contrast, full siblings, who share the same genetic mother *and* father, share 25% of their alleles through the mother *and* 25% of their alleles through the father, for a total, on average, of 25% + 25% = 50% of their alleles. In other words, the degree of relatedness for full siblings is 0.50. The degree of relatedness between a nephew or niece and an aunt or uncle is 0.25, and the degree of relatedness between first cousins is 0.125. Thus, individuals should be more likely to help close relatives because, by increasing a close relative's fitness, the individual is helping to propagate some of its own alleles.

Although altruistic behavior can reduce the relative fitness of an individual through its own reproduction, it may increase the altruist's inclusive fitness through reproduction by its relatives because they share some of the altruist's alleles. This form of natural selection is called **kin selection.**

For example, suppose a male wolf helps his parents rear four pups that would have died without his assistance ($B = 4$). If the pups are his full siblings, they share 50% of his genes ($r = 0.5$). Thus, on average, the helper wolf has generated "by proxy" two ($r \times B = 0.50 \times 4 = 2$) copies of any allele that contributed to his altruistic behavior. If he had not acted altruistically, the helper wolf might have raised, say, two surviving offspring of his own. Each of his offspring would carry half of his alleles ($r = 0.5$), preserving just one ($0.50 \times 2 = 1$) copy of a given allele. Under these hypothetical circumstances, reproducing on his own would have resulted in lower inclusive fitness (that is, fewer copies of his alleles in the next generation) than helping to raise his siblings.

Although our example of the altruistic wolf is hypothetical, sibling helpers are common in many birds and mammals, especially when inexperienced young adults are unable to control sufficient resources to reproduce successfully on their own. Their altruistic behavior not only assists reproduction by close relatives, but it provides useful practice for rearing their own future offspring.

RECIPROCAL ALTRUISM Hamilton's kin-selection hypothesis explains altruistic behavior between closely related individuals, but behavioral biologists have also observed examples of altruism between nonrelatives. For example, the common vampire bat *(Desmodus rotundus),* which feeds on the blood of sleeping mammals, must consume a meal every two days to avoid starving to death. Bats that have consumed a large meal often share their bounty with unrelated members of their group. Why would one bat share its resources with a nonrelative? Robert Trivers, then of Harvard University, proposed that individuals will help nonrelatives if they are likely to return the favor in the future. Trivers called this form of altruistic behavior **reciprocal altruism,** because each member of the partnership can potentially benefit from the relationship. Trivers hypothesized that reciprocal altruism would be favored by natural selection as long as individuals that do *not* reciprocate—called "cheaters" by behavioral biologists—are denied future aid. Observations of vampire bats and some other animals have confirmed Trivers' hypothesis: when a vampire bat accepts a "blood donation" from another bat, but then refuses to share food that it has collected, the other bats refuse to share their food with it in the future.

Haplodiploidy May Contribute to Eusociality in Some Insects

Hamilton's insights lead to a critical prediction about self-sacrificing behavior: altruism should usually be directed to close relatives. Evidence from many animal species overwhelmingly supports this prediction, but some species of ants, bees, and wasps as well as termites may provide a truly remarkable example. In these **eusocial** insects, thousands of related individuals—most of them sterile females—live and work together in a colony for the reproductive benefit of a single queen and her mate(s). How did this self-sacrificing social behavior evolve, and why does it persist over time? The failure of altruistic workers to reproduce should doom any alleles that promote altruism to extinction.

For example, in a honeybee *(Apis mellifera)* colony, which may contain 30,000 to 50,000 related individuals, the only fertile female is the queen bee; all of the workers are her daughters **(Figure 56.28).** The queen's role in the colony is to reproduce. Workers perform all other tasks in the colony, feeding the queen and her larvae, constructing new honeycomb, foraging for nectar and pollen, sharing food, and guarding the entrance to the hive. Some pay the ultimate sacrifice when they sting intruders: this act of defense tears open the bee's abdomen, leaving the stinger and the poison sac behind in the intruder's skin, but killing the bee.

Why do bees and other eusocial insects devote their entire lives to helping their mother produce hundreds of thousands of eggs? One contributing factor may be an unusual pattern of sex determination called **haplodiploidy (Figure 56.29).** Female honeybees are diploid, receiving one set of chromosomes from each parent, but male bees are haploid because they hatch from unfertilized eggs. When a queen bee mates with one drone (a male), all of the sperm he delivers are genetically identical because males have only one set of chromosomes. Thus, all workers inherit the same set of alleles from their male parent,

FIGURE 56.28 Life in a honeybee *(Apis mellifera)* colony. A court of sterile worker daughters surrounds a queen bee, the only female of the colony that reproduces.

HAPLODIPLOIDY IN SOME EUSOCIAL INSECTS

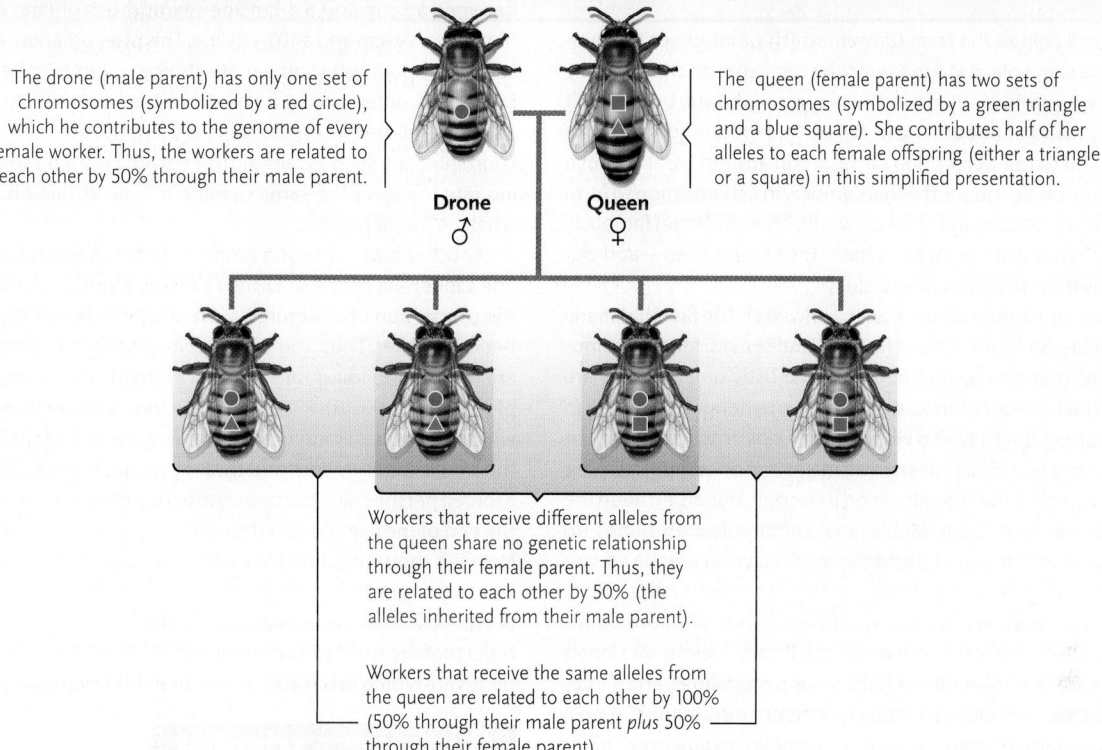

The drone (male parent) has only one set of chromosomes (symbolized by a red circle), which he contributes to the genome of every female worker. Thus, the workers are related to each other by 50% through their male parent.

Drone ♂

The queen (female parent) has two sets of chromosomes (symbolized by a green triangle and a blue square). She contributes half of her alleles to each female offspring (either a triangle or a square) in this simplified presentation.

Queen ♀

Workers that receive different alleles from the queen share no genetic relationship through their female parent. Thus, they are related to each other by 50% (the alleles inherited from their male parent).

Workers that receive the same alleles from the queen are related to each other by 100% (50% through their male parent *plus* 50% through their female parent).

SUMMARY The genetic system of many eusocial insects produces full siblings that have an exceptionally high degree of relatedness. Although this simplified model ignores recombination between the queen's two sets of chromosomes, it demonstrates how half the workers are related to each other by 50% and half are related to each other by 100%. Thus, the average degree of relatedness among workers is 75%. Including recombination would complicate the illustration, but the conclusion would be the same.

think like a scientist How would the average degree of relatedness among workers in a bee colony change if the queen mated with two drones instead of just one?

producing a 50% degree of relatedness among them. Like other diploid organisms, the workers are also related to each other by an average of 25% through their female parent. Adding these two components of relatedness, we see that *on average* workers are related to each other by 75%, a higher coefficient of relatedness than they would have to any offspring they produced if they were fertile. Thus, helping their mother produce additional offspring increases their inclusive fitness more than reproducing themselves.

This extremely high degree of relatedness among the workers in some eusocial insect colonies may explain their exceptional level of cooperation. When Hamilton first worked out this explanation of social behavior in these insects, he suggested that the workers devote their lives to caring for their siblings—the queen's other offspring—because a few of those siblings, which carry 75% of the workers' alleles, may become future queens and produce enormous numbers of offspring themselves.

Some researchers maintain that haplodiploidy and inclusive fitness do not fully explain the evolution of eusociality. They note that some eusocial species do not have haplodiploid sex determination; both males and females are diploid. Moreover, in some eusocial insects, a queen may mate with multiple males. Both of these factors reduce the coefficient of relatedness among sterile workers. Moreover, some evidence suggests that in some species worker sterility is not altruistic because the queen suppresses reproduction by workers through the released of pheromones. Thus, although altruistic behavior in eusocial insects is consistent with the predictions of kin selection theory, other factors may have contributed to its evolution.

STUDY BREAK 56.7

1. What do the social behaviors of musk oxen and sawfly larvae have in common?
2. Which animals in a dominance hierarchy are most likely to reproduce?
3. Why might the genetic system of many eusocial insects promote altruistic behavior?

Why do animals communicate in more than one sensory modality?

Jacob von Uexkell coined the term *Umwelt,* which pointed out that animals living in the same place at the same time can experience drastically different sensory worlds. Humans and bats can cohabitate, but we split up the acoustic world at about 20 kHz. Below that are the sonic sounds we tap into to add to our perception of the world around us, and above 20 kHz are the ultrasonic sounds that bats employ in echolocation to form a very different "acoustic image" of their world. Thomas Nagel famously asked in 1974, "What is it like to be a bat?" His answer then—and our answer now—is that we do not have a clue.

To complicate our appreciation of alien Umwelts is the fact that many animals are relying on sensory input from several senses. Many communication systems that we thought operated in mostly one modality are in fact multimodal. Spiders thrust their front appendages in the air in conspicuous display, but they also vibrate their webs; frogs call, but their vocal sacs are important visual cues to rival males and discerning females; and elephants trumpet, but they also send infrasonic signals through the ground for miles around them. Multimodal communication brings up questions about both the signaler and the receiver. What are the advantages of using such signals: redundancy, a safeguard against habituation, targeting multiple receivers, encoding different kinds of information? From the vantage of the receiver, we can ask if they weight all signals equally, if there are nonlinear interactions among modalities, and if integrating information provides qualitatively different information. A possible advantage relative to the last point is that male túngara frogs might be able to compare the time of arrival differences between a rival male's call and the water surface vibrations that emanate from the rival while calling. If the receiver can do the math, he can figure out the distance of his rival, much as we can decipher the distance of a storm by determining the time between when we hear the thunder clap and when the lightening flashes. All animals are probably detecting their world through multiple senses, but we have little notion of how often this happens and what advantages it might accrue.

How important are biases in stimulus perception?

Technology allows extremely accurate measures of the world around us, but we do not always perceive the world just as it is. When we compare stimulus strength, for example, our comparisons are usually based on proportional rather than absolute differences. Thus, the difference between a 2-cm and a 4-cm line seems much greater than the distance between a 98-cm and a 102-cm line. This phenomenon, known as Weber's Law, also operates in some mate choice systems. To be perceived as different, the difference in a courtship trait between males needs to be absolutely larger when the traits are larger. Thus, as the magnitude of a courtship trait, such as tail length, evolves it must do so at an ever increasing rate to derive the same benefit in mate attraction. Or at least this is what we might predict.

Another bias in making judgments about size is forced perspective. The Cinderella Castle at Disney's Magic Kingdom looks larger because the proportion of structures, such as windows, get smaller as the castle becomes taller. Thus, the windows appear farther away than they really are and the building taller than it really is. The greater bowerbird has picked up on this trick and creates a long avenue that leads to the area where he courts, lining the avenue with rocks and shells, always placing the smaller ones in front and the larger ones in back. This pattern creates a forced perspective, but opposite to that of the Cinderella Castle, making the end of the avenue and the court seem smaller than they actually are. Now, it is thought, when the male strolls out into his court he is perceived as much larger. In how many other instances do Weber's Law and forced perspective bias the animal's view of its world and drive the evolution of traits that are used to communicate about it? And what other cognitive biases are hiding in our and others' brains that we have yet to think about?

think like a scientist Group decision making is one example of an emergent group property, in which the decision is not a property of any one individual. Think about social groups of organisms—such as packs of hyenas, schools of fishes, colonies of bees, and groups of slime molds—and list the types of "decisions" these groups need to make and how they might make them together.

Michael J. Ryan

Michael J. Ryan is the Clark Hubbs Regents Professor in Zoology at The University of Texas at Austin, where he studies the evolution and function of animal behavior. Most of his work has addressed sexual selection and communication in frogs and fish. To learn more about Dr. Ryan's research, go to http://www.sbs.utexas.edu/ryan/.

REVIEW KEY CONCEPTS

For access to MindTap and additional study materials visit www.cengagebrain.com.

56.1 Instinctive and Learned Behaviors

- Although most behaviors have both instinctive and learned components, many are either largely instinctive or largely learned (Table 56.1). Some behaviors are produced only if the animal's nervous system acquires inputs from specific experiences during a critical stage of its development (Figure 56.1).

- Instinctive behaviors are those that an animal performs completely the first time it is presented with a stimulus.

- Fixed action patterns are highly stereotyped behaviors that animals exhibit in response to simple cues called sign stimuli (Figures 56.2 and 56.3). Fixed action patterns often change through time in response to an animal's experiences.

- Behavioral differences between individuals often reflect underlying genetic differences. Research on garter snakes suggests that certain food preferences are genetically based (Figure 56.4).

- Learned behaviors develop only after an animal has had certain experiences in its environment. The different forms of learning include imprinting (Figure 56.5), classical conditioning, operant conditioning, and cognition (Figure 56.6). Habituation is a learned loss of responsiveness to specific stimuli.

56.2 Neurophysiological and Endocrine Control of Behavior

- Animal behavior requires an anatomical, physiological, and biochemical foundation based in the nervous system. An individual's experience alters cells of the nervous system in ways that produce particular patterns of behavior.
- The physiological basis of bird singing behavior resides in specific neuron clusters, called nuclei, that communicate with each other in the bird's brain.
- Bird song and some other behaviors develop only after specific genes are activated within the neurons that produce the behavior (Figure 56.7).
- Hormones can mediate the expression of specific behaviors by activating genes that change the biochemistry, morphology, and number of neurons in specific nuclei. Estrogen stimulates the production of neurons in the higher vocal center of male zebra finches (Figure 56.8).
- Age-related changes in hormone levels can alter the behavior of animals over the course of their lives. Changes in juvenile hormone concentration are correlated with changes in task specialization in honeybees (Figure 56.9).
- Behavioral interactions with other individuals can alter an animal's hormone levels, inducing changes in its behavior. Research on male cichlid fishes suggests that variations in coloration and aggressive behavior associated with territorial status and social interactions are mediated by the production of certain hormones (Figure 56.10).

56.3 Migration and Wayfinding

- Some animals migrate seasonally, traveling from their birthplace to a distant locality and back again (Figures 56.11 and 56.12).
- Migrating animals use various behaviors to find their way. In piloting, animals use familiar landmarks to guide their journey. In compass orientation, animals use the position of the Sun or stars, polarized light, or the Earth's magnetic field as a guide (Figure 56.13). In navigation, animals use mental maps of their position to find their destination.
- Biologists often interpret migratory behavior as an adaptive response to changing food supplies. Some animals breed in northern habitats when food is plentiful during the spring and summer. They generally head south to seasonally more productive habitats before the onset of winter (Figure 56.14).

56.4 Habitat Selection and Territoriality

- Animals use multiple criteria to select their habitats.
- Kineses and taxes help animals orient to appropriate portions of the habitats they occupy.
- Habitat selection often has a largely genetic basis, but learning and prior experience influence habitat selection in some species (Figure 56.15).

- Animals may maintain territories to gain exclusive use of defendable resources that are in short supply. The costs of territoriality include the time and energy devoted to territory defense, the risk of injury from fights, and exposure to predators.

56.5 The Evolution of Communication

- Animal communication occurs between a signaler, which sends a message, and a signal receiver, which receives and interprets the message.
- Animals communicate using acoustical, visual, chemical, tactile, or electrical signals (Figures 56.16–56.18). Each sensory channel provides specific advantages. Animals may use more than one channel simultaneously.
- Honeybees use a combination of tactile, acoustical, and chemical channels to share information about the location of food sources (Figure 56.19).

56.6 The Evolution of Reproductive Behavior and Mating Systems

- Males and females exhibit different reproductive strategies. Males can increase their reproductive success by inseminating the eggs of many females. Females generally seek mates that provide successful alleles to offspring, have access to abundant resources, or help care for young.
- Males often compete for access to females (Figure 56.20). Sexual selection has produced elaborate structures that males use for displays to females and for aggressive interactions with other males (Figures 56.21 and 56.22). Females may prefer males that have showy structures and great stamina, which are signs that they possess successful alleles.
- The type of mating system a species uses is tied to its pattern of territoriality and the amount of parental care the male parent provides.
- The comparative method allows biologists to trace the evolution of specific behaviors, such as nest building, by superimposing the behaviors on a phylogenetic tree for the lineage (Figure 56.23).

56.7 The Evolution of Social Behavior

- Social interactions between individuals have both benefits and costs. Group living may provide better protection from predators, more efficient feeding, and communal care of young (Figures 56.24 and 56.26). The costs of living in a group include increased competition for scarce resources and an increase in the spread of contagious diseases (Figure 56.25).
- Dominance hierarchies are highly structured societies in which some individuals have high status and first access to resources.
- Altruistic behavior appears to contradict Darwinian evolutionary theory, because altruistic individuals sacrifice their own fitness for the benefit of others. However, kin selection may ensure that individuals display altruistic behavior to close relatives that share some of their alleles (Figure 56.27).
- An unusual mechanism of sex determination, haplodiploidy, makes the workers in some eusocial insect colonies more closely related to each other than siblings in most species are (Figures 56.28 and 56.29). Haplodiploidy may have fostered the evolution of highly altruistic behavior.

TEST YOUR KNOWLEDGE

Remember/Understand

1. Marler concluded that white-crowned sparrows can learn their species' song only:
   a. after receiving hormone treatments.
   b. during a critical period of their development.
   c. under natural conditions.
   d. from their genetic father.
   e. if they are reared in isolation cages.

2. A stimulus that always causes an animal to behave in a highly stereotyped way is called:
   a. fixed action pattern.
   b. an instinct.
   c. habituation.
   d. a sign stimulus.
   e. a reinforcement.

3. Arnold's experiments on the feeding preferences of garter snakes demonstrated that food choice is largely governed by a snake's:
   a. early experiences.
   b. genetics.
   c. size and color.
   d. diet while it was developing inside its mother.
   e. trial-and-error learning.

4. The development of the song system in male songbirds depends on:
   a. direct connections between sensory neurons and motor neurons.
   b. a decrease in the number of neurons in the song system.
   c. the behaviors of females, which stimulate hormone production.
   d. the successful defense of a territory.
   e. the production of estrogen early in life.

5. In cichlid fishes, high levels of the hormone GnRH:
   a. make females more receptive to male attention.
   b. cause males to be sexually aggressive but not territorial.
   c. stimulate a male to defend its territory.
   d. cause males to abandon their territories.
   e. cause males to lose their bright colors.

6. In comparison to males, the females of many animal species:
   a. compete for mates.
   b. choose mates that are well camouflaged in their habitats.
   c. choose to mate with many partners.
   d. are always monogamous.
   e. choose their mates carefully.

Apply/Analyze

7. Which of the following statements about animal migration is true?
   a. Piloting animals use the position of the Sun to acquire information about their direction of travel.
   b. Animals migrating by compass orientation use mental maps of their position in space.
   c. Navigating animals use familiar landmarks to guide their journey.
   d. Navigating animals use a compass and a mental map of their position to reach a destination.
   e. Most migrating birds use olfactory cues to return to the place where they hatched from eggs.

8. Which signal type would provide the fastest communication between bats flying in a dark forest?
   a. chemical signals
   b. acoustical signals
   c. visual signals
   d. tactile signals
   e. electrical signals

9. Social behavior:
   a. is exhibited *only* by animals that live in groups with close relatives.
   b. cannot evolve in animals that maintain territories.
   c. evolved because group living provides benefits to individuals in the group.
   d. is never observed in insects and other invertebrate animals.
   e. can be explained only by the hypothesis of kin selection.

10. The degree of relatedness between a parent and its biological offspring:
    a. is the same as that between full siblings.
    b. is less than that between brother and sister.
    c. depends on how many siblings the parent has.
    d. promotes an individual's reproductive success.
    e. is the same as between first cousins.

11. **Discuss Concepts** Is learning always superior to instinctive behavior? If you think so, why do so many animals react instinctively to certain stimuli? Are there some environmental circumstances in which being able to respond "correctly" the first time would have a big payoff?

12. **Discuss Concepts** In Chapters 51, 54, and 55, you learned about some of the environmental changes associated with global warming. What effects might global warming have on animal species that undertake seasonal migrations?

Evaluate/Create

13. **Discuss Concepts** Using an example from your own experience, explain why habituation to a frequent stimulus might be beneficial. Also describe an example in which habituation might be harmful or even dangerous.

14. **Discuss Concepts** Although females provide parental care far more often than males in the animal kingdom as a whole, exceptions exist, especially among birds and fishes. Develop three evolutionary hypotheses to explain why male birds are so likely to involve themselves in caring for their offspring.

15. **Design an Experiment** You find that some fruit flies in your lab are quick to come to a dish containing citrus oils, but others are not as responsive. How could you test whether these behavioral differences are caused by genetic differences among the flies or environmental differences in their prior experience?

16. **Apply Evolutionary Thinking** Some birds that are frequently kept as pets, such as parrots and myna birds, have the ability to imitate human speech faithfully. Develop a hypothesis that explains why the ability to be a good mimic might have evolved in these species. What features of the birds' brains might be involved in this behavior?

For selected answers, see Appendix A.

INTERPRET THE DATA

A female Seychelles warbler (*Acrocephalus sechellensis*) typically produces just one egg per nesting event. A male of this species will guard his mate from the attentions of neighboring males until she lays that egg, thereby increasing the likelihood that he fathered the offspring inside it. Jan Komdeur of the University of Groningen, The Netherlands, studied the behavior of male Seychelles warblers—before females had laid eggs—in relation to the number of other males that lived nearby. First, Komdeur watched focal males (that is, the ones being studied) for 30-second time periods and recorded the percentage of time periods in which they fed or guarded mates. Then, for focal males with more than three male neighbors, Komdeur experimentally reduced the number of neighbors to three and surveyed the behavior of the focal males again to see if it changed. His results are illustrated in the graphs reproduced below. (Because males sometimes performed both behaviors during the same observation period, some observation periods counted toward both behaviors; thus, the percentages of time periods spent foraging or mate guarding sum to more than 100%.) If a male warbler has many male neighbors, what cost is associated with ensuring his paternity of his mate's egg?

Source: Adapted from J. Komdeur. 2001. Mate guarding the Seychelles warbler is energetically costly and adjusted to paternity risk. *Proceedings of the Royal Society of London B* 268:2103–2111.

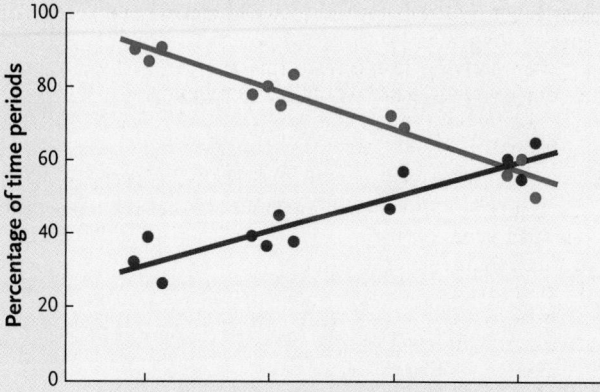

A. Before removal of some male neighbors

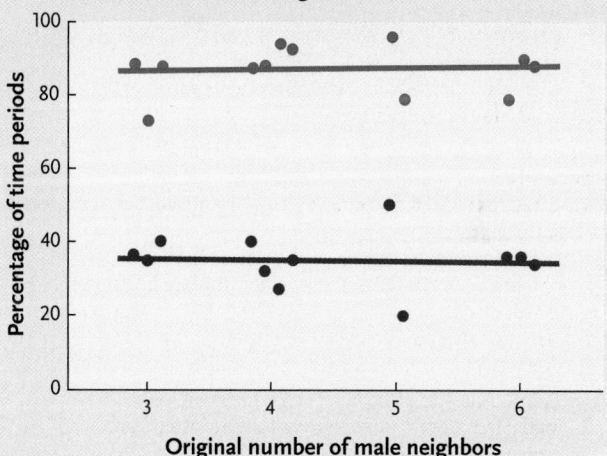

B. After removal of some neighbors

Original number of male neighbors

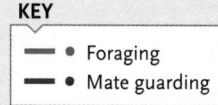

KEY
- Foraging
- Mate guarding

Appendix A: Answers

Chapter 1

Study Break

STUDY BREAK 1.1

1. The major levels in the hierarchy of life and some of their emergent properties are: cells—life; organisms—learning; populations—birth and death rates; communities, ecosystems, biosphere—diversity and stability.
2. Organisms use energy collected from the external environment for growth (including the production of new molecules and cells), maintenance and repair of body parts, and reproduction.
3. A life cycle is the series of structurally and functionally distinct developmental stages through which organisms pass.

STUDY BREAK 1.2

1. In artificial selection, humans selectively breed individuals with desirable heritable characteristics to enhance those traits in the next generation. In natural selection, genetically-based characteristics that increase survival and reproduction become more common in the next generation.
2. Random changes in DNA—mutations—may change the structure of proteins that contribute to the physical appearance and internal functions of an organism.
3. Being camouflaged may make an animal less likely to be noticed by a predator.

STUDY BREAK 1.3

1. In the cells of prokaryotic organisms, DNA is not separated from other parts of the cell. In the cells of eukaryotic organisms, DNA is enclosed within a nucleus.
2. Humans are classified in Domain Eukarya and Kingdom Animalia.
3. Model organisms are usually easy to maintain and study, and they have been so well studied that researchers already know a lot about their biology.

STUDY BREAK 1.4

1. A scientific hypothesis must be falsifiable. In other words, we must be able to imagine what sort of data would demonstrate that the hypothesis is incorrect.
2. The copper lizard models told the researchers how frequently lizards would perch in the sun just by chance and what the temperatures of nonthermoregulating lizards would be.
3. When scientists describe a set of ideas as a "theory," they recognize that the ideas have already withstood many scientific tests.

Think Like a Scientist

FIGURE 1.11

Animals are more closely related to fungi than they are to plants. According to the phylogenetic tree, animals and fungi shared a common ancestor more recently than did animals and plants.

FIGURE 1.15

Assignable activity in MindTap. Answers provided to instructors.

FIGURE 1.16

Assignable activity in MindTap. Answers provided to instructors.

Test Your Knowledge

1. c 2. b 3. d 4. b 5. d 7. c 8. d 9. a 10. e 12. c

Interpret the Data

Assignable activity in MindTap. Answers provided to instructors.

Chapter 2

Study Break

STUDY BREAK 2.1

An element is a pure substance that consists of one type of atom. An atom is the smallest unit of an element that retains its chemical and physical properties. A molecule is a collection of atoms chemically combined in fixed numbers and ratios. Molecules can consist of the same atoms, as is seen for the two oxygen atoms in the oxygen molecule, or of different atoms, as in the combination of two hydrogen atoms and one oxygen atom in a molecule of water. Molecules with component atoms that are different, such as water, are compounds.

STUDY BREAK 2.2

1. Protons and neutrons are found in the nucleus of an atom. Electrons are found in orbitals located in energy levels (shells) that surround the nucleus.
2. Carbon-11 has six protons and five neutrons. Oxygen-15 has eight protons and seven neutrons.
3. The number of valence electrons—the electrons in the outermost shell of an atom—determines its chemical reactivity. If the outermost shell is not completely filled with electrons, the atom tends to be chemically reactive, whereas if that shell is completely filled, the atom is nonreactive.

STUDY BREAK 2.3

1. An ionic bond forms between atoms when those atoms gain or lose electrons completely. An example is the ionic bond in NaCl.
2. A covalent bond forms when atoms share a pair of valence electrons rather than gaining or losing them completely.
3. Electronegativity is a measure of an atom's attraction for the electrons it shares in a chemical bond with another atom. When electrons are shared equally, the atoms remain uncharged and the result is a nonpolar covalent bond. When electrons are shared unequally, one atom carries a partial negative charge and the other atom carries a partial positive charge. The molecule then has polarity, and the bond is a polar covalent bond.
4. In a chemical reaction, atoms or molecules interact to form new chemical bonds or break old ones. Atoms are added to or removed from molecules, or linkages of atoms in molecules are rearranged as a result of bond formation.

STUDY BREAK 2.4

1. Hydrogen bonds between neighboring water molecules produce a water lattice. The constant breakage and reformation of hydrogen bonds in the lattice allows water to flow easily. The polarity of water molecules also contributes to the properties of water; that is, in liquid water, the lattice resists invasion by other molecules unless the invading molecules also contain polar regions that can form competing attractions with water molecules. In that case, the water lattice opens, forming a cavity in which the polar or charged molecule can move. However, nonpolar molecules are unable to affect the water lattice. Hydrogen bonds also give water its unusual ability to resist changes in temperature by absorbing or releasing heat energy, its unusually high boiling point, and its unusually high internal cohesion and surface tension.
2. A solute is a dissolved substance. A solvent is a substance capable of dissolving another substance. A solution is a solute dissolved in a solvent. For example, salt (NaCl) is a solute that can dissolve in the solvent water.

STUDY BREAK 2.5

1. Acids are hydrogen ion (proton, H^+) donors; bases are proton acceptors. An acid dissociates in water to produce a hydrogen ion and an anion. Most bases dissociate in water to give hydroxide ions, which then accept protons to produce water.
2. Buffers act to control the pH of a solution. In living organisms, buffers keep the pH of body and cell fluids within a narrow range, enabling normal cell and body functions to occur. Outside of the normal pH range, the functions of proteins can be affected, thereby adversely affecting the functions of the organism.

Think Like a Scientist

UNANSWERED QUESTIONS

Local strategies such as reducing pollution, overfishing, coastal development, and destructive fishing practices, often by setting up networks of well-managed marine protected areas, have been shown to make marine ecosystems more resilient to global stressors such as climate change. Global strategies to reduce impacts of ocean acidification and climate change should be focused on reducing CO_2 levels in the atmosphere, which is a major global political and social challenge. Additional global strategies could include large multinational networks of marine reserves, bioengineering strategies for CO_2 storage or sequestration, and global treaties to ensure that global CO_2 levels are reduced quickly enough to avoid major ecosystem-level damage. Experiments and studies to track the effectiveness of these efforts could include: (1) rapid, global monitoring studies to study the health of marine ecosystems at a rate sufficient for adaptive management; (2) multistressor, multiscale experiments so that we can better understand impacts from the molecular to the ecosystem level; and (3) molecular and physiological studies so that we can determine the mechanisms of the impacts and develop strategies to reduce impacts.

Test Your Knowledge

1. d 2. e 3. a 4. b 5. a 6. d 7. c 9. e 10. b

Interpret the Data

Assignable activity in MindTap. Answers provided to instructors.

Chapter 3

Study Break

STUDY BREAK 3.1

1. Organic molecules are molecules based on carbon. Hydrocarbons are a type of organic molecule that consists of carbon linked only to hydrogen atoms.
2. The maximum number of bonds that a carbon atom can form is four.
3. Carboxyl groups donate a hydrogen ion in water and therefore act as acids. Amino groups accept a hydrogen ion in water and therefore act as bases. Phosphate groups donate hydrogen ions in water and therefore act as acids.
4. In a dehydration synthesis reaction, components of water (—H and —OH) are removed. In hydrolysis, the components of a water molecule are added to functional groups as molecules are broken down into smaller subunits.

STUDY BREAK 3.2

A monosaccharide is the structural unit of carbohydrate molecules. Monosaccharides are simple sugars such as trioses, pentoses, and hexoses. Glucose, galactose, and fructose are hexoses. A disaccharide is a molecule assembled from two monosaccharides linked by a dehydration synthesis reaction. Lactose, sucrose, and maltose are disaccharides. A polysaccharide is a polymer of monosaccharide subunits. The subunits are identical or different, depending on the particular polysaccharide. Glycogen, starch, cellulose, and chitin are polysaccharides.

STUDY BREAK 3.3

The three most common lipids found in living organisms are neutral lipids, phospholipids, and steroids. Most neutral lipids consist of a three-carbon backbone chain formed from glycerol, with each carbon linked to a fatty acid side chain. In the most common phospholipids, glycerol is the backbone, with two of its binding sites linked to fatty acids. The third binding site is linked to a polar phosphate group. Steroids have structures based on a framework of four carbon rings. Differences in side groups attached to the rings distinguish the different types of steroids.

STUDY BREAK 3.4

1. Differences in the side groups (R in the figures) give the amino acids their individual properties.
2. A peptide bond is the bond between the C of the carboxyl group of one amino acid and the N of the amino group of the adjacent amino acid (see Figure 3.16). The bond is formed in a dehydration synthesis reaction between an amino group of one amino acid and a carboxyl group of another amino acid.
3. Domains are distinct structural subdivisions in the final folded forms of proteins. They are the result of the amino acid sequence of the protein (the primary structure of the protein) and the secondary, tertiary, and quaternary (if more than one polypeptide is involved) structures of the protein.

STUDY BREAK 3.5

1. Nucleic acids are formed from nucleotide monomers. A nucleotide consists of a nitrogenous base, a five-carbon sugar, and a phosphate.
2. In DNA, the five-carbon sugar is deoxyribose; in RNA it is ribose. DNA has the pyrimidine nitrogenous base T (thymine), and RNA has U (uracil).

Think Like a Scientist

FIGURE 3.21

Assignable activity in MindTap. Answers provided to instructors.

MOLECULAR INSIGHTS

1. The enzyme that assembles DNA molecules in the bacterium *Escherichia coli* (see Figure 3.23). The enzyme has two domains, one to assemble the DNA molecules (the polymerase domain) and the other to correct mistakes during DNA assembly (the exonuclease domain).
2. The method of analysis described in *Molecular Insights* was the comparison of domain structure and organization. Fundamentally, the domains consist of particular sequences of amino acids. The function of a domain depends on the sequence of amino acids plus the higher-order structures it forms within the protein. As a domain diverges through evolutionary time, core functional elements must stay the same or highly similar in order that the function of the domain be retained. At the same time, other parts of the domain can experience changes in amino acid sequence. Therefore, through bioinformatics-based computer analysis of proteins sharing a particular function, researchers can determine how the domain has changed over time, and thereby construct a phylogenetic tree based on the changes.

UNANSWERED QUESTIONS

In type 2 diabetes, SREBPs in the liver are activated, and this leads to overproduction of fatty acids and triglycerides. Inhibitors of SREBP processing might reduce plasma triglycerides and prevent their toxic effects on blood vessels.

Test Your Knowledge

1. a 2. d 3. c 4. e 5. c 6. d 7. b 8. d 9. b 10. a

Interpret the Data

Assignable activity in MindTap. Answers provided to instructors.

Chapter 4

Study Break

STUDY BREAK 4.1

The plasma membrane is a bilayer of lipid and suspended protein molecules that bounds the cytoplasm of a cell. The plasma membrane is a selective barrier for the passage of water, ions, nutrients, and waste. It maintains the specialized internal ionic and molecular environments required for cellular life.

STUDY BREAK 4.2

The DNA of a prokaryotic cell is located in the nucleoid region, a central area of the cell that has no membrane around it to separate it from the cytoplasm. In most prokaryotes, the DNA is a folded mass. In its unfolded state, it is a circular DNA molecule.

STUDY BREAK 4.3

1. Most of the DNA of a eukaryotic cell is found within a nucleus located roughly in the center of the cell. The nucleus is bounded by the nuclear envelope, a membrane that separates its contents from the cytoplasm. The DNA is complexed with proteins and organized into several linear chromosomes.
2. The nucleolus is an area within the nucleus. It forms around the genes for rRNA in the chromosomes and is the location in which the information in those genes is copied into rRNA. The rRNA combines with proteins in the nucleolus to form the large and small ribosomal subunits. A large and a small ribosomal subunit function together in the cytoplasm to synthesize proteins.
3. The endomembrane system is a collection of organelles, membranous channels, and vesicles that form a major traffic network for the synthesis, distribution, and storage of proteins and other molecules. Its structure includes the endoplasmic reticulum (ER) and the Golgi complex. The rough ER has ribosomes on its outer surface. Proteins synthesized by those ribosomes enter the ER lumen, where they fold into their final shape and may be modified chemically. Then they are delivered to other regions of the cell within vesicles that pinch off from the rough ER. Most of the proteins made on rough ER go to the Golgi complex.

   The smooth ER consists of membranes that lack ribosomes. Functions of the smooth ER include synthesis of lipids that become parts of cell membranes.

   The Golgi complex is a stack of flattened, membranous sacs. The *cis* face of this organelle receives vesicles containing proteins released from the rough ER and continues their chemical modifications. The proteins are then sorted into vesicles that pinch off from the *trans* face of the Golgi complex, the side that faces the plasma membrane. Some of the released vesicles remain in the cytoplasm as storage vesicles of various types, whereas others, called secretory vesicles, release their contents to the outside of the cell. In addition, some vesicles bud from the *cis* face of the Golgi complex to carry certain proteins that function in the ER to the ER.
4. A mitochondrion is an organelle enclosed by two membranes. The outer mitochondrial membrane is smooth and covers the outside of the organelle. The inner mitochondrial membrane is highly folded into cristae. Within the inner membrane is the mitochondrial matrix, which contains DNA and ribosomes. Most of the energy required for eukaryotic cellular activities is generated by reactions in the cristae and the matrix. Those reactions break down sugars, fats, and other fuel molecules into water and carbon dioxide, releasing energy mostly in the form of ATP.
5. The cytoskeleton is an internal cytoplasmic network of filaments and tubules composed mainly of actin and tubulin proteins. The function of the cytoskeleton is to maintain the shape of the cell, reinforce the plasma membrane, and organize internal structures. Changes in the cytoskeleton are responsible for movements of cell organelles, movements of parts of the cell, or movements of the whole cell.

STUDY BREAK 4.4

1. A chloroplast has two membranes: an outer boundary membrane and an inner boundary membrane. The latter, similar to the cristae of the mitochondrion, is highly folded. The two membranes enclose an inner compartment known as the stroma. Within the stroma is a membrane system composed of flattened, closed sacs called thylakoids. Chloroplasts are the sites of photosynthesis in plant cells. The thylakoid membranes contain chlorophyll and other molecules that absorb light energy and convert it into chemical energy. Enzymes in the stroma use the chemical energy to make carbohydrates and other complex organic molecules from water, carbon dioxide, and other simple inorganic precursors.
2. The tonoplast, the membrane surrounding the central vacuole, contains transport proteins that move substances into and out of the vacuole. Central vacuoles also store organic and inorganic salts, organic acids, sugars, storage proteins, pigments, and, in some cells, waste products. Chemical defense molecules are found in the central vacuoles of some plants. Enzymes capable of breaking down biological molecules are present in some central vacuoles, supporting the view of some scientists that they may have some of the properties of animal lysosomes.

STUDY BREAK 4.5

1. Anchoring junctions are spots or belts that run entirely around cells, effectively sticking adjacent cells together. Microfilaments (in adherens junctions) or intermediate filaments (in desmosomes) anchor the junction in the underlying cytoplasm. Tight junctions involve fusion of a network of junction proteins in the outer halves of the plasma membranes of adjacent cells, forming a tight seal that can keep even ions from moving between the cells. Gap junctions open direct channels between adjacent cells through which ions and small molecules can pass directly. The gap junctions are formed by aligned hollow protein cylinders in the plasma membranes of two cells.
2. The extracellular matrix (ECM) is a complex of proteins and polysaccharides secreted by the cells that it surrounds. Depending on the nature of the network of proteoglycans (carbohydrate-rich glycoproteins) in the ECM, the consistency ranges from soft and jellylike to hard and elastic.

Think Like a Scientist

FIGURE 4.11

Assignable activity in MindTap. Answers provided to instructors.

FIGURE 4.17

The chemical is inhibiting a step in the pathway for protein secretion from the cell. Possible steps affected are the transport of the protein in vesicles from the ER to the Golgi complex, the production of secretory vesicles by budding from the Golgi membranes, or the release of the protein at the cell surface by exocytosis.

MOLECULAR INSIGHTS

When two species have genes with similar functions that have similar sequences, the interpretation is that the genes in the two species have evolved from a common ancestral gene as the species themselves evolved. Hence, 60% of the *Drosophila* genes identified have common ancestry with genes in humans. If a gene of one species with a particular function is not similar in sequence to any of the genes with similar functions in another species, the interpretation is that the second species does not contain a gene that is evolutionarily related to the gene in the first species. Hence,

40% of the *Drosophila* genes identified do not have common ancestry with genes in humans.

UNANSWERED QUESTIONS

Perturbing the activity of individual genes and proteins allows cell biologists to determine whether each plays a necessary role in a given cellular process, but does not indicate what mechanistic role the protein performs or how it cooperates with other proteins to perform its functions. On the other hand, reconstituting a process in a test tube allows cell biologists to build a functioning system from individual components and understand how each contributes to the system, but does not address the functional importance of each component in the context of the cell. Both approaches can be technically challenging, but reconstitution is often more challenging because it requires both identification of all of the important players and the isolation of the players in a functional form.

Test Your Knowledge

1. e 2. c 3. a 4. b 5. c 6. b 7. b 8. d 10. b 11. b

Interpret the Data

Assignable activity in MindTap. Answers provided to instructors.

Chapter 5

Study Break

STUDY BREAK 5.1

1. The fluid mosaic model proposes that the membrane consists of a fluid phospholipid bilayer in which proteins are embedded and float freely.
2. Integral proteins include transport proteins, receptor proteins, recognition proteins, and cell adhesion proteins. Peripheral proteins include microtubules, microfilaments, intermediate filaments, and proteins that link the cytoskeleton together.

STUDY BREAK 5.2

1. In passive transport, molecules and ions move across the membrane from the side with the higher concentration to the side with the lower concentration; that is, the difference in concentration provides the energy for passive transport. In active transport, molecules and ions move across the membrane from the side with the lower concentration to the side with the higher concentration—that is, against the concentration gradient. The energy for active transport comes from hydrolysis of ATP.
2. The transport of substances through membranes based solely on molecular size and lipid solubility is simple diffusion. The diffusion of polar and charged molecules across membranes with the help of transport proteins is facilitated diffusion.

STUDY BREAK 5.3

1. Osmosis is the passive transport of water across a membrane. This movement follows concentration gradients. For osmosis to occur, there must be a selectively permeable membrane—that is, a membrane that will allow water molecules, but not molecules of the solute, to pass. As long as the solute is at different concentrations on the two sides of the membrane, water movement will occur; that is, it is not necessary for pure water to be present on one side of the membrane.
2. If animal cells are in a hypertonic solution, water molecules will move by osmosis from within the cells to the surrounding solution. If the outward movement of water exceeds the capacity of the cells to replace the lost water, the cells will shrink.

STUDY BREAK 5.4

1. Active transport is the movement of substances across membranes against their concentration gradients by pumps; the energy for active transport comes from ATP hydrolysis. In primary active transport, ATP hydrolysis directly drives the process; that is, the same protein that

transports a substance also hydrolyzes ATP. In secondary active transport, ATP hydrolysis indirectly drives the process; that is, the transport proteins themselves do not hydrolyze ATP. Rather, the transporters use a favorable concentration gradient of ions, generated by primary active transport (when ATP hydrolysis is used), as their energy source for active transport of a different ion or molecule.
2. A membrane potential is a voltage difference across a membrane. Ion transport by membrane pumps contributes to this voltage difference. The sodium-potassium pump in the plasma membrane pushes three sodium ions out of the cell and two potassium ions into the cell with each turn of the pump. This leads to an accumulation of positive charges outside the membrane, causing the inside of the cell to become negatively charged with respect to the outside of the cell. In addition, an unequal distribution of ions across the membrane is created by passive transport. The electrical potential difference (voltage) across the plasma membrane is the membrane potential.

STUDY BREAK 5.5

1. In exocytosis, secretory vesicles in the cytoplasm contact and fuse with the plasma membrane, releasing their contents to the outside of the cell.
2. Endocytosis is a mechanism by which substances are brought into the cell from the exterior. The substances become trapped in pitlike depressions that bulge inward from the plasma membrane. The depression pinches off as an endocytic vesicle. In bulk endocytosis, no binding by surface receptors is involved. Extracellular water is taken in together with any other molecules that are in solution in the water. This is the simplest form of endocytosis. In receptor-mediated endocytosis, molecules to be taken in become bound to the outer cell surface by receptor proteins. The receptor proteins are specific in that they recognize and bind only certain molecules from the solution that surrounds the cell. The molecules recognized are mostly proteins or other molecules carried by proteins. Once the receptors have bound their target molecules, the receptors collect into a coated pit, a depression in the plasma membrane. The pits, with the contained target molecules, pinch off from the plasma membrane to form endocytic vesicles.

Think Like a Scientist

FIGURE 5.6

Assignable activity in MindTap. Answers provided to instructors.

FIGURE 5.11

The hydrolysis of ATP to ADP plus phosphate is key to the ability of the channel protein to transport ions. If the rate of hydrolysis of ATP was reduced drastically, the primary active transport pump would only be able to move the ion it transports at a much lower rate.

MOLECULAR INSIGHTS

The result showed that unlabeled LDL competed with labeled LDL for binding to the cells. The interpretation was that there are specific binding sites for LDL on the cell surface and that their number is limited. The result supported the researchers' overall conclusion because it indicated that specific binding was occurring in the experiments rather than nonspecific binding; that is, there is a specific receptor for LDL.

UNANSWERED QUESTIONS

Restricting the passage of protons aids in conservation of the membrane's electrochemical potential.

Test Your Knowledge

1. d 2. b 3. a 4. e 5. b 6. a 7. c 8. c 9. e 10. d

Interpret the Data

Assignable activity in MindTap. Answers provided to instructors.

Chapter 6

Study Break

STUDY BREAK 6.1

1. Kinetic energy is the energy of motion, whereas potential energy is stored energy.
2. An isolated system exchanges neither matter nor energy with its environment. A closed system can exchange energy, but not matter, with its environment. An open system can exchange both energy and matter with its environment.

STUDY BREAK 6.2

1. The two factors that must be considered are the change in the energy content of a system and its change in entropy. Reactions tend to be spontaneous when the products have less potential energy than the reactants, and when the products are less ordered than the reactants.
2. The greater the negative value of ΔG, the further a reaction will proceed toward completion, and therefore the greater the concentration of product molecules versus reactant molecules.
3. An exergonic reaction releases free energy—ΔG is negative because the products contain less free energy than the reactants. An endergonic reaction requires free energy from the surroundings to run—ΔG is positive because the products contain more free energy than the reactants.

   In a catabolic reaction, energy is released during the breakdown of a complex reactant molecule to a simpler product molecule, whereas in an anabolic reaction, energy is used to create a product molecule that is more complex than the reactant molecule.

   Individual reactions may be exergonic or endergonic. When a series of reactions (each of which may have a positive or a negative ΔG value) forms a metabolic pathway, the pathway is catabolic if the overall sum of individual reaction ΔG values is negative, and it is anabolic if the overall sum of individual reaction ΔG values is positive.

STUDY BREAK 6.3

1. ATP contains the five-carbon sugar ribose, with the nitrogenous base adenine linked to one of the carbons and a chain of three phosphate bonds linked to another carbon. The phosphate groups are closely associated with each other and their negative charges strongly repel each other, making the bonding arrangements unstable and storing potential energy. Hydrolysis of ATP to remove one or two of the three phosphates is a spontaneous reaction that relieves the repulsion and releases large amounts of free energy.
2. Many individual reactions found in living cells are not spontaneous because they have a positive ΔG. By joining such a reaction to another reaction with a large negative ΔG, the reaction can be completed. The combined reaction is called a coupled reaction. ATP participates in coupled reactions in the enzyme-driven process of energy coupling; that is, ATP comes into close contact with a reactant molecule in an endergonic reaction. When ATP is hydrolyzed, the terminal phosphate group is transferred to the reactant molecule, which makes that molecule less stable so that the reaction continues spontaneously.

STUDY BREAK 6.4

1. Enzymes accelerate reactions by reducing the activation energy of a reaction, the initial input of energy required to start a reaction. Enzymes lower activation energy by rearranging the atoms and bonds of the reacting molecules into the transition state, an activated state that is highly unstable. With relatively little change in energy, the transition state can move forward toward products or backward toward reactants.
2. No.

1. There is a maximum rate at which an enzyme can combine with substrates and release products. Beyond that point, increasing substrate concentration does not increase the reaction rate any further.
2. In competitive inhibition, the inhibitor competes with the normal substrate molecule for binding to the active site of the enzyme, whereas in noncompetitive inhibition, the inhibitor does not compete directly with the substrate for binding to the active site.
3. As the temperature increases, the increasing kinetic motions of the enzyme's amino acid chains eventually disrupt the enzyme's three-dimensional structure, causing it to unfold and denature. At that point, there is no enzyme activity.

STUDY BREAK 6.6

A ribozyme is an RNA molecule that accelerates the rate of a biological reaction. A ribozyme qualifies as an enzyme because it remains unchanged after the reaction is complete; that is, it is a true catalyst.

Think Like a Scientist

MOLECULAR INSIGHTS

Both the 5′ and 3′ constant regions are necessary for the active ribozyme structure to form.

UNANSWERED QUESTIONS

The problem is that each nucleotide in the ribozyme would need to serve as template for polymerization. This means that the ribozyme would have to open up its own structure and inactivate its own catalytic site. Therefore, the existence of such a molecule seems very unlikely. However, it may be possible to design such a molecule if it has two catalytic centers, or if the crucial parts of the ribozyme do not need to template because they exist twice in the molecule.

Test Your Knowledge

1. c 2. d 3. b 4. c 5. e 6. a 7. a 8. b 9. d 10. e

Interpret the Data

Assignable activity in MindTap. Answers provided to instructors.

Chapter 7

Study Break

STUDY BREAK 7.1

1. Oxidation is the removal of electrons from a substance; reduction is the addition of electrons to a substance.
2. Cellular respiration refers to the reactions in which oxygen is used as final electron acceptor; it includes the reactions that transfer electrons from organic molecules to oxygen and the reactions that make ATP. Oxidative phosphorylation is the process by which ATP is synthesized using the energy released by electrons as they are transferred to oxygen.

STUDY BREAK 7.2

1. The initial steps of glycolysis, which require 2 ATP, convert glucose to a phosphorylated derivative. The later steps, which release 4 ATP, remove electrons from the glucose derivatives and generate two molecules of pyruvate.
2. The redox reaction in glycolysis is the glyceraldehyde-3-phosphate (G3P) to 1,3-bisphosphoglycerate reaction (see Figure 7.8, step 6).
3. ATP is synthesized in glycolysis by substrate-level phosphorylation occurring in the 1,3-bisphosphoglycerate to 3-phosphoglycerate reaction (see Figure 7.8, step 7), and in the phosphoenolpyruvate (PEP) to pyruvate reaction (see Figure 7.8, step 10).

STUDY BREAK 7.3

The three-carbon pyruvate molecules are transported from the cytosol into the mitochondria, where they are converted into two-carbon acetyl units through pyruvate oxidation. The citric acid cycle oxidizes the acetyl units completely to carbon dioxide with the transfer of electrons to NAD^+ or FAD.

STUDY BREAK 7.4

1. Each complex contains a unique combination of nonprotein carriers that pick up and release electrons.
2. The proton pumps push protons (H^+) from the mitochondrial matrix to the intermembrane compartment, increasing the proton concentration there. The resulting proton gradient produces an electrical gradient across the inner mitochondrial membrane with the matrix negatively charged with respect to the intermembrane compartment. The charge and proton concentration differences together provide energy for ATP synthesis in what is called proton-motive force.

STUDY BREAK 7.5

1. Anaerobic respiration and fermentation are both processes in which ATP is generated by the catabolism of food molecules in the absence of oxygen. Anaerobic respiration uses a molecule other than oxygen as the final electron acceptor of an electron transfer system, whereas in fermentation electrons are transferred to an organic molecule and no electron transfer system is involved.
2. The two types of fermentation are lactate fermentation and alcoholic fermentation. In lactate fermentation, the end product of glycolysis, pyruvate, is converted to lactate. The lactate stores electrons temporarily, transferring them to the mitochondrial electron transfer system when the oxygen content of cells returns to normal. In alcoholic fermentation, pyruvate is converted into ethyl alcohol.

STUDY BREAK 7.6

1. Acetyl–CoA.
2. Amino acids, fatty acids, triglycerides, sugars of nucleotides, glucose-1-phosphate.
3. If excess ATP is present in the cytosol, it binds to and inhibits the activity of phosphofructokinase. As a result, the concentration of fructose-1,6-bisphosphate, the product of the phosphofructokinase reaction, decreases, and the subsequent reactions of glycolysis are slowed or stopped. This is reversed when the ATP level in the cytosol decreases. In the end, this control mechanism helps prevent the needless oxidation of fuel molecules when the cell has an adequate supply of ATP. Phosphofructokinase is also inhibited by citrate, and stimulated by AMP.

Think Like a Scientist

FIGURE 7.11

The H^+ concentration will increase in the intermembrane compartment. The increase will lead to an increase in ATP production.

FIGURE 7.13

Assignable activity in MindTap. Answers provided to instructors.

MOLECULAR INSIGHTS

Knowing the core proteome for mitochondria could lead to the development of diagnostic tools for mitochondrial disorders related to the core proteins, and potentially then for the identification of mitochondrial protein targets for the development of therapeutic drugs.

UNANSWERED QUESTIONS

One possible hypothesis: there is a direct cause and effect relationship between altered regulation of mitochondrial expression and AD. If a significant number of mice whose genes were not genetically altered were to develop AD, this hypothesis would have to be rejected.

Test Your Knowledge

1. a 2. b 3. b 4. e 5. a 6. d 7. a 8. c 9. e 10. a

Interpret the Data

Assignable activity in MindTap. Answers provided to instructors.

Chapter 8

Study Break

STUDY BREAK 8.1

1. The two stages of photosynthesis are: (1) the light-dependent reactions, in which the energy of sunlight is absorbed and converted into chemical energy in the form of ATP and NADPH; and (2) the light-independent (dark) reactions, in which electrons carried by NADPH are used as a source of energy to convert carbon dioxide from inorganic to organic form.
2. In plants, photosynthesis takes place in the chloroplast. The light-dependent reactions are carried out on the thylakoid membranes and stromal lamellae. The light-independent reactions are carried out in the stroma.

STUDY BREAK 8.2

1. The chlorophyll *a* molecules in the antenna complexes are normal molecules of the pigment, consisting of a carbon ring structure with a magnesium atom bound at the center and an attached hydrophobic side chain. These chlorophyll *a* pigments absorb light. The chlorophyll *a* molecules in the reaction centers have modified light absorption properties that result from interactions with particular proteins of the photosystems. The two special chlorophyll *a* molecules of photosystem II are P680; those of photosystem I are P700. These pigment molecules capture light energy from the antenna complex pigments in the form of an excited electron that is passed to a primary acceptor molecule. That electron is passed to the electron transfer system.
2. The making of NADPH begins when electrons derived from water splitting are pushed to higher energies by light absorption in photosystem II. The high-energy electrons pass to a primary acceptor in photosystem II and then down an electron transfer system to P700 in photosystem I, losing energy along the way. Light energy absorbed by photosystem I again excites the electrons, which pass to different electron carriers, ending with ferredoxin. The ferredoxin transfers high-energy electrons to $NADP^+$, which is reduced to NADPH by $NADP^+$ reductase.
3. In the linear electron flow pathway, electrons run through the entire set of photosystems and electron carriers, producing both NADPH and ATP. In the cyclic electron flow pathway, electrons flow cyclically around photosystem I; photosystem II is not involved. The cycle of electrons is through the cytochrome complex and plastocyanin to photosystem I, to ferredoxin, but then back to the cytochrome complex rather than on to $NADP^+$ reductase. Only ATP is produced by this pathway.

STUDY BREAK 8.3

1. Rubisco catalyzes a reaction combining carbon dioxide with RuBP to form two molecules of 3-phosphoglycerate (3PGA). Rubisco, an enzyme unique to photosynthetic organisms, is the key enzyme for producing the world's food because it is responsible for carbon dioxide fixation, a process that ultimately provides organic molecules for most of the world's organisms. Rubisco is the key regulatory site of the Calvin cycle for the following reason: During the daytime, sunlight powers the light-dependent reactions, and the NADPH and ATP produced by those reactions stimulate rubisco, which, in turn, keeps the Calvin cycle running. In darkness, however, NADPH and ATP levels are low, and as a result, rubisco's activity is inhibited and the Calvin cycle slows down or stops.
2. For each carbon atom that is released from the Calvin cycle in a carbohydrate molecule, one carbon dioxide molecule must enter the cycle. Therefore, to produce a molecule containing 12 carbon atoms, 12 molecules of carbon dioxide must enter the cycle.

1. Photorespiration uses oxygen and releases CO_2. It occurs when oxygen concentrations are high relative to CO_2 concentrations. In that condition, rubisco acts as an oxygenase rather than a carboxylase, catalyzing the combination of RuBP with O_2 rather than CO_2. The toxic products formed by this reaction cannot be used in photosynthesis and are eliminated from the plant as CO_2. Photorespiration uses energy to salvage the carbons from phosphoglycolate, which greatly reduces the efficiency of energy use in photosynthesis. This can be seen in the reduced growth of plants grown under photorespiration conditions.

2. In the C_4 pathway, carbon fixation involves the reaction of CO_2 with phosphoenolpyruvate (PEP) to produce a four-carbon molecule, oxaloacetate. The oxaloacetate is reduced to malate by electrons transferred from NADPH, and malate then is oxidized to pyruvate in a reaction releasing CO_2, which is used in the rubisco-catalyzed first step of the Calvin cycle. In C_4 plants, carbon fixation and the Calvin cycle occur in different cell types: carbon fixation in mesophyll cells, and the Calvin cycle in bundle sheath cells. This alternative method of carbon fixation minimizes photorespiration.

3. In C_4 plants, carbon fixation and the Calvin cycle occur in different cell types, mesophyll cells and bundle sheath cells, respectively. In CAM plants, carbon fixation and the Calvin cycle occur at different times, at night and during the day, respectively.

STUDY BREAK 8.5

The reactions of photosynthesis and cellular respiration are essentially the reverse of one another, with CO_2 and H_2O being the reactants of photosynthesis and the products of cellular respiration. Phosphorylation reactions involving electron transfer systems are part of each process, namely, photophosphorylation in photosynthesis and oxidative phosphorylation in cellular respiration. G3P is an intermediate in both pathways: in photosynthesis it is a product of the Calvin cycle, and in cellular respiration it is generated in glycolysis in the conversion of glucose to pyruvate. In photosynthesis, G3P is used for the synthesis of sugars and other fuel molecules, and in cellular respiration, it is part of the catabolism of sugars to simpler organic molecules.

Think Like a Scientist

FIGURE 8.4

Assignable activity in MindTap. Answers provided to instructors.

FIGURE 8.10

Two.

MOLECULAR INSIGHTS

Perform essentially the same experiment described except, instead of using water deficit for the experimental plants, irrigate them with water containing a high salt concentration. Study shoot elongation and photosynthesis, and collect shoot tip samples for proteomic analysis over the time course of the experiment. Analyze the proteomic data to determine which proteins change significantly in abundance and compare them with the set of proteins that changed in the water deficit study.

UNANSWERED QUESTIONS

Imagine two types of desert plants. Some, like the wild watermelon plants of the Kalahari, grow very rapidly, with very robust photosynthesis, as soon as rain falls. They produce as many seeds as they can, quickly, before severe drought sets in. This strategy requires rapid photosynthesis, even at the risk of photodamage. Others, such as desert scrub plants like sage, persist throughout the drought. They may grow slowly, with highly protected photosynthesis, even during times of rain. If they grew too fast, their large leaf surface areas would result in high water loss during drought. In some invasive plant species, aggressive

photosynthesis and high growth rates have a selective advantage, allowing them to outcompete their rivals for resources such as sunlight or growth space. In these cases, high rates of photosynthesis may be an advantage even as they risk photodamage. Other plants may invest for the longer term, building long-lived resilient structures (leaves, wood, etc.). The large investment may render risky photosynthetic strategies less viable, instead preferring more "conservative" down-regulatory strategies. Still other plants may have to contend with lack of key nutrients, in which high photosynthetic rates or the need for rapid repair of the photosynthetic apparatus would be too "expensive" in terms of resources. Finally, humans have selected for traits in crop plants that are not related to photosynthetic yield—for example, tasty fruits, disease resistance, or short growing season. These traits often take precedence over photosynthetic efficiency.

Test Your Knowledge

1. d 2. b 3. a 4. c 5. e 6. d 7. d 8. b 9. a 10. b

Interpret the Data

Assignable activity in MindTap. Answers provided to instructors.

Chapter 9

Study Break

STUDY BREAK 9.1

The specificity of a cellular response depends on the signaling molecule–receptor interaction. Specificity starts with the signaling molecule; that is, the specific signaling molecule is the messenger that elicits a specific cellular response. For example, the hormone epinephrine causes glucose to be released into the bloodstream. Specificity also depends on the target cells; that is, only target cells respond to the signaling molecule because they exclusively have receptors for the signaling molecule.

STUDY BREAK 9.2

1. Protein kinases are enzymes that add phosphate groups to other proteins. The result of phosphorylation is that the protein will be either stimulated or inhibited in its activity. The cellular responses of signal transduction pathways are produced through the actions of protein kinases.

2. Amplification is the phenomenon of an increase in the magnitude of each step of a signal transduction pathway. Amplification typically occurs because the proteins conducting each step of the pathway are enzymes; that is, each enzyme, when it becomes activated, activates large numbers of molecules entering the next step of the pathway.

STUDY BREAK 9.3

1. For a receptor tyrosine kinase to become activated, the signaling molecule first binds to the receptor, which then assembles into a dimer. The receptor adds phosphate groups to tyrosines on the cytoplasmic side of itself, which activates the receptor.

2. A fully activated receptor tyrosine kinase has phosphorylated tyrosines on each of its two monomers. A signaling protein that recognizes a phosphorylated tyrosine as well as surrounding parts of the polypeptide binds to the receptor. Depending on the signaling protein, the binding itself may activate the signaling protein, or it is activated by tyrosine phosphorylation catalyzed by the receptor. In its activated form, the signaling protein initiates a transduction pathway leading to a cellular response. A given receptor can initiate different responses because different combinations of signaling proteins can bind to the receptor.

3. The first messenger in a G-protein–coupled receptor-controlled pathway is the extracellular signaling molecule. When it binds to the G-protein–coupled receptor, it activates a site on the cytoplasmic side of the receptor;

the activated receptor, in turn, activates the G protein next to it.

4. The effector is activated by the G protein. The effector is a plasma membrane-associated enzyme that generates a nonprotein signaling molecule called the second messenger. The second messenger leads to the activation of protein kinases, leading to the cellular responses triggered by the signaling molecule.

5. A main way the pathway is turned off is by the conversion of cAMP to 5′-AMP by phosphodiesterase. As long as the receptor is bound by the signaling molecule, cAMP is being generated by the activated effector. The continued synthesis of cAMP balances the degradation of cAMP by phosphodiesterase, ensuring that the pathway continues to run. However, if the signaling molecule no longer is bound to the receptor, the effector again becomes inactive, cAMP therefore is not generated, and existing cAMP is rapidly degraded by phosphodiesterase. As a result, the protein kinase cascade is shut down and no cellular responses occur.

6. When a ligand binds to a ligand-gated ion channel receptor, the conformation of the receptor channel changes, opening or closing an ion channel and thereby controlling movement of ions into or out of the cell.

STUDY BREAK 9.4

1. The steroid receptor is within the cell, whereas the receptor tyrosine kinase and G-protein–coupled receptors are in the membrane or associated with the membrane. Also, the activated steroid receptor directly activates genes, whereas the other two receptors, when active, are just the first steps in pathways that may or may not activate genes.

2. A steroid hormone brings about a specific cellular response because whether a cell responds to a steroid hormone depends on whether it has the internal receptor for the hormone. Then, within the cells with the receptor, the specific genes that are controlled are those with regulatory sequences that are recognized by the activated receptor.

STUDY BREAK 9.5

Signal transduction pathways, cellular response systems triggered by cell adhesion molecules, and communication pathways that involve gap junctions between adjacent cells might be integrated in a cross-talk network.

Think Like a Scientist

FIGURE 9.2

Assignable activity in MindTap. Answers provided to instructors.

FIGURE 9.7

The transduction pathways controlled by the receptor tyrosine kinase would be active all the time, producing cellular responses in an uncontrolled manner—that is, in the absence of the signaling molecule. Mutations such as this are known to be associated with some cancers.

FIGURE 9.9

There are two main possibilities. One is that the receptor would no longer recognize the first messenger signaling molecule. In this case, the cellular response(s) controlled by the signal molecule would not occur. The other is that the receptor would be active even in the absence of the first messenger signaling molecule. In this case, cellular response(s) would occur in an uncontrolled way. As you learned in the text, there are a very large number of G-protein–coupled receptors, and they control a great many cellular processes. Thus, malfunctions of these receptors can have serious consequences. In fact, mutations in genes for G-protein–coupled receptors have been shown to be responsible for more than 30 human diseases.

MOLECULAR INSIGHTS

Binding of ERβ to the mitochondrial genome suggests that this receptor has some involvement in mitochondrial genome function.

UNANSWERED QUESTIONS

1. Perhaps the behavior of the male induces the release of dopamine or another relevant neurotransmitter onto specific neurons in the brain of the female, which contain progestin receptors and are involved in regulation of sexual behavior. The dopamine, in turn, may activate those receptors, leading to neuronal changes resulting in the expression of sexual behavior.

2. To answer this question, you have to think of other situations in which the environment might cause the release of a neurotransmitter that then, via cross-talk, activates a steroid receptor resulting in changes in behavior. Perhaps stimulation from pups activates neuronal steroid hormone receptors, resulting in changes in maternal behavior in a lactating mother. There are many other examples in which stimulation from the environment or another animal might, via neurotransmitter release, influence the function of a particular steroid hormone receptor (by activating it).

Test Your Knowledge

1. b 2. a 3. d 4. d 5. c 6. c 7. e 8. b 9. c 10. e

Interpret the Data

Assignable activity in MindTap. Answers provided to instructors.

Chapter 10

Study Break

STUDY BREAK 10.1

1. (1) An elaborate master program of molecular checks and balances ensures an orderly and timely progression through the cell cycle. (2) The process of DNA synthesis replicates each DNA chromosome into two copies with almost perfect fidelity. (3) A structural and mechanical web of interwoven "cables" and "motors" of the mitotic cytoskeleton separates the DNA copies precisely into the daughter cells.

2. A linear DNA molecule complexed with proteins.

3. The nucleosome consists of two molecules each of histones H2A, H2B, H3, and H4 assembled into a nucleosome core particle wrapped with almost two turns of DNA. The diameter of the nucleosome is 10 nm.

4. Histone H1 is responsible for the next level of chromosome packing above the nucleosome. H1 binds to the exit/entry point of DNA on the nucleosome and to the linker DNA and brings about a coiling of the chromatin into the 30-nm chromatin fiber. The coiled structure is called the solenoid.

STUDY BREAK 10.2

1. Each daughter cell has the same number and types of chromosomes, and contains the same genetic information, as the parent cell before its chromosomes were duplicated.

2. In order, the stages of mitosis are prophase, prometaphase, metaphase, anaphase, and telophase.

3. Each eukaryotic chromosome has a specialized region known as a centromere. The centromere is where a complex of several proteins, called a kinetochore, forms. During mitosis, some spindle microtubules attach to each kinetochore. These connections determine the outcome of mitosis because they attach the sister chromatids of each chromosome to microtubules leading to the opposite spindle poles; during anaphase, the spindle separates sister chromatids and pulls them to opposite spindle poles. In brief, the centromeres are key to chromosome segregation during mitosis. Although not mentioned in the chapter, this is also apparent when problems occur in which a chromosome fragment without a centromere breaks off from a chromosome. The fragment without a centromere cannot connect to the spindle, and hence is not segregated properly.

STUDY BREAK 10.3

1. In animal cells, all of which have centrosomes, the spindle forms through division of the cell center. As the dividing centrosome separates into two parts, the microtubules of the spindle form between them. Plant cells lack centrosomes. In plant cells, the spindle microtubules simply assemble around the nucleus. In either case, the microtubules assemble in a parallel array that creates two poles in the dividing cell.

2. Chromosomes (sister chromatids) move apart during anaphase. During the anaphase movements, the kinetochores move along the kinetochore microtubules, which become shorter as anaphase progresses. The nonkinetochore microtubules slide over each other, decreasing the degree of overlap and pushing the poles farther apart. The total distance traveled by the chromosomes is the sum of the two movements.

STUDY BREAK 10.4

1. A Cdk can become active only once it has complexed with a cyclin protein. Each cyclin is present only during a particular segment of the cell cycle, controlled by when it is synthesized and degraded. Thus, the period in the cell cycle when a particular Cdk is active depends on when its activating cyclin is present.

2. When the kinase of the Cdk is activated upon binding to a cyclin, it phosphorylates target proteins in the cell, regulating their activities. Those proteins play roles in initiating or regulating key events of the cell cycle, namely, DNA replication, mitosis, and cytokinesis. The progression through the cell cycle is regulated, then, by a succession of cyclin–Cdk complexes, each of which has specific regulatory effects.

3. An oncogene is an altered gene in an organism that contributes to the development of cancer—that is, uncontrolled cell division. Some of the genes that become oncogenes encode components of the cyclin–Cdk system that regulates cell division, whereas others encode proteins that regulate gene activity, form cell-surface receptors, or make up elements of the systems controlled by the receptors.

4. Metastasis is when cells break loose from a tumor, spread throughout the body, and grow into new tumors in other body regions.

STUDY BREAK 10.5

1. Bacterial cell division begins with replication of the bacterial chromosome, starting with duplication of the origin of replication. Once the origin of replication is duplicated, the two origins actively migrate to the two ends of the cells, a process that separates the two replicating chromosomes in the cell. Division of the cytoplasm then occurs by means of a partition of cell wall material that grows inward until the cell is separated into two parts. The cytoplasmic division divides the replicated DNA molecules and cytoplasmic structures between the daughter cells.

2. Present in eukaryotic cell division, but absent from bacterial cell division, are the following: the process of mitosis; any form of microtubules for chromosome segregation; a spindle apparatus; cyclin/CDK control proteins.

Think Like a Scientist

FIGURE 10.4

Joined sister chromatid pairs attach to kinetochore microtubules during prometaphase and begin their migration to the metaphase plate. The spindle microtubules are also necessary for segregating the sister chromatids to opposite poles of the cell during anaphase. Colchicine would block spindle formation and no kinetochore microtubules would be present to attach to the sister chromatids. As a result, the cells would not be able to complete mitosis. In fact, colchicine is used to get cells to arrest in metaphase when the chromosomes are at their most condensed. At that stage, the chromosomes are the easiest to visualize by microscopy (see Figure 10.8).

FIGURE 10.13

Assignable activity in MindTap. Answers provided to instructors.

FIGURE 10.15

Assignable activity in MindTap. Answers provided to instructors.

MOLECULAR INSIGHTS

The cell cycle proceeds the same way in all eukaryotes, with some variation in details. Likely the proteins involved have been conserved during evolution with perhaps some species-specific modifications. Thus, you would predict that many, if not most, of the proteins identified in the study would be shown to be cell cycle-dependent in other eukaryotes. Some proteins could be specific to humans, and some proteins could be specific to other eukaryotes. For instance, we know that the details of cell cycle regulation differ in mammals versus yeast, with some differences in the sets of cyclins and Cdks in each.

UNANSWERED QUESTIONS

Although *E. coli* is easy to culture, there is no known prokaryotic equivalent of mitosis; mammalian cells are hard to culture and cannot be kept in long-term culture (see *Focus on Research: Basic Research*). *Focus on Research: Model Organisms* provides abundant rationale for *Saccharomyces* as a model organism for this type of research. Ubiquitin is given this name because it is indeed ubiquitous—present in almost the same form in essentially all eukaryotes (see Chapter 16).

Test Your Knowledge

1. c 2. e 3. e 4. b 5. d 6. d 7. a 8. b 9. b 10. b 11. c

Interpret the Data

Assignable activity in MindTap. Answers provided to instructors.

Chapter 11

Study Break

STUDY BREAK 11.1

1. Mitosis produces daughter cells that are genetically identical to the parent cell. Either a haploid or a diploid cell can undergo mitosis. Meiosis starts with a diploid cell. There is one round of DNA replication but two rounds of cell division, with the result that four haploid cells are produced from the parent diploid cell.

2. Recombination is the physical exchange of segments between the chromatids of homologous chromosomes. Recombination occurs in prophase I, when homologous chromosomes have each duplicated to produce sister chromatids and are aligned fully in an organization called a tetrad.

3. Meiosis II is the meiotic division that is similar to a mitotic division.

STUDY BREAK 11.2

1. There are three ways in which sexual reproduction generates genetic variability. First, recombination, which involves the physical exchange of segments between homologous chromatids in prophase I of meiosis, generates new combinations of alleles. Second, the random separation of homologous chromosomes during meiosis generates genetic variability; that is, in metaphase I, for each homologous pair of chromosomes, one chromosome makes spindle connections leading to one pole and the other chromosome connects to the opposite pole. This process operates independently for each homologous pair of chromosomes; thus, for each meiosis random combinations of maternal and paternal chromosomes move to the poles during anaphase I. Third, the random joining of male and female gametes produces additional genetic variability.

2. The proportion of gametes that will have chromosomes that originate from the animal's female parent is: $(1/2)^6 = 1/64$.

STUDY BREAK 11.3

In animals, the diploid phase dominates the life cycle; mitotic divisions occur only in this phase. Meiosis in the diploid phase gives rise to products that develop directly into egg and sperm cells without undergoing mitosis.

In most plants, the life cycle alternates between haploid and diploid generations, both of which grow by mitotic divisions. Fertilization produces the diploid sporophyte generation; after growth by mitotic divisions, cells of the sporophyte undergo meiosis and produce haploid spores. The spores germinate and grow by mitotic divisions into the gametophyte generation. After growth of the gametophyte, cells develop directly into egg or sperm nuclei, which fuse in fertilization to produce the diploid sporophyte generation again.

Think Like a Scientist

MOLECULAR INSIGHTS
The result suggests that CYP26B1 plays an essential role in germ cell development perhaps among all vertebrates, not just mammals.

UNANSWERED QUESTIONS
The outcome depends on a variety of factors, such as the extent of the rearrangement, its location in the genome, and the particular organism or species being analyzed. However, very frequently, a phenomenon referred to as "synaptic adjustment" is observed. This allows the duplications and/or inversions to be accommodated (in many cases these just "loop out" from the fully paired homologs) and the homologs still succeed in synapsing. Occasionally, these rearrangements can also lead to nonhomologous synapsis and asynapsis.

Test Your Knowledge
1. b 2. a 3. d 4. c 5. b 6. b 7. d 8. a 9. b 10. c

Interpret the Data
Assignable activity in MindTap. Answers provided to instructors.

Chapter 12

Study Break

STUDY BREAK 12.1
1. The numerical results approximate a 9:3:3:1 ratio. Therefore, both parents must be heterozygous for both genes. If we designate the alleles for one of the pairs of traits as *A* and *a*, and the alleles for the other pair of traits as *B* and *b*, each parent has the genotype *Aa Bb*.
2. A ratio of 1:1:1:1 is the typical outcome of a testcross involving a parent who is heterozygous for the alleles of two genes. Using the same allele symbols as in (1), the genotypes of the parents are *Aa Bb* and *aa bb*; this is a testcross.

STUDY BREAK 12.2
1. The color pattern involved is an incompletely dominant trait.
2. The fur colors here involve multiple alleles of a single gene. The allele symbols are *C* for wild type, c^{ch} for chinchilla, c^h for Himalayan, and *c* for albino, with dominance in the order $C \to c^{ch} \to c^h \to c$: that is, the *C* allele is completely dominant to the c^{ch} allele, the c^{ch} allele is completely dominant to the c^h allele, and so on. Therefore, we have these genotypes and phenotypes:

   C with c^{ch}, c^h, or *c* = agouti (*CC*, Cc^{ch}, Cc^h, *Cc*)
   c^{ch} with c^{ch}, c^h, or *c* = chinchilla ($c^{ch}c^{ch}$, $c^{ch}c^h$, $c^{ch}c$)
   c^h with c^h or *c* = Himalayan (c^hc^h, c^hc)
   c with *c* = albino (*cc*)

Think Like a Scientist

FIGURE 12.5
Assignable activity in MindTap. Answers provided to instructors.

FIGURE 12.8
Assignable activity in MindTap. Answers provided to instructors.

FIGURE 12.9
Assignable activity in MindTap. Answers provided to instructors.

FIGURE 12.13
Assignable activity in MindTap. Answers provided to instructors.

MOLECULAR INSIGHTS
In Chapter 6 you learned that the substrate or substrates for an enzyme interact with a small region of an enzyme called the *active site*. It is at the active site where enzyme catalysis occurs. The active site of an enzyme is formed by a particular array of amino acids in the polypeptide chain. Those amino acids have evolved to be optimal for enzyme activity. A mutation that alters an amino acid in or near the active site can affect the structure of the active site, and therefore adversely affect the enzyme's catalytic activity. The extent of the deleterious effect depends on the particular amino acid substitution and the chemical changes brought about in the protein by the substitution.

UNANSWERED QUESTIONS
If you sequence lots of species you will identify lots of regions of strong evolutionary constraint. If you look at many individuals within a species, you can look at the rate of fast-evolving alleles that might have responded to population-specific pressures or drift and start asking questions pertaining to the genome of the human. Ultimately, your choice will depend on the question you are trying to answer. If you are trying to understand developmental processes, cross-species comparisons are extremely useful, but if you are looking at immunity, perhaps the within-species comparison is better.

Test Your Knowledge
1. (a) The *CC* parent produces all *C* gametes, and the *Cc* parent produces 1/2 *C* and 1/2 *c* gametes. All offspring would have colored seeds—half homozygous *CC* and half heterozygous *Cc*. (b) The *Cc* parent produces 1/2 *C* gametes and 1/2 *c* gametes, and the *cc* parent produces all *c* gametes. Half of the offspring are colored (1/2 *Cc*) and half are colorless (1/2 *cc*). (c) Both parents produce 1/2 *C* and 1/2 *c* gametes. Of the offspring, three-fourths would have colored seeds (1/4 *CC* + 1/2 *Cc*) and one-fourth would have colorless seeds (1/4 *cc*).
2. The genotypes of the parents are *Lele* and *lele*.
3. (a) All *A B*. (b) 1/2 *A B* + 1/2 *a B*. (c) 1/2 *A b* + 1/2 *a b*. (d) 1/4 *A B* + 1/4 *A b* + 1/4 *a B* + 1/4 *a b*.
4. (a) All *Aa Bb*. (b) 1/4 *Aa Bb* + 1/4 *aa Bb* + 1/4 *aa bb*. (c) 1/4 *AA BB* + 1/4 *AA Bb* + 1/4 *Aa BB* + 1/4 *Aa Bb*. (d) 1/4 *Aa Bb* + 1/8 *AA Bb* + 1/8 *Aa BB* + 1/8 *Aa bb* + 1/8 *aa Bb* + 1/16 *AA BB* + 1/16 *AA bb* + 1/16 *aa BB* + 1/16 *aa bb*.
5. (a) All *A B C*. (b) 1/4 *A B C* + 1/4 *A B c* + 1/4 *a B C* + 1/4 *a B c*. (c) 1/2 *A B C* + 1/2 *a B C*. (d) 1/8 *A B C* + 1/8 *A B c* + 1/8 *A b C* + 1/8 *A b c* + 1/8 *a B C* + 1/8 *a B c* + 1/8 *a b C* + 1/8 *a b c*.
6. The parental cross is *GG LeLe RR* × *gg lele rr*. All offspring of this cross are expected to be tall plants with green pods and round seeds, or *Gg Lele Rr*. When crossed, this heterozygous F₁ generation is expected to produce eight different phenotypes among the offspring— green–tall–round, green–dwarf–round, yellow–tall–round, green–tall–wrinkled, yellow–dwarf–round, green–dwarf–wrinkled, yellow–tall–wrinkled, yellow–dwarf–wrinkled—in a 27:9:9:9:3:3:3:1 ratio.
7. The cross is expected to produce white, tabby, and black kittens in a 12:3:1 ratio.
9. The taster parents could have a nontaster child, but nontaster parents are not expected to have a child who can taste PTC. The chance that they might have a taster child is 3/4. The chance of a nontaster child being born to the taster couple would be 1/4. Because each combination of gametes is an independent event, the chance of the couple having a second child, or any child, who cannot taste PTC is expected to be 1/4.
10. Because the man can produce only 1 type of allele for each of the 10 genes, he can produce only 1 type of sperm cell with respect to these genes. The woman can produce 2 types of alleles for each of her 2 heterozygous genes, so she can produce 2 × 2 = 4 different types of eggs with respect to the 10 genes. In general, as the number of heterozygous genes increases, the number of possible types of gametes increases as 2^n, where *n* = the number of heterozygous genes.
11. Use a standard testcross; that is, cross the guinea pig with rough, black fur with a double recessive individual, *rr bb* (smooth, white fur). If your animal is homozygous *RR BB*, you would expect all of the offspring to have rough, black fur.
12. One gene probably controls pod color. One allele, for green pods, is dominant; the other allele, for yellow pods, is recessive.
13. The cross *RR* × *Rr* will produce 1/2 *RR* and 1/2 *Rr* offspring. The cross *Rr* × *Rr* will produce 1/4 *RR*, 1/2 *Rr*, and 1/4 *rr* as combinations of alleles. However, the 1/4 *rr* combination is lethal, so it does not appear among the offspring. Therefore, the offspring will be born with only two types, *RR* and *Rr*, with twice as many *Rr* as *RR* in a 1:2 ratio (or 1/3 *RR* + 2/3 *Rr*).
14. The genotypes are: bird 1, *Ff Pp*; bird 2, *FF PP*; bird 3, *Ff PP*; bird 4, *Ff Pp*.
15. Yes, it can be determined that the child is not hers, because the father must be AB to have both an A and B child with a type O wife; none of the woman's children could have type O blood with an AB father.
16. The mother is homozygous recessive for both genes, and the father must be heterozygous for both genes. The child is homozygous recessive for both genes. The chance of having a child with normal hands is 1/2, and that of having a child with woolly hair is 1/2. Using the product rule of probability, the probability of having a child with normal hands and woolly hair is 1/2 × 1/2 = 1/4.

Interpret the Data
Assignable activity in MindTap. Answers provided to instructors.

Chapter 13

Study Break

STUDY BREAK 13.1
The cross to use is the testcross. Here, the testcross would be *Aa Bb* × *aa bb*. A testcross is used so that you can follow the meiotic events in the dihybrid parent (including the consequences of crossing over between linked genes) because all of the gametes from the testcross parent carry recessive alleles for the genes in the cross. A testcross shows linkage when the ratio of 1:1:1:1 for the four possible phenotypes is not seen; that is, the 1:1:1:1 ratio result occurs when two genes assort independently. However, if two genes are linked, there will be excess of the two parental classes of progeny compared with the two recombinant classes.

STUDY BREAK 13.2
The differences between sex-linked inheritance and autosomal inheritance are seen clearly when reciprocal crosses are made and followed through to the F₂ generation. If sex-linked inheritance is involved, a cross of miniature-winged female × normal-winged male flies will give an F₁ generation of all normal-winged female and all miniature-winged male flies. (This result would not be found with autosomal inheritance. Instead, all F₁ flies—both males and females— would have normal wings.) Selfing the F₁ flies will give an F₂ generation with 1:1 normal-winged: miniature-winged flies in both sexes. (For autosomal inheritance, you would see a 3:1 ratio of normal-winged: miniature-winged flies in both sexes.)

In the reciprocal cross of true-breeding normal-winged female × miniature-winged male, the F₁ flies will all have normal wings if sex-linked inheritance is involved. (This result is the same as for autosomal inheritance.) Selfing the F₁ flies will give an F₂ generation in which all females will have normal wings and the males will be 1/2 normal-winged and 1/2 miniature-winged. (For autosomal inheritance, you would see a 3:1 ratio of normal-winged: miniature-winged flies in both sexes.)

In summary, reciprocal crosses show different segregation patterns of phenotypes for sex-linked inheritance and autosomal inheritance. The two modes of inheritance are easiest to distinguish in a cross of a mutant female × wild-type male because then, in the F₁ generation, all males show the mutant phenotype when sex-linked inheritance is involved.

STUDY BREAK 13.3

(a) Duplication of a chromosome segment occurs when a segment breaks from one chromosome and is inserted into its homolog.

(b) An individual with Down syndrome results when non-disjunction of chromosome 21 during meiosis occurs (usually in females), producing gametes with two copies of chromosome 21 and one copy of every other chromosome. When such a gamete fuses with a normal gamete, the result is a zygote with three copies of chromosome 21 and two copies of the other chromosomes. This individual will have Down syndrome. Aneuploidy is the term for the condition of extra or missing chromosomes.

(c) A translocation occurs when a broken segment of a chromosome becomes attached to a different, nonhomologous chromosome.

(d) Polyploidy means that there are more sets of chromosomes than the typical diploid set. Polyploidy may result if the spindle fails to function properly in mitosis of cell lines leading to gametes. Cells affected in this way will have twice the normal number of sets of chromosomes. When meiosis subsequently occurs, the gametes produced will have two sets of chromosomes. Fusion of these gametes with, for instance, a gamete with one set of chromosomes will produce a zygote with three sets of chromosomes—a triploid cell.

STUDY BREAK 13.4

1. Autosomal recessive inheritance: For a child to exhibit an autosomal recessive inheritance, he or she must inherit one recessive allele from each parent. For autosomal recessive inheritance to explain Simpson syndrome in the family, the father must be homozygous for the Simpson syndrome allele, ss, and the mother must be heterozygous, Ss. The expectation would be that 1/2 of the children would be Ss and 1/2 would be ss, regardless of sex, and that is what is found. Therefore, on the assumption that the mother is heterozygous, the syndrome could be an autosomal recessive trait. Sex-linked recessive inheritance: One characteristic of sex-linked recessive inheritance is that affected females pass on the trait to all of their sons. Here, we start with an affected male, and he would have to be X^sY if it is a sex-linked recessive trait. To explain the children, we would have to assume that the mother is heterozygous, X^SX^s. The cross of $X^SX^s \times X^sY$ is expected to give 1/2 females with the syndrome and 1/2 males with the syndrome, which is what is described. Therefore, the syndrome could be a sex-linked recessive trait.

2. Autosomal recessive inheritance: In pedigrees of autosomal recessive traits, the appearance of progeny with a trait when both parents do not have the trait is one common feature. If we assume that both parents are heterozygous, Ww, then the children can be explained; that is, you can get both wiggly-eared children (ww, expected frequency 1/4) and nonwiggly-eared children (WW or Ww, combined expected frequency 3/4), regardless of sex. Therefore, wiggly ears could be an autosomal recessive trait based on this family. Sex-linked recessive inheritance: Because a male individual has only one X chromosome, it is not possible for two nonwiggler parents to produce a wiggler daughter. To get a wiggler daughter, the male would have to be X^WY (wiggler) and the female would have to be X^WX^w (nonwiggler), which is not the case here. Therefore, we cannot conclude that ear wiggling is a sex-linked recessive trait based on this family.

STUDY BREAK 13.5

A mutant trait that shows cytoplasmic inheritance is caused by an alteration in the DNA of an organelle, either the mitochondrion or the chloroplast. A key property of cytoplasmic inheritance is that a trait is transmitted by a parent to all offspring, regardless of sex. The most common form of this is maternal inheritance, in which the progeny inherit the trait from their mother, paralleling the inheritance of mitochondria and mitochondrial DNA from the female parent and not from the male parent. This pattern of inheritance would not be seen for genes on chromosomes in the nucleus (as explained in Chapter 11).

Think Like a Scientist

FIGURE 13.2

Assignable activity in MindTap. Answers provided to instructors.

FIGURE 13.8

Assignable activity in MindTap. Answers provided to instructors.

MOLECULAR INSIGHTS

You would predict that expression of the *Xist* gene would lead to inactivation of the autosome on which the gene is located as a result of Xist lncRNA coating that chromosome and recruiting the protein machinery to alter the chromosomes' structure and silence its genes. In fact, this experiment has been done and that is what happens.

UNANSWERED QUESTIONS

No, the genes in the del(5q) region still have some activity. In these patients, only one copy of the gene is lacking due to a chromosomal deletion. The other copy, or allele, is present and appears to have its normal function. Because these patients now only have one copy of the genes in this deleted region, there is a lowered gene dosage, referred to as "haploinsufficiency." A lowered gene dosage (and resultant production of lower protein levels) can have profound effects on cell growth. Scientists currently believe that the lowered gene expression of *multiple* genes in the del(5q) region work together to play a role in disease development.

Test Your Knowledge

1. All sons will be color-blind, but none of the daughters will be. However, all daughters will be heterozygous carriers of the trait.

2. The chance that her son will be color-blind is 1/2, regardless of whether she marries a normal or color-blind male.

3. All of these questions can be answered from the pedigree. Polydactyly is caused by a dominant allele, and it is not a sex-linked trait. The genotypes of each person are:

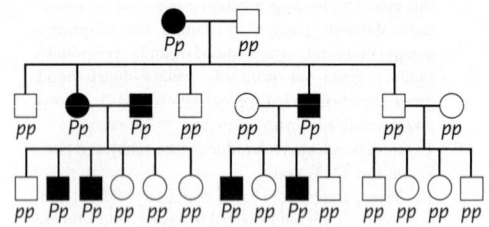

4. The sequence of the genes is *ADBC*.

5. The initial cross is X^wX^w white female × $X^{w^+}Y$ red male. The F₁ females produced are $X^{w^+}X^w$, with red eyes, and the F₁ males are X^wY, with white eyes.

   The cross of an F₁ female with a parental male, therefore, is $X^{w^+}X^w$ (red) × $X^{w^+}Y$ (red). The offspring are $X^{w^+}X^{w^+}$ and $X^{w^+}X^w$ females, both of which have red eyes, and $X^{w^+}Y$ (red) and X^wY (white) males in equal proportions. Thus the phenotypic ratio for females is 2 red: 0 white, and that for males is 1 red: 1 white.

   The cross of an F₁ male with a parental female is X^wY (white) × X^wX^w (white). All progeny of both sexes will be white-eyed.

7. Let the allele for wild-type gray body color = b^+, and the allele for black body = b. Let the allele for wild-type red eye color = p^+, and the allele for purple eyes = p. Then the parents are:

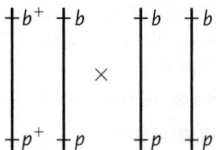

The F₁ flies with black bodies and red eyes are:

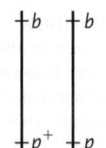

and the flies with gray bodies and purple eyes are:

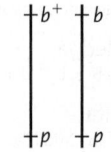

8. The genes are linked by their presence on the same chromosome (an autosome), but they are not sex-linked. Because the F₁ females must have produced 600 gametes to give these 600 progeny, and because 42 + 30 of these were recombinant, the percentage of recombinant gametes is 72/600, or 12%, which implies that 12 map units separate the two genes.

10. You might suspect that a recessive allele is sex-linked and is carried on one of the two X chromosomes of the female parent in the cross. When present on the single X of the male (or if present on both X chromosomes of a female), the gene is lethal.

Interpret the Data

Assignable activity in MindTap. Answers provided to instructors.

Chapter 14

Study Break

STUDY BREAK 14.1

³⁵S-labeled phages in this scenario will have labeled protein coats and labeled DNA. When these phages infect bacteria, radioactivity enters the cell and is found in the progeny phages. In addition, radioactivity is found in the phage material removed by the blender. ³²P-labeled phages in this scenario are like the phages in Hershey and Chase's experiment; they have labeled DNA but unlabeled protein. When these phages infect bacteria, radioactivity enters the cell and is found in the progeny phages. No radioactivity is found in the phage material removed by the blender.

STUDY BREAK 14.2

1. Adenine and guanine are purines. Thymine and cytosine are pyrimidines.

2. Complementary base pairs are held together by hydrogen bonds. Each base is attached to the deoxyribose sugar by a covalent bond.

3. Watson and Crick described the right-handed double helix that consists of two sugar–phosphate backbones on the outside and complementary base pairs between the two backbones. A complementary base pair is a purine paired with a pyrimidine, more specifically, an A with a T, and a G with a C. The two strands of DNA are antiparallel. The key dimensions of the molecule are:

diameter = 2 nm; 1 base pair = 0.34 nm; 1 turn of the helix = 10 base pairs = 3.4 nm.

4. The question focuses on the complementary base-pairing rules: A = T and G = C. If A = 20%, then T = 20%, giving 40% of the DNA as A–T base pairs. Therefore, 60% of the base pairs in this DNA molecule are G–C, and the percentage of C is 30%.

STUDY BREAK 14.3

1. Complementary base pairing ensures that the new DNA double helix is a faithful copy of the parental DNA double helix. For whatever base is exposed on the template strand, the DNA polymerase inserts the nucleotide with the complementary base.

2. DNA polymerases cannot initiate a DNA strand; they can add DNA nucleotides only to the 3′ end of an existing strand. The primer serves to provide a short stretch of nucleic acid that can be extended by DNA polymerase. The primer consists of RNA, rather than DNA, and is made by primase.

3. DNA polymerase III is the main DNA polymerase for replication in *E. coli*. This enzyme extends each primer that is synthesized on the lagging strand template, and synthesizes the leading strand. DNA polymerase I is used in lagging strand DNA synthesis. This enzyme replaces DNA polymerase when a new DNA fragment reaches the 5′ end of the Okazaki fragment that was made previously. With its 5′→3′ exonuclease activity, DNA polymerase I removes the RNA primer of that Okazaki fragment, and with its 5′→3′ polymerizing activity, it replaces that primer with DNA nucleotides. In this way the Okazaki fragment is converted from an RNA–DNA hybrid into a DNA fragment.

4. Telomeres are buffers against the progressive loss of the ends of chromosomes by repeated rounds of replication. Only when the hundreds to thousands of copies of the telomere repeats have been lost are genes exposed. When those genes are lost by continued chromosome shortening and/or when chromosomes break down in the absence of telomeres, the cell is severely damaged. Telomeres also act as caps on the ends of chromosomes. The capping prevents the staggered ends from being recognized by the cellular machinery that detects broken DNA in need of repair.

STUDY BREAK 14.4

Proofreading prevents errors from being introduced into the DNA sequence. The DNA sequence in an organism's genome specifies everything about that organism—most notably, its function and reproduction. If significant errors occur during replication, gene sequences could be changed and the function of the organism could be adversely affected. Particularly if there is a high rate of errors, as there would be in the absence of proofreading, these errors would have potentially lethal consequences.

As part of the proofreading process, DNA polymerase reverses and removes the mispaired nucleotide. The enzyme then continues to move forward, inserting the correct nucleotide.

Error-correcting mechanisms are mismatch repair, which corrects replication errors that escape proofreading, and excision repair mechanisms that correct various kinds of DNA damage. The mechanisms differ in how the error is recognized and removed, but then further steps are similar in each case; that is, synthesis of a replacement DNA segment by a repair DNA polymerase, and sealing the new segment to the adjacent DNA by DNA ligase.

Think Like a Scientist

FIGURE 14.2
Assignable activity in MindTap. Answers provided to instructors.

FIGURE 14.3
Assignable activity in MindTap. Answers provided to instructors.

FIGURE 14.9
Assignable activity in MindTap. Answers provided to instructors.

MOLECULAR INSIGHTS
It indicates that the structures arose early in evolution and that the molecules are fundamentally important for efficient replication. Further, it indicates that the mechanism of DNA replication has been conserved throughout evolution.

UNANSWERED QUESTIONS
Telomerase alone is not the sole determinant of telomere length. The complex of proteins at the telomere regulates access of telomerase to the telomere. Thus, even if the amount of telomerase in the cell is unchanged, certain conditions could alter the function of one or more proteins at the telomere to increase the ability of telomerase to elongate the chromosome ends.

Test Your Knowledge
1. b 2. d 3. a 4. a 5. d 6. c 7. b 8. a 9. c

Interpret the Data
Assignable activity in MindTap. Answers provided to instructors.

Chapter 15

Study Break

STUDY BREAK 15.1

1. Although most enzymes are proteins, not all proteins are enzymes. And, some proteins consist of more than one polypeptide subunit. Each different polypeptide is encoded by a different gene, hence the one gene–one polypeptide hypothesis.

2. There are four different letters in the code (A, U, G, C), so a five-letter code would have 4^5 possible combinations = 1,024 codons.

STUDY BREAK 15.2
1. 5′-GUUUAACCGAAUAAUGGCCUAC-3′
2. The promoter determines where transcription of a gene will begin. In prokaryotes, RNA polymerase binds to the nucleotide sequence of the promoter and orients in the correct way to transcribe the associated gene. In eukaryotes, transcription factors bind to the promoter and then recruit RNA polymerase, which then orients properly for transcription from the transcription start point.

STUDY BREAK 15.3
1. Both pre-mRNAs and mRNAs have a 5′ cap, exons, and a 3′ poly(A) tail. Only pre-mRNAs have introns, which are removed from pre-mRNAs to produce mRNAs.
2. Particular snRNPs bind to the ends of an intron using their contained RNAs to recognize the boundary sequences of the intron. Other snRNPs then bind, causing the intron to loop out, completing the active spliceosome. Cleavage at each intron–exon junction, looping back of the intron on itself, and joining the two exons together completes the splicing event; the intron and snRNPs are then released.

STUDY BREAK 15.4
1. In eukaryotes, a complex of the small ribosomal subunit, initiator tRNA, initiation factors, and GTP binds to the 5′ cap of the mRNA and scans along the mRNA until it reaches the first AUG codon, which is the start codon. The anticodon of the initiator tRNA binds to the start codon, the large ribosomal subunit binds, and the initiation factors are released when GTP is hydrolyzed.

In prokaryotes, a complex of the small ribosomal subunit, initiator tRNA, initiator factors, and GTP binds to the region of the mRNA where the AUG start codon is located, directed by a specific RNA sequence upstream of the start codon. The other steps are the same as those in eukaryotes.

2. The P site is where the tRNA with the growing polypeptide is located. Downstream of the P site is the A site.

An incoming aminoacyl–tRNA enters the A site and its anticodon base pairs with the codon of the mRNA in that site. When the polypeptide is transferred to the amino acid on the tRNA in the A site, the ribosome translocates one codon along the mRNA. As translocation takes place, the empty tRNA that was in the P site is moved to the E site. It remains there, blocking a new aminoacyl–tRNA entering the A site until translocation is finished. Then the empty tRNA is released from the ribosome.

3. Proteins found in the cytosol are made on free ribosomes. Proteins are sorted to the endomembrane system by cotranslational import. The proteins begin their synthesis on free ribosomes. These proteins have signal sequences at their N-terminal ends that direct them and the ribosome to dock with a receptor on the rough ER membrane. Continued translation inserts the growing polypeptide into the lumen of the ER, and the signal sequence is removed by signal peptidase. The proteins are then tagged to target them for sorting to their final destinations. Some proteins remain in the ER, whereas others are transported via the Golgi to vesicles for sorting to lysosomes, secreting them from the cell, or depositing them in the plasma membrane.

Proteins are sorted to mitochondria, chloroplasts, microbodies, and the nucleus by posttranslational import. Proteins destined for the mitochondria, chloroplasts, and microbodies have targeting signals that target them to the organelle. The targeting signals are recognized by specific receptors on the organelle surface, triggering the import of the protein into the organelle by specific pathways. Proteins destined for the nucleus have a nuclear localization signal that is bound by a cytosolic protein. The complex then interacts with the nuclear pore complex and the polypeptide is then transported into the nucleus through the pore.

STUDY BREAK 15.5
1. A missense mutation involves a change from a sense codon to another sense codon that specifies a different amino acid. A silent mutation involves a change from one sense codon to another sense codon, but where both codons specify the same amino acid.

2. Genetic recombination occurs by crossing over between two homologous sequences. TE transposition occurs by integration into a new location with which the TE has no sequence homology.

3. A DNA transposon is a transposable element (TE) that moves from one location to another in the genome using a DNA intermediate.

A retrotransposon is a transposable element that moves from one location to another in the genome using an RNA intermediate; that is, the integrated DNA element is transcribed to produce an RNA copy. The RNA copy is reverse-transcribed into DNA, which then integrates at a new location in that genome.

4. Both: (1) have inverted repeats at their ends; (2) contain a transposase gene; and (3) integrate into target sites and cause a duplication of the target site.

Think Like a Scientist

FIGURE 15.1
Assignable activity in MindTap. Answers provided to instructors.

FIGURE 15.5
No. Although transcription occurs in highly similar ways in eukaryotes and prokaryotes, the initiation processes in particular are specific; that is, different promoter sequences are used, and eukaryotic RNA polymerases typically recognize only eukaryotic promoter sequences, and prokaryotic RNA polymerases typically recognize only prokaryotic promoter sequences. Therefore, *E. coli* RNA polymerase is unlikely to be able to recognize the promoter of the eukaryotic protein-coding gene in the figure.

MOLECULAR INSIGHTS

A reasonable hypothesis would be that sleep deprivation causes a modest loss in the number of transcripts that show circadian rhythmicity. The hypothesis could be tested by using transcriptome analysis to compare groups of students with normal and restricted sleep times. The time course of any changes seen could be studied in volunteers who are transitioned from normal sleep to restricted sleep.

UNANSWERED QUESTIONS

1. The first ribosomes were made only of RNA.
2. Not to make proteins, but to make simple protein fragments (peptides) that bind to RNA to help RNA carry out its biological functions before the evolution of proteins.

Test Your Knowledge

1. a 2. e 3. d 4. b 5. d 6. b 7. e 8. d 9. b 10. a

Interpret the Data

Assignable activity in MindTap. Answers provided to instructors.

Chapter 16

Study Break

STUDY BREAK 16.1

1. The Lac repressor is active when it is made. In normal cells, in the absence of lactose, the Lac repressor binds to the operator, blocking transcription. In the mutant, the Lac repressor is not made, so transcription can never be blocked because no repressor is available to bind to the operator. As a consequence, the lactose-metabolizing enzymes will be made both in the presence and absence of lactose in the medium. The mutation involved is an example of a regulatory mutant. Mutations such as this were valuable to Jacob and Monod in developing the operon model for gene regulation.
2. The Trp repressor is inactive when it is made. In normal cells, in the presence of tryptophan, the Trp repressor is activated, binding to the operator and blocking transcription. In the mutant, the Trp repressor is not made, so the operon cannot be turned off when tryptophan is present. This means that the tryptophan biosynthesis enzymes will be produced both in the presence and absence of tryptophan.

STUDY BREAK 16.2

1. Histones are general negative regulators of gene expression. When DNA is complexed with histones in normal chromatin, gene promoters typically are not very accessible to the transcription machinery. By acetylating histones, the chromatin is remodeled, making the promoter now accessible to the transcription machinery. Acetylation of histones occurs in response to the binding of an activator to a regulatory sequence associated with the gene.
2. General transcription factors bind to the promoter and recruit RNA polymerase II, orienting the enzyme so that it will begin transcription at the beginning of the gene. Activators bind to regulatory sequences associated with genes and increase the rate of transcription. Activators that bind to regulatory sequences in the proximal promoter region interact directly with the general transcription factors at the promoter to exert their action. Activators that bind to regulatory sequences in the enhancer stimulate transcription indirectly. These latter activators bind to a coactivator that also binds to the complex of proteins at the promoter, and transcription then occurs at the maximal possible rate.

STUDY BREAK 16.3

1. A microRNA (miRNA), in a complex with particular proteins, binds to an mRNA by complementary base pairing. Either the proteins cut the mRNA in the region of pairing, thereby destroying that molecule, or the double-stranded RNA region blocks translation.
2. Removal of the poly(A) tail would result in the mRNA not being translated.

STUDY BREAK 16.4

1. Determination is the process by which the developmental fate of a cell is set. Differentiation is the establishment of a cell-specific developmental program in cells. Determination and differentiation are under molecular control. Regulatory genes encode regulatory proteins that bind to promoters of the genes they control, switching the genes on or off depending on the interaction.
2. The segmentation genes subdivide the embryo progressively into regions, thereby determining the segments of the embryo and the adult. In essence, they organize the embryo into segments. The *Hox* genes specify the identity of each segment with respect to the body part it will become.

STUDY BREAK 16.5

1. A driver mutation is a mutation in a gene that confers a selective advantage to the cell in which it occurs. The gene in which the driver mutation occurs is called a driver gene.
2. A tumor suppressor gene encodes a product that has an inhibitory role in the cell division cycle. If a tumor suppressor gene is mutated so that its product is nonfunctional or its function is significantly diminished, then the gene's inhibitory control of cell division is lost or reduced. As a result, the cell may have a selective growth advantage over its neighbors and may progress toward becoming a cancer cell.
3. A proto-oncogene encodes a product that has a stimulatory role in cell division. Mutations that lead to increased levels of that product have converted a proto-oncogene to an oncogene. The increased amount of product is stimulatory to cell division and will confer a selective growth advantage on the cell.
4. Some miRNAs have been shown to regulate the expression of mRNA transcripts of particular tumor suppressor genes. Overexpression of those miRNAs can abnormally inhibit the activity of those tumor suppressor genes, thereby removing or decreasing inhibitory signals for cell proliferation. Other miRNAs have been shown to regulate the expression of mRNA transcripts of particular proto-oncogenes. Inactivation of those miRNAs means that the proto-oncogenes are expressed at higher-than-normal levels, which can stimulate cell proliferation.

Think Like a Scientist

FIGURE 16.2

If the Lac repressor cannot bind to the operator, then the *lac* operon structural genes will be transcribed even in the absence of lactose because there is no block to the binding of RNA polymerase to the promoter. By contrast, in a normal cell, the Lac repressor binds to the operator and blocks RNA polymerase from transcribing the *lac* structural genes as long as lactose is absent from the medium.

FIGURE 16.4

If the Trp repressor cannot bind tryptophan, then the repressor could not be activated when tryptophan is present in the medium. Without an active repressor in this condition, RNA polymerase cannot be blocked from binding to the promoter. Therefore, the Trp structural genes will be transcribed. By contrast, in a normal cell, the presence of tryptophan in the medium activates the Trp repressor, which then binds to the operator and blocks transcription of the *trp* structural genes by RNA polymerase.

MOLECULAR INSIGHTS

Some questions might be:

1. What is the molecular mechanism that *Braveheart* uses to regulate *MesP1* expression? Some experiments to address this question are discussed in the paper.
2. Is there an equivalent lncRNA that is critical for cardiac development in humans? It is a good assumption that there is. However, sequence comparisons indicate that there is not a human analog of *Braveheart*. That may not be surprising, because lncRNAs are known to evolve more rapidly than genes that code for proteins.

3. Are there other noncoding RNAs that play a role in cardiac development in the mouse? The answer is not known at present, but it is a research direction the MIT scientists are pursuing.

UNANSWERED QUESTIONS

Dr. Kay's laboratory uses mice to test RNAi therapies against hepatitis C virus. You could test various RNAi agents via an intravenous infusion to find a dose where the viral load (or level) is reduced substantially without causing any dangerous toxicities that have been determined by standard laboratory testing of blood samples. Kay's lab aims to reduce viral load by at least 100-fold.

Test Your Knowledge

1. d 2. a 3. b 4. c 5. b 6. b 7. d 8. a 9. b 10. a

Interpret the Data

Assignable activity in MindTap. Answers provided to instructors.

Chapter 17

Study Break

STUDY BREAK 17.1

1. An F^+ cell contains the F factor plasmid in addition to the chromosomal DNA. The F factor enables an F^+ cell to conjugate with an F^- cell, which lacks the F factor. By a special replication mechanism, a copy of the F factor is transferred to the recipient F^- cell in an $F^+ \times F^-$ conjugation, so the recipient is converted to an F^+ cell.

   An Hfr cell has the F factor integrated into the chromosomal DNA. When an Hfr cell conjugates with an F^- cell, the F factor begins its replicative transfer into the recipient as in an $F^+ \times F^-$ conjugation and, by that transfer mechanism, brings in chromosomal genes from the Hfr donor. Those chromosomal genes can recombine with the genes in the recipient. Because replication of the F factor begins in the middle of the plasmid, the entire F factor cannot be transferred to the recipient unless the entire chromosome is transferred to the recipient, which occurs only rarely. Hence, the recipient remains F^- in this conjugation experiment.
2. In gene segregation in sexually reproducing organisms, genetic material moves from one generation to the next, basically by descent. In horizontal gene transfer, the movement of genetic material between organisms is other than by descent. For instance, in transformation, a DNA molecule in the environment is taken up by a recipient cell—the DNA has moved horizontally from a donor cell (from which the DNA was released) to the recipient cell. Genetic recombination in the recipient can alter the phenotype of that cell.

STUDY BREAK 17.2

1. A virulent phage always enters the lytic cycle when it infects a bacterial cell. The end result is the assembly of many progeny phages and their release into the surroundings when the cell breaks open.

   A temperate phage enters either the lytic cycle or the lysogenic cycle when it infects a cell. The lytic cycle is the same as that for a virulent phage. The lysogenic cycle involves integration of the phage's chromosome into the bacterial chromosome. In the integrated state the phage—now called the prophage—is inactive and replicates only when the bacterial chromosome replicates. In response to an adverse environmental signal to the cell, the phage chromosome can excise itself from the bacterial chromosome and enter the lytic cycle.
2. Animal cells: Viruses without an envelope bind by their recognition proteins to receptor proteins in the host cell's plasma membrane and are then taken into the cell by receptor-mediated endocytosis. For some enveloped viruses, the genome-containing capsid enters the cell when the envelope fuses with the host cell's plasma

membrane. For other enveloped viruses, the complete virus, with the envelope, enters the cell by endocytosis.

Plant cells: All plant viruses lack envelopes. They enter cells either through mechanical injuries to leaves and stems or by the action of biting and feeding insects.

3. Non-retroviral RNA viruses have RNA genomes that are replicated in an RNA-to-RNA manner. Retroviruses have RNA genomes that are replicated via a DNA intermediate; that is, the RNA genome is copied to double-stranded DNA by reverse transcriptase and the DNA molecule integrates into the host cell's nuclear chromosomes, from which location new RNA viral genomes are transcribed.

STUDY BREAK 17.3

Although all are infectious agents, viruses have protein coats, whereas viroids and prions do not. A virus consists of a nucleic acid genome surrounded by a protein coat and, in some cases an envelope. Viroids are naked, single-stranded RNA circular molecules. Prions are proteins.

Think Like a Scientist

FIGURE 17.1

Assignable activity in MindTap. Answers provided to instructors.

FIGURE 17.4

The transfer of chromosomal genes by conjugation is dependent on the F factor. Genes on the F factor direct the one-way transfer of a copy of the F factor to a recipient in $F^+ \times F^-$ conjugations. When the F factor is integrated into the chromosome to produce an Hfr, again the F factor genes direct a one-way transfer of DNA from donor to recipient, in this case involving the F factor attached to the bacterial chromosome. There is no mechanism, therefore, for transfer of chromosome genes in the opposite direction—that is, from recipient to donor.

FIGURE 17.8

As you learned in the text, the prophage state is maintained by a phage-encoded repressor molecule that prevents expression of phage genes for the lytic cycle. When the prophage is transferred into a recipient cell, that cell will not contain repressor proteins, and because it takes time for new repressor proteins to be synthesized, the result will be that the λ lytic cycle genes will be expressed and the recipient will be killed by phage release. If the recipient already contained a λ prophage, then repressor proteins would already be present and the transferred prophage would remain repressed with respect to the lytic cycle genes.

MOLECULAR INSIGHTS

The researchers demonstrated that a >30,000-year-old virus from the permafrost was capable of infecting and killing present-day *Acanthamoeba* cells. Perhaps there are other ancient pathogenic viruses lying dormant and undiscovered in the permafrost that would be released upon thawing. If such viruses were pathogenic to animals or even humans, that could be a serious issue.

UNANSWERED QUESTIONS

Researchers can further study the interactions between normal prion protein and prion aggregates and the effect of these interactions on neurons. One possible function for the normal prion protein is to act as a signaling molecule to regulate apoptosis. Interactions with prion particles may induce aberrant signaling through the normal prion protein. Because the normal prion protein also binds to many other molecules (for example, extracellular matrix components), one could also investigate the effects of the conversion process on the interactions of normal prion protein with other molecules.

Test Your Knowledge

1. e 2. b 3. c 4. d 5. a 6. c 7. e 8. a 9. b 10. a

Interpret the Data

Assignable activity in MindTap. Answers provided to instructors.

Chapter 18

Study Break

STUDY BREAK 18.1

1. Each restriction enzyme recognizes a specific sequence in DNA, typically in the range of 4 to 6 bp, and cuts both strands of the DNA within the sequence. Restriction enzymes differ with respect to the DNA sequences—the restriction sites—they recognize and cut. The enzymes most useful for cloning produce sticky ends.

2. Useful features include the following: replication origin so that the plasmid will replicate in *E. coli*, an antibiotic resistance gene to allow selection of bacteria containing the plasmid, a cluster of restriction sites (a multiple cloning site) to provide choices for inserting fragments of DNA, and the *lacZ*$^+$ gene to use in blue–white screening, in which colonies containing recombinant plasmids (white) are distinguished from colonies with vectors lacking inserted DNA (blue).

3. PCR is a method to amplify a specific segment of DNA. The amplification process depends on DNA replication, and therefore requires primers. A limitation of PCR, then, is that DNA sequence information must be available for the segment of DNA to be amplified. Otherwise, the primers cannot be synthesized. The ingredients of a PCR are the template DNA containing the sequence to be amplified, the two primers, a buffer, and a DNA polymerase that is tolerant to the high temperature used to denature the DNA repeatedly during the cycles of the reaction.

STUDY BREAK 18.2

1. A transgenic organism is one into which a gene or genes from an external source have been introduced as a means to modify the organism genetically.

2. Germline cells develop into reproductive cells, so modifying this cell type genetically will lead to the genetic modification being passed to offspring. In somatic cell gene therapy, germ-line cells and their products are not involved, so the inserted genes remain with the individual and are not passed to offspring.

STUDY BREAK 18.3

1. Restriction fragment length polymorphisms (RFLPs) are restriction enzyme-generated DNA fragments of different lengths from the same region of the genome. They are the result of base-pair differences in the region that affect restriction sites.

2. A single-nucleotide polymorphism (SNP) is a base-pair mutation at a particular site (SNP locus) in the genome. A SNP locus typically has two alleles, with a common allele and a rare allele found in the population with a frequency of at least 1%.

3. Each human (or other organism) has unique combinations and variations of DNA sequences. DNA fingerprinting exploits these combinations and variations to distinguish between different individuals (with the exception of identical twins). By using DNA technologies to analyze the particular regions of the genome showing sequence variation, it is possible to compare two DNA samples to see if they are from the same person or not. DNA fingerprinting is used in several ways, including forensics, paternity testing, and basic research.

Think Like a Scientist

FIGURE 18.10

Assignable activity in MindTap. Answers provided to instructors.

MOLECULAR INSIGHTS

There are many possibilities, including: How does the metabolome of an animal cell change in response to treatment of an organism with a drug? How does the metabolome of an organism change in response to a specific mutation that alters a phenotype? How does the metabolome of an animal cell change in response to an extracellular signaling molecule? How does the metabolome of an animal cell change when it becomes a cancer cell? How does the metabolome change during the embryonic development of an animal? How does the cellular metabolome change when a cell is infected with a virus? Is the metabolome a of cloned animal or plant significantly different from its nontransgenic parent?

UNANSWERED QUESTIONS

Tissue-specific stem cell niches serve to both protect stem cells against environmental insults and control stem cell responses. Cellular and neuronal signals that control stem cell dormancy and proliferation are often uniquely adapted to these microenvironments. This ensures that tissue-specific stem cells receive signals in a tightly controlled manner.

Test Your Knowledge

1. a 2. c 3. e 4. d 5. e 6. b 7. e 8. b 9. a 10. b

Interpret the Data

Assignable activity in MindTap. Answers provided to instructors.

Chapter 19

Study Break

STUDY BREAK 19.1

Having the complete sequence of a genome enables a researcher to ask questions about the organization of genes and of other sequences of importance in the genome. The most important question is how the functions of all of the genes in a genome direct that organism's life. It also allows researchers to compare gene content and gene order between organisms, which provides insight into how genomes evolve.

STUDY BREAK 19.2

1. In whole-genome shotgun genome sequencing, the entire genome is broken into thousands to millions of random, overlapping fragments, each fragment is cloned and sequenced, and then the genome sequence is assembled by computer on the basis of sequence overlaps between fragments.

2. The key sequences are protein-coding genes, noncoding RNA genes, regulatory sequences associated with genes, origins of replication, transposable elements and related sequences, pseudo-genes, and short repetitive sequences.

3. To find protein-coding genes in a bacterium, computer algorithms search a genome sequence for open reading frames (ORFs), which are defined as the DNA equivalent of a start codon (ATG) in frame (a multiple of three away) from a stop codon (TAG, TAA, or TGA). Such a segment of DNA could potentially produce an mRNA that could be translated into a protein. Other analyses would be required to show if any given ORF identified in this way is an actual protein-coding gene.

To find protein-coding genes in a mammal, one also needs to find ORFs. However, the presence of introns in eukaryotic protein-coding genes complicates matters. Therefore, more sophisticated computer algorithms are needed—in this case ones that attempt to locate exon–intron boundaries while they search for ORFs.

4. Prokaryotic genomes vary widely in size. Nonetheless, their genes are densely packed in their genomes (a fairly consistent density of 900 genes per Mb), with little noncoding space between them. Genomes of eukaryotes also show a great range of sizes; overall, their genomes are much larger than those of prokaryotes. Gene density also varies considerably in eukaryotes, ranging from 500 per Mb for the single-celled yeast to fewer than 10 per Mb in mammals. There is no uniformity to the packing of genes, however.

Prokaryotic genes are arranged either singly or in operons, depending on the genes. All eukaryotic genes are arranged singly; no operons are found in eukaryotes. The number of regulatory sequences typically is much higher for eukaryotic protein-coding genes than for prokaryotic protein-coding genes. There are no introns in prokaryotic protein-coding genes, whereas most eukaryotic protein-coding genes contain introns. Introns can make up significant proportions of eukaryotic genomes—for example, about 20% in humans. In contrast to prokaryotes, eukaryotes contain many pseudogenes, many transposable element sequences, and a variety of short repeated sequences. A prokaryotic chromosome has one origin of replication, whereas a eukaryotic chromosome may have hundreds to thousands of origins of replication.

STUDY BREAK 19.3

1. The function of a gene identified in a genome sequence may be assigned by a sequence similarity search of sequence databases to find a match with a gene of known function, by comparing the structure of the protein encoded by the gene with the structures of proteins with known function, and by experiments that alter the expression of the gene to see how the phenotype of a cell or organism is affected.
2. A genome-wide analysis of gene expression here could involve DNA microarray assays. The mRNAs could be isolated from untreated tissue culture cells and from cells treated with a steroid hormone. Convert each mRNA preparation to cDNAs using reverse transcriptase, using nucleotide precursors with different fluorescent labels for the two batches—for example, green for the untreated (reference) sample and red for the treated (experimental) sample. Mix the two cDNAs and pump them through a DNA chip prepared to have spots with DNA representing every gene in the genome. Allow hybridization to occur, and analyze the hybridization by laser detection. The colors of the spots indicate which genes are affected by the steroid hormone. Purely red spots represent genes that are active only in hormone-treated cells. Purely green spots represent genes that are active only in untreated cells. Spots that are a mixture of green and red represent genes that are active in both types of cells. Based on controls, it would be possible to see if any of the genes have higher or lower expression levels in the treated cells.
3. The proteome is complete set of proteins that can be expressed by an organism's genome. Proteomics has the three major goals of determining the number and structures and functions of proteins of the proteome, determining the location of each protein within or outside the cell, and determining the physical interactions between proteins.

STUDY BREAK 19.4

1. Tandem duplication happens when homologous pairs of chromosomes do not line up properly in meiosis. As a result of unequal crossing over, one gamete inherits a chromosome with two copies of the stretch of DNA between the two points of crossing over, and therefore duplicate copies of any genes on that stretch of DNA. Dispersed duplication happens when transposable elements copy the DNA for a gene and insert the new copy somewhere else in the genome.
2. Gene duplication produces two identical copies of a gene that only evolve distinct functions as different mutations occur in each copy. Exon shuffling, by contrast, combines parts of two or more genes in an evolutionarily novel way. The protein coded for by the new gene contains a combination of domains that may never before have existed in one protein. Because a protein's domains usually each perform a different molecular function, the newly evolved protein may have a combination of functionalities unlike any previously existing protein.

Think Like a Scientist

UNANSWERED QUESTIONS

Knowledge that an individual is genetically predisposed to developing a disease or is at greater risk for the aftereffects of disease based on genotype may cause emotional distress, and fear of discrimination by employers or insurance companies.

Similar concerns may exist if genetic information suggests that disease treatment options for an individual are limited. The Genetic Information Nondiscrimination Act (GINA) was recently enacted to help address these concerns.

Test Your Knowledge

1. b 2. a 3. a 4. a 5. d 6. d 7. d 8. a 9. d 10. d

Interpret the Data

Assignable activity in MindTap. Answers provided to instructors.

Chapter 20

Study Break

STUDY BREAK 20.1

1. Buffon did not understand how "anatomically perfect" animals could have useless structures.
2. Lamarck proposed that all species change through time, the changes are inherited by the next generation, the changes arise in response to environmental conditions, and specific mechanisms caused the changes.
3. The concepts of gradualism and uniformitarianism suggest that Earth's major geological features were produced by the very slow action of geological processes observed today. Thus, it must have taken more than 6,000 years for these features to assume their present forms.

STUDY BREAK 20.2

1. Darwin observed that living organisms often resemble fossils found in the same area; that organisms found in South America resembled one another, even if they occupied different environments; and that many species found on the Galápagos Islands resembled species from the South American mainland.
2. Darwin realized that the effects of competition for resources in nature were similar to the action of a plant or animal breeder who used only certain individuals as parents to produce the next generation.
3. Darwin's theory relied on physical explanations for the origin of biodiversity: he recognized that evolutionary change takes place within a population rather than in individuals; he recognized that natural selection is a multistage process; and he emphasized the importance of environmental conditions to the process of natural selection.

STUDY BREAK 20.3

1. Two problems that slowed the acceptance of Darwin's theory are that Mendel's genetic studies focused on simple traits and that these traits often changed in just a few generations.
2. Microevolution refers to small, genetically based changes within populations. Macroevolution refers to larger-scale evolutionary changes observed in species and more inclusive groups. Current research suggests that microevolutionary change and macroevolutionary change result from the same evolutionary processes.
3. Evidence for evolution comes from studies of adaptation, the fossil record, historical biogeography, comparative morphology, and comparative molecular biology.

Think Like a Scientist

FIGURE 20.12

Assignable activity in MindTap. Answers provided to instructors.

FIGURE 20.14

Most vertebrate biologists agree that the loss of legs in snakes was probably favored by natural selection as a mechanism to facilitate locomotion through environments with dense vegetation or to facilitate burrowing underground.

MOLECULAR INSIGHTS

If dogs were domesticated before the rise of agriculture, they probably fed on scraps of meat and bones near human encampments. If dogs were domesticated after the rise of agriculture, they may have eaten more vegetable matter left behind by humans.

UNANSWERED QUESTIONS

If climate change rapidly disrupts physical environments everywhere on Earth, Arctic species are unlikely to adapt to those changes. Given the frequency of extinction in the fossil record, one might assume that most of the species living today will become extinct eventually, and rapid climate change is likely to speed that process.

Test Your Knowledge

1. d 2. b 3. b 4. d 5. c 6. e 7. a 8. d 10. c 11. b

Interpret the Data

Assignable activity in MindTap. Answers provided to instructors.

Chapter 21

Study Break

STUDY BREAK 21.1

1. The variation in skunks is qualitative.
2. The researchers used artificial selection to change the activity levels of the mice.
3. Genetic variation, differing environmental effects on individuals, and interactions between genes and the environment affect phenotypic variation in a population.

STUDY BREAK 21.2

1. Genotype frequencies specify how alleles are combined in individuals, and allele frequencies specify how common the alleles are.
2. The Hardy–Weinberg principle is a null model because it identifies the conditions under which evolution will *not* occur.
3. If genotype frequencies are not already in equilibrium, they will stop changing after one generation of random mating.

STUDY BREAK 21.3

1. Mutation and gene flow tend to increase genetic variation within populations, and natural selection and genetic drift tend to decrease it.
2. Stabilizing selection increases the representation of the average phenotype in a population.
3. Sexual selection, like directional selection, favors extreme phenotypes.

STUDY BREAK 21.4

1. Diploidy protects harmful recessive alleles from natural selection because dominant alleles mask their effects in heterozygotes.
2. A balanced polymorphism is one in which two or more phenotypes are maintained in fairly stable proportions over many generations.
3. The sickle-cell allele is rare in Northern Europe because, in the absence of the malarial parasite, it confers no advantage on individuals that carry it.

STUDY BREAK 21.5

1. The adaptive value of a trait can be evaluated by comparing closely related species that live in different environments.
2. Natural selection preserves traits that were useful when the organisms subject to selection were alive and reproducing.

Think Like a Scientist

FIGURE 21.6

Assignable activity in MindTap. Answers provided to instructors.

FIGURE 21.9

Dachshunds are the product of directional selection on two traits: body length was increased and leg length was decreased.

FIGURE 21.10

Assignable activity in MindTap. Answers provided to instructors.

FIGURE 21.11
Assignable activity in MindTap. Answers provided to instructors.

FIGURE 21.13
Assignable activity in MindTap. Answers provided to instructors.

FIGURE 21.15
Assignable activity in MindTap. Answers provided to instructors.

FIGURE 21.16
Assignable activity in MindTap. Answers provided to instructors.

MOLECULAR INSIGHTS
Sexual selection apparently increased the sensitivity of horn-producing cells to the presence of insulin/IGF.

UNANSWERED QUESTIONS
If you choose two plants, consider how they are pollinated. How might the pollen of one species land on the stigma of another species? What might be all of the possible consequences? Climate change is altering the conditions under which plants and pollinators survive, and students might enjoy doing their own research into this question.

Test Your Knowledge
1. d 2. b 3. b 4. c 5. d 6. c 7. b 8. c 9. e 10. a

Interpret the Data
Assignable activity in MindTap. Answers provided to instructors.

Chapter 22

Study Break

STUDY BREAK 22.1
1. The morphological species concept defines species based on morphological differences among them. The biological species concept defines species as populations that can successfully interbreed under natural conditions. The phylogenetic species concept defines a species as a cluster of populations with a recent shared evolutionary history.
2. Clinal variation is a pattern of smooth variation along a geographical gradient.

STUDY BREAK 22.2
1. Prezygotic isolating mechanisms either prevent individuals of different species from mating or prevent sperm of one species from fertilizing the eggs of another. Postzygotic isolating mechanisms limit the survivorship or reproductive capability of hybrid individuals.
2. The scenario illustrates a behavioral isolating mechanism.

STUDY BREAK 22.3
1. In the first stage of allopatric speciation, populations become geographically separated. In the second stage, they become reproductively isolated.
2. Insects from different host races spend most of their time on different host plant species. Thus, they rarely encounter each other and would be unlikely to mate under natural conditions.

STUDY BREAK 22.4
1. Natural selection cannot promote reproductive isolation in allopatric populations directly, but it can lead to genetic divergence, which results in reproductive isolation.
2. Polyploidy changes the number of chromosome sets in the cells of an organism. Polyploid individuals are often reproductively isolated from their parent species because their gametes contain different numbers of chromosomes. When a gamete from a polyploid individual fuses with a gamete from a nonpolyploid individual, the resulting offspring is usually sterile because the odd number of chromosomes cannot segregate properly during meiosis.

Think Like a Scientist

FIGURE 22.15
After a spontaneous doubling of the chromosomes in the hybrid offspring, the allopolyploid would have $2n = 20$.

FIGURE 22.17
Assignable activity in MindTap. Answers provided to instructors.

MOLECULAR INSIGHTS
After the loss of mbCHC production in *D. birchii*, the presence of mbCHC would signal to a female *D. serrata* that a potential mate is of her own species, rather than *D. birchii*.

UNANSWERED QUESTIONS
Most of the asexual organisms for which the biological species definition is not a good match are bacteria. Because these organisms do not exhibit as much morphological distinctness as many animals do, biologists rely on comparisons of their DNA or RNA sequences to analyze their evolutionary relationships. These data allow researchers to construct phylogenetic trees for the organisms they study, indicating that any new species definition that applies to them would probably resemble the phylogenetic species definition more than the morphological species definition.

Test Your Knowledge
1. a 2. e 3. c 4. b 5. e 6. b 7. d 8. b 9. c 12. a

Interpret the Data
Assignable activity in MindTap. Answers provided to instructors.

Chapter 23

Study Break

STUDY BREAK 23.1
1. Hard parts, such as the shells or bones of animals, are the materials most likely to fossilize.
2. The fossil record provides an incomplete portrait of life in the past because not all organisms are equally likely to form fossils, fossils do not form in all types of habitats, and fossils are often destroyed by geological processes and erosion.
3. The fossil record provides information about the morphology of ancient organisms, how structures changed over time, and the proliferation and extinction of evolutionary lineages. It also offers indirect evidence about the behavior, ecology, and physiology of organisms that lived in the past.

STUDY BREAK 23.2
1. Continental drift caused large-scale geographical separation of populations and lineages that subsequently evolved in isolation.
2. Sea levels fell whenever a large proportion of Earth's water was incorporated into glaciers.

STUDY BREAK 23.3
1. Continuous distribution requires no special explanations. Biologists infer that lineages with a continuous distribution simply occupy all or one part of their historical range.
2. Distantly related species that live in widely separated parts of the world may resemble each other because convergent evolution fosters similar adaptations to the environments they occupy.

STUDY BREAK 23.4
1. A population of organisms may occupy a new adaptive zone after the evolution of a key morphological innovation that allows it to use the environment in a unique way or after a once-successful group of organisms declines.
2. Huge volcanic eruptions triggered a chain of events that led to the mass extinction at the end of the Permian period. These eruptions warmed the atmosphere, acidified rainwater, and may have ignited massive underground coal deposits.

3. The first major adaptive radiation of animals took place in the Cambrian era.

STUDY BREAK 23.5
1. The horse lineage was highly branched, and it included some species that were larger and some that were smaller than their ancestors.
2. Phyletic gradualism predicts that morphological change is slow and steady, producing fossils with morphologies that are intermediate between the structures found in lower and higher strata. The punctuated equilibrium hypothesis predicts that morphological changes occur rapidly as new species form and that most species experience little morphological change for long periods of time.

STUDY BREAK 23.6
1. Allometry refers to the differential growth of body parts. Heterochrony refers to changes in the timing of developmental events.
2. Similar developmental control genes are present in a wide variety of animals, plants, fungi, and prokaryotes. Their widespread distribution suggests that they were present in the ancestor of all of these organisms and have been conserved through countless generations.
3. The *Pitx1* gene is expressed in fin buds that later produce spines, and it is not expressed in those that fail to produce spines.

Think Like a Scientist

FIGURE 23.17
No, artificial selection has not reversed the many evolutionary trends observed in the fossil record for horses. A reduction in body size appeared multiple times in horse evolution.

FIGURE 23.18
Assignable activity in MindTap. Answers provided to instructors.

FIGURE 23.19
Assignable activity in MindTap. Answers provided to instructors.

FIGURE 23.22
Assignable activity in MindTap. Answers provided to instructors.

MOLECULAR INSIGHTS
The prolonged expression of *Hox* genes in the developing limbs of tetrapods is an example of heterochrony, a change in the timing of developmental events.

UNANSWERED QUESTIONS
For genes that are very pleiotropic—that is, have many diverse functions—regulatory evolution may be more likely. Shifts in gene expression domains can produce rapid morphological change while leaving other functions intact. However, we can never rule out coding changes, especially if gene duplication has reduced pleiotropy.

Test Your Knowledge
1. a 2. e 3. d 4. c 5. c 6. a 7. d 8. e 9. b 10. b

Interpret the Data
Assignable activity in MindTap. Answers provided to instructors.

Chapter 24

Study Break

STUDY BREAK 24.1
1. The system of binomial nomenclature avoids ambiguity in the naming of species because it assigns a unique two-part name to each species.
2. The taxonomic hierarchy helps biologists organize information about different species because it categorizes them into increasingly inclusive groups. Species that are included in a lower taxonomic category share many characteristics, whereas those included only in the same higher taxonomic category share fewer characteristics.

STUDY BREAK 24.2

1. A phylogenetic tree is a formal hypothesis about the evolutionary relationships among species. A classification is an arrangement of organisms into hierarchical groups that reflect their relatedness.
2. A monophyletic taxon contains an ancestor and all of its descendants. A polyphyletic taxon includes species from different clades. A paraphyletic taxon includes an ancestor and some, but not all, of its descendants.

STUDY BREAK 24.3

1. Systematists use homologous characters in their analyses because similarities in homologous characters indicate genetic relatedness and shared ancestry.
2. Morphological traits are often useful for tracing the evolutionary relationships within a group of organisms because they can be observed and measured in fossils as well as in living organisms.
3. Molecular characters provide several advantages in systematic analyses: (1) they provide abundant data; (2) molecular sequences can be compared between distantly related organisms that share no organismal characteristics or between closely related species with only minor morphological differences; and (3) proteins and nucleic acids are not directly affected by developmental or environmental factors that cause nongenetic morphological variation.

STUDY BREAK 24.4

Traditional systematics emphasizes the divergence of groups as well as branching evolution. Groups that have evolved very different morphology are sometimes placed in a taxon that is different from the one to which the rest of their clade is assigned.

STUDY BREAK 24.5

1. Outgroup comparison is a technique that compares the group under study—the ingroup—to more distantly related organisms—the outgroup—to identify the ancestral and derived versions of characters. Character states observed in the outgroup are considered ancestral.
2. In a cladistic analysis, organisms with synapomorphies (shared derived characters) are grouped together within a clade.
3. The principle of parsimony suggests that the cladogram or phylogenetic tree with the fewest number of hypothesized evolutionary changes is the best current hypothesis about the phylogenetic history of a group.

STUDY BREAK 24.6

1. An assumption that underlies the use of genetic sequence differences as molecular clocks is that mutations arise at a relatively constant rate.
2. Birds are more closely related to nonavian dinosaurs than they are to crocodilians.

STUDY BREAK 24.7

1. Different strains of HIV were acquired from chimpanzees and from sooty mangabey monkeys.
2. Phylogenetic analyses of prokaryotes based on morphological data were not very successful because prokaryotes do not have many morphological features. Analyses based on molecular sequence data have been more successful because researchers can identify many molecular differences among the lineages of prokaryotes.

Think Like a Scientist

FIGURE 24.2

This arrangement does not imply that humans are more closely related to orangutans than to gorillas because the phylogenetic tree shows that humans and gorillas shared a more recent common ancestor than did humans and orangutans.

FIGURE 24.10

No, the data in the table do not contain any information that would allow the development of an hypothesis about the evolutionary relationship of cycads and ginkgophytes. For the five characters listed, these groups exhibit identical character states. The phylogenetic trees illustrated in the figure are actually based on data about many additional characters in all of the plant groups listed. The analysis in the figure is simplified to illustrate the parsimony method for evaluating phylogenetic trees.

FIGURE 24.12

Assignable activity in MindTap. Answers provided to instructors.

MOLECULAR INSIGHTS

Even after discovering that shared genetic mechanisms are responsible for producing electrocytes from cells that might have become skeletal muscle, the structural and functional differences in the electrocytes in these distantly related groups of fishes are examples of homoplasy. The absence of electric organs in all of the other vertebrates in the phylogenetic tree suggest that electrocytes and electric organs were produced by convergent evolution.

UNANSWERED QUESTIONS

The Linnaean system of classification cannot easily incorporate the phenomenon of horizontal gene transfer because it pigeonholes species into a nested hierarchy of rigidly defined taxonomic categories. Although a phylogenetic view of biodiversity also depends on the vertical transmission of derived traits from one generation to the next, horizontal gene transfer can at least be illustrated on phylogenetic trees.

Test Your Knowledge

1. c 2. b 3. e 4. a 5. a 6. e 7. c 8. a 9. b 10. d

Interpret the Data

Assignable activity in MindTap. Answers provided to instructors.

Chapter 25

Study Break

STUDY BREAK 25.1

1. Only in a reducing atmosphere can amino acids be produced from simpler chemicals and energy. Without amino acids there can be no life. Thus, any theory for the origin of life must consider the necessity of reducing conditions.
2. Current thinking is that early Earth's atmosphere *per se* was not reducing; scientists have looked for localized regions where conditions were of a reducing nature, such as ocean floor hydrothermal vents. The recently discovered alkaline hydrothermal vents have the most promising set of environmental conditions.

STUDY BREAK 25.2

1. Membranes could have first formed within microscopic pores throughout vent chimneys, and chemical reactions taking place in the vents produce nonequilibrium conditions that could have been exploited as energy sources to support early metabolism. For example, because the vent water is alkaline and the surrounding ocean water is more acidic, protocells could have made use of energy released as H^+ ions diffuse from ocean water outside the chimney walls to the vent water within the chimneys.
2. RNA is unique among biological molecules in that it contains genetic information (like DNA) and can fold into ribozymes that catalyze chemical reactions (like proteins). In protocells, replicase ribozymes may have facilitated the replication of other RNA molecules having the same nucleotide sequence.
3. Protocells are hypothesized to have been surrounded by a lipid-based membrane and to contain populations of some form of self-replicating polymer such as RNA. Unlike modern-day cellular organisms, protocells are not thought to have had the machinery for protein synthesis or DNA-based genomes, and their membranes were more likely to have been composed of fatty acids than more complex lipids such as phospholipids.

STUDY BREAK 25.3

The basic tenet of the theory of endosymbiont origins for mitochondria and chloroplasts is that organelles such as mitochondria and chloroplasts originated from symbiotic relationships between two prokaryotic organisms. An anaerobic prokaryote is proposed to have ingested an aerobic prokaryote, which persisted in the cytoplasm, continuing to respire aerobically. A gradual process of mutual adaptation transformed the cytoplasmic aerobes into mitochondria.

The same basic mechanism is believed to have led to the appearance of membrane-bound plastids (including chloroplasts) at a later time. In this case, nonphotosynthetic aerobic cells with mitochondria are proposed to have ingested photosynthetic prokaryotes that were perhaps similar to present-day cyanobacteria. Again, through mutual adaptation, the photosynthetic prokaryotes changed into plastids.

Think Like a Scientist

FIGURE 25.2

Assignable activity in MindTap. Answers provided to instructors.

MOLECULAR INSIGHTS

The RNA world model states that the first genes and enzymes were RNA molecules—that is, ribozymes. Ribozymes, which are RNA molecules capable of catalyzing biochemical reactions, may have functioned both as informational molecules and catalysts in protocells, without requiring protein enzymes. Thus, the observation that a ribozyme could become activated within the protocell-like vesicles supports the RNA world model and the possibility that living cells may have developed from protocells.

UNANSWERED QUESTIONS

Two possible related explanations might be summarized as "division of labor" and "efficiency" In an "RNA world," RNA would have a dual function, both as the transcriptional unit (RNA copy of DNA) and the molecule that carries out the activity of catalyzing biochemical reactions. With a division of labor between proteins, the work unit, and RNA, the informational unit, RNA becomes more specialized and efficient, and hence the whole process becomes more efficient. Proteins then evolve specialization to catalyze biochemical reactions.

Test Your Knowledge

1. d 2. e 3. e 4. b 5. b 6. c 7. c 8. d 9. d 10. e

Interpret the Data

Assignable activity in MindTap. Answers provided to instructors.

Chapter 26

Study Break

STUDY BREAK 26.1

1. Prokaryotes have no major cytoplasmic organelles that are equivalent to the endoplasmic reticulum, the Golgi complex, mitochondria, or chloroplasts of eukaryotes, nor do they have lysosomes.

   The genetic material of a prokaryote generally is a single, circular DNA molecule localized in a nonmembrane-bound central region of the cell called a nucleoid. By contrast, the genetic material of a eukaryote is distributed among a number of linear chromosomes, which consist of DNA complexed with basic proteins known as histones.

   Most prokaryotes are surrounded by a cell wall located outside the plasma membrane. Animal cells do not have a cell wall, but plants, fungi, and some other eukaryotes do. The compositions of the eukaryotic cell walls are chemically different from bacterial cell walls.
2. A chemoheterotroph oxidizes organic molecules as its energy source and obtains carbon in organic form. A photoautotroph uses light as its energy source and carbon dioxide as its carbon source.

3. Obligate anaerobes are poisoned by oxygen. They survive either by fermentation, or by a form of respiration in which inorganic molecules are used as final electron acceptors. Facultative anaerobes use oxygen when it is present but live by fermentation when conditions are anaerobic.

4. Nitrogen fixation is the reduction of atmospheric nitrogen (N_2) to ammonia (NH_3). Nitrification is the conversion of ammonium (NH_4^+) to nitrate (NO_3^-).

   Nitrogen fixation, an exclusively prokaryotic process, is the only means of replenishing the nitrogen sources used by most microorganisms, and by all animals and plants.

STUDY BREAK 26.2

1. A biofilm is a complex aggregation of microorganisms (either one or multiple microorganisms, many or all of them prokaryotes) attached, in most cases, to a surface. Biofilms are used in a variety of beneficial applications, including bioremediation of toxic organic chemicals contaminating groundwater. On the other hand, biofilms can have adverse effects on human health. For instance, biofilms can result in antibiotic-resistant infections if they adhere to surgical materials, such as catheters and implants.

2. An exotoxin is a toxic protein that is secreted from the bacterium that makes it, or is released when that bacterium lyses. An endotoxin is a normal lipopolysaccharide component of the outer membrane of Gram-negative bacteria; it is released when the bacteria die and lyse. An exoenzyme is an enzymatic protein that is released from cells.

   An exotoxin interferes with biochemical processes of body cells. An endotoxin overstimulates the immune system, often causing inflammation. Depending on the bacterium, the endotoxin release has different effects, which may include organ failure and death. An exoenzyme digests plasma membranes, causing cells of the infected host to rupture and die. Exoenzymes may also digest extracellular materials and red and white blood cells.

3. Many pathogenic bacteria develop resistance to antibiotics as a result of mutations. The mutations allow the bacteria to break down the antibiotics or otherwise counteract their effects. Pathogenic bacteria may also develop resistance to antibiotics as a result of horizontal gene transfer when antibiotic resistance genes on plasmids are transferred between bacteria by conjugation, or when DNA containing an antibiotic resistance gene is introduced into the pathogen in other ways, such as by transformation or transduction.

STUDY BREAK 26.3

1. See Table 25.1 for a comparison of the properties of organisms in each of the three domains. The classification of Archaea as a distinct domain was based on comparisons of DNA and rRNA sequences.

2. A methanogen lives in reducing environments, generating energy by converting substrates such as carbon dioxide, hydrogen gas, methanol, or acetate into methane gas. All known methanogens belong to the Euryarchaeota.

3. Extreme halophilic archaeans live in high-salt environments, requiring at least 1.5 M NaCl in order to live. Most of these organisms are aerobic chemoheterotrophs, obtaining energy from sugars, alcohols, and amino acids using pathways similar to those of bacteria. All known extreme halophilic archaeans belong to the Euryarchaeota.

4. Extreme thermophiles live in extremely hot environments such as thermal hot springs and hydrothermal ocean floor vents. Psychrophiles grow optimally at temperatures in the range −10°C to −20°C, such as in the Antarctic and Arctic oceans.

Think Like a Scientist

FIGURE 26.12

Assignable activity in MindTap. Answers provided to instructors.

MOLECULAR INSIGHTS

Regulation of gene expression in bacteria was discussed in Section 16.1. In some way the bacteria must "sense" their density and then trigger the transition from planktonic to biofilm growth. Changes in gene expression must involve molecular signals, probably involving operons (see Section 16.1). Thus, down-regulation of a gene or of genes in an operon would require repression of those genes or the removal of a gene activator, whereas up-regulation would require removal of repression or the production of a gene activator.

UNANSWERED QUESTIONS

One possibility is that the last universal common ancestor (LUCA) of all life on the planet had a genome composed not of DNA but of RNA. If the switch to DNA-based genomes occurred after LUCA, once in the lineage that led to Bacteria and once entirely independently in the lineage that led to Archaea and Eukarya, then the two machineries for replicating DNA need not be related to one another. A second possibility is that LUCA had a DNA-based genome but that the cellular replication machinery was displaced by that of a virus in either the bacterial or archaeal/eukaryal lineages.

Test Your Knowledge

1. d 2. a 3. c 4. b 5. e 6. a 7. d 8. b 9. c 10. d

Interpret the Data

Assignable activity in MindTap. Answers provided to instructors.

Chapter 27

Study Break

STUDY BREAK 27.1

A protist is distinguished from a prokaryote by having typical eukaryotic cell features such as a nuclear envelope surrounding its genetic material, and cell organelles such as mitochondria (in most protists), chloroplasts (in some protists), endoplasmic reticulum, Golgi complex, and so on.

Distinguishing protists from fungi, animals, and plants is more blurry. Fungi are nonmotile at all stages of their life cycles, whereas most protists are motile or have motile stages in their life cycles. Cell wall structure is also different from fungi, and from plants. Protists differ from both animals and plants by lacking highly differentiated structures and by not having complex developmental stages. Collagen, the extracellular support protein of animals, is absent in protists.

STUDY BREAK 27.2

1. Metamonads' nuclear genomes contain genes that are of mitochondrial origin, suggesting that they once had mitochondria.

2. The chloroplast will have two membranes: one derived from the plasma membrane of the engulfing eukaryote and the other from the plasma membrane of the cyanobacterium.

Think Like a Scientist

FIGURE 27.11

Assignable activity in MindTap. Answers provided to instructors.

MOLECULAR INSIGHTS

The genome sequencing data had provided a correlation of a mutation in the *gacA* gene with food bacteria and the absence of that mutation with nonfood bacteria. By knocking out the *gacA* gene experimentally in nonfood bacteria, the researchers could test directly the hypothesis that a *gacA* gene mutation (or loss) is causal for converting bacteria from a nonfood to a food phenotype. Their metabolite and edibility results supported that hypothesis.

UNANSWERED QUESTIONS

1. Evolution is a dynamic field of study utilizing many tools, from fossils to gene sequence data to complicated mathematical models. New ideas about the first eukaryotes and the significance of endosymbioses will continue to emerge by using the scientific method.

2. Anything is possible, but no current evidence suggests that prokaryotes evolved from a single-celled eukaryote. Fossil evidence and evidence for the endosymbiotic theory support a prokaryote-first world. Unlike *Giardia* that has mitochondrial genes, for example, prokaryotes do not have "eukaryotic" genes that appear to have no (or lost) function.

Test Your Knowledge

1. e 2. d 3. d 4. b 5. b 6. c 7. b 8. a 9. b 10. c

Interpret the Data

Assignable activity in MindTap. Answers provided to instructors.

Chapter 28

Study Break

STUDY BREAK 28.1

1. Evolution of a root system gave land plants access to minerals and water in soil and provided physical support for aerial parts. The evolving shoot system of land plants, including lignified tissues in stems, allowed vascular plants to grow taller and stay erect, thereby gaining better access to sunlight for photosynthesis. Reproductive structures borne on aerial stems (such as flowers) might serve as platforms for more efficient dispersal of spores from the parent plant. Vascular tissues were innovations for distributing water (xylem) and sugars (phloem) throughout the plant body.

2. Homosporous plants produce a single type of sexual spore and are in effect bisexual, with each gametophyte capable of producing both sperm and eggs. Heterosporous species, including angiosperms and gymnosperms, produce two types of spores, which develop into sexually different gametophytes that produce either sperm or eggs. Plant scientists associate the evolution of heterospory with several key reproductive innovations in land plant evolution, including the protection of male gametes inside pollen grains and the protection of plant embryos inside seeds.

STUDY BREAK 28.2

1. Like aquatic plants, bryophytes produce flagellated sperm that must swim through water to reach eggs, and they lack a complex vascular system (although some have a primitive type of conducting tissue). Bryophytes do have parts that are rootlike, stemlike, and leaflike, although the "roots" are rhizoids, and bryophyte "stems" and "leaves" did not evolve from the same structures that vascular plant stems and leaves did. Sporophytes of some species have a water-conserving cuticle and stomata. Like most plants, bryophytes also have both sexual and asexual reproductive modes.

2. In general, mosses are the bryophytes that most closely resemble vascular plants. Some species produce structurally complex gametophytes that have a central strand of primitive water-conducting tissue that resembles the xylem of vascular plants, and in a few species the water-conducting cells are surrounded by sugar-conducting tissue resembling the phloem of vascular plants.

STUDY BREAK 28.3

1. In bryophytes, the gametophyte is much larger than the sporophyte and obtains its nutrition independently. The comparatively tiny sporophyte remains attached to the gametophyte and depends on the gametophyte for much of its nutrition. In modern lycophytes (club mosses and their close relatives), the gametophyte is free-living—although it is nourished by mycorrhizae instead of carrying out photosynthesis—and it is smaller than the sporophyte, which is a photosynthetic autotroph.

2. Fern leaves often take the form of feathery fronds, and roots extend from underground stems called rhizomes. Whisk ferns lack true leaves and roots; instead, small leaflike scales dot an upright, green, branching stem, which arises from a horizontal rhizome system anchored

by rhizoids. Horsetail sporophytes typically have underground rhizomes and roots that anchor the rhizome to the soil. The scalelike leaves are arranged in whorls around a photosynthetic stem.

3. In horsetails, the sporangia that produce spores are borne in strobili, and spores are carried away from the plant by air currents. In ferns, sporangia are produced on the lower surface or margin of leaves, and spores are forcefully dispersed from the parent plant when contraction of a beltlike annulus rips open the sporangium and ejects the spores.

STUDY BREAK 28.4

1. Seedless plants play various important roles in ecosystems. For example, when bryophytes colonize bare land, their bodies trap organic and inorganic matter, helping to build soil on bare rock and stabilizing soil surfaces. Certain hornworts harbor mutualistic nitrogen-fixing cyanobacteria, thereby increasing the amount of nitrogen available to other plants. In arctic tundra, mosses and other bryophytes make up as much as half the biomass and are crucial components of the food web.

As the source of the world's coal, oil, and natural gas deposits, seedless vascular plants have been and continue to be crucial to human economies.

2. Ferns and some other seedless vascular plants are used in research on plant growth and development. The recently sequenced genomes of the lycophyte *Selaginella moellendorffii* and the moss *Physcomitrella patens* are proving extremely useful in elucidating the evolutionary pathways of groups throughout the plant kingdom.

Think Like a Scientist

MOLECULAR INSIGHTS

Given that land plants apparently arose from an ancient charophyte alga, genomic data from a living charophyte would have provided the researchers with deeper insight into the genetic basis of adaptations crucial for the initial transition to land. Genomic data from a conifer would have shed light on the divergence of seed plants from a seedless lineage.

UNANSWERED QUESTIONS

If the number and density of genes remain largely unchanged, increases in chromosome number should generally increase independent assortment because genes will be distributed on more independent linkage groups. The result is a higher effective recombination rate. Such changes in chromosome number may happen because of fission of existing chromosomes into more than one chromosome. In most eukaryotes, one obstacle that chromosomes produced by fission must overcome is the development of a centromere in which spindle fibers link sister chromatids. Chromosomes produced by duplication do not generally need a new centromere, but must also overcome being the odd chromosome in the population. The likelihood that a chromosome will be duplicated is at least partly based on the dosage sensitivity of the genes located on it. For example, the human genetic disorder Down syndrome is caused by the presence of an extra copy of chromosome 21 (trisomy 21). Thus, the consequences of gene duplication to phenotype and fitness play a significant role in shaping which regions of the genome may duplicate or not—in plants, animals, and all other eukaryotes.

Test Your Knowledge

1. b 2. d 3. e 4. a 5. a 6. b 7. c 8. b 9. a 10. a

Interpret the Data

Assignable activity in MindTap. Answers provided to instructors.

Chapter 29

Study Break

STUDY BREAK 29.1

1. Four major adaptations facilitated the evolution of the seed: the trend toward reduced gametophytes; the emergence of sexually different spores ("male" microspores and "female" megaspores); protection of female spores and gametophytes within the body of the sporophyte; and the evolution of the ovule, the developmental precursor of the seed.

2. The evolution of pollen and pollination permitted sexual reproduction in the absence of environmental water, allowing seed plant lineages to radiate and diversify into a wide range of new, drier habitats. The survival advantages of seeds—including protection and nourishment of embryos—facilitated this diversification.

STUDY BREAK 29.2

1. Gymnosperm seeds are "naked" because they are not enclosed within floral structures but instead develop on the shelflike scales of ovulate (female) cones.

2. Cycads, the sole extant *Ginkgo* species, the maidenhair tree *Ginkgo biloba*, and gnetophytes represent ancient plant lineages. Extant cycads superficially resemble small palms and are remnants of a lineage that likely arose from a seed fern ancestor. All are native to tropical or subtropical habitats. Leaves and other parts of *Ginkgo biloba* closely resemble fossils going back at least 200 million years. Gnetophytes are structurally and functionally simpler than most other seed plants. Species in the three genera (*Gnetum*, *Ephedra*, and *Welwitschia*) are native to tropical regions or arid deserts.

3. Conifers are long-lived trees or shrubs—the mature sporophytes. Most produce pollen in clusters of small pollen cones and ovules in larger, woody ovulate cones.

A pollen cone consists of small scales (sporophylls). In the pine life cycle, haploid microspores develop by meiosis in two microsporangia on the underside of the scales. Further mitotic development yields winged pollen grains, each an immature male gametophyte that consists of four cells. Two later degenerate, leaving cells known as the tube cell and the generative cell. Similarly, cone scales on ovulate cones bear two megasporangia (ovules) in which meiosis yields four haploid megaspores. Three of these degenerate. Following pollination, the surviving megaspore develops into a mature female gametophyte in which eggs develop. At pollination, one pollen grain develops into a pollen tube that is the mature male gametophyte. The generative cell gives rise to two unflagellated sperm—essentially, sperm nuclei. Growth of the pollen tube stimulates maturation of the female gametophyte and the production of eggs. At fertilization, one sperm nucleus fertilizes an egg, forming a zygote that is the first cell of the new individual. A seed forms as the zygote develops into an embryo. Eventual germination of the seed and growth of the embryo into a mature sporophyte completes the life cycle.

STUDY BREAK 29.3

1. Seeds leaves (cotyledons) and pollen morphology are two major features used to distinguish monocots and eudicots. Monocots have a single seed leaf and pollen grain with a single groove. Eudicots have two seed leaves and pollen grains with three grooves. Monocots (such as grasses, lilies, and palms) also generally have fibrous root systems, leaves with parallel veins, flower parts in multiples of three, and scattered vascular bundles in stems. Eudicots (most flowering trees and shrubs, roses, sunflowers, and beans) usually have netlike leaf venation, a primary taproot, flower parts in fours or fives, and vascular tissues arranged in a ring.

2. Adaptations that have contributed to the evolutionary success of angiosperms include vascular tissue modifications that make transport of water and nutrients more efficient; double fertilization, which results in enhanced nutrition (endosperm) for embryos; physical protection of embryos within ovaries and seeds; and coevolution with animal pollinators, which increases the likelihood that pollination will occur.

STUDY BREAK 29.4

1. Polyploidy results from the duplication of an organism's genome. In the course of angiosperm evolution, polyploidy has been an important factor in speciation and probably was a major factor in the rapid diversification of flowering plants. Many of these angiosperms originated as *allopolyploids*—hybrids of a mating between two closely related species. If such hybrid offspring can interbreed but are reproductively isolated from their parents, a new species has come about within a single generation.

2. Among other benefits, having the entire sequences of a variety of plant genomes allows researchers to use the resulting to elucidate the relationship between large-scale shifts in genomes and the evolution of fundamental characteristics of major lineages. For example, all land plants carry the *LFY* gene, which encodes the regulatory protein LEAFY. Researchers studying the protein's effects in groups as different as mosses and angiosperms have found that, although the phenotypic effects—and resulting adaptations—vary markedly in different plant groups, only a few changes in the base sequence of the *LFY* gene have been associated with the shifts.

STUDY BREAK 29.5

1. Human societies rely on seed plants for food, lumber, fibers used to manufacture products such as paper and textiles, and for medicines and other drugs—among many other uses.

2. The global food supply depends heavily on species of grasses (the Poaceae), including wheat, corn, and rice. The pea family (Fabaceae) runs a close second as the source of soybeans and other beans. Many other food plants are in the potato family (Solanaceae) and the rose family (Rosaceae).

Think Like a Scientist

FIGURE 29.19

Assignable activity in MindTap. Answers provided to instructors.

MOLECULAR INSIGHTS

Forest managers have eagerly awaited the sequencing of conifer genomes, such as that of *P. taeda*. For example, having a fuller understanding of the genetic basis of traits such as wood formation may lead to strategies for selectively breeding pines that grow to maturity more quickly or that develop wood having enhanced, commercially useful properties. Enhancing pines' natural resistance to damaging insects such as pine bark beetles may be another practical application of genomic research.

UNANSWERED QUESTIONS

One explanation hinges on the theory that all apical meristems (except root meristems), whether vegetative or reproductive, produce lateral organs as an innate function. Vegetative meristems produce leaves that arise laterally on the stem, and flower meristems produce sepals that arise laterally on the floral axis. According to this theory, once a flower meristem is formed, it will automatically produce sepals with no new gene functions needed. Then, if B-gene function is added to the baseline gene activity in the meristem (which includes E function), petals will be produced—so instead of requiring A + B function to produce petals, only B function is required. Stamen and carpel formation is exactly the same as in the ABC model: B + C = stamens, and C = carpels. Thus, only two gene functions, B and C, in addition to E, are required for the formation of all four floral organ types.

Test Your Knowledge

1. d 2. b 3. a 4. d 5. a, c, d, e 6. a, b, d 7. b, c, d 8. a 9. b, c, d 10. b, c, d

Interpret the Data

Assignable activity in MindTap. Answers provided to instructors.

Chapter 30

Study Break

STUDY BREAK 30.1

1. Some fungi are multicellular, whereas others, the yeasts, are single cells. (Some species alternate between these two forms at different life-cycle stages.) The cells of all fungi are surrounded by a hardened wall; in most cases the hardener is the polysaccharide chitin. The body of a multicellular fungus consists of a dense mesh of filaments called hyphae, which in some groups are separated into cell-like compartments by cross walls (septa). Aggregations of hyphae are the structural foundation for all other parts that develop as part of a multicellular fungus. For example, in some species, modified hyphae form rhizoids that anchor the fungus to its substrate.

2. Fungal spores are microscopic, usually nonmotile reproductive cells in which haploid nuclei are surrounded by a tough outer wall. They are produced sexually or asexually. Sexual spores are produced by genetically different parent fungi and may unite in a sexual process that gives rise to a diploid life stage. Asexual spores are genetically identical to the parent fungus and may give rise to a new, haploid individual.

3. Many fungal species have a life-cycle stage called a dikaryon, which contains two haploid nuclei (a condition expressed as $n + n$). A dikaryon forms as the result of plasmogamy, a sexual stage in which the cytoplasms of two genetically different partners fuse. This fusion ensures genetic diversity in new individuals. At some point after a dikaryon forms, the nuclei fuse (karyogamy) to form a short-lived zygote. Meiosis in the zygote produces haploid nuclei that become packaged into sexual spores.

STUDY BREAK 30.2

1. The main phyla of fungi are the Chytridiomycota (chytrids), Zygomycota (zygospore-forming fungi), Glomeromycota, Ascomycota (sac fungi), and Basidiomycota (club fungi). A proposed new phylum, the Cryptomycota, includes single-celled parasitic fungi formerly in the Chytridiomycota, as well as microsporidia. Chytrids are the only fungi that produce motile, flagellated spores. Zygomycetes often reproduce asexually, but sometimes reproduce sexually by way of hyphae that occur in + and − mating strains; haploid nuclei in the hyphae function as gametes. Following plasmogamy, further development produces zygospores in which karyogamy gives rise to diploid zygotes ($2n$ nuclei), which then undergo meiosis as sexual spores form. Glomeromycetes reproduce asexually, by way of spores that form at the tips of hyphae. All are specialized to form mycorrhizae with plant roots. In ascomycetes, chains of asexual spores called conidia, each containing a haploid nucleus, develop during asexual reproduction. Ascomycetes produce haploid sexual spores in pouch-like cells called asci. Most basidiomycetes reproduce only sexually: club-shaped basidia develop on a basidiocarp (for example, a "mushroom") and bear sexual spores on their outer surface. When dispersed, the spores may germinate and give rise to a haploid mycelium. Cytoplasmic fusion may occur between hyphae of two compatible mating strains, producing a dikaryotic mycelium from which basidiocarps may grow. Microsporidia, single-celled parasites in the Cryptomycota, resemble spores but lack mitochondria. Fungi for which no sexual life stage has been identified are placed in a convenience grouping called "conidial fungi."

2. Anatomically, the simplest fungi are the microscopic microsporidia and chytrids, and zygomycetes, which have aseptate hyphae. Ascomycetes, basidiomycetes, and glomeromycetes all form septate (walled) hyphae.

STUDY BREAK 30.3

1. A lichen is a communal life form representing a symbiosis between a photosynthetic green alga or species of cyanobacteria (the photobiont) and a nonphotosynthetic fungus (the mycobiont). The algal cells supply the lichen's carbohydrates, most of which are absorbed by the fungus. In some cases, the alga is protected from desiccation or some other environmental threat.

2. A mycorrhiza is a symbiotic association between a fungus and plant roots. The fungal hyphae make mineral ions and sometimes water available to the plant's roots, and in exchange the fungus absorbs carbohydrates, amino acids, and possibly other growth-enhancing substances provided by the plant. Mycorrhizae greatly enhance the plant's ability to extract various nutrients, especially phosphorus and nitrogen, from soil, and they are crucial to the survival of many plant species.

STUDY BREAK 30.4

1. Mycorrhizal fungi form symbiotic associations with plant roots, whereas endophytic fungi live within leaves or other plant tissues. Most such associations are mutualisms, in which the fungus feeds on the host's carbohydrates and other substances, while providing the host with some benefit. Mycorrhizae typically provide plants with enhanced access to nitrogen or other nutrients. Many endophytic fungi synthesize compounds thought to deter predation, enhance resistance to abiotic stresses such as drought, or otherwise enhance the plant's survival. Parasitic fungi feed on a living host without providing a reciprocal benefit.

2. Humans use fungi as food and to make antibiotics and fermented products such as beer and cheeses. Some fungi have been exploited as poisons and recreational hallucinogens. The yeast *Saccharomyces cerevisiae* and the mold *Neurospora crassa* both are pivotal model organisms in studies of DNA structure and function; *S. cerevisiae* has also been important in the development of genetic engineering methods.

Think Like a Scientist

MOLECULAR INSIGHTS

Staining provided three important pieces of information: (1) It demonstrated that the genes governing chitin synthesis in cell walls of other fungi have the same role in *Rozella*; (2) It also showed that chitin is present in two *Rozella* life stages; and (3) Identifying a concentrated region of chitin in cysts suggests a possible function—strengthening the cyst wall region that must penetrate a host cell wall, which also is reinforced with chitin. Further studies may provide deeper understanding of the infection process in *Rozella* and similar fungal parasites.

UNANSWERED QUESTIONS

1. Ecological theory predicts that species with complete overlap in the way they use resources and interact with other species would not coexist in a community. Therefore, it is likely that ectomycorrhizal fungi that coexist on a single tree differ in factors such as how they gain nutrients from the tree, how they gain nutrients from the surrounding soil, how they interact with the tree (for example, what parts of the root system they colonize, what types and amounts of nutrients they provide, or what additional benefits they confer to the tree), where they occur in space (for example, at what soil depth they occur), or how they respond to different environmental conditions (such as rainfall and temperature).

2. Because mycorrhizal symbiosis may be controlled by the plant symbiont, the fungal symbiont, or both, the dominance of fungal symbionts within a community may result from interactions between different fungal symbionts (for example, direct competition between mycelia or indirect competition for nutrients) or from differences in the way they interact with the tree (such as the "selection" by the tree of one symbiont over another).

3. Understanding the nature of interactions between plants, fungi, and the environment can be beneficial for protecting rare fungal or plant species, maintaining forest diversity, predicting the effects of climate or land use change, and improving the outcome of restoration efforts.

Test Your Knowledge

1. a, c 2. b, e 3. a 4. d 5. e 6. c 7. a 8. b, e 9. c, d 10. d

Interpret the Data

Assignable activity in MindTap. Answers provided to instructors.

Chapter 31

Study Break

STUDY BREAK 31.1

1. Several characteristics distinguish animals from plants: plant cells have cell walls, but animal cells do not; almost all plants are autotrophic, whereas all animals are heterotrophic; and plants are usually sessile, but animals are motile at some stage of their life cycle. Animals differ from fungi in that the cells of fungi have cell walls, and most fungi are sessile.

2. The ability of animals to move through the environment allows them to search for and pursue the food items that supply them with nutrients and energy.

STUDY BREAK 31.2

1. A tissue is a group of cells that share a common structure and function. The three primary tissue layers that contribute to the bodies of most animals are endoderm, mesoderm, and ectoderm.

2. Humans are bilaterally symmetrical.

3. The coelom is a space within which internal organs can move independently of the body wall muscles. The fluid within it provides protection for internal organs. In some animals the coelom functions as a hydrostatic skeleton.

4. Having a segmented body may allow an animal to survive damage to some parts of its body segments and may allow improved control over body movements.

STUDY BREAK 31.3

1. Molecular sequence studies have confirmed the distinctions between the Parazoa and the Eumetazoa, between the Radiata and the Bilateria, and between the Protostomia and the Deuterostomia.

2. An enterocoelom appears to be the ancestral body cavity among the deuterostomes.

STUDY BREAK 31.4

1. Sponges do not exhibit any kind of body symmetry.

2. A sponge gathers food from its environment by drawing water into its body through numerous small pores and harvesting particulate matter from the water with its choanocytes, or collar cells.

STUDY BREAK 31.5

1. Cnidarians capture animal prey by stinging it with their nematocysts and using their tentacles to pull it into their mouths.

2. The anthozoans, including sea anemones and corals, have only a polyp stage in their life cycle.

3. Ctenophores capture microscopic plankton in sticky filaments on their two tentacles, which are then drawn across the mouth.

STUDY BREAK 31.6

1. Free-living flatworms have digestive, excretory, nervous, and reproductive systems. Tapeworms lack a digestive system.

2. Ectoprocts, brachiopods, and phoronid worms all have a circular or U-shaped feeding structure called a lophophore, a characteristic that reveals their close evolutionary relationship.

3. The anatomical and physiological systems that allow squids and other cephalopods to be more active than other types of mollusks include a closed circulatory system with accessory hearts and a complex nervous system with giant nerve fibers. Many cephalopods use their excurrent siphon to expel jets of water, allowing them to move rapidly through the environment.

4. The organ systems that exhibit segmentation in most annelid worms include: respiratory surfaces; parts of the nervous, circulatory, and excretory systems; and the body wall and coelom.

STUDY BREAK 31.7

1. The cuticle protects a nematode from the digestive enzymes of its host.
2. Although the rigid exoskeletons of arthropods do not expand, these animals grow a new, soft exoskeleton inside the existing one. After shedding the old exoskeleton, they grow to a larger size by expanding the new exoskeleton with either water or air before it hardens.
3. The body regions of the four living subphyla of arthropods differ in how they have become fused. Chelicerates have a fused cephalothorax and an abdomen. Crustaceans show variable patterns, but many have a fused cephalothorax and an abdomen. Myriapods have a head and a trunk. Hexapods have a separate head, thorax, and abdomen.
4. Insects with incomplete metamorphosis hatch from their eggs as wingless nymphs, which vary in how closely they resemble adults; nymphs then undergo metamorphosis into the adult form. Insects with complete metamorphosis hatch from eggs as larvae, which are always very different from adults. After becoming a pupa, their cells and tissues are reorganized into the adult form.

Think Like a Scientist

MOLECULAR INSIGHTS

Given their early appearance in the fossil record, the clade of trilobites would probably branch off the root of the tree earlier than and to the left of Chelicerata.

Unanswered Questions

1. Recent genetic comparisons of invertebrate taxa confirm the distinction between Parazoa and Eumetazoa and the distinction between Radiata and Bilateria.
2. The nucleotide sequences most likely to shed light on the diversity of living invertebrates are those in small subunit ribosomal RNA, mitochondrial DNA, and *Hox* genes of species within the protostome and deuterostome lineages.

Test Your Knowledge

1. b 2. d 3. c 4. a 5. c 6. e 7. a 8. d 9. b 10. e

Interpret the Data

Assignable activity in MindTap. Answers provided to instructors.

Chapter 32

Study Break

STUDY BREAK 32.1

1. Echinoderms have a water vascular system, which operates the tube feet that are used for locomotion and/or feeding.
2. Water enters the pharynx of a hemichordate through its mouth, and it exits the pharynx through the branchial slits. The animal extracts oxygen and particulate food from the water as it passes through the pharynx.

STUDY BREAK 32.2

1. The animal is not a chordate, because chordates have a dorsal nerve cord.
2. Vertebrates have an internal bony skeleton, including a cranium and vertebral column in most groups, as well as structures derived from neural crest cells.

STUDY BREAK 32.3

1. Vertebrates have multiple *Hox* gene complexes, which provide them with several copies of each *Hox* gene. Cephalochordates have just one *Hox* gene complex.
2. Of the three groups listed, Gnathostomata has the most species, and Amniota has the fewest.

STUDY BREAK 32.4

1. Hagfishes lack bone, paired fins, and scales in their skin. Hagfishes have neither a cranium nor a vertebral column, and lampreys have only rudimentary traces of vertebrae. These observations suggest that their lineages arose before these structures appeared in vertebrates.

2. The derived traits possessed by conodonts and ostracoderms include structures made of bone or a bonelike material and, in some ostracoderms, a brain divided into three regions.

STUDY BREAK 32.5

1. Sharks are more efficient predators than acanthodians or placoderms were because they have well-developed sensory systems to detect prey; their lightweight skeletons and absence of heavy body armor allow them to pursue prey rapidly; and they have numerous teeth that are replaced when damaged or worn, as well as loosely attached jaws that permit them to suck in large chunks of food.
2. The air bladders of ray-finned bony fishes increase their locomotor abilities by allowing them to rise or sink easily in the water column. Their fin rays allow them to engage in precise movements during locomotion.
3. The lungs of lungfish allow them to survive in environments with low oxygen content because they can acquire oxygen from the air.

STUDY BREAK 32.6

1. For the first tetrapods, the advantages of moving onto land included abundant food resources, the rarity of predators, and readily available oxygen. The disadvantages included the need for more skeletal support against gravity, mechanisms to prevent dehydration in air, and modifications of sensory systems so that they would function in air.
2. The parts of the amphibian life cycle that are most dependent on water are the egg and larval stages.

STUDY BREAK 32.7

1. The amniote egg freed amniotes from a dependence on standing water because the egg can survive on land. The shells of amniote eggs mediate gas exchange and water exchange with the environment.
2. The two major amniote lineages are the Synapsida, which includes the mammals, and the Reptilia, which includes turtles, lizards, snakes, crocodilians, and birds.
3. Because lizards are lepidosaurs and both birds and crocodilians are archosaurs, crocodilians are more closely related to birds than they are to lizards.

STUDY BREAK 32.8

1. Besides their loss of legs, snakes differ from their lizard ancestors in having smaller skull bones with more elastic connections between them.
2. Various species of lizards feed on vegetation, insects, or larger animals. Virtually all snakes are carnivores, and they swallow their prey whole.

STUDY BREAK 32.9

1. The overall structure of turtles differs from that of other amniotes in that their bodies are enclosed within a bony, keratin-covered shell.
2. Several characteristics reveal the close evolutionary relationship of crocodilians and birds, including a four-chambered heart and maternal care of offspring.
3. The specific adaptations that allow birds to fly either reduce their weight or increase their muscle power. Weight-reducing adaptations include a lightweight skeleton, the absence of teeth and a urinary bladder, and the habit of laying an egg as soon as it has a shell. Power-promoting adaptations include large wing muscles; efficient digestive, respiratory, and circulatory systems; and a high metabolic rate.
4. The structure of a bird's bill reflects its diet. For example, hummingbirds that drink nectar have long, thin bills, and parrots that eat hard nuts have stout, sharp bills. Wings and feet are adapted to birds' flying habits and habitats. For example, ducks have webbed feet that allow them to paddle in water, and albatrosses have long, thin wings that work efficiently for long distance flight.

STUDY BREAK 32.10

1. Most mammals were probably active at night during the Mesozoic era to avoid competition with and predation by dinosaurs, most of which were active during the day.

2. The key adaptations that allow mammals to be active under many types of environmental conditions include insulating fur and fat and a high metabolic rate that generates lots of body heat.
3. The three major groups of living mammals are distinguished on the basis of their reproductive habits. Monotremes lay eggs. Marsupials give birth to relatively undeveloped young after a short period of gestation. Placentals give birth to more developed young after a long period of gestation.

STUDY BREAK 32.11

1. The characteristics that allow many species of primates to spend a lot of time in trees include flexible shoulder and hip joints, grasping hands, and excellent depth perception.
2. The lowest taxonomic group that includes monkeys, apes, and humans is the Anthropoidea. The lowest taxonomic group that includes only apes and humans is the Hominoidea.
3. Gorillas, chimpanzees, and bonobos are the apes that spend the most time on the ground.

STUDY BREAK 32.12

1. Researchers usually use the criterion of bipedal locomotion to distinguish between humans and apes. Humans (that is, hominins) are bipedal, and apes are not.
2. The strongest evidence suggesting that Neanderthals and modern humans belong to different species comes from mtDNA sequence data: the differences between gene sequences of Neanderthals and humans are much greater than the differences between any two modern humans.
3. Independent genetic studies of mitochondrial DNA and the Y chromosome indicate that all human populations are descended from a common ancestor that originated in Africa and then migrated to various regions on Earth.

Think Like a Scientist

FIGURE 32.8

Based on information in the figure, a change from radial to bilateral symmetry accompanied the evolution of an increased number of *Hox* genes in the *Hox* gene complex.

MOLECULAR INSIGHTS

If early primates were leaping or climbing through trees, increased visual acuity and better depth perception would allow them to estimate the distances between branches, decreasing their likelihood of falling. Increased olfactory function would not contribute to better arboreal locomotion.

UNANSWERED QUESTIONS

Viviparity could provide several benefits. By carrying her developing young, the mother can protect them from some detrimental environmental factors; she can regulate the temperature at which they develop by regulating her own body temperature; she can provide them with nutrients and energy throughout their development; and she can select a favorable environment into which she releases them when they are born.

Test Your Knowledge

1. a 2. c 3. d 4. b 5. e 6. d 7. a 8. c 9. e 10. b

Interpret the Data

Assignable activity in MindTap. Answers provided to instructors.

Chapter 33

Study Break

STUDY BREAK 33.1

1. A land plant's shoot system consists of its photosynthetic tissues and organs—stems, leaves, and buds. Stems are frameworks for upright growth and favorably position leaves for light exposure and flowers for pollination. Leaves increase a plant's surface area and thus its exposure to sunlight. Buds eventually extend the shoot or

give rise to a new, branching shoot. The shoot system of a flowering plant also includes flowers and fruits. Parts of the shoot system store carbohydrates manufactured during photosynthesis.

The root system usually grows below ground. It anchors the plant, and sometimes structurally supports its upright parts. It also absorbs water and dissolved minerals from soil and stores carbohydrates.

2. Meristem tissue is self-perpetuating embryonic tissue. Apical meristems, at the tips of shoots and roots, gives rise to a young plant's stems, buds, roots, and other primary tissues. In plants that show secondary growth, cylinders of lateral meristem tissue give rise to (often woody) secondary tissues that increase the diameter of older stems and roots.

STUDY BREAK 33.2

1. The ground tissue system makes up most of the plant body. It includes three types of structurally simple tissues—parenchyma, collenchyma, and sclerenchyma—each of which is composed mainly of one type of cell. Parenchyma makes up most of a plant's primary tissue and typically has air spaces between its cells, which are alive at maturity and can continue to divide. Subgroups of parenchyma cells are specialized for photosynthesis, secretion, and storage (of starch). Collenchyma is flexible ground tissue that contains cellulose. Its cells remain alive and metabolically active at maturity. They provide mechanical support for parenchyma and often collectively form strands or a sheathlike cylinder under the dermal tissue of growing shoot regions and leaf stalks. Cells of sclerenchyma are dead at maturity, but they develop thick secondary walls while alive that typically are lignified and provide additional support and protection in mature plant parts.

2. Xylem and phloem are the tissues of the vascular tissue system. The two types of xylem cells, called tracheids and vessel members, both develop thick, lignified secondary cell walls and die at maturity. The empty cell walls of abutting cells serve as pipelines for water and minerals. The conducting cells of phloem, called sieve tube members, form sieve tubes that conduct solutes, mainly sugars made during photosynthesis, throughout a plant.

3. The dermal tissue system serves as a skinlike protective covering for the plant body. Cells of the epidermis are tightly packed and cover the primary plant body. They secrete a cuticle that coats all plant parts except the very tips of the shoot and most absorptive parts of roots. Some epidermal cells become modified for specialized functions. Examples include guard cells, which form stomata; root hairs, which absorb water and minerals; and hairlike trichomes, which function in defense against herbivory or secrete sugars that attract pollinators.

STUDY BREAK 33.3

1. Most eudicots have a taproot system—a single main root, or taproot, that is adapted for storage and smaller branching lateral roots. As the main root grows downward, its diameter increases, and the lateral roots emerge along the length of its older, differentiated regions. Grasses and many other monocots develop a fibrous root system in which several main roots branch to form a dense mass of smaller roots. Fibrous root systems are adapted to absorb water and nutrients from the upper layers of soil, and tend to spread out laterally from the base of the stem.

2. The root apical meristem and the actively dividing cells behind it form the zone of cell division. Cells in the center of the root tip become the procambium; those just outside the procambium become ground meristem; and those on the periphery of the apical meristem become protoderm. The zone of cell division merges into the zone of elongation, where most of the increase in a root's length occurs. Above the zone of elongation, cells may differentiate further and take on specialized roles in the zone of maturation.

3. Primary root growth produces a system of vascular pipelines extending from root tip to shoot tip. The root procambium produces cells that mature into the root's xylem and phloem. Ground meristem gives rise to the root's cortex, its ground tissue of starch-storing parenchyma cells that surround the stele. In many flowering plants, the outer root cortex cells give rise to an exodermis, a thin band of cells that may limit water losses from roots and help regulate the absorption of ions. The innermost layer of the root cortex is the thin endodermis, which helps control the movement of water and dissolved minerals into the stele. Between the stele and the endodermis is the pericycle, which gives rise to lateral roots. In some cells in the developing root epidermis, the outer surface extends into root hairs.

STUDY BREAK 33.4

1. Stems have four main functions: (1) they provide mechanical support for body parts involved in growth, photosynthesis, and reproduction; (2) they house the vascular tissues (xylem and phloem), which transport products of photosynthesis, water and dissolved minerals, hormones, and other substances throughout the plant; (3) they often are modified to store water and food; and (4) they have specific stem regions that contain meristematic tissue, which gives rise to new cells of the shoot.

A plant stem is divided into modules, each consisting of a node, where leaves are attached, and an internode, the space between nodes. New primary growth occurs in buds—a terminal bud at the apex of the main shoot, and lateral buds, which produce branches (lateral shoots), in the leaf axils. Meristem tissue in buds gives rise to leaves, flowers, or both.

In eudicots, most primary growth in a stem's length occurs directly below the shoot apical meristem. When a meristematic cell divides, one of its daughter cells becomes an initial, a cell that remains as part of the meristem. The other daughter cell becomes a derivative, which typically divides once or twice and then enters on the path to differentiation. As derivatives differentiate, they give rise to three primary meristems: protoderm, procambium, and ground meristem. These primary meristems produce cells that differentiate into specialized cells and tissues. In eudicots, the primary meristems are also responsible for elongation of the plant body. Each primary meristem occupies a different position in the shoot tip. Outermost is protoderm, which gives rise to the stem's epidermis. Inward from the protoderm, the ground meristem gives rise to ground tissue (mostly parenchyma). Procambium, which produces the primary vascular tissues, is sandwiched between ground meristem layers. In most plants, inner procambial cells give rise to xylem and outer procambial cells to phloem. The developing vascular tissues become organized into vascular bundles that are wrapped in sclerenchyma and thread lengthwise through the parenchyma. In the stems and roots of most eudicots and some conifers, the vascular bundles form a stele (vascular cylinder) that vertically divides the column of ground tissue into an outer cortex and an inner pith.

2. Leaves are organs specialized for photosynthesis. In both eudicots and monocots, the leaf blade provides a large surface area for absorbing sunlight and carbon dioxide. Many eudicot leaves have a broad, flat blade attached to the stem by a petiole. Unless a petiole is very short, it holds a leaf away from the stem and helps prevent individual leaves from shading one another. In most monocot leaves, such as those of rye grass or corn, the blade is longer and narrower and its base simply forms a sheath around the stem.

3. Leaves develop on the sides of the shoot apical meristem. Initially, meristem cells near the apex divide and their derivatives elongate. The resulting bulge enlarges into a thin, rudimentary leaf, or leaf primordium. As the plant grows and internodes elongate, the leaves

become spaced at intervals along the length of the stem or its branches. Leaf tissues typically form several layers. Uppermost is epidermis, with cuticle covering its outer surface. Just beneath the epidermis is mesophyll, which is composed of loosely packed parenchyma cells that contain chloroplasts. Leaves of many plants, especially eudicots, contain two layers of mesophyll. Palisade mesophyll cells contain more chloroplasts and are arranged in compact columns with smaller air spaces between them, typically toward the upper leaf surface. Spongy mesophyll, which tends to be located toward the underside of a leaf, consists of irregularly arranged cells with a network of air spaces that enhance the uptake of carbon dioxide and release of oxygen during photosynthesis and account for 15% to 50% of a leaf's volume. Below the mesophyll is another cuticle-covered epidermal layer. Except in grasses and a few other plants, this layer contains most of the stomata through which water vapor exits the leaf and gas exchange occurs. Vascular bundles form a network of veins throughout the leaf.

4. Plants that live for many years may spend part of their lives in a juvenile phase, and then shift to a mature, or adult phase. The differences between juveniles and adults often are reflected in leaf size and shape, in the arrangement of leaves on the stem, or in a change from vegetative growth to a reproductive stage. Most woody plants must attain a certain size before their meristem tissue can respond to the hormonal signals that govern flower development.

STUDY BREAK 33.5

1. Secondary growth processes add girth to roots and stems over two or more growing seasons. In plant species that have secondary growth, older stems and roots become more massive and woody through the activity of two types of lateral meristems. One of these meristems, the vascular cambium, produces secondary xylem and phloem. The other, the cork cambium, produces cork, a secondary epidermis that is one element of bark.

2. Vascular cambium consists of two types of cells—fusiform initials and ray initials. Fusiform initials are derived from cambium inside the vascular bundles and give rise to secondary xylem and phloem cells. Secondary xylem forms on the inner face of the vascular cambium, and secondary phloem forms on the outer face. Ray initials are derived from the parenchyma cells between vascular bundles. Their descendants form spokelike rays of parenchyma cells—horizontal channels that carry water sideways through the stem. As the mass of secondary xylem inside the ring of vascular cambium increases, it forms hard tissue known as wood. Bark encompasses all of the living and nonliving tissues between the vascular cambium and the stem surface. It includes the secondary phloem and the periderm, the outermost portion of bark that consists of cork, cork cambium, and secondary cortex.

Think Like a Scientist

FIGURE 33.3

If part of an apical meristem were to be removed, hormonal signals would prompt the remaining meristematic cells to give rise to replacement cells.

MOLECULAR INSIGHTS

An obvious way to test this hypothesis is by elucidating and comparing the gene regulatory networks involved in xylem development in major lineages of vascular plants, including gymnosperms, basal angiosperms, and a range of eudicots and monocots, as well as seedless vascular plants. Research to date indicates that genes responsible for xylem development vary little (they are highly conserved) among conifers, and vary somewhat more among angiosperms. Genomic studies of lycophytes and mosses will help complete the picture.

UNANSWERED QUESTIONS

Despite producing progeny cells that adopt different fates, both shoot and root stem cells are pluripotent cells that share the properties of continuous division and self-renewal, and thus their molecular signatures are likely to be more similar than different.

Test Your Knowledge

1. d 2. c 3. b 4. a 5. c 6. d 7. e 8. b 9. a 10. c

Interpret the Data

Assignable activity in MindTap. Answers provided to instructors.

Chapter 34

Study Break

STUDY BREAK 34.1

1. Water potential is potential energy stored in water. It is the driving force for osmosis, which in turn is responsible for the movement of water into and out of plant cells, including root cells.

2. Solute potential reflects the presence (if any) of dissolved substances in a solution. It is always a negative value. Pressure potential reflects the effects of physical pressure. It can be either a negative or positive value. Solute potential and pressure potential combine to yield water potential—a statement represented by the equation $\Psi = \Psi_S + \Psi_P$.

3. The apoplastic pathway is an extracellular route that passes through a continuous network of adjoining cell walls and air spaces. The symplastic pathway is a plant-wide compartment that consists of the protoplasts of living plant cells, which are connected by plasmodesmata. Water in the transmembrane pathway diffuses short distances, moving across plasma membranes or entering cells through membrane aquaporins.

STUDY BREAK 34.2

1. In the apoplastic pathway, water and dissolved substances do not pass through living root cells but instead move through the continuous network of adjoining cell walls and air spaces. When apoplastic water and solutes reach the endodermis, however, they must detour around the impermeable Casparian strip and pass through cells to move into the stele. The symplastic pathway passes through living cells. Water that diffuses into root cells moves in this pathway from cell to cell through plasmodesmata.

2. Epidermal cells of root hairs actively transport most mineral ions into root epidermal cells. These ions travel inward via the transmembrane pathway. Other ions may be dissolved in apoplastic water. They ultimately travel to the xylem in the symplast after crossing into and through endodermal cells of the Casparian strip. Once an ion reaches the stele, it enters the xylem.

STUDY BREAK 34.3

1. In the cohesion–tension mechanism, water transport begins as water evaporates from the walls of mesophyll cells inside leaves and into the intercellular spaces. This water vapor escapes by transpiration through open stomata. As water molecules exit the leaf, they are replaced by others from the mesophyll cell cytoplasm. The water loss gradually reduces the water potential in a transpiring cell below the water potential in the leaf xylem. Water from the xylem in the leaf veins then follows the gradient into cells, replacing the water lost in transpiration.

2. As water evaporates from mesophyll cell walls into leaf air spaces, surface tension in the water film remaining at air–water interfaces pulls the film inward, producing a curved *meniscus*. This curved shape signals the development of a negative pressure potential—and, hence, a lower water potential—in a cell's water. Because the water potential in neighboring cells now is comparatively higher, some of water moves out by osmosis, replacing that lost at the start of transpiration. As the loss and

replacement of water continues from cell to cell in mesophyll, eventually replacement water is pulled osmotically out of small xylem veins in the leaf, which in turn withdraw water from larger veins, and so on. The cohesion of water molecules maintains the pull ever farther from the air–water interface, a pull that extends through the xylem all the way down to a plant's roots.

3. Humidity, air temperature, and wind all affect transpiration. Relative humidity is a measure of the amount of water vapor in air. The less water vapor in the air, the more evaporates from leaves because the water potential is higher in the leaves than in the dry air. Rising air temperature at the leaf surface also speeds evaporation and hence transpiration. Although evaporation cools the leaf a little, the amount of water lost can double for each 10°C rise in air temperature. Air moving across the leaf surface carries water vapor away from the surface and so makes a steeper gradient. Together these factors explain why on extremely hot, dry, breezy days, the leaves of some plants must completely replace their water each hour.

STUDY BREAK 34.4

1. Stomata open and close in response to changing environmental cues, such as light levels (detected via blue-light receptors), CO_2 concentration in the air spaces inside leaves, and the amount of water available to the plant. Stomata open when hydrogen ions are pumped out of guard cells, setting up the symport of H^+ and K^+ into the guard cells through ion channels. Water then follows by osmosis. Stomata close when H^+ pumping in guard cells ceases and K^+ moves out of guard cells, with water again following by osmosis.

2. Through their ability to open and close, stomata help regulate water loss by plants and the uptake of carbon dioxide for photosynthesis.

STUDY BREAK 34.5

1. Translocation is the long-distance transport of substances in plants. The term generally applies to the transport of organic compounds, mainly sucrose, in phloem. Transpiration is the evaporation of water from a plant's aerial parts, mainly leaves. This water moves from roots upward to aerial parts in the xylem.

2. The mechanism of pressure flow moves sucrose from a source (such as a leaf or stem) into sieve tubes. Pressure builds up at the source end of a sieve tube system as sucrose enters sieve tubes at sources and water follows by osmosis. Under high pressure, sucrose moves by bulk flow toward a sink (plant parts that take up sucrose as metabolic fuel), where the sugar is unloaded.

Think Like a Scientist

FIGURE 34.9

It would reduce transpiration by increasing the relative humidity of the air.

FIGURE 34.15

Assignable activity in MindTap. Answers provided to instructors.

FIGURE 34.16

The main known sucrose channels are sucrose/H^+ antiporters and symporters. Other phloem nutrients enter sink cells by other means, including nonselective cation channels, aquaporins, H^+/ATPases, and amino acid/H^+ symporters.

MOLECULAR INSIGHTS

Autolysis makes sense because we might expect external enzymatic destruction of the cell to harm the walls of sieve-tube elements. And as you now know, those walls, although not living, are essential to the physiological function of sieve-tube elements.

UNANSWERED QUESTIONS

The results support a hypothesis that active transport may sometimes move substances through plasmodesmata. Active transport is unidirectional—a transported solute moves either into or out of a cell—and the tracer moved

unidirectionally in the experiments. The halt in tracer movement when cell metabolism was inhibited by sodium azide also is consistent with an active transport mechanism, because active transport requires ATP produced by the cell's metabolism.

Test Your Knowledge

1. a 2. d 3. e 4. d 5. a 6. e 7. c 8. c 9. b

Interpret the Data

Assignable activity in MindTap. Answers provided to instructors.

Chapter 35

Study Break

STUDY BREAK 35.1

1. Plants require relatively large amounts of macronutrients, such as nitrogen, sulfur, potassium, and calcium, and trace amounts of micronutrients, such as iron, chlorine, zinc, nickel, and copper.

2. Plants vary in their nutritional requirements. For example, leafy plants require more nitrogen and magnesium than other plant types do, and alfalfa, a grass, requires significantly more potassium than lawn grasses do. An adequate amount of an essential element for one plant also may be toxic for another. For these reasons, the nutrient content of soils is an important factor determining which plants grow well in a given location.

STUDY BREAK 35.2

1. Humus is important in soil because it generally contains nutrient-rich organic material and because it absorbs water, which contributes to the water-holding capacity of soil.

2. The amount of water that is available in soil to be taken up by plant roots depends primarily on the relative proportions of different soil components. Water moves quickly through sandy soils, whereas soils rich in clay and humus tend to hold the most water.

3. A plant's ability to absorb soil minerals depends partly on cation exchange, in which one cation, usually H^+, replaces a soil cation. As H^+ enters the soil solution, it displaces adsorbed mineral cations attached to clay and humus, freeing them to move into roots. Anions in the soil solution, such as nitrate (NO_3^-), sulfate (SO_4^{2-}), and phosphate (PO_4^-), generally move more readily into root hairs. Soil pH also affects the availability of some mineral ions because chemical reactions in very acidic (low pH) soils can trigger chemical reactions that bind various mineral cations in compounds that are insoluble in soil water.

STUDY BREAK 35.3

1. As described in this chapter and in Section 28.3, a mycorrhiza is a symbiotic association between a fungus and plant roots. Most plants form mycorrhizal associations, which facilitate the plant's ability to extract soil nutrients such as nitrogen and phosphorus. As with plant roots, mineral ions enter fungal hyphae by way of transport proteins. Some of the plant's sugars and nitrogenous compounds nourish the fungus, and as the root grows, it takes up a portion of the minerals that the fungus has secured. In some types of mycorrhizae the fungus actually lives inside cells of the root cortex.

2. Nitrogen fixation refers to the incorporation of atmospheric nitrogen into compounds, especially nitrate (NO_3^-), which plants can readily take up. Ammonification is a process in which soil bacteria known as ammonifying bacteria break down decaying organic matter and convert it to ammonium (NH_4^+). In nitrification, nitrifying bacteria oxidize NH_4^+ to NO_3^-. Inside root cells, absorbed NO_3^- is converted by a multistep process back to NH_4^+. In this form, it is rapidly used to synthesize organic molecules, mainly amino acids. In denitrification, soil bacteria convert

nitrites or nitrates first into nitrous oxide (N_2O) and then into molecular nitrogen (N_2) that escapes to the atmosphere. Because denitrification removes nitrates from the soil, it reduces the amount of nitrogen available to plant roots.

3. Associations with bacteria supply nitrogen to certain types of plants, such as legumes. The host plant provides organic molecules that the bacteria use for cellular respiration, and the bacteria supply NH_4^+ that the plant uses to produce nitrogenous molecules. In legumes the nitrogen-fixing bacteria reside in root nodules. Usually, a single species of nitrogen-fixing bacteria colonizes a single legume species, drawn to the plant's roots by chemical attractants (mainly flavonoids) that the roots secrete. By way of exchanged molecular signals, bacteria then are able to penetrate a root hair and form a colony inside the root cortex. Each cell in a root nodule may contain several thousand bacteria (now called bacteroids). The plant takes up some of the nitrogen fixed by the bacteroids, and the bacteroids use some compounds produced by the plant.

Think Like a Scientist

FIGURE 35.10

Apparently not. Recent studies have identified a few species of nodule-forming bacteria (although not in the genera *Rhizobium* and *Bradyrhizobium*) that lack *nod* genes. The bacteria may enter root cortex cells by infiltrating cracks in the epidermis where lateral roots arise.

MOLECULAR INSIGHTS

These lines of research may lead to methods for reducing the amount of phosphate fertilizers added to crop plants by identifying mycorrhizal fungi providing the most efficient phosphate uptake—or by engineering crop plants with an improved capacity to take in this essential nutrient.

UNANSWERED QUESTIONS

Even after microorganisms satisfy their N requirements, they continue to consume nitrogen-containing organic compounds because they need the carbon in these compounds to fuel metabolism and growth.

Test Your Knowledge

1. e 2. c 3. a, c, d 4. c 5. b 6. c 7. a 8. d 9. a 10. e

Interpret the Data

Assignable activity in MindTap. Answers provided to instructors.

Chapter 36

Study Break

STUDY BREAK 36.1

1. The two alternating generations of plants are the sporophyte (spore-producing) and gametophyte (gamete-producing) generations.
2. Sporophytes produce spores that give rise to gametophytes. Gametophytes then may produce gametes; male gametophytes produce sperm, and female gametophytes produce eggs. In all seed plants the sporophyte is much larger and longer-lived than the gametophyte, and the gametophyte is protected within sporophyte tissues for all or part of its life. Gametophytes also are dependent upon the sporophyte for their nutrition.

STUDY BREAK 36.2

1. Flowers are specialized for reproduction. Before an angiosperm can produce a flower, biochemical signals (triggered in part by environmental cues such as day length and temperature) travel to the apical meristem of a shoot. In response, cells there change their activity: instead of continuing vegetative growth, the shoot is modified into a floral shoot that will give rise to floral organs.
2. Pollen grains are the mature male gametophytes. They arise by the following steps:
   a. Spores that give rise to male gametophytes are produced in a flower bud's anthers.

b. Diploid microsporocytes inside an anther's pollen sacs undergo meiosis; eventually each one produces four small haploid microspores.
c. Microspores then divide by mitosis.
d. One of the two resulting nuclei divides again by mitosis, yielding a three-celled immature gametophyte: two haploid sperm cells and a third cell that will control the development of a pollen tube after pollen lands on a receptive stigma.
e. A mature male gametophyte consists of the pollen tube and sperm cells.

3. Female gametophytes develop inside ovules in a flower's carpels. They arise by the following steps:
   a. In an ovule, a diploid megasporocyte divides by meiosis, forming four haploid megaspores.
   b. Three (usually) of these megaspores disintegrate.
   c. The remaining megaspore undergoes three rounds of mitosis without cytokinesis. The result is a single large cell with eight nuclei arranged in two groups of four.
   d. One nucleus in each group migrates to the center of the cell.
   e. After the cell undergoes cytokinesis, a cell wall forms around these two polar nuclei, forming a large "central cell."
   f. A wall also forms around each of the remaining nuclei, and three of them, including an egg cell, cluster near the micropyle.
   g. The result is an embryo sac containing seven cells and eight nuclei. This sac is the mature female gametophyte.

STUDY BREAK 36.3

1. A pollen grain that lands on a compatible stigma absorbs moisture and germinates a pollen tube, which burrows through the stigma and style toward an ovule. Chemical cues from the two synergid cells help guide the pollen tube toward the egg. Before or during these events, the pollen grain's haploid sperm-producing cell divides by mitosis, forming two haploid sperm. When the pollen tube reaches the ovule, it enters through the micropyle and an opening forms in its tip. The two sperm are released into the cytoplasm of a disintegrating synergid. Next double fertilization occurs: typically, one sperm nucleus fuses with the egg to form a diploid ($2n$) zygote. The other sperm nucleus fuses with the central cell, forming a cell with a triploid ($3n$) nucleus. Tissues derived from the $3n$ cell are called endosperm.

2. As a seed matures, the embryo inside it develops a root-shoot axis with a root apical meristem at one end and a shoot apical meristem at the other end. Depending on the plant group, one or two cotyledons also develop. In monocots, a single large cotyledon develops and stores endosperm; protective tissues arise around the root and shoot apical meristems. In eudicots, two endosperm-storing cotyledons form. Near the micropyle, the radicle (embryonic root) attaches to the cotyledon at a region called the hypocotyl. Beyond the hypocotyl is the epicotyl, which has the shoot apical meristem at its tip and often bears a cluster of tiny foliage leaves, the plumule. At germination, when the root and shoot first elongate and emerge from the seed, the cotyledons are positioned at the first stem node with the epicotyl above them and the hypocotyl below them.

3. Imbibition causes the seed coat to split, and water and oxygen move more easily into the seed. Metabolism switches into high gear as cells divide and elongate to produce the seedling. Enzymes that were synthesized before dormancy become active; other enzymes are produced as the genes encoding them begin to be expressed. The increased gene activity and enzyme production mobilize the seed's food reserves in cotyledons or endosperm. Nutrients released by the enzymes sustain the developing seedling sporophyte until its root and shoot systems are established.

STUDY BREAK 36.4

1. Flowering plants may reproduce asexually (vegetatively) by fragmentation, in which cells in a piece of the parent plant dedifferentiate and then regenerate a whole plant; by apomixis, in which a diploid embryo develops from an unfertilized egg or from diploid cells in ovule tissue; or by the production of structures such as rhizomes or suckers from a nonreproductive plant part, typically meristem tissues in a bud on a root or stem.
2. Totipotency is the capacity of fully differentiated cells to dedifferentiate, return to an unspecialized embryonic state, and then develop into a fully functional mature plant. Plant tissue culture procedures trigger the development of a mass of dedifferentiated cells (a callus), some of which regain totipotency and develop into plantlets.

STUDY BREAK 36.5

1. A homeobox gene is a regulatory gene in the genome of an organism that encodes a transcription factor. In *A. thaliana*, homeobox genes govern the development of the root and shoot tissue systems, as well as of floral organs.
2. The two basic mechanisms of plant morphogenesis are oriented cell division and cell expansion. Oriented cell division establishes the general shape of a plant organ, and cell expansion enlarges the cells in a developing organ in particular directions. They increase in circumference (girth) when new cell walls form parallel to the nearest plant surface (such as the surface of a stem or tree trunk) or when cell walls form at right angles both to the nearest surface and to the transverse plane.
3. Leaves arise through a developmental program that begins with gene-regulated activity in meristematic tissue. Hormones or other signals may arrive at target cells via the stem's vascular tissue, activating genes that regulate development. Small phloem vessels penetrate a young leaf primordium almost immediately after it begins to bulge out from the underlying meristematic tissue, followed by xylem. A growing primordium becomes cone-shaped (wider at its base than at its tip). Rapid mitosis in a particular plane in cells along the flanks of the cone (perpendicular to the surface in eudicots and parallel to the surface in monocots) produce the leaf blade that is characteristic of the particular species. Leaf tip cells typically are the first to stop dividing. By the time a leaf has expanded to its mature size, mitosis has ended and the leaf is a fully functional photosynthetic organ.

Think Like a Scientist

FIGURE 36.5

Flowers of self-fertilizing angiosperms—estimated at about 10–15% of angiosperms—typically are inconspicuous and lack nectar and a pronounced scent because they are not adapted for pollination by animals or other external means. Examples include the flowers of some grasses (oat, wheat, rice), some legumes (peanut), some orchids, tomato, apricot, and peach. In the flowers of self-fertilizing species, anthers are situated so that pollen is shed directly onto stigmas.

FIGURE 36.9

Researchers may use several strategies for identifying the signaling events that guide pollination. One approach is to identify mutants that show particular defects in the guidance steps, and then analyze the defects to shed light on what kind of signal(s) operate in normal plants. Some investigators also employ microscopy-based real-time assays to directly observe pollen-tube behavior in various mutants.

FIGURE 36.25

Assignable activity in MindTap. Answers provided to instructors.

MOLECULAR INSIGHTS

Because trichomes are accessible and transparent, different mutants can (and do) serve as cellular "laboratories" for studying a range of subcellular processes, including interactions among organelles and the behavior of the actin and microtubule cytoskeletons under varying conditions.

1. Pollen-tube guidance signals are likely to be species specific. Because they are involved in mating, it is likely that these signals ensure that successful pollination, and consequently fertilization, occur only within compatible species. It is also likely that the later the guidance event, the more specific is the guidance signal. For example, the guidance signal from an ovule is likely more selective and specific than those that mediate early interactions between a pollen tube and stigma.
2. If pollen-tube growth and guidance are defective, seeds are not produced. Around 80% of the world's staple food is derived from seeds of crop plants, so studying this process is very important agriculturally. By understanding pollen-tube guidance signals, we are equipping ourselves with knowledge that could help us to: (a) improve seed yield; (b) regulate interspecies hybridizations, and thereby generate novel plant hybrids; and (c) contain pollen spreading from genetically modified crops—a possibility that will reassure concerned public and regulatory agencies.

Test Your Knowledge
1. c 2. a 3. b 4. e 5. d 6. c 7. d 8. c 9. c 10. e

Interpret the Data
Assignable activity in MindTap. Answers provided to instructors.

Chapter 37

Study Break

STUDY BREAK 37.1
1. Auxins, gibberellins, cytokinins, and brassinosteroids all promote the growth of plant parts, and ethylene stimulates cell division in seedlings. Strigolactones interact with other hormones, including auxin and cytokinins, to modify the development of roots and shoots as environmental conditions shift. In particular, strigolactones regulate lateral branching in shoots and roots. Abscisic acid is the major growth-inhibiting plant hormone. ABA also triggers plant responses to various abiotic environmental stresses, including cold snaps and drought.
2. Ethylene is a good example of a hormone that can stimulate or inhibit growth at various stages of the plant life cycle. In seedlings, it simultaneously slows elongation of the stem and stimulates cell divisions that increase stem girth. In mature plants of deciduous species, it governs senescence (including fruit ripening) and the abscission of flowers, fruits, and leaves. Studies of brassinosteroids have revealed that this family of steroid hormones has different effects in different tissues—for example, promoting the elongation of vascular tissue and pollen tubes, but inhibiting elongation in roots.
3. Two general types of signal response pathways convert chemical signals into altered plant cell activity. Some plant hormones and growth factors bind to receptors at the target cell's plasma membrane. Others cross the plasma membrane and bind to receptors inside the cell. Second messengers enhance the cellular response.

STUDY BREAK 37.2
1. General plant responses to attack include mobilization of jasmonates and salicylic acid, systemin (in tomato), the hypersensitive response, PR proteins, and bioactive compounds called phytoalexins. Defenses against feeding herbivores include structural features such as spines and leathery leaves and compounds such as alkaloids and phenolics. Plants respond to specific threats by way of pattern recognition receptors and effector-triggered immunity. The former include receptors that can recognize PAMPs—pathogen-associated molecular patterns, which often are proteins associated with an invader's cell wall or flagellum. Detection launches a complex web of signals that culminate in the cell's response, such as changes in gene expression that lead to the synthesis of phytoalexins. Some plant immune responses are triggered by bacterial virulence effectors—proteins produced by pathogenic bacteria that can bypass a plant cell's pattern recognition receptors. Effector-triggered immunity is an inducible mechanism in which resistance proteins encoded by so-called R genes (for "resistance") trigger an immediate defense response in the plant.
2. Salicylic acid (SA) is considered to be a general systemic response to damage because experiments show that when a plant is wounded, soon thereafter SA can be detected in a variety of its tissues.
3. While the hypersensitive response is under way, SA also is synthesized and operates in other defensive chemical pathways in a plant. This effect includes the synthesis of PR (pathogenesis-related) proteins that attack pathogenic cells.

STUDY BREAK 37.3
1. Directional light of blue wavelengths is the direct stimulus for phototropism (an example of photomorphogenesis). The most widely accepted scientific explanation for gravitropism is the sinking of amyloplasts in cells surrounding vascular bundles in response to gravity. Sinking amyloplasts may provide a mechanical stimulus that triggers a gene-guided redistribution of IAA. The changing auxin gradient in turn adjusts a plant's growth pattern.
2. Unlike tropisms, nastic movements occur in response to nondirectional stimuli, such as mechanical pressure resulting from an insect brushing against hairlike sensory structures in the leaves of a Venus flytrap plant.

STUDY BREAK 37.4
1. Plant responses to changes in photoperiod rely on different chemical forms of the blue-green pigment phytochrome. Daylight converts the inactive phytochrome (P_r) to an active form (P_{fr}). When light levels fall, P_{fr} reverts to P_r. This switching mechanism helps regulate light-related processes such as photosynthesis.
2. Photoperiod length is a factor in the seasonal flowering of many angiosperms. Spinach and irises are examples of long-day plants, which flower in spring when the period of daylight extends at least 9 to 16 hours. Chrysanthemums and potatoes are examples of short-day plants, which flower as day length becomes shorter than some critical period, a condition that occurs naturally in the fall.
3. Dormancy is an adaptive response because it attunes a plant's growth to the most favorable environmental conditions for survival.

Think Like a Scientist

FIGURE 37.3
Assignable activity in MindTap. Answers provided to instructors.

MOLECULAR INSIGHTS
The only other seed plants are gymnosperms. Therefore, one logical next step for testing the authors' proposed scenario for the emergence of the major groups of plant hormones would be to perform the same analysis but include data from one or more of the available conifer genomes. This approach would be particularly interesting with respect to the current study's conclusions about ethylene because ethylene is known to regulate several aspects of growth and development in conifers.

UNANSWERED QUESTIONS
1. The genome reorganization occurs in a specific subset of the genome and reproducibly occurs under the same stress conditions. Therefore, it is not random, but the mechanism by which these regions of the genome are recognized is currently unknown. Most mutation systems are assumed to be random within the genome, although there is evidence that transposons do have preferential sites.
2. The data indicate that the environment can direct higher rates of mutations to particular regions of the genome, and therefore mutating and selecting plants simultaneously might provide a better frequency of adaptive mutations. Because plants are sessile, do they need mechanisms that enable the generation of diversity in the face of chronic stress growth conditions for evolution? If so, partitioning of the plant genome into regions that are protected from change (because they are essential) and variable regions (in which changes have phenotypic effects but are not lethal) is important. Using the information might permit the more rapid breeding of stress tolerant crops.

Test Your Knowledge
1. e 2. a 3. c 4. d 5. a 6. b, c 7. e 8. a, d 9. e 10. b 11. a

Interpret the Data
Assignable activity in MindTap. Answers provided to instructors.

Chapter 38

Study Break

STUDY BREAK 38.1
1. Multicellularity made it possible for animals to create an internal fluid environment for fulfilling the need for nutrient supply, waste removal, and osmotic balance in individual cells. As a result, multicellular organisms could evolve to occupy a variety of habitats, including dry terrestrial environments, in which single cells cannot survive. Multicellularity also allowed major life functions to be distributed among specialized groups of cells, with each group having a single activity. The specialized groups of cells are typically organized into tissues, the tissues into organs, and the organs into organ systems.
2. A tissue is a group of cells with the same structure and function. Cells in the tissue work together to perform one or more activities. An organ integrates two or more different tissues into a structure that carries out a specific function. An organ system coordinates the activities of two or more organs to carry out a major body function.

STUDY BREAK 38.2
1. Exocrine and endocrine glands are formed by epithelia. An exocrine gland remains connected to the epithelium by a duct, whereas an endocrine gland is suspended in connective tissue underlying the epithelium, with no ducts leading to the epithelial surface.
2. Loose connective tissue, fibrous connective tissue, cartilage, bone, adipose tissue, and blood.
3. Skeletal, cardiac, and smooth.

STUDY BREAK 38.3
See Figure 38.7.

STUDY BREAK 38.4
A stimulus—a change in the external or internal environment—starts the homeostatic control system. The stimulus is detected by a *sensor*, an *integrator* compares the environmental change with a set point, and the *effector* or effectors become activated by the integrator and function to return the environmental parameter to the set point.

Think Like a Scientist

MOLECULAR INSIGHTS
The observation suggests that miR-4423 is an miRNA that has evolved recently.

UNANSWERED QUESTIONS
Some organisms needed to anticipate light in order to harvest its energy, whereas other organisms needed to avoid its damaging ultraviolet radiation. Thus, anticipation, via circadian rhythm, was and is crucial for survival.

Test Your Knowledge
1. b 2. a 3. e 4. c 5. b 6. a 7. c 8. e 9. e 10. d

Interpret the Data

Assignable activity in MindTap. Answers provided to instructors.

Chapter 39

Study Break

STUDY BREAK 39.1

1. A dendrite receives signals and conducts them toward the cell body. An axon conducts signals away from the cell body toward another neuron or effector.

2. An afferent neuron conducts information from its sensory receptors to interneurons. Efferent neurons conduct signals from interneuron networks to effectors, the muscles and glands that carry out the response. The afferent and efferent neurons constitute the peripheral nervous system (PNS). Interneurons process information from afferent neurons and send a response to the efferent neurons. Interneurons form the brain and spinal cord, the central nervous system (CNS).

3. In an electrical synapse, the plasma membrane of the axon terminal of the presynaptic cell is in direct contact with the postsynaptic cell, allowing ions to pass directly between the cells when an electrical impulse arrives. In a chemical synapse, the presynaptic and postsynaptic cells are separated by a small gap. Neurotransmitters released from the presynaptic cell diffuse across the gap and bind to receptors in the plasma membrane of the postsynaptic cell. If enough neurotransmitter molecules bind to those receptors, the postsynaptic cell generates a new electrical impulse, which travels along its axon to the next neuron or effector in the circuit.

STUDY BREAK 39.2

1. At the peak of an action potential, the plasma membrane of the neuron enters a short refractory period, in which the threshold for generation of an action potential is much higher than normal. Only the region in front of the action potential can fire, meaning that the impulse can only move in one direction—that is, toward the axon tip. The refractory period remains in effect until the membrane again reaches the resting potential. By that time, the action potential has moved too far away to cause a second action potential in the same region.

2. Neurons insulated with myelin sheaths have gaps in the sheaths called nodes of Ranvier, where the axon membrane is exposed to extracellular fluids. The inward movement of sodium ions at a node produces depolarization and an action potential, but the adjacent myelin sheath prevents the excess positive ions from exiting through the membrane. Instead, they diffuse rapidly to the next node, where they cause depolarization, inducing an action potential there. Continuation of this process allows the action potential to jump rapidly along the axon from node to node, at a faster rate than a nonmyelinated neuron of the same diameter.

STUDY BREAK 39.3

1. A neurotransmitter is synthesized in a neuron, is released into the synaptic cleft from a presynaptic axon terminal, and binds to receptors in the plasma membrane of the postsynaptic cell. Depending on the type of receptor, a neurotransmitter either stimulates or inhibits the generation of action potentials.

2. A neurotransmitter that binds to an ionotropic neurotransmitter receptor, like all neurotransmitters, is stored in synaptic vesicles in the cytoplasm at an axon terminal. When an action potential arrives at the terminal, the change in membrane potential opens voltage-gated Ca^{2+} channels in the axon terminal, allowing Ca^{2+} to flow back into the cytoplasm. The rise in Ca^{2+} concentration triggers the release of the neurotransmitters into the synaptic cleft by exocytosis. Neurotransmitters that bind to an ionotropic neurotransmitter receptor diffuse across the synaptic cleft and open or close ligand-gated ion channels in the postsynaptic neuron's membrane.

Most of the channels regulate Na^+ or K^+ movement through the membrane, although some regulate Cl^-. Depending on the ion flow, action potentials are either stimulated or inhibited in the postsynaptic neuron.

STUDY BREAK 39.4

One way a postsynaptic neuron integrates signals is through summation of EPSPs and IPSPs that alter the neuron's membrane potential. EPSPs move the membrane potential toward the threshold for an action potential, whereas IPSPs move the membrane potential away from the threshold for an action potential. The final change in membrane potential depends on the particular array of EPSPs and IPSPs received. The patterns of synaptic connections made by a neuron also contribute to integration.

Think Like a Scientist

FIGURE 39.7

Assignable activity in MindTap. Answers provided to instructors.

FIGURE 39.11

The activity of the K^+ channel is responsible for causing the membrane potential to fall. Inhibiting the K^+ channel would decrease significantly the outflow of K^+, and therefore slow the fall in membrane potential. In other words, it would take much longer than normal for the membrane potential to reach its resting value.

MOLECULAR INSIGHTS

The chimeric proteins must fold in an appropriate way in order to function. The hydrophilic domain must have a three-dimensional structure that can recognize the neurotransmitter, and the membrane-embedded hydrophobic domain must have a three-dimensional structure that can carry out the ion conduction function. The simplest interpretation of the results is that only two of the five artificial receptors had hydrophobic domains that were folded properly for conducting ions.

UNANSWERED QUESTIONS

If we understood the biological nature of conscious thought, then perhaps we could design a more rational judicial system, for example.

Test Your Knowledge

1. d 2. b 3. a 4. c 5. e 6. a 7. c 8. b 9. e 10. a

Interpret the Data

Assignable activity in MindTap. Answers provided to instructors.

Chapter 40

Study Break

STUDY BREAK 40.1

1. A nerve net is a loose meshwork of neurons organized in a radial pattern that reflects the radial symmetry of the animal in which it is found. Nerves are bundles of axons surrounded by connective tissue. Nerve cords are bundles of nerves.

2. Cephalization is the formation of a distinct head region containing ganglia, which form a major central control center or brain, and major sensory structures. Cephalization is found in more complex invertebrates and in all vertebrates.

STUDY BREAK 40.2

(a) Sympathetic nervous system; this is the classic "fight or flight" scenario for this autonomic nervous system division.

(b) Parasympathetic nervous system.

STUDY BREAK 40.3

1. The blood–brain barrier is a distinct separation between the blood and the brain resulting from tight junctions between the epithelial cells forming the walls of the capillaries in the brain. The blood–brain barrier regulates exchanges between blood and the brain, in particular

preventing most dissolved substances from entering the cerebrospinal fluid, thereby protecting the brain and spinal cord from viruses, bacteria, and toxic substances that may be in the blood.

2. Gray matter consists of nerve cell bodies and dendrites; white matter consists of axons, many of them surrounded by myelin sheaths.

3. The brain stem regulates a number of the body's vital housekeeping functions without conscious involvement or control by the cerebrum, including heart and respiration rates, blood pressure, constriction and dilation of blood vessels, coughing, and reflex activities of the digestive system.

4. The cerebellum is an outgrowth of the pons, but it is structurally and functionally separate from the brain stem. The cerebellum has extensive connections with other parts of the brain, through which it receives sensory inputs from receptors in muscles and joints, from balance receptors in the inner ear, and from the receptors of touch, vision, and hearing. The cerebellum integrates the various sensory signals and compares them with signals from the cerebrum that control voluntary body movements. Information flow from the cerebellum to the cerebrum, brain stem, and spinal cord modifies and fine-tunes movements of the body. In humans, the cerebellum also is involved in the learning and memory of motor skills.

The cerebral cortex is a thin, folded layer of gray matter that forms the surface of the cerebrum. The cerebral cortex contains sensory areas that receive and integrate sensory information of many kinds, including touch, pain, temperature, pressure, hearing, vision, smell, and taste. It also contains motor areas that are involved in controlling body movements and position, and association areas, which integrate information from the sensory areas and send responses to the motor area. Most higher functions of the human brain, including critical thinking, abstract thought, musical ability, and aspects of personality, involve activities of many regions of the cerebral cortex.

STUDY BREAK 40.4

Short-term memory involves transient changes in neurons. Short-term memory loss resulting from aging may be caused by a loss of control of the mechanisms involved, or, more likely, by a loss or degeneration of the neurons that constitute the short-term memory system.

Learning first involves storing memories. If the short-term memory system is faulty, perhaps due to loss or degeneration of neurons, then the transfer of short-term memories to the long-term memory system will be impaired. Another possibility is a loss or degeneration of the neurons responsible for long-term memory.

Think Like a Scientist

FIGURE 40.7

A slow response to the knee-jerk reflex test might indicate a defect in nerve conduction.

FIGURE 40.13

Assignable activity in MindTap. Answers provided to instructors.

MOLECULAR INSIGHTS

There are many possible experiments, including:

- Analyze the connectome at earlier ages to determine what sex differences exist at birth and the earliest times they are observable.
- Analyze the connectome with respect to aging.
- Analyze the connectomes of patients with diseases such as Alzheimer disease.
- Analyze the connectomes of patients who show abnormal behaviors.
- Analyze the connectomes of identical twins.

UNANSWERED QUESTIONS

A good guess is that the generalist will show better learning. The reason is that it normally must find food across a wide variety of environmental conditions that might vary in predictability. Therefore, learning ability would be highly adaptive. The specialist could adapt quite well by "hard wiring" an attraction to all the specific chemical cues that are normally associated with its single food source. Of course, only an experimental test would verify this hypothesis.

Test Your Knowledge

1. d 2. b 3. b 4. b 5. e 6. c 7. a 8. c 9. c 10. a

Interpret the Data

Assignable activity in MindTap. Answers provided to instructors.

Chapter 41

Study Break

STUDY BREAK 41.1

1. Sensory transduction is the conversion of a stimulus into a change in membrane potential.
2. The signals from particular sensory receptors are routed by afferent neurons to specific regions of the CNS. Processing of the incoming signals by those regions gives the "sense" of the stimulus—a smell, pain, and so forth.

STUDY BREAK 41.2

1. Proprioceptors detect stimuli that are processed to provide the animal with information about movements and position of the body.
2. All proprioceptors are mechanoreceptors because they are stimulated by a mechanical force.

STUDY BREAK 41.3

1. An octopus has mechanoreceptors on its head and tentacles, similar to the lateral line of fishes. These mechanoreceptors detect vibrations in the water. Many insects have sensory receptors in the form of hairs or bristles that vibrate in response to sound waves, whereas some insects (for example, moths, grasshoppers, and crickets) have auditory organs on either side of the abdomen or on the first pair of walking legs.
2. In humans, vibrations representing sound frequencies are transmitted into the fluid-filled inner ear. They travel through the inner ear and cause the basilar membrane to vibrate in response, bending the sensory hair cells and stimulating them to release a neurotransmitter that triggers action potentials in afferent neurons leading from the inner ear. The vibrations from a particular sound frequency cause the basilar membrane to vibrate maximally at one particular location, stimulating the hair cells in that region to initiate action potentials. That information is sent to the brain, which integrates it into a perception of the sound stimulus.

STUDY BREAK 41.4

(a) A photopigment is an association of retinal with one of several different opsin proteins. (b) A photoreceptor is a receptor specialized for detection of colors (different wavelengths of light). (c) A ganglion cell's receptive field is the specific set of photoreceptors that sends signals to that cell. Receptive fields are usually circular and vary in size; the smaller the receptive field, the more precise the information sent to the brain and the sharper the image.

STUDY BREAK 41.5

1. Most odor perceptions arise from a combination of different olfactory receptors, which are located in the nasal cavities in humans. We have about 1,000 different olfactory receptor types, each of which responds to a different class of chemicals. The stimulated olfactory receptors send signals via the olfactory bulbs to the olfactory centers of the cerebral cortex, where the signals are interpreted as particular smells.
2. Chemicals bind to taste receptors and generate signals. Signals from taste receptors are relayed to the thalamus. From there, some signals go to gustatory centers in the cerebral cortex, where they are integrated to produce taste perception. Other signals go to the brain stem and limbic system, which links tastes to involuntary visceral and emotional responses, such as sensations of pleasure or revulsion.

STUDY BREAK 41.6

The other sensory receptors involve specialized receptor structures. Afferent neurons synapse with the receptors. When a receptor is stimulated, the change in membrane potential (sensory transduction) is transmitted to the afferent neuron, which transmits the signal to the interneuron networks of the CNS. By contrast, both thermoreceptors and nociceptors consist of free nerve endings formed by the dendrites of afferent neurons, with no specialized receptor structures involved.

STUDY BREAK 41.7

Electroreceptors are used in aquatic vertebrates for electrolocation (locating other animals such as prey), electrocommunication (communicating with other members of the same species), and killing prey (involving high-voltage discharge).

Think Like a Scientist

MOLECULAR INSIGHTS

The mutation or mutations produced a new phenotype rather than a loss of function, so the mutation or mutations were of the gain-of-function type.

Base-pair changes for loss-of-function mutation: Missense mutation that causes loss of or significant decrease in protein function; nonsense mutation; frameshift mutation.

Base-pair changes for gain-of-function mutation: Missense mutation that alters protein function.

UNANSWERED QUESTIONS

1. Photoreceptors (they will clearly "detect" that the color of the eggs is odd); chemoreceptors (although the taste of the eggs is not altered, the person eating the green eggs may make a facial expression that leads you to believe that he or she has detected an altered taste). Combining an MRI and a PET scan might be a good method for testing synesthesia.
2. Detection and response to light are logically a key sensory system for any living organism—from prokaryotes to plants to animals—but sensing light with photoreceptors does not have to involve "vision." Yet, the genes for photoreceptor proteins would logically be of early evolutionary origin.

Test Your Knowledge

1. c 2. a 3. c 4. b 5. e 6. b 7. d 8. a 9. e 10. d

Interpret the Data

Assignable activity in MindTap. Answers provided to instructors.

Chapter 42

Study Break

STUDY BREAK 42.1

1. A *hormone* is a signaling molecule secreted by one group of cells and transported through the circulatory system to other, target cells, whose activities they change. Specific target cells react to the hormone because they carry receptors for the hormone. The best-known hormones are secreted by cells of the endocrine system and elicit a response in target cells that have receptors for the hormone. An *exocrine secretion* is a molecule released from an exocrine gland. Such molecules are secreted into ducts that lead outside the body or into the cavities of the digestive tract.
2. In *endocrine signaling,* a hormone secreted by an endocrine gland into the ECF is transported in the blood and causes a response in a target cell that has a receptor for the hormone. In *autocrine signaling,* a local regulatory molecule is released from a cell and binds to cell surface receptors on the same cells that produce the molecule, triggering a response.

STUDY BREAK 42.2

1. Glucagon, a peptide hormone, triggers a response by binding to a surface receptor. Glucagon binding activates the receptor, which triggers a signal transduction pathway inside the cell, leading to phosphorylation of target proteins. The altered activities of the phosphorylated proteins produce responses in the target cells, in this case the breakdown of glycogen in liver cells to glucose.

    Aldosterone, a steroid hormone, passes though the plasma membrane and binds to an internal receptor in the cytoplasm or nucleus, activating it. The hormone-activated receptor complex binds to control sequences in the DNA that the receptor recognizes and either activates or inhibits transcription of the associated target genes. Through binding to a receptor, the hormone affects transcription of specific genes in target cells.
2. A target cell could respond to different hormones if it carries receptors for those different hormones. Turning this around, a target cell will respond only to one hormone if it just has the receptor for that hormone. The same hormone could produce different effects in different cells if there are different receptors for that hormone, each triggering a distinct response pathway.

STUDY BREAK 42.3

1. The hypothalamus and anterior pituitary gland are connected by nerve and vascular tissues. The hypothalamus releases peptide neurosecretory hormones into the linking blood vessels. These hormones regulate the secretion of peptide hormones by the anterior pituitary gland. The pituitary hormones regulate several key body systems.
2. Secretion of hormones from the anterior pituitary is controlled by releasing hormones (RHs) and inhibiting hormones (IHs), which are released by the hypothalamus. For the posterior pituitary, neurosecretory neurons directly secrete the hormones into the circulatory system.

STUDY BREAK 42.4

1. Parathyroid hormone stimulates the dissolution of calcium and phosphate ions from bone and their release into the bloodstream.
2. The adrenal medulla secretes epinephrine and norepinephrine. Epinephrine prepares the body for handling stress or physical activity by, among other actions: (1) increasing heart rate; (2) breaking down glycogen and fats, thereby releasing glucose and fatty acids into the blood for fuel; (3) dilating blood vessels in the heart, skeletal muscles, and lungs to increase blood flow; (4) constricting blood vessels elsewhere, thereby raising blood pressure, reducing blood flow to the intestine and kidneys, and inhibiting smooth muscle contraction, which reduces water loss and slows down the digestive system; and (5) dilating airways in the lungs, thereby increasing airflow.

    Norepinephrine has similar effects to epinephrine on heart rate, blood pressure, and blood flow to the heart muscle. In contrast to epinephrine, norepinephrine causes blood vessels in skeletal muscles to contract.
3. Glucocorticoids and mineralocorticoids are the two major classes of steroid hormones secreted by the adrenal cortex. Glucocorticoids help maintain the concentration of glucose and other fuel molecules in the blood, and mineralocorticoids regulate the levels of Na^+ and K^+ in blood and the ECF.
4. Estradiol is an estrogen and progesterone is a progestogen; both are steroid hormones. Estradiol produced by the ovaries stimulates maturation of sex organs at puberty and development of secondary sexual characteristics. Progesterone, also produced by the ovaries, prepares and maintains the uterus for implantation of a fertilized egg and the growth and development of an embryo.

STUDY BREAK 42.5

In general, invertebrates have fewer hormones regulating fewer body processes and responses than vertebrates do. Peptide and steroid hormones are produced in both invertebrates and vertebrates. However, most of those hormones are different in structure and molecular function in the two groups, even though the reaction pathways stimulated by the hormones are the same.

Think Like a Scientist
FIGURE 42.5
Assignable activity in MindTap. Answers provided to instructors.

MOLECULAR INSIGHTS
The observation mentioned at the beginning was that oxytocin may cause increased recollection of stressful events. Given the fact that humans and mice are both mammals, there would seem to be a strong likelihood that a similar mechanism of fear enhancement involving oxytocin receptors would occur in humans. However, it cannot be ruled out that humans and mice are different in this neuroendocrine system.

UNANSWERED QUESTIONS
Although no animal model perfectly mimics human breast cancer, animals are useful in understanding the effect of specific exposures on breast cancer risk and the risk of recurrence, without having to worry that some unintended factors explain the findings. Further, it is far easier to study the effects of timing of exposure to estrogenic compounds, and multigenerational effects, on breast cancer in animal models than it is in humans. The benefits of cell culture studies include the ability to perform highly mechanistic studies to understand causal effects of estrogens and their biological targets in affecting breast cancer growth. The main weakness of cell culture studies is that breast cancer cells are studied in isolation from other cells in the breast, which are known to interact with the cancer cells to impact their growth.

Test Your Knowledge
1. b 2. d 3. a 4. c 5. c 6. e 7. a 8. b 9. d 10. b

Interpret the Data
Assignable activity in MindTap. Answers provided to instructors.

Chapter 43
Study Break
STUDY BREAK 43.1
1. An axon terminal of an efferent neuron makes a synapse called a neuromuscular junction with a muscle fiber. When an action potential arrives at that junction, the axon terminal releases acetylcholine, which triggers an action potential in the muscle fiber that moves in all directions over its surface and penetrates to the interior of the fiber through T tubules. When the action potential reaches the end of the T tubules, ion channels are opened in the sarcoplasmic reticulum that allow Ca^{2+} ions to flow from the sarcoplasmic reticulum into the cytosol. Troponin then binds the ion and undergoes a conformational change that allows tropomyosin to enter the grooves in the actin helix of the thin filaments. As a result, the myosin-binding sites are uncovered, and the crossbridge cycle is turned on, leading to muscle contraction.
2. In the sliding filament mechanism of muscle contraction, the thin filaments on each side of a sarcomere slide over the thick filaments toward the center of the A band, thereby bringing the Z lines closer together. A crossbridge cycle is responsible for the contraction. At the beginning of the cycle, the myosin crossbridge has an ATP bound to it and is not in contact with the actin of the thin filament. The ATP is hydrolyzed, causing the myosin crossbridge to bend away from the tail and bind to a myosin-binding site on the thin filament that was uncovered in response to the release of Ca^{2+} ions into the cytosol from the sarcoplasmic reticulum. When the myosin crossbridge binds to the actin, the crossbridge snaps back toward the myosin tail to produce the power stroke that pulls the thin filament over the thick filament. ADP and phosphate are released from the

crossbridge in this step. A new molecule of ATP now binds to the crossbridge, causing the myosin to detach from the actin. The cycle then repeats.
3. Contraction intensity: low in slow aerobic fibers, intermediate in fast aerobic fibers, and high in fast anaerobic fibers.
   Fatigue resistance: high in slow aerobic fibers, intermediate in fast aerobic fibers, and low in fast anaerobic fibers.
   Oxidative phosphorylation capacity: high in slow aerobic fibers, high in fast aerobic fibers, and low in fast anaerobic fibers.
   Number of mitochondria: many in slow aerobic fibers, many in fast aerobic fibers, and few in fast anaerobic fibers.
   Myoglobin content: high in slow aerobic, high in fast aerobic, and low in fast anaerobic.

STUDY BREAK 43.2
1. A hydrostatic skeleton provides support to the body or body part through muscles acting on compartments filled with fluid under pressure. In vertebrates, the penis is a hydrostatic skeletal structure.
   An exoskeleton is a rigid external body covering. The force of muscle contraction against the covering provides support to the body or part of the body. In vertebrates, the shell of a turtle or tortoise and the bony plates in the skin, the abdominal ribs, the collarbones, and most of the bony skull of the American alligator are exoskeletal structures.
   An endoskeleton consists of internal body structures such as bones. The force of muscle contraction against the internal body structures provides support. In vertebrates, the endoskeleton is the primary skeletal system.
2. The bones of the vertebrate endoskeleton provide support for the body and body parts, protect key internal organs, store calcium and phosphate ions, and are the sites where new blood cells form.

STUDY BREAK 43.3
1. Synovial joints consist of the ends of two bones that are enclosed by a fluid-filled capsule of connective tissue. Fluid within the capsule and a smooth layer of cartilage over the ends of the bones enable the bones to slide easily as the joint moves. In these joints, ligaments extend across the joints across the capsule. The ligaments confine the motion of the joint and protect it, to an extent, from the deleterious effects of heavy loads.
   Cartilaginous joints have no fluid-filled capsule surrounding them, and the bones involved are covered with cartilage. The bones of the joint are covered by and connected with fibrous connective tissue. Cartilaginous joints are less movable than are synovial joints.
2. Antagonistic muscle pairs are muscles arranged so that bones can be extended, flexed, or rotated in opposite directions around a joint. For example, in humans, the biceps and triceps muscles are an antagonistic muscle pair.

Think Like a Scientist
FIGURE 43.4
Assignable activity in MindTap. Answers provided to instructors.

FIGURE 43.5
Inhibiting myosin ATPase would reduce ATP hydrolysis. As the figure shows, the energy of ATP hydrolysis is used in the crossbridge cycle to "drive" muscle fiber contraction. Inhibiting ATPase would reduce, or perhaps abolish, the rate of movement of the thin filament over the thick filament (the power stroke). Overall, the force of muscle fiber contraction would be reduced.

MOLECULAR INSIGHTS
The key result came from the experiment knocking down gene expression of an SCPP gene family member in zebrafish. Treated zebrafish developed with significantly reduced endoskeletal bone formation. This result showed causation—that

is, a direct link between an SCPP gene family member's expression and an effect on bone formation. By contrast, the original discovery of the absence of the SCPP gene family in the elephant shark genome was only a correlation with the lack of bone formation.

UNANSWERED QUESTIONS
Skeletal muscle consumes metabolites, including fatty acids and carbohydrates, both for movement and for building new muscle. Thus, maintaining and using new muscle could increase fat metabolism and its breakdown and oxidation for energy, while also diverting carbohydrates, which are often converted to fat, away from fat cells and to muscle cells.

Test Your Knowledge
1. c 2. e 3. a 4. d 5. b 6. b 7. c 8. a 9. d 10. b

Interpret the Data
Assignable activity in MindTap. Answers provided to instructors.

Chapter 44
Study Break
STUDY BREAK 44.1
1. The three basic features of animal circulatory systems are: (1) a specialized fluid medium for transporting molecules, exemplified by the blood of vertebrates; (2) a muscular heart for pumping the fluid; and (3) tubular vessels for distributing the fluid pumped by the heart.
2. In an open circulatory system, there is no distinction between blood and interstitial fluid. Vessels from the heart release hemolymph directly into body spaces and the fluid is subsequently collected and reenters the heart. In a closed circulatory system, blood is channeled in blood vessels leading to and from the heart and is distinct from the interstitial fluid.
   In an open circulatory system, most of the fluid pressure generated by the heart dissipates when the blood is released into the body spaces. Consequently, blood flows relatively slowly. Humans could not function as we do with such a system, because we would not be able to distribute oxygen efficiently throughout the body, nor would we be able to eliminate the wastes we produce.
3. Atria are chambers of the heart that receive blood returning to the heart. Ventricles are chambers of the heart that pump blood from the heart. Arteries are vessels of circulatory systems that conduct blood away from the heart at relatively high pressure. Veins are vessels of circulatory systems that carry blood back to the heart.
4. The systemic circuit of the circulatory system is the circuit from the heart to most of the tissues and cells of the body and back to the heart. The pulmocutaneous circuit goes from the heart to the skin and lungs or gills in amphibians and back to the heart. The pulmonary circuit goes from the heart to the lungs and back to the heart.

STUDY BREAK 44.2
1. Erythrocytes originate as pluripotent stem cells in the red bone marrow. In humans and other mammals, they lose their nucleus, cytoplasmic organelles, and ribosomes as they mature, becoming essentially a membrane-bound hemoglobin reservoir that is not capable of protein synthesis. At the end of their lifespan—about 4 months—erythrocytes are engulfed by macrophages, a type of leukocyte, in the spleen, liver, and bone marrow.
2. Low oxygen content triggers a negative feedback mechanism to increase erythrocytes in the blood. The kidneys are stimulated to synthesize the hormone erythropoietin (EPO), which stimulates stem cells in the bone marrow to increase erythrocyte production. EPO synthesis is stopped when the oxygen content of the blood rises above normal levels.
3. Leukocytes act as the first line of defense against invading organisms, eliminate dead and dying cells from the body, and remove cellular debris. Platelets assist in blood clotting. When blood vessels are damaged, platelets stick

to the collagen fibers exposed to leaking blood and recruit other platelets to the site. Eventually a plug forms at the site, sealing off the damaged area.

STUDY BREAK 44.3

1. The right atrium receives blood in the systemic circuit returning to the heart in vessels coming from the entire body, with the exception of the lungs. The right atrium pumps blood into the right ventricle.

    The right ventricle receives blood from the right atrium, and pumps it into the pulmonary arteries going to the lungs, beginning the pulmonary circuit.

    The left atrium receives blood returning from the lungs in the pulmonary veins, completing the pulmonary circuit. The left atrium pumps this blood into the left ventricle.

    The left ventricle pumps the blood received from the left atrium into the aorta, where the blood begins its path in the systemic circuit.

2. Systole is the period of contraction and emptying of the heart. Diastole is the period of relaxation and filling of the heart between contractions.

3. Neurogenic hearts beat under the control of signals from the nervous system. If the signals cease, this type of heart stops beating. Myogenic hearts maintain a contraction rhythm without signals from the nervous system. In the event of a serious trauma to the nervous system, this type of heart keeps beating.

4. Pacemaker cells of the SA node undergo regularly timed depolarizations, which initiate waves of contraction that travel over the heart. The waves stop before they reach the bottom part of the heart as they encounter an insulating layer of connective tissue. The contraction signals at this point excite cells of the AV node, and then pass to the bottom of the heart along Purkinje fibers. There they induce waves of contraction that begin at the bottom of the heart and move upward, expelling blood from the ventricles into the aorta and pulmonary arteries.

STUDY BREAK 44.4

1. Blood flow through capillary networks is controlled by contraction of smooth muscles in arterioles and by contraction of precapillary sphincters at the junctions of capillaries and arterioles.

2. In most body tissues, there are small spaces between capillary endothelial cells, but in the brain, the capillary endothelial cells are tightly sealed together, forming a blood–brain barrier.

3. The two major mechanisms are diffusion along concentration gradients and bulk flow. Diffusion along concentration gradients occurs both through the spaces between the capillary endothelial cells and through the plasma membranes. The direction of movement of the molecule or ion depends on the concentration gradient. Thus, oxygen and glucose, which are at higher concentrations in the capillaries, diffuse into the interstitial fluid and then into the cells of the tissues. Carbon dioxide, which is at a higher concentration in the interstitial fluid, diffuses into the capillaries.

    Bulk flow carries water, ions, and molecules out of the capillaries. Driven by the pressure of the blood, which is higher than the pressure of the interstitial fluid, bulk flow occurs through the spaces between capillary endothelial cells.

STUDY BREAK 44.5

1. Arterial blood pressure must be regulated within limits to provide sufficient blood flow for the brain and other tissues and to prevent damage to blood vessels, tissues, and organs that would occur at high blood pressures.

2. The three main mechanisms for regulating blood pressure are: (1) controlling cardiac output (the pressure and amount of blood pumped by the ventricles); (2) controlling the degree of constriction of the blood vessels (mostly the arterioles); and (3) controlling the total blood volume.

3. The hormone epinephrine raises blood pressure by increasing the strength and rate of the heart rate, and by

stimulating vasoconstriction of arterioles in certain parts of the body.

STUDY BREAK 44.6

Lymph is interstitial fluid, an aqueous solution containing molecules, ions, infecting bacteria, damaged cells, cellular debris, and lymphocytes. Because the lymph capillaries consist of a single layer of endothelial cells, the lymphatic fluid can enter them at sites where the endothelial cells overlap when the pressure of the interstitial fluid forces the flaps open. The openings produced are large enough for cells to enter.

Think Like a Scientist

FIGURE 44.13

An SA node dysfunction would involve nonnormal electrical signals. Because the SA node is the pacemaker for cardiac contraction, SA node dysfunction would result in an abnormal heart rhythm. The abnormality could be a rate of heart contraction that is too fast or too slow, depending on the dysfunction.

FIGURE 44.19

Assignable activity in MindTap. Answers provided to instructors.

MOLECULAR INSIGHTS

The key element of the experimental design was that the specific knockout of both α_{1D}-adrenergic receptor genes in the mice enabled the researchers to study just one variable—the loss of the receptor. Studies with drugs could not lead to conclusions specifically about this receptor because no drug selectively affects just one subtype of adrenergic receptors.

UNANSWERED QUESTIONS

1. Probably not. The most common form of hemophilia is a result of a mutation in clotting factor VIII (see Section 42.2), and lampreys lack that clotting factor.

2. When a blood vessel is injured during surgery or extended bed rest, the body uses platelets and fibrin to form a blood clot. The coagulation process is similar in lampreys and mammals, so the molecular basis of drugs used to prevent DVT could be elucidated by using DVT preventive drugs on lampreys.

Test Your Knowledge

1. d 2. e 3. b 4. b 5. c 6. a 7. e 8. e 9. d 10. a

Interpret the Data

Assignable activity in MindTap. Answers provided to instructors.

Chapter 45

Study Break

STUDY BREAK 45.1

1. Mucous layers are physical barriers to pathogens. Mucous layers may contain toxins and other chemicals that kill pathogens. The aqueous environment in contact with an epithelial surface may inhibit or kill pathogens by being acidic, or by containing enzymes or bile juices.

2. Innate immunity provides a nonspecific, immediate response to a pathogen. It is the first response system that comes into play when a pathogen is encountered, but it retains no memory of the encounter. The innate immune responses include inflammation and specialized cells that attack pathogens.

    Adaptive immunity, by contrast, provides a specific response to a pathogen, and it retains a memory of exposure to the pathogen so that it can respond more quickly to future attacks. The adaptive immune response takes several days to become protective, in contrast to minutes for an innate immune response.

STUDY BREAK 45.2

1. The inflammatory response is characterized by heat, pain, redness, and swelling at the infection site.

2. Inflammation-mediating molecules released at the infection site cause the dilation of local blood vessels and increase their permeability. As a result, blood flow

increases and fluid leaks from the blood vessels. Heat, redness, and swelling are direct consequences of these effects. Pain is caused by the migration of macrophages and neutrophils to the infection site, and their activities at the site.

3. The complement system is a nonspecific defense mechanism in which more than 30 interacting plasma proteins are activated by molecules on the surface of pathogens. Activated complement system proteins produce membrane attack complexes that insert into the plasma membrane of many types of bacterial cells and coat other types of cells with fragments of the complement proteins. Membrane-attack complexes create pores that lead to loss of osmotic balance, which causes the pathogens to lyse. Pathogens coated with complement protein fragments are recognized by phagocytes that then engulf and destroy the pathogens.

4. Viral pathogens do not have surface molecules that are distinct from those of the host. The main strategies used by the host in innate immunity are interferon and natural killer cells.

STUDY BREAK 45.3

1. The antibody-mediated immune response uses antibodies secreted by plasma cells (differentiated from activated B cells) to target antigens in various body fluids. The cell-mediated immune response uses cytotoxic T cells to target and kill cells infected with a pathogen.

2. An antibody is a member of the immunoglobulin (Ig) family of proteins. It consists of four polypeptides, two of which are identical heavy chains and two of which are identical light chains. Structurally, an antibody looks like a Y. Each of the two arms of the Y involves pairing of a light chain with part of the heavy chain. The tail of the Y consists of the remaining parts of the heavy chains paired together. Each heavy chain and each light chain has a variable region and a constant region. The variable regions, which occur at the top half of the arms of the Y, form two identical binding sites for the antigen with which the antibody can react.

3. DNA rearrangements of genes during differentiation are responsible for generating diverse light-chain genes and heavy-chain genes for the B-cell receptor and for the T-cell receptor. In brief, there are many different V DNA segments, one C DNA segment, and a few different J (joining) DNA segments. During differentiation, a light-chain gene is assembled from one V segment, one J segment, and the C segment, creating a gene that can be transcribed. Transcription of the gene produces a pre-mRNA that is spliced to remove introns and generate the translatable mRNA. The various combinations of segments plus imprecision in the joining at the DNA level of the V and J segments produces many types of light-chain genes. A similar DNA rearrangement process occurs for heavy-chain genes. Together, this results in tremendous variability in the antibody molecules that can be made.

4. Clonal selection is the process by which an antigen stimulates the production of a clone of B-cell-derived plasma cells that secrete antibodies against that antigen. Clonal selection accounts for the rapid response, specific action, and diversity of antibodies seen for the adaptive immune system. It also accounts for immunological memory through the production of memory B cells and memory T cells.

5. Immunological memory is the aspect of adaptive immunity that allows for a rapid, intense immune reaction to develop if the body encounters an antigen it has seen before.

STUDY BREAK 45.4

1. Immunological tolerance, a feature of the adaptive immune system, protects the body's own molecules from attack by its immune system. B cells and T cells are involved in the development of immunological tolerance, but the exact process is not understood.

2. Basically, an allergy results from an overreaction of the immune system to a particular antigen. The substances responsible for allergic reactions are a class of antigens known as allergens. Allergens stimulate B cells to secrete IgE antibodies in high amounts. IgE binds to mast cells, which are then induced to oversecrete histamine. Histamine contributes to inflammation, and at high amounts, the histamine-induced inflammation can be severe and even life threatening.

STUDY BREAK 45.5

Invertebrates lack the adaptive immune system seen in mammals; they have no B or T cells, for instance. Invertebrate immune defenses, like the innate immune system of mammals, are nonspecific. For example, all invertebrates have phagocytic cells that engulf pathogens, and many invertebrates produce antimicrobial proteins such as lysozyme. Also, many invertebrates produce proteins of the immunoglobulin family. Although these are not antibody molecules, in some invertebrates they provide a protective function against some pathogens.

Think Like a Scientist

FIGURE 45.6

The key cellular event is the interaction of an antigen-presenting B cell with a helper T cell that differentiated as a result of encountering and recognizing the same antigen as did the B cell.

FIGURE 45.11

Memory cytotoxic T cells in the cell-mediated response are cells that have been programmed to recognize a particular antigen of a pathogen. Similar to the memory cells for the antibody-mediated immune response, memory cytotoxic T cells are long-lived cells and serve to mount a rapid response should the same antigen be detected on the surface of an infected host cell.

Molecular Insights

Using cloned genes for class II MHC, CD4, and the other protein, you could analyze the transcriptomes of other cod family members. The absence of any mRNAs for those proteins would indicate either that the genes were present but silent, or that the genes were absent. PCR could be used to try to amplify the genes from the genomes of the other cod family members, assuming conservation of gene sequence, which is likely. An absence of an amplified product would indicate absence of the genes from the genomes of the family members.

UNANSWERED QUESTIONS

MHC-I genes are highly polymorphic. The large number of MHC-I variants expands the number of antigens that can potentially be targeted by the immune system. A new MHC-I molecule like HLA-C with unique characteristics might be selected for in a population if it allowed the host immune system to better control a pathogen that would otherwise evade the immune system. In fact, some investigators have hypothesized that pathogens put pressure on the evolution of MHC-I sequences.

Test Your Knowledge

1. a 2. c 3. b 4. e 5. e 6. d 7. b 8. c 9. a 10. d

Interpret the Data

Assignable activity in MindTap. Answers provided to instructors.

Chapter 46

Study Break

STUDY BREAK 46.1

1. The respiratory medium is the environmental source of oxygen and the repository for released CO_2. For aquatic animals, water is the respiratory medium; for terrestrial animals, it is air. The respiratory surface is the layer of epithelial cells between the body and the respiratory medium. Gas exchange occurs across the respiratory surface—O_2 in and CO_2 out of the body. In certain small animals, the body surface itself is the respiratory surface. Among larger animals, aquatic animals use gills, insects use tracheal systems, and terrestrial animals use lungs as the respiratory surface.

2. The advantage of water over air as a respiratory medium is that it readily enables the respiratory surface to remain wet at all times. Two key advantages of air over water as a respiratory medium are: (1) there is much more oxygen in air than in water; and (2) air is much less dense and less viscous than water, so significantly less energy is needed to move air over the respiratory surface than to move water.

STUDY BREAK 46.2

1. The advantages of gills to water-breathing animals over skin breathing are greater efficiency of gas exchange, the ability to live in more diverse habitats, and the potential to achieve a greater body mass.

2. In countercurrent exchange, the respiratory medium flows in the opposite direction of the blood flow under the respiratory surface. Examples are the flow of water over the gills in sharks, fishes, and some crabs; all these are opposite to the flow of blood. The advantage of this mechanism is that it maximizes the amounts of O_2 and CO_2 exchanged with the respiratory medium.

3. In the tracheal system, fine branches called tracheoles end in tips that are in contact with body cells. The tracheole tips are filled with fluid, and gas exchange occurs through the fluid and the plasma membrane of the body cells in contact with the tips. Air enters the tracheal system through spiracles on the body surface. Movement of air through the tracheal system occurs by muscle-driven contractions and expansions of air sacs within the system.

4. In positive pressure breathing, gulping, swallowing, or pumping action forces air into the lungs. In negative pressure breathing, muscular activity expands the lungs, lowering the pressure of air in the lungs and thereby causing air to be pulled inward.

STUDY BREAK 46.3

1. Contraction of the diaphragm and the external intercostal muscles pulls the ribs upward and outward, expanding the chest and lungs. By this negative pressure mechanism, the air pressure within the lungs is lower than that outside of the body. The higher outside pressure drives air into the lungs, expanding and filling them. Relaxation of the diaphragm and the intercostal muscles reverses the pressure condition, and the elastic recoil of lungs expels the air.

2. The most important feedback stimulus for breathing is the level of CO_2 in the blood.

3. The chemoreceptors in the medulla respond to the CO_2 level in the blood. If it rises, the medulla triggers an increase in the rate and depth of breathing. If it falls, the medulla triggers a decrease in the rate and depth of breathing.

STUDY BREAK 46.4

1. Hemoglobin is present in red blood cells. O_2 diffuses from alveoli into the plasma solution in the capillaries and then into the red blood cells. In the red blood cells, the O_2 binds to hemoglobin, thereby lowering the partial pressure of O_2 in the plasma. This leads to more O_2 molecules diffusing down the oxygen concentration gradient from alveolar air to blood.

The binding of O_2 to hemoglobin is reversible. Further, the affinity of hemoglobin for O_2 increases as the partial pressure of O_2 increases, and vice versa. This property is important for determining the release of O_2 from hemoglobin keyed to tissue requirements.

2. CO has much greater affinity for hemoglobin than does O_2, and so it displaces O_2 from hemoglobin, reducing the amount of O_2 carried in the blood. Because the brain does not monitor O_2 levels, and other receptors do not respond until blood O_2 levels are critically low, victims can easily lapse into unconsciousness and death.

STUDY BREAK 46.5

- More blood per unit body weight than land animals
- Additional red blood cells, many stored in the spleen and released during a dive
- More myoglobin in muscles than land animals
- Slowing of the heart by as much as 90%
- Reduction of blood circulation to internal organs and muscles by up to 95%
- Retention of lactic acid in the muscles with no release into the blood until the animal returns to the surface

Think Like a Scientist

FIGURE 46.14

Assignable activity in MindTap. Answers provided to instructors.

MOLECULAR INSIGHTS

One explanation is that most Han Chinese lost the Denisovan-derived *EPAS1* gene variant because of natural selection. That is, the Han Chinese mostly do not live at high altitudes, and hence retaining the variant would not confer any selective advantage, and so the gene variant was lost in most Han Chinese. The opposite is the case, of course, with the Tibetans: having the gene variant aided them in adapting to life on the Tibetan plateau.

It is also possible that interbreeding between Han Chinese and Tibetans contributes to the presence of the variant in present-day Han Chinese.

UNANSWERED QUESTIONS

1. Carbon dioxide level.
2. Contraction of smooth muscle in the heart is under the control of the autonomic nervous system. Nicotine stimulates the postganglionic neurons of both the sympathetic (increases heart rate) and parasympathetic (decreases heart rate) divisions and disrupts heart rate rhythm. Therefore, research is focused on both divisions, as nicotine affects the opposing actions of these divisions.

Test Your Knowledge

1. e 2. d 3. a 4. c 5. a 6. d 7. b 8. d 9. e 10. a

Interpret the Data

Assignable activity in MindTap. Answers provided to instructors.

Chapter 47

Study Break

STUDY BREAK 47.1

1. Carnivores primarily eat other animals as their source of organic materials. Herbivores primarily eat plants as their source of organic materials. Omnivores obtain their organic materials from any digestible organic source.

2. Essential nutrients are the amino acids, fatty acids, vitamins, and minerals that the animal cannot make itself and must obtain from its diet. The list of essential nutrients varies among animal types.

3. Deposit feeders pick up or scrape particles of organic matter from solid material they live in or on. Suspension feeders ingest small organisms that are suspended in water. Depending on the animal, the ingested organisms may be bacteria, protozoa, algae, or small crustaceans, or fragments of those organisms.

STUDY BREAK 47.2

1. Intracellular digestion takes place within body cells, whereas extracellular digestion takes place outside the cells, either in a pouch or a tube enclosed by the body.

2. (1) Mechanical processing, to break up the food; (2) secretion of enzymes and other substances that aid digestion; (3) enzymatic hydrolysis of food molecules into simpler molecular subunits; (4) absorption of the molecular subunits into body fluids and cells; and (5) elimination of undigested materials.

1. The two classes of vitamins are the water-soluble vitamins and the fat-soluble vitamins. Fat-soluble vitamins ingested in the diet in excess of bodily needs can be stored in adipose tissues. In contrast, excess water-soluble vitamins in the diet are excreted in the urine. Hence, it is critical that humans meet their daily requirements for the water-soluble vitamins.

2. The four layers of the gut are the mucosa, the submucosa, the muscularis, and the serosa. The muscularis layer, which is formed from two smooth muscle layers, is responsible for peristalsis.

3. Pepsin is secreted from chief cells into the stomach lumen as an inactive precursor molecule, pepsinogen. The pepsinogen is converted to pepsin by the highly acid conditions of the stomach. Pepsin is a digestive enzyme that begins the digestion of proteins by making breaks in polypeptide chains.

4. In the small intestine, the digestion of macromolecules into their molecular subunits occurs, and those subunits are absorbed. In the large intestine, water and mineral ions are absorbed from the remaining digestive contents, leaving the undigested remnants, the feces, which are expelled from the body.

STUDY BREAK 47.4

1. Gastrin stimulates the secretion of pepsinogen and HCl in the stomach. Pepsin generated by cleavage of pepsinogen digests protein in the swallowed food. Gastrin also stimulates contraction of the stomach and the intestines. Cholecystokinin (CCK), released in the duodenum in response largely to fat content in chyme, inhibits gastric activity; that is, CCK has the opposite effect of gastrin.

2. The hypothalamus has two interneuron centers that play a central role in regulating the digestive processes. One center stimulates appetite and reduces oxidative metabolism. The other center works in opposition to the first center by stimulating the release of a peptide hormone that inhibits appetite.

STUDY BREAK 47.5

1. Incisors nip or cut food. Canines bite and pierce food. Premolars and molars crush, grind, and shear food.

2. Symbiotic microorganisms aid digestion in many herbivores by assisting in the breakdown of plant material. The microorganisms synthesize cellulase, an enzyme vertebrates cannot make, which hydrolyzes the cellulose of plant cell walls into glucose subunits. Mutualistic microorganisms that synthesize essential amino acids and vitamins, and digest particular components of food that otherwise are indigestible, are found in the gut microbiomes of all mammals.

Think Like a Scientist

MOLECULAR INSIGHTS

One could colonize the intestines of the germ-free normal mice with bacteria harvested from the intestines of obese (*ob/ob*) and (as a control) wild-type (*ob⁺/ob⁺*) mice and determine the effect on body fat. If the intestinal bacteria directly contribute to production of body fat, the prediction is that the percentage of body fat in obese mice would increase significantly, whereas it would not in wild-type mice. Gordon's research group did this experiment and observed the result of the prediction.

UNANSWERED QUESTIONS

1. Tortoises are ectotherms and allocate energy to activity associated with maintaining body temperature; hares are endotherms and allocate energy to physiological processes to maintain body temperature.

2. They are used widely as stored energy. In fact, fats are an excellent source of energy in the diet.

Test Your Knowledge

1. b 2. b 3. a 4. d 5. c 6. e 7. c 8. d 9. e 10. a

Interpret the Data

Assignable activity in MindTap. Answers provided to instructors.

Chapter 48

Study Break

STUDY BREAK 48.1

Osmosis is a process in which water molecules move across a selectively permeable membrane from a region where they are more highly concentrated to one where they are less highly concentrated.

Osmolarity is the total solute concentration of a solution. Osmolarity is measured in osmoles per liter of solution, where an osmole is the number of solute molecules and ions (in moles).

A solution that is hypoosmotic has lower osmolarity than the solution on the other side of a selectively permeable membrane. The solution with the higher osmolarity is said to be hyperosmotic.

An osmoregulator is an animal that uses an active control mechanism to keep the osmolarity of cellular and extracellular fluids the same. This osmolarity value may differ from the osmolarity of the surroundings.

Tubules that carry out osmoregulation and excretion are formed from transport epithelium, a layer of cells with specialized transport proteins in their plasma membranes that move specific molecules and ions into and out of the tubule.

STUDY BREAK 48.2

Protonephridia are found in flatworms and larval mollusks. They are the simplest invertebrate excretory tubules. Body fluids enter the end of protonephridia and are moved through the tubule by movement of cilia on the flame cells. As the fluids move through the protonephridia, some molecules and ions are reabsorbed, and others, including nitrogenous wastes, are secreted into the tubules. Excess fluid is released through pores connecting the network of protonephridia to the body surface.

Metanephridia are found in annelids and most adult mollusks. Body fluids enter the funnel-like proximal end, driven by cilia surrounding that end. Some molecules and ions are reabsorbed as the fluids move through the tubule, and other ions and nitrogenous wastes are secreted into the tubule and excreted from the body surface.

Malpighian tubules are found in insects and other arthropods. Body fluids enter the tubules through spaces between the tubule cells. The distal ends of the tubules empty into the gut. Pressure is not used in the filtration step in Malpighian tubules. The secretion of K^+ from the hemolymph into the lumen draws in Cl^-. The accumulation of KCl causes water to enter the tubule from the hemolymph by osmosis. Organic wastes now are secreted into the tubule. When the fluids reach the hindgut and rectum, K^+ and Cl^- are reabsorbed, followed by water. As water leaves, uric acid precipitates as crystals that are excreted from the rectum in the feces.

STUDY BREAK 48.3

1. At the proximal end, a human nephron forms a cuplike region called Bowman's capsule. Bowman's capsule surrounds the glomerulus, a complex of blood capillaries. The capsule and the glomerulus are located in the renal cortex. The proximal convoluted tubule descends in a U-shaped bend called the loop of Henle into the renal medulla, and ascends again to form the distal convoluted tubule. Up to eight distal convoluted tubules drain into one collecting duct.

2. The collecting ducts are permeable to water but not to salt ions. The ducts, which begin in the cortex and descend into the medulla of the kidney, become surrounded by an ever-increasing solute concentration. As the urine passes down the collecting ducts, water moves osmotically out of the ducts, causing the concentration of the urine to increase. At the bottom of the collecting ducts, the urine is about four times more concentrated than other body fluids.

STUDY BREAK 48.4

The RAAS compensates for excessive loss of salt and body fluids. The ADH system compensates for excessive water intake or loss. Combined, the two systems play an important role in regulating the interactions between the kidneys and the rest of the body.

The RAAS is activated when the Na^+ concentration of body fluids falls, causing the volume of extracellular fluids and blood pressure also to drop. Cells in the juxtaglomerular apparatus secrete renin, which activates angiotensinogen to produce angiotensin I. Angiotensin-converting enzyme converts angiotensin I to angiotensin II, which has three effects: (1) It constricts arterioles around the body, thereby quickly increasing blood pressure; (2) It stimulates the secretion of aldosterone from the adrenal cortex, which increases Na^+ reabsorption in the kidneys and raises the osmolarity of body fluids; as a result, water moves from the tubules into the interstitial fluid, conserving water; and (3) It stimulates thirst so that more water is brought into the body. The RAAS is suppressed if the NaCl concentration in body fluids is higher than normal.

In the ADH system, ADH is released from the posterior pituitary when osmoreceptors in the hypothalamus detect an increase in the osmolarity of body fluids. ADH increases water reabsorption from the distal convoluted tubules and collecting ducts of the kidney; as a result, urinary output is reduced and water is conserved. In the case of a decrease in osmolarity, ADH release from the pituitary is inhibited, thereby decreasing water reabsorption in the kidney.

STUDY BREAK 48.5

1. Marine teleosts live in seawater, which is hyperosmotic to their body fluids. They drink seawater continuously to replace water lost to the environment by osmosis. Na^+, K^+, and Cl^- ions from the seawater they drink are excreted by the gills. Nitrogenous wastes are released from the gills, primarily as ammonia, by simple diffusion.

   Freshwater fishes live in fresh water, which is hypoosmotic to their body fluids. They take in water by osmosis and excrete excess water. Salts needed for bodily functions are obtained from food and by active transport through the gills from the water. Nitrogenous wastes are excreted from the gills as ammonia.

2. The excretion of urea in mammals involves expelling the urea in solution—urine. Thus, there is loss of water with excretion of nitrogenous waste as urea. Excretion of nitrogenous waste in the form of uric acid conserves water because uric acid crystals are almost water-free.

STUDY BREAK 48.6

Ectothermy applies to animals that obtain heat energy primarily from the environment; endothermy applies to animals that obtain heat energy primarily from internal reactions. Generally speaking, ectotherms are highly successful in warm environments, but their bodily functions slow down as the temperature drops. Endotherms remain active over a broader range of environmental temperatures than ectotherms do, but they require an almost constant supply of energy to maintain their body temperature.

STUDY BREAK 48.7

1. The thermoregulatory responses shown by ectotherms can be physiological, such as regulating blood flow to internal organs or to the skin, or behavioral, such as physically moving to a location in the environment suitable for their heat energy needs at the time.

2. Thermal acclimatization refers to the physiological changes that many ectotherms make to compensate for seasonal shifts in environmental temperature.

STUDY BREAK 48.8

Temperature regulation in endothermic animals involves mechanisms that balance internal heat production against heat loss from the body. Internal heat production is controlled by negative feedback pathways triggered by thermoreceptors in the skin, the hypothalamus, and the spinal cord. When a deviation from the set point occurs, this system operates to return the core temperature to that set point through changes in metabolic processes, behavioral activities, and control of heat loss at the skin surface.

Think Like a Scientist

FIGURE 48.11
Assignable activity in MindTap. Answers provided to instructors.

MOLECULAR INSIGHTS
The fact that the miR-17~92 miRNA cluster is directly implicated in the pathogenesis of PKD suggests that the miRNAs in the cluster might be targets for a therapeutic approach.

UNANSWERED QUESTIONS
Alleles can be deleterious in some environments and beneficial in others, and deleterious in, say, the homozygous state, and beneficial in the heterozygous state. The best know example of this is sickle-cell disease, in which two hemoglobin gene mutations cause disease, but heterozygosity for a mutation provides some protection against malaria. Something similar is thought to be the case with *APOL1*. Heterozygosity appears to protect people from some forms of African sleeping sickness. These are both examples of "balancing selection."

Test Your Knowledge
1. d 2. c 3. a 4. b 5. d 6. e 7. c 8. b 9. d 10. e

Interpret the Data
Assignable activity in MindTap. Answers provided to instructors.

Chapter 49

Study Break

STUDY BREAK 49.1
Recall that asexual reproduction produces offspring with genes from only one parent. In most animals that undergo asexual reproduction, the offspring are produced by mitosis, and therefore are genetically identical to one another and to the parent. Such genetic homogeneity can be advantageous in stable and uniform environments. Another advantage is that individuals do not need to use energy to produce gametes or to find and select a mate. Further, for individuals in sparse populations, or for sessile animals, asexual reproduction can be an advantage. A disadvantage is that a genetically homogeneous population may not adapt readily to new environments.

By contrast, sexual reproduction always generates genetic diversity among offspring. This provides a population the opportunity to adapt to changing environments, and perhaps to move to and colonize new environments. Disadvantages of sexual reproduction include the expenditure of energy and raw materials to produce gametes and to find and select mates. Another disadvantage of sexual reproduction is that it produces fewer offspring per parent than does asexual reproduction; that is, asexual reproduction of a female produces all-female progeny, each of which can produce offspring for the next generation. However, in sexual reproduction, only one half of the progeny of a mating are females, and only they can produce the offspring of the next generation.

STUDY BREAK 49.2
1. Egg coats are surface coats around the egg. They are added during oocyte development or fertilization. Egg coats provide protection against mechanical injury, infection by microorganisms, and, in some species, loss of water.

   Mammalian eggs have an egg coat called the zona pellucida immediately surrounding the egg. This gel-like coating is called the vitelline coat in other organisms. In birds, a thick solution of proteins—the egg white—surrounds a vitelline coat. Bounding the egg white is a hard shell.

2. The slow block to polyspermy occurs in many organisms, including mammals. The fusion of egg and sperm triggers an increase in calcium ions in the cytosol. The calcium ions activate proteins that initiate a high level of metabolic activity in the egg. The released calcium ions also cause the cortical granules to fuse with the plasma membrane of the egg and release their contents to the outside. Enzymes released from those granules alter the egg coats in a matter of minutes after fertilization, and this blocks further sperm from attaching to and penetrating the egg.

STUDY BREAK 49.3
- FSH stimulates a number of oocytes to begin meiosis. A follicle forms around each oocyte.
- FSH and LH interact to stimulate follicle cells to secrete estrogens.
- Later in the cycle, an increased level of estrogen leads to a new burst of release of FSH and LH. This new LH stimulates follicle cells to release enzymes that digest away the follicle wall, leading to egg release.
- The new LH also causes the remaining follicle cells to grow into the corpus luteum.
- FSH and LH levels decrease. This removes the stimulatory signal for follicular growth, and no new follicles grow in the ovary.

STUDY BREAK 49.4
The oral contraceptive pill inhibits the secretion of FSH and LH by the pituitary. FSH and LH stimulate oocytes in the ovaries to begin meiosis. They also stimulate follicle cells that surround the oocytes to secrete estrogens. Later, LH causes ovulation—the release of an egg. Therefore, without FSH and LH, ovulation cannot occur.

Think Like a Scientist

FIGURE 49.12
The menstrual cycle would not proceed normally; that is, menstruation would not occur.

FIGURE 49.13
Assignable activity in MindTap. Answers provided to instructors.

MOLECULAR INSIGHTS
The study showed clearly a human male superiority in olfactory sensitivity for bourgeonal. (The authors stated that it was the first study to show a male superiority in olfactory sensitivity.) Potentially, this might be due to differences in its behavioral relevance for males and females, although that remains to be shown.

UNANSWERED QUESTIONS
It could be used in cases of a low sperm count (few sperm produced), poorly motile or immotile sperm, incomplete sperm maturation, or inability to produce sperm by ejaculation (sperm could be collected surgically from the epididymis).

Test Your Knowledge
1. b 2. a 3. d 4. a 5. e 6. c 7. b 8. e 9. c 10. d

Interpret the Data
Assignable activity in MindTap. Answers provided to instructors.

Chapter 50

Study Break

STUDY BREAK 50.1
1. In cleavage divisions, cycles of DNA replication and division occur without the production of new cytoplasm. Therefore, the cytoplasm of the egg partitions into many cells without increasing in overall size or mass. In cell division in an adult organism, the mitotic cell cycle involves cycles of DNA replication and division interspersed with cell growth; that is, the cell grows, then divides, then the two progeny cells grow, then divide, and so on. New cytoplasm is produced during these cell cycles so that, overall, there is an increase in mass of the cells.

2. The three primary cell layers of the embryo are produced by gastrulation. They are ectoderm, mesoderm, and endoderm.

STUDY BREAK 50.2
1. The gray crescent establishes the dorsal–ventral axis, with the gray crescent marking the future dorsal side.
2. If cells of the dorsal lip of the blastopore are transplanted to another location in the egg, they cause a second blastopore, and subsequently a second embryo, to form in that region.
3. The extraembryonic membranes in birds and their functions are as follows:
   - Yolk sac: Surrounds the yolk, which provides nutrients to the embryo.
   - Chorion: Exchanges oxygen and carbon dioxide with the environment through the egg shell.
   - Amnion: Encloses the embryo and secretes amniotic fluid into the space that provides an aquatic environment in which the embryo can develop.
   - Allantois: Stores nitrogenous wastes, primarily in the form of uric acid, that are derived from the embryo. Part of the allantoic membrane forms a bed of capillaries that is connected to the embryo and that delivers carbon dioxide from the embryo to the chorion and picks up oxygen absorbed through the shell.

STUDY BREAK 50.3
1. The three primary tissues, ectoderm, mesoderm, and endoderm, give rise to the tissues and organs of the embryo.
2. The central nervous system, notably the brain and spinal cord, develops from the neural tube. Neural crest cells give rise to parts of the nervous system (including cranial nerves, ganglia of the autonomic nervous system, peripheral nerves from the spinal cord to body structures, and nerves of the developing gut) and contribute to a variety of other body structures (for example, bones of the inner ear and skull, cartilage of facial structures, teeth, pigment cells in the skin, and the adrenal medulla).

STUDY BREAK 50.4
1. Cells of the trophoblast are responsible for implantation of the blastocyst in the endometrium (uterine lining). When the blastocyst is ready for implantation, trophoblast cells secrete proteases that digest pathways between endometrial cells. Dividing trophoblast cells fill the spaces and, through continued digestion and division, the blastocyst burrows into the endometrium and eventually becomes covered by a layer of endometrial cells. From then on, cells originating in the trophoblast support the development of the embryo/fetus in the uterus through the production of the chorion.

   The inner cell mass becomes the embryo/fetus itself. During implantation, the inner cell mass separates into the epiblast and the hypoblast. The epiblast produces the ectoderm, mesoderm, and endoderm of the developing embryo. The hypoblast gives rise to part of the extraembryonic membranes.
2. Oxytocin. This hormone is responsible for stimulating contractions of the uterus.

STUDY BREAK 50.5
1. Cell division, cell movement, and cell adhesion.
2. Induction is the process whereby a region of the embryo acts on other cells to alter the course of development. Induction is brought about by proteins; hence, induction is under genetic control.
3. In the absence of a death signal, CED-9 is active. Active CED-9 inactivates CED-4 and, as a result, CED-3 protease is inactive. Therefore, the cascade of activations necessary for apoptosis do not occur. If a death signal binds to the surface receptor and activates it, the result is to

inactivate CED-9. Therefore, CED-9 is not available to inactivate CED-4, and active CED-4 activates CED-3. The cascade of activations necessary for apoptosis to occur then takes place.

Think Like a Scientist

FIGURE 50.12
The embryo/fetus and mother are different genetically Therefore, the cells of the embryo/fetus will have surface antigens recognized as foreign by the mother. This will induce an immune response by the mother (see Chapter 45) in which antibodies will be produced that will attack the cells of the embryo/fetus and likely kill it.

FIGURE 50.17
Assignable activity in MindTap. Answers provided to instructors.

FIGURE 50.19
Assignable activity in MindTap. Answers provided to instructors.

MOLECULAR INSIGHTS
This is a Mendelian cross, essentially (see Chapter 12). A cross of two heterozygotes of genotype $Pomt2^{+/-}$ is predicted to produce progeny in a ratio of 1 $Pomt2^{+/+}$: 2 $Pomt2^{+/-}$: 1 $Pomt2^{-/-}$. However, the $Pomt2^{-/-}$ do not survive beyond the embryo stage, so the predicted ratio of genotypic classes among living progeny is 1 $Pomt2^{+/+}$: 2 $Pomt2^{+/-}$.

UNANSWERED QUESTIONS
No (and, in fact, there are not any). This can be inferred from the description of sex determination in mammals and from the list of gynandromorph animals. The significance of the research is unchanged.

Much can be learned about sex determination in mammals by studying sex determination in other organisms, and one could argue that understanding why it is different in other animals only provides greater insight into sex determination in mammals.

Test Your Knowledge
1. d 2. e 3. a 4. a 5. e 6. c 7. d 8. d 9. b 10. c

Interpret the Data
Assignable activity in MindTap. Answers provided to instructors.

Chapter 51

Study Break

STUDY BREAK 51.1
1. Studies of ecosystems are more "inclusive" than studies of populations because ecosystems include the populations of many different species.
2. Mathematical models are useful in ecological research because they help scientists formalize hypotheses about the relationships between variables and because they allow researchers to simulate the effects of changing variables before investing time and resources in experiments or observational studies.

STUDY BREAK 51.2
1. Because of Earth's spherical shape, sunlight striking the planet's surface is more concentrated near the equator than at the poles. As a result, temperatures are higher at low latitudes. The concentrated sunlight near the equator heats the atmosphere, causing air masses near the equator to rise, establishing three circulation cells in the Northern Hemisphere and three in the Southern Hemisphere.
2. Earth's fixed tilt on its axis causes seasonal variation in the amount of sunlight striking the temperate zone as the planet orbits the Sun.
3. Dry conditions prevail at 30° N and S latitudes because sinking air masses warm as they descend, causing them to absorb water from the land.
4. Mountains affect local precipitation because rising air masses on the windward side of a mountain cool

adiabatically and release moisture. When the air masses descend on the leeward side of a mountain, they warm and absorb moisture, causing a rain shadow.

STUDY BREAK 51.3
1. *Anolis* lizards in the Dominican Republic bask more frequently at high elevation than they do at low elevation.
2. Global warming will likely cause the geographical distributions of species to shift or expand to higher latitudes and to higher elevations.

STUDY BREAK 51.4
1. Tropical rainforest and temperate rainforest are the terrestrial biomes that receive the most rainfall.
2. Savannas, chaparral, and temperate grasslands are renewed by periodic fires.
3. Tropical rainforests and temperate rainforests have the tallest vegetation. Arctic and alpine tundra have the shortest vegetation.
4. Trees are usually evergreen in tropical rainforest, temperate rainforest, and taiga.

STUDY BREAK 51.5
1. Dissolved oxygen concentration is usually high in the headwaters of a stream, gradually diminishing as water flows into a river.
2. The factors that cause seasonal overturns in lakes include seasonal changes in environmental temperature, variations in wind velocity, and the fact that water is densest at 4°C.
3. Oligotrophic lakes are better than eutrophic lakes for recreational purposes because the water in oligotrophic lakes is clear, whereas the water in eutrophic lakes is often clogged with strands of algae and cyanobacteria.

STUDY BREAK 51.6
1. The benthic province of the ocean includes all of the bottom sediments. The pelagic province includes all of the water.
2. Estuaries experience the largest fluctuations in salinity over time.
3. Estuaries, the intertidal zone, and the upper layer of the oceanic pelagic zone receive substantial energy inputs from sunlight.
4. The benthos of the oceanic zone receives nutrients and energy from the detritus sinking from the upper layers of water.
5. Chemoautotrophic bacteria are the primary producers of hydrothermal vent communities and cold seep communities. Unlike photosynthetic organisms of the photic zone, they use hydrogen sulfide and other molecules, instead of sunlight, as an energy source for their chemosynthetic activity.

Think Like a Scientist

FIGURE 51.9
Assignable activity in MindTap. Answers provided to instructors.

FIGURE 51.11
Climate warming will not change the distribution of biomes on the climograph, because the biomes will still occur at particular combinations of temperature and precipitation. However, global warming will probably change the distribution of biomes on the Earth's surface, as reflected in the map in Figure 51.12. Biomes may shift toward the north and south as the climate warms at higher latitudes.

FIGURE 51.25
Assignable activity in MindTap. Answers provided to instructors.

MOLECULAR INSIGHTS
Yes, differences in the voltage gated K^+ channels in the neurons of different octopus species are subject to natural selection. Although selection may not act on the DNA sequences directly, it can act on the genes coding for the RNA editing

machinery, favoring higher levels of editing in species that live in very cold ocean waters.

UNANSWERED QUESTIONS
We should expect to see the loss of species in mountaintop habitats because the species have no cooler habitats into which they can migrate. We would also expect to see the loss of polar species, especially those that depend on sea ice, the extent of which is shrinking rapidly.

Test Your Knowledge
1. c 2. e 3. b 4. a 5. d 6. a 7. a 8. d 9. a 10. d

Interpret the Data
Assignable activity in MindTap. Answers provided to instructors.

Chapter 52

Study Break

STUDY BREAK 52.1
1. A population's size is simply the number of individuals it contains. Its density is the number of individuals per area or volume of habitat occupied.
2. A clumped pattern of dispersion implies that individuals in the population help each other or that some vital resource in the environment also has a clumped distribution. A uniform pattern of dispersion implies that individuals in the population repel each other. A random pattern of dispersion does not imply either positive or negative interactions among individuals in the population.

STUDY BREAK 52.2
1. A life table usually summarizes statistics about the age-specific survival rates, age-specific mortality rates, and age-specific fecundity of a population.
2. Humans in the industrialized countries exhibit Type I survivorship curves because they provide lots of care to their offspring, thus reducing infant and childhood mortality to low levels.

STUDY BREAK 52.3
1. Children spend most of their energy on growth and maintenance.
2. Fecundity and the amount of parental care devoted to each offspring exhibit an inverse relationship because organisms that produce few offspring can devote substantial time and energy to each, whereas those that produce many offspring can devote only minimal time and energy to each.

STUDY BREAK 52.4
1. The model of exponential population growth predicts unlimited population growth over time, generating a J-shaped curve of population size versus time. The logistic model predicts that population growth slows down as the population approaches its carrying capacity, generating an S-shaped curve of population size versus time.
2. Carrying capacity is the maximum number of individuals in a population that an environment can support. The carrying capacity is thus a property of the environment with reference to a particular population.
3. A time lag is a delay in a population's response to a changing environment. It may cause a population's size to oscillate around its carrying capacity.

STUDY BREAK 52.5
1. The effects of density-dependent factors get stronger (that is, they affect a larger percentage of the individuals in the population) as the population's density increases. The effects of density-independent factors do not change (that is, they affect the same percentage of the individuals in a population) as the population's density changes.
2. The effects of infectious diseases are usually density-dependent effects because disease-causing pathogens spread more quickly through dense populations of the organisms they infect.

STUDY BREAK 52.6

1. Humans have sidestepped the controls that regulate the populations of other organisms by expanding their geographical range to include a wide variety of habitats, by increasing the carrying capacity through agricultural production, and by decreasing their death rate through the introduction of medical care and improved sanitation.
2. The age structure of a population influences its future population growth by determining how many individuals will reach reproductive age in the future. Populations with a bottom-heavy age structure (that is, with many young children) will experience rapid population growth when children alive today reach sexual maturity. Populations with a more even age structure will not experience a dramatic future increase in population size.

Think Like a Scientist

FIGURE 52.17

Assignable activity in MindTap. Answers provided to instructors.

FIGURE 52.19

Assignable activity in MindTap. Answers provided to instructors.

FIGURE 52.20

The lynx and hare appear to be on a 10-year population cycle. From the data presented in the graph, the lynx population increase followed the hare population increase almost instantaneously in 1865, 1885, and 1905. The lynx population increased more slowly after increases in the hare population in 1855, 1875, 1895, 1915, and 1925.

MOLECULAR INSIGHTS

Colonies that are four or five years old contribute the most to the next generation. Even though fecundity is steady throughout a colony's lifetime, the steady rate of mortality in colonies after they first reproduce reduces the overall reproductive contribution of colonies that are older than four or five years.

UNANSWERED QUESTIONS

1. The "ideal" conditions under which the equations predict growth simply never exist, as is clear, for example, from the growth of populations illustrated in the chapter (see growth of *Thrips imaginis*). Therefore, unlike a law such as the law of gravity, population growth cannot simply be predicted by applying the logistic or exponential equation for population growth. The mathematical growth models are constructed after basic observations define the variables of the models.
2. The time scales for most population studies are too short to consider selection, genetic drift, or mutation. Molecular studies allow population biologists to measure changes in gene frequencies at the time scales of single generations and before those changes are manifested in noticeable changes in fecundity or survival (or morphology and behavior).

Test Your Knowledge

1. b 2. c 3. d 4. a 5. d 6. b 7. b 8. e 9. e 10. d

Interpret the Data

Assignable activity in MindTap. Answers provided to instructors.

Chapter 53

Study Break

STUDY BREAK 53.1

1. Some carnivores spend more time and energy capturing large prey than small prey because large prey provides a larger return on their investment of time and energy in the hunt.
2. Cryptic coloration makes an organism inconspicuous, allowing it to blend in with its surroundings. Aposematic coloration makes an organism highly conspicuous, advertising its unpalatability. Mimicry allows one organism, the mimic, to resemble another species, the model; models are usually unpalatable or poisonous. A mimic will have aposematic coloration if it resembles an aposematic model.
3. Field experiments can demonstrate that two species are competing for limiting resources if the removal of one species increases population size or density in the other or if the addition of a potential competitor decreases the population size or density of the other.

STUDY BREAK 53.2

1. Gleason's individualistic view of communities suggests that they are just chance assemblages of species that happen to be adapted to similar abiotic environmental conditions.
2. Ecologists find more species living in an ecotone than in the communities on either side of it because ecotones contain species from both neighboring communities as well as species that are adapted to transitional environmental conditions.

STUDY BREAK 53.3

1. The plant growth forms found in tropical forests include a canopy of tall trees, an understory of shorter trees and shrubs, an herb layer, vinelike lianas, and epiphytes.
2. The species richness of a community is the number of species it contains. Relative abundance refers to the commonness or rarity of species in the community.
3. Pigeons, which eat grain and other vegetable matter, are included in the second trophic level, primary consumers. Peregrine falcons, which feed on pigeons and other birds, are in the third trophic level, secondary consumers.

STUDY BREAK 53.4

1. On the one hand, the ecological literature may overestimate the importance of competition because ecologists are more likely to study and publish papers on interactions in which competition is important than on interactions in which it is not. On the other hand, the literature may underestimate the importance of competition because, if strong competition between species cannot persist for long periods of time, we are unlikely to find populations competing strongly in nature.
2. Keystone species are those that have a substantial effect on community structure even if their populations are not very dense. Keystone species may either increase or decrease species richness in the communities they occupy.

STUDY BREAK 53.5

1. Strong storms allow coral communities to be rejuvenated through the recruitment of new individuals because they scour the seafloor, removing existing coral colonies from the community. These openings provide spaces where coral larvae may settle and initiate the growth of new colonies.
2. Moderately severe and moderately frequent disturbances increase a community's species richness by creating opportunities for *r*-selected species to colonize the habitat while allowing populations of *K*-selected species to persist.

STUDY BREAK 53.6

1. Primary succession occurs in places without soil; secondary succession occurs after a disturbance has destroyed vegetation.
2. A climax community differs from earlier successional stages in having taller, longer-lived vegetation, generally higher species richness, and a buffered physical environment under the vegetation.
3. The three hypotheses about the underlying causes of succession differ in how they view the role of population interactions. The facilitation hypothesis specifies no particular role for population interactions. The inhibition hypothesis suggests that the species already present prevent other species from joining a community. The tolerance hypothesis suggests that as environmental conditions within the community change during succession, only species that can compete strongly under the changing conditions will persist.

STUDY BREAK 53.7

1. Some explanations of the high species richness in the tropics suggest that the benign climate and historically low levels of severe disturbance have fostered more rapid rates of speciation. Other explanations suggest that the year-round availability of food resources and complex food webs allow more species to coexist in tropical regions.
2. According to the equilibrium theory of island biogeography, large islands will harbor more species than small islands, and islands that are close to the mainland will harbor more species than those that are farther away.

Think Like a Scientist

FIGURE 53.8

Assignable activity in MindTap. Answers provided to instructors.

FIGURE 53.12

Assignable activity in MindTap. Answers provided to instructors.

FIGURE 53.24

Assignable activity in MindTap. Answers provided to instructors.

FIGURE 53.25

Assignable activity in MindTap. Answers provided to instructors.

FIGURE 53.28

Primary succession after the retreat of a glacier in the very far north would be similar to that observed at Glacier Bay, Alaska. However, the succession would produce a climax community that included abundant lichens and a relatively small number of very short woody plants.

MOLECULAR INSIGHTS

The communities of organisms living in bromeliad cups are analogous to those living in British treeholes. Thus, we would expect the species richness of microorganisms and invertebrates living in bromeliad cup communities to exhibit species-area functions similar to those observed in communities living on islands.

UNANSWERED QUESTIONS

1. Many, but certainly not all, mutualisms involve partners "taking" resources from each other. In pollination, for example, bees take pollen, and plants "take" the bees' time and direct their movements.
2. Yes, because of shared traits among close relatives. This is one reason cannibalism may be rare.

Test Your Knowledge

1. c 2. c 3. a 4. b 5. b 6. d 7. d 8. b 9. b 11. e

Interpret the Data

Assignable activity in MindTap. Answers provided to instructors.

Chapter 54

Study Break

STUDY BREAK 54.1

1. In the generalized compartment model of biogeochemical cycling, nutrient pools are classified as available or unavailable and as organic or inorganic.
2. The advantage of using conceptual models of ecosystem function is that they are a simplification of the processes that determine ecosystem function in nature. The disadvantage of these models is that they do not include data on the processes that carry nutrients and energy out of one ecosystem into another; neither do they include the details of exactly how specific ecosystems function. Thus, conceptual models do not provide precise predictions about potential changes in ecosystem function.

3. Before constructing a simulation model of an ecosystem, ecologists must collect data about the population sizes of important species, the average energy and nutrient content of each, the food webs in which they participate, the quantity of food each species consumes, and the productivity of each population; the ecosystem's energy and nutrient gains and losses caused by erosion, weathering, precipitation, and runoff; and seasonal and annual variations in these factors.

STUDY BREAK 54.2

1. Gross primary productivity is a measure of the total amount of solar energy converted into chemical energy by the producers in an ecosystem. Net primary productivity is the amount of chemical energy that remains after deducting the producers' maintenance costs from the gross primary productivity.
2. In terrestrial ecosystems, primary productivity may be influenced by the availability of light, water, and nutrients and by the environmental temperature. In aquatic ecosystems, primary productivity is often limited by the joint availability of light and nutrients in the same place.
3. Energy is lost from an ecosystem at every transfer between trophic levels because some of the energy is not assimilated, because organisms use some of the energy they assimilate for maintenance costs, and because biological processes are never 100% efficient.
4. The presence of predators can influence an ecosystem's productivity by consuming herbivores and thereby changing the population dynamics of the herbivores and the plants they eat. These effects can reverberate through an ecosystem in a trophic cascade.

STUDY BREAK 54.3

1. The global hydrologic cycle maintains its balance because the amount of water returned to the atmosphere by evaporation and transpiration is equal to the amount that falls as precipitation. Runoff from the land maintains the balance between terrestrial and marine components of the cycle.
2. Respiration, excretion, leaching, and the burning of fossil fuels move large quantities of carbon from an organic compartment to an inorganic compartment of an ecosystem.
3. Bacteria, cyanobacteria, and fungi drive the global nitrogen cycle through their activities in nitrogen fixation, ammonification, nitrification, and denitrification.
4. Marine sediments are Earth's main reservoir for phosphorus, which is recycled slowly after geological uplifting and erosion make it available to producers.

STUDY BREAK 54.4

1. The greenhouse effect refers to the tendency of certain gases, called greenhouse gases, to foster the accumulation of heat in the lower atmosphere by hindering the escape of infrared heat into space. CO_2 enhances the greenhouse effect because it is one of the greenhouse gases.
2. The combustion of fossil fuels and wood is the human activity that releases the most CO_2 into the atmosphere.
3. Irrigation, the use of synthetic nitrogen-containing fertilizers, the harvesting of crops, and the rearing of livestock contribute to the disruption of the nitrogen cycle.

Think Like a Scientist

FIGURE 54.7
Assignable activity in MindTap. Answers provided to instructors.

FIGURE 54.10
Assignable activity in MindTap. Answers provided to instructors.

MOLECULAR INSIGHTS
Moderate acidification of ocean water appears to affect the transcription of genes that influence biomineralization. Thus, the production of tests (that is, shells) by sea urchins might be adversely affected.

UNANSWERED QUESTIONS
Large cities release substantial quantities of airborne pollutants that can be carried by prevailing winds even to distant forests. These materials can have a profound influence on the health and biological activity of primary producers in the forest.

Test Your Knowledge
1. c 2. d 3. c 4. a 5. a 6. d 7. a 8. e 9. e 10. c

Interpret the Data
Assignable activity in MindTap. Answers provided to instructors.

Chapter 55

Study Break

STUDY BREAK 55.1

1. The growing human population in south Florida has increased the likelihood of desertification there by withdrawing groundwater for agricultural, recreational, and residential uses faster than it is replenished.
2. Overexploitation of fish populations typically causes fishes to reach reproductive maturity at a smaller size and younger age, decreases population sizes, and sometimes leads to the extinction of populations.
3. The construction of dams changes the water flow in rivers, leading to changes in the physical structure of the river and the species that inhabit it; disrupting the environmental cues that trigger successful reproduction of organisms; inhibiting the free movement of migrating animals through all parts of the river system; and opening the habitat to invasion by species that originated elsewhere.

STUDY BREAK 55.2

1. The use of the anti-inflammatory drug diclofenac on livestock has decimated vulture populations in South Asia. Diclofenac is highly toxic to birds, and the vultures ingest it when they feed on the carcasses of livestock.
2. By one estimate, extinction rates today may be 1,000 times greater than the background extinction rate evident in the fossil record.

STUDY BREAK 55.3

1. Living systems are a storehouse of potentially useful genetic information because naturally occurring compounds may prove to be useful in the treatment of disease, in the manufacture of new products, or in agriculture.
2. Naturally occurring organisms provide many ecosystem services, such as the sequestration of carbon dioxide, fixation of nitrogen into forms that plants can absorb, recycling of nutrients within ecosystems, and the retention of water in ecosystems.

STUDY BREAK 55.4

1. The IUCN designates the conservation status of species and higher taxa in the following categories: least concern, near threatened, vulnerable, endangered, critically endangered, extinct in the wild, and extinct. Some species and higher taxa are omitted from consideration because of insufficient data.
2. Conservation biologists identify a site as one where extinction is imminent if it houses 95% or more of the individuals in an endangered species and it has definable boundaries that encompass distinctive habitats.
3. Conservation biologists are especially alarmed about deforestation in the New World tropics because these forests harbor many species in imminent danger of becoming extinct.

STUDY BREAK 55.5

1. Population bottlenecks—large, temporary reductions in a population's size—inevitably foster genetic drift, thereby reducing a population's genetic variability, which increases its likelihood of becoming extinct.
2. A population viability analysis allows a conservation biologist to identify the minimum population size that

is likely to survive both predictable and unpredictable environmental change. It therefore specifies how many individuals must be conserved for the continued survival of the population and species.
3. Several small preserves would collectively experience more edge effects than one large preserve of the same total size.

STUDY BREAK 55.6

1. The Pine Bush habitat in the state of New York is an example of conservation through preservation.
2. The Royal Chitwan National Park has been judged a success because local residents benefit from the park's existence and therefore support it, and because populations of many animals, including tigers and rhinoceroses, have increased within its borders.
3. Economists can determine the economic values of specific ecosystem services and convince local governments that it is economically beneficial to preserve ecosystems and the services they provide.

Think Like a Scientist

FIGURE 55.6
Assignable activity in MindTap. Answers provided to instructors.

FIGURE 55.19
Assignable activity in MindTap. Answers provided to instructors.

MOLECULAR INSIGHTS
DNA barcoding might be useful in the effort to track the spread of Asian carp throughout North American rivers and lakes because biological samples can be subjected to DNA barcoding analysis to detect the presence of these species in habitats in which they had not previously been found.

UNANSWERED QUESTIONS
Student answers may vary. Because students are asked to come up with a list of unanswered questions on their own, we can't provide an answer to this question.

Test Your Knowledge
1. e 2. b 3. b 4. b 5. a 6. d 7. a 8. c 9. d 10. b

Interpret the Data
Assignable activity in MindTap. Answers provided to instructors.

Chapter 56

Study Break

STUDY BREAK 56.1

1. An instinctive behavior is a genetically or developmentally programmed response that appears in complete and functional form the first time it is used. A learned behavior is one that is dependent on having a particular kind of experience during development.
2. Arnold demonstrated that the receptiveness of garter snakes to a meal of banana slugs had a genetic basis by breeding snakes that almost always eat banana slugs with snakes that rarely eat them. The behavior of the hybrid offspring was intermediate between the behaviors of the two parent populations.
3. Tail wagging by a dog when it sees its owner pick up a leash is an example of classical conditioning.
4. Sleeping through an alarm clock is an example of habituation.

STUDY BREAK 56.2

1. The conclusion that certain neurons in a young male bird's brain are influenced only by acoustical signals from its own species and only during a critical period is supported by two observations: (1) young birds that did not hear taped songs during the critical period never produced the full song of their species; and (2) young birds that heard recordings of *other* bird species' songs during the critical period never generated replicas of those songs as they matured.

2. The role of the ZENK enzyme in song learning is to program the nerve cells of the bird's brain to anticipate key acoustical events of potential biological importance.
3. A high estrogen concentration in the brains of young male zebra finches stimulates the production of more neurons in the higher vocal center.
4. The loss of a territory by a male African cichlid fish causes its brain to produce less GnRH.

STUDY BREAK 56.3

1. When piloting, animals use familiar landmarks to guide their journey. In compass orientation, animals use external environmental cues such as the position of the Sun or stars as a compass to move in a particular direction, often over a specific distance or for a prescribed length of time. When navigating, animals use a compass as well as a mental map of their position in relation to their destination.
2. Seasonal changes in temperature and food availability are the most likely selection pressures responsible for the evolution of migratory behavior in birds.

STUDY BREAK 56.4

1. Wood lice accumulate in moist habitats because they move much less in moist habitats than they do in dry habitats.
2. The costs of maintaining a territory include the time and energy needed to defend territory borders, the possibility of being injured or killed during a territorial encounter, and the increased likelihood of being noticed and captured by a predator. The benefits of holding a territory include access to all of the resources within the territory.

STUDY BREAK 56.5

1. Humans consciously use acoustical, visual, and tactile channels to communicate with each other.

2. A honeybee uses the waggle dance to communicate the location of a distant food source to its hive-mates.
3. A young wandering raven that discovers a food source in the territory of other ravens will call vigorously to attract other wandering ravens from outside the territory. Collectively, the nonterritorial birds can overwhelm the defenses of the resident birds and then consume the food.

STUDY BREAK 56.6

1. For monogamous species, the males that are most attractive to females are those that can demonstrate their good genes with large showy morphological characteristics and elaborate behavioral displays and those that hold territories rich in resources.
2. Male sage grouse perform displays at a lek, trying to attract the attention of females. Female sage grouse go to a lek to evaluate the qualities of the males that are displaying there.
3. A female red-winged blackbird might settle in the territory of a male even in the presence of other resident females if the male's territory is very rich in resources.

STUDY BREAK 56.7

1. The social behavior of musk oxen and sawfly larvae includes cooperative defense of the group against predators.
2. Among the animals in a dominance hierarchy, the most dominant individuals are the most likely to reproduce.
3. The genetic system in eusocial insects may promote altruistic behavior because the workers in a colony are more closely related to each other than are siblings in most other animal species.

Think Like a Scientist

FIGURE 56.2

Assignable activity in MindTap. Answers provided to instructors.

FIGURE 56.10

Assignable activity in MindTap. Answers provided to instructors.

FIGURE 56.13

Assignable activity in MindTap. Answers provided to instructors.

FIGURE 56.29

If the queen in a bee colony mated with two drones instead of just one, the average degree of relatedness among the workers would decrease. Although their relatedness through their shared mother would not change, some workers would be unrelated to each other through their different fathers.

MOLECULAR INSIGHTS

As described in Section 21.3, a recessive mutation is not easily eliminated from a population because recessive mutations have no impact on heterozygotes. Some form of genetic screening would be necessary to identify people who carry the mutation and counsel them about the possible consequences of having children together. Unidentified heterozygotes could produce offspring who are homozygous for the mutation and would therefore exhibit the condition that it causes.

UNANSWERED QUESTIONS

Answers will vary, but some decisions might include: Where do worker bees forage to find that patch of nectar-rich plants one of their sisters just found? How close should members of a fish community come in their approach to a novel species to determine if it is a predator? At what point are relatives closely enough related that they should share food?

Test Your Knowledge

1. b 2. d 3. b 4. e 5. c 6. e 7. d 8. b 9. c 10. a

Interpret the Data

Assignable activity in MindTap. Answers provided to instructors.

Appendix B: Classification System

The classification system presented here is based on a combination of organismal and molecular characters and is a composite of several systems developed by microbiologists, botanists, and zoologists. This classification reflects current trends toward a phylogenetic approach to taxonomy, one that incorporates the ever more detailed information about the relationships of monophyletic lineages provided by new molecular sequence data. In keeping with these trends, we have omitted reference to the traditional taxonomic categories, such as "class" and "order." Instead, we present the major monophyletic lineages in each of the three domains, and we indicate their relationships within a nested hierarchy that parallels that of traditional Linnaean classification.

Although researchers generally agree on the identity of the major monophyletic lineages, the biologists who study different groups have not established universal criteria for identifying the somewhat arbitrary taxonomic categories included in the traditional Linnaean hierarchy. As a result, a "class" or an "order" of flowering plants may not be the equivalent of a "class" or an "order" of animals. In fact, as described in *Unanswered Questions* at the end of Chapter 24, systematic biologists are shifting toward a more phylogenetic approach to taxonomy and classification, such as the one presented here.

Bear in mind that we include this appendix to introduce the diversity of life and illustrate many of the evolutionary relationships that link monophyletic groups. Like all phylogenetic hypotheses, this classification is open to revision as new information becomes available. Moreover, the classification is incomplete because it includes only those lineages that are described in Unit Four.

Organisms fall into two groups, prokaryotes and eukaryotes, based on the organization of their cells. Prokaryotes consist of the Domains Bacteria and Archaea and are characterized by a central region, the nucleoid, which has no boundary membrane separating it from the cytoplasm, and by membranes typically limited to the plasma membrane. Most prokaryotes are single-celled species, although some are found in simple associations. All other organisms are eukaryotes, which make up the Domain Eukarya. Eukaryotes are characterized by cells with a central, membrane-bound nucleus, and an extensive membrane system. Some eukaryotes are single-celled species, whereas others are multicellular.

Domain Bacteria

The largest and most diverse group of prokaryotes. Includes photoautotrophs, chemoautotrophs, and heterotrophs.

PROTEOBACTERIA Gram-negative bacteria that include chemoautotraphs, chemoheterotrophs, and photoautotrophs. Some are aerobic and some are anaerobic. Five subgroups of proteobacteria are recognized.

SPIROCHETES Helically spiraled bacteria that move by twisting in a corkscrew pattern

CHLAMYDIAS Gram-negative intracellular parasites of animals, with cell walls that lack peptidoglycans

GRAM-POSITIVE BACTERIA Chemoheterotrophic bacteria with thick cell walls

CYANOBACTERIA Photoautotrophic Gram-negative bacteria that carry out photosynthesis using the same chlorophyll as in plants and release oxygen as a by-product

Domain Archaea

Prokaryotes that are evolutionarily between eukaryotic cells and the bacteria. Most are chemoautotrophs. None is photosynthetic. Originally discovered in extreme habitats, they are now known to be widely dispersed. Compared with bacteria, the Archaea have a distinctive cell wall structure and unique membrane lipids, ribosomes, and RNA sequences. Some are symbiotic with animals, but none is known to be pathogenic.

EURYARCHAEOTA Includes methanogens, extreme halophiles, and some extreme thermophiles

NANOARCHAEOTA Extreme thermophile (only one living member known, which lives as a symbiont of a thermophilic member of the Crenarchaeota). Other members represented by sequences detected in environmental samples.

CRENARCHAEOTA Includes most of the archaean extreme thermophiles, as well as psychrophiles; mesophilic species comprise a large part of plankton in cool, marine waters

THAUMOARCHAEOTA All living species (four) are mesophilic chemoautotrophs. Other members are represented by DNA sequences in environmental samples.

KORARCHAEOTA Only living member is a thermophile. Other members are represented by DNA sequences in environmental samples.

Domain Eukarya

PROTISTS A collection of single-celled and multicelled lineages, which are almost certainly not a monophyletic group.

Excavata (excavates) Single-celled animal parasites that have greatly reduced mitochondria, or organelles derived from mitochondria, and move by means of flagella; most have a scooped out (excavated) feeding apparatus on the ventral surface of the cell

  Metamonada (metamonads)—consist of the Diplomonadida and the Parabasala

    Diplomonadida (diplomonads)—cells have two nuclei; move by multiple freely beating flagella

    Parabasala (parasabala)—move by freely beating flagella and an undulating membrane formed by a flagellum buried in a fold of cytoplasm

  Euglenozoa (euglenozoans)—mostly single-celled, highly motile cells that swim using flagella and have disc-shaped inner mitochondrial membranes

    Euglenids—free-living with anterior flagella; most are photosynthetic autotrophs

    Kinetoplastids—nonphotosynthetic heterotrophs that live as animal parasites. Their single mitochondrion contains a large DNA–protein deposit called a kinetoplast.

Chromalveolata (chromalveolates) Heterogeneous group with a range of forms and lifestyles

  Alveolata (alveolates)—characterized by small membrane-bound vesicles called alveoli in a layer under the plasma membrane

    Ciliophora (ciliates)—motile, primarily single-celled, highly complex heterotrophs; swim by means of cilia

Dinoflagellata (dinoflagellates)—nonmotile single-celled marine heterotrophs or autotrophs; shell formed from cellulose plates

Apicomplexa (apicomplexans)—nonmotile parasites of animals with apical complex for attachment and invasion of host cells

Stramenopila (stramenopiles)—characterized by two different flagella, one with hollow tripartite projections, and the other plain

Oomycota (oomycetes)—water molds, white rusts, and mildews; all are funguslike heterotrophs lacking chloroplasts

Bacillariophyta (diatoms)—single-celled autotrophs that carry out photosynthesis by pathways similar to those of plants; covered by a glassy silica shell; most are free-living, whereas some are symbionts

Chrysophyta (golden algae)—colonial; each cell of the colony has a pair of flagella and is covered by a glassy shell consisting of plates or scales; most are autotrophs that carry out photosynthesis by pathways similar to those of plants

Phaeophyta (brown algae)—photosynthetic autotrophs

Rhizaria (rhizarians) Amoebas with stiff, filamentous pseudopodia; some with outer shells

Radiolaria (radiolarians)—marine heterotrophs; glassy internal skeleton with projecting axopods—raylike strands of cytoplasm supported internally by long bundles of microtubules

Foraminifera (forams)—marine heterotrophs; shells, most of which are chambered, spiral structures, consist of organic matter reinforced by calcium carbonate

Chlorarachniophyta (chlorarachniophytes)—green, photosynthetic amoebas

Archaeplastida (archaeplastids) Red algae, green algae, and land plants (*viridaeplantae*), photosynthesizers with a common evolutionary origin

Rhodophyta (red algae)—marine seaweeds, typically multicellular with plantlike bodies; most are reddish in color

Chlorophyta (green algae)—green single-celled, colonial, and multicellular autotrophs that carry out photosynthesis using the same pigments as plants; likely ancestor of land plants

Amoebozoa (amoebozoans) Includes most of the amoebas as well as the slime molds. All use pseudopods for locomotion and for feeding.

Amoebas—single-celled; use lobose (nonstiffened) pseudopods for locomotion and feeding

Cellular slime molds—heterotrophs; primarily individual cells; move by amoeboid motion, or as a multicellular mass

Plasmodial slime molds—heterotrophs; live as plasmodium, a large composite mass with nuclei in a common cytoplasm, that moves and feeds like a giant amoeba

Opisthokonta (opisthokonts) A single posterior flagellum at some stage in the life cycle; consist of the fungi, animals, and two protist groups, the choanoflagellates and the nucleariids.

Choanoflagellata (choanoflagellates)—motile protists with a single flagellum surrounded by collar of closely packed microvilli; likely ancestor of animals and fungi

Nucleariidae (nucleariids)—heterotrophic, mainly spherical amoebae with radiating, fine pseudopods that are not supported by microtubules; more closely related to fungi than to animals

PLANTAE Multicellular autotrophs, mostly terrestrial, and most of which gain energy via photosynthesis; life cycle characterized by alternation of a gametophyte (gamete-producing) generation and sporophyte (spore-producing) generation.

Bryophytes (seedless plants) Nonvascular plants and seedless vascular plants

Hepatophyta (liverworts)—leafy or simple flattened thallus with rhizoids; no true leaves, stems, roots, or stomata (porelike openings for gas exchange); spores in capsules[1]

Anthocerophyta (hornworts)—simple flattened thallus, hornlike sporangia[1]

Bryophyta (mosses)—feathery or cushiony thallus; some with hydroids; spores in capsules[1]

Lycophyta (club mosses)—simple leaves, cuticle, stomata, true roots; most species have sporangia on sporophylls; fertilization by swimming sperm[2]

Monilophyta (ferns, whisk ferns, horsetails)—Ferns: Finely divided leaves; sporangia in sori. *Whisk ferns:* Branching stem from rhizomes; sporangia on stem scales. *Horsetails:* hollow stem, scalelike leaves, sporangia in strobili.[2]

Spermatophyta (seed plants) Vascular plants in which embryos develop within seeds

Gymnosperms—seeds born on stems, on leaves, or under scales

Cycadophyta (cycads)—shrubby or treelike with palmlike leaves; male and female strobili on separate plants

Ginkgophyta (ginkgos)—lineage with a single living species (*Ginkgo biloba);* tree with deciduous, fan-shaped leaves; male and female reproductive structures on separate plants

Gnetophyta (gnetophytes)—shrubs or woody vinelike plants; male and female strobili on separate plants

Coniferophyta (conifers)—predominant extant gymnosperm group; mostly evergreen trees and shrubs with needlelike or scalelike leaves; male and female cones usually on the same plant

Anthophyta (angiosperms/flowering plants)—reproductive structures in flowers

Monocotyledones (monocots)—grasses, palms, lilies, orchids and their relatives; a single cotyledon (seed leaf); pollen grains have one groove

Eudicotyledones (eudicots)—roses, melons, beans, potatoes, most fruit trees, others; two cotyledons; pollen grains have three grooves

Other major angiosperm lineages: magnoliids (Family Magnoliaceae); star anise (Family Illicium); water lilies (Family Nymphaeaceae); *Amborella* (Family Amborellaceae)

FUNGI Heterotrophic, mostly multicellular organisms with cell wall containing chitin and cell nuclei occurring in threadlike hyphae; life cycle typically includes both asexual and sexual phases, with sexual structures used as the basis for phylum-level classification. Single-celled species are known as yeasts.

Chytridiomycota (chytrids)—mostly aquatic; asexual reproduction by way of motile zoospores; sexual reproduction via gametes produced in gametangia; hyphae mostly aseptate

Zygomycota (zygomycetes)—terrestrial; asexual reproduction via nonmotile haploid spores formed in sporangia; sexual spores (zygospores) form in zygosporangia; aseptate hyphae

Glomeromycota (glomeromycetes)—terrestrial; asexual reproduction via spores at the tips of hyphae; form mycorrhizal associations with plant roots

Ascomycota (ascomycetes/sac fungi)—terrestrial and aquatic; sexual spores form in asci; asexual reproduction occurs via conidia (nonmotile spores); septate hyphae

Basidiomycota (basidiomycetes)—terrestrial; reproduction usually via sexual basidiospores produced by basidia; septate hyphae

Basidiomycetes: mushroom-forming fungi and relatives

Teliomycetes: rusts

Ustomycetes: smuts

Cryptomycota—terrestrial and aquatic; single-celled sporelike parasites of animals; include *Rozella* species and microsporidia, and may represent the most basal clade of fungi

Conidial fungi—not a true phylum but a convenience grouping of species for which no sexual phase is known

---

[1]Nonvascular bryophytes)—a polyphyletic group of seedless plants with no specialized structures for transporting water and nutrients; swimming sperm require liquid water for sexual reproduction
[2]Seedless vascular plants—a polyphyletic group of plants in which embryos are not housed inside seeds

ANIMALIA Multicellular heterotrophs; nearly all with tissues, organs, and organ systems; motile during at least part of the life cycle; sexual reproduction in most; embryos develop through a series of stages; many with larval and adult stages in life cycle

Parazoa Animals lacking tissues and body symmetry

Porifera (sponges)—multicellular; extract oxygen and particulate food from water drawn into a central cavity

Eumetazoa Animals with tissues and either radial or bilateral symmetry

Radiata—acoelomate animals with radial symmetry and two tissue layers

Cnidaria (cnidarians)—two tissue layers; single opening into gastrovascular cavity; nerve net; nematocysts for defense and predation; some sessile, some motile; most are predatory, some with photosynthetic endosymbionts; freshwater and marine

Hydrozoa: hydrozoans

Scyphozoa: jellyfishes

Cubozoa: box jellyfishes

Anthozoa: sea anemones, corals

Ctenophora (comb jellies)—two (possibly three) tissue layers; feeding tentacles capture particulate food; beating cilia provide weak locomotion; marine

Bilateria—animals with bilateral symmetry and three tissue layers

Protostomia—acoelomate, pseudocoelomate, or schizocoelomate; many with spiral, indeterminate cleavage; blastopore forms mouth; nervous system on ventral side

Lophotrochozoa—many with either a lophophore for feeding and gas exchange or a trochophore larva

Ectoprocta (bryozoans)—coelomate; colonial; secrete hard covering over soft tissues; lophophore; sessile; particulate feeders; marine

Brachiopoda (lamp shells)—coelomate; dorsal and ventral shells; lophophore; sessile; particulate feeders; marine

Phoronida (phoronid worms)—coelomate; secrete tubes around soft tissues; lophophore; sessile; particulate feeders; marine

Platyhelminthes (flatworms)—acoelomate; dorsoventrally flattened; complex reproductive, excretory, and nervous systems; gastrovascular cavity in many; free-living or parasitic, often with multiple hosts; terrestrial, freshwater, and marine

Turbellaria: free-living flatworms

Trematoda: flukes

Monogenoidea: flukes

Cestoda: tapeworms

Rotifera (wheel animals)—pseudocoelomate; microscopic; complete digestive system; well-developed reproductive, excretory, and nervous systems; particulate feeders; major components of marine and freshwater plankton

Nemertea (ribbon worms)—schizocoelomate; proboscis housed within rhynchocoel; complete digestive tract; circulatory system; predatory; mostly marine

Mollusca (mollusks)—schizocoelomate; many with trochophore larva; many with shell secreted by mantle; body divided into head-foot, visceral mass, and mantle; well-developed organ systems; variable locomotion; herbivorous or predatory; terrestrial, freshwater, and marine

Polyplacophora: chitons

Gastropoda: snails, sea slugs, land slugs

Bivalvia: clams, mussels, scallops, oysters

Cephalopoda: squids, octopuses, cuttlefish, nautiluses

Annelida (segmented worms)—schizocoelomate; many with trochophore larva; segmented body and organ systems; well-developed organ systems; many use hydrostatic skeleton for locomotion; some predatory, some particulate feeders, some detritivores; terrestrial, freshwater, and marine

"Polychaeta": marine worms

Clitellata: freshwater and terrestrial worms, leeches

Ecdysozoa—cuticle or exoskeleton is shed periodically

Nematoda (roundworms)—pseudocoelomate; body covered with tough cuticle that is shed periodically; well-developed organ systems; thrashing locomotion; many are parasitic on plants or animals; mostly terrestrial

Onychophora (velvet worms)—schizocoelomate; segmented body covered with cuticle; locomotion by many unjointed legs; complex organ systems; predatory; terrestrial

Arthropoda (arthropods)—schizocoelomate; jointed exoskeleton made of chitin; segmented body, some with fusion of segments in head, thorax, or abdomen; complex organ systems; variable modes of locomotion, including flight; specialization of numerous appendages; herbivorous, predatory, or parasitic; terrestrial, freshwater, and marine

Trilobita: trilobites (extinct)

Chelicerata: horseshoe crabs, spiders, scorpions, ticks, mites

Mandibulata

Myriapoda: centipedes, millipedes

Pancrustacea: shrimps, crayfishes, lobsters, crabs, barnacles, copepods, isopods, springtails, insects

Deuterostomia—enterocoelomate; many with radial, determinate cleavage; blastopore forms anus; nervous system on dorsal side in many

Echinodermata (echinoderms)—secondary radial symmetry, often organized around five radii; hard internal skeleton; unique water vascular system with tube feet; complete digestive system; simple nervous system; no circulatory or respiratory system; generally slow locomotion using tube feet; predatory, herbivorous, particulate feeders, detritivores; exclusively marine

Asteroidea: sea stars

Ophiuroidea: brittle stars

Echinoidea: sea urchins, sand dollars

Holothuroidea: sea cucumbers

Crinoidea: feather stars, sea lilies

Concentricycloidea: sea daisies

Hemichordata (acorn worms)—pharynx perforated with branchial slits; proboscis; complex organ systems; tube-dwelling in soft sediments; particulate or deposit feeders; exclusively marine

Chordata (chordates)—notochord; segmental body wall and tail muscles; dorsal hollow nerve chord; perforated pharynx; complex organ systems; variable modes of locomotion; extremely varied diets; terrestrial, freshwater, and marine

Cephalochordata: lancelets

Urochordata: tunicates, sea squirts

Vertebrata: vertebrates

Myxinoidea: hagfishes

Petromyzontoidea: lampreys

Placodermi: placoderms (extinct)

Chondrichthyes: sharks, skates, and rays

Acanthodii: acanthodians

Osteichthyes: bony fishes

Amphibia: salamanders, frogs, caecilians

Synapsida: mammals

Reptilia: turtles, sphenodontids, lizards, snakes, crocodilians, birds

Glossary

3′ end The end of a polynucleotide chain at which a hydroxyl group is bonded to the 3′ carbon of a deoxyribose sugar.

3′ untranslated region (3′ UTR) The part of an mRNA between the stop codon and the 3′ end of the molecule; this region does not code for amino acids.

5′ cap In eukaryotes, a guanine-containing nucleotide attached in a reverse orientation to the 5′ end of pre-mRNA and retained in the mRNA produced from it. The 5′ cap on an mRNA is the site where ribosomes attach to initiate translation.

5′ end The end of a polynucleotide chain at which a phosphate group is bound to the 5′ carbon of a deoxyribose sugar.

5′ untranslated region (5′ UTR) The part of an mRNA between the 5′ end of the molecule and the start codon; this region does not code for amino acids.

10-nm chromatin fiber The most fundamental level of chromatin packing of a eukaryotic chromosome in which DNA winds for almost two turns around an eight-protein nucleosome core particle to form a nucleosome and linker DNA extends between adjacent nucleosomes. The result is a beads-on-a-string type of structure with a 10-nm diameter.

30-nm chromatin fiber Level of chromatin packing of a eukaryotic chromosome in which histone H1 binds to the 10-nm chromatin fiber causing it to package into a coiled structure about 30 nm in diameter and with about six nucleosomes per turn. Also referred to as a *solenoid*.

A site The site where the incoming aminoacyl–tRNA carrying the next amino acid to be added to the polypeptide chain binds to the mRNA.

abdomen The region of the body that contains much of the digestive tract and sometimes part of the reproductive system; in insects, the region behind the thorax.

abiotic Nonbiological, often in reference to physical factors in the environment.

abscisic acid (ABA) A plant hormone involved in the abscission of leaves, flowers, and fruits, dormancy of buds and seeds, and closing of stomata.

abscission In plants, the dropping of flowers, fruits, and leaves in response to environmental signals.

absorption Within the digestive tube, the process in which molecular subunits are absorbed from the digestive contents into body fluids and cells.

absorption spectrum Curve representing the amount of light absorbed at each wavelength.

abyssal zone The bottom sediments that lie permanently below deep ocean water.

accessory fruits A fruit in which floral parts in addition to each ovary are incorporated as the fruit develops.

acetylcholine A neurotransmitter that is released by axon terminals at chemical synapses. It has mostly excitatory effects.

acetyl–CoA (acetyl–coenzyme A) An acetyl group bonded to coenzyme A. In cellular respiration, acetyl–CoA generated by pyruvate oxidation carries acetyl groups to the citric acid cycle.

acclimation When a set point for a homeostatically controlled factor changes artificially in a laboratory setting.

acclimatization When a set point for a homeostatically controlled factor changes naturally because of an alteration in environmental conditions.

accommodation A process by which the lens changes to enable the eye to focus on objects at different distances.

acid Proton donor that releases H^+ (and anions) when dissolved in water.

acid precipitation Rainfall with low pH, primarily created when gaseous sulfur dioxide (SO_2) dissolves in water vapor in the atmosphere, forming sulfuric acid.

acid-growth hypothesis A hypothesis to explain how the hormone auxin promotes growth of plant cells; it suggests that auxin stimulates H^+ pumps in the plasma membrane to move H^+ from the cell interior into the cell wall, which increases wall acidity, making the wall expandable.

acidity The concentration of H^+ in a water solution, as compared with the concentration of OH^-.

acoelomate A body plan of bilaterally symmetrical animals that lack a body cavity between the gut and the body wall.

acoustical signal A means of animal communication in which a signaler produces a sound that is heard by a signal receiver.

acquired immunity *See* **adaptive (acquired) immunity.**

acquired immunodeficiency syndrome (AIDS) A constellation of disorders that follows infection by the human immunodeficiency virus (HIV).

acrosome A specialized secretory vesicle on the head of an animal sperm, which helps the sperm penetrate the egg.

acrosomal reaction The process in which enzymes contained in the acrosome are released from an animal sperm and digest a path through the egg coats.

actin A protein that, in an interaction with the protein myosin, causes muscle contraction.

Actinopterygii A monophyletic lineage of bony fishes that includes ray-finned fishes.

action potential The rapid and transient change in membrane potential that occurs when a neuron conducts an electrical impulse.

action spectrum Graph produced by plotting the effectiveness of light at each wavelength in driving photosynthesis.

activation energy The initial input of energy required to start a reaction.

activator A regulatory protein that controls the expression of one or more genes.

active immunity The production of antibodies in the body in response to exposure to a foreign antigen.

active parental care Parents' investment of time and energy in caring for offspring after they are born or hatched.

active site The region of an enzyme to which substrate(s) bind and where catalysis occurs.

active transport The mechanism by which ions and molecules move against the concentration gradient across a membrane, from the side with the lower concentration to the side with the higher concentration.

adaptation Characteristic that helps an organism survive longer or reproduce more under a particular set of environmental conditions.

adaptation, evolutionary The accumulation of adaptive traits over time.

adaptation, sensory *See* **sensory adaptation.**

adaptive (acquired) immunity A vertebrate-specific defense mechanism that recognizes specific molecules (free, or on the surface of pathogens or foreign cells) as foreign and clears those molecules from the body.

adaptive immune system The inherited mechanisms leading to the synthesis of molecules that target pathogens in a specific way. The adaptive immune system is one of two components of the immune system in vertebrates. It is not found in invertebrates.

adaptive radiation A cluster of closely related species that are each adaptively specialized to a specific habitat or food source.

adaptive trait A genetically based characteristic, preserved by natural selection, that increases an organism's likelihood of survival or its reproductive output.

adaptive zone An environment or part of an environment that may be occupied by a group of species exploiting resources in a similar manner.

adductor muscle A muscle that pulls inward toward the median line of the body; in bivalve mollusks, it pulls the shell closed.

adenine A purine that base-pairs with either thymine in DNA or uracil in RNA.

adherens junction Animal cell junction in which intermediate filaments are the anchoring cytoskeletal component.

adhesion The adherence of molecules to the walls of conducting tubes, as in plants.

adiabatic cooling A decrease in temperature without the actual loss of heat energy, occurring in air masses that expand as they rise in the atmosphere.

adipocytes Densely clustered cells within adipose tissue that are specialized for fat storage.

adipose tissue Connective tissue containing large, densely clustered cells called *adipocytes* that are specialized for fat storage.

adrenal cortex The outer region of the adrenal glands, which contains endocrine cells that secrete two major types of steroid hormones, the glucocorticoids and the mineralocorticoids.

adrenal medulla The central region of the adrenal glands, which contains neurosecretory neurons that secrete the catecholamine hormones epinephrine and norepinephrine.

adrenergic receptors G-protein–coupled receptors to which hormones originating in the adrenal gland can bind.

adrenocorticotropic hormone (ACTH) A hormone that triggers hormone secretion by cells in the adrenal cortex.

adventitious root A root that develops from the stem or leaves of a plant.

aerobe An organism that requires oxygen for cellular respiration.

aerobic glycolysis The process in most cancer cells in which energy production is dependent on glycolysis. Glycolysis rates are higher than in normal cells, and substrate-level phosphorylation reactions of glycolysis are used to produce ATP regardless of oxygen present.

aerobic respiration The form of cellular respiration found in eukaryotes and many prokaryotes in which oxygen is a reactant in the ATP-producing process.

afferent arteriole The vessel that delivers blood to the glomerulus of the kidney.

afferent division One of two divisions of the peripheral nervous system (PNS). The afferent division of the PNS carries signals to the central nervous system (CNS) and includes all the neurons that transmit sensory information from their receptors.

afferent neuron A neuron that transmits stimuli collected by a sensory receptor to an interneuron.

afterbirth The placenta and any remnants of the umbilical cord and embryonic membranes expelled in the third stage of labor which follows the birth of the baby.

agar A gelatinous product extracted from certain red algae or seaweed used as a culture medium in the laboratory and as a gelling or stabilizing agent in foods.

agarose gel electrophoresis Technique by which DNA, RNA, or protein molecules are separated in an agarose gel subjected to an electric field.

age structure A statistical description or graph of the relative numbers of individuals in each age class in a population.

age-specific fecundity The average number of offspring produced by surviving females of a particular age.

age-specific mortality The proportion of individuals alive at the start of an age interval that died during that age interval.

age-specific survivorship The proportion of individuals alive at the start of an age interval that survived until the start of the next age interval.

agglutination One of two important mechanisms to clear foreign antigens from the body, the immobilization of pathogens by antibodies.

aggregate fruit A fruit that develops from multiple separate carpels of a single flower, such as a raspberry or strawberry.

agonist A muscle that causes movement in a joint when it contracts.

albumin The most abundant protein in blood plasma, important for osmotic balance and pH buffering; also, the portion of an egg that serves as the main source of nutrients and water for the embryo.

alcohol A molecule of the form R—OH in which R is a chain of one or more carbon atoms, each of which is linked to hydrogen atoms.

alcoholic fermentation Reaction in which pyruvate is converted into ethyl alcohol and CO_2 in a two-step series that also converts NADH into NAD^+.

aldehyde Molecule in which the carbonyl group is linked to a carbon atom at the end of a carbon chain, along with a hydrogen atom.

aldosterone A mineralocorticoid hormone released from the adrenal cortex that increases the amount of Na^+ reabsorbed from the urine in the kidneys and absorbed from foods in the intestine, reduces the amount of Na^+ secreted by salivary and sweat glands, and increases the rate of K^+ excretion by the kidneys, keeping Na^+ and K^+ balanced at the levels required for normal cellular function.

aleurone The thin layer of cells that separates the endosperm of a seed from the pericarp.

algin Alginic acid, found in the cell walls of brown algae.

alkaloids A class of secondary metabolites produced by plant cells and that serve as defenses against herbivores. Examples include caffeine and strychnine

allantois In an amniote egg, an extraembryonic membrane sac that fills much of the space between the chorion and the yolk sac and store's the embryo's nitrogenous wastes.

allele One of two or more versions of a gene.

allele frequency The abundance of one allele relative to others at the same gene locus in individuals of a population.

allergen A type of antigen responsible for allergic reactions, which induces B cells to secrete an overabundance of IgE antibodies.

allolactose An isomer of lactose that is the inducer for the *E. coli lac* operon, meaning that it induces the transcription of the operon's structural genes. Allolactose works by inactivating the Lac repressor.

allometric growth A pattern of postembryonic development in which parts of the same organism grow at different rates.

allopatric speciation The evolution of reproductive isolating mechanisms between two populations that are geographically separated.

allopolyploidy The genetic condition of having two or more complete sets of chromosomes from different parent species.

all-or-nothing principle The principle that an action potential is produced only if the stimulus is strong enough to cause depolarization to reach the threshold.

allosteric activator Molecule that converts an enzyme with an allosteric site, a regulatory site outside the active site, from the inactive form to the active form.

allosteric inhibitor Molecule that converts an enzyme with an allosteric site, a regulatory site outside the active site, from the active form to the inactive form.

allosteric regulation Specialized control mechanism for enzymes with an allosteric site, a regulatory site outside the active site, that may either slow or accelerate activity depending on the enzyme.

allosteric site A regulatory site outside the active site.

alpha (α) helix A type of secondary structure of a polypeptide in which the amino acid chain is twisted into a regular, right-hand spiral.

α-melanocyte-stimulating hormone (alpha-MSH) A peptide hormone released from the hypothalamus that inhibits appetite.

alphaproteobacteria One of five subgroups of the proteobacteria group of bacteria. Alphaproteobacteria species show considerable metabolic diversity, and they include many species that form symbiotic relationships with plants as animals, as well as some pathogens.

alpine tundra A biome that occurs on high mountaintops throughout the world, in which dominant plants form cushions and mats.

alternation of generations The regular alternation of mode of reproduction in the life cycle of an organism, such as the alternation between diploid (sporophyte) and haploid (gametophyte) phases in plants.

alternative hypothesis An explanation of an observed phenomenon that is different from the explanation being tested.

alternative splicing Mechanism by which a pre-mRNA in a eukaryotic cell is processed by reactions that join exons in different combinations to produce different mRNAs from a single gene.

altruism A behavioral phenomenon in which individuals appear to sacrifice their own reproductive success to help other individuals.

Alveolata (alveolates) A subgroup of the Chromalveolata protist evolutionary group; characterized by small, flattened, membrane-bound vesicles called alveoli in a layer just under the plasma membrane.

alveolus (plural, *alveoli*) One of the millions of tiny air pockets in mammalian lungs, each surrounded by dense capillary networks.

amacrine cell A type of neuron that forms lateral connections in the retina of the eye, connecting bipolar cells and ganglion cells.

amine hormone A hormone that is an amine. Most amine hormones are based on tyrosine and, with one major exception, they are hydrophilic molecules. Amine hormones are involved in classical endocrine and neuroendocrine signaling.

amino acid A molecule that contains both an amino and a carboxyl group.

amino group Group that acts as an organic base, consisting of a nitrogen atom bonded on one side to two hydrogen atoms and on the other side to a carbon chain.

aminoacylation The process of adding an amino acid to a tRNA. Also referred to as *charging*.

aminoacyl–tRNA A tRNA linked to its "correct" amino acid, which is the finished product of aminoacylation.

aminoacyl–tRNA synthetase An enzyme that catalyzes aminoacylation.

aminopeptidase An enzyme that hydrolyzes small peptides to produce individual amino acids.

ammonification A metabolic process in which bacteria convert organic nitrogen compounds into ammonia and ammonium ions; part of the nitrogen cycle.

amniocentesis Technique of prenatal diagnosis in which cells are obtained from the amniotic fluid for DNA testing, for biochemical analysis, or to test for the presence of chromosomal mutations.

amnion In an amniote egg, an extraembryonic membrane that encloses the embryo, forming the amniotic cavity and secreting amniotic fluid, which provides an aquatic environment in which the embryo develops.

Amniota The monophyletic group of vertebrates that have an amnion during embryonic development.

amniote egg A shelled egg that can survive and develop on land.

amoeba A descriptive term for a single-celled protist that moves by means of temporary cellular projections called *pseudopods*.

Amoebozoa (amoebozoans) A protist evolutionary group that includes most of the amoebas as well as the cellular and plasmodial slime molds; characterized by the use of pseudopods for locomotion and feeding for all or part of their life cycles.

amphibians A monophyletic group of tetrapods that generally have moist skin and unshelled eggs.

amplification An increase in the magnitude of each step as a signal transduction pathway proceeds.

amygdala A gray-matter center of the brain that works as a switchboard, routing information about experiences that have an emotional component through the limbic system.

amylase An enzyme that catalyzes the hydrolysis of starches.

amyloplast Colorless plastid that stores starch in plants.

anabolic-androgenic steroid (AAS) A synthetic derivative of androgens that mimics their effects. Also called an *anabolic steroid*.

anabolic pathway A metabolic pathway in which energy is used to build complicated molecules from simpler ones; also called a *biosynthetic pathway*. An individual reaction in an anabolic pathway is an anabolic reaction, also called a *biosynthetic reaction*.

anabolic reaction Metabolic reaction that requires energy to assemble simple substances into more complex molecules.

anabolic steroid *See* **anabolic-androgenic steroid (AAS).**

anaerobe An organism that does not require oxygen to live.

anaerobic respiration The production of ATP in a number of types of prokaryotes using an electron transfer system in which the final electron acceptor is an inorganic molecule rather than oxygen.

anagenesis The slow accumulation of evolutionary changes in a lineage over time.

anal sphincter A muscular ring that controls the opening and closing of the anus.

anaphase The phase of mitosis during which the spindle separates sister chromatids and pulls them to opposite spindle poles.

anaphase-promoting complex (APC) An enzyme complex activated by M phase–promoting factor that controls the separation of sister chromatids and the onset of daughter chromosome separation in anaphase of mitosis.

anaphylactic shock A severe inflammation stimulated by an allergen, involving extreme swelling of air passages in the lungs that interferes with breathing, and massive leakage of fluid from capillaries that causes blood pressure to drop precipitously.

anatomy The study of the structures of organisms.

ancestral character state A trait that was present in a distant common ancestor.

anchoring junction Cell junction that forms belts that run entirely around cells, "welding" adjacent cells together.

androgen One of a family of hormones that promote the development and maintenance of sex characteristics.

aneuploid An individual with extra or missing chromosomes.

angiosperm A flowering plant. Its egg-containing ovules mature into seeds within protected chambers called *ovaries*.

angiotensin A peptide hormone that raises blood pressure quickly by constricting arterioles in most parts of the body; it also stimulates release of the steroid hormone aldosterone.

angiotensin I In the renin–angiotensin-aldosterone system (RAAS), the molecule produced by cleavage of the plasma protein angiotensinogen.

angiotensin II In the renin–angiotensin-aldosterone system (RAAS), the molecule converted from angiotensin I by angiotensin-converting enzyme (ACE); angiotensin II is a hormone that constricts arterioles to raise blood pressure, stimulates synthesis of aldosterone and its secretion from the adrenal cortex, and stimulates thirst.

angiotensin-converting enzyme (ACE) In the renin–angiotensin–aldosterone system (RAAS), the enzyme that converts angiotensin I to angiotensin II.

animal behavior The responses of animals to specific internal and external stimuli.

animal pole The end of the egg where the egg nucleus is located, which typically gives rise to surface structures and the anterior end of the embryo.

Animalia The taxonomic kingdom that includes all living and extinct animals.

anion A negatively charged ion.

annual An herbaceous plant that completes its life cycle in one growing season and then dies.

annulus In ferns, a ring of thick-walled cells that nearly encircles the sporangium and functions in spore release.

antagonistic pair Two skeletal muscles, one of which flexes as the other extends to move joints.

antenna A chemosensory appendage attached to the head of some adult arthropods.

antenna complex (light-harvesting complex) In photosystems, the sites at which light is absorbed and converted into chemical energy during photosynthesis, an aggregate of many chlorophyll pigments and a number of carotenoid pigments that serves as the primary site of absorbing light energy in the form of photons.

anterior Indicating the head end of an animal.

anterior pituitary The glandular part of the pituitary, composed of endocrine cells that synthesize and secrete several tropic and nontropic hormones.

anther The pollen-bearing part of a stamen.

antheridium (plural, *antheridia*) In plants, a structure in which sperm are produced.

Anthocerophyta The phylum comprising hornworts.

Anthophyta The phylum comprising flowering plants.

Anthropoidea The monophyletic lineage of primates comprising monkeys, apes, and humans.

antibody A highly specific soluble protein molecule that circulates in the blood and lymph, recognizing and binding to antigens and clearing them from the body.

antibody class The molecular type of an antibody determined by the constant regions of the heavy chains of the molecule.

antibody-mediated immunity Adaptive immune response in which plasma cells secrete antibodies. Also called *humoral immunity*.

anticodon The three-nucleotide segment in a tRNA that pairs with a codon in an mRNA.

antidiuretic hormone (ADH) A hormone secreted by the posterior pituitary that increases water absorption in the kidneys, thereby increasing the volume of the blood.

antigen A foreign molecule that triggers an adaptive immunity response.

antigen-binding site The region at one end of an antibody molecule, T-cell receptor (TCR), or B-cell receptor (BCR) that binds to a specific antigen.

antigenic determinants *See* **epitope.**

antigen-presenting cell (APC) A cell that presents an antigen to T cells in antibody-mediated immunity and cell-mediated immunity.

antihistamines Substances that block histamine receptors thereby reducing symptoms of an allergic reaction.

antiparallel Strands of double-stranded DNA that run in opposite directions with the 3′ end of one strand opposite the 5′ end of the other strand.

antiport A secondary active transport mechanism in which a molecule moves through a membrane channel into a cell and powers the active transport of a second molecule out of the cell. Antiport is a type of cotransport.

aorta A large artery from the heart that branches into arteries leading to all body regions except the lungs.

aortic body One of several small clusters of chemoreceptors, baroreceptors, and supporting cells located along the aortic arch that measures changes in blood pressure and the composition of arterial blood flowing past it.

aphotic zone Deeper water of a lake or ocean where sunlight does not penetrate.

apical dominance Inhibition of the growth of lateral buds in plants due to auxin diffusing down a shoot tip from the terminal bud.

apical meristem A region of unspecialized dividing cells at shoot tips and root tips of a plant.

apical surface The outer surface of epithelial cells.

Apicomplexa (apicomplexans) A lineage within the Alveolata subgroup of the Chromalveolata protist evolutionary group; nonmotile parasites of animals characterized by the apical complex, a special grouping of fibrils, microtubules, and organelles at one end of the cell that functions in attachment and invasion of host cells.

apicoplast A remnant plastid found in most apicomplexan parasites. Apicomplexans are protists in the Chromalveolate group.

apomixis In plants, the production of offspring without meiosis or formation of gametes.

apomorphy A derived character state.

apoplast The compartment made up of plant cell walls and spaces between cells.

apoplastic pathway The route followed by water moving through plant cell walls and intercellular spaces (the apoplast). *Compare* **symplastic pathway.**

apoptosis A type of programmed cell death.

aposematic coloration Bright, contrasting patterns that advertise the unpalatability of poisonous or repellant species.

appendicular skeleton The bones comprising the pectoral (shoulder) and pelvic (hip) girdles and limbs of a vertebrate.

appendix A fingerlike sac that extends from the cecum of the large intestine.

applied research Research conducted with the goal of solving specific practical problems.

aquaporin A specialized protein channel that facilitates diffusion of water through cell membranes.

aquatic succession A process in which debris from rivers and runoff accumulates in a body of fresh water, causing it to fill in at the margins.

aqueous humor A clear fluid that fills the space between the cornea and lens of the eye.

arbuscules The branched hyphae of endomycorrhizae.

Archaea One of two domains of prokaryotes; archaeans have some unique molecular and biochemical traits, but they also share some traits with Bacteria and other traits with Eukarya.

archaeal chromosome DNA molecule in archaeans in which hereditary information is encoded.

Archaeplastida (archaeplastids) An evolutionary group consisting of the Rhodophyta and Chlorophyta protists, and the land plants; all are photosynthesizers.

archegonium The flask-shaped structure in which bryophyte eggs form.

Archelosauria A monophyletic group of Reptilia that includes turtles, crocodilians, nonavian dinosaurs, and birds.

archenteron The central endoderm-lined cavity of an embryo at the gastrula stage, which forms the primitive gut.

arctic tundra A treeless biome that stretches from the boreal forests to the polar ice cap in Europe, Asia, and North America.

arteriole A branch from a small at the point where it reaches the organ it supplies.

artery A vessel that conducts blood away from the heart at relatively high pressure.

arthropods A monophyletic lineage of ecdysozoans, including myriapods, chelicerates, drustaceans, and insects.

artificial selection Selective breeding of organisms to ensure that certain desirable traits appear at higher frequency in successive generations.

ascending reticular formation One of two parts of the reticular formation; it contains neurons that convey stimulatory signals via the thalamus to arouse and activate the cerebral cortex. It is responsible for the sleep–wake cycle. Also called the *reticular activating system.*

ascocarp In ascomycete (sac) fungi, a reproductive body that bears or contains asci.

ascus (plural, *asci*) A saclike cell in ascomycetes (sac fungi) in which meiosis gives rise to haploid sexual spores (meiospores).

asexual reproduction Any mode of reproduction in which a single individual gives rise to offspring without fusion of gametes; that is, without genetic input from another individual. *See also* **vegetative reproduction.**

association area One of several areas surrounding the sensory and motor areas of the cerebral cortex that integrate information from the sensory areas, formulate responses, and pass them on to the primary motor area.

aster Radiating array produced as microtubules extending from the centrosomes of cells grow in length and extent.

asthma In an allergic reaction, a severe response to allergens that involves constriction of airways in the lungs.

astrocyte A star-shaped glial cell that provides support to neurons in the vertebrate central nervous system.

asymmetrical Characterized by a lack of proportion in the spatial arrangement or placement of parts.

atmosphere The component of the biosphere that includes the gases and airborne particles enveloping the planet.

atom The smallest unit that retains the chemical and physical properties of an element.

atomic nucleus The nucleus of an atom, which consists of subatomic particles called protons and neutrons.

atomic number The number of protons in the nucleus of each kind of atom.

atomic weight The weight of an element in grams, equal to the mass number.

ATP (adenosine triphosphate) The primary agent that couples exergonic and endergonic reactions.

ATP/ADP cycle The continual hydrolysis and resynthesis of ATP in living cells.

ATP synthase A membrane-spanning protein complex that couples the energetically favorable transport of protons across a membrane to the synthesis of ATP.

atrial natriuretic factor (ANF) A peptide hormone that inhibits renin release and increases the filtration rate by dilating the arterioles that deliver blood to glomeruli and by inhibiting aldosterone release.

atrial siphon A tube through which invertebrate chordates expel digestive and metabolic wastes.

atriopore The hole in the body wall of a cephalochordate through which water is expelled from the body.

atrioventricular node (AV node) A region of the heart wall that receives signals from the sinoatrial node and conducts them to the ventricle.

atrioventricular valve (AV valve) The valve between each atrium and ventricle in the heart.

atrium (plural, *atria*) A body cavity or chamber surrounding the perforated pharynx of invertebrate chordates; also, one of the chambers that receive blood returning to the heart.

austrobaileyales The star anise group of flowering plants.

autocrine regulation A type of cell signaling in the endocrine system in which a local regulator acts on the same cells that produced it.

autoimmune disease A malfunction of the immune system in which the body reacts against its own proteins or cells.

autoimmune reaction The production of antibodies against molecules of the body.

autonomic nervous system A subdivision of the peripheral nervous system that consists of nerves to smooth muscle, cardiac muscle, exocrine glands, and some endocrine glands. It controls largely involuntary processes including digestion, secretion by sweat glands, circulation of the blood, many functions of the reproductive and excretory systems, and contraction of smooth muscles in all parts of the body.

autopolyploidy The genetic condition of having more than two sets of chromosomes from the same parent species.

autosomal dominant inheritance Pattern in which the allele that causes a trait is dominant, and only homozygous recessives are unaffected.

autosomal recessive inheritance Pattern in which individuals with a trait are homozygous for a recessive allele.

autosome Chromosome other than a sex chromosome.

autotroph An organism that produces its own food using CO_2 and other simple inorganic compounds from its environment and energy from the Sun or from oxidation of inorganic substances.

autumn overturn A process in which winds mix the water in a lake vertically, equalizing the concentrations of dissolved gases and nutrients at all depths.

auxin Any of a family of plant hormones that stimulate growth by promoting cell elongation in stems and coleoptiles, inhibit abscission, govern responses to light and gravity, and have other developmental effects.

auxotroph A mutant strain that requires for its growth a nutrient supplement that is not needed by the wild-type strain.

Avogadro's number The number 6.022×10^{23}, derived by dividing the atomic weight of any element by the weight of an atom of that element.

***Avr* gene** A gene in certain plant pathogens that encodes a product triggering a defensive response in the plant.

axial skeleton The bones comprising the head and trunk of a vertebrate: the cranium, vertebral column, ribs, and sternum (breastbone).

axil The upper angle between the stem and an attached leaf.

axon The single elongated extension of a neuron that conducts signals away from the cell body to another neuron or an effector.

axon hillock A junction with the cell body of a neuron from which the axon arises.

axon terminal A branch at the tip of an axon that ends as a small, buttonlike swelling.

B cell A lymphocyte that recognizes antigens in the body.

Bacillariophyta (diatoms) A lineage of the Stramenopila subgroup of the Chromalveolata protist evolutionary group; single-celled organisms covered by an intricate, glassy silica shell.

bacillus (plural, *bacilli*) A cylindrical or rod-shaped prokaryote.

background extinction rate The average rate of extinction of taxa through time.

Bacteria One of the two domains of prokaryotes; collectively, bacteria are the most metabolically diverse organisms.

bacterial chromosome DNA molecule in bacteria in which hereditary information is encoded.

bacterial flagellum *See* **flagellum.**

bacteriophage A virus that infects bacteria. Also referred to as a *phage.*

bacteroid A rod-shaped or branched bacterium in the root nodules of nitrogen-fixing plants.

balanced polymorphism The maintenance of two or more phenotypes in fairly stable proportions over many generations.

ball-and-socket joint Type of joint in the body that can rotate about its axis.

bark The tough outer covering of woody stems and roots, composed of all the living and nonliving tissues between the vascular cambium and the stem surface.

Barr body The inactive, condensed X chromosome seen in the nucleus of female placental mammals.

basal angiosperm Any of the earliest branches of the flowering plant lineage; includes the star anise group and water lilies.

basal body Structure that anchors cilia and flagella to the surface of a cell.

basal lamina The membrane that fixes the epithelium to underlying tissues (also called the *basement membrane*).

basal nucleus One of several gray-matter centers that surround the thalamus on both sides of the brain and moderate voluntary movements directed by motor centers in the cerebrum.

basal surface The inner surface of epithelial cells.

base-excision repair An excision repair mechanism that repairs nonbulky damage to DNA bases by removing the erroneous base and replacing it with the correct one complementary to the base on the other DNA strand.

base Proton acceptor that reduces the H^+ concentration of a solution.

base-pair mismatch An error in the assembly of a new nucleotide chain in which bases other than the correct ones pair together.

base-pair mutation A type of mutation that involves a change of a single base pair in the genetic material.

basic research Research conducted to search for explanations about natural phenomena in order to satisfy curiosity and to advance collective knowledge of living systems.

basidiocarp A fruiting body of a basidiomycete; mushrooms are examples.

basidiomycetes Club fungi, all of which are in the phylum Basidiomycota. Examples include the mushroom-forming species, shelf fungi, smuts, rusts, and puffballs.

basidiospore A haploid sexual spore produced by basidiomycete fungi.

basidium (plural, *basidia*) A small, club-shaped structure in which sexual spores of basidiomycetes arise.

basilar membrane A part of the floor of the cochlear duct, which anchors sensory hair cells in the organ of Conti and which vibrates in response to vibrations moving through the inner ear.

basophil A type of leukocyte located in blood that responds to IgE antibodies in an allergic response by secreting histamine, which stimulates inflammation.

Batesian mimicry The form of defense in which a palatable or harmless species resembles an unpalatable or poisonous one.

B-cell receptor (BCR) The receptor on B cells that is specific for a particular antigen.

behavioral ecology The set of actions that an individual or population can perform to interact with stimuli in its environment.

behavioral isolation A prezygotic reproductive isolating mechanism in which two species do not mate because of differences in courtship behavior; also known as ethological isolation.

behavioral repertoire The set of actions that an animal can perform in response to stimuli in its environment.

benign tumor A type of tumor that is not invasive, meaning that it does not spread to other tissues.

benthic province The bottom sediments in the ocean.

benthos Species living in and on the bottom sediments of the ocean.

betaproteobacteria One of five subgroups of the proteobacteria group of bacteria. Betaproteobacteria are metabolically diverse and include a number of important pathogens.

beta (β) sheet A type of primary structure in a polypeptide in which the amino acid chain zigzags in a flat plane to form a beta strand, and beta strands then align side by side in the same or opposite direction.

biennial A plant that completes its life cycle in two growing seasons and then dies; limited secondary growth occurs in some biennials.

Bikonta (bikonts) Eukaryotes with two flagella.

bilateral symmetry The body plan of animals in which the body can be divided into mirror image right and left halves by a plane passing through the midline of the body.

Bilateria A branch of the Eumetazoa, comprising animals with bilateral body symmetry.

bilayer A membrane with two molecular layers.

bile A mixture of substances including bile salts, cholesterol, and bilirubin that is made in the liver, stored in the gallbladder, and used in the digestion of fats.

bile salts A component of bile, consisting of derivatives of cholesterol and amino acids, that aids fat digestion.

bilirubin A component of bile consisting of a waste product derived from worn-out red blood cells.

binary fission The process of splitting or dividing a prokaryotic cell into two parts.

binomial Relating to or consisting of two names or terms.

binomial nomenclature The naming of species with a two-part scientific name, the first indicating the genus and the second indicating the species.

biodiversity The richness of living systems as reflected in genetic variability within and among species, the number of species living on Earth, and the variety of communities and ecosystems.

biodiversity hotspot An area where biodiversity is both highly concentrated and endangered.

biofilm A prokaryotic multicellular association consisting of a complex aggregation of microorganisms (either one or multiple species) attached, in most cases, to a surface.

biofuel A fuel produced from a living organism.

biogeochemical cycle Any of several global processes in which a nutrient circulates between the abiotic environment and living organisms.

biogeographical realm A major region of Earth that is occupied by distinct evolutionary lineages of plants and animals.

biogeography The study of the geographical distributions of plants and animals.

bioinformatics The application of mathematics and computer science to extract information from biological data, including those related to genome structure and function.

biological clock An internal time-measuring mechanism that adapts an organism to recurring environmental changes.

biological evolution The process by which some individuals in a population experience changes in their DNA and pass those modified instructions to their offspring.

biological lineage An evolutionary sequence of ancestral organisms and their descendants.

biological magnification The increasing concentration of nondegradable poisons in the tissues of animals at higher trophic levels.

biological research The collective effort of individuals who have worked to understand how living systems function.

biological rhythm (biorhythm) The change of a regulated factor in a predictable and cycling pattern.

biological species concept The definition of species based on the ability of populations to interbreed and produce fertile offspring.

bioluminescent An organism that glows or releases a flash of light, particularly when disturbed.

biomass The dry weight of biological material per unit area or volume of habitat.

biome A large scale vegetation type and its associated microorganisms, fungi, and animals.

biosignature Particular organic molecules in sedimentary rocks that could only have been formed by cellular activity.

biosphere All regions of Earth's crust, waters, and atmosphere that sustain life.

biosynthetic reaction An individual reaction in an anabolic pathway (biosynthetic pathway).

biota The total collection of organisms in a geographical region.

biotechnology The manipulation of living organisms to produce useful products.

biotic Biological, often in reference to living components of the environment.

bipedalism The habit in animals of walking upright on two legs.

bipolar cell A type of neuron in the retina of the eye that connects the rods and cones with the ganglion cells.

blade The expanded part of a leaf that provides a large surface area for absorbing sunlight and carbon dioxide.

blastocoel A fluid-filled cavity in the blastula embryo.

blastocyst An embryonic stage in mammals; a single-cell-layered hollow ball of about 120 cells with a fluid-filled blastocoel in which a dense mass of cells is localized to one side.

blastodisc A disclike layer of cells at the surface of the yolk produced by early cleavage divisions.

blastomere A small cell formed during cleavage of the embryo.

blastopore The opening at one end of the archenteron in the gastrula that gives rise to the mouth in protostomes and the anus in deuterostomes.

blastula The hollow ball of cells that is the result of cleavage divisions in an early embryo.

blending theory of inheritance Theory suggesting that hereditary traits blend evenly in offspring through mixing of the blood of the two parents.

blood A fluid connective tissue composed of blood cells suspended in a fluid extracellular matrix called **plasma.**

blood–brain barrier A specialized arrangement of capillaries in the brain that prevents most substances dissolved in the blood from entering the cerebrospinal fluid and thus protects the brain and spinal cord from viruses, bacteria, and toxic substances that may circulate in the blood.

blood pressure The measurement of the hydrostatic pressure on the walls of the arteries as the heart pumps blood through the body.

blue-light receptors In plants, light-absorbing proteins, including certain pigment molecules and cryptochrome, that are sensitive to blue light wavelengths and operate in various light-based growth responses.

body system *See* **organ system.**

Bohr effect The reduction in affinity of hemoglobin for O_2 that results from a conformational change caused by lowered pH.

bolting Rapid formation of a floral shoot in plant species that form rosettes, such as lettuce.

bolus The food mass after chewing.

bone The densest form of connective tissue, in which living cells secrete the mineralized matrix of collagen and calcium salts that surrounds them; forms the skeleton.

bony temporal arches Bones that border the temporal fenestrae in the skulls of some amniotes.

book lungs Pocketlike respiratory organs found in some arachnids consisting of several parallel membrane folds arranged like the pages of a book.

boreal forest A biome that is a circumpolar expanse of evergreen coniferous trees in Europe, Asia, and North America.

Bowman's capsule An infolded region at the proximal end of a nephron that cups around the glomerulus and collects the water and solutes filtered out of the blood.

brachiation A pattern of locomotion among primates in which an individual swings below branches from one handhold to another.

brain A single, organized collection of nervous tissue in an organism's head that forms the control center of the nervous system that integrates major sensory input with motor output.

brain stem A stalklike structure formed by the pons and medulla, along with the midbrain, which connects the forebrain with the spinal cord.

branchial slits Openings in the walls of the pharynx that allow water to exit the pharynx.

brassinosteroid Any of a family of plant hormones that stimulate cell division and elongation and differentiation of vascular tissue.

breathing The exchange of gases with the respiratory medium by animals.

Broca's area A language area in the frontal lobe of the cerebral cortex of the brain. The motor programs for coordination of the lips, tongue, jaws, and other structures producing the sounds of speech are generated by Broca's area.

bronchiole One of the small, branching airways in the lungs that lead into the alveoli.

bronchus (plural, *bronchi*) An airway that leads from the trachea to the lungs.

brown adipose tissue A specialized tissue in which the most intense heat generation by nonshivering thermogenesis takes place.

brown algae *See* **Phaeophyta.**

brown fat *See* **brown adipose tissue.**

brush border The fingerlike projections of the plasma membrane of the epithelial cells covering the intestinal villi.

Bryophyta The phylum of seedless nonvascular plants plants to which mosses are assigned.

bryophyte A general term for plants (such as mosses) that lack internal transport vessels.

budding A mode of asexual reproduction in which a new individual grows and develops while attached to the parent.

buffer Substance that compensates for pH changes by absorbing or releasing H^+.

bulbourethral gland One of two pea-sized glands on either side of the prostate gland, which secrete a mucous fluid that is added to semen.

bulk feeder An animal that consumes sizeable food items whole or in large chunks.

bulk flow The group movement of molecules in response to a difference in pressure between two locations.

bulk-phase endocytosis Mechanism by which extracellular water is taken into a cell together with any molecules that happen to be in solution in the water. Also referred to as *pinocytosis*.

C_3 pathway *See* **light-independent reaction**; also referred to as the *Calvin cycle*.

C_3 plant A plant that initially fixes carbon into a three-carbon molecule using the *C_3 pathway*.

C_4 pathway In C_4 plants the pathway to fix CO_2 into oxaloacetate in mesophyll cells and then produce CO_2 for the Calvin cycle in bundle sheath cells.

C_4 plant A plant that initially fixes carbon into a four-carbon molecule using the *C_4 pathway*.

Ca^{2+} pump (calcium pump) Pump that pushes Ca^{2+} from the cytoplasm to the cell exterior, and also from the cytosol into the vesicles of the endoplasmic reticulum.

cadherins (calcium-dependent adhesion molecules) The major class of cell adhesion molecules. Cadherins are necessary for establishing and maintaining connections between cells, and they play a critical role in the organization of cells in the body.

calcitonin A nontropic peptide hormone that lowers the level of Ca^{2+} in the blood by inhibiting the ongoing dissolution of calcium from bone.

callus An undifferentiated tissue that develops on or around a cut plant surface or in tissue culture.

calorie (cal) The amount of heat required to raise 1 g of water by 1°C; known as a "small" calorie; when capitalized, a unit equal to 1,000 small calories.

Calvin cycle *See* **light-independent reaction**.

calyx The outermost whorl of a flower, made up of sepals; early in the development of a flower, it encloses all the other parts, as in an unopened bud.

CAM pathway In CAM plants the pathway to fix CO_2 into oxaloacetate and then produce CO_2 for the Calvin cycle, both occurring in mesophyll cells, but separated by time of day. CAM stands for "crassulacean acid metabolism."

cancer When a tumor becomes malignant and its cells invade and disrupt surrounding tissues.

cancer genomics The study of the cancer genome to identify genetic changes that cause a cell to become a cancer, and to distinguish one type or subtype of cancer from another.

canines Pointed, conical teeth of a mammal, located between the incisors and the first premolars, that are specialized for biting and piercing.

CAP (catabolite activator protein) Key regulatory molecule involved in positive gene regulation of the *lac* operon.

CAP site Region in the promoter of the *lac* operon and in the promoters of a large number of other operons that control the catabolism of many sugars to which activated catabolite activator protein (CAP) binds, thereby enabling RNA polymerase to bind and transcribe the operon's structural genes.

capillary The smallest diameter blood vessel, with a wall that is one cell thick, which forms highly branched networks well adapted for diffusion of substances.

capsid The protective layer of protein that surrounds the nucleic acid core of a virus in free form; also known as a *coat*.

capsule An external layer of sticky or slimy polysaccharides coating the cell wall in many prokaryotes.

carapace A protective outer covering that extends backward behind the head on the dorsal side of an animal, such as the shell of a turtle or lobster.

carbon cycle The global circulation of carbon atoms, especially via the processes of photosynthesis and respiration.

carbonyl group The reactive part of aldehydes and ketones, consisting of an oxygen atom linked to a carbon atom by a double bond.

carboxyl group The characteristic functional group of organic acids, formed by the combination of carbonyl and hydroxyl groups.

carboxypeptidase An enzyme that digests peptides to produce amino acids.

cardiac cycle The systole-diastole sequence of the heart.

cardiac muscle The contractile tissue of the heart.

carnivore An animal that primarily eats other animals.

carotenoid Molecule of yellow-orange pigment by which light is absorbed in photosynthesis.

carotid body A small cluster of chemoreceptors and supporting cells located near the bifurcation of the carotid artery that measures changes in the composition of arterial blood flowing through it.

carpel The reproductive organ of a flower that houses an ovule and its associated structures.

carrageenan A chemical extracted from the red alga *Eucheuma* that is used to thicken and stabilize paints, dairy products such as pudding and ice cream, and many other creams and emulsions.

carrier A heterozygote—an individual who carries a recessive mutant allele and could pass it on to offspring, but does not display its symptoms.

carrier protein Transport protein that binds a specific single solute and transports it across the lipid bilayer.

carrying capacity The maximum size of a population that an environment can support indefinitely.

Cartagena Protocol on Biosafety An international agreement that promotes biosafety as it relates to the handling and use of genetically modified organisms.

cartilage A tissue composed of sparsely distributed chondrocytes surrounded by networks of collagen fibers embedded in a tough but elastic matrix of the glycoprotein.

cartilaginous joint A joint between bones that is not very movable and in which the ends of the bones are covered with layers of cartilage but with no fluid-filled capsule surrounding them.

Casparian strip A thin, waxy impermeable band that seals abutting cell walls in roots; the strip helps control the type and amount of solutes that enter the stele by blocking the apoplastic pathway at the endodermis and forcing substances to pass through cells (the symplast).

catabolic pathway A metabolic pathway in which energy is released by the breakdown of complex molecules to simpler compounds. An individual reaction in a catabolic pathway is a catabolic reaction.

catabolic reaction Cellular reaction that breaks down complex molecules such as sugar to make their energy available for cellular work.

catalysis The process of accelerating a chemical reaction with a catalyst.

catalyst Substance with the ability to accelerate a spontaneous reaction without being changed by the reaction.

catastrophism The theory that Earth has been affected by sudden, violent events that were sometimes worldwide in scope.

catecholamine Any of a class of compounds derived from the amino acid tyrosine that circulates in the bloodstream, including epinephrine and norepinephrine.

cation A positively charged ion.

cation exchange Replacement of one cation with another, as on a soil particle.

$CD4^+$ T cell A type of T cell in the lymphatic system that has CD4 receptors on its surface. This type of T cell binds to an antigen-presenting cell in antibody-mediated immunity.

$CD8^+$ T cell A type of T cell in the lymphatic system that has CD8 receptors on its surface. This type of T cell binds to an antigen-presenting cell in cell-mediated immunity.

cecum A blind pouch formed at the junction of the large and small intestine.

cell Smallest unit with the capacity to live and reproduce.

cell adhesion molecule (CAM) A membrane protein that mediates the selective binding of cells together.

cell body The portion of the neuron containing genetic material and cellular organelles.

cell–cell recognition A form of direct cell communication in which animal cells with particular membrane-bound cell surface molecules dock with one another, initiating communication between the cells.

cell–cell recognition protein A protein in the plasma membrane that identifies a cell as part of the same individual or as foreign, facilitates cell–cell linking, binds cells to the extracellular matrix, or links the extracellular matrix to the cytoskeleton.

cell center *See* **centrosome**.

cell culture Living cells growing in a growth medium in a laboratory vessel.

cell cycle The sequence of events during which a cell experiences a period of growth followed by nuclear division and cytokinesis.

cell differentiation A process in which changes in gene expression establish cells with specialized structure and function.

cell division The process whereby a preexisting cell divides to form two new cells. Cell division enables an organism to grow, reproduce, and repair damaged tissues and organs.

cell expansion A mechanism that enlarges the cells in specific directions in a developing organ.

cell fractionation Technique that divides cells into fractions containing a single cell component.

cell junction Junction that seals the spaces between cells and provides direct communication between cells.

cell lineage Cell derivation from the undifferentiated tissues of the embryo.

cell-mediated immunity An adaptive immune response in which a subclass of T cells—cytotoxic T cells—becomes activated and, with other cells of the immune system, attacks host cells infected by pathogens, particularly those infected by a virus.

cell plate In cytokinesis in plants, a new cell wall that forms between the daughter nuclei and grows laterally until it divides the cytoplasm.

cell signaling The system of communication between cells through signaling pathways.

cell theory Three generalizations yielded by microscopic observations: all organisms are composed of one or more cells; the cell is the smallest unit that has the properties of life; and cells arise only from the growth and division of preexisting cells.

cell wall A rigid external layer of material surrounding the plasma membrane of cells in plants, fungi, bacteria, and some protists, providing cell protection and support.

cellular respiration The process by which energy-rich molecules are broken down to produce energy in the form of ATP.

cellular slime mold Any of a variety of primitive organisms of the phylum Acrasiomycota, especially of the genus *Dictyostelium*; the life cycle is characterized by a slimelike amoeboid stage and a multicellular reproductive stage.

cellulose One of the primary constituents of plant cell walls, formed by chains of carbohydrate subunits.

centimorgan *See* **map unit.**

central canal The central portion of the vertebral column in which the spinal cord is found.

central dogma The name given by Francis Crick to the flow of information from DNA to RNA to protein.

central nervous system (CNS) One of the two major divisions of the nervous system containing the brain and spinal cord.

central vacuole A large, water-filled organelle in plant cells that maintains the turgor of the cell and controls movement of molecules between the cytosol and sap.

centriole A cylindrical structure consisting of nine triplets of microtubules in the centrosomes of most animal cells.

centromere A specialized chromosomal region that connects sister chromatids and attaches them to the mitotic spindle.

centrosome (cell center) The main microtubule organizing center of a cell, which organizes the microtubule cytoskeleton during interphase and positions many of the cytoplasmic organelles.

cephalization The development of an anterior end with a high concentration of sensory organs and nervous system tissue.

cephalochordate A monophyletic group of chordates, including invertebrates known as lancelets.

cerebellum The portion of the brain that receives sensory input from receptors in muscles and joints, from balance receptors in the inner ear, and from the receptors of touch, vision, and hearing.

cerebral cortex A thin outer shell of gray matter covering a thick core of white matter within each hemisphere of the brain; the part of the forebrain responsible for information processing and learning.

cerebrospinal fluid Fluid that circulates through the central canal of the spinal cord and the ventricles of the brain, cushioning the brain and spinal cord from jarring movements and impacts, as well as nourishing the CNS and protecting it from toxic substances.

cerebrovascular accident *See* **stroke.**

cervix The lower end of the uterus.

channel protein Transport protein that forms a hydrophilic channel in a cell membrane through which water, ions, or other molecules can pass, depending on the protein.

chaparral A biome comprising a scrubby mix of short trees and shrubs that dominates coastal land between 30° and 40° latitude, where winters are cool and wet and summers hot and dry.

chaperone protein (chaperonin) "Guide" protein that binds temporarily with newly synthesized proteins, directing their conformation toward the correct tertiary structure and inhibiting incorrect arrangements as the new proteins fold.

character A specific heritable attribute or property of an organism.

character differences Alternative forms of characters. Also called a *trait.*

character displacement The phenomenon in which allopatric populations are morphologically similar and use similar resources, but sympatric populations are morphologically different and use different resources; may also apply to characters influencing mate choice.

character states One or more forms of a character used in a phylogenetic analysis.

Chargaff's rules The findings that, in DNA, the amount of purines equals the amount of pyrimidines and, more specifically, that the amount of adenine equals the amount of thymine, and the amount of guanine equals the amount of cytosine.

charging *See* **aminoacylation.**

Charophyta (charophytes) The phylum of green algae from which the first land plants arose.

charophyte A member of the group of green algae most similar to the algal ancestors of land plants.

checkpoint Internal control of the cell cycle that prevents a critical phase from beginning until the previous phase is complete.

chelicerae The first pair of fanglike appendages near the mouth of an arachnid, used for biting prey and often modified for grasping and piercing.

Chelicerata A clade within Arthropoda that includes spiders, ticks, mites, scorpions, and horse-shoe crabs.

chemical bond Link formed when atoms of reactive elements combine into molecules.

chemical energy In biological systems, the potential energy that can be released in a chemical reaction. Chemical energy is stored in the bonds between atoms.

chemical equation A chemical reaction written in balanced form.

chemical evolution The formation of the organic molecules that allowed the first forms of life on Earth to originate.

chemical reaction A reaction that occurs when atoms or molecules interact to form new chemical bonds or break old ones.

chemical signal Any secretion from one cell type that can alter the behavior of a different cell that bears a receptor for it; a means of cell communication. In animal behavior, a mechanism of communication in which a signal is received by an olfactory receptor.

chemical synapse In the nervous system where the plasma membranes of the presynaptic and postsynaptic cells are separated by a narrow gap. Communication across such a synapse occurs by means of a neurotransmitter; that is, an electrical impulse arriving at an axon terminal of the presynaptic cell triggers release of a neurotransmitter that crosses the gap and binds to a receptor on the postsynaptic cell, triggering an electrical impulse in that cell.

chemiosmosis In oxidative phosphorylation, the process of generating an H^+ gradient from free energy released by electron flow through the electron transfer system.

chemiosmotic hypothesis Model proposing that mitochondrial electron transfer produces an H^+ gradient and that the gradient powers ATP synthesis by ATP synthase.

chemoautotroph An organism that obtains energy by oxidizing inorganic substances such as hydrogen, iron, sulfur, ammonia, nitrites, and nitrates and uses carbon dioxide as a carbon source.

chemoheterotroph An organism that oxidizes organic molecules as an energy source and obtains carbon in organic form.

chemokine A protein secreted by activated macrophages that attracts other cells, such as neutrophils.

chemoreceptor A sensory receptor that detects specific molecules, or chemical conditions such as acidity.

chemotroph An organism that obtains energy by oxidizing inorganic or organic substances.

chewing In mammals with teeth, the first step in the digestive process in which ingested food is sliced, torn, and ground into small pieces. Also called mastication.

chiasmata *See* **crossover.**

chief cells Cells in the gastric glands of the stomach lining that secrete pepsinogen.

chitin A polysaccharide that contains nitrogen and is present in the cell walls of fungi and the exoskeletons of arthropods.

Chlorarachniophyta (chlorarachniophytes) A lineage within the Rhizaria protist evolutionary group; green, photosynthetic amoebas that also engulf food.

chlorophyll Molecule of green pigment that absorbs photons of light in photosynthesis.

Chlorophyta (green algae) A lineage within the Archaeplastida protist evolutionary group; autotrophs that carry out photosynthesis using the same pigments as plants.

chloroplast The site of photosynthesis in plant cells.

chlorosis An abnormal yellowing of plant tissues due to lack of chlorophyll; a sign of nutrient deficiency or infection by a pathogen.

choanocyte One of the inner layer of flagellated cells lining the body cavity of a sponge.

Choanoflagellata (choanoflagellates) A group of minute, single-celled protists found in water; the flask-shaped body has a collar of closely packed microvilli that surrounds the single flagellum by which it moves and takes in food. Choanoflagellates are a subgroup of the opisthokonts.

cholecystokinin (CCK) Hormone released in response to fat, and to a lesser extent protein, in the chyme that enters the duodenum and inhibits gastric activity.

cholesterol The predominant sterol of animal cell membranes.

Chondrichthyes A monophyletic group of fishes, including sharks, skates, and rays, that lack all traces of bone in their skeletons.

chondrocyte A cartilage-producing cell.

Chordata A monophyletic group of deuterostomes that includes vertebrates and their closest invertebrate relatives.

chorion In an amniote egg, an extraembryonic membrane that surrounds the embryo and yolk sac completely and exchanges oxygen and carbon dioxide with the environment; becomes part of the placenta in mammals.

chorionic villus (plural, *villi*) One of many treelike extensions from the chorion, which greatly increase the surface area of the chorion.

chorionic villus sampling Technique of prenatal diagnosis in which cells are obtained from portions of the placenta that develop from tissues of the embryo for DNA testing, for biochemical analysis, or to test for the presence of chromosomal mutations.

Chromalveolata (chromalveolates) A protist evolutionary group consisting of the Alveolata and the Stramenopila.

chromatin A complex of DNA and proteins in a eukaryotic nucleus.

chromatin remodeling The type of change in chromatin structure in which (typically) large multiprotein complexes displace acetylated nucleosomes in the promoter region from the DNA, or move them along the DNA away from the promoter as part of the process of transcription initiation in eukaryotes.

chromoplast Plastid containing red and yellow pigments.

chromosomal mutation A variation from the normal condition in chromosome structure or chromosome number.

chromosomal protein A histone and nonhistone protein associated with DNA in a eukaryotic nuclear chromosome.

chromosome In eukaryotic cells, a linear structure composed of a single DNA molecule complexed with protein. Each eukaryotic species has a characteristic number of chromosomes in the nucleus. Most prokaryotes have a single, usually circular chromosome with few or no associated proteins.

chromosome segregation The equal distribution of daughter chromosomes to each of the two cells that result from cell division.

chromosome theory of inheritance The principle that genes and their alleles are carried on the chromosomes.

Chrysophyta (golden algae) A lineage of the Stramenopila subgroup of the Chromalveolata protist evolutionary group; mostly colonial protists in which each cell of a colony has a pair of flagella and is surrounded by a glassy shell.

chylomicron A small triglyceride droplet coated with a layer of lipoprotein from the endoplasmic reticulum.

chyme Digested content of the stomach released for further digestion in the small intestine.

chymotrypsin An enzyme that digests peptides to produce small peptides.

chytrids Fungi in the phylum Chytridiomycota; all have flagella and nearly all are microscopic aquatic organisms.

ciliary body A fine ligament in the eye that anchors the lens to a surrounding layer of connective tissue and muscle.

Ciliophora (ciliates) A lineage of the Alveolata subgroup of the Chromalveolata protist evolutionary group.

cilium (plural, *cilia*) Motile structure, extending from a cell surface, that moves a cell through fluid or fluid over a cell.

circadian rhythm Any biological activity that is repeated in cycles, each about 24 hours long, independently of any shifts in environmental conditions.

circulatory system An organ system consisting of a fluid, a heart, and vessels for moving important molecules, and often cells, from one tissue to another.

circulatory vessel An element of the circulatory system through which fluid flows and carries nutrients and oxygen to tissues and remove wastes.

circumcision Removal of the prepuce for religious, cultural, or hygienic reasons.

cisternae (singular, *cisterna*) Membranous channels and vesicles that make up the endoplasmic reticulum.

citric acid cycle Series of reactions in which acetyl groups are oxidized completely to carbon dioxide and some ATP molecules are synthesized. Also referred to as *Krebs cycle* and *tricarboxylic acid cycle.*

clade A monophyletic group of organisms that share homologous features derived from a common ancestor.

cladistics An approach to systematics that uses shared derived characters to infer the phylogenetic relationships and evolutionary history of groups of organisms.

claspers A pair of organs on the pelvic fins of male crustaceans and sharks, which help transfer sperm into the reproductive tract of the female.

class A Linnaean taxonomic category that ranks below a phylum and above an order.

class II major histocompatibility complex (MHC) A collection of proteins that present antigens on the cell surface of an antigen-presenting cell in an antibody-mediated immune response.

classical conditioning A type of learning in which an animal develops a mental association between two phenomena that are usually unrelated.

classical endocrine signaling A type of cell signaling in the endocrine system in which hormones are secreted into the extracellular fluid by the cells of endocrine glands.

classification An arrangement of organisms into hierarchical groups that reflect their relatedness.

clathrin The network of proteins that coat and reinforce the cytoplasmic surface of cell membranes.

cleavage Mitotic cell divisions of the zygote that produce a blastula from a fertilized ovum.

climate The weather conditions prevailing over an extended period of time.

climax community A relatively stable, late successional stage in which the dominant vegetation replaces itself and persists until an environmental disturbance eliminates it, allowing other species to invade.

climograph A graph that portrays the particular combination of temperature and rainfall conditions where each terrestrial biome occurs.

cline A pattern of smooth variation in a characteristic along a geographical gradient.

clitoris The structure at the junction of the labia minora in front of the vulva, homologous to the penis in the male.

cloaca The cavity in reptiles, birds, amphibians, and many fishes into which both the intestinal and genital tracts empty.

clonal analysis In plants, a method of culturing meristematic tissue that contains a mutated embryonic cell having a readily observable trait, such as the absence of normal pigment.

clonal expansion In antibody-mediated immunity, the proliferation of an activated CD4$^+$ T cell by cell division to produce a clone of cells.

clonal reproduction The type of asexual reproduction in which the parent and offspring are genetically identical to one another.

clonal selection The process by which a lymphocyte is specifically selected for cloning when it encounters a foreign antigen from among a randomly generated, enormous diversity of lymphocytes with receptors that specifically recognize the antigen.

clone An individual genetically identical to an original cell from which it descended.

cloning vector DNA molecule into which a DNA fragment can be inserted to form a recombinant DNA molecule for the purpose of cloning.

closed circulatory system A circulatory system in which the fluid, blood, is confined in blood vessels and is distinct from the interstitial fluid.

clumped dispersion A pattern of distribution in which individuals in a population are grouped together.

cnidocyte A prey-capturing and defensive cell in the epidermis of cnidarians.

CO$_2$ fixation Process in which electrons are used as a source of energy to convert inorganic CO$_2$ to an organic form.

coactivator (mediator) In eukaryotes, a large multiprotein complex that bridges between activators at an enhancer and proteins at the promoter and promoter proximal region to stimulate transcription.

coat *See* **capsid.**

coated pit A depression in the plasma membrane that contains receptors for macromolecules to be taken up by endocytosis.

coccus (plural, *cocci*) A spherical prokaryote.

cochlea A snail-shaped structure in the inner ear containing the organ of hearing.

codominance Condition in which alleles have approximately equal effects in individuals, making the alleles equally detectable in heterozygotes.

codon Each three-letter word (triplet) of the genetic code.

coefficient of relatedness A calculation of the average proportion of alleles that two relatives are likely to share.

coelom A fluid-filled body cavity in bilaterally symmetrical animals that is completely lined with derivatives of mesoderm.

coelomate A body plan of bilaterally symmetrical animals that have a coelom.

coenocytic Condition in which a single cell has many nuclei.

coenzymes Organic cofactors that include complex chemical groups of various kinds.

coevolution The evolution of genetically based, reciprocal adaptations in two or more species that interact closely in the same ecological setting.

cofactor An inorganic or organic nonprotein group that is necessary for catalysis to take place.

cognition A form of learning in which animals use insight to solve a novel problem; sometimes called trial-and-error learning.

cohesins Proteins that encircle newly formed sister chromatids along their length to hold them tightly together, a process called sister chromatid cohesion.

cohesion The high resistance of water molecules to separation.

cohesion–tension mechanism of water transport The mechanism that transports water upward from roots to shoot parts in the xylem of vascular plants; due to cohesion of water molecules, the evaporation (transpiration) of water from shoot parts such as leaves creates a continuous negative pressure (tension) that draws water upward from roots.

cohort A group of individuals of similar age.

coleoptile A protective sheath that covers the shoot apical meristem and plumule of the embryo in monocots, such as grasses, as it pushes up through soil.

coleorhiza A sheath that encloses the radicle of an embryo until it breaks out of the seed coat and enters the soil as the primary root.

collagen Fibrous glycoprotein—very rich in carbohydrates—embedded in a network of proteoglycans.

collecting duct A location where urine leaving individual nephrons is processed further.

collenchyma A ground tissue that flexibly supports rapidly growing plant parts. Its elongated cells are alive at maturity and collectively often form strands or a sheathlike cylinder under the dermal tissue of growing shoot regions and leaf stalks.

colon The main part of the large intestine.

colony Multiple individual organisms of the same species living in a group.

combinatorial gene regulation The combining of a few regulatory proteins in particular ways so that the transcription of a wide array of genes can be controlled and a large number of cell types can be specified.

commaless The sequential nature of the words of the nucleic acid code, with no indicators such as commas or spaces to mark the end of one codon and the beginning of the next.

commensalism A symbiotic interaction in which one species benefits and the other is unaffected.

common pili (singular, *pilus*) The common form of pili among bacteria and archaeans. Common pili are relatively short, there are many per cell, and there are many subvarieties. Also called *fimbriae.*

communication The conveyance of information to other individuals through one or more sensory channels.

community Populations of all species that occupy the same area.

community ecology The ecological discipline that examines groups of populations occurring together in one area.

companion cell A specialized parenchyma cell that is connected to a mature sieve tube member by plasmodesmata and assists sieve tube members both with the uptake of sugars and with the unloading of sugars in tissues.

comparative genomics Comparison of the sequences of entire genomes (or extensive portions of them) to understand evolutionary relationships and the basic biological similarities and differences among species.

comparative morphology Analysis of the structure of living and extinct organisms.

compartment model A graphical depiction of the pathways through which nutrients and energy move between the living and nonliving components of an ecosystem.

compass orientation A wayfinding mechanism that allows animals to move in a particular direction, often over a specific distance or for a prescribed length of time.

competition The use of limiting resources by two or more individuals or populations.

competitive exclusion principle The ecological principle stating that populations of two or more species cannot coexist indefinitely if they rely on the same limiting resources and exploit them in the same way.

competitive inhibition Inhibition of an enzyme reaction by an inhibitor molecule that resembles the normal substrate closely enough so that it fits into the active site of the enzyme.

complement system A nonspecific defense mechanism activated by invading pathogens, made up of more than 30 interacting soluble plasma proteins circulating in the blood and interstitial fluid.

complementary base pairing Feature of DNA in which the specific purine–pyrimidine base pairs A–T (adenine–thymine) and G–C (guanine–cytosine) occur to bridge the two sugar–phosphate backbones.

complementary DNA (cDNA) A DNA molecule that is complementary to an mRNA molecule, synthesized by reverse transcriptase.

complete digestive system (or tract) A digestive system having a mouth at one end, through which food enters, and an anus at the other end, through which undigested waste is voided.

complete flower A flower in which all four whorls (sepals, petals, stamens, carpels) are present.

complete medium A growth medium containing a full complement of nutrient substances that a wild-type microorganism can make for itself.

complete metamorphosis The form of metamorphosis in which an insect passes through four separate stages of growth: egg, larva, pupa, and adult.

complex virus A bacteriophage with a DNA genome that has a tail attached at one side of a polyhedral head.

compound A molecule whose component atoms are different.

compound eye The eye of most insects and some crustaceans, composed of many faceted, light-sensitive units called ommatidia fitted closely together, each having its own refractive system and each forming a portion of an image.

concentration The number of molecules or ions of a substance in a unit volume of space.

concentration gradient A difference in concentration of molecules or ions between two areas.

condensation reaction Reaction during which the components of a water molecule are removed, usually as part of the assembly of a larger molecule from smaller subunits. Also referred to as *dehydration synthesis reaction.*

conduction The flow of heat between atoms or molecules in direct contact.

cone In the vertebrate eye, a photoreceptor in the retina that is specialized for detection of different wavelengths (colors). In cone-bearing plants, a cluster of sporophylls.

conformation The overall three-dimensional shape of a protein.

conformational change Alteration in the three-dimensional shape of a protein.

conformers Animals having internal environments that change as the external environment changes.

conidiophore In ascomycete fungi, a modified hyphal branch that gives rise to conidia.

conidium (plural, *conidia*) An asexual spore produced by many species of ascomycetes.

Coniferophyta The major phylum of cone-bearing gymnosperms, most of which are substantial trees; includes pines, firs, and other conifers.

conjugation In bacteria, the process by which DNA of the donor cell moves through the cytoplasmic bridge into the recipient cell. With some types of donor cells, this can lead to genetic recombination in the recipient cell. In ciliate protists, a process of sexual reproduction in which individuals of the same species temporarily couple and exchange genetic material.

connective tissue Tissue having cells scattered through an extracellular matrix; forms layers in and around body structures that support other body tissues, transmit mechanical and other forces, and in some cases act as filters.

connectome The entire network of neural connections in the brain.

conodont A soft-bodied vertebrate dating from the early Paleozoic era through the early Mesozoic era.

consciousness Awareness of oneself, identity, and surroundings, with understanding of the significance and likely consequences of events.

consensus sequence The series of nucleotides found most frequently at the particular sites in the different sequences which occur in nature.

conservation biology An interdisciplinary science that focuses on the maintenance and preservation of biodiversity.

constant (C) region For the light and heavy polypeptides of a particular type of antibody molecule, the regions that have the same amino acid sequences for all molecules.

constitutive defenses In plants, built-in barriers to threats, such as bark, the cuticle covering dermal tissue, and spines or hairs.

consumer An organism that consumes other organisms in a community or ecosystem.

contact inhibition The inhibition of movement or proliferation of normal cells that results from cell–cell contact.

continental climate Climate not moderated by the distant ocean.

continental drift The long-term movement of continents as a result of plate tectonics.

continuous distribution A geographical distribution in which a species lives in suitable habitats throughout a geographical area.

contraception The prevention of pregnancy.

contractile vacuole A specialized cytoplasmic organelle that pumps fluid in a cyclical manner from within the cell to the outside by alternately filling and then contracting to release its contents at various points on the surface of the cell.

control Treatment that tells what would be seen in the absence of the experimental manipulation.

convection The transfer of heat from a body to a fluid, such as air or water, that passes over its surface.

convergent evolution The evolution of similar adaptations in distantly related organisms that occupy similar environments.

copulation The physical act involving the introduction of the accessory sex organ of a male into the accessory sex organ of a female to accomplish internal fertilization.

coral reef A structure made from the hard skeletons of coral animals or polyps; found largely in shallow tropical and subtropical marine environments.

corepressor In the regulation of gene expression in bacteria, a regulatory molecule that combines with a repressor to activate it and shut off an operon.

core temperature The temperature within the central core of the body consisting of the abdominal and thoracic organs, the CNS, and the skeletal muscle.

cork A nonliving, impermeable secondary tissue that is one element of bark.

cork cambium A lateral meristem in plants that forms periderm, which in turn produces cork.

cornea The transparent layer that forms the front wall of the eye, covering the iris.

corolla The structure formed collectively by the petals of a flower.

coronary arteries The arteries from the aorta that branch extensively over the heart, supplying blood to the cardiac muscle cells.

corpus callosum A structure formed of thick axon bundles that connect the two cerebral hemispheres and coordinate their functions.

corpus luteum Cells remaining at the surface of the ovary during the luteal phase; the structure acts as an endocrine gland, secreting several hormones: estrogens, large quantities of progesterone, and inhibin.

cortex Generally, an outer, rindlike layer. In mammals, the outer layer of the brain, the kidneys, or the adrenal glands. In plants, the outer region of tissue in a root or stem lying between the epidermis and the vascular tissue, composed mainly of parenchyma.

cortical granule A secretory vesicle just under the plasma membrane of an egg cell.

cortical reaction The reaction in which cortical granules fuse with the plasma membrane of the egg and release their contents to the outside.

cortisol The major glucocorticoid steroid hormone secreted by the adrenal cortex, which increases blood glucose by promoting breakdown of proteins and fats.

cotranslational import A mechanism by which a polypeptide being sorted via the endomembrane system in a eukaryotic cell begins its import into the endoplasmic reticulum simultaneously with translation of the mRNA encoding the polypeptide.

cotransport Secondary active transport mechanisms that move both ions and organic molecules across membranes. The two types of cotransport are *symport* and *antiport.*

cotyledon A seed leaf.

countercurrent exchange A mechanism in which the water flowing over the gills moves in a direction opposite to the flow of blood under the respiratory surface.

coupled reaction Reaction that occurs when an exergonic reaction is joined to an endergonic reaction, producing an overall reaction that is exergonic.

courtship display A behavior performed by males to attract potential mates or to reinforce the bond between a male and a female.

covalent bond Bond formed by electron sharing between atoms.

cranial nerve A nerve that connects the brain directly to the head, neck, and body trunk.

cranium The part of the skull that encloses the brain.

crassulacean acid metabolism (CAM) A biochemical variation of photosynthesis that was discovered in a member of the plant family Crassulaceae. Carbon dioxide is taken up and stored during the night to allow the stomata to remain closed during the daytime, decreasing water loss.

Crenarchaeota A major group of the domain Archaea that contains most of the extreme thermophiles. The group also includes psychrophiles and mesophiles. Members include obligate anaerobes, facultative anaerobes, and aerobes.

CRISPR (clusters of regularly interspersed short palindromic repeats) A locus in sequenced bacterial and archaeal genomes that, with *cas* genes, encode an immune system against foreign bacteriophages and plasmids. Each CRISPR locus consists of repeated sequences about 40 bp long that have palindromic regions, and that are interspersed with unique sequences of about the same length.

criss-cross inheritance The transmission pattern characteristic of an X-linked allele from a parent of one sex to a "child" of the opposite sex to a "grandchild" of the first sex.

crista (plural, *cristae*) Fold that expands the surface area of the inner mitochondrial membrane.

critical period A restricted stage of development early in life during which an animal has the capacity to respond to specific environmental stimuli.

crop Of birds, an enlargement of the digestive tube where the digestive contents are stored and mixed with lubricating mucus.

cross-fertilization Fertilization of one plant by a different plant.

crossing-over The recombination process in meiosis, in which chromatids exchange segments.

crossover Site of recombination during meiosis. Also referred to as a *chiasmata*.

cross-pollination *See* **cross-fertilization.**

cross-talk Interaction by which cell signaling pathways communicate with one another to integrate their responses to cellular signals.

cryptic coloration Coloration that allows an organism to match its background and hence become less vulnerable to predation or recognition by prey.

cryptochrome A light-absorbing protein that is sensitive to blue light and that may also be an important early step in various light-based growth responses.

C-terminal end The end of an amino acid chain with a —COO⁻ group.

cupula In certain mechanoceptors, a gelatinous structure with stereocilia extending into it that moves with pressure changes in the surrounding water; movement of the cupula bends the stereocilia, which triggers release of neurotransmitters.

cuticle The outer layer of plants and some animals, which helps prevent desiccation by slowing water loss and may serve as a barrier to infection or other threats.

Cycadophyta A phylum of palmlike gymnosperms known as cycads; the pollen-bearing and seed-bearing cones (strobili) occur on separate plants.

cyclic AMP (cAMP) In particular signal transduction pathways, a second messenger that activates protein kinases, which elicit the cellular response by adding phosphate groups to specific target proteins. cAMP functions in one of two major G-protein–coupled receptor response pathways.

cyclic electron flow An electron transport pathway associated with photosystem I in photosynthesis that produces ATP without the synthesis of NADPH.

cyclin In eukaryotes, protein that regulates the activity of CDK (cyclin-dependent kinase) and controls progression through the cell cycle.

cyclin-dependent kinase (CDK) A protein kinase that controls the cell cycle in eukaryotes.

cytochrome Protein with a heme prosthetic group that contains an iron atom.

cytokine A molecule secreted by one cell type that binds to receptors on other cells and, through signal transduction pathways, triggers a response. In innate immunity, cytokines are secreted by activated macrophages.

cytokinesis Division of the cytoplasm into two daughter cells following nuclear division in mitosis or meiosis.

cytokinin A hormone that promotes and controls growth responses of plants.

cytoplasm All the parts of the cell that surround the central nucleus (eukaryotes) or nucleoid region (prokaryotes).

cytoplasmic determinants The mRNA and proteins stored in the egg cytoplasm that direct the early stages of animal development in the period before genes of the zygote become active.

cytoplasmic inheritance Pattern in which inheritance follows that of genes in the cytoplasmic organelles, mitochondria, or chloroplasts.

cytoplasmic streaming Intracellular movement of cytoplasm.

cytosine A pyrimidine that base-pairs with guanine in nucleic acids.

cytoskeleton The interconnected system of protein fibers and tubes that extends throughout the cytoplasm of a eukaryotic cell.

cytosol Aqueous solution in the cytoplasm containing ions and various organic molecules.

cytotoxic T cell A T lymphocyte that functions in cell-mediated immunity to kill body cells infected by viruses or transformed by cancer.

daily torpor A period of inactivity and lowered metabolic rate that allows an endotherm to conserve energy when environmental temperatures are low.

dalton A standard unit of mass, about 1.66×10^{-24} grams.

daughter chromosomes The chromosomes that result after sister chromatids separate.

day-neutral plant A plant that flowers without regard to photoperiod.

decomposer A small organism, such as a bacterium or fungus, that feeds on the remains of dead organisms, breaking down complex biological molecules or structures into simpler raw materials.

dedifferentiation A process in which cells lose their normal regulatory controls and revert partially or completely to an embryonic developmental state.

deep vein thrombosis A medical condition that results when sitting for a long period of time causes blood to pool in the veins of the body below the heart and then clot, particularly in the legs.

defecation reflex The opening of the anal sphincter and expulsion of feces in response to feces entering the rectum and stretching the rectal wall.

defensins In innate immunity, antimicrobial peptides that protect the epithelial surfaces against invading pathogens.

degeneracy (redundancy) The feature of the genetic code in which, with two exceptions, more than one codon represents each amino acid.

dehydration synthesis reaction *See* **condensation reaction.**

deletion Chromosomal alteration that occurs if a broken segment is lost from a chromosome.

deltaproteobacteria One of five subgroups of the proteobacteria group of bacteria. Deltaproteobacteria consist of only a few species; they include a group of anaerobes involved in the cyclic interchanges of sulfur-containing compouns in the biosphere, and other species with complex life cycles.

demographic transition model A graphical depiction of the historical relationship between a country's economic development and its birth and death rates.

demography The statistical study of the processes that change a population's size and density through time.

denaturation A loss of both the structure and function of a protein due to extreme conditions that unfold it from its normal conformation.

dendrite The branched extension of the nerve cell body that receives signals from other nerve cells.

dendritic cell A type of phagocyte, so called because it has many surface projections that resemble dendrites of neurons, which engulfs a bacterium in infected tissue by phagocytosis.

denitrification A metabolic process in which certain bacteria convert nitrites or nitrates into nitrous oxide and then into molecular nitrogen, which enters the atmosphere.

dense connective tissue A type of connective tissue in which fibroblasts are distributed sparsely among dense masses of collagen and elastin fibers that are arranged to resist stretch and provide strength.

density-dependent Description of environmental factors for which the strength of their effect on a population varies with the population's density.

density-independent Description of environmental factors for which the strength of their effect on a population does not vary with the population's density.

deoxyribonucleic acid *See* **DNA.**

deoxyribonucleotide Nucleotide containing deoxyribose as the sugar; deoxyribonucleotides are components of DNA.

deoxyribose A 5-carbon sugar to which a nitrogenous base and a phosphate group link covalently in a nucleotide of DNA.

depolarized State of the membrane (which was polarized at rest) as the membrane potential becomes less negative.

deposit feeder An animal that consumes particles of organic matter from the solid substrate on which it lives.

derivative One of the daughter cells produced when a plant cell divides; it typically divides once or twice and then enters on the path to differentiation.

derived character state A new version of a trait found in the most recent common ancestor of a group.

dermal tissue system The plant tissue system that comprises the outer tissues of the plant body, including the epidermis and periderm; it serves as a protective covering for the plant body.

dermis The skin layer below the epidermis; it is packed with connective tissue fibers such as collagen, which resist compression, tearing, or puncture of the skin.

descending reticular formation One of two parts of the reticular formation; it receives information from the hypothalamus and connects with interneurons in the spinal cord that control skeletal muscle contraction, thereby controlling muscle movement and posture.

descent with modification Biological evolution.

desert A sparsely vegetated biome that forms where rainfall averages less than 25 cm per year.

desertification A process in which large tracts of subtropical forest are cleared and overused, the groundwater table recedes to deeper levels, less surface water is available for plants, soil accumulates high concentrations of salts, and topsoil is eroded by wind and water.

desmosome Anchoring junction for which microfilaments anchor the junction in the underlying cytoplasm.

determinate cleavage A type of cleavage in protostomes in which each cell's developmental path is determined as the cell is produced.

determinate growth The pattern of growth in most animals in which individuals grow to a certain size and then their growth slows dramatically or stops.

determination Mechanism in which the developmental fate of a cell is set.

detritivore An organism that extracts energy from the organic detritus (refuse) produced at other trophic levels.

deuterostome A group of animals in which cleavage usually follows a radial pattern, mesoderm arises from cells in the roof of the archenteron, the coelom forms in a space pinched off by the developing mesoderm, and the blastopore forms the anus.

development A series of programmed changes encoded in DNA, through which a fertilized egg divides into many cells that ultimately are transformed into an adult, which is itself capable of reproduction.

diabetes mellitus A disease that results from problems with insulin production or action.

diacylglycerol (DAG) In particular signal transduction pathways, a second messenger that activates protein kinases, which elicit the cellular response by adding phosphate groups to specific target proteins. DAG is involved in one of two major G-protein–coupled receptor–response pathways.

diapsid A member of a group within the amniote vertebrates having a skull with two temporal arches. Their living descendants include lizards and snakes, crocodilians, and birds.

diastole The period of relaxation and filling of the heart between contractions.

diastolic blood pressure The low point of the arterial blood pressure in the cardiac cycle that occurs between ventricular contractions.

diatoms *See* **Bacillariophyta.**

differentiation Process by which cells that have been committed to a particular developmental fate by the determination process now develop into specialized cell types with distinct structures and functions.

diffusion The net movement of ions or molecules from a region of higher concentration to a region of lower concentration.

digestion The splitting of carbohydrates, proteins, lipids, and nucleic acids in foods into chemical subunits small enough to be absorbed into the body fluids and cells of an animal.

digestive tract Part of an extracellular digestive system; it is tubelike with two openings that form a separate mouth and anus. Digestive contents move in one direction through specialized regions of the tube, from the mouth to the anus.

dihybrid A zygote produced from a cross that involves two characters.

dihybrid cross A cross between two individuals that are heterozygous for two pairs of alleles.

dikaryon The life stage in certain fungi in which a cell contains two genetically distinct haploid nuclei.

Dinoflagellata (dinoflagellates) A lineage of the Alveolata subgroup of the Chromalveolata protist evolutionary group; mostly single-celled marine phytoplankton that live as heterotrophs or autotrophs or sometimes using both modes of nutrition.

dinosaurs A monophyletic group of archosauromorpha that includes ornithischians, nonavian saurischians, and birds.

dioecious Having male flowers and female flowers on different plants of the same species.

dipeptidase An enzyme that splits dipeptides (two amino acids joined together) into individual amino acids.

diploblastic An animal body plan in which adult structures arise from only two cell layers, the ectoderm and the endoderm.

diploid An organism or cell with two copies of each type of chromosome in its nucleus.

directional selection A type of selection in which individuals near one end of the phenotypic spectrum have the highest relative fitness.

disaccharide A carbohydrate consisting of two monosaccharides bonded together.

disaccharidase An enzyme that splits disaccharides (two sugar molecules joined together) into individual monosaccharides (single sugar molecules).

disclimax community An ecological community in which regular disturbance inhibits successional change.

discontinuous replication Replication in which a DNA strand is formed in short lengths that are synthesized in the direction opposite of DNA unwinding.

disjunct distribution A geographical distribution in which populations of the same species or closely related species live in widely separated locations.

dispersal The movement of organisms away from their place of origin.

dispersed duplication Gene copies that are found in different places in the genome, often on two different chromosomes.

dispersion The spatial distribution of individuals within a population's geographical range.

disruptive selection A type of natural selection in which extreme phenotypes have higher relative fitness than intermediate phenotypes.

dissociation The separation of water to produce positively charged hydrogen ions and hydroxide ions.

distal convoluted tubule The tubule in the human nephron that drains urine into a collecting duct that leads to the renal pelvis.

disulfide linkage Linkage that occurs when two sulfhydryl groups interact during a linking reaction.

DNA (deoxyribonucleic acid) The large, double-stranded, helical molecule that is the genetic material of all living organisms.

DNA chip *See* **DNA microarray.**

DNA amplification *In vitro* technique that produces an amount of DNA to the point where it can be analyzed or manipulated easily. (Also see **PCR.**)

DNA fingerprinting Technique in which DNA samples are used to distinguish between individuals of the same species.

DNA helicase An enzyme that catalyzes the unwinding of DNA template strands.

DNA ligase In DNA replication, an enzyme that seals the nicks left after RNA primers are replaced with DNA.

DNA marker A site or region in the genome that is polymorphic. A DNA marker is a type of *genetic marker.*

DNA methylation Process in which a methyl group is added enzymatically to cytosine bases in the DNA.

DNA microarray A solid surface divided into a microscopic grid of thousands of spaces each containing thousands of copies of a DNA probe. DNA microarrays are used commonly for analysis of gene activity and for detecting differences between cell types. Also referred to as a *DNA chip.*

DNA polymerase I In *E. coli*, the replication enzyme that replaces the RNA primer at the start of a new DNA segment with DNA.

DNA polymerase III The principal replication polymerase in *E. coli* that synthesizes the majority of the new DNA.

DNA profiling *See* **DNA fingerprinting.**

DNA repair mechanism Mechanism to correct base-pair mismatches that escape proofreading.

DNA technologies Techniques to isolate, purify, analyze, and manipulate DNA sequences.

DNA transposon A transposable element that transposes using a DNA intermediate.

domain In protein structure, a distinct, large structural subdivision produced in many proteins by the folding of the amino acid chain. In systematics, the highest taxonomic category; a group of cellular organisms with characteristics that set it apart as a major trunk of the Tree of Life.

dominance The masking effect of one allele over another.

dominance hierarchy A social system in which the behavior of each individual is constrained by that individual's status in a highly structured social ranking.

dominant The allele expressed when paired with a recessive allele.

dominant species The species that is represented by a large proportion of the individuals present in an ecological community.

dormancy A period in the life cycle in which biological activity is suspended.

dorsal Indicating the back side of an animal.

dorsal lip of the blastopore A crescent-shaped depression rotated clockwise 90° on the embryo surface that marks the region derived from the gray crescent, to which cells from the animal pole move as gastrulation begins.

dosage compensation mechanism Mechanism in placental mammals by which the effects of most genes carried on the X chromosome in females are equalized in females (who have two X chromosomes) and males (who have one X chromosome).

double fertilization A characteristic feature of sexual reproduction in flowering plants. In the embryo sac, one sperm nucleus unites with the egg to form a diploid zygote from which the embryo develops, and another unites with two polar nuclei to form the primary endosperm nucleus.

double helix Two nucleotide chains wrapped around each other in a spiral that resembles a twisted ladder.

double-helix model Model of DNA consisting of two polynucleotide strands twisted around each other.

downy mildews One of the subgroups of the Oomycota; parasites of plants.

driver gene In the context of cancer, a gene that contains a *driver mutation.*

driver mutation In the context of cancer, a mutation that confers a selective growth advantage to the cell in which it occurs.

duodenum A short region of the small intestine where secretions from the pancreas and liver enter a common duct.

duplication Chromosomal alteration that occurs if a segment is broken from one chromosome and inserted into its homolog.

E site The site where an exiting tRNA binds before its release from the ribosome in translation.

ecdysis Shedding of the cuticle, exoskeleton, or skin; molting.

ecdysone A steroid hormone secreted by the prothoracic glands of insects.

Ecdysozoa A monophyletic group of protostomes that periodically shed their exoskeleton or cuticle.

echolocation A technique for locating prey by making squeaking or clicking noises, and then listening for the echoes that bounce back from objects in their environment.

Echinodermata A monophyletic group of deuterostomes characterized by secondary radial symmetry, shell like elements in their skeletons, and a water vascular system.

ecological community An assemblage of species living in the same place.

ecological efficiency The ratio of net productivity at one trophic level to net productivity at the trophic level below it.

ecological footprint The sum total of all resources used by an individual or a population.

ecological isolation A prezygotic reproductive isolating mechanism in which species that live in the same geographical region occupy different habitats.

ecological niche The resources a population uses and the environmental conditions it requires over its lifetime.

ecological pyramid A diagram illustrating the effects of energy transfer from one trophic level to the next.

ecological succession A somewhat predictable series of changes in the species composition of a community over time.

ecology The study of the interactions between organisms and their environments.

ecosystem Group of biological communities interacting with their shared physical environment.

ecosystem services The ecological processes on which all life depends, which include decomposition of wastes, nutrient recycling, oxygen production, maintenance of fertile topsoil, and air and water purification.

ecosystem valuation A process in which ecosystem services are assigned an economic value.

ecotone A wide transition zone between adjacent communities.

ecotourism An activity in which visitors, often from wealthy countries, pay a fee to visit a nature preserve.

ectoderm The outermost of the three primary germ layers of an embryo, which develops into epidermis and nervous tissue.

ectomycorrhizae Mycorrhizae that grow between and around the roots of trees and shrubs but do not enter root cells. Compare to endomycorrhizae.

ectoparasite A parasite that lives on the exterior of its host organism.

ectotherm An animal that obtains its body heat primarily from the external environment.

edge effect A phenomenon in which the removal of natural vegetation disrupts the local physical environment, exposing the borders of the remaining habitat to additional sunlight, wind, and rainfall.

effector In signal transduction, a plasma membrane–associated enzyme, activated by a G protein, that generates one or more second messengers. In

homeostatic feedback, the system that returns the condition to the set point if it has strayed away.

effector T cell A cell involved in effecting—bringing about—the specific immune response to an antigen.

effector-triggered immunity An inducible defense mechanism in which proteins produced by a pathogen trigger an immediate defense response in an affected plant; so-called *resistance proteins* carry out the response.

efferent arteriole The arteriole that receives blood from the glomerulus.

efferent division One of two divisions of the peripheral nervous system (PNS). The efferent division of the PNS carries signals from the CNS to the muscles and glands, which act as effectors to bring about the desired response.

efferent neuron A neuron that carries the signals indicating a response away from the interneuron networks to the effectors.

egg cell The female reproductive cell.

elastin A rubbery protein in some connective tissues that adds elasticity to the extracellular matrix—it is able to return to its original shape after being stretched, bent, or compressed.

electrical signal A means of animal communication in which a signaler emits an electric discharge that can be received by another individual.

electrical synapse In the nervous system, where plasma membranes of two connecting neurons are in direct contact and communication across such a synapse occurs by the direct flow of an electrical signal.

electrocardiogram (ECG) Graphic representation of the electrical activity within the heart, detected by electrodes placed on the body.

electrochemical gradient A difference in ion concentration and electric charge difference across a membrane.

electromagnetic spectrum The range of wavelengths or frequencies of electromagnetic radiation extending from gamma rays to the longest radio waves and including visible light.

electronegativity The measure of the tendency of an atom to attract electrons to itself in a chemical bond (that is, to become negative).

electron Negatively charged particle outside the nucleus of an atom.

electron microscope Microscope that uses electrons to illuminate the specimen.

electron transfer system Stage of cellular respiration in which high-energy electrons produced from glycolysis, pyruvate oxidation, and the citric acid cycle are delivered to oxygen by a sequence of electron carriers.

electroreceptor A specialized sensory receptor that detects electrical fields.

element A pure substance that cannot be broken down into simpler substances by ordinary chemical or physical techniques.

elimination Referring to the digestive tube, the process in which undigested materials are expelled through the anus.

elongation In transcription, the step in which RNA polymerase (RNA polymerase II in eukaryotes) moves along the gene extending the RNA chain, with the DNA continuing to unwind ahead of the enzyme. In translation, the step in which the assembled translation complex reads the string of codons in the mRNA one at a time while joining the specified amino acids into the polypeptide.

elongation factor (EF) A protein that aids in an elongation step of translation.

embryo An organism in its early stage of development, beginning in the first moments after fertilization.

embryo sac The female gametophyte of angiosperms, within which the embryo develops; it usually consists of seven cells: an egg cell, an endosperm mother cell, and five other cells with fleeting reproductive roles.

embryogenesis Stages of development from a fertilized egg to an embryo.

embryonic stem cell (ES cell) A type of stem cell in mammals that is pluripotent, meaning that it is capable of differentiating into many of the different cell types of the body, but not all of them. ES cells are found in a mass of cells inside an early-stage embryo.

embryophyte Any plant in which the embryo is retained within maternal tissue.

emergent property Characteristic that depends on the level of organization of matter, but does not exist at lower levels of organization.

emigration The movement of individuals out of a population.

endangered species A species in danger of extinction throughout all or a significant portion of its range.

endemic species A species that occurs in only one place on Earth.

endergonic reaction Reaction that can proceed only if free energy is supplied.

endocrine gland Any of several ductless secretory organs that secrete hormones into the blood or extracellular fluid.

endocrine system The system of glands that release their secretions (hormones) directly into the circulatory system.

endocytic vesicle Vesicle that carries proteins and other molecules from the plasma membrane to destinations within the cell.

endocytosis In eukaryotes, the process by which molecules are brought into the cell from the exterior involving a bulging in of the plasma membrane that pinches off to form an endocytic vesicle.

endoderm The innermost of the three primary germ layers of an embryo, which develops into the gastrointestinal tract and, in some animals, the respiratory organs.

endodermis The innermost layer of the root cortex; a selectively permeable barrier that helps control the movement of water and dissolved minerals into the stele.

endomembrane system In eukaryotes, a collection of interrelated internal membranous sacs that divide a cell into functional and structural compartments.

endometrium Lining of the uterus formed by layers of connective tissue with embedded glands and supplied by extensive blood vessels.

endomycorrhiza A mycorrhiza in which the fungal hyphae penetrate into cells of the root.

endoparasite A parasite that lives in the internal organs of its host organism.

endophyte A fungus that lives within plant tissues in a symbiotic relationship with the host.

endoplasmic reticulum (ER) In eukaryotes, an extensive interconnected network of cisternae that is responsible for the synthesis, transport, and initial modification of proteins and lipids.

endorphin One of a group of small proteins occurring naturally in the brain and around nerve endings that bind to opiate receptors and thus can raise the pain threshold.

endoskeleton An internal body structure, such as bones, that provides support.

endosperm Nutritive tissue inside the seeds of flowering plants.

endospore A small, metabolically inactive, asexual spore that develops within some bacterial cells when environmental conditions become unfavorable.

endosymbiont Organism that lives symbiotically within a host cell.

endosymbiotic theory The proposal that the membranous organelles of eukaryotic cells (mitochondria and chloroplasts) may have originated from mutualistic relationships between two prokaryotic cells.

endothelium A specialized type of simple squamous epithelial tissue that lines the entire circulatory system.

endotherm An animal that obtains most of its body heat from internal physiological sources.

endothermic Referring to a reaction that absorbs energy, that is, a reaction in which the products have more potential energy than the reactants.

endotoxin A lipopolysaccharide released from the outer membrane of the cell wall when a bacterium dies and lyses.

end-product inhibition *See* **feedback inhibition**.

energy The capacity to do work.

energy budget The total amount of energy that an organism can accumulate and use to fuel its activities.

energy coupling The process in living cells by which the hydrolysis of ATP is coupled to an endergonic reaction so that energy is not wasted as heat.

energy levels Regions of space within an atom where electrons are found. Also referred to as *shells*.

enhancer In eukaryotes, a sequence in the genome that increases transcription of a gene independently of its position, orientation, and distance from a promoter.

enteric nervous system A network of nerves in the wall of the digestive tract.

enterocoelom In deuterostomes, the body cavity pinched off by outpocketings of the archenteron.

enthalpy The potential energy in a system.

entropy Disorder, in thermodynamics.

enveloped virus A virus that has a surface membrane derived from its host cell.

enzymatic hydrolysis A process in which chemical bonds are broken by the addition of H^+ and OH^-, the components of a molecule of water.

enzymatic protein A membrane protein that confers a specific property on the membrane.

enzyme Protein that accelerates the rate of a cellular reaction.

enzyme specificity The ability of an enzyme to catalyze the reaction of only a single type of molecule or group of closely related molecules.

eosinophil A type of leukocyte that secretes substances that kill eukaryotic parasites such as worms.

epiblast The top layer of the blastodisc.

epicotyl The upper part of the axis of an early plant embryo, located between the cotyledons and the first true leaves.

epidermis A complex tissue that covers an organism's body in a single, continuous layer or sometimes in multiple layers of tightly packed cells.

epididymis A coiled storage tubule attached to the surface of each testis.

epigenetics A phenomenon in which a change in gene expression does not involve a change in the DNA sequence of the gene or of the genome.

epiglottis A flaplike valve at the top of the trachea.

epilimnion The top layer of the limnetic zone in a lake.

epinephrine A nontropic amine hormone secreted by the adrenal medulla.

epiphyte A plant that grows independently on other plants and obtains nutrients and water from the air.

epistasis Interaction of genes, with one or more alleles of a gene at one locus inhibiting or masking the effects of one or more alleles of a gene at a different locus.

epithelium (plural, *epithelia*) A tissue consisting of sheetlike layers of cells with little extracellular matrix material between them that protects the internal environment of the body and regulates the exchange of material between the internal and external environments.

epithelial tissue Tissue formed of sheetlike layers of cells that are usually joined tightly together, with little extracellular matrix material between them. They protect body surfaces from invasion by bacteria and viruses, and secrete or absorb substances.

epitope The small region of an antigen molecule to which BCRs or TCRs bind.

epsilonproteobacteria One of five subgroups of the proteobacteria group of bacteria. Epsilonproteobacteria consist of a small group of bacteria that inhabit the animal digestive tract.

equilibrium potential The electrical potential necessary to balance the diffusional potential of an ion at the plasma membrane of an axon.

equilibrium theory of island biogeography A hypothesis suggesting that the number of species on an island is governed by a give and take between the immigration of new species to the island and the extinction of species already there.

ER (endoplasmic reticulum) lumen The enclosed space surrounded by a cisterna.

erythrocyte A red blood cell, which contains hemoglobin, a protein that transports O_2 in blood.

erythropoietin (EPO) A hormone that stimulates stem cells in bone marrow to increase erythrocyte production.

esophagus A connecting passage of the digestive tube.

essential amino acid Any amino acid that is not made by the human body but must be taken in as part of the diet.

essential element Any of a number of elements required by living organisms to ensure normal reproduction, growth, development, and maintenance.

essential fatty acid Any fatty acid that the body cannot synthesize but needs for normal metabolism.

essential mineral Any inorganic element such as calcium, iron, or magnesium that is required in the diet of an animal.

essential nutrient Any of the essential amino acids, fatty acids, vitamins, and minerals required in the diet of an animal.

estivation Seasonal torpor in an animal that occurs in summer.

estradiol A form of estrogen.

estrogen Any of the group of female sex hormones.

estuary A habitat tidal seawater mixes with fresh water from rivers, streams, and runoff.

ethology A discipline that focuses on how animals behave in their natural environments.

ethylene A plant hormone that helps regulate seedling growth, stem elongation, the ripening of fruit, and the abscission of fruits, leaves, and flowers.

euchromatin The loosely packed regions of chromosomes that have a high level of chromosome packing.

etiolation A plant growth response to darkness, in which the plant's metabolic resources are channeled into elongation of the stem rather than leaf expansion and root growth.

eudicot A plant belonging to the Eudicotyledones, one of the two major classes of angiosperms; their embryos generally have two seed leaves (cotyledons), and their pollen grains have three grooves.

Euglenids A lineage of the Euglenozoa subgroup of the Excavata protist evolutionary group; free-living with anterior flagella.

Euglenozoa (euglenozoans) A subgroup of the Excavata protist evolutionary group; single-celled, highly motile cells that swim by means of flagella and that contain mitochondria characterized by disc-shaped cristae.

Eukarya The domain that includes all eukaryotes, organisms that contain a membrane-bound nucleus

within each of their cells; all protists, plants, fungi, and animals.

eukaryote Organism in which the DNA is enclosed in a nucleus.

eukaryotic chromosome A DNA molecule, with its associated proteins, in the nucleus of a eukaryotic cell.

Eumetazoa A subgroup of Animalia, comprising species that have tissues.

euploid An individual with a normal set of chromosomes.

Euryarchaeota A major group of the domain Archaea, members of which are found in different extreme environments. They include methanogens, extreme halophiles, and some extreme thermophiles.

eusocial A form of social organization, observed in some insect species, in which numerous related individuals—a large percentage of them sterile female workers—live and work together in a colony for the reproductive benefit of a single queen and her mate(s).

Eutheria A monophyletic group of mammals in which embryonic development is completed within the uterus of the mother.

eutrophic lake A lake that is rich in nutrients and organic matter.

evaporation Heat transfer through the energy required to change a liquid to a gas.

evolutionary developmental biology A field of biology that compares the genes controlling the developmental processes of different animals to determine the evolutionary origin of morphological novelties and developmental processes.

evolutionary divergence A process whereby natural selection or genetic drift causes populations to become more different over time.

exaptation A trait that is adaptive in a context different from the context in which it originally evolved.

Excavata A protist evolutionary group: most have a scooped-out feeding apparatus on the ventral surface of the cell; some have flagella; some are anaerobes; some are photosynthetic; and some are parasites.

excision repair Mechanism for correcting various kinds of DNA damage by removing erroneous or damaged bases and replacing them with the correct ones.

excitatory postsynaptic potential (EPSP) The change in membrane potential caused when a neurotransmitter opens a ligand-gated Na^+ channel and Na^+ enters the cell, making it more likely that the postsynaptic neuron will generate an action potential.

excretion The process by which the fluid containing waste materials—urine—is released from the body into the environment from the distal end of an excretory tubule.

excretory tubules Minute tubular structures that carry out osmoregulation and excretion.

exergonic reaction Reaction that has a negative ΔG because it releases free energy.

exocrine gland A gland that is connected to the epithelium by a duct and that empties its secretion at the epithelial surface.

exocytosis In eukaryotes, the process by which a secretory vesicle fuses with the plasma membrane and releases the vesicle contents to the exterior.

exodermis In the roots of some plants, an outer layer of root cortex that may limit water losses from roots and help regulate the absorption of ions.

exoenzyme An enzymatic protein released by a bacterium that digests plasma membranes and causes cells of the infected host to rupture and die.

exon An amino acid-coding sequence present in pre-mRNA that is retained in a spliced mRNA that is translated to produce a polypeptide.

exon shuffling Molecular evolutionary process that combines exons of two or more existing genes to produce a gene that encodes a protein with an unprecedented function.

exoskeleton A hard external covering of an animal's body that blocks the passage of water and provides support and protection.

exothermic Referring to a reaction that releases energy, that is, a reaction in which the products have less potential energy than the reactants.

exotoxin A toxic protein that leaks from or is secreted from a bacterium and interferes with the biochemical processes of body cells in various ways.

experimental data Information that describes the result of a careful manipulation of the system under study.

experimental variable The variable in a scientific study that is manipulated by the experimenter.

exploitative competition Form of competition in which two or more individuals or populations use the same limiting resources.

exponential model of population growth Mathematical model that describes unlimited population growth.

expression vector Cloning vector that, in addition to normal features, contains regulatory sequences for transcription and translation of a gene.

extensor muscle A muscle that extends a joint, thereby increasing the angle between the two bones; with a flexor muscle, constitutes the antagonistic pair of muscles that control a joint's movement.

external environment The environment outside of the bodies of multicellular organisms.

external fertilization The process in which sperm and eggs are shed into the surrounding water, occurring in most aquatic invertebrates, bony fishes, and amphibians.

external gill A gill that extends out from the body and lacks a protective covering.

extinction The death of the last individual in a species or the last species in a lineage.

extracellular digestion Digestion that takes place outside body cells, in a pouch or tube enclosed within the body.

extracellular fluid The fluid occupying the spaces between cells in multicellular animals.

extracellular matrix (ECM) A molecular system that supports and protects cells and provides mechanical linkages.

extraembryonic membrane A primary tissue layer extended outside the embryo that conducts nutrients from the yolk to the embryo, exchanges gases with the environment outside the egg, or stores metabolic wastes removed from the embryo.

F factor Plasmid in a donor bacterial cell that confers on that cell the ability to conjugate with a recipient bacterial cell.

F pilus Structure on the cell surface that allows an F^+ donor bacterial cell (a cell containing an F factor) to attach to an F^- recipient bacterial cell (a cell lacking an F factor). Also referred to as a *sex pilus*.

F^- cell Recipient cell in conjugation between bacteria; it lacks an F factor.

F^+ cell Donor cell in conjugation between bacteria; it contains an F factor.

F_1 generation The first generation of offspring from a genetic cross.

F_2 generation The second generation of offspring from a genetic cross produced by interbreeding F_1 individuals.

facilitated diffusion Mechanism by which polar and charged molecules diffuse across membranes with the help of transport proteins.

facilitation hypothesis A hypothesis that explains ecological succession, suggesting that species modify the local environment in ways that make it less suitable for themselves but more suitable for colonization by species typical of the next successional stage.

facultative anaerobe An organism that can live in the presence or absence of oxygen, using oxygen when it is present and living by fermentation under anaerobic conditions.

familial (hereditary) cancer Cancer that runs in a family.

family A Linnaean taxonomic category that ranks below an order and above a genus.

family planning program A program that educates people about ways to produce an optimal family size on an economically feasible schedule.

fast aerobic fiber A type of *fast muscle fiber* that contracts relatively quickly and powerfully; it has abundant mitochondria, a rich blood supply, and a high concentration of myoglobin.

fast anaerobic fiber A type of *fast muscle fiber* that contracts relatively quickly and powerfully; it typically contains high concentrations of glycogen, relatively few mitochondria, and a more limited blood supply than fast aerobic fibers.

fast block to polyspermy The barrier set up by the wave of depolarization triggered when sperm and egg fuse, making it impossible for other sperm to enter the egg.

fast muscle fiber A muscle fiber that contracts relatively quickly and powerfully.

fat Neutral lipid that is semisolid at biological temperatures.

fate map Mapping of adult or larval structures onto the region of the embryo from which each structure developed.

fat-soluble vitamin A vitamin that dissolves in liquid fat or fatty oils, in addition to water.

fatty acid One of two components of a neutral lipid, containing a single hydrocarbon chain with a carboxyl group linked at one end.

fatty acid-derived molecules One of four molecular classes into which most hormones and local regulators fall; involved in paracrine and autocrine regulation.

feather A sturdy, lightweight structure of birds, derived from scales in the skin of their ancestors.

feces Condensed and compacted digestive contents in the large intestine.

feedback inhibition In enzyme reactions, regulation in which the product of a reaction acts as a regulator of the reaction. Also referred to as *end-product inhibition*.

feeding The uptake of food from the surroundings.

fermentation Process in which electrons carried by NADH are transferred to an organic acceptor molecule rather than to an electron transfer system.

fertilization The fusion of the nuclei of an egg and sperm cell, which initiates development of a new individual.

fetus A developing human from the eighth week of gestation onward, at which point the major organs and organ systems have formed.

fever A condition characterized by a rise in body temperature above the normal range.

fiber In sclerenchyma, an elongated, tapered, thick-walled cell that gives plant tissue its flexible strength.

fibrin A protein necessary for blood clotting; fibrin forms a weblike mesh that traps platelets and red blood cells and holds a clot together.

fibrinogen A plasma protein that plays a central role in the blood-clotting mechanism.

fibroblast The type of cell that secretes most of the collagen and other proteins in the loose connective tissue.

fibronectin A class of glycoproteins that aids in the attachment of cells to the extracellular matrix and helps hold the cells in position.

fibrous root system A root system that consists of branching roots rather than a main taproot; roots tend to spread laterally from the base of the stem.

filament In flowers, the stalk of a stamen, which supports the anther.

filtration The nonselective movement of some water and a number of solutes—ions and small molecules, but not large molecules such as proteins—into the proximal end of the renal tubules through spaces between cells.

first law of thermodynamics The principle that energy can be transferred and transformed but it cannot be created or destroyed.

first messenger The extracellular signal molecule in signal transduction pathways controlled by G-protein–coupled receptors.

fission The mode of asexual reproduction in which the parent separates into two or more offspring of approximately equal size.

fixed action pattern A highly stereotyped instinctive behavior; when triggered by a specific cue, it is performed over and over in almost exactly the same way.

flagellum (plural, *flagella*) A long, threadlike cellular appendage responsible for movement; found in both prokaryotes and eukaryotes, but with different structures and modes of locomotion.

flatulence Gas expelled through the anus.

flexor muscle A muscle that decreases the angle between the two bones at a joint, thereby performing the opposite action to an extensor muscle; with an extensor muscle, constitutes the antagonistic pair of muscles that control a joint's movement.

floral shoot A reproductive shoot that gives rise to one or more flowers.

florigen The name conferred in the 1930s on the hormone hypothesized to regulate flowering.

flower The reproductive structure of angiosperms, consisting of floral parts grouped on a stem; the structure in which seeds develop.

fluid feeder An animal that obtains nourishment by ingesting liquids that contain organic molecules in solution.

fluid mosaic model Model proposing that the membrane consists of a fluid phospholipid bilayer in which proteins are embedded and float freely.

fMRI *See* **functional magnetic resonance imaging.**

follicle cell A cell that grows from ovarian tissue and nourishes the developing egg.

follicle-stimulating hormone (FSH) The pituitary hormone that stimulates oocytes in the ovaries to continue meiosis and become follicles. During follicle enlargement, FSH interacts with luteinizing hormone to stimulate follicular cells to secrete estrogens.

food chain A depiction of the trophic structure of a community; a portrait of who eats whom.

food vacuole A membrane-bound sac used for digestion.

food web A set of interconnected food chains with multiple links.

foot The muscular structure in the bodies of many mollusks that is responsible for crawling or burrowing locomotion.

Foraminifera (forams) A lineage of the Rhizaria protist evolutionary group; heterotrophic marine organisms with shells consisting of organic matter reinforced by calcium carbonate.

forebrain The largest division of the brain, which includes the cerebral cortex and basal ganglia. It is credited with the highest intellectual functions.

foreskin A loose fold of skin that covers the glans of the penis.

formula The name of a molecule written in chemical shorthand.

fossil The remains or traces of an organism of a past geologic age embedded and preserved in Earth's crust.

foundation species A species that defines the nature of a community by creating locally stable environmental conditions.

founder effect An evolutionary phenomenon in which a population that was established by just a few colonizing individuals has only a fraction of the genetic diversity seen in the population from which it was derived.

fovea The small region of the retina around which cones are concentrated in mammals and birds with eyes specialized for daytime vision.

fragmentation A type of vegetative reproduction in plants in which cells or a piece of the parent break off, then develop into new individuals.

frameshift mutation Mutation in a protein-coding gene that causes the reading frame of an mRNA transcribed from the gene to be altered, resulting in the production of a different, and nonfunctional, amino acid sequence in the polypeptide.

free energy The energy in a system that is available to do work.

freeze–fracture technique Technique in which experimenters freeze a block of cells rapidly, then fracture the block to split the lipid bilayer and expose the hydrophobic membrane interior.

frequency-dependent selection A form of natural selection in which rare phenotypes have a selective advantage simply because they are rare.

frontal lobe Division of the cerebral hemisphere of the brain responsible for voluntary motor activity, expressing language, and elaboration of thought.

fruit A mature ovary, often with accessory parts, from a flower.

fruiting body In some fungi, a stalked, spore-producing structure such as a mushroom.

functional genomics The study of the functions of genes and of other parts of the genome.

functional groups Small, reactive groups of atoms.

functional magnetic resonance imaging (fMRI) A scanning technique for studying activities of the cerebral cortex.

fundamental niche The range of conditions and resources that a population can possibly tolerate and use.

fungal spore In fungi, a haploid reproductive cell that can germinate and produce a mycelium.

furrow In cytokinesis, a groove that girdles the cell and gradually deepens until it cuts the cytoplasm into two parts.

G_0 phase The phase of the cell cycle in eukaryotes in which many cell types stop dividing.

G_1 phase The initial growth stage of the cell cycle in eukaryotes, during which the cell makes proteins and other types of cellular molecules, but not nuclear DNA.

G_2 phase The phase of the cell cycle in eukaryotes during which the cell continues to synthesize proteins and to grow, completing interphase.

gallbladder The organ that stores bile between meals, when no digestion is occurring.

gametangium In certain plants and fungi, a cell or organ in which gametes are produced.

gamete A haploid cell, an egg or sperm. Haploid cells fuse during sexual reproduction to form a diploid zygote.

gametic isolation A prezygotic reproductive isolating mechanism caused by incompatibility between the sperm of one species and the eggs of another; may prevent fertilization.

gametogenesis The formation of male and female gametes.

gametophyte In organisms in which alternation of generations occurs, notably plants and green algae, the multicellular haploid generation that produces gametes. *Compare* **sporophyte.**

gammaproteobacteria One of five subgroups of the proteobacteria group of bacteria. Gammaproteobacteria are the largest subgroup of the proteobacteria; they are metabolically diverse, and are known for facultative anaerobes that inhabit mammalian colons.

ganglion A functional concentration of nervous system tissue composed principally of nerve-cell bodies, usually lying outside the central nervous system.

ganglion cell A type of neuron in the retina of the eye that receives visual information from photoreceptors via various intermediate cells such as bipolar cells, amacrine cells, and horizontal cells.

gap gene In *Drosophila* embryonic development, one of the first activated set of segmentation genes that progressively subdivide the embryo into regions, determining the segments of the embryo and the adult. Gap genes subdivide the embryo along the anterior–posterior axis into broad regions that later develop into several distinct segments.

gap junction Junction that opens direct channels allowing ions and small molecules to pass directly from one cell to another.

gastric glands Glands in the stomach wall that contain cells which secrete some of the products needed to digest food.

gastric juice A substance secreted by the stomach that contains the digestive enzyme pepsin.

gastric pits Entrances to gastric glands.

gastrin A hormone secreted in response to stimulation of chemoreceptors in the stomach by the presence of food molecules, particularly proteins.

gastroesophageal sphincter Ring of muscle at the junction of the esophagus and the stomach that controls the movement of a bolus of food into the stomach.

gastrovascular cavity A saclike body cavity with a single opening, a mouth, which serves both digestive and circulatory functions.

gastrula The developmental stage resulting when the cells of the blastula migrate and divide once cleavage is complete.

gastrulation The second major process of early development in most animals, which produces an embryo with three distinct primary tissue layers.

gated channel Ion transporter in a membrane that switches between open, closed, or intermediate states.

gemma (plural, *gemmae*) Small cell mass that forms in cuplike growths on a thallus.

gene A specific sequence of base pairs in a genome containing the code for a protein molecule or one of its parts, or for functioning RNA molecules such as tRNA and rRNA.

gene cloning DNA cloning when it involves a gene.

gene duplication Any process that produces two identical copies of a gene in an organism's genome. It is the simplest and most common mechanism for the evolution of new genes.

gene expression The process by which information encoded in genes guides the production of RNA molecules and proteins.

gene flow The transfer of genes from one population to another through the movement of individuals or their gametes.

gene isoform One of the different forms of mRNAs produced by transcription of particular protein-coding genes.

gene pool The sum of all alleles at all gene loci in all individuals in a population.

gene targeting The knocking out, replacement, or addition of a gene in a genome.

gene therapy Correction of genetic disorders using genetic engineering techniques.

general transcription factor (basal transcription factor) In eukaryotes, a protein that binds to the promoter of a gene and recruits and orients RNA polymerase II to initiate transcription at the correct place.

generalized transduction Transfer of bacterial genes between bacteria using virulent phages that have incorporated random DNA fragments of the bacterial genome.

generation time The average time between the birth of an organism and the birth of its offspring.

genetic code The nucleotide information that specifies the amino acid sequence of a polypeptide.

genetic counseling Counseling that allows prospective parents to assess the possibility that they might have a child affected by a genetic disorder.

genetic distance method An approach to phylogenetic analysis that calculates the overall proportion of nucleotide bases that differ among species.

genetic drift Random fluctuations in allele frequencies as a result of chance events; usually reduces genetic variation in a population.

genetic engineering The use of DNA technologies to alter genes for practical purposes.

genetic equilibrium The point at which neither the allele frequencies nor the genotype frequencies in a population change in succeeding generations.

genetic marker A mutation or genetic variant that gives a distinguishable phenotype. The two types of genetic markers are *gene markers* and *DNA markers*.

genetic recombinants Nonparental combinations of alleles. In eukaryotes they result from crossing-over in meiosis.

genetic recombination The process by which the combinations of alleles for different genes in two parental individuals become shuffled into new combinations in offspring individuals.

genetic screen A technique to search through a mutagenized population of organisms to find individuals with the mutant phenotypes of interest.

genetic screening Biochemical or molecular tests for identifying inherited disorders applied to children and adults, or to newborn infants in hospitals.

genetically altered organism An organism that has its genome altered to change a genetic trait or traits.

genetically modified organism (GMO) A term for a genetically altered organism whose genome has been engineered to introduce or change a genetically controlled trait.

genomic imprinting Pattern of inheritance in which the expression of a nuclear gene is based on whether an individual organism inherits the gene from the male or female parent.

genome All of the DNA in the cells of a living organism, or all of the DNA or RNA comprising the genetic material of a virus.

genomics The branch of biology that characterizes entire genomes, including sequencing genomes and identifying likely protein coding and noncoding segments of DNA, determining the specific functions of protein-coding genes and noncoding DNA sequences, and comparing genomes of different organisms to see how the genomes have evolved.

genotype The genetic constitution of an organism in terms of its genes and alleles.

genotype frequency The percentage of individuals in a population possessing a particular genotype.

genus A Linnaean taxonomic category ranking below a family and above a species.

geographical range The overall spatial boundaries within which a population lives.

germ cell An animal cell that is set aside early in embryonic development and gives rise to the gametes.

germination The onset of the growth of the embryo enclosed within a plant seed.

germ layers The three primary cell layers of the embryo: the outer ectoderm, the inner endoderm, and the mesoderm between the ectoderm and the endoderm.

germ-line cells Cells that develop into sperm or eggs.

germ-line gene therapy Experiment in which a gene is introduced into germ-line cells of an animal to correct a genetic disorder..

gestation The period of mammalian development in which the embryo develops in the uterus of the mother.

gibberellin Any of a large family of plant hormones that regulate aspects of growth, including cell elongation.

gill A respiratory organ formed as evagination of the body that extends outward into the respiratory medium.

gill arch One of the series of curved supporting structures between the slits in the pharynx of a chordate.

gill slit One of the openings in the pharynx of a chordate through which water passes out of the pharynx.

Ginkgophyta A plant phylum with a single living species, the ginkgo (or maiden-hair) tree.

gizzard The part of the digestive tube that grinds ingested material into fine particles by muscular contractions of the wall.

gland A cell or group of cells that produces and releases substances nearby, in another part of the body, or to the outside.

glans A soft, caplike structure at the end of the penis, containing most of the nerve endings producing erotic sensations.

glial cell A nonneuronal cell contained in the nervous tissue that physically supports and provides nutrients to neurons, provides electrical insulation between them, and scavenges cellular debris and foreign matter.

globulin A plasma protein that transports lipids (including cholesterol) and fat-soluble vitamins; a specialized subgroup of globulins, the immunoglobulins, constitute antibodies and other molecules contributing to the immune response.

glomerular filtration The process by which plasma filters through the glomerular capillaries into Bowman's capsule as blood flows through the glomerulus; it is the first step of urine formation.

glomerulus A ball of blood capillaries surrounded by Bowman's capsule in the human nephron.

glucagon A pancreatic hormone with effects opposite to those of insulin: it stimulates glycogen, fat, and protein degradation.

glucocorticoid A steroid hormone secreted by the adrenal cortex that helps maintain the blood concentration of glucose and other fuel molecules.

gluconeogenesis Glucose biosynthesis from intermediates of the glycolysis and citric acid pathways, as well as from molecules derived from those pathways. The gluconeogenesis reactions are the reverse of glycolysis.

glucose-dependent insulinotropic peptide (GIP) Hormone released in response to a meal entering the digestive tract that triggers insulin release from the pancreas.

glycocalyx A carbohydrate coat covering the cell surface.

glycogen Energy-providing carbohydrates stored in animal cells.

glycolipid A lipid molecule with carbohydrate groups attached.

glycolysis Stage of cellular respiration in which sugars such as glucose are partially oxidized and broken down into smaller molecules.

glycoprotein A protein with carbohydrate groups attached.

glycosidic bond Bond formed by the linkage of two α-glucose molecules with oxygen as a bridge between a carbon of the first glucose unit and a carbon of the second glucose unit.

Gnathostomata A monophyletic group of vertebrates that use jaws for feeding.

golden algae *See* **Chrysophyta.**

Golgi complex In eukaryotes, the organelle responsible for the final modification, sorting, and distribution of proteins and lipids.

Golgi tendon organ A proprioceptor of tendons.

gonad A specialized gamete-producing organ in which the germ cells collect. Gonads are the primary source of sex hormones in vertebrates: ovaries in the female and testes in the male.

gonadotropin A hormone that regulates the activity of the gonads (ovaries and testes).

gonadotropin-releasing hormone (GnRH) A tropic hormone secreted by the hypothalamus that causes the pituitary to make luteinizing hormone (LH) and follicle-stimulating hormone (FSH).

G-protein–coupled receptor In signal transduction, a surface receptor that responds to a signal by activating a G protein.

graded potential A change in membrane potential that does not necessarily trigger an action potential.

gradualism The view that Earth and its living systems changed slowly over its history.

gradualist hypothesis The hypothesis that large changes in either geological features or biological lineages result from the slow, continuous accumulation of small changes over time.

Gram stain technique A technique of staining bacteria to distinguish between types of bacteria with different cell wall compositions.

Gram-negative Describing bacteria that appear pink when stained using the Gram stain technique.

Gram-positive Describing bacteria that appear purple when stained using the Gram stain technique.

granum Structure in the chloroplasts of higher plants formed by thylakoids stacked one on top of another.

gravitropism A directional growth response to Earth's gravitational pull that is induced by mechanical and hormonal influences.

gray crescent A crescent-shaped region of the underlying cytoplasm at the side opposite the point of sperm entry exposed after fertilization when the pigmented layer of cytoplasm rotates toward the site of sperm entry.

gray matter Areas of densely packed nerve cell bodies and dendrites in the brain and spinal cord.

Great Oxygen Event (GOE) The dramatic increase in atmospheric oxygen level around 2.5 to 2.3 billion years ago. GOE has been linked traditionally to the evolution of oxygenic photosynthesis.

greater vestibular gland One of two glands located slightly below and to the left and right of the opening of the vagina in women. They secrete mucus that provides lubrication, especially when the woman is sexually aroused.

green algae *See* **Chlorophyta.**

greenhouse effect A phenomenon in which certain gases foster the accumulation of heat in the lower atmosphere, maintaining warm temperatures on Earth.

gross primary productivity The rate at which producers convert solar energy into chemical energy.

ground meristem The primary meristematic tissue in plants that gives rise to ground tissues, mostly parenchyma.

ground tissue system One of the three basic tissue systems in plants; includes all tissues other than dermal and vascular tissues.

growth factor A molecule (typically a peptide hormone) that stimulates cell division of a target cell.

growth hormone (GH) A hormone that stimulates cell division, protein synthesis, and bone growth in children and adolescents, thereby causing body growth.

growth-inhibiting factor A molecule that inhibits cell division of a target cell.

guanine A purine that base-pairs with cytosine in nucleic acids.

guard cell Either of a pair of specialized crescent-shaped cells that control the opening and closing of stomata in plant tissue.

gut microbiome The microbiome—complete collection of microorganisms—associated with a digestive tract.

guttation The exudation of water from leaves as a result of strong root pressure.

gymnosperm A seed plant that produces "naked" seeds not enclosed in an ovary.

H⁺ pump *See* **proton pump.**

habitat The specific environment in which a population lives, as characterized by its biotic and abiotic features.

habitat fragmentation A process in which remaining areas of intact habitat are reduced to small, isolated patches.

habituation The learned loss of responsiveness to stimuli.

half-life The time it takes for half of a given amount of a radioisotope to decay.

Hamilton's rule A cost-benefit analysis that predicts when an individual should exhibit altruistic behavior to another individual.

haplodiploidy A pattern of sex determination in insects in which females are diploid and males are haploid.

haploid An organism or cell with only one copy of each type of chromosome in its nuclei.

Haplorhini A monophyletic group of Primates that includes animals with derived primate characters.

Hardy–Weinberg principle An evolutionary rule of thumb that specifies the conditions under which a population of diploid organisms achieves genetic equilibrium.

haustorium (plural, *haustoria*) The hyphal tip of a parasitic fungus that penetrates a host plant and absorbs nutrients from it; likewise in parasitic flowering plants, a root that can penetrate a host's tissues and absorb nutrients.

head The anteriormost part of an organism's body, containing the brain, sensory structures, and feeding apparatus. For a bacteriophage, the usually polyhedral part of the virus containing the genetic material.

head–foot In mollusks, the region of the body that provides the major means of locomotion and contains concentrations of nervous system tissues and sense organs.

heart The muscular organ that pumps fluid through the circulatory system.

heartwood The inner core of a woody stem; composed of dry tissue and nonliving cells that no longer transport water and solutes and may store resins, tannins, and other defensive compounds.

heat of vaporization The heat required to give water molecules enough energy of motion to break loose from liquid water and form a gas.

heat-shock protein (HSP) Any of a group of chaperone proteins that are present in all cells in all life forms. They are induced when a cell undergoes various types of environmental stresses such as heat, cold, and oxygen deprivation.

heavy chain The heavier of the two types of polypeptide chains that are found in immunoglobulin and antibody molecules.

helical virus A virus in which the protein subunits of the coat assemble in a rodlike spiral around the genome.

helper T cell A clonal cell that assists with the activation of B cells.

hemolymph The circulatory fluid of invertebrates with open circulatory systems, including mollusks and arthropods.

hepatic portal vein The blood vessel that leads to capillary networks in the liver.

Hepatophyta The phylum that includes liverworts and their bryophyte relatives.

herbicide A compound that, at proper concentration, kills plants.

herbivore An animal that obtains energy and nutrients primarily by eating plants.

herbivory The process in which herbivores consume plants.

hermaphroditism The mechanism in which both mature egg-producing and mature sperm-producing tissue are present in the same individual.

heterochromatin The densely packed regions of chromosomes that have a high level of chromosome packing.

heterochrony Changes in the relative rate of development of morphological characters.

heterogametic sex The sex that produces two types of gametes with respect to the sex chromosomes.

heterokaryon A fungal mycelium in which two or more genetically different nuclei are present.

heterosporous Producing two types of spores, "male" microspores and "female" megaspores.

heterotroph An organism that acquires energy and nutrients by eating other organisms or their remains.

heterozygote An individual with two different alleles of a gene.

heterozygote advantage An evolutionary circumstance in which individuals that are heterozygous at a particular locus have higher relative fitness than either homozygote.

heterozygous State of possessing two different alleles of a gene.

Hexapoda A clade within the Pancrustacea that includes insects and their closest relatives.

Hfr cell A special donor cell that can transfer genes on a bacterial chromosome to a recipient bacterium.

hibernation Extended torpor during winter.

high-throughput technique A method that facilitates the handling of many samples simultaneously, such as DNA molecules for sequencing or genes for analysis.

hindbrain The lower area of the brain that includes the brain stem, medulla oblongata, and pons.

hinge joint A type of joint that can move only in one direction to open or close the angle between the bones flanking it.

hippocampus A gray-matter center that is involved in sending information.

histamine An inflammatory signaling molecule.

histone acetylation The acetylation of histone tails in nucleosomes in eukaryotes, one mechanism that plays an important role in determining whether chromatin is transcriptionally inactive or active.

histone A small, positively charged (basic) protein that is complexed with DNA in the chromosomes of eukaryotes.

histone code A regulatory mechanism for altering eukaryotic chromatin structure and, therefore, gene activity, based on signals in histone tails represented by chemical modification patterns.

historical biogeography The study of the geographical distributions of plants and animals in relation to their evolutionary history.

homeobox A region of a homeotic gene that corresponds to an amino acid section of the homeodomain.

homeobox gene A gene containing the highly conserved homeobox sequence. Homeobox genes regulate the formation of many body structures during early embryonic development of animals, and play key roles in the development of plants and fungi. Many homeobox genes encode transcription factors that regulate the expression of other genes.

homeodomain An encoded transcription factor of each protein that binds to a region in the promoters of the genes whose transcription it regulates.

homeostasis The regulation of the internal environment to maintain it in a relatively stable state.

homeostatic control systems The processes and activities responsible for homeostasis.

homeotic transformation A major change in the developmental fate of a segment in the *Drosophila* embryo caused by a mutation in a *Hox* gene.

hominin A member of a monophyletic group of primates, characterized by an erect bipedal stance, which includes modern humans and their recent ancestors.

Hominoidea The monophyletic group of primates that includes apes and humans.

homogametic sex The sex that produces only one type of gamete with respect to the sex chromosomes.

homologies Characteristics shared by a set of species because they inherited them from their common ancestor.

homologous chromosomes The two chromosomes of each pair in a diploid cell—one of the pair derives from the maternal parent and the other derives from the paternal parent. Homologous chromosomes have the same genes, in the same order, in their DNA.

homologous genes Genes that have highly conserved sequences because they have evolved from a gene in a common ancestor.

homologous traits Characteristics that are similar in two species because they inherited the genetic basis of the trait from their common ancestor.

homoplasies Characteristics shared by a set of species, often because they live in similar environments, but not present in their common ancestor; often the product of convergent evolution.

homosporous Producing only one type of spore.

homozygote An individual with two copies of the same allele.

homozygous State of possessing two copies of the same allele.

horizon A noticeable layer of soil, such as topsoil, having a distinct texture and composition that varies with soil type.

horizontal cell A type of neuron that forms lateral connections among photoreceptor cells in the retina of the eye.

horizontal gene transfer Movement of genetic material between organisms other than by descent.

hormone A signaling molecule secreted by a cell that can alter the activities of any cell with receptors for it; in animals, typically a molecule produced by one tissue and transported via the bloodstream to another specific tissue to alter its physiological activity.

host A species that is fed on by a parasite.

host race A population of insects that may be reproductively isolated from other populations of the same species as a consequence of their adaptation to feed on a specific host plant species.

housekeeping gene A gene that is expressed in almost all cell types.

Hox **gene** A gene that specifies what each segment of the *Drosophila* embryo will become after metamorphosis.

human chorionic gonadotropin (hCG) A hormone that keeps the corpus luteum in the ovary from breaking down.

human immunodeficiency virus (HIV) A retrovirus that causes acquired immunodeficiency syndrome (AIDS).

humoral immunity *See* **antibody-mediated immunity.**

humus The organic component of soil remaining after decomposition of plants and animals, animal droppings, and other organic matter.

hybrid An organism produced by a mating between parents of different species or subspecies.

hybrid breakdown A postzygotic reproductive isolating mechanism in which hybrids are capable of reproducing, but their offspring have either reduced fertility or reduced viability.

hybrid inviability A postzygotic reproductive isolating mechanism in which a hybrid individual has a low probability of survival to reproductive age.

hybrid sterility A postzygotic reproductive isolating mechanism in which hybrid offspring cannot form functional gametes.

hybrid zone A geographical area where the hybrid offspring of two divergent populations or species are common.

hybridoma A B cell that has been induced to fuse with a cancerous lymphocyte called a *myeloma cell,* forming a single, composite cell.

hydration layer A surface coat of water molecules that covers other polar and charged molecules and ions.

hydrocarbon Molecule consisting of carbon linked only to hydrogen atoms.

hydrogen bond Noncovalent bond formed by unequal electron sharing between hydrogen atoms and oxygen, nitrogen, or sulfur atoms.

hydrologic alterations Human-induced changes in the pathways through which water moves in the hydrologic cycle.

hydrologic cycle The global cycling of water between the ocean, the atmosphere, land, freshwater ecosystems, and living organisms.

hydrolysis Reaction in which the components of a water molecule are added to functional groups as molecules are broken into smaller subunits.

hydrophilic In chemistry and biology, referring to polar molecules that associate readily with water.

hydrophobic In chemistry and biology, referring to nonpolar substances that are excluded by water and other polar molecules.

hydroponic culture A method of growing plants not in soil but with the roots bathed in a solution that contains water and mineral nutrients.

hydrosphere The component of the biosphere that encompasses all the waters on Earth, including oceans, rivers, and polar ice caps.

hydrostatic skeleton A structure consisting of muscles and fluid that, by themselves, provide support for the animal or part of the animal; no rigid support, such as a bone, is involved.

hydroxyl group Group consisting of an oxygen atom linked to a hydrogen atom on one side and to a carbon chain on the other side.

hymen A thin flap of tissue that partially covers the opening of the vagina.

hyperpolarized The condition of a neuron when its membrane potential is more negative than the resting value.

hypersensitive response A plant defense that physically cordons off an infection site by surrounding it with dead cells.

hypertension Commonly called *high blood pressure,* a medical condition in which blood pressure is chronically elevated above normal values.

hyperthermia The condition resulting when the heat gain of the body is too great to be counteracted by physiological responses.

hypertonic Solution containing dissolved substances at higher concentrations than the cells it surrounds.

hypha (plural, *hyphae*) Any of the threadlike filaments that form the mycelium of a fungus.

hypoblast The bottom layer of a blastodisc.

hypocotyl The region of a plant embryo's vertical axis between the cotyledons and the radicle.

hypodermis The innermost layer of the skin that contains larger blood vessels and additional reinforcing connective tissue.

hypolimnion The deep water of the profundal zone of a lake.

hypothalamus The portion of the brain that contains centers regulating basic homeostatic functions of the body and contributing to the release of hormones.

hypothermia A condition in which the core temperature falls below normal for a prolonged period.

hypothesis A tentative explanation for an observation, phenomenon, or scientific problem that can be tested by further investigation.

hypotonic Solution containing dissolved substances at lower concentrations than the cells it surrounds.

ice lattice A rigid, crystalline structure formed when a water molecule in ice forms four hydrogen bonds with neighboring molecules.

IgA The class of antibodies found mainly in secretions at particular locations in the body; IgA functions to bind to surface groups on pathogens and block their attachment to body surfaces.

IgD The class of antibodies that occurs with IgM as a receptor on the surfaces of B cells; the function of IgD is uncertain.

IgE The class of antibodies secreted by plasma cells of the skin and tissues lining the gastrointestinal tract and respiratory tract that binds to basophils and mast cells to trigger the release of histamine, thereby causing an inflammatory response.

IGF *See* **insulin-like growth factor.**

IgG The most abundant antibody type circulating in the blood and lymphatic system that is involved in primary and secondary immune responses.

IgM The first antibody type secreted by B cells in a primary immune response.

imbibition The movement of water into a seed as the water molecules are attracted to hydrophilic groups of stored proteins; the first step in germination.

immigration Movement of organisms into a population.

immune response The defensive reactions of the immune system.

immune system The body's system of defenses against disease, composed of certain white blood cells and antibodies.

immunoglobulin A specific protein substance produced by plasma cells to aid in fighting infection.

immunological memory The capacity of the immune system to respond more rapidly and vigorously to the second contact with a specific antigen than to the primary contact.

immunological tolerance The process that protects the body's own molecules from attack by the immune system.

imperfect flower A type of incomplete flower that has stamens or carpels, but not both.

imprinting The process of learning the identity of a caretaker and potential future mate during a critical period early in life.

inbreeding A special form of nonrandom mating in which genetically related individuals mate with each other.

incisors Flattened, chisel-shaped teeth of mammals, located at the front of the mouth, that are used to nip or cut food.

inclusive fitness The sum of individual fitness and the fitness gained through the reproduction by relatives.

incomplete dominance Condition in which the effects of recessive alleles can be detected to some extent in heterozygotes.

incomplete flower Flower lacking one or more of the four floral whorls.

incomplete metamorphosis In certain insects, a life cycle characterized by the absence of a pupal stage between the immature and adult stages.

incurrent siphon A muscular tube that brings water containing oxygen and food into the body of an invertebrate.

incus The second of the three sound-conducting middle ear bones in vertebrates, located between the malleus and the stapes.

independent assortment Mendel's principle that the alleles of the genes that govern two characters assort independently during formation of gametes. Mechanistically this is the case because any combination of chromosomes may be segregated to the spindle poles during meiosis I.

indeterminate cleavage A type of cleavage, observed in many deuterostomes, in which the developmental fates of the first few cells produced by mitosis are not determined as soon as cells are produced.

indeterminate growth Growth that is not limited by an organism's genetic program, so that the organism grows for as long as it lives; typical of many plants. *Compare* **determinate growth.**

induced mutation A mutation that occurs when an organism is exposed either deliberately or accidentally to a physical or chemical mutagen.

inducer Concerning regulation of gene expression in bacteria, a molecule that turns on the transcription of the genes in an operon.

inducible defense In plants, a defense response that arises as the result of contact with a threat, such as a pathogenic virus or bacterium.

inducible operon Operon whose expression is increased by an inducer molecule.

induction A mechanism in which one group of cells (the inducer cells) causes or influences another nearby group of cells (the responder cells) to follow a particular developmental pathway.

infection thread In the formation of root nodules on nitrogen-fixing plants, the tube formed by the plasma membrane of root hair cells as bacteria enter the cell.

infertility The inability for the female of a couple to get pregnant after 12 months of frequent, contraceptive-free intercourse.

inflammation The heat, pain, redness, and swelling that occur at the site of an infection.

inflorescence A cluster of flowers.

ingestion The feeding methods used to take food into the digestive cavity.

inheritance The transmission of DNA (that is, genetic information) from one generation to the next.

inhibin A peptide that, in females, is an inhibitor of FSH secretion from the pituitary thereby diminishing the signal for follicular growth. In males, inhibin inhibits FSH secretion from the pituitary, thereby decreasing spermatogenesis.

inhibiting hormone (IH) A hormone released by the hypothalamus that inhibits the secretion of a particular anterior pituitary hormone.

inhibition hypothesis A hypothesis suggesting that new species are prevented from occupying a community by whatever species are already present.

inhibitory postsynaptic potential (IPSP) A change in membrane potential caused when hyperpolarization occurs, pushing the neuron farther from threshold.

initial A plant cell that remains permanently as part of a meristem and gives rise to daughter cells that differentiate into specialized cell types.

initiation In transcription, the step in which the molecular machinery that carries out transcription assembles at the promoter and begins synthesizing an RNA copy of the gene. In translation, the step in which the translation components assemble on the start codon of the mRNA.

initiation factor (IF) A protein that aids an initiation step of translation.

initiator codon *See* **start codon.**

innate immune system A nonspecific line of defense against pathogens that includes inflammation, which creates internal conditions that inhibit or kill many pathogens, and specialized cells that engulf or kill pathogens or infected body cells. The innate system is found in all animals and is one of two components of the immune system in vertebrates.

innate immunity The initial response by the body to eliminate cellular pathogens that involves mechanisms of the innate immune system.

inner boundary membrane Membrane lying just inside the outer boundary membrane of a chloroplast, enclosing the stroma.

inner cell mass The dense mass of cells within the blastocyst that will become the embryo.

inner ear That part of the ear, particularly the cochlea, that converts mechanical vibrations (sound) into neural messages that are sent to the brain.

inner mitochondrial membrane Membrane surrounding the mitochondrial matrix.

inorganic molecule Molecule without carbon atoms in its structure.

inositol triphosphate (IP₃) In particular signal transduction pathways, a second messenger that activates transport proteins in the endoplasmic reticulum to release Ca^{2+} into the cytoplasm. IP_3 is involved in one of two major G-protein–coupled receptor–response pathways.

insertional mutagenesis A transposon insertion event that disrupts a gene and, thereby, prevents the synthesis of its protein.

insertion sequence A transposable element that contains only genes for its transposition.

instinctive behavior A genetically "programmed" response that appears in complete and functional form the first time it is used.

insulin A hormone secreted by beta cells in the islets, acting mainly on cells of nonworking skeletal muscles, liver cells, and adipose tissue (fat) to lower blood glucose, fatty acid, and amino acids levels, and promote the storage of those molecules.

insulin-like growth factor (IGF) A peptide released by target tissues in response to binding by growth hormone that directly stimulates growth processes.

integral protein Protein embedded in a phospholipid bilayer.

integration The third component of neural signaling involving the sorting and interpretation of neural messages and the determination of the appropriate response(s).

integrator In homeostatic feedback, the control center that compares a detected environmental change with a set point.

integument Skin.

intercalated disk The joining point between two cells in cardiac muscle.

interference competition Form of competition in which individuals fight over resources or otherwise harm each other directly.

interferon A cytokine produced by infected host cells affected by viral dsRNA, which acts both on the infected cell that produces it, an autocrine effect, and on neighboring uninfected cells, a paracrine effect.

interkinesis A brief interphase separating the two meiotic divisions.

interleukin In the antibody-mediated immune response, the molecule secreted by an antigen-presenting cell that activates the associated T cell.

intermediate disturbance hypothesis Hypothesis proposing that species richness is greatest in communities that experience fairly frequent disturbances of moderate intensity.

intermediate filament A cytoskeletal filament about 10 nm in diameter that provides mechanical strength to cells in tissues.

intermediate-day plant A plant that flowers only when day length falls between the values for long-day and short-day plants.

internal environment The fluid within an organism that supplies all the needs of individual cells.

internal fertilization The process in which sperm are released by the male close to or inside the entrance of the reproductive tract of the female.

internal gill A gill located within the body that has a cover providing physical protection for the gills. Water must be brought to internal gills.

interneuron A neuron that integrates information to formulate an appropriate response.

internode The region between two nodes on a plant stem.

interphase The first stage of the mitotic cell cycle, during which the cell grows and replicates its DNA before undergoing mitosis and cytokinesis.

interspecific competition The competition for resources between species.

interstitial fluid The fluid occupying the spaces between cells in multicellular animals.

intertidal zone The shoreline that is alternately submerged and exposed by tides.

intestinal villus A microscopic, fingerlike extension in the lining of the small intestine.

intestine The portion of digestive system where organic matter is hydrolyzed by enzymes secreted into the digestive tube. As muscular contractions of the intestinal wall move the mixture along, cells lining the intestine absorb the molecular subunits produced by digestion.

intracellular fluid (ICF) The fluid within cells.

intracellular digestion The process in which cells take in food particles by endocytosis.

intraspecific competition The dependence of two or more individuals in a population on the same limiting resource.

intrinsic rate of increase The maximum possible per capita population growth rate in a population living under ideal conditions.

intron A non-protein-coding sequence that interrupts the protein-coding sequence in a eukaryotic gene. Introns are removed by splicing in the processing of pre-mRNA to mRNA.

invagination The process in which cells changing shape and pushing inward from the surface produce an indentation, such as the dorsal lip of the blastopore.

invasive species A nonnative species that can occupy a wide range of habitats and competitively exclude native species from those habitats.

inversion Chromosomal alteration that occurs if a broken segment reattaches to the same

chromosome from which it was lost, but in reversed orientation, so that the order of genes in the segment is reversed with respect to the other genes of the chromosome.

invertebrate An animal without a vertebral column.

involution The process by which cells migrate into the blastopore.

ion A positively or negatively charged atom.

ion channel A channel protein that facilitates the transport of ions such as sodium, potassium, calcium, and chlorine ions. Ion channels are found in all eukaryotes.

ionic bond A bond that results from electrical attractions between atoms that gain or lose valence electrons completely.

ionotropic neurotransmitter receptor A receptor in the postsynaptic membrane to which a neurotransmitter binds directly, causing an associated ion channel gate to open or close thereby altering the flow of a specific ion or ions in the postsynaptic cell.

iris Of the eye, the colored muscular membrane that lies behind the cornea and in front of the lens, which by opening or closing determines the size of the pupil and hence the amount of light entering the eye.

islets of Langerhans Endocrine cells of the pancreas that secrete the peptide hormones insulin and glucagon into the bloodstream.

isomers Two or more molecules with the same chemical formula but different molecular structures.

isotonic Equal concentration of water inside and outside cells.

isotope A distinct form of the atoms of an element, all with the same number of protons but different number of neutrons.

iteroparity An evolved life history characteristic in which individuals reproduce multiple times during their lifetimes.

jasmonate Any of a group of plant hormones that help regulate aspects of growth and responses to stress, including attacks by predators and pathogens.

juvenile hormone (JH) A peptide hormone secreted by the corpora allata, a pair of glands just behind the brain in insects.

juxtaglomerular apparatus A structure in the kidney near a point where the distal convoluted tubule contacts the afferent arteriole carrying blood to the glomerulus; specialized tubule cells in the juxtaglomerular apparatus monitor the salt level of the fluid flowing past them in the tubule.

karyogamy In plants, the fusion of two sexually compatible haploid nuclei after cell fusion (plasmogamy).

karyotype The complete set of metaphase chromosomes of a species, arranged according to size and centromere position.

keeled sternum The ventrally extended breastbone of a bird to which the flight muscles attach.

ketone Molecule in which the carbonyl group is linked to a carbon atom in the interior of a carbon chain.

keystone species A species that has a greater effect on community structure than its numbers might suggest.

kilocalorie (kcal) The scientific unit equivalent to a Calorie and equal to 1,000 small calories.

kin selection An explanation for altruistic behavior to close relatives, allowing them to produce proportionately more surviving copies of the altruist's genes than the altruist might otherwise have produced on its own.

kinesis A change in the rate of movement or the frequency of turning movements in response to environmental stimuli.

kinetic energy The energy of motion.

kinetochore A specialized structure consisting of proteins attached to a centromere that mediates the attachment and movement of chromosomes along the mitotic spindle.

kinetochore microtubule In mitotic and meiotic spindles, a microtubule originating from a spindle pole and binding to a kinetochore of a chromosome.

Kinetoplastids A lineage of the Euglenozoa subgroup of the Excavata protist evolutionary group; nonphotosynthetic, heterotrophic cells that live as animal parasites and that contain a single mitochondrion with a large DNA–protein deposit called a kinetoplast.

kingdom A Linnaean taxonomic category that ranks below a domain and above a phylum.

Kingdom Animalia The taxonomic kingdom that includes all living and extinct animals.

Kingdom Fungi The taxonomic kingdom that includes all living or extinct fungi.

Kingdom Plantae The taxonomic kingdom encompassing all living or extinct plants.

knee-jerk reflex *See* **patellar tendon reflex.**

knockout mouse A mouse in which a gene in the genome has been knocked out so that its function is lost. A knockout mouse is made by a gene targeting method.

Korarchaeota A group of the domain Archaea consisting of one living species, with other members represented in sequences in environmental samples. This group diverged very early in the evolution of the Archaea.

Krebs cycle *See* **citric acid cycle.**

K-selected species Long-lived species that thrive in more stable environments.

labia majora A pair of fleshy, fat-padded folds that partially cover the labia minora.

labia minora Two folds of tissue that run from front to rear on either side of the opening to the vagina.

lac operon In *E. coli*, the cluster of genes for the catabolism of lactose, and adjacent sequences that control the expression of those genes.

Lac repressor The regulatory protein responsible for controlling gene expression of the negatively regulated *lac* operon.

lactate fermentation Reaction in which pyruvate is converted into lactate.

lactose intolerance The unpleasant intestinal symptoms that result when an individual loses the capacity to synthesize lactase, the enzyme that breaks down the milk sugar.

lagging strand The new DNA strand synthesized discontinuously during replication in the direction opposite to that of DNA unwinding.

lagging strand template The DNA template strand for the lagging strand.

landscape ecology The field that examines how large-scale ecological factors—such as the distribution of plants, topography, and human activity—influence local populations and communities.

language A written or spoken form of communication used to convey ideas.

larva A sexually immature stage in the life cycle of many animals that is morphologically distinct from the adult.

larynx The voice box.

latent state The time during which an animal virus (such as a pathogen) remains in an infected cell in an inactive form and cannot be isolated and identified.

lateral bud A bud on the side of a plant stem from which a branch may grow.

lateral geniculate nuclei Clusters of neurons located in the thalamus that receive visual information from the optic nerves and send it on to the visual cortex.

lateral inhibition Visual processing in which lateral movement of signals from a rod or cone proceeds to a horizontal cell and continues to bipolar cells with which the horizontal cell makes inhibitory connections, serving both to sharpen the edges of objects and enhance contrast in an image.

lateral line system The complex of mechanoreceptors along the sides of some fishes and aquatic amphibians that detect vibrations in the water.

lateral meristem A plant meristem that gives rise to secondary tissue growth. *Compare* **primary meristem.**

lateral root A root that extends away from the main root (or taproot).

lateralization A phenomenon in which some brain functions are more localized in one of the two hemispheres.

lateral-line system The complex of organs and sensory receptors along the sides of many fishes and amphibians that detects vibrations in water.

leaching The process by which soluble materials in soil are washed into a lower layer of soil or are dissolved and carried away by water.

leading strand The new DNA strand synthesized during replication in the direction of DNA unwinding.

leading strand template The DNA template strand for the leading strand.

leaf primordium A lateral outgrowth from the apical meristem that develops into a young leaf.

learned behavior A response of an animal that is dependent on having a particular kind of experience during development.

learning A process in which experiences stored in memory change the behavioral responses of an animal.

leghemoglobin An iron-containing, red-pigmented protein produced in root nodules during the symbiotic association between *Bradyrhizobium* or *Rhizobium* and legumes.

lek A display ground where males each possess a small territory from which they court attentive females.

lens The transparent, biconvex intraocular tissue that helps bring rays of light to a focus on the retina.

Lepidosauria A monophyletic group within Reptilia, including lizards, snakes, and sphenodontids.

leukocyte A white blood cell, which eliminates dead and dying cells from the body, removes cellular debris, and participates in defending the body against invading organisms.

Leydig cell A cell that produces the male sex hormones.

lichen A single vegetative body that is the result of an association between a fungus and a photosynthetic partner, often an alga.

life cycle The sequential stages through which individuals develop, grow, maintain themselves, and reproduce.

life history The lifetime pattern of growth, maturation, and reproduction that is characteristic of a population or species.

life table A chart that summarizes the demographic characteristics of a population.

ligament A fibrous connective tissue that connects bones to each other at a joint.

ligand A molecule that binds to another molecule.

ligand-gated ion channel A type of cell surface receptor that, in response to a ligand binding, changes conformation, opening or closing an ion channel, thereby allowing or blocking ion movement through the channel.

ligation The process of sealing two DNA molecules together with DNA ligase to produce a recombinant DNA molecule.

light chain The lighter of the two types of polypeptide chains found in immunoglobulin and antibody molecules.

light microscope Microscope that uses light to illuminate the specimen.

light-dependent reactions The first stage of photosynthesis, in which the energy of sunlight is absorbed and converted into chemical energy in the form of ATP and NADPH.

light-independent reactions The second stage of photosynthesis, in which electrons are used as a source of energy to convert inorganic CO_2 to an organic form. Also referred to as the *Calvin cycle.*

lignin A tough, rather inert polymer that strengthens the secondary walls of various plant cells and thus helps vascular plants to grow taller and stay erect on land.

limbic association cortex A region of the temporal lobe of the brain that is important for motivation, emotion, and memory.

limbic system A functional network formed by parts of the thalamus, hypothalamus, and basal nuclei, along with other nearby gray-matter centers—the amygdala, hippocampus, and olfactory bulbs—sometimes called the "emotional brain."

limiting nutrient An element in short supply within an ecosystem, the shortage of which limits productivity.

limnetic zone The sunlit, open water in a lake, beyond the zone where plants rooted in the bottom can grow.

linear electron flow In the light-dependent reactions of photosynthesis, the one-way movement of electrons from H_2O to $NADP^+$ via photosystem II, the plastoquinone pool, the cytochrome complex, plastocyanin, photosystem I, and ferredoxin.

linkage The phenomenon of genes being located on the same chromosome.

linkage map Map of a chromosome showing the relative locations of genes based on recombination frequencies.

linked genes Genes on the same chromosome.

linker A short segment of DNA extending between one nucleosome and the next in a eukaryotic chromosome.

lipase A pancreatic enzyme that hydrolyzes fats.

lipid One of a diverse group of water-insoluble, primarily nonpolar biological molecules composed mainly of hydrocarbons.

lipopolysaccharide A large molecule that consists of a lipid and a carbohydrate joined by a covalent bond.

lithosphere The component of the biosphere that includes the rocks, sediments, and soils of the crust.

littoral zone The shallow, sunlit water near the shore of a lake or pond.

liver A large organ whose many functions include aiding in digestion, removing toxins from the body, and regulating the chemicals in the blood.

lncRNA *See* **long noncoding RNA.**

loam Any well-aerated soil composed of a mixture of sand, clay, silt, and organic matter.

local homeostatic controls Controls that operate only within an organ where a change in the internal environment needs to be addressed.

local inflammation Inflammation that occurs at the site of an infection.

locus The particular site on a chromosome at which a gene is located.

logistic model of population growth Model of population growth that assumes that a population's per capita growth rate decreases as the population gets larger.

Lokiarchaeota A candidate group of the domain Archaea, members of which contain more eukaryotic-like genes than any of the other known archaeal species.

long-day plant A plant that flowers in spring when dark periods become shorter and day length becomes longer.

long noncoding RNA (lncRNA) A noncoding RNA—RNA that does not code for a protein—that is greater than 200 nucleotides long.

long-term memory Memory that stores information from days to years or even for life.

long-term potentiation A long-lasting increase in the strength of synaptic connections in activated neural pathways following brief periods of repeated stimulation.

loop of Henle A U-shaped bend of the proximal convoluted tubule.

loose connective tissue A tissue formed of sparsely distributed cells surrounded by a more or less open network of collagen and other glycoprotein fibers.

lophophore The circular or U-shaped fold with one or two rows of hollow, ciliated tentacles that surrounds the mouth of brachiopods, bryozoans, and phoronids and is used to gather food.

Lophotrochozoa A lineage of protostomes that includes phyla that do not shed their exoskeleton or cuticle.

lumen The inside of the digestive tube.

lung One of a pair of invaginated respiratory surfaces, buried in the body interior where they are less susceptible to drying out; the organs of respiration in mammals, birds, reptiles, and most amphibians.

luteinizing hormone (LH) A hormone secreted by the pituitary that stimulates the growth and maturation of eggs in females and the secretion of testosterone in males.

Lycophyta The plant phylum that includes club mosses and their close relatives.

lymph The interstitial fluid picked up by the lymphatic system.

lymph node One of many small, bean-shaped organs spaced along the lymph vessels that contain macrophages and other leukocytes that attack invading disease organisms.

lymphatic system An accessory system of vessels and organs that helps balance the fluid content of the blood and surrounding tissues and participates in the body's defenses against invading disease organisms.

lymphocyte A leukocyte that carries out most of its activities in tissues and organs of the lymphatic system. The main subtypes of lymphocytes play major roles in innate and adaptive immunity.

lysogenic cycle Cycle in which the DNA of the bacteriophage is integrated into the DNA of the host bacterial cell and may remain for many generations.

lysosome Membrane-bound vesicle containing hydrolytic enzymes for the digestion of many complex molecules.

lytic cycle The series of events from infection of one bacterial cell by a phage through the release of progeny phages from lysed cells.

M phase-promoting factor (MPF) A complex of M cyclin and cyclin-dependent kinase 1 (Cdk1).

The complex initiates mitosis and orchestrates some of its key events.

macroevolution Large-scale evolutionary patterns in the history of life, producing major changes in species and higher taxonomic groups.

macromolecule A single polymer molecular with a mass of 1,000 Da (daltons) or more.

macronucleus In ciliophorans, a single large nucleus that develops from a micronucleus but loses all genes except those required for basic "housekeeping" functions of the cell and for ribosomal RNAs.

macronutrient In humans, a mineral required in amounts ranging from 50 mg to more than 1 gram per day. In plants, a nutrient needed in large amounts for normal growth and development.

macrophage A type of phagocytic leukocyte that engulfs infected cells, pathogens, and cellular debris in damaged tissues, and helps activate lymphocytes carrying out an immune response.

magnetoreceptor A receptor found in some animals that navigate long distances which allows them to detect and use Earth's magnetic field as a source of directional information.

magnification The ratio of an object as viewed to its real size.

magnoliids An angiosperm group that includes magnolias, laurels, and avocados; they are more closely related to monocots than to eudicots.

major histocompatibility complex A large cluster of genes encoding the MHC proteins.

malignant tumor A tumor that has invaded and disrupted surrounding tissues. The tissue is now called a **cancer.**

malleus The outermost of the sound-conducting bones of the middle ear in vertebrates.

malnutrition A condition resulting from a diet in which intake of organic fuels is inadequate or whose assimilation of such fuels is abnormal (= **undernutrition**), or in which there is excessive intake of specific nutrients (= **overnutrition**).

Malpighian tubule The main organ of excretion and osmoregulation in insects, helping them to maintain water and electrolyte balance.

mammals A monophyletic group of tetrapods characterized by hair or fur, a high body temperature, and the production of milk to feed their young.

mammary glands Specialized organs of female mammals that produce energy-rich milk, a watery mixture of fats, sugars, proteins, vitamins, and minerals.

Mandibulata A clade within Arthropoda that includes millipedes, centipedes, crustaceans, and insects.

mantle One or two folds of the body wall that lines the shell and secretes the substance that forms the shell in mollusks.

map unit The unit of a linkage map, equivalent to a recombination frequency of 1%. Also referred to as a *centimorgan.*

maritime climate Climate tempered by ocean winds.

marsupium An external pouch on the abdomen of many female marsupials, containing the mammary glands, and within which the young continue to develop after birth.

mass The amount of matter in an object.

mass extinctions The disappearance of a large number of species in a relatively short period of geological time.

mass number The total number of protons and neutrons in the atomic nucleus.

mast cell A type of cell dispersed through connective tissue that releases histamine when activated by the death of cells, caused by a pathogen at an infection site.

master regulatory gene A gene encoding a transcription factor that controls genes for developmental events.

mastication (chewing) In mammals with teeth, the first step in the digestive process in which ingested food is sliced, torn, and ground into small pieces.

maternal chromosome The chromosome derived from the female parent of an organism.

maternal inheritance A type of uniparental inheritance in which all progeny (both males and females) have the phenotype of the female parent.

maternal-effect gene One of a class of genes that regulate the expression of other genes expressed by the mother during oogenesis and that control the polarity of the egg and, therefore, of the embryo.

mating The pairing of a male and a female for the purpose of sexual reproduction.

mating systems The social systems describing how males and females pair up.

mating type A genetically defined strain of an organism (such as a fungus) that can only mate with an organism of the opposite mating type; mating types are often designated + and −.

matter Anything that occupies space and has mass.

maximum likelihood methods A statistical technique that compares alternative phylogenetic trees with specific models of evolutionary change.

mechanical isolation A prezygotic reproductive isolating mechanism caused by differences in the structure of reproductive organs or other body parts.

mechanical processing In the digestive tube, the process of chewing, grinding, and tearing food chunks into smaller pieces.

mechanoreceptor A sensory receptor that detects mechanical energy, such as changes in pressure, body position, or acceleration. The auditory receptors in the ears are examples of mechanoreceptors.

medusa The tentacled, usually bell-shaped, free-swimming sexual stage in the life cycle of a cnidarian.

megapascal A unit of pressure used to measure water potential.

megaphyll A broad leaf having multiple veins.

megasporangia Sporangia in which in which megaspores form and give rise to female gametophytes that produce eggs.

megaspore A plant spore that develops into a female gametophyte; usually larger than a microspore.

meiosis The division of diploid cells to haploid progeny, consisting of two sequential rounds of nuclear and cellular division.

meiosis I The first division of the meiotic cell cycle in which homologous chromosomes pair and undergo an exchange of chromosome segments, and then the homologous chromosomes separate, resulting in two cells, each with the haploid number of chromosomes and with each chromosome still consisting of two chromatids.

meiosis II The second division of the meiotic cell cycle in which the sister chromatids in each of the two cells produced by meiosis I separate and segregate into different cells, resulting in four cells each with the haploid number of chromosomes.

melanocyte-stimulating hormone (MSH) A hormone secreted by the anterior pituitary that controls the degree of pigmentation in melanocytes.

melatonin A peptide hormone secreted by the pineal gland that helps maintain daily biorhythms.

membrane attack complexes (MAC) An abnormal activation of the complement (protein) portion of the blood, forming a cascade reaction that brings blood proteins together, binds them to the cell wall, and then inserts them through the cell membrane.

membrane potential An electrical voltage that measures the potential inside a cell membrane relative to the fluid just outside; it is negative under resting conditions and becomes positive during an action potential.

memory The storage and retrieval of a sensory or motor experience, or a thought.

memory B cell In antibody-mediated immunity, a long-lived cell expressing an antibody on its surface that can bind to a specific antigen. A memory B cell is activated the next time the antigen is encountered, producing a rapid secondary immune response.

memory cell An activated lymphocyte that circulates in the blood and lymph, ready to initiate a rapid immune response on subsequent exposure to the same antigen.

memory cytotoxic T cells Cells differentiated from cytotoxic T cells that are stored for subsequent cell-mediated immunity involving the same antigens.

memory helper T cell In cell-mediated immunity, a long-lived cell differentiated from a helper T cell, which remains in an inactive state in the lymphatic system after an immune reaction has run its course and ready to be activated on subsequent exposure to the same antigen.

meninges Three layers of connective tissue that surround and protect the spinal cord and brain.

menstrual cycle A cycle of approximately 1 month in the human female during which an egg is released from an ovary and the uterus is prepared to receive the fertilized egg; if fertilization does not occur, the endometrium breaks down, which releases blood and tissue breakdown products from the uterus to the outside through the vagina.

meristem An undifferentiated, permanently embryonic plant tissue that gives rise to new cells forming tissues and organs.

mesenteries Sheets of loose connective tissue, covered on both surfaces with epithelial cells, which suspend the abdominal organs in the coelom and provide lubricated, smooth surfaces that prevent chafing or abrasion between adjacent structures as the body moves.

mesoderm The middle layer of the three primary germ layers of an animal embryo, from which the muscular, skeletal, vascular, and connective tissues develop.

mesohyl The gelatinous middle layer of cells lining the body cavity of a sponge.

mesophyll The ground tissue located between the two outer leaf tissues, composed of loosely packed parenchyma cells that contain chloroplasts.

messenger RNA (mRNA) The RNA transcribed from a protein-coding gene. Translation of an mRNA produces a polypeptide.

metabolism The biochemical reactions that allow a cell or organism to extract energy from its surroundings and use that energy to maintain itself, grow, and reproduce.

metabotropic neurotransmitter receptor A receptor in the postsynaptic membrane, typically of the G-protein–coupled receptor type, to which a neurotransmitter binds as a first messenger, activating the receptor and triggering generation of a second messenger that brings about changes in ion-conducting channels in the membrane.

Metamonada (metamonads) A subgroup of the Excavata protist evolutionary group containing the Diplomonadida and the Parabasala. Metamonads have multiple flagella and lack mitochondria.

metamorphosis A reorganization of the form of certain animals during postembryonic development.

metanephridia The excretory tubules of most annelids and mollusks.

metaphase The phase of mitosis during which the spindle reaches its final form and the spindle microtubules move the chromosomes into alignment at the spindle midpoint.

metapopulation A group of neighboring populations that exchange individuals.

metastasis The spreading of a malignant tumor through the blood system or lymphatic system, forming new tumors at other locations in the body.

Metatheria A monophyletic group of mammals in which the gestation of young is brief, and development is completed in a marsupium.

micelles The tiny droplets produced when fats are emulsified.

microbiome The complete collection of microorganisms associated with an organism or part of an organism.

microbody Small, membrane-bound organelle that carries out vital reactions linking metabolic pathways.

microclimate The abiotic conditions immediately surrounding an organism.

microevolution Small-scale genetic changes within populations, often in response to shifting environmental circumstances or chance events.

microfilament A cytoskeletal filament composed of actin.

microfossil The remains of a cell that has decayed and been filled in by calcium carbonate or silica.

micronucleus In ciliophorans, one or more diploid nuclei that contains a complete complement of genes, functioning primarily in cellular reproduction.

micronutrient Any mineral required by an organism only in trace amounts.

microphyll A narrow leaf having a single strand of vascular tissue (a vein).

micropyle A small opening at one end of an ovule through which the pollen tube passes before fertilization.

microRNA (miRNA) One of the major types of small regulatory RNAs in eukaryotes. miRNAs are involved in RNA interference (RNAi).

microsatellite *See* **short tandem repeat (STR)**.

microscope Instrument of microscopy with different magnifications and resolutions of specimens.

microscopy Technique for producing visible images of objects that are too small to be seen by the human eye.

microsporangia Sporangia in which microspores form and five rise to male gametophytes that produce sperm.

microspore A plant spore from which a male gametophyte develops; usually smaller than a megaspore.

microsporidium (plural, *microsporidia*) A fungal parasite of animals; many mycologists believe they make up a possible sixth phylum within the Kingdom Fungi.

microtubule A cytoskeletal component formed by the polymerization of tubulin into rigid, hollow rods about 25 nm in diameter.

microtubule organizing center (MTOC) An anchoring point near the center of a eukaryotic cell from which most microtubules extend outward.

microvilli Fingerlike projections forming a brush border in epithelial cells that cover the villi.

midbrain The uppermost of the three segments of the brainstem, serving primarily as an intermediary between the rest of the brain and the spinal cord.

middle ear The air-filled cavity containing three small, interconnected bones: the malleus, incus, and stapes.

middle lamella Layer of gel-like polysaccharides that holds together walls of adjacent plant cells.

migration The predictable seasonal movement of animals from the area where they are born to a distant and initially unfamiliar destination, returning to their birth site later.

mimic The species in Batesian mimicry that resembles the model.

mimicry A form of defense in which one species evolves an appearance resembling that of another.

mineralocorticoid A steroid hormone secreted by the adrenal cortex that regulates the levels of Na^+ and K^+ in the blood and extracellular fluid.

minimal medium A growth medium containing the minimal ingredients that enable a nonmutant organism, such as *E. coli* or *Neurospora crassa,* to grow.

minimum viable population size The smallest population size that is likely to survive both predictable and unpredictable environmental variation.

miRNA *See* **microRNA.**

miRNA-induced silencing complex (miRISC) Protein complex containing an miRNA that binds to sequences in the 3′ UTRs of target mRNAs, resulting in either inhibition of translation of the mRNAs or their degradation.

mismatch repair Repair mechanism that removes a segment of a newly synthesized DNA strand containing a base-pair mismatch, and replacing it with a new segment complementary to the template strand.

missense mutation A base-pair substitution mutation in a protein-coding gene that results in a different amino acid in the encoded polypeptide than the normal one.

mitochondrial electron transfer system Series of electron carriers that alternately pick up and release electrons, ultimately transferring them to their final acceptor, oxygen.

mitochondrial matrix The innermost compartment of the mitochondrion.

mitochondrion (plural, *mitochondria*) Membrane-bound organelle found in the cells of all eukaryotic groups of organism that is responsible for synthesis of most of the ATP in eukaryotic cells.

mitosis Nuclear division that produces daughter nuclei that are exact genetic copies of the parental nucleus.

model The species in Batesian mimicry that is resembled by the mimic.

model organism An organism with characteristics that make it a particularly useful subject of research because it is likely to produce results widely applicable to other organisms.

modern synthesis A unified theory of evolution developed in the middle of the twentieth century.

molarity (*M*) The number of moles of a substance dissolved in 1 L of solution.

molars Posteriormost teeth of mammals, with a broad chewing surface for grinding food.

mold Asexual, spore-producing stage of many multicellular fungi.

mole (mol) Amount of substance that contains as many atoms or molecules as there are atoms in exactly 12 g of carbon-12, which is 6.022×10^{23}.

molecular clock A technique for dating the time of divergence of two species or lineages, based on the number of molecular sequence differences between them.

molecular geometry The three-deimensional arrangement of the atoms in a molecule; in other words, its shape.

molecular phylogenetics Approach of using DNA or amino acid sequence comparisons to determine evolutionary relationships among organisms.

molecular replicator A molecule that is able to store and reproduce genetic information.

molecular weight The weight of a molecule in grams, equal to the total mass number of its atoms.

molecule A unit composed of atoms combined chemically in fixed numbers and ratios.

molt-inhibiting hormone (MIH) A peptide neurohormone secreted by a gland in the eye stalks of crustaceans that inhibits ecdysone secretion.

monilophyta The plant phylum of ferns and their close relatives.

monoclonal antibody An antibody that reacts only against the same segment (epitope) of a single antigen.

monocot A plant belonging to the Monocotyledones, one of the two major classes of angiosperms; monocot embryos have a single seed leaf (cotyledon) and pollen grains with a single groove.

monocyte A type of leukocyte that enters damaged tissue from the bloodstream through the endothelial wall of the blood vessel and then differentiates into a macrophage.

monoecious Having both "male" flowers (which possess only stamens) and "female" flowers (which possess only carpels).

monogamous Mating behavior in which one male and one female form a long-term association.

monohybrid An F_1 heterozygote produced from a genetic cross that involves a single character.

monohybrid cross A genetic cross between two individuals that are each heterozygous for the same pair of alleles.

monophyletic taxon A group of organisms that includes a single ancestral species and all of its descendants.

monoploid An individual with one set of chromosomes instead of the usual two.

monosaccharides The smallest carbohydrates, containing three to seven carbon atoms.

monotreme A lineage of mammals that lay eggs instead of bearing live young.

monounsaturated Fatty acids with one double bond.

monsoon cycle A wind pattern that brings seasonally heavy rain to a region by blowing moisture-laden air from the sea to the land.

morphogenesis Orderly, genetically programmed changes in the size, shape, and proportion of body parts of an organism; the process by which specialized tissues and organs form.

morphological species concept The concept that all individuals of a species share measurable traits that distinguish them from individuals of other species.

morphology The form or shape of an organism, or of a part of an organism.

morula The first stage of animal development, a solid ball or layer of blastomeres.

motile Capable of self-propelled movement.

motor neuron An efferent neuron that carries signals to skeletal muscle.

motor unit A block of muscle fibers that is controlled by branches of the axon of a single efferent neuron.

mRNA *See* **messenger RNA.**

mRNA splicing Process that removes introns from pre-mRNAs and joins exons together.

mucosa The lining of the gut that contains epithelial and glandular cells.

mucous cells Cells of gastric pits that secrete alkaline mucus, which protects the mucosal layer of the stomach.

Müllerian duct The bipotential primitive duct associated with the gonads that leads to a cloaca.

Müllerian mimicry A form of defense in which two or more unpalatable species share a similar appearance.

multicellular organism Individual consisting of interdependent cells.

multigene family A family of homologous genes in a genome. The members of a multigene family have all evolved from one ancestral gene, and therefore have similar DNA sequences and produce proteins with similar structures and functions.

multiple alleles More than two different alleles of a gene.

multiple fruit A fruit that develops from the enlarged ovaries of several flowers clustered in an inflorescence. Pineapple is an example.

multiple sclerosis An autoimmune disease resulting from an attack against a protein of the myelin sheaths insulating the surfaces of neurons.

multipotent In development, referring to cells derived from pluripotent cells; multipotent cells give rise to cells with particular functions.

muscle fiber A bundle of elongated, cylindrical cells that make up skeletal muscle.

muscle spindle A bundle of small, specialized muscle cells wrapped with the dendrites of afferent neurons.

muscle tissue Cells that have the ability to contract (shorten) forcibly.

muscle twitch A single, weak contraction of a muscle fiber.

muscularis The muscular coat of a hollow organ or tubular structure.

mutagen A physical or chemical agent that produces a mutation.

mutagenesis The production of mutations in the laboratory by exposure to a mutagen.

mutation A spontaneous and heritable change in DNA.

mutualism A symbiotic interaction between species in which both partners benefit.

mycelium A network of branching hyphae that constitutes the body of a multicellular fungus.

mycobiont The fungal component of a lichen.

mycorrhiza A mutualistic symbiosis in which fungal hyphae associate intimately with plant roots.

myelin sheath Glial cells wrapping around axons in a jellyroll fashion; the myelin sheath acts as an electrical insulator.

myofibril A cylindrical contractile element about 1 mm in diameter that run lengthwise inside the muscle fiber cell.

myogenic heart A heart that maintains its contraction rhythm with no requirement for signals from the nervous system.

myoglobin An oxygen-storing protein closely related to hemoglobin.

myosin A protein that interacts with the protein actin to cause muscle contraction.

Myriapoda A group within Mandibulata that includes centipedes and millipedes.

Myxinoidea An early group of vertebrates, including hagfishes.

Na^+/K^+-ATPase *See* Na^+/K^+ pump.

Na^+/K^+ pump Pump that pushes 3 Na^+ out of the cell and 2 K^+ into the cell in the same pumping cycle. Also referred to as the *sodium–potassium pump* or as *Na^+/K^+-ATPase.*

NAD^+ *See* **nicotinamide adenine dinucleotide.**

NADPH *See* **nicotinamide adenine dinucleotide phosphate.**

Nanoarchaeota A group of the domain Archaea consisting of one living species that lives as a symbiont and species represented by DNA samples.

nastic movement In plants, a reversible response to nondirectional stimuli, such as mechanical pressure or humidity.

natural history The branch of biology that examines the form and variety of organisms in their natural environments.

natural killer (NK) cell A type of lymphocyte that destroys virus-infected cells.

natural selection The evolutionary process by which alleles that increase the likelihood of survival and the reproductive output of the individuals that carry them become more common in subsequent generations.

natural theology A belief that knowledge of God may be acquired through the study of natural phenomena.

navigation A wayfinding mechanism in which an animal moves toward a specific destination, using both a compass and a "mental map" of where it is in relation to the destination.

negative control In the regulation of gene expression, when genes are expressed unless they are switched off by a repressor regulatory protein.

negative feedback The primary mechanism of homeostasis, in which a stimulus—a change in the external or internal environment—triggers a response that compensates for the environmental change.

negative pressure breathing Muscular contractions that expand the lungs, lowering the pressure of the air in the lungs and causing air to be pulled inward.

nekton Animals that can actively swim against water currents.

nematocyst A coiled thread, encapsulated in a cnidocyte, that cnidarians fire at prey or predators, sometimes releasing a toxin through its tip.

nephron A specialized excretory tubule that contributes to osmoregulation and carries out excretion, found in all vertebrates.

neritic zone The shallow water of the oceans above the continental shelves.

Nernst equation An equation to calculate the equilibrium potential for a single ion with differing concentrations across a membrane.

nerve A bundle of axons enclosed in connective tissue and all following the same pathway.

nerve cord A bundle of nerves that extends from the central ganglia to the rest of the body, connected to smaller nerves.

nerve net A simple nervous system that coordinates responses to stimuli but has no central control organ, or brain.

nervous tissue Tissue that contains neurons, which serve as lines of communication and control between body parts.

net primary productivity The chemical energy remaining in an ecosystem after a producer's cellular respiration is deducted from the gross primary productivity.

neural circuit A connection between axon terminals of one neuron and the dendrites or cell body of a second neuron; typically a neural circuit consists of an afferent (sensory) neuron connected to one or more interneurons, and an efferent neuron.

neural crest A band of cells that arises early in the embryonic development of vertebrates near the region where the neural tube pinches off from the ectoderm; later, the cells migrate and develop into unique structures.

neural plate Ectoderm thickened and flattened into a longitudinal band, induced by notochord cells.

neural signaling Communication by neurons; that is, the process by which an animal responds appropriately to a stimulus.

neural tube A hollow tube in vertebrate embryos that develops into the brain, spinal cord, spinal nerves, and spinal column.

neuroendocrine signaling A type of cell signaling in the endocrine system in which specialized neurosecretory neurons release neurohormones into the circulatory system when stimulated appropriately.

neurogenic heart A heart that beats under the control of signals from the nervous system.

neurohormone The type of hormone secreted by a neurosecretory neuron.

neuromuscular junction The junction between a nerve fiber and the muscle it supplies.

neuron An electrically active cell of the nervous system responsible for controlling behavior and body functions.

neuroscience The integrated study of the structure, function, and development of the nervous system.

neurosecretory neuron A neuron that releases a neurohormone into the circulatory system when appropriately stimulated.

neurotransmitter A chemical released by an axon terminal at a chemical synapse.

neurulation The process in vertebrates by which organogenesis begins with development of the nervous system from ectoderm.

neutral lipid Energy-storing molecule consisting of a glycerol backbone and three fatty acid chains.

neutral variation hypothesis An evolutionary hypothesis that some variation at gene loci coding for enzymes and other soluble proteins is neither favored nor eliminated by natural selection.

neutralization One of two important mechanisms that clear foreign antigens from the body; toxins produced by an invading pathogen are inactivated (neutralized) by antibodies.

neutron Uncharged subatomic particle in the nucleus of an atom.

neutrophil A type of phagocytic leukocyte that engulfs pathogens and tissue debris in damaged tissues.

nicotinamide adenine dinucleotide (NAD^+) A coenzyme that serves as an electron carrier.

nicotinamide adenine dinucleotide phosphate (NADPH) In photosynthesis, the molecule that carries electrons that are pushed to high energy levels by absorbed light.

nitrification A metabolic process in which certain soil bacteria convert ammonia or ammonium ions into nitrites that are then converted by other bacteria to nitrates, a form usable by plants.

nitrogenase The enzyme that catalyzes the reduction of gaseous nitrogen (N_2) to ammonium (NH_4^+) in nitrogen fixation.

nitrogen cycle A biogeochemical cycle that moves nitrogen between the huge atmospheric pool of gaseous molecular nitrogen and several much smaller pools of nitrogen-containing compounds in soils, marine and freshwater ecosystems, and living organisms.

nitrogen fixation A metabolic process in which certain bacteria and cyanobacteria convert molecular nitrogen into ammonia and ammonium ions, forms usable by plants.

nitrogenous base A nitrogen-containing molecule that accepts protons.

nociceptor A sensory receptor that detects tissue damage or noxious chemicals; their activity registers as pain.

node In phylogenetic trees, a node is a branching point from which two descendants emerge. In plants, the point on a stem where one or more leaves are attached. In systematics, the point of intersection of two branches of a phylogenetic tree.

node of Ranvier The gap between two Schwann cells, which exposes the axon membrane directly to extracellular fluids.

noncoding RNA An RNA molecule that is not translated to produce a polypeptide. Noncoding RNAs are transcribed from noncoding RNA genes.

noncoding RNA gene A gene encoding an RNA that is not translated; that is, a gene other than a protein-coding gene.

noncompetitive inhibition Inhibition of an enzyme reaction by an inhibitor molecule that binds to the enzyme at a site other than the active site and, therefore, does not compete directly with the substrate for binding to the active site.

noncyclic electron flow Pathway in photosynthesis in which electrons travel in a one-way direction from H_2O to $NADP^+$.

nondisjunction The failure of homologous pairs to separate during the first meiotic division or of chromatids to separate during the second meiotic division.

nonhistone protein All the proteins associated with DNA in a eukaryotic chromosome that are not histones.

nonkinetochore microtubule In mitotic and meiotic spindles, a microtubule originating from a spindle pole that does not bind to a kinetochore of a chromosome.

nonpolar association Association that occurs when nonpolar molecules clump together.

nonpolar covalent bond Bond in which electrons are shared equally or nearly equally; the atoms have no charge.

nonsense codon *See* **stop codon.**

nonsense mutation A base-pair substitution mutation in a gene in which the base-pair change results in a change from a sense codon to a nonsense codon in the mRNA. The polypeptide translated from the mRNA is shorter than the normal polypeptide because of the mutation.

nonshivering thermogenesis The generation of heat by oxidative mechanisms in nonmuscle tissue throughout the body.

nonvascular plant *See* **bryophyte.**

norepinephrine A nontropic amine hormone secreted by the adrenal medulla.

no-till farming The planting of fields using techniques that are less physically destructive of the soil and retain the nutrient-dense remains of a previous season's crop.

notochord A flexible rodlike structure, constructed of fluid-filled cells surrounded by tough connective tissue, which supports a chordate embryo from head to tail.

N-terminal end The end of a polypeptide chain with an $—NH_3^+$ group.

nuclear envelope In eukaryotes, membranes separating the nucleus from the cytoplasm.

Nucleariidae (nucleariids) A protist group consisting of heterotrophic, predominantly spherical amoebae with radiating, fine pseudopods that are not supported by microtubules. Nucleariids are a subgroup of the opisthokonts and are more closely related to fungi than to animals.

nuclear localization signal A short amino acid sequence in a protein that directs the protein to the nucleus.

nuclear pore complex A large, octagonally symmetrical, cylindrical structure that functions to exchange molecules between the nucleus and cytoplasm and prevents the transport of material not meant to cross the nuclear membrane. A nuclear pore—a channel through the complex—is the path for the exchange of molecules.

nuclease An enzyme that digests a nucleic acid molecule.

nucleoid The central region of a prokaryotic cell with no boundary membrane separating it from the cytoplasm, where DNA replication and RNA transcription occur.

nucleolus The nuclear site of rRNA transcription, processing, and ribosome assembly in eukaryotes.

nucleoplasm The liquid or semiliquid substance within the nucleus.

nucleoporin A type of protein that makes up a nuclear pore complex.

nucleoside Molecule containing only a nitrogenous base and a five-carbon sugar.

nucleosidase An enzyme that, with a phosphatase enzyme, breaks down nucleosides into the component nitrogenous bases, five-carbon sugars, and phosphates.

nucleosome The basic structural unit of chromatin in eukaryotes, consisting of DNA wrapped around a histone core.

nucleosome core particle An eight-protein particle formed by the combination of two molecules each of H2A, H2B, H3, and H4, around which DNA winds for almost two turns.

nucleotide The monomer of nucleic acids, consisting of a five-carbon sugar, a nitrogenous base, and a phosphate.

nucleotidase An enzyme that breaks down nucleotides into nucleosides.

nucleotide-excision repair A type of excision repair mechanism for repairing DNA damage involving bulky distortion to the DNA. Nucleotide-excision repair involves recognition of the damage, removing the DNA segment with the damage, and replacing the removed DNA with new DNA.

nucleus The central region of eukaryotic cells, separated by membranes from the surrounding cytoplasm, where DNA replication and messenger RNA transcription occur. In the nervous system, a nucleus is a concentration of nerve cells within the central nervous system that have related functions.

null hypothesis A statement of what might be seen if the hypothesis being tested is not correct.

null model A conceptual model that predicts what one would see if a particular factor had no effect.

nutrition The processes by which an organism takes in, digests, absorbs, and converts food into organic compounds.

Nymphaeales The basal angiosperm lineage that includes water lilies.

obligate aerobe A microorganism that uses oxygen for cellular respiration and requires oxygen in its surroundings to support growth.

obligate anaerobe A microorganism that cannot use oxygen and can grow only in the absence of oxygen.

observational data Basic information on biological structures or the details of biological processes.

oceanic zone The deep ocean water beyond the continental shelves.

occipital lobe Division of the cerebral hemisphere of the brain responsible for initial processing of visual input.

ocellus (plural, *ocelli*) The simplest eye, which detects light but does not form an image.

oil Neutral lipid that is liquid at biological temperatures.

Okazaki fragments The short lengths of lagging strand DNA produced by discontinuous replication.

olfactory bulb A gray-matter center that relay inputs from odor receptors to both the cerebral cortex and the limbic system.

oligodendrocyte A type of glial cell that populates the CNS and is responsible for producing myelin.

oligotrophic lake A lake that is poor in nutrients and organic matter, but rich in oxygen.

omasum One of the four stomach chambers of a ruminant; in this chamber, water is absorbed from the mass of digesting material.

ommatidium (plural, *ommatidia*) A faceted visual unit of a compound eye.

omnivore An animal that feeds at several trophic levels, consuming plants, animals, and other sources of organic matter.

oncogene A gene capable of inducing one or more characteristics of cancer cells.

one gene–one enzyme hypothesis Hypothesis showing the direct relationship between genes and enzymes.

one gene–one polypeptide hypothesis Restatement of the one gene–one enzyme hypothesis, taking into account that some proteins consist of more than one polypeptide and not all proteins are enzymes.

oocyte A developing gamete that becomes an ootid at the end of meiosis.

oogenesis The process of producing eggs.

oogonium A cell that enters meiosis and gives rise to gametes, produced by mitotic divisions of the germ cells in females.

Oomycota (water molds, white rusts, and downy mildews) A lineage of the Stramenopila subgroup of the Chromalveolata protist evolutionary group; funguslike organisms that lack chloroplasts and live as heterotrophs.

open circulatory system An arrangement of the circulatory system in some invertebrates in which, when the heart contracts, arteries leaving the heart release a bloodlike fluid, hemolymph, directly into body spaces called sinuses that surround organs.

open reading frame (ORF) Segment of a protein-coding gene or an mRNA transcribed from such a gene that involves a start codon separated by a multiple of three nucleotides from one of the stop codons. The ORF is a potential polypeptide-coding sequence.

operant conditioning A form of associative learning in which animals learn to link a voluntary activity, an operant, with its favorable consequences, the reinforcement.

operator A DNA regulatory sequence that controls transcription of an operon.

operculum A lid or flap of bone serving as the gill cover in some fishes.

operon A cluster of prokaryotic genes organized into a single transcription unit and their associated regulatory sequences.

Opisthokonta (opisthokonts) A protist evolutionary group; a broad group of eukaryotes in which a single posterior flagellum is found at some stage in the life cycle.

opisthosoma The rear end of a chelicerate's body, derived from the abdomen in ancestral arthropods.

opsin One of several different proteins that bond covalently with the light-absorbing pigment of rods and cones (retinal).

optic chiasm Location just behind the eyes where the optic nerves converge before entering the base of the brain, a portion of each optic nerve crossing over to the opposite side.

optimal foraging theory A set of mathematical models that predict the diet choices of animals as they encounter a range of potential food items.

oral hood Soft fleshy structure at the anterior end of a cephalochordate that frames the opening of the mouth.

orbital The region of space where the electron "lives" most of the time.

order A Linnaean taxonomic category of organisms that ranks above a family and below a class.

organ Two or more different tissues integrated into a structure that carries out a specific function.

organ of Corti An organ within the cochlear duct that contains the sensory hair cells detecting sound vibrations transmitted to the inner ear.

organ system The coordinated activities of two or more organs to carry out a major body function such as movement, digestion, or reproduction.

organelles In eukaryotic cells, the nucleus and other specialized structures important for cell function.

organic acid (carboxylic acid) Acid for which the characteristic functional group is a carboxyl group (—COOH).

organic molecule Molecule based on carbon.

organismal ecology An ecological discipline in which researchers study the genetic, biochemical, physiological, morphological, and behavioral adaptations of organisms to their abiotic environments.

organogenesis Rearrangements of the three germ layers during animal development to produce tissues and organs.

orgasm A sensation of intense physical pleasure that is the peak of excitement for sexual intercourse.

oriented cell division Cell division in different planes; establishes the overall shape of a plant organ.

origin of replication A specific region at which DNA replication commences. Bacterial chromosomes have single origins of replication (*ori*)

whereas eukaryotic chromosomes have multiple origins.

orthogenesis An obsolete theory that evolution is goal oriented, striving to perfect organisms.

orthologous group The collection of all of the orthologs of a gene in all species.

orthologs Genes in two or more different organisms that are closely related to one another evolutionarily, and perform the same function in every organism.

oscula One or more openings in a sponge through which water is expelled.

osmoconformer An animal in which the osmolarity of the cellular and extracellular solutions matches the osmolarity of the environment.

osmolarity The total solute concentration of a solution, measured in osmoles—the number of solute molecules and ions (in moles)—per liter of solution.

osmoreceptor A sensory receptor in the hypothalamus that responds to changes in the osmolarity of the fluid surrounding it, which reflects the osmolarity generally of the body fluids.

osmoregulation The regulation of water and ion balance.

osmoregulator An animal that uses control mechanisms to keep the osmolarity of cellular and extracellular fluids the same, but at levels that may differ from the osmolarity of the surroundings.

osmosis The passive transport of water across a selectively permeable membrane in response to solute concentration gradients, a pressure gradient, or both.

osmotic pressure The pressure that must be applied to a solution to prevent water movement across a membrane.

Osteichthyes A monophyletic group of vertebrates that includes bony fishes and their descendants.

osteoblast A cell that produces the collagen and mineral of bone.

osteoclast A cell that removes bone minerals and recycles them through the bloodstream.

osteocyte A mature bone cell.

osteon The structural unit of bone, consisting of a minute central canal surrounded by osteocytes embedded in concentric layers of mineral matter.

ostracoderm One of an assortment of extinct, jawless fishes that were covered with bony armor.

otolith One of many small crystals of calcium carbonate embedded in the otolithic membrane of the hair cells.

outer boundary membrane A smooth membrane that surrounds a chloroplast, enclosing the stroma.

outer ear The external structure of the ear, consisting of the pinna and meatus.

outer membrane In Gram-negative bacteria, an additional boundary membrane that covers the peptidoglycan layer of the cell wall.

outer mitochondrial membrane The smooth membrane covering the outside of a mitochondrion.

outgroup comparison A technique used to identify ancestral and derived characters by comparing the group under study to more distantly related species that are not otherwise included in the analysis.

ova *See* **ovum.**

oval window An opening in the bony wall that separates the middle ear from the inner ear.

ovarian cycle The cyclic events in the ovary leading to ovulation.

ovary In animals, the female gonad, which produces female gametes and reproductive hormones. In flowering plants, the enlarged base of a carpel in which one or more ovules develop into seeds.

overexploitation The excessive harvesting of an animal or plant species, potentially leading to its extinction.

overnutrition A form of malnutrition. The condition caused by excessive intake of specific nutrients.

oviduct The tube through which the egg moves from the ovary to the outside of the body.

oviparous Referring to animals that lay eggs containing the nutrients needed for development of the embryo outside the mother's body.

ovoviviparous Referring to animals in which fertilized eggs are retained within the body and the embryo develops using nutrients provided by the egg; eggs hatch inside the mother.

ovulation The process in which oocytes are released into the oviducts as immature eggs.

ovule In plants, the structure in a carpel in which a female gametophyte develops and fertilization takes place.

ovum (plural, *ova*) A female sex cell, or egg.

oxidation The partial or full loss of electrons from a substance.

oxidative phosphorylation Synthesis of ATP in which ATP synthase uses an H^+ gradient built by the electron transfer system as the energy source to make the ATP.

oxidized Substance—the electron donor—from which the electrons are lost during oxidation.

oxytocin A hormone that stimulates the ejection of milk from the mammary glands of a nursing mother.

P generation The parental individuals used in an initial genetic cross.

P site The site in the ribosome where the tRNA carrying the growing polypeptide chain is bound during translation.

pacemaker cell A specialized cardiac muscle cell in the upper wall of the right atrium that sets the rate of contraction in the heart.

paedomorphosis A common form of heterochrony in which juvenile characteristics are retained in a reproductive adult.

pairing *See* **synapsis.**

pair-rule gene In *Drosophila* embryonic development, one of the second activated set of segmentation genes that progressively subdivide the embryo into regions, determining the segments of the embryo and the adult. Pair-rule genes control the division of the embryo into units of two segments each.

paleobiology The study of ancient organisms.

PAMPs Acronym for pathogen-associated molecular patterns, pathogen-specific molecules that are recognized by the innate immune system. In plants, PAMPs launch specific inducible defenses.

pancreas A mixed gland composed of an exocrine portion that secretes digestive enzymes into the small intestine and an endocrine portion, the islets of Langerhans, that secretes insulin and glucagon.

pancreatic enzymes Digestive enzymes secreted by exocrine cells in the pancreas into ducts that empty into the lumen of the duodenum.

Pancrustacea A clade within the Mandibulata that includes crustaceans and insects.

papilla (plural, *papillae*) An outgrowth on the tongue. Taste buds are embedded in the papillae.

parabronchi In bird lungs, an array of fine, parallel tubes through which air flows, and across which a capillary network crosses in a perpendicular direction to produce a crosscurrent pattern of blood flow relative to the air flow.

paracrine regulation A type of cell signaling in the endocrine system in which a cell releases a signaling molecule that diffuses through the ECF and acts on nearby cells, resulting in local regulation.

paraphyletic taxon A group of organisms that includes an ancestral species and some, but not all, of its descendants.

parapodia Fleshy lateral extensions of the body wall of aquatic annelids, used for locomotion and gas exchange.

parasite An organism that feeds on the tissues of or otherwise exploits its host.

parasitism A symbiotic interaction in which one species, the parasite, uses another, the host, in a way that is harmful to the host.

parasitoid An insect species in which a female lays eggs in the larva or pupa of another insect species, and her young consume the tissues of the living host.

parasympathetic division The division of the autonomic nervous system that predominates during quiet, low-stress situations.

parathyroid gland One of a pair of glands that produce parathyroid hormone (PTH) (found only in tetrapod vertebrates).

parathyroid hormone (PTH) The hormone secreted by the parathyroid glands in response to a fall in blood Ca^{2+} levels.

Parazoa A subgroup of Animalia, comprising species that lack tissues.

parenchyma A ground tissue with cells having a thin primary wall, which is pliable and permeable to water. Parenchyma cells may be specialized for photosynthesis, storage, secretion, or other tasks.

parental Phenotypes identical to the original parental individuals.

parental chromosomes In genetics, the chromosomes with parental combinations of alleles.

parental investment The time and energy devoted to the production and rearing of offspring.

parietal cells Cells of gastric glands that secrete H^+ and Cl^-, which combine to form HCl in the lumen of the stomach.

parietal lobe Division of the cerebral hemisphere of the brain mainly responsible for receiving and processing sensory input.

parthenogenesis A mode of asexual reproduction in which animals produce offspring by the growth and development of an egg without fertilization.

partial diploid A condition in which part of the genome of a haploid organism is diploid. Recipients in bacterial conjugation between an Hfr and an F$^-$ cell become partial diploids for part of the Hfr bacterial chromosome.

partial pressure In a mixture of gases, the pressure of each individual gas.

parturition The process of giving birth.

passenger mutation In a typical tumor, the many mutations that do not have any effect on cancer progression.

passive immunity The acquisition of antibodies as a result of direct transfer from another person.

passive parental care The amount of energy invested in offspring—in the form of the energy stored in eggs or seeds or energy transferred to developing young through a placenta—before they are born.

passive transport The transport of substances across cell membranes without expenditure of energy, as in diffusion.

patch-clamp technique An experimental method to study how ions flowing through channels can change membrane potential.

patellar tendon reflex (knee-jerk reflex) A programmed movement that takes place without conscious effort in which a tap to the tendon just below the knee cap causes the leg to kick.

paternal chromosome The chromosome derived from the male parent of an organism.

pathogen-associated molecular patterns *See* **PAMPs.**

pathogenesis-related (PR) protein A hydrolytic enzyme that breaks down components of a pathogen's cell wall.

pattern formation The arrangement of organs and body structures in their proper three-dimensional relationships.

PCR (polymerase chain reaction) Process that amplifies a specific DNA sequence from a DNA mixture to an extremely large number of copies.

pectoral girdle A bony or cartilaginous structure in vertebrates that supports and is attached to the forelimbs.

pedicellariae Small pincers at the base of short spines in starfishes and sea urchins.

pedigree Chart that shows all parents and offspring for as many generations as possible, the sex of individuals in the different generations, and the presence or absence of a trait of interest.

pedigree analysis The study of a pedigree.

pelagic province The open water in the oceans.

pellicle A layer of supportive protein fibers located inside the cell, just under the plasma membrane, providing strength and flexibility instead of a cell wall.

pelvic girdle A bony or cartilaginous structure in vertebrates that supports and is attached to the hind limbs.

PEP carboxylase The enzyme in the C$_4$ pathway of plants that catalyzes the carbon fixation reaction producing oxaloacetate.

pepsin An enzyme made in the stomach that breaks down proteins.

pepsinogen The inactive precursor molecule for pepsin.

peptic ulcer Lesion in the stomach wall resulting from attack by HCl and pepsin.

peptide bond A link formed by a dehydration synthesis reaction between the amino group of one amino acid and the carboxyl group of a second.

peptide hormone A hormone consisting of a chain of 3 to more than 200 amino acids. Peptide hormones are released into the ECF and then enter the blood; they are involved in classical endocrine signaling and neuroendocrine signaling.

peptidoglycan A polymeric substance formed from a polysaccharide backbone tied together by short polypeptides, which is the primary structural molecule of bacterial cell walls.

peptidyl transferase An enzyme that catalyzes the reaction in which an amino acid is cleaved from the tRNA in the P site of the ribosome and forms a peptide bond with the amino acid on the tRNA in the A site of the ribosome.

peptidyl–tRNA A tRNA linked to a growing polypeptide chain containing two or more amino acids.

per capita growth rate The difference between the per capita birth rate and the per capita death rate of a population.

perception The conscious awareness of our external and internal environments derived from the processing of sensory input.

perennial A plant in which vegetative growth and reproduction continue year after year.

perfect flower A flower that has both male (stamen) and female (carpel) sexual organs.

perforin A protein secreted by natural killer cells of the immune system that creates pores in a virus-infected cell's membrane.

perfusion The flow of blood or other body fluids on the internal side of the respiratory surface.

pericarp The fruit wall.

pericycle A tissue of plant roots, located between the endodermis and the phloem, which gives rise to lateral roots.

periderm The outermost portion of bark; consists of cork, cork cambium, and secondary cortex.

peripheral nervous system (PNS) All nerve roots and nerves (motor and sensory) that supply the muscles of the body and transmit information about sensation (including pain) to the central nervous system.

peripheral protein Protein held to membrane surfaces by noncovalent bonds formed with the polar parts of integral membrane proteins or membrane lipids.

peristalsis The rippling motion of muscles in the intestine or other tubular organs characterized by the alternate contraction and relaxation of the muscles that propel the contents onward.

peritoneum The thin tissue derived from mesoderm that lines the abdominal wall and covers most of the organs in the abdomen.

peritubular capillary A capillary of the network surrounding the glomerulus.

permafrost Perpetually frozen ground below the topsoil.

peroxisome Microbody that produces hydrogen peroxide as a by-product.

PET *See* **positron emission tomography.**

petal Part of the corolla of a flower, often brightly colored.

petiole The stalk by which a leaf is attached to a stem.

Petromyzontoidea A clade of primitive vertebrates, including lampreys.

pH scale The numerical scale from 0 to 14 used by scientists to measure acidity.

Phaeophyta (brown algae) A lineage of the Stramenopila subgroup of the Chromalveolata evolutionary subgroup; photosynthetic autotrophs that range from microscopic forms to giant kelps.

phage *See* **bacteriophage.**

phagocyte A white blood cell that engulfs bacteria or other cellular debris by the process of phagocytosis.

phagocytosis Process in which some types of cells engulf bacteria or other cellular debris to break them down.

pharmacogenomics The study of how genes affect individual responses to drug therapy.

pharyngeal arches Embryonic features of all vertebrates; they contribute to the formation of the face, neck, mouth, larynx, and pharynx.

pharynx The throat. In some invertebrates, a protrusible tube used to bring food into the mouth for passage to the gastrovascular cavity; in mammals, the common pathway for air entering the larynx and food entering the esophagus.

phenolics Tannins and other chemically similar compounds that serve as defensive chemicals in plants. *See also* **alkaloids.**

phenomics The study of gene function by identifying changes in phenotypes.

phenotype The observable or measurable (biochemical, molecular) characteristics of an organism that are produced by an interaction between the genotype and the environment.

phenotypic variation Differences in appearance or function between individual organisms.

pheromone A distinctive volatile chemical released in minute amounts to influence the behavior of members of the same species.

phloem The food-conducting tissue of a vascular plant.

phloem sap The solution of water and organic compounds that flows rapidly through the sieve tubes of plants.

phosphatase An enzyme that, with a nucleosidase, breaks down a nucleoside to a nitrogenous base, a five-carbon sugar, and a phosphate. In general, a phosphatase removes a phosphate group from a molecule.

phosphate group Group consisting of a central phosphorus atom held in four linkages: two that bind —OH groups to the central phosphorus atom, a third that binds an oxygen atom to the central phosphorus atom, and a fourth that links the phosphate group to an oxygen atom.

phosphodiester bond The linkage of nucleotides in polynucleotide chains by a bridging phosphate group between the 5′ carbon of one sugar and the 3′ carbon of the next sugar in line.

phospholipid A phosphate-containing lipid.

phosphorus cycle A biogeochemical cycle in which weathering and erosion carry phosphate ions from rocks to soil and into streams and rivers, which eventually transport them to the ocean, where they are slowly incorporated into rocks.

phosphorylation The addition of a phosphate group to a molecule.

phosphorylation cascade A series of phosphorylation reactions catalyzed by a series of protein kinases.

photic zone Surface water of a lake or ocean that sunlight penetrates.

photoautotroph Photosynthetic organism that uses light as its energy source and carbon dioxide as its carbon source.

photobiont The photosynthetic component of a lichen.

photoheterotroph An organism that uses light as the ultimate energy source but obtains carbon in organic form rather than as carbon dioxide.

photoperiodism The response of plants to changes in the relative lengths of light and dark periods in their environment during each 24-hour period.

photophosphorylation The synthesis of ATP coupled to the transfer of electrons energized by photons of light.

photopigment Light-absorbing pigment.

photopsin One of three photopigments in which retinal is combined with different opsins.

photoreceptor A sensory receptor that detects the energy of light.

photorespiration A process that metabolizes a by-product of photosynthesis.

photoreversible The property of certain light-sensitive pigment proteins (such as phytochromes) in which light absorption triggers a shift from one conformation to another.

photosynthesis The conversion of light energy to chemical energy in the form of sugar and other organic molecules.

photosystem A large complex into which the light-absorbing pigments for photosynthesis are organized with proteins and other molecules.

photosystem I In photosynthesis, a protein complex in the thylakoid membrane that uses energy absorbed from sunlight to synthesize NADPH.

photosystem II In photosynthesis, a protein complex in the thylakoid membrane that uses energy absorbed from sunlight to synthesize ATP.

phototroph An organism that obtains energy from light.

phototropism The tendency of a plant shoot to bend toward a source of light.

phyletic gradualism hypothesis The hypothesis that most morphological change occurs gradually over long periods of time.

PhyloCode A formal set of rules governing phylogenetic nomenclature.

phylogenetic species concept A concept that seeks to delineate species as the smallest group of populations that can be united by shared derived characters.

phylogenetic tree A branching diagram depicting the evolutionary relationships of groups of organisms.

phylogeny The evolutionary history of a group of organisms.

phylum (plural, *phyla*) A major Linnaean division of a kingdom, ranking above a class.

physical barriers The first of three lines of defense that humans and other mammals have against threats of pathogens; parts of the organism that physically block access to a pathogen. Examples are skin and epithelial surfaces covering internal body cavities and ducts. Physical barriers are not part of the immune system.

physiological respiration The process by which animals exchange gases with their surroundings—how they take in oxygen from the outside environment and deliver it to body cells, and remove carbon dioxide from body cells and deliver it to the environment.

physiology The study of the functions of organisms—the physico-chemical processes of organisms.

phytoalexin A biochemical that functions as an antibiotic in plants.

phytochrome A blue-green pigmented plant chromoprotein involved in the regulation of light-dependent growth processes.

phytoplankton Microscopic, free-flowing aquatic plants and protists.

phytoremediation The use of plants to remove pollutants from the environment.

phytosterol A sterol that occurs in plant cell membranes.

piloting A wayfinding mechanism in which animals use familiar landmarks to guide their journey.

pilus (plural, *pili*) A hair or hairlike appendage on the surface of a prokaryote.

pinacoderm In sponges, an unstratified outer layer of cells.

pineal gland A light-sensitive, melatonin-secreting gland that regulates some biological rhythms.

pinna The external structure of the outer ear, which concentrates and focuses sound waves.

pinocytosis *See* bulk-phase endocytosis.

pith The soft, spongelike, central cylinder of plant stems, composed mainly of parenchyma.

pituitary gland A gland consisting mostly of two fused lobes suspended just below the hypothalamus by a slender stalk of tissue that contains both neurons and blood vessels; it interacts with the hypothalamus to control many physiological functions, including the activity of some other glands.

placenta A specialized temporary organ that connects the embryo and fetus with the uterus in mammals, mediating the delivery of oxygen and nutrients.

plasma The clear, yellowish fluid portion of the blood in which cells are suspended. Plasma consists of water, glucose and other sugars, amino acids, plasma proteins, dissolved gases, ions, lipids, vitamins, hormones and other signal molecules, and metabolic wastes.

plasma cell A large antibody-producing cell that develops from B cells.

plasma membrane The membrane surrounding the cytoplasm of all cells that is responsible for the regulation of substances moving into and out of cells. The plasma membrane is a lipid bilayer with embedded proteins.

plasmid A DNA molecule in the cytoplasm of certain prokaryotes, which often contains genes with functions that supplement those in the nucleoid and which can replicate independently of the nucleoid DNA and be passed along during cell division.

plasmodesma (plural, *plasmodesmata*) A minute channel that perforates a cell wall and contains extensions of the cytoplasm that directly connect adjacent plant cells.

plasmodial slime mold A slime mold of the class Myxomycetes.

plasmodium The composite mass of plasmodial slime molds consisting of individual nuclei suspended in a common cytoplasm surrounded by a single plasma membrane.

plasmogamy The sexual stage of fungi during which the cytoplasms of two genetically different partners fuse.

plasmolysis Condition due to outward osmotic movement of water, in which plant cells shrink so much that they retract from their walls.

plasticity In animals, with respect to the nervous system, the experience-dependent change in structure and function.

plastids A family of plant organelles that includes chloroplasts, amyloplasts, and chromoplasts.

plastron The ventral part of the shell of a turtle.

plate tectonics The geological theory describing how Earth's crust is broken into irregularly shaped plates of rock that float on its semisolid mantle.

platelet An oval or rounded cell fragment enclosed in its own plasma membrane, which is found in the blood; they are produced in red bone marrow by the division of stem cells and contain enzymes and other factors that take part in blood clotting.

pleiotropy Condition in which single genes affect more than one character of an organism.

pleura The double layer of epithelial tissue covering the lungs.

ploidy The number of chromosome sets of a cell or species.

plumule The rudimentary terminal bud of a plant embryo located at the end of the hypocotyl, consisting of the epicotyl and a cluster of tiny foliage leaves.

pluripotent In development, referring to cells derived from totipotent cells; pluripotent cells can give rise to most adult cell types.

polar association Association that occurs when polar molecules attract and align themselves with other polar molecules and with charged ions and molecules.

polar body A nonfunctional cell produced in oogenesis.

polar covalent bond Bond in which electrons are shared unequally.

polar nucleus In the embryo sac of a flowering plant, one of two nuclei that migrate into the center of the sac, become housed in a central cell, and eventually give rise to endosperm.

polar transport Unidirectional movement of a substance from one end of a cell (or other structure) to the other.

polarity The unequal distribution of yolk and other components in a mature egg.

pollen grain The male gametophyte of a seed plant.

pollen sac The microsporangium of a seed plant, in which pollen grains develop.

pollen tube A tube that grows from a germinating pollen grain and carries sperm cells to the ovary.

pollination The transfer of pollen to a seed plant's female reproductive parts typically by air current or on the bodies of animal pollinators.

pollutant Materials or energy in a form or quantity that organisms do not usually encounter.

poly(A) polymerase The enzyme in eukaryotes that adds a chain of adenine nucleotides—the poly(A) tail—to the cleaved 3′ end of a pre-mRNA.

poly(A) tail The string of A nucleotides added posttranscriptionally to the 3′ end of a cleaved pre-mRNA molecule and retained in the mRNA produced from it that enables the mRNA to be translated efficiently and protects it from attack by RNA-digesting enzymes in the cytoplasm.

polyadenylation signal Sequence near the 3′ end of a eukaryotic gene which, in the pre-mRNA transcript of the gene, specifies where the transcript should be cleaved. Once cleaved, a poly(A) tail is added to the 3′ end of the RNA.

polyandry A polygamous mating system in which one female mates with multiple males.

polygamous Mating behavior in which either males or females may have many mating partners.

polygenic inheritance Inheritance in which several to many different genes contribute to the same character.

polygyny A polygamous mating system in which one male mates with many females.

polyhedral virus A virus in which the coat proteins form triangular units that fit together like the parts of a geodesic sphere.

polymer A molecule assembled from subunit monomer molecules into a chain by covalent bonds.

polymerase chain reaction *See* **PCR**.

polymerization The process of assembly of a polymer from monomers.

polymorphism The existence of discrete variants of a character among individuals in a population.

polyp The tentacled, usually sessile stage in the life cycle of a cnidarian.

polypeptide The chain of amino acids formed by sequential peptide bonds.

polyphyletic taxon A group of organisms that belong to different evolutionary lineages and do not share a recent common ancestor.

polyploid An individual with one or more extra copies of the entire haploid complement of chromosomes.

polyploidy The condition of having one or more extra copies of the entire haploid complement of chromosomes.

polysaccharide Carbohydrate polymers with more than 10 linked monosaccharide monomers.

polysome The entire structure of an mRNA molecule and the multiple associated ribosomes that are translating it simultaneously.

polyspermy In animal reproduction, the phenomenon in which more than one sperm fertilizes an egg.

polyunsaturated Fatty acid with more than one double bond.

population A group of organisms of the same kind that live together in the same place.

population bottleneck An evolutionary event that occurs when a stressful factor reduces population size greatly and eliminates some alleles from a population.

population density The number of individuals per unit area or per unit volume of habitat.

population dynamics Changes in the characteristics of populations through time or over space.

population ecology The ecological discipline that focuses on how a population's size and other characteristics change in space and time.

population genetics The branch of science that studies the prevalence and variation in genes among populations of individuals.

population size The number of individuals in a population at a specified time.

population viability analysis A mathematical analysis used by conservation biologists to determine the minimum viable population size for threatened or endangered species.

portal vein A vein that connects two tissues or organs, such as the hypothalamus and the anterior pituitary of the brain, or the small intestine and liver (hepatic portal vein).

positive control A gene regulatory mechanism in which genes are expressed (switched on) only when an activator regulatory protein is present.

positive feedback A mechanism that intensifies or adds to a change in internal or external environmental condition.

positive pressure breathing A gulping or swallowing motion that forces air into the lungs.

positron emission tomography (PET) A scanning technique for studying activities of the cerebral cortex.

posterior Indicating the tail end of an animal.

posterior parietal cortex A region of the parietal lobe of the cerebral cortex of the brain that receives and processes sensory input and guides the premotor complex in controlling the coordination of complex movements.

posterior pituitary The neural portion of the pituitary, which stores and releases two hormones made by the hypothalamus, antidiuretic hormone and oxytocin.

postganglionic neuron The second neuron in an autonomic nervous system pathway; the dendrites and cell body of this neuron are in the PNS, and its axon extends to the effector carrying out the autonomic response.

postsynaptic cell The neuron or the surface of an effector after a synapse that receives the signal from the presynaptic cell.

posttranslational import A mechanism by which proteins are sorted to their final cellular locations in a eukaryotic cell after they have been made on free ribosomes in the cytosol.

postzygotic isolating mechanism A reproductive isolating mechanism that acts after zygote formation.

potential energy Stored energy.

predation The interaction between animals and the animal prey they consume.

prediction A statement about what the researcher expects to happen to one variable if another variable changes.

prefrontal association cortex A region of the frontal lobe that is the key part of the brain involved in thinking, such as planning for voluntary activity, decision making, and creativity, as well as for personality traits.

preganglionic neuron The first neuron in an autonomic nervous system pathway; it has its dendrites and cell body in the CNS, and its axon extends to a ganglion in the PNS.

pregnancy The period of mammalian development in which the embryo develops in the uterus of the mother.

premolars Teeth located in pairs on each side of the upper and lower jaws of mammals, positioned behind the canines and in front of the molars.

premotor cortex The region of the frontal lobe of the brain that controls skeletal muscles in coordinating complex movements.

pre-mRNA (precursor-mRNA) The primary transcript of a eukaryotic protein-coding gene, which is processed to form messenger RNA.

prenatal diagnosis Techniques in which cells derived from a developing embryo or its surrounding tissues or fluids are tested for the presence of mutant alleles or chromosomal alterations.

prepuce Foreskin; a loose fold of skin that covers the glans of the penis.

pressure flow mechanism In vascular plants, pressure that builds up at the source end of a sieve tube system and pushes solutes by bulk flow toward a sink, where they are removed.

pressure potential The influence of physical pressure on water potential (the potential energy of water).

presynaptic cell The neuron with an axon terminal on one side of the synapse that transmits the signal across the synapse to the dendrite or cell body of the postsynaptic cell.

prezygotic isolating mechanism A reproductive isolating mechanism that acts before the production of a zygote, or fertilized egg.

primary acceptor An electron-accepting molecule that receives excited electrons from chlorophylls in photosynthesis.

primary active transport Transport in which the same protein that transports a substance also hydrolyzes ATP to power the transport directly.

primary cell layers The ectoderm, mesoderm, and endoderm layers that form the embryonic tissues.

primary cell wall The initial cell wall laid down by a plant cell.

primary consumer An herbivore, a member of the second trophic level.

primary endosymbiosis In the model for the origin of plastids in eukaryotes, the first event in which a eukaryotic cell engulfed a photosynthetic cyanobacterium.

primary growth The growth of plant tissues derived from apical meristems. *Compare* **secondary growth.**

primary immune response The response of the immune system to the first challenge by an antigen.

primary meristem Root and shoot apical meristems, from which a plant's primary tissues develop. *Compare* **lateral meristem.**

primary motor cortex The area of the cerebral cortex that runs in a band just in front of the primary somatosensory area and is involved in coordinating skeletal muscles for voluntary movement.

primary plant body The portion of a plant that is made up of primary tissues.

primary producer An organism that uses light energy or chemical energy to convert simple inorganic molecules into organic molecules.

primary structure The particular and unique linear sequence of amino acids linked to each other by peptide bonds to form a polypeptide.

primary succession Predictable change in species composition of an ecological community that develops on bare ground.

primary tissue A plant tissue that develops from an apical meristem.

primase An enzyme that assembles the primer for a new DNA strand during DNA replication.

Primates A monophyletic group of eutherians that includes lemurs, lorises, monekys, apes, and humans.

primer A short nucleotide chain made of RNA that is laid down as the first series of nucleotides in a new DNA strand, or made of DNA for use in the polymerase chain reaction (PCR).

primitive groove In development of birds, the sunken midline of the primitive streak that acts as a conduit for migrating cells to move into the blastocoel.

primitive streak In development of birds, the thickened region of the embryo produced by cells of the epiblast streaming toward the midline of the blastodisc.

Principle of Independent Assortment Mendel's principle that the alleles of the genes that govern two characters assort independently during formation of gametes.

principle of parsimony A principle of systematic biology that states that a particular trait is unlikely to evolve independently in separate evolutionary lineages.

Principle of Segregation Mendel's principle that the pairs of alleles that control a character segregate as gametes are formed, with half the gametes carrying one allele, and the other half carrying the other allele.

prion An infectious agent that contains only protein and does not include a nucleic acid molecule.

probability The possibility that an outcome will occur if it is a matter of chance.

procambium The primary meristem of a plant that develops into primary vascular tissue.

product An atom or molecule leaving a chemical reaction.

product rule Mathematical rule in which the final probability is found by multiplying individual probabilities.

profundal zone The perpetually dark layer below the limnetic zone in a lake.

progesterone A female sex hormone that stimulates growth of the uterine lining and inhibits contractions of the uterus.

progestin A class of sex hormones synthesized by the gonads of vertebrates and active predominantly in females.

programmable RNA-guided genome editing system A molecular technique based on the CRISPR-Cas system that can cut and edit any sequence of DNA in a genome.

progymnosperms Lineages ancestral to gymnosperms, which evolved from seedless vascular plants.

prokaryote Organism in which the central DNA-containing region of the cell has no boundary membrane separating it from the cytoplasm. Prokaryotes make up two domains of organisms, the Bacteria and the Archaea.

prokaryotic chromosome The genetic material of prokaryotes, in most cases a single, circular DNA molecule.

prolactin (PRL) A peptide hormone secreted by the anterior pituitary that stimulates breast development and milk secretion in mammals.

prometaphase A transition period between prophase and metaphase during which the microtubules of the mitotic spindle attach to the kinetochores and the chromosomes shuffle until they align in the center of the cell.

promiscuous Mating behavior in which individuals do not form close pair bonds, and both males and females mate with multiple partners.

promoter The site to which RNA polymerase binds (prokaryotes) or to which general transcription factors bind and recruit RNA polymerase (eukaryotes) for initiating transcription of a gene.

promoter proximal element Regulatory sequence within the promoter proximal region, a region upstream of the promoter of a eukaryotic protein-coding gene. Regulatory proteins bind to promoter proximal elements and stimulate or inhibit the rate of transcription initiation.

promoter proximal region Upstream of a eukaryotic gene, a region containing regulatory sequences—promoter proximal elements—for transcription called promoter proximal elements.

proofreading A mechanism for correcting base-pair mismatch errors made by DNA polymerase during replication.

prophage A viral genome inserted in the host cell DNA.

prophase The beginning phase of mitosis during which the duplicated chromosomes within the nucleus condense from a greatly extended state into compact, rodlike structures.

proprioceptor A mechanoreceptor that detects stimuli used in the CNS to maintain body balance and equilibrium and to monitor the position of the head and limbs.

prosoma The fused head and thorax of chelicerates.

prostaglandin One of a group of local regulators derived from fatty acids that are involved in paracrine and autocrine regulation.

prostate gland An accessory sex gland in males that adds a thin, milky fluid to the semen and adjusts the pH of the semen to the level of acidity best tolerated by sperm.

protease An enzyme that hydrolyses proteins.

proteasome Large cytoplasmic protein complex in eukaryotic cells that degrades ubiquitinylated proteins.

protein Molecules that carry out most of the activities of life, including the synthesis of all other biological molecules. A protein consists of one or more polypeptides depending on the protein.

protein chip *See* **protein microarray.**

protein-coding gene A gene encoding a protein.

protein kinase Enzyme that transfers a phosphate group from ATP to one or more sites on particular proteins.

protein phosphatase Enzyme that removes phosphate groups from target proteins.

proteoglycans Glycoproteins that consist of small proteins noncovalently attached to long polysaccharide molecules.

proteome The complete set of proteins that can be expressed by the genome of an organism.

proteomics The study the proteome, the complete set of proteins that can be produced by a genome. Proteomics involves characterizing the structures and functions of all expressed proteins of an organism, the localization of proteins within the cell, and the interactions among proteins in the cell.

prothoracicotropic hormone (PTTH) A peptide hormone secreted by neurosecretory neurons in the brains of insects that is one of the hormones regulating molting and metamorphosis.

protists A diverse and polyphyletic group of single-celled and multicellular eukaryotic species.

protocell A primitive cell-like structure that has some of the properties of life and that might have been the precursor of cells.

protoderm The primary meristem that will produce stem epidermis.

proton A subatomic particle in the nucleus of an atom that carries one unit of charge.

proton pump Pump that moves hydrogen ions across membranes and pushes hydrogen ions across the plasma membrane from the cytoplasm to the cell exterior. Also referred to as H^+ *pump.*

protonema The structure that arises when a liverwort or moss spore germinates and eventually gives rise to a mature gametophyte.

protonephridium The simplest form of invertebrate excretory tubule.

proton-motive force Stored energy that contributes to ATP synthesis, as well as to the cotransport of substances to and from mitochondria.

proto-oncogene A gene that encodes various kinds of proteins that stimulate cell division. Mutated proto-oncogenes—oncogenes—contribute to the development of cancer.

protoplast The cytoplasm, organelles, and plasma membrane of a plant cell.

protoplast fusion A plant breeding process in which protoplasts are fused into a single cell.

protostomes A lineage of animals, in many of which cleavage follows a spiral pattern; mesoderm arises from cells near the blastopore, the coelom arises from a split in mesoderm tissue, and the blastopore forms the mouth.

provirus DNA copy of a retrovirus RNA genome that becomes inserted into host DNA.

prototheria A monophyletic group of mammals in which females reproduce by laying eggs.

proximal convoluted tubule The tubule between the Bowman's capsule and the loop of Henle in the nephron of the kidney, which carries and processes the filtrate.

proximate causes The ways in which genetic, cellular, physiological, and anatomical mechanisms underlie an animal's behavior.

pseudocoelom A fluid- or organ-filled body cavity between the gut (a derivative of endoderm) and the muscles of the body wall (a derivative of mesoderm).

pseudocoelomate A body plan of bilaterally symmetrical animals with a body cavity that lacks a complete lining derived from mesoderm.

pseudopod (plural, *pseudopodia*) A temporary cytoplasmic extension of a cell.

pseudogene A gene that is very similar to a functional gene at the DNA sequence level but that has one or more inactivating mutations that prevent it from producing a functional gene product.

psychrophile An archaean or bacterium that grows optimally at temperatures in the range −10°C to −20°C.

pulmocutaneous circuit In amphibians, the branch of a double blood circuit that receives deoxygenated blood and moves it to the skin and lungs or gills.

pulmonary arteries The arteries from the right ventricle of the heart that lead to the lungs.

pulmonary circuit The circuit of the cardiovascular system that supplies the lungs.

pulmonary veins The veins that carry oxygenated blood from the lungs to the right atrium of the heart.

pulvinus (plural, *pulvini*) A jointlike, thickened pad of tissue at the base of a leaf or petiole; flexes when the leaf makes nastic movements.

punctuated equilibrium hypothesis The evolutionary hypothesis that most morphological variation arises rapidly during speciation events in isolated populations at the edge of a species' geographical distribution.

Punnett square Method for determining the genotypes and phenotypes of offspring and their expected proportions by combining gametes and their probabilities of occurrence in a matrix table.

pupa The nonfeeding stage between the larva and adult in the complete metamorphosis of some insects, during which the larval tissues are completely reorganized within a protective cocoon or hardened case.

pupil The dark center in the middle of the iris through which light passes to the back of the eye.

purine A type of nitrogenous base with two carbon–nitrogen rings.

Purkinje fibers Fibers that carry an electrical signal from the atrioventricular node (AV node) to the bottom of the heart.

pyloric sphincter The muscular ring at the junction between the stomach and small intestine that controls the flow of materials from the stomach.

pyramid of biomass A diagram that illustrates differences in standing crop biomass in a series of trophic levels.

pyramid of energy A diagram that illustrates the amount of energy that flows through a series of trophic levels.

pyramid of numbers A diagram that illustrates the number of individual organisms present in a series of trophic levels.

pyrimidine A type of nitrogenous base with one carbon–nitrogen ring.

pyrogens Chemicals released by macrophages in response to infection that stimulate prostaglandin release from the hypothalamus thereby leading to an increase in body temperature (fever).

pyruvate oxidation (pyruvic acid oxidation) Stage of cellular respiration in which the three-carbon molecule pyruvate is converted into a two-carbon acetyl group that is completely oxidized to carbon dioxide.

qualitative variation Variation that exists in two or more discrete states, with intermediate forms often being absent.

quantitative trait A character that displays a continuous distribution of the phenotype involved, typically resulting from several to many contributing genes.

quantitative trait loci (QTLs) The individual genes that contribute to a quantitative trait.

quantitative variation Variation that is measured on a continuum (such as height in human beings) rather than in discrete units or categories.

quaternary structure The arrangement of bonded polypeptide chains in a protein that contains more than one chain.

quiescent center A region in a root apical meristem where there is no cell division.

quorum sensing Communication between bacteria in a population by sending and receiving chemical signals.

R gene A resistance gene in a plant; dominant *R* alleles confer enhanced resistance to plant pathogens.

R plasmid A bacterial plasmid containing genes that provide resistance to unfavorable conditions.

RAAS *See* **renin–angiotensin–aldosterone system.**

radial cleavage A cleavage pattern in deuterostomes in which newly formed cells lie directly above and below other cells of the embryo.

radial symmetry A body plan of organisms in which structures are arranged regularly around a central axis, like spokes radiating out from the center of a wheel.

Radiata A branch of the Eumetazoa comprising animals with radial body symmetry.

radiation The transfer of heat energy as electromagnetic radiation.

radicle The rudimentary root of a plant embryo.

radioactivity The giving off of particles of matter and energy by decaying nuclei.

radioisotope An unstable, radioactive isotope.

Radiolaria (radiolarians) A lineage of the Rhizaria protist evolutionary group; characterized by axopods, slender, raylike strands of cytoplasm supported internally by long bundles of microtubules.

radiometric dating A dating technique that uses the clockwork decay of unstable isotopes to estimate the age of organic material, rocks, or fossils that contain them.

radula The tooth-lined "tongue" of mollusks that scrapes food into small particles or drills through the shells of prey.

rain shadow An area of reduced precipitation on the leeward side of a mountain.

random dispersion A pattern of distribution in which the individuals in a population are distributed unpredictably in their habitat.

rapid eye movement (REM) sleep The period during deep sleep when the delta wave pattern is replaced by rapid, irregular beta waves characteristic of the waking state. The person's heartbeat and breathing rate increase, the limbs twitch, and the eyes move rapidly behind the closed eyelids.

reactants The atoms or molecules entering a chemical reaction.

reaction center Part of photosystems I and II in chloroplasts of plants. In the light-dependent reactions of photosynthesis, the reaction center receives light energy absorbed by the antenna complex in the same photosystem.

reading frame The series of codons for a polypeptide encoded by the mRNA.

realized niche The range of resources and environmental conditions actually used by a population in nature.

receptacle The expanded tip of a flower stalk that bears floral organs.

reception In signal transduction, the binding of a signal molecule with a specific receptor in a target cell. In neural signaling, the first of four components in which a stimulus is detected by specialized sensory receptors.

receptive field The region surrounding a receptor within which the receptor responds to a stimulus.

receptor potential The change in membrane potential exhibited by a sensory receptor in response to a stimulus in its receptive field.

receptor protein Protein that recognizes and binds molecules from other cells that act as chemical signals.

receptor tyrosine kinase In signal transduction, a surface receptor with built-in protein kinase activity.

receptor-mediated endocytosis The selective uptake of macromolecules that bind to cell surface receptors concentrated in clathrin-coated pits.

recessive An allele that is masked by a dominant allele.

reciprocal altruism Form of altruistic behavior in which individuals help nonrelatives if they are likely to return the favor in the future.

reciprocal cross A genetic cross in which the two parents are switched with respect to which trait is associated with each sex.

recombinant Phenotype with a different combination of traits from those of the original parents.

recombinant chromosomes Chromosomes that contain nonparental combinations of alleles. In eukaryotes they are generated by crossing-over in meiosis.

recombinant DNA DNA from two or more different sources joined together.

recombination *See* **genetic recombination.**

recombination frequency In the construction of linkage maps of diploid eukaryotic organisms, the percentage of testcross progeny that are recombinants.

rectum The final segment of the large intestine.

red tide A growth in dinoflagellate populations that causes red, orange, or brown discoloration of coastal ocean waters.

redox reaction Coupled oxidation–reduction reaction in which electrons are removed from a donor molecule and simultaneously added to an acceptor molecule.

reduced Substance—the electron acceptor—that gains electrons during reduction.

reduction The partial or full gain of electrons to a substance.

reflex A programmed movement that takes place without conscious effort, such as the sudden withdrawal of a hand from a hot surface.

refractory period A period that begins at the peak of an action potential and lasts a few milliseconds, during which the threshold required for generation of an action potential is much higher than normal.

regeneration In the light-independent reactions of photosynthesis, the process by which glyceraldehyde-3-phosphate (G3P) molecules generated by three turns of the Calvin cycle that do not exit the cycle are used to produce ribulose 1,5-bisphosphate (RuBP).

regulated gene A gene that is expressed in a controlled way, meaning that it may or may not be expressed at any given time or in a given cell type.

regulators Animals that maintain factors of the internal environment in a relatively constant state.

regulatory gene Gene that encodes a protein that regulates the expression of a structural gene or genes.

regulatory protein DNA-binding protein that binds to a regulatory sequence and affects the expression of an associated gene or genes.

regulatory sequence DNA sequence involved in the regulation of a gene or genes to which a regulatory protein binds to control the transcription of the gene or genes.

reinforcement The enhancement of reproductive isolation that had begun to develop while populations were geographically separated.

relative abundance The relative commonness of populations within a community.

relative fitness The number of surviving offspring that an individual produces compared with the number left by others in the population.

release factor A protein that recognizes stop codons in the A site of a ribosome translating an mRNA and terminates translation. Also referred to as the *termination factor.*

releasing hormone (RH) A peptide neurohormone that control the secretion of hormones from the anterior pituitary.

renal artery An artery that carries bodily fluids into the kidney.

renal cortex The outer region of the mammalian kidney that surrounds the renal medulla.

renal medulla The inner region of the mammalian kidney.

renal pelvis The central cavity in the kidney where urine drains from collecting ducts.

renal vein The vein that routes filtered blood away from the kidney.

renaturation The reformation of a denatured protein into its folded, functional state.

renin An enzyme secreted by cells in the juxtaglomerular apparatus into the bloodstream that converts a blood protein into the peptide hormone angiotensin.

renin–angiotensin–aldosterone system (RAAS) The most important hormonal system involved in regulation of Na^+ in mammals.

replica plating Technique for identifying and counting genetic recombinants in conjugation, transformation, or transduction experiments in which the colony pattern on a plate containing solid growth medium is pressed onto sterile velveteen and transferred to other plates containing different combinations of nutrients.

replicates Multiple subjects that receive either the same experimental treatment or the same control treatment.

replication bubble The two Y-shaped replication forks joined together at the tops of the Ys after DNA is unwound at an origin of replication.

replication fork The region of DNA synthesis where the parental strands separate and two new daughter strands elongate.

replisome The key proteins and enzymes for replication assembled into a DNA replication complex.

repressible operon Operon whose expression is prevented by a repressor molecule.

repressor A regulatory protein that prevents the operon genes from being expressed.

reproduction The process by which parents produce offspring.

reproductive isolating mechanism A biological characteristic that prevents the gene pools of two species from mixing.

reproductive strategy One of a set of behaviors that lead to reproductive success.

Reptilia A monophyletic group of amniotes with a dry, scaly, nonglandular skin.

residual volume The air that remains in lungs after exhalation.

resilin An elastic protein related to elastin, found only in insects and some crustaceans.

resistance proteins Plant defensive proteins that are synthesized as part of effector-triggered immunity.

resolution The minimum distance two points in a specimen can be separated and still be seen as two points.

resource partitioning The use of different resources or the use of resources in different ways by species living in the same place.

respiratory medium The environmental source of O_2 and the "sink" for released CO_2. For aquatic animals, the respiratory medium is water; for terrestrial animals, it is air.

respiratory surface A layer of epithelial cells that provides the interface between the body and the respiratory medium.

respiratory system All the parts of the body involved in exchanging air between the external environment and the blood.

response In signal transduction, the last stage in which the transduced signal causes the cell to change according to the signal and to the receptors on the cell. In neural signaling, the fourth and last component involving the action resulting from the integration of neural messages.

resting membrane potential A steady negative membrane potential exhibited by the membrane of a neuron that is not stimulated—that is, not conducting an impulse.

restriction endonuclease (restriction enzyme) An enzyme that cuts DNA at a specific sequence.

restriction fragment A DNA fragment produced by cutting a long DNA molecule with a restriction enzyme.

restriction fragment length polymorphisms When comparing different individuals, restriction enzyme-generated DNA fragments of different lengths from the same region of the genome.

reticular formation A complex network of interconnected neurons that runs through the length of the brain stem, connecting to the thalamus at the anterior end and to the spinal cord at the posterior end.

reticulum One of the four chambers of the stomach of a ruminant; in this chamber, fermentation reactions by symbiotic microorganisms begin digesting boluses of swallowed plant matter.

retina A light-sensitive membrane lining the posterior part of the inside of the eye.

retrotransposon A transposable element that transposes via an intermediate RNA copy of the transposable element.

retrovirus A virus with an RNA genome that replicates via a DNA intermediate.

reverse transcriptase An enzyme that uses RNA as a template to make a DNA copy of the retrotransposon. Reverse transcriptase is used to make DNA copies of RNA in test tube reactions.

reverse transcriptase–PCR (RT–PCR) A method that couples reverse transcription with PCR to convert an RNA molecule to double-stranded DNA.

reverse transcription The synthesis of a double-stranded DNA molecule from a single-strandaed RNA molecule catalyzed by reverse transcriptase.

reversible The term indicating that a reaction may go from left to right or from right to left, depending on conditions.

rheumatoid arthritis An autoimmune disease that results by a self-attack on connective tissues, particularly in the joints, causing pain and inflammation.

Rhizaria (rhizarians) A protist evolutionary group; amoebas with filamentous pseudopods.

rhizome A horizontal, modified stem that can penetrate a substrate and anchor the plant.

Rhodophyta (red algae) A lineage of the Archaeplastida protist evolutionary group; mostly freeliving autotrophic marine seaweeds.

rhodopsin A photopigment consisting of retinal bound covalently to an opsin protein.

ribonucleic acid *See* **RNA**.

ribonucleotide Nucleotide containing ribose as the sugar; ribonucleotides are components of RNA.

ribose A five-carbon sugar to which the nitrogenous bases in nucleotides link covalently.

ribosomal RNA (rRNA) The RNA component of ribosomes.

ribosome A ribonucleoprotein particle that carries out protein synthesis by translating mRNA into chains of amino acids.

ribosome binding site In translation initiation in prokaryotes, a sequence just upstream of the start codon that directs the small ribosomal subunit to bind and orient correctly for the complete ribosome to assemble and start translating in the correct spot.

ribozyme An RNA-based catalyst that is part of the biochemical machinery of all cells.

ring species A species with a geographical distribution that forms a ring around uninhabitable terrain.

RNA A polymer assembled from repeating nucleotide monomers in which the five-carbon sugar is ribose. Cellular RNAs include mRNA (which is translated to produce a polypeptide), tRNA (which brings an amino acid to the ribosome for assembly into a polypeptide during translation), and rRNA (which is a structural component of ribosomes). The genetic material of some viruses is RNA.

RNA interference (RNAi) The phenomenon of silencing a gene posttranscriptionally by a small, single-stranded RNA that is complementary to part of an mRNA.

RNA polymerase An enzyme that catalyzes the assembly of ribonucleotides into an RNA strand.

RNA-seq Whole-transcriptome sequencing; a technique using high-throughput sequencing of cDNAs to identify and quantify RNA transcripts in a sample.

RNA world model A model which states that the first genes and enzymes were RNA molecules.

rod In the vertebrate eye, a type of photoreceptor in the retina that is specialized for detection of light at low intensities.

root An anchoring structure in land plants that also absorbs water and nutrients and (in some plant species) stores food. In phylogenetic trees, the root of the tree represents the ancestor of all species included within the tree. In systematics, the common ancestor of all species depicted in a phylogenetic tree.

root cap A dome-shaped cell mass that forms a protective covering over the apical meristem in the tip of a plant root.

root hair A tubular outgrowth of the outer wall of a root epidermal cell; root hairs absorb much of a plant's water and minerals from the soil.

root nodule A localized swelling on a root in which symbiotic nitrogen-fixing bacteria reside.

root pressure The pressure that develops in plant roots as the result of osmosis, forcing xylem sap upward and out through leaves. *See also* **guttation.**

root primordium A rudimentary root.

root system An underground (or submerged) network of roots with a large surface area that favors the rapid uptake of soil water and dissolved mineral ions.

rough ER Endoplasmic reticulum with many ribosomes studding its outer surface.

round window A thin membrane that faces the middle ear.

rRNA *See* **ribosomal RNA.**

r-selected species A short-lived species adapted to function well in a rapidly changing environment.

rubisco *See* **RuBP carboxylase/oxygenase.**

RuBP carboxylase/oxygenase (rubisco) An enzyme that catalyzes the key reaction of the Calvin cycle, carbon fixation, in which CO_2 combines with RuBP (ribulose 1,5-bisphosphate) to form 3-phosphoglycerate.

rumen One of the four chambers of the stomach of a ruminant; in this chamber, fermentation reactions by symbiotic microorganisms begin digesting boluses of swallowed plant matter.

ruminant An animal that has a complex, four-chambered stomach.

S phase The phase of the eukaryotic cell cycle during which DNA replication occurs.

saccule Saclike, fluid-filled structure that is part of the vestibular apparatus. With the utricle, provides information about the up–down position of the head, as well as changes in the rate of linear movement of the head.

salicylic acid (SA) In plants, a chemical synthesized following a wound that has multiple roles in plant defenses, including interaction with jasmonates in signaling cascades.

saliva In humans, a secretion of three pairs of salivary glands in the mouth.

salivary amylase A substance that hydrolyzes starches to the disaccharide maltose.

salivary gland A gland that secretes saliva through a duct on the inside of the cheek or under the tongue; the saliva lubricates food and begins digestion.

salt marsh A tidal wetland dominated by emergent grasses and reeds.

saltatory conduction A mechanism that allows small-diameter axons to conduct impulses rapidly.

saprobe An organism nourished by dead or decaying organic matter.

sapwood The newly formed outer wood located between heartwood and the vascular cambium. Compared with heartwood, it is wet, lighter in color, and not as strong.

sarcomere The basic unit of contraction in a myofibril.

sarcoplasmic reticulum In vertebrate muscle fibers, a complex system of vesicles modified from the smooth endoplasmic reticulum that encircles the sarcomeres. The sarcoplasmic reticulum is part of the pathway for the stimulation of muscle contraction by neural signals.

Sarcopterygii A monophyletic group of boney fishes that includes the fleshy-finned fishes, lungfishes, and their descendants.

saturated With respect to a fatty acid, if only single bonds link the carbon atoms. With respect to an enzyme reaction, when the enzyme is cycling as rapidly as possible so that further increases in substrate concentration have no effect on the reaction rate.

savanna A biome comprising grasslands with few trees, which grows in areas adjacent to tropical deciduous forests.

schizocoelom In protostomes, the body cavity that develops as inner and outer layers of mesoderm separate.

Schwann cell A type of glial cell in the PNS that wraps nerve fibers with myelin and also secretes regulatory factors.

scientific method An investigative approach in which scientists make observations about the natural world, develop working explanations about what they observe, and then test those explanations by collecting more information.

scientific name A two-part name identifying the genus to which a species belongs and designating a particular species within that genus.

scientific theory A broadly applicable idea or hypothesis that has been confirmed by every conceivable test.

scion A bud or branch from a plant with desirable fruit traits, which is grafted to a root or stem of a plant having desirable root traits.

sclereid A type of sclerenchyma cell; sclereids typically are short and have thick, lignified walls.

sclerenchyma A ground tissue in which cells develop thick secondary walls, which commonly are lignified and perforated by pits through which water can pass.

scrotum The baglike sac in which the testes are suspended in many mammals.

scutellum The shield-shaped cotyledon of a grass.

second law of thermodynamics Principle that for any process in which a system changes from an initial to a final state, the total disorder of the system and its surroundings always increases.

second messenger In particular signal transduction pathways, an internal, nonprotein signal molecule that directly or indirectly activates protein kinases, which elicit the cellular response.

secondary active transport Transport indirectly driven by ATP hydrolysis.

secondary cell wall A layer added to the cell wall of plants that is more rigid and may become many times thicker than the primary cell wall.

secondary consumer A carnivore that feeds on herbivores, a member of the third trophic level.

secondary contact The reestablishment of geographical overlap after a barrier that separated allopatric populations no longer keeps them apart.

secondary endosymbiosis In the model for the origin of plastids in eukaryotes, the second event, in which a nonphotosynthetic eukaryote engulfed a photosynthetic eukaryote.

secondary growth Plant growth that originates at lateral meristems and increases the diameter of older roots and stems. *Compare* **primary growth.**

secondary immune response The rapid immune response that occurs during the second (and subsequent) encounters of the immune system of a mammal with a specific antigen.

secondary plant body The part of a plant made up of tissues that develop from lateral meristems.

secondary productivity Energy stored in new consumer biomass as energy is transferred from producers to consumers.

secondary structure The coiling or folding of a segment of a polypeptide chain produced by hydrogen bonding between different amino acids in the segment. Alpha helices and beta strands are examples of secondary structure.

secondary succession Predictable changes in species composition in an ecological community that develops after existing vegetation is destroyed or disrupted by an environmental disturbance.

secondary tissue In plants, the tissue that develops from lateral meristems.

secretin Hormone released into the bloodstream from glandular cells in the small intestine in response to stimulation by acidic chyme.

secretion A selective process in which specific small molecules and ions are transported from the body fluids (in animals with open circulatory systems) or blood (in animals with closed circulatory systems) into the excretory tubules. In animal nutrition, secretion is one of the steps of the digestive process; it involves the release into the digestive tract of enzymes and other substances that aid the process of digestion, such as acids, emulsifiers, and lubricating mucus.

secretory vesicle Vesicle that transports proteins to the plasma membrane.

seed The structure that forms when an ovule matures after a pollen grain reaches it and a sperm fertilizes the egg.

seed coat The outer protective covering of a seed.

seed ferns An extinct, paraphyletic group of gymnosperms that produced seeds and had leaves resembling fern fronds.

segment polarity gene In *Drosophila* embryonic development, one of the third activated set of segmentation genes that progressively subdivide the embryo into regions, determining the segments of the embryo and the adult. Segment polarity genes set the boundaries and anterior–posterior axis of each segment in the embryo, thereby determining the regions that become segments of larvae and adults.

segmentation The production of body parts and some organ systems in repeating units.

segmentation genes Genes that work sequentially, progressively subdividing the embryo into regions, determining the segments of the embryo and the adult.

selective cell adhesion A mechanism in which cells make and break specific connections to other cells or to the extracellular matrix.

selectively neutral *See* **neutral variation hypothesis.**

selectively permeable Membranes that selectively allow, impede, or block the passage of atoms and molecules.

self-fertilization (self-pollination) Fertilization in which sperm nuclei in pollen fertilize egg cells of the same plant.

self-incompatibility In plants, the inability of a plant's pollen to fertilize ovules of the same plant.

semelparity An evolved life history characteristic in which individuals reproduce only once in their lifetimes.

semen The secretions of several accessory glands in which sperm are mixed before ejaculation.

semicircular canals Fluid-filled structures that are part of the vestibular apparatus. The three semicircular canals are positioned at angles corresponding to the three planes of space and detect rotational (spinning) motions.

semiconservative replication The process of DNA replication in which the two parental strands separate and each serves as a template for the synthesis of new progeny double-stranded DNA molecules.

seminal fluid Fluid secreted by the seminal vesicles that contains prostaglandins, which when ejaculated into the female trigger contractions of the female reproductive tract that help move the sperm into and through the uterus.

seminal vesicle A vesicle that secretes seminal fluid.

seminiferous tubule One of the tiny tubes in the testes where sperm cells are produced, grow, and mature.

senescence The biologically complex process of aging in mature organisms that leads to the death of cells and eventually the whole organism.

sense codon A codon that specifies an amino acid.

sensillum (plural, *sensilla*) A hollow sensory bristle of an insect that contains a taste receptor.

sensitization Increased responsiveness to mild stimuli after experiencing a strong stimulus; one of the simplest forms of memory.

sensor A tissue or organ that detects a change in an external or internal factor such as pH, temperature, or the concentration of a molecule such as glucose.

sensory adaptation A condition in which the effect of a stimulus is reduced if it continues at a constant level.

sensory neuron A neuron that transmits stimuli collected by the sensory receptors to interneurons.

sensory receptor A receptor formed by the dendrites of afferent neurons, or by specialized receptor cells making synapses with afferent neurons that pick up information about the external and internal environments of the animal.

sensory transduction The conversion of a stimulus into a change in membrane potential.

sepal One of the separate, usually green, parts forming the calyx of a flower.

septum (plural, *septa*) A thin partition or cross wall that separates body segments.

sequential hermaphroditism The form of hermaphroditism in which individuals change from one sex to the other.

serosa The serous membrane: a thin membrane lining the closed cavities of the body; has two layers with a space between that is filled with serous fluid.

Sertoli cell One of the supportive cells that completely surrounds developing spermatocytes in the seminiferous tubules. Follicle-stimulating hormone stimulates Sertoli cells to secrete a protein and other molecules that are required for spermatogenesis.

sessile Unable to move from one place to another.

set point The level at which the condition controlled by a homeostatic pathway is to be maintained.

seta (plural, *setae*) A chitin-reinforced bristle that protrudes outward from the body wall in some annelid worms.

sex chromosomes Chromosomes that are different in male and female individuals of the same species.

sex pilus *See* **F pilus.**

sex ratio The relative proportions of males and females in a population.

sex-linked gene Gene located on a sex chromosome.

sexual dimorphism Differences in the size or appearance of males and females.

sexual reproduction The mode of reproduction in which male and female parents produce offspring through the union of egg and sperm generated by meiosis.

sexual selection A form of natural selection established by male competition for access to females and by the females' choice of mates.

shade avoidance An adaptive response in which a plant's metabolic resources are directed into unusually rapid vertical growth; it may occur when photosynthetic parts of a plant adapted to grow in full sunlight are shaded by the leaves of other, nearby plants.

shells *See* **energy levels.**

shoot system The stems and leaves of a plant.

short tandem repeat (STR) Short 2- to 6-bp DNA sequence repeated in series.

short-day plant A plant that flowers in late summer or early autumn when dark periods become longer and light periods become shorter.

short-term memory Memory that stores information for seconds.

sieve tube A series of phloem cells joined end to end, forming a long tube through which nutrients are transported; a feature mainly of flowering plants.

sieve tube element Any of the main conducting cells of phloem that connect end to end, forming a sieve tube.

sign stimulus A simple cue that triggers a fixed action pattern.

signaling cascade The cascade of reactions that include several different molecules that constitute transduction, the second step in the processing of a signal by a target cell.

signal peptide A short segment of amino acids to which the signal recognition particle binds, temporarily blocking further translation. A signal peptide is found on polypeptides that are sorted to the endoplasmic reticulum. Also referred to as *signal sequence.*

signal recognition particle (SRP) Protein-RNA complex that binds to signal sequences and targets polypeptide chains to the endoplasmic reticulum.

signal sequence *See* **signal peptide.**

signal transduction pathway With respect to cell communication by long-distance signaling, the process of changing a signal into the form necessary to cause the cellular response in the transduction stage; the pathway typically involves a cascade of reactions that include several different molecules.

signaling pathway The series of events by which a signal molecule released from a controlling cell causes a response (affects the function) of target cells with receptors for the signal. Target cells process the signal in the three sequential steps of reception, transduction, and response.

signature gene A key gene whose expression changes in a way that correlates with a normal cell becoming a cancer cell.

silencing Phenomenon in which methylation of cytosines in eukaryotic promoters inhibits transcription and turns the genes off.

silent mutation A base-pair substitution mutation in a protein-coding gene that does not alter the amino acid specified by the gene.

simple diffusion Mechanism by which certain small substances diffuse through the lipid part of a biological membrane.

simple fruit A fruit that develops from a single ovary; in many of them at least one layer of the pericarp is fleshy.

simulation modeling An analytical method in which researchers gather detailed information about a system and then create a series of mathematical equations that predict how the components of the system interact and respond to change.

simultaneous hermaphroditism A form of hermaphroditism in which individuals develop functional ovaries and testes at the same time.

single-lens eye An eye type that works by changing the amount of light allowed to enter into the eye and by focusing this incoming light with a lens.

single-nucleotide polymorphism (SNP) A single base-pair mutation at a site (locus) in DNA. A SNP locus typically has two alleles, with one allele more frequent that the other in the population. To be a SNP, the frequency of the rarer allele must be at least 1%.

single-stranded binding protein (SSB) Protein that coats single-stranded segments of DNA, stabilizing the DNA for the replication process.

sink Any region of a plant where organic substances are being unloaded from the phloem and used or stored.

sink population In metapopulation analysis, a population that routinely declines in size after being replenished by immigrants from a source population.

sinoatrial node (SA node) The region of the heart that controls the rate and timing of cardiac muscle cell contraction.

sinus A body space that surrounds an organ.

siRNA-induced silencing complex (siRISC) Protein complex containing an siRNA that binds to a sequence in a target RNA resulting in cleavage of that RNA.

sister-chromatid cohesion The process by which newly formed sister chromatids are held together tightly by cohesin proteins encircling them along their length.

sister chromatid One of two exact copies of a chromosome duplicated during replication.

sister clades Two evolutionary lineages (that is, clades) that emerge from the same node in a phylogenetic tree.

sister species Two species that are descended from the same recent ancestral species.

skeletal muscle A muscle that connects to bones of the skeleton, typically made up of long and cylindrical cells that contain many nuclei.

sliding clamp A protein that encircles the DNA and binds to the DNA polymerase to tether the enzyme to the template, thereby making replication more efficient.

sliding filament model A model for muscle fiber contraction in which the thin filaments on each side of a sarcomere slide over the thick filaments toward the center of the A band, which brings the Z lines closer together, shortening the sarcomeres and contracting the muscle.

slime layer A coat typically composed of polysaccharides that is loosely associated with bacterial cells.

slime molds Members of the Amoebozoa; heterotrophic protists that, at some stage of their life cycle, exist as individuals that move by amoeboid motion but that exist in more complex forms the remainder of the time.

slow block to polyspermy The process in which enzymes released from cortical granules alter the egg coats within minutes after fertilization, so that no other sperm can attach and penetrate to the egg.

slow muscle fiber A muscle fiber that contracts relatively slowly and with low intensity.

small interfering RNA (siRNA) One of the major types of small, single-stranded regulatory RNAs in eukaryotes involved in RNA interference (RNAi).

small nuclear ribonucleoprotein particle (snRNP) A complex of RNA and proteins in the nucleus that is involved in mRNA splicing in eukaryotes.

smooth ER Endoplasmic reticulum with no ribosomes attached to its membrane surfaces. Smooth ER has various functions, including synthesis of lipids that become part of cell membranes.

smooth muscle A relatively small and spindle-shaped muscle cell in which actin and myosin molecules are arranged in a loose network rather than in bundles.

SNP *See* **single-nucleotide polymorphism.**

snRNP *See* **small nuclear ribonucleoprotein particle.**

social behavior The interactions that animals have with other members of their species.

sodium–potassium pump *See* **Na$^+$/K$^+$ pump.**

soil solution A combination of water and dissolved substances that coats soil particles and partially fills pore spaces.

solenoid model *See* **30-nm chromatin fiber.**

solute The molecules of a substance dissolved in water.

solute potential The effect of dissolved solutes on water potential (the potential energy of water).

solution Substance formed when molecules and ions separate and are suspended individually, surrounded by water molecules.

solvent The water in a solution in which the hydration layer prevents polar molecules or ions from reassociating.

somaclonal selection A procedure in which somatic embryos derived from tissue culture are screened to identify those having desired characteristics, such as disease resistance.

somatic cell Any of the cells of an organism's body other than reproductive cells.

somatic embryo A plant embryo that is genetically identical to the parent because it arose through asexual means.

somatic gene therapy Gene therapy in which genes are introduced into somatic cells.

somatic nervous system A subdivision of the peripheral nervous system controlling body movements that are primarily conscious and voluntary.

somatosensory cortex The area of the cerebral cortex that runs in a band across the parietal lobes of the brain and registers information on touch, pain, temperature, and pressure.

somites Paired blocks of mesoderm cells along the vertebrate body axis that form during early vertebrate development and differentiate into dermal skin, bone, and muscle.

soredium (plural, *soredia*) A specialized cell cluster produced by lichens, consisting of a mass of algal cells surrounded by fungal hyphae; soredia function like reproductive spores and can give rise to a new lichen.

sorus (plural, *sori*) A cluster of sporangia on the underside of a fern frond; reproductive spores arise by meiosis inside each sporangium.

source In plants, any region (such as a leaf) where organic substances are being loaded into the sieve tube system of phloem.

source population In metapopulation analyses, a population that is either stable or increasing in size.

spatial summation The summation of EPSPs produced by firing of different presynaptic neurons.

specialized transduction Transfer of bacterial genes between bacteria using temperate phages that have incorporated fragments of the bacterial genome as they make the transition from the lysogenic cycle to the lytic cycle.

speciation The process of species formation.

species A group of populations in which the individuals are so similar in structure, biochemistry, and behavior that they can successfully interbreed.

species cluster A group of closely related species recently descended from a common ancestor.

species composition The particular combination of species that occupy a site.

species diversity A community characteristic defined by species richness and the relative abundance of species.

species fusion Merger of two populations into one after the establishment of secondary contact.

species richness The number of species that live within an ecological community.

specific epithet The species name in a binomial.

specific heat The amount of heat required to increase the temperature of a given quantity of water.

spermatocyte A developing gamete that becomes a spermatid at the end of meiosis.

spermatogenesis The process of producing sperm.

spermatogonium (plural, *spermatogonia*) A cell that enters meiosis and gives rise to gametes, produced by mitotic divisions of the germ cells in males.

spermatozoan Also called *sperm*; a haploid cell that develops into a mature sperm cell when meiosis is complete.

sphincter A powerful ring of smooth muscle that forms a valve between major regions of the digestive tract.

spinal cord A column of nervous tissue located within the vertebral column and directly connected to the brain.

spinal nerve A nerve that carries signals between the spinal cord and the body trunk and limbs.

spindle The structure that separates sister chromatids and moves them to opposite spindle poles.

spindle pole One of the pair of centrosomes in a cell undergoing mitosis from which bundles of microtubules radiate to form the part of the spindle from that pole.

spinneret A modified opisthosomal appendage from which spiders secrete silk threads.

spiracle An opening in the chitinous exoskeleton of an insect through which air enters and leaves the tracheal system.

spiral cleavage The cleavage pattern in many protostomes in which newly produced cells lie in the space between the two cells immediately below them.

spiral organ *See* **organ of Corti.**

spiral valve A corkscrew-shaped fold of mucous membrane in the digestive system of elasmobranchs, which slows the passage of material and increases the surface area available for digestion and absorption.

spirillum (plural, *spirilla*) Any flagellated aerobic bacterium twisted helically like a corkscrew.

spliceosome A complex formed between the pre-mRNA and small ribonucleoprotein particles, in which mRNA splicing takes place.

spongocoel The central cavity in a sponge.

spontaneous mutation A mutation that occurs naturally within a cell.

spontaneous reaction Chemical or physical reaction that occurs without outside help.

sporadic (nonhereditary) cancer Cancer that is not inherited.

sporangium (plural, *sporangia*) A single-celled or multicellular structure in fungi and plants in which spores are produced.

spore A haploid reproductive structure, usually a single cell, that can develop into a new individual without fusing with another cell; found in plants, fungi, and certain protists.

sporophyll A specialized leaf that bears sporangia (spore-producing structures).

sporophyte An individual of the diploid generation produced through fertilization in organisms that undergo alternation of generations; it produces haploid spores.

sporopollenin A tough polymer in the walls of spores and pollen grains, the presence of which helps such structures resist decay.

spring overturn The mixing of surface water with deep water in a lake or pond, causing oxygen at the surface to move to the bottom, and nutrients from the bottom to move to the surface.

SRP (signal recognition particle) receptor A protein on the membrane of the endoplasmic reticulum that binds the signal recognition particle.

stability The ability of a community to maintain its species composition and relative abundances when environmental disturbances eliminate some species from the community.

stabilizing selection A type of natural selection in which individuals expressing intermediate phenotypes have the highest relative fitness.

stamen A "male" reproductive organ in flowers, consisting of an anther (pollen producer) and a slender filament.

standing crop biomass The total dry weight of plants present in an ecosystem at a given time.

stapes The smallest of three sound-conducting bones in the middle ear of tetrapod vertebrates.

starch A storage polysaccharide in plants consisting of branched or unbranched chains of glucose subunits.

start codon The first codon read in an mRNA in translation—AUG. Also referred to as the *initiator codon.*

statocyst A mechanoreceptor in invertebrates that senses gravity and motion using statoliths.

statolith A movable starch- or carbonate-containing stonelike body involved in sensing gravitational pull.

steady state When a homeostatically regulated factor does not vary much from a steady level.

stele The central core of vascular tissue in roots and shoots of vascular plants; it consists of the xylem and phloem together with supporting tissues.

stem cell A cell capable of undergoing many divisions in an unspecialized, undifferentiated state and that also has the ability to differentiate into specialized cell types.

stereocilia Microvilli covering the surface of hair cells clustered in the base of neuromasts.

stereoisomers Molecules that are mirror images of one another are an example of stereoisomers.

steroid A type of lipid derived from cholesterol.

steroid hormone receptor Internal receptor that turns on specific genes when it is activated by binding a signal molecule.

steroid hormone response element The DNA sequence to which the hormone-receptor complex binds.

steroid hormone A hydrophobic molecule derived from cholesterol that functions as a hormone in classical endocrine signaling.

sterol Steroid with a single polar —OH group linked to one end of the ring framework and a complex, nonpolar hydrocarbon chain at the other end.

sticky end End of a DNA fragment generated by digestion with a restriction enzyme, with a single-stranded structure that can form hydrogen bonds with a complementary sticky end on any other DNA molecule cut with the same enzyme.

stigma The receptive end of a carpel where deposited pollen germinates.

stimulus A component of a negative feedback control system maintaining homeostasis, specifically an environmental change that triggers a response.

stoma (plural, *stomata*) The opening between a pair of guard cells in the epidermis of a plant leaf or stem, through which gases and water vapor pass.

stomach The portion of the digestive system in which food is stored and digestion begins.

stomach ulcer *See* **peptic ulcer.**

stop codon A codon that does not specify amino acids. The three nonsense codons are UAG, UAA, and UGA. Also referred to as the *nonsense codon* and *termination codon.*

STR *See* **short tandem repeat.**

Stramenopila (stramenopiles) A subgroup of the Chromalveolata protist evolutionary group; protists with two different flagella, one with hollow tripartite projections, and one that is plain.

stratification Horizontal layering of sedimentary rocks beneath the soil surface.

Strepsirhini A monophyletic group of Primates that includes ancestral primate characters.

stretch reflex A programmed movement that takes place without conscious effort when a muscle is stretched.

strict aerobe Cell with an absolute requirement for oxygen to survive, unable to live solely by fermentations.

strict anaerobe Organism in which fermentation is the only source of ATP.

strigolactones Plant hormones that adjust growth, notably lateral branching in shoots and roots, in ways that foster the availability of adequate nutritional support for the whole plant.

strobilus *See* **cone.**

stroke (cerebrovascular accident) A loss of critical brain functions because of the death of nerve cells in the brain.

stroma The fluid within the compartment formed by the inner membrane of a chloroplast.

stock A root or stem having desired traits, to which a bud or branch from another plant is grafted. *Compare* **scion.**

structural gene Gene that encodes a protein that has a function other than gene regulation.

structural isomers Two molecules with the same chemical formula but atoms that are arranged in different ways.

style The slender stalk of a carpel situated between the ovary and the stigma in plants.

submucosa A thick layer of elastic connective tissue that contains neuron networks and blood and lymph vessels.

subsoil The region of soil beneath topsoil, which contains relatively little organic matter.

subspecies A taxonomic subdivision of a species.

substrate The particular reacting molecule or molecular group that an enzyme catalyzes.

substrate-level phosphorylation An enzyme-catalyzed reaction that transfers a phosphate group from a substrate to ADP.

sugar–phosphate backbone Structure in a polynucleotide chain that is formed when deoxyribose sugars (in DNA) or ribose sugars (in RNA) are linked by phosphate groups in an alternating sugar–phosphate–sugar–phosphate pattern.

sulfhydryl group Group that works as a molecular fastener, consisting of a sulfur atom linked on one side to a hydrogen atom and on the other side to a carbon chain.

sum rule Mathematical rule in which final probability is found by summing individual probabilities.

surface tension The force that places surface water molecules under tension, making them more resistant to separation than the underlying water molecules.

survivorship curve Graphic display of the rate of survival of individuals over a species' lifespan.

suspension feeder An animal that ingests small food items suspended in water.

suspensor In seed plants, a stalklike row of cells that develops from a zygote and helps position the embryo close to the nourishing endosperm.

sustainable agriculture Farming methods in which fields are replanted using techniques such as no-till farming that maintain soil structure and fertility.

swallowing reflex The involuntary action produced by contractions of muscles in the walls of the pharynx that direct food into the esophagus.

swim bladder A gas-filled internal organ that helps bony fishes maintain buoyancy.

symbiosis An interspecific interaction in which the ecological relations of two or more species are intimately tied together.

symmetry Exact correspondence of form and constituent configuration on opposite sides of a dividing line or plane.

sympathetic nervous system The division of the autonomic nervous system that predominates in situations involving stress, danger, excitement, or strenuous physical activity.

sympatric speciation Speciation that occurs without the geographical isolation of populations.

symplast The compartment represented by a plant's living parts; collectively, the interior of cells. *Compare* **apoplast**.

symplastic pathway The route taken by water that moves through the cytoplasm of plant cells (the symplast). *Compare* **apoplastic pathway**.

symport The transport of two molecules in the same direction across a membrane. Also referred to as *cotransport*.

synapomorphy A derived character state found in two or more species.

synapse A site where a neuron makes a communicating connection with another neuron or an effector such as a muscle fiber or gland.

Synapsida A monophyletic lineage of amniotes that includes mammals.

synapsis Process in meiosis in which homologous chromosomes come together and pair. Also referred to as *pairing*.

synaptic cleft A narrow gap that separates the plasma membranes of the presynaptic and postsynaptic cells.

synaptic plasticity The process by which a neuron's dendrites can change in shape and extent, thereby establishing new synaptic connections with other neurons.

synaptic signaling Communication across a synapse between neurons.

synaptic vesicle Membrane-bound vesicle located near the axon terminal of a neuron which contains neurotransmitter molecules.

synaptonemal complex A protein framework that tightly holds together homologous chromosomes as they pair.

synovial joint The most movable joint of a vertebrate skeleton consisting of two bones enclosed by a fluid-filled capsule of connective tissue.

systematics The branch of biology that studies the diversity of life and its evolutionary relationships.

systemic acquired resistance A plant defense response to microbial invasion; defensive chemicals including salicylic acid may spread throughout a plant, rendering healthy tissues less vulnerable to infection.

systemic circuit In amphibians, the branch of a double blood circuit that receives oxygenated blood and provides the blood supply for most of the tissues and cells of a body.

systemic homeostatic controls Controls that are initiated outside of an organ or organ system to control that organ's or organ system's activity.

systemic inflammation Inflammation that occurs throughout the body.

systemic lupus erythematosus (lupus) An autoimmune disease caused by production of a wide variety of anti-self antibodies against blood cells, blood platelets, and internal cell structures and molecules such as mitochondria and proteins associated with the DNA in the cell nucleus; characterized by anemia and problems with blood circulation and kidney function.

systemin A plant peptide hormone that functions in defense responses to wounds.

systems biology An area of biology that studies the organism as a whole to unravel the integrated and interacting network of genes, proteins, and biochemical reactions responsible for life.

systole The period of contraction and emptying of the heart.

systolic blood pressure The peak of high blood pressure in the cardiac cycle that occurs as the heart ventricles contract and move blood outward through the arteries.

T (transverse) tubule The tubule that passes in a transverse manner from the sarcolemma across a myofibril of striated muscle.

T cell A lymphocyte produced by the division of stem cells in the bone marrow and then released into the blood and carried to the thymus. T cells participate in adaptive immunity.

tactile signal A means of animal communication in which the signaler uses touch to convey a message to the signal receiver.

taiga *See* **boreal forest**.

tail With respect to bacteriophages, the structure attached to one side of the polyhedral head.

tandem duplication Duplicate genes clustered together in the same region of the same chromosome. Unequal crossing-over produces such an arrangement.

taproot system A root system consisting of a single main root from which lateral roots can extend; often stores starch.

target protein In a signal transduction pathway, the last protein in a phosphorylation cascade, the phosphorylation of which stimulates or inhibits its activity depending on the particular protein to bring about the cellular response.

targeting signal Sequences on proteins that are destined for the mitochondria, chloroplasts, and microbodies. Receptors on the organelle surface recognize the targeting signal triggering the import of the protein into the organelle by specific pathways.

target site The location in a genome to which a transposable element moves when it transposes.

taste bud A small, pear-shaped capsule with a pore at the top that opens to the exterior that contains the taste receptors. Found in most vertebrates.

TATA box An initiator sequence in DNA for transcription in eukaryotes.

taxis A behavioral response that is directed either toward or away from a specific stimulus.

taxon (plural, *taxa*) A name designating a group of organisms included within a category in the Linnaean taxonomic hierarchy.

taxonomic hierarchy A system of classification based on arranging organisms into ever more inclusive categories.

taxonomy The science of the classification of organisms into an ordered system that indicates natural relationships.

T-cell receptor (TCR) A receptor that covers the plasma membrane of a T cell, specific for a particular antigen.

tectorial membrane A membrane that extends the length of the cochlear canal of the inner ear in which the stereocilia of the sensory hair cells of the organ of Corti are embedded.

telomerase An enzyme that adds telomere repeats to chromosome ends.

telomeres Repeats of simple-sequence DNA that maintain the ends of linear chromosomes.

telophase The final phase of mitosis, during which the spindle disassembles, the chromosomes decondense, and the nuclei reform.

temperate bacteriophage Bacteriophage that may enter an inactive phase (lysogenic cycle) in which the host cell replicates and passes on the bacteriophage DNA for generations before the phage becomes active and kills the host (lytic cycle).

temperate deciduous forest A forested biome found at low to middle altitudes at temperate latitudes, with warm summers, cold winters, and annual precipitation between 75 and 250 cm.

temperate grassland A nonforested biome that stretches across the interiors of most continents, where winters are cold and snowy and summers are warm and fairly dry.

temperate rainforest A coniferous forest biome supported by heavy rain and fog, which grows where winters are mild and wet and the summers are cool.

template A nucleotide chain used in DNA replication for the assembly of a complementary chain.

template strand The DNA strand that is copied into an RNA molecule during gene transcription.

temporal isolation A prezygotic reproductive isolating mechanism in which species live in the same habitat but breed at different times of day or different times of year.

temporal lobe Division of the cerebral hemisphere of the brain responsible for receiving auditory input.

temporal summation The summation of several EPSPs produced by successive firing of a single presynaptic neuron over a short period of time.

tendon A type of fibrous connective tissue that attaches muscles to bones.

terminal bud A bud that develops at the apex of a shoot.

termination In transcription, the step in which transcription ends and the RNA transcript and RNA polymerase (RNA polymerase II in the case of eukaryotes) are released from the DNA template. In translation, the step in which the translation complex disassembles after the last amino acid of the polypeptide specified by the mRNA has been added to the polypeptide.

termination codon *See* **stop codon.**

termination factor *See* **release factor.**

terminator Specific DNA sequence for a gene that signals the end of transcription of a gene. Terminators are common for prokaryotic genes.

territory A plot of habitat, defended by an individual male or a breeding pair of animals, within which the territory holders have exclusive access to food and other necessary resources.

tertiary consumer A carnivore that feeds on other carnivores, a member of the fourth trophic level.

tertiary structure The folding of the complete amino acid sequence of a polypeptide chain, with its secondary structures, into the overall three-dimensional shape.

testcross A genetic cross between an individual with the dominant phenotype and a homozygous recessive individual.

testis (plural, *testes*) The male gonad. In male vertebrates, they secrete androgens and steroid hormones that stimulate and control the development and maintenance of male reproductive systems.

testosterone A hormone produced by the testes, responsible for the development of male secondary sex characteristics and the functioning of the male reproductive organs.

tetanus A situation in which a muscle fiber cannot relax at all between stimuli, and twitch summation produces a peak level of continuous contraction.

Tetrapoda A monophyletic lineage of vertebrates that includes animals having four feet, legs, or leglike appendages.

T-even bacteriophage Virulent bacteriophages, T2, T4, and T6, that have been valuable for genetic studies of bacteriophage structure and function.

thalamus A major switchboard of the brain that receives sensory information and relays it to the regions of the cerebral cortex concerned with motor responses to sensory information of that type.

thallus A plant body not differentiated into stems, roots, or leaves.

Thaumarchaeota A group of the domain Archaea consisting of four living species, all of which are mesophilic, chemoautotrophic ammonia oxidizers, and other members represented by DNA sequences in environmental samples.

thermal acclimatization A set of physiological changes in ectotherms in response to seasonal shifts in environmental temperature, allowing the animals to attain good physiological performance at both winter and summer temperatures.

thermocline The narrow depth range in a lake where water temperature changes abruptly at the boundary between the epilimnion and the hypolimnion.

thermodynamics The study of the energy flow during chemical and physical reactions.

thermoreceptor A sensory receptor that detects the flow of heat energy.

thermoregulation The control of body temperature.

thick filament A type of filament in striated muscle composed of myosin molecules; they interact with thin filaments to shorten muscle fibers during contraction.

thigmomorphogenesis A plant response to a mechanical disturbance, such as frequent strong winds; includes inhibition of cellular elongation and production of thick-walled supportive tissue.

thigmotropism Growth in response to contact with a solid object.

thin filament A type of filament in striated muscle composed of actin, tropomyosin, and troponin molecules; they interact with thick filaments to shorten muscle fibers during contraction.

thorax The central part of an animal's body, between the head and the abdomen.

thorn forest A forested biome that grows at the arid borders of true savanna, where large mammals are less abundant.

thorn A sharp, pointed modified plant stem that may serve to deter predation.

threshold potential In signal conduction by neurons, the membrane potential at which the action potential fires.

thylakoid A flattened, closed sac within the stroma of a chloroplast.

thymine A pyrimidine that base-pairs with adenine.

thymus An organ of the lymphatic system that plays a role in filtering viruses, bacteria, damaged cells, and cellular debris from the lymph and bloodstream, and in defending the body against infection and cancer.

thyroid gland A gland located beneath the voice box (larynx) that secretes hormones regulating growth and metabolism.

thyroid-stimulating hormone (TSH) A hormone that stimulates the thyroid gland to grow in size and secrete thyroid hormones.

thyroxine (T$_4$) The main hormone of the thyroid gland, responsible for controlling the rate of metabolism in the body.

Ti (tumor-inducing) plasmid A plasmid used to make transgenic plants.

tidal volume The volume of air entering and leaving the lungs during inhalation and exhalation.

tight junction Region of tight connection between membranes of adjacent cells.

time lag The delayed response of organisms to changes in environmental conditions.

tissue A group of cells and intercellular substances with the same structure that function as a unit to carry out one or more specialized tasks.

tolerance hypothesis Hypothesis asserting that ecological succession proceeds because competitively superior species replace competitively inferior ones.

toll-like receptor A pattern recognition receptor in innate immunity that is found on the cell surface and within the cell on various membrane-bound compartments. Each type of toll-like receptor recognizes a different, specific set of molecular patterns on pathogens.

tonicity The effect a solution has on a cell when the solution surrounds it.

tonoplast The membrane that surrounds the central vacuole in a plant cell.

topoisomerase An enzyme that relieves the overtwisting and strain of DNA ahead of the replication fork.

topsoil The rich upper layer of soil where most plant roots are located; it generally consists of sand, clay particles, and humus.

torpor A sleeplike state produced when a lowered set point greatly reduces the energy required to maintain body temperature, accompanied by reductions in metabolic, nervous, and physical activity.

torsion The realignment of body parts in gastropod mollusks that is independent of shell coiling.

totipotent Having the capacity to produce cells that can develop into or generate a new organism or body part.

trace element An element that occurs in organisms in very small quantities (less than 0.01%); in nutrition, a mineral required by organisms only in small amounts.

tracer Isotope used to label molecules so that they can be tracked as they pass through biochemical reactions.

trachea In insects, an extensively branched, air-conducting tube formed by invagination of the outer epidermis of the animal, and reinforced by rings of chitin. In vertebrates, the windpipe, which branches into the bronchi.

tracheal system A branching network of tubes that carries air from small openings in the exoskeleton of an insect to tissues throughout its body.

tracheid A conducting cell of xylem, usually elongated and tapered.

tracheophyte A plant with xylem, phloem, and usually well-developed roots, stems, and leaves.

traditional systematics An approach to systematics that uses phenotypic similarities and differences to infer evolutionary relationships, grouping together species that share both ancestral and derived characters.

trait One of the forms of a genetic character.

transcription The mechanism by which the information encoded in DNA is made into a complementary RNA copy.

transcription factor (TF) In eukaryotes, the proteins required for RNA polymerase to initiate transcription or that regulate that process. One class of transcription factors recognizes and binds to the promoter in the area of the TATA box and then recruit RNA polymerase.

transcription initiation complex Combination of general transcription factors with RNA polymerase II.

transcription unit A region of DNA that transcribes a single primary transcript.

transcriptional regulation The fundamental level of control of gene expression that determines which genes are transcribed into mRNA.

transcriptome The complete set of transcripts in a cell, tissue, or organism.

transcriptomics The study of a transcriptome qualitatively and quantitatively.

transduction In cell signaling, the process of changing a signal into the form necessary to cause the cellular response. In prokaryotes, the process in which DNA is transferred from donor to recipient bacterial cells by an infecting bacteriophage.

transfer RNA (tRNA) The RNA that brings amino acids to the ribosome for addition to the polypeptide chain.

transformation The conversion of the hereditary type of a cell by the uptake of DNA released by the breakdown of another cell.

transgene A gene introduced into an organism by genetic manipulation to alter its genotype.

transgenic An organism that has been modified to contain genetic information from an external source.

translation The use of the information encoded in mRNA to assemble amino acids into a polypeptide.

translocation In genetics, a chromosomal alteration that occurs if a broken segment is attached to a different, nonhomologous chromosome. In vascular plants, the long-distance transport of substances by xylem and phloem.

transmembrane pathway The path followed by water when it enters root cells by crossing the cells' plasma membranes or enters cells through membrane aquaporins.

transmembrane protein An integral membrane protein that extends entirely through the plasma membrane.

transmission The second component of neural signaling involving the sending of a message along a neuron, and then to another neuron or to a muscle or gland.

transpiration The evaporation of water from a plant, principally from the leaves.

transport The controlled movement of ions and molecules from one side of a membrane to the other.

transport epithelium A layer of cells with specialized transport proteins in their plasma membranes.

transport protein A protein embedded in the cell membrane that forms a channel allowing selected polar molecules and ions to pass across the membrane.

transposable element (TE) A sequence of DNA that can move from one place to another within the genome of a cell.

transposase An enzyme that catalyzes some of the reactions inserting or removing the transposable element from the DNA.

transposition The movement of a transposable element from one site to another in a genome.

transposon A bacterial transposable element with an inverted repeat sequence at each end enclosing a central region with one or more genes.

tricarboxylic acid cycle *See* **citric acid cycle.**

trichocyst A dartlike protein thread that can be discharged from a surface organelle for defense or to capture prey.

trichome A single-celled or multicellular outgrowth of the plant epidermis; examples include root hairs and leaf trichomes that resemble hairs.

triglyceride A nonpolar compound produced when a fatty acid binds by a dehydration synthesis reaction at each of glycerol's three —OH-bearing sites.

triiodothyronine (T_3) A hormone secreted by the thyroid gland that regulates metabolism.

trilobites An ancient group of marine arthropods, now extinct.

trimester A division of human gestation, three months in length.

triple response Set of growth changes in plants that include slowed stem elongation, stem thickening, and a shift to horizontal growth. The response is triggered by ethylene as an adaptation to mechanical stress.

triploblastic An animal body plan in which adult structures arise from three primary germ layers, endoderm, mesoderm, and ectoderm.

trochophore The small, free-swimming, ciliated aquatic larva of various invertebrates, including certain mollusks and annelids.

trophic cascade The effects of predator–prey interactions that reverberate through other population interactions at two or more trophic levels in an ecosystem.

trophic level A position in a food chain or web that defines the feeding habits of organisms.

trophoblast The outer single layer of cells of the blastocyst.

tropical deciduous forest A tropical forest biome that occurs where winter drought reduces photosynthesis, and most trees drop their leaves seasonally.

tropical forest Any forest that grows between the Tropics of Capricorn and Cancer, a region characterized by high temperature and rainfall and thin, nutrient-poor topsoil.

tropical montane forest A tropical forest biome of short trees, which are frequently enveloped in mist; also known as a "cloud forest."

tropical rainforest A dense tropical forest biome that grows where some rain falls every month, mean annual rainfall exceeds 250 cm, mean annual temperature is at least 25°C, and humidity is above 80%.

tropics The latitudes between 23.5° N and 23.5° S, the Tropics of Cancer and Capricorn.

tropism The turning or bending of an organism or one of its parts toward or away from an external stimulus, such as light, heat, or gravity.

***trp* operon** In *E. coli*, the cluster of five genes for the biosynthesis of the amino acid, tryptophan, and adjacent sequences that control the expression of those genes.

tRNA *See* **transfer RNA.**

true-breeding Individual that passes traits without change from one generation to the next.

trypsin A digestive enzyme that digests peptides to produce small peptides.

tube feet Parts of the water vascular system of Echnodermata that allow them to grasp surfaces and food items.

tubular reabsorption The selective process in which some molecules and ions are transported by the transport epithelium from the lumen of the tubule back into the extracellular fluid and eventually into the blood as the solution produced by filtration moves through an excretory tubule.

tubular secretion The selective process in which specific small molecules and ions are transported from the extracellular fluid and blood into an excretory tubule.

tumor Tissue mass that results with differentiated cells of complex multicellular organisms deviate from their normal genetic program and begin to grow and divide.

tumor-suppressor gene A gene that encodes a protein that inhibits cell division.

turgid Osmotic condition of a plant cell when it contains sufficient water to press the plasma membrane against the cell wall and prevent further wall expansion.

turgor pressure The internal hydrostatic pressure within plant cells.

turnover rate The rate at which one generation of producers in an ecosystem is replaced by the next.

tympanic membrane Also called the *eardrum,* a sheet of tissue forming the boundary between the outer ear and the middle ear that vibrates in response to sound waves that move through the auditory canal.

tympanum A thin membrane in the auditory canal that vibrates back and forth when struck by sound waves.

ultimate causes The reasons why behaviors have adaptive value to the organisms displaying them.

umbilical cord A long tissue with blood vessels linking the embryo and the placenta.

umbilicus Navel; the scar left when the short length of umbilical cord still attached to the infant after birth dries and shrivels within a few days.

undernutrition A form of malnutrition. A condition in animals in which intake of organic fuels is inadequate, or whose assimilation of such fuels is abnormal.

undulating membrane In parabasalid protists, a finlike structure formed by a flagellum buried in a fold of the cytoplasm that facilitates movement through thick and viscous fluids; an expansion of the plasma membrane in some flagellates that is usually associated with a flagellum.

unequal crossing-over The rare phenomenon in meiosis in which, instead of crossing-over occurring at the exact same point on each homolog of a homologous pair of chromosomes, crossing-over occurs at different points.

unicellular organism Individual consisting of a single cell.

uniform dispersion A pattern of distribution in which the individuals in a population are evenly spaced in their habitat.

uniformitarianism The concept that the geological processes that sculpted Earth's surface over long periods of time—such as volcanic eruptions, earthquakes, erosion, and the formation and movement of glaciers—are exactly the same as the processes observed today.

Unikonta (unikonts) Eukaryotes with a single flagellum.

uniparental inheritance A pattern of inheritance in which all progeny (both males and females) have the phenotype of only one of the parents.

universal A feature of the nucleic acid code, with the same codons specifying the same amino acids in all living organisms.

unreduced gamete A gamete that contains the same number of chromosomes as a somatic cell.

unsaturated With respect to a fatty acid, if one or more double bonds link the carbon atoms.

ureter The tube through which urine flows from the renal pelvis to the urinary bladder.

urethra The tube through which urine leaves the bladder. In most animals, the urethra opens to the outside.

urinary bladder A storage sac located outside the kidneys.

Urochordata A monophyletic group of chordates, including tunicates and sea squirts.

uterine cycle The menstrual cycle.

uterus A specialized saclike organ, in which the embryo develops in viviparous animals.

utricle Saclike, fluid-filled structure that is part of the vestibular apparatus. With the saccule, provides information about the up–down position of the head, as well as changes in the rate of linear movement of the head.

vaccination The process of administering a weakened form of a disease to patients as a means of giving them immunity to a more serious form of the disease.

vagina The muscular canal that leads from the cervix to the exterior.

valence electron An electron in the outermost energy level of an atom.

Van der Waals forces Weak molecular attractions over short distances.

variable An environmental factor that may differ among places or an organismal characteristic that may differ among individuals.

variable (V) region For the light and heavy polypeptides of a particular type of antibody, the regions that have different amino acids sequences from molecule to molecule.

vas deferens The tube through which sperm travel from the epididymis to the urethra in the male reproductive system.

vascular bundle A cord of plant vascular tissue; often multistranded with both xylem and phloem.

vascular cambium A lateral meristem that produces secondary vascular tissues in plants.

vascular plant *See* **tracheophyte.**

vascular tissue system One of the three tissue systems in plants that provide the foundation for plant organs; it consists of transport tubes for water and nutrients.

vegetal pole The end of the egg opposite the animal pole, which typically gives rise to internal structures such as the gut and the posterior end of the embryo.

vegetative reproduction Asexual reproduction in plants by which new individuals arise (or are created) without seeds or spores; examples include fragmentation from the parent plant or the use of cuttings by gardeners.

vein In a plant, a vascular bundle that forms part of the branching network of conducting and supporting tissues in a leaf or other expanded plant organ. In an animal, a vessel that carries the blood back to the heart.

ventilation The flow of the respiratory medium (air or water, depending on the animal) over the respiratory surface.

ventral Indicating the lower or "belly" side of an animal.

ventricle In the brain, an irregularly shaped cavity containing cerebrospinal fluid. In the heart, a chamber that pumps blood out of the heart.

venule A capillary that merges into the small veins leaving an organ.

vernalization The stimulation of flowering by a period of low temperature.

vertebrae The series of bones that form the vertebral column of vertebrate animals.

vertebral column The series of vertebrae that surrounds and protects the dorsal nerve cord and forms the supporting axis of the body.

vertebrate A member of the monophyletic group of chordate that possess a vertebral column.

vesicle A small, membrane-bound compartment that transfers substances between parts of the endomembrane system.

vessel In plants, one of the tubular conducting structures of xylem, typically several centimeters long; most angiosperms and some other vascular plants have xylem vessels.

vessel element Any of the short cells joined end to end in tubelike columns in xylem.

vestibular apparatus The specialized sensory structure of the inner ear of most terrestrial vertebrates that is responsible for detecting rotational movement of the head and provides information about the up–down positioning of the head as well as changes in the rate of linear movement of the head. The vestibular apparatus consists of three semicircular canals, the utricle, and the saccule.

vestigial structure An anatomical feature of living organisms that no longer retains its ancestral function.

vibrio Any of various short, motile, S-shaped or comma-shaped bacteria of the genus *Vibrio*.

vicariance The fragmentation of a continuous geographical distribution by nonbiological factors.

virion A complete virus particle.

viroid A plant pathogen that consists of strands or circles of RNA, smaller than any viral DNA or RNA molecule, that have no protein coat.

virulent bacteriophage Bacteriophage that kills its host bacterial cells during each cycle of infection.

virus An infectious agent that contains either DNA or RNA surrounded by a protein coat.

visceral mass In mollusks, the region of the body containing the internal organs.

visual signal A means of communication in which animals use facial expressions or body language to send messages to other individuals.

vital capacity The maximum tidal volume of air that an individual can inhale and exhale.

vitamin An organic molecule required in small quantities that the animal cannot synthesize for itself.

vitamin D A steroidlike molecule that increases the absorption of Ca^{2+} and phosphates from ingested food by promoting the synthesis of a calcium-binding protein in the intestine; it also increases the release of Ca^{2+} from bone in response to PTH.

vitelline coat A gel-like matrix of proteins, glycoproteins, or polysaccharides immediately outside the plasma membrane of an egg cell.

vitreous humor The jellylike substance that fills the main chamber of the eye, between the lens and the retina.

viviparous Referring to animals that retain the embryo within the mother's body and nourish it during at least early embryo development.

voltage An electric potential difference.

voltage-gated ion channel A membrane-embedded protein that opens and closes as the membrane potential changes.

vulva The external female sex organs.

water lattice An arrangement formed when a water molecule in liquid water establishes an average of 3.4 hydrogen bonds with its neighbors.

water molds *See* **Oomycota.**

water potential The potential energy of water, representing the difference in free energy between pure water and water in cells and solutions; it is the driving force for osmosis.

watershed An area of land from which precipitation drains into a single stream or river.

water-soluble vitamin A vitamin with a high proportion of oxygen and nitrogen able to form hydrogen bonds with water.

water vascular system A locomotor system, including internal canals and tube feet, unique to Echinodermata.

wavelength In electromagnetic radiation, the horizontal distance between the crests of successive waves.

wax A substance insoluble in water that is formed when fatty acids combine with long-chain alcohols or hydrocarbon structures.

weathering Physical, chemical, or biological events that break down rock and inorganic particles, forming soil.

weight A measure of the pull of gravity on an object.

Wernicke's area A language area in the frontal lobe of the cerebral cortex of the brain. Comprehension of spoken and written language depends on coordination of inputs from the visual, auditory, and general sensory association areas by Wernicke's area.

wetland A highly productive ecotone often at the border between a freshwater biome and a terrestrial biome.

white blood cell *See* **leukocyte.**

white matter The myelinated axons that surround the gray matter of the central nervous system.

white rusts *See* **Oomycota.**

whole genome shotgun sequencing A strategy for obtaining the complete DNA sequence of a genome in which DNA is isolated and purified, that DNA is broken into many random, overlapping fragments each of which is amplified and sequenced, and the entire genome sequence is assembled from those sequences using computer algorithms.

wilting The drooping of leaves and stems caused by a loss of turgor.

wobble hypothesis Hypothesis stating that the complete set of 61 sense codons can be read by fewer than 61 distinct tRNAs because of particular pairing properties of the bases in the anticodons.

Wolffian duct A bipotential primitive duct associated with the gonads that leads to a cloaca.

wood The secondary xylem of trees and shrubs, lying under the bark and consisting largely of cellulose and lignin.

X chromosome Sex chromosome that occurs paired in female cells and single in male cells.

X-linked gene A gene on the X chromosome.

X-linked inheritance The pattern of inheritance of an X-linked gene.

X-linked recessive inheritance Pattern in which displayed traits are due to inheritance of recessive alleles carried on the X chromosome.

X-ray diffraction Method for deducing the position of atoms in a molecule.

xylem The plant vascular tissue that distributes water and nutrients.

xylem sap The dilute solution of water and solutes that flows in the xylem.

Y chromosome Sex chromosome that is paired with an X chromosome in male cells.

yeast A single-celled fungus that reproduces by budding or fission.

yolk The portion of an egg that serves as the main energy source for the embryo.

yolk sac In an amniote egg, an extraembryonic membrane that encloses the yolk.

zero population growth A circumstance in which the birth rate of a population equals the death rate.

zona pellucida A gel-like matrix of proteins, glycoproteins, or polysaccharides immediately outside the plasma membrane of the egg cell.

zone of cell division The region in a growing root that consists of the root apical meristem and the actively dividing cells behind it.

zone of elongation The region in a root where newly formed cells grow and elongate.

zone of maturation The region in a root above the zone of elongation where cells do not increase in length but may differentiate further and take on specialized roles.

zoogeographical realms Contiguous regions on Earth that are occupied by animals with a shared evolutionary history.

zooplankton Small, usually microscopic, animals that float in aquatic habitats.

zygomycetes Fungi in the phylum Zygomycota, which reproduce sexually by way of zygospores. Many species are saprobes that live in soil, feeding on plant debris.

zygospore A multinucleate, thick-walled sexual spore formed in zygomycete fungi (Zygomycota).

zygote A fertilized egg.

Index

Myxinoidea (hagfishes), **752**–753, 753i
Myxobacteria, 602, 602i
Myxoma virus, 1222

Nachman, Michael W., 8–9
NAC proteins, 826
NAD+ (nicotinamide adenine dinucleotide), **149**
in glycolysis, 149–151, 151i, 152i, 160
in oxidative phosphorylation, 149, 149i
in pyruvate oxidation, 153–155
NADH (nicotinamide adenine dinucleotide)
in glycolysis, 149–150, 149i, 151, 151i
in oxidative phosphorylation, 160, 160i, 162
in pyruvate oxidation, 153–154, 154i, 155
NADP+ (nicotinamide adenine dinucleotide phosphate), in photosynthesis, **170**, 176–177
NADPH (nicotinamide adenine dinucleotide phosphate), in photosynthesis, **170**, 170i, 176, 177i–179i, 178i, 180, 181i, 183, 184
Nagel, Thomas, 1360
Na+/K+ pump (sodium-potassium pump), **116**, 117i, 118, 935, 937, 1126
Nanoarchaeota, 610
Nanoarchaeum equitans, 610
Nanoplankton, 624
Nanos gene, 374
Narwhal (*Monodon monoceros*), 774i
Nastic movements, **902**–903, 902i, 903i
National Academy of Sciences, 20
National Center for Biotechnology Information, 439
National Human Genome Research Institute, 336
National Institutes of Health (NIH), 433
National Science Foundation, 12
Natrosobacterium, 610
Natural experiments, 1195
Natural gas deposits, 656
Natural history, **458**
Natural killer (NK) cells, 1054t, **1057**
Natural selection
adaptation and, 9, 467, 472, 473
adaptive traits favored by, 464
balanced polymorphisms, mainte-nance of, 492–495, 493i, 494i
Darwin and, 7–8, 464
defined, **7, 464, 484**
genetic variability shaped by, 484–488, 485i
kin selection, 1356–1358, 1357i
recessive alleles masked by dip-loidy, 492, 493t
Natural theology, **458, 459**
Nature Conservancy, 1328
Nature preserves, 1326–1327, 1329–1330. *See also* Conserva-tion biology
Nautiluses, 726, 726i

Nautilus macromphalus, 726i
Navarro, Arcadi, 517
Navigation, **1346**
Neale, David B., 678
Neanderthals, 780, 781
Necrotizing fasciitis, 608
Nectar, 677
Nef (negative factor) protein, 1071
Negative control, gene regulation, **355**, 356–357, 356i, 357i, 359, 360i
Negative feedback control systems
in endocrine system, 996, 997, 997i, 1004, 1007i, 1010
homeostatis and, **924**–926, 924i, 925i, 929
in human reproduction, 1153, 1157i
stretch reflexes, 958
thermoregulation, 1132
Negative pressure breathing, **1079**, 1080, 1081
Negative pressure. *See* Surface tension
Neisseria gonorrhoeae, 598, 602
Neisseria meningitidis, 602
Nekton, **1216**
Neljubov, Dimitry, 894
Nematocyst, **716**, 716i
Nematode, 729–730, 729i. *See also Caenorhabditis elegans*
parasitic, 722, 730
Nematode-trapping fungi, 694i, 695
Nemertea (ribbon worms), 723–724, 723i
Nemhauser, Jennifer, 911
Neofunctionalization, 681
Neogene period, 524t–525t, 530i, 536i
Neon (element), 28i
Neoplasms, 296
Neotiella (fungus), 240i
Nephron, **1123**, 1124i, 1125t, 1126–1127, 1126i
Nephrotic syndrome, 1141
Nereis, 728i
Neritic zone, **1213**, 1214i, 1216, 1216i
Nernst equation, **936**
Nerve cell, 921, 932i. *See also* Neuron
Nerve cord, 747, 747i, 952i, **953**
Nerve fiber, 931
Nerve growth factor, 200
Nerve net, **716, 952,** 952i
Nerve ring, 952, 952i
Nerves, **952**
Nervous system, 922i, 951–971
central nervous system, 953, 957–966 (*See also* Central nervous system (CNS))
compared to endocrine system, 994–995
development, 1173–1174, 1174i
of invertebrates, 952–953, 952i
memory, learning, and conscious-ness, 966–967, 967i
nerve nets, 952, 952i
neuron structure and organization in, 930, 931–935
peripheral nervous system, 932, 953, 955–957, 955i, 956i
of protostomes, 720, 726, 736
structure of, 953–954, 954i
of vertebrates, 953–954, 954i

Nervous tissue, 914i, 920i, **921**
Nest-building behavior, evolution of, 1354–1355, 1354i
Net primary productivity (NPP), **1287**, 1288i, 1288t, 1289, 1289i, 1306
Neural circuit, **932**, 948, 968, 1341
Neural crest, 748–**749, 1173**, 1174i, 1181, 1184
Neural ectoderm, 1173, 1175
Neural plate, **1173**, 1174i, 1181, 1183
Neural signaling, 210, **931**, 931i
Neural tube, **953, 1173**, 1174i, 1184, 1186i
Neuroendocrine signaling, **995**, 995i, 996
Neurogenic heart, **1040**
Neurohormone, **995**–996, 1004, 1012
Neuromasts, 976
Neuromuscular junction, **1018**, 1018i, 1020i, 1023i
Neuron, 930–950
afferent neurons, 931–932, 973, 973i, 974
in autonomic nervous system, 956, 956i
in cell communications, 193, 194i
cells in retina, 983–984, 984i
chemical synapses, transmission, 941–944, 943i, 944i
defined, **921**
efferent neurons, 931–932, 931i
glial cell support, 932, 932i
integration of signals, 944–947, 947i
chemical synapses, summation of, 944–947
neurosecretory neurons, 995, 1004, 1011
number in human brain, 947, 948, 957
organization in nervous systems, 931–935
patterns of synaptic connections, 947
respiratory, 1089
signaling by, 935–941
action potential production, 937–938, 938i–939i
membrane potential during action potential, 936–937, 937i
membrane potential production, 935
neural impulse propagation, 939–941, 940i
resting membrane potential, 935–936, 937i
saltatory conduction, 941, 942i
structure, 920i, 921, 931–932, 932i
synapse communication, 932–935, 933i
Neuropeptide, 945t
Neuropeptide Y, 1114
Neurophysiology, of behavior, 1341–1342
Neuroscience, **1335**

Neurosecretory neuron, **995**, 1004, 1011
Neurospora crassa (orange bread mold)
cell cycle of, 215
gene–enzyme relationship, 324, 325i
genetic research, 686, 693
genome, 440t
life cycle, 694, 695i
as model organism, 702
Neuroteratogens, 1089
Neurotransmitter, **934**
action of, 934–935
in cell communications, 193, 197, 201, 204, 205, 210
compared to neurohormones, 995–996
ion flow, 944
receptors, 941–943, 943i, 946
release by exocytosis, 943–944
types, 944, 945t
Neurulation, 1176–1177, 1182
Neutralization, by antibodies, **1063**, 1063i
Neutral lipid, 52–54
Neutral variation hypothesis, **495**
Neutron, 26–27
Neutrophil, 1054i, 1054t, **1055**–1056, 1056i
New Guinea impatiens (*Impatiens hawkeri*), 823i
Newt, 761, 761i
Newton, Sir Isaac, 458
Next-generation DNA sequencing, 208, 916
Ngorongoro Conservation Area, 1328–1329
Niacin, 1100t
Niche, ecological, **1256**–1259, 1257i, 1258i
Niche cells, 809
Nicholas II (Czar), 283, 428
Nickel, plant micronutrient, 835t, 836
Nicolson, Garth L., 107
Nicotinamide adenine dinucleotide. *See* NAD+; NADH
Nicotinamide adenine dinucleotide phosphate. *See* NADP+; NADPH
Nicotine, 680, 945t, 946, 1004, 1089, 1161
Nicotine addiction, 206
Nicotinic acetylcholine receptor (nAChR), 946
Niedergerke, Ralph, 1019i
Nifedipine, 1051, 1051i
NIH (National Institutes of Health), 433
Niimura, Yoshihito, 772
Nirenberg, Marshall, 327, 328
Nitella, 630
Nitrate, 161, 600, 839, 843, 844–845, 845i, 848, 1300
Nitric oxide (NO), 207, 451, 1046, 1046i, 1303
Nitric oxide synthase, 451
Nitrification, **600, 844,** 845i, 1298t, 1299i, **1300**
Nitrifying bacteria, 600
Nitrite, 600, 1300
Nitrobacter, 600

genomes sequenced, 438t
habitat, 617–618
metabolism, 616–617
plastid evolution from endosymbionts, 633–635, 634i, 635–336
reproduction, 615, 617, 620, 621i
structure, 617, 617i
"Protista" kingdom, 570, 572
Protist groups, 618–635
Archaeplastida, 627–630
Chlorophyta (green algae), 628–630, 628i–630i
Rhodophyta (red algae), 628, 628i
Chromalveolata, 620–626
Alveolata, 620–622, 621i
Stramenopila, 622–626, 622i–626i
Excavata, 618–619
Euglenozoans, 619, 619i
Metamonada, 618, 618i
Rhizarian amoebas, 626–627
Chlorarachniophyta, 627
Foraminifera (forams), 627, 627i
Radiolaria, 627, 627i
Unikonta, 630–632
Amoebozoa, 630–631
Opisthokonta, 631–632, 632i
Protocell, 55–56, 578, 582–583, 582i
Protoderm, 797, 797i, 800, 800i
Proton, 26–27
gradients in photosynthesis, 176–177, 178i
hydrogen ions in water ionization, 38–39
Protonema, 649, 650i
Protonephridium, 1121–1122, 1122i
Proton-motive force, 158, 159i, 176–177
Proton pump, 118, 159i, 815i
Proto-oncogene, 379
Protoplast, 789
Protostome
blastopore development in, 1170
developmental patterns, 710–711, 710i
Ecdysozoa, 712 (See also Ecdysozoan protostomes)
Lophotrochozoa, 711 (See also Lophotrochozoan protostomes)
Prototheria, 771
Protozoa, 12, 13i, 616
Proventriculus, 1097, 1097i
Provirus, 400, 401, 401i, 402i
Provisioning ecosystem services, 1318–1319
Proximal convoluted tubule, 1123, 1125i, 1125t, 1126–1127, 1126i
Proximal end, of excretory tubule, 1119
Proximate causes of behavior, 1335
PR proteins. See Pathogenesis-related proteins
Prusiner, Stanley, 402
Pseudoceros dimidiatus, 721i
Pseudocoelom, 709i, 710
Pseudocoelomate animal, 709i, 710, 721, 739
Pseudogene, 438, 439, 442
Pseudohermaphrodite, 1151
Pseudomonas, 613
Pseudomonas fluorescens, 75i, 633
Pseudomyrmex ferruginea, 1259–1260, 1261i

Pseudo-nitzschia, 624
Pseudopodia, 617, 626–627, 630
Psilotum, 654, 655i
P site (peptidyl site, in ribosome), 338, 339i–343i, 340, 341
Psychrophile, 610
Pterocorys, 627i
Pterosaurs, 567, 764
PTH. See Parathyroid hormone
PTTH. See Prothoracicotropic hormone
Puberty, 1181
Public health, 1244
Pufferfish (Takifugu rubripes), 440t, 441, 941, 1048
Pulmocutaneous circuit, 1034, 1034i
Pulmonary artery, 1038i, 1039, 1039i
Pulmonary circuit, 1034, 1034i, 1038i, 1039, 1042
Pulmonary vein, 1038i, 1039
Pulvini, 902
Punctuated equilibrium hypothesis, 539–540, 540i, 547
Punnett, Reginald, 258
Punnett square, 258, 258i, 261i
Pupa, 739, 739i, 1011, 1012i
Pupil, eye, 981, 981i
Pure-breeding, 253
Purine, 65, 66i, 304, 304i, 305
Purkinje fiber, 1040, 1041i
Purple martin (Progne subis), 501i
PVA. See Population viability analysis
Pyloric sphincter, 1101i, 1105
Pyramid of biomass, 1291–1292, 1292i
Pyramid of numbers, 1292, 1292i
Pyramids of energy, 1291, 1292i
Pyranose, 590
Pyrenoid protein bodies, 650
Pyrimidine, 65, 66i, 304, 304i, 305
Pyrobolus, 610
Pyrococcus, 13i, 610
Pyrogen, 1056
Pyrophilus, 610
Pyrophosphate, 310i
Pyrrolnitrin, 633
Pyruvate
in C4 pathway, 185, 186i
fermentation reactions and, 162, 162i
in glycolysis, 151, 151i, 152i
Pyruvate dehydrogenase complex (PDC), 155
Pyruvate kinase, 152i, 165
Pyruvate oxidation, 149, 149i, 153–155, 153i, 160, 160i, 163i
Pyruvic acid oxidation, 153
Python, 469, 470i, 988

QTLs. See Quantitative trait loci
Qualitative variation, 478, 478i
Quantitative comparative proteomic profiling, 166
Quantitative trait loci (QTLs), 268
Quantitative traits, 268
Quantitative variation, 477, 478i
Quaternary structure of protein, 60, 60i, 63–64
Quiescent center, 797, 797i
Quillwort, 653
Quinine, 680
Quinoa (Chenopodium quinoa), 1240i
Quorum sensing, 195, 606

RAAS (renin–angiotensin–aldosterone system), 1129–1130, 1129i
Rabbit, 251, 773, 1222–1223
Rabies, 1316
Race (subspecies), 503
Racker, Efraim, 159i
Radial canals, 745, 745i
Radial cleavage, 710, 710i, 745
Radial nerve, 952, 952i
Radial symmetry
of animals, 709, 709i
of flowers, 857
echinoderms, secondary, 744i, 745
eumetazoans with, 715–718
Radiation, 172, 172i, 236
evolutionary, 533-537, 561
Radiation, of heat, 1134–1135, 1134i
Radiation mutagen, 347, 378
Radiation therapy, 27
Radicle, 796, 861, 863i, 866, 866i
Radioactivity, 27
Radioisotope, 27–28
Radiolaria (radiolarians), 627, 627i
Radiometric dating, 28, 524, 526i, 584–585
Radium, 236
Radula, 724, 725i
Radulovic, Jelena, 1006
Rainforest
deforestation, 1310–1311, 1310i
temperate, 1209
tropical, 1205, 1205i, 1206
Rain shadow, 1199, 1200i, 1207
Ramalina lichens, 699
Random coil, 60–61, 64
Random dispersion, 1224, 1225i
Rao, Potu N., 226, 227i
Raphanobrassica, 515
Rapid-eye movement (REM) sleep, 967
Ras G protein, 204–205, 205i
Raspberry (Rubus), 863
Rat, 1314i, 1315
Rat snake (Elaphe obsoleta), 503i
Rattlesnake, 765, 765i, 988, 988i, 1253
Raup, David, 535
Raven (Corvus corax), 1340, 1340i, 1351–1352
Ray, plant secondary growth, 807
Ray-finned fishes. See Actinopterygii (ray-finned fishes)
Rays, manta, 756, 756i, 919, 990, 1070, 1131
Reabsorption
excretory tubular, 1120, 1120i, 1126–1127
of sodium, 1129–1130
of water, 1130, 1132i–1133i
Reactant, 34
Reaction. See Chemical reaction
Reaction center, 175–176, 176i
Reactive oxygen species, 179, 189
Reading frame (see also open reading frame), 329, 334, 340, 346
Realized niche, 1256–1259, 1257i, 1258i
Realms, biogeographical and zoogeographical, 531–532, 532i
Receptacle, flower, 854, 854i
Reception, in cell communication, 195, 197i, 198i, 199i, 201i–205i, 207i
Reception, in neural signaling, 931, 931i
Receptive field, 973, 975, 984

Receptor, 6
hormones binding to, 995, 997–999, 998i, 999i, 1000i
internal, 206–207
for peptide hormone, 197
in respiration, 1082–1083
sensory (See Sensory receptors)
steroid hormone (See Steroid hormone receptors)
surface (See Surface receptors, cell communication)
Receptor-mediated endocytosis, 119, 120i, 121, 397
Receptor potential, 973
Receptor protein, 57t, 106, 110, 995
Receptor tyrosine kinase (RTK), 199–201, 200i, 204–205, 205i
Recessive allele, 255, 492, 493t
Recessive trait, 291i
autosomal, 290–291, 319
X-linked, 292
Reciprocal altruism, 1358
Reciprocal cross, 254, 281
Reciprocal translocation, 286, 286i
Recognition protein, 106
Recombinant chromosomes, 245
Recombinant DNA, 409, 410, 414, 424
Recombinant phenotypes, 276i, 277–279, 277i
Recombinant plasmids, 409, 409i, 410, 410i, 411i
Recombinants, 244i, 245
Recombination. See Genetic recombination
Recombination frequency, 278–279, 278i
Recruitment, population, 1268
Rectal salt gland, 1131
Rectum, 1099i, 1108–1109, 1108i
Red algae, 628, 628i
Red blood cell. See Erythrocyte
Red buckeye (Aesculus), 803i
Red deer (Cervus elaphus), 1228, 1229i, 1230
Red fir (Abies magnifica), 1238i
Red-green color blindness, 281, 283
Red harvester ants (Pogonomyrmex barbatus), 1228–1229
Red-legged frog (Rana aurora), 1349
Red marrow, 1025, 1025i
Red oak (Quercus rubra), 793i, 841
Redox reaction, 148, 148i
Red-spotted newt (Notophthalmus viridescens), 761i
Red tides, 621
Reduced Emissions from Deforestation and Degradation (REDD), 1331
Reduced substance, 147
Reduction, in Calvin cycle, 180, 181i
Reduction, redox reactions, 147–148, 148i
Reductionist approach in systems biology, 19, 100
Redundancy, in genetic code, 328, 681
Red-winged blackbirds (Agelaius phoeniceus), 1353–1354
Redwoods, 13i, 668, 821, 823i
Reference genome sequences, 678
Reflexes, 958, 959i
Refractory period, 937, 937i, 940, 1158
Regeneration, 180, 1145
Regeneration cells, 214
Regier, Jerome C., 732

Stimulus
   in classical conditioning, 1340
   defined, **924, 973**
   homeostatis and negative feedback, 924, 924i, 925i
   in neural signaling, 931i, 941
   perception biases, 1360
   sensory receptors and, 973–975, 973i, 989
Stipes, of kelp, 624
Stock, grafting, **869,** 869i
Stoddart, Angela, 296
Stoeckenius, Walther, 159i
Stolon, 800, 802i
Stoma. *See* Stomata (stoma)
Stomach, 1104–1105
   defined, **1097**
   of human, 1099i, 1101i
   of insect and bird, 1097, 1097i
   of mammal, 1104–1105, 1104i
   as pathogen destroyer, 1053
   ruminant, 1111–1112, 1112i
   structure and function, 1101i, 1104–1105, 1104i
   wall layers, 920, 1100, 1101i, 1102, 1105
Stomata (stoma), **172, 641, 794**
   adaptation, 641, 641i, 644
   in bryophytes, 648, 649, 650
   carbon dioxide and, 795
   genetics of, 795–796
   guard cells, 794
   location, 805i
   opening and closing, 823–824, 823i
   in photosynthesis, 171i, **172,** 184, 186i
   response to environmental stress, 895
   structure, 795i
   transpiration and, 820, 820i, 822i
   water loss, regulation of, 823–825, 823i, 825i
Stone, Anne, 780
Stonecrops, 186
Stoneking, Mark, 780, 781
Stop codons, **329,** 342, 342i, 346
Storage proteins, 57t
Storage root, 796i, 797
Stork, Nigel, 1322
Storz, Jay, 1088i
STR. *See* Short tandem repeat
Strahl, Sabine, 1184
Stramenopila (stramenopiles), 622–626, 635–636
   Bacillariophyta (diatoms), 624, 624i, 625i
   Chrysophyta (golden algae), 624, 625i
   Oomycota, 622–624, 622i
   Phaeophyta (brown algae), 615i, 624, 625i, 626, 626i
Strata, in fossils, 522–523, 522i, 523–524
Stratified epithelium, 915, 915i
Strawberry (*Fragaria ananassa*), 800, 802i, 823i, 863, 868, 898, 898i
Streams, 1211, 1211i, 1270–1271
Strepsirhini, **775,** 775i
Streptococcus pneumoniae, 301, 302i, 390, 597, 608
Streptococcus pyogenes, 604, 604i, 608
Streptomycin, 338, 349, 608
Stretch reflex, **958**
Striated muscle, 335, 1017–1018, 1023

Strict aerobe, **163**
Strict anaerobe, **162**
Strigolactones (SLs), 882t, **889–891,** 890i, 892
Strobilus
   club moss, **653**
   horsetail, 655i
Stroke, 926–927, **1045,** 1161
Stroma in chloroplast, 96, 96i, **171,** 172i, 177
Stromal cells, 296
Stromal lamellae, 171–172, 171i, 174
Stromatolites, **585,** 586, 586i
Structural gene, **356,** 356i, 357i, 360i
Structural isomer, **47,** 48i
Structural protein, 57t
Strychnine, 898
Sturgeon, 757, 757i
Sturtevant, Alfred, 278
Stylosphaera, 627i
Styracosaurus, 764
Style, flower part, **675, 854,** 854i
Suberin, 806, 818
Subfunctionalization, 681
Submucosa, 1101i, **1102**
Subsoil, **838,** 838i
Subspecies, **503,** 503i, 504, 504i
Substance P, 945t, 989
Substrate-level phosphorylation, **149,** 150i, 152i, 153, 154i, 155, 160i
Substrate molecule, **135,** 136
Succession. *See* Ecological succession
Succinate, 154i, 155
Succinate dehydrogenase, 154i, 155, 157
Succinyl–CoA, 154i
Succinyl–CoA synthetase, 154i
Sucrose
   breakdown to glucose and fructose, 133, 134
   as disaccharide, 50, 50i
   in photosynthesis, 182–183
   transport in plants, 826–829, 828i
Sudden infant death syndrome (SIDS), 1089
Sudden oak death, 703
Sugar, transport in phloem, 794, 794i
Sugarcane (*Saccharum*), 679, 1259
Sugar–phosphate backbone, **304,** 304i, 305, 306i
Sulfate, 161, 839
Sulfhydryl group, **46,** 47t, 59
Sulfolobus, 75i, 611
Sulfur
   essential mineral nutrient for humans, 1101t
   plant macronutrient, 835, 835t
   radioisotopes, 28
Sulfur cycle, 161
Sulson, John, 1187
Summation, twitch, 1021, 1021i
Sum rule in probability, **257,** 258i
Sundew, English (*Drosera anglica*), 126i
Sunflower (*Helianthus annuus*), 791i, 861, 881i
Sunlight. *See* Solar energy/sunlight
Superior vena cava, 1038i, 1039
Superorganisms, 1261
Support ecosystem services, 1319
Suprachiasmatic nucleus, 960, 1011
Surface area-to-volume ratio, 75, 77, 77i

Surface receptors, cell communication, 197–206
   G-protein–coupled receptors, 201–205, 201i, 202i–205i
   as integral membrane glycoproteins, 197, 198i
   ligand-gated ion channels, 205–206, 206i
   off switches, 199
   signal transduction pathways, 198–199
   on target cell, 192
   tyrosine kinases, 199–201, 200i, 204–205, 205i
Surface tension of water, 35, **36**–37, 36i, 820–821, 820i
Surfactant, 1081
Surfactant-A (SP-A), 1189
Surgeonfish (*Acanthurus lineatus*), 1349
Surroundings, 127, 128–129, 128i
Survivorship, age-specific, 1226, 1226t
Survivorship curve, **1227,** 1227i, 1250
Suspension feeder, 747–748, **1095,** 1095i
Suspensor, **861,** 862i
Sustainable agriculture, **840**
Sutherland, Earl, 195, 196i
Sutherland, John, 142
Sutton, Walter, 263–264
Swallow, John G., 479
Swallowing reflex, **1103,** 1103i
Swallows, 501i, 1354i, 1355, 1356
Swamp rose-mallow (*Hibiscus pallustris*), 855i
Swamp sparrow (*Melospiza georgiana*), 1334, 1335i
Swan, Shanna, 1165
Sweat gland, 285, 1138
Sweet pea (*Lathyrus odoratus*), 804i
Swell shark (*Cephaloscylium ventricosum*), 756i
Swim bladder, **757,** 758i
Swimmerets, 735i, 736
Switches, developmental, 545
Swordtail (*Xiphophorus helleri*), 512
Symbiosis, 1259–1261. *See also* Endosymbiosis
   commensalism, 1259, 1259i
   defined, **699, 1259**
   fungal, for nutrition, 642
   herbivore digestive system microorganisms, 1111–1112, 1112i
   lichen, 699–700, 700i
   mutualism, 1259–1260, 1260i, 1261i
   mycorrhizae, 700, 701i
   parasitism, 1260–1261
Symmetrical body plan, **709**
Sympathetic nervous system, 955, 955i, **956–957,** 956i
Sympatric speciation, **509,** 511, 511i, 512–516, 514i, 517, 520
Symplast, **817–819,** 829
Symplastic pathway, **817–819,** 818i
Symport, **118,** 118i, 814, 815i, 824, 829
Synapomorphy, **562–563**
Synapse, **193,** 932–935
   chemical (*See* Chemical synapse)
   electrical, 193, 932–933, 933i
Synapsida (synapsids), 561i, 563, **763,** 770

Synapsis of homologous chromosomes, 240i, 242i
Synaptic cleft, 933i, **934,** 943–944, 943i
Synaptic plasticity, nervous system, **963,** 966
Synaptic signaling, **193,** 194i
Synaptic vesicle, 933i, **934**–935, 943, 943i
Synaptonemal complex, 240i, 248
Synaptosome, 935
Synergid, 858i, 859, 859i, 860, 878
Synesthesia, 990
Synovial joint, **1025**–1026, 1026i
Synura, 625i
Syphilis, 602
System, in thermodynamics, **127,** 128–129, 128i
Systematic biologists (systematists), 551
Systematics, 550–576
   binomial nomenclature and taxonomy, 551–552, 573
   cladistic revolution, 561–566
   conservation biology and, 1322
   defined, **551**
   molecular phylogenetics, 569–573
   phylogenetic analyses, sources of data, 556–560
   phylogenetic trees (*See also* Phylogenetic tree)
   evolutionary history and classification, 552–556
   as research tools, 567–569
   traditional classification and paraphyletic groups, 560–561
Systematics and the Origin of Species (Mayr), 501
Systemic acquired resistance, 896t, **899,** 899i
Systemic circuit, **1034,** 1034i, 1038i, 1039, 1042
Systemic homeostatic controls, **924**
Systemic inflammation, **1055**–1056
Systemic lupus erythematosus (lupus), **1068**
Systemin, 882t, 896t, **897,** 897i
Systems biology, **19**–20, 20i, 100
Systole, **1039,** 1039i
Systolic blood pressure, 1040–1042, 1051i
Szostak, Jack, 317, 583

T2 phage, 301–302, 303i, 395
T$_3$. *See* Triiodothyronine (T$_3$)
T$_4$. *See* Thyroxine (T$_4$)
Tactile senses. *See* Mechanoreceptors
Tactile signals, **1350,** 1350i
Taiga, **1209**
Tail
   chordate, 747, 747i
   of virus, 394i, **395,** 395i, 396, 397i
Talley, Brooke L., 1318
Tallgrass prairie, 1208, 1208i
Talmage, David, 1063
Tamoxifen, 1012
Tandem duplication, **450**
Tang, Yiwei, 1192
Tannin, 893, 898
Tanzania, conservation area, 1328–1329, 1329i
Tapeworms, 721, 721i, 1260
Taproot system, **796,** 796i
Taq polymerase, 412
Tarantula, Texas brown (*Aphonopelma hentzi*), 2i